Correlation Guide For

Stewart *College Physics for the AP® Physics 1 & 2 Courses*

Third Edition

Find a complete correlation guide to the AP® Physics 1 & 2 Curriculum Frameworks in your online resources and at www.bfwpub.com/stewart3e

College Physics

for the AP® Physics 1 & 2 Courses

Third Edition

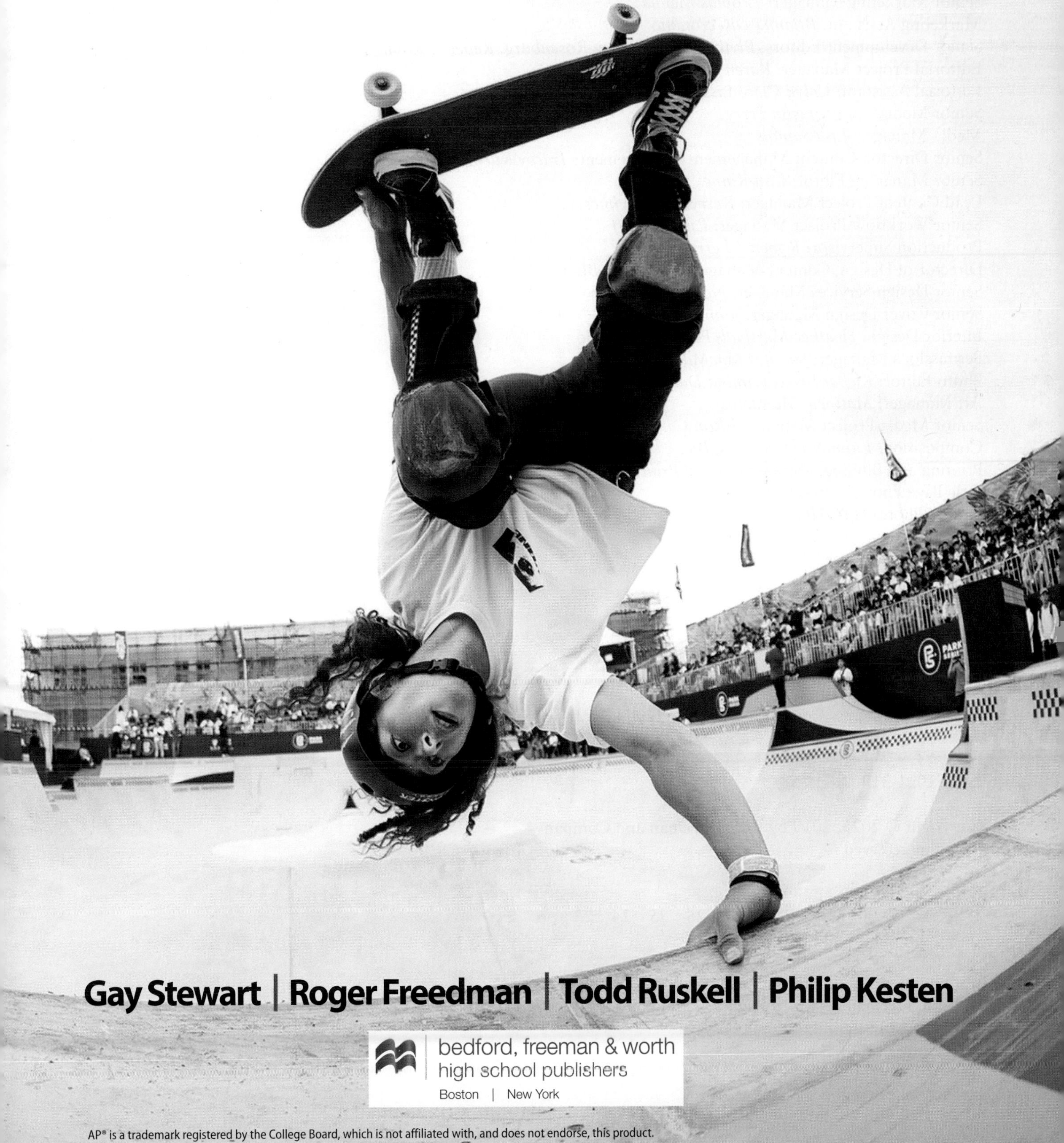

Gay Stewart | Roger Freedman | Todd Ruskell | Philip Kesten

bedford, freeman & worth
high school publishers
Boston | New York

Executive Vice President & General Manager: *Charles Linsmeier*
Vice President, Social Sciences & High School: *Shani Fisher*
Executive Program Director, High School: *Ann Heath*
Senior Executive Program Manager, High School STEM: *Yolanda Cossio*
Senior Marketing Manager: *Thomas Menna*
Marketing Assistant: *Brianna DiGeronimo*
Senior Development Editors: *Blythe Robbins, Meg Rosenburg, Rebecca Kohn*
Editorial Project Manager: *Karen Misler*
Editorial Assistant: *Calyn Clare Liss*
Senior Media Editor: *Justin Perry*
Media Manager: *Lisa Samols*
Senior Director, Content Management Enhancement: *Tracey Kuehn*
Senior Managing Editor: *Lisa Kinne*
Lead Content Project Manager: *Kerry O'Shaughnessy*
Senior Workflow Project Manager: *Paul Rohloff*
Production Supervisor: *Robert Cherry*
Director of Design, Content Management: *Diana Blume*
Senior Design Services Manager: *Natasha Wolfe*
Senior Cover Design Manager: *John Callahan*
Interior Design: *Heather Marshall, Periwinkle Studio*
Permissions Manager: *Jennifer MacMillan*
Photo Editor: *Richard Fox, Lumina Datamatics, Inc.*
Art Manager: *Matthew McAdams*
Senior Media Project Manager: *Elton Carter*
Composition: *Lumina Datamatics, Inc.*
Printing and Binding: *Transcontinental Printing*
Title Page Photo: *Anthony Acosta*
Banner Photo: *HYONGGAAA/Shutterstock*

Library of Congress Control Number: 2022944275

ISBN-13: 978-1-319-48621-1
ISBN-10: 1-319-48621-5

Printed in Canada
1 2 3 4 5 6 28 27 26 25 24 23

Bedford, Freeman & Worth High School Publishers
120 Broadway
25th Floor
New York, NY 10271
highschool.bfwpub.com/catalog

GAY

As always, for anything important I do in life, to John, my husband and the rock I stand on; to our awesome children, Bernadette and Katherine, who were proud of me for trying to make a difference in the world, even if it meant I made more mistakes as a mom. For this project, I must also include all my students and the AP® teachers (some are now the same people) who have taught me so much over the years. I couldn't have done it without all of your love and support.

ROGER

To the memory of S/Sgt Ann Kazmierczak Freedman, WAC, and Pvt. Richard Freedman, AUS.

TODD

To Susan and Allison, whose never-ending patience, love, and support made it possible.
And to my parents, from whom I learned so much—especially my father, who so effectively demonstrates what it means to be an effective teacher both in and out of the classroom.

PHIL

To my parents for instilling in me a love of learning, to my wife for her unconditional support, and to my children for letting their kooky dad infuse so much of their lives with science.

Brief Contents

*Enrichment chapter

Contents

Unit 10 Electric Force, Field, and Potential 778

Unit 11 DC Circuits 872

Unit 12 Magnetism and Electromagnetism 930

*Enrichment chapter

About the Authors

GAY STEWART received her PhD in physics from University of Illinois, Urbana-Champaign in 1994. She then accepted a faculty position at University of Arkansas (UA), where she focused on three interrelated issues: improving the introductory sequence to better prepare students to succeed in science and engineering programs; improving the preparation of physics majors for the variety of career options open to them; and the preparation of future faculty, for both high school and professoriate. The undergraduate program saw dramatic improvement, with a 10-fold increase in number of graduates. Dr. Stewart led UA's efforts as one of the first six primary program institutions in the Physics Teacher Education Coalition, PhysTEC, which now has more than 300 members. Dr. Stewart first received NSF support for her work in 1995. As a teaching assistant mentor, she developed a preparation program that grew into one of four sites for the NSF/AAPT "Shaping the Preparation of Future Science Faculty." She was co-PI of an NSF GK-12 project that placed fellows in middle school mathematics and science classrooms. The results were so favorable that helping math and science teachers to work together was a component of the $7.3M NSF-MSP (the College Ready in Mathematics and Physics Partnership). Through the Noyce Program she received support for future and master physics teachers. She chaired the College Board's Science Academic Advisory Committee, co-chaired the Advanced Placement® Physics Redesign Commission, responsible for AP® Physics 1 and 2, and the AP® Physics 2 Development Committee. In 2014, Dr. Stewart transitioned to West Virginia University, where she is Eberly Professor of STEM Education and professor of physics, and the founding director of the WVU Center for Excellence in STEM Education. The transdisciplinary Center works with faculty across STEM (Science, Technology, Engineering, and Mathematics) and related disciplines at WVU, partner programs, and the WV Department of Education to enhance STEM education and STEM education opportunities in West Virginia, grades K-20. She is former president of the American Association of Physics Teachers (AAPT) and former member of the Board of Directors, Council of Representatives, and the Committee on Education of the American Physical Society (APS). She is a Fellow of both the AAPT and the APS and received the Oersted Medal, the AAPT's highest honor, which "recognizes those who have had an outstanding, widespread, and lasting impact on the teaching of physics" in 2019.

Roger Freedman is an accomplished textbook author of such best-selling titles as *Universe* (W. H. Freeman), *Investigating Astronomy* (W. H. Freeman), and *University Physics* (Pearson). Dr. Freedman is a lecturer in physics at the University of California, Santa Barbara. He was an undergraduate at the University of California campuses in San Diego and Los Angeles, and did his doctoral research in theoretical nuclear physics at Stanford University. He came to UCSB in 1981 after 3 years of teaching and doing research at the University of Washington. At UCSB, Dr. Freedman has taught in both the Department of Physics and the College of Creative Studies, a branch of the university intended for highly gifted and motivated undergraduates. In recent years, he has helped to develop computer-based tools for learning introductory physics and astronomy and has been a pioneer in the use of classroom response systems and the "flipped" classroom model at UCSB. Roger holds a commercial pilot's license and was an early organizer of the San Diego Comic-Con, now the world's largest popular culture convention.

Todd Ruskell As a Teaching Professor of Physics at the Colorado School of Mines, Todd Ruskell focuses on teaching at the introductory level and continually develops more effective ways to help students learn. One method used in large enrollment introductory courses is Studio Physics. This collaborative, hands-on environment helps students develop better intuition about, and conceptual models of, physical phenomena through an active learning approach. Dr. Ruskell brings his experience in improving students' conceptual understanding to the text, as well as a strong liberal arts perspective. Dr. Ruskell's love of physics began with a BA in physics from Lawrence University in Appleton, Wisconsin. He went on to receive an MS and a PhD in optical sciences from the University of Arizona. He has received awards for teaching excellence, including Colorado School of Mines' Alumni Teaching Award.

Philip Kesten, Associate Professor of Physics at Santa Clara University, holds a BS in physics from the Massachusetts Institute of Technology and received his PhD in high-energy particle physics from the University of Michigan. Since joining the Santa Clara faculty in 1990, Dr. Kesten has also served as Chair of Physics, Faculty Director of the ATOM and da Vinci Residential Learning Communities, and Director of the Ricard Memorial Observatory. He has received awards for teaching excellence and curriculum innovation, was Santa Clara's Faculty Development Professor for 2004–2005, and was named the California Professor of the Year in 2005 by the Carnegie Foundation for the Advancement of Education. Dr. Kesten is co-founder of Docutek (a SirsiDynix Company), an Internet software company, and has served as the Senior Editor for *Modern Dad,* a newsstand magazine.

Acknowledgments

THANK YOU FROM GAY STEWART

Writing this book was an amazing undertaking and updating it has been a chance to think even more deeply about how to potentially help out some awesome physics teachers, better supporting them in teaching AP® Physics 1 and 2. None of this would have been possible without the love and support of my husband John, who tolerated a whole lot of late nights and busy weekends while doing his best to keep me healthy and sane. Our daughters Bernadette and Katherine continually served as moral support, letting me bounce ideas off of them and telling me how proud they were of my efforts.

Creating a textbook is a large undertaking that would not be possible without a dedicated, supportive team. I would like to thank Blythe Robbins for convincing me to take on this project and for her continued assistance in this edition, as well as Senior Executive Program Manager Yolanda Cossio who has remained deeply committed to the quality of the book throughout the process to make sure students get what they need to succeed in the AP® Physics 1 and 2 courses. I would also like to thank Developmental Editors Meg Rosenburg and Rebecca Kohn, Editorial Project Manager Karen Misler, and Lead Content Project Manager Kerry O'Shaughnessy for their patience and attention to detail, and for keeping the book on track and on time. Also a big thank you to Media Manager Lisa Samols, Media Editor Justin Perry, Editorial Assistant Calyn Clare Liss, and Permissions Manager Jennifer MacMillan. Of course, none of this would be possible without the support of Executive Program Director of High School Ann Heath, VP of Social Sciences & High School Shani Fisher, and Executive Vice President & General Manager Charles Linsmeier. Special thanks also go to our talented marketing team, Senior Marketing Manager Thomas Menna and Marketing Assistant Brianna DiGeronimo.

Friends and Family

Roger thanks his wife, Caroline Robillard, for her patience with the seemingly endless hours that went into preparing this textbook. I also thank my students at the University of California, Santa Barbara, for giving me the opportunity to test and refine new ideas for making physics more accessible.

Todd thanks his wife, Susan, and daughter, Allison, for their limitless patience and understanding with the countless hours spent working on this book. I also thank my parents who showed me how to live a balanced life.

Philip would like to acknowledge valuable and insightful conversations on physics and physics teaching with Richard Barber, John Birmingham, and J. Patrick Dishaw of Santa Clara University, and to offer these colleagues my gratitude. Finally, I offer my gratitude to my wife Kathy and my children Sam and Chloe for their unflagging support during the arduous process that led to the book you hold in your hands.

AP® Reviewers

We would also like to thank the many colleagues who carefully reviewed chapters for us. Their insightful comments significantly improved the book.

Claire Anton, *McLean High School*
Brian Boone, *Fairless High School*
Elizabeth Carpenter, *Chittenango High School*
Eleanor Dorso, *Brentwood High School*
Bart Frey, *Stark County High School*
Catherine Garland, *Uncommon Charter High School*
Marla Glover, *Rossville High School*
Frederick Heyler, *Iolani School*
Roxana Invernizzi, *Pembroke Pines Charter High School*
Leonard Itzkowitz, *Schechter School of Long Island*
Martin Kirby, *Retired, Hart High School*
Jay Kurima, *Texas Academy of Biomedical Sciences*
Martha Lietz, *Niles West High School*
Dan Marek, *Grapevine High School*
Elmarie Mortimer, *Trinity Preparatory School*
Travis Martin, *William G Enloe High School*
Timothy Nawrocki, *West Catholic High School*
Joel C. Palmer, *Ed.D.*
Barry Panas, *St. John's-Ravenscourt School*
Kim Quire, *Academy of the Holy Names High School*
Marc Reif, *Fayetteville High School*
Jack Replinger, *The Soulsville Charter School*
James Ringlein, *Lancaster Country Day School*
Mark Schweizer, *Winslow Township High School*
Joe Stieve, *Retired*
Tiffany Taylor, *Rogers Heritage High School*
Joshua Winter, *BASIS Independent Brooklyn*
Patricia Zober, *Goucher College*

College Reviewers

We would also like to thank the many colleagues who carefully reviewed chapters for us. Their insightful comments significantly improved our book.

Don Abernathy, *North Central Texas College*
Elise Adamson, *Wayland Baptist University*
Miah Muhammad Adel, *University of Arkansas at Pine Bluff*

Mikhail M. Agrest, *The Citadel*
Ricardo Alarcon, *Arizona State University*
Z. Altounian, *McGill University*
Abu Amin, *Riverland Community College*
Vasudeva Rao Aravind, *Clarion University of Pennsylvania*
Sanjeev Arora, *Fort Valley State University*
Llani Attygalle, *Bowling Green State University*
Yiyan Bai, *Houston Community College*
Michael Bates, *Moraine Valley Community College*
Luc Beaulieu, *Memorial University*
Jeff J. Bechtold, *Austin Community College*
E. C. Behrman, *Wichita State University*
Antonia Bennie-George, *Green River College*
David Bennum, *University of Nevada*
Satinder Bhagat, *University of Maryland*
Ken Bolland, *The Ohio State University*
Dan Boye, *Davidson College*
Matthew Joseph Bradley, *Santa Rosa Junior College*
Matteo Broccio, *University of Pittsburgh*
Jeff Bronson, *Blinn College*
Douglas Brumm, *Florida State College at Jacksonville*
Mark S. Bruno, *Gateway Community College*
Brian K. Bucklein, *Missouri Western State University*
Michaela Burkardt, *New Mexico State University*
Kris Byboth, *Blinn College*
Joel W. Cannon, *Washington & Jefferson College*
Kapila Clara Castoldi, *Oakland University*
Paola M. Cereghetti, *Lehigh University*
Hong Chen, *University of North Florida*
Zengjun Chen, *Tuskegee University*
Uma Choppali, *Dallas County Community College*
Todd Coleman, *Century College*
Daniel J. Costantino, *The Pennsylvania State University*
José D'Arruda, *University of North Carolina, Pembroke*
Tinanjan Datta, *Georgia Regents University*
Chad L. Davies, *Gordon College*
Adam Davis, *Wayne State College*
Brett DePaola, *Kansas State University*
Sharvil Desai, *The Ohio State University*
Eric Deyo, *Fort Hays State University*
Sandra Doty, *Ohio University*
James Dove, *Metro Community College*
Carl T. Drake, *Jackson State University*
Diana I. Driscoll, *Case Western Reserve University*
Rodney Dunning, *Longwood University*
Vernessa M. Edwards, *Alabama A & M University*
Davene Eyres, *North Seattle College*
Hasan Fakhruddin, *Ball State University*
William Falls, *Erie Community College*
Paul Fields, *Pima Community College*
Lewis Ford, *Texas A & M University*
J.A. Forrest, *University of Waterloo*
Scott Freedman, *Philadelphia Academy Charter High School*
Tim French, *Harvard University*
James Friedrichsen III, *Austin Community College*
Sambandamurthy Ganapathy, *SUNY Buffalo*
J. William Gary, *University of California, Riverside*
L. Gasparov, *University of North Florida*
Vladimir Gasparyan, *California State University, Bakersfield*
Brian Geislinger, *Gadsden State Community College*
Oommen George, *San Jacinto College*
Frank Gerlitz, *Washtenaw Community College*
Anindita Ghosh, *Suffolk County Community College*
Svetlana Gladycheva, *Towson University*
Romulus Godang, *University of South Alabama*
Alan I. Goldman, *Iowa State University*
Javier Gomez, *The Citadel*
Richard Goulding, *Memorial University of Newfoundland*
Morris C. Greenwood, *San Jacinto College Central*
Thomas P. Guella, *Worcester State University*
Alec Habig, *University of Minnesota, Duluth*
Edward Hamilton, *Gonzaga University*
Ania Harlick, *Memorial University of Newfoundland*
C. A. Haselwandter, *University of Southern California*
Erik Helgren, *California State University, East Bay*
Perry G. Hillburn, *Gannon University*
Zvonko Hlousek, *California State University, Long Beach*
Micky Holcomb, *West Virginia University*
Kevin M. Hope, *University of Montevallo*
J. Johanna Hopp, *University of Wisconsin, Stout*
Zdeslav Hrepic, *Columbus State University*
Leon Hsu, *University of Minnesota*
Olenka Hubickyj *Cabot, San Jose State University*
Patrick Huth, *Community College of Allegheny County*
Richard Ignace, *East Tennessee State University*
Elizabeth Jeffery, *James Madison University*
Matthew Jewell, *University of Wisconsin, Eau Claire*
Yong Joe, *Ball State University*
Darrin Eric Johnson, *University of Minnesota, Duluth*
David Kardelis, *Utah State University, College of Eastern Utah*
Wafaa Khattou, *Valencia College*
Agnes Kim, *Georgia State College*
Ju H. Kim, *University of North Dakota*
Seong-Gon Kim, *Mississippi State University*
Seth T. King, *University of Wisconsin, La Crosse*
Patrick Koehn, *Eastern Michigan University*
Kathleen Koenig, *University of Cincinnati*
Ameya S. Kolarkar, *Auburn University*

Olga Korotkova, *University of Miami*
Minjoon Kouh, *Drew University*
Maja Krcmar, *Grand Valley State University*
Elena Kuchina, *Thomas Nelson Community College*
Tatiana Krivosheev, *Clayton State University*
Michael Kruger, *University of Missouri, Kansas City*
Avishek Kumar, *Arizona State University*
Jessica C. Lair, *Eastern Kentucky University*
Josephine M. Lamela, *Middlesex County College*
Patrick M. Len, *Cuesta College*
Shelly R. Lesher, *University of Wisconsin, La Crosse*
Chunfei Li, *Clarion University of Pennsylvania*
Bruce W. Liby, *Manhattan College*
David M. Lind, *Florida State University*
Jeff Loats, *Metropolitan State College of Denver*
Susannah E. Lomant, *Georgia Perimeter College*
Jose Lozano, *Bradley University*
Jia Grace Lu, *University of Southern California*
Mark Lucas, *Ohio University*
Lianxi Ma, *Blinn College*
Aklilu Maasho, *Dyersburg State Community College*
Dan MacIsaac, *SUNY Buffalo State College*
Ron MacTaylor, *Salem State University*
Eric Mandell, *Bowling Green State University*
Maxim Marienko, *Hofstra University*
Mark Matlin, *Bryn Mawr College*
Dan Mattern, *Butler Community College*
Mark E. Mattson, *James Madison University*
Linda McDonald, *North Park University*
Francis Mensah, *Virginia Union University*
Jo Ann Merrell, *Saddleback College*
Michael R. Meyer, *Michigan Technological University*
Karie A. Meyers, *Pima Community College*
Andrew Meyertholen, *University of Redlands*
Victor Migenes, *Texas Southern University*
John H. Miller Jr., *University of Houston*
Ronald C. Miller, *University of Central Oklahoma*
Ronald Miller, *Texas State Technical College System*
Hector Mireles, *California State Polytechnic University, Pomona*
Ted Monchesky, *Dalhousie University*
Steven W. Moore, *California State University, Monterey Bay*
Mark Morgan-Tracy, *University of New Mexico*
Krishna Mukherjee, *Slippery Rock University of Pennsylvania*
Dennis Nemeschansky, *University of Southern California*
Rumiana Nikolova-Genov, *College of DuPage*
Moses Ntam, *Tuskegee University*
Terry F. O'Dwyer, *Nassau Community College*
John S. Ochab, *J. Sargeant Reynolds Community College*
Martin O. Okafor, *Georgia Perimeter College*
Umesh C. Pandey, *Central New Mexico Community College*
Archie Paulson, *Madison Area Technical College*
Gabriela Petculescu, *University of Louisiana at Lafayette*
Yuriy Pinelis, *University of Houston, Downtown*
Sulakshana Plumley, *Community College of Allegheny County*
Christian Poppeliers, *Augusta State University*
James R. Powell, *University of Texas, San Antonio*
Michael Pravica, *University of Nevada, Las Vegas*
Kenneth M. Purcell, *University of Southern Indiana*
Kenneth Ragan, *McGill University*
Milun Rakovic, *Grand Valley State University*
Jyothi Raman, *Oakland University*
Natarajan Ravi, *Spelman College*
Lawrence Rees, *Brigham Young University*
Lou Reinisch, *Jacksonville State University*
David Richardson, *Northwest Missouri State University*
Sandra J. Rhoades, *Kennesaw State University*
Carlos Roldan, *Central Piedmont Community College*
John Rollino, *Rutgers University, Newark*
Rodney Rossow, *Tarrant County College*
Larry Rowan, *University of North Carolina*
Jeffrey Sabby, *Southern Illinois University Edwardsville*
Arun Saha, *Albany State University*
Michael Sampogna, *Pima Community College*
Haiduke Sarafian, *Pennsylvania State University*
Tumer Sayman, *Eastern Michigan University*
Jim Scheidhauer, *DePaul University*
Katrin Schenk, *Randolph College*
Paul Schmidt, *Ball State University*
Morton Seitelman, *Farmingdale State College*
Surajit Sen, *SUNY Buffalo*
Saeed Shadfar, *Oklahoma City University*
Jerry Shakov, *Tulane University*
Douglas Sharman, *San Jose State University*
Ananda Shastri, *Minnesota State University, Moorhead*
Weidian Shen, *Eastern Michigan University*
Jason Shulman, *Richard Stockton College of New Jersey*
Michael J. Shumila, *Mercer County Community College*
Marllin L. Simon, *Auburn University*
R. Seth Smith, *Francis Marion University*
Stanley J. Sobolewski, *Indiana University of Pennsylvania*
Frank Somer, *Columbia College*
Chad Sosolik, *Clemson University*
Brian Steinkamp, *University of Southern Indiana*

Narasimhan Sujatha, *Wake Technical Community College*
Maxim Sukharev, *Arizona State University*
Erin C. Sutherland, *Kennesaw State University*
James H. Taylor, *University of Central Missouri*
Richard Taylor, *University of Oregon*
Sarah F. Tebbens, *Wright State University*
Fiorella Terenzi, *Florida International University*
E. Tetteh-Lartey, *Blinn Community College*
Gregory B. Thompson, *Adrian College*
Marshall Thomsen, *Eastern Michigan University*
Dmitri Tsybychev, *Stony Brook University*
Som Tyagi, *Drexel University*
Vijayalakshmi Varadarajan, *Des Moines Area Community College*
John Vasut, *Baylor University*
Dimitrios Vavylonis, *Lehigh University*
Kendra L. Wallis, *Eastfield College*
Laura Weinkauf, *Jacksonville State University*
Laura Whitlock, *Georgia Perimeter College*
Heather M. Whitney, *Wheaton College*
Pushpa Wijesinghe, *Arizona State University*
Fengyuan Yang, *The Ohio State University*
Capp Yess, *Morehead State University*
Garett Yoder, *Eastern Kentucky University*
Chadwick Young, *Nicholls State University*
Jiang Yu, *Fitchburg State University*
Yifu Zhu, *Florida International University*
Raymond L. Zich, *Illinois State University*
Ulrich Zurcher, *Cleveland State University*

To the Student
A Note from Gay Stewart

As you begin your study of AP® Physics 1 and 2, I want to welcome you to the subject that I love. Physics is a study of how the universe works. Yes, you will learn how to solve big problems and get good grades on exams, but you will also learn things that will help you in your daily life, such as how to be a safer driver! Physics also helps lay the foundation for successful advanced study in the other sciences. While much of this material was worked out a long time ago, there is still ongoing research related to even these basic ideas.

This book was created with you in mind. If you read to prepare for class, pay attention in class, keep up with your homework, and engage in laboratory work with your classmates, you will build the knowledge and skills needed to be successful on the AP® Physics exams. Research indicates that if you learn by engaging in active problem solving and discussion, as scientists do, you will learn better, so active participation is important.

The AP® Physics 1 and 2 courses took years to design and refine, and they represent a unique opportunity to delve into the world of physics to gain a coherent, deeper level of understanding by asking questions, doing experiments, and learning more about the world around you.

This textbook is different from others you might have seen because not only is it a college-level text, but it also offers you support and practice in preparing for the AP® Physics exams. This book models essential best practices to help you become a competent conceptual learner and problem solver, while it allows you to apply this knowledge to actual AP® practice problems and exams throughout the text. You will be able to see the direct connection between the content you're learning daily to how you will be assessed on the AP® Physics exams. This will help you arrive for your AP® Physics exam confident and ready to take the test!

Here are some ways to use the book to your benefit.

Read Carefully

Read *before* class, *not* after. Years ago, I switched from conducting physics research to studying how people learn physics, and something that our research has supported is that reading *before* class sets you up for success. This book introduces the concepts you need to comprehend to be successful on the AP® Physics exams, and it also includes a large number of examples with clear explanations that help you learn how to logically solve problems. This process will allow you to develop the habit of reflecting on your answers, which is a necessary skill for success in the course and on the exam. Reading the book with a pencil in hand, taking notes on important ideas, and writing down any questions you have, particularly while doing the worked examples, will help you develop the deeper understanding that the AP® Physics exams are designed to measure.

When your teacher works a problem, it may look like it all makes sense, but then when you try to do a similar problem, you may realize there was some step you did not quite understand. Even though I have tried to anticipate many of them, it is harder to ask a book questions! If you read the book first, you can jot down questions about anything you don't understand; then, when your teacher discusses the content, you can see if your questions are answered (you will pay more attention, since you have questions!) or ask your teacher to explain anything that you still don't understand. You'll be amazed at how much more you'll get out of class if you read first. This is a major success strategy for learning in general.

Getting a 5 on the exam requires deep understanding of the concepts. If something doesn't seem easy the first time you try it that does not mean you are not good at physics, it just means you need to practice. This is true for almost any interesting activity, such as solving physics problems or competing in the Olympics as a skate boarder! Practice is the way people learn. At the end of each example, you'll find NOW WORK cross-reference problems in the end-of-section problems. Doing these suggested problems after reading the examples, and solving the rest of the practice problems in the chapter, will allow you to gain the skills and habits you need for success.

In the back of the book, you'll find answers to the odd questions. We have created mini-solutions for these odd problems so you can really see how to do them, not just check your answer. If these don't get you over wherever you are stuck, be sure to seek help quickly since physics concepts build upon each other.

About the AP® Physics Exams

It's important to understand how the AP® Physics exams are set up, so that you know how best to study for them. The AP® exams require you to make scientific arguments, justify solutions, decide what will happen when something changes, and design experiments as well as do the appropriate mathematics. You will find many opportunities to practice solving AP® questions throughout the text. There are AP® Practice Problems at the end of every chapter, and three complete AP® Practice Exams. The first exam appears after Chapter 9 and tests the content of Chapters 1–9. The second exam appears after Chapter 13 and tests all the content covered on the AP® Physics 1 Exam. The third exam appears after Chapter 25 and covers all the content on the AP® Physics 2 Exam. It's important to practice taking the AP®-style questions so that on the day of the test, you feel comfortable and ready to take the exam. If you take advantage of the features, study consistently, and practice solving the problems, you will be able to develop the understanding that leads to a successful performance on the AP® exams. But, more importantly, you will be able to use the important problem-solving skills that you will learn along the way in your other studies and in your daily life! Please do not hesitate to contact me with any questions, suggestions, or comments that might improve this text. I look forward to hearing from you. I hope that you enjoy learning physics and do well not only in AP® Physics 1 and AP® Physics 2, but in all your studies.

Getting the Most from This Book

Learn the concepts and realize success on the AP® Physics exams by knowing how to use your text effectively.

YOU WILL LEARN TO:

- Recognize the differences among one-, two-, and three-dimensional motion.
- Describe the properties of a vector and how to find the sum or difference of two vectors.
- Describe the motion of an object in two dimensions using quantities such as displacement, distance, velocity, speed, and acceleration, in an appropriate way for a chosen coordinate system.
- Explain how the displacement, velocity, and acceleration of an object or the center of mass of a system are described in terms of vectors, and can be modeled mathematically by the kinematic equations when acceleration is constant.
- Recognize that the velocity observed for an object depends on the reference frame in which it is observed, and be able to describe the reference frames of a given observer from a description of the physical situation.
- Be able to convert the velocity of an object relative to one reference frame into the velocity of that same object observed from a different reference frame.
- Identify the key features of projectile motion and how to interpret this kind of motion.
- Solve problems involving projectile motion.

Each chapter begins with a **You Will Learn To** box that clearly outlines the chapter learning goals.

The **You Will Learn To** boxes at the beginning of the chapter directly correspond to the **What Did You Learn?** box at the end of the chapter.

WHAT DID YOU LEARN?

Chapter learning goals	Section(s)	Related example(s)	Prep for the AP Exam: Relevant section review exercises
Recognize the differences among one-, two-, and three-dimensional motion.	3-1		1
	3-2		3
	3-3	3-1	
Describe the properties of a vector and how to find the sum or difference of two vectors.	3-2		4
	3-3	3-1, 3-2	3, 4, 5
Describe the motion of an object in two dimensions using quantities such as displacement, distance, velocity, speed, and acceleration, in an appropriate way for a chosen coordinate system.	3-2		2
	3-3	3-2	6
	3-4	3-3	4, 5
	3-5		1, 2, 3, 4, 5
Explain how the displacement, velocity, and acceleration of an object or the center of mass of a system are described in terms of vectors, and can be modeled mathematically by the kinematic equations when acceleration is constant.	3-4	3-3	1, 9
	3-5		
	3-6	3-4	
Recognize that the velocity observed for an object depends on the reference frame in which it is observed, and be able to describe the reference frames of a given observer from a description of the physical situation.	3-4	3-3	1, 2, 4, 5
Be able to convert the velocity of an object relative to one reference frame into the velocity of that same object observed from a different reference frame.	3-4	3-3	1, 2, 4, 5
Identify the key features of projectile motion and how to interpret this kind of motion.	3-5		1, 3
	3-6	3-4, 3-5	
Solve problems involving projectile motion.	3-6	3-5, 3-6, 3-7	1, 2, 9

Each chapter ends with a **What Did You Learn?** box. This is a great study tool that connects the chapter learning goals to each section, example, and related section review exercises.

Learn with Less Friction

The book's features will help you grasp the big concepts in AP® Physics 1 and AP® Physics 2!

WATCH OUT !

Projectile motion means constant vertical acceleration.

Just as for vertical free fall, it's important to remember that the vertical, y, component of acceleration is the same ($a_y = -g$, for positive y defined upward) at every point in the motion of the projectile.

Watch Out! boxes draw attention to important ideas that you need to remember as you read through the chapters.

NEED TO REVIEW?

To review a coordinate system, turn to Chapter 2, Section 2-2.

Need to Review? boxes provide on-the-spot notes that explain where to find concepts to review or to study.

Equation in Words boxes translate complex equations into everyday language that everyone can understand.

Displacement (change in position) of the object during the time interval from time t_1 to a later time t_2

Position of the object at later time t_2

$$\Delta\vec{r} = \vec{r}_2 - \vec{r}_1 \quad (3\text{-}6)$$

Position of the object at earlier time t_1

To calculate the displacement, **subtract** the earlier position from the later position (just as for motion in a straight line).

EQUATION IN WORDS
Displacement equals change in position during a time interval

The most general kind of motion is three dimensional. In many cases, however, the motion is in a plane and so is two dimensional. In figures (b) and (c) any side-to-side motion can be neglected; all motion is up and down and left and right.

This aircraft follows a complicated, three-dimensional path.

(a)

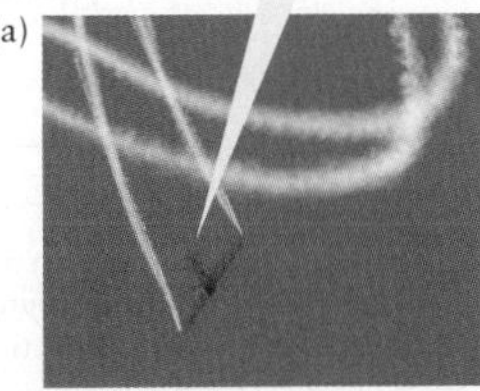

Design Pics Inc/Alamy

This motorcycle's path through the air is two dimensional (in a plane): It moves left to right as well as up and down.

(b)

Stephen Barnes/Science/Alamy

Each car on this Ferris wheel moves in the plane of the wheel, so its path is two dimensional.

(c)

Holly Kuchera/Getty Images

Figure 3-1 Motion in three and two dimensions: three examples of motion These figures depict motion in two or three dimensions. In this course you will be asked to solve problems in only two dimensions, although you should be able to describe what you would need to do in a three-dimensional motion situation.

Art designed to teach uses word bubbles to break down key physics concepts in a visual narrative.

Powerful AP® Practice

Be confident! The study and review features in the book are specifically designed to help you prepare for the AP® exam.

 **Exam Tip**

On the AP® Physics 2 Exam, you will only be asked to calculate the magnetic force on a charged object when the field and velocity are at angles of 0, 90, or 180 degrees. You can be asked for qualitative descriptions for the force or subsequent behavior of the object for other angles.

 **Exam Tip**

On the AP® Physics 1 exam, air resistance is assumed to be negligible unless stated otherwise. If air resistance were not negligible, the acceleration's magnitude in the vertical direction would be smaller than the acceleration due to gravity, and there would also be an acceleration in the horizontal direction, opposing the motion of the object. We will consider the effects of air resistance in Chapter 5.

AP **Exam Tips** offer hints and tips on how to succeed on both AP® Physics exams.

An AP **Group Work question** appears in the end-of-chapter content, providing a question designed to be solved as a group work activity.

Prep for the AP Exam

AP Group Work

Directions: The problem below is designed to be done as group work in class.

Data were collected on the motion of a projectile launched with an initial speed of 70.7 m/s at an angle of 45.0°. The data (labeled as "observed") were not in agreement with the trajectory that was predicted in the figure. The prediction was based on the equations of motion for a projectile where air resistance is neglected.

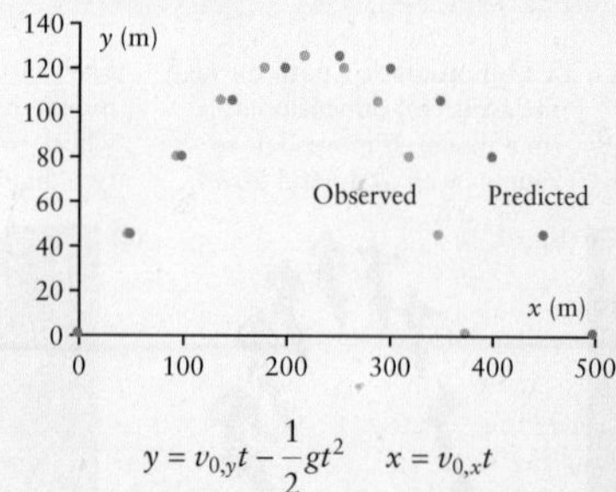

$$y = v_{0,y}t - \frac{1}{2}gt^2 \qquad x = v_{0,x}t$$

(a) Based on the evidence provided by these data, justify the claim that the observed trajectory does not exhibit an acceleration in the y direction different from the acceleration due to gravity.

(b) Based on the evidence provided by these data, justify the claim that the observed trajectory does exhibit an acceleration in the x direction.

The numerical values of the observed and predicted trajectories are shown in the table.

$x_{predicted}$	$y_{predicted}$	$y_{observed}$	$y_{predicted}$
0	0	0	0
50	45	49	45
100	80	96	80
150	105	140	105
200	120	182	120
250	125	222	125
300	120	260	120
350	105	295	105
400	80	328	80
450	45	359	45
500	0	388	0

(c) Design a strategy for the analysis of these data to determine the magnitude of the acceleration in the x direction.

(d) Apply your strategy to calculate the acceleration in the x direction.

Every chapter section ends with The Takeaway, which provides key opportunities to practice the problem-solving skills presented throughout the chapter in a unique three-tiered approach.

AP® **Building Blocks** (tier one) test students' knowledge of concepts by putting them into practice. Easy cross-references to examples in the chapter provide extra support when solving problems.

Prep for the AP® Exam

AP® Building Blocks

1. A bumblebee has a mass of about 0.250 g. Calculate its kinetic energy when it is moving with a speed of 10.0 m/s.

EX 7-3 2. A man rides a scooter on a level road. The interaction between the scooter and the road causes the road to exert a constant net force, F_1, on the scooter in the forward direction over a distance, d. The combined mass of the man and his scooter is m. Suddenly, the wind picks up and air resistance pushes against him in the direction opposite his motion over this same distance with a constant force F_2.
   (a) If he is initially moving at a speed v_i on a level road, how will you determine the man's speed after the scooter has moved a distance d?
   (b) How will you determine whether the speed of the man will increase or decrease?
   (c) Use your methods to predict the speed of the man after moving a distance d given these values: $F_1 = 1200$ N, $F_2 = 800$ N, $m = 90$ kg, $v_i = 5$ m/s, and $d = 20$ m.

EX 7-3 3. A small truck has a mass of 2100 kg. What is the work required to decrease the speed of the vehicle from 22.0 to 12.0 m/s on a level road?

AP® Skill Builders

4. A 1000-kg car moves along a straight, level road. Initially the car's velocity is 60 mph due east and later the car's velocity is 60 mph due west.
   (a) Create a mathematical model of the change in kinetic energy as a function of time, $\Delta K(t)$, of the car as time passes. Assume that the acceleration is constant so that the change in velocity is a linear function of time.
   (b) Construct a diagram of the function $\Delta K(t)$ you have created for the time interval over which the velocity changes from 60 mph due east to 60 mph due west.

5. Explain how the kinematic equation, $v_f^2 = v_i^2 + 2a(x_f - x_i)$; Newton's second law, $\sum F_{ext,x} = ma_x$; and the definition of work lead to the work-energy theorem for an object.

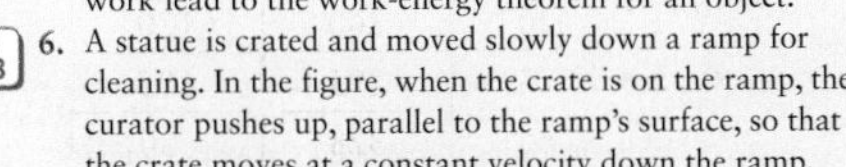

EX 7-3 6. A statue is crated and moved slowly down a ramp for cleaning. In the figure, when the crate is on the ramp, the curator pushes up, parallel to the ramp's surface, so that the crate moves at a constant velocity down the ramp.

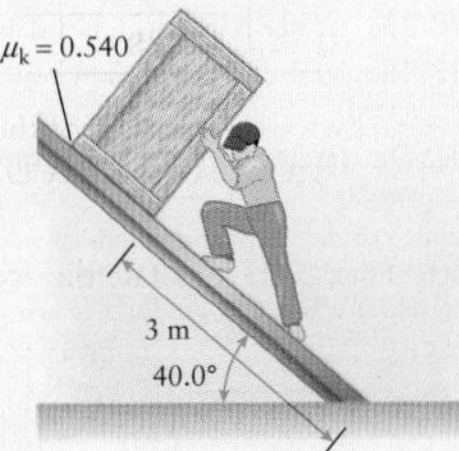

   (a) Construct a free-body diagram of the crate, in which the magnitudes of the forces are consistent with the crate having a constant velocity as it moves down the ramp.
   (b) Use the free-body diagram representation, by identifying the forces that do work on the crate as it slides down the ramp, to predict the sign of the work done on the crate by those forces.
   (c) Construct mathematical representations of each contribution to the total work done on the crate using variables defined in your free-body diagram, and needed given quantities and constants.
   (d) Calculate each contribution to the total work if the statue slides 3.00 m down the ramp with no acceleration and the coefficient of kinetic friction between the crate and the ramp is 0.540. The masses of the statue and the crate are each 150 kg. The ramp is inclined at 40.0°.
   (e) Explain why the sum of work done by the forces exerted on the crate is zero.

AP® Skills in Action

7. A horizontal snow surface with a length Δx at the base of a slope allows a sled and rider to come to a stop after descending the slope. Beyond the length Δx is a sheer cliff. As the Sun sets and the temperature falls the coefficient of kinetic friction is reduced.
   (a) Explain why a reduction in the coefficient of kinetic friction is a cause for concern.
   (b) Express the minimum coefficient of kinetic friction in terms of the velocity of the sled and rider at the base of the slope and the length Δx.

8. A worker exerts a force, F, on a box with mass m initially at rest on a level floor. The force exerted by the worker is directed downward at angle θ with the floor. The floor exerts a friction force on the box.
   (a) The box does not move. Explain why changing the angle that the external force makes with the floor might get the box moving.
   (b) The strategy in part (a) works and the box starts moving. Express the kinetic energy of the box in terms of the force exerted by the worker, F; the mass, m; the coefficient of kinetic friction, μ_k; and the displacement of the box, Δx.
   (c) If the worker continues to push on the box at the same downward angle θ, express how he should reduce the force he exerts on the box, relative to the initial force, $F/F_{initial}$, to displace the box with the least amount of work.

AP® **Skills in Action** (tier three) problems provide a scaffolded opportunity to practice valuable reasoning skills vital to success on the AP® Physics exams.

AP® **Skill Builders** (tier two) include problems that guide students toward developing the reasoning skills they need to be successful physics students.

The Practice Problems You Want

AP® PRACTICE PROBLEMS

Prep for the AP® Exam

Multiple-Choice Questions

Directions: The following questions have a single correct answer.

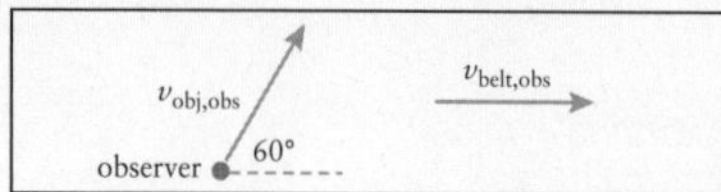

1. As shown in the figure, a conveyer belt is moving to the right at a speed v. A stationary observer sees an object that is moving with the belt, but in addition is sliding relative to the belt, moving away from the observer at the same speed v but at an angle of 60° with respect to the edge of the belt. What is the speed of the object relative to the belt?
   (A) $0.5\,v$ (B) v
   (C) $\frac{\sqrt{3}}{2}v$ (D) $\left(1+\frac{\sqrt{3}}{2}\right)v$
2. A ball rolls off the edge of a tall horizontal ledg a speed v_0. Assuming it does not strike the grou beforehand, at what time after it leaves the surf its velocity be in a direction 45° below the horiz
   (A) $\frac{v_0}{g}$ (B) $\frac{2v_0}{g}$
   (C) $\frac{v_0}{2g}$ (D) $\frac{\sqrt{2}v_0}{2g}$
3. An object is accelerated horizontally by a sm propeller. As the object leaves the supporting face it has an initial horizontal velocity v_0 = (has a constant horizontal acceleration a = +0.
   Taking the origin as the point of departure an to the right and +y upward, the path of the ol it falls will be described by which of the follov
   (A) $y = -\frac{1}{2}x$ (B) $y = -2x$
   (C) $y = -\frac{1}{2}x^2$ (D) $y = -2x^2$
4. A student is on a platform that is moving in th x direction with a constant speed v_0. The stude tosses a ball straight up, providing an addition tial velocity in the y direction (also of magnitu Relative to the ground, when the ball reaches i greatest height its speed will be what fraction o speed as it left the student's hand?
   (A) $\frac{1}{4}$ (B) $\frac{1}{2}$
   (C) $\frac{3}{4}$ (D) $\frac{\sqrt{2}}{2}$

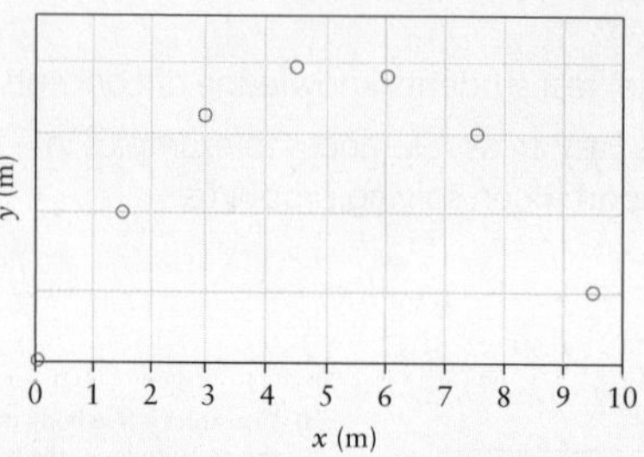

5. According to a laboratory report the positions of a projectile were measured at equal time intervals and recorded on the graph of the x and y positions as shown. The expected uncertainty in measurement is

AP® Practice Problems at the end of each chapter provide practice solving the types of problems found on the AP® Physics exams.

Black and white art mimics the type of art that appears on the AP® exams.

AP® Physics 1 Practice Exam 1

Prep for the AP® Exam

Multiple-Choice Questions

Directions: In the following questions, select the best answer choice from the four options listed.

Questions 1–2 refer to the following diagram.

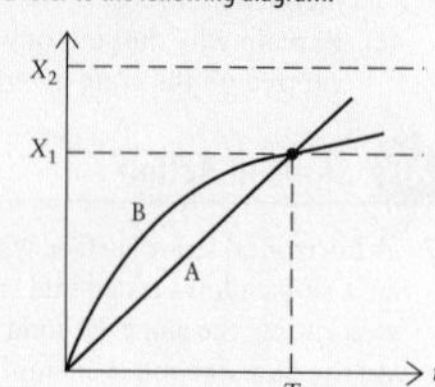

1. Two trains, A and B, begin moving on parallel tracks from the same starting position at a train station. The graph provided shows their respective displacements from the starting point as functions of time. Which of the following is true at time T?
   (A) Train A is moving faster than train B.
   (B) Train B is moving faster than train A.
   (C) Trains A and B are moving at the same speed.
   (D) Train A is ahead of train B.
2. If the trains continue in the same motion as shown in the graph, which of the following statements is true?
   (A) Train B moves faster than train A at all points on the graph.
   (B) Train A moves faster than train B at all points on the graph.
   (C) Train A will reach the position X_2 before train B.
   (D) Train B will reach the position X_2 before train A.
3. Which of the following graphs of velocity as a function of time represents the motion of a ball that is thrown vertically upward and moves under the influence of gravity until it returns to its initial position?

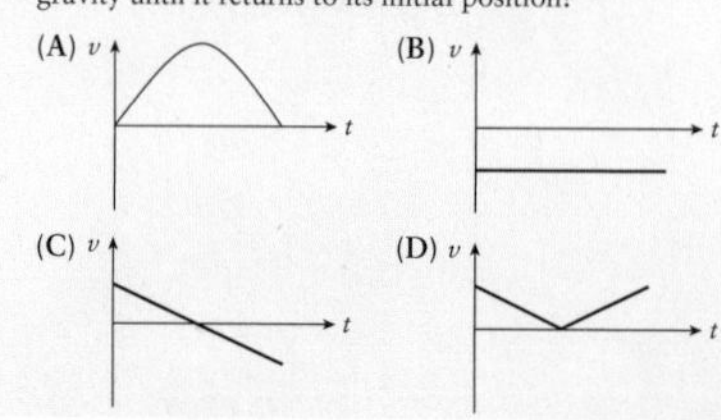

4. A projectile is launched at a speed of v_0 and at an angle θ above the horizontal. Which of the following are the correct values of the projectile's speed and the magnitude of its acceleration when it is at the highest point in its path?

	Speed	Acceleration
(A)	0	0
(B)	$v_o\cos(\theta)$	0
(C)	0	9.8 m/s^2
(D)	$v_o\cos(\theta)$	9.8 m/s^2

5. A block of mass m is at rest on a rough inclined plane. Which of the following diagrams best represents the correct directions for the normal force ($\vec{F}_n$), weight ($\vec{F}_g$), and friction force ($\vec{F}_f$) being exerted on the block?

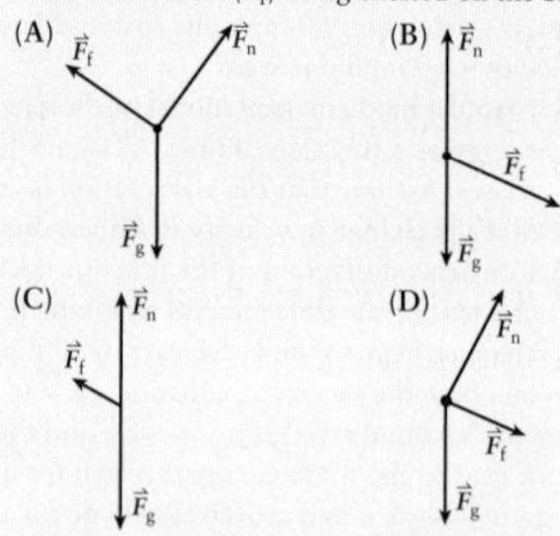

6. A person of weight 800 N is riding in an elevator while standing on a scale. The scale reading is 700 N. Which of the following is true of the elevator's motion?
   (A) The elevator is moving upward at constant speed.
   (B) The elevator is moving upward at increasing speed.
   (C) The elevator is moving downward at constant speed.
   (D) The elevator is moving downward at increasing speed.
7. Which of the following graphs of velocity as a function of time could represent the motion of an object on which a constant net force is being exerted?

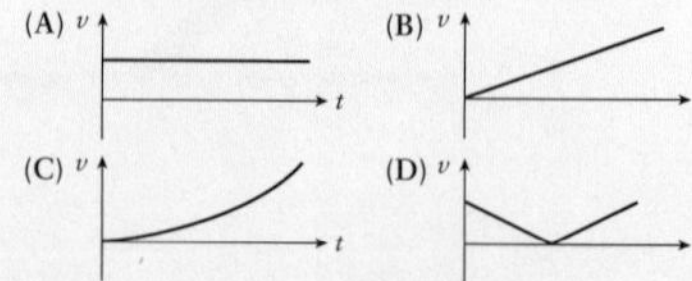

Full-length AP® Practice Exams appear after Chapters 9, 13, and 25 to further help prepare for test day.

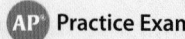

At the end of every chapter you will find a cumulative review of the material. Answers to odd problems in the back of the book are mini-solutions designed to clarify problem-solving steps.

Chapter 3 Review Problems

1. During the motion of a projectile, which of the following quantities are constant during the flight: x, y, v_x, v_y, a_x, a_y? (Neglect any effects due to air resistance.)
2. For a given fixed launch speed, at what angle should you launch a projectile to achieve
   (a) the longest range,
   (b) the longest time of flight, and
   (c) the greatest height? (Neglect any effects due to air resistance.)
3. Evaluate your ability to calculate the orientation, magnitude, and components of a vector.
   (a) Find the x and y components of a velocity with a magnitude of 34.00 m/s at 210.0°.
   (b) A rocket is launched with a speed of 1200.0 m/s at an angle measured relative to the ground of 43.0°. What are the x and y components of the launch velocity?
   (c) From the x and y components of a displacement, $R_y = 120$ m and $R_x = 345$ m, find the magnitude of the vector.
   (d) From the x and y components of a displacement, $R_x = 345$ m and $R_y = 120$ m, find the direction of the vector.
   (e) Find the x and y components of a velocity with a magnitude of 15.0 m/s at 12.0°.

Review Problems at the end of every chapter provide a focused practice of chapter concepts.

Answers to Odd Problems are included in the back of the book. These short solutions highlight key steps in the problem-solving process so that it's easy to see how to correctly solve the problem.

Chapter 3 Review Problems

1. During the motion of a projectile, only v_x, $a_x = 0$, and $a_y = -g$ (if up is positsive) are constant. The position (x, y) of the projectile is constantly changing while it is moving. Acceleration in the y direction is nonzero, which means v_y also changes.
3. (a) $v_y = (34.00 \text{ m/s}) \sin 210.0° = -17.00 \text{ m/s}$
   $v_x = (34.00 \text{ m/s}) \cos 210.0° = -29.44 \text{ m/s}$
   (b) $v_x = (1200.0 \text{ m/s}) \cos(43.0°) = 878 \text{ m/s}$
   $v_y = (1200.0 \text{ m/s}) \sin(43.0°) = 818 \text{ m/s}$
   (c) $R = \sqrt{R_x^2 + R_y^2} = \sqrt{120^2 \text{ m}^2 + 345^2 \text{ m}^2} = 365 \text{ m}$
   (d) $\alpha = \tan^{-1}\left(\dfrac{120 \text{ m}}{345 \text{ m}}\right) = 19.2°$
   (e) $v_x = (15.0 \text{ m/s}) \cos(12.0°) = 14.7 \text{ m/s}$
   $v_y = (15.0 \text{ m/s}) \sin(12.0°) = 3.12 \text{ m/s}$

Studying Made Easier

Look for these useful features when studying to help foster your reasoning and analytical skills, which are key for success in the course!

THE TAKEAWAY for Section 3-4

✔ If an object moves during a time interval, its displacement $\Delta\vec{r}$ is a vector that points from its position at the beginning of the interval to its position at the end of the interval.

✔ The velocity $\vec{v}$ of an object is a vector. It equals the displacement $\Delta\vec{r}$ for a very short time interval divided by the duration of the interval. The velocity always points along the object's trajectory.

✔ The reference frame chosen determines the direction and the magnitude of quantities measured by an observer in that reference frame.

✔ The observed velocity of an object results from the combination of the object's velocity and the velocity of the observer's reference frame.

The Takeaway summary highlights important information needed to "take away" from each chapter section to ensure understanding of the content.

Chapter 3 Review

Key Terms

All the Key Terms can be found in the Glossary/Glosario on page G1 in the back of the book.

component 95
component method 95
motion in a plane 103
parabola 117
projectile 114
projectile motion 114
reference frame 103
relative velocity 105
trajectory 103
vector addition 91
vector difference 92

Chapter Summary

Topic	Equation or Figure
Vectors and scalars: A vector quantity such as displacement or velocity has both a magnitude and a direction. By contrast, a scalar quantity such as temperature or time has no direction. We write a vector in boldface italic with an arrow above it. The magnitude of a vector $\vec{A}$ is written as A.	(Figure 3-2d)

(d)
Vectors $\vec{H}$ and $\vec{J}$ have different magnitudes but the same direction.
Vectors $\vec{J}$ and $\vec{K}$ have the same magnitude but different directions.
$\vec{H}$
$\vec{J}$
$\vec{K}$
Vectors $\vec{H}$ and $\vec{K}$ have different magnitudes and different directions.
A vector has only two attributes: magnitude and direction.

Key Terms/Chapter Summary synthesize ideas after reading or class time. This is a valuable study tool.

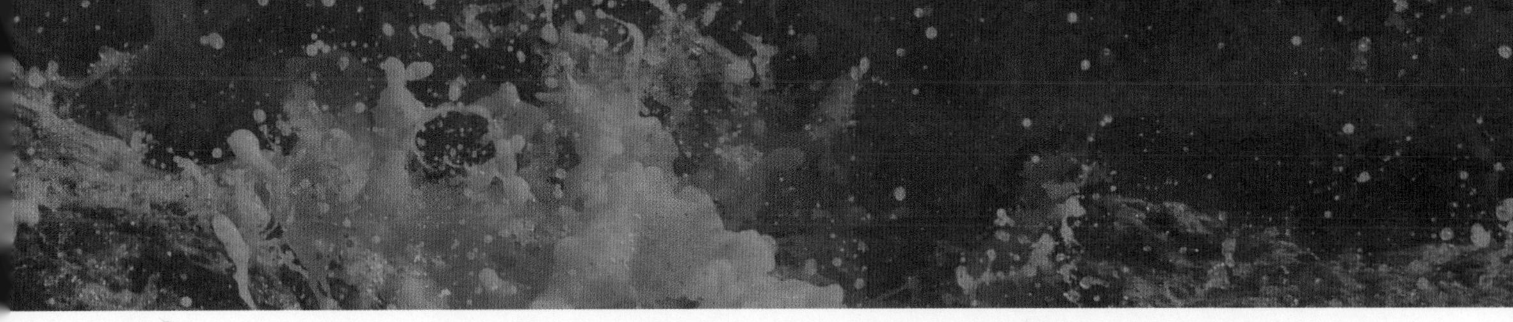

Examples mirror the approach scientists take to solve problems by developing reasoning and analysis skills.

Set Up. The first step in each problem is to determine an overall approach and to gather the necessary pieces of information needed to solve it. These might include sketches, equations related to the physics, and concepts.

Solve. Rather than simply summarizing the mathematical manipulations required to move from first principles to the final answer, the authors show many intermediate steps in working out solutions to the sample problems, highlighting a crucial part of the problem-solving process.

Reflect. An important part of the process of solving a problem is to reflect on the meaning, implications, and validity of the answer. Is it physically reasonable? Do the units make sense? Is there a deeper or wider understanding that can be drawn from the result? The authors address these and related questions when appropriate. We will, in some examples, *Extend* the example, to provide connections to other sorts of problems, or to show how the concepts used could be taken further.

EXAMPLE 2-7 Motion with Constant Acceleration II: Which Solution Is Correct?

Consider again the jet takeoff described in Example 2-6. There is a marker alongside the runway 328 m from the point where the jet starts to move. How long after beginning its takeoff roll does the jet pass the marker?

Set Up

We use the same x axis and choice of origin as in Example 2-6. Again we have $x_0 = 0$ and $v_{0x} = 0$, and we know the value of a_x from our work in Example 2-6. Our goal is to calculate the time t at which $x = 328$ m. We don't know the jet's velocity at this point, nor are we asked for its value, so we put a red "X" next to v_x in the table of knowns and unknowns.

Table 2-3 tells us that the equation we need to find t from the given information (x, x_0, v_{0x} and a_x) is Equation 2-9.

Position, acceleration, and time for constant acceleration only:

$$x = x_0 + v_{0x}t + \frac{1}{2}a_xt^2 \quad (2\text{-}9)$$

Quantity	Know/Don't Know
t	?
x_0	0
x	328 m
v_{0x}	0
v_x	X
a_x	1.64 m/s^2

Solve

First let's substitute the known values $x = 328$ m, $x_0 = 0$, $v_{0x} = 0$, and $a_x = 1.64$ m/s^2 into Equation 2-9.

Substitute $x_0 = 0$ and $v_{0x} = 0$, then simplify:

$$x = 0 + (0)t + \frac{1}{2}a_xt^2 = \frac{1}{2}a_xt^2$$

Then substitute $x = 328$ m, $a_x = 1.64$ m/s^2:

$$328\text{ m} = \frac{1}{2}(1.64\text{ m/s}^2)t^2$$

Rearrange the equation to solve for t.

$$t^2 = \frac{2(328\text{ m})}{1.64\text{ m/s}^2} = 4.00 \times 10^2\text{ s}^2$$

$$t = \pm\sqrt{4.00 \times 10^2\text{ s}^2} = \pm 20.0\text{ s}$$

The $\pm$ symbol in front of 20.0 s means that the square of either $t = +20.0$ s or $t = -20.0$ s is equal to $t^2 = 4.00 \times 10^2$ s^2. In this case we clearly want a positive value of time: The jet must reach the marker at $x = 328$ m some time *after* beginning its takeoff roll at $t = 0$. So the correct answer is $t = 20.0$ s.

Choose the value that corresponds to a time after the jet started moving:

$$t = 20.0\text{ s}$$

Reflect

In Example 2-6 we found that the jet takes 38.2 s to travel 1.20 km. Our answer says that it takes less time (20.0 s) to travel a shorter distance (328 m), which is reasonable.

What may not seem reasonable is that we got *two* answers to the problem, only one of which made sense. To see the meaning of the other answer, $t = -20.0$ s, remember that we assumed that the jet has *constant* acceleration $a_x = 1.64$ m/s^2. As far as the equations are concerned, the jet has *always* had this acceleration, even before $t = 0$! According to these equations, before $t = 0$ the jet had the same positive value of a_x but had a *negative* velocity v_x and was moving *backward* along the runway in the negative x direction. The answer $t = -20.0$ s refers to the time during this fictitious motion when the backward-moving jet passed through the point $x = 328$ m.

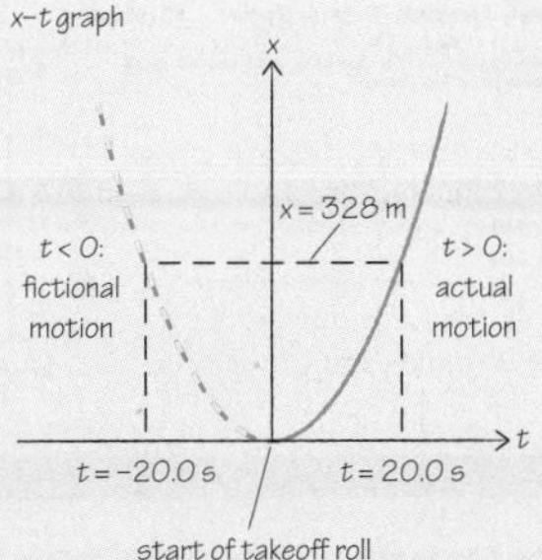

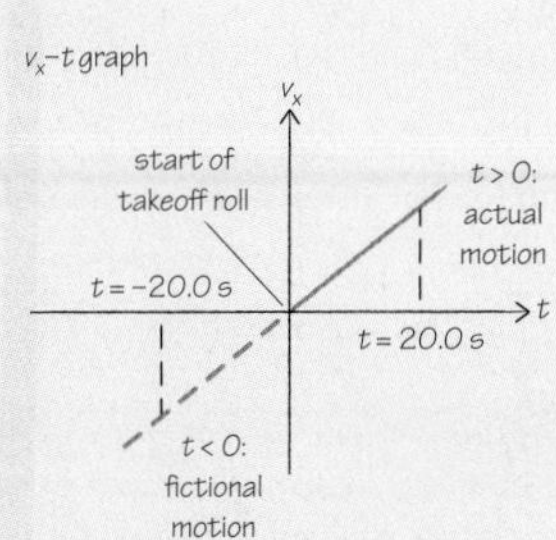

NOW WORK Problem 3 from The Takeaway 2-5.

NOW WORK notes after every example reference related problems in The Takeaway.

Our Digital Platform Encourages Efficient Learning

Everything you need in one place!

e-book

Alem Omerovic/EyeEm/Getty Images

Lemberg Vector studio/Shutterstock

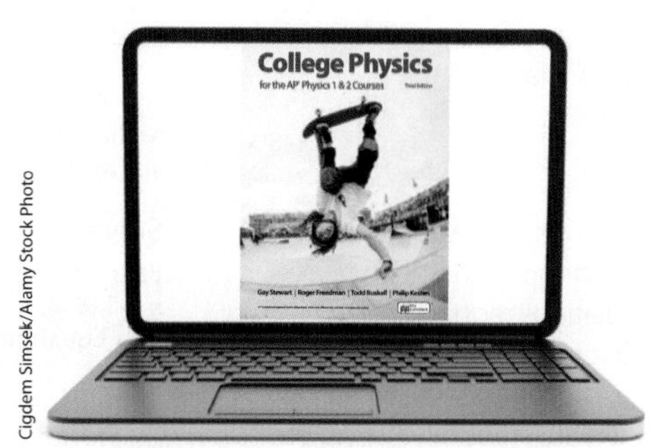

Cigdem Simsek/Alamy Stock Photo

The **interactive, mobile-ready e-book** allows you to read and reference the text when you are working online, or to download it to read when an internet connection is not available. All offline highlights and notes sync when you connect to the internet.

Online Homework System

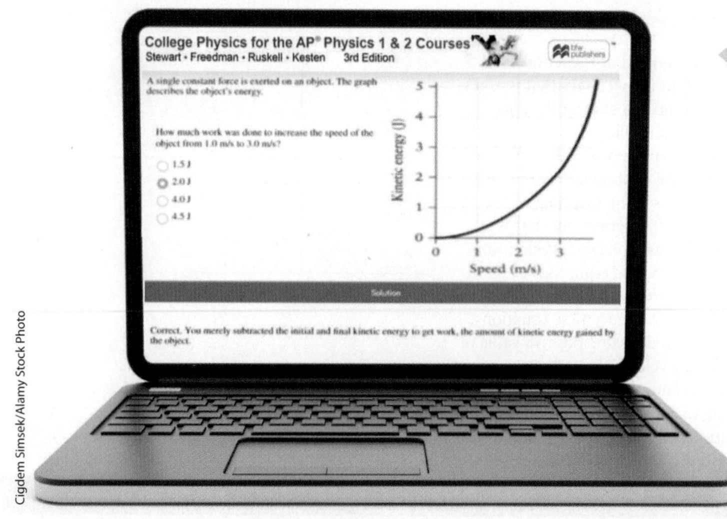

Cigdem Simsek/Alamy Stock Photo

Created and supported by educators, the **online homework system** includes all the resources you need in one convenient place.

Hundreds of **online homework questions** allow you to do your homework efficiently. The program offers hints and guidance to help you learn even when the answer you select is wrong.

College Physics

for the AP® Physics 1 & 2 Courses

Introduction

Chapter 1 Introduction to Physics

Cameron Spencer/Getty Images

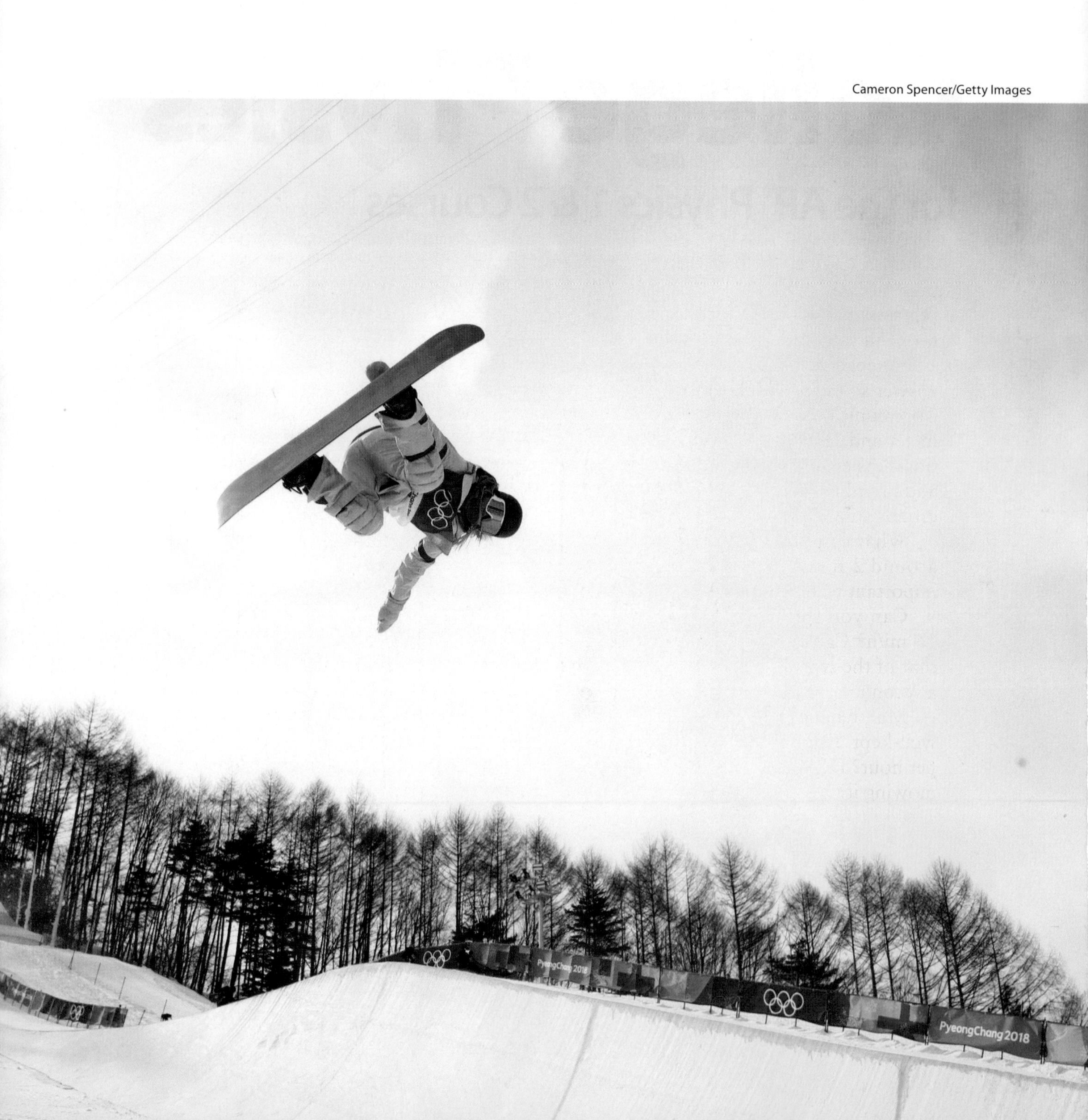

Case Study: How fast is a snowboarder?

American Chloe Kim got some "big air" on her way to winning a gold medal in the halfpipe competition in the 2018 Winter Olympics. Zooming back and forth across the downward-sloped run, she covered the distance across the run and back—somewhat under 40 m—in about one-tenth of a minute. She was moving pretty fast! But how does her speed compare to that of an Olympic marathon runner, who can run a bit more than 26 miles in just over 2 hours?

A friend suggests you can find Kim's speed by multiplying 40 m by one-tenth of a minute...but that can't be right! Speed, the rate at which position changes per a set time interval, must have dimensions of distance *divided by* time. Chloe Kim's speed is (approximately) 40 m divided by 0.1 min, or 400 m/min. One valuable approach in problem solving is to do a dimensional analysis.

A speed in units of m/min? These units aren't conventional, which makes the answer sound strange. It will also make it harder for others to understand your result. Communication is important in science. But you know that there are 1000 m in a kilometer and 1609 m in a mile, and also that there are 60 minutes in an hour. Those relationships enable you to convert Kim's speed to about 24 km/h and 15 mi/h. In science and in ordinary life, a variety of units are used; it's straightforward to convert between different units for the same quantity.

What can you learn from this answer? The marathon runner runs 26 miles in around 2 hours, or 13 mi/h. That is slow for a car, but still quite fast. It's always important to reflect on the results of a problem to see that they make sense.

Can you show that a speed of 400 m/min is equivalent to about 24 km/h and 15 mi/h? Can you convert this speed to units that are appropriate to compare it to that of the world's fastest snail? The common garden snail can travel nearly 3 mm in a second.

Maintenance needs to mow the local park's lawn once a week to keep it looking well-kept. Estimate the rate of growth of the grass in m/s. What is that rate in miles per hour? Do you think the maintenance crew can see the grass growing while they are mowing it?

By the end of this chapter, you should be able to identify the fundamental units used for measuring physical quantities, convert from one set of units to another, and use dimensional analysis to check your results.

1 Introduction to Physics

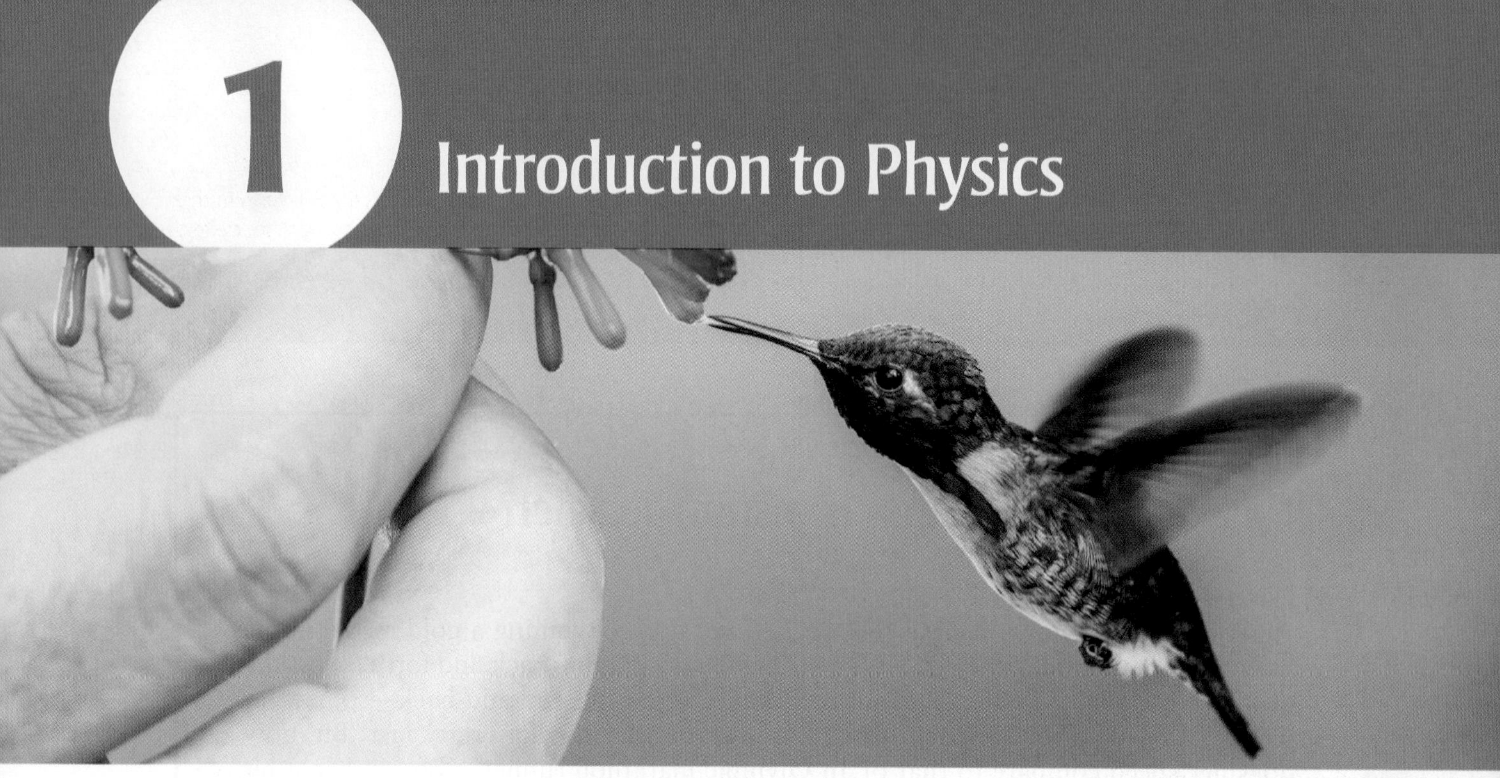

robertharding/Alamy

YOU WILL LEARN TO:

- Explain the roles that concepts and scientific practices play in physics.
- Describe three key steps in solving any physics problem.
- Identify the fundamental units used for measuring physical quantities and convert from one set of units to another.
- Use significant digits in calculations.
- Use dimensional analysis to check algebraic results.

1-1 Scientists use special practices to understand and describe the natural world

A hummingbird is a beautiful sight to anyone. To a physicist, however, a hummingbird is also an illustration of several physical principles. As they beat, the hummingbird's wings push downward on the air, which causes the air to push upward in response and keep the bird aloft. The flow of blood through the hummingbird's body is governed by the properties of moving fluids, including the interplay of pressure and friction in the narrowest blood vessels. The bending of light rays as they enter the bird's eyes forms an image on the retina, providing the hummingbird with vision. And the subtle interference of light waves reflecting off the feathers at the bird's throat gives them their characteristic iridescence. To truly understand and appreciate hummingbirds or any other living organism, or the world around us, or the physical universe, we need to understand the principles of physics. The purpose of this book is to help you gain that understanding.

Learning physics is somewhat like learning a foreign language. If you're paying a brief visit to a foreign country, you may just memorize a few key phrases such as "Where is the hotel?" in the local language. But if you want to be fluent in the language, you don't memorize millions of phrases for the countless situations you might encounter. Instead, you learn a vocabulary of useful words and the grammar that allows you to combine those words into meaningful sentences. Using those basic tools, you can express an enormous variety of ideas, even those you haven't heard before. In the same way, it may seem at the outset that physics is a long compilation of rules, equations, definitions, and concepts. But we will discover that this long list of physics "phrases" is created from a much smaller set of physics "grammar" and the key

AP Exam Tip

To be successful on the AP® Physics 1 exam, you need to be able to extract both quantitative and qualitative information from equations and other representations and develop logical arguments. It is helpful to practice explaining your answers with words—both orally and in writing.

vocabulary of physics. Your goal is to become fluent in physics by understanding how to use the rules of physics grammar to connect the vocabulary of physics.

The vocabulary of physics is often expressed in terms of mathematical quantities, and these quantities, along with the grammar that relates them, are the equations of physics. Like most physics books, this one contains many equations. Not all equations are equally important, however. A much smaller set of *fundamental* equations expresses the key ideas of physics. You will find the ones related to the content of this book at the back of the book. By starting with these fundamental equations and combining them using the rules of algebra, you'll be able to solve all of the quantitative physics problems that you encounter in this book. You will need to be able to describe the relationship of the equations you choose to the physical quantities you are studying, select and follow a logical pathway to a solution or estimate, and assess the reasonableness of your results.

NEED TO REVIEW? Turn to the **Glossary** in the back of the book for definitions of bolded Key Terms.

Other times you will be asked purely qualitative questions, which will require you to apply your knowledge of theories and concepts to describe or explain phenomena or observations, compare results for different systems, or predict future outcomes. Sometimes you will be asked questions that require you to mix qualitative and quantitative approaches and describe the relationship of features of a model or representation to physical quantities using both approaches to demonstrate your understanding.

As you read through this textbook, each equation you will see is a mathematical representation for a concept that can be much more profound than the equation might suggest. One example is Albert Einstein's famous equation

$$E = mc^2$$

While this equation is not one you will need to master in this course, it is one many people have seen, so it is a good example to launch our discussion. In this equation E represents energy, m represents mass, and c represents the speed at which light travels in a vacuum. But what the equation *means* is something deeper. In this book we'll write often-used equations with explanations attached, such as this one, which we refer to as an "Equation in Words":

AP® Exam Tip

Einstein's rest energy is an example of refining a theory. In nuclear reactions, the mass changes are measurable. In chemical reactions, it is customary to say mass is conserved. This is a very good approximation, but the energy change is always equal to a change in the rest mass of the reactants. Because c is so big, this change is too small to routinely measure in chemical reactions.

Rest energy of an object; **Mass** of the object; **Speed of light** in a vacuum

$$E_0 = m_0c^2 \quad (1\text{-}1)$$

The greater the mass of an object when it is at rest, the greater its rest energy.

EQUATION IN WORDS
Einstein's equation for rest energy

The **rest energy** of an object is the amount of energy it possesses because it has mass. (Einstein was the first physicist to introduce this as a general principle, and created this famous equation that describes it.) Relativity, which is introduced in Physics 2, describes how an object's mass changes as its speed increases, although this isn't noticeable unless it is moving very fast. For rest energy, we use the mass the object would have if it were not moving, called its rest mass. In most situations the rest mass is indistinguishable from the mass of the object, so we will just talk about mass. Equation 1-1 shows that rest energy is *directly proportional* to mass: The greater the mass of an object, the greater its rest energy and vice versa. If the rest energy of an object decreases, its mass must decrease as well. You can find an example of this in many kitchens (**Figure 1-1**). In a gas range, natural gas (mostly methane, CH_4) combines with oxygen to produce carbon dioxide (CO_2) and water (H_2O). The chemical reaction that takes place is

$$CH_4 + 2O_2 \rightarrow CO_2 + 2H_2O$$

This process releases energy in the form of warmth and light. In Chapter 7 we will discuss the source of this "chemical energy" as the breaking of chemical bonds. The actual source of the energy in these chemical bonds is quite complicated (building on concepts introduced in Physics 2), however, Equation 1-1 tells us that the combined rest energy of the molecules present after the reaction (a CO_2 molecule and two H_2O molecules) is less than the rest energy of the reactants (a CH_4 molecule and two O_2 molecules). While this is a remarkable statement, experimental evidence supports the mass-energy equivalence. When you look at a flame, Equation 1-1 itself isn't the important part of the story: It's the concept behind an equation that's truly important. Concepts are the keys

Figure 1-1 $E = mc^2$ When you burn natural gas in a kitchen range, the energy released as light and warmth is equal to a decrease in the reactants' rest energy.

WATCH OUT !

This is important.

When we come across a situation that requires special attention, we'll put it in a *Watch Out!* box like this one. For instance, in this textbook we will avoid using the word "heat" as it is commonly used in everyday language. Heat has a very specific meaning in physics, as you will find if you continue your studies: Heat is the amount of energy transferred through **thermal processes,** *not* a type of energy itself. So we will instead use the term "warmth" to convey the energy being transferred in a thermal process, or "heating" to refer to the process. This may take a while to get used to, but it's important to use the correct terms as you move through the course to make sure you are building your understanding.

 **Exam Tip**

The Free-Response section of the AP® Physics 1 exam will always include a question that requires you to demonstrate your understanding, connecting across multiple representations, both qualitative and quantitative, including equations, words, graphs, or diagrams.

to understanding scientific principles. Two related scientific terms are important in the comprehension of concepts: *theory* and *law.* A scientific **theory** or **law** expresses a relationship of an aspect of the natural world and universe that has been repeatedly tested and verified in accordance with the scientific method, using accepted protocols of observation, measurement, and evaluation of results. Whereas a theory includes an explanation of natural phenomena, a law simply summarizes the relationship between variables.

Equations are just one representation of the concepts underlying physics. To describe and explain the real world, physicists often simplify it in the form of equations (which are mathematical models) and conceptual **models** to make analysis possible. Conceptual or mathematical models identify the most important characteristics of a phenomenon or system to simplify analysis. This simplification makes it possible to find an answer or make a prediction. Models can be used to predict how new phenomena will occur.

Communication is very important in science. You have to be able to describe the relationship between an equation and the natural phenomenon you are investigating as well as correctly manipulate that equation to understand what it describes about natural phenomena, or use it to make a numerical estimate. Various representations beyond equations are used to communicate models in introductory physics. These other representations include verbal descriptions and pictures, graphs, or diagrams (referred to as visual representations). It's essential to learn how to create these representations and also how to use one representation to generate another one (to translate between them) to define and solve problems. One example is our Equations in Words feature, where we translate between a verbal conceptual description and an equation. In Chapter 2, you will see examples of going from a picture to a sketch where we can emphasize important details. Then we will go from those sketches to useful diagrams and from the diagrams to graphs that will allow us to answer different questions. In Chapter 4, you will see many examples of translating between a diagram and an equation: A diagram lets us simply and methodically capture the information we need to solve the problem, and then we translate it into an equation to "do the math" and get a numerical answer. Creating and using models and representations is a **scientific practice** in which scientists routinely engage. Such communications skills are very important, as you must be able to communicate your process in answering a question or solving a problem, as well as your results. You sometimes need to collaborate with others to solve larger, harder problems. You will also need to be able to communicate to describe phenomena. In order to be successful in physics, and in science in general, it is crucial to use multiple modes to communicate and this scientific practice supports you in engaging in other practices of science.

You also need to engage in **scientific questioning** to extend your thinking or guide investigations within the context of this course. In your laboratory experience and in the problems, you will create scientific experimental procedures that are aligned to your question. You will analyze data and evaluate the evidence it provides in relation to your question. An important feature of this scientific practice is refining your observations or measurements as needed. You will work with scientific explanations and theories, learning how scientists argue from evidence to make predictions and justify solutions. This scientific practice is at least as important as finding mathematical solutions! This suite of scientific practices will allow you to treat problems as a scientist would and better understand and describe the world around you. Learning to carefully reason from evidence, and to evaluate the quality of that evidence before you make decisions based upon it and develop a logical argument, will allow you to better engage as a citizen in our increasingly complex, information-rich society.

 **Exam Tip**

On the AP® Physics 1 exam, you will be asked to write responses in the Free-Response Questions that demonstrate your ability to communicate your understanding of a physical situation, using several of these scientific practices. It is helpful to practice creating a coherent, organized, and sequential description of the analysis of a situation where you argue from evidence, cite physical principles, and clearly present your thinking without repeating the question or other extraneous information. It will support your understanding if you can develop a logical argument that uses a single line of reasoning to explain, justify, or predict the behavior of a system. You will be given some practice in producing such responses in this text as you are learning new content.

THE TAKEAWAY for Section 1-1

✔ Like all other sciences, physics is a collection of interlinked concepts. Paired with scientific practices, these concepts allow us to understand and describe the physical universe.

✔ Equations, graphs, diagrams, and other representations help us summarize, communicate, and use complex models.

✔ It's essential to understand the concepts behind any equation you use.

Prep for the  Exam

AP® Building Blocks

1. AP® Physics requires that the student be able to justify the use of a mathematical equation to solve a problem. A carpenter might use screws and a screwdriver or nails and a hammer to join two boards.
   (a) In your own words, describe how the carpenter's selection of the correct combination of fastener and tool is similar to the requirement that you be able to select and justify the correct data and mathematical equations.
   (b) Explain how participating in a conversation in a foreign language rather than using a memorized phrase is similar to recognizing and responding to the limitations of a mathematical equation.
2. Antoine Lavoisier wrote a book on chemistry in 1789 that presented as a law the idea that, in a chemical reaction, mass is conserved. The *Oxford English Dictionary* defines a law as "a statement of fact, deduced from observation, to the effect that a particular natural or scientific phenomenon always occurs if certain conditions are present." In your own words, explain how Lavoisier's law of mass conservation can be consistent with Einstein's law of mass-energy conservation in chemical reactions.

1-2 Success in physics requires well-developed problem solving using mathematical, graphical, and reasoning skills

More than just broadening our understanding of the world, the process of learning physics involves learning to solve problems. Physics problems usually can't be done simply by selecting an equation and plugging in values; using mathematics appropriately is just one of the skills you need. Solving physics problems is easier if you build up a set of tools and techniques and then apply a consistent strategy. Our strategy for solving problems involves three steps that we refer to as ***Set Up, Solve,*** and ***Reflect.*** We'll use these steps in the worked examples in this book. You won't have to extend your reflections to consider other examples, but in this textbook we will sometimes extend ours in the worked examples to help you make connections to other problems.

AP® Exam Tip

Every AP® Physics 1 exam includes a question where you are asked to predict what will happen if something about the problem were to be changed. The *Extend* step we use in some of our worked examples will help you with this.

EXAMPLE 1-1 Strategy for Solving Problems

Set Up

At the beginning of any problem, you must first ask yourself: What is the problem asking you to find? With the goal in mind, carefully reading the problem will help you determine which quantities are known, and which are unknowns whose values you need to determine to solve the problem. Then you must ask yourself these two questions: What concepts are appropriate and relevant, and which models and representations should you use to apply them? The answers to these questions are the primary focus of this textbook. We will find there are multiple ways to solve some problems. We will consider not just what concepts apply, but which will help us most efficiently get to the answer.

Set Up the problem.

Solve for the desired quantities.

Reflect on the answer.

Once you've answered these questions, you'll have a good idea of how to get started solving the problem. Carefully read the problem, one sentence at a time. As you read each sentence, begin to use the following important strategies that you can use to set up any problem's solution. You may need to add to and refine these strategies as you learn more information about the problem.

Draw a picture. Physicists rely heavily on diagrams. You'll find one or more in many of the worked examples in this text. A good problem-solving picture should capture the motion, the process, the geometry, or whatever else defines the problem. You don't need an artistic masterpiece; it's preferable to represent objects as dots. In addition to a sketch of the physical situation, you will also learn to draw new kinds of diagrams that have formal rules that then make them valuable, even required, parts of a solution. We will show you plenty of examples and explain the rules as we get to the appropriate content in this textbook.

Label all quantities. It makes sense to label all quantities such as length, speed, and position right on the picture you've drawn. But be careful! First, more than one of a given kind of variable is often in a problem. For example, if two objects are moving at different speeds, but you use the same symbol v for the speeds of both objects, confusion is bound to result. Subscripts are useful in such a case, for example, v_A and v_B for the speeds of the two objects. Also, don't hesitate to label lengths, velocities, and other quantities even if you haven't been given a value for them. You might be able to find the value of one of these quantities and use it later in the problem, or you might gain an insight that will guide you to a solution.

Exam Tip

Variables and subscripts are often defined for you in a problem on an AP® Physics exam, so take care to use the variables and subscripts given, and carefully define any needed variables not already provided.

Look for connections. Often the best strategy for solving a physics problem is to look for connections between quantities. These connections could be relationships between two or more variables implied by the description in the problem and physics concepts, or an actual statement such as that a certain parameter has the same value for two objects. Furthermore, these connections will often enable you either to eliminate variables or to find a value for an unknown variable. As a result, for quantitative problems, you sometimes will be able to quickly simplify to an equation that defines the variable of interest only in terms of variables that are known. Always be on the lookout for connections such as these when starting a problem.

Solve

Once you've sized up the situation and selected the concepts and models to use, finding a quantitative answer is often just an exercise in mathematics. During this step in the worked examples, we'll break down and show many intermediate steps as we describe our reasoning and thought processes. In this way, we'll demonstrate good problem solving, which will lead you to develop your own problem-solving skills and habits. For questions on an AP® exam, or those a researcher might approach in her or his own work in a lab, describing and justifying your reasoning may be the most important part of the problem! In purely qualitative problems, you may not have any mathematical support to fall back on. For such problems, being able to make a scientific claim about what will happen and justify that claim based on physical principles, laws, or models will generally provide a good solution. Sometimes, you will be required to understand what is wrong with an explanation and improve upon it. Often, you will find it helps you to approach a problem from both the qualitative and the quantitative perspectives, and to compare those answers and ensure they agree and make sense. When you can do this, you know you truly understand the problem, and you can build on that understanding to solve more complicated ones! Even if the problem doesn't ask for it, we will often use this strategy in the Reflect step to check our solution.

Reflect

For many problems, this last step will be the most important. Supporting claims with evidence from experimental data, physical principles, laws, or models, and assessing the reasonableness of your results or solutions not only demonstrates your understanding, but allows you to build upon it.

At the first level, you always need to review your answer to see what it tells you. If the problem calls for a numerical answer, it's important to ask, "Do the number and the units make sense?" This is often a good way for you to check the accuracy of your calculations. For example, if you've been asked to calculate the diameter of Earth and you come up with 40 centimeters, or if you come up with an answer in kilograms (the unit of mass), it's time to go back and check your calculations. If your answer is a formula rather than a single number, check that what the formula says is reasonable. As an example, if you've found a formula that relates the weight of a cat to its volume, your answer should show that a heavier cat has a larger volume than a lightweight one. If it doesn't, there's an error in your calculations that you need to find and fix.

If your answer is qualitative, checking your answer is just as important. If you are asked to predict the motion of an object as a certain process is occurring, you will need to think about not just what seems like common sense, but what you learn of how motion works, including details that you may not have been previously prepared to see and consider. For example, if you say that an object that you are pushing will stop sliding once you stop pushing, are you considering how much friction there is between the object and the surface you are pushing it along? If there were no friction (spoiler alert, you will find out about this in Chapter 4), it would never stop!

Extend: Additionally, some Reflect sections will include an "Extend" paragraph that shows you how to extend your learning to real-life applications, or consider the effect of changes in the problem.

NOW WORK Problems 1 and 2 from The Takeaway 1-2.

Learning physics will be easier if you study the many worked examples in this book and get in the habit of following the same steps of ***Set Up, Solve,*** and ***Reflect*** in your own problem solving. Including a written explanation for the decisions you make in a problem will ensure mastery and allow you to connect the vocabulary of physics into a rich, and useful, language for solving problems and understanding your world.

THE TAKEAWAY for Section 1-2

✔ Success in physics requires having a methodical approach to problem solving.

✔ When approaching any problem, follow these three steps: ***Set Up*** a problem by selecting the appropriate concepts, models, and representations; ***Solve*** by working through the mathematics or making a prediction and justifying that prediction based on physical principles, laws, or representations; and ***Reflect*** on the results to see that they make sense.

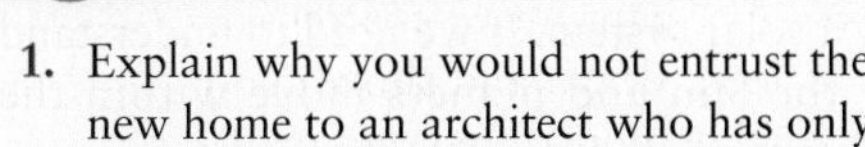

1. Explain why you would not entrust the construction of a new home to an architect who has only a "mental model" of the plan. In your own words connect this expectation to the use of a drawing in the solution of a physics problem.

EX 1-1 2. Scientific theories must be testable. This means that a theory should lead to a prediction for the behavior of an object or system under given circumstances. In your own words connect the last step in problem solving, ***Reflect***, to the role of experimentation in science.

1-3 Scientists use simplifying models to make it possible to solve problems; an "object" will be an important model in your studies

The real world is complex. To describe and explain phenomena, it is often necessary to simplify real objects, systems, and processes. As discussed in Section 1-1, these simplifications are models, and these models are tested by using them to predict how new phenomena may occur. For some uses, a simplified model may give results that are accurate enough, such as when we predict where a baseball will land given a certain pitch by neglecting any effects of the air on the motion of the object. This would lead to a simple equation we will learn in Chapter 3. In other cases, models must be refined: If the person who threw the ball was a professional pitcher and it went very fast, your answer would be quite a bit off if you neglected the effect of the air, so you would have to add air resistance to your model, and consider the spinning motion of the ball as it traveled. This would lead to a very complicated equation, beyond the scope of this book. So simple models need to be refined as we need more precise answers. Models that seem to be complete are replaced when better models are discovered. For instance, people used to think mass was constant in any process ($m_{\text{final}} = m_{\text{initial}}$), but then as we saw in Equation 1-1, Einstein showed that mass could be converted into energy and so change! In this book, you will learn how to describe, use, refine, and create several types of important models, such as objects or rigid bodies, pure rolling motion, or neglecting friction or the mass of a string. You will express key elements of models of natural phenomena using multiple representations (for instance, you might convert a graphical representation into an equation, and also explain it in words).

The most common model we will use in this textbook is the **object model.** Using the object model for something means that you can ignore its size and shape, anything going on inside it, or any internal structure, and treat it like a point moving through space. The object model is used consistently in AP® Physics to mean you can neglect internal structure. In other textbooks you may have seen the word *particle* used instead. We have avoided using that word, since in physics, we use the term **fundamental particles** to refer to things like electrons and quarks that we believe really have no internal structure.

In contrast, we use the **system model** when we cannot ignore internal structure. Once we have determined if we can use a **system** or an **object** model to describe what we are interested in, we refer to it as a system or an object.

Systems may be built from fundamental particles (this is true of systems like protons and neutrons, which are composed of fundamental particles called quarks), objects, or even other systems. According to the problem you are solving, or the process you are trying to describe, it may be appropriate to describe an atom as a system, paying attention to the protons and neutrons and electrons inside. But other times, you may not need to know what is going on inside the atom, and it would then be perfectly fine to model the atom as an object, ignoring its internal structure.

Often, you will be working with something that seems simple, like a ball. However, while a ball is simple, it does really have internal structure. For instance, it may be squeezed or spun. If the question you are trying to answer depends on the fact that not all the parts of the ball move exactly the same distance when you do something to the ball, you will have to use the system model (and refer to the ball as a system). If, however, the properties you are trying to study do not depend on the internal structure of the ball, such as predicting how far it goes when you throw it gently, you can ignore its internal structure and treat it as an object (use the object model).

A **system** can be complex, like a galaxy or a solar system. If we need to understand how the Sun and planets move within the solar system, then we must treat it as a system. However, even with something as large and complex as a solar system, if we look at how solar systems move in the galaxy, a good first approximation could be found by treating the solar system as an object (**Figure 1-2**). Living things are some of the most complex systems around. For instance, we can move our bodies, and we digest food to convert it into energy to help us do so. But sometimes you might just want to know how far you can jump. In this case, you can get a pretty good first approximation by treating yourself as an object once you know how fast and in what direction you are going when you leave the ground.

So when we can ignore the internal structure of a system, we can model it as an object. Choosing whether to model something as an object or a system is a fundamental step in determining how to describe and analyze a physical situation.

Figure 1-2 Systems and objects can be very large Astronomers have shown that the Milky Way (our galaxy) is on a collision course with the neighboring Andromeda spiral galaxy. In 2007, refined models allowed new calculations to show how that would affect our solar system. The calculations treat the solar system as an object and show it will be exiled to the outer reaches of the merged galaxy. The collision is predicted to take place before the Sun becomes a burned-out white dwarf star (in just a few billion years). This image is of a collision of galaxies already in progress.

WATCH OUT!

Which Model: System or Object?

In AP® Physics, use of the *object* model implies that you can neglect the internal structure of the system you are studying. To determine if you can use the object model, you must make sure you do not need to attribute anything to it that requires internal structure. You can use an object model as long as all points on the system move together in exactly the same way. In other words, you can treat it as an object if you can completely describe the system by a single point in space and its motion with the motion of that point. When you cannot neglect internal structure (the system's shape changes, or it spins, or there are internal interactions you cannot ignore) you must use a *system model. Particle* will be used in AP® Physics only to refer to *fundamental particles*.

THE TAKEAWAY for Section 1-3

✔ The universe contains fundamental particles that have no internal structure and systems built up from these, such as protons and neutrons, from which all other systems are built.

✔ If the properties of interest in a problem do not depend on the internal structure of a system, you may model that system as an object.

✔ If internal structure is necessary to solve the problem, you must use a system model (such as when something changes shape or spins).

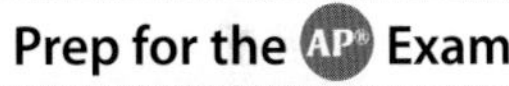
Prep for the AP® Exam

AP® Building Blocks

1. An object model can be used when it is not necessary to consider internal structure. When internal structure plays a role in behavior we must use a system model defined in terms of the composing objects. In one or more sentences, justify your claim that an object model is or is not appropriate for the bolded word(s) in each of the following contexts. After responding to parts (a)–(g), reflect on your answers and in one or more sentences connect the roles played by internal structure in your claims when a system description is needed.
   (a) A description of the motion of **Earth** as it orbits the Sun
   (b) An explanation for the motion of **Earth as it orbits the Sun**
   (c) A description of the motion of a **girl on a bike** coasting down a hill toward an intersection
   (d) The mass of the **water** in a goldfish bowl
   (e) **Water** poured into a goldfish bowl, assuming the shape of the bowl
   (f) An **ice skater** gliding in a line across the ice
   (g) An **ice skater** spinning in place on the ice
2. Someone may be praised by calling them a "systems thinker." In your own words describe why this may be considered praiseworthy and pose a question that initiates systems thinking.
3. On a long journey you rely on an airplane to take off and land. In scheduling a flight you expect the departure and arrival times to be well-defined. Explain why treating the airplane as an object simplifies your thinking about these events, but may fail to describe the actual outcome.

1-4 Measurements in physics are based on standard units of time, length, and mass

To the average person, physics calls to mind speculative theories about the nature of matter and the origins of space and time. Someone a little more familiar with physics may think of space ships or advanced computers. But at its heart physics is an *experimental* science based on measurements of the natural world. Over a century ago the renowned Scottish physicist Lord Kelvin (1824–1907) expressed this viewpoint: "When you can measure what you are speaking about and express it in numbers, you know something about it; but when you cannot measure it...your knowledge is of a meager and unsatisfactory kind."

Any measurement requires a standard of comparison. If you want to compare the heights of two children, for example, you might stand them side by side and use one child's height as a reference for the other. A better and more reliable method would be to use a vertical rod with equally spaced markings (in other words, a ruler) that acts as a standard of length.

TABLE 1-1 Fundamental Quantities and Their SI Units

Quantity	Unit	Abbreviation
time	second	s
length	meter	m
mass	kilogram	kg
temperature	kelvin	K
electric current	ampere	A
amount of substance	mole	mol
luminous intensity	candela	cd

In physics, three fundamental quantities for which it is essential to have standards are *time, length,* and *mass.* Common standards, or **units**, for these three quantities are the *second*, the *meter*, and the *kilogram*, respectively, and the system of units based on these quantities is called the **Système International**, or **SI** for short. By understanding how these standards are defined, we'll get a sense of the precision toward which physicists strive in their measurements. **Table 1-1** lists the fundamental SI quantities. The definition of the second (abbreviated "s") is based on the properties of the cesium atom. When such atoms are excited, they emit radio waves of a very definite frequency. The second is defined to be the amount of time required for cesium to emit 9,192,631,770 complete cycles of these waves. This may seem like a bizarre way to measure time, but modern radio technology is highly refined, and it's relatively easy to prepare a gas of cesium atoms with the right properties. Hence a physicist anywhere in the world could readily repeat these measurements with fairly common equipment.

In the past, metal bars were used as standards of length. These were not very good standards, however, because their length changes slightly but measurably when the temperature changes. Today light waves are used to define the meter (abbreviated "m"), the SI standard of length. The speed of light in a vacuum is defined to be precisely 299,792,458 meters per second, so the meter is the distance that light travels in 1/299,792,458 of a second. This is a robust and repeatable standard because light in a vacuum travels at the same speed anywhere on Earth and, as best we can tell, everywhere in the universe. (We specify a vacuum because light travels more slowly in a material such as glass, water, or even air.)

Mass is a measure of the amount of material in an object or system. A feather has less mass than a baseball, which in turn has less mass than a cinder block. The ideal standard for mass would be one that is, like the standards for time and length, based on something that is the same throughout the universe—for example, the mass of a hydrogen atom. Unfortunately, physicists' present-day ability to measure mass on the atomic scale is not yet precise enough to give a satisfactory standard of mass. From 1889 to 2019, the kilogram (abbreviated "kg") was defined to be the mass of a special cylinder of platinum-iridium alloy kept in a repository near Paris, France (**Figure 1-3**). (The *gram*, equal to 1/1000 of a kilogram, is not a fundamental SI unit.) The kilogram is now defined in terms of the meter, the second, and fundamental constants, based on the use of a very precise instrument called the Kibble balance. This, like the frequency definition of a second, allows a properly equipped laboratory to calibrate a measurement instrument directly by measuring a natural phenomenon instead of by using an artifact such as the cylinder. The actual quantities measured to allow this calibration are current and potential difference, which you will learn about in Physics 2. They are used to determine a force. We will see how a precise force could get us to a definition of mass in Chapter 4.

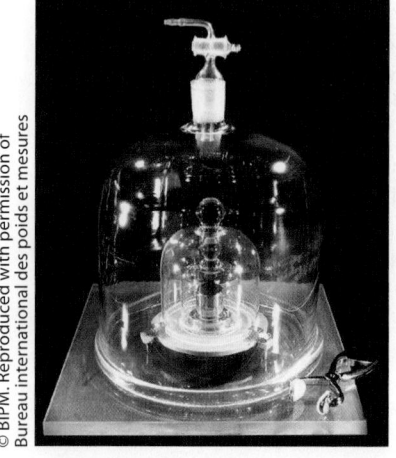

Figure 1-3 The international mass standard for more than 100 years This cylinder was defined to have a mass of precisely 1 kilogram. The cylinder was treated very carefully in an attempt to prevent even a small number of atoms from being scraped off. Even more problematic was the addition of mass through surface contamination. This cylinder and its replicas had come to differ by as much as 50 micrograms, motivating a change to a new fundamental definition of the kilogram. These cylinders still serve as secondary standards.

WATCH OUT !

Mass and weight are different quantities.

It's important not to confuse *mass* and *weight*. While an object's mass tells you how much material that object contains, its weight tells you how strongly gravity pulls on that object. Consider a person who weighs 110 pounds on Earth, corresponding to a mass of 50 kilograms. Gravity is only about one-sixth as strong on the Moon as it is on Earth, so on the Moon this person would weigh only one-sixth of 110 pounds, or about 18 pounds. But that person's mass of 50 kilograms is the same on the Moon; wherever you go in the universe, you take all of your material along with you. We'll explore the relationship between mass and weight in Chapter 4.

Units for many other physical quantities can be derived from the fundamental units of time, length, and mass. For example, speed is measured in meters per second (m/s), and weight is measured in kilograms times meters per second per second, or kilogram-meters per second squared ($kg \cdot m/s^2$). Some derived quantities are used so frequently that physicists have named a new unit to represent them. For example, the SI unit of weight is the newton (N), equal to $1\ kg \cdot m/s^2$, or about 0.225 pound.

If you continue your study of physics you will be introduced to the other **fundamental units** (see Table 1-1), such as the kelvin, which is the fundamental unit of temperature and the ampere, the fundamental unit for electric current. Another derived unit called the coulomb, a unit of electric charge, is derived from the ampere. Just as we will find that gravity is an interaction between objects with mass, the electric force is an interaction between objects with charge.

In this course, the units we'll work with will be based on seconds, meters, and kilograms.

Unit Conversions

In science and in ordinary life, a variety of other units are used for time, length, and mass. As we'll see, it's straightforward to convert between different units for the same quantity.

In addition to the second, some other common units of time include the following:

$$1 \text{ minute (min)} = 60 \text{ s}$$
$$1 \text{ hour (h)} = 3600 \text{ s}$$
$$1 \text{ day (d)} = 86{,}400 \text{ s}$$

Some other units of length that are handy when discussing sizes and distances include millimeters (mm), centimeters (cm), and kilometers (km). These units of length are related to the meter as follows:

$$1 \text{ mm} = 0.001 \text{ m}$$
$$1 \text{ cm} = 0.01 \text{ m}$$
$$1 \text{ km} = 1000 \text{ m}$$

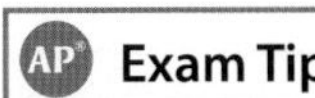
Exam Tip

Of the fundamental quantities, only time, length, and mass are part of the AP® Physics 1 curriculum.

Although the English system of inches (in), feet (ft), and miles (mi) is much older than the SI, today the English system is actually based on the SI system: The inch is *defined* to be exactly 2.54 cm. A useful set of conversions is as follows:

$$1 \text{ in} = 2.54 \text{ cm}$$
$$1 \text{ ft} = 0.3048 \text{ m}$$
$$1 \text{ mi} = 1.609 \text{ km}$$

You'll find more conversion factors in Appendix A.

We can use conversions such as these to convert any quantity from one set of units to another. For example, the glass skyscraper in London, England, called the Shard is 1016 feet tall (**Figure 1-4**). **Figure 1-5** shows the general technique for converting units, using the Shard as an example. When doing unit conversions or other multiplication and division calculations involving units, you can treat the unit associated with the number as a variable. Units can get multiplied together, and they can also cancel if the same unit appears in the numerator and denominator. Making sure all of the units but the one you want have canceled is the best way to know you have converted units correctly!

You can multiply by 1 as many times as necessary to get the units you need. Notice we have to use 1 three times to get from days to seconds, since 1 day (d) equals 24 hours (h), 1 h equals 60 minutes (min), and 1 min equals 60 seconds (s):

$$1 \text{ d} = 1\cancel{\text{d}}\left(\frac{24\cancel{\text{h}}}{1\cancel{\text{d}}}\right)\left(\frac{60\cancel{\text{min}}}{1\cancel{\text{h}}}\right)\left(\frac{60 \text{ s}}{1\cancel{\text{min}}}\right) = 86{,}400 \text{ s}$$

Notice that we were careful to check the cancellation of each pair of like units (d and d, h and h, min and min). You should carry out conversions such as this one in an equally methodical way and always be careful to ensure the unwanted units do cancel. For instance, if you had multiplied $1016 \text{ ft} = 1016 \text{ ft} \times 1 = 1016 \text{ ft} \times \dfrac{1 \text{ ft}}{0.3048 \text{ m}} = 3333\dfrac{\text{ft}^2}{\text{m}}$ your units would not be the ones you wanted and your answer would be meaningless.

Allan Baxter/Getty Images

Figure 1-4 From feet to meters The tallest building in the United Kingdom, the Shard, stands 1016 feet tall. How tall is this in meters?

(1) To convert units—for example, to convert a distance in feet (ft) to a distance in meters (m)—multiply by 1 in the appropriate form. Multiplying a number by 1 does not change the number.

$$1016 \text{ ft} = 1016 \text{ ft} \times 1$$

(2) To find the appropriate form of 1, begin with the conversion between the two units...

(3) ...and divide one side by the other. (If you divide a number by an equal number, the result is always equal to 1.) This gives two ways to express the conversion between feet and meters.

$$1 \text{ ft} = 0.3048 \text{ m, so } \frac{1 \text{ ft}}{0.3048 \text{ m}} = 1 \text{ and } \frac{0.3048 \text{ m}}{1 \text{ ft}} = 1$$

(4) Multiply the quantity you want to convert by 1 (equal to the ratio of the two units) that enables the unwanted units to cancel out: Multiply 1016 ft by the second form of 1...

(5) ...the units of ft cancel and the result is in m, as desired.

$$1016 \text{ ft} = 1016 \text{ ft} \times 1 = 1016 \cancel{\text{ft}} \times \frac{0.3048 \text{ m}}{1 \cancel{\text{ft}}} = 309.7 \text{ m}$$

Figure 1-5 Converting units To convert units, remember that a quantity does not change if you multiply by 1. The key to converting units is to write 1 in an appropriate way.

AP Exam Tip

Some basic conversions are included on the AP® Physics 1 equation sheet. Other necessary conversions will be stated in specific problems. It is helpful to have a sense of the relative size of units, to see if mathematical answers are sensible.

EXAMPLE 1-2 Converting Units: Speed

The world's fastest bird, the white-throated needletail (*Hirundapus caudacutus*), can move at speeds up to 47.0 m/s in level flight. (Falcons can go even faster, but they must dive to do so.) What is this speed in km/h? in mi/h? Note that because both desired conversions include hours, we can find one, and then just do the conversion between kilometers and miles to find the second.

Set Up

This is a problem in converting units. To find the speed in km/h, we will have to convert meters (m) to kilometers (km) and seconds (s) to hours (h). To then find the speed in mi/h, we will have to convert km to miles (mi). In each stage of the conversion, we'll follow the procedure shown in Figure 1-5.

Conversions:

1 h = 3600 s

1 km = 1000 m

1 mi = 1.609 km

Solve

We write the first two conversions above as a ratio equal to 1, such as (1 km)/(1000 m) = 1. We then multiply each of these by 47.0 m/s to find the speed in km/h. Note that we must set up the ratios so that the units of meters (m) and seconds (s) cancel as shown.

$$47.0 \text{ m/s} = \left(47.0\frac{\cancel{\text{m}}}{\cancel{\text{s}}}\right)\left(\frac{3600 \cancel{\text{s}}}{1 \text{ h}}\right)\left(\frac{1 \text{ km}}{1000 \cancel{\text{m}}}\right)$$
$$= 169 \text{ km/h}$$

To convert this speed to mi/h, we use the ratio (1 mi)/(1.609 km) = 1.

$$169 \text{ km/h} = \left(169\frac{\cancel{\text{km}}}{\text{h}}\right)\left(\frac{1 \text{ mi}}{1.609 \cancel{\text{km}}}\right)$$
$$= 105 \text{ mi/h}$$

Reflect

Because there are 3600 s in an hour, an object that moves at 1 m/s will travel 3600 m, or 3.6 km, in 1 h. So 1 m/s = 3.6 km/h. This says that *any* speed in km/h will be 3.6 times greater than the speed in m/s. You can check our first result by verifying that 169 is 3.6 times 47.0.

To check our second result, note that a mile is a greater distance than a kilometer. Hence our fast-flying bird will travel fewer miles than kilometers in a 1-hour time interval, and the speed in mi/h will be less than in km/h. This is just what we found.

NOW WORK Problem 3 from The Takeaway 1-4.

Quantities Versus Units

In this book we'll often use symbols to denote the values of physical quantities, by which we mean any property of an object or system that can be measured using numbers, such as mass or temperature. We'll always write these symbols using *italics*. Some examples include the symbols v for speed and E for energy. (As these examples show, the symbol is not always the first letter of the name of the quantity.) By contrast, the letters that we use to denote units, such as m (meters), s (seconds), and kg (kilograms), are not italicized. If the unit is named for a person, the letter in the unit abbreviation is capitalized. An example is the unit of power, the watt (abbreviated W), named for the Scottish inventor James Watt.

In some cases the same symbol is used for more than one thing. One example is the symbol T, which can denote a temperature or a period of time. The letter T (not italicized, so not a symbol) is also used as an abbreviation for the unit of magnetic field, the tesla (named for the Serbian-American physicist Nikola Tesla). You'll often need to pay careful attention to the context in which a symbol is used to discern the quantity or unit it represents.

Scientific Notation

Physicists investigate objects and systems that vary in size from the largest structures in the universe, including galaxies and clusters of galaxies, down to atomic nuclei and the fundamental particles, quarks, found within nuclei. The time intervals that they analyze range from the age of the universe to a tiny fraction of a second. To describe such a wide range of phenomena, we need an equally wide range of both large and small numbers.

To avoid such confusing terms as "a million billion billion," physicists use a standard shorthand system called **scientific notation**. All the cumbersome zeros that accompany a large number are consolidated into one term consisting of 10 followed by an exponent, which is written as a superscript. The **exponent** indicates the number of times you need to multiply (if the exponent is positive) or divide (if the exponent is negative) by 10 to write out the long form of the number. This is why the exponent is also called the power of ten. For example, ten thousand can be written as 10^4 ("ten to the fourth" or "ten to the fourth power") because $10^4 = 10 \times 10 \times 10 \times 10 = 10{,}000$. Some powers of ten are so common that we have created abbreviations for them that appear as a prefix to the actual unit. For example, rather than writing 10^9 watts (10^9 W), we can write 1 gigawatt, or 1 GW. A negative exponent tells you to *divide* by the appropriate number of tens. For example, 10^{-2} ("ten to the minus two") means to divide by 10 twice, so $10^{-2} = (1/10) \times (1/10) = 1/100 = 0.01$. **Table 1-2** lists the prefix and its abbreviation for each of several common positive and negative powers of ten.

In scientific notation, numbers (the first factor) are usually written as a number between 1 and 10 multiplied by the appropriate power of ten (the second factor). The approximate distance between Earth and the Sun, for example, can be written as 1.5×10^8 km. This is far more convenient than writing 150,000,000 km or one hundred and fifty million kilometers. A power output of 2500 watts (2500 W) can be written in scientific notation as 2.5×10^3 W. (Alternatively, we can write this as 2.5 kW, where 1 kW = 1 kilowatt = 10^3 W.) You can also use scientific notation to express a number such as 0.00245: $0.00245 = 2.45 \times 0.001 = 2.45 \times 10^{-3}$. (Again, in scientific notation the first factor, a number between 1 and 10, contains the measured or calculated digits while the second factor merely places the decimal.) This notation is particularly useful when dealing with very small numbers. As an example, the diameter of a hydrogen atom is much more convenient to state in scientific notation (1.1×10^{-10} m) than as a decimal (0.00000000011 m) or a fraction (110 trillionths of a meter).

Note that most electronic calculators use a shorthand for scientific notation. To enter the number 1.5×10^8 you first enter 1.5; then press a key labeled EXP, EE, or 10^x; then enter the exponent 8. (The EXP, EE, or 10^x key takes care of the $\times 10$ part of the expression.) Note that for a number like 10^4, you would need to enter 1, then the appropriate key, then the exponent 4.

Exam Tip

The prefixes from *pico-* to *tera-* are included on the AP® Physics 1 equation sheet.

TABLE 1-2 Prefixes

Factor	Prefix	Symbol
10^{-15}	femto	f
10^{-12}	pico	p
10^{-9}	nano	n
10^{-6}	micro	μ
10^{-3}	milli	m
10^{-2}	centi	c
10^{-1}	deci	d
10^{1}	deka	da
10^{2}	hecto	h
10^{3}	kilo	k
10^{6}	mega	M
10^{9}	giga	G
10^{12}	tera	T

WATCH OUT !

Reading scientific notation on calculators.

Confusion can result from the way calculators display scientific notation. For example, the number 1.5×10^8 will appear on your calculator's display as 1.5 E 8, 1.5 8, or some variation of these; typically the $\times 10$ is not displayed as such. It is not uncommon to think that 1.5×10^8 is the same as 1.5^8. That is not correct, however; 1.5^8 is equal to 1.5 multiplied by itself eight times, or 25.63, which is not even close to $150{,}000{,}000 = 1.5 \times 10^8$. There are some variations from one calculator to another, so you should spend time making sure you know the correct procedure for working with numbers in scientific notation. You will be using this notation continually in your study of physics, so this is time well spent. Reading over the manual for your calculator will help you to avoid common errors.

EXAMPLE 1-3 Unit Conversion and Scientific Notation: Hair Growth

The hair on a typical person's head grows at an average rate of 1.5 cm per month (**Figure 1-6**). Approximately how far (in meters) will the ends of your hair move during a 50-min physics class? Express the answer using scientific notation.

Figure 1-6 Hair growth How much does a person's hair grow during a 50-minute class period?

Set Up

Like Example 1-2, this is a problem in converting units. We'll again follow the procedure shown in Figure 1-5. Not all months have the same number of days, so we'll assume an average 30-day month. We'll use the idea that, because speed equals distance per time, the distance the hair moves equals the growth speed multiplied by the time in one class period.

Conversions:

$$1 \text{ min} = 60 \text{ s}$$
$$1 \text{ d} = 86{,}400 \text{ s}$$
$$1 \text{ month (average)} = 30 \text{ d}$$
$$1 \text{ cm} = 0.01 \text{ m}$$
$$\text{speed} = \frac{\text{distance}}{\text{time}}$$
$$\text{so distance} = \text{speed} \times \text{time}$$

Solve

Let's first express the speed at which hair grows in SI units (m/s). We use the conversion factors relating meters to centimeters, months to days, and days to seconds, and cancel units. We must move the decimal point to the right by nine steps to put the result in scientific notation, so the exponent is –9.

$$1.5 \text{ cm/month}$$
$$= \left(1.5\frac{\cancel{\text{cm}}}{\cancel{\text{month}}}\right)\left(\frac{0.01 \text{ m}}{1 \cancel{\text{cm}}}\right)\left(\frac{1 \cancel{\text{month}}}{30 \cancel{\text{d}}}\right)\left(\frac{1 \cancel{\text{d}}}{86{,}400 \text{ s}}\right)$$
$$= 0.0000000058 \text{ m/s} = 5.8 \times 10^{-9} \text{ m/s}$$

To find the distance that hair grows in 50 minutes, we multiply this speed by 50 min and include the conversion factor relating seconds and minutes. We then put the result into scientific notation. Note that the first number in the result is between 1 and 10. Remember that when multiplying together powers of ten, we simply add the exponents.

You can also express this result in micrometers (μm). Note that when dividing powers of ten, we subtract the exponent of the denominator from that of the numerator. A distance of 17 μm is about 1/60 mm.

$$\text{distance} = (5.8 \times 10^{-9})\frac{\text{m}}{\cancel{\text{s}}} \times (50 \cancel{\text{min}})\left(\frac{60 \cancel{\text{s}}}{1 \cancel{\text{min}}}\right)$$
$$= (5.8 \times 10^{-9})(3 \times 10^{3}) \text{ m}$$
$$= 5.8 \times 3 \times 10^{-9} \times 10^{3} \text{ m}$$
$$= 17 \times 10^{-9} \times 10^{3} \text{ m}$$
$$= 1.7 \times 10^{1} \times 10^{-9} \times 10^{3} \text{ m}$$
$$= 1.7 \times 10^{1-9+3} \text{ m}$$
$$= 1.7 \times 10^{-5} \text{ m}$$

Reflect

You would not be able to notice this change in length with the unaided eye, which makes sense, as you don't expect to be able to watch someone's hair grow!

$$1.7\times10^{-5}\text{ m} = (1.7\times10^{-5}\text{ m})\left(\frac{1\ \mu\text{m}}{10^{-6}\text{ m}}\right)$$
$$= 1.7\times\frac{10^{-5}}{10^{-6}}\ \mu\text{m}$$
$$= 1.7\times10^{-5-(-6)}\ \mu\text{m}$$
$$= 1.7\times10^{1}\ \mu\text{m} = 17\ \mu\text{m}$$

NOW WORK Problems 4, 5, and 6 from The Takeaway 1-4.

THE TAKEAWAY for Section 1-4

✔ Many physical quantities can be measured in units that are combinations of the SI units seconds, meters, and kilograms.

✔ To convert a physical quantity from one set of units to another, multiply it by an appropriate conversion factor equivalent to 1.

✔ In scientific notation, quantities are expressed as a number between 1 and 10 multiplied by 10 raised to a power.

✔ Prefixes can be used instead of scientific notation to indicate the power to which the 10 would be raised in scientific notation.

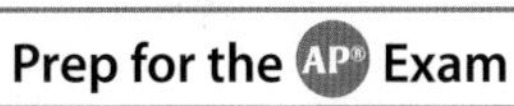

Prep for the AP® Exam

AP® Building Blocks

1. Complete this table without reference materials.

Quantity	SI unit
time	
length	
mass	

2. Complete this table without reference materials.

Prefix	Power of ten
micro	
mega	
centi	
kilo	
nano	
milli	
giga	

EX 1-2 3. The distance from New York City, United States, to Lisbon, Portugal, is 5420 km. A flight takes 6 h and 40 min. What is the speed of the plane in meters per second?

EX 1-3 4. Galileo used his pulse to measure time. He also used a barrel with a small hole that produced drips of water that he could hear.

Suppose that he measured 95 drips in a minute. He rolled a ball on a track that was 12 braccia in length in a time of 34 drips. A braccia is 0.7 m. What is the speed of the ball in meters per second?

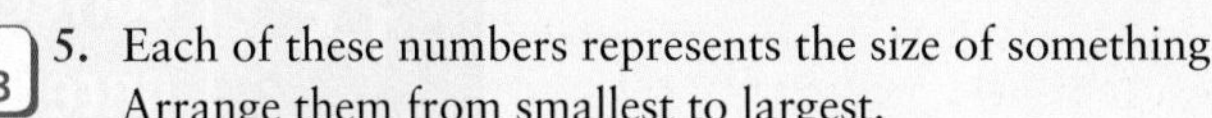

EX 1-3 5. Each of these numbers represents the size of something. Arrange them from smallest to largest.
(A) 0.1 mm
(B) 7 μm
(C) 6380 km
(D) 165 cm
(E) 200 nm

EX 1-3 6. For each pair, determine which quantity is bigger and find the ratio of the larger quantity to the smaller quantity.
(a) 1 mg, 1 kg
(b) 1 mm, 1 cm
(c) 1 MW, 1 kW
(d) 10^{-10} m, 10^{-14} m
(e) 10^{10} m, 10^{14} m

AP® Skill Builders

7. The thin, roughly 2-mm-thick outer layer of the brain is called the cerebral cortex. This cortex is highly convoluted (which just means folded in a complex way). If the cortex were spread out it would cover an area of roughly 1 m^2. When researchers use a microscope to look at the cortex, they find between 120,000 and 3,200,000 neurons (brain cells) per cm^2 on its surface. Neurons are entangled and difficult to count. Also, the number of neurons in a small piece of the cortex is different in different areas of the cortex. Using their findings, calculate roughly how many "brain cells" you have on the surface of your cerebral cortex.

Solve this problem using each of the steps described in the text in **Example 1-1** of Section 1-2.

- **Determine** what the problem is asking for, which quantities you know or need, and what will help you solve the problem.
- **Draw** a picture.
- **Label** each variable in the picture and choose a symbol to represent each variable.
- **Construct** a representation of the unknown variable in terms of known variables.
- **Solve** for the unknown variable.
- **Reflect** on your answer.

Show each unit conversion factor used and express values using scientific notation.

8. Betelgeuse is a red supergiant in the constellation Orion and is one of the few stars visible in the night sky even in urban areas. It is a rapidly changing star. It was observed to shrink in size by 15% between 1993 and 2009. It is a large star with a diameter of approximately 1.4×10^{12} m. This is greater than the distance from Earth to the Sun, which is approximately 1.5×10^{11} m.

Measurement of length in space is often made in parsecs. [Interesting background information: 1 parsec (pc) is the distance to a star whose parallax is 1 s of arc (or 1000 milliarcseconds, 1000 mas). Parallax is the shift in apparent position when viewed from two different observation points. In this particular case, the two points of observation are opposite positions in Earth's orbit.] One parsec is 3.26 light years (ly). The distance from Earth to Betelgeuse is 197 pc.

(a) What is the distance from Earth to Betelgeuse in light years? Write the answer using scientific notation. A light year is the distance light travels (at a speed of approximately 3.0×10^8 m/s) in one Earth year.

(b) What is the length of a light year in meters? Write each unit conversion factor in scientific notation.

(c) What is the distance from Earth to Betelgeuse in meters?

1-5 Correct use of significant digits helps keep track of uncertainties in numerical values

NEED TO REVIEW?

Turn to page M1 in the Math Tutorial in the back of the book for more information on significant digits (also called significant figures).

Although physicists strive to make their measurements as precise as possible, the unavoidable fact is that there is always some uncertainty in every measurement. This is not a result of sloppiness or of a lack of care on the part of the person doing the measurement; rather, it is due to fundamental limitations on how well measurements can be carried out.

If you try to measure the length of your thumbnail using a ruler with markings in millimeters, you might find that it's between 15 and 16 mm. With keen eyes you might be able to tell that the length is very nearly halfway between those two markings on the ruler, so you might say that the length is approximately 15.5 mm. However, armed only with your ruler it would be impossible to say whether the length was 15.47, 15.51, or 15.54 mm. We would say that the fourth digit is *not significant*, that is, it does not contain meaningful information. Hence this measurement has only three **significant digits**. We specify this implicitly by giving the measured value as 15.5 mm, by which we mean that it's between 15.4 and 15.6 mm. More precise measurements have more significant digits; less precise measurements have fewer significant digits.

WATCH OUT !

Sometimes you need to be able to estimate uncertainty in a calculation.

While you are not required to do propagation of error on the AP® Physics 1 exam, sometimes it is still useful to examine if a calculated value makes sense. To determine the impact of uncertainties in measurements of quantities on values calculated from those quantities, an estimate can be made by using all of the measured quantities with the limit that gives the lowest value to calculate your lower bound, and all of the measured quantities with the limit that gives the highest value to calculate your upper bound. For instance, the lower bound on speed would be found by dividing the measured value of distance minus the uncertainty in the measurement, by the measured value of elapsed time plus the uncertainty in that measurement. Since uncertainty usually partially cancels, this technique overestimates your uncertainty, so you can be reasonably confident the value falls within the range you calculate.

An alternative way to present the number would be to give the length of your thumbnail as 15.5 ± 0.1 mm, with the number after the ± sign representing the uncertainty. This kind of presentation is very common in experimental work. In this text, however, we'll usually use significant digits to represent uncertainty and you will practice using the ± sign representation in your laboratory work.

Note that some numbers are *exact*, with no uncertainty whatsoever. Three examples are the number of meters in a kilometer (exactly 1000), the speed of light (defined to be exactly 299,792,458 m/s, as we learned in Section 1-4), and the square root of 4. There might also be no uncertainty when you are counting objects, such as "there are four chairs at the table." The value of any measured quantity, however, will have some uncertainty.

Scientific notation is helpful in representing the number of significant digits in a measurement. When we say that the diameter of the Moon at its equator is 3476 km, we imply that there are four significant digits in the value and that the uncertainty in the diameter is only about ±1 km. Since there are 1000 m in 1 km, we could also express the diameter as 3,476,000 m. This is misleading, however, since it gives the impression that there are *seven* significant digits in the value. That isn't correct; we do *not* know the value of the Moon's equatorial radius to within 1 m. A better way to express the diameter in meters is as 3.476×10^6 m, which makes it explicit that there are only four significant digits in the value. The radius of the minor planet Huya is less well known and is about 460 km; only the first two digits are significant, so it's more accurate to express this dimension as 4.6×10^2 km.

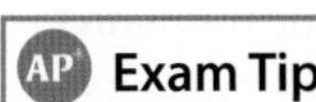

Exam Tip

On the AP® Physics 1 exam you will not need to calculate uncertainty. However, you will be required to show that you understand the *principles* of uncertainty. You will be able to practice these types of problems throughout this textbook. Additionally, you'll gain practice in the lab accompanying this course. This is useful because on the AP® exam you may be required to assess an experimental design or procedure and decide whether the conclusions can be supported based on the evidence. It's important to remember that you'll always need to be able to discuss how the results of an experiment may be affected by modifications to an experimental procedure or sources of experimental uncertainty, as well as be able to report which measurement or variable in a procedure contributes most to overall uncertainty in a final result. In fact, on the AP® exam, it's expected that you'll be able to explain the effects of error and error propagation on conclusions that you draw from a given data set, and how the change in measurement (either in technique or precision) would affect the overall outcome. Additionally, while the experiment or data analysis questions do not ask you to carry out propagation of error or linear regression on the exam, it's expected that you'll be able to linearize data and to estimate a line of best fit to given data or data that you plot. You may also be asked to identify that there is no significant difference between two reported measurements if they differ by less than the uncertainty in the measurements. And, last but not least, you should be able to report results of calculations to an appropriate number of significant digits, and reason in terms of percent error.

AP **Exam Tip**

Percent error (relative error × 100) is one of the simplest types of uncertainty to calculate and is needed to reason if there is a difference between measured and theoretical or predicted (expected) values in experimental situations on the AP® exam. We define absolute error as the difference between our measured and expected values. Relative error is the ratio of the absolute error to the expected value:

$$\frac{(\Delta x)}{x} = \frac{(x_{\text{measured}} - x_{\text{expected}})}{x_{\text{expected}}}$$

This, multiplied by 100, gives you the percent error. If it is negative, your measured value was smaller than your expected value.

There are special rules to help determine whether zeroes in a number are significant. *Leading zeroes* are not significant, so the number 0.0035 has only two significant digits. *Trailing zeroes* to the right of the decimal point are used only if they are significant, so the number 0.3500 has four significant digits. When we have trailing zeroes to the left of the decimal point, as in 350, things get ambiguous. To make it clear that the trailing zero is significant, it is best to write 3.50×10^2.

However, there are times when it's more expedient to write exact numbers without scientific notation. For example, there are exactly 60 s in 1 min, so the 60 has as many significant digits as we need, and we won't bother writing 6.0000×10^1 s or 60.000 to indicate that the value should have five significant digits.

In most problems in this book, we'll use numbers with two, three, or four significant digits. For example, you might be asked about the motion of an ostrich that runs 43.9 m (three significant digits) in 2.3 s (two significant digits). If you use your calculator to find the speed of this ostrich (the distance traveled divided by the elapsed time), the result will be

$$\frac{43.9 \text{ m}}{2.3 \text{ s}} = 19.08695652 \text{ m/s} \left(\text{incorrect}\right)$$

This answer is incorrect because there are *not* 10 significant digits in the answer. You do not know the result that precisely. The general rule for *multiplying or dividing* numbers is that the result has the same number of significant digits as the input number with the *fewest* significant digits. In our example the elapsed time of 2.3 s has fewer significant digits (two) than the distance of 43.9 m, so the result has only two significant digits:

$$\frac{43.9 \text{ m}}{2.3 \text{ s}} = 19 \text{ m/s} \left(\text{correct}\right)$$

Following the same rule, if a strip of cloth is 35.65 cm long and 2.49 cm wide, the area (length times width) has only three significant digits: $(35.65 \text{ cm}) \times (2.49 \text{ cm}) = 88.7685 \text{ cm}^2$, which rounds to 88.8 cm^2.

There's a slightly different rule for *adding or subtracting* numbers. Suppose that you drive 44.3 km (according to the odometer of your car) from Central City to the parking lot at the Metropolis shopping mall, then walk an additional 108 m, or 0.108 km (according to your pedometer), to your favorite clothing store. The total distance you have traveled is $(44.3 \text{ km}) + (0.108 \text{ km})$, which according to your calculator is 44.408 km. This isn't quite right, however, because you know your driving distance only to the nearest tenth of a kilometer. As a result, you need to round off the answer to the nearest tenth of a kilometer, or 44.4 km.

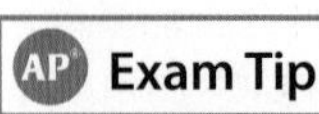

Exam Tip

The AP® Physics 1 exam does not require the use of the strict rules of significant digits, but the number of digits should be reasonable. Usually two to four digits are acceptable.

One digit to the right of the decimal point | Three digits to the right of the decimal point

$$44.3 \text{ km} + 0.108 \text{ km} = 44.4 \text{ km}$$

Result has only one digit to the right of the decimal point

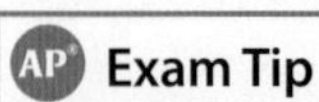

Exam Tip

On the AP® Physics 1 exam, you may be asked to recognize if data agree with a prediction. For instance, if you would expect a result to be the same for multiple trials, the numbers given do not all have to be exactly the same, but within either significant digits or experimental uncertainty.

Thus the general rule for adding or subtracting numbers is that the result can have no less uncertainty than the most uncertain input number. In our example that's the 44.3-km distance, which has an uncertainty of about 0.1 km. Stated another way, the answer may have no more digits to the right of the decimal point than the input number with the fewest digits to the right of the decimal point.

Let's summarize our general rules for working with significant digits when numbers are appropriately expressed in scientific notation:

Rule 1 When multiplying or dividing numbers, the result should have the same number of significant digits as the input number with the fewest significant digits. The same rule applies to squaring or taking the square root. When calculating something where a function changes shape, such as sines and cosines, this is only approximate, but for our purposes, we will be safe in doing so.

Rule 2 When adding or subtracting numbers, when all numbers are expressed in scientific notation with the same exponent the result should have the same number of digits to the right of the decimal point as the input number with the fewest digits to the right of the decimal point. Remember that you should write numbers so that the rightmost number is significant (not 350 but 3.5×10^2, for instance, if we don't know the ones place exactly).

We'll use these rules for working with significant digits in all of the worked examples in this text. (If you look back, you'll see that we used them in Examples 1-2 and 1-3 in Section 1-3.) Make sure that you follow these same rules in your own problem solving.

WATCH OUT !

Be careful with significant digits and scientific notation.

For a number given in scientific notation, the exponent of 10 is known exactly; it doesn't affect the number of significant digits. So 1.45×10^{-2} has three significant digits, and 2.2×10^{-4} has two significant digits. If you *multiply* these together, by Rule 1 their product has only two significant digits: $(1.45 \times 10^{-2})(2.2 \times 10^{-4}) = 3.19 \times 10^{-2+(-4)} = 3.19 \times 10^{-6}$, which rounds to 3.2×10^{-6}. To *add* these two numbers together, first write them using the same exponent of 10: $1.45 \times 10^{-2} + 2.2 \times 10^{-4} = 1.45 \times 10^{-2} + 0.022 \times 10^{-2} = (1.45 + 0.022) \times 10^{-2}$.

In this sum, there are two significant digits after the decimal point for the first number versus three significant digits for the second number. By Rule 2 the sum can have only two significant digits after the decimal point, so we must round $1.45 + 0.022 = 1.472$ to 1.47, and the final answer is 1.47×10^{-2}.

WATCH OUT !

Be mindful of significant digits when rounding numbers.

If you're doing a calculation with several steps, it's best to retain *all* of the digits in your calculation until the very end. Then you can round off the result as necessary to give the answer to the correct number of significant digits. For example, suppose you want to calculate $32.1 \times 4.998 \times 4.87$. Rule 1 tells us that the final answer has only three significant digits. The product of the first two numbers is $32.1 \times 4.998 = 160.4358$, which if you round off at this point becomes 160. If you multiply this rounded number by 4.87, you get 779.2, which rounds to 779. If, however, you calculate $32.1 \times 4.998 \times 4.87$ directly using your calculator *without* rounding the intermediate answer, you get 781.322346. Rounding this to three significant digits gives the correct answer, which is 781. Moral: Don't round until the end of your calculation!

EXAMPLE 1-4 Significant Digits: Combining Volumes

What is the volume (in cubic kilometers) of Earth? What is the volume of the Moon? What is the combined volume of the two worlds together? The radius of Earth is 6371 km, and the radius of the Moon is 1737 km.

Set Up

From the Math Tutorial the volume of a sphere of radius R is $4\pi R^3/3$. Because we multiply the radius by itself, the answer can have no more significant digits than the value of the radius.

Radius of Earth $R_E = 6371$ km
Radius of Moon $R_M = 1737$ km

Solve

The radius of Earth, R_E has four significant digits, so the volume has only four significant digits.

Volume of Earth:

$$V_E = \frac{4\pi(6371\text{ km})^3}{3}$$

$$= 1.083206917 \times 10^{12}\text{ km}^3$$

Round to four significant digits:

$$V_E = 1.083 \times 10^{12}\text{ km}^3$$

The radius of the Moon also has four significant digits, so its volume does as well.

Volume of Moon:

$$V_M = \frac{4\pi(1737\ \text{km})^3}{3}$$
$$= 2.195270618 \times 10^{10}\ \text{km}^3$$

Round to four significant digits:

$$V_M = 2.195 \times 10^{10}\ \text{km}^3$$
$$V_E = 1.083 \times 10^{12}\ \text{km}^3$$
$$V_M = 2.195 \times 10^{10}\ \text{km}^3$$
$$= 0.02195 \times 10^{12}\ \text{km}^3$$

To find the combined volume we add V_E and V_M. To make sure we retain the correct number of significant digits in our answer, we express both numbers in scientific notation with the same exponent. (Remember to write both numbers to the same power of ten before adding.) According to Rule 2 for addition, our answer must have only three significant digits to the right of the decimal point.

$$V_E + V_M = \left(1.083206917 \times 10^{12}\ \text{km}^3\right) + \left(0.02195270618 \times 10^{12}\ \text{km}^3\right)$$
$$= (1.083206917 + 0.02195270618) \times 10^{12}\ \text{km}^3$$
$$= 1.10515962318 \times 10^{12}\ \text{km}^3$$

To avoid intermediate rounding errors we will use the unrounded values and then round the result to three digits to the right of the decimal point at the end of the calculation.

This has too many significant digits, so we round to the final answer of

$$V_E + V_M = 1.105 \times 10^{12}\ \text{km}^3$$

Reflect

In this problem we had to use both the significant digit rule for multiplication and the significant digit rule for addition. You should be prepared to use both of these rules when solving problems on your own. In this example, if we had used the rounded values of the volumes, we would have gotten $V_E + V_M = 1.083 \times 10^{12}\ \text{km}^3 + 0.02195 \times 10^{12}\ \text{km}^3 = 1.10495 \times 10^{12}\ \text{km}^3$, which still rounds to $1.105 \times 10^{12}\ \text{km}^3$. But this is not always the case, so don't round until you have finished all your calculations. You don't necessarily have much experience thinking about numbers as large as these, but we can see that our answer is reasonable: Each radius was between 10^3 and 10^4, so by the time we cube them and add them together, 10^{12} is reasonable.

NOW WORK Problem 1 from The Takeaway 1-5.

THE TAKEAWAY for Section 1-5

✔ To use significant digits correctly, numbers should be expressed in scientific notation.

✔ Numbers that are known exactly, such as 60 s in a minute, have unlimited significant digits.

✔ Significant digits characterize the precision of a measurement or the uncertainty of a numerical value.

✔ When multiplying or dividing numbers the result has the same number of significant digits as the input number with the fewest significant digits.

✔ When adding or subtracting numbers, the result has the same number of digits to the right of the decimal point as the input number with the fewest number of digits to the right of the decimal point when all numbers are expressed in scientific notation with the same exponent.

✔ When writing a number in scientific notation all trailing zeroes to the right of the decimal point are significant digits.

Prep for the AP® Exam

AP® Building Blocks

EX 1-4 1. A pair of stars in the constellation Cygnus comprise an eclipsing binary system. This means that from Earth we observe one star pass in front of the other as the two orbit about a point between them. This allows a direct observation of the time it takes for each orbit, the orbital period. The orbital period of this system has been measured for many years and is decreasing. The explanation is that the distance between these stars is decreasing. They are predicted to merge with a brilliant explosion that will be visible from Earth without a telescope in 2022.

Relative to the mass of the Sun, $M_{Sun} = 1.989 \times 10^{30}$ kg, these stars, labeled KIC 9832227 A and KIC 9832227 B, have masses $m_A = 1.395M_{Sun}$ and $m_B = 0.3180M_{Sun}$. Express each numerical value using scientific notation and apply the rules for significant digits to write your answer.
(a) What is the mass of KIC 9832227 A?
(b) What is the mass of KIC 9832227 B?
(c) What is the combined mass of this system?

2. Apply the rules of significant digits to write the results of these calculations. Record the entire number produced by your calculator. Cross out the numbers that should not be included in the answer. If you are using a calculator with the EE button, then use it rather than the 10^x key.
(a) $3411/62 =$ ________
(b) $3411/62.0 =$ ________
(c) $3411.0/62.0 =$ ________
(d) $6.05 \times 10^5 / 1.0 \times 10^{11} =$ ________
(e) $6.05 \times 10^5 / 1.00 \times 10^{11} =$ ________
(f) $6.05 \times 10^5 / 0.1 =$ ________
(g) $6.05 \times 10^5 \times 1 \times 10^{-1} =$ ________

3. Simple rules for the determination of the number of significant digits apply for addition, subtraction, multiplication, and division. Other functions may not be so simple. There are two significant digits in $x = 1.0 \times 10^1$. This is equivalent to the statement that $9.5 < x < 10.5$. Let's calculate the number of significant digits in e^x, where x has two significant digits. If you plug values into your calculator, you will find $e^{9.5} = 13359.7$, $e^{10} = 22026.5$, and $e^{10.5} = 36315.5$.

What is the number of significant digits in e^{10}?

4. Which of the following gives the product $1.4 \times 15 \times 7.15 \times 8.003$ using the proper number of significant digits?
(A) 1201.65045
(B) 1201.7
(C) 1202
(D) 1200

5. Which of the following gives the result of 0.0688/0.028 using the proper number of significant digits?
(A) 2
(B) 2.5
(C) 2.46
(D) 2.457

AP® Skill Builders

6. Explain why the product of the measured numbers 83 m and 1.2 m cannot be represented as 100.0 m^2. Rather than count the number of significant digits in each number and apply a rule to infer the significance of each number in the product, describe the reasoning underlying the rule.

7. The mass of an object is measured on a pan balance with a precision of 0.005 g and a recorded value of 128.01 g. A mass of a second object is measured on a pan balance with a precision of 0.005 kg and a recorded value of 0.13 kg. Evaluate the claim that these two objects have different masses.

8. A large rectangular sheet of steel plate is measured with a ruler whose smallest increment is 0.1 cm. The length and width of the sheet are 26.55 cm and 19.20 cm, respectively. Design specifications require the area of the plate to be less than 510.0 cm^2. **Justify the claim** that this sheet of steel does not meet specifications.

9. Predict the heel-to-heel length of your unhurried step to the nearest inch. Convert that length to meters. Your step can provide a useful estimate of meter-scale lengths. To get a feel for the accuracy of this measurement scale, you measure a box and find it to be one step in length. For a "theoretical" (expected) value, you use a meter stick to measure the length of the box, given the meter stick is more precise than your step. Assume you predicted your step to be 30 ± 1 inches, and measured the box to be one step. You then use the meter stick and find it to be 71 cm. Calculate the percent error of your measurement, defined as

$$\frac{(x_{measured} - x_{expected})}{x_{expected}} \times 100$$

10. Prediction of the width of the fingernail on the index finger of your left hand, using the widest part of the nail might be 1 cm, 10 mm. This provides a useful personal reference for the estimate of centimeter-scale lengths. Measure the width of your fingernail, and using this as the expected value, calculate the percent error as defined in the preceding question using the 10-mm estimate as your measured value.

11. Predict the arc subtended by your closed fist held with your arm extended. One way to do this would be to estimate how many widths of your fist you move your hand as you go the 90 degrees from holding your arm straight out to your side to straight out in front of your shoulder. This provides a useful personal reference for the estimate of celestial angles. Reflect on the measurement. Each time that you positioned your hand it was positioned relative to the last position for which you had only a mental construction—a memory. How accurate is that location? Assume you counted between 11 and 12 fists in the 90° arc. Report the interval of degrees/fist this uncertainty represents.

12. The rest mass of a proton is a fundamental quantity from which other atomic-scale quantities are often inferred, such as the rest mass of an electron. Heiβe and co-workers have measured the rest mass of a proton with the greatest precision yet attained. They report the mass using a common notation in which a space is left after every three digits to the right of the decimal point and the uncertainty in the rightmost digits is expressed with parentheses:

$$m_p = 1.007\ 276\ 466\ 583(29)\ \text{u}$$

where u is the unified atomic mass unit which is one-twelfth of the mass of a carbon-12 atom. The number in parentheses is the uncertainty, ±0.000 000 000 029.

Heiβe and co-workers were also able to measure the ratio of the proton (p) and electron (e) rest masses:

$$m_p / m_e = 1836.152\ 673\ 346\ (81)$$

(a) Express the electron rest mass, m_e, in terms of the proton rest mass, m_p, and the mass ratio, m_p/m_e.
(b) Using their results, calculate the electron rest mass, m_e, in units of atomic mass units, u, and report the result using ± notation.
(c) Using the conversion factor u = 1.660 539 040 × 10^{-27} kg, calculate the electron rest mass in kg to nine significant digits. Show each factor used in unit conversion.
(d) With one or more sentences, compare your calculated value with that reported by the National Institute of Science and Technology of $9.109\ 383\ 56(11) \times 10^{-31}$ kg.

Skills in Action

13. Considering the thumbnail example from the beginning of this section (p. 16), in your own words summarize the reasoning upon which the convention that the precision of measurement with a linear scale is equal to one-half of the smallest increment on which the scale is based.

14. Galileo used his pulse for the measurement of time. Measure the time between your heartbeats after sitting quietly for a few minutes to the nearest 0.1 s. You can obtain that level of precision with an ordinary analog clock by counting several pulses and dividing the elapsed time by the number of pulses. Speed is the distance you traveled divided by the time that elapsed while you traveled that distance.
(a) Predict your unhurried walking speed. Count off 10 steps, counting your heartbeats as you walk, to measure your walking speed in paces per beat and test your prediction.
(b) The length of your step contained some uncertainty when you measured it. Your measurement of time contains some uncertainty. Use these to estimate how much uncertainty there is in your measurement of your walking speed. Report your measurement in meters per second and include the uncertainty.

1-6 Dimensional analysis is a powerful way to check the results of a physics calculation

Exam Tip

You should always practice any calculations in AP® Physics starting from the equation, then substituting in numbers with units. You can cancel the units as we did with unit conversions in Section 1-4. Getting the answer with the correct units helps ensure you did not make an easy-to-correct math error!

Once you've solved a problem, it's useful to be able to quickly determine how likely it is that you've solved it correctly. One valuable approach is to do a dimensional analysis. This approach is particularly useful when you calculate an algebraic result and want to check it before substituting numerical values.

To understand what we mean by dimensional analysis, first note that any physical quantity has a **dimension**, which describes the base quantity and to which we apply units of measure. Three physical quantities that we introduced in Section 1-4 are the base quantities: mass, length, and time. (Table 1-1 in Section 1-4 lists additional physical quantities.) We say that the dimension of the mass of a hydrogen atom is mass, the dimension of the diameter of the atom is length, and the dimension of the amount of time it takes light to travel that diameter is time. Mass, length, and time are *fundamental* dimensions because they cannot be expressed in terms of other, more fundamental quantities. (You will encounter a few other fundamental dimensions in Physics 2.)

Dimensions are not units themselves. Units are used to help express the numerical value of a dimension. A dimension of length or distance may be expressed in several different units including inches, miles, feet, or meters; for example, the length of a typical house cat is 0.65 m. The dimension of time might be expressed in units of nanoseconds, and the dimension of mass might be expressed in units of kilograms.

The dimensions of many quantities are made up of a combination of fundamental dimensions. One example is speed, which has dimensions of distance per time, or (distance)/(time) for short. Another is volume, which has dimensions of distance cubed, or $(\text{distance})^3$. Volume may be expressed in any number of different units such as m^3, cm^3, mm^3, or km^3. But no matter what units are used, volume *always* has dimensions of $(\text{distance})^3$.

Here's the key to using **dimensional analysis:** *In any equation, the dimensions must be the same on both sides of the equation.* As an example, suppose you hold a ball at height h above the ground and then let it fall. The ball takes a time t to reach the ground. Suppose you calculate that the relationship between the height h and the time t is

$$h = vt^2 \qquad (1\text{-}2)$$

where v is the speed of the ball just before it hits the ground. Is Equation 1-2 dimensionally correct? To find out, replace each symbol by the dimensions of the quantity that it represents. The height h has dimensions of length or distance, speed v has dimensions of (distance)/(time), and time t has dimensions of time. Dimensions cancel just like algebraic quantities, so if we simplify this, we get

$$\begin{aligned} \text{distance} &= \frac{\text{distance}}{\text{time}} \times (\text{time})^2 \\ &= \frac{\text{distance}}{\cancel{\text{time}}} \times \cancel{\text{time}} \times \text{time} \\ &= \text{distance} \times \text{time} \end{aligned}$$

This is *inconsistent:* The left-hand side of Equation 1-2 (h) has dimensions of distance, while the right-hand side (vt^2) has dimensions of distance multiplied by time. So Equation 1-2 *cannot* be correct.

It turns out that the correct relationship between the height h from which the ball is dropped and the time t it takes to reach the ground is

$$h = \frac{1}{2} g t^2 \tag{1-3}$$

where g is a constant. Let's use dimensional analysis to figure out the dimensions of g. Again h has units of distance and t has units of time. The number 1/2 has no dimensions at all, so it can be ignored (it is a pure number with no units, so it is *dimensionless*). We rewrite Equation 1-3 in terms of dimensions and solve for the dimensions of g:

$$\text{distance} = (\text{dimensions of } g) \times (\text{time})^2$$

$$(\text{dimensions of } g) = \frac{\text{distance}}{(\text{time})^2}$$

The SI units of g must then be m/s^2 (meters per second squared, or meters per second per second). We'll learn the significance of the quantity g (called the *acceleration due to gravity*) in Chapter 2.

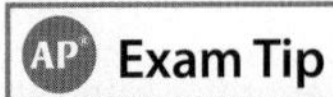

AP® Exam Tip

Performing dimensional analysis can tell you whether an answer is wrong, but even if the dimensions of your answer are correct, the numerical value might still be wrong.

AP® Exam Tip

When performing dimensional analysis on an assessment, it is common practice to place the quantity you are exploring in square brackets. For example, instead of stating (dimensions of g), you would use [g]. It is also common to use L for length or distance, M for mass, and T for time.

THE TAKEAWAY for Section 1-6

✔ Physical quantities are measured in terms of dimensions, which have no numerical value. Units allow us to assign a number to a dimension.

✔ Units can be converted with multiplication by factors that are equal to one.

✔ Dimensions of any quantity can be expressed in terms of a few fundamental dimensions, which include mass, length, and time.

✔ In any equation the dimensions must be the same on both sides of the equation. Doing this dimensional analysis is a technique of checking the correctness of your algebraic results.

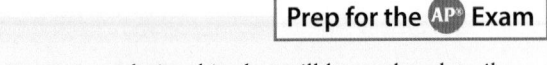

Prep for the AP® Exam

AP® Building Blocks

1. Using dimensional analysis and the definition of the prefix "centi-," confirm that there are 1 million cubic centimeters in a cubic meter.
2. Using dimensional analysis and the definition of the prefix "centi-," confirm that there are 10 thousand square centimeters in a square meter.
3. Velocity has dimensions L/T, where L is length and T is time (as with units, dimensions are not italicized). Acceleration is the time rate of change of velocity. In other words acceleration is the change in velocity divided by the time during which the change occurs. What are the dimensions of acceleration? What are the SI units of acceleration?
4. An important relationship that will be used to describe motion in a circle, such as the orbit of a satellite, is

$$a = \frac{v^2}{r}$$

Show that the dimensions on both sides of this equation are the same.

AP® Skill Builders

5. In Section 1-1 Einstein's statement of mass-energy equivalence was used:

$$E = mc^2$$

where E is the energy of an object in units of joules, m is the mass of the object in units of kg, and c is the speed of light in units of m/s.

(a) Show that for this relationship to be dimensionally consistent energy must have dimensions

$$[E] = \text{mass}\frac{L^2}{T^2}$$

(b) Define the joule in terms of the fundamental SI units for mass, length, and time.

6. Which of the following could be correct based on a dimensional analysis?
(A) The volume flow rate is 64 m^3/s.
(B) The height of the Transamerica Pyramid is 332 m^2.
(C) The time required for a fortnight is 66 m/s.
(D) The speed of the train is 9.8 m/s^2.
(E) The mass of a bar is 2.2 kg.
(F) The density of gold is 19.3 kg/m^2.

7. A block of mass m oscillates back and forth on the end of a spring. You'll find that the time T that the block takes for one full back-and-forth cycle depends on a constant k that has dimensions of mass divided by time squared according to $T = 2\pi\sqrt{\frac{m}{k}}$. Use dimensional analysis to determine whether this result could be correct.

WHAT DID YOU LEARN?

Prep for the AP® Exam

Chapter learning goals	Section(s)	Related example(s)	Relevant section review exercises
Explain the roles that concepts and scientific practices play in physics.	1-1	1-1	1
	1-2		1
	1-3		1
Describe three key steps in solving any physics problem.	1-2	1-1	1, 2
Identify the fundamental units used for measuring physical quantities and convert from one set of units to another.	1-4	1-2, 1-3	3, 4, 5, 6
Use significant digits in calculations.	1-5	1-4	1
Use dimensional analysis to check algebraic results.	1-6		4, 5, 6

Chapter 1 Review

Key Terms

All the Key Terms can be found in the Glossary/Glosario on page G1 in the back of the book.

dimension 22
dimensional analysis 22
exponent 13
fundamental particles 8
fundamental units 11
law 4
mass 10
model 4
object 8
object model 8
rest energy 3
scientific practice 4
scientific notation 13
scientific questioning 4
significant digits 16
system 8
Système International (SI) 10
system model 8
theory 4
thermal process 4
unit 10

Chapter Summary

Topic	Equation or Figure
Using physics to describe the world and solve problems involves using concepts and scientific practices: These are what we will spend the rest of this text developing.	
Problem-solving strategy: We'll solve all problems using a strategy that incorporates three steps.	***Set Up*** the problem. ***Solve*** for the desired quantities. ***Reflect*** on the answer.
Unit conversion: Units can be converted by multiplying any value by an appropriate representation of 1.	④ Multiply the quantity you want to convert by 1 (equal to the ratio of the two units) that enables the unwanted units to cancel out: Multiply 1016 ft by the second form of 1... ⑤ ...the units of ft cancel and the result is in m, as desired. $1016\text{ ft} = 1016\text{ ft} \times 1 = 1016\,\not{\text{ft}} \times \frac{0.3048\text{ m}}{1\,\not{\text{ft}}} = 309.7\text{ m}$ (Figure 1-5)
Scientific notation: Any number can be written in scientific notation, which usually consists of a number between 1 and 10 multiplied by 10 raised to some power. Prefixes can be used as abbreviations for many powers of ten.	$150{,}000{,}000\text{ km} = 1.5 \times 10^8\text{ km}$ $2500\text{ watts} = 2.5 \times 10^3\text{ W}$ $= 2.5\text{ kilowatts} = 2.5\text{ kW}$
Significant digits: Any measured quantity has a finite number of significant digits. You need to keep track of the number of significant digits in all your calculations. There are different rules for addition/subtraction and multiplication/division. **Rule 1** When multiplying or dividing numbers, the result should have the same number of significant digits as the input number with the fewest significant digits. For our purposes we can apply the same rule to squaring, taking the square root, and so on. **Rule 2** When adding or subtracting numbers expressed in scientific notation to the same exponent, the result should have the same number of digits to the right of the decimal point as the input number with the fewest digits to the right of the decimal point.	One digit to the right of the decimal point Three digits to the right of the decimal point $44.3\text{ km} + 0.108\text{ km} = 44.4\text{ km}$ Result has only one digit to the right of the decimal point
Dimensional analysis: Analyzing the dimensions of an equation can provide a clue as to the correctness of that equation. Dimensions and units are not the same thing.	$h = \frac{1}{2} g t^2$ $\text{distance} = (\text{dimensions of } g) \times (\text{time})^2$ $(\text{dimensions of } g) = \frac{\text{distance}}{(\text{time})^2}$ (1-3)

Chapter 1 Review Problems

1. Rank the following lengths from largest to smallest. If any are equal indicate it.
(A) 10 nm
(B) 10 cm
(C) 10^2 mm
(D) 10^{-2} m
(E) 1 m

2. Which of the following is the sum of 1.4 + 15 + 7.15 + 8.003 using the proper number of significant digits?
(A) 31.553
(B) 31.550
(C) 31.55
(D) 31.6
(E) 32

3. Which of the following is the quotient of 0.688/0.28 using the proper number of significant digits?
(A) 2.4571
(B) 2.457
(C) 2.46
(D) 2.5
(E) 2

4. Which of the following relationships is dimensionally consistent with a value for acceleration that has dimensions of distance per time per time? In these equations x is distance, t is time, and v is speed.
(A) v^2/t
(B) v/t
(C) v/t^2
(D) v/x^2
(E) v^2/x^2

5. Which of the following is the product of 25.8 × 70.0 using the proper number of significant digits?
(A) 1806.0
(B) 1806
(C) 1810
(D) 1800
(E) 2000

6. Write the following numbers in scientific notation:
(a) 237
(b) 0.00223
(c) 45.1
(d) 1115
(e) 14,870
(f) 214.78
(g) 0.00000442
(h) 12,345,678

7. Write the following quantities in scientific notation without prefixes: (a) 300 km, (b) 33.7 μm, and (c) 77.5 GW. You should eventually be able to do this without reference materials.

8. Write the following with prefixes and not using scientific notation: (a) 3.45×10^3 s, (b) 2×10^6 W, (c) 7.5×10^{-9} g, and (d) 6.54×10^{-2} m.

9. Write the following numbers using decimals:
(a) 4.42×10^{-3}
(b) 7.09×10^{-6}
(c) 8.28×10^2
(d) 6.02×10^6
(e) 456×10^3
(f) 22.4×10^{-3}
(g) 0.375×10^{-4}
(h) 138×10^{-6}

10. Complete the following conversions:
(a) 125 cm = ______ m
(b) 233 g = ______ kg
(c) 786 ms = ______ s
(d) 454 kg = ______ mg
(e) 208 cm^2 = ______ m^2
(f) 444 m^2 = ______ cm^2
(g) 12.5 cm^3 = ______ m^3
(h) 144 m^3 = ______ cm^3

11. The United States is about the only country that still uses the units feet, miles, and gallons. However, you might see some car specifications that give fuel efficiency as 7.9 km per kilogram of fuel. Given that a mile is 1.609 km, a gallon is 3.785 liters, and a liter of gasoline has a mass of 0.729 kg, what is the car's fuel efficiency in miles per gallon?

12. Complete the following operations using the correct number of significant digits:
(a) $5.26 \times 2.0 =$ ________
(b) $\frac{14.2}{2} =$ ________
(c) $2 \times 3.14159 =$ ________
(d) $4.040 \times 5.55 =$ ________
(e) $4.444 \times 3.33 =$ ________
(f) $\frac{1000}{333.3} =$ ________
(g) $2244 \times 88.66 =$ ________
(h) $133 \times 2.000 =$ ________

13. One equation that describes motion of an object under certain conditions is

$$x = at^2/2$$

where x is the position of the object with units of meters, a is the acceleration with units of meters per second squared, and t is time with units of seconds. Show that the dimensions in the equation are consistent.

14. Give the number of significant digits in each of the following numbers:
    (a) 112.4
    (b) 10
    (c) 3.14159
    (d) 700
    (e) 1204.0
    (f) 0.0030
    (g) 9.33×10^3
    (h) 0.02240
15. One equation that describes motion of an object is $x = vt + x_0$, where x is the position of the object, v is its speed, t is time, and x_0 is the initial position. Show that the dimensions in the equation are consistent.
16. The kinetic energy of a particle is $K = \frac{1}{2}mv^2$, where m is the mass of the particle and v is its speed. Show that 1 joule (J), the SI unit of energy, is equivalent to 1 Nm.
17. The period T of a simple pendulum, the time for one complete oscillation, is given by $T = 2\pi\sqrt{(L/g)}$, where L is the length of the pendulum and g is the acceleration due to gravity. Show that the dimensions in the equation are consistent.
18. A student measures the period of a pendulum of length 30.0 cm ten times and obtains an average period of 1.14 seconds. The theoretical period for a simple pendulum of this length is calculated using the equation $T = 2\pi\sqrt{\frac{L}{g}}$, where L is in meters and $g = 9.80\ \text{m/s}^2$.
    (a) Calculate the theoretical value of the period of a pendulum of this length.
    (b) Calculate the percent error of the value obtained by the student.
19. Convert 30.0 miles per hour to meters per second.
20. The acceleration g (units m/s^2) of a falling object near a planet is given by the following equation $g = GM/R^2$. If the planet's mass M is expressed in kg and the distance of the object from the planet's center R is expressed in m, determine the units of the gravitational constant G.
21. During a certain experiment, light is found to take 37.1 µs to traverse a measured distance of 11.12 km. Determine the speed of light from the data. Express your answer in SI units and in scientific notation, using the appropriate number of significant digits.
22. The body mass index (BMI) estimates the amount of fat in a person's body. It is defined as the person's mass m in kg divided by the square of the person's height h in m.
    (a) Write the formula for BMI in terms of m and h.
    (b) In the United States, most people measure weight in pounds and height in feet and inches. Show that with weight W in pounds and height h in inches, the BMI formula is BMI = $703\ W/h^2$.
    (c) A person with a BMI between 25.0 and 30.0 kg/m^2 is considered overweight. If a person is 5'11" tall, for what range of mass will he be considered overweight?
23. The cost of gas in France is currently 1.33 euros per liter. Currently, the U.S. dollar is worth 0.84 euro. Calculate the current cost of gas in France in U.S. dollars per gallon.
24. Most of the world measures fuel economy in liters per 100 kilometers. A certain Peugeot is listed at 4.6 liters per 100 kilometers. Calculate the fuel economy of this car in miles per gallon.

Prep for the AP® Exam

AP® Group Work

Directions: The problem below is designed to be done as group work in class.

You hire a printer to print concert tickets. He delivers them in circular rolls labeled as 1000 tickets each. You want to check the number of tickets in each roll without counting thousands of tickets. You decide to do it by measuring the diameter of the rolls. The tickets are 4.0 cm long, 1.5 cm wide, and 0.22 mm thick and are rolled on a circular cardboard core 3.0 cm in diameter. What should be the diameter of a roll of 1000 tickets?

Apply mathematical reasoning using each of the steps described in the text:

(a) Set Up
   i. Draw a picture of the coiled roll of tickets lying flat.
   ii. Label each variable in the picture and choose a symbol to represent each variable—area, thickness per ticket, length, length per ticket, number of tickets.
   iii. Look for connections:
   The area of the coiled tickets is hard to calculate because each ring of tickets in the coil has a different radius. What area is easy to calculate and how is that area connected to the area of the coil?

(b) Solve
   i. Construct a representation of the unknown variable in terms of known variables.
   ii. Solve for the unknown variable.
   iii. Your model predicts the size of a roll of 1000 tickets. To count the tickets you can now measure the area. You have a ruler with a smallest increment of 1 mm. How many tickets could there be, given a measurement of the area that agrees with the expected area, but includes the measurement uncertainty?

iv. As an alternative method of indirectly counting tickets you obtain a balance with a precision of ±0.05 g. The mass density, ρ, of the paper used to make the tickets is 250.0 kg/m^3. Calculate the mass of one ticket to determine the uncertainty in the number of tickets counted in this way. Use that to determine the uncertainty in the number of tickets with this method.

(c) Reflect

i. Do your answers seem reasonable?

ii. Is there any reason you can identify that one method would be more reliable than the other?

AP® PRACTICE PROBLEMS

Prep for the AP® Exam

Multiple-Choice Questions

Directions: This is an introductory chapter, so the following questions practice skills for the AP® exam, but do not likely address actual concepts to be directly tested on the exam. The following questions have a single correct answer.

1. Suppose there is a system of units in which the length is the standard thumb, and three thumbs are equal to two clumsys. Which of the following would be the correct volume of a box that is 9.0 thumbs long, 6.0 thumbs wide, and 1.0 thumb tall?
   (A) 16 clumsys^3
   (B) 24 clumsys^3
   (C) 36 clumsys^3
   (D) 81 clumsys^3

2. A student makes the following measurements to determine the density, mass/volume, of a cube.

	Trial 1	Trial 2	Trial 3	Average
Mass (kg)	0.0305	0.0303	0.0303	0.0304
Length of side (m)	0.020	0.021	0.019	0.020

   Which of the following would be the best report of the density?
   (A) 4×10^3 kg/m^3
   (B) 3.8×10^3 kg/m^3
   (C) 3.80×10^3 kg/m^3
   (D) 1.50×10^3 kg/m^3

3. An object's position is given as $x = pt + qt^{-1} + rt^2$, where x is measured in m and t in seconds. The proper dimensions of p, q, and r should be
   (A) p in m/s, q in m/s, r in m/s^2.
   (B) p in m·s, q in m/s^2, r in m·s^2.
   (C) p in m/s, q in m·s, r in m·s^2.
   (D) p in m/s, q in m·s, r in m/s^2.

4. A Newton (N) is defined as a kg·m/s^2. Which of the following would also be true?
   (A) 1 N = 100,000 g·cm/s^2
   (B) 1 N = 1,000 kg·km/s^2
   (C) 1 N = 1,000 kg·cm/ms^2
   (D) 1 N = 10^{-4} kg·m/s^2

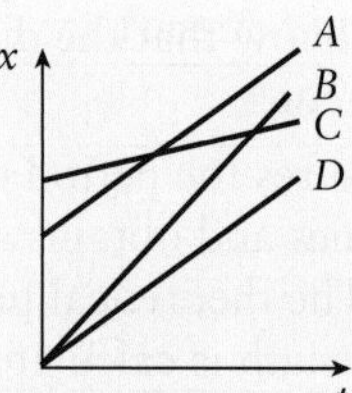

5. Consider this graph for the position as a function of time of an object. The motion of four objects, A, B, C, and D, is represented in the graph. We will find that an important property of the objects' motion is represented by the slope of the line representing their motion in this type of graph. Which of the following ranks this property of the objects' motion from greatest to least based on the slope of the line that represents their motion?
   (A) $A > B > C > D$
   (B) $C > A > (B = D)$
   (C) $(A = D) > B > C$
   (D) $B > (A = D) > C$

Free-Response Question

1. Students are given a series of circular lids from food containers of different sizes. Their teacher asks them to develop a mathematical relationship between the diameter of a circle and the circumference.

(a) Describe an experimental procedure that students could use to collect data that would allow them to plot a relationship between the diameter and the circumference. Assume students have access to the equipment available in a typical physics classroom. Include a diagram showing how they will make the measurements, as well as a list of what equipment they will use to make these measurements.

Another group of students is asked to find the mathematical relationship between the diameter of a circle and its area. Students are given graph paper where each square is 1.0 cm on each side. They use the paper to measure the area of the lids. They obtain the following data.

Diameter d (cm)	Area A (cm^2)	
4.3	14	
7.5	44	
9.5	71	
11.9	111	
15.6	186	

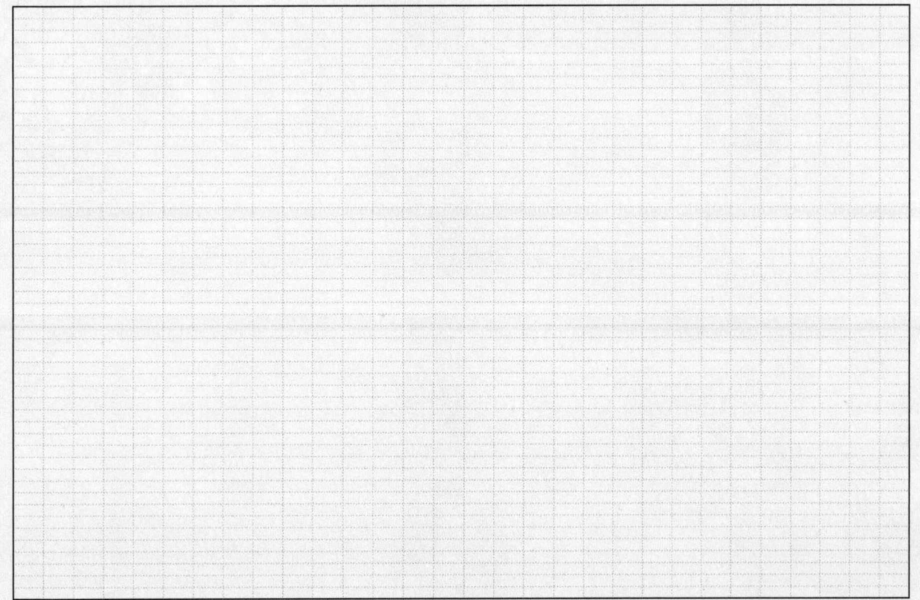

(b) Given that the area of a circle is πr^2, where r is the radius of the circle, sketch the shape of the graph that would be obtained if the area is plotted as a function of diameter.

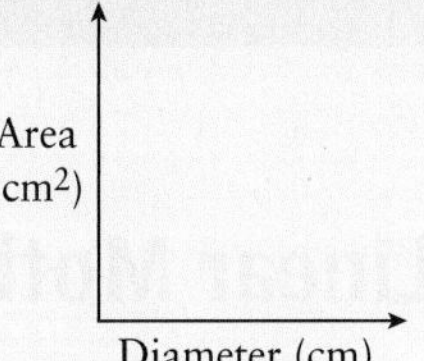

(c) Which variables should students plot on each axis to obtain a linear graph?

(d) Use the empty spaces in the preceding table to calculate the variables you listed in part (c). Plot the data on the axes below and draw a best fit line to the data.

(e) Compare the slope of the best fit line to the experimental data to the accepted value for that slope.

Kinematics

Chapter 2 Linear Motion

Chapter 3 Motion in Two or Three Dimensions

Case Study: How can fundamental physics help us understand baseball?

In the first game of the 1954 World Series between the Cleveland Indians and New York Giants, a Cleveland batter smashed a pitch into deep center field, well over the head of baseball legend Willie Mays. But at the crack of the bat, Mays turned and sprinted toward the center field wall, catching the ball over his shoulder while running at top speed. This play, often cited as one of most iconic moments in sports history, was a result of spectacular athleticism combined with fundamental physics.

Willie Mays was standing still when the ball was struck. To meet the ball at the end of its flight, he had to first accelerate and then run at top speed. Acceleration is the rate of change of velocity. Speed is a measure of how fast something is moving. At any instant, the speed, together with the direction that an object is moving, defines that object's velocity; velocity is the rate of change of position.

The force of Earth's gravity governed the motion of the ball after it was hit. The ball rocketed upward, but gravity caused its vertical speed to decrease as the ball went up. After rising to its peak, the tug of gravity brought the ball back down, faster and faster. But the force of gravity is only exerted in the vertical direction, so the horizontal speed of the ball remained essentially the same throughout its flight. (Not exactly the same, because the moving ball had to push air molecules out of its way, which slowed it down.) It is the combination of the vertical and horizontal components of the ball's motion that resulted in its graceful, parabolic trajectory. However, each component of an object's motion is just a one-dimensional problem that can be treated separately.

How does the angle at which a baseball is launched off a bat affect the distance it travels? Consider a steep angle, a 45° angle, and a shallow angle.

In the famous play from that 1954 World Series game, the ball was in the air about 6 seconds. Mays had to run about 10 m to catch it. At top speed, Willie Mays could cover about 7 m per second—why do you think he was still running hard when he made his spectacular catch?

By the end of these chapters, you should be able to define and explain the relationships among distance, displacement, instantaneous velocity, average velocity, constant velocity, and speed. You should also be able to define acceleration and explain its relationship to velocity, solve constant-acceleration linear motion problems, and solve problems involving projectile motion, such as how fast and in what direction the ball left the bat in 1954!

2 Linear Motion

Patrik Giardino/Getty Images

YOU WILL LEARN TO:

- Define the linear motion of a system and when and how it can be simplified to an object model.
- Define and explain the relationships among distance, displacement, instantaneous velocity, average velocity, constant velocity, and speed.
- Define acceleration, explain its relationship to velocity, and interpret graphs of velocity versus time.
- Create, use, and interpret narrative (written), mathematical, and graphical representations for linear motion with constant acceleration.
- Solve constant-acceleration linear motion problems when given a description of the motion of an object. The description could be in the form of words, graphs, or experimental data.
- Explain when objects are in free fall (constant acceleration due to gravity).

2-1 Studying linear motion is the first step in understanding physics

NEED TO REVIEW?
Turn to the **Glossary** in the back of the book for definitions of bolded terms.

We live in a universe of motion. We are surrounded by speeding cars, scampering animals, and moving currents of air. Our planet is in motion as it spins on its axis and orbits the Sun. There is motion, too, within our bodies, including the beating of the heart and the flow of blood through the circulatory system. One of the principal tasks of physics is to describe motion in all of its variety, and we begin our study of physics with the description of motion, a subject called **kinematics**.

In general, objects can move in all three dimensions—forward and back, left and right, and up and down—and we have to describe the motion in each dimension. In this chapter, however, we'll concentrate on the simpler case of **linear motion** (also called *one-dimensional motion*): motion along a line. Although by their very definition lines are always straight, it is common to see linear motion referred to as straight-line motion. Examples of linear motion include an airliner accelerating down the runway before it takes off, a peregrine falcon diving from height toward its potential prey, and a rocket climbing upward from the launch pad.

These examples of what is traveling in linear motion can each be described as a system. A system, as you might recall from Chapter 1, is a collection of objects that may have internal interactions. The individual objects within a system may not all move in the same way (the wheels on the airliner are rotating about their axles while their axles move in a line along with the airliner, for instance). Often when we describe motion, we don't care about the details of all of the pieces of the system. When we care only about the general direction the system is moving in, and not how its pieces act within it (and only small things we can ignore such as the wheels on the airliner are rotating—we will discuss rotational motion later in this book), it is a great time to use our object model to approach problem solving.

NEED TO REVIEW?
The object model is discussed in Chapter 1, Section 1-3.

When we approach kinematics problems, it's important for us to first consider if the system is rigid. If it is rigid, the parts don't move with respect to one another, such as with the body of a car. This is an important consideration because if the system *is* rigid, we can pick any point in that rigid system as a reference for the location of our *object* model. For example, if we pick the rear-view mirror as the point of reference in our car, this means that all points in the car around the mirror move the same distance when any point in the system moves, so we can describe the motion of the car by describing how the position of the mirror moves. However, not every system is rigid. An example of a nonrigid system is the jumping young woman in the chapter opening photo. In this instance, if we chose a point—say the jumper's hand—it would not always be in the same place with respect to her knee when the system (the jumper) moves. For cases like this, physicists use the concept of *center of mass*. The *center of mass* can be considered as a balance point: for example, as the point at which you could balance a tray with several objects on it on a single finger. While you will practice calculating the center of mass later in Chapter 9 to develop a deeper familiarity with it, you primarily need to understand the concept, which we will develop more in Chapter 3. So we can treat a system as a single object if we don't care how its individual parts are moving, and we can assign a single point to represent its location. If our system is rigid, that single point can be any point, but if the individual parts are moving with respect to each other, that single point should be the center of mass.

Exam Tip

The AP® Physics 1 exam requires a good conceptual understanding of center of mass. You will only be expected to calculate it for relatively simple configurations of mass, but you will need to understand qualitatively how it changes to make predictions for more complex configurations.

All of the concepts that we'll study in this chapter also apply directly to motion in two or three dimensions, which we'll study in Chapter 3. We will also see that no matter how complicated the system, the object model will help us solve many problems. You'll use the ideas of this chapter and the next throughout your study of physics, so the time you spend understanding them now will be an excellent investment for the future.

THE TAKEAWAY for Section 2-1

✔ Linear motion (also called *one-dimensional motion*) is motion along a line.

✔ For many problems we can model systems as objects: For rigid systems we can use any point to represent the motion of the object; for systems with relative motion, we use the center of mass.

✔ We'll build on the ideas of linear motion in later chapters to understand more complex kinds of motion, such as motion in two or three dimensions.

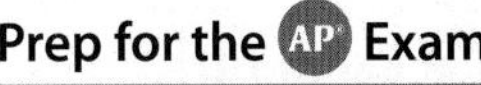

AP® Building Blocks

1. Imagine two methods of pulling a wagon behind a bicycle. In the first method, the wagon can be attached to the bicycle by a long rope. In the second method, the wagon can be attached to the bicycle by an aluminum bar. In which of these methods can the motion of the system, consisting of the wagon and the bicycle, be described using an object model for each of these two different motions?
   (a) While the bicycle is moving along a line without changing speed
   (b) When the bicycle slows down to a stop

2-2 Constant velocity means moving at a constant speed without changing direction

Figure 2-1 shows objects in linear motion: three swimmers traveling along the lanes of a swimming pool. Consider the swimmers as objects, ignoring how their arms and legs are moving relative to the motions of their centers of mass. Their motion is particularly simple because each swimmer has a **constant velocity**: For every equal unit of time, the

Each swimmer moves with a constant speed and always moves in the same direction. Hence, each swimmer has a **constant velocity**.

Fuse/Getty Images

Figure 2-1 Constant velocity An object has a constant velocity when its speed remains the same (it neither speeds up nor slows down) and its direction of motion remains the same. These three swimmers all move in the same direction yet have different constant velocities because their speeds are different. The middle swimmer (who reaches the end of the pool first) is the fastest.

swimmer moves the same distance and always in the same direction. We introduced units in Section 1-4. When we make an argument like this it doesn't matter which unit for time you choose, the definition is the same. You could think in terms of the fundamental unit for time, seconds: Constant velocity means that in each second, you move exactly the same distance, always in the same direction.

To understand constant velocity, we must compare it to something more familiar, **speed.** Just as you might guess, speed is a measure of how fast each swimmer is moving. *Velocity* is not just another word for speed, even though they have the same fundamental units of meters per second. **Velocity** is a measure of how fast position (including direction) is changing. If the velocity is constant, it is the change in position divided by the time it took to make that change. The dimension of length means something different in speed and velocity. In speed, the quantity that has the dimension of length is the distance traveled: How far did the point you were tracking go? In velocity, the quantity that has the dimension of length is how far and in what direction that point ended up from where it started (this quantity is called **displacement**). The remainder of this section digs deeper into these definitions.

Position, Displacement, and Average Velocity

AP Exam Tip

While x is defined as position on the AP® Physics 1 equation sheet, position can also be defined as any other letter: It is also often denoted by y or z in one dimension, or r in three dimensions (Chapter 3).

To be able to mathematically define the changing position of a moving object (such as a swimmer) we must define a **coordinate system**, as shown in **Figure 2-2a**. Since the swimmers are moving in just a single direction, we don't have to worry about the width of the pool. We can define position as where the swimmer, our object, is along the length of the pool. We will use the symbol x to denote the position of the object along the length of the pool at a given time t. The value of x can be positive or negative, depending on where the object is relative to an **origin**, a location we choose that we define to be $x = 0$. We can choose the origin to be anywhere that's convenient for the motion we are trying to describe. For example, if x represents the possible locations of an object along the length of a swimming pool, then we could choose $x = 0$ to be the point where we're sitting beside the pool. If our seat is 10.0 meters (10.0 m for short) from one end of a swimming pool with a length of 50.0 m, then one end of the pool is at $x = -10.0$ m and the other end is at $x = +40.0$ m. We are free to choose the positive x direction (the direction in which x increases) as we wish; in Figure 2-2a, we've chosen it to be to the right.

The **position** of an object relative to an origin is what is called a vector quantity. **Vectors** are special quantities that contain two kinds of information: **magnitude** and **direction**. The magnitude of a vector is its size. For position vectors, the magnitude is the length of the position vector, and is numerically equal to the distance of the object from the origin. For instance, from $x = 0$ the magnitude of the position of the right-hand end of the pool in Figure 2-2a is 40.0 m, and to get there from $x = 0$

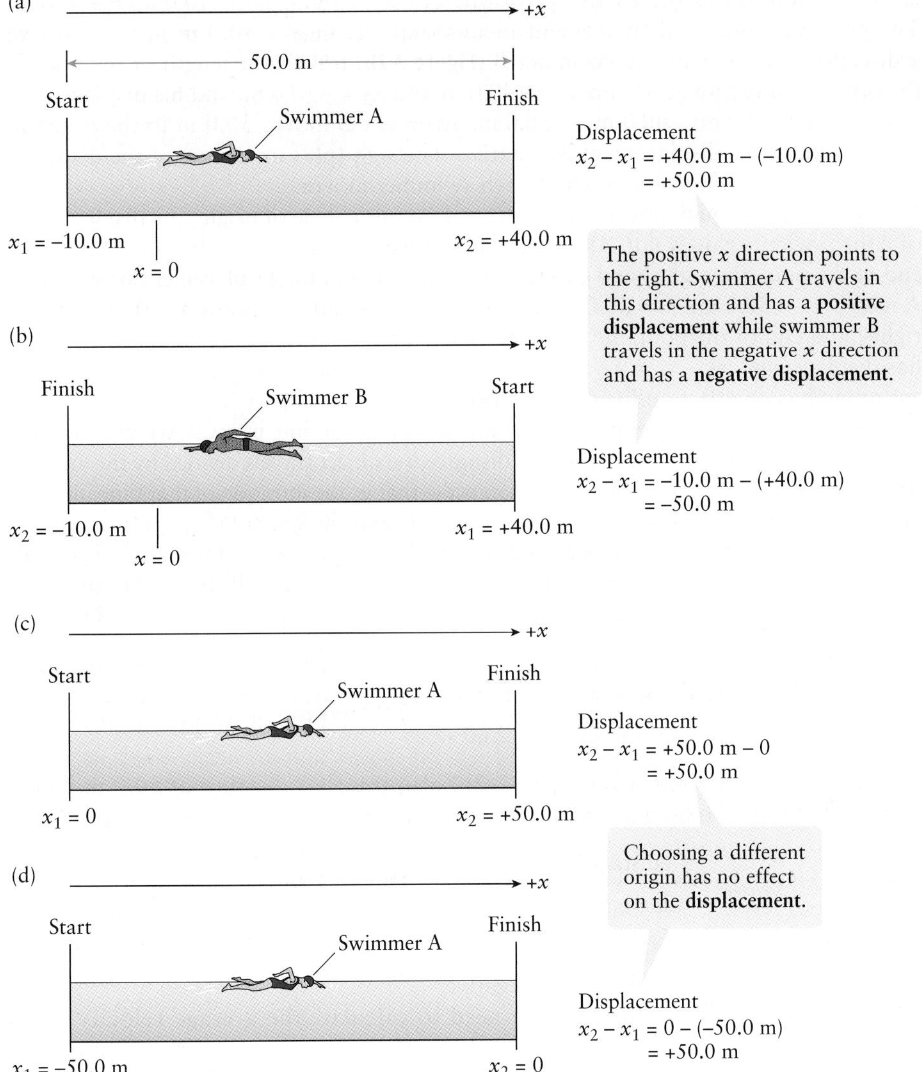

Figure 2-2 Coordinate systems and displacement (a) With our choice of coordinate system, swimmer A has a positive displacement and (b) swimmer B has a negative displacement. (c) and (d) show that displacement is independent of the coordinate system.

you must travel in the positive x direction. We summarize these statements by saying that the right-hand end of the pool is at $x = +40.0$ m. The left-hand end of the pool in Figure 2-2a is 10.0 m away from $x = 0$, and to get there you must travel in the negative x direction (the direction in which x decreases). So the position of this end in our chosen coordinate system is $x = -10.0$ m. We will discuss vectors in much more depth in Chapter 3, when we start looking at motion in more than one direction. In this chapter, we will develop our understanding of vectors in just one dimension. Then we will be able to treat motion in two or three dimensions as just two or three one-dimensional motions!

Distance, by comparison, is a **scalar** quantity: A scalar does not contain information about direction. (If you say something is 50.0 m away, you're stating its distance; if you say it's 50.0 m *to the east*, you're stating its position.) That's why distance, unlike position, is always given as a positive number. If you were asked, "How far is it from New York City to Boston?" you would never answer, "Negative 225 miles!"

The displacement of an object is the difference between its positions at two different times. It tells us how far and in what direction the object moves, so displacement, like position, is a vector. The positive x direction as we defined it in this example points to the right. Swimmer A travels in this direction and has a *positive displacement* while swimmer B travels in the opposite direction and so has a *negative displacement*. For example, if swimmer A travels the length of the swimming pool from left to right (Figure 2-2a), she starts at $x_1 = -10.0$ m and ends at $x_2 = +40.0$ m. (The subscripts "1" and "2" help us keep track of which position is which.) Her displacement is the

WATCH OUT !

A quantity is a vector only if the positive or negative denotes direction.

Some scalar quantities can be either positive or negative. For instance, the temperature might be +35°C at noon in midsummer but –10°C on a cold winter night. But there is no direction such as up or left associated with temperature, so it is a scalar, not a vector, quantity. You need to rely on the context to decide whether the sign of a quantity refers to a direction.

later position minus the *earlier* position: $x_2 - x_1 = +40.0\text{ m} - (-10.0\text{ m}) = +50.0\text{ m}$. The positive value of displacement means that she moves 50.0 m in the positive x direction. By comparison, swimmer B (**Figure 2-2b**) travels the length of the pool in the opposite direction. For him $x_1 = +40.0$ m and $x_2 = -10.0$ m, and his displacement is $x_2 - x_1 = (-10.0\text{ m}) - 40.0\text{ m} = -50.0\text{ m}$. Swimmer B moves 50.0 m in the negative x direction, so his displacement is negative. The *sign* (plus or minus) of the displacement tells you the *direction* in which each swimmer moves.

Note that although position does depend on our choice of origin, the displacement of either swimmer does *not*. As an example, if we choose $x = 0$ to be at the left-hand end of the pool, the right-hand end is at $x = 50.0$ m, and the displacement of swimmer A is $x_2 - x_1 = 50.0\text{ m} - 0 = 50.0\text{ m}$ (**Figure 2-2c**). If instead we choose $x = 0$ to be at the right-hand end of the pool, the left-hand end is at $x = -50.0$ m, and swimmer A again has displacement $x_2 - x_1 = 0 - (-50.0\text{ m}) = 50.0\text{ m}$ (**Figure 2-2d**). So, choosing a different origin has no effect on the displacement.

WATCH OUT !

Displacement and distance are not the same thing.

Remember that distance has no direction and is always a positive number. But displacement, like position, is a vector quantity that has both a magnitude and a direction. Swimmer A in Figure 2-2a and swimmer B in Figure 2-2b both travel the same distance, 50.0 m, but their displacements are different (+50.0 m for swimmer A, −50.0 m for swimmer B) because the *directions* of their motion are different. Either swimmer when they swim back to where they started has a displacement of zero, no matter how far they swam in between, since their initial and final positions are the same when they return to where they started.

To determine how quickly an object moves in a given time interval, we define a new quantity, **average speed.** This is the total distance the object travels divided by the amount of time it takes the object to travel that distance—that is, the duration of that time interval. We use the symbol t to denote a value of the time, and the symbol v_{average} ("v-average") for average speed. For example, suppose you start timing swimmer A just as she starts off at $x_1 = -10.0$ m, so the time at this instant is $t_1 = 0$ (Figure 2-2a). If she reaches the other end of the pool at time $t_2 = 25.0$ s, the time interval is $t_2 - t_1 = 25.0\text{ s} - 0 = 25.0\text{ s}$. She traveled a distance of 50.0 m in that time interval, so her average speed is

$$v_{\text{average}} = \frac{\text{distance}}{\text{time interval}} = \frac{50.0\text{ m}}{25.0\text{ s}} = 2.00\text{ m/s for swimmer A}$$

Slow-moving swimmer B (Figure 2-2b) also travels a distance of 50.0 m, but it takes him 100.0 s. So his average speed is

$$v_{\text{average}} = \frac{\text{distance}}{\text{time interval}} = \frac{50.0\text{ m}}{100.0\text{ s}} = 0.500\text{ m/s for swimmer B}$$

Note that average speed, like distance, tells you nothing about the direction of motion. Like distance, average speed is never negative.

To get direction information, we need to calculate the **average velocity** of an object, which is a vector quantity. We use the symbol $v_{\text{average},x}$ ("v-average-x") for average velocity in the x direction. This is very close to the symbol for average speed but includes the subscript "x" to remind us that the object is moving in either the positive or negative x direction. The average velocity equals the displacement divided by the time interval for that displacement. For swimmer A in Figure 2-2a, who starts at position $x = -10.0$ m and ends at position $x = +40.0$ m in 25.0 s, the average velocity is

$$v_{\text{average},x} = \frac{x_2 - x_1}{t_2 - t_1} = \frac{40.0\text{ m} - (-10.0\text{ m})}{25.0\text{ s} - 0}$$
$$= \frac{50.0\text{ m}}{25.0\text{ s}} = +2.00\text{ m/s for swimmer A}$$

WATCH OUT !

Speed and velocity are not the same thing.

Average speed and average velocity are *not* the same quantity. Average speed is equal to the *distance* traveled divided by the time interval to travel that distance. Distances are always positive, so average speed is always positive. Average velocity is a vector equal to the displacement divided by the time interval for that displacement.

So swimmer A moves in the positive x direction (as shown by the positive sign of $v_{\text{average},x}$), or from left to right in Figure 2-2a, at an average *speed* of 2.00 m/s. Swimmer B in Figure 2-2b starts at $x_1 = +40.0$ m at $t_1 = 0$, and at $t_2 = 100.0$ s he reaches $x_2 = -10.0$ m. Hence the average velocity of swimmer B is

$$v_{\text{average},x} = \frac{x_2 - x_1}{t_2 - t_1} = \frac{-10.0\text{ m} - 40.0\text{ m}}{100.0\text{ s} - 0}$$
$$= \frac{-50.0\text{ m}}{100.0\text{ s}} = -0.500\text{ m/s for swimmer B}$$

Swimmer B moves at only 0.500 m/s. The minus sign of $v_{\text{average},x}$ says that he travels in the negative x direction, or from right to left in Figure 2-2b. Equation 2-1 summarizes these ideas about average velocity:

Average velocity of an object in linear motion

Displacement (change in position) of the object over a certain time interval: The object moves from x_1 to x_2, so $\Delta x = x_2 - x_1$.

$$v_{\text{average},x} = \frac{x_2 - x_1}{t_2 - t_1} = \frac{\Delta x}{\Delta t} \quad (2\text{-}1)$$

For both the displacement and the time interval, subtract the earlier value (x_1 or t_1) from the later value (x_2 or t_2).

Time interval for the motion: The object is at x_1 at time t_1 and x_2 at time t_2, so the time interval is $\Delta t = t_2 - t_1$.

EQUATION IN WORDS
Average velocity for linear motion equals displacement divided by time interval

In the example above, average speed and average velocity are very similar because they have the same magnitude. However, let's imagine the case where swimmer A travels back to her starting point, both starting and ending at position $x = -10.0$ m, in a period of 50.0 s. The total distance traveled in this 50.0 s would be 100 m, two times the length of the pool, so the average speed for swimmer A would be

$$v_{\text{average}} = \frac{\text{distance}}{\text{time interval}} = \frac{100 \text{ m}}{50 \text{ s}} = 2.00 \text{ m/s for swimmer A}$$

On the other hand, the average velocity of swimmer A is

$$v_{\text{average},x} = \frac{\text{displacement}}{\text{time interval}} = \frac{-10 \text{ m} - (-10 \text{ m})}{50 \text{ s}} = 0.00 \text{ m/s for swimmer A}$$

Average speed and average velocity have the same magnitude for a given motion only when the direction of that motion does not change. The standard SI units for average velocity and average speed are meters per second (m/s). **Table 2-1** lists some other common units for these quantities.

AP Exam Tip

Be sure you understand the meaning of Δ. The symbol Δx means the *change* in the value of x. In Equation 2-1 Δx ("delta-x") is an abbreviation for the change $x_2 - x_1$ in the position x. In the same way, Δt ("delta-t") means the change in the value of time, $t_2 - t_1$, that is, the time interval. Throughout our study of physics we'll use the symbol Δ (the capital Greek letter "delta") to represent the change in a quantity.

TABLE 2-1 Units of Velocity and Speed

The SI unit of velocity and speed is meters per second (m/s). Other common units are:
1 kilometer per hour (km/h) = 0.2778 m/s
1 mile per hour (mi/h) = 0.4470 m/s = 1.609 km/h
1 foot per second (ft/s) = 0.3048 m/s = 1.097 km/h = 0.6818 mi/h

Constant Velocity, Motion Diagrams, and *x*–*t* Graphs

We've calculated the average velocity for each swimmer's entire trip from one end of the pool to the other. But we can also calculate the average velocity for any *segment* of a swimmer's trip. A swimmer has *constant velocity* if the average velocity calculated for *any* segment of the trip has the same value as for any other segment. Consider the motion diagram in **Figure 2-3a,** which represents the position of swimmer A at moments in time separated by equal time intervals. Swimmer A is at $x_1 = 0.0$ m at $t_1 = 5.0$ s and at $x_2 = 30.0$ m at $t_2 = 20.0$ s. Her average velocity for this part of her motion is

$$v_{\text{average},x} = \frac{x_2 - x_1}{t_2 - t_1} = \frac{30.0 \text{ m} - 0.0 \text{ m}}{20.0 \text{ s} - 5.0 \text{ s}}$$

$$= \frac{30.0 \text{ m}}{15.0 \text{ s}} = +2.00 \text{ m/s for swimmer A}$$

That's the same value we calculated above for the entire trip from one end of the pool to the other. If the average velocity has the same value for all segments, we sometimes

Figure 2-3 Motion diagrams for constant velocity A series of dots shows the positions of (a) swimmer A and (b) swimmer B at equal time intervals.

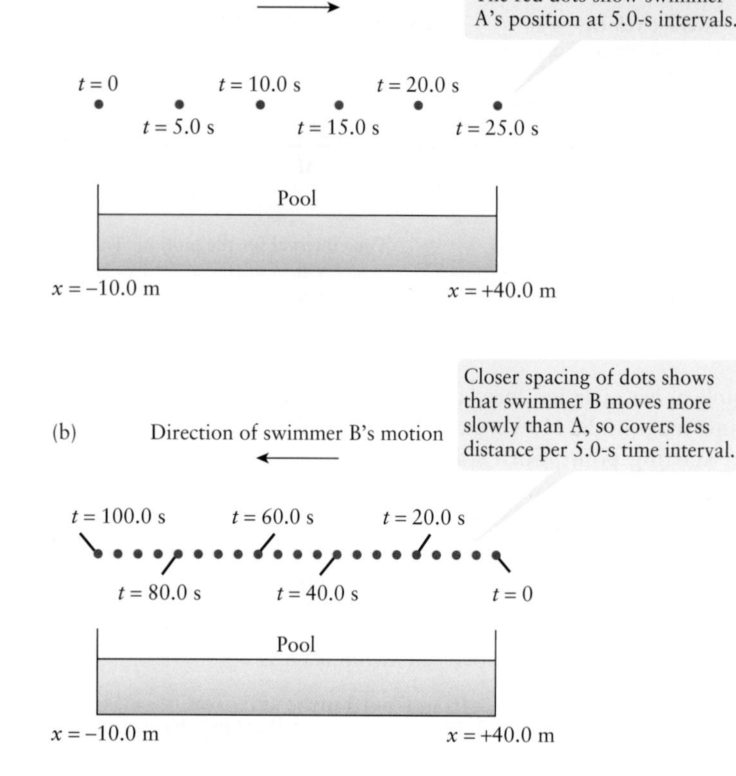

Exam Tip

Representing data in diagrams and on graphs is a very important skill to develop to prepare for the AP® Physics 1 exam.

WATCH OUT !

The *x* axis can be the vertical axis.

When we want to show how a quantity varies with time, we always put time on the horizontal axis. Then, whatever quantity we are interested in, even if it is a horizontal position, goes on the vertical axis. That's why we put the position x on the vertical axis, no matter how the object's motion is oriented in space. Another example is a graph of how a stock mutual fund performs. Time (that is, dates) is on the horizontal axis and the value of the fund on those dates is on the vertical axis.

Exam Tip

Be careful when you are asked to describe motion. For instance, a line has a constant slope. Describing a line as *straight* is not the same as describing it as *horizontal*.

omit the word "constant" and call it simply the *velocity*. We give it the symbol v_x ("v-x"), omitting the subscript "average" because it represents the speed at any moment during the constant velocity motion.

Figure 2-3a shows how to depict swimmer A's entire motion in a way that makes it clear that her velocity is constant. We model the swimmer as an object, using just one point to represent her location, and draw dots to represent her positions at equal time intervals (say, $t = 0$, $t = 5.0$ s, $t = 10.0$ s, $t = 15.0$ s, and so on). Taken together, these dots make up a motion diagram. The spacing between adjacent dots shows you the distance that the object traveled during the corresponding time interval. The spacing is the same between any two adjacent dots in Figure 2-3a, which tells you that swimmer A travels equal distances in the $+x$ direction—that is, has equal displacement—in equal time intervals. That's just what we mean by saying that swimmer A has constant velocity.

Figure 2-3b shows a motion diagram for swimmer B. Since he moves more slowly than swimmer A (speed 0.5 m/s compared to 2.00 m/s), he travels a smaller distance in the same time interval. Hence the dots in swimmer B's motion diagram are more closely spaced than those for swimmer A.

Yet another useful way to depict motion is in terms of a graph of position versus time, also known as an x–t graph. In **Figure 2-4a,** we've turned swimmer A's motion diagram on its side so that the x values are now represented on the vertical axis. Because each dot in the diagram corresponds to a specific time t, we've moved each dot to the right by an amount that corresponds to the value of t. We add a horizontal axis for time to make it easy to read the value of t that corresponds to each dot. The result is an x–t graph, a graph of coordinate x of the swimmer's position versus the time t. To finish the graph, we draw a smooth curve connecting the dots. For this constant-velocity motion, the graph is a single line. Although the swimmer's motion in the x direction in Figure 2-3a is *horizontal*, we've drawn the x–t graph of Figure 2-4a with the x axis *vertical*. We do this because in graphs that show how a quantity (the dependent variable) varies with time, we always put time (the independent variable) on the horizontal axis.

Figure 2-4b shows the x–t graph for swimmer B. This swimmer also has a constant velocity, so the x–t graph is again a straight line. However, this graph differs from the x–t graph for swimmer A (Figure 2-4a) in two ways: where the line touches the vertical axis, and the *slope* of the line (how steep it is and whether it slopes up or down).

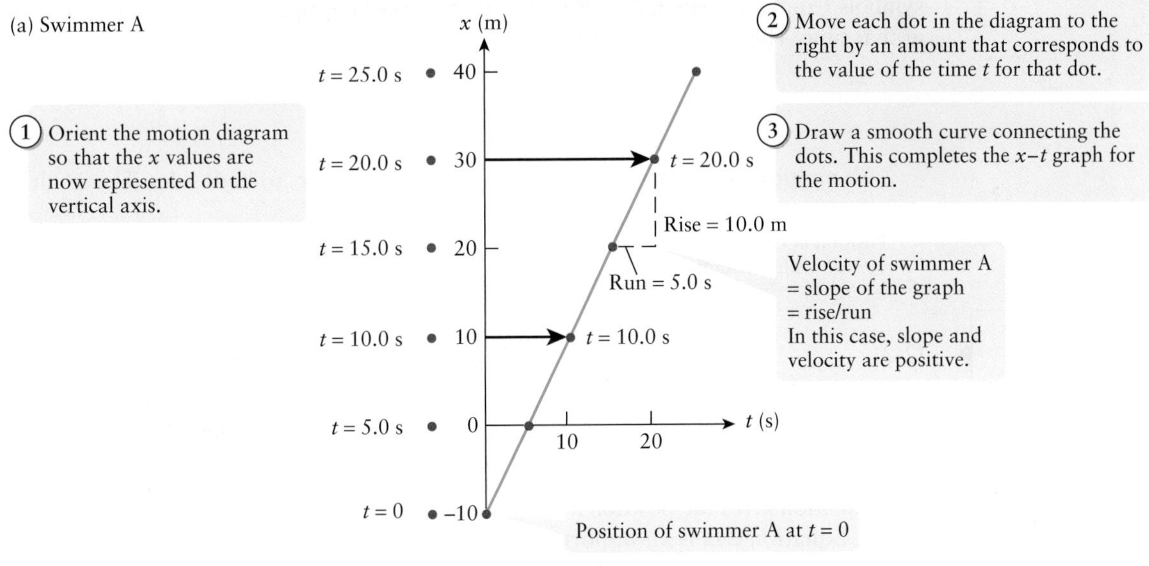

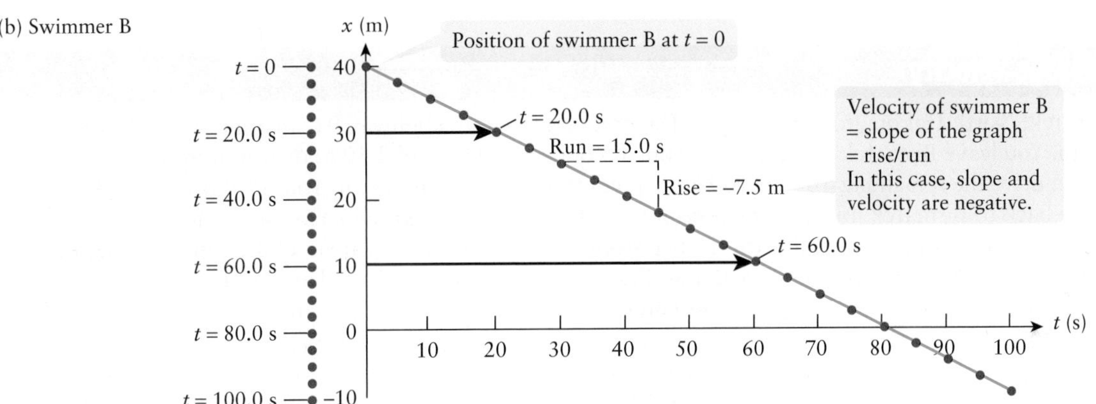

Figure 2-4 From motion diagrams to *x–t* graphs: constant velocity How to convert the motion diagrams from Figure 2-3 for (a) swimmer A and (b) swimmer B into graphs of position *x* versus time *t*. If the velocity is constant, the *x–t* graph is a line. The slope (rise divided by run) of the line equals the velocity.

The point on the graph where the line touches the vertical (x) axis shows you the position x of the object at $t = 0$. We give this the symbol x_0 (typically spoken as "x-naught," where naught means nothing, another way to say zero). If the motion starts at $t = 0$, then x_0 represents the *initial position* of the object. The slope of the line tells you the value of the object's *velocity*. If you consider two points on the line, corresponding to times t_1 and t_2 and positions x_1 and x_2, the slope is the vertical difference $x_2 - x_1$ between those two points (the "rise" of the graph) divided by the horizontal difference $t_2 - t_1$ (the "run" of the graph). But that's just our definition of the average velocity, Equation 2-1. The x–t graph in Figure 2-4a has a steep slope because swimmer A is moving rapidly, and slopes upward because she is moving in the positive x direction. In Figure 2-4b the x–t graph has a shallow slope because swimmer B is moving slowly, and slopes downward because he is moving in the negative x direction.

The Equation for Constant-Velocity Linear Motion

We can write a simple and useful equation for constant-velocity motion by starting with Equation 2-1, $v_{\text{average},x} = (x_2 - x_1)/(t_2 - t_1)$. Since the velocity is constant, we replace $v_{\text{average},x}$ by v_x. We choose the earlier time t_1 in this equation to be $t = 0$ and use x_0 instead of x_1 as the symbol for the object's position at $t = 0$. We also use the symbols t and x instead of t_2 and x_2 to represent the later time and the object's position at that time. Then Equation 2-1 becomes

$$v_x = \frac{x - x_0}{t - 0} = \frac{x - x_0}{t}$$

We multiply both sides of this equation by t then add x_0 to both sides. The result is

EQUATION IN WORDS
Position versus time for linear motion with constant velocity

$$x = x_0 + v_x t \qquad (2\text{-}2)$$

- x: **Position at time t** of an object in linear motion with constant velocity
- x_0: **Position at time $t = 0$** of the object
- v_x: **Constant velocity** of the object
- t: **Time** at which the object is at position x

This equation gives the position x of the object as a function of the time t. As we'll see below, Equation 2-2 is an important tool for solving problems about linear motion with constant velocity. But even with this equation, an essential part of solving any such problem is drawing x–t graphs like those shown in Figure 2-4.

EXAMPLE 2-1 Average Velocity

You drive from Bismarck, North Dakota, to Fargo, North Dakota, on Interstate Highway 94, an approximately straight road 315 km in length. You leave Bismarck and travel 210 km at constant velocity in 2.50 h, then stop for 0.50 h at a rest area. You then drive the remaining 105 km to Fargo at constant velocity in 1.00 h. You then turn around immediately and drive back to Bismarck nonstop at constant velocity in 2.75 h. (a) Draw an x–t graph for the entire round trip. Then calculate the average velocity for (b) the trip from Bismarck to the rest area, (c) the trip from the rest area to Fargo, (d) the entire outbound trip from Bismarck to Fargo, (e) the return trip from Fargo to Bismarck, and (f) the round trip from Bismarck to Fargo and back to Bismarck.

Set Up

We can use linear motion because we are told the motion is along an approximately straight road. We choose to set up our coordinates as shown, with $x = 0$ at the starting position (Bismarck) and the positive x direction from Bismarck toward Fargo. We also choose $t = 0$ to be when the car leaves Bismarck. Using Equation 2-1 will tell us the average velocity for each portion of the trip. Because we only care about when the car is at specific locations, we don't need to worry about the relative motion of the components of the car, so it is fine to use the object model.

Average velocity for linear motion:

$$v_{\text{average},x} = \frac{x_2 - x_1}{t_2 - t_1} = \frac{\Delta x}{\Delta t} \qquad (2\text{-}1)$$

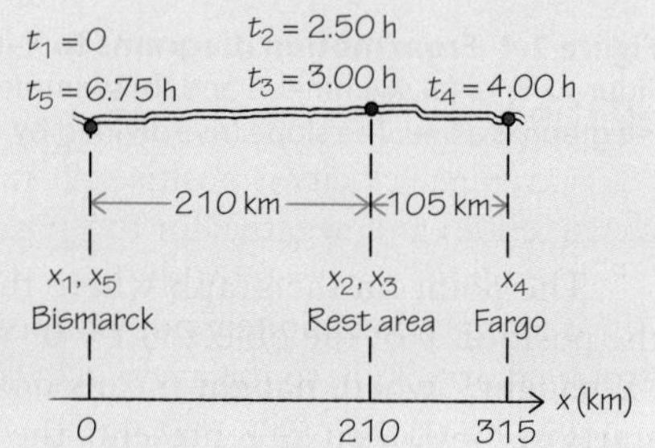

Solve

(a) To draw the x–t graph for the car's motion, we first put a dot on the graph for the car's positions and the corresponding times at the beginning and end of each leg of the trip. The velocity is constant on each leg, so we draw lines connecting successive points. The slope of the graph is positive (upward) when the car is moving in the positive x direction toward Fargo and negative (downward) when the car is moving in the negative x direction back toward Bismarck. The slope is *zero* (the line is horizontal) while the car is at the rest area and not moving.

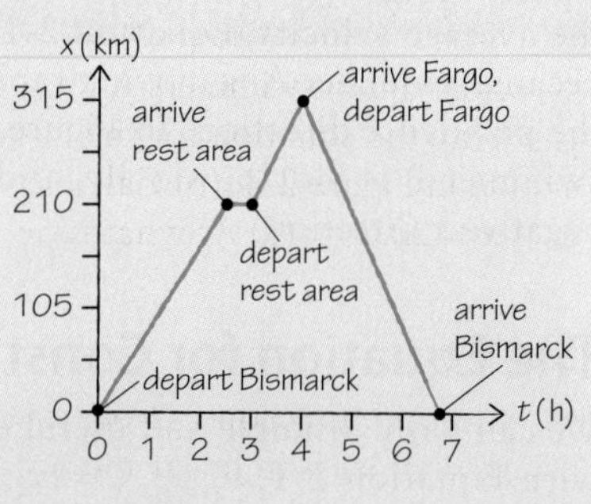

(b) Calculate the average velocity for the trip from Bismarck (position $x_1 = 0$ at time $t_1 = 0$) to the rest area (position $x_2 = 210$ km at time $t_2 = 2.50$ h).

Bismarck to rest area:

$$v_{\text{average},x} = \frac{\Delta x}{\Delta t} = \frac{x_2 - x_1}{t_2 - t_1} = \frac{210 \text{ km} - 0}{2.50 \text{ h} - 0} = +84.0 \text{ km/h}$$

(c) Calculate the average velocity for the trip from the rest area (position $x_3 = 210$ km at time $t_3 = 2.50 \text{ h} + 0.50 \text{ h} = 3.00$ h) to Fargo (position $x_4 = 315$ km at time $t_4 = 3.00 \text{ h} + 1.00 \text{ h} = 4.00$ h).

Rest area to Fargo:

$$v_{\text{average},x} = \frac{\Delta x}{\Delta t} = \frac{x_4 - x_3}{t_4 - t_3} = \frac{315 \text{ km} - 210 \text{ km}}{4.00 \text{ h} - 3.00 \text{ h}} = +105 \text{ km/h}$$

(d) Calculate the average velocity for the entire outbound trip from Bismarck (position $x_1 = 0$ at time $t_1 = 0$) to Fargo (position $x_4 = 315$ km at time $t_4 = 4.00$ h).

Bismarck to Fargo:

$$v_{\text{average},x} = \frac{\Delta x}{\Delta t} = \frac{x_4 - x_1}{t_4 - t_1} = \frac{315 \text{ km} - 0}{4.00 \text{ h} - 0} = +78.8 \text{ km/h}$$

(e) Calculate the average velocity for the return trip from Fargo (position $x_4 = 315$ km at time $t_4 = 4.00$ h) to Bismarck (position $x_5 = 0$ at time $t_5 = 4.00 \text{ h} + 2.75 \text{ h} = 6.75$ h).

Fargo back to Bismarck:

$$v_{\text{average},x} = \frac{\Delta x}{\Delta t} = \frac{x_5 - x_4}{t_5 - t_4} = \frac{0 - 315 \text{ km}}{6.75 \text{ h} - 4.00 \text{ h}} = -115 \text{ km/h}$$

(f) For the round trip from Bismarck (position $x_1 = 0$ at time $t_1 = 0$) to Fargo and back to Bismarck (position $x_5 = 0$ at time $t_5 = 6.75$ h), the net displacement $\Delta x = x_5 - x_1$ is *zero*: The car ends up back where it started. Hence the average velocity for the round trip is zero, no matter how long it took.

Bismarck to Fargo and back to Bismarck:

$$v_{\text{average},x} = \frac{\Delta x}{\Delta t} = \frac{x_5 - x_1}{t_5 - t_1} = \frac{0 - 0}{6.75 \text{ h} - 0} = 0$$

Reflect

The average velocity is positive for the trips from Bismarck toward Fargo (because the car moves in the positive x direction) and negative for the trip from Fargo back to Bismarck (because the car moves in the negative x direction). The result $v_{\text{average},x} = 0$ for the round trip means that the net displacement for the round trip was zero. The average *speed* for the trip was definitely not zero, however: The car traveled a total distance of 2(315 km) = 630 km in a total time interval of 6.75 h, so the average speed was (630 km)/(6.75 h) = 93.3 km/h. Displacement and distance are not the same thing; likewise, average velocity and average speed are not the same thing any time the direction of travel changes.

NOW WORK Problem 6 from The Takeaway 2-2.

EXAMPLE 2-2 When Swimmers Pass

Consider again swimmers A and B, whose motion is depicted in Figure 2-3. Both of them begin swimming at $t = 0$. (a) Write the equation for the position as a function of time, Equation 2-2, for each swimmer. Use the coordinate system shown in Figure 2-3. (b) Calculate the position of each swimmer at $t = 10.0$ s. (c) Calculate the time when each swimmer is at $x = 20.0$ m. (d) At what time and position do the two swimmers pass each other?

Set Up

We can use Equation 2-2 because each swimmer is in linear motion with constant velocity. Figure 2-3 shows where each swimmer starts and each swimmer's constant

$$x = x_0 + v_x t \quad (2\text{-}2)$$

Swimmer A starts at $x = -10.0$ m at $t = 0$ and has velocity +2.00 m/s, so

$$x_{A0} = -10.0 \text{ m}$$

$$v_{Ax} = +2.00 \text{ m/s}$$

velocity, which are just the quantities we need to substitute into Equation 2-2. (We use subscripts "A" and "B" to denote the quantities that pertain to each swimmer.)

Swimmer B starts at $x = +40.0$ m at $t = 0$ and has velocity -0.500 m/s, so

$$x_{B0} = +40.0 \text{ m}$$
$$v_{Bx} = -0.500 \text{ m/s}$$

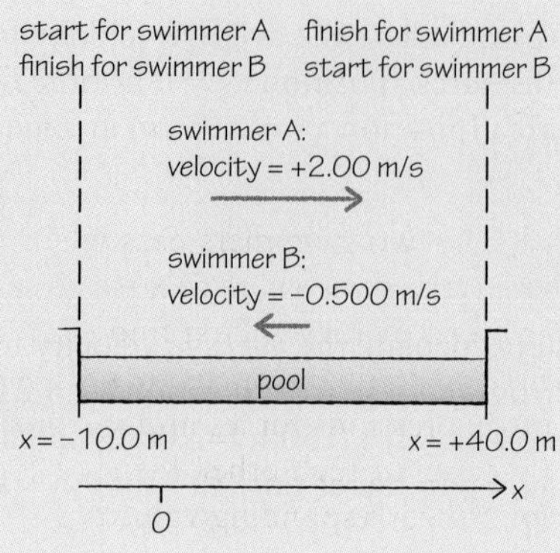

Solve

(a) Use the given information to write Equation 2-2 for each swimmer. It's useful to draw the x–t graphs for both swimmers on the same graph (see Figure 2-4). This will allow us to check that the times and positions we calculate (using mathematical representations) match the ones we get from the graphical representation. We can draw the graphs because we know where each swimmer starts, their x_0, which will be the location where the line touches the vertical axis, and their velocity, which is the slope of the line.

Swimmer A:

$$x_A = x_{A0} + v_{Ax}t \text{ so}$$
$$x_A = -10.0 \text{ m} + (2.00 \text{ m/s})t$$

Swimmer B:

$$x_B = x_{B0} + v_{Bx}t \text{ so}$$
$$x_B = +40.0 \text{ m} + (-0.500 \text{ m/s})t$$

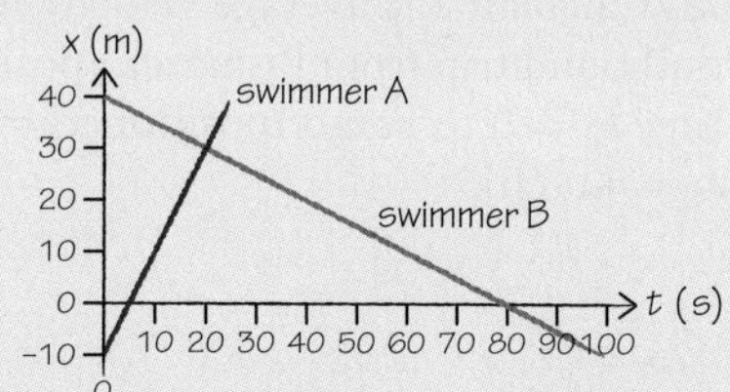

(b) To find the position of each swimmer at $t = 10.0$ s, just substitute this value of t into the equations for x_A and x_B from part (a).

Swimmer A:

$$x_A = -10.0 \text{ m} + (2.00 \text{ m/s})(10.0 \text{ s})$$
$$= -10.0 \text{ m} + 20.0 \text{ m}$$
$$= +10.0 \text{ m}$$

Swimmer B:

$$x_B = +40.0 \text{ m} + (-0.500 \text{ m/s})(10.0 \text{ s})$$
$$= +40.0 \text{ m} + (-5.00 \text{ m})$$
$$= +35.0 \text{ m}$$

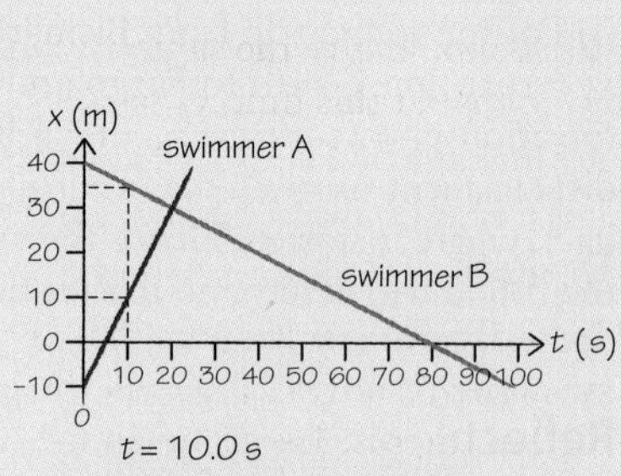

(c) To solve for the times when the two swimmers are at $x = 20.0$ m, we must first rearrange the equations from (a) so that t (the quantity we are trying to find) is by itself on one side of each equation. We then substitute 20.0 m for both x_A and x_B.

Swimmer A:

$$x_A = -10.0 \text{ m} + (2.00 \text{ m/s})t \text{ so}$$
$$x_A + 10.0 \text{ m} = (2.00 \text{ m/s})t$$
$$t = \frac{x_A + 10.0 \text{ m}}{2.00 \text{ m/s}}$$

Substitute $x_A = 20.0$ m:

$$t = \frac{20.0 \text{ m} + 10.0 \text{ m}}{2.00 \text{ m/s}} = 15.0 \text{ s for A}$$

Swimmer B:

$$x_B = +40.0 \text{ m} + (-0.500 \text{ m/s})t \text{ so}$$
$$x_B - 40.0 \text{ m} = (-0.500 \text{ m/s})t$$
$$t = \frac{x_B - 40.0 \text{ m}}{(-0.500 \text{ m/s})}$$

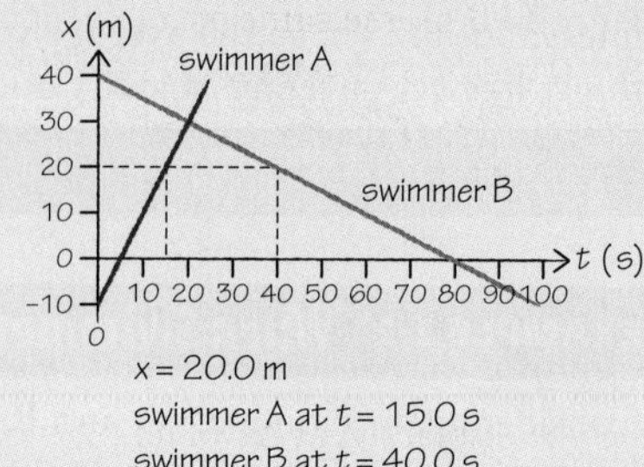

Substitute $x_B = 20.0$ m:

$$t = \frac{20.0 \text{ m} - 40.0 \text{ m}}{(-0.500 \text{ m/s})} = 40.0 \text{ s for B}$$

(d) The two swimmers pass when they are at the *same* value of x (so $x_A = x_B$) at the *same* time t. First find the time *when* this happens by setting the expressions for x_A and x_B from (a) equal to each other, then solving for the corresponding value of t.

Swimmer A:

$$x_A = -10.0 \text{ m} + (2.00 \text{ m/s})t$$

Swimmer B:

$$x_B = +40.0 \text{ m} + (-0.500 \text{ m/s})t$$

When the two swimmers pass, $x_A = x_B$

$$-10.0 \text{ m} + (2.00 \text{ m/s})t = 40.0 \text{ m} + (-0.500 \text{ m/s})t$$

$$(2.00 \text{ m/s})t - (-0.500 \text{ m/s})t = 40.0 \text{ m} - (-10.0 \text{ m})$$

$$(2.50 \text{ m/s})t = 50.0 \text{ m}$$

$$t = \frac{50.0 \text{ m}}{2.50 \text{ m/s}} = 20.0 \text{ s}$$

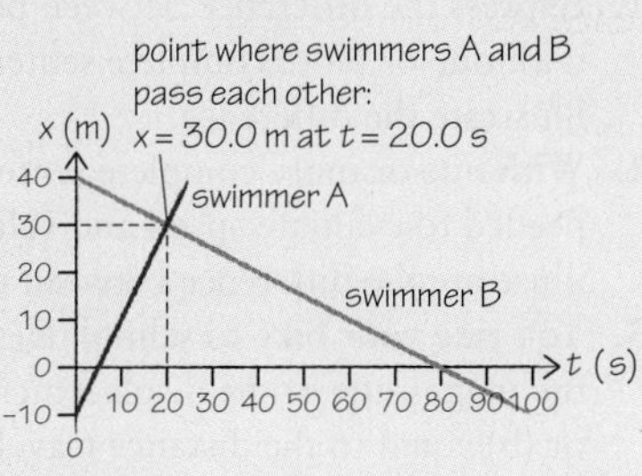

To find *where* the two swimmers pass, substitute the value of t for when they pass into either the equation for x_A or the equation for x_B.

At $t = 20.0$ s swimmer A's position is

$$x_A = -10.0 \text{ m} + (2.00 \text{ m/s})t = -10.0 \text{ m} + (2.00 \text{ m/s})(20.0 \text{ s}) = 30.0 \text{ m}$$

(Both should give the same answer, given that at this time $x_A = x_B$.)

$t = 20.0$ s is the time when the two swimmers pass, so the position where they pass is $x = 30.0$ m.

Reflect

To verify our results, note that swimmer B moves 1/4 as fast as swimmer A. This checks out for part (b): In 10.0 s, swimmer A moves 20.0 m in the positive x direction from her starting point at $x_{A0} = -10.0$ m, and swimmer B moves 5.00 m (that is, 1/4 as far) in the negative x direction from his starting point at $x_{B0} = +40.0$ m. It also checks out for part (c): Swimmer A travels 30.0 m in 15.0 s, while slow-moving swimmer B travels a shorter distance (20.0 m) in a longer time (40.0 s). And in part (d), the two swimmers meet 10.0 m from where swimmer B started but 40.0 m (four times farther) from where swimmer A started. This also matches up with the position and time where the two lines we drew cross. You can also check the answer to part (d) by substituting $t = 20.0$ s into the equation for x_B. You should find $x_B = 30.0$ m, the same as x_A (because $t = 20.0$ s is when the two swimmers pass). Do you?

NOW WORK **Problem 7 from The Takeaway 2-2.**

THE TAKEAWAY for Section 2-2

✔ The displacement Δx of an object in linear motion is the difference between the object's position x at two different times (displacement Δx is the *later* position minus the *earlier* position).

✔ The average velocity of the object between those two different times, $v_{\text{average},x}$, is the displacement Δx (the change in position) divided by the time interval Δt (the time interval to travel that displacement).

✔ An object has constant velocity if $v_{\text{average},x}$ has the same value for any time interval. If the velocity is constant, the object always moves in the same direction and maintains a steady speed.

✔ Motion diagrams and x–t graphs are important tools for interpreting what happens during linear motion. The slope of an x–t graph equals the object's velocity.

✔ You can solve many problems that involve linear motion by using Equation 2-1, which relates average velocity, displacement, and time interval.

✔ If the velocity is constant, you can use Equation 2-2 to relate velocity, position, and time.

Prep for the AP® Exam

AP® Building Blocks

1. Express the difference between position and displacement with one or more complete sentences. Give an example to illustrate the difference.
2. With one or more complete sentences, identify the data needed to evaluate speed and velocity. Give an example to illustrate the difference between these physical properties.
3. You ride your bike to school. Describe situations in which the magnitude of the displacement could be (a) less than or (b) equal to the distance traveled. (c) Could the magnitude of the displacement ever be greater than the distance traveled? Explain why or why not.
4. In a graph of position as a function of time, when velocity is constant:
   (a) Describe how the graphical information could be used to determine the speed.
   (b) Describe how the graphical information could be used to determine the initial and final positions.
5. In the United States, transportation rates are described in units of miles per hour. Elsewhere the conventional unit is kilometers per hour. The SI unit for speed is meters per second. Calculate the numerical values for the conversion factors for converting from miles per hour to meters per second, and kilometers per hour to meters per second.

EX 2-1

6. Traveling 221 miles from Boston Back Bay Station to New York Penn Station takes 3 hours and 40 minutes on a train. Calculate the average speed of the train in units of m/s.

EX 2-2

7. Traveling 221 miles from the Boston Back Bay Station to New York City Penn Station takes 3 hours and 40 minutes on a train. A train traveling from Penn Station to Back Bay Station takes 4 hours and 5 minutes. If both trains depart at the same time and move at their average speed, what is the displacement of each train when they pass? Assume the rails are parallel and indicate which direction you choose as positive.
8. Consider the v–t graph.

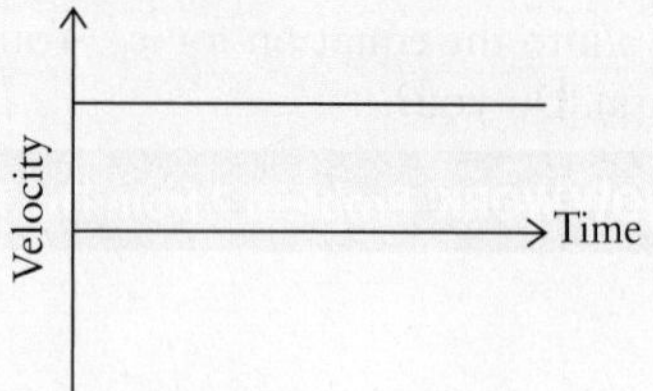

   (a) Give an example of an object with a motion that is represented by this v–t graph.
   (b) Consider how Equation 2-1 relates to this graph and the fact that, for a rectangle, area is the product of length and width. Using these ideas, describe why the displacement during a certain time interval is equal to the area under the graph during that time interval.

AP® Skill Builders

9. Two hikers complete a 1.0-km trail in two segments of length 500 m. One hiker measures the speeds, v_1 and v_2, for each segment and from these measurements calculates the average speed for the complete hike. The second hiker measures the elapsed times for each segment, Δt_1 and Δt_2, and from these data calculates the average speed for the complete hike.
   (a) Construct a diagram representing the trail for each hiker, the segments, and the variables representing the data that each hiker has collected.
   (b) Derive an algebraic expression (an equation) for the total time that elapsed for each hiker.
   (c) Express the average speed of each hiker in terms of the total distance and the variables representing the data collected.
10. What kind of data are needed to calculate the speed of an object? What kind of data are needed to calculate the velocity of an object? Justify your selections.
11. A motion detector emits a high-pitched sound wave that bounces off a surface of an object in motion. This reflected sound wave is received by the detector and, using a set of emitted and received signals, the position of the object is calculated for a series of closely spaced times so that it approximates measuring position as a function of time. A motion detector was used to collect the data shown in the table below.

Time (s)	0	0.5	1.0	1.5	2.0	2.5	3.0	3.5	4.0	4.5	5.0
Position (m)	0	0.9	1.8	2.6	3.6	4.3	5.3	6.1	6.9	7.9	8.7

   (a) Construct a graph of the data.
   (b) Analyze the data to determine if the motion occurred at constant velocity.
   (c) If the motion is a constant velocity motion, calculate the value of the velocity; if not, calculate the average velocity and describe the motion.
12. You happen to live on the same street as the grocery store. For a trip by car from your home to the grocery store, justify the selection of data needed to calculate the average speed and the average velocity.
13. Suppose that you left home on a short trip by car and that you wore a blindfold during the trip. Pose one or more scientific questions to the driver involving average speed and average velocity that could differentiate between a trip to your school and a round trip from your home to your school and back to your home again.
14. Joe Hardy asks his brother Frank, "How far is it from home to the Ghost Dude Ranch?" Describe how Frank can refine Joe's question so that it has an unambiguous answer.
15. Measured values of the change in position of an object and the elapsed time for that change in position of the object are 3409 ± 1 m and 69.0 ± 0.5 s, respectively. If you use a calculator to evaluate the average velocity, you will get something like 49.4057971 m/s. (Exactly what you get will depend on your calculator.) Explain why the scientist reports the average velocity as 49.4 ± 0.4 m/s.

AP® Skills in Action

16. A person with his back to a wall moves in the positive direction for 3.0 s at an average speed of 2.5 m/s. The person then pauses for 2.0 s before moving in the negative direction at an average speed of 2.0 m/s until he returns to the wall. Assume velocity is constant and equal to the

average velocity over each interval and choose the wall as the origin.

(a) Construct a position-versus-time graph of this motion.
(b) Predict the position of the person at 5.0 s.
(c) As the motion continues after 5.0 s predict the time that elapses in moving back to the wall from the position predicted in part (b).
(d) Construct a graph of velocity versus time. (Velocity versus time means place velocity on the y axis, and time on the x axis.) Constant values of the y coordinate are represented by horizontal lines. When the quantity being graphed changes in a smaller time than you are keeping track of, such as the implied quick changes in this problem, represent the ends of the horizontal lines with small open circles. Do not connect them with vertical lines, which would imply that the velocity had all of those values at the same time.

17. A trainer times her racehorse as it completes a workout on a long, straight track. The position-versus-time data are given in the table. Based on these data and the average speed of the horse between (a) 0 and 10 s, (b) 10 and 30 s, and (c) 40 and 50 s, construct a claim, supported by reasoning, regarding the strategy that the jockey can use to improve the horse's competitiveness in a race.

x (m)	t (s)	x (m)	t (s)
0	0	500	30
90	5	550	35
180	10	600	40
270	15	650	45
360	20	700	50

2-3 Velocity is the rate of change of position, and acceleration is the rate of change of velocity

In Section 2-2 we considered linear motion with constant velocity: an object that moves in a line with constant speed and always in the same direction. But in many important situations the speed, the direction of motion, or both, can change as an object moves. Some examples include a car speeding up to get through an intersection before the light turns red; the same car slowing down as a police officer signals the driver to pull over; or a dog that runs away from you to pick up a thrown stick, then changes direction and returns the stick to you.

Whenever an object's speed or direction of motion changes—that is, whenever its velocity changes—we say that it *accelerates* or undergoes an **acceleration**. Before we study acceleration, however, we first need to take a closer look at velocity.

Instantaneous Velocity

Figure 2-5 is a motion diagram of an object that moves with a changing velocity: a jogger who starts at rest, speeds up, then slows down toward the end of a 100-m run that takes him 38.0 s. The dots show his position at equal time intervals of 2.0 s. Because he speeds up and slows down, these dots are *not* equally spaced: His displacement, and so his average velocity, is *not* the same for all 2.0-s intervals.

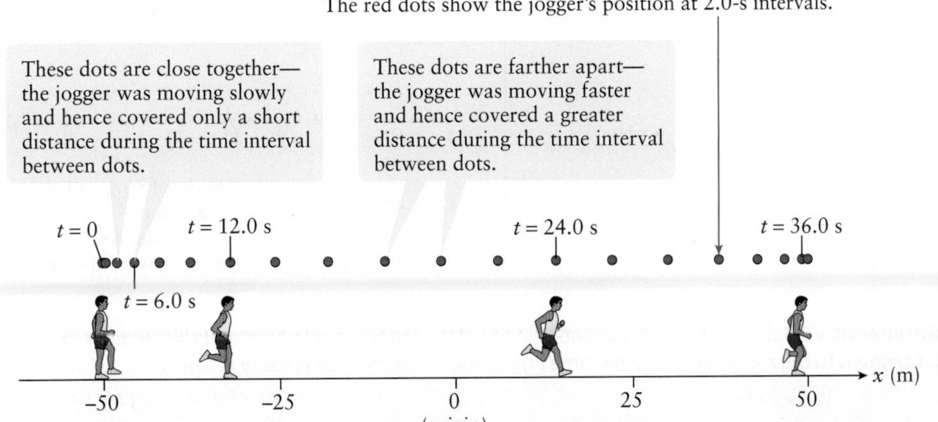
The red dots show the jogger's position at 2.0-s intervals.
These dots are close together—the jogger was moving slowly and hence covered only a short distance during the time interval between dots.
These dots are farther apart—the jogger was moving faster and hence covered a greater distance during the time interval between dots.
t = 0
t = 12.0 s
t = 24.0 s
t = 36.0 s
t = 6.0 s
x (m)
−50
−25
0
(origin)
25
50

Figure 2-5 Motion with varying velocity A jogger's velocity changes as he speeds up and slows down. This motion diagram shows these velocity changes.

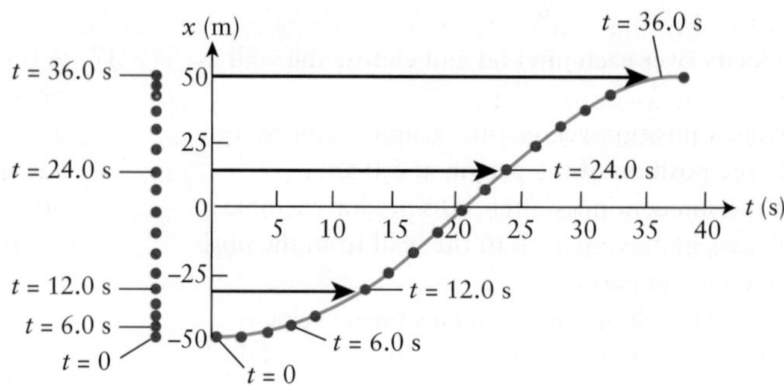

Figure 2-6 From motion diagram to *x–t* graph: varying velocity Converting the jogger's motion diagram from Figure 2-5 to an *x–t* graph. Compare to Figure 2-4.

> **AP Exam Tip**
>
> An important skill for the AP® Physics 1 Exam is describing how a graph would change if the physical situation it was describing were altered. You should practice this both by making sketches of the new graph and describing the physical situation in words.

Figure 2-6 shows the x–t graph for the jogger. The graph is not straight because the velocity is not constant. In Section 2-2 we saw that if the velocity *is* constant, as for the two swimmers whose motion is graphed in Figure 2-4, the slope of the x–t graph tells us the value of the velocity. We'll now show that the same idea is true when the velocity is not constant.

In **Figure 2-7a,** we've picked out two points during the jogger's motion, representing a time interval from $t = 6.0$ s to $t = 24.0$ s. The rise of a line connecting these points on the x–t graph equals the displacement Δx during the time interval, and the run of this line equals the duration Δt of the time interval. From Equation 2-1, the average velocity for this time interval is $v_{\text{average},x} = \Delta x/\Delta t$; this is the rise divided by the run, or the *slope* of this line. For the case shown in Figure 2-7a, $\Delta x = +59.5$ m, and $\Delta t = 18.0$ s, so $v_{\text{average},x} = (59.5\text{ m})/(18.0\text{ s}) = +3.31$ m/s. The plus sign means that the line slopes upward, which tells us that x increased during the time interval, and the jogger moved in the positive x direction.

The line in Figure 2-7a doesn't exactly match the shape of the x–t curve between the two points. However, if we decrease the time interval from $t = 6.0$ s to $t = 12.0$ s,

(a) Average x velocity on an x–t graph

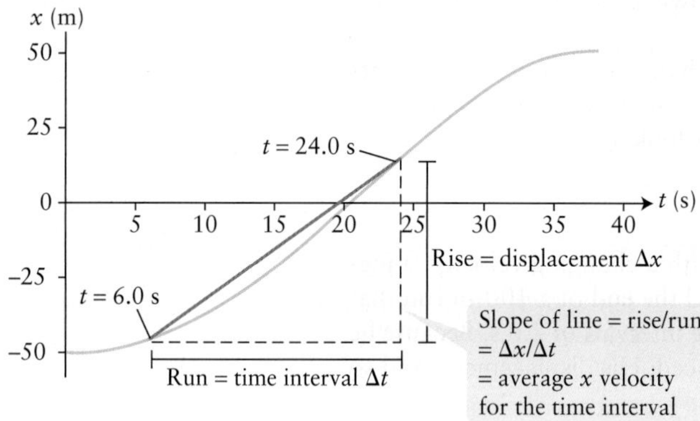

(b) Average x velocity for a shorter time interval

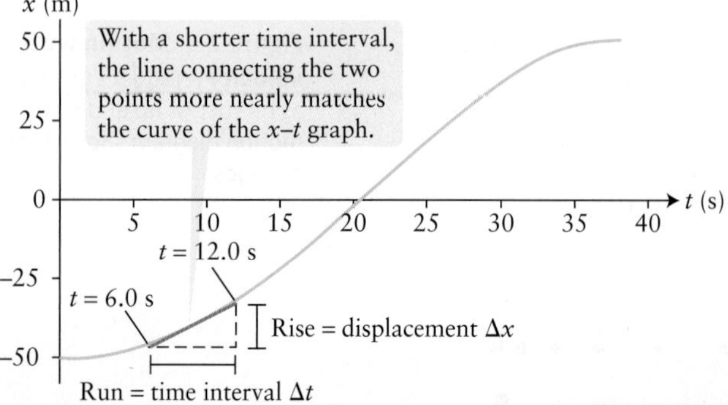

(c) Instantaneous x velocity on an x–t graph

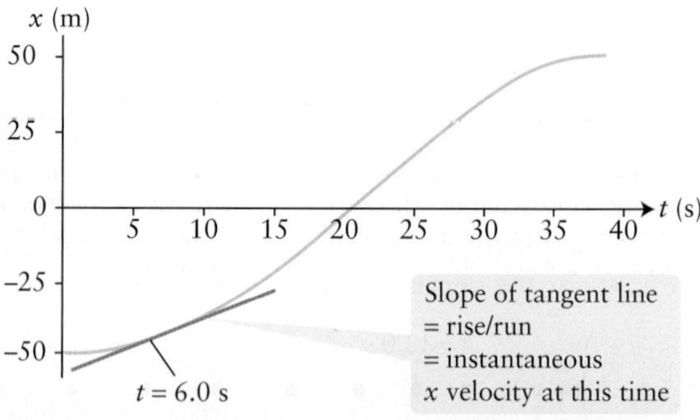

Figure 2-7 Average and instantaneous velocity Finding the average velocity of the jogger over shorter and shorter time intervals. The slope of a line that is tangent to the *x–t* graph at a given time *t* gives the instantaneous velocity at time *t*.

as in **Figure 2-7b**, the line is a better match to the x–t curve. And if we make the time interval *very* short—say, from $t = 6.000$ s to 6.001 s—the line becomes a nearly perfect match to the x–t curve during that time interval. Indeed, if we make the time interval infinitesimally small in duration, the line has the same slope as a line tangent to the x–t curve, as in **Figure 2-7c.** This slope equals the average velocity during an infinitesimally brief interval around that point in time (instant). We call this quantity v_x (with no "average"), or the **instantaneous** velocity at the instant in question. (We used the same symbol v_x in Section 2-2 when we discussed objects that move with constant velocity. If the velocity is constant, v_x has the same value at every instant.)

The sign of the instantaneous velocity tells us what direction the object is moving at the instant in question: v_x is positive if the object is moving in the positive x direction, and v_x is negative if the object is moving in the negative x direction. In a single instant, the only way the direction is changing is if the object is instantaneously at rest while it is turning around. In that case, the magnitude of the velocity and the speed are both zero. The *instantaneous* speed is always equal to the magnitude of the *instantaneous* velocity. We denote the instantaneous speed of the object by the symbol v without a subscript, since speed has no direction. Instantaneous speed is always positive or zero, never negative. A car's speedometer is a familiar device for measuring instantaneous speed.

As an example, consider the jogger's x–t graph in Figure 2-7c and the tangent line at the point for $t = 6.0$ s. If you carefully measure the rise and the run of this tangent line, then take their quotient, you'll find that the slope of this line is +1.50 m/s. So at $t = 6.0$ s the jogger's instantaneous velocity is $v_x = +1.50$ m/s; at this instant he is moving in the positive x direction at speed $v = 1.50$ m/s. Henceforth we'll use the terms "velocity" to refer to instantaneous velocity (which we'll use much more often than average velocity) and "speed" to refer to instantaneous speed.

WATCH OUT !

Instantaneous speed is the magnitude of instantaneous velocity.

Unlike average speed, which is generally not the absolute value of the average velocity, the *instantaneous* speed is always equal to the absolute value of the *instantaneous* velocity since in general the direction is not changing in a single instant. The only time the direction is changing in a single instant is when something turns around. In that instant, the magnitude of the velocity and the speed are both zero.

AP Exam Tip

Students frequently confuse velocity (instantaneous velocity) with average speed or average velocity. The definitions and differences between these three quantities are very important to understand for the AP® exam.

WATCH OUT !

Precise meanings are different for similar terms.

When we use just the word "velocity" we are referring to the instantaneous velocity. Similarly, when we use just the word "speed" we are referring to the instantaneous speed. When we are referring to the average speed or average velocity, we will always include the word "average," unless the speed or velocity is constant, in which case the average and instantaneous quantities are the same.

Mathematically, an object's velocity v_x is the *rate of change* of the object's position as given by the coordinate x with respect to time. The faster the position changes, the greater the magnitude of the velocity v_x and the greater the speed of the object. Positive v_x means that x is increasing, the object is moving in the positive x direction; negative v_x means that x is decreasing, the object is moving in the negative x direction.

EXAMPLE 2-3 Decoding an x–t Graph

Figure 2-8 shows an x–t graph for the motion of an object. (a) Determine the direction in which the object is moving at t = 0, 10 s, 20 s, 30 s, and 40 s. (b) Describe the object's motion in words and draw a motion diagram for the object for the period from $t = 0$ to $t = 40$ s.

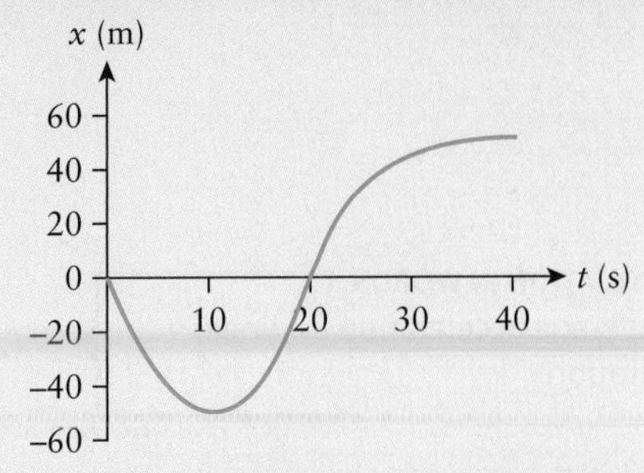

Figure 2-8 An x–t graph to interpret How must an object move to produce this x–t graph?

Set Up

We'll use the idea that the slope of the x–t graph for an object tells us the object's velocity v_x. The algebraic sign of the velocity (plus or minus) tells us whether the object is moving in the positive or negative x direction.

$$v_x = \text{slope of } x\text{–}t \text{ graph}$$

Solve

(a) On a copy of Figure 2-8, we've drawn green lines tangent to the blue x–t curve at the five times of interest. The slope is negative at $t = 0$ (point 1 in the figure), so $v_x < 0$, and the object is moving in the negative x direction. At $t = 10$ seconds (point 2 in the figure), the slope of the tangent line is zero, thus the object is momentarily at rest ($v_x = 0$) at that time, so it is not moving in either direction. At $t = 20$ s and 30 s (points 3 and 4 in the figure), the slope is positive, indicating the object is moving in the positive x direction ($v_x > 0$); at $t = 40$ s (point 5 in the figure) the object is again at rest ($v_x = 0$).

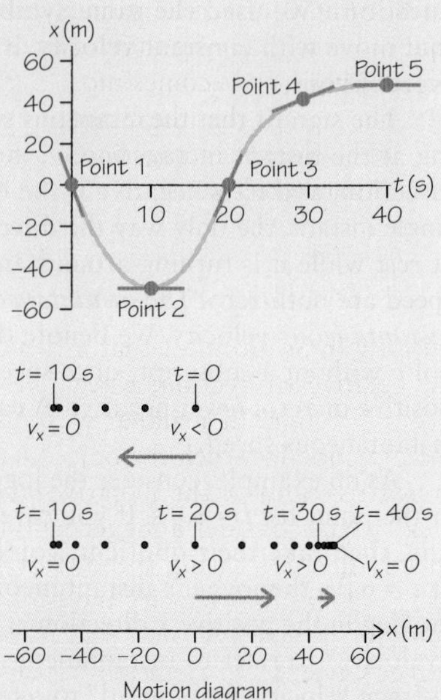

(b) At time $t = 0$ the object is at $x = 0$ and moving in the negative x direction. The x–t graph has a negative slope until $t = 10$ s, so the object continues to move in the negative x direction; because the slope of the x–t graph becomes shallower, the object is slowing down, so the dots are closer together. At $t = 10$ s, the object comes to rest momentarily at $x = -50$ m. After $t = 10$ s, the x–t graph has a positive slope, so the object moves in the positive x direction. From $t = 10$ s to 20 s, the object is speeding up because the slope of the x–t graph is increasing. The object has the fastest speed (because the graph has the greatest slope) at $t = 20$ s when the object again passes through $x = 0$. After $t = 20$ s the slope of the x–t graph is still positive but getting shallower; so the object is still moving in the positive x direction, but its speed is decreasing such that the dots again get closer together. The object finally comes to rest at $x = +50$ m at $t = 40$ s.

Reflect

The vertical *value* of an x–t graph for any time t shows you the object's *position* at that time (in other words, where the object is), while the *slope* of the x–t graph tells you its *velocity* (that is, how fast and in which direction the object is moving). You can read the positions off for the times of interest and they make sense. The object's position gets more negative from $t = 0$ to $t = 10$ s. After this value of time, the curve goes upward, which means the direction changes. After $t = 10$ s the position gets more positive as time increases. The x–t graph has zero *slope* at $t = 10$ s, so at this instant the coordinate x is neither increasing nor decreasing, and the velocity is *zero*. It also makes sense that the velocity would need to go through zero as the direction of motion changes.

NOW WORK Problem 2 from The Takeaway 2-3.

Average Acceleration and Instantaneous Acceleration

Just as velocity is how rapidly and in what direction position changes, acceleration is how rapidly and in what direction velocity changes. Suppose an object like the jogger in Figure 2-7 has velocity v_{1x} at time t_1 and velocity v_{2x} at time t_2. We define the object's average acceleration $a_{\text{average},x}$ over this time interval to be the change in velocity $\Delta v_x = v_{2x} - v_{1x}$ divided by the time interval $\Delta t = t_2 - t_1$:

Average acceleration of an object in linear motion

Change in **velocity** of the object over a certain time interval: The velocity changes from v_{1x} to v_{2x}.

EQUATION IN WORDS
Average acceleration for linear motion

$$a_{\text{average},x} = \frac{\Delta v_x}{\Delta t} = \frac{v_{2x} - v_{1x}}{t_2 - t_1} \quad (2\text{-}3)$$

The time interval: The object has velocity v_{1x} at time t_1, and has velocity v_{2x} at time t_2.

If a racecar initially at rest blasts away from the starting line and rapidly comes to its top speed, its velocity changes substantially (Δv_x is large) in a short time (Δt is small), and its average acceleration $a_{\text{average},x}$ has a large magnitude. If, instead, an elderly horse starts from rest and gradually speeds up to a slow amble, the horse's velocity changes only a little (Δv_x is small) and takes a long time to change (Δt is large). In this case the horse's average acceleration $a_{\text{average},x}$ has only a small magnitude.

Velocity has units of meters per second (m/s) and time has units of seconds (s), so the SI units of acceleration are meters per second per second, or meters per second squared (m/s^2). Saying that an object has an acceleration of 2.0 m/s^2 means that the object's velocity becomes more positive by 2.0 m/s every second; an acceleration of –4.5 m/s^2 means that the change in velocity is –4.5 m/s every second.

Like velocity, average acceleration is a vector and contains direction information, as we explain below.

Interpreting Positive and Negative Acceleration

We saw in Section 2-2 that a positive velocity means that the object is moving in the positive x direction; a negative velocity means the object is moving in the negative x direction. In a similar way, the sign of the average acceleration $a_{\text{average},x}$ tells us the direction in which the object is *accelerating*: in the positive x direction if $a_{\text{average},x}$ is positive and in the negative direction if $a_{\text{average},x}$ is negative. To see what this means, let's consider a jogger who moves in the positive x direction, first increasing speed, then decreasing speed (**Figure 2-9a**). During the time interval from $t_1 = 12.0$ s to $t_2 = 24.0$ s, his velocity increases from +3.00 to +4.00 m/s, and the change in his velocity is positive: $\Delta v_x = v_{2x} - v_{1x} = (+4.00 \text{ m/s}) - (+3.00 \text{ m/s}) = +1.00$ m/s. The jogger's average acceleration for this time interval is also positive: $a_{\text{average},x} = \Delta v_x/\Delta t = (+1.00 \text{ m/s})/(24.0 \text{ s} - 12.0 \text{ s}) = +0.08$ m/s^2.

(a)

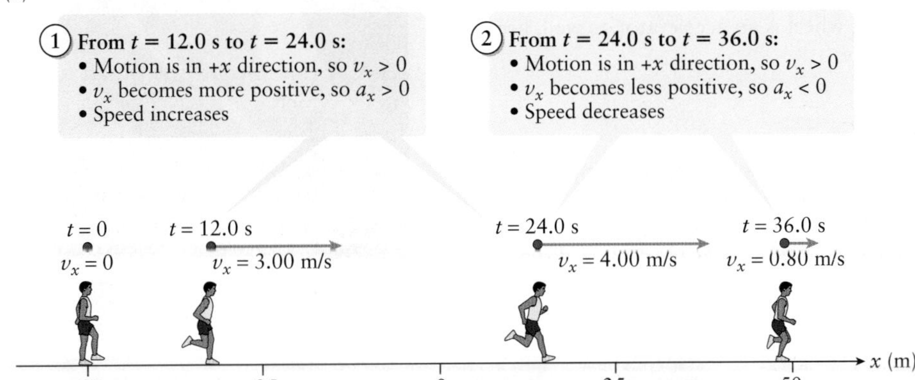

Figure 2-9 Interpreting positive and negative acceleration You must consider the algebraic signs (plus or minus) of acceleration *and* velocity to determine whether an object is slowing down or speeding up.

(b)

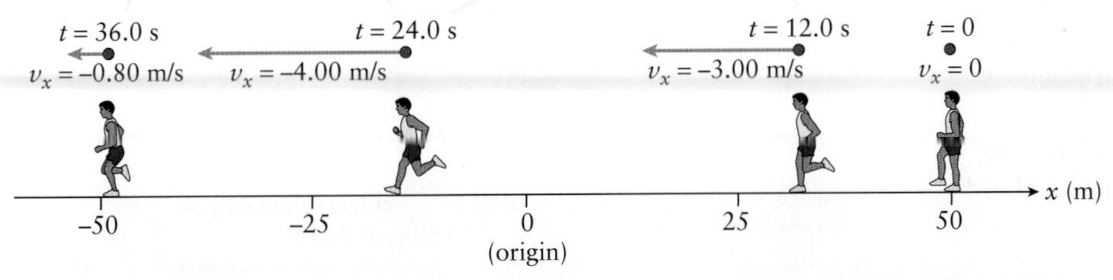

- If velocity v_x and acceleration a_x have the same sign (both positive as in 1 or both negative as in 3), the jogger speeds up.
- If velocity v_x and acceleration a_x have the opposite signs (one positive and one negative as in 2 or 4), the jogger slows down.

We say that he accelerates in the positive x direction. The jogger's acceleration is in the *same direction* as his velocity, and he *speeds up*.

By contrast, for the time interval from $t_2 = 24.0$ s to $t_3 = 36.0$ s the jogger's velocity decreases from +4.00 to +0.80 m/s, and the change in his velocity is *negative*: $\Delta v_x = v_{3x} - v_{2x} = (+0.80\text{ m/s}) - (+4.00\text{ m/s}) = -3.20\text{ m/s}$. This means that the average acceleration for this time interval is also negative: $a_{\text{average},x} = \Delta v_x/\Delta t = (-3.20\text{ m/s})/(36.0\text{ s} - 24.0\text{ s}) = -0.27\text{ m/s}^2$. We say that the jogger accelerates in the negative x direction. In this case the jogger's acceleration is *opposite* to his velocity, and he *slows down*.

It's a common mistake to think that positive acceleration always corresponds to speeding up and that negative acceleration always corresponds to slowing down. That's *not* true! **Figure 2-9b** shows an example: The jogger from Figure 2-9a again speeds up and then slows down, but in the negative x direction, so his velocity is negative at all times. The figure shows that the jogger has *negative* average acceleration when he is speeding up $a_{\text{average},x} = \Delta v_x/\Delta t = (-3\text{ m/s} - 0)/(12\text{ s} - 0) = -0.25\text{ m/s}^2$ and *positive* acceleration when he is slowing down $a_{\text{average},x} = \Delta v_x/\Delta t = (-0.80\text{ m/s} - 4\text{ m/s})/(36\text{ s} - 24\text{ s}) = 3.2\text{ m/s}^2$. Here's a simple rule to help you understand when an object is speeding up and when it is slowing down:

> *When an object moving along a line speeds up, its velocity and acceleration have the same sign (both positive or both negative). When an object moving along a line slows down, its velocity and acceleration have opposite signs (one positive and the other negative).*

WATCH OUT !

Acceleration doesn't have to mean increasing speed.

In everyday language, "acceleration" is used to mean "speeding up" and "deceleration" is used to mean "slowing down." In physics, however, acceleration refers to *any* change in velocity and so includes *both* speeding up and slowing down (**Figure 2-10**). Speeding up is accelerating in the direction of motion and slowing down is accelerating opposite to the direction of motion. We will never use the word "deceleration"; we use the word "acceleration" to mean any change in velocity. In the next chapter, we will see that the steering wheel is an accelerator as well, because changing direction is also a change in velocity.

SuperStock/AGE Fotostock

Figure 2-10 Two pedals that cause acceleration The right-hand pedal in a car is called the accelerator because it's used to make the car go faster. But the left-hand pedal, which controls the brakes, is also an accelerator pedal because it's used to make the car slow down.

EXAMPLE 2-4 Calculating Average Acceleration

The object described in Example 2-3 has velocity $v_x = -10$ m/s at $t = 0$, $v_x = 0$ at $t = 10$ s, $v_x = +10$ m/s at $t = 20$ s, $v_x = +1.4$ m/s at $t = 30$ s, and $v_x = 0$ at $t = 40$ s. Find the object's average acceleration for the time intervals (a) $t = 0$ to 10 s, (b) $t = 10$ s to 20 s, (c) $t = 20$ s to 30 s, and (d) $t = 30$ s to 40 s.

Set Up

The figures in Example 2-3 show the x–t graph and motion diagram for this object. For each time interval we'll use the definition of average acceleration, Equation 2-3. We know from the slopes of the tangent lines drawn in Example 2-3 that the instantaneous velocity is 0 at points 2 and 5 and can estimate it to be about −10 m/s at point 1, 10 m/s at point 3, and 1.5 m/s at point 4. These estimates are not far off the given values!

Average acceleration for linear motion:

$$a_{\text{average},x} = \frac{\Delta v_x}{\Delta t} = \frac{v_{2x} - v_{1x}}{t_2 - t_1} \quad (2\text{-}3)$$

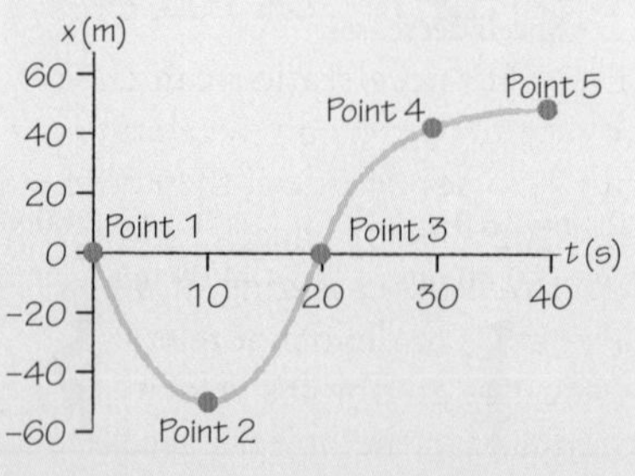

Solve

(a) For the time interval $t_1 = 0$ to $t_2 = 10$ s (from point 1 to point 2 in the x–t graph), we have $v_{x1} = -10$ m/s and $v_{x2} = 0$. Because the change in velocity is positive (it goes from negative to zero), the average x acceleration is positive.

$$a_{\text{average},x} = \frac{v_{2x} - v_{1x}}{t_2 - t_1} = \frac{0 - (-10 \text{ m/s})}{10 \text{ s} - 0} = \frac{+10 \text{ m/s}}{10 \text{ s}} = +1.0 \text{ m/s}^2$$

(b) For the time interval $t_2 = 10$ s to $t_3 = 20$ s (from point 2 to point 3 in the x–t graph), we have $v_{x2} = 0$ and $v_{x3} = +10$ m/s. The change in velocity is positive (it goes from zero to a positive value), so the average x acceleration is again positive.

$$a_{\text{average},x} = \frac{v_{3x} - v_{2x}}{t_3 - t_2} = \frac{(+10 \text{ m/s}) - 0}{20 \text{ s} - 10 \text{ s}} = \frac{+10 \text{ m/s}}{10 \text{ s}} = +1.0 \text{ m/s}^2$$

(c) For the time interval $t_3 = 20$ s to $t_4 = 30$ s (from point 3 to point 4 in the x–t graph), we have $v_{x3} = +10$ m/s and $v_{x4} = +1.4$ m/s. The change in velocity is negative, so the average x acceleration is negative.

$$a_{\text{average},x} = \frac{v_{4x} - v_{3x}}{t_4 - t_3} = \frac{(+1.4 \text{ m/s}) - (+10 \text{ m/s})}{30 \text{ s} - 20 \text{ s}} = \frac{-8.6 \text{ m/s}}{10 \text{ s}} = -0.86 \text{ m/s}^2$$

(d) For the time interval $t_4 = 30$ s to $t_5 = 40$ s (from point 4 to point 5 in the x–t graph), we have $v_{x4} = +1.4$ m/s and $v_{x5} = 0$. Again the change in velocity is negative and the average acceleration is negative.

$$a_{\text{average},x} = \frac{v_{5x} - v_{4x}}{t_5 - t_4} = \frac{0 - (+1.4 \text{ m/s})}{40 \text{ s} - 30 \text{ s}} = \frac{-1.4 \text{ m/s}}{10 \text{ s}} = -0.14 \text{ m/s}^2$$

Reflect

Let's check these results to see how they agree with the general rule about the algebraic signs of velocity v_x and average acceleration $a_{\text{average},x}$. From $t_1 = 0$ to $t_2 = 10$ s, v_x is negative (the x–t graph slopes downward), but $a_{\text{average},x} = +1.0$ m/s^2 is positive. Because the signs are different, the speed must decrease during this interval, and indeed it does (from 10 m/s to zero). From $t_2 = 10$ s to $t_3 = 20$ s, v_x is positive (the x–t graph slopes upward), and $a_{\text{average},x} = +1.0$ m/s^2 is also positive. During this interval the signs of v_x and $a_{\text{average},x}$ are the same, and the speed increases (from 0 to 10 m/s), in accordance with the rule. During the intervals from $t_3 = 20$ s to $t_4 = 30$ s (for which $a_{\text{average},x} = -0.86$ m/s^2) and from $t_4 = 30$ s to $t_5 = 40$ s (for which $a_{\text{average},x} = -0.14$ m/s^2), v_x and $a_{\text{average},x}$ have opposite signs (v_x is positive and $a_{\text{average},x}$ is negative), and the speed decreases (from 10 m/s to 1.4 m/s to zero) as it should. When looking at changes in speed, the sign of the acceleration alone isn't what's important—what matters is how the signs of velocity and acceleration compare.

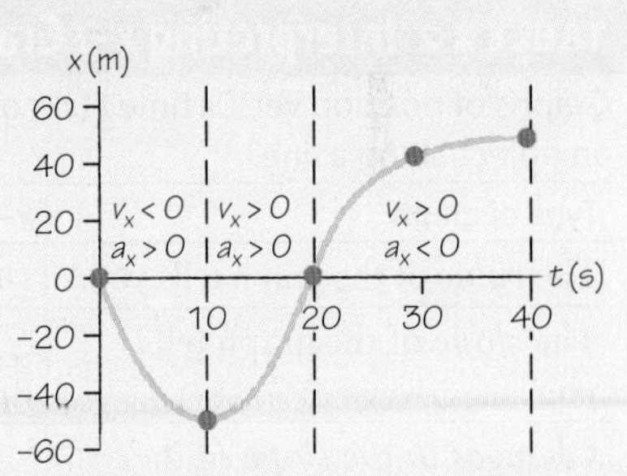

NOW WORK Problem 3 from The Takeaway 2-3.

Instantaneous Acceleration and the v_x–t Graph

An object's acceleration can change from one moment to the next. For example, a car moving forward can accelerate forward and gain speed, cruise at a steady velocity with zero acceleration, then accelerate backward when the driver steps on the brakes. The instantaneous acceleration, which we will call simply acceleration, describes how the velocity is changing at a given instant. We use the symbol a_x (without the word "average") to denote acceleration.

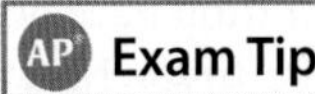

Creating graphs of motion is a very important skill that you need to develop for the AP® Physics exam.

Just as an object's velocity v_x is the rate of change of its position as expressed by its coordinate x, the object's acceleration a_x is the rate of change of its velocity v_x. In other words, *acceleration is to velocity as velocity is to position*. The slope of a graph of an object's

(a) Jogger's x–t graph

At each point on an x–t curve, the slope of the tangent to the curve at that point equals the instantaneous velocity v_x.

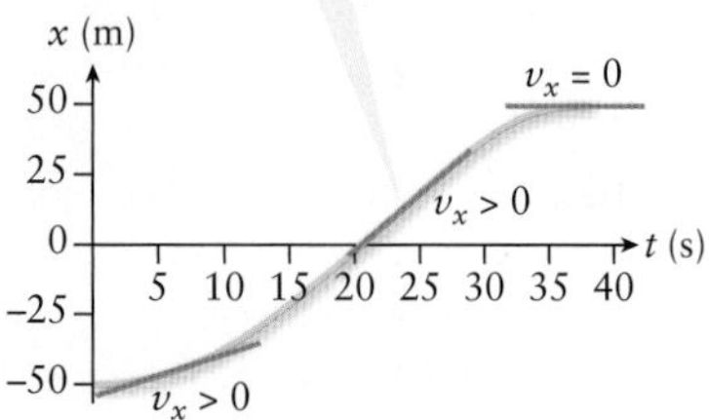

(b) Jogger's v_x–t graph

At each point on a v_x–t curve, the slope of the tangent to the curve at that point equals the instantaneous acceleration a_x.

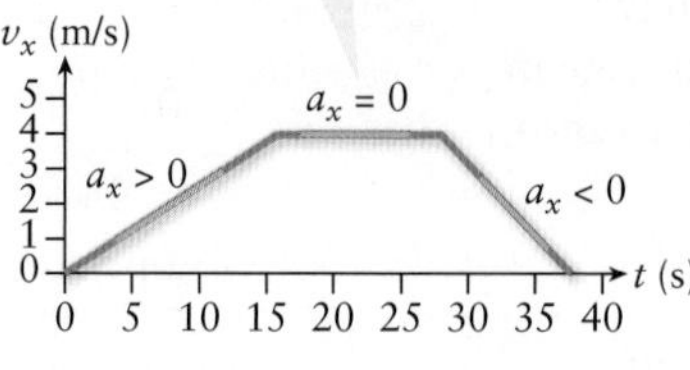

(c) Jogger's v_x–t graph

Because the slope of the v_x–t graph is constant over three time intervals, the acceleration is constant in each time interval, and found from the change in velocity divided by the time interval, for t = 0–16 s, t = 16–28 s, and t = 28–38 s.

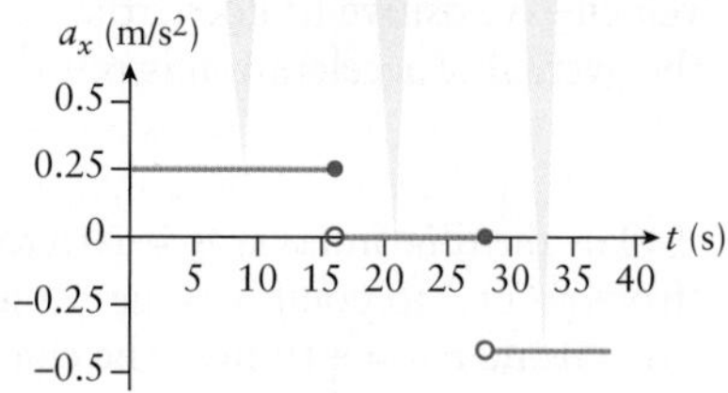

Figure 2-11 From x–t graph to v_x–t graph to a_x–t graph Graphs of a jogger's (a) position versus time (b) velocity versus time and (c) acceleration versus time.

position versus time (an x–t graph) indicates its velocity v_x. In the same way, the slope of a graph of the object's velocity versus time—that is, a v_x–t graph—tells us its acceleration a_x. If the v_x–t graph has an upward (positive) slope, the change in velocity is positive, and the acceleration is positive; if the v_x–t graph has a downward (negative) slope, the change in velocity is negative, and the acceleration is negative. If the velocity is not changing, so that the acceleration is zero, the v_x–t graph is horizontal and has zero slope.

You can see these properties of the v_x–t graph by inspecting **Figure 2-11,** which shows both the x–t graph and the v_x–t graph for the motion of the jogger depicted in Figures 2-5, 2-6, and 2-9a. From $t = 0$ to $t = 16$ s his velocity is positive and increasing, which we can see in two ways: First, the x–t graph has an *increasing*, positive slope, and, second, the v_x–t graph has a *positive* slope (which means a_x is positive). From $t = 16$ s to $t = 28$ s, the jogger has a constant velocity, which is why the x–t graph has a *constant* slope (which means v_x is not changing), and the v_x–t graph is a horizontal line with *zero* slope (which means a_x is zero) during this time interval. Finally, from $t = 28$ s to $t = 38$ s, the change in the jogger's velocity is negative, so the x–t graph has a *decreasing* slope (which means v_x is decreasing in magnitude), and the v_x–t graph has a *negative* slope (which means a_x is negative). **Table 2-2** summarizes these key features of x–t and v_x–t graphs.

TABLE 2-2 Interpreting x–t Graphs and v_x–t Graphs

Graphs of position versus time t (x–t graph) and velocity v_x versus time t (v_x–t graph) are different ways to depict the motion of an object along a line.

Type of graph	x–t graph	v_x–t graph
The *value* of the graph tells you...	...the position of the object at a given time t.	...the velocity v_x of the object at a given time t.
The *slope* of the graph tells you...	...the velocity v_x of the object at a given time t.	...the acceleration a_x of the object at a given time t.
Changes in the slope of the graph tell you...	...the acceleration a_x of the object at a given time t.	...whether the acceleration is changing.

EXAMPLE 2-5 Decoding a v_x–t Graph

Figure 2-12 shows both the x–t graph and the v_x–t graph for the motion of the same object that we examined in Examples 2-3 and 2-4. Using these graphs alone, determine whether the object's acceleration a_x is positive, negative, or zero at (a) $t = 0$, (b) $t = 10$ s, (c) $t = 20$ s, and (d) $t = 30$ s.

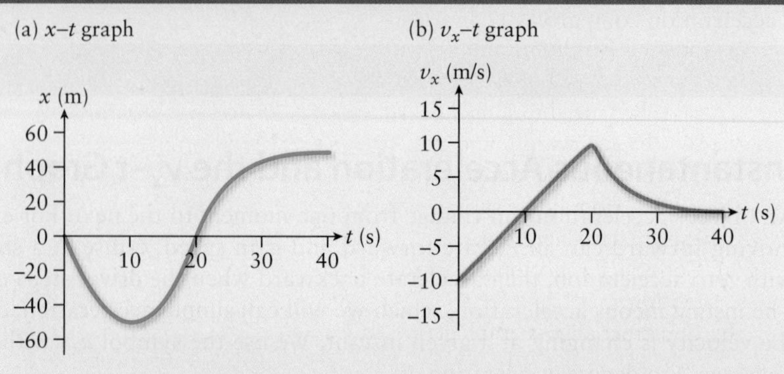

Figure 2-12 Two graphs to interpret The x–t graph and v_x–t graph for the object from Examples 2-3 and 2-4.

Set Up

We use two ideas: (i) acceleration is the rate of change of velocity, and (ii) the value of the acceleration equals the slope of the v_x–t graph (Table 2-2).

Solve

(a) At $t = 0$ the v_x–t graph has a positive (upward) slope, which means that v_x is increasing and hence that a_x is positive. The x–t graph shows the same thing. This graph has a steep negative slope at $t = 0$, which means v_x is negative, but as time increases the slope becomes shallower as the slope and v_x become closer to zero. Because v_x is changing from a negative value toward zero, the change in velocity is positive, and a_x is positive.

(b) At $t = 10$ s the v_x–t graph has the same upward slope as at $t = 0$, which means that the acceleration a_x is again positive. The velocity is zero at this instant, so the object is momentarily at rest. It's still accelerating, however; the object has a negative velocity just before $t = 10$ s, and the velocity is positive just after $t = 10$ s. This is the time at which the object turns around. You can also see this from the x–t graph, the slope of which is changing from negative to zero to positive around $t = 10$ s.

(c) At $t = 20$ s the v_x–t graph has zero slope, so the acceleration is zero. At this instant the velocity is neither increasing nor decreasing. That's why the x–t graph at this time closely approximates a line (that is, with nearly constant slope), indicating that the velocity isn't changing at this instant.

(d) At $t = 30$ s the v_x–t graph has a negative (downward) slope, so a_x is negative. The x–t graph has a positive slope at this instant (v_x is positive) but is flattening out, so the slope and v_x are both becoming less positive, which is another way of saying that the acceleration is negative.

Reflect

We can check our answers by seeing what they tell us about how the *speed* of the object is changing. At $t = 0$ the object has negative velocity and positive acceleration. Because v_x and a_x have opposite signs, the object is slowing down. At $t = 20$ s the velocity is positive but the acceleration is zero; the velocity is not changing at this instant, so the object is neither speeding up nor slowing down. Finally, at $t = 30$ s, the velocity is positive and the acceleration is negative, so the object is once again slowing down. You could further confirm these conclusions by comparing them with the motion diagram in Example 2-3 to show they are consistent. You could also use Figure 2-12 to show that at $t = 40$ s the object is at rest (it has zero velocity) and is remaining at rest (its acceleration is zero, so the velocity remains equal to zero).

NOW WORK Problems 4 and 5 from The Takeaway 2-3.

THE TAKEAWAY for Section 2-3

✔ Instantaneous velocity v_x (or *velocity* for short) tells you an object's speed and direction of motion at a given instant. It is equal to the rate of change of the object's position and the slope of the object's x–t graph at that instant.

✔ The average acceleration of an object, $a_{\text{average},x}$, is the change in its velocity v_x that occurs over a given time interval Δt divided by that time interval.

✔ Instantaneous acceleration a_x (or *acceleration* for short) equals the rate of change of an object's velocity at a given instant. It also equals the slope of the object's v_x–t graph at that instant.

✔ An object speeds up when its velocity and acceleration have the same sign and slows down when they have opposite signs.

Prep for the AP® Exam

AP® Building Blocks

1. Identify the units of (a) the slope of a graph of displacement (y axis) versus time (x axis), and (b) the slope of a graph of velocity (y axis) versus time (x axis).

EX 2-3
2. A woman leaves home and jogs along a straight road at a constant speed of 4 m/s. After 180 s she suddenly increases her speed to 8 m/s. When she is 1 km from home she runs back toward home at a constant speed of 4 m/s for 200 m. Then she increases her speed to 8 m/s. When she is 80 m from home she reduces her speed to 2 m/s until she reaches her door.
   (a) Construct a motion diagram representing her motion.
   (b) Construct an x–t graph representing her motion.

EX 2-4
3. Calculate the acceleration of the motion represented by the v–t graph shown.

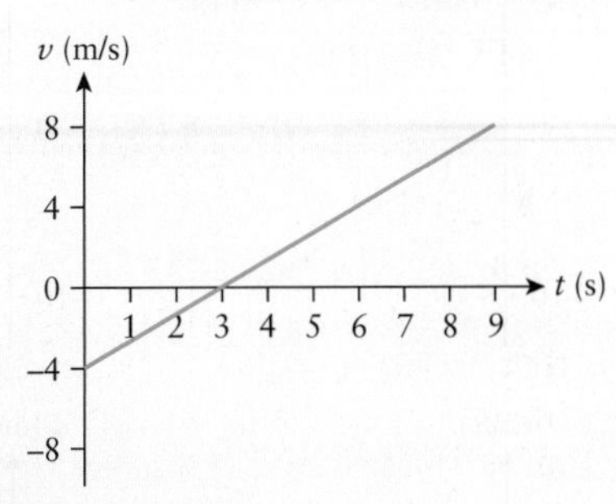

EX 2-5
4. The v–t graph represents the motion of a car on a toy rollercoaster.

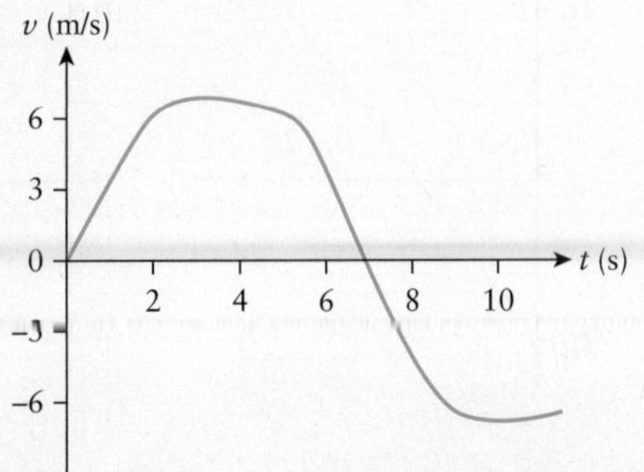

(a) Write a brief description of the motion represented in this graph. In your description include phrases "slowing down" and "speeding up" and the direction of the velocity during the motion of the car.
(b) Identify the interval(s) of time when the average acceleration is (i) positive, (ii) negative, and (iii) approximately zero.
(c) Identify time(s) at which the instantaneous acceleration is zero.
(d) Use the scenario of the car on the rollercoaster as evidence to support your answers to these questions:
 i. Can an object have zero velocity and still be accelerating?
 ii. Can an object have constant velocity and still have a varying speed?
 iii. Can the velocity of an object change sign when the object's acceleration is constant?
 iv. Can an object have increasing speed if its acceleration is negative?

EX 2-5 **5.** At $t = 0$, an object moving along an x axis is at position $x_0 = -20$ m. The signs of the object's initial velocity v_0 (at time t_0) and constant acceleration, a, are, respectively, for four situations: (i) +, +; (ii) +, –; (iii) –, +; (iv) –, –. In which situation will the object
(a) speed up?
(b) slow down?
(c) come to a momentary stop?
(d) definitely pass through the origin (given enough time)?
(e) definitely not pass through the origin (given enough time)?

AP® Skill Builders

6. Using the following graphs of velocity versus time, construct the corresponding graph of acceleration versus time.

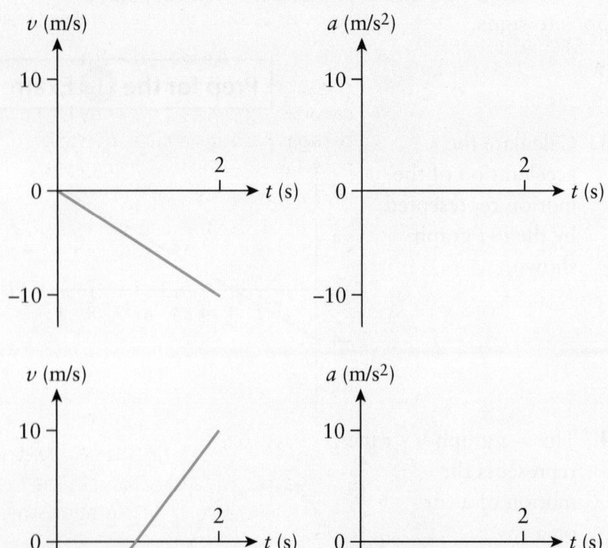

7. Measurements of the velocity of an object are made at three times separated at equal intervals. Identify and justify a mathematical procedure, using the definition of acceleration, to test the claim that the object moved with a constant velocity.

8. The following table displays measured values of position and time for an object in motion.
(a) Use the values in the first two columns to calculate the minimum and maximum estimated values of the average velocities v_a and v_b, indicated by blanks in the table below.

Time (s ± 0.05 s)	Position (m ± 0.1 m)	Velocity (m/s)
0	1	
0.8		____ $\le v_{a,\,average} \le$ ____
1.6	5.8	
2.4		____ $\le v_{b,\,average} \le$ ____
3.2	9.4	

(b) Evaluate the claim that the object is moving with a constant velocity.

9. The distance from Dubai to Abu Dhabi, two cities in the United Arab Emirates, is 139 km. Hyperloop One is building a high-speed transportation system between these cities. The goal is to make the trip in 13 minutes.
(a) Connect the targeted average speed of the Hyperloop One system with your experience by comparing the time required for you to drive to a nearby city in a car with an average speed of 45 mi/h and the time it would take for a Hyperloop One to drive to the same city from your starting position operating at the targeted speed. In making your comparison, use the method of dimensional analysis that was described in the last chapter.
(b) There is a problem with the Hyperloop One design. Pilots of flights involving maneuvers such as sudden changes in direction or lift-off into space must train to build their capacity to endure relatively high accelerations of as much as 80 m/s². If Hyperloop One passengers could become unconscious at accelerations as small as 30 m/s², predict whether the maximum speed of the Hyperloop will need to be greater or smaller than the targeted speed in part (a) to achieve a travel time of 13 minutes. Justify your prediction.

10. Humans have two gaits: walking and running. Quadruped mammals have three gaits: walking, trotting, and running. Students found a claim on the Internet that there is a transition for the preferred gait in humans that occurs at about 2.0 m/s (4.5 mi/hour). The students decide to test this claim of a transition speed where humans switch from walking to running. They will use a treadmill on which the speed of the track can be preprogrammed and which displays the elapsed time and track speed in miles per hour to the runner. The runner will not be able to see the speed of the treadmill.
(a) Pose the test of this claim as a scientific question.
(b) Justify the decision to blindfold the subject.
(c) Design a plan for collecting data that answers the question posed.

The data collected for 10 subjects in their high school physics class are:

Student	1	2	3	4	5	6	7	8	9	10
Transition speed (mi/hour)	4.2	4.0	4.5	4.3	3.8	4.6	4.0	4.7	3.9	4.1

(d) Evaluate the evidence from these data in relation to the question posed.

At the time when each subject passed the threshold speed they were asked, "Why did you choose to begin running?" Some responses were:

- "I was afraid of falling off the track."
- "My hips were hurting."

(e) The students discussed an extension of their investigation in which explanations for the transition threshold speed could be tested. Construct possible explanations for each of the responses shown above.

Skills in Action

11. Consider the motion of an object described by the data displayed in the graph.

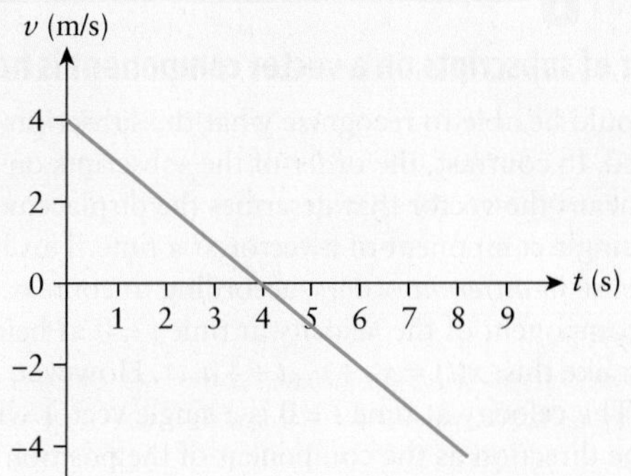

(a) Describe a scenario in which the motion will result in a velocity graph like the one shown. In your description include the phrases "slows down" and "speeds up" and the direction of the velocity during the motion of the object.

(b) Calculate the acceleration of the object.

(c) Calculate the average velocity in the time interval between 0 and 4 s.

(d) Describe how the displacement of the object between 0 and 4 s can be expressed in terms of the average velocity.

2-4 Tools for describing constant acceleration motion

When a basketball falls from a player's hand or a car comes to a halt with the brakes applied, the acceleration of the object turns out to be nearly *constant*—that is, the velocity of the object changes at a steady rate (**Figure 2-13**). In addition to using graphs, we can represent this motion using a set of kinematic equations applicable to linear motion with constant acceleration. These equations turn out to be useful for making predictions even in situations where the acceleration isn't strictly constant.

If the acceleration is constant, the *instantaneous* acceleration a_x at any instant of time is the same as the *average* acceleration $a_{\text{average},x}$ for any time interval. Thus we can use a_x in place of $a_{\text{average},x}$ in any equation. In particular let's consider a time interval from $t = 0$ to some other time t. We use the symbol v_{0x} for the velocity at $t = 0$, which we call the initial velocity, and the symbol v_x (with no zero) for the velocity at time t. From the definition of average acceleration given in Equation 2-3,

Yauhen_D/Shutterstock

Figure 2-13 An object with constant acceleration As this car brakes to a halt, its velocity is changing at a steady rate. Hence its acceleration—the rate of change of velocity—remains constant.

$$a_x = \frac{v_x - v_{0x}}{t - 0} \tag{2-4}$$

We can rewrite Equation 2-4 to get an expression for the velocity v_x at any time t. Multiply both sides of the equation by t, add v_{0x} to both sides, and rearrange:

Velocity at time t of an object in linear motion with constant acceleration

Velocity at time $t = 0$ of the object

$$v_x = v_{0x} + a_x t \tag{2-5}$$

Constant acceleration of the object

Time at which the object has velocity v_x

EQUATION IN WORDS
Velocity as a function of time for constant acceleration only

Equation 2-5 represents the object's velocity v_x as a function of time. If $a_x = 0$ so that the object is not accelerating, v_x at any time t is the same as the velocity v_{0x} at $t = 0$; in

WATCH OUT!

The order of subscripts on a vector component is not fixed.

You should be able to recognize what the subscripts on a *component of a vector* describe, in whichever order the subscripts are listed. In contrast, the order of the subscripts on a symbol for a *vector* often has a specific meaning. For instance, $\vec{r}_{12}$ most likely means the vector that describes the displacement from point 1 to point 2 in space. In AP® Physics 1, we will only work with a single component of a vector at a time. This has less accepted formalism, and the order of subscripts even seems to make more sense in *different* orders, according to context. When we think of just a single component, it might make sense to think of the component of the velocity at time $t = 0$ as being the most important feature, and so we might write the subscripts on the velocity like this: $x(t) = x_0 + v_{x0}t + \frac{1}{2}a_x t^2$. However, the important thing to remember is x is a single component of a position vector. The velocity at time $t = 0$ is a single vector with as many as three components, and we are just using the component in the same direction as the component of the position in which we are interested. Normally, $\vec{v}_0$ is written in terms of v_{0x}, v_{0y}, and v_{0z}. In this context, we then write $x(t) = x_0 + v_{0x}t + \frac{1}{2}a_x t^2$. To keep you thinking in a way that will make it easy to move forward in your studies, this is the notation we will most often use in this book, but the important point is that components may have a subscript that denotes *when*: 0, i, or f; or that denotes the component x, y, or z. Of course, they may also have a subscript that denotes *which*, like 1 or 2 or the name of the object. You need to be able to pick out the important information, no matter in what order the subscripts are listed.

other words, the velocity is constant. If a_x is not zero and the object is accelerating, the term $a_x t$ in this equation indicates how much the velocity changes in time t.

We'd also like to have an expression that shows how the object's x coordinate varies with time. To obtain such an expression first note that if the object is at coordinate x_0 at $t = 0$ and at coordinate x at time t, its x displacement during the time interval from 0 to t is $\Delta x = x - x_0$. From Equation 2-1, Δx divided by the duration $\Delta t = t - 0$ of the time interval equals the average velocity for this time interval:

$$v_{\text{average},x} = \frac{\Delta x}{\Delta t} = \frac{x - x_0}{t - 0}$$

We can rewrite this expression using the same steps we used in writing Equation 2-5:

$$x = x_0 + v_{\text{average},x}t \tag{2-6}$$

Equation 2-6 tells us that to determine where the object is at time t—that is, what its x coordinate is—we need to know the average velocity $v_{\text{average},x}$ for the time interval from 0 to t.

The average velocity is easy to find if the acceleration is constant. For example, suppose that the initial velocity is $v_{0x} = 3.0$ m/s and the acceleration is $a_x = 2.0$ m/s². From Equation 2-5, the velocity at the beginning, middle, and end of the time total time interval is

$$v_x = (3.0\text{ m/s}) + (2.0\text{ m/s}^2)(0\text{ s}) = 3.0\text{ m/s at } t = 0$$
$$v_x = (3.0\text{ m/s}) + (2.0\text{ m/s}^2)(1.0\text{ s}) = 5.0\text{ m/s at } t = 1.0\text{ s}$$
$$v_x = (3.0\text{ m/s}) + (2.0\text{ m/s}^2)(2.0\text{ s}) = 7.0\text{ m/s at } t = 2.0\text{ s}$$

To find the average of these three velocities—that is, the average velocity $v_{\text{average},x}$ for the time interval from $t = 0$ to $t = 2.0$ s—we add them together and divide by 3:

$$v_{\text{average},x} = \frac{3.0\text{ m/s} + 5.0\text{ m/s} + 7.0\text{ m/s}}{3} = \frac{15.0\text{ m/s}}{3} = 5.0\text{ m/s}$$

Notice that $v_{\text{average},x} = 5.0$ m/s is also the value of the instantaneous velocity v_x at the midpoint of the interval ($t = 1.0$ s). Furthermore, $v_{\text{average},x}$ is equal to the average of the instantaneous velocities at the *beginning and end* of the time interval [$v_x = 3.0$ m/s at $t = 0$ and $v_x = 7.0$ m/s at $t = 2.0$ s; the average of these is (3.0 m/s + 7.0 m/s)/2 = (10.0 m/s)/2 = 5.0 m/s]. You will *always* (and *only*) get this result if the acceleration is constant. So for constant a_x, the average velocity for the time interval from time 0 to time t is the average of v_{0x} and v_x:

$$v_{\text{average},x} = \frac{v_{0x} + v_x}{2} \tag{2-7}$$

If we substitute the expression for v_x from Equation 2-5 into Equation 2-7, we get

$$v_{\text{average},x} = \frac{v_{0x} + (v_{0x} + a_x t)}{2} = \frac{2v_{0x} + a_x t}{2} = v_{0x} + \frac{1}{2}a_x t \tag{2-8}$$

WATCH OUT!

Instantaneous velocity and average velocity are different quantities.

Equations 2-5 and 2-8 look quite similar, but they describe two very different things. The first of these, $v_x = v_{0x} + a_x t$, tells us the *instantaneous velocity* at the *end* of the time interval from 0 to t. The second of these, $v_{\text{average},x} = v_{0x} + \frac{1}{2}a_x t$, is a formula for the *average velocity during* this time interval.

Equation 2-8 gives the average velocity of an object moving with constant acceleration for the interval from 0 to t.

If we substitute the expression for $v_{\text{average},x}$ from Equation 2-8 into Equation 2-6, we get an equation for the coordinate of the object x at time t:

$$x = x_0 + v_{\text{average},x} t = x_0 + \left(v_{0x} + \frac{1}{2} a_x t \right) t$$

or, simplifying,

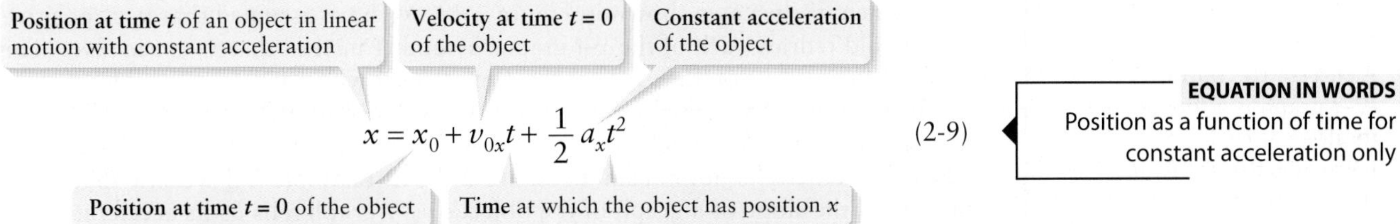

$$x = x_0 + v_{0x} t + \frac{1}{2} a_x t^2 \tag{2-9}$$

EQUATION IN WORDS
Position as a function of time for constant acceleration only

If we know the object's initial position x_0, its initial velocity v_{0x}, and its constant acceleration a_x, Equation 2-9 tells us its position x at any time t.

A good way to check Equation 2-9 is to consider a couple of special cases. If $v_{0x} = 0$ (the object is initially at rest) and $a_x = 0$ (the object's velocity doesn't change), Equation 2-9 says that $x = x_0$ at any time t. In other words, the object doesn't move! If v_{0x} is not zero (the object is moving initially) but $a_x = 0$ (the acceleration is zero), the object has a constant velocity $v_x = v_{0x}$ and its position at time t is $x = x_0 + v_{0x}t = x_0 + v_x t$. This relationship is the same as Equation 2-2 for linear motion with constant velocity, which we found in Section 2-2. Note that if there is a nonzero acceleration a_x, Equation 2-9 shows that there is an additional displacement of $\frac{1}{2} a_x t^2$ during the time interval from 0 to t compared to the case of constant velocity. Below, we will see a graphical way to find this same relationship.

Graphing Motion with Constant Acceleration

A good way to interpret Equations 2-5 and 2-9 is to draw the associated motion diagram, x–t graph, and v_x–t graph. **Figure 2-14** shows three examples of these motion diagrams and graphs. In each case the spacing between the dots of the motion diagram changes as the object's velocity changes. In cases (a) and (b) the acceleration is positive,

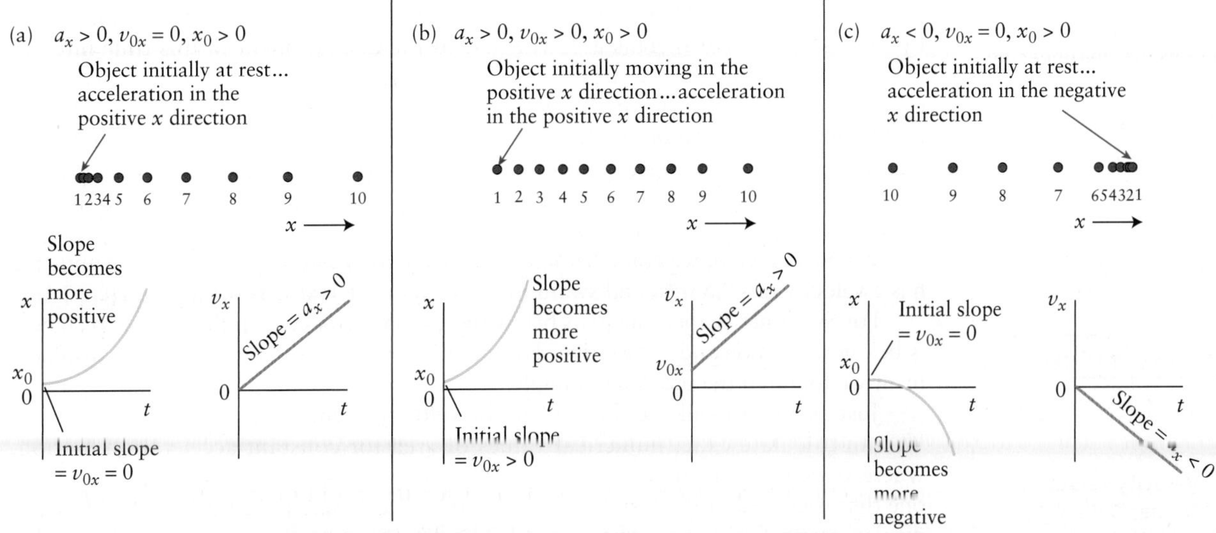

Figure 2-14 Motion with constant acceleration: three examples Each of these three sets of an x–t graph and a v_x–t graph depicts a different example of linear motion with constant acceleration.

WATCH OUT!

Know your graphs.

Be careful not to confuse the x–t graph and the v_x–t graph for constant acceleration. Remember that acceleration means a change in velocity, so constant acceleration means that the *rate* at which position is changing is constantly changing, so the x–t graph is *curved*. By contrast, the v_x–t graph is a *line* because its slope at each point represents the *acceleration* at the corresponding time. The acceleration is constant, so the v_x–t graph has a constant slope (which makes the graph a line).

so the velocity is becoming more positive. Whenever the acceleration is a positive constant, the x–t graph is a special curve called a parabola that curves upward (the slope, which denotes v_x, is increasing), and the v_x–t graph is a line that slants upward (the slope, which denotes a_x, is positive). In case (c) the acceleration is negative. Whenever the acceleration is a negative constant, the x–t graph is a parabola that curves downward (because v_x is getting more negative, the slope of this curve is also becoming more negative), and the v_x–t graph is a line that slants downward (because a_x, and thus the slope of this curve, is negative). For any nonzero acceleration, the x–t graph is *curved* because its slope at each point represents the velocity at the corresponding time t. The velocity is continually changing, so the slope is different at each point and the graph is curved.

In each of the cases shown in Figure 2-14, we have assumed that x_0 is positive. You should redraw each of the x–t graphs assuming instead that x_0 is negative. (This change has no effect on the v_x–t graphs. Can you see why not?)

In Section 2-2 we saw that if the velocity is constant, the slope of the x–t graph tells us the value of the velocity. In Section 2-3 we saw that the same idea is true when the velocity is not constant (Figure 2-7). In Example 2-5 we saw that in the same way, the slope of a graph of the object's v_x–t graph tells us its acceleration a_x. This is not an accident. The slope of a line indicates the rate that the quantity being graphed on the y axis varies as the quantity being graphed on the x axis varies. Velocity is the rate of change of position with respect to time, so velocity is the slope of a graph of position versus time. Acceleration is the rate of change of velocity with respect to time, so acceleration is the slope of the velocity versus time graph.

It turns out that there is another fundamental relationship between quantities that depends on finding the area *under* the graph! For instance, when we plot the quantity v_x that would be the slope of the x–t graph versus t, we can find out how much x would change in any interval of time by finding the area *under* the graph during that interval. *Under* means between the graph line and the horizontal axis. If the velocity is negative, then the relevant area is between that negative line and the horizontal axis, so technically *above* the graph line. But physicists (and mathematicians) still refer to this as the area under the graph. If we look at the mathematical relationship $v = \Delta x/\Delta t$, we see that $\Delta x = v\Delta t$, the product of v and Δt. The "area rule" is a mathematical fact that works for *any* product: When you plot the multiplicand and multiplier on the axes of a graph, their product is the area under the graph. If the area extends below the x axis, the product is negative. Let's calculate this for some examples with which we have already been working.

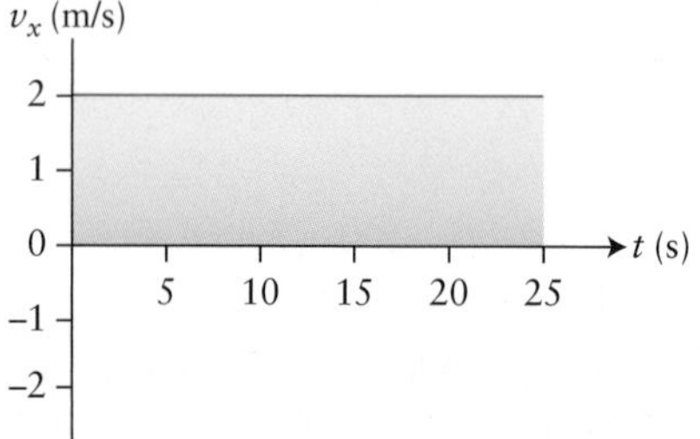

Figure 2-15 Velocity versus time graph for swimmer A The shaded section represents the area under the graph between 0 and 25 s.

In Figure 2-4 (and then further explored in Example 2-2) we drew the position versus time graphs for two swimmers, A and B. We can take that data and create velocity versus time graphs. We know that Swimmer A swam for 25 seconds, and had a velocity of +2 m/s, as shown in **Figure 2-15.**

The area under the graph of velocity versus time equals the area of the colored rectangle in Figure 2-15; that is, the height of the rectangle (the velocity v) multiplied by its base (the time t). This area is equal to the displacement in this time interval (notice that the units for the area are correct for displacement, m/s · s = m). *On a graph of velocity versus time, the displacement in any time interval equals the area under the graph during that time interval.* To find the position at the end of this time interval, you would add this displacement to the position at the beginning of the time interval. For Swimmer A, this area is 2 m/s · 25 s = 50 m, the displacement we found in the previous example.

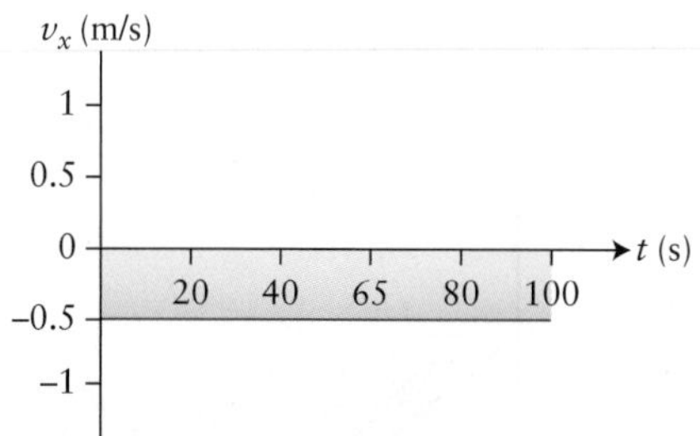

Figure 2-16 Velocity versus time graph for swimmer B The shaded section represents the area under the graph between 0 and 100 s.

For Swimmer B, we come back to what we said about negative areas. Swimmer B has a velocity of –0.5 m/s, and swam for an interval of 100 s, as shown in **Figure 2-16.**

For Swimmer B, the velocity is below the x axis, so the area in the colored rectangle is below the x axis and is negative. This area is –0.5 m/s · 100 s = –50 m, the displacement we found in the previous example.

Just as we can still use slope when the velocity is changing, we can still use area. We can calculate exact numerical values of area for constant acceleration motion, where velocity versus time is a line, but given any graph of velocity versus time, even one that is curved, it is possible to use this rule to estimate the area, and therefore the displacement, during any time interval. The displacement during any time interval is *always* given by the area under the velocity–time graph.

Figure 2-17 shows a velocity versus time graph for an object undergoing a constant acceleration. Since the acceleration is constant, the graph of velocity as a function of time is a line. To find the displacement between two times, t_f and t_i, we can divide up the regions into a triangle and a rectangle (I and II in Figure 2-17b). Then we can see how that compares to Equation 2-9 relating velocity, acceleration, and position, $x = x_0 + v_{0,x}t + \frac{1}{2}a_x t^2$.

Next, let's rewrite the equation to find the change in position in terms of the initial and final times given in Figure 2-17b: $x_f = x_i + v_{i,x}(t_f - t_i) + \frac{1}{2}a_x(t_f - t_i)^2$. If we rearrange it just slightly, we can get the displacement, $\Delta x = x_f - x_i = v_{i,x}(t_f - t_i) + \frac{1}{2}a_x(t_f - t_i)^2$. The first term is just the area of the rectangle, area II: $v_{i,x}(t_f - t_i)$ since $v_{i,x}$ is the height of the rectangle, and $t_f - t_i$ is its width. The second term is a little less straightforward to see, but it is in fact the area of the triangle, I. The area of a triangle is ½ its width times its height. Just like the rectangle, the width of the triangle is $t_f - t_i$. Its height is $v_{f,x} - v_{i,x}$, but we need to get that in terms of acceleration and time. From Equation 2-3, $a_{\text{average},x} = \dfrac{\Delta v_x}{\Delta t} = \dfrac{v_{2x} - v_{1x}}{t_2 - t_1}$, so for our values, $a_x = (v_{f,x} - v_{i,x})/(t_f - t_i)$. Rearranging this gives us $v_{f,x} - v_{i,x} = a_x(t_f - t_i)$. Substituting this in for the height, we have the area of the triangle $= \frac{1}{2}\cdot(t_f - t_i)\cdot a_x(t_f - t_i) = \frac{1}{2}a_x(t_f - t_i)^2$. This is exactly the second term in our equation for displacement, so we can see that the area method of the equation must give the same results. Having two ways to solve a problem always gives us a good way to check our answers. Additionally, we could use the area method to estimate the displacement for an object even if its acceleration was not constant, in which case we would not have an equation, and area would provide our only solution tool. We will find other cases where the area rule is useful in later chapters.

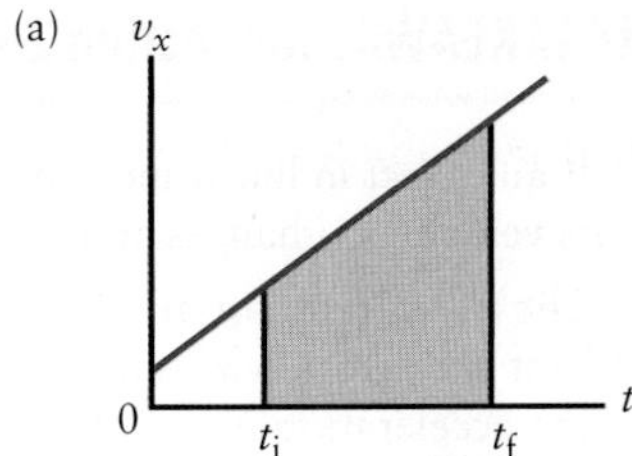

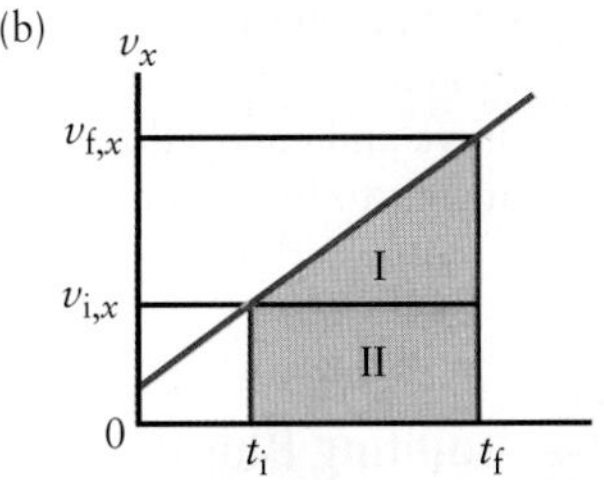

Figure 2-17 Finding the area under a *velocity–time* graph (a) The shaded area represents the area under the *velocity–time* graph for the time interval $t_f - t_i$. (b) For constant acceleration, the area breaks into a rectangle and a triangle, which give the displacement if the initial velocity had been constant, and the change in displacement due to the acceleration.

The Kinematic Equations for Motion with Constant Acceleration

Equations 2-5 and 2-9 give the object's x coordinate and velocity at a given time t. There is also a third equation that is often useful in analyzing linear motion with constant acceleration. This equation relates the object's displacement and acceleration to its initial and final velocities, without reference to time. To create this equation, first solve Equation 2-5 for the time at which the object has velocity v_x:

$$t = \frac{v_x - v_{0x}}{a_x}$$

If you substitute this expression into Equation 2-9, $x = x_0 + v_{0x}t + \frac{1}{2}a_x t^2$, every place you see the quantity t, you get:

$$x = x_0 + v_{0x}\left(\frac{v_x - v_{0x}}{a_x}\right) + \frac{1}{2}a_x\left(\frac{v_x - v_{0x}}{a_x}\right)^2 \qquad (2\text{-}10)$$

NEED TO REVIEW?
See the Math Tutorial for more information on solving simultaneous equations (M-4) and quadratic equations (M-5).

You should fill in the algebraic steps needed to rewrite Equation 2-10 as

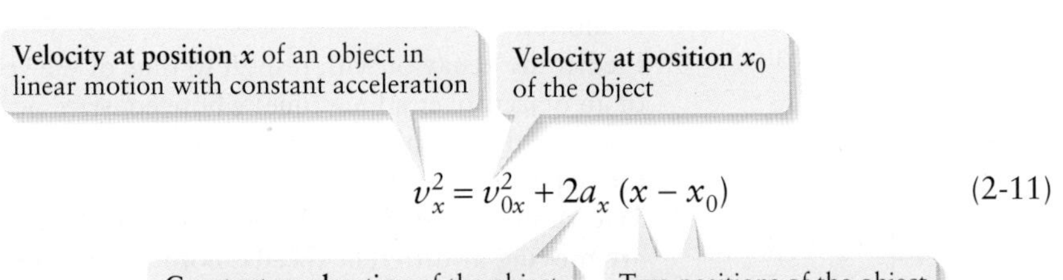

(2-11)

EQUATION IN WORDS
Velocity as a function of position for constant acceleration only

Equations 2-5, 2-9, and 2-11 are tremendously useful because they allow you to solve *any* problem in kinematics that involves linear motion with constant acceleration. Situations with constant or nearly constant acceleration are quite common in nature and technology, so these equations will be useful to us throughout our study of physics. In the following section we'll see how to tackle problems of this kind.

THE TAKEAWAY for Section 2-4

✔ If an object in linear motion has a constant acceleration a_x, its velocity v_x changes at a steady rate.

✔ The v_x–t graph for an object in linear motion with constant acceleration a_x is a line. The slope of the line tells you the acceleration of the object.

✔ The x–t graph for an object in linear motion with constant acceleration a_x is a parabola. The parabola curves upward if a_x is positive and curves downward if a_x is negative. The slope of the tangent to the parabola where it meets the vertical axis tells you the object's velocity at $t = 0$.

✔ The slope of a line always can be used to find the rate of change of the quantity on the y axis with respect to the quantity graphed on the x axis. Since velocity is rate of change of position with respect to time, it can always be found from the slope of a position versus time graph.

✔ The area under the graph of velocity versus time during a given time interval equals the displacement in this time interval.

Prep for the AP® Exam

AP® Building Blocks

1. (a) Acceleration is defined as
   (A) the change in displacement.
   (B) rate of change of displacement with respect to time.
   (C) the change in velocity.
   (D) the rate of change of velocity with respect to time.
   (b) Velocity is defined as
   (A) the change in displacement.
   (B) the rate of change of displacement with respect to time.
   (C) the change in distance.
   (D) the rate of change of distance with respect to time.
2. The graph shows the velocity of an object moving along the x axis. What are the (a) initial and (b) final directions of travel in terms of x? (c) Does the object stop momentarily? (d) Is the acceleration constant or does it vary?

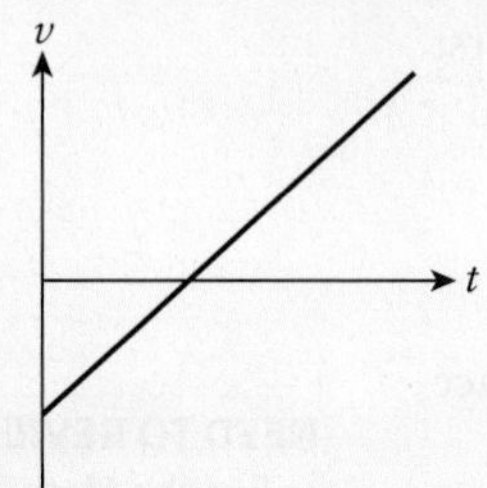

3. Explain the difference between "going slow" and "slowing down" with a complete sentence. Give an example to illustrate the difference.
4. The one-dimensional position of an object is measured in increments of 0.5 s, as shown in the table.

t (s)	x (m)	v_{average} (m/s)
0	0.00	
0.5	0.81	1.62
1.0	2.25	
1.5	4.31	
2.0	7.00	
2.5	10.31	
3.0	14.25	
3.5	18.81	
4.0	24.00	

(a) Construct an x–t graph using these data.
(b) Complete the table by calculating the average speeds in each of the 0.5 second intervals. For example, in the first 0.5 second interval

$$v_{\text{average}} = \frac{0.81 \text{ m}}{0.5 \text{ s}}$$

(c) Construct a graph of these values. Draw a best-fit line and find an equation for the line. You may use a calculator or computer.
(d) Construct a motion diagram from these data.

AP® Skill Builders

5. Leon claims that in the movie *2 Fast 2 Furious* the Mitsubishi Eclipse could have beaten the Charger in a drag race had the Eclipse not exploded. In a drag race two cars are driven 1320 ft (a quarter of a mile) starting from rest. The shortest time interval wins the race. Leon supports his claim by stating that the maximum speed is higher for the Eclipse than for the Charger. Pose a question that another student could ask Leon to challenge Leon's claim.
6. Many physical and social behaviors are characterized by rates of change. Many of these are time rates of change (the variable of interest changes with time). Connect a familiar physical or social behavior to the concepts of velocity and acceleration by summarizing an analogy between these changes and the properties of motion.
7. Predict the behavior with respect to time of the velocity of an object (over a long time) whose constant acceleration is in the (a) same direction as the velocity and (b) in the opposite direction.
8. Two runners in different lanes of a running track are moving at different speeds. Both speeds are constant. Predict whether or not the motion of the faster runner is accelerated from the perspective of the slower runner. As always, a prediction must be justified.

AP® Skills in Action

9. A car is coasting forward on a straight, horizontal road. The car's acceleration is constant and in the direction opposite to the direction of the car's motion. Which of the following correctly predicts the car's motion?
   (A) The velocity of the car will continue to be in the same direction while the speed does not change.
   (B) The velocity of the car will continue to be in the same direction while the speed increases.
   (C) The car will continue to roll in its original direction until the speed is zero and then the car will stop rolling and remain at rest.
   (D) The car will continue to roll in its original direction until the speed is zero and then the car will roll in the same direction as the acceleration and the speed will increase.

10. If you know the constant acceleration of a car, its initial velocity, and the time interval, which of the following can you predict?
   (A) The direction of the car's final velocity
   (B) The magnitude of the car's final velocity
   (C) The displacement of the car
   (D) All of the above

2-5 Solving linear motion problems: constant acceleration

Table 2-3 summarizes the properties of Equations 2-5, 2-9, and 2-11 for linear motion with constant acceleration. The three equations in Table 2-3 involve six physical quantities: the time t, the object's position x_0 at time 0, the object's position x at time t, its velocity v_{0x} at time 0, its velocity v_x at time t, and its constant acceleration a_x. However, not all of these quantities appear in any one of the three constant-acceleration equations. In the following worked examples, we'll show how to decide which equation or equations are the proper ones to use for a given problem.

TABLE 2-3 Equations for Linear Motion with Constant Acceleration

By using one or two of these equations, you can solve any problem involving linear motion for which the acceleration is constant. The check marks indicate which of the quantities t (time), x_0 (the x coordinate of position at time 0), x (the x coordinate of position at time t), v_{0x} (the velocity at time 0), v_x (the velocity at time t), and a_x (the constant acceleration) appear in each equation.

Equation		Which quantities the equation includes					
		t	x	x_0	v_{0x}	v_x	a_x
$v_x = v_{0x} + a_x t$	(2-5)	✓			✓	✓	✓
$x = x_0 + v_{0x}t + \frac{1}{2}a_x t^2$	(2-9)	✓	✓	✓	✓		✓
$v_x^2 = v_{0x}^2 + 2a_x(x - x_0)$	(2-11)		✓	✓	✓	✓	✓

AP® Exam Tip

You may learn other useful kinematic equations for particular constant acceleration situations, but they can all be derived from Equations 2-5 and 2-9. It's important to note that only these three kinematic equations are given on the AP® Physics 1 exam equation sheet.

There are situations where you'll need to use two of the equations in Table 2-3 to solve the problem, because you are solving for two unknown quantities (see the rest of the chapter for examples). In these situations, you'll need to use your skills at solving simultaneous equations. In addition, Equation 2-9 is a *quadratic* equation: It involves the square of the time t as well as t. This quadratic equation can have two, one, or no solutions—this will depend on the particular problem you are working on and you will have to choose the answer that makes sense given the problem situation. Similarly, Equation 2-11 is quadratic in velocity, so if you are solving for velocity you will have to use other things you know from the problem to determine the sign of the velocity.

NEED TO REVIEW?

See Math Tutorials to review how to work with simultaneous equations (M-4) and with quadratic equations (M-5).

WATCH OUT !

You will never use all three equations at once.

You will never use all three equations for linear motion with constant acceleration at once because we derived Equation 2-11 from Equations 2-5 and 2-9, so only two of the three equations can give you independent information. You can always use any two of the equations, but what information you are given can make it easier to use specific equations (for instance, 2-11 does not involve time, so if you don't have time this one is easier, which is why we derived it for you). If you believe you have three unknown quantities in a problem involving just one object, you should probably reread it carefully to find any relationships you missed. If you have two objects, each will have its own two equations, so you could solve for four unknowns!

Exam Tip

An illustrative part of solving any quantitative, one-dimensional kinematics problem is identifying the known and unknown quantities. Write down each of the six variables (from the three kinematics equations) and its value if known, or indicate that it is unknown. This will help you select which one or two of the equations will help you find the desired quantity of interest, and will help anyone grading your work understand what you are doing.

Exam Tip

When solving kinematics problems, we are often only interested in the displacement, which is represented by $(x - x_0)$. If this is the case, this reduces the number of unknowns.

EXAMPLE 2-6 Motion with Constant Acceleration I: Cleared for Takeoff!

After winning the lottery you are shopping for a corporate jet. During a demonstration flight with a salesperson, you notice that a particular type of jet takes off after traveling 1.20 km down the runway and after reaching a speed of 226 km/h (140 mi/h). (a) Sketch a motion diagram, an x–t graph, and a v_x–t graph for the jet's motion as it rolls down the runway. (b) What is the magnitude of the airplane's acceleration (assumed constant) during the takeoff roll? (c) How long does it take the airplane to reach takeoff speed?

Set Up

We take the positive x axis to point along the runway in the direction that the jet travels, choose the origin at the point where the jet starts to move, and choose $t = 0$ to be when it starts to move. We are told the jet's initial velocity ($v_{0x} = 0$ because the jet starts at rest) as well as its velocity $v_x = 226$ km/h at position $x = 1.20$ km. Our goals in parts (b) and (c) are to find the jet's acceleration a_x and the time t when the jet attains $v_x = 226$ km/h. To summarize what we know and what we don't, we make a table of the six physical quantities involved in this problem, with a question mark for each of the quantities we want to find.

To find a_x from x_0, x, v_{0x}, and v_x:

$$v_x^2 = v_{0x}^2 + 2a_x(x - x_0) \qquad (2\text{-}11)$$

To find t from v_{0x}, v_x, and a_x:

$$v_x = v_{0x} + a_x t \qquad (2\text{-}5)$$

start of takeoff roll
$v_{0x} = 0$
takeoff
$v_x = 226$ km/h
0
1.20 km
x

Because we want to determine the values of *two* unknown quantities, a_x and t, we need two equations. Table 2-3 tells us that to find a_x from x_0, x, v_x, and v_{0x}, we should use Equation 2-11. Once we know the value of a_x, we can use Equation 2-5 to find the time t at which $v_x = 226$ km/h.

Quantity	Know/Don't Know
t	?
x_0	0
x	1.20 km
v_{0x}	0
v_x	226 km/h
a_x	?

Solve

(a) Because the jet speeds up as it rolls down the runway, successive dots in its motion diagram are spaced increasingly far apart. The x–t graph is a parabola (because a_x is constant) that starts at $x_0 = 0$ at $t = 0$ (the jet is initially at the origin), has zero slope at $t = 0$ (the initial velocity is zero), and curves upward (the slope increases as v_x increases). The v_x–t graph starts at $v_{0x} = 0$ at $t = 0$ (the initial velocity is zero) and is a line that slopes upward (the acceleration is constant and positive). (Compare Figure 2-14a.) We draw the graphs only to the point of takeoff because after that point the jet is flying and no longer simply moving along the x axis.

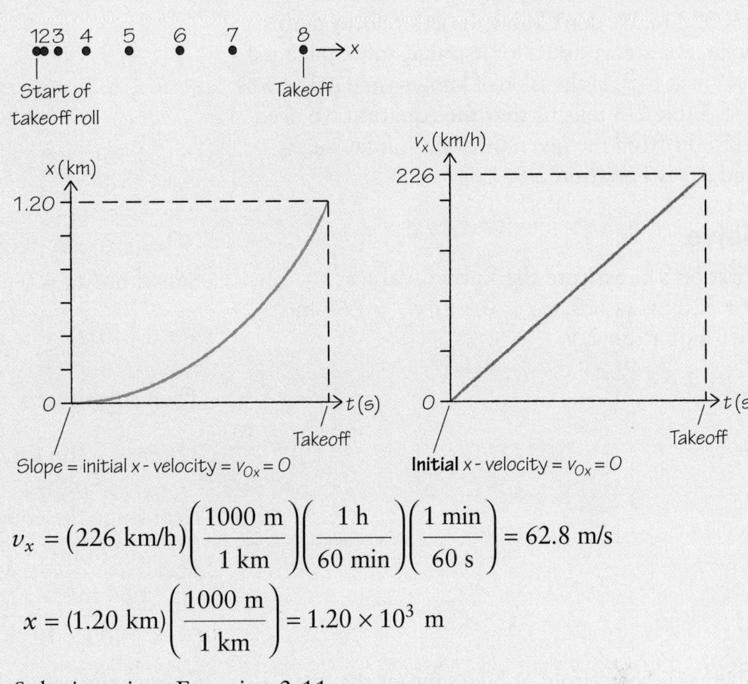

(b) We first convert $v_x = 226$ km/h to m/s and convert $x = 1.20$ km to m. We then solve Equation 2-11 for a_x and substitute in the known values.

$$v_x = (226 \text{ km/h})\left(\frac{1000 \text{ m}}{1 \text{ km}}\right)\left(\frac{1 \text{ h}}{60 \text{ min}}\right)\left(\frac{1 \text{ min}}{60 \text{ s}}\right) = 62.8 \text{ m/s}$$

$$x = (1.20 \text{ km})\left(\frac{1000 \text{ m}}{1 \text{ km}}\right) = 1.20 \times 10^3 \text{ m}$$

Substitute into Equation 2-11:

$$a_x = \frac{v_x^2 - v_{0x}^2}{2(x - x_0)} = \frac{(62.8 \text{ m/s})^2 - (0 \text{ m/s})^2}{2(1.20 \times 10^3 \text{ m} - 0 \text{ m})} = 1.64 \text{ m/s}^2$$

(c) We now know the value of a_x from part (b), so we can solve Equation 2-5 for the time t at which $v_x = 226$ km/h $= 62.8$ m/s. When you enter the values for v_x and a_x into your calculator, retain the extra digits from your calculations above and round the value for the time t only at the end.

Substitute a_x from part (b) into Equation 2-5:

$$t = \frac{v_x - v_{0x}}{a_x} = \frac{(62.8 \text{ m/s}) - (0 \text{ m/s})}{1.64 \text{ m/s}^2} = 38.2 \text{ s}$$

Reflect

Both of our numerical answers have the correct units: a_x is in m/s^2 and t is in s. We can check our answers to parts (b) and (c) by substituting $x = 0$, $v_{0x} = 0$, $a_x = 1.64$ m/s^2, and $t = 38.2$ s into Equation 2-9, the one constant-acceleration equation that we haven't used yet. We get $x = 1.20$ km, which means that when the jet reaches takeoff speed, it is 1.20 km down the runway—just as we were told in the statement of the problem. That tells us our results are consistent.

Substitute a_x from part (b) and t from part (c) into Equation 2-9:

$$x = x_0 + v_{0x}t + \frac{1}{2}a_x t^2$$

$$= 0 + (0)(38.2 \text{ s}) + \frac{1}{2}(1.64 \text{ m/s}^2)(38.2 \text{ s})^2$$

$$= 1.20 \times 10^3 \text{ m} = 1.20 \text{ km}$$

NOW WORK Problems 1 and 2 from The Takeaway 2-5.

EXAMPLE 2-7 Motion with Constant Acceleration II: Which Solution Is Correct?

Consider again the jet takeoff described in Example 2-6. There is a marker alongside the runway 328 m from the point where the jet starts to move. How long after beginning its takeoff roll does the jet pass the marker?

Set Up

We use the same x axis and choice of origin as in Example 2-6. Again we have $x_0 = 0$ and $v_{0x} = 0$, and we know the value of a_x from our work in Example 2-6. Our goal is to calculate the time t at which $x = 328$ m. We don't know the jet's velocity at this point, nor are we asked for its value, so we put a red "X" next to v_x in the table of knowns and unknowns.

Table 2-3 tells us that the equation we need to find t from the given information (x, x_0, v_{0x} and a_x) is Equation 2-9.

Position, acceleration, and time for constant acceleration only:

$$x = x_0 + v_{0x}t + \frac{1}{2}a_x t^2 \qquad (2\text{-}9)$$

Quantity	Know/Don't Know
t	?
x_0	0
x	328 m
v_{0x}	0
v_x	X
a_x	1.64 m/s^2

Solve

First let's substitute the known values $x = 328$ m, $x_0 = 0$, $v_{0x} = 0$, and $a_x = 1.64$ m/s^2 into Equation 2-9.

Substitute $x_0 = 0$ and $v_{0x} = 0$, then simplify:

$$x = 0 + (0)t + \frac{1}{2}a_x t^2 = \frac{1}{2}a_x t^2$$

Then substitute $x = 328$ m, $a_x = 1.64$ m/s^2:

$$328 \text{ m} = \frac{1}{2}(1.64 \text{ m/s}^2)t^2$$

Rearrange the equation to solve for t.

$$t^2 = \frac{2(328 \text{ m})}{1.64 \text{ m/s}^2} = 4.00 \times 10^2 \text{ s}^2$$

$$t = \pm\sqrt{4.00 \times 10^2 \text{ s}^2} = \pm 20.0 \text{ s}$$

The $\pm$ symbol in front of 20.0 s means that the square of either $t = +20.0$ s or $t = -20.0$ s is equal to $t^2 = 4.00 \times 10^2$ s^2. In this case we clearly want a positive value of time: The jet must reach the marker at $x = 328$ m some time *after* beginning its takeoff roll at $t = 0$. So the correct answer is $t = 20.0$ s.

Choose the value that corresponds to a time after the jet started moving:

$$t = 20.0 \text{ s}$$

Reflect

In Example 2-6 we found that the jet takes 38.2 s to travel 1.20 km. Our answer says that it takes less time (20.0 s) to travel a shorter distance (328 m), which is reasonable.

What may not seem reasonable is that we got *two* answers to the problem, only one of which made sense. To see the meaning of the other answer, $t = -20.0$ s, remember that we assumed that the jet has *constant* acceleration $a_x = 1.64$ m/s^2. As far as the equations are concerned, the jet has *always* had this acceleration, even before $t = 0$! According to these equations, before $t = 0$ the jet had the same positive value of a_x but had a *negative* velocity v_x and was moving *backward* along the runway in the negative x direction. The answer $t = -20.0$ s refers to the time during this fictitious motion when the backward-moving jet passed through the point $x = 328$ m.

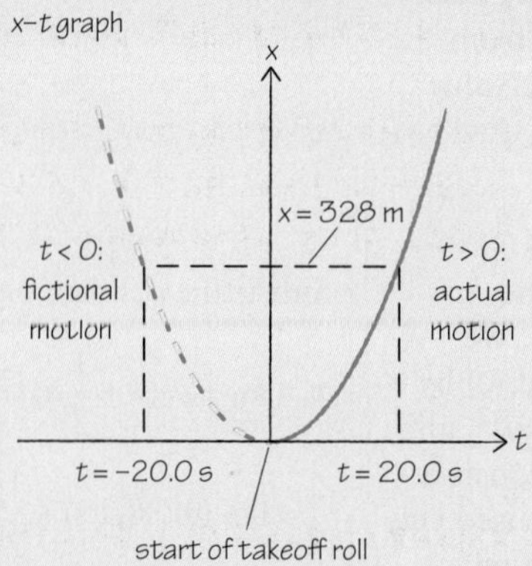

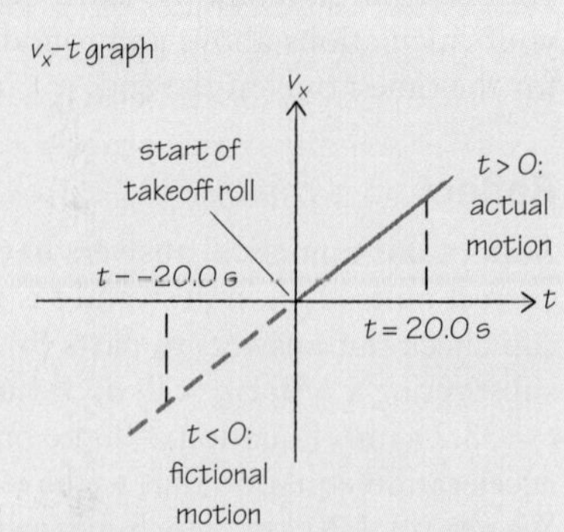

NOW WORK Problem 3 from The Takeaway 2-5.

EXAMPLE 2-8 Motion with Constant Acceleration III: Demolition Derby

As a stunt for a movie, two cars are to collide with each other head-on (**Figure 2-18**). The two cars initially are 125 m apart, car A is heading straight for car B at 30.0 m/s, and car B is at rest. Car A maintains the same velocity, while car B accelerates toward car A at a constant 4.00 m/s². (a) When do the cars collide? (b) Where do the cars collide? (c) How fast is car B moving at the moment of collision?

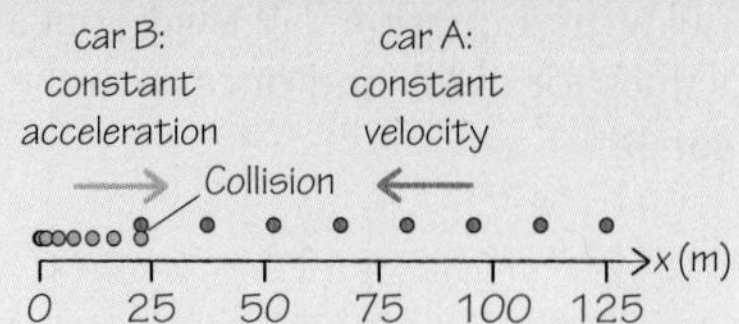

Figure 2-18 A head-on collision When and where will these two cars collide?

Set Up

This situation is similar to the two swimmers in Example 2-2 in Section 2-2. The two cars collide when they are at the same place—in other words, at the time t when they have the same x coordinate for position, as the motion diagram shows. We choose the x axis so that car B is initially at the origin and car A is initially at $x = x_{A0} = 125$ m. Then car A has initial velocity $v_{A0x} = -30.0$ m/s (it is moving in the negative x direction, toward the origin) and zero acceleration, while car B has zero initial velocity (it starts at rest) and constant acceleration $a_{Bx} = +4.00$ m/s² (it accelerates in the positive x direction).

The two cars collide at the time t when their x–t graphs cross. You can also graph this on your calculator and find the intersection to help you be more confident of your calculated answer. Note from the v_x–t graph that the two cars never have the same *velocity* because they move in opposite directions.

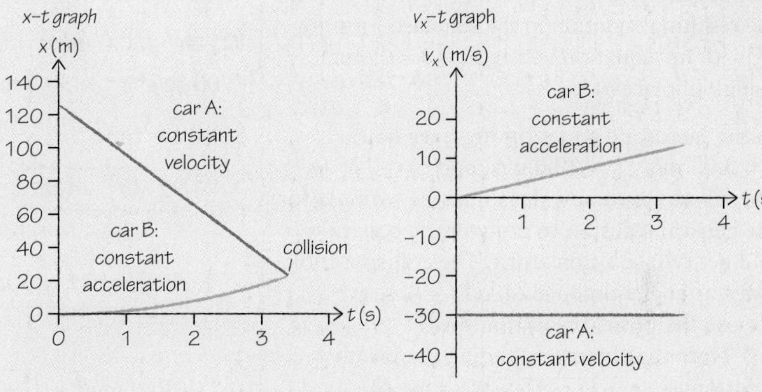

Our goals are to find the value of t when the two cars collide, the common value of x for the two cars at this time, and the velocity of car B at this time. (The tables of knowns and unknowns for the two cars seem to show *five* unknowns. But remember that the two cars have the same values of t and the same values of x when they collide.) Because time is an essential quantity in this problem, we use the two equations from Table 2-3 that involve t.

Position, acceleration, and time for constant acceleration only:

$$x = x_0 + v_{0x}t + \frac{1}{2}a_x t^2 \quad (2\text{-}9)$$

Velocity, acceleration, and time for constant acceleration only:

$$v_x = v_{0x} + a_x t \quad (2\text{-}5)$$

For car A:

Quantity	Know/Don't know
t	?
x_0	125 m
x	?
v_{0x}	−30.0 m/s
v_x	−30.0 m/s
a_x	0

For car B:

Quantity	Know/Don't know
t	?
x_0	0
x	?
v_{0x}	0
v_x	?
a_x	4.00 m/s²

Solve

(a) Write Equation 2-9 twice, once with the values for car A and once with the values for car B.

Car A ($x_{A0} = 125$ m, $v_{A0x} = -30.0$ m/s, $a_{Ax} = 0$):

$$x_A = 125\text{ m} + (-30.0\text{ m/s})t + \frac{1}{2}(0)t^2$$
$$= 125\text{ m} - (30.0\text{ m/s})t$$

Car B ($x_{B0} = 0$, $v_{B0x} = 0$, $a_{Bx} = 4.00\text{ m/s}^2$):

$$x_B = 0 + (0)t + \frac{1}{2}(4.00\text{ m/s}^2)t^2 = (2.00\text{ m/s}^2)t^2$$

The collision occurs when $x_A = x_B$. We rewrite the resulting equation in the standard form for a quadratic equation, $ax^2 + bx + c = 0$, but using t in place of x.

When the cars collide, $x_A = x_B$, so

$$125\text{ m} - (30.0\text{ m/s})t = (2.00\text{ m/s}^2)t^2 \text{ or}$$
$$(2.00\text{ m/s}^2)t^2 + (30.0\text{ m/s})t - 125\text{ m} = 0$$

In the quadratic equation for t, we have $a = 2.00\text{ m/s}^2$, $b = 30.0$ m/s, and $c = -125$ m. We substitute these values into the formula for the general solution to a quadratic equation and get two solutions for t. The collision took place after the time we called $t = 0$, so we choose the positive solution for t.

Note that before rounding the positive solution is $t = 3.397247$ s. We'll use this unrounded result in part (b).

Solve for time t:

$$t = \frac{-b \pm \sqrt{b^2 - 4ac}}{2a}$$
$$= \frac{-(30.0\text{ m/s}) \pm \sqrt{(30.0\text{ m/s})^2 - 4(2.00\text{ m/s}^2)(-125\text{ m})}}{2(2.00\text{ m/s}^2)}$$
$$= \frac{-(30.0\text{ m/s}) \pm \sqrt{1900\text{ m}^2/\text{s}^2}}{2(2.00\text{ m/s}^2)}$$
$$= \frac{-(30.0\text{ m/s}) \pm (43.6\text{ m/s})}{4.00\text{ m/s}^2}$$
$$= 3.40\text{ s or } -18.4\text{ s}$$

The answer must be positive, so the desired answer is $t = 3.40$ s.

(b) To find where the collision took place, we can plug t from part (a) into either the equation for x_A or the equation for x_B because at this time x_A and x_B are equal. Let's choose the first of these. Make sure to use the unrounded value of $t = 3.397247$ s in your calculation and round only at the end.

Solve for the position of car A at $t = 3.40$ s:

$$x_A = 125\text{ m} - (30.0\text{ m/s})t$$
$$= 125\text{ m} - (30.0\text{ m/s})(3.397247\text{ s})$$
$$= 23.1\text{ m}$$

So the collision takes place 23.1 m from the origin (the place where car B started).

(c) To calculate how fast car B was moving when the collision occurred, we will use Equation 2-5, now that we know from part (a) the time at which the collision occurred.

Solve for the velocity of car B at $t = 3.40$ s. Its initial velocity is 0, and its acceleration is 4.00 m/s^2.

$$v_x = v_{0x} + a_x t = 0 + (4.00\text{ m/s}^2)(3.397247\text{ s})$$
$$= 13.6\text{ m/s}$$

Reflect

You can check the answer for part (b) by verifying that you get the same result if you substitute t from part (a) into the equation for x_B. Be sure to try this!

Here's another way to look at the answer to part (b). When the two cars collide, car B has moved 23.1 m (from the origin to $x = 23.1$ m) while car A has moved more than four times as far (from $x = 125$ m to $x = 23.1$ m). These results make sense: Car B started at rest and reached a speed of only 13.6 m/s, while car A moved at a faster steady speed of 30.0 m/s and so covered more distance in the same amount of time.

NOW WORK Problem 4 from The Takeaway 2-5.

THE TAKEAWAY for Section 2-5

✔ To mathematically solve problems that involve linear motion with constant acceleration, you need just one or two of the three equations (see Table 2-3) that involve six quantities: time t, initial position x_0, position x at time t, velocity v_{0x} at time $t = 0$, velocity v_x at time t, and (constant) acceleration a_x.

Prep for the AP® Exam

AP® Building Blocks

EX 2-6

1. A car is able to stop with an acceleration of -3 m/s^2. Justify the mathematical routine used to calculate the distance required to stop from a velocity of 100 km/h by choosing the correct answer below.
(A) 16.7 m because the average velocity is 50 km/h, the change in velocity divided by the acceleration is the time, and the time multiplied by the average velocity is the distance
(B) 64.4 m because the average velocity is 13.9 m/s, the average velocity divided by the acceleration is the time, and the time multiplied by the average velocity is the distance
(C) 129 m because the average velocity is 13.9 m/s, the change in velocity divided by the acceleration is the time, and the time multiplied by the average velocity is the distance
(D) 257 m because the initial velocity is 27.8 m/s, the initial velocity divided by the acceleration is the time, and the time multiplied by the initial velocity is the distance

EX 2-6

2. The Bugatti Veyron is able to accelerate at 12.3 m/s^2. Calculate the time required to reach 100 km/h from rest if both maximum acceleration and velocity are given; justify the mathematical routine used.
(A) 8.13 s because change in velocity divided by acceleration is time
(B) 4.07 s because average speed divided by acceleration is time
(C) 2.26 s because change in velocity divided by acceleration is time
(D) 1.13 s because average speed divided by acceleration is time

EX 2-7

3. An object is moving at a constant velocity of -10 m/s when something happens, at time $t = 0$ s, to accelerate the object with a constant acceleration of 10 m/s^2.
(a) Express the displacement as a function of time after $t = 0$ s.
(b) Calculate the time when the displacement of the object relative to its position at $t = 0$ s is equal to 15 m using (i) the quadratic equation and (ii) factoring the expression from part (a).

EX 2-8

4. Two cars emerge side by side from the end of a tunnel on a four-lane road. Car A is traveling with a speed of 17 m/s and has an acceleration of 1.9 m/s^2. Car B has a speed of 11 m/s and has an acceleration of 3.8 m/s^2.
(a) What is the displacement of each car relative to the end of the tunnel 1 second before, at $t = -1$s?
(b) Which car is passing and which car is being passed as they come out of the tunnel?
(c) The two cars will later have the same displacement if they both continue with the same velocity and same acceleration. At what time after leaving the tunnel will one car pass the other?

AP® Skill Builders

5. A train of the Japanese high-speed rail system has a maximum acceleration of 0.72 m/s^2. Calculate the minimum time for the train to reach a speed of 270 km/h from rest.
6. A bus driver sees a wounded muskrat on the road just 50 m ahead and at that instant the speed of the bus is 13 m/s (30 mi/h). The maximum magnitude of the safe stopping acceleration of the bus is 1.8 m/s^2. Predict if the bus can stop before hitting the muskrat. Justify your prediction using the following three mathematical routines: (a) Equation 2-6, (b) Equation 2-9, and (c) Equation 2-11.
7. An object moves along a linear path. The initial velocity of the object is 20 m/s west and there is a constant acceleration of 2 m/s^2 east.
(a) Predict the time, T, at which the velocity of the object changes sign.
(b) Predict the velocity of the object at times of $T/2$ and $2T$. In your justification include your evaluation of the sign of the initial velocity and the acceleration.
8. A driver in a car moving at $v_0 = 20$ m/s (45 mi/h) sees the traffic light $\Delta x = 85$ m ahead turn from green to yellow. The conventional yellow time interval for traffic lights in cities is 3.5 s. The response time of the driver is 0.25 s. Assume that the driver responds immediately so that $t = 3.25$ s. The magnitude of the braking acceleration of the car is $a_{braking} = 5$ m/s^2 and to increase speed the car has a maximum magnitude of acceleration of 3.7 m/s^2.
(a) Construct inequalities for Δx in terms of v_0, a, $a_{braking}$, and t as needed that will let you determine the safety for each of the following strategies the driver might use: (i) maintain constant speed, (ii) press the accelerator pedal, or (iii) press the brake.
(b) Make a claim regarding the best strategy for the situation described above.
(c) If the brakes have not been maintained on this automobile the braking acceleration may be less than 5 m/s^2. Predict the minimum acceleration that will stop the car before reaching the intersection as the light changes.
(d) An internal memo circulates within a transportation engineering company raising the question "Is 3.5 s the best yellow time interval for a traffic light?" Refine this question and select the types of data needed to answer the question.
9. *Daphnia*, also known as water fleas, are just visible to unaided human eyes. They live in the upper layer of water, always below the surface, where they feed on algae. These animals have the record for the greatest acceleration as

a multiple of body length. They have jointed legs, which they use to stroke the water to accelerate.

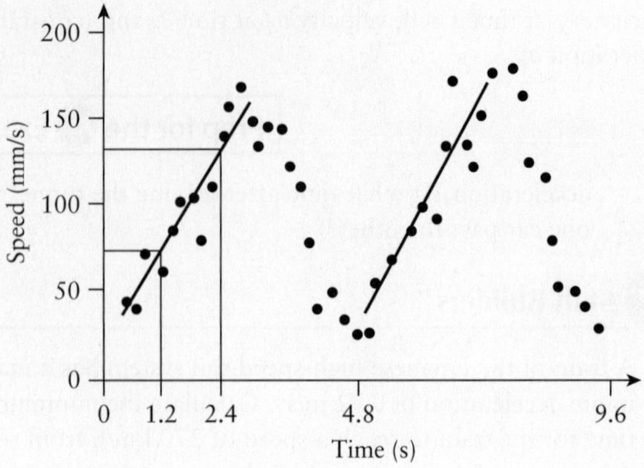

(a) Use the graph shown to analyze the data to determine the acceleration for the two strokes of the animal's legs. The species of *Daphnia* from which these data were taken have a body length of approximately 0.3 mm.

(b) From these data express the acceleration in units of body lengths per second squared and the average speed in body lengths per second.

(c) The record time for the 100-m race is 9.6 s, set by Usain Bolt in 2009. Analysis of his performance shows that the greatest acceleration was between 20 and 40 m from the starting line, where his speeds were 5.6 and 11.4 m/s, respectively.

i. Calculate Bolt's average speed.

ii. Calculate Bolt's average acceleration during his period of maximum acceleration.

iii. Express the average speed and maximum acceleration in units of body lengths per second and body lengths per second squared. Bolt's body length is 2.0 m.

(d) Compare the results of your analysis from parts (b) and (c) and pose two or three scientific questions whose pursuit might allow you to connect the properties of living systems over this large span of spatial domains.

AP® Skills in Action

10. The Bugatti Veyron is able to accelerate at 12.3 m/s^2. The Ford Fusion is able to accelerate at 4.2 m/s^2. Which of the following statements best calculates the additional time required for the Ford to reach 1 km from rest if both are given maximum acceleration and best describes the relationship between the mathematical model used and physical quantities?

(A) 9.0 s because the time elapsed is equal to the change in speed divided by the acceleration

(B) 157 s because the time elapsed is equal to the displacement divided by the change in speed

(C) 157 s because the square of the final speed is proportional to the product of displacement and acceleration

(D) 9.0 s because the time elapsed is equal to the change in speed divided by the acceleration and the displacement is proportional to the elapsed time squared

11. A lab cart is given a push upward along a low-friction track. The initial speed of the cart is 4.0 m/s. If the acceleration of the cart is –3.0 m/s^2, calculate the time until its instantaneous velocity is zero and calculate the maximum displacement of the cart from its starting point.

12. Analyze the data below graphically. Using the graph you construct, identify time intervals of approximately constant velocity and acceleration. Calculate the values of velocity and the signs of the acceleration in these intervals. Annotate the graph to display your calculated values. Explain why you were not asked to calculate values for acceleration.

t (s)	x (m)	t (s)	x (m)	t (s)	x (m)	t (s)	x (m)	t (s)	x (m)
0	–12	5	12	10	70	15	60	20	30
1	–4	6	24	11	80	16	50	21	30
2	0	7	36	12	78	17	42	22	32
3	0	8	48	13	74	18	36	23	34
4	4	9	60	14	68	19	32	24	36

2-6 Objects falling freely near Earth's surface have constant acceleration

Perhaps the most important case of constant acceleration is the motion of falling objects near the surface of Earth (**Figure 2-19**). Experiment shows that if air resistance isn't important, an object falling near Earth's surface has a constant, downward acceleration. We will learn later that this acceleration is due to a gravitational force exerted by Earth on the object.

When is air resistance important? To help understand this, drop a sheet of notebook paper. It speeds up for a second or so then wafts downward at a slow, roughly constant speed. The speed doesn't keep increasing, so this is definitely *not* a case of constant acceleration. But now take a second, identical piece of notebook paper and wad it into a ball. If you drop the first sheet and the second, wadded sheet side by side, you'll find that the wadded one falls more rapidly and *does* keep on gaining speed as it falls. The difference is that by wadding up the second sheet, you've given it a smaller cross-sectional area and thus made it less susceptible to air resistance. Competition bicyclists use this same trick when they bend low over the handlebars. This reduces their cross section and thus the effect of air resistance so that the bicyclists can go faster for the same effort.

But even the wadded-up paper experiences some air resistance. Imagine, though, that you could somehow crush the paper to an infinitesimally small sphere. There would be essentially no air resistance on this tiny sphere, and only the pull of gravity would affect its fall. In this idealized situation, called **free fall**, the sphere would have a constant downward acceleration. The acceleration for objects in free fall is the same whether the object is moving up or down or momentarily at rest while it switches between going up and down, or what the size of the object is. In free fall near the surface of Earth, the acceleration always has magnitude g and is always directed downward.

Free fall is an idealization, but many real-life situations come very close to this ideal. Some examples of situations that can be described as approximately free fall include the basketball and players shown in Figure 2-19, a high diver descending toward the water, and a leaping frog in midair. In each of these cases, the falling object experiences minimal air resistance because it has a relatively slow speed, a small cross section, or both, and so falls with constant downward acceleration.

If we can ignore the effects of the air, the players and the ball are all in **free fall:** Their acceleration is constant and downward.

• An object in free fall has a constant downward acceleration.
• The magnitude g of the acceleration is the same no matter what the size of the object.
• The acceleration is the same whether the object is moving upward, moving downward, or momentarily at rest.

Goodshoot/Getty Images

Figure 2-19 Free fall If an airborne object is moving relatively slowly, the effects of the air on the object can be ignored. Then the object moves with a constant acceleration due to gravity.

WATCH OUT !

Be careful when falling isn't free.

If an object is falling at high speed, or has a relatively large cross section, we *cannot* ignore air resistance and so the motion is *not* free fall since its acceleration is not constant. Some examples include a sky diver descending under a parachute, a falling leaf, and a hawk with folded wings plummeting at high speed downward onto its prey. In each of these cases, the falling object reaches a maximum speed, or *terminal speed*, at which air resistance pushes upward exactly as strongly as gravity is pulling downward on the object. We'll return to situations of this kind in the next unit, where we learn how to mathematically describe these pushes and pulls.

Measuring Free-Fall Acceleration

You might be wondering how we can make the claim that the acceleration of a freely falling object near the surface of Earth is constant. The answer is simple—by experiment! **Figure 2-20** shows a strobe photograph of a billiard ball dropped from rest. The image of the ball is recorded at intervals of 0.10 s. For this example, we will call the vertical axis the y axis (and choose the positive y direction to point upward) so that increasing values of y correspond to increasing height. After the ball is dropped from rest, it accelerates downward (in the negative y direction for our choice of coordinate system) and acquires a negative velocity v_y (we use the subscript "y" since we defined the motion to be along the y axis). Let's choose the initial position of the ball (its maximum height) to be $y_0 = 0$ so that as the ball descends, its y coordinate becomes an increasingly negative number.

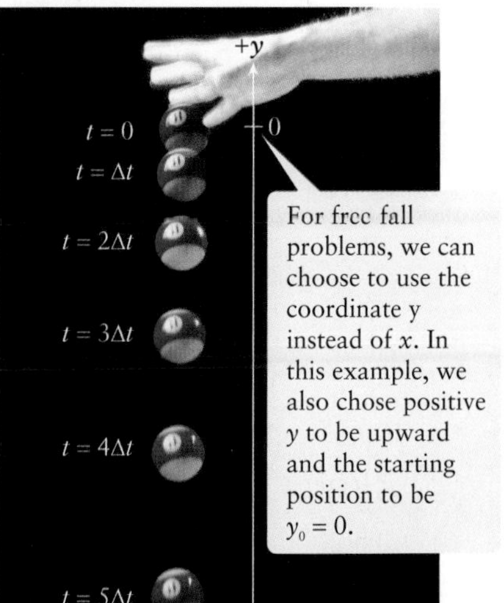

+y
0
t = 0
t = Δt
t = 2Δt
t = 3Δt
t = 4Δt
t = 5Δt
For free fall problems, we can choose to use the coordinate y instead of x. In this example, we also chose positive y to be upward and the starting position to be y₀ = 0.

Richard Megna/Fundamental Photographs

Figure 2-20 Analyzing a freely falling object When an object is dropped from rest and falls freely, it travels a greater distance in successive time intervals: It accelerates downward. Measurements of this falling ball confirm that it has a constant acceleration.

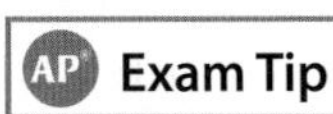

Exam Tip

The three kinematic equations (2-5, 2-9, 2-11) from Section 2-5, and the ones on the AP® Physics 1 equation sheet, are written for motion along the x axis. They are clearly just as valid if we call the axis "y" or "z" instead—you can just replace "x" everywhere in these equations with the appropriate letter. But note that you cannot mix the axes—the equations are only for motion along a single line. When solving problems, be sure to clearly define the axis, and only use quantities along that axis to plug into the three kinematic equations for motion along that axis.

Careful measurements of the y coordinate and velocity of the ball yield the values shown in **Table 2-4**. We plot the values of v_y and y versus time in **Figure 2-21**. We learned in Section 2-3 that the slope of an object's velocity–time graph represents that object's acceleration. Because a line has a constant slope, we can conclude from

TABLE 2-4 Motion of a Falling Ball

This table lists the y coordinate of the position and the y velocity at 0.10-s intervals for the falling ball shown in Figure 2-20. Initially the ball is at rest at $y = 0$. We defined the positive y direction as upward, so as the ball descends it has a negative y velocity and it moves to increasingly negative values of y. Uncertainty is implied by the last significant figure.

Time t (s)	y coordinate (m)	y velocity (m/s)
0	0.00	0.00
0.10	−0.05	−0.95
0.20	−0.20	−1.96
0.30	−0.44	−2.97
0.40	−0.78	−3.92
0.50	−1.23	−4.86
0.60	−1.76	−5.94

Figure 2-21a that the acceleration of the falling ball is constant. We can determine the value of this acceleration by measuring the slope of the v_y–t graph. We would not normally want to use data points to directly calculate acceleration since each individual point in an experiment has some error, but instead we would use the slope of a graph of all of the data, since the line we draw helps minimize the impact of that error. You should get practice in this in the lab. When we carefully plot the data above and draw a best fit line, we find two of the data points lie on the line, so we can use them. These data points are velocity $v_{1y} = -1.96$ m/s at $t_1 = 0.20$ s, and velocity $v_{2y} = -3.92$ m/s at $t_2 = 0.40$ s, so using these points on the line we find the acceleration a_y of the falling ball is

$$a_y = \text{slope of } v_y - t \text{ graph} = \frac{v_{2y} - v_{1y}}{t_2 - t_1} = \frac{-3.92 \text{ m/s} - (-1.96 \text{ m/s})}{0.40 \text{ s} - 0.20 \text{ s}} = -9.8 \text{ m/s}^2$$

We can confirm this value of acceleration in free fall by comparing the graph of position versus time in **Figure 2-21b** with the positions we can calculate by using the value $a_y = -9.8$ m/s^2. To make these calculations, we use Equation 2-9 for motion with constant acceleration, $x = x_0 + v_{0x}t + \frac{1}{2}a_xt^2$. Since we are going to do a lot of examples of free fall, we will stick with the new coordinate system we just defined, so we need to change every x in this equation to a y. This does not change its meaning, it is just for convenience. Using the initial velocity $v_{0y} = 0$ since it started at rest, and initial y coordinate we chose as $y_0 = 0$, this equation becomes $y = \frac{1}{2}a_yt^2$. If we substitute $t = 0.30$ s and $a_y = -9.8$ m/s^2, our predicted position for the falling ball is

$$y = \frac{1}{2}(-9.8 \text{ m/s}^2)(0.30 \text{ s})^2 = -0.44 \text{ m}$$

which agrees within uncertainty with the value in Table 2-4.

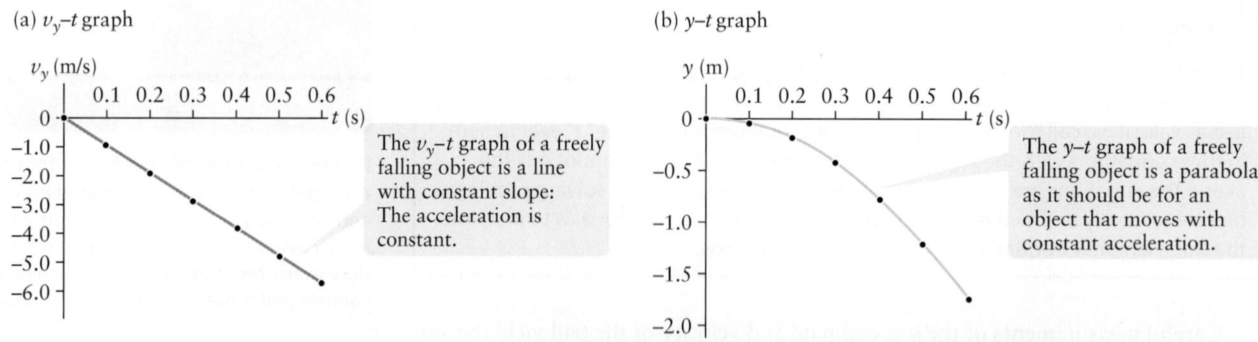
(a) v_y–t graph
v_y (m/s)
0.1 0.2 0.3 0.4 0.5 0.6
t (s)
0
-1.0
-2.0
-3.0
-4.0
-5.0
-6.0
The v_y–t graph of a freely falling object is a line with constant slope: The acceleration is constant.
(b) y–t graph
y (m)
0.1 0.2 0.3 0.4 0.5 0.6
t (s)
0
-0.5
-1.0
-1.5
-2.0
The y–t graph of a freely falling object is a parabola, as it should be for an object that moves with constant acceleration.

Figure 2-21 Graphing free fall Graphs of velocity versus time and position versus time for the freely falling object shown in Figure 2-20.

Acceleration Due to Gravity

Experiments such as that depicted in Figure 2-20 show that an object in free fall near Earth's surface has a downward acceleration of approximate magnitude 9.8 m/s^2. (There are slight variations in this value from one place to another, related to differences in elevation and in the density of material beneath the surface. We will use 9.80 m/s^2 as an *average* value, stated to three significant figures for convenience in calculation.) Physicists refer to this magnitude as the **acceleration due to gravity** and denote it by the symbol g:

$$g = 9.80 \text{ m/s}^2 \tag{2-12}$$

The value of g decreases very slightly with altitude: It is about 0.3% less at 10,000 m (32,800 ft) above sea level than at sea level. As we will learn in Chapter 6, a spacecraft journeying far beyond Earth encounters very different values of g. The value of g is also totally different on the surfaces of the Moon and other planets.

The downward force of gravity also causes objects that are moving upward or are even momentarily stationary to accelerate downward. If a ball is tossed straight up, as it ascends its velocity is positive (upward), and its acceleration is negative (downward). Since the velocity and acceleration have opposite signs, the ball slows down (see our discussion in Section 2-3). At the highest point, the velocity is zero because the ball is momentarily at rest. The velocity is changing, however; the ball's velocity was upward a moment before reaching the high point, and the ball's velocity is downward a moment after. If the acceleration at the high point of the motion actually *were* zero, the tossed ball would reach the high point of its motion and simply stop in midair. That doesn't happen in the real world, which tells us that the acceleration can't be zero at the highest point. After the ball has reached the highest point of its motion and is descending, its velocity and acceleration are both negative (downward). The signs of velocity and acceleration are the same, so the ball speeds up as it falls. At all times that the ball is in free fall, it has the *same* magnitude and direction of acceleration (**Figure 2-22**).

AP Exam Tip

While the acceleration due to gravity near Earth's surface is given on the AP® Physics 1 equation sheet as 9.8 m/s^2, you *may* use $g = 10$ m/s^2 as a simple approximation in your calculations. But why not be more accurate and use 9.8 m/s^2?

WATCH OUT !

***g* is always positive, never negative.**

Note that g is a *positive* number because it is the *magnitude* (absolute value) of the acceleration due to gravity. If we take the positive y direction to point upward, as we did in Figure 2-20, the acceleration due to gravity is negative: $a_y = -g = -9.80$ m/s^2. But no matter whether you take the positive direction to be upward or downward, the quantity g is always positive. Acceleration is a vector, so its sign is always determined by its direction in your chosen coordinate system.

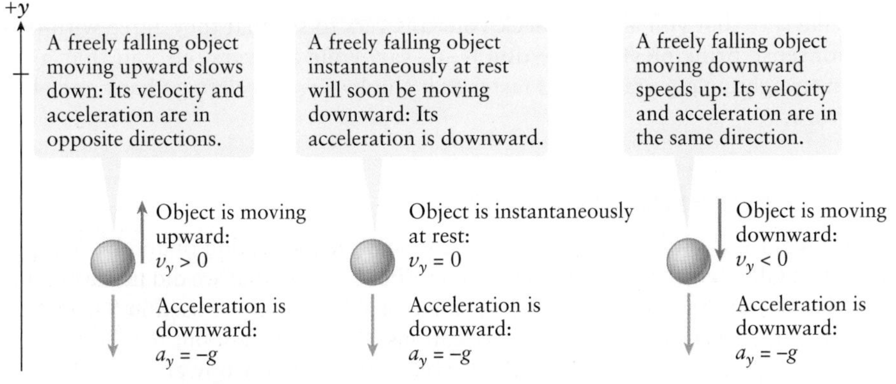

+y
A freely falling object moving upward slows down: Its velocity and acceleration are in opposite directions.
A freely falling object instantaneously at rest will soon be moving downward: Its acceleration is downward.
A freely falling object moving downward speeds up: Its velocity and acceleration are in the same direction.
Object is moving upward: $v_y > 0$
Object is instantaneously at rest: $v_y = 0$
Object is moving downward: $v_y < 0$
Acceleration is downward: $a_y = -g$
Acceleration is downward: $a_y = -g$
Acceleration is downward: $a_y = -g$

No matter how it is moving, an object in free fall has a constant downward acceleration of magnitude g.

Figure 2-22 Free-fall acceleration The acceleration due to gravity does not depend on how the object is moving.

The Equations of Free Fall

Let's see how to describe free fall that involves motion that is straight up and down—that is, *one-dimensional* free fall or *linear* free fall. As we did in Figures 2-20, 2-21, and 2-22, we will call the vertical axis the y axis and choose the positive y direction to point upward. Then the acceleration in free fall points in the negative y direction, so the acceleration is $a_y = -g$. (Remember from Equation 2-12 that g is a *positive* number.) The equations for free fall are just Equations 2-5, 2-9, and 2-11, by replacing every x in these equations with a y and using $-g$ for the acceleration a_y:

EQUATION IN WORDS
Velocity as a function of time for constant acceleration modified for free fall

$$v_y = v_{0y} - gt \tag{2-13}$$

- v_y: **Velocity at time** t of an object in free fall
- v_{0y}: **Velocity at time** $t = 0$ of the object
- g: **Acceleration due to gravity** (g is positive)
- t: **Time** at which the object has velocity v_y

EQUATION IN WORDS
Position as a function of time for constant acceleration modified for free fall

$$y = y_0 + v_{0y}t - \frac{1}{2}gt^2 \tag{2-14}$$

- y: **Position at time** t of an object in free fall
- v_{0y}: **Velocity at time** $t = 0$ of the object
- g: **Acceleration due to gravity** (g is positive)
- y_0: **Position at time** $t = 0$ of the object
- t: **Time** at which the object is at position y

EQUATION IN WORDS
Velocity as a function of position for constant acceleration modified for free fall

$$v_y^2 = v_{0y}^2 - 2g\,(y - y_0) \tag{2-15}$$

- v_y: **Velocity at position** y of an object in free fall
- v_{0y}: **Velocity at position** y_0 of the object
- y_0: **Position** of object when it has velocity v_{0y}
- g: **Acceleration due to gravity** (g is positive)
- y: **Position** of object when it has velocity v_y

WATCH OUT !

Up or down, it's still free fall.

In many free-fall problems an object like a tossed ball first moves upward then falls back downward (see Example 2-10). Many students make such problems more complicated by writing separate sets of equations for the upward and downward parts of the motion. But remember that the acceleration is the same, $a_y = -g$, the whole time the object is in free fall. One set of equations is all you need for both the upward and downward motion, as Example 2-10 shows.

Note the minus sign in front of each term containing g, which is a reminder that the acceleration is always *downward*, in the negative y direction for the coordinate system we have defined. If a freely falling object is rising, it slows down; if the object is descending, it speeds up (Figure 2-22). When solving problems that involve free fall, make sure that you always check your answers to see that they agree with these commonsense principles! If they don't—if, say, your answers tell you that a ball thrown straight up goes faster and faster as it climbs—you need to go back and check your work.

Solving Free-Fall Problems

Equations 2-13, 2-14, and 2-15 can be used to analyze the motion of an object experiencing free fall if we make the choices for coordinate system that we did in the previous examples. We use the same approach to solve problems that we used in Section 2-5. After all, these equations are the same equations as in Table 2-3, but for the choice of coordinate system of motion along the y axis, with positive y upward, and therefore $a_y = -g$. Remember that understanding the process we use to solve each problem is no less important than the final answer.

EXAMPLE 2-9 Free Fall I: A Falling Monkey

A monkey drops from a tree limb to grab a piece of fruit on the ground 1.80 m below (**Figure 2-23**). (a) How long does it take the monkey to reach the ground? (b) How fast is the monkey moving just before it reaches the ground? Neglect the effects of air resistance, so you can treat the monkey as being in free fall.

Figure 2-23 A monkey in free fall What are the monkey's position and velocity as functions of time?

Set Up

To describe our freely falling monkey, we choose the positive y direction to be upward (see Figure 2-23). Then the monkey's constant downward acceleration is $a_y = -g = -9.80\ \text{m/s}^2$ and we can use the equations in Table 2-3 as needed. Our goals are to find the time when he reaches the ground and the magnitude of his velocity just *before* reaching the ground. (Once the monkey is on the ground, his speed is zero.)

As Figure 2-23 shows, we choose the origin ($y = 0$) to be on the ground. Then the monkey's initial y coordinate is $y_0 = 1.80$ m, his y coordinate when he reaches the ground is zero, and his initial velocity is $v_{0y} = 0$ (because he starts at rest).

We've drawn the y–t graph and v_y–t graph for the monkey's motion using these values of y_0 and v_{0y}. Because the acceleration is negative, the y–t graph curves downward (the slope, which represents the velocity v_y, becomes increasingly negative as time goes by), and the v_y–t graph has a downward (negative) slope.

As we did for the examples in Section 2-5, we summarize the problem with a table of known and unknown quantities. We know the values of y_0, y, v_{0y}, and a_y, and want to find the values of (a) t and (b) v_y. To solve for t from the known information, we use Equation 2-14. Once we know the value of t, we can substitute it into Equation 2-13 to solve for the monkey's velocity v_y.

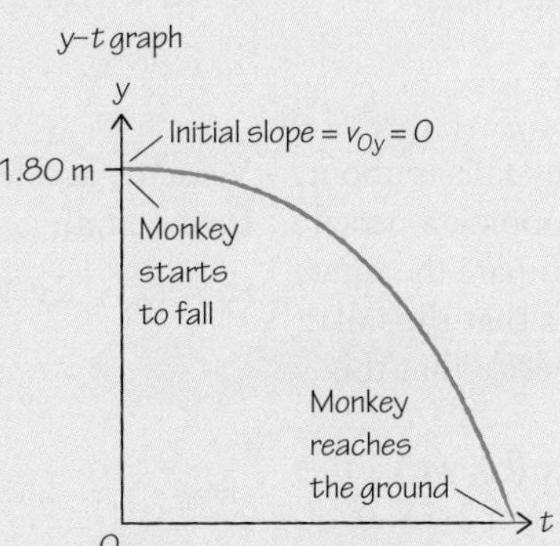

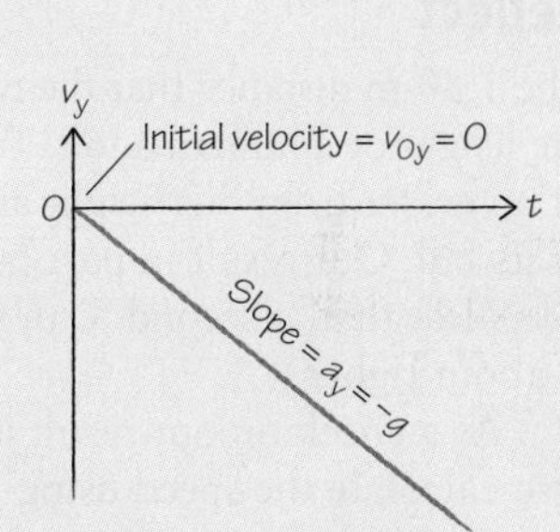

Position, acceleration, and time for free fall:

$$y = y_0 + v_{0y}t - \frac{1}{2}gt^2 \qquad (2\text{-}14)$$

Velocity, acceleration, and time for free fall:

$$v_y = v_{0y} - gt \qquad (2\text{-}13)$$

Quantity	Know/Don't know
t	?
y_0	1.80 m
y	0
v_{0y}	0
v_y	?
$a_y = -g$	$-9.80\ \text{m/s}^2$

Solve

(a) Substitute the known values from our table into Equation 2-14. Then solve for t by rearranging the equation so that t^2 is by itself on the left-hand side of the equation and by taking the square root. The equation for t has both a positive solution and a negative solution. The monkey started to fall at $t = 0$, and we are looking for a time after this, so we must choose the positive value of t.

To find the time t when the monkey reaches the ground, substitute values into Equation 2-14:

$$0 = 1.80\ \text{m} + (0)t - \frac{1}{2}(9.80\ \text{m/s}^2)t^2$$

Solve for t:

$$\frac{1}{2}(9.80\ \text{m/s}^2)t^2 = 1.80\ \text{m}$$

$$t^2 = \frac{1.80\ \text{m}}{(1/2)(9.80\ \text{m/s}^2)} = 0.367\ \text{s}^2$$

$$t = \pm\sqrt{0.367\ \text{s}^2} = \pm 0.606\ \text{s}$$

Choose the positive value of t:
$t = 0.606$ s

(b) To find the monkey's velocity v_y just before it reaches the ground, substitute the known values and $t = 0.606$ s from part (a) into Equation 2-13. We want the monkey's *speed*, which is a positive number and is just the absolute value of v_y.

To find the velocity just before reaching the ground, substitute into Equation 2-13:

$$v_y = 0 - (9.80\ \text{m/s}^2)(0.606\ \text{s})$$
$$= -5.94\ \text{m/s, so}$$

speed = 5.94 m/s

Reflect

The 1.80-m distance that the monkey falls is about the height of a human adult. Try dropping a pencil or an eraser from this height and estimate the time of its fall. Our result in part (a) says that the fall takes less than a second. Can you verify that this is about right?

As a check on our result in part (b), we can also calculate the speed using Equation 2-15. If you substitute $v_{0y} = 0$, $g = 9.80\ \text{m/s}^2$, $y_0 = 1.80$ m, and $y = 0$ into this equation, you should again find that the monkey's speed is 5.94 m/s. Do you? Check and find out!

Velocity, acceleration, and position for free fall:

$$v_y^2 = v_{0y}^2 - 2g(y - y_0) \qquad (2\text{-}15)$$

NOW WORK Problem 4 from The Takeaway 2-6.

EXAMPLE 2-10 Free Fall II: A Pronking Springbok

When startled, a springbok (*Antidorcas marsupialis*) like the one shown in **Figure 2-24** leaps upward several times in succession, reaching a height of several meters. (This behavior, called *pronking*, may inform predators that the springbok knows of their presence.) During one such jump, a springbok leaves the ground at a speed of 6.00 m/s. (a) What maximum height above the ground does the springbok attain? (b) How long does it take the springbok to attain this height? (c) What is the springbok's velocity when it is 1.25 m above the ground? (d) At what times is the springbok 1.25 m above the ground?

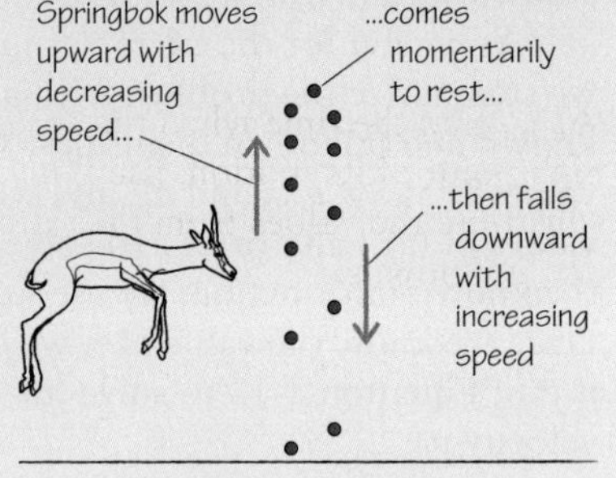

Figure 2-24 Pronking When a springbok jumps into the air, it is in free fall. The green arrows represent its directions of motion and the red dots are the motion diagram, showing how its speed changes throughout the motion.

Set Up

Like the monkey in Example 2-9, the springbok is in free fall once it leaves the ground. Unlike the monkey, however, the springbok is moving initially and so v_{0y} is *not* zero.

At the moment when the springbok attains its maximum height, its velocity is zero. So parts (a) and (b) are really asking for the springbok's vertical coordinate and the time t when $v_y = 0$. In part (c) our goal is to find the springbok's

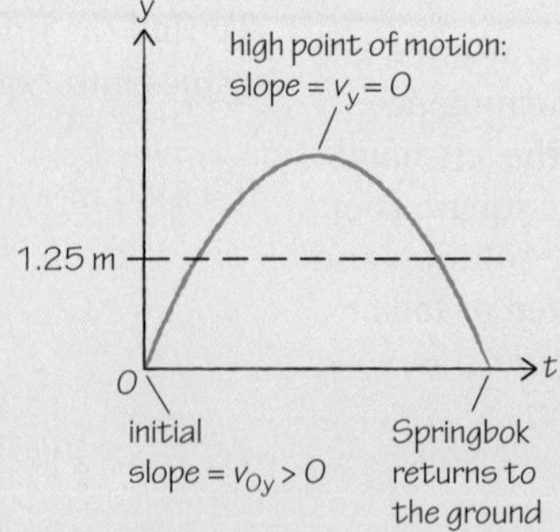

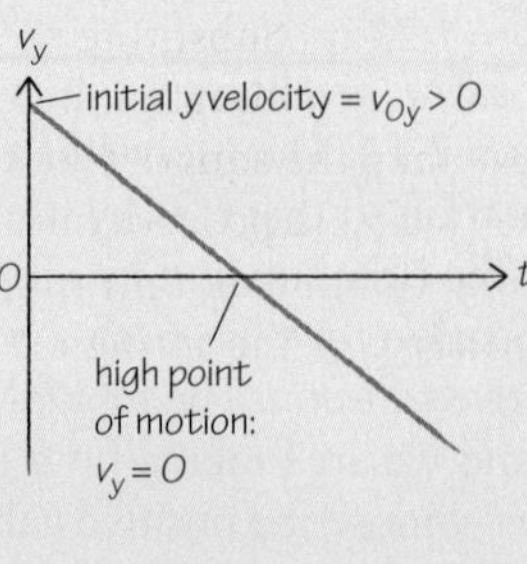

velocity v_y at a certain vertical position, and in part (d) we wish to find the times t at which the springbok is located at this position.

We again choose the y axis to point upward and again place the origin at ground level. Then the springbok has initial y position $y_0 = 0$ and initial velocity $v_{0y} = +6.00$ m/s (positive because it's moving upward). The springbok's y–t graph and v_y–t graph are sketched above. Because of the initial upward velocity, the y–t graph has an upward slope at $t = 0$, and the v_y–t graph starts with a positive value at $t = 0$.

The tables of known and unknown quantities tell us that in parts (a) and (b) we want to calculate the value of y and the time t when the springbok reaches its highest point. We can use Equation 2-15 to find the position when the velocity is zero and Equation 2-13 to determine how much time is required for the velocity to equal zero. In part (c) we want the value of v_y at a height corresponding to $y = 1.25$ m. We can again use Equation 2-15. In part (d) we'll use Equation 2-13 to find the time t when the springbok has this value of y. We expect the springbok to pass through this height twice, once moving upward and once moving back downward.

For parts (a) and (c):

$$v_y^2 = v_{0y}^2 - 2g(y - y_0) \qquad (2\text{-}15)$$

For parts (b) and (d):

$$v_y = v_{0y} - gt \qquad (2\text{-}13)$$

For parts (a) and (b)

Quantity	Know/Don't know
t	?
y_0	0
y	?
v_{0y}	+6.00 m/s
v_y	0
$a_y = -g$	−9.80 m/s²

For parts (c) and (d)

Quantity	Know/Don't know
t	?
y_0	0
y	1.25 m
v_{0y}	+6.00 m/s
v_y	?
$a_y = -g$	−9.80 m/s²

Solve

(a) To determine the value of y corresponding to the high point of the motion, we substitute the values from our first table into Equation 2-15. Then we solve this equation for y, the height of the springbok above its launch point ($y_0 = 0$).

Substitute into Equation 2-15:

$$0^2 = (6.00 \text{ m/s})^2 - 2(9.80 \text{ m/s}^2)(y - 0)$$

Solve for y at the high point of the motion:

$$2(9.80 \text{ m/s}^2)y = (6.00 \text{ m/s})^2$$

$$y = \frac{(6.00 \text{ m/s})^2}{2(9.80 \text{ m/s}^2)} = 1.84 \text{ m}$$

(b) To find the time when the springbok is at the high point of its motion, use Equation 2-13 and substitute the values from the table of knowns and unknowns.

Substitute into Equation 2-13:

$$0 = 6.00 \text{ m/s} - (9.80 \text{ m/s}^2)t$$

Solve for t at the high point of the motion:

$$(9.80 \text{ m/s}^2)t = 6.00 \text{ m/s}$$

$$t = \frac{6.00 \text{ m/s}}{9.80 \text{ m/s}^2} = 0.612 \text{ s}$$

(c) Now we want to determine the value of v_y when $y = 1.25$ m. Substitute the values from the second table into Equation 2-15 to solve for v_y. When we take the square root of v_y^2, we get two solutions for v_y: a positive one corresponding to the springbok moving upward and a negative one corresponding to the springbok moving downward.

Substitute into Equation 2-15:

$$v_y^2 = (6.00 \text{ m/s})^2 - 2(9.80 \text{ m/s}^2)(1.25 \text{ m} - 0)$$
$$= 36.0 \text{ m}^2/\text{s}^2 - 24.5 \text{ m}^2/\text{s}^2$$
$$= 11.5 \text{ m}^2/\text{s}^2$$

Solve for v_y when $y = 1.25$ m:

$$v_y = \pm\sqrt{11.5 \text{ m}^2/\text{s}^2}$$

Solutions:

$v_y = +3.39$ m/s (springbok moving up)

$v_y = -3.39$ m/s (springbok moving down)

(d) From part (c) we know the two values of v_y for when the springbok passes through $y = 1.25$ m. To find the times t when it passes through this height, we substitute each value of v_y into Equation 2-13, along with $v_{0y} = 6.00$ m/s and $g = 9.80$ m/s^2.

Substitute into Equation 2-13:

Springbok moving up ($v_y = +3.39$ m/s):

$$+3.39 \text{ m/s} = +6.00 \text{ m/s} - (9.80 \text{ m/s}^2)t$$

Solve for t:

$$(9.80 \text{ m/s}^2)t = 6.00 \text{ m/s} - 3.39 \text{ m/s}$$

$$t = \frac{6.00 \text{ m/s} - 3.39 \text{ m/s}}{9.80 \text{ m/s}^2} = \frac{2.61 \text{ m/s}}{9.80 \text{ m/s}^2}$$

$$= 0.266 \text{ s}$$

Springbok moving down ($v_y = -3.39$ m/s):

$$-3.39 \text{ m/s} = +6.00 \text{ m/s} - (9.80 \text{ m/s}^2)t$$

Solve for t:

$$(9.80 \text{ m/s}^2)t = 6.00 \text{ m/s} - (-3.39 \text{ m/s})$$

$$t = \frac{6.00 \text{ m/s} - (-3.39 \text{ m/s})}{9.80 \text{ m/s}^2} = \frac{9.39 \text{ m/s}}{9.80 \text{ m/s}^2}$$

$$= 0.958 \text{ s}$$

Reflect

A professional basketball player can jump to a height of about 1 m; our result in part (a) says a springbok can jump even higher. This makes sense because a springbok has four powerful legs to propel it upward rather than just two!

We can check our result in part (c) by noticing that 3.39 m/s is less than 6.00 m/s. In other words, when the springbok is rising through $y = 1.25$ m, it is going more slowly than when it left the ground—as it should be because gravity slows the springbok as it ascends. Its speed is the same at that height when it is going up or down, which makes sense, because it is the same distance below the point where its velocity is zero and the acceleration is constant.

You can also check the times that we calculated in part (d) by solving for them directly without first calculating the values of v_y. You can do this with Equation 2-14: Substitute $y = 1.25$ m, $y_0 = 0$, $v_{0y} = 6.00$ m/s, and $g = 9.80$ m/s^2 and solve for t in the same way that we did in Example 2-8 in Section 2-5. Try this! You should get the same two answers for t, 0.266 s and 0.958 s.

NOW WORK Problem 5 from The Takeaway 2-6.

THE TAKEAWAY for Section 2-6

✔ An object is in free fall if its motion is influenced by Earth's gravity alone (that is, if air resistance is so small that it can be ignored). Free fall includes moving upward, moving downward, and the high point of the motion.

✔ Free-fall motion near the surface of Earth is constant-acceleration motion, with a downward acceleration of magnitude $g = 9.80$ m/s^2. The magnitude g is always a positive number.

✔ When discussing free fall, the acceleration is always downward. If we choose our coordinate system such that the positive y direction is upward, the acceleration of an object in free fall is $a_y = -g$. For an object in free fall for this choice of coordinate system, the y–t graph is a parabola with a downward curvature, and the v_y–t graph is a line with downward (negative) slope.

✔ You can use the three equations in Table 2-3 to solve any problems involving linear motion in free fall.

✔ Free-fall problems are just one special (but very common) case of what we learned in Section 2-4 for general constant-acceleration problems.

Prep for the AP® Exam

AP® Building Blocks

1. A ball is thrown upward from the ground. Describe the speed, velocity, and acceleration of the ball until it has returned to the ground in terms of these conditions: (a) increasing, (b) decreasing, (c) constant, (d) zero, and for vectors include (e) direction.

2. Objects moving downward in free fall move
   (A) with increasing speed because the velocity increases as the distance to Earth decreases.
   (B) with increasing speed because the acceleration increases as the distance to Earth decreases.

(C) at a constant velocity because the gravitational acceleration is constant.
(D) with increasing speed because the gravitational acceleration is constant.

3. A baseball catcher throws a ball vertically upward and catches it in the same spot as it returns to the catcher's mitt. At what point in the path is the velocity of the ball zero, whereas at the same time acceleration is nonzero?
(A) One-half the distance to the top of the path
(B) At the top of its path
(C) The instant it leaves the catcher's hand
(D) As it arrives in the catcher's mitt

EX 2-9

4. A ball is thrown straight down from the edge of a roof 10.0 m above the ground with a speed of 8.0 m/s. Take the positive y direction to be downward.
(a) Calculate the time for the ball to reach the ground, using the equation for y as a function of t.
(b) What is the speed of the ball when it strikes the ground?
(c) Repeat the calculations of parts (a) and (b), taking the positive y direction to be upward.
(d) Repeat the calculation of parts (a) and (b) using:

$$v^2 = v_0^2 + 2g\Delta y$$

and

$$\Delta v = g\Delta t$$

EX 2-10

5. A ball is thrown straight up from the ground with a speed of 18.0 m/s. Take the positive y direction to be downward.
(a) Calculate the time it takes for the ball to reach its greatest height using Equation 2-5. Using the symmetry of the motion, predict the time that it takes the ball to return to the point from which it was thrown (neglect air resistance).
(b) Calculate the maximum height of the ball using Equation 2-9.
(c) Repeat the calculations of part (a) and (b), taking the positive y direction to be upward.
(d) Repeat the calculations of parts (a) and (b) using

$$\Delta v = g\Delta t$$

and

$$v^2 = v_0^2 + 2g\Delta y$$

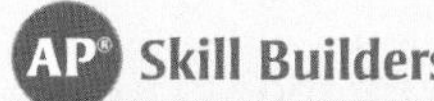

AP® Skill Builders

6. The upper limit on the magnitude of the braking acceleration for cars is about the same as the acceleration of a boulder falling from a cliff. Which of these statements might connect these? Choose all that you think are possible.
(A) The car and the boulder have about the same mass.
(B) All falling bodies have about the same acceleration.
(C) Falling boulders have about the same speed as fast cars.
(D) Engineers design brakes to produce accelerations that are comfortable for humans.
(E) This is just a coincidence and the two have no connection.

7. A video is made of a ball being thrown up into the air and then falling. Is there any way to tell whether the video is being played backward? Justify your claim.

8. Katie serves a ping-pong ball by tossing the ball straight up and hitting the ball once it returns to the same point where it was tossed. If Katie neglects the effects of air resistance she knows that she will be able to predict the initial speed of the ping-pong ball when it leaves her hand.
(a) Select the data that are needed for the prediction and justify your selection.
(b) To refine her model of the motion of the ping-pong ball she wants to include the effect of air resistance. She expects air resistance to slow the ball. She claims that the initial speed predicted when air resistance is neglected will be smaller than the actual initial speed. Justify her claim with a v–t graph, taking the upward direction to be positive. On the graph include annotations indicating the measured time, lines for velocity with and without the neglect of resistance (assuming an average acceleration), and actual and predicted initial speeds.

AP® Skill Building

9. A device launches a ball straight up from the edge of a cliff so that the ball falls and hits the ground at the base of the cliff. The device is then turned so that a second, identical ball is launched straight down from the same height. Predict, with justification, the relative velocities of the first and second launches just before the time of impact with the ground.

10. Construct a displacement–time graph of a ball thrown straight up into the air with an initial speed of 30 m/s near the surface of Earth, assuming magnitude of acceleration of 10 m/s^2. Calculate values of the instantaneous velocity at 1.5 s, 3 s, 4.5 s, and 6 s. Add annotations to your graph showing these values of the instantaneous velocity.

WHAT DID YOU LEARN?

Prep for the AP® Exam

Chapter Learning Goals	Section(s)	Related example(s)	Relevant section review exercises
Define the linear motion of a system and when and how it can be simplified to an object model.	2-1	2-1	6
Define and explain the relationships among distance, displacement, instantaneous velocity, average velocity, constant velocity, and speed.	2-2	2-1, 2-2	6, 7
Define acceleration, explain its relationship to velocity, and interpret graphs of velocity versus time.	2-3	2-3, 2-4, 2-5	2, 3, 5
Create, use, and interpret narrative (written), mathematical, and graphical representations for linear motion with constant acceleration.	2-4 2-5 2-6	 2-6, 2-7, 2-8 2-10	1, 2, 3, 4 5
Solve constant-acceleration linear motion problems when given a description of the motion of an object. The description could be in the form of words, graphs, or experimental data.	2-4 2-5 2-6	 2-6, 2-7, 2-8 2-10	1, 2, 3, 4 5
Explain when objects are in free fall (constant acceleration due to gravity).	2-6	2-9	4

Chapter 2 Review

Key Terms

All the Key Terms can be found in the Glossary/Glosario on page G1 in the back of the book.

acceleration 45
acceleration due to gravity 71
average velocity 36
constant velocity 33
coordinate system 34
direction 34
displacement 34
distance 35
free fall 69
instantaneous 47
kinematics 32
linear motion 32
magnitude 34
origin 34
position 34
scalar 35
speed 34
vector 34
velocity 34

Chapter Summary

Topic	Equation or Figure
Linear motion: We describe the motion of an object along a line by stating how its position x changes with the passage of time t. The change in the object's position during a time interval is called the displacement. Displacement is different than distance, which contains no information about direction and is always positive. Displacement and distance traveled during any time will only be the same if the direction of motion never changes.	(a) Direction of swimmer A's motion 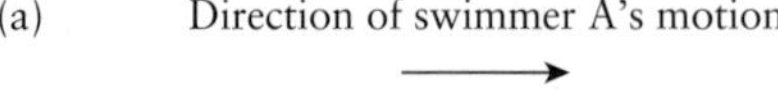 The red dots show swimmer A's position at 5.0-s intervals. 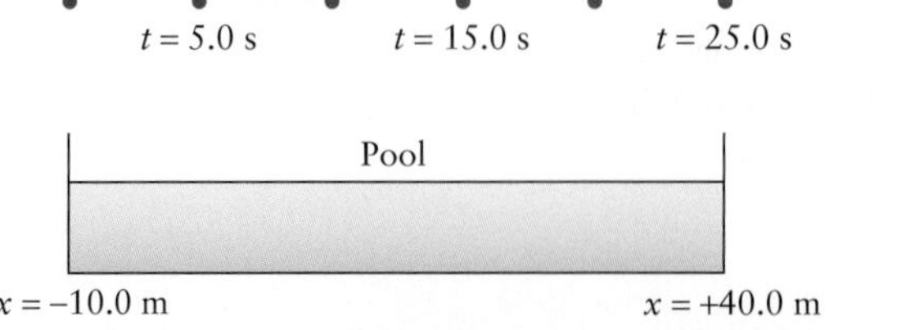 The change in position between any two red dots is the displacement of the swimmer over that 5.0-s time interval. (Figure 2-3a)

Average velocity and instantaneous velocity: The average velocity of an object during a time interval describes how the displacement of the object is changing with time during that time interval. If the time interval is very short, the average velocity becomes the instantaneous velocity v_x (*velocity* for short). Speed and velocity are not the same quantities. Average speed is the total distance the object travels divided by the amount of time it takes the object to travel that distance. Like distance, average speed tells you nothing about the direction of motion and therefore is never negative. Instantaneous speed will be the same as the magnitude of instantaneous velocity, but average speed will be different than the magnitude of average velocity if the direction of motion changes.

Average velocity of an object in linear motion

Displacement (change in position) of the object over a certain time interval: The object moves from x_1 to x_2, so $\Delta x = x_2 - x_1$.

$$v_{\text{average},x} = \frac{x_2 - x_1}{t_2 - t_1} = \frac{\Delta x}{\Delta t} \tag{2-1}$$

For both the displacement and the time interval, subtract the earlier value (x_1 or t_1) from the later value (x_2 or t_2).

Time interval for the motion: The object is at x_1 at time t_1 and x_2 at time t_2, so the time interval is $\Delta t = t_2 - t_1$.

Constant-velocity linear motion: If an object has the same velocity v_x at all times, it maintains a steady speed and always moves in the same direction. A graph of position x versus time t for constant-velocity motion is a line with slope v_x.

Position at time t of an object in linear motion with constant velocity

Position at time $t = 0$ of the object

$$x = x_0 + v_x t \tag{2-2}$$

Constant velocity of the object

Time at which the object is at position x

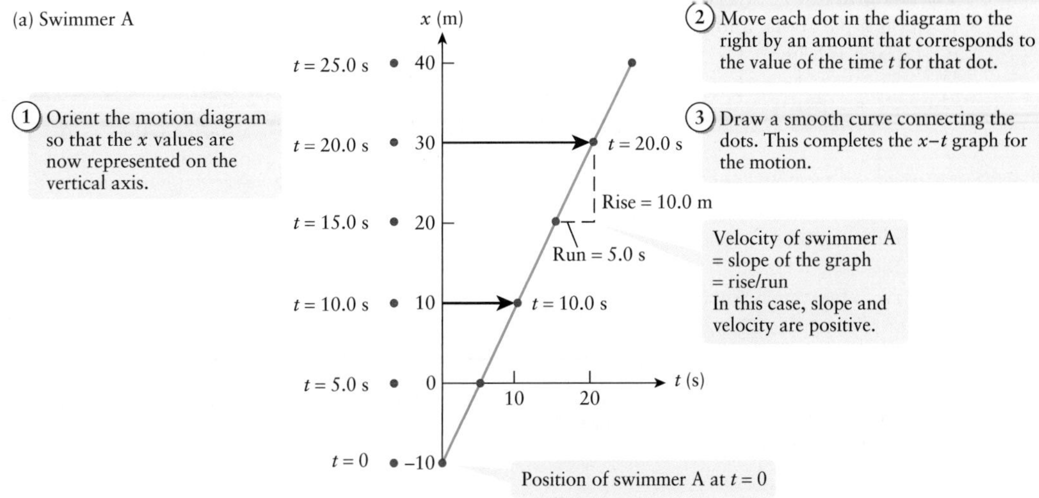

(a) Swimmer A
x (m)
t (s)
t = 25.0 s
t = 20.0 s
t = 15.0 s
t = 10.0 s
t = 5.0 s
t = 0
40
30
20
10
0
–10
10
20
1 Orient the motion diagram so that the x values are now represented on the vertical axis.
2 Move each dot in the diagram to the right by an amount that corresponds to the value of the time t for that dot.
3 Draw a smooth curve connecting the dots. This completes the x–t graph for the motion.
t = 20.0 s
Rise = 10.0 m
Run = 5.0 s
t = 10.0 s
Velocity of swimmer A = slope of the graph = rise/run In this case, slope and velocity are positive.
Position of swimmer A at t = 0

(Figure 2-4a)

Average acceleration and instantaneous acceleration: An object's average acceleration over a time interval equals the average rate of change of the object's velocity during that time interval. If the time interval is very short, average acceleration becomes instantaneous acceleration a_x (or *acceleration* for short).

Average acceleration of an object in linear motion

Change in **velocity** of the object over a certain time interval: The velocity changes from v_{1x} to v_{2x}.

$$a_{\text{average},x} = \frac{\Delta v_x}{\Delta t} = \frac{v_{2x} - v_{1x}}{t_2 - t_1} \qquad (2\text{-}3)$$

The time interval: The object has velocity v_{1x} at time t_1, and has velocity v_{2x} at time t_2.

Interpreting velocity and acceleration: The slope of an object's x–t graph (graph of position versus time) is its velocity v_x. The slope of its v_x–t graph (graph of velocity versus time) is its acceleration a_x. An object speeds up if v_x and a_x have the same sign, and slows down if v_x and a_x have opposite signs. The displacement during any time interval is always given by the area under the velocity–time curve. It is positive if the area is above the time axis and negative if the area is below the time axis.

(a) x–t graph

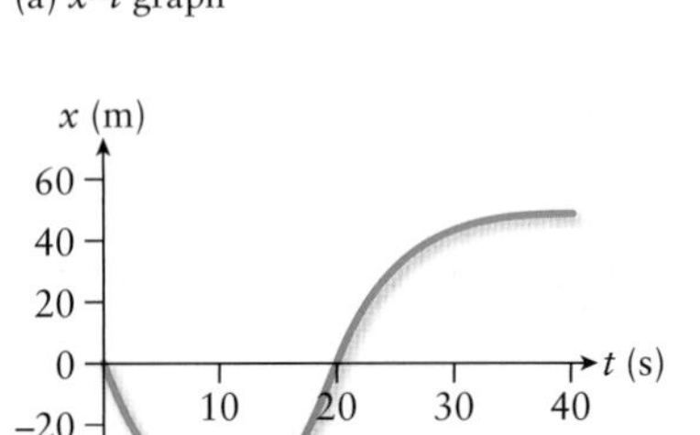

(b) v_x–t graph

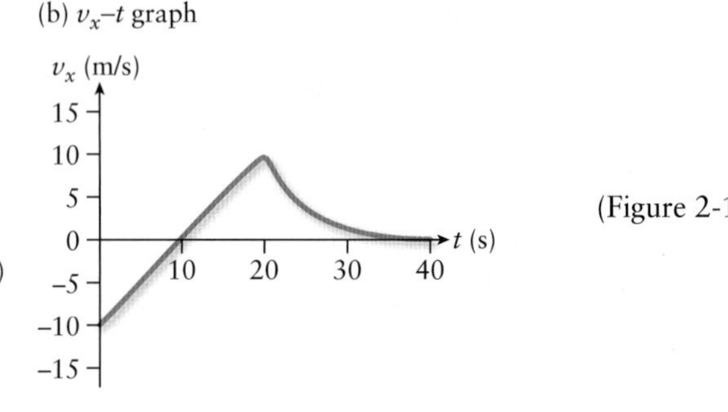

(Figure 2-12)

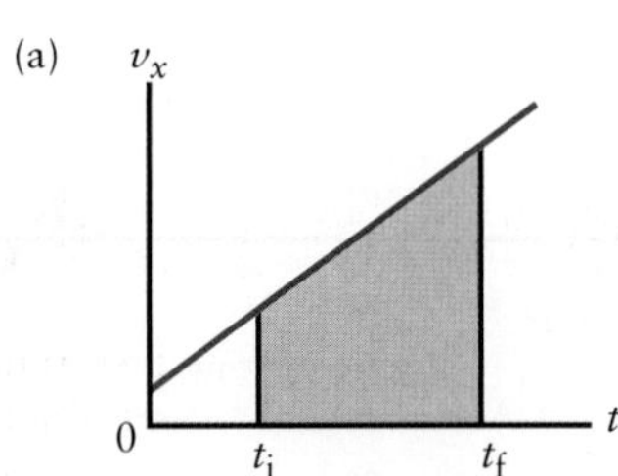

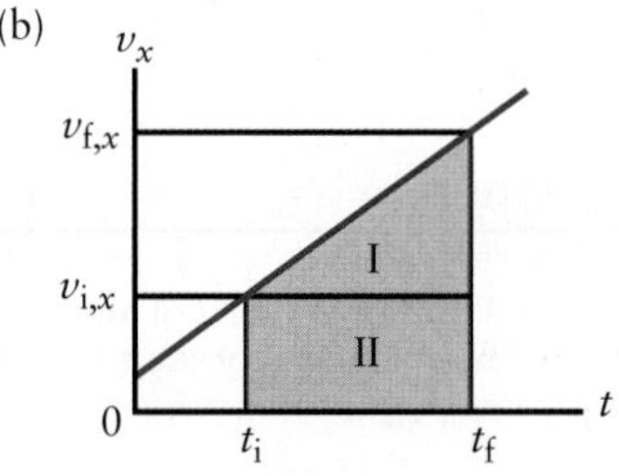

(Figure 2-17)

Motion with constant acceleration: The velocity of an object in constant acceleration motion changes at the same rate at all times. The x–t graph for such an object is a parabola, and the v_x–t graph is a line.

Velocity at time *t* of an object in linear motion with constant acceleration

Velocity at time *t* = 0 of the object

$$v_x = v_{0x} + a_x t \qquad (2\text{-}5)$$

Constant acceleration of the object

Time at which the object has velocity v_x

Position at time *t* of an object in linear motion with constant acceleration

Velocity at time *t* = 0 of the object

Constant acceleration of the object

$$x = x_0 + v_{0x}t + \frac{1}{2}a_x t^2 \qquad (2\text{-}9)$$

Position at time *t* = 0 of the object

Time at which the object has position x

Velocity at position *x* of an object in linear motion with constant acceleration

Velocity at position x_0 of the object

$$v_x^2 = v_{0x}^2 + 2a_x(x - x_0) \qquad (2\text{-}11)$$

Constant acceleration of the object

Two positions of the object

Free fall: A special case of constant acceleration, an object that moves upward or downward under the influence of gravity alone is in free fall. It has a constant downward acceleration of magnitude g. Near the surface of Earth, $g = 9.80$ m/s^2.

$$v_y = v_{0y} - gt \tag{2-13}$$

Velocity at time t of an object in free fall; **Velocity at time $t = 0$** of the object; **Acceleration due to gravity** (g is positive); **Time** at which the object has velocity v_y

$$y = y_0 + v_{0y}t - \frac{1}{2}gt^2 \tag{2-14}$$

Position at time t of an object in free fall; **Velocity at time $t = 0$** of the object; **Acceleration due to gravity** (g is positive); **Position at time $t = 0$** of the object; **Time** at which the object is at position y

$$v_y^2 = v_{0y}^2 - 2g(y - y_0) \tag{2-15}$$

Velocity at position y of an object in free fall; **Velocity at position y_0** of the object; **Position** of object when it has velocity v_{0y}; **Acceleration due to gravity** (g is positive); **Position** of object when it has velocity v_y

Chapter 2 Review Problems

1. The figure shows a position versus time graph for a moving object. At which lettered point does the object have the least speed?

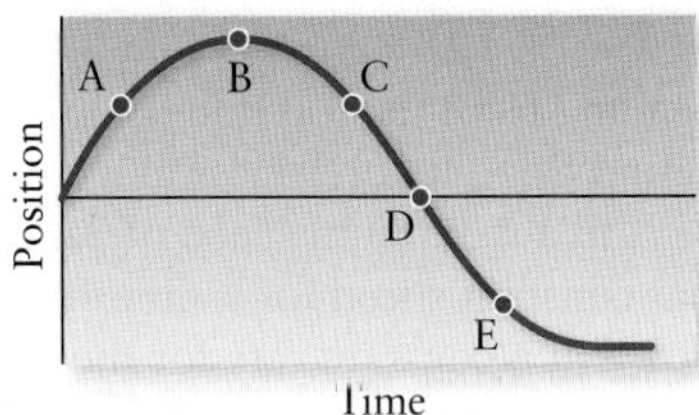

(A) A (B) B (C) C (D) D (E) E

2. The figure shows a position versus time graph for a moving object. At which lettered point does the object have the greatest speed?

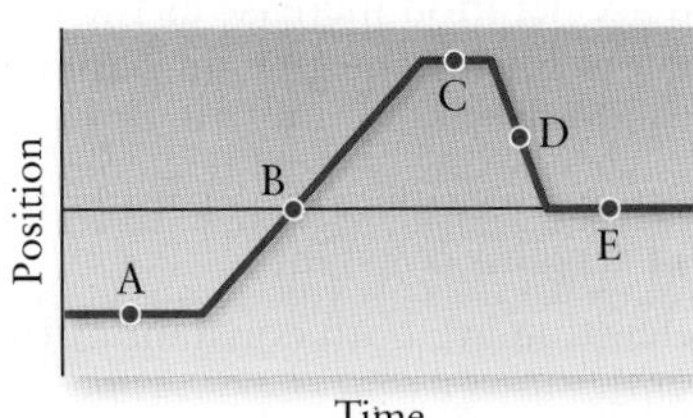

(A) A (B) B (C) C (D) D (E) E

3. A 1-kg ball and a 10-kg ball are dropped from a height of 10 m at the same time. In the absence of air resistance
   (A) the 1-kg ball hits the ground first.
   (B) the 10-kg ball hits the ground first.
   (C) the two balls hit the ground at the same time.
   (D) the 10-kg ball will take 10 times the amount of time to reach the ground.
   (E) there is not enough information to determine which ball hits the ground first.

4. The Olympic record for the men's marathon set in 2008 is 2 h, 6 min, 32 s. The marathon distance is 26.2 mi. What was the average speed of the record-setting runner in km/h?

5. Kevin completes his morning workout at the pool. He swims 4.00×10^3 m (80 laps in the 50.0-m-long pool) in 1.00 h.
   (a) What is the average velocity of Kevin during his workout?
   (b) What is his average speed?
   (c) With a burst of speed, Kevin swims one 25.0-m stretch in 9.27 s. What is Kevin's average speed over those 25 m?

6. Select the kind of data needed to estimate the average speed of a marathon runner.

7. Calculate the displacement and distance of a swimmer completing a typical workout of 20 laps (a complete lap is twice the length of the pool, returning to the starting position) in a pool with a length of 25 m.

8. The 100-m dash record time for women, 10.49 s, was set by Florence Griffith-Joyner in 1988.
   (a) Calculate the average speed for this record time.
   (b) If the average speed was reached after 2.0 s calculate the average acceleration for this 2.0-s segment.

(c) If the average speed is not achieved until 2 s into the race, what does this imply about the actual average speed for the rest of the race, and the value of the acceleration you calculated?

9. Alcohol consumption slows people's reaction times. In a controlled government test, it takes a certain driver 0.320 s to hit the brakes in a crisis when unimpaired and 1.00 s when drunk. When the car is initially traveling at 90.0 km/h, how much farther does the car travel before coming to a stop when the person is drunk compared to sober?

10. Select the type of data needed to estimate the acceleration of a car as it is on the entrance ramp to a highway. Explain how you would use the data to calculate the acceleration.

11. The figure shows a v–t graph for some object.

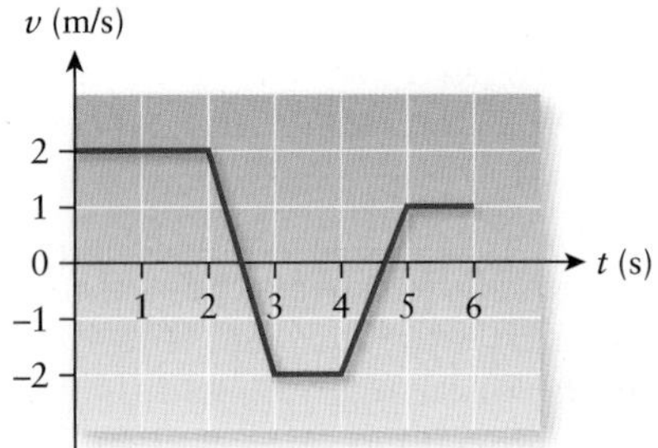

(a) What is the object's maximum velocity?

(b) How many times does the object change direction?

(c) What is the acceleration of the object between $t = 2$ s and $t = 3$ s?

(d) What is the object's average acceleration between $t = 0$ and $t = 6$ s?

12. Consider the x–t plot in the figure. For each defined interval identify the *signs* (that is, + or −) of velocity v and acceleration a.

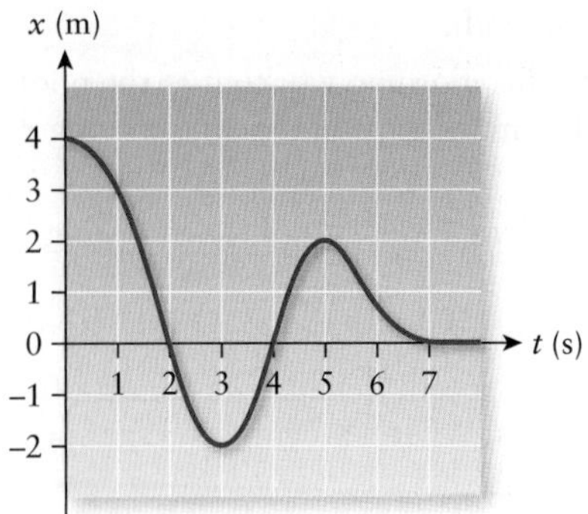

(a) $0 \text{ s} \le t \le 2 \text{ s}$

(b) $2 \text{ s} \le t \le 3 \text{ s}$

(c) $3 \text{ s} \le t \le 4 \text{ s}$

(d) $4 \text{ s} \le t \le 5 \text{ s}$

(e) $5 \text{ s} \le t \le 5.8 \text{ s}$

13. A car is stopped at a traffic light, defined as position $x = 0$. At $t = 0$ the light turns green, and the car accelerates constantly at 3 m/s^2 until it reaches 15 m/s at $t = 5$ s, at which time it continues on at that velocity. At $t = 2$ s a speeding truck passes through the traffic light in the same direction traveling at a constant velocity of 20 m/s. On the *same* set of axes, draw the position versus time graphs for both the car and the truck out to $t = 6$ s.

14. Paola can flex from a bent position to a position in which her legs are straight such that the displacement of her hips in a line straight upward is 20.0 cm. Paola leaves the ground when her legs are straight, at a speed of 4.43 m/s. Calculate the magnitude of her average acceleration.

15. A car traveling at 35.0 km/h speeds up to 45.0 km/h in a time of 5.00 s. The same car later speeds up from 65.0 km/h to 75.0 km/h in a time of 5.00 s.

(a) Calculate the magnitude of the constant acceleration for each of these intervals.

(b) Determine the distance traveled by the car during each of these time intervals.

16. Consider the graph below of an object's velocity as a function of time. Take the initial position of the object to be at the origin.

(a) Calculate the position of the object at $t = 3.0$ s and $t = 6.0$ s.

(b) Calculate the acceleration of the object at $t = 3.0$ s and $t = 6.0$ s.

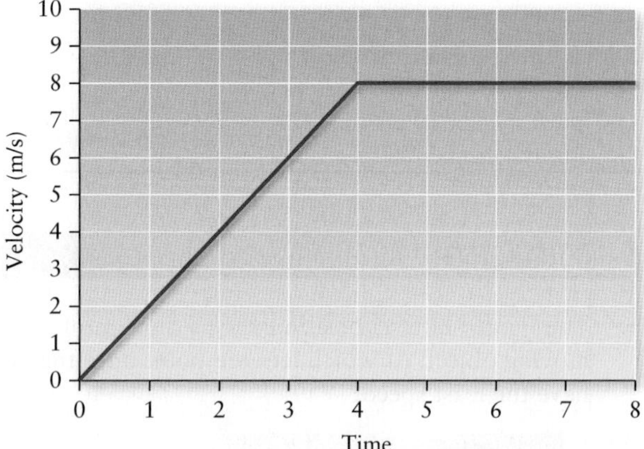

17. An elevator is descending from the top floor of a building down to the ground floor. It accelerates at a constant rate for a time t_1 to a maximum speed of 8.00 m/s. It then travels at that constant speed for a time $4t_1$, and then it slows down at a constant rate for the same time t_1 before coming to rest a distance of 400 m below its starting point. What is the total time for the descent?

18. Choose the statement below that best predicts the properties of the speed and acceleration of a painter who loses his balance and falls from a step stool.

(A) Both speed and acceleration increase as the painter falls and just before hitting the ground both have their largest magnitudes.

(B) Speed becomes more negative as the painter falls and acceleration remains constant.

(C) Speed and acceleration are both constant as the painter falls.

(D) Speed increases as the painter falls and acceleration remains constant.

Briefly justify your selection.

19. Select the type of data needed to estimate the time in the air of a professional basketball player making a dunk shot.

20. Birds that prey on flying insects, such as swifts, swallows, and purple martins, have forms that permit very high accelerations during turns (as much as 6 g has

been reported) at relatively low speeds. Birds who attack prey by diving, such as falcons and osprey, have forms that allow nearly 1 g over a very long dive.

(a) Predict the maximum speed of a falcon at the end of a dive of 1200 m. Compare your prediction to the reported speed of 157 m/s from that height.

(b) Using the speed you predicted in part (a), predict the minimum stopping time if the largest sustainable acceleration while stopping is 6 g.

21. Velocity, acceleration, and displacement are vectors. Explain why each of the following pairs of vectors is parallel, perpendicular, or antiparallel in terms of the mathematical meanings of "speeding up" and "slowing down."

(a) Acceleration and velocity in one-dimension motion for an object that is speeding up

(b) Acceleration and velocity in one-dimension motion for an object that is slowing down

(c) Acceleration and displacement in one-dimension motion for an object that is slowing down

(d) Acceleration and displacement in one-dimension motion for an object that is speeding up

22. A pinecone falls from the top of a tall tree. Complete the following table to predict the properties of the motion as it falls freely. For convenience use $g = 10\ \text{m/s}^2$ and avoid the use of a calculator or spreadsheet.

Time (s)	Displacement (m)	Speed (m/s)	Acceleration (m/s^2)
0.5			
1.0			
1.5			
2.0			

23. A student builds a small racing cart that is propelled by a fan with two blades. The rotation of the blades is driven by a rubber band that is wound by turning the blades with a finger. The finger remains in place as the cart is placed at the starting line. When the finger is released the cart rolls on a track toward a motion sensor that accurately measures velocity. Design an experiment to determine the relationship between the number of rotations through which the blade is wound and the acceleration of the cart as it leaves the starting position. In the description of your design:

- Identify measured variables as dependent or independent.
- Create a diagram of the experiment.
- Describe a strategy that can be used to obtain the desired relationship from the measured variables.

24. Four cars have leaky oil pans that leave drops of oil at a rate of 1 drop every 1 s, creating a motion diagram on the street, as shown in the figure. The x axis measures distance in units of meters with a smallest increment of 0.2 m.

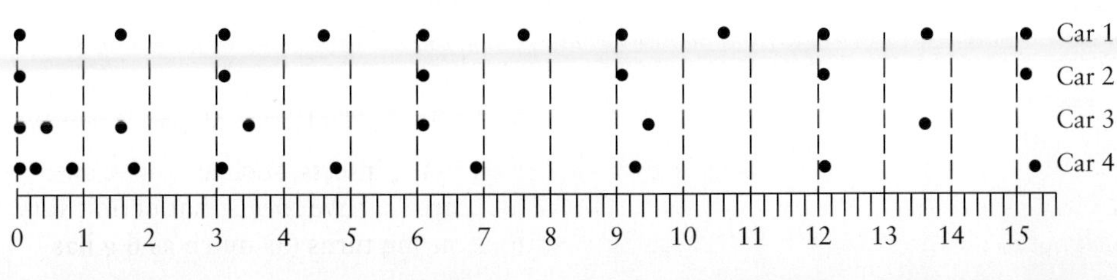

Standing at the side of the street at a fixed position of 10 m from the origin, describe in complete sentences what you observe about the motion of the four cars. Include the label of the cars, the sequence in which they pass your position, the time intervals during which they pass, and any prominent features of their motion.

25. A ball is dropped from rest at a height of 25.0 m above the ground.

(a) How fast is the ball moving when it is 10.0 m above the ground?

(b) How much time is required for it to reach the ground level? Ignore the effects of air resistance.

26. Alex climbs to the top of a tall tree while his friend Gary waits on the ground below. Gary throws a ball up to Alex, who allows the ball to go past him on its way up before catching the ball on its way down. The ball has an initial speed of 10 m/s and is caught 3.5 m above where it was thrown. How long after the ball was thrown does Alex catch it? Ignore the effects of air resistance.

27. A fox locates its prey, usually a mouse, under the snow by slight sounds the rodent makes. The fox then leaps straight into the air and burrows its nose into the snow to catch its next meal. The fox jumps to a maximum height of 85.0 cm.

(a) Calculate the velocity at which the fox leaves the snow and

(b) Calculate the total time the fox is in the air. Ignore the effects of air resistance.

28. A tennis ball is hit straight up at 20.0 m/s from the edge of a sheer cliff. Sometime later, the ball passes the original height from which it was hit. Ignore the effects of air resistance.

(a) How fast is the ball moving at that time?

(b) If the cliff is 30.0 m high, how long will it take the ball to reach the ground level?

(c) What total distance did the ball travel?

29. A ball is thrown straight up at 18.0 m/s. How fast is the ball moving after 1.00 s? After 2.00 s? After 5.00 s? When does the ball reach its maximum height? Ignore the effects of air resistance.

30. Julia runs the 100-meter dash by accelerating at a constant rate for 3.0 seconds and then maintaining a constant speed for the remainder of the race. She finishes in a total time of 15.0 seconds. Calculate Julia's maximum speed for this race.

31. A stone is dropped from a bridge that is 50.0 m above the surface of a river. A second stone is thrown downward 2.00 seconds after the first. Both stones strike the water surface at the same time. What is the initial speed of the second stone?

32. Blythe and Geoff compete in a 1.00-km race. Blythe's strategy is to run the first 600 m of the race at a constant speed of 4.00 m/s, and then accelerate to her maximum speed of 7.50 m/s, which takes her 1.00 min, and then finish the race at that speed. Geoff decides to accelerate to his maximum speed of 8.00 m/s at the start of the race and to maintain that speed throughout the rest of the race. It takes Geoff 3.00 min to reach his maximum speed. Assuming all accelerations are constant, who wins the race?

33. A ball is dropped from an upper floor, some unknown distance above your apartment. As you look out of your window, which is 1.50 m tall, you observe that it takes the ball 0.180 s to traverse the length of the window. Determine how high above the top of your window the ball was dropped. Ignore the effects of air resistance.

34. A ball is thrown straight up at 15.0 m/s.
    (a) How much time does it take for the ball to be 5.00 m above its release point?
    (b) How fast is the ball moving when it is 7.00 m above its release point?
    (c) How much time is required for the ball to reach a point that is 7.00 m above its release point? Why are there two answers to part (c)?

35. Kharissia wants to complete a 1000-m race with an average speed of 8.00 m/s. After 750 m, she has averaged 7.20 m/s. What average speed must she maintain for the remainder of the race in order to attain her goal?

36. A ball is thrown straight down at 1.50 m/s from a tall tower. Two seconds (2.00 s) later a second ball is thrown straight up at 4.00 m/s. How far apart are the two balls 4.00 s after the second ball is thrown? Ignore the effects of air resistance.

37. A two-stage model rocket blasts off vertically from rest on a launch pad. During the first stage, which lasts for 15.0 s, the rocket's acceleration is a constant 2.00 m/s^2 upward. At the end of the first stage, the second stage engine fires, giving the rocket an upward acceleration of 3.00 m/s^2 that lasts for 12.0 s. At the end of the second stage, the engines no longer fire and therefore cause no acceleration, so the rocket coasts to its maximum altitude.
    (a) What is the maximum altitude of the rocket?
    (b) Over the time interval from blastoff at the launch pad to the instant that the rocket falls back to the launch pad, assuming no parachute, what are its (i) average speed and (ii) average velocity? Ignore the effects of air resistance.

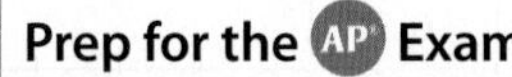

AP® Group Work

Directions: The problem below is designed to be done as group work in class.

An interchange, as shown in the figure, allowing access between a small road, r_1, and a large road, R_1, is designed to allow cars to reach the speed of cars in the traffic that they merge with.

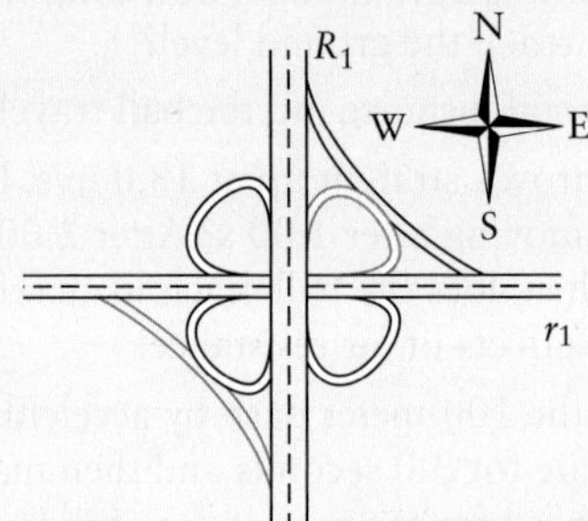

(a) The speed limit on R_1 is 110 km/h (30.6 m/s) and on r_1 is 40 km/h (11.1 m/s). The section of the on-ramp colored red, connecting road r_1 in the east direction to road R_1 in the south direction, has a length, d, of approximately 150 m. Assume for your calculation that it is straight. Calculate the magnitude of the acceleration, a, in units of m/s^2, of a car along this on-ramp using

$$v_{R1}^2 = v_{r1}^2 + 2ad$$

(b) Headed north on R_1 a motorist uses the cloverleaf off-ramp shown in green to turn onto road r_1 in the west direction. As you will see in the next chapter, it is safer to take the curved part of the ramp at a relatively constant speed, and that the larger the speed, the larger the radius of the curved part of the ramp should be. This curved ramp is designed to be traveled safely at a speed of 11.1 m/s, and so is about 50 m in radius. In order to fit, the length of the acceleration lane where you can change speed before entering the curved part of the ramp is about 100 m. Predict whether the magnitude of the acceleration in the acceleration lane is larger than, smaller than, or approximately the same as the acceleration estimated in part (a). Support your prediction with reasoning. Again, assume for your calculation that the acceleration lane is straight.

(c) Suppose that there is a stop sign at the end of the off-ramp in part (a). Predict whether the magnitude of the acceleration on the off-ramp is larger than, smaller than, or approximately the same as the acceleration you calculated in part (a). Support your prediction with reasoning.

AP® PRACTICE PROBLEMS

Prep for the AP® Exam

Multiple-Choice Questions

Directions: The following questions have a single correct answer.

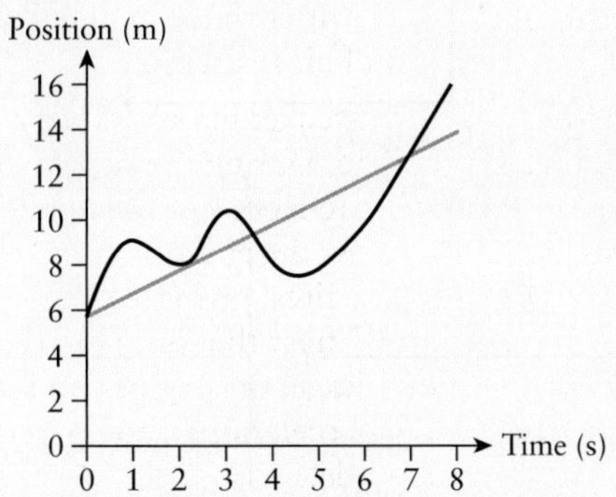

1. The position graphs of two objects are shown above. For which of the following time intervals do the two objects have approximately the same average velocity?
   (A) 0–2 s and 4 s–6 s
   (B) 2 s–4 s and 6 s–8 s
   (C) 2 s–4 s and 4 s–6 s
   (D) 0–2 s and 6 s–8 s

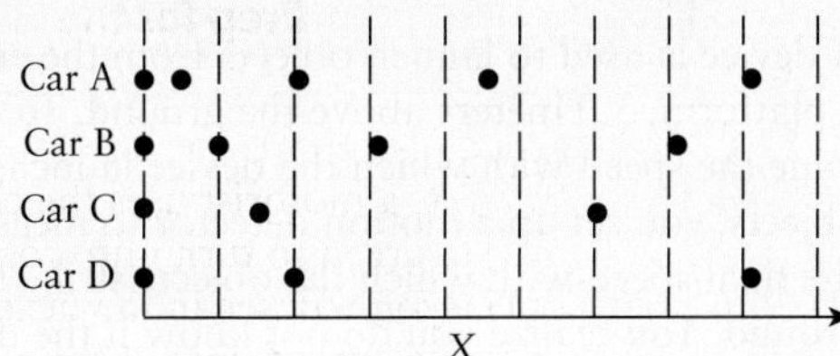

2. Four toy cars start from rest at the same location and move in the x-direction, as shown above. In each case the time intervals between the recorded positions are the same. For which car is the value of the acceleration *certainly not* constant?
   (A) Car A
   (B) Car B
   (C) Car C
   (D) Car D

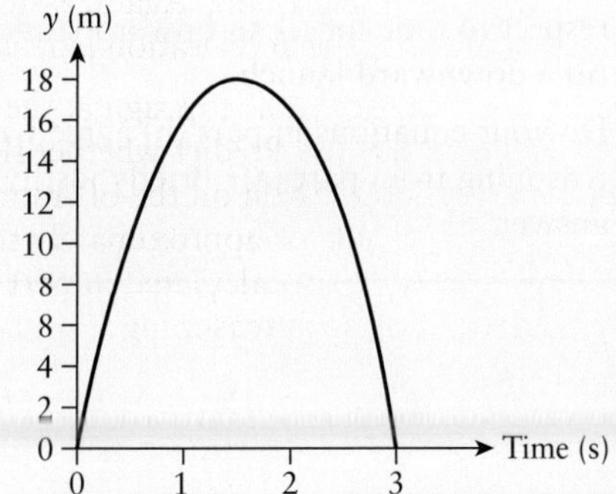

3. Suppose that a projectile on a different planet is shot directly upward. If a graph of its position *vs.* time is that shown above, what is the magnitude of the acceleration due to gravity near the surface of that planet?
   (A) 2 m/s^2
   (B) 4 m/s^2
   (C) 8 m/s^2
   (D) 16 m/s^2

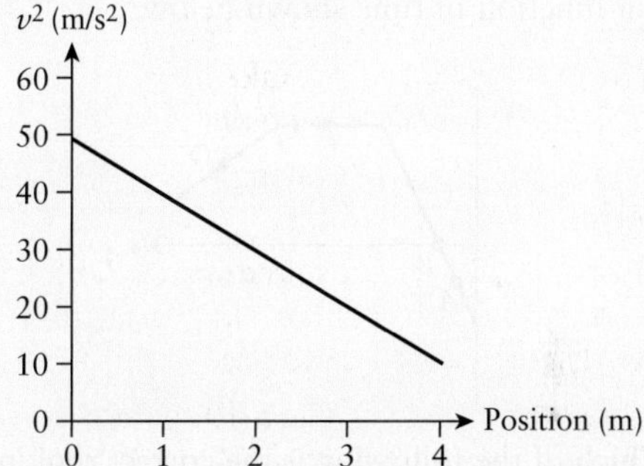

4. The graph above shows the velocity squared as a function of position for a given object. What is the acceleration of the object?
   (A) −5 m/s^2
   (B) +5 m/s^2
   (C) −10 m/s^2
   (D) +10 m/s^2

Questions 5 and 6 refer to the following material.

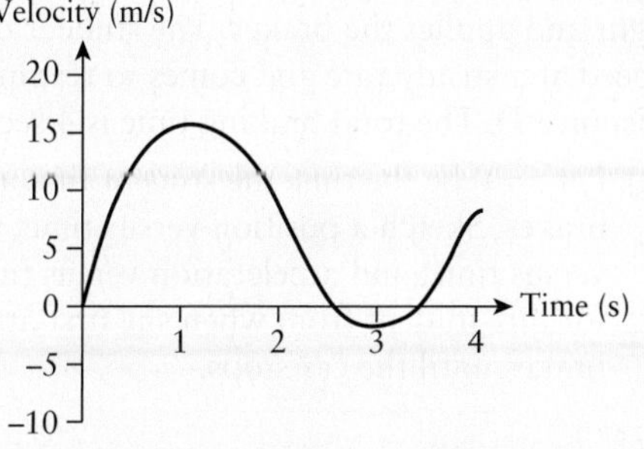

The graph above shows velocity as a function of time for a given object.

5. At which of the following times does the object have the greatest displacement from the start?
   (A) 1 s (B) 2 s
   (C) 3 s (D) 4 s

6. During which time interval does the object have the greatest acceleration?
   (A) 0 s–1 s (B) 1 s–2 s
   (C) 2 s–3 s (D) 3 s–4 s

(Continued)

7. Three objects all start from rest. Object A has an acceleration of a for a time interval t. Object B has an acceleration of $a/2$ for a time interval $2t$. Object C has an acceleration of $a/3$ for a time interval $3t$. Which object will travel the farthest during the time interval in which its acceleration is nonzero?
   (A) Object A
   (B) Object B
   (C) Object C
   (D) None; the objects all travel the same distance.
8. An object moves according to the graph of velocity as a function of time shown below.

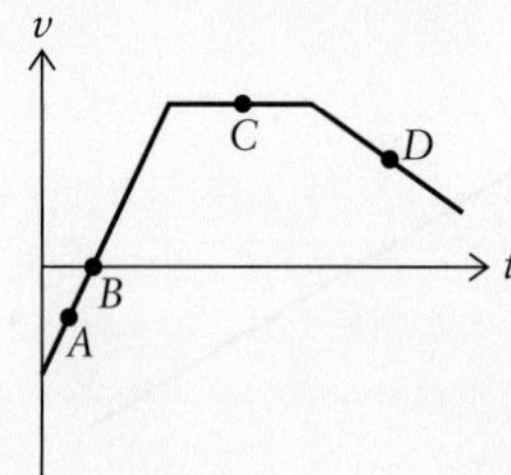

Which of the following is the correct ranking of the four points labeled in the graph based on the magnitude of the acceleration of the object at that moment?
   (A) $C > D > B > A$
   (B) $D > A > C > B$
   (C) $(A = B) > D > C$
   (D) $D > C > (A = B)$

Free-Response Questions

1. A student is driving at speed v_o when she sees a red light and applies the brakes. The student decreases speed at a steady rate and comes to rest in a distance D. The total braking time is 4 seconds.
   (a) Let $t = 0$ be the time the student first applies the brakes. Sketch a position versus time, velocity versus time, and acceleration versus time graph for this student from when she first applies the brakes until the car stops.

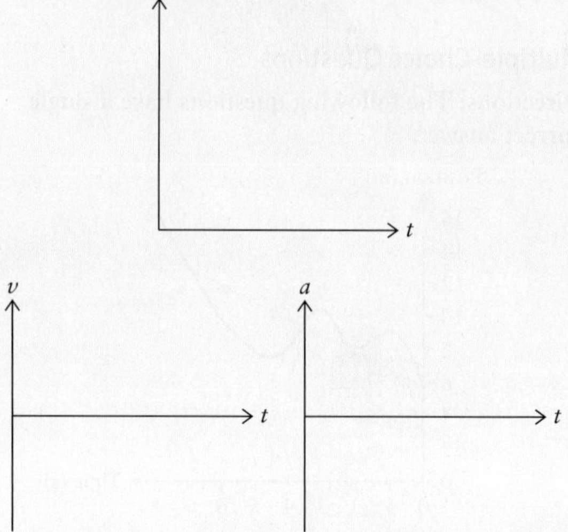

   (b) Write the equations for position versus time, velocity versus time, and acceleration versus time in terms of the variables given and any necessary fundamental constants.
   (c) Does the car travel a greater distance in the first two seconds of braking, or in the last two seconds? Briefly explain your reasoning, referring to your graphs and your equations as evidence to support your choice.
2. A device is used to launch objects from the edge of a platform, 5.0 meters above the ground. To determine the speed with which the device launches the objects, you set up a motion detector to measure the final speed with which the objects strike the ground. You realize you do not know if the device is set to launch the objects vertically upward or vertically downward. A student argues that in order to find the launch direction you would need to observe the initial direction of motion of the object.
   (a) Without citing or manipulating equations, describe why the student may believe this, and if you agree with the student or believe there is another option. (Define your system and include in your description the motion of a launched object for both an upward and a downward launch.)
   (b) Write an equation for the vertical position with respect to time for (i) an upward launch and (ii) a downward launch.
   (c) Do your equations in part (b) agree with your reasoning from part (a)? Briefly justify your answer.

3 Motion in Two or Three Dimensions

John Bingham/Alamy

YOU WILL LEARN TO:

- Recognize the differences among one-, two-, and three-dimensional motion.
- Describe the properties of a vector and how to find the sum or difference of two vectors.
- Describe the motion of an object in two dimensions using quantities such as displacement, distance, velocity, speed, and acceleration, in an appropriate way for a chosen coordinate system.
- Explain how the displacement, velocity, and acceleration of an object or the center of mass of a system are described in terms of vectors, and can be modeled mathematically by the kinematic equations when acceleration is constant.
- Recognize that the velocity observed for an object depends on the reference frame in which it is observed, and be able to describe the reference frames of a given observer from a description of the physical situation.
- Be able to convert the velocity of an object relative to one reference frame into the velocity of that same object observed from a different reference frame.
- Identify the key features of projectile motion and how to interpret this kind of motion.
- Solve problems involving projectile motion.

3-1 The ideas of linear motion help us understand motion in two or three dimensions

The high jumper shown in the above photo follows a *curved* path as she passes over the crossbar. That means that we can't describe her path simply by using the ideas of Chapter 2 for one-dimensional motion. But we can extend the same concepts of displacement, velocity, and acceleration to motion along curved paths in two or three dimensions—that is, to motion in a plane or to generalized motion in space (**Figure 3-1a**).

We'll look in detail at an important kind of two-dimensional motion called *projectile motion*. This kind of motion in a plane occurs when the only source of an object's acceleration is the downward pull of gravity (**Figure 3-1b**). Projectile motion is not a perfect description of the flight of batted baseballs or kicked footballs, for which air resistance can be important, but it can still give useful approximate results.

Exam Tip

On the AP® Physics 1 exam, you may assume that air resistance is negligible unless otherwise stated.

The most general kind of motion is three dimensional. In many cases, however, the motion is in a plane and so is two dimensional. In figures (b) and (c) any side-to-side motion can be neglected; all motion is up and down and left and right.

This aircraft follows a complicated, three-dimensional path.

(a)

Design Pics Inc/Alamy

This motorcycle's path through the air is two dimensional (in a plane): It moves left to right as well as up and down.

(b)

Stephen Barnes/Science/Alamy

Each car on this Ferris wheel moves in the plane of the wheel, so its path is two dimensional.

(c)

Holly Kuchera/Getty Images

Figure 3-1 Motion in three and two dimensions: three examples of motion These figures depict motion in two or three dimensions. In this course you will be asked to solve problems in only two dimensions, although you should be able to describe what you would need to do in a three-dimensional motion situation.

Another important example of motion in a plane is motion in a circle (**Figure 3-1c**). Even if an object travels around a circle at a constant speed, it is nonetheless accelerating. That's because the object is *turning*—the direction of its motion is continuously changing. We'll use this observation to help explain why birds bank their wings when they turn, why cars sometimes skid when turning on a rain-slicked road, and other aspects of the natural and technological world. In later chapters, we will dig more deeply into the important case of uniform circular motion.

You might worry that motion in two or three dimensions is much harder to analyze than one-dimensional motion along a straight line. Not so! We introduced vectors in Section 2-2 in studying one-dimensional motion. The only difference is that in two- or three-dimensional motion, the vector quantities position, displacement, velocity, and acceleration do *not* always point along the same line, and we have to deal with each component separately. The good news is that each component is just a one-dimensional problem that we already know how to do. In the following section we'll see how to work with vectors in two and three dimensions.

THE TAKEAWAY for Section 3-1

✔ Motion in two or three dimensions uses the same ideas of velocity and acceleration as linear motion.

✔ In two- and three-dimensional motion, position, velocity, and acceleration have to be addressed along each dimension separately, treated as separate one-dimensional problems.

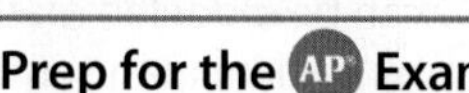

Prep for the AP Exam

AP® Building Blocks

1. Angelo lives in a building at the corner of a square block that has an unusual feature: a diagonal alley that starts behind his apartment building. Not everyone knows about the alley and Angelo uses it as a secret resource when racing against those who do not. When his opponent quickly races ahead along the streets, Angelo pretends to tie his shoelace. Angelo knows that it takes him 24 s to run from corner to corner, along a side of the block, a distance L.
   (a) If Angelo and his opponent run at the same constant speed, v, and race to the far corner of the block, a distance $2L$ away along the streets, how long can Angelo pretend to tie his shoe and still tie his opponent?
   (b) Explain how this two-dimensional problem was solved with one-dimensional thinking.

3-2 A vector quantity has both a magnitude and a direction

Imagine that one afternoon you decide to take your textbook to study at your favorite coffee house. As **Figure 3-2a** shows, you could take a number of different paths to the coffee house. No matter which path you choose, however, the net result is that you

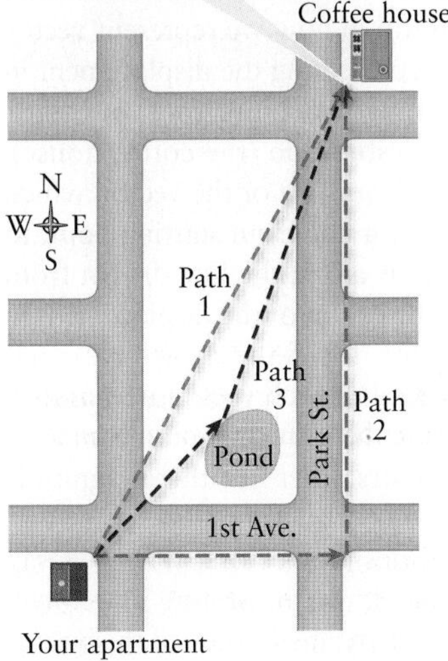

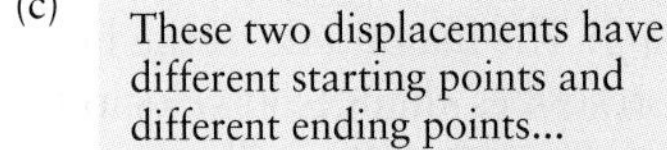

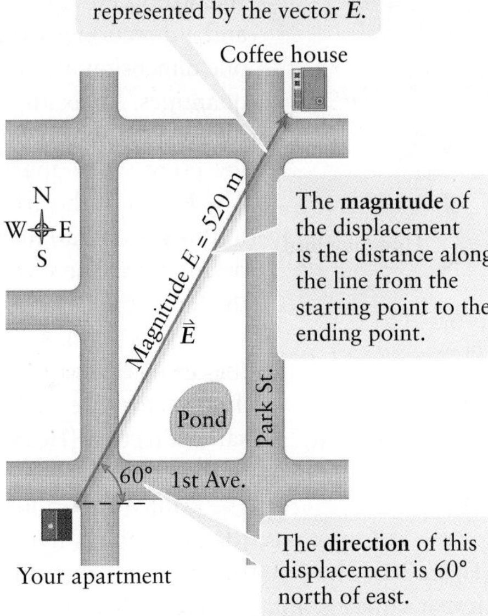

Figure 3-2 **Vectors** (a) Three paths from your apartment to your favorite coffee house. (b) Each path has the same displacement: from apartment to coffee house. (c) Two equal vectors. (d) Three unequal vectors.

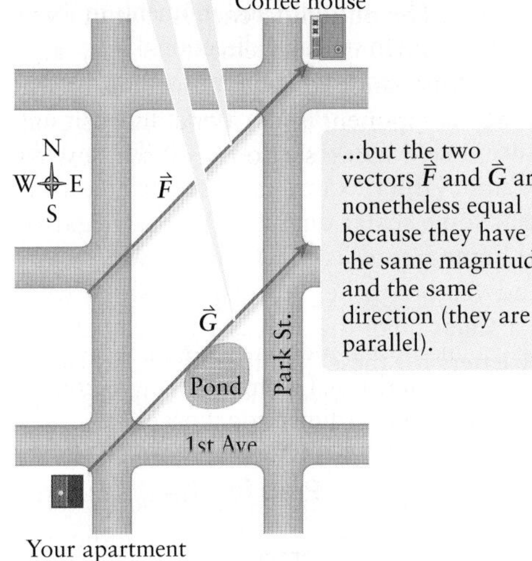

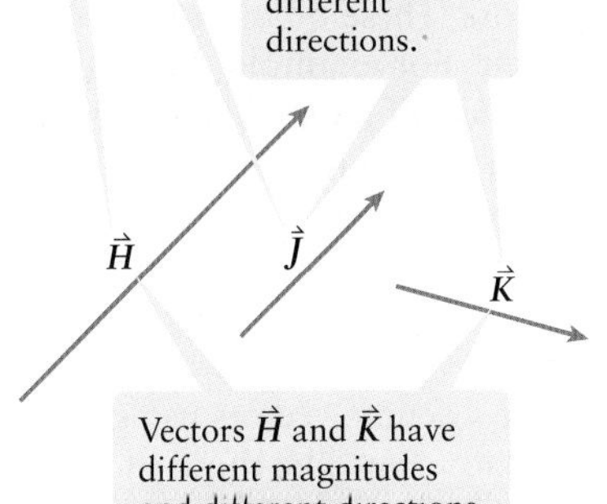

(d)
Vectors $\vec{H}$ and $\vec{J}$ have different magnitudes but the same direction.
Vectors $\vec{J}$ and $\vec{K}$ have the same magnitude but different directions.
$\vec{H}$
$\vec{J}$
$\vec{K}$
Vectors $\vec{H}$ and $\vec{K}$ have different magnitudes and different directions.

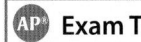

A vector has only two attributes: magnitude and direction.

end up 520 m from your apartment in a direction 60° north of east (**Figure 3-2b**). This change in your position is your *displacement*, and it has both a *magnitude* (520 m) and a *direction* (60° north of east). This is an extension of the idea of displacement that we introduced in Chapter 2 for motion along a straight line. In that case, the sign of the displacement (positive or negative) told us whether the displacement was in the positive or negative x direction. Because your motion from apartment to coffee house is in a *plane*, however, we need to expand our definition of direction.

AP® Exam Tip

The AP® Physics 1 equation sheet has the constant acceleration equations for the x direction, with the subscript x, but the same three equations can be used just as well for the y or the z direction—just make sure to use them independently in each direction. The only thing that is common about the three directions is the time, t.

WATCH OUT !

You need a plus or minus sign to indicate positive or negative along each dimension.

In two- or three-dimensional motion, we will use the subscripts "x," "y," and "z," as needed, to indicate the coordinate axis that goes with the variable. For constant acceleration problems, we will use our same one-dimensional equations, but one set along each axis, so we will need to define which direction is positive along each axis.

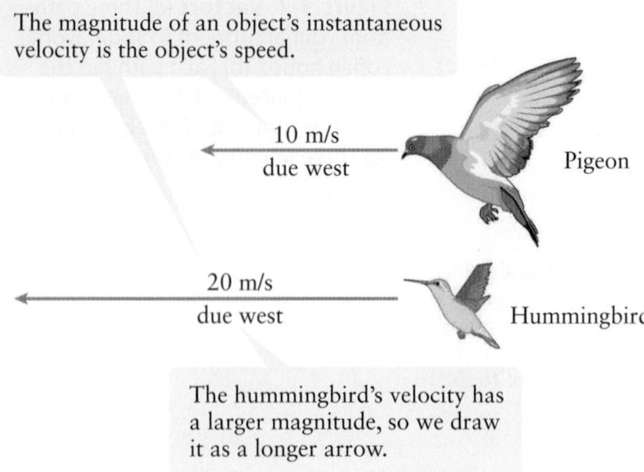

The magnitude of an object's instantaneous velocity is the object's speed.
10 m/s
due west
Pigeon
20 m/s
due west
Hummingbird
The hummingbird's velocity has a larger magnitude, so we draw it as a longer arrow.

Figure 3-3 Velocity The velocity of a flying bird is a vector: It has both a magnitude and a direction (the direction of the bird's travel).

Vectors and Scalars

As introduced in Chapter 2, any quantity that has both a magnitude and a direction is called a *vector*. As we learn to work in more than one dimension, we must be more careful in how we represent vector quantities. An example is the vector representing the displacement in Figure 3-2b, which we depict as an arrow that points from your starting point (your apartment) to your destination (the coffee house). The length of the arrow denotes the *magnitude* of the vector, which in this case is 520 m, the distance along a line from starting point to destination. The direction of a vector is along the line drawn from the tail to the head of the vector. The direction of the displacement in this case is 60° north of east. Two vectors are in the same direction if the lines drawn from the tail to the head of each vector are parallel. Two vectors are equal *only* if they have the same magnitude *and* the same direction (**Figure 3-2c**). If two vectors differ in either magnitude or direction, they are not equal to each other (**Figure 3-2d**).

Many important physical quantities are vectors. When we say that a pigeon is flying at 10 m/s due west, we are stating its *velocity* (**Figure 3-3**). The pigeon's *speed* at a particular instant in time is the magnitude of this vector at that same instant. Just as for the displacement, we use an arrow to denote the vector, and we use the length of the arrow to denote the magnitude. That's why we draw the arrow representing the velocity for a hummingbird flying at 20 m/s twice as long as the arrow representing the velocity for the 10-m/s pigeon in Figure 3-3: The magnitude of the hummingbird's velocity is twice as great. Acceleration (which describes changes in velocity) is also a vector. We have worked with these vectors in a single dimension. We will encounter other vectors in our study of physics such as force (a push or pull), and momentum. Now we will add the mathematical tools to work with vectors in multiple directions.

Other physical quantities have *no* direction associated with them, so they are *not* vectors. These quantities are called scalars and can be described by just a number and a unit. For example, the average temperature inside a home is 20°C (68°F), and the duration of a typical university lecture is 50 min. Other examples of scalar quantities are area, volume, mass, and density. Note that some scalar quantities can be negative; for example, the temperature inside a typical home freezer is −18°C or below.

We use different symbols for scalars and vectors to help us distinguish between them. We always use an *italic* letter to denote a scalar quantity, such as T for temperature or t for time. In handwriting you use ordinary letters for these quantities. By contrast, we always denote vectors using ***boldface italic*** letters with an arrow on top. Thus we write displacements like those shown in Figure 3-2 as $\vec{E}$, $\vec{F}$, $\vec{G}$, and so on. In handwriting you should *always* draw an arrow over the symbol for a vector quantity.

The symbol for the magnitude of a vector is the same symbol as for the vector itself, but *without* the arrow over the symbol and in italic rather than boldface italic. For example, the magnitude of the displacement $\vec{E}$ shown in Figure 3-2b is $E = 520$ m. We also sometimes use absolute value signs to denote the magnitude of a vector: $|\vec{E}| = E = 520$ m.

Like a scalar, the magnitude of a vector is given by a number and a unit. The unit is meters (m) for the displacement shown in Figure 3-2b. The magnitude of a vector can *never* be a negative number. If asked the distance from your apartment to the coffee house in Figure 3-2b, you would never say, "It's negative 520 m from here"! In the same way, speed (the magnitude of velocity) is never negative, which is why there are no negative numbers on a speedometer. Remember, when we use the word "speed" or "velocity" without a modifier, we are referring to the instantaneous value.

WATCH OUT !

Always determine if you are working with a scalar or vector quantity.

In the previous chapter, we reminded you that some scalar quantities could be expressed as positive or negative, like temperature, but that the sign had to be associated with direction for the quantity to be a vector. In two- or three-dimensional motion, a single "+" or "−" sign is not enough to express direction. You need one along each separate dimension. Sometimes vectors are expressed in terms of a magnitude and an angle direction instead, like in the coffee house example. So you must read carefully to determine if you are dealing with a vector or a scalar quantity; you cannot just look for signs!

WATCH OUT !

The vector arrow symbol always means the same thing.

The arrow symbol drawn over letters always indicates a vector; all mean the same thing, no matter the length of arrow.

WATCH OUT !

Use vector notation correctly.

Note that it is *never* correct to write an equation such as $\vec{E} = 520$ m. A vector is not simply equal to its magnitude; its direction is just as important. You must instead say, "$\vec{E}$ has magnitude 520 m and points 60° north of east." (Later in this chapter we'll see a different, shorthand way to represent a vector mathematically.)

Adding Vectors

Scalars add together according to the rules of ordinary arithmetic: If you have a cup of coffee that is at 45°C and you make it 5 degrees hotter, it will always end up at 50°C. Vectors behave differently, however. If you start out 45 m from your door and walk 5 more meters, but in different directions, you do not end up at the same place. Or, you could walk different individual displacements but still get to the same place. For instance, suppose you walk due east from your apartment to the corner of 1st Avenue and Park Street, a displacement $\vec{A}$, then walk due north to the coffee house, a displacement $\vec{B}$ (**Figure 3-4a**). Your net displacement is from your apartment to the coffee house; the combined effect of these two displacements is the same as the single displacement $\vec{E}$. Combining $\vec{A}$ and $\vec{B}$ to get $\vec{E}$ is an example of doing **vector addition**. The vector $\vec{E}$ is the *vector sum* of $\vec{A}$ and $\vec{B}$, $\vec{E} = \vec{A} + \vec{B}$. Figure 3-4a shows that another way to get to the coffee house is to travel first from your apartment to the duck pond (displacement $\vec{C}$) and then from the duck pond to the coffee house (displacement $\vec{D}$). So it's also true that $\vec{E} = \vec{C} + \vec{D}$.

Figure 3-4b illustrates that to perform vector addition we draw the two vectors in sequence, with the tail of the second vector up against the tip of the first vector. The sum of these two vectors then points from the tail of the first vector to the tip of

NEED TO REVIEW?
Turn to the **Glossary** in the back of the book for definitions of bolded Key Terms.

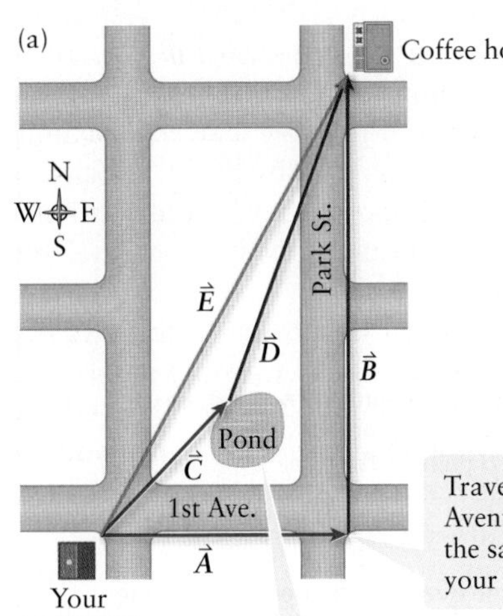
(a)
Coffee house
N
W
E
S
Park St.
$\vec{E}$
$\vec{D}$
$\vec{B}$
Pond
$\vec{C}$
1st Ave.
$\vec{A}$
Your apartment
Traveling from your apartment to the corner of 1st Avenue and Park Street, then to the coffee house, gives you the same net displacement as traveling in a single line from your apartment to the coffee house. So $\vec{E} = \vec{A} + \vec{B}$.
A trip via the duck pond also gives you the same net displacement as a single-line trip. So $\vec{E} = \vec{C} + \vec{D}$.

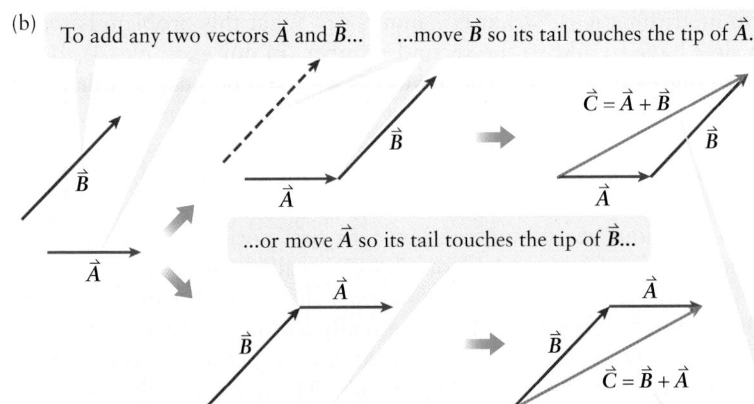
(b)
To add any two vectors $\vec{A}$ and $\vec{B}$...
...move $\vec{B}$ so its tail touches the tip of $\vec{A}$...
$\vec{B}$
$\vec{A}$
$\vec{C} = \vec{A} + \vec{B}$
...or move $\vec{A}$ so its tail touches the tip of $\vec{B}$...
$\vec{C} = \vec{B} + \vec{A}$
...and the sum $\vec{C}$ of the two vectors runs from the tail of the first vector to the tip of the second vector. You get the same result no matter in which order you add the vectors: $\vec{C} = \vec{A} + \vec{B} = \vec{B} + \vec{A}$.

Figure 3-4 Vector addition (a) All three paths from your apartment to the coffee house give the same total displacement: $\vec{E} = \vec{A} + \vec{B} = \vec{C} + \vec{D}$. (b) How to place the vectors $\vec{A}$ and $\vec{B}$ so they can be added.

When adding two vectors, place the tip of the first vector so it touches the tail of the second vector. The vector sum then extends from the tail of the first vector to the tip of the second vector.

the second vector. Even if the two vectors are not originally tip to tail, you can make them that way by moving or *translating* one vector's tail to the other vector's tip while keeping the directions of the vectors the same. Figure 3-4b also shows that the order in which you add vectors doesn't matter, so $\vec{A} + \vec{B} = \vec{B} + \vec{A}$. Ordinary scalar addition behaves the same way; for example, $3 + 4 = 4 + 3$.

WATCH OUT !

Vector addition is not the same as ordinary addition.

The magnitude of $\vec{A} + \vec{B}$, the sum of two vectors, is equal to the sum of the magnitude of $\vec{A}$ and the magnitude of $\vec{B}$ only if they point in the same direction (**Figure 3-5a**). If there is any difference in their direction this is not the case. In **Figure 3-5b** the vectors $\vec{A}$ and $\vec{B}$ are perpendicular, so these two vectors and their sum $\vec{C} = \vec{A} + \vec{B}$ make up three sides of a right triangle. In this case the Pythagorean theorem tells us that the magnitude of $\vec{C}$ is given by $C^2 = A^2 + B^2$, so $C = \sqrt{A^2 + B^2}$. This is less than $A + B$. **Figure 3-5c** shows that if $\vec{A}$ and $\vec{B}$ point in *opposite* directions, the magnitude $|\vec{A} + \vec{B}|$ is equal to the *difference* of A and B. In Section 3-3 we'll learn a mathematical technique for calculating the magnitude and direction of a vector sum.

(a) $\vec{A}$ Magnitude $A = 4$ m $\vec{B}$ Magnitude $B = 3$ m

$\vec{C} = \vec{A} + \vec{B}$

Magnitude of this vector sum:
$C = A + B$
$= 4\text{ m} + 3\text{ m}$
$= 7\text{ m}$

(b)

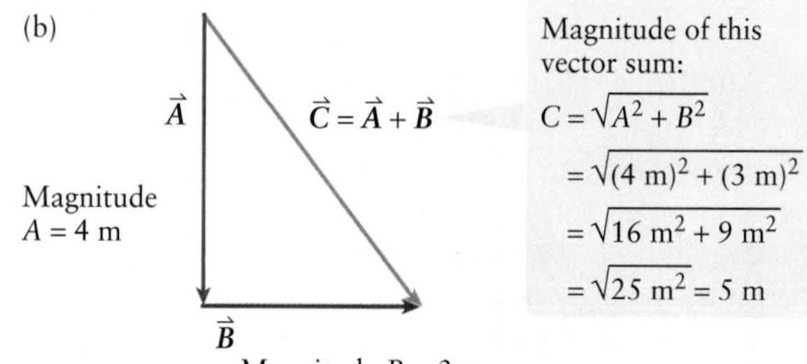

(c)

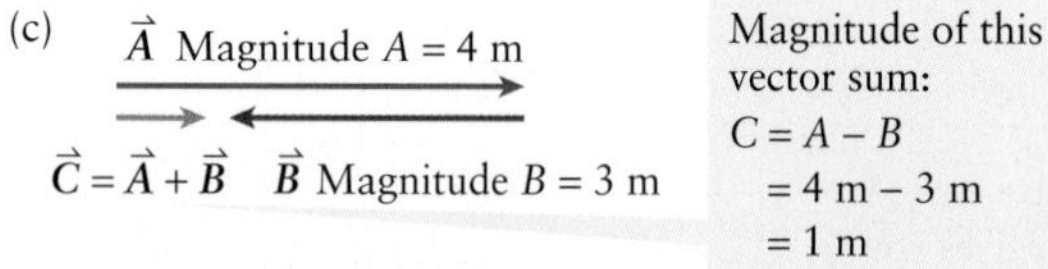

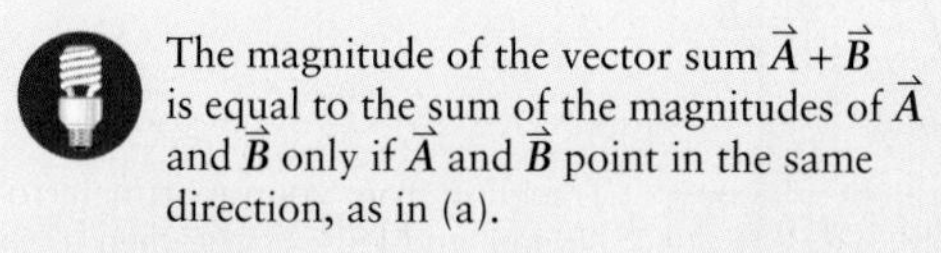

Figure 3-5 Special cases of vector addition The sum $\vec{A} + \vec{B}$ of two vectors that point (a) in the same direction, (b) perpendicular to each other, and (c) in opposite directions.

Subtracting Vectors

A simple problem in arithmetic is "What is 7 minus 4?" What this problem is really asking is "What do I have to add to the second number (in our example, 4) to get the first number (in our example, 7)?" The answer is $7 - 4 = 3$ because 3 added to 4 gives 7. Subtracting vectors (*vector subtraction*) works in much the same way. When we say that $\vec{D} - \vec{E} = \vec{F}$, we mean that the vector $\vec{F}$ is what we would have to add to the vector $\vec{E}$ to get the vector $\vec{D}$. **Figure 3-6a** shows how the vectors $\vec{D}$, $\vec{E}$, and $\vec{F}$ are related.

To see how to carry out vector subtraction, let's first notice something about ordinary subtraction: When we *subtract* 4 from 7, we're really *adding* –4 to 7, so that $7 - 4 = 7 + (-4) = 3$. In the same way, when we *subtract* the vector $\vec{E}$ from the vector $\vec{D}$ to calculate the **vector difference** $\vec{D} - \vec{E}$, we're really *adding* the vector $-\vec{E}$ to the vector $\vec{D}$; that is, $\vec{D} - \vec{E} = \vec{D} + (-\vec{E})$. To take the *negative of a vector*, we keep its magnitude the same and reverse its direction (**Figure 3-6b**). Thus, if $\vec{E}$ is a displacement of magnitude 600 m that points due *east*, the vector $-\vec{E}$ also has magnitude 600 m but points due *west*. Figure 3-6b shows how to carry out the subtraction $\vec{D} - \vec{E}$ by adding $\vec{D}$ and $-\vec{E}$. If you know how to add two vectors, you also know how to subtract them!

(a) Subtracting vectors:

The difference 7 – 4 = 3 is what you must add to 4 to get to 7.

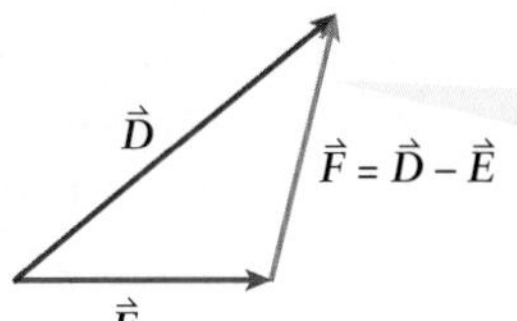

Similarly, the vector difference $\vec{F} = \vec{D} - \vec{E}$ is what you must add to $\vec{E}$ to get $\vec{D}$.

(b) How to calculate $\vec{F} = \vec{D} - \vec{E}$:

(1) Form the vector $-\vec{E}$ (with the same magnitude as $\vec{E}$ but the opposite direction).

(2) Add $\vec{D}$ and $-\vec{E}$ by placing them tip to tail. Their sum is $\vec{D} + (-\vec{E}) = \vec{D} - \vec{E}$.

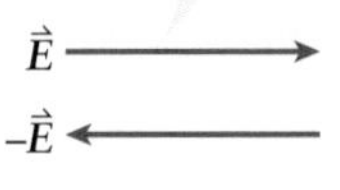

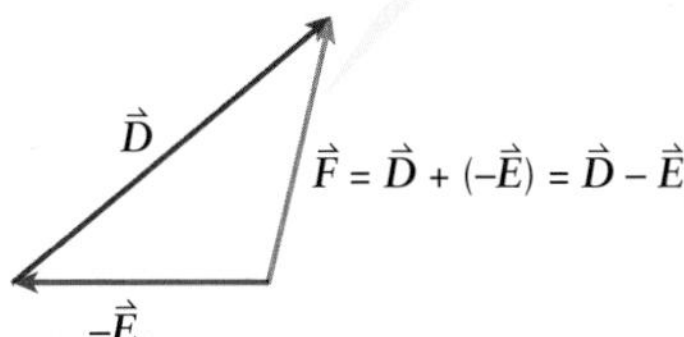

Figure 3-6 Vector subtraction Subtracting $\vec{E}$ from $\vec{D}$ is the same as adding $-\vec{E}$ to $\vec{D}$. (a) Subtracting one vector from another. (b) How to place the vectors $\vec{D}$ and $\vec{E}$ so they can be subtracted.

A number of different animal species use vectors to navigate. As an example, honeybees (**Figure 3-7a**) fly along a jagged path from flower to flower in search of nectar, but they fly straight back—on a beeline—to their hive. Although most models of bee navigation include a map-like spacial memory and the potential of using landmarks, all models include bees referencing the vector components of their paths and doing vector addition and subtraction to return to the hive (**Figure 3-7b**). There is evidence that fiddler crabs and desert ants count their steps and use directional cues. We'll use vector subtraction in Section 3-4 to help us extend the ideas of velocity and acceleration to motion in two or three dimensions.

Exam Tip

A common student error on the AP® Physics 1 exam is to add vectors by adding their magnitudes. This is only correct if the vectors being added point in the same direction.

(a)

Rob Flynn/USDA ARS

(b)

Flower
$\vec{A}$
$\vec{B}$
Flower
Hive
$\vec{C}$
$-(\vec{A} + \vec{B} + \vec{C})$
Flower

Figure 3-7 Honeybee vectors (a) A honeybee (genus *Apis*) visiting a flower. (b) Honeybees keep track of their displacements $\vec{A}, \vec{B}, \vec{C}, \ldots$ and use vector subtraction to calculate the direct route back to the hive.

Multiplying a Vector by a Scalar

A third useful bit of mathematics involving vectors is **multiplying a vector by a scalar. Figure 3-8** shows the simple rules for this. If c is a *positive* scalar, then the product of c and a vector $\vec{A}$ is a new vector $c\vec{A}$ that points in the same direction as $\vec{A}$ but has a different magnitude, equal to cA (the product of c and the magnitude of $\vec{A}$). If c is a *negative* scalar, then $c\vec{A}$ points in the direction opposite to $\vec{A}$ and has magnitude $|c|A$ (the absolute value of c multiplied by the magnitude of $\vec{A}$). We actually used this idea already when we defined the negative of a vector (Figure 3-6b): The vector $-\vec{E}$ equals the product of the negative scalar -1 and the vector $\vec{E}$, so $-\vec{E}$ points in the direction opposite to $\vec{E}$ and has magnitude $|-1|E = (1)E = E$ (that is, the same magnitude as $\vec{E}$).

We can also *divide* a vector by a scalar. This is really the same thing as multiplying by a scalar. As an example, dividing the vector $\vec{A}$ by 2 is the same as multiplying $\vec{A}$ by 1/2: $\vec{A}/2 = (1/2)\vec{A}$. So $\vec{A}/2$ is a vector that points in the same direction as $\vec{A}$ and has one-half the magnitude (Figure 3-8).

We'll multiply and divide a vector by a scalar frequently in our study of physics. In Section 3-4 we'll use these ideas to help define the velocity and acceleration vectors.

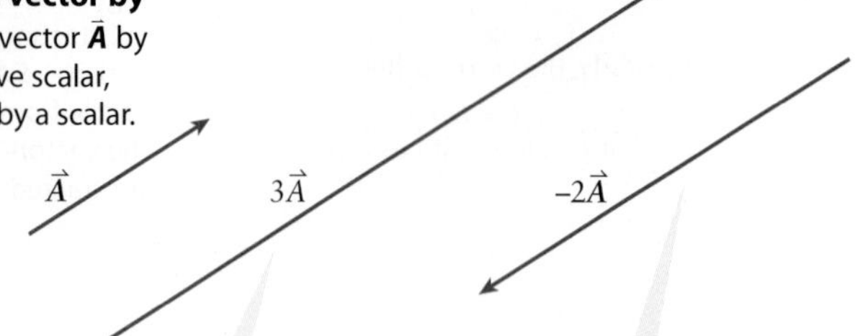

$\vec{A}$
$3\vec{A}$
$-2\vec{A}$
$\frac{\vec{A}}{2} = \frac{1}{2}\vec{A}$
The number 3 is positive, so the vector $3\vec{A}$ points in the same direction as $\vec{A}$ but has 3 times the magnitude.
The number –2 is negative, so the vector $-2\vec{A}$ points in the direction opposite to $\vec{A}$ and has 2 times the magnitude.
Dividing a vector by 2 is the same as multiplying it by 1/2, so the vector $\vec{A}/2$ points in the same direction as $\vec{A}$ (since 1/2 is positive) and has 1/2 the magnitude.

Figure 3-8 Multiplying a vector by a scalar How to multiply a vector $\vec{A}$ by a positive scalar or a negative scalar, and how to divide a vector by a scalar.

THE TAKEAWAY for Section 3-2

✔ Vectors are quantities that have both a magnitude and a direction. Scalar quantities have only magnitudes.

✔ To find the vector sum $\vec{A} + \vec{B}$ of two vectors, place the tail of the second vector ($\vec{B}$) against the tip of the first vector ($\vec{A}$). The vector sum points from the tail of the first vector to the tip of the second vector.

✔ The vector difference of two vectors, $\vec{A} - \vec{B}$, is the sum of $\vec{A}$ and $-\vec{B}$ (the negative of vector $\vec{B}$, which has the same magnitude as $\vec{B}$ but points in the opposite direction).

✔ The product of a scalar c and a vector $\vec{A}$ is a new vector that has magnitude $|c|A$ (the absolute value of c multiplied by the magnitude of $\vec{A}$). This vector points in the same direction as $\vec{A}$ if c is positive but in the opposite direction if c is negative.

Prep for the AP® Exam

AP® Building Blocks

1. What is the difference between a scalar and a vector? Give an example of a scalar and an example of a vector.
2. Which of the following is not a vector? Justify your choice.
   (A) average velocity
   (B) instantaneous velocity
   (C) distance
   (D) displacement
   (E) acceleration
3. Using a sheet of graph paper, construct vectors that represent the following excursion:
   (a) Dora starts at her tent and walks 10 km north on plains near the Nazca Lines. Then she turns and walks 15 km east. On your representation of her journey label the northward path as $\vec{A}$ and the eastward path as $\vec{B}$. Draw the two individual motions to scale (in other words, to the correct relative lengths).
   (b) Using a ruler, measure the length of each line segment, and use these lengths to determine the scale of your map in km per cm.
   (c) Draw a line connecting her starting point and her ending point. Measure the length of the line segment and, using your map scale, convert the magnitude of the displacement in cm to a magnitude of the displacement in km.
   (d) Label the line connecting her starting point and her ending point as $\vec{C}$. Express the vector $\vec{C}$ in terms of the vectors $\vec{A}$ and $\vec{B}$.
   (e) Confirm that the Pythagorean theorem works to predict the magnitude of the displacement of Dora.
   (f) Dora accidently dropped her canteen when she turned east. She calls in to base camp on her cell phone for a resupply mission. Base camp can get her location using GPS. How would Dora express the location of her canteen from her current location using vector addition?
   (g) Is the scale of your map a scalar or a vector?
4. (a) Can the sum of two vectors that have different magnitudes ever be equal to zero? If so, give an example. If not, explain why the sum of two vectors with different magnitudes cannot be equal to zero.
   (b) Can the sum of three vectors that have different magnitudes ever be equal to zero?

AP® Skill Builders

5. Before constructing a machine, an engineer decides to build a model so that specific parts of the machine do not accidentally come into contact with each other when the configuration of the machine changes during operation. Parts of the final machine are then fabricated by "scaling up" each part of the model, once the model has been shown to work. Explain, using a simple triangle as an example of a set of three contacting vectors, why scaling up, that is, multiplying all parts by the same scalar, does not change the shape of the machine.

AP® Skills in Action

6. All cars have speedometers in the dashboard. Many cars have compasses. Design a device that displays both speed and direction in a single window and give the device a name similar to *speedometer* so that the public will understand it.

3-3 Vectors can be described in terms of components

We saw in Figure 3-5 that adding two vectors $\vec{A}$ and $\vec{B}$ is straightforward if the vectors are perpendicular, point in the same direction, or point in opposite directions. But what can we do in situations in which vectors are not so conveniently arranged?

A powerful technique that we can use for all kinds of calculations with vectors, no matter how they're oriented, is the **component method**. As an application, consider the arc of a kicked football in the absence of air resistance. We know the acceleration due to gravity in the downward vertical direction. As a result, it does not affect the football's horizontal motion. Likewise, wind blowing horizontally does not affect the vertical motion of the ball. As a result we can analyze the two-dimensional motion of the football as two separate, one-dimensional motions—vertical motion and horizontal motion. We must describe the football's velocity as a vector, so it makes sense to break that vector into horizontal and vertical components.

As we have defined our axes in **Figure 3-9,** this vector, $\vec{A}$, has a positive x component A_x and a positive y component A_y.

Figure 3-9a shows a pair of mutually perpendicular coordinate axes labeled x and y. We will refer to such a set of perpendicular coordinate axes as a coordinate system. These two axes define a plane, and any point in the plane can be identified by its x and y coordinates. For instance, a point located at $x = 3$ and $y = 5$ has coordinates (3, 5) in the units of the vector being represented. The figure also shows a vector $\vec{A}$ that lies in the x–y plane. If we place the tail of this vector at the *origin*—that is, at the point $x = 0$, $y = 0$—the coordinates of the tip of the vector $\vec{A}$ are (A_x, A_y), as shown in Figure 3-9a. The quantities A_x (we say "A-sub-x") and A_y (we say "A-sub-y") are called the components of the vector $\vec{A}$: A_x is the x **component**, and A_y is the y component.

NEED TO REVIEW?
To review a coordinate system, turn to Chapter 2, Section 2-2.

As Figure 3-9a shows, to find the x component A_x we first draw the vector $\vec{A}$ with its tail at the origin, then draw a line perpendicular to the x axis from the tip of $\vec{A}$ to

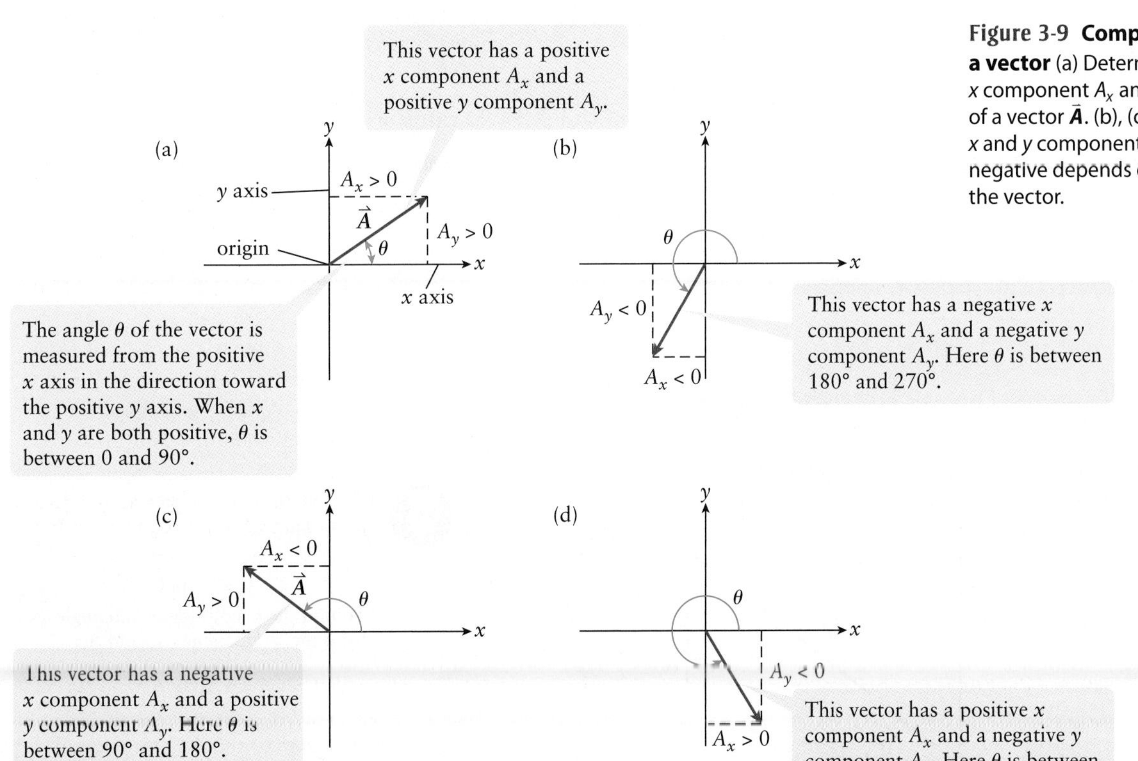

(a)
This vector has a positive x component A_x and a positive y component A_y.
y
y axis
$A_x > 0$
$\vec{A}$
$A_y > 0$
origin
θ
x
x axis
The angle θ of the vector is measured from the positive x axis in the direction toward the positive y axis. When x and y are both positive, θ is between 0 and 90°.
(b)
y
θ
x
$A_y < 0$
$A_x < 0$
This vector has a negative x component A_x and a negative y component A_y. Here θ is between 180° and 270°.
(c)
y
$A_x < 0$
$\vec{A}$
$A_y > 0$
θ
x
This vector has a negative x component A_x and a positive y component A_y. Here θ is between 90° and 180°.
(d)
y
θ
x
$A_y < 0$
$A_x > 0$
This vector has a positive x component A_x and a negative y component A_y. Here θ is between 270° and 360°.

Figure 3-9 Components of a vector (a) Determining the x component A_x and y component A_y of a vector $\vec{A}$. (b), (c), (d) Whether the x and y components are positive or negative depends on the direction of the vector.

the x axis. Where this line intersects the x axis tells you the value of A_x. In a similar way, to find the value of A_y, we draw a line perpendicular to the y axis from the tip of $\vec{A}$ to the y axis. Depending on the direction of the vector, the components can be both positive (Figure 3-9a), both negative (**Figure 3-9b**), or of different signs (**Figures 3-9c** and **3-9d**). Once you have stated the x component and y component of a vector in the x–y plane, you have defined the vector completely.

Two Common Ways to Describe Vector Components

You can describe a vector such as $\vec{A}$ in terms of either (1) its magnitude A and direction, in terms of the angle θ or (2) its components A_x and A_y. You can use whichever description is more convenient. **Figure 3-10** shows how to convert between these two descriptions. Note that the vector $\vec{A}$ of magnitude A, its x component A_x, and its y component A_y form three sides of a right triangle. From trigonometry the cosine of the angle θ equals the adjacent side A_x divided by the hypotenuse A, the sine of θ equals the opposite side A_y divided by the hypotenuse A, and the tangent of θ equals the opposite side A_y divided by the adjacent side A_x. Furthermore, the Pythagorean theorem tells us that $A^2 = A_x^2 + A_y^2$.

Rewriting these relationships, we get two equations that tell us how to find the components from the magnitude and direction, and two that tell us how to find the magnitude and direction from the components when the angle θ is measured from the positive x axis toward the positive y axis:

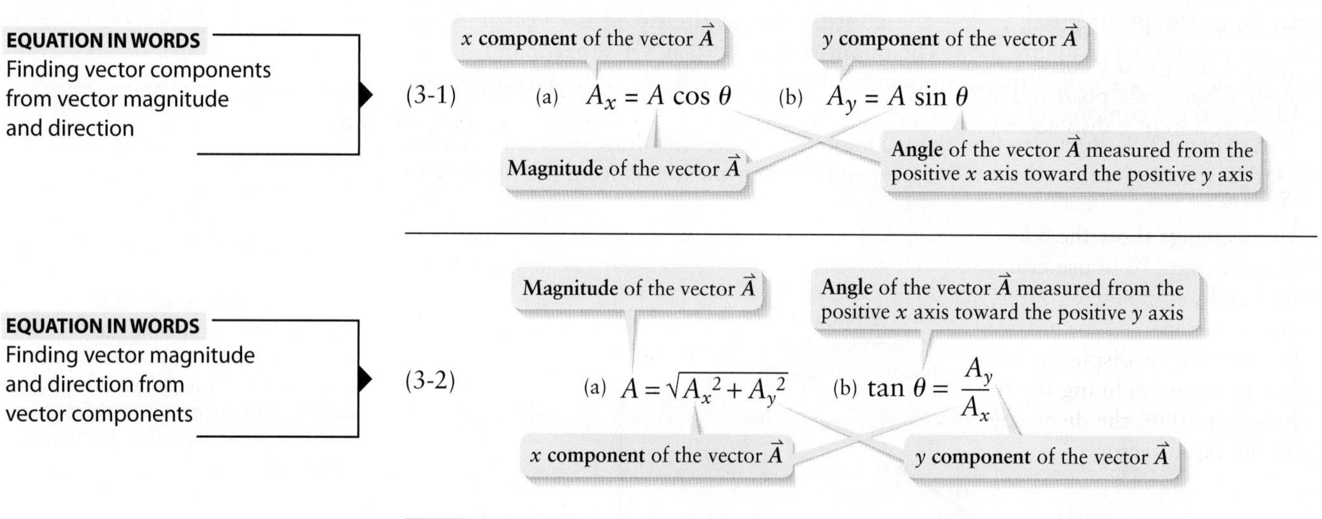

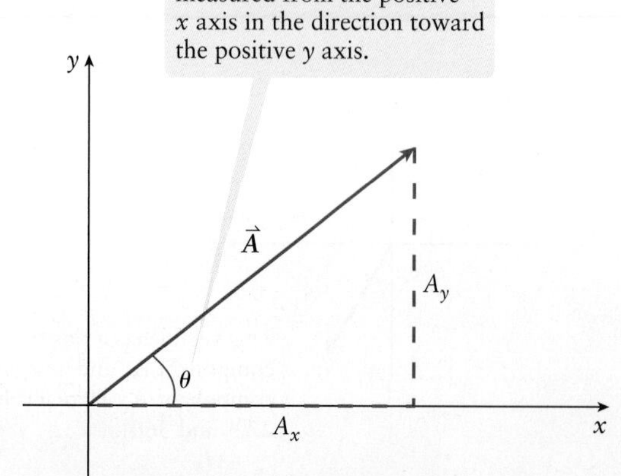

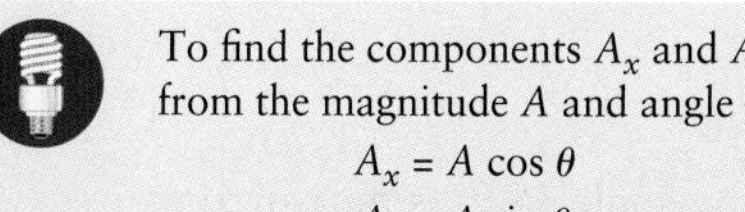

To find the components A_x and A_y from the magnitude A and angle θ:
$A_x = A\cos\theta$
$A_y = A\sin\theta$
To find the magnitude A and angle θ from the components A_x and A_y:
$A = \sqrt{A_x^2 + A_y^2}$
$\tan\theta = \frac{A_y}{A_x}$

Figure 3-10 Two ways to describe a vector You can describe a vector $\vec{A}$ in terms of its magnitude A and direction θ or in terms of its components A_x and A_y.

WATCH OUT !

Be careful with angles when doing calculations with vector components.

Often in the real world it is just not physically possible with the equipment you have to measure the angle from the positive x axis toward the positive y axis. In such a situation, you will have to use the information about right triangles to apply the trigonometric relations carefully to properly relate the magnitude, x component, and y component to the given angle, and not just rely on Equations 3-1 and 3-2. For instance, in the triangle below, where O is the origin, and $\vec{A}$ is our vector, if we measure the angle θ, we use Equations 3-1 and 3-2, no problem! If, however, we had to measure the angle ϕ, then our x component, A_x would be $A \sin \phi$ and our y component, A_y would be $A \cos \phi$. Also, given that the sum of all the interior angles of a triangle is 180 degrees, if one is able to measure one of the two acute angles in a right triangle, one could always calculate the other, and then employ Equations 3-1 and 3-2, as before.

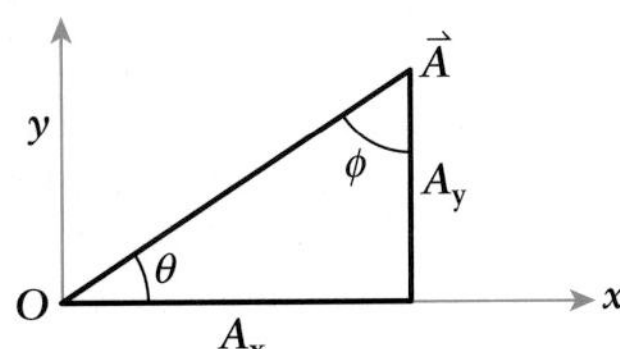

You also need to be careful if you're given an angle θ in degrees, or are trying to calculate an angle in degrees, to always check that your calculator is in "degrees" mode.

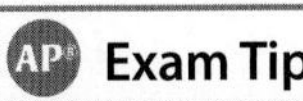

Exam Tip

The equations relating the angle and the sides of a right triangle are given on the AP® Physics 1 equation sheet. Be sure to understand how that translates to vector components and the equations in this section. Note that on the equation sheet, if the angle θ is measured relative to the x axis, then b is the x component and a is the y component; and if theta is measured relative to the y axis, then b is the y component and a is the x component.

EXAMPLE 3-1 Different Descriptions of a Vector

(a) Find the x and y components of a velocity $\vec{v}$ that has magnitude 35.0 m/s and points in a direction 36.9° west of north. Choose the positive x direction to point east and the positive y direction to point north. (b) Find the magnitude and direction of a displacement $\vec{r}$ that has x component –24.0 m and y component –11.0 m.

Set Up

Our drawings show the two vectors and the x- and y-axes. We'll use Equations 3-1 to find the x and y components of the velocity $\vec{v}$ in part (a) and Equations 3-2 to find the magnitude and direction of the displacement $\vec{r}$ in part (b). Note that as we are defining the start of the vector as the origin (0,0), the displacement and position are the same vector.

Finding vector components from vector magnitude and direction:

$$A_x = A \cos \theta \quad (3\text{-}1a)$$

$$A_y = A \sin \theta \quad (3\text{-}1b)$$

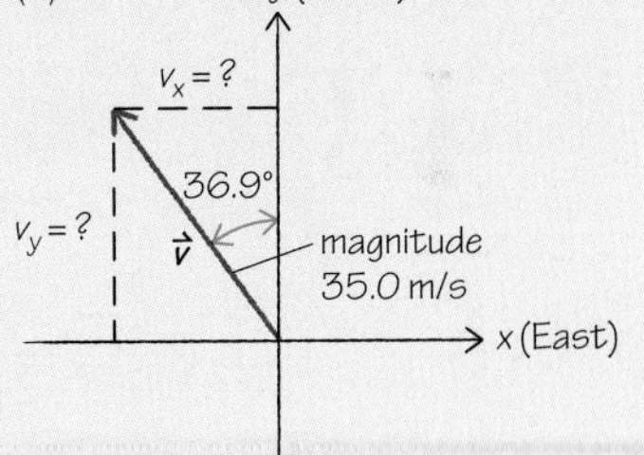

Finding vector magnitude and direction from vector components:

$$A = \sqrt{A_x^2 + A_y^2} \quad (3\text{-}2a)$$

$$\tan \theta = \frac{A_y}{A_x} \quad (3\text{-}2b)$$

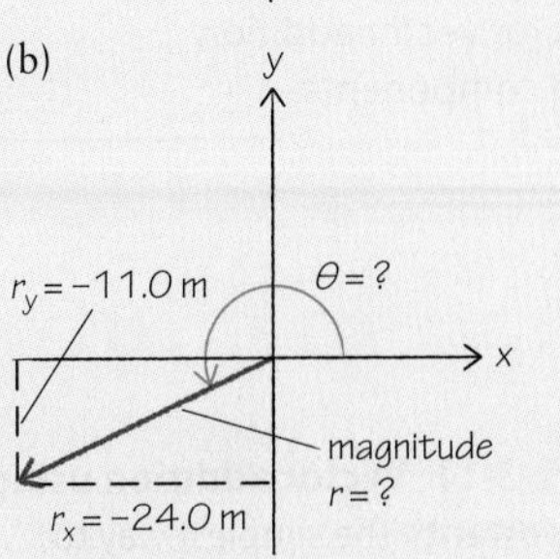

Solve

(a) The angle 36.9° is measured from the positive y axis. To use Equations 3-1, the angle of the vector $\vec{v}$ must be measured from the positive x axis. The figure shows that the angle we need is 90.0° + 36.9° = 126.9°. Use this angle in Equations 3-1 along with the magnitude v = 35.0 m/s of $\vec{v}$.

$$\begin{aligned} v_x &= (35.0 \text{ m/s}) \cos 126.9° \\ &= (35.0 \text{ m/s})(-0.600) \\ &= -21.0 \text{ m/s} \end{aligned}$$

$$\begin{aligned} v_y &= (35.0 \text{ m/s}) \sin 126.9° \\ &= (35.0 \text{ m/s})(0.800) \\ &= +28.0 \text{ m/s} \end{aligned}$$

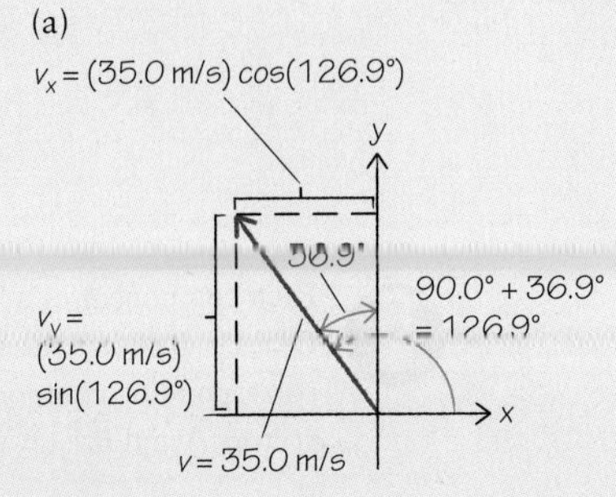

(b) We are given the x and y components of the displacement $\vec{r}$, so we can use Equation 3-2a to calculate the magnitude r.

$$r = \sqrt{r_x^2 + r_y^2}$$
$$= \sqrt{(-24.0 \text{ m})^2 + (-11.0 \text{ m})^2}$$
$$\text{so } r = 26.4 \text{ m}$$

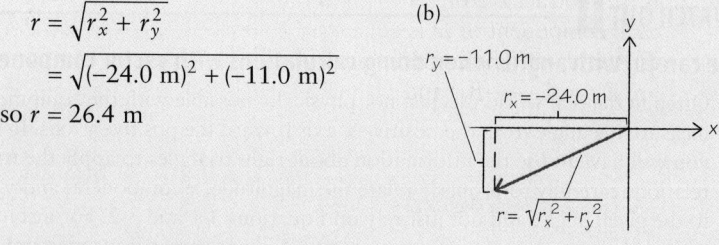

From Equation 3-2b, the tangent of the angle θ shown in the figure equals r_y divided by r_x, or 0.458. Use your calculator to find the inverse tangent of 0.458. The result is 24.6°. However, the tangent function has the property that $\tan \varphi = \tan(180° + \varphi)$. Hence 0.458 is equal to *both* tan 24.6° and tan (180.0° + 24.6°) = tan 204.6°. The figure shows that the angle of $\vec{r}$ measured from the positive x axis is greater than 180°, so the answer we want is $\theta = 204.6°$.

$$\tan \theta = \frac{r_y}{r_x} = \frac{-11.0 \text{ m}}{-24.0 \text{ m}} = 0.458$$
$$\theta = \tan^{-1} 0.458 = 24.6° \text{ or } = 204.6°$$

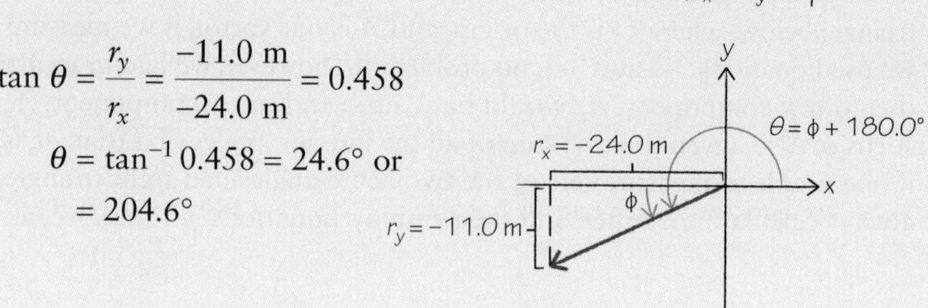

Reflect

The drawing of $\vec{v}$ in part (a) shows that this vector points in the negative x direction and the positive y direction. This agrees with our results, which show that v_x is negative and v_y is positive. The drawing of $\vec{r}$ in part (b) helped us decide which value of θ was the correct one. This shows why it's important to *always* draw a picture for any problem that involves vectors.

NOW WORK Problems 1 and 2 from The Takeaway 3-3.

Vector Arithmetic with Components

Describing vectors in terms of their components greatly simplifies the vector arithmetic that we described in Section 3-2. Here are the rules:

1. **Figure 3-11** shows that if you add the vectors $\vec{A}$ and $\vec{B}$ to form the vector sum $\vec{C} = \vec{A} + \vec{B}$, each component of $\vec{C}$ is just the sum of the corresponding components of $\vec{A}$ and $\vec{B}$:

EQUATION IN WORDS
Rules for vector addition using components

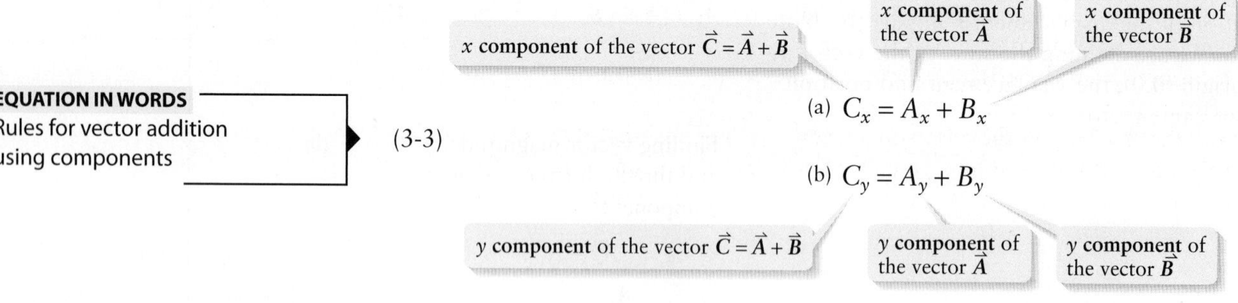

The y component of the vector $\vec{C} = \vec{A} + \vec{B}$ equals the sum of the y components of $\vec{A}$ and $\vec{B}$: $C_y = A_y + B_y$

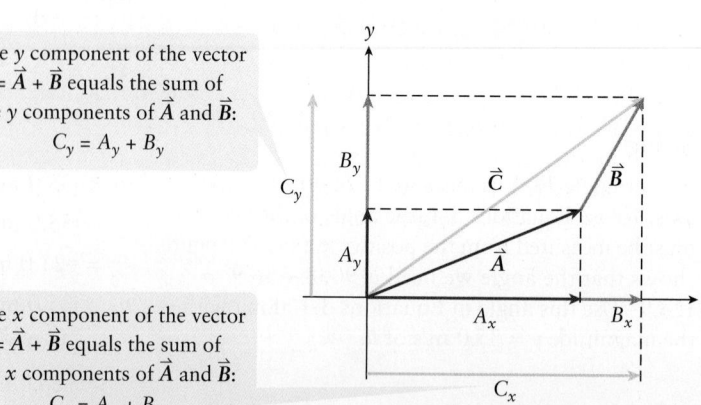

The x component of the vector $\vec{C} = \vec{A} + \vec{B}$ equals the sum of the x components of $\vec{A}$ and $\vec{B}$: $C_x = A_x + B_x$

y, x, B_y, C_y, A_y, $\vec{C}$, $\vec{B}$, $\vec{A}$, A_x, B_x, C_x

Figure 3-11 Vector addition using components The simplest way to add two vectors is in terms of their components.

Each component of a vector sum $\vec{C} = \vec{A} + \vec{B}$ equals the sum of the corresponding components of the vectors $\vec{A}$ and $\vec{B}$.

2. If we subtract $\vec{B}$ from $\vec{A}$ to form the vector difference $\vec{D} = \vec{A} - \vec{B} = \vec{A} + (-\vec{B})$, each component of $\vec{D}$ is equal to the sum of the corresponding components of $\vec{A}$ and $-\vec{B}$. The components of $-\vec{B}$ are $-B_x$ and $-B_y$, so the components of $\vec{D}$ are $D_x = A_x + (-B_x)$ and $D_y = A_y + (-B_y)$, or

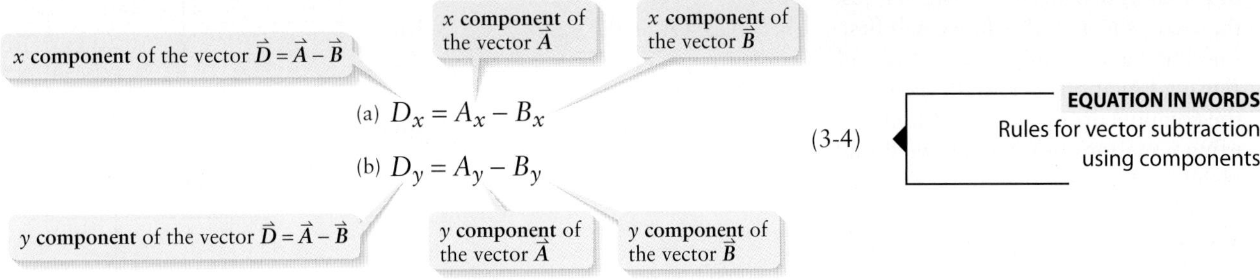

$$\text{(a) } D_x = A_x - B_x \qquad \text{(b) } D_y = A_y - B_y \tag{3-4}$$

EQUATION IN WORDS
Rules for vector subtraction using components

That is, each component of $\vec{D} = \vec{A} - \vec{B}$ equals the difference between the corresponding components of $\vec{A}$ and $\vec{B}$.

3. Finally, multiplying a vector $\vec{A}$ by a scalar c gives a new vector $\vec{E} = c\vec{A}$ whose components are just the components of $\vec{A}$ multiplied by c:

x **component** of the vector $\vec{E} = c\vec{A}$ — **Scalar** c — *x* **component** of the vector $\vec{A}$

$$\text{(a) } E_x = cA_x \qquad \text{(b) } E_y = cA_y \tag{3-5}$$

y **component** of the vector $\vec{E} = c\vec{A}$ — **Scalar** c — *y* **component** of the vector $\vec{A}$

EQUATION IN WORDS
Rules for multiplying a vector by a scalar using components

Note that if c is negative, the sign of each vector component is reversed—which means that the direction of the vector as a whole is reversed. That's just what we saw in Figure 3-8.

To use these rules you need to know the vectors $\vec{A}$ and $\vec{B}$ in terms of their components. If instead you're given $\vec{A}$ and $\vec{B}$ in terms of their magnitudes and directions, you can find their components using Equations 3-1, making sure the angle you use is measured from the positive x axis toward the positive y axis. If you are given a different angle, you can use the appropriate components for the angle given, or the basic facts that the sum of the angles in a triangle is 180° and that right angles are 90° to find the angle you need. Once you've found the components of the desired result (such as a vector sum or a vector difference), you can determine its magnitude and direction using Equations 3-2.

EXAMPLE 3-2 Adding Vectors Using Components

You travel 62.0 km in a direction 60.0° east of north, then turn to a direction 50.0° south of east and travel an additional 23.0 km (**Figure 3-12**). How far and in what direction are you from your starting point (your net displacement for the trip)? Note that because the initial starting point is (0,0), your final position and the net displacement for the trip are the same.

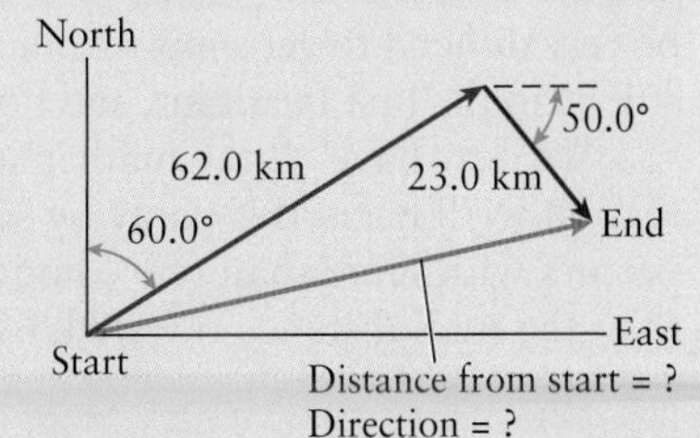

Figure 3-12 A problem in vector addition What is your total (net) displacement?

Set Up

Each leg of the trip is a displacement, which we show as $\vec{A}$ and $\vec{B}$ in the sketch. To find the magnitude and direction of the total displacement, which is just the vector sum $\vec{C} = \vec{A} + \vec{B}$, we will first find the x and y components of $\vec{A}$ and $\vec{B}$ using Equations 3-1. We'll then find the components of the vector sum $\vec{C}$ using Equations 3-3. Finally, we'll find the magnitude and direction of $\vec{C}$ using Equations 3-2. It is best to calculate angles from ratios directly instead of calculating intermediate numbers, given that tangents vary quickly.

$$A_x = A\cos\theta \quad (3\text{-}1a)$$

$$A_y = A\sin\theta \quad (3\text{-}1b)$$

$$C_x = A_x + B_x \quad (3\text{-}3a)$$

$$C_y = A_y + B_y \quad (3\text{-}3b)$$

$$A = \sqrt{A_x^2 + A_y^2} \quad (3\text{-}2a)$$

$$\tan\theta = \frac{A_y}{A_x} \quad (3\text{-}2b)$$

Solve

We choose the positive x axis to point east and the positive y axis to point north, and draw both $\vec{A}$ and $\vec{B}$ with their tails at the origin. The vector $\vec{A}$ points east of north at an angle of 60.0°. Because north and east are perpendicular to each other, the angle vector $\vec{A}$ makes with the x axis is 30.0°. The vector $\vec{B}$ points 50.0° south of east, which makes this angle −50.0° (negative, because we need the angle toward the positive y axis and the 50.0° angle is measured below the positive x axis in the direction *away from* the positive y axis). We use Equations 3-1 to find their components.

$$A_x = (62.0 \text{ km})\cos 30.0° = 53.7 \text{ km}$$

$$A_y = (62.0 \text{ km})\sin 30.0° = 31.0 \text{ km}$$

$$B_x = (23.0 \text{ km})\cos(-50.0°) = 14.8 \text{ km}$$

$$B_y = (23.0 \text{ km})\sin(-50.0°) = -17.6 \text{ km}$$

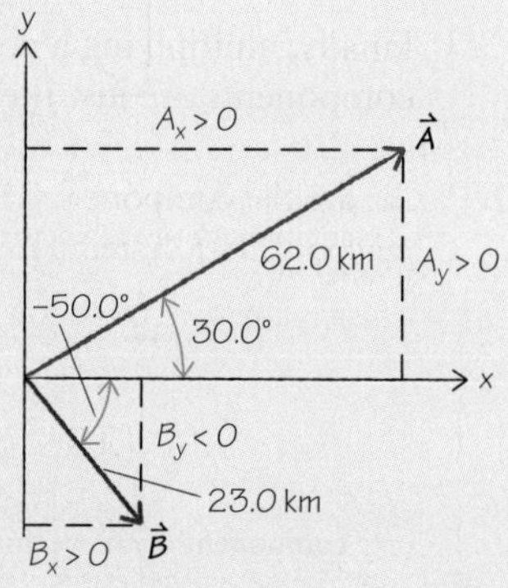

Given the components of $\vec{A}$ and $\vec{B}$, we can now find the components of $\vec{C} = \vec{A} + \vec{B}$ using Equations 3-3.

$$C_x = A_x + B_x = 53.7 \text{ km} + 14.8 \text{ km} = 68.5 \text{ km}$$

$$C_y = A_y + B_y = 31.0 \text{ km} + (-17.6 \text{ km}) = 13.4 \text{ km}$$

Now calculate the magnitude and direction of $\vec{C}$ using Equations 3-2. Both components of $\vec{C}$ are positive, so the vector points north of east, and the angle θ_C is between 0 and 90°. Hence the desired angle is 11.1°, not 180° + 11.1° = 191.1°.

$$C = \sqrt{C_x^2 + C_y^2} = \sqrt{(68.5 \text{ km})^2 + (13.4 \text{ km})^2}$$

$$C = 69.8 \text{ km}$$

$$\tan\theta_C = \frac{C_y}{C_x} = \frac{13.4 \text{ km}}{68.5 \text{ km}}$$

$$\theta_C = \tan^{-1} 13.4/68.5 = 11.1°$$

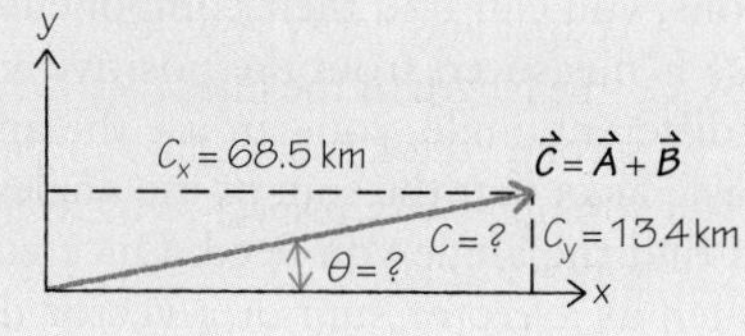

Reflect

The net displacement takes you 69.8 km from the starting point at an angle of 11.1° north of east. These numbers would be very difficult to get simply from the drawing of the two vectors shown above, although it was clear the answer should still be in the first quadrant, and not too much different in magnitude than A.

We now have all the mathematical tools that we need to study motion in two or three dimensions. In the next section we'll begin this study by seeing how to relate the position, velocity, and acceleration of an object using vectors with more than one component. Note that the length of this vector, the distance from the start, is different than the total distance traveled, 62 m + 23 m = 85 m.

NOW WORK Problems 3, 4, and 5 from The Takeaway 3-3.

THE TAKEAWAY for Section 3-3

✔ The easiest way to do arithmetic with vectors (adding, subtracting, and multiplying by a scalar) is to use components. Equations 3-1 let you find the components of a vector from its magnitude and direction, and Equations 3-2 let you find the magnitude and direction of a vector from its components. Furthermore, equations 3-3, 3-4, and 3-5 show how to add vectors, subtract vectors, and multiply a vector by a scalar using components.

✔ For any problem involving vectors, it's essential to draw a sketch of the situation and to measure angles from the positive x direction toward the positive y direction.

✔ Be cautious using Equation 3-2b, $\tan\theta = A_y/A_x$. This equation actually gives two answers for the direction θ of a vector $\vec{A}$; your sketch will help you decide which answer is correct.

Prep for the AP® Exam

AP® Building Blocks

EX 3-1 **1.** Calculate the magnitude of the vector in the figure.

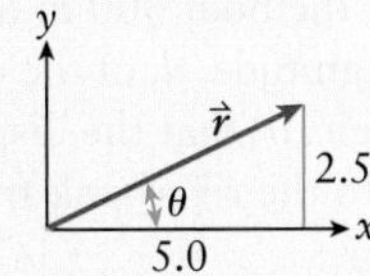

EX 3-1 **2.** What are the components A_x and A_y of the vector in the three coordinate systems shown in the figure?

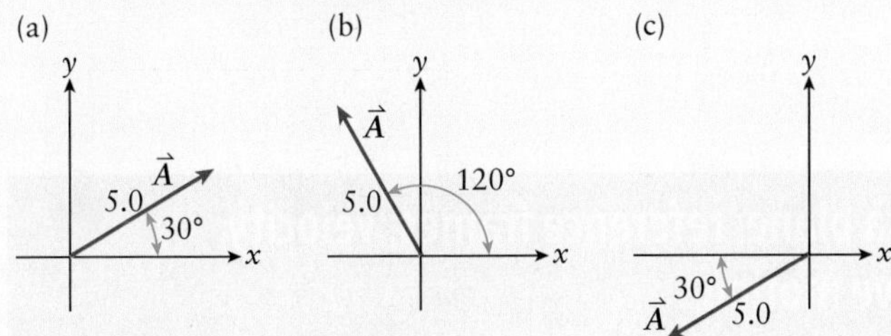

EX 3-2 **3.** If $\vec{r}$ has a magnitude of 6.0 units and makes an angle of 36° with respect to the negative x axis, below the axis:

(a) Find the vector components of $\vec{r}$.

(b) Use a protractor to construct the vector $\vec{r}$ on graph paper and verify your answer in part (a) by measuring the length of the lines representing its components.

A second vector, $\vec{s}$, has a magnitude of 8.0 units and makes a 30° angle with respect to the positive x axis toward the positive y axis.

(c) Find the vector components of $\vec{s}$.

(d) Use a protractor to construct the vector $\vec{s}$ on the same grid used in part (b) and verify your answer in part (c) by measuring the length of the lines representing its components.

(e) Construct the vectors $\vec{s}+\vec{r}$ and $\vec{s}-\vec{r}$ on the same graph used in parts (b) and (d). Use a different color to show copies of the vectors $\vec{r}$ and $-\vec{r}$ and to show the head-to-tail method of vector addition.

(f) Confirm with values taken from your graph that $(\vec{s}+\vec{r})_y = s_y + r_y$ and $(\vec{s}-\vec{r})_y = s_y - r_y$.

EX 3-2 **4.** Given a vector with components $A_x = 2.00$ and $A_y = 6.00$, and a vector with components $B_x = 3.00$ and $B_y = 22.00$, calculate the magnitude and angle with respect to the x axis of the vector sum, assuming both vectors have the same units.

EX 3-2 **5.** Three vectors all represent the same type of quantity, in the same units, so they may be added and subtracted. In these units, vector $\vec{A}$ has components $A_x = 6$ and $A_y = 9$. Vector $\vec{B}$ has components $B_x = 7$, $B_y = 23$, and vector $\vec{C}$ has components $C_x = 0$, $C_y = 26$. Determine the components of the following vector sums: a. $\vec{A}+\vec{B}$, b. $\vec{A}-2\vec{C}$, c. $\vec{A}+\vec{B}-\vec{C}$, and d. $\vec{A}+\frac{1}{2}\vec{B}-3\vec{C}$.

6. An object is displaced twice, first along the vector $\vec{A}$ and then along the vector $\vec{B}$ as shown in the figure. The units of the displacement are centimeters.

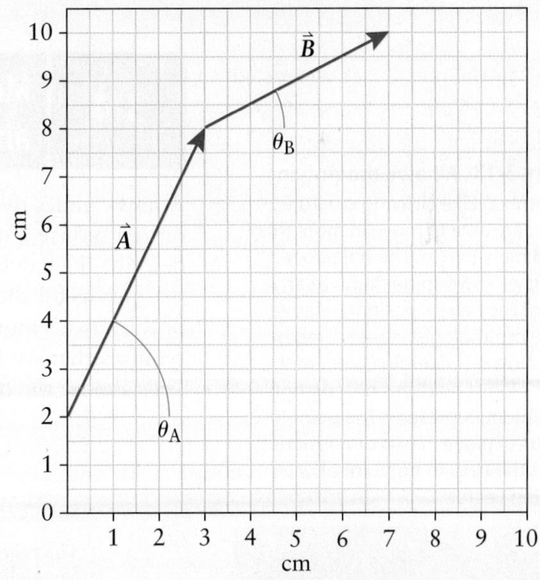

(a) What is the magnitude of the displacement along $\vec{A}$?

(b) What is the magnitude of the displacement along $\vec{B}$?

(c) What is the direction of the displacement along $\vec{A}$ in terms of the angle θ_A?

(d) What is the direction of the displacement along $\vec{B}$ in terms of the angle θ_B?

(e) What is the magnitude of the net displacement?

(f) What is the direction of the net displacement in terms of the angle that the net displacement makes with the x axis?

AP® Skill Builders

7. A unit vector has a magnitude of 1.
   (a) Using graph paper and a protractor, construct a graph of two unit vectors in the first quadrant making angles of 30° and 60° with the positive x axis toward the positive y axis.
   (b) Determine the x and y components of the vectors from your vector drawings.
   (c) Calculate the sum of the squares of the components of each vector.
   (d) Explain in terms of ratios of the x and y components to the hypotenuse and trigonometric identities, why the value of this sum makes sense.

AP® Skills in Action

8. Consider the set of vectors in the figure, representing the same type of quantity in the same units. Nathan says the magnitude of the resultant vector is 7.0, and the resultant vector points in a direction 37° in the northeasterly direction. What, if anything, is wrong with his statement? If something is wrong, explain the error(s) and how to correct it (them).

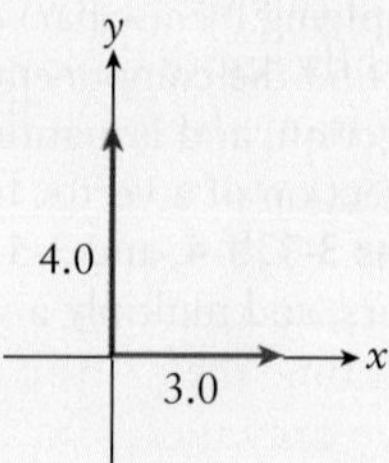
y
4.0
x
3.0

9. A boat is steered in a direction perpendicular to the bank of a river and its motor provides a velocity of 2.0 km/h in that direction. But in the time that the boat moves 200 m toward the other side of the river the current of the river has taken the boat 300 m downstream.
   (a) What is the magnitude, d, of the displacement?
   (b) What is the angle, θ, that the displacement makes perpendicular to the riverbank from which the boat was launched?
   (c) What is the speed, V, of the river?
   (d) If the river is 400 m wide, how long, T, does it take the boat to cross the river?

3-4 Motion in a plane: reference frames, velocity, and relative motion

A hawk glides over a meadow at a shallow angle, then steepens its flight path to dive onto a mouse before sailing upward again (**Figure 3-13a**). How can an ornithologist describe the hawk's motion using the language of physics?

Just as for the linear motion that we discussed in Chapter 2, a complete description of the hawk's motion describes its position at every instant of time. Such a description requires that we have a reference point, or origin, from which to measure positions. Let's choose the origin to be at the point where the ornithologist holds his observation

(a)

James Little/Alamy

(b)

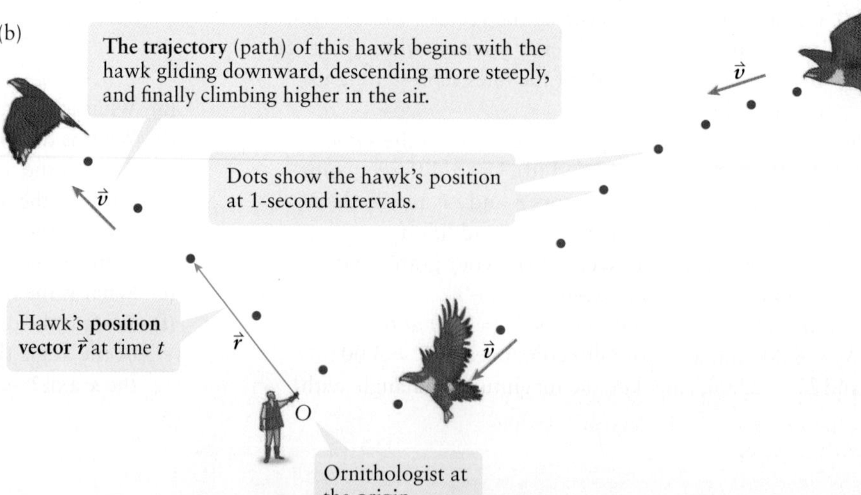
The trajectory (path) of this hawk begins with the hawk gliding downward, descending more steeply, and finally climbing higher in the air.
$\vec{v}$
Dots show the hawk's position at 1-second intervals.
$\vec{v}$
Hawk's position vector $\vec{r}$ at time t
$\vec{r}$
$\vec{v}$
O
Ornithologist at the origin

Figure 3-13 A hawk moving in a plane (a) Like all living creatures, hawks are able to move in more than one dimension. (b) The trajectory of a hawk's swooping flight. At any instant the hawk's position relative to the ornithologist is measured by a position $\vec{r}$. The origin O, where the ornithologist holds his observation instrument, represents the point (0, 0) in the x–y plane in two dimensions. This is the beginning point of each position vector.

instrument. We call an origin and the attached coordinate system (which we defined in Section 2-1) a **reference frame**. So far, and in this example, our reference frames have been attached to an origin that is not moving, which are sometimes called *fixed* or *stationary*. Later in this section, we will learn how to describe reference frames for which the origin, and therefore the coordinate system, is moving.

In this reference frame fixed on the ornithologist's instrument, at each instant we can visualize a position vector $\vec{r}$ that extends from the origin to the hawk's position at that instant. **Figure 3-13b** shows one such position vector. If we know what $\vec{r}$ is at each instant of time t, we know the path or **trajectory** that the hawk follows through space as well as what time the hawk passes through each point along that trajectory. In Figure 3-13b we have drawn dots to show the hawk's position at 1-s intervals, so this figure also shows the hawk's motion diagram.

For simplicity let's assume that our hawk moves in only two of the three dimensions of space, so it is in *two-dimensional motion*. (Motion in all three dimensions is *three-dimensional motion*.) This means that the hawk may move up and down as well as forward and back, but doesn't turn left or right. One key reason for making this simplifying assumption is that two-dimensional motion is a *lot* easier to draw than three-dimensional motion since a piece of paper is two dimensional. More importantly, many real-life motions are two-dimensional—for example, the flight of a thrown baseball and Earth's orbit around the Sun. In these situations we need only two coordinate axes, which we typically call x and y. In mathematics two such axes define a plane, so another name for two-dimensional motion is **motion in a plane**. If we use x and y to define the two perpendicular directions within the plane, we'll refer to this plane as the x–y plane.

Velocity in Two Dimensions

Just as for the one-dimensional motion that we studied in Chapter 2, understanding two- or three-dimensional motion means knowing how rapidly and in what direction the object moves along its trajectory. To see how to describe this type of trajectory, consider how the hawk shown in Figure 3-13b moves during a time interval between two instants of time that we call t_1 and t_2 (see **Figure 3-14**).

The change of the hawk's position during that interval is its displacement vector $\Delta\vec{r}$ for that time interval. This vector points from the object's position at t_1 to its position at t_2 and is the difference between the object's position vector $\vec{r}_2$ at t_2 and its position vector $\vec{r}_1$ at t_1:

$$\Delta\vec{r} = \vec{r}_2 - \vec{r}_1 \qquad (3\text{-}6)$$

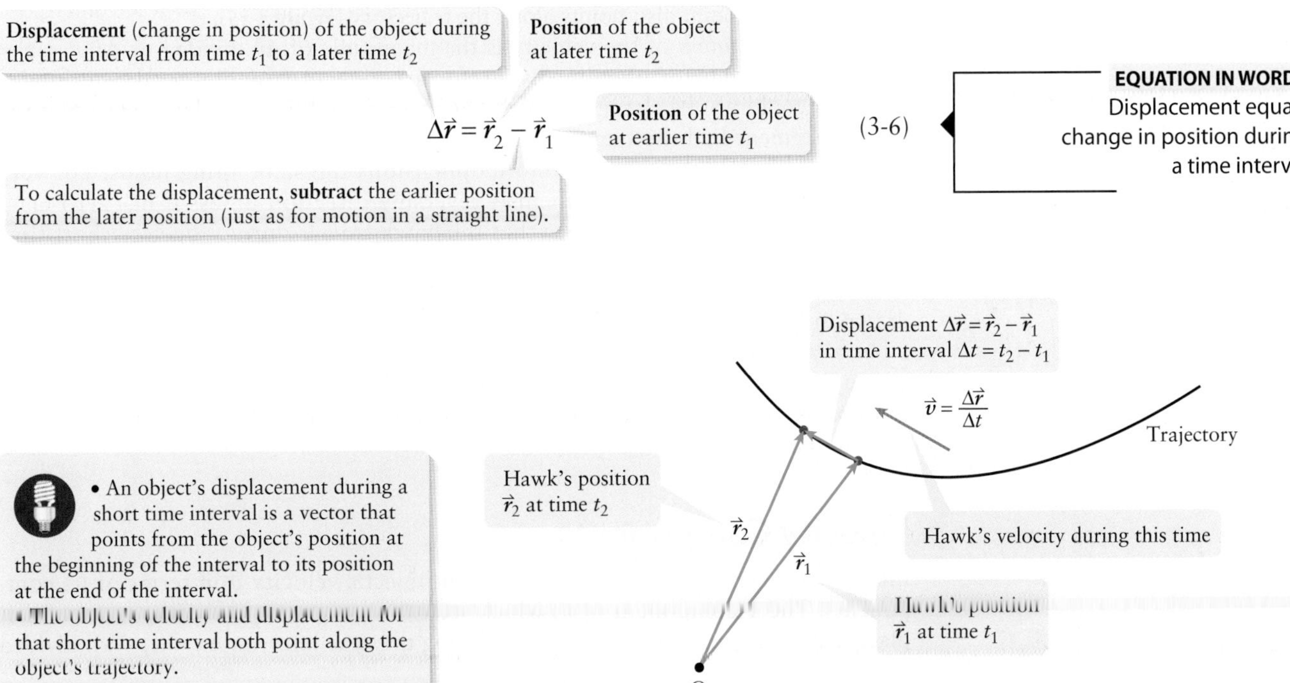

Displacement (change in position) of the object during the time interval from time t_1 to a later time t_2
Position of the object at later time t_2
Position of the object at earlier time t_1
To calculate the displacement, subtract the earlier position from the later position (just as for motion in a straight line).
EQUATION IN WORDS
Displacement equals change in position during a time interval
Displacement $\Delta\vec{r} = \vec{r}_2 - \vec{r}_1$ in time interval $\Delta t = t_2 - t_1$
$\vec{v} = \frac{\Delta\vec{r}}{\Delta t}$
Trajectory
Hawk's position $\vec{r}_2$ at time t_2
Hawk's velocity during this time
Hawk's position $\vec{r}_1$ at time t_1
$\vec{r}_2$
$\vec{r}_1$
O
• An object's displacement during a short time interval is a vector that points from the object's position at the beginning of the interval to its position at the end of the interval.
• The object's velocity and displacement for that short time interval both point along the object's trajectory.

Figure 3-14 The displacement vector and velocity vector The displacement and velocity vectors both point along the trajectory.

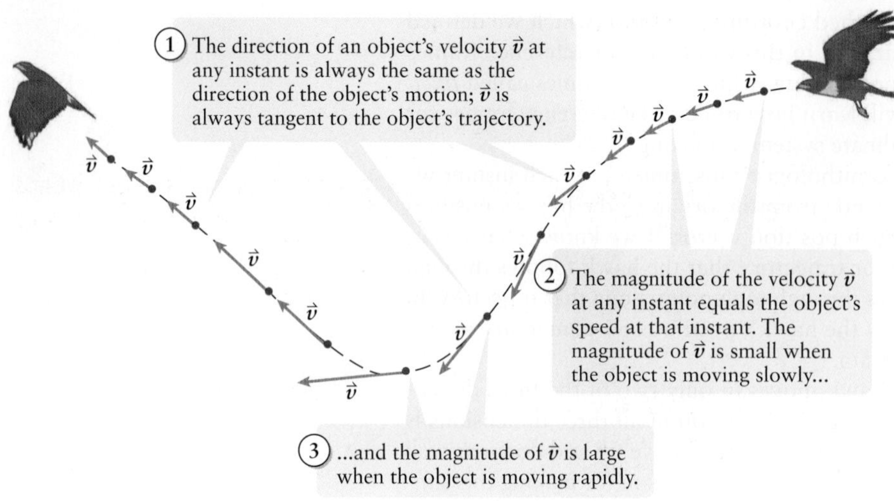

1 The direction of an object's velocity $\vec{v}$ at any instant is always the same as the direction of the object's motion; $\vec{v}$ is always tangent to the object's trajectory.
2 The magnitude of the velocity $\vec{v}$ at any instant equals the object's speed at that instant. The magnitude of $\vec{v}$ is small when the object is moving slowly...
3 ...and the magnitude of $\vec{v}$ is large when the object is moving rapidly.

Figure 3-15 Velocities at different instants along a curved trajectory The varying velocity of the swooping hawk depicted in Figure 3-13.

The direction and magnitude at any instant of an object's velocity $\vec{v}$ tells you the direction of the object's motion and the object's speed at that instant.

If the time interval $t_2 - t_1$ between these two points is very small, $\Delta\vec{r}$ in Figure 3-14 is a very small vector and points very nearly along the hawk's trajectory, even if the trajectory is curved. By analogy to what we did for one-dimensional motion in Chapter 2, we define the hawk's *instantaneous* velocity $\vec{v}$ (or just velocity for short) at a given point along the trajectory as follows:

EQUATION IN WORDS
Velocity equals displacement divided by time interval

$$\vec{v} = \frac{\Delta\vec{r}}{\Delta t} = \frac{\vec{r}_2 - \vec{r}_1}{t_2 - t_1} \tag{3-7}$$

Velocity for the object over a very short time interval from time t_1 to a later time t_2

Displacement (change in position) of the object during the short time interval

Elapsed time for the time interval

For both the displacement and the elapsed time, **subtract** the earlier value from the later value.

Equation 3-7 tells us to multiply the displacement $\Delta\vec{r}$ by $1/\Delta t$ (see Figure 3-8). Since $1/\Delta t$ is positive, the velocity $\vec{v}$ defined by Equation 3-7 points in the same direction as $\Delta\vec{r}$. So $\vec{v}$ at each point also points along the trajectory (**Figure 3-15**).

The *magnitude* v of the velocity is the magnitude of $\Delta\vec{r}$ divided by Δt: $v = |\vec{v}| = |\Delta\vec{r}|/\Delta t$. Recall that for one dimension, for a very short time interval, we showed direction could not be changing unless the magnitude of the displacement was zero, so displacement had the same magnitude as distance and thus instantaneous speed and velocity were the same. In more dimensions the same thing holds: For very short time intervals no component is reversing direction unless it has zero magnitude, so $|\Delta\vec{r}|$ is just the distance that the hawk travels during the very short time interval of duration Δt. This means v is the instantaneous speed (distance per time) of the hawk at the instant in question. That's why the arrows that represent velocity, drawn in Figure 3-15, have different lengths at different points along the trajectory: $\vec{v}$ has a large magnitude where the hawk is moving rapidly and a small magnitude where the hawk is moving slowly (compare to Figure 3-3). Like velocity and speed for motion in a straight line, the magnitude of the velocity $\vec{v}$ has units of meters per second (m/s).

Velocity Components

Like any other vector, we can describe an object's velocity $\vec{v}$ in terms of its components. The x component of $\vec{v}$, which we call v_x, equals the x component of displacement (the change in x coordinate during the time interval, $\Delta x = x_2 - x_1$) divided by the time interval $\Delta t = t_2 - t_1$; we define v_y, the y component of $\vec{v}$, in a similar way. Thus

$$\text{(a) } v_x = \frac{\Delta x}{\Delta t} = \frac{x_2 - x_1}{t_2 - t_1}$$
$$\text{(b) } v_y = \frac{\Delta y}{\Delta t} = \frac{y_2 - y_1}{t_2 - t_1} \quad (3\text{-}8)$$

x **component** of the **velocity** $\vec{v}$

y **component** of the **velocity** $\vec{v}$

x **component** of the **displacement** during a short time interval

y **component** of the **displacement** during a short time interval

Elapsed time for the time interval

EQUATION IN WORDS
Components of velocity

Equation 3-8a is the same expression that we wrote in Section 2-2 for the average velocity along the x axis in linear motion. When the time interval is short—or when the velocity is constant—this expression also gives the x component of the instantaneous velocity. Equation 3-8b is the corresponding equation for linear motion along the y axis. So our two-dimensional vector definition of velocity is a natural extension of our definition of linear motion.

AP® Exam Tip

The definitions of average or instantaneous velocity are not on the AP® Physics 1 equation sheet, but the instantaneous velocity can be determined from the equations of motion on the equation sheet.

Reference Frames and Relative Motion

At the beginning of this section we said that to describe the motion of the hawk, we had to define the reference frame we use to describe the hawk's motion. The velocities we have worked with so far have been measured with respect to the ground or some fixed coordinate system; our examples have reference frames fixed to a stationary origin.

Velocities, however, can be expressed relative to any observer. If that observer is moving, then to understand what that observer would measure, we can consider that observer to be the origin of a coordinate system that moves with that observer. This would be a moving reference frame. Every moving observer has its own reference frame that is moving at the same velocity the observer is. Thus, as measured in the reference frame of the observer, anything moving with the same velocity as the observer has a zero velocity. In other words, in your own reference frame that moves with you, you always measure your own position and velocity to be zero!

We discussed how changing our choice of the origin could change the position, but not the displacement, that describes the motion of an object. Since velocities depend on displacement, we can see that the velocity will not depend on the location of the origin. But someone in a reference frame where the origin is moving will measure a different velocity than someone in a reference frame where the origin is fixed. We can use the concept of relative motion to convert a velocity we measure or calculate from one observer's reference frame to the velocity *relative to* another observer's reference frame. We will only do this for cases where the observer is not accelerating. This velocity of the object relative to the observer is what we call the **relative velocity.**

When finding relative velocities, the easiest way to make a mistake is to get confused as to which velocity is which. We will use a common notation that helps us keep track of the velocities and ensures we are using the relationship between the velocities to find the one we want. We will introduce it in the context of a child on a train, and the observers on the platform who are watching the train pass by, as illustrated in **Figure 3-16.** First, we identify the object we are interested in and determine the possible observers. Each observer is treated as the origin of a reference frame. We pick a symbol for the object being observed, and each of the reference frames. In Figure 3-16, we are going to be measuring the child's velocity, so she is the object being observed and we will label her "C." There are two reference frames from which she could be observed, that of the platform, which we will label "P," and that of the train, which we will label "T." In **Figure 3-16a,** the velocity of the train relative to the platform (the velocity the people on the platform would measure for the train) is shown. We will label this $\vec{v}_{TP}$, using the subscripts to show this is the velocity of the train relative to the platform by the order we have chosen for the subscripts. In **Figure 3-16b,** the velocity of the child relative to the platform (the velocity people on the platform would measure for the child) is shown. We will label

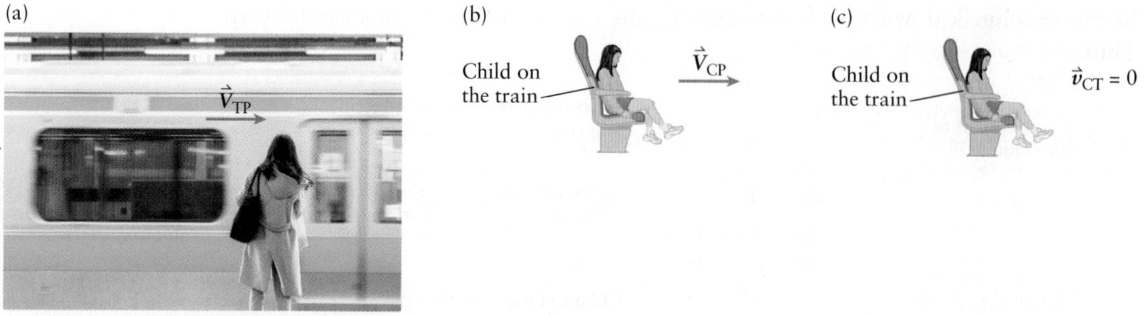

KO HONG-WEI/Alamy

Figure 3-16 Observing the motion of a child on a train In this example of relative velocity, we introduce a notation system that will help keep track of which vectors you need to add. Each velocity is denoted by two subscripts. The first subscript denotes the object for which the velocity is being measured. The second subscript is for the reference frame from which that velocity is being measured. For this example, the subscript "T" stands for train, "C" for child, and "P" for platform. (a) Someone standing on the platform would measure the speed of the train as $\vec{v}_{TP}$. (b) Someone standing on the platform would measure the speed of the child as $\vec{v}_{CP}$. (c) Someone standing on the train would measure the speed of the child as $\vec{v}_{CT}$.

this $\vec{v}_{CP}$, using the same convention for the order of the subscripts. In **Figure 3-16c,** the velocity of the child relative to the train (the velocity the people in the train would measure for the child) is shown. We will label this $\vec{v}_{CT}$. Once we have labeled our vectors in this way, we can write the relationships between them in the form

EQUATION IN WORDS
Relative Velocity

The velocity of the object, here labeled "first," is relative to another reference frame of interest, here labeled "other."

$$\vec{v}_{\text{first, last}} = \vec{v}_{\text{first,other}} + \vec{v}_{\text{other,last}} \tag{3-9}$$

The velocity of the object, here labeled "first," is relative to the reference frame, here labeled "last."

The velocity of the other reference frame of interest, here labeled "other," is relative to the reference frame labeled "last" here and in the first term.

The important thing to keep straight is that the first subscript on the velocity you want to know must match the first subscript in the series of velocities you are summing, the last subscript on the velocity you want to know must match the last subscript on the last velocity you are summing, and the subscripts in the middle must match each other. If needed, you can extend this approach to more than two reference frames, as long as you make sure the subscripts match up in this way, first and last consistent with the subscripts on the one you are trying to find, and the ones in the middle matching. If the velocities of the observer and the object they are observing fall along the same line and are thus the same vector component, then the techniques we learned in Chapter 2 are sufficient. In one dimension, the addition or subtraction of the components completely describes the direction. However, if there is an angle between the velocities then we need to apply the techniques we learned in Section 3-3 to the velocity components we defined earlier in this section. Although the example in Figure 3-16 is fairly simple, it allows us to explore several different ideas in a way that is a bit easier to visualize and comprehend. In Example 3-3, we will first practice with these basic concepts, and will then apply what we know to a more challenging problem.

Let's rewrite Equation 3-9 in terms of the symbols we chose for the object and reference frames in Figure 3-16. Start by assuming that we want to know the velocity of the child with respect to the platform. Then, our "first" and "last" subscripts are "C" and "P" and the other subscript in this case is "T." This gives us $\vec{v}_{CP} = \vec{v}_{CT} + \vec{v}_{TP}$. This is a one-dimensional problem. Designate positive to the right. If the train is traveling 10 m/s, then we get $v_{CP,x} = v_{CT,x} + v_{TP,x} = 0 + 10$ m/s $= 10$ m/s (where positive is to the right). This is pretty simple because the child is moving with the train so she has the same velocity as the train with respect to the platform. We can also imagine what we would find if the child were to stand and start walking toward the right end of the train car. If she were walking at a constant 1 m/s relative to the train, from the platform her velocity would appear to be $v_{CP,x} = v_{CT,x} + v_{TP,x} = 1$ m/s $+ 10$ m/s $= 11$ m/s. If on the other hand she were to walk toward the left of the train car she would appear to be passing the platform more slowly, and we get $v_{CP,x} = v_{CT,x} + v_{TP,x} = -1\ m/s + 10\ m/s = 9\ m/s$.

What if we want to know the velocity of the platform with respect to the child? It turns out any time we switch the order of the subscripts we are multiplying the vector by -1 ($\vec{v}_{CP} = -\vec{v}_{PC}$)! As the train passes the platform at 10 m/s to the right, to someone on the train it looks as though the platform is moving to the left at 10 m/s, $v_{PT,x} = -10$ m/s. This important relationship is also true for motion in two or three dimensions. If you take the next physics class, you will learn a more rigorous way to think about this sort of *relativity* as a prelude to learning how this way of calculating relative velocities breaks down for objects traveling near the speed of light. Even at our everyday sorts of velocities, there is in fact a relativistic correction, it is just much too small to measure! For a Global Positioning System (GPS) to be accurate, relativistic corrections, due to the satellite's motion, must be included. So, many of us rely on relativity quite often, even if we are unaware of it.

Now, let's apply this technique to determine relative velocity in two dimensions.

EXAMPLE 3-3 Relative Motion in Two Dimensions

Britta swims from a dock to a small island that is home to turtles. Her map of the lake shows that the island is 225 m from the dock, as shown in **Figure 3-17**.

When the water is calm at sunrise she can swim to the island and back in 15 min. Later, the wind usually becomes strong, causing a current that runs from left to right. In a strong wind the velocity of the current reaches 1.0 m/s. With this current, it takes Britta 15 min just to arrive at the island. (a) At what angle with respect to her path in calm water and how much faster does she need to swim to reach the island in 15 minutes when there is a current? (b) She is tired when she reaches the island and decides to swim aiming straight for the dock as she would in calm water. She just lets the current carry her along as she swims, thinking it will be easier to just walk back to the dock along the shore instead of fighting the current. How long will it take her to reach the shore and how far from the dock does she finally come ashore? Assume all velocities are constant.

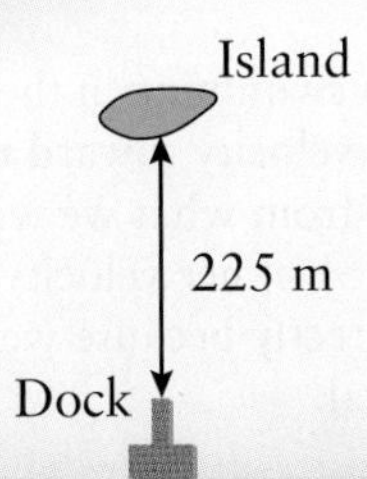

Figure 3-17 A map of Britta's lake Not drawn to scale.

Set Up

(a) In getting to the island, Britta's velocity relative to the dock when the wind is blowing is less than her velocity relative to the dock when the water is calm (it takes her twice as long to move through the displacement from the dock to the island). She must also swim in a different direction relative to the water to reach the island because of the motion of the water. To figure out her new velocity, we must first figure out her velocity in calm water, $v_{\text{calm water}}$. We will assume her speed was constant when she makes a round trip, so we can find her velocity by the displacement to the island, divided by half the time interval for the round trip. Her velocity in calm water can also be thought of as the sum of her velocity relative to the water and the water's velocity relative to the dock, because in this first case the water is not moving, so the water's velocity relative to the dock is zero.

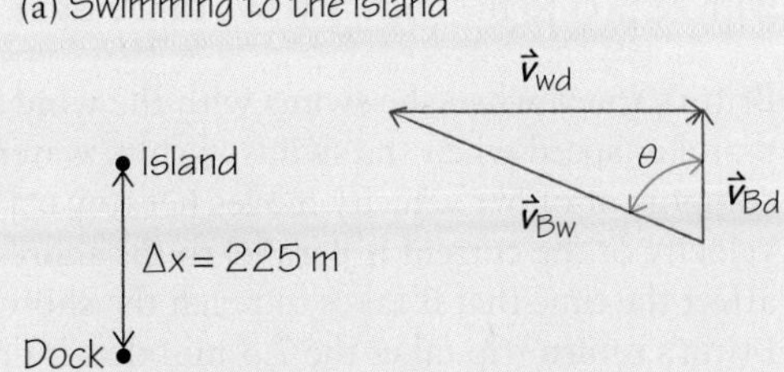

Her velocity with respect to the dock when the wind is blowing is half this size (because it takes her twice as long to get to the island), and we can unpack it to find Britta's velocity relative to the water. The vector representation of Britta's velocity relative to the dock (v_{Bd}) is the vector sum of the velocity of the water relative to the dock (v_{wd}) and Britta's velocity relative to the water (v_{Bw}), as she swims from the dock to the island, as shown in the figure.

(b) On the return trip Britta actually swims always pointed toward the shore and so her velocity relative to the water ($v_{\text{Bw, return}}$) has the same magnitude as that of her velocity in calm water. The time for Britta to reach the shore is independent of the velocity of the water relative to the dock, v_{wd}, because the directions of the shore and the current are perpendicular. Britta's velocity relative to the dock on the return is represented as $v_{\text{Bd, return}}$ in the diagram. It is the resultant of the vector addition of the velocity at which she is swimming with respect to the water ($v_{\text{calm water}}$) and the velocity of the water with respect to the dock, 1.0 m/s to the right (v_{wd}), as shown in the figure. We can predict that the angle between the vectors, θ, will not be as large because $v_{\text{Bw, return}}$, the vector adjacent to the angle, is now twice as large. The distance she must walk back along the shore to get to the dock is the magnitude of $d_{\text{walk back}}$.

(b) Swimming back from the island

Solve

(a) First, we calculate her speed in calm water.

$$v_{\text{calm water}} = \frac{450\text{ m}}{15\text{ min}}\frac{1\text{ min}}{60\text{ s}} = 0.50\text{ m/s}$$

The vector sum of her velocity when swimming in the wind and the velocity of the current is the velocity toward the island, v_{Bd} in the diagram. We know from what we were told in the problem statement that this is ½ of her velocity in calm water, but we can also calculate it directly because we know the displacement and the time interval.

$$v_{\text{Bd}} = \frac{225\text{ m}}{15\text{ min}}\frac{1\text{ min}}{60\text{ s}} = 0.25\text{ m/s}$$

We get the same answer either way, so we are in good shape to keep going! The magnitude and direction of her swim in the wind can then be calculated from this net velocity.

$$v_{\text{Bw}} = \sqrt{v_{\text{Bd}}^2 + v_{\text{wd}}^2} = \sqrt{1.0625\text{ m}^2/\text{s}^2}$$
$$= 1.0\text{ m/s}$$

To reach the island she must swim more than twice as fast as her speed in calm water and almost directly into the current.

$$\text{and}\quad \theta = \tan^{-1}\frac{1.0\text{ m/s}}{0.25\text{ m/s}} = 76°$$

(b) As she returns to the shore, her velocity with respect to the dock is again the resultant of the vector addition of the velocity at which she is swimming with respect to the water (0.50 m/s toward the shore), and the velocity of the current with respect to the dock, 1.0 m/s to the right.

$$v_{\text{Bd, return}} = \sqrt{v_{\text{Bw, return}}^2 + v_{\text{wd}}^2} = \sqrt{1.25\text{ m}^2/\text{s}^2}$$
$$= 1.1\text{ m/s}$$

$$\text{and}\quad \theta = \tan^{-1}\frac{1.0\text{ m/s}}{0.50\text{ m/s}} = 63°$$

Britta's speed when she swims with the wind is much larger than her speed when she swims in calm water. But only one component of her velocity moves her toward the shore. The velocity of the current is parallel to the shore and does not affect the time that it takes to reach the shore. This means Britta's return trip takes the 7.5 min that it did in calm water, but the point where she reaches the shore has changed.

Her displacement, $\vec{d}_{\text{swim with current}}$, is the vector sum of the displacement from the island to the dock, 225 m toward the shore, and the displacement that she must walk to return home, $d_{\text{walk back}}$, to the left along the shore as drawn. The magnitude of the net displacement vector $d_{\text{swim with current}}$, is the product of the magnitude of Britta's net velocity, $v_{\text{Bd, return}}$, and the swimming time.

$$d_{\text{swim with the current}} = (v_{\text{Bd, return}})(7.5\text{ min})\left(\frac{60\text{ s}}{\text{min}}\right)$$
$$= (1.1\text{ m/s})(450\text{ s}) = 500\text{ m}$$

Using vector subtraction or the Pythagorean theorem (because the vectors are perpendicular), we can solve for the missing leg of the trip. If we use the Pythagorean theorem, $d_{\text{walk back}}$ and d_{dock} are the sides of the triangle, and $d_{\text{swim with current}}$ is the hypotenuse.

$$d_{\text{walk back}} = \sqrt{(d_{\text{swim with current}})^2 - (d_{\text{dock}})^2}$$

$$d_{\text{walk back}} = \sqrt{(500)^2\text{ m}^2 - (225)^2\text{ m}^2} = 450\text{ m}$$

Britta decides that from now on she will swim in the morning and go sailing in the afternoon!

Reflect

In order to see if our answer makes sense, in this case we have a relatively easy check we can make for the distance she would walk back. The displacement along the shore (along the direction of the current) has to be the component of the velocity in that direction (1.0 m/s) multiplied by the time over which the motion took place (7.5 min). If you do the math, you see you do get the same answer. For the direction of the swim coming back, we mentioned that the angle between the vectors should be less because the vector next to the angle was now twice as large, and the vector opposite the angle had not changed. We did get a smaller angle, so that looks good.

Finally, our first answer just makes sense if we think about it. If you are trying to go the exact same direction as you usually would and something keeps trying to carry you sideways much faster than you would normally move, you would have to move faster and keep moving against the direction that something is trying to move you.

Extend: In the second scenario, the time for the swimmer to reach the shore was independent of the velocity of the current; the direction from the dock to the shore and direction of the current were perpendicular. In the analysis of projectile motion that we consider in Section 3.5, we will see that motion in two dimensions can always be treated as two one-dimensional motions along perpendicular directions.

NOW WORK Problems 1, 2, 4, and 5 from The Takeaway 3-4.

THE TAKEAWAY for Section 3-4

✔ If an object moves during a time interval, its displacement $\Delta\vec{r}$ is a vector that points from its position at the beginning of the interval to its position at the end of the interval.

✔ The velocity $\vec{v}$ of an object is a vector. It equals the displacement $\Delta\vec{r}$ for a very short time interval divided by the duration of the interval. The velocity always points along the object's trajectory.

✔ The reference frame chosen determines the direction and the magnitude of quantities measured by an observer in that reference frame.

✔ The observed velocity of an object results from the combination of the object's velocity and the velocity of the observer's reference frame.

Prep for the AP® Exam

AP® Building Blocks

EX 3-3

1. A small motorboat moves at a constant speed of 8.00 m/s relative to still water. The driver of the boat takes a trip in a river that has a steady current of 3.00 m/s.
   (a) How long does it take the boat to travel a distance of 1.00 km upstream?
   (b) How long does it take the boat to travel a distance of 1.00 km downstream?

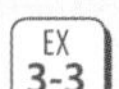
EX 3-3

2. A boat can travel at 2.50 m/s in still water. It heads directly across a river 200 m wide, with a current that is 1.35 m/s.
   (a) How far downstream is the boat pulled by the current as it crosses the river?
   (b) What is the magnitude of the displacement of the boat from its initial point on one shore to its final point on the other shore?
   (c) What direction upstream must the boat head in order to reach a point directly across from the one from which it started?
3. Cody starts at a point 6.0 km to the east and 4.0 km to the south of a location that represents the origin of a coordinate system for a map. He ends up at a point 10.0 km to the west and 6.0 km to the north of the map origin.
   (a) What was his average velocity if the trip took him 4.0 h to complete?
   (b) What is his average speed? Please explain your answer.
   (c) Cody walks to his destination at a constant rate. His friend Marcus covers the distance with a combination of jogging, walking, running, and resting so that the total trip time is also 4.0 h. How do their average velocities compare?

EX 3-3

4. A Chevy is moving at a constant velocity of 50 km/h in the positive x direction behind a Ford that is moving in the same direction at a constant velocity of 40 km/h. These velocities are both defined in a stationary reference frame, a coordinate system with an origin at the position of a particular road sign.

   In the reference frame of the Ford:
   (a) What is the velocity of the Ford?
   (b) What is the velocity of the Chevy?
   In the frame of reference of the Chevy (c) what is the velocity of the Chevy and (d) what is the velocity of the Ford?

   The cars are initially 200 m apart and both cars continue to move at constant speeds.
   (e) In the reference frame of the stationary sign how long will it be before they collide?
   (f) In the reference frame of the Ford how long will it be before they collide?
   (g) In the reference frame of the Chevy how long will it be before they collide?

5. A spider is on a floating leaf that is moving at 0.5 m/s with the river current parallel to the riverbank. The spider walks at a speed of 0.1 m/s across the leaf toward the riverbank. What is the spider's velocity relative to the riverbank? What is the spider's velocity relative to the leaf? What is the magnitude of the spider's velocity relative to the riverbank? Define y to be in the direction of the river current and x to be toward the riverbank.

AP® Skill Builders

6. A boy on a bicycle moving with a speed of 6.0 m/s throws a ball with an initial speed of 4.0 m/s in a direction to the right and perpendicular to the path of the bicycle that is moving toward a girl standing by the road. Because the boy is carrying the ball, it has the same velocity he does until he throws it relative to the bicycle and himself.
   (a) In the reference frame of the bicycle, what are the components of the initial velocity of the ball parallel and perpendicular to the path of the bicycle?
   (b) In the stationary reference frame of the girl, what are the components of the initial velocity of the ball parallel and perpendicular to the path of the bicycle?
   (c) In the stationary reference frame of the girl, determine the magnitude and the direction of the initial velocity of the ball.
   (d) In the moving reference frame of the bicycle, calculate the direction of the initial velocity of the ball.
7. When there is no wind, an airplane is flying 70 m/s with its nose pointed due north. A wind begins blowing toward the west with a constant speed relative to the ground of 18 m/s. What are the magnitude and direction of the velocity of the plane relative to the ground if the pilot does not adjust for the wind?

AP® Skills in Action

8. A skydiver has deployed her parachute and is descending with a constant speed at an angle of 30.0° relative to the horizontal. The chase team drives the pickup vehicle along a road parallel to the plane of the skydiver's motion at a speed of 50.0 km/h. From the reference frame of the pickup vehicle the velocity of the skydiver has only a single nonzero component.
   (a) Construct a vector diagram of the motion of the skydiver and the pickup vehicle in the plane of their motions.
   (b) Construct a symbolic representation of the relationship between the x and y components of the velocity of the skydiver relative to the ground, $v_{sg,x}$ and $v_{sg,y}$; the velocity of the truck relative to the ground, $v_{tg,x}$; the velocity of the skydiver relative to the truck, $v_{st,y}$; and the angle of descent, 30.0°.
   (c) Evaluate the expressions in part (b) and calculate the speed of the skydiver relative to the ground.
9. In a ride-by fruiting, a tomato was thrown by a biker at an innocent jogger who has paused to catch his breath. The biker, riding parallel to the sidewalk at a speed of 6.0 m/s relative to the ground, launched the fruit horizontally from a height of 1.2 m at an angle of 45° relative to the edge of the sidewalk, as shown in the figure from the investigative report. When the tomato was thrown, the victim was 3.0 m away along a line perpendicular to the path of the biker. The initial speed of the fruit in the reference frame of the biker, as shown in the figure, was 4.0 m/s. We know that motion in each dimension is independent, so in parts (a)–(c), just consider the horizontal motion of the tomato.

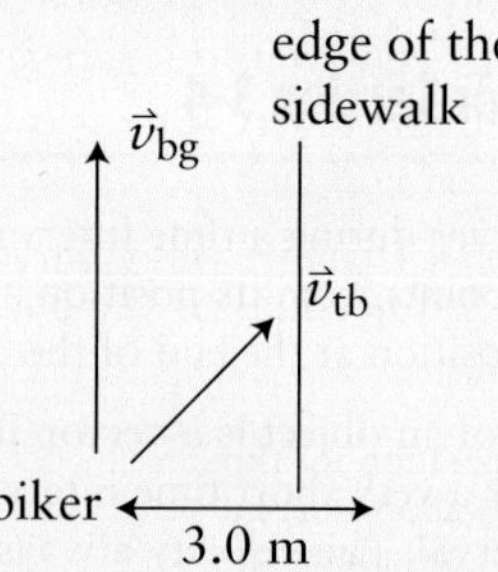

   (a) Calculate the velocity of the fruit in the frame of reference of the ground.
   (b) Calculate the time of flight.
   (c) Calculate the angle, relative to the edge of the sidewalk, of the velocity of the fruit in the x–y plane when it has traveled the 3 m in the x direction toward the jogger.
   (d) It is very unlikely that this throw will strike the intended target. Discuss two additional quantities you could calculate that would help determine if the tomato strikes the jogger.
10. A boat moves at a steady speed relative to the current. It takes the boat 200 seconds to move a distance 800 m upstream, and a total of 100 seconds to move the same distance downstream.
   (a) What is the speed of the boat relative to still water?
   (b) What is the speed of the current relative to the shore?

3-5 Motion in a plane: acceleration and projectile motion

We now have almost all the tools we need to analyze the kind of motion shown in the photograph that opens this chapter, in which an athlete leaps over a crossbar. The curved path followed by the athlete is the same as those followed by streams of water from a fountain and by clumps of molten lava (**Figure 3-18**). The velocity of an object that follows such a curved path is constantly changing. Since the velocity is changing, the object must be *accelerating*. Remember from Section 2-3 that an object is accelerating whenever its velocity is changing. In one-dimensional motion, this acceleration caused the object to either speed up or slow down, but now, we are learning to describe motion in more than one dimension. Because velocity is a vector, this means that an object accelerates if there is a change in *any* aspect of the velocity. Imagine driving a car along a country road. You speed up along a straight part of the road, then slow down when you see a curve up ahead, and finally go around the curve at a constant speed (**Figure 3-19**). In each of these three cases you are accelerating: speeding up, slowing down, and *turning*.

When your car speeds up (**Figure 3-20a**) or slows down (**Figure 3-20b**) while moving in a line, the magnitude of the velocity changes. When your car is turning (**Figures 3-20c** and **3-20d**), the *direction* of the velocity changes and so there is an acceleration, even if the car's speed doesn't change. So, to complete our ability to describe motion in a plane, we must learn to describe acceleration in more than one dimension.

(a)

Backyard Productions/Alamy

(b)

Tom Pfeiffer/VolcanoDiscovery/Getty Images

Figure 3-18 Two examples of projectile motion (a) Water in a fountain. (b) Clumps of molten lava ejected from a volcano.

The Acceleration Vector and Its Direction

The direction of a car's acceleration depends on whether it is speeding up, slowing down, turning, or a combination of these. To explore this, first recall from Section 2-3 our definition of the acceleration at a given time: the change in velocity during a very short time interval around that time divided by the duration of the time interval. For an object that moves in a plane, we have to consider the change in each component of its velocity. We again consider an infinitesimally short time interval from time t_1, when the object's velocity is $\vec{v}_1$, to a later time t_2, when the object has velocity $\vec{v}_2$. The object's acceleration $\vec{a}$ (short for *instantaneous* acceleration) at the time of this infinitesimal time interval is equal to the change in velocity divided by the duration of the interval:

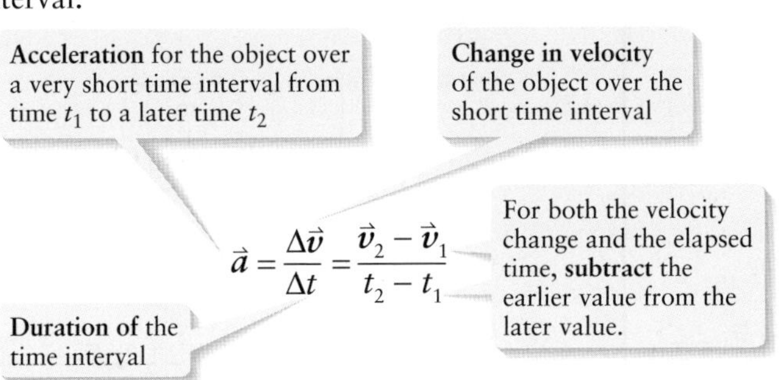
Acceleration for the object over a very short time interval from time t_1 to a later time t_2
Change in velocity of the object over the short time interval
For both the velocity change and the elapsed time, **subtract** the earlier value from the later value.
Duration of the time interval

$$\vec{a} = \frac{\Delta\vec{v}}{\Delta t} = \frac{\vec{v}_2 - \vec{v}_1}{t_2 - t_1} \qquad (3\text{-}10)$$

EQUATION IN WORDS
Acceleration equals change in velocity divided by time interval

The acceleration $\vec{a}$ points in the same direction as $\Delta\vec{v}$, and its magnitude equals the magnitude of $\Delta\vec{v}$ divided by Δt: $|\vec{a}| = |\Delta\vec{v}|/\Delta t$. Just like acceleration for linear motion, the magnitude of the acceleration has units of meters per second squared (m/s^2).

Whereas an object's velocity $\vec{v}$ always points along its trajectory, its acceleration may not. **Figure 3-21** shows how to find the direction of the acceleration $\vec{a} = \Delta\vec{v}/\Delta t$ for each situation shown in Figure 3-20. In **Figure 3-21a** a car is speeding up as it moves in one dimension. The velocity change $\Delta\vec{v}$ points in the same direction as the car's motion, and so its acceleration is likewise in the direction of motion. If instead the car slows down as it moves in one dimension, as shown in **Figure 3-21b,** $\Delta\vec{v}$ and the acceleration both point opposite to the car's motion. In **Figures 3-21c** and **3-21d** the car is going around a curve at a constant speed. In this case the acceleration is *perpendicular* to the car's trajectory and points toward the *inside* of the curve. If the car is turning to the right, as in Figure 3-21c, the acceleration is to the right; if it is turning to the left, as in Figure 3-21d, the acceleration is to the left. If an object goes around a curve at constant speed the acceleration must always be perpendicular to the velocity at any instant.

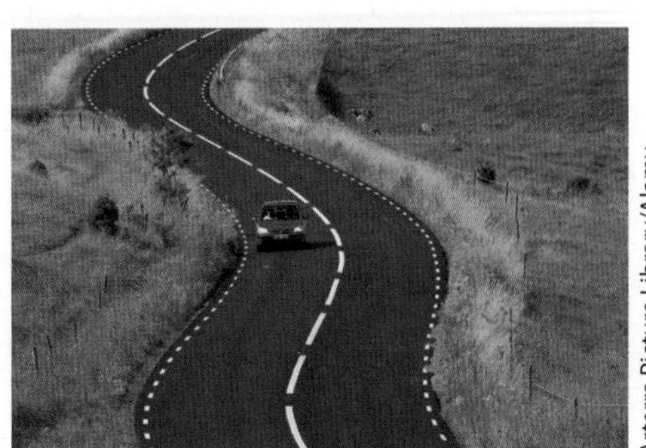

Arterra Picture Library/Alamy

Figure 3-19 An accelerating car Although the gas pedal is commonly called the accelerator, a car really has *three* "accelerators": the gas pedal, which makes the car speed up; the brake pedal, which makes it slow down; and the steering wheel, which changes its direction of motion. All three change the car's velocity $\vec{v}$.

To convince yourself that Figures 3-21c and 3-21d correctly show the direction of acceleration while turning, note that *when you are riding in a vehicle that accelerates, you feel a push in the direction opposite the vehicle's acceleration*. (What we feel when we are inside a system that is moving is

(a) Speeding up in linear motion

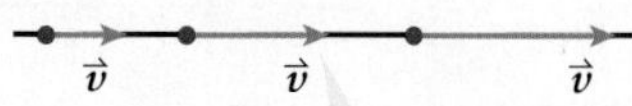

The magnitude of the velocity is changing, so the object is accelerating.

(b) Slowing down in linear motion

The magnitude of the velocity is changing, so the object is accelerating.

(c) Turning right at a steady speed

(d) Turning left at a steady speed

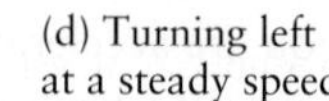

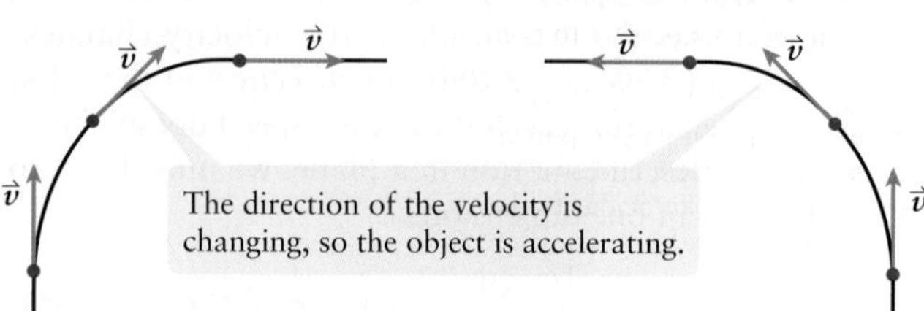

The direction of the velocity is changing, so the object is accelerating.

An object accelerates if it speeds up, slows down, or changes direction. That's because in each case the velocity changes in either its magnitude or its direction.

Figure 3-20 Four different examples of acceleration An object is accelerating if *any* aspect of its velocity is changing.

not what we think it is! We'll see why this is so in Chapter 4.) If you're riding in a car that speeds up, you feel pushed *backward* into your seat. This push is opposite to the car's *forward* acceleration (see Figure 3-21a). If the car brakes suddenly, you feel thrown *forward* in a direction opposite to the *backward* acceleration of the car (see

(a) Speeding up in linear motion

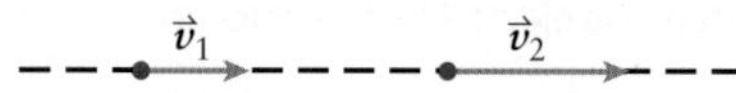

1. Add $\vec{v}_2$ and $-\vec{v}_1$ to form $\Delta\vec{v} = \vec{v}_2 - \vec{v}_1$:

$\vec{v}_2$

$\Delta\vec{v}$ $-\vec{v}_1$

2. Average acceleration $\vec{a}_{average}$ is in the same direction as $\Delta\vec{v}$, in this case, in the direction of motion.

(b) Slowing down in linear motion

1. Add $\vec{v}_2$ and $-\vec{v}_1$ to form $\Delta\vec{v} = \vec{v}_2 - \vec{v}_1$:

$\Delta\vec{v}$ $\vec{v}_2$

$-\vec{v}_1$

2. Average acceleration $\vec{a}_{average}$ is in the same direction as $\Delta\vec{v}$—in this case, opposite to the direction of motion.

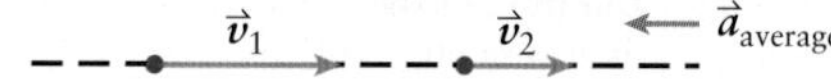

(c) Turning right at a steady speed

1. Add $\vec{v}_2$ and $-\vec{v}_1$ to form $\Delta\vec{v} = \vec{v}_2 - \vec{v}_1$:

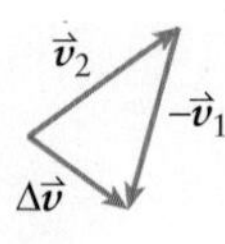

$\Delta\vec{v} = \vec{v}_2 - \vec{v}_1$

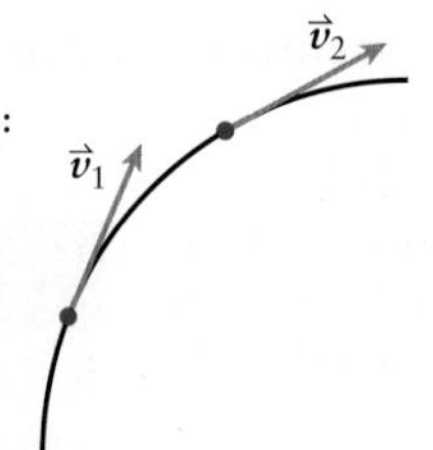

2. Average acceleration $\vec{a}_{average}$ is in the same direction as $\Delta\vec{v}$—in this case, toward the inside of the turn.

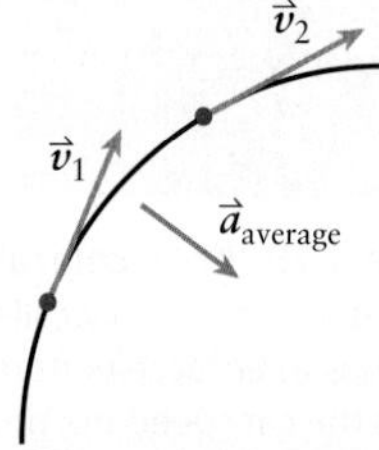

(d) Turning left at a steady speed

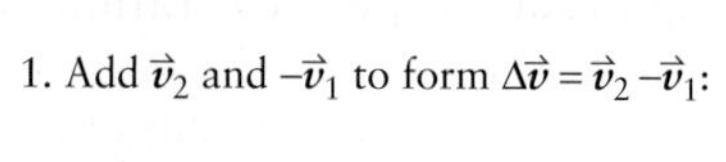

1. Add $\vec{v}_2$ and $-\vec{v}_1$ to form $\Delta\vec{v} = \vec{v}_2 - \vec{v}_1$:

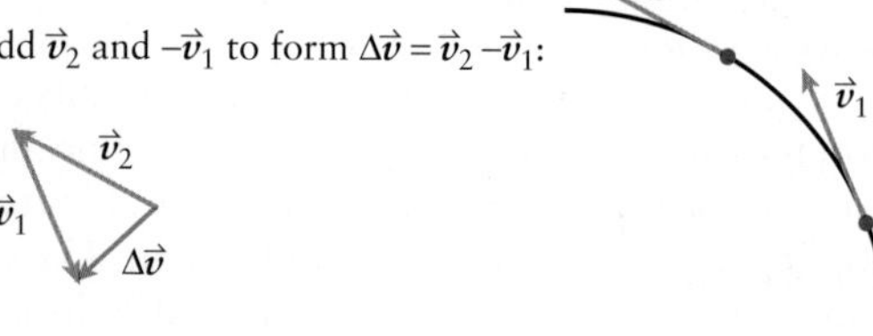

2. Average acceleration $\vec{a}_{average}$ is in the same direction as $\Delta\vec{v}$—in this case, toward the inside of the turn.

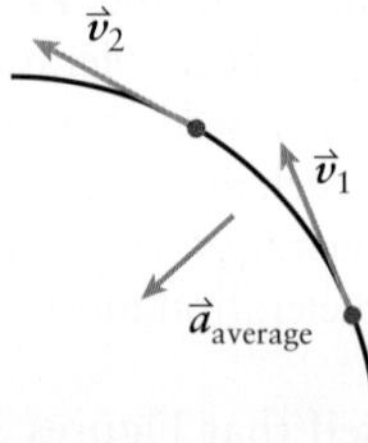

Figure 3-21 Finding the direction of the acceleration An object's acceleration $\vec{a}$ for a given brief time interval points in the same direction as the difference $\Delta\vec{v} = \vec{v}_2 - \vec{v}_1$ between the object's velocities at the end and beginning of the interval. The four examples shown here correspond to the four situations in Figure 3-20.

Figure 3-22 Finding the acceleration of a swooping hawk How to find the acceleration of an object from its change in velocity.

(1) Find the direction of $\vec{a}$ for (a) the interval from point 1 to point 2 and for (b) the interval from point 2 to point 3.

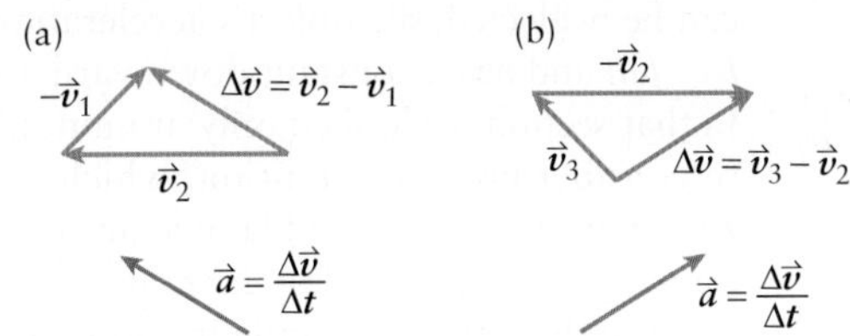

A hawk speeds up as it moves from point 1 to point 2 and slows down as it moves from point 2 to point 3.

For each interval we can break the acceleration $\vec{a}$ into a component along the trajectory and a component perpendicular to the trajectory. It is safer to do this for smaller intervals of time. For larger time intervals this is just an approximation; it gives the average acceleration over the time interval.

(2) For both intervals there is a component of acceleration that points perpendicular to the trajectory (toward the inside of the turn). This describes changes in the hawk's direction of motion.

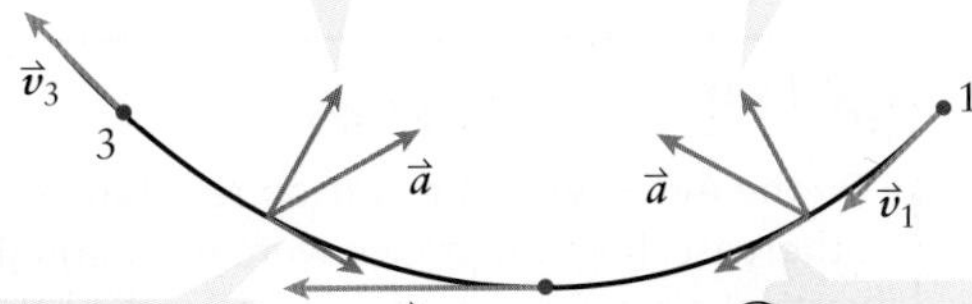

(4) For the interval from point 2 to point 3, there is a component of acceleration opposite to the direction of the hawk's motion. This describes the decrease in the hawk's speed.

(3) For the interval from point 1 to point 2, there is a component of acceleration in the same direction as the hawk's motion. This describes the increase in the hawk's speed.

If an object's speed is increasing, $\vec{a}$ has a component in the direction of motion.
If an object's speed is decreasing, $\vec{a}$ has a component opposite to the direction of motion.
If an object's direction of motion is changing (the trajectory is curved), $\vec{a}$ has a component perpendicular to its trajectory that points *toward* the inside of the turn.

Figure 3-21b). And if the car makes a turn at constant speed, you feel pushed to the *outside* of the turn. The acceleration is opposite to this direction and hence must point to the *inside* of the turn, as Figures 3-21c and 3-21d show.

Figure 3-22 shows another useful way to think about the acceleration. At a given point along an object's trajectory, $\vec{a}$ has two components: one that is perpendicular to the trajectory and one that points along the trajectory in the direction that the object is moving. The acceleration component perpendicular to the trajectory tells us about the change in the object's *direction*, and the component along the trajectory tells us about the change in the object's *speed*.

It's also often useful to express the acceleration of an object in terms of its x and y components. Use the symbols v_{1x} and v_{1y} for the velocity components of the object at time t_1 and the symbols v_{2x} and v_{2y} for the velocity components at time t_2. Then for the average acceleration, we can find a_x, the x component of $\vec{a}$, by calculating the x component of velocity change ($\Delta v_x = v_{x2} - v_{x1}$) divided by the time interval $\Delta t = t_2 - t_1$, and the y component of $\vec{a}$ equals the y component of velocity change ($\Delta v_y = v_{y2} - v_{y1}$) divided by Δt:

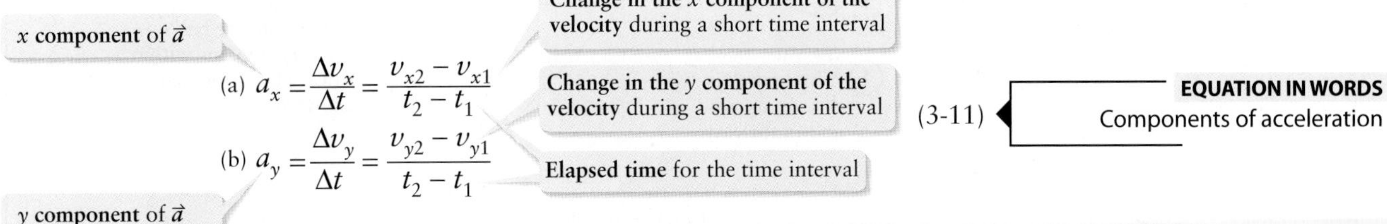

$$\text{(a)}\ a_x = \frac{\Delta v_x}{\Delta t} = \frac{v_{x2} - v_{x1}}{t_2 - t_1}$$

$$\text{(b)}\ a_y = \frac{\Delta v_y}{\Delta t} = \frac{v_{y2} - v_{y1}}{t_2 - t_1} \qquad (3\text{-}11)$$

EQUATION IN WORDS
Components of acceleration

In the next section we'll use the component description of velocity and acceleration to help us analyze the motion of a *projectile*—an object moving through the air, which we neglect, treating the object as if it is accelerated by gravity alone.

Defining Projectile Motion

If the object is moving through the air at a relatively slow speed so that air resistance can be neglected, the object's acceleration is due to gravity alone. Then the object is in *free fall* and has a constant downward acceleration, just as we discussed in Section 2-6. In that section we looked only at motion that was straight up and down; now we want to consider **projectile motion**, which is free fall with both vertical motion *and* horizontal motion. You could launch an object straight up, so the free-fall motion could be thought of as the simplest example of projectile motion. We will just call that free fall, and use the term "projectile motion" to describe free-fall motion with an initial horizontal velocity component. (We often refer to an object undergoing this kind of motion as a **projectile**.) Just as for vertical free fall, the downward acceleration of a projectile has a magnitude g, and is always directed downward. We'll continue to use the value $g = 9.80 \text{ m/s}^2$.

AP® Exam Tip

On the AP® Physics 1 exam, air resistance is assumed to be negligible unless stated otherwise. If air resistance were not negligible, the acceleration's magnitude in the vertical direction would be smaller than the acceleration due to gravity, and there would also be an acceleration in the horizontal direction, opposing the motion of the object. We will consider the effects of air resistance in Chapter 5.

For projectile motion, we will always place one axis of our coordinate system along the direction of the acceleration due to gravity. We can always make the direction of the component perpendicular to gravity the other axis of our coordinate system. So projectile motion is *two*-dimensional: The motion lies in a plane. To see why, think about tossing a basketball upward and toward the east (**Figure 3-23**). When the ball leaves your hand, its velocity vector has a horizontal component toward the east, a vertical component that points upward, and zero components along the north–south direction. Because the acceleration vector points straight downward, there are no north–south or east–west components of acceleration. Therefore, the north–south velocity doesn't change—this component of velocity starts at zero and remains at zero.

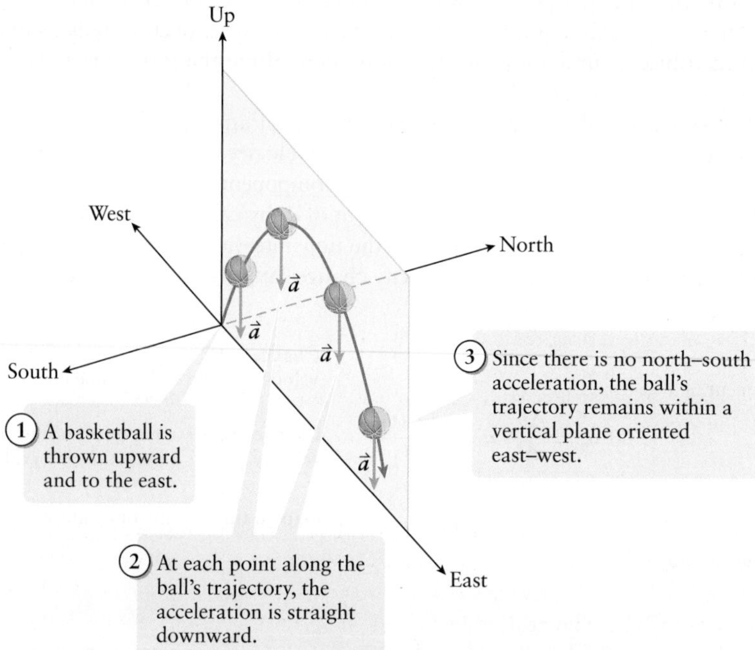

Up
West
North
South
East
$\vec{a}$
1 A basketball is thrown upward and to the east.
2 At each point along the ball's trajectory, the acceleration is straight downward.
3 Since there is no north–south acceleration, the ball's trajectory remains within a vertical plane oriented east–west.

Figure 3-23 Why projectile motion is two-dimensional motion The trajectory of a projectile is in a vertical plane because the acceleration due to gravity points downward.

Figure 3-24 Projectile motion
Acceleration components along and perpendicular to the trajectory.

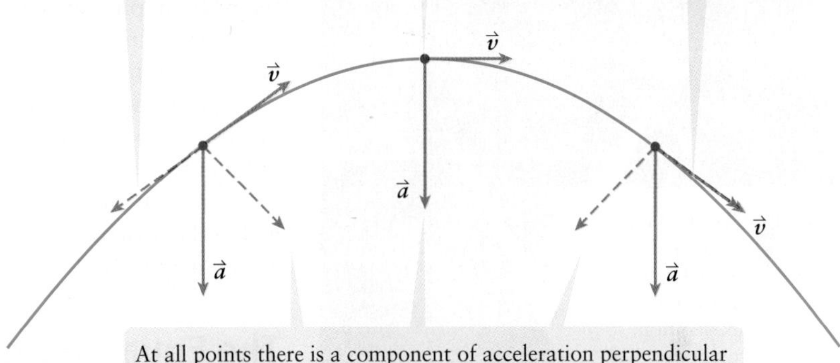
There is no component of acceleration along the direction of motion: The speed of the projectile has its minimum value (its velocity is purely horizontal), only its vertical direction of motion is changing.
There is a component of acceleration opposite to the direction of motion: The projectile is slowing down.
There is a component of acceleration in the direction of motion: The projectile is speeding up.
$\vec{v}$
$\vec{v}$
$\vec{v}$
$\vec{a}$
$\vec{a}$
$\vec{a}$
At all points there is a component of acceleration perpendicular to the trajectory: The direction of motion is changing.

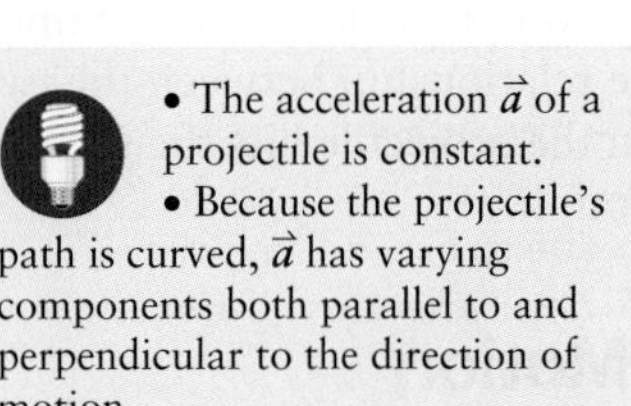

• The acceleration $\vec{a}$ of a projectile is constant.
• Because the projectile's path is curved, $\vec{a}$ has varying components both parallel to and perpendicular to the direction of motion.

The *east* component also stays constant. So the basketball always remains in a vertical plane that's oriented east–west, as Figure 3-23 shows.

Figure 3-24 shows the acceleration $\vec{a}$ at several points along a projectile's trajectory as well as the components of $\vec{a}$ along and perpendicular to the trajectory. There is a perpendicular component of $\vec{a}$ at *all* points, so the projectile's direction of motion is changing at all points, and the trajectory is curved all along its length. When the projectile is ascending there is a component of acceleration opposite to its motion, so the projectile is slowing down. When the projectile is descending there is an acceleration component in the direction of motion, so the projectile is speeding up. The only point at which the speed is instantaneously *not* changing is at the highest point of the trajectory, where the acceleration vector is entirely perpendicular to the trajectory. At this point the projectile has its *minimum* speed—its velocity in the vertical direction is zero, and the downward acceleration is changing the projectile's vertical direction of motion from upward to downward.

WATCH OUT !

A projectile is a special case.

Not every object in flight can be regarded as a projectile. As an example, a gliding bird is definitely *not* a projectile: The air exerts lift and drag forces on the bird that can't be ignored because these forces are comparable to the force of gravity. A batted baseball also feels a substantial drag force exerted by the air because of its high speed (much greater than the speed of the basketball in Figure 3-23), so this object isn't really a projectile either. Nonetheless, studying projectile motion is an important first step toward understanding these more complex kinds of motion.

Remember that to be considered a projectile, an object must be in free fall (accelerating only due to gravity). We can see that the lava chunks and the water droplets in Figure 3-18a are relatively slow-moving objects that cannot power themselves once they are in flight. The athlete in the opening photograph of this chapter is more complicated. First, an athlete can move herself; however, once she has left the ground,

WATCH OUT !

What you see depends on your own motion.

When you are a passenger in a car and you drop an object, you notice that it falls straight into your lap. You might not have time to notice that it is accelerating downward; but it is, because it's in free fall. However, if you were observing from outside the car, you would see a very different trajectory: The object would be in projectile motion. Even though it was not "launched," it has the initial horizontal velocity you had when you let it go. It makes sense that it must be moving forward, or it would not stay in front of you as it fell, since you are moving forward. As we saw in the last section, observing from the ground outside the car or from inside the car is considered observing from a particular reference frame. The reference frame changes the description of the motion of an object or system, but not its actual motion, as you can recover that information by including information about the motion of the reference frame itself.

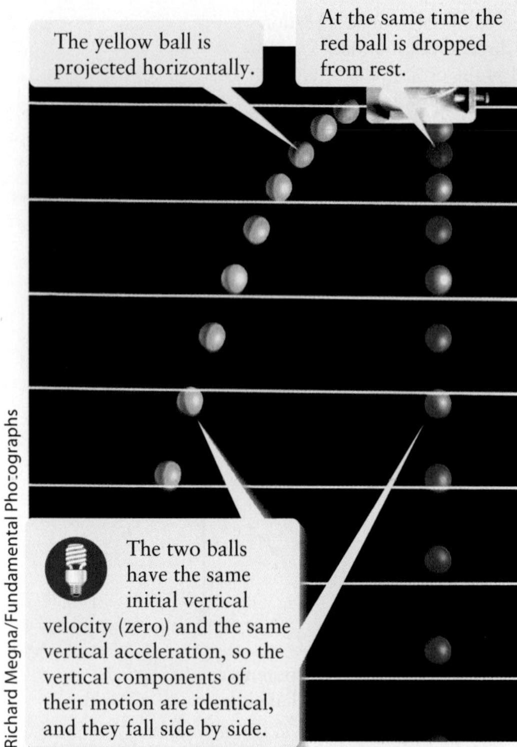

The yellow ball is projected horizontally.
At the same time the red ball is dropped from rest.

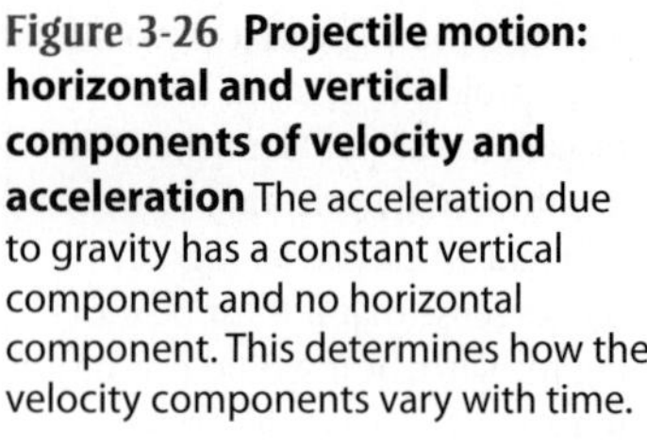

The two balls have the same initial vertical velocity (zero) and the same vertical acceleration, so the vertical components of their motion are identical, and they fall side by side.

Richard Megna/Fundamental Photographs

Figure 3-25 Projectile motion: The vertical and horizontal components of motion are independent Although these two balls have different initial horizontal velocities, their initial vertical velocities—and hence their subsequent vertical motions—are identical.

she no longer can push on the ground to cause herself to accelerate, so gravity is the only cause of her acceleration. The athlete is obviously not a good fit for the object model because her limbs are moving in different trajectories. The athlete's center of mass is fairly close to her navel. As she jumps she moves her arms and legs such that as her center of mass moves as a projectile her body clears the bar. You can see this by looking at her shape in each instant captured in the photograph. We will learn more about center of mass in the next chapter, but we can see that the center of mass of the athlete does follow the same trajectory as any projectile, so, if we are interested only in the motion of her center of mass, then we can treat her as an object. As we begin our study of the causes of motion in the next chapter, we will find this is generally true.

The multiflash photograph in **Figure 3-25** helps to show the horizontal and vertical motions of a projectile are independent.

As we mentioned in Section 3-1, we can treat two-dimensional motion as just two one-dimensional problems that we learned to solve in Chapter 2, so we can treat projectile motion problems as two simpler problems we already know how to do. The relationship between the two is that its horizontal motion also ends when the projectile hits the ground, a time we can find from just the vertical motion.

The Equations of Projectile Motion

Let's be more quantitative about how a projectile's velocity changes during its flight. As we develop some equations, let's use the x and y axes to define the plane in which the projectile moves, with the positive x axis pointing in a horizontal direction, and the positive y axis pointing vertically upward (**Figure 3-26**).

We could label our axes differently using different symbols to denote the horizontal and vertical axes, but that would just change some labels and not the overall coordinate system. The important point is that we choose one axis to be vertically aligned with the acceleration due to gravity, and one to be horizontal, aligned with the horizontal component of the motion. We can pick either direction on either axis to be positive. As long as we choose our coordinate system and carefully follow it, we will ultimately get the same results. With

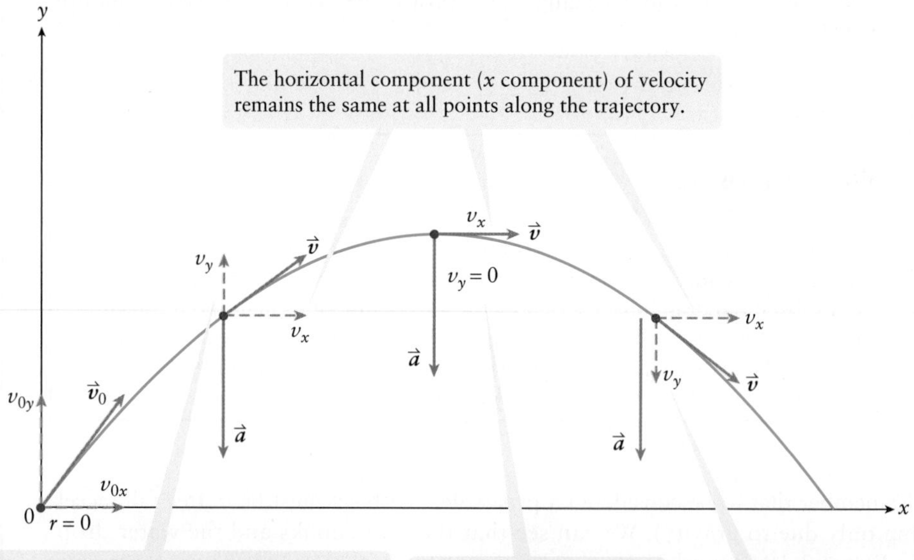

Figure 3-26 Projectile motion: horizontal and vertical components of velocity and acceleration The acceleration due to gravity has a constant vertical component and no horizontal component. This determines how the velocity components vary with time.

the choice of axes described, the components of the constant, vertically downward acceleration $\vec{a}$ are

x **component** of the **acceleration** $\vec{a}$ for **projectile motion**

Zero value means that the *x* component of the velocity is constant.

$$\text{(a)}\quad a_x = 0$$
$$\text{(b)}\quad a_y = -g \qquad (3\text{-}12)$$

y **component** of the **acceleration** $\vec{a}$ for **projectile motion**

Constant negative value means that the change in the *y* component of the velocity is always becoming more downward, and that change happens at a steady rate.

EQUATION IN WORDS
Components of the acceleration in projectile motion for a coordinate system in which up is defined as positive

WATCH OUT !

Projectile motion means constant vertical acceleration.

Just as for vertical free fall, it's important to remember that the vertical, *y*, component of acceleration is the same ($a_y = -g$, for positive *y* defined upward) at every point in the motion of the projectile.

Because both the *x* and *y* components of acceleration are constant, we can use our results from Chapter 2 to easily write the equations that show how the *x* and *y* components of the projectile's velocity $\vec{v}$ vary with time. We'll let $t = 0$ represent the instant when the projectile begins its flight. At this moment the projectile has *x* component of velocity v_{0x} and *y* component of velocity v_{0y}. (The 0 in the subscripts reminds us that these are the components at $t = 0$.) We found in Chapter 2 that if the *x* component of acceleration is constant, the *x* component of velocity v_x as a function of time is $v_x = v_{0x} + a_x t$. The same equation rewritten for the *y* direction is $v_y = v_{0y} + a_y t$. If we substitute the values $a_x = 0$ and $a_y = -g$ for the chosen coordinate system, we get

x **component** of an object's **velocity** $\vec{v}$ at time *t* for **projectile motion**

x component of the object's velocity at $t = 0$

$$\text{(a)}\quad v_x = v_{0x}$$
$$\text{(b)}\quad v_y = v_{0y} - gt \qquad (3\text{-}13)$$

Acceleration due to gravity (*g* is positive)

Time at which the object has velocity components v_x and v_y

y **component** of an object's **velocity** $\vec{v}$ at time *t* for **projectile motion**

y component of the object's velocity at $t = 0$

EQUATION IN WORDS
Velocity, acceleration, and time for projectile motion

WATCH OUT !

Different symbols can be used to indicate the *initial* state.

It is not unusual to see the subscript "i" for "initial" used instead of 0. If you are just given an initial value and Δt and have not defined $t = 0$, i is actually more precise. However, you will never be tested on this distinction. It is just important that you be able to recognize the appropriate variable for any calculation you wish to make.

Equation 3-13a tells us that *in projectile motion the horizontal component of velocity doesn't change*. The vertical component of velocity *does* change with time, however; as Equation 3-13b shows, the *y* component of velocity decreases at a steady rate. If v_{0y} is positive, v_y decreases to zero (at the high point of the trajectory) and then, as shown in Figure 3-26, becomes increasingly negative (that is, the projectile moves downward at an ever-faster rate).

We can use another equation from Chapter 2 to write the equations for the projectile's *x* and *y* coordinates at any time *t*. We let x_0 and y_0 be the projectile's coordinates at time $t = 0$. Because the *x* component of acceleration a_x is constant, we know that $x = x_0 + v_{0x}t + (1/2)a_x t^2$ from Equation 2-9 in Section 2-4. The same equation rewritten for the *y* direction is $y = y_0 + v_{0y}t + (1/2)a_y t^2$. For our chosen coordinate system $a_x = 0$ and $a_y = -g$, so these equations become

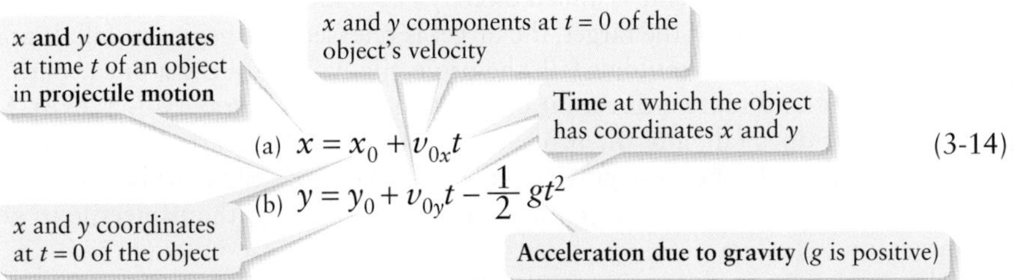

(3-14)

EQUATION IN WORDS
Position, acceleration, and time for projectile motion

The curved path that the projectile follows in accordance with Equations 3-14 is called a **parabola**. You can see this shape in the photograph that opens this chapter and in Figure 3-18.

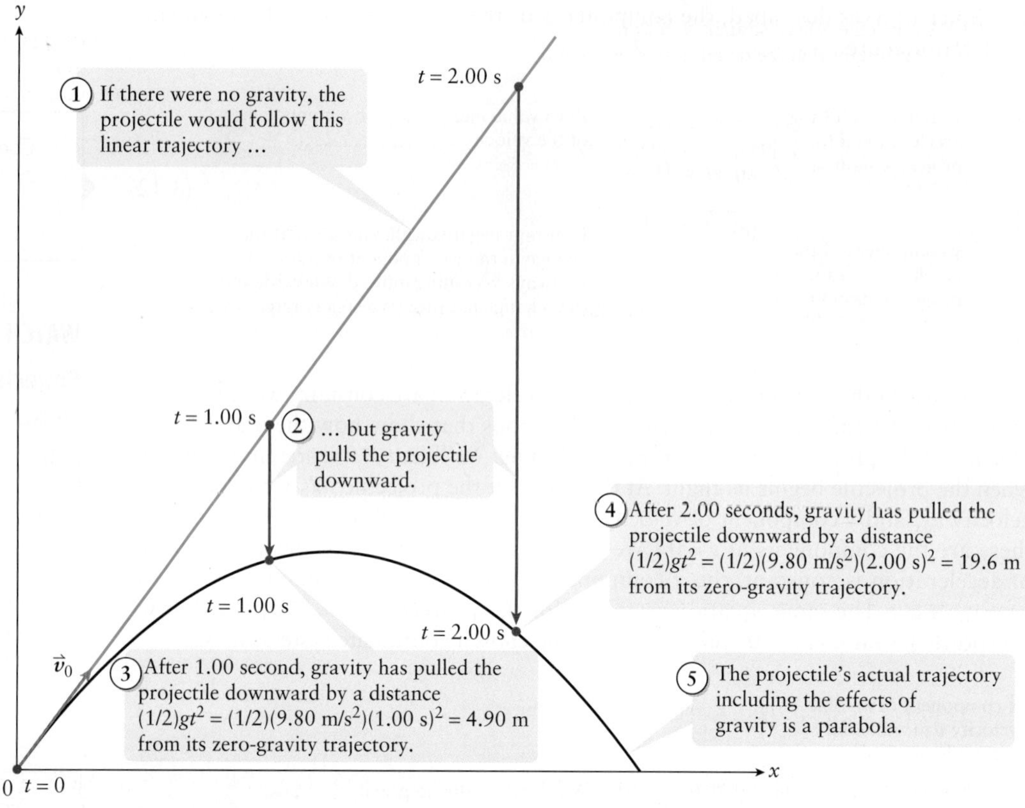

y
x
0 t = 0
$\vec{v}_0$
t = 2.00 s
t = 1.00 s
t = 1.00 s
t = 2.00 s
1 If there were no gravity, the projectile would follow this linear trajectory ...
2 ... but gravity pulls the projectile downward.
3 After 1.00 second, gravity has pulled the projectile downward by a distance $(1/2)gt^2 = (1/2)(9.80\ \text{m/s}^2)(1.00\ \text{s})^2 = 4.90$ m from its zero-gravity trajectory.
4 After 2.00 seconds, gravity has pulled the projectile downward by a distance $(1/2)gt^2 = (1/2)(9.80\ \text{m/s}^2)(2.00\ \text{s})^2 = 19.6$ m from its zero-gravity trajectory.
5 The projectile's actual trajectory including the effects of gravity is a parabola.

Figure 3-27 Projectile motion: with and without gravity If there were no gravity, a projectile would follow a linear path. The effect of gravity is to pull a projectile downward from this straight line by a distance $(1/2)gt^2$ in a time *t*.

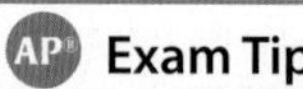
Exam Tip

A common student error is to confuse quantities in the horizontal direction with quantities in the vertical direction. Be careful to treat the two directions independently, with the only common variable being time *t*.

You probably recognize Equations 3-13b and 3-14b from our discussion of free fall in Section 2-6; they're exactly the same as Equations 2-13 and 2-14, respectively. So we can think of projectile motion as a combination of horizontal motion with constant velocity and vertical free fall. As we promised, the horizontal and vertical motions of a projectile are independent: The expressions for v_x and x in Equations 3-13 and 3-14 involve only x quantities, while the expressions for y and v_y involve only y quantities. This was illustrated in Figure 3-25. The relationship is that a projectile's horizontal motion also ends when the projectile hits the ground, a time we can find using just the projectile's vertical motion.

Another way to think about projectile motion is to first imagine what would happen if we could turn gravity off so that $g = 0$. Then a thrown projectile would have zero acceleration, its velocity would remain constant, and its trajectory would be a straight line. However, a projectile's trajectory curves downward because gravity *does* exist. Equation 3-14b tells us that after a time t, the projectile has been dragged downward by a distance $(1/2)gt^2$ from the linear path it would follow in the absence of gravity (**Figure 3-27**).

The demonstration depicted in **Figure 3-28** illustrates this way of thinking about projectile motion. A projectile is aimed directly at a hanging target. Just as the projectile starts its flight toward the target, the target is released and begins to fall. If gravity did not exist, the target wouldn't fall, the projectile would follow a linear path, and the projectile would score a direct hit on the target. Because gravity does exist, however, the target does fall, and the projectile follows a curved path. But the projectile still scores a direct hit because gravity pulls *both* the projectile and the target downward by the *same* distance $(1/2)gt^2$ in the time t it takes for the projectile to get to the target.

Thus we have two alternative ways of thinking about projectile motion: (1) a combination of horizontal, constant-velocity motion and vertical free fall or (2) straight-line motion in the direction of the initial velocity, but pulled downward by a distance $(1/2)gt^2$ in time t. As we'll see in the next section, both of these interpretations can be useful when solving problems with projectile motion.

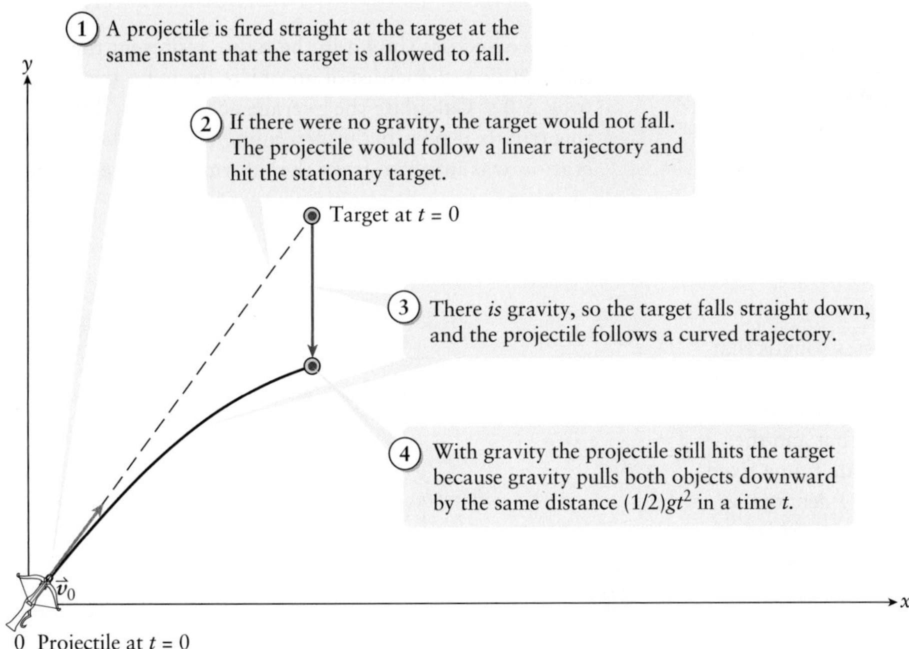

1 A projectile is fired straight at the target at the same instant that the target is allowed to fall.
2 If there were no gravity, the target would not fall. The projectile would follow a linear trajectory and hit the stationary target.
Target at $t = 0$
3 There *is* gravity, so the target falls straight down, and the projectile follows a curved trajectory.
4 With gravity the projectile still hits the target because gravity pulls both objects downward by the same distance $(1/2)gt^2$ in a time t.
y
x
$\vec{v}_0$
0 Projectile at $t = 0$

Figure 3-28 Hitting a dropped target This experiment demonstrates that all objects in free fall have the same downward acceleration.

THE TAKEAWAY for Section 3-5

✔ The acceleration $\vec{a}$ of an object is also a vector. It equals the change in velocity $\Delta\vec{v}$ for a very short time interval divided by the duration of the interval. The direction of the acceleration depends on how the object's speed is changing and on whether the object is changing direction.

✔ A projectile is an object that is launched and then falls freely (that is, does not have a motor and moves without air resistance).

✔ Only gravity accelerates a projectile, so the acceleration is downward and has the same magnitude throughout the motion.

✔ Projectile motion can be thought of as a combination of horizontal motion with constant velocity and vertical free fall. These components of motion, like any two-dimensional motion, are independent, related only by time.

✔ Projectile motion can also be thought of as linear motion in the direction of the initial velocity with an additional downward motion due to gravity.

Prep for the AP® Exam

AP® Building Blocks

1. Jack drops a ball from rest from the top floor of a building and, at the same time, Jill throws a ball horizontally from the same location. Explain why both balls hit the level ground at the same instant. (Neglect any effects due to air resistance.)
2. An object travels with a constant acceleration for 10.0 s. The vectors in the figure represent the final and initial velocities.

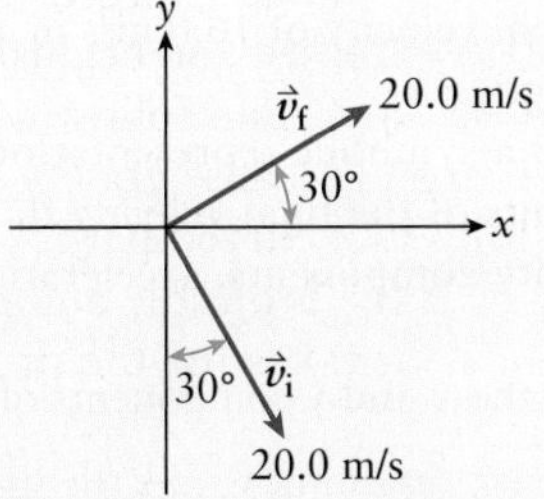

(a) Calculate the x and y components of the acceleration.
(b) Express the x and y components of the velocity as a function of time during this time interval.
(c) Calculate the times at which the object momentarily stops (if any) in its motion in the x direction and in the y direction.
(d) Explain why the initial values of the x and y components of the velocity and the times calculated in part (c), or given, are sufficient to construct graphs of the horizontal and vertical components of the velocity.
(e) On the same set of axes carefully plot the x component of the velocity versus time and the y component of the velocity versus time.

3. An object is undergoing projectile motion as shown from the side in the figure below. Assume the object starts its motion at ground level. For the five positions shown, draw to scale vectors representing the magnitudes of (a) the x component of the velocity, (b) the y component of the velocity, and (c) the acceleration. Neglect any effects due to air resistance.

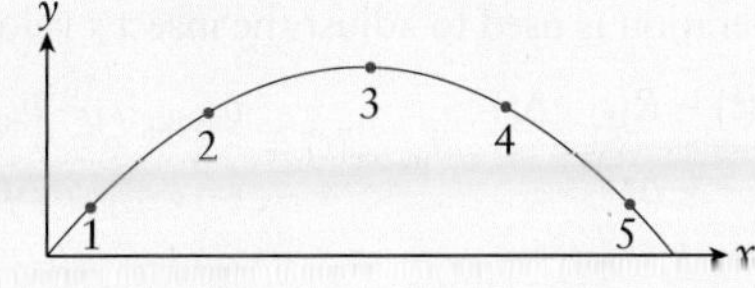

4. 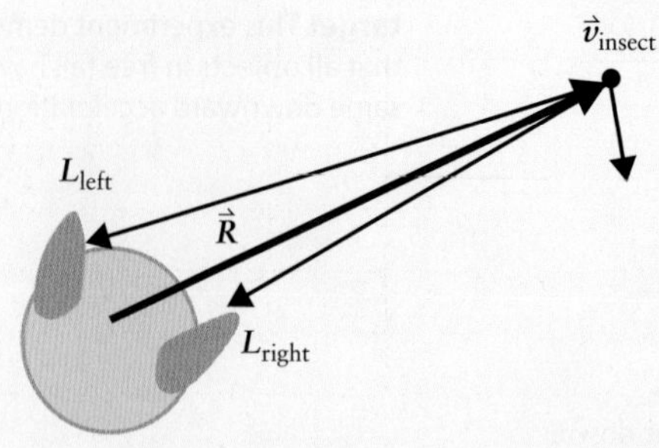

To navigate by echolocation, a bat processes sounds that it transmits that are reflected back to it from surrounding objects. At the instant shown in the figure, an insect is located at $\vec{R}(t)$ relative to the bat. Both the bat and the insect are in motion. Relative to a coordinate frame fixed on the bat, the insect has a velocity at time t denoted by $\vec{v}_{insect}(t)$. Models that are consistent with observed behavior claim that movement of the bat in the plane formed by the target and the two ears of the bat is guided by the loudness, L, of signals received at the left and right ears. The bat emits sound in pulses with time intervals Δt that decrease as the bat nears the insect. As each reflected pulse is received the sonic landscape is updated. Four students present possible models to predict the trajectory that intercepts the insect as a repetition of three steps. For each of the students' models below (one is correct), in a sentence explain why you believe the model is correct or incorrect.

(A) Current and last insect positions are used to calculate the bat's velocity. Current and last insect velocities are used to calculate the insect's acceleration. The insect acceleration is then used to adjust the bat's velocity.

$$\vec{v}_{bat}(t) = \frac{\vec{R}(t) - \vec{R}(t - \Delta t)}{\Delta t}; \ \vec{a}_{insect}(t) = \frac{\vec{v}_{insect}(t) - \vec{v}_{insect}(t - \Delta t)}{\Delta t}$$

(B) Current and last insect positions are used to calculate the bat's velocity. Current and last bat velocities are used to calculate the bat's acceleration. The bat acceleration is then used to adjust the bat's velocity.

$$\vec{v}_{bat}(t) = \frac{\vec{R}(t) - \vec{R}(t - \Delta t)}{\Delta t}; \ \vec{a}_{bat}(t) = \frac{\vec{v}_{bat}(t) - \vec{v}_{bat}(t - \Delta t)}{\Delta t}$$

(C) Current and last insect positions are used to calculate the insect's velocity. Current and last insect velocities are used to calculate the insect's acceleration. The insect's acceleration is used to adjust the bat's velocity.

$$\vec{v}_{insect}(t) = \frac{\vec{R}(t) - \vec{R}(t - \Delta t)}{\Delta t}; \ \vec{a}_{insect}(t) = \frac{\vec{v}_{insect}(t) - \vec{v}_{insect}(t - \Delta t)}{\Delta t}$$

(D) Current and last insect positions are used to calculate the insect's velocity. Current and last insect velocities are used to calculate the insect's acceleration. The insect's acceleration is used to adjust the insect's velocity.

$$\vec{v}_{insect}(t) = \frac{\vec{R}(t) - \vec{R}(t - \Delta t)}{\Delta t}; \ \vec{a}_{insect}(t) = \frac{\vec{v}_{insect}(t) - \vec{v}_{insect}(t - \Delta t)}{\Delta t}$$

5. The two vectors shown in the figure represent the initial and final velocities of an object during a trip that took 5.0 s. Calculate the average acceleration during this trip. Is it possible to determine whether the acceleration was uniform from the information given in the problem?

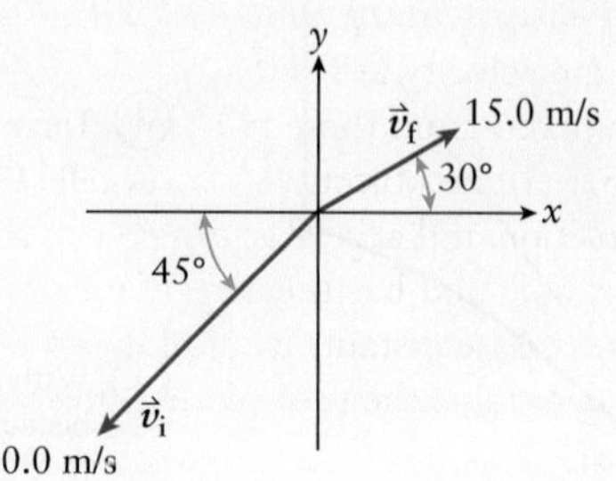

AP® Skill Builders

6. A spear is thrown with an initial velocity of 20.0 m/s at an angle of 30.0° relative to the horizontal, and with an acceleration in the vertical direction of $-10.0\ \text{m/s}^2$. Treat the spear as an object, only worrying about the motion of its center of mass.
   (a) Calculate the horizontal and vertical components of the initial velocity.
   (b) Express the horizontal and vertical positions of the spear as functions of time.
   (c) Express the vertical position of the spear as a function of horizontal position.
   (d) Construct a diagram in which the x–y coordinate system is rotated so that the x axis lies along the initial velocity vector, 30.0° relative to the horizontal.
   (e) Calculate the components of the acceleration vector in the rotated coordinate system.
   (f) Express the x and y components of the spear as functions of time in the new coordinate system.
   (g) Calculate a few pairs of x and y values in the new coordinate system and plot these points on the diagram constructed in part (d).
   (h) Use your result in part (g) to justify a claim regarding the dependence of the motion on the coordinate frame used to describe it.
7. An object initially moving in the x–y plane has a constant acceleration of $2.00\ \text{m/s}^2$ in the $-x$ direction for 2.70 s, attaining a final velocity of 16.0 m/s in a direction 45.0° from the $+x$ axis.
   (a) Construct a symbolic representation of the x and y components of the final velocity in terms of the initial velocity components, acceleration components, and time.
   (b) Calculate the x and y components of the initial velocity.
   (c) Calculate the initial speed of the object.

AP® Skills in Action

8. The equation of a parabola is $y = dx^2 + cx + b$, where d, c, and b are constants. In your study of linear motion in Chapter 2 you investigated two types of motion: without acceleration, $x = x_0 + v_{0x}t$ and with constant acceleration, a, $y = y_0 + v_{0y}t + at^2/2$, where v_{0x} and v_{0y} are the components of the velocity at $t = 0$.
   (a) Eliminate t from these two equations and show that the path of an object with no acceleration along the x direction and acceleration along the y direction is a parabola and has the form $y = cx + dx^2$.
   (b) Express the constants c and d in terms of the factors that determine the properties of the motion, v_{0x}, v_{0y}, and a.
   (c) Using graph paper or a sketch of the graph displayed on your calculator, interpret the dependence of the shape of the parabola in terms of y_{max} and x_{max} by comparing curves with these parameters. Complete the table below. Label each curve (A–F) in the sketch or graph. For each use $a = -g = -10$ m/s^2.

Curve	v_{0x}	v_{0y}	y_{max}	x_{max}	Equation of parabola
A	1	1			
B	2	1			
C	1	5			
D	2	5			
E	1	10			
F	2	10			

   (d) Summarize your examination with brief answers to these questions:
      i. As v_{0y} increases with fixed v_{0x} what happens to y_{max} and x_{max}?
      ii. As v_{0x} increases with fixed v_{0y} what happens to y_{max} and x_{max}?
   (e) Predict, with evidence, the change in shape of the parabola with fixed values of v_{0x} and v_{0y} if:
      i. The magnitude of a is increased by a factor of 2.
      ii. The magnitude of a is decreased by a factor of 2.

3-6 You can solve projectile motion problems using techniques learned for linear motion

Equations 3-13 and 3-14 are the principal equations we'll use for solving problems in projectile motion. Since these equations are identical to those for constant-velocity motion (along the x direction) and free fall (along the y direction) from Chapter 2, we'll be able to use the same problem-solving techniques learned in that chapter.

In **Table 3-1** we've collected Equations 3-13 and 3-14 along with information about which quantities are involved in each equation. In the same way that you used Table 2-3 in Section 2-5, you can use this table to decide which equations to use for a given projectile problem.

TABLE 3-1 Equations for Projectile Motion

By using one or more of these equations, you can solve any problem involving projectile motion for which you have defined the coordinate system as x horizontal, and positive y upward. The check marks indicate which of the quantities t (time), x_0 (the x component of the position at time 0), x (the x component of the position at time t), v_{0x} (the x component of the velocity at time 0), v_x (the x component of the velocity at time t), y_0 (the y component of the position at time 0), y (the y component of the position at time t), v_{0y} (the y component of the velocity at time 0), and v_y (the y component of the velocity at time t) are present in each equation. Note that g, the magnitude of the acceleration due to gravity, is a *positive* quantity.

Equation		Which quantities the equation includes								
		t	x_0	x	v_{0x}	v_x	y_0	y	v_{0y}	v_y
$v_x = v_{0x}$	(3-13a)				✓	✓				
$v_y = v_{0y} - gt$	(3-13b)	✓							✓	✓
$x = x_0 + v_{0x}t$	(3-14a)	✓	✓	✓	✓					
$y = y_0 + v_{0y}t - \frac{1}{2}gt^2$	(3-14b)	✓					✓	✓	✓	

EXAMPLE 3-4 Cliff Diving I

A diver leaps from a cliff in La Quebrada (Acapulco), Mexico. The cliff is 30.0 m above the surface of the water. He leaves the cliff moving at 2.00 m/s at 40.0° above the horizontal. If air resistance can be neglected, (a) how long does it take him to hit the water, and (b) how far does he travel horizontally before reaching the ocean?

Set Up

We can treat the diver as a projectile because air resistance is negligible and we are analyzing only the trajectory of his center of mass. We choose the origin ($x_0 = 0$, $y_0 = 0$) to be where the diver leaves the cliff and starts to move as a projectile. The diver's initial velocity (of magnitude $v_0 = 2.00$ m/s and angle $\theta_0 = 40.0°$) has positive x and y components that we can find using trigonometry. The time that the highest point in projectile motion is reached (y_{max} in the figure) is always straightforward to find, because it is when the y component of the velocity goes to zero. In part (a) we'll find the time t_t when he reaches his highest point, and then the time, t_l, from the highest point until he reaches the water at $y = -30.0$ m. We'll use the total time, $t = t_t + t_l$ in part (b) to find his x position when he hits the water. This tells us how far he traveled horizontally during the dive. Table 3-1 tells us which equations to use for each part.

$$v_{0x} = v_0 \cos\theta_0$$

$$v_{0y} = v_0 \sin\theta_0 \quad (3\text{-}1)$$

For part (a):

$$v_y = v_{0y} - gt \quad (3\text{-}13b)$$

$$y = y_0 + v_{0y}t - \frac{1}{2}gt^2 \quad (3\text{-}14b)$$

For part (b):

$$x = x_0 + v_{0x}t \quad (3\text{-}14a)$$

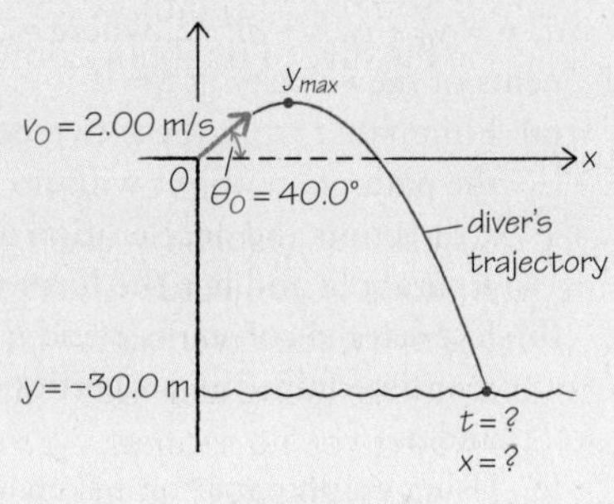

Solve

(a) First find the x and y components of the diver's initial velocity. These early calculations are not directly part of the answer we need to give, just steps along the way, so we will not round until we are done with our calculations. I am not going to write down all the digits though.

$$v_{0x} = v_0 \cos\theta_0$$
$$= (2.00 \text{ m/s}) \cos 40.0°$$
$$= 1.532 \text{ m/s}$$
$$v_{0y} = v_0 \sin\theta_0$$
$$= (2.00 \text{ m/s}) \sin 40.0°$$
$$= 1.286 \text{ m/s}$$

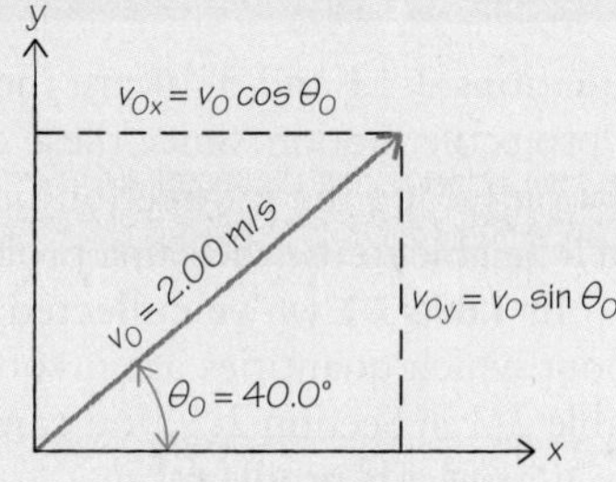

We could cut straight to solving this with a quadratic equation, but let's instead step through it, physically reasoning through the motion. First, we know a projectile launched upward goes upward until its y component of velocity goes to zero. So first we solve Equation 3-13b for time t_t when the projectile is at y_{max} and insert the known values $v_{0y} = 1.286$ m/s, and $g = 9.80$ m/s².

$$v_y = v_{0y} - gt_t$$
$$0 = 1.286 \text{ m/s} - 9.80 \text{ m/s}^2\, t_t$$
$$t_t = 0.131 \text{ s}$$

Next, solve Equation 3-14b for y_{max}, the y-component of position at this time, using the known values $v_{0y} = 1.286$ m/s, and $g = 9.80$ m/s².

$$y_{max} = y_0 + v_{0y}\, t_t - \frac{1}{2}gt_t^2$$
$$y_{max} = 0 + (1.286 \text{ m/s})(0.131 \text{ s}) - \frac{1}{2}(9.80 \text{ m/s}^2)(0.131 \text{ s})^2$$
$$= 0.084 \text{ m}$$

Then solve Equation 3-14b for time t_l (the time from when the projectile is at this value of y at the top of its motion until when the projectile lands) using the known values, $y = -30.0$ m, $v_{0y} = 0$ m/s, and $g = 9.80$ m/s².

$$-30.0 \text{ m} = y_{max} + 0t_l - \frac{1}{2}gt_l^2$$
$$-30.0 \text{ m} = 0.0843 \text{ m} + 0 - \frac{1}{2}gt_l^2$$
$$t_l^2 = \frac{2(30.0843 \text{ m})}{9.80 \text{ m/s}^2} \Rightarrow t_l = 2.478 \text{ s}$$

The diver reaches the water after leaving the cliff at $t = 0$, so the value of t we want is the sum of how long it took to go up and to come back down, t_t and t_l.

$$t = t_t + t_l = 0.131 \text{ s} + 2.478 \text{ s} = 2.61 \text{ s}$$

(b) We substitute the diver's constant x component of velocity $v_{0x} = 1.53$ m/s along with $t = 2.61$ s from part (a) into Equation 3-14a to determine the diver's final horizontal position.

$$x - x_0 = v_{0x}t = (1.532 \text{ m/s})(2.608 \text{ s}) = 4.00 \text{ m}$$

Reflect

We can check our results by using one of our interpretations of projectile motion. If there were no gravity, the diver wouldn't fall but would continue along a straight line at an angle of 40.0° above the horizontal. Then in 2.61 s the diver would rise 3.37 m above the cliff. In fact the diver falls a distance $\frac{1}{2}gt^2 = 33.4$ m below his zero-gravity trajectory in 2.61 s. The net result is that the diver goes through a vertical displacement of −30.0 m—exactly the displacement from the height of the cliff to sea level.

If gravity did not exist, in time $t = 2.61$ s the diver would travel horizontally: $x - x_0 = v_{0x}t$

$$= (1.53 \text{ m/s})(2.61 \text{ s}) = 4.00 \text{ m}$$

vertically: $y - y_0 = v_{0y}t$

$$= (1.286 \text{ m/s})(2.61 \text{ s}) = 3.37 \text{ m}$$

The distance the diver falls when just gravity is considered would be

$$\frac{1}{2}gt^2 = \frac{1}{2}(9.80 \text{ m/s}^2)(2.61 \text{ s})^2 = 33.4 \text{ m}$$

so the diver's net vertical displacement in 2.61 s is 3.37 m − 33.4 m = −30.0 m.

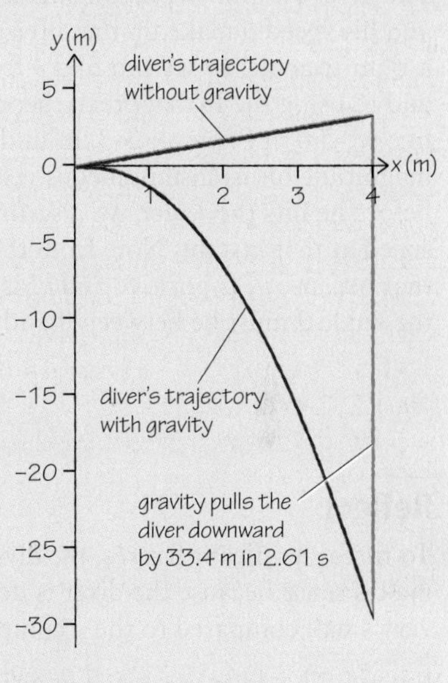

NOW WORK Problem 1(b) from The Takeaway 3-6.

EXAMPLE 3-5 Cliff Diving II

What is the velocity of the diver in Example 3-4 when he enters the water?

Set Up

From Example 3-4 we know the diver's initial ($t = 0$) velocity components, $v_{0x} = 1.53$ m/s and $v_{0y} = 1.286$ m/s, as well as the time $t = 2.61$ s that it takes for him to fall to the surface of the water. We'll use this information and Equations 3-13 to calculate his velocity components v_x and v_y when he reaches the water. We could then use trigonometry to find the magnitude and direction of his velocity at that point. Either expression—components, or magnitude and direction—describes the velocity as a vector.

Velocity, acceleration, and time for projectile motion:

$$v_x = v_{0x} \quad (3\text{-}13a)$$

$$v_y = v_{0y} - gt \quad (3\text{-}13b)$$

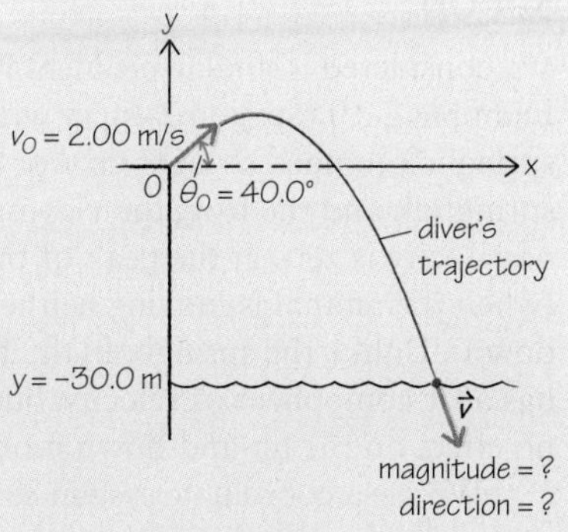

Solve

Equation 3-13a tells us that v_x doesn't change as the diver plummets toward the water, so it has the same value as when he left the cliff at $t = 0$. By contrast, Equation 3-13b tells us that v_y decreases by an amount gt.

$$v_x = v_{0x} = 1.53 \text{ m/s}$$

$$v_y = v_{0y} - gt$$
$$= 1.286 \text{ m/s} - (9.80 \text{ m/s}^2)(2.61 \text{ s})$$
$$= -24.3 \text{ m/s}$$

While these components describe the diver's velocity, we can take this a step further. The diver's x and y components of velocity and his speed v make up the three sides of a right triangle, so we can find v from v_x and v_y using the Pythagorean theorem just like we did in Example 3-3. In finding the magnitude of his instantaneous velocity just before he hits the water, we also find his speed at that instant. Note from the sketch that because v_x is positive and v_y is negative, the angle θ must be between 0 and –90°.

Magnitude of $\vec{v}$:

$$v = \sqrt{v_x^2 + v_y^2}$$
$$= \sqrt{(1.53 \text{ m/s})^2 + (-24.3 \text{ m/s})^2}$$
$$= 24.3 \text{ m/s}$$

Direction of $\vec{v}$:

$$\tan\theta = \frac{v_y}{v_x}$$
$$= \frac{-24.3 \text{ m/s}}{1.53 \text{ m/s}} = -15.9$$
$$\theta = \tan^{-1}(-15.9)$$
$$= -86.4°$$

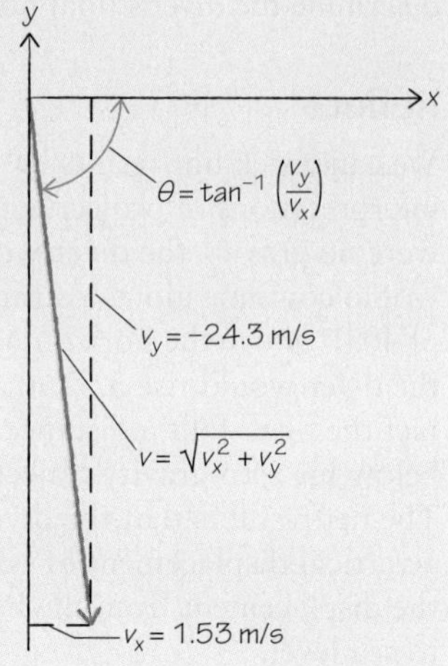

Reflect

To three significant figures, the diver's speed v is the same as the absolute value of his y component of velocity v_y. This makes sense because the diver is descending at a very steep angle of almost –90°, so the x component of his velocity is very small compared to the y component.

Extend: The diver is moving at a substantial speed when he enters the water. You can show that $v = 87.7$ km/h or 54.5 mi/h, which is comparable to the highway driving speed of an automobile. (In fact the diver's speed will be much less due to the effects of air resistance.) It's essential that the diver enter the water with his body oriented at the correct angle to avoid a painful "belly flop," or an even more serious injury, especially because this is an incredibly high dive.

NOW WORK Problems 1(a) and 9 from The Takeaway 3-6.

EXAMPLE 3-6 How High Does It Go?

A frog hops so that it leaves the ground at speed v_0 and at an angle θ_0 above the horizontal. What is the maximum height that the frog reaches during its leap? Ignore air resistance.

Set Up

We considered a similar problem in Example 2-10 (Section 2-6), in which a springbok jumped straight up. For both the springbok and the frog, the y component of velocity is zero at the peak of the motion (when the animal is moving neither up nor down). Unlike the springbok, the frog also has an x component of velocity, but this has no effect on the up-and-down motion.

We use a coordinate system as shown in the sketch, with the origin at the frog's starting point so that its initial coordinates are $x_0 = 0$, $y_0 = 0$. We'll use Equation 3-13b to find the time t_1 when the frog's y component of velocity is zero so that it is at the highest point of its leap. We'll then use Equation 3-14b to find the frog's y coordinate y_1 at this time.

Velocity, acceleration, and time for projectile motion:

$$v_y = v_{0y} - gt \quad (3\text{-}13b)$$

Position, acceleration, and time for projectile motion:

$$y = y_0 + v_{0y}t - \frac{1}{2}gt^2 \quad (3\text{-}14b)$$

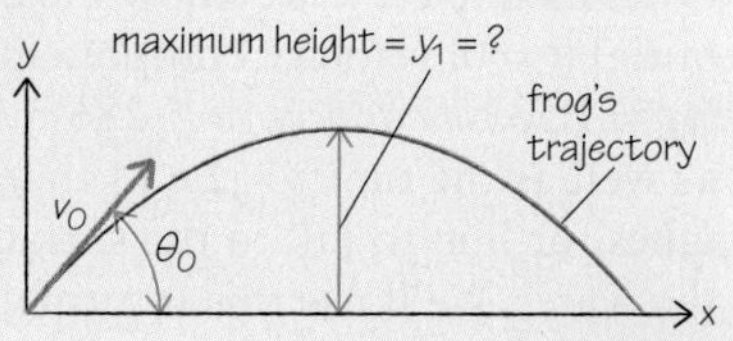

Solve

First express the frog's initial y component of velocity v_{0y} in terms of its initial speed v_0 and launch angle θ_0. Then, substitute this expression into Equation 3-13b. Set $v_y = 0$ (corresponding to the high point of the motion) at time $t = t_1$, then solve for t_1.

$$v_{0y} = v_0 \sin\theta_0 \text{ so}$$
$$v_y = v_0 \sin\theta_0 - gt$$
$$0 = v_0 \sin\theta_0 - gt_1$$
$$gt_1 = v_0 \sin\theta_0$$
$$t_1 = \frac{v_0 \sin\theta_0}{g}$$

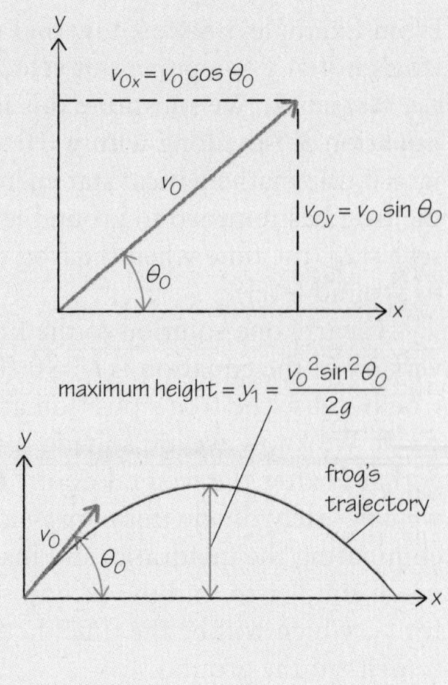

Now use Equation 3-14b to solve for the frog's y coordinate at time t_1. Substitute $y_0 = 0$, $v_{0y} = v_0 \sin\theta_0$, and $t_1 = (v_0 \sin\theta_0)/g$ from above.

$$y_1 = 0 + (v_0 \sin\theta_0)\left(\frac{v_0 \sin\theta_0}{g}\right) - \frac{1}{2}g\left(\frac{v_0 \sin\theta_0}{g}\right)^2$$
$$= \frac{v_0^2 \sin^2\theta_0}{g} - \frac{v_0^2 \sin^2\theta_0}{2g} \text{ so}$$
$$y_1 = \frac{v_0^2 \sin^2\theta_0}{2g}$$

Reflect

In this example our result is an algebraic expression rather than a number. You can verify that our answer for y_1 has the correct units.

Next we need to interpret what this equation tells us, and see if it makes sense. Let's look at how the maximum height y_1 depends on the values of the frog's initial speed v_0 and launch angle θ_0 as well as on the value of g. Because y_1 is proportional to the square of v_0, doubling the frog's launch speed makes it go *four* times as high. Note also that y_1 is proportional to $\sin^2\theta_0$. The sine function has its maximum value for $\theta_0 = 90°$, which corresponds to the frog jumping straight up. Both the dependence on v_0 and the dependence on θ_0 make sense; the frog leaps higher if its launch speed is greater or if it jumps at a more vertical angle.

Finally, note that y_1 is *inversely* proportional to g (the factor of g is in the denominator). Hence, if the value of g were smaller, you would divide by a smaller number to calculate y_1 and you would get a larger answer. This result also makes sense: If gravity were weaker, the frog would reach a greater height.

NOW WORK Problem 1(c) from The Takeaway 3-6.

EXAMPLE 3-7 How Far Does It Go?

When the frog in Example 3-6 returns to the ground, how far from its launch point does it land?

Set Up

The frog started its flight at $x_0 = 0$, $y_0 = 0$. When the frog lands, it returns to the same height $y = 0$ at horizontal coordinate x_2. The frog's distance from its launch point when it lands is the x component of its displacement, $x_2 - x_0$. To determine x_2, we'll first use Equation 3-14b to calculate the time t_2 when the frog lands and returns to $y = 0$. We'll then substitute this time into Equation 3-14a to calculate the frog's x coordinate when it lands, x_2.

Position, acceleration, and time for projectile motion:

$$x = x_0 + v_{0x}t \qquad (3\text{-}14a)$$

$$y = y_0 + v_{0y}t - \frac{1}{2}gt^2 \qquad (3\text{-}14b)$$

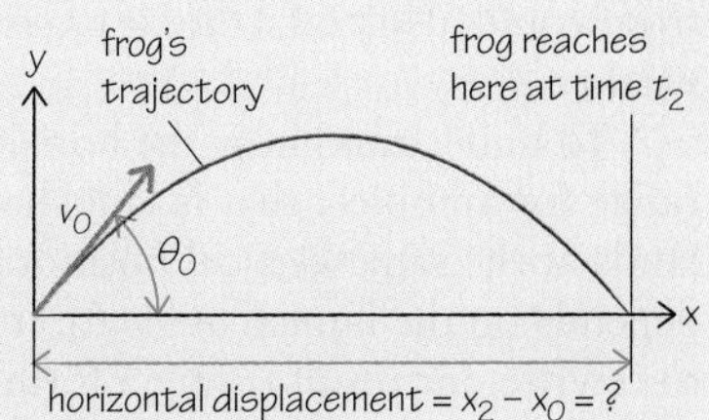

Solve

From Example 3-6 we know that the frog's initial y component of velocity is $v_{0y} = v_0 \sin\theta_0$. We substitute this into Equation 3-14b along with $y = 0$ and $y_0 = 0$, the mathematical statement that the frog has returned to ground level, and set $t = t_2$ (the time when the frog returns to ground level).

$v_{0y} = v_0 \sin\theta_0$

Equation 3-14b becomes

$$0 = 0 + (v_0 \sin\theta_0)t_2 - \frac{1}{2}gt_2^2$$

$$0 = (v_0 \sin\theta_0) - \frac{1}{2}gt_2 \text{ or}$$

$$t_2 = \frac{2v_0 \sin\theta_0}{g}$$

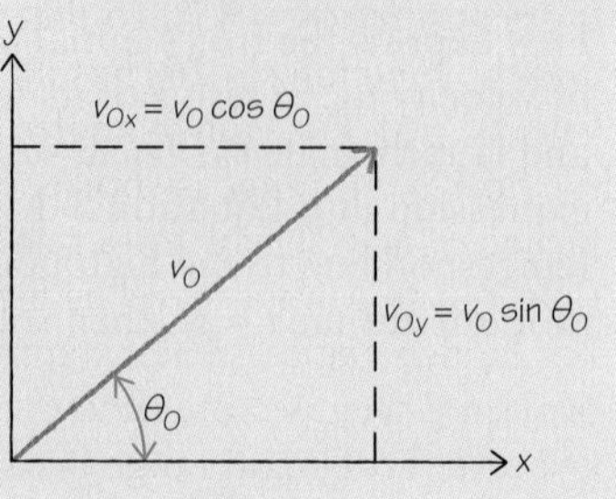

Clearly one solution to the first version of the equation is $t_2 = 0$. This is true because the frog's position at $t = 0$ is $y = 0$. We know we are solving for a time t_2 that is after the frog takes off. Thus, we can safely divide this expression by t_2, eliminating the quadratic, and making it much simpler to find the second solution for t_2, which will be the time the frog *returns* to the ground.

Note that the time $t_2 = (2v_0 \sin\theta_0)/g$ for the frog to return to its starting height is exactly *twice* the time $t_1 = (v_0 \sin\theta_0)/g$ that we calculated in Example 3-6 for the frog to reach its maximum height. In other words, a projectile takes just as long to ascend from its starting height to its maximum height as it does to descend from its maximum height back to its starting height.

To find the frog's x component of displacement $x_2 - x_0$ when it returns to the ground, substitute $t = t_2$ into Equation 3-14a. We also need the frog's initial x component of velocity, which from Example 3-6 is $v_{0x} = v_0 \cos\theta_0$. The distance $x_2 - x_0$ that the frog travels horizontally during its projectile motion is called the *horizontal range*.

$$x_2 - x_0 = v_{0x}t_2$$

At time $t_2 = (2v_0 \sin\theta_0)/g$, find the displacement $x_2 - x_0$:

Horizontal range for an object that starts and lands at the same height:

$$x_2 - x_0 = (v_0 \cos\theta_0)\left(\frac{2v_0 \sin\theta_0}{g}\right) = \frac{2v_0^2 \sin\theta_0 \cos\theta_0}{g}$$

Reflect

The horizontal range is proportional to v_0^2 (which says that increasing the launch speed makes the frog go farther) and inversely proportional to g (which says that the frog would also go farther if gravity were weaker).

To understand how the horizontal range for an object that is launched and lands at the same vertical position (height) depends on the launch angle θ_0, it's helpful to rewrite our result (using a formula from trigonometry) as $v_0^2 \sin 2\theta_0/g$.

$$R = \frac{2v_0^2 \sin\theta_0 \cos\theta_0}{g}$$

From trigonometry,

$$2 \sin\theta_0 \cos\theta_0 = \sin 2\theta_0$$

so

$$R = \frac{v_0^2 \sin 2\theta_0}{g}$$

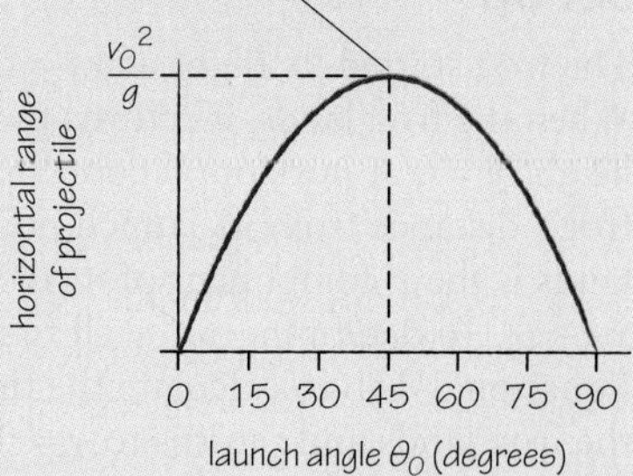

This expression shows that horizontal range is greatest for $\theta_0 = 45°$, so that $2\theta_0 = 90°$ and the function $\sin 2\theta_0$ has its greatest value ($\sin 90° = 1$). For launch angles less than 45°, the horizontal range is shorter because the projectile is in the air for a relatively short time and so cannot travel as far horizontally. For launch angles greater than 45°, the projectile spends more time in the air yet has a shorter horizontal range because its initial velocity is mostly upward (see the sketch).

Notice that for $\theta_0 = 90°$ the horizontal range is *zero*: The projectile lands right back where it started. However, with this launch angle the projectile reaches the greatest possible height, as we discussed in Example 3-6. R = horizontal range for an object that starts and lands at the same height.

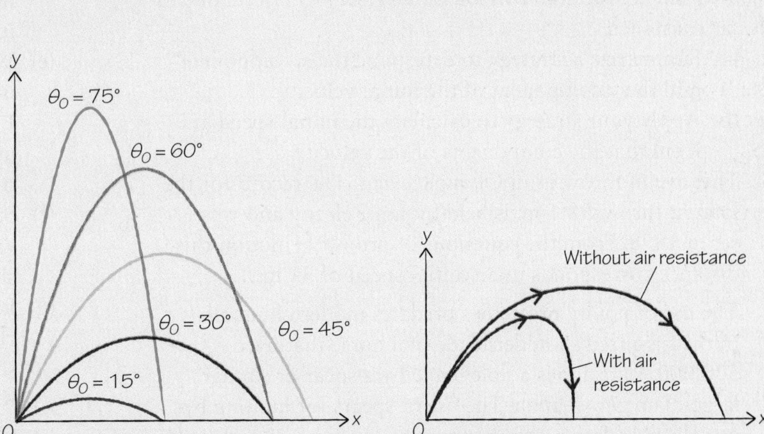

Extend: Because of the quadratic dependence on t, projectile motion problems can appear mathematically to have more than one solution, just like problems in linear motion with constant acceleration (see Example 2-8 in Section 2-5) or free-fall problems (see Example 2-10 in Section 2-6). Be careful and remember to use your common sense to choose the right root!

NOW WORK Problems 1(d) and 2 from The Takeaway 3-6.

WATCH OUT !

Air resistance affects range.

The rule that a launch angle of 45° for a projectile launched from the same height it will land gives the greatest horizontal range doesn't work for objects such as kicked footballs or batted baseballs. The reason is that air resistance plays a significant role in the flight of these objects, so their acceleration is not constant, their trajectories are not parabolic, and the formulas we've developed in this section don't apply. Detailed calculations that account for air resistance show that a batted baseball has the greatest horizontal range for a launch angle of about 35° above the horizontal. The difference in height can be ignored assuming the ball is caught at the same height from which it is launched. If not, that would add a further error.

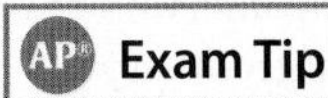

AP® Exam Tip

Since the kinematic equations are written assuming the starting time is $t = 0$, for simplicity, it is usually easiest to consider the start of the motion $t = 0$ unless specifically told otherwise. If you do need to use multiple start times, you just need to keep track of the initial values for each calculation, and use Δt.

THE TAKEAWAY for Section 3-6

✔ Projectile motion problems are easiest to solve when the components are chosen as the horizontal and vertical directions.

✔ In solving problems that involve projectile motion, you can use the same techniques that we used in Section 2-6 to solve one-dimensional free-fall problems and Section 2-2 to solve constant-velocity motion problems.

Prep for the AP® Exam

AP® Building Blocks

1. An object is thrown from the roof of a building with a height of 30.0 m near the surface of Earth with an initial speed $v_0 = 14.2$ m/s at an angle with the horizontal of 60.0°.
   - EX 3-5 (a) Calculate the x and y components of the initial velocity.
   - EX 3-4 (b) Express the x and y components of the displacement as functions of time.
   - EX 3-6 (c) Using the quadratic formula, calculate the time that the object is in the air.
   - EX 3-7 (d) Calculate the horizontal distance traveled by the object.

   The quadratic formula approach can be avoided or checked using a graphical solution or with an alternative mathematical solution strategy:
   - (e) Calculate the time it takes for the object to reach the highest point in its flight.
   - (f) Calculate the maximum height of the object in its flight.
   - (g) Calculate the time it takes for the object to fall to the surface from the maximum height.
   - (h) Calculate the total time of flight of the object.
   - (i) Calculate the horizontal distance traveled by the object.

EX 3-7

2. An object undergoing projectile motion travels 1.00×10^2 m in the horizontal direction before returning to its initial height. The object is thrown initially at a 30.0° angle from the horizontal. Neglect any effects due to air resistance.
   (a) Summarize a strategy to determine the x component and the y component of the initial velocity.
   (b) Apply your strategy to calculate the initial speed and from that the components of the velocity.
3. The javelin throw is an Olympic event. The record for the longest throw, 98.5 m, is held by Jan Zelezny and was set in 1998. From the equations of projectile motion this distance corresponds to an initial speed of 31 m/s.

   The use of pointy projectiles predates modern humans. In the fossilized shoulder blade of a horse that lived 500,000 years ago is a hole made by a spear or similar large-diameter weapon. The use of spears for hunting by our closest relative, the chimpanzee, has been documented. Persistence hunting with a spear is a strategy of pursuing prey until they tire and are within range of the spear throw. When ungulates, such as gazelles and antelopes, are pursued their evasion strategy is to run in a zigzag path. Unless the hunter is very close, the position of the prey after the time of flight of the spear must be anticipated.
   (a) Given a time interval between direction changes (zigs and zags) of from 1 to 2.5 s, predict the interval of initial throwing speeds from this model of time interval for the change in direction of the fleeing prey. Assume that the throw is made with an angle to the horizontal of 45°, and that the projectile strikes its prey at the same height from which it was launched.
   (b) Calculate the interval of distances for a throw that strikes the prey.
   (c) Based on a survey of all anthropological data, the maximum effective range of a hand-held spear is 7.8 m. As a test of the model, compare this effective range with the predicted range and compare the predicted interval of initial speeds with the Olympic record.

AP® Skill Builders

4. The equations describing the x and y components of the displacement of an object in parabolic motion are not provided on the AP® exam formula sheet. It is assumed that if you understand this type of motion then you will see it as a combination of the one-dimension kinematics equations that are provided on the formula sheet. In the following use 10 m/s^2 for the magnitude of the gravitational acceleration. Neglect air resistance.
   (a) An object is launched with an initial speed, $v_{i,y}$, straight upward near the surface of Earth. Express the vertical velocity and vertical displacement of the object using one-dimension kinematic equations.
   (b) Predict the time for the object to reach its greatest height and the time that the object is in the air if the initial velocity is 100 m/s.
   (c) An object is launched with an initial velocity $\vec{v}_i$, at an angle θ, relative to the surface of Earth. Express the vertical velocity and vertical displacement of the object using one-dimension kinematic equations.
   (d) Predict the time for the object to reach its greatest height and the time that the object is in the air if the initial velocity is 100 m/s and the launch angle is 30°.
   (e) While the object in part (c) is rising and falling it is also moving horizontally at a constant speed. Express the horizontal displacement of the object, using one-dimension kinematic equations, in terms of v_i, θ, and g.
   (f) Summarize in your own words the sequence of steps, with the key idea at each step, leading to the relationship in part (e).
   (g) In the *Discourse on Two New Sciences*, Galileo claims that "for [launch angles] which exceed or fall short of 45° by equal amounts, the ranges are equal." Knowing that 45° gives the maximum horizontal distance for a projectile that is launched and lands at the same height (neglecting air resistance) greatly simplifies the construction of a cannon. Examine your result in part (e) to justify the claim that 45° gives a maximum range by analyzing the x and y components of the projectile motion equations.
5. Select a logical computational pathway to calculate, from known quantities, the horizontal displacement of an object, when it strikes the ground, that was launched upward from the edge of a cliff with a height $h = 100$ m at an angle $\alpha = 30°$ and initial speed $v = 30$ m/s. The launch occurs near the surface of Earth where $g = 9.80$ m/s^2. Choose the correct answer below.
   (A) The equation $x_{max} = v_{i,x}t_{flight} = v_i \cos\alpha \dfrac{2v_i \sin\alpha}{g}$ can be applied because α, v, and g have known values.
   (B) The time of flight, t_{flight}, can be calculated using $-h = v_i \sin\alpha\, t_{flight} - \dfrac{1}{2}gt_{flight}^2$ and used to calculate the maximum distance from $x_{max} = v_i \cos\alpha\, t_{flight}$.
   (C) The time of flight, t_{flight}, can be calculated using $h = v_i \sin\alpha\, t_{flight} + \dfrac{1}{2}gt_{flight}^2$ and used to calculate the maximum distance from $x_{max} = v_i \cos\alpha\, t_{flight}$.
   (D) The equation $x_{max} = h + v_{i,x}t_{flight} = h + v_i \cos\alpha \dfrac{2v_i \sin\alpha}{g}$ can be applied because h, α, v, and g have known values.
6. The table shows trajectory information for the two Boston Red Sox players who hit the most home runs in the month of September 2017 (ESPN). Also tabulated are predicted values of the maximum height, y_{max}, and horizontal distance of the trajectory in vacuum, which represents their predicted values if air resistance were neglected. The relative error represents the percentage difference of the predicted from the actual values. Evaluate the data to determine whether projectile motion is a good model for these trajectories. Describe the direction of any additional acceleration vectors needed. If you choose to ignore any of the data, explain your choice.

Player	Launch angle	Initial speed (m/s)	Actual y_{max} (m)	Vacuum y_{max} (m)	Relative error	Actual distance (m)	Vacuum distance (m)	Relative error
Mookie Betts	25.6	42.8	16.2	17.4	0.07	109.7	145.5	0.33
Mookie Betts	27.3	47.4	24.4	24.1	−0.01	137.3	186.7	0.36
Mookie Betts	25.9	48.6	24.4	23.0	−0.06	141.8	189.2	0.33
Mookie Betts	28.3	47.6	25.0	26.0	0.04	135.3	192.8	0.42
Mookie Betts	23.5	46.4	18.9	17.4	−0.08	128.1	160.5	0.25
Mitch Moreland	28.8	44.2	27.4	23.1	−0.16	131.5	168.1	0.28
Mitch Moreland	23.2	46.6	21.9	17.2	−0.21	140.8	160.3	0.14
Mitch Moreland	31.2	45.7	26.8	28.6	0.07	130.2	188.7	0.45
Mitch Moreland	22.1	52.5	18.9	19.9	0.05	140.4	195.9	0.40

7. A zookeeper with a tranquilizer dart gun is trying to shoot a monkey sitting at the top of a tree, so that it may be taken in for needed medical care. The monkey, attempting to dodge the dart, drops from the tree at the same instant the zookeeper fires the dart.
   (a) Choose the correct claim below about where the zookeeper should aim to hit the monkey. (Neglect any effects due to air resistance.)
      (A) Aim straight at the monkey because the dart is moving very quickly while the velocity of the monkey changes very slowly during the time it takes for the dart to arrive.
      (B) Aim lower than the monkey because that is where the monkey will be when the dart arrives and how much lower depends on the speed of the dart.
      (C) Aim at a point one-half of the height of the tree because the time of flight of the dart, which first rises and then falls, is twice the time it takes the monkey to fall.
      (D) Aim straight at the monkey because the dart and the monkey are both in free fall with the same acceleration.

   A common demonstration tests the correct claim from part (a). This demonstration may have been conducted in your class or you may have seen the demonstration in the MIT lecture-demo series, or even on YouTube. In the MIT device, the fall of the monkey and the shot are synchronized when the projectile blocks a photogate that opens a circuit with an electromagnet suspending the monkey. That the objects collide should be surprising and requires an explanation.
   (b) The demonstration described suggests that a successful collision requires that the projectile motion of the dart and the one-dimensional falling motion of the monkey need to start at the same instant. But there is no discussion of the speed of the dart. **Pose a question** that could be pursued to develop a model of the relationship between the velocity of the dart (magnitude and direction) and the height of the collision.

   Let the distance from the zookeeper to the point directly below the monkey be labeled d. The components of the initial velocity of the projectile are v_{0x} and v_{0y}.
   (c) Construct a diagram of the scenario that includes d, v_{0x}, v_{0y}, the initial height of the monkey, h, and the angle θ that the initial projectile path makes with the horizontal.
   (d) Express the time, T, for the projectile to move a horizontal distance d in terms of d and v_{0x}.
   (e) Express the vertical position of the projectile at time T, $y_p(T)$, in terms of d, g, v_{0x}, and v_{0y}.
   (f) Express the vertical position of the monkey at the time T, $y_m(T)$, in terms of d, g, and v_{0x}.
   (g) Analyze your expressions to obtain a condition that is sufficient to require that the projectile and the monkey collide.
   (h) Explain, using the diagram constructed in part (c), why the condition of part (g) is satisfied if the zookeeper aims at the monkey.
   (i) Analyze the expression for the vertical position of the monkey at time T, $y_m(T)$, obtained in part (f):
      i. Locate the point where the collision occurs when the x component of the projectile velocity is very large.
      ii. Determine the value for the x component of the projectile velocity for a collision that occurs at the same height as the launch in terms of g, d, and h. Discuss how you could find the minimum horizontal velocity that would allow the projectile to strike the monkey.

AP® Skills in Action

8. Two golf balls are hit from the same point on a flat field. Both are hit at an angle of 30° above the horizontal. Ball 2 has twice the initial velocity of ball 1. Balls 1 and 2 land a displacement d_1 and d_2, respectively, from the initial point. Predict the relationship between the displacements d_1 and d_2, neglecting any effects due to air resistance, by choosing the best answer below.
   (A) $d_2 = 2d_1$ because the time of flight is proportional to the y component of the initial velocity.
   (B) $d_2 = 2d_1$ because the range is proportional to the x component of the initial velocity.
   (C) $d_2 = 4d_1$ because both time of flight and range are proportional to the initial velocity.
   (D) $d_2 = 4d_1$ because time of flight is proportional to the initial velocity, and range is proportional to both initial velocity and time of flight.

EX 3-5 9. You toss a ball into the air at an initial angle 40° from the horizontal. Neglecting any effects due to air resistance, predict the point during motion that the ball has the smallest speed and choose the correct answer below.
(A) at the highest point in its flight because both the horizontal and vertical components of the velocity are zero
(B) at the highest point in its flight because the vertical component of the velocity is zero
(C) when it is first thrown because at that time the ball has both horizontal and vertical components of acceleration
(D) over the entire range because it moves at a constant horizontal velocity

10. The acceleration due to gravity on another planet, g_P, is larger than on Earth, g_E. Predict the properties of a launch on the other planet that achieves the same horizontal distance as the trajectory of a projectile launched on Earth, by choosing the correct answer below.
(A) The launch speed should by increased by a factor of g_P/g_E so that the time of flight will be the same.
(B) The sine of the launch angle should be increased by a factor of g_P/g_E so that the time of flight will be the same.
(C) The launch speed should by increased by a factor of $\sqrt{\frac{g_P}{g_E}}$ so that the time of flight will be the same while the horizontal component of the velocity increases.
(D) The launch speed should be increased by a factor of $\sqrt{\frac{g_P}{g_E}}$ so that, although the time of flight decreases some, the increase in the horizontal component of the velocity will compensate.

WHAT DID YOU LEARN?

Chapter learning goals	Section(s)	Related example(s)	Prep for the AP Exam: Relevant section review exercises
Recognize the differences among one-, two-, and three-dimensional motion.	3-1		1
	3-2		3
	3-3	3-1	
Describe the properties of a vector and how to find the sum or difference of two vectors.	3-2		4
	3-3	3-1, 3-2	3, 4, 5
Describe the motion of an object in two dimensions using quantities such as displacement, distance, velocity, speed, and acceleration, in an appropriate way for a chosen coordinate system.	3-2		2
	3-3	3-2	6
	3-4	3-3	4, 5
	3-5		1, 2, 3, 4, 5
Explain how the displacement, velocity, and acceleration of an object or the center of mass of a system are described in terms of vectors, and can be modeled mathematically by the kinematic equations when acceleration is constant.	3-4	3-3	1, 9
	3-5		
	3-6	3-4	
Recognize that the velocity observed for an object depends on the reference frame in which it is observed, and be able to describe the reference frames of a given observer from a description of the physical situation.	3-4	3-3	1, 2, 4, 5
Be able to convert the velocity of an object relative to one reference frame into the velocity of that same object observed from a different reference frame.	3-4	3-3	1, 2, 4, 5
Identify the key features of projectile motion and how to interpret this kind of motion.	3-5		1, 3
	3-6	3-4, 3-5	
Solve problems involving projectile motion.	3-6	3-5, 3-6, 3-7	1, 2, 9

Chapter 3 Review

Key Terms

All the Key Terms can be found in the Glossary/Glosario on page G1 in the back of the book.

component 95
component method 95
motion in a plane 103
parabola 117
projectile 114
projectile motion 114
reference frame 103
relative velocity 105
trajectory 103
vector addition 91
vector difference 92

Chapter Summary

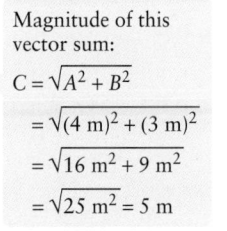

Topic	Equation or Figure
Vectors and scalars: A vector quantity such as displacement or velocity has both a magnitude and a direction. By contrast, a scalar quantity such as temperature or time has no direction. We write a vector in boldface italic with an arrow above it. The magnitude of a vector $\vec{A}$ is written as A.	(d) Vectors $\vec{H}$ and $\vec{J}$ have different magnitudes but the same direction. Vectors $\vec{J}$ and $\vec{K}$ have the same magnitude but different directions. 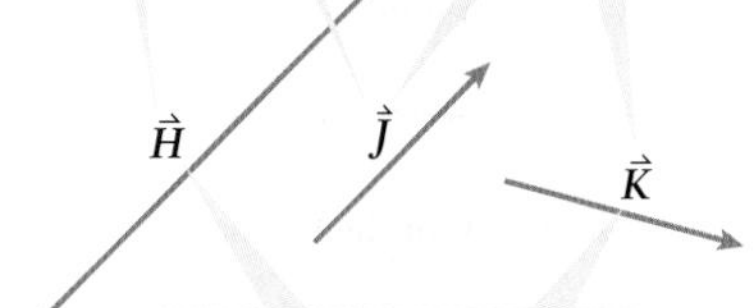 Vectors $\vec{H}$ and $\vec{K}$ have different magnitudes and different directions.  A vector has only two attributes: magnitude and direction. (Figure 3-2d)
Vector arithmetic: To find the sum of two vectors $\vec{A}$ and $\vec{B}$, place them tip to tail. Then $\vec{A}+\vec{B}$ points from the tail of the first vector to the tip of the second vector. To find the difference $\vec{A}-\vec{B}$ of two vectors, add $\vec{A}$ and $-\vec{B}$ (a vector with the same magnitude as $\vec{B}$ that points opposite to $\vec{B}$). Multiplying a vector by a scalar can change the vector's magnitude, direction, or both.	(b) 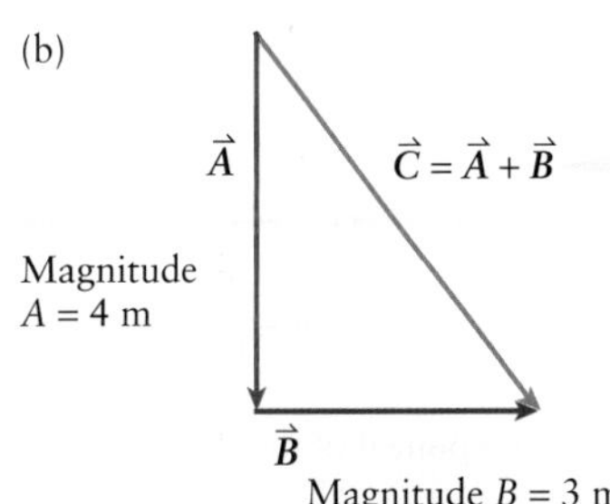 Magnitude of this vector sum: $C=\sqrt{A^2+B^2}$ $=\sqrt{(4\text{ m})^2+(3\text{ m})^2}$ $=\sqrt{16\text{ m}^2+9\text{ m}^2}$ $=\sqrt{25\text{ m}^2}=5\text{ m}$ (Figure 3-5b)  The magnitude of the vector sum $\vec{A}+\vec{B}$ is equal to the sum of the magnitudes of $\vec{A}$ and $\vec{B}$ only if $\vec{A}$ and $\vec{B}$ point in the same direction, as in (a).

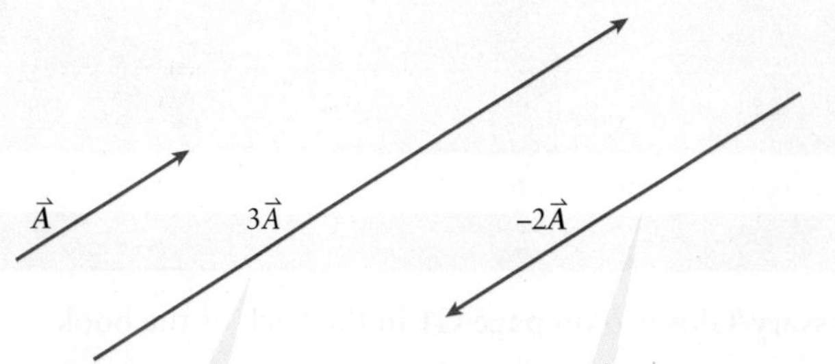

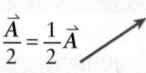

(Figure 3-8)

The number 3 is positive, so the vector $3\vec{A}$ points in the same direction as $\vec{A}$ but has 3 times the magnitude.

The number −2 is negative, so the vector $-2\vec{A}$ points in the direction opposite to $\vec{A}$ and has 2 times the magnitude.

Dividing a vector by 2 is the same as multiplying it by 1/2, so the vector $\vec{A}/2$ points in the same direction as $\vec{A}$ (since 1/2 is positive) and has 1/2 the magnitude.

Vector components: A vector can be expressed in terms of either its magnitude and direction or its x and y components. The components can be calculated from the magnitude and direction, and the magnitude and direction can be calculated from the components.

This vector has a positive x component A_x and a positive y component A_y.

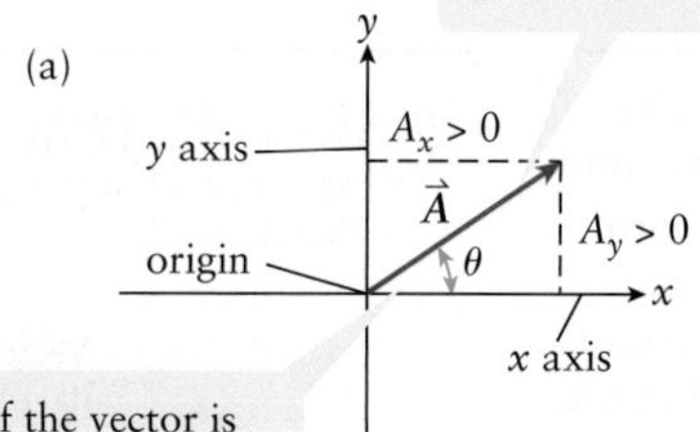

(Figure 3-9a)

The angle θ of the vector is measured from the positive x axis in the direction toward the positive y axis. When x and y are both positive, θ is between 0 and 90°.

---

x component of the vector $\vec{A}$

y component of the vector $\vec{A}$

$$\text{(a)}\quad A_x = A\cos\theta \qquad \text{(b)}\quad A_y = A\sin\theta \tag{3-1}$$

Magnitude of the vector $\vec{A}$

Angle of the vector $\vec{A}$ measured from the positive x axis toward the positive y axis

---

Magnitude of the vector $\vec{A}$

Angle of the vector $\vec{A}$ measured from the positive x axis toward the positive y axis

$$\text{(a)}\quad A = \sqrt{A_x^2 + A_y^2} \qquad \text{(b)}\quad \tan\theta = \frac{A_y}{A_x} \tag{3-2}$$

x component of the vector $\vec{A}$

y component of the vector $\vec{A}$

Vector arithmetic with components: To find the components of a vector sum $\vec{A}+\vec{B}$, add the respective components. To find the components of a vector difference $\vec{A}-\vec{B}$, subtract the respective components. To find the components of a vector $\vec{A}$ multiplied by a scalar c, multiply each component of $\vec{A}$ by c.

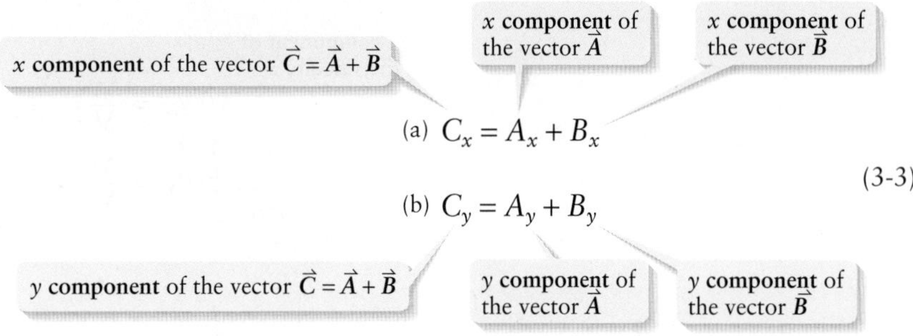

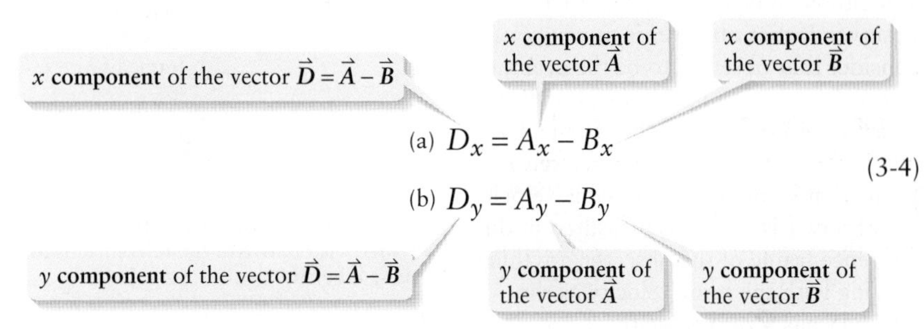

x component of the vector $\vec{E}=c\vec{A}$ | **Scalar** c | **x component** of the vector $\vec{A}$

$$\text{(a) } E_x = cA_x$$

$$\text{(b) } E_y = cA_y \qquad (3\text{-}5)$$

y component of the vector $\vec{E}=c\vec{A}$ | **Scalar** c | **y component** of the vector $\vec{A}$

Displacement and velocity in more than one dimension: The displacement $\Delta\vec{r}$ of an object during a time interval is a vector that points from the object's position at the earlier time t_1 to its position at the later time t_2. The velocity $\vec{v}$ at a given time points along the object's trajectory; its magnitude is the object's speed. The individual vector components of $\vec{v}$ are given by the same expressions we used in Chapter 2 for velocity in linear motion.

Displacement (change in position) of the object during the time interval from time t_1 to a later time t_2

Position of the object at later time t_2

$$\Delta\vec{r} = \vec{r}_2 - \vec{r}_1 \qquad (3\text{-}6)$$

Position of the object at earlier time t_1

To calculate the displacement, **subtract** the earlier position from the later position (just as for motion in a straight line).

Velocity for the object over a very short time interval from time t_1 to a later time t_2

Displacement (change in position) of the object during the short time interval

$$\vec{v} = \frac{\Delta\vec{r}}{\Delta t} = \frac{\vec{r}_2 - \vec{r}_1}{t_2 - t_1} \qquad (3\text{-}7)$$

For both the displacement and the elapsed time, **subtract** the earlier value from the later value.

Elapsed time for the time interval

x component of the **velocity** $\vec{v}$

y component of the **velocity** $\vec{v}$

$$\text{(a) } v_x = \frac{\Delta x}{\Delta t} = \frac{x_2 - x_1}{t_2 - t_1}$$

$$\text{(b) } v_y = \frac{\Delta y}{\Delta t} = \frac{y_2 - y_1}{t_2 - t_1}$$

x **component** of the **displacement** during a short time interval

y **component** of the **displacement** during a short time interval

Elapsed time for the time interval

(3-8)

Velocities can be expressed relative to any observer. If that observer is moving, we consider that observer to be the origin of a coordinate system that moves with that observer, a moving reference frame. Every moving observer has its own reference frame that is moving at the same velocity the observer is. Thus, as measured in the reference frame of the observer, anything moving with the same velocity as the observer has a zero velocity.

The velocity of the object, here labeled "first," is relative to another reference frame of interest, here labeled "other."

$$\vec{v}_{\text{first, last}} = \vec{v}_{\text{first,other}} + \vec{v}_{\text{other,last}} \qquad (3\text{-}9)$$

The velocity of the object, here labeled "first," is relative to the reference frame, here labeled "last."

The velocity of the other reference frame of interest, here labeled "other," is relative to the reference frame labeled "last" here and in the first term.

Acceleration in more than one dimension: An object has a nonzero acceleration whenever it is speeding up, slowing down, *or* changing direction. The *x* and *y* components of $\vec{a}$ are given by the same expressions we used in Chapter 2 for acceleration in linear motion. Alternatively, we can express $\vec{a}$ in terms of its component along the direction of motion (which describes changes in speed) and its component perpendicular to the direction of motion (which describes changes in direction).

Acceleration for the object over a very short time interval from time t_1 to a later time t_2

Change in velocity of the object over the short time interval

$$\vec{a} = \frac{\Delta \vec{v}}{\Delta t} = \frac{\vec{v}_2 - \vec{v}_1}{t_2 - t_1} \qquad (3\text{-}10)$$

Duration of the time interval

For both the velocity change and the elapsed time, **subtract** the earlier value from the later value.

x **component** of $\vec{a}$

y **component** of $\vec{a}$

$$\text{(a) } a_x = \frac{\Delta v_x}{\Delta t} = \frac{v_{x2} - v_{x1}}{t_2 - t_1}$$

$$\text{(b) } a_y = \frac{\Delta v_y}{\Delta t} = \frac{v_{y2} - v_{y1}}{t_2 - t_1}$$

Change in the *x* **component of the velocity** during a short time interval

Change in the *y* **component of the velocity** during a short time interval

Elapsed time for the time interval

(3-11)

Projectile motion: A projectile is a moving object accelerated by gravity alone. It has a constant downward acceleration of magnitude g. In the equations for the motion of a projectile, we have chosen the x axis to be horizontal and the y axis to point vertically upward.

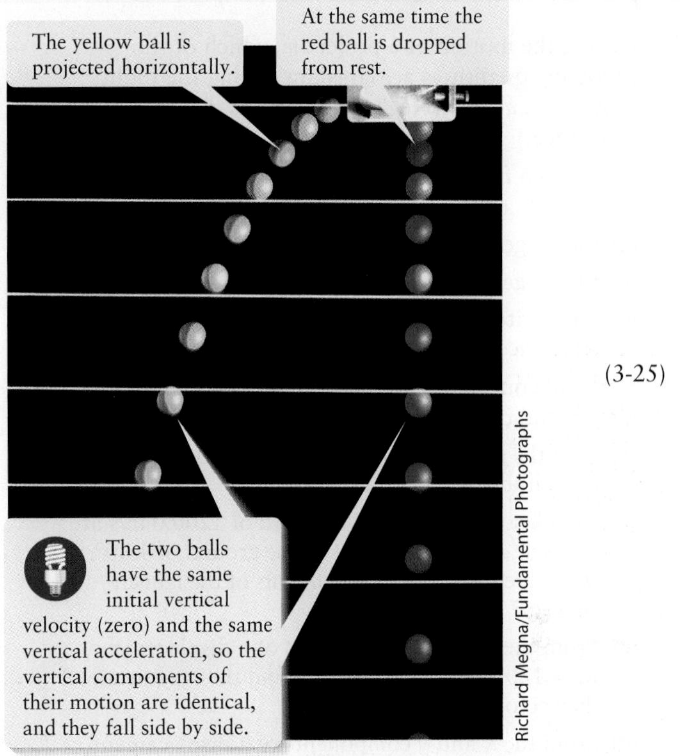

The yellow ball is projected horizontally.
At the same time the red ball is dropped from rest.
The two balls have the same initial vertical velocity (zero) and the same vertical acceleration, so the vertical components of their motion are identical, and they fall side by side.
Richard Megna/Fundamental Photographs

(3-25)

x component of an object's **velocity** $\vec{v}$ at time t for **projectile motion**

x component of the object's velocity at $t = 0$

$$\text{(a)}\ v_x = v_{0x}$$

$$\text{(b)}\ v_y = v_{0y} - gt \qquad (3\text{-}13)$$

Acceleration due to gravity (g is positive)

Time at which the object has velocity components v_x and v_y

y component of an object's **velocity** $\vec{v}$ at time t for **projectile motion**

y component of the object's velocity at $t = 0$

x and y coordinates at time t of an object in **projectile motion**

x and y components at $t = 0$ of the object's velocity

Time at which the object has coordinates x and y

$$\text{(a)}\ x = x_0 + v_{0x}t$$

$$\text{(b)}\ y = y_0 + v_{0y}t - \frac{1}{2}gt^2 \qquad (3\text{-}14)$$

x and y coordinates at $t = 0$ of the object

Acceleration due to gravity (g is positive)

Chapter 3 Review Problems

1. During the motion of a projectile, which of the following quantities are constant during the flight: x, y, v_x, v_y, a_x, a_y? (Neglect any effects due to air resistance.)
2. For a given fixed launch speed, at what angle should you launch a projectile to achieve
   (a) the longest range,
   (b) the longest time of flight, and
   (c) the greatest height? (Neglect any effects due to air resistance.)
3. Evaluate your ability to calculate the orientation, magnitude, and components of a vector.
   (a) Find the x and y components of a velocity with a magnitude of 34.00 m/s at 210.0°.
   (b) A rocket is launched with a speed of 1200.0 m/s at an angle measured relative to the ground of 43.0°. What are the x and y components of the launch velocity?
   (c) From the x and y components of a displacement, $R_y = 120$ m and $R_x = 345$ m, find the magnitude of the vector.
   (d) From the x and y components of a displacement, $R_x = 345$ m and $R_y = 120$ m, find the direction of the vector.
   (e) Find the x and y components of a velocity with a magnitude of 15.0 m/s at 12.0°.
4. The vector in the figure makes a 30.0° angle with respect to the y axis as shown. Which pair of x and y components of the vector are possible?

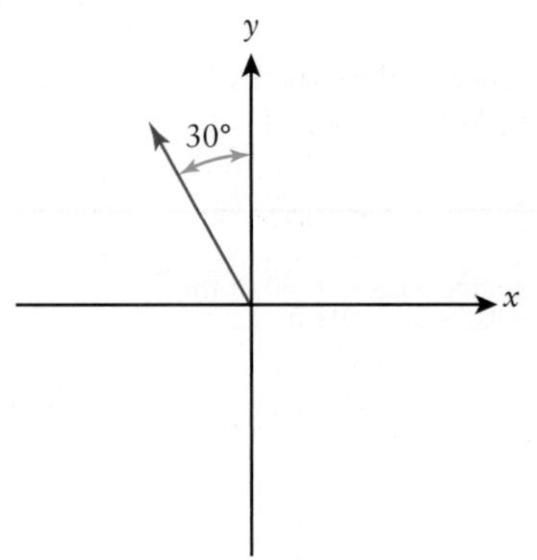

   (A) 3.46, 2.00
   (B) −2.00, 3.46
   (C) −3.46, 2.00
   (D) 2.00, −3.46
5. (a) Find the x and y components of a velocity with a magnitude of 240.00 m/s at 330.0°.
   (A) −237.89 m/s and −31.77 m/s
   (B) 207.85 m/s and −120.00 m/s
   (C) 120.00 m/s and 207.85 m/s
   (D) 31.77 m/s and −237.89 m/s

   (b) Find the x and y components of a displacement with a magnitude of 15.00 m at 12.0°.
   (A) −8.05 m and 12.66 m
   (B) 14.67 m and 3.12 m
   (C) 3.12 m and 14.67 m
   (D) 12.66 m and −8.05 m

   (c) A rocket is launched with x (horizontal) and y (vertical) components of the initial velocity, $v_x = 120.00$ m/s and $v_y = 345.00$ m/s. Find the speed.
   (A) 133,425 m/s
   (B) 365.27 m/s
   (C) 465.00 m/s
   (D) 225.00 m/s

   (d) From the x (horizontal) and y (vertical) components of velocity, $v_y = 31.0$ m/s and $v_x = 8.0$ m/s, find the direction of the vector.
   (A) 76° (B) 39° (C) 14.5° (D) 15°

   (e) Use the x and y components of an acceleration, $a_x = -15\ \text{m/s}^2$, $a_y = 12\ \text{m/s}^2$, to find the direction of the vector.
   (A) −53° (B) 321° (C) 141° (D) 309°

   (f) From the x (horizontal) and y (vertical) components of a velocity, $v_y = 31.0$ m/s and $v_x = 8.0$ m/s, find the speed.
   (A) 39.0 m/s
   (B) 32.0 m/s
   (C) 30.0 m/s
   (D) 23.0 m/s
6. A vector has an x component and a y component that are equal in magnitude. Which of the following angles could the vector make with the x axis in the same x–y coordinate system?
   (A) 0° (B) 45° (C) 60° (D) 90°
   (E) 120° (F) 135° (G) 180° (H) 210°
   (I) 225° (J) 245° (K) 270°
7. Each of the following vectors is given in terms of its x and y components. Find the magnitude of each vector and the angle it makes with respect to the x axis.
   (a) $A_x = 3.0$, $A_y = -2.0$
   (b) $A_x = -2.0$, $A_y = 2.0$
   (c) $A_x = 0$, $A_y = -2.0$
8. $\vec{A}$ is 66.0 m long at a 28° angle with respect to the x axis. $\vec{B}$ is 40.0 m long at a 56° angle with respect to the x axis. What is the sum of the vectors (magnitude and angle with the x axis)?

9. Given the vector $\vec{A}$ with components $A_x = 2.00$, $A_y = 6.00$, the vector $\vec{B}$ with components $B_x = 2.00$, $B_y = 22.00$, and the vector $\vec{D} = \vec{A} - \vec{B}$, calculate the magnitude and angle with the x axis of the vector $\vec{D}$.

10. Two velocities are given as follows: $\vec{A} = 30$ m/s, 45° north of east and $\vec{B} = 40$ m/s, due north. Calculate the components of each vector sum and the direction of the vector: (a) $\vec{A} + \vec{B}$, (b) $\vec{A} - \vec{B}$, and (c) $2\vec{A} - \vec{B}$.

11. A rock is thrown from a bridge at an angle 20° below horizontal. At the instant of impact, is the rock's speed greater than, less than, or equal to the speed with which it was thrown? Explain your answer. (Neglect any effects due to air resistance.)

12. In the 1970 National Basketball Association championship, Jerry West made a 60-ft shot from beyond half court to lead the Los Angeles Lakers to an improbable tie at the buzzer with the New York Knicks. Neglecting air resistance, estimate the initial speed of the ball. (The Knicks won the game in overtime.)

13. An object with a constant acceleration has an initial velocity of 30 m/s south. After 10 s, the object has a final velocity of 40 m/s west.
    (a) Construct a vector diagram with the vectors $\vec{v}_i$, $\vec{v}_f$, and $\vec{a}t$.
    (b) Calculate the magnitude of the acceleration.
    (c) Calculate the components of the acceleration.

14. A potato gun fires a spud with a muzzle velocity of 15.0 m/s. What is the maximum range for a root vegetable launched from this weapon? Assume the magnitude of g is 10.0 m/s^2.

15. A soccer ball is kicked with a horizontal velocity of 11.3 m/s and a vertical velocity of 3.5 m/s. What are the magnitude and direction of the ball's initial velocity?

16. A professional quarterback needs to be able to throw a football the entire length of the field. What is the minimum initial velocity needed to get the ball 91 m down field? Assume that air resistance is negligible, and assume that the ball lands at the height from which it is launched. This will lead to a slightly high estimate, which is safe, given that you are looking for a minimum.

17. An architect is planning a golf course with these vectors connecting tee-off locations and holes:

$$L_a = 450 \text{ m at } 20°, L_b = 250 \text{ m at } 270°, L_c = 630 \text{ m at } 70°$$

What are the length and direction of a vector connecting the tee-off for hole a with hole c? The measurements given assume the angles are measured with respect to due east at each hole, and holes a and b are the tee-offs for holes b and c, respectively.

18. A plane moving 200.0 km/h at an altitude of 1000.0 m drops a package. What is the horizontal displacement of the package from when it is released until it strikes the ground below?

19. An object moves in a plane with a constant velocity of 2 m/s in the positive x direction and a constant acceleration in the negative y direction with a magnitude of 4 m/s^2. The initial velocity of the object in the y direction is 0.
    (a) Complete the following table. Your teacher suggests that given the precision to which the time, initial velocity, and acceleration were given, that you report all calculated values to one decimal place.

t (s)	v_x (m/s)	v_y (m/s)	x (m)	y (m)
0.0				
0.2				
0.4				
0.6				
0.8				
1.0				

    (b) Construct a graph in which x and y are plotted versus time on the same graph.
    (c) Construct a graph in which y is plotted versus x.

20. A tiger leaps horizontally out of a tree that is 4.00 m high. If he lands 5.00 m from the base of the tree, calculate his initial speed. (Neglect any effects due to air resistance.)

21. A football is punted at 25.0 m/s at an angle of 30.0° above the horizon. What is the velocity of the ball when it is 5.00 m above ground level? Assume it starts 1.00 m above ground level. (Neglect any effects due to air resistance.)

22. A diver leaps into the air from a platform 5.0 m above the water's surface. The diver's initial speed is 4.4 m/s and her launch angle is 70°. (a) How long is the diver in the air? (b) What is her horizontal displacement from when she leaves the board until she strikes the water's surface?

23. An electron is shot from an electron gun with a velocity of 3.0×10^6 m/s parallel to the surface of Earth. How far will it fall as it travels a horizontal distance of 1.0 m? The only acceleration is that due to gravity.

24. Two trajectories of projectile motions are shown in the figure. For each of the statements below, make a claim, and support your claim with evidence from the figure and physical principles using one or more complete sentences.

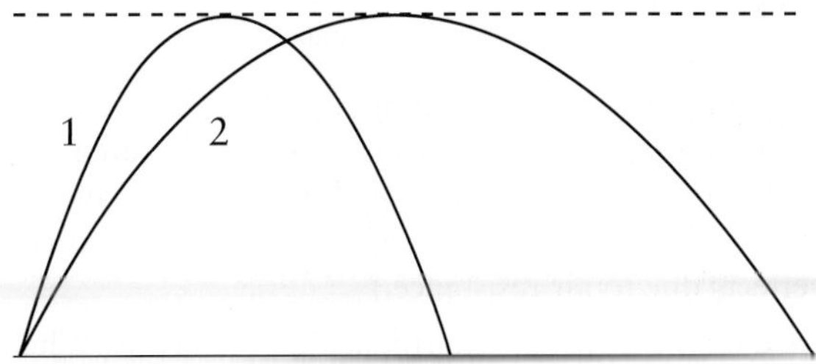

(a) Compare the y components of the initial velocities.

(b) Compare the x components of the initial veocities.

(c) Compare the times of flight for the motions.

(d) Compare the magnitudes of the initial velocities.

25. Two armies are launching missiles at their enemies. One has the high ground with an elevation $h = 400$ m. Both weapons were purchased from the same manufacturer and so they have the same initial speed for the projectile of 420 m/s. Both have initial launch velocities with the same launch angle of 60.0° relative to the horizontal. The figures below are given to aid interpretation and are not drawn to scale.

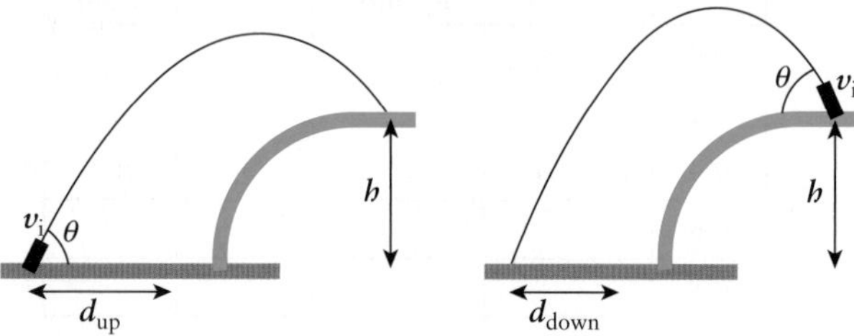

(a) Calculate the maximum horizontal distance that each army can shoot their missiles. It may be easier to work this problem by breaking the motion into steps.

(b) In part (a), the advantage for the army on the high ground was demonstrated for a particular initial velocity. Justify the claim that the additional horizontal distance is

$$\Delta x = v_{i,x}\left[\sqrt{\frac{2(h + y_{max})}{g}} - \sqrt{\frac{2(y_{max} - h)}{g}}\right]$$

26. In Detroit in 1971, Reggie Jackson hit one of the most memorable home runs in the history of the Major League Baseball All-Star Game. The approximate trajectory is plotted in the figure. (The asymmetry is due to air resistance.) Using the information in the graph, estimate the initial speed of the ball as it left Reggie's bat.

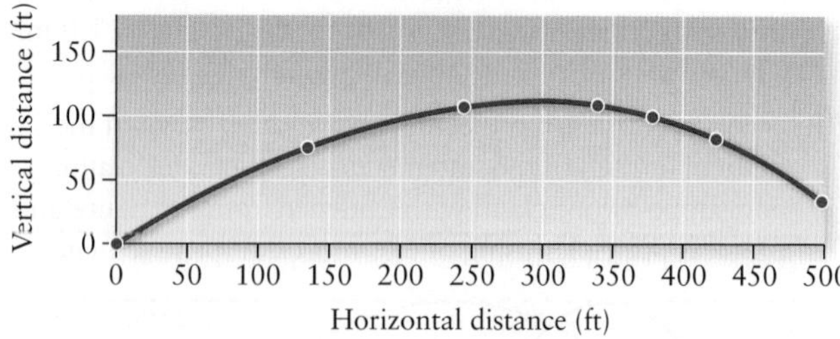

27. A Chinook salmon can jump out of water with a speed of 6.30 m/s. How far horizontally can a Chinook salmon travel through the air if it leaves the water with an initial angle of 40.0°? (Neglect any effects due to air resistance.)

28. A dart is thrown at a dartboard 2.37 m away. When the dart is released at the same height as the center of the dartboard, it hits the center in 0.447 s. At what angle above the horizontal was the dart thrown? (Neglect any effects due to air resistance.)

29. Marcus and Cassie want to hike to a destination directly 12.0 km due north of their starting point. Before heading to the destination, Marcus walks 10.0 km in a direction that is 30.0° north of east and Cassie walks 15.0 km in a direction that is 45.0° north of west. How much farther must each hike on the second part of the trip?

30. Nathan walks due east a certain distance and then walks due south twice that distance. He finds himself 15.0 km from his starting position. How far east and how far south does Nathan walk?

31. A water balloon is thrown horizontally at a speed of 2.00 m/s from the roof of a building that is 6.00 m above the ground. At the same instant the balloon is released, a second balloon is thrown straight down at 2.00 m/s from the same height. Determine which balloon hits the ground first and how much sooner it hits the ground than the other balloon. (Neglect any effects due to air resistance.)

32. An airplane releases a ball as it flies parallel to the ground at a height of 255 m, as shown in the figure. If the ball lands on the ground at a horizontal displacement of exactly 255 m from the release point, calculate the speed of the plane. (Neglect any effects due to air resistance.)

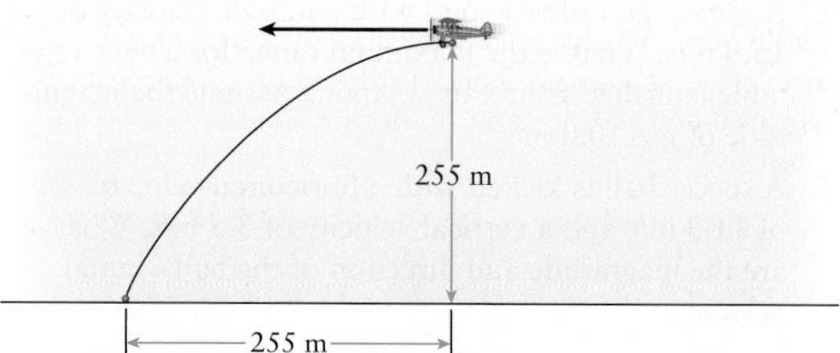

33. A boy runs horizontally off the end of a diving platform at a speed of 5.00 m/s. The platform is 10.0 m above the surface of the water. (Neglect any effects due to air resistance.)

(a) Calculate the boy's speed when he hits the water.

(b) How much time is required for the boy to reach the water?

(c) How far horizontally will the boy travel before he hits the water?

34. Tim wants to make a solo flight from Chicago to Milwaukee, about 100 miles north of Chicago. He needs to make the flight in 50 minutes, but there is a wind blowing toward the east at 40 mi/h. In what direction and at what speed should he fly in order to make it to Milwaukee in the allotted time?

35. Your teacher wants you to determine how fast you can throw a baseball. With a tape measure for measuring

distances, you set to work. You throw the ball horizontally from shoulder height, which is about 1.5 m. After five trials, you determine the ball lands an average distance of 15 m from the point on the ground directly below where you released it.

(a) What was your throwing speed?

(b) You send your friend down the field to catch the ball as you attempt to throw it as far as you can. Assuming the same speed you found in (a), what is the maximum horizontal distance from you that your friend should stand? Your friend also catches the ball at shoulder height (1.5 m above the ground).

36. A pilot needs to travel due south a distance of 800 miles in 1.5 hours. She discovers that she must fly the plane at a speed (relative to the air) of 570 miles per hour at an angle of 14 degrees west of south in order to arrive on time at her destination. Calculate the speed and direction of the air flow along her flight path (that is, the wind velocity).

37. The speed of a boat relative to the water is v_{bw}. The boat is to make a round trip in a river with a current that travels at speed v_{wg}. (Assume $v_{wg} < v_{bw}$.)

(a) Derive an expression for the time needed to make a round trip of total distance D if the boat makes the round trip by moving

i. upstream and back downstream.

ii. directly across the river (perpendicular to the direction of motion of the water) and back.

(b) Which of these two times (i or ii) is greater? Explain your reasoning.

38. A pilot wishes to fly due north and arrive at a location 1000.0 miles to the north in exactly 2 hours. There is a wind that is blowing 75.0 miles per hour directed southeast. In what direction should the pilot head and with what speed should she fly the plane (relative to the air) in order to reach her destination on time?

39. Anish throws a ball such that its horizontal range (to the same level) is five times its maximum height. What is the angle at which he threw the ball?

Prep for the AP Exam

Group Work

Directions: The problem below is designed to be done as group work in class.

Data were collected on the motion of a projectile launched with an initial speed of 70.7 m/s at an angle of 45.0°. The data (labeled as "observed") were not in agreement with the trajectory that was predicted in the figure. The prediction was based on the equations of motion for a projectile where air resistance is neglected.

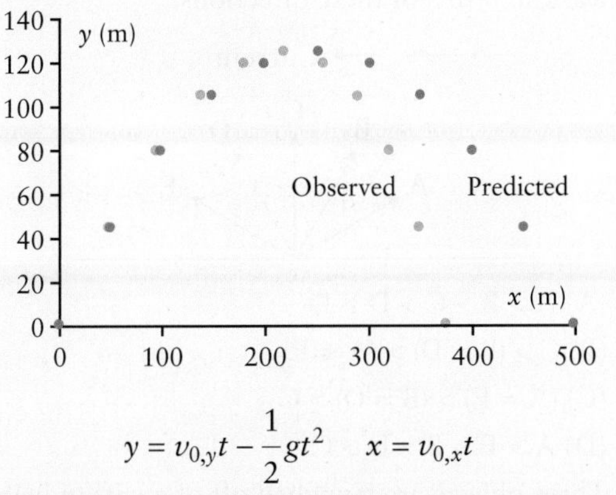

$$y = v_{0,y}t - \frac{1}{2}gt^2 \qquad x = v_{0,x}t$$

(a) Based on the evidence provided by these data, justify the claim that the observed trajectory does not exhibit an acceleration in the y direction different from the acceleration due to gravity.

(b) Based on the evidence provided by these data, justify the claim that the observed trajectory does exhibit an acceleration in the x direction.

The numerical values of the observed and predicted trajectories are shown in the table.

$x_{predicted}$	$y_{predicted}$	$y_{observed}$	$y_{predicted}$
0	0	0	0
50	45	49	45
100	80	96	80
150	105	140	105
200	120	182	120
250	125	222	125
300	120	260	120
350	105	295	105
400	80	328	80
450	45	359	45
500	0	388	0

(c) Design a strategy for the analysis of these data to determine the magnitude of the acceleration in the x direction.

(d) Apply your strategy to calculate the acceleration in the x direction.

AP® PRACTICE PROBLEMS

Prep for the AP® Exam

Multiple-Choice Questions

Directions: The following questions have a single correct answer.

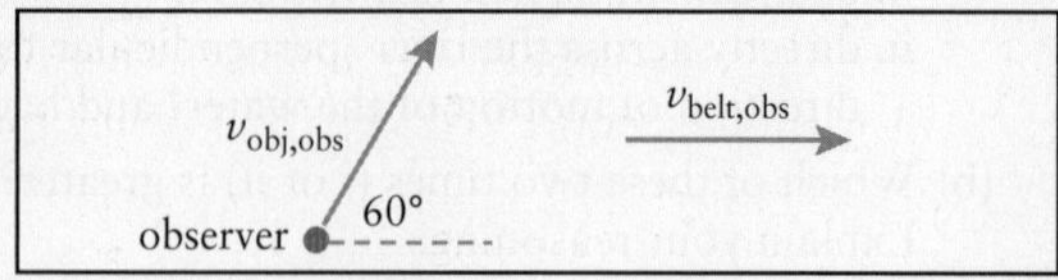

1. As shown in the figure, a conveyer belt is moving to the right at a speed v. A stationary observer sees an object that is moving with the belt, but in addition is sliding relative to the belt, moving away from the observer at the same speed v but at an angle of 60° with respect to the edge of the belt. What is the speed of the object relative to the belt?
(A) $0.5\ v$ (B) v
(C) $\frac{\sqrt{3}}{2}v$ (D) $\left(1+\frac{\sqrt{3}}{2}\right)v$

2. A ball rolls off the edge of a tall horizontal ledge with a speed v_0. Assuming it does not strike the ground beforehand, at what time after it leaves the surface will its velocity be in a direction 45° below the horizontal?
(A) $\frac{v_0}{g}$ (B) $\frac{2v_0}{g}$
(C) $\frac{v_0}{2g}$ (D) $\frac{\sqrt{2}v_0}{2g}$

3. An object is accelerated horizontally by a small propeller. As the object leaves the supporting surface it has an initial horizontal velocity $v_0 = 0$ and has a constant horizontal acceleration $a = +0.5g$. Taking the origin as the point of departure and $+x$ to the right and $+y$ upward, the path of the object as it falls will be described by which of the following?
(A) $y = -\frac{1}{2}x$ (B) $y = -2x$
(C) $y = -\frac{1}{2}x^2$ (D) $y = -2x^2$

4. A student is on a platform that is moving in the x direction with a constant speed v_0. The student tosses a ball straight up, providing an additional initial velocity in the y direction (also of magnitude v_0). Relative to the ground, when the ball reaches its greatest height its speed will be what fraction of its speed as it left the student's hand?
(A) $\frac{1}{4}$ (B) $\frac{1}{2}$
(C) $\frac{3}{4}$ (D) $\frac{\sqrt{2}}{2}$

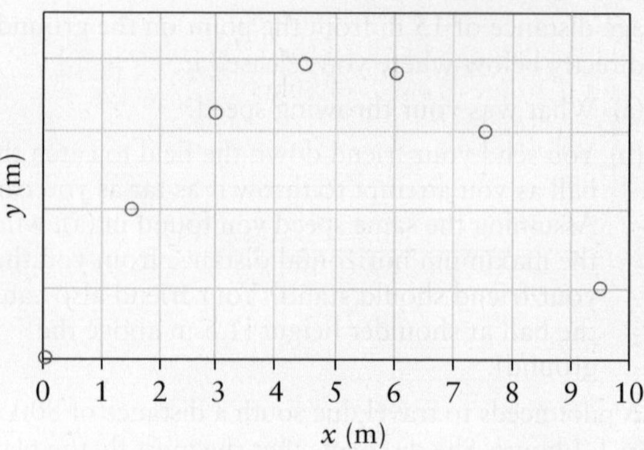

5. According to a laboratory report the positions of a projectile were measured at equal time intervals and recorded on the graph of the x and y positions as shown. The expected uncertainty in measurement is represented by the size of the circles plotted. During which portion of the trajectory does there appear to have been a mistake in the recorded time interval?
(A) 0 m – 2.5 m (B) 2.5 m – 5.0 m
(C) 5.0 m – 7.5 m (D) 7.5 m – 10.0 m

6. A student wants to swim from one side of a river to the other side. The picture below shows several directions in which she could aim herself (relative to the water). Directions B and D each make an angle of 60° with the shore, directions A and E each make an angle of 30° with the shore, and C is perpendicular to the shore. Assume she swims at the same speed relative to the water in whichever direction she aims. Which of the following is the correct ranking of the time to cross the river, greatest to least, in terms of these directions?

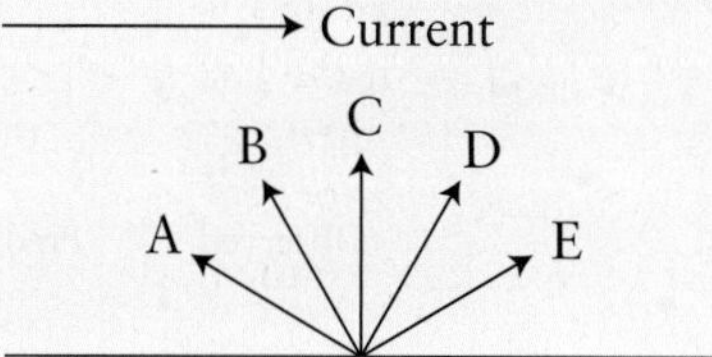

(A) A > B > C > D > E
(B) C > (B = D) > (A = E)
(C) (A = E) > (B = D) > C
(D) A > E > B > D > C

7. Three objects are launched off of a cliff of height H. Object A is launched at an angle θ above the horizontal. Object B is launched horizontally. Object C is launched at an angle θ below the horizontal. Which of the following ranks the objects based on the time they are in the air before they strike the ground below the cliff?
(A) A > B > C (B) A > C > B
(C) (A = C) > B (D) A = B = C

(Continued)

Free-Response Questions

1. The following graphs represent the horizontal and vertical components of the velocity of a ball thrown off the edge of a cliff of unknown height, H. Neglect air resistance.

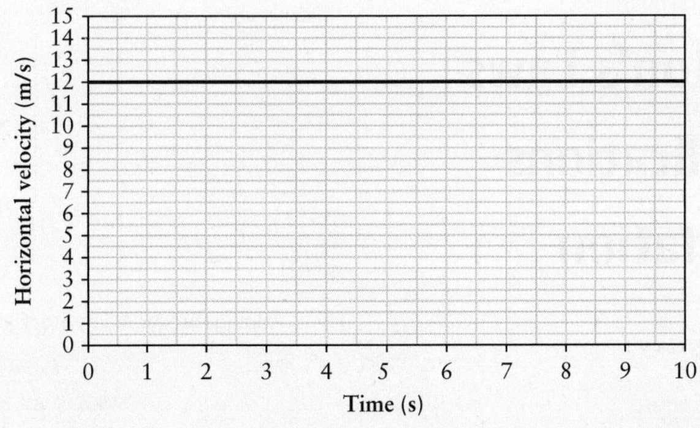

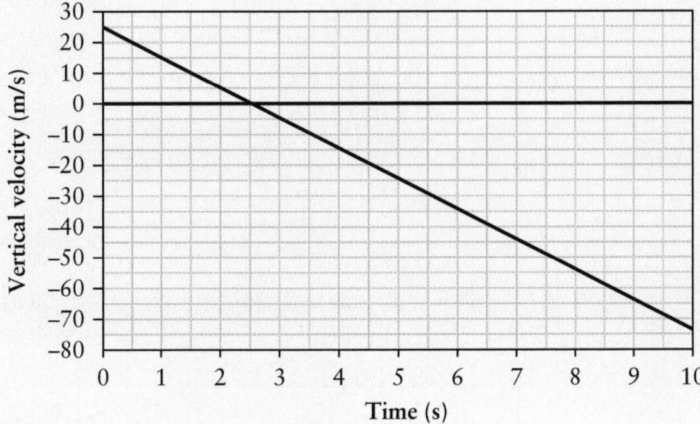

The projectile lands on the ground below the cliff at $t = 10$ seconds.

(a) Calculate how far from the base of the cliff the ball strikes the ground.

(b) Calculate the height, H, of the cliff.

(c) Draw the horizontal and vertical components of the final velocity of the ball just before it strikes the ground. Draw them tip to tail and draw the resultant final velocity vector.

(d) i. Write equations for the horizontal and vertical components of the velocity as a function of time for this ball.

ii. Write equations of motion for horizontal and vertical components of the position of the ball as a function of time.

A second, identical ball is released from rest at the top of the cliff.

(e) In a few sentences, describe how the velocity-versus-time graphs for this second ball would compare to the graphs shown above.

2. Two projectiles are launched from ground level with a speed v_0. Projectile A is launched at an angle θ above a horizontal line and projectile B is launched at an angle $90° - \theta$ above the ground, which may be represented by a horizontal line.

(a) In the space below, create a sketch to model the two-dimensional motion of the two projectiles, with the horizontal line representing the ground.

(b) Derive expressions for the horizontal displacements of both projectiles.

(c) On the following figures, sketch graphs of the vertical and horizontal components of the velocity for both projectiles. Label the graphs clearly to indicate which projectile is represented. The graph should only include the time that the projectile is in flight (that is, before it strikes the ground).

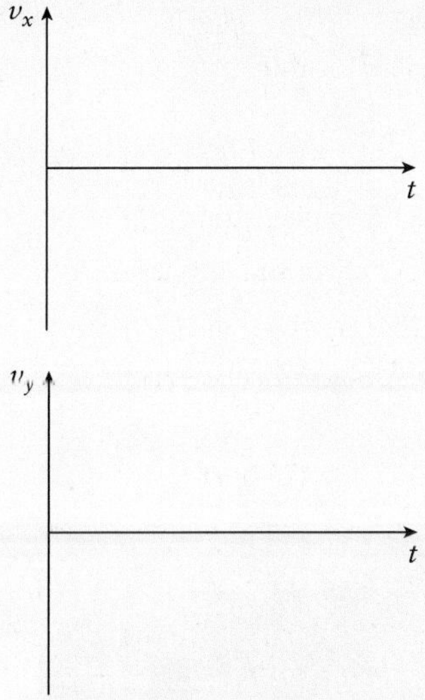

Force and Translational Dynamics

Imagebroker/Alamy Stock Photo

Case Study: Who wins in a tug-of-war contest?

Rex enjoys a game of tug-of-war by pulling on one end of a rope in the direction opposite to his friend Alice. Each exerts a force—the interaction of one object with another—on the rope. When one of them pulls harder than the other on the rope, the rope begins to move. A nonzero net force on the rope results in an acceleration of the rope.

The magnitude of an object's acceleration depends on the magnitude of the net force exerted on the object and the object's mass. Mass is a measure of the quantity of material in the object. In particular, acceleration depends on the object's inertial mass, the property of an object or system that determines how its motion changes when a net force is exerted on it. *The greater the mass, the smaller the acceleration.*

When Rex and Alice pull equally hard, the stationary rope does not begin to move. Force is a vector—it has direction as well as magnitude. Rex and Alice exert forces on the rope in opposite directions, so if the magnitudes of those forces are equal the two forces cancel. The rope's motion doesn't change.

In the tug-of-war contest, Rex exerts a force on the rope and the rope exerts a force on him. This pair of forces are sometimes referred to as an action-reaction force pair; they are always equal in magnitude and opposite in direction. Two forces can be a force pair only if they are exerted between the same two objects. The force of Rex on the rope is not, then, paired with the force that the rope exerts on Alice. As we will see, Rex and Alice do not directly interact; neither exerts a force on the other.

Earth's gravity exerts a force on Rex. If we consider this the action force, what force (exerted by what and exerted on what) is the reaction paired with this force? Careful—remember that force pairs are between the same two objects. There is at least one other force pair (in addition to that involving Rex and Earth) in evidence in the photo. Can you identify it?

Friction forces resist the motion of one object relative to another. How would the contest between Rex and Alice be different if there were no friction between their feet and the ground?

By the end of these chapters, you should be able to use the relationship between force and acceleration to predict the motion of an object; construct explanations of physical situations involving the interaction of objects using forces, accelerations, and the representation of action-reaction pairs of forces; construct a graphical representation of all the external forces exerted on an object; and use that representation to determine unknowns about the motion of the object.

4 Forces and Motion I: Newton's Laws

Ceri Breeze/Alamy

YOU WILL LEARN TO:

- Describe a force as an interaction between two objects or systems and identify both objects or systems for any force.
- Analyze a scenario and make claims about the forces exerted on an object by other objects or systems for different types of interactions.
- Use Newton's second law to predict the motion of an object subject to forces exerted by other objects in a variety of physical situations.
- Categorize forces as long-range or contact forces, and make claims about contact forces due to the microscopic cause of those forces.
- Recognize the distinctions among mass, weight, and inertia.
- Draw and use free-body diagrams in problems to analyze situations involving multiple forces exerted on an object.
- Construct explanations of physical situations involving the interaction of objects using Newton's third law and the representation of action–reaction pairs of forces.
- Use a free-body diagram and Newton's second law to construct a mathematical representation relating the acceleration of an object to its mass, and to solve that equation for an unknown quantity.

4-1 How objects move is determined by their interactions with other objects, which can be described by forces

NEED TO REVIEW?
Turn to the **Glossary** in the back of the book for definitions of bolded Key Terms.

In Chapter 1, we defined "object" and "system." In Unit 1, we learned how to describe motion, kinematics. The object and system models will be crucial to our ability to describe the causes of motion, the focus of this unit. Dynamics is sometimes referred to as the branch of physics dealing with forces and their effects on motion. When one object or system interacts externally with another, we can describe that interaction as a **force**. Forces allow us to predict the future motion of objects or systems. However, if the interaction between objects is internal to a system, we don't want to use a force description for that interaction because there are other ways to predict motion. We will see examples of this beginning in Chapter 7. To begin our exploration of forces as a way to describe interactions, we will look at interactions between objects, or systems that we can model as objects.

When two objects touch, it is easy to see the contact interaction, especially if it is an obvious push or pull. Pushes and pulls are common ways to describe forces. To push an object means to exert a force on the object directed away from yourself, and to pull an object means to exert a force directed toward yourself. An example of this is demonstrated in the chapter opening photo of the airliner. In the language of physics, the tug (a heavy vehicle used to move another) shown in the photo exerts a force on the airliner. Force is a vector: It has both magnitude and direction. For example, in the photo, the tug exerts a force on the airliner that points in the direction of the airliner's motion. Additionally, as we saw in the previous chapter, air can resist motion through itself and a surface resists things being dragged over itself. This resistance to relative motion is called a kinetic friction interaction, a kinetic friction force, or just **kinetic friction.** The wheels of the airliner slide as well as roll; the ground therefore exerts a kinetic friction force on the airliner that points backward (opposite to the airliner's motion). At low speeds the kinetic friction with the ground is much greater than air resistance so we will neglect the effect of the air. The magnitude of each force is a measure of how strong it pushes or pulls on the airliner.

What determines how the airliner moves is not any one single force but rather the combined effect of *all* the forces exerted on it, also called the **net external force** on the object (or, for short, the **net force**). To initially get the plane rolling, the tug must supply a force in the direction of desired motion. If, once it gets rolling, the forward force from the tug has the same magnitude as the backward-directed kinetic friction force, these two oppositely directed forces cancel and so the net force is zero. We will soon see that acceleration and force are directly related (proportional), so if the net force is zero, so is the acceleration. As we learned in Chapter 3, if the acceleration of an object is zero, its velocity, which could be zero, will not change. So since the airliner is already moving, and the net force on it is zero, it continues to move with constant velocity. But if the tug pulls harder than the force of kinetic friction, the forces do not cancel, and there is a net force in the forward direction. This means that the airliner accelerates. Remember that when the acceleration is in the direction of motion, speed will increase.

Exam Tip

Remember that "no net force" does not mean that there are no forces on the system, but it means that the forces on the system add as vectors to equal zero. It is impossible on Earth—or any planet—to have no forces on a system.

If there were no kinetic friction, then once the tug got the plane moving, you could detach the tug and the plane would just keep moving. It is the fact that we don't "see" friction that often leads us to incorrectly predict what will happen. Experiment confirms that an object in linear motion will remain in that same motion if there is no net force exerted on it.

Experiment also shows that no matter how hard the tug pulls on the airliner, the airliner pulls back on the tug with an equally strong force. (This force on the tug is why the tug's engine has to work harder with the airliner attached.) This turns out to be true for all interactions, no matter what kind: If one object exerts a force on a second object, the second object necessarily exerts a force on the first one.

These observations are at the heart of *Newton's laws of motion*, a time-tested set of physical principles that have a tremendous range of applicability. In later physics courses you will find that more sophisticated models are needed when you try to predict the motion of objects or systems that move very fast or that are very tiny. However, for almost all phenomena we directly observe daily, Newton's laws will allow us to make valid predictions. We'll use Newton's three laws of motion throughout much of the remainder of this book. We'll devote this chapter and the next to understanding these laws and some of their most important applications.

THE TAKEAWAY for Section 4-1

✔ What determines how an object moves is the combined effect of all of the forces (pushes or pulls) that are exerted on the object, the net force.

✔ If one object exerts a force on a second object, the second object must also exert a force on the first object.

Prep for the 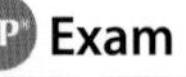 **Exam**

AP® Building Blocks

1. Students are in the hall outside of the physics classroom getting a feel for force. A pair of students stand with brooms a few paces from another pair of students who have a collection of different balls: a bowling ball, a basketball, a tennis ball, and a ping-pong ball. When each of these balls is rolled forward, the students with brooms try to change its velocity—stopping it or changing its direction. In your own words, describe what you expect

the students to observe. In particular, think about how interactions between the broom and the balls will differ and what causes this difference.

2. The movement of a car involves an interaction between the car and the road. In your own words describe forces within the car–road system that push and pull the car.

4-2 If a net external force is exerted on an object, the object accelerates

If you want to start a soccer ball rolling along the ground, you have to give it a push or kick. If you're a hockey goalie and want to deflect a puck away from the goal that you're defending, you have to hit the puck with your hockey stick. And if your dog runs in pursuit of the neighbor's cat, you have to pull on the dog's leash to slow it down and bring it to a halt. In each of these cases you're changing the velocity of an object, making it speed up (for the soccer ball), changing the direction of its motion (for the hockey puck), or making it slow down (for your dog). And in each case, to cause the object's velocity to change, you have to exert a *force*—that is, a push or a pull—on that object.

Forces are *vectors* because they have both magnitude and direction. (You might pull a flower gently toward you to take a closer look or push a plate of cafeteria food strongly away from you if you don't like something on the plate.) We saw in the previous section that when more than one force is exerted on an object, what determines how the object moves is the *vector sum* of all the forces exerted on that object: the net external force. If the individual forces exerted on an object are $\vec{F}_1$, $\vec{F}_2$, $\vec{F}_3$, and so on, we can write the net force on that object as

EQUATION IN WORDS
Net external force on an object

The **net external force** exerted on an object...

...equals the **vector sum** of all of the individual forces that are exerted on the object.

$$\sum \vec{F}_{ext} = \vec{F}_1 + \vec{F}_2 + \vec{F}_3 + \ldots \quad (4\text{-}1)$$

The sum includes only external forces (forces exerted on the object by other objects).

For example, there are three interactions experienced by a softball player as she slides to get safely to base (**Figure 4-1**). Each of these interactions can be described by a force. One of these interactions is the downward pull of Earth's gravity, called the **gravitational force** (denoted by $\vec{F}_g$). This is a long-range interaction, which means Earth doesn't have to be touching you for you to experience this force. The gravitational force is one of the fundamental forces that can be used to describe our universe. The fundamental force

(a)

Moodboard Stock Photography/Alamy Stock Photo

(b)

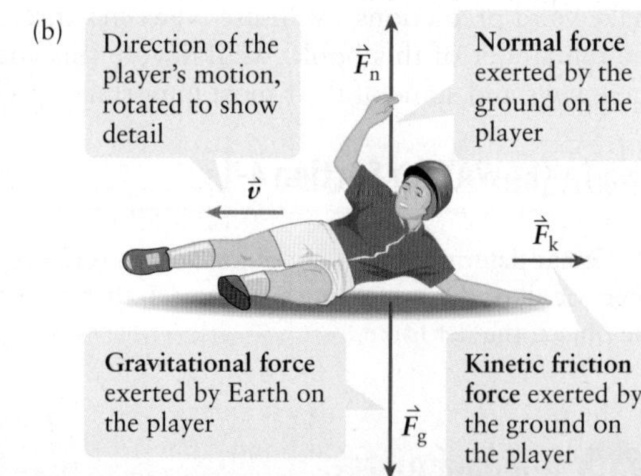

Figure 4-1 Forces on a softball player (a) A softball player sliding to get safely on base. (b) The red vectors indicate the direction of each external force exerted on the player.

The only forces that affect the player's motion are *external* forces (forces exerted on the player by other objects). These include Earth's gravitational force (the player's weight) and forces from objects that the player is touching (in this case the normal force and kinetic friction force exerted by the ground).

that dominates the properties of the objects in our everyday experiences is actually the *electric force*, which you will learn about in your next physics course. We will occasionally mention it, because it is what is actually responsible for a lot of the forces we regularly experience. However, to use this description, we would have to consider every atom! So it is more convenient to describe such everyday forces as nonfundamental contact forces. The normal force and kinetic friction are two such **contact forces**, called that because they are most easily described mathematically as being exerted through contact with a surface, such as the ground on which the player slides.

If you stand on any surface and it does not break, the force that keeps you from falling through the surface is called a **normal force** (denoted by F_n). On the macroscopic level, the normal force is represented as an arrow pointing directly away from the surface. **Figure 4-2** shows both how the normal force is usually represented and a way to think about what is actually causing the normal force. The normal force gets its name from the mathematical definition for normal, which is "meets at a 90° angle." For two lines we usually call this "perpendicular." "Normal" is commonly used when referring to a surface. The normal force is always exerted by a surface on an object in contact with the surface in a direction away from and perpendicular to the surface. If that surface is a plane the normal is perpendicular to any line in the plane. If the surface is curved, you would draw a tangent to the surface at the point of interest, and the line perpendicular to that tangent is normal to the surface at that point.

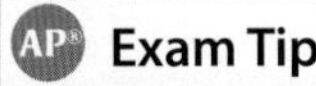

Exam Tip

While the friction force and gravitational force are on the equation sheet, the normal force is not. This is because the former two forces have a single equation that defines them, which the normal force does not. The normal force can only ever be determined by an application of Newton's first or second law to the system of interest.

WATCH OUT!

Any solid surface will exert some normal force on an object coming into contact with it, no matter how slippery the surface is.

On a micro level, this contact force is actually due to other fundamental forces exerted between atomic components (primarily electrons) of the object and the surface. Unlike the fundamental force of gravity, which always attracts, the fundamental force between charged objects can attract *or* repel. When you push on a surface, the electrons in your hand get repelled by the electrons in the surface. The electrons inside the surface get squeezed together, sort of like the springs of a mattress when you push on it (Figure 4-2). Eventually the repelling force exerted by your hand balances the repelling force from squeezing the electrons together and you cannot push into the surface any further. If the surface is "hard" this happens quite rapidly. There are many, many electrons, so any sideways forces cancel, and the force is always directly away from the surface. Since we just want to use the object model and not worry about what is going on inside, on a macro level we simply refer to it as a *normal force*.

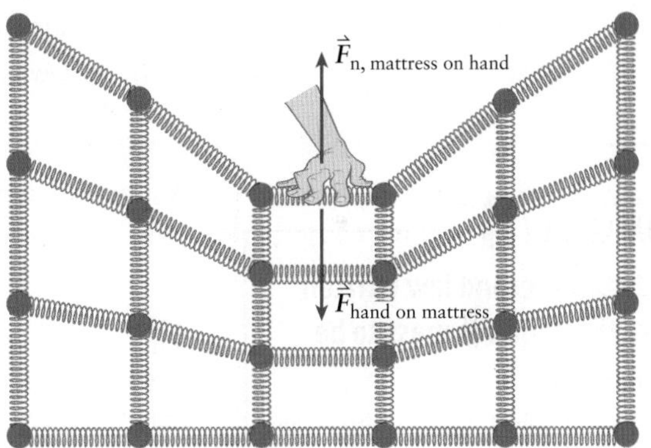

Figure 4-2 Springs The picture represents the way that springs in a mattress interact with each other as you compress them. As a mattress compresses it pushes back on your hand as hard as your hand pushes down on the mattress. How far the mattress compresses depends on how hard you push and how firm the mattress is.

Although contact forces may make calculations easier, to understand the world around you, it's important that you know that contact forces all arise from the electric force exerted between microscopic objects at the interatomic or intermolecular level. The electric force is a component of the fundamental electroweak force. The normal force prevents the softball player from falling through the ground below her. If you're sitting in a chair right now, you're feeling an upward normal force exerted by the chair on your rear end; if you're standing, you're feeling an upward normal force exerted on your feet by the ground. You don't have to think about electrons to describe these forces.

The **kinetic friction force** on the softball player (which we denote by $\vec{F}_k$) is exerted in the direction parallel to the surface and opposite to the player's motion. The friction force arises from chemical bonds that form between atoms in the ground and atoms in the player's uniform, as well as surface roughness. These bonds resist being broken, so a gentle push may not begin to move an object that is initially at rest. When this occurs, it's called a **static friction force** (static, for standing still). Once an object begins to slide, however, the bonds begin to be broken. The sliding object, in this case the softball player, feels the kinetic friction force, which is smaller than the static friction force,

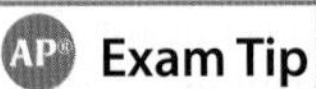

Exam Tip

The three fundamental forces of nature are the gravitational force, the electroweak force, and the strong force. Knowing the existence of these three fundamental forces is required for the course, but on the AP® Physics 1 exam, you are only expected to do calculations using the gravitational force and the various contact forces described in Unit 2.

slowing her down as she slides. On a micro level, chemical bonds are also primarily a result of the electric force, but unlike the repulsive interaction that causes the normal force, this one is an attractive interaction between electrons and nuclei. However, for our purposes, the macro description of the contact force gives us all the information we need to solve most problems.

The three forces that are exerted on the softball player in Figure 4-1 are called **external forces** (denoted as F_{ext}) because they are exerted by other objects outside of the player's body. These are Earth's gravitational force and the normal force and kinetic friction force exerted by the ground. **Internal forces** are also exerted by one part of the player's body on another. One example is the force that the player's shoulder exerts on her left arm to keep it from being pulled off as she slides. While internal forces are important for understanding the interactions of one part of the player's body with another, these forces do not affect the motion of the player and so are *not* included in the sum in Equation 4-1. That's why the left-hand side of that equation has the subscript "ext" for "external." In a moment we'll see why it's appropriate to include only external forces in Equation 4-1. We will not discuss internal forces again until we begin to use another description for interactions in Chapter 7.

Newton's Second Law

In the late seventeenth century the English physicist and mathematician Sir Isaac Newton (1642–1727) published a treatise in which he described three fundamental relationships between force and motion. These relationships, called **Newton's laws of motion**, explain nearly all physical phenomena in our everyday experience. One of these relationships, commonly known as **Newton's second law**, is the most commonly used. It states:

If a net external force is exerted on an object, the object accelerates in the direction of the force. The net external force is equal to the product of the object's mass and acceleration.

Mathematically this is expressed directly as (as long as all quantities are in appropriate units)

$$\sum \vec{F}_{\text{ext}} = m\vec{a}$$

WATCH OUT !

Newton's second law written like this requires mass to be constant!

Fortunately, for many situations we encounter, the mass of the object in which we are interested does not change, and so we can use this form of Newton's second law. In Chapter 9, we will learn a more general relationship that derives from a conservation principle that simplifies to this one for constant-mass situations.

The magnitude of acceleration that the net external force causes depends on the mass m of the object (which depends on the quantity of material in the object and that material's density). The greater the mass, the smaller the acceleration.

We have spent the last two chapters developing the skills necessary to model and describe motion. We found that acceleration is always a description of the change of the motion of an object or system and our kinematic equations describe the motion of objects with constant (possibly zero) acceleration. Since constant accelerations are common in our everyday existence, these equations give us a powerful tool for predicting motion. Forces are just one way to describe the interactions that can cause changes in motion (although very handy ones), so it is appropriate to make acceleration the focus of our equation, allowing us to use it to predict motion through the kinematic equations we have already learned.

This also makes sure we don't fall into the trap of thinking that just because we know an object's acceleration, we know the force exerted on the object. Remember, the acceleration is due to the sum of all the forces exerted on the object. So we are less likely to think about this the wrong way if instead we choose to write Newton's second law as:

EQUATION IN WORDS
Newton's second law of motion

The acceleration of an object...

...is proportional to and in the same direction as the net external force exerted on the object...

$$\vec{a} = \frac{\sum \vec{F}_{\text{ext}}}{m} \quad (4\text{-}2)$$

...and inversely proportional to the mass of the object.

Newton's second law is simple to state but can be challenging to fully understand. Here are its three essential features:

1. *A net force in a certain direction causes acceleration in that direction.* The softball player shown in Figure 4-1 has a backward acceleration (she slows down) because the net force exerted on her points opposite to the direction of her motion (**Figure 4-3a**). If we ignore air resistance, only a single force is exerted on a falling basketball, the downward gravitational force. So the net force on the ball is downward, and it accelerates downward (**Figure 4-3b**).
2. *The magnitude of acceleration caused by the net force depends on the object's inertial mass.* **Inertia** means resistance to change in motion. Recall that a change in motion refers to a change in velocity, so this means inertia is resistance to acceleration. **Inertial mass** is the property of an object or system that determines how its motion changes when it interacts with other objects and systems. The **gravitational mass** of an object determines the amount of force exerted on an object through a gravitational interaction. From these definitions, it would not seem obvious that these two quantities, gravitational and inertial mass, would be identical. Einstein's theory of general relativity was based on his assumption that these two quantities were in fact the same thing. General relativity is not a part of this course, but impacts it. Einstein's assumption was experimentally verified so we simply refer to both quantities as **mass**, a measure of how much matter an object has. Thus mass quantifies both the inertia of an object and the gravitational force that Earth exerts on it. The SI unit of mass is the **kilogram**, abbreviated kg. To provide some scale, a liter of water has a mass of 1 kg. Newton's second law says that the product of an object's mass and the object's acceleration equals the net external force on the object. So if you deliver identically strong kicks to a tennis ball (mass $m = 0.058$ kg) and to a soccer ball (mass $m = 0.43$ kg), the more massive soccer ball will experience less acceleration while it's in contact with your foot and will fly off with a smaller speed than the tennis ball will.
3. *Only external forces exerted on an object affect that object's acceleration.* As an example, sit in your chair with your feet off the ground and pull upward on your belt with both hands. No matter how hard you try, you can't lift yourself out of the chair! Your body has to accelerate to rise out of the chair (it has to go from being stationary to being in motion), but the force of your hands on your belt is an *internal* force (one part of your body pulls on another part).

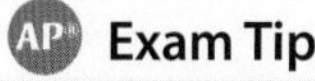
Exam Tip

Remember that the direction of acceleration is not necessarily the direction of motion of an object or a system; a common error is to assume that those are always the same. Acceleration is always in the direction of change of motion. Acceleration is only in the direction of motion for objects that are speeding up along a line or starting from rest.

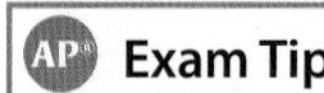
Exam Tip

When determining quantities on the AP® Physics 1 exam, you are not expected to differentiate between inertial or gravitational mass, but you are expected to cite the difference between the two, and you might be asked a qualitative question to elicit this understanding.

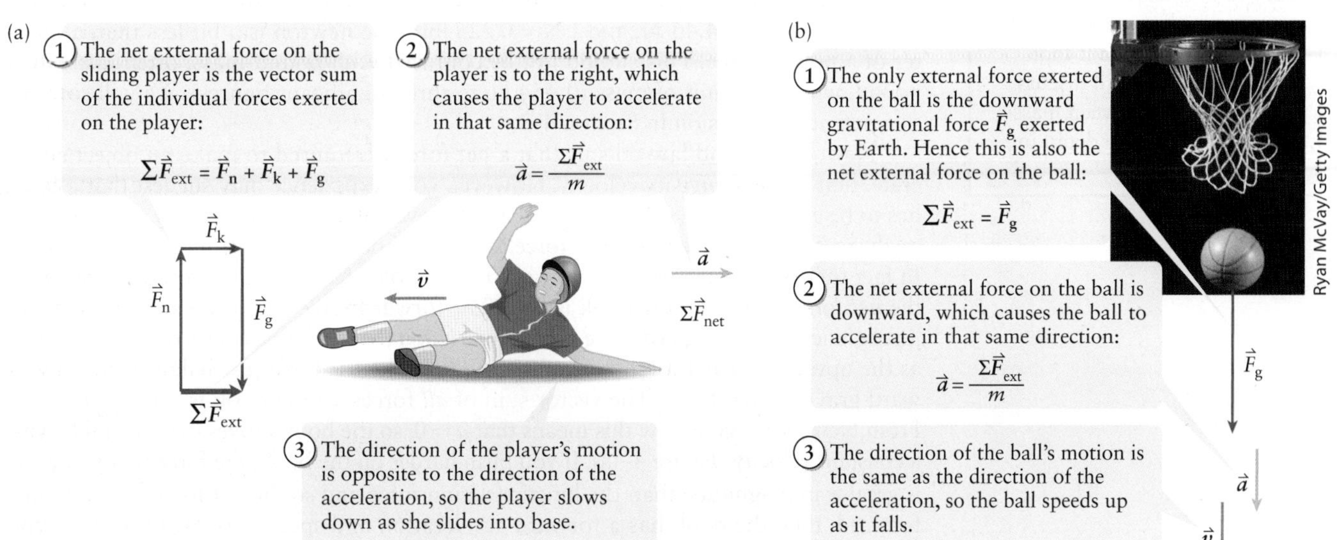

(a)
1 The net external force on the sliding player is the vector sum of the individual forces exerted on the player:
$\Sigma\vec{F}_{ext} = \vec{F}_n + \vec{F}_k + \vec{F}_g$
2 The net external force on the player is to the right, which causes the player to accelerate in that same direction:
$\vec{a} = \frac{\Sigma\vec{F}_{ext}}{m}$
$\vec{F}_k$
$\vec{F}_n$
$\vec{F}_g$
$\Sigma\vec{F}_{ext}$
$\vec{v}$
$\vec{a}$
$\Sigma\vec{F}_{net}$
3 The direction of the player's motion is opposite to the direction of the acceleration, so the player slows down as she slides into base.
(b)
1 The only external force exerted on the ball is the downward gravitational force $\vec{F}_g$ exerted by Earth. Hence this is also the net external force on the ball:
$\Sigma\vec{F}_{ext} = \vec{F}_g$
2 The net external force on the ball is downward, which causes the ball to accelerate in that same direction:
$\vec{a} = \frac{\Sigma\vec{F}_{ext}}{m}$
$\vec{F}_g$
3 The direction of the ball's motion is the same as the direction of the acceleration, so the ball speeds up as it falls.
$\vec{a}$
$\vec{v}$
Ryan McVay/Getty Images

Figure 4-3 The net external force determines acceleration The net external force on an object is the vector sum $\sum \vec{F}_{ext}$ of the individual forces exerted on that object. The acceleration $\boldsymbol{a}$ produced by the net external force is in the same direction as $\sum \vec{F}_{ext}$.

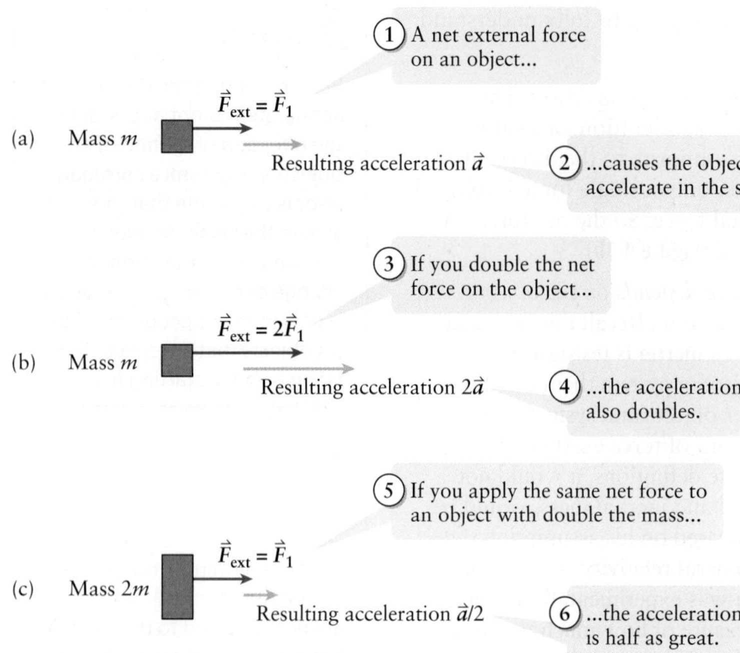

(a) Mass m
$\vec{F}_{ext} = \vec{F}_1$
Resulting acceleration $\vec{a}$
(1) A net external force on an object...
(2) ...causes the object to accelerate in the same direction.
(3) If you double the net force on the object...
(b) Mass m
$\vec{F}_{ext} = 2\vec{F}_1$
Resulting acceleration $2\vec{a}$
(4) ...the acceleration also doubles.
(5) If you apply the same net force to an object with double the mass...
(c) Mass $2m$
$\vec{F}_{ext} = \vec{F}_1$
Resulting acceleration $\vec{a}/2$
(6) ...the acceleration is half as great.

Figure 4-4 Net force, mass, and acceleration Newton's second law, $\vec{a} = \frac{\sum \vec{F}_{ext}}{m}$, tells us how much acceleration results when a net external force $\sum \vec{F}_{ext}$ is exerted on an object of mass m.

The acceleration of an object is directly proportional to the net external force that is exerted on an object. The acceleration of an object is inversely proportional to the mass m of the object.

$$\vec{a} = \frac{\Sigma \vec{F}_{ext}}{m}$$

WATCH OUT !

Mass and weight are not the same thing.

An object's mass is independent of the object's location; for example, an astronaut walking on the Moon has the same mass as she has on Earth because the amount of matter in her body is exactly the same on either world. However, the astronaut has less *weight* on the Moon because the Moon exerts a smaller gravitational force on the astronaut than the Earth does. We'll discuss the distinction between mass and weight in more detail in Section 4-3.

Thus this force can't produce an acceleration. A helpful friend could lift you out of your chair, but that would happen because your friend exerts an *external* force (one that originates outside your body).

Figure 4-4 illustrates the ideas of Equation 4-2.

Units of Force

The SI unit of force is the **newton**, abbreviated N. (An uppercase abbreviation is used for units that bear a person's name.) A net force of one newton applied to an object with a mass of one kilogram gives the object an acceleration of 1 m/s^2. From Equation 4-2, we can see that $\sum \vec{F}_{ext} = m\vec{a}$ so it follows that

$$1 \text{ N} = 1 \text{ kg} \times 1 \text{ m/s}^2 = 1 \text{ kg} \cdot \text{m/s}^2$$

The English unit of force is the **pound-force** (abbreviated lbf). To three significant digits, 1 lbf equals 4.45 N, and 1 N = 0.225 lbf. One newton is a bit less than a quarter of a pound-force. This should not be confused with pound-mass, often just called pound, which is a unit of mass (0.454 kg to three significant figures). We will sort out this potential confusion in Section 4-3.

Newton's second law tells us that a net force is required to make an object accelerate, that is, to *change* its velocity. However, your experience may suggest that a force has to be exerted on an object to make it move even at a *constant* velocity. After all, you might reason, you have to exert a force to make a book slide across a table (**Figure 4-5a**). In fact there's no contradiction between these two statements! The explanation is that the *net* force on the sliding book is zero. The forward force that you exert on the book just balances the backward force of kinetic friction that the table exerts on the book, just as the upward normal force exerted on the book by the table just balances the downward gravitational force. The vector sum of *all* forces on the book is zero: $\sum \vec{F}_{ext} = 0$. From Newton's second law this means that $\vec{a} = 0$, so the book moves over the table with a constant velocity (Figure 4-4a). If you push harder on the book, the force that you exert is greater in magnitude than the kinetic friction force and so the net force $\sum \vec{F}_{ext}$ points forward; then the book has a forward acceleration and speeds up (**Figure 4-5b**). If you stop pushing the sliding book, the only force left along the surface of the table is the kinetic friction force, so the net force $\sum \vec{F}_{ext}$ points backward, giving the book a backward acceleration until it slows to a stop (**Figure 4-5c**). Just remember that it's the *net* force exerted on an object that determines its acceleration, not any one particular force.

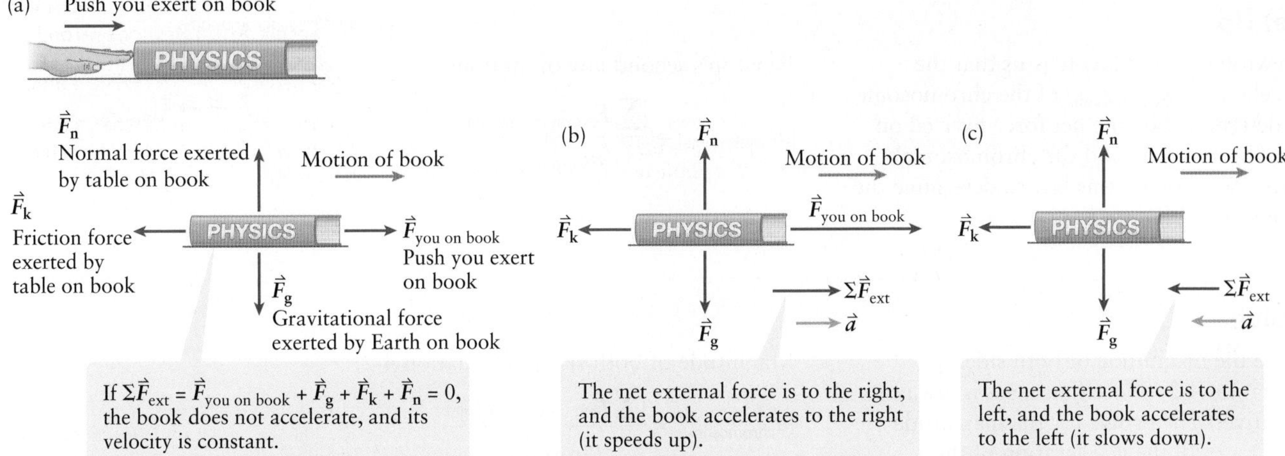

Figure 4-5 It's the net external force that matters (a) If you consider only the force that *you* apply to this book, you might think that a force is required to maintain constant velocity. If instead you consider *all* of the forces exerted on the book, you'll see that a zero *net* external force keeps the book's velocity constant. If the net external force on the book is not zero, the book accelerates in the direction of the net force.

The sum $\sum \vec{F}_{ext}$ in Newton's second law involves vector addition. We saw in Chapter 3 that it's usually easiest to add vectors if we use components, so we'll often use Equation 4-2 in component form:

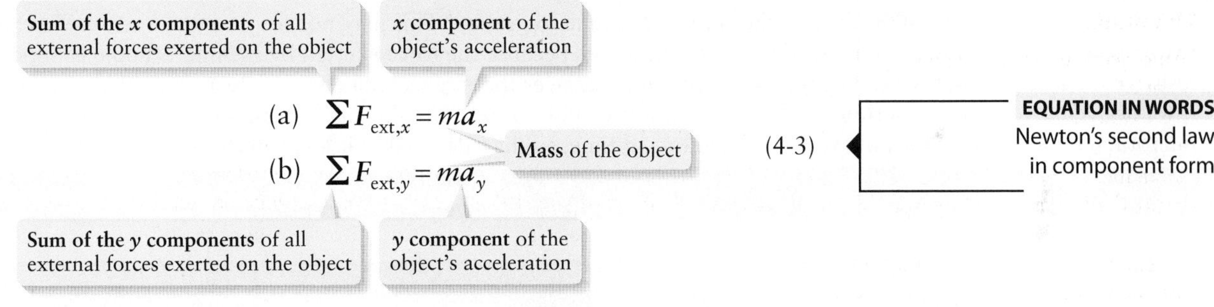

$$\text{(a)}\quad \sum F_{ext,x} = ma_x \qquad \text{(4-3)}$$
$$\text{(b)}\quad \sum F_{ext,y} = ma_y$$

The two examples below illustrate how to use Newton's second law to relate net force, mass, and acceleration.

EXAMPLE 4-1 Small but Forceful

Microtubules are assembled from protein molecules (**Figure 4-6**). Microtubules help cells maintain their shape and are responsible for various kinds of movements within cells, such as pulling apart chromosomes during cell division. Measurements show that microtubules can exert forces from a few pN ($1\ \text{pN} = 1\ \text{piconewton} = 10^{-12}\ \text{N}$) up to hundreds of nN ($1\ \text{nN} = 1\ \text{nanonewton} = 10^{-9}\ \text{N}$). A particular bacterial chromosome has a mass of 2.00×10^{-17} kg. If a microtubule exerts a force of 1.00 pN on the chromosome, what is the magnitude of the chromosome's acceleration? (Ignore any other forces that might be exerted on the chromosome.)

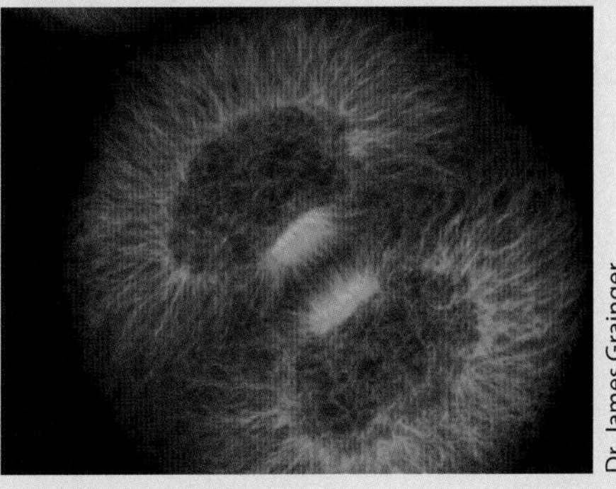

Figure 4-6 Microtubules Microtubules (shown here in green within a fertilized sea urchin egg undergoing division) are protein molecules found within cells. They help cells hold their shape and are responsible for various kinds of movements within cells.

Set Up

Newton's second law tells us that the acceleration $\vec{a}_{chromosome}$ of the chromosome is determined by the net force exerted on the chromosome and the chromosome's mass. We will use this law to determine the magnitude of $\vec{a}_{chromosome}$.

Newton's second law of motion:

$$|\vec{a}_{chromosome}| = \frac{\left|\sum \vec{F}_{ext\ on\ chromosome}\right|}{m_{chromosome}} \qquad (4\text{-}2)$$

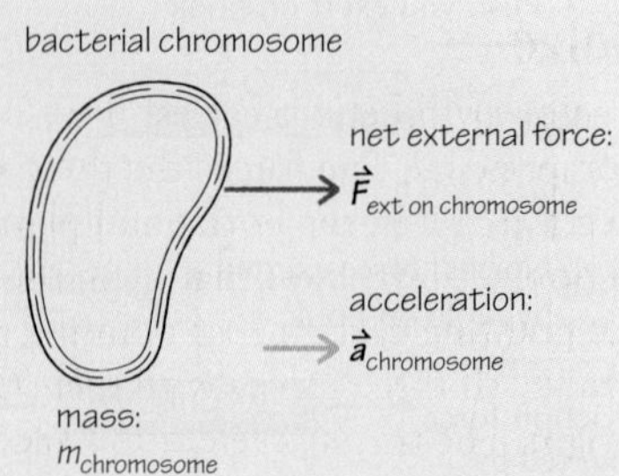

Solve

Take the magnitude of both sides of Equation 4-2. (Note that mass is always positive.) Then solve for the magnitude $a_{chromosome}$ of the acceleration of the chromosome.

Magnitude of both sides of Equation 4-2:

$$|\vec{a}_{chromosome}| = \frac{\left|\sum \vec{F}_{ext\ on\ chromosome}\right|}{m_{chromosome}}$$

$$a_{chromosome} = \frac{1.00\ \text{pN}}{2.00 \times 10^{-17}\ \text{kg}} = \frac{1.00 \times 10^{-12}\ \text{N}}{2.00 \times 10^{-17}\ \text{kg}}$$
$$= 5.00 \times 10^{4}\ \text{N/kg}$$

Use the definition $1\ \text{N} = 1\ \text{kg} \cdot \text{m/s}^2$:

$$a_{chromosome} = (5.00 \times 10^{4}\ \text{N/kg})\left(\frac{1\ \text{kg} \cdot \text{m/s}^2}{1\ \text{N}}\right)$$
$$= 5.00 \times 10^{4}\ \text{m/s}^2$$

Reflect

This acceleration is about 5000 times g, the acceleration due to gravity! This may not seem to make sense. We cannot really use our common sense to judge this answer, however, because the force exerted by the microtubule is opposed by resistive forces that we chose to neglect. When we get an answer that seems unlikely, it is always important to consider if we neglected anything important. So the acceleration would not be quite this big in reality. The forces exerted by microtubules are exerted for only very short periods of time, so that also makes it less scary!

NOW WORK Problems 5, 6, and 7 from The Takeaway 4-2.

EXAMPLE 4-2 A Three-Person Tug-of-War

Jesse, Karim, and Luis, three hungry students, are each pulling on a cafeteria tray laden with desserts. The mass of the tray and its contents is 2.50 kg. The tray sits on a horizontal dining table, and each of the students pulls horizontally on the tray. Jesse pulls due south with a force of magnitude 85.0 N, and Karim pulls with a force of magnitude 90.0 N in a direction 35.0° north of east. If the tray briefly accelerates due north at 20.0 m/s^2, how hard and in what direction does Luis pull on the tray during that time? Ignore the effects of friction. Using complete sentences, justify how you have identified all of the forces exerted on the tray.

Set Up

The tray accelerates due to the net force that is exerted on it, which is the vector sum of the individual forces exerted by Jesse, Karim, and Luis. We've drawn these individual force vectors with their tails together. (We're told to ignore friction, so no other horizontal forces are exerted on the tray. The tray doesn't accelerate vertically, so there's no net vertical force—the upward normal force exerted by the table top balances the downward gravitational force.) Because we know the mass and acceleration of the tray, we can use Newton's second law to calculate the net force on the tray. We'll then use vector addition and subtraction to determine the unknown force exerted by Luis.

$$\vec{a}_{tray} = \sum \vec{F}_{ext\ on\ tray} / m_{tray} \qquad (4\text{-}2)$$

In component form:

$$\sum F_{ext\ on\ tray,x} = m_{tray}\, a_{tray,x} \qquad (4\text{-}3a)$$
$$\sum F_{ext\ on\ tray,y} = m_{tray}\, a_{tray,y} \qquad (4\text{-}3b)$$

Net force on tray = sum of forces exerted by Jesse, Karim, and Luis:

$$\sum \vec{F}_{ext\ on\ tray} = \vec{F}_{Jesse\ on\ tray} + \vec{F}_{Karim\ on\ tray} + \vec{F}_{Luis\ on\ tray}$$

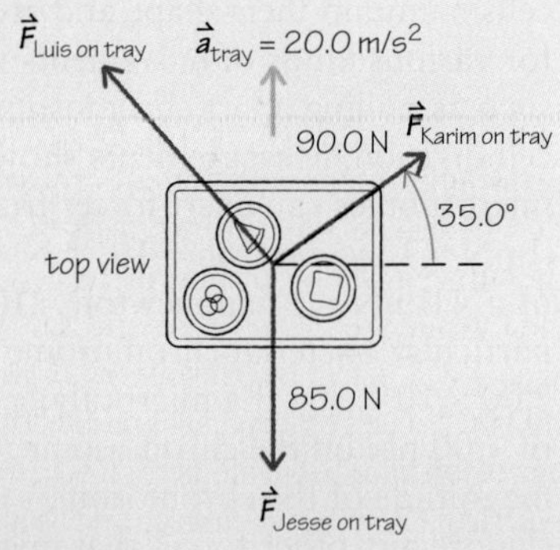

Solve

Vector arithmetic is easiest if we use vector components. The forces the three students exert are all in the horizontal plane, so we choose x and y axes that lie in this plane with $+x$ pointing east and $+y$ pointing north. We've drawn all three forces with their tails at the origin. The tray's acceleration vector $\vec{a}_{tray}$ points in the positive y direction.

$$a_{tray,x} = 0$$
$$a_{tray,y} = 20.0\ \text{m/s}^2$$

Use Newton's second law in component form, Equations 4-3a and 4-3b, to calculate the x and y components of the net force on the tray. Recall that $1\ \text{kg} \cdot \text{m/s}^2 = 1\ \text{N}$.

$$\sum F_{\text{ext on tray},x} = m_{tray} a_{tray,x} = m_{tray}(0) = 0$$
$$\sum F_{\text{ext on tray},y} = m_{tray} a_{tray,y} = (2.50\ \text{kg})(20.0\ \text{m/s}^2)$$
$$= 50.0\ \text{kg} \cdot \text{m/s}^2 = 50.0\ \text{N}$$

The x component of the net force on the tray is the sum of the x components of the individual forces on the tray, and similarly for the y component.

$$\sum F_{\text{ext on tray},x} = 0$$
$$= F_{\text{Jesse on tray},x} + F_{\text{Karim on tray},x} + F_{\text{Luis on tray},x}$$
$$\sum F_{\text{ext on tray},y} = 50.0\ \text{N}$$
$$= F_{\text{Jesse on tray},y} + F_{\text{Karim on tray},y} + F_{\text{Luis on tray},y}$$

Solve for the components of the force that Luis exerts on the tray:

$$F_{\text{Luis on tray},x} = -F_{\text{Jesse on tray},x} - F_{\text{Karim on tray},x}$$
$$F_{\text{Luis on tray},y} = 50.0\ \text{N} - F_{\text{Jesse on tray},y} - F_{\text{Karim on tray},y}$$

To proceed, we need to know the x and y components of the forces that Jesse and Karim exert on the tray. Once we find the forces, we will write our justification that we have identified all forces exerted on the tray.

$\vec{F}_{\text{Jesse on tray}}$:
magnitude 85.0 N, points due south (in the negative y direction), so

$$F_{\text{Jesse on tray},x} = 0$$
$$F_{\text{Jesse on tray},y} = -85.0\ \text{N}$$

$\vec{F}_{\text{Karim on tray}}$:

magnitude 90.0 N, points 35.0° north of east, so $F_{\text{Karim on tray},x}$ and $F_{\text{Karim on tray},y}$ are both positive.

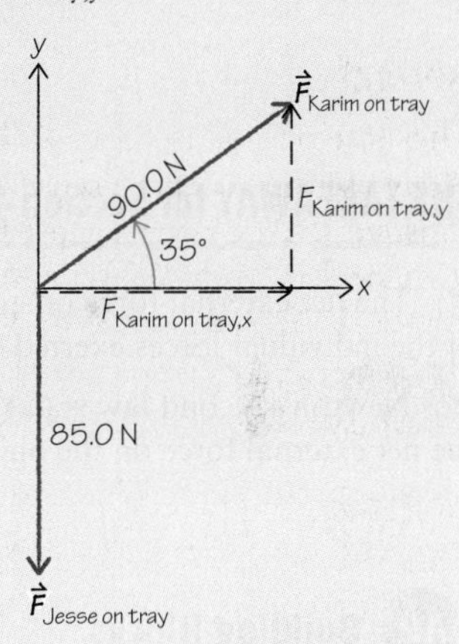

$$F_{\text{Karim on tray},x} = (90.0\text{N}) \cos 35.0°$$
$$= 73.7\ \text{N}$$
$$F_{\text{Karim on tray},y} = (90.0\ \text{N}) \sin 35.0°$$
$$= 51.6\ \text{N}$$

Substitute these values into the expressions for the components $F_{\text{Luis on tray},x}$ and $F_{\text{Luis on tray},y}$ and solve.

$$F_{\text{Luis on tray},x} = -(0) - (73.7\ \text{N})$$
$$= -73.7\ \text{N}$$
$$F_{\text{Luis on tray},y} = 50.0\ \text{N} - (-85.0\ \text{N}) - (51.6\ \text{N})$$
$$= 83.4\ \text{N}$$

The force that Luis exerts, $\vec{F}_{\text{Luis on tray}}$, has a negative x component and positive y component, so it points west of north. Calculate the magnitude of $\vec{F}_{\text{Luis on tray}}$ using the Pythagorean theorem and the direction of $\vec{F}_{\text{Luis on tray}}$ using trigonometry. The students are touching the tray, so are exerting contact forces on it. The table is also touching the tray, exerting the normal force, another contact force. The action is going on near Earth's surface, so the gravitational force, a long-range fundamental force, must also be included.

Magnitude of $\vec{F}_{\text{Luis on tray}}$:

$$F_{\text{Luis on tray}} = \sqrt{(F_{\text{Luis on tray},x})^2 + (F_{\text{Luis on tray},y})^2}$$
$$= \sqrt{(-73.7\ \text{N})^2 + (83.4\ \text{N})^2}$$
$$= 111\ \text{N}$$

Angle θ of $\vec{F}_{\text{Luis on tray}}$ measured west of north:

$$\tan\theta = \frac{73.7\ \text{N}}{83.4\ \text{N}} = 0.884$$
$$\theta = \tan^{-1} 0.884 = 41.5°$$

Reflect

Luis pulls on the tray with a force of 111 N at an angle of 41.5° west of north. In terms of components, Luis pulls west (in the negative x direction) with a force of 73.7 N, thus canceling Karim's 73.7-N pull to the east. Luis also pulls north (in the positive y direction) with a force of 83.4 N; combined with Karim's 51.6-N northward pull, Luis's pull overwhelms Jesse's 85.0-N pull toward the south. Hence the net force on the tray and its delicious contents is northward, and its acceleration is northward as well. The sizes we calculated seemed to make sense given the size of the other students' exerted forces, and the directions came out as expected.

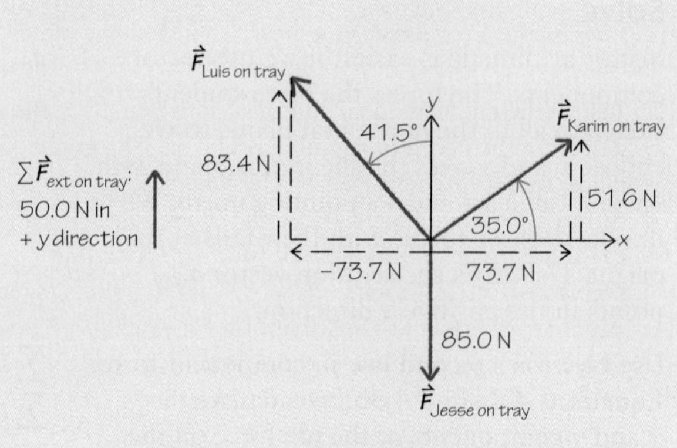

NOW WORK Problems 8 and 9 from The Takeaway 4-2.

Example 4-2 illustrates an important point: *It's always wise to choose coordinate axes that align with one or more of the vectors.* In this example we chose the axes to be east–west and north–south so that the tray's acceleration $\vec{a}_{\text{tray}}$ and the force $\vec{F}_{\text{Jesse on tray}}$ exerted by Jesse were both along one of the coordinate axes. This choice made it easy to express these vectors in terms of components and so simplified the calculation. Using a different choice of coordinate axes (say, one in which the positive x axis pointed 30° north of west and the positive y axis pointed 30° east of north), you would have ended up with the same result, but with a good deal more effort.

THE TAKEAWAY for Section 4-2

✔ The net external force on an object is the vector sum of all of the individual forces exerted on it by other objects.

✔ Newton's second law states that an object accelerates if the net external force on the object is not zero.

✔ The acceleration is in the same direction as the net force, is directly proportional to the magnitude of the net force, and is inversely proportional to the mass of the object (a measure of the quantity and type of material in the object).

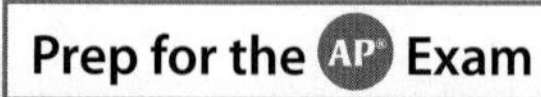

AP Building Blocks

1. What are the basic SI units (kg, m, s) for acceleration? What are the basic SI units for force according to Newton's second law $\sum \vec{F}_{\text{ext}} = m\vec{a}$?
2. If the sum of the forces exerted on an object equals zero, does this imply that the object is at rest?
3. The net force on a moving object suddenly becomes zero and remains zero. The object will
   (A) stop abruptly.
   (B) reduce speed abruptly.
   (C) reduce speed gradually.
   (D) continue to move at constant velocity.
4. In the absence of a net force, an object cannot
   (A) remain at rest.
   (B) be in motion with a constant velocity.
   (C) be accelerating.
   (D) be displaced.
5. EX 4-1 According to Newton's second law of motion, when a net force is exerted on an object, the acceleration is
   (A) inversely proportional to the object's mass.
   (B) independent of the object's mass.
   (C) inversely proportional to the net force exerted on the object.
   (D) directly proportional to the object's mass.
6. EX 4-1 What is the acceleration of a 2.00×10^3-kg car if the net force on the car is 4.00×10^3 N?
7. EX 4-1 What net force would need to be exerted on a 2.00×10^3-kg car to accelerate it at 4.00 m/s^2?
8. EX 4-2 Two forces of magnitudes 30 N and 70 N are exerted on an object. What are the minimum and maximum values for the magnitude of the sum of these two forces?
9. EX 4-2 A block on the ground is initially at rest. Two forces are exerted on the block as shown in the drawing. To move the block in the positive x direction, what minimum external force, F_1, must be exerted? Express your answer in terms of F_1 and the angle, θ. Draw a diagram of the right-angle triangle that justifies your expression. For this problem, do not consider acceleration in the y direction.

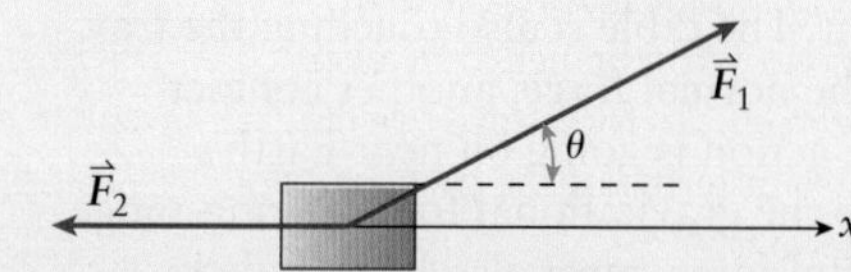

AP® Skill Builders

10. Justify the mathematical routine you would use to calculate the acceleration of a 1-kg box, given that the net external force exerted on the box is 2 N. Justify two different but equivalent routines that you could use to calculate the new net external force needed to accelerate the same box at 4 m/s^2.
11. A person exerts a 60-N horizontal force on a box that causes the box to move across the floor at constant velocity. Predict the mathematical dependence of velocity on time in a graph of the velocity of the box versus time. Predict the way the graph of this relationship would change if the force the person exerted on the box were suddenly doubled.
12. A marble is at rest in the aisle of a train car. Qualitatively predict the displacement of the marble relative to the smooth, horizontal floor of the car as the train enters and leaves a station.
13. During the sudden impact of a car accident, a person's neck can experience abnormally large forces, resulting in an injury commonly known as whiplash. From the relative motion of the head and the neck, as shown in the figure, predict the direction from which the car in which the person is riding was struck in the collision.

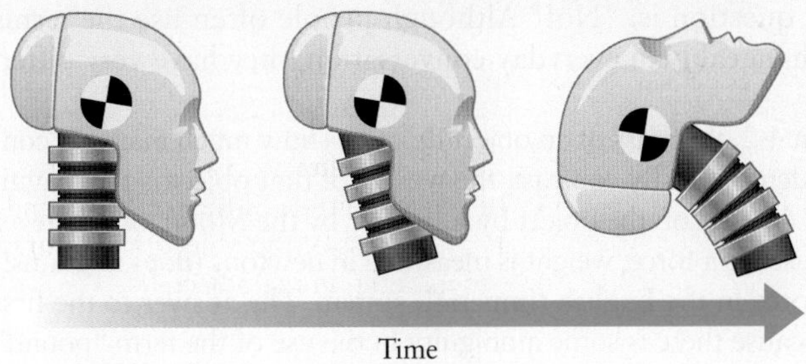

14. Three forces are exerted on an object at angles θ_1, θ_2, and θ_3, as shown in the figure.

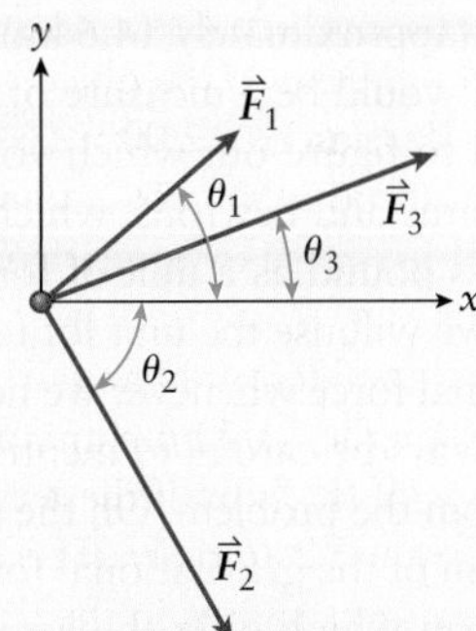

(a) Construct a representation of the sums of the components of the forces, F_{1x}, F_{1y}, F_{2x}, F_{2y}, F_{3x}, and F_{3y}.
(b) Refine the representation to express these sums in terms of the magnitudes of the forces F_1, F_2, and F_3, and the angles, θ_1, θ_2, and θ_3.
(c) Explain why the net force exerted on the object cannot be zero if the forces are oriented as shown in the figure in terms of the properties of the cosine and sine functions.
(d) Refine the representation if the object, with mass m, is accelerating in the $+x$ direction.
(e) Justify the mathematical routine needed to obtain the magnitudes F_2 and F_3 under these circumstances: $m = 2.00$ kg, $F_1 = 1.00$ N, $\theta_1 = 40.0°$, $\theta_2 = 60.0°$, $\theta_3 = 20.0°$, and the acceleration of the object in the x direction is 1.5 m/s^2.

15. A current exerts a force on the hull of a boat of mass m along the vector $\vec{F}_2$ in the figure. The wind exerts a force along the vector $\vec{F}_1$. The view in the figure is from above the boat.

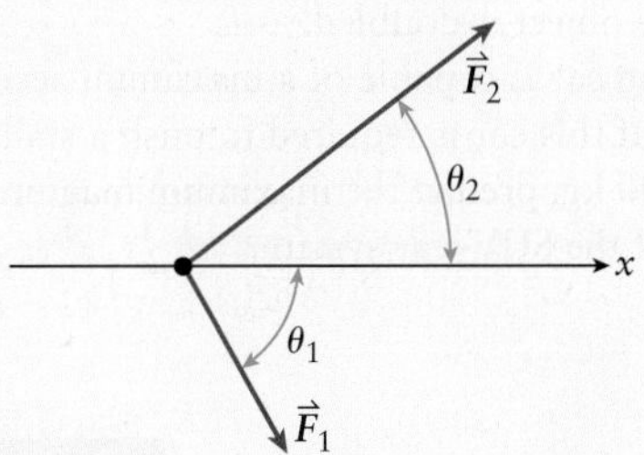

(a) Justify a mathematical routine that can be used to choose the magnitude and direction of a third force exerted by a propeller that would hold the boat in a fixed position.
(b) Redraw the diagram showing the boat and the vectors $\vec{F}_1$ and $\vec{F}_2$. Refine this representation by constructing, using vector addition, the force exerted by the propeller $\vec{F}_3$.
(c) Continue to set up the quantitative solution to this problem by expressing the relationships among the components of the vectors $\vec{F}_1$, $\vec{F}_2$, and $\vec{F}_3$ parallel and perpendicular to the $+x$ direction in terms of the angles θ_1 and θ_2 and the magnitudes $\vec{F}_1$ and $\vec{F}_2$.

The force exerted by the wind has a magnitude of 1.00 kN at an angle θ_1 equal to 60.0° with respect to the $+x$ direction. The force exerted by the current has a magnitude of 2.00 kN at an angle θ_2 equal to 37.0° with respect to the $+x$ direction. The boat and sailor have a combined mass of 2250 kg.

(d) Evaluate the expressions from part (c) using these data to obtain a pair of equations.
(e) Calculate the numerical values of the magnitude and direction of $\vec{F}_3$.

The diesel fuel tank for the boat engine needs to be switched over to the reserve tank. It will take 10 s to make the change. During that time, the engine and propeller will be idle.

(f) Predict the direction and magnitude of the displacement of the boat while the tank is being changed.

AP® Skills in Action

16. Predict the change in the acceleration of the object under the following scenarios (evidence is always needed to support a prediction):
(a) The force exerted on the object is doubled while the mass of the object is fixed.
(b) The mass of the object is halved while the force exerted on the object is fixed.

(c) The force exerted on the object is doubled and the mass of the object is doubled.
(d) The force exerted on the object is doubled and the mass of the object is halved.
(e) The force exerted on the object is halved while the mass of the object is fixed.
(f) The mass of the object is doubled while the force exerted on the object is fixed.
(g) The force exerted on the object is halved and the mass of the object is halved.
(h) The force exerted on the object is halved and the mass of the object is doubled.

17. A 1300-kg car is capable of a maximum acceleration of 5.0 m/s^2. If this car is required to push a stalled SUV of mass 1700 kg, predict the maximum magnitude of acceleration of the SUV–car system.

18. Zu Yang draws a motion diagram of the horizontal component of the motion of an arrow shot from a bow; it shows displacements during equal intervals of time. The diagram shows that as the arrow gets farther from the bow, displacements decrease with distance. His partner Jia claims that an alternative motion diagram in which the displacements during equal intervals of time are nearly identical is supported by reasoning based on the theory of motion and the relationship between force and acceleration.
(a) Evaluate Jia's reasoning.
(b) Explain what Jia may be considering when she says that the horizontal displacements are "nearly" identical. Zu's representation may be more accurate, but not for the right reason. What should he have said to justify his diagram appropriately?

4-3 Mass and weight are distinct but related concepts

How much do you weigh? If you grew up in the United States, Liberia, or the Sultantate of Brunei, chances are your answer will be in pounds. If not, you'll probably use kilograms. We've already declared that the SI units of mass are kilograms. Does that mean that pounds and kilograms are both units of the same quantity? Are weight and mass the same?

The answer to this last question is: "No!" Although people often use the terms "mass" and "weight" interchangeably in everyday conversation, they have very different meanings in science.

As we discussed in Section 4-2, the mass of an object describes how much matter is contained in the object and how dense it is. By contrast, the **weight** of that object is the magnitude of the gravitational force exerted on the object by Earth (or by the Moon, or whatever the object is located on). Because it's a force, weight is measured in newtons (*not* kilograms!) in the SI system and pound-force in the English (Imperial) system. The answer to the first question above is not clear because there is some ambiguity in the use of the term "pound" in the English or U.S. customary systems of units. In some contexts, the term "pound" is used almost exclusively to refer to the unit of force, which technically is the "pound-force" or "pound of force" (lbf) and not the unit of mass (lb). In those contexts, the preferred unit of mass is the slug, which is defined as 1 lbf · s^2/ft (approximately 14.6 kg). In other contexts, the unit "pound" refers to a unit of mass, which would be a measure of the same quantity as kilograms. Since you need to be very careful to figure out which you should be using from the context, it is much easier to use kilograms and newtons, which always represent mass and force respectively. When we do discuss pound as a unit of force (which is what you are measuring when you measure weight), we will use the unit lbf for clarity. We'll use the symbol F_g for the magnitude of the gravitational force whenever we need to consider the weight of an object in a problem. We will not always be careful to mention what is causing the gravitational force, as it is usually obvious from the problem. On the surface of Earth, it is always Earth. We will learn a more general form of the gravitational force in Chapter 6.

Consider an object that has mass m and on which only the gravitational force is exerted. The object is a ball falling in a **vacuum**, a space from which all the air has been removed, so there is no air resistance (**Figure 4-7**). Newton's second law, Equation 4-2, tells us that acceleration a is the gravitational force $\vec{F}_g$ divided by the object's mass:

$$a = F_g/m$$

The magnitude of the gravitational force is the weight F_g, so, rearranging and taking magnitudes, we get

$$F_g = ma$$

where a is the magnitude of the object's acceleration. We know from Section 2-6 that if only the force of gravity is exerted on an object, the object's acceleration

① If an object falls without air resistance near the surface of Earth, the net external force on the object equals the downward gravitational force exerted on that object by Earth, which has magnitude F_g.

② An object falling without air resistance accelerates downward. The magnitude of the acceleration is g.

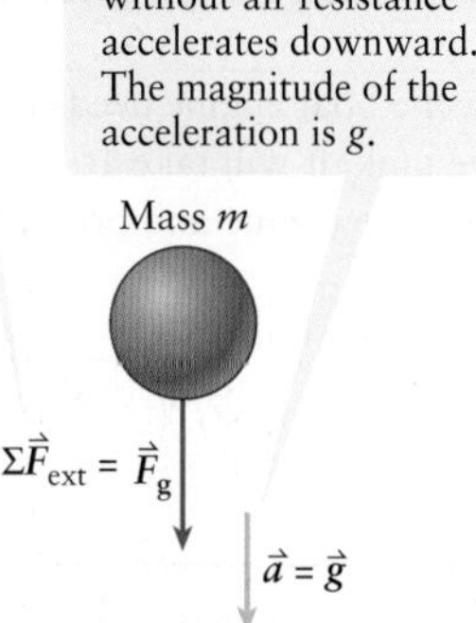

③ Newton's second law tells us that $\vec{F}_g = m\vec{g}$ and so $F_g = mg$.

Figure 4-7 Gravitational force Applying Newton's second law to a freely falling object shows that the magnitude of the gravitational force is $F_g = mg$.

has magnitude g. So we get the following expression for the weight of an object of mass m:

Weight of an object (equal to the magnitude of the gravitational force on that object near the surface of Earth)

$$F_g = mg \tag{4-4}$$

Mass of the object

Magnitude of the acceleration due to gravity near the surface of Earth

EQUATION IN WORDS
Weight of an object of mass m

That is, *the weight of an object near Earth's surface is equal to the object's mass multiplied by g, the acceleration due to gravity near Earth's surface.* This statement makes it clear that the units of weight cannot be the same as the units of mass. In Chapters 2 and 3 we used an average value of $g = 9.80 \text{ m/s}^2$ but recognized that the exact value of g varies slightly from place to place on Earth. Therefore, the weight of an object must change as it is moved from place to place. *Hence weight is not an intrinsic property of an object.* Not only does the weight of an object change depending on where you are on Earth, but as we will see in Chapter 6, it can be significantly different on other worlds such as the Moon and Mars.

AP Exam Tip

Notice that on the AP® Physics 1 equation sheet, $\vec{F}_g$ (vector) is mentioned in two different ways, and neither looks exactly like the equation above. The equation sheet shows a more general form of the gravitational force—the force between any two masses at a distance, not just between Earth and a mass at its surface—and then shows how the gravitational acceleration vector $\vec{g}$ would be calculated from that. The direction of $\vec{g}$ for Earth is always toward its center (toward the ground). We will introduce this more general equation for gravitational force in Chapter 6.

WATCH OUT !

Objects have weight even if the net force exerted on them is zero!

We arrived at Equation 4-4 for the weight of an object by assuming that the object was falling freely. However, Equation 4-4 is true even if the object is not accelerating. A 10.0-kg object near Earth's surface has a weight of 98.0 N whether it's sitting on a table, falling off the table, or being pulled across the floor.

Consider dropping two different objects side by side in a vacuum, so that the only force exerted on either object is the gravitational force due to Earth (**Figure 4-8**). The heavier object has twice the mass, so twice as much gravitational force is exerted on it, and since that is the only force, that net force is twice as much as the one exerted on the lighter object.

So you might expect that the heavier object would have a greater downward acceleration $\vec{a}$. But the heavier object also has a greater mass and Equation 4-2, $\vec{a} = \frac{1}{m}\sum \vec{F}_{ext}$, tells us that acceleration and mass are inversely proportional. The heavier object has twice the gravitational force exerted on it; it also has twice the mass. Hence the quotient $\frac{1}{m}\sum \vec{F}_{ext}$ has the *same* value for both objects, and they both fall with the same acceleration in a vacuum.

Can you lift an object that weighs 10 N? A 1000-N object? In Section 4-2 we mentioned the following conversions, which are valid to three significant digits:

$$1.00 \text{ N} = 0.225 \text{ lbf}$$

$$1.00 \text{ lbf} = 4.45 \text{ N}$$

(You can find more precise conversion factors in Appendix A.) According to Equation 4-4 the mass of an object that weighs 10.0 N is $m = F_g/g = (10.0 \text{ N})/(9.80 \text{ m/s}^2) = 1.02 \text{ kg}$. One liter of water has a mass of 1 kg; that is, it weighs about 10 N. In pound-force, that's 10 multiplied by 0.225 lbf or about 2.2 lbf. You'd have no problem lifting that amount. Many people can lift an object that weighs 1000 N (225 lbf); that's a mass of about 100 kg. Remembering that 1 L of water has a mass of 1 kg and a weight of 10 N can help you get an intuitive feel for the numerical values of masses and forces.

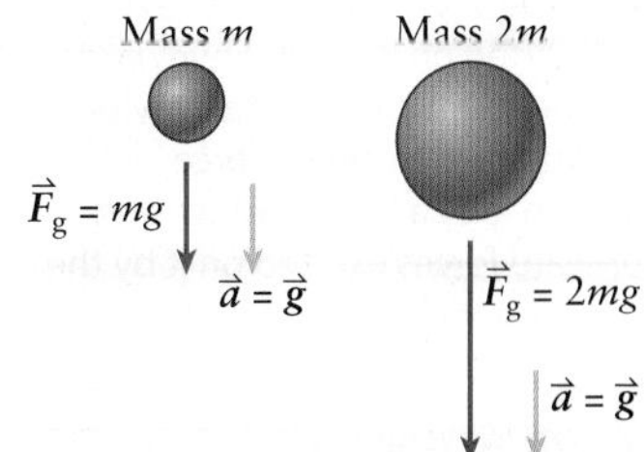

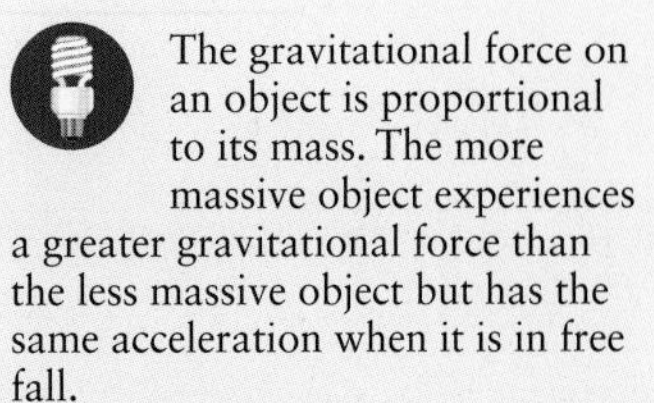

The gravitational force on an object is proportional to its mass. The more massive object experiences a greater gravitational force than the less massive object but has the same acceleration when it is in free fall.

Figure 4-8 Greater mass means greater gravitational force Both of these objects are falling in a vacuum, so the only force exerted on each object is the gravitational force of Earth.

Mass, Inertia, and Newton's First Law

As we discussed in Section 4-2, the concept of mass is intertwined with the observation that all objects *resist changes* in their state of motion. We use the term "inertia" to refer to the tendency of an object to resist a change in motion. For example, a *stationary* object (such as a rock lying on the ground or a roommate sleeping on the couch) will remain stationary unless an external force is exerted on it by another object. An object *in motion*, like a fast-moving hockey puck sliding on ice, also keeps moving in the same direction with the same speed unless an external force is exerted on it. Note that in both situations (stationary and constant velocity), the acceleration is zero. While we mentioned that this has been experimentally verified, you also have experienced this yourself. If you are driving down the road at a constant 55 mi/h, then everything in your car is also moving at 55 mi/h. Because your cup and soda are traveling with you, you can drink your cold soda with confidence as long as you don't hit a bump or put on the brakes. If you do put on the brakes, there is an external force on you (your seat and seat belt) and your cup (you are holding it), but your soda is free to keep moving. This can be a bit cold!

WATCH OUT !

Don't confuse mass, weight, and inertia.

Note the differences between mass, weight, and inertia. Mass is the *quantity of material* in an object; no matter where in the universe you take the object, its mass remains the same. Weight is the *magnitude of the gravitational force* on an object; it is proportional to the mass but depends on the value of g at the object's location. Inertia is the tendency of an object to maintain the same motion and is also a property of the object. We quantify the inertia of an object by the object's mass.

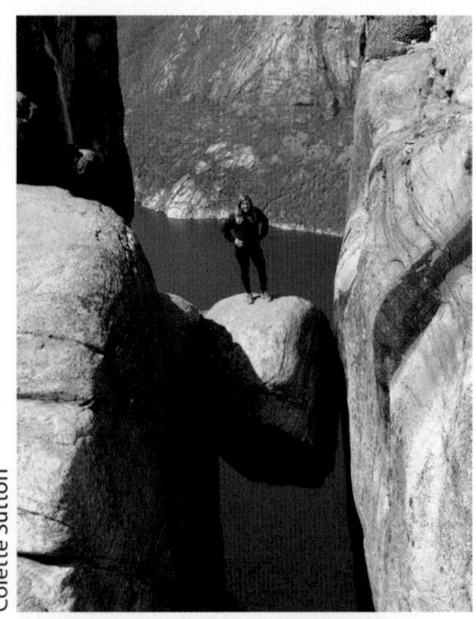

Colette Sutton

Figure 4-9 An object at rest Kjeragbolten is a boulder suspended above a 984-m-deep (3228-ft-deep) abyss in Norway. It remains at rest because the vector sum of the external forces exerted on it—the downward force exerted on it by Earth's gravity, the additional downward force exerted on it by a brave tourist, and the upward forces exerted on it by the surrounding rock—is zero.

Inertia is an intrinsic property of an object: an object's amount of resistance to change in velocity. Inertia is quantified by the object's mass. Just saying "inertia" is shorthand for the "principle of inertia" which is stated in **Newton's first law:**

An object at rest tends to stay at rest, and an object in uniform motion tends to stay in motion with the same speed and in the same direction unless a net external force is exerted upon the object.

Newton's first law contradicts the older theories of motion developed by the ancient Greeks, principally by the philosopher Aristotle (384–322 B.C.). Aristotle believed that every object had a natural place in the world. For example, heavy objects, such as rocks, are naturally at rest on Earth, and light objects, such as clouds, are naturally at rest in the sky. In Aristotle's view, a moving object tended to stop when it found its natural rest position. To keep an object moving, reasoned Aristotle, a force had to be exerted on it; to make it move faster, a greater force was required. Newton saw more deeply. He identified the concept of the *net* force on an object and realized that the net force on an object determines its acceleration rather than its speed. (This realization is Newton's second law.) If the net force on an object is zero, the object has zero acceleration and thus moves with a constant velocity—that is, it keeps the same speed in the same direction. (You should review the discussion of Figure 4-5 in Section 4-2.)

For objects, you can think of Newton's first law as a special case of the *second* law that applies when the net force on an object is zero. When we apply these laws to more complex systems we will see that the second law has a larger meaning that we will discuss briefly at the end of Chapter 9. In equation form we can write the first law as follows:

If the net external force on an object is zero... ...the object does not accelerate...

$$\text{If } \sum \vec{F}_{\text{ext}} = 0, \text{ then } \vec{a} = 0 \text{ and } \vec{v} = \text{constant} \tag{4-5}$$

...and the velocity of the object remains constant. If the object is at rest, it remains at rest; if it is in motion, it continues in linear motion at a constant speed.

EQUATION IN WORDS
Newton's first law of motion

We say that an object is in **equilibrium** if the net external force on it is zero (**Figure 4-9**). A stationary chandelier hanging from the ceiling is in equilibrium: The chain from which the chandelier is suspended exerts an upward force that exactly balances the downward gravitational force on the chandelier. The chandelier's velocity is zero and remains zero. But a moving airliner is also in equilibrium if it flies in a line at a constant speed. The forward force provided by the airliner's engines balances the backward drag force that the air exerts on the airliner, and the upward lift force exerted by air flowing around the wings balances the downward gravitational force. The net force on it is zero. The airliner has a nonzero velocity, but its *acceleration* is zero and so the airliner is in equilibrium.

The following example illustrates how to use Newton's first law to analyze an object at rest.

EXAMPLE 4-3 Let Sleeping Cats Lie

A 40.0-N cat is asleep on a ramp that is tilted by an angle of 15.0° from the horizontal. Three forces are exerted on the cat: the downward gravitational force, a normal force perpendicular to the ramp, and a static friction force directed uphill parallel to the ramp. The cat remains at rest while sleeping. Determine the magnitude of each force exerted on the cat.

Set Up

As in Example 4-2, we've drawn all of the external forces on the cat with their tails touching. The cat's weight is $F_{g,cat} = 40.0$ N; this is just the magnitude of the gravitational force on the cat ($\vec{F}_{g,cat}$), and the direction is straight down. We know the directions of the normal force $\vec{F}_n$ (perpendicular to the ramp) and the friction force $\vec{F}_S$ (uphill parallel to the ramp). Our task is to find the magnitudes of $\vec{F}_n$ and $\vec{F}_S$. We'll do this using Newton's first law: Because the cat remains at rest, the sum of $\vec{F}_{g,cat}$, $\vec{F}_n$, and $\vec{F}_S$ must be zero. So we can solve this problem by using vector addition.

Newton's first law of motion:

$$\sum \vec{F}_{\text{ext on cat}} = \vec{F}_{g,cat} + \vec{F}_n + \vec{F}_S = 0 \quad (4\text{-}5)$$

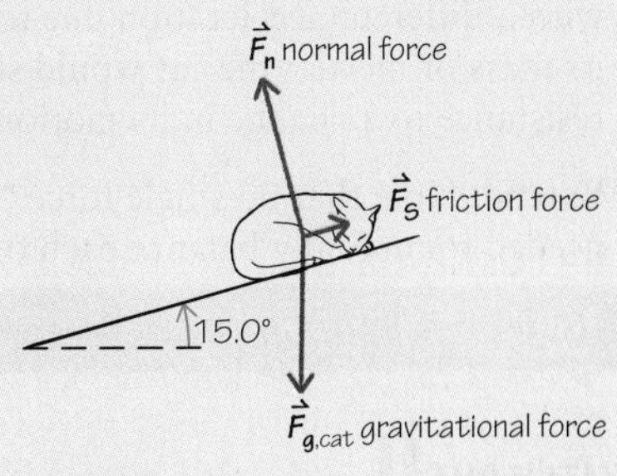

Solve

Like most vector addition problems, this one is most easily solved using components. If $\sum \vec{F}_{\text{ext on cat}} = 0$, then each of the components of $\sum \vec{F}_{\text{ext on cat}}$ must also be equal to zero. It's convenient to choose the x axis to be along the tilted ramp and the y axis to be perpendicular to the ramp. (The x and y axes must be perpendicular to each other but don't have to be horizontal or vertical.) Using this choice, two of the vectors, $\vec{F}_n$ and $\vec{F}_S$, lie either directly along or directly opposite to one of the axes.

Newton's first law in component form:

$$\sum F_{\text{ext on cat},x} = F_{g,cat,x} + F_{n,x} + F_{S,x} = 0$$

$$\sum F_{\text{ext on cat},y} = F_{g,cat,y} + F_{n,y} + F_{S,y} = 0$$

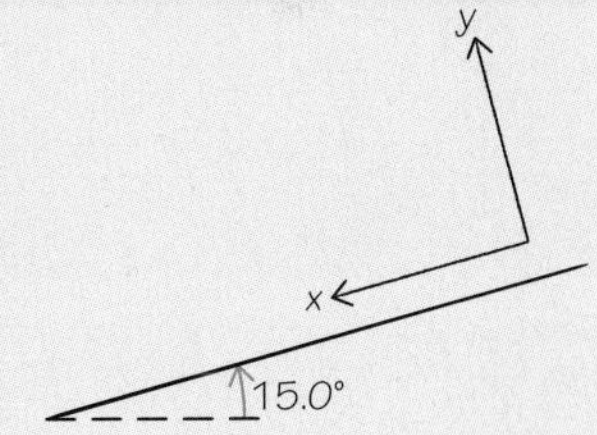

Write the x and y components of each of the external forces in terms of their magnitudes $F_{g,cat}$, F_n, and F_S and the angle $\theta = 15.0°$ of the ramp. Note that $\vec{F}_{g,cat}$ has a positive x component (down the ramp) and a negative y component (into the ramp); $\vec{F}_n$ has only a positive y component (perpendicular to the ramp), and $\vec{F}_S$ has only a negative x component (up the ramp).

Components of gravitational force $\vec{F}_{g,cat}$:

$$F_{g,cat,x} = +F_{g,cat} \sin\theta$$

$$F_{g,cat,y} = -F_{g,cat} \cos\theta$$

Components of normal force $\vec{F}_n$:

$$F_{n,x} = 0$$

$$F_{n,y} = +F_n$$

Components of $\vec{F}_S$:

$$F_{S,x} = -F_S$$

$$F_{S,y} = 0$$

Substitute the expressions for $F_{g,cat,x}$, $F_{g,cat,y}$, $F_{n,x}$, $F_{n,y}$, $F_{S,x}$, and $F_{S,y}$ into Newton's first law and solve for the magnitudes F_n and F_S.

Substitute into Newton's first law in component form:

$$\sum F_{\text{ext on cat},x} = F_{g,cat} \sin\theta + 0 + (-F_S) = 0$$

$$\sum F_{\text{ext on cat},y} = -F_{g,cat} \cos\theta + F_n + 0 = 0$$

From the y equation:

$$F_n = F_{g,cat} \cos\theta = (40.0 \text{ N}) \cos 15.0° = 38.6 \text{ N}$$

From the x equation:

$$F_S = F_{g,cat} \sin\theta = (40.0 \text{ N}) \sin 15.0° = 10.4 \text{ N}$$

Reflect

The downward gravitational force has magnitude 40.0 N, the normal force exerted perpendicular to the ramp has magnitude 38.6 N, and the friction force exerted up the ramp has magnitude 10.4 N. The magnitudes of the last two forces are less than the magnitude of the gravitational force exerted on the cat by Earth, which makes sense. Each just balances a single component of the gravitational force. If this were to be repeated on a different planet with a different acceleration due to gravity, the weight of the cat would be changed, but not its mass or inertia (the cat would still be made up of the same stuff, and still have the same resistance to a change in its motion).

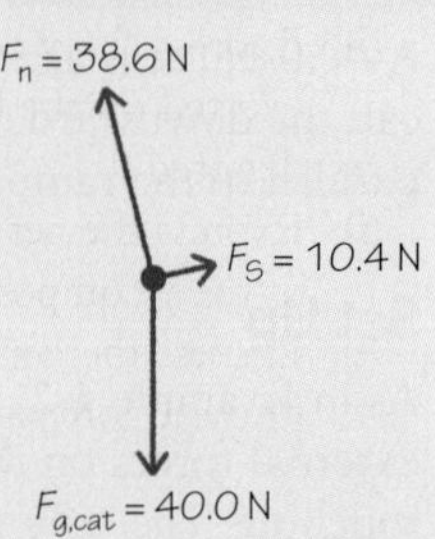

Extend: Note that no single force "balances" any of the other forces: All three forces are needed to mutually balance each other and keep the cat at rest.

NOW WORK Problem 6 from The Takeaway 4-3.

WATCH OUT 

The normal force isn't always equal to the weight.

It's a common mistake to assume that the normal force on an object (in this case, the sleeping cat) has the same magnitude as the object's weight. That's clearly not true in this case: The normal-force magnitude $F_n = 38.6$ N is less than the cat's 40.0-N weight. As a general rule it's always safest to treat the magnitude of the normal force as an unknown, as we have done in this example.

AP® Exam Tip

There is no equation defining the normal force. The only way to calculate it is to apply Newton's first or second law, whichever is appropriate to the problem. In order to calculate it correctly, you need a good understanding of vector components and vector addition.

THE TAKEAWAY for Section 4-3

✔ The weight of an object is the magnitude of the gravitational force exerted on it.

✔ Newton's first law states that if the external forces exerted on an object sum to zero, the object does not accelerate. If the object is stationary, it remains at rest, and if it is in motion, it continues to move in a line with constant velocity.

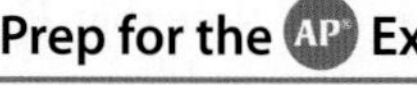

AP® Building Blocks

1. What is the net force on a bathroom scale when a 75-kg person stands on it? What is the normal force that the scale exerts on the person?
2. Why would it be easier to lift a truck on the Moon's surface than it is on Earth?
3. What is the weight on Earth of a wrestler who has a mass of 120 kg?
4. A bluefin tuna has a mass of 252 kg. What is its weight on Earth?
5. An astronaut has a mass of 80.0 kg. How much would the astronaut weigh on Mars, where the acceleration due to gravity at the surface is 38.0% of that on Earth?

6. A brick with a mass of 2.5 kg is at rest on a board that has an inclination of 15° from the horizontal as shown.

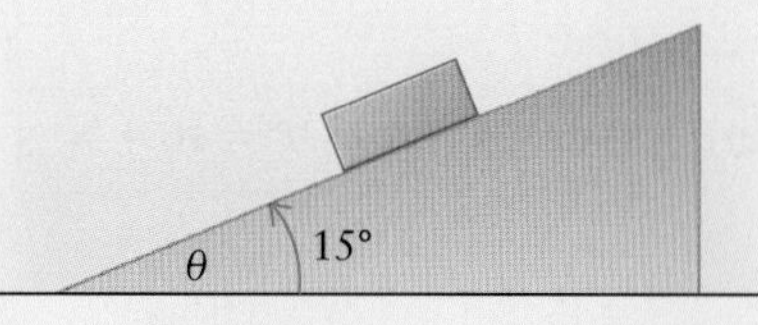

(a) Identify each of the forces exerted on the brick.
(b) Make a drawing of the system and add to your drawing an x–y coordinate system where the origin is at the center of the brick.
(c) In the coordinate system just drawn, identify each of the x or y components of the forces identified in part (a) that are zero.
(d) In the coordinate system just drawn, identify each of the x or y components of the forces identified in part (a) that are nonzero.
(e) Add a vector to your drawing to represent the gravitational force exerted on the brick.
(f) Add a triangle to your drawing in which the hypotenuse is the magnitude of the gravitational force exerted on the brick and the acute angle at the upper vertex of the triangle is equal to the angle of inclination of the ramp.
(g) Express the x component of the gravitational force exerted on the brick in terms of the inclination angle denoted θ.

(h) Express the y component of the gravitational force exerted on the brick in terms of the inclination angle denoted θ.

(i) Express the net force exerted in the y direction in terms of the y component of each force exerted on the brick.

(j) Express the net force exerted in the x direction in terms of the x component of each force exerted on the brick.

AP® Skill Builders

7. Evaluate evidence from your experience supporting the claim that an upward force is exerted on an object at rest on a horizontal surface.
8. Two teams of three people play tug-of-war with a rope. When the rope is taut an object with a mass of 5 kg is hung at the midpoint between the teams and the rope sags. Construct an argument that, no matter how hard the teams pull, the sag cannot be removed so that the rope is straight again.
9. A new nanotech material in the form of conductive molybdenum sulfide sheets has set a weightlifting record. By exciting the flexible sheet with a small electric potential difference, the 1.6-mg device is able to exert a force sufficient to lift an object with a mass of 265 mg through a distance of several millimeters. The sheet can raise an object 166 times its own mass. This is like a 60-kg person being able to lift 10,000 kg, or 10 metric tons. Cycles of expansion and contraction of a sheet can be rapidly repeated by changing the electric potential difference that is applied across the sheet.

   These sheets can be stacked. These sheets can be shaped. Applications will be developed by engineers who pose questions about how these materials will be used. To help you frame such a question, identify a potential application for these sheets.

AP® Skills in Action

10. Predict the relative monetary value of a newton of gold on Earth and a newton of gold on the Moon by choosing the best answer.
    (A) The value is the same regardless of location because the value of gold is determined by its weight.
    (B) The newton of gold on Earth has more value because the value of gold is determined by its mass. A newton of gold on Earth has a greater mass because the acceleration due to gravity is greater on Earth.
    (C) The newton of gold on the Moon has more value because the value of gold is determined by its mass. A newton of gold on the Moon has a greater mass because the acceleration due to gravity is greater on Earth.
    (D) The value is the same regardless of location because the mass of gold does not change when it is taken from Earth to the Moon.
11. 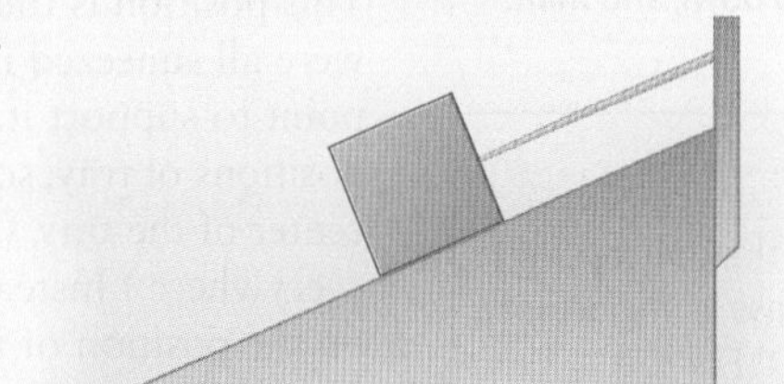

    Consider a stationary block on an incline, attached by a rope to the top of the incline, as shown in the figure. How does the magnitude of the normal force exerted on the block by the ramp compare to the weight of the block? The normal force is
    (A) equal to the weight of the block.
    (B) greater than the weight of the block.
    (C) less than the weight of the block.
    (D) possibly equal to or less than the weight of the block, depending on whether or not the ramp surface is smooth.

4-4 A free-body diagram is essential in solving any problem involving forces

As we solved the problems in Examples 4-2 and 4-3, we started by drawing vectors representing the forces on the object in the problem in a specific way. This is part of an important technique needed to solve physics problems involving forces. There are some important concepts we need to allow us to fully understand how to construct such a representation, which is called a *free-body diagram*.

A **free-body diagram** is a graphical representation of all the external forces exerted on an object. It's useful because Newton's second law and Newton's first law both involve the sum of all external forces on an object, $\sum \vec{F}_{ext}$. The term "free body" means that we draw only the object on which the forces are exerted, *not* the other objects that exert those forces.

Exam Tip

A free-body diagram is a basic skill that you should expect to be tested on during the AP® Physics 1 exam. To succeed at this skill, it's important to get the features of the diagram correct as well as understand the nature of each of the possible forces exerted on an object. Figure 4-11 illustrates the steps in drawing a free-body diagram.

The Center of Mass of a System Moves as Though All the System's Mass Were Concentrated There

We learned in Section 4-2 that the acceleration of an object is determined by the external forces on the object as described by Newton's second law:

$$\vec{a} = \frac{\sum \vec{F}_{ext}}{m} \qquad (4\text{-}2)$$

Figure 4-10 Center of mass at the restaurant The tray balances if it is supported at the center of mass of the system of tray, soup bowl, and mint.

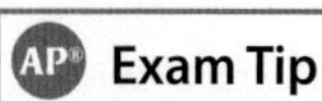

Exam Tip

Understanding the concept of the center of mass is important for AP® Physics 1. For the AP® Physics 1 exam, you will only be required to calculate it for relatively simple situations. You will learn how to calculate it later in this course.

But if we cannot use the object model, such as if the system is rotating like a gymnast tumbling in midair or a barrel rolling downhill, or if we are squeezing a system such as if you exert a force on something soft like a pillow, different parts of the object have different accelerations so the equation doesn't seem to make sense. The quantity $\vec{a}$ in Equation 4-2 must refer to the acceleration of a specific point on the object. But what point is that?

The point we are looking for is called the **center of mass** of the object. As we will see, the center of mass moves as though all of the object's mass were squeezed into that point and all of the external forces are exerted on that point. In Chapter 9 we will learn how to calculate the position of the center of mass; then we'll use ideas about momentum to justify why the center of mass behaves as it does. For now, we just need to have a feel for this quantity. If every object in a system moves with the same acceleration as the center of mass, then that system can always be modeled as an object. When we have to worry about parts of the system not moving with the same acceleration, then we cannot use the object model. The idea of center of mass lets us see how to decide which model to use, so it is pretty important! When we can use the object model, as for the cat in Example 4-3, we need to represent it as a single point to use Newton's laws. How do we find that point?

Imagine a waiter who supports a tray that has a bowl of soup on one side and a small mint on the other (**Figure 4-10**). To balance the combination of tray, soup bowl, and mint, the waiter supports the tray at a position much closer to the soup than the mint. This position is the center of mass of the three objects: If the tray, soup bowl, and mint were all squeezed into a single point, the waiter would put his hand directly under that point to support it. The position of the center of mass isn't the ordinary average of the positions of tray, soup bowl, and mint, which would be a point closer to the geometrical center of the tray. (If the waiter tried to balance the tray there, the result would be soup everywhere.) Instead, the center of mass is a *weighted* average that puts more emphasis on the position of the high-mass soup bowl than on the position of the low-mass mint. Most of us can almost instinctively feel the point we need to use to balance such a system, and for now we just need that instinctive understanding of center of mass.

Making a Free-Body Diagram

When we draw a free-body diagram, we are declaring that we are using an object model. Either the object model holds (every point on an extended object moves the same way) or we are only interested in the motion of the center of mass of a system. We already drew something very close to free-body diagrams for the tray in Example 4-2 and the cat in Example 4-3 (see the Set Up step in each example). We were just too artistic in these cases. The first rule of free-body diagrams is to keep them as clean and simple as possible, so you can clearly read the information you need from them. As an example, **Figure 4-11a** shows a block on a horizontal table. A wind is blowing from left to right,

(a) A block on a horizontal table

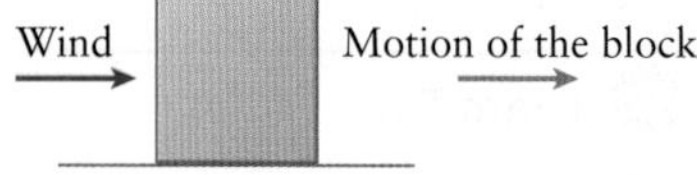

(b) Drawing the free-body diagram for the block

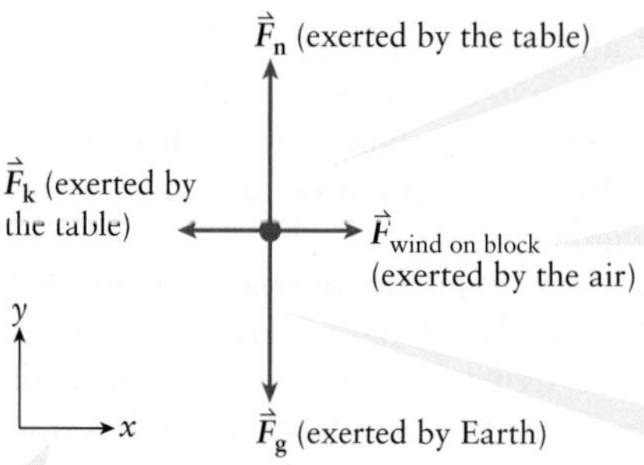

$\vec{F}_n$ (exerted by the table)
$\vec{F}_k$ (exerted by the table)
$\vec{F}_{\text{wind on block}}$ (exerted by the air)
$\vec{F}_g$ (exerted by Earth)
y
x

(1) Sketch a simplified version of the object on which the forces are exerted. A dot, or a small circle or square is fine: something easy to draw where you can place the tails of your vectors. Remember you are representing the center of mass if it is an extended object, so you don't want to think about the shape too much.
(2) Identify what other objects exert forces on this object. This includes Earth (which exerts a gravitational force) and anything that touches the object.
(3) For each force that is exerted on the object, draw the force with its tail at the center of the object. Make sure each vector points in the correct direction. Do not include forces that the object exerts on the other objects.
(4) Label each force with its symbol. Be certain that you can identify what other object exerts each force. If you're not sure what exerts a force, it's probably not a real force!
(5) Choose, draw, and label the directions of the positive x and y axes. Choose the axes so that as many forces as possible lie along one of the axes.

Figure 4-11 Constructing a free-body diagram (a) The wind pushes to the right on this block, which slides on a horizontal table. (b) How to create a free-body diagram that depicts all of the external forces exerted on the block.

and the wind pushes on the block and makes it slide at constant velocity. **Figure 4-11b** lists the steps involved in constructing this block's free-body diagram. The dot indicates the object, or the system's center of mass.

Often we'll label a force to indicate what object is exerting the force and what object the force is exerted on. For example, if a girl pushes on a book, we will label that force $\vec{F}_{\text{girl on book}}$. (We used a label of this kind in Figure 4-11b for the force of the wind blowing on the block.) However, we'll label some commonly encountered forces in a different way. We'll use $\vec{F}_g$ for the gravitational force that Earth exerts on an object. If the object is in contact with a surface, we'll use $\vec{F}_s$ for the static friction force on an object at rest or $\vec{F}_k$ for the kinetic friction force that is exerted by the surface on the object parallel to the surface as the object slides relative to the surface, and $\vec{F}_n$ for the normal force exerted on the object by the surface, which is always perpendicular to the surface (see Figure 4-11b). As we mentioned in Section 4-2, we'll refer to friction forces and the normal force as contact forces: They arise only when two objects are in contact with each other and are caused by electric forces on smaller particles within the objects that we don't really want to consider when we are using the object model. Remember that the gravitational force is a long-range force, one of the fundamental forces: A basketball in midair is pulled downward by the gravitational force even though it's not in contact with the ground.

Examining Free-Body Diagrams

Let's look at some examples of free-body diagrams and what we can learn from them. In **Figure 4-12a**, a stationary box rests on a table with a horizontal top. The free-body diagram in Figure 4-11b shows the two forces exerted on the box: the downward gravitational force $\vec{F}_g$ and the upward normal force $\vec{F}_n$ exerted by the table.

The box exerts a downward force on the surface of the table, but we don't include this force in the free-body diagram of the box since it is not a force exerted *on* the box.

We choose the y axis to be vertical; both forces lie along this axis. The sum of $\vec{F}_g$ and $\vec{F}_n$ must be zero in order for the box to remain at rest, so the free-body diagram tells us that $\vec{F}_g$ and $\vec{F}_n$ must be opposite in direction and have the same magnitude.

In **Figure 4-13a** a box is placed on an inclined ramp with a rough surface. There is friction between the box and the ramp, so the free-body diagram in **Figure 4-13b** includes a static friction force $\vec{F}_s$ (since the box is not moving) as well as the gravitational force $\vec{F}_g$ and the normal force $\vec{F}_n$. Because the ramp is tilted by an angle θ from the horizontal, the normal force (which is perpendicular to the ramp) is tilted from the vertical by the same angle θ. We've chosen the x axis to be parallel to the ramp and the y axis to be perpendicular to the ramp. Choosing these as our axes means any acceleration is along one of the axes, and that $\vec{F}_s$ has only an x component, and $\vec{F}_n$ has only a y component. So, this choice of axes simplifies the vector math we will have to do to solve the problem. If the block remains at rest, the net external force $\sum \vec{F}_{\text{ext}}$ on it—that is, the vector sum of $\vec{F}_g$, $\vec{F}_n$, and $\vec{F}_s$—must be zero. So the friction force $\vec{F}_s$ must point in the negative x direction in order to cancel the (positive) x component of the gravitational force $\vec{F}_g$; the normal force $\vec{F}_n$ points in the positive y direction and must cancel the (negative) y component of $\vec{F}_g$. (This is the same situation as in Example 4-3 in Section 4-3.) Note that the x and y components of the gravitational force $\vec{F}_g$ are less than the magnitude of the

(a)

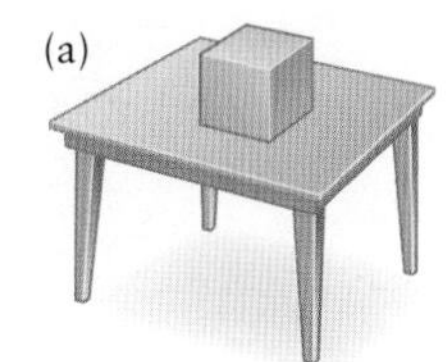

(b)

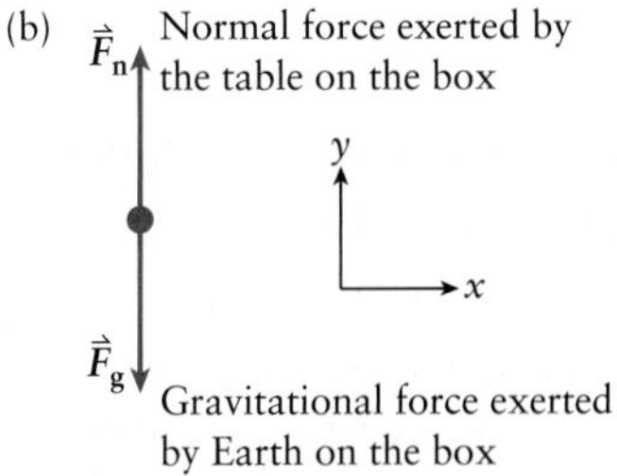

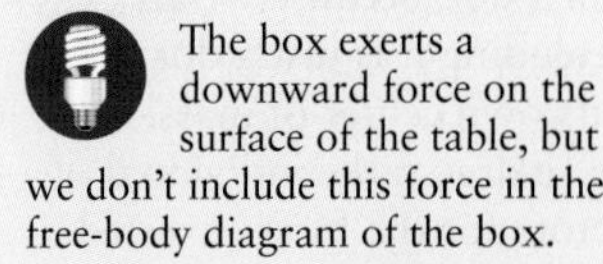

Figure 4-12 A free-body diagram: a stationary box atop a table
(a) A box rests on a horizontal tabletop. (b) The free-body diagram for the box.

Free-body diagrams on the AP® Physics 1 exam are not required to be drawn to scale unless explicitly requested. Always drawing free-body diagrams to scale prepares you for questions where it is requested and makes them a better problem-solving tool for you.

(b)

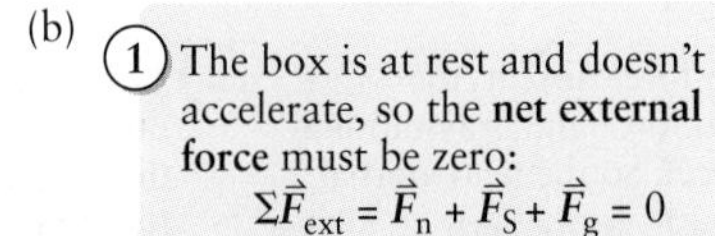

1 The box is at rest and doesn't accelerate, so the **net external force** must be zero:
$\Sigma\vec{F}_{\text{ext}} = \vec{F}_n + \vec{F}_s + \vec{F}_g = 0$

2 With this choice of axes, the normal force has only a y component, and the friction force has only an x component. (The gravitational force has both an x component and a y component.)

(a)

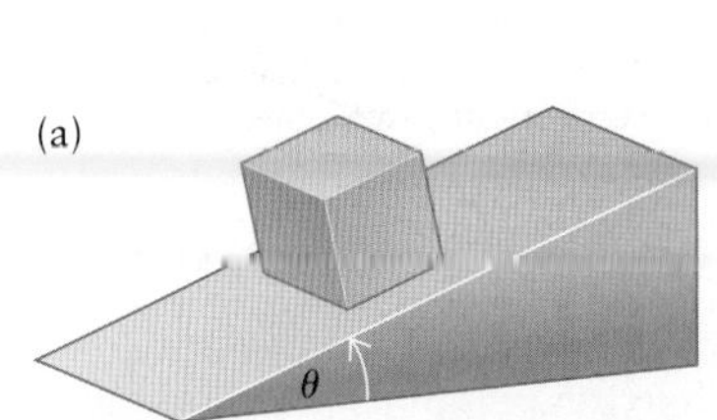

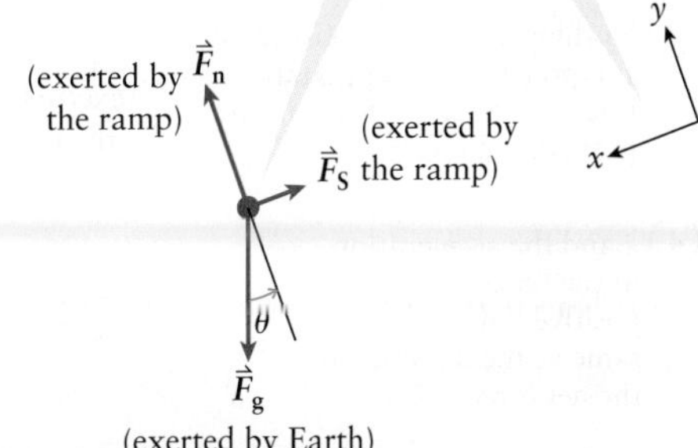

Figure 4-13 A free-body diagram: a stationary box on a ramp
(a) Friction keeps the box from sliding down the ramp. (b) The free-body diagram for the box.

force, which is the box's weight $\vec{F}_g$. The free-body diagram therefore tells us something useful: The magnitudes of the friction force and normal force are both *less* than the weight of the box. Compare this situation to the box on a horizontal surface shown in Figure 4-12, for which the magnitude of the normal force is exactly equal to the weight.

What happens if we grease the surface of the ramp so that friction is negligible? In this case the friction force $\vec{F}_S$ goes to zero, and the free-body diagram is as shown in **Figure 4-14.** Now the vector sum $\sum \vec{F}_{ext}$ of the external forces on the box cannot be zero: For the normal force and gravitational forces to cancel each other, they would have to point in opposite directions, which the free-body diagram in Figure 4-14 shows they do not. Instead the net external force points down the ramp in the positive x direction. According to Newton's second law, $\vec{a} = \sum \vec{F}_{ext}/m$ (Equation 4-2), the box must therefore have a nonzero acceleration $\vec{a}$ down the ramp. We learn this simply by examining the free-body diagram for the box.

Note that neither the normal force nor the gravitational force in Figure 4-14 depends on how the box is moving. Whether the box is sliding up the ramp, momentarily at rest, or sliding down the ramp, these forces are the same. This means the acceleration is also the same in each of these cases. If the box is moving up the ramp, it slows down; if it is momentarily at rest, the box starts moving down the ramp (changing direction); and if it is moving down the ramp, it speeds up. So the free-body diagram also tells us that the box shown in Figure 4-14 moves in a line with a constant acceleration—which means we can analyze its motion using the techniques that we learned in Chapter 2.

Figure 4-14 also includes vectors for the net external force $\sum \vec{F}_{ext}$ and the acceleration $\vec{a}$. If you add vectors like this to your free-body diagram, draw them *alongside* (not touching) the object, and clearly labeled as separate from your representation of the individual forces, as in Figure 4-14. In your diagram only the individual forces that are exerted on the object should be touching it.

We'll use free-body diagrams throughout our study of forces and motion, and you should as well. It is very important to be able to translate between multiple representations to be able to successfully solve physics problems. Before we delve more deeply into how to solve problems using Newton's first and second laws, we'll introduce Newton's *third* law. This will give us deeper insight into the way in which objects exert forces on each other.

WATCH OUT !

Draw the vectors for the forces in a free-body diagram with their tails touching.

In Figures 4-11, 4-12, 4-13, and 4-14, we have represented the object as something small, even a dot, because we are pretending it is just a dot at its own center of mass. We've drawn the individual vectors for the forces exerted on an object so that their tails all touch at the center of mass. (We also did this in Examples 4-2 and 4-3.) This makes it easier to calculate the x and y components of these vectors. It is also more accurate because the object acts as if all the forces were exerted on its center of mass.

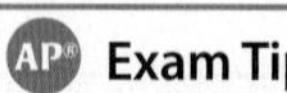

Exam Tip

On the AP® Physics exams, there are rules for free-body diagrams that must be followed although you may do them differently, but still correctly, elsewhere. These rules produce diagrams like those expected in most college classes and modeled in this book. Some important rules to remember are:

- All vectors representing forces exerted on the object must originate on and *touch* the dot representing your object (vectors should never be added head-to-tail).
- You should never draw vector components on free-body diagrams.
- Every vector should be labeled to denote clearly that it is a force and what kind of force it is, and thus the gravitational force should never be labeled "g" or "G," since those have other meanings in physics.

If you are asked to draw a free-body diagram on the AP® exam, follow these rules. If you need to break a vector into components to use Newton's second law, draw another set of coordinate axes off to the side and use that to find the components of the vector.

Figure 4-14 A free-body diagram: a box sliding on a ramp The free-body diagram for a box sliding on a ramp with negligible friction.

(1) With the friction force absent, the remaining normal force and gravitational force do not cancel. So the net external force $\Sigma\vec{F}_{ext} = \vec{F}_n + \vec{F}_g$ is not zero.

(2) The box neither leaps off the ramp nor sinks down into it, so it doesn't accelerate in the y direction. So the normal force must cancel the (negative) y component of the gravitational force.

(3) Nothing cancels the positive x component of the gravitational force, so the net force is in the positive x direction...

(4) ...and the acceleration of the block is in the positive x direction, the same as the direction of the net force.

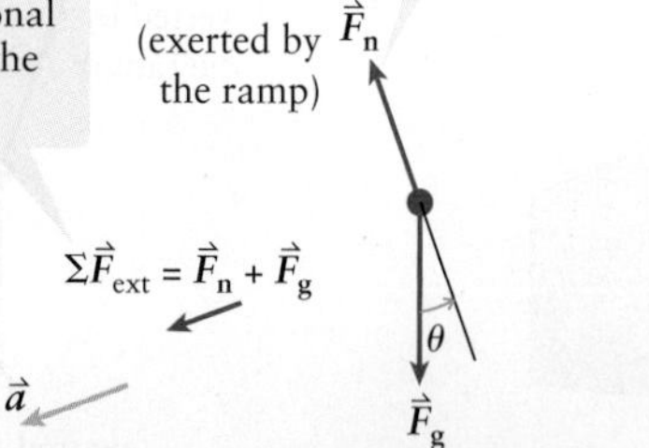

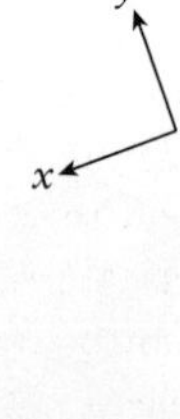

THE TAKEAWAY for Section 4-4

✔ The center of mass of a system moves as if all the system's mass were located at that point.

✔ To use the object model, you must identify the point that represents the system you are modeling as an object. This point is the center of mass.

✔ In order to solve a problem about how an object moves and the forces that are exerted on it, you must draw a free-body diagram for the object. This diagram depicts all of the external forces that are exerted on the object.

✔ Choose the x and y axes in a free-body diagram so that most of the forces point along one or the other of the axes and, if at all possible, one of the axes is in the direction of the acceleration.

Prep for the AP® Exam

AP® Building Blocks

1. Draw a free-body diagram for a heavy crate being lowered by a steel cable straight down at a constant speed.
2. Draw a free-body diagram for a bicycle rolling at constant speed down a hill. Ignore the friction between the bicycle wheels and the hill, but consider any air resistance.
3. Two children, one large and one small, sit on a seesaw (also called a teeter-totter). One type of seesaw is a long board resting on a pivot at its center around which the board can rotate. Describe, in general, how the two children should arrange themselves on the board so that the gravitational force on the center of mass of the system of the two children and the board is along the same line as the upward force of the pivot on the board.

AP® Skill Builders

4. Explain why, in a free-body diagram, the lengths of the arrows representing the forces exerted on an apple at rest in your hand should be equal.
5. Explain why an analysis of a free-body diagram:
   (a) of an object moving on an inclined plane is simplified by choosing the rotated coordinate system A rather than coordinate system B.
   (b) of the block suspended by two cables is simplified by choosing coordinate system C rather than coordinate system B.

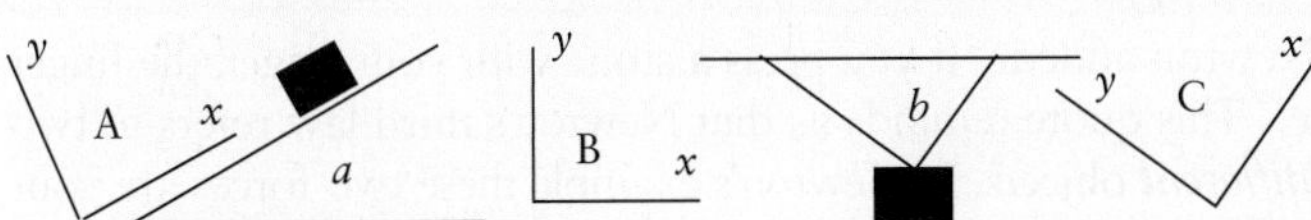

6. A sign is supported by two ropes, as shown in the figure.
   (a) In a coordinate system that is rotated so that $\vec{F}_{T,R}$ lies along the x axis, construct a free-body diagram showing the angle that $\vec{F}_{T,L}$ makes with the y axis of the rotated coordinate system.

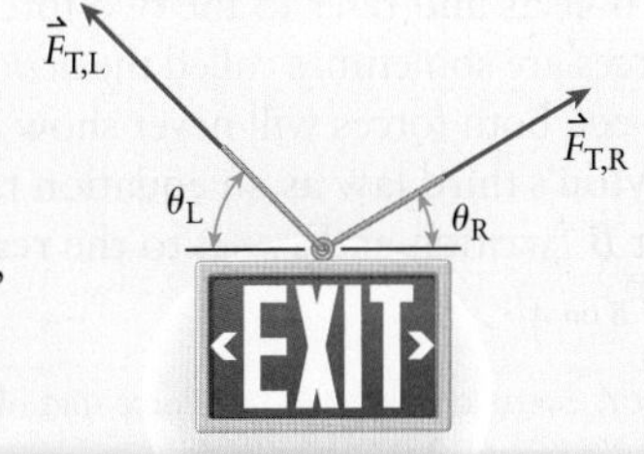

   (b) Refine the representation from part (a) by adding components of the weight in the rotated coordinate system, represented as clearly labeled dashed or differently colored lines, so that they are not confused with the forces on the diagram.
      i. Re-express the diagram in part (b) as a pair of mathematical equations for the object when the sign is at rest.
      ii. Justify a mathematical routine that could be used to determine $F_{T,R}$ and $F_{T,L}$, given the mass of the sign and the angles θ_R and θ_L.
      iii. Express the relationships among the tension forces in the two ropes, angles, and weight of the sign in a coordinate system with the x axis parallel to the bottom edge of the sign.
7. A sheet of cardboard can be treated like a two-dimensional object. An irregular shape is cut from a sheet of cardboard.
   (a) Design a procedure using a sharpened pencil to locate the center of mass of the shape.
   (b) Suppose the object is shaped like a typical bagel, with a hole at its center. Evaluate the claim that any procedure developed in part (a) will locate the center of mass of this object.
   (c) Refine the procedure in part (a) so that the center of mass of a bagel-shaped object can be located.

AP® Skills in Action

8. Draw a free-body diagram for a box being pushed horizontally to the left by a person across a smooth floor (negligible friction) at a steadily increasing velocity.
9. A salvage boat uses its winch to pull up a sunken treasure with an upward force of 4000 N. The mass of the treasure is 200 kg and the water exerts a force of 1500 N on the treasure, resisting the force exerted by the winch. The boat is directly above the treasure. (a) Draw a free-body diagram for the treasure. (b) Calculate the acceleration, and write an equation for position as a function of time for the treasure, assuming it is initially at rest and the winch starts at $t = 0$.
10. Consider a standard chair with four legs. Draw a picture of the chair balanced on the back two of its four legs. Explain where the center of mass of the chair must be for the chair to be balanced in this position.

Figure 4-15 Comparing the forces that objects exert on each other (a) Two spring scales connected together. Each scale measures the force exerted on it. (b) One scale is connected to a wall, and you pull on the other end. (c) The force you exert on the wall with the scales is equal in magnitude to the force that the wall exerts back on you.

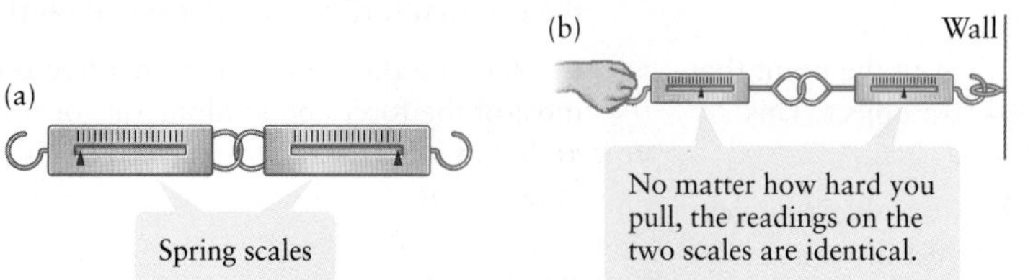
(b)
Wall
No matter how hard you pull, the readings on the two scales are identical.

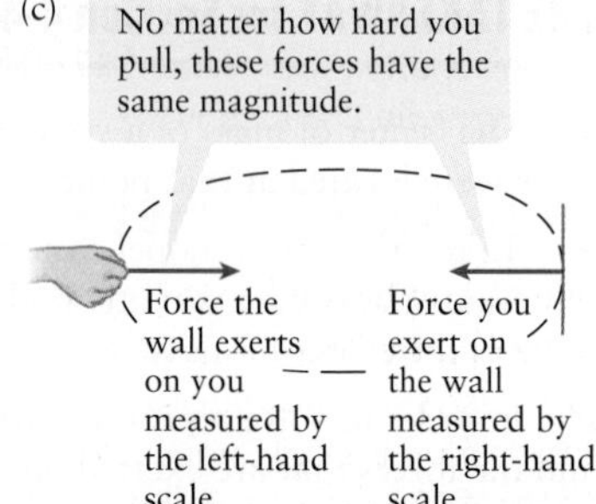
(c)
No matter how hard you pull, these forces have the same magnitude.
Force the wall exerts on you measured by the left-hand scale
Force you exert on the wall measured by the right-hand scale

(a)

Spring scales

4-5 Newton's third law relates the forces that two objects exert on each other

If you hold a book in your palm or dribble a basketball, you're exerting a force on an object. However, you know from experience that the *object* exerts a force on *you* as well. You can feel the book pushing back against your palm, and you can feel the slap of the basketball against your hand as you push down on it. What is the connection between the force that you exert on an object and the force that the object exerts back on you?

Here's an experiment that offers an answer to this question. Take two spring scales (such as you might use for weighing a fish) and connect their upper ends together, then turn them apart sideways (**Figure 4-15a**). You hold the free end of one scale, and you attach the free end of the other scale to a hook mounted to the wall (**Figure 4-15b**). When you pull on the free end of the first scale, the wall pulls on the free end of the other scale. Remarkably, no matter how hard you pull, the reading on the scale that you hold (which measures the amount of force you exert) is *identical* to the reading on the scale attached to the wall (which measures the amount of force that the wall exerts). In other words, the force that you exert on the wall has the *same* magnitude as the force that the wall exerts back on you (**Figure 4-15c**).

In Figure 4-15 both you and the wall are stationary. We can try the same experiment with two objects in motion—say, a car and a trailer connected together by the same arrangement of two (much stronger) spring scales (**Figure 4-16**). The readings on both scales depend on whether the car and trailer are moving at a constant speed, speeding up, or slowing down. In all of these cases, however, experiment shows that the spring scale attached to the car reads *exactly* the same as the spring scale attached to the trailer. So the force that the car exerts on the trailer always has the same magnitude as the force that the trailer exerts on the car. The only difference between the two forces is that they are in *opposite* directions: The car pulls forward on the trailer, while the trailer pulls backward on the car.

The car and trailer are linked together by two spring scales (like those shown in Figure 4-15) connected end to end.

Both spring scales have the same reading: The force that the car exerts on the trailer always has the same magnitude as the force that the trailer exerts on the car. This is true no matter how the car and trailer move (and is also true if they aren't moving at all).

Figure 4-16 Comparing the forces that moving objects exert on each other A pair of spring scales like those shown in Figure 4-15a are used to connect a car and trailer.

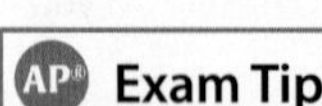

Exam Tip

On the AP® Physics 1 Exam, "force meters" or "force probes" are sometimes discussed in questions. A force meter or probe functions like one of these spring scales—measuring the amount of force exerted on it.

We can summarize these observations in **Newton's third law:**

If object A exerts a force on object B, object B exerts a force on object A that has the same magnitude but is in the opposite direction.

In declaring this law Newton added: "If you press a stone with your finger, the finger is also pressed by the stone." This quote reminds us that Newton's third law refers to two forces that are exerted on *different* objects. In Newton's example these two forces are your finger pressing on the stone and the stone pressing on your finger. Physicists speak of the *interaction* between two objects and refer to the two forces involved in an interaction as a **force pair.** (These two forces are sometimes called the *action* and the *reaction.*) Because they are never on the same object, both forces will never show up in a single free-body diagram!

We can express Newton's third law as an equation relating the force that an object A exerts on another object B (written as $\vec{F}_{A \text{ on } B}$) to the **reaction force** that object B exerts on object A (written as $\vec{F}_{B \text{ on } A}$):

EQUATION IN WORDS
Newton's third law of motion

Force that object A exerts on object B | Force that object B exerts on object A

$$\vec{F}_{A \text{ on } B} = -\vec{F}_{B \text{ on } A} \quad (4\text{-}6)$$

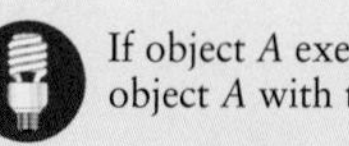

If object A exerts a force on object B, object B must exert a force on object A with the same magnitude but opposite direction.

The minus sign means that $\vec{F}_{B \text{ on } A}$ has the same magnitude as $\vec{F}_{A \text{ on } B}$ but is in the opposite direction (**Figure 4-17**). The two forces in a force pair are sometimes called "equal and opposite," which means that they are equal *in magnitude* and opposite *in direction.*

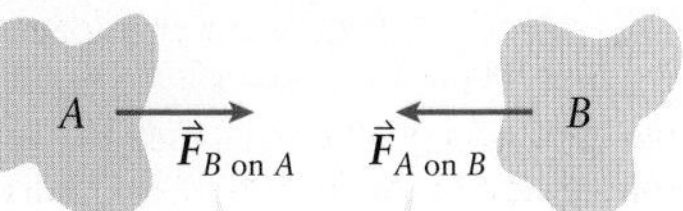
A
B
$\vec{F}_{B \text{ on } A}$
$\vec{F}_{A \text{ on } B}$
These two forces are equal in magnitude but opposite in direction.

Figure 4-17 Newton's third law If object A exerts a force $\vec{F}_{A \text{ on } B}$ on object B, then object B necessarily exerts a force $\vec{F}_{B \text{ on } A}$ on object A with the same magnitude but the opposite direction.

Comparing Newton's Three Laws

We've emphasized that Newton's third law is fundamentally different from the first and second laws because it tells us about forces that are exerted on *two different* objects. By contrast, Newton's first and second laws tell us about the effect of the net external force exerted on a *single* object. Here's an example that illustrates this important distinction. **Figure 4-18a** shows a ball held at rest in a person's hand. The drawing shows three forces: the gravitational force of Earth pulling down on the ball (the ball's weight), the normal force of the person's hand on the ball, and the normal force of the ball on the person's hand. All three forces have equal magnitudes but for different reasons.

- When the ball is at rest, the force of Earth on the ball and the force of the person's hand on the ball have equal magnitudes and opposite directions according to Newton's *first* law. These are the only two external forces exerted *on* the ball, so their vector sum $\sum \vec{F}_{\text{ext}}$ must be zero (that is, the two forces must balance each other) in order for the ball to remain at rest.
- The force of the person's hand on the ball and the force of the ball on the person's hand have equal magnitudes and opposite directions according to Newton's *third* law. These two forces are a force pair: If A represents the hand and B represents the ball, these are the forces of A on B and of B on A.

Figure 4-18b shows the situation when the person's hand is accelerating the ball upward. Now the vector sum $\sum \vec{F}_{\text{ext}}$ of external forces exerted on the ball is not zero, so the force of Earth on the ball does *not* have the same magnitude as the force of the person's hand on the ball. However, the force of the person's hand on the ball still *does* have the same magnitude of the force of the ball on the hand because Newton's third law works at all times, even when the objects are accelerating.

Here's a general rule to keep in mind:

Use Newton's first law or second law when dealing with the forces exerted on a given object. Use Newton's third law when relating the forces that two objects exert on each other.

WATCH OUT !

Newton's third law involves only two objects.

Two forces can be a force pair only if they are exerted between the *same* two objects. Consider the situation in which you are holding a rock in your hand. The gravitational force exerted by Earth on the rock cannot ever be a force pair to the normal force of your hand on the same rock. These two forces involve three objects: Earth, rock, and hand. So these two forces cannot be a force pair.

(a) Holding a ball at rest

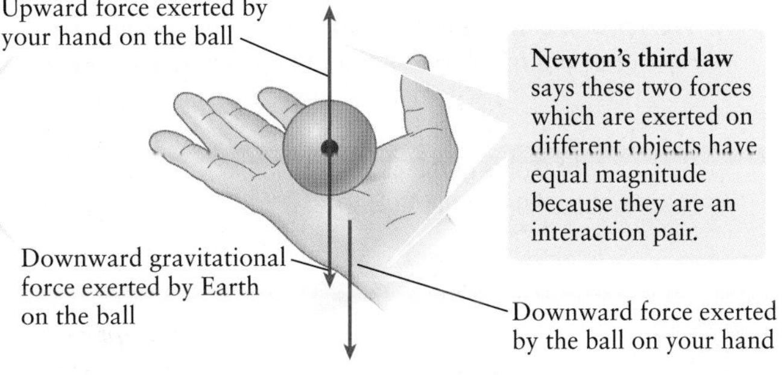
Upward force exerted by your hand on the ball
Newton's first and second laws say these two forces exerted on the same object (the ball) have equal magnitude since the ball is not accelerating.
Newton's third law says these two forces which are exerted on different objects have equal magnitude because they are an interaction pair.
Downward gravitational force exerted by Earth on the ball
Downward force exerted by the ball on your hand

(b) Accelerating a ball upward

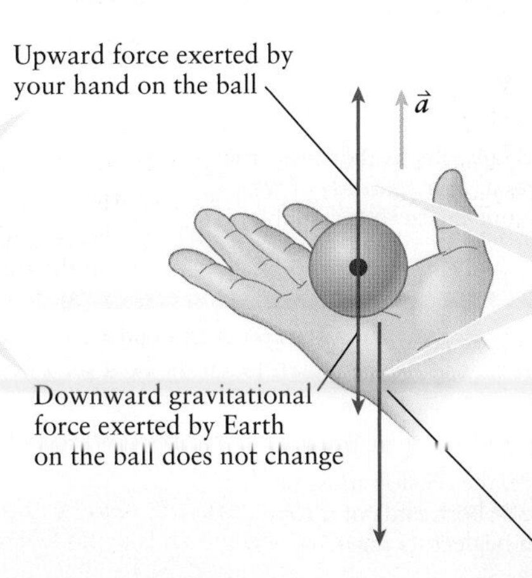
Upward force exerted by your hand on the ball
$\vec{a}$
Newton's second law says these two forces exerted on the ball do not have equal magnitudes. Their sum must be upward (in the direction of the acceleration). Since acceleration is not zero, we cannot use Newton's first law.
Newton's third law says these two forces have equal magnitudes, whether the ball is accelerating or not. Notice in this case (b) the force the ball exerts on the hand is greater than it was in (a). In fact, if this was you accelerating the ball upward, you would notice that it felt heavier to you, since it would be pushing down on you harder.
Downward gravitational force exerted by Earth on the ball does not change
Downward force exerted by the ball on your hand

Figure 4-18 Comparing Newton's three laws (a) Newton's first and third laws applied to a ball at rest in your palm. (b) Newton's second and third laws applied to a ball that you accelerate upward.

Newton's Third Law and Tension

We can use Newton's second and third laws to help us understand another important type of contact force. **Tension** is the force exerted by a rope (or thread, string, cable, or wire) on an object to which the rope is connected. As an example, the rope shown in **Figure 4-19a** exerts a tension force on the sailor who holds on to the left-hand end of the rope and exerts a tension force on the boat tied to its right-hand end. We'll use the symbol $\vec{F}_T$ for a tension force.

Here's an important fact about tension: If a rope has sufficiently small mass, the tension forces that the rope exerts at its two ends have the *same* magnitude. To see why these forces have the same magnitude, consider a sailor who docks a boat by pulling on one end of a rope whose other end is tied to the boat (Figure 4-19a). If he pulls with a given force—say, 25 N—on his end of the rope, Newton's third law tells us that the rope pulls back on him with a force of 25 N. How much force does the *other* end of the rope exert on the boat?

To see the answer, consider **Figure 4-19b,** which shows the forces that are exerted *on* the rope. The weight of the rope causes the rope to sag so that the angle of the rope is different at the two ends. However, in many situations the weight of the rope is very small compared to the pulling forces exerted on the rope by the objects attached to its ends. That's the case in Figure 4-19a and expanded in **Figure 4-19c.** So we can safely approximate the rope as having *zero* mass and hence *zero* weight. Using this approximation, the rope won't sag at all under its own weight (because it has none) and will be perfectly straight. Then the only forces exerted on the rope are those at either end. Because the rope is straight, these two forces point in opposite directions (Figure 4-19c).

Newton's *second* law states that the sum of external forces exerted on the rope, $\sum \vec{F}_{\text{ext}}$, is equal to $m\vec{a}$, the product of the rope's mass m and its acceleration $\vec{a}$. However, if the rope has zero mass ($m = 0$), the product $m\vec{a}$ is always equal to zero, even if the rope is accelerating. Therefore, the sum of external forces on the rope must be zero $\left(\sum \vec{F}_{\text{ext}} = 0\right)$ under all circumstances. For this to be the case, the oppositely directed forces exerted on either end of the rope (Figure 4-19c) must have the same magnitude whether the rope is accelerating or not.

Newton's *third* law tells us that the force exerted *on* each end of the rope has the same magnitude as the tension forces exerted *by* each end of the rope. Because equal-magnitude forces are exerted on each end of rope of negligible mass, it follows that the tension is also the same at *both* ends: If the tension is 25 N at the left-hand end, it is 25 N at the right-hand end (Figure 4-19c). So we are left with the following general rule:

The tension is the same at both ends of a rope or string, provided the weight of the rope or string is much less than the forces exerted on the ends of the rope or string.

WATCH OUT !

Newton's law still works!

If the mass of the rope were not negligible, we could still use Newton's laws. The force of the boat on the rope and the rope on the boat would still be equal. In the case where we cannot neglect the mass, we would have to think of the rope as a string of small objects each exerting forces on each other. The forces between each pair of objects would be the same, but the net force on each object would have to be nonzero. If an object's mass is nonzero, there must be a net force to accelerate it.

Figure 4-19 Comparing the tensions at the two ends of a rope (a) A sailor docking a boat pulls on one end of a rope. (b) The forces exerted on the rope. (c) The forces exerted on a rope of negligible mass. In this case the tension force must have the same magnitude at either end of the rope if we can neglect its mass.

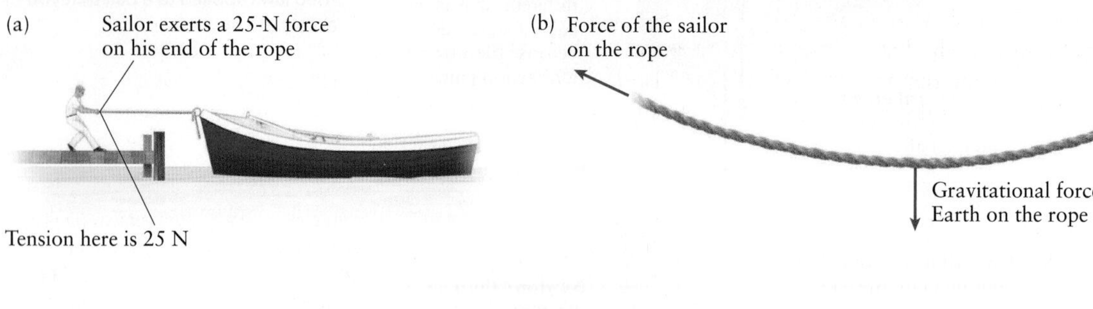

(c)

1 Newton's second law tells us that these two forces must be equal in magnitude, if we assume that the rope has zero (negligible) mass.

Force of the sailor on the rope = 25 N

Force of the boat on the rope = 25 N

Tension at this end = force of the rope on the sailor = 25 N

Tension at this end = force of the rope on the boat = 25 N

2 Newton's third law tells us that these two forces must be equal in magnitude...

3 ...and Newton's third law tells us that these two forces must be equal in magnitude.

4 Therefore, the tension must be the same at both ends of a rope if we can neglect its mass.

Newton's Third Law and Propulsion

Newton's third law plays an essential role in biology: All living systems (including you) use it for *propulsion*. As an example, suppose you start running from a standing start. It's common to say that you propel yourself forward, but this isn't strictly correct: It takes an external force to accelerate you forward from rest, and an external force has to come from outside your body. What happens is that to start running, you use your feet to exert a *backward* force on the surface of the running track. By Newton's third law, the track exerts a *forward* force of the same magnitude on your feet. It's this force exerted by the track that's responsible for propelling you forward.

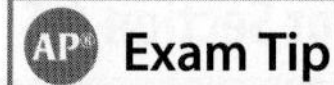

Exam Tip

On the AP® Physics 1 equation sheet, there is no symbol for F_T (tension) because it can only be determined by an application of Newton's laws.

Bird flight also depends on Newton's third law. As they flap, a bird's wings push air both backward and downward. Newton's third law tells us that the air therefore exerts forward and upward components of force on the bird's wings. For a bird flying without changing direction at constant speed (zero acceleration) we know each component of the net force on the bird is zero. The forward component of force exerted on the bird's wings by the air balances the backward force of air resistance exerted on the bird's body. The upward component of force that the air exerts on the bird's wings balances the downward gravitational force on the bird's body. If the bird wished to accelerate it would need to change how hard it was pushing on the air, so the resultant force of the air on the bird would cause an unbalanced force in the desired direction. The forces of the air on the wing and the wing on the air would still be equal in magnitude and opposite in direction. For instance, if the bird wished to accelerate downward, it could decrease the force its wings exert on the air in the downward direction, which would reduce the upward force the air exerts on the bird's wings. This new upward force would no longer balance gravity, and the bird would begin to accelerate downward.

Fish typically have a number of fins of various shapes and sizes (**Figure 4-20a**) that they use to take full advantage of Newton's third law. In almost all species of fish, the primary means of forward propulsion is the side-to-side motion of the caudal fin and the rear part of the fish's body. **Figure 4-20b** shows a fish from above as it sweeps its caudal fin to the right. In doing so the fish exerts a force $\vec{F}_{\text{fish on water}}$ on the surrounding water in a direction perpendicular to the fish's body, in a manner similar to a normal force. In accordance with Newton's third law, the water exerts a force on the fish, $\vec{F}_{\text{water on fish}}$, of equal magnitude in the opposite direction. The forward component of this force on the fish, labeled F_{thrust} in Figure 4-20b, propels the fish forward against the backward resistance of water on the front of the fish. In some species the motion of the dorsal fin (Figure 4-20a) also contributes to forward propulsion.

Exam Tip

Since the action and reaction forces are always on two different objects (for example, the fish and the water) they are never drawn on the same free-body diagram.

Figure 4-20b shows that there is also a force component F_{lateral} on the fish that is perpendicular to the net motion of the fish. These lateral forces contribute to the stability of the fish and help it turn and maneuver. Most fish can also make subtle adjustments to the orientation of the pectoral fins on their sides, enabling them to stop, start, and change direction quickly.

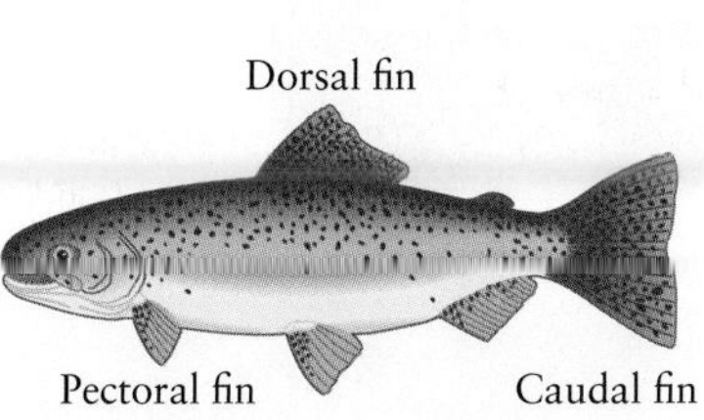

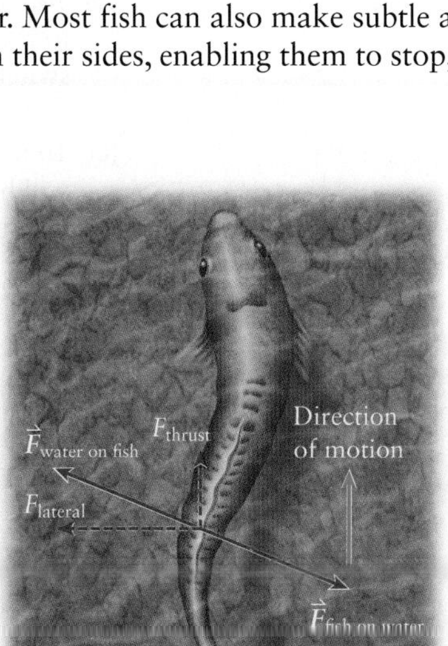

Figure 4-20 Fish propulsion (a) Fish have a variety of fins for propulsion and directional control. The caudal fin is primarily responsible for pushing the fish forward through the water. (b) When a fish swishes its caudal fin back and forth, it exerts a force $\vec{F}_{\text{fish on water}}$ on the water. The reaction to this is the force $\vec{F}_{\text{water on fish}}$ that the water exerts on the fish; the forward component of this force, F_{thrust}, propels the fish forward. Note that F_{thrust} and F_{lateral} are components of $\vec{F}_{\text{water on fish}}$, and would not be drawn on a free body diagram.

THE TAKEAWAY for Section 4-5

✔ Newton's third law tells us that if object A exerts a force on object B, object B has to exert a force on object A of the same magnitude and in the opposite direction.

✔ A rope exerts the same tension at both ends, provided the mass of the rope can be ignored.

AP® Building Blocks

1. We know that Earth exerts a force on the Moon, holding it in orbit around Earth. What evidence do we have that the Moon also exerts a force on Earth?
2. Use Newton's third law to explain how birds are able to fly forward.

AP® Skill Builders

3. A boxer claims that Newton's third law helps him while boxing. He says that during a boxing match the force that his jaw feels is the same as the force that his opponent's fist feels (when the opponent is doing the punching). Therefore, his opponent will feel the same force as he feels and he will be able to fight as long as his opponent can, no matter how many punches he receives or gives. It will always be an "even fight." Evaluate the boxer's claim where the boxer and his opponent are both treated as objects and where the boxer and his opponent are systems with internal structure.
4. A chair rests on a scale inside an elevator as shown.
   (a) Construct separate free-body diagrams for the chair, the scale, and the elevator using the labeled forces shown in the figure. Assume that the elevator has just begun to descend from rest.
   (b) Beneath each free-body diagram use Newton's second law to construct mathematical equations that relate the forces to the motion of the chair.
   (c) Construct a free-body diagram and the equivalent mathematical equation for the system composed of the elevator, the scale, and the chair.
   (d) Analyze the representations from parts (b) and (c) to obtain a condition that the four normal forces must satisfy.
   (e) Justify the claim that the condition in part (d) is the sum of two third-law conditions.

$F_{n,chair\ on\ scale}$

$F_{n,scale\ on\ chair}$

$F_{n,floor\ on\ scale}$

$F_{n,scale\ on\ floor}$

$F_{g,elevator}$

$F_{g,scale}$

$F_{g,chair}$

$F_{cable\ on\ elevator}$

5. Consider the object below, hanging at rest from two ropes, one of which is horizontal and the other that makes an angle of 60° with the horizontal.

Prep for the AP® Exam

   (a) Draw a free-body diagram of the object.
   (b) Calculate the value of the tension, $\vec{F}_{T1}$, in the diagonal rope.
   (c) Calculate the value of the tension, $\vec{F}_{T2}$, in the horizontal rope.
   (d) Explain how each of these tensions would change if the angle with the horizontal were decreased to a value less than 60°. Justify your answer based on physics principles.

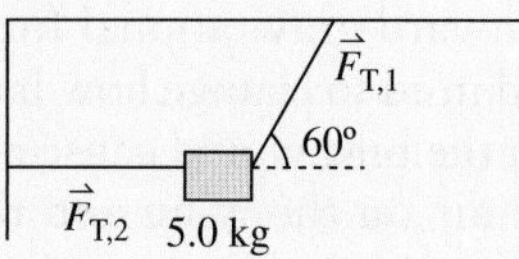

AP® Skills in Action

6. A horse is attached to a cart but refuses to pull the cart. "According to Newton's third law, it will do no good," says the horse, "no matter how hard I pull on the cart, the cart will pull on me with an equal and opposite force, and so the forces will cancel." Construct a claim supported by evidence that should convince the horse that pulling on the cart will make the cart move.
7. Alvin claims that the speed of a cart on which a fan is mounted can be increased by attaching a sail to the cart that catches the air as it passes through the fan. Amy challenges Alvin's claim by using a free-body diagram and Newton's laws of motion to support her claim.
   (a) What claim is Amy making? Answer in complete sentences and with a free-body diagram.
   (b) Still unconvinced, Alvin tapes a light sheet of paper to the face of the fan and, when the cart is at rest, switches the fan on. Predict what happens to the motion of the cart.
   (c) To test her claim Amy removes the tape. Predict what will happen and explain why this action is a test of her claim.

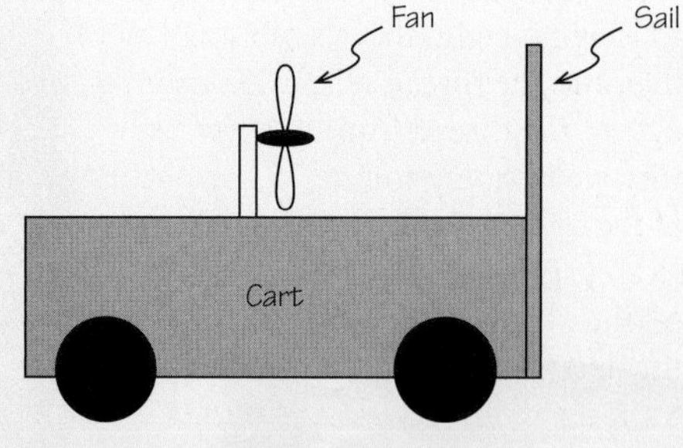

4-6 All problems involving forces can be solved using the same series of steps

In this section we'll use Newton's laws to solve a variety of problems that involve forces and acceleration. We can set up *any* force problem using just two steps.

1. Make a free-body diagram for the object (or for each of the objects) of interest using the rules that we outlined in Section 4-4. Choose the coordinate axes and label the positive directions. (As we've mentioned before, it's best to choose these axes so that as many as possible of the vectors in the problem—the acceleration and the individual forces—lie along the axes.)
2. Write Newton's second law for each object and in each direction separately. Use the corresponding free-body diagram to make sure that for each object, all of the forces exerted on that object are included in the sum of external forces. Then solve for the unknowns.

NEED TO REVIEW?
See the Math Tutorial, Section M8, in the back of the book for more information on trigonometry.

Problems Involving a Single Object

Our first few examples involve just a single object. In some problems the goal is to find the object's acceleration; in other cases you need to determine one or more of the forces exerted on the object. Sometimes determining the acceleration or unknown force is not the goal itself but just a necessary step to arrive at the final answer.

One thing to think about in these problems is the choice of x and y axes for your force diagram. If the object is on an incline or otherwise has a known direction of acceleration, choose one of the axes to point along the direction of the acceleration, and make the other axis perpendicular to it. If the object is moving in a known direction, it's best to choose one of the axes to point in that direction. If you're not sure which way the object will accelerate (say, whether to the left or to the right), don't worry about it—choose an axis to point in one direction or the other. If it turns out that the acceleration is in the direction opposite to the one you chose, you'll know because the acceleration will turn out to be negative.

EXAMPLE 4-4 What's the Angle?

A pair of fuzzy dice hangs on a lightweight thread from the rearview mirror of your high-performance sports car. At what angle from the vertical do the dice hang when the car is accelerating forward at a constant 6.50 m/s^2?

Set Up

At first the dice appear to swing backward on the thread, but really, they are trying to stay still as the thread tries to bring them along with the car. Once they are hanging at a constant angle, the dice are moving along with the car and so have the same forward acceleration of 6.50 m/s^2. This means there must be a net forward force on each of the dice. The individual forces on each of the dice are the gravitational force $\vec{F}_{g,dice}$ exerted by Earth and the tension force $\vec{F}_T$ exerted by the thread. Because the thread is lightweight, the gravitational force on it is negligible, and the thread will be straight when taut. So the angle of the thread from the vertical—which is the quantity we're trying to find—is the same as the angle θ of the tension force $\vec{F}_T$. Draw the free-body diagram to represent the forces you identified that are exerted on the object. Choose the coordinate system to have the direction of the known acceleration and the known force of gravity along the axes. Then we will only have to take components of one vector, the tension force. We'll find this angle by applying Newton's second law to the dice. Because the dice

Newton's second law in component form:

$$\sum F_{\text{ext on dice},x} = m_{\text{dice}} a_{\text{dice},x} \quad (4\text{-}3a)$$

$$\sum F_{\text{ext on dice},y} = m_{\text{dice}} a_{\text{dice},y} \quad (4\text{-}3b)$$

Net force on dice = sum of tension force and gravitational force:

$$\sum \vec{F}_{\text{ext on dice}} = \vec{F}_T + \vec{F}_{g,\text{dice}}$$

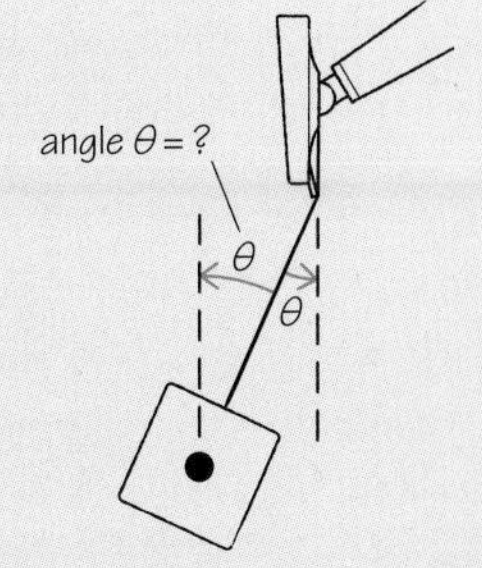

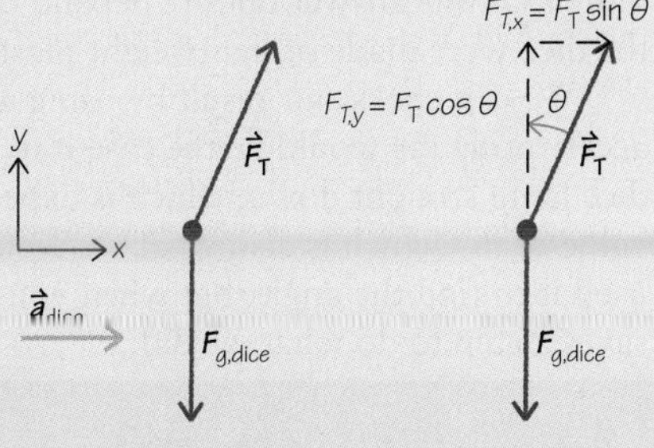

are identical, we only have to do this for one. Note that we are adding in our solution the components of the tension vector in a different line style and clearly indicating they are components. We place the components in a second version of the diagram used for working the problem out.

Solve

Use the free-body diagram to determine how to write the x and y components of the external forces exerted on the dice and the acceleration of the dice.

Components of tension force $\vec{F}_T$:

$F_{T,x} = F_T \sin\theta$

$F_{T,y} = F_T \cos\theta$

Components of gravitational force $\vec{F}_{g,dice}$ with magnitude $F_{g,dice} = m_{dice}g$:

$F_{g,dice,x} = 0$

$F_{g,dice,y} = -F_{g,dice} = -m_{dice}g$

Components of acceleration $\vec{a}_{dice}$:

$a_{dice,x} = a_{dice} = 6.50 \text{ m/s}^2$

$a_{dice,y} = 0$

Substitute the expressions for the components of $\vec{F}_g$, $\vec{F}_T$, and $\vec{a}_{dice}$ into Newton's second law. This gives two equations, one from the x components and one from the y components.

Newton's second law in component form:

$x: \sum F_{\text{ext on dice},x} = F_{T,x} + F_{g,dice,x} = m_{dice}\,a_{dice,x}$, so

$F_T \sin\theta + 0 = m_{dice}a_{dice}$

$y: \sum F_{\text{ext on dice},y} = F_{T,y} + F_{g,dice,y} = m_{dice}a_{dice,y}$, so

$F_T \cos\theta + (-m_{dice}g) = 0$

We don't know either the magnitude F_T or the angle θ of the tension force. All we are asked to solve for is the value of θ, so we eliminate F_T from the two equations above. We also use the definition of the tangent function, $\tan\theta = (\sin\theta)/(\cos\theta)$.

From the y equation:

$F_T \cos\theta = m_{dice}g$, so

$$F_T = \frac{m_{dice}g}{\cos\theta}$$

Substitute this into the x equation:

$F_T \sin\theta = m_{dice}a_{dice}$, so

$$\frac{m_{dice}g}{\cos\theta}\sin\theta = m_{dice}a_{dice}$$

Divide through by m_{dice} and g and apply the definition of tangent:

$$\frac{\sin\theta}{\cos\theta} = \tan\theta = \frac{a_{dice}}{g}$$

Solve for θ:

$$\theta = \tan^{-1}\left(\frac{a_{dice}}{g}\right) = \tan^{-1}\left(\frac{6.50 \text{ m/s}^2}{9.80 \text{ m/s}^2}\right) = \tan^{-1}(0.663)$$

$$= 33.6°$$

Reflect

Note that our answer doesn't depend on the mass m_{dice}. The angle θ would be the same whether the dice were made of lightweight plastic or solid gold.

We can check our result by using a range of values of acceleration. When the car is not accelerating (as would be the case if its velocity is constant), $a_{dice} = 0$ and $\theta = \tan^{-1} 0 = 0$. The dice hang straight down, which is expected. The greater the acceleration a_{dice}, the greater the value of θ—which is also what we would expect. We could have solved for the tension and then used it to find the angle, but when you have two unknowns and you only need one of them, it saves a step to do what we did.

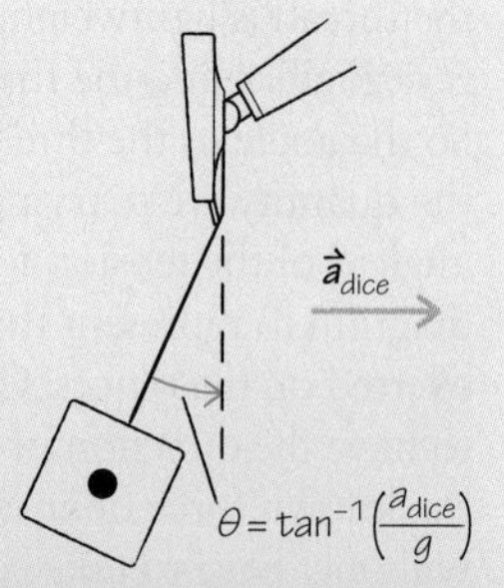

NOW WORK Problems 1 and 2 from The Takeaway 4-6.

WATCH OUT !

Newton's laws apply only in inertial frames of reference.

You might argue that the dice in the preceding example accelerate *backward* rather than forward. After all, from your perspective as the driver, the dice swing backward when you step on the accelerator pedal. The driver's seat isn't a good place from which to apply Newton's laws, however, because you are yourself accelerating! Until the string is tight, the dice are accelerating more slowly than you are, so *comparatively* they seem to accelerate backward. It turns out that Newton's laws work *only* if you observe from a vantage point that is *not* accelerating. Such a vantage point is called an **inertial frame of reference.** Newton's first law is considered to define inertial frames of reference. One such frame of reference would be a person standing still by the road. Although the dice appear to move backward to the driver, they accelerate forward with the car, just not as quickly, to the observer standing by the road. Once the dice have stabilized, they move with the same acceleration as the car, so to the driver they do not seem to be accelerating. Once you reach a constant velocity, neither you nor the dice are accelerating, and they hang straight down. The same inertial effect explains why you feel pushed back in your seat when the car accelerates forward, and you feel thrown forward against your safety belt if the car brakes rapidly (accelerates in the direction opposite its motion). No force is exerted on you in the direction you *feel* accelerated in either case: It's your inertia resisting a change in your motion. (If you think inertia *is* a real force, ask yourself what could be exerting it. If you think it's your own body, remember that only external forces affect motion.)

EXAMPLE 4-5 Slipping and Sliding I

One wintry day you accidentally drop your physics book (mass 2.50 kg) on an icy sidewalk (a surface that exerts negligible friction) that tilts at an angle of 10.0° with the horizontal. What is the book's acceleration as it slides downhill?

Set Up

We'll find the book's acceleration using Newton's second law. Because we can assume there is no friction in this situation, the only forces exerted on the book are the gravitational force $\vec{F}_{g,book}$ (which points straight down) and the normal force $\vec{F}_n$ (which points perpendicular to the surface of the inclined sidewalk). Draw the free-body diagram to represent the forces you identified that are exerted on the object. This situation is similar to the block shown in Figure 4-13 (Section 4-4). Because the motion is down the ramp, we choose the x axis to be in that direction.

Newton's second law in component form applied to the book:

$$\sum F_{ext\ on\ book,x} = m_{book}\, a_{book,x} \quad (4\text{-}3a)$$

$$\sum F_{ext\ on\ book,y} = m_{book}\, a_{book,y} \quad (4\text{-}3b)$$

Net force on book = sum of normal force and gravitational force:

$$\sum \vec{F}_{ext\ on\ book} = \vec{F}_n + \vec{F}_{g,book}$$

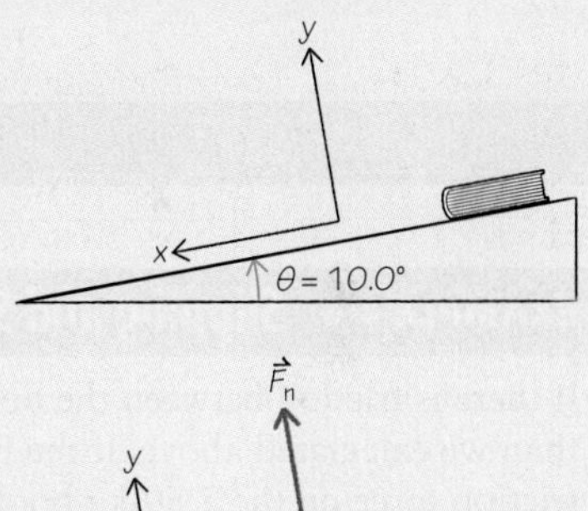

Solve

Reproduce a working copy of the free-body diagram to determine how to write the forces and the acceleration in component form. The book doesn't move or change its motion perpendicular to the ramp, so the y component of its acceleration is zero.

Components of normal force $\vec{F}_n$:

$$F_{n,x} = 0$$

$$F_{n,y} = F_n$$

Components of gravitational force $\vec{F}_{g,book}$:

$$\text{magnitude } F_{g,book} = m_{book}g$$

$$F_{g,book,x} = F_{g,book} \sin\theta = m_{book}g \sin\theta$$

$$F_{g,book,y} = -F_{g,book} \cos\theta = -m_{book}g \cos\theta$$

Components of acceleration $\vec{a}_{book}$:

$a_{book,x}$: to be determined

$$a_{book,y} = 0$$

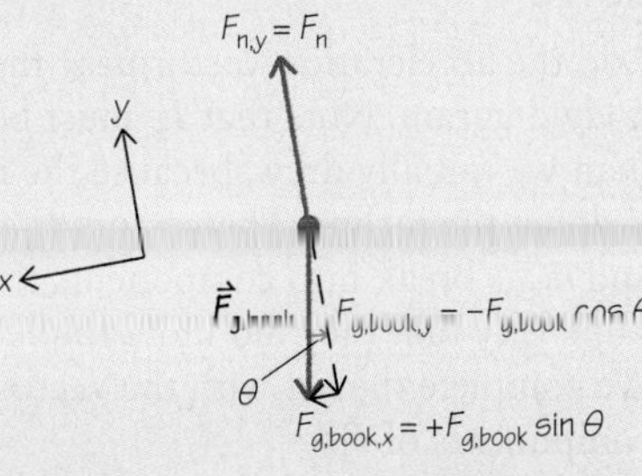

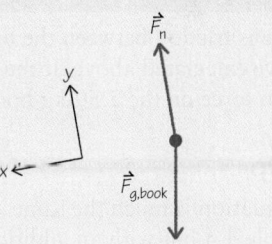

Use the components to write Newton's second law in component form for the book.

Newton's second law in component form:

$$x: \sum F_{\text{ext on book},x} = F_{n,x} + F_{g,x} = 0 + m_{\text{book}} g \sin\theta$$
$$= m_{\text{book}} a_{\text{book},x}, \text{ so}$$
$$m_{\text{book}} g \sin\theta = m_{\text{book}} a_{\text{book},x}$$
$$y: \sum F_{\text{ext on book},y} = F_{n,y} + F_{g,y} = F_n - m_{\text{book}} g \cos\theta = m_{\text{book}} a_{\text{book},y}, \text{ so}$$
$$F_n - m_{\text{book}} g \cos\theta = 0$$

All we are asked to find is the acceleration of the book, so we need only the x equation.

Dividing the x equation by m_{book}:

$$a_{\text{book},x} = g \sin\theta = (9.80 \text{ m/s}^2) \sin 10.0°$$
$$= 1.70 \text{ m/s}^2$$

Reflect

Our result tells us that the book accelerates in the positive x direction (downhill). This result makes sense: When no friction opposes its motion, the book should gain speed as it slides downhill. If the slope were steeper, θ and $\sin\theta$ would be greater, and the acceleration $a_{\text{book},x} = g \sin\theta$ would be greater as well. Just as for an object in free fall, the acceleration doesn't depend on the book's mass.

Extend: Note also that our analysis would be exactly the same if you put the book near the bottom of the incline and gave it an upward push to start it moving. The gravitational force and normal force would be unchanged, so the magnitude and direction of the acceleration would likewise be unchanged. Because the acceleration is downhill and the velocity uphill, the book would slow down as it moved up the incline—again, just as we would expect.

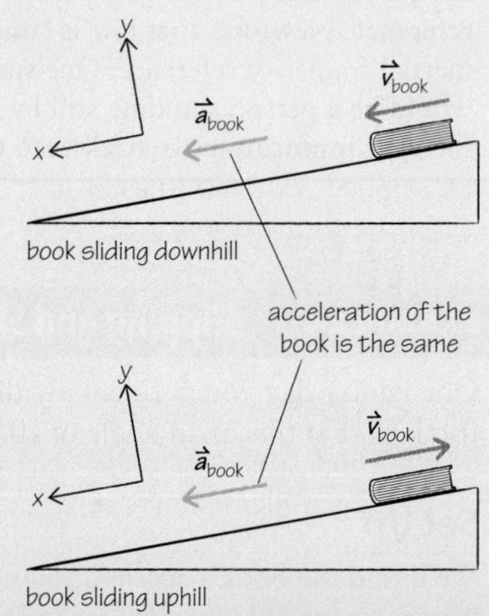

NOW WORK Problem 3 from The Takeaway 4-6.

EXAMPLE 4-6 Slipping and Sliding II

If there is friction between the book and the incline in the previous example, the book's downhill acceleration will be less than we calculated above. If the book gains speed at only 0.200 m/s² as it slides downhill, what is the magnitude of the friction force on the 2.50-kg book?

Set Up

The situation is much the same as in Example 4-5 but with an additional kinetic friction force $\vec{F}_k$ which points up the incline, opposing the book's motion. Draw the free-body diagram to represent the forces you identified that are exerted on the object. We are given the book's acceleration, so we'll use Newton's second law to determine the magnitude of the friction force.

Newton's second law in component form applied to the book:

$$\sum F_{\text{ext on book},x} = m_{\text{book}} a_{\text{book},x} \quad (4\text{-}3a)$$
$$\sum F_{\text{ext on book},y} = m_{\text{book}} a_{\text{book},y} \quad (4\text{-}3b)$$

Net force on book = sum of normal force, gravitational force, and friction force:

$$\sum \vec{F}_{\text{ext on book}} = \vec{F}_n + \vec{F}_{g,\text{book}} + \vec{F}_k$$

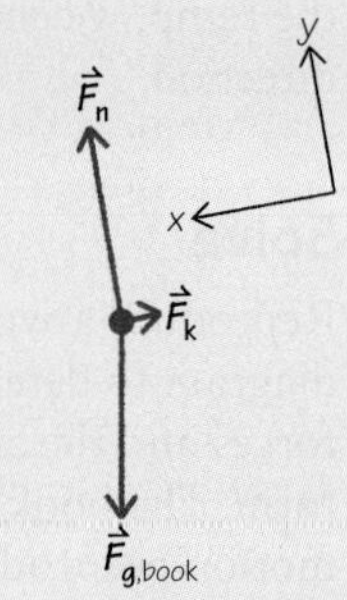

Solve

Add the acceleration vector near the free-body diagram. Note that $\vec{F}_k$ must be shorter than we initially drew, because the net force is down the incline. The vectors $\vec{F}_n$, $\vec{F}_{g,\text{book}}$, and $\vec{a}_{\text{book}}$ break into components the same way that they did in Example 4-5. We complete the list with the vector components of $\vec{F}_k$.

Components of friction force $\vec{F}_k$:

$$F_{k,x} = -F_k$$
$$F_{k,y} = 0$$

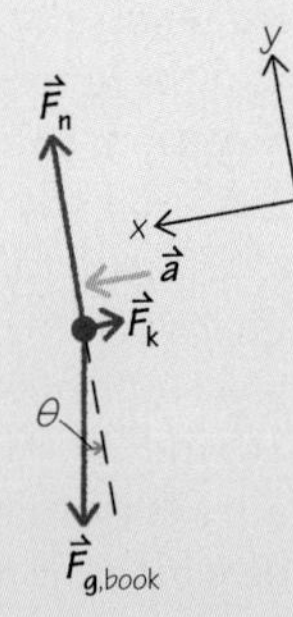

Use the components to write Newton's second law in component form for the book.

Newton's second law in component form:

$x: \sum F_{\text{ext on book},x} = F_{n,x} + F_{g,x} + F_{k,x} = 0 + m_{\text{book}} g \sin\theta - F_k = m_{\text{book}} a_{\text{book},x}$, so

$m_{\text{book}} g \sin\theta - F_k = m_{\text{book}} a_{\text{book},x}$

$y: \sum F_{\text{ext on book},y} = F_{n,y} + F_{g,y} + F_{k,y} = F_n - m_{\text{book}} g \cos\theta + 0 = m_{\text{book}} a_{\text{book},y}$, so

$F_n - m_{\text{book}} g \cos\theta = 0$

Solve the x equation for the magnitude F_k of the friction force.

$m_{\text{book}} g \sin\theta - F_k = m_{\text{book}} a_{\text{book},x}$ so

$$\begin{aligned} F_k &= m_{\text{book}} g \sin\theta - m_{\text{book}} a_{\text{book},x} \\ &= m_{\text{book}}(g \sin\theta - a_{\text{book},x}) \\ &= (2.50\ \text{kg})[(9.80\ \text{m/s}^2)\sin 10.0° - 0.200\ \text{m/s}^2] \\ &= 3.75\ \text{kg}\cdot\text{m/s}^2 = 3.75\ \text{N} \end{aligned}$$

Reflect

This example is very much like Example 4-3 in Section 4-3. The only difference is that the net force on the cat in Example 4-3 is zero (the cat is at rest), whereas the net force on the sliding book in this example is *not* zero.

NOW WORK Problem 4 from The Takeaway 4-6.

EXAMPLE 4-7 Apparent (Effective) Weight: When You Don't "Weigh" *mg*

A 70.0-kg student stands on a bathroom scale in an elevator. What is the reading on the scale (a) when the elevator is moving upward at a constant 0.500 m/s? (b) When the elevator is accelerating downward at 1.00 m/s^2?

Set Up

Remember from our discussion of spring scales in Section 4.5 that the spring scale measures the force exerted on it. Your bathroom scale operates in this same fashion, so when you stand on the bathroom scale, the reading is how hard the scale is having to push up on you to cause your upward acceleration. When you are at rest, your net vertical acceleration is zero, and the upward normal force from the scale just balances the downward force of gravity, your weight. However, in other situations where there is an acceleration, the reading is different. (Try pushing on a bathroom scale with your hand. The reading measures how strong you are, not how much you weigh.) Newton's third law tells us that the downward normal force of the student on the scale is equal in magnitude to the upward normal force exerted by the scale on the student, $\vec{F}_{\text{n,scale on student}}$. So what we want to know is the magnitude of $\vec{F}_{\text{n,scale on student}}$ in each situation, which we'll find by applying Newton's second law to the student. The only other force exerted on the student is the gravitational force $\vec{F}_{\text{g,student}}$. Both forces and the acceleration in part (b) are in the vertical direction, so we need only the y component. Draw the free-body diagrams to represent the forces you identified as being exerted on the object. Because there are two parts you will need two figures, so we will place our free-body diagrams in the Solve parts for (a) and (b). The gravitational force on the student is always the same, because the actual amount of matter of which the student is made does not change!

y component of Newton's second law applied to the student:

$$\sum F_{\text{ext on student},y} = m_{\text{student}} a_{\text{student},y} \qquad \text{(4-3b)}$$

Net force on student = sum of normal force and gravitational force:

$$\sum \vec{F}_{\text{ext on student}} = \vec{F}_{\text{n,scale on student}} + \vec{F}_{\text{g,student}}$$

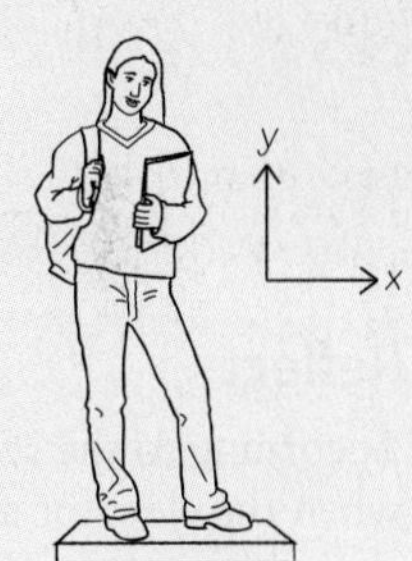

Solve

Draw the free-body diagram to represent the forces you identified that are exerted on the object. (a) If the elevator is moving at a constant velocity, the student's acceleration is zero: The sum of the external forces on the student must be zero. We then solve for the magnitude of the normal force that the scale exerts on the student.

Student's y velocity is constant, so

$a_{\text{student},y} = 0$

Newton's second law in component form:

$y{:}\sum F_{\text{ext on student},y}$
$= F_{\text{n,scale on student}} + (-F_{\text{g,student}})$
$= m_{\text{student}} a_{\text{student},y} = 0$ so

$F_{\text{n,scale on student}}$
$= F_{\text{g,student}} = m_{\text{student}}\, g$
$= (70.0 \text{ kg})(9.80 \text{ m/s}^2)$
$= 686 \text{ N}$

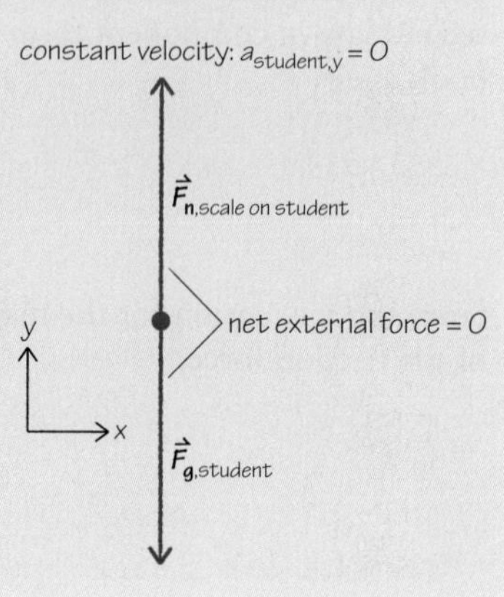

(b) If the elevator, and so also the student, are accelerating downward (in the negative y direction) at 1.00 m/s^2, the net force on the student must be downward. The net force can be downward only if the upward normal force exerted by the scale is smaller than the downward gravitational force. Again we solve for the magnitude $F_{\text{n,scale on student}}$ using Newton's second law.

The student's acceleration has a downward (negative) y component:

$a_{\text{student},y} = -1.00 \text{ m/s}^2$

Newton's second law in component form:

$y{:}\sum F_{\text{ext on student},y}$
$= F_{\text{n,scale on student}} + (-F_{\text{g,student}})$
$= m_{\text{student}} a_{\text{student},y}$

$F_{\text{n,scale on student}}$
$= F_{\text{g,student}} + m_{\text{student}} a_{\text{student},y}$
$= 686 \text{ N} + (70.0 \text{ kg})(-1.00 \text{ m/s}^2) = 616 \text{ N}$

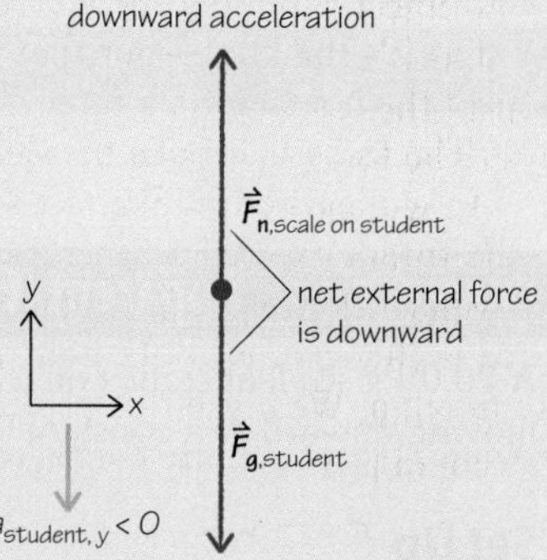

Reflect

According to the scale the student "weighs" 70 N less when the elevator accelerates downward. You've probably experienced this phenomenon in an elevator when it starts moving downward from rest or when it comes to a halt when moving upward, so this makes sense. If the elevator were in free fall, such as briefly in some amusement park rides, you would experience "weightlessness"; the normal force exerted on you by the scale would be zero!

If the elevator is in free fall,

$a_{\text{student},y} = -g$

$F_{\text{n,scale on student}}$
$= F_{\text{g,student}} + m_{\text{student}} a_{\text{student},y}$
$= m_{\text{student}} g + m_{\text{student}}(-g)$
$= 0$

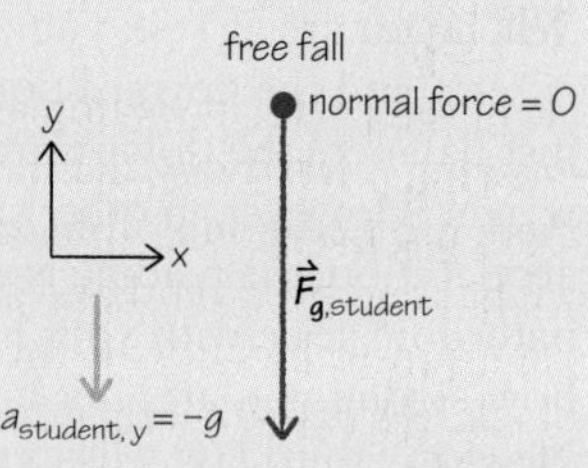

Extend: A similar phenomenon occurs in an elevator that accelerates *upward*. Then $a_{\text{student},y}$ is positive, $F_{\text{n,scale on student}}$ is greater than the student's weight $F_{\text{g,student}}$, and the student feels heavier than normal.

NOW WORK **Problem 5 from The Takeaway 4-6.**

In Example 4-7, we showed that in free fall, the scale reads zero, and the student will feel weightless! This situation is called *apparent weightlessness*. The student isn't truly weightless because Earth still exerts a gravitational force on her. We'll see in Chapter 6 that astronauts in Earth orbit are also in free fall, so they *feel* weightless just like the student in the free-falling elevator.

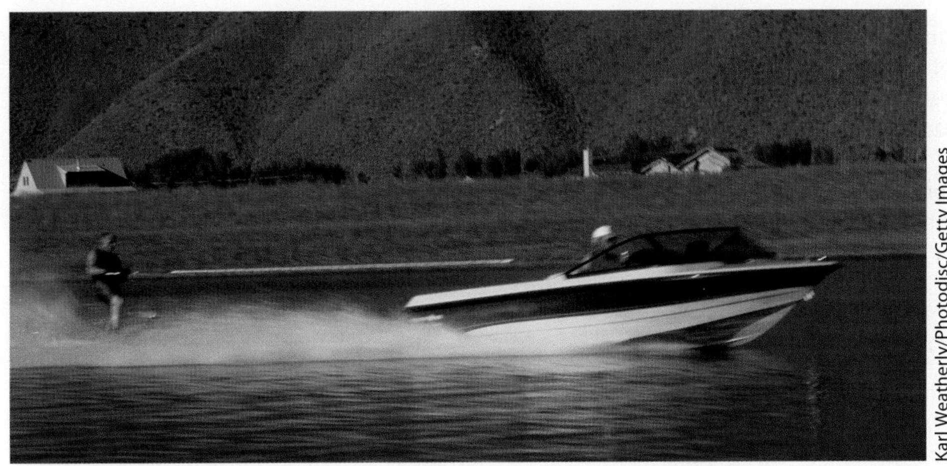

Karl Weatherly/Photodisc/Getty Images

Figure 4-21 Towing a water skier The boat exerts a force on the tow rope, and the tow rope pulls on the skier. The rope appears straight so we can safely neglect its mass.

Problems Involving Ropes and Tension

What object is pulling the water skier shown in **Figure 4-21**? You might be tempted to say that it's the boat—but that can't be correct because the boat doesn't touch him. Instead the boat exerts a force on the tow rope, and it's the tow rope that pulls on the skier. The force that pulls the skier is therefore a *tension* force.

As we learned in Section 4-5, if the weight of the tow rope is small compared to the forces exerted by the objects to which it's attached (in this case, the boat and the water skier), we can ignore its mass altogether. Then the magnitude of the tension force is the *same* at both ends of the rope no matter how the boat, skier, and rope are moving. We'll use this principle in the next two examples, both of which involve moving objects that are connected by a lightweight string.

> **AP® Exam Tip**
>
> A pulley is used to change the direction of the tension force in a string or rope. We use the model *ideal pulley* when the pulley has small enough mass to ignore compared to other objects in the system and causes negligible friction in the system. On the AP® Physics 1 exam, unless you are given a mass of the pulley or told there is friction you may assume any pulleys are ideal. Nonideal pulleys will be introduced later in the course.

EXAMPLE 4-8 Tension in a String

You pull two blocks connected by a lightweight string across a horizontal table. Block 1 has a mass of 1.00 kg, and block 2 has a mass of 2.00 kg. There is negligible friction between the blocks and the table. The string will break if the tension is greater than 6.00 N. What is the maximum force with which you can pull block 1 without breaking the string?

Set Up

As in the previous examples we start by drawing a free-body diagram. Note that here there are *two* bodies (block 1 and block 2), so we have to be careful and draw *two* free-body diagrams to include the forces exerted on each block. We can label them m_1 and m_2 given that we will be using those symbols to express their masses. We choose the positive x axis to be in the direction of the blocks' horizontal motion and choose the positive y axis to be vertically upward.

$$\vec{a}_1 = \sum \vec{F}_{\text{ext on }1}/m_1$$

$$\vec{a}_2 = \sum \vec{F}_{\text{ext on }2}/m_2 \qquad (4\text{–}2)$$

Net force on block 1 = sum of force you exert, normal force, gravitational force, and tension:

$$\sum \vec{F}_{\text{ext on }1} = \vec{F}_{\text{you on }1} + \vec{F}_{\text{n},1} + \vec{F}_{\text{g},1} + \vec{F}_{\text{T,string on }1}$$

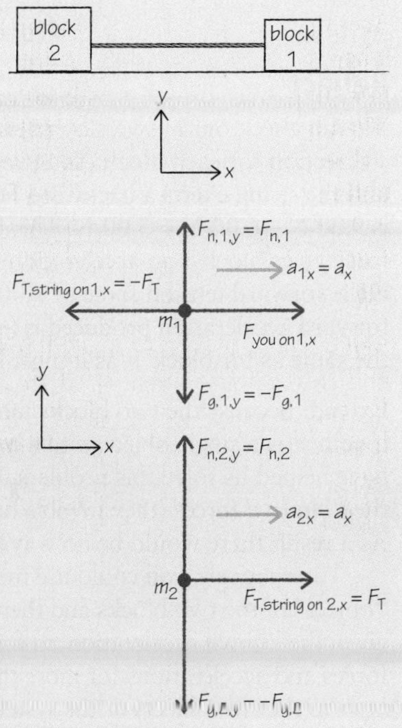

We are told that the string connecting the blocks is lightweight, so applying Newton's third law as we did in Figure 4-19, the tension at either end of the string has the same magnitude: $F_{T,\text{string on }1} = F_{T,\text{string on }2}$. We'll call this magnitude F_T for short. Our goal is to find the maximum value of $F_{\text{you on }1}$ (the magnitude of the force that you exert on block 1) so that F_T is no greater than 6.00 N.

Net force on block 2 = sum of normal force, gravitational force, and tension:

$$\sum \vec{F}_{\text{ext on }2} = \vec{F}_{n,2} + \vec{F}_{g,2} + \vec{F}_{T,\text{string on }2}$$

The magnitude of the tension is the same at both ends of the string:

$$F_{T,\text{string on }1} = F_{T,\text{string on }2} = F_T$$

Solve

We have already figured out the components on the free-body diagrams above. We then write Newton's second law in component form for each block. If we assume that the string doesn't stretch, the two blocks move together and have the same x component of acceleration, which we call a_x.

Both blocks have the same x acceleration:

$$a_{1x} = a_{2x} = a_x$$

Both blocks have zero y acceleration.

Newton's second law in component form for block 1:

$$x: \sum F_{\text{ext on }1,x} = F_{\text{you on }1} + (-F_T) = m_1 a_x$$

$$y: \sum F_{\text{ext on }1,y} = F_{n,1} + (-F_{g,1}) = 0$$

Newton's second law in component form for block 2:

$$x: \sum F_{\text{ext on }2,x} = F_T = m_2 a_x$$

$$y: \sum F_{\text{ext on }2,y} = F_{n,2} + (-F_{g,2}) = 0$$

We want to find the value of $F_{\text{you on }1}$ when $F_T = 6.00$ N (the maximum tension before the string breaks). We use the x equation for block 2 to find the x acceleration a_x in this situation and then we use this value of a_x in the x equation for block 1 to solve for $F_{\text{you on }1}$.

First solve for the value of a_x when $F_T = 6.00$ N. Use the x equation for block 2:

$$F_T = m_2 a_x$$

$$a_x = \frac{F_T}{m_2} = \frac{6.00\text{ N}}{2.00\text{ kg}} = 3.00\text{ N/kg} = 3.00\text{ m/s}^2$$

Now substitute $a_x = 3.00\text{ m/s}^2$ into the x equation for block 1 and solve for $F_{\text{you on }1}$:

$$F_{\text{you on }1} - F_T = m_1 a_x$$

$$F_{\text{you on }1} = F_T + m_1 a_x = 6.00\text{ N} + (1.00\text{ kg})(3.00\text{ m/s}^2)$$
$$= 6.00\text{ N} + 3.00\text{ kg}\cdot\text{m/s}^2 = 6.00\text{ N} + 3.00\text{ N} = 9.00\text{ N}$$

Reflect

We can check our answers by referring back to Newton's second law in the x direction for each block. You exert a forward force $F_{\text{you on }1} = 9.00$ N on block 1, and the string exerts a backward force of 6.00 N. So the net forward force on block 1 is 9.00 N − 6.00 N = 3.00 N. This block has a mass of 1.00 kg, so the 3.00-N net force gives block 1 an acceleration of 3.00 m/s². The only horizontal force on block 2 is the forward tension force of 6.00 N; because block 2 has a mass of 2.00 kg, the forward acceleration produced is (6.00 N)/(2.00 kg) = 3.00 m/s². This acceleration is the same as for block 1, as it must be because the two blocks move together.

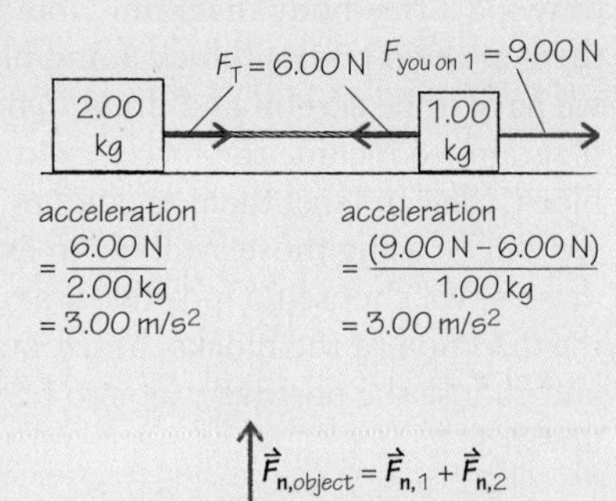

Extend: Because the two blocks move together, we could have treated the two blocks together as a single object with a mass equal to $m_1 + m_2 = 3.00$ kg. That wouldn't have helped us solve this problem, however. The reason is that the tension forces are then *internal* forces (they involve one part of the "object" pulling on another part). As a result there would be no way to apply the condition on the tension.

Alternatively, you could use the object model to find the acceleration of the "object" of the two blocks and then use the free-body diagram for the 2.00-kg block, given the known acceleration, to find the tension. When solving problems that involve forces and accelerations for more than one object, you will need to apply Newton's laws to the two objects or the system and one of the objects separately whenever you need to solve for a force that involves an interaction between the objects.

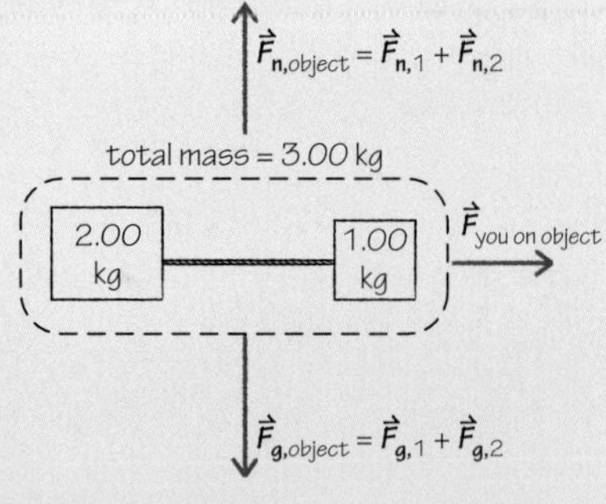

Can you solve the equations for the two blocks as an object to show that this acceleration is 3.00 m/s^2?

Newton's second law in component form for the two blocks as a single object:

$x: \sum F_{\text{ext on object},x}$
$= F_{\text{you on 1}}$
$= (m_1 + m_2)a_{\text{object},x}$

$y: \sum F_{\text{ext on 1},y} = F_n + (-F_g) = 0$

We can't solve for $F_{\text{you on 1}}$ because F_T does not appear in these equations, but once you know $F_{\text{you on 1}}$, you can use it to confirm a_x and use the acceleration and a free-body diagram of the 2.00-kg block to find the tension.

NOW WORK Problem 6 from The Takeaway 4-6.

EXAMPLE 4-9 Atwood's Machine

Two blocks are connected by a lightweight string that passes over an ideal pulley, as shown in **Figure 4-22.** (This arrangement is known as *Atwood's machine* after Rev. George Atwood, who devised it in 1784 as a way to demonstrate motion with constant acceleration.) Block 1 is more massive than block 2, so when released block 1 accelerates downward, and block 2 accelerates upward. Because the pulley is ideal, the mass of the pulley is so small that we can neglect it. Derive expressions for (a) the acceleration of each block and (b) the tension in the string.

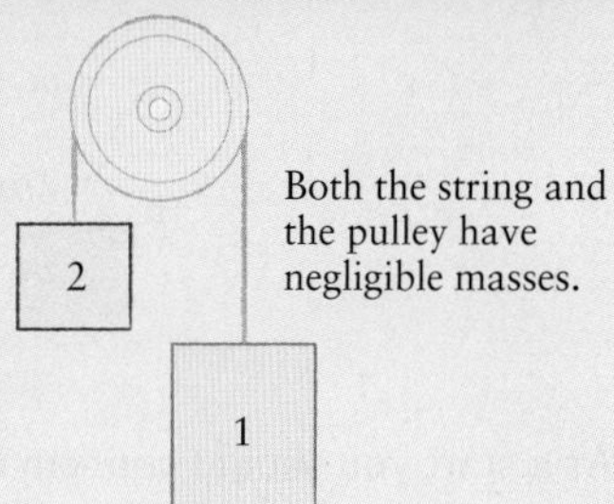

2
1
Both the string and the pulley have negligible masses.

Figure 4-22 Atwood's machine When the blocks are released from rest, block 1 falls and block 2 rises. What are their accelerations, and what is the tension in the string?

Set Up

Each block has only two forces exerted on it: a downward gravitational force and an upward tension force. We choose the positive y direction for block 1 to be downward and the positive y direction for block 2 to be upward. Using this choice, both blocks accelerate in the positive y direction and both blocks have the same y component of acceleration (because we assume that the string doesn't stretch). One of our goals is to calculate this acceleration, which we call a_y.

Because the pulley is ideal, the magnitude of the tension force has the same value F_T at both ends of the lightweight string. Our other goal is to calculate the value of F_T. We draw separate free-body diagrams for each block. We can again label them m_1 and m_2 because we will be using those symbols to express their masses.

Newton's second law in the y direction applied to each block:

$$\sum F_{\text{ext on 1},y} = m_1 a_{1y}$$
$$\sum F_{\text{ext on 2},y} = m_2 a_{2y} \qquad (4\text{-}3b)$$

Net force on block 1 = sum of gravitational force and tension:

$$\sum \vec{F}_{\text{ext on 1}} = \vec{F}_{g,1} + \vec{F}_{\text{T,string on 1}}$$

Net force on block 2 = sum of gravitational force and tension:

$$\sum \vec{F}_{\text{ext on 2}} = \vec{F}_{g,2} + \vec{F}_{\text{T,string on 2}}$$

Both blocks have the same acceleration, a y component in the direction chosen to be positive:

$$a_{1y} = a_{2y} = a_y$$

Magnitude of tension is the same at both ends of the string:

$$F_{\text{T,string on 1}} = F_{\text{T,string on 2}} = F_T$$

Solve

Write the y component of Newton's second law for each block.

Newton's second law in component form for block 1:

$$y: \sum F_{\text{ext on 1},y} = F_{g,1} + (-F_T) = m_1 a_y \text{ or } m_1 g - F_T = m_1 a_y$$

Newton's second law in component form for block 2:

$$y: \sum F_{\text{ext on 2},y} = F_T + (-F_{g,2}) = m_2 a_y \text{ or } F_T - m_2 g = m_2 a_y$$

We have two equations (one for block 1 and one for block 2) and two quantities that we want to find (a_y and F_T). One way to proceed is to first solve for one unknown in terms of the other. Let's solve the block 1 equation for F_T in terms of a_y. Then substitute this expression for F_T into the block 2 equation and solve for a_y.

Solve the block 1 equation for F_T:

$m_1 g - F_T = m_1 a_y$ so $F_T = m_1 g - m_1 a_y$

Substitute this into the block 2 equation:

$(m_1 g - m_1 a_y) - m_2 g = m_2 a_y$

Solve for a_y:

$m_1 g - m_2 g = m_1 a_y + m_2 a_y$

$(m_1 - m_2)g = (m_1 + m_2)a_y$ or

$$a_y = \left(\frac{m_1 - m_2}{m_1 + m_2}\right)g$$

Now that we have an expression for a_y, substitute it back into the equation for F_T that we found from the block 1 equation. This resulting expression gives us our final answer for F_T.

From the block 1 equation for F_T:

$F_T = m_1 g - m_1 a_y$

Substitute our expression for a_y and simplify:

$$F_T = m_1 g - m_1\left(\frac{m_1 - m_2}{m_1 + m_2}\right)g = m_1 g\left(1 - \frac{m_1 - m_2}{m_1 + m_2}\right)$$

$$= m_1 g\left(\frac{m_1 + m_2}{m_1 + m_2} - \frac{m_1 - m_2}{m_1 + m_2}\right) = m_1 g\left(\frac{m_1 + m_2 - m_1 + m_2}{m_1 + m_2}\right) \text{ or}$$

$$F_T = \frac{2m_1 m_2 g}{m_1 + m_2}$$

Reflect

Do these results make sense? As a start you should confirm that our expression for a_y (an acceleration) has units of m/s^2, and our expression for F_T (a force magnitude) has units of newtons. Do they?

To check our expression for acceleration a_y in more detail, notice that it depends on the difference $m_1 - m_2$ between the two masses. If $m_1 > m_2$ as we assumed here, a_y is positive; then both blocks accelerate in the positive y direction, so block 1 (the more massive and heavier one) accelerates downward, and block 2 accelerates upward.

If instead $m_1 < m_2$ so that block 2 is more massive and heavier, the acceleration a_y is negative. Then block 1 (which is now the lighter one) accelerates upward and block 2 downward. Finally, if the two blocks have equal mass so that $m_1 = m_2$, our expression says that $a_y = 0$, or the blocks don't accelerate at all. If the two equal-mass blocks begin at rest, they will remain at rest, just as we would expect.

What about our expression for the tension F_T? To check this out let's choose some specific values of m_1 and m_2. With $m_1 = 4.00$ kg and $m_2 = 2.00$ kg, the tension is $F_T = 26.1$ N. That's less than the 39.2-N weight of block 1 but greater than the 19.6-N weight of block 2. This result makes sense because for block 1 the downward gravitational force exceeds the upward tension force and the block accelerates downward, while for block 2 the upward tension force exceeds gravity, and the block accelerates upward. Note that because the tension force "holds back" the downward acceleration of block 1, the value of a_y is less than g.

Let $m_1 = 4.00$ kg and $m_2 = 2.00$ kg

Weight of block 1 $= m_1 g$

$= (4.00 \text{ kg})(9.80 \text{ m/s}^2)$

$= 39.2$ N

Weight of block 2 $= m_2 g$

$= (2.00 \text{ kg})(9.80 \text{ m/s}^2)$

$= 19.6$ N

Tension in string:

$$F_T = \frac{2m_1 m_2 g}{m_1 + m_2}$$

$$= \frac{2(4.00 \text{ kg})(2.00 \text{ kg})(9.80 \text{ m/s}^2)}{4.00 \text{ kg} + 2.00 \text{ kg}}$$

$$= 26.1 \text{ N}$$

Extend: You should repeat these calculations for different values of the two masses. Try $m_1 = 1.00$ kg and $m_2 = 3.00$ kg: How does the tension compare to the weights of the two blocks in this case? Is the acceleration positive or negative in this case?

Acceleration of either block:

$$a_y = \left(\frac{m_1 - m_2}{m_1 + m_2}\right)g$$

$$= \left(\frac{4.00 \text{ kg} - 2.00 \text{ kg}}{4.00 \text{ kg} + 2.00 \text{ kg}}\right)(9.80 \text{ m/s}^2)$$

$$= 3.27 \text{ m/s}^2$$

NOW WORK Problem 7(b) from The Takeaway 4-6.

Kinematics and Newton's Laws

All of the objects that we considered in Examples 4-4 through 4-9 move linearly. What's more, the forces exerted on each object are constant. So the *accelerations* that these forces produce are constant as well. That means we can apply all of the kinematic formulas for linear motion with constant acceleration that we learned in Chapter 2. If you can solve problems that use these formulas, and can also solve problems that involve Newton's laws, it's straightforward to solve problems that involve both of these. Here's an example.

EXAMPLE 4-10 Down the Slopes

A 60.0-kg skier starts from rest at the top of a hill with a 30.0° slope. She reaches the bottom of the slope 4.00 s later. If there is a constant 72.0-N kinetic friction force that resists her motion, how long is the hill?

Set Up

All three of the forces exerted on the skier—the normal force $\vec{F}_n$, the gravitational force $\vec{F}_{g,\text{skier}}$, and the kinetic friction force $\vec{F}_k$—are constant. So the net force on the skier is constant, as is her acceleration. We can therefore use a constant-acceleration equation to find the length of the hill. We'll find the skier's downhill acceleration using Newton's second law. We start by drawing a free-body diagram, and we add the dashed lines to work out the components. Components are not shown as part of a free-body diagram, but only added to a working diagram as part of a solution. Because we were not asked to produce a free-body diagram as part of the solution, this is okay.

$$x = x_0 + v_{0x}t + \frac{1}{2}a_x t^2 \quad (2\text{-}9)$$

Newton's second law in component form for the skier:

$$x: \sum F_{\text{ext on skier},x} = m_{\text{skier}} a_x \quad (4\text{-}3\text{a})$$

$$y: \sum F_{\text{ext on skier},y} = m_{\text{skier}} a_y \quad (4\text{-}3\text{b})$$

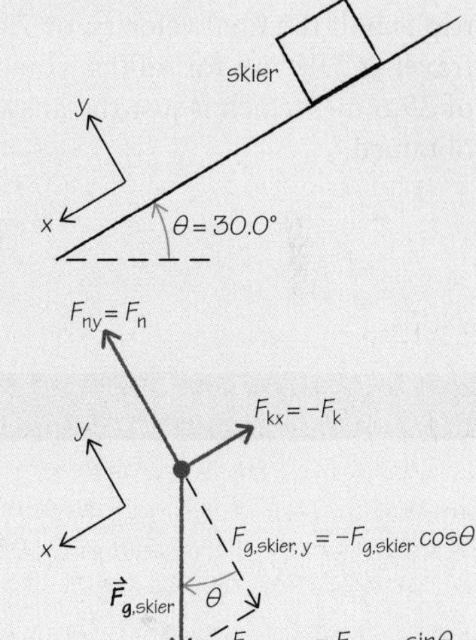

We choose the positive x axis to point down the slope and the positive y axis to point perpendicular to the slope as shown. (Compare Example 4-6.) Then the skier's acceleration points along the x axis only.

Net force on skier = sum of normal force, gravitational force, and friction force:

$$\sum \vec{F}_{\text{ext on skier}} = \vec{F}_n + \vec{F}_{g,\text{skier}} + \vec{F}_k$$

Solve

The normal force $\vec{F}_n$ has only a positive y component. The gravitational force $\vec{F}_{g,\text{skier}}$ has a positive (downhill) x component and a negative y component, whereas the friction force has only a negative (uphill) x component. Use these components and Newton's second law to solve for the skier's x component of acceleration a_x.

Find the skier's x component of acceleration from the x component of Newton's second law:

$$\sum F_{\text{ext on skier},x} = F_{g,\text{skier}} \sin\theta + (-F_k) = m_{\text{skier}} a_x$$

Weight of skier:

$F_{g,\text{skier}} = m_{\text{skier}} g$, subbing this into the equation we just found

$$m_{\text{skier}} a_x = m_{\text{skier}} g \sin\theta - F_k$$

$$a_x = g \sin\theta - \frac{F_k}{m_{\text{skier}}}$$

$$= (9.80 \text{ m/s}^2)\sin 30.0° - \frac{72.0 \text{ N}}{60.0 \text{ kg}}$$

$$= 4.90 \text{ m/s}^2 - 1.20 \text{ m/s}^2$$

$$= 3.70 \text{ m/s}^2$$

Now that we know the skier's acceleration in the x direction, we can use the kinematic equation to solve for her final position, which will be equal to the length of the hill as long as she starts at $x_0 = 0$. Because the skier starts at rest, her initial velocity is $v_{0x} = 0$.

Find the skier's final position after 4.00 s, assuming she starts at $x_0 = 0$.

$$x = x_0 + v_{0x}t + \frac{1}{2}a_x t^2$$

$$= 0 + 0 + \frac{1}{2}(3.70 \text{ m/s}^2)(4.00 \text{ s})^2$$

$$= 29.6 \text{ m}$$

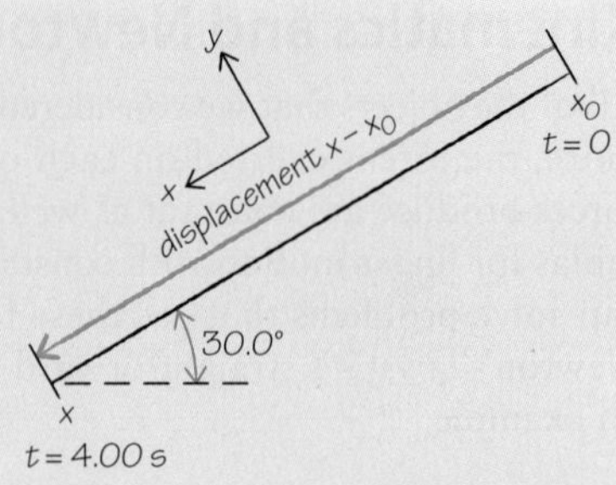

Reflect

Is our answer consistent? With an acceleration of 3.70 m/s², our skier reaches a velocity of 14.8 m/s after 4.00 s. Because she started at rest, her average x component of velocity for the trip is half the final velocity, or 7.40 m/s. If you travel at 7.40 m/s for 4.00 s, you cover a distance of 29.6 m—which is just the answer that we obtained.

Skier's velocity at $t = 4.00$ s:

$$v_x = v_{0x} + a_x t \qquad (2\text{-}5)$$

The skier's initial velocity is $v_{0x} = 0$, so

$$v_x = 0 + (3.70 \text{ m/s}^2)(4.00 \text{ s}) = 14.8 \text{ m/s}$$

Average velocity for the skier's motion with constant acceleration:

$$v_{\text{average},x} = \frac{v_{0x} + v_x}{2} = \frac{0 + 14.8 \text{ m/s}}{2} = 7.40 \text{ m/s} \qquad (2\text{-}7)$$

Distance traveled in 4.00 s:

$$x - x_0 = v_{\text{average},x}t = 0 + (7.40 \text{ m/s})(4.00 \text{ s}) = 29.6 \text{ m}$$

NOW WORK Problem 7(d) and (e) from The Takeaway 4-6.

By now you should have a pretty good sense of how to attack problems that involve force and acceleration. In the rest of this unit we'll use the same techniques to tackle problems involving a more detailed description of friction as well as motion in a circle.

 **Exam Tip**

Since you will never be asked to solve a problem in linear motion where the acceleration is changing, if you always choose your axes for your free-body diagram so that one axis is the direction of the acceleration, then the acceleration along the other axis is zero. This means that if you then apply kinematics to the problem, it is just one-dimensional kinematics, and you only need to solve for variables involving acceleration in one direction of motion. Along the other axis will always be the simpler constant-velocity (including $v = 0$) motion.

THE TAKEAWAY for Section 4-6

✔ To solve a problem that involves forces, begin by making a free-body diagram for the object (or one for each object) of interest, choosing coordinate axes as aligned with forces as possible.

✔ Once you have drawn the free-body diagram(s), write Newton's second law for each coordinate axis for each object, and solve for the unknowns.

 **Building Blocks**

1. Frying pans coated with polytetrafluoroethylene (PTFE), also known as Teflon, exert very small friction forces on grilled cheese sandwiches. A 200-g sourdough and melted raclette cheese sandwich is sliding along the bottom of a nonstick frying pan that exerts a negligible friction force on the sandwich and is held at an inclination of 20° relative to the plate.
   (a) Construct a free-body diagram of the sandwich.
   (b) Calculate the normal force exerted by the sandwich on the pan.
   (c) Calculate the acceleration of the sandwich.

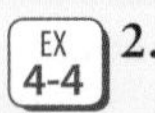

2. Smartphones and tablets detect screen orientation relative to the gravitational force. A micromechanical sensor converts the acceleration of a micron-scale bead into an electrical signal. The mechanism is similar to the fuzzy die that hangs on the rear-view mirror of the car in the figure. In the micromechanical accelerometer, the micron-scale object is supported by a cantilevered shelf of silicon that is anchored to a circuit board. When the board is accelerated the shelf deforms and the bead is accelerated in the same direction, but lagging behind because of its inertia.

(a) Construct a free-body diagram to represent the forces exerted on the fuzzy die. Assume that the string has negligible mass.

(b) In the reference frame of an observer standing on the ground, the fuzzy die is accelerating. Apply Newton's second law to re-express the x and y components of the free-body diagram in the frame of reference of the observer.

(c) Analyze the forces exerted on the die to predict the dependence of the angle θ on the acceleration of the car and the acceleration due to gravity, g.

When a smartphone is not accelerating in the reference frame of Earth's surface, the situation for the micromechanical sensor is the same as that for the die on the rearview mirror when $\theta = 0$. The accelerometer in your phone also has to make it possible for your phone to maintain a single screen orientation relative to the user. Whichever direction the cantilever is deflected is detected as "downward" by the phone. An illustration of the cantilevered bead surrounded by a case is shown.

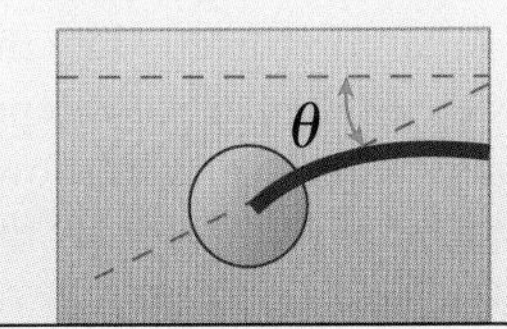

(d) When you turn your phone on its side the screen rotates. How many accelerometers must be in your smartphone for it to correctly sense this motion? Justify your answer. Use complete sentences.

EX 4-5 **3.** Some structures are supported by cantilevered beams, such as the balconies on the face of apartment buildings that are supported by horizontal bars embedded in concrete. The bars must bend because there is a large, downward gravitational force.

(a) Construct a free-body diagram representing the forces exerted on the center of mass of the balcony slab. The building must hold the slab up (consider this a tension force in the bar, which we will learn more about in a later chapter). We can model this by assuming a tension force exerted at a small angle upward, straight out from the bar, when the bar is modeled as being tilted downward slightly, instead of bending. Because this tension force is at a small angle upward, the bar pushes horizontally on the face of the apartment building as well. By Newton's third law, the building must exert an equal magnitude horizontal force to keep the bar from collapsing into the building, $F_{\text{building on bar}}$.

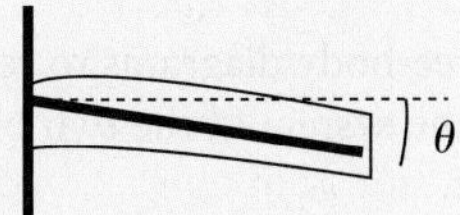

(b) Apply the second law to express the force exerted on the bar by the building in terms of m, θ, and $F_{\text{building on bar}}$.

(c) A small balcony is 3 m wide, 4 m long, and 0.25 m thick. The density of concrete is 2500 kg/m^3. The deflection is 3°. Calculate $F_{\text{building on bar}}$ in this case.

EX 4-6 **4.** A flat-bed truck is unloading a crate by elevating the front of the bed until the 1200-N crate just begins to slide.

(a) Construct a free-body diagram for the crate when the inclination of the bed relative to the horizontal is θ, the angle at which the crate begins to slide.

(b) The crate just begins to slide when $\theta = 19°$. Calculate the friction and normal forces exerted on the crate at that inclination of the bed.

(c) The driver is impatient and raises the bed to $\theta = 23°$ so that the acceleration is 0.80 m/s^2. Calculate the friction and normal forces exerted on the crate at that inclination of the bed.

EX 4-7 **5.** A small mob of zombies enters a freight elevator at the top, 13th, floor of an apartment building. None speak. All stare toward the back of the elevator car. When the door closes they all randomly choose the 5th floor. The elevator has a weight of 1700 N and the zombies have a weight of 8900 N. The cable that raises the elevator can safely support forces less than 12,000 N; beyond this limit the cable snaps. The building is fairly old and there are no safety systems installed in the elevator shaft.

(a) Construct a free-body diagram of the elevator and its contents.

(b) The elevator begins to descend with a sudden acceleration of magnitude $g/4$. Determine if the cable is at risk of snapping.

(c) The elevator comes to a constant speed, then begins to come to a stop. The zombies are nonplussed when the cable snaps. Explain why the cable snaps near the 5th floor by calculating the acceleration at which the cable snaps.

(d) What is the apparent weight (the normal force the elevator exerts on the zombies) of the zombies after the cable snaps?

EX 4-8 **6.** Two blocks with masses m_1 and m_2, as shown, are connected by a light string. A second string is also connected to the block of mass m_2. The string is used to pull the blocks up a ramp with a tension F directed parallel to the ramp. The ramp is inclined at an angle θ to the horizontal, as shown in the diagram below.

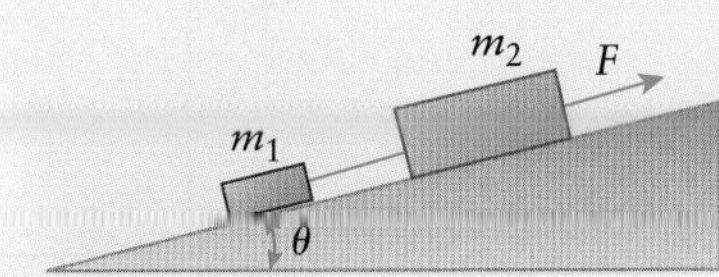

(a) Construct free-body diagrams to represent each of the blocks and the system of the two blocks treated as a single object.
(b) Apply Newton's second law to create a mathematical representation of the relations between the forces exerted on each block.
(c) Express the acceleration of the two blocks in terms of m_1, m_2, F, and θ assuming that the strings do not stretch.
(d) Express the tension, F_T, in the string connecting the two blocks in terms of m_1, m_2, F, θ, and the acceleration of the system, a.
(e) Calculate the numerical value of the tension in the string connecting the two blocks if $m_1 = 0.12$ kg, $m_2 = 0.24$ kg, $F = 3.1$ N, and $\theta = 10°$.

7. A Benedictine abbot at Westminster Abbey once was training a novice to ring the bell. Practicing without speaking, they grasped ends of a rope dangling over an ideal pulley from a machine left by Rev. George Atwood. On one occasion, the abbot stepped off a platform, grabbing his end of the rope, 2.5 m above the floor.

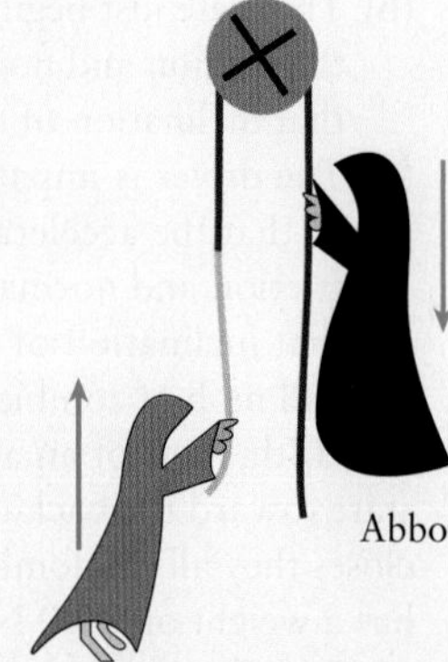

(a) Construct free-body diagrams for the abbot and the novice.

(b) Apply Newton's second law to express the acceleration of the system in terms of the masses of the abbot and the novice, and fundamental constants.
(c) Assume the abbot weighed 200 lbf (890 N) and the novice weighed 100 lbf. This event occurred before the British adopted SI in 1965. Calculate the acceleration of the novice as the abbot descended.

(d) Calculate the time interval for the descent of the abbot.

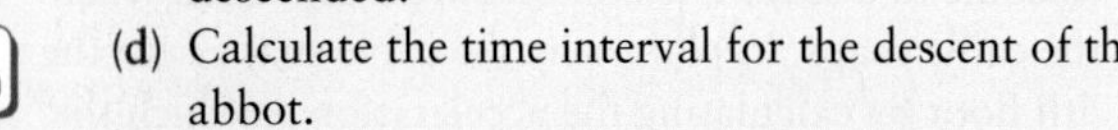

(e) When the abbot reached the floor, he released his end of the rope. Calculate the time interval for the descent of the novice.

AP® Skill Builders

8. Box A rests on a table as shown. A rope that connects boxes A and B drapes over a pulley so that box B hangs above the table, as shown. Assume that the pulley and rope have negligible mass and that the pulley turns with negligible friction.

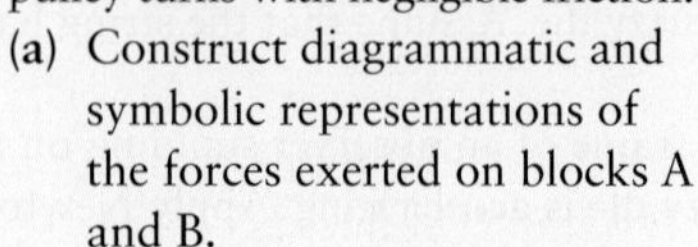

(a) Construct diagrammatic and symbolic representations of the forces exerted on blocks A and B.
(b) Add your equations from part (a) to construct a representation of the net force exerted on the system of two objects.
(c) Re-express the symbolic representations of part (b) as an expression for the expected normal force $F_{n,A}$ in terms of the weights of the blocks.

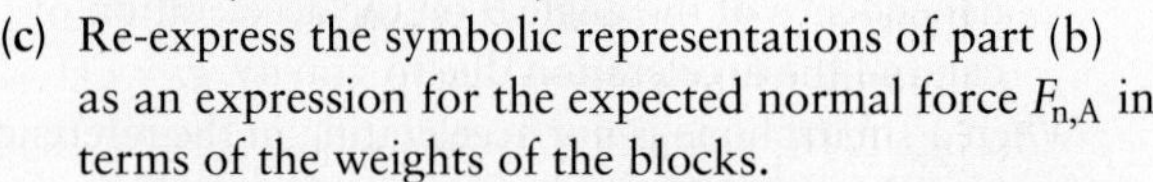

(d) A pan balance is used to separately determine the weights of blocks A and B. When the same pan balance is placed beneath block A with blocks A and B connected as shown in the figure and stationary, the magnitude of the normal force is larger than predicted by the expression obtained in part (c). Consider the assumptions made and evaluate the completeness of the identification of forces within the system.

9. A weightlifter stands on a weight scale while holding a barbell. The normal force exerted by the scale on the lifter–barbell system as the barbell is repetitively raised and lowered is shown in the graph below.
(a) Construct a free-body diagram for the lifter and another for the barbell.
(b) Use your diagrams to create a mathematical representation of the relationship between the normal force exerted by the scale on the lifter, the masses, g, and the acceleration of the barbell, assuming the acceleration of the lifter is 0.
(c) Use the equation from part (b) to explain the motion of the barbell causing the graph and why the display on the scale decreases between the points labeled b and c.
(d) If the mass of the barbell is 25 kg, analyze the data from the graph to determine the mass of the lifter.

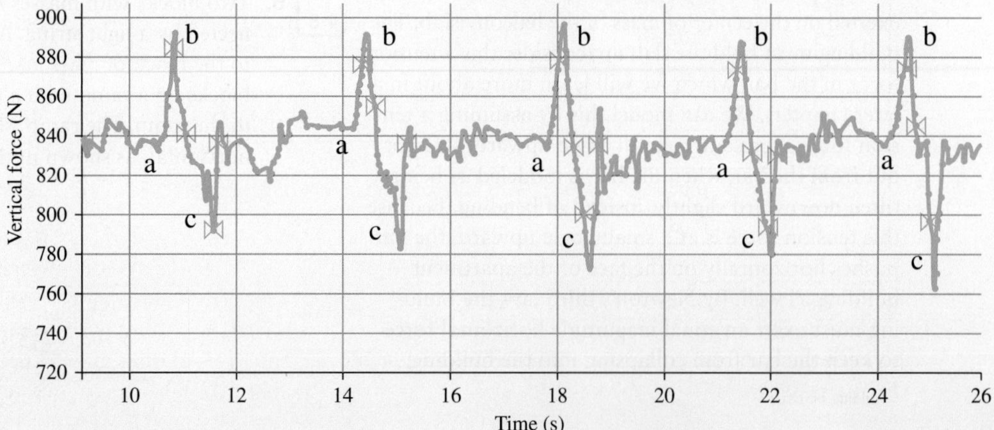

10. A person pulls three crates along a smooth horizontal floor, as shown. The pulling force is constant. The crates are connected to each other by identical horizontal strings A and B. Each string has a negligible mass and can support a maximum tension of 45.0 N before breaking.

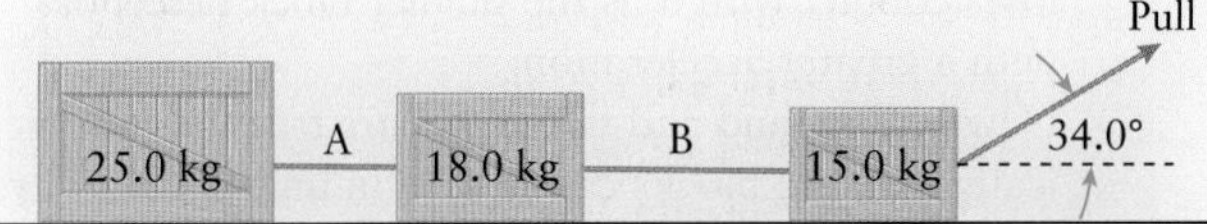

(a) Construct free-body diagrams for each of the crates and for the overall system of three crates treated as a single object.
(b) Predict the maximum force that can be exerted by the pull on the 15.0-kg crate without causing it to lose contact with the floor.
(c) Express the acceleration of the system of three creates in terms of the external force.
(d) Justify the claim that, if the pulling force is slowly increased, string B breaks first and calculate the pulling force that causes the string to break.

11. On a climb one partner lowers the other over the edge of a cliff. The rope being used will break under any tension greater than 800 N.
(a) Predict the minimum acceleration of an object weighing 850 N lowered over the edge of the cliff.
(b) Predict the most hazardous time in the process and explain a strategy to reduce the risk at that time.

12. A low-friction cart with mass $m_{\text{cart}} = 226$ g rests on a flat, horizontal track, and is attached to a horizontal string. The string runs over a pulley and is tied to some washers. The mass of each washer is 7.2 ± 0.2 g. The mass of the string is negligible. The masses of the cart and the washers were obtained using a pan balance. Assume the mass of the pulley can be neglected and that it exerts negligible friction on the string.

The number of washers, N_w, and their total mass, M, are shown in the table. The slope of the velocity–time graph for each value of N_w was fit to a line to obtain acceleration.

N_w	M (kg)	a (N/kg)
1	0.0072	0.33
2	0.0144	0.50
3	0.0216	0.78
4	0.0288	1.06
5	0.0360	1.42
6	0.0432	1.44
7	0.0504	1.74
8	0.0576	1.76
9	0.0648	2.14
10	0.0720	2.38

(a) Construct free-body diagrams for the cart and for the hanging washers.
(b) Re-express these free-body diagrams as a mathematical representation of the acceleration.
(c) Explain the distinction between the meaning of the mass, M, as it appears in the numerator and denominator of the mathematical representation for the acceleration.
(d) With the help of the equation(s) you developed in part (b), design a plan to analyze these data based on a *linear* relationship between a and a variable defined in terms of M and m_{cart}.
(e) Construct a table to aid in analyzing the data in which the independent variable that linearizes the data is displayed in one column.
(f) Construct a graph of the dependent variable, a, and independent variable.
(g) Calculate the slope from a best-fit line drawn through the data.
(h) Evaluate the results of the experiment in terms of the slope obtained from the best-fit line.

AP® Skills in Action

13.

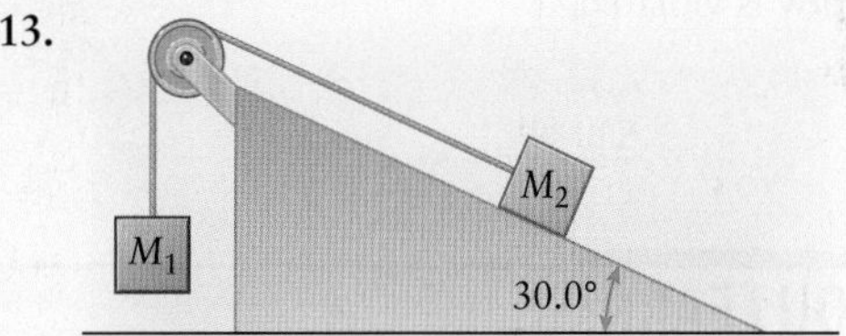

Two blocks with masses M_1 and M_2 are connected by a string of negligible mass that passes over an ideal pulley, which only changes the direction of the string. The block M_2 rests on a long ramp with inclination $\theta = 30.0°$. The friction force that the ramp exerts on the block is negligible.
(a) Construct free-body diagrams for each block and re-express the free-body diagrams as mathematical representations.
(b) The net force on each block is zero. Calculate the mass M_1 in terms of the mass M_2.
(c) The inclination of the ramp is reduced to 20.0° and block M_2 is released from rest. Predict the time dependence of the displacement of block M_2 while the string attached to the block remains taut.

14. A woman in an elevator stands on a weight scale that displays the normal force that the scale exerts on the woman. She presses the "down" button in the elevator, rides down, realizes she has forgotten something and rides back up. Consider the reading on the scale during each of the motions listed below. Rank these motions based on the scale reading, from greatest to least. Assume that the elevator has the same magnitude of acceleration any time there is an acceleration.
(A) The elevator starts descending with increasing speed.
(B) The elevator descends at a constant speed.
(C) The elevator descends with decreasing speed.
(D) The elevator is stopped.
(E) The elevator ascends with increasing speed.
(F) The elevator ascends with constant speed.
(G) The elevator ascends with decreasing speed.

15.

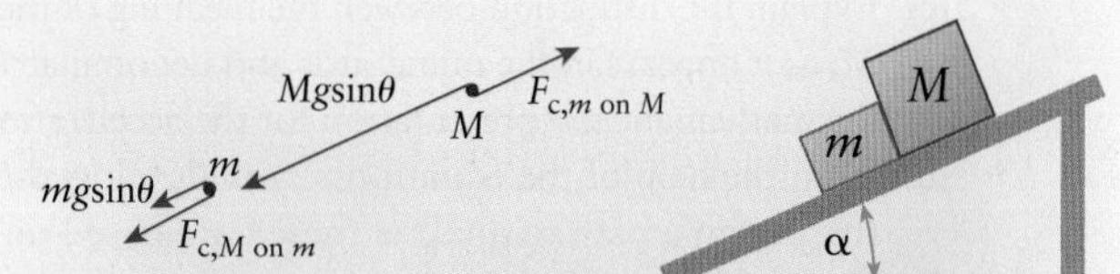

Two blocks are placed on an inclined plane with negligible friction and released. One of the blocks, labeled M, has a larger mass. Newton's third law implies that the contact force that block m exerts on block M must be equal and opposite to the contact force that block M exerts on block m. The vector diagrams drawn for the forces along the direction of motion show that the net force on each block is not proportional to the block's mass, but you know that both blocks have an acceleration equal to $g \sin \theta$. How can both conditions be true? Choose the best answer below.

(A) The contact force between the two blocks is 0, then neither law is violated.

(B) The third law is violated by this system because the material in the smaller block is more compressed by the larger block, but just enough so that they do slide down the ramp together.

(C) The second law is violated by this system because the larger block gives the smaller block a push and then loses contact with it as the smaller block descends with a greater acceleration.

(D) Newton's second and third laws are true only theoretically and real blocks can behave in unexpected ways, such as this.

16. You are in an inertial frame of reference and you measure the net force on a system and find that it is zero. From this measurement, the most complete description you can make of the system is that the system

(A) has no linear velocity in the frame of reference in which the measurement was made.

(B) has no linear velocity in any frame of reference.

(C) has no linear acceleration in the frame of reference in which the measurement was made.

(D) has no linear acceleration in any inertial frame of reference.

WHAT DID YOU LEARN?

Chapter learning goals	Section(s)	Related example(s)	Prep for the AP Exam: Relevant section review exercise(s)
Describe a force as an interaction between two objects and identify both objects for any force.	4-1		1, 2
	4-6	4-10	
Analyze a scenario and make claims about the forces exerted on an object by other objects for different types of interactions.	4-2	4-1, 4-2	5, 13
Use Newton's second law to predict the motion of an object subject to forces exerted by other objects in a variety of physical situations.	4-2	4-2	9, 15
Categorize forces as long-range or contact forces, and make claims about contact forces due to the microscopic cause of those forces.	4-2	4-2	
	4-4		
	4-5		
	4-6		1
Recognize the distinctions among mass, weight, and inertia.	4-3	4-3	5, 10
	4-6	4-7	
Draw and use free-body diagrams in problems to analyze situations involving interactions among several objects.	4-4		1
	4-6	4-4, 4-9	7
Construct explanations of physical situations involving the interaction of objects using Newton's third law and the representation of action-reaction pairs of forces.	4-5		1, 3
	4-6	4-8	
Use a free-body diagram and Newton's second law to construct a mathematical representation relating the acceleration of an object to its mass, and to solve that equation for an unknown quantity.	4-4		
	4-6	4-5, 4-6	3, 4

Chapter 4 Review

Key Terms

All the Key Terms can be found in the Glossary/Glosario on page G1 in the back of the book.

apparent weight (effective weight) 175
center of mass 162
contact forces 147
equilibrium 158
external forces 148
force 144
force pair 166
frame of reference 173
free-body diagram 161
gravitational force 146
gravitational mass 149
inertia 149
inertial frame of reference 173
inertial mass 149
internal forces 148
kilogram (kg) 149
kinetic friction 145
kinetic friction force 147
mass 149
net external force 145
net force 145
newton (N) 150
Newton's first law 158
Newton's laws of motion 148
Newton's second law 148
Newton's third law 166
normal force 147
pound-force (lbf) 150
reaction force 166
static friction force 147
vacuum 156
weight 156

Chapter Summary

Topic	Equation or Figure	
Forces and Newton's second law of motion: The net external force on an object is the vector sum of all forces exerted on that object by other objects. Common forces include one of the fundamental long-range forces, the gravitational force, and contact forces that are caused by the electric force: the normal force, static and kinetic friction forces, and the tension force. Newton's second law states that if the net external force on an object is not zero, the object must accelerate.	The **net external force** exerted on an object... ...equals the **vector sum** of all of the individual forces that are exerted on the object. $$\sum \vec{F}_{ext} = \vec{F}_1 + \vec{F}_2 + \vec{F}_3 + \ldots$$ The sum includes only external forces (forces exerted on the object by other objects).	(4-1)
	The acceleration of an object... ...is proportional to and in the same direction as the net external force exerted on the object... $$\vec{a} = \frac{\sum \vec{F}_{ext}}{m}$$ ...and inversely proportional to the mass of the object.	(4-2)
	Sum of the x components of all external forces exerted on the object; **x component** of the object's acceleration; **Mass** of the object $$\text{(a)}\ \sum F_{ext,x} = ma_x$$ $$\text{(b)}\ \sum F_{ext,y} = ma_y$$ **Sum of the y components** of all external forces exerted on the object; **y component** of the object's acceleration	(4-3)

Mass, weight, inertia, and Newton's first law of motion: The mass of an object describes the quantity and density of material that it contains. Its weight is the magnitude of the gravitational force on the object. Inertia is the tendency of an object to maintain the same motion. Newton's first law states that if the net external force on an object is zero, it maintains a constant velocity (zero acceleration).

Weight of an object (equal to the magnitude of the gravitational force on that object near the surface of Earth)

$$F_g = mg \qquad (4\text{-}4)$$

Mass of the object

Magnitude of the acceleration due to gravity near the surface of Earth

If the net external force on an object is zero...

...the object does not accelerate...

$$\text{If } \sum \vec{F}_{ext} = 0, \text{ then } \vec{a} = 0 \text{ and } \vec{v} = \text{constant} \qquad (4\text{-}5)$$

...and the velocity of the object remains constant. If the object is at rest, it remains at rest; if it is in motion, it continues in linear motion at a constant speed.

Free-body diagrams: A free-body diagram depicts all of the external forces that are exerted on an object. The coordinate axes are chosen so that as many of the vectors as possible (force and acceleration) point along one of the axes.

(a) A block on a horizontal table

Wind

Motion of the block

(b) Drawing the free-body diagram for the block

$\vec{F}_n$ (exerted by the table)

$\vec{F}_k$ (exerted by the table)

$\vec{F}_{\text{wind on block}}$ (exerted by the air)

$\vec{F}_g$ (exerted by Earth)

y

x

(1) Sketch a simplified version of the object on which the forces are exerted. A dot, or a small circle or square is fine: something easy to draw where you can place the tails of your vectors. Remember you are representing the center of mass if it is an extended object, so you don't want to think about the shape too much.

(2) Identify what other objects exert forces on this object. This includes Earth (which exerts a gravitational force) and anything that touches the object.

(3) For each force that is exerted on the object, draw the force with its tail at the center of the object. Make sure each vector points in the correct direction. Do not include forces that the object exerts on the other objects.

(4) Label each force with its symbol. Be certain that you can identify what other object exerts each force. If you're not sure what exerts a force, it's probably not a real force!

(5) Choose, draw, and label the directions of the positive x and y axes. Choose the axes so that as many forces as possible lie along one of the axes.

(Figure 4-11)

Center of mass: The center of mass of a system of objects is a point that represents the weighted average of the positions of the individual objects in the system. The system's center of mass acts as if all of the mass were located at that point. If all points in a system move with the center of mass (or you only need to know the center-of-mass motion to answer the problem), it is safe to use the object model.

Lightweight mint
Heavier soup bowl
The system behaves as though all of its mass were concentrated at the center of mass. That's where the waiter supports the tray.

(Figure 4-10)

Newton's third law of motion: The third law relates the forces that two objects exert on each other. This law is valid no matter how the objects move.

Force that object A exerts on object B

Force that object B exerts on object A

$$\vec{F}_{A \text{ on } B} = -\vec{F}_{B \text{ on } A} \qquad (4\text{-}6)$$

If object A exerts a force on object B, object B must exert a force on object A with the same magnitude but opposite direction.

Solving force problems: All problems that involve forces can be solved by (1) first drawing a free-body diagram for the object in question and (2) then writing out Newton's second law in component form and solving for the unknowns.

$$\text{(a)}\ \sum F_{\text{ext},x} = ma_x \qquad \text{(b)}\ \sum F_{\text{ext},y} = ma_y \tag{4-3}$$

Sum of the x components of all external forces exerted on the object

x component of the object's acceleration

Mass of the object

Sum of the y components of all external forces exerted on the object

y component of the object's acceleration

Chapter 4 Review Problems

1. A fisherman has a fishing line that is rated at 4 lbf. Tensions greater than this will cause the line to break.
   (a) How can the fisherman reel in a fish that weighs 5 pounds?
   (b) Experienced fishermen scoop the fish into a net before lifting it out of the water. Why does this increase their chance of getting the fish into the boat without snapping the line?
2. List all the forces exerted on a bottle of water if it were sitting on your desk, first thinking of the bottle and the water it contains as a single object, but then thinking of the bottle and the water it contains as a system, so including internal interactions.
3. Tension is a very common force in everyday life. Identify five ordinary situations that involve the force of tension.
4. Consider a person walking along a floor: Use Newton's third law to relate the forces exerted by the feet and by the floor as the person walks. Include both vertical and horizontal components.
5. A large block is hung by a rope from a beam. A smaller block hangs below the first, connected by a second rope. The rope connected to the beam breaks. Evaluate the claim that the larger block will collide with the smaller block. Air resistance is negligible.
6. A solid cube hangs by a thread from a beam. A second identical thread hangs below the cube. You exert a force on the lower thread. Evaluate these claims: If the force is exerted quickly, the lower thread is more likely to break. If the force is exerted slowly, the upper thread breaks first.
7. In an attempt to define Newton's third law, a student states that action and reaction forces are equal in magnitude and opposite in direction to each other. If this is the case, the student claims, there can never be a nonzero net force on an object. Evaluate the claim.
8. You and a friend are ice skating. Standing in the middle of the rink, you give your 85-kg friend a push with a force of 3.0×10^2 N. Assuming that the ice surface exerts negligible friction on your friend, what magnitude of acceleration does your friend experience?
9. Draw and describe a physical situation to which the following equation could apply.

$$a = \frac{40\ \text{N} - (4.0\ \text{kg})\left(9.8\dfrac{\text{N}}{\text{kg}}\right)\sin(20°)}{4.0\ \text{kg}}$$

10. A spaceship takes off vertically from rest with an acceleration of 29.0 m/s^2. What net force is exerted on a 75.0-kg astronaut during takeoff? What is the normal force exerted on the astronaut by the spaceship? Express this normal force in newtons and also as a multiple of the astronaut's weight on Earth.
11. A car is proceeding at a speed of 14.0 m/s when it collides with a stationary car in front of it. During the collision, the first car moves a distance of 3.00 m as it comes to a stop. The driver is wearing her seat belt, so she remains in her seat during the collision. If the driver's mass is 52.0 kg, how much force does the belt exert on her during the collision? Neglect any friction between the driver and the seat, or any change in the shape of the cars.
12. 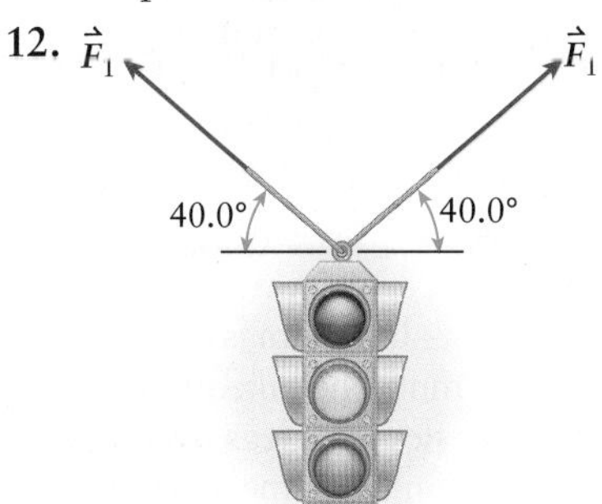

A traffic light is supported equally by two cables as shown. One cable pulls up and to the right, 40.0° above the horizontal; the other cable pulls up and to the left, 40.0° above the horizontal. The two forces exerted on the light by the cables have equal magnitude F_1. Predict the force, F_2, exerted by the cables, in terms of F_T, if both angles are decreased to 30.0°.

13. The distance between two telephone poles is 50.0 m. A 0.500-kg bird lands on the telephone wire midway between the poles and when the bird comes to rest,

the wire sags 0.15 m, measured from the middle of the initially horizontal phone line.

(a) Construct a free-body representation of the bird on the point on the wire as described.

(b) Construct a mathematical representation of the forces exerted on the bird. Using Newton's second law, derive an expression for the force of tension in the wire in terms of the vertical sag Δy, the length of the wire L, and the weight of the bird $m_{\text{bird}}g$. You may find the small-angle approximation $\sin(\theta) \sim \tan(\theta)$ useful.

(c) For the given values above, calculate a numerical value for the tension in the wire, neglecting the mass of the wire.

14. A locomotive pulls 10 identical freight cars with a constant acceleration. Construct a mathematical representation for the coupling force exerted by each car on the one following it, and use it to predict the dependence of this force on the number of the pulling car from the locomotive. You may notice a relationship that allows you not to write out the relationship separately for each pair of cars.

15.

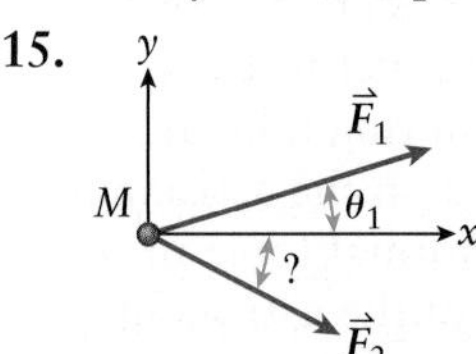

Two forces are exerted on an object of mass M as shown. The direction and magnitude of the force F_1 are known. The magnitude of the force F_2 is not known, and neither is the angle, θ_2, that the force F_2 makes with the x axis.

(a) Represent the x and y components of the acceleration symbolically in terms of the magnitudes F_1 and F_2 and the angles θ_1 and θ_2.

(b) The unknown direction of F_2 is to be determined by adjusting the orientation and magnitude of the force F_1 until the object has only an x component of the acceleration. Analyze these data to obtain the magnitude and direction of F_2: $M = 2.4 \times 10^{-8}$ kg, $a_x = 1.7 \times 10^3$ m/s^2, $F_1 = 4.8 \times 10^{-5}$ N, and $\theta_1 = 17°$.

16. On average, froghopper insects have a mass of 12.3 mg and jump to a height of 428 mm. The takeoff velocity is achieved as the little critter flexes its legs over a distance of approximately 2.00 mm. Assume a vertical jump with constant acceleration.

(a) For how much time is the froghopper pushing off the ground, and what is the froghopper's acceleration during that time?

(b) Make a free-body diagram of the froghopper during the time it is pushing off the ground, but has not yet left the ground.

(c) Calculate the magnitude of the force exerted by the ground on the froghopper during the time the froghopper was pushing off the ground. Express your answer in millinewtons and as a multiple of the insect's weight.

17. A car traveling at 28.0 m/s hits a bridge abutment. A 45-kg passenger moves forward 50 cm while being brought to rest by an inflated air bag. The car is designed to crumple so that the passenger moves another 1.0 m (so 1.5 m total) in coming to rest. Assuming the force that stops the passenger is constant, find its magnitude and direction.

18.

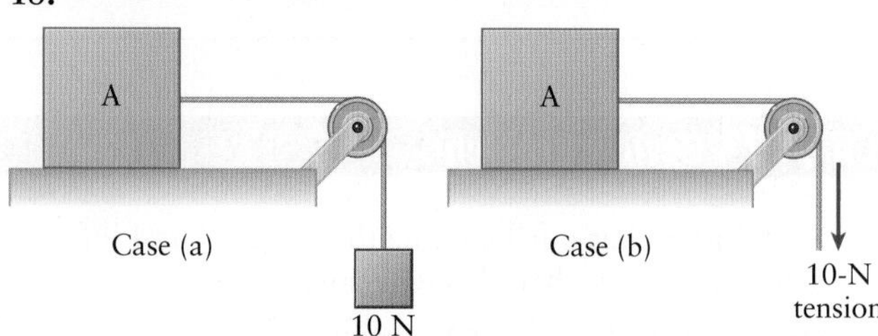

Case (a) in the figure shows block A accelerated across a table with negligible friction by the tension in a string created by a hanging 10-N block. In case (b) the same block A is accelerated by a steady 10-N tension in the string. Treat the masses of the strings, as well as the masses and friction of the pulleys, as negligible. The acceleration of block A in case (b) is

(A) greater than its acceleration in case (a).

(B) less than its acceleration in case (a).

(C) equal to its acceleration in case (a). Justify your choice.

19. Biomimetic engineers are working on a model of fish propulsion with the goal of building an autonomous robotic mackerel. The first step in the project is to locate the center of mass of a frozen mackerel. Shown are two configurations of the same stiff fish hanging from a cord attached to the ceiling. A second cord is tied to the point of the attachment of the fish to the cord attached to the ceiling. At the other end of the second cord is an object with mass.

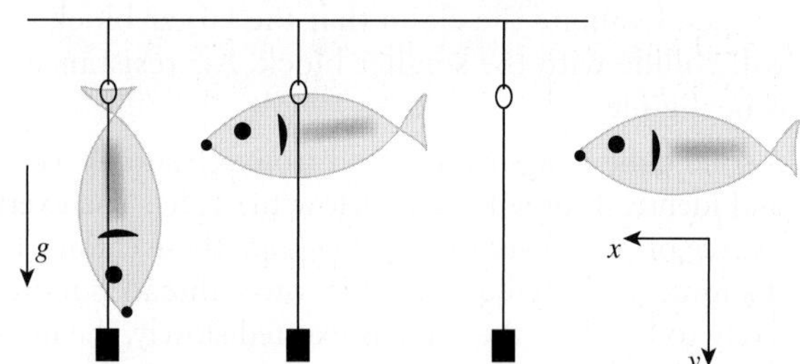

(a) Explain how evidence from the diagram can be used to locate the center of mass of the frozen mackerel in the x–y plane.

(b) Justify the selection of a third orientation of the mackerel that can be used to locate the center of mass of the fish.

20. In yet another ill-advised attempt to catch the elusive roadrunner, a reckless coyote engineer straps on a pair of ice skates and attaches a rocket capable of producing 5.0×10^3 N of thrust to his back. Together, the coyote and the rocket have a mass of 120 kg. If the coyote starts at rest on a level, icy surface that exerts negligible friction on him and bends over such that the rocket thrust is directed parallel to the ice, what is his final speed if the rocket burns for 5.0 s?

21. You stand at the base of a 4.00-m-long ramp that is inclined at an angle of 9.00°. You want to slide a 2.00-kg object up the ramp so that it stops *just* as it reaches the top. What initial velocity must you give the object if friction between the object and the ramp is negligible?

22.

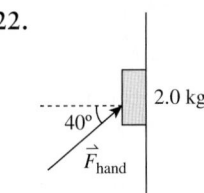

A 2.0-kg box is held against a vertical wall, as shown above, by an external force, F_{hand}, exerted on the box at an angle of 40° with the horizontal. Assume friction between the box and the wall is negligible.

(a) Draw a free-body diagram of the box.

(b) What is the value of F_{hand} such that the box remains at rest?

(c) What is the magnitude of the force the wall exerts on the box for this value of F_{hand}?

(d) Suppose friction is present between the wall and the box. If static friction is directed upward parallel to the wall, what does this mean about the force of the hand in part (b) above?

(e) If static friction is directed downward parallel to the wall, what does that mean about the force of the hand in part (b) above?

23. A skier starts at rest atop a smooth ski slope that has a slope of exactly 18.0°. Her total mass including her gear is 72.0 kg. Assuming the slope exerts negligible friction on the skier and that the skier skis straight down the slope, what will her speed be after 25.0 s?

24. A 30.0-kg golden retriever stands on a scale in an elevator in one of the world's tallest buildings. The scale is calibrated to measure the force exerted on it in newtons. Calculate the reading on the scale when the elevator (a) accelerates at 3.50 m/s^2 downward, (b) cruises down at a steady speed, and (c) accelerates at 4.00 m/s^2 upward.

25.

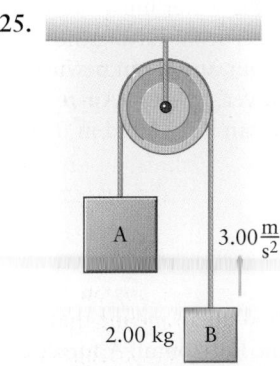

The picture shows an Atwood's machine. Box A of unknown mass is attached to box B that has a mass of 2.00 kg. The two boxes are attached by a rope of negligible mass that hangs over a pulley; the pulley's mass and friction force exerted on the rope are negligible. The pulley is attached to the ceiling. When in motion, it is found that box B moves upward with an acceleration of 3.00 m/s^2.

(a) Construct free-body diagrams to represent the forces exerted on each block and on the pulley.

(b) Calculate the tension in the rope.

(c) Calculate the mass of box A.

(d) Calculate the force exerted by the pulley on the ceiling.

26. A rider in an elevator weighs 700 N. If this person stands on a scale in the elevator, (a) find a general expression for the reading on the scale in terms of the mass or weight of the rider, g, and the acceleration of the elevator a_y. Calculate the values of the reading on the scale as the elevator, (b) accelerates upward at 3.00 m/s^2, (c) cruises upward at 4.00 m/s, (d) slows to a stop at 2.00 m/s^2, then (e) free-falls toward the bottom of the elevator shaft before safety measures bring the elevator to a safe stop.

27. A person weighs 588 N. If she stands on a scale while riding on the Inclinator (the lift at the Luxor Hotel in Las Vegas), what will be the reading on the scale? Assume the Inclinator moves at a constant acceleration of 1.25 m/s^2, in a direction 39.0° above the horizontal, and that the floor of the elevator car is level and horizontal.

28.

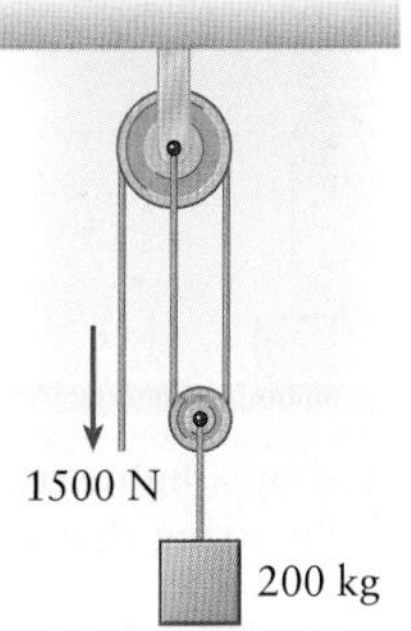

A 2.00×10^2-kg block is hoisted by two pulleys, one fixed pulley attached to the ceiling and one moveable (suspended) pulley, as shown. Neglect the masses of the pulleys and friction forces exerted by the pulleys. A force of 1.50×10^3 N is applied to the rope, which also has a negligible mass and does not stretch.

(a) Construct two free-body diagrams: one representing the forces exerted on the 200-kg block, and the other representing the forces exerted on the suspended pulley.

(b) Justify the claims that the acceleration of the block is the same as the acceleration of the suspended pulley and that the suspended pulley and block can be treated as a single object.

(c) Calculate the acceleration of the block.

29. A window washer sits in a bosun's chair that dangles from a rope of negligible mass that runs over a pulley of negligible mass and friction and back down to the man's hand. The other end of the rope is attached to his chair. The pulley is at a fixed position, attached to a scaffold that is extended outward from the roof of the building. The combined mass of the man, chair, and a bucket is 95.0 kg. With how much force must he pull downward on the rope to raise himself (a) at constant speed and (b) with an upward acceleration of 1.50 m/s^2?

30.

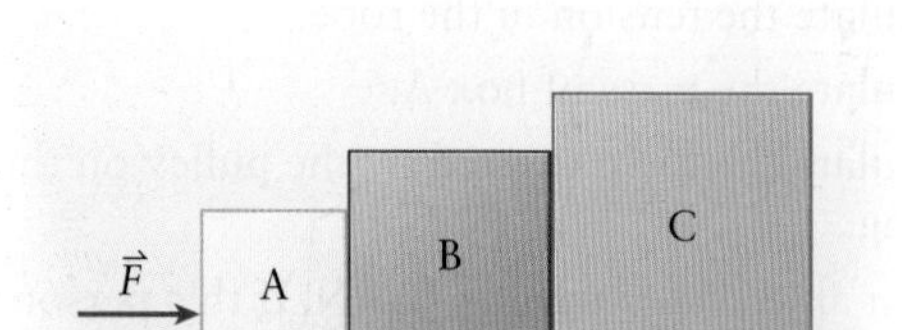

Three boxes with masses m_A, m_B, and m_C are lined up so that they are in contact with each other, as shown in the figure. An external force ($\vec{F}$) pushes on box A toward the right. All blocks remain in contact. Friction can be considered to be negligible.

(a) Construct three free-body diagrams, one for each block, representing the forces exerted on each block individually.

(b) Qualitatively predict the relative magnitudes of forces F, $F_{\text{A on B}}$, and $F_{\text{B on C}}$ using the fact that the system accelerates to the right.

(c) Quantitatively predict the dependence of the acceleration on F by treating the system as a single object.

31.

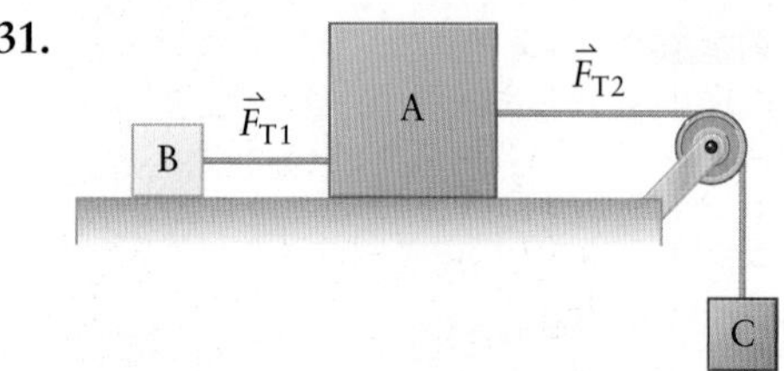

Two blocks, A and B, are connected by a light string. Block A is connected to another light string that passes over a pulley and is connected to block C, as shown in the figure above. The horizontal tabletop exerts negligible friction on the blocks. The masses of blocks A, B, and C are m_A, m_B, and m_C, respectively. The strings remain taut at all times. Neglect the mass of and friction forces exerted by the pulley.

(a) Construct free-body diagrams to represent the forces exerted on each of the three blocks.

(b) Re-express the diagrams in part (a) as symbolic representations by applying Newton's second law.

(c) Qualitatively predict the relative magnitudes of tensions F_{T1} and F_{T2} using the fact that the system accelerates.

(d) Derive algebraic expressions for the magnitudes of tensions F_{T1} and F_{T2} by application of Newton's second law. Treating the three blocks as a single object may be helpful as part of the solution.

32. A 2.00-kg object A is connected to a light string that passes over a pulley of negligible mass and friction to a 3.00-kg object B. The smaller object rests on a smooth plane, which is tilted at an angle of $\theta = 40.0°$. (a) Draw a diagram of the system. (b) Calculate the acceleration of each object and the tension in the string.

33.

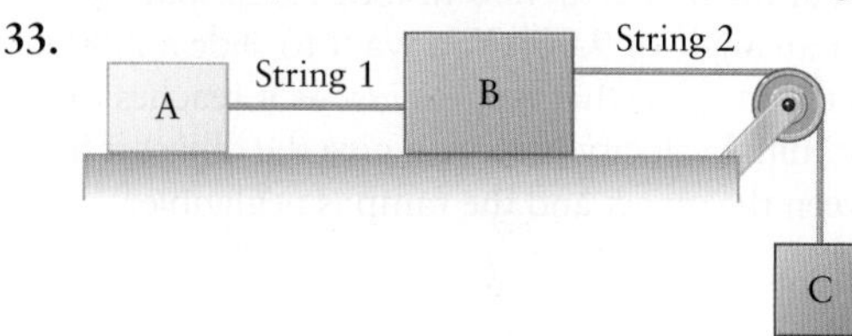

A 1.00-kg object A is connected with a string to a 2.00-kg object B, which is connected with a second string that passes over a pulley to a 4.00-kg object C. Neglect the mass of and friction forces exerted by the pulley. The strings have negligible mass and do not stretch, and the level tabletop exerts negligible friction forces.

(a) Calculate the acceleration of each object in terms of g.

(b) Calculate the tension in string 2 in terms of g.

(c) Calculate the tension in string 1 in terms of g.

34.

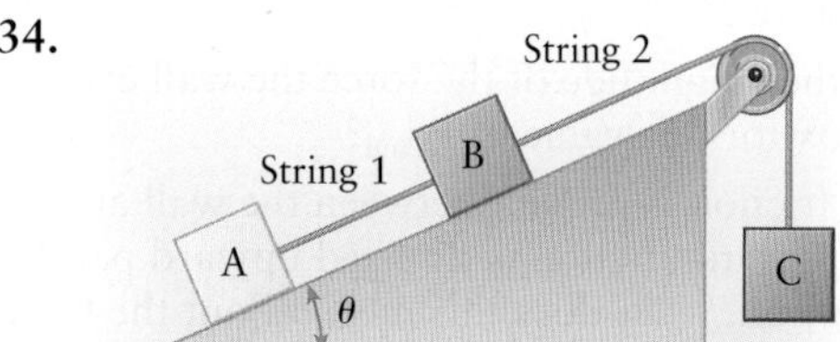

An object A is connected with a string to an object B, which is connected with a second string over a pulley with negligible mass and friction to an object C. The string does not stretch. The first two objects are placed onto an inclined plane that makes an angle θ with the horizontal, as shown. The inclined plane along which objects A and B move is fixed in place and exerts a negligible friction force on the objects.

(a) Construct free-body diagrams for each object.

(b) Express the net force on the system of three objects, A, B, and C in terms of m_A, m_B, m_C, and g.

35. An athlete drops from rest from a platform 10.0 m above the surface of a 5.00-m-deep pool. Assuming that the athlete enters the water vertically and moves through the water with constant acceleration, what is the minimum average force the water must exert on a 62.0-kg diver to prevent her from hitting the bottom of the pool? Express your answer in newtons and also as a multiple of the diver's weight. Air resistance during the athlete's dive can be ignored in this problem.

36.

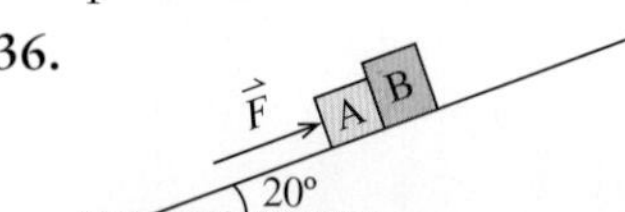

Two boxes, A and B, are pushed up an incline that makes an angle of 20° with the horizontal. A force, $\vec{F}$, is exerted on box A in a direction parallel to the incline, and friction is negligible.

(a) Draw a free-body diagram for box A, for box B, and for the system of two boxes.

(b) If the masses of boxes A and B are 1.0 kg and 2.0 kg, respectively, what is the magnitude of the force F required to push the blocks up the incline at a constant speed?

(c) What are the magnitude and direction of the force that B exerts on A when they are moving up the incline at constant speed?

Prep for the AP® Exam

AP® Group Work

Directions: The problem below is designed to be done as group work in class.

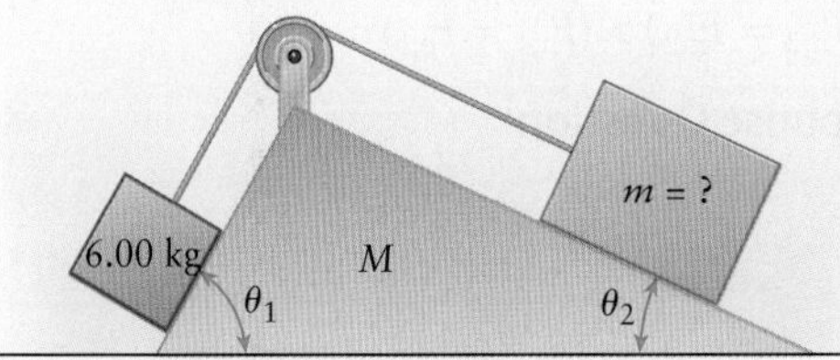

In the figure, two blocks are connected by a string with negligible mass running over a pulley with negligible mass, which exerts negligible friction on the string. The supporting structure has a triangular cross section so that it forms two ramps with inclinations θ_1 and θ_2. The ramps exert negligible friction forces on the blocks. The block on the left ramp has a mass of 6.00 kg. The supporting structure has mass $M = 21.7$ kg and rests on a horizontal surface that also has negligible friction. The angles θ_1 and θ_2 are 60° and 25°, respectively.

(a) Construct free-body diagrams for each of the three system components and a free-body diagram treating the entire system as a single object.

(b) Calculate the mass, m, for which the blocks are at rest in the reference frame of the supporting structure with a triangular cross section.

(c) Calculate the acceleration of the system composed of the blocks and supporting structure in a reference frame fixed on the horizontal surface.

The blocks are held in place while the ramps are adjusted so that θ_1 and θ_2 are 60° and 30°, respectively. The blocks are then released.

(d) Qualitatively predict the response of the system in the stationary reference frame of the horizontal surface.

(e) If the supporting structure were not free to move on the horizontal surface, find the acceleration of the two blocks.

(f) In the case of negligible friction, would it be valid to use Newton's second law to calculate the acceleration of the blocks in the reference frame of the supporting structure? Why or why not?

AP® PRACTICE PROBLEMS

Prep for the AP® Exam

Multiple-Choice Questions

Directions: The following questions have a single correct answer.

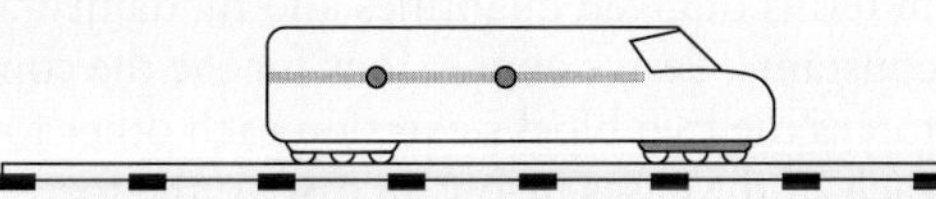

1. A train is moving with constant velocity on a level track. Which of the following are action-reaction pairs?
   (A) The force the engine exerts on the train to keep it moving forward, and the force of friction of the track on the wheels of the train holding it back
   (B) The force of friction of the track on the wheels and the force of friction of the wheels on the track
   (C) The force of gravity on the train and the normal force of the track on the wheels
   (D) The normal force the train exerts on the track and the gravitational pull of the train on Earth

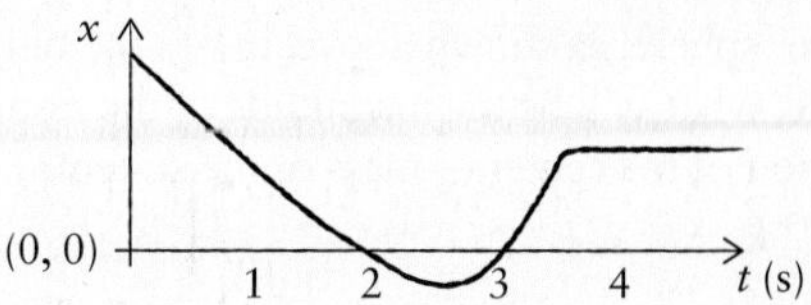

2. The figure shows a position vs. time graph for an object. At which of the following times shown on the graph is the sum of the forces on the object zero?
   (A) 1 s and 4 s (B) 2 s and 3 s
   (C) 2.5 s only (D) 4 s only

$m = 1$ kg
$a = 2.00\ \text{m/s}^2$ $a = 0.50\ \text{m/s}^2$
M F_x m M F_x

3. A student exerts a horizontal force F_x on an object of unknown mass, M, that is resting on a horizontal surface of negligible friction. The object accelerates at

(Continued)

2.0 m/s^2. The student attaches a known mass of 1.0 kg on top of the object of mass M and finds that when the same force, F_x, is exerted on the combination the acceleration is 0.50 m/s^2. What is the mass M?

(A) 0.12 kg (B) 0.25 kg
(C) 0.33 kg (D) 0.50 kg

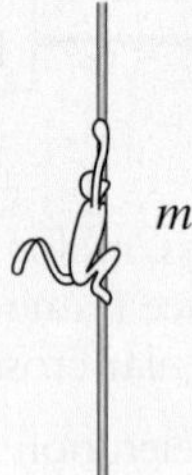

4. A monkey of mass m slides down a rope of negligible mass with a decreasing speed. Rank the magnitudes of the forces.
(A) $mg > F_{\text{monkey on rope}} > F_{\text{tension}}$
(B) $mg > F_{\text{monkey on rope}} = F_{\text{tension}}$
(C) $F_{\text{tension}} > F_{\text{monkey on rope}} = mg$
(D) $F_{\text{tension}} = F_{\text{monkey on rope}} > mg$

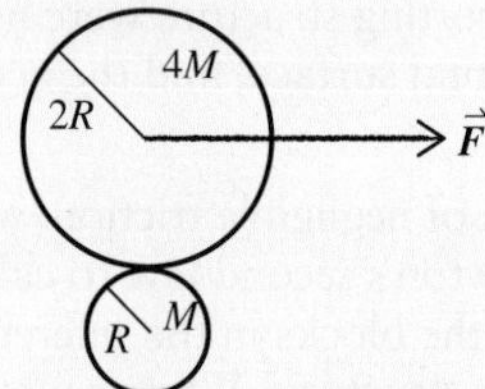

5. Suppose a system in space consists of two spheres joined at their surface. One sphere has a radius R and a mass M. The other sphere has a radius $2R$ and a mass $4M$. A force of magnitude F is exerted on the larger sphere, perpendicular to a line joining the centers of the two spheres, and in line with the center of the larger sphere, as shown above. If $\vec{F}$ is the only external force, what will be the magnitude of the resulting acceleration of the center of mass of the system?

(A) $\frac{F}{5M}$ (B) $\frac{F}{4M}$
(C) $\frac{2F}{5M}$ (D) $\frac{4F}{5M}$

6. A block of mass m is given an initial velocity v_o up a plane that is inclined at an angle θ with the horizontal. It comes to a stop in a distance d. Which of the following correctly expresses the magnitude of the friction force that the surface exerts on the block as the block moves up the incline?

(A) $mg \sin \theta$ (B) $m\frac{v_0^2}{2d}$
(C) $mg \sin \theta + m\frac{v_0^2}{2d}$ (D) $m\left(\frac{v_0^2}{2d} - g \sin \theta\right)$

7. Two boxes of masses m_1 and m_2 ($m_1 < m_2$) are at rest on a horizontal surface, as shown below. Let $F_{g,1}$ and $F_{g,2}$ represent the gravitational forces on each box; let $F_{n,1}$ represent the force exerted upward by the floor on box 1 and let $F_{n,2}$ represent the force exerted upward by box 1 on box 2. Which of the following is the correct ranking of these forces?

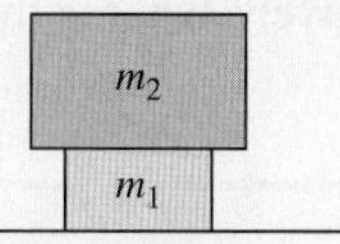

(A) $F_{n,1} > (F_{n,2} = F_{g,2}) > F_{g,1}$
(B) $(F_{n,1} = F_{g,1}) > (F_{n,2} = F_{g,2})$
(C) $F_{n,1} > F_{n,2} > F_{g,2} > F_{g,1}$
(D) $(F_{n,2} = F_{g,2}) > (F_{n,1} = F_{g,1})$

Free-Response Question

1. Two blocks, of mass M and $2M$, respectively, are being pushed along a smooth horizontal surface by a force $\vec{F}_0$, as shown above.
(a) The two dots, below, represent each of the blocks shown above. On each dot, draw a free-body diagram of the corresponding block. Each force should be represented by a single arrow starting on the dot.

Block of mass M Block of mass $2M$

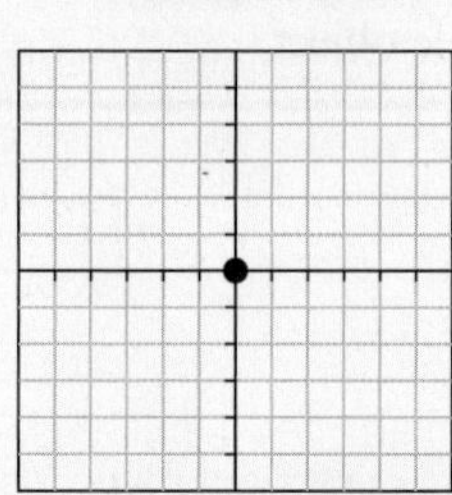

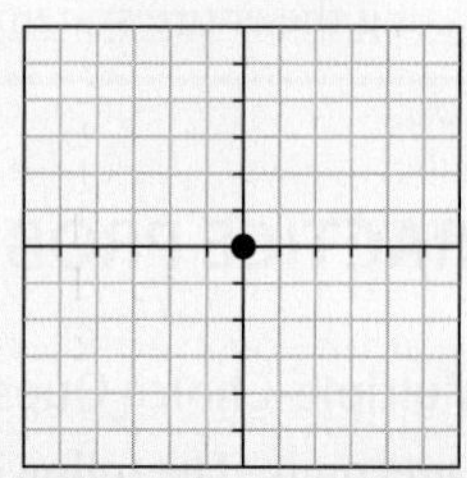

(b) In terms of given quantities and fundamental constants, derive expressions for the the contact forces the two blocks exert on each other for each of the cases below. In case A, the force $\vec{F}_0$, is exerted on the left side of the smaller block and directed to the right. In case B, the force $\vec{F}_0$, is exerted on the right side of the larger block and directed to the left.

Case A Case B

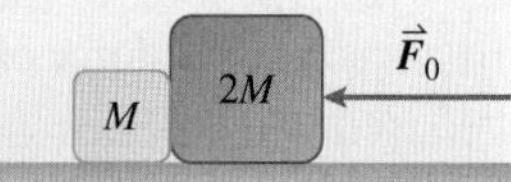

(c) Indicate whether the acceleration of the blocks is greater in case A or case B, or the same in both cases, and explain your reasoning, referring to your equations and free-body diagrams above.

5 Forces and Motion II: Applications

Derek Darke/Alamy

YOU WILL LEARN TO:

- Recognize what determines the magnitude of the static friction force and find the magnitude and direction of the maximum static friction force exerted on an object by a surface.
- Find the magnitude and direction of the force of kinetic friction exerted on an object by a surface.
- Explain contact forces on an object as arising from interatomic electric interactions and determine the direction in which these forces must therefore be exerted on the object.
- Construct free-body diagrams, and extract quantitative or qualitative information from them to solve for properties of the motion of an object when the forces on an object include static or kinetic friction.
- Analyze situations in which fluid resistance is important to determine when application of constant acceleration equations would give poor results.
- Analyze equilibrium force problems involving the spring force to solve for properties of the system such as mass, spring constant, or compression or extension of the spring.

5-1 We can use Newton's laws in situations beyond those we have already studied

In Chapter 4 we learned the basic laws that describe the motions of objects, and we saw how external forces exerted on an object by other objects or systems determine the magnitude and direction of the object's acceleration. In this chapter we'll see how to apply these laws to a variety of very common situations. In particular, we'll take a much closer look at situations involving friction forces and drag forces—two forces that resist the motion of one object relative to another.

The force of *friction* plays an important role in many physical situations. There is friction in almost every case when two surfaces are in contact, and it can be present whether or not the two surfaces are sliding past each other, as long as there is some force exerted trying to move one surface with respect to the other. In Chapter 4 we used the friction force in some examples; in this chapter we'll learn more about the nature of friction and see how to solve more complex problems that involve friction. Another

WATCH OUT !

Sketches are used to identify forces and their directions, but not where they are exerted!

For sketches of systems we can treat as objects, the arrows only represent the direction of forces on the objects. Sometimes we draw the arrows as emanating from the center of mass because that is the point we use to represent our system as an object to describe its motion. Although gravity is always exerted at the center of mass, we will find that the normal and friction forces are not. When we begin to study motion for systems in which the object model does not hold, we will modify our procedure and have to concern ourselves with where the forces are being exerted on the system.

important situation arises when objects move through a fluid (a gas or liquid), such as a football flying through the air or a submarine cruising beneath the surface of the ocean. The fluid exerts a *drag force* on each of these objects. The faster an object travels, the more difficult the drag force makes it for the object to move through the fluid.

THE TAKEAWAY for Section 5-1

✔ Newton's laws also apply in situations that involve friction and drag forces.

Prep for the AP® Exam

Building Blocks

1. Describe two situations in which the normal force exerted on an object is not equal to the object's weight.

Skill Builders

2. When you observe a leaf flutter as it falls, you do not see the forces of gravity or the wind. You infer the presence of these invisible forces exerted on the leaf from the accelerations that they produce. A brick is placed on one end of a board that is initially horizontal, but then one end is lifted, increasing the angle θ, as shown. As the incline is increased slowly the brick does not slide.

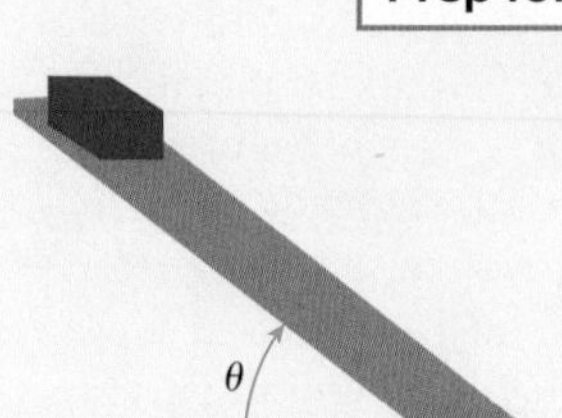

(a) Predict the directions of forces exerted on the brick.
(b) Justify the claim that no component of forces exerted on the brick parallel or perpendicular to the board remains constant as the angle of inclination increases while the brick does not slide.
(c) Explain the differences and similarities of the forces that cause and resist the motion of the brick along the board and the forces that we infer to be present as we watch a leaf flutter and fall.

(a)

The ground exerts a static friction force on the stationary children's feet as they are pulled on by the rope. This force prevents their feet from sliding.

Tanya Constantine/Getty Images

(b)

The water slide exerts a kinetic friction force on the moving child that slows her down. The water reduces the friction but doesn't eliminate it.

JGI/Jamie Grill/Getty Images

Figure 5-1 Static friction versus kinetic friction Both (a) static friction and (b) kinetic friction arise from the interaction between two surfaces in contact.

5-2 The static friction force changes magnitude to offset other forces being exerted on a system

Friction is a force that resists the sliding of one object past another. If you push or pull on an object to make it slide but it doesn't move, we call the force that is preventing sliding a *static friction* force (**Figure 5-1a**). If the object is moving, we call the force that opposes this motion a *kinetic friction force* (**Figure 5-1b**). Both static and kinetic friction forces can result from interactions between the microscopic irregularities (bumps) of the two surfaces in contact. Even surfaces that appear or feel smooth can be rough at the microscopic level. Irregularities on one surface can catch on or be impeded by irregularities on the other surface. This is a convenient way to think of the interactions of surfaces of objects or systems in terms of our everyday experiences, at a **macroscopic** level. Macroscopic means the physical objects can be seen by the unaided eye.

As we mentioned in the last chapter, on a microscopic level friction (and drag) forces result from electrostatic interactions. All of the forces that provide resistance to motion are models that greatly simplify calculating the effect of these microscopic interactions on motion. On a small enough length scale, matter is mostly empty space between atoms, with electrons moving through this space. The actual size of the atoms would not be enough to keep one object from simply falling through another object, if it were not for the electrostatic forces between electrons that push them apart. So both the force of friction and the normal force exerted by any surface are caused by electric interactions. These electric interactions are fundamental forces; in this case, it is a repulsive force between electrons. Because the contact forces of friction and the normal force are a result of these electric interactions, the direction of these contact forces resists relative motion between the objects or systems (sometimes referred to as surfaces, depending on the interaction) involved. We will see that the force of friction resists one object or system sliding with respect to another and that the normal force resists one object or system from sinking into another. Friction will always depend on the surfaces of the systems that are interacting, and not on a single point.

However, as long as for all other interactions we can treat the system on which the friction force is exerted as an object, we can still use the object model, which greatly simplifies solving problems. In Chapter 8 we will learn how to approach one application where the object model is not completely adequate.

We will spend the rest of this chapter exploring the models for friction forces that we will most often use to make predictions of motion. We'll concentrate on static friction in this section and analyze kinetic friction in Section 5-3.

Properties of Static Friction

To demonstrate some properties of static friction, gently rest your open palm and fingertips against a wall and then try to drag your hand downward while maintaining contact with the wall. You'll find that it doesn't take much effort to make your hand slide. If you repeat the same experiment, but push a bit harder against the wall, you'll notice that it takes a bit more effort to get your hand to move. If you push as hard as you can on the wall, you'll find it very challenging to slide your hand down the wall. If the wall has a smooth surface, however, you'll find that your hand slips more easily than on a rough-surfaced wall.

Here's what we learn from this simple experiment:

1. In each case the force that tries to keep your stationary hand from sliding is the force of static friction. From now on, we'll use the symbol $\vec{F}_S$ ("*F*-sub-S") for this force. (The subscript "S" reminds us that we're talking about *static* friction.) The static friction force $\vec{F}_S$ exerted on your hand by the wall is exerted parallel to your hand and parallel to the wall, that is, *parallel to the surfaces in contact.*
2. You can overcome static friction by pushing against it with sufficient force. This says that in a given situation, the static friction force has a maximum magnitude $F_{S,max}$. The force of static friction is zero until you try to slide your hand, and is just balancing whatever force you are exerting until you exceed the maximum static friction force. You overcome the static friction force and start your hand sliding if you exert a force on your hand in a direction parallel to the surfaces whose magnitude is greater than $F_{S,max}$.
3. The value of $F_{S,max}$ depends on how hard the two surfaces are pushed together—that is, on the *normal force* that is exerted *perpendicular to the two surfaces*. The greater the magnitude F_n of the normal force one surface exerts on the other, the greater the maximum static friction force exerted by that surface and the more difficult it is to make the surfaces slide past each other. In this example your hand was the object you were trying to move, and the two surfaces were the wall and your hand. As you pushed harder on the wall, the wall exerted a greater normal force on your hand.
4. The force of static friction depends on the properties of the surfaces in contact.

We can combine these observations into simple mathematical relationships:

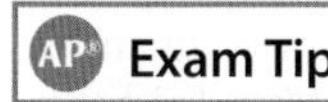

Exam Tip

The AP® Physics 1 equation sheet combines these two equations describing friction into one equation that expresses both ideas at once. The single friction equation on the equation sheet also includes information about kinetic friction, so the subscript "f" is used. On the exam, you'll need to recognize when it applies to static friction and when it applies to kinetic friction. We will spend a full section on each type of friction force to ensure you understand the similarities and differences. In Chapter 7, we will introduce another force, the spring force, which will also be represented by $\vec{F}_S$. The AP® Physics 1 equation sheet uses $\vec{F}_S$ for the spring force and $\vec{F}_f$ for the combined expression for the friction force. It is not unusual for scientists to use the same symbol for different variables in different contexts, so it is very important to understand the meaning, so you can always choose the right equation.

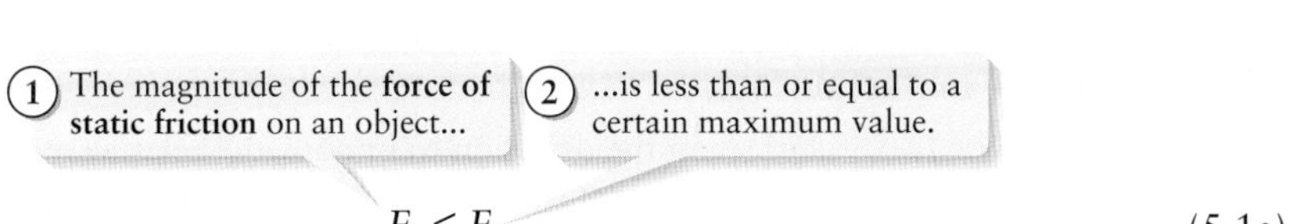

(1) The magnitude of the **force of static friction** on an object...
(2) ...is less than or equal to a certain maximum value.

$$F_S \le F_{S,max} \quad (5\text{-}1a)$$

$$F_{S,max} = \mu_S F_n \quad (5\text{-}1b)$$

(3) The **maximum force of static friction** depends on...

(4) ...the **coefficient of static friction** (which depends on the properties of the two surfaces in contact)...

(5) ...and the **normal force** pressing the object against a surface.

EQUATION IN WORDS
Magnitude of the static friction force

The quantity μ_S in Equation 5-1b is called the **coefficient of static friction.** Its value depends on the properties of both surfaces in contact; for example, μ_S has different values for steel on steel, for steel on glass, and for glass on glass. Because both F_S and F_n have units of force, Equation 5-1b tells us that μ_S must be dimensionless.

Typical values of the coefficient of static friction are between about 0.1 and 0.8. For example, μ_S is approximately 0.1 for a ski on snow, approximately 0.5 for two pieces of wood in contact, and approximately 0.7 for two pieces of metal in contact.

Lubrication reduces the friction force between surfaces and hence the value of μ_S. In human joints, such as the elbow and knee, the presence of lubricating fluid results in a coefficient of static friction of about 0.01. This relatively low value of μ_S minimizes sticking and reduces wear and tear of the cartilage that supports these joints.

WATCH OUT !

Static friction is described by an inequality, not an equality.

All of the mathematical relationships we've seen so far in our study of physics have been equalities, with an equal sign relating two quantities. But Equation 5-1a is an *inequality*. **Figure 5-2** helps show why this is so. Imagine that you apply a horizontal force to a stationary block on a horizontal table. You find that if you apply a 20.0-N force, the block remains at rest. Therefore, the static friction force must have the same 20.0-N magnitude to balance the force that you apply (Figure 5-2a). What happens if you exert a lesser force, say only 10.0 N? The block would still remain at rest, so the force of static friction is now only 10.0 N in magnitude to exactly cancel the force you exert (Figure 5-2b). The block also remains at rest if you don't push on it at all, so in this case the static friction force is zero (Figure 5-2c). If you push enough to overcome the maximum force of static friction given by Equation 5-1b, the object will slide. In Section 5-3 we'll discuss what happens then.

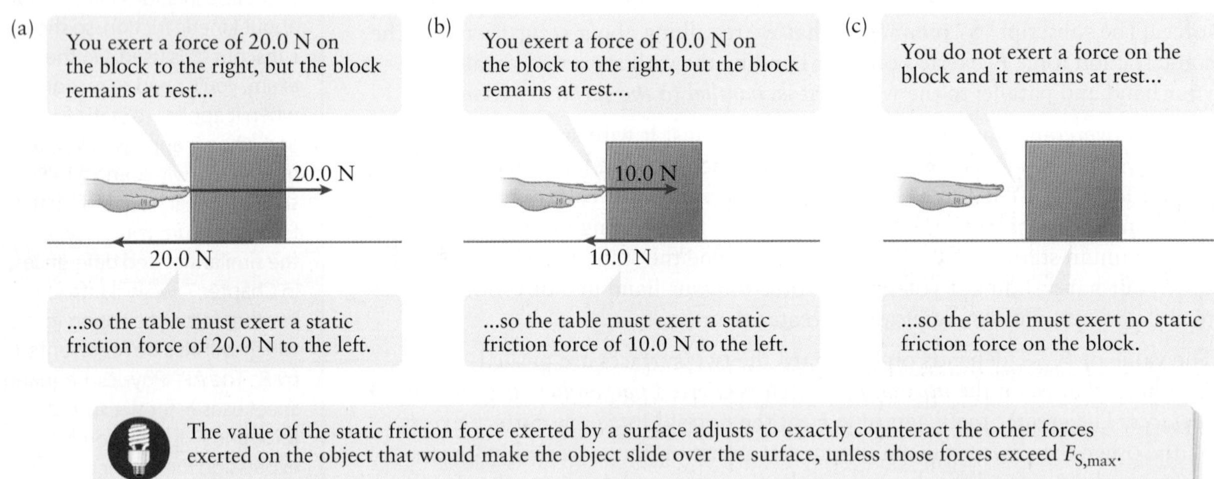

(a)
You exert a force of 20.0 N on the block to the right, but the block remains at rest...
20.0 N
20.0 N
...so the table must exert a static friction force of 20.0 N to the left.
(b)
You exert a force of 10.0 N on the block to the right, but the block remains at rest...
10.0 N
10.0 N
...so the table must exert a static friction force of 10.0 N to the left.
(c)
You do not exert a force on the block and it remains at rest...
...so the table must exert no static friction force on the block.
The value of the static friction force exerted by a surface adjusts to exactly counteract the other forces exerted on the object that would make the object slide over the surface, unless those forces exceed $F_{S,max}$.

Figure 5-2 The magnitude of the static friction force For a given object at rest on a given surface, the magnitude of the static friction force is *not* always the same: It depends on the other forces exerted on the object trying to move it relative to the surface.

WATCH OUT !

The direction of the static friction force is parallel to the contact surface.

Equations 5-1 relate only the *magnitudes* of the static friction force and the normal force. These forces are *not* in the same direction! The friction force exerted by a surface on an object is always parallel to the surface in the direction to oppose any relative motion, while the normal force is always perpendicular to the surface. (Remember that *normal* is another word for *perpendicular*.)

Static friction explains why a nail does not slide out of the wood into which it is hammered, even if the nail is upside down (**Figure 5-3a**). The wood fits tightly around the nail and so exerts strong normal forces on the sides of the nail (**Figure 5-3b**). Equation 5-1b tells us that the maximum static friction force is therefore quite large, more than enough to keep the nail in place. If you live in a wood-frame house, static friction deserves much of the credit for holding it together! The same effect explains why it is difficult to take off a tight-fitting pair of boots or to pull an overstuffed pillow out of its pillowcase.

Geckos (tropical lizards) use static friction to walk on vertical surfaces, even on a smooth plate of glass (**Figure 5-4**). At first glance this seems impossible: On a vertical surface there is nothing pushing the gecko into the surface, so the normal force would be zero, and so there would be zero static friction force (see Equation 5-1b). However, at the end of each of a gecko's toes are hundreds of thousands of tiny *setae* (hairs) each of which splits into hundreds of even smaller, flat pads. Each pad or *spatula* is only about 0.2 μm (2×10^{-7} m) across—so small that it can nestle in one of the tiny crevices on the surface of a pane of glass. Like what happens to the nail in Figure 5-3, the normal force between a spatula and the walls of its crevice gives rise to a static friction force. Multiply that tiny force by the millions of spatulae on a gecko's four feet, and the result is a maximum force of friction as great as 20 N. This is more than adequate to support the weight of an average gecko, which is less than 0.5 N, so the gecko can accelerate up the wall as well as simply hang on.

(a)

Mike Flippo/Getty Images

(b)

Figure 5-3 Static friction force on a nail (a) Hammering a nail upward into a piece of wood. (b) Forces on the nail. (Please note that (b) is not a free-body diagram, which would show all the force arrows drawn with their tails on the object, which would be represented by a dot.)

Measuring the Coefficient of Static Friction

Here's a simple experiment for measuring the coefficient of static friction. Place a block of weight F_g on a ramp whose angle θ can be varied (**Figure 5-5a**). If the angle θ is small enough, the block will remain at rest. However, the block will start to slide if the angle is greater than a certain critical value. Let's see how to determine the value of μ_S from this critical angle, which we'll call θ_{slip}.

We'll follow the steps for solving problems involving forces that we used in Section 4-6. (You should review that strategy because we'll be using it a lot in this chapter.) **Figure 5-5b** shows the free-body diagram for the block when $\theta = \theta_{slip}$. Three forces are exerted on the block: the normal force of magnitude F_n, which is perpendicular to the surface of the ramp; the gravitational force of magnitude F_g, which points downward; and the static friction force of magnitude F_S, which points opposite to the direction the block would move if it could. At this angle the block is just about to slip, which means the friction force must be at its maximum value. So from Equations 5-1, $F_S = F_{S,max} = \mu_S F_n$.

We set the sum of the external forces on the block in each direction equal to its mass multiplied by its acceleration in that direction (Newton's second law). To simplify our equations as much as possible, we will take the positive x direction to point down the ramp and the positive y direction to point perpendicular to the ramp, as Figure 5-5b shows. The block remains at rest, so $a_x = a_y = 0$. Then the equations of Newton's second law are

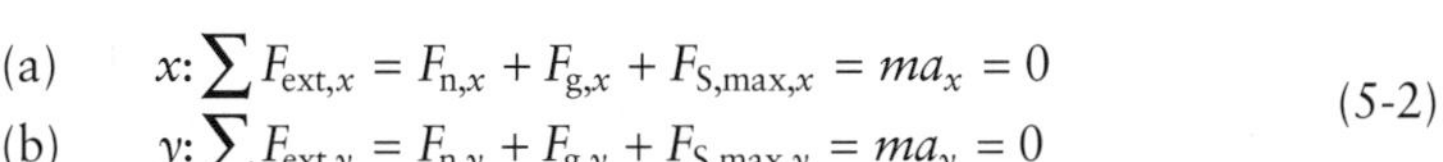

$$\text{(a)} \quad x: \sum F_{ext,x} = F_{n,x} + F_{g,x} + F_{S,max,x} = ma_x = 0 \qquad (5\text{-}2)$$
$$\text{(b)} \quad y: \sum F_{ext,y} = F_{n,y} + F_{g,y} + F_{S,max,y} = ma_y = 0$$

Figure 5-4 Walking on a glass wall A gecko uses static friction to support its weight while walking on a vertical surface. It can do this thanks to the specialized architecture of its feet.

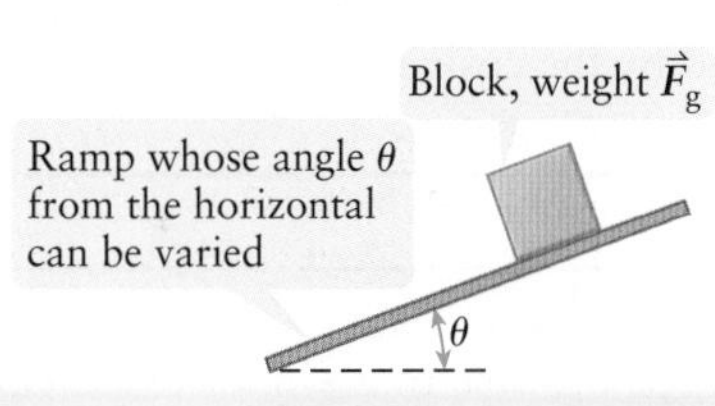

(a) A block at rest on an inclined ramp

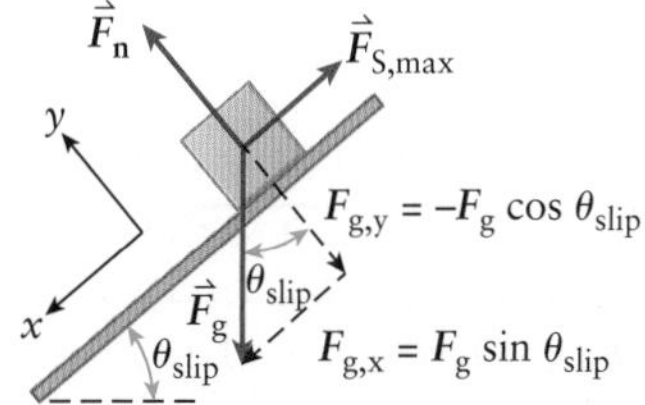

(b) Free-body diagram and force components for the block at the angle where it is just about to slip

Figure 5-5 Measuring the coefficient of static friction The angle at which a block on an incline just begins to slip tells you the coefficient of static friction μ_S.

Figure 5-5b shows that the gravitational force has a positive (downhill) x component $F_{g,x} = F_g \sin \theta_{slip}$ and a negative y component $F_{g,y} = -F_g \cos \theta_{slip}$. The normal force points in the positive y direction (so $F_{n,x} = 0$ and $F_{n,y} = F_n$), and the static friction force points in the negative x direction (so $F_{S,max,x} = -F_{S,max}$ and $F_{S,max,y} = 0$). So Equations 5-2 become

$$\begin{aligned} &\text{(a)} \quad x: \sum F_{ext,x} = 0 + F_g \sin \theta_{slip} + (-F_{S,max}) = 0 \\ &\text{(b)} \quad y: \sum F_{ext,y} = F_n + (-F_g \cos \theta_{slip}) + 0 = 0 \end{aligned} \qquad (5\text{-}3)$$

Our goal is to find the coefficient of static friction μ_S and so we need to incorporate this quantity into our equations. We do this by substituting Equation 5-1b, $F_{S,max} = \mu_S F_n$, into Equation 5-3a. Then, by rearranging Equations 5-3, we get the following expressions:

$$(5\text{-}4) \qquad \begin{aligned} &\text{(a)} \quad \mu_S F_n = F_g \sin \theta_{slip} \\ &\text{(b)} \quad F_n = F_g \cos \theta_{slip} \end{aligned}$$

Equation 5-4b tells us the value of F_n, the magnitude of the normal force, so we can substitute this quantity into Equation 5-4a and solve for μ_S:

$$\mu_S (F_g \cos \theta_{slip}) = F_g \sin \theta_{slip}$$

$$\mu_S \cos \theta_{slip} = \sin \theta_{slip}$$

Or, using $\tan \theta = (\sin \theta)/(\cos \theta)$,

EQUATION IN WORDS
Angle at which an object just begins to slip on an incline

Coefficient of static friction for an object at rest on an incline

Angle of the surface at which the object just **begins to slip**

$$(5\text{-}5) \qquad \mu_S = \tan \theta_{slip}$$

NEED TO REVIEW?
Turn to the Math Tutorial, Section M8, in the back of the book for more information on trigonometry.

The greater the angle θ_{slip} at which the box begins to slide, the greater the value of $\tan \theta_{slip}$ and hence the greater the coefficient of static friction must be. Note that Equation 5-5 does *not* involve the weight of the block. A plastic block on a wooden ramp will begin to slide at the same angle θ independent of its weight.

WATCH OUT!

The coefficient of static friction can be greater than 1.

Because typical values of the coefficient of static friction are less than 1, it is tempting to assume that μ_S is *always* less than 1. Not so! **Figure 5-6** shows a wooden block that rests without slipping on a ramp covered with sandpaper. The ramp is set at a 50° angle; if this is the angle at which the block just begins to slip, then from Equation 5-5 we find that $\mu_S = \tan 50° = 1.2$. Whenever θ_{slip} is greater than 45°, the coefficient of static friction is greater than $\tan 45° = 1$.

David Tauck

Figure 5-6 A block on a sandpaper surface This block can remain at rest at an angle of up to 50°, so the coefficient of static friction for this block on sandpaper is tan 50° = 1.2 (see Equation 5-5).

WATCH OUT!

The normal force exerted on an object is not always equal to the object's weight.

Since the maximum force of static friction a surface can exert on an object is related to the normal force that a surface exerts on the object, it's important to remember that the normal force exerted on an object by a surface is *not* always equal to the object's weight. (We cautioned you about this in Section 4-3.) The situation shown in Figure 5-5 is an example of this caveat. Equation 5-4b shows that the normal force on this object is the object's weight multiplied by the cosine of the angle of the incline.

AP Exam Tip

The normal force is not just determined by the weight and angle of the surface an object rests on; it might also be affected by any other forces on the object that are exerted perpendicular to the surface. The normal force on an object can only ever be determined from a correct free-body diagram and the application of Newton's second law to an object.

EXAMPLE 5-1 Friction in Joints

The wrist is made up of eight small bones called *carpals* that glide back and forth as you wave your hand from side to side. A thin layer of cartilage covers the surfaces of the carpals, making them smooth and slippery. In addition, the spaces between the bones contain synovial fluid, which provides lubrication. During a laboratory experiment, a physiologist applies a compression force to squeeze the bones together along their nearly planar bone surfaces. She then measures the force that must be exerted parallel to the surface of contact to make them move. (**Figure 5-7** shows the contact region between these two carpal bone surfaces.) When the compression force is 11.2 N, the minimum force required to move the bones is 0.135 N. What is the coefficient of static friction in the joint?

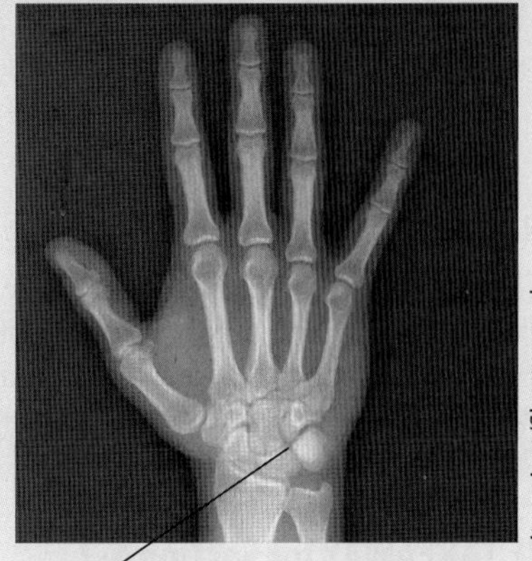

Figure 5-7 Static friction in the wrist This x-ray image shows the carpal bones of the wrist.

Set Up

We begin by sketching the situation to identify all the forces exerted on one of the two carpals in question, which we call carpal A, when it is just about to slip relative to carpal B. This sketch is just to help us identify the forces and their directions. The physiologist exerts two of the four forces exerted on carpal A: the compression force $\vec{F}_{\text{compression}}$ (of magnitude 11.2 N) and the force $\vec{F}_{\text{slide}}$ exerted parallel to the contact surface (of magnitude 0.135 N) to make bone A slide relative to bone B. The other two forces exerted on carpal A are the normal force $\vec{F}_{\text{n}}$ and the static friction force $\vec{F}_{\text{S,max}}$, both exerted by carpal B on carpal A. (This friction force is the *maximum* force of static friction because the carpals are just about to slip.) We then draw the free-body diagram, using a dot to represent carpal A. As always, in a free-body diagram we carefully place the forces with their tails on the dot representing the object. Our goal is to find the value of μ_{S}, so we'll also use the relationship between $F_{\text{S,max}}$ and the normal force given by Equation 5-1b.

Newton's second law for carpal A:

$$\sum \vec{F}_{\text{ext on } A} = \vec{F}_{\text{compression}} + \vec{F}_{\text{slide}} + \vec{F}_{\text{n}} + \vec{F}_{\text{S,max}} = m_A \vec{a}_A$$

Magnitude of the static friction force:

$$F_{\text{S,max}} = \mu_{\text{S}} F_{\text{n}} \quad (5\text{-}1\text{b})$$

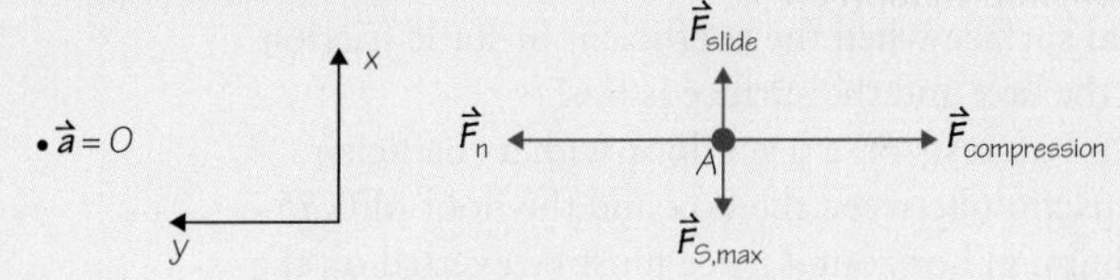

Solve

Make sure we have included the axes in our free-body diagram, so that we can translate from the visual representation to a mathematical one. Let the positive x axis point up and the positive y axis point to the left. Because we are considering the situation in which the carpals have not yet begun to slide, carpal A is at rest and $a_{A,x} = a_{A,y} = 0$. We use this information to write Newton's second law in component form for carpal A.

Newton's second law in component form for carpal A:

$$x{:} \sum F_{\text{ext on } A,x} = 0 + F_{\text{slide}} + 0 + (-F_{\text{S,max}}) = m_A a_{A,x} = 0$$

$$y{:} \sum F_{\text{ext on } A,y} = (-F_{\text{compression}}) + 0 + F_{\text{n}} + 0 = m_A a_{A,y} = 0$$

To find the value of μ_{S} using Equation 5-1b, we need to know the values of F_{n} and $F_{\text{S,max}}$. We find these quantities by using Newton's second law equations and the known magnitudes of the forces $F_{\text{compression}} = 11.2$ N and $F_{\text{slide}} = 0.135$ N.

Solve Equation 5-1b for μ_{S}:

$$\mu_{\text{S}} = \frac{F_{\text{S,max}}}{F_{\text{n}}}$$

Solve Newton's second law equations for $F_{\text{S,max}}$ and F_{n}. From the x component equation:

$$F_{\text{S,max}} = F_{\text{slide}} = 0.135 \text{ N}$$

From the y component equation:

$$F_n = F_{compression} = 11.2 \text{ N so}$$

$$\mu_S = \frac{0.135 \text{ N}}{11.2 \text{ N}} = 0.0121 = 1.21 \times 10^{-2}$$

Reflect

This coefficient of static friction is very small, which means that very little effort is required to make your wrist move. The lubricating fluid between the bones is the key to keeping the friction so small. The fluid also helps reduce wear on the joints. If the bones or even the cartilage surfaces were in direct contact, our joints would wear down relatively quickly.

NOW WORK Problems 1 and 2 in The Takeaway 5-2.

THE TAKEAWAY for Section 5-2

✔ If two objects are at rest with respect to each other and an external force is exerted on one of the objects to try to make it slide past the other, the other object exerts a static friction force on it to oppose that external force.

✔ In any situation, the magnitude of the static friction force on an object adjusts to balance the other forces exerted on the object. However, there is a maximum possible magnitude of the static friction force, given by Equation 5-1b.

✔ If an object is placed on a tilted ramp, it will slide if the angle of the ramp is greater than the slip angle given by Equation 5-5.

Prep for the AP® Exam

AP® Building Blocks

EX 5-1 1. What is the minimum horizontal force that must be exerted on a 5.00-kg box to make it begin to slide on a horizontal surface when the coefficient of static friction between the box and the surface is 0.67?

EX 5-1 2. A 7.60-kg box rests on a level floor with a coefficient of static friction between the box and the floor of 0.55. What minimum horizontal force must be exerted on the box to start it sliding?

3. External forces exerted on objects A and B are directed to make them slide past each other if they were to begin to move, yet they remain at rest. Which of the following is true about the friction force exerted on object A by object B?
   (A) The friction force is proportional and parallel to the normal force that A exerts on B.
   (B) The friction force has a magnitude that is proportional to the magnitude of the normal force that object A exerts on object B, and is in the direction opposing other forces that would accelerate object B relative to object A.
   (C) The friction force is proportional and parallel to the normal force that B exerts on A.
   (D) The friction force has a magnitude that is proportional to the magnitude of the normal force that object B exerts on object A and is in the direction opposing other forces that would accelerate object A relative to object B.

4. If the force of friction always opposes the sliding of an object, how then can a friction force cause an object to accelerate in the direction of the motion?

AP® Skill Builders

5. A person, with a mass m_P, exerts a force on the edge of a table, which has a mass m_T. The force is directed at a clockwise angle θ relative to the horizontal. The floor exerts friction forces on the table and the person with coefficients of static friction $\mu_{S,table}$ and $\mu_{S,person}$, respectively. As the force exerted by the person increases, either the person or the table will slip on the floor.

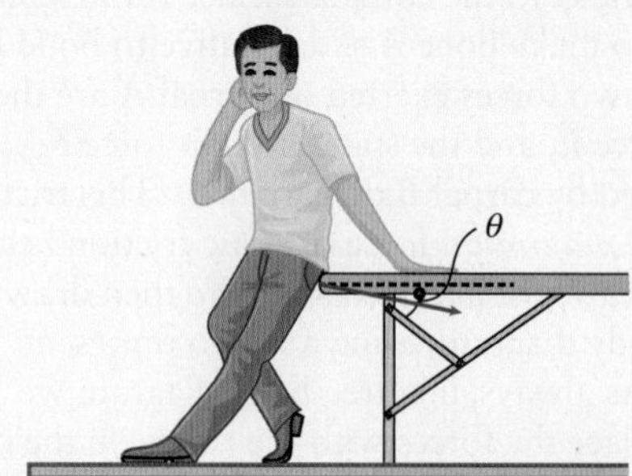

   (a) Suppose m_P and m_T are equal and that the person's shoes are slippery ($\mu_{S,person} \sim 0.1$) and $\mu_{S,table} \sim 0.5$. In a clear, coherent paragraph-length response, evaluate the claim that for some angle θ (without lifting the table off the floor) the table will slide first at some threshold force.
   (b) For the same situation as in part (a), $\mu_{S,person} \sim 0.1$ and $\mu_{S,table} \sim 0.5$, quantitatively predict the ratio of masses m_P/m_T when the threshold force, exerted along $\theta = 0$, causes the table and the person to slide simultaneously. Using complete sentences, briefly explain your reasoning without manipulating equations.
   (c) Construct a free-body diagram of the table. Justify in words the relative sizes of the vectors in your representation.
   (d) Construct a free-body diagram of the person. Justify in words the relative sizes of the vectors in your representation.
   (e) Using your free-body diagrams express $F_{person\ on\ table}$ and $F_{table\ on\ person}$ as inequalities in terms of the static friction coefficients, the angle θ, and the weights of the person and the table.

(f) Remember that the contact forces exerted by the table on the person, and vice versa, are always equal in magnitude, so we can set the expressions you got in (e) equal to each other. The resulting ratio ($F_{g,person}/F_{g,table}$) is graphed versus θ in the figure. Consider your answers to parts (a) and (b). Analyze these data for the situation with $\mu_{S,person} = \mu_{S,table} = 0.5$ in terms of regions in the graph in which the threshold force causes (i) the person to slip first, (ii) the table to slip first, and (iii) both to slip simultaneously.

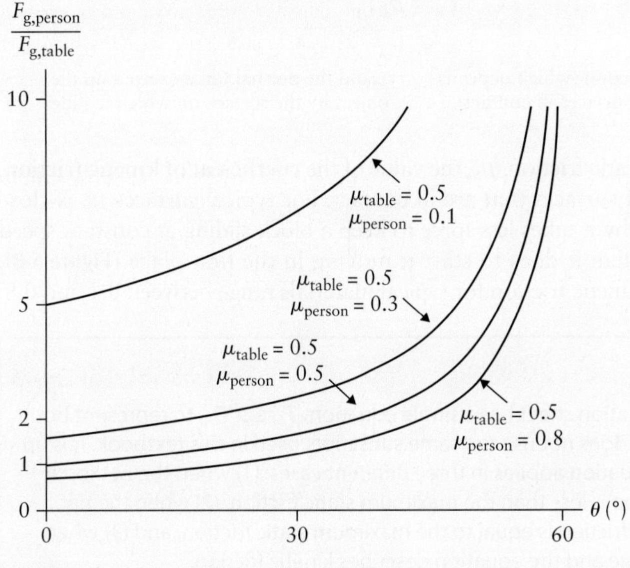

AP® Skills in Action

6. Construct a free-body diagram for the situation shown in the figure. An object of mass M rests on a ramp; there is friction between the object and the ramp.

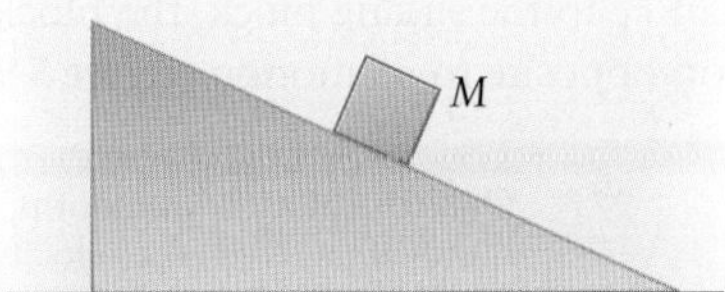

7. Tribology is the name given to the scientific study of friction. The significance of tribology is comparable to the search for energy resources that are not carbon based: One-third of world energy resources are used to overcome friction forces. It is an ancient field. Leonardo da Vinci introduced the idea that friction force is proportional to weight and through experimentation estimated the constant of proportionality as 0.25.
(a) Using da Vinci's model, in a clear, coherent paragraph-length response, explain how the coefficient of static friction between two materials can be measured by slowly elevating an inclined plane made from one of the materials upon which rests a block made from the other material.
(b) Da Vinci built machines and used olive oil to lubricate the junction between wooden shafts and metal hubs supporting the shaft. The coefficient of static friction between oak and bronze is 0.19 when the wooden surface is lubricated by olive oil.
Calculate the inclination at which a bronze block just begins to slide on an oak board coated with olive oil.
(c) After the Second World War, researchers at the National Bureau of Standards (today known as the National Institute of Standards and Technology) evaluated synthetic lubricants to replace petroleum-based materials. Some of the measurements for the static friction coefficient between surfaces of stainless steel with several lubricants are shown in the table.

	Lubricant	μ_{min}	μ_{max}	$\mu_{average}$
A	Navy 2110 oil	0.158	0.188	0.166
B	Grease G (23% sodium soap)	0.060	0.090	0.068
C	Grease G + 5% molybdenum disulfide	0.038	0.053	0.047
D	Fluorinated hydrocarbon	0.184	0.195	0.188
E	Fluorinated hydrocarbon + 1% molybdenum disulfide	0.054	0.070	0.067

Identify pairs of lubricants in these measurements that could be difficult to distinguish in such an experiment. Identify the best lubricant.

5-3 The kinetic friction force on a sliding object has a constant magnitude

In the previous section, we used a block on a tilted ramp (Figure 5-5) to help us explore the force of static friction. If the ramp is inclined by more than a certain critical angle, not even the maximum force of static friction is sufficient to prevent the block from sliding. The block still experiences a resistive force as it slides. However, this force is no longer static friction but *kinetic friction*, the friction force between two surfaces that move relative to each other.

The Magnitude of the Kinetic Friction Force

Like static friction, the force of kinetic friction, $\vec{f}_k$ ("*f*-sub-k"), between two objects is proportional to the normal force that one object exerts on the other perpendicular to their surface of contact—in other words, it is proportional to how hard the two surfaces are pushed together. However, the kinetic friction force is less than the maximum force of static friction. Imagine for a moment if this weren't the case: If we tilted a surface just enough for an object on it to start sliding, it would start and then stop again due to kinetic friction. However, then it would be stopped and static friction wouldn't be big enough to hold it, so it would start sliding again and would just continue to stop and start!

We'll model the kinetic friction force as being independent of how fast the two objects (or surfaces or an object on a surface) slide past each other, as well as independent of the area of contact between the objects. (This model works well in many situations.) Then we can write an expression for the magnitude F_k of the kinetic friction force:

EQUATION IN WORDS
Magnitude of the kinetic friction force

The force of kinetic friction exerted on an object by a surface depends on...

$$F_k = \mu_k F_n \quad (5\text{-}6)$$

...the **coefficient of kinetic friction** (which depends on the properties of the two surfaces in contact)...

...and the **normal force** exerted on the object by the surface on which it slides.

Like the coefficient of static friction, μ_S, the value of the **coefficient of kinetic friction**, μ_k, depends on the kinds of surfaces that are in contact. For typical surfaces μ_k is close to but less than μ_S. That's why it takes less force to keep a block sliding at constant speed over a horizontal surface than it does to start it moving in the first place (**Figure 5-8**). Values of the coefficient of kinetic friction for typical materials range between 0.1 and 0.8.

AP® Exam Tip

Since the AP® Physics 1 equation sheet has a single equation, $|\vec{F}_f| \le \mu|\vec{F}_n|$, to represent both static and kinetic friction, it does not use the same subscripts used in this textbook. It is up to you to know how the equation applies in three different cases: (1) when the net force exerted against static friction is less than the maximum static friction, (2) when the net force exerted against static friction is equal to the maximum static friction, and (3) when the static friction is overcome and the equation describes kinetic friction.

In **Figure 5-9a** the block is moving at a constant velocity, so its acceleration is zero and the net force exerted on the block is zero as well. In this case, the horizontal force that you apply just balances the force of kinetic friction of magnitude F_k opposing the motion. If the horizontal force that you apply is greater in magnitude than F_k, the net force points in the direction of the block's motion and the block speeds up (**Figure 5-9b**). If you apply a horizontal force with a magnitude less than F_k to the sliding block, the block slows down because the net force points in the direction opposite to its motion (**Figure 5-9c**).

(a) You push a stationary block with a horizontal force less than the maximum force of static friction.

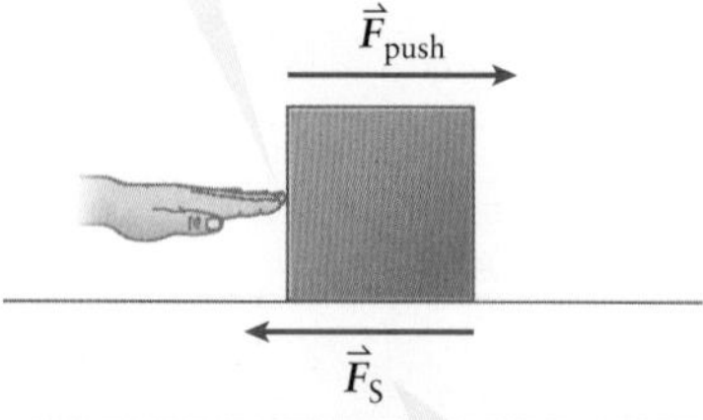

Static friction balances applied force: block remains at rest.

(b) You push a stationary block with a horizontal force equal to the maximum force of static friction.

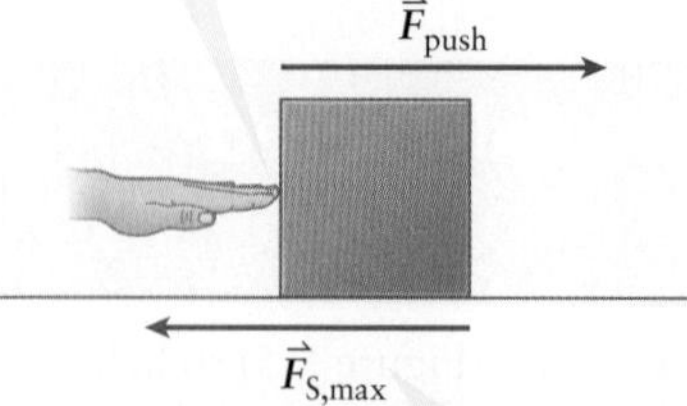

Static friction just balances applied force: block is at rest but just about to slip.

(c) Once the block is in motion, you can keep it moving at constant velocity by exerting a horizontal force equal to the force of kinetic friction.

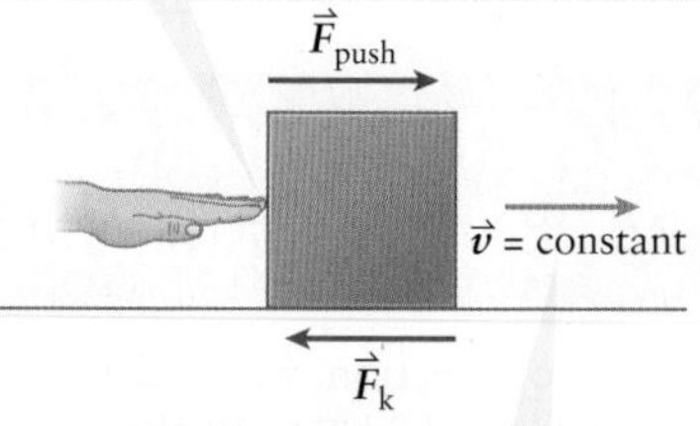

Block moves at a constant velocity.

If you push hard enough on an object at rest, you can overcome the force of static friction.

Figure 5-8 From static friction to kinetic friction If you push hard enough on an object to overcome static friction and start the object sliding, the force of friction becomes the kinetic friction force.

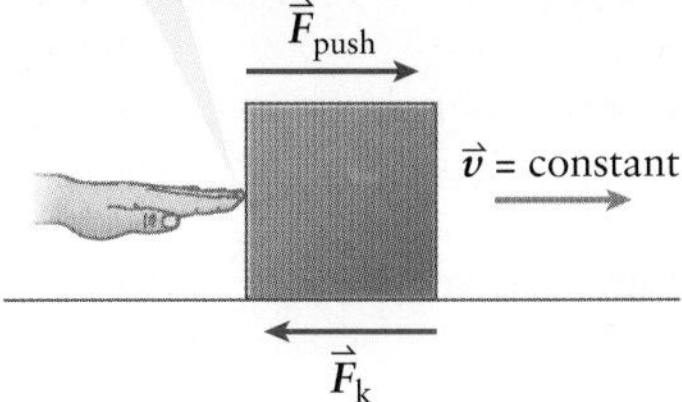
(a)
You can keep a block sliding with constant velocity over a horizontal surface by exerting a horizontal force equal to the force of kinetic friction exerted on the block by the surface.
$\vec{F}_{push}$
$\vec{v}$ = constant
$\vec{F}_k$

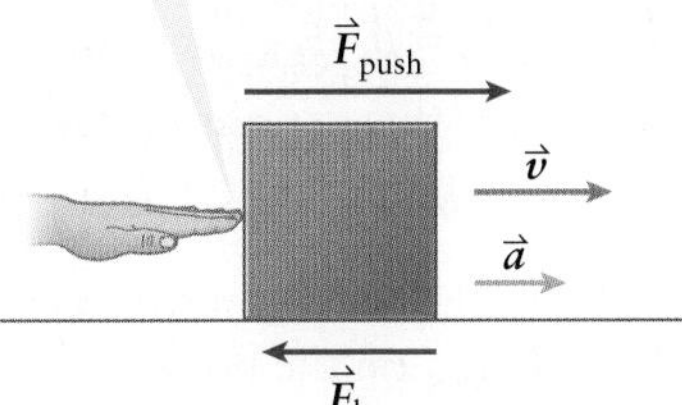
(b)
If you exert a horizontal force on the moving block greater than the force of kinetic friction exerted on the block by the surface, it speeds up.
$\vec{F}_{push}$
$\vec{v}$
$\vec{a}$
$\vec{F}_k$

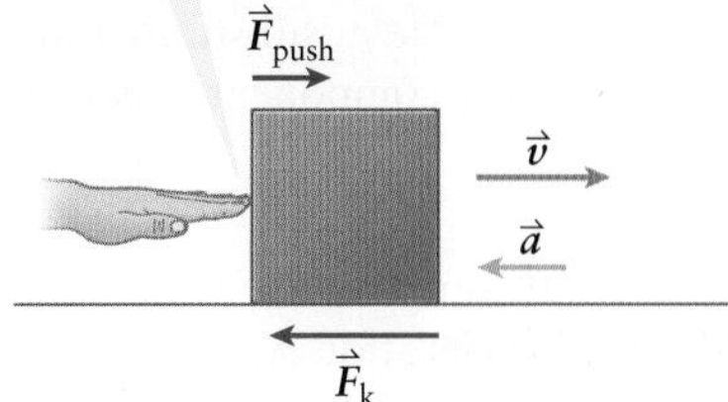
(c)
If you exert a horizontal force on the moving block less than the force of kinetic friction exerted on the block by the surface, it slows down.
$\vec{F}_{push}$
$\vec{v}$
$\vec{a}$
$\vec{F}_k$

The kinetic friction force exerted on the box by the surface does not depend on how hard you push the object, nor does it depend on the speed of the object.

Figure 5-9 Kinetic friction: three examples Depending on how hard you push on an object sliding with kinetic friction, it will (a) maintain a constant velocity, (b) speed up, or (c) slow down. The size of the kinetic friction force on a sliding object does not depend on how hard you push the object, nor does it depend on the speed of the object, but it goes to zero when the object stops moving.

EXAMPLE 5-2 Sliding with Kinetic Friction

In Example 4-6 we considered a book sliding down an incline with friction. In a variation of this problem, suppose the coefficient of kinetic friction between the incline and the sliding 2.50-kg book is 0.350. (a) What is the downhill acceleration of the sliding book if the incline is at an angle from the horizontal of 30.0°? (b) What is the acceleration of the sliding book if the angle is 10.0°?

Set Up

The book is moving down the incline of angle θ, so the kinetic friction force (which always opposes sliding) must point up the incline. To find the net external force on the book and hence the book's acceleration, we need to determine the magnitude of the friction force by using Equation 5-6. Once we've found a general expression for the book's acceleration in terms of θ, we'll substitute the values $\theta = 30.0°$ and $\theta = 10.0°$.

Newton's second law applied to the book:

$$\sum \vec{F}_{\text{ext on book}} = \vec{F}_n + \vec{F}_{g,\text{book}} + \vec{F}_k = m_{\text{book}}\vec{a}_{\text{book}}$$

Magnitude of the kinetic friction force:

$$F_k = \mu_k F_n \tag{5-6}$$

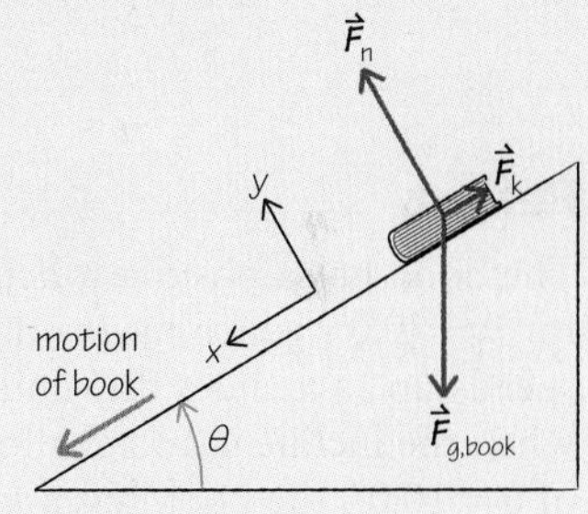
$\vec{F}_n$
$\vec{F}_k$
y
x
motion of book
θ
$\vec{F}_{g,book}$

Solve

Note that in the sketch above, we placed the forces as if they were all exerted on the center of mass of the book, which is what the free-body diagram represents. Although the force of friction and the normal force are not exerted at the center of mass, the object's motion is the same as if they were. We use the sketch to draw the free-body diagram. Because the free-body diagram was not requested, we can include the components as dashed arrows. Use this diagram to write Newton's second law in component form for the book. Because the book's acceleration is along the incline let's choose that for the positive x direction. Choose the positive y direction perpendicular to and away from the plane; then it follows that $a_{\text{book},y} = 0$.

Newton's second law in component form for the book:

$$x: \sum F_{\text{ext on book},x} = 0 + F_{g,\text{book}} \sin\theta + (-F_k) = m_{\text{book}} a_{\text{book},x}$$

$$y: \sum F_{\text{ext on book},y} = F_n - F_{g,\text{book}} \cos\theta + 0 = m_{\text{book}} a_{\text{book},y} = 0$$

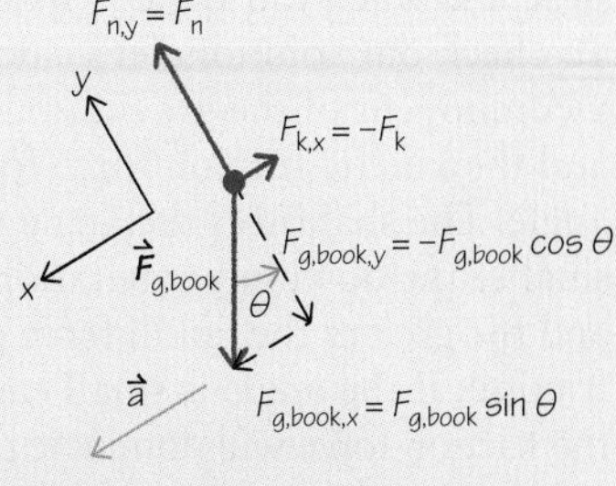
$F_{n,y} = F_n$
y
$F_{k,x} = -F_k$
x
$\vec{F}_{g,book}$
$F_{g,book,y} = -F_{g,book} \cos\theta$
θ
$\vec{a}$
$F_{g,book,x} = F_{g,book} \sin\theta$

To find the magnitude of the kinetic friction force, solve the y component equation for the magnitude of the normal force F_n and then substitute this expression into the equation for F_k. Next, substitute our result for F_k into the x component equation.

From the y component equation:

$F_n = F_{g,book} \cos\theta$

Substitute this into Equation 5-6:

$F_k = \mu_k F_n = \mu_k (F_{g,book} \cos\theta)$

Substitute this expression for F_k into the x equation:

$F_{g,book} \sin\theta - \mu_k F_{g,book} \cos\theta = m_{book} a_{book,x}$

Now we can complete our solution for the book's acceleration $a_{book,x}$. Use the relationship between gravitational force and mass.

Relationship between the mass m_{book} of the book and the gravitational force $F_{g,book}$ on the book:

$F_{g,book} = m_{book} g$

Substitute into the x equation and solve for $a_{book,x}$:

$m_{book} g \sin\theta - \mu_k m_{book} g \cos\theta = m_{book} a_{book,x}$

The factor of m_{book} cancels, so

$a_{book,x} = g \sin\theta - \mu_k g \cos\theta$ or

$a_{book,x} = g(\sin\theta - \mu_k \cos\theta)$

Use this general equation for $a_{book,x}$ for the given value of μ_k and the two given values of θ.

(a) Substitute $\theta = 30.0°$:

$$\begin{aligned} a_{book,x} &= g(\sin\theta - \mu_k \cos\theta) \\ &= (9.80\text{ m/s}^2)[\sin 30.0° - (0.350)(\cos 30.0°)] \\ &= (9.80\text{ m/s}^2)[0.500 - (0.350)(0.866)] \\ &= 1.93\text{ m/s}^2 \end{aligned}$$

(b) Substitute $\theta = 10.0°$:

$$\begin{aligned} a_{book,x} &= g(\sin\theta - \mu_k \cos\theta) \\ &= (9.80\text{ m/s}^2)[\sin 10.0° - (0.350)(\cos 10.0°)] \\ &= (9.80\text{ m/s}^2)[0.174 - (0.350)(0.985)] \\ &= -1.68\text{ m/s}^2 \end{aligned}$$

Reflect

The book has a positive x acceleration for $\theta = 30.0°$, which means the book speeds up as it slides downhill. However, when the incline is at a shallower angle of $\theta = 10.0°$, the book has a *negative* (uphill) acceleration: Its speed decreases as it moves down the incline. This makes sense because gravity is "helping" it slide less when the angle is shallower. To check our solution, let's calculate the x component of the gravitational force and the kinetic friction force for each angle. The shallower the angle θ, the smaller the downhill gravitational force and the greater the uphill force of kinetic friction. If the angle is small enough, the friction force will dominate over the gravitational force—in which case the net force and acceleration point up the incline.

Downhill component of gravitational force:

$F_{g,book,x} = F_{g,book} \sin\theta = m_{book} g \sin\theta$

Uphill component of kinetic friction force:

$F_k = \mu_k F_n = \mu_k m_{book} g \cos\theta$

If $\theta = 30.0°$,

$$\begin{aligned} F_{g,book,x} &= (2.50\text{ kg})(9.80\text{ m/s}^2) \sin 30.0° \\ &= 12.3\text{ N} \end{aligned}$$

$$\begin{aligned} F_k &= (0.350)(2.50\text{ kg})(9.80\text{ m/s}^2) \cos 30.0° \\ &= 7.43\text{ N} \end{aligned}$$

so gravity dominates over kinetic friction, and the net force is downhill.

If $\theta = 10.0°$,

$$\begin{aligned} F_{g,book,x} &= (2.50\text{ kg})(9.80\text{ m/s}^2) \sin 10.0° \\ &= 4.25\text{ N} \end{aligned}$$

$$\begin{aligned} F_k &= (0.350)(2.50\text{ kg})(9.80\text{ m/s}^2) \cos 10.0° \\ &= 8.44\text{ N} \end{aligned}$$

Thus, kinetic friction dominates over gravity and so the net force is uphill. If the book comes to a stop, the acceleration will also stop, so a book that starts out moving downhill will never turn around and move uphill due to friction! You would just have a case of static friction less than the maximum value.

Extend: Can you show that the acceleration of the book will be zero (that is, it will slide down the incline with constant velocity) if $\theta = 19.3°$?

NOW WORK Problems 1–4 in The Takeaway 5-3.

EXAMPLE 5-3 How Far Up the Incline?

You set the incline described in Example 5-2 at an angle of 30.0° from the horizontal and then push the 2.50-kg book up the incline with an initial speed of 4.20 m/s. The coefficient of kinetic friction between book and ramp is again 0.350. How far up the incline does the book travel before coming to a stop?

Set Up

As in Example 5-2, the three forces exerted on the sliding book are the normal force, gravitational force, and kinetic friction force and again we just sketched them at the center of mass. The difference here is that the friction force points downhill rather than uphill because the book is now sliding uphill. (Remember that the kinetic friction force always opposes the direction of the relative sliding of the object over the surface.)

This problem asks for a distance, so it's really a kinematics problem. Because all three forces exerted on the book are constant, the net external force and acceleration are constant as well. Therefore, we can use the formulas from Section 2-4 for linear motion with constant acceleration. While we could use any of these equations, given that we don't know time, it is easiest to choose the equation that relates the book's initial velocity v_{0x}, its final velocity v_x, and its acceleration $a_{\text{book},x}$ to its displacement $x - x_0$. We are asked to find distance, which will be the magnitude of the displacement, given that the motion is in one direction.

Newton's second law applied to the book:

$$\sum \vec{F}_{\text{ext on book}} = \vec{F}_{\text{n}} + \vec{F}_{\text{g,book}} + \vec{F}_{\text{k}} = m_{\text{book}}\vec{a}_{\text{book}}$$

Magnitude of the kinetic friction force:

$$F_{\text{k}} = \mu_{\text{k}} F_{\text{n}} \quad (5\text{-}6)$$

Velocity, acceleration, and position for constant acceleration only:

$$v_x^2 = v_{0x}^2 + 2a_{\text{book},x}(x - x_0) \quad (2\text{-}11)$$

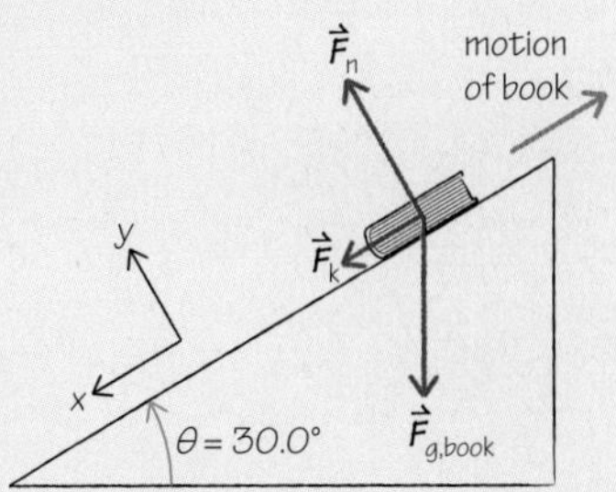

Solve

Draw the free-body diagram. Because the free-body diagram was not requested as part of the solution, we can include the components as dashed arrows. Use this diagram to write Newton's second law in component form for the book. Choose the same coordinate system as in the last example. The equations are the same as in Example 5-2 except that the friction force now has a positive (downhill) x component. The book's acceleration is along the x axis only, so $a_{\text{book},y} = 0$.

Newton's second law in component form for the book:

$$x: \sum F_{\text{ext on book},x} = 0 + F_{\text{g,book}} \sin\theta + F_{\text{k}} = m_{\text{book}} a_{\text{book},x}$$

$$y: \sum F_{\text{ext on book},y} = F_{\text{n}} - F_{\text{g,book}} \cos\theta + 0 = m_{\text{book}} a_{\text{book},y} = 0$$

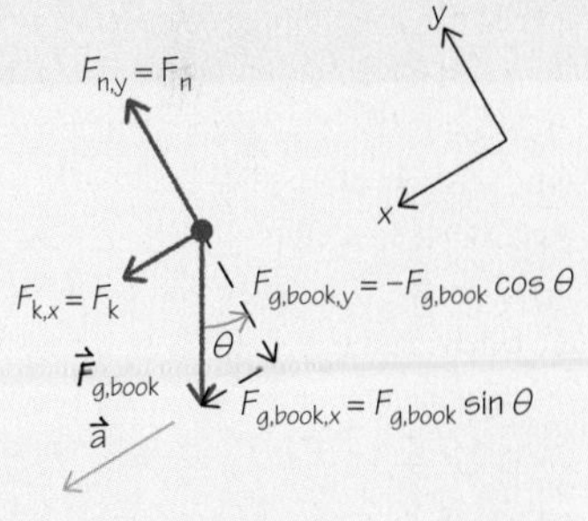

Use these equations and Equation 5-6 to solve for the acceleration of the book.

From the y component equation and $F_{\text{g,book}} = m_{\text{book}} g$:

$$F_{\text{n}} = F_{\text{g,book}} \cos\theta = m_{\text{book}} g \cos\theta$$

Substitute into Equation 5-6 and find the magnitude of the kinetic friction force:

$$F_{\text{k}} = \mu_{\text{k}} F_{\text{n}} = \mu_{\text{k}} m_{\text{book}} g \cos\theta$$

Substitute into x component equation and solve for $a_{\text{book},x}$:

$$m_{\text{book}} g \sin\theta + \mu_{\text{k}} m_{\text{book}} g \cos\theta = m_{\text{book}} a_{\text{book},x}$$

$$g \sin\theta + \mu_{\text{k}} g \cos\theta = a_{\text{book},x}$$

$$a_{\text{book},x} = g(\sin\theta + \mu_{\text{k}} \cos\theta)$$

$$= (9.80 \text{ m/s}^2)[\sin 30.0° + (0.350)(\cos 30.0°)]$$

$$= (9.80 \text{ m/s}^2)[0.500 + (0.350)(0.866)]$$

$$= 7.87 \text{ m/s}^2$$

(Acceleration is positive because it points downhill.)

Now we can solve for the displacement of the book from where it starts to where it stops. Its initial velocity v_{0x} is negative because the book starts by moving uphill. Its final velocity v_x is zero because it is at rest after traveling the maximum distance up the incline. Use Equation 2-11 to find $x - x_0$.

Equation 2-11:

$$v_x^2 = v_{0x}^2 + 2a_{\text{book},x}(x - x_0)$$

with $v_{0x} = -4.20$ m/s, $v_x = 0$, and $a_{\text{book},x} = 7.87$ m/s^2

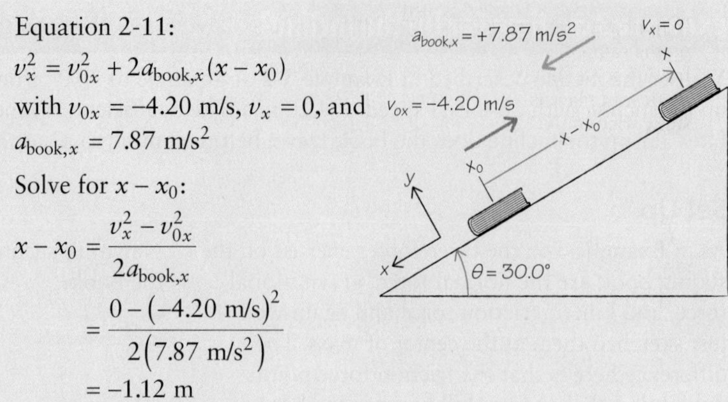

Solve for $x - x_0$:

$$x - x_0 = \frac{v_x^2 - v_{0x}^2}{2a_{\text{book},x}}$$

$$= \frac{0 - (-4.20 \text{ m/s})^2}{2(7.87 \text{ m/s}^2)}$$

$$= -1.12 \text{ m}$$

(Displacement is negative because the book moves uphill.) So the book travels 1.12 m.

Reflect

The downhill acceleration of the book (7.87 m/s^2) is much greater than in Example 5-2 (1.93 m/s^2). This makes sense because the force of kinetic friction now points downhill and so augments the downhill component of the gravitational force.

We can check our result by seeing how far the book would travel if the surface were horizontal, so $\theta = 0$. In this case the book travels 2.57 m, which is more than twice the distance we found for the book on the 30.0° incline. It travels a greater distance if the surface is horizontal because the force of gravity is no longer helping to slow the book down. This result is physically reasonable and gives us confidence that our answer is correct.

If the surface is horizontal ($\theta = 0$):

$$a_{\text{book},x} = g(\sin 0 + \mu_k \cos 0)$$
$$= g[0 + \mu_k(1)]$$
$$= \mu_k g$$
$$= (0.350)(9.80 \text{ m/s}^2)$$
$$= 3.43 \text{ m/s}^2$$

horizontal surface

$a_{book,x}$ = +3.43 m/s²

V_{ox} = −4.20 m/s

V_x = 0

Then the displacement of the book from where it starts to where it stops is

$$x - x_0 = \frac{v_x^2 - v_{0x}^2}{2a_{\text{book},x}} = \frac{0 - (-4.20 \text{ m/s})^2}{2(3.43 \text{ m/s}^2)}$$
$$= -2.57 \text{ m}$$

So the book travels 2.57 m.

NOW WORK Problems 5 and 6 in The Takeaway 5-3.

Rolling Friction

In addition to static friction and kinetic friction, a third important type of friction is **rolling friction**. If you start a billiard ball rolling across a horizontal surface, it will roll for quite a distance but will eventually come to a stop. A force must therefore be exerted on the ball opposite to its motion, and it's this force that we call rolling friction (**Figure 5-10**). This force always opposes the motion of a rolling object, whether it's rolling uphill, downhill, or on a horizontal surface.

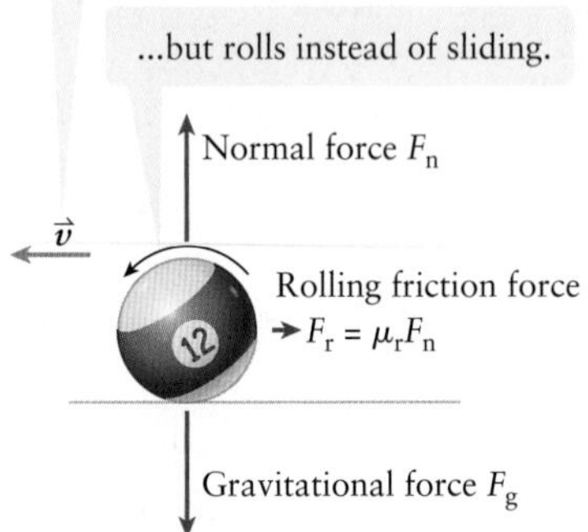

Figure 5-10 A rolling billiard ball A ball can roll much farther on a horizontal surface than a block can slide on the same surface. The reason is that rolling friction is a much weaker force than kinetic friction.

Just as for static and kinetic friction, the force of rolling friction is proportional to the normal force exerted on the rolling object by the surface over which it rolls. So we can express the magnitude F_r of the rolling friction force as

$$F_r = \mu_r F_n \text{ (force of rolling friction)} \quad (5\text{-}7)$$

The **coefficient of rolling friction** μ_r can be a very small number—about 0.02 for rubber automobile tires on concrete and about 0.002 for the steel wheels of a railroad car on steel rails. As a result, rolling friction is generally much smaller than kinetic friction, so a round object will roll much farther than a brick-shaped object of the same mass will slide. In general, the more rigid the rolling object and the more rigid the surface over which it rolls, the smaller the value of μ_r. The tires of your car are an example: If you don't keep them properly inflated, the coefficient of rolling friction for your tires will be greater than it should be, and your fuel consumption will be greater as well. If the surfaces are totally rigid, then you would get no loss of speed at all as long as the object was just rolling and not sliding because, as we will see later in this text, the point in contact would always be at rest (Weird, huh? Stick with me!). The problem is, if there is no friction, you also lose your ability to steer or brake because there is

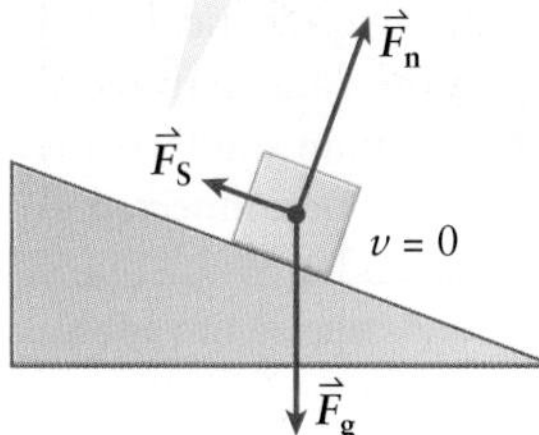
(a)
If an object is at rest on the ramp, the friction force that holds it in place is **static friction.**
$\vec{F}_n$
$\vec{F}_S$
$v = 0$
$\vec{F}_g$

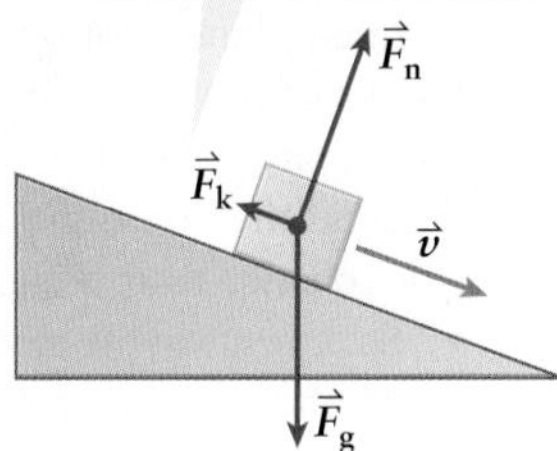
(b)
If the object slides on the ramp, the friction force that the ramp exerts on it is **kinetic friction.**
$\vec{F}_n$
$\vec{F}_k$
$\vec{v}$
$\vec{F}_g$

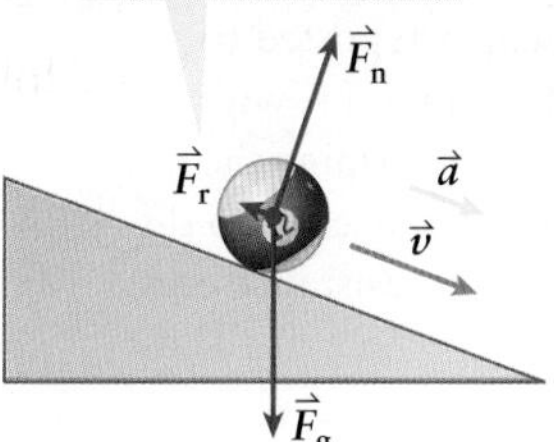
(c)
If the object rolls on the ramp, the friction force that the ramp exerts on it is **rolling friction.**
$\vec{F}_n$
$\vec{F}_r$
$\vec{a}$
$\vec{v}$
$\vec{F}_g$

Figure 5-11 Three kinds of friction Comparing (a) static friction, (b) kinetic friction, and (c) rolling friction. In each case, we have drawn the forces as if they are exerted on the center of mass, since we are using an object model.

no force exerted along the direction of motion, which we need to change the speed of something. Rolling friction occurs because the surfaces do deform and rub across each other, so you get something like kinetic friction, just a much reduced amount. This is why trains are so much more fuel efficient than cars—train wheels are made out of metal and so they do not deform, but they have to run on tracks in order to be steered.

Technically, when something is rolling, not all of the points on the rolling item are moving in exactly the same way, so the object model does not fit. However, since we are worrying just about the straight-line motion of the center of the item that is rolling, we can still use the object model. We will learn how to consider other aspects of such motion when we discuss rotational motion.

Figure 5-11 summarizes the three varieties of friction force.

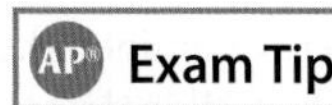
AP Exam Tip

While it is important to learn about rolling friction to completely understand the concept of friction, rolling friction is not a topic on the AP® Physics 1 exam.

THE TAKEAWAY for Section 5-3

✔ If an object is sliding over a surface, the kinetic friction force exerted on the object by the surface is always in the direction that opposes the sliding.

✔ The magnitude of the kinetic friction force is proportional to the normal force exerted on the object by the surface on which it is sliding (Equation 5-6).

✔ Like kinetic friction, rolling friction opposes motion and is proportional to the normal force, but the coefficient of friction for rolling is much smaller than that for sliding.

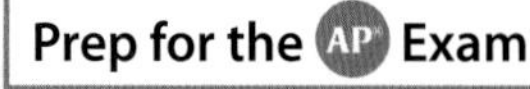
Prep for the AP Exam

AP® Building Blocks

EX 5-2

1. A friction force of 12.7 N is exerted on an object in motion relative to a level surface. If the coefficient of kinetic friction between the object and the surface is 0.37, what is the mass of the object?

EX 5-2

2. A 25.0-kg crate rests on a level floor. A horizontal force of 50.0 N is exerted on the crate, accelerating it at 1.0 m/s^2. Calculate
   (a) the normal force exerted by the floor on the crate;
   (b) the friction force exerted by the floor on the crate; and
   (c) the coefficient of kinetic friction between the crate and the floor.

EX 5-2

3. A mop is pushed across the floor by a handle that exerts a force $\vec{F}$ of 50.0 N at an angle of $\theta = 50.0°$ on the mop head. The mass of the mop head is 3.75 kg. Identify all of the forces exerted on the mop head and specify their directions, then calculate the acceleration of the mop head if the coefficient of kinetic friction between the mop head and the floor is 0.40.

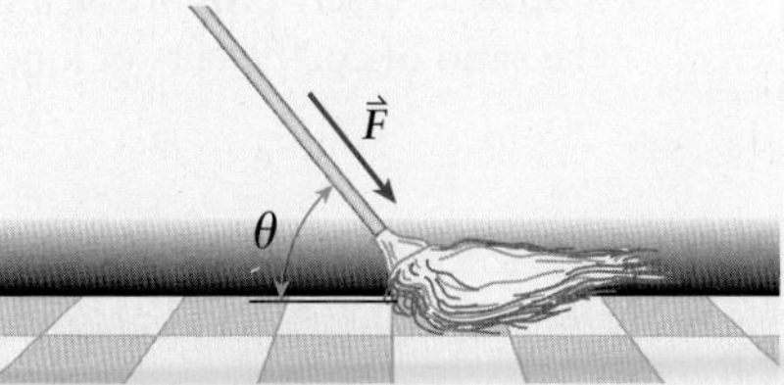
$\vec{F}$
θ

4. EX 5-2 If the coefficient of kinetic friction between an object with mass $M = 3.0$ kg and a flat surface is 0.40, calculate the magnitude of force that must be exerted on the object to cause it to accelerate at 2.5 m/s^2, if the force is exerted at an angle of $\theta = 30.0°$ below the horizontal, as shown.

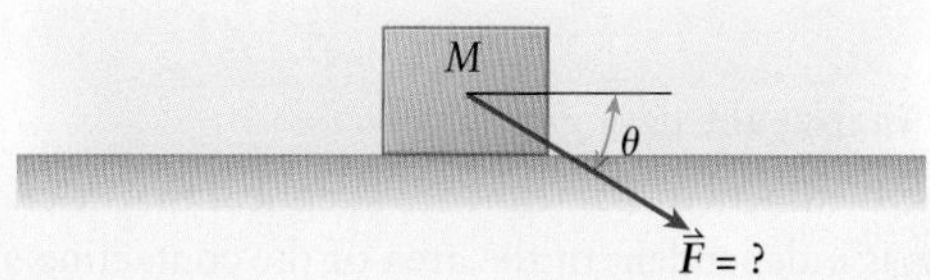
M
θ
$\vec{F} = ?$

5. EX 5-3 A spring-loaded launcher is used to propel a small steel block upward along a polystyrene ramp. A website states that the kinetic friction coefficient for this pair of materials is between 0.3 and 0.35. The spring in the launcher can be compressed to vary the initial velocity of the block. The displacement of the steel block is measured. A trial run is made to assess the validity of the reported friction coefficient.

The following data were obtained: initial velocity as the block left the launcher = 4.5 m/s, distance the block travelled up the ramp = 2.1 m, and ramp inclination = 10°. Evaluate the kinetic friction coefficient from these data.

EX 5-3

6. A reasonably priced 1000-kg sportscar is reported to accelerate from 0 to 58 mph (26.4 m/s) in 5.5 s, with some uncertainty in the measurement. Determine the precision of measurement of the time interval needed to differentiate between a coefficient of rolling resistance equal to 0.02 and one equal to 0.

AP® Skill Builders

7. A team of students obtained the following data by allowing a block to slide down an inclined board. The inclination angle was defined relative to the horizontal. A motion sensor was used to measure the velocity for each angle and accelerations were calculated from linear best-fits of the velocity versus time graphs obtained for each angle. The following data were obtained.

Inclination angle (°)	Acceleration (m/s^2) Run 1	Acceleration (m/s^2) Run 2
20	0.4	0.3
23	1.0	0.9
29	2.1	2.3
35	2.8	3.2
39	3.7	3.6

(a) Construct a free-body diagram to represent the forces exerted on the block.
(b) Equation 5-6 claims that the friction force exerted on the block is proportional to the normal force exerted by the ramp on the block. As the inclination of the ramp increases, the acceleration increases. Evaluate the evidence for the validity of Equation 5-6 provided by these data.
(c) Evaluate the components of the net force exerted on the block to express the acceleration in terms of the inclination angle.
(d) The inclination angle is the independent variable in this experiment. But analysis of the data is complicated by the dependence of the acceleration a on θ. Re-express the mathematical representation obtained in part (c) by defining new variables x and y so that

$$y = g(1 - \mu_k x)$$

(e) Analyze the data graphically.

Some members of the team are skeptical that friction force is independent of the area of the contacting surfaces. It is claimed that friction forces are electrostatic. So if the area increases, the number of atoms interacting increases.

(f) Design an experiment to test the claim that the friction force is independent of the surface area. The inclined plane, motion sensor, and two identical blocks of the type used to obtain the data analyzed in part (e) are available. The blocks remain stable as they descend the board if they are stacked or connected side to side. Your design should include:
- Statements of the independent and dependent variables
- Methods of data collection
- Methods of data analysis

8. You want to push a heavy box across a rough floor. You know that the coefficient of static friction (μ_S) is larger than the coefficient of kinetic friction (μ_k). Predict which of the following processes are easier based on how hard you must push the box:
(A) Push the box for a short distance, rest, push the box another short distance, and then repeat the process until the box is where you want it; or
(B) push the box and when it is moving keep pushing the box across the floor.

Skills in Action

9. A block of mass m slides down a rough incline with constant speed. If a similar block that has a mass of $4m$ were sliding down the same incline, it would
(A) slide down at a constant speed.
(B) accelerate down the incline.
(C) slowly slide down the incline and then stop.
(D) slide down with an increasing acceleration.

10. Wax is applied by a skier to the bottom surface of a cross-country ski to optimize performance for different snow surfaces. Kicking or poling up even a gentle slope can be arduous with the wrong wax on the surface of the ski. To make a quick estimate of the reduction in the kinetic friction coefficient from an application of wax to her skis, the skier pushes the waxed and unwaxed skis forward over a horizontal snow surface with the same initial velocity. Which of the following best describes her method of estimation?
(A) The ratio of distances where each ski stops is proportional to the ratio of coefficients of kinetic friction.
(B) The ratio of distances where each ski stops is inversely proportional to the ratio of coefficients of kinetic friction.
(C) The square root of the ratio of distances until the skis stop is proportional to the square root of the ratio of coefficients of kinetic friction.
(D) The square root of the ratio of distances until the skis stop is inversely proportional to the square root of the ratio of coefficients of kinetic friction.

5-4 Problems involving friction are solved like any other force problems

In this section we'll look at a number of problems that involve static friction, kinetic friction, or both. We'll solve these problems using the same two-step approach that we introduced in Section 4-6. First, draw a free-body diagram for each object of interest;

second, use the diagram to write down a Newton's second law equation for each object and solve for the unknowns. The only new wrinkle for problems that involve friction is now you must identify the appropriate friction force and choose the correct expressions for either the magnitude of the static friction force ($F_S \le F_{S,max} = \mu_S F_n$) or the kinetic friction force ($F_k = \mu_k F_n$).

Exam Tip

It is very important during the AP® Physics 1 exam to be able to differentiate between situations when (1) static friction is exerted on an object: when an object is not moving relative to another or when we are interested in what it would take to get it moving; and (2) when kinetic friction is exerted on an object: when an object is sliding relative to another.

EXAMPLE 5-4 Two Boxes, a String, and Friction

Two boxes are connected by a lightweight string that passes over a pulley of negligible mass and friction (**Figure 5-12**). Box 1 (mass $m_1 = 2.00$ kg) sits on a ramp inclined at 30.0° from the horizontal, while box 2 (mass $m_2 = 4.00$ kg) hangs from the other end of the string. When the boxes are released, the string remains taut, box 1 accelerates up the ramp, and box 2 accelerates downward. The string does not stretch. The coefficient of kinetic friction for box 1 sliding on the ramp is $\mu_k = 0.250$. What is the acceleration of each box?

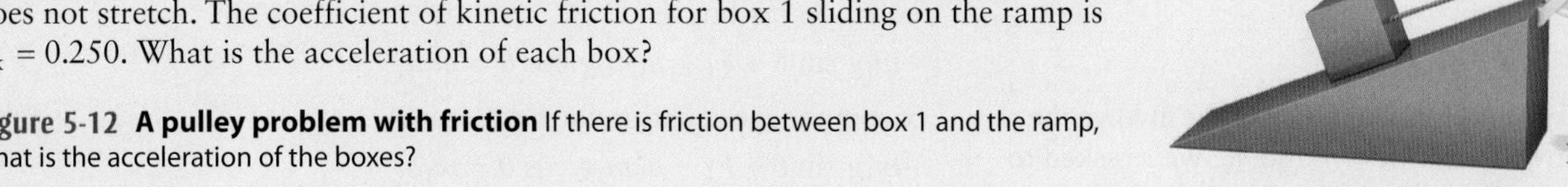

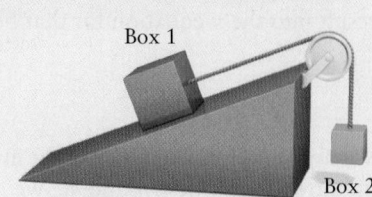

Figure 5-12 A pulley problem with friction If there is friction between box 1 and the ramp, what is the acceleration of the boxes?

Set Up

Because the direction of the friction force is dependent on the direction of motion, it is helpful to ask ourselves some questions to ensure that we get the direction of motion correct. We start by considering in which direction each nonfriction force would cause the system to accelerate. The force of gravity would pull box 2 straight downward. The component of the force of gravity along the ramp would pull box 1 down the ramp. (The forces perpendicular to the ramp must cancel.) The tension forces mean the blocks move together. Because box 2 has a mass greater than box 1 by a larger factor than m_1 plus the coefficient of friction times m_1, the boxes will move in the direction determined by the force of gravity on box 2. So we expect that box 2 will fall, accelerating downward and pulling box 1 up the ramp. Let's, then, choose to have the positive x axis point up the ramp for box 1 and straight down for box 2. Box 1 has four forces exerted on it: a normal force $\vec{F}_{n,1}$ from the surface of the ramp, its weight $\vec{F}_{g,1}$, the tension force from the string $\vec{F}_{T,string\ on\ 1}$, and the kinetic friction force from the ramp $\vec{F}_{k,1}$. Two forces are exerted on box 2: its weight $\vec{F}_{g,2}$ and the tension force from the string $\vec{F}_{T,string\ on\ 2}$. Because the string is lightweight and the pulley is of negligible mass and friction, the tensions $\vec{F}_{T,string\ on\ 1}$ and $\vec{F}_{T,string\ on\ 2}$ at either end have the same magnitude F_T. If the string doesn't stretch, the two boxes move together and their accelerations $\vec{a}_1$ and $\vec{a}_2$ have the same x component a_x. (Note that the positive x direction for each box was chosen to be the same as the direction that the box moves.) Our goal is to calculate a_x.

Newton's second law applied to box 1:

$$\sum \vec{F}_{ext\ on\ 1} = \vec{F}_{n,1} + \vec{F}_{g,1} + \vec{F}_{T,string\ on\ 1} + \vec{F}_{k,1} = m_1\vec{a}_1$$

Magnitude of the kinetic friction force on box 1:

$$F_{k1} = \mu_k F_{n,1} \quad (5\text{-}6)$$

Newton's second law applied to box 2:

$$\sum \vec{F}_{ext\ on\ 2} = \vec{F}_{g,2} + \vec{F}_{T,string\ on\ 2} = m_2\vec{a}_2$$

Tensions have the same magnitude:

$$F_{T,string\ on\ 1} = F_{T,string\ on\ 2} = F_T$$

Accelerations have the same x component:

$$a_{1x} = a_{2x} = a_x$$

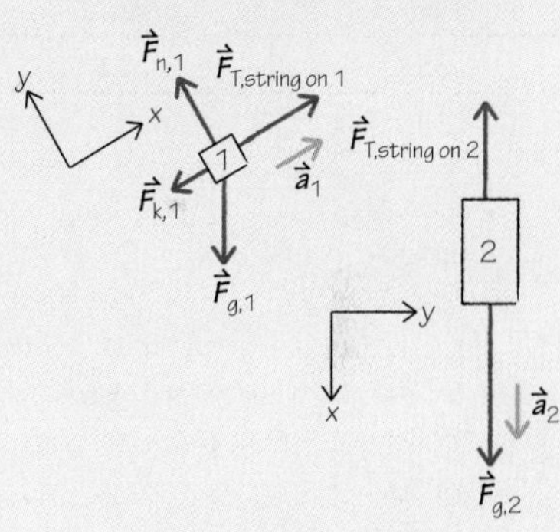

Solve

Draw the free-body diagrams for each box and then sketch in the components as dashed arrows to use them to write Newton's second law in component form for each box. Use F_T for the magnitude of each tension and a_x for the x component of each acceleration. Box 1 doesn't move in the y direction, so its y component of acceleration is zero and the net external force on box 1 has a zero y component. It helps to redraw the free-body diagram for box 1, so we can find the components of the gravitational force, which does not lie along either of our axes.

Newton's second law in component form for box 1:

$$x: 0 + (-F_{g,1} \sin \theta) + F_T + (-F_k) = m_1 a_x$$

$$y: F_{n,1} + (-F_{g,1} \cos \theta) + 0 + 0 = 0$$

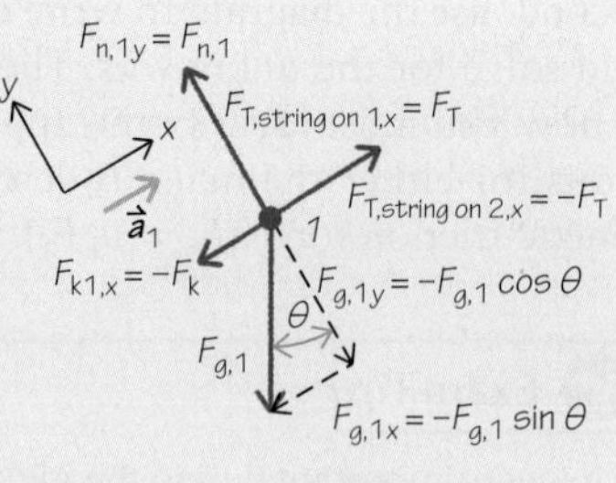

Newton's second law in component form for box 2:

$$x: F_{g,2} + (-F_T) = m_2 g - F_T = m_2 a_x$$

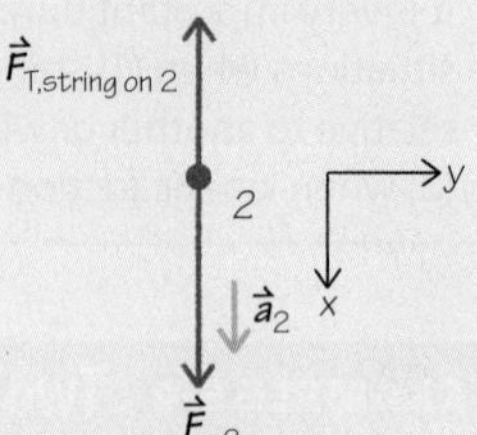

Calculate the normal force $F_{n,1}$ exerted on box 1 from the y component equation for that box. Use this expression to find the magnitude of the kinetic friction force on box 1 and substitute the result into the x equation for that box.

From the y component equation for box 1:

$$F_{n,1} = F_{g,1} \cos \theta = m_1 g \cos \theta$$

So the kinetic friction force on box 1 has magnitude

$$F_{k,1} = \mu_k F_{n,1} = \mu_k m_1 g \cos \theta$$

Substitute this expression and $F_{g,1} = m_1 g$ into the x component equation for box 1:

$$-m_1 g \sin \theta + F_T - \mu_k m_1 g \cos \theta = m_1 a_x$$

We now have two equations that involve the unknown quantities F_T and a_x. We're asked to find only the value of a_x, so we eliminate F_T between the two equations.

x component equation for box 1:

$$-m_1 g \sin \theta + F_T - \mu_k m_1 g \cos \theta = m_1 a_x$$

x component equation for box 2:

$$m_2 g - F_T = m_2 a_x$$

Solve for F_T using the x component equation for box 1:

$$F_T = m_1 g \sin \theta + \mu_k m_1 g \cos \theta + m_1 a_x$$

Substitute this into the x component equation for box 2:

$$m_2 g - m_1 g \sin \theta - \mu_k m_1 g \cos \theta - m_1 a_x = m_2 a_x$$

Rearrange and solve for a_x:

$$m_2 g - m_1 g \sin \theta - \mu_k m_1 g \cos \theta = m_1 a_x + m_2 a_x$$

$$(m_2 - m_1 \sin \theta - \mu_k m_1 \cos \theta) g = (m_1 + m_2) a_x$$

$$a_x = \frac{(m_2 - m_1 \sin \theta - \mu_k m_1 \cos \theta)}{(m_1 + m_2)} g$$

$$= \frac{[4.00 \text{ kg} - (2.00 \text{ kg}) \sin 30.0° - (0.250)(2.00 \text{ kg}) \cos 30.0°]}{(2.00 \text{ kg} + 4.00 \text{ kg})}$$

$$\times (9.80 \text{ m/s}^2)$$

$$= 4.19 \text{ m/s}^2$$

Reflect

If there were no friction to slow the boxes down, we would expect the acceleration to be greater. We can check this conclusion by looking at the case $\mu_k = 0$ (which implies zero kinetic friction). Sure enough, a_x is greater in this case.

If there were no friction between box 1 and the ramp:

$\mu_k = 0$ so

$$a_x = \frac{(m_2 - m_1 \sin \theta)}{(m_1 + m_2)} g$$

$$= \frac{[4.00 \text{ kg} - (2.00 \text{ kg}) \sin 30.0°]}{(2.00 \text{ kg} + 4.00 \text{ kg})} (9.80 \text{ m/s}^2)$$

$$= 4.90 \text{ m/s}^2$$

Extend: Returning to the case where friction is present, you can imagine that if box 1 were sufficiently more massive than box 2, the kinetic friction force would be so great that the boxes would slow down rather than speed up as they move and would eventually come to a stop because the static friction force is larger. The mass would have to be significantly greater to cause the system to move in the other direction. In that case a_x would be negative. Using $m_2 = 4.00$ kg and $\mu_k = 0.250$, can you show that a_x would be negative if m_1 were greater than 5.58 kg?

NOW WORK Problems 1–4 in The Takeaway 5-4.

EXAMPLE 5-5 Pinned Against a Wall

You place a block of plastic that weighs 33.0 N against a vertical wall and push it toward the wall with a force of 55.0 N (**Figure 5-13**). The coefficients of static and kinetic friction between the plastic and the wall are $\mu_S = 0.420$ and $\mu_k = 0.400$, respectively. Does the block remain at rest? If not, with what acceleration does it slip down the wall? Neglect any friction between you and the block (you don't impede its downward motion; you only are pushing it toward the wall).

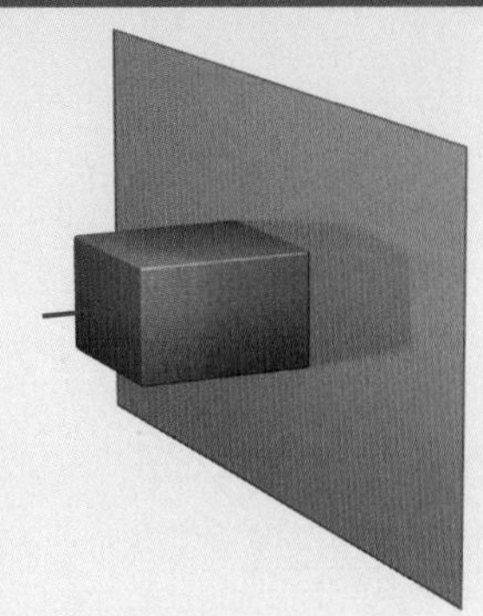

Figure 5-13 Pushing a block against a wall If you push a block against a wall with a force of a given magnitude, will the block remain at rest or slide down the wall?

Set Up

Four forces are exerted on the block: the gravitational force $\vec{F}_{g,block}$, the force $\vec{F}_{you\ on\ block}$ you exert to push it against the wall, the normal force $\vec{F}_{n,block}$ that the wall exerts, and the upward friction force $\vec{F}_f$ (which opposes the downward gravitational force). The block will remain at rest only if the force of static friction is large enough to balance the 33.0-N gravitational force.

We'll use Equation 5-1b to determine the maximum force of static friction, which will help us decide whether the block will slip. If the block slips, we'll use Equation 5-6 for the force of kinetic friction to determine its acceleration. Draw the free-body diagram and use it to write Newton's second law for the block.

Newton's second law applied to the block:

$$\sum \vec{F}_{ext\ on\ block} = \vec{F}_{g,block} + \vec{F}_{you\ on\ block} + \vec{F}_{n,block} + \vec{F}_f = m_{block}\vec{a}_{block}$$

Magnitude of the static friction force:

$$F_{S,max} = \mu_S F_n \qquad (5\text{-}1b)$$

Magnitude of the kinetic friction force:

$$F_k = \mu_k F_n \qquad (5\text{-}6)$$

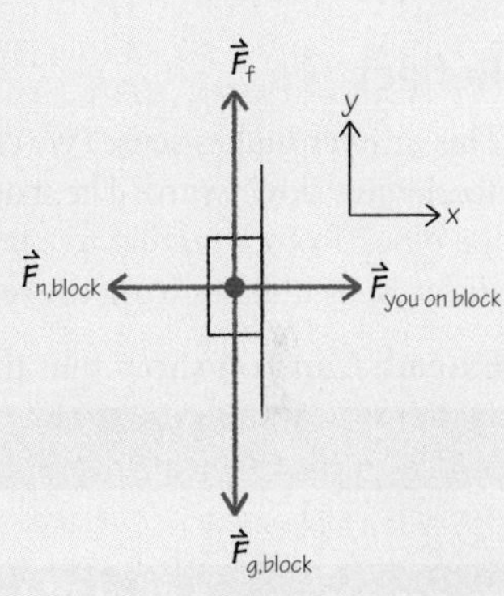

Solve

We use Newton's second law to determine the normal force that the wall exerts on the block and Equation 5-1b to find the maximum static friction force available. Only 23.1 N of static friction is available, which is less than the gravitational force on the block, so we conclude that the block will slip.

Newton's second law in component form for the block, assuming the block remains at rest (so the friction force is static friction):

x: $F_{you\ on\ block} + (-F_{n,block}) = 0$

y: $F_S + (-F_{g,block}) = 0$

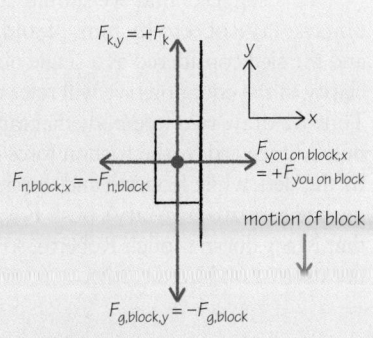

From the x component equation:

$$F_{n,block} = F_{you\ on\ block} = 55.0\text{ N}$$

From Equation 5-1b, the maximum static friction force available is

$$F_{S,max} = \mu_S F_{n,block} = (0.420)(55.0\text{ N}) = 23.1\text{ N}$$

From the y component equation, the required static friction force is

$$F_S = F_{g,block} = 33.0\text{ N}$$

This force is more than the maximum static friction force available, so the block can't remain at rest.

We again apply Newton's second law to the block to find its acceleration as it slides down the wall. Again we find the normal force that the wall exerts on the block, and we use this value to find the kinetic friction force. Note that we are given the block's weight, not its mass, so we have to calculate its mass m_{block}.

Newton's second law in component form for the block, assuming the block is sliding downward (so the friction force is kinetic friction):

x: $F_{you\ on\ block} + (-F_{n,block}) = 0$

y: $F_k + (-F_{g,block}) = m_{block}a_{block,y}$

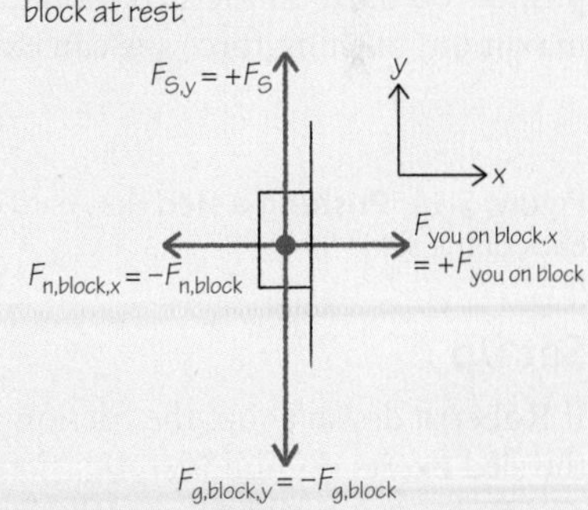

From the x component equation:

$$F_{\text{n,block}} = F_{\text{you on block}} = 55.0 \text{ N}$$

From Equation 5-6:

$$F_{\text{k}} = \mu_{\text{k}} F_{\text{n,block}} = (0.400)(55.0 \text{ N}) = 22.0 \text{ N}$$

Find the mass of the block:

$$F_{\text{g,block}} = m_{\text{block}} g$$

$$m_{\text{block}} = \frac{F_{\text{g,block}}}{g} = \frac{33.0 \text{ N}}{9.80 \text{ m/s}^2} = 3.37 \text{ kg}$$

Substitute into the y equation and solve for $a_{\text{block},y}$:

$$a_{\text{block},y} = \frac{F_{\text{k}} - F_{\text{g,block}}}{m_{\text{block}}} = \frac{22.0 \text{ N} - 33.0 \text{ N}}{3.37 \text{ kg}} = -3.27 \text{ m/s}^2$$

Reflect

This answer makes sense. We chose the positive y direction to be upward, so the negative value of $a_{\text{block},y}$ means that the block accelerates downward. The magnitude of the acceleration is somewhat less than $g = 9.80 \text{ m/s}^2$ because kinetic friction prevents the block from attaining free fall. The direction of the kinetic friction force is always directed to resist the relative motion of the surfaces, as it arises from the repulsive electric force between electrons. It doesn't matter if the surface is horizontal or vertical.

Extend: Can you show that the block would remain at rest if you pushed with a force of at least 78.6 N?

NOW WORK Problem 5 in The Takeaway 5-4.

EXAMPLE 5-6 A Sled Ride

Two children, Roberto (mass 35.0 kg) and Mary (mass 30.0 kg), go out sledding one winter day. Roberto sits on the sled of mass 5.00 kg, and Mary gives the sled a forward push (**Figure 5-14**). The coefficients of friction between Roberto and the upper surface of the sled are $\mu_{\text{S}} = 0.300$ and $\mu_{\text{k}} = 0.200$; the friction force between the sled and the icy ground is so small that we can ignore it. Mary finds that if she pushes too hard on the sled, Roberto slides toward the back of the sled. What is the maximum pushing force she can exert without this happening?

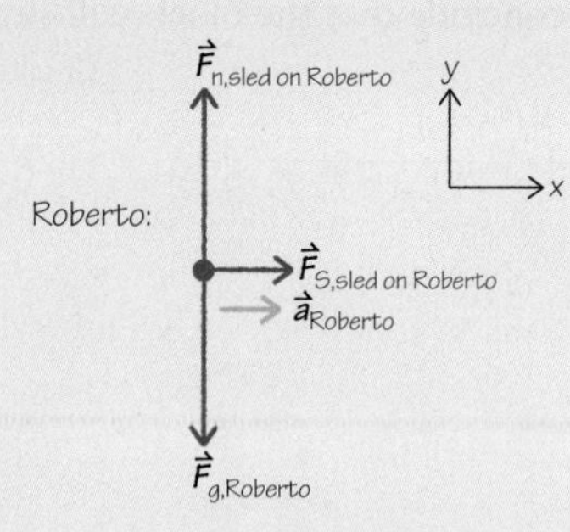

Figure 5-14 Pushing a sled How hard can Mary push without making Roberto slide toward the back of the sled?

Set Up

If Roberto doesn't slip, the friction force that the sled exerts on him is *static* friction. (The sled and Roberto are sliding over the ice, but they're not sliding relative to each other.) The maximum force of static friction sets a limit on how great the forward acceleration of Roberto and the sled together can be.

This suggests that we should consider two objects: (a) Roberto by himself and (b) Roberto and the sled considered as a single object, which for clarity in the equations we will refer to as a "unit." Thus we draw two free-body diagrams. Roberto is pushed forward by the friction force exerted on him by the sled, while Roberto and the sled together are pushed forward by Mary's push $\vec{F}_{\text{Mary on unit}}$. (Note that Mary doesn't touch Roberto, so her push is not exerted on him directly.)

Newton's second law equation for Roberto:

$$\sum \vec{F}_{\text{ext on Roberto}} = \vec{F}_{\text{n,sled on Roberto}} + \vec{F}_{\text{g,Roberto}} + \vec{F}_{\text{S,sled on Roberto}} = m_{\text{Roberto}} \vec{a}_{\text{Roberto}}$$

Newton's second law equation for Roberto and the sled considered as a single object (unit):

$$\sum \vec{F}_{\text{ext on unit}} = \vec{F}_{\text{n,ground on unit}} + \vec{F}_{\text{g,unit}} + \vec{F}_{\text{Mary on unit}} = m_{\text{unit}} \vec{a}_{\text{unit}}$$

Roberto:
$\vec{F}_{\text{n,sled on Roberto}}$
$\vec{F}_{\text{S,sled on Roberto}}$
$\vec{a}_{\text{Roberto}}$
$\vec{F}_{\text{g,Roberto}}$
y
x

We'll determine the maximum acceleration that will prevent Roberto from slipping and then use the result to find the maximum forward force that Mary can exert. As long as the pushing force she exerts on the Roberto–sled object is less than this maximum, Roberto and the sled will move together through the same displacement as they are pushed, so treating them as a single object is valid. If Mary pushes harder they will no longer move together exactly and we would have to treat them as a system.

Magnitude of the static friction force:

$$F_{S,max} = \mu_S F_n \qquad (5\text{-}1b)$$

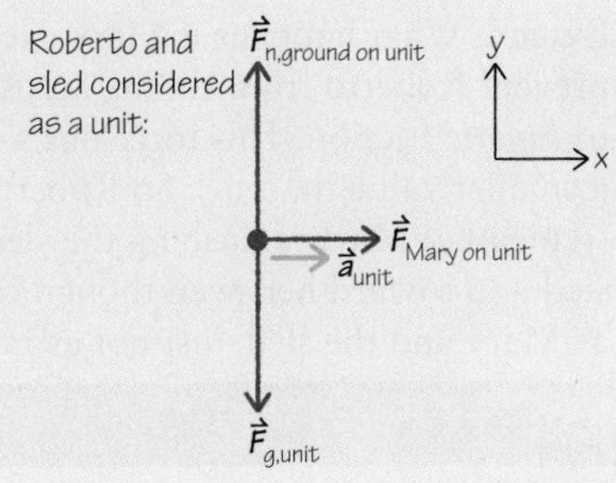

Solve

Let's take positive x to be in the direction of motion, and positive y upward, as shown near the free-body diagrams above. First let's consider just Roberto and the forces that are exerted on him. He is pushed forward by the static friction force that the sled exerts on him. If Roberto is just about to slip, the static friction force has its maximum value. To determine this value we first find the normal force that the sled exerts on Roberto.

Newton's second law in component form for Roberto, assuming that the static friction force has its maximum value:

x: $F_{S,max\ of\ sled\ on\ Roberto} = m_{Roberto} a_{Roberto,x}$

y: $F_{n,sled\ on\ Roberto} + (-F_{g,Roberto}) = 0$

From the y component equation:

$$F_{n,sled\ on\ Roberto} = F_{g,Roberto} = m_{Roberto} g$$

So the maximum force of static friction that the sled exerts on Roberto is

$$F_{S,max\ of\ sled\ on\ Roberto} = \mu_S F_{n,sled\ on\ Roberto} = \mu_S m_{Roberto} g$$

So from the x component equation, Roberto's maximum acceleration is

$$a_{Roberto,x} = \frac{F_{S,max\ of\ sled\ on\ Roberto}}{m_{Roberto}} = \frac{\mu_S m_{Roberto} g}{m_{Roberto}} = \mu_S g$$

As long as the acceleration does not exceed this value Roberto does not slip, which means he and the sled have the same acceleration. So we can treat Roberto and the sled as a single object that accelerates forward due to Mary's push. (We don't have to consider the forces that Roberto and the sled exert on each other. That's because these are *internal* forces.) Because we know Roberto's (and the sled's) maximum acceleration that will prevent slipping, we can calculate the maximum force that Mary can exert on the Roberto–sled object, which we will again refer to as a "unit" for the equations and diagrams.

Newton's second law in component form applied to the Roberto–sled unit:

x: $F_{Mary\ on\ unit} = m_{unit} a_{unit,x}$

y: $F_{n,ground\ on\ unit} + (-F_{g,unit}) = 0$

If Roberto has his maximum acceleration and does not slip on the sled,

$$a_{unit,x} = a_{Roberto,x} = \mu_S g$$

So the maximum force that Mary can exert without causing Roberto to slip is

$$\begin{aligned} F_{Mary\ on\ unit} &= m_{unit}\mu_S g \\ &= (m_{Roberto} + m_{sled})\mu_S g \\ &= [(35.0\ \text{kg}) + (5.00\ \text{kg})](0.300)(9.80\ \text{m/s}^2) \\ &= 118\ \text{kg}\cdot\text{m/s}^2 = 118\ \text{N} \end{aligned}$$

Reflect

This force is based on an acceleration of about one-third that of gravity. The magnitude of the force would be about that required to lift a 12-kg (26-lb) object against gravity, which seems like something a relatively large child could do with some effort. This may be a larger push than she would really give him as she is trying to keep her own footing as she runs through the snow, so he would move along with the sled fine, which is what usually happens if you have ever tried this or seen it in a movie!

Extend: What happens if Mary exerts a force greater than 118 N on the sled? There will not be enough static friction to prevent Roberto from sliding across the top of the sled. The net force on him will still be forward, but now it will be due to *kinetic* friction. This force has a smaller magnitude than the maximum force of static friction (we are told that μ_k has a smaller value than μ_S). So Roberto's forward acceleration will be less than that of the sled. From Mary's perspective (remember Mary is pushing the sled so she must be running along with it), Roberto will slide backward relative to the sled and toward her, even though to an observer on the ground, Roberto would still be accelerating in the same direction as Mary and the sled, just not as fast.

NOW WORK Problems 6 and 7 in The Takeaway 5-4.

Exam Tip

The examples in this section should be solved by starting with a free-body diagram. Drawing a free-body diagram is a basic skill you should expect to be tested on during the AP® Physics 1 exam. It is important to get the mechanics of the diagram correct as well as understand the nature of each of the possible forces on an object. Note that if the friction changes from static to kinetic within a problem, then that would be represented by two different free-body diagrams, with the kinetic friction force represented by a shorter arrow, since it is always smaller.

THE TAKEAWAY for Section 5-4

✔ If friction has to be included in a problem that involves forces, approach the problem using the same steps as described in Section 4-6: Draw a free-body diagram for each object of interest; then use the diagram to write a Newton's second law equation for each component of the motion for each object.

✔ If an object is not sliding, use Equations 5-1 to describe the static friction force. If the problem asks for a limiting case such as "when the object just begins to slip," assume the static friction force is maximum.

✔ If an object is sliding, use Equation 5-6 to describe the kinetic friction force.

Prep for the AP Exam

AP® Building Blocks

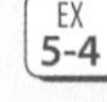
EX 5-4 1. A student exerts a horizontal force $\vec{F}$ on a stationary block with a mass M of 2.0 kg as shown. The coefficient of static friction between the block and the floor is 0.75; the coefficient of kinetic friction is 0.45.

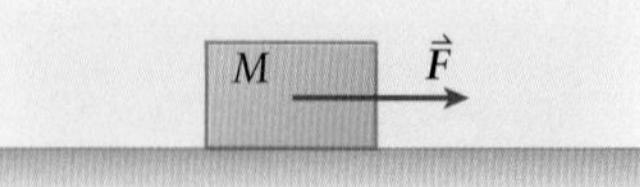

(a) If the magnitude of the horizontal force, $\vec{F}$, is 10.0 N, what is the acceleration of the box?
(A) 0 N/kg (B) 2.5 N/kg
(C) 3.8 N/kg (D) 5.6 N/kg

(b) If the magnitude of the horizontal force, $\vec{F}$, is 20.0 N, what is the acceleration of the box?
(A) 0 N/kg (B) 2.5 N/kg
(C) 3.8 N/kg (D) 5.6 N/kg

EX 5-4 2. Two blocks are connected over a pulley. Neglect the mass of the string and friction in the pulley. The mass of block 2 is 8.00 kg, and the coefficient of kinetic friction between block 2 and the incline is 0.22. The angle θ of the incline is 28.0°. Block 2 slides up the incline at constant speed.

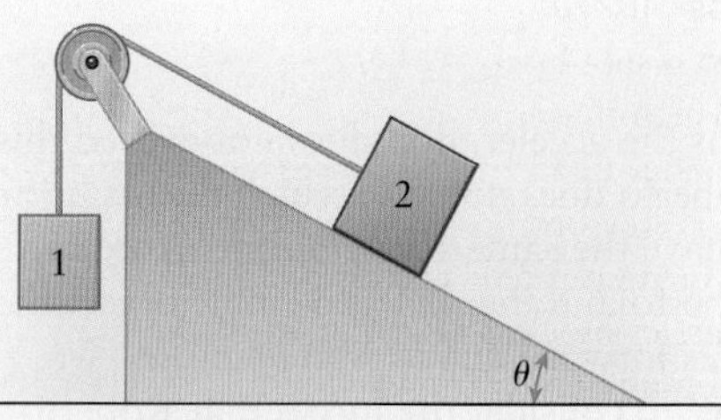

(a) Construct a free-body diagram for each block.
(b) Using the free-body diagrams you created in part (a), write equations for the acceleration of each block based on Newton's second law. Add an appropriate coordinate system near each free-body diagram.
(c) Solve the system of equations in part (b) for the mass of block 1.

EX 5-4 3. A taut string of negligible mass connects a crate with mass M_1 = 5.0 kg to a crate with mass M_2 = 12.0 kg. The coefficient of static friction between the smaller crate and the floor is 0.57; the coefficient of static friction between the larger crate and the floor is 0.43.

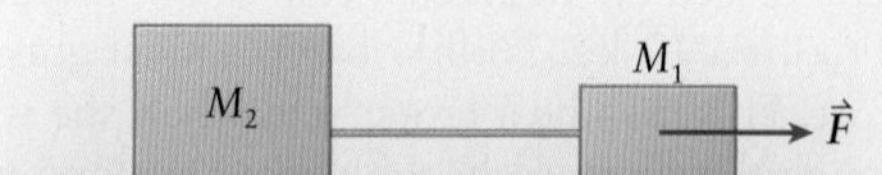

(a) Construct a free-body diagram for each crate and for the model in which the system consisting of both blocks and the string is treated as a single object.

(b) Write equations based on Newton's second law for the horizontal forces on each crate.
(c) Calculate the minimum tension force in the string and the minimum force, F, required to slide the crate with mass M_2.

4. A block with mass M_1 of 2.85 kg moves along an inclined plane that makes an angle θ of 40.0° with the horizontal, and is connected by a string that passes over a pulley to a second block that has a mass M_2 of 4.75 kg. Both blocks are composed of the same material. The string and pulley have negligible mass and the pulley exerts no friction force on the string. The coefficient of kinetic friction between the plane and the block is 0.55.

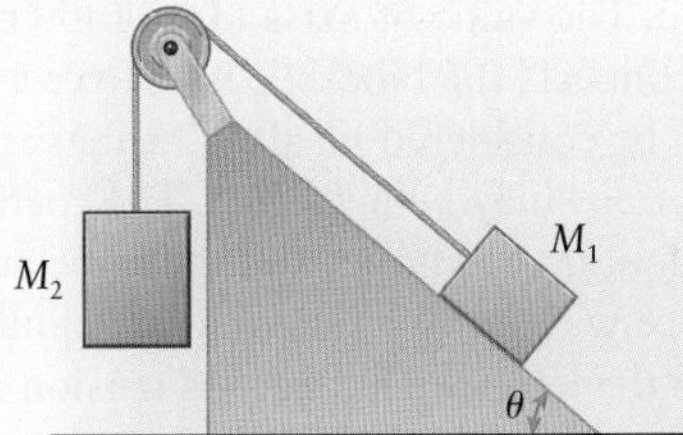

(a) Construct a free-body diagram for each block.
(b) Write equations based on Newton's second law for each block.
(c) Calculate the magnitude and direction of the acceleration of the blocks.
(d) Calculate the magnitude of the tension in the string.
(e) The blocks are swapped. Repeat parts (a)–(c) for the new configuration. If anything unusual happens in your calculations, explain why it may make sense.

5. A large block is being pushed against a smaller block such that the smaller block remains elevated while being pushed, as shown in the figure. The mass of the smaller block is $m = 0.75$ kg. The blocks need to have a minimum acceleration of $a = 15 \text{ m/s}^2$ in order for the smaller block to remain elevated and not slide down.

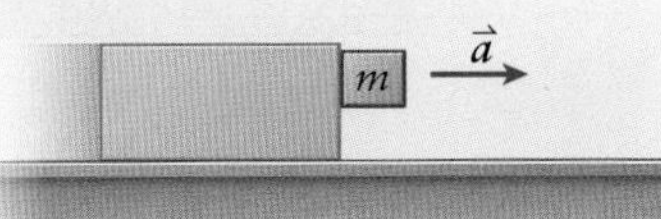

(a) Express the condition that the friction force on the small block must satisfy.
(b) Calculate the coefficient of static friction between the two blocks.

6. A box of mass $M_{box} = 2.00$ kg rests on top of a crate with mass $M_{crate} = 5.00$ kg. The coefficients of static and kinetic friction between the box and the crate are 0.67 and 0.50, respectively. The coefficients of static and kinetic friction between the crate and the floor are 0.40 and 0.30, respectively. An external force is exerted on the crate. A rope of negligible mass connects the box to a stationary wall. Assume the rope can be spooled out from a pulley to keep the tension it exerts on the box constant.

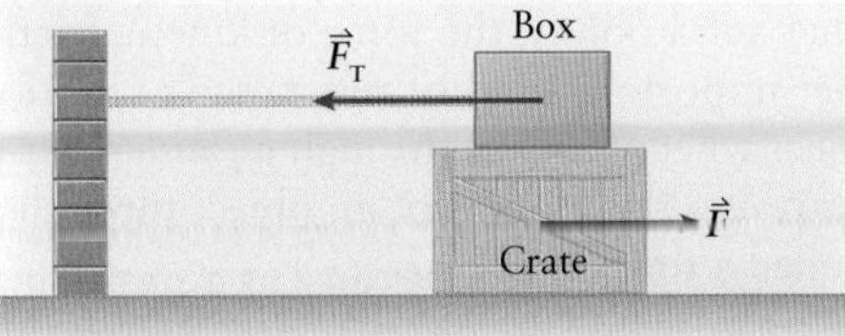

(a) Construct free-body diagrams for the box and the crate modeled as separate objects, and for the model in which the system of box/crate is treated as a single object.
(b) Apply Newton's second law for the horizontal forces on each object.
(c) Express the condition that must be met to justify the claim that the crate and box can be treated as a single object.
(d) Calculate the largest magnitude for $\vec{F}_T$ that will satisfy the condition identified in part (c), and the value of $\vec{F}$ for that value of $\vec{F}_T$.
(e) How would the tension have to change to allow a larger pulling force $\vec{F}$?

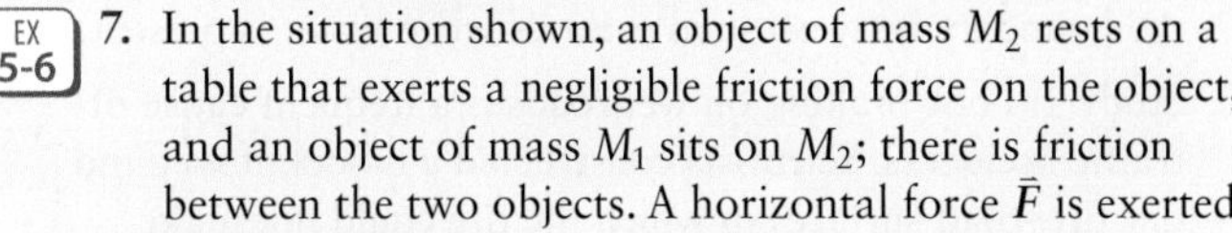

7. In the situation shown, an object of mass M_2 rests on a table that exerts a negligible friction force on the object, and an object of mass M_1 sits on M_2; there is friction between the two objects. A horizontal force $\vec{F}$ is exerted on the lower object.

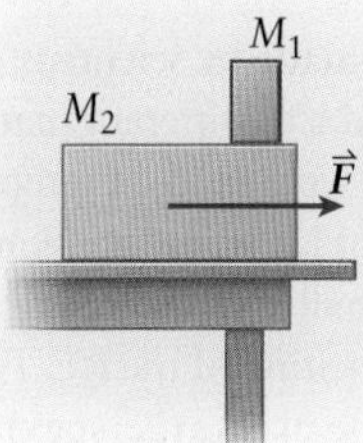

(a) Express the conditions that must be satisfied by accelerations and the force $\vec{F}$ if the system of two blocks is to be treated as a single object.
(b) Assuming that the condition identified in part (a) is met, express the acceleration of the object.
(c) Assuming that the condition identified in part (a) is not met, construct a free-body diagram for each block.
(d) Apply Newton's second law to each block and Newton's third law for the horizontal pair forces on blocks 1 and 2 to derive the accelerations of the two blocks.
(e) Assuming that the condition identified in part (a) is not met, qualitatively predict the relative acceleration, $a_2 - a_1$, when $M_1 \ll M_2$ and the force exerted on the block 2 is very large.

Skill Builders

8. The data shown were collected with a force probe attached to a 458-g copper block initially at rest on a horizontal wooden surface for three different trials (open and filled-in circles and dots).

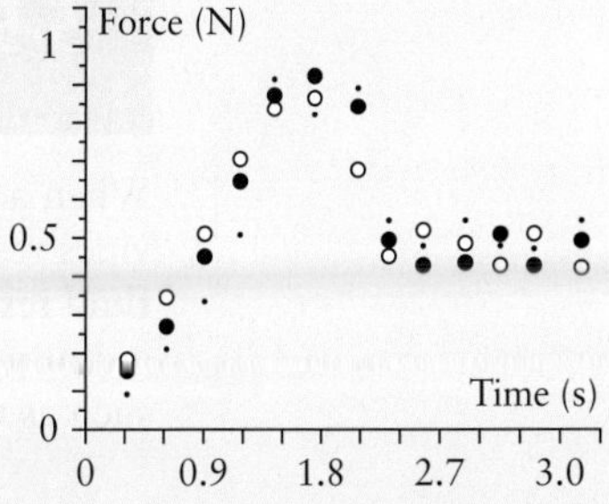

(a) Interpret the representation to determine the time intervals during which the block was (i) at rest and (ii) sliding. Report intervals in the form $[t_{min}, t_{max}]$.
(b) Using the data, calculate a numerical value for the coefficient of static friction and the coefficient of kinetic friction for the copper–wood interface. Report values in the form average ± uncertainty.
(c) Describe a change in the design of this data collection that would improve the precision of measurement for each trial.

9. As you read the following two studies of friction forces, pose three scientific questions whose pursuit might connect these studies and lead to explanations of friction forces at the boundary between solid surfaces when water is present.

Study 1: Poor braking on wet roads is a frequent cause of traffic accidents. Scientists constructed a model of wet and dry tire–road surfaces in which the tire could not make contact with the road because of pools of water filling small pits in the road and smoothing the surface.

Study 2: A study was made by scientists of the friction forces between ice skates and ice. Different materials from which skates could be made were used, at temperatures below the melting temperature of ice where ice skates are used. The pressure due to the small surface area of the skates touching the ice creates some melting of the ice. Two metals (copper and mild steel) and one plastic (Perspex) were considered. Mild steel is a less expensive steel alloy with a small amount of carbon. Data in the graph were presented.

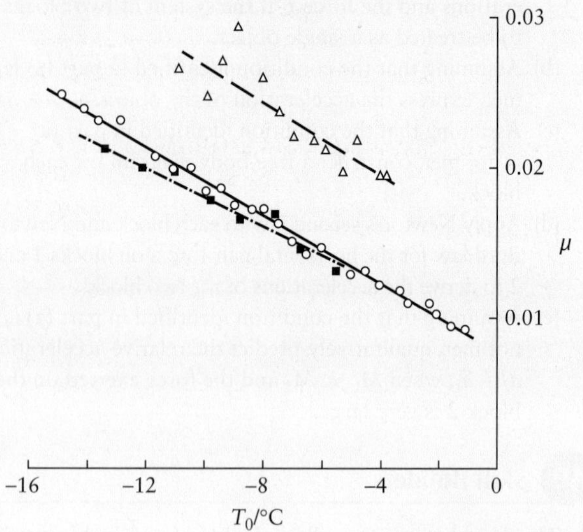

The graph shows the variation of coefficient of friction with air temperature for various rod materials. Velocity is 3.16 m/s and total load is 45.5 N. (Δ) copper rods, (■) Perspex rods, and (o) mild steel rods.

AP® Skills in Action

10. A 10-kg crate is placed on a horizontal conveyor belt moving with a constant speed. The crate does not slip. If the coefficients of friction between the crate and the belt are $\mu_s = 0.50$ and $\mu_k = 0.30$, what is the magnitude of the friction force exerted on the crate?

11. A block of mass M rests on a block of mass M_1, which is on a tabletop. The value of mass M_1 is known. A light string, which connects the blocks, passes over a pulley. The pulley can be considered ideal, so its mass or any friction interactions may be neglected. The coefficient of kinetic friction, μ_k, is known and the same at both surfaces. A force with known magnitude, F, pulls the upper block to the left, which creates a tension in the string, which then pulls the lower block to the right. The blocks are moving at a constant speed. Derive an expression for the mass M in terms of the given values of M_1, F, and μ_k.

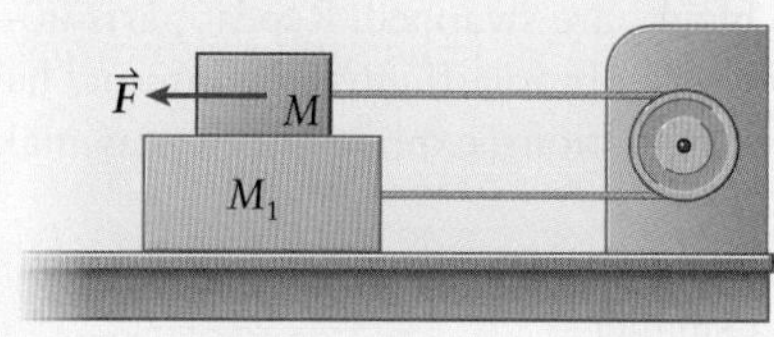

5-5 An object moving through air or water experiences a drag force

When a solid object slides over another solid object, the force of kinetic friction provides resistance to its motion. Another important kind of resistance to motion is the **fluid resistance** experienced by an object when it moves through a *fluid*—that is, a liquid or gas (**Figure 5-15**). The force that resists the motion of an object through a liquid such as water or a gas such as air is called a **drag force.**

(a)

This sea turtle feels a backward force of fluid resistance (drag) as it moves through the water.

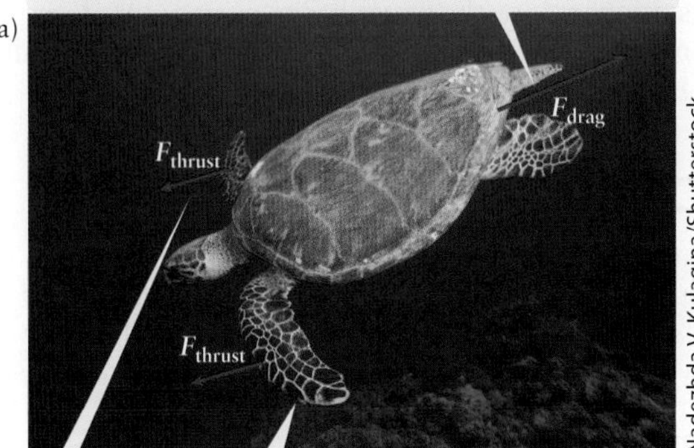

To maintain a constant forward velocity, there must be a balancing forward force exerted on the turtle. To create this forward force, the turtle pushes backward on the water with its flippers. The water responds by exerting a forward force on the flippers (Newton's third law).

(b)

These skydivers feel an upward force of fluid resistance (drag) as they move downward through the air.

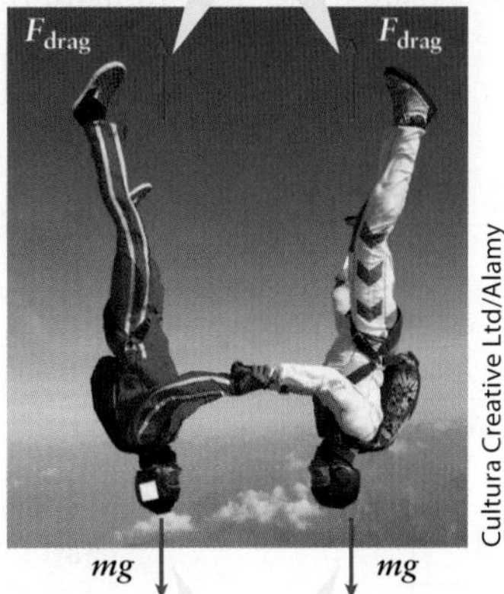

In this case, the fluid resistance opposes the force due to gravity. This slows their fall, giving them more time to enjoy the descent. If the drag force exactly balances the force of gravity (it cannot be bigger), the skydivers stop accelerating and move at a constant velocity.

Figure 5-15 Fluid resistance When an object moves through either (a) a liquid such as water or (b) a gas such as air, it experiences fluid resistance (a drag force).

Drag Force on Small, Slow-Moving Objects

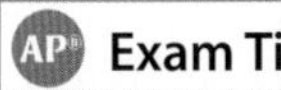

Exam Tip

On the AP® Physics 1 exam, it is important to understand when any force, including the drag force, might apply, but the nature and quantitative application of the drag force is not content that is tested.

Unlike the force of kinetic friction, the drag force depends on the object's speed v as it moves through the fluid. The faster the object's speed, the greater the magnitude of the drag force. For a very slow-moving object that is very small, such as a living cell, a dust particle, or pollen, the magnitude of the drag force is directly proportional to the speed v:

Magnitude of the **drag force** on a **small object** moving at a **low speed**

$$F_{drag} = bv \quad (5\text{-}8)$$

Constant that depends on the properties of the object and of the fluid

Speed of the object relative to the fluid

EQUATION IN WORDS
Drag force for small objects at low speeds

The value of b in Equation 5-8 depends on the size, shape, and surface characteristics of the object and the properties of the liquid or gas through which the object is moving. As an example, for an algal spore (which has a spherical shape) of radius 0.020 mm moving in water, the coefficient b is 3.8×10^{-7} Ns/m. At a speed of 5.0×10^{-5} m/s, the drag force exerted on this spore by the water has magnitude

$$F_{drag} = bv = (3.8 \times 10^{-7}\ \text{Ns/m})(5.0 \times 10^{-5}\ \text{m/s}) = 1.9 \times 10^{-11}\ \text{N}$$

This force is so small that you might think it would have no effect on the spore at all. However, the mass of a spore with this size is only 3.6×10^{-11} kg. If the only force exerted on the spore were the drag force (**Figure 5-16**), the resulting acceleration would be

$$a_{spore,x} = \frac{F_{drag,x}}{m_{spore}} = \frac{-1.9 \times 10^{-11}\ \text{N}}{3.6 \times 10^{-11}\ \text{kg}} = -0.53\ \text{m/s}^2$$

The negative values of $F_{drag,x}$ and $a_{spore,x}$ mean that the drag force opposes the spore's motion, as in Figure 5-16. This substantial acceleration brings the slow-moving spore nearly to a halt relative to the water in a fraction of a second. Since such microscopic spores move very little *relative* to the water around them, they move readily from place to place if the water is itself in motion; the drag force exerted by the moving

Algal spore moving in water

Velocity of spore through water

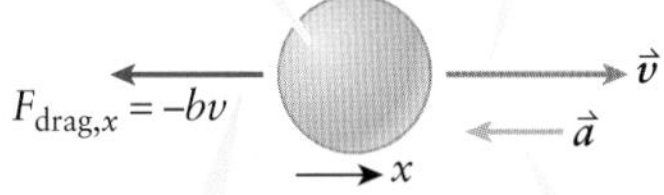

The drag force on the spore is opposite to its velocity relative to the water...

...and therefore the acceleration of the spore is also opposite to the velocity, so the spore slows down.

Figure 5-16 Drag on a microscopic spore A spherical algal spore moves through water in the positive x direction.

water carries the spores along with it (since according to Newton's first law, the spore would just remain still—which would be moving opposite the direction of the water's flow, if we thought of the water as stationary—so the water has to exert a drag force to carry it along). Thus, due to fluid resistance, the location where an algal spore finally germinates and produces a new organism depends crucially on water currents.

Drag Force on Larger Objects

For a larger object moving at a faster speed, such as the sea turtle or skydivers in Figure 5-15, the drag force has a different dependence on the speed v: Its magnitude is approximately proportional to the *square* of v. In this case we can write

EQUATION IN WORDS
Drag force for larger objects at faster speeds

Magnitude of the **drag force** on a **larger object** moving at a **faster speed**

$$F_{\text{drag}} = cv^2 \tag{5-9}$$

Constant that depends on the properties of the object and of the fluid

Speed of the object relative to the fluid

Like the coefficient b in Equation 5-8, the quantity c in Equation 5-9 depends on the fluid through which the object moves and the size, shape, and surface properties of the object. For a baseball flying through the air, $c = 1.3 \times 10^{-3}\ \text{Ns}^2/\text{m}^2$. If the baseball is traveling at 25 m/s (a relatively slow pitching speed for a professional player), the magnitude of the drag force on the ball is

$$F_{\text{drag}} = (1.3 \times 10^{-3}\ \text{Ns}^2/\text{m}^2)(25\ \text{m/s})^2 = 0.81\ \text{N}$$

This drag force is substantial: F_{drag} is a bit more than half the magnitude of the gravitational force on a regulation baseball, which weighs 1.4 N (5 oz). You can see that the drag force plays an important role in baseball and other ball sports.

Giving an object a streamlined shape can lower its value of c. For example, a dolphin (which has a streamlined body) moving in water has a value of c that is only about 1% as large as that of a sphere of the same cross-sectional area. Furthermore, because air is less dense than water, the value of c for a dolphin moving through air is only about 10^{-3} as great as for a dolphin moving through water. Some scientists believe this is one reason why dolphins jump out of the water: The dramatic reduction in drag force that they experience while airborne more than compensates for the effort required to leap clear of the water.

Another species that minimizes its value of the quantity c in Equation 5-9 in order to achieve high speed is the peregrine falcon. This species of predatory bird preys on other, smaller birds, and attacks its prey by diving on it from high altitude. To achieve maximum speed in a dive, which maximizes the impact force it can exert when it strikes its prey, a peregrine falcon streamlines its shape by folding back its wings and tail and tucking in its feet. As a result a diving falcon can attain remarkable speeds, as the following example shows.

EXAMPLE 5-7 Terminal Speed

When it is diving straight down toward its prey, two forces are exerted on a peregrine falcon: a downward gravitational force, and a drag force of magnitude F_{drag} given by Equation 5-9 directed vertically upward (opposite to the direction of the falcon's motion through the air). As the falcon falls and its speed v increases, the value of F_{drag} also increases. When the drag force becomes equal in magnitude to the gravitational force, the net force on the falcon is zero and the falcon ceases to accelerate. It has reached its *terminal speed*, so it no longer speeds up nor does it slow down. Find the terminal speed of a female peregrine falcon of mass 1.2 kg, for which the value of c is $1.6 \times 10^{-3}\ \text{Ns}^2/\text{m}^2$.

Set Up

Draw a free-body diagram, taking positive y to be upward. The sketch shows there are two forces exerted on the falcon. We use Equation 5-9 to find the value of the speed v at which the sum of these forces is zero so that the acceleration is zero and the downward velocity is constant.

$$\sum \vec{F}_{\text{ext on falcon}} = \vec{F}_{\text{drag on falcon}} + \vec{F}_{\text{g,falcon}} = m\vec{a}_{\text{falcon}} = 0$$

Drag force for larger objects at faster speeds:

$$F_{\text{drag}} = cv^2 \quad (5\text{-}9)$$

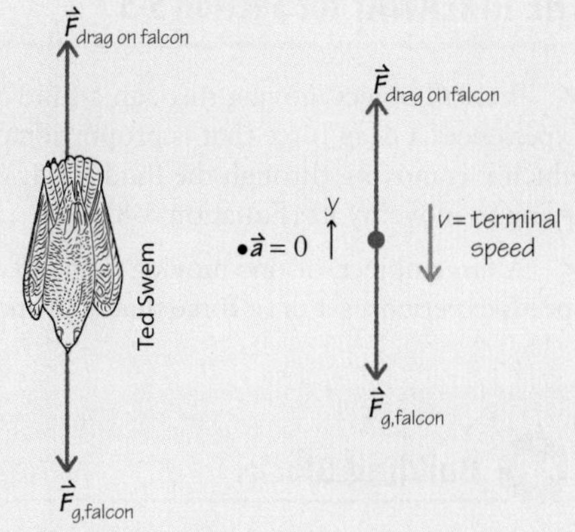

Solve

Using your free-body diagram, write Newton's second law in component form and solve for the terminal speed v_{term}.

Newton's second law in component form applied to the falcon:

$$y: F_{\text{drag on falcon}} + (-F_{\text{g,falcon}}) = 0$$

At the terminal speed v_{term},

$$F_{\text{drag on falcon}} = cv_{\text{term}}^2 \text{ so}$$

$$cv_{\text{term}}^2 - F_{\text{g,falcon}} = 0$$

$$cv_{\text{term}}^2 = F_{\text{g,falcon}} = mg$$

$$v_{\text{term}}^2 = \frac{mg}{c}$$

$$v_{\text{term}} = \sqrt{\frac{mg}{c}}$$

Substitute the numerical values of m and c for the falcon.

Using $m = 1.2$ kg and $c = 1.6 \times 10^{-3}\ \text{Ns}^2/\text{m}^2$,

$$v_{\text{term}} = \sqrt{\frac{(1.2\ \text{kg})(9.80\ \text{m/s}^2)}{1.6 \times 10^{-3}\ \text{Ns}^2/\text{m}^2}}$$

$$= 86\ \text{m/s} = 310\ \text{km/h} = 190\ \text{mi/h}$$

Reflect

The high diving speed attained by the peregrine falcon seems astonishing, but is correct; this type of falcon is the fastest member of the animal kingdom.

Extend: The relationship $v_{\text{term}} = \sqrt{\frac{mg}{c}}$ explains the common notion that heavier objects fall faster. A baseball and an iron ball of the same radius falling side by side in air have the same value of c (because they have the same shape and size), but the iron ball will have a greater terminal speed because its mass m is greater. So a heavier object *does* fall faster if we take the drag force into account. For objects that are not the same shape, the issue is significant because c depends on shape. If you and a mouse dropped from the same height and landed on soft ground, the mouse would scamper away, having fallen much more slowly, whereas you might be hurt. If the baseball and iron ball were dropped side by side in a vacuum, however, they would have the same acceleration of magnitude g and so would always have the same speed.

NOW WORK Problems 1–4 in The Takeaway 5-5.

WATCH OUT !

The drag force is not a constant force.

Unlike other forces we've considered, the magnitude of the drag force on an object changes as the object's speed changes. So the drag force exerted by the fluid on the object and hence the object's acceleration are *not* constant. As an example, a skydiver who falls without opening her parachute reaches half of her terminal speed ($v = v_{\text{term}}/2$) in 3 to 5 s. To achieve nearly terminal speed takes more than twice as long, about 15 to 20 s after beginning the fall. So her acceleration from rest to $v = v_{\text{term}}/2$ is much greater than that from $v = v_{\text{term}}/2$ to $v = v_{\text{term}}$. For variable forces, we cannot apply the constant-acceleration formulas from Chapters 2 and 3.

THE TAKEAWAY for Section 5-5

✔ A small object moving through a fluid at low speed experiences a drag force that is proportional to the speed at which it is moving through the fluid (or at which the fluid is trying to move by it) (Equation 5-8).

✔ A large object, or one moving through a fluid at a fast speed, experiences a drag force that is proportional to the square of the speed at which it is moving through the fluid (or at which the fluid is trying to move by it) (Equation 5-9).

✔ A falling object reaches its terminal speed when the upward drag force exerted on the object by the fluid just balances the downward gravitational force exerted on the object by Earth.

Prep for the AP® Exam

AP® Building Blocks

EX 5-7 1. A single-celled organism called a paramecium propels itself quite rapidly through water by using its hair-like cilia. A certain paramecium experiences a drag force of magnitude $F_{\text{drag}} = cv^2$ in water, where the drag coefficient c is approximately $0.31\ \text{Ns}^2/\text{m}^2$. What propulsion force does this paramecium generate when moving at a constant (terminal) speed v of 1.5×10^{-4} m/s?

2. A 1-kg balsa wood ball and a 1-kg lead ball are dropped simultaneously from a tall tower. Air resistance is present.

(a) Suppose that the two balls have identical sizes, shapes, and surface characteristics. Assume that the drag force is proportional to v^2 with a coefficient of $1.0 \times 10^{-3}\ \text{Ns}^2/\text{m}^2$. The density of balsa wood is $0.1\ \text{g/cm}^3$. The density of lead is about $11\ \text{g/cm}^3$. So, the lead ball must be hollow. Qualitatively predict the relative magnitudes of the accelerations during the fall.
- (A) The accelerations are initially the same and later the lead ball has the larger acceleration.
- (B) The accelerations are initially the same and later the wood ball has the larger acceleration.
- (C) The acceleration of each ball remains constant for the entire fall.
- (D) The accelerations of both balls change but both hit the ground simultaneously.

(b) Suppose that the lead ball is solid and that the two balls have different sizes, but still have the same mass.
- (A) The accelerations are initially the same and later the lead ball has the larger acceleration.
- (B) The accelerations are initially the same and later the wood ball has the larger acceleration.
- (C) The acceleration of each ball remains constant for the entire fall.
- (D) The accelerations of both balls change but both hit the ground simultaneously.

EX 5-7 3. As a skydiver falls faster and faster through the air, does the magnitude of his acceleration increase, decrease, or remain the same?
- (A) The magnitude of the acceleration increases because the drag force increases with speed squared.
- (B) The magnitude of the acceleration decreases because the drag force increases until it is equal to the gravitational force.
- (C) The magnitude of the acceleration remains constant because the skydiver is a falling object.
- (D) The magnitude of the acceleration remains constant because the drag force is equal and opposite to the gravitational force.

EX 5-7 4. A skydiver is falling at her terminal speed. Immediately after she opens her parachute
- (A) the magnitude of the drag force on the skydiver will decrease.
- (B) the net force on the skydiver is in the downward direction.
- (C) the magnitude of the drag force is larger than the skydiver's weight.
- (D) the net force on the skydiver is zero.

AP® Skill Builders

5. The *Escherichia coli* bacterium propels itself slowly through water by means of a long, thin structure called a flagellum. Predict, by choosing the correct answer below, changes in the motion of a bacterium moving at a constant speed if the force exerted by the flagella on the surrounding fluid doubles.
- (A) The acceleration will immediately double because the acceleration is proportional to force.
- (B) The velocity will immediately double because the force on the fluid is equal and opposite to the force on the bacterium and this is a situation in which the bv model of drag force is appropriate.
- (C) The velocity will eventually double because the force on the fluid is equal and opposite to the force on the bacterium and this is a situation in which the bv model of drag force is appropriate.
- (D) The velocity will eventually increase by a factor of 4 because the force on the fluid is equal and opposite to the force on the bacterium and this is a situation in which the cv^2 model of drag force is appropriate.

6. A wind tunnel is used to study the drag forces exerted on a model airplane wing as a function of attack angle, α, in the figure. The scale model wing is mounted on rods, the deflection of which is used to measure the horizontal component of the drag forces, which is what we typically call drag, F_D, exerted on the wing in the direction of the constant velocity stream of air. Tension in the rods can be used to measure the vertical component of the drag forces, which is commonly called lift, F_L, exerted on the wing in a direction perpendicular to the air velocity due to the attack angle of the wing. While both of these forces we are calling lift and drag depend on the drag forces exerted on the wing by the air, the horizontal component of the drag forces resist the forward motion of the plane, whereas the vertical component makes it possible for the plane to stay up. (Lift also depends on other properties of airflow over the wing that are beyond our coverage in this course.

However, the source of these other lift forces does not affect how you answer this question.)

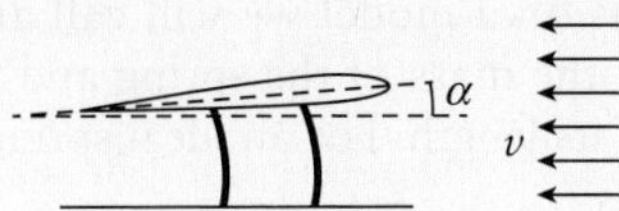

$\alpha°$	c_L	c_D	c_L/c_D
0	0.18	0.033	5.4
0.5	0.25	0.035	7.1
1	0.32	0.037	8.7
1.5	0.37	0.038	9.7
2	0.43	0.040	10.7
2.5	0.49	0.042	11.8
3	0.53	0.044	11.9
3.5	0.58	0.046	12.6
4	0.64	0.048	13.1
4.5	0.66	0.052	12.8
5	0.71	0.054	13.2
5.5	0.74	0.057	13.0
6	0.77	0.060	12.9
6.5	0.82	0.063	13.0
7	0.84	0.066	12.7
7.5	0.87	0.070	12.4
8	0.91	0.072	12.6

When α is equal to zero there is still a nonzero drag force parallel and a lift force perpendicular to the air velocity. As α increases, the cross-sectional area exposed to the stream of air increases. As the cross-sectional area increases, so do the drag and lift forces. A simple model for the dependence of drag and lift forces on the attack angle for small angles is

$$F_x = F_{x,\alpha=0} + c_D v^2 \alpha \qquad F_y = F_{y,\alpha=0} + c_L v^2 \alpha$$

where c_D and c_L are coefficients of drag and lift, respectively.

(a) Analyze these experimental data for a model airfoil in the table to test the prediction of a linear relationship between the coefficients of lift and drag and the attack angle.

(b) Using test data justify the claim that attack angles between 5° and 6° result in the best fuel efficiency for a plane with this wing style.

7. The terminal speed of a raindrop that is 4.00 mm in diameter is approximately 8.50 m/s under controlled, windless conditions. The density of water is 1.00×10^3 kg/m^3. Recall that the density of an object is its mass divided by its volume and model the raindrops as spheres.
   (a) If we model the air drag as being proportional to the square of the speed, $F_{drag} = cv^2$, **calculate** the value of c.
   (b) Predict the terminal speed of a raindrop that is 8.0 mm in diameter. Use the fact that for spherical objects, the value of c depends on the object's cross-sectional area.

AP® Skills in Action

8. Golfers and baseball batters know that when a ball is struck on the lower half, the ball has "lift" and "carries" farther. A ball struck on the top half "dies" in flight. A sinker in baseball is a pitch that falls much more than expected as it flies toward the batter.

   The diagrams show a spinning ball moving through air with a velocity v.

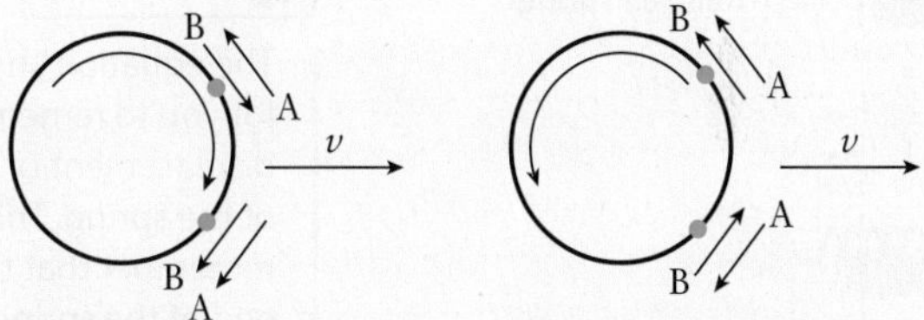

   Also shown for four example points on the surface of the ball are vectors describing the velocity of that point, labeled B, and the velocity of the air, labeled A, as it moves by that point.

   Consider the relative velocities of the air and the surface of the ball and use the model of drag force described in this section to explain how a pitcher makes a ball "sink."
9. A skeptical student reads that at low speeds the drag force on an object moving through a fluid is proportional to its velocity. The student claims that this is a violation of Newton's second law. Evaluate this claim.
10. We model the drag force of the atmosphere exerted on a 70.0-kg person falling through the air as $F_{drag} = cv^2$, where the value of c is 18.0 kg/m.
    (a) Calculate the person's terminal speed.
    (b) With a parachute, the constant c for the same person is a factor of 8 larger than if the person had no parachute. Predict the factor by which the terminal speed is decreased for a person using a parachute.
11. A common scene in action films involves a flailing person falling from a plane who is pursued by a savior carrying a parachute. Explain how this scene can have a happy ending.

5-6 An ideal spring force can be used to model many interactions

We used a picture of springs in Figure 4-2 to introduce the normal force. We also considered springs when we discussed how to measure forces in Section 4-5. Many of us are familiar with what we think of as a spring, a system that when you pull on it pulls back on you. When you try to squeeze it, compressing it to make it shorter, it pushes back. In reality many things behave like springs, although the amount they change shape may be very tiny. Experiment shows that if you stretch or compress a spring (by a relatively small amount compared to its size), the force that the *spring* exerts on you is directly proportional to the amount of stretch (**Figure 5-17**).

Unsurprisingly, we are going to focus on a model we will call an **ideal spring**. In the ideal spring model, we can neglect the mass of the spring and we assume the amount we stretch it is small compared to its length. For an ideal spring, we can write this relationship, known as **Hooke's law**, as

EQUATION IN WORDS
Hooke's law for the force exerted by an ideal spring

Force exerted by an **ideal spring** | **Spring constant** of the spring (a measure of its stiffness)

(5-10) $$F_{s,x} = -k\Delta x$$

The force exerted by the spring is always in the opposite direction to the displacement of the end of the spring.

Displacement of the end of the spring from its equilibrium (unstretched) position

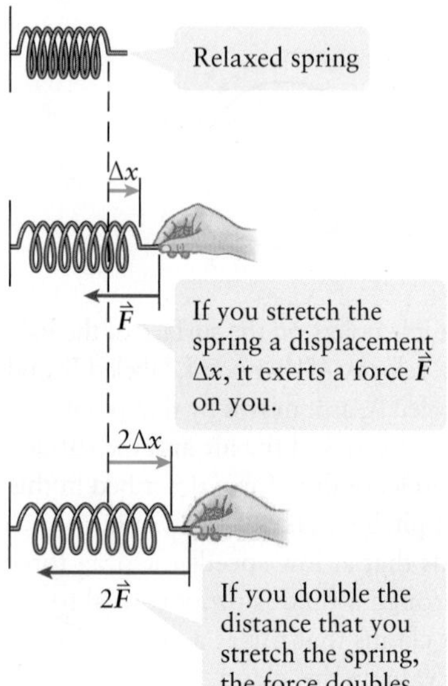

Relaxed spring
Δx
$\vec{F}$
If you stretch the spring a displacement Δx, it exerts a force $\vec{F}$ on you.
$2\Delta x$
$2\vec{F}$
If you double the distance that you stretch the spring, the force doubles.

Figure 5-17 Hooke's law If you stretch an ideal spring, the force that it exerts on you is directly proportional to its extension.

AP Exam Tip

The equation sheet for the AP® exam lists the force of a spring as $|F_s| = k|x|$. It is important for you to remember that this x is not measured from an arbitrary origin, but is the displacement of the end of the spring measured from the relaxed equilibrium position of the spring. This equation also just calculates the size of the force, and requires you to remember that the direction of the spring force is always opposite the displacement of the end of the spring from the equilibrium. So remember the specific definition of x for this equation on the exam. To help you, we will continue to stress in this textbook that this is the displacement from the relaxed equilibrium position of the end of the spring.

The minus sign in Equation 5-10 means that the force that the spring exerts on you is in the direction *opposite* to the stretch. If you pull one end of the spring in the positive x direction, $x_f - x_i > 0$ then $\Delta x > 0$ and the spring pulls back on you in the negative x direction. (As we will see below, Hooke's law also describes situations in which the spring is compressed rather than stretched.) The quantity k, called the **spring constant**, depends on the stiffness of the spring: The greater the value of k, the stiffer the spring. If the force is measured in newtons and the extension in meters, the spring constant has units of N/m. Because this force depends on the stretch of the spring, we will only be able to use it with our current techniques to solve problems for which the stretch does not change. In Chapter 7, we will learn new techniques that will allow us to consider springs for a broader range of problems. We will learn about the origin of Hooke's law and k later in this book. This important part of the ideal spring model, that the stretch (or compression) is directly proportional to the force, is valid for many materials in nature over small enough stretches or compressions, which is why this model and law are so important. Real springs can be stretched or compressed too far for this law to continue to hold, so when we use real springs in the lab, we will have to be careful over what ranges we use this model.

Equation 5-10 gives the force that the *spring* exerts on *you* as you stretch it. By Newton's third law, the force that *you* exert on the *spring* has the same magnitude but the opposite direction. Thus, the force you exert has the opposite sign of the force in Equation 5-10:

(5-11) $$F_{\text{you on spring},x} = +k\Delta x \text{ (force that you exert on the spring)}$$

Stretching the spring in the positive x direction means $\Delta x > 0$ and to do this you must exert a force in the same direction. So $F_{\text{you on spring},x} > 0$ if $\Delta x > 0$ which is just what Equation 5-11 tells us. If you double the amount you stretch the spring from its equilibrium length, the size of the force you must exert also doubles, as Equation 5-11 tells us, and as we can see in Figure 5-17.

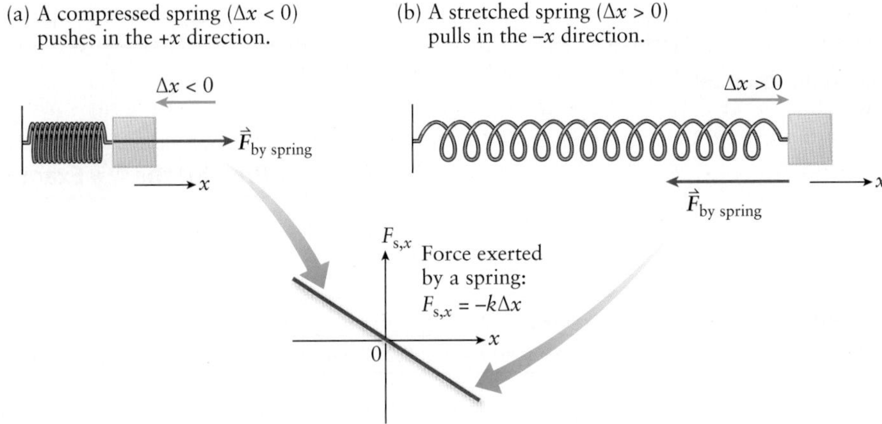

(a) A compressed spring ($\Delta x < 0$) pushes in the $+x$ direction.
(b) A stretched spring ($\Delta x > 0$) pulls in the $-x$ direction.
$\Delta x < 0$
$\Delta x > 0$
$\vec{F}_{\text{by spring}}$
x
$F_{s,x}$
Force exerted by a spring: $F_{s,x} = -k\Delta x$
0

Figure 5-18 Compressing and stretching an ideal spring Hooke's law, $F_{s,x} = -k\Delta x$ can be written as $F_s = -kx$ when $x = 0$ is defined as the position of the end of the spring when it is at relaxed equilibrium. This equation applies equally well to an ideal spring whether it is (a) compressed or (b) stretched.

The spring in Figure 5-17 exerts a force when it is stretched. A spring also exerts a force when it is *compressed* (**Figure 5-18a**). One example is a car's suspension, whose springs compress as you load passengers and luggage into the car. The force that an ideal spring exerts when compressed is given by Equation 5-10, $F_{s,x} = -k\Delta x$, the *same* equation that describes the force exerted by a *stretched* spring (**Figure 5-18b**). The only difference is that Δx is negative if the spring is compressed. Hence if you compress the spring (push its end in the negative x direction), the spring pushes back on you in the positive x direction.

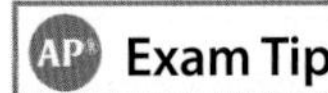

Exam Tip

The AP® exam does not refer to Hooke's law, but only refers to it as "the spring force."

As mentioned above, Hooke's law and Equations 5-10 and 5-11 are only *approximate* descriptions of how real springs, elastic cords, and tendons behave. As an example, **Figure 5-19** is a graph of the force needed to stretch a human patellar tendon (which connects the kneecap to the shin). The curve isn't a straight line, which means that the force isn't directly proportional to the amount of stretch. The force you have to apply to the tendon is also greater when you stretch than when you let it relax. What's more, the tendon can change its properties: The two sets of curves in Figure 5-19 are for males in their 70s before and after a 14-week course of physical training, which caused the patellar tendon to become much stronger and stiffer. Even when something is not an ideal spring, there may be ranges over which the ideal spring equations are good approximations. For instance, as shown in Figure 5-19, the post-training tendons show an approximately linear relationship between force and stretch from about 1.5 to 3 mm. So since this linear relationship exists for this range, it would be safe to use Hooke's law, although you would have different values of k for stretching and compressing. Even regular springs have ranges over which you can't use Hooke's law. If you compress a spring too much, the coils pack together and it can no longer compress like a spring. It turns out a solid tube of metal also has a spring constant, but it is much different than wound wire, so the behavior is much different. Stretch a spring too far, and the coils come out of shape, again changing its behavior. These are things you need to consider when using springs in the lab.

Nonetheless, Hooke's law is a very useful approximation for the behavior of many materials, which exert an almost linear force as a function of stretch when the amount of stretch or compression is small. This type of force relationship is very important in nature. We'll use Hooke's law repeatedly in our study of physics.

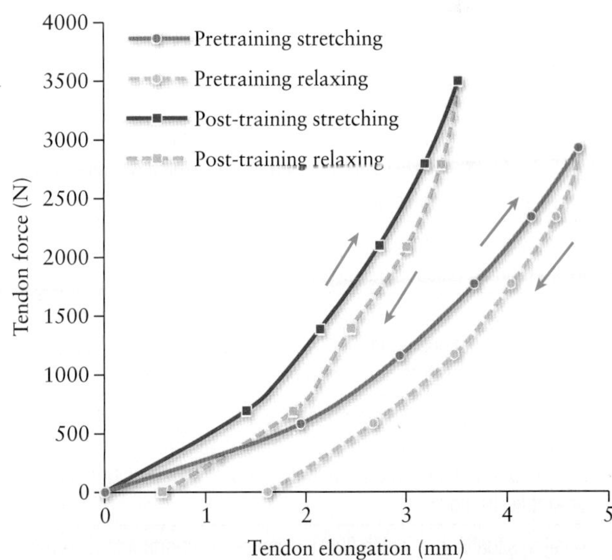

Figure 5-19 Tendons are not ideal springs This graph of force versus extension for a human patellar tendon is very different from that for an ideal spring (compare to Figure 5-18). The graph is not a straight line; there is a different graph for relaxing than for stretching the tendon; and the graph for tendons that have undergone exercise (square, red data points) is different from that for tendons that have not (circular, green data points).

THE TAKEAWAY for Section 5-6

✔ An ideal spring exerts a force proportional to the distance that it is stretched or compressed (Hooke's law).

Prep for the AP Exam

AP Building Blocks

1. An ideal spring with spring constant $k = 2.00 \times 10^2$ N/m is oriented vertically with one end on a flat solid table. What distance will the spring compress to reach equilibrium if a 1.20-kg object is placed on its upper end?
2. You use a spring, held horizontally, to pull a 3.0-kg box across the floor at constant velocity. The spring is stretched 5.0 cm at this velocity. If the coefficient of kinetic friction with the floor is 0.25, what is the spring constant of the spring, assuming it is ideal?

AP Skill Builders

3. A gravitational force exerted on an object of mass m attached so that it is hanging from a vertical spring will cause the spring to stretch. When the object is in equilibrium, the gravitational force exerted on the object is equal in magnitude to the spring force exerted on the object: $mg = k\Delta x$. This table provides data for a spring for eight different values of mass.

Mass, m (kg)	0.10	0.20	0.30	0.40	0.50	0.60	0.70	0.80
Stretch, Δx (cm)	1.0	2.2	3.4	4.6	5.2	6.2	7.2	8.4

   (a) Make a graph of mg versus Δx.
   (b) Use your graph to calculate the value of the spring constant, k, using all of the data values.
4. An object of mass m_2 is attached to a spring with negligible mass and spring constant k, and the equilibrium position of the spring when it is supporting m_2 is measured to be x_0. The object is removed from the spring, and an object with mass m_1 is attached to the spring. In each case, the spring and object hang at rest vertically. What is the position of m_1 in terms of m_1, m_2, x_0, and any other necessary constants? Assume x_0 is measured from the relaxed equilibrium of the spring when it is not supporting any object, which we will define as zero.

WHAT DID YOU LEARN?

Chapter learning goals	Section(s)	Related example(s)	Prep for the AP Exam: Relevant section review exercise(s)
Recognize what determines the magnitude of the static friction force and find the magnitude and direction of the maximum static friction force exerted on an object by a surface.	5-2 5-4	5-1	1 4
Find the magnitude and direction of the force of kinetic friction exerted on an object by a surface.	5-3	5-2, 5-3	3, 4, 5, 6
Explain contact forces on an object as arising from interatomic electric interactions and determine the direction in which these forces must therefore be exerted on an object.	5-2 5-4	5-5	3 5
Construct free-body diagrams and extract quantitative or qualitative information from them to solve for properties of the motion of an object when the forces on an object include static or kinetic friction.	5-4	5-4, 5-6	3, 4, 5, 6, 7
Analyze situations in which fluid resistance is important to determine when application of constant acceleration equations would give poor results.	5-5	5-7	2
Analyze equilibrium force problems involving the spring force to solve for properties of the system such as mass, spring constant, or compression or extension of the spring.	5-6		1, 2, 3, 4

Chapter 5 Review

Key Terms

All the Key Terms can be found in the Glossary/Glosario on page G1 in the back of the book.

drag force 218
coefficient of kinetic friction μ_k 204
coefficient of rolling friction μ_r 208
coefficient of static friction μ_S 197
fluid resistance 218
Hooke's law 224
ideal spring 224
macroscopic 196
rolling friction 208
spring constant 224

Chapter Summary

Topic	Equation or Figure
Static friction: If an object is at rest on a surface, a static friction force F_S is exerted by the surface to oppose other forces trying to make the object move relative to the surface. The static friction force can have any magnitude up to a maximum value.	(Figure 5-2); (5-1a); (5-1b)
An object at rest on an incline: Static friction will keep an object at rest on an inclined surface provided the angle θ of the incline is less than a critical value θ_{slip}.	(Figure 5-5a); (5-5)

(a) You exert a force of 20.0 N on the block to the right, but the block remains at rest...

20.0 N

20.0 N

...so the table must exert a static friction force of 20.0 N to the left.

(b) You exert a force of 10.0 N on the block to the right, but the block remains at rest...

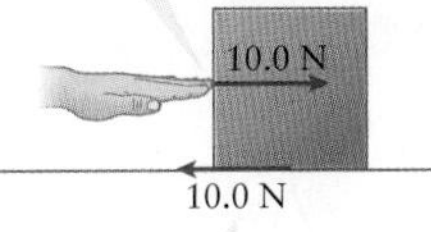

...so the table must exert a static friction force of 10.0 N to the left.

(c) You do not exert a force on the block and it remains at rest...

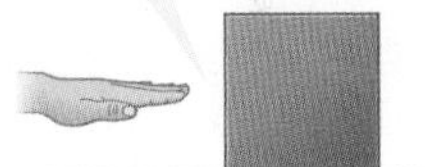

...so the table must exert no static friction force on the block.

The value of the static friction force exerted by a surface adjusts to exactly counteract the other forces exerted on the object that would make the object slide over the surface, unless those forces exceed $F_{S,max}$.

(Figure 5-2)

(1) The magnitude of the **force of static friction** on an object...

(2) ...is less than or equal to a certain maximum value.

$$F_S \leq F_{S,max} \qquad (5\text{-}1a)$$

(3) The **maximum force of static friction** depends on...

$$F_{S,max} = \mu_S F_n \qquad (5\text{-}1b)$$

(4) ...the **coefficient of static friction** (which depends on the properties of the two surfaces in contact)...

(5) ...and the **normal force** pressing the object against a surface.

Block, weight $\vec{F}_g$

Ramp whose angle θ from the horizontal can be varied

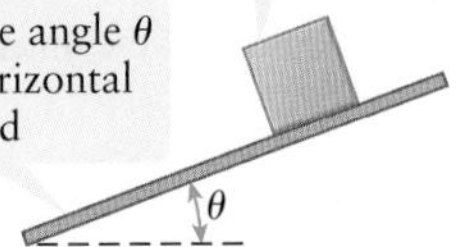

(a) A block at rest on an inclined ramp

(Figure 5-5a)

Coefficient of static friction for an object at rest on an incline

Angle of the surface at which the object just **begins to slip**

$$\mu_S = \tan \theta_{slip} \qquad (5\text{-}5)$$

Kinetic friction: If an object is sliding over a surface, that surface exerts a kinetic friction force F_k that opposes the relative motion of the object and the surface. We model this force as being independent of the object's speed. There is a similar expression for the force of rolling friction; the difference is that the coefficient for rolling friction has a much smaller value than that for sliding friction.

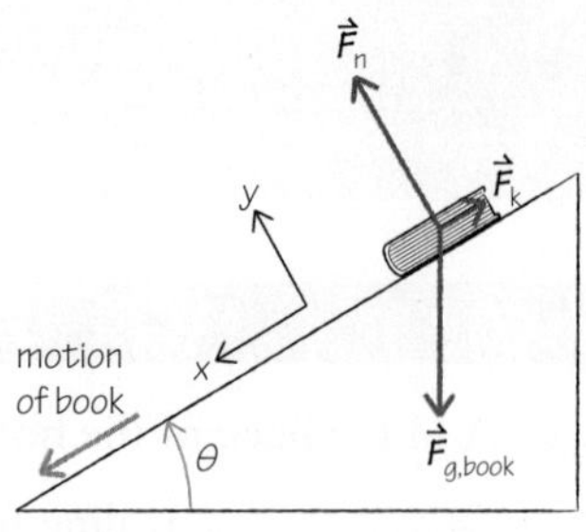

(Example 5-2)

The force of kinetic friction exerted on an object by a surface depends on...

$$F_k = \mu_k F_n \tag{5-6}$$

...the **coefficient of kinetic friction** (which depends on the properties of the two surfaces in contact)...

...and the **normal force** exerted on the object by the surface on which it slides.

(b) If the object slides on the ramp, the friction force that the ramp exerts on it is **kinetic friction.**

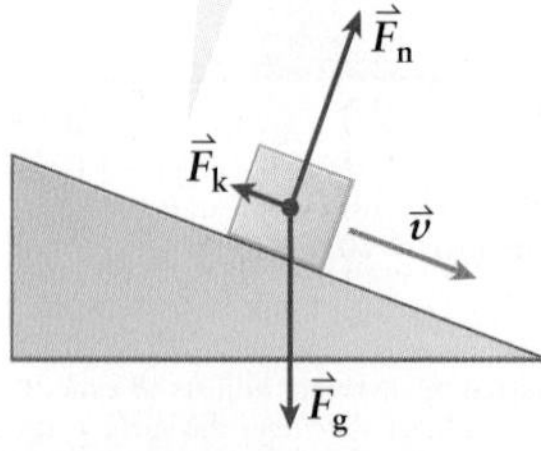

(c) If the object rolls on the ramp, the friction force that the ramp exerts on it is **rolling friction.**

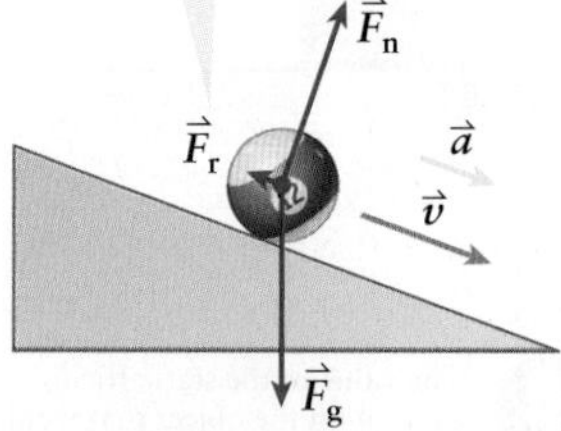

(Figure 5-11b,c)

Drag force: A resistive force that opposes motion is exerted on an object moving through a fluid (gas or liquid). If the object is small and moving at a low speed v relative to the fluid, the magnitude of the drag force exerted on the object by the fluid is proportional to v. For larger objects moving at faster speeds, the drag force is proportional to the square of v. Since these forces are not constant, if a drag force on an object is not negligible, the equations from Chapters 2 and 3 can no longer be used to describe its motion.

Magnitude of the **drag force** on a **small object** moving at a **low speed**

$$F_{drag} = bv \tag{5-8}$$

Constant that depends on the properties of the object and of the fluid

Speed of the object relative to the fluid

Magnitude of the **drag force** on a **larger object** moving at a **faster speed**

$$F_{drag} = cv^2 \tag{5-9}$$

Constant that depends on the properties of the object and of the fluid

Speed of the object relative to the fluid

The spring force and Hooke's law: An ideal spring exerts a force that is proportional to how far it is stretched or compressed from its relaxed equilibrium length, and it is always directed opposite to the stretch or compression. The spring force is not constant, so the work needed to stretch or compress a spring is not simply the magnitude of the exerted force multiplied by the displacement. Since this is always in one dimension (along the length of the spring), positive and negative signs are enough to determine the direction of $\vec{F}_S$.

Force exerted by an **ideal spring**

Spring constant of the spring (a measure of its stiffness)

$$F_{s,x} = -k\Delta x \qquad (5\text{-}10)$$

The force exerted by the spring is always in the opposite direction to the displacement of the end of the spring.

Displacement of the end of the spring from its equilibrium (unstretched) position

Chapter 5 Review Problems

1. If a sport utility vehicle (SUV) is being designed for rugged exploration that would require it to accelerate up a slope of 45°, justify the selection of the condition that must be satisfied by the relevant coefficient of friction between the SUV's tires and the surface. Assume that the tires are rolling and not sliding on the surface, as discussed in Section 5-3.
   (A) $\mu_S > 1$ because there must be a net force in the direction of motion and the point of contact between the tire and the surface is at rest
   (B) $\mu_S < 1$ because there must be a net force in the direction of motion and the point of contact between the tire and the surface is at rest
   (C) $\mu_k > 1$ because there must be a net force in the direction of motion and the tire and the surface are in relative motion
   (D) $\mu_k < 1$ because there must be a net force in the direction of motion and the tire and the surface are in relative motion
2. The green algae *Chlamydomonas reinhardtii* propels itself with two long, thin structures called flagella that extend from one end of the cell. When the two flagella exert a force of 3.50×10^{-14} N, the alga swims through water at a constant speed of 210.0 μm/s. Find the constant speed of the alga in water when the force exerted by its flagella is 7.00×10^{-14} N.
3. A book is pushed across a horizontal table at a constant speed. The horizontal force applied to the book is equal to one-half of the book's weight. Calculate the coefficient of kinetic friction between the book and the tabletop.
4. A solid rectangular block made out of a uniform material has sides of three different areas. You may choose to rest any of the sides on the floor as you apply a horizontal force to the block. Justify the claim that the choice of side on the floor does not affect how hard it is to push the block.
5. You press a book against the wall with the tip of your finger. You may assume your finger exerts no friction force on the book, only a normal force. As you get tired you exert less force, but the book remains at the same spot on the wall. Predict whether each of the following forces increases, decreases, or does not change in magnitude when you reduce the force you are applying to the book. Briefly justify your predictions.
   (a) The weight of the book
   (b) The normal force exerted by the wall on the book
   (c) The friction force exerted by the wall on the book and
   (d) The maximum static friction force that can be exerted by the wall on the book
6. You find yourself pushing a 42.0-kg box up a 6.00° metal ramp into a moving truck. The coefficient of kinetic friction between the box and the ramp is 0.320. To save energy, you only push hard enough to move the box up the ramp at constant velocity. Assuming that your pushing force is directed parallel to the ramp, with what force do you push?
7. While playing tug-of-war with your dog, you decide to pull him across the carpeted floor. Your dog has a mass of 18.5 kg, the rope toy you are pulling him with

makes an angle of 23.0° with the floor, and the coefficient of kinetic friction between your dog's feet and the floor is 0.367. If you pull your dog at a constant velocity, with what force are you pulling on the rope?

8. A runaway ski slides down a 250-m-long slope inclined at 37.0° with the horizontal. If the initial speed of the ski is 10.0 m/s, obtain an expression for the time it takes the ski to reach the bottom of the incline in terms of the coefficient of kinetic friction, μ_k, between the ski and snow. Evaluate that expression when μ_k is (a) 0.10 and (b) 0.15.

9. Coefficients of friction are not necessarily constant; they may depend on temperature and on the relative speed of the surfaces. Cross-country skiers apply wax to their skis in response to changes in the friction coefficient between ski and snow. A friction force is exerted on the bottom of an unwaxed cross-country ski by the film of ice that forms on the ski surface. Scientists reported measurements of the kinetic friction coefficient between ice and ice. They measured the force, F, exerted on a fixed slab of ice resting on a rotating sheet of ice held at uniform temperatures of −15°C, −5°C, and −1°C. Four values of the speed of the rotating ice sheet at the point where the force was measured were considered: 0.5 m/s, 1.0 m/s, 2.0 m/s, and 3.0 m/s. Also, the normal force exerted by the sheet on the slab, F_n, was varied by adjusting the external force, F, used to push the slab onto the sheet. (The recording instrument was calibrated so that the force recorded as F included the weight of the slab.) The coefficient of friction was calculated as F/F_n.

Results below show how the measured force varied with temperature for two values of F_n and two speeds of the sheet. Results in the bottom graph show how the force varied with speed for two values of F_n and the three values of temperature. Select the claim that is supported by these data. Neglect drag force due to air resistance.

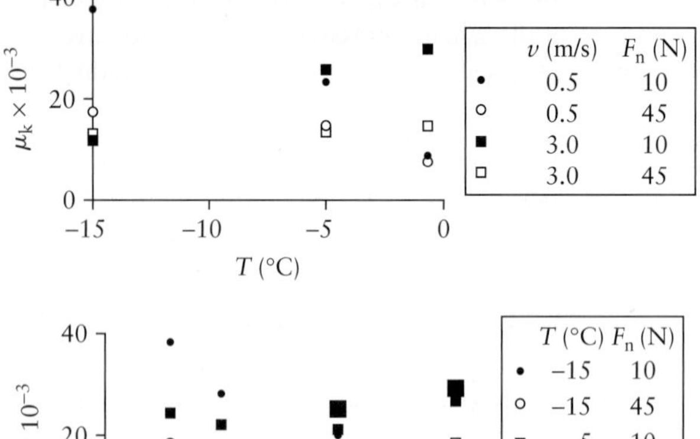

(A) The friction force decreases as the normal force decreases at all temperatures, so on the same downhill slope a skier with a smaller mass slides faster than a skier with a larger mass.

(B) The friction force increases as the normal force decreases at all temperatures, so on the same downhill slope a skier with a larger mass slides faster than a skier with a smaller mass.

(C) As the temperature decreases, more force must be exerted on the snow surface by the skier's poles to climb a slope at low speeds because the friction force decreases as temperature decreases.

(D) The friction force exerted on a skier near the melting temperature is a drag force exerted by a lubricating layer of liquid water on the surface of the ski because the friction force increases with increasing speed, whereas at other temperatures the friction force decreases or is nearly constant as speed increases.

10. Luis absentmindedly leaves his physics textbook on the horizontal top of his 1825-kg car while getting in. The coefficients of friction between the 1.84-kg textbook and the top of the car are $\mu_s = 0.180$ and $\mu_k = 0.113$. If the car is on level ground, what is the magnitude of the maximum acceleration the car can have before Luis's textbook slides off the car?

11. A librarian moonlighting as a magician wants to pull one book out of a stack of books. The stack is seven books high and all the books are of roughly equal mass M, of comparable size, and the coefficients of friction between the books are μ_s and μ_k. The librarian needs the third book in the stack, counting from the bottom. The librarian gives the third book a hard, straight yank with a force of magnitude F perpendicular to the stack. In terms of M, g, and the appropriate coefficient of friction, what is the minimum value of the horizontal force, F, with which she must yank the book?

12. Two blocks are in contact. A horizontal force is applied to the smaller block which has mass m. The surface between the smaller block and larger block with mass M has a coefficient of static friction, μ_s. The larger block slides over a surface with negligible friction.

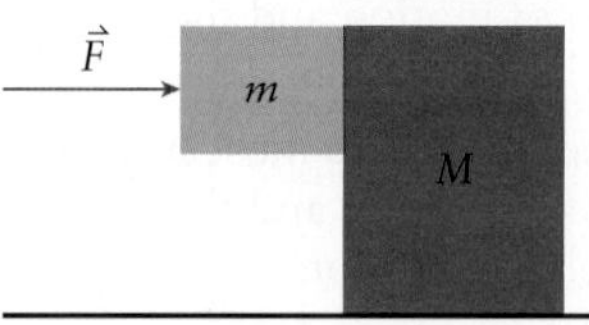

(a) Construct a free-body diagram for each block.

(b) Use the free-body diagrams in part (a) to express the normal force exerted by the larger block on the smaller block, $F_{M \text{ on } m}$ in terms of the applied force, F, M, and m, and any necessary constants.

(c) Predict the minimum horizontal force that must be exerted on the smaller block to keep it from sliding down.

13. You are sliding a piece of furniture across the floor at constant velocity when a friend jumps on top of it.

Assume that the force that you exert on the piece of furniture remains constant. Predict what happens.

14. A 1-kg wood ball and a 5-kg lead ball have identical sizes, shapes, and surface characteristics. They are dropped simultaneously from a tall tower. Air resistance is present. How do their accelerations compare?
   (A) The 1-kg wood ball has the larger acceleration, but the accelerations cannot be compared more precisely without more information.
   (B) The 5-kg lead ball has the larger acceleration, but the accelerations cannot be compared more precisely without more information.
   (C) The accelerations are the same.
   (D) The 5-kg ball accelerates at five times the acceleration of the 1-kg ball.
   (E) The 1-kg ball accelerates at five times the acceleration of the 5-kg ball.

15. A girl rides her scooter on a hill that is inclined at 10.0° with the horizontal. The combined mass of the girl and scooter is 50.0 kg. On the way down, she coasts at a constant speed of 12.0 m/s, while experiencing a drag force that is proportional to the square of her velocity. What force, parallel to the surface of the hill, is required to increase her speed to a constant 20.0 m/s? Neglect any other resistive forces, including the friction between the scooter and the hill.

16. A 150-kg crate rests in the bed of a truck that slows from 50.0 km/h to a stop in 12.0 s. The coefficient of static friction between the crate and the truck bed is 0.655.
   (a) Will the crate slide during the braking period? Explain your answer.
   (b) What is the minimum stopping time for the truck that prevents the crate from sliding?

17. A 2.50-kg package slides down a 12.0-m-long inclined plane that makes an angle of 20.0° with the horizontal. The package has an initial speed of 2.00 m/s at the top of the incline. What must the coefficient of kinetic friction between the package and the inclined plane be so that the package comes to rest as it reaches the bottom of the incline?

18. Synovial fluid lubricates the surfaces where bones meet in joints, making the coefficient of friction between bones very small. A condition of aging and many diseases is that synovial fluid degrades; investigations leading to synthetic replacements are ongoing. The synovial fluid contains lubricin, a small protein strand whose ends are weakly attracted to the cartilage that covers the bones separated by the joint, as shown in this drawing. The weak attraction between the ends of the lubricin polymer and the cartilage covering the bone are shown in the drawing as dots. The drawing also shows that some strands of lubricin span the space between the bone surfaces and in some both ends of the strand attach to the same surface. Synovial fluid is also composed of another polymer, hyaluronan, in a matrix of water.

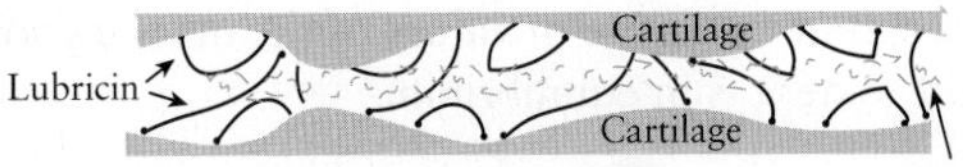

Without physical activity, joints become "stiff." This is very much like in mechanical systems such as a bearing. The lubricating oil in a bearing does not cover the internal metal or plastic surfaces uniformly when it is not moving. The film drains in response to gravitational forces. Only when the bearing begins to turn does the lubricating film become distributed. The fluid must be drawn in by the motion of internal surfaces.
   (a) One reason that athletes perform stretching exercises is to increase blood flow to muscle tissue. Explain why these exercises might also increase flexibility of the skeleton.
   (b) Sitting for a long time can make your back "stiff." The weight of your body causes compression of the facet joints between adjacent vertebrae. You might relieve pain by touching your toes and twisting your back. Predict how continued compression could affect the attachment of lubricin polymers to cartilage surfaces and how repeated extension, compression, and twisting the vertebrae could rearrange strands of lubricin in the facet joints.

19. Two blocks are connected by a light string. Assume the string does not stretch as it passes over a pulley with negligible mass and friction, as shown in the figure. Block 1 has a mass of 1.00 kg and block 2 has a mass of 0.400 kg. The angle θ of the incline is 30.0°. The coefficients of static friction and kinetic friction between block 1 and the incline are $\mu_S = 0.500$ and $\mu_k = 0.400$, respectively. What is the magnitude of the tension in the string?

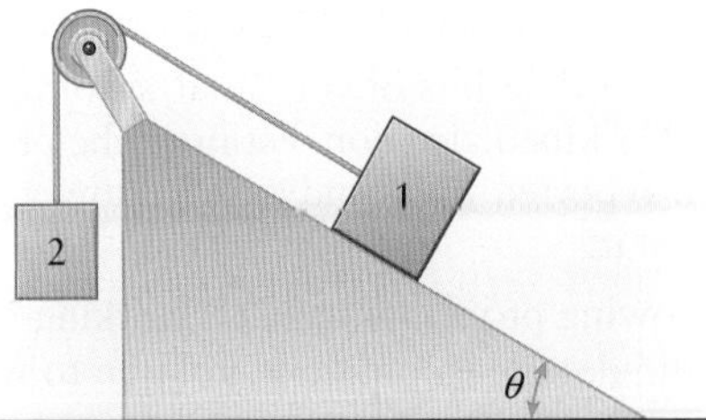

20. Two blocks are connected by a light string. Assume the string does not stretch as it passes over a pulley with negligible mass and friction, as shown in the figure from problem 19. Block 1 has a mass m_1 of 1.00 kg and block 2 has a mass m_2 of 2.00 kg. The angle θ of the incline is 30.0°. The coefficients of static friction and kinetic friction between block 1 and the incline are $\mu_S = 0.500$ and $\mu_k = 0.400$. What is the acceleration of block 1?

21. In the figure, two blocks are connected to each other by a string of negligible mass that passes over a pulley with negligible mass and friction. The string does not stretch. The mass of block 1 is $m_1 = 6.00$ kg. Assuming the coefficient of static friction $\mu_S = 0.542$ for all

surfaces, find the range of values of the mass m_2 so that the system is in equilibrium.

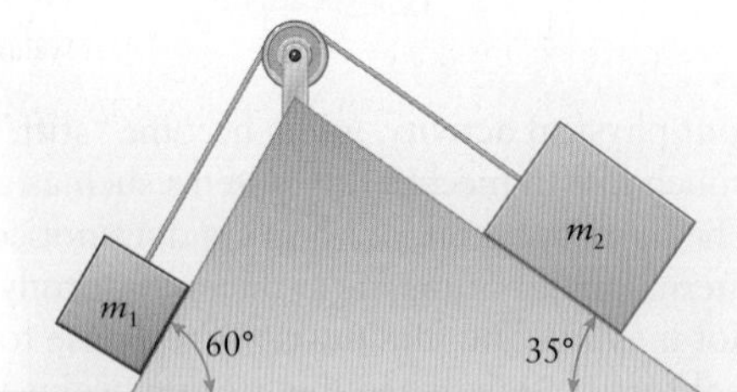

22. The following data were collected for the acceleration of a 342-g object down an inclined plane. Using these data, determine the coefficient of kinetic friction between the object and the plane.

Inclination (degrees)	Acceleration (m/s^2)	y	x
22.0	1.56 ± 0.03		
27.0	2.45 ± 0.03		
32.0	3.25 ± 0.03		
37.0	4.14 ± 0.03		

(a) Using a free-body diagram express the acceleration of the object as a function of the acceleration due to gravity, g, and the coefficient of kinetic friction.

(b) Define variables y and x that linearize the relationship so that it can be expressed in the form $y = 1 - \mu_k x$.

(c) Complete the table for each ramp inclination.

(d) Using a spreadsheet or your graphing calculator graph your values of x and y.

(e) Obtain a best fit line for y versus x and display the equation as an annotation on your graph.

(f) What should be the y-intercept of your graph? What does the slope of your graph represent?

(g) Report the results of your analysis of the coefficient of kinetic friction. Estimate the precision of your reported value and explain how you estimated it.

23. The following problem involves "working backward." Draw and describe a physical situation to which the following equation could apply:

$$a = \frac{20\text{ N} - 0.40(2.0\text{ kg})\left(9.8\dfrac{\text{N}}{\text{kg}}\right)\cos(30°) - (2.0\text{ kg})\left(9.8\dfrac{\text{N}}{\text{kg}}\right)\sin(30°)}{2.0\text{ kg}}$$

24. (a) A block of mass 2.3 kg is initially at rest on an incline that makes an angle of 30° with the horizontal. The coefficient of kinetic friction between the block and the incline is 0.35 and the coefficient of static friction is 0.74. Calculate the magnitude of the force of friction on this block.

(b) The upper end of the incline is raised until the incline makes an angle of 45° with the horizontal. Calculate the magnitude of friction on the block in this case.

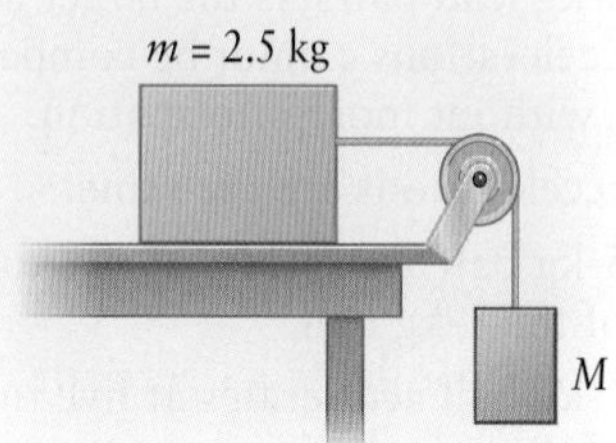

25. A block of mass $m = 2.5$ kg is on a horizontal table and connected by a light string that passes over a pulley of negligible mass to block of mass M. The coefficient of kinetic friction between the block and the table is 0.32 and the coefficient of static friction is 0.52.

(a) Calculate the maximum value of M such that the system remains at rest.

(b) The block of mass M is removed and replaced with a block of mass 1.5 kg. Calculate the acceleration of the system and the tension in the string in this case.

26. A block of mass 0.500 kg is initially moving at a speed v_0 when it is 1.00 m from the edge of a table. The coefficient of kinetic friction between the table and the block is 0.270. The table has a height of 1.10 m above the floor. The block is observed to slide across the table and then off the edge of the table. The block strikes the floor 0.800 m from the base of the table. Calculate the initial speed v_0 of the block.

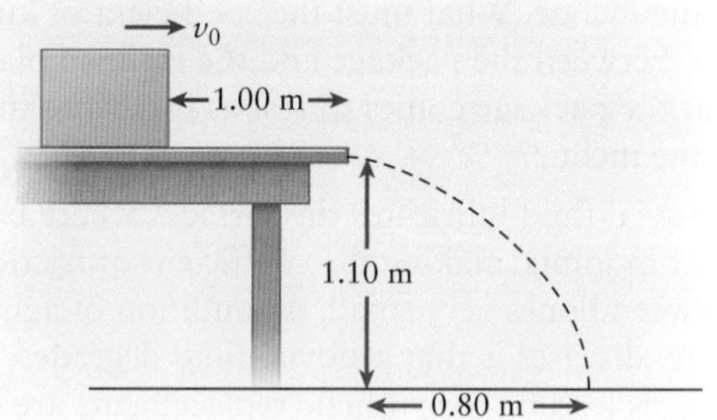

Prep for the AP Exam

AP® Group Work

Directions: This problem is designed to be done as group work in class.

It is claimed that a new tire material will reduce the tendency of automobile tires to slip on wet surfaces. Equipment to test this claim includes the following:

- A moving track whose variable speed can be measured in meters per minute
- Sheets that can be fixed to the track surface; these sheets are composed of the new material and the current material that it may replace
- Disks with different diameters whose rims are surfaced with the asphalt used on road surfaces
- A tachometer attached to the axle of the disk that measures the number of rotations per minute

- A nozzle that sprays films of water of varying thickness on the track surface

(a) Define a particular scientific question that can be addressed with this equipment and that will test the claim regarding the tendency of the new material to reduce slipping.

(b) Select the independent and dependent variables that will be collected in your design to test the claim.

(c) Design a plan that describes the measurements to be made.

(d) Describe how the evidence provided by these data could be used to **evaluate** the question defined in part (a).

AP® PRACTICE PROBLEMS

Prep for the AP® Exam

Multiple-Choice Questions

Directions: The following questions have a single correct answer.

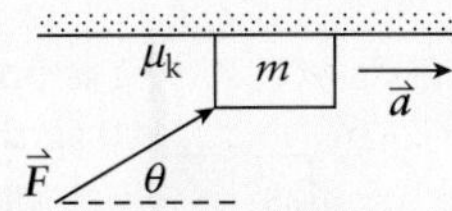

1. A block is being pushed against a ceiling with a force of magnitude F that makes an angle θ with the horizontal. The block remains in contact with the ceiling and slides with an acceleration, a, to the right. The coefficient of kinetic friction between the surfaces is μ_k. Of the following, which equation correctly describes the situation?

(A) $F\cos\theta - \mu_k(mg + F\sin\theta) = ma$

(B) $F\cos\theta - \mu_k mg = ma$

(C) $F\cos\theta + \mu_k(mg + F\sin\theta) - ma = 0$

(D) $F\cos\theta - \mu_k(F\sin\theta - mg) - ma = 0$

Questions 2–4 refer to the following material.

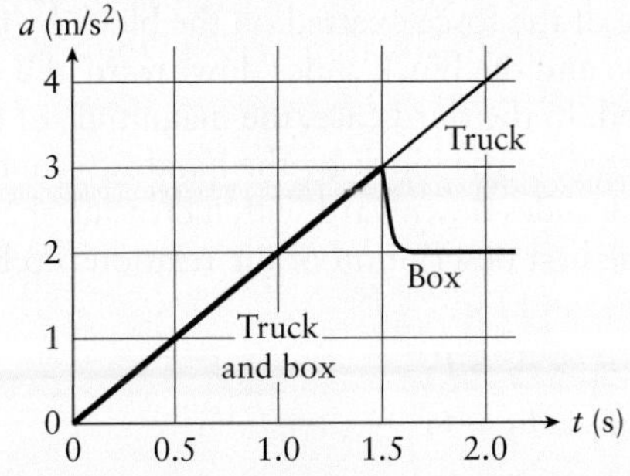

A 0.50-kg box rests on the bed of a truck on a horizontal surface. At $t = 0$ s the truck begins to accelerate, as shown in the graph. At some time the box begins to slide on the bed of the truck.

2. Of the following, which one is the closest value for the coefficient of static friction between the box and the surface of the bed of the truck?

(A) 0.5 (B) 0.4 (C) 0.3 (D) 0.2

3. Of the following, which one is the closest value for the coefficient of kinetic friction between the box and the surface of the bed of the truck?

(A) 0.5 (B) 0.4 (C) 0.3 (D) 0.2

4. At the time $t = 2.0$ s, what is the difference in the velocities of the truck and the box?

(A) 0.75 m/s

(B) 1.0 m/s

(C) 1.5 m/s

(D) 2.0 m/s

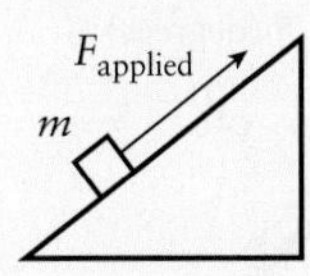

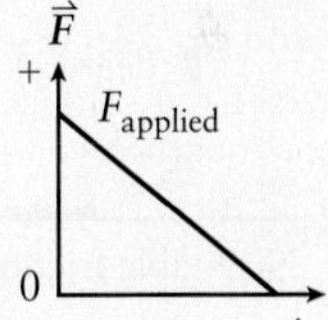

5. A force, $F_{applied}$, is exerted on a block of mass m to pull it up a rough incline. The force is decreased over time, as illustrated by the above graph. Eventually the block begins to slide down the incline. Of the following graphs, which one most nearly illustrates the size and the direction of the friction force, $F_{friction}$, exerted on the block?

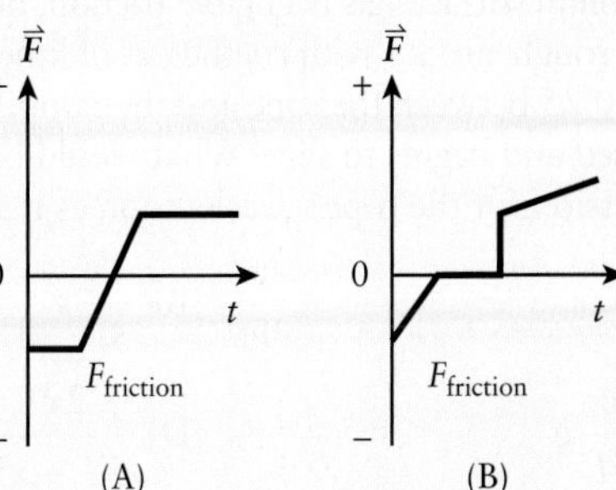

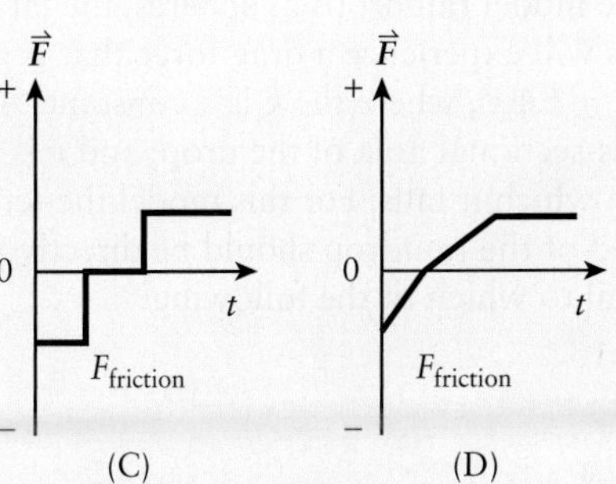

(*Continued*)

θ

6. A block is sitting on a board that can gradually be raised through angles from zero degrees to 90 degrees. Which of the following graphs best represent the friction force on the block as a function of the angle θ?

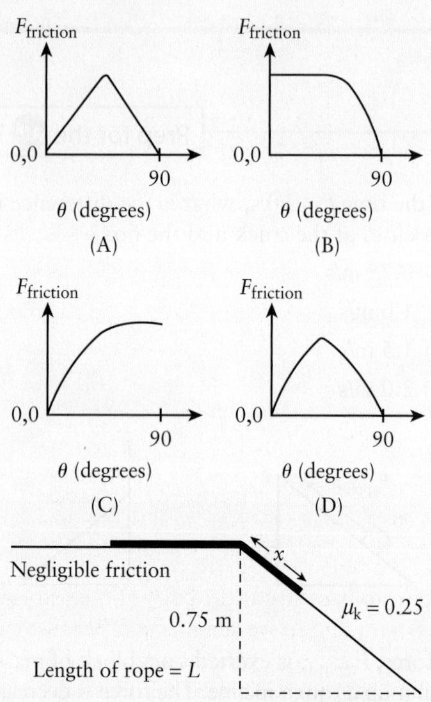

7. A rope of length L and mass m is held resting on a surface with a portion of its length, x, hanging on a downward ramp. As shown, the ramp is 0.75 m high and extends a horizontal distance of 1.00 m. The top horizontal surface has negligible friction, but the ramp has a rough surface with coefficient of kinetic friction $\mu_k = 0.25$ between the rope and the ramp. The rope is released and begins to slide. What would be the proper expression for the rope's acceleration as it slides?

(A) 0
(B) $\frac{2x}{5L}g$
(C) $\frac{3x}{5L}g$
(D) $\frac{2x}{5(L-x)}g$

8. If we model raindrops as spheres, the larger ones will experience a drag force that is given by $F_{drag} = kAv^2$, where the k is a constant, A is the cross-sectional area of the drop, and v is the speed with which it falls. For this model the terminal speed of the raindrop should be directly proportional to which of the following?

(A) $\frac{1}{r}$
(B) r
(C) r^2
(D) $\sqrt{r}$

9. A student learning to drive quickly applies the brakes when she is 20 m from an intersection and sees the light turn yellow. It takes her 2.5 seconds to come to rest at the intersection. Can the speed of the car just before she applied the brakes be determined directly from this information, assuming constant acceleration?

(A) No, because the fundamental definition of velocity contains acceleration.
(B) No, because acceleration is used in the kinematic equation $\Delta x = v_{0x}t + 1/2a_xt^2$.
(C) Yes, by dividing the distance to stop (20 m) by the time it took to stop (2.5 s).
(D) Yes, by finding the average speed during braking and doubling it.

10. F_1 F_2 F_3

m m m

$v = 0$ v constant v increasing with time

A block of mass m is pressed against a vertical wall, as shown above. There is friction between the wall and the block. Assume the friction between the hand and the block is negligible (the hand exerts a horizontal force only on the block). Consider three cases. In the first case, the magnitude of the force exerted on the block by the hand is F_1, and the block does not move. In the second case, the magnitude of the force exerted on the block by the hand is F_2, and the block slides downward at a constant speed. In the third case, the magnitude of the force exerted on the block by the hand is F_3, and the block slides downward with increasing speed. What is the best description of the relationship between F_1, F_2, and F_3?

(A) $F_1 > F_2 > F_3$
(B) $F_1 > F_2 = F_3$
(C) $F_3 < F_1 = F_2$
(D) $F_3 < F_1 < F_2$

11. Consider the following scenario: a 200-N block is placed on a horizontal surface. The coefficient of kinetic friction between the block and the table is 0.30 and the coefficient of static friction is 0.50. A horizontal force of magnitude F is exerted on the block. Which of the following is the correct ranking of the cases shown below in terms of the magnitude of the friction force exerted on the block 5 seconds after the force is applied?

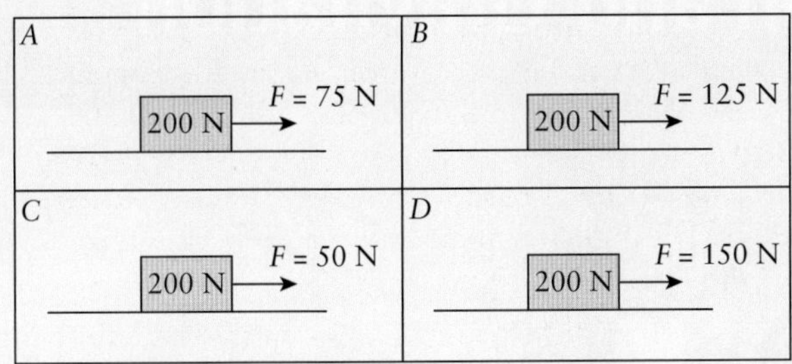

(A) $D > B > A > C$

(B) $(A = C) > (B = D)$

(C) $A > (B = D) > C$

(D) $(B = D) > (A = C)$

Free-Response Questions

1.

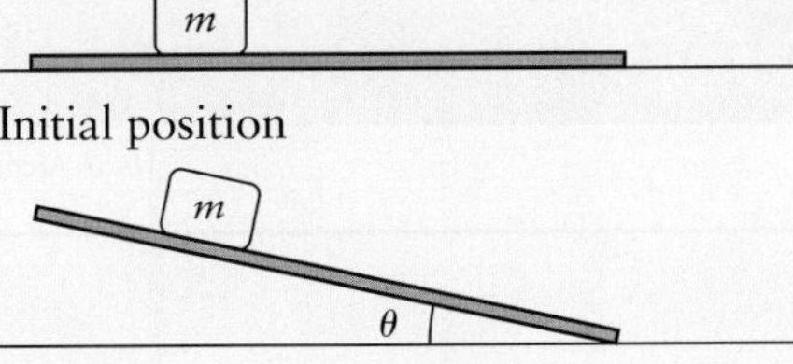

Incline raised to $\theta < \theta_{max}$

A block of mass m is initially at rest on a rough board, which is initially horizontal on a tabletop. The left end of the board is slowly raised until the board makes an angle θ with the horizontal. The block remains at rest.

(a) On the dot, which represents the block, draw vectors representing the forces on the block when the incline makes an angle θ with the horizontal. The vectors should start on the dot and point away from the dot. The vectors should be drawn to an appropriate length consistent with the motion of the block.

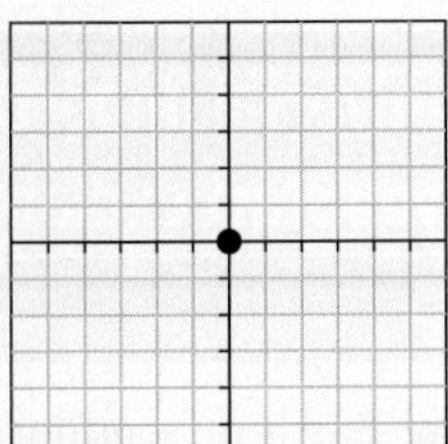

When the block reaches an angle θ_{max}, the block breaks free and slides down the board.

(b) At all angles $\theta < \theta_{max}$, the block remains at rest on the incline. Derive an expression for the friction force on the block as a function of θ.

(c) Derive an expression for the coefficient of static friction, μ_S, between the block and the incline in terms of θ_{max}, m, and physical constants, as appropriate.

Assume the coefficient of kinetic friction between the block and the board is less than the coefficient of static friction.

(d) Describe the motion of the block as it slides down the incline. Explain your reasoning.

2.

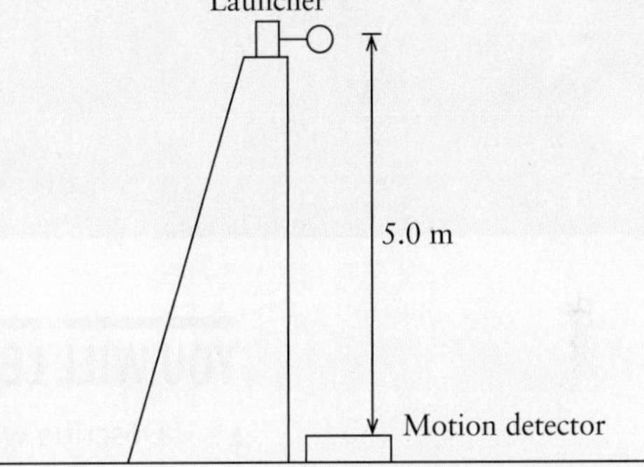

Your friend builds a device that can be used to launch objects from the edge of a platform, $\Delta y = 5.0$ meters above the ground. Because you study physics, your friend asks you to come up with a method to determine the speed with which the device launches the objects. You lend your friend an instrument that records the final velocity of the objects. You are sent the data to analyze, but you realize you do not know if the launcher was set to launch the objects vertically downward or horizontally. You may neglect air resistance.

(a) Do you need to know the initial direction of motion of the object to know in which direction it was launched? Define your system and support your claim, without using equations, by describing the motion of a launched object for both a downward and a horizontal launch.

(b) Derive expressions for the final speed of the object in terms of given quantities and necessary constants for each of the two launch directions.

(c) Do your equations in part (b) match your claim in part (a)? Justify your answer.

6 Circular Motion and Gravitation

NASA Archive/Alamy

YOU WILL LEARN TO:

- Describe why an object moving in a circle is always accelerating even when its speed is not changing.
- Apply Newton's laws to objects in uniform circular motion.
- Recognize what it means to say that gravitation is universal, and articulate when the gravitational force is the dominant force between objects or systems.
- Apply Newton's law of universal gravitation to describe or calculate the gravitational force any two objects exert on each other, and use that force in contexts other than orbital motion.
- Apply the law of universal gravitation to analyze circular orbits of satellites and planets.
- Relate the gravitational field at a point in space to the gravitational force exerted on an object at that point in space.
- Explain the origin of apparent weightlessness.

6-1 Gravitation is a force of universal importance; add circular motion and you start explaining the motion of the planets

Eighty years ago, the idea of humans orbiting Earth or sending spacecraft to other worlds was regarded as science fiction. Today these ideas have become commonplace reality. Humans live and work in Earth orbit aboard the International Space Station (see photograph at the start of this chapter), have ventured as far as the Moon, and have sent robotic spacecraft to explore all the planets of the solar system.

While we think of space flight as an innovation of the twentieth century, we can trace its origins to Isaac Newton's revolutionary seventeenth-century **law of universal gravitation**—the idea that all objects in the universe exert gravitational forces on each other. In Chapters 4 and 5 we learned that Earth exerts a gravitational force on all objects near our planet's surface. In this chapter we'll extend our discussion to consider the gravitational force (sometimes referred to as gravitational attraction, since the gravitational force is always attractive) exerted by any object with mass on any other object with mass. Gravitational forces between the components of our planet are

NEED TO REVIEW?
Turn to the **Glossary** in the back of the book for definitions of bolded Key Terms.

(a) Earth

NASA/NOAA
Earth is held together by the mutual gravitational attraction of all its parts.

(b) Saturn, its rings, and three of its moons

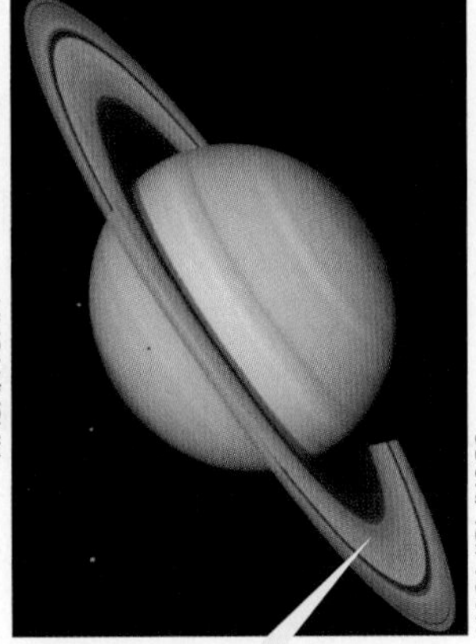

NASA/JPL/USGS
Saturn's rings are made of countless small objects that, like Saturn's moons, are held in orbit by the planet's gravitational force.

(c) Galaxy NGC 6744

ESO
This galaxy (a near-twin of our Milky Way) contains more than 10^{11} stars. The stars' gravitational attraction for each other holds the galaxy together.

Figure 6-1 Gravitation is universal The force of gravitation is responsible for (a) keeping Earth from flying apart, (b) keeping moons and rings in orbit around their planets, and (c) holding galaxies—among the largest structures in the universe—together.

partly responsible for holding Earth together (**Figure 6-1a**); gravitational forces exerted by Saturn keep its moons and other small objects in orbit (**Figure 6-1b**); and gravitational forces between stars and the material between the stars hold entire galaxies of stars together (**Figure 6-1c**).

In this chapter we will learn about the properties of the gravitational force, including how Newton deduced that this force must grow weaker as objects move farther apart. We will study circular motion, and in conjunction with gravitation, begin to understand the basics of motion that keep a car on a road as it travels around a curve—whether on a flat or inclined road—or a space station in its correct orbital revolution. We'll also use the idea of universal gravitation to understand why astronauts in orbit feel weightless, and learn about some of the physiological challenges that astronauts face as a result.

While we will begin to see how to use these ideas to understand the motions of satellites and orbits, we won't complete our understanding of gravitation until we have also learned about gravitational potential energy and angular momentum. In a later chapter, we will come back to gravitation to examine more complicated motions than we can now tackle.

 **Exam Tip**

Force is a vector, so it has both magnitude and direction. The AP® Physics 1 equation sheet gives only the magnitude of the gravitational force between two objects with mass, so you need to know that it is always attractive to be able to specify its direction.

THE TAKEAWAY for Section 6-1

✔ Isaac Newton deduced that gravitation is universal—it is a force that is exerted by any object with mass on every other object with mass throughout the universe.

✔ The ideas of universal gravitation and circular motion will help us understand the motions of satellites around Earth and of planets around the Sun.

Prep for the AP® Exam

AP® Building Blocks

1. In the model of gravitational force that Newton developed, the magnitude of the force grows weaker as the distance increases between objects with mass. Qualitatively compare this to the dependence of the strength of a signal with distance from the source: (i) your ability to hear a sound, and (ii) how bright a light appears to you.

AP® Skill Builder

2. The gravitational force on a basketball with mass m near Earth's surface may be modeled as a constant force mg. As the ball rolls from one end of a basketball court to the other, the direction and magnitude of the gravitational force doesn't change. But if the basketball court is stretched until it is the size of Kansas, you will see that the direction of the force is not constant. This is modeled in the accompanying figure. The vectors used to represent gravitational force at any two locations are parallel when a plane is used to model an area on Earth and not parallel when a smooth sphere is used to model an area on Earth.

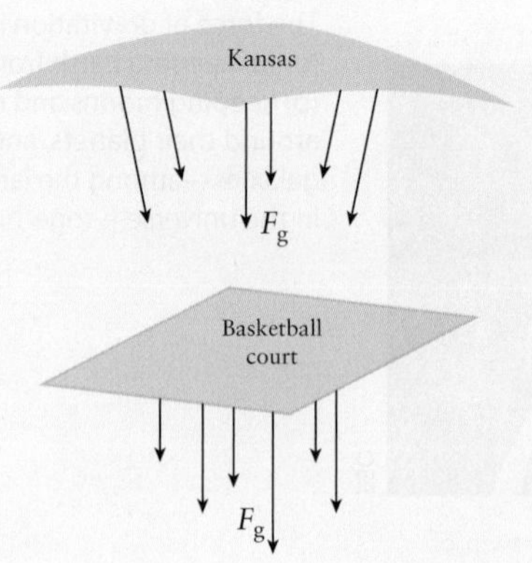

(a) Describe how the direction of the gravitational force exerted by Earth on an object near the Earth's surface is defined such that it is consistent with both the planar and spherical models.

On a featureless area of Earth's surface, for example, just south of Liberal, Kansas, where Dorothy may have lived before visiting Oz, the Oklahoma border is at the horizon, roughly 5000 m away.

(b) Calculate the angle between the gravitational force at Liberal and at the Oklahoma border. Model Earth as a smooth sphere with a radius of 6.37×10^6 m.

6-2 An object moving in a circle is accelerating even if its speed is constant

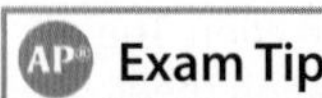

Whenever the velocity of an object changes direction, the object accelerates, even if the speed is constant, since a changing direction means a changing velocity. Uniform circular motion means that an object is moving at constant speed in a circle, so therefore it must be accelerating toward the center of the circle.

We begin our study of circular motion by thinking about the motion of a car. There is an assembly that is responsible for translating the rotation of the steering wheel into left or right motion of two of the car's wheels in parallel. Exactly what this assembly is depends on what sort of steering the car has. How hard you turn the steering wheel affects how quickly the car's wheels begin to change the car's direction of motion, and the Newton's third law reaction to the friction force the tires exert on the road provides the force to accelerate the car through the change in direction. To round a tight corner, you have to turn more than when you are rounding a gentle curve, and your tires have to exert a larger friction force to keep you turning more tightly. If you round a gentle curve faster, your tires have to exert a larger friction force on the road than if you round that same curve at low speed. You may have seen someone going around a curve at a speed that would normally be fine on a dry road lose control of their car when the road was wet and friction was reduced, or if their tires were worn. We saw in Section 3-5 that a car following a curved path experiences an acceleration directed toward the inside of the curve. So the car's acceleration must be greater for a high-speed, tight turn than for a low-speed, gentle turn. But *how much* greater? What is the relationship between the car's acceleration, its speed, and the radius of its circular path?

To see the answer let's look at the case of an object going around a circular path at a constant speed, also called **uniform circular motion**. The time for an object to travel a complete circle is the **period**, T. Its speed is constant, the circumference of the circle/T. Because the object's speed is constant, its acceleration $\vec{a}$ has no component tangent to its trajectory: It is perpendicular to the trajectory. We can see it must point toward the *center* of the circle if we just do a quick comparison of the direction of the velocity before and after any point on the circle (**Figure 6-2**). An acceleration that points

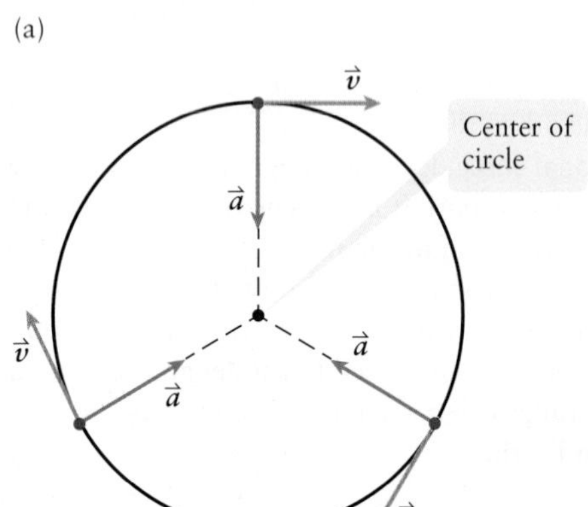

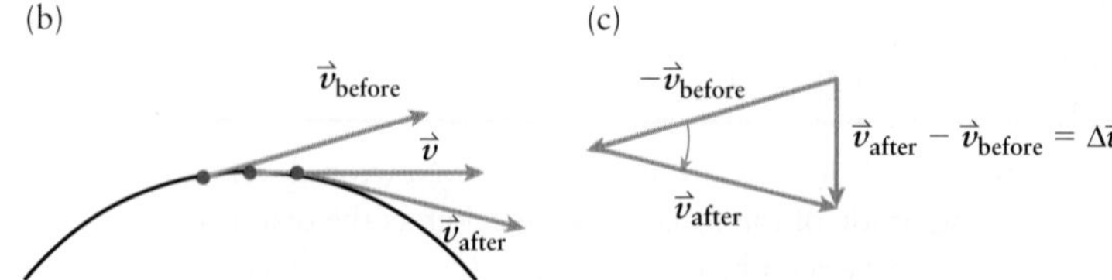

Figure 6-2 Uniform circular motion (a) An object moving around a circle at a constant speed. (b) A closer view of the motion at the top of the circle, showing the velocities before and after the point right at the top of the circle. (c) Since time is a scalar, $\vec{a} = \Delta\vec{v}/\Delta t$ must be in the direction of $\Delta\vec{v}$, the change in velocity, calculated here.

in this direction is called a **centripetal acceleration**, which comes from a Latin term meaning "seeking the center." In the next section, we will see how to calculate the magnitude of the centripetal acceleration and calculate its direction more carefully.

Analyzing Motion in a Circle

Figure 6-3a shows two points along the trajectory of an object in uniform circular motion around a circular path of radius r. From point 1 to point 2, the object travels a distance Δs around the circle, and an imaginary line drawn from the center of the circle to the object rotates through an angle $\Delta\theta$. We can write a simple relationship between Δs and $\Delta\theta$ provided that we express $\Delta\theta$ in *radians*:

$$\Delta\theta = \frac{\Delta s}{r} \quad (\Delta\theta \text{ in radians}) \tag{6-1}$$

You can check this formula by thinking about the case in which the object goes all the way around the circle so that the distance Δs equals $2\pi r$, the circumference of the circle. Then Equation 6-1 says that $\Delta\theta = (2\pi r)/r = 2\pi$, which is just the number of radians that make up a complete circle.

If the time interval between points 1 and 2 is very short, the angle $\Delta\theta$ is small, and the curved arc of length Δs is nearly linear. Then the lines from the center of the circle to points 1 and 2 and the curved arc form a triangle with two sides of length r separated by an angle $\Delta\theta$ and a third side of length Δs (see **Figure 6-3b**). Thus Equation 6-1 is a relationship between the length Δs of the short side of the triangle, the length r of the two long sides, and the angle $\Delta\theta$.

Figure 6-3c shows why Equation 6-1 is useful: The velocity rotates through the same angle $\Delta\theta$ as the object moves from point 1 (velocity $\vec{v}_1$) to point 2 (velocity $\vec{v}_2$). Because the speed is constant, $\vec{v}_1$ and $\vec{v}_2$ have the same magnitude v. In **Figure 6-3d,** we've converted $\vec{v}_1$ into $-\vec{v}_1$ by reversing its direction and placed the head of $-\vec{v}_1$ against the tail of $\vec{v}_2$ to form the vector difference $\Delta\vec{v} = \vec{v}_2 - \vec{v}_1$. This vector diagram is equivalent to a triangle with two sides of length v separated by an angle $\Delta\theta$ and a third side of length Δv, which is the magnitude of the velocity change between points 1 and 2 (**Figure 6-3e**).

You can see that the triangle in Figure 6-3e is *similar* to the triangle in Figure 6-3b. Both triangles have two equal sides separated by the same angle, so the relationship between the lengths of their sides is the same. Comparing the two figures shows that

(a)

An object in uniform circular motion moves a distance Δs from point 1 to point 2.

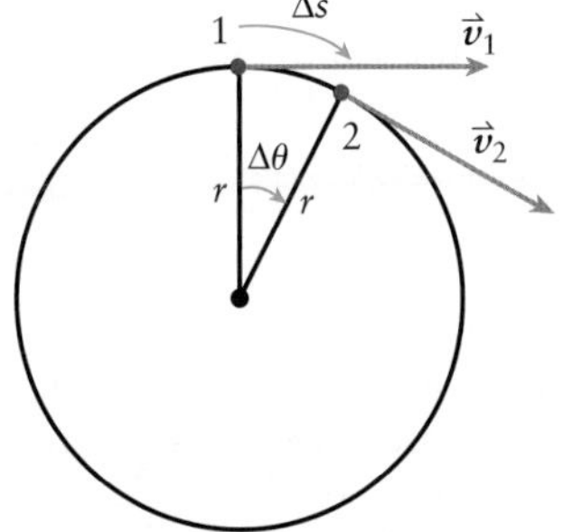

(b)

The three lengths r, r, and Δs approximate the three sides of a triangle.

(c)

The velocity of the object rotates through the same angle $\Delta\theta$.

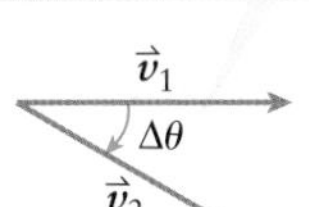

(d)

The vectors $-\vec{v}_1$, $\vec{v}_2$, and $\Delta\vec{v}$ form the three sides of a triangle... $\Delta\vec{v}$ is in the direction of the object's acceleration.

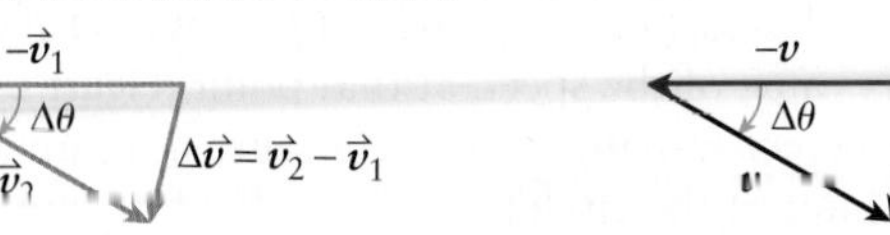

(e)

In uniform circular motion, speed is constant, which means both $-v_1$ and v_2 have the same magnitude v. This triangle is similar to that in part (b).

$-v$ $\Delta\theta$ Δv

Figure 6-3 Analyzing uniform circular motion (a) An object's circular path between points 1 and 2. (b) For a small enough angle, Δs is approximately a line, so a triangle relating the radius r of the circle, the distance traveled Δs, and the angle $\Delta\theta$ can be drawn. (c) Relating the velocities at points 1 and 2. (d) Calculating the velocity change between points 1 and 2. (e) A triangle relating the speed v, the magnitude Δv of the velocity change, and the angle $\Delta\theta$. The angle is again assumed to be small, so the vector drawn approximates the direction of Δv at the center of the angle $\Delta\theta$.

Δv and v in Figure 6-3e correspond to Δs and r, respectively, in Figure 6-3b. So we can write an equation involving Δv and v that's the equivalent of Equation 6-1:

$$\Delta\theta = \frac{\Delta v}{v} \tag{6-2}$$

We'd like to calculate the magnitude of the object's acceleration, which is equal to $a_{\text{cent}} = \Delta v/\Delta t$. (We use the subscript *cent* to remind us that the acceleration is *centripetal.*) To do this calculation we first set our two expressions for $\Delta\theta$, Equations 6-1 and 6-2, equal to each other, which we can do as long as $\Delta\theta$ is very small, and the magnitude of v does not change. This gives a new equation that relates Δv, v, Δs, and r:

$$\frac{\Delta v}{v} = \frac{\Delta s}{r}$$

To get an expression for $a_{\text{cent}} = \Delta v/\Delta t$, multiply both sides of this equation by v and divide both sides by Δt. Then we get

$$a_{\text{cent}} = \frac{\Delta v}{\Delta t} = \left(\frac{\Delta s}{r}\right)\left(\frac{v}{\Delta t}\right) = \left(\frac{\Delta s}{\Delta t}\right)\left(\frac{v}{r}\right) \tag{6-3}$$

Now the object's speed v is equal to the distance Δs that it travels divided by the time interval Δt required to travel this distance: $v = \Delta s/\Delta t$. Therefore, we can replace $\Delta s/\Delta t$ in Equation 6-3 by v, and our expression for the object's centripetal acceleration becomes $a_{\text{cent}} = v(v/r)$, or

EQUATION IN WORDS
Centripetal acceleration for motion in a circle

$$a_{\text{cent}} = \frac{v^2}{r} \tag{6-4}$$

Centripetal acceleration: magnitude of the acceleration of an object in uniform circular motion

Speed of the object as it moves around the circle

Radius of the circle

The units of v and r are m/s and m, respectively. Hence v^2/r in Equation 6-4 has units $(\text{m/s})^2/\text{m}$ or m/s^2, which are the correct units for acceleration.

Equation 6-4 tells us that the magnitude of the centripetal acceleration is proportional to the square of the speed and inversely proportional to the radius of the circle. So there is greater centripetal acceleration the greater the object's speed and the smaller the radius (the tighter the turn). That's entirely consistent with our discussion at the beginning of this section.

Although we derived Equation 6-4 for the case of an object moving around a circle at constant speed, the same equation applies to an object that travels at constant speed around any *portion* of a circle. As an example, think of a bird initially flying north that makes a 90° right turn along a semicircular arc at constant speed until it is heading east. The part of the bird's flight during which it is turning constitutes uniform circular motion, with an acceleration directed toward the center of the circle of which the arc is part (**Figure 6-4**).

Equation 6-4 is also valid when the object's speed changes (a case called *nonuniform* circular motion). In this case, as we saw in Section 3-5, the acceleration has both a component along the direction of motion (which describes changes in speed) and a component perpendicular to the motion (which describes changes in direction). It turns out that the magnitude of this perpendicular component at each point around the circular path is given by Equation 6-4, $a_{\text{cent}} = v^2/r$, provided that we set v equal to the instantaneous value of the speed at that point (**Figure 6-5**). No matter what, an object moving in a circle must have an acceleration toward the center of the circle!

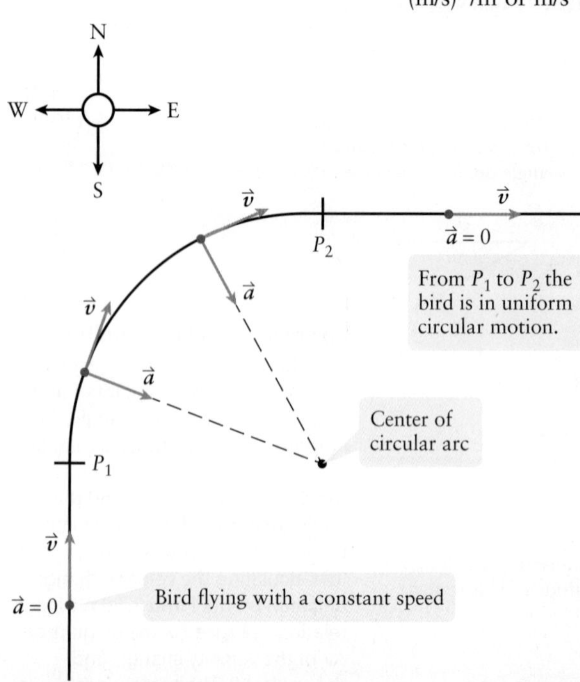

Figure 6-4 Uniform circular motion along an arc An object doesn't have to travel all the way around a complete circle to be in uniform circular motion.

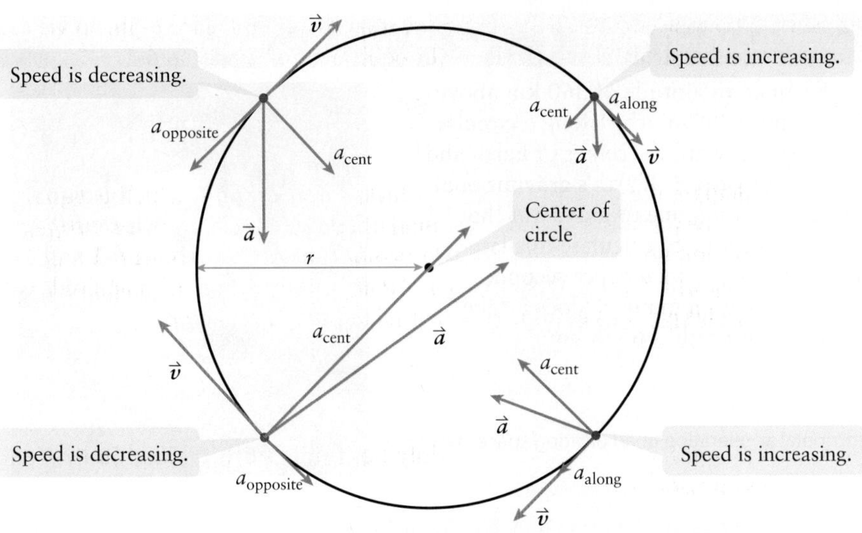

• If an object moves around a circle with varying speed, the acceleration $\vec{a}$ has a component along or opposite the direction of motion.

• Whether the speed is increasing, decreasing, or remaining constant, the acceleration always has a component $a_{cent} = v^2/r$ directed toward the center of the circle.

Figure 6-5 Nonuniform circular motion If the speed varies in circular motion, the acceleration is not directed toward the center of the circle, although a component of it always must be. The component toward the center of the circle is the centripetal acceleration. The component along or opposite the direction of motion is referred to as tangential acceleration, as this will always be tangential to the circular path.

Exam Tip

The magnitude of the centripetal acceleration is labeled as a_c on the AP® Physics 1 equation sheet. You need to know that this is the component of the acceleration of any object moving in a circle pointing toward the center of the circle. If the object is changing speed as well, it will also have a component of acceleration that is tangential to the circle.

EXAMPLE 6-1 Around the Bend

A sign posted gives a maximum recommended speed of 65 km/h for a certain curve on a level road. The curve is a circular arc with a radius of 95 m. What is the centripetal acceleration of a car that takes this curve at the maximum recommended speed?

Set Up

We are given the car's speed v = 65 km/h and the radius r = 95 m of its trajectory, so we can calculate its centripetal acceleration a_{cent} using Equation 6-4.

Centripetal acceleration:

$$a_{cent} = \frac{v^2}{r} \qquad (6\text{-}4)$$

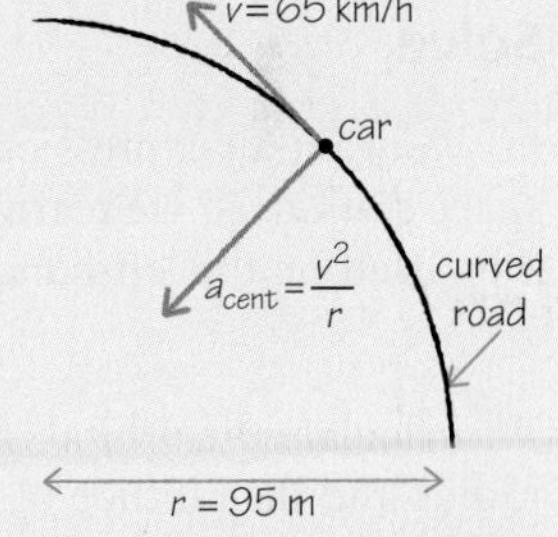

Solve

We are given r in m and v in km/h. To calculate the car's acceleration in m/s², we must first convert v to m/s.

$$v = 65\text{ km/h} \times \left(\frac{1000\text{ m}}{1\text{ km}}\right) \times \left(\frac{1\text{ h}}{3600\text{ s}}\right)$$

$$= 18\text{ m/s}$$

Given the value of v we calculate a_{cent} using Equation 6-4.

$$a_{cent} = \frac{v^2}{r} = \frac{(18\text{ m/s})^2}{95\text{ m}} = 3.4\text{ m/s}^2$$

Reflect

The magnitude of the centripetal acceleration is just over one-third of the magnitude of the acceleration due to gravity during free fall. We know that to change direction takes an acceleration, just like to change speed, given that acceleration is the change in the velocity of an object. To see if this acceleration makes sense, let's compare it to the size of a typical linear acceleration. We hear of high-performance cars that can go from 0 to 60 mph in about 6 seconds. Converting to m/s, this would be about (27 m/s – 0)/6 s = 4.5 m/s². The maximum recommended acceleration for the curve we found is about the same size.

NOW WORK Problems 2 and 4 from The Takeaway 6-2.

EXAMPLE 6-2 Orbital Speed

The International Space Station (ISS) orbits Earth at an altitude of 360 km above Earth's surface. The acceleration required to keep the ISS in orbit along a circular path is provided by Earth's gravity, which is directed toward the center of Earth and hence toward the center of the orbit, as shown in **Figure 6-6.** (Earth's gravitational pull decreases gradually with altitude, so the acceleration due to gravity at the altitude of the ISS is only 8.78 m/s^2. We will find out how to calculate this later in this chapter.) (a) What is the orbital speed of the ISS in meters per second, kilometers per hour, and miles per hour? (b) How long (in minutes) does it take the ISS to complete an orbit? Note that the radius of Earth is 6378 km.

Figure 6-6 An orbiting space station The centripetal acceleration of an orbiting space station or other satellite is provided by Earth's gravity.

Set Up

We are given the ISS's centripetal acceleration a_{cent} and orbital radius r (the sum of Earth's radius and the altitude of the ISS, as the sketch shows). Using this information, we can calculate the speed v using Equation 6-4. Once we know v we can calculate the time T for the ISS to complete one orbit (that is, to travel a distance equal to the circumference of the orbit) using the relationship (speed) = (distance)/(time).

Centripetal acceleration:

$$a_{cent} = \frac{v^2}{r} \qquad (6\text{-}4)$$

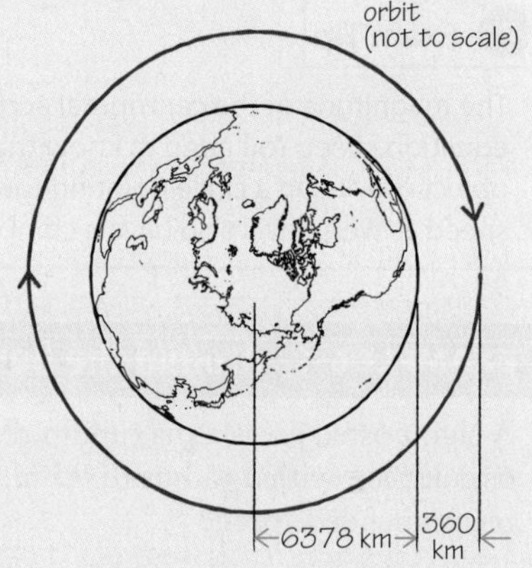

Solve

(a) We want an expression for the speed v of the space station, so we rearrange Equation 6-4 to put v by itself on the left-hand side.

$$a_{cent} = \frac{v^2}{r}$$
$$v^2 = ra_{cent}$$
$$v = \sqrt{ra_{cent}}$$

We calculate the orbital radius r by adding Earth's radius and the altitude of the ISS, then convert the result to meters.

$$r = 6378 \text{ km} + 360 \text{ km}$$
$$= 6738 \text{ km} \times \left(\frac{10^3 \text{ m}}{1 \text{ km}}\right) = 6.738 \times 10^6 \text{ m}$$

Substitute our calculated value of r into the expression for v:

$$v = \sqrt{(6.738 \times 10^6 \text{ m})(8.78 \text{ m/s}^2)}$$
$$= \sqrt{5.92 \times 10^7 \text{ m}^2/\text{s}^2} = 7.69 \times 10^3 \text{ m/s}$$

Convert the speed to km/h and mi/h:

$$v = 7.69 \times 10^3 \text{ m/s} \times \left(\frac{1 \text{ km}}{10^3 \text{ m}}\right) \times \left(\frac{3600 \text{ s}}{1 \text{ h}}\right)$$
$$= 2.77 \times 10^4 \text{ km/h}$$
$$v = 2.77 \times 10^4 \text{ km/h} \times \left(\frac{1 \text{ mi}}{1.609 \text{ km}}\right)$$
$$= 1.72 \times 10^4 \text{ mi/h}$$

(b) The distance that the ISS travels in one orbit is equal to the orbit circumference $2\pi r$. Because (speed) = (distance)/(time), the time T for one orbit equals the orbit circumference divided by the speed.

We convert this answer to minutes:

$$\text{distance} = 2\pi(6.738 \times 10^6 \text{ m})$$
$$= 4.234 \times 10^7 \text{ m}$$

$$\text{time } T = \frac{\text{distance}}{\text{speed}} = \frac{4.234 \times 10^7 \text{ m}}{7.69 \times 10^3 \text{ m/s}}$$
$$= 5.50 \times 10^3 \text{ s} = (5.50 \times 10^3 \text{ s}) \times \left(\frac{1 \text{ min}}{60 \text{ s}}\right)$$
$$= 91.7 \text{ min}$$

Reflect

Let's think about what we know to determine if the idea that there is one particular speed for a circular orbit of a given altitude makes sense. The sketch helps show this. If you drop a ball, it falls straight downward. If you throw it horizontally, it follows a curved path before hitting the ground. The faster you throw the ball, the farther it travels prior to impact, and the more gradually curved its trajectory. If you throw the ball with a sufficiently great speed, the curvature of its trajectory precisely matches the curvature of Earth. Then the ball remains at the same height above Earth's surface and never hits the ground. Putting a satellite such as the International Space Station into orbit works the same way, except that a rocket rather than a strong pitching arm is used to provide the necessary speed. Gravity is making the ISS "fall," but because of its high speed, it falls *around* Earth rather than *toward* Earth. As we calculated in part (b), the ISS makes a complete orbit about every hour and a half. This might not be something you could say makes sense from what you know, but the reason for calculating so many different versions of numbers in this example was to help you start to get a feel for the size of numbers in this type of motion. It is the fact that the astronauts and everything in the station are all accelerating toward Earth at the local gravitational acceleration that causes the astronauts' sense of weightlessness. See the Watch Out! box below for more on weightlessness.

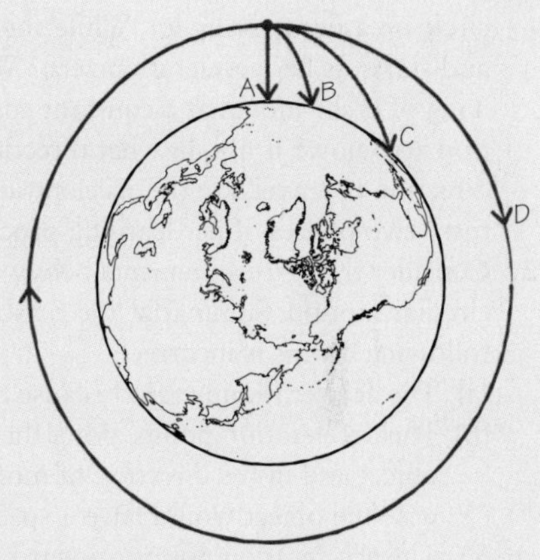

NOW WORK Problem 5 from The Takeaway 6-2.

WATCH OUT!

What causes weightlessness—and what doesn't.

It's a common misconception that an object in orbit is "beyond the pull of Earth's gravity." It's true that an astronaut on board the International Space Station floats around the cabin *as though* she were weightless and unaffected by gravity. But if gravity weren't exerted by Earth on an orbiting satellite or its occupants, the satellite would have *zero* acceleration and hence couldn't stay in orbit: It would maintain a constant velocity and simply fly off into space in a linear path. In fact a satellite such as the ISS remains in orbit precisely because it *does* feel the pull of gravity, which provides the acceleration necessary to make the satellite follow a circular trajectory around Earth. An astronaut on board the ISS *seems* to be weightless because she is really an independent satellite of Earth following the same orbit as the ISS does. Because the astronaut and the ISS are both following the same trajectory, there is nothing pushing the astronaut toward any of the station's walls. As a result, even though gravity is exerted on the astronaut, she *feels* weightless. We will return to this idea in Section 6-6.

 **Exam Tip**

The gravitational force on an object is never zero except in the idealized situation when we are ignoring the gravitational forces of other objects because they are small compared to other forces being considered.

 **Exam Tip**

The feeling of weightlessness means that any surface you touch has the same acceleration as you, so that the surface exerts no normal force on you. It is the normal force exerted on you that gives you the perception of weight.

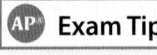

THE TAKEAWAY for Section 6-2

✔ If an object follows a curved trajectory, it has an acceleration, even if its speed remains constant. Motion in a circle is an important example of motion of this kind.

✔ Uniform circular motion is motion around a circle at a constant speed. The acceleration of an object in uniform circular motion is centripetal: It points toward the center of the circle.

✔ Problems in uniform circular motion involve the relationships among centripetal acceleration, speed, and radius expressed by Equation 6-4.

Prep for the AP® Exam

AP® Building Blocks

1. A child who is learning to ride a bike pedals around in a circle on a flat parking lot. While she is learning she stops and starts. Is her acceleration zero? When she gets the hang of it she moves at a constant speed. Is her acceleration zero now? If not, in what directions, relative to her direction of travel, are the accelerations pointed in each of these two phases of her learning process?

EX 6-1 2. Consider the three statements below regarding uniform circular motion. Summarize the reason that each of the following claims is incorrect.
   (a) The acceleration is zero because the speed is constant.
   (b) The acceleration points along the circular path of the object and in the direction of motion, because otherwise the object would have a speed of zero.
   (c) The acceleration points outward from the center of the circle, because otherwise the object would spiral toward the center of the circular path.

3. A car is moving in a circular path on flat ground with a constant speed. Describe, with a complete sentence and with a drawing, the direction of the acceleration at the instant that the car is pointed north and is turning to the right. In your drawing, add vectors displaying the direction of the velocity and acceleration as the car travels through one quarter of the circle.

4. A ball is twirled on a 0.870-m-long string with a constant speed of 3.36 m/s. Calculate the acceleration of the ball. Be sure to specify the direction of the acceleration.

5. A ball attached to a string is twirled in a circle of radius 1.25 m with a constant speed of 2.25 m/s. Calculate the time required to travel once around the circle.

AP® Skill Builders

6. A document intended to support quality control for laboratory analyses specifies that centrifuges used to separate components of whole blood must have a relative centrifugal force (RCF) of between 1000 and 1200 g. The document provides the following formula to calculate RCF.

$$\text{RCF} = 1.119 \times 10^{-5} \times r \times n^2$$

In this formula, the document claims that "r is the rotating radius (cm) (that is, the radius in millimeters measured from the center of rotation) and n is the rotation rate (rpm)."

   (a) Evaluate these claims, assuming that the document can be interpreted to mean the following:
   - "Relative centrifugal force" is not actually a force but the centripetal acceleration.
   - The radius is measured in centimeters not millimeters.
   - The units of the calculated RCF are g, the acceleration due to gravity.

   The lab director has unpacked a new centrifuge and needs to write brief instructions for technicians who will use the centrifuge to do analyses of whole blood. She measures the radius of rotation to be 10.0 ± 0.05 cm. On the face of the centrifuge is a switch that is used to select rotation rates. The switch can be set at any one of the eight positions on the dial from 500 to 4000 rpm in increments of 500 rpm. She makes the following drawing of the dial:

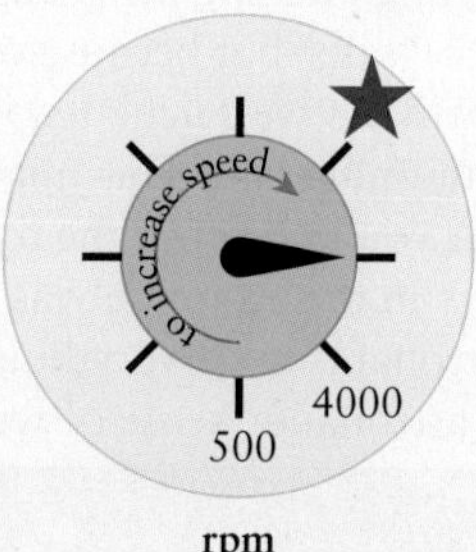

   (b) In her drawing, the star is above the setting that lab technicians must use to centrifuge whole blood. Justify the claim that the location of the star is consistent with $1000\ g \leq \text{RCF} \leq 1200\ g$.
   (c) The radius of rotation was difficult to measure and the lab director is concerned that the uncertainty in the measurement might be greater than ±0.05 cm. Justify the claim that if the uncertainty is ±0.1 cm the centrifuge cannot be used for the analysis of whole blood when at this setting.

7. An object shot into the air from a launcher on Earth's surface moves along a curved path in a projectile motion. An object launched from Earth's surface with a sufficiently large initial speed moves in an orbit around Earth that may be a uniform circular motion. The same event with two different initial speeds leads to two motions that we describe differently as projectile motion or as circular motion.
   (a) Summarize in complete sentences the differences and similarities between the projectile motion and circular motion paths in terms of the vector components of the displacement, velocity, and acceleration parallel and perpendicular to the surface of Earth.
   (b) Summarize the differences and similarities by completing the following table.

Vector components	Projectile motion	Uniform circular motion
Displacement perpendicular to the surface		
Displacement parallel to the surface		
Velocity perpendicular to the surface		
Velocity parallel to the surface		
Acceleration perpendicular to the surface		
Acceleration parallel to the surface		

AP® Skills in Action

8. In a vertical dive, a peregrine falcon can accelerate at 0.60 times the free-fall acceleration (that is, at 0.60 *g*) to reach a speed of about 1.0×10^2 m/s.
   (a) Express the minimum vertical height from which the dive begins.
   (b) If a falcon pulls out of a dive into a circular arc at this speed and can sustain a centripetal acceleration of 6 *g*, determine the minimum radius of the turn.

9. Mary and Kelly are running side by side on the 200-m track shown. As they enter one of the semicircular turns, Mary is running on the inside lane, and Kelly is running in the outer lane. The lanes both have a width of 1.50 m and are divided by a chalk line with a radius of 20.0 m on the semicircular segments. They both enter and leave the semicircular turn at the same time.

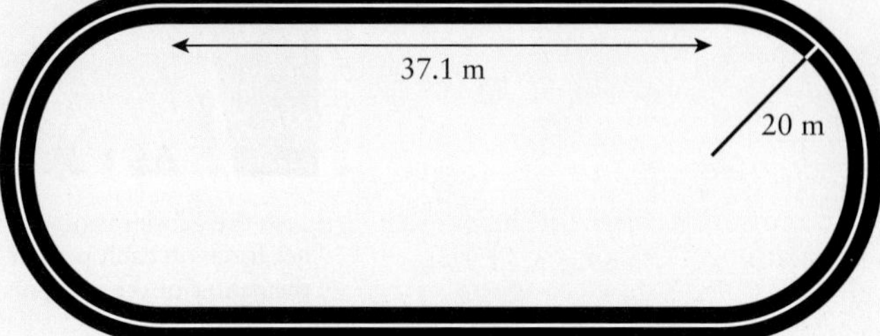

   (a) Express the relative centripetal accelerations of the two runners in terms of the radius of Mary's uniform circular motion, r_{Mary}, and the difference between the radii of Kelly's and Mary's motion, Δr.
   (b) Calculate the numerical value of the expression in part (a).
   (c) Mary's speed when she enters the turn is 8.33 m/s. Calculate Kelly's speed if they both maintain constant speeds.

6-3 For an object in uniform circular motion, the net force exerted on the object points toward the center of the circle

Let's apply the concept of circular motion to something a little closer to home. Think of thrill seekers on fast-moving amusement park rides or a motorcycle making a sharp turn (**Figure 6-7**). In both cases an object is following a curved path, so the direction of its velocity is changing. Because the direction of the velocity is changing, the object is *accelerating* (the object's speed can be constant or changing) with a component of the acceleration directed toward the inside of the curve. In the previous section we learned that this acceleration, called the *centripetal acceleration*, depends on the radius of the curve and on the object's speed. From Section 4-2 we know that for an object to accelerate there must be a net external force on the object that points in the direction of the acceleration $\vec{a}$. In this section we'll put these ideas together and complete our analysis of objects that move along curved paths.

What kinds of force can provide the acceleration needed to make an object follow a curved path? The answer is simple: *any kind of force*, provided it has a component perpendicular to the object's trajectory. In **Figure 6-7a,** for example, each person's acceleration arises from the normal force exerted on each person by the seat on which that person sits. For the motorcycle in **Figure 6-7b,** the acceleration of the motorcycle is provided by the friction force that the ground exerts on the motorcycle's tires. If the road is banked, then the normal force exerted by the road also has a component inward. The force on the motorcycle rider is provided by a combination of normal and friction forces exerted on him by the motorcycle. Since they are traveling together, we can just think of them as a single object, though. Whatever force the motorcycle is exerting on the rider, the rider is exerting an equal and opposite force back on the motorcycle.

One of the simplest situations in which an object follows a curved trajectory is that which we introduced in the previous section, *uniform circular motion*. Because

 **Exam Tip**

For an object in uniform circular motion, the speed with which it travels can be related to properties of the motion. The object travels a distance equal to the circumference of the circular path or radius *R* during the time it takes to complete one full circle, the period *T*. So, the speed for uniform circular motion can be written as $(2\pi R)/T$.

Figure 6-7 Forces exerted on an object in uniform circular motion An object moving on a circular path at a constant speed is in uniform circular motion. Since the object's direction of motion is continuously changing, the object is accelerating and a net external force must be exerted on it.

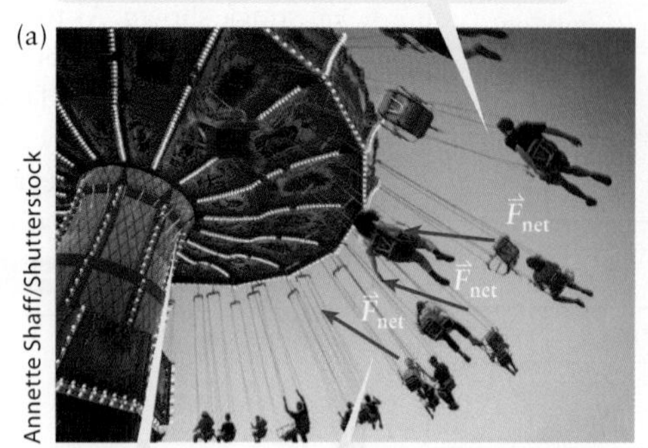
Each person on this amusement park ride moves in a circle at constant speed...
(a)
$\vec{F}_{net}$
$\vec{F}_{net}$
$\vec{F}_{net}$
...so the acceleration of each person and the net force on each person both point toward the center of the person's circular path.
Annette Shaff/Shutterstock

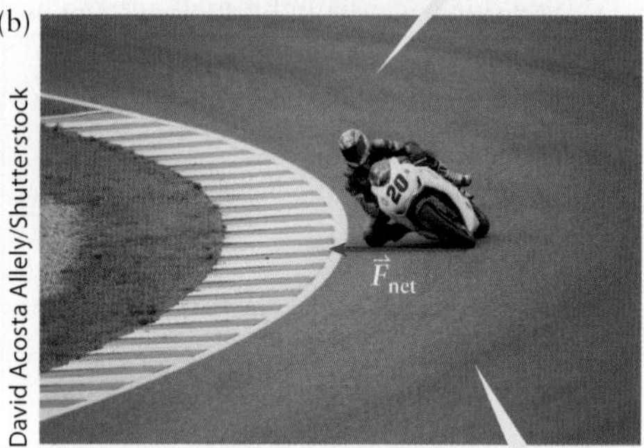
This motorcycle rider also moves along a circular path at constant speed...
(b)
$\vec{F}_{net}$
...so his acceleration and the net force on him both point toward the center of the circular path.
David Acosta Allely/Shutterstock

the acceleration is always toward the center of the circle, it follows from Newton's second law that at any instant the net external force exerted on the object of mass m must likewise be directed toward the center of the circle. Furthermore, the magnitude of the net external force must be equal to ma_{cent}. We can write this statement in equation form as

EQUATION IN WORDS
Newton's second law equation for uniform circular motion

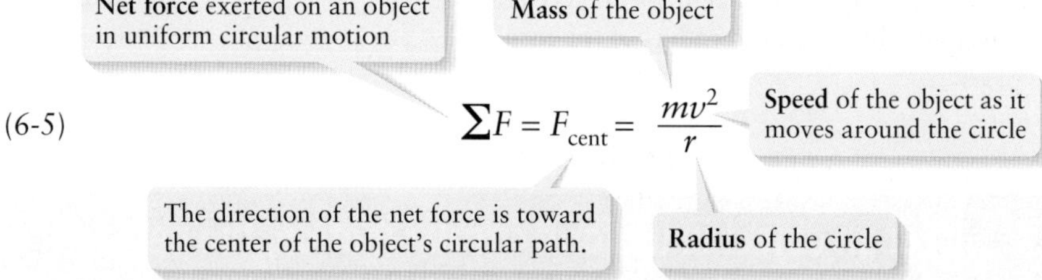
Net force exerted on an object in uniform circular motion
Mass of the object
Speed of the object as it moves around the circle
The direction of the net force is toward the center of the object's circular path.
Radius of the circle

(6-5) $$\sum F = F_{cent} = \frac{mv^2}{r}$$

This equation says that in uniform circular motion the magnitude of the net external force exerted on the object is equal to the object's mass m multiplied by the magnitude of the centripetal acceleration $a_{cent} = v^2/r$.

Let's look at several examples of how to apply these ideas to problems in which an object moves around a circle at constant speed.

WATCH OUT !

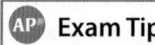

"Centripetal force" is not a special *kind* of force—it describes the *direction* of the net force.

The force that points toward the inside of an object's curving trajectory and produces the centripetal acceleration is sometimes called the **centripetal force**. Keep in mind, however, that this is not a separate kind of force that you need to include in a free-body diagram. The word *centripetal* simply describes the direction of the force. Just as we might write F_x to denote the x component of a force, we write F_{cent} to denote the component of a force directed toward the center of an object's circular trajectory (although sometimes we will just define toward the center at some instant x, and still call it F_x). As we'll see in the following examples, in different circumstances different forces play the role of the centripetal force.

AP® Exam Tip

Since the free-body diagram is supposed to show all the individual forces on an object, and not the components or sums, it should never have a force labeled as the centripetal force. You will then use the free-body diagram to determine what the centripetal force is by identifying the net force—or the component of the net force—toward the center of the circular motion.

EXAMPLE 6-3 A Rock on a String

You tie one end of a lightweight string around a rock of mass m so that the distance from the point at which you hold the string to the center of the rock (which you can treat as an object) is a length, L. You make the rock swing in a horizontal circle at constant speed. As you swing the rock the string makes an angle θ with the vertical (**Figure 6-8**). Derive an expression for the speed at which the rock moves around the circle in terms of m, L, θ, and relevant physical constants. Ignore the effect of drag forces.

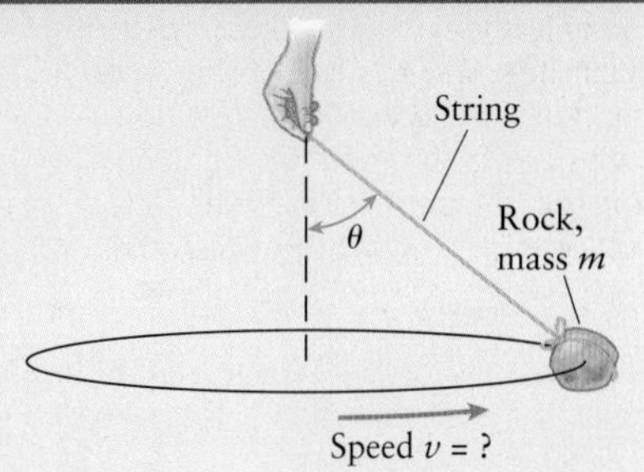

Figure 6-8 Moving in a circle How fast must the rock move to make the string hang at an angle θ from the vertical?

Set Up

Because the rock is not too large compared to the length of the string, we can treat it as an object, and use the length, L, from the hand to the center of the rock. Because the rock moves at constant speed in a horizontal circle, its acceleration points horizontally toward the center of the circle. We choose the positive x axis to point in this direction for one point in time, and the positive y axis to point upward.

Only two forces are exerted on the rock: the tension force $\vec{F}_T$ exerted by the string and the downward gravitational force $\vec{F}_{g,rock}$. The tension force is directed along the string and so points at an angle θ to the vertical. The vertical component of $\vec{F}_T$ balances the downward force of gravity, while the horizontal component of $\vec{F}_T$ provides the centripetal acceleration. Our goal is to solve for the rock's speed v. First we draw a free-body diagram with this information to make it easier to translate Newton's second law into equations we can solve. We also draw a sketch to help us determine the radius of the rock's circular path.

Newton's second law equation for the rock:

$$\sum \vec{F}_{\text{ext on rock}} = \vec{F}_T + \vec{F}_{g,rock} = m_{rock}\vec{a}_{rock}$$

Centripetal acceleration on rock:

$$a_{cent} = \frac{v^2}{r} \quad (6\text{-}4)$$

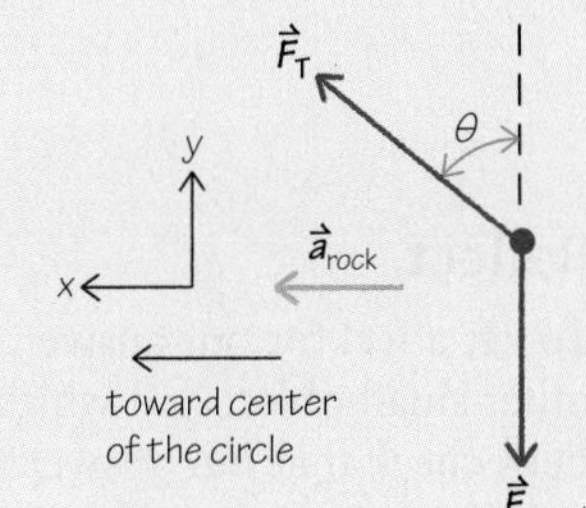

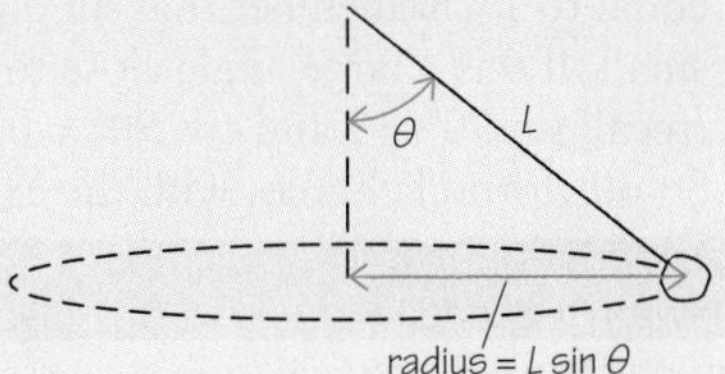

Solve

Write Newton's second law in component form for the rock. Note that the quantity v, which is what we're trying to find, doesn't appear in these equations. We'll introduce it in the next step through the expression for the rock's centripetal acceleration.

Newton's second law in component form applied to the rock:

x: $F_T \sin\theta = m_{rock}a_{cent}$

y: $F_T \cos\theta + (-F_{g,rock}) = F_T \cos\theta - m_{rock}g = 0$

Use Equation 6-4 to express the acceleration of the rock. Note that because the string is at an angle, the radius of the rock's circular path is *not* equal to the length, L.

Radius of the rock's circular path:

$$r = L\sin\theta$$

so the rock's centripetal acceleration is

$$a_{cent} = \frac{v^2}{r} = \frac{v^2}{L\sin\theta}$$

Substitute the expression for centripetal acceleration into the Newton's second law equation for the x direction and solve for v.

Newton's second law equations for the rock become

x: $F_T \sin\theta = m_{rock}a_{cent} = \dfrac{m_{rock}v^2}{L\sin\theta}$

y: $F_T \cos\theta = m_{rock}g$

Solve the y component equation for F_T:

$$F_T = \frac{m_{rock}g}{\cos\theta}$$

Substitute this expression into the x component equation and solve for v:

$$\frac{m_{rock}g}{\cos\theta}\sin\theta = \frac{m_{rock}v^2}{L\sin\theta}$$

$$\frac{g\sin\theta}{\cos\theta} = \frac{v^2}{L\sin\theta}$$

$$v^2 = \frac{gL\sin^2\theta}{\cos\theta}$$

$$v = \sqrt{\frac{gL\sin^2\theta}{\cos\theta}}$$

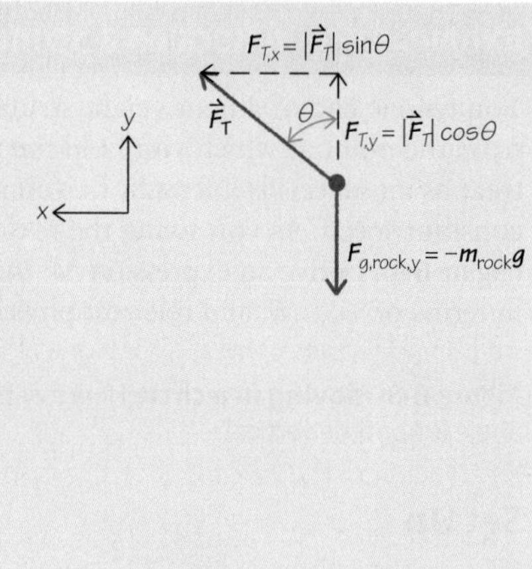

Reflect

To get a feel for our answer, you can experiment with this motion yourself by tying a piece of string around an eraser or other small object and whirling it in a horizontal circle. You'll find that the faster the object moves, the greater the angle θ. Let's check that our answer agrees with this conclusion.

If θ is a small angle close to zero so that the string hangs down almost vertically, $\sin\theta$ is small and $\cos\theta$ is nearly equal to 1. (Remember that $\sin 0 = 0$ and $\cos 0 = 1$.) So the ratio $(\sin^2\theta)/(\cos\theta)$ is a small number, and the speed v will be small. If θ is a large angle close to 90°, so that the string is nearly horizontal, $\sin\theta$ is a little less than 1 and $\cos\theta$ is small (recall $\sin 90° = 1$ and $\cos 90° = 0$). In this case the ratio $(\sin^2\theta)/(\cos\theta)$ is a large number, and the speed v will be large. So our formula agrees with the experiment!

NOW WORK Problems 1 and 5 from The Takeaway 6-3.

WATCH OUT !

No "centrifugal force," please.

In Example 6-3 you may have been tempted to add a force that points toward the *outside* of the rock's curved trajectory. Don't do it! If you felt this temptation, it's because you're thinking of the so-called centrifugal force you feel when riding in a car that makes a sharp turn. This "centrifugal force" seems to push you toward the outside of the turn (*centrifugal* means "fleeing the center"). This force is *fictitious*, however; nothing is exerting an outward force on you. Rather, you're just feeling your body's inertia, which is the tendency of all objects to resist changing direction. When you are in a car going around a curve, the seat belt, friction with the car seat, and maybe a normal force from the door all help to push you into the curve. By Newton's third law, you are pushing outward with an equal magnitude force, but the friction force exerted by the road on the tires keeps the car going on its curved path. Both you and the car are experiencing a net centripetal force. The law of inertia, Newton's first law, explains what would happen if you let go of the string attached to the rock: The rock would fly off in a line as seen from above (**Figure 6-9**), and so it would move away from the center of its circular path because of the *absence* of the inward tension force. An object going around a curve is *not* in balance: The object is accelerating, so the net external force on it is *not* zero!

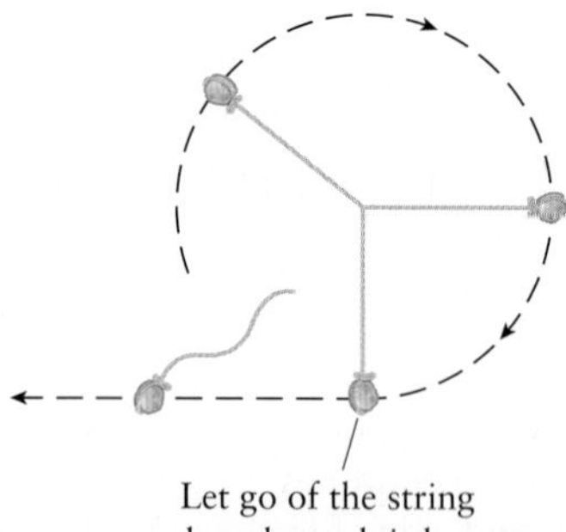

Figure 6-9 Force is required to keep direction changing In a top view of the rock's trajectory, we see that if the string is let go or breaks, the rock flies off in a line due to its inertia—because there is no longer a centripetal force (the horizontal component of the string's tension in this case) to keep it moving in a circle.

EXAMPLE 6-4 Making an Airplane Turn

An airplane banks (dips its wing) to one side to turn in that direction. By banking the plane, the *lift force*, a force exerted perpendicular to the direction of flight due to the differences in the motion of air over and under the airplane's wings—ends up with both a vertical component and a horizontal component (**Figure 6-10**). The horizontal component of this force is directed toward the center of the desired circular path (centripetally), causing the airplane to move in a circular path. (a) If the airplane of mass m_{airplane} is traveling at speed v and is banked by an angle θ derive an expression for the radius of the turn. (b) The pilot of the airplane has mass m_{pilot} and weight $F_{\text{g,pilot}} = m_{\text{pilot}}g$. Derive an expression for the force that the seat exerts on the pilot as the airplane makes its banked turn.

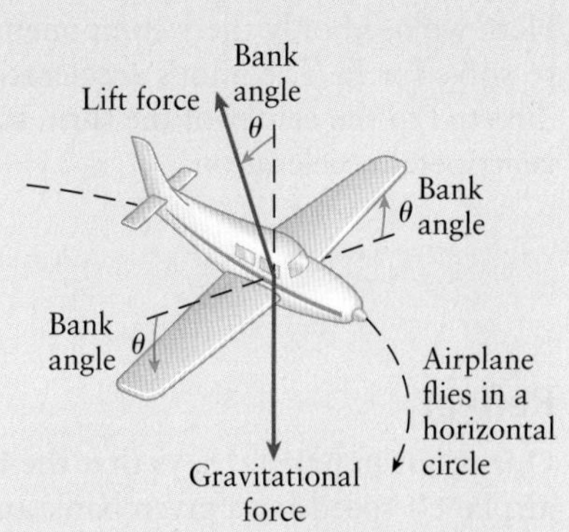

Figure 6-10 An airplane in a banked turn What mathematical expression describes the radius of the airplane's turn? The forward force on the plane due to the interaction of the engine and the air and the backward drag force on the plane due to the air cancel because the plane's speed is not changing, and are not shown.

Set Up

We draw two free-body diagrams, one for the airplane of mass m_{airplane} and one for the pilot of mass m_{pilot}. (In both cases the airplane and pilot are coming at us head on.) Both of these diagrams are almost identical to the free-body diagram for the rock in Example 6-3. Besides the gravitational force, a second force has a vertical component that balances the gravitational force (so that the airplane doesn't accelerate up or down) and a horizontal component that gives the object a centripetal acceleration.

In part (a) we want to find the radius r of the turn; in part (b) our goal is to determine the value of the normal force F_n (the force that the seat exerts on the pilot). The value of F_n is the pilot's *apparent weight* (also referred to as *effective weight*), or how much she *feels* that she weighs during the turn.

Newton's second law equation for the airplane:

$$\sum \vec{F}_{\text{ext on airplane}} = \vec{F}_{\text{L}} + \vec{F}_{\text{g,airplane}} = m_{\text{airplane}}\vec{a}_{\text{airplane}}$$

Newton's second law equation for the pilot:

$$\sum \vec{F}_{\text{ext on pilot}} = \vec{F}_{\text{n}} + \vec{F}_{\text{g,pilot}} = m_{\text{pilot}}\vec{a}_{\text{pilot}}$$

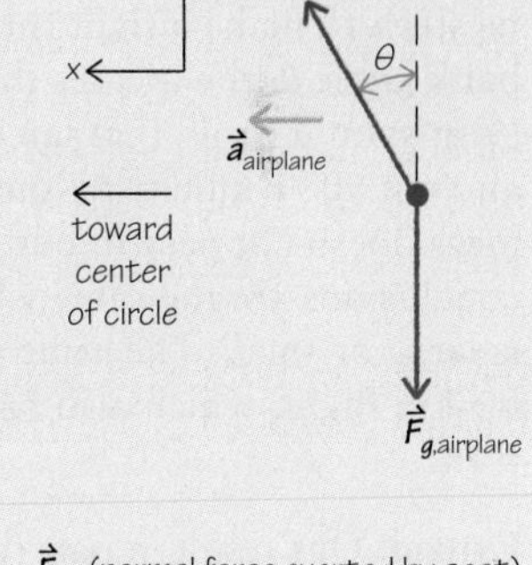

Centripetal acceleration:

$$a_{\text{airplane}} = a_{\text{pilot}} = a_{\text{cent}} = \frac{v^2}{r} \quad (6\text{-}4)$$

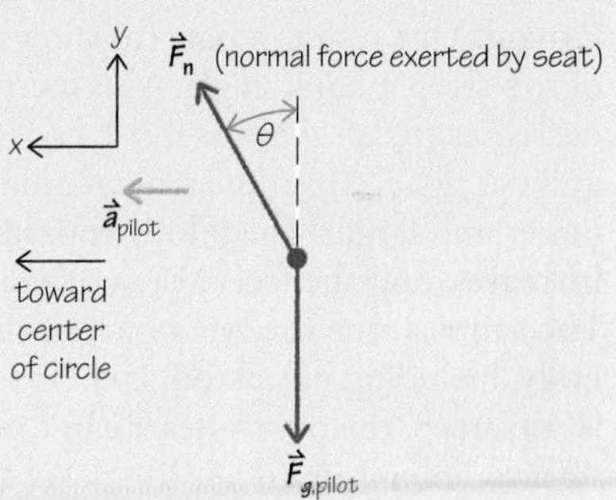

Solve

(a) Once again, we begin by writing Newton's second law in component form. The x component equation involves the radius r of the turn, but we don't know the value of the lift force F_L. We find this force from the y component equation, and we then substitute the value of F_L into the x equation and solve for r. The airplane's acceleration is directed toward the center of the turn, so it is a centripetal acceleration.

Newton's second law in component form applied to the airplane:

$$x\text{: } F_L \sin\theta = m_{\text{airplane}}a_{\text{cent}} = m_{\text{airplane}}\frac{v^2}{r}$$

$$y\text{: } F_L \cos\theta + (-F_{\text{g,airplane}}) = F_L \cos\theta - m_{\text{airplane}}g = 0$$

From the y equation,

$$F_L \cos\theta = m_{\text{airplane}}g$$

$$F_L = \frac{m_{\text{airplane}}g}{\cos\theta}$$

Substitute this expression for F_L into the x equation and solve for r:

$$m_{\text{airplane}}\frac{v^2}{r} = F_L \sin\theta = \frac{m_{\text{airplane}}g}{\cos\theta}\sin\theta = m_{\text{airplane}}g\tan\theta$$

$$\frac{v^2}{r} = g\tan\theta$$

$$r = \frac{v^2}{g\tan\theta}$$

(b) The component form of Newton's second law for the pilot is very similar to that for the airplane—the main difference is that the lift force F_L is replaced by the normal force F_n. Here we need only the y component equation to solve for F_n. The pilot's acceleration is directed to the center of the turn, so it is a centripetal acceleration.

Newton's second law in component form applied to the pilot:

$$x: F_n \sin\theta = m_{pilot}a_{cent} = m_{pilot}\frac{v^2}{r}$$

$$y: F_n \cos\theta + (-F_{g,pilot}) = F_n \cos\theta - m_{pilot}g = 0$$

Solve the y component equation for F_n:

$$F_n \cos\theta = m_{pilot}g$$

$$F_n = \frac{m_{pilot}g}{\cos\theta}$$

Reflect

Our result in part (a) says that the faster the airplane's speed for a given bank angle θ, the larger the radius r of the turn. To make a tight turn at high speed (that is, to have r be small even though v is large), the quantity $\tan\theta$ has to be as large as possible. So the bank has to be steep to make a tight turn: The greater the bank angle θ, the greater the value of $\tan\theta$. (You should recall that $\tan 0 = 0$, $\tan 45° = 1$, and $\tan 90° =$ infinity.) Note that the airplane's mass doesn't appear in our result, so our conclusions are the same whether the airplane is large or small. The same result applies to a bird in flight, which also banks its wings to turn.

The greater the bank angle θ, the smaller the turning radius r of the airplane:

$$r = \frac{v^2}{g\tan\theta}$$

For $\theta = 45°$: $\tan 45° = 1$

$$r = \frac{v^2}{g}$$

For $\theta = 60°$: $\tan 60° = 1.73$

$$r = \frac{v^2}{1.73g} = 0.577\frac{v^2}{g}$$

For $\theta = 80°$: $\tan 80° = 5.67$

$$r = \frac{v^2}{5.67g} = 0.176\frac{v^2}{g}$$

Extend: Our result in part (b) shows the danger of too steep a bank angle. As θ increases, $\cos\theta$ decreases (recall that $\cos 0 = 1$, $\cos 60° = 1/2$, and $\cos 90° = 0$), so the scale reading becomes larger and larger. The pilot's apparent weight increases, and she feels heavier than normal. The same is true for every part of the pilot's body, including her blood. If the bank angle is too steep, the pilot's heart can't pump this "heavy" blood up to her brain. As a result, a typical pilot will lose consciousness after 5 to 10 s at an apparent weight of 4 to 5 times $m_{pilot}g$.

To counter these effects, fighter pilots (who often make very sharp turns at steep bank angles) wear *g-suits*. These are garments that apply pressure to the legs and lower abdomen, thus preventing blood from pooling in the lower body, even during extreme aerial maneuvers.

The greater the bank angle θ, the greater the pilot's scale reading and the heavier she feels:

actual weight $= m_{pilot}g$

apparent weight $= F_n = \dfrac{m_{pilot}g}{\cos\theta}$

For $\theta = 45°$: $\cos 45° = 0.707$

$$F_n = \frac{m_{pilot}g}{0.707} = 1.41\, m_{pilot}g$$

(The pilot feels "1.41 g.")

For $\theta = 60°$: $\cos 60° = 0.500$

$$F_n = \frac{m_{pilot}g}{0.500} = 2.00\, m_{pilot}g$$

(The pilot feels "2.00 g.")

For $\theta = 80°$: $\cos 80° = 0.174$

$$F_n = \frac{m_{pilot}g}{0.174} = 5.76\, m_{pilot}g$$

(The pilot feels "5.76 g" and might pass out.)

NOW WORK Problem 2 from The Takeaway 6-3.

EXAMPLE 6-5 A Turn on a Level Road

A car of mass m_{car} exits the freeway and travels around a circular off-ramp at a constant speed of 15.7 m/s (56.5 km/h, or 35.1 mi/h). The ramp is level (it is not banked), so the centripetal force has to be provided by the force of friction exerted by the road on the tires. If the coefficient of static friction between the car's tires and the road is 0.550 (a typical value when the road surface is wet), what radius of curvature is required for the off-ramp so that the car does not skid?

Set Up

This example is very similar to Example 5-6 in Section 5-4. The only difference is that in Example 5-6 the static friction force gave Roberto a *forward* acceleration, while in this example the static friction force causes a *centripetal* acceleration.

To begin, we draw the free-body diagram, showing the car head-on. Note that even though the car is rolling on its tires, the force that provides the centripetal acceleration is *static* friction rather than rolling friction. That's because the force of interest is the one that prevents the car from sliding sideways away from the center of its circular path. (By contrast, rolling friction is exerted opposite to the car's forward motion and so doesn't contribute to the centripetal acceleration.) There would be only a sideways kinetic friction if the car had already started to skid, and that would be smaller than the static friction, so let's make sure that doesn't happen! The minimum value of the radius r corresponds to the maximum centripetal acceleration $a_{cent} = v^2/r$, which in turn corresponds to the maximum value $F_{S,max}$ of the static friction force.

Newton's second law equation for the car:

$$\sum \vec{F}_{\text{ext on car}} = \vec{F}_n + \vec{F}_{g,car} + \vec{F}_{S,max} = m_{car}\vec{a}_{car}$$

Centripetal acceleration:

$$a_{car} = a_{cent} = \frac{v^2}{r} \quad (6\text{-}4)$$

Magnitude of the static friction force:

$$F_{S,max} = \mu_S F_n \quad (5\text{-}1b)$$

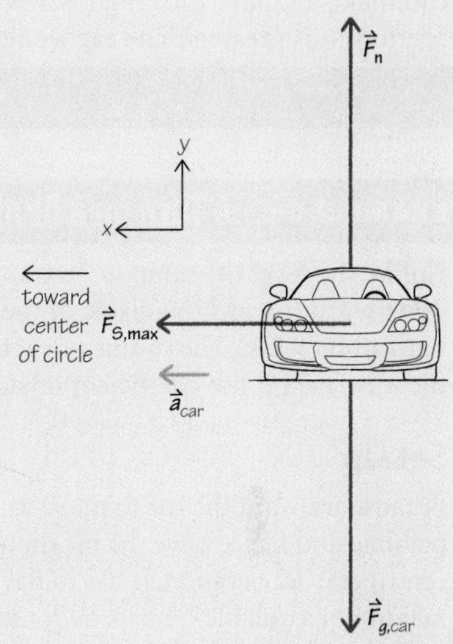

Solve

We write Newton's second law in component form for the car, with the static friction force set equal to its maximum value $F_{S,max}$, and use our free-body diagram to get the right forces and directions for our equation. The horizontal (x) acceleration of the car is the centripetal acceleration; there is zero net vertical (y) acceleration. We then solve for the radius r of the car's circular path.

Newton's second law in component form for the car, assuming that the static friction force has its maximum value:

x: $F_{S,max} = m_{car}a_{cent}$

y: $F_n + (-F_{g,car}) = 0$

From the y component equation,

$$F_n = F_{g,car} = m_{car}g$$

So the maximum force of static friction of the tires is

$$F_{S,max} = \mu_S F_n = \mu_S m_{car} g$$

Substitute this expression and $a_{cent} = v^2/r$ into the x component equation, then solve for r:

$$\mu_S m_{car} g = m_{car}\frac{v^2}{r}$$

$$r = \frac{v^2}{\mu_S g}$$

Using $\mu_S = 0.550$ and $v = 15.7$ m/s,

$$r = \frac{(15.7 \text{ m/s})^2}{(0.550)(9.80 \text{ m/s}^2)} = 45.7 \text{ m}$$

$F_{n,y} = +F_n$
y
x
$F_{S,max,x} = +F_{S,max}$
$F_{g,car,y} = -F_{g,car}$

Reflect

Our answer is the *minimum* radius of the off-ramp. If the radius is larger, the car will have a smaller centripetal acceleration $a_{cent} = v^2/r$ and the static friction force required will be less than the maximum force available. This would give a diameter of about one football field, which seems a little large, but that is why ramps are usually banked at an angle to utilize a component of the normal force exerted on the car as a centripetal force. We will look at that in detail in the next example. So this answer is reasonable. Looking at our equation, the radius gets larger as the speed increases, which makes sense. It gets smaller as the coefficient of static friction gets bigger for a given speed. This also makes sense: The car can make a tighter turn (that is, r will be smaller) if the tires have a better "grip" on the road (so the coefficient of static friction μ_S is greater). The car would be able to make a tighter turn if the road were dry, in which case μ_S is about 1.0.

NOW WORK Problem 3 from The Takeaway 6-3.

EXAMPLE 6-6 A Turn on a Banked Road

Unlike the level off-ramp in Example 6-5, many freeway off-ramps, as well as corners at racetracks, are banked (angled downward toward the inside of the curved path). As an example, the corners at the Indianapolis Motor Speedway are banked by 9.20°. The radius of each corner is 256 m. What is the fastest speed at which a race car could take one of these corners if the coefficient of static friction is $\mu_S = 0.550$, as in Example 6-5?

Set Up

Because we want the car to move as quickly as possible and hence have the maximum possible centripetal acceleration, it has to have the maximum available centripetal force exerted on it. In this extreme case the static friction force will be at its maximum value. Our goal is to find the value of v in this situation. We start, as with all force problems, by drawing the free-body diagram. In this problem, given that the acceleration is directed horizontally toward the center of the circle, we choose the x axis in that direction (instead of parallel to the ramp). In contrast to the situation in Example 6-5, the net centripetal force on the car comes from *two* forces exerted by the banked road: the horizontal component of the normal force and the horizontal component of the static friction force.

Newton's second law equation for the car:

$$\sum \vec{F}_{\text{ext on car}} = \vec{F}_n + \vec{F}_{g,car} + \vec{F}_{S,max} = m_{car}\vec{a}_{car}$$

Centripetal acceleration:

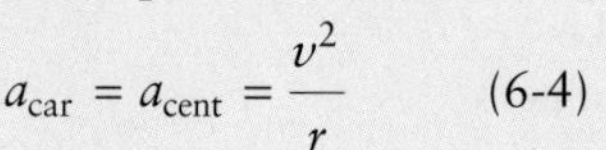

$$a_{car} = a_{cent} = \frac{v^2}{r} \qquad (6\text{-}4)$$

Magnitude of the static friction force:

$$F_{S,max} = \mu_S F_n \qquad (5\text{-}1b)$$

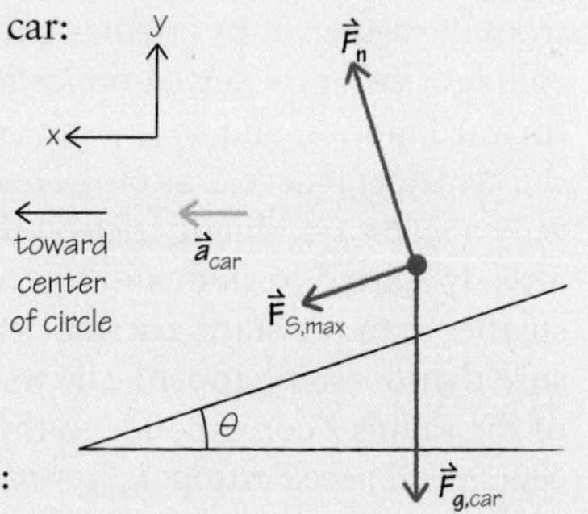

Solve

As in Example 6-5, we write Newton's second law in component form for the car, with the static friction force set equal to its maximum value $F_{S,max}$ and use our free-body diagram to identify the forces and their directions. Again the horizontal (x) acceleration equals the centripetal acceleration; there is zero vertical (y) acceleration. The difference is that now the normal force and friction force have both x and y components.

Newton's second law in component form for the car, assuming that the static friction force has its maximum value:

$$x\text{: } F_n \sin\theta + F_{S,max}\cos\theta = m_{car}a_{cent}$$

$$y\text{: } F_n \cos\theta + (-F_{g,car}) + (-F_{S,max}\sin\theta) = 0$$

Substitute the expressions for the maximum force of static friction, the centripetal acceleration, and the weight of the car:

$$F_{S,max} = \mu_S F_n;\ a_{cent} = \frac{v^2}{r};\ F_{g,car} = m_{car}g$$

Then the Newton's law equations become

$$x\text{: } F_n \sin\theta + \mu_S F_n \cos\theta = m_{car}\frac{v^2}{r}$$

$$y\text{: } F_n \cos\theta - m_{car}g - \mu_S F_n \sin\theta = 0$$

Solve the y component equation for the normal force F_n.

Rearrange the y component equation to find F_n:

$$F_n \cos\theta - \mu_S F_n \sin\theta = m_{car}g$$

$$F_n(\cos\theta - \mu_S \sin\theta) = m_{car}g$$

$$F_n = \frac{m_{car}g}{\cos\theta - \mu_S \sin\theta}$$

Substitute this expression for F_n into the x component equation and solve for the speed v.

Rearrange the x component equation:

$$F_n \sin\theta + \mu_S F_n \cos\theta = m_{car}\frac{v^2}{r}$$

$$F_n(\sin\theta + \mu_S \cos\theta) = m_{car}\frac{v^2}{r}$$

Substitute the expression for F_n and solve for v:

$$\frac{m_{car}g}{\cos\theta - \mu_S \sin\theta}(\sin\theta + \mu_S\cos\theta) = m_{car}\frac{v^2}{r}$$

$$g\left(\frac{\sin\theta + \mu_S\cos\theta}{\cos\theta - \mu_S\sin\theta}\right) = \frac{v^2}{r}$$

$$v^2 = gr\left(\frac{\sin\theta + \mu_S\cos\theta}{\cos\theta - \mu_S\sin\theta}\right)$$

$$v = \sqrt{gr\left(\frac{\sin\theta + \mu_S\cos\theta}{\cos\theta - \mu_S\sin\theta}\right)}$$

Use the values of r, θ, and μ_S given for the Indianapolis Motor Speedway.

Using $r = 256$ m, $\theta = 9.20°$, and $\mu_S = 0.550$,

$\sin\theta = \sin 9.20° = 0.160$

$\cos\theta = \cos 9.20° = 0.987$

$$v = \sqrt{(9.80\ \text{m/s}^2)(256\ \text{m})\left[\frac{0.160 + (0.550)(0.987)}{0.987 - (0.550)(0.160)}\right]}$$

$$= 44.3\ \text{m/s}$$

$$= (44.3\ \text{m/s})\left(\frac{1\ \text{km}}{1000\ \text{m}}\right)\left(\frac{3600\ \text{s}}{1\ \text{h}}\right) = 159\ \text{km/h}$$

$$= (159\ \text{km/h})\left(\frac{1\ \text{mi}}{1.61\ \text{km}}\right) = 99.0\ \text{mi/h}$$

Reflect

This speed is quite a bit less than that reached by race cars on the straightaway (200 to 300 km/h), so it's necessary to slow down before turning this corner.

We can check our result by looking at two special cases: (a) If the curve is *not* banked, then $\theta = 0$, $\sin\theta = 0$, and $\cos\theta = 1$. In this case our result reduces to what we found for the flat road in Example 6-5. (b) If the curve is banked but there is no friction, then $\mu_S = 0$ (because then the maximum force of friction is $F_{S,max} = \mu_S F_n = 0$). In this situation our result simplifies to what we found for the airplane making a level turn in Example 6-4. (The airplane isn't on a road, but the lift force $\vec{F}_L$ on a turning airplane plays exactly the same role as the normal force $\vec{F}_n$ on a car on a banked curve when friction is negligible.)

Extend: You can show that in case (a) the car's maximum speed would be 37.1 m/s, while in case (b) it would be only 20.2 m/s. A banked curve with friction allows the highest speeds!

(a) If the curve is not banked (so $\theta = 0$):

$$v = \sqrt{gr\left(\frac{\sin 0 + \mu_S\cos 0}{\cos 0 - \mu_S\sin 0}\right)}$$

$$= \sqrt{gr\left(\frac{0 + \mu_S(1)}{1 - \mu_S(0)}\right)}$$

$$= \sqrt{\mu_S g r}$$

This equation is equivalent to what we found in Example 6-5 for a flat road:

$$r = \frac{v^2}{\mu_S g},\ \text{so}\ v^2 = \mu_S g r\ \text{and}\ v = \sqrt{\mu_S g r}$$

(b) If the curve is banked but there is no friction (so $\mu_S = 0$):

$$v = \sqrt{gr\left(\frac{\sin\theta + (0)\cos\theta}{\cos\theta - (0)\sin\theta}\right)}$$

$$= \sqrt{gr\left(\frac{\sin\theta}{\cos\theta}\right)}$$

$$= \sqrt{gr\tan\theta}$$

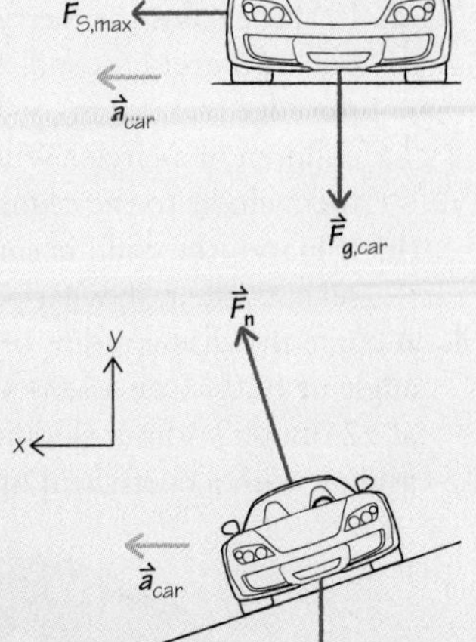

This equation is equivalent to what we found in Example 6-4 for an airplane making a banked turn:

$$r = \frac{v^2}{g\tan\theta},\ \text{so}\ v^2 = gr\tan\theta\ \text{and}$$

$$v = \sqrt{gr\tan\theta}$$

NOW WORK Problem 4 from The Takeaway 6-3.

THE TAKEAWAY for Section 6-3

✔ The net force exerted on an object in uniform circular motion must be directed toward the center of the circle to result in the object's acceleration (Equation 6-5). The net force in this case is called the centripetal force.

✔ Any type of force (or combination of forces) can provide the centripetal force on an object in uniform circular motion. *Centripetal* is just a word indicating the direction of that force.

Prep for the AP® Exam

AP® Building Blocks

EX 6-3 1. A small 25.0-g metal washer is tied to a 60.0-cm-long string.
- (a) The washer on the string is first whirled around in a circle on a horizontal surface with negligible friction with a constant speed of 6.0 m/s. Calculate the tension in the string.
- (b) The washer on the string is then whirled around in a vertical circle. If the speed is 6.0 m/s at the highest point in its path calculate the tension in the string.
- (c) The washer on the string is again whirled around in a vertical circle. If the speed is 6.0 m/s at the lowest point in its path calculate the tension in the string.

2. An airplane is traveling due north and makes a turn to head due east, completing the turn in a uniform circular motion with a radius of 200 m and a centripetal acceleration of 1 *g*.
- (a) Calculate the speed of the plane.
- (b) Calculate the time required to make this 90° turn.

3. Marie cares for twin 4-year-olds when she is not studying physics. She is taking the kids to the park and has just noticed that Helen is wearing sneakers ($\mu_S = 0.80$) and Hazel is in socks ($\mu_S = 0.10$). They want to stand while they ride on the merry-go-round. Before it starts rotating she thinks about where they should stand to have the most fun while still being safe, even if they let go of the railing.
- (a) If Marie spins the merry-go-round so that it completes one turn every second, and stands the children near the edge ($r = 2.0$ m), will the friction force hold both children in a circle? Will they be safe if she stands them closer to the center ($r = 0.50$ m)?
- (b) If Hazel (the child in socks) stands near the edge and releases the rail, what period for a turn is safe?

EX 6-6 4. A curve that has a radius of 1.00×10^2 m is banked at an angle of 10.0°. Can a 1.00×10^3-kg car navigate the curve at 125 km/h without skidding? The road surface is dry asphalt with a coefficient of static friction equal to 0.850.

- (a) Draw a free-body diagram of the car while it is making the turn.
- (b) The circular path of the car lies in a horizontal plane. Draw a diagram to represent (i) the horizontal and vertical components of the friction force exerted on the car and (ii) the horizontal and vertical components of the normal force exerted on the car.
- (c) Apply Newton's second law for the horizontal and vertical components of the net force exerted on the car in uniform circular motion for the maximum friction force.
- (d) Calculate the magnitudes of the normal force and the maximum friction force exerted on the car.
- (e) Calculate the maximum safe speed of the car and predict what happens if the car exceeds this speed.

EX 6-3 5. A centrifuge spins small tubes in a circle of radius 25 cm at a rate of 1.20×10^3 rev/min. Calculate the centripetal acceleration.

AP® Skill Builders

6. When a roller coaster car passes over a crest (A) or through a dip (B), as shown in the figure, there is an acceleration due to the changing direction of the velocity. The radius of a circle that fits the curve at each of these points (sketched in for A and B) defines the radius of curvature, r_A and r_B, respectively.

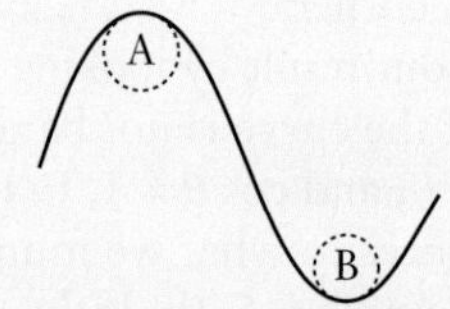

- (a) Construct a free-body diagram of a passenger in the roller coaster car at points A and B.
- (b) Find the normal force on the passenger at points A and B in terms of the radius of curvature at that point, r_A or r_B; the speed, v_A or v_B; the mass of the passenger, *m*; and any necessary constants.
- (c) At sufficiently high speed the passenger will lose contact with the seat at point A. Qualitatively predict the dependence of the speed at which the passenger loses contact on the radius of curvature at the crest.
- (d) Safety standards for roller coasters specify a maximum centripetal acceleration for a brief period as 4 *g*. A very tall roller coaster can reach speeds as high as 50 m/s. Calculate the minimum radius of curvature at point B that satisfies this safety standard on such a tall roller coaster.

7. The condition for the maximum acceleration on a banked turn can be expressed in several ways. Here are four equivalent statements.

(A) $\dfrac{v^2}{r} = \dfrac{g(\sin\theta + \mu_S \cos\theta)}{\cos\theta - \mu_S \sin\theta}$

(B) $\dfrac{v^2}{r} = \dfrac{g(\tan\theta + \mu_S)}{1 - \mu_S \tan\theta}$

(C) $r = \dfrac{(1 - \mu_S \tan\theta)}{(\mu_S + \tan\theta)g} v^2$

(D) $\tan\theta = \dfrac{v^2 - \mu_S rg}{\mu_S v^2 + rg}$

While they are all equivalent, different ones are most useful for different situations, in terms of what is given and what needs to be found. For the following situations, use complete sentences to justify the selection of the most appropriate expression(s).

(a) A civil engineer is determining the elevation for the outer edge of a road, relative to the inner edge of a banked turn, for slippery road conditions.

(b) A state department of transportation is evaluating the speed limit for an existing banked turn.

(c) A construction firm is preparing a proposal for the construction of a new highway that will require land acquisition.

AP® Skills in Action

8. A pail of water is tied to a rope and whirled around in a vertical circle with a constant speed of 4.00 m/s. Which of the statements below best describes and justifies the tension in the rope?

(A) The magnitude of the tension in the rope is constant over the entire circular path because the tension always points to the center and the magnitude of the centripetal acceleration is constant in uniform circular motion.

(B) The magnitude of the tension in the rope is greatest at the lowest point because the gravitational force exerted on the pail of water is downward, whereas the tension force is upward and the magnitude of the centripetal acceleration is constant in uniform circular motion.

(C) The magnitude of tension in the rope is greatest at the highest point in the circular motion because the tension force and gravitational force exerted on the pail of water are both downward.

(D) The magnitude of the tension in the rope is greatest at the lowest point in the circular motion because the gravitational force exerted on the pail of water is downward while the tension force is upward.

9. An object of mass $m_1 = 0.125$ kg undergoes uniform circular motion and is connected by a string with negligible mass through a hole in a table to a larger object of mass $m_2 = 0.225$ kg. The hanging object is stationary and the radius R of the circular path is equal to 1.0 m. Neglect friction forces.

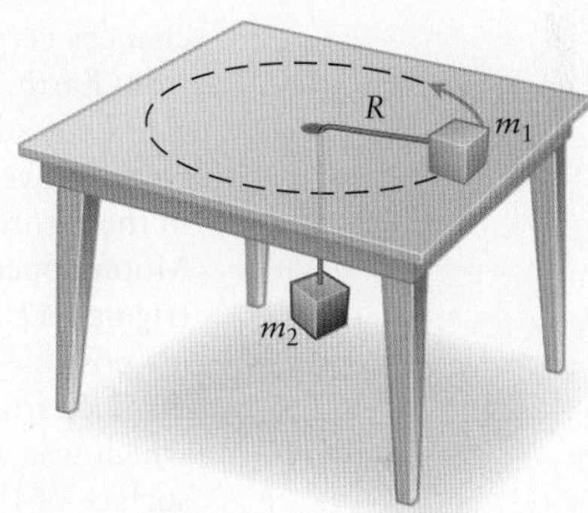

(a) Draw free-body diagrams for both blocks.

(b) Calculate the tension in the string from which the object of mass m_2 is hanging.

(c) Calculate the speed of the block of mass m_1.

6-4 Newton's law of universal gravitation explains the orbit of the Moon, and introduces us to the concept of field

Newton determined much of what we now know about gravitation by considering the motion of Earth's moon (**Figure 6-11**). It had been understood for centuries that the Moon orbits around Earth. Newton knew from his laws of motion that there must therefore be a force exerted on the Moon to keep it from flying off into space, and that Earth must exert this force on the Moon. This is just the force of Earth's gravitation. But how strong is that force? Is this force the same, stronger, or weaker than it would be if the Moon were closer? Let's see how Newton answered these questions by using the ideas of circular motion.

Newton knew that the Moon orbits around Earth at a nearly constant speed in a nearly circular path. The evidence for this is that the Moon changes its position against the background stars. It moves against this background at a nearly constant rate (which tells us that its speed is nearly constant), and as it moves its apparent size

Donna Pagakis/Alamy

Figure 6-11 The Moon Isaac Newton used the known distance to the Moon and the Moon's speed to determine its acceleration, which he concluded was caused by Earth's gravitational attraction. This gave him an important clue about how the gravitational force depends on distance.

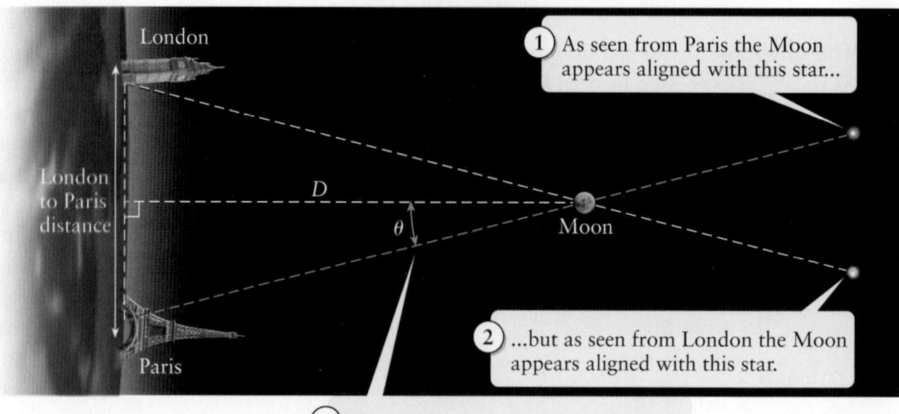

London
1 As seen from Paris the Moon appears aligned with this star...
London to Paris distance
D
θ
Moon
2 ...but as seen from London the Moon appears aligned with this star.
Paris
3 Given the distance from London to Paris and the angle θ, the distance D to the Moon can be calculated.

Figure 6-12 Measuring the distance to the Moon The Moon is close enough that it appears to be in slightly different positions against the backdrop of distant stars when viewed from different locations on Earth. This made it possible to calculate the distance to the Moon many centuries before humans were able to send a spacecraft there.

changes very little (which tells us that the Moon maintains a roughly constant distance from Earth, and so its orbit is nearly circular).

The radius of the Moon's orbit—that is, the distance from Earth to the Moon—was also well known in Newton's time, even though travel to the Moon was centuries in the future. This was possible because as seen from different locations on Earth, the Moon appears to be in slightly different positions relative to the background of stars (**Figure 6-12**). By measuring the differences in apparent position and using trigonometry, it's possible to measure the distance D in Figure 6-12. From these measurements, the distance from the center of Earth to the Moon (equal to Earth's radius $R = 6370$ km, which was also known in Newton's time, plus the distance D in Figure 6-12 from the surface of Earth to the Moon) turns out to be 384,000 km $= 3.84 \times 10^8$ m. This is the radius r of the Moon's orbit.

The Moon's acceleration is due to the gravitational pull of Earth on the Moon, so a_{cent} is equal to the value of g at the position of the Moon. Equation 6-4 shows that to determine the Moon's acceleration Newton had to know the values of the radius r of the Moon's orbit—that is, the distance from the center of Earth to the Moon—and the speed v of the Moon in its orbit.

From observations of the Moon's apparent motion relative to the stars, it was known in Newton's time that the Moon makes one complete orbit around Earth in a time $T = 27.3$ days, or 2.36×10^6 s. The speed v of the Moon in its orbit is therefore the circumference of its orbit, equal to 2π times the radius $r = 3.84 \times 10^8$ m of its orbit, divided by the time $T = 2.36 \times 10^6$ s to complete an orbit. The result is $v = 1.02 \times 10^3$ m/s (about 3670 km/h or 2280 mi/h). If you substitute these values for v and r into Equation 6-4, you'll find that the Moon's acceleration is, to two significant digits,

$$a_{\text{cent}} = \frac{v^2}{r} = \frac{(1.02 \times 10^3 \text{ m/s})^2}{3.84 \times 10^8 \text{ m}} = 2.7 \times 10^{-3} \text{ m/s}^2$$

Note that a_{cent} is *very* much less than the value $g = 9.8$ m/s^2 at the surface of Earth. This implies that Earth's gravitational attraction—which is what exerts a force on the Moon to give it its centripetal acceleration—gets weaker with increasing distance. In fact, the Moon's centripetal acceleration in its orbit around Earth is less than the value of g at Earth's surface by a factor of

$$\frac{a_{\text{cent, Moon}}}{g_{\text{surface of Earth}}} = \frac{2.7 \times 10^{-3} \text{ m/s}^2}{9.8 \text{ m/s}^2} = 2.8 \times 10^{-4} = \frac{1}{3600} = \frac{1}{60^2}$$

As it happens, the distance from the center of Earth to Earth's surface is smaller than the distance from the center of Earth to the Moon by a factor of

$$\frac{r_{\text{Earth center to surface}}}{r_{\text{Earth center to Moon}}} = \frac{6370 \text{ km}}{384{,}000 \text{ km}} = \frac{1}{60}$$

From this Newton deduced that

the acceleration due to Earth's gravity decreases in proportion to the square of the distance from Earth's center.

If you could stand atop a tower 6370 km tall—that is, a tower whose height equals the radius of Earth—you would be twice as far from Earth's center as a person at sea level, and the acceleration due to gravity would be $1/(2)^2 = 1/4$ as great as at sea level. By contrast if you were to stand atop Earth's tallest mountain, Mount Everest, you would be a little less than 9 km above sea level. The distance from Earth's center to the peak of Mount Everest is 6379 km as compared to 6370 km from Earth's center to sea level, which is greater by a factor of (6379 km)/(6370 km) = 1.0014. So the value of g atop Mount Everest is about $1/(1.0014)^2 = 0.997$ of the value at sea level, or about 0.3% less than the sea-level value. This shows why there is only a small difference in the value of g with elevation on Earth.

WATCH OUT !

The object model gets a bit stretched.

When working with the fundamental gravitational force beyond that at the surface of Earth, it is necessary to consider the radius of the "objects" which exert the force and on which the force is exerted. We can still treat these as being located at their centers of mass, and the force is exerted at the center of mass, but we do have to consider the radius of a planet, moon, or star to determine how close these centers of mass can get. For things like people, rocks, boxes, and cars, we can ignore their size. We will see some special considerations for the use of this model in the next Watch Out!

The Law of Universal Gravitation

The following equation expresses Newton's observation that the value of the acceleration g due to Earth's gravitation decreases in proportion to the square of the distance r from the center of Earth:

$$g = \frac{\text{(a constant)}}{r^2}$$

The value of the constant in this equation is chosen so that at a point on Earth's surface, where r equals Earth's radius of 6370 km, the value of g equals 9.8 m/s^2. We learned in Section 4-3 that the magnitude of Earth's gravitational force on an object of mass m is mg. So if an object of mass m is located a distance r from Earth's center, the magnitude of the gravitational force of Earth on this object is

$$F_{\text{Earth on object}} = \frac{\text{(a constant)}m}{r^2}$$

That is, Earth's gravitational force on an object of mass m is directly proportional to the mass m of the object that experiences the force.

Newton generalized this to say that *any* object of mass m_1 exerts a gravitational force on a second object of mass m_2, that this force is directly proportional to the mass m_2 of the object that experiences the force, and that this force is inversely proportional to the square of the distance r between the centers of the objects (**Figure 6-13a**):

$$F_{1 \text{ on } 2} = \frac{\text{(a constant)}m_2}{r^2} \tag{6-6}$$

For the force of Earth on the Moon, m_2 is the Moon's mass, and r is the Earth–Moon center-to-center distance. If Equation 6-6 is true for *any* two objects, then it must also be true that the object of mass m_2 exerts a gravitational force on the object of mass m_1, and this force is directly proportional to the mass m_1. But Newton's third law (Section 4-5) tells us that the forces that the two objects exert on each other have opposite directions and the same magnitude: $F_{2 \text{ on } 1} = F_{1 \text{ on } 2}$. Hence the gravitational force that each object exerts on the other must be directly proportional to *both* m_1 and m_2. The mathematical relationship that gives a result directly proportional to

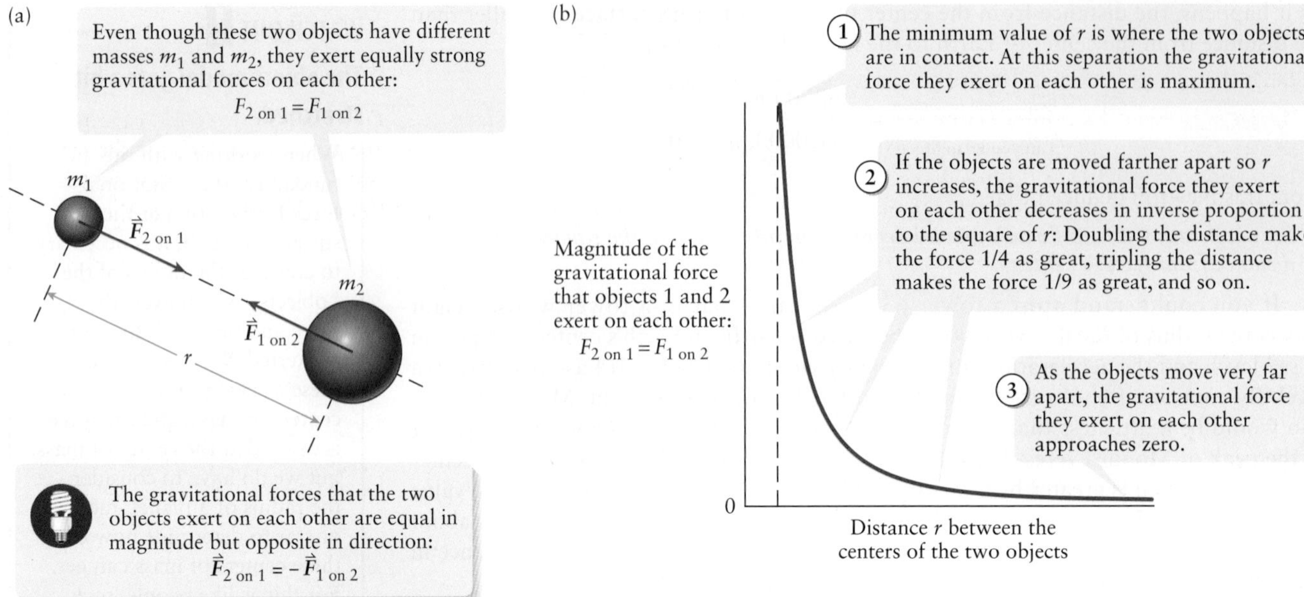

(a)
Even though these two objects have different masses m_1 and m_2, they exert equally strong gravitational forces on each other:
$F_{2 \text{ on } 1} = F_{1 \text{ on } 2}$
m_1
$\vec{F}_{2 \text{ on } 1}$
m_2
$\vec{F}_{1 \text{ on } 2}$
r
The gravitational forces that the two objects exert on each other are equal in magnitude but opposite in direction:
$\vec{F}_{2 \text{ on } 1} = -\vec{F}_{1 \text{ on } 2}$
(b)
1 The minimum value of r is where the two objects are in contact. At this separation the gravitational force they exert on each other is maximum.
2 If the objects are moved farther apart so r increases, the gravitational force they exert on each other decreases in inverse proportion to the square of r: Doubling the distance makes the force 1/4 as great, tripling the distance makes the force 1/9 as great, and so on.
Magnitude of the gravitational force that objects 1 and 2 exert on each other:
$F_{2 \text{ on } 1} = F_{1 \text{ on } 2}$
3 As the objects move very far apart, the gravitational force they exert on each other approaches zero.
0
Distance r between the centers of the two objects

Figure 6-13 The law of universal gravitation (a) Any two objects exert gravitational forces on each other. These forces are equal in magnitude but opposite in direction: $\vec{F}_{2 \text{ on } 1} = -\vec{F}_{1 \text{ on } 2}$. (b) The magnitude of the gravitational force two objects exert on each other is inversely proportional to the square of the center-to-center distance r between the objects.

both is a product, so we must multiply m_1 and m_2. This leads us to Newton's law of universal gravitation:

EQUATION IN WORDS
Newton's law of universal gravitation

(6-7)

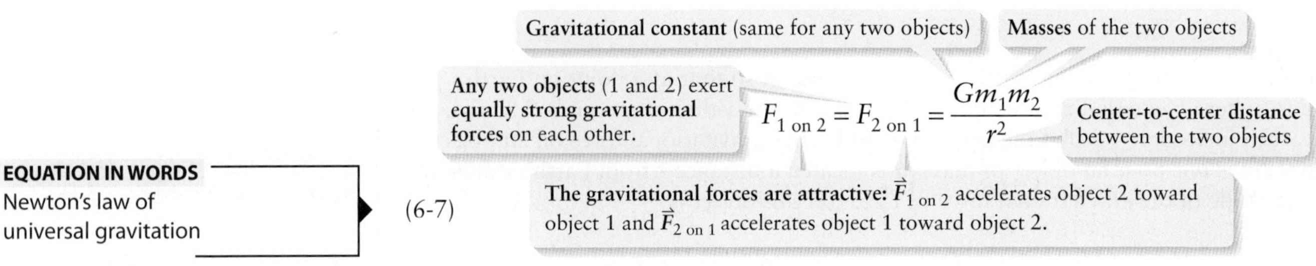

WATCH OUT !

Be sure to understand the limitations of Equation 6-7.

Strictly speaking, Equation 6-7 applies to two infinitesimally small objects with masses m_1 and m_2. It's much more challenging to calculate the gravitational force between two extended objects with complicated shapes, such as between Earth and the Moon (neither of which is a perfect sphere) or between Earth and the International Space Station (which is cylindrical with protrusions, one of which you can see in the photograph that opens this chapter). However, in practice we can safely use Equation 6-7 if the *distance* between the two objects is large compared to the *size* of either object. That's the case for Earth and the Moon, which are separated by a distance of 384,000 km—much larger than the radius of either Earth (6370 km) or the Moon (1740 km).

It's also safe to use Equation 6-7 if one object is very much smaller than the other, as is the case for a satellite (like the International Space Station) orbiting Earth, as long as we can approximate the large object (in this case Earth) as being roughly uniform. In other situations, such as calculating the force between the two galaxies shown in Figure 6-1c (which are comparable in size and whose separation is comparable to the size of either galaxy), Equation 6-7 still gives useful estimates of the magnitude of the gravitational force. Note also that Equation 6-7 is not valid if one object is *inside* the other. For example, we can't use this equation to find the gravitational force on you when you're in a tunnel deep inside Earth's interior unless we are careful about what r means and we know how the mass of Earth is distributed.

WATCH OUT !

The distance r in Equation 6-7 is the center-to-center distance.

Always remember that r in Equation 6-7 is the distance between the *centers* of the two objects. If you're calculating the gravitational force that Earth exerts on a person standing on Earth's surface, the distance r equals the radius of Earth—that is, the distance from the center of Earth to that person. For nearly symmetrical objects like Earth or the Moon, the geometrical center is the same as the center of mass, so we'll often use the terms *center* and *center of mass* interchangeably when discussing gravitational forces.

As Figure 6-13a shows, the forces that the two objects exert on each other are attractive and are directed along the line that connects the centers of the two objects. The proportionality constant G is called the **gravitational constant**. We'll see later in this section how the value of G (which Newton did not know) is determined. Its currently accepted value, to three significant digits, is

$$G = 6.67 \times 10^{-11}\ \text{Nm}^2/\text{kg}^2$$

Figure 6-13b graphs the magnitude of the gravitational force between the two objects as a function of the distance r between their centers.

As Equation 6-7 shows, the gravitational force that one object exerts on another is proportional to $1/r^2$, which is the reciprocal of the square of the distance r or "the inverse square" of r for short. That's why the law of universal gravitation is sometimes called *the inverse-square law for gravitation*. If you continue to study physics, you will learn that the *electric* force that one charged object exerts on another (for instance, the force between an electron and a proton) also obeys an inverse-square law.

Because the gravitational constant has such a small value, gravitational forces between two objects also tend to be pretty small unless at least one of the objects is relatively large (that is, planet-sized). The following examples illustrate this.

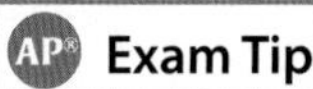

Exam Tip

A common error that students make on the AP® Physics 1 exam is leaving the units off constants, or assuming constants have no units. You should note that constants in equations can have units or not: The units of proportionality constants are determined by solving for that constant and doing a dimensional analysis.

EXAMPLE 6-7 You and Your Backpack

While you sit studying, what is the magnitude of the gravitational force that your 10.0-kg backpack resting on your desk exerts on you? Assume that your mass is 70.0 kg and that your center of mass is 0.60 m from the center of your backpack.

Set Up

We'll use Newton's law of universal gravitation, Equation 6-7, to find the magnitude of the gravitational force. Because we are not worried about direction and are just calculating the size of one force, we can skip the free-body diagram.

Newton's law of universal gravitation:

$$F_{\text{bag on you}} = \frac{G m_{\text{bag}} m_{\text{you}}}{r^2} \qquad (6\text{-}7)$$

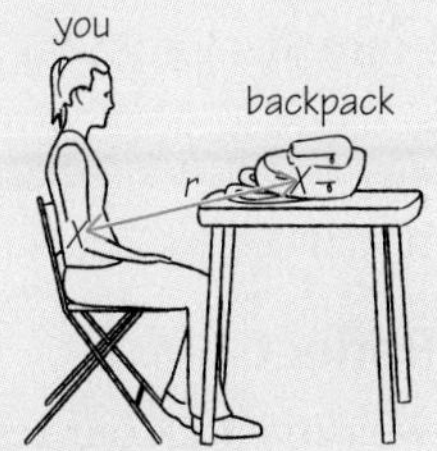

Solve

Substitute the given values of $G, m_{\text{bag}} = 10.0$ kg, $m_{\text{you}} = 70.0$ kg, and $r = 0.60$ m into the expression for gravitational force. The distance is given to only two significant digits, so our result has only two significant digits.

$$F_{\text{bag on you}} = \frac{G m_{\text{bag}} m_{\text{you}}}{r^2}$$

$$= \frac{(6.67 \times 10^{-11}\ \text{Nm}^2/\text{kg}^2)(10.0\ \text{kg})(70.0\ \text{kg})}{(0.60\ \text{m})^2}$$

$$= 1.3 \times 10^{-7}\ \text{N}$$

Reflect

The gravitational force your bag exerts on you is *very* small, about one ten-millionth of a newton. This is equivalent to the weight of a few specks of dust. This force is far smaller than the gravitational force Earth exerts on you (see the next example). It's also far smaller than the force of friction your chair exerts on you should you start to slide. This makes sense, because we generally neglect the gravitational force that everyday objects exert on each other. The gravitational force is a weaker force than the other fundamental forces. However, whereas the gravitational force that everyday objects exert on each other is small, the gravitational force dominates on large scales because the gravitational force, unlike the other fundamental forces, is always attractive.

NOW WORK Problems 1 and 2 from The Takeaway 6-4.

EXAMPLE 6-8 Earth's Gravitational Force on You

Calculate the magnitude of the gravitational force that Earth exerts on you, again assuming you have a mass of 70.0 kg. Earth has a mass of 5.97×10^{24} kg and radius of 6370 km.

Set Up

Again we'll find the magnitude of the gravitational force using Newton's law of universal gravitation, Equation 6-7. The distance r between the centers of the two objects (Earth and you) equals Earth's radius, 6370 km. The extra meter or so distance from Earth's surface to the center of your body is insignificant compared to the size of Earth's radius, so we ignore it.

Newton's law of universal gravitation:

$$F_{\text{Earth on you}} = \frac{Gm_{\text{Earth}}m_{\text{you}}}{r^2} \quad (6\text{-}7)$$

r = distance from center of Earth to you

$= R_{\text{Earth}}$ = radius of Earth

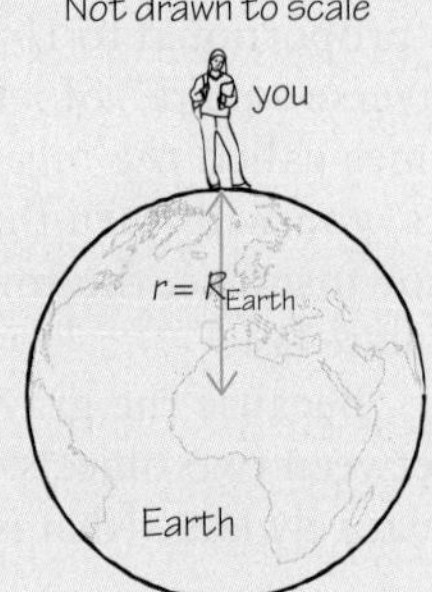

Solve

To use the distance $r = R_{\text{Earth}} = 6370$ km in Equation 6-7, we first convert it to meters.

$$r = R_{\text{Earth}}$$
$$= (6370 \text{ km}) \times \frac{1000 \text{ m}}{1 \text{ km}} = 6.37 \times 10^6 \text{ m}$$

Substitute the values of G, m_{Earth}, m_{you}, and r into Equation 6-7.

$$F_{\text{Earth on you}} = \frac{Gm_{\text{Earth}}m_{\text{you}}}{r^2} = \frac{Gm_{\text{Earth}}m_{\text{you}}}{R_{\text{Earth}}^2}$$
$$= \frac{(6.67 \times 10^{-11} \text{ Nm}^2/\text{kg}^2)(5.97 \times 10^{24} \text{ kg})(70.0 \text{ kg})}{(6.37 \times 10^6 \text{ m})^2}$$
$$= 687 \text{ N} = 690 \text{ N to two significant digits}$$

Reflect

We can check our result by using the expression $F_g = mg$ for gravitational force that we introduced in Section 4-3. We get the same result as above to two significant digits, which shows that the values of G and m_{Earth} are compatible with the value of g at Earth's surface.

Using the expression $F_g = mg$ for gravitational force,

$$F_{\text{Earth on you}} = m_{\text{you}}g$$
$$= (70.0 \text{ kg})(9.80 \text{ m/s}^2)$$
$$= 686 \text{ N}$$
$$= 690 \text{ N to two significant digits}$$

NOW WORK Problem 3 from The Takeaway 6-4.

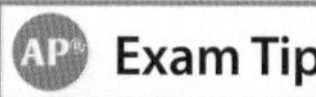
Exam Tip

If Earth exerts a 690-N gravitational force on you, then by Newton's third law, you also must exert a 690-N gravitational force on Earth. This can also be seen by applying Newton's law of universal gravitation to determine what force you exert on Earth. Sometimes you will see a force like this called the force *between* two objects. This does not imply that a single force exists at some location "between" the two objects: There are always two forces, each of which is exerted on different objects, and equal in magnitude and opposite in direction.

Finding the Value of *G* and the Mass of Earth

Example 6-8 shows that there is a connection between the value of g at Earth's surface and the values of G and m_{Earth}, the gravitational constant and Earth's mass, respectively. To see this more clearly take the two expressions for the gravitational force on you and set them equal to each other:

$$F_{\text{Earth on you}} = m_{\text{you}}g = \frac{Gm_{\text{Earth}}m_{\text{you}}}{R_{\text{Earth}}^2}$$

Then divide both sides of the equation by your mass m_{you} to get an expression for g:

Gravitational constant | Mass of Earth

Acceleration due to gravity at Earth's surface

$$g = \frac{Gm_{\text{Earth}}}{R_{\text{Earth}}^2} \qquad (6\text{-}8)$$

Radius of Earth

EQUATION IN WORDS
Value of g at Earth's surface

This equation tells us that whenever we calculate a weight or an acceleration using g, we are simply using a shorthand form of the law of universal gravitation applied near Earth's surface. So the familiar expression $F_g = mg$ is just a special form of the general law of universal gravitation, Equation 6-7.

Equation 6-8 relates the value of g (which we can measure by observing a freely falling object) to the values of Earth's radius R_{Earth}, Earth's mass m_{Earth}, and the gravitational constant G. The value of R_{Earth} has been known since the third century B.C., when Eratosthenes, a scholar from Cyrene in modern-day Libya, first determined its value. (**Figure 6-14** shows how Eratosthenes was able to do this.) But that leaves two unknown values in Equation 6-8, G and m_{Earth}. With only one equation to relate them, it seems impossible to determine the values of both of these unknowns.

(1) As the story is told, on the day of the summer solstice (when the Sun is highest in the sky as seen from the northern hemisphere), Eratosthenes observed the shadow cast by a column in the city of Alexandria (Not drawn to scale.)

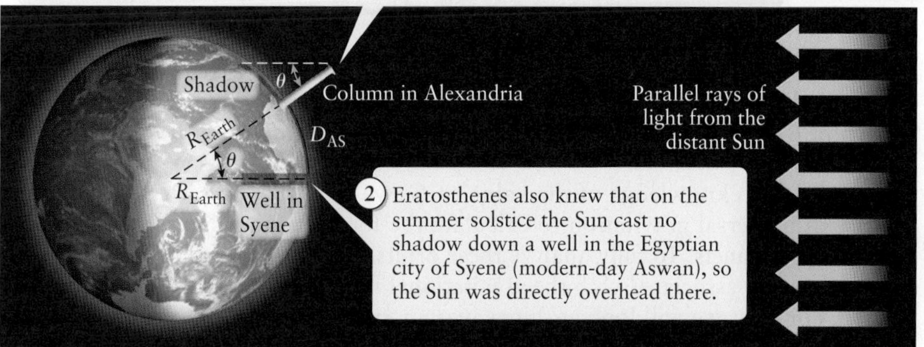
Shadow
θ
Column in Alexandria
Parallel rays of light from the distant Sun
R_{Earth}
D_{AS}
θ
R_{Earth}
Well in Syene
(2) Eratosthenes also knew that on the summer solstice the Sun cast no shadow down a well in the Egyptian city of Syene (modern-day Aswan), so the Sun was directly overhead there.

(3) Given the known distance D_{AS} from Alexandria to Syene and his measurement of the angle θ, Eratosthenes was able to calculate the radius R_{Earth}.

Figure 6-14 How Eratosthenes measured the radius of Earth Around 240 B.C. the scholar Eratosthenes measured Earth's radius by analyzing the shadows cast by the Sun at two different locations.

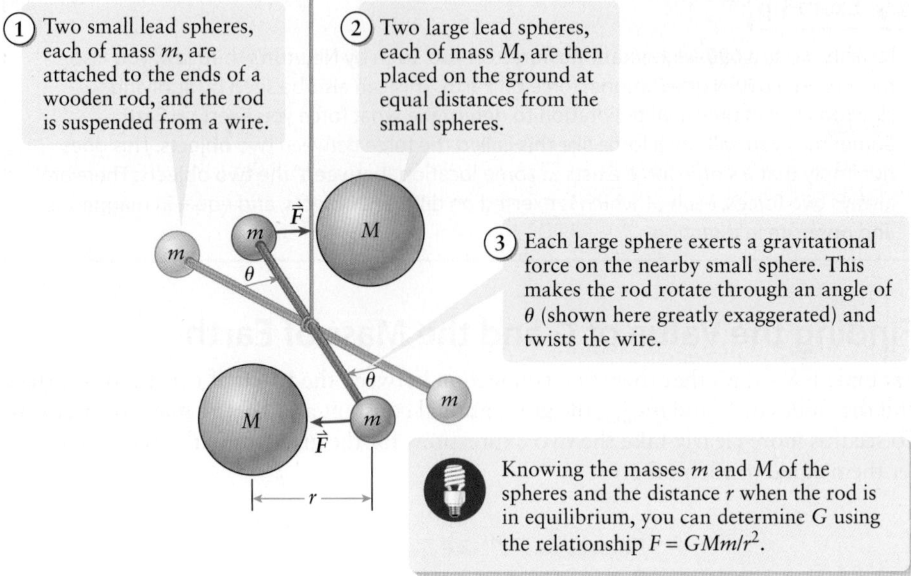

1 Two small lead spheres, each of mass m, are attached to the ends of a wooden rod, and the rod is suspended from a wire.
2 Two large lead spheres, each of mass M, are then placed on the ground at equal distances from the small spheres.
3 Each large sphere exerts a gravitational force on the nearby small sphere. This makes the rod rotate through an angle of θ (shown here greatly exaggerated) and twists the wire.
Knowing the masses m and M of the spheres and the distance r when the rod is in equilibrium, you can determine G using the relationship $F = GMm/r^2$.
m
M
F
θ
r

Figure 6-15 The Cavendish experiment This experiment is used to determine the value of G. In the original (1798) version of the experiment, the wooden rod was 1.8 m in length, each small lead sphere had a mass of 0.73 kg, each large lead sphere had a mass of 158 kg, and the distance r was about 23 cm.

The solution to this problem was provided by the British scientist Henry Cavendish in 1798, more than a century after Newton published the law of universal gravitation. Cavendish conducted the elegant experiment shown schematically in **Figure 6-15**, now known simply as the **Cavendish experiment**. The gravitational force of each large sphere on the nearby small sphere makes the wooden rod rotate, which twists the wire from which the rod is suspended. The wire resists being twisted and exerts a torque (we will learn about torque in a later chapter; for now you can think of it as the "twist" equivalent of force) on the rod that tries to return the wire to its relaxed state, much as a stretched spring exerts a force that tries to return the spring to its unstretched length. In equilibrium the torque exerted by the wire just balances the gravitational torque. By measuring the wire beforehand Cavendish knew how much torque the wire exerted for a given rotation angle, and thus he could determine the equal-magnitude gravitational torque. He could then determine the gravitational force on each small sphere required to cause this torque on the rod. With this technique Cavendish could measure extremely small forces on the order of 10^{-7} N, which made his experiment possible.

Given the magnitude of the force F; the masses m and M of the small and large spheres, respectively; and the distance r between neighboring large and small spheres, we can use Newton's law of universal gravitation to solve for the value of G:

$$F = \frac{GMm}{r^2} \quad \text{so} \quad G = \frac{Fr^2}{Mm}$$

Modern experiments to measure G to great precision still use variations on Cavendish's original apparatus. Once the value of G is known, we can use Equation 6-8 to determine the value of Earth's mass m_{Earth} (now known to be 5.97×10^{24} kg).

Armed with knowledge of the value of G, we have the tools we need to apply the law of universal gravitation to calculate the forces responsible for the orbits of satellites and planets. To fully analyze these orbits, it's important to first consider the potential energy associated with the gravitational force, which is discussed in the next chapter. For now, we will look at things we can understand making use of the law of universal gravitation.

EXAMPLE 6-9 At What Altitude Is *g* Cut in Half?

At what height above Earth's surface is the value of g equal to one-half the value at the surface?

Set Up

Equation 6-8 tells us the value of g at Earth's surface, which we'll call $g_{surface}$. Our goal is to find the height h above the surface at which g has half this value. We'll use the law of universal gravitation and the relationship between the gravitational force on an object and the value of g. Remember that the distance r in the law of universal gravitation is *not* the height above the surface but rather the distance to Earth's center. This equals the radius of Earth plus the height of the object.

Value of g at Earth's surface:

$$g_{\text{surface}} = \frac{Gm_{\text{Earth}}}{R_{\text{Earth}}^2} \quad (6\text{-}8)$$

Newton's law of universal gravitation:

$$F_{\text{Earth on object}} = \frac{Gm_{\text{Earth}}m_{\text{object}}}{r^2} \quad (6\text{-}7)$$

Gravitational force on an object in terms of g:

$$F_{\text{Earth on object}} = m_{\text{object}}g$$

Distance from Earth's center to a height h above the surface:

$$r = R_{\text{Earth}} + h$$

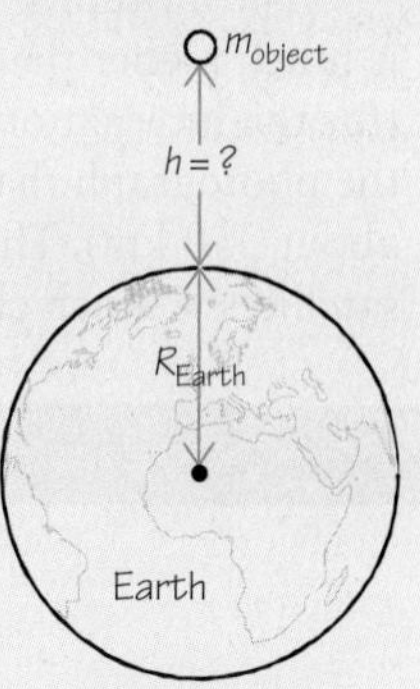

Solve

Use the two expressions for the gravitational force on an object to solve for g at a height h above the surface.

Set the two expressions for $F_{\text{Earth on object}}$ equal to each other:

$$F_{\text{Earth on object}} = m_{\text{object}}g = \frac{Gm_{\text{Earth}}m_{\text{object}}}{r^2}$$

To solve for g divide through by m_{object}:

$$g = \frac{Gm_{\text{Earth}}}{r^2}$$

Substitute $r = R_{\text{Earth}} + h$ in the expression for g:

$$g = \frac{Gm_{\text{Earth}}}{(R_{\text{Earth}} + h)^2}$$

We want to find the value of h at which g equals one-half of $g_{surface}$.

The value of g at height h equals 1/2 the value of g_{surface}:

$$g = \frac{1}{2}g_{\text{surface}}$$

Substitute the expression for g we found above and the expression for g_{surface} from Equation 6-8:

$$\frac{Gm_{\text{Earth}}}{(R_{\text{Earth}} + h)^2} = \frac{1}{2}\frac{Gm_{\text{Earth}}}{R_{\text{Earth}}^2}$$

Note that Gm_{Earth} cancels, leaving

$$\frac{1}{(R_{\text{Earth}} + h)^2} = \frac{1}{2R_{\text{Earth}}^2}$$

Take the reciprocal of both sides of this equation:

$$(R_{\text{Earth}} + h)^2 = 2R_{\text{Earth}}^2$$

To eliminate the squares take the square root of both sides:

$$R_{\text{Earth}} + h = \sqrt{2}R_{\text{Earth}}$$

To get an expression for h, subtract R_{Earth} from both sides of this equation:

$$h = \sqrt{2}R_{\text{Earth}} - R_{\text{Earth}}$$
$$= (\sqrt{2} - 1)R_{\text{Earth}}$$

Substitute $R_{\text{Earth}} = 6370$ km:

$$h = (\sqrt{2} - 1)(6370\ \text{km}) = (1.414 - 1)(6370\ \text{km})$$
$$= 2640\ \text{km}$$

Reflect

Our calculation shows that g, like the gravitational force, is inversely proportional to the square of the distance r from Earth's center, which makes sense.

The altitude $h = (\sqrt{2} - 1)R_{Earth} = 2640$ km is much higher than that of most Earth satellites (for the International Space Station shown in the photograph that opens this chapter, h is only about 350 km). That's why the value of g for these satellites is much closer to $g_{surface}$ than to $g_{surface}/2$.

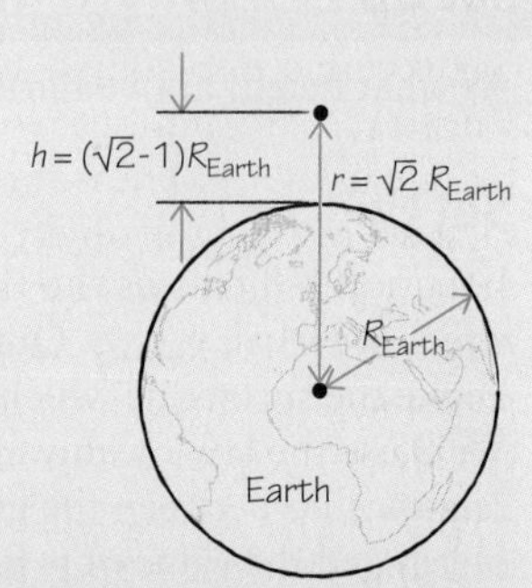

NOW WORK Problem 4 from The Takeaway 6-4.

Fields Are a Useful Way to Model Fundamental Forces

The previous example gives us insight into another way to think about the gravitational force. The gravitational force, like the other fundamental forces, is exerted "at a distance." This means that the object upon which the force is exerted does not need to be in contact with the object exerting the force. To understand and to calculate such forces, it is often useful to model them in terms of **fields**, which associate a value of some physical quantity with every point in space. Forces are vectors, and so the associated fields are also vectors, having a magnitude and a direction assigned to each point in space. The source of a gravitational field is an object with mass. When more than one source is present, the net field value can be determined by vector addition. The gravitational field at a point in space is calculated by dividing the gravitational force exerted by the field on a test object by the mass of that test object. The direction of the field is the same as the direction of the force. So the magnitude of a gravitational field outside of an object of mass M is $\frac{GM}{r^2}$, where r is the distance from the center of mass of the object to the point of interest. The units of the gravitational field are N/kg. The units of a field can always be expressed as units of force divided by the property that relates to that force, in this case mass. At the surface of Earth, this is just Equation 6-8, the expression for g. So our equation for gravitational field is just $\vec{g} = \frac{GM}{r^2}$ directed toward the center of the object causing the field. A gravitational field $\vec{g}$ at the location of an object with mass m causes a gravitational force of magnitude mg to be exerted on the object in the direction of the field. On Earth, we substitute the mass and radius of Earth into our equation for $\vec{g}$ and this gravitational force is called weight. For any location in space, if the gravitational force is the only force exerted on an object, then the observed free-fall acceleration of the object is equal to the magnitude of the gravitational field at that location.

AP® Exam Tip

On the AP® Physics 1 equation sheet, the equation $\vec{g} = \frac{\vec{F}_g}{m}$ indicates how the gravitational field at a point in space can be calculated from the gravitational force on an object of mass m at that point. The magnitude of this value is also the magnitude of the gravitational acceleration g at that point.

EXAMPLE 6-10 The Moons of Saturn

Figure 6-1b shows Saturn's moons Tethys, Dione, and Rhea. (Saturn's moons, of which 62 were known as of this writing, are named for mythological figures associated with the Roman god Saturn.) Their masses are 6.2×10^{20} kg for Tethys, 1.1×10^{21} kg for Dione, and 2.3×10^{21} kg for Rhea. All three of these moons are quite small compared to Earth's moon (mass 7.36×10^{22} kg). Tethys, Dione, and Rhea move in different orbits around Saturn and at different speeds, and on occasion they form a right triangle (**Figure 6-16**). Calculate the *net* gravitational force on Tethys due to Dione and Rhea when the three moons are in this configuration. Compare this to the force that Saturn (mass 5.7×10^{26} kg) exerts on Tethys, which orbits 2.95×10^5 km from Saturn's center.

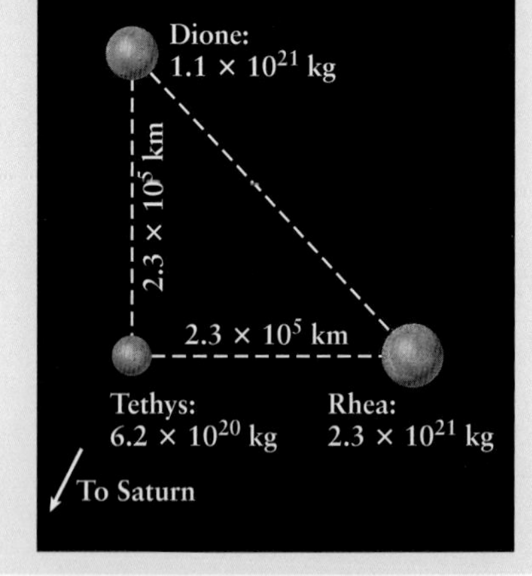

Figure 6-16 Three moons of Saturn What is the net gravitational force on Saturn's moon Tethys due to the moons Dione and Rhea?

Set Up

Begin with a free-body diagram. Both Dione and Rhea exert gravitational forces on Tethys as given by Equation 6-7. These forces are attractive and are directed toward the object causing the force, so they are at right angles to each other. The *net* gravitational force on Tethys due to the two other moons is the vector sum of these two individual forces. We'll use the ideas of vector addition and vector components from Sections 3-2 and 3-3 to find this vector sum and determine the magnitude and direction of the net force.

Newton's law of universal gravitation:

$$F_{\text{moon on Tethys}} = \frac{Gm_{\text{moon}}m_{\text{Tethys}}}{r^2_{\text{moon to Tethys}}} \qquad (6\text{-}7)$$

$\vec{F}_{\text{Dione on Tethys}}$

Tethys

$\vec{F}_{\text{Rhea on Tethys}}$

y

x

Solve

The easiest way to add vectors is by using components, so we choose x and y axes as shown. With this choice of axes, the force of Dione on Tethys has only a y component, and the force of Rhea on Tethys has only an x component. Calculate these components using the values of r shown in Figure 6-16.

Convert the Dione–Tethys distance and Rhea–Tethys distance from kilometers to meters:

$$r_{\text{Dione to Tethys}} = r_{\text{Rhea to Tethys}} = 2.3 \times 10^5 \text{ km}$$
$$= (2.3 \times 10^5 \text{ km})\left(\frac{1000 \text{ m}}{1 \text{ km}}\right) = 2.3 \times 10^8 \text{ m}$$

The force of Dione on Tethys has zero x component and a positive y component:

$$F_{\text{Dione on Tethys},x} = 0$$
$$F_{\text{Dione on Tethys},y} = +\frac{Gm_{\text{Dione}}m_{\text{Tethys}}}{r^2_{\text{Dione to Tethys}}}$$
$$= +\frac{(6.67 \times 10^{-11} \text{ Nm}^2/\text{kg}^2)(1.1 \times 10^{21} \text{ kg})(6.2 \times 10^{20} \text{ kg})}{(2.3 \times 10^8 \text{ m})^2}$$
$$= +8.6 \times 10^{14} \text{ N}$$

The force of Rhea on Tethys has a positive x component and zero y component:

$$F_{\text{Rhea on Tethys},x} = +\frac{Gm_{\text{Rhea}}m_{\text{Tethys}}}{r^2_{\text{Rhea to Tethys}}}$$
$$= +\frac{(6.67 \times 10^{-11} \text{ Nm}^2/\text{kg}^2)(2.3 \times 10^{21} \text{ kg})(6.2 \times 10^{20} \text{ kg})}{(2.3 \times 10^8 \text{ m})^2}$$
$$= +1.8 \times 10^{15} \text{ N}$$
$$F_{\text{Rhea on Tethys},y} = 0$$

The net force on Tethys is the vector sum of the two individual forces. Find the components of the net force.

The x component of the net force on Tethys is the sum of the x components of the individual forces:

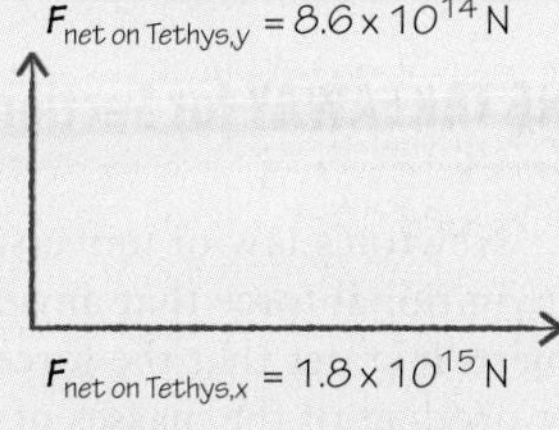

$$F_{\text{net on Tethys},x}$$
$$= F_{\text{Dione on Tethys},x} + F_{\text{Rhea on Tethys},x}$$
$$= 0 \text{ N} + 1.8 \times 10^{15} \text{ N}$$
$$= +1.8 \times 10^{15} \text{ N}$$

Similarly, the y component of the net force on Tethys is the sum of the y components of the individual forces:

$$F_{\text{net on Tethys},y} = F_{\text{Dione on Tethys},y} + F_{\text{Rhea on Tethys},y}$$
$$= +8.6 \times 10^{14} \text{ N} + 0 \text{ N}$$
$$= +8.6 \times 10^{14} \text{ N}$$

From the components of the net force, find its magnitude and direction using the techniques learned in Section 3-3.

Find the magnitude of the net force:

$$F_{\text{net on Tethys}} = \sqrt{(F_{\text{net on Tethys},x})^2 + (F_{\text{net on Tethys},y})^2}$$
$$= \sqrt{(+1.8 \times 10^{15} \text{ N})^2 + (+8.6 \times 10^{14} \text{ N})^2}$$
$$= 2.0 \times 10^{15} \text{ N}$$

Find the angle of the net force relative to the x axis:

$$\tan\theta = \frac{F_{\text{net on Tethys},y}}{F_{\text{net on Tethys},x}} = \frac{+8.6\times10^{14}\ \text{N}}{+1.8\times10^{15}\ \text{N}} = 0.48$$

so

$$\theta = \tan^{-1} 0.48 = 26°$$

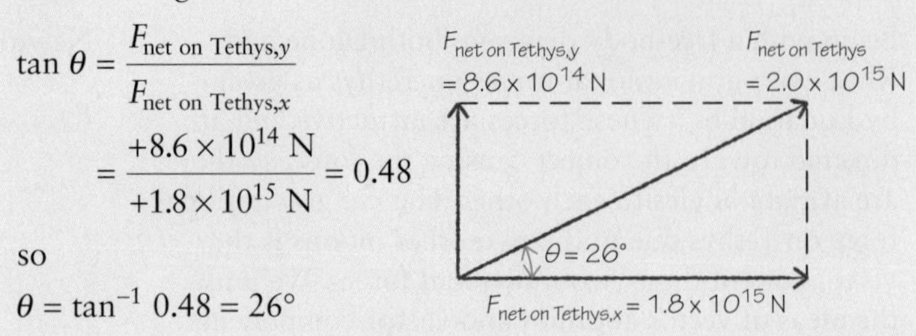

Reflect

The net gravitational force on Tethys from Dione and Rhea is 2.0×10^{15} N, which seems tremendous. To see if this is reasonable, we need to see if it is smaller than the gravitational force that Saturn exerts on Tethys. When we do the calculation we find the gravitational force due to Saturn on Tethys is about 1.4×10^{5} (140,000) times greater than the net force exerted on Tethys by the other moons. This means the orbit of Tethys around Saturn is affected hardly at all by the presence of the other moons, so this answer is plausible.

Note that the net force on Tethys from the other two moons is not directed at either of those moons but somewhere between the two. This isn't surprising: The sum of two vectors doesn't point in the direction of either individual vector unless they are pointing along the same line, or one is so large we can neglect the other.

Convert the distance from the center of Saturn to Tethys from kilometers to meters:

$$r_{\text{Saturn to Tethys}} = (2.95\times10^{5}\ \text{km})\left(\frac{1000\ \text{m}}{1\ \text{km}}\right) = 2.95\times10^{8}\ \text{m}$$

Then the gravitational force that Saturn exerts on Tethys is

$$F_{\text{Saturn on Tethys}} = \frac{Gm_{\text{Saturn}}m_{\text{Tethys}}}{r^2_{\text{Saturn to Tethys}}}$$

$$= \frac{(6.67\times10^{-11}\ \text{Nm}^2/\text{kg}^2)(5.7\times10^{26}\ \text{kg})(6.2\times10^{20}\ \text{kg})}{(2.95\times10^{8}\ \text{m})^2}$$

$$= 2.7\times10^{20}\ \text{N}$$

This force is greater than the net force that Dione and Rhea together exert on Tethys by a factor of

$$\frac{F_{\text{Saturn on Tethys}}}{F_{\text{net on Tethys}}} = \frac{2.7\times10^{20}\ \text{N}}{2.0\times10^{15}\ \text{N}} = 1.4\times10^{5}$$

Note also that we could have simplified our calculation somewhat by finding the gravitational field at the location of Tethys due to the other two moons. This would have involved the same vectors, but not multiplying by Tethys's mass until we had a final value. When you are just finding the force on one object, this may not save a lot of time; but if we wanted to routinely find the gravitational force on a variety of objects at that point (like we do on the surface of Earth), it would save us a lot of time!

The field concept also gives us a way to further check our answer for reasonableness. The net force on Tethys from Saturn is 2.7×10^{20} N. If we divide this by the mass of Tethys, we get a number that is smaller than g. Saturn is larger than Earth, and Saturn and Tethys are closer than Earth and the Moon, so this being larger than the gravitational field of Earth at the Moon, but smaller than at Earth's surface, is reasonable.

NOW WORK Problem 5 from The Takeaway 6-4.

THE TAKEAWAY for Section 6-4

✔ Newton's law of universal gravitation describes the gravitational force that any object exerts on another object. It states that the force is directly proportional to the product of the masses of two objects and inversely proportional to the square of the distance between their centers.

✔ The relationship we use for weight near Earth's surface, $F_g = mg$, is simply an application of the law of universal gravitation.

✔ A gravitational field $\vec{g}$ at the location of an object with mass m causes a force of magnitude mg to be exerted on the object in the direction of the field.

AP® Building Blocks

1. Would the magnitude of the acceleration due to gravity near Earth's surface increase more if Earth's mass were doubled and its radius kept the same, or if Earth's radius were cut in half and its mass kept the same? Because forces always come in pairs, just as Earth causes an object on its surface to accelerate toward it, that object must also cause Earth to accelerate toward it. Why don't you notice Earth's acceleration? Justify your answers.

EX 6-7

2. A 5.00×10^{2}-kg tree stump is located 1.00×10^{3} m from a 12,000-kg boulder. Determine the magnitude and direction of the gravitational force exerted by the tree stump on the boulder.

3. Compare the magnitude of the gravitational force on an apple on Earth's surface exerted by Earth with the gravitational force exerted on the same apple by the Moon when the Moon is overhead. Assume that Earth and the Moon are spherical and that both have their masses concentrated at their respective centers. The Earth–Moon distance is 3.844×10^8 m, and the radii of Earth and the Moon are 6.371×10^6 m and 1.737×10^6 m, respectively. The masses of Earth and the Moon are 5.972×10^{24} kg and 7.348×10^{22} kg, respectively.

4. No net force is exerted on a space probe located at some distance $r_{\text{Earth–probe}}$ along a line joining the center of Earth and the center of the Moon. Representing the center–center Earth–Moon distance as $r_{\text{Earth–Moon}}$, express the condition of no net force as a quadratic equation whose solution is the distance $r_{\text{Earth–probe}}$.

5. The exoplanet Kepler 452b in the constellation Cygnus orbits a star similar to Sol, Earth's star. Kepler 452b has a radius 1.6 times larger than the radius of Earth. The mass has been estimated as 5 times the mass of Earth.
   (a) Calculate the gravitational field on the surface of Kepler 452b.
   (b) For an object that weighs 10 N on Earth, calculate its weight when it is on the surface of Kepler 452b.

AP® Skill Builder

6. Explanations of the fact that people perceive the Moon to be larger when it is near the horizon than when it is at a higher elevation have been recorded since the time of Ptolemy (150 B.C.). From the following statements, identify the best procedure for testing the claim that the radius of the Moon is independent of position in the sky relative to an observer on the surface of Earth.
   (A) Numerical data of the position of the Moon as a function of time as seen from Earth are analyzed to confirm that the orbital speed of the Moon does not depend on distance to the horizon.
   (B) A series of photographs with time stamps of the Moon are taken while the Moon rises over the horizon.
   (C) Two photographs of the Moon with the same magnification are compared at two different times during the night when the Moon is at different elevations.
   (D) A small disk is positioned so that it appears to cover the Moon and the distance of the disk to the eye of the observer is measured.

AP® Skills in Action

7. We observe that the Moon revolves around Earth once every 27.3 days. The average distance from the center of Earth to the center of the Moon is 3.84×10^8 m. Determine the acceleration of the Moon due to its motion around Earth assuming that Earth is stationary and the Moon is in uniform circular motion.
8. Commercial ultracentrifuges can rotate at rates of 100,000 rpm (revolutions per minute). A particular centrifuge produces a centripetal acceleration of 800,000 g. Calculate the distance from the rotation axis to the sample chamber in this device.
9. The radius of Earth is 6.371×10^6 m, and in 1 day it completes one rotation in a counterclockwise direction when viewed from above the north geographic pole.
   (a) Draw a diagram that shows the dependence of the rotational radius on latitude θ.
   (b) Calculate the centripetal acceleration of an object moving with the rotation of Earth when it is located (i) on the equator (latitude 0°) and (ii) at latitude 40.0° north.

6-5 Newton's law of universal gravitation begins to explain the orbits of planets and satellites

We saw in Section 6-4 that Newton deduced the law of universal gravitation from the properties of the Moon's orbit around Earth. Because gravitation is universal, the same principles apply to the orbit of *any* celestial object around another. We will begin to explore this concept for circular orbits, but we will need information on gravitational potential energy and rotational motion when we return to this topic for *all* orbits.

Circular Orbits: Orbital Speed

The simplest type of orbit to analyze is a circular orbit. Many Earth satellites, including the International Space Station (see the photograph that opens this chapter) and the satellites that provide signals for Global Positioning System (GPS) navigation, are in circular or nearly circular orbits.

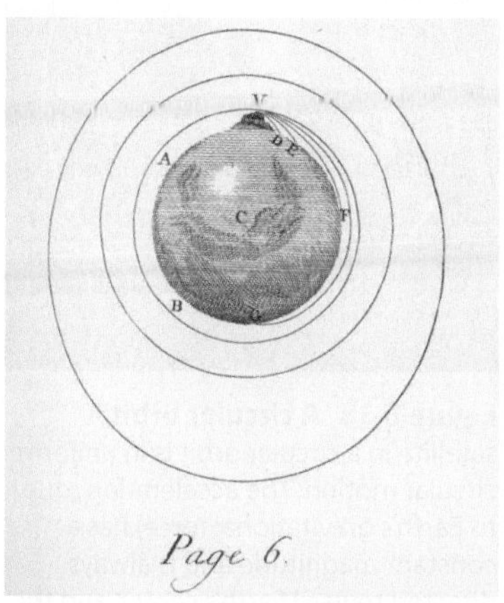

Photo by Christina Micek

Figure 6-17 Newton's recipe for an Earth satellite Isaac Newton imagined that if a cannonball were fired horizontally from the top of a mountain with sufficient speed, the rate at which it fell could be made to match the rate at which Earth's surface fell away. The cannonball would therefore end up orbiting Earth. Newton created this illustration for the same 1687 book in which he presented the law of universal gravitation.

Although the first Earth satellite was put in orbit in 1957, Newton understood what was required to put a satellite in a circular orbit almost three centuries earlier (**Figure 6-17**). If a cannonball is dropped from a great height, such as the top of a mountain, it will fall straight down. If it is thrown horizontally at a moderate speed, the cannonball will follow a curved arc before hitting the ground. But if it is thrown horizontally with just the right speed, the surface of Earth will fall away below the cannonball so that the cannonball always remains at the same height above the surface. Put another way, Earth's gravitational attraction will cause the cannonball to accelerate

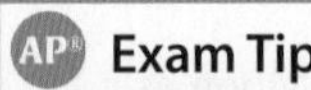
Exam Tip

In AP® Physics 1, you only need to know how to apply Newton's law of gravitation to circular orbits.

toward Earth's center. If the cannonball's speed is just right, the result will be uniform circular motion of the sort that we studied in Section 6-2, in which the acceleration is always directed toward the center of the cannonball's circular path (**Figure 6-18**).

We can find the speed v required for a circular orbit of radius r from Equation 6-4 for the acceleration in uniform circular motion: $a_{cent} = v^2/r$. For a satellite of mass m orbiting Earth, the acceleration is provided by Earth's gravitational force. From Newton's second law, this says

$$F_{\text{Earth on satellite}} = ma_{\text{cent}}$$

or, from Equation 6-4 for uniform circular motion and Equation 6-7 for the law of universal gravitation,

$$\frac{Gm_{\text{Earth}}m}{r^2} = m\frac{v^2}{r}$$

To solve for v, multiply both sides of this equation by r/m

$$\frac{Gm_{\text{Earth}}}{r} = v^2$$

and take the square root:

EQUATION IN WORDS
Speed of an Earth satellite in a circular orbit

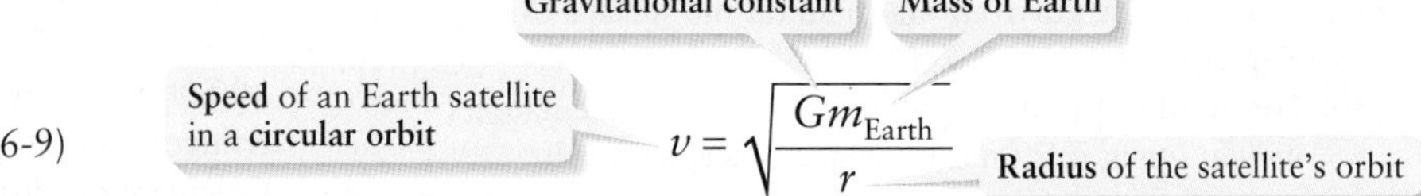

$$v = \sqrt{\frac{Gm_{\text{Earth}}}{r}} \tag{6-9}$$

Many satellites, including the International Space Station, are in *low Earth orbit*: Their height above the surface is a few hundred kilometers, which is a short distance compared to Earth's radius of 6370 km. So we can find the approximate speed of a satellite in low Earth orbit by replacing r in Equation 6-9 with the radius of Earth and substituting the values $G = 6.67 \times 10^{-11}$ Nm^2/kg^2, $m_{\text{Earth}} = 5.97 \times 10^{24}$ kg, and $R_{\text{Earth}} = 6.37 \times 10^6$ m:

$$v = \sqrt{\frac{Gm_{\text{Earth}}}{R_{\text{Earth}}}} = 7.91 \times 10^3 \text{ m/s} = 7.91 \text{ km/s}$$

(orbital speed for low Earth orbit)

This speed is about 28,500 km/h or 17,700 mi/h.

Equation 6-9 shows that increasing the orbital radius r decreases the speed necessary to maintain a stable circular orbit. We saw an example of this in Section 6-4: The Moon orbits Earth at an average distance of 3.84×10^8 m, about 60 times Earth's radius, and its orbital speed is only 1.02×10^3 m/s = 1.02 km/s. You can understand why the speed decreases with increasing orbital radius by considering a ball on the end of a string. Imagine whirling the ball on a string around your hand so that the ball makes a circular orbit around your hand. To make the ball move at high speed around a small circle, you must exert a substantial pull on the string. But if you lengthen the string and make the same ball move at low speed around a large circle, much less pull is required. Similarly, a satellite that orbits close to Earth experiences a substantial gravitational pull and so moves at high speed, while that same satellite in a larger orbit experiences less gravitational "pull" and so moves at a lower speed.

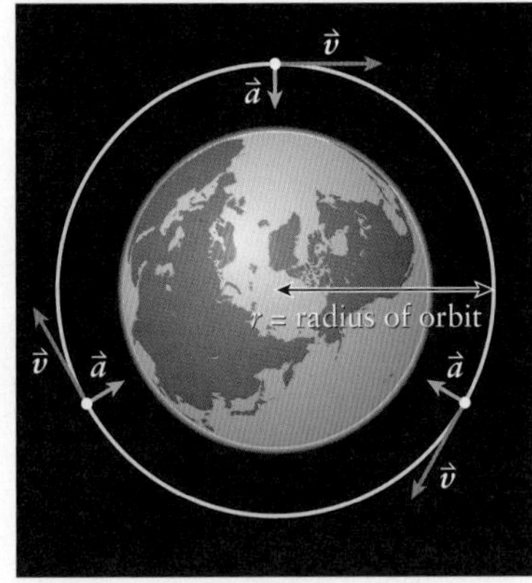

Figure 6-18 A circular orbit A satellite in a circular orbit is in uniform circular motion: The acceleration (due to Earth's gravitational force) has a constant magnitude and is always directed toward Earth's center, and the velocity has a constant magnitude.

Circular Orbits: Orbital Period

Another useful way to describe how rapidly a satellite moves around its circular orbit is in terms of the **orbital period**, which is the time required to complete one orbit. In uniform circular motion the speed v is constant, so the orbital period T is just the circumference $2\pi r$ of the orbit (the distance around the circular orbit of radius r) divided by the speed: $T = 2\pi r/v$. Using Equation 6-9 for v, this becomes

$$T = \frac{2\pi r}{v} = 2\pi r\sqrt{\frac{r}{Gm_{\text{Earth}}}}$$

It's convenient to square both sides of this equation to eliminate the square root. Note that the square of r is r^2 and the square of $\sqrt{r}$ is r, so we end up with a factor of $r^2 \times r = r^3$ on the right-hand side of the equation:

$$T^2 = \frac{4\pi^2}{Gm_{\text{Earth}}} r^3 \tag{6-10}$$

Period of an Earth satellite in a circular orbit (T^2)

Radius of the satellite's orbit (r^3)

Gravitational constant (G)

Mass of Earth (m_{Earth})

EQUATION IN WORDS
Relationship between orbital period and radius for a circular orbit

For an object in low Earth orbit, for which r equals R_{Earth}, this tells us that $T = 5.06 \times 10^3$ s = 84.4 min, or a little less than an hour and a half.

In words, Equation 6-10 says that the *square* of the orbital period T is directly proportional to the *cube* of the radius r of the orbit. By comparison, Equation 6-9 tells us that the orbital speed v is inversely proportional to the *square root* of the orbital radius r. As an example, consider a satellite that orbits at a distance $r = 4R_{\text{Earth}}$ from Earth's center (that is, at an altitude of $3R_{\text{Earth}}$ above Earth's surface). Since r is four times greater than for a satellite in low Earth orbit, the quantity r^3 is $4^3 = 4 \times 4 \times 4 = 64$ times greater. So the quantity T^2 is also 64 times greater than for an object in low Earth orbit, which means the period T is $\sqrt{64} = 8$ times longer. The orbital speed is proportional to $1/\sqrt{r}$, which is $1/\sqrt{4} = 1/2$ as great as for a satellite in low Earth orbit. These proportionality rules therefore tell us that for a satellite with $r = 4R_{\text{Earth}}$, the orbital period is 8×84.4 min = 675 min (11.2 hours), and the orbital speed is $(1/2) \times 7.91$ km/s = 3.95 km/s.

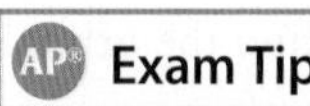
Exam Tip

Arguing results from proportionalities is an important skill in AP® Physics 1. Be sure, also, to understand the difference between a proportional relationship and other types of relationships such as *linear* and *exponential* to ensure you will be successful on the exam.

WATCH OUT!

When a satellite orbits Earth, Earth moves too.

One aspect of orbital motion that we've ignored in this discussion is Newton's third law: If Earth exerts a gravitational force on a satellite, the satellite exerts an equally strong force on Earth. So Earth also moves in a small orbit. Strictly speaking Earth and the satellite both orbit around the center of mass of the Earth–satellite system. However, even the largest satellite humans have ever placed in orbit has a tiny mass compared to the mass of Earth. So for all practical purposes the center of mass of the Earth–satellite system is at Earth's center. By contrast the Moon's mass is a reasonable fraction of Earth's mass (about 1.2%), so for detailed calculations this effect must be taken into account (**Figure 6-19**).

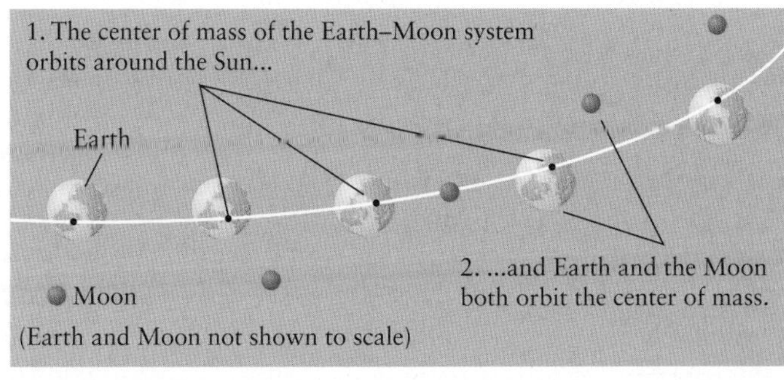

1. The center of mass of the Earth–Moon system orbits around the Sun...
Earth
2. ...and Earth and the Moon both orbit the center of mass.
Moon
(Earth and Moon not shown to scale)

Figure 6-19 The motions of Earth and the Moon Because Earth and the Moon both exert gravitational forces on each other, we are using a simplified model when we say that the Moon orbits Earth. In reality, both objects move around their common center of mass. However, you can see from the figure that this does not cause a large change in Earth's path. For the level of precision we need for many problems, such as the ones we will do in this textbook, using the model that the Moon orbits Earth gives sufficient detail.

WATCH OUT!

When a satellite falls to Earth, it's not just gravity's fault.

Orbiting satellites do sometimes fall out of orbit and crash back to Earth. When this happens, however, the real culprit is not gravity but air resistance. A satellite in a relatively low orbit is actually flying through the tenuous outer wisps of Earth's atmosphere. Air resistance, drag, as we saw in Chapter 5 reduces the speed of the satellite. When the speed decreases below that required for a circular orbit, the satellite doesn't go fast enough to keep its same distance from Earth and the orbital radius r becomes smaller, and the satellite sinks to a lower altitude, where it encounters more air resistance and sinks even lower. Eventually the satellite either strikes Earth or burns up in flight due to air friction. By contrast, the Moon and planets orbit in the near-vacuum of interplanetary space. Hence, they are unaffected by this kind of air resistance, and they have remained in orbit around Earth or the Sun since the solar system formed 4.56 billion years ago.

EXAMPLE 6-11 A Satellite for Satellite Television

The broadcasting satellites used in a satellite television system orbit Earth's equator with a period of 24 hours, the same as Earth's rotation period. As a result, these satellites are *geostationary*: They always remain over the same spot on the equator, so they always appear to be in the same position in the sky as seen from anywhere on Earth's surface. (A satellite TV receiver "dish" is aimed to receive the signal broadcast from one of these satellites.) (a) At what distance from Earth's center must a television satellite orbit? (b) What must be its orbital speed?

Set Up

A geostationary satellite has $T = 24$ h; we'll use this information and Equation 6-10 to find the radius r of the orbit. Once we find the value of r, we'll use Equation 6-9 to find the orbital speed of the satellite.

Relationship between period and radius for a circular orbit:

$$T^2 = \frac{4\pi^2}{Gm_{\text{Earth}}} r^3 \qquad (6\text{-}10)$$

Speed of an Earth satellite in a circular orbit:

$$v = \sqrt{\frac{Gm_{\text{Earth}}}{r}} \qquad (6\text{-}9)$$

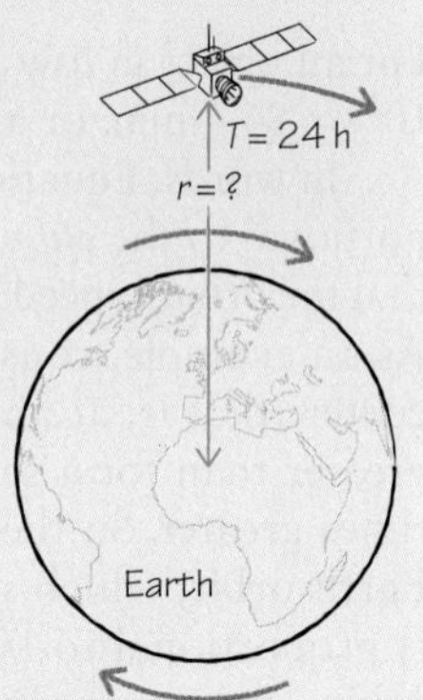

Solve

(a) To use Equation 6-10 to find the orbital radius r, first convert the orbital period T to seconds.

Period of the satellite orbit:

$$T = 24\ \text{h}\left(\frac{60\ \text{min}}{1\ \text{h}}\right)\left(\frac{60\ \text{s}}{1\ \text{min}}\right) = 8.64 \times 10^4\ \text{s}$$

Solve Equation 6-10 for r^3:

$$r^3 = \frac{Gm_{\text{Earth}}T^2}{4\pi^2}$$

$$= \frac{(6.67 \times 10^{-11}\ \text{Nm}^2/\text{kg}^2)(5.97 \times 10^{24}\ \text{kg})(8.64 \times 10^4\ \text{s})^2}{4\pi^2}$$

$$= 7.53 \times 10^{22}\ \frac{\text{Nm}^2 \cdot \text{s}^2}{\text{kg}} = 7.53 \times 10^{22}\ \text{m}^3$$

(We used $1\ \text{N} = 1\ \text{kg} \cdot \text{m/s}^2$.)

Take the cube root to find r:

$$r = \sqrt[3]{r^3} = \sqrt[3]{7.53 \times 10^{22}\ \text{m}^3}$$

$$= 4.22 \times 10^7\ \text{m} = 4.22 \times 10^4\ \text{km}$$

This is 6.63 times Earth's radius ($R_{\text{Earth}} = 6.37 \times 10^3$ km).

(b) Then use Equation 6-9 to find the orbital speed of the satellite.

Orbital speed:

$$v = \sqrt{\frac{Gm_{\text{Earth}}}{r}}$$

$$= \sqrt{\frac{(6.67 \times 10^{-11}\ \text{Nm}^2/\text{kg}^2)(5.97 \times 10^{24}\ \text{kg})}{4.22 \times 10^7\ \text{m}}}$$

$$= \sqrt{9.43 \times 10^6\ \frac{\text{Nm}}{\text{kg}}} = \sqrt{9.43 \times 10^6\ \frac{\text{kg} \cdot \text{m}}{\text{s}^2}\frac{\text{m}}{\text{kg}}}$$

$$= \sqrt{9.43 \times 10^6\ \frac{\text{m}^2}{\text{s}^2}}$$

$$= 3.07 \times 10^3\ \text{m/s} = 1.11 \times 10^4\ \text{km/h} = 6.87 \times 10^3\ \text{mi/h}$$

(Again we used $1\ \text{N} = 1\ \text{kg} \cdot \text{m/s}^2$.)

Reflect

Our answer to part (a) shows that geostationary broadcast satellites orbit at a tremendous distance from Earth. You can check the result for part (b) by confirming that the satellite's orbital speed v equals $2\pi r$ (the circumference of the circular orbit of radius r) divided by T (the time to complete one orbit). Does it?

NOW WORK Problems 3 and 4 from The Takeaway 6-5.

THE TAKEAWAY for Section 6-5

✔ Newton's law of universal gravitation explains how planets and satellites move.

✔ For an object in a circular orbit, the orbital speed is inversely proportional to the square root of the orbital radius.

✔ For an object in a circular orbit, the square of the orbital period is directly proportional to the cube of the orbital radius.

Prep for the AP® Exam

AP® Skill Builders

1. The Ferris wheel at the Santa Monica Pier has a diameter of 26 m and has a minimum rotation period of 26 s. The much larger High Roller Ferris wheel in Las Vegas has a diameter of 158.5 m.
   (a) Calculate the average speed of a cabin on the Santa Monica Ferris wheel for the minimum rotation period.
   (b) Predict the speed of a cabin on the High Roller if it had the same rotation period of 26 s.
   (c) The actual maximum speed of the High Roller is 48 km/h. Determine the minimum period.
2. The gravitational force between two objects with mass is always attractive. In terms of the relationship between the force, acceleration, and velocity, explain why a satellite orbits Earth.
3. EX 6-11 A satellite is to be raised from one circular orbit to one farther from Earth's surface. What will happen to its period?
4. EX 6-11 A geostationary satellite orbits in the equatorial plane of Earth and has a unique orbital radius and speed.
   (a) Describe in complete sentences why the radius and speed are unique.
   (b) Imagine that a geosynchronous satellite orbits in a plane intersecting Earth's surface at a latitude θ (the angle above Earth's equator). Construct a diagram showing the plane and radius of the satellite from Earth's center, r, and the radius of its orbital path if it were to travel in a circle.
   (c) It turns out that a geosynchronous satellite's orbit must lie in a plane through Earth's center. Write down the equation for gravitational force and that for mass times centripetal acceleration, and identify what the r's are and why the relationships previously used don't work if this isn't the case.

Skills in Action

5. According to Newton's universal law of gravitation, if the masses of both the object exerting the force and the object on which the force is exerted are doubled, the force is
   (A) four times as large as the original value.
   (B) twice as large as the original value.
   (C) one-half of the original value.
   (D) one-fourth of the original value.
6. Eratosthenes (276–196 B.C.E.) measured the size of Earth by measuring the length of a shadow cast at noon at two points on Earth's surface in Egypt: Aswan and Alexandria.
   (a) The caption in Figure 6-14 describes the method Eratosthenes used. The angle θ was measured directly from the length of a shadow cast by a pillar whose height can easily be determined by direct measurement. The distance, D, from Alexandria to Aswan by road today is 1076.4 km. Eratosthenes must have addressed the measurement of this distance to complete his measurement of the size of Earth. Although the distance D had probably not been measured directly in 240 B.C.E., caravans of camels laden with commercial products regularly traveled from Aswan to Alexandria. Ripe dates are time-sensitive and where there is commerce there is information. Design a strategy that Eratosthenes could have used to determine D.
   (b) Eratosthenes's measurement gave a circumference of 250,000 stadia. This is not the usual stadion (600 Greek feet, each of which is 31.6 cm in modern units). This is the Egyptian surveyor's stadion (300 royal cubits). Each royal cubit is 0.524 m in length in modern units. Construct a representation of the unit conversion (see Example 1-3) to express 250,000 stadia in meters.
   (c) Modern measurements of the radius of Earth give a value of 6371 km. Evaluate the quality of Eratosthenes's measurements by calculating the relative error, taking the modern value as the "theoretical" or accepted value. Remember that the percent error =
   $$\frac{\text{measured value} - \text{theoretical value}}{\text{theoretical value}} \times 100$$
7. Two planets, A and B, with the same mass are in circular orbits around the same star. The orbital radius of A is four times larger than the orbital radius of B. The period of planet B is
   (A) eight times the period of planet A.
   (B) four times the period of planet A.
   (C) one-fourth of the period of planet A.
   (D) one-eighth of the period of planet A.

8. According to Newton's universal law of gravitation, if the distance r between two objects is doubled, the force exerted by one of these objects on the other is
(A) four times as large as the original value.
(B) twice as large as the original value.
(C) one-half of the original value.
(D) one-fourth of the original value.

9. Consider the figure below. The net gravitational force on a satellite placed at point L_1 or L_2 causes the satellite to orbit the Sun with the same period as Earth, T_{Earth}. You may assume the satellites have just two forces exerted on them: one by Earth and one by the Sun.

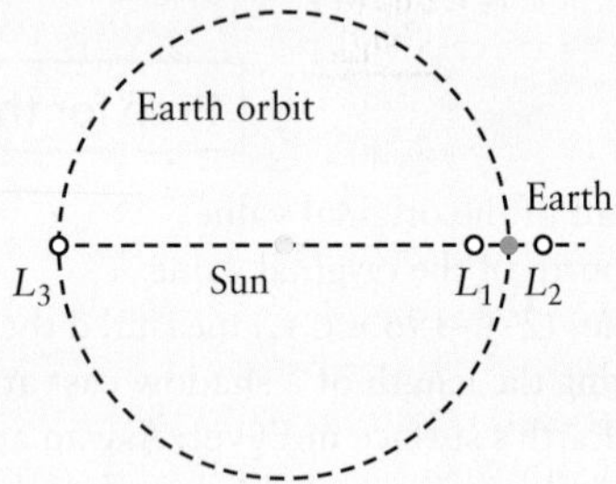

(a) Using free-body diagrams and Newton's second law, show that the centripetal acceleration, a_{L1} of an object at the point L_1 and the centripetal acceleration, a_{L2} of an object at the point L_2 can be expressed as

$$-\frac{GM_{Sun}}{r^2_{Sun-L1}} + \frac{GM_{Earth}}{r^2_{Earth-L1}} = a_{L1}$$

and

$$-\frac{GM_{Sun}}{r^2_{Sun-L2}} - \frac{GM_{Earth}}{r^2_{Earth-L2}} = a_{L2}$$

(b) Explain why the orbital period of L_1 and Earth will be equal if

$$-\frac{M_{Sun}}{r^3_{L1-Sun}} + \frac{M_{Earth}}{r^2_{Earth-L1} r_{Sun-L1}} = -\frac{M_{Sun}}{r^3_{Earth-Sun}}$$

This expression can be used to calculate the distance from Earth to L_1. The distance is approximately 1.5×10^9 m, to be compared with $r_{Earth-Sun} = 1.5 \times 10^{11}$ m.

(c) Euler described $r_{L1-Earth}$ and $r_{L2-Earth}$ in 1765. He also described a third point, L_3, with the same orbital period as Earth, T_{Earth}. Explain why the periods of L_3 and Earth must be the same.

6-6 Apparent weight and what it means to be "weightless"

Figure 6-20 shows astronauts Shane Kimbrough and Sandra Magnus as they watch a basket's worth of fruit floating freely about the cabin of the Space Shuttle *Endeavour*. The astronauts *feel* as though they are weightless, but this is a misnomer: Earth exerts a gravitational force on them just as it does on their spacecraft. As **Figure 6-21** shows, because the astronauts are the same distance from Earth as their spacecraft, the shuttle, astronauts, and fruit are all in free fall, experiencing the same acceleration due to Earth's gravity. This acceleration is only about 10% less than at the surface of Earth. Since they fall at the same rate, they stay in the same relative position. Since the astronauts and fruit have no tendency to move toward the floor, ceiling, or any other part of the spacecraft, they are what some would call "floating." This is **apparent weightlessness** (or **effective weightlessness**).

Figure 6-20 Astronauts in orbit In the apparent weightlessness aboard an orbiting spacecraft, there is no up or down.

The primary way we perceive our weight is through the force the floor exerts on our feet. If you ride in an elevator that accelerates rapidly upward, you'll feel the floor pushing upward harder on your feet as though your weight had increased. If instead the elevator accelerates rapidly downward, the force that the floor exerts on your feet is reduced, and you feel as though your weight had decreased. In the extreme case where the elevator is falling freely with a downward acceleration of magnitude $g = 9.80$ m/s^2, the force of the floor on your feet goes to zero and you have a sense of apparent weightlessness, even though gravity is still exerted on you. As long as both you and your surroundings experience the same acceleration, you will feel "weightless."

So it's not necessary for a spacecraft to be in orbit around a planet for the occupants to experience apparent weightlessness. All that's needed is for the spacecraft to be falling freely, so the only external force on the spacecraft is the gravitational force. Astronauts will feel weightless whether they are orbiting Earth aboard the International Space Station, on a mission from Earth to the Moon, or on an interplanetary voyage to Mars. For a mission to Mars or any other planet, the spacecraft follows an elliptical orbit chosen so that it intersects the orbit of

Figure 6-21 The origin of apparent weightlessness If the only external force exerted on a spacecraft is the gravitational force, then the spacecraft and its occupant have the same acceleration $a = F/m$. As a result the spacecraft and the astronaut fall freely along the same trajectory, and the astronaut feels "weightless."

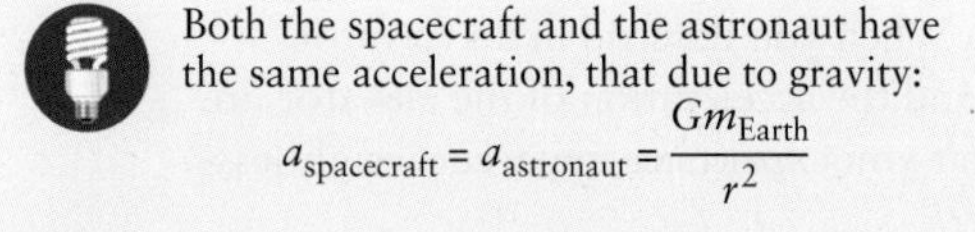

the destination planet. The only times during the flight when the astronauts do not feel weightless are when the rockets are firing, either to put the spacecraft into the desired orbit or to take it out of orbit for landing.

Both the spacecraft and the astronaut have the same acceleration, that due to gravity:
$a_{\text{spacecraft}} = a_{\text{astronaut}} = \dfrac{Gm_{\text{Earth}}}{r^2}$

At first glance it might seem that being in a state of apparent weightlessness would be relaxing. In fact there are many very negative effects on an astronaut's body. One common problem is *space adaptation syndrome* (also called *space sickness*). About half of all astronauts suffer from this condition during their first few days of spaceflight. The symptoms include motion sickness and disorientation.

Two other problems that bedevil all astronauts are *loss of muscle mass* and *blood loss*. You exercise the muscles in your calves and along your spine simply by walking and standing. But in apparent weightlessness these muscles are not exercised and can lose as much as 5% of their mass per week. During a very long-duration spaceflight, such as a mission to Mars (which could take 10 months each way), astronauts might lose up to 40% of their capacity to do physical work, equivalent to a 40-year-old astronaut's muscles deteriorating to those of an 80-year-old. For this reason astronauts exercise as much as possible in space using special equipment designed to mimic gravity. (Building muscle mass before going into space does not appear to help: Astronauts who begin their missions with the greatest muscle mass also show the greatest decline while in space.)

The cause of blood loss is more subtle. When you stand upright on Earth, blood pools in your feet due to gravity, so your blood pressure is higher in your feet than in your brain. In apparent weightlessness, however, blood distributes itself more evenly through the body. (This gives astronauts in space a characteristic puffy-faced appearance.) This initiates a reflex to reduce blood volume. As a result, astronauts can lose up to 22% of their blood volume within 3 days of apparent weightlessness. The weightless environment also reduces demand on the heart because it does not have to overcome the effects of gravity to provide the brain with blood, so the heart muscle begins to atrophy as well. This is another reason astronauts do cardiovascular exercise while in space.

Astronauts recover muscle strength relatively quickly after returning from space (at a rate of about a day of recovery per day in space), and blood volume can be restored within a few days by drinking fluids. More problematic is recovery from the *bone loss* that occurs in space. Astronauts' bones atrophy at a rate of about 1% per month in apparent weightlessness, and recovery from a 6-month mission may require 2 or 3 years back on Earth coupled with a program of strenuous exercise.

Another threat to astronauts' well-being is that while human muscle and bone tend to atrophy in apparent weightlessness, microbes such as *Escherichia coli* and *Staphylococcus* actually reproduce more rapidly under these conditions. This leads to increased risk for contamination and serious infection during a long-duration spaceflight. Such are the biological challenges that confront the future of human exploration of space.

THE TAKEAWAY for Section 6-6

✔ Astronauts in space feel apparent weightlessness because their acceleration is the same as that of the spacecraft.

✔ The physiological challenges of apparent weightlessness include space sickness; loss of muscle mass, blood, and bone; and increased risk of microbial infection.

Prep for the AP Exam

 Skill Builders

1. Astronauts and the ISS (International Space Station) in which they travel have the same acceleration as each other, and the astronauts and the ISS experience apparent weightlessness. When you descend in an elevator, your acceleration and the acceleration of the elevator are identical. Why don't you experience apparent weightlessness in an elevator?
2. Predict whether the normal force exerted on a person standing on a bathroom scale on the floor of an elevator increases or decreases when the elevator starts upward from rest.

AP Skills in Action

3. A mountain climber ascends a very tall mountain. How does her weight change? Choose the correct answer.
   (A) She weighed more at the bottom of the mountain because the gravitational field is increasing as she ascends.
   (B) She weighed less at the bottom of the mountain because the gravitational field is decreasing as she ascends.
   (C) She weighed more at the bottom of the mountain because the gravitational field is decreasing as she ascends.
   (D) Her weight does not change because the gravitational field is constant.

WHAT DID YOU LEARN?

Chapter learning goals	Section(s)	Related example(s)	Prep for the AP Exam: Relevant section review exercises
Describe why an object moving in a circle is always accelerating even when its speed is not changing.	6-2	6-1	1
Apply Newton's laws to objects in uniform circular motion.	6-3	6-3 6-4 6-5 6-6	1, 2, 3, 4, 5
Recognize what it means to say that gravitation is universal, and articulate when the gravitational force is the dominant force between objects or systems.	6-4	6-7	1
Apply Newton's law of universal gravitation to describe or calculate the gravitational force any two objects exert on each other, and use that force in contexts other than orbital motion.	6-4	6-7 6-8 6-9	2, 3, 5
Apply the law of universal gravitation to analyze circular orbits of satellites and planets.	6-5	6-11	3, 4
Relate the gravitational field at a point in space to the gravitational force exerted on an object at that point in space.	6-4	6-9 6-10	4 5
Explain the origin of apparent weightlessness.	6-2 6-6	6-2	 1

Chapter 6 Review

Key Terms

All the Key Terms can be found in the Glossary/Glosario on page G1 in the back of the book.

apparent weightlessness (effective weightlessness) 272
Cavendish experiment 262
centripetal acceleration 239
centripetal force 246
field 264
gravitational constant 259
law of universal gravitation 236
orbital period 268
period 238
uniform circular motion 238

Chapter Summary

Topic	Equation or Figure
Centripetal acceleration: In order for an object to travel in a path that is not straight, it must be accelerating even if its speed is not changing, because its velocity is changing direction. For an object going in a circle, this change in direction is perpendicular to the direction of motion at any instant, toward the center of the circular path, and is called centripetal acceleration. If an object's speed is not changing it must not have an acceleration in its direction of motion.	$a_{cent} = \frac{v^2}{r}$ (6-4) **Centripetal acceleration:** magnitude of the acceleration of an object in uniform circular motion **Speed** of the object as it moves around the circle **Radius** of the circle
Centripetal force: To cause an object to accelerate so that it moves in a circle, there must be a net force on the object that causes this acceleration, so directed toward the center of the circular path. To describe the direction, this net force is referred to as centripetal force, and can be caused by any type of force exerted on the object.	$\sum F = F_{cent} = \frac{mv^2}{r}$ (6-5) **Net force** exerted on an object in uniform circular motion **Mass** of the object **Speed** of the object as it moves around the circle The direction of the net force is toward the center of the object's circular path. **Radius** of the circle
Newton's law of universal gravitation: Any two objects attract each other with a gravitational force. This force is proportional to the product of the masses of the objects and inversely proportional to the square of the distance between the centers of the two objects.	$F_{1\text{ on }2} = F_{2\text{ on }1} = \frac{Gm_1m_2}{r^2}$ (6-7) **Gravitational constant** (same for any two objects) **Masses** of the two objects **Any two objects** (1 and 2) exert **equally strong gravitational forces** on each other. **Center-to-center distance** between the two objects **The gravitational forces are attractive:** $\vec{F}_{1\text{ on }2}$ accelerates object 2 toward object 1 and $\vec{F}_{2\text{ on }1}$ accelerates object 1 toward object 2.

Value of g at Earth's surface: The acceleration due to gravity at our planet's surface is related to Earth's mass and radius. At greater distances r from Earth's center, the value of g decreases in inverse proportion to the square of r.

Gravitational constant | Mass of Earth

Acceleration due to gravity at Earth's surface

$$g = \frac{Gm_{\text{Earth}}}{R_{\text{Earth}}^2} \qquad (6\text{-}8)$$

Radius of Earth

The Cavendish experiment: The value of the gravitational constant G in the law of universal gravitation can be determined by measuring the gravitational attraction between objects of known mass.

(1) Two small lead spheres, each of mass m, are attached to the ends of a wooden rod, and the rod is suspended from a wire.

(2) Two large lead spheres, each of mass M, are then placed on the ground at equal distances from the small spheres.

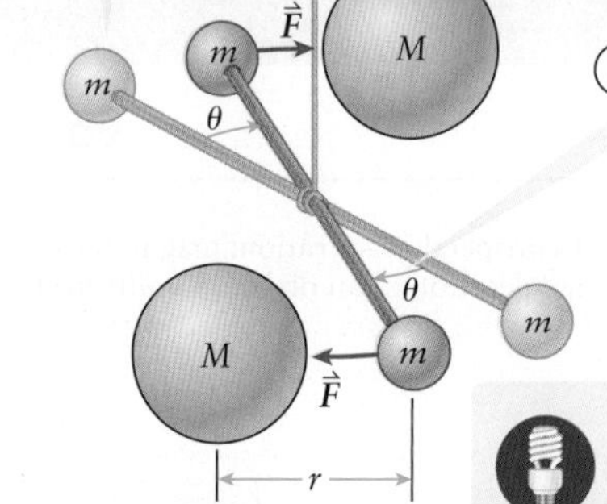

(3) Each large sphere exerts a gravitational force on the nearby small sphere. This makes the rod rotate through an angle of θ (shown here greatly exaggerated) and twists the wire.

Knowing the masses m and M of the spheres and the distance r when the rod is in equilibrium, you can determine G using the relationship $F = GMm/r^2$.

(Figure 6-15)

Circular orbits: A satellite in a circular orbit around Earth moves with a constant speed v. The orbital period T in a circular orbit increases with increasing orbital radius r.

Period of an Earth satellite in a circular orbit | **Radius** of the satellite's orbit

Gravitational constant

$$T^2 = \frac{4\pi^2}{Gm_{\text{Earth}}} r^3 \qquad (6\text{-}10)$$

Mass of Earth

Apparent weightlessness: Because gravitation is universal, astronauts riding in a spacecraft (with the rockets off) have the same acceleration as the spacecraft. As a result they "fall" along with the spacecraft and feel as though they are weightless.

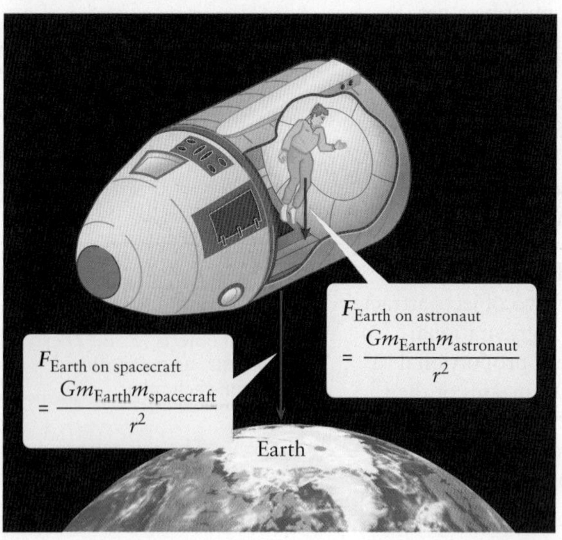

$F_{\text{Earth on astronaut}} = \dfrac{Gm_{\text{Earth}}m_{\text{astronaut}}}{r^2}$
$F_{\text{Earth on spacecraft}} = \dfrac{Gm_{\text{Earth}}m_{\text{spacecraft}}}{r^2}$
Earth

(Figure 6-21)

Both the spacecraft and the astronaut have the same acceleration, that due to gravity:

$$a_{\text{spacecraft}} = a_{\text{astronaut}} = \frac{Gm_{\text{Earth}}}{r^2}$$

Fields: For forces that are exerted at a distance, it is useful to model them as fields, which associate a value of some physical quantity with every point in space. Forces are vectors, and so the associated fields are also vectors, having a magnitude and a direction assigned to each point in space. Since mass is the source of a gravitational force, a gravitational field and the acceleration due to gravity are identical at any point in space since the gravitational field is the gravitational force that would be exerted on a test mass at some point in space divided by the test mass. So, Equation 6-8 represents acceleration or gravitational field, with the understanding that either is a vector directed toward the center of the object causing the force. If there are multiple sources, then the total field is the vector sum of each of the individual fields.

Gravitational constant; Mass of Earth

Acceleration due to gravity at Earth's surface

$$g = \frac{Gm_{\text{Earth}}}{R_{\text{Earth}}^2} \quad (6\text{-}8)$$

Radius of Earth

Chapter 6 Review Problems

1. In 1892 George W. G. Ferris designed a carnival ride in the shape of a large wheel. This Ferris wheel had a diameter of 76 m and completed one revolution every 20 min. Calculate the magnitude of the centripetal acceleration and the speed of one of the cars, which were located on the circumference of the wheel.
2. A car races at a constant speed of 330 km/h around a flat, circular track 1.00 km in diameter. What is the car's centripetal acceleration in m/s^2?
3. Fidget spinners rotate at about 4000 rpm. Calculate the speed and acceleration at the edge of a fidget spinner that has a diameter of 6.0 cm.
4. Anne is working on a research project that involves the use of a centrifuge. Her samples must first experience an acceleration of 100 *g*, but then the acceleration must increase by a factor of 8. By how much will the rotation speed have to increase? Express your answer as a multiple of the initial rotation rate.
5. A fast-pitch softball player does a windmill pitch, moving her hand through a circular arc with her arm straight. She releases the ball at a speed of 34.3 m/s. Just before the ball leaves her hand, the ball's centripetal acceleration is 1960 m/s^2. What is the length of her arm from the pivot point at her shoulder?
6. Very high-speed ultracentrifuges are useful devices to sediment materials quickly or for separating materials. An ultracentrifuge spins a small tube in a circle of radius 10.0 cm at 6.00×10^4 rpm. Calculate the force exerted on an object that has a mass of 3.00 g rotating in the centrifuge.
7. In the Large Hadron Collider near Geneva, Switzerland, protons travel in a circular orbit with a circumference of 27 km at a speed of nearly 3.0×10^8 m/s (0.999999991 *c*). The mass of a single proton is approximately 1.673×10^{-27} kg. Superconducting magnets constrain the motion in the circular orbit by exerting a centripetal force on the protons. What is the magnitude of this force (neglecting relativistic effects, even though they are sizable in this case)?
8. What is the magnitude of the force exerted on a jet pilot by her seat as she completes a vertical loop that is 5.00×10^2 m in radius at a speed of 2.00×10^2 m/s? Assume her mass is 70.0 kg and that she is located at the bottom of the loop.
9. A coin that has a mass of 25.0 g rests on a horizontal record player's turntable that rotates at 78.0 rpm. The center of the coin is 12.0 cm from the turntable axis. If the coin does not slip, what is the minimum value of the coefficient of static friction between the coin and the turntable surface?
10. A 25.0-g metal washer is tied to a 60.0-cm-long string and whirled around in a vertical circle at a constant speed of 6.00 m/s. Calculate the tension in the string (**a**) when the washer is at the bottom of the circular path and (**b**) when it is at the top of the path.
11. In executing a windmill pitch, a fast-pitch softball player moves her hand through a circular arc of radius 0.60 m. She begins with her arm high behind her and rotates her arm down and then forward. The 0.19-kg ball leaves her hand at 33.0 m/s. (**a**) When in the circular motion would she want to release the ball so that it has the maximum forward velocity? (**b**) What is the magnitude of the upward force exerted on the ball by her hand immediately before she releases it?

12. Two stars are separated by a center-to-center distance of 7.50×10^9 km. The first star has a mass equal to the mass of our Sun, $M_1 = M_{Sun}$. The second star has a mass that is one-half of the Sun's mass, $M_2 = M_{Sun}/2$. Let r_1 be the distance from the center of the first star to a point along the line between the stars, as shown.

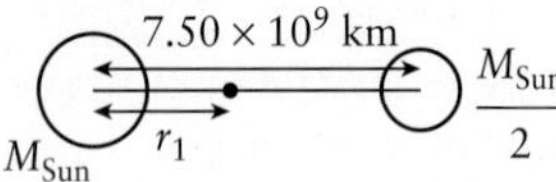

(a) Construct a free-body diagram for an object with mass m instantaneously at rest at point r_1.

(b) Construct but do not solve a mathematical equation in terms of M_{Sun}, r_1, and any necessary constants that you could use to solve for r_1, a distance such that the object remains at rest at that point.

13. Occupants of cars hit from behind often suffer serious neck injury from whiplash. During a low-speed rear-end collision on a level road, the car suddenly jerks forward while the head of a person tries to maintain its previous motion because of inertia. In this accelerating reference frame, the person's experience is that the head suddenly pivots backward about the base of the neck. Assume for one particular collision this rotation is through a 60.0° angle and that the motion lasts 250 ms. The distance from the base of the neck to the center of the head is typically about 0.20 m, and the head normally comprises about 6.0% of a person's body weight. Model the motion of the head as uniform speed over the course of its pivot.

(a) Construct a diagram representing the model.

(b) Calculate the centripetal acceleration of the head (modeled as an object at the center of the head) about the pivot point of the neck during the collision.

(c) Calculate the force required to cause this acceleration for the head of a 75-kg person. Neglect other forces exerted on the head, including the gravitational force. Although this rotation is caused by inertia, it effectively exerts this force on the neck, damaging the disks between the vertebrae as well as the muscles and tendons in the neck.

14. Apply the universal law of gravitation and the relationship between the gravitational field and weight on the surface of a planet to obtain an expression for the field in terms of the mass and radius of the planet.

15. Determine the average magnitude of the force of gravity exerted by the Sun on Earth.

16. Determine the net force of gravity exerted on the Moon during an eclipse when it is directly between Earth and the Sun.

17. Earth orbits the Sun with an average orbital radius of 1.5×10^{11} m and a period of 365.25 days. The mass of Earth is 5.97×10^{24} kg.

(a) Calculate the average velocity of Earth in its orbit.

(b) Calculate the mass of the Sun.

18. We observe the radius of the orbit of the Moon around Earth, 3.84×10^8 m, and the period of the orbit, 27.4 days. The mass of Earth can be determined with measurements of the acceleration due to gravity on the surface of Earth, and the radius of Earth. Earth's radius, if we approximate Earth as a sphere, can be determined from the measurement of the distance between two points on the surface. Explain why this information is enough to calculate the average speed of the Moon in its orbit, but is not enough to calculate the mass of the Moon.

19. A ball swings from a cord attached to the ceiling, as shown in the figure.

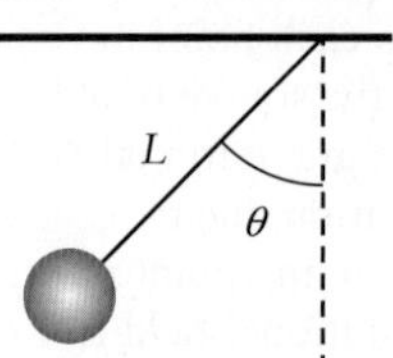

(a) Construct a free-body diagram representing the forces exerted on the ball the moment it is released from rest in the position shown.

(b) Draw a free-body diagram representing the forces exerted on the ball when the string is vertical and the ball has reached its maximum speed and is just beginning to swing upward.

(c) Apply Newton's second law to express the mathematical relationship between the horizontal and vertical components of the acceleration and the corresponding components of the net force at positions described in parts (a) and (b).

(d) Explain why the equation describing circular motion caused by a central force, $F = mv^2/r$, is consistent with the instant in the motion of the swinging ball described in part (b).

20. In the game of tetherball, a 1.25-m rope connects a 0.750-kg ball to the top of a vertical pole, so that the ball can revolve around the pole, as shown in the figure.

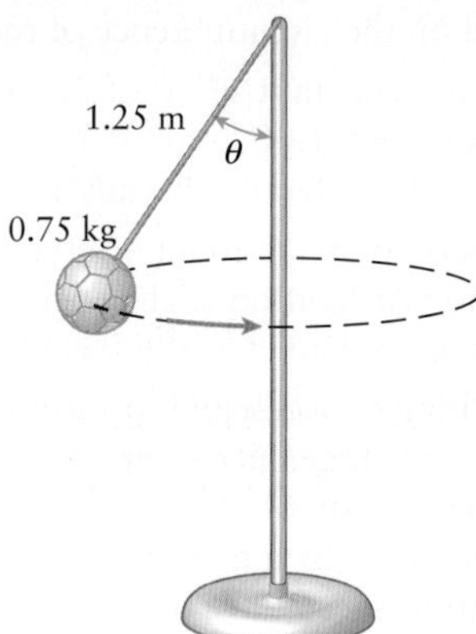

(a) Draw a free-body diagram of the tetherball when the angle θ of the rope is 35.0° with the vertical.

(b) Apply Newton's second law to calculate the tension in the rope.

(c) Calculate the speed of the ball as it rotates around the pole.

(d) As the ball passes one of the players the player slaps the ball to exert a force in the direction in which the ball was moving. Explain why the angle between the rope and the vertical increases.

21. In a laboratory test of tolerance for high centripetal acceleration, pilots were swung in a circle 13.4 m in diameter. It was found that they blacked out when they were spun at 30.6 rpm (rev/min).

(a) At what acceleration (in SI units and in multiples of g) did the pilots black out?
(b) If you want to decrease the acceleration by 25.0% without changing the diameter of the circle, by what percent must you change the time for the pilots to make one spin?

22. Modern pilots with specialized flight suits can survive centripetal accelerations up to 9 g (88 m/s^2). Can a fighter pilot flying at a constant speed of 500 m/s and in a circle that has a diameter of 8800 m survive to tell about his experience?

23. A 1.50×10^3-kg truck rounds an unbanked curve on the highway at a speed of 20.0 m/s. If the maximum friction force between the surface of the road and all four of the tires is 8.00×10^3 N, calculate the minimum radius of curvature for the curve to prevent the truck from skidding off the road.

24. Biomedical laboratories routinely use ultracentrifuges, some of which are able to spin at 1.00×10^5 rev/min about the central axis. The turning rotor in certain models is about 20.0 cm in diameter. At its top spin speed, what force does the rotor exert on a 2.00-g sample that is positioned at the greatest distance from the spin axis to keep it moving in the circle? Would the force needed be appreciably different if the sample were spun in a vertical or a horizontal circle? Why or why not?

25. An amusement park ride called the Rotor debuted in 1955 in Germany. Passengers stand in the cylindrical drum of the Rotor as it rotates around its axis. Once the Rotor reaches its operating speed, the floor drops but the riders remain pinned against the wall of the cylinder. Suppose the cylinder makes 25.0 rev/min and has a radius of 3.50 m. What is the minimum coefficient of static friction between the wall of the cylinder and the backs of the riders?

26. The four largest of Jupiter's moons are listed in the table below. Using these data, Equation 6-10, and the law of universal gravitation, (a) complete the table and (b) calculate the mass of Jupiter. Assume circular orbits, and that the semimajor axis is the radius of the circular orbit.

Moon	Semimajor axis (km)	Orbital period (days)
Io	421,700	1.769
Europa	671,034	?
Ganymede	?	7.155
Callisto	?	16.689

27. The former Soviet Union launched the first artificial Earth satellite, *Sputnik*, in 1957. Its mass was 83.6 kg, and it made one orbit every 96.2 min.
(a) Calculate the altitude of *Sputnik*'s orbit above Earth's surface, assuming circular orbits.
(b) What was *Sputnik*'s weight in orbit and at Earth's surface?

28. Locate the point(s) along the line AB where a small, 1.00-kg object could be placed such that the net gravitational force on it due to the two objects shown is exactly zero.

0.400 m
A B
10.0 kg 3.00 kg

29. The Ferris wheel at the Santa Monica pier has a diameter of 26.0 m and a period of 26.0 s and rotates at a constant angular speed. Consider a student riding the Ferris wheel who has a mass of 71.0 kg. The student remains upright for the entire ride. Calculate the apparent weight of the student at the (a) top of the ride and (b) the bottom of the ride.

30. Draw and describe a physical situation to which the following equation could apply.

$$\frac{\left(8.5\,\frac{\text{m}}{\text{s}}\right)^2}{5.0\text{ m}} = \frac{F_n + (61.0\text{ kg})\left(9.8\,\frac{\text{N}}{\text{kg}}\right)}{61.0\text{ kg}}$$

31. A satellite passes over Earth's north pole twice in 24 hours. Calculate the altitude of this satellite.

32. Two satellites are in stable, circular orbits about a planet. One is at a radius R_A and the other is at a radius $R_B = 2R_A$.
(a) Calculate the ratio of the speeds of these satellites in their orbit, $\frac{v_B}{v_A}$.
(b) Calculate the ratio of the periods of these satellites, $\frac{T_B}{T_A}$.

33. A large passenger airplane flying north approaches an airport for landing at 75 m/s when the pilot receives instructions from the control tower to change runways by turning east along a circular path. Assume that the plane maintains the same speed and remains at the same altitude, and that the ratio of the apparent weight to the true weight of a passenger does not exceed 1.2.
(a) Using the notation in Example 6-4, construct free-body diagrams of the airplane and a passenger while the plane is making the turn.
(b) Apply Newton's second law to construct a mathematical representation of the relationships between the horizontal and vertical components of the net force exerted on the plane and the centripetal acceleration in terms of the bank angle (the angle between the lift vector and the negative of the gravitational force).
(c) Apply Newton's second law to construct a mathematical representation of the relationships between vertical components of the net force exerted on the passenger in terms of the bank angle.
(d) Calculate the maximum bank angle for the turn.
(e) Calculate the time required to complete the turn through 90° at the maximum bank angle.

The control tower adds another restriction: The turn must be completed within 15 s.
(f) Calculate the minimum bank angle that meets the control tower restriction while maintaining the same velocity and calculate the ratio of apparent and actual weights of passengers during the turn.
(g) Explain why the bank angle must be modified to meet the additional restriction if the speed does not change.

Prep for the AP® Exam

AP® Group Work

Directions: This problem is designed to be done as group work in class.

A string is threaded through a plastic tube and tied at one end to a rubber stopper. At the other end a paper clip is bent to support washers. Each washer has a mass of 25.0 g. By holding the plastic tube and rotating her wrist without touching the string, a student can swing the stopper in a nearly horizontal circle while the washers remain at constant distance below the tube. With a stopwatch her partner measures the period of 20 complete revolutions of the stopper. The radius, r, of the circular motion is also measured.

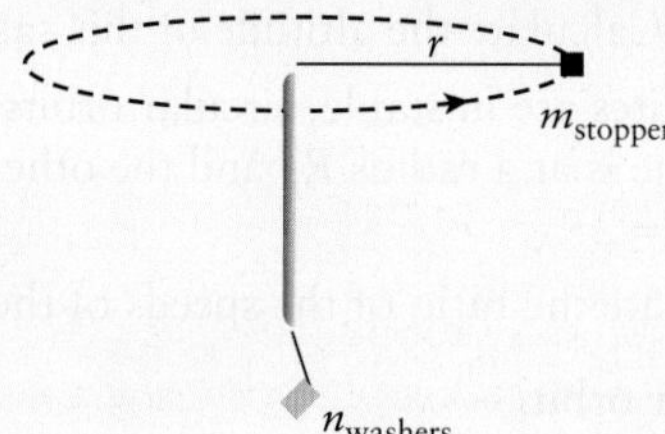

The data in the following table were collected with this procedure.

Number of washers	String length (m)	Period for 20 revolutions (s)
5	1.1	19.0
6	0.75	13.5
7	0.5	11.0
8	0.45	9.5
9	0.4	8.5
10	0.35	7.5

(a) Apply Newton's second law to express the relationship among the length of the string extended above the tube, r, the period of one revolution of the stopper T, the mass of the washers $m_{washers}$, the mass of the stopper $m_{stopper}$, and the gravitational acceleration g.

(b) Analyze the data graphically in terms of the independent variable, $m_{washers}g$, and the dependent variable, r/T^2.

(c) Apply the results of your analysis of the data to predict the mass of the stopper.

(d) Using a pan balance the mass of the stopper was measured as 23.2 g. Construct a claim regarding the accuracy of the prediction in part (c) using percent relative error.

(e) Construct free-body diagrams of the washers and of the stopper.

(f) Using the free-body diagrams in part (e) explain why the circular path of the stopper is only approximately horizontal and, given the experimental mass of the stopper, evaluate the bias in the measured radius of the circular path in terms of the range of the percent relative error in use of measured radius or string lengths.

AP® PRACTICE PROBLEMS

Prep for the AP® Exam

Multiple-Choice Questions

Directions: The following questions have a single correct answer.

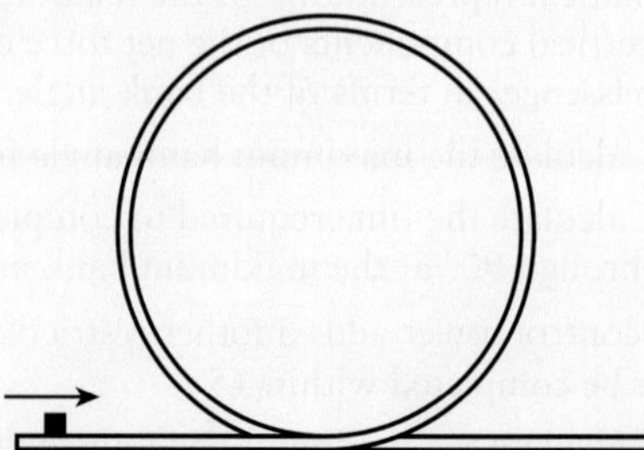

1. A small object is projected horizontally along a track that exerts negligible friction on the object. It then enters a vertical circular section, completes the loop, and continues on. Of the following, which shows the correct direction of the object's acceleration at each of the four points: initially at the bottom, on the right side, at the top, and finally as it comes down the left side?

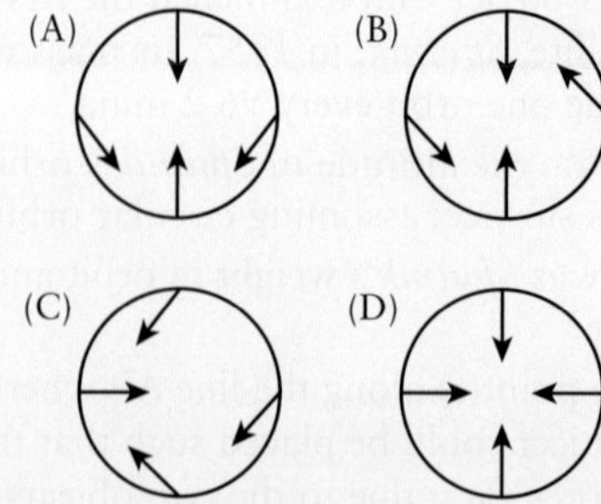

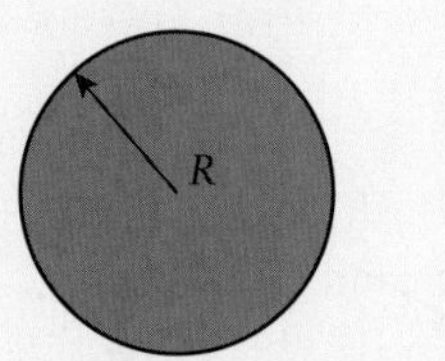

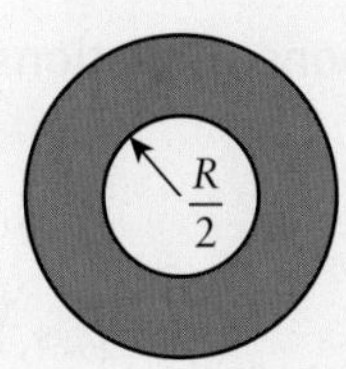

2. Suppose a planet of uniform density throughout has a radius R and an acceleration of gravity at its surface g. If a second planet were of the same radius R and made of the same material but had a hollow center of radius $0.50R$, what would be the acceleration of gravity at its surface?

(A) $\frac{1}{8}g$ (B) $\frac{1}{4}g$ (C) $\frac{1}{2}g$ (D) $\frac{7}{8}g$

3. Consider a nonrotating planet of radius R that has no atmosphere. On Earth the atmosphere helps keep everything on the surface in place as the planet rotates. Because it has no atmosphere, on this other planet just the gravitational force exerted by the planet keeps things on the surface. Take the acceleration of gravity on the surface of that planet as g_P. Now consider a mass m located on the surface at the planet's equator. How fast would the planet have to rotate at its surface for the object's apparent weight to be reduced to three-quarters of that for the nonrotating planet?

(A) $v = \frac{\sqrt{g_P R}}{2}$ (B) $v = \frac{\sqrt{3g_P R}}{2}$

(C) $v = 2\sqrt{g_P R}$ (D) $v = \frac{3\sqrt{g_P R}}{4}$

4. Consider an object of mass m located at an angle θ from the axis of rotation of a planet. It is hanging from a string. Of the following, which diagram best shows the relationship between the force of gravity, mg, and the tension in the string, F_T, on that rotating planet?

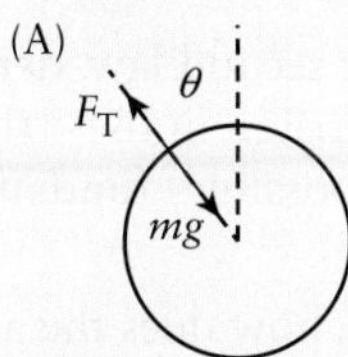

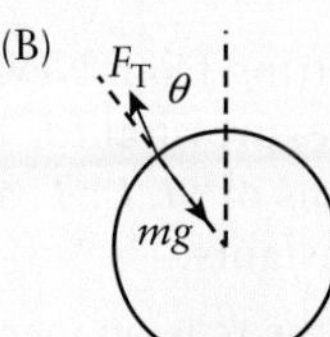

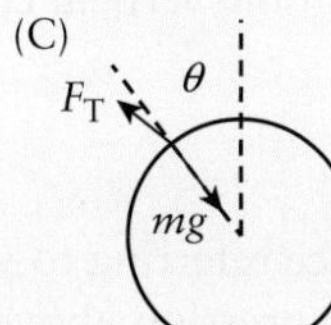

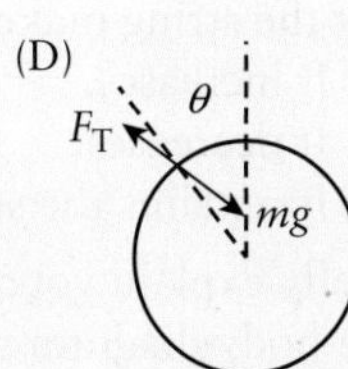

5. Two objects are in circular orbits around the same planet. One is traveling at a speed v and makes a complete orbit in time T. If the other is traveling at twice the speed, the time for it to orbit will be which of the following?

(A) $\frac{1}{8}T$ (B) $\frac{1}{4}T$ (C) $\frac{1}{2}T$ (D) $2T$

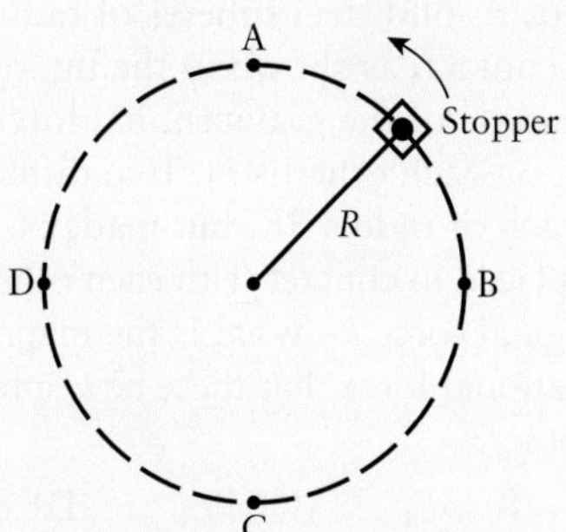

6. Consider a rubber stopper at the end of a string twirled in a vertical circle at constant speed, as shown above. Four points, A through D, are labeled on the circular path. Which of the following is the correct ranking of these points based on the tension in the string at that point?

(A) (D = B) > C > A
(B) A > (B = D) > C
(C) C > (B = D) > A
(D) C > A > (B = D)

7. The net gravitational force exerted on an interstellar traveler by two stars is zero. What conclusion can be drawn regarding the location of the traveler relative to the two stars? Choose the best answer.

(A) The two stars are the same distance from the traveler.
(B) The ratios of the mass of each star to the distance from the star to the traveler are the same.
(C) The ratios of the mass of each star to the distance squared from the star to the traveler are the same.
(D) The ratios of the mass of each star to the distance squared from the star to the traveler are the same and the traveler is located on a line connecting the centers of the two stars.

(Continued)

Questions 8 and 9 refer to the following material.

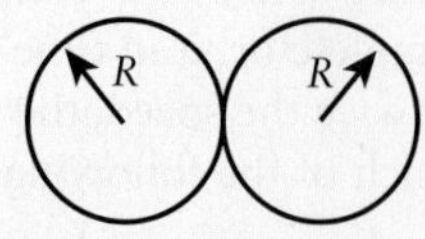

Cross section

8. Two identical solid steel spheres of radius R are placed in contact, as shown in the figure above. The magnitude of the gravitational force that they exert on each other is F_1. Two different solid spheres, each of radius $3R$, but made of this same steel, are placed in contact with each other. In terms of the original force F_1, what is the magnitude of the gravitational force that these new spheres exert on each other?

(A) F_1 (B) $3F_1$ (C) $9F_1$ (D) $81F_1$

9. Two solid identical steel spheres of radius R are placed in contact, as shown in the figure above. The magnitude of the gravitational force that they exert on each other is F_1. Two solid identical brass spheres of the same radius R (brass is denser than steel) are placed in contact. The magnitude of the gravitational force that they exert on each other is F_2. Two solid identical aluminum spheres of the same radius R (aluminum is less dense than steel) are placed in contact. The magnitude of the gravitational force that they exert on each other is F_3. How do the magnitudes of the gravitational forces compare?

(A) $F_1 = F_2 = F_3$

(B) $F_1 > F_2 > F_3$

(C) $F_2 > F_1 > F_3$

(D) $F_3 > F_1 > F_2$

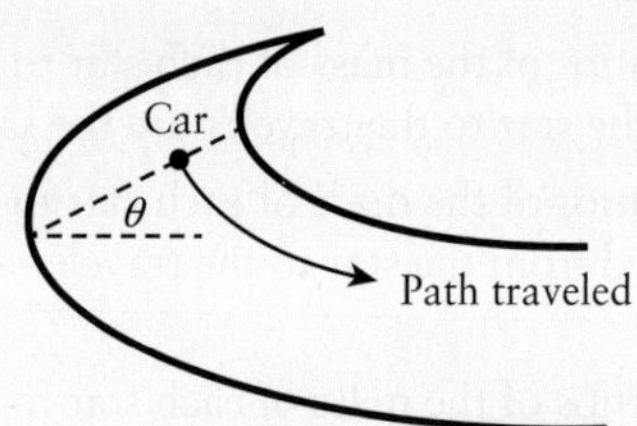

10. As shown above, a car is traveling around a curve that has been improperly banked. The surface of the road makes an angle of θ with the horizontal, and as it proceeds around the curve the car does not move up or down the incline. Which of the following expressions is correct?

(A) $F_{S,max} > mg \sin \theta$

(B) $mg = F_{S,max} \sin \theta + F_n \cos \theta$

(C) $F_{S,max} \cos \theta = F_n \sin \theta$

(D) $F_n \cos \theta = mg$

Free-Response Questions

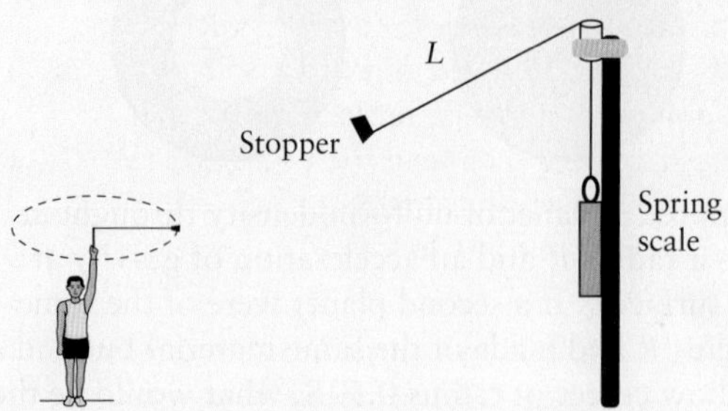

1. A student swings a rubber stopper with mass m in a horizontal circle overhead (see above figure) using a spring scale to measure the tension in the string. The string is attached to the stopper, passes through a small glass tube, and is tied to the spring scale, as shown in the figure. The student varies the tension, F_T, and measures the period, T, of the circular motion and the length of the string. The length, L, of the string is kept constant, but the angle that the string makes with the vertical varies with the speed of the stopper. The students performing the experiment have no method of measuring this angle.

(a) On the dot below, draw a free-body diagram of the stopper in the position shown in the figure.

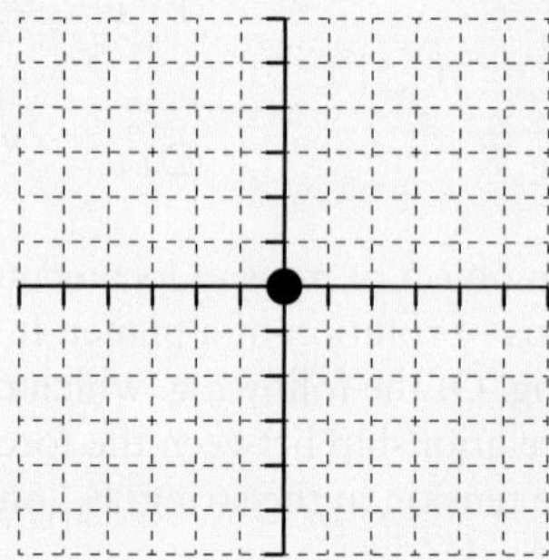

(b) Starting from Newton's second law, derive an expression for the tension in the string in terms of m, L, T, and (possibly) fundamental constants.

(c) As the tension increases, how does the angle that the string makes with the vertical change?

(A) It increases.

(B) It decreases.

(C) It remains the same.

Briefly explain your choice referring to your free-body diagram and expression above.

2. Students are to carry out a series of experiments with blocks in a lab. For the first experiment, some students decide to use a launcher to give both blocks the same initial velocity, v_0, and launch them across a table. The coefficient of kinetic friction between the table and the blocks is μ_k. The blocks are made of the same material, but the mass of block A (m_A) is twice the mass of block B (m_B).

(a) i. Student 1 says block A will slide farther along the table before coming to rest because it has more inertia. What is correct or incorrect about this statement? Explain briefly. (Do not use an equation; state your reasoning conceptually.)

ii. Student 2 says block B will slide farther because the floor exerts a smaller normal force on it. What is correct or incorrect about this statement? Explain briefly. (Do not use an equation; state your reasoning conceptually.)

(b) Derive an expression for the distance traveled by block A after it leaves the launcher, in terms of given quantities and physical constants.

(c) Use the expression you derived in part (b) to evaluate the correctness of the student statements above. Explain briefly.

Work, Energy, and Power

Chapter 7 Conservation of Energy and an Introduction to Energy and Work

Chapter 8 Application of Conservation Principles to Energy, Work, and Power

Pictorial Press Ltd/Alamy Stock Photo

Case Study: How do we determine the energy of a roller coaster?

A flume ride at Universal Studios Hollywood carries a boat filled with thrill-seekers through scenes of life-like dinosaurs from the movie *Jurassic Park*. Near the end of the ride the boat gently dips over the edge of a 33-m-long downhill track filled with rushing water, and accelerates down the 51° slope before entering a pond of still water. The boat glides about 15 m before coming to a near stop due to drag forces.

Kinetic energy is associated with motion. Conservative interactions within a system give rise to potential energy; for example, the gravitational interaction between Earth and the boat and riders can be described as gravitational potential energy of the Earth–boat/rider system. Energy can be converted from one type into the other, but total energy is always conserved.

That is certainly the case for our intrepid riders and their boat: Energy is neither created nor destroyed; once the boat starts its descent, we can use the resulting relationship to determine the properties of its motion. After you learn the physics of the conservation of energy and the relationship between work and energy, you'll be able to answer many questions about the science that underlies the *Jurassic Park* ride. For example...

The speed of the boat at the top of the final drop is close to zero. Estimate the speed of the boat at the bottom of the final drop. Ignore any nonconservative forces, which are relatively small. Does the mass of the boat or the riders affect your answer?

Estimate the average drag force that the water in the pond exerts on the boat. Do you think you yourself could exert a force of that magnitude?

By the end of these chapters, you should be able to apply conservation principles and the work-energy theorem to real-world problems. For instance, another amusement park wants to build a replica of this ride, but in order to slow down the boats—and to save space!—it needs to replace the pond at the end of the ride and consider whether they can use a horizontal spring. You will be able to help them decide.

7 Conservation of Energy and an Introduction to Energy and Work

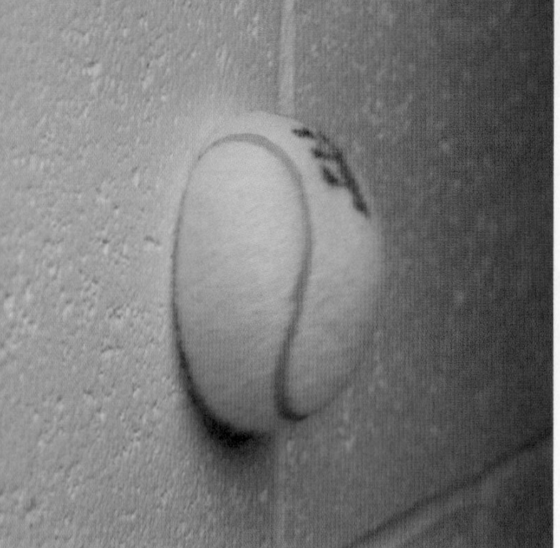

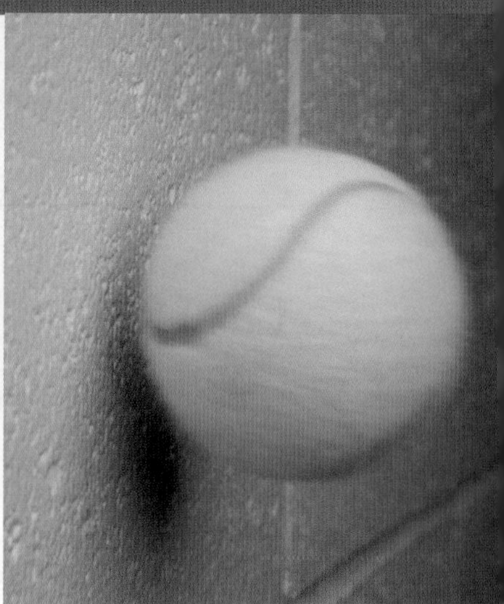

Ted Kinsman/Science Source

YOU WILL LEARN TO:

- Explain what it means for a quantity to be conserved.
- Describe what conditions must be met for work to be done on an object, for both positive and negative work.
- Explain the relationship between work and kinetic energy for an object.
- Explain why something modeled as an object can have only kinetic energy, and why a system can have other types of energy.
- Describe how the work-energy theorem relates to conservation of energy for an object or a system.
- Explain the meaning of potential energy and how conservative interactions, such as those described by the gravitational and spring forces, give rise to potential energy.
- Recognize why the work-energy theorem applies even for curved paths and varying forces such as the spring force.

7-1 The ideas of work and energy are intimately related, and this relationship is based on a conservation principle

In Chapter 1 we introduced *system* and *object*; the distinction between these will be foundational to our understanding of energy. In Chapters 2 and 3, we learned to describe the motion of an object, with some guidance on how that description is a simplification for the motion of a system. In Chapters 4 and 5, we explored how to use the concept of force to describe the interactions between objects and systems. In this chapter, we will begin applying one of the most fundamental ideas (and one of the most useful tools) in physics: The changes that occur as a result of interactions between objects and systems are constrained by conservation laws.

NEED TO REVIEW?
Turn to the **Glossary** in the back of the book for definitions of bolded Key Terms.

What does this mean? **Conservation** is often a poorly understood concept because of the use of the word in everyday life. We all try to conserve energy, switching off the lights when we leave a room, or turning down the thermostat in the winter. When we use the word *conservation* in science, however, we mean something much more profound than "using less." In science, a conserved quantity is a quantity that can be transferred between objects or systems, or converted from one type to another, but is neither created nor destroyed. When quantities are not created or destroyed, the amount of that quantity does not change, and this gives rise to some of the most powerful, fundamental laws in physics: conservation laws. Conservation laws constrain the possible motions of the objects in a system, or the outcome of an interaction or process. A **conservation law** is a statement that a measurable physical quantity of a system does not change as the system evolves over time. This physical quantity can be used to characterize a system.

We use the word *energy* almost every day, but most people would still find it hard to define. In science, **energy** is defined as a scalar quantity used to measure the state or motion of an object or a system (we will see some examples to help make this clearer). Energy is always conserved, but not all energy is equally useful. We will continue to explore this idea throughout this chapter and the next. So when we say we want to "conserve energy" in our everyday lives, we really mean that we want to not waste the energy that is most useful to us.

Because conservation also considers transfer of a quantity, we need to define a system to know how to apply a conservation law. This is a little tricky, as different sciences, and even different fields within a single science, define the types of systems differently. In chemistry, a closed system is a system where no matter can escape, but energy can be exchanged freely through heating. A closed system can be used when conducting chemical experiments where temperature is not a factor. In nonrelativistic classical mechanics, a closed system is a physical system that doesn't exchange any matter with its surroundings and isn't subject to any external forces (in quantum mechanics you have to add information to the list of things that cannot be exchanged). A closed system in classical mechanics would be considered an isolated system in thermodynamics. Closed systems are often used to limit the factors that can affect the results of a specific problem or experiment. In thermodynamics, a closed system can exchange energy (as heating or work), but not matter, with its surroundings. An isolated system cannot exchange energy or matter with the surroundings, while an open system can exchange both energy and matter. So just to make sure we don't miss anything because of a previous definition you know, we will define **closed, isolated systems** as those where no energy or matter is transferred to or from the system and there are no interactions between objects in the system and objects outside of the system. The definition for this combination is consistent, if a bit repetitive in cases, with all of the other definitions that exist.

We saw in Chapter 4 that a force is one way to describe the interaction between two objects, so another way to define an isolated system is a system for which no forces are exerted on objects inside the system by objects outside the system. We will see in this chapter how to use interactions that can be described by forces to transfer energy, but there are other processes (such as heating) to transfer energy that are beyond the scope of this book. The total amount of energy in a closed, isolated system cannot change, and all interactions and processes in the system are constrained by that fact. In contrast, in a system that is open, energy can cross the boundary of the system, in which case conservation no longer means the energy in the system is constant! It means that the changes in energy in the system are equal to the transfer of energy into or out of the system by interactions with other systems or processes.

We begin our study of conservation laws with the law of conservation of energy, the most pervasive conservation law across all areas of physics, and all sciences in general. Energy is used in nearly every living process—moving, breathing, circulating blood, digesting food, and absorbing nutrients. Within these processes, there are different types of energy, including kinetic, potential, and internal. The **law of conservation of energy** states that energy can be converted from one type into another but never destroyed. We have defined energy as a scalar quantity used to measure the state or motion of an object or a system; but what exactly does that mean? It may be easiest to begin to develop our definition of energy by considering one of the ways to transfer it.

To delve into how energy is transferred, we must first explore the concept of work. *Work* is another word with many different meanings in everyday life. In the language of physics, however, work has a very specific definition: **Work** is the transfer of energy from one object or system to another through a *mechanical* process that happens when an external force is exerted on an object or a system along its direction of motion as the point of contact where the force is exerted on the object or system moves. Remember, *mechanics* is the area of physics concerned with motion due to interactions (including in the special case

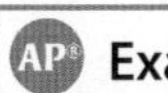 **Exam Tip**

On the AP® Physics 1 exam, when you are asked to justify a result, you often need to relate your argument to a fundamental principle such as this conservation law. For example: If asked to compare two projectiles on the exam, it would not be enough to say that if they each move through the same height and start with the same speed, they would therefore have the same final speed. For a complete answer, you would need to relate your response to the fundamental principle of energy conservation.

WATCH OUT !

***Closed system* means something different in mechanics and chemistry.**

Chemists don't worry about interactions that can be described by forces in their experiments (they have a lot more heating), so their definition of closed does not exclude such interactions. For mechanics this is a very important thing to exclude, so we will add the word *isolated* to our description of systems, emphasizing that for such systems no interactions *or* matter cross the system boundaries.

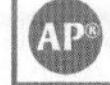 **Exam Tip**

Although thermodynamics is not part of AP® Physics 1, it is important to understand that mechanical energy can go into warming up systems. We will learn about the type of energy associated with temperature change in Chapter 8.

of interactions that result in zero motion). One example of doing work is lifting a book from your desk to a high bookshelf; another is a football player pushing a blocking sled across a field (**Figure 7-1a**). In each of these cases, the point of contact of where the force is exerted on the object moves. The definitions of work and energy reference each other, but things will become clearer as we move through this chapter and the next. We will see that if an object or a system has energy, then that object or system may be able to do work.

One type of energy is **kinetic energy**, which is the energy that an object has due to its motion (**Figure 7-1b**). An object with kinetic energy has the ability to do work; for example, a moving ball has the ability to displace objects in its path (**Figure 7-1c**). It's important to remember that for anything for which we use the object model, the only type of energy that it can have is kinetic energy. This is due to the fact that by the definition of an object, we cannot change its shape, or the way its internal components are moving relative to each other, because an object has no internal structure. Conversely, systems, because they have internal structure, can have internal and potential energy, which we will define shortly.

Kinetic energy can be converted to other types of energy that can't be used to do work—such as when an egg thrown at high speed splatters against a wall. Before the egg hits the wall, it can be modeled as an object because every point on the egg is moving in the same trajectory. However, once the egg hits the wall, the points on the now-broken egg no longer travel together as the egg changes shape. The egg can no longer, therefore, be modeled as an object to describe it as it breaks. Breaking apart on the wall is an *irreversible change* in shape, a change that disrupts the arrangement of the system in such a way that it cannot simply return to its initial shape. For example, you could not easily reassemble the egg, putting the yolk back in its sack, the egg white back into a smooth shield, and the shell all into one piece. Such disordered changes dissipate energy into a type that is no longer useful: The egg no longer has kinetic energy after it breaks. All the bits come to rest. The kinetic energy went into breaking the egg. Conversely, if that egg is replaced with a rubber ball, the ball bounces off the wall with nearly the same speed at which it hit the wall and so we recover most of the ball's kinetic energy. We say that as the ball compresses against the wall, its kinetic energy is converted into **potential energy**—energy associated not with the ball's motion but with a *reversible change* in its shape, a change that will allow the relative position, the structural arrangement, of its parts to return to their original condition. This potential energy is again converted into kinetic energy as the ball bounces back to its original shape (Figure 7-1c or the photographs opening this chapter). Potential energy is associated with a reversible change in shape of a system, which can be described as a reversible change in the configuration of a system of objects. (We cannot use the object model for the ball striking the wall during this process, because the various points on the ball do not all move together at the same speed as the center of mass. The point in contact with the wall has stopped, but the center of mass of the ball moves toward the wall as the ball slows down and then away from the wall as the ball speeds back up.)

The football player exerts a force of magnitude F on the sled in its direction of motion while the point on which he is exerting the force moves a distance d. Hence he does an amount of work on the sled equal to Fd.

(a)

AP Photo/Kevin Wolf

The girl does work on the basketball: She exerts a force on the ball as she pushes it away from her. As a result the ball acquires kinetic energy (energy of motion).

(b)

Jens Karlsson/Getty Images

(c)

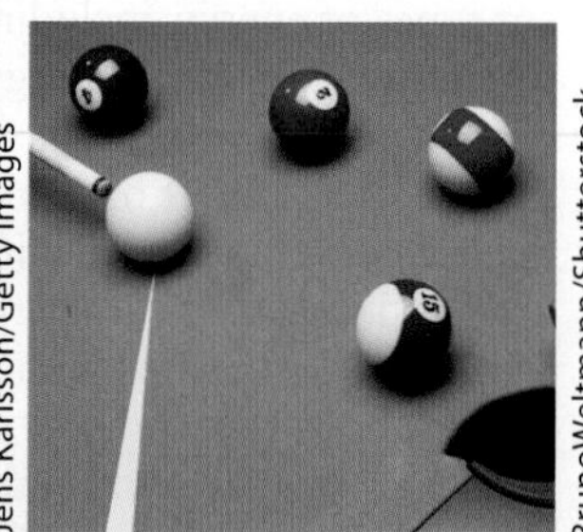

BrunoWeltmann/Shutterstock

As the cue ball strikes a second pool ball, it stops and the second pool ball moves off at nearly identical speed to that of the cue ball before the collision. The cue ball exerts a force on the second ball as it comes to a stop.

Figure 7-1 Work and energy In this chapter we'll explore the ideas of work and energy.

In this chapter we'll look at the ways in which kinetic energy, potential energy, and other types of **internal energy**—energy stored within a system, sometimes in a way that can't easily be extracted—can be converted from one type of energy into another or transferred into or out of a system. As an example, the energy you need to make it through today is extracted from energy stored in food that you consumed earlier. This sort of internal energy is stored in chemical bonds, related to the structural arrangement of atoms and molecules. Internal energy can also be due to motion: When the molecules in water start moving more quickly, the water warms up. Temperature is directly related to this sort of internal energy. Although it sounds like kinetic energy because it involves motion, the motions are of microscopic objects inside the system and are random and internal, not causing the system to move. In this course, we will focus on *work* as the method of energy transfer through mechanical processes, but when you continue your studies you will also learn how to calculate energy transfer due to *heating*, which is transferring energy by thermal processes (and does not require motion of the system, but just causes these random internal motions). In this course, you just need to know this can happen, like when you boil water!

THE TAKEAWAY for Section 7-1

✔ Conservation laws constrain the possible behaviors of objects or systems.

✔ Conservation of a quantity does not mean that the quantity in a system cannot change, but that any change in that quantity in a system must equal transfers into or out of the system. Conservation means the quantity is neither created nor destroyed.

✔ An object's or system's energy is related to its ability to do work.

✔ There are different types of energy, including kinetic, potential, and internal. Energy can be converted from one of these types into another.

Prep for the AP® Exam

AP® Building Blocks

1. Describe how you know winds blowing across a field of tall grass have a property that we call energy and that work is being done by the wind on the grass.
2. In everyday language, energy is consumed for transportation. In this language, if you say that energy should be conserved, you might mean that you should drive a smaller car or drive less. In physics, we look more deeply and see that the internal energy of the molecules of fuel is converted to the motion of a car (the car's kinetic energy) and into heating of the molecules of exhaust and of the engine. How does this conversion satisfy energy conservation? In your own words compare these two energy conservation statements. For example, can they be violated?

AP® Skill Builder

3. Missing from the definitions of open, closed, and isolated systems given in Section 7-1 is a system that cannot exchange energy with its surroundings but can exchange matter. Explain why this category is not possible in terms of possible exchange processes.

AP® Skills in Action

4. An open system can exchange both energy and matter with its surroundings. A closed, isolated system cannot exchange energy or matter with its surroundings. Categorize each of the following systems (a, b, c) as open or closed, isolated and describe evidence from your own experience to support your categorization.
(a) Earth
(b) You
(c) A very good ice chest while the lid is shut

7-2 The work done by a constant force exerted on a moving object depends on the magnitude of the force and the distance the object moves in the direction of the force

The man depicted in **Figure 7-2** is doing work as he pushes a crate up a ramp. The amount of work that he does depends not only on how hard he pushes on the crate (that is, on the magnitude of the force that he exerts) but also on the distance over which he moves the point on which he is pushing the crate (and thus the crate since it

The man exerts a constant force $\vec{F}$ on the crate. The direction of the force is parallel to the ramp.

As he exerts the force, the crate moves through a displacement $\vec{d}$ up the ramp.

stevecoleimages/E+/Getty Images

Figure 7-2 Work depends on force and displacement If the force $\vec{F}$ the man exerts on the crate is in the same direction as the displacement $\vec{d}$ of the point on which he is exerting the force on the crate, the work W that he does on the crate is the product of that force and displacement: $W = Fd$.

can be treated as an object) in the direction in which the force is exerted. In a similar way, the football players shown in Figure 7-1a will be more exhausted if the coach asks them to push the blocking sled all the way down the field rather than a short distance, if they exert the same force on it over the longer distance. In Figure 7-1c, kinetic energy gets transferred from one pool ball to the next as the two balls collide without friction or breaking. Because of the large resistance to motion provided by friction between the ground and the sled, the kinetic energy of the blocking sled in Figure 7-1a does not continue to increase as the football player pushes on it.

These examples suggest how we should define the work done on an object or a system by a force exerted on the object or system. Let us start by thinking about an object. Suppose a constant force $\vec{F}$ is exerted on an object as it moves through a displacement $\vec{d}$, and the force $\vec{F}$ is in the same direction as $\vec{d}$. Then the work done by the force equals the product of the magnitude of the force F and the magnitude of the displacement, d, over which the point of contact where the force is exerted on the object moves:

Work done on an object by a constant force $\vec{F}$ exerted on the object in the same direction as the object's displacement $\vec{d}$

Magnitude of the constant force $\vec{F}$

$$W = Fd \qquad (7\text{-}1)$$

Magnitude of the displacement $\vec{d}$

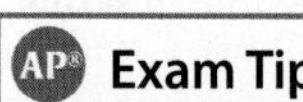

EQUATION IN WORDS
Work done by a constant force exerted on an object in the same direction as the object's displacement

Note that Equation 7-1 refers only to situations in which the force is exerted on the object in the *same* direction as the object's displacement. You've already seen two situations of this sort: The football players in Figure 7-1a push the sled backward as it moves backward, and the man in Figure 7-2 pushes the crate uphill as it moves uphill. Later we'll consider the case in which force and displacement are *not* in the same direction.

We saw in Chapters 4 and 5 that it's important to keep track of which object *exerts* a given force and on which object that force is exerted. It's equally important to keep track of both the object which exerts a force and the object on which the force does work (and these are the same observations!). For example, in Figure 7-2, the object exerting a force is the man, and the object on which the force is exerted and on which work is done is the crate. Just like a force must be exerted by something external to the object or the system, work is done on an object or a system by an external force. Work is the first way we will explore how to *transfer* energy.

We know that the unit of force is the newton and the unit of distance is the meter. Therefore, the unit of work is the newton · meter, or Nm. This unit is also called the **joule** (J), named after the nineteenth-century English physicist James Joule, who did fundamental research on the relationship between motion and work. From Equation 7-1,

$$1\ \text{J} = (1\ \text{N})(1\ \text{m})$$

You do 1 J of work when you exert a 1-N push on an object in the direction it is moving as it moves through a distance of 1 m.

AP® Exam Tip

You always need to put units on a numerical answer, but Nm and J are both acceptable SI units for energy.

AP® Exam Tip

There is not a symbol called weight on the AP® equation sheet, but the force of gravity between two objects with mass is given, and then is used in the definition of the gravitational field. The force of gravity near the surface of Earth is weight. While weight is not defined on the equation sheet, a symbol representing weight may be defined for you in AP® problems. Always carefully note the definition of all symbols introduced in a problem statement, and be sure to carefully define any symbols you introduce.

WATCH OUT!

Don't use *w* as a symbol for weight.

Because *work* and *weight* begin with the same letter, it's important to use different symbols to represent them in equations. We'll use an uppercase W for work and F_g or mg for weight (the magnitude of the gravitational force on an object near the surface of Earth), and we recommend that you do the same to prevent confusion.

EXAMPLE 7-1 Lifting a Book

How much work must you do to lift a textbook with a mass of 2.00 kg a vertical distance of 5.00 cm? You lift the book at a constant speed.

Set Up

Newton's first law tells us that the net force on the book must be zero if it is to move upward at a constant speed. Hence the upward force you exert must be constant and equal in magnitude to the gravitational force on the book. Because the force you exert is constant and in the direction in which the book moves, we can use Equation 7-1 to calculate the work done on the book by that force.

Work done on an object, force in the same direction as displacement:

$$W = Fd \qquad (7\text{-}1)$$

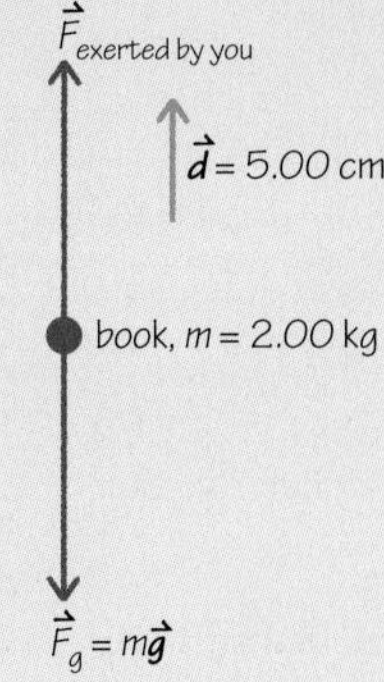

Solve

Calculate the magnitude F of the force that you exert on the book, then substitute this and the displacement $d = 5.00$ cm into Equation 7-1 to determine the work that you do on the book.

$$\begin{aligned} F = mg &= (2.00 \text{ kg})(9.80 \text{ m/s}^2) \\ &= 19.6 \text{ N} \\ W = Fd &= (19.6 \text{ N})(5.00 \text{ cm}) \\ &= (19.6 \text{ N})(0.0500 \text{ m}) \\ &= 0.980 \text{ Nm} \\ &= 0.980 \text{ J} \end{aligned}$$

Reflect

The work that you do in lifting the book is almost exactly 1 joule. We don't yet have much practice with units of energy and work, but consider the basic definition of a joule. You exerted nearly 20 N of force on the book, but only for 5 cm of motion, 1/20th of a meter. The product then should be about 1 Nm, a little lower because the force you exerted wasn't quite 20 N. That is what we got! You may be worried because you did work on the book and did not change its kinetic energy and we said that work was a way to transfer energy. We will find that this was because your lift was not the only force exerted on the book. The force of gravity was pulling exactly opposite you. We will learn how to calculate the work done by a force in the direction opposite to displacement after the next example.

Extend: The actual amount of energy that you would need to *expend* is several times more than 0.980 J. That's because your body isn't 100% efficient at converting energy into work. Some of the energy goes into lifting your arm, and some into heating your muscles. In fact, your muscles can consume energy even when they do *no* work, as we describe below. The value of 0.980 J that we calculated is just the amount of work that you do *on the book*.

NOW WORK Problem 1 from The Takeaway 7-2.

Muscles and Doing Work

Pick up a heavy object and hold it in your hand at arm's length (**Figure 7-3**). After a while you'll notice your arm getting tired: It feels like you're doing work to hold the object in midair. But Equation 7-1 says that you're doing *no* work on the object because it isn't moving (its displacement d is zero). So why does your arm feel tired?

The reason your arm tires while holding a heavy object is that whether you are moving the object or not, you still have to exert a force to *balance* that of gravity (it is the same size force, and in the opposite direction) to hold the object at a constant height. Your body has to convert energy stored in the chemical bonds of fat and sugar (chemical energy) to a type that your muscle cells can use to exert this force, but these reactions generate waste products that change the conditions in the muscle, including warming it up. This process of changing energy from one type into another leads to the feeling that you've been doing work—even though you're doing no work on the *object* you're holding.

Figure 7-3 Getting tired while doing zero work Weights that you hold stationary in your outstretched arms undergo no displacement, so you do zero work on them. Why, then, do your arms get tired?

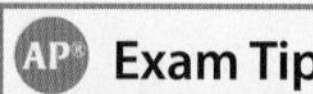

Exam Tip

Chemical energy is not a topic on the AP® Physics 1 exam.

If you again hold the object in one hand, but now rest your elbow on a table, you'll be able to hold the object for a longer time with far less fatigue. That's because in this case you're using only the muscles of your hand and forearm, not those of your upper arm. Fewer muscles are involved, so you need to convert less energy. The table does not have to use chemical energy to support you or the book.

Work by Forces Not Parallel to Displacement, and Negative Work

How can we calculate the work done by a constant force that is *not* exerted in the direction of the object's motion? As an example, in **Figure 7-4a** a groundskeeper is using a rope to pull a screen across a baseball diamond to smooth out the dirt. The force $\vec{F}$ that the man exerts on the screen is at an angle θ with respect to the displacement $\vec{d}$ of the screen, so only the *component* of the force along the displacement does work on the screen (**Figure 7-4b**). This component is $F\cos\theta$, so the amount of work done by the force is

EQUATION IN WORDS
Calculating the work done by a constant force exerted at an angle θ to the displacement

Work done on an object by a constant force $\vec{F}$ that points at an angle θ to the object's displacement $\vec{d}$

Magnitude of the constant force $\vec{F}$

(7-2)
$$W = Fd\cos\theta$$

Magnitude of the object's displacement $\vec{d}$

Angle between the directions of $\vec{F}$ and $\vec{d}$

Exam Tip

$F\cos\theta$ is the component of the force that is parallel to the displacement, so is also called $F_{\parallel}$ on the equation sheet.

To understand Equation 7-2, let's look at some special cases.

- If $\vec{F}$ is in the same direction as the motion, $\theta = 0$ and $\cos\theta = \cos 0 = 1$. This is the same situation as shown in Figure 7-2, and in this case Equation 7-2 gives the same result as Equation 7-1: $W = Fd$.
- If the angle θ between the force and displacement is more than 0 but less than 90°, as in Figure 7-4b, then $\cos\theta$ is less than 1 but still positive. The work W done by the force $\vec{F}$ on the object is positive but less than Fd.
- If $\vec{F}$ is perpendicular to the direction of motion, $\theta = 90°$ and $\cos\theta = \cos 90° = 0$. In this case Equation 7-2 tells us that force $\vec{F}$ does *zero* work on the object. An example is the force exerted by a lazy dog lying on top of the screen from Figure 7-4a (**Figure 7-5**). The dog exerts a downward force on the screen as the screen moves horizontally, so $\theta = 90°$ and the lazy dog does no work at all.

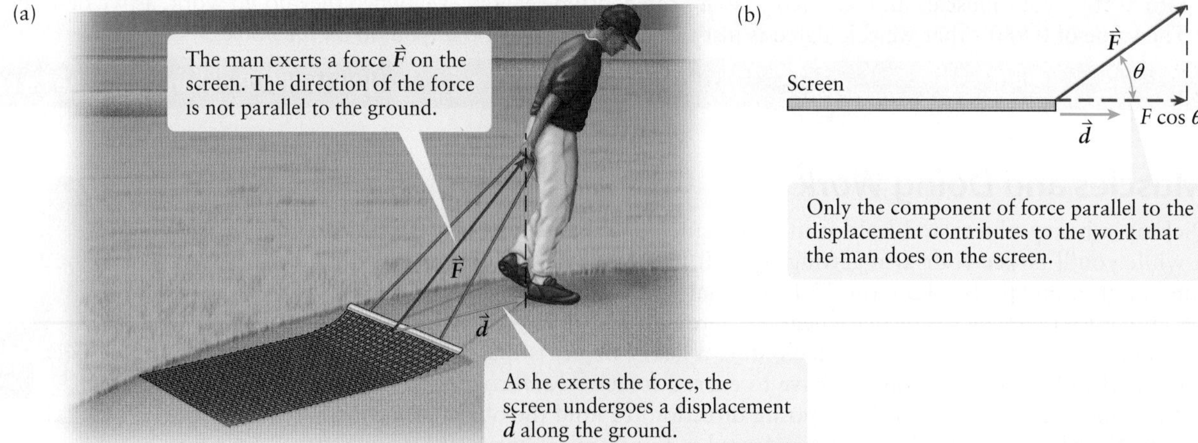

Figure 7-4 When force is not aligned with displacement (a) A man exerts a force on a screen to pull it across a baseball field. (b) Finding the work that the man does on the screen (seen from the side). If a constant force $\vec{F}$ is exerted on an object at an angle θ to the object's displacement $\vec{d}$ the work done on the object by the force equals the force component $F\cos\theta$ parallel to the displacement multiplied by the object's displacement d: $W = Fd\cos\theta$.

- If the angle θ in Equation 7-2 is greater than 90°, the value of $\cos\theta$ is negative. In this case the work done by force $\vec{F}$ on the object is *negative*: $W < 0$. As an example, **Figure 7-6** shows a cart rolling along the floor as a person tries to slow it down. The force $\vec{F}$ that the person exerts on the cart is directed opposite to the cart's displacement $\vec{d}$, so the angle θ in Equation 7-2 is 180°. Because $\cos 180° = -1$, this means that the work that the person does on the cart is $W_{\text{person on cart}} = -Fd$, which is negative.

What does it mean to do negative work on an object? Remember that when the force exerted on an object (and hence the object's acceleration) points in the opposite direction of the object's velocity and displacement, the object slows down (see Figure 2-9). So doing negative work on an object slows it down. Conversely, if you do positive work on an object, you can speed it up. This makes sense: Because work is a way to transfer energy, if you make a positive energy transfer, you increase the energy of the system receiving it. If you make a negative energy transfer, you decrease the system's energy.

What else happens when you do negative work on an object? Newton's third law provides the answer when two objects in contact exert forces on each other: If the person in Figure 7-6 exerts a force $\vec{F}$ on the cart, the cart exerts a force $-\vec{F}$ on her (the same magnitude of force but in the opposite direction). The hands of the person have the same displacement $\vec{d}$ as the cart, and the force the cart exerts on her hands is in the *same* direction as her hands' displacement ($\theta = 0$ and $\cos\theta = +1$ in Equation 7-2), so the work that the *cart* does on the *person* is $W_{\text{cart on person}} = +Fd$. *For objects in contact, if object A does negative work on object B, then object B does an equal amount of positive work on object A.* For example, as in Figure 7-1c, when a moving cue ball hits a stationary pool ball on a pool table, the cue ball does *positive* work on the pool ball: It pushes the pool ball forward (in the direction that the pool ball moves) and makes the pool ball speed up. The pool ball does *negative* work on the cue ball: It pushes back on the cue ball (in the direction opposite to the cue ball's motion) and makes the cue ball slow down. With the pool balls, the energy transfer is almost all kinetic. They are modeled very well as objects. For the person pushing the box, her hands actually don't necessarily move the same distance as the rest of her body, and her muscles warm up, so the object model would not be the right model for the person. In this case, her kinetic energy doesn't change much because of internal motion but primarily because she is also pushing against the floor so (more Newton's third law) the floor is pushing against her, helping her stay put.

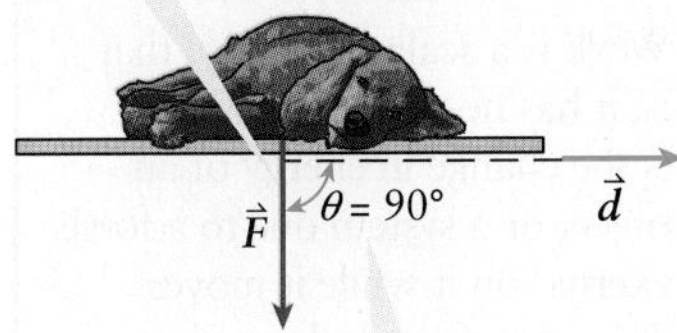

A lazy dog exerts a downward force on the screen.
$\theta = 90°$
$\vec{F}$
$\vec{d}$
The angle between the force and the displacement $\vec{d}$ of the screen is 90° (the force and displacement are perpendicular). Hence the force has zero component in the direction of $\vec{d}$ and does zero work.

Figure 7-5 Doing zero work A dog resting atop the screen exerts a force on the screen as it slides but does zero work on the screen.

WATCH OUT !

Not all objects exert contact forces on each other!

While Newton's third law is always true, when objects interact by forces that do not require contact, sometimes the objects do not move through the same displacement. That is why action-reaction pair forces do not always do equal but opposite work.

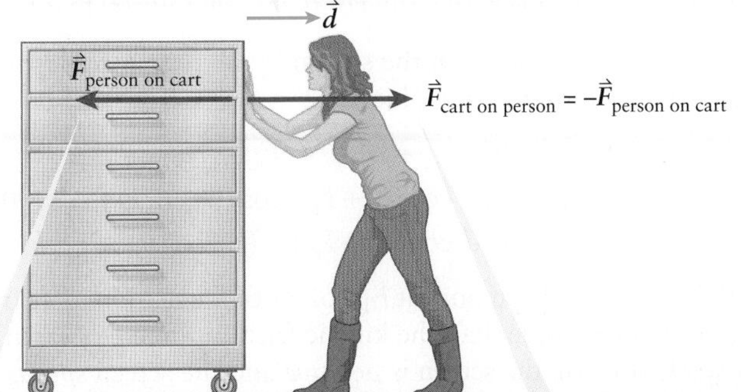

(1) As the person tries to make the cart slow down, the cart and the person's hands move together to the right (they have the same displacement since her hands and the cart are in contact).
$\vec{d}$
$\vec{F}_{\text{person on cart}}$
$\vec{F}_{\text{cart on person}} = -\vec{F}_{\text{person on cart}}$
(2) The force of the person on the cart is opposite to the cart's displacement. Hence $\theta = 180°$, $\cos\theta = -1$, and the person does negative work on the cart.
(3) By Newton's third law, the cart exerts an equally strong force on the person, but in the opposite direction—that is, in the same direction of the displacement of the person's hands. So the cart does positive work on the person.

Figure 7-6 Doing negative work As the cart rolls to the right, the person pushes on the cart to the left in order to make it slow down. As a result, she does negative work on the cart.

For objects in contact, if one object (such as the person) does negative work on a second object (such as the cart), the second object does an equal amount of positive work on the first object.

WATCH OUT !

Don't let the signs fool you!

Work is a scalar quantity; that is, it has no direction. Work is the change in energy of an object or a system due to a force exerted on it while it moves. The sign of work determines if the work is increasing (+) or decreasing (–) the energy of the object or system. Add work terms to find the total change in energy of the object or system.

 **Exam Tip**

It is important to be able to distinguish conceptually between when a particular force is doing positive or negative work. Note that an object can still be speeding up while a particular force is doing negative work on it; in which case, there must be another force doing positive work on the object.

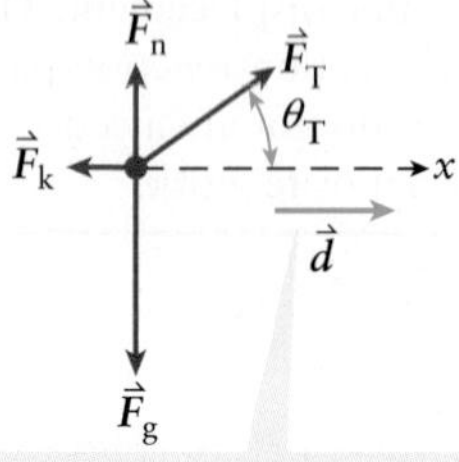

Always draw the displacement $\vec{d}$ to one side so you don't confuse it with the forces.

Figure 7-7 Calculating the work done by multiple forces The free-body diagram for the screen shown in Figure 7-4a. To find the work done by each force exerted on the object, draw the displacement $\vec{d}$ on the diagram, and label angles to match forces.

WATCH OUT !

Forces are not what do work!

Forces are just a way to describe an interaction. When we talk about a force doing work, we are using a sort of abbreviation for the work done by whatever is exerting the force.

Table 7-1 summarizes the relationship between the angle at which a force is exerted on an object relative to its displacement and the work done on the object. It only considers an angle between 0 and 180°.

TABLE 7-1 Work Done when the Force Exerted on an Object Is Constant

If the angle between a force $\vec{F}$ exerted on an object and the displacement $\vec{d}$ of the object is...	...then the work done on the object is...
0° to 90°	positive
90°	zero
90° to 180°	negative

Calculating Work Done when Multiple Forces Are Exerted on an Object

What if more than one force is exerted on an object as it moves? Then the *total* work done on the object is the *sum* of the work done on the object by each object or system exerting a force on it individually. We simplify this language by saying it is the work done by the forces, but remember that the energy has to come from whatever is exerting the force. For example, four forces are exerted on the screen in Figure 7-4a: the tension in the rope, the gravitational force, the normal force, and the force of friction. The free-body diagram in **Figure 7-7** shows all of these forces, as well as the displacement $\vec{d}$. Note that we draw $\vec{d}$ to one side in the diagram to avoid confusing it with the forces, and that the direction of motion is in the positive x direction as indicated by the dashed arrow. Because there are several forces exerted on the object, add subscripts to your angles so you can tell which angle goes with which force.

We use Equation 7-2 to determine the work that each force does on the screen:

Tension force: This force $\vec{F}_T$ (which we labeled $\vec{F}$ in Figure 7-4a) is exerted at an angle θ_T with respect to the direction of motion, so $W_{tension} = F_T d \cos \theta_T$.

Gravitational force: The gravitational force $\vec{F}_g$ is perpendicular to the direction of motion, so the angle θ in Equation 7-2 is 90° for this force. Because $\cos 90° = 0$, the work done by the gravitational force is $W_{grav} = 0$.

Normal force: The normal force $\vec{F}_n$ is also perpendicular to the direction of motion. So like the gravitational force, the normal force does no work: $W_{normal} = 0$.

Kinetic friction force: Because friction opposes sliding, the kinetic friction force $\vec{F}_k$ points in the direction opposite to the motion. The angle to use in Equation 7-2 is therefore 180°, and $W_{friction} = F_k d \cos 180° = -F_k d$. *When a force is exerted to oppose the motion of the object, the value of the work done by that force is negative.*

The total work done on the screen by all four forces is the sum of the work done by each force exerted on the object:

$$\begin{aligned} W_{total} &= W_{tension} + W_{grav} + W_{normal} + W_{friction} \\ &= F_T d \cos \theta_T + F_g d \cos (90°) + F_n d \cos (90°) + (-F_k d) \\ &= F_T d \cos \theta_T - F_k d = (F_T \cos \theta_T - F_k) d \end{aligned}$$

If the horizontal component $F_T \cos \theta_T$ of the tension force (which does positive work) is greater in magnitude than the kinetic friction force F_k (which does negative work), the total work done on the screen is *positive* and the screen *speeds up*. But if the magnitude of $F_T \cos \theta_T$ is less than F_k, the total work done on the screen is *negative* and the screen *slows down*. Note that the kinetic friction force cannot do any work on an object unless that object is already in motion, or the surface is in motion, or some other force is causing the surface or object to move. This makes sense: Because the ground doesn't move, you can stand on the ground all day and if you don't try to move, it isn't going to make you.

While the dog lying on the screen in Figure 7-5 would not do any work itself, the presence of the dog would increase the normal force and therefore the friction force, and so the negative work done by the friction force. So the tension force needed to drag the screen increases, making the person do more work to pull the screen the same distance.

EXAMPLE 7-2 Up the Hill

You need to push a box of supplies (weight 225 N) from your car to your campsite, a distance of 6.00 m up a 5.00° incline. You exert a force of 85.0 N parallel to the incline, and a 56.0-N kinetic friction force is exerted on the box by the ground. Calculate (a) how much work you do on the box, (b) how much work the force of gravity does on the box, (c) how much work the friction force does on the box, and (d) the net work done on the box by all forces exerted on it as it moves up the incline.

Set Up

Because this problem involves forces, we draw a free-body diagram for the box. The forces do not all point along the direction of the box's displacement $\vec{d}$, so we'll have to use Equation 7-2 to calculate the work done by each force.

Work done by a constant force, displacement:

$$W = Fd \cos \theta \qquad (7\text{-}2)$$

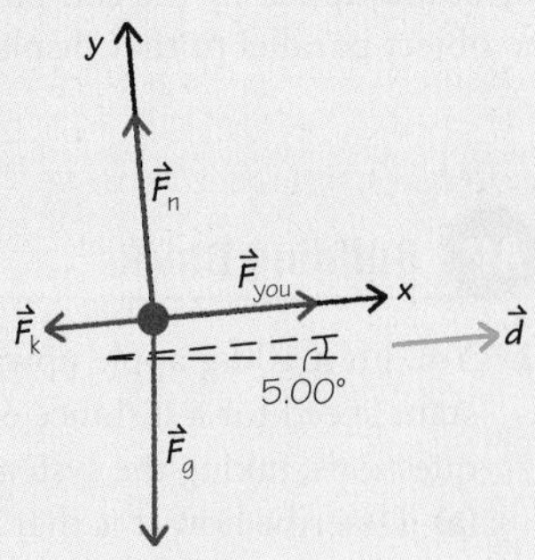

Solve

(a) The force you exert on the box is in the same direction as the box's displacement, so $\theta = 0°$ in Equation 7-2.

$$W_{you} = F_{you}\, d \cos 0° = (85.0\ \text{N})(6.00\ \text{m})(1) = 5.10 \times 10^2\ \text{Nm} = 5.10 \times 10^2\ \text{J}$$

(b) The angle between the gravitational force and the displacement is $\theta = 90.0° + 5.00° = 95.0°$. Because this is more than 90°, $\cos \theta$ is negative and the gravitational force does negative work on the box.

$$W_{grav} = F_g d \cos 95.0° = (225\ \text{N})(6.00\ \text{m})(-0.0872) = -1.18 \times 10^2\ \text{J}$$

(c) The friction force points opposite to the displacement, so for this force $\theta = 180°$ and $\cos \theta = -1$.

$$W_{friction} = F_k d \cos 180° = (56.0\ \text{N})(6.00\ \text{m})(-1) = -3.36 \times 10^2\ \text{J}$$

(d) The net work done on the box is the sum of the work done by all four forces that are exerted on the box. The normal force points perpendicular to the displacement, so it does zero work (for this force, $\theta = 90°$ and $\cos \theta = 0°$).

$$W_{total} = W_{you} + W_{grav} + W_{friction} + W_n = (5.10 \times 10^2\ \text{J}) + (-1.18 \times 10^2\ \text{J}) + (-3.36 \times 10^2\ \text{J}) + 0 = 0.56 \times 10^2\ \text{J} = 56\ \text{J}$$

Reflect

To check our result, let's calculate how much work is done by the *net* force that is exerted on the box. Our discussion of Equation 7-2 tells us that we need only the component of the net force in the direction of the displacement, which in our figure is the x component. So the work done by the net force is just $\sum F_x$ multiplied by d.

There are two observations we can use to ensure this makes sense. First, the work done by the net force has the *same* value as the sum of the work done by the individual forces. It has to, because this is just another way to describe the exact same situation! Second, the net force is in the direction of the displacement, so the box picks up speed as it moves *and* the net work done on the box is positive.

Net force up the incline:

$$\sum F_x = F_{you} - F_k - F_g \sin 5.00° = 85.0\ \text{N} - 56.0\ \text{N} - (225\ \text{N})(0.0872) = 9.4\ \text{N}$$

(positive, so the net force is uphill)

Work done by net force:

$$W_{net} = \left(\sum F_x\right) d = (9.4\ \text{N})(6.00\ \text{m}) = 56\ \text{J}$$

This equals the sum of the work done by the four forces individually.

Extend: We'll use both of these observations in the next section to relate the net work done on an object to the change in its speed.

NOW WORK Problem 3 from The Takeaway 7-2.

THE TAKEAWAY for Section 7-2

✔ If a force is exerted on an object that undergoes a displacement, the object or system exerting that force can do work on that object.

✔ For a constant force and linear displacement, the amount of work done on an object equals the displacement of the object multiplied by the component of the force exerted on the object parallel to that displacement.

✔ Whether the work done is positive, negative, or zero depends on the angle between the direction in which the force is exerted on an object and the direction of the object's displacement.

✔ For objects in contact, if one object does negative work on a second object, the second object must do an equal amount of positive work on the first object.

Prep for the AP® Exam

AP® Building Blocks

EX 7-1 1. You lift a 100-g apple upward from the ground at a constant speed for a distance of 1 m. Answer the following questions, taking the system to be the apple alone.
(a) Describe how it is that you know the size of the force that you exert on the apple.
(b) Describe how you can determine the quantity of work done by you on the apple.
(c) Describe how you can determine the sign of the work done by you on the apple.
(d) You release the apple from rest and it falls 1 m. Describe how you know that work is done on the apple as it falls.
(e) Identify the force that does the work on the apple as it falls and describe how you would determine the quantity of work done by that force on the apple. Neglect air resistance.
(f) Compare the quantities of work done on the apple for the processes in parts (a) and (d) and any assumptions that are useful in making this comparison.
(g) Compare accelerations and net forces for the processes in parts (b) and (d). Describe how your answer to part (b) would change if work done by the net force exerted on the apple was considered, instead of only considering the force exerted by you.
(h) When the apple fell, energy was transferred. Describe how the energy of the apple changed during the process from when the apple started falling to when it came to rest on the ground. Also describe the sizes and directions of the forces exerted by the apple on the ground and the ground on the apple, and justify your description in terms of the change in the apple's energy.

2. A crane lifts a 2.00×10^2-kg crate a vertical distance of 15.0 m at a slow, constant speed. How much work in joules does the crane do on the crate?

3. A constant force, $\vec{F}$, is exerted on a box, dragging the box a distance d across a floor. $\vec{F}$ makes an angle θ with the horizontal as shown in the figure. This is repeated several times; each time the angle θ is increased by a few degrees, but the magnitude of the force $\vec{F}$ remains unchanged. Describe how the work done by the force $\vec{F}$ in dragging the box changes.

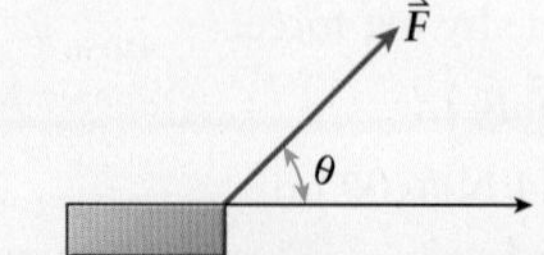

4. The figure shows four situations in which a box slides to the right a distance d across a floor as a result of forces exerted on the box, *one* of which is shown. Friction between the box and the floor is negligible. The magnitudes of the forces shown in each situation are identical. Rank the four situations in order of increasing work done on the box by the force shown as the box goes through the displacement, d.

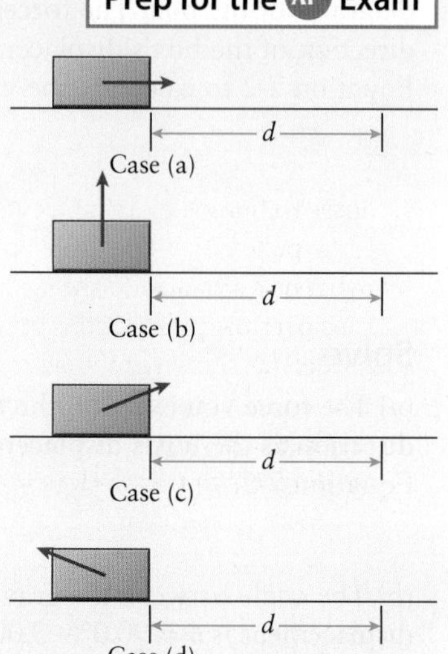

AP® Skill Builders

5. In the men's weight-lifting competition at the 2008 Beijing Olympics, Matthias Steiner made his record lift of 446 kg from the floor to over his head (2.0 m). How much work did Steiner do on the weights during the lift?

6. The work done on an object is $W = Fd \cos \theta$, where the magnitude of the displacement of the object is d, the magnitude of the force exerted on the object is F, and the angle between the direction of displacement and the direction of the force is θ. Evaluate this relationship to identify all the possible conditions where the work done by this force is equal to zero.

7. You (i) lift a 10-N box to a height of 1 m from the floor at a constant vertical speed, (ii) carry it horizontally 3 m across the room at constant speed, and (iii) set it back down on the floor, also at constant speed. (In this question, neglect the brief changes in speed as you first pick it up and then bring it to rest vertically at the end of the lift and start moving it horizontally, and similarly as you change its direction to put it back down.)
(a) Construct a free-body diagram on the box for each straight segment of the path (i–iii), clearly labeling each force exerted on the box.
(b) Construct representations of the displacement for each straight segment of the overall path of the box.
(c) Predict the work done *by you on the box* along each straight segment of the overall path of the box and the total work done by you on the box during the overall path.

(d) On the horizontal segment **ii.** you exert a force on the box as you walk but the kinetic energy of the box does not change. On this path the box exerts an equal and opposite force on you but there is no work done by that force on you. If you walked 3 km rather than 3 m you would probably notice that you had used a lot more of your internal energy. Many students claim that the extra energy you used is transferred to the box. Evaluate this claim and justify your answer.

(e) How would your answer to part (c) change if you did not neglect when you were changing the velocity of the box?

AP® Skills in Action

8. Some technologies are ancient. One example is the pulley. To lift a burden above the floor, a pulley is attached to the ceiling and a rope runs over the pulley and is attached to the burden (as shown in the figure). A force F is exerted on the rope and the burden is then displaced upward by Δx at a constant speed.

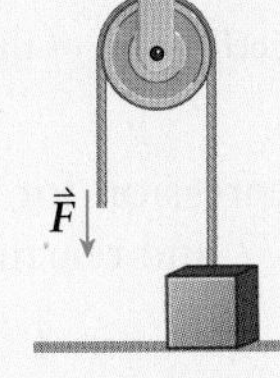

(a) Express the work done on the burden by the external force in terms of F and Δx.

Hero of Alexandria wrote about what we now call simple machines in the first century, C.E.

"Whenever we want to move some weight, if we tie a rope to this weight we pull with as much force as is equal to the burden. But if we untie the rope from the weight, and tie one of its ends to a stationary point and pass its other end over a pulley fastened to the burden and draw on the rope, we will more easily move the weight."

In Hero's machine a rope runs over a pulley connected to a stationary beam (A), then runs through a second pulley (B) that is attached to the "burden," then runs back to the beam where it is tied as shown in the diagram.

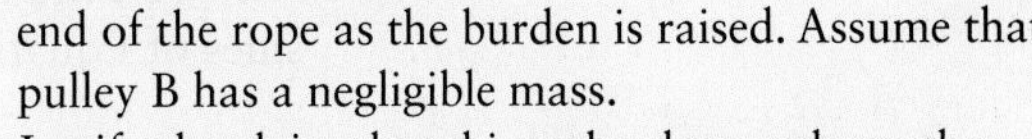

(b) This machine is used to lift the burden to the same height, Δx. Predict the work done on the burden, the external force that must be exerted to do this work, and the displacement of the free end of the rope as the burden is raised. Assume that pulley B has a negligible mass.

(c) Justify the claim that this technology reduces the effort required to achieve the same outcome required to lift the burden to the same height.

7-3 Newton's second law applied to an object allows us to determine a formula for kinetic energy and state the work-energy theorem for an object

In Example 7-2 we considered a box being pushed uphill that gained speed as it moved. We found that the total amount of work being done on this box was positive. Let's now show that there's a *general* relationship between the total amount of work done on an *object* and the change in that object's speed. To find this important relationship, which defines kinetic energy, we'll combine our definition of work from Section 7-2 with what we learned about one-dimensional motion in Chapter 2 and our knowledge of Newton's laws from Chapter 4. Remember that for something that can be modeled as an object, all points on that object must move the same distance in any motion, so, the point of contact of a force exerted on an object and the center of mass of the object always have the same displacement.

Let's begin by considering an object that moves along a line that we'll call the x axis, traveling from initial position x_i to final position x_f with a constant acceleration a_x (**Figure 7-8**).

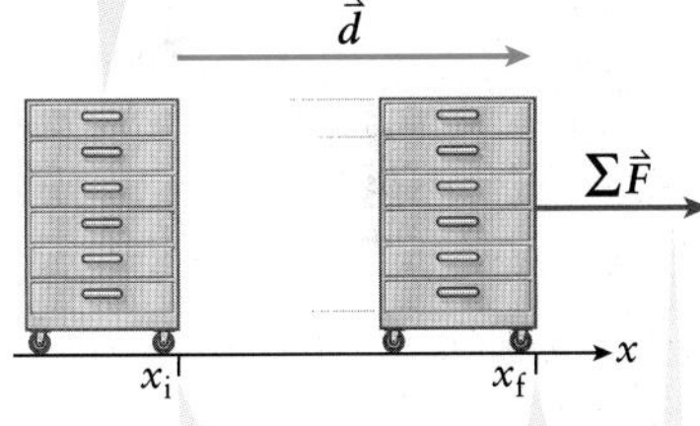

① A cart of mass m moves with negligible friction along a line that we'll call the x axis.
$\vec{d}$
$\Sigma\vec{F}$
x
x_i
x_f
② The cart moves from initial position x_i to final position x_f.
③ As the cart moves, a constant net force in the x direction is exerted on the cart.

Figure 7-8 Deriving the work-energy theorem for an object A cart moves along a line, traveling from initial position x_i to final position x_f. The net force exerted on the cart is constant, so the cart has a constant acceleration a_x.

Equation 2-11, reintroduced here as Equation 7-3, gives us a relation between the object's velocity v_{ix} at x_i and its velocity v_{fx} at x_f:

$$v_{fx}^2 = v_{ix}^2 + 2a_x(x_f - x_i) \quad (7\text{-}3)$$

In Equation 7-3 v_{fx}^2 is the square of the velocity at x_f, but it also equals the square of the object's *speed* v_f at x_f. That's because v_{fx} is equal to $+v_f$ if the object is moving in the positive x direction and equal to $-v_f$ if the object is moving in the negative x direction. In either case, $v_{fx}^2 = v_f^2$. For the same reason $v_{ix}^2 = v_i^2$, where v_i is the object's speed at x_i. So we can rewrite Equation 7-3 as

Speed at position x_f of an object in linear motion with constant acceleration

Speed at position x_i of the object

$$v_f^2 = v_i^2 + 2a_x(x_f - x_i) \quad (7\text{-}4)$$

Constant acceleration of the object

Two positions of the object

EQUATION IN WORDS
Relating speed, acceleration, and position for linear motion with constant acceleration

In Equation 7-4 $x_f - x_i$ is the displacement of the object along the x axis. Using Equation 7-2, we multiply this by the x component of the net force exerted on the object, ΣF_x—that is, by the net force component in the direction of the displacement and get a slightly different expression for W_{net}, the work done by the net force on the object. (This assumes that the net force on the object is constant, which is consistent with our assumption that the object's acceleration is constant.) That is,

$$W_{net} = \left(\sum F_x\right)(x_f - x_i) \quad (7\text{-}5)$$

Newton's second law tells us that $\Sigma F_x = ma_x$, where m is the mass of the object. So we can rewrite Equation 7-5 as

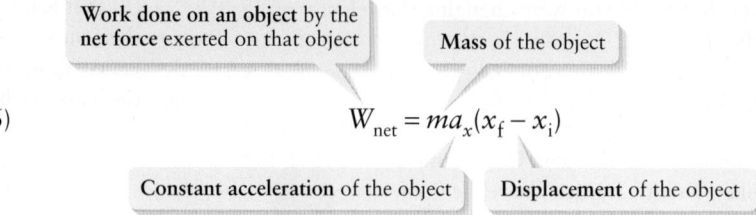

EQUATION IN WORDS
Calculating the work done on an object by a constant net force, linear motion

$$W_{net} = ma_x(x_f - x_i) \quad (7\text{-}6)$$

Work done on an object by the **net force** exerted on that object
Mass of the object
Constant acceleration of the object
Displacement of the object

We can get another expression for the quantity $ma_x(x_f - x_i)$ by multiplying both sides of Equation 7-4 by $m/2$ and rearranging:

$$\frac{1}{2}mv_f^2 = \frac{1}{2}mv_i^2 + ma_x(x_f - x_i)$$

$$ma_x(x_f - x_i) = \frac{1}{2}mv_f^2 - \frac{1}{2}mv_i^2$$

The left-hand side of this equation is the work done on the object by the net force (see Equation 7-6). So

$$W_{net} = \frac{1}{2}mv_f^2 - \frac{1}{2}mv_i^2 \quad (7\text{-}7)$$

The right-hand side of Equation 7-7 is the *change* in the quantity $\frac{1}{2}mv^2$ over the course of the displacement (the value at the end of the displacement, where the speed is v_f, minus the value at the beginning where the speed is v_i). We call this quantity, $\frac{1}{2}mv^2$, the kinetic energy K of the object:

EQUATION IN WORDS
Kinetic energy of an object

$$K = \frac{1}{2}mv^2 \quad (7\text{-}8)$$

Kinetic energy of an object
Mass of the object
Speed of the object

The units of kinetic energy are $\text{kg} \cdot \text{m}^2/\text{s}^2$. Because $1 \text{ J} = 1 \text{ Nm}$ and $1 \text{ N} = 1 \text{ kg} \cdot \text{m/s}^2$, you can see that kinetic energy is measured in joules, the same as work. Using the definition of kinetic energy given in Equation 7-8, we can rewrite Equation 7-7 as

EQUATION IN WORDS
The work-energy theorem for an object

$$W_{net} = K_f - K_i \quad (7\text{-}9)$$

Work done on an object by the **net force** exerted on that object
Kinetic energy of the object **after** the work is done on it
Kinetic energy of the object **before** the work is done on it

When an object undergoes a displacement, the work done on it by the net force equals the change in the object's kinetic energy (its kinetic energy at the end of the displacement minus its kinetic energy at the beginning of the displacement). This statement is called the **work-energy theorem** for an object. It is valid as long as the object model is appropriate, such as when a system is rigid—that is, it doesn't deform like a rubber ball might. Although we have derived this theorem for the special case of linear motion with constant forces, it turns out to be valid even if the object follows a curved path and the forces exerted on it vary. (We'll justify this claim in Section 7-5.) The *net* in *net force* is essential to this definition. Remember, the net force exerted on an object is the sum of all of the forces exerted on the object. If there is more than one force exerted on an object, the work done by each individual force does not tell you the change in kinetic energy of the object. You have to add them all.

The Meaning of the Work-Energy Theorem for an Object

What is kinetic energy, and how does the work-energy theorem for an object help us solve physics problems? To answer these questions let's first return to the cart from Figure 7-6 and imagine that it starts at rest on a horizontal floor and that its interaction with the floor is such that friction with the floor can be neglected. If you give the cart a push as in **Figure 7-9a,** the net force on the cart equals the force that you exert (the upward normal force on the cart cancels the downward gravitational force), so the work that you do is the net work W_{net}. The cart starts at rest, so $v_i = 0$ and so the cart's initial kinetic energy $K_i = \frac{1}{2}mv_i^2$ is zero. After you've finished the push, the cart has final speed $v_f = v$ and kinetic energy $K_f = \frac{1}{2}mv^2$. So the work-energy theorem for an object states that

$$W_{\text{you on cart}} = K_f - K_i = \frac{1}{2}mv^2 - 0 = \frac{1}{2}mv^2$$

In words, this special case gives us our first interpretation of kinetic energy: *An object's kinetic energy equals the work that was done to accelerate it from rest to its present speed.*

Now suppose your friend stands in front of the moving cart and brings it to a halt (**Figure 7-9b**). The cart's initial kinetic energy is $K_i = \frac{1}{2}mv^2$, and its final kinetic energy is $K_f = 0$ (the cart ends up at rest). The net force on the cart is the force exerted by your

(a) Making the cart speed up

(1) The cart slides without friction and the normal force balances the gravitational force.

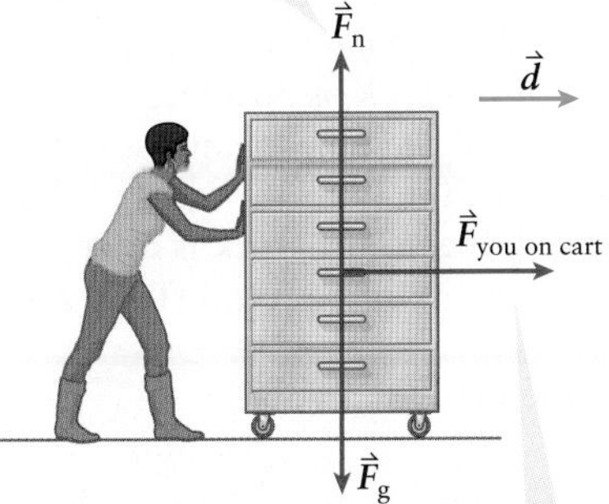

(2) The force you exert therefore equals the net force on the cart. You do positive work on the cart, and the cart gains speed and kinetic energy.

An object's kinetic energy equals the work that would need to be done on the object to accelerate it from rest to its current speed.

(b) Making the cart slow down

(1) The cart slides without friction and the normal force balances the gravitational force.

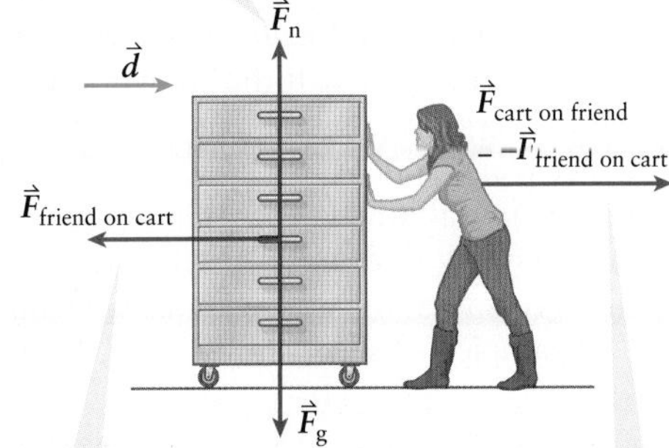

(2) The force your friend exerts therefore equals the net force on the cart. She does negative work on the cart, and the cart loses speed and kinetic energy.

(3) As the cart loses kinetic energy, it pushes on your friend and does positive work on her.

An object's kinetic energy equals the work it could do on another object in coming to rest from its current speed.

Figure 7-9 The meaning of kinetic energy We can interpret kinetic energy in terms of the amount of energy (a) that must be transferred to an object to accelerate it from rest to a given speed or (b) the amount of energy that can be transferred by the object as it slows to a halt.

WATCH OUT !

Kinetic energy is a scalar.

Although you have become comfortable breaking an object's motion into components, remember that kinetic energy depends on *speed*, not velocity. Because speed is not a vector, kinetic energy is a *scalar* quantity and *not* a vector. It wouldn't be meaningful to set up components of kinetic energy in different directions. Note also that we've only discussed the kinetic energy associated with the motion of an object where every point must have exactly the same speed, in the same direction, which we call **translational kinetic energy**. Kinetic energy is also associated with the rotation of a system around its own axis where different points in the system have different speeds and directions. We'll return to this *rotational kinetic energy* in later in the text.

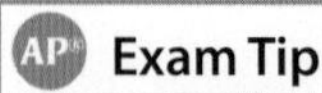

Exam Tip

The table of equations given with the exam does not differentiate between the total work and the work done by a single force. It is important to understand in which situations you would use either one, and that just the total work—or work done by the net force—is equal to the change in kinetic energy of an object.

friend, so the work she does on the cart equals the net work on the cart. If we apply the work-energy theorem for an object to the part of the motion where she exerts a force on the cart,

$$W_{\text{friend on cart}} = K_f - K_i = 0 - \frac{1}{2}mv^2 = -\frac{1}{2}mv^2$$

Our discussion in Section 7-2 tells us that the *cart* does an amount of work on your *friend* that's just the negative of the work that your friend does on the cart. So we can write

$$W_{\text{cart on friend}} = -W_{\text{friend on cart}} = \frac{1}{2}mv^2$$

This gives us a second interpretation of kinetic energy: *An object's kinetic energy equals the amount of work it can do in the process of coming to a halt from its present speed.* Your friend is interacting with the floor through friction, and her hands can move a different distance than her center of mass so she cannot be modeled as an object so she does not end up with this kinetic energy!

This should remind you of the general description of energy as the ability to do work, which we introduced in Section 7-1.

You already have a good understanding of this interpretation of kinetic energy. If you see a thrown basketball coming toward your head, you know intuitively that due to its mass and speed it has a pretty good amount of kinetic energy $K = \frac{1}{2}mv^2$, so it can do a pretty good amount of work on you (it can exert a force on your nose that pushes it inward a painful distance). You don't want this to happen, which is why you duck!

Net Work and Net Force

In Equation 7-9 we interpret W_{net} as the work done by the net force exerted on an object. But we saw in Example 7-2 that the work done by the net force equals the *sum* of the amount of work done by each individual force exerted on the object. So we can also think of W_{net} in Equation 7-9 as the *net work* done by all of the individual forces. It turns out that this statement is true not just for the situation in Example 7-2; it applies to all situations.

This gives us a simplified statement of the work-energy theorem for an object: *The net work done by all forces exerted on an object, W_{net}, equals the change in its kinetic energy $K_f - K_i$.* If the net work done is positive, then $K_f - K_i$ is positive, the final kinetic energy is greater than the initial kinetic energy, and the object gains speed. If the net work done is negative, then $K_f - K_i$ is negative, the final kinetic energy is less than the initial kinetic energy, and the object loses speed. If zero net work is done, the kinetic energy does not change, and the object maintains the same speed (**Table 7-2**). This agrees with the observations we made in Section 7-1.

Example 7-3 shows how to use the work-energy theorem for an object.

TABLE 7-2 The Work-Energy Theorem for an Object

If the net work done on an object is…	…then the change in the object's kinetic energy is $K_f - K_i$…	…and the speed of the object…
$W_{\text{net}} > 0$ (positive net work)	$K_f - K_i > 0$ (kinetic energy increases)	increases (object speeds up)
$W_{\text{net}} < 0$ (negative net work)	$K_f - K_i < 0$ (kinetic energy decreases)	decreases (object slows down)
$W_{\text{net}} = 0$ (zero net work)	$K_f - K_i = 0$ (kinetic energy stays the same)	is unchanged (object maintains the same speed)

EXAMPLE 7-3 Up the Hill Revisited

Consider again the box of supplies that you pushed up the incline in Example 7-2. Use the work-energy theorem for an object to find how fast the box is moving when it reaches your campsite, assuming it is moving uphill at 0.75 m/s at the bottom of the incline.

Set Up

Our goal is to find the final speed v_f of the box. In Example 7-2 we were given the weight (and hence the mass m) of the box, and we found that the net work done on the box is $W_{net} = 56$ J. Because we're given $v_i = 0.75$ m/s, we can use Equations 7-8 and 7-9 to solve for v_f.

Work-energy theorem for an object:

$$W_{net} = K_f - K_i \quad (7\text{-}9)$$

Kinetic energy:

$$K = \frac{1}{2}mv^2 \quad (7\text{-}8)$$

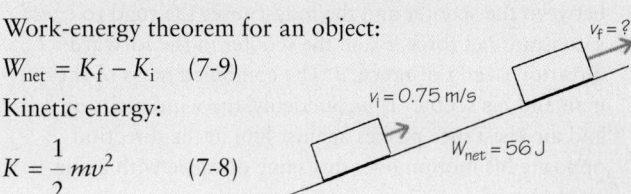

Solve

Use the work-energy theorem for an object to solve for v_f in terms of the initial speed, v_i, the net work done, W_{net}, and the mass m.

Isolate the $\frac{1}{2}mv_f^2$ term on one side of the equation, then solve for v_f by multiplying through by $2/m$ and taking the square root.

Combine Equations 7-8 and 7-9:

$$W_{net} = \frac{1}{2}mv_f^2 - \frac{1}{2}mv_i^2$$

$$\frac{1}{2}mv_f^2 = \frac{1}{2}mv_i^2 + W_{net}$$

$$v_f^2 = v_i^2 + \frac{2W_{net}}{m}$$

$$v_f = \sqrt{v_i^2 + \frac{2W_{net}}{m}}$$

From Example 7-2 we know that $W_{net} = 56$ J. We find the mass m of the box from its known weight $F_g = 225$ N and the relationship $F_g = mg$. Use these to find the value of v_f.

$$m = \frac{F_g}{g} = \frac{225 \text{ N}}{9.80 \text{ m/s}^2} = 23.0 \text{ kg}$$

$$v_f = \sqrt{(0.75 \text{ m/s})^2 + \frac{2(56 \text{ J})}{23.0 \text{ kg}}}$$

$$= 2.3 \text{ m/s}$$

Reflect

We *could* have solved this problem by first finding the net force on the box, then calculating its acceleration from Newton's second law, and finally solving for v_f by using one of the kinematic equations from Chapter 2. This would result in the correct answer, but using the work-energy theorem for an object is easier and gives fewer opportunities to make a mistake.

As a check on our result, note that we found in Example 7-2 that the net force on the box is 9.4 N uphill. We can use Equation 2-11 to find the final velocity after moving 6.00 m up the incline and it agrees, as it must if you did not make a math mistake.

NOW WORK Problems 2, 3, and 6 from The Takeaway 7-3.

Example 7-3 illustrates how using the work-energy theorem can simplify solving problems. In the following section we'll see more examples of how to apply this powerful theorem to various cases of linear motion. In Section 7-5 we'll see how the work-energy theorem can be applied to problems in which the motion is along a curved path and in which the forces are not constant.

THE TAKEAWAY for Section 7-3

✔ The net work done on an object (the sum of the work done on it by all forces exerted on it) as it undergoes a displacement is equal to the change in the object's kinetic energy during that displacement.

✔ The formula for the kinetic energy of an object of mass m and speed v is $K = \frac{1}{2}mv^2$. The kinetic energy of an object is equal to the amount of work that would be needed to accelerate the object from rest to its present speed.

✔ The kinetic energy of an object is also equal to the maximum amount of work the object can do in the process of coming to a halt from its present speed.

Prep for the AP® Exam

AP® Building Blocks

1. A bumblebee has a mass of about 0.250 g. Calculate its kinetic energy when it is moving with a speed of 10.0 m/s.

EX 7-3

2. A man rides a scooter on a level road. The interaction between the scooter and the road causes the road to exert a constant net force, F_1, on the scooter in the forward direction over a distance, d. The combined mass of the man and his scooter is m. Suddenly, the wind picks up and air resistance pushes against him in the direction opposite his motion over this same distance with a constant force F_2.
   (a) If he is initially moving at a speed v_i on a level road, how will you determine the man's speed after the scooter has moved a distance d?
   (b) How will you determine whether the speed of the man will increase or decrease?
   (c) Use your methods to predict the speed of the man after moving a distance d given these values: $F_1 = 1200$ N, $F_2 = 800$ N, $m = 90$ kg, $v_i = 5$ m/s, and $d = 20$ m.

3. A small truck has a mass of 2100 kg. What is the work required to decrease the speed of the vehicle from 22.0 to 12.0 m/s on a level road?

AP® Skill Builders

4. A 1000-kg car moves along a straight, level road. Initially the car's velocity is 60 mph due east and later the car's velocity is 60 mph due west.
   (a) Create a mathematical model of the change in kinetic energy as a function of time, $\Delta K(t)$, of the car as time passes. Assume that the acceleration is constant so that the change in velocity is a linear function of time.
   (b) Construct a diagram of the function $\Delta K(t)$ you have created for the time interval over which the velocity changes from 60 mph due east to 60 mph due west.
5. Explain how the kinematic equation, $v_f^2 = v_i^2 + 2a(x_f - x_i)$; Newton's second law, $\sum F_{ext,x} = ma_x$; and the definition of work lead to the work-energy theorem for an object.

EX 7-3

6. A statue is crated and moved slowly down a ramp for cleaning. In the figure, when the crate is on the ramp, the curator pushes up, parallel to the ramp's surface, so that the crate moves at a constant velocity down the ramp.

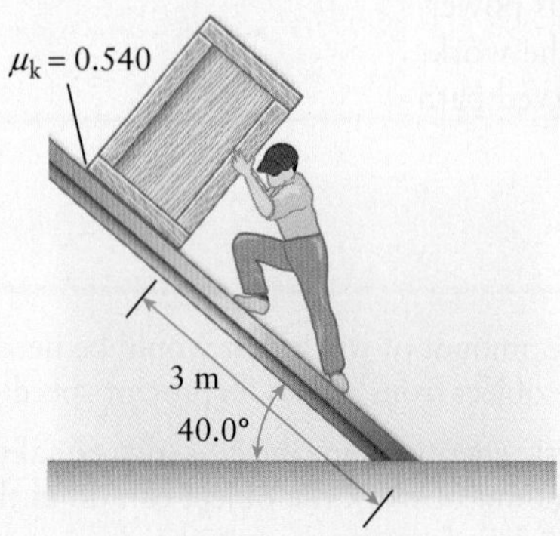

   (a) Construct a free-body diagram of the crate, in which the magnitudes of the forces are consistent with the crate having a constant velocity as it moves down the ramp.
   (b) Use the free-body diagram representation, by identifying the forces that do work on the crate as it slides down the ramp, to predict the sign of the work done on the crate by those forces.
   (c) Construct mathematical representations of each contribution to the total work done on the crate using variables defined in your free-body diagram, and needed given quantities and constants.
   (d) Calculate each contribution to the total work if the statue slides 3.00 m down the ramp with no acceleration and the coefficient of kinetic friction between the crate and the ramp is 0.540. The masses of the statue and the crate are each 150 kg. The ramp is inclined at 40.0°.
   (e) Explain why the sum of work done by the forces exerted on the crate is zero.

AP® Skills in Action

7. A horizontal snow surface with a length Δx at the base of a slope allows a sled and rider to come to a stop after descending the slope. Beyond the length Δx is a sheer cliff. As the Sun sets and the temperature falls the coefficient of kinetic friction is reduced.
   (a) Explain why a reduction in the coefficient of kinetic friction is a cause for concern.
   (b) Express the minimum coefficient of kinetic friction in terms of the velocity of the sled and rider at the base of the slope and the length Δx.
8. A worker exerts a force, F, on a box with mass m initially at rest on a level floor. The force exerted by the worker is directed downward at angle θ with the floor. The floor exerts a friction force on the box.
   (a) The box does not move. Explain why changing the angle that the external force makes with the floor might get the box moving.
   (b) The strategy in part (a) works and the box starts moving. Express the kinetic energy of the box in terms of the force exerted by the worker, F; the mass, m; the coefficient of kinetic friction, μ_k; and the displacement of the box, Δx.
   (c) If the worker continues to push on the box at the same downward angle θ, express how he should reduce the force he exerts on the box, relative to the initial force, $F/F_{initial}$, to displace the box with the least amount of work.

7-4 The work-energy theorem can simplify many physics problems

In this section we'll explore the relationships among work, force, and speed by applying the definitions of work and kinetic energy (Equations 7-2 and 7-8) and the work-energy theorem for an object (Equation 7-9) to a variety of physical situations. Even if the problem could be solved using Newton's second law, the work-energy theorem often makes the solution easier as well as gives additional insight.

Shortly, we will find out there are many more ways to use the work-energy theorem for an object, but we will start with the simplest case: problems that involve an object that moves a distance along a line while constant forces are being exerted on it. You can use this theorem to relate the forces exerted on the object, the displacement of the object, and the speed of the object at the beginning and end of the displacement.

Note that the work-energy theorem for an object makes no reference to the *time* it takes the object to move through this displacement. If the problem requires you to use or find this time, you should use a different approach, such as using Newton's laws in conjunction with kinematics.

Just as we described in Example 1-1, a strategy for solving problems involves three steps, which below we tie directly to problems using the work-energy theorem for an object:

Set Up

Always draw a picture of the situation that shows the object's displacement. Include a free-body diagram, showing all the forces exerted on the object. Draw the direction of each force carefully, because the direction is crucial for determining how much work each force exerted on the object does on the object. Decide what unknown quantity the problem is asking you to determine (for example, the object's final speed or the magnitude of one of the forces).

Exam Tip

Practice drawing free-body diagrams and sketches of setups; they are often required on the AP® Physics exam.

Solve

Use Equation 7-2 to find expressions for the work done by each force exerted on the object. It can be helpful to create secondary diagrams on which you place the vector representing the force tail-to-tail with the vector representing displacement to determine the angle between them. This can be repeated for each force for which you want to calculate the work done. The sum of the work done by each force is the net work done on the object, W_{net}. Then use Equations 7-8 and 7-9 to relate this to the object's initial and final kinetic energies. Solve the resulting equations for the desired unknown.

Reflect

Always check whether the numbers have reasonable values and that each quantity has the correct units.

EXAMPLE 7-4 Work and Kinetic Energy: Force at an Angle

At the start of a race, a four-man bobsleigh crew pushes their sleigh as fast as they can down the 50.0-m straight, relatively horizontal starting stretch (**Figure 7-10**). The force that the four men together exert on the 210-kg sleigh has magnitude 285 N and is directed at an angle of 20.0° below the horizontal. As they push, a 60.0-N net kinetic friction and air drag force is also exerted on the sleigh. What is the speed of the sleigh right before the crew jumps in at the end of the starting stretch?

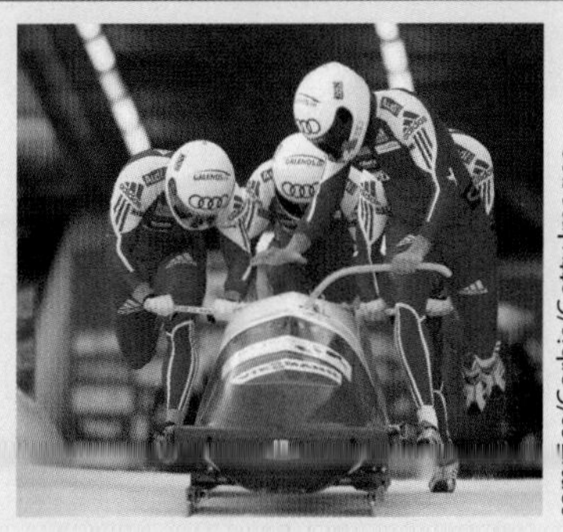

samsics/Corbis/Getty Images

Figure 7-10 Bobsleigh start The success of a bobsleigh team depends on the team members giving the sleigh a competitive starting speed.

Set Up

Again we'll use the work-energy theorem for an object, this time to determine the final speed v_f of the bobsleigh (which starts at rest, so its initial speed is $v_i = 0$). The normal force and gravitational force do no work on the sleigh because they are exerted on the sleigh perpendicular to its displacement. The four men of the crew do positive work, while the kinetic friction and drag forces do negative work (we will just refer to this combined force as F_k). The new wrinkle is that the force exerted by the crew points at an angle to the sleigh's displacement. We'll deal with this using Equation 7-2. Note that the horizontal and vertical forces are drawn to different scales.

Work done by a constant force, linear displacement:

$$W = Fd\cos\theta \quad (7\text{-}2)$$

Kinetic energy:

$$K = \frac{1}{2}mv^2 \quad (7\text{-}8)$$

Work-energy theorem for an object:

$$W_{net} = K_f - K_i \quad (7\text{-}9)$$

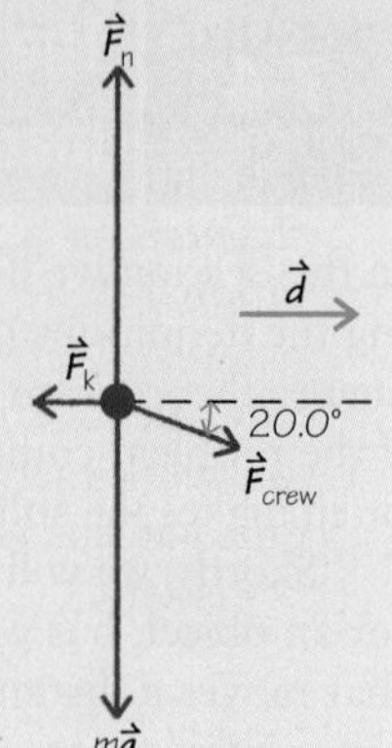

Solve

The sleigh starts at rest, so its initial kinetic energy is zero. From Equation 7-9 the net work done on the sleigh is therefore equal to its final kinetic energy, which is related to the final speed v_f.

Work-energy theorem for an object with initial kinetic energy equal to zero:

$\vec{d} = 50.0$ m

$v_i = 0$

v_f

$$W_{net} = K_f - 0 = \frac{1}{2}mv_f^2$$

The force of the crew is exerted at $\theta = 20.0°$ to the displacement, and the friction force is at $\theta = 180°$. Use these in Equation 7-2 to find the net work done on the sleigh.

net work = work done by the crew plus work done by friction

$$\begin{aligned} W_{net} &= W_{crew} + W_{friction} \\ &= F_{crew}d\cos 20.0° + F_k d\cos 180° \\ &= (285\text{ N})(50.0\text{ m})(0.940) \\ &\quad + (60.0\text{ N})(50.0\text{ m})(-1) \\ &= 1.04 \times 10^4\text{ J} \end{aligned}$$

Substitute W_{net} into the work-energy theorem for an object and solve for the final speed v_f.

Work-energy theorem for an object: $W_{net} = \frac{1}{2}mv_f^2$

Solve for v_f: $v_f = \sqrt{\frac{2W_{net}}{m}} = \sqrt{\frac{2(1.04 \times 10^4\text{ J})}{210\text{ kg}}}$

$= 9.95$ m/s

Reflect

The sleigh gains speed and kinetic energy during the motion, so the net work done on the sleigh by the crew and friction must be positive (even though individually one was negative), which is what we got! The answer for v_f is relatively close to the speed of world-class sprinters, so it seems reasonable. In reality, the start of an Olympic four-man bobsleigh race is slightly downhill, so the sleigh's speed at the end of the starting stretch is typically even faster (11 or 12 m/s) because a component of the force of gravity exerted on the sleigh by Earth is in the direction of motion, adding to the work done.

NOW WORK Problems 2 and 3 from The Takeaway 7-4.

EXAMPLE 7-5 Find the Work Done by an Unknown Force

An adventurous parachutist of mass 70.0 kg drops from the top of Angel Falls in Venezuela, the world's highest waterfall. The waterfall is 979 m (3212 ft) tall and the parachutist deploys her chute after falling 295 m, at which point her speed is 54.0 m/s. During the 295-m drop, (a) what was the net work done on her and (b) what was the work done on her by the force of air resistance?

Set Up

We are given the parachutist's mass and the distance that she falls, so we can find the gravitational force that Earth exerts on her and the work done by that force using Equation 7-2. Air resistance also exerts a force on the parachutist. We don't know its magnitude, so we can't directly calculate the work done by this force. Instead, we'll use the work-energy theorem for an object. We'll assume that the parachutist starts at rest, so her initial kinetic energy is zero.

Work done by a constant force, linear displacement:

$$W = Fd\cos\theta \quad (7\text{-}2)$$

Work-energy theorem for an object:

$$W_{net} = K_f - K_i \quad (7\text{-}9)$$

Kinetic energy:

$$K = \frac{1}{2}mv^2 \quad (7\text{-}8)$$

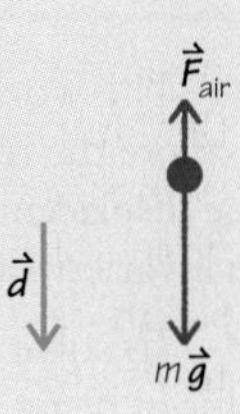

Solve

(a) We can model the parachutist as an object well, because the relative motion of her limbs does not affect what we are trying to find, which is just how much work is done by the various forces in this situation. (How the parachutist holds herself does slightly impact the effects of air resistance, but we aren't trying to change the size of that.) Because $K_i = 0$, Equation 7-9 states that the net work done on the parachutist is just equal to her final kinetic energy.

$$W_{net} = K_f = \frac{1}{2}mv_f^2$$

$$= \frac{1}{2}(70.0\text{ kg})(54.0\text{ m/s})^2$$

$$= 1.02 \times 10^5\text{ J}$$

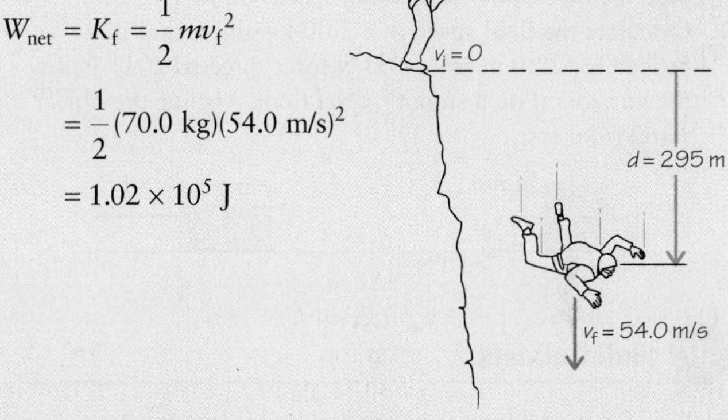

(b) We know that total energy must be conserved. This does not mean that total energy is constant (that is only true for a closed, isolated system, which our parachutist is not), but that any change in the energy of the system must be equal to transfers of energy into or out of the system by external interactions. Any time we use work to determine changes in energy, we are building from this conservation principle. The net work on the parachutist is the sum of the work done by the forces exerted on her; the work done by the gravitational force, W_{grav}; and the work done by the force of air resistance, W_{air}. Hence W_{air} is the difference between W_{net} and W_{grav}.

Work done by the gravitational force: Displacement is in the same direction (downward) as the force, so

$$W_{grav} = mgd$$

$$= (70.0\text{ kg})(9.80\text{ m/s}^2)(295\text{ m})$$

$$= 2.02 \times 10^5\text{ J}$$

Substitute and solve: The work done by the force of air resistance is

$$W_{air} = W_{net} - W_{grav}$$

$$= 1.02 \times 10^5\text{ J} - 2.02 \times 10^5\text{ J}$$

$$= -1.00 \times 10^5\text{ J}$$

Reflect

The work done by the force of air resistance is negative because the force of air resistance is directed upward, opposite to the downward displacement of the parachutist. We can use the calculated value of W_{air} to find the average value of the force of air resistance. This is only an *average* value because this force is *not* constant: The faster the parachutist falls, the greater the force of air resistance. This is less than the gravitational force on the parachutist, as it needs to be, in order for the parachutist to fall and gain speed.

Average upward force of air resistance:

$$W_{air} = F_{air}d\cos 180° = -F_{air}d$$

Solving:

$$F_{air} = -\frac{W_{air}}{d} = -\frac{(-1.00 \times 10^5\text{ J})}{295\text{ m}}$$

$$= 339\text{ J/m} = 339\text{ N}$$

NOW WORK Problem 4 from The Takeaway 7-4.

THE TAKEAWAY for Section 7-4

✔ The work-energy theorem for an object allows us to explore the relationship between the force exerted on an object, the displacement through which the object travels, and its speed in a variety of physical situations.

✔ To solve problems using the work-energy theorem for an object, begin by drawing a diagram that shows all the forces that are exerted on the object in question, and indicates the object's displacement. Then write expressions for the kinetic energies at the beginning and end of the displacement and for the net work done on the object (the sum of the work done by each force). Relate these using the work-energy theorem for an object and solve for the unknown quantity.

Prep for the AP® Exam

AP® Building Blocks

1. Two model racing cars are propelled by identical launchers. One car has twice the mass of the other. Qualitatively describe the relative maximum speeds of the two cars.

EX 7-4 2. Calculate the final speed of a 2.00-kg object that is pushed for 22.0 m by a 40.0-N force directed 20.0° below the horizontal on a smooth, level floor. Assume the object starts from rest.

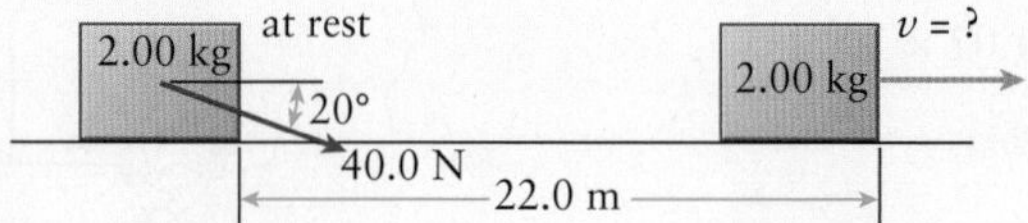

AP® Skill Builders

EX 7-4 3. A 325-g model boat is facing east and is floating on a pond. Because it is floating there is no net force exerted on the boat perpendicular to the water's surface. However, there are forces exerted on the boat in the plane of the water's surface. The wind in its sail provides a force of 1.85 N that points 25.0° north of east. The average force on its keel (a part of the boat's hull that lies beneath the water and helps prevent the boat from slipping sideways) is 0.782 N pointing south. The average drag force of the water on the boat is 0.750 N toward the west.

(a) Construct a free-body diagram of the forces exerted on the model boat in the plane of the water's surface.

(b) The boat starts from rest and heads east. Using the work-energy theorem for an object, predict the speed of the boat after it has moved through a displacement of 3.55 m.

(c) Using the same initial velocity and forces, use Newton's second law to predict the speed of the boat as a function of time.

(d) Compare your results from parts (b) and (c). Evaluate the second law and work-energy theorem for an object as alternative scientific explanations in terms of the role of time.

EX 7-5 4. In a baseball game, assume the ball is thrown from the pitcher to the catcher at a constant velocity, v. We can make this assumption, that the ball travels in a horizontal line from the pitcher to the catcher, if we neglect gravity, which over the time of flight is reasonable, given the speed delivered by a professional pitcher. After contacting the glove, the ball travels only a small distance, d, before coming to a complete stop.

(a) If the mass of the ball is m, construct a symbolic representation of the average horizontal force that the glove exerts on the ball during the catch in terms of v, d, and m, using the work-energy theorem for an object.

(b) Calculate the value of the average force for the glove on the ball found in part (a) using $v = 50$ m/s, $d = 10$ cm, and $m = 0.145$ kg.

(c) The catcher remains in position crouching behind the batter. The catcher's mitt is held vertically and the ball is caught in a padded pocket in the mitt. Explain why the baseball gloves used by other players have much less padding than the catcher's mitt.

AP® Skills in Action

5. A 65.0-kg woman steps off a 10.0-m diving platform and drops straight down into the water. After entering the water she continues falling through the water until she reaches a depth of 4.00 m.

(a) Calculate the work done on the woman by gravity and by the water. Ignore air resistance.

(b) What is the average net force exerted on her over the 4.00-m displacement in the water?

(c) How does the average net force exerted on the woman while she is moving downward in the water compare to the force exerted on her just by the water during this displacement?

7-5 The work-energy theorem is also valid for curved paths and varying forces, and, with a little more information, systems as well as objects

We've derived the work-energy theorem only for the special case of an object in linear motion with constant forces. Because we already know how to solve problems for that kind of motion, you may wonder what good this theorem is. Here's the answer: The work-energy theorem also works for motion along a *curved* path and in cases where

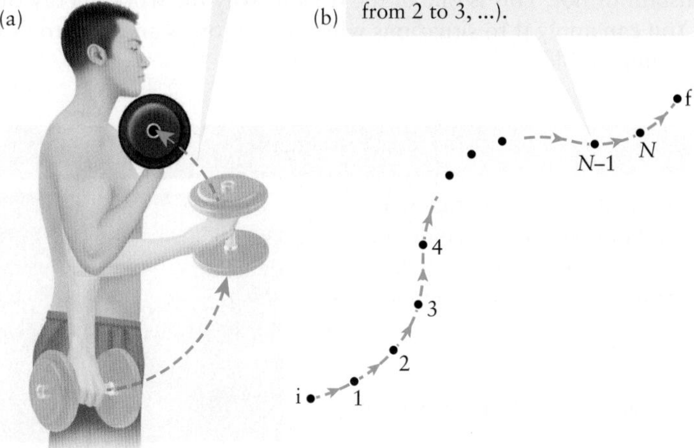
A weight lifter doing bicep curls makes the dumbbell move through a curved path. The force that he exerts on the dumbbell varies during its travel.
(a)
An object's curved path from an initial point i to a final point f. We mark a large number N of equally spaced intermediate points 1, 2, 3, ..., N along the path and imagine breaking the path into short segments (from i to 1, from 1 to 2, from 2 to 3, ...).
(b)
f
N
N–1
4
3
2
i
1

Figure 7-11 Motion along a curved path (a) The dumbbell travels in a curved path during the bicep curl. (b) Analyzing motion along a general curved path.

the forces are *not* constant (**Figure 7-11a**) and can be extended to describe systems as well as objects.

To see that the work-energy theorem for an object is valid for curved paths and varying forces, consider **Figure 7-11b,** which shows an object's curved path from an initial point i to a final point f. We mark a large number N of equally spaced intermediate points 1, 2, 3,..., N along the path and imagine breaking the path into short segments (from i to 1, from 1 to 2, from 2 to 3, and so on). If each segment is sufficiently short, we can treat it as a line. Furthermore, the forces exerted on the object don't change very much during the brief time the object traverses one of these segments, so we can treat the forces as constant over that segment. If we really want to be sure, we could make the segments even shorter. It would not change this analysis, it would just get very tedious! The forces exerted on the object may be different from one segment to the next, but all that matters for this analysis is that the segments are small enough that the forces exerted on the object are relatively constant over each segment. Because each segment is a line with constant forces, we can safely apply the work-energy theorem for an object in Equation 7-9 to every segment:

$$\begin{aligned} W_{\text{net,i to 1}} &= K_1 - K_{\text{i}} \\ W_{\text{net,1 to 2}} &= K_2 - K_1 \\ W_{\text{net,2 to 3}} &= K_3 - K_2 \\ &\cdots \\ W_{\text{net},N \text{ to f}} &= K_{\text{f}} - K_N \end{aligned}$$

Now add all of these equations together:

$$\begin{aligned} &W_{\text{net,i to 1}} + W_{\text{net,1 to 2}} + W_{\text{net,2 to 3}} + \cdots + W_{\text{net},N \text{ to f}} \\ &\quad = (K_1 - K_{\text{i}}) + (K_2 - K_1) + (K_3 - K_2) + \cdots + (K_{\text{f}} - K_{\text{N}}) \end{aligned} \tag{7-10}$$

The left-hand side of Equation 7-10 is the sum of the amounts of work done by the net force on the object as it moves through each segment. This sum equals the *net* work done by the net force along the entire path from i to f, which we call simply W_{net}. Remember that although force is a vector, work is a scalar so these quantities simply add algebraically! On the right-hand side of the equation, $-K_1$ in the second term cancels K_1 in the first term, $-K_2$ in the third term cancels K_2 in the second term, and so on. The only quantities that survive on the right-hand side are $-K_i$ and K_f, so we are left with

$$W_{net} = K_f - K_i$$

This is *exactly* the same statement of the work-energy theorem for an object as shown in Equation 7-9. So the work-energy theorem is valid for *any* path and for *any* forces, whether constant or not. This is one of the reasons why the work-energy theorem is so important: You can apply it to situations where using forces and Newton's laws would be difficult or impossible.

EXAMPLE 7-6 A Swinging Spider

Figure 7-12 shows a South African kite spider (*Gasteracantha*) swinging on a strand of spider silk. Suppose a momentary gust of wind blows on a spider of mass 1.00×10^{-4} kg hanging on such a strand. As a result, the spider acquires a horizontal velocity of 1.3 m/s when it is hanging straight down. How high will the spider swing? Ignore air resistance.

Figure 7-12 A spider swinging on silk If we know the spider's speed at the low point of its arc, how do we determine how high it swings?

Emil von Maltitz/Getty Images

Set Up

Because the spider follows a curved path, this would be a very difficult problem to solve using Newton's laws directly. Instead, we can use the work-energy theorem for an object. We are given the spider's mass and initial speed $v_i = 1.3$ m/s, which allows us to calculate its initial kinetic energy. We want to find its maximum height h at the point where the spider comes momentarily to rest—so its speed and kinetic energy are zero—before it begins swinging back downward.

The diagram shows that only two forces are exerted on the swinging spider: the gravitational force and the tension force exerted by the silk. All we have to do is calculate the work done on the spider by these forces along this curved path. We'll do this by breaking the path into a large number of segments as in Figure 7-11. We need to be able to determine the displacement (change in position) for each segment, so let's choose the horizontal position to be x and the vertical position to be y. Because the path is a circle, we need the segments to be very tiny and to be approximated by lines. This also lets us approximate the tension force as a constant over each segment.

Work-energy theorem for an object:

$$W_{net} = K_f - K_i \qquad (7\text{-}9)$$

Kinetic energy:

$$K = \frac{1}{2}mv^2 \qquad (7\text{-}8)$$

Work done by a constant force, linear displacement:

$$W = Fd\cos\theta \qquad (7\text{-}2)$$

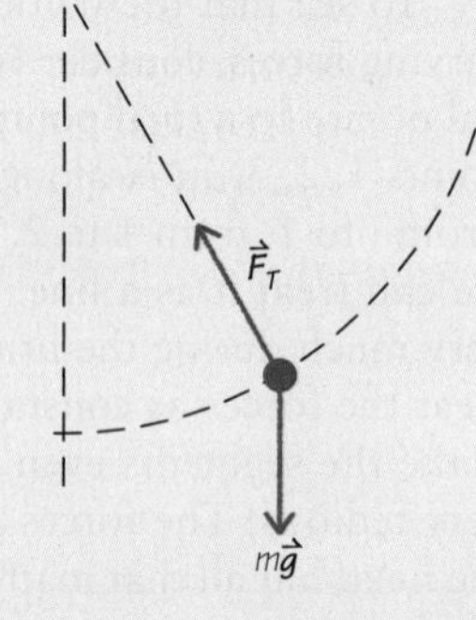

Solve

Rewrite the work-energy theorem for an object in terms of the initial and final speeds of the spider and the work done by each force.

Combine Equations 7-9 and 7-8:

$$W_{net} = \frac{1}{2}mv_f^2 - \frac{1}{2}mv_i^2$$

Net work is the sum of the work done by the tension force F_T and the work done by the gravitational force:

$$W_{net} = W_{grav} + W_T$$

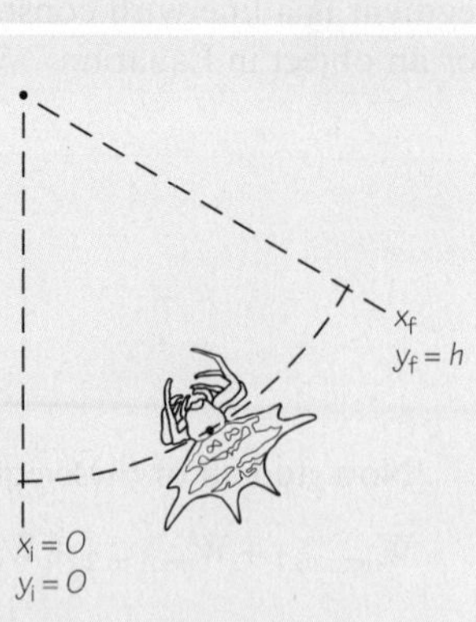

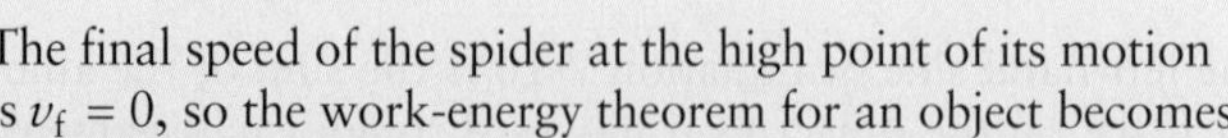

The final speed of the spider at the high point of its motion is $v_f = 0$, so the work-energy theorem for an object becomes

$$W_{grav} + W_T = -\frac{1}{2}mv_i^2$$

To calculate the work done by each force along the spider's curved path, break the path up into segments so short that each one can be considered as a line tangent to the path. On each segment the spider has a displacement $\vec{d}$ with horizontal component Δx and vertical component Δy. These are different for each individual segment. In the figure, we exaggerated the length of the segment to make it easier to draw.

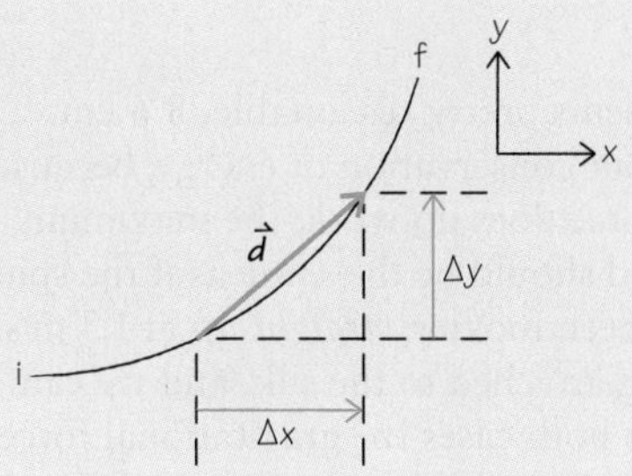

On each segment of the path the tension force $\vec{F}_T$ points radially inward, perpendicular to the path, and because of our very short segment approximation, perpendicular to the displacement $\vec{d}$. Hence $\theta = 90°$ in Equation 7-2, so the tension force does *no* work during this displacement.

Work done by the tension force as the spider moves through a short segment of its path:

$$W_{\text{T,segment}} = F_T d \cos 90° = 0$$

The gravitational force $\vec{F}_g$ always points straight downward and has constant magnitude $F_g = mg$. We apply Equation 7-2 to the displacement and gravitational force drawn as shown. We are also going to use the identity $\cos(90° + \phi) = -\sin\phi$. Don't be scared by this, even if you haven't seen it before. (If you look at graphs of the cosine and sine of an angle versus the angle you can see that if you add 90° to any angle the cosine of the sum is the same as the negative of the sine of the angle!)

Work done by the gravitational force as the spider moves through a short segment of its path:

$$W_{\text{grav,segment}} = F_g d \cos\theta$$

Because the angle between the displacement and the gravitational force is greater than 90°, the gravitational force does negative work. Our trig identity gave us this sign, so the equation we came up with looks correct.

The angle between the gravitational force and the displacement is $\theta = 90° + \phi$ so

$$W_{\text{grav,segment}} = F_g d \cos(90° + \phi) = -F_g d \sin\phi$$

The figure shows

$d \sin\phi = \Delta y$, so

$$W_{\text{grav,segment}} = -F_g \Delta y = -mg\Delta y$$

Now we can calculate the *net* work done on the spider by each force as it moves along its curved path. This is just the sum of the individual bits of work done on it along the short segments.

The tension force does zero work on each short segment, so when you add them all up

$$W_T = 0$$

Work done by the gravitational force over the entire path:

$$W_{\text{grav}} = (-mg\Delta y_1) + (-mg\Delta y_2) + (-mg\Delta y_3) + \cdots = -mg(\Delta y_1 + \Delta y_2 + \Delta y_3 + \cdots)$$

The sum of all the Δy terms is the net vertical displacement $y_f - y_i = h - 0 = h$. So $W_{\text{grav}} = -mgh$.

Substitute the expressions for W_T and W_{grav} into the work-energy theorem for an object and solve for h. Note that we didn't need the length of the silk for this problem.

Work-energy theorem for an object:

$$W_{\text{grav}} + W_T = -mgh + 0 = -\frac{1}{2}mv_i^2$$

$$h = \frac{v_i^2}{2g} = \frac{(1.3\ \text{m/s})^2}{2(9.80\ \text{m/s}^2)} = 0.086\ \text{m}$$

Reflect

The height seems pretty reasonable, 8.6 cm. Thinking about conservation of energy, because the tension force does no work, the maximum height reached should be the *same* as if the spider had initially been moving *straight up* at 1.3 m/s without being attached to the silk, and we can check that. In both cases the gravitational force does the same amount of (negative) work to reduce the spider's kinetic energy to zero, because the tension does no work. Because the tension is always perpendicular to the motion, we know it can cause only the direction of motion to change, not the speed.

The tension force in this problem is complicated because its magnitude and direction change as the spider moves through its swing.

But in the work-energy approach we don't have to worry about the tension force at all because it does no work on the spider.

If the spider was initially moving straight up:

$$W_{\text{net}} = \frac{1}{2}mv_{\text{f}}^2 - \frac{1}{2}mv_{\text{i}}^2$$

$$W_{\text{grav}} = F_{\text{g}}h \cos 180° = -mgh$$

$$W_{\text{net}} = W_{\text{grav}} = -mgh = 0 - \frac{1}{2}mv_{\text{i}}^2$$

$$h = \frac{v_{\text{i}}^2}{2g} = \frac{(1.3 \text{ m/s})^2}{2 \times 9.8 \text{ m/s}^2}$$

$$h = 0.086 \text{ m}$$

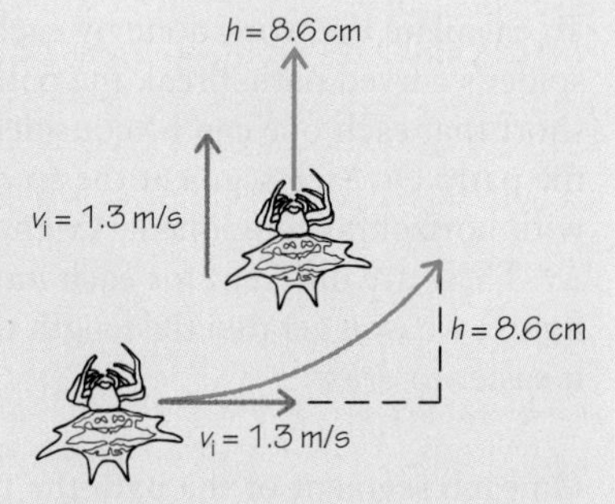

NOW WORK Problems 3 and 7 from The Takeaway 7-5.

Example 7-6 illustrates an important point: *If a force is always exerted perpendicular to an object's path, it does zero work on the object.* This enabled us to ignore the effects of the tension force in this problem. We'll use this same idea in many contexts!

The Work-Energy Theorem Is Not Just for Objects

When you look at the chapter-opening sequence of photographs, you will see, given our discussion in Section 1-3, that we cannot treat the ball as an object during this time; we must use a system description. Additionally, we discussed the need to describe the motion of one point in a system to represent the whole system if we want to use the object model. For the object model, we normally choose this point to be the center of mass. In Section 4-4 we discussed how to qualitatively locate the center of mass, which will be sufficient for what we are doing now. (In Section 9-7 we will learn how to calculate the location of the center of mass, when it becomes clearer in the context we will be studying.) Examining the chapter-opening sequence of photographs, you see the ball expand as it moves away from the wall. Because it was at rest when it was completely against the wall, it is speeding up in this process. You can imagine what this sequence of photographs would look like if it were longer and involved the ball initially striking the wall. The ball would compress as it slowed to a stop.

Given what we have been discussing so far in this chapter, our first reaction might be to think the wall has done work on the ball. After all, the ball's kinetic energy has to go to zero at the instant it is fully compressed against the wall, as it changes direction. The wall does no work on the ball, however. Our definition of work says that the force must be exerted on the object as the point of contact of the force on the object *moves* for there to be work done on the object. In this example, the point of contact of the ball and the wall does not move, so the force exerted by the wall does not do work. Actually, this makes a lot of sense if you think about it: You can lean against a wall all day, and it is not going to make you start moving!

The shortcut we took in deriving the work-energy theorem for an object (Equation 7-9) is clear in its name: We used the object model. By definition, for the object model to be valid, the center of mass and any point on the object must move the same distance. That is not true for a system such as a compressible ball. In the chapter-opening photographs, we see that the point of contact for the force does not move, but using our qualitative understanding of the center of mass, we see the center of mass clearly does get further from the wall as the ball expands. Work is determined by the displacement of the point of contact while the force doing the work is being exerted, but Equation 7-9 is derived from an equation that describes the motion of the center of mass. This means the displacement of the center of mass while the force is being exerted describes change in the kinetic energy of the system. For a system, the displacement of the center of mass and the point of contact where a force is being exerted can be different, so types of energy other than kinetic can change.

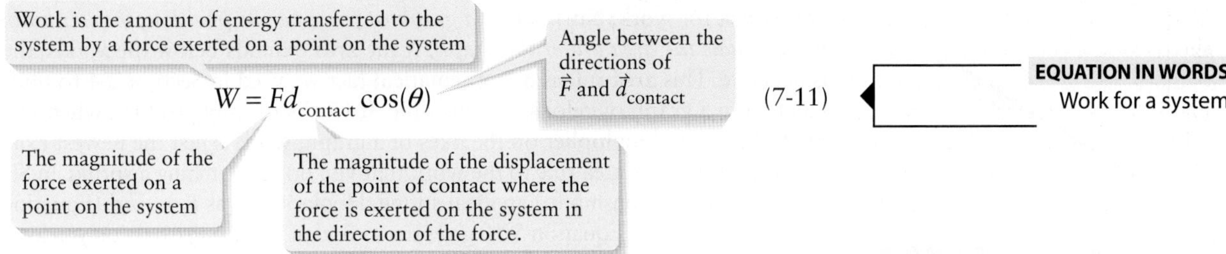

Work is the amount of energy transferred to the system by a force exerted on a point on the system
Angle between the directions of $\vec{F}$ and $\vec{d}_{contact}$
The magnitude of the force exerted on a point on the system
The magnitude of the displacement of the point of contact where the force is exerted on the system in the direction of the force.

$$W = Fd_{contact}\cos(\theta) \quad (7\text{-}11)$$

EQUATION IN WORDS
Work for a system

AP® Exam Tip

A stationary object, such as a wall, exerting a normal force, cannot do work: It cannot add or remove energy. However, the normal force exerted by the stationary object on a system allows the system to convert energy from one type to another without changing the system's total energy.

For instance, in the series of chapter-opening photos, the wall is exerting a normal force on the ball, but the point of contact does not move, so the total work done by the wall on the ball is zero. The wall transfers no energy to the ball. However, the center of mass of the ball does move while the wall is exerting a force on the ball, so the ball's kinetic energy does change.

When we discuss energy, we will be most focused on the changes in energy. In Equation 7-9 we wrote out K_f and K_i. It is easier to use the notation for changes we introduced in Chapter 2, $\Delta K = K_f - K_i$. Using this notation, we can now write the full work-energy theorem:

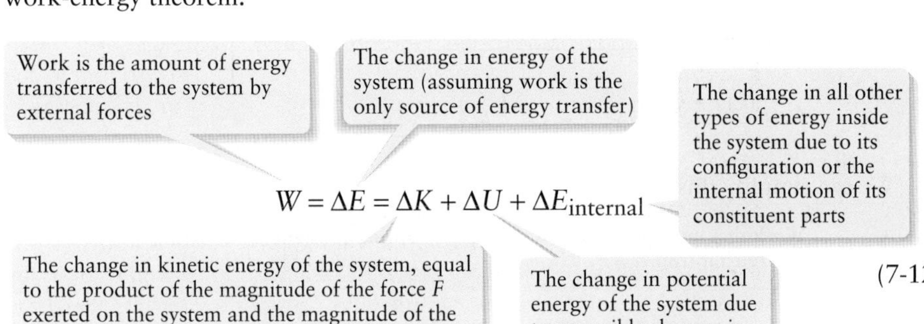

Work is the amount of energy transferred to the system by external forces
The change in energy of the system (assuming work is the only source of energy transfer)
The change in all other types of energy inside the system due to its configuration or the internal motion of its constituent parts
The change in kinetic energy of the system, equal to the product of the magnitude of the force F exerted on the system and the magnitude of the displacement of the center of mass of the system in the direction of the force.
The change in potential energy of the system due to reversible changes in its configuration

$$W = \Delta E = \Delta K + \Delta U + \Delta E_{internal} \quad (7\text{-}12)$$

EQUATION IN WORDS
Work-Energy Theorem

AP® Exam Tip

You are likely to need the *area rule* on an AP® exam: Be ready to estimate the area under a force versus displacement graph, but also be sure to distinguish that from a force versus time graph (the area of which gives impulse), which you will learn about in Chapter 9.

Both Equations 7-11 and 7-12 simplify our earlier equations if the displacement of the point of contact of the force on the system is the same as the displacement of the center of mass, which is a requirement for the object model to be valid. This is true even if the center of mass is not in the same place as the point of contact; it just requires the system to be rigid so that all points in the system move the same distance and direction. When things are rotating, we will need additional tools to solve such problems. Those tools will be provided later in the text. These equations also do not consider other ways of adding energy to a system, such as heating. You will get to the full expression if you take the next course. For now, just remember this equation doesn't include energy added to or removed from the system by thermal processes so it doesn't work in those contexts!

Work Done by a Varying Force

Up until now, we have always assumed a constant force when calculating work done. In Example 7-6 we saw how we might cope with a varying force, but because the tension force exerted on the spider did no work, it might not have seemed very important. In many situations, however, a force of variable magnitude *does* do work on an object or a system. As an example, you must do work to stretch a spring. The force you exert on the spring to stretch it is not constant: The farther you stretch the spring, the greater the magnitude of the force you must exert on the spring. How can we calculate the amount of work that you do while stretching the spring?

To see the answer let's first consider a *constant* force F that is exerted on an object in the direction of its linear motion. Because we are working in one dimension, we are just using one component, so positive and negative are enough to determine direction and we don't need arrows to denote vectors. **Figure 7-13** shows a graph of this force versus position as the object undergoes a displacement d. The *area* under the graph of force

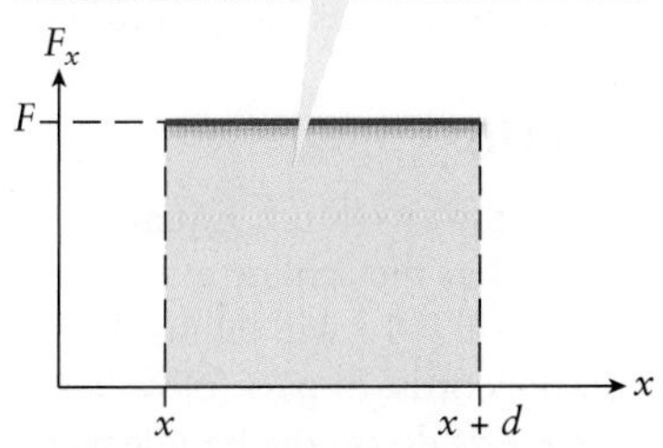

Figure 7-13 The area rule for work
Finding the work done by a constant force exerted on an object or a system using a graph of force versus position.

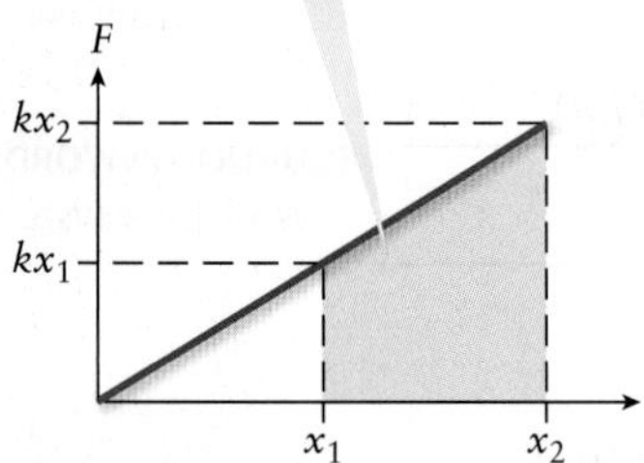

Figure 7-14 Applying the area rule to an ideal spring We can use the same technique as in Figure 7-13 to find the work required to stretch a spring.

versus position equals the area of the colored rectangle in Figure 7-13; that is, the height of the rectangle (the force $\vec{F}$) multiplied by its width (the displacement $\vec{d}$). But the area Fd is just equal to the work done by the constant force (notice that the units for the area are correct for work, Nm). So *on a graph of force versus position, the work done by the force equals the area under the graph*. If the area extends below the x axis, the product is negative. This area rule is a mathematical fact we used in Section 2-4 to find displacement from a graph of velocity versus time. It works for *any* product, when you plot the multiplicand and multiplier on the axes of a graph; work is just the newest example.

Let's apply this area rule to the work that you do in stretching a spring. In Section 5-6, we wrote the relationship for an ideal spring, Hooke's law, as $F_s = -kd$ (Equation 5-10).

Remember that Equation 5-10 gives the force that the *spring* exerts on *you* as you stretch it. By Newton's third law, the force that *you* exert on the *spring* has the same magnitude but is exerted in the opposite direction. We wrote this as Equation 5-11: $F_{\text{you on spring},x} = +kd$.

Figure 7-14 graphs the force that you exert as a function of the distance that the spring has been stretched. *Each position can be thought of as an incredibly tiny displacement, over which the force is a constant, just like we did for the forces on the spider in Example 7-6.* If the spring is initially stretched to x_1 from its equilibrium position, which we call $x = 0$, and you stretch it further to x_2, the work W that you do is equal to the area of the colored trapezoid in Figure 7-14.

From geometry, this area is equal to the *average* height of the graph multiplied by the width $x_2 - x_1$. Using Equation 5-11, we find

$$W = \left[\frac{(\text{force you exert at } x_1) + (\text{force you exert at } x_2)}{2}\right](x_2 - x_1)$$

$$= \frac{(kx_1 + kx_2)}{2}(x_2 - x_1) = \frac{1}{2}k(x_1 + x_2)(x_2 - x_1)$$

We can simplify this to

WATCH OUT!

Who's doing the work?

Note that Equation 7-13 tells us the work that *you* do on the *spring*. Because this is a contact force, the work that the *spring* does on *you* is equal to the negative of Equation 7-13.

EQUATION IN WORDS
Work done to stretch a spring

$$W = \frac{1}{2}kx_2^2 - \frac{1}{2}kx_1^2 \quad (7\text{-}13)$$

Work that must be done on a spring to stretch it from $x = x_1$ to $x = x_2$

Spring constant of the spring (a measure of its stiffness)

x_1 = **initial stretch** of the spring
x_2 = **final stretch** of the spring
x_1 and x_2 are measured from $x = 0$, the location of the end of the spring when it is relaxed

Example 7-7 shows how to use Equation 7-13 to attack a problem that would have been impossible to solve with the force techniques from Chapters 4 and 5.

EXAMPLE 7-7 Work Those Muscles!

An athlete stretches a set of exercise cords 47 cm from their unstretched length. The cords behave like a spring with a spring constant of 86 N/m. (a) How much force does the athlete exert to hold the cords in this stretched position? (b) How much work did he do to stretch them? (c) The athlete loses his grip on the cords. If the mass of the handle is 0.25 kg, how fast is it moving when it hits the wall to which the other end of the cords is attached? You can ignore gravity and assume that the cords themselves have a negligible mass.

Set Up

In part (a) we'll use Equation 5-11 to find the force that the athlete exerts, and in part (b) Equation 7-13 will tell us the work that he does on the cords. In part (c) we'll see how much work the *cords* do on the handle as they go from being stretched to relaxed. This work goes into the kinetic energy of the handle. We're ignoring the mass of the cords and therefore assuming that they have no kinetic energy of their own.

$$F_{\text{athlete on cords},x} = +kd \quad (5\text{-}11)$$

$$W = \frac{1}{2}kx_2^2 - \frac{1}{2}kx_1^2 \quad (7\text{-}13)$$

$$W_{\text{net}} = K_f - K_i \quad (7\text{-}9)$$

$$K = \frac{1}{2}mv^2 \quad (7\text{-}8)$$

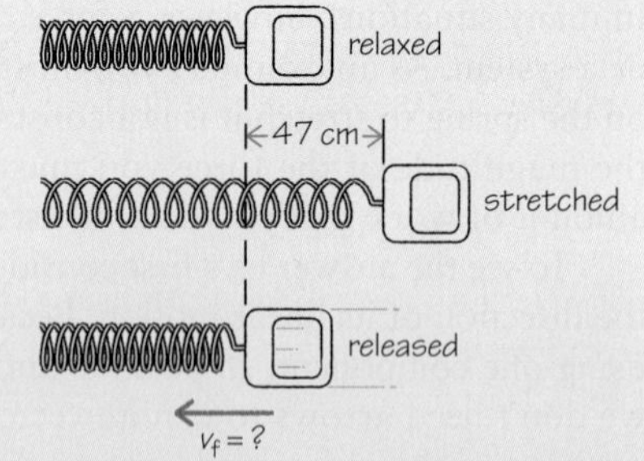

Solve

(a) The force that the athlete exerts is proportional to the distance $x = 47$ cm that the cords are stretched (assuming the end of the cords when they are not stretched is at $x = 0$).

Force exerted by the athlete to hold the cords stretched:

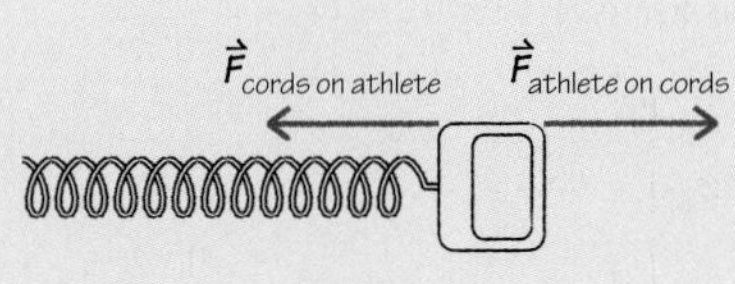

$$F_{\text{athlete on cords},x} = +kd$$

$$= (86 \text{ N/m})(47 \text{ cm})\left(\frac{1 \text{ m}}{100 \text{ cm}}\right)$$

$$= 4.0 \times 10^1 \text{ N}$$

(b) The cords are not stretched at all to start (so $x_1 = 0$) and end up stretched by 47 cm (so $x_2 = 47 \text{ cm} = 0.47$ m).

Work done by the athlete to stretch the cords:

$$W_{\text{athlete on cords}} = \frac{1}{2}kx_2^2 - \frac{1}{2}kx_1^2$$

$$= \frac{1}{2}(86 \text{ N/m})(0.47 \text{ m})^2 - \frac{1}{2}(86 \text{ N/m})(0 \text{ m})^2$$

$$= 9.5 \text{ Nm} = 9.5 \text{ J}$$

(c) When the athlete releases the handle, the cords relax from their new starting position ($x_1 = 0.47$ m) to their final, unstretched position ($x_2 = 0$). Unlike an actual spring, the cords don't push back as the handle passes through the equilibrium position. So, the work that the cords do on the handle is given by the negative of Equation 7-13.

Work done by the cords on the handle as they relax:

$$W_{\text{cords on handle}} = -\left(\frac{1}{2}kx_2^2 - \frac{1}{2}kx_1^2\right)$$

$$= -\left(\frac{1}{2}(86 \text{ N/m})(0 \text{ m})^2 - \frac{1}{2}(86 \text{ N/m})(0.47 \text{ m})^2\right)$$

$$= -(-9.5 \text{ Nm}) = 9.5 \text{ J}$$

We'll ignore the force of gravity (that is, we assume the handle flies back horizontally and doesn't fall). Then the net work done on the handle equals the work done by the cords. This is equal to the change in the handle's kinetic energy. Use this to find the handle's speed when the cords are fully relaxed.

Net work done on handle:

$$W_{\text{net}} = W_{\text{cords on handle}} = 9.5 \text{ J}$$

Work-energy theorem applied to handle:

$$W_{\text{net}} = K_f - K_i = \frac{1}{2}mv_f^2 - \frac{1}{2}mv_i^2$$

Handle is initially at rest, so $v_i = 0$. Solve for final speed of the handle:

$$\frac{1}{2}mv_f^2 = W_{\text{net}} = 9.5 \text{ J}$$

$$v_f^2 = \frac{2W_{\text{net}}}{m}$$

$$v_f = \sqrt{\frac{2W_{\text{net}}}{m}} = \sqrt{\frac{2(9.5 \text{ J})}{0.25 \text{ kg}}} = 8.7 \text{ m/s}$$

Reflect

The spring force is not a *constant* force, so the work done must be less than you would calculate if you multiplied the force needed to hold the cords fully stretched by the cords' displacement, because the force exerted is smaller than that except at the maximum stretch. So we expect our answer to be less than $F_s d = (4.0 \times 10^1 \text{ N})(0.47 \text{ m}) = 19$ J, which it is.

Extend: Notice that the amount of work that the athlete does to stretch the cords (9.5 J) is the *same* as the amount of work that the cords do on the handle when they relax. This suggests that the athlete stores energy in the cords by stretching them. We'll explore this idea of *storing energy* in Section 7-6 as we explore potential energy.

NOW WORK Problem 5 from The Takeaway 7-5.

Like Equation 5-10, Equation 7-13 is also valid when a spring is compressed. If a spring with spring constant $k = 1000$ N/m is initially relaxed (so $x_1 = 0$) and you stretch it by 20 cm (so $x_2 = +20 \text{ cm} = +0.20$ m), the work that you do is

$$W = \frac{1}{2}kx_2^2 - \frac{1}{2}k(0)^2 = \frac{1}{2}(1000 \text{ N/m})(0.20 \text{ m})^2 = 20 \text{ J}$$

The work that you do to *compress* the same spring by 20 cm (so $x_2 = -20$ cm = -0.20 m) is

$$W = \frac{1}{2}kx_2^2 - \frac{1}{2}k(0)^2 = \frac{1}{2}(1000\text{ N/m})(-0.20\text{ m})^2 = 20\text{ J}$$

So you have to do positive work to compress a relaxed spring as well as to stretch one.

THE TAKEAWAY for Section 7-5

✔ The work-energy theorem for an object applies even when the object follows a curved path or the forces that are exerted on the object are not constant.

✔ A force that is always exerted perpendicular to an object's path does zero work on the object.

✔ The work-energy theorem for a system that cannot be modeled as an object gives the total energy change of the system on which the force is exerted, including internal and potential energies. Objects can only have kinetic energy.

✔ If the point of contact where a force is exerted on an object or a system does not move, there is no work done by that force on that object or system.

✔ An ideal spring exerts a force proportional to the distance that it is stretched or compressed (Hooke's law). The work required to stretch or compress a spring by a given distance from its equilibrium length is proportional to the square of that distance.

Prep for the AP® Exam

AP® Building Blocks

1. A gravitational force exerted on an object of mass m attached to a vertical spring will cause the spring to stretch. When the object is in equilibrium, the gravitational force exerted on the object is equal in magnitude to the spring force exerted on the object: $mg = k\Delta x$. This table provides data for a spring for eight different values of mass.

Mass, m (g)	1.0	2.0	3.0	4.0	5.0	6.0	7.0	8.0
Stretch, Δx (cm)	0.5	1.1	1.7	2.3	2.6	3.1	3.6	4.2

(a) Make a graph of mg versus Δx.
(b) Use your graph to calculate the value of the spring constant, k, using all of the data values.
(c) Shade the area in the graph that shows the work done to stretch the spring by 2.5 cm and calculate the work done by the object on the spring resulting in that stretch. Justify your use of the graph to calculate the work.

2. A sled of mass $m = 22.0$ kg is accelerated from rest on a horizontal surface to a velocity of $v = 12.5$ m/s, as it travels from $x = 0$ to $x = 30.0$ m. The net force in the direction of the displacement, in terms of F_{max}, is graphed below as a function of displacement for the sled.

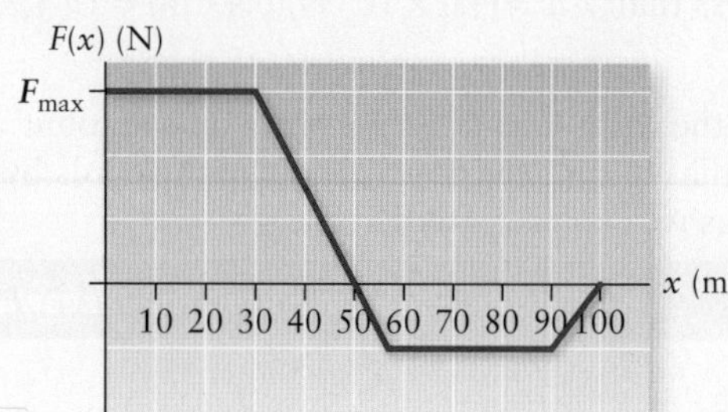

(a) Use the graph and the values given to find the value of the work done on the sled as it moves from $x = 0$ to $x = 30.0$ m, and use the work to calculate the value of the maximum net force exerted on the sled, F_{max}.
(b) Does the speed of the sled increase or decrease as it travels from $x = 30.0$ m to $x = 50.0$ m? Your prediction must be based on scientific theories.
(c) Calculate the velocity of the sled at $x = 100.0$ m.

EX 7-6 3. A boy swings a ball with a 6-N weight on a string at constant speed in a horizontal circle that has a diameter equal to 1 m. What is the work done on the ball by the 10-N tension force in the string as the ball travels through one-half of the circular path?
(A) 31.4 J (B) 10 J (C) 6 J (D) 0 J

4. An ideal spring with a constant $k = 1$ N/m is attached to a stationary beam so that it hangs vertically. At the lower end of the spring a 100-g object is attached. The object is supported by a hand so that the spring force exerted on the object is initially zero (the spring is at its equilibrium length). The object is then released and as it falls it passes through a position where the extension of the spring is 0.5 m from its initial equilibrium length. Take the value of the gravitational field strength to be 10 N/kg.
(a) What is the work done on the object by the gravitational force as it travels through this 0.5-m displacement?
(A) 1 J (B) 0.5 J (C) −0.5 J (D) −1 J
(b) What is the work done on the object by the spring force as it travels through this 0.5-m displacement?
(A) 1/8 J (B) 1/10 J (C) −1/8 J (D) −1/10 J

EX 7-7 5. An object resting on a smooth table is attached to the free end of a horizontal spring that has a spring constant equal to 450 N/m. The object is initially held at rest 12 cm beyond the equilibrium length of the spring, and is then pulled to a position 18 cm beyond equilibrium, where it is again held at rest. Calculate the work the spring does on the object between these two positions, both algebraically and graphically.

AP® Skill Builders

6. Two forces are simultaneously exerted on an object over the range of positions, x, $0 \le x \le 2$ m. The forces are

represented by $F_1(1-x)$ and $F_2(x-1)$, where F_1 and F_2 are constants.

(a) On a single graph construct a representation of both forces as a function of position.

(b) Refine the representation by shading the area that represents the work done on the object by these forces from $x = 0$ to $x = 2$ m.

(c) Apply mathematical reasoning to support the claim that the net work done on the object as the object moves from $x = 0$ to $x = 2$ m is zero for any values of F_1 and F_2.

EX 7-6 7. Before giving a slick ice cube at the bottom of a large, smooth ceramic bowl a flick with your finger, you want to calculate the height to which the flick will cause the ice cube to rise. You estimate the mass of the ice cube to be between 25 and 50 g and the initial velocity of the ice cube after it leaves your finger at the bottom of the bowl to be between 1 and 2 m/s.

(a) Justify the selection of the use of the energy conservation principle rather than Newton's second law to make this prediction.

(b) Construct a free-body diagram of the ice cube at some point as it rises along the inner surface of the bowl and annotate the diagram with a vector representing the displacement of the ice cube at that instant.

(c) Using the diagram constructed in part (b), explain why the normal force exerted by the wall of the bowl on the ice cube does not do work on the ice cube.

(d) Let a vector along a short segment of the path of the ice cube as it rises in the bowl be $\vec{d}$. Construct a diagram in which the displacement $\vec{d}$ is the hypotenuse of a triangle with one component that is antiparallel to the gravitational force. In the diagram, label the angle between the displacement and this component as θ. Using this diagram explain why the sum of displacements along the field is equal to the height the ice cube goes up the bowl.

(e) Using the diagram constructed in part (d) explain how the height to which the ice cube rises can be predicted.

(f) Explain why the predicted height is insensitive to your uncertainty in the mass of the ice cube but very sensitive to your uncertainty in the initial velocity.

AP® Skills in Action

8. An object with a mass of 0.5 kg and a velocity of $\sqrt{10}$ m/s enters a region in which a changing force F in the x direction is exerted on the object. The object moves from an initial position $x = 0$.

(a) Analyze the data in the table to express F/m as a function of x.

x (m)	0	0.2	0.4	0.6	0.8	1.0	1.2	1.4	1.6	1.8
F/m (N/kg)	−10	−9	−8	−7	−6	−5	−4	−3	−2	−1

(b) Calculate the initial kinetic energy of the object.

(c) Predict the point, x_s, where the object momentarily comes to a stop and begins to move in the negative x direction. *Hint*: Apply the work-energy theorem to find the point where the object stops (the kinetic energy of the object is zero).

(d) Justify the claim that the kinetic energy of the object will have the value calculated in part (b) when, after stopping, it returns to the position $x = 0$.

9. An ideal spring is used to stop blocks as they slide along a track with negligible friction, as shown in the figure. Measurements are made of the maximum displacement of the spring when struck by a 0.20-kg block for different velocities. Use these data to determine the spring constant graphically.

Velocity (m/s)	Displacement (cm)
0.8	3.5
1.8	7.2
2.8	11
3.7	14.5
4.3	17.5
5.3	22

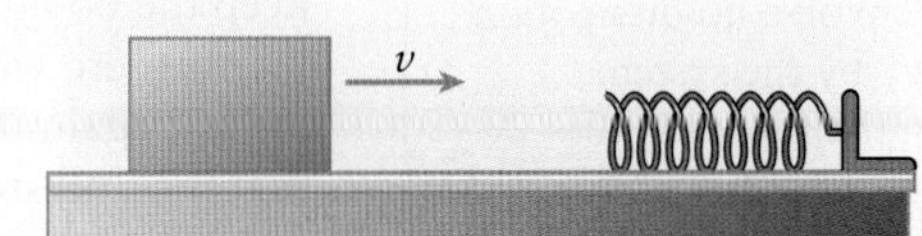

7-6 Potential energy is energy related to reversible changes in a system's configuration

When we do work on a spring to compress it, we don't give it kinetic energy, but we could use that compressed spring to give kinetic energy to a ball, for instance, by using the spring as a launcher. Let's go back to our photographs at the beginning of the chapter. When the ball comes to rest compressed against the wall some of its kinetic energy gets stored in the compression of the ball, just like the energy stored in springs as we saw in the last section. The more elastic the ball is, the greater the proportion of the kinetic energy that gets converted into energy stored in the compression of the ball. Some of the energy goes into warming up the ball, and some into making noise. The energy that goes into these types is said to be *dissipated*. Dissipated energy is energy that has been converted into a type other than kinetic or potential energy, that cannot ever be fully converted back into kinetic or potential energy. In many cases the amount of dissipated energy will be small enough that we will choose to ignore it. If all of the kinetic energy got stored in the compression of the ball, then the ball would have the *potential* to come flying off the wall with the same speed and exactly the same shape it had when it first struck the wall. We call this energy stored in the reversible changes in a configuration of the system (in this case the compression of the ball) potential energy.

Conservative and Nonconservative Forces

To use potential energy instead of force to describe and predict motion for a system, the force must be like that exerted by an ideal spring or the gravitational force: The work done by the force does not depend on the path taken (the total distance), just the displacement of the point of contact of the force. Such a force is called a **conservative force.** Another way this is often stated is that if a force exerted on an object does no net work during any closed loop (zero displacement but not zero distance), then the force is said to be conservative. If the net work is not zero, the force is said to be nonconservative. (In Chapter 8 we'll more fully develop this term.) By contrast, the kinetic friction force is an example of a **nonconservative force** for which we *cannot* use the concept of potential energy. The reason is that unlike the force exerted by an ideal spring, the "work" (the quotation marks are intentional and will be explained soon!) done by the kinetic friction force *does* depend on the path taken from the initial point to the final point.

In **Figure 7-15a** we slide a book across a tabletop from an initial point to a final point along two different paths. Along either path the kinetic friction force has the same magnitude and points opposite to the direction of motion. Hence the kinetic friction force does more (negative) "work" along the curved path than along the straight path. Because the "work" done by kinetic friction depends on more than just the initial and final positions, we can't write it in terms of a change in potential energy. That's why there's no such thing as "friction potential energy."

WATCH OUT!

Conservative forces have two defining properties.

For a force to be conservative, it must (1) depend on a reversible change in configuration, such as a change in length of a spring or a change in separation between Earth and an object, and (2) involve no dissipation of energy by the system exerting the force.

Remember, potential energy is associated with reversible changes in a system's configuration, so there is an equivalent way to decide whether a certain kind of force is conservative: If the work done by the force on a *round trip* (that is, one where the initial and final positions are the same) is *zero*, the force is conservative and we can use the idea of potential energy instead of using the force to describe or predict motion. This is the case for the gravitational force. If you toss a ball straight up, the gravitational force does negative work on the ball as it rises and positive work on the ball as it falls. If $y_f = y_i$, then $W_{grav} = -mgy_f + mgy_i = 0$. The same is true for the spring force: If $x_f = x_i$, then $W_{spring} = \frac{1}{2}kx_f^2 - \frac{1}{2}kx_i^2 = 0$. But if you slide a book on a round trip on a tabletop, the total amount of work that you do on the book is *not* zero (**Figure 7-15b**). To keep the book moving, you have to do positive work on the book as you push it.

If there were no kinetic friction, you would have to push the book to get it going, push it again if you wanted to change its direction, and push it again to stop it. Getting it going and stopping it in each direction would require equal magnitude and opposite amounts of work (pushing in the direction of motion, and then opposite the direction

(a) Friction dissipated more energy along path 2 than along path 1. Since the "work" done (energy dissipated) by friction depends on the path, we conclude that friction is **not** a conservative force.

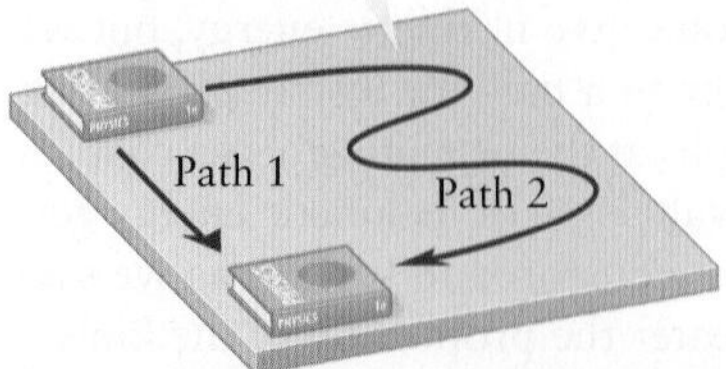

(b) The book makes a round trip that begins and ends at this point. As you push the book around the path, you decide where to push the book and where to stop and start it, and you have to use your own internal energy. So we can conclude that force exerted by a person is **not** a conservative force.

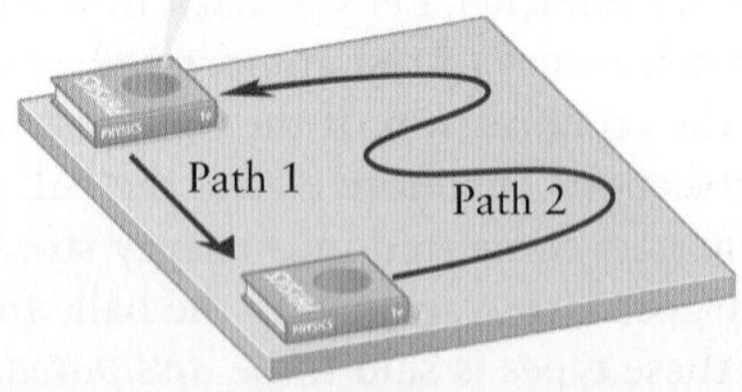

Figure 7-15 Nonconservative forces Because the amount of energy dissipated by kinetic friction depends on the path, it is a nonconservative force. Any force that does nonzero net work on a round trip or requires internal energy is nonconservative.

of motion), so the total amount of work you did to return the book to its initial position would be zero. This looks like it might satisfy our statement that "if a force exerted on an object does no net work during any closed loop then the force is said to be conservative," but it really points out something we need to be careful about when we see this common statement. As a system with internal energy (a human being), we decide where we push the book and when to stop it. We could have just as easily done the same zero net work on the book in the case with no friction but not returned it to its original position, either stopping it sooner or letting it go farther. So there is no special relationship between where the book is and the force you are exerting on it. As you do this zero net work in either case, you have to convert your own internal energy to do so, and not just count on your body returning to the shape it prefers!

Even for nonconservative forces, kinetic friction is weird. You may have noticed the quotes around "work" when we were talking about friction above. You can calculate Fd for a kinetic friction force, but sometimes it doesn't meet our definition of an energy transfer into or out of a system. When you push on an object or a system, the force you exert on the object times the displacement of the point you are pushing on tells you how much energy you transferred to the system. This is not the case for kinetic friction because the "point of contact" is not simply a point, but the entire contact area. As we discussed in Chapter 4, the two surfaces in contact stretch and break. What you end up with is some warming up of the object and some of the surface, and some damage (maybe microscopic) to both. So you cannot say how much of that energy is going into the system. You can say only that much energy is being removed from the potential and kinetic energy in the system and cannot be recovered into those types (the energy is dissipated). So for kinetic friction, while you can call it work you must remember that it is the friction force times the total distance traveled, because the kinetic friction force always opposes the direction of motion, no matter how you change that direction, and that you cannot account for exactly where it went, just that it is dissipated. We will look at this more in Chapter 8.

We saw in Section 5-6 that the force exerted by a human tendon is also nonconservative. The tendon exerts more force on its end while it is being stretched than when it is relaxing to its original length, so it does more negative work on the muscle attached to its end as it stretches than it does positive work as it relaxes. So the tendon does a nonzero (and negative) amount of work on this "round trip." To make the tendon go through a complete cycle, the muscle has to do a nonzero amount of *positive* work on the tendon, just like the positive work you must do to push the book around the path in Figure 7-15b.

There are many examples of potential energy in your environment. The spring in a mousetrap has the potential to do very destructive work on any mouse unlucky enough to release the trap. There are also many other sources of energy that drive the world around you. For now we will just consider these sources of internal energy and we will learn how to include them in our solutions in the next chapter. If you study physics further, you will learn more about them. For example, in Chapter 1, we saw Einstein's famous equation relating mass to energy, $E = mc^2$. The positively charged protons in a uranium nucleus exert an electric force on each other, pushing each other apart, but are prevented from doing so by the strong force between the constituents of the nucleus. If the nucleus is broken apart, the resultant pieces have less mass than the uranium nucleus; that missing mass is converted to energy. This is the process that provides the energy released in a nuclear reactor.

Quantifying Potential Energy

An object near Earth's surface gains kinetic energy when it is dropped. An example is the barbell in **Figure 7-16.** When held at rest above the weight lifter's head, the barbell has no kinetic energy. If the weight lifter should drop the barbell, however, it will fall to the ground gaining kinetic energy as it does. It can also leave a dent in the floor as it comes to rest on the floor, because it does work on the floor by exerting a large downward force on the floor over a very small distance. If we treat the barbell as an object, then Earth exerts an external force of gravity that does work on the barbell as it falls to the floor, increasing the barbell's kinetic energy. Because the force of gravity is conservative, we can instead think of a change in potential energy associated with this

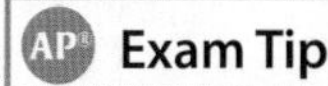

Exam Tip

Remember, a *force* is a way to describe an interaction. Only forces exerted by something external to a system may do work on (change the total energy of) a system. While it may seem handy to think of interactions inside a system in terms of forces, remember that internal interactions cannot change the total energy of the system, but only allow energy to be converted from one type into another. Always carefully define your system for any problem or pay attention to how a system is defined for you in the problem. If a system consists of only one object (rather than, say, the object and Earth), it can have only kinetic energy and no potential energy.

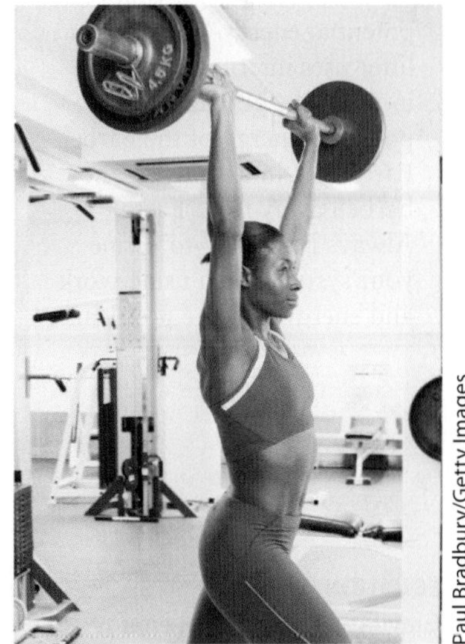

Paul Bradbury/Getty Images

Figure 7-16 Potential energy The barbell in this photo is at rest and so has zero kinetic energy. It would acquire kinetic energy if the weight lifter should drop it. How we describe that gain in kinetic energy depends on our definition of the system.

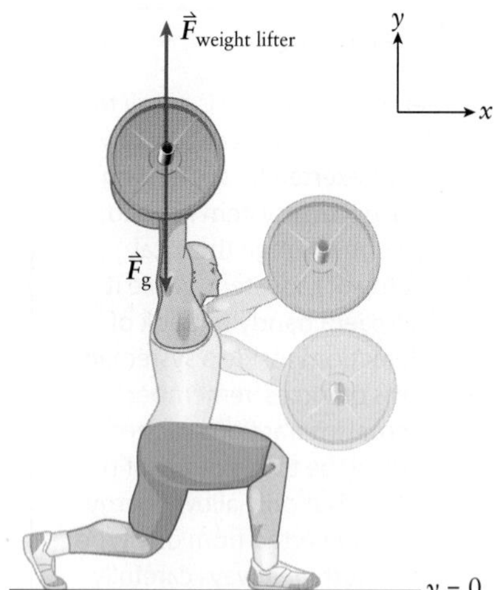

Figure 7-17 Adding potential energy The change in the kinetic energy of the barbell as it is raised by the weight lifter is zero. The work that the weight lifter does to lift the barbell goes into increasing the potential energy of the Earth–barbell system.

change in location of the barbell. Potential energy changes depend on changes in the configuration of the system. The shape of the barbell does not change, but the distance between the barbell and Earth's surface does change. If we choose our system to be the barbell and Earth, we can no longer use work done by the force of gravity, because gravity is no longer an external force! Instead we talk about the gravitational potential energy of the system. How we choose our system completely determines whether we use work or potential energy to describe the effects of conservative interactions.

Gravitational Potential Energy

There are two ways to quantify the amount of potential energy stored in the Earth–barbell system in Figure 7-16. First, we compare the potential energy due to a reversible change in the configuration of a system to the work that would be done if the object were isolated. If the system is just the barbell, and it has mass m and is initially a height h above the floor (we are going to consider the floor as $h = 0$), as it falls, Earth (because Earth is not in the system, the gravitational force on the barbell in its direction of motion is an external force) would do an amount of work $W_{grav} = F_g d \cos\theta = F_g h \cos 0° = F_g h = mgh$ on the barbell. From the work-energy theorem this is equal to the kinetic energy that the dropped barbell has just before it hits the floor. We're ignoring any effects of air resistance. We could instead say that the gravitational potential energy (symbol U_{grav}) of the Earth–barbell system before the barbell was dropped was $U_{grav} = mgh$, and that this potential energy was converted to kinetic energy as the barbell fell, so the change in gravitational potential energy is the negative of the change in kinetic energy $\Delta U_{grav} = -mgh$. Note that potential energy has units of joules, the same as work and kinetic energy.

This is one of the reasons we like to focus on changes in energy. Notice that if we had decided to call $h = 0$ the height from which the weight lifter dropped the barbell, then the final position of the barbell would be $-h$. The displacement would still be $h_f - h_i = 0 - h = -h$, and so the changes in energy would be exactly the same, as they must be.

A second way to quantify the amount of potential energy is to consider where the initial potential energy came from for the Earth–barbell system. To see this, consider what happens as a weight lifter *raises* the barbell from the floor to a height h (**Figure 7-17**). During the lifting, the barbell begins with zero kinetic energy (sitting on the floor) and ends up with zero kinetic energy (at rest above the weight lifter's head), and the weight lifter exerts an average force mg upward over the displacement h, to keep the barbell moving upward at constant speed. The net *change* in its kinetic energy is zero and the force of gravity does no work on the system, because it is internal to the system. Hence the positive work that the weight lifter did to raise the barbell must provide the potential energy mgh of the Earth–barbell system when the barbell is at height h (again assuming U_{grav} is zero at the floor). The potential energy stored in the configuration of a system because of a gravitational interaction (mgh near the surface of Earth) is called **gravitational potential energy**. The gravitational potential energy mgh would be the same if the barbell remained at rest but Earth as a whole were pushed down a distance h. The weight lifter converted internal energy (that snack eaten earlier) into the work done on the barbell. Any time energy comes from changing the configuration of chemicals in food, some of that internal energy goes into warming the weight lifter up. We will discuss this more in the next chapter.

WATCH OUT!

Objects can have only kinetic energy; it takes a system to have potential energy.

Never be sloppy and say, "The barbell *has* gravitational potential energy." The weight lifter uses internal energy to increase the gravitational potential energy of the barbell–Earth system by moving the barbell away from Earth. Always be careful to define your system when using work and energy to solve problems.

In general, if an object of mass m is at a vertical coordinate y above the surface of Earth, the gravitational potential energy of the Earth–object system is

EQUATION IN WORDS
Gravitational potential energy stored in an Earth–object system near the surface of Earth

$$U_{grav} = mgy \tag{7-14}$$

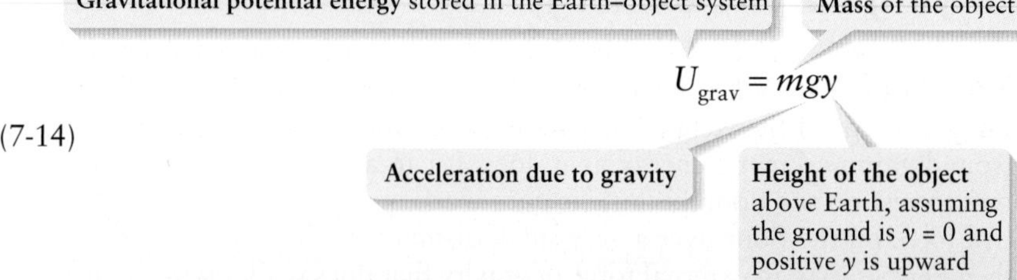

When the weight lifter in Figure 7-16 raises the barbell (so y increases), the gravitational potential energy of the Earth–barbell system increases; when she lowers or drops the barbell, y decreases, and so the gravitational potential energy of the system decreases.

Potential Energy for a Curved Path

Here's how the work-energy theorem describes what happens to a dropped isolated barbell (just the object, not the Earth–barbell system) in the absence of air resistance: As the barbell falls, the gravitational force does work on it, and this work goes into changing the barbell's kinetic energy.

Now let's generalize to the case in which the barbell follows a curved path (as would be the case if the barbell was lifted at an angle upward, rather than dropped). The barbell in **Figure 7-18** moves along a curve from an initial y coordinate y_i to a final coordinate y_f. As we did for the swinging spider in Example 7-6 (Section 7-5), we divide the path into a large number of short segments, each of which is small enough that we can treat it as a line. The displacement along such a segment is $\vec{d}$, with horizontal component Δx and vertical component Δy. The work done along each segment by the force of gravity $\vec{F}_g$ is exactly the same as it was in Example 7-6 (because the barbell is moving upward and the gravitational force is downward):

AP® Exam Tip

When solving problems, it is important to distinguish clearly between the potential energy of a system when it is in a particular configuration (for a certain value of y, the separation of Earth and an object, for instance) and the change in potential energy that comes from a change in that configuration.

$$W_{grav} = -mg\ \Delta y$$

The total work W_{grav} done by the force of gravity along the entire curved path is the sum of the ΔW_{grav} terms for each segment. The sum of all the Δy terms is the total change in the y coordinate, $y_f - y_i$, so

$$\begin{aligned} W_{grav} &= (-mg\ \Delta y_1) + (-mg\ \Delta y_2) + (-mg\ \Delta y_3) + \cdots \\ &= -mg(\Delta y_1 + \Delta y_2 + \Delta y_3 + \cdots) \\ &= -mg(y_f - y_i) = -mgy_f + mgy_i \end{aligned}$$

We can *reinterpret* this statement in terms of the gravitational potential energy as given by Equation 7-14, $U_{grav} = mgy$:

$$\text{Earth external: } W_{grav} = -mgy_f + mgy_i = -(mgy_f - mgy_i)$$
$$\text{Earth in system: } -(U_{grav,f} - U_{grav,i}) = \Delta U_{grav} \qquad (7\text{-}15)$$

In other words, *the work done by gravity on an object if we choose our system to be the isolated object is equal to the negative of the change in gravitational potential energy of the Earth–object system if we choose our system to be Earth and the object.* If an object descends, we can think of it as the downward gravitational force doing positive work on the object or as the gravitational potential energy of the Earth–object system decreasing (its change is negative). If an object rises, we can describe it as the downward gravitational force doing negative work on it or as the gravitational potential energy of the Earth–object system increasing (its change is positive). If an object begins and ends its motion at the same height, we can describe it as the gravitational force doing zero net work on it or as there being zero net change in the Earth–object system's gravitational potential energy. Remember, we have to define our system, so we must choose only one of these interpretations: either Earth is in (use potential energy) or out (use work) of the system. However we define our system, we will get the same results, as we must. We usually choose our system to make the problem we are interested in easier to solve.

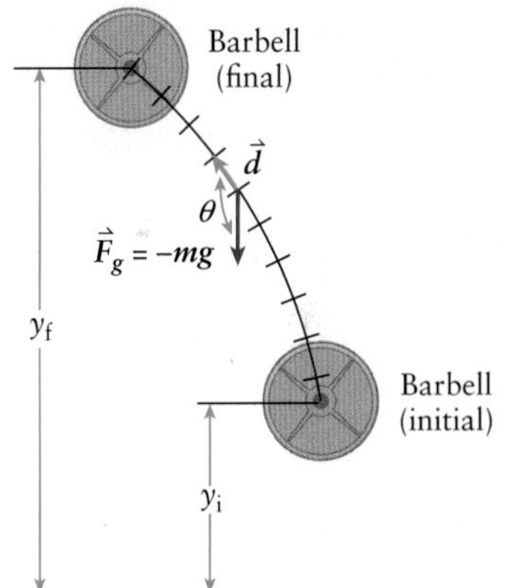

Figure 7-18 Raising a barbell along a curved path The work done on the barbell by the gravitational force $\vec{F}_g$ along a short segment $\vec{d}$ of this curved path is $W_{grav} = F_g d \cos\theta$. Adding up the work done along all such segments gives the net work done by the gravitational force between height y_i and height y_f.

WATCH OUT !

The choice of $y = 0$ for gravitational potential energy doesn't affect what happens.

The value of $U_{grav} = mgy$ depends on what height you choose to be $y = 0$. But Equation 7-15 shows that what matters is the *change* in gravitational potential energy, and that does *not* depend on your choice of $y = 0$. That's because the change in gravitational potential energy depends only on the *difference* between the initial and final heights, not the heights themselves: $\Delta U_{grav} = U_f - U_i = mgy_f - mgy_i = mg(y_f - y_i)$. Example 7-8 illustrates this important point.

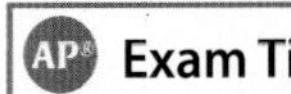

AP® Exam Tip

Be sure to understand how the W done by a force external to the system is related to the change in potential energy when the interaction instead is defined as internal to the system.

EXAMPLE 7-8 A Ski Jump

A skier of mass m starts at rest at the top of a ski jump ramp (**Figure 7-19**). The vertical distance from the top of the ramp to the lowest point is H, and the vertical distance from the lowest point to where the skier leaves the ramp is D. The first part of the ramp is at an angle θ from the horizontal, and the second part is at an angle ϕ. Derive an expression for the speed of the skier when she leaves the ramp in terms of the variables given in the problem and appropriate physical constants. Assume that her skis are well waxed, so that there is negligible friction between the skis and the ramp, and ignore air resistance.

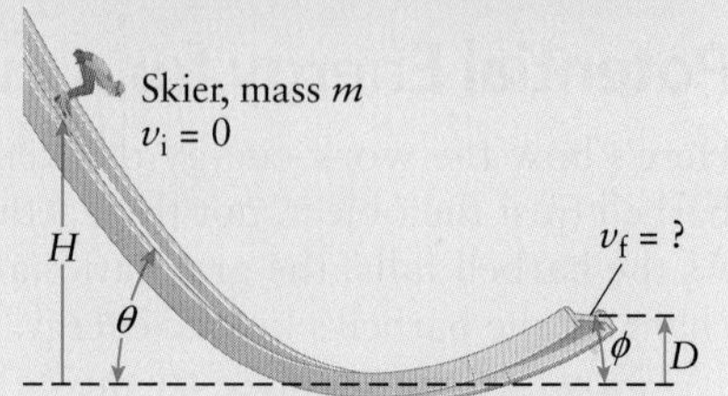

Figure 7-19 Flying off the ramp What is the skier's speed when she leaves the ramp, neglecting friction?

Set Up

If we neglect friction, the only forces that are exerted on the skier are the gravitational force and the normal force exerted by the ramp. The normal force does no work on the skier because it is always exerted perpendicular to the ramp and so is perpendicular to her direction of motion. We are not worried about the skier's own internal energy to keep herself warm, and so on; just the energy that affects her motion, and she just slides down the slope, so we can neglect her internal energy. The gravitational force does no work if we choose our system to be the Earth–skier system. So we can use Equation 7-12, with $W = 0$ and $\Delta E_{\text{internal}} = 0$ to calculate the change in the skier's kinetic energy and hence her final speed v_f.

Work-energy theorem:

$$W = \Delta K + \Delta U_{\text{grav}} + \Delta E_{\text{internal}} \quad (7\text{-}12)$$

Gravitational potential energy:

$$U_{\text{grav}} = mgy \quad (7\text{-}14)$$

Kinetic energy:

$$K = \frac{1}{2}mv^2 \quad (7\text{-}8)$$

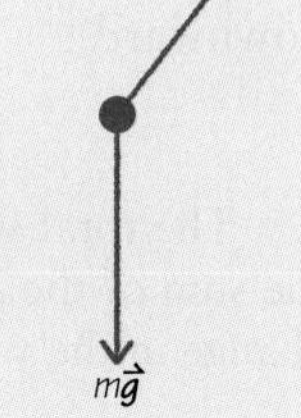

Solve

Let's take $y = 0$ to be at the low point of the ramp. The skier then begins at rest ($v_i = 0$) at $y_i = H$ and is at $y_f = D$ when she leaves the ramp. Use Equations 7–8, 7-12, and 7-14 to solve for v_f.

At starting point:

$$K_i = \frac{1}{2}mv_i^2 = 0$$

$$U_{\text{grav,i}} = mgy_i = mgH$$

At the point where the skier leaves the ramp:

$$K_f = \frac{1}{2}mv_f^2$$

$$U_{\text{grav,f}} = mgy_f = mgD$$

Use Equation 7-12:

$$W = \Delta K + \Delta U_{\text{grav}} + \Delta E_{\text{internal}}$$

$$W = 0 \text{ and } \Delta E_{\text{internal}} = 0$$

$$0 = \Delta K + \Delta U_{\text{grav}}$$

$$-\Delta U_{\text{grav}} = \Delta K, \text{ here}$$

$$\Delta U_{\text{grav}} = U_{\text{grav,f}} - U_{\text{grav,i}} = mgD - mgH$$

$$= mg(D - H) = -mg(H - D)$$

(negative, because $D < H$) and

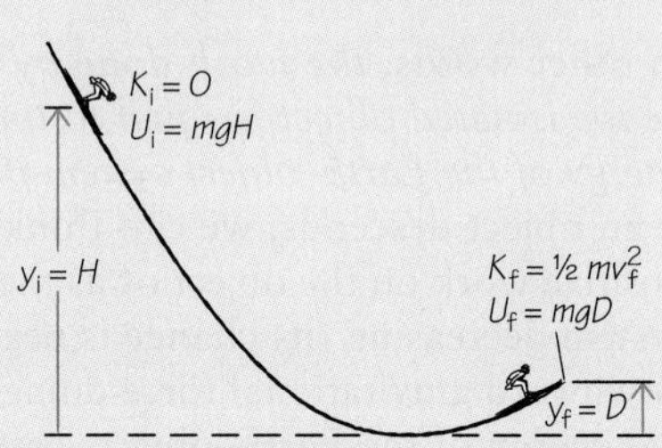

$$\Delta K = K_f - K_i$$
$$= \frac{1}{2}mv_f^2 - 0 = \frac{1}{2}mv_f^2$$

(positive, because the skier is moving faster at the end of the ski jump than at the beginning) So

$$+mg(H - D) = \frac{1}{2}mv_f^2$$
$$v_f^2 = 2g(H - D)$$
$$v_f = \sqrt{2g(H - D)}$$

Reflect

The answer does *not* involve the angles θ or ϕ, or any other aspect of the ramp's shape. All that matters is the difference between the skier's initial and final heights. If the ramp had a different shape, the final speed v_f would be exactly the same, as long as the difference in heights remained the same.

As we mentioned in the **Watch Out!** feature just before this example, your answer also shouldn't depend on our having chosen $y = 0$ to be at the low point of the ramp. For example, instead choose $y = 0$ to be where the skier leaves the ramp, and the result for v_f should be the same. If $y = 0$ is the skier's starting point, the end of the ramp is at $y = -(H - D)$ and you get the same ΔU_{grav}.

At starting point:

$$U_{grav,i} = mgy_i = 0$$

At the point where skier leaves the ramp:

$$U_{grav,f} = mgy_f = mg(H - D), \text{ so}$$
$$\Delta U_{grav} = U_{grav,f} - U_{grav,i}$$
$$= -mg(H - D) - 0$$
$$= -mg(H - D)$$

This is the same as with our previous choice of $y = 0$, so we'll find the same value of v_f.

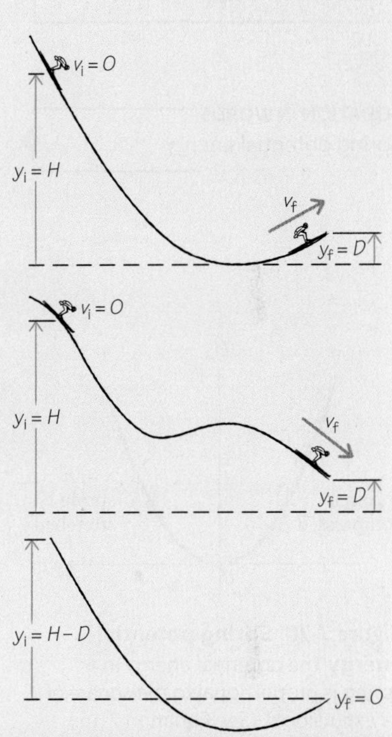

NOW WORK Problems 2 and 5 in The Takeaway 7-6.

Spring Potential Energy

Our definition of potential energy is that a change in such energy must depend only on the system's final and initial configurations. For instance, the gravitational potential energy cannot depend on the path taken between the initial and final positions of the object above Earth. Using gravitational potential energy, we saw how to quantify the potential energy by comparing the change in potential energy when Earth was in the system to the change in kinetic energy of an object due to work done by Earth when it is outside of the system. We also saw that the amount of work that we would have to do to lift an object between those two points was ΔU_{grav}, the change in the gravitational potential energy values for the Earth–object system as the object moved between the two points. The same is true for the work done by an ideal spring. We saw in Section 7-5 that the work you must do to stretch a spring from an extension x_i to an extension x_f is $\frac{1}{2}kx_f^2 - \frac{1}{2}kx_i^2$ (this is Equation 7-13, with 1 replaced by i and 2 replaced by f). Note that this also depends on only the initial and final extensions of the spring, not on the details of how you got the end of the spring from one position to the other. In this case, the spring is our system. If you fix one end of the spring, and pull on the other end, it stretches and if you hold it at a given stretch it has no kinetic

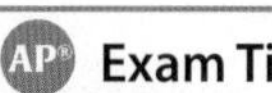

Exam Tip

The equation sheet on the AP® Physics 1 exam lists the spring potential energy as an extension *x* from its relaxed equilibrium, but does not list the change in potential energy. Understand why the change in potential energy when you are stretching a spring is as given in the equation to the right and cannot be written in terms of Δx, unless x_i is the equilibrium length of the spring, $x = 0$.

energy, so we can use the work-energy theorem to write the change in potential energy of an ideal spring in terms of the change in its length. We are assuming that it is an ideal spring, which means no energy is going into heating or deforming it and that its mass is negligible, so the only type of energy it has due to it being a system is potential energy, so we can rewrite Equation 7-12 as $W = \Delta K + \Delta U$. ΔK is zero, because it is at rest at both the initial and final lengths. W is the amount of work you do on the spring, $\frac{1}{2}kx_f^2 - \frac{1}{2}kx_i^2$. Substituting, $\frac{1}{2}kx_f^2 - \frac{1}{2}kx_i^2 = 0 + \Delta U_s$.

$$\Delta U_s = \tfrac{1}{2}kx_f^2 - \tfrac{1}{2}kx_i^2$$

We can rewrite this:

$$\Delta U_s = U_{s,f} - U_{s,i} = \tfrac{1}{2}kx_f^2 - \tfrac{1}{2}kx_i^2$$

Because $U_{s,f}$ is the value of U_s at x_f, if we set $x_i = 0$ we can just call x_f, x. In this case x represents the extension of the spring from its relaxed equilibrium length ($x_i = 0$) and the quantity U_s is the **spring potential energy:**

EQUATION IN WORDS
Spring potential energy

Spring potential energy of a stretched or compressed spring

Spring constant of the spring

$$U_s = \frac{1}{2}kx^2 \qquad (7\text{-}16)$$

Extension of the spring, when the equilibrium position of the end of the spring when it is relaxed is defined as $x = 0$ ($x > 0$ if spring is stretched, $x < 0$ if spring is compressed)

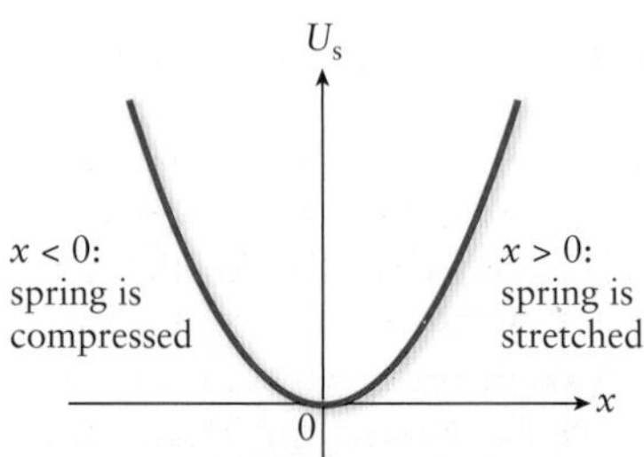

Figure 7-20 Spring potential energy The potential energy in a spring is proportional to the square of its extension *d*. (See Equation 7-16.)

The spring potential energy is zero if the spring is relaxed ($x = 0$) and positive if the spring is stretched ($x > 0$) or compressed ($x < 0$) (**Figure 7-20**). This says that we have to do work to either stretch the spring or compress it, and the work that we do goes into the spring potential energy.

While a human tendon is not an ideal spring, we can think of it as storing spring potential energy when it is stretched. When you are running as you move over your foot in contact with the ground, the Achilles tendon at the back of that leg stretches. The spring potential energy stored in that tendon is part of "springing" you back in the air, helping you to sustain your running pace.

Yes, you still get tired. Remember, the human tendon is not an ideal spring, so you had to put more energy into stretching it out from equilibrium than you get back when it relaxes back to equilibrium. Even though the approximation is not perfect, it still gives us insight into how such a complicated system, the human body, functions!

THE TAKEAWAY for Section 7-6

✔ The work done by a conservative force depends only on the initial and final positions of the object, not on the path the object followed from one position to the other. The gravitational force and the force exerted by an ideal spring are examples of conservative forces.

✔ Potential energy is used to describe conservative interactions inside a system; a conservative force is a way to describe conservative interactions with objects or systems outside the system of interest.

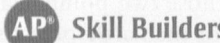

AP® Building Blocks

1. What does it mean when a force is referred to as conservative?

EX 7-8 2. Over 630 m in height, the Burj Khalifa is the world's tallest skyscraper. Calculate the change in gravitational potential energy of the coin–Earth system when a 1-dirham coin (6.1 g) is carried from ground level to the top of the Burj Khalifa. Calculate the speed of the coin just before it hits the ground if it is dropped from rest at the top of the skyscraper and air resistance is neglected. Based on the speed you calculated, do you feel it was a good idea to neglect air resistance?

3. A spring that is compressed by 12.5 cm stores 3.33 J of potential energy. Calculate the value of the spring constant.

AP® Skill Builders

4. A 350-kg box initially at rest is pulled 7.00 m up a 30.0° inclined plane by an external force, $F_{ext} = 5.00 \times 10^3$ N, that is exerted parallel to the plane. Assume friction is negligible.
   (a) Construct a representation of the scenario that includes the ramp, the inclination of the ramp, the external force exerted on the box, the displacement, the box, and the coordinate system to be used for the

analysis of the problem. Represent each characteristic with a label that will be used in an analysis.

(b) Construct a free-body diagram for the box.

(c) Construct a diagram representing the decomposition of the gravitational force into components parallel and perpendicular to the ramp. One of the angles in your vector diagram will be the same as the angle of inclination of the ramp. Mark this angle on your vector diagram. Add the displacement to this diagram and mark the angle, taken to be positive in the counterclockwise direction, between the displacement and the gravitational force, to your diagram.

(d) Express the components of the gravitational force along the ramp and perpendicular to the ramp in terms of F_g and θ.

(e) Express the work done on the box by the gravitational force in terms of θ. *Hint*: To be sure of your expression you can use the trigonometric identity

$$\cos(\alpha - \beta) = \cos(\alpha)\cos(\beta) + \sin(\alpha)\sin(\beta)$$

to express the work done on the box by the gravitational force.

(f) Apply the object model to the box and express the final kinetic energy of the box in terms of the work done on the box by the external force and the gravitational field.

(g) Consider the Earth–box system and apply energy conservation to express the final kinetic energy in terms of the work done by the external force and the change in gravitational potential energy of the Earth–box system.

(h) Compare the expressions from parts (f) and (g) and explain similarities and differences.

(i) Calculate the final speed of the box.

EX 7-8 5. A 12.0-kg block (M) is released from rest on an incline that makes an angle of 28.0° with the horizontal. Below the block is a spring that has a spring constant of 13,500 N/m. The figure shows the initial arrangement. The friction between the block and incline is negligible. The block momentarily stops when it compresses the spring by 5.50 cm. You will analyze the dynamics of the block with the goal of predicting the displacement of the block from its initial position at the instant that it momentarily stops while compressing the spring.

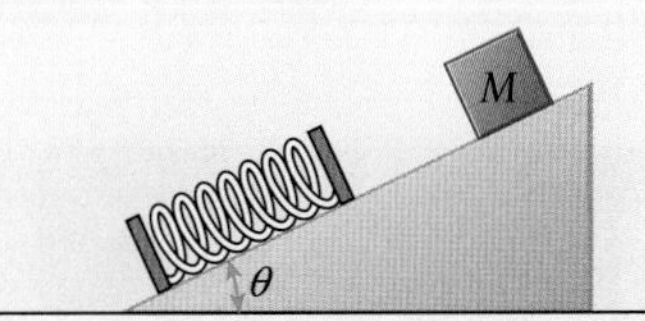

(a) Construct a diagram for your analysis of the scenario that includes initial (i) and final (f) states, displacement prior to collision with the spring, displacement after collision with the spring, and the angle of inclination of the ramp with labels for each variable to be used in your analysis.

(b) Justify your selection of the system to be analyzed.

(c) Express potential and kinetic energies at the initial and final states mathematically. If work is done to your system express the work mathematically.

(d) Apply the principle of energy conservation and evaluate the distance that the block moves down the incline from its release point to the stopping point.

(e) Analyze the dynamics of the block using an alternative choice of system to that used in your analysis in part (d).

6. Ben says to Jerry that the potential energy change of the Ben–Earth system as he climbs a ladder from the first rung to the fourth rung is smaller than the potential energy change of the same system as he climbs from the second rung to the fifth rung because gravitational potential energy is proportional to height and the height of the fifth rung is greater than the height of the fourth rung. In response, Jerry draws a diagram with two ladders with one in the basement and one on the first floor. Construct Jerry's diagram and annotate the diagram to support his evaluation of Ben's claim.

7. Carrie's group is analyzing the motion of a cart that moves up a ramp. Carrie claims that when the cart slows down the kinetic energy is negative. She justifies this claim by stating that "the work-energy theorem for an object says that when a force pushes an object the kinetic energy increases but now the force is pulling the object." Andrea disagrees and claims that the kinetic energy of the cart is never negative anywhere on the ramp but doesn't provide reasoning to support her claim.

(a) Treating the cart as an object upon which Earth exerts a gravitational force, justify Andrea's claim.

(b) Carrie responds by claiming that by treating the cart–Earth pair as an isolated system, no work is done and so according to the work-energy theorem the kinetic energy of the system cannot change. And because the total kinetic energy of the system is the sum of the kinetic energies of the components, for the sum to be zero one of the terms in this sum must be negative. Evaluate Carrie's explanation of why kinetic energy can be negative.

AP® Skills in Action

8. Several identical springs, with spring constant k, are attached end-to-end, as shown in the first diagram below. An external force is exerted on the first spring with the response shown in the second diagram. The springs are stretched as shown.

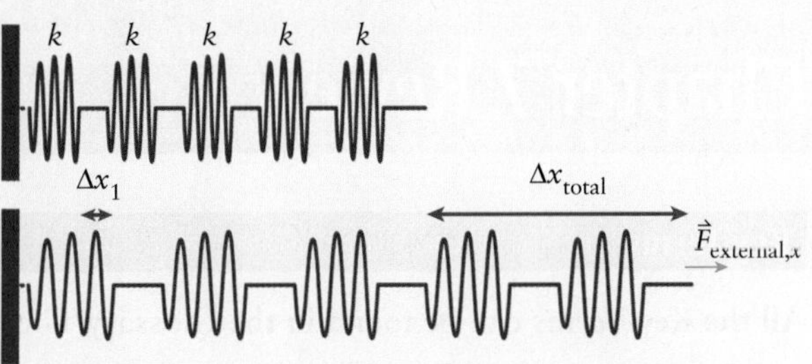

(a) Justify the claim that

$$\Delta x_{\text{total}} = n\frac{F_{\text{external},x}}{k}$$

(b) Explain why the work done by the external force on the springs attached in series is the sum of work done on each of the springs.

You are provided with several weights with known mass, a ruler, and several identical springs with unknown spring constant that can be connected end-to-end, as shown.

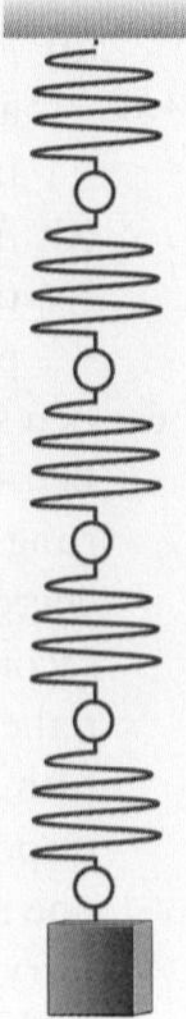

(c) Design procedures to collect and analyze data to test the claim in part (a). Include a careful description of methods to be used to measure length.

WHAT DID YOU LEARN?

Prep for the AP Exam

Chapter learning goals	Section(s)	Related example(s)	Relevant section review exercises
Explain what it means for a quantity to be conserved.	7-1 7-2	7-1	1
Describe what conditions must be met for work to be done on an object, for both positive and negative work.	7-2	7-1, 7-2	1, 3
Explain the relationship between work and kinetic energy for an object.	7-3	7-3	2, 3, 6
Explain why something modeled as an object can have only kinetic energy, and why a system can have other types of energy.	7-4	7-4	2, 3, 4
Describe how the work-energy theorem relates to conservation of energy for an object or a system.	7-5	7-6, 7-7	3, 5, 7
Explain the meaning of potential energy and how conservative interactions, such as those described by the gravitational and spring forces, give rise to potential energy.	7-6	7-8	2, 5
Recognize why the work-energy theorem applies even for curved paths and varying forces such as the spring force.	7-5	7-6, 7-7	1, 2, 3, 7

Chapter 7 Review

Key Terms

All the Key Terms can be found in the Glossary/Glosario on page G1 in the back of the book.

Chapter Summary

Topic	Equation or Figure	
Work done on an object by a constant force that points in the same direction as the displacement: If an object moves in a line while a constant force is exerted on the object in the same direction as the displacement, the work is equal to the magnitude of the force times the magnitude of the displacement.	**Work done on an object** by a constant force $\vec{F}$ exerted on the object in the **same direction** as the object's displacement $\vec{d}$ **Magnitude of the constant force** $\vec{F}$ $W = Fd$ **Magnitude of the displacement** $\vec{d}$	(7-1)
Work done on an object by a constant force exerted at an angle θ to the displacement: If an object moves in a line while a constant force is applied at some angle θ to the displacement, the work is equal to the magnitude of the force times the magnitude of the displacement multiplied by the cosine of the angle between the force and displacement.	**Work done on an object** by a constant force $\vec{F}$ that **points at an angle** θ to the object's displacement $\vec{d}$ **Magnitude of the constant force** $\vec{F}$ $W = Fd\cos\theta$ **Magnitude of the object's displacement** $\vec{d}$ **Angle** between the directions of $\vec{F}$ and $\vec{d}$	(7-2)
Relating speed, acceleration, and position for linear motion with constant acceleration: If an object moves in a line with a constant acceleration, its displacement can be related to the change in the square of its velocity.	**Speed** at position x_f of an object in linear motion with constant acceleration **Speed** at position x_i of the object $v_f^2 = v_i^2 + 2a_x\,(x_f - x_i)$ **Constant acceleration** of the object **Two positions** of the object	(7-4)
Calculating the work done on an object by the net force exerted on that object in linear motion: If an object moves in a line with a constant acceleration due to the net force exerted on the object, the work done on the object can be related to its displacement and the acceleration.	**Work done on an object** by the **net force** exerted on that object **Mass** of the object $W_{net} = ma_x(x_f - x_i)$ **Constant acceleration** of the object **Displacement** of the object	(7-6)
Kinetic energy: An object's kinetic energy is the energy associated with its motion. Because objects are defined as having no internal structure and all points on the object as moving in exactly the same way as its center of mass, this is the only type of energy an object can have.	**Kinetic energy** of an object **Mass** of the object $K = \frac{1}{2}mv^2$ **Speed** of the object	(7-8)

The work-energy theorem for an object: The work-energy theorem for an object states that the net work done as an object moves through a displacement—the sum of the work done on the object by individual forces exerted on the object—is equal to the change in the object's kinetic energy during that displacement. This theorem is valid whether the path is curved or straight and whether the forces are constant or varying.

Work done on an object by the **net force** exerted on that object

$$W_{\text{net}} = K_{\text{f}} - K_{\text{i}} \tag{7-9}$$

Kinetic energy of the object **after** the work is done on it

Kinetic energy of the object **before** the work is done on it

Work for a system: Work is the amount of energy transferred to the system by a force exerted on a point on the system. The net work on a system is equal to the sum of the work due to each external force exerted on the system, which for each force is the magnitude of the external force times the magnitude of the displacement of the point of contact of that force in the direction of the force.

Work is the amount of energy transferred to the system by a force exerted on a point on the system

$$W = Fd_{\text{contact}}\cos(\theta) \tag{7-11}$$

Angle between the directions of $\vec{F}$ and $\vec{d}_{\text{contact}}$

The magnitude of the force exerted on a point on the system

The magnitude of the displacement of the point of contact where the force is exerted on the system in the direction of the force.

Work-energy theorem: Work is the amount of energy transferred to the system by external forces exerted on the system. If the system is not an object (the point of contact can move a different distance than the center of mass of the system) this work can go into more than one type of energy. If there is no other source of energy transfer, such as heating, then the total change in energy of the system is equal to the work done on the system. If the point of contact moves a different distance than the center of mass of the system then energy can be converted from one type to another within the system, even if the net work on the system is zero.

Work is the amount of energy transferred to the system by external forces

The change in energy of the system (assuming work is the only source of energy transfer)

The change in all other types of energy inside the system due to its configuration or the internal motion of its constituent parts

$$W = \Delta E = \Delta K + \Delta U + \Delta E_{\text{internal}} \tag{7-12}$$

The change in kinetic energy of the system, equal to the product of the magnitude of the force F exerted on the system and the magnitude of the displacement of the center of mass of the system in the direction of the force.

The change in potential energy of the system due to reversible changes in its configuration

Work can be found from the area under a graph of the force exerted along the direction of motion versus the position of the point at which that force is exerted: For a constant force, the area is just the same as the multiplication given in the definition. When the force is not constant this still works. This area rule is a mathematical fact that works for any product, when you plot the multiplicand and multiplier on the axes of a graph; work is just one example.

Work done by a constant force along direction of motion = Fd = area under graph of force versus position

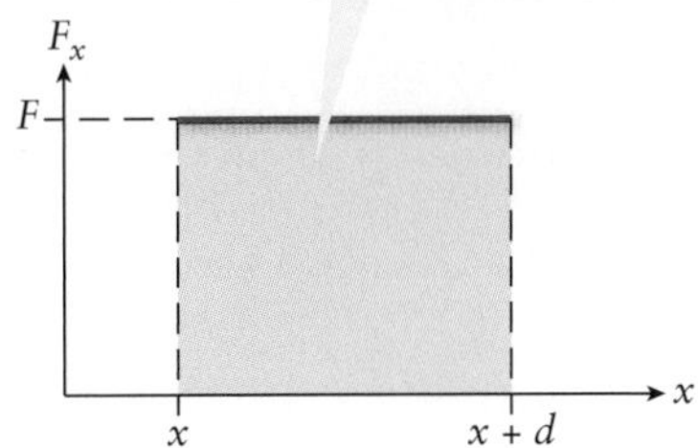

(Figure 7-13)

The spring force and work: The spring force is not constant, so the work needed to stretch or compress a spring is not simply the magnitude of the exerted force multiplied by the displacement, but can be found by the area under the graph of force versus position.

Work that must be done on a spring to stretch it from $x = x_1$ to $x = x_2$

Spring constant of the spring (a measure of its stiffness)

$$W = \frac{1}{2}kx_2^2 - \frac{1}{2}kx_1^2 \quad (7\text{-}13)$$

x_1 = **initial stretch** of the spring
x_2 = **final stretch** of the spring
x_1 and x_2 are measured from $x = 0$, the location of the end of the spring when it is relaxed

Area under curve = work you do to stretch spring from x_1 to x_2

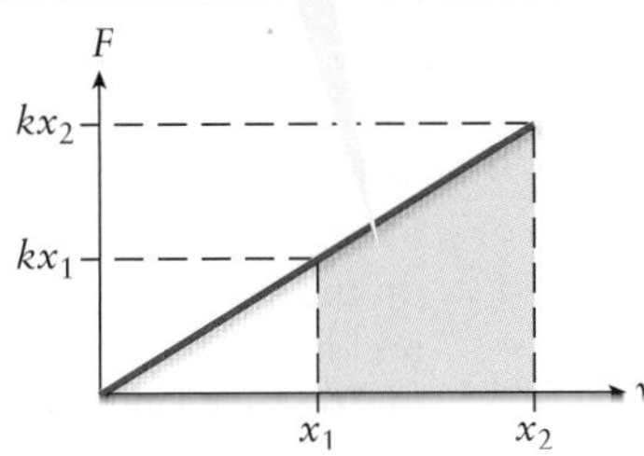

(Figure 7-14)

Potential energy: Unlike kinetic energy, which is associated with a property of a single object (its speed), potential energy is associated with the configuration of a system. Gravitational potential energy near the surface of Earth increases with separation of the object and Earth, measured in terms of the height of the object relative to the surface of Earth. The potential energy of a spring increases with the stretch or compression of the spring. Only conservative interactions are associated with a potential energy.

Gravitational potential energy stored in the Earth–object system

Mass of the object

$$U_{grav} = mgy \quad (7\text{-}14)$$

Acceleration due to gravity

Height of the object above Earth, assuming the ground is $y = 0$ and positive y is upward

Spring potential energy of a stretched or compressed spring

Spring constant of the spring

$$U_s = \frac{1}{2}kx^2 \quad (7\text{-}16)$$

Extension of the spring, when the equilibrium position of the end of the spring when it is relaxed is defined as $x = 0$ ($x > 0$ if spring is stretched, $x < 0$ if spring is compressed)

How you define your system matters: Only forces exerted by an object or a system external to a system can do work on that system. Conservative interactions can be described by work if exerted by an object or a system external to the system, or as potential energy if internal. These two descriptions will always be the same magnitude but opposite signs.

Earth external: $W_{grav} = -mgy_f + mgy_i = -(mgy_f - mgy_i)$

Earth in system: $-(U_{grav,f} - U_{grav,i}) = \Delta U_{grav}$ (7-15)

Chapter 7 Review Problems

1. A satellite orbits Earth in a circular path at a high altitude. Explain why the gravitational force does zero work on the satellite.

2. An assistant for the football team carries a 30.0-kg cooler of water from the top row of the stadium, which is 20.0 m above the field level, down to the bench area on the field.
   (a) If the speed of the cooler is constant throughout the trip, calculate the work done by the assistant on the cooler of water.
   (b) How much work is done by the force of gravity on the cooler of water?

3. A 1250-kg car moves at 20.0 m/s. How much work must be done on the car to increase its speed to 30.0 m/s?

4. Bicycling to the top of a hill is much harder than coasting down. Does the person–bike–Earth system have more potential energy at the top of the hill or at the bottom? Explain your answer.

5. An 8500-metric-ton freight train is out of control and moving at 90 km/h on level track. How much work must a superhero do on the train to bring it to a halt?

6. A 375-g toy boat is floating in a swimming pool. One child exerts a force of 1.85 N in the direction 30° north of east on the boat. Another child exerts a force on the boat of 0.75 N south. The average drag force of the water on the boat is approximately one-half the net force (due to the children) in the direction of motion, and is in the opposite direction of this net force. If the boat starts from rest, how fast is it moving after it travels a distance of 5.0 m?

7. A catcher in a baseball game stops a pitched ball originally moving at 44.0 m/s at the moment it first came in contact with the catcher's glove. After contacting the glove, the ball traveled an additional 12.5 cm before coming to a complete stop. The mass of the ball is 0.145 kg. What is the average force that the glove imparts to the ball during the catch? Comment on the force exerted on the catcher's hand during the catch.

8. The graph below shows the net force exerted on a low-friction cart of mass 500 grams as it moves along a horizontal track. Assume that the speed of the cart at position $x = 0$ is zero.
   (a) Calculate the velocity of the cart after 1.0 m.
   (b) Calculate the velocity of the cart after 3.0 m.

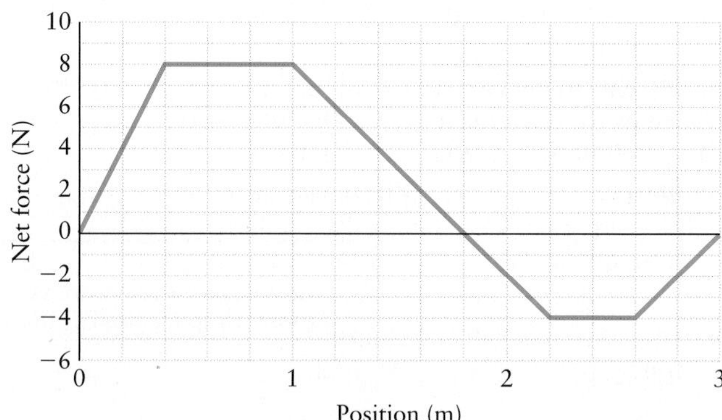

9. You are pulling your younger sister on a sled up a hill inclined at 10° with the horizontal. You exert a force on the sled that makes an angle of 20° with the hill, and has a magnitude of 120 N. You walk a distance of 20.0 m along the hill. The mass of the sister–sled system is 18.0 kg. Assume the friction between the hill and the sled is negligible.
   (a) Calculate the work done on the sister–sled system by your pulling force as you move 20.0 m along the hill.
   (b) Calculate the work done on the sister–sled system by gravity during this motion.
   (c) Calculate the work done by the normal force on the sister–sled system during this motion.
   (d) Calculate the total work done on the sister–sled system during this motion.

10. A 5.00-kg object is attached to one end of a horizontal spring that has a negligible mass and a spring constant of 250 N/m. The other end of the spring is fixed to a wall. The spring is compressed by 10.0 cm from its equilibrium position and released from rest.
   (a) What is the speed of the object when it is 8.00 cm from equilibrium?

(b) What is the speed when the object is 5.00 cm from equilibrium?

(c) What is the speed when the object is at the equilibrium position?

11. Wei drags a piece of driftwood for 910 m along an irregular path. If Wei ends 750 m from where he started and exerted a force of constant magnitude 625 N, at all times directed parallel to his path on the piece of driftwood, how much work did he do on the driftwood?

12. Earth orbits the Sun at a radius of about 1.5×10^8 km. At this distance the force of gravity on Earth due to the Sun is 3.6×10^{22} N. Assuming Earth's orbit to be perfectly circular, how much work does the Sun's gravity do on Earth in one year?

13. An ideal spring of constant 8.0 N/m is fixed at one end to the wall, and at the other end to a low-friction cart of mass 750 g that is initially at rest on a horizontal table. A student moves the cart a distance of 0.50 m, stretching the spring and holds it at rest at that position.

(a) Sketch a graph of the force exerted on the cart by the spring as a function of the distance the cart is moved, from 0 to 0.50 m.

(b) Use your graph to calculate the work done on the cart by the spring, and the potential energy stored in the spring when it is stretched 0.50 m.

(c) The cart is released from rest at 0.50 m and the spring pulls it back to the equilibrium position. Calculate the speed of the cart as it passes through the equilibrium position.

14. Draw and describe a physical situation to which the following equation could apply:

$$\frac{1}{2}\left(20\frac{\text{N}}{\text{m}}\right)(0.10\text{ m})^2 - \frac{1}{2}\left(20\frac{\text{N}}{\text{m}}\right)(0)^2 = \frac{1}{2}(5\text{ kg})(v_f)^2 - \frac{1}{2}(5\text{ kg})(0)^2$$

15. A 40.0-kg boy steps on a skateboard and pushes off from the top of a hill. What is the change in the potential energy of the Earth–boy system as the boy glides down to the bottom of the hill, 4.35 m below the starting level?

16. How much additional potential energy is stored in a spring that has a spring constant of 15.5 N/m if the spring is stretched so that the displacement of the end of the spring moves from 10.0 cm to 15.0 cm, measured from its relaxed equilibrium position?

17. An ideal spring with spring constant $k = 2.00 \times 10^2$ N/m is oriented vertically with one end on the ground.

(a) What distance must the spring compress for a 2.00-kg object placed on its upper end to reach equilibrium?

(b) By how much does the potential energy stored in the spring increase during the compression? (Later in this book, we will see why this answer makes sense.)

18. A 0.145-kg baseball rebounds off of a wall. The rebound speed is one-third of the original speed. By what percent does the kinetic energy of the baseball change in the collision with the wall? Where does the energy go?

19. Gravel-filled runaway truck lanes are designed to stop trucks that have lost their brakes on mountain grades. Such a lane is horizontal or uphill and about 35.0 m long. We can think of the ground as exerting a friction force on the truck. If a truck enters the gravel lane with a speed of 55.0 mph (24.6 m/s), use the work-energy theorem to find the minimum coefficient of kinetic friction between the truck and the lane to be able to stop the truck if the lane is horizontal.

20. Three clowns try to move a 3.00×10^2-kg crate 12.0 m to the right across a smooth, low-friction floor. Moe pushes to the right with a force of 5.00×10^2 N, Larry pushes to the left with 3.00×10^2 N, and Curly pushes straight down with 6.00×10^2 N. Calculate the work done by each of the clowns.

21. The 2010 Americans with Disabilities Act established a standard for wheelchair ramp inclinations. Assuming the ramp is a right triangle with the inclined surface as its hypotenuse, the horizontal base of the ramp must have a length 12 times greater than the vertical height of the ramp. Justify the selection of data that guided engineers in the creation of this standard by describing the data they must have sought and the forces exerted on the wheelchair.

22. A child slides down a slide that is 4 m in length and has an inclination, relative to the horizontal, of 28°. Describe in words how you will determine the change in potential energy of the child–Earth system as she descends the slide. In your description include your choice of the origin for the definition of initial and final positions, the algebraic expression for potential energy, and any assumptions about friction forces exerted on the child.

23. A boy ties a string to a horizontal pipe and attaches a ball with a weight of 6 N to the other end of the string so that the ball is 1 m from the horizontal pipe. He holds the ball so that the string is taut and level with the pipe. He then releases the ball. The complete path of the ball is a nearly semicircular arc followed by a return swing that nearly reaches the boy, who then grabs the ball. Use the ball as your system in answering the following questions.

(a) What is the work done on the ball by the gravitational force and the tension force in the string during the complete path of the ball?
(A) 31.4 J (B) 10 J (C) 6 J (D) 0 J

(b) If the ball is allowed to continue to swing until it comes to rest, what is the work done by the gravitational force?
(A) 31.4 J (B) 10 J (C) 6 J (D) 0 J

(c) If the ball is allowed to continue to swing until it comes to rest, what is the work done by the tension force?
(A) 31.4 J (B) 10 J (C) 6 J (D) 0 J

24. A common classroom demonstration involves holding a bowling ball attached by a rope to the ceiling close

to your face and releasing it from rest, as shown in part (a) of the figure.

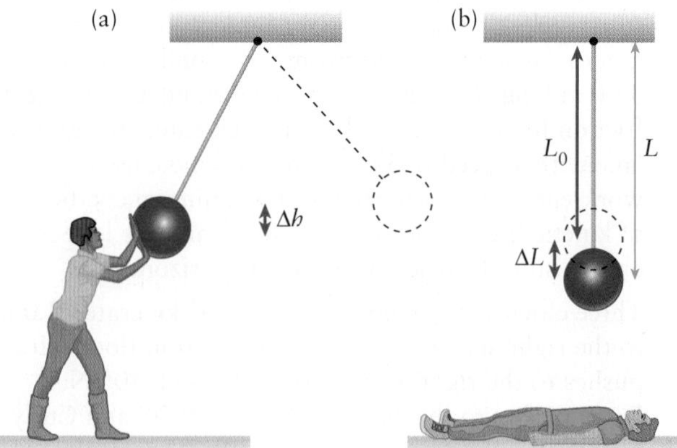

Note: Figure not drawn to scale.

(a) Predict what will happen to the relative height, Δh, to which the ball rises if the ball is pushed rather than released from rest in terms of the work done, $W_{\text{on ball}}$, by the push.

A teacher tires of this demonstration and attaches the bowling ball to an ideal spring with constant k. As shown in part (b) of the figure, the equilibrium length of the spring hanging vertically when the ball is attached is L_0. Lying on the floor with the bowling ball suspended at rest above his chest he pulls down slightly on the ball a distance ΔL, and releases it.

(b) Take the ball to be initially at rest above the teacher's chest with a spring length $L_0 + \Delta L$. What are the forces exerted on the ball at this time?

(c) Using the conservation of energy, describe the motion of the ball after the teacher releases it.

(d) Which demonstration is safer, do you think? Justify your answer in terms of the subsequent motion of the ball in each case, (a) and (c).

25. As you pedal a bike up a hill, predict

(a) the sign of the work done by the gravitational force on you,

(b) the sign of the work done by your foot on the bike pedal as it pushes the pedal down, and

(c) the sign of the work done by the pedal on your foot as the pedal moves downward.

Hint: Remember that predictions are always supported by reasoning and that coordinate systems are chosen for convenience.

26. A 1.0-kg object is at rest when a force is exerted on the object. The positions at several times are shown in the figure.

$t = 0$ s, $t = 0.5$ s, $t = 1$ s, $t = 1.5$ s, $t = 2$ s

0.0 0.2 0.4 0.6 0.8 x (m)

(a) Use mathematical reasoning to express displacement as a function of time for the object by evaluating the acceleration.

(b) Calculate the numerical value of the force exerted on the object.

(c) Express work done by the force on the object and kinetic energy of the object as functions of time.

The external force is switched off and the object is returned to the original initial position. At the instant the force is switched on again a second object with a mass of 2.0 kg enters along a parallel path with an initial position $x = 0$ and an initial speed of $v = 1$ m/s. Interactions between the two objects are negligible.

At $t = 1$ s the external force is switched off.

(d) Express the displacements of both objects as a function of time.

(e) Express the kinetic energy of the two-object system as a function of time.

(f) Explain why the kinetic energy of the two-object system is constant for $t > 1$ s.

Prep for the AP® Exam

AP® Group Work

Directions: The following problem is designed to be done as group work in class.

As part of a lab experiment that uses a horizontal air track, Allison pushes an air-track cart of mass m up against a spring to compress the spring with spring constant k by an amount Δx from its equilibrium length. The air track has negligible friction. When Allison lets go, the spring launches the cart.

Allison's lab partner Gwen predicts that because the force increases linearly as the spring is compressed, the square of the velocity when the spring is released should increase in proportion to the compressed length. So the velocity should increase in proportion to the square root of the displacement. To support her claim she refers to the kinematic equation

$$v_f^2 = v_i^2 + 2a\Delta x$$

and she draws the graph of force versus spring compression as shown in the following graph. Gwen argues that the energy added to the spring is the area under the force-displacement graph, and points out that the two shaded triangles have the same area. So a second compression by a distance d beyond a first compression by a distance d doubles the area and therefore doubles the energy. The energy is proportional to the velocity squared. So, Gwen argues, the square of the velocity should be proportional to the compression.

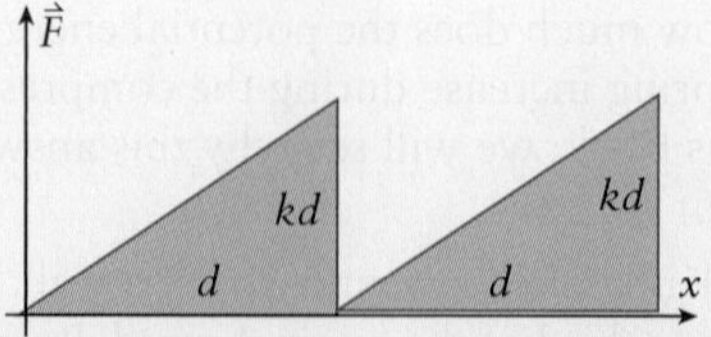

(a) Refine Gwen's symbolic and graphical representations of the relationships among force, spring compression, and speed to critically evaluate Gwen's explanation.

(b) Allison has persuaded Gwen that the speed is proportional to the compression length but she can't remember how the speed depends on m and k. Construct the representation using dimensional analysis to confirm that the dependence is $\sqrt{\frac{k}{m}}$.

To test the prediction, Allison and Gwen determined the mass of the cart (m = 0.25 kg) and placed it on the air track. A photogate timer was placed in the middle of the air track. A piece of cardboard with a length of 1 cm was mounted on top of the cart to interrupt the light beam of the photogate. The following data were obtained where Δx is the spring compression from the equilibrium position and Δt is the time that elapses as the cart breaks the photogate beam.

Δx (cm)	Δt (ms)
2	40.0
4	16.7
6	10.9
8	9.6

(c) Construct from these data a graph that tests the prediction in part (b).

(d) Analyze the data to obtain the value of the spring constant in units of N/m.

(e) Evaluate the evidence provided by the data in terms of both Allison's model and Gwen's original model and evaluate the reliability of the data.

AP® PRACTICE PROBLEMS

Multiple-Choice Questions

Directions: The following questions have a single correct answer.

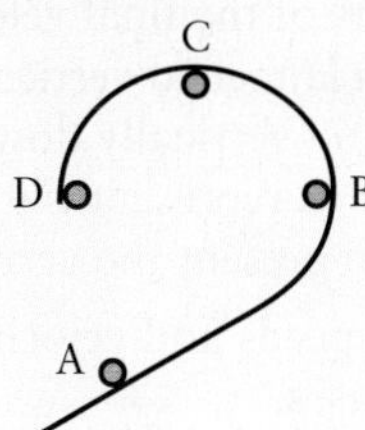

1. A ball is projected up an incline and rolls around a track as shown above. At which labeled point is the work being done on the ball by the gravitational force as the ball moves through that point zero?

(A) A (B) B (C) C (D) D

Force (N)
3.0
2.0
1.0
0
0 10 20 30
Displacement (m)

2. A single force is exerted on an object in the x direction. Shown is a graph of that force and the displacement of the object in that direction. What amount of work is done for the 30 m shown?

(A) 30 J (B) 45 J (C) 50 J (D) 60 J

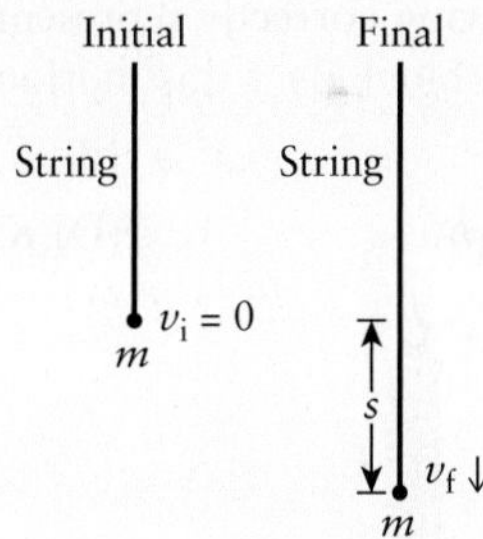

3. An object of mass m hanging from the end of a light string is initially at rest. It is lowered by the string a distance s and reaches a speed v_f. How much work was done on the object by the string?

(A) $\frac{1}{2}mv_f^2$

(B) $-\frac{1}{2}mv_f^2$

(C) $\frac{1}{2}mv_f^2 - mgs$

(D) $-\frac{1}{2}mv_f^2 - mgs$

(Continued)

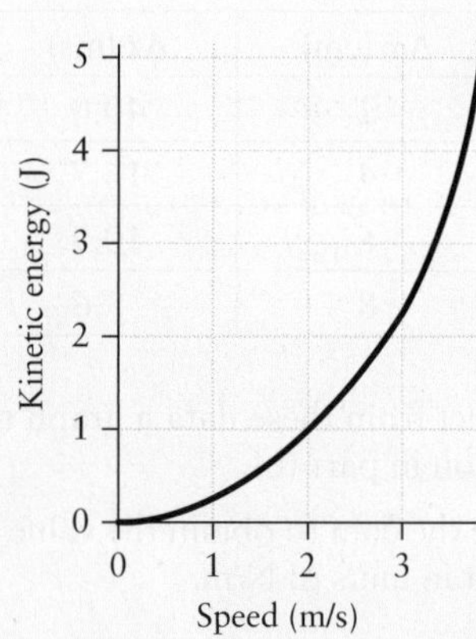

4. A single constant force is exerted on an object. The graph describes the object's energy. How much work was done to increase the speed of the object from 1.0 to 3.0 m/s?

(A) 1.5 J (B) 2.0 J (C) 4.0 J (D) 4.5 J

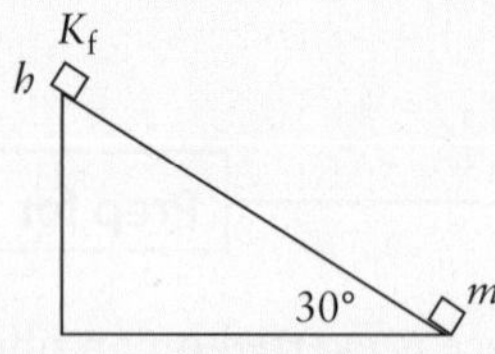

5. A block of mass m is initially at rest at the bottom of a plane inclined at 30° with the horizontal, as shown in the figure. The block is pulled up the plane. At a height h it has kinetic energy K_f. The coefficient of friction with the incline is μ. Which of the following correctly represents the net work done on the block over this motion?

(A) K_f
(B) $K_f + mgh$
(C) $K_f - mgh$
(D) $K_f + \mu mgh$

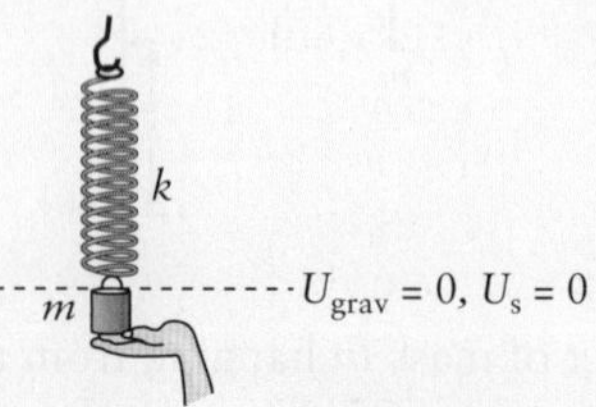

6. An object of mass m is attached to a spring hanging from the ceiling as shown. It is initially supported so the spring is unstretched. The object is released and falls. Which of the following correctly describes the kinetic energy of the object when it has fallen a distance h below the starting point?

(A) $K = mgh - \frac{1}{2}kh^2$

(B) $K = mgh + \frac{1}{2}kh^2$

(C) $K = -mgh + \frac{1}{2}kh^2$

(D) $K = -mgh - \frac{1}{2}kh^2$

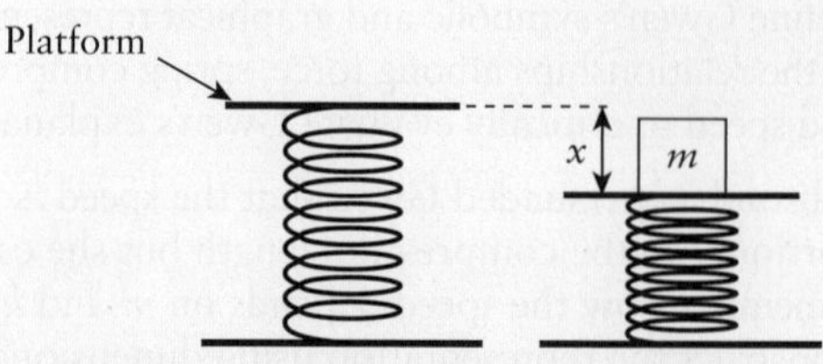

7. A small platform is attached to the top end of a spring of spring constant k. The spring is oriented vertically, as shown in the figure. A block of mass m is placed on the platform, and the spring is compressed a distance x. The spring is then released, and the block travels upward a distance H above its starting point. The spring is then replaced with one of spring constant $4k$, and the procedure is repeated with the same compression, x, as before. To what distance above its starting point will the block now travel?

(A) $H/4$ (B) H (C) $2H$ (D) $4H$

8. A device is used to launch objects from the edge of a platform 5.0 meters above the ground. To determine the speed with which the device launches the objects, you set up a motion detector to measure the final speed with which the objects strike the ground. The device can be set to launch the objects vertically upward, vertically downward, or horizontally from the same initial position, and with the same magnitude of initial velocity. How do the components of the final velocities of the object compare when launched vertically upward, v_U, horizontally, v_H, and vertically downward, v_D? We use the subscript x to represent the horizontal component and y to represent the vertical component.

(A) The final speeds and velocities are the same in all three cases.

(B) $v_{Hy} = v_{Dy} = v_{Uy}$; $v_{Dx} = v_{Ux} < v_{Hx}$

(C) $v_{Hy} < v_{Dy} = v_{Uy}$; $v_{Dx} = v_{Ux} < v_{Hx}$

(D) $v_{Hy} < v_{Dy} < v_{Uy}$; $v_{Dx} = v_{Ux} < v_{Hx}$

9. Various boxes are being pushed along a horizontal surface by a constant, external force exerted on them horizontally, as shown in the figure. The magnitude of the force and the mass of the block are given in each case. The blocks and the surfaces are the same materials in each case, and friction is not negligible. Which of the following is the correct ranking of the cases in terms of the kinetic energy of the block after being pushed a distance d?

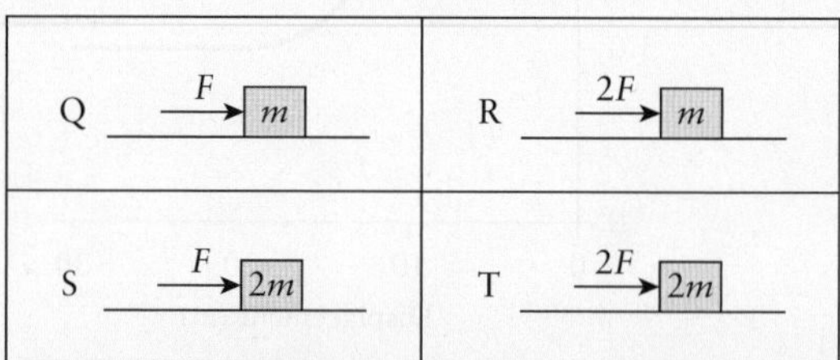

(A) R > (Q = T) > S
(B) (R = T) > (Q = S)
(C) S > (Q = T) > R
(D) R > T > Q > S

Free-Response Questions

1. Consider a tennis ball of mass m that has been thrown upward from ground level with an initial speed v_0 and then rises to a maximum height H above ground level before falling back down to the ground. Air resistance can be considered negligible. In analyzing this situation, we can apply energy principles to at least two different systems: the system that includes just the ball, and the ball–Earth system.

   (a) i. Write a sentence that describes the energy conversions within and the energy transfers into and out of the system consisting of just the ball as the ball rises to H.

   ii. Write a sentence that describes the energy conversions within and the energy transfers into and out of the system consisting of the ball and Earth as the ball rises to H.

   (b) Apply the work-energy theorem to the system consisting of just the ball to derive an expression for the maximum height H in terms of the initial speed, v_0.

   (c) Sketch a graph of kinetic energy of the ball as a function of its height above the ground from the time it leaves the ground until the time it reaches its maximum height.

   (d) Another tennis ball is thrown vertically upward with a speed v_2 and it reaches a height $h_2 = 2H$. Calculate the ratio v_2/v_0.

   The tennis ball is replaced by a softball that has approximately three times the mass of the tennis ball ($M = 3m$). Another student claims that because the kinetic energy depends on the mass of the ball, the softball has more kinetic energy if it leaves the ground with the same speed. Thus it will reach a maximum height that is three times the height that the tennis ball reached.

   (e) Do you agree with this student or not? Justify your choice by reasoning with the work-energy theorem.

2. Students are to conduct a series of experiments in their physics classroom with two blocks, A and B. For the first experiment, some students decide to use a launcher to give both blocks the same initial velocity, v_0, and launch them across a table. The coefficient of kinetic friction between the table and each block is μ_k. The blocks are made of the same material, but the mass of block A (m_A) is twice the mass of block B (m_B).

   (a) i. Student 1 says block A will slide farther along the table before coming to rest than block B will. What reasoning would support this prediction? Explain your answer using work and energy concepts and without using equations.

   ii. Student 2 says block B will slide farther along the table than block A will. What reasoning would support this prediction? Explain your answer using work and energy concepts and without using equations.

   (b) What forces are exerted (and by what other objects are they exerted) on block A as it slides to rest? Describe and justify their relative magnitudes and directions and how they compare to the forces on block B as it slides to rest.

   (c) Another student in your class derives a mathematical expression, $\Delta x = \frac{1}{2}v_0^2/\mu g$, for the distance traveled by block A after it leaves the launcher and slides to rest. Does the expression support the reasoning you used in part (a) to support either student 1 or student 2, or both? Explain briefly, again using work and energy concepts.

   (d) In a second experiment, students set up a single launcher, in which a block is placed at the end of a compressed spring. The block is then released from rest and loses contact with the spring when the spring reaches its equilibrium length. Explain why compressing the spring farther might increase the distance the block slides before coming to rest. Include a discussion of how the friction force affects the increased slide distance.

8 Application of Conservation Principles to Energy, Work, and Power

engel.ac/Shutterstock

YOU WILL LEARN TO:

- Identify which kinds of problems are best solved with energy conservation and the steps to follow in solving these problems.
- Describe how changing the way you identify a system changes the description of conservation of energy (but not the results).
- Identify the types of energy involved in an interaction between objects and systems.
- Describe what power is and its relationship to work and energy.
- Describe the general expression for gravitational potential energy and how to relate it to the expression used near Earth's surface.

8-1 Total energy is always conserved, but it is constant only for a closed, isolated system

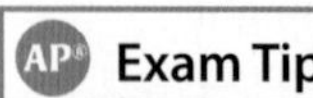

Exam Tip

Recall that when defining the gravitational potential energy at a height y, it is defined relative to where you have defined $y = 0$, which is where $U = 0$. The AP® Physics 1 equation sheet instead has $\Delta U = mg\Delta y$, which does not require the definition of where the potential energy equals zero.

NEED TO REVIEW?

Turn to the **Glossary** in the back of the book for definitions of bolded Key Terms.

Take a pencil in your hand and toss it upward. Consider its motion after it has left your hand. In this scenario, we can define the system we are examining as the Earth–pencil system. As the pencil ascends, it loses speed so its kinetic energy $K = \frac{1}{2}mv^2$ decreases. At the same time the pencil gains height so that the gravitational potential energy of the Earth–pencil system $U_{\text{grav}} = mgy$ increases until the pencil comes to a stop and all the energy of the system is gravitational potential energy.

As the pencil falls downward, its kinetic energy increases and the gravitational potential energy of the system decreases as the pencil loses height. This way of thinking about the pencil's up-and-down motion suggests that energy is converted from one type (kinetic) into a different type (gravitational potential) as the pencil rises and is converted back as the pencil descends.

We learned that mechanics is the branch of physics concerned with motion. Similarly, energy related to motion is called **mechanical energy**. It is easy to see that kinetic energy fits this definition since it depends on speed. Even though a system can have potential energy when nothing is moving (consider when the pencil is at the top of the flight), potential energy is also considered a type of mechanical energy because all the potential energy in a system can become kinetic energy again (such as when the pencil descends). Another way to say this is that mechanical energy of a system is the sum of the kinetic energy of the objects in the system and the energy stored in the system due to the configuration of (the position of the objects within) the system.

Since you are not in the system, when you toss the pencil, the upward force you exert on the pencil as you push it upward provides the positive work that gives the pencil the kinetic energy it has when it leaves your hand. Work requires motion, so is considered a **mechanical energy transfer.**

Any system where there are no transfers of energy of any type into or out of the system (such as the Earth–pencil system's up-and-down motion while the pencil is in flight) is what we defined in Chapter 7 to be a *closed, isolated system*. While you were throwing the pencil upward, there was a transfer of energy to the Earth–pencil system, so the system was not closed and isolated. You would have to start your constant-energy calculation after the pencil left your hand. We will return to this example in the next section.

 **Exam Tip**

Interactions between the Earth–pencil system and your hand result in changes in the energy of the system. You always need to define your system carefully as it determines how you describe the changes in energy.

In the real world, truly closed, isolated systems are not common, but are often an excellent first approximation. They allow you to start thinking about how to solve a problem, and can give you a good approximate answer. As you conduct laboratory investigations you will often make modifications to an experiment to try to make the system as closed and isolated as possible, in order to better test predictions. When we discussed Equation 7-12, the work-energy theorem $W = \Delta K + \Delta U + \Delta E_{\text{internal}}$, we can see it simplifies in a closed, isolated system. Energy is always conserved, but in an open system (one that is not closed or isolated) the amount transferred into or out of the system equals the sum of the changes of the energy inside the system. So, in an open system the amount of energy in the system is not constant; rather, energy is being transferred in or out by the work done on the system. In a closed, isolated system, $W = 0$, so the total energy is constant.

One important note to remember: For a closed, isolated system that does not involve just mechanical energy, mechanical energy *can* be converted into internal energy and warm things up. Total energy stays constant; it just doesn't all remain in a form that is mechanical. Typically, when energy is converted into a type that is not mechanical, we say that energy has been *dissipated*, whether that energy is transferred out of the system or stays in the system. If, however, your system involves only mechanical energy (all the interactions can be described by potential energy), mechanical energy stays constant, just converting back and forth between potential and kinetic energy. Sometimes texts will discuss mechanical energy as being more *useful* because all that energy remains available to do work. Practically, though, when it is cold outside, warmth seems like a pretty useful type of energy!

WATCH OUT !

A closed, isolated system means the total amount of energy stays constant, not that all the energy stays mechanical.

In a closed, isolated system, no energy can enter or leave the system. This does not mean that the mechanical energy in the system cannot be converted into other types of energy. For example, if you threw a very bouncy ball into an absolutely insulating large box (meaning that no energy entered or exited) and closed the lid, the ball would bounce around, stretch and compress, and warm up. The air around it would also warm up. Eventually the ball would come to rest, and all the kinetic energy it originally had would transfer into increasing the temperature of the ball, the inner wall of the box, and the air, which would all be at the same temperature at that point. If there had been no air in the box, the bouncing would have gone on a bit longer, and the ball and inner wall of the box would end up just a little warmer, since all the energy would be in the temperature of the ball and box wall. No energy had to leave the box, no work was done on the ball—energy just changed type.

 **Exam Tip**

If anything in the system can change shape irreversibly, or convert internal energy, the total energy of the system must still stay constant. But mechanical energy can be converted into other types of energy, and other types of energy can be converted into mechanical energy although never completely. Some energy is always dissipated when there is a conversion between mechanical and nonmechanical energy.

In closed, isolated systems made up of objects experiencing only conservative interactions, the *sum* of kinetic energy and potential energy—a sum that we call the **total mechanical energy** of the system—keeps the same value, it stays *constant*. This is a special case of the conservation of energy. When the system's kinetic energy K increases, its potential energy U decreases, and when K decreases, U increases—just as in the up-and-down motion of the Earth–pencil system. This is a direct result of the law of conservation of energy that tells us the amount of energy in a system can change only as a result of transfers of energy into or out of the system. No transfers means the

 **Exam Tip**

Examples of conservative interactions that are illustrated on the AP® Physics 1 equation sheet are the gravitational and spring forces.

total stays constant. We will start by looking at systems with just potential and kinetic energy, but this is true for any closed, isolated system; we would just have to keep track of each of the other types of energy in the system. In the next section we'll see how to express these ideas in equation form, building from our general equation for conservation of energy, the work-energy theorem from simple to more complex situations.

THE TAKEAWAY for Section 8-1

✔ Energy is conserved in any system. Energy transferred into or out of a system (when the system is not closed and isolated) can change the kinetic, potential, and internal energies of the system.

✔ For a closed, isolated system, conservation of energy means total energy stays constant, since no energy can enter or leave such a system.

✔ If the only interactions inside a closed, isolated system are conservative, the system's total mechanical energy will remain constant.

Prep for the AP® Exam

AP® Building Blocks

1. Enrique is skeptical of the claim that energy is always conserved. He draws a picture of a block with an initial speed sliding over the surface of a table and coming to a stop.

To support his claim that energy is not always conserved, he argues that: (1) there is no change in gravitational potential energy in the Earth–block system, (2) the change in the total energy of the system is the sum of the change in the kinetic energy plus the change in the potential energy, and (3) the kinetic energy has changed. This shows that energy is not conserved. Identify and explain the flaw in Enrique's argument.

AP® Skill Builder

2. Three balls are thrown off a tall building with the same speed but in different directions. Ball A is thrown in the horizontal direction, ball B starts out at 45° above the horizontal, and ball C begins its flight at 45° below the horizontal. Which of the following both correctly ranks the speed of each ball just before it strikes the ground and best justifies that ranking? Ignore any effects due to air resistance.
(A) B ≥ A ≥ C because the increase in gravitational potential energy as ball B rises is converted to additional kinetic energy as the ball falls, while the direction in which ball A is thrown does not cause an increase in gravitational potential energy and the direction for ball C causes gravitational potential energy to be lost.
(B) B ≥ A ≥ C because the time of flight of each ball increases in this order and the longer a ball is in the air the more kinetic energy it acquires.
(C) C ≥ A ≥ B because the kinetic energy in the downward direction is greatest for ball C, zero for ball A, and negative for ball B.
(D) A = B = C because in each case, the ball–Earth system has the same initial kinetic and potential energies, and the same final potential energy so energy conservation requires that they have the same final kinetic energy.

AP® Skills in Action

3. A 1-kg rock is dropped from a cliff 100 m above the sea. The rock strikes the sea and continues to fall through the water until coming to rest on the sea floor. Neglect air resistance but consider drag force as the rock falls through the water. Which of the following best describes the rock–Earth system and its energy as the rock is falling?
(A) The system is closed and isolated, and all internal forces are conservative, so energy is conserved and mechanical energy is constant.
(B) The system is closed and isolated until the rock strikes the sea surface, so mechanical energy is constant, but as the rock falls through the water the system is no longer isolated and energy is not conserved.
(C) The system is closed and isolated with a constant mechanical energy until the rock strikes the water, when a nonconservative force is exerted on the rock, so mechanical energy is not constant although total energy is conserved.
(D) The system is closed and isolated with a constant mechanical energy until the rock strikes the water, when a nonconservative force is exerted on the rock, so energy is transferred out of the system and total energy is no longer conserved.

8-2 Choosing systems and considering multiple interactions, including nonconservative ones

Suppose that as an object moves from an initial position to a final position, the only force that does work on the object is the gravitational force exerted by Earth. An example is the skier on a ski jump in Example 7-8 (Section 7-6). In addition to the gravitational force on the skier, there is a normal force exerted by the ramp, but it does no work on the skier because this force is always perpendicular to the skier's path.

If we instead include Earth in our system, then Earth is no longer external and we can no longer consider the work done by the gravitational force because an interaction must be external to the system in order to transfer energy into or out of the system. We can treat the skier in this example as an object, because we are neglecting anything the skier does, except slide down the ramp. When we consider this Earth–skier system, the work-energy theorem, $W = \Delta K + \Delta U + \Delta E_{\text{internal}}$, simplifies since both W and $\Delta E_{\text{internal}}$ are zero (no external are forces exerted on the system in the direction of motion and there are no nonconservative internal interactions). The potential energy of the system is gravitational potential energy stored in the separation of the skier and Earth. Substituting this information into the work-energy theorem, we find the change in the object's kinetic energy is equal to the negative of the change in the potential energy of the Earth–object system:

$$\Delta K = -\Delta U_{\text{grav}} \tag{8-1}$$

This equation is a mathematical expression of the constant mechanical energy we discussed in Section 8.1. The change in kinetic energy equals the final value minus the initial value, and likewise for the gravitational potential energy. Using this expansion, we can rewrite Equation 8-1 as

$$K_{\text{f}} - K_{\text{i}} = -(U_{\text{grav,f}} - U_{\text{grav,i}}) = -U_{\text{grav,f}} + U_{\text{grav,i}}$$

Let's rearrange this equation so that all the terms involving the initial situation are on the left-hand side of the equal sign and all the terms involving the final situation are on the right-hand side. Since energy is conserved, the sum of the initial energies and the sum of the final energies is each equal to the total energy, E. We get

$$K_{\text{i}} + U_{\text{grav,i}} = K_{\text{f}} + U_{\text{grav,f}} = E \tag{8-2}$$

If the system includes Earth and the object, there are no nonconservative interactions in the system, so no external force does work on the object as it moves. Its speed v and kinetic energy $K = \frac{1}{2}mv^2$ can change, and its height y and the gravitational potential energy of the Earth–object system $U_{\text{grav}} = mgy$ can change. But the *sum* of K and U_{grav} has the same value at the end of the motion as at the beginning.

The work-energy theorem includes all types of potential energy, so we know we can include all types of potential energy in Equation 8-2 as well. Our results from Section 7-6 show that the work done by *any* conservative force can be written as the negative of the change in the associated potential energy when you redefine the system to include the object or system causing the force. We also found change in potential energy by looking at the work that needed to be done on the system to increase its potential energy, such as for a spring. We did this in developing Equation 7-16, for the case of the work done to stretch or compress an ideal spring: $W_{\text{on spring}} = \Delta \frac{1}{2}kx^2$, where k is the spring constant and x is the extension or compression of the spring. We know that the force the spring exerts on you as you extend the spring is equal and opposite to the force you must exert on the spring to extend it, so $W_{\text{by spring}} = -\Delta \frac{1}{2}kx^2$. This is the negative of the change we would find if we were discussing potential energy instead of work, just as for gravity. This is true for any conservative force (you will learn about others if you continue your study of physics). So if a number of conservative forces ($\vec{F}_{\text{A}}, \vec{F}_{\text{B}}, \vec{F}_{\text{C}},\ldots$) do work on an object as it moves, and the potential energies associated with each of these forces are ($U_{\text{A}}, U_{\text{B}}, U_{\text{C}},\ldots$), then the total work done by each of these forces, when the source of the force is considered external to the system, can be used to calculate the potential energy when the source of the force is instead included as an interaction internal to the system (just as we did for gravity or springs).

This means the quantity U_{grav} in Equation 8-2 can be replaced by the *total* potential energy, which is just the sum of the individual potential energies $U_{\text{A}}, U_{\text{B}}, U_{\text{C}},\ldots$ So we can generalize Equation 8-2 for the total potential energy due to all conservative interactions within the system:

Values of the **kinetic energy** K of an object at two points (i and f) during its motion

$$K_{\text{i}} + U_{\text{i}} = K_{\text{f}} + U_{\text{f}} \tag{8-3}$$

Values of the **potential energy** of the system U at the same two points (i and f)

EQUATION IN WORDS
Mechanical energy is constant if there are only conservative interactions in the system and no external forces do work on the system.

You can see that this is a more general form of Equation 8-2, which refers to the case in which only one conservative interaction—the gravitational interaction—exists in the system. Note that we can take the initial point i and the final point f to be *any* two points during the motion, as long as only conservative interactions exist.

The sum of kinetic energy K and potential energy U is the total mechanical energy of the system. Equation 8-3 tells us that if a system is defined to include the sources of the conservative forces (the examples we have worked with so far are Earth's gravity and springs), mechanical energy is constant; that is, it maintains the same value during the motion. You can now see the origin of the term *conservative* for forces like the gravitational force and the spring force: If only interactions of this kind are present, the system can be defined so that we need to use only energy to solve problems (instead of dealing with vector forces and work). To summarize, energy is always conserved and in closed, isolated systems, constant. When only conservative interactions are present, the energy remains all mechanical; it just slops back and forth between kinetic and potential.

It is often easier to consider changes in energy rather than totals. If we do the same substitution for all forms of potential energy that we did to get from Equation 8-2 to 8-3, we can rewrite Equation 8-1 as

EQUATION IN WORDS
Constant mechanical energy—alternative version

Change in the kinetic energy K of an object during its motion

Change in the potential energy U of a system containing the object during the object's motion

(8-4) $$\Delta K + \Delta U = 0$$

Equation 8-4, which is a restatement of the work-energy theorem for the special case of no work and no internal energy, refers to the change in kinetic energy ΔK of an object and the change in potential energy ΔU of the system containing that object and the sources of the conservative interactions during the motion. Any increase in kinetic energy comes with an equal decrease in potential energy, and vice versa. You can get to any of these equations from the work-energy theorem if you understand what the terms mean.

Exam Tip

Be sure to differentiate between the potential energy of the system or the kinetic energy of the object at a particular instant, and changes in potential or kinetic energy between two instants. Note that if mechanical energy is conserved, $\Delta K = -\Delta U$ always, but $K_i = U_f$ is not generally true.

WATCH OUT !

Total mechanical energy is a property of a system, not a single object.

We saw in Section 7-6 that gravitational potential energy is a property of a *system* of an object and Earth. So the total mechanical energy for a baseball in flight when you include gravitational potential energy is a property of the ball–Earth system. Likewise, if an object is attached to a horizontal spring and allowed to oscillate back and forth, the total mechanical energy is a shared property of the object (which has kinetic energy) and the spring (which has spring potential energy). Whenever you think about total mechanical energy, you should always be able to identify the system to which that mechanical energy belongs.

In Section 8-3 we'll use the idea of conservation of energy to solve a number of physics problems. The easiest ones are when total mechanical energy is constant, but those are not the only problems we will be able to solve. In many situations, total mechanical energy is *not* constant, even though energy is always conserved. The system might not be closed and isolated. As an example, at the beginning of the previous section we asked you to toss a pencil into the air. During the toss you had to do *positive* work on the pencil: You exerted an upward force on the pencil as you pushed it upward. As a result the total mechanical energy of the pencil increased by the amount of mechanical energy you transferred into the system by doing work on the pencil. Another way that the mechanical energy of a system can change, and change even in the case of the system being closed and isolated, is if there are interactions within the system that could be described by *nonconservative* forces. With the number of variables to consider, we will find it useful to develop a diagram representation of conservation of energy to help us keep track of all the energy types. In the remainder of this section we will develop one such representation and work through our pencil tossing example in a number of ways to explore the differences caused by choice of system.

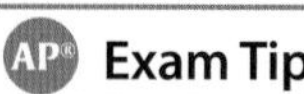
Exam Tip

If you define your system as a single object, it can only have kinetic energy. Any other object outside of the system, such as Earth or a spring, could do work on your system to change its kinetic energy, but there cannot be potential energy because there is no way to change the configuration of a single object.

Using Conservation of Energy Can Simplify Problems; Work-Energy Bar Diagrams Help

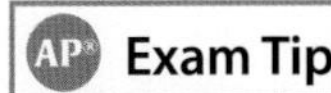

Exam Tip

Since there is no one accepted format for work-energy bar charts, you will not be asked to present them on the AP Physics 1 exam. However, students report much greater success in solving problems when they use some form of these diagrams. Your teacher may have a favorite method that you will be expected to use in class. If not, the method described here is an effective alternative.

In solving Newton's second law problems, we found free-body diagrams to be extremely valuable in keeping track of what our object was, and what the forces exerted on it were, so that we could write equations that would allow us to solve the given problem. Work-energy bar diagram is a type of representation that serves a similar purpose in solving problems using conservation of energy. However, unlike free-body diagrams, there is no one common form of work-energy bar diagrams. All forms allow us to carefully track the types of energy, making it much easier to come up with an equation that you can use to solve the problem. They always require you to identify what is in and out of your system and so all the energy types you need to consider and what can do work on the system. We will develop a basic form of work-energy bar diagrams that we will use to track energy through several examples, starting with a relatively simple system. We will conclude with the pencil toss we discussed in the previous section, including internal energy, which shows the potential complications.

Our first step, as in any approach to a physics problem, is to identify the system. Once the system is identified we can identify any types of energy in the system or external forces that can do work on the system. We are actually drawing a bar diagram that represents the energies in the work-energy theorem. We can do this for different times, tracking how energy changes in our system. Most commonly, we will use our diagram to write the work-energy theorem for the situation, so we will have before and after diagrams, including one bar for each type of energy that we need to describe in both the before and after diagrams. We break each bar into vertical segments. The exact amount of energy we represent in each bar is not that important, but we need to be able to compare the totals. If an external force does work on the system, we need to add a bar to represent that between our initial and final cases so that we can use the diagram to generate a complete equation.

Let's start out with a situation where we don't have any work. Consider a skier moving down a smooth, snow-covered ramp, so we can ignore friction. To not have work, the skier and Earth must both be in the system. The potential energy of the Earth–skier system decreases as the skier gets closer to Earth's surface. Since there is no external force, this decrease in potential energy of the system becomes an increase in kinetic energy of the skier—at each stage of the motion the total energy E is constant. Since there are no nonconservative forces, E is all mechanical energy. While the exact height of each bar is not important, the total height of all the bars in each diagram must be equal when there is no work. Diagrams showing this constant mechanical energy are shown for four configurations of the Earth–skier system in **Figure 8-1.**

The diagrams should carefully account for the changes in energy between types. Even if we are just estimating these changes, our diagrams show good approximations, and would be useful in checking numerical answers because they would ensure we had the right types of energy, and that the changes we calculated were the right sign. We can tell when some types of energy are zero, including when there is no motion (kinetic energy), when a spring is relaxed (spring potential energy), or when an object is at the height at which we defined the zero of gravitational potential energy. The example of tossing a pencil we opened this chapter with is represented in **Figure 8-2** for the pencil–Earth system. The total energy is not the same for each bar diagram because the total mechanical energy is increased by the amount of energy you transfer into the system by doing work on the pencil. You then do *negative* work on the pencil when you catch it because you exert an upward force on it as the pencil moves downward. If you track the energy of the system only after the pencil leaves your hand, the diagrams in Figure 8-2 are complete, since the Earth–pencil system has an almost constant amount of mechanical energy (the energy lost warming up the air by the system is very small).

As Figure 8-2 shows, the total mechanical energy increases during the throw and decreases during the catch. These changes in energy in the system are equal to the amount of work you did on the pencil, and therefore the system. So, to show the whole motion, we have to

The kinetic energy K and the gravitational potential energy U_{grav} both change, but their sum E remains the same when there are no external or dissipative forces.

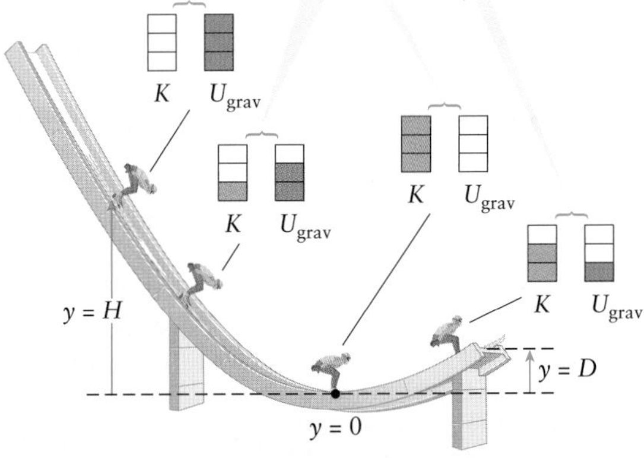

Figure 8-1 **Total mechanical energy is constant** As the skier moves on a ramp where friction can be neglected, the skier's kinetic energy K and the Earth–skier system's gravitational potential energy U_{grav} both change, but their sum $E = K + U_{grav}$—the total mechanical energy—always has the same value.

Because you do work on the pencil, the total mechanical energy E—the sum of kinetic energy K and gravitational energy U_{grav}— of the pencil–Earth system is not constant throughout this process.

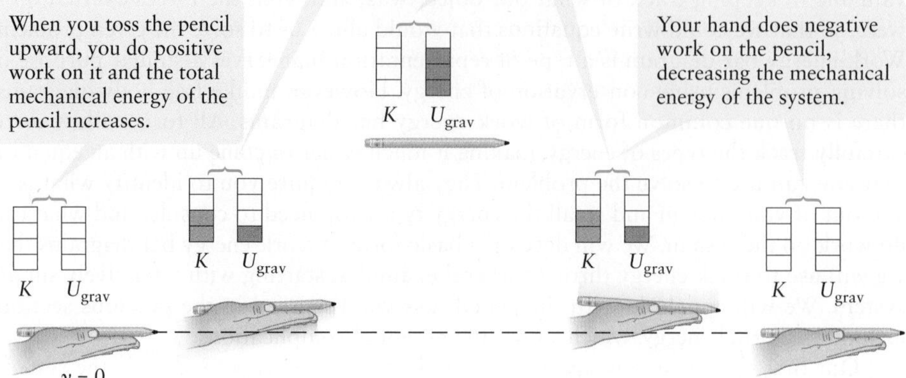

Figure 8-2 Tossing and catching a pencil The total mechanical energy of the pencil–Earth system increases as you toss the pencil upward, remains constant while the pencil is in flight, then decreases as you catch it.

account for the work done. We will also include a single separate column to represent the work done on the system. We will know we have it right when the size of the bars representing the initial configuration of our system and the work together are equal to the total of the bars representing the final configuration of our system. Since work can be positive or negative, we will sometimes put a zero at the top of the column and measure the negative work downward. Work exists in our work-energy bar diagrams only when there is an external interaction, and we should always label the source. Let's look at Figure 8-2 again. There are five stages to the motion depicted by the bar diagrams in Figure 8-2. These diagrams let us set up the mathematical statement of the work-energy theorem for each stage.

The first stage of motion for tossing the pencil is the throw. To create a work-energy bar diagram for this first stage, where there is as yet no mechanical energy, we add a column representing the work done by the external force exerted on the system (that is, by *you*) between two individual bar diagrams representing the initial energy and the final energy. We treat these two bar diagrams as being the initial and final states for this process of you adding work (you both lift the pencil and speed it up as you do work on it). Because the system only has mechanical energy after it leaves your hand, the total mechanical energy it has (the pencil–Earth potential energy plus the pencil's kinetic energy) must equal the work you did:

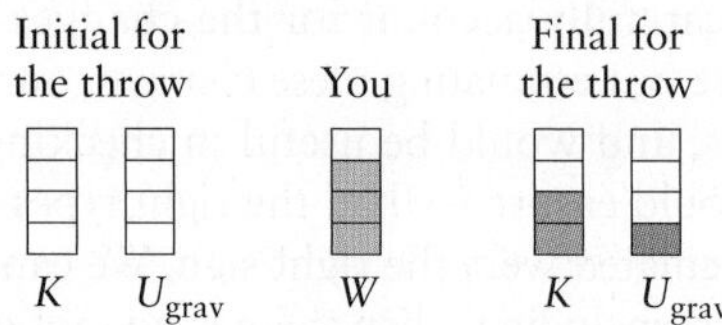

The next stage of the motion is the ascent and top of the flight of the pencil. The top of the flight is called out separately because of the change in motion there.

In work-energy bar diagram for the top of the flight, the initial state shown in the individual bar diagram is the final state from the throw. The final state, represented in the second individual bar diagram, is the top of the flight, where the pencil is momentarily at rest, so all the energy is the gravitational potential energy of the Earth–pencil system. The system has only mechanical energy after it leaves your hand, and there is no work done (external force exerted on the system = none), so the total mechanical energy must stay the same. This means that the work column tying the two individual bar diagrams into a single work-energy bar diagram will be empty.

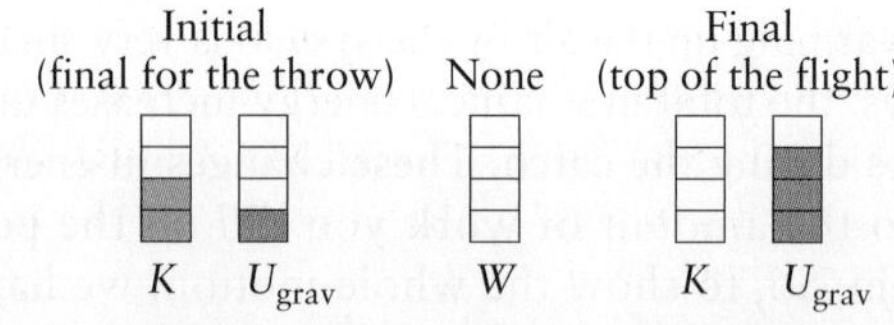

The next stage of motion is the pencil's descent. The changes in types of energy of the Earth–pencil system as the pencil begins its descent are the opposite of those as it rose. The system has only mechanical energy, and there is no external force so no work done, so the total mechanical energy must stay the same. The initial state here is our final state at the top of the flight. The final state is the fourth bar diagram from Figure 8-2, where the pencil is moving downward. If we choose this final position to be at the same height as when we finished the throw, the gravitational potential energy stored in the Earth–pencil system must be the same.

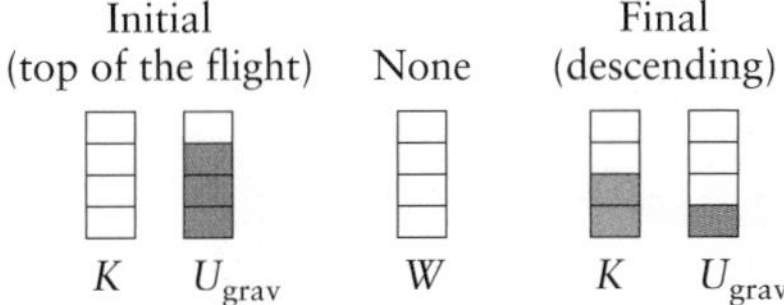

The last stage of the motion is as the pencil comes to rest when you catch it. Here, again work is being done (you are external to the system and exerting a force on it), but this time the work is negative, and in our final state, the pencil ends at rest (no kinetic energy) at the height $y = 0$, which we defined as a system's gravitational potential energy of $U_{grav} = 0$ in our first step. Our final state from the last step (descending) is our initial state, the same as the fifth bar diagram in Figure 8-2. We define the zero as the top of the bar graph for work in this case and draw a downward arrow to show it is negative. It is drawn in red to emphasize this point.

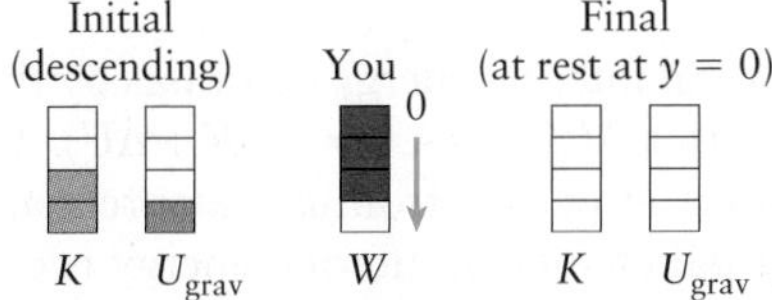

In viewing the work-energy diagrams, there are three important points to notice. First, there is always a column for each type of energy we are tracking. Second, if there is something external to the system interacting with the system, there will be work. Third, the total number of bars in all of the columns in the initial state diagram plus the number of bars in the work column must equal the number of bars in all of the columns of the final state diagram. As long as these three points are correctly represented in your diagram, the diagram should successfully guide your solution to a problem.

Generalizing the Idea of Conservation of Energy: Nonconservative Forces

In the example above of the Earth–pencil system you were outside of the system, and did work on the system. If the force of your hand were a conservative force, all the energy you expended to toss the pencil would be returned to your body when you caught the pencil; you could toss the pencil up and down all day and you would never get tired. This is *not* the case, however, and you *do* get tired. (If you don't believe this, try it with a book rather than a pencil and try tossing the book up and catching it a few dozen times.) Because the energy you expend in the toss is *not* returned to you in the catch, the force of your hand is nonconservative, so energy changes due to this interaction cannot be described as a type of potential energy even if you redefine the system so that you are part of it.

To toss the pencil upward you use energy stored in the chemical bonds of adenosine triphosphate (ATP), carbohydrates, and fat cells. As we discussed in the Reflection section of Example 7-1, when this internal energy conversion occurs your muscles get warm and waste products are produced as your body converts chemical energy. So things start to get more complicated if you choose to put yourself in the system. As long as you are *outside* of the system all that matters is the work you do on the pencil. If you are *inside* the system $W = 0$, and we have to consider changes to the internal energy of the system $E_{internal}$ due to the inclusion of your body into the system, even if you cannot exactly quantify them.

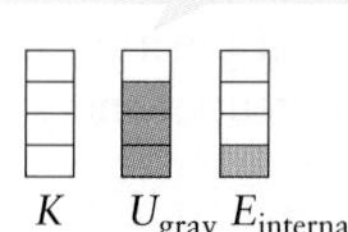

Total energy—the sum of total mechanical energy (K plus U grav) and internal energy E internal—is conserved in this process.

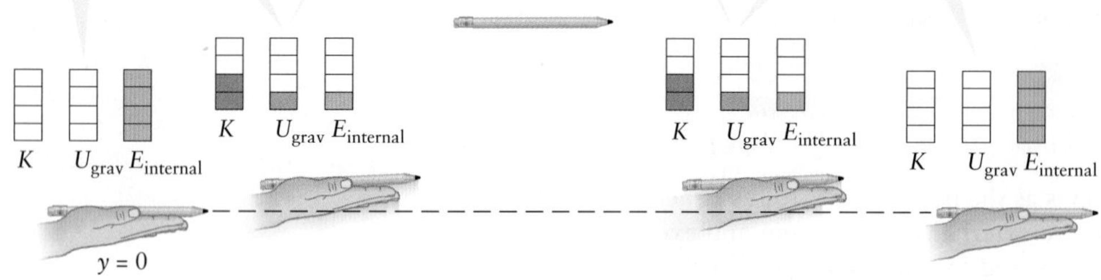

Before you toss the pencil, all the energy is your internal energy. After you toss the pencil some of your internal energy has been converted to mechanical energy of the pencil–Earth system.
Before you catch the pencil, the pencil–Earth system has mechanical energy. This mechanical energy is converted to internal energy of your arm and the pencil (thermal energy), raising their temperatures.
K
U grav
E internal
y = 0

Figure 8-3 Tossing and catching a pencil, revisited When you toss the pencil, some of your internal energy (stored in the arrangement of atoms and molecules: chemical bonds, $E_{internal}$) is converted to mechanical energy of the pencil. When you catch and stop the pencil this added mechanical energy is converted into a different type of internal energy. While the pencil is in free fall, the mechanical energy of the Earth–pencil system remains constant, shifting from kinetic to potential and back to kinetic.

AP® Exam Tip

Although thermodynamics and chemical energy are not part of AP® Physics 1, it is important to understand that mechanical energy can convert to (dissipate as) thermal and other energies.

During the toss, your total internal energy decreases by the amount the mechanical energy of the system increases $\Delta E_{internal} < 0 = -(\Delta K + \Delta U)$. This doesn't look so bad: We can calculate this change in energy from a measurement of how high the pencil goes (**Figure 8-3**). But the total change in internal energy is only part of the story. Not all of the released chemical energy goes into moving your muscles; some goes into warming your muscles (which is why you get warm when you exercise) and hence into **thermal energy**, which is energy associated with the random motion of atoms and molecules in an object or system (in this case inside your arm). Thermal energy is also considered part of $E_{internal}$. So during the toss, some chemical energy is converted into thermal energy as well as the energy that goes into creating the motion of the pencil (**Figure 8-4**). If we unpack the changes in $E_{internal}$ we find a decrease in chemical energy that is bigger than the change in mechanical energy of the system, because internal thermal energy (your muscles warmed up and you warmed up the pencil slightly) increased, and that was also supplied by your internal chemical energy.

When you start to catch the pencil, it is initially moving but you decrease its kinetic energy to zero and it drops a little lower, so $\Delta E_{internal} = -(\Delta K + \Delta U)$ is positive. You again use your muscles and so you use more of the energy stored in chemical bonds. In the catch, however, both your arm and the pencil warm more, and although this warming is a slight amount it is actually measurable if you have sensitive enough equipment. This increase in thermal energy is greater than the loss of chemical energy (Figure 8-4) since during the catch, $\Delta E_{internal} > 0$ because the pencil is transferring its kinetic energy to your hand. This increase in your internal energy due to the decrease in the pencil's kinetic energy cannot go back into your chemical energy; it all gets dissipated as thermal energy. Even though we do not yet know how to calculate such types of internal energy changes, we can still determine some of their properties, using conservation of energy.

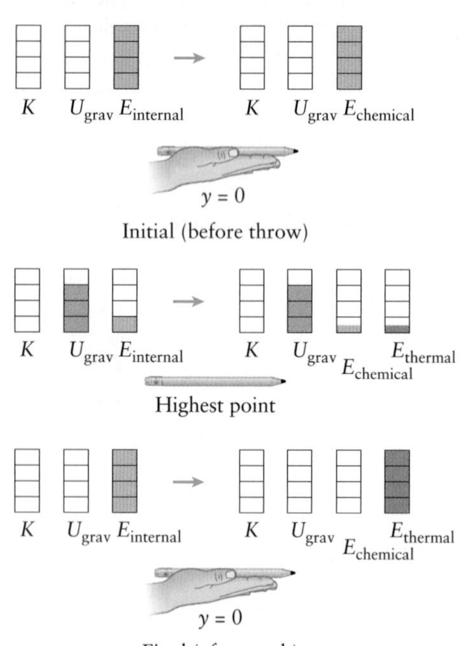

Figure 8-4 Tossing and catching a pencil, revisited and expanded For three points in the motion, we expand the internal energy into its components, chemical and thermal energy. Even though we do not know exact details of the internal energy conversion, we can estimate the amount of internal energy of each type at each stage of a process. We start with chemical energy. We use that chemical energy to add mechanical energy to the pencil and Earth, and also generate a little thermal energy in the process. To catch the pencil and bring it to a stop takes a little bit more chemical energy, and heats up the pencil a little more.

These observations suggest that changes in internal energy are caused by including the source of nonconservative forces within a system. Sometimes, you will want to calculate the size of a non-conservative force, for instance, the force of kinetic friction, as we will see in Example 8-1. Remember that in Chapter 7, for this type of nonconservative force, we put work in quotation marks because we could not be quite sure where all the energy went, just that it was dissipated from the mechanical energy of the system. If we are considering the force exerted by the person on the pencil as we did in our discussion of Figures 8-2 through 8-4, we are pretty safe in considering that most of the change in internal energy required to generate that force is in the person, so we can neglect changes in internal energy when the person is out of the system (although the pencil does warm up just slightly when you catch it). If we put the boundary between the two systems undergoing a kinetic friction interaction, as we must to talk about a force exerted on one of them by the other, not all of the dissipated energy is transferred out of the system we are examining. Both the surface and the things sliding on the surface heat up significantly, so some of the energy goes into each system. If there are many types of energy changes occurring, the safest way to solve such a problem is to find the change in internal energy for the case where you put everything that is interacting into the system. Then we can work our way backward, using this change in internal energy of the system to find the force we know caused it.

In the common case of a friction interaction, we will call this "work" something different: $W_{\text{nonconservative}}$. In the simplest case of a single object sliding on a rough surface, this $W_{\text{nonconservative}}$ is just the force of kinetic friction times the displacement of the object in the direction of the force (Equation 7-2). Since the direction of the force of kinetic friction is always opposite the direction of motion, this $W_{\text{nonconservative}}$ is always negative, equal to the amount of the kinetic energy of the object dissipated by the kinetic friction interaction. We don't know where exactly the energy is, just that it has dissipated from the initial kinetic energy of the object.

Indeed, many careful measurements show that in *all* situations involving a friction interaction of some type, such as kinetic friction or air resistance, if you have correctly identified your system, the amount of energy transfer you would calculate as $W_{\text{nonconservative}}$ is *exactly* equal to the negative of $\Delta E_{\text{internal}}$ of all the interacting objects and the air around them—if you put the nonconservative interaction inside the system. In other words, instead of talking about mechanical energy removed from the system due to a friction force, you could talk about the increase in the internal energy of the system as it heats up, which must be equal to the decrease in the mechanical energy of the system. We will show you how to do this in Example 8-1. This is mathematically identical to how we found that potential energy for an interaction internal to the system was the negative of the work done for the same interaction when the system is chosen so that the interaction is external. However, unlike conservative interactions, we cannot tell exactly where the energy went. This dissipated energy goes not only into warming but into damage (nonreversible changes in configuration, such as scratches or breaks) and sound.

The **change in internal energy** of a system when the frictional interaction is included in the system

Magnitude of the friction force exerted on an object

$$-\Delta E_{\text{internal}} = W_{\text{nonconservative}} = -F_{\text{friction}}d \qquad (8\text{-}5)$$

"Work" done by the friction force if it was external to the system

Distance object moved while the friction force was exerted on it

When we expand a system to include not only the object we are interested in and the sources of the conservative forces that produce ΔU but also the sources of the nonconservative forces (such as your hand in the case of the thrown pencil), work goes to zero. As an object in the system moves, K, U, and E_{internal} can all change values, but the sum of these changes is zero: One kind of energy can be converted into another, but the total amount of energy remains the same. When any of the sources of these interactions is outside the system, then we have to include work, since energy is transferred into and out of the system. For mechanical energy transfers, the work-energy theorem is the most general statement of the law of conservation of energy and any of the other statements we have used are simplifications for specific cases. If you continue your study of physics, you will find there are other ways to transfer energy to a system, resulting in an even more generalized law of conservation of energy. Scientists have made an exhaustive search for situations in which conservation of energy is not observed. Classically, no such situation has

WATCH OUT !

We can only estimate internal energy unless we are given more information.

As we qualitatively reason our way through the changes that must happen in a system, the work-energy bar diagrams we produce can tell us what types of energy exist in the system we have defined and the changes in those types of energy. But unless we are given some amount for thermal or chemical energy, the amounts we put in for both of those are estimates. In Example 8-4, you will see how this information can be given.

 **Exam Tip**

Just like doing work on a system is a mechanical transfer of energy into or out of a system, heating (or cooling) is a thermal transfer of energy into (or out of) a system. For AP Physics 1, it is only necessary to realize it exists. Thermal energy transfers are a topic of AP Physics 2.

EQUATION IN WORDS

The change in internal energy for a closed, isolated system containing a friction interaction is equal to the "work" done by the interaction if it were instead treated like an external force.

WATCH OUT !

How you define your system determines how you solve your problem.

Whenever you want to use a potential energy approach, you must ensure that both objects in the interaction are in your system. Anything causing an interaction you wish to consider as a source of work should be placed outside the system so that it is an external force.

ever been found, and the law of conservation of energy has been used to predict new phenomena that were later observed. From these results, we conclude that conservation of energy is a law of nature. In the following section we'll see how to apply this law to solve physics problems. As long as you understand all the terms in the work-energy theorem, you can solve any conservation of energy problem you will be asked in AP® Physics 1.

WATCH OUT !

Things get weird when the system gets very small.

In quantum mechanics, there are uncertainty principles that tell us we cannot know two related quantities exactly. One such relationship is that of energy and time. For a short enough time period, a little extra energy can be produced. These numbers are tiny ($\Delta E \times \Delta t < 10^{-34}$ J × s)! Scientists know this occurs because it does have effects that can be observed, although only in events that occur on a smaller scale than an atom. For any of the macroscopic systems we work with in this text, the law of conservation of energy is absolute, and even on those tiny scales, the energy before and after the interaction is the same; the fluctuations occur for only tiny periods of time during the interaction.

THE TAKEAWAY for Section 8-2

✔ Total energy—the sum of kinetic, potential, and internal energies—is always conserved.

✔ For an open system, energy transfers into or out of the system are equal to the change in energy of the system.

✔ If the only interactions in a closed, isolated system are conservative, then the total mechanical energy E (the sum of the kinetic energy K and the potential energy U) of the system is constant. The values of K and U may change, but the value of E remains the same.

✔ If nonconservative interactions occur inside a closed, isolated system, the total mechanical energy of the system changes. The change in total mechanical energy is equal and opposite to the change in internal energy, since the total energy of a closed, isolated system remains constant.

Prep for the AP® Exam

AP® Building Blocks

1. Adam and Bobby are twins who have the same weight. Adam drops to the ground from a tree at the same time that Bobby begins his descent down a slide. The slide exerts a negligible friction force on Bobby. Initially both boys are at rest and are at the same height above the ground. Neglect air resistance and compare their kinetic energies at the instant just before they strike the ground. Which of the following best describes the kinetic energies of each of the twins? For each statement explain why you agree or disagree.
   (A) Adam's kinetic energy is greater because, having traveled straight down, the time interval of his fall is shorter and therefore his velocity is greater.
   (B) Adam's kinetic energy is greater because, having traveled only in the direction of the gravitational force, the work done on him by the force is greater.
   (C) Adam and Bobby have the same kinetic energy because the changes in potential energy of the Earth–boy systems are the same and mechanical energy is constant.
   (D) Bobby's velocity is greater because as he descends down the slide, the slide exerts a contact force that does work on him and so increases his kinetic energy.

AP® Skills in Action

2. Jack and Jill, starting from rest, ascended to an elevation of 300 m to fetch a pail of water. When they arrived at the well their speeds were the same. Jack's path was a climb up a sheer rock face to reach the top, while Jill followed a meandering path 835 m long. Jill was ready to hurry home but Jack needed to rest. Treat each child as an object in a child–Earth system and explain how it can be that $\Delta E_{\text{Jack}} - \Delta E_{\text{Jill}} \neq 0$, even though $\Delta K_{\text{Jack}} - \Delta K_{\text{Jill}} = 0$ and $\Delta U_{\text{Jack}} - \Delta U_{\text{Jill}} = 0$.
3. A box with mass m_2 is pulled upward along a ramp by a falling box with a mass m_1, as shown in the figure. A friction force, $F_k = \mu_k F_n$ is exerted on the box with mass m_2.

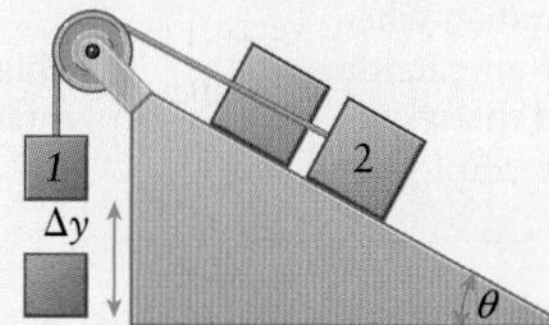

   (a) Identify the free-body diagram that best represents the forces exerted on the box with mass m_2.

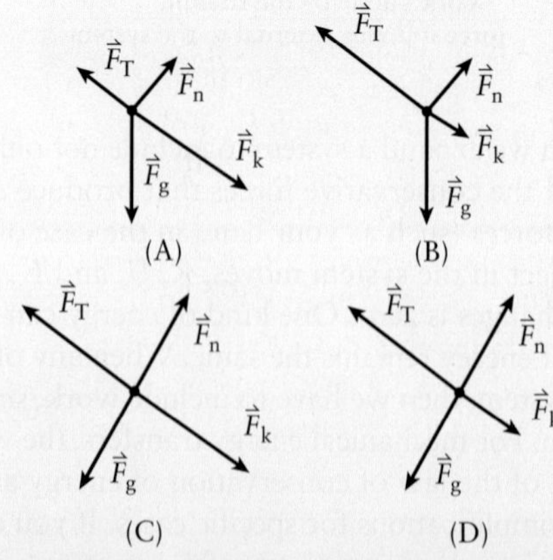

   (b) Describe three choices you could make for your system of interest for this problem to calculate the speed of the box with mass m_2. For each choice, identify what is interacting with the system (external forces). Remember that free-body diagrams are used for systems that can be treated as objects, and that the forces drawn on the diagram are forces external to that system.

(c) Predict if energy will be transferred out of the system composed of the two boxes shown in the figure and Earth while the box with mass m_1 falls a distance Δy. And, if energy is transferred, predict the sign of that energy transfer. Make sure you use the convention that energy transferred into a system is a positive energy transfer.

4. An ice cube starts at rest at point A, and slides down a track with negligible friction as shown in the figure. Assume melting of the ice cube is negligible along this track.

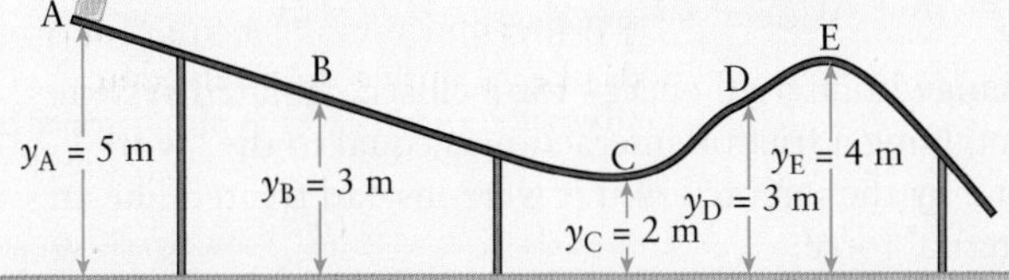

(a) Qualitatively predict the relative potential energies of the ice cube–Earth system at the five labeled points along the path of the ice cube. Each potential energy is measured relative to the shared horizontal plane from which each height is measured. Express your prediction by ranking U_A, U_B, U_C, U_D, and U_E, from greatest to least.

(b) Qualitatively predict the kinetic energies of the ice cube at the labeled points. Express your prediction by ranking K_A, K_B, K_C, K_D, and K_E, from greatest to least.

5. Consider a system composed of a plastic ball and Earth. External to this system is a student who cocks a spring-loaded launcher located on the ground and then pulls the trigger to launch the plastic ball, which is in the launcher. The ball slides along the walls of the launcher as it is being launched. Consider three stages, beginning at the point at which the student cocks the launcher until the ball strikes the ground:

- Stage 1: student cocks the launcher
- Stage 2: student pulls the trigger
- Stage 3: ball flies, initially upward

For each stage summarize

(a) the forces on the ball,

(b) the total mechanical energy and changes in kinetic and potential energies, and

(c) any changes in the total mechanical energy and the relationships between the work done on the system by the student and each of these changes.

8-3 Energy conservation is an important tool for solving a wide variety of problems

Now let's see how to use energy conservation to solve problems. As a rule, if a problem involves a conserved quantity (that is, a quantity whose value remains unchanged), it simplifies solving the problem tremendously to use a conservation law. For instance, when we use conservation of energy, we can use scalar calculations, instead of finding vector components of forces. Often a conservation equation, which relates the value of the conserved quantity at one position in an object's motion to the value at a different position, is all you need to solve for the desired unknown. We'll see several problems of that kind in this section.

EXAMPLE 8-1 A Ski Jump, Revisited

As in Example 7-8 (Section 7-6), a skier of mass m starts at rest at the top of a ski jump ramp (**Figure 8-5**). (a) Use energy conservation to find an expression for the skier's speed as she flies off the ramp. Assume there is negligible friction between the skis and the ramp and you can ignore air resistance. (b) After the skier reaches the ground at point P in Figure 8-5, she begins braking and comes to a halt a distance L later, at point Q. Find an expression for the magnitude of the constant friction force that is exerted on her between points P and Q.

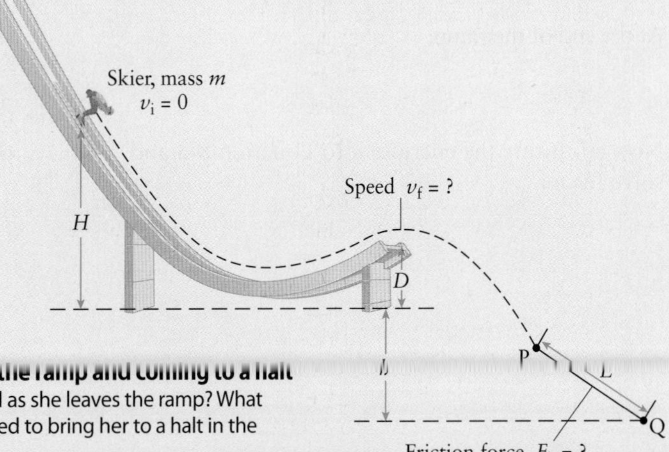

Figure 8-5 Flying off the ramp and coming to a halt
What is the skier's speed as she leaves the ramp? What force of friction is required to bring her to a halt in the distance specified?

Set Up

In part (a) of this problem the skier travels from an initial point (quantities from this point are designated by the subscript "i") at the top of the ramp to a final point (quantities from this point are designated by the subscript "f") at the end of the ramp. During this motion only the conservative interaction of gravity exists. (The ramp exerts a normal force on her, but this force does no work because it is always perpendicular to her motion.) There are no springs, or anything else, exerting a force on the skier. Our goal for this part is to find the skier's speed v_f as she leaves the ramp. Because we want a speed, the best approach is to use the Earth–skier system and apply conservation of energy. Remember that if the work and internal energy terms can be set to zero in the work-energy theorem, mechanical energy must be constant. Then, after rearranging, Equation 8-3 is obtained.

Constant mechanical energy:

$$K_i + U_i = K_f + U_f \qquad (8\text{-}3)$$

Gravitational potential energy:

$$U_{grav} = mgy \qquad (7\text{-}14)$$

Kinetic energy:

$$K = \frac{1}{2}mv^2 \qquad (7\text{-}8)$$

Work-energy theorem:

$$W = \Delta K + \Delta U_{grav} + \Delta E_{internal} \qquad (7\text{-}12)$$

Change in internal energy for a closed, isolated system containing a friction interaction is equal to the "work" done by the interaction if it were instead treated like an external force:

$$-\Delta E_{internal} = W_{nonconservative} = -F_{nonconservative}d \qquad (8\text{-}5)$$

For part (b) we will consider the skier's entire motion from the top of the ramp (which we again call point i) to where she finally comes to rest at point Q. A normal force is also exerted on her when she reaches the ground, but again this force does no work on her because it is perpendicular to her motion. Between points P and Q, a nonconservative *friction* force is exerted on her. If we consider a closed, isolated system that contains the friction interaction for the motion as a whole, we can use the work-energy theorem to find the change in internal energy for the system, and then use Equation 8-5 to find the nonconservative force that does the "work" that is equivalent to this change in internal energy. Although we are not using an object model for the skier, we can draw a free-body diagram to help us find the relationship of the friction force to the normal force and the skier's weight. We don't really need the normal force, because our goal for part (b) is to find the magnitude F_k of the friction force, but we would need it to find the coefficient of friction.

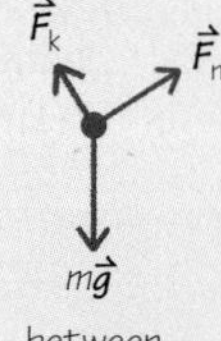

Solve

(a) Let's take $y = 0$ as the location of the lowest point of the ramp, given that both H and D are measured from there in the diagram. Write expressions for the skier's kinetic energy K and the gravitational potential energy U_{grav} of the Earth–skier system when the skier is at the top of the ramp (at $y_i = H$, where the speed is $v_i = 0$) and the end of the ramp (at $y_f = D$, where the speed is v_f).

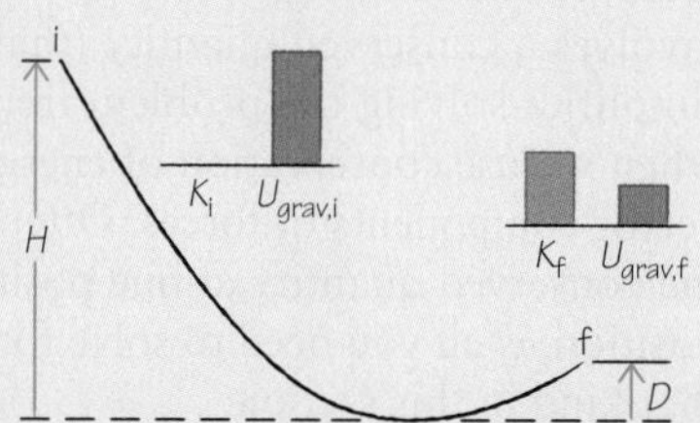

At the top of the ramp:

$$K_i = \frac{1}{2}mv_i^2 = 0$$

$$U_{grav,i} = mgy_i = mgH$$

At the end of the ramp:

$$K_f = \frac{1}{2}mv_f^2$$

$$U_{grav,f} = mgy_f = mgD$$

Now substitute the energies into Equation 8-3 and solve for v_f:

$$K_i + U_i = K_f + U_f \qquad (8\text{-}3)$$

$$0 + mgH = \frac{1}{2}mv_f^2 + mgD$$

$$\frac{1}{2}mv_f^2 = mgH - mgD = mg(H - D)$$

$$v_f^2 = 2g(H - D)$$

$$v_f = \sqrt{2g(H - D)}$$

(b) Now the final position is at point Q, a distance h below the base of the ramp, so $y_Q = -h$. At this point the skier is again at rest, so $v_Q = 0$. Our system is still the Earth–skier system. The friction force F_k is internal to the system, so it can do no work on the system, but the interaction causes a change in the internal energy of the system. There is no work done on the system by external forces, so the work-energy theorem becomes $0 = \Delta K + \Delta U_{grav} + \Delta E_{internal}$. We can conserve energy between any two points, so we could use the position at the bottom of the ramp as the skier's initial position for the conservation of energy equation, given that we know values of speed and height. However, if we choose our initial position as the skier's initial starting point, her speed there was zero, so we have one less number to calculate. Because we are not going to use the details in the middle of the path, we can just indicate Q on our drawing and not the specific path, so just drawing a dashed line is okay.

At the top of the ramp:

$$K_i = \frac{1}{2}mv_i^2 = 0$$

$$U_{grav,i} = mgy_i = mgH$$

At point Q:

$$K_Q = \frac{1}{2}mv_Q^2 = 0$$

$$U_{grav,Q} = mgy_Q = mg(-h) = -mgh$$

$$\Delta U_{grav} = \Delta U_{grav,f} - \Delta U_{grav,i} = -mgh - mgH = -mg(h + H)$$

Conserving energy:

$$W = \Delta K + \Delta U_{grav} + \Delta E_{internal}$$

$$0 = 0 + \Delta U_{grav} + \Delta E_{internal}$$

$$-\Delta U_{grav} = \Delta E_{internal}$$

$$\Delta E_{internal} = +mg(h + H)$$

To consider friction, we need to know distance she traveled along the ramp, L. Now use Equation 8-5 to solve for F_k. "Work" done by the friction force must be the negative of $\Delta E_{internal}$:

$$W_{nonconservative} = -mg(h + H) = F_k L \cos\theta$$

Because the kinetic friction force is opposite the direction of motion, θ is 180°, so $-mg(h + H) = F_k L(-1)$.

$$F_k = mg\left(\frac{H + h}{L}\right)$$

Reflect

Our answer for part (a) is the same one that we found in Example 7-8.

Our result for F_k in part (b) has the correct dimensions of force, the same as mg, because the dimensions of $H + h$ (the total vertical distance that the skier descends) cancel those of L (the skier's stopping distance). Note that the smaller the stopping distance L compared to $H + h$, the greater the friction force required to bring the skier to a halt, which also makes sense.

Extend: An alternative way to solve part (b) would be to treat the second part of the skier's motion separately, with the initial point at the end of the ramp [at height D, where the skier's speed is $\sqrt{2g(H - D)}$ as found in part (a)] and the final point at point Q (at height $-h$, where the skier's speed is zero). You should try solving the problem this way using the same equations. You should get the same answer for F_k. Do you?

NOW WORK Problem 1 from The Takeaway 8-3.

EXAMPLE 8-2 Warming Skis and Snow

When the skier in Example 8-1 brakes to a halt, her muscles, her skis, and the snow over which she slides all warm up. What is the increase in the internal energy of the system in this process?

Set Up

We could do exactly the same calculation that we did in (b) in the previous example, or we could answer this without considering the nonconservative work by just changing our system. Again, we choose to just look at the initial and final states, so we do not need to worry about the location of point P or the exact shape of the path taken. During the skier's motion (from the top of the ramp to when she finally comes to rest at point Q), if we put the skier, Earth, and the surface between them all in the system, there is no work done on the system, so the sum of the changes in kinetic energy K, potential energy U, and internal energy E_{internal} must be zero.

Work-energy theorem:

$$W = \Delta K + \Delta U + \Delta E_{\text{other}} = 0 \qquad (7\text{-}12)$$

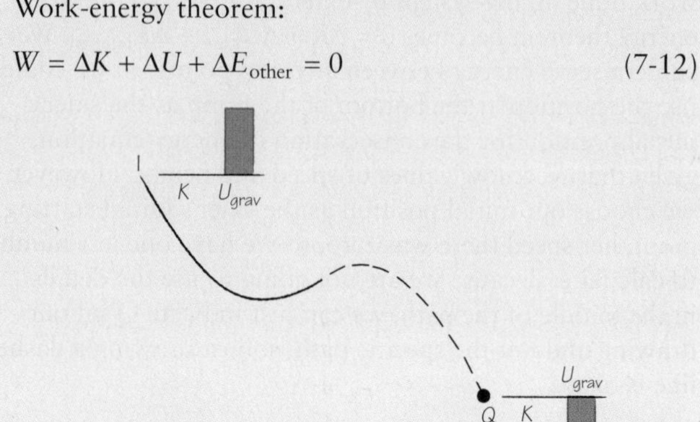

Solve

The change in kinetic energy ΔK is zero, because she starts and stops at rest. The change in gravitational potential energy ΔU, then, must all go into the increase in internal energy.

Change in potential energy:

$$\Delta U = U_{\text{grav,Q}} - U_{\text{grav,i}}$$
$$= -mgh - mgH = -mg(H + h)$$

The work-energy theorem with ΔK and $W = 0$:

$$0 + \Delta U + \Delta E_{\text{internal}} = 0$$
$$0 + [-mg(H + h)] + \Delta E_{\text{internal}} = 0$$
$$\Delta E_{\text{internal}} = mg(H + h)$$

Reflect

The overall motion of the skier begins and ends with zero kinetic energy. The net result is that the (gravitational) potential energy decreases by an amount $mg(H + h)$ and the internal energy increases by the same amount. This primarily goes into the thermal energy of the skier's skis and the snow over which she slides as she brakes to a halt.

The net change in internal energy is caused by the nonconservative friction interaction, and we get the same answer no matter which system we choose, as we must.

NOW WORK Problem 2 from The Takeaway 8-3.

EXAMPLE 8-3 How Far to Stop?

The U.S. National Highway Traffic Safety Administration lists the minimum braking distance for a car traveling at 40.0 mi/h to be 101 ft. If the braking force is the same at all speeds, what is the minimum braking distance for a car traveling at 65.0 mi/h?

Note: The braking force is a model. This model allows us to ignore many complications but still solve the problem. This is one of the powers of models. We know an object cannot exert a force on itself, and even if we were to consider the car a system instead of an object, internal interactions cannot change the total energy of systems. Yet the brake pads squeeze on things around the wheels, which are also part of the car. So that alone cannot stop your car. Your car stops because as your wheels are affected by the brakes and begin to slow their rotation, the friction between your wheels and the surface of the road makes this slower rotation result in a reduced speed for your car. It is the friction force exerted by the road on your car that allows your brakes to bring your car to rest. We will unpack this model more when we get to rotational motion and understand how wheels work later in the course!

Set Up

If we neglect air resistance, only three forces are exerted on the car—the gravitational force, the normal force exerted by the road, and the braking force exerted by the road—but if the road is level, only the braking force does work on the car. (The other two forces are perpendicular to the displacement, so $Fd \cos 90° = 0$.) We'll use Equations 7-2, 7-8, and 7-12 to find the relationship among the initial speed v_i, the braking force, and the displacement of the car while this force is exerted to result in the final speed of the car, being $v_f = 0$.

Work done by a constant force, linear displacement:

$$W = Fd \cos \theta \qquad (7\text{-}2)$$

Kinetic energy:

$$K = \frac{1}{2}mv^2 \qquad (7\text{-}8)$$

Work-energy theorem:

$$W = \Delta K + \Delta U + \Delta E_{other} = 0 \qquad (7\text{-}12)$$

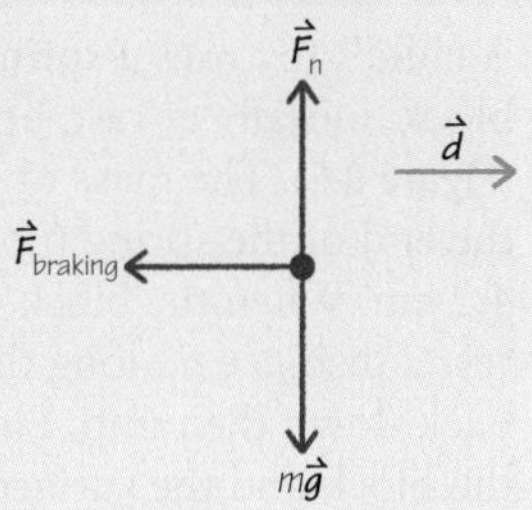

Solve

The car ends up at rest, so its final kinetic energy is zero. We included gravity as an external force, so we don't include Earth in the system. Our braking force model hides any internal energy changes, so we have just kinetic energy left. The work-energy theorem tells us that the net work done on the car (equal to the work done on it by the braking force) equals the negative of the initial kinetic energy.

Work-energy theorem with final kinetic energy equal to zero, no change in potential energy, and no consideration of internal energy:

$$W_{net} = 0 - K_i = -\frac{1}{2}mv_i^2$$

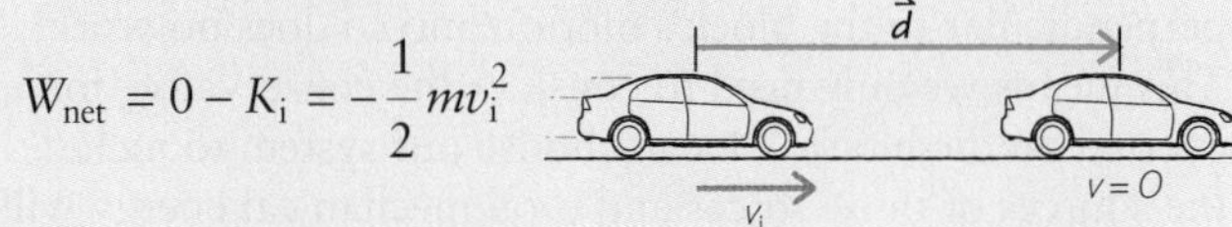

The braking force of magnitude $F_{braking}$ is directed opposite to the displacement, so $\theta = 180°$ in the expression for the work done by this force.

Net work = work done by the braking force:

$$W_{net} = F_{braking}d \cos 180° = -F_{braking}d$$

Substitute $W_{net} = -F_{braking}d$ into the work-energy theorem and solve for the braking distance d. The result tells us that d is proportional to the square of the initial speed v_i.

Setting the two expressions for the net work equal gives

$$-F_{braking}d = -\frac{1}{2}mv_i^2$$

Solve for d:

$$d = \frac{mv_i^2}{2F_{braking}}$$

We aren't given the mass m of the car or the magnitude $F_{braking}$ of the braking force, but we can set up a ratio between the values of distance for $v_{i1} = 40.0$ mi/h and $v_{i2} = 65.0$ mi/h. Because this calculation is based on the ratio of two speeds, any speed unit would work here. This is a rare case in which there is no need to convert to SI units! The mass of the car and the braking force are the same for both initial speeds, so they cancel.

Use this expression to set up a ratio for the value of d at two different initial speeds:

$$\frac{d_2}{d_1} = \frac{v_{i2}^2}{v_{i1}^2}$$

Solve for the stopping distance d_2 corresponding to $v_{i2} = 65.0$ mi/h:

$$d_2 = d_1\frac{v_{i2}^2}{v_{i1}^2} = (101 \text{ ft})\left(\frac{65.0 \text{ mi/h}}{40.0 \text{ mi/h}}\right)^2$$
$$= 267 \text{ ft}$$

Reflect

Stopping distance is proportional to kinetic energy, which is proportional to the *square* of the speed, so the answer makes sense. This example demonstrates what every driver training course stresses: The faster you travel, the more distance you should leave between you and the car ahead in case an emergency stop is needed. Increasing your speed by 62.5% (from 40.0 to 65.0 mi/h) increases your stopping distance by 164% (from 101 to 267 ft).

NOW WORK Problem 8 from The Takeaway 8-3.

EXAMPLE 8-4 A Ramp and a Spring

A child's toy uses a spring with spring constant $k = 36$ N/m to shoot a block, initially at rest, up a ramp inclined at $\theta = 30°$ from the horizontal (**Figure 8-6**). The mass of the block is $m = 8.0$ g. The block is held against the end of the spring (not attached to it). The spring is compressed 4.2 cm. When the block is released it slides up the ramp and comes to rest a distance d along the ramp from where it started, before sliding back down the ramp. Find the value of d. Neglect friction between the block and the surface of the ramp. Use the subscript "i" to denote quantities at the initial point, when the spring is compressed, and the subscript "f" to denote quantities for the final point of interest, when the block reaches its highest point, at d.

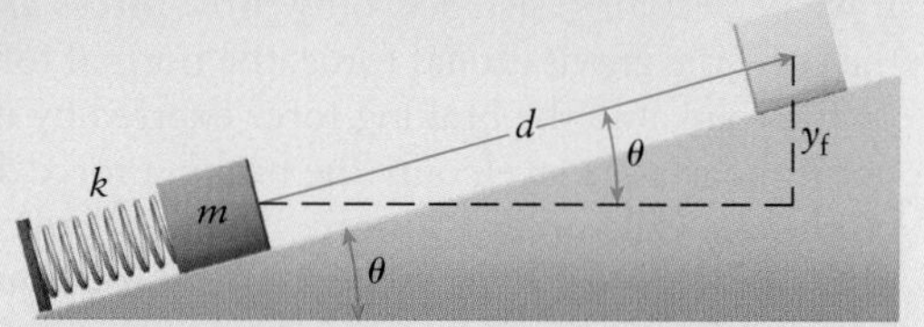

Figure 8-6 Spring potential energy and gravitational potential energy If the spring is compressed a certain distance, how far up the ramp will the block go before it slides back down, if we can neglect friction?

Set Up

The forces on the block are a normal force, the gravitational force, and a spring force exerted on the block while the block is in contact with the spring. The normal force is always perpendicular to the block's motion and so does no work. This means we only need to consider the conservative forces exerted on the block. We can choose our system to include the sources of those forces and total mechanical energy will be constant. This will make things much easier, because we won't have to take components of forces as we would if we used kinematics (gravity is not along the direction of motion) and we don't have to deal with a variable force (the spring force is not constant). This problem would require a lot more mathematical sophistication to solve using the techniques in Chapters 4 and 5. Instead, we'll use Equation 8-3 and a little trigonometry to solve for the distance along the ramp, d.

$$K_i + U_i = K_f + U_f \quad (8\text{-}3)$$

$$U_{grav} = mgy \quad (7\text{-}14)$$

$$U_s = \frac{1}{2}kx^2 \quad (7\text{-}16)$$

$$K = \frac{1}{2}mv^2 \quad (7\text{-}8)$$

The dotted line is parallel to the base of the ramp from which θ is measured, so the angle we want will be the same angle (see figure in Example M-11 in Math Tutorial M-10 at the back of the book).

Solve

The block is at rest with zero kinetic energy at both the initial point and the final point. If we take $y = 0$ to be the height of the block where it is released, then $y_i = 0$; the final height y_f is related to the distance d by trigonometry. We use Equation 7-14 to write the initial and final gravitational potential energies. The initial spring potential energy is given by Equation 7-16, with $k = 36$ N/m and $x_i = -4.2$ cm; the final spring potential energy is zero because the block is not attached to the spring, so the block cannot extend the spring beyond the point where it is relaxed (has zero potential energy). We will have to double-check this assumption if the block doesn't go far enough for the spring to come to equilibrium.

Before block is released:

$$K_i = \frac{1}{2}mv_i^2 = 0$$

$$U_{grav,i} = mgy_i = 0$$

$$U_{s,i} = \frac{1}{2}kx_i^2$$

After block has traveled a distance d up the ramp:

$$K_f = \frac{1}{2}mv_f^2 = 0$$

From Figure 8-6,

$y_f = d \sin\theta$, so

$$U_{grav,f} = mgy_f = mgd \sin\theta$$

$$U_{s,f} = \frac{1}{2}kx_f^2 = 0$$

Substitute the energies into Equation 8-3. (Remember that the *total* potential energy U is the sum of the gravitational and spring potential energies.) Then solve for the distance along the ramp, d.

Equation 8-3:

$K_i + U_i = K_f + U_f$, with two types of potential energy:

$$K_i + U_{grav,i} + U_{s,i} = K_f + U_{grav,f} + U_{s,f},$$

so

$$0+0+\frac{1}{2}kx_i^2 = 0+mgd\sin\theta+0$$

$$d = \frac{kx_i^2}{2mg\sin\theta}$$

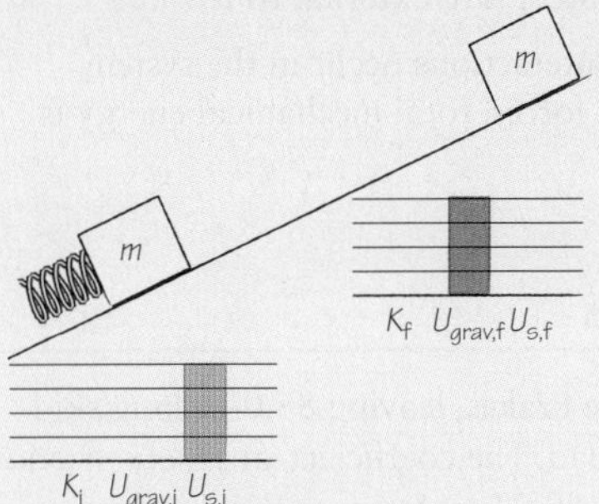

Insert numerical values (being careful to express distances in meters and masses in kilograms) and find the value of d.

Recall that $1\text{ N} = 1\text{ kg}\cdot\text{m/s}^2$.

$$x_i = (-4.2\text{ cm})\left(\frac{1\text{ m}}{100\text{ cm}}\right) = -0.042\text{ m}$$

$$m = (8.0\text{ g})\left(\frac{1\text{ kg}}{1000\text{ g}}\right) = 0.0080\text{ kg}$$

$$d = \frac{(36\text{ N/m})(-0.042\text{ m})^2}{2(0.0080\text{ kg})(9.80\text{ m/s}^2)\sin 30°}$$

$$= 0.81\frac{\text{N}\cdot\text{s}^2}{\text{kg}} = 0.81\frac{\text{kg}\cdot\text{m s}^2}{\text{s}^2\text{ kg}}$$

$$= 0.81\text{ m} = 81\text{ cm}$$

Reflect

In our solution we assumed that the final spring potential energy is zero, which means that the spring ends up neither compressed nor stretched. In other words, we assumed that the block moves so far up the ramp that it loses contact with the spring. That's consistent with our result: The block moves 81 cm up the ramp, much greater than the 4.2-cm distance by which the spring was originally compressed.

As expected, the net effect of the block's motion is to convert the system's initial spring potential energy to gravitational potential energy at the final position. The block begins and ends with zero kinetic energy, just as the skier does in Examples 7-8 and 8-2. Although the block *does* have kinetic energy at intermediate points during the motion, when it's moving up the ramp, we don't need to worry about these intermediate stages of the motion when using conservation of energy. Of course, if there really were no friction, the block would slide back down, compressing the spring by the same amount and starting the cycle of motion over. If there is friction, the block will go a little less high each time, and compress the spring a little less.

NOW WORK Problem 3 from The Takeaway 8-3.

WATCH OUT!

Interactions can always be described as forces, even if they cannot do work, so it is very important to discriminate between internal and external forces.

In Chapters 7 and 8 so far we have been very careful to just call the interactions inside a system interactions. *Force* is another way to describe an interaction, but only forces exerted by something external to a system can do work on a system, changing its total energy. Internal interactions can only change energy from one type to another, as we have seen in two cases (potential energy from conservative interactions and internal energy from friction interactions). You may see these internal interactions referred to simply as forces. You must always define your system, and identify if the interaction being described by a force is internal (exerted by something in the system) or external (exerted by something outside the system). Only if it is an external force can it be used to calculate work, the amount of energy transferred to the system by mechanical processes. Now that you have had a chance to develop your understanding of the difference, we will begin to use the descriptions *internal forces* and *external forces*.

THE TAKEAWAY for Section 8-3

✔ The law of conservation of energy simplifies solving problems because energy is a scalar quantity; forces are vectors and require more information (direction) and calculations to solve a problem.

✔ Identifying the system is essential for deciding which forces do work on the system (are external to it).

✔ If only conservative interactions occur in the system and there are no external forces, total mechanical energy is constant. If there are nonconservative interactions or external forces do work, you must use the more general statement of energy conservation.

✔ Interactions can be described by forces. It is very important to distinguish between internal and external forces, as only external forces can do work, changing the total energy of an object or system.

AP® Building Blocks

1. A driver slams on the brakes, leaving 88.0-m-long skid marks on the level road. The coefficient of kinetic friction is estimated to be 0.480. How fast was the car moving when the driver hit the brakes?

2. An 18.0-kg suitcase falls from a hot-air balloon that is at rest at a height of 385 m above the surface of Earth. The suitcase reaches a speed of 30.0 m/s just before it hits the ground. Calculate the change in internal energy of the suitcase–Earth–air system. Where does this energy go?

3. A block with mass m is released from rest at the top of a ramp with a length L and inclination θ. It slides down the ramp and is stopped by a spring that is compressed by a length Δd from its equilibrium length, d, measured from the bottom of the ramp. Write each of the relevant terms in the equation for energy conservation for the block–spring–Earth system, indicating which are final and which are initial. Express these in terms of the given values and any necessary constants. Assume that the mass of the spring and friction are negligible.

AP® Skill Builders

4. An external force causes a 3.50-kg box to move at a constant speed up a ramp. The friction force between the ramp and the box is negligible. The external force is exerted on the box in a direction parallel to the ramp.

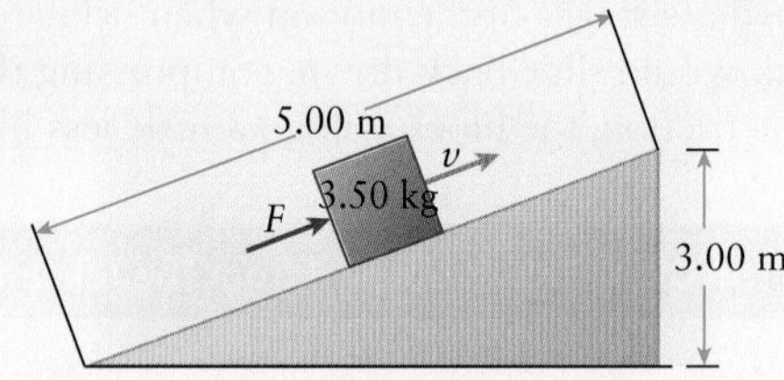

(a) Construct a free-body diagram for the forces exerted on the box.

(b) Treat the box as an object upon which the external force is exerted over a displacement $d = 5$ m, while velocity remains constant, and apply the work-energy theorem to evaluate the numerical value of the work, W, done by this force. What is the total work done on the box by the net force?

(c) Treat the box and Earth as a system upon which the external force is exerted and apply the principle of energy conservation to evaluate the numerical value of the work, W, done on the system.

Prep for the AP Exam

5. A box with mass m_2 (box 2) is pulled upward along a ramp by a falling box with mass m_1 (box 1), as shown in the figure. The inclination of the ramp, θ, can be adjusted. A friction force, $F_k = \mu_k F_n$, is exerted on box 2 by the ramp. Box 2 rises at a constant velocity while box 1 falls a distance Δy.

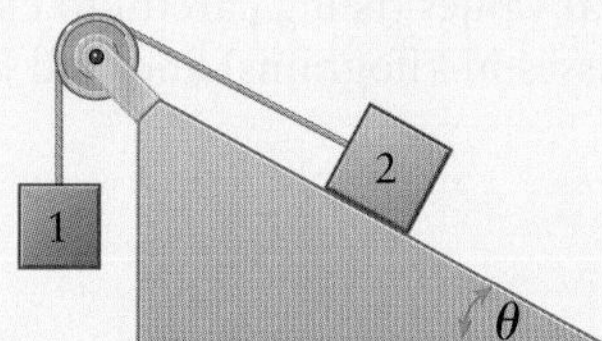

(a) Taking the system to be composed of Earth and the two boxes shown in the figure, apply energy conservation to obtain an expression that can be used to predict the energy transferred from the system to the environment in terms of m_1, m_2, Δy, and θ.

(b) Construct a mathematical representation of the condition that must be met by m_1, m_2, μ_k, and θ if the boxes move with constant velocity.

(c) Design an experiment in which m_1 and m_2 are specified but the value of the maximum static friction is unknown, to determine the value of $\mu_{S,max}$ using only the materials as they are described adjacent to the figure above. Include in your design the method of analysis to be used to determine the value of $\mu_{S,max}$.

6. An object with mass m is released from rest on a ramp of negligible friction with an angle of inclination θ_1 from a height H_1 above the base of the ramp in the figure. The bottom end of the ramp merges smoothly with a second ramp that rises at angle θ_2. From the lowest point in the path onward the second ramp exerts a friction force, F_k, on the object. When released from the height H_1 the object ascends the second ramp to the height H_2. Measurements are to be conducted with the goal of characterizing the friction force, F_k.

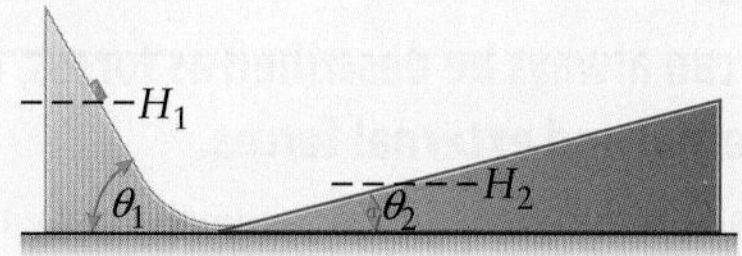

(a) Explain why the principle of energy conservation, rather than Newton's second law, is the appropriate approach to constructing a model of this system.

(b) A linear scale is etched on the second ramp along the surface on which the object moves a distance d before coming to rest. Re-express H_2 in terms of d and θ_2.

(c) Construct a representation of the work done on the block, W, by the ramp in terms of H_1, H_2, m, g, and θ_2, as needed.

(d) The work done on the block by the ramp, W, can be expressed in terms of the distance d and the force exerted on the block by the ramp. Use this relationship to express W in terms of the mass of the object, m, the angle θ_2, and the kinetic coefficient of friction, μ_k.

(e) Analyze the measured data in the table to obtain an estimate of the friction coefficient if the angle of inclination, θ_2, is equal to 7.0°.

H_1 (cm)	d (cm)
5	29.6
10	59.1
15	88.7
20	118.3
25	147.8

AP® Skills in Action

7.

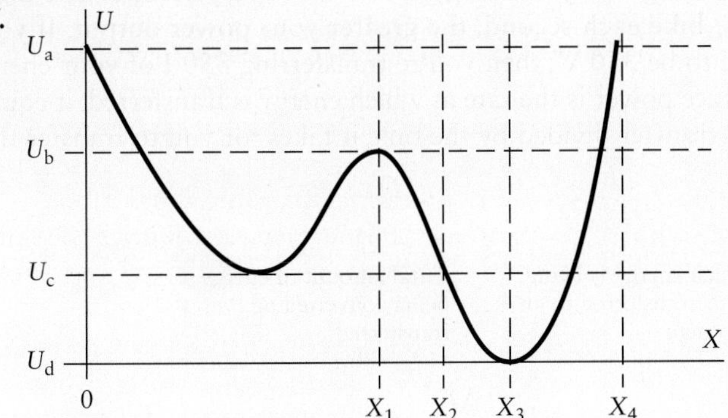

(a) The graph above represents the potential energy of a closed, isolated system of an object and Earth. Assume there are no other internal interactions. If initially the object is at rest at $X = 0$, then which of the following statements is correct?

(A) The object will remain at $X = 0$ because the system has no kinetic energy.

(B) The system will eventually lose energy in the amount $U_a - U_d$ and the object will come to rest at $X = X_3$.

(C) The object will be in motion with a kinetic energy of $U_c - U_d$ when it is at $X = X_3$.

(D) The object will move back and forth along X between $X = X_4$ and $X = 0$.

(b) If the graph above represents the potential energy of an open system of an object and Earth and initially the object is at rest at $X = 0$ and work in the amount of $W = -U_b$ is done on the system, then which of the following statements could be correct?

(A) The object will remain at $X = 0$ because the system has no kinetic energy.

(B) The system will eventually lose energy in the amount $U_a - U_d$ and the object will come to rest at $X = X_3$.

(C) The object will be in motion when it is at $X = X_3$.

(D) The object will move back and forth along X between $X = X_4$ and $X = 0$.

(c) For the system described in part (a), if the object is initially at rest at $X = X_2$ then which of the following statements is correct?

(A) The object will remain at $X = X_2$ because the system has no kinetic energy.

(B) The system will eventually lose energy in the amount $U_c - U_d$ and the object will come to rest at $X = X_3$.

(C) The object will be in motion with a kinetic energy of $U_c - U_d$ when it is at $X = X_3$.

(D) The system will gain potential energy until it reaches $X = X_4$ and then the object will move back and forth along X between $X = X_4$ and $X = 0$.

EX 8-3 **8.** As part of a lab experiment, Meagen uses an air-track cart of mass m to compress a spring of constant k by an amount x from its equilibrium length. The air track has negligible friction. When Meagen lets go, the spring launches the cart.

(a) Draw a diagram of the track and in the diagram identify the most useful initial (i) and final (f) positions for the analysis of the spring–cart system to predict the maximum speed of the cart.

(b) For these positions express K_i, K_f, U_i, and U_f in terms of m, k, x_i, x_f, v_i, and v_f.

(c) Why is it possible to assume that mechanical energy is constant in this system?

(d) Express the constancy of total mechanical energy in terms of m, k, x_i, x_f, v_i, and v_f.

9. A pendulum is constructed by attaching a small metal ball with mass m to one end of a very light string whose length, L, is 1.25 m. The string hangs from the ceiling in the figure. The ball is released when it is raised high enough for the string to make an angle of $\theta = 30.0°$ with the vertical.

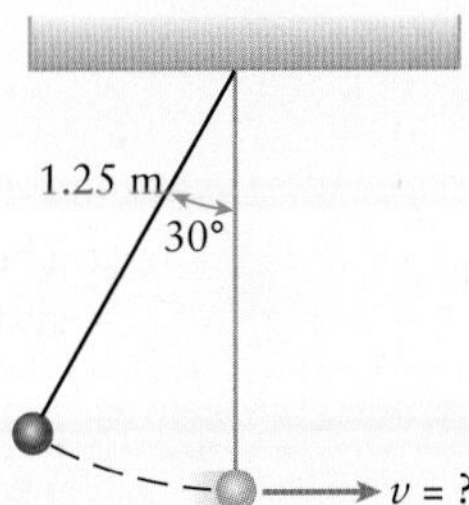

(a) Redraw the diagram and in it identify the most useful initial (i) and final (f) configurations for the analysis of the ball–string–Earth system to predict the maximum speed of the ball.

(b) For these configurations express K_i, K_f, U_i, and U_f in terms of θ, L, g, m, v_i, and v_f.

(c) What must we assume to use total mechanical energy as a constant for this system?

(d) Express the constancy of total mechanical energy in terms of θ, L, g, m, v_i, and v_f.

(e) Evaluate the expression in part (d) to obtain its speed when the ball is at the lowest point in its motion.

8-4 Power is the rate at which energy is transferred into or out of a system or converted within a system

The AGE/Fairfax Media/Getty Images

Figure 8-7 Running versus walking Both running and walking involve converting internal energy to mechanical energy and other types of internal energy; the difference is the rate at which that energy has to be converted.

If you walk for a kilometer, your heart rate will increase above its resting value. But if you run for a kilometer, your heart rate will increase to an even higher value (**Figure 8-7**). Now that we've learned about energy, this may seem paradoxical: You covered the same distance and converted roughly the same amount of energy in both cases, so why is a higher heart rate needed for running? The answer, as we will see, is related to the *rate* at which energy is converted from one type to another in these two cases.

The rate at which energy is transferred from one system to another, or converted from one type of energy to another, is called **power**. The unit of power is the joule per second, or **watt** (abbreviated W): 1 W = 1 J/s. As an example, a 50-W light bulb is designed so that when it is in operation in an electric circuit, 50 J of energy is delivered to it every second by that circuit. This energy is converted into warmth and light. Other common units of power are the *kilowatt* (1 kW = 1000 W) and *horsepower* (1 hp = 746 W, a number related to the rate at which a horse does work by pulling on a plow). We'll denote power by the uppercase symbol P.

Let's apply the concept of power to the rate of *transferring energy*. If you are using a stationary exercise bike at the gym, and your power output is measured to be 200 W, that means you are transferring 200 J of energy to the bike every second. The more energy you transfer to the bike each second, the greater your power output. If your power output is measured to be 350 W, then you're transferring 350 J of your energy to the bike each second. Since power is the rate at which energy is transferred, it equals the amount of energy you transfer divided by the time it takes for you to transfer that energy:

EQUATION IN WORDS
Total power is the energy transferred into or out of a system or converted from one type to another per unit time.

Rate at which energy is being converted or transferred to an object or system

Total amount of energy being converted or transferred

$$P = \frac{\Delta E}{\Delta t} \tag{8-6}$$

Time over which that conversion or transfer occurs

The quantity P in this equation is sometimes called the power *delivered* to the object or system to which energy is being transferred.

To help interpret this equation, let's think again about the groundskeeper pulling a screen across a baseball diamond that we described in Section 7-2 (Figure 7-4). The groundskeeper exerted a force of magnitude F on the screen at an angle θ to the screen's displacement. So to move the screen a distance d over the ground at a steady speed, the groundskeeper did an amount of work on the screen–ground system, $W = Fd\cos\theta$ (Equation 7-2). Remember our definition of work: Work is the transfer of energy from one object or system to another through a *mechanical* process. If the groundskeeper took a time t to pull the screen this distance, t was the time taken to transfer this amount of energy. The average speed of the screen was $v = d/t$. Substituting Equation 7-2 for the amount of energy transferred into Equation 8-6 and using the relationship $v = d/t$ tells us that the average power output that the groundskeeper delivered to the screen–ground system was

$$P = \frac{W}{\Delta t} = \frac{(F\cos\theta)d}{\Delta t} = (F\cos\theta)v \qquad (8\text{-}7)$$

- $W = \Delta E$, Work is the amount of energy transferred by a mechanical process.
- Magnitude of the constant force $\vec{F}$
- Rate at which energy is being transferred to an object or system by a force exerted on the object or system
- Angle between the directions of $\vec{F}$ and $\vec{d}$
- Time over which that transfer occurs
- Magnitude of the displacement $\vec{d}$
- Magnitude of the velocity $\vec{v}$ in the direction $\vec{d}$

EQUATION IN WORDS
Average power associated with a single force exerted on an object or a system moving at constant speed

This equation says that the greater the force F that the groundskeeper exerts and the faster the speed v at which he makes the screen move, the greater the power he delivers to the screen–ground system if the angle is held constant.

While we calculated this for an average rate of energy transfer, we can imagine that if we look at very small intervals of time, we can find the instantaneous power using this same equation, if we use the value of force and velocity at particular instants in time. This is necessary if either vector is changing (like an object speeding up because you are exerting a force on it). The same equation also applies to the power that you deliver to the pedals of a bicycle or a stationary exercise bike, where the direction between force and displacement is changing. At the instant your foot pushes straight down on one of the pedals, as in **Figure 8-8a,** the angle θ in Equation 8-7 is zero (the force is in the direction that the pedal is moving). Since $\cos\theta = 1$, assuming the pedal is moving at a constant speed, the power you deliver is then $P = Fv$. But if the force is at an angle of $\theta = 45°$, as in **Figure 8-8b,** the power you deliver is smaller: $P = (F\cos 45°)v = 0.707\ Fv$. So the power you deliver to the pedal is maximized when you push straight down on a downward-moving pedal. (It's also maximized when you pull straight up on an upward-moving pedal, so again $\theta = 0$, which is why many serious cyclists use toe clips or straps to lock their feet onto the pedals: You cannot do this work if you don't have a way to pull on the pedal as well as push.) You can notice this more easily if you stand up on the pedals; you can feel the greater power being delivered when your feet move vertically.

We can now use the concept of power to explain the difference between walking and running that we mentioned at the beginning of this section. Whether you walk or run, you expend some of the internal energy $E_{internal}$ stored in your body to exert a static friction force backward on the ground. Because of Newton's third law, the ground exerts a forward static friction force on you (static as long as your feet don't slide). The friction force of the ground is in the direction you are going, but the point of contact of this force doesn't move with you, so no work is done on you. This force lets you convert your own internal energy into kinetic energy, because your center of mass does move while you are pushing on the ground and the ground is pushing back. Your total change in kinetic energy from when you start to when you stop is zero, so directly applying conservation of energy is not very helpful in this case, but we can get at the answer qualitatively using what we know so far. (Presenting a coherent qualitative solution to a problem is an important skill to acquire in this course!)

Whether you are walking or running, you have to keep pushing backward on the ground to keep going. Every time you push, you should speed up, so if your speed is constant, there must be a force exerted on you resisting your motion. There is some kinetic friction (as your shoes slip a little) with the ground, but a large resistance is due to

(a)

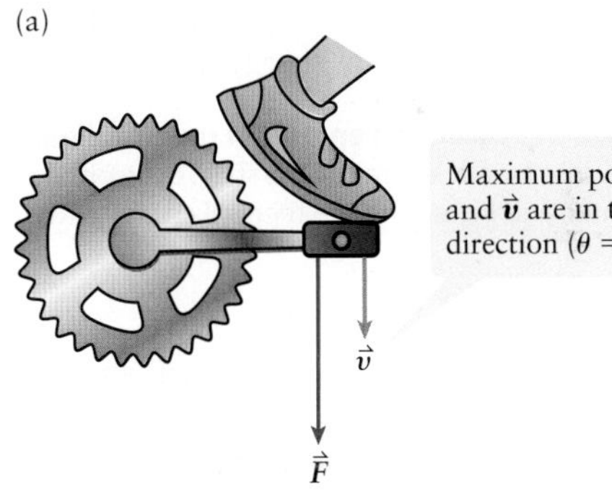

Maximum power when $\vec{F}$ and $\vec{v}$ are in the same direction ($\theta = 0$)

(b)

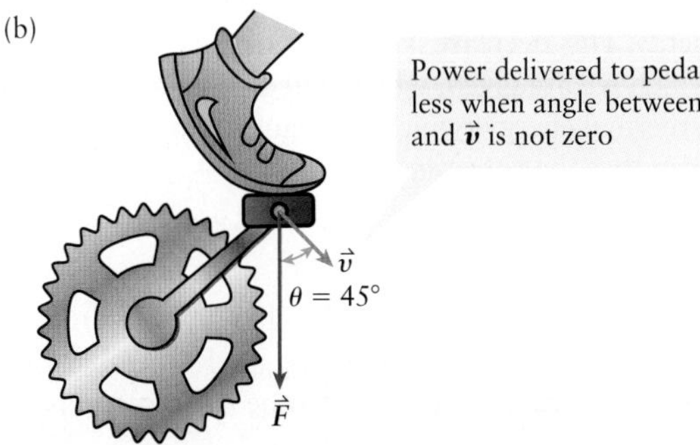

Power delivered to pedal is less when angle between $\vec{F}$ and $\vec{v}$ is not zero

Figure 8-8 Power in cycling The power you deliver to a bicycle pedal depends on the angle between the force you exert on the pedal and the direction of the pedal's motion.

the air around you (this resistance increases as you go faster, but we are going to ignore that for now). You are exerting a force on the air as you move forward, and you exert this force over the same distance whether you walk or run, so you convert at least the same amount of energy when you run as when you walk. Since you cover the same distance in less time when you run, you need to divide the amount of energy you converted by the smaller time. This means you converted more energy per unit time to overcome the forces pushing back on you as you moved forward when you were running. *Converting more energy per unit time means you deliver more power*. If we did consider that the resistance is greater the faster we move, that would mean we would need to exert an even larger force on the air when we went faster, increasing the total energy we needed to convert as well as decreasing the time, which makes the power delivered even greater.

When you convert energy more rapidly your muscles consume internal energy more rapidly and also require more oxygen per second. You see this when your respiration rate goes up while running in order to inhale more oxygen per second, and your heart rate goes up in order to more rapidly supply your muscles with oxygenated blood. This elevated heart rate is why running is more demanding than walking as a cardiovascular exercise.

In walking, running, and all other forms of exercise, only part of the chemical energy that you expend goes into changing your kinetic energy and maintaining your speed in the presence of various types of friction interactions. The rest of the chemical energy goes into warming your body and your surroundings. The faster the rate at which you convert energy—that is, the greater your power output—the more rapidly you use up your chemical energy and the more rapidly your thermal energy increases (the more rapidly you warm up). Example 8-5 explores this for one particular type of vigorous exercise.

EXAMPLE 8-5 Power in a Rowing Race

Rowing demands one of the highest average power outputs of all competitive sports (**Figure 8-9**). In a typical rowing race a racing shell (a racing row boat) travels 2000 m in 6.00 min, during which time each rower has an average power output of 4.00×10^2 W and expends 7.20×10^5 J = 720 kJ of chemical energy. (a) How much internal energy does each rower provide to propel the shell? (b) How much thermal energy does each rower produce?

Figure 8-9 Power and energy in rowing These athletes are converting chemical energy into kinetic energy of the shell, overcoming friction between the shell and the water, and increasing the thermal energy of themselves, the shell, and the water.

Set Up

We are given the rower's power output P and the time Δt over which that power is delivered, so we can use Equation 8-6 to find the total energy that the rower provides. The rower has no gravitational potential energy change, so all the energy the rower delivers must come from a decrease in the rower's internal energy. This decrease is the sum of the change in chemical energy (which we are given, which will also be negative) and the change in thermal energy (which we want to find, which will be positive, given that the rower gets warmer).

$$P = \frac{\Delta E}{\Delta t} \quad (8\text{-}6)$$

Solve

(a) Solve Equation 8-6 for the change in internal energy of the rower. Remember E here is E_{internal}. Remember that 1 W = 1 J/s, and be careful to express the time Δt in seconds.

Equation 8-6:

$$P = \frac{\Delta E}{\Delta t}, \text{ so}$$

$$\Delta E = P\Delta t$$

$$P = 4.00 \times 10^2 \text{ W} = 4.00 \times 10^2 \text{ J/s}$$

$$\Delta t = (6.00 \text{ min})\left(\frac{60 \text{ s}}{1 \text{ min}}\right) = 3.60 \times 10^2 \text{ s}$$

$$\Delta E_{\text{internal}} = (4.00 \times 10^2 \text{ W})(3.60 \times 10^2 \text{ s}) = 1.44 \times 10^5 \text{ J} = 144 \text{ kJ}$$

(b) We can now use the fact that the total internal energy change $\Delta E_{\text{internal}}$ is the negative of the energy that the rower delivered. We know the change in the rower's chemical energy is $\Delta E_{\text{chemical}} = -7.20 \times 10^5$ J (this is negative because the rower's chemical energy decreases). So we can find the change in thermal energy $\Delta E_{\text{thermal}}$.

$-\Delta E = \Delta E_{\text{internal}}$

$\Delta E_{\text{internal}} = -\Delta E = -1.44 \times 10^5$ J

This is the sum of $\Delta E_{\text{chemical}}$ and $\Delta E_{\text{thermal}}$:

$\Delta E_{\text{internal}} = \Delta E_{\text{chemical}} + \Delta E_{\text{thermal}}$, so

$$\begin{aligned}\Delta E_{\text{thermal}} &= \Delta E_{\text{internal}} - \Delta E_{\text{chemical}} \\ &= (-1.44 \times 10^5\ \text{J}) - (-7.20 \times 10^5\ \text{J}) \\ &= +5.76 \times 10^5\ \text{J} = +576\ \text{kJ}\end{aligned}$$

Reflect

Our answer makes sense. The sign worked out positive and the size seems appropriate. In most of the world chemical energy or food energy is expressed in kilojoules, or kJ. In the United States it is expressed in food calories (symbol Cal, also called kilocalories), where 1 Cal = 4.186 kJ. The unit calorie (not capitalized) is 1/1000th of a food calorie, just 4.186 J. So our rower "burned" 720 kJ or 172 food calories during the race. This sounds reasonable, being about the number of calories in a typical granola bar.

We can also see that, as expected, much less than 100% of the chemical energy used was provided to propel the shell, only 0.200 or 20.0%. The other 80.0% went into warming the rower and his surroundings. Fortunately, much of the thermal energy that rowers produce gets transferred to the air around them, or they would all get quite sick! The kinetic energy of the boat will end up dissipated, primarily increasing the thermal energy of the water as the boat comes to rest.

Chemical energy used:

$$(720\ \text{kJ})\left(\frac{1\ \text{Cal}}{4.186\ \text{kJ}}\right) = 172\ \text{Cal}$$

Fraction of chemical energy that was transferred to the shell:

$$\frac{144\ \text{kJ}}{720\ \text{kJ}} = 0.200$$

Fraction of chemical energy that went into thermal energy:

$$\frac{576\ \text{kJ}}{720\ \text{kJ}} = 0.800$$

NOW WORK Problems 1, 2, 7, and 8 from The Takeaway 8-4.

The wind turbine in the opening picture of this chapter is used to generate electricity commercially. Such wind turbines convert the kinetic energy of the air into the motion of electrical charge in a process you will learn about in AP® Physics 2. The rate at which large commercial turbines convert wind energy into electrical energy, their power output, is as much as 3,600,000 W under ideal conditions. Turbines come in a range of sizes, from the commercial ones at around 100 m in diameter (that cost millions of dollars) down to 20-cm ones you can buy at a local outdoor store for your own use (around a hundred dollars) that generate as much as 400 J each second (400 W).

Exam Tip

Be sure that you carry units through your calculations, and have units with all of your answers. Be careful to make the appropriate substitution if there is a prefix (*kilo-* for example), or carry it along. Always check that your answer appears to be the right order of magnitude, because if it is not, you might have dropped a unit prefix, such as accidentally using meters instead of millimeters.

THE TAKEAWAY for Section 8-4

✔ Power is the amount of energy transferred from one object or system to another, or converted from one type into another, per unit of time.

✔ The power delivered to an object is related to the speed of the object and the magnitude and direction of the force exerted on the object that is doing the work causing the energy transfer.

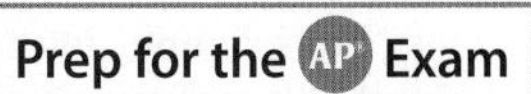

AP® Building Blocks

1 The power output of professional cyclists averages about 350 W.
(a) How much energy does a typical 70-kg pro cyclist (70 kg is the mass of both cyclist and bicycle) expend during a 12.3-km-long ride at an average speed of 22.5 km/h on a flat road?
(b) How much energy does the cyclist expend if she maintains the same average speed during a 12.3-km-long climb on a mountain road with a 3.9° average slope?

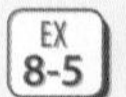

2. Neglecting extraneous factors such as wind resistance, and given that the human body is only about 25% efficient, how much power must an 80-kg runner produce to run at a constant speed of 3.75 m/s up an 8° incline?

Skill Builders

3. Approximately 1.7×10^{17} watts of solar power have been transferred to Earth continuously for the last 3.8 billion years with very little warming. Explain why the rate of energy transfer out of the Earth system must be approximately equal to the solar power received.
4. Photovoltaic cells convert radiant energy to electric potential energy that can be used to provide power to electric devices. The maximum theoretical efficiency (energy output/energy input) for photovoltaic cells is 0.86. The actual efficiency in 2017 was closer to 0.2.
   (a) Justify the selection of data needed to estimate the minimum area of photocells needed to capture the equivalent of the total energy consumption in your state.

State	P/m^2	GJ/person
Alaska	0.7 kW/m²	900
New York	1.0 kW/m²	200
Texas	1.2 kW/m²	500

   The table shows the approximate solar power received per square meter averaged over the year, and the yearly average energy consumption per person for three states.
   (b) Estimate the area of photocells needed to supply the total energy consumption for a person in each of these states and compare that area with the typical studio apartment (50 m²) and the median area of a home (250 m²) in the United States.
5. In many regions of the United States wind power production is increasingly important. In 2017 Texas had 21 GW of wind power generation. The turbines in wind farms are typically more than 70 m tall and have three blades, some greater than 50 m in length, completing a full rotation every 3 or 4 s. It was once common to see much shorter, typically 10-m, towers with 12-blade disks throughout the Midwest and Great Plains used to pump water. In any given region, wind speeds tend to increase with height. A perpendicular surface projecting behind the blade array kept the disk of the array facing into the wind. The force of the wind, F, struck each tilted blade, causing the disk to rotate as much as one full rotation per second.
   (a) Designs of the shape and number of blades are determined experimentally and guided by design equations. For example, in the design of the short-tower multi-blade turbines used to pump water for wind speeds greater than 5 m/s, the average power output for a 12-blade array with a radius of 1 m was found to be $P = 120\ \text{kg} \cdot \text{m/s}^2 \cdot v$, where v is the wind speed. Predict the power output at a wind speed of 10 m/s.

   The design equation for power in part (a) for a turbine was for a fixed orientation, θ, of 25° for each of the blades, as shown in the figure. With the wind perpendicular to the array, the component of the wind's force, F, that causes a rotation of the blades is $F \sin\theta \cos\theta$. A design equation for the dependence of power output on θ was developed,

Orientation of the blade with respect to the plane of the array

Top view

Front view

Wind direction

θ

Orientation of plane of the array

   $P = c \sin\theta \cos\theta\, v$, where c is a constant to be determined.
   (b) Calculate the value of c for a 12-blade array with a radius of 1 m and a fixed blade orientation of 25° using the two forms of the design equation.

   In later versions of these short-tower water pumps, the rotation of the blades was used to drive an electric motor. An electric motor operates best when the power delivered to it is constant. To maintain the rotation speed appropriate for the motor, the tilt angle θ of the blade was mechanically adjusted to a value that depended on the wind speed, θ_{match}.
   (c) Construct a mathematical representation from which the angle θ_{match} could be calculated as a function of wind speed, v, that is designed for a $\frac{1}{2}$-horsepower (0.373-kW) motor.
   (d) Explain the design differences between the turbines used in modern wind farms and the earlier turbines used to pump water. Include height, rotational speed, and number of blades.
6. Lithium-ion batteries are primarily intended for mobile storage of power supplied by large-scale wind, solar, and conventional power plants. Several companies are in competition with each other to manufacture and sell such lithium-ion batteries. By 2021 it is expected that yearly production of batteries will reach a storage capacity of 1 terajoule (10^{15} J). Tesla manufactures battery electric vehicles (BEVs) with battery capacities between 250 and 360 MJ with a range of up to 100 km. The current generation of BEVs can recharge their batteries at home or at charging stations at rates up to 7 kW.
   (a) As an illustration of the challenges involved in large-scale adoption of this technology, calculate the charging time for a completely discharged 360-MJ battery.
   (b) Pose three scientific questions that will need to be addressed by the design of a transportation system based on BEVs that includes distributed charging stations.

AP® Skills in Action

EX 8-5 7. Two of the fastest elevators in the world are found in the Taipei 101 building. They ascend at 16.83 m/s, rising 382.2 m to the building's 89th floor. Each of these elevators has a maximum load capacity, including the cabin and passengers, of 1600 kg. What power is required as a maximum load is being transported from the lowest level to the top floor at this constant speed?

EX 8-5 8. A tower crane has a hoist motor rated at 125 kW. Assuming the crane is limited to using 70% of its maximum hoisting power for safety reasons, what is the shortest time in which the crane can lift a 5700-kg load over a distance of 85 m?

8-5 Gravitational potential energy is much more general, and profound, than our near-Earth approximation

In Chapter 7 we introduced the ideas of kinetic energy (Section 7-3) and potential energy (Section 7-6). We found that if a force is *conservative*—the work done by that force on an object as it moves depends only on the object's starting point and ending point, not on the path the object takes between them—then we can associate that force with potential energy by changing the system to include the source of the force.

One particularly important conservative force is the gravitational force on an object of mass m that is near Earth's surface, where since g is approximately constant, $F_g = mg$. We found that the potential energy associated with this force when we place Earth in the system is $U_{grav} = mgy$, where y is the height of the object above some level that we choose to call $y = 0$. But because this simple expression for U_{grav} is based on the expression $F_g = mg$ for gravitational force, it *cannot* be correct for the more general form of the gravitational force between two objects $F = Gm_1m_2/r^2$ we introduced in Chapter 6. What, then, is the correct expression for the gravitational potential energy of two objects of mass m_1 and m_2 separated by a distance r?

Deriving this expression from the law of universal gravitation is a task that requires calculus, which is beyond our scope. Instead we'll just present the answer, and then examine it to see why it makes sense:

Gravitational constant (same for any two objects) | **Masses** of the two objects

Gravitational potential energy of a system of two objects (1 and 2)

$$U_{grav} = -\frac{Gm_1m_2}{r} \quad (8\text{-}8)$$

Center-to-center distance between the two objects

The gravitational potential energy of a system of two objects is zero when those objects are infinitely far apart. If the objects are brought closer together (so r is made smaller), U_{grav} decreases (it becomes more negative).

EQUATION IN WORDS
Gravitational potential energy

The expression in Equation 8-8 looks nothing like the familiar formula $U_{grav} = mgy$. To understand the difference let's consider two different cases in which gravitational potential energy changes (**Figure 8-10**).

In Figure 8-10a a woman on Earth's surface raises a book of mass m_{book} a vertical displacement d. Earth exerts a gravitational force on the book of magnitude $m_{book}g$, so she must exert an upward force of the same magnitude. She does an amount of work $m_{book}gd$ in the process. The book begins and ends at rest, so the work $m_{book}gd$ that she does goes into increasing the gravitational potential energy of the Earth–book system: $\Delta U_{grav} = m_{book}gd$. If the woman stands on a ladder and once again lifts the book by a vertical displacement d, the gravitational force she works against is still $m_{book}g$ since the gravitational force exerted by Earth changes hardly at all near Earth's surface. So she again does work $m_{book}gd$, and the gravitational potential energy increases by the same amount $m_{book}gd$ as before. As Figure 8-10b shows, the gravitational potential energy increases in direct proportion to the object's height y, so the graph of U_{grav} versus height is a line: $U_{grav} = m_{book}gy$.

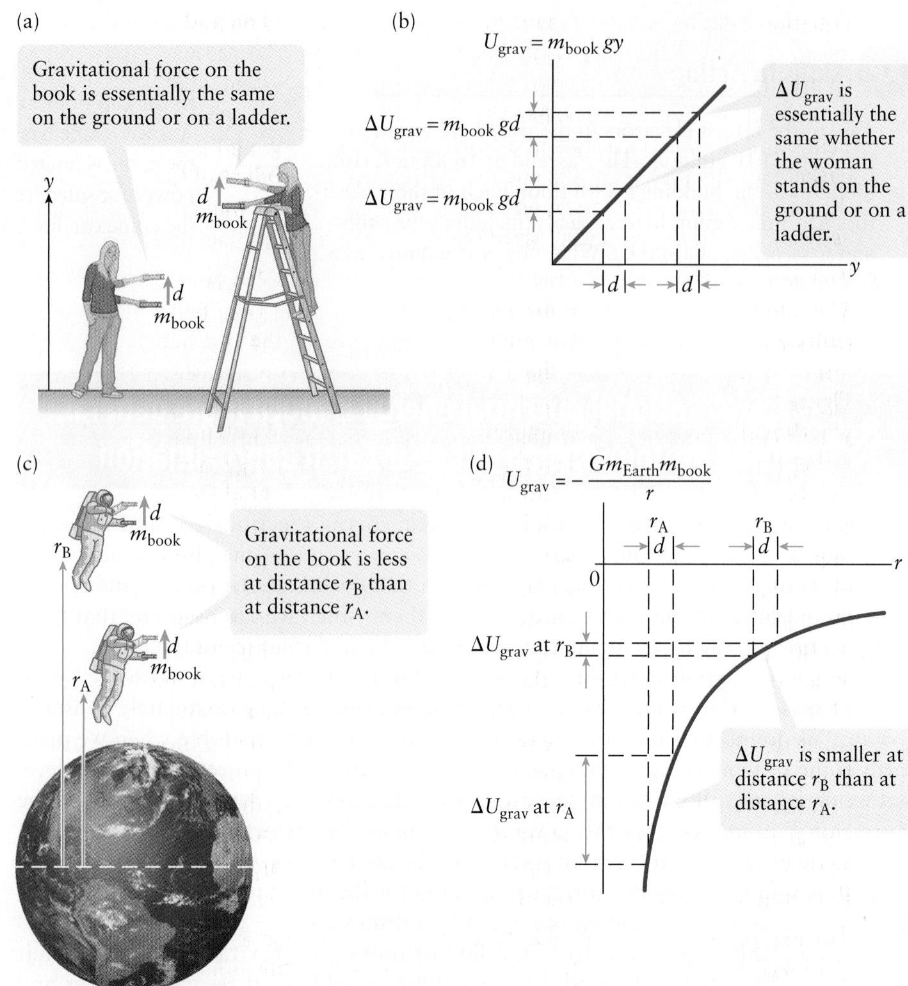

Figure 8-10 Gravitational potential energy near Earth's surface and far away (a) A woman lifts a book of mass m_{book} a vertical displacement d while standing on the ground. She then repeats the process while standing on a ladder. (b) The gravitational force, and so the change in gravitational potential energy, is basically the same in both cases, so the graph of U_{grav} versus height is a line. (c) The woman again moves the book a displacement d away from Earth, first at distance r_A from Earth's center, then at distance r_B. (d) The gravitational force and the change in gravitational potential energy are less at distance r_B than at distance r_A, so the graph of U_{grav} versus distance is a curve that gets shallower as r increases.

Now imagine that this same person dons a spacesuit and brings the book with her to a point in space a distance r_A from the center of Earth (Figure 8-10c). At this point the gravitational force that Earth exerts on the book has magnitude $F_A = Gm_{\text{Earth}}m_{\text{book}}/r_A{}^2$. If she moves the book a small displacement d away from Earth, she has to exert a force of approximately that magnitude on the book and so does an amount of work $F_A d = (Gm_{\text{Earth}}m_{\text{book}}/r_A^2)d$. (The force actually changes with distance, but the change will be very slight if we keep d very small.) Just as on Earth, this work goes into increasing the gravitational potential energy of the Earth–book system. If she now moves to a greater initial distance r_B from the center of Earth and repeats the process, Earth's gravitational force has a smaller magnitude $F_B = Gm_{\text{Earth}}m_{\text{book}}/r_B{}^2$. So at this greater distance she does a smaller amount of work to move the book a displacement d away from Earth, $F_B d = (Gm_{\text{Earth}}m_{\text{book}}/r_B{}^2)d$, and so adds a smaller amount of gravitational potential energy. This means that the gravitational potential energy does *not* increase in direct proportion to the distance from the center of the Earth, r, but by a smaller value for the same displacement of an object, d, the larger r gets. The graph of U_{grav} versus distance is not a line, but a curve that becomes shallower with increasing values of r (Figure 8-10d). This is just what Equation 8-8 tells us, which helps to justify this expression for gravitational potential energy.

More on Gravitational Potential Energy

Here are three important attributes of Equation 8-8 for gravitational potential energy.

1. *The gravitational potential energy of two interacting objects is a property of the system of the two objects.* We learned in Section 7-6 that the gravitational potential energy $U_{\text{grav}} = mgy$ is the potential energy of the Earth–object system for an object of mass m near Earth's surface. In $U_{\text{grav}} = -Gm_1m_2/r$, as given by

Equation 8-8, the *system* is made up of an object of mass m_1 and an object of mass m_2. Earth is not necessarily one of those objects: This same equation can be used for a system of any two spherical objects or a system where one of the two objects is much smaller than the other, even if it is not spherical, such as Venus and the Sun, two stars in a binary system, a rocket and the Moon, or a basketball and a fly. Since the energy depends on the product of the masses of the two objects in the system, we can see that the gravitational potential energy associated with two everyday-sized objects such as a basketball and a fly is tiny!

2. *The gravitational potential energy as given by Equation 8-8 is never positive.* The minus sign in Equation 8-8 means that the value of U_{grav} given by this expression is always negative, approaching zero when the two objects are infinitely far apart (because the denominator depends on r, as shown in Figure 8-10d). Don't be concerned by this! We learned in Section 7-6 that what's really physically meaningful are not the values of potential energy, but rather the *differences* between the values of potential energy at two different points. That means we can choose the point where potential energy equals zero to be wherever it's convenient. Our choice is that $U_{grav} = 0$ when the two objects are infinitely far apart. With this choice, U_{grav} as given by Equation 8-8 decreases and becomes more and more negative as an object of mass m_1 and an object of mass m_2 move toward each other, just as gravitational potential energy decreases as a ball drops toward the ground (that is, toward Earth). When we consider things on the cosmic scale this choice of zero makes a lot of sense. Two objects extremely far apart with no kinetic energy, if there is no other type of potential energy or any external forces exerted upon them, would begin to fall toward each other because of the attractive gravitational interaction between them. As they begin to fall toward one another, their potential energy decreases (becomes more negative) and their kinetic energies increase as they get closer together, keeping their mechanical energy constant, as in Equation 8-2. In this case, that total energy is approximately 0.

3. *The gravitational potential energy as given by Equation 8-8 is consistent with the formula* $U_{grav} = mgy$. To see how this can be, imagine an object of mass m a height y_i above Earth's surface—that is, at a distance $r = R_{Earth} + y_i$ from Earth's center. If you lift this object to a new height y_f above the surface, its new distance from Earth's center is $r = R_{Earth} + y_f$. The change in the gravitational potential energy in this process is

$$\Delta U_{grav} = U_{grav,f} - U_{grav,i} = \left(-\frac{Gm_{Earth}m}{R_{Earth} + y_f}\right) - \left(-\frac{Gm_{Earth}m}{R_{Earth} + y_i}\right)$$
$$= Gm_{Earth}m\left(\frac{1}{R_{Earth} + y_i} - \frac{1}{R_{Earth} + y_f}\right)$$

If you express this in terms of a common denominator, you should be able to show that you get

$$\Delta U_{grav} = m\left[\frac{Gm_{Earth}}{(R_{Earth} + y_i)(R_{Earth} + y_f)}\right](y_f - y_i) \tag{8-9}$$

This looks like a horrible mess! But if the heights y_i and y_f are both small compared to Earth's radius R_{Earth}—that is, if the object of mass m remains close to Earth's surface—then the quantities $R_{Earth} + y_i$ and $R_{Earth} + y_f$ are both approximately equal to R_{Earth}. In that case the quantity in square brackets in Equation 8-9 becomes Gm_{Earth}/R_{Earth}^2, which we found in Chapter 6 is equal to the value of g at Earth's surface (see Equation 6-8). Then Equation 8-9 becomes

$$\Delta U_{grav} = U_{grav,f} - U_{grav,i}$$
$$= mg(y_f - y_i) = mgy_f - mgy_i$$

That's precisely the result we would have obtained using $U_{grav} = mgy$! In other words if we restrict ourselves to looking at the motion of the object near the surface of Earth, we can safely use the expression from Section 7-6 for gravitational potential energy.

 **Exam Tip**

The potential energy of a spring U_s and gravitational potential energy of an object-Earth system U_G are given for a position (x and r, respectively) that represents a particular configuration of the system on the equation sheet. This is in contrast to the change in potential energy of an object-Earth system when the object is near Earth's surface, ΔU_g, which is given for a change in position of the object with respect to the surface of Earth. It is important to note that this is a hint about how they are defined. U_s and U_G, as given, are defined as zero at specific positions (U_s is zero when the spring is at its relaxed equilibrium length, U_G is zero when the center-to-center separation r = infinity) while U_g can be defined as zero at any position, and the form given on the equation sheet will still work.

Gravitational Potential Energy and Constant Total Mechanical Energy

If both objects interacting through the gravitational force are in a described system, we discuss only their gravitational potential energy; if no other objects or systems are exerting a force on that system, then the total energy E is all mechanical energy—the sum of the kinetic energy K and the gravitational potential energy U_{grav} given by Equation 8-8—and is *constant*, just as it was when we were using the expression for gravitational potential energy near Earth's surface. An example is a projectile that's launched from Earth's surface with a certain initial speed. Earth is so massive compared to anything we might launch from its surface that we can assume that there is no effect on our planet. (In Chapter 9 we will prove this is a great approximation!) Then, if we ignore the effects of air resistance during the projectile's initial climb through the atmosphere, the total mechanical energy of the system is constant. The following example demonstrates how to use this idea.

WATCH OUT !

Infinity is pretty big!

We use the term *infinity* a lot in this chapter. Whatever the true value of the initial potential energy you are neglecting by saying they are infinitely far apart is the size of the error you are introducing. So, according to how precise you want your answer, infinite separation can really just mean that the two interacting objects are very far apart compared to the size of either one. Two atoms across a living room are practically infinitely far apart for most calculations. When it comes to rocket ships and planets and stars, the distances get a lot larger, but for most purposes it just means really big.

EXAMPLE 8-6 How High Does It Go?

You launch a rocket of mass 2.40×10^4 kg straight upward from Earth's surface. The rocket's engines burn for a short time and a very short distance, giving the rocket an initial speed of 9.00 km/s (32,400 km/h or 20,100 mi/h), then shut off (**Figure 8-11**). To what maximum height above Earth's surface will this rocket rise? Earth's mass is 5.97×10^{24} kg, and its radius is 6370 km = 6.37×10^6 m. Ignore the effect of air resistance on the rocket as it ascends through the atmosphere.

U.S. Air Force/Science Source

Figure 8-11 A rocket launch What is the relationship between the initial speed of the rocket after the launch and the maximum height that the rocket reaches?

Set Up

Because we are neglecting air resistance, after the launch the rocket only interacts with Earth, so if we put Earth in the system we can use potential energy of the system, and total mechanical energy is constant.

With such a tremendous launch speed, we expect the rocket to reach a very high altitude. Because the rocket will not remain close to Earth's surface, we must use the more general form for gravitational potential energy given by Equation 8-8.

Initially the rocket has speed v_i = 9.00 km/s = 9.00×10^3 m/s and given that we are told it reaches this speed over a very

Total mechanical energy is constant:

$$K_i + U_{grav,i} = K_f + U_{grav,f} \quad (8\text{-}2)$$

Gravitational potential energy:

$$U_{grav} = -\frac{Gm_{Earth}m_{rocket}}{r} \quad (8\text{-}8)$$

Kinetic energy of rocket:

$$K = \frac{1}{2}m_{rocket}v^2 \quad (7\text{-}8)$$

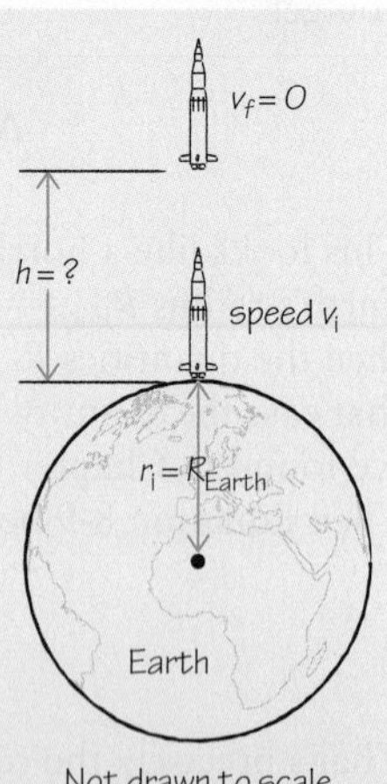

short distance, we will assume it is initially at a distance r_i from Earth's center equal to the radius of Earth. At its maximum height the rocket is at rest ($v_f = 0$) and is a distance h above the surface. Our goal is to determine h. (We know that the rocket must rise some while it is picking up this speed, but as long as h is large, neglecting the initial height above Earth's surface is a reasonable first approximation. We just need to make sure we are aware of what our approximations do to our answer.)

Solve

We first determine the total mechanical energy of the system using the given values for the rocket's initial speed and position.

When the rocket is initially launched approximately at the surface, its kinetic energy is

$$K_i = \frac{1}{2}m_{\text{rocket}}v_i^2 = \frac{1}{2}(2.40 \times 10^4 \text{ kg})(9.00 \times 10^3 \text{ m/s})^2$$
$$= 9.72 \times 10^{11} \text{ kg} \cdot \text{m}^2/\text{s}^2 = 9.72 \times 10^{11} \text{ J}$$

(Recall that $1 \text{ J} = 1 \text{ kg} \cdot \text{m}^2/\text{s}^2$.)

The gravitational potential energy of the Earth–rocket system is

$$U_{\text{grav,i}} = -\frac{Gm_{\text{Earth}}m_{\text{rocket}}}{r_i} = -\frac{Gm_{\text{Earth}}m_{\text{rocket}}}{R_{\text{Earth}}}$$
$$= -\frac{(6.67 \times 10^{-11} \text{ Nm}^2/\text{kg}^2)(5.97 \times 10^{24} \text{ kg})(2.40 \times 10^4 \text{ kg})}{6.37 \times 10^6 \text{ m}}$$
$$= -1.50 \times 10^{12} \text{ Nm} = -1.50 \times 10^{12} \text{ J}$$

(Recall that $1 \text{ J} = 1 \text{ Nm}$.)

The total mechanical energy is

$$E_i = K_i + U_{\text{grav,i}} = (9.72 \times 10^{11} \text{ J}) + (-1.50 \times 10^{12} \text{ J}) = -5.28 \times 10^{11} \text{ J}$$

When the rocket is at the high point of its trajectory, it is momentarily at rest, and its kinetic energy is zero. At this point the total mechanical energy is equal to the gravitational potential energy. Use this to solve for the rocket's distance r_f from Earth's center at its high point.

At the high point of the trajectory, the rocket's speed is $v_f = 0$, and its kinetic energy is

$$K_f = \frac{1}{2}m_{\text{rocket}}v_f^2 = 0$$

The total mechanical energy has the same value as when the rocket was launched:

$$E_f = K_f + U_{\text{grav,f}} = E_i = -5.28 \times 10^{11} \text{ J}$$

Because $K_f = 0$, this means that

$$U_{\text{grav,f}} = -\frac{Gm_{\text{Earth}}m_{\text{rocket}}}{r_f} = E_i = -5.28 \times 10^{11} \text{ J}$$

Solve for r_f:

$$r_f = -\frac{Gm_{\text{Earth}}m_{\text{rocket}}}{E_i}$$
$$= -\frac{(6.67 \times 10^{-11} \text{ Nm}^2/\text{kg}^2)(5.97 \times 10^{24} \text{ kg})(2.40 \times 10^4 \text{ kg})}{(-5.28 \times 10^{11} \text{ J})}$$
$$= 1.81 \times 10^7 \text{ m} = 1.81 \times 10^4 \text{ km}$$

Subtract Earth's radius from r_f to find the rocket's final height above the surface. The fact that the height when the initial velocity is achieved is actually away from the surface of Earth makes our answer a lower approximation; neglecting air resistance makes our answer a high approximation.

The rocket's final distance from Earth's center, r_f, equals the radius of Earth (R_{Earth}) plus the rocket's final height h above Earth's surface:

$$r_f = R_{Earth} + h$$

Solve for h:

$$h = r_f - R_{Earth} = 1.81 \times 10^4 \text{ km} - 6.37 \times 10^3 \text{ km} = 1.17 \times 10^4 \text{ km}$$

Reflect

The final height is nearly twice the radius of Earth *above the surface* of Earth! This justifies our decision to use Equation 8-8 for gravitational potential energy and to neglect the actual height it had achieved when it reached its maximum velocity. Had we used the expression $U_{grav} = m_{rocket}gy$, we would have gotten an incorrect answer because that expression assumes that g has the same value at all heights. In fact g decreases substantially in value at greater distances from Earth, which means that the rocket climbs much higher than the expression $U_{grav} = m_{rocket}gy$ would predict.

Extend: From Example 6-9 (Section 6-4), the value of g at a distance r from Earth's center is

$$g = \frac{Gm_{Earth}}{r^2}$$

At the rocket's maximum distance

$$g = \frac{Gm_{Earth}}{r_f^2} = \frac{(6.67 \times 10^{-11} \text{ Nm}^2/\text{kg}^2)(5.97 \times 10^{24} \text{ kg})}{(1.81 \times 10^7 \text{ m})^2} = 1.22 \text{ m/s}^2$$

which is *much* less than $g = 9.80 \text{ m/s}^2$ (remember m/s^2 is the same unit as N/kg) at Earth's surface.

$g = 1.22 \text{ m/s}^2$

$h \approx 1.17 \times 10^4 \text{ km}$

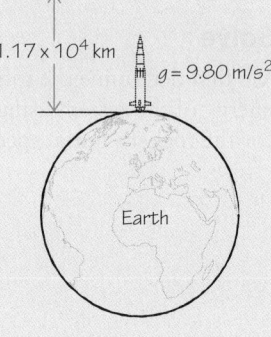

$g = 9.80 \text{ m/s}^2$

Earth

NOW WORK Problem 1 from The Takeaway 8-5.

WATCH OUT !

***g* is used in a variety of ways, but the units always reduce to the same thing.**

We first met g as the acceleration due to gravity at Earth's surface in Chapter 2. In Chapter 6 we found out that this is also the value of the gravitational field at the surface of Earth. The symbol g will, in the context of gravitational field, be used for other planets as well. In any situation, a gravitational field at some point in space is the gravitational force exerted on an object at that point in space, divided by the object's mass. Since, in general, $F = ma$, we can see that for gravitational interactions, gravitational field and acceleration due to gravity are always going to be the same. The value of g on the AP® Physics 1 equation sheet is the value of the acceleration due to gravity at Earth's surface, so in this text we will generally use the units given there, m/s^2.

Escape Speed

Like a ball thrown vertically upward, the rocket in Example 8-6 reaches a peak height and eventually returns to the surface. If the rocket is launched with a great enough speed, however, it will *never* return. That's because the gravitational force that Earth exerts on the rocket decreases with increasing distance from Earth and approaches zero as the rocket moves toward being an infinite distance from Earth (see Equation 6-7 and Figure 6-13). As in Example 8-6, the rocket engines give it a certain initial speed, but then shut off. The rocket then coasts upward and loses speed due to Earth's gravity. As the rocket climbs, the gravitational force on the rocket decreases, so the rocket loses an ever-smaller amount of speed per second. If the initial speed of the rocket is great enough, there will still be some speed remaining when the rocket is very far from Earth, and it will keep on going forever. In such a case, we say that the rocket has *escaped* from Earth. Alternatively, we could discuss this in terms of gravitational potential energy, which reaches a maximum of zero when the interacting objects are infinitely far apart. If the initial kinetic energy of the rocket is large enough so that its total energy is greater than zero when it launches, it will still be moving when the gravitational potential energy is zero. We will actually use this idea to find escape speed.

We can find the minimum speed necessary for a projectile launched from Earth's surface to escape from Earth by using Equation 8-8 for gravitational potential energy. As in Example 8-6, we'll ignore the effects of air resistance during the time that the projectile is passing through Earth's atmosphere, so the total energy of the projectile–Earth system will be the mechanical energy and it will be constant. The projectile (mass m) is launched from Earth's surface (that is, from a distance $r_i = R_{Earth}$ from Earth's center) with a speed v_i. The total energy just as the projectile is launched is found by substituting these values into the initial energy part of Equation 8-2:

$$E_i = K_i + U_{grav,i} = \frac{1}{2}mv_i^2 + \left(-\frac{Gm_{Earth}m}{R_{Earth}}\right)$$

When the projectile is a distance r from Earth's center and moving at a speed v, the total energy is found by substituting in the values to the final energy part of Equation 8-2. Since we are doing this for a general distance from the center of Earth, r, we leave off the subscript for final:

$$E = K + U_{grav} = \frac{1}{2}mv^2 + \left(-\frac{Gm_{Earth}m}{r}\right)$$

Since total energy is constant, E_i is equal to E so we can restate Equation 8-2 for a problem involving a launch from the surface of Earth as

$$\frac{1}{2}mv_i^2 + \left(-\frac{Gm_{Earth}m}{R_{Earth}}\right) = \frac{1}{2}mv^2 + \left(-\frac{Gm_{Earth}m}{r}\right) \quad (8\text{-}10)$$

Now let's suppose that the projectile just barely escapes Earth, so its speed is zero when it is infinitely far away. That corresponds to replacing v with zero and r with infinity in Equation 8-10. The reciprocal of infinity is approximately zero, so in this case both terms on the right-hand side of Equation 8-10 are zero: The projectile ends up with zero kinetic energy (it is at rest), and the gravitational potential energy ends up with a zero value (the projectile is infinitely far from Earth; see Figure 8-10d). Then Equation 8-10 becomes

$$\frac{1}{2}mv_i^2 + \left(-\frac{Gm_{Earth}m}{R_{Earth}}\right) = 0$$

The initial speed v_i in this case is called the **escape speed**, which we denote by the symbol v_{escape}. You may hear the term *escape velocity*, but, since we don't care about the direction, we will refer to this as escape speed. It is the minimum speed at which an object must be launched from Earth's surface to escape to infinity. To find its value, replace v_i with v_{escape}, and solve for v_{escape}. To do this, add $Gm_{Earth}m/R_{Earth}$ to both sides, and multiply both sides by $2/m$:

$$\frac{1}{2}mv_{escape}^2 = \frac{Gm_{Earth}m}{R_{Earth}}$$

$$v_{escape}^2 = \frac{2Gm_{Earth}}{R_{Earth}}$$

Finally, take the square root of both sides:

$$v_{escape} = \sqrt{\frac{2Gm_{Earth}}{R_{Earth}}} \quad (8\text{-}11)$$

Speed that a projectile must have at Earth's surface in order to **escape Earth's gravitational pull**

Gravitational constant

Mass of Earth

Radius of Earth

EQUATION IN WORDS
Escape speed

If you substitute $G = 6.67 \times 10^{-11}$ Nm^2/kg^2, $m_{Earth} = 5.97 \times 10^{24}$ kg, and $R_{Earth} = 6.37 \times 10^6$ m into Equation 8-11, you'll find that $v_{escape} = 1.12 \times 10^4$ m/s = 11.2 km/s (about 40,300 km/h or 25,000 mi/h). Rockets designed to send spacecraft to other planets must have powerful rocket engines (see Figure 8-11) to accelerate their

WATCH OUT !

You can never be "beyond" the pull of Earth's gravity.

There's a common misconception that if a rocket travels far enough from Earth, it reaches a point where it's "beyond" the pull of our planet's gravity. But Equation 6-7 shows that in fact the gravitational force that Earth exerts on an object of mass m, $F = Gm_{Earth}m/r^2$, becomes zero only when the distance r from Earth to the object is *infinity*. The force gets progressively weaker as the object moves farther away from Earth, but it never entirely disappears. A rocket can escape from Earth if it has enough kinetic energy to travel to infinity, but the rocket still feels that attraction at all points along its infinite voyage. As it moves away from Earth, the many other massive objects in space will also exert gravitational forces on it, so this is a simplification.

payload to such high speeds. To find the escape speed for an object launched from some height above Earth, you would need to replace R_{Earth} with the distance from the center of Earth to the initial launching point.

Notice that the launch speed of the rocket in Example 8-6 was only 9.00 km/s, which is less than escape speed. That's why the rocket in that example did *not* escape but reached a maximum distance from Earth before falling back. To have the rocket escape, it would have to be launched at 11.2 km/s or faster.

We can also use Equation 8-11 to find the escape speed from the surface of *any* planet or satellite: Just replace m_{Earth} and R_{Earth} with the mass and radius of the object from which the projectile is launched.

WATCH OUT !

Space travel is actually more complicated.

You may have noticed making calculations like this is a lot more stressful in science fiction movies. This is because Earth is not the only object exerting a gravitational force on your rocket. A rocket doesn't have to get very far away for other objects in our solar system to be exerting significant forces on it. There are also tricks you will learn with a little more physics that can help effectively increase the launch speed for long trips. You may have heard of the term *launch window*, implying a small period of time when the launch of a rocket or satellite will be successful. This is because Earth itself is moving through space. If we can line up the orbital velocity of Earth with the direction the rocket has to leave Earth's surface to have the trajectory it needs to get where we are sending it, Earth itself serves as a booster rocket for the launch. (At 30 km/s, this speed is enough to travel from Earth to the Moon in 4 hours, so it is quite a boost!) In AP® Physics 1, we will generally deal with simplified systems like the examples in this chapter—these form a solid foundation for learning the more complicated stuff!

EXAMPLE 8-7 Escaping from a Martian Moon

You have landed your spacecraft on Phobos, the larger of the two airless moons of Mars (**Figure 8-12**). Phobos is roughly spherical with an average radius of 11.1 km = 1.11×10^4 m. By dropping a baseball while standing on Phobos, you find that g has a very small value at the moon's surface, only 0.00580 m/s^2. At what speed would you have to throw the baseball to have it escape Phobos?

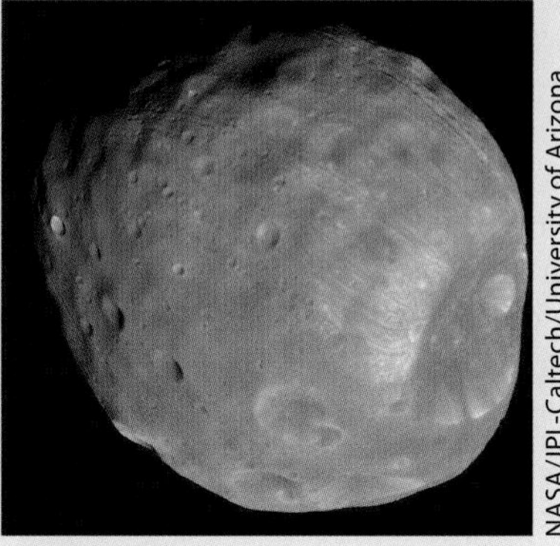

NASA/JPL-Caltech/University of Arizona

Figure 8-12 Phobos What is the escape speed from this miniature moon of Mars, just 11.1 km in radius?

Set Up

To find the escape speed v_{escape} using Equation 8-11, we need the radius of Phobos, which we are given, and the mass of Phobos, which we are *not* given. However, we recall from Equation 6-8 in Section 6-4 that the value of g on Earth's surface is related to Earth's mass and radius. We can write this same equation for Phobos, which will let us use the measured value of g on Phobos to determine its mass.

Escape speed from Phobos:

$$v_{\text{escape}} = \sqrt{\frac{2Gm_{\text{Phobos}}}{R_{\text{Phobos}}}} \quad (8\text{-}11)$$

Value of g at the surface of Phobos:

$$g = \frac{Gm_{\text{Phobos}}}{R^2_{\text{Phobos}}} \quad (6\text{-}8)$$

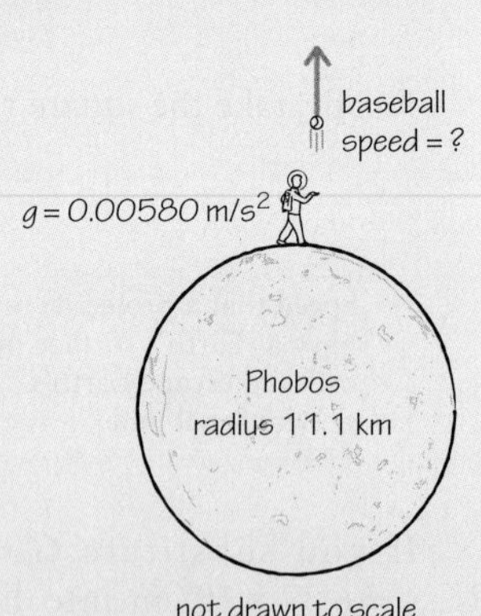

Solve

First use Equation 6-8 to determine the mass of Phobos.

From Equation 6-8:

$$g = \frac{Gm_{\text{Phobos}}}{R_{\text{Phobos}}^2}$$

(Recall that $1\ \text{N} = 1\ \text{kg} \cdot \text{m/s}^2$.)

Solve this for the mass of Phobos:

$$m_{\text{Phobos}} = \frac{gR_{\text{Phobos}}^2}{G} = \frac{(0.00580\ \text{m/s}^2)\,(1.11 \times 10^4\ \text{m})^2}{6.67 \times 10^{-11}\ \text{Nm}^2/\text{kg}^2}$$

$$= 1.07 \times 10^{16}\,\frac{\text{kg}^2 \cdot \text{m/s}^2}{\text{N}} = 1.07 \times 10^{16}\ \text{kg}$$

Then use Equation 8-11 to find the escape speed from Phobos.

(We once again use $1\ \text{N} = 1\ \text{kg} \cdot \text{m/s}^2$.)

Escape speed:

$$v_{\text{escape}} = \sqrt{\frac{2Gm_{\text{Phobos}}}{R_{\text{Phobos}}}}$$

$$= \sqrt{\frac{2(6.67 \times 10^{-11}\ \text{Nm}^2/\text{kg}^2)\,(1.07 \times 10^{16}\ \text{kg})}{1.11 \times 10^4\ \text{m}}}$$

$$= \sqrt{129\,\frac{\text{Nm}}{\text{kg}}} = \sqrt{129\,\frac{\text{kg} \cdot \text{m}}{\text{s}^2}\,\frac{\text{m}}{\text{kg}}} = \sqrt{129\,\frac{\text{m}^2}{\text{s}^2}}$$

$$= 11.3\ \text{m/s} = 40.8\ \text{km/h} = 25.4\ \text{mi/h}$$

Reflect

The escape speed on Phobos is almost exactly 1/1000th the escape speed on Earth! In general, the smaller the planet or moon, the lower the escape speed.

A typical high school baseball pitcher can easily throw a baseball at more than 25 m/s (90 km/h or 56 mi/h), so a well-trained astronaut should have no problem throwing a baseball at the 11.3-m/s escape speed on Phobos, even encumbered by a space suit.

NOW WORK Problem 3 from The Takeaway 8-5.

WATCH OUT !

Direction doesn't matter for escape.

Suppose you are on Phobos and throw the baseball with a speed faster than 11.3 m/s. **Figure 8-13** shows four possible directions in which you could throw the ball at that speed. Let's think about in which of these directions the ball would escape from Phobos. From the conservation principle we know so far, conservation of energy, it should not matter! If the ball has escape speed or greater, the total mechanical energy is enough for the ball to travel to infinity. The total mechanical energy is a scalar, not a vector, so the direction in which the ball is thrown doesn't matter as long as it does not hit the planet. Changing the direction in which the ball is thrown simply affects the path that the ball follows to infinity. As we study rotation later in this book, we will introduce another conservation law that will put additional constraints on this problem.

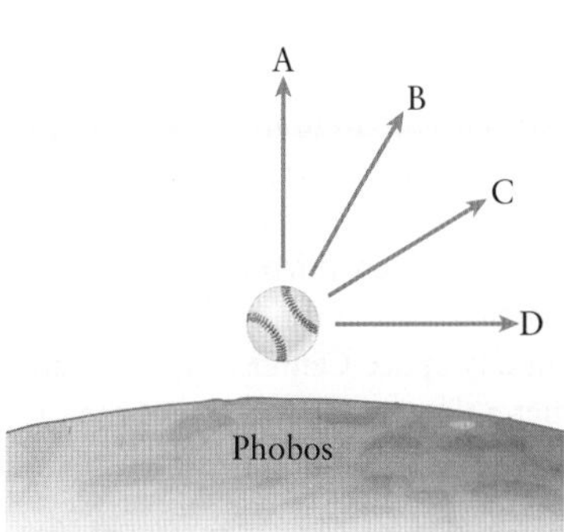

Figure 8-13 Four directions for escaping Phobos If a ball is thrown at escape speed, in what direction(s) can it be thrown so that it *will* escape?

THE TAKEAWAY for Section 8-5

✔ When analyzing situations in which an object does not remain close to the surface of Earth, you must use the expression for gravitational potential energy that is derived from Newton's law of universal gravitation to correctly use the conservation of energy.

✔ Using Newton's law of universal gravitation, the gravitational potential energy of two objects is always negative or zero. It is zero when the two objects are infinitely far apart, and it decreases (becomes more and more negative) as the objects move closer together.

✔ Escape speed is the speed at which a projectile must be launched from a planet or moon so that it never falls back to the world from which it was launched.

Prep for the AP® Exam

AP® Building Blocks

EX 8-6

1. A change in the position of an object near the surface of Earth from r_{Earth} to $r_{\text{Earth}} + h$ results in a change in the gravitational potential energy of the Earth–object system. Use Equation 8-8 to show that, although the gravitational potential energy of the system at both positions is negative, the change in gravitational potential energy of the system is positive.
2. A rocket with mass m is launched from the surface of Mars (m_{Mars}, r_{Mars}) with a speed v. In order to express the energy conservation principle in a form that can be used to predict the maximum height above the planet surface r that the rocket reaches, write the initial and final potential and kinetic energies in terms of m, m_{Mars}, r_{Mars}, and v.

EX 8-7

3. A satellite with a mass of 1.00×10^3 kg is orbiting Earth at 5.00×10^2 km above Earth's surface with a period of 1.50 h. If its orbital height changes to 5.40×10^2 km without an energy transfer, calculate its final kinetic energy.

AP® Skill Builders

4. (a) Explain the following properties of the gravitational potential energy of a pair of objects with mass.
   i. Approaching zero as the separation between the objects approaches infinity
   ii. Being negative

   (b) Construct a graphical representation of the gravitational potential energy of a pair of objects with mass as a function of separation and use the graph to provide evidence in support of your explanations above.
5. The surface of Earth is in motion as it completes one rotation every day. This means a point on Earth must move far enough to move in one complete circle every 24 hours. Thus rockets are launched from a moving platform—the surface of Earth.

   (a) The Kennedy Space Center is at a latitude of 28.6° N. Construct a diagram that displays the radius needed to calculate the speed of the launch site and calculate that speed.

   (b) Explain why the escape speed of a rocket that is launched from the surface of Earth is equal to $v_{\text{escape}} = \sqrt{\dfrac{2Gm_{\text{Earth}}}{r_{\text{Earth}}}}$ by identifying the initial and final points to which energy conservation is applied to obtain this result and assumptions about energies at these points.

   (c) Estimate the relative contribution to the minimum kinetic energy of escape of the motion of the rocket while still on the launch pad.

   (d) Justify the claim that the contribution of the motion of the site to the initial energy at launch is negligible, even when the launch is not made at escape speed, by calculating the initial kinetic energy per unit mass of a launch that results in an orbit similar to that of the ISS: $r = r_{\text{Earth}} + 400$ km and $T = 90$ min. *Note*: The real boosting speed Earth provides (and that determines "launch windows") comes from the motion of Earth around the Sun, not from the rotation of Earth.
6. At great expense, large rockets place satellites in orbit around Earth. Once the satellite is in orbit, the amount of work required to maintain the orbit is negligible.

   (a) Treating (i) the satellite as an object in the gravitational field of Earth and (ii) the satellite–Earth pair as a system, explain why the orbital motion requires no additional energy input.

   (b) Pose a question to which seeking an answer could lead to an important and lucrative engineering project that addresses the consequences of there being no financial penalty for leaving nonfunctioning satellites in orbit.
7. Rocket Raccoon needs to guide his ship through a binary star system and is running low on fuel. One is a main sequence star with a mass of 3×10^{30} kg. The other is a red giant with a mass that is 6 times larger. The stars are separated by a distance of 2×10^{11} m. A sketch of the gravitational potential energy of the two stars and the ship as a function of position along the line that passes through the centers of the two stars is shown in the figure.

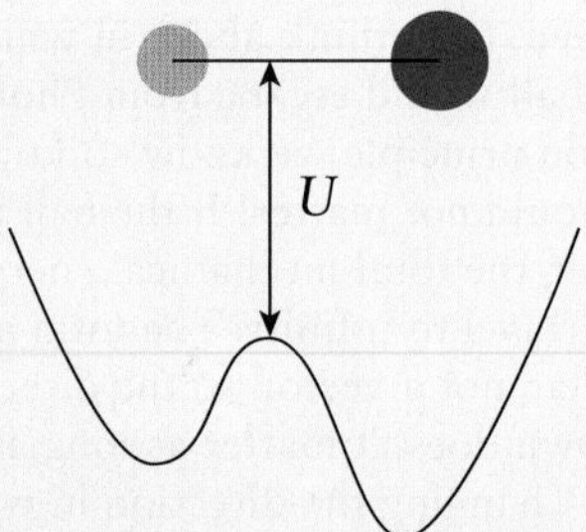

   (a) In terms of changes in the potential energy of the system composed of the two stars and the ship, why should Rocket Raccoon choose to travel between the stars at the point where the potential energy is maximum, shown in the diagram?

   (b) The maximum as shown is closer to the smaller star. Describe qualitatively why that is.

(c) We have defined the gravitational potential energy of a system of two objects to be zero when they are infinitely far apart, the r between them being very, very large. This is the point where, if we were instead talking about the gravitational force exerted on an object by a star, that force would be zero. Predict the location of the maximum potential energy of the stars–ship system if the two stars have the same mass and justify your prediction mathematically, using this concept.

8. How the Moon formed remains an unanswered question. Energy conservation principles have contributed to several possible answers.
   (a) One idea is that the Moon split from Earth. Evaluate gravitational potential energy stored in the two-object system when an object with mass m is separated into two objects with masses m_{Earth} and m_{Moon}, by a final distance $r_{\text{Earth–Moon}}$. Use these values: $m_{\text{Earth}} = 5.97 \times 10^{24}$ kg, $m_{\text{Moon}} = 7.35 \times 10^{22}$ kg, and $r_{\text{Earth–Moon,f}} = 3.84 \times 10^{8}$ m. The full calculation is actually pretty complicated because originally the objects were not separated, so we don't exactly know how to calculate how far apart they were to find the initial gravitational potential energy. We will simplify this by assuming they were already two objects, located right next to each other. This gives the original separation as just the sum of their radii ($r_E = 6.37 \times 10^{6}$ m, $r_M = 1.74 \times 10^{6}$ m) so $r_{\text{Earth–Moon,i}} = 8.11 \times 10^{6}$ m.
   (b) Some think that the splitting of Earth from the Moon could have been caused by an asteroid. Asteroids in our solar system have been observed with speeds between 1.50×10^{4} and 3.00×10^{4} m/s. Nearer to the Sun, where Earth's orbit lies, asteroids have speeds at the high end of this range. If the Moon were formed by a collision of an asteroid with an early Earth, predict the range of possible masses of the asteroid that would be needed to create the energy you found in part (a).
   (c) What type of data would be needed to test the prediction in part (b)?
   (d) What scientific questions could be pursued from the analysis of lunar samples that might help answer if the Moon was split from Earth?

WHAT DID YOU LEARN?

Prep for the AP Exam

Chapter learning goals	Section(s)	Related example(s)	Relevant section review exercises
Identify which kinds of problems are best solved with energy conservation and the steps to follow in solving these problems.	8-1		
	8-2		
	8-3	8-1, 8-2, 8-3, 8-4	1, 2, 3
Describe how changing the way you identify a system changes the description of conservation of energy (but not the results).	8-2		
	8-3	8-4	3
Identify the types of energy involved in an interaction between objects and systems.	8-2		
	8-3	8-1, 8-2, 8-3, 8-4	1, 2, 3
	8-5	8-6, 8-7	1, 3
Describe what power is and its relationship to work and energy.	8-4	8-5	1, 2, 7, 8
Describe the general expression for gravitational potential energy and how to relate it to the expression used near Earth's surface.	8-5	8-6, 8-7	1, 3

Chapter 8 Review

Key Terms

All the Key Terms can be found in the Glossary/Glosario on page G1 in the back of the book.

escape speed 365
mechanical energy 334
mechanical energy transfer 335
power 354
thermal energy 342
total mechanical energy 335
watt 354

Chapter Summary

Topic	Equation or Figure
Total mechanical energy is a special case of conservation of energy: In closed, isolated systems made up of objects experiencing only conservative interactions, the *sum* of kinetic energy and potential energy—a sum that we call the total mechanical energy of the system—keeps the same value; it stays *constant*. This is a special case of the conservation of energy.	Values of the **kinetic energy** K of an object at two points (i and f) during its motion $$K_i + U_i = K_f + U_f \quad (8\text{-}3)$$ Values of the **potential energy** of the system U at the same two points (i and f) **Change in the kinetic energy** K of an object during its motion **Change in the potential energy** U of a system containing the object during the object's motion $$\Delta K + \Delta U = 0 \quad (8\text{-}4)$$
Nonconservative forces can be found from applying the conservation of energy: The amount of energy transferred by a nonconservative external force exerted on a system $W_{\text{nonconservative}}$ is exactly equal to the negative of the change in internal energy $\Delta E_{\text{internal}}$ of all the interacting objects and the air around them if the nonconservative interaction were instead inside the defined system.	The **change in internal energy** of a system when the frictional interaction is included in the system **Magnitude of the friction** force exerted on an object $$-\Delta E_{\text{internal}} = W_{\text{nonconservative}} = -F_{\text{friction}}d \quad (8\text{-}5)$$ **"Work" done by the friction** force if it was external to the system **Distance object moved** while the friction force was exerted on it
Power: Power is the rate at which energy is transferred into or out of a system or converted from one type to another with respect to time. If work is being done on an object, the power delivered equals the rate at which work is done. This rate can change. When it does, Equation 8-6 can be used to calculate the average power. Equation 8-7 gives the instantaneous power, with F and v the force and velocity at that instant. In the case of constant force, Equation 8-7 can also be used to find average power, by substituting the average value of the velocity over the interval of time for which the average power is being calculated for v.	**Rate** at which energy is being converted or transferred to an object or system **Total amount of energy** being converted or transferred $$P = \frac{\Delta E}{\Delta t} \quad (8\text{-}6)$$ **Time** over which that conversion or transfer occurs $W = \Delta E$, Work is the amount of energy transferred by a mechanical process. Magnitude of the constant force $\vec{F}$ Rate at which energy is being transferred to an object or system by a force exerted on the object or system Angle between the directions of $\vec{F}$ and $\vec{d}$ $$P = \frac{W}{\Delta t} = \frac{(F\cos\theta)d}{\Delta t} = (F\cos\theta)v \quad (8\text{-}7)$$ Time over which that transfer occurs Magnitude of the displacement $\vec{d}$ Magnitude of the velocity $\vec{v}$ in the direction $\vec{d}$

Gravitational potential energy: The general expression for the gravitational potential energy of two objects follows from the law of universal gravitation. In this expression the potential energy is zero when the two objects are infinitely far apart; for any finite separation, U_{grav} is negative.

$$U_{grav} = -\frac{Gm_1m_2}{r} \quad (8\text{-}8)$$

- **Gravitational constant** (same for any two objects)
- **Masses** of the two objects
- **Gravitational potential energy** of a system of two objects (1 and 2)
- **Center-to-center distance** between the two objects
- The gravitational potential energy of a system of two objects is zero when those objects are infinitely far apart. If the objects are brought closer together (so r is made smaller), U_{grav} decreases (it becomes more negative).

Escape speed: If an object is launched from Earth's surface at the escape speed or faster, it will never fall back to Earth but will escape to infinity (neglecting air resistance).

$$v_{escape} = \sqrt{\frac{2Gm_{Earth}}{R_{Earth}}} \quad (8\text{-}11)$$

- **Gravitational constant**
- **Mass of Earth**
- **Speed** that a projectile must have at Earth's surface in order to **escape Earth's gravitational pull**
- **Radius of Earth**

Chapter 8 Review Problems

1. When does the kinetic energy of a rock that is dropped from the edge of a high cliff reach its maximum value? Answer the question for:

   (a) when the air resistance is negligible; and

   (b) when there is significant air resistance.

2. Starting from rest, a 75.0-kg skier skis down a slope 8.00×10^2 m long that has an average incline of 40.0°. The slope is covered with about a foot of loose powdery snow, so friction cannot be neglected. If the speed of the skier at the bottom of the slope is 20.2 m/s, how much work was done by nonconservative forces?

3. A book slides across a level, carpeted floor at an initial speed of 4.00 m/s and comes to rest after 3.25 m. Calculate the coefficient of kinetic friction between the book and the carpet. Assume the only forces exerted on the book are friction, gravity, and the normal force.

4. A ball is thrown straight up with an initial speed of 15.0 m/s. How far will the ball have traveled when it has one-half of its initial speed?

5. A water balloon is thrown straight down with an initial speed of 12.0 m/s from a second-floor window, 5.00 m above ground level. How fast is the balloon moving just before it hits the ground?

6. Starting from rest, a 30.0-kg child rides a 9.00-kg sled down a ski slope with negligible friction. At the bottom of the hill, her speed is 7.00 m/s. If the slope makes an angle of 15.0° with the horizontal, how far did she slide on the sled?

7. A skier leaves the starting gate at the top of a ski jump with an initial speed of 4.00 m/s as shown in the figure. The starting position is 120 m higher than the end of the ramp, which is 3.00 m above the snow. Find the final speed of the skier if he lands 145 m down the 20.0° slope. Assume the friction exerted by the ramp is negligible, but air resistance causes a 50% loss in the final kinetic energy. The GPS reading of the elevation of the skier is 4212 m at the top of the jump and 4039 m at the landing point.

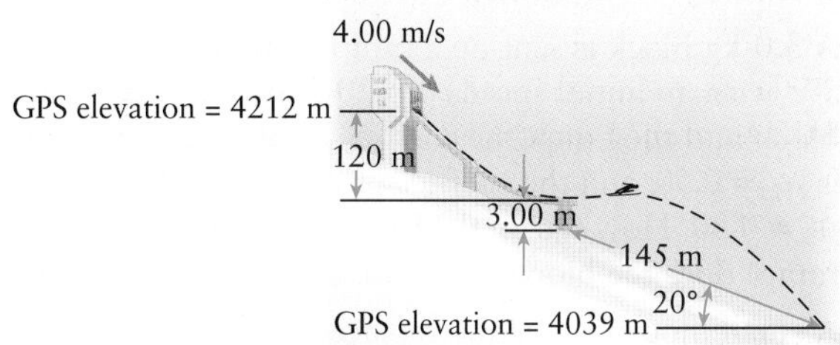

8. A child slides down a snow-covered slope on a sled. At the top of the hill, her mother gives her a push to start her off with a speed of 1.00 m/s. The friction force exerted on the sled is one-fifth of the combined weight of the child and the sled. If she travels for a distance of 25.0 m and her speed at the bottom is 4.00 m/s, calculate the angle that the hill makes with the horizontal.

9. Neil and Gus are having a competition to see who can launch a marble higher in the air using their own spring. Neil has a firm spring (k_{Neil} = 50.8 N/m), but it can be compressed only a maximum of $y_{Neil,max}$ = 14.0 cm. Gus's spring is less firm than Neil's (k_{Gus} = 12.7 N/m), but its maximum compression is greater ($y_{Gus,max}$ = 27.0 cm). Given that the mass of the marble is 4.50 g and ignoring the effects of air resistance, who can launch the marble to the higher position, and what is the winning height? Assume the marble starts from the same height in both cases, when the springs are at maximum compression.

10. Continuing with the previous problem, what is the speed of each marble as it leaves its spring?

11. An adult male dolphin is about 5.00 m long and weighs about 1600 N. How fast must he be moving as he leaves the water in order to jump to a height of 2.50 m? Ignore any effects due to air resistance.

12. A 3.00-kg block is placed at the top of a track consisting of two frictionless quarter circles of radius $R = 2.00$ m connected by a 7.00-m-long, straight, horizontal surface, and negligible friction, as shown in the figure. The coefficient of kinetic friction between the block and the horizontal surface is $\mu_k = 0.100$. The block is released from rest. What maximum vertical height does the block reach on the right-hand section of the track?

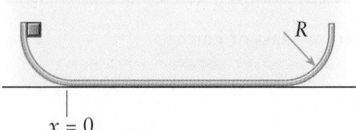

13. An average froghopper insect has a mass of 12.3 mg and reaches a maximum height of 290 mm when its takeoff angle is 58.0° above the horizontal. What is the takeoff speed of the froghopper?

14. A 1.00-kg object is attached by a thread of negligible mass, which passes over a pulley of negligible mass, to a 2.00-kg object. The objects are positioned so that they are the same height from the floor and then released from rest. What are the speeds of the objects when they are separated vertically by 1.00 m?

15. A 3.0-kg block is sent up a ramp of angle θ equal to 37° with an initial speed $v_0 = 20.0$ m/s. Between the block and the ramp, the coefficient of kinetic friction is $\mu_k = 0.50$, and the coefficient of static friction is $\mu_s = 0.80$. How far up the ramp (measured along the ramp) does the block go before it comes to a stop?

16. A small block of mass M is placed halfway up on the inside of a circular loop of radius R, as shown in the figure. The size of the block is very small compared to the radius of the loop. Determine an expression for the minimum downward speed with which the block must be launched in order to guarantee that it will make the full loop.

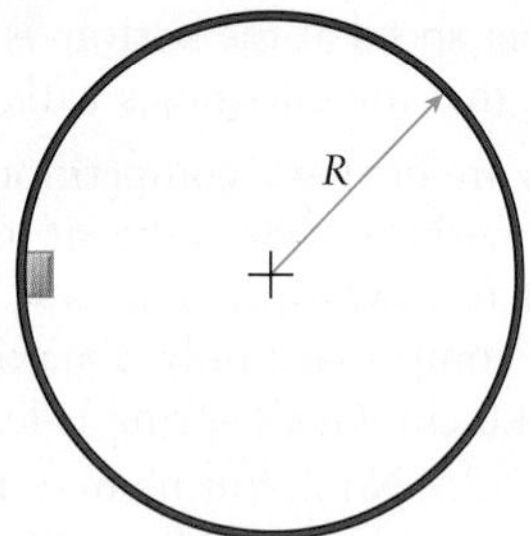

17. When a pitcher throws a fastball, the 145-g baseball goes from rest to a speed of 4.0×10^1 m/s.
    (a) Calculate the work done on the ball by the pitcher.
    (b) After the windup in the throwing phase of the pitch, the pitcher's arm extends behind the pitcher. At the instant of furthest extension the ball is nearly at rest. The body rotates, the arm moves through an arc, and the ball is released. The time interval of the throwing phase is only about 5.0×10^{-2} s. Calculate the average power of the pitch during the throwing phase.

18. A small asteroid that has a mass of 100 kg is moving at 200 m/s when it is 1000 km above the surface of the Moon.
    (a) How fast will the meteorite be traveling when it impacts the lunar surface if it is heading straight toward the center of the Moon?
    (b) The atmosphere of the Moon is negligible. How much work does the Moon do in stopping the asteroid? The radius and mass of the Moon are 1.737×10^6 m and 7.35×10^{22} kg, respectively.

19. Amit claims that if the force required to compress a spring increases as the length of the spring decreases, then the force should be inversely proportional to the length. Pose a scientific question that Kamala, Amit's sister, could ask to help Amit better understand how the spring force depends on the spring length.

20. A team of students is building a model roller coaster. They find that when the cart is released from the height h_1, as shown in the diagram, the cart stops rolling at height h_4. Sofia claims that it is a drag force that is responsible for the failure of the cart to maintain a constant mechanical energy. Jackie claims that it is friction between the cart and the track or friction between the axles and the wheels. The track is composed of three sections (blue, green, and red in the diagram). Each section is the same length, l. They have placed sensors at positions S_1 and S_2 that can be used to measure the speed of the cart at these points. The body of the cart is an open box into which they can place small metal spheres. The box can be removed, leaving just the wheels, and there is a second shorter box that can be mounted on the wheels.

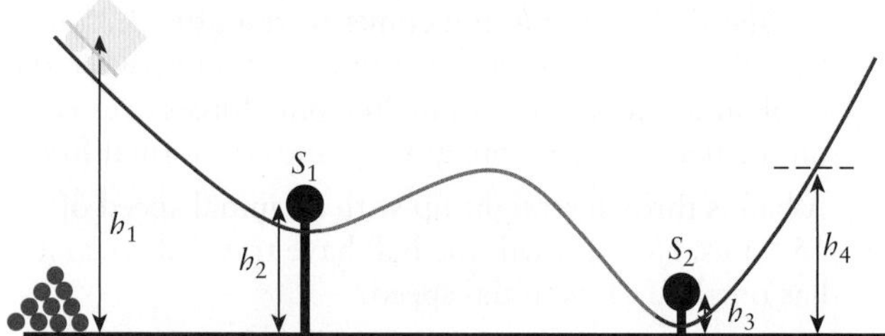

    (a) Express Sofia's claim in terms of mechanical energy and the drag force on the cart and identify the possible independent and dependent variables in this expression.
    (b) Express Jackie's claim in terms of mechanical energy and the friction force on the cart and identify the possible independent and dependent variables in this expression.
    (c) Describe two different designs that Sofia can use to collect the evidence needed to test her claim.
    (d) Describe two different designs that Jackie can use to collect the evidence needed to test her claim.

21. An object with mass m is placed against the free end of a spring (with spring constant k) that is compressed a distance Δx from the equilibrium length, which is a

distance of 1.25 m from the edge of the tabletop. Once released, the object slides across the tabletop and eventually lands a distance d from the edge of the table onto the floor, as shown.

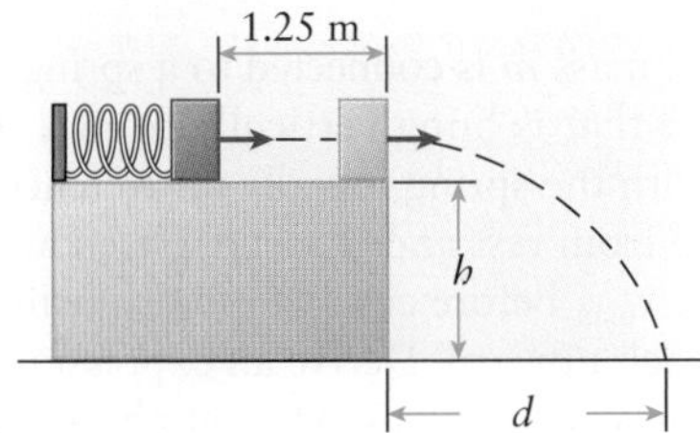

(a) Define your system and construct a diagram for the analysis, including the labeling of variables.

(b) Express the velocity of the block at the instant that it leaves the table in terms of k, m, and Δx.

(c) Express the velocity of the block at the instant that it leaves the table in terms of d and h.

(d) Define values of system variables as follows:

$d = 1.60$ m; $h = 1.0$ m; $m = 0.10$ kg; $k = 75$ N/m; and $\Delta x = 0.15$ m

Justify the claim that the table exerts a friction force on the block.

(e) Calculate the coefficient of kinetic friction for the interaction of the tabletop and the block.

22. Logan, with mass $m = 75$ kg, is standing at the edge of a dock holding onto a rope, which is suspended from a tree branch above. The distance along the rope from Logan's hands to the point of attachment to the tree is $L = 4.0$ m. The rope is taut and makes an angle, θ, of 30° with the vertical direction. Logan very gently steps off of the dock and swings in a circular arc until he releases the rope when it makes an angle, ϕ, of 12° with the vertical, but on the other side of the branch.

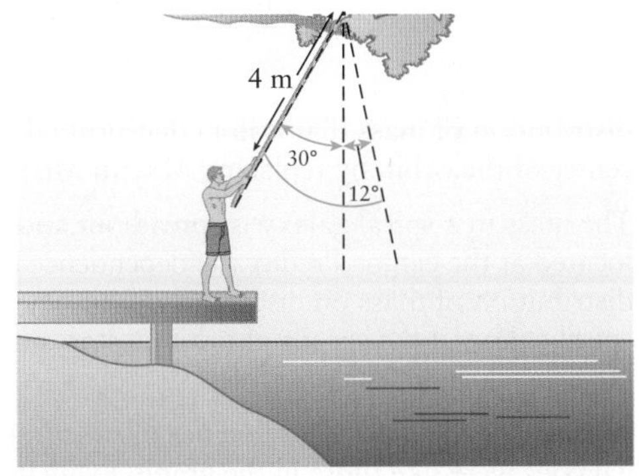

(a) Define your system so that you may use it to complete the rest of this problem, and construct a diagram showing the initial (i) and final (f) configurations of Logan's swing.

(b) Apply the principle of energy conservation to the system in terms of K_i, K_f, U_i, and U_f in terms of θ, ϕ, L, g, m, v_i, and v_f.

(c) Evaluate the expression in part (b) to determine Logan's speed when he releases the rope.

(d) Justify the selection of additional data required to determine the horizontal displacement of Logan from the edge of the dock to when he strikes the water and summarize the mathematical model from which the displacement can be calculated.

23. The unknown spring constant of a spring is determined by measurement of the maximum compression, x, when the spring is struck by a 100-g block with a known velocity. The surface over which the block travels exerts a negligible friction force on the block. The data obtained from the measurements for a particular spring are shown in the table.

v_i (m/s)	x (cm)
5	4.5
10	9
15	14
20	18
25	22

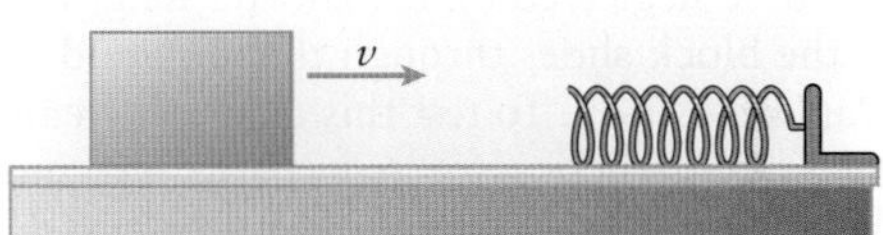

(a) Express the compression of the spring in terms of the velocity of the block using the principle of energy conservation.

(b) Analyze the data to determine the spring constant in units of N/m.

24. In the apparatus shown, a block with mass m is released from height h_1, descends along a track with an inclination α, does a loop with radius R and minimum height h_2, and is launched horizontally from a height $h_3 = h_2 + 2R$. The team of students who designed and built the track predicted x_2 as the distance from the launch position to landing. When they tested the prediction they observed the distance x_1. They knew that there were friction forces because they heard the block sliding, so they want to add a frictional work term to their energy conservation analysis. They have placed three photogates at positions at the top and bottom of the circular loop in the track and at the point where the block leaves the track.

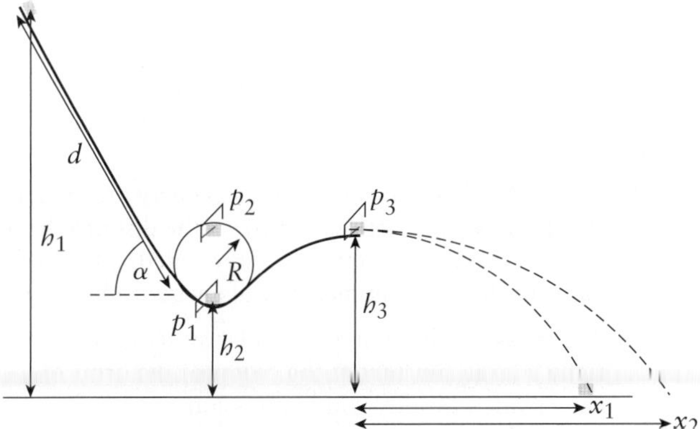

(a) First apply energy conservation for a closed system to obtain their prediction x_2.

The team believed that most of the frictional work occurs on the descent before the loop, given that $d > R$.

(b) Apply energy conservation to obtain the relationship that the team used to measure this work in terms of the speed measured at photogate p_1.

(c) Explain why the team predicted that the speeds at photogates p_2 and p_3 would be the same.

In fact, the team discovered that the speeds at photogates p_2 and p_3 were significantly different. So they decided that if they were going to succeed in understanding the system they would need to estimate the coefficient of kinetic friction.

(d) Apply energy conservation and Newton's second law to express the friction coefficient, μ_k, in terms of g, d, α, and the velocity at photogate p_1, v_1, before the block passes through the loop.

(e) The team then assumed that the frictional work could be neglected on the shorter lengths of track as the block slides through the loop and up to the launch position. To test this claim the team modified the prediction for x_2 in part (a) to include the friction force along the track before the loop. Obtain their expression.

(f) Is there a problem with their assumption in part (e)?

25. A new planet is discovered that has twice the radius of Earth and, given that it has the same density, eight times the mass of Earth. Let v_E represent the escape velocity from the surface of Earth. Calculate the escape velocity from the new planet, v_P, in terms of v_E.

26. Draw and describe a physical situation to which the following equation could apply.

$$(1.30\text{ kg})\left(9.8\frac{\text{m}}{\text{s}^2}\right)(0.50\text{ m}) + (0.25)(1.30\text{ kg})\left(9.8\frac{\text{m}}{\text{s}^2}\right)(d)\cos(180°)$$

27. A block of mass m is connected to a spring of unknown constant k that is hung vertically from a support, as shown. With the spring initially unstretched, the block is released from rest and descends, stretching the spring a distance x_{max} before coming momentarily to rest and moving back upward. Derive an expression for k in terms of m, x_{max}, and any relevant physical constants.

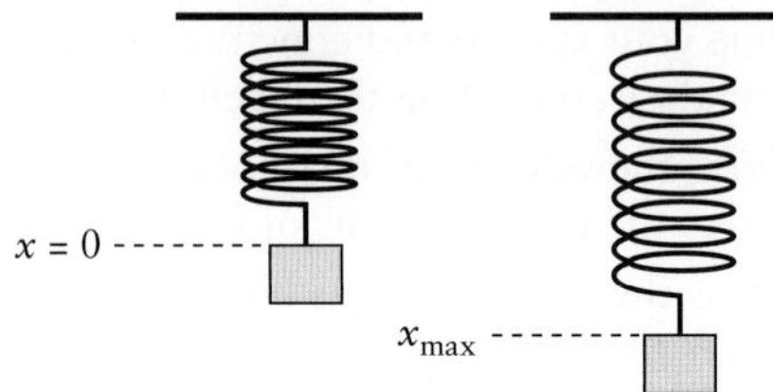

28. A block of mass m is compressed a distance x_0 against a spring of constant k and released from rest. The block is not connected to the spring. The block then slides with negligible friction along a track and around a vertical circle of radius R, as shown. What is the minimum compression, x_0, such that the block will slide completely around the loop without losing contact at point A?

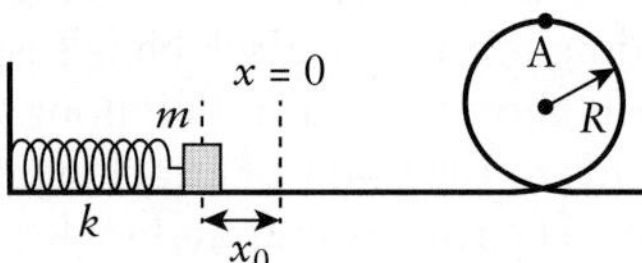

AP® Group Work

Directions: The problem below is designed to be done as group work in class.

Evidence for the existence of dark matter (matter that is nonluminous but exerts gravitational force) was obtained by the astronomer Vera Rubin from an analysis of the speed of stars in spiral galaxies. Although we do not yet know how to calculate orbits (we will learn about that later in the course), it turns out that the dependence on distance for a stable circular orbit is the same as that for escape velocity, but smaller by a fixed constant, which will not affect your argument.

(a) Apply the constancy of mechanical energy to obtain the dependence of orbital speed on the distance from and the mass of the object that is orbited, taking the value of the constant mechanical energy to be zero.

(b) The mass distribution in our solar system is essentially a large constant at the center of the orbit of each planet. Modify your expression to allow a distribution of mass that is not concentrated at the center of the orbit, by replacing M with $M(r)$.

(c) The mass in a spiral galaxy is spread out and not all located at the center. Predict the dependence of the distribution of mass on distance from the center of a spiral galaxy if the speeds of orbiting stars are independent of the orbital distance.

(d) Rubin observed that the velocities of stars in spiral galaxies were like those in the graph. Evaluate the evidence provided by the typical behavior of the speed of stars in a spiral galaxy as a function of distance from the center.

The light from a spiral galaxy is much brighter near the center. The number of stars and their masses were estimated from the observed light that came from the stars, which was confirmed experimentally.

(e) Describe how evidence supported Rubin's claim that not all the mass is visible.

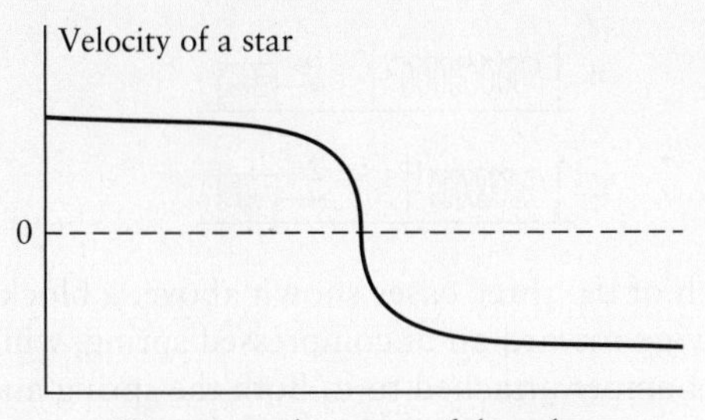

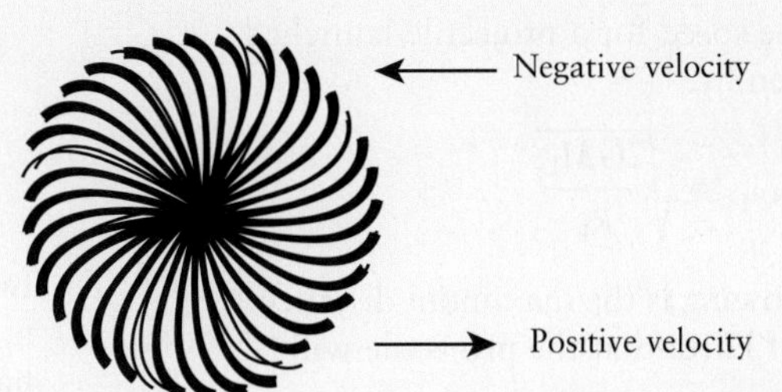

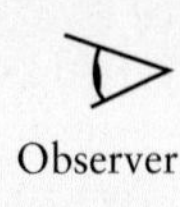

AP® PRACTICE PROBLEMS

Prep for the AP® Exam

Multiple-Choice Questions

Directions: The following questions have a single correct answer.

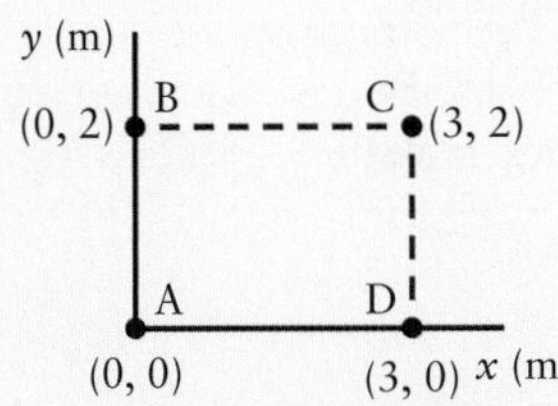

1. A force is exerted on an object in the x–y plane. The force has x and y components given by:

$$F_x = \left(2\frac{\text{N}}{\text{m}}\right)y$$

$$F_y = \left(2\frac{\text{N}}{\text{m}^2}\right)x^2$$

Of the following, which one correctly describes the work and nature of the force over the two given paths ABC and ADC?

(A) $W_{ABC} = W_{ADC}$ and the force is conservative
(B) $W_{ABC} = W_{ADC}$ and the force is nonconservative
(C) $W_{ABC} < W_{ADC}$ and the force is conservative
(D) $W_{ABC} < W_{ADC}$ and the force is nonconservative

2. A rough model of falling raindrops as spheres leads to the following two equations for the drag force and the terminal velocity:

$$F_{drag} = CAv^2 \qquad v_{term} = \sqrt{kr}$$

The C and k constants depend on such things as the density of air, the density of water, and the acceleration of gravity.

For raindrops falling at terminal velocity, the rate at which mechanical energy is dissipated is then proportional to which of the following?

(A) r (B) r^2 (C) $\sqrt{r^3}$ (D) $\sqrt{r^7}$

3. A 2.0-kg crate starting at rest slides down a rough 4.0-m-long ramp, inclined 30° with the horizontal. The force of friction between crate and ramp is 1.0 N. Which one of the following will be closest to the speed of the crate at the bottom of the ramp?

(A) 4 m/s (B) 6 m/s (C) 16 m/s (D) 18 m/s

4. A person jumps up and grabs a rope hanging from a tree. At the moment they grab the rope they are moving upward with an initial velocity v_0. They proceed to climb up the rope and stop when they reach the top. Graphs of the kinetic energy of the person and the potential energy of the person–Earth system are represented below. The shaded bars represent energy at the moment the person catches the rope, and the transparent bars represent the energy when the limb is reached. In each case the final kinetic energy is zero. Which set of graphs could be a correct match for this situation?

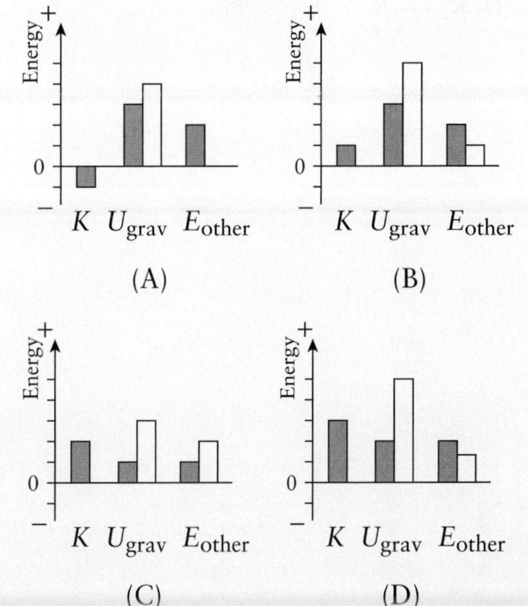

(*Continued*)

5. A projectile is launched from Earth with a speed that is just one-half of the speed it will take to escape. The escape speed for a projectile launched from Earth is given by:

$$v_{\text{escape}} = \sqrt{\frac{2GM_E}{R_E}}$$

Which of the following is the maximum distance from the center of Earth that the projectile will reach?

(A) $R = \frac{4}{3}R_E$

(B) $R = \frac{3}{2}R_E$

(C) $R = 2R_E$

(D) $R = 4R_E$

6. An object of mass m is moved from the surface of one planet of mass M_1 and radius R_1 to that of another planet of mass M_2 and radius R_2. The potential energy of the object–planet system has twice the magnitude on the second planet. Which of the following could contribute to this change?

	Radius of planet	**Density of planet**
(A)	$R_2 = \frac{1}{2}R_1$	$\rho_2 = 2\rho_1$
(B)	$R_2 = \frac{1}{2}R_1$	$\rho_2 = 4\rho_1$
(C)	$R_2 = 2R_1$	$\rho_2 = \frac{1}{2}\rho_1$
(D)	$R_2 = \frac{1}{3}R_1$	$\rho_2 = 9\rho_1$

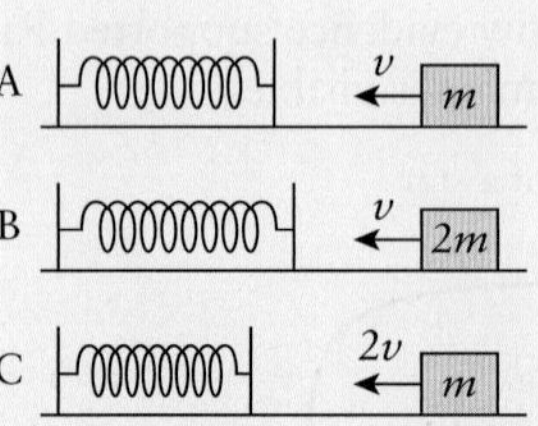

7. In each of the three cases shown above, a block is moving toward an uncompressed spring, which has a bumper attached to it. Both the spring and the bumper are of negligible mass. The block strikes and sticks to the bumper and compresses the spring a distance x before coming momentarily to rest. All three springs have the same spring constant, but the masses and initial speeds of the blocks vary. Which of the following is the correct ranking of the distances the spring is compressed in each case?

(A) $x_A > x_B > x_C$

(B) $x_C > x_B > x_A$

(C) $(x_B = x_C) > x_A$

(D) $x_C > (x_A = x_B)$

Free-Response Question

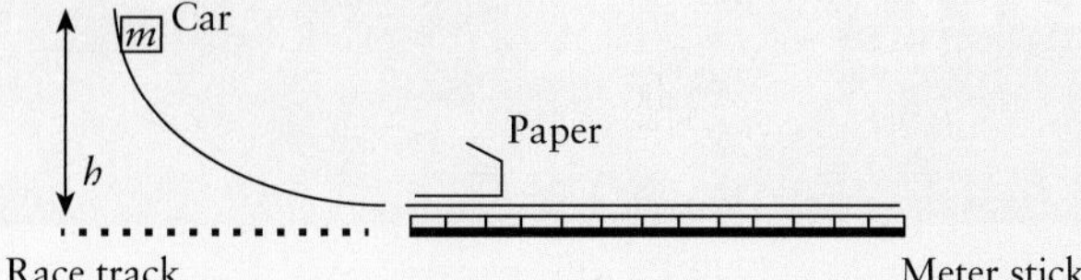

1. Students are performing an experiment in which they release a small toy car of mass m from rest at the top of an incline, and it rolls with negligible friction until it strikes a paper catcher of negligible mass at the bottom of the incline. The car and the catcher slide across the horizontal lab table and come to rest. The car is released at height h and the car and catcher slide a distance D across the table.

   (a) Derive an expression for the coefficient of kinetic friction, μ_k, between the paper and the table in terms of the given variables and fundamental constants.

   The students take the average data, as shown in the table, releasing the small toy car from rest at height h.

h (m)	D (m)		
0.08	0.22		
0.12	0.30		
0.16	0.44		
0.20	0.58		
0.24	0.67		
0.28	0.78		
0.32	0.90		

   (b) i. Indicate two quantities you would plot to obtain a linear graph.

   ii. Use the remaining columns in the table above, as needed, to record any quantities that you indicated that are not given.

   iii. On the axes on the accompanying graph, graph the data from the table that will produce a linear relationship. Clearly mark scales and label all axes, including units if appropriate. Draw a line of best fit.

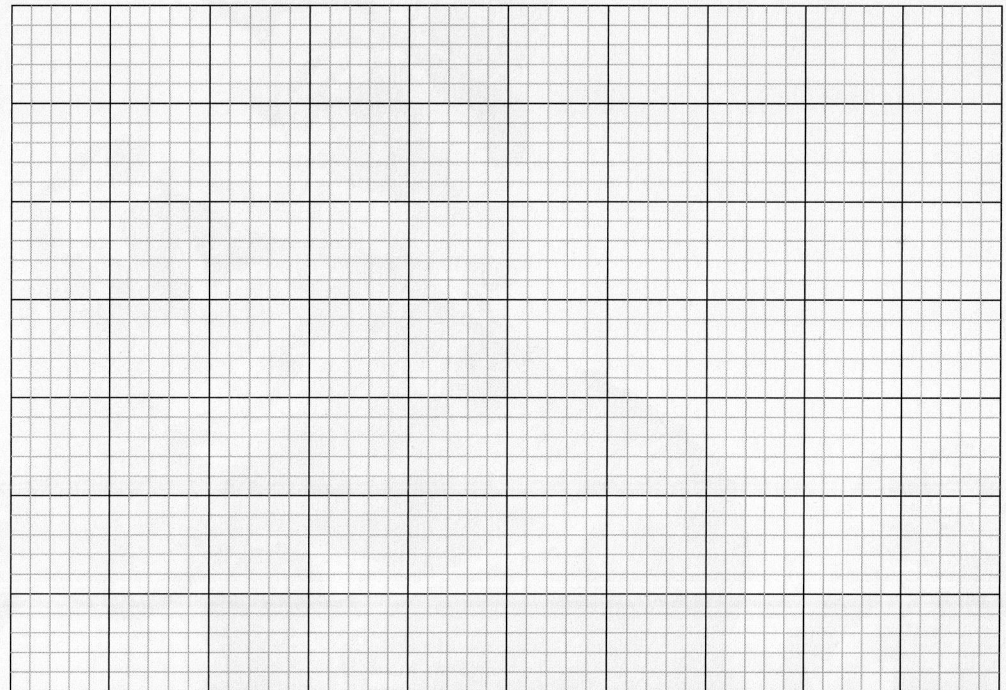

   (c) The car that students use has a mass of 0.030 kg. Use this information and your graph to calculate a numerical value for μ_k.

   (d) The students repeat the experiment, this time taping an additional 20 grams to the top of the car. Predict how the line of best fit for this new set of data would compare to the line of best fit drawn above. Justify your answer.

   (e) The teacher asks students to perform another experiment to confirm their result for the coefficient of kinetic friction between the paper and the table. Assume the students have access to equipment usually available in a physics classroom. Briefly describe an experiment that could be performed and describe how the data taken could be used to calculate μ_k.

Linear Momentum

Chapter 9 Momentum, Collisions, and the Center of Mass

Westend61/Getty Images

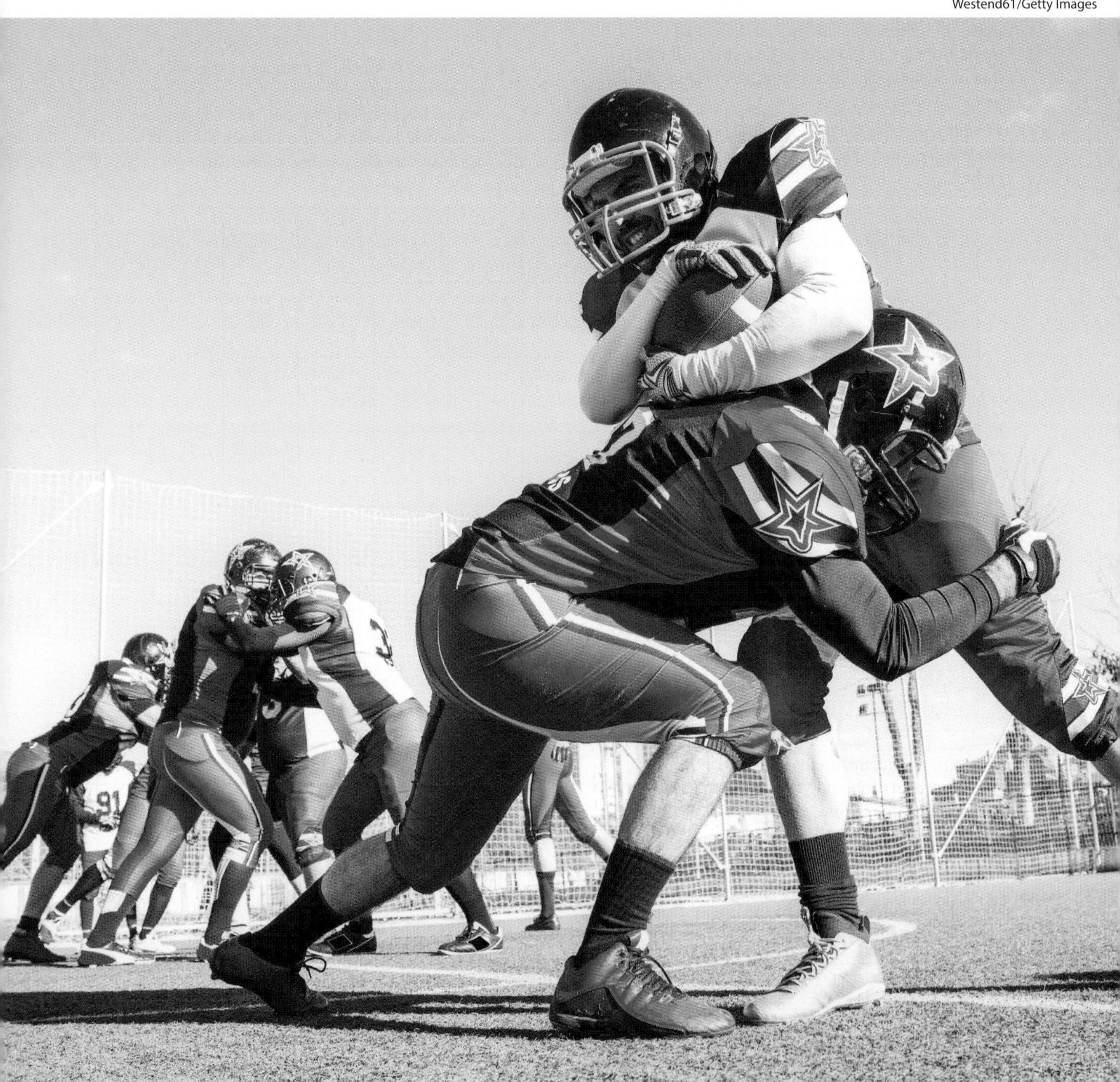

Case Study: When two football players collide... who wins?

If you've ever watched football, you know that collisions between opposing players are essential to winning a game. But just what does a player need to do to "win" a collision—to knock over another player? To answer this, we need a new concept: momentum.

Although the total energy of a closed, isolated system is always constant, the kinetic energy isn't unchanged when one player tackles another. How do we know? Although both players were moving just before the collision, the two might come to a stop. So what happens? Energy is not lost, but some is converted to different types—sound for example—that cannot be recovered. So what *does* remain the same? If they are only interacting with each other, total momentum, the sum of the products of mass and velocity for each player, remains unchanged. *The total momentum of a system is always conserved.*

We'll discover that momentum is a vector. When two players of about equal mass are running toward each other at about the same speed, their momenta (the plural of *momentum*) are similar in magnitude but opposite in direction. That makes the vector sum of their momenta before the collision about zero. If at the instant they collide they are not in contact with the ground, then immediately after the tackle, both stop moving...yes, the total momentum is still zero. So to "win" the collision, the magnitude of one player's momentum must be larger than the other's! (This is only considering linear motion. Rotation, which we will begin to discuss in the next chapter, can impact who falls down, even though momentum is constant.)

We've seen that the center of mass of a system moves as though all the mass of the system were concentrated there. The center of mass of the two-person system, assuming their masses are similar, is close to midway between them. With no external forces on the system, if their momenta are equal in magnitude and opposite in direction, the center of mass doesn't move. As they collide, the system's center of mass remains in the same place. After the tackle is complete, both players are in a heap with their center of mass unmoved. If one player had a much larger momentum, then the center of mass would continue to move in the direction of that player's initial motion. Does the momentum of each individual football player remain constant as the collision proceeds?

By the end of this chapter, you should be able to use the concepts of momentum, center of mass, and system to predict the behavior of objects in everyday situations, define the linear momentum of an object, and apply momentum conservation to problems about collisions.

9 Momentum, Collisions, and the Center of Mass

IgorNP/Shutterstock

YOU WILL LEARN TO:

- Use the concepts of momentum, center of mass, and system to predict the behavior of objects in everyday situations.
- Define the linear momentum of an object and explain how it differs from kinetic energy.
- Explain the conditions under which the total momentum of a system is constant and why total momentum is constant in a collision.
- Identify the differences and similarities between elastic, inelastic, and *completely* inelastic collisions.
- Apply conservation of momentum and mechanical energy to problems involving elastic collisions.
- Relate the momentum change of an object, the force that causes the change, and the time over which the force is exerted on the object.
- Find the center of mass of a system and describe how the net force on a system affects the motion of the system's center of mass.

9-1 Newton's third law will help lead us to the idea of momentum

The ideas of kinetic energy and the work-energy theorem that we introduced in Chapters 7 and 8 gave us new ways to think about motion. These ideas relate fundamentally to Newton's second law, which states that the net external force on an object determines the object's acceleration. In this chapter we'll extend this idea by reconsidering Newton's third law: When two objects interact with each other, they exert forces of equal magnitude but opposite direction on each other. These forces can change the velocities of the objects. An important way to express such a change is in terms of the **momentum** of an object (written as $\vec{p}$), which is the product of an object's mass and its velocity. Since velocity is a vector and mass is a scalar, momentum is also a vector. We'll see that we can describe the behavior of two interacting objects or systems, such as a baseball bat and a ball (**Figure 9-1a**) or a squid and the water it ejects to propel itself (**Figure 9-1b**), in terms of momentum: Each object in the pair undergoes a change in velocity and hence in momentum, and the momentum changes are equal in magnitude but opposite in direction. This important observation will lead us to a physical principle called the **law of conservation of momentum.** This law, which we will discuss in more detail in Section 9-3, will turn out to be essential for

NEED TO REVIEW?
Turn to the **Glossary** in the back of the book for definitions of bolded Key Terms.

(a) (b) (c)

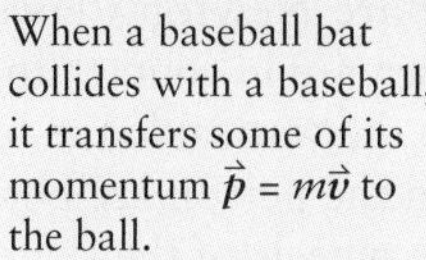

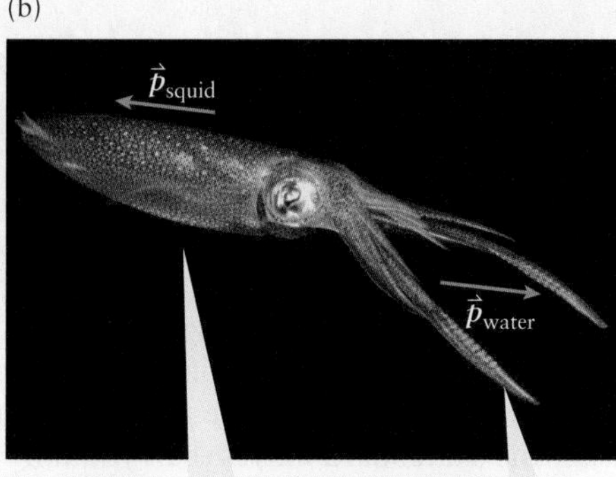

When a baseball bat collides with a baseball, it transfers some of its momentum $\vec{p} = m\vec{v}$ to the ball.

A squid propels itself by ejecting water at high speed. The water acquires momentum in one direction, and the squid acquires momentum of the same magnitude in the opposite direction.

The center of mass (labeled with a yellow × in the figure) of a typical woman is just below the navel. (For men, it is just above the navel.) You can change exactly where your center of mass is located by changing the relative positions of your arms and legs.

In order to balance on one foot, this woman must place her center of mass directly above that foot. By raising her other leg and shifting her arms slightly, she has moved her center of mass slightly to her left from where it would be if she were relaxed.

(a) Stephen Mcsweeny/Shutterstock; (b) David Fleetham/Alamy; (c) Gino's Premium Images/Alamy

Figure 9-1 Momentum and center of mass In this chapter we'll explore (a) momentum and (b) conservation of momentum, and learn how to quantify (c) the center of mass of a system.

analyzing what happens when two objects or systems interact with each other. In fact it turns out that this law is the fundamental physical principle from which Newton's three laws arise, and it can be used in more general situations than those in which we are able to apply Newton's laws, including cases when mass changes instead of velocity!

To discuss forces on systems and the work-energy theorem, we had to understand the notion of center of mass. We found that the foundation of the object model was whether every point on the system we were considering could be approximated to move the same displacement as the center of mass of the system, and if that was not true, other types of energy besides kinetic energy could result from work. This special point associated with a system moves as though all the mass of the system were concentrated there. We will further explore this idea and expand upon it because we will see systems behave as if all the forces on the system are exerted at that point. You've used the idea of the center of mass if you've ever balanced on one foot (**Figure 9-1c**). If all the mass of your body were concentrated into a small blob placed on the ground, that blob would remain at rest because it would be supported from below. In the same way, to have your body remain at rest while standing on the ground on one foot, you lean so that your center of mass (located near your navel) is above your foot and so is supported from below. We'll finally be able to fully develop the idea of the center of mass by expanding on what we'll learn about momentum.

AP® Exam Tip

Momentum is typically represented by a lowercase $\vec{\boldsymbol{p}}$, and is a vector. Power is typically represented by a capital *P* and is a scalar. If you make sure to understand what each equation means, you will not confuse these two from the AP® Physics 1 equation sheet when solving problems.

NEED TO REVIEW?

Turn to Chapter 4, Section 4-4, for a review of center of mass.

THE TAKEAWAY for Section 9-1

✔ The idea of momentum is useful in situations where two or more objects or systems interact with each other.

Prep for the AP® Exam

AP® Building Blocks

1. You are seated in a wooden chair with a rigid back. With all four legs of the chair on the ground you are stable and the center of mass of the system is near your navel, approximately at the base of the arrow shown in the figure.

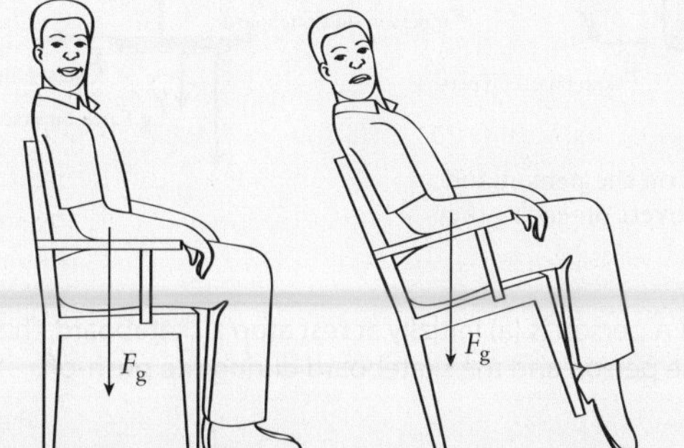

(a) Explain why there is a large angle that the seat can make with the horizontal without toppling and describe that angle in terms of the displacement of your navel.
(b) Design a method to measure the angle at which you would no longer be stable.

9-2 Momentum is a vector that depends on an object's mass, speed, and direction of motion

We'll begin our exploration of the momentum concept by considering a simple system: a person standing on a skateboard who decides to jump off. A person is standing atop a stationary skateboard (**Figure 9-2a**). He then jumps straight to the left off the skateboard, and the skateboard rolls away to the right (**Figure 9-2b**). Imagine yourself in this person's place. If you were to try this move, you'd find that the skateboard rolls to the right much more quickly than you fly through the air to the left. Why is this? What determines how much faster the skateboard moves than you do?

The tools we have so far to begin to answer this question are Newton's laws. **Figure 9-2c** shows the forces on the person and skateboard while they are in contact and he is pushing off. The vertical forces on the skateboard cancel, as do the vertical forces on the person, so the net force on the skateboard and the net force on the person are both horizontal. There is negligible friction between the ground and the wheels of the skateboard, so the only horizontal forces we need to consider are the forces that the person and the skateboard exert on each other. Again, imagine you are the person on the skateboard. If we apply Newton's second law to the skateboard and to you, we get the equations

$$\begin{aligned}\sum \vec{F}_{\text{ext on skateboard}} &= \vec{F}_{\text{you on skateboard}} = m_{\text{skateboard}}\vec{a}_{\text{skateboard}} \\ \sum \vec{F}_{\text{ext on you}} &= \vec{F}_{\text{skateboard on you}} = m_{\text{you}}\vec{a}_{\text{you}}\end{aligned} \tag{9-1}$$

The vectors $\vec{a}_{\text{skateboard}}$ and $\vec{a}_{\text{you}}$ are the accelerations of the skateboard and you, respectively, during the push-off.

Equations 9-1 involve the *accelerations* of the skateboard and you, but we're interested in the *velocities* of the skateboard and you after the push-off. To see how to get these, let's go back to the definition of acceleration in Section 3-5:

EQUATION IN WORDS
Acceleration equals change in velocity divided by time interval.

Acceleration for the object over a very short time interval from time t_1 to a later time t_2

Change in velocity of the object over the short time interval

$$\vec{a} = \frac{\Delta\vec{v}}{\Delta t} = \frac{\vec{v}_2 - \vec{v}_1}{t_2 - t_1} \tag{3-10}$$

Duration of the time interval

For both the velocity change and the elapsed time, subtract the earlier value from the later value.

In Equation 3-10 the time interval Δt is how long the push-off lasts. The initial velocity before the push-off is $\vec{v}_1 = 0$ for both the skateboard and you, since both begin at rest. We'll use the symbols $\vec{v}_{\text{skateboard}}$ and $\vec{v}_{\text{you}}$ for the velocities of the skateboard and you just after the push-off (corresponding to $\vec{v}_2$ in Equation 3-10). When we substitute

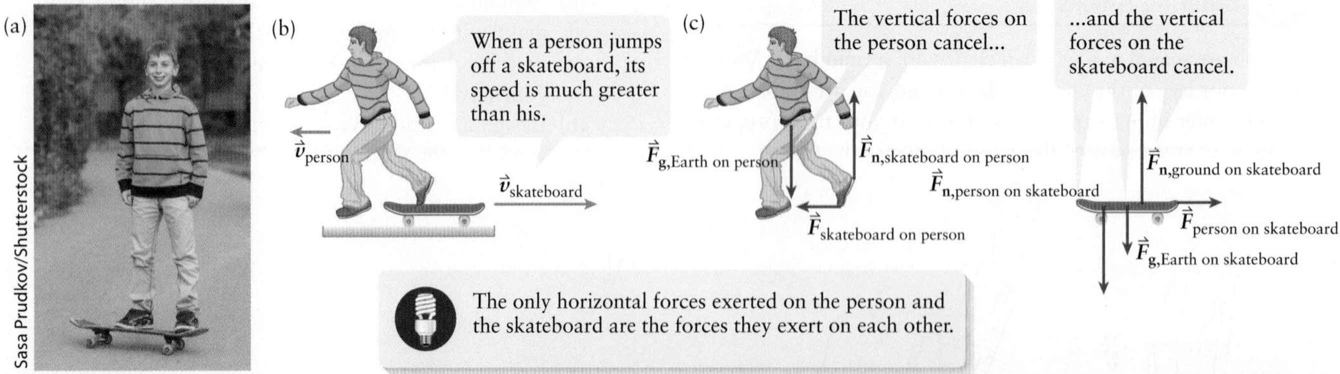

(a)
Sasa Prudkov/Shutterstock
(b)
When a person jumps off a skateboard, its speed is much greater than his.
$\vec{v}_{\text{person}}$
$\vec{v}_{\text{skateboard}}$
(c)
The vertical forces on the person cancel...
...and the vertical forces on the skateboard cancel.
$\vec{F}_{\text{g,Earth on person}}$
$\vec{F}_{\text{n,skateboard on person}}$
$\vec{F}_{\text{n,person on skateboard}}$
$\vec{F}_{\text{n,ground on skateboard}}$
$\vec{F}_{\text{skateboard on person}}$
$\vec{F}_{\text{person on skateboard}}$
$\vec{F}_{\text{g,Earth on skateboard}}$
The only horizontal forces exerted on the person and the skateboard are the forces they exert on each other.

Figure 9-2 Jumping off a skateboard A person is (a) initially at rest atop a skateboard, then (b) jumps off horizontally. (c) Forces on the person and the skateboard during the push-off.

these values into Equations 9-1, then multiply both sides of these equations by Δt (the duration of the push-off), we get

$$\vec{F}_{\text{you on skateboard}}\Delta t = m_{\text{skateboard}}\vec{v}_{\text{skateboard}}$$
$$\vec{F}_{\text{skateboard on you}}\Delta t = m_{\text{you}}\vec{v}_{\text{you}} \quad (9\text{-}2)$$

Now remember that from Newton's third law, the force that the skateboard exerts on you has the same magnitude as the force that you exert on the skateboard but points in the opposite direction (Figure 9-2c): $\vec{F}_{\text{skateboard on you}} = -\vec{F}_{\text{you on skateboard}}$. If we substitute this into Equations 9-2, you can see that the quantities $m_{\text{skateboard}}\vec{v}_{\text{skateboard}}$ and $m_{\text{you}}\vec{v}_{\text{you}}$ likewise have the same magnitude but point in opposite directions:

$$m_{\text{skateboard}}\vec{v}_{\text{skateboard}} = -m_{\text{you}}\vec{v}_{\text{you}} \quad (9\text{-}3)$$

Here's how to interpret Equation 9-3:

- The minus sign in Equation 9-3 tells us that $\vec{v}_{\text{skateboard}}$ and $\vec{v}_{\text{you}}$ are in opposite directions. As Figure 9-2b shows, if you push the skateboard to the right, you must move to the left.
- If we take the magnitude of both sides of this equation, we get

$$m_{\text{skateboard}}v_{\text{skateboard}} = m_{\text{you}}v_{\text{you}} \quad \text{or} \quad v_{\text{skateboard}} = \left(\frac{m_{\text{you}}}{m_{\text{skateboard}}}\right)v_{\text{you}} \quad (9\text{-}4)$$

  The speed of the skateboard is equal to your speed multiplied by the ratio $m_{\text{you}}/m_{\text{skateboard}}$. You are much more massive than the skateboard, so this ratio is a large number and the skateboard moves much more quickly than you do.
- If you push off from the skateboard with greater force, both the skateboard and you will fly off at faster speeds. But this force doesn't appear in either Equation 9-3 or 9-4. That means the *ratio* of the skateboard's speed to yours will be the same no matter how hard you push.

Our analysis of the skateboard problem, and especially Equation 9-3, suggests that we think about a quantity that's equal to the product of an object's mass m and its velocity $\vec{v}$. We'll call this quantity the object's **linear momentum**:

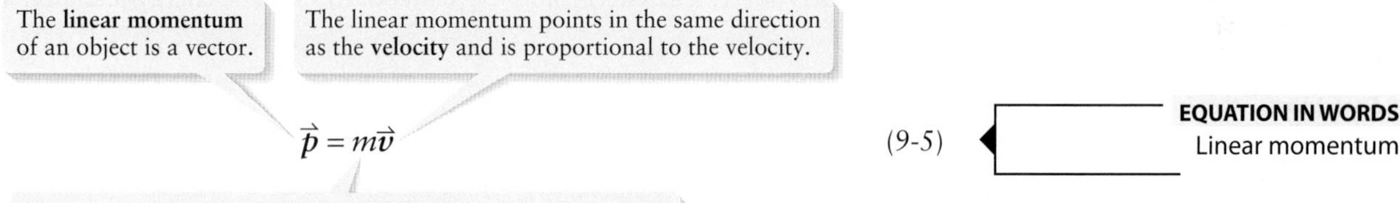

The **linear momentum** of an object is a vector.
The linear momentum points in the same direction as the **velocity** and is proportional to the velocity.
The linear momentum is also proportional to the object's **mass**.

$$\vec{p} = m\vec{v} \quad (9\text{-}5)$$

EQUATION IN WORDS
Linear momentum

Unless we are dealing with a system that is also rotating (we will discuss this in a later chapter), we'll usually call $\vec{p}$ simply the momentum of the object. Note that since the mass m is a positive scalar quantity, the momentum $\vec{p}$ points in the same direction as the velocity $\vec{v}$. **Figure 9-3** compares the momenta of three different objects.

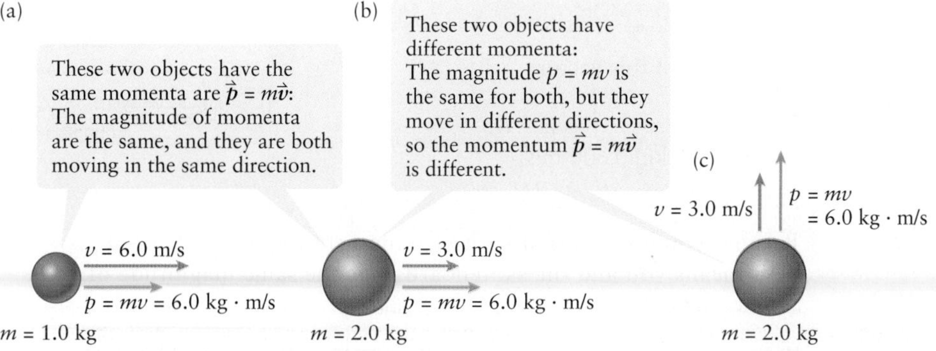

Figure 9-3 Momentum is a vector Comparing the momenta of three different objects.

Equation 9-5 shows that the units of momentum are the units of mass multiplied by the units of velocity, or kg · m/s. This combination of units doesn't have a special name in the SI system, so we simply say that the momentum of a 1000-kg car driving at 20 m/s has magnitude $p = mv = (1000 \text{ kg})(20 \text{ m/s}) = 20{,}000 \text{ kg} \cdot \text{m/s}$, or "twenty thousand kilogram-meters per second."

We can use the definition of momentum, Equation 9-5, to rewrite Equation 9-3 for the skateboarder and you when you push off from rest: $m_{\text{skateboard}}\vec{v}_{\text{skateboard}} = -m_{\text{you}}\vec{v}_{\text{you}}$ becomes

$$\vec{p}_{\text{skateboard}} = -\vec{p}_{\text{you}} \tag{9-6}$$

Just after the push-off, if you both start at rest, you and the skateboard each have the same *magnitude* of momentum, but your momentum is directly opposite to the skateboard's momentum. We'll use this observation in Example 9-1. In the next section we will see that it is always the changes in momentum of two interacting objects that are of the same magnitude and in opposite directions.

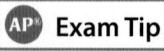

AP® Exam Tip

Note that kinetic energy only depends on speed, while momentum depends on velocity.

WATCH OUT !

Don't confuse momentum and kinetic energy.

Like momentum, the kinetic energy K of an object (introduced in Chapter 7) depends on its mass m and its speed v: $K = \frac{1}{2}mv^2$. But kinetic energy is a *scalar* quantity (it has no direction), whereas momentum $\vec{p} = m\vec{v}$ is a *vector* quantity. Furthermore, two objects with the same magnitude of momentum can have very different kinetic energies. Example 9-1 explores this further.

EXAMPLE 9-1 Momentum and Kinetic Energy

Suppose that you have a mass of 50.0 kg. Your skateboard has a mass of 2.50 kg. Initially, you and the skateboard are at rest. Just after you push off, as in Figure 9-2, you are moving to the left at 0.600 m/s. (a) What is the magnitude of your momentum and the magnitude of the skateboard's momentum just after you push off? (b) What is the speed of the skateboard just after you push off? (c) What are your kinetic energy and the skateboard's kinetic energy just after you push off?

Set Up

We use the definitions of linear momentum and kinetic energy. Our discussion so far and Equation 9-6 tell us how to find the speed of the skateboard from your speed if you both start at rest.

$$\vec{p} = m\vec{v} \tag{9-5}$$

$$K = \frac{1}{2}mv^2 \tag{7-8}$$

$$\vec{p}_{\text{skateboard}} = -\vec{p}_{\text{you}} \tag{9-6}$$

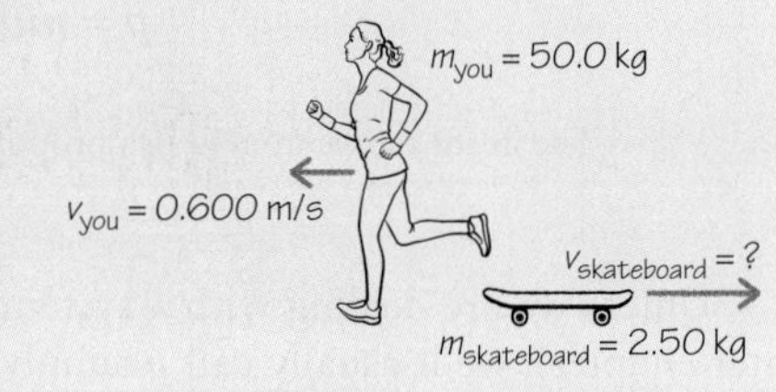

Solve

(a) Equation 9-5 gives us the magnitude of your momentum. Equation 9-6 tells us that after the interaction, because both started from rest, the skateboard has as much momentum to the right as you have to the left.

Your momentum:

$$\vec{p}_{\text{you}} = m_{\text{you}}\vec{v}_{\text{you}}$$

Magnitude of this vector:

$$p_{\text{you}} = m_{\text{you}}v_{\text{you}} = (50.0 \text{ kg})(0.600 \text{ m/s}) = 30.0 \text{ kg} \cdot \text{m/s}$$

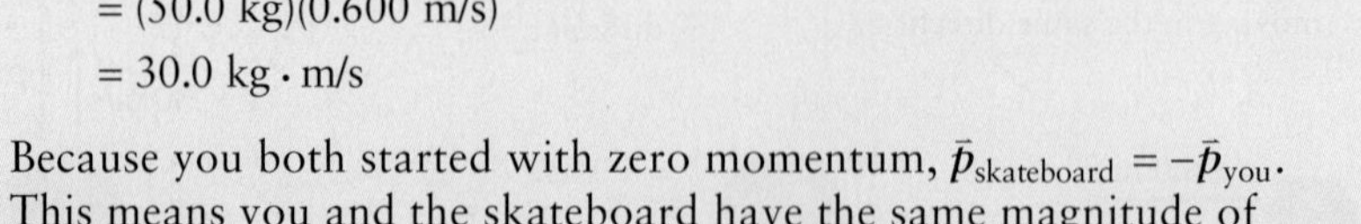

Because you both started with zero momentum, $\vec{p}_{\text{skateboard}} = -\vec{p}_{\text{you}}$. This means you and the skateboard have the same magnitude of momentum:

$$|\vec{p}_{\text{skateboard}}| = |m_{\text{skateboard}}\vec{v}_{\text{skateboard}}| = |\vec{p}_{\text{you}}| = |m_{\text{you}}\vec{v}_{\text{you}}| = 30.0 \text{ kg} \cdot \text{m/s}$$

(b) We know the skateboard's mass and momentum, so we can solve for its speed.

$$\vec{p}_{\text{skateboard}} = m_{\text{skateboard}}\vec{v}_{\text{skateboard}} \text{ so}$$

$$v_{\text{skateboard}} = \frac{p_{\text{skateboard}}}{m_{\text{skateboard}}} = \frac{30.0 \text{ kg} \cdot \text{m/s}}{2.50 \text{ kg}} = 12.0 \text{ m/s}$$

$v_{you} = 0.600$ m/s

$v_{skateboard} = 12.0$ m/s
(20.0 times faster than you)

(c) Using Equation 7-8, we find the kinetic energies from the masses and speeds of the two objects.

Your kinetic energy:

$$K_{\text{you}} = \frac{1}{2}m_{\text{you}}v_{\text{you}}^2 = \frac{1}{2}(50.0 \text{ kg})(0.600 \text{ m/s})^2 = 9.00 \text{ kg} \cdot \text{m}^2/\text{s}^2 = 9.00 \text{ J}$$

$$K_{\text{skateboard}} = \frac{1}{2}m_{\text{skateboard}}v_{\text{skateboard}}^2 = \frac{1}{2}(2.50 \text{ kg})(12.0 \text{ m/s})^2 = 180 \text{ J}$$

$K_{you} = 9.00$ J

$K_{skateboard} = 180$ J
(20.0 times more than you)

Reflect

The skateboard's speed is 12.0 m/s, about the same as the speed limit for cars in a residential neighborhood. It's not difficult to shove a skateboard hard enough to give it that speed.

As we discussed earlier, the skateboard moves much more quickly than you do after the push-off because it has much less mass. The ratio of your mass to that of the skateboard is 20.0 to 1: This same factor of 20.0 appears in the ratio of speeds (the skateboard is faster). It also appears in the ratio of kinetic energies: The skateboard has 20.0 times more kinetic energy than you do. If two objects have the same magnitude of momentum, the less massive one *always* has more kinetic energy.

$$\text{Skateboard speed: } v_{\text{skateboard}} = \left(12.0\,\frac{\text{m}}{\text{s}}\right)\left(\frac{1 \text{ km}}{1000 \text{ m}}\right)\left(\frac{3600 \text{ s}}{1 \text{ h}}\right) = 43.2 \text{ km/h} = 26.8 \text{ mi/h}$$

$$\text{Ratio of masses: } \frac{m_{\text{you}}}{m_{\text{skateboard}}} = \frac{50.0 \text{ kg}}{2.50 \text{ kg}} = 20.0$$

$$\text{Ratio of speeds: } \frac{v_{\text{you}}}{v_{\text{skateboard}}} = \frac{0.600 \text{ m/s}}{12.0 \text{ m/s}} = 0.0500 = \frac{1}{20.0}$$

$$\text{Ratio of kinetic energies: } \frac{K_{\text{you}}}{K_{\text{skateboard}}} = \frac{9.00 \text{ J}}{180 \text{ J}} = 0.0500 = \frac{1}{20.0}$$

NOW WORK Problems 1–4 from The Takeaway 9-2.

We can get more insight into why the two objects in Example 9-1 have the same magnitude of momentum but different kinetic energies. Let's combine Equations 9-2 and 9-3 with the definition $\vec{p} = m\vec{v}$ for momentum:

$$\vec{F}_{\text{you on skateboard}}\Delta t = m_{\text{skateboard}}\vec{v}_{\text{skateboard}} = \vec{p}_{\text{skateboard}}$$
$$\vec{F}_{\text{skateboard on you}}\Delta t = m_{\text{you}}\vec{v}_{\text{you}} = \vec{p}_{\text{you}} \quad (9\text{-}7)$$

Equations 9-7 say that the momentum that each object acquires during the push-off from rest is equal to the net force exerted on it multiplied by the *time* over which the force is exerted. Newton's third law tells us that you and the skateboard exert forces of equal magnitude on each other and that when one force is present the other one must be as well. Hence both forces are exerted for the same time Δt, and since you both started from rest, the skateboard and you end up with equal magnitudes of momentum (although in opposite directions). The kinetic energy that each object acquires, however, is equal to the work done on that object during the push-off, and work equals force multiplied by the displacement of the object in the direction in which the force is exerted. If we treat you and the skateboard as individual objects, the forces you and the skateboard exert on each other are external forces and can do work. The skateboard moves more quickly than you do during the push-off, so it travels a greater distance than you do and so has more work done on it by the same magnitude of force.

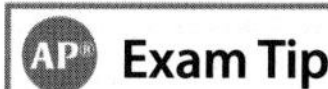

Note that the magnitude of momentum for an object is mass times the magnitude of the object's velocity. At any instant, the magnitude of the object's velocity is its speed. Since kinetic energy depends on speed, we can use this relationship to find the kinetic energy in terms of the magnitude of the momentum:

$K = \frac{1}{2}mv^2 = \frac{1}{2}m(p/m)^2 = p^2/2m$

Figure 9-4 Same momentum but different kinetic energy During the push-off you and the skateboard acquire the same magnitude of momentum, but the skateboard acquires more kinetic energy.

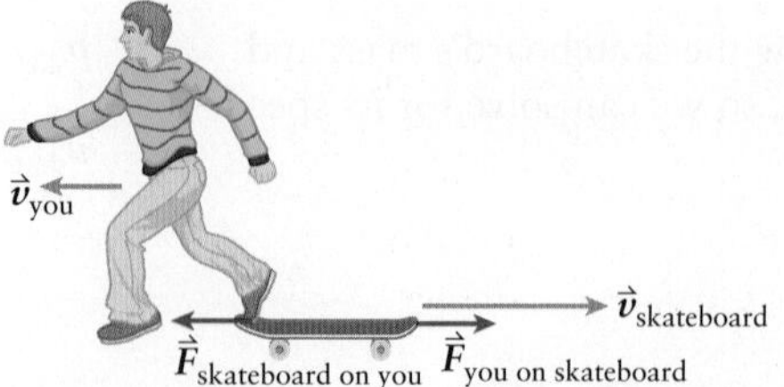

(1) $\vec{F}_{\text{skateboard on you}}$ and $\vec{F}_{\text{you on skateboard}}$ have the same magnitude.

(2) These forces are exerted on you and the skateboard for the same amount of time, so you and the skateboard acquire the same magnitude of momentum.

(3) The skateboard moves more quickly and covers a greater distance while the forces are exerted, so more work is done on the skateboard and it acquires more kinetic energy.

Therefore, the skateboard ends up with more kinetic energy (**Figure 9-4**). Another way to see this without considering work is that if the momenta of two objects have the same magnitude, the lighter object must have a bigger speed by a factor of the mass, and kinetic energy depends on speed squared. The kinetic energy of an object can be written $1/2mv^2 = 1/2(mv)^2/m = |p|^2/2m$, so for the same momentum, whichever object has the smaller mass must have the greater kinetic energy.

We've seen how the idea of momentum is useful for the special case of a person pushing off a skateboard from rest. In the next section we'll see how to apply this idea to the general case in which any two objects interact with each other over a short period of time. We will call this sort of interaction a **collision**, which we define as a short, strong interaction between two objects or systems. The "colliding" objects can be interacting through forces that do not require contact, so sometimes colliding objects do not touch or move through the same displacement during the collision. In the next section, we will find that what happens *during* the interaction determines the type of interaction that occurs.

WATCH OUT !

Nothing is truly modeled properly as an object during a collision.

Throughout Example 9-1, you and the skateboard are discussed as two objects. When you and the skateboard interact, however, you are pushing off from the skateboard, so you must be using internal energy to do so. Before and after this interaction, however, the object model works fine because you are doing nothing that affects the motion of you or the skateboard (and the skateboard is well modeled as an object throughout). We will find that we can use momentum to understand how the motion will change from *before* to *after* an interaction. The object model works well for this type of interaction because we will be using mathematics only to model the motion before and after the interaction, not during the interaction.

THE TAKEAWAY for Section 9-2

✔ The linear momentum of an object or a system is a vector. For an object, it is equal to the product of the object's mass and its velocity.

✔ If two objects or systems at rest push away from each other, each acquires an equal magnitude of momentum in opposite directions (if no net external forces are exerted on the objects or systems other than the ones they exert on each other).

✔ If two objects of different masses have the same magnitude of momentum, the object with the smaller mass has more kinetic energy than the other object does.

✔ If two interacting systems are well modeled as objects before and after the interaction, we can model them as objects for momentum calculations.

✔ Collisions are short, strong interactions between two objects or systems.

Prep for the AP® Exam

AP® Building Blocks

EX 9-1 1. A 1.00×10^4-kg train car moves east at 15 m/s. Determine the magnitude of the momentum and the kinetic energy of the train car.

EX 9-1 2. The magnitude of the momentum of a 57-g tennis ball is 2.6 kg · m/s. What is its speed?

EX 9-1 3. Determine the following vectors, taking the forward direction to be the positive x direction, for a car with a mass of 1250 kg:

(a) Initial momentum while backing up at 5.00 m/s

(b) Final momentum while moving forward at 14.0 m/s

(c) Change in momentum between these two times

4. Blythe and Geoff are ice skating together. Blythe has a mass of 50.0 kg, and Geoff has a mass of 80.0 kg. Blythe pushes Geoff in the chest when both are at rest, causing him to move away at a speed of 4.00 m/s. Determine Blythe's speed after she pushes Geoff.

AP® Skill Builders

5. If the mass of a basketball is 18 times that of a tennis ball, can they ever have the same momentum? Explain your answer.
6. The magnitudes of the momenta of two objects, labeled A and B, are the same. Object B has a larger kinetic energy. Which of the following statements most accurately relates the masses and velocities of objects A and B? Explain your choice.
   (A) The magnitude of the velocity of object A is larger than the magnitude of the velocity of object B, and the mass of object B is larger than the mass of object A.
   (B) The magnitude of the velocity of object B is larger than the magnitude of the velocity of object A, and the mass of object A is larger than the mass of object B.
   (C) The magnitude of the velocity of object A is larger than the magnitude of the velocity of object B, and the ratio of the masses m_B/m_A is equal to the ratio of velocities v_A/v_B.
   (D) The magnitude of the velocity of object B is larger than the magnitude of the velocity of object A, and the ratio of the masses m_B/m_A is equal to the ratio of velocities v_A/v_B.

AP® Skills in Action

7. A major league baseball has a mass of 0.145 kg. A ball leaves the bat of a slugger at an angle, θ, of 37° above the horizontal with a speed of 34.9 m/s. Take away from the bat and upward to be the positive x and y directions, respectively.
   (a) For the instant when the ball leaves the bat, calculate:
      i. The magnitude of the ball's momentum, p
      ii. The horizontal component of the momentum, p_x
      iii. The vertical component of the momentum, p_y
   (b) The wall at the baseball field is 130 m from home plate. If friction forces are negligible, and we neglect the difference between the height at which the ball was hit and the ground, the time of flight of the ball is $2v_y/g$. Does the ball reach the wall? Use $g = 9.8$ m/s^2.

9-3 The total momentum of a system is always conserved; it is constant for systems that are closed and isolated

In Section 9-2 we found that just after you pushed off a skateboard when you both started at rest, your momentum and the momentum of the skateboard had the same magnitude but were in opposite directions. In terms of vectors (Equation 9-6), we wrote this as

$$\vec{p}_{\text{skateboard}} = -\vec{p}_{\text{you}}$$

Let's rearrange this equation so that both momenta are on the same side of the equal sign:

$$\vec{p}_{\text{you}} + \vec{p}_{\text{skateboard}} = 0 \qquad (9\text{-}8)$$

The left-hand side of Equation 9-8 is the **total momentum**, the momentum of a system of objects, just after the push-off of you and the skateboard together. Since momentum is a vector, we have to use the rules of vector arithmetic to add them together. Since these two vectors have equal magnitudes but opposite directions, they add to zero. So the system of you and the skateboard has zero total momentum just after the push-off. During the interaction of you pushing off the skateboard, nothing else is interacting with you or the skateboard with a force that is even close to the same magnitude as you exert on the skateboard. Gravity and the normal force balance each other out, so exert a zero net force on you or the skateboard. Even though there may be friction with the ground, it is small compared to your push-off, so the system of you and the skateboard is well approximated as closed and isolated.

Note also that the total momentum just *before* the push-off is also zero: Initially neither you nor the skateboard is moving, so both objects have zero momentum and the sum of these is likewise zero. So for this special case of a short, strong interaction, the total momentum of the system of you and the skateboard is *constant*: It has the same value (in this case zero) before and after the push-off (**Figure 9-5**).

We saw in Chapter 7 that the *total energy E* of a system is always conserved and could be constant under special circumstances (when the system is a closed, isolated system). Is something similar true for the total momentum of a system? And if so, what

(1) Before the push-off: No motion, so total momentum of the system of you and skateboard is zero:

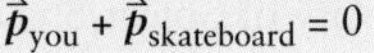

$\vec{p}_{\text{you}} + \vec{p}_{\text{skateboard}} = 0$

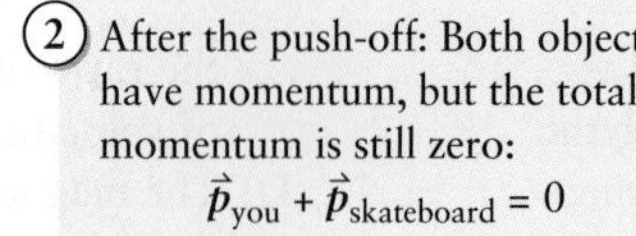

(2) After the push-off: Both objects have momentum, but the total momentum is still zero:

$\vec{p}_{\text{you}} + \vec{p}_{\text{skateboard}} = 0$

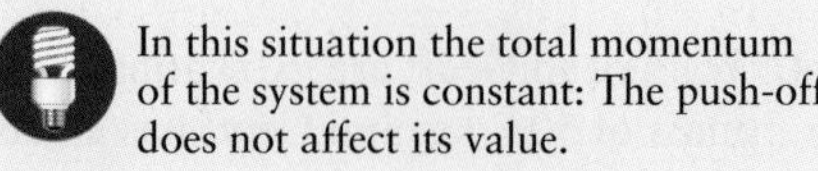

In this situation the total momentum of the system is constant: The push-off does not affect its value.

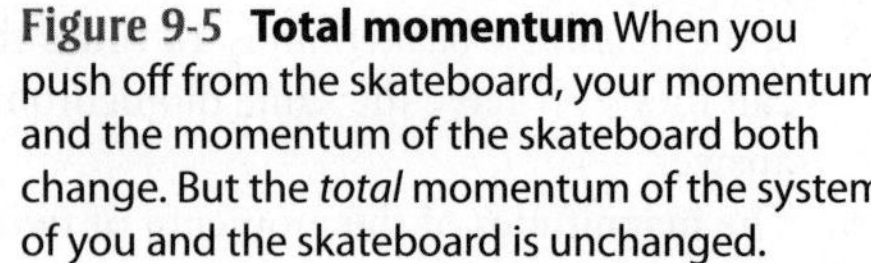

Figure 9-5 Total momentum When you push off from the skateboard, your momentum and the momentum of the skateboard both change. But the *total* momentum of the system of you and the skateboard is unchanged.

are the special circumstances under which total momentum is constant? Let's find the answers to these questions, using our discussion from Section 9-2 as a guide.

A System of Two Objects: Internal and External Forces

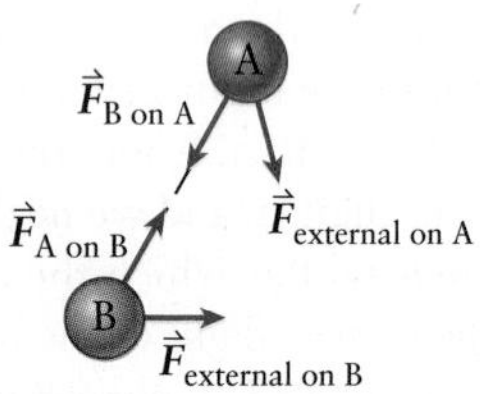

Figure 9-6 Internal forces and external forces If we treat the two objects or systems A and B as a single system, the interaction between A and B is most easily described as equal and opposite *internal* forces. Forces exerted on A and B by other objects outside the system are *external* forces. Recall that external forces can do work on a system, changing its total energy, but internal interactions (which we could describe as internal forces) cannot change the energy of the system, only rearrange it.

Figure 9-6 shows a system of two objects. These objects could be billiard balls on a billiard table, oxygen molecules in the air around you, or a planet and its moon. As noted in the previous section, the object model will fit well *before* and *after* a collision, but not during the collision itself. We will therefore discuss the participants in a collision as objects unless we are looking at the details of the interaction.

Two kinds of forces can be exerted on each object: forces exerted by other members of the system, which we call internal forces, and forces exerted by objects outside the system, which we call external forces. For two billiard balls the internal forces are the forces that each ball exerts on the other when they collide; the external forces are the normal and friction forces exerted by the billiard table, the gravitational force exerted by Earth, and the force you exert on the cue ball with the cue. Internal forces can either pull the components of a system together or push them apart, and can allow the objects or systems that make up the system of interest to convert energy from one type to another, while keeping the total energy constant. (We can see this in our skateboard example: To jump off the skateboard, you had to convert some of your internal energy to kinetic energy.) In order for two true objects to interact, one of them would already have to be moving, since objects can have only kinetic energy! For the two objects A and B shown in Figure 9-6, we can write Newton's second law as

$$\begin{aligned} \sum \vec{F}_{\text{on A}} &= \sum \vec{F}_{\text{external on A}} + \vec{F}_{\text{B on A}} = m_{\text{A}}\vec{a}_{\text{A}} \\ \sum \vec{F}_{\text{on B}} &= \sum \vec{F}_{\text{external on B}} + \vec{F}_{\text{A on B}} = m_{\text{B}}\vec{a}_{\text{B}} \end{aligned} \tag{9-9}$$

We can reduce Equations 9-9 to a single equation by adding them together. This has several advantages. First, it leaves us with a single equation to analyze. Second, from Newton's third law, we know that $\vec{F}_{\text{B on A}} = -\vec{F}_{\text{A on B}}$. So when we add Equations 9-9, these two terms cancel each other out. Mathematically, we end up with

$$\sum \vec{F}_{\text{external on A}} + \sum \vec{F}_{\text{external on B}} = m_{\text{A}}\vec{a}_{\text{A}} + m_{\text{B}}\vec{a}_{\text{B}} \tag{9-10}$$

The internal forces that one object in our system exerts on another have disappeared from Equation 9-10. This will prove to be incredibly important for analyzing collisions: We'll be able to calculate the effects of a collision without knowing the details of those internal forces. It doesn't even matter if they are contact forces.

Over a time interval Δt the velocity of A changes from $\vec{v}_{\text{Ai}}$ to $\vec{v}_{\text{Af}}$ ("i" for initial, "f" for final), and the velocity of B changes from $\vec{v}_{\text{Bi}}$ to $\vec{v}_{\text{Bf}}$. Using $\vec{a} = \Delta\vec{v}/\Delta t$, we can rewrite each of the accelerations in Equation 9-10 in terms of the changes in velocity:

$$\sum \vec{F}_{\text{external on A}} + \sum \vec{F}_{\text{external on B}} = \frac{m_{\text{A}}(\vec{v}_{\text{Af}} - \vec{v}_{\text{Ai}})}{\Delta t} + \frac{m_{\text{B}}(\vec{v}_{\text{Bf}} - \vec{v}_{\text{Bi}})}{\Delta t} \tag{9-11}$$

We can further rewrite Equation 9-11 by multiplying both sides of each equation by Δt, using the definition of momentum $\vec{p} = m\vec{v}$, and rearranging terms on the right-hand side:

$$\begin{aligned} \left(\sum \vec{F}_{\text{external on A}} + \sum \vec{F}_{\text{external on B}}\right)\Delta t &= m_{\text{A}}\vec{v}_{\text{Af}} - m_{\text{A}}\vec{v}_{\text{Ai}} + m_{\text{B}}\vec{v}_{\text{Bf}} - m_{\text{B}}\vec{v}_{\text{Bi}} \\ \left(\sum \vec{F}_{\text{external on A}} + \sum \vec{F}_{\text{external on B}}\right)\Delta t &= (\vec{p}_{\text{Af}} + \vec{p}_{\text{Bf}}) - (\vec{p}_{\text{Ai}} + \vec{p}_{\text{Bi}}) \end{aligned} \tag{9-12}$$

In Equation 9-12 $\vec{p}_{Ai}$ and $\vec{p}_{Bi}$ are the momenta of A and B, respectively, at the beginning of the time interval Δt, and $\vec{p}_{Af}$ and $\vec{p}_{Bf}$ are the momenta of A and B, respectively, at the end of the time interval.

The quantity in parentheses on the left-hand side of Equation 9-12 is the sum of all the *external* forces that are exerted on the *system* of objects A and B. We'll call this $\sum \vec{F}_{\text{external on system}}$ for short. The right-hand side of the equation is the difference between the *total* momentum of the system after the time interval Δt, $\vec{p}_{\text{total,f}} = \vec{p}_{Af} + \vec{p}_{Bf}$, and the *total* momentum of the system before the time interval, $\vec{p}_{\text{total,i}} = \vec{p}_{Ai} + \vec{p}_{Bi}$. So we can rewrite Equation 9-12 as

The **sum of all external forces** exerted on a system of objects

Duration of a time interval over which the external forces are exerted

$$\left(\sum \vec{F}_{\text{external on system}}\right)\Delta t = \vec{p}_{\text{total,f}} - \vec{p}_{\text{total,i}} = \Delta\vec{p}_{\text{total}} \tag{9-13}$$

Change during that time interval **in the total momentum** of the objects that make up the system

EQUATION IN WORDS
External force and total momentum change for a system of objects

Equation 9-13 indicates something quite remarkable: *Only the external forces exerted on a system can affect the system's total momentum.* The internal forces of one object on another allow momentum to be transferred between the objects (for example, when a moving cue ball hits an 8-ball at rest and sends the 8-ball flying), but they don't affect the value of the *total* momentum. This parallels our result that internal forces only allow energy to be converted or transferred between objects in a system, and cannot change the total energy of a system. This is the most general expression of the conservation of momentum. Just like the work-energy theorem told us that any change in energy of a system must equal the transfer of energy into or out of the system, any change in momentum of a system must equal the transfer of momentum, $F_{\text{ext}}\Delta t$. Just as $F_{\text{ext}}\Delta x \cos\theta$ has a name (work), $F_{\text{ext}}\Delta t$ also has a name, which we will learn later in this chapter.

There is something else profound in this statement. Although we used mathematical reasoning to get to Equation 9-13, instead of advanced mathematics, it turns out that in our process we passed through the most fundamental expression of Newton's second law! $\sum \vec{F}_{\text{external on system}} = \dfrac{\Delta\vec{p}_{\text{total}}}{\Delta t}$ shows an external force exerted on a system results in a change in that system's momentum. In most cases we will deal with, the mass of that system will remain constant. Because momentum is $\vec{p} = m\vec{v}$, if the mass does not change only the velocity does, $\dfrac{\Delta\vec{p}_{\text{total}}}{\Delta t}$ becomes $m\dfrac{\Delta\vec{v}_{\text{total}}}{\Delta t}$ and so reduces to Equation 4-3, Newton's second law. However, there are important cases in nature and in human-made design where the mass changes, like a squid using a water jet ejected from its body to accelerate or a rocket ship burning fuel. If you continue your study of physics and mathematics, you will gain the tools necessary to solve these problems as well. An understanding of Newton's second law as a condition on momentum allows you to make good predictions in most situations.

Exam Tip

On the AP® Physics 1 equation sheet, the conservation of momentum equation is listed as $\Delta\vec{p} = \vec{F}\Delta t$; it's important to understand for the exam that the force in this equation refers to the sum of the external forces on the system of interest.

We now have the answer to the question, "When is momentum constant?" Equation 9-13 says that the total momentum does not change over the time interval Δt if the net external force on the system is zero. Then the left-hand side of Equation 9-13 is zero and so there is zero difference between the final total momentum $\vec{p}_{\text{total,f}} = \vec{p}_{Af} + \vec{p}_{Bf}$ and the initial total momentum $\vec{p}_{\text{total,i}} = \vec{p}_{Ai} + \vec{p}_{Bi}$. This shortened law of conservation of momentum for the special case of no external forces can also be used in all collisions, so is remarkably useful:

If the net external force exerted on a system of objects is zero, **the total momentum of the system** is constant.

$$\vec{p}_{\text{total,f}} = \vec{p}_{\text{total,i}} \tag{9-14}$$

Then the total momentum of the system at the end of a time interval...

...is equal to the total momentum of the system at the beginning of that time interval.

EQUATION IN WORDS
Law of conservation of momentum for closed, isolated systems and collisions

Equation 9-14 is the law of conservation of momentum for a closed, isolated system or for a collision. It explains why the total momentum was constant for the system of you and the skateboard in Section 9-2: There were external forces exerted on you and on the skateboard, but, except for friction, the vector sum of these external forces was zero. The friction force was much smaller than the force you exerted on the skateboard and could be considered negligible in this case. Hence the total momentum of you and the skateboard had the same value (zero) immediately before and after the push-off.

The law of conservation of momentum for closed, isolated systems turns out to be useful even when the net external force on a system is *not* small. One example is a collision between two automobiles (**Figure 9-7**). During the collision the net vertical force on each car is zero: The upward normal force exerted by the ground balances the downward gravitational force. But the ground and air also exert fairly large horizontal friction forces on each car, and there is nothing to balance the friction forces. So there is a net external force on the system of two cars. To see why we can still ignore this net external force over the collision, let's rewrite Equation 9-12 the way it would appear if we didn't immediately cancel the internal forces when we added Equations 9-9 (the internal forces still appear on the left-hand side of the equation):

$$\left(\sum \vec{F}_{\text{external on A}} + \vec{F}_{\text{B on A}}\right)\Delta t + \left(\sum \vec{F}_{\text{external on B}} + \vec{F}_{\text{A on B}}\right)\Delta t = \vec{p}_{\text{Af}} + \vec{p}_{\text{Bf}} - \vec{p}_{\text{Ai}} - \vec{p}_{\text{Bi}} \quad (9\text{-}15)$$

The forces $\vec{F}_{\text{B on A}}$ and $\vec{F}_{\text{A on B}}$ that the cars exert on each other are *very* large: They can deform metal and shatter both plastic and glass. By comparison, even though much larger than the force of friction on your skateboard, the external friction forces are much smaller in magnitude than the internal forces. So for each object, it's a very good approximation to ignore the external friction forces during the brief duration Δt of the collision and set the terms $\sum \vec{F}_{\text{external on A}}$ and $\sum \vec{F}_{\text{external on B}}$ in Equation 9-15 equal to zero:

$$\left(\vec{F}_{\text{B on A}} + \vec{F}_{\text{A on B}}\right)\Delta t = \vec{p}_{\text{Af}} + \vec{p}_{\text{Bf}} - \vec{p}_{\text{Ai}} - \vec{p}_{\text{Bi}} \quad (9\text{-}16)$$

But by Newton's third law we still have $\vec{F}_{\text{B on A}} = -\vec{F}_{\text{A on B}}$, so the left-hand side of Equation 9-16 for the two colliding cars is zero. Thus the change in total momentum during the collision is still zero, or

$$\vec{p}_{\text{Af}} + \vec{p}_{\text{Bf}} = \vec{p}_{\text{Ai}} + \vec{p}_{\text{Bi}} \quad (9\text{-}17)$$

This is just a different way to write Equation 9-14, the law of conservation of momentum for a closed, isolated system. So we conclude that

If the internal forces during an interaction are much greater in magnitude than the external forces, the total momentum of the interacting objects has the same value just before and just after the interaction.

NEED TO REVIEW?
Turn to Chapter 4, Section 4-2, for a review of internal forces.

For *most* real-life collisions (things running into each other) the internal forces are much larger in magnitude than the external forces. For example, when two hockey

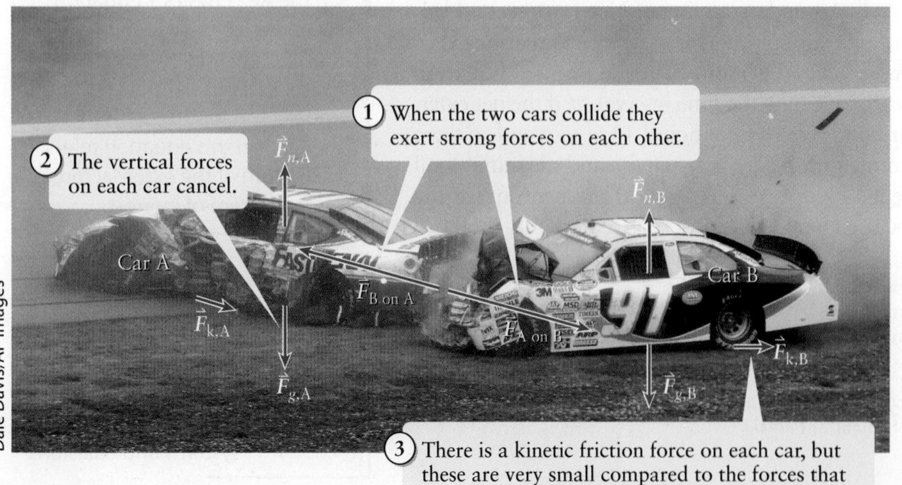

1 When the two cars collide they exert strong forces on each other.
2 The vertical forces on each car cancel.
3 There is a kinetic friction force on each car, but these are very small compared to the forces that car A exerts on car B and car B exerts on car A.
Car A
Car B
$\vec{F}_{n,A}$
$\vec{F}_{n,B}$
$\vec{F}_{\text{B on A}}$
$\vec{F}_{\text{A on B}}$
$\vec{F}_{k,A}$
$\vec{F}_{k,B}$
$\vec{F}_{g,A}$
$\vec{F}_{g,B}$

Dale Davis/AP Images

Figure 9-7 Forces in a collision During a collision between two cars, the internal forces (the forces of one car on another) are so great that we can ignore the external forces on the system.

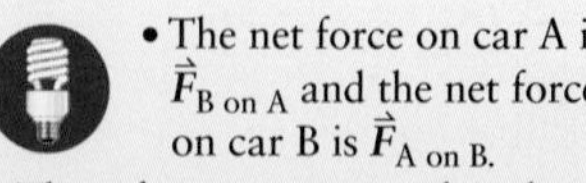

- The net force on car A is $\vec{F}_{\text{B on A}}$ and the net force on car B is $\vec{F}_{\text{A on B}}$.
- These forces are internal to the system of two cars.

players collide on the ice, the internal forces of one player on the other are strong enough to knock the wind out of the players or even cause injury. These internal forces are much greater than the friction forces that the ice exerts on the players. Similarly, the forces that are exerted between a tennis ball and tennis racquet in a serve are so large that the ball distorts noticeably. The external forces that are exerted on the ball and racquet during the time they are in contact—the gravitational forces and the force of the player's hand on the racquet—are very feeble by comparison. So it's almost always safe to say that the total momentum of a system of colliding objects just *after* the collision is the same as just *before* the collision. That is why in physics the definition of collision is expanded to include any brief interaction where the internal forces dominate, such as when you jumped off your skateboard in Example 9-1. Even though Equation 9-14 is a special case of the law of conservation of momentum, it is not uncommon to see it called just the law of conservation of momentum. We will do this, and learn a different name for the more general law, which includes the transfer term, in Section 9-6. Conservation of momentum, like conservation of energy, is one of the fundamental laws of our universe that governs the interactions of all objects and systems.

WATCH OUT !

In a collision, momentum is constant only *during* the collision.

The general rule that we have discovered says that the total momentum of a system of colliding objects is constant *only* during the very brief period during which the collision takes place. Once the collision is over, the strong internal forces are no longer exerted, and the external forces become dominant, so the total momentum is no longer constant. For example, the two colliding cars shown in Figure 9-7 continued to move after the collision (with a total momentum just after the collision that equaled their total momentum just before the collision). Once the collision ended, however, friction with the track and grass soon caused the cars to come to a halt, transferring all their momentum to Earth. The mass of Earth is so great that it is very hard to see this effect, but the most careful experiments have never failed to confirm that total momentum is conserved!

WATCH OUT !

Remember, momentum is a vector.

It's important to note that Equation 9-14, the law of conservation of momentum, is a *vector* equation. If the vectors $\vec{p}_{\text{total,f}}$ and $\vec{p}_{\text{total,i}}$ are equal, it must be that the x components of the two vectors are equal *and* the y components of the two vectors are equal. This is true even in just one dimension. You must know in which direction things are moving. Examples 9-2, 9-3, and 9-4 illustrate how to solve problems that involve the vector nature of momentum.

EXAMPLE 9-2 Conservation of Momentum: A Collision on the Ice

Gordie, a 100-kg hockey player, is initially moving to the right at 5.00 m/s directly toward Mario, a stationary 80.0-kg player. After the two players collide head-on, Mario is moving to the right at 3.75 m/s. (a) In what direction and at what speed is Gordie moving after the collision? (b) What was the change in Gordie's momentum in the collision? What was the change in Mario's momentum?

Set Up

The system that we're considering is made up of the two players. The vertical forces on each player (the normal force exerted by the ice and the gravitational force) cancel each other, so there is no net external force in the vertical direction. The friction forces between the players and the ice are small compared to the forces that

Momentum conservation:

$$\vec{p}_{\text{total,f}} = \vec{p}_{\text{total,i}} \quad (9\text{-}14)$$

Definition of momentum:

$$\vec{p} = m\vec{v} \quad (9\text{-}5)$$

Gordie
$m_G = 100$ kg
Mario (at rest)
$m_M = 80.0$ kg
$v_{Gi} = 5.00$ m/s
$v_{Mi} = 0$
+x
before the collision

Gordie and Mario exert on each other. So we can treat the total momentum of the system as constant during the collision. We'll use this to find Gordie's final velocity and the changes in momentum of each player. Because momentum is a vector we must define our coordinate system. We choose the positive x direction to be to the right, as shown.

Solve

(a) Write the equation of momentum conservation, using the subscript "i" (for initial) for values just before the collision and subscript "f" (for final) for values just after. Also, we use the subscripts "G" for Gordie and "M" for Mario.

Just before the collision the total momentum is

$$\vec{p}_{\text{total,i}} = \vec{p}_{\text{Gi}} + \vec{p}_{\text{Mi}} = m_{\text{G}}\vec{v}_{\text{Gi}} + m_{\text{M}}\vec{v}_{\text{Mi}}$$

Gordie

Mario

$v_{\text{Gf}} = ?$

$v_{\text{Mf}} = 3.75$ m/s

$+x$

after the collision

Just after the collision the total momentum is

$$\vec{p}_{\text{total,f}} = \vec{p}_{\text{Gf}} + \vec{p}_{\text{Mf}} = m_{\text{G}}\vec{v}_{\text{Gf}} + m_{\text{M}}\vec{v}_{\text{Mf}}$$

Momentum is constant:

$$\vec{p}_{\text{total,f}} = \vec{p}_{\text{total,i}}$$

$$m_{\text{G}}\vec{v}_{\text{Gf}} + m_{\text{M}}\vec{v}_{\text{Mf}} = m_{\text{G}}\vec{v}_{\text{Gi}} + m_{\text{M}}\vec{v}_{\text{Mi}}$$

Note that $\vec{v}_{\text{Mi}} = 0$, because Mario is originally at rest. We need to find $\vec{v}_{\text{Gf}}$ (Gordie's velocity just after the collision).

Because the motion is entirely along the x axis, we need only the x component of the momentum conservation equation. Solve for Gordie's final x velocity, $v_{\text{Gf}x}$; then substitute the values of the players' masses, Gordie's velocity before the collision, and Mario's velocity after the collision.

$$m_{\text{G}}v_{\text{Gf}x} + m_{\text{M}}v_{\text{Mf}x} = m_{\text{G}}v_{\text{Gi}x} + m_{\text{M}}v_{\text{Mi}x}$$

Mario is originally at rest, so $v_{\text{Mi}x} = 0$ and

$$m_{\text{G}}v_{\text{Gf}x} + m_{\text{M}}v_{\text{Mf}x} = m_{\text{G}}v_{\text{Gi}x}$$

$$m_{\text{G}}v_{\text{Gf}x} = m_{\text{G}}v_{\text{Gi}x} - m_{\text{M}}v_{\text{Mf}x}$$

$$v_{\text{Gf}x} = v_{\text{Gi}x} - \frac{m_{\text{M}}v_{\text{Mf}x}}{m_{\text{G}}}$$

Players' masses: $m_{\text{G}} = 100$ kg, $m_{\text{M}} = 80.0$ kg

Gordie's initial x velocity: $v_{\text{Gi}x} = +5.00$ m/s

Mario's final x velocity: $v_{\text{Mf}x} = +3.75$ m/s

$$\begin{aligned} v_{\text{Gf}x} &= (+5.00 \text{ m/s}) - \frac{(80.0 \text{ kg})(+3.75 \text{ m/s})}{100 \text{ kg}} \\ &= +5.00 \text{ m/s} - 3.00 \text{ m/s} \\ &= +2.00 \text{ m/s} \end{aligned}$$

Gordie ends up moving at 2.00 m/s to the right (in the positive x direction).

(b) The change in each player's momentum equals his momentum after the collision minus his momentum before the collision.

$$\begin{aligned} \text{Gordie: } \Delta p_{\text{G}x} &= m_{\text{G}}v_{\text{Gf}x} - m_{\text{G}}v_{\text{Gi}x} \\ &= (100 \text{ kg})(+2.00 \text{ m/s}) - (100 \text{ kg})(+5.00 \text{ m/s}) \\ &= 200 \text{ kg} \cdot \text{m/s} - 500 \text{ kg} \cdot \text{m/s} \\ &= -300 \text{ kg} \cdot \text{m/s} \end{aligned}$$

$$\begin{aligned} \text{Mario: } \Delta p_{\text{M}x} &= m_{\text{M}}v_{\text{Mf}x} - m_{\text{M}}v_{\text{Mi}x} \\ &= (80.0 \text{ kg})(+3.75 \text{ m/s}) - (80.0 \text{ kg})(0 \text{ m/s}) \\ &= +300 \text{ kg} \cdot \text{m/s} \end{aligned}$$

Reflect

Our answer to part (a) tells us that after the collision Gordie is still moving in the positive x direction but with reduced speed: He has lost momentum in the x direction, while Mario (who was originally at rest) has gained momentum in the x direction. This answer makes sense because we know that the momentum of the system is constant. In fact, as part (b) confirms, the amount of momentum in the x direction that Gordie loses (300 kg · m/s) is exactly the same as the amount of momentum in the x direction that Mario gains. Thus, we can think of the collision between the two players as a *transfer* of momentum between Gordie and Mario. Momentum is not changed by internal forces; it simply is transferred from one player (part of the system) to the other.

Notice that in this problem we needed to know the x component of Mario's final velocity $v_{\text{Mf}x}$ to find the x component of Gordie's final velocity $v_{\text{Gf}x}$. That's because the statement that momentum is conserved in the x direction gave us only one equation that relates $v_{\text{Mf}x}$ and $v_{\text{Gf}x}$, so we were able to solve for only one unknown quantity. If we didn't know either of the players' final velocities, we would have needed additional information—such as the duration of the collision and the magnitude of the force that one player exerted on the other during the collision, or a description of how energy changed during the collision—which we unfortunately do not have. The law of conservation of momentum is a great tool, but by itself it can't tell you everything! Fortunately, we will learn some more things about collisions that will help give us the extra information needed to solve a wider variety of problems.

Extend: Note that neither player's *individual* momentum is constant. Gordie's momentum after the collision is different than before the collision and likewise for Mario's momentum. It's only the *total* momentum of the Gordie–Mario system that is constant. If we look at either Gordie or Mario, we can see that we would not be able to consider either as a closed, isolated system because the forces exerted between them would be external forces, and those forces are large. *In general, the momentum of any particular object within a system will not be constant.*

NOW WORK Problems 1 and 2 from The Takeaway 9-3.

EXAMPLE 9-3 A Collision at the Bowl-a-Rama

The sequence of images in **Figure 9-8** shows a bowling ball striking a stationary pin. The second image shows that when the collision occurs the ball strikes a bit to the left of the horizontal center of the pin. The third image shows what happens after the collision: The ball moves on a path to the left of its original direction, and the pin shoots off to the right. Just after one such collision, the ball is moving off horizontally at a 10.0° angle and the pin is moving off horizontally at a 60.0° angle (measured relative to the original direction of motion of the ball). The mass of the ball is 3.50 times greater than the mass of the pin. Just after the collision is the pin moving more quickly or more slowly than the ball, and by what factor?

Figure 9-8 Hitting a bowling pin off-center After the bowling ball strikes the left-hand side of the pin, the ball deflects to the left and the pin moves off to the right. How do their speeds compare?

Set Up

The forces that the ball and pin exert on each other are very strong (imagine what it would feel like if your finger were between the ball and pin when they hit!). Compared to these, we can ignore the external forces exerted on the system of ball and pin, so the total momentum of this system is constant. The collision is two-dimensional, so we must account for both the x and y components of momentum.

Momentum conservation:

$$\vec{p}_{\text{total,f}} = \vec{p}_{\text{total,i}} \qquad (9\text{-}14)$$

Linear momentum:

$$\vec{p} = m\vec{v} \qquad (9\text{-}5)$$

("B" for ball, "P" for pin)

We are given the directions of motion of the ball and pin before and after the collision, and we want to find the *ratio* of the pin's final speed v_{Pf} to the ball's final speed v_{Bf}. We don't know the mass of either the ball or the pin, but we do know the ratio of their masses.

Ratio of the mass of the ball to the mass of the pin:

$$\frac{m_B}{m_P} = 3.50$$

Solve

The total momentum before the collision (subscript "i") equals the total momentum after the collision (subscript "f"). Write this for both the x component and the y component of momentum.

Linear momentum:

$$\vec{p}_{total} = \vec{p}_B + \vec{p}_P$$

Momentum conservation:

$$\vec{p}_{Bf} + \vec{p}_{Pf} = \vec{p}_{Bi} + \vec{p}_{Pi}, \text{ or}$$

$$m_B\vec{v}_{Bf} + m_P\vec{v}_{Pf} = m_B\vec{v}_{Bi} + m_P\vec{v}_{Pi}, \text{ or}$$

$$m_B v_{Bfx} + m_P v_{Pfx} = m_B v_{Bix} + m_P v_{Pix}$$

(total momentum in the x direction is constant)

$$m_B v_{Bfy} + m_P v_{Pfy} = m_B v_{Biy} + m_P v_{Piy}$$

(total y momentum is constant)

Just prior to the collision, the ball has no component of velocity in the y direction (so $v_{Biy} = 0$) and the pin is at rest (so $v_{Pix} = v_{Piy} = 0$). Just after the collision the ball is moving at a 10.0° angle with positive x and y components of velocity, while the pin is moving at a 60.0° angle with a positive x velocity and a negative y velocity. Use these facts to rewrite the equations for the conservation of x and y momentum.

x equation:

$$m_B v_{Bf} \cos 10.0° + m_P v_{Pf} \cos 60.0° = m_B v_{Bi}$$

y equation:

$$m_B v_{Bf} \sin 10.0° + (-m_P v_{Pf} \sin 60.0°) = 0$$

Our goal is to find the ratio of the speed of the pin just after the collision (v_{Pf}) to the speed of the ball just after the collision (v_{Bf}). The equation for momentum in the x direction isn't useful for this because we aren't given the value of the ball's speed v_{Bi} before the collision. Instead, we use the equation for momentum in the y direction to solve for v_{Pf}/v_{Bf}, and then substitute the ratio of the masses of the two objects.

$$m_B v_{Bf} \sin 10.0° = m_P v_{Pf} \sin 60.0°$$

$$\frac{v_{Pf}}{v_{Bf}} = \frac{m_B}{m_P}\frac{\sin 10.0}{\sin 60.0} = (3.50)\frac{0.174}{0.866} = 0.702$$

After the collision the speed of the pin is 0.702 times that of the ball.

Reflect

With a more head-on impact the pin could end up moving away from the collision at a much faster speed than the ball. Here, the pin moves relatively slowly. This makes sense given that it received only a glancing blow from the ball.

NOW WORK Problem 3 from The Takeaway 9-3.

Momentum Conservation Applies Even When There Are More Than Two Interacting Objects

There are three objects in the collision in upcoming Example 9-4, rather than two as in Examples 9-2 and 9-3. But the same ideas of momentum conservation apply, no matter how many objects are in the system, as long as the internal forces are the dominant interactions in the system (we can consider the interaction a collision). The third-law pairs of forces will still cancel when all the forces on each of the objects in the system are summed, leaving only the net external force on the system.

EXAMPLE 9-4 A Tricky Billiards Shot

On a billiards table the 7-ball and 8-ball are initially at rest and touching each other. You hit the cue ball in such a way that it acquires a speed of 1.7 m/s before hitting the 7-ball and 8-ball simultaneously. After the collision the cue ball is at rest, and the other two balls are each moving at 45° from the direction that the cue ball was moving. What are the speeds of the 7-ball and 8-ball immediately after the collision? Each billiard ball has a mass of 0.16 kg.

Set Up

This is a more complicated collision than in the previous two examples, but the fundamental principle is the same: Total momentum is constant because the internal forces between the pool balls are much larger than any net external forces exerted on the pool balls during the collision.

Initially all the momentum is that of the cue ball and the momentum points along the direction of its motion. After the collision the 7-ball and 8-ball must travel on opposite sides of the cue ball's initial path as shown. (If they were on the same side, there would be no chance for their momenta to cancel, so there would be a nonzero total momentum perpendicular to the cue ball's initial path, and momentum would not be constant!)

As in Example 9-2 we know the directions of motion of the balls before and after the collision. We know the initial speed $v_{\text{cue,i}} = 1.7$ m/s and final speed $v_{\text{cue,f}} = 0$ of the cue ball, and we want to find the final speeds, $v_{7\text{f}}$ and $v_{8\text{f}}$, of the 7-ball and 8-ball, respectively.

Momentum conservation

$$\vec{p}_{\text{total,f}} = \vec{p}_{\text{total,i}} \qquad (9\text{-}14)$$

(cue ball, 7-ball, and 8-ball) has the same value just before and just after the collision

Linear momentum:

$$\vec{p} = m\vec{v} \qquad (9\text{-}5)$$

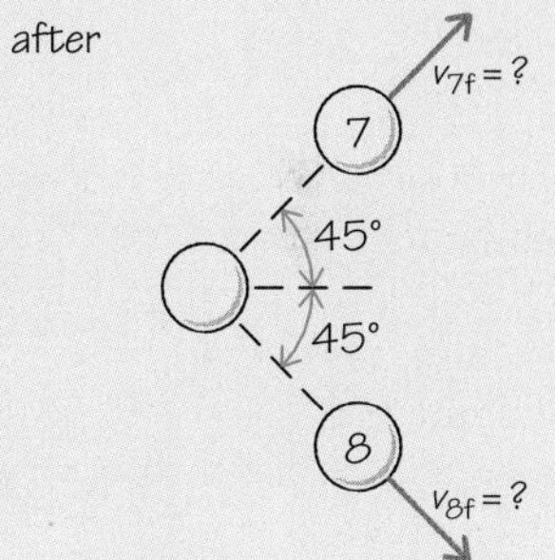

Solve

Write in component form the momentum conservation equation, which says that the total momentum before the collision (subscript "i") equals the total momentum after the collision (subscript "f"). Take the positive x axis to be in the direction that the cue ball was moving before the collision. Note that all three balls have the same mass $m = 0.16$ kg.

Linear momentum:

$$\vec{p}_{\text{total}} = \vec{p}_{\text{cue}} + \vec{p}_7 + \vec{p}_8$$

Momentum conservation:

$$\vec{p}_{\text{cue,f}} + \vec{p}_{7\text{f}} + \vec{p}_{8\text{f}} = \vec{p}_{\text{cue,i}} + \vec{p}_{7\text{i}} + \vec{p}_{8\text{i}}$$

The 7-ball and 8-ball are not moving before the collision, so $\vec{p}_{7\text{i}} = \vec{p}_{8\text{i}} = 0$, and the cue ball is not moving after the collision, so $\vec{p}_{\text{cue,f}} = 0$. So the momentum conservation equation becomes

$$\vec{p}_{7\text{f}} + \vec{p}_{8\text{f}} = \vec{p}_{\text{cue,i}}$$

$$m\vec{v}_{7\text{f}} + m\vec{v}_{8\text{f}} = m\vec{v}_{\text{cue,i}}$$

or in component form

$$mv_{7\text{f}x} + mv_{8\text{f}x} = mv_{\text{cue,i}x}$$

(total momentum in the x direction is constant)

$$mv_{7\text{f}y} + mv_{8\text{f}y} = mv_{\text{cue,i}y}$$

(total y momentum is constant)

The cue ball's initial velocity had zero y component (so $v_{\text{cue,i}y} = 0$). Of the 7-ball and 8-ball, one must have positive final y component of velocity and the other negative final y component of velocity to satisfy the condition that total y component of momentum is constant. Use these observations to simplify the momentum equations.

x equation (the x component is adjacent to the angle, so we use cosine):

$$mv_{7\text{f}} \cos 45° + mv_{8\text{f}} \cos 45° = mv_{\text{cue,i}}$$

y equation (the y component is opposite the angle, so we use sine):

$$mv_{7\text{f}} \sin 45° + (-mv_{8\text{f}} \sin 45°) = 0$$

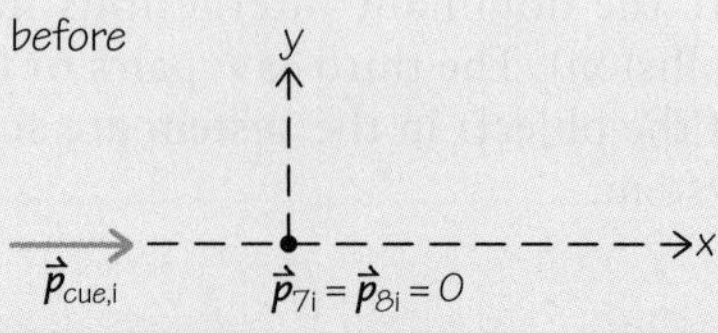

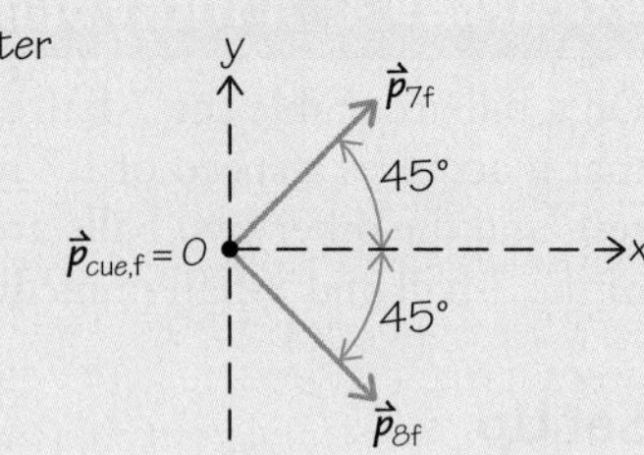

Solve the momentum conservation equations for the final speeds of the 7-ball and 8-ball.

From the y equation

$mv_{7\text{f}} \sin 45° = mv_{8\text{f}} \sin 45°$ so

$$v_{7\text{f}} = v_{8\text{f}}$$

After the collision, for the momentum in the y direction to remain zero, the 7-ball and the 8-ball must move with the same speed because they are moving at the same angle and have the same mass. To find this common speed, replace $v_{8\text{f}}$ in the x equation with $v_{7\text{f}}$ and then solve for $v_{7\text{f}}$:

$$mv_{7\text{f}} \cos 45° + mv_{7\text{f}} \cos 45° = mv_{\text{cue,i}}$$

$$2mv_{7\text{f}} \cos 45° = mv_{\text{cue,i}}$$

$$v_{7\text{f}} = \frac{v_{\text{cue,i}}}{2 \cos 45°} = \frac{1.7 \text{ m/s}}{2 \cos 45°} = 1.2 \text{ m/s}$$

After the collision, both the 7-ball and the 8-ball are moving at 1.2 m/s.

Reflect

The speeds we solved for are reasonable, a bit smaller than the initial ball's speed, but there are two of them. (Note that we didn't need the value of the billiard ball mass. That's because all three balls have the same mass m, so m canceled out in the equations. That didn't happen in Example 9-3 because the bowling ball and pin had *different* masses.)

NOW WORK Problem 4 from The Takeaway 9-3.

Exam Tip

AP® Physics 1 includes a quantitative treatment of momentum conservation in one dimension, but only a semi-quantitative treatment of momentum conservation in two dimensions: On the exam, you'll need to be able to set up momentum conservation equations and reason qualitatively, but you will not be asked to solve simultaneous equations.

THE TAKEAWAY for Section 9-3

✔ In a system of objects, internal forces are forces that one object in the system exerts on another object in the system. External forces are forces exerted on objects in the system by objects outside of the system.

✔ If the net external force on a system of objects is zero, the total momentum of the system is constant. Objects within the system can exchange momentum with each other, but the vector sum of the momentum of all members of the system remains constant.

✔ During a collision the internal forces are typically much larger in magnitude than any external forces. Hence the external forces can be ignored, and the total momentum just after the collision equals the total momentum just before the collision.

Prep for the AP® Exam

AP® Building Blocks

EX -2 1. A 2.00-kg object is moving east at 4.00 m/s when it collides with a 6.00-kg object that is initially at rest. After the collision the larger object moves east at 2.00 m/s. What is the final magnitude and direction of the velocity of the smaller object after the collision? What assumption must you make to solve this problem?

EX -2 2. An 8000.0-kg gondola, a railroad freight car that is open at the top, is decoupled from the train and rolling at a constant speed of 20.0 m/s when it begins to rain heavily. The rain falls straight down. After water has collected in the car, it slows to 19.0 m/s. What mass of water has collected in the car? For simplicity, assume that the water collects quickly enough that you can neglect any air resistance on the freight car and that the train tracks exert negligible friction on the freight car.

EX 9-3 3. In a game of pool, the cue ball is rolling at 2.0 m/s in a direction 30.0° north of east when it collides with the 8-ball (initially at rest). The mass of the cue ball is 170 g, but the mass of the 8-ball is only 156 g. After the collision, the cue ball heads off at 10.0° north of east, and the 8-ball moves due north. What are the final magnitudes of the velocity of each ball after the collision? What assumptions do you need to make?

EX 9-4 4. An object of mass $3M$, moving in the $+x$ direction at speed v_0, breaks into two pieces of mass M and $2M$ as shown. If $\theta_1 = 45.0°$ and $\theta_2 = 30.0°$, determine the final magnitudes and directions of the velocities of the resulting pieces in terms of v_0. What assumptions do you need to make?

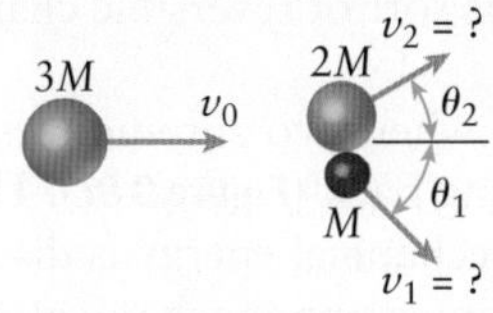

AP® Skill Builders

5. Starting from Newton's second law, explain how a system upon which no external forces are exerted and in which two objects collide has a constant momentum. In other words, explain how the momentum of the system remains constant during the collision.

6. A tennis player smashes a ball with a mass m horizontally at a vertical wall. The ball rebounds at the same speed v with which it struck the wall. Choose the correct claim below, and justify the relationship between external force and change in momentum.

(A) Because the ball exerts an external force on the wall the momentum of the ball changes. And because the magnitude of the velocity before and after collision with the wall is the same,

$$\vec{F}_{\text{ball on wall}}\Delta t = \Delta\vec{p}_{\text{ball}} = m_{\text{ball}}\vec{v}_{\text{ball,i}} - m_{\text{ball}}\vec{v}_{\text{ball,f}} = 2m_{\text{ball}}\vec{v}_{\text{ball,i}}$$

(B) Because the wall exerts an external force on the ball the momentum of the ball changes. And because the magnitude of the velocity before and after collision with the wall is the same,

$$\vec{F}_{\text{wall on ball}}\Delta t = \Delta\vec{p}_{\text{ball}} = m_{\text{ball}}\vec{v}_{\text{ball,f}} - m_{\text{ball}}\vec{v}_{\text{ball,i}} = -2m_{\text{ball}}\vec{v}_{\text{ball,i}}$$

(C) Because the ball exerts an external force on the wall the momentum of the ball changes. And because the magnitude of the velocity before and after collision with the wall is the same,

$$\vec{F}_{\text{ball on wall}}\Delta t = \Delta\vec{p}_{\text{ball}} = m_{\text{ball}}\vec{v}_{\text{ball,f}} - m_{\text{ball}}\vec{v}_{\text{ball,i}} = 0$$

(D) Because the wall exerts an external force on the ball the momentum of the ball changes. And because the magnitude of the velocity before and after collision with the wall is the same,

$$\vec{F}_{\text{wall on ball}}\Delta t = \Delta\vec{p}_{\text{ball}} = m_{\text{ball}}\vec{v}_{\text{ball,f}} - m_{\text{ball}}\vec{v}_{\text{ball,i}} = 0$$

7. Pairs of objects i and j in a system interact with pair forces $F_{\text{i on j}}$ and $F_{\text{j on i}}$.

(a) For a system composed of three objects labeled 1, 2, and 3, justify the claim that the sum of internal forces within a system is zero by writing out and rearranging all the pair forces in this sum:

$$\vec{F}_{\text{internal}} = \vec{F}_{1\text{ on }2} + \vec{F}_{1\text{ on }3} + \vec{F}_{2\text{ on }1} + \vec{F}_{2\text{ on }3} + \vec{F}_{3\text{ on }1} + \vec{F}_{3\text{ on }2}$$

(b) Explain why the sum doesn't include $F_{\text{j on j}}$.

AP® Skills in Action

8. Two ice skaters, Twyla and Ani, face each other while stationary and push against each other's hands. Twyla's mass is two times larger than Ani's mass.

(a) Calculate their relative speeds after they lose contact. Assume that friction forces are negligible.

(b) In fact, forces resisting the motion are only negligible during the interaction. After the interaction you can see that there are forces exerted on the skaters. Drag forces exerted by air and water at the skate–ice boundary and kinetic friction forces are present. Explain how you can tell such forces exist and pose questions regarding the dependence of these forces on mass and speed that would need to be addressed to refine the calculation in part (a).

9. During football season concussions frequently occur on the field. Two players—for example, a pass receiver with a smaller mass and a larger speed and an outside linebacker with a larger mass and smaller speed—might collide, exerting very large forces on each other over a short period of time. The pass receiver is seriously injured more frequently.

(a) Explain why the greater effect on the receiver is not due to the magnitudes of the forces exerted.

(b) Explain why the greater effect is not because the momentum change of the receiver is greater.

(c) If neither force nor momentum differences can account for the increased likelihood of concussion for the receiver, how is this difference explained?

9-4 In an inelastic collision some of the mechanical energy is dissipated

We've learned that momentum is constant in collisions of all kinds. We also learned in Chapter 7 that the mechanical energy of a system is constant under certain circumstances. Now we ask: Are *both* mechanical energy and momentum constant in collisions?

The answer to this question is "sometimes." In Section 7-6 we learned that the mechanical energy of a closed, isolated system is constant if there are only conservative interactions inside the system. An example of this is a system made up of two smaller systems that behave like ideal springs when they collide: They compress, converting some of the kinetic energy of the colliding objects into spring potential energy; then they relax and convert all of their potential energy back into kinetic energy (**Figure 9-9a**). A collision of this kind, in which the internal forces between the colliding objects are conservative, is called an **elastic collision.** In an elastic collision both total momentum *and* total mechanical energy are constant. In elastic collisions, during the interaction, all the initial kinetic energy not required to conserve momentum is converted into potential energy. After the interaction, all the energy converted into potential energy is restored to the kinetic energy. If the initial momentum of a system is not zero, then not all the kinetic energy can be stored because enough kinetic energy must remain to keep momentum constant.

Elastic collisions are happening all around you: When oxygen and nitrogen molecules in the air collide with each other, the collisions are almost always elastic. That's because the forces between molecules (which are fundamentally electric in nature) are not large enough to break any bonds (cause a change in configuration that cannot be reversed) within the molecules. The molecules compress slightly upon the collision, a reversible change in configuration, storing energy as potential energy. After the collision, the molecules return to their original configuration, restoring all the stored potential energy as kinetic. On a larger scale, a collision between billiard balls is very nearly elastic (**Figure 9-9b**). A billiard ball deforms slightly like a very strong spring when hit, but immediately springs back to its original shape with very little of the energy being dissipated. We will learn more about this sort of reversible change in later in this course.

Something very different usually happens when two automobiles collide: The bodies of the automobiles deform and do *not* spring back (**Figure 9-9c**). The forces involved in this deformation are *nonconservative* so mechanical energy is dissipated (converted to internal energy). (This is by design. By converting mechanical energy to internal

Because the ideal springs between these colliding objects change shape in a reversible way, mechanical energy is conserved in this **elastic collision.**

(a)

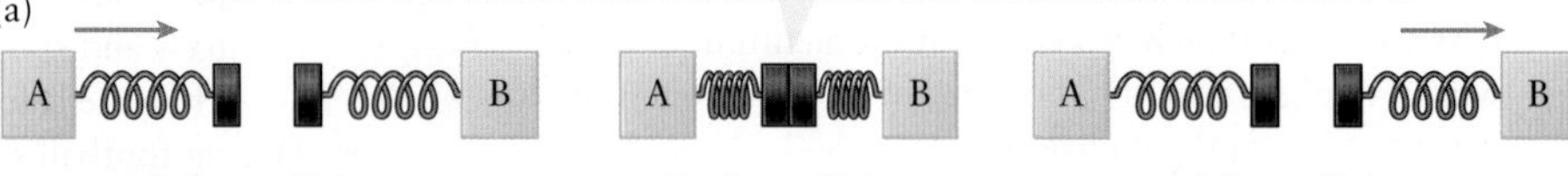

(b) The collision of the billiard balls is elastic: It does not cause any permanent deformation.

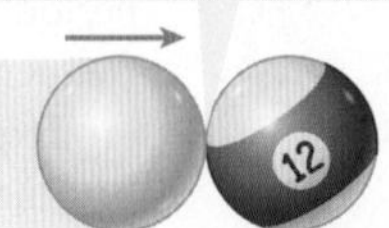

The forces between colliding ideal billiard balls are conservative. No mechanical energy is lost in this **elastic collision.**

(c) The collision of the two cars is inelastic: It causes a permanent deformation.

The forces between colliding cars are not conservative. Mechanical energy is lost in this **inelastic collision.**

Figure 9-9 Collision variations (a) The ideal springs connected to these colliding objects exert conservative internal forces. Mechanical energy is constant in this elastic collision. (b) The forces between colliding billiard balls are very nearly conservative (although we treat them as objects, billiard balls actually flex a tiny bit, like very strong springs), and the collision is very nearly elastic. (c) The forces between colliding cars are not conservative. Mechanical energy is dissipated, and the collision is inelastic.

In elastic collisions, mechanical energy is conserved. In inelastic collisions, mechanical energy is converted to other types of energy.

energy as it deforms, the structure of an automobile prevents that energy from being used to do potentially harmful work on the automobile's occupants.) A collision in which mechanical energy is *not* constant is called an **inelastic collision.**

In some inelastic collisions, mechanical energy is actually put into the system by the interaction. An example of this is Example 9-1. When you push off the skateboard, you convert some of your own internal energy into kinetic energy. We will refer to an inelastic collision in which mechanical energy is increased as an **explosive collision** and reserve the term *inelastic collision* for collisions where mechanical energy is decreased.

In inelastic collisions, during the interaction, all the initial kinetic energy not required to conserve momentum is converted into internal forms, but not all of it is converted into potential energy. Some is converted directly into other forms of internal energy. After the interaction, any energy converted into potential energy is restored to the kinetic energy, but the internal energy is considered dissipated, since it cannot result in motion.

Returning to Figure 9-9, if we know the masses of the colliding objects and their velocities before and after the collision, it's straightforward to determine whether the collision is elastic or inelastic. Here's the idea: By analogy to a collision between ideal springs (Figure 9-9a), there is zero potential energy just before and just after the collision (corresponding to the springs being relaxed). So just before and just after the collision, the total mechanical energy of the system is equal to the sum of the *kinetic energies* of the colliding objects. If the total kinetic energy has the same value before and after the collision, the collision is elastic; if there is less total kinetic energy after the collision, the collision is inelastic (or if there is more total kinetic energy after the collision, the collision is explosive). Example 9-5 illustrates this technique for analyzing collisions.

EXAMPLE 9-5 Elastic or Inelastic?

Determine whether the following collisions are elastic or inelastic: (a) the collision of two hockey players in Example 9-2 (Section 9-3); (b) the collision of three billiard balls in Example 9-4 (Section 9-3).

Set Up

For each collision we calculate the total kinetic energy (the sum of the individual kinetic energies of the colliding objects) just before and just after the collision. If these are equal, total mechanical energy is conserved and the collision is elastic; if they are not equal, the collision is inelastic.

Kinetic energy:

$$K = \frac{1}{2}mv^2 \qquad (7\text{-}8)$$

Solve

(a) The collision between the two hockey players in Example 9-2 is one-dimensional (that is, along a line). So the square of each player's speed v is the same as the square of each player's x velocity v_x.

Players' masses:

$$m_G = 100 \text{ kg}, \; m_M = 80.0 \text{ kg}$$

Before the collision, Gordie has x velocity $v_{Gix} = +5.00$ m/s and Mario is at rest. So the initial total kinetic energy is

$$K_{Gi} + K_{Mi} = \frac{1}{2}m_G v_{Gix}^2 + \frac{1}{2}m_M v_{Mix}^2$$

$$= \frac{1}{2}(100 \text{ kg})(5.00 \text{ m/s})^2$$

$$+ \frac{1}{2}(80.0 \text{ kg})(0)^2$$

$$= 1.25 \times 10^3 \text{ J}$$

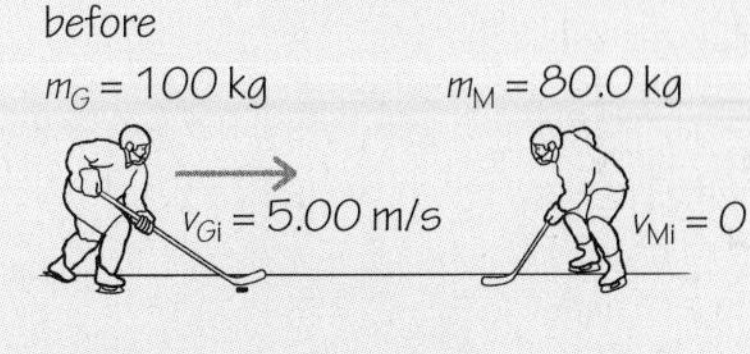

After the collision, Gordie has x velocity $v_{Gfx} = +2.00$ m/s and Mario has x velocity $v_{Mfx} = +3.75$ m/s. So the final total kinetic energy is

$$K_{Gf} + K_{Mf} = \frac{1}{2} m_G v_{Gfx}^2 + \frac{1}{2} m_M v_{Mfx}^2$$
$$= \frac{1}{2}(100 \text{ kg})(2.00 \text{ m/s})^2 + \frac{1}{2}(80.0 \text{ kg})(3.75 \text{ m/s})^2$$
$$= 763 \text{ J}$$

The total kinetic energy is less after the collision than before, so this collision is inelastic.

(b) Repeat the calculation for the three billiard balls in Example 9-4, for which we know the masses, the speeds before the collision, and the speeds after the collision.

Each ball has mass $m = 0.16$ kg. Before the collision, the cue ball is moving at speed $v_{cue,i} = 1.7$ m/s, and both the 7-ball and 8-ball are at rest. So the initial total kinetic energy is

$$K_{cue,i} + K_{7i} + K_{8i}$$
$$= \frac{1}{2} m v_{cue,i}^2 + \frac{1}{2} m v_{7i}^2 + \frac{1}{2} m v_{8i}^2$$
$$= \frac{1}{2}(0.16 \text{ kg})(1.7 \text{ m/s})^2$$
$$+ \frac{1}{2}(0.16 \text{ kg})(0)^2$$
$$+ \frac{1}{2}(0.16 \text{ kg})(0)^2 = 0.23 \text{ J}$$

After the collision, the cue ball is at rest, the 7-ball is moving at 1.2 m/s, and the 8-ball is also moving at 1.2 m/s. So the final total kinetic energy is

$$K_{cue,f} + K_{7f} + K_{8f} = \frac{1}{2} m v_{cue,f}^2 + \frac{1}{2} m v_{7f}^2 + \frac{1}{2} m v_{8f}^2$$
$$= \frac{1}{2}(0.16 \text{ kg})(0)^2 + \frac{1}{2}(0.16 \text{ kg})(1.2 \text{ m/s})^2$$
$$+ \frac{1}{2}(0.16 \text{ kg})(1.2 \text{ m/s})^2$$
$$= 0.23 \text{ J}$$

The total kinetic energy is unchanged by the collision, so this collision is elastic.

Reflect

The human body is not as "springy" as a billiard ball, so the forces between Gordie and Mario when they collide are nonconservative. Some of the mechanical energy was dissipated, increasing the two players' internal energies, which means that their temperatures increased slightly as a result of the collision; maybe there was some bruising, certainly some "wind" was knocked out of them. Some of the energy went into compressing and slightly warming the padding they wore (hence the padding!).

NOW WORK Problem 1 from The Takeaway 9-4.

Completely Inelastic Collisions

The type of collision in which the *most* mechanical energy is dissipated is a **completely inelastic collision**, in which two objects stick together after they collide. A collision between two cars is completely inelastic if the cars lock together and don't separate. (If the cars deform and then bounce apart, the collision is inelastic but not *completely* inelastic.) When a bird like the Cooper's hawk in **Figure 9-10** catches its prey and carries it off, this is a completely inelastic collision.

In a completely inelastic collision, none of the initial kinetic energy is stored as potential energy. The amount of kinetic energy needed to keep the momentum constant remains, but all the rest of the initial kinetic energy is converted to internal energy. When an object of mass m_A and velocity $\vec{v}_{Ai}$ undergoes a completely inelastic collision with a second object of mass m_B and velocity $\vec{v}_{Bi}$, we can regard what remains after the collision as a single object of mass $m_A + m_B$. Momentum conservation then gives us an equation for the velocity $\vec{v}_f$ of this combined object just after the collision:

Jim Zipp/Science Source

Figure 9-10 A completely inelastic collision A Cooper's hawk attacking a pigeon from behind will then move off with the pigeon in its talons.

Velocities of objects A and B **before** they undergo a **completely inelastic collision** (they stick together)

$$m_A\vec{v}_{Ai} + m_B\vec{v}_{Bi} = (m_A + m_B)\vec{v}_f \quad (9\text{-}18)$$

Masses of objects A and B

Velocity of the two objects moving together **after the collision**

EQUATION IN WORDS
Momentum conservation in a completely inelastic collision

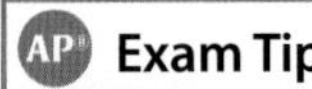

Exam Tip

On the AP® Physics 1 exam, you will need to recognize that there is a common final velocity in a total inelastic collision, but this is not an equation given on the equation sheet. You will be expected to be able to determine this velocity from momentum conservation.

The completely inelastic collision can help give us insight into any collision. Because all the mass of the system travels off together after a completely inelastic collision, the final velocity in such a collision is the velocity of the center of mass of the system! From this velocity we can calculate the minimum amount of kinetic energy required to conserve momentum. Examples 9-6 and 9-7 illustrate how to use Equation 9-18.

EXAMPLE 9-6 A Head-On Collision

In a scene from an action movie, a 1.50×10^3-kg car moving north at 35.0 m/s collides head-on with a 7.50×10^3-kg truck moving south at 25.0 m/s. The car and truck stick together after the collision. (a) How fast and in what direction is the wreckage traveling just after the collision? (b) How much mechanical energy is dissipated in the collision?

Set Up

Because the two vehicles stick together, this is a completely inelastic collision. We'll use Equation 9-18 to find the final velocity $\vec{v}_f$ of the wreckage. We'll then compare the final kinetic energy of the wreckage to the combined kinetic energies of the car and truck before the collision; the difference is the amount of mechanical energy that's dissipated in the collision.

Momentum conservation in a completely inelastic collision:

$$m_{car}\vec{v}_{car,i} + m_{truck}\vec{v}_{truck,i} = (m_{car} + m_{truck})\vec{v}_f \quad (9\text{-}18)$$

Kinetic energy:

$$K = \frac{1}{2}mv^2 \quad (7\text{-}8)$$

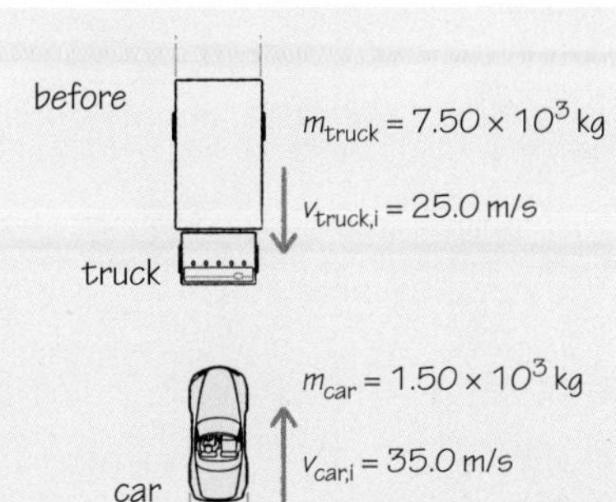

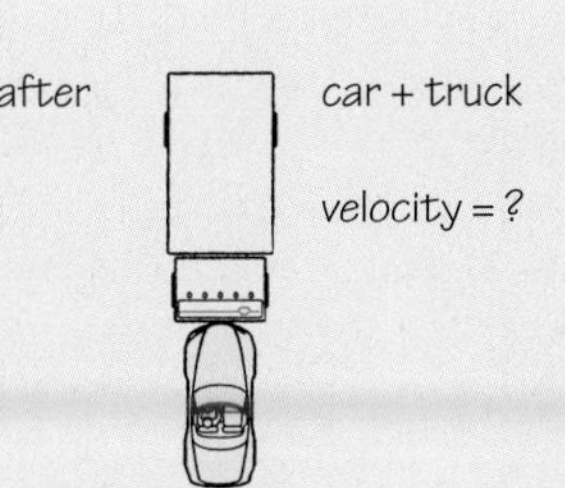

Solve

(a) The collision is along a line, which we call the x axis. We take the positive x axis to be to the north, in the direction of the car's motion before the collision. We use the x component of Equation 9-18 to solve for the final velocity.

Equation for conservation of momentum in the x direction in a completely inelastic collision:

$$m_{\text{car}}v_{\text{car},ix} + m_{\text{truck}}v_{\text{truck},ix} = (m_{\text{car}} + m_{\text{truck}})v_{fx}$$

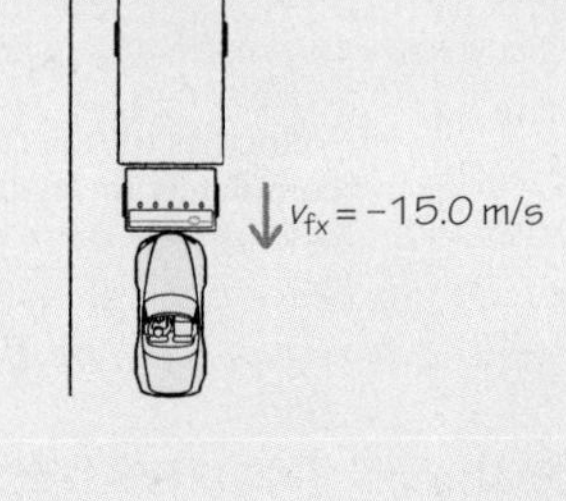

Solve for final velocity of wreckage:

$$v_{fx} = \frac{m_{\text{car}}v_{\text{car},ix} + m_{\text{truck}}v_{\text{truck},ix}}{m_{\text{car}} + m_{\text{truck}}}$$

With our choice of x axis, we have

$$v_{\text{car},ix} = +35.0 \text{ m/s}$$

$$v_{\text{truck},ix} = -25.0 \text{ m/s}$$

Substitute values:

$$v_{fx} = \frac{\left[\begin{array}{c}(1.50 \times 10^3 \text{ kg})(+35.0 \text{ m/s}) \\ +(7.50 \times 10^3 \text{ kg})(-25.0 \text{ m/s})\end{array}\right]}{1.50 \times 10^3 \text{ kg} + 7.50 \times 10^3 \text{ kg}} = -15.0 \text{ m/s}$$

The wreckage moves at 15.0 m/s to the south (in the negative x direction).

(b) Calculate the kinetic energies before and after the collision and compare. Because the two objects move off with no motion relative to each other, the final kinetic energy is the minimum amount of kinetic energy required to keep the momentum of the system constant.

Total kinetic energy before the collision:

$$\begin{aligned} K_{\text{car,i}} + K_{\text{truck,i}} &= \frac{1}{2}m_{\text{car}}v_{\text{car,i}}^2 + \frac{1}{2}m_{\text{truck}}v_{\text{truck,i}}^2 \\ &= \frac{1}{2}(1.50 \times 10^3 \text{ kg})(35.0 \text{ m/s})^2 \\ &\quad + \frac{1}{2}(7.50 \times 10^3 \text{ kg})(25.0 \text{ m/s})^2 \\ &= 9.19 \times 10^5 \text{ J} + 2.34 \times 10^6 \text{ J} \\ &= 3.26 \times 10^6 \text{ J} \end{aligned}$$

Total kinetic energy after the collision:

$$\begin{aligned} K_{\text{car,f}} + K_{\text{truck,f}} &= \frac{1}{2}(m_{\text{car}} + m_{\text{truck}})v_{\text{f}}^2 \\ &= \frac{1}{2}(1.50 \times 10^3 \text{ kg} + 7.50 \times 10^3 \text{ kg})(15.0 \text{ m/s})^2 \\ &= 1.01 \times 10^6 \text{ J} \end{aligned}$$

The amount of mechanical energy dissipated in the collision is the difference between the initial and final energies:

$$\begin{aligned} (K_{\text{car,i}} + K_{\text{truck,i}}) - (K_{\text{car,f}} + K_{\text{truck,f}}) &= 3.26 \times 10^6 \text{ J} - 1.01 \times 10^6 \text{ J} \\ &= 2.25 \times 10^6 \text{ J} \end{aligned}$$

Reflect

It makes sense that the wreckage moves in the direction of the truck's initial motion. Before the collision the magnitude of the southbound truck's momentum was much greater than that of the northbound car, so the total momentum was to the south.

Our result in part (b) shows that more than two-thirds of the total mechanical energy is dissipated in the collision. This dissipated energy goes into internal energy, heating up and deforming the two vehicles, which makes for a very impressive collision scene in the movie. (Personally, I hope never to see a collision like this in real life.) Although the internal forces of the car and the truck are equal and opposite, the acceleration of the car would be so large (because its mass is so much less than the truck) that the car (and most probably its occupants) would be destroyed.

Momentum of car before the collision:

$$p_{\text{car},ix} = m_{\text{car}}v_{\text{car},ix}$$
$$= (1.50 \times 10^3\ \text{kg})(+35.0\ \text{m/s})$$
$$= +5.25 \times 10^4\ \text{kg} \cdot \text{m/s}$$

Momentum of truck before the collision:

$$p_{\text{truck},ix} = m_{\text{truck}}v_{\text{truck},ix}$$
$$= (7.50 \times 10^3\ \text{kg})(-25.0\ \text{m/s})$$
$$= -1.88 \times 10^5\ \text{kg} \cdot \text{m/s}$$

Total momentum before the collision:

$$p_{\text{total},ix} = p_{\text{car},ix} + p_{\text{truck},ix}$$
$$= +5.25 \times 10^4\ \text{kg} \cdot \text{m/s} + (-1.88 \times 10^5\ \text{kg} \cdot \text{m/s})$$
$$= -1.35 \times 10^5\ \text{kg} \cdot \text{m/s}$$

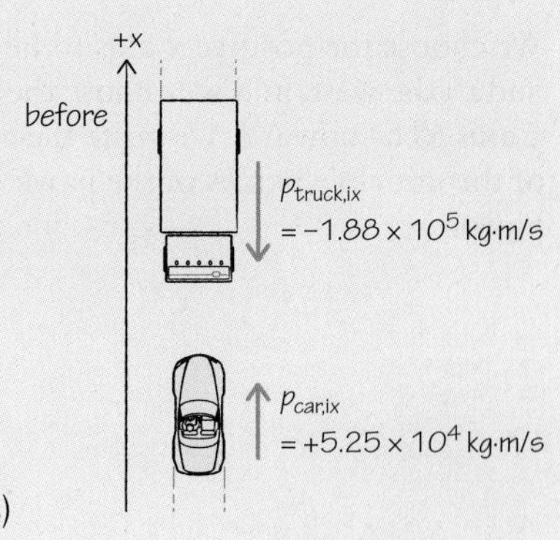

NOW WORK Problems 2 and 3 from The Takeaway 9-4.

EXAMPLE 9-7 Drama in the Skies

In Figure 9-10, the pigeon is gliding due west at 8.00 m/s at a constant altitude, and the Cooper's hawk is diving on the pigeon at a speed of 12.0 m/s at an angle of 30.0° below the horizontal. Each bird has a mass of 0.600 kg. Just after the Cooper's hawk grabs the pigeon in its talons, how fast and in what direction are the two birds moving?

Set Up

This is a completely inelastic collision because the two birds move together afterward. Unlike Example 9-6 this collision takes place in two dimensions, so we will need to consider more than one component of Equation 9-18 to find the final velocity $\vec{v}_f$ of the two birds after the collision. We won't get scared though, because this is just like solving two one-dimensional problems, which we now know how to do!

Momentum conservation in a completely inelastic collision:

$$m_{\text{pigeon}}\vec{v}_{\text{pigeon,i}} + m_{\text{hawk}}\vec{v}_{\text{hawk,i}} = (m_{\text{pigeon}} + m_{\text{hawk}})\vec{v}_{\text{f}} \quad (9\text{-}18)$$

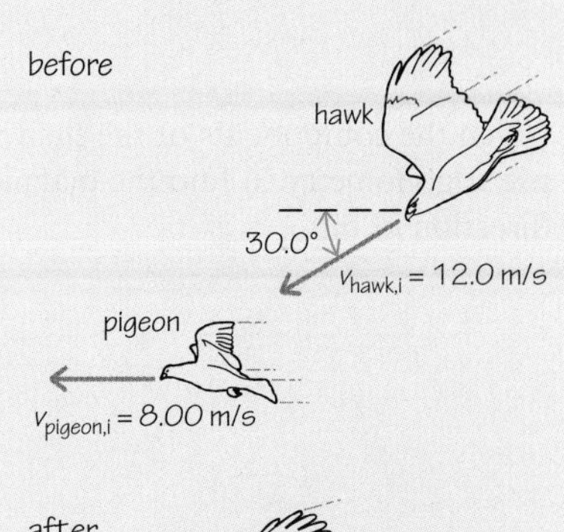

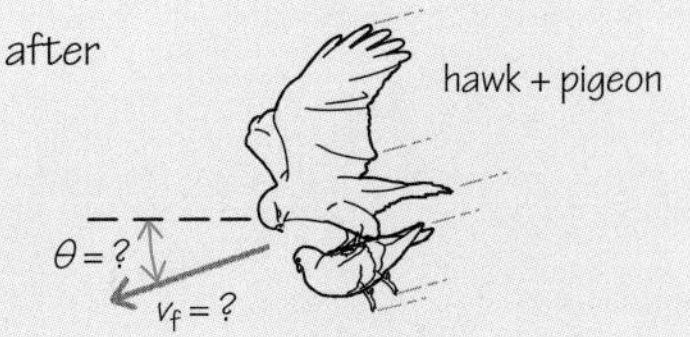

Solve

We choose the positive x axis to be horizontal and to the west, and we choose the positive y axis to be upward. We write the components of the initial velocities of the hawk and pigeon.

Pigeon is initially moving in the positive x direction, so

$$v_{\text{pigeon},ix} = v_{\text{pigeon},i} = +8.00 \text{ m/s}$$

$$v_{\text{pigeon},iy} = 0$$

Hawk is initially moving in the positive x direction and the negative y direction, so

$$v_{\text{hawk},ix} = v_{\text{hawk},i} \cos 30.0° = (12.0 \text{ m/s}) \cos 30.0° = +10.4 \text{ m/s}$$

$$v_{\text{hawk},iy} = -v_{\text{hawk},i} \sin 30.0° = -(12.0 \text{ m/s}) \sin 30.0° = -6.00 \text{ m/s}$$

Use conservation of momentum in component form to find the components of the final velocity of the two birds. Equation for conservation of momentum in the x direction for the collision:

$$m_{\text{pigeon}}v_{\text{pigeon},ix} + m_{\text{hawk}}v_{\text{hawk},ix} = (m_{\text{pigeon}} + m_{\text{hawk}})v_{fx}$$

Solve for the final x component of the velocity of the two birds:

$$v_{fx} = \frac{m_{\text{pigeon}}v_{\text{pigeon},ix} + m_{\text{hawk}}v_{\text{hawk},ix}}{m_{\text{pigeon}} + m_{\text{hawk}}} = \frac{(0.600 \text{ kg})(+8.00 \text{ m/s}) + (0.600 \text{ kg})(+10.4 \text{ m/s})}{0.600 \text{ kg} + 0.600 \text{ kg}} = +9.20 \text{ m/s}$$

Equation for conservation of momentum in the y direction for the collision:

$$m_{\text{pigeon}}v_{\text{pigeon},iy} + m_{\text{hawk}}v_{\text{hawk},iy} = (m_{\text{pigeon}} + m_{\text{hawk}})v_{fy}$$

Solve for the final y component of the velocity of the two birds:

$$v_{fy} = \frac{m_{\text{pigeon}}v_{\text{pigeon},iy} + m_{\text{hawk}}v_{\text{hawk},iy}}{m_{\text{pigeon}} + m_{\text{hawk}}} = \frac{(0.600 \text{ kg})(0) + (0.600 \text{ kg})(-6.00 \text{ m/s})}{0.600 \text{ kg} + 0.600 \text{ kg}} = -3.00 \text{ m/s}$$

Given the components of the final velocity $\vec{v}_f$, use trigonometry to find the magnitude and direction of $\vec{v}_f$.

Speed after collision = magnitude of final velocity:

$$v_f = \sqrt{v_{fx}^2 + v_{fy}^2} = \sqrt{(+9.20 \text{ m/s})^2 + (-3.00 \text{ m/s})^2} = 9.67 \text{ m/s}$$

Direction of final velocity:

$$\tan \theta = \frac{v_{fy}}{v_{fx}} = \frac{-3.00 \text{ m/s}}{+9.20 \text{ m/s}} = -0.326$$

$$\theta = \arctan(-0.326) = -18.1°$$

After the collision, the two birds move at 9.67 m/s in a direction of 18.1° below the horizontal.

Reflect

The impact causes the hawk to slow down from 12.0 to 9.67 m/s and causes the pigeon to speed up from 8.00 to 9.67 m/s. This is just what we would expect when a slow-moving pigeon is hit from behind by a fast-moving hawk.

NOW WORK Problems 4 and 5 from The Takeaway 9-4.

WATCH OUT !

Kinetic energy depends on the observer.

The only type of energy an object can have is kinetic energy, but that does not mean that the kinetic energy measured for an object only depends on the object. Kinetic energy depends on the object's speed, so the kinetic energy of an object measured by any observer depends on the reference frame of the observer. For example, if you were to throw a ball in the direction a train was going while you were riding in the train, the kinetic energy you would measure for the ball would be less than that measured by someone standing on the ground as the train went by. This does not change the amount of energy that can be dissipated when objects interact, though. That really depends on how fast the objects that interact are moving with respect to each other. Imagine riding in a bumper car. If you are at rest, and your friend runs into you at 5 mph, you would get shaken. But if you and your friend were both moving in the same direction at 5 mph (you a *little* slower, so she could run into you), you would barely notice it. In these two cases, from your frame of reference, your friend has a different speed and so a different kinetic energy, although her speed with respect to the ground did not change. Whenever two objects stick together, if you were to measure their kinetic energy after the collision from the frame of reference of either object, you would always measure zero.

Three Special Cases of Completely Inelastic Collisions

Figure 9-11a shows an object of mass m_A moving with velocity $\vec{v}_{Ai}$ and about to collide with a second object of mass m_B that is initially at rest, so m_B. Remember, *during* a collision, A and B are not well modeled as objects, but we call them that since they do act as objects before and after the collision. You could imagine the two objects have some Velcro on them, causing them to attach when they collide, because in order to stick together, there must be some way to dissipate mechanical energy, and objects cannot have internal energy! (We know Velcro dissipates energy because you have to pull pieces of Velcro apart once they are stuck together.) If the collision is completely inelastic, we can use Equation 9-18 to find the velocity $\vec{v}_f$ after the collision. Since $\vec{v}_{Bi} = 0$,

$$m_A \vec{v}_{Ai} = (m_A + m_B)\vec{v}_f$$

and so

$$\vec{v}_f = \frac{m_A \vec{v}_{Ai}}{m_A + m_B} \tag{9-19}$$

(completely inelastic collision, object B initially at rest)

Let's consider three important special cases of this equation.

1. *The moving object has much greater mass than the object at rest* (**Figure 9-11b**). Whenever one quantity in an expression is significantly larger than another, it's a good approximation to ignore the smaller quantity if the two quantities are added or subtracted. So when the mass of object A is much larger than the mass of object B, we can safely replace $m_A + m_B$ in Equation 9-19 with m_A:

$$\vec{v}_f \approx \frac{m_A \vec{v}_{Ai}}{m_A} = \vec{v}_{Ai}$$

WATCH OUT !

An inelastic collision doesn't have to be *completely* inelastic.

A common misconception is that a collision is inelastic if two objects come together and stick, and it is elastic if they bounce off each other. Only the first part of this statement is true! Part (a) of Example 9-5 describes a collision between two hockey players that is inelastic even though the players bounce off each other. For a collision to be inelastic, some mechanical energy has to be dissipated in the collision, and this can happen even if the colliding objects don't stick together.

(a) A completely inelastic collision

Before

A $\vec{v}_{Ai}$ B $\vec{v}_{Bi} = 0$

Mass m_A Mass m_B

After

A B $\vec{v}_f$

Mass $m_A + m_B$

(b) Special case: mass m_A much greater than mass m_B

Before

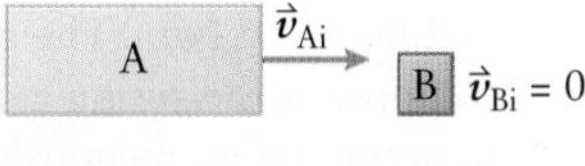

After

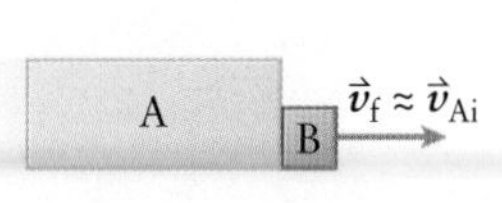

(c) Special case: mass m_A equals mass m_B

Before

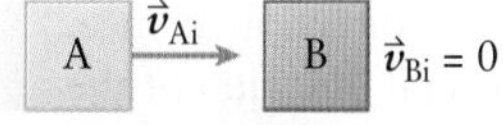

After

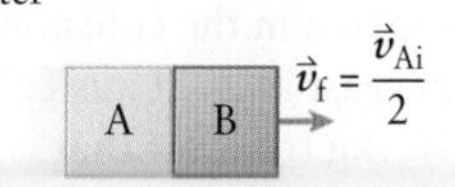

(d) Special case: mass m_A much less than mass m_B

Before

After

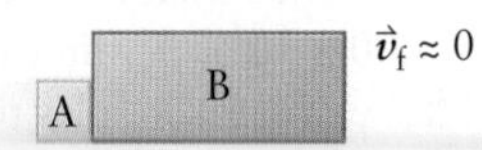

Figure 9-11 Analyzing a completely inelastic collision when one object is initially at rest (a) Object A has a completely inelastic collision with object B (initially at rest). (b) If A is much more massive than B, the final velocity is almost the same as the initial velocity of A. (c) If A and B have the same mass, the final velocity is one-half the initial velocity of A. (d) If A is much less massive than B, the final velocity is nearly zero.

Oleksandr Kulichenko/Getty Images

Figure 9-12 Firewood physics In splitting this log, why does the axe blade get hot as it repeatedly strikes the log?

In this case the stationary object is so small that the moving object is essentially unaffected by the collision, and its velocity remains the same. If a fast-moving car runs into a mosquito, the car slows down only imperceptibly.

2. *The two objects have the same mass* (**Figure 9-11c**). If $m_A = m_B$, Equation 9-19 becomes

$$\bar{v}_f = \frac{\bar{v}_{Ai}}{2}$$

This says that when a moving object collides with and sticks to a stationary one of the same mass, the combined system moves at half the initial speed.

3. *The moving object has much less mass than the object at rest* (**Figure 9-11d**). In this case the mass of object A is much less than the mass of object B, so the quantity m_A in the numerator of Equation 9-19 is very much less than the sum $m_A + m_B$ in the denominator. Therefore $m_A/(m_A + m_B)$ is close to zero, and in this approximation Equation 9-19 becomes

$$\bar{v}_f \approx 0$$

This says that because object A has so little mass, object B hardly moves at all when it is struck. A real-world example of this is using an axe to split logs (systems, not objects) for the fireplace (**Figure 9-12**). Each impact is a completely inelastic collision between the axe (object A) and a large, massive, stationary system B made up of the log, the stump on which the log rests, and the ground below the stump. The final velocity of the axe and log is zero, which means *all* the kinetic energy of the axe is dissipated. This dissipated energy goes into heating the axe and heating and ripping apart the log. Momentum is still conserved, it is just the change in velocity of the ground below the stump would not be something that you could measure, even if you got everyone else on the planet to stand still while you tried, it is so close to zero.

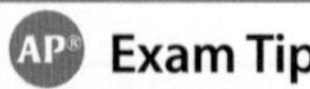

On the AP® Physics 1 exam, students need to carefully distinguish between situations in which momentum or mechanical energy is constant; the two are often confused. Particularly for collisions, momentum is always constant, but only use the conservation of mechanical energy if you know it is an elastic collision.

THE TAKEAWAY for Section 9-4

✔ In an elastic collision the forces the colliding objects exert on each other (the internal forces) are conservative. Both momentum and mechanical energy are constant in an elastic collision (some of the initial kinetic energy is converted to potential energy during the collision and that potential energy is restored to kinetic energy after the collision).

✔ In an inelastic collision the forces the colliding objects exert on each other are nonconservative forces. In this case, some of the initial kinetic energy is turned directly into internal energy, so while momentum is constant in the collision, mechanical energy is not.

✔ In a completely inelastic collision the colliding objects stick together. Momentum is constant, but the maximum amount of kinetic energy possible is converted to internal energy.

✔ If the initial momentum of a system is not zero, then not all the kinetic energy can be dissipated, no matter what kind of collision. Enough kinetic energy to keep momentum constant will remain during and after the collision.

✔ The final kinetic energy in a completely inelastic collision is the minimum kinetic energy required to conserve momentum for any type of collision that system may undergo.

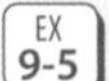

1. An object of mass $3M$, moving in the $+x$ direction at speed v_0, collides with an object with mass M moving in the $-x$ direction with speed $2v_0$. After the collision, the object with mass $3M$ moves in the $-x$ direction at speed $v_0/5$. Calculate the final magnitude and direction of the velocity of the object with a mass M. Determine if the collision was elastic, and if it was inelastic calculate the fraction of the initial kinetic energy dissipated in the collision.

EX 9-6

2. A 1.00×10^4-kg train car moving due east at 20.0 m/s collides with and couples to a 2.00×10^4-kg train car that is initially at rest.
   (a) What is the common velocity of the two-car train after the collision?
   (b) What is the total kinetic energy of the two train cars before and after the collision?

EX 9-6

3. A 12.0-g bullet is fired with a speed of 2.00×10^3 m/s into a block of wood, which is initially at rest. The block is attached to an ideal spring that has a spring constant of 2.00×10^3 N/m. The block with the embedded bullet compresses the spring a distance $\Delta x = 30.0$ cm to the right, before momentarily coming to a stop. Calculate the mass of the wooden block.

EX 9-7 **4.** A 1200-kg car is moving at 20.0 m/s due north. A 1500-kg car is moving at 18.0 m/s due east. The two cars simultaneously approach an icy intersection where, with no brakes or steering, they collide and stick together. Calculate the magnitude and direction of the momentum of the combined two-car wreck immediately after the collision.

EX 9-7 **5.** Two hockey players collide on the ice and go down together in a tangled heap. Player 1 has a mass of 105 kg, and player 2 has a mass of 92 kg. Before the collision, player 1 had a velocity of $v_1 = 6.3$ m/s in the $+x$ direction, and player 2 had a velocity of $v_2 = 5.6$ m/s at an angle of 72° with respect to the $+x$ axis as shown. At what speed and in what direction do they slide together on the ice immediately after the collision?

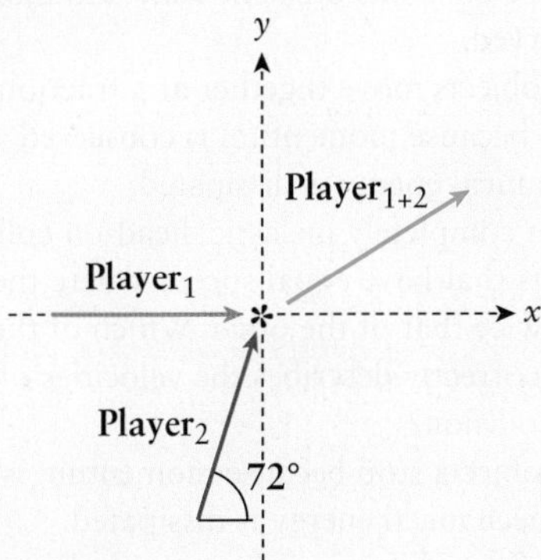

AP® Skill Builders

6. An object moves at constant speed directly toward a solid barrier, as shown in (A) in the figure below. It strikes the barrier in (B) and is reflected in (C), as shown.

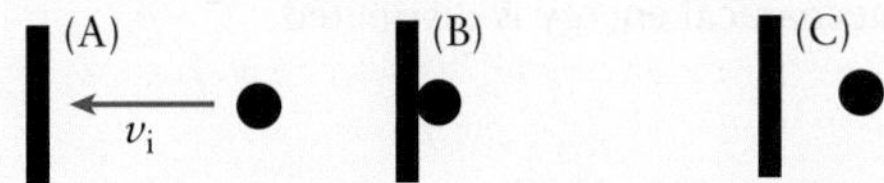

(a) Construct a diagram representing the scenario, including the elements shown in the figure. Also include the following:

- A choice for the positive direction in this one-dimensional motion
- A representation of the initial and final momenta
- A representation of the force exerted by the barrier on the object and the force exerted by the object on the barrier, with variable labels for each
- Labels for vectors that are positive in the coordinate system chosen and vectors that are negative in that coordinate system
- An expression for the change in momentum of the system, in terms of the contact time, assuming the contact time during the collision is Δt
- The sign for each momentum and the change in momentum for the coordinate system chosen

(b) Predict the sign of any change in kinetic energy during the collision and explain why the sign does not depend on your choice of coordinate frame.

(c) Explain why the use of an object model to describe both the object in motion and the barrier cannot account for the sign predicted in part (b).

7. Momentum and mechanical energy changes can be used to model the motion of objects within a system and also their external interactions. These conditions might be observed for the vector momentum and scalar mechanical energy.

(A) The momentum and mechanical energy changes are both zero.

(B) The momentum does not change and the change in mechanical energy is negative.

(C) The momentum does not change and the change in mechanical energy is positive.

(D) The momentum changes but mechanical energy does not.

(E) The momentum changes and the change in mechanical energy is positive.

(F) The momentum changes and the change in mechanical energy is negative.

Predict which of the conditions listed above would be observed in the following scenarios. If the scenario involves an interaction that causes the change, predict the change in terms of the value prior to the interaction subtracted from the value following the interaction. If a change in either momentum or energy is predicted to be negligible, include that expectation in your reasoning. Define your system in each situation.

(a) A golf ball initially at rest on the grass is struck by a golf club.

(b) Two hockey pucks slide over ice and collide.

(c) A bullet strikes a block of wood suspended by a rope.

(d) After two cars with the same mass approach each other at constant, but different, speeds, they collide and fuse, sliding together until they come to a stop.

(e) A balloon pops.

(f) A cart glides over a horizontal air track and bounces off the stationary rubber band stretched over the end of the track. A negligible friction force is exerted by the track on the cart and the collision of the cart with the rubber band is elastic.

(g) A cart glides repeatedly over a horizontal air track and bounces off stationary rubber bands stretched over the ends of the track until it finally comes to rest.

(h) A cart glides up an inclined air track, comes instantaneously to rest and then glides back down the track. The track exerts a negligible friction force on the cart.

8. A female siamang, a type of ape, in a forest in Indonesia is perched on a limb at a height H above the forest floor. She grabs a vertically hanging vine with a length L, pushes horizontally off from the limb with a speed V in the direction of her infant, and swings in a smooth arc on the vine. Her mass M is greater than the mass m of her infant, who is running horizontally away from the mother with a speed v on a limb at a height h above the forest floor. The mother swings upward, clutches the infant, and the two continue to swing on the mother's vine.

(a) Construct a representation of the path of the mother between her initial position and her position when she catches the infant. In your diagram draw and label the distances H, L, and h.

(b) Refine your representation to show that the relationship of the angle between the initial, vertical orientation of her vine and the orientation of her vine when she reaches the infant is

$$\cos\theta = \frac{L-(h-H)}{L}$$

(c) Construct a representation of an x–y coordinate system with the velocity of the infant at the origin and the velocity of the mother when she reaches the infant along the y axis.

(d) Refine your representations in parts (a) and (c) to show that the angle between the infant's velocity and the mother's velocity when she reaches the infant is θ.

(e) Apply the constancy of the mechanical energy of the mother to express her speed, V', when she reaches the infant in terms of h, H, and V.

(f) Apply the conservation of momentum to express the speed of the mother and infant, $V_{\text{with infant}}$, in terms of V, H, h, and the masses of the two siamangs.

(g) In the preceding analysis, the siamangs were represented as objects without internal structure. However, siamangs move by brachiating, arms swinging about their shoulders, which clearly shows that this is an approximation. Predict the effect of internal structure in the mother on the difference between the calculated and actual velocities at the position where the infant is captured.

 Skills in Action

9. You throw a bouncy rubber ball and a wet lump of clay, both of mass m, at a wall. Both strike the wall at speed v, but while the ball bounces off with no loss of speed, the clay sticks. What is the change in momentum of the clay and ball, respectively, assuming that toward the wall is the positive x direction?
(A) $\Delta p_{\text{clay},x} = mv$; $\Delta p_{\text{ball},x} = 0$
(B) $\Delta p_{\text{clay},x} = -mv$; $\Delta p_{\text{ball},x} = 0$
(C) $\Delta p_{\text{clay},x} = -mv$; $\Delta p_{\text{ball},x} = mv$
(D) $\Delta p_{\text{clay},x} = -mv$; $\Delta p_{\text{ball},x} = -2mv$

10. Consider a completely inelastic, head-on collision between two objects that have equal masses and equal speeds. Which of the following sentences correctly describes the velocities of the objects after the collision?
(A) Both objects stop because momentum is conserved and mechanical energy is dissipated.
(B) One of the objects continues with the same velocity and the other comes to rest because the collision is inelastic and momentum is conserved.
(C) One of the objects continues with the same velocity, and the other reverses direction at twice the speed because the collision is inelastic and momentum is conserved.
(D) Both objects move together at a fraction of the initial speed because momentum is conserved and some mechanical energy is dissipated.

11. Consider a completely inelastic, head-on collision between two objects that have equal speeds where the mass of one object is twice that of the other. Which of the following sentences correctly describes the velocities of the objects after the collision?
(A) Both objects stop because momentum is conserved and mechanical energy is dissipated.
(B) One of the objects continues with the same velocity and the other comes to rest because the collision is inelastic and momentum is conserved.
(C) One of the objects continues with the same velocity, and the other reverses direction at twice the speed because the collision is inelastic and momentum is conserved.
(D) Both objects move together at a fraction of the initial speed because momentum is conserved and some mechanical energy is dissipated.

9-5 In an elastic collision both momentum and mechanical energy are constant

In Section 9-4 we introduced the idea of an *elastic* collision in which the forces between colliding objects are conservative and no mechanical energy is lost in the collision. Hence *both* total momentum (a vector) and total mechanical energy (a scalar) are constant in an elastic collision. Since all the potential energy is restored to kinetic energy after the interaction, this means that the final kinetic energy must equal the initial kinetic energy. Call the two colliding objects A and B, with masses m_A and m_B. We then have two conservation equations that relate the velocities $\vec{v}_{Af}$ and $\vec{v}_{Bf}$ of the two objects in the system after the collision to the velocities $\vec{v}_{Ai}$ and $\vec{v}_{Bi}$ before the collision:

(9-20)

(a) $$m_A\vec{v}_{Af} + m_B\vec{v}_{Bf} = m_A\vec{v}_{Ai} + m_B\vec{v}_{Bi}$$

(Total momentum is constant in an elastic collision.)

(b) $$\frac{1}{2}m_Av_{Af}^2 + \frac{1}{2}m_Bv_{Bf}^2 = \frac{1}{2}m_Av_{Ai}^2 + \frac{1}{2}m_Bv_{Bi}^2$$

(Kinetic energy is restored in an elastic collision.)

Equations 9-20 require some effort to solve, and even after that they often don't tell the whole story of an elastic collision. As an example, consider a collision on a pool table between a moving cue ball (A) with known initial velocity $\vec{v}_{Ai}$ and an 8-ball (B) that is initially at rest so $\vec{v}_{Bi} = 0$. Such a collision is almost perfectly elastic. If you're a pool player, you would like to know how fast and in what direction each ball will go after the collision, which means that you want to know the x and y components of their final velocities. So there are four unknowns in this problem: v_{Afx}, v_{Afy}, v_{Bfx}, and v_{Bfy}. But Equations 9-20 give us only three equations for these four unknowns: two from Equation 9-20a for the x and y components of momentum and one from Equation 9-20b. So these equations by themselves aren't enough to solve the problem.

Equations 9-20 allow us to solve any two-dimensional elastic collision problem for three unknowns. In one dimension, we would only have one component equation for momentum, so only two equations, allowing us to solve for two unknowns. With two unknowns, we can solve these equations completely for the final velocities in a *one-dimensional* elastic collision, so all the motions are along a line. This is the only type of problem you will be asked to solve mathematically on the AP® Physics 1 exam. For more than one dimension, you may need to set up a system of equations, but primarily you will be responsible for knowing the key physics of an elastic collision: Momentum is constant and kinetic energy is restored. In the following example, we explore in detail an elastic collision for the case when an object collides with another object with less mass that is initially at rest and consider the cases of identical masses or the moving object having less mass.

EXAMPLE 9-8 When Molecules Collide Elastically

In air, at room temperature, a typical molecule of oxygen (O_2) moves at about 500 m/s. (In Physics 2, you will learn the relationship between the temperature of a gas and the speeds of molecules in that gas. The warmer something is, the faster the molecules that make it up are moving.) The mass of an O_2 molecule is 32.0 u, where 1 u = 1 atomic mass unit = 1.66×10^{-27} kg. Suppose an O_2 molecule is initially moving at 5.00×10^2 m/s in the positive x direction and has a head-on elastic collision with a molecule of hydrogen (H_2, of mass 2.02 u) that is at rest. This is the same situation shown in **Figure 9-13.** Determine the final velocity of each molecule, the initial and final momentum of each molecule, and the initial and final kinetic energy of each molecule.

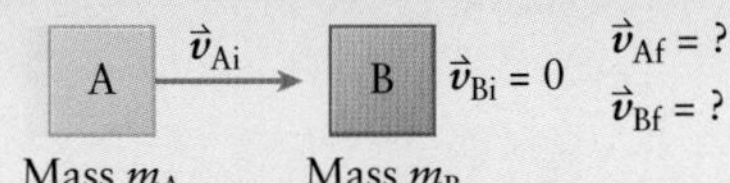

Figure 9-13 A head-on elastic collision If momentum and mechanical energy are both conserved in this head-on collision, what are the final velocities of the two objects?

Set Up

This is a one-dimensional elastic collision in which one object is initially at rest. We know that, in addition to momentum being conserved, the final kinetic energy is equal to the initial kinetic energy for the collision. This means we have two equations, enough to find two unknowns. Once we determine the two final velocities we can calculate the final momenta and kinetic energies from those. Use A to represent the moving oxygen molecule and B to represent the hydrogen molecule. We can then use the definitions of momentum and kinetic energy to find the initial and final values of these quantities.

Final velocities in a head-on elastic collision, object B initially at rest:

$$m_A\vec{v}_{Af} + m_B\vec{v}_{Bf} = m_A\vec{v}_{Ai} + m_B\vec{v}_{Bi}$$

$$\frac{1}{2}m_Av_{Af}^2 + \frac{1}{2}m_Bv_{Bf}^2 = \frac{1}{2}m_Av_{Ai}^2 + \frac{1}{2}m_Bv_{Bi}^2 \quad (9\text{-}20)$$

$$\vec{p} = m\vec{v} \quad (9\text{-}5)$$

$$K = \frac{1}{2}mv^2 \quad (7\text{-}8)$$

before

$v_{Aix} = +5.00 \times 10^2$ m/s

$v_{Bix} = 0$

A B x

$m_A = m_{O_2} = 32.0$ u

$m_B = m_{H_2} = 2.02$ u

after

A B x

$v_{Afx} = ?$ $v_{Bfx} = ?$

Solve

Calculate the final velocities using Equations 9-20. With two equations and two unknowns, one technique is to solve one equation for one of the unknowns in terms of the other, and then substitute that into the second equation to get it in terms of only one unknown, then you can solve for it. If we divide all terms by one of the masses, then these equations will only involve the ratios of the masses, and we can use the mass values in atomic mass units without needing to convert to kilograms.

Hydrogen molecule B is initially at rest. Oxygen molecule A is initially moving in the positive x direction, so

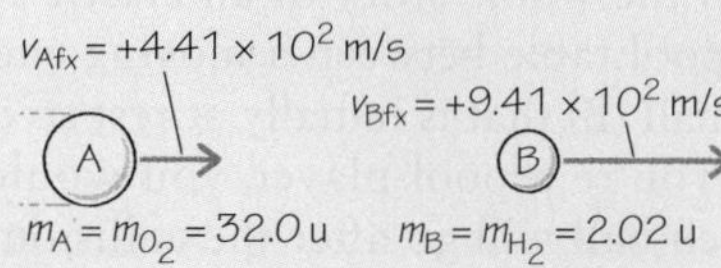

$v_{Aix} = +5.00 \times 10^2$ m/s

Mass of molecule A(O_2): $m_A = 32.0$ u

Mass of molecule B (H_2): $m_B = 2.02$ u

Substitute values into the momentum part of Equation 9-20 (We are only going to put in the 0 before the last step, given that the symbols are easier to write out than the numbers):

$$m_A v_{Aix} + 0 = m_A v_{Afx} + m_B v_{Bfx}$$

Divide through by m_B and solve for v_{Bfx} in terms of knowns and the unknown:

$$v_{Bfx} = (m_A/m_B)(v_{Aix} - v_{Afx})$$

Now, substitute $v_{Bi} = 0$ and our expression for v_{Bfx} into Equation 9-20 for kinetic energy:

$$\frac{1}{2}m_A v_{Aix}^2 = \frac{1}{2}m_A v_{Afx}^2 + \frac{1}{2}\frac{m_A^2}{m_B}(v_{Aix} - v_{Afx})^2$$

Subtract the initial kinetic energy term from both sides. There is a $1/2m_A$ in every term; divide it out.

$$0 = v_{Afx}^2 + \frac{m_A}{m_B}(v_{Aix} - v_{Afx})^2 - v_{Aix}^2$$

$$= v_{Afx}^2 + \frac{m_A}{m_B}\left(v_{Afx}^2 - 2v_{Afx}v_{Aix} + v_{Aix}^2\right) - v_{Aix}^2$$

Because we know the initial velocity and the ratio of the masses we can insert those numbers and group our resulting quadratic in terms to get the a, b, and c for our solution:

$$ax^2 + bx + c = 0$$

So, with v_{Afx} as our x, $x = \dfrac{-b \pm \sqrt{b^2 - 4ac}}{2a}$ for a quadratic expression of the form

$$0 = \left(\frac{32.0 \text{ u}}{2.02 \text{ u}} + 1\right) \times v_{Afx}^2 - 2\left(\frac{32.0 \text{ u}}{2.02 \text{ u}}\right)(v_{Aix}) \times v_{Afx} + \left(\frac{32.0 \text{ u}}{2.02 \text{ u}} - 1\right)(v_{Aix})^2$$

Comparing, we see that $a = \left(\dfrac{32.0 \text{ u}}{2.02 \text{ u}} + 1\right)$;

$b = -2\left(\dfrac{32.0 \text{ u}}{2.02 \text{ u}}\right)(v_{Aix})$; and

$c = \left(\dfrac{32.0 \text{ u}}{2.02 \text{ u}} - 1\right)(v_{Aix})^2$

Final velocity of the O_2 molecule (the u's cancel), v_{Afx}:

$$= \frac{\left(2\left(\frac{32.0}{2.02}\right)(v_{Aix})\right) \pm \sqrt{\left(-2\left(\frac{32.0}{2.02}\right)(v_{Aix})\right)^2 - 4\left(\left(\frac{32.0}{2.02} + 1\right)\right)\left(\left(\frac{32.0}{2.02} - 1\right)(v_{Aix})^2\right)}}{2\left(\left(\frac{32.0}{2.02} + 1\right)\right)}$$

$= +4.41 \times 10^2$ m/s (one solution to the quadratic gives the initial velocity, which we know has to change because there is something in its way, so we take the other solution).

Final velocity of the H_2 molecule:

$$v_{Bfx} = (m_A/m_B)(v_{Aix} - v_{Afx})$$
$$= +9.41 \times 10^2 \text{ m/s}$$

Both final velocities are positive, so after the collision both molecules are moving in the same direction as the initial motion of the O_2 molecule. The H_2 molecule is moving almost twice as fast as the initial velocity of the O_2 molecule.

We now know both the initial and final velocities, so we can calculate the initial and final values of momentum and kinetic energy for both molecules. For these calculations we need the masses in kilograms.

Masses of the molecules in kilograms:

$$m_A = \text{mass of } O_2 \text{ molecule}$$
$$= (32.0 \text{ u})(1.66 \times 10^{-27} \text{ kg/u}) = 5.31 \times 10^{-26} \text{ kg}$$
$$m_B = \text{mass of } H_2 \text{ molecule}$$
$$= (2.02 \text{ u})(1.66 \times 10^{-27} \text{ kg/u}) = 3.35 \times 10^{-27} \text{ kg}$$

Calculate the initial and final values of momentum in the x direction:
For molecule A (O_2):

$$p_{Aix} = m_A v_{Aix} = (5.31 \times 10^{-26} \text{ kg})(+5.00 \times 10^2 \text{ m/s})$$
$$= +2.66 \times 10^{-23} \text{ kg} \cdot \text{m/s}$$
$$p_{Afx} = m_A v_{Afx} = (5.31 \times 10^{-26} \text{ kg})(+4.41 \times 10^2 \text{ m/s})$$
$$= +2.34 \times 10^{-23} \text{ kg} \cdot \text{m/s} = 0.881 p_{Aix}$$

For molecule B (H_2):

$$p_{Bix} = m_B v_{Bix} = 0 \text{ (molecule B initially at rest)}$$
$$p_{Bfx} = m_B v_{Bfx} = (3.35 \times 10^{-27} \text{ kg})(+9.41 \times 10^2 \text{ m/s})$$
$$= +3.15 \times 10^{-24} \text{ kg} \cdot \text{m/s} = 0.119 p_{Aix}$$

In the collision, the O_2 molecule transfers 0.119 of its initial momentum to the H_2 molecule, leaving 0.881 for itself.

Calculate the initial and final values of kinetic energy.
For molecule A (O_2):

$$K_{Ai} = \frac{1}{2} m_A v_{Aix}^2 = \frac{1}{2}(5.31 \times 10^{-26} \text{ kg})(+5.00 \times 10^2 \text{ m/s})^2$$
$$= 6.64 \times 10^{-21} \text{ J}$$
$$K_{Af} = \frac{1}{2} m_A v_{Afx}^2 = \frac{1}{2}(5.31 \times 10^{-26} \text{ kg})(+4.41 \times 10^2 \text{ m/s})^2$$
$$= 5.16 \times 10^{-21} \text{ J} = 0.777\, K_{Ai}$$

For molecule B (H_2):

$$K_{Bi} = \frac{1}{2} m_B v_{Bix}^2 = 0 \text{ (molecule B initially at rest)}$$
$$K_{Bf} = \frac{1}{2} m_B v_{Bfx}^2 = \frac{1}{2}(3.35 \times 10^{-27} \text{ kg})(9.41 \times 10^2 \text{ m/s})^2$$
$$= 1.48 \times 10^{-21} \text{ J} = 0.223\, K_{Ai}$$

In the collision, the O_2 molecule transfers 0.223 of its initial kinetic energy to the H_2 molecule, leaving 0.777 for itself.

Reflect

Our numerical answers work out to show what we expected: Momentum is constant and the kinetic energy is restored for the system of two molecules.

In the case of a much more massive object (A) striking a stationary object (B), we found that the much more massive object slows down only slightly and its momentum and kinetic energy decrease very little. The lighter object flies off moving faster than the original speed of the more massive object.

Extend: You can repeat these calculations for the case in which the moving O_2 molecule collides with a second O_2 molecule that is at rest. You'll find that the moving O_2 molecule ends up at rest and the stationary O_2 molecule ends up moving with the original velocity of the first molecule, 5.00×10^2 m/s. In this case the first O_2 molecule transfers all its momentum and all its kinetic energy to the second O_2 molecule. This is always true in elastic collisions between identical objects—they actually completely trade momentum.

You could also repeat these calculations for the case in which the moving O_2 molecule collides with a protein molecule of mass 3.20×10^4 u that is at rest. In this case you'll find that the O_2 molecule has a final velocity of -4.99×10^2 m/s. The negative sign means that the O_2 molecule *bounces back*, and the magnitude means that the O_2 molecule ends up with nearly the same speed as it had originally (5.00×10^2 m/s). You'll also find that the massive protein molecule ends up with a final velocity of just +0.999 m/s, so it moves very slowly in the initial direction of motion of the O_2 molecule. The protein molecule ends up with about twice the initial momentum of the O_2 molecule, but has little kinetic energy.

NOW WORK Problems 1 and 2 from The Takeaway 9-5.

Example 9-8 suggests that just as we did in our exploration of completely inelastic collisions (Section 9-4), we can gain insight into the physics of elastic collisions by considering three special cases involving the relative masses of the two objects. Again, consider a moving object A that has a head-on elastic collision with a stationary object B as is shown in **Figure 9-14**.

1. *The moving object has much greater mass than the object at rest* (**Figure 9-14a**). In this case $m_A/m_B \gg 1$, so we can neglect the two terms in the quadratic that are not multiplied by this ratio, $0 = v_{Afx}^2 + \frac{m_A}{m_B}\left(v_{Afx}^2 - 2v_{Afx}v_{Aix} + v_{Aix}^2\right)^2 - v_{Aix}^2$, leaving just $0 = \frac{m_A}{m_B}\left(v_{Afx}^2 - 2v_{Afx}v_{Aix} + v_{Aix}^2\right)^2$, so $v_{Af}x \approx v_{Ai}x$

   In this extreme case the motion of the massive object A is almost unaffected by the elastic collision, but the second, much smaller object B flies off more quickly than the initial speed of object A. To find the final velocity of B takes some algebra to prove in general, so we are just going to point out that it makes sense. A slows down a tiny bit, B is much lighter, so to conserve momentum it must go much faster to make up this small decrease in the momentum of A. That's what happens in the nearly elastic collision between a bowling ball and a stationary bowling pin, which has only about one-quarter the mass of the ball: The ball continues with nearly the same speed, while the pin flies off at a greater speed.

2. *The two objects have the same mass* (**Figure 9-14b**). If $m_A = m_B$, we can replace the ratio of masses in our quadratic equation in Example 9-8 with 1. This simplifies the equation and we can see that object A comes to a stop during the elastic collision and object B leaves the collision with the initial velocity of A. If a cue ball hits an 8-ball head-on, the result of this nearly elastic collision is that the cue ball stops and the 8-ball rolls away with the velocity that the cue ball had before the collision.

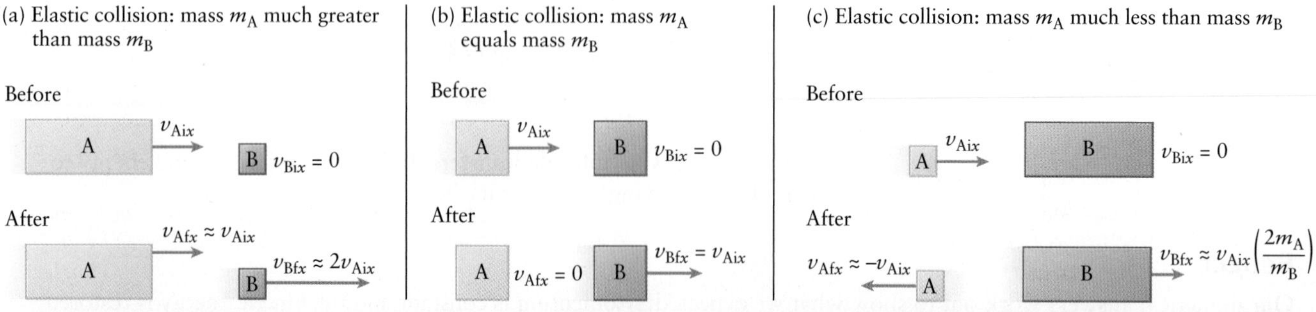

(a) Elastic collision: mass m_A much greater than mass m_B
Before: A v_{Aix}; B $v_{Bix} = 0$
After: A $v_{Afx} \approx v_{Aix}$; B $v_{Bfx} \approx 2v_{Aix}$

(b) Elastic collision: mass m_A equals mass m_B
Before: A v_{Aix}; B $v_{Bix} = 0$
After: A $v_{Afx} = 0$; B $v_{Bfx} = v_{Aix}$

(c) Elastic collision: mass m_A much less than mass m_B
Before: A v_{Aix}; B $v_{Bix} = 0$
After: A $v_{Afx} \approx -v_{Aix}$; B $v_{Bfx} \approx v_{Aix}\left(\frac{2m_A}{m_B}\right)$

Figure 9-14 Analyzing an elastic collision Object A has a head-on elastic collision with object B (initially at rest). (a) If A is much more massive than B, A is almost unaffected and B moves off with twice the initial velocity of A. (b) If A and B have the same mass, A comes to rest and B moves off with the initial velocity of A. (c) If A is much less massive than B, A bounces back with nearly its initial speed and B moves forward very slowly.

(a) Elastic collision: momentum constant; mechanical energy constant; kinetic energy restored

Before

$\vec{v}_{Ai}$ $\vec{v}_{Bi}$ A B

After

$\vec{v}_{Af}$ A B $\vec{v}_{Bf}$

(b) Inelastic collision: momentum constant; mechanical energy not constant; internal energy increased

Before

$\vec{v}_{Ai}$ $\vec{v}_{Bi}$ A B

After

$\vec{v}_{Af}$ A B $\vec{v}_{Bf}$

(c) Completely inelastic collision: momentum constant; mechanical energy not constant; colliding objects stick together and internal energy increased as much as possible

Before

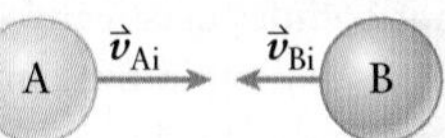

After

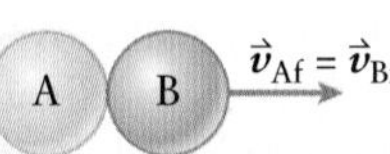

Figure 9-15 Three types of collision Comparing (a) elastic, (b) inelastic, and (c) completely inelastic collisions.

3. *The moving object has much less mass than the object at rest* (**Figure 9-14c**). In this case the mass of object A is small compared to the mass of object B, so we will approximate m_A/m_B as equal to 0. Putting this into our quadratic, we find that object A is reflected back from the elastic collision, moving with almost the same speed it had initially but in the opposite direction. Hence object A has lost almost no kinetic energy, but its momentum has changed from $p_{Aix} = +m_A v_{Aix}$ to $p_{Afx} = -m_A v_{Aix}$. Again the algebra is complicated, but we can see that object B must move very slowly since object A kept most of its kinetic energy. (It actually looks like it would be zero if we solve our equations but is not really zero, since m_A/m_B is just much less than 1, not really zero). But because its mass is so great, even though it is moving slowly object B turns out to have *twice* as much momentum as object A had before the collision since object A's momentum reversed direction and momentum must still be conserved,

$$p_{Afx} + p_{Bfx} = m_A v_{Aix} = -m_A v_{Aix} + p_{Bfx}$$
$$p_{Bfx} = 2m_A v_{Aix}$$

which is the same as the momentum of object A before the collision. So total momentum is constant in this case, as it must be. An example is a pebble hitting a boulder: The pebble bounces back, while the recoil of the boulder is so little as to be imperceptible. These results reiterate the reflection from Example 9-8, which describes what happens when a moving O_2 molecule collides elastically with a stationary molecule of smaller mass (an H_2 molecule), the same mass (another O_2 molecule), or greater mass (a protein molecule).

Figure 9-15 summarizes the differences between elastic collisions, inelastic collisions, and completely inelastic collisions.

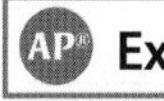

Exam Tip

Although you will not be asked to solve complicated simultaneous equations for the AP® Physics 1 exam, you should be able to reason the qualitative results for collisions of objects with different masses from momentum and energy considerations on the exam.

THE TAKEAWAY for Section 9-5

✔ A collision is elastic if kinetic energy is completely restored and linear momentum is constant for the system of colliding objects.

✔ If a moving object has a head-on elastic collision with a second, stationary object, how the objects move after the collision depends on the ratio of masses of the two objects.

Prep for the AP Exam

AP® Building Blocks

EX 9-8 1. One-dimensional, head-on elastic collisions between an object with mass m_1 and initial speed v_{1ix} and an object with mass m_2 and initial speed v_{2ix} result in a final velocity v_{1fx} for a system satisfying both momentum conservation and conservation of mechanical energy.

$$v_{1fx} = \frac{m_1 - m_2}{m_1 + m_2} v_{1ix} + \frac{2m_2}{m_1 + m_2} v_{2ix}$$

(a) The labels 1 and 2 are arbitrary. Write the result for final velocity v_{2fx} satisfying both momentum conservation and conservation of mechanical energy by exchanging the labels.

(b) Show for the case of equal masses that these equations give the same results as Equations 9-20 when one of the objects is initially at rest.

(c) A 2.00-kg ball moves at 3.00 m/s toward the right. It collides elastically with a 4.00-kg ball that is initially at rest. Determine the velocities of the balls after the collision.

(d) A 10.0-kg block of ice is sliding due east at 8.00 m/s when it collides elastically with a 6.00-kg block of ice that is sliding in the same direction at 4.00 m/s. Determine the velocities of the blocks of ice after the collision.

(e) A 0.170-kg ball moves at 4.00 m/s toward the right. It collides elastically with a 0.155-kg ball moving at 2.00 m/s toward the left. Determine the final velocities of the balls after the collision.

2. One way that scientists measure the mass of an unknown particle is to create a collision with a particle whose mass is known, such as a proton or an electron.

(a) A head-on collision occurs between a proton and a stationary neutron. The mass of the proton, m_p, is 1.672622×10^{-27} kg. The mass of the neutron, m_n, is 1.674927×10^{-27} kg. The initial speed of the proton is v_{pi}. Express the recoil speed of the proton, v_{pf}, and the final speed of the neutron, v_{nf}, in terms of the masses and the initial speed of the proton, assuming the collision is elastic.

(b) Confirm that momentum is constant and kinetic energy is restored (the same before and after the collision).

 Skill Builders

3. When one of the rigid balls suspended in a well-balanced and stable Newton's cradle (as shown) is raised and released, the cascade of collisions through the system from right to left and back again repeats for a long time.

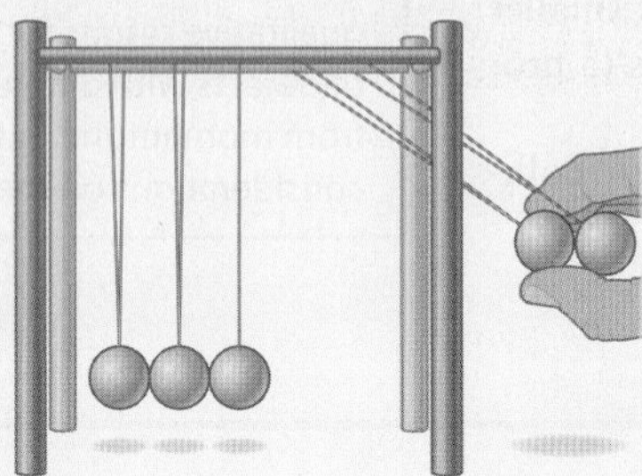

(a) Explain why the system behaves in this manner.

(b) Consider a Newton's cradle consisting of just two balls. One ball is raised and released. Justify the claim that the height after collision of the ball initially at rest can be no greater than the height from which the first ball was released.

(c) Assume the direction of motion of the ball is positive as it is about to collide with the ball initially at rest. Given your conclusion in part (b), justify the claim that the velocity of the ball colliding with the ball initially at rest cannot be negative after the collision.

(d) Given your conclusions in parts (b) and (c), justify the claim that all the momentum of the ball initially raised must be transferred to the ball initially at rest.

(e) Predict what happens when two balls are raised and released together in a Newton's cradle with at least three balls.

4. Two blocks are released from rest on either side of a half-pipe with negligible friction and whose maximum height on either side is h. Labeling the blocks on the left and right as L and R, respectively, the masses of blocks and the initial heights are shown in the figure: $m_L = m$, $m_R = m/2$, $h_{Li} = h$, $h_{Ri} = h$

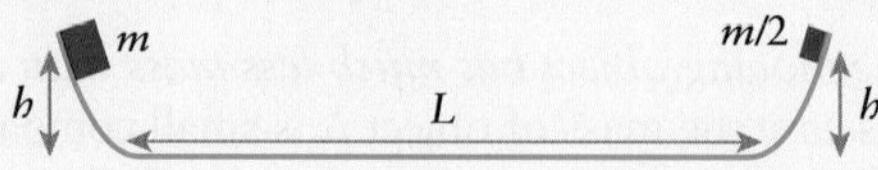

(a) Express the velocities v_{Li} and v_{Ri} of each block just before they collide, taking the positive direction to be to the right in the figure.

(b) Assume that the collision is elastic. Using momentum conservation and the conservation of mechanical energy, justify the claim that after they initially collide the momenta of the blocks are

$$p_{L,f} = -\frac{m}{3}\sqrt{2gh} \quad p_{R,f} = \frac{5m}{6}\sqrt{2gh}$$

(c) Predict the subsequent motion of the blocks.

 Skills in Action

5. The classic collision problem involves two hard spheres colliding. When you define the system as both spheres and there are no external forces exerted on the spheres, the momentum of the system is constant. When you define the system as only one of the spheres, the momentum is not constant. Explain why the momentum is conserved in both cases but the momentum of the system is constant only in the first case.

6. Two identical spheres are moving toward each other with the same speed, v_i, and collide elastically, head-on. Justify your evaluation of the validity of each of the following claims for the system of two spheres.

After the collision:

(A) Both spheres move in opposite directions with final speeds that are greater than v_i, because each sphere is accelerated by the force exerted by the other sphere.

(B) Both spheres move in opposite directions with the same speed v_i, because otherwise momentum would not be constant.

(C) One sphere moves off with all the system's initial mechanical energy because otherwise kinetic energy would not be restored.

(D) Both spheres move in opposite directions with the same speed v_i, because otherwise mechanical energy and momentum will not remain constant.

9-6 What happens in a collision is related to the time the colliding objects are in contact

During a collision that brings a car to a sudden stop, the air bags installed in most cars inflate (**Figure 9-16**). Their purpose is to prevent serious injury by cushioning the impact between the car's occupants and hard surfaces in the vehicle.

The principle of the air bag is to minimize the *force* on the occupants by maximizing the *time* that it takes to bring the occupants to rest. To see how this works, let's look at the relationship between the net force on an object, the time that the net force is exerted on the object, and the change in the object's momentum caused by the net force.

Exam Tip

The change in momentum of an object or system is equal to the net impulse, which must therefore be due to the net external force exerted on the system. Note that any force can contribute to the impulse. On the AP® exam, you may be asked to calculate the impulse due to a single force, which can be calculated by $\vec{J} = \vec{F}_{avg}\,\Delta t$. The contribution to the impulse by any single force is not equal to the change in momentum of the system unless it is the only force exerted on the system.

Collision Force, Contact Time, and Momentum Change

We learned in Section 9-3 that if external forces are exerted on a system for a time Δt, the result is a change in the momentum of the system from an initial value $\vec{p}_{\text{total,i}}$ to a final value $\vec{p}_{\text{total,f}}$. This change is given by Equation 9-13.

The left-hand side of Equation 9-13 is called the **impulse** J exerted on the system. This impulse can be the result of a large net external force $\sum \vec{F}_{\text{external on system}}$ exerted for a short time Δt or a smaller net external force that is exerted for a longer time. Just like the work-energy theorem, the **impulse-momentum theorem** is a statement of conservation. Any change in the momentum of a system has to be due to the transfer of momentum into or out of the system by a net external force. The system could be made up of only a single object. Then we replace the symbol $\vec{p}_{\text{total}}$ (for the momentum of a system) by $\vec{p}$ (for the momentum of a single object), and Equation 9-13 becomes

The **sum of all external forces** exerted on an object

Duration of the time interval over which the external forces are exerted

Impulse

$$\vec{J} = \left(\sum \vec{F}_{\text{external on object}}\right)\Delta t = \vec{p}_{\text{f}} - \vec{p}_{\text{i}} = \Delta\vec{p} \qquad (9\text{-}21)$$

Change in the momentum $\vec{p}$ of the object during that time interval

EQUATION IN WORDS
Impulse-momentum theorem

This equation tells us that to change the momentum of an *object* from $\vec{p}_{\text{i}}$ to $\vec{p}_{\text{f}}$, a certain amount of impulse is required. Equations 9-13 and 9-21 both reflect the same concept, that any change in momentum of an object or a system is due to an impulse. The system in Equation 9-13 could be a single object, but Equation 9-21 explicitly deals with a single object.

If the momentum change occurs as a result of a collision, the net external force on an object is predominantly due to the other object with which it is colliding; any other forces are very weak by comparison (see Section 9-3). So for the special case of a collision, Equation 9-21 becomes

fStop Images GmbH/Alamy

Figure 9-16 A sudden stop with air bag How does a deployed air bag help prevent serious injury?

Force exerted on an object during a collision

Duration of the collision = **contact time**

$$\vec{F}_{\text{collision}}\,\Delta t = \vec{p}_{\text{f}} - \vec{p}_{\text{i}} = \Delta\vec{p} \qquad (9\text{-}22)$$

Change in the momentum $\vec{p}$ of the object during the collision

EQUATION IN WORDS
Impulse-momentum theorem for an object for a collision

The **contact time** in Equation 9-22 is the amount of time that the colliding objects interact and hence the amount of time that the objects exert forces on each other.

Equation 9-22 explains the principle of the air bag shown in Figure 9-16. As a car comes to a sudden stop, the momentum of one of its occupants changes from a large initial value $\vec{p}_{\text{i}}$ to a final value of zero ($\vec{p}_{\text{f}} = 0$). So the momentum change $\vec{p}_{\text{f}} - \vec{p}_{\text{i}}$ is large, which means the product $\vec{F}_{\text{collision}}\Delta t$ also has a large value (**Figure 9-17a**). If there are no air bags, the occupant has a "hard" collision with the structure of the car, coming to a sudden stop as the occupant hits the hard surface. In this case the contact time Δt is very short

(a)

$\vec{p}_i$ $\quad$ $\vec{p}_f = 0$ $\quad$ $\vec{F}_{collision}\,\Delta t = \vec{p}_f - \vec{p}_i = -\vec{p}_i$

As the crash test dummy comes to rest, its momentum changes from an initial value $\vec{p}_i$ to a final value $\vec{p}_f = 0$ (zero momentum). The product $\vec{F}_{collision}\,\Delta t$ equals the negative of $\vec{p}_i$.

(b)

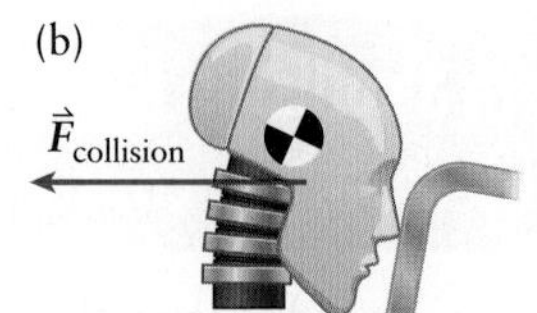

If the crash test dummy strikes a hard surface, Δt is very short and $\vec{F}_{collision}$ is very large.

(c)

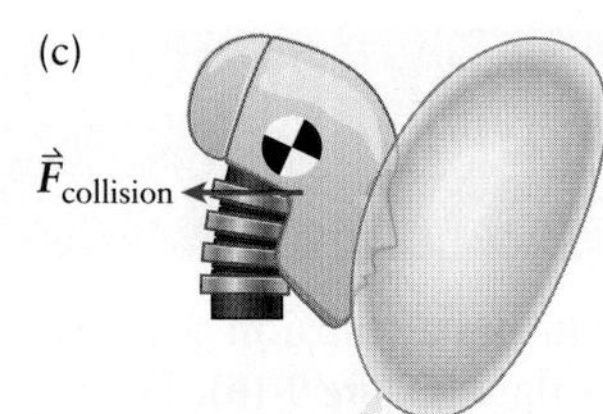

If the crash test dummy strikes an air bag, Δt is longer and $\vec{F}_{collision}$ is reduced.

Figure 9-17 Contact time (a) To bring a crash test dummy to rest requires that $\vec{F}_{collision}\,\Delta t$ have a certain large value. (b) A short contact time Δt means that the force $\vec{F}_{collision}$ will be large. (c) A longer contact time Δt reduces the magnitude of $\vec{F}_{collision}$.

and so $\vec{F}_{collision}$ exerted on the occupant by the surface must have a *very* large value: The car's structure exerts a tremendous force on the occupant that is likely to cause injury, as **Figure 9-17b** shows. But if the car is equipped with air bags, the air bag compresses much more than the hard surface as the occupant strikes it, so the occupant covers a greater distance while coming to rest and so is in contact with the air bag for a longer time Δt (**Figure 9-17c**). In this case $\vec{F}_{collision}$, the force exerted on the occupant by the air bag, is greatly reduced, and injury is avoided. For the same reason, if you jump down from a table it's less painful if you bend your knees when landing. Flexing your legs during the collision between your feet and the ground maximizes the contact time Δt that it takes for your center of mass to come to a stop. If you are flexing your legs, you need to use your center of mass to describe your motion since you are not moving as an object.

In Chapter 2 (Figure 2-17), we saw that we can use the area under a velocity-time graph to find displacement (change in position) over a particular time interval Δt. (On a graph of velocity versus time, the displacement in any time interval equals the area under the graph during that time interval.) In Chapter 7, we stated that this "area" rule is a mathematical fact that works for any product whenever we draw a plot with the multiplicand and multiplier on the axes of the graph, and gave work as another example. We can use that same relationship to find impulse (change in momentum) from the area under a force-time graph, since impulse is the product of force and time.

WATCH OUT !

Using constant force or average force

In deriving Equations 9-21 and 9-22, we have assumed that the forces exerted on the object are *constant* forces that do not change during the time Δt. We can still use these equations to get a very good approximation even if the forces are not constant, however: Treat $\sum \vec{F}_{external\ on\ object}$ in Equation 9-22 and $\vec{F}_{collision}$ in Equation 9-24 as the *average* values of these forces over the time interval Δt.

Example 9-9 shows how to apply the idea of contact time to a collision in which an object speeds up rather than slows down.

EXAMPLE 9-9 Follow-through

Tennis players are taught to *follow through* as they serve the ball—that is, to continue to swing the racket after first striking the ball. The rationale is that by increasing the time that the ball and racket are in contact, the ball might have a larger speed as it leaves the racket. Using her racket, a player can exert an average force of 560 N on a 57-g tennis ball during an overhand serve (**Figure 9-18**). At what speed does the ball leave the player's racket if the contact time between them is 0.0050 s? What would the speed of the ball be if the player improved her follow-through and increased the contact time by a factor of 1.3 to 0.0065 s? During an overhand serve, the tennis ball is struck when it is essentially motionless.

Figure 9-18 Tennis physics How does an improved follow-through affect the speed of a tennis ball?

Set Up

We are given the force that the racket exerts on the ball and the contact time (the time that the ball is touching the racket). We also know that the ball starts at rest and so has zero momentum. We can use Equation 9-22 to calculate the final momentum of the ball, and then use Equation 9-5 to determine the final speed.

Collision force, contact time, and momentum change:

$$F_{collision}\Delta t = \vec{p}_f - \vec{p}_i \quad (9\text{-}22)$$

$$\vec{p} = m\vec{v} \quad (9\text{-}5)$$

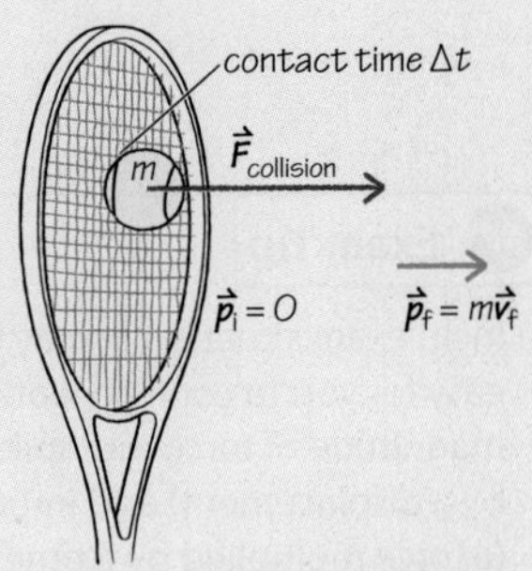

Solve

The racket follows a curved path during the serve. But because the collision between racket and ball lasts such a short time, the ball is in contact with the racket for just a short segment of that path, which we can treat as a line. We choose the positive x axis to be in the direction of the motion of the ball.

The collision is one-dimensional, so we just need the x component of Equation 9-22:

$$F_{collision,x}\Delta t = p_{fx} - p_{ix}$$

The ball is initially at rest, so $v_{ix} = 0$ and $p_{ix} = 0$.

After the collision with the racket, the ball has (unknown) velocity v_{fx}, so

$$p_{fx} = mv_{fx}$$

Then the x component of Equation 9-22 becomes

$$F_{collision,x}\Delta t = mv_{fx} - 0 = mv_{fx}$$

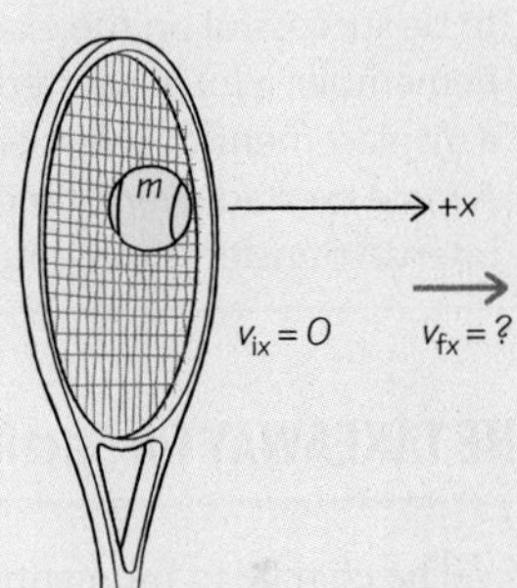

Solve for the final velocity of the ball in each case.

Rewrite the x component of Equation 9-22 to solve for v_{fx}:

$$v_{fx} = \frac{F_{collision,x}\Delta t}{m}$$

In both cases $F_{collision,x} = 560$ N

$$m = 57\text{ g}\left(\frac{1\text{ kg}}{1000\text{ g}}\right) = 0.057\text{ kg}$$

Calculate v_{fx} if the contact time is 0.0050 s: Recall that $1\text{ N} = 1\text{ kg}\cdot\text{m/s}^2$:

$$v_{fx} = \frac{F_{collision,x}\Delta t}{m} = \frac{(560\text{ N})(0.0050\text{ s})}{0.057\text{ kg}}$$

$$= 49\,\frac{\text{N}\cdot\text{s}}{\text{kg}}\left(\frac{1\text{ kg}\cdot\text{m/s}^2}{1\text{ N}}\right) = 49\text{ m/s}$$

With improved follow-through and a contact time of 0.0065 s,

$$v_{fx} = \frac{F_{collision,x}\Delta t}{m} = \frac{(560\text{ N})(0.0065\text{ s})}{0.057\text{ kg}} = 64\text{ m/s}$$

Reflect

For a given force, the change in momentum and therefore the change in velocity during a collision are directly proportional to the contact time. Increasing the contact time Δt from 0.0050 s to (1.3)(0.0050 s) = 0.0065 s results in an increase in the final speed of the tennis ball from 49 m/s to (1.3)(49 m/s) = 64 m/s, so our answer makes sense.

NOW WORK Problems 1–4 from The Takeaway 9-6.

Impulse and Momentum Versus Work and Kinetic Energy

AP® Exam Tip

In an exam situation, it may be easy for you to confuse work (a magnitude of force multiplied by a displacement) and impulse (a force multiplied by a time); it's important to carefully distinguish between the two to be successful on the exam. Remember, a force exerted over a displacement transfers energy. A force exerted over a time interval transfers momentum.

Equation 9-21 says that the product of the net force on an object and the time that the force is exerted equals the change in momentum of the object. This should remind you of the work-energy theorem for an object from Section 7-3, which states that the net work on an object—which is the product of the net force on an object parallel to its displacement and the displacement of the object during the time that the force is exerted—equals the change in kinetic energy of the object. This theorem gives us an alternative way to understand the effect of the air bag shown in Figure 9-16. To bring a moving crash test dummy to rest, its kinetic energy $K = \frac{1}{2}mv^2$ must change from a large initial value K_i to a final value $K_f = 0$, so the required change in kinetic energy is $K_f - K_i = 0 - K_i = -K_i$. Hence a certain amount of (negative) work has to be done on the dummy to bring it to rest. Without an air bag, the dummy stops in a very short distance when it strikes the hard surfaces of the car, so the magnitude of the displacement d is small and the force F on the dummy must be large. Since the air bag expands to meet the dummy, then compresses as the dummy moves into it, the dummy travels a larger distance as it slows to a stop. In this case the magnitude of the displacement d is larger, and the magnitude of the force F is smaller and less injurious.

THE TAKEAWAY for Section 9-6

✔ The change in momentum of an object during a collision equals the force exerted on the object during the collision multiplied by the contact time—the length of time objects interact during the collision.

✔ The same change in momentum can be produced by a large force exerted for a short time or by a small force exerted for a longer time.

Prep for the AP Exam

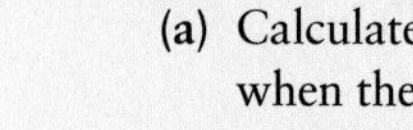

AP® Building Blocks

1. A sudden gust of wind exerts a force of 10.0 N for 1.20 s in a direction opposite to the flight of a red-tailed hawk, whose speed was 5.00 m/s before the gust of wind. As a result, the bird ends up moving in the opposite direction at 7.00 m/s. What is the mass of the hawk?

2. Determine the magnitude and direction of the average force exerted on your hand as you catch a 0.200-kg ball moving at 20.0 m/s. Assume the time of contact is 0.0250 s.

3. A baseball of mass 0.145 kg is thrown at a speed of 40.0 m/s. The batter strikes the ball with a force of 2.5×10^4 N; the bat and ball are in contact for 0.500 ms. Assuming that the force is aligned exactly opposite to the original direction of the ball's motion, calculate the final speed of the ball.

4. A 0.0750-kg ball is thrown so that it strikes a brick wall at 25.0 m/s along a line perpendicular to the wall.
   (a) Calculate the impulse that the ball imparts to the wall when the ball rebounds at 25.0 m/s in the direction opposite its original motion.
   (b) Assuming the ball is in contact with the wall for 0.0100 s, determine the magnitude and direction of the average force that the wall exerts on the ball. Compare this to the force of gravity exerted by Earth on the ball during its interaction with the wall.

AP® Skill Builders

5. An arrow strikes a straw target with a speed v and penetrates a distance d. Predict the penetration distance if an identical arrow strikes the same target with twice the momentum.

6. A ball strikes a smooth wall with a velocity v_i at an angle, as shown in the figure. The angle, θ_i, is measured relative to a line perpendicular to the wall. The wall exerts no force on the ball in a direction in the plane of the wall and does exert a normal force on the ball. The ball is reflected with a velocity, v_f, that makes an angle θ_f with the perpendicular, as shown in the figure.

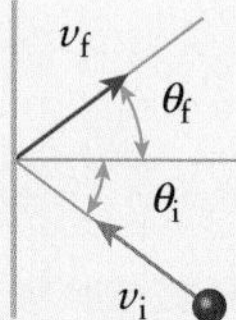

   (a) Apply momentum conservation to this collision in two dimensions to obtain the relationship between the components of the impulse, $F_{\text{wall on ball},x}\Delta t$ and $F_{\text{wall on ball},y}\Delta t$, and components of the initial and final momentum of the ball in terms of the angles θ_i and θ_f.
   (b) Interpret the representation of the wall as smooth as a mathematical condition on the impulse.
   (c) Assume that the collision between the wall and the ball is elastic, and apply the conservation of mechanical energy to the collision to obtain a condition on the magnitude of the initial and final momenta.
   (d) Based on the evidence provided by parts (b) and (c), justify the claim that the initial and final angles θ_i and θ_f are equal.

7. One smartphone manufacturer reported in 2015 that 50% of users have cracked a screen. Dropping a phone onto one of its corners is the most common cause of cracked

screens. Protective cases made from deformable materials are intended to protect the corners and sides during impact. The consensus among phone manufacturers is that cases should be used and 75% of owners do use them.

(a) Explain how cases can reduce cell phone screen damage.

Compliance standards based on the U.S. Military Standard MIL-STD-810 include Method 522.1 (Ballistic Shock) involving "momentum exchange caused by an inelastic collision of two elastic bodies during a time interval of less than 180 msec." Method 522.1 includes a drop test from 48 inches (1.22 m) onto hardwood over concrete with success indicated by 26 drops without failure.

(b) Calculate the magnitude of the force corresponding to a 180-ms collision duration for a 174-g phone dropped from 1.22 m. Assume that the collision is completely inelastic so that the final momentum of the phone is equal to zero (because the phone–Earth system would have negligible speed). Is this an upper or a lower limit on the force that you find?

(c) The protective case of a smartphone provides a layer of deformable materials with a thickness d. The purpose of this material is to spread out the time over which the collision occurs, to decrease the force on the phone to a survivable level. Predict the minimum time over which the phone must come to rest for a phone that drops from a height of 1.22 m and has a mass of 174 grams if the largest safe force exerted on the phone is 4 N. How far would the phone travel in this time if we assume the material only increases the distance over which the momentum is decreased?

(d) The distance the phone travels as it comes to a stop could be an approximation for the thickness of the case wall. If the necessary thickness of the case wall calculated in part (c) does not seem appropriate for a mobile device, consider the limitations of the model used to describe the falling phone. Also, explain why the formation of cracks in the screen indicates that the phone has an internal structure necessitating the use of a system model to describe the phone.

8. A force probe is mounted at the end of a horizontal air track over which a cart moves with negligible friction force. The probe reports the magnitude of the force exerted on it as a function of time. At the other end of the air track a spring is mounted that can be compressed by the cart and then released, giving the cart an initial velocity toward the force probe.

The spring constant is not known. The apparatus will be used to experimentally determine the spring constant. The mass of the cart is measured as 137.5 g. When the spring is compressed by 2 cm the force probe reports the data shown.

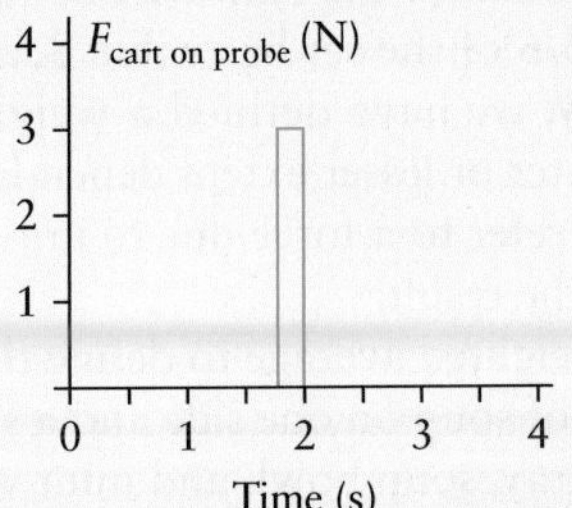

(a) Explain how the area of the rectangle shown in the data can be used to determine the speed of the cart when it loses contact with the spring. Include any assumptions.

(b) In the collected data shown, Δx is the compression of the spring from its rest length and F is the force exerted on the probe by the cart. The time interval of contact between the cart and the force probe is the same for each collision. Analyze these data to obtain a linear relationship, the slope of which is equal to the spring constant. Construct a graph of the data, and evaluate the spring constant.

Δx (cm)	F (N)
0.5	0.7
1.0	1.5
2.0	3.0
3.0	4.4

(c) A photogate timer is placed alongside the track and used to measure the time interval as a 0.5-cm flag mounted on top of the cart blocks the beam of the photogate. Use just these two values of the time interval at the photogate timer to calculate the energy dissipated during an interval from before (i) to after (f) the collision, including any assumptions: $\Delta t_i = 1.67$ ms, $\Delta t_f = 1.72$ ms.

(d) Using the data provided in part (c), evaluate the uncertainty in the measured energy dissipated. Use uncertainties in the measurements of the flag (±0.05 cm), time interval (±0.005 ms), and mass (±0.05 g) and design procedures, possibly involving alternative or additional measurement tools that can be used to increase the precision of the measurement.

AP® Skills in Action

9. The data shown were obtained from a force probe attached to the front edge of a cart rolling over a track that exerted negligible friction force on the cart. The force exerted by the cart on the probe is displayed. This is nonzero whenever the probe comes in contact with another object that changes the probe's motion, and through the probe, that of the cart. The mass of the cart is 187.4 g. Equations for the best-fit lines through the data in the interval between 1.0 and 6.0 ms and in the interval between 12 and 18 ms are displayed.

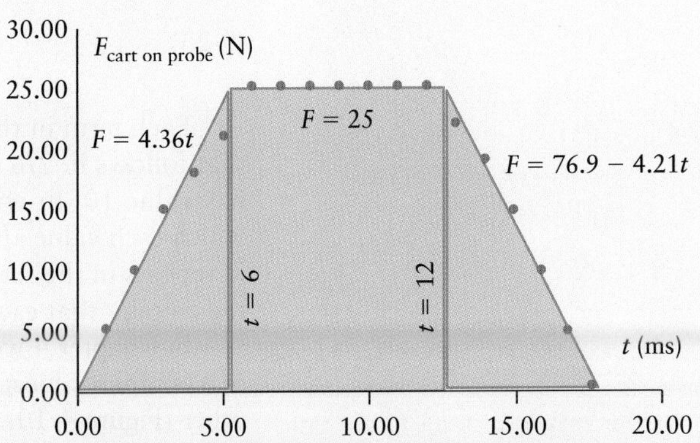

(a) Before collision with a stationary block at the end of the track, the cart was moving with a speed of 0.850 m/s. Calculate the speed and direction of the cart 18 ms after collision with the wall.

(b) Calculate the change in kinetic energy of the cart.

10. A golf ball appears to be rigid; deformation is negligible when the ball is squeezed in your hand. But when struck at high speed the ball compresses in the direction of the force exerted by the rigid head of a golf club at the end of a flexible shaft.

(a) Explain why a flexible shaft increases the displacement of the golf ball.

(b) If the duration of contact between the head and the ball is 470 μs, what is the force exerted by the head on the ball in a 275-m drive along a trajectory that maximizes horizontal displacement, assuming that air resistance is negligible? A golf ball has a mass of 46 g.

9-7 The center of mass of a system moves as though all the system's mass were concentrated there

NEED TO REVIEW?
Turn to Chapter 4, Section 4-2, for a review of external forces.

The concept of momentum can help us quantify an idea we have been using since Chapter 4. We learned in that chapter that the acceleration of an object is determined by the external forces on the object as described by Newton's second law: $\vec{a} = \frac{\sum \vec{F}_{ext}}{m}$. But if a system is rotating, like a gymnast tumbling in midair or a barrel rolling downhill, or if it is changing shape, then the object model no longer holds, since different points have different accelerations. So the quantity $\vec{a}$ in $\vec{a} = \frac{\sum \vec{F}_{ext}}{m}$ refers to the acceleration of a specific point, the center of mass of the system.

As we will see, the center of mass moves as though all the object's mass were squeezed into a tiny blob at that point and all the external forces are exerted on that blob. We'll learn how to calculate the position of the center of mass; then we'll use ideas about momentum to justify why the center of mass behaves as it does.

Averages and Weighted Averages

The position of a system's center of mass is a kind of *average* that takes into account the masses and positions of all the objects that make up that system. If the system is a car, you can think of these pieces as the various components of a car, or the individual atoms that comprise the car. To see how this average is defined, let's first remind ourselves how to calculate an ordinary average.

To find the average of a set of N numbers, add them together and divide by N. For example, the average of the three numbers ($N = 3$) 4, 8, and 18 is $(4+8+18)/3 = 30/3 = 10$. The average of the set of six numbers $n_1 = 4$, $n_2 = 4$, $n_3 = 8$, $n_4 = 8$, $n_5 = 8$, $n_6 = 10$ is

$$\bar{n} = \frac{1}{6}\sum_{i=1}^{6} n_i = \frac{4+4+8+8+8+10}{6} = \frac{42}{6} = 7$$

Here n_i represents the ith value in the set, and the symbol $\bar{n}$ (n with a bar on top) denotes the average value of the numbers n_i. Notice that because some of the values in the set occur more than once, we could write this average as

$$\bar{n} = \frac{2(4)+3(8)+10}{6} = \frac{2}{6}(4) + \frac{3}{6}(8) + \frac{1}{6}(10) = 7 \quad (9\text{-}23)$$

Each term in this sum includes the fraction of the entire set comprising the value that follows it: 2/6 of the set has value 4, 3/6 of the set has value 8, and 1/6 of the set has value 10. In writing the sum this way, we have defined a **weighted average**, in which each value affects the result to a greater or lesser extent depending on how often it appears in the set. Here *weight* does not refer to a force due to gravity but rather to the amount that each value contributes to the result.

In Chapter 4 we qualitatively used a weighted average to define the center of mass of a complex object, a tray that has a bowl of soup on one side and a small mint on the other (Figure 4-10). We argued that if the tray, soup bowl, and mint were all squeezed into a single blob, the waiter would put his hand directly under the blob to support it, so that the point where he would put his hand would be the center of mass for the system of the tray, soup, and mint.

Equation 9-23 provides some guidance as to how to write such a weighted average. Suppose we have a system of N objects with different masses $m_1, m_2, m_3, \ldots, m_N$. The total mass M_{tot} of the system is the sum of these N individual masses:

$$M_{tot} = m_1 + m_2 + m_3 + \ldots + m_N = \sum_{i=1}^{N} m_i \qquad (9\text{-}24)$$

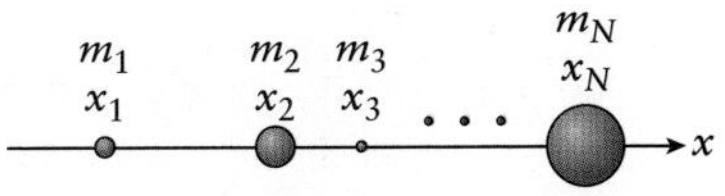

The mass of each individual object represents a fraction m_i/M of the total mass. Now suppose the N objects are at different positions along the x axis: $x_1, x_2, x_3, \ldots, x_N$ (**Figure 9-19**).

Figure 9-19 A system of objects The position of the center of mass of this collection of objects depends on the masses and positions of the individual objects (Equation 9-25).

By analogy to Equation 9-23 we write the position of the center of mass as a weighted average of these positions:

Position of the center of mass of a system of N objects

m_i = masses of the individual objects

$$x_{CM} = \frac{m_1}{M_{tot}}x_1 + \frac{m_2}{M_{tot}}x_2 + \frac{m_3}{M_{tot}}x_3 + \ldots + \frac{m_N}{M_{tot}}x_N = \frac{1}{M_{tot}}\sum_{i=1}^{N} m_i x_i \qquad (9\text{-}25)$$

Total mass of all N objects together

x_i = **positions** of the individual objects

EQUATION IN WORDS
Position of the center of mass of a system

The more massive a given object, the greater the ratio m_i/M_{tot} and the greater the importance of that object's position x_i in the sum given by Equation 9-25. Example 9-10 illustrates this property of the position of the center of mass.

EXAMPLE 9-10 Locating the Center of Mass

Two objects are located on the x axis at the positions $x_1 = 2.0$ m and $x_2 = 8.0$ m. For the system made up of these two objects, find the position of the center of mass (a) if m_1 and m_2 are both equal to 3.0 kg and (b) if $m_1 = 3.0$ kg and $m_2 = 33$ kg.

Set Up

Because there are just two objects, the total mass of the system is just the sum of their two masses. We use Equation 9-25 in each case to locate the center of mass.

Position of the center of mass of a system:

$$x_{CM} = \frac{m_1}{M_{tot}}x_1 + \frac{m_2}{M_{tot}}x_2 + \frac{m_3}{M_{tot}}x_3 + \ldots + \frac{m_N}{M_{tot}}x_N = \frac{1}{M_{tot}}\sum_{i=1}^{N} m_i x_i \qquad (9\text{-}25)$$

Total mass of the system of two objects:

$$M_{tot} = m_1 + m_2 \qquad (9\text{-}24)$$

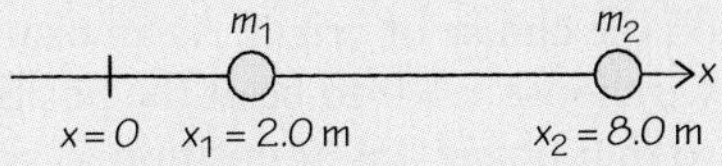

Solve

(a) Calculate the position of the center of mass for the case in which both objects have the same mass.

The two objects have equal masses:

$$m_1 = m_2 = 3.0 \text{ kg}$$

Total mass of the system:

$$M_{tot} = m_1 + m_2 = 3.0 \text{ kg} + 3.0 \text{ kg} = 6.0 \text{ kg}$$

Position of the center of mass:

$$x_{CM} = \frac{m_1}{M_{tot}}x_1 + \frac{m_2}{M_{tot}}x_2$$

$$= \frac{3.0 \text{ kg}}{6.0 \text{ kg}}(2.0 \text{ m}) + \frac{3.0 \text{ kg}}{6.0 \text{ kg}}(8.0 \text{ m}) = 5.0 \text{ m}$$

(b) Repeat part (a) for the case in which m_2 is much larger than m_1.

The two objects have different masses:

$m_1 = 3.0$ kg, $m_2 = 33$ kg

Total mass of the system:

$M_{tot} = m_1 + m_2$

$3.0 \text{ kg} + 33 \text{ kg} = 36 \text{ kg}$

Position of the center of mass:

$$x_{CM} = \frac{m_1}{M_{tot}}x_1 + \frac{m_2}{M_{tot}}x_2$$

$$= \frac{3.0 \text{ kg}}{36 \text{ kg}}(2.0 \text{ m}) + \frac{33 \text{ kg}}{36 \text{ kg}}(8.0 \text{ m})$$

$$= 7.5 \text{ m}$$

Reflect

In part (a) the position $x_{CM} = 5.0$ m is 3.0 m from object 1 and 3.0 m from object 2—in other words exactly halfway between the two identical objects. In part (b), by contrast, the center of mass is 5.5 m from the less massive object 1 ($m_1 = 3.0$ kg) and only 0.5 m from the more massive object 2 ($m_2 = 33$ kg). If one object in the system is much more massive than the others, the center of mass is closest to that more massive object. In both cases, if the two objects were connected by a rigid lightweight rod, you could balance the system in your hand if you supported it at the center of mass, just like the tray shown in Figure 4-10. The center of mass of the system of two objects, if they were constrained to move together by this rod, would move as if all their mass were located at this point.

Extend: It's important to note that the relative position of the center of mass does *not* depend on the choice of origin. As an example, suppose we choose $x = 0$ to be at the position of object 1 so that object 2 is at position $x_2 = 6.0$ m. In this case we find the center of mass in case (b) to be at $x_{CM} = 5.5$ m. That looks different from the value of 7.5 m that we found above but is still 5.5 m from object 1 and 0.5 m from object 2—exactly as we found before.

Find the center of mass in case (b), with two objects of different mass, but now with the origin chosen to be at the position of object 1:

$m_1 = 3.0$ kg, $m_2 = 33$ kg so $M_{tot} = 36$ kg

$x_1 = 0$, $x_2 = 6.0$ m

Position of the center of mass:

$$x_{CM} = \frac{m_1}{M_{tot}}x_1 + \frac{m_2}{M_{tot}}x_2$$

$$= \frac{3.0 \text{ kg}}{36 \text{ kg}}(0) + \frac{33 \text{ kg}}{36 \text{ kg}}(6.0 \text{ m})$$

$$= 5.5 \text{ m}$$

NOW WORK Problem 1 from The Takeaway 9-7.

More on the Position of the Center of Mass

In writing Equation 9-25 we assumed that all the objects are arranged along a line that we call the x axis. If the objects are arranged in a plane defined by the x and y axes, we need to specify the x and y coordinates of the position of the center of mass. Equation 9-25 gives the x coordinate, and a very similar equation gives us the y coordinate:

$$y_{CM} = \frac{m_1}{M_{tot}}y_1 + \frac{m_2}{M_{tot}}y_2 + \frac{m_3}{M_{tot}}y_3 + \ldots + \frac{m_N}{M_{tot}}y_N = \frac{1}{M_{tot}}\sum_{i=1}^{N} m_i y_i \qquad (9\text{-}26)$$

(y coordinate of the position of the center of mass)

If the objects are arranged in three dimensions, you would also need an equation similar to Equations 9-25 and 9-26 for the z coordinate, z_{CM}, of the center of mass.

The center of mass of the system in part (a) of Example 9-10 is midway between the two objects, at the *geometrical* center of the system. This is true whenever the system is *symmetrical*, so that the mass is distributed in the same way on both sides of the center of mass. For example, the center of mass of a billiard ball is at its geometrical center, as is the center of mass of a uniform metal rod. In part (b) of Example 9-10 the center of mass is *not* at the geometrical center because the system is not symmetrical: Masses m_1 and m_2 have very different values.

WATCH OUT !

There doesn't have to be any mass at the center of mass.

Note that in both parts (a) and (b) of Example 9-10, there's nothing physically located *at* the center of mass. The same is true for any symmetrical object with a hole at its center, like a Blu-ray disc, a doughnut, or a ping-pong ball: In each case the center of mass is located in the center of the empty hole if the mass is distributed uniformly.

Momentum, Force, and the Motion of the Center of Mass

It's now time to justify the statement that the center of mass moves as though all the mass of the system were concentrated there. To begin, let's combine Equations 9-25 and 9-26 into a single equation for the position of the center of mass:

$$\vec{r}_{CM} = \frac{m_1}{M_{tot}}\vec{r}_1 + \frac{m_2}{M_{tot}}\vec{r}_2 + \frac{m_3}{M_{tot}}\vec{r}_3 + \ldots + \frac{m_N}{M_{tot}}\vec{r}_N = \frac{1}{M_{tot}}\sum_{i=1}^{N} m_i\vec{r}_i \qquad (9\text{-}27)$$

(position of the center of mass)

In Equation 9-27 $\vec{r}_{CM}$ is the position of the center of mass, and the vectors $\vec{r}_1$, $\vec{r}_2$, $\vec{r}_3, \ldots, \vec{r}_N$ represent the positions of the N individual objects that make up the system. Each vector $\vec{r}_i$ has components x_i and y_i (and, if a third dimension is required, z_i), and $\vec{r}_{CM}$ has components x_{CM} and y_{CM} (and z_{CM} if required).

Let's now allow the objects that make up the system to move. (The objects need not be connected, so they may or may not move together.) During a time interval Δt each of the vectors $\vec{r}_i$ in Equation 9-27 changes by an amount $\Delta\vec{r}_i$, which can have a different value for each object. From Equation 9-27 the change $\Delta\vec{r}_{CM}$ in the position of the center of mass during Δt is

$$\Delta\vec{r}_{CM} = \frac{m_1}{M_{tot}}\Delta\vec{r}_1 + \frac{m_2}{M_{tot}}\Delta\vec{r}_2 + \frac{m_3}{M_{tot}}\Delta\vec{r}_3 + \ldots + \frac{m_N}{M_{tot}}\Delta\vec{r}_N = \frac{1}{M_{tot}}\sum_{i=1}^{N} m_i\Delta\vec{r}_i \qquad (9\text{-}28)$$

(change in position of the center of mass)

Equation 3-7 in Section 3-4 tells us that if an object changes its position by $\Delta\vec{r}$ during a time Δt, the velocity of the object is $\vec{v} = \Delta\vec{r}/\Delta t$. If we divide Equation 9-31 through by Δt, we get an expression for the velocity of the center of mass:

$$\begin{aligned}\vec{v}_{CM} &= \frac{\Delta\vec{r}_{CM}}{\Delta t} \qquad (9\text{-}29)\\ &= \frac{1}{\Delta t}\left(\frac{m_1}{M_{tot}}\Delta\vec{r}_1 + \frac{m_2}{M_{tot}}\Delta\vec{r}_2 + \frac{m_3}{M_{tot}}\Delta\vec{r}_3 + \ldots + \frac{m_N}{M_{tot}}\Delta\vec{r}_N\right)\\ &= \frac{m_1}{M_{tot}}\vec{v}_1 + \frac{m_2}{M_{tot}}\vec{v}_2 + \frac{m_3}{M_{tot}}\vec{v}_3 + \ldots + \frac{m_N}{M_{tot}}\vec{v}_N = \frac{1}{M_{tot}}\sum_{i=1}^{N} m_i\vec{v}_i\end{aligned}$$

(velocity of the center of mass)

Just as the position of the center of mass is a weighted average of the positions of the objects that make up the system, the *velocity* of the center of mass is a weighted average of the *velocities* of the objects.

The quantity $m_i\vec{v}_i$ on the right-hand side of Equation 9-29 is just the *momentum* of the ith object in the system. So $\sum_{i=1}^{N} m_i\vec{v}_i$ is the vector sum of the momentum of all objects that make up the system. This is just the *total* momentum of the system, which we denote as $\vec{p}_{\text{total}}$. If we multiply Equation 9-32 by the total mass of the system M_{tot}, we get

EQUATION IN WORDS
Total momentum and the velocity of the center of mass

The **total momentum** of a system... ...equals the **vector sum** of the **momentum of all objects** in the system...

$$\vec{p}_{\text{total}} = \sum_{i=1}^{N} m_i\vec{v}_i = M_{\text{tot}}\vec{v}_{\text{CM}} \qquad (9\text{-}30)$$

...and also equals the **total mass of the system** multiplied by the **velocity of the center of mass.**

The total momentum of the system is $\vec{p}_{\text{total}} = m_1\vec{v}_1 + m_2\vec{v}_2 + m_3\vec{v}_3 ...$

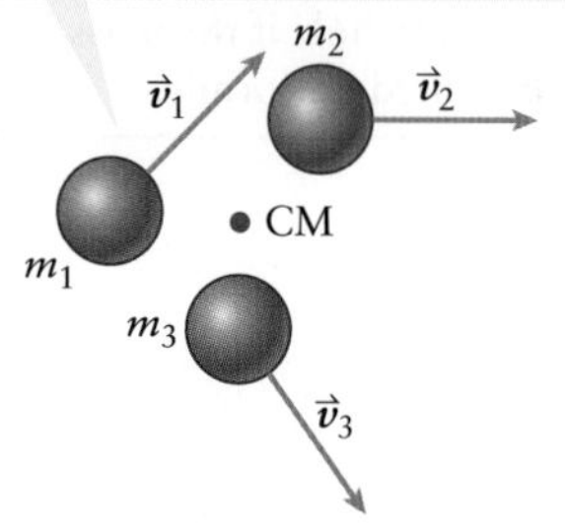

$M_{\text{tot}} = m_1 + m_2 + m_3 + ...$

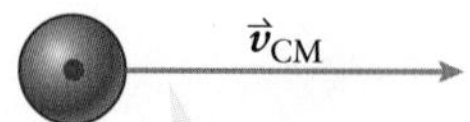

...which is the same as the momentum of all the mass moving together at the velocity of the center of mass: $\vec{p}_{\text{total}} = M_{\text{tot}}\vec{v}_{\text{CM}}$.

Figure 9-20 Total momentum of a system Two ways to represent the total momentum of a system of objects.

In other words the total momentum of a system of objects of total mass M_{tot} is the *same* as if all the objects were squeezed into a single blob moving at the velocity of the center of mass (**Figure 9-20**). This means that the total linear momentum of a football doesn't depend on how fast the football is spinning but only on the velocity of the center of mass of the football (which is at the football's geometrical center because the football is symmetrical).

We can now use Equation 9-30 to see what affects the motion of the center of mass. This equation shows that in order for the velocity of the center of mass to change by an amount $\Delta\vec{v}_{\text{CM}}$, there must be a change in the total momentum of the system:

$$\Delta\vec{p}_{\text{total}} = M_{\text{tot}}\Delta\vec{v}_{\text{CM}}$$

(The mass M_{tot} doesn't change since the system always includes the same set of objects.) But Equation 9-13 in Section 9-3 tells us that the total momentum can change during a time Δt *only* if a net external force is exerted on the system during that time:

$$\left(\sum \vec{F}_{\text{external on system}}\right)\Delta t = \Delta\vec{p}_{\text{total}}$$

Combining these two equations, we get

$$\left(\sum \vec{F}_{\text{external on system}}\right)\Delta t = M_{\text{tot}}\Delta\vec{v}_{\text{CM}}$$

Divide both sides of this equation by Δt, and recall that the acceleration $\vec{a}$ of an object equals the change in its velocity divided by the time over which the velocity changes: $\vec{a} = \Delta\vec{v}/\Delta t$ (Equation 3-10). The result is

EQUATION IN WORDS
How the net external force on a system affects the center of mass

The **net external force** on a system... ...causes the **center of mass** of the system to **accelerate.**

$$\sum \vec{F}_{\text{external on system}} = M_{\text{tot}}\frac{\Delta\vec{v}_{\text{CM}}}{\Delta t} = M_{\text{tot}}\vec{a}_{\text{CM}} \qquad (9\text{-}31)$$

Equation 9-31 looks almost exactly like Newton's second law for a single object of mass m, when we rearrange it to isolate the force term:

$$\sum \vec{F}_{\text{ext}} = m\vec{a}$$

So what Equation 9-31 tells us is that *the center of mass of a system of objects moves exactly as if the entire mass* M_{tot} *of the system were concentrated at the center of mass, and all the external forces on the system were exerted on that concentrated mass*. This justifies the statements we made at the beginning of this section about the significance of the center of mass.

Equation 9-31 also says that only *external* forces affect the motion of the center of mass. If various parts of the system exert forces on each other, these forces can affect how those parts move relative to each other, but they have *no* effect on the motion of the center of mass. As an example, consider the snowboarder shown in **Figure 9-21.** As she moves from right to left, her arms, legs, head, and torso move in complicated ways relative to each other. But her center of mass, shown by the red x's, moves in a simple parabolic trajectory. This is the same trajectory that would be followed by a blob with the same mass as the snowboarder with the same launch velocity on which the only external force exerted was gravity. If we only care about the motion of the center of

Exam Tip

Equation 9-30 is not on the AP® Physics 1 equation sheet because you are expected to understand that $\vec{p} = m\vec{v}$ represents both the momentum for an individual object moving at velocity $\vec{v}$, and the total momentum for a system of objects with total mass m in which the velocity of the center of mass is $\vec{v}$.

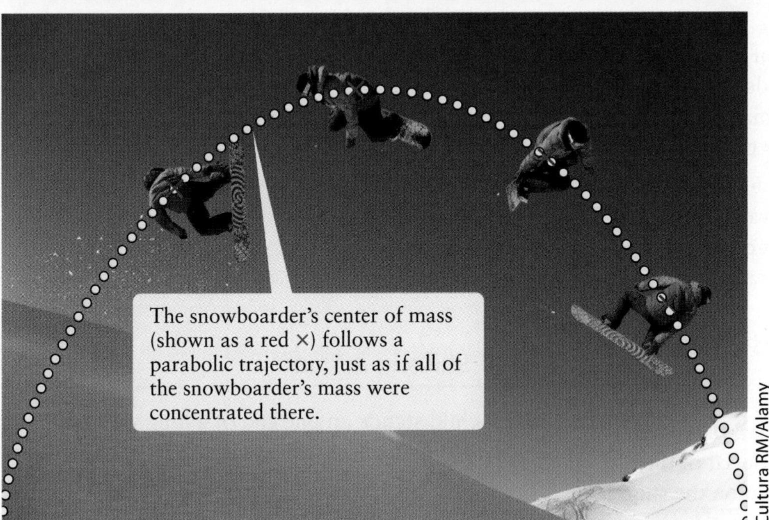

The snowboarder's center of mass (shown as a red ×) follows a parabolic trajectory, just as if all of the snowboarder's mass were concentrated there.
Cultura RM/Alamy

Figure 9-21 Center of mass of a snowboarder The motion of the snowboarder's body is complex, but the motion of the snowboarder's center of mass is described by an object in projectile motion.

mass, we can consider the snowboarder to be an object; but if we are interested in the complicated motion about the center of mass, we must consider her to be a system.

If the net external force exerted on a system is zero, Equation 9-31 says that the center of mass cannot accelerate even if the various parts of the system accelerate relative to each other. As an example, if the snowboarder in Figure 9-21 were in gravity-free outer space, her center of mass would continue moving in a line at a constant speed no matter how she moved the various parts of her body.

The center of mass will play an important role when we discuss rotational motion in the next chapter. We'll see that if a system is rotating as it moves through space, like a spinning football in flight or a spinning tire on a moving car, its motion can naturally be thought of as being in two pieces: the motion of the system's center of mass, plus rotation of the system around its center of mass.

THE TAKEAWAY for Section 9-7

✔ The center of mass of a system of objects is a weighted average of the positions of the individual objects.

✔ The total momentum of the system equals the system's total mass multiplied by the velocity of the center of mass.

✔ If there is a net external force on the system, the center of mass accelerates as though all the mass of the system were concentrated into a blob at the center of mass. If there is zero net external force on the system, the center of mass does not accelerate.

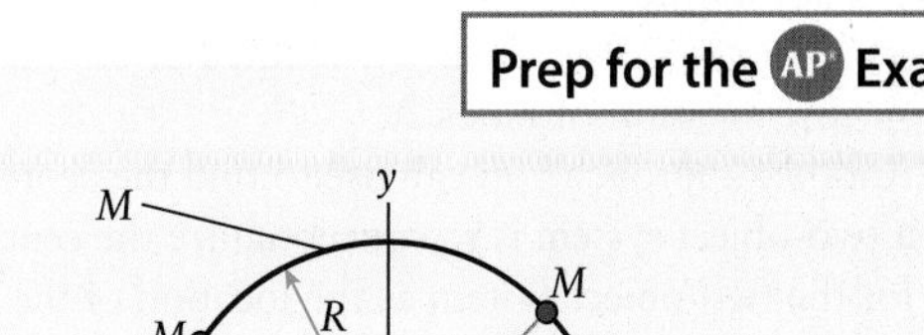

Prep for the AP Exam

AP® Building Blocks

EX 10

1. Find the coordinates of the center of mass of the three objects shown in the figure if $m_1 = 4.00$ kg, $m_2 = 2.00$ kg, and $m_3 = 3.00$ kg. Distances are in meters.

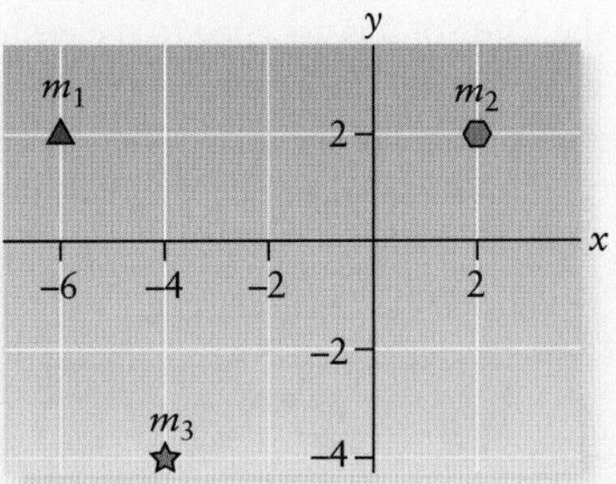

2. Four beads, each of mass M, are attached at various locations to a hoop of mass M and radius R. Find the center of mass of the hoop and beads.

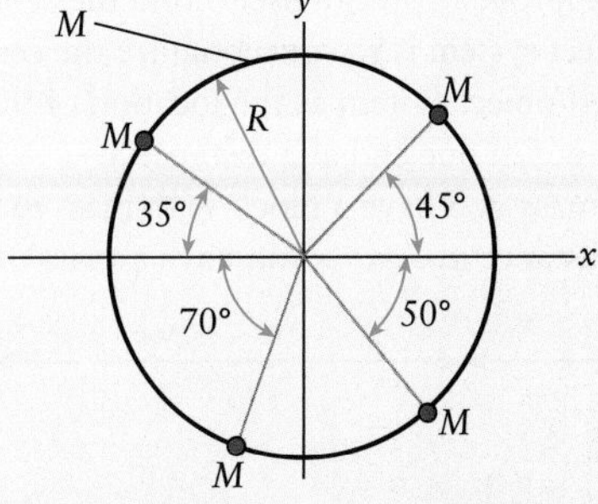

3. A man and his dog sit at opposite ends of a 3-m-long flat-bottom boat on which there are negligible horizontal forces exerted by the surface of the still pond. The boat is pointed toward shore with the dog in the bow at the front of the boat and the man in the stern at the rear of the boat. The bow is 6 m from the shore. The masses of the man and dog are 65 and 20 kg, respectively. The mass of the boat is 135 kg. Assume the center of mass of the boat

is located at its center. Neglect the extent of the man and dog (assume they are all the way at the ends to start).

(a) Describe the location of the center of mass of the dog–man–boat system relative to the center of the boat.

The dog suddenly walks rapidly toward the man.

(b) Justify the claim that in a coordinate frame that moves with the boat momentum is not constant.

(c) Justify the claim that in a coordinate frame fixed on the shore momentum is constant.

(d) Predict the direction that the boat moves as the dog is moving.

(e) Construct a diagram representing the system, including the masses of the dog, man, and boat; the speed of the dog relative to the boat and the speed of the boat relative to a coordinate frame fixed on the shore. Label each of these variables.

AP® Skill Builders

4. A 3.00-kg block (A) is attached to a 1.00-kg block (B) by a spring of negligible mass that is compressed and locked in place, as shown. The blocks are sliding with negligible friction along the x direction at an initial constant speed of 2.00 m/s.

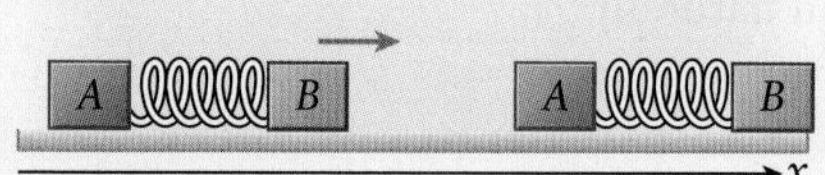

(a) At time $t = 0$ the positions of blocks A and B are $x = 1.00$ m and $x = 1.20$ m, respectively. Describe the location of the center of mass in the mass–spring system at time $t = 0$.

(b) At time $t = 0$ a mechanism releases the spring, and the blocks begin to oscillate as they slide. Predict the location of the center of mass 2.00 s later.

(c) If at $t = 2.00$ s block B is located at 6.00 m, describe the location of block A.

5. Another approach you can use to find the center of mass of a two-object system is to conceptualize the center of mass for the two-object system as the location of the fulcrum on a seesaw. On one end at a distance $x_{CM} - x_A$ from the center of mass you have a block with mass m_A. At the other end at a distance $x_B - x_{CM}$ you have a mass m_B.

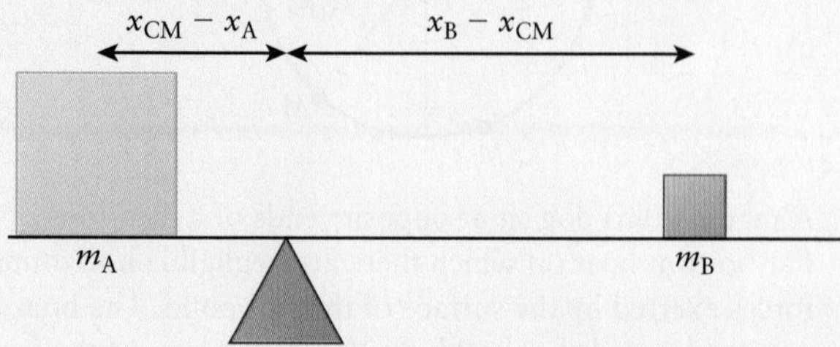

(a) Using Equation 9-25, justify the claim that the ratio of distances is inversely proportional to the ratio of masses.

$$\frac{(x_{CM} - x_A)}{(x_B - x_{CM})} = \frac{m_B}{m_A}$$

(b) The center of mass of a system doesn't depend on the coordinate frame that is used to locate it. Justify the claim using Equation 9-25 that when the coordinate frame is chosen so that object B is at the origin, $x_B = 0$, and so with x_A defined relative to x_B, the center of mass is located at

$$x_{CM} = \frac{m_A x_A}{m_A + m_B}$$

AP® Skills in Action

6. A child stands on one end of a long wooden plank with a length L that rests on an icy surface. The ice exerts a negligible friction force on the plank. The plank and the child have the same mass, m. The child's mother stands to the side of the plank, marking the center of mass.

(a) Describe the location of the center of mass of the child–plank system relative to the end of the plank where the child is standing.

(b) The child runs to the other end of the plank with a speed v_{child} while the mother maintains her position on the ice. Predict how the center of mass of the system moves while the child runs.

(c) Predict the location of the middle of the plank and the location of the child relative to the mother when the child has reached the other end of the plank.

(d) From the child's point of view she has been displaced by a distance L. Explain why her displacement from her mother's point of view is not equal to L.

7. Two laboratory carts are at rest on a track that exerts negligible friction on the carts. They are connected by a spring, with equilibrium length d_0 and negligible mass, that is locked in a compressed state until unlocked by a remote trigger. Justify your answers to each of these questions. Neglect the effects of air resistance.

(a) Is momentum constant during the expansion of the spring?

(b) Is mechanical energy constant during the expansion of the spring?

The two carts have the same mass, M. Ten blocks are available, each with a mass $M/10$. Each cart can hold up to all 10 blocks. Data will be collected with n blocks in cart 1 and m blocks in cart 2 using all 10 blocks in each trial: $n + m = 10$. A scale is marked on a tape adjacent to the track on which the value of 0 is assigned to the center of the track, as shown in the figure. The carts are initially positioned so that the center of mass of the system is at 0. The length of each cart is L and the compressed length of the spring is d.

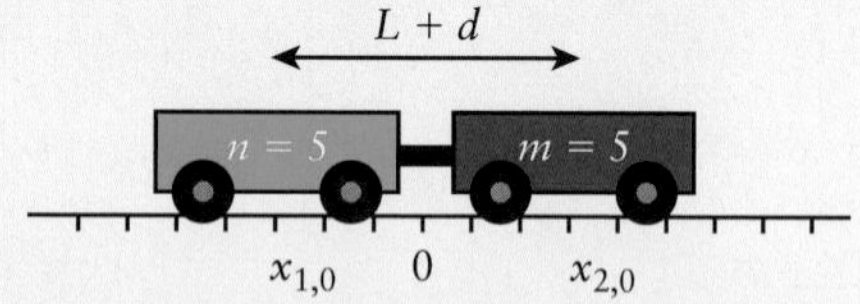

(c) Confirm that the figure shows a starting configuration with the system center of mass located at the origin with $n = m = 5$.

(d) Predict a starting configuration, $x_{1,0}$ and $x_{2,0}$, that will give the center of mass at the same location when $n = 10$ and $m = 0$.

(e) Express momentum conservation in terms of m, n, v_1, and v_2.

(f) With a spring constant, k, express the total mechanical energy in terms of n, m, M, k, d, d_0, v_1, and v_2.

(g) Using photogate timers, the speeds of the two carts after the spring has been triggered are measured. Describe a test of the prediction in part (e).

(h) By removing the spring and applying a variable force the spring constant is determined. Describe how this new information can be used to test the prediction in part (f).

WHAT DID YOU LEARN?

Prep for the AP Exam

Chapter learning goals	Section(s)	Related example(s)	Relevant section review exercises
Use the concepts of momentum, center of mass, and system to predict the behavior of objects in everyday situations.	9-1 9-2	9-1	4
Define the linear momentum of an object and explain how it differs from kinetic energy.	9-7	9-1	1
Explain the conditions under which the total momentum of a system is constant and why total momentum is constant in a collision.	9-3	9-2, 9-3, 9-4	1, 2, 4, 5
Identify the differences and similarities between elastic, inelastic, and *completely* inelastic collisions.	9-4	9-5, 9-6	1, 3
Apply conservation of momentum and mechanical energy to problems involving elastic collisions.	9-5	9-8	1, 2
Relate the momentum change of an object, the force that causes the change, and the time over which the force is exerted on the object.	9-6	9-9	1, 2, 3, 4
Find the center of mass of a system and describe how the net force on a system affects the motion of the system's center of mass.	9-7	9-10	1, 2, 3

Chapter 9 Review

Key Terms

All the Key Terms can be found in the Glossary/Glosario on page G1 in the back of the book.

collision 386
completely inelastic collision 401
contact time 415
elastic collision 398
explosive collision 399
impulse 415
impulse-momentum theorem 415
inelastic collision 399
law of conservation of momentum 380
linear momentum (momentum) 383
momentum (linear momentum) 380
total momentum 387
weighted average 420

Chapter Summary

Topic	Equation or Figure

Linear momentum: The momentum of an object is a vector that points in the same direction as its velocity. It depends on both the mass and velocity of the object. Do not confuse momentum with kinetic energy, which is a scalar quantity.

The **linear momentum** of an object is a vector.

The linear momentum points in the same direction as the **velocity** and is proportional to the velocity.

$$\vec{p} = m\vec{v} \qquad (9\text{-}5)$$

The linear momentum is also proportional to the object's **mass.**

External force and total momentum change for a system: The total momentum of a system of objects is the vector sum of the individual momentum of each object in the system. Only external forces (exerted by objects outside the system) can change the total momentum of a system. In contrast, internal forces (exerted by one member of the system on another) do not because for each internal force there is another force that is equal in magnitude and opposite in direction, by Newton's third law.

The **sum of all external forces** exerted on a system of objects

Duration of a time interval over which the external forces are exerted

$$\left(\sum \vec{F}_{\text{external on system}}\right)\Delta t = \vec{p}_{\text{total,f}} - \vec{p}_{\text{total,i}} = \Delta\vec{p}_{\text{total}} \qquad (9\text{-}13)$$

Change during that time interval **in the total momentum** of the objects that make up the system

Collisions and the law of conservation of momentum: The total momentum of a system is constant (maintains the same value) if there is zero net external force exerted on the system. Because collisions are short-time events, with the collision itself being the dominant interaction, external forces can be neglected and the total momentum of the colliding objects is constant during a collision.

If the net external force exerted on a system of objects is zero, **the total momentum of the system** is constant.

$$\vec{p}_{\text{total,f}} = \vec{p}_{\text{total,i}} \qquad (9\text{-}14)$$

Then the total momentum of the system at the end of a time interval...

...is equal to the total momentum of the system at the beginning of that time interval.

Types of collisions: Momentum is constant in collisions of all types. In an elastic collision mechanical energy and momentum are constant. Kinetic energy gets converted only to potential energy during the collision and is then restored to kinetic energy after the collision. In an inelastic collision, kinetic energy gets converted into internal as well as potential energy. In a completely inelastic collision, the colliding objects stick together after the collision and kinetic energy is converted only to internal energy. In each of these types of collisions, the amount of kinetic energy that is not converted during the collision depends on the center-of-mass velocity of the system and is the amount required to conserve momentum.

(a) Elastic collision: momentum constant; mechanical energy constant; kinetic energy restored

Before

A $\vec{v}_{Ai}$ $\vec{v}_{Bi}$ B

After

$\vec{v}_{Af}$ A B $\vec{v}_{Bf}$

(b) Inelastic collision: momentum constant; mechanical energy not constant; internal energy increased

Before

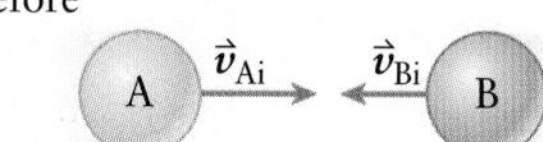

After

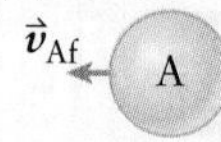

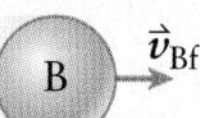

(c) Completely inelastic collision: momentum constant; mechanical energy not constant; colliding objects stick together and internal energy increased as much as possible

Before

A $\vec{v}_{Ai}$ $\vec{v}_{Bi}$ B

After

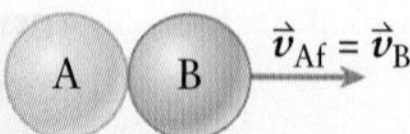

(Figure 9-15)

Completely inelastic collisions: The final velocity in a completely inelastic collision is common, all colliding objects stick together, so it depends only on the masses and initial velocities of the colliding objects.

Velocities of objects A and B **before** they undergo a **completely inelastic collision** (they stick together)

$$m_A\vec{v}_{Ai} + m_B\vec{v}_{Bi} = (m_A + m_B)\vec{v}_f \quad (9\text{-}18)$$

Masses of objects A and B

Velocity of the two objects moving together **after the collision**

Impulse and external force: To cause a certain change in an object's or a system's momentum requires a certain impulse (the product of the net external force on the object or system and the time over which the net force is exerted on the object or system). For a system, impulse is given by the left term of Equation 9-13.

The **sum of all external forces** exerted on an object

Duration of the time interval over which the external forces are exerted

Impulse

$$\vec{J} = \left(\sum \vec{F}_{\text{external on object}}\right)\Delta t = \vec{p}_f - \vec{p}_i = \Delta\vec{p} \quad (9\text{-}21)$$

Change in the momentum $\vec{p}$ of the object during that time interval

Contact time: The impulse in a collision is determined by the force exerted by one of the objects on the other in the collision and the contact time (the time that the colliding objects interact with each other). The same momentum change can be caused by a strong force with a short contact time or a weak force with a long contact time.

Force exerted on an object during a collision

Duration of the collision = **contact time**

$$\vec{F}_{\text{collision}}\Delta t = \vec{p}_f - \vec{p}_i = \Delta\vec{p} \quad (9\text{-}22)$$

Change in the momentum $\vec{p}$ of the object during the collision

Center of mass: The center of mass of a system of objects is a weighted average of the positions of the individual objects.

Position of the center of mass of a system of N objects

m_i = **masses** of the individual objects

$$x_{CM} = \frac{m_1}{M_{tot}}x_1 + \frac{m_2}{M_{tot}}x_2 + \frac{m_3}{M_{tot}}x_3 + \ldots + \frac{m_N}{M_{tot}}x_N = \frac{1}{M_{tot}}\sum_{i=1}^{N} m_i x_i \quad (9\text{-}25)$$

Total mass of all N objects together

x_i = **positions** of the individual objects

Total momentum and the center of mass: The total momentum of the system is the same as if all the mass were concentrated at the center of mass and moving with the center-of-mass velocity.

The **total momentum** of a system...

...equals the **vector sum** of the **momentum of all objects** in the system...

$$\vec{p}_{\text{total}} = \sum_{i=1}^{N} m_i\vec{v}_i = M_{tot}\vec{v}_{CM} \quad (9\text{-}30)$$

...and also equals the **total mass of the system** multiplied by the **velocity of the center of mass.**

Motion of the center of mass: The center of mass of a system behaves as though all the mass of the system were concentrated there and all the external forces exerted at that point. If there is no net external force exerted on the system, the center of mass does not accelerate.

The **net external force** on a system...

...causes the **center of mass** of the system to **accelerate.**

$$\sum \vec{F}_{\text{external on system}} = M_{tot}\frac{\Delta\vec{v}_{CM}}{\Delta t} = M_{tot}\vec{a}_{CM} \quad (9\text{-}31)$$

Chapter 9 Review Problems

1. One ball has four times the mass and twice the speed of another.
   (a) How does the magnitude of momentum of the more massive ball compare to the magnitude of momentum of the less massive one?
   (b) How does the kinetic energy of the more massive ball compare to the kinetic energy of the less massive one?
2. A large fish has a mass of 25.0 kg and swims at 1.00 m/s toward and then swallows a smaller fish that is not moving. If the smaller fish has a mass of 1.00 kg, what is the speed of the larger fish immediately after it finishes lunch?
3. What is the speed of a 75-kg soccer player with a momentum of 550 kg · m/s?
4. Two small, identical steel balls collide completely elastically. Initially, ball 1 is moving with velocity v_1, and ball 2 is stationary. After the collision, the final velocities of ball 1 and ball 2 are
   (A) $1/2v_1$; $1/2v_1$ (B) 0; v_1 (C) $-v_1$; 0 (D) $-v_1$; $2v_1$
5. An expert boxer delivers a 2.8×10^3-N punch to the head of his opponent. If the punch contacts for 0.10 s, and the unfortunate opponent's head has a mass of 5.0 kg, what is the speed of his head when the punch loses contact? Assume no other forces are exerted on the opponent's head. (In reality, the force of the neck on the head plays an important role.)
6. Bean bag rounds are nonlethal projectiles fired from shotguns. If a 40.0-g round strikes at 70.0 m/s and delivers all its momentum over a time of 0.15 s, what is the average force of impact?
7. The kinetic energy of an object increases by a factor of 9. By what factor does its momentum increase?
8. A girl kayaks on choppy water at a constant speed of 4.5 m/s. The mass of the girl is 35.2 kg and the mass of the kayak is 42.7 kg.
   (a) Calculate the magnitude of the momentum of the girl–kayak system.
   (b) To maintain her speed, the average force that she exerts on the water with her paddles is 13.7 N. Calculate the power she delivers.
   (c) The water then smooths out and the girl and kayak are flowing with the water at this same speed without her needing to paddle. To avoid a rock, she exerts a force with the same magnitude she exerted before in a direction opposing the motion of the kayak for 3.0 s. Calculate her new speed.
   (d) Calculate the additional time she would need to exert this force to stop the forward motion of the kayak.
9. Two objects have equal kinetic energies. Justify the claim that the magnitudes of their momenta are equal if and only if they have the same mass.
10. A friend throws a heavy ball toward you while you are standing on smooth ice. You can either catch the ball or deflect it back toward your friend. Rank your speed right after your interaction with the ball for these options, with greatest speed first. Use the letters (A), (B), (C), (D), and (E) to represent speed for that option and use > or = to compare the speeds as you place them in order.
    (A) You catch the ball and hold it.
    (B) You catch the ball and then immediately drop it.
    (C) You deflect the ball back toward your friend at the same speed with which it hit your hand.
    (D) You deflect the ball back toward your friend at half the speed with which it hit your hand.
    (E) You deflect the ball back toward your friend at twice the speed with which it hit your hand.
11. A 3.00-kg object is moving toward the right at 6.00 m/s. A 5.00-kg object moves to the left at 4.00 m/s. After the two objects collide the 3.00-kg object moves toward the left at 2.00 m/s. What is the final velocity of the 5.00-kg object? Assume no external forces are exerted on the objects.
12. Consider a completely elastic head-on collision between two objects that have the same mass and the same speed. Explain why each of these statements describing the state of the system after the collision is correct or incorrect.
    (a) Both total momentum and total mechanical energy are zero.
    (b) The magnitudes of the velocities are the same, but the directions are reversed.
    (c) Both of the objects continue with the same velocity.
    (d) One of the objects continues with the same velocity, and the other reverses direction at twice the speed.
13. An object is traveling in the positive x direction with speed v. A second object that has half the mass of the first is traveling in the opposite direction with the same speed. The two experience a completely inelastic collision. The final velocity in the x direction is
    (A) $v/2$ (B) $v/3$ (C) v (D) $2v/3$
14. A 5.0-km-wide asteroid with a speed of 2.8×10^4 m/s collides with Mars, whose orbital speed is 2.4×10^4 m/s and whose diameter is 6.8×10^6 m and mass is 6.4×10^{23} kg. Assume that the density of the iron-rich asteroid is twice the density of Mars.
    (a) Estimate the change in Mars's orbital speed if **i.** the collision is a head-on collision and **ii.** the asteroid has a path perpendicular to the orbital path.
    (b) Pose questions needed to estimate the long-term effect of asteroid collisions on the orbital speed of Mars.
15. An 85.0-kg linebacker is running at 8.00 m/s directly toward the sideline of a football field. He tackles a 75.0-kg running back moving at 9.00 m/s straight toward the goal line (perpendicular to the original

direction of motion of the linebacker). Determine their common velocity immediately after they collide.

16. A 2.00-kg ball moves at 3.00 m/s toward the right. It collides elastically with a 4.00-kg ball that is initially at rest. Write, including substitutions of quantities given and defining your coordinate system, but do not solve, the equations you could use to determine the velocities of the balls after the collision.

17. A baseball bat strikes a ball when both are moving at 31.3 m/s (relative to the ground) toward each other. The bat and ball are in contact for 1.20 ms, after which the ball is traveling opposite its initial direction of motion at a speed of 42.5 m/s. The mass of the bat and the ball are 850 and 145 g, respectively. Calculate the magnitude and direction of the impulse given to **(a)** the ball by the bat and **(b)** the bat by the ball. **(c)** What average force does the bat exert on the ball? **(d)** Why doesn't the force shatter the bat?

18. Sally finds herself stranded on a frozen pond so slippery that she can't stand up or walk on it. To save herself, she throws one of her heavy boots horizontally, directly away from the closest shore. Sally's mass is 50 kg, the boot's mass is 5.0 kg, and Sally throws the boot with speed equal to 5.0 m/s.
    (a) What is Sally's speed immediately after throwing the boot?
    (b) Where is the center of mass of the Sally–boot system, relative to where she threw the boot, after 10.0 s?
    (c) How long does it take Sally to reach the shore, a distance of 10.0 m away from where she threw the boot? For all parts, neglect friction between Sally and the ice.

19. You have been called to testify as an expert witness in a trial involving a head-on collision. Car A has a mass 680 kg and was traveling eastward. Car B has a mass 500 kg and was traveling westward at 72.0 km/h. The cars locked bumpers and slid eastward with their wheels locked for 6.00 m before stopping. You have measured the coefficient of kinetic friction between the tires and the pavement to be 0.750. How fast (in miles per hour) was car A traveling just before the collision?

20. The sport of curling is quite popular in Canada. A curler slides a 19.1-kg stone so that it strikes a competitor's stationary stone at 6.40 m/s before moving at an angle of 120° from its initial direction. The competitor's stone moves off at 5.60 m/s. Set up but do not solve the equations you could use to determine the final speed of the first stone and the final direction of the second one. State the assumption about the collision you need to make.

21. In a ballistic pendulum experiment, a small marble is fired into a cup attached to the end of a pendulum. If the mass of the marble is 0.00750 kg and the mass of the pendulum is 0.250 kg, how high will the pendulum swing if the marble has an initial speed of 6.00 m/s? Assume that the mass of the pendulum is concentrated at its end and the velocity of the marble as it enters the cup is purely horizontal.

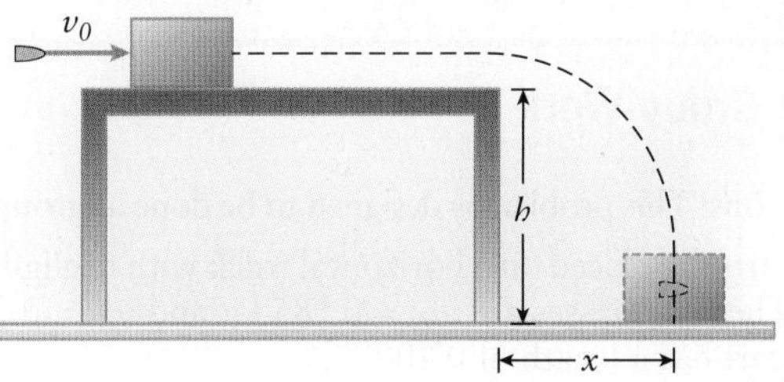

22. A bullet of mass 0.020 kg is moving horizontally with speed v_0 when it hits a block of mass 1.50 kg that is at rest on a horizontal table, as shown above. The surface of the table exerts negligible friction on the block and is at a height $h = 0.94$ m above the floor. After the impact, the bullet and the block slide off the table and strike the floor a distance $x = 1.60$ m from the edge of the table. Calculate the following quantities:
    (a) The speed of the block (with the bullet in it) as it leaves the table
    (b) The speed of the bullet before it strikes the block

23. An explosion breaks an object into two pieces, one of which has twice the mass of the other. The original object was initially at rest. If the total energy released in the explosion is 12,000 J, and half of that goes into the kinetic energy of the two pieces, how much energy does each piece acquire?

24. A block of mass M, initially at rest, is attached to an ideal spring of constant k, which is attached to a wall. A bullet of mass m is fired into the block and the block moves, compressing the spring a distance x_{max} before coming to rest. Assume that friction between the block and the floor is negligible.

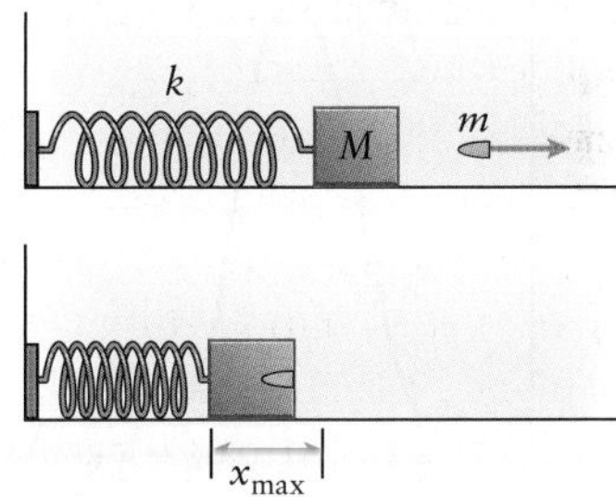

    (a) Derive an expression for the speed of the bullet just before it strikes the block, in terms of the given quantities and physical constants.
    (b) If the bullet instead passed through the block, would the compression of the spring, x_{max2}, be greater than, less than, or the same as for when the bullet remains embedded? Justify your answer.

25. A bullet of mass 0.025 kg is fired at a speed of 250 m/s from below into a block of unknown mass, which is initially at rest on a horizontal surface. The block with the bullet embedded in it rises to a maximum height of 1.30 m before falling back downward. Calculate the mass of the block.

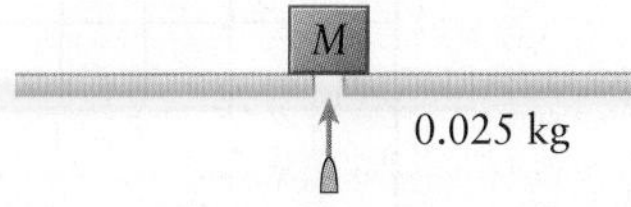

Prep for the AP® Exam

AP® Group Work

Directions: This problem is designed to be done as group work.

Two carts are placed on a horizontal track with negligible friction. The cart masses are $m_A = 0.185$ kg and $m_B = 0.370$ kg. Each cart has a length of 0.100 m.

(a) A ruler that runs along the side of the 3.0-m track shows that the centers of mass of cart A and cart B are initially at $x_{i,A} = 0.20$ m and $x_{i,B} = 2.80$ m. Describe the center of mass of the system of the two carts.

Motion sensors that measure the dependence of position and velocity on time are placed at each end of the track. The motion sensor at 3.00 m will record position and velocity for cart B and the motion sensor at 0.00 m will record position and velocity for cart A. On the front of each cart is a force probe that can be used to measure impulse on the cart.

A team of students will use this apparatus to collect data to confirm momentum conservation. The students will be able to analyze data files with values of x and v in increments of 0.05 s for each cart and for force data received from each force probe.

For a preliminary test of the system, the students push carts A and B with these initial velocities: $v_{i,A} = 1.00$ m/s and $v_{i,B} = -0.50$ m/s from the initial positions defined in part (a). Data collected from the force probe fixed on cart B are shown in the graph (points) with a best-fit curve (line).

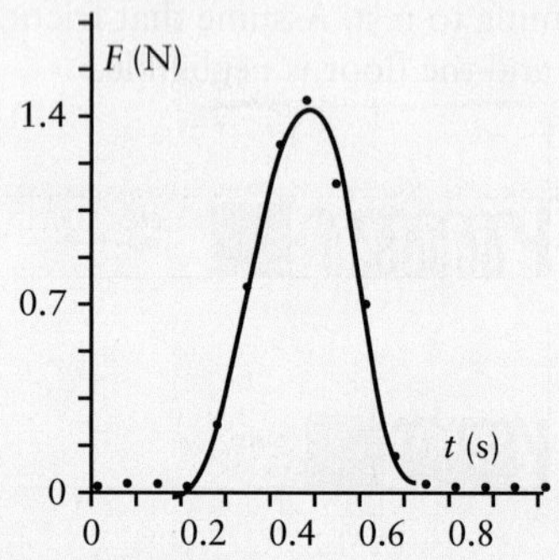

(b) Using these data predict

i. The time at which the collision occurs, $t_{collision}$

ii. The position of each cart when the collision occurs

iii. The position of the center of mass at the time of collision

iv. The speed of each cart after the collision

v. The position of each cart at a time twice the time at which the collision occurs, $2t_{collision}$

vi. The position of the center of mass at $2t_{collision}$

(c) Using the axes provided, construct graphs that predict the time dependence of x_{CM} and p for each cart.

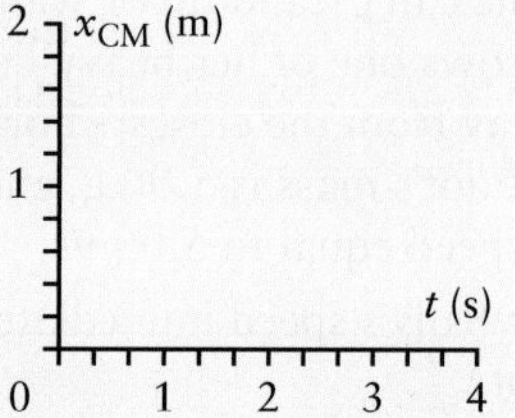

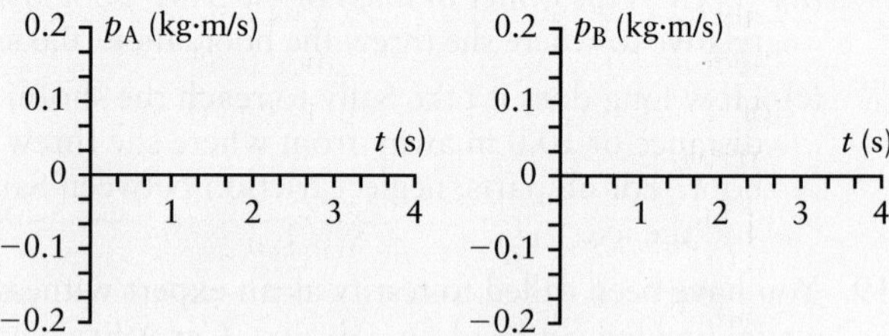

(d) Identify the features of the graphs that could be measured experimentally to test the claim that momentum is constant.

(e) Describe a procedure to use this apparatus to obtain the data identified in part (d).

AP® PRACTICE PROBLEMS

Prep for the AP® Exam

Multiple-Choice Questions

Directions: The following questions have a single correct answer.

Questions 1–2 refer to the following material.

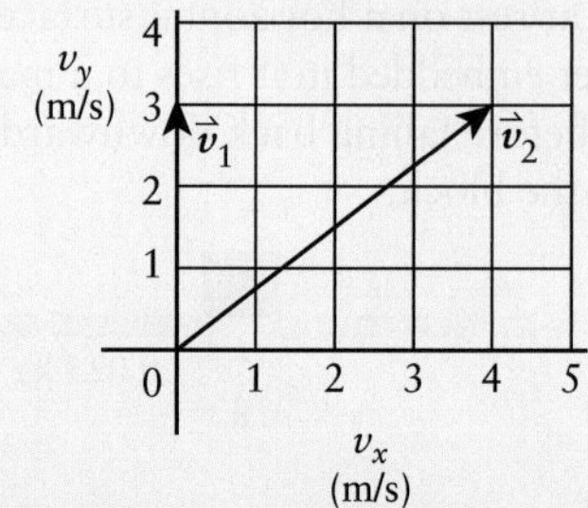

1. An object of mass of 4.0 kg is initially moving with a velocity $\vec{v}_1$ as shown. A force is then exerted on it, which changes its velocity to $\vec{v}_2$, as shown. The object's change in momentum has a magnitude of which of the following?

(A) 4.0 kg · m/s
(B) 8.0 kg · m/s
(C) 12 kg · m/s
(D) 16 kg · m/s

2. What would be the change in kinetic energy of the object?

(A) 16 J (B) 18 J (C) 32 J (D) 64 J

t(s)	0.00	0.50	1.00	1.50	2.00	2.50	3.00	3.50	4.00
F(N)	2.01	2.00	2.01	3.00	4.02	5.00	4.02	2.99	2.00

3. The table above presents measurements of the net force exerted on a 0.50-kg object moving along a line during a 4.00-second time period. Of the following, which would be the best estimate of the change in velocity for the object?
(A) 27 m/s
(B) 19 m/s
(C) 13 m/s
(D) 10 m/s

4. Two lumps of clay moving in opposite directions collide with each other and move off as one with no external forces exerted on the system. The first lump has a mass of 0.40 kg and a velocity of +2.0 m/s and the other a mass of 1.6 kg and a velocity of –1.0 m/s. What is the velocity of the center of mass before the collision?
(A) –0.40 m/s
(B) +0.40 m/s
(C) –1.2 m/s
(D) 1.2 m/s

5. Labeling two colliding objects as I and II, which of the following correctly describes collisions in general for unequal masses in which no external forces are exerted on the system?

Change in momentum of each	Work done on each by the other
(A) $\Delta\vec{p}_{\text{I}} = -\Delta\vec{p}_{\text{II}}$	W_{I} in general $\neq W_{\text{II}}$
(B) $\Delta\vec{p}_{\text{I}} = \Delta\vec{p}_{\text{II}}$	W_{I} in general $\neq W_{\text{II}}$
(C) $\Delta\vec{p}_{\text{I}} = -\Delta\vec{p}_{\text{II}}$	$W_{\text{I}} = -W_{\text{II}}$
(D) $\Delta\vec{p}_{\text{I}} = \Delta\vec{p}_{\text{II}}$	$W_{\text{I}} = W_{\text{II}}$

6. In a one-dimensional collision, two objects with masses M and m collide head-on with no external forces exerted. Both the mass of M and its speed are greater than those of m, that is, $M > m$ and the speed $v_M > v_m$. The two objects bounce off each other when they collide. Which of the following bar graphs might correctly represent this collision?

(A)

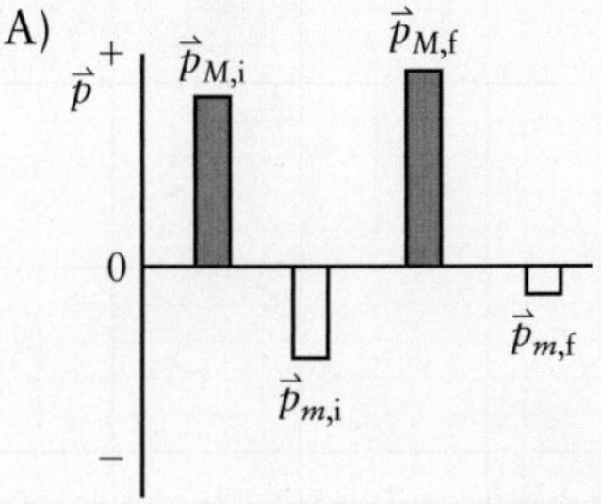

(B)

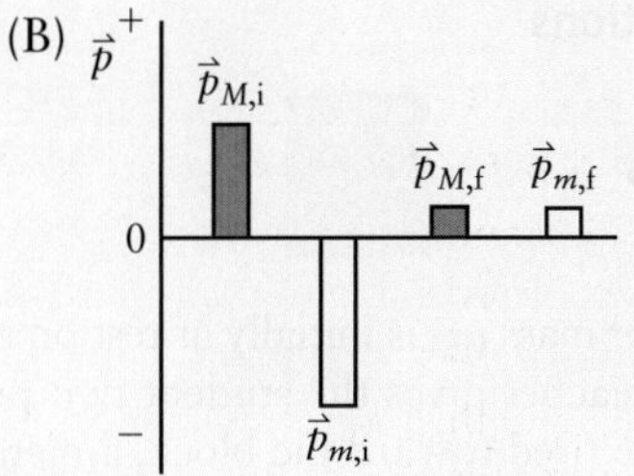

(C)

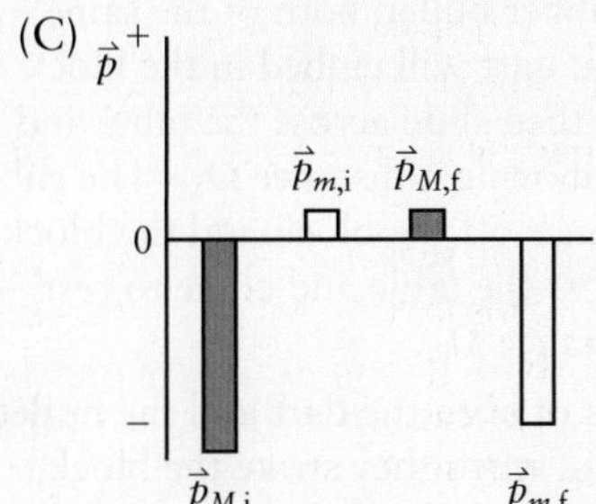

(D)

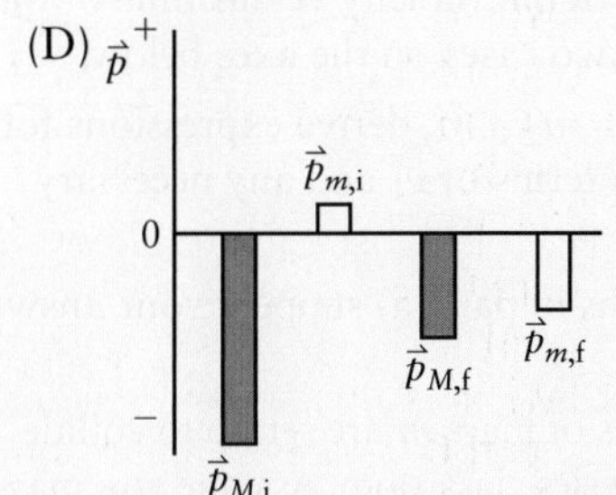

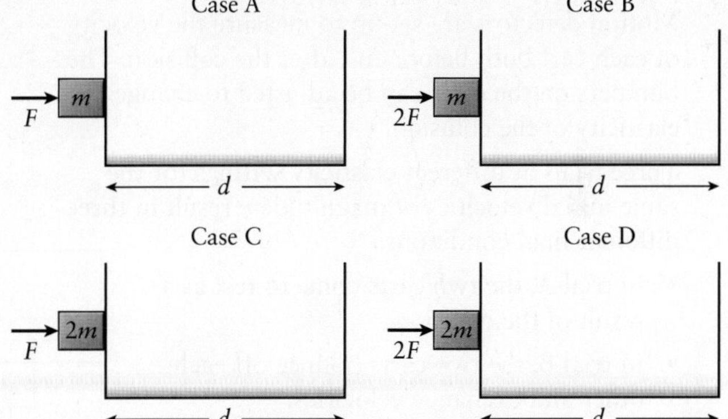

7. A block is pushed a distance d across the floor in the four cases shown above. The block and the constant pushing force vary in value in each case as shown. The friction between the floor and the block is negligible in each case. Which of the following is the correct ranking of the cases based on the momentum of the block after having been pushed a distance d?
(A) (B = D) > (C = D)
(B) (A = B) > (C = D)
(C) B > (A = D) > C
(D) D > (B = C) > A

(Continued)

Free-Response Questions

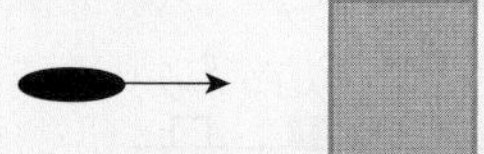

1. A wooden block of mass m_2 is initially at rest on a rough table. The teacher gives the student two projectiles that can be fired toward the block, a metal dart and a hard rubber bullet, both of the same mass m_1. The metal dart will embed in the block and the block will then slide across the table and come to rest after moving a distance D_M. The rubber bullet will bounce off the block and the block will then slide across the table and come to rest after moving a distance D_R.
   (a) If the velocities of both the dart and the bullet are the same, v_1, when they strike the block, sketch graphs of the velocity versus time of the block for the two cases on the axes below.
   (b) Assuming $m_1 = m_2 = M$, derive expressions for D_M and D_R in terms of v_1 and any necessary constants.
   (c) Do your graphs in part (a) support your answer in part (b)?

2. Two identical carts of mass m are set up to collide on a track in a physics classroom. Assume you may neglect friction between the carts and the track. Motion detectors are set up to measure the velocity of each cart both before and after the collision. The bumpers on the carts can be adjusted to change the elasticity of the collision.

   Three trials at different elasticity settings for the same initial velocities of magnitude v result in three different final conditions.

   - In trial A, the two carts come to rest as a result of the collision.
   - In trial B, the two carts bounce off each other and exchange velocities.
   - In trial C, the two carts bounce off each other and move with speeds that are 2/3 the magnitude of the speeds they had before the collision.

   (a) For each trial, identify the type of collision and describe the center-of-mass velocity before, during, and after the collision. Also describe any energy changes during the interaction, clearly defining your system.
   (b) For each trial, calculate the change in kinetic energy of the two carts for the collision.
   (c) Do your calculations in (b) support your claims in (a)? Briefly justify your answer.

3. Two students are playing a game called tetherball, where a ball of mass M_b is connected (tethered) by a rope to the top of a vertical pole, so that it can move around the pole, as shown in the figure.

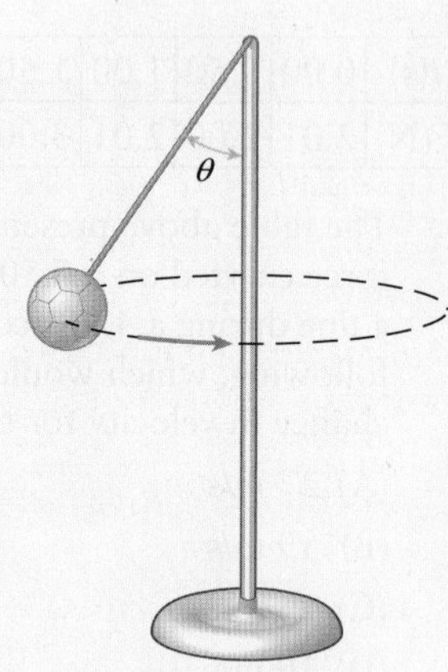

   As the tethered ball passes directly in front of the pole, such that it is moving to the right with speed v_0, it collides head-on with, and sticks to, a small lump of clay of mass M_c. The clay is thrown at it softly by a student, such that its velocity upon impact can be approximated as 0. The rope breaks at the same instant the clay collides with the tethered ball.
   (a) On the axes below, sketch a graph of the speed of the tetherball from just before the collision until the tetherball–clay system strikes the ground.
   (b) The tetherball is at a height H when the rope breaks. In terms of the given and fundamental constants:
      i. Derive an expression for the displacement of the center of mass of the system as a function of time from immediately after the collision until just before the system strikes the ground.
      ii. Derive an expression for the displacement of the center of mass of the system as the system strikes the ground.
   (c) Relate your expression in (b) to your graph in (a).

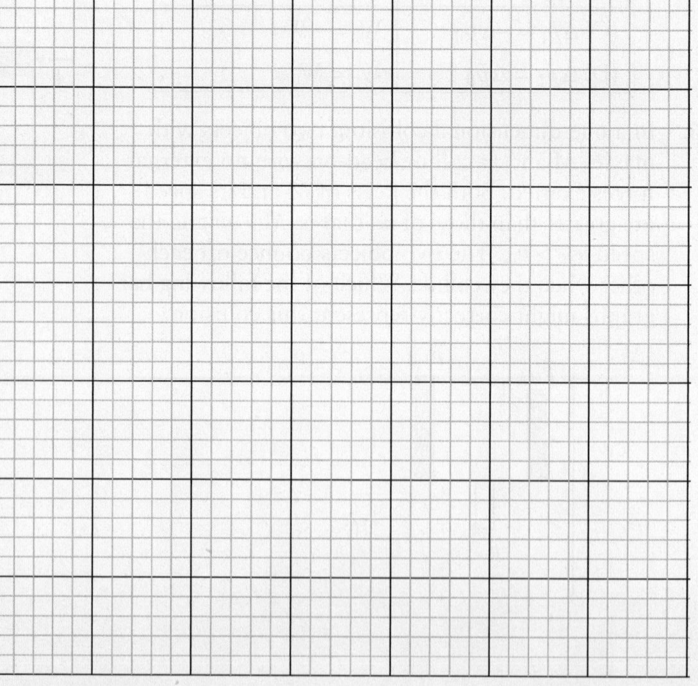

AP® Physics 1 Practice Exam 1

Prep for the AP® Exam

Multiple-Choice Questions

Directions: In the following questions, select the best answer choice from the four options listed.

Questions 1–2 refer to the following diagram.

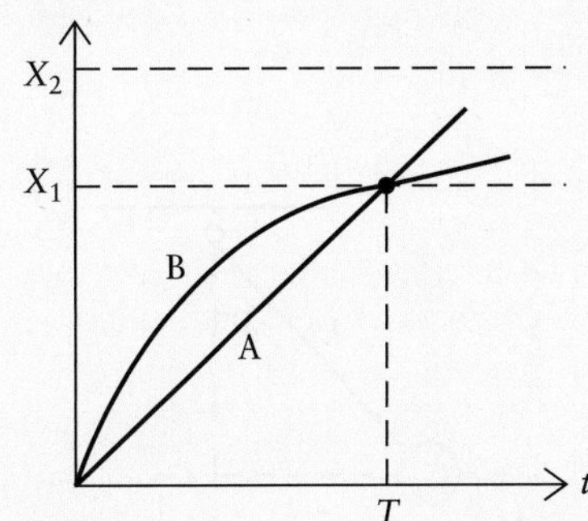

1. Two trains, A and B, begin moving on parallel tracks from the same starting position at a train station. The graph provided shows their respective displacements from the starting point as functions of time. Which of the following is true at time T?
   (A) Train A is moving faster than train B.
   (B) Train B is moving faster than train A.
   (C) Trains A and B are moving at the same speed.
   (D) Train A is ahead of train B.

2. If the trains continue in the same motion as shown in the graph, which of the following statements is true?
   (A) Train B moves faster than train A at all points on the graph.
   (B) Train A moves faster than train B at all points on the graph.
   (C) Train A will reach the position X_2 before train B.
   (D) Train B will reach the position X_2 before train A.

3. Which of the following graphs of velocity as a function of time represents the motion of a ball that is thrown vertically upward and moves under the influence of gravity until it returns to its initial position?

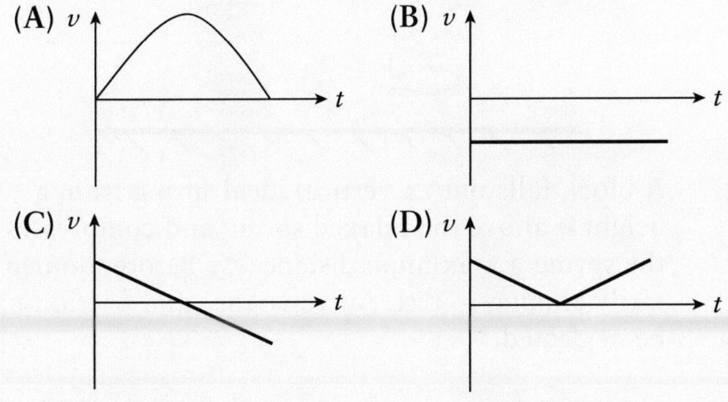

4. A projectile is launched at a speed of v_0 and at an angle θ above the horizontal. Which of the following are the correct values of the projectile's speed and the magnitude of its acceleration when it is at the highest point in its path?

	Speed	Acceleration
(A)	0	0
(B)	$v_o \cos(\theta)$	0
(C)	0	9.8 m/s^2
(D)	$v_o \cos(\theta)$	9.8 m/s^2

5. A block of mass m is at rest on a rough inclined plane. Which of the following diagrams best represents the correct directions for the normal force ($\vec{F}_n$), weight ($\vec{F}_g$), and friction force ($\vec{F}_f$) being exerted on the block?

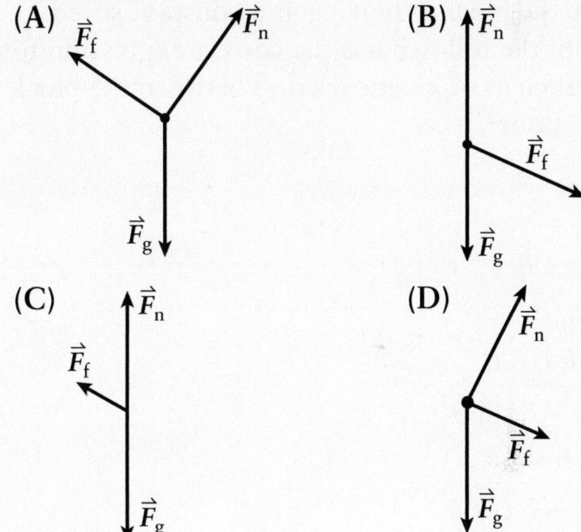

6. A person of weight 800 N is riding in an elevator while standing on a scale. The scale reading is 700 N. Which of the following is true of the elevator's motion?
   (A) The elevator is moving upward at constant speed.
   (B) The elevator is moving upward at increasing speed.
   (C) The elevator is moving downward at constant speed.
   (D) The elevator is moving downward at increasing speed.

7. Which of the following graphs of velocity as a function of time could represent the motion of an object on which a constant net force is being exerted?

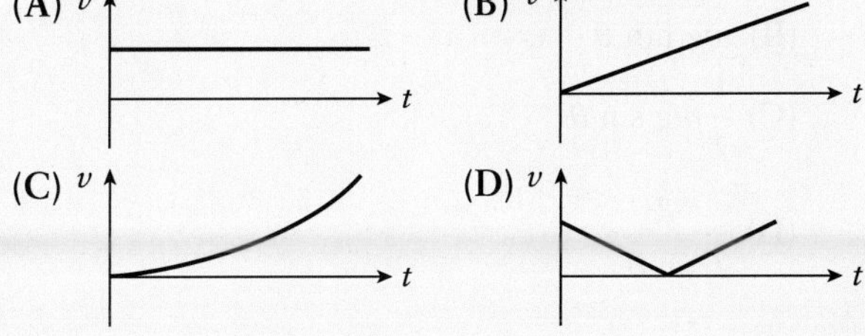

Questions 8–9 refer to the following diagram.

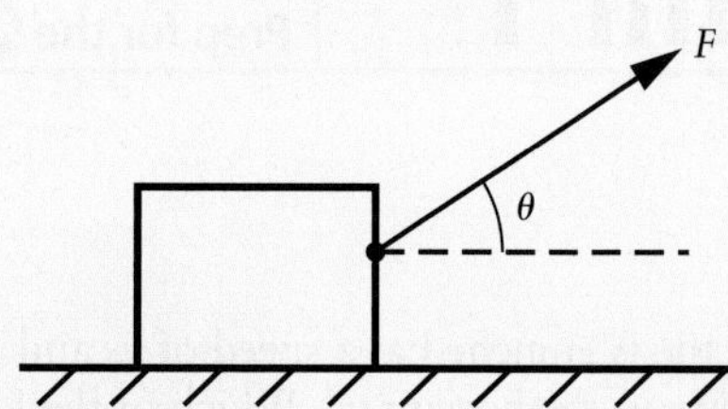

A force of magnitude F is exerted at an angle θ from the horizontal on a block of mass m and pulls it across a level floor through a displacement Δx.

8. The work done on the block by the force during this displacement is which of the following?

(A) $\frac{F\Delta x}{\sin\theta}$

(B) $\frac{F\Delta x}{\cos\theta}$

(C) $F\Delta x\cos\theta$

(D) $F\Delta x\sin\theta$

9. Assume the block is moving at a constant speed. Which of the following is the correct expression for the coefficient of kinetic friction between the block and the floor?

(A) $\frac{F}{mg}$

(B) $\frac{F\cos\theta}{mg}$

(C) $\frac{F\cos\theta}{mg - F\sin\theta}$

(D) $\frac{F\sin\theta}{mg - F\cos\theta}$

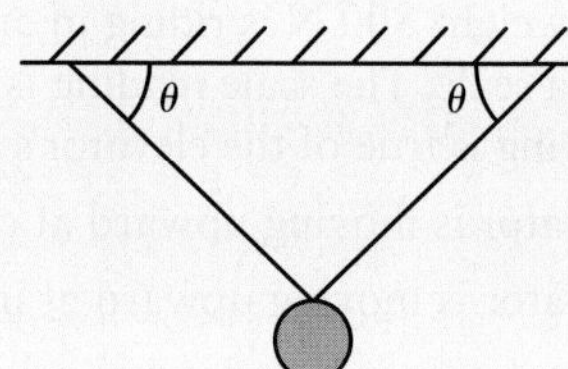

10. An object of mass m is hung from two light strings, as shown. Both strings make an angle θ with the ceiling. Which of the following is the correct expression for the tension in each string?

(A) $\frac{mg}{2\cos\theta}$

(B) $mg\tan\theta$

(C) $\frac{1}{2}mg\sin\theta$

(D) $\frac{mg}{2\sin\theta}$

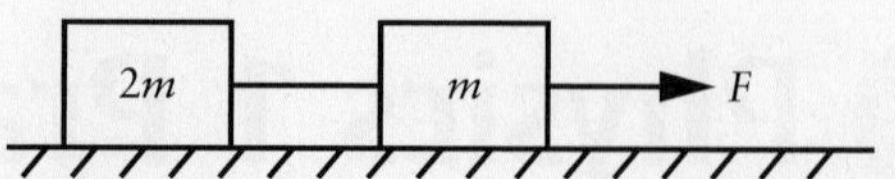

11. Two blocks are connected by a string of negligible mass. Block 1 has mass m and is pulled by a force of magnitude F that accelerates both blocks along a surface of negligible friction. Block 2 has mass $2m$. Which of the following is the correct expression for the tension in the string between the two blocks?

(A) $2F$

(B) $2F/3$

(C) $F/2$

(D) $F/3$

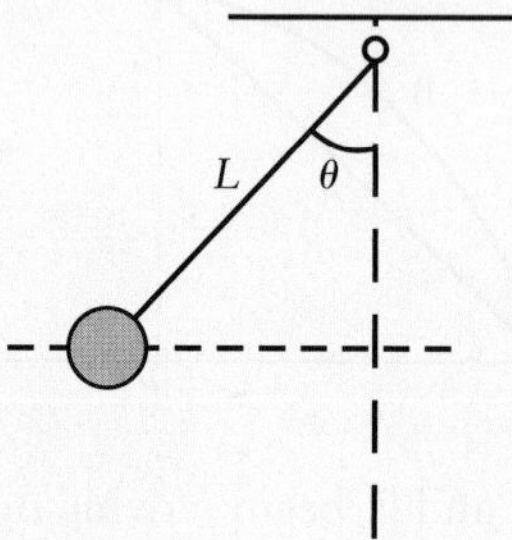

12. A pendulum of length L is pulled to the side at an angle θ measured from the vertical, as shown. The pendulum is released from rest. Which of the following is the correct expression for the speed of the pendulum at the lowest point in the swing?

(A) $\sqrt{2gL}$

(B) $\sqrt{2gL\cos\theta}$

(C) $\sqrt{2g(L - L\sin\theta)}$

(D) $\sqrt{2g(L - L\cos\theta)}$

Questions 13–14 refer to the following diagram.

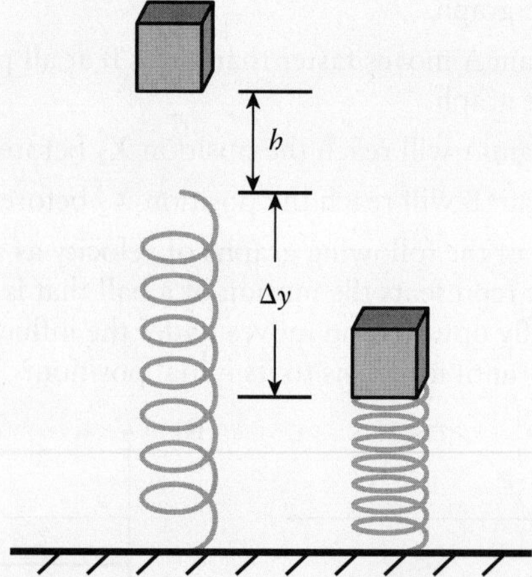

A block falls onto a vertical ideal spring from a height h above the relaxed spring and compresses the spring a maximum distance Δy before momentarily coming to rest. Air resistance and friction may be neglected.

13. Consider the motion of the block while it is in contact with the spring. At the point of maximum compression, which of the following must be true?
   (A) The speed of the block is at its maximum.
   (B) The acceleration of the block is at its maximum.
   (C) The elastic potential energy of the spring is at its minimum.
   (D) The kinetic energy of the block is at its maximum.

14. The spring expands upward and launches the block vertically back into the air. Which of the following is true after the block leaves the spring?
   (A) The block has no kinetic energy.
   (B) The block–Earth system has no gravitational potential energy.
   (C) The block will rise to exactly half the height from which it was dropped.
   (D) The block will rise to the same height from which it was dropped.

15. Student A lifts a 60-kg crate onto a truck bed 1 m high in 3 s. Student B lifts sixty 1-kg boxes onto the same truck in a time of 2 min. Which of the following statements is true?
   (A) Student A does more work than student B does.
   (B) Student B does more work than student A does.
   (C) The two students do the same amount of work, but the power generated by student A is greater.
   (D) The two students do the same amount of work, but the power generated by student B is greater.

16. A box of mass m and initial velocity v slides along a level floor. The box comes to rest after a time t. There is a constant friction force exerted on the box as it slides. The average rate at which the kinetic energy of the box is dissipated is given by which of the following expressions?
   (A) mvt
   (B) $\frac{mv^2}{2t}$
   (C) $\frac{mv}{t}$
   (D) $\frac{mv^2}{t}$

17. Two students can move a cart to the left or to the right on a long horizontal track with negligible friction. During the time intervals when the cart is accelerating, the acceleration is constant. Data are taken on the motion of the cart and recorded in the following table.

Velocity v (m/s)	Time t (s)
−4	0
−2	1
−2	2
−2	3
1	6
1	7
0	9
0	10

In which time interval is the magnitude of the acceleration the greatest?
   (A) 0 to 1 s
   (B) 3 to 6 s
   (C) 6 to 7 s
   (D) 7 to 9 s

18. Consider the following graphs of position x as a function of time t and velocity v as a function of time t.

In which of the graphs does the moving object reverse its direction?

(A) x
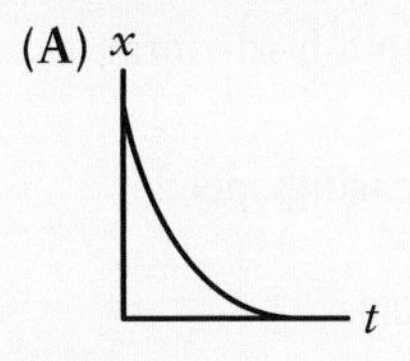

(B) v
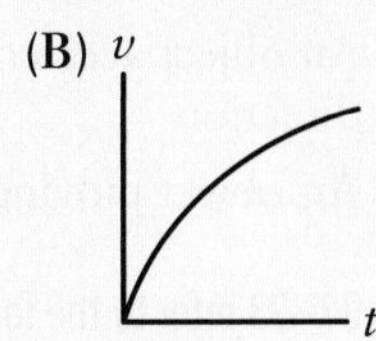

(C) v
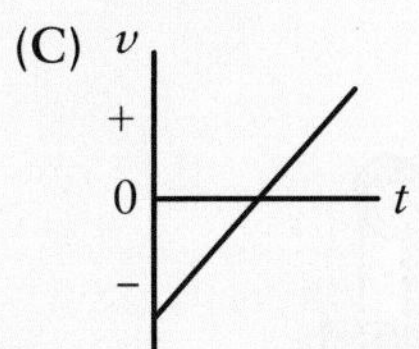

(D) x
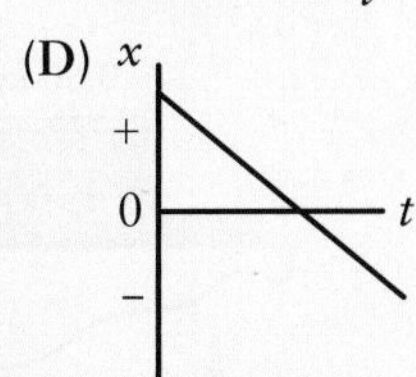

19. A ball launched horizontally with speed v from the top of a tower lands a distance D from the base of the tower. A second identical ball is launched horizontally from the top of the same tower with speed $2v$. How far does it land from the base of the tower?
   (A) D
   (B) $\sqrt{2}D$
   (C) $2D$
   (D) $4D$

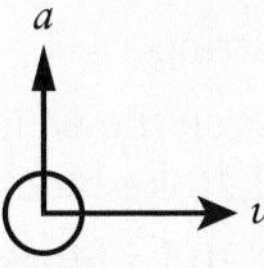

20. The velocity and acceleration vectors associated with the motion of an object are shown in the diagram. For the object under consideration, the upward direction is toward the top of the page. Which of the following could be true about the motion of the object?
   (A) The object could be traveling in a circular path at a constant speed.
   (B) The speed of the object must be increasing.
   (C) The object could be a projectile at the highest point in its trajectory.
   (D) The net force exerted on the object must be zero.

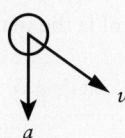

21. The velocity and acceleration vectors associated with the motion of an object are shown in the diagram. Which of the following could be represented by the diagram?

(A) An object moving at constant speed

(B) A projectile following a parabolic path

(C) An object accelerating only in a horizontal direction

(D) An object moving with decreasing speed

Questions 22–23 refer to the following diagram.

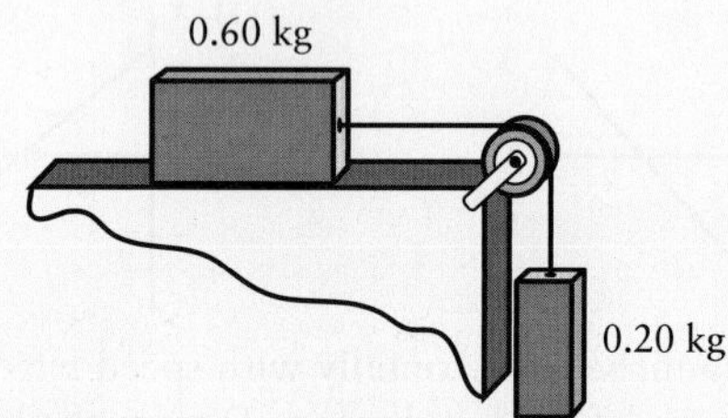

A 0.60-kg block on a surface of negligible friction is connected to a string that passes over a pulley of negligible mass and friction. The other end of the string is connected to a 0.20-kg block. The system is released from rest.

22. The net force exerted on the 0.60-kg block after it is released from rest is equal to which of the following?

(A) Weight of the 0.60-kg block

(B) Tension in the string

(C) Difference between the weights of the 0.60-kg block and the 0.20-kg block

(D) Weight of the 0.20-kg block

23. In terms of the acceleration due to gravity g, the magnitude of the acceleration of each block is given by which of the following?

(A) $\frac{g}{5}$ (B) $\frac{g}{4}$

(C) $\frac{g}{3}$ (D) g

Questions 24–25 refer to a student who is swinging a ball of mass *m* in a horizontal circle of radius *r* with a constant period *T* and a constant speed *v*.

24. During one complete period of the motion, the work done by the net force exerted on the ball is given by which of the following?

(A) $\frac{mv^2}{r}$ (B) $2\pi mv^2$

(C) $\frac{1}{2}mv^2$ (D) Zero

25. The student notices that he cannot keep the string perfectly horizontal as the ball moves in a circle. Which of the following is the best explanation for why the string cannot be perfectly horizontal?

(A) The student is not exerting enough centripetal force on the ball to keep the string horizontal.

(B) The centripetal force exerted on the ball can never be larger than the force of gravity exerted on the ball.

(C) The force of gravity exerted on the ball must be balanced by a vertical component of the tension for the vertical acceleration to be zero.

(D) The student is not able to swing the ball at a high enough speed to keep the string horizontal.

26. Two students want to determine the coefficient of static friction between a wooden block and a wooden board. They place the block in the middle of the board and begin slowly lifting one end of the board. The block just begins to slide when the board makes an angle θ with the horizontal. Which of the following would the students need to measure directly in order to calculate the coefficient of static friction between the block and the board?

(A) The mass of the block, the height of the block at the instant it begins to slide, and the length of the board

(B) Both the mass of the block and the length of the board

(C) Both the mass of the block and the angle

(D) Only the angle θ

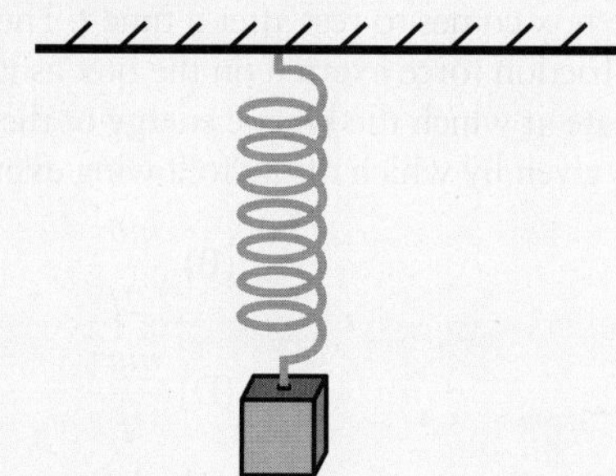

27. A block of mass m is attached to the end of a vertical spring at its equilibrium length, as shown. The spring is then compressed upward and then the block is released from rest. Let U_{grav} represent the potential energy of the block–Earth system, U_{spring} represent the elastic potential energy of the spring, and K represent the kinetic energy of the block. Immediately after the block is released, which of the following are increasing?

(A) K, U_{grav}, and U_{spring}

(B) K and U_{grav}

(C) K only

(D) U_{grav} and U_{spring}

28. A ball is released from rest and falls a distance h from a tower. Assume air resistance is negligible. Which of the following statements is true as the ball falls?

(A) The potential energy of the ball–Earth system is conserved.

(B) The kinetic energy of the ball is conserved.

(C) The difference between the potential energy of the ball–Earth system and kinetic energy of the ball is a constant.

(D) The sum of the kinetic energy of the ball and potential energy of the ball–Earth system is a constant.

29. Three students, A, B, and C, have designed a projectile launcher to hit a target on the ground. They launch the projectile from level ground at a speed of 10 m/s at an angle of 30° to the horizontal, but the projectile travels only halfway to the target before hitting the ground. Which of the students is correct on how they should adjust their projectile launcher to hit the target?

(A) Student A says that they should increase the launch speed to 20 m/s and keep the angle the same.

(B) Student B says they should increase the angle of launch to 60° and keep the launch speed the same.

(C) Student C says they should increase the launch speed to 14 m/s and keep the angle the same.

(D) None of the students is correct.

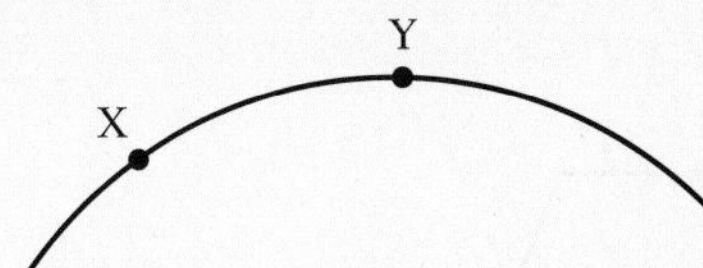

30. A ball is projected at an angle above the horizontal and follows a parabolic path, as shown in the figure. Point Y is the highest point on the path, and air resistance is negligible. Which of the following vectors best shows the direction of the acceleration of the ball at point X?

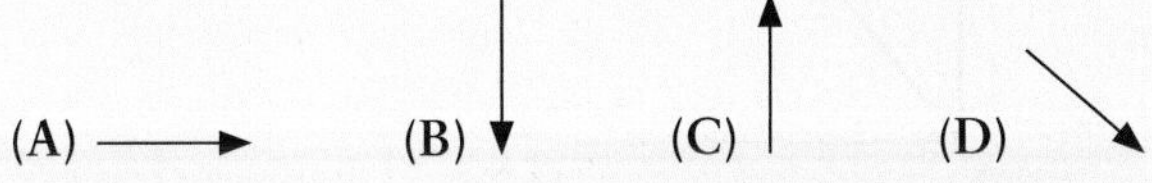

31. A satellite moves in a stable circular orbit with speed v_0 at a distance R from the center of a planet. For this satellite to move in a stable circular orbit a distance $2R$ from the center of the planet, the speed of the satellite must be

(A) $2v_0$

(B) $\sqrt{2}v_0$

(C) $\dfrac{v_0}{\sqrt{2}}$

(D) $\dfrac{v_0}{2}$

32. A ball with a mass of 0.500 kg and a speed of 6.0 m/s collides perpendicularly with a wall and bounces off with a speed of 4.0 m/s in the opposite direction. What is the magnitude of the impulse exerted on the ball?

(A) 1.0 Ns

(B) 2.0 Ns

(C) 5.0 Ns

(D) 10.0 Ns

33. An object starts with a velocity of 10 m/s and accelerates at $-2.0\ \text{m/s}^2$ for 3.0 s, then accelerates at $-1.0\ \text{m/s}^2$. What is the total distance the object travels before coming to rest?

(A) 11 m

(B) 29 m

(C) 52 m

(D) 75 m

Questions 34–35 refer to the following diagram.

A ball on the end of a string is being swung in a vertical circle at constant speed, as shown in the figure. Points I, II, and III are labeled on the circle.

34. Which of the following statements is true?

(A) The tension in the string is greater at point III than at point I.

(B) The tension in the string is greater at point I than at point III.

(C) The tension at point I is equal to the tension at point III.

(D) The tension in the string is greatest at point II.

35. If the string were to break at point II, what would be the initial direction of travel of the ball?

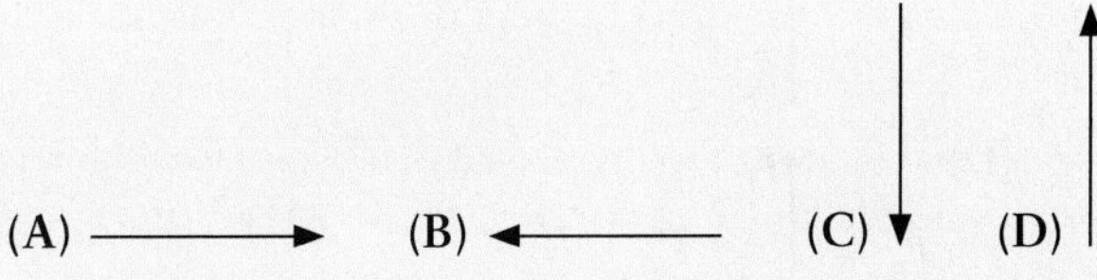

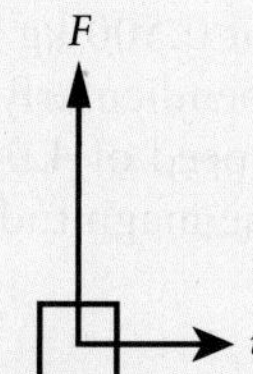

36. An object is moving to the right with a velocity $\boldsymbol{v}$ across a surface with negligible friction. A net force $\boldsymbol{F}$ is exerted on the object perpendicular to its direction of motion. Which of the following statements is true at the moment shown in the diagram?
(A) The object is moving with a constant velocity.
(B) The object is moving with a constant speed.
(C) The force $\boldsymbol{F}$ is doing positive work on the object.
(D) The force $\boldsymbol{F}$ is increasing the kinetic energy of the object.

Questions 37–38 refer to the following diagram.

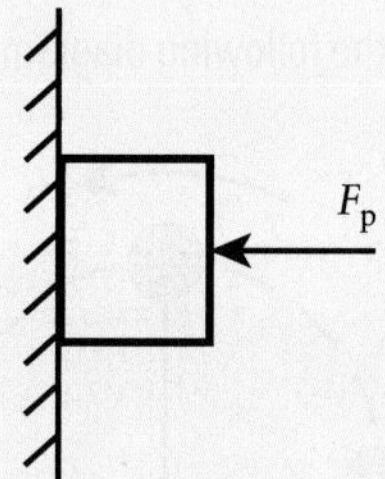

A force of magnitude F_p is exerted on a block of mass m, pushing it against a rough, fixed wall, as shown in the figure. The magnitude of the friction force between the block and the surface is F_f, the weight of the block is F_g, and the magnitude of the normal force exerted on the block is F_n.

37. Which of the following diagrams correctly shows all the forces exerted on the block?

(A)
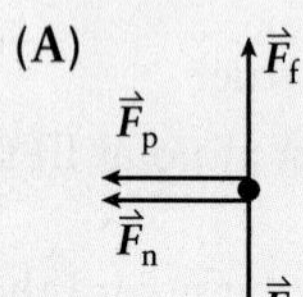

(B)
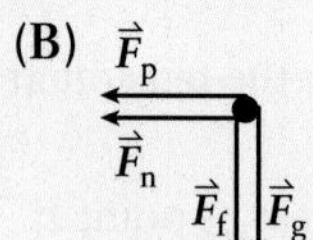

(C)
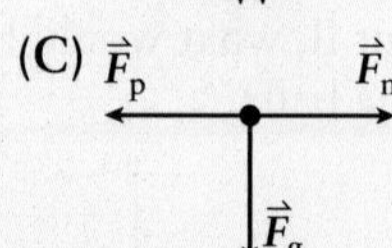

(D)
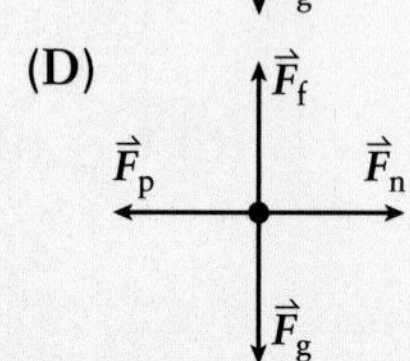

38. The force F_p is adjusted so that it is the minimum value required to keep the block in static equilibrium. In this situation, the magnitude of the force of friction exerted on the block is equal to the magnitude of the force
(A) F_p
(B) F_g
(C) F_n
(D) $F_n + F_p$

39. A block of unknown mass is initially at rest on a floor that exerts negligible friction on the block. A force $\vec{F}$ is exerted on the block for a time t and moves the block a horizontal displacement Δx. Which of the following graphs would allow you to determine the change in the kinetic energy of the block?

(A)
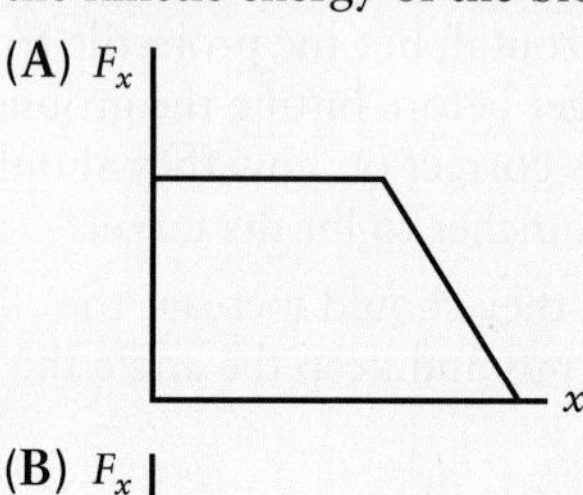

(B)
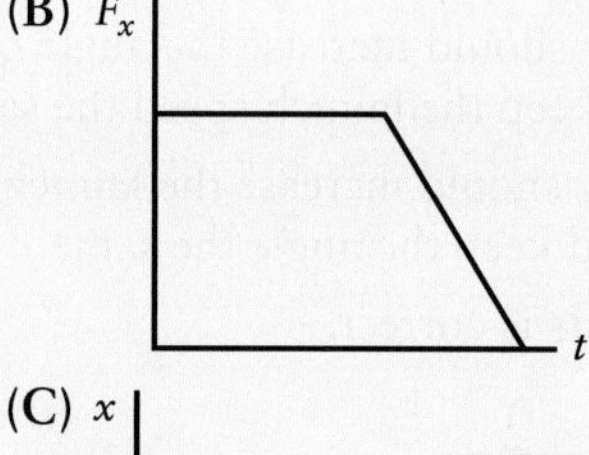

(C)
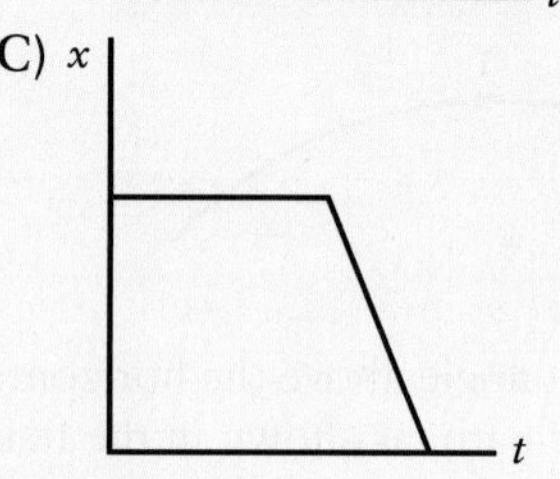

(D)
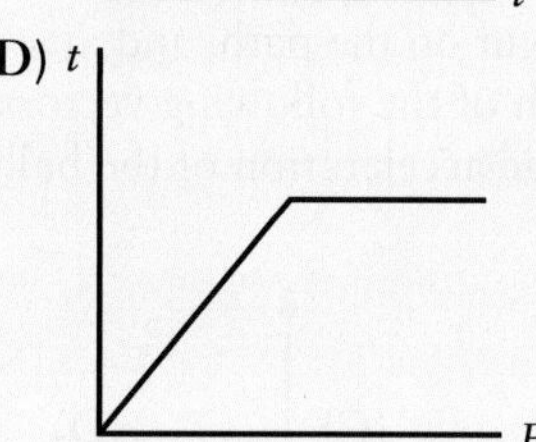

40. A dart of mass m moving with speed v_0 is fired into a foam block on top of a cart that can move with negligible friction. The foam–cart system has a mass M. Which of the following is a correct expression for the speed of the dart–foam–cart system after the collision?
(A) $\frac{M}{m}v_0$
(B) $\frac{m}{M+m}v_0$
(C) $\frac{M+m}{m}v_0$
(D) $\frac{m}{M}v_0$

41. A rubber ball and a ball of clay, of equal mass, are released from rest at the same height. They both strike the ground. The ball bounces up to half its initial height and the clay sticks to the ground. Which of the following statements is true?

(A) The clay had a greater change in momentum than the ball.

(B) The clay had a greater impulse exerted on it than the ball.

(C) The ball and the clay had the same change in momentum.

(D) The ball had a greater impulse exerted on it than the clay.

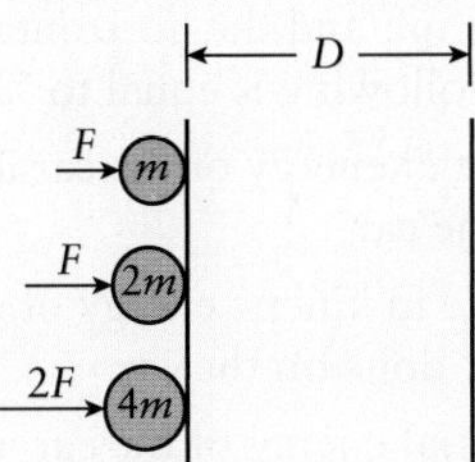

42. Three pucks are pushed by constant forces a distance D across the ice, as shown in the top view diagram. The pucks have masses m, $2m$, and $4m$, as shown, and are pushed by a force F or $2F$, as shown. All the pucks start from rest. Which puck has the greatest momentum after it has been pushed a distance D?

(A) The puck of mass m

(B) The puck of mass $2m$

(C) The puck of mass $4m$

(D) All the pucks have the same momentum.

43. The escape speed for a planet of mass M and radius R is v_{esc}. What is the escape speed for a planet of mass $3M$ and radius R?

(A) $\frac{v_{esc}}{3}$

(B) v_{esc}

(C) $\sqrt{3}v_{esc}$

(D) $3v_{esc}$

44. Two metal spheres attract each other with a gravitational force F. If the mass of one sphere is doubled, and the distance between the two spheres is doubled, what is the new force the two spheres exert on each other?

(A) $\frac{F}{2}$

(B) F

(C) $2F$

(D) $\frac{F}{4}$

45. A block of mass M is moving with speed v when it encounters a rough patch and stops in a time t_1. An identical block is moving with speed $2v$ when it encounters the same rough patch. What is the time it takes for this second block to stop?

(A) t_1

(B) $\sqrt{2}t_1$

(C) $2t_1$

(D) $4t_1$

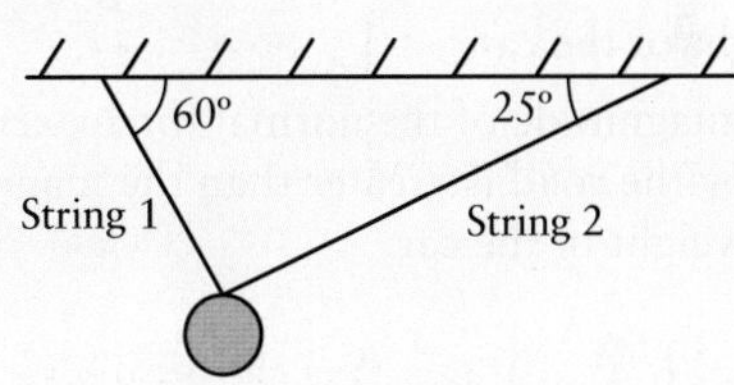

46. A ball hangs in equilibrium from two strings at the angles shown in the figure. String 1 is on the left and string 2 is on the right. Which of the following statements is true?

(A) The tension in string 1 is equal to the tension in string 2.

(B) The tension in string 2 is greater than the tension in string 1.

(C) The sum of the magnitudes of the two tensions is equal to the weight of the ball.

(D) The sum of the vertical components of the tensions in the strings is equal to the weight of the ball.

View of car from behind

47. A car is traveling on a rough, banked road in a circular path at a constant speed. A view from behind the car is shown. If the car were to go any faster, it would begin to slide outward from its circular path. Which of the following statements is correct?

(A) The net centripetal force exerted on the car is provided by only the normal force exerted on the car by the road.

(B) The centripetal force exerted on the car is provided by only the friction force exerted on the car by the road.

(C) The magnitude of the normal force exerted by the car on the road is equal to the magnitude of the weight of the car.

(D) The magnitude of the normal force exerted on the car by the road is greater than the magnitude of the weight of the car.

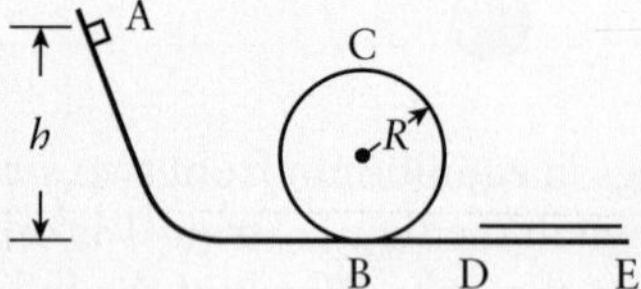

48. A small block of mass m begins at rest at the top of a curved track at height h and travels around a circular loop of radius R, as shown in the diagram. The track exerts negligible friction on the block from point A to point D, but there is friction on the track from point D to point E. The block comes to rest at point E. The track is level from point B to point E. Take point B as the zero of gravitational potential energy of the block–Earth system. Which of the following statements is true?

(A) The gravitational potential energy of the block–Earth system at point A is equal to the kinetic energy of the block at point D.

(B) The gravitational potential energy of the block–Earth system at point C is equal to the kinetic energy at point B.

(C) The magnitude of the work done by friction is less than the kinetic energy at point D.

(D) The centripetal force at point C is provided only by the normal force exerted by the track on the block.

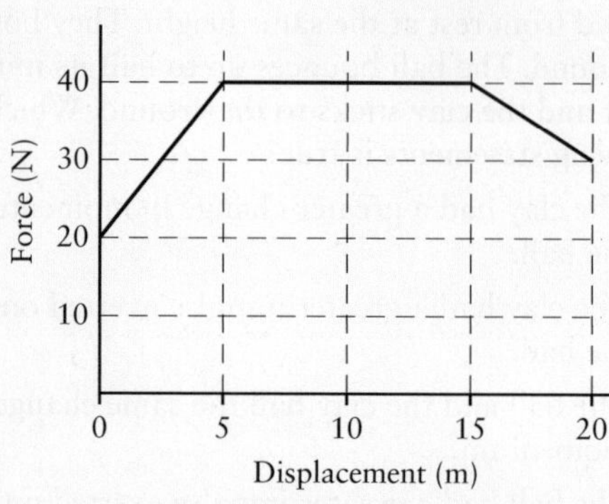

49. A variable horizontal net force is exerted on a toy car on a level track, as shown in the graph. The area between the graph and the horizontal axis is 725 Nm. Which of the following is equal to 725 Nm?

(A) The potential energy of the car and the total work done on the car

(B) The change in kinetic energy of the car and the total work done on the car

(C) The potential energy of the car and the change in kinetic energy of the car

(D) The difference in the work done by the force and the change in kinetic energy of the car

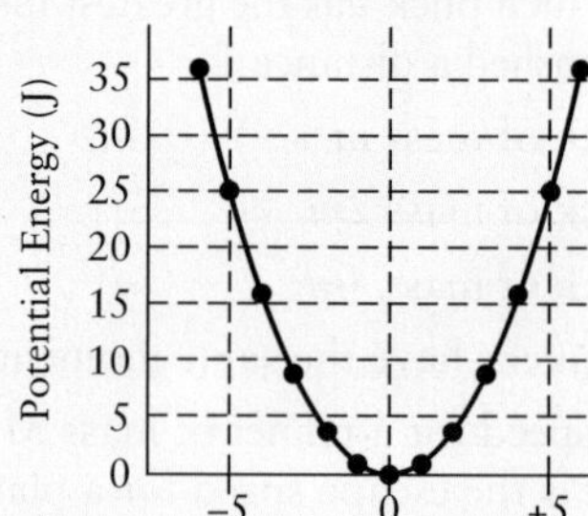

50. The elastic potential energy versus displacement x is shown for the system of a block oscillating at the end of an ideal horizontal spring. The maximum kinetic energy of the oscillating block is 25 J. Which of the following statements is true?

(A) The potential energy of the spring–block system at $x = -2$ m is 6 J.

(B) The block does not have enough energy to reach the position $x = 6$ m.

(C) The kinetic energy of the block at $x = -5$ m is 25 J.

(D) The maximum kinetic energy of the block occurs at $x = 5$ m.

Free-Response Questions

Question 1

Block 1 of mass m_1 is connected by a light string that passes over a pulley of mass m_1 and radius r, to block 2 of mass $m_2 (m_2 > m_1)$, as shown. The system is released from rest and the blocks are allowed to accelerate.

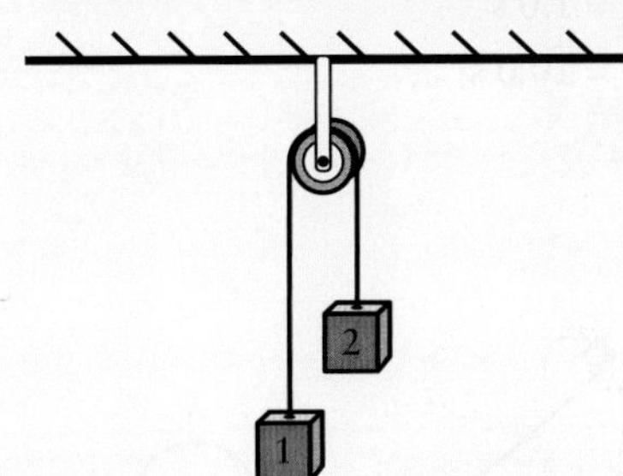

(a) The dots drawn in the following diagram represent blocks 1 and 2. Draw free-body diagrams showing and labeling forces (not components) exerted on each block as it accelerates. Draw the relative lengths of all vectors to reflect the relative magnitudes of all forces.

Block 1 Block 2

(b) Derive an expression for the acceleration of block 1 in terms of the given quantities and fundamental constants, as appropriate.

(c) Consider the system including the two blocks, the string, and the pulley. Briefly describe the work done on the system and the changes in energy of the system as block 2 descends. In your description, justify the sign of work and why the acceleration of the two blocks must be less in this case than if the pulley had negligible mass.

Question 2

Students are asked to use circular motion to measure the coefficient of static friction between two materials. They have a round turntable with a surface made from one of the materials, for which they can vary the speed of rotation. They also have a small block of mass m made from the second material. A rough sketch of the apparatus is shown in the figure below. Additionally they have equipment normally found in a physics classroom.

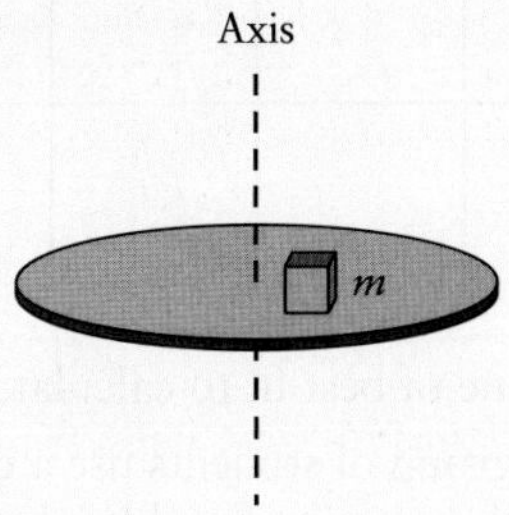

(a) Briefly describe a procedure that would allow you to use this apparatus to calculate the coefficient of static friction, μ.

(b) Based on your procedure, determine how to analyze the data collected to calculate the coefficient of friction.

(c) One group of students collects the following data.

r (m)	f_m (rev/s)		
0.050	1.30		
0.10	0.88		
0.15	0.74		
0.20	0.61		
0.25	0.58		

i. Use the empty spaces in the table as needed to calculate quantities that would allow you to use the slope of a line graph to calculate the coefficient of friction, providing labels with units, as in the first two columns.

ii. On the axes below, plot the data you indicated, and draw a line of best fit.

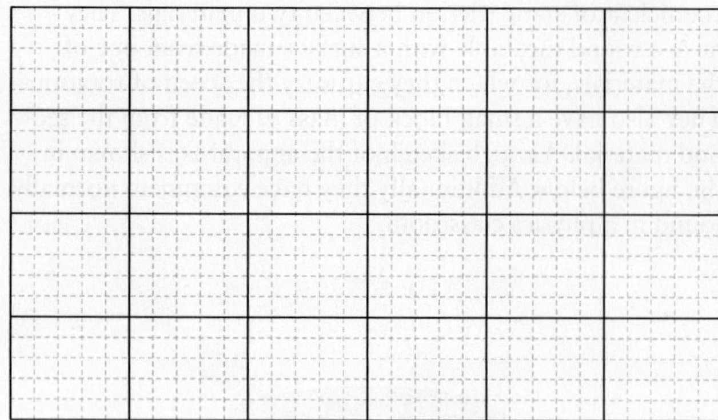

(d) Use the line of best fit to calculate the value of μ.

(e) Another group of students use a different block made of the same material but having a greater mass than that of the first group. What are the differences, if any, between the new graph and its best-fit line, and the first group's? Briefly justify your answer.

Question 3

The velocity of an elevator is given by the graph shown. Assume the positive direction is upward.

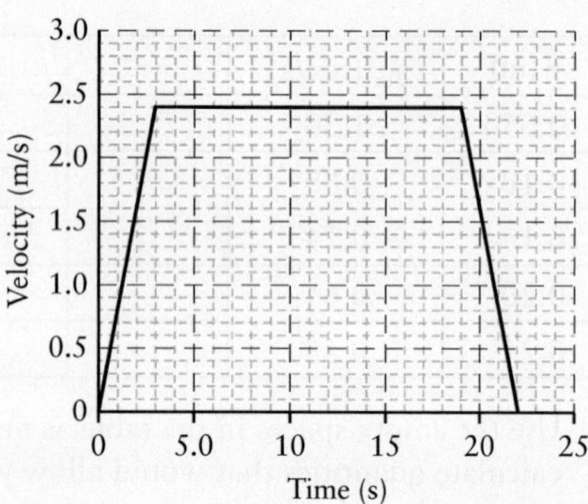

(a) Briefly describe the motion of the elevator. Justify your description with reference to the graph.

(b) Assume the elevator starts from an initial position of $y = 0$ at $t = 0$. Deriving any numerical values you need from the graph:

i. Write an equation for the position as a function of time for the elevator from $t = 0$ to $t = 3.0$ seconds.

ii. Write an equation for the position as a function of time for the elevator from $t = 3.0$ seconds to $t = 19$ seconds.

(c) A student of weight mg gets on the elevator and rides the elevator during the time interval shown in the graph. Consider the force of contact, F_n, between the floor and the student. How does F_n compare to mg at the following times? Justify your answer with reference to the graph and your equations above.

i. $t = 1.0$ s

ii. $t = 10.0$ s

Question 4

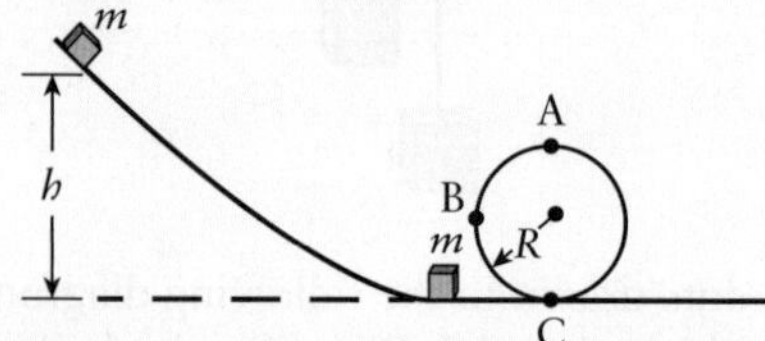

A block of mass m slides down a ramp of height h and collides with an identical block that is initially at rest. The two blocks stick together and travel around a loop of radius R without losing contact with the track. Point A is at the top of the loop, point B is at the end of a horizontal diameter, and point C is at the bottom of the loop, as shown in the figure above. Assume that friction between the track and blocks is negligible.

(a) The dots below represent the two connected blocks at points A, B, and C. Draw free-body diagrams showing and labeling the forces (not components) exerted on the blocks at each position. Draw the relative lengths of all vectors to reflect the relative magnitude of the forces.

Point A | Point B | Point C

(b) For each of the following, derive an expression in terms of m, h, R, and fundamental constants.

i. The speed of moving block at the bottom of the ramp, just before it contacts the stationary block

ii. The speed of the two blocks immediately after the collision

iii. The speed of the two blocks at the top of the loop (point A)

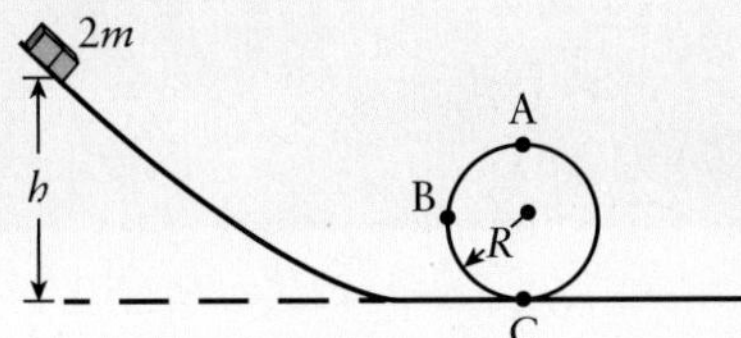

(c) In the scenario shown above, the blocks are connected and the combination of two blocks is released from rest at the top of the incline. The incline can be modeled as a ramp of length L at an angle θ from the horizontal, although the curve still exists to ensure speed only changes direction at the bottom of the incline. On the axis below, sketch a graph of v_C as a function of h. Label limiting values on the horizontal axis.

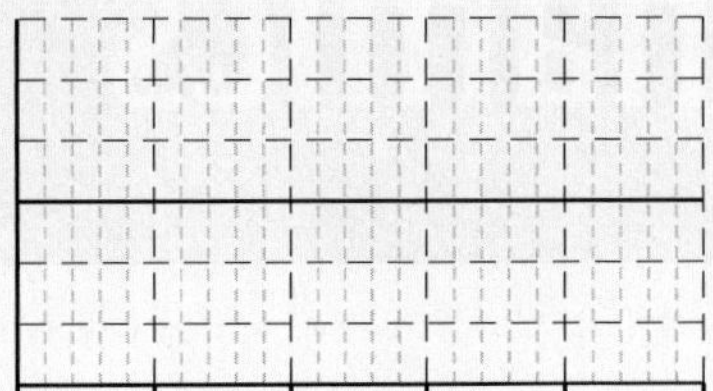

(d) In which scenario—the colliding blocks in (b) or the connected blocks in (c)—is the minimum initial height required for the blocks to complete the circle least? Briefly justify your answer referring to your free-body diagrams and equations as needed.

Torque and Rotational Dynamics

Chapter 10 Rotational Motion I: A New Kind of Motion

Anna Jackowska/Shutterstock

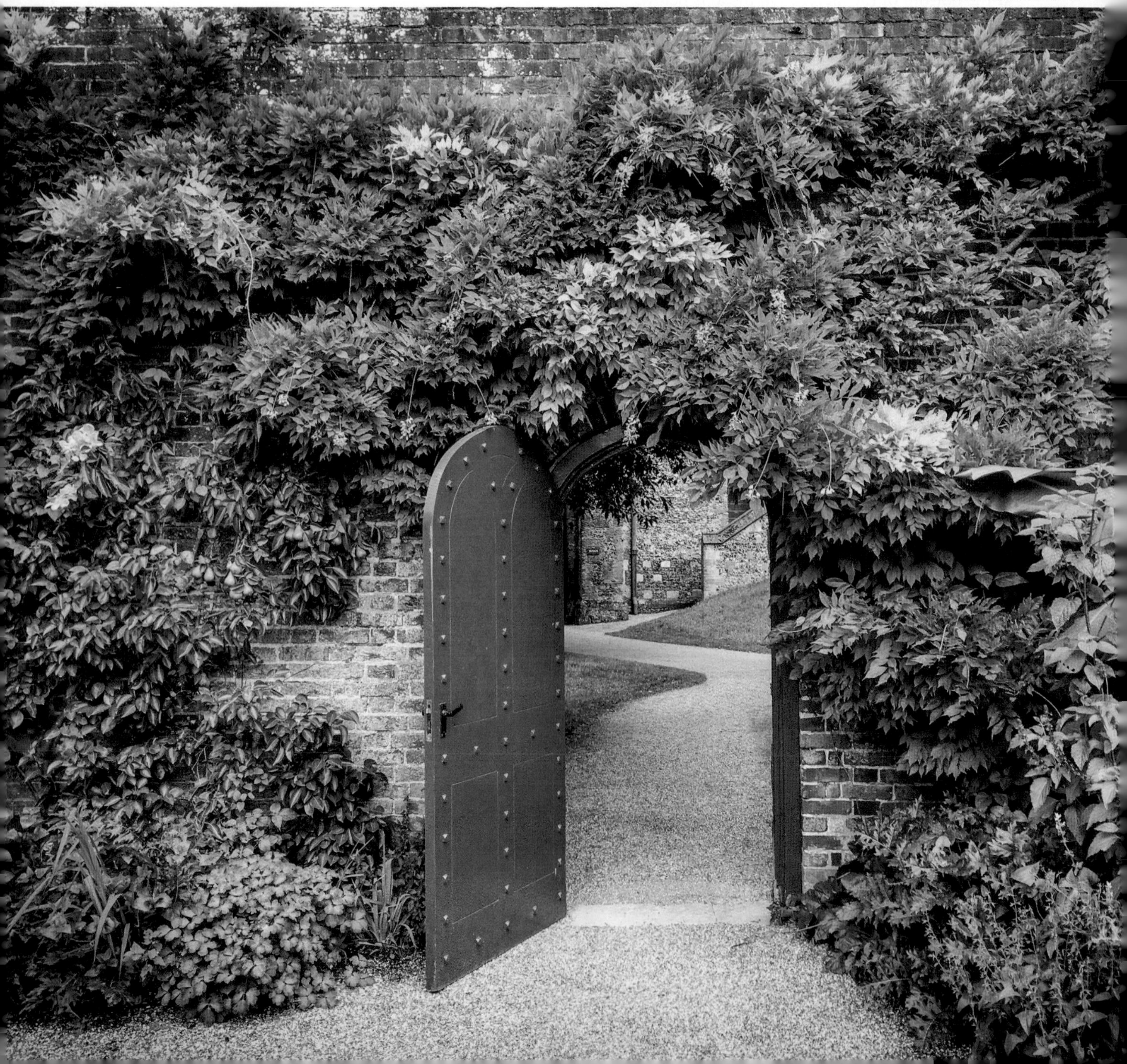

Case study: How to make it easier to open a new door?

To this point in the book, we have focused on things we could model as an object, at least for most of their motion. We saw in a collision that we had to relax the model during the interaction, but it was fine to use it before and after the interaction. Even for something traveling in a circle, we found this model was satisfactory, as long as the only important consideration was where the center of mass was. For example, if we are interested in Earth traveling around the Sun, this approach works well. However, if we want to understand day and night, we would be sunk! So, as we start our study of rotation, things are going to be different. Even in our everyday activities, there are many situations in which we have to move beyond our object model. With this unit, we are opening a new door!

When you push on a door and it swings open you might notice that it doesn't bend. The whole thing moves together rigidly, holding its shape. However, the edge of the door closest to the hinges barely moves, while the edge of the door opposite the hinges moves rapidly. This observation makes sense if we recall Chapter 6, where we saw that the arc length traveled depended on the angle swept out and the radius of the path. That is, if the angle swept out was the same (the whole door moves together), the farther you are from the hinge, the larger the radius and the faster the segment under consideration must move to go the same angle in the same amount of time. This also means that it is not possible to use the object model, which only holds for situations for which every point in the system has a motion that is the same as that of the center of mass. But, since the door swings as one unit, it seems like we should be able to find a simple way to describe it, and we will in this chapter.

Let's think about what happens when you push on a door. The door handle is usually on the side of the door opposite the hinges. Why is that true? If you push on the door with exactly the same force at different distances from the hinges, it responds differently. How can we describe that?

By the end of this chapter, you should be able to explain the connections between rotational and linear motion. You should be able to define and explain the relationships among angular displacement, angular velocity, angular acceleration, torque, and rotational inertia. Additionally, you should be able to use these quantities to describe the motion of a system, and to calculate the rotational motion of objects traveling in curved paths or that of systems that hold their shape, like the door pictured. You will be able to use these concepts to explain why most doors are designed with the handle farther from the hinges!

10 Rotational Motion I: A New Kind of Motion

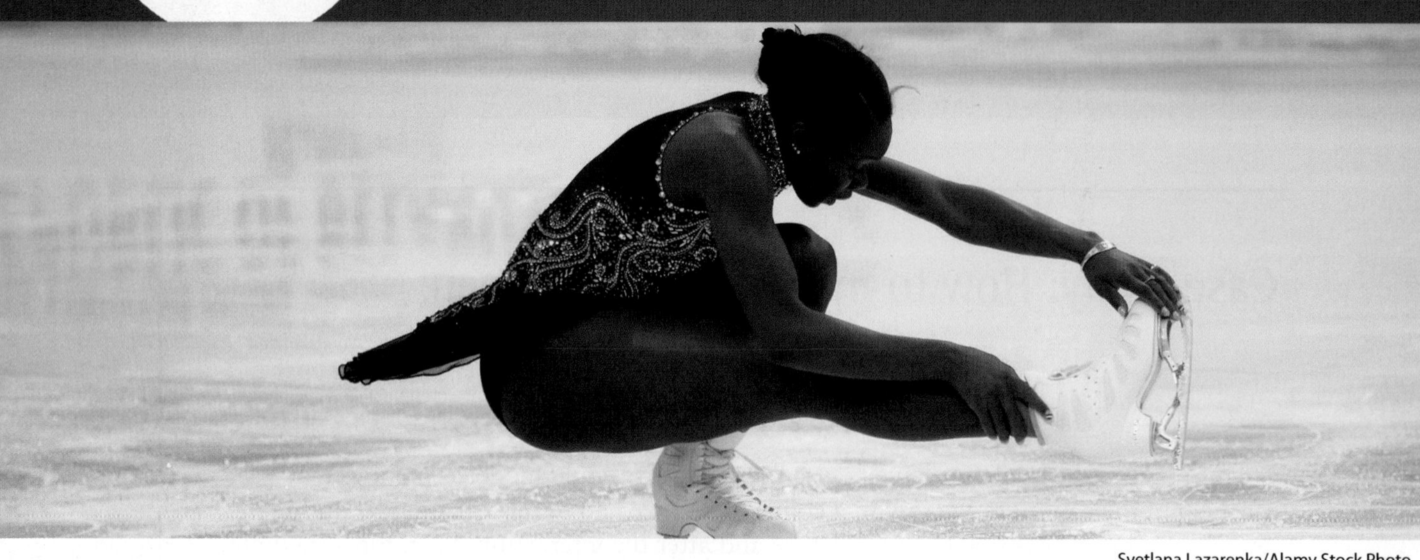

Svetlana Lazarenka/Alamy Stock Photo

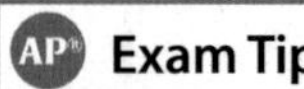

Exam Tip

The coverage of rotation is compact! Because of the similarities between rotational kinematics and dynamics and their linear counterparts, these topics have been condensed into a single chapter. Building on the similarities to linear motion will improve your understanding of rotational motion, and you must be able to explain these similarities on the exam. We defined kinematics for linear motion in Unit 1 and dynamics for linear motion in Unit 2.

YOU WILL LEARN TO:

- Define the rotational motion of a system and describe when and how a system can be simplified to an object model.
- Define and explain the relationships among angular displacement, angular velocity, and angular acceleration and use these quantities to describe the rotation of an object or a system of rigidly connected objects, which is a rigid body.
- Explain what is meant by the rotational inertia of a system and how to use it to calculate the rotational kinetic energy of that system.
- Describe the techniques for finding the rotational inertia of a rigid body, including use of the parallel-axis theorem.
- Explain the similarities between rotational and linear motion, and describe how the two forms of motion correspond for a point on a rotating rigid system.
- Solve constant-acceleration angular motion problems when given a description of the motion of an object or a system. The description could be in the form of words, graphs, or experimental data.
- Define the concept of lever arm and know how to use it to calculate the torque caused by a force exerted on a rigid body.
- Define rotational equilibrium and calculate it for a rigid body.
- Explain how to find the direction of angular velocity and torque.

10-1 Rotation is an important and ubiquitous kind of motion

NEED TO REVIEW?

Turn to the **Glossary** in the back of the book for definitions of bolded Key Terms.

Up to this point we've considered the motions of objects such as a thrown baseball, a speeding car, and a soaring bird. We use the word **translation** to refer to these kinds of motion, in which an object or some rigid system that could be treated as an object (all points on it move in the same way as its center of mass) moved from one point in space to another. However, we haven't addressed the kind of motion in which a system spins around an axis so that not all points move through the same distance and in the same direction. One example of such a system is the figure skater in the chapter-opening photo. We will refer to such a system, in which the pieces are connected but do not

(a)

A Ferris wheel rotates around a fixed (stationary) axis.

(b)

This mother-of-pearl moth caterpillar (*Pleurotya ruralis*) wraps itself into a circle for self-defense: It can roll away from predators 40 times faster than it can walk. As it rolls, the caterpillar moves as a whole (it translates) as well as rotates.

(c)

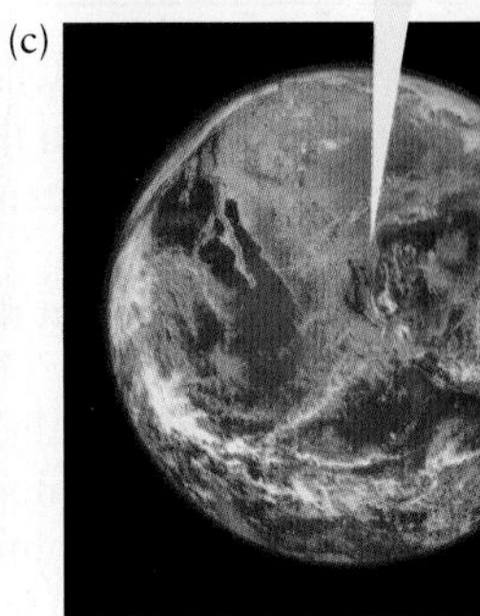

Our planet rotates to the east once per day around an axis that extends through the planet from the north pole to the south pole. Earth also revolves around an external axis, held in a nearly circular path by the force of gravity exerted on it by the Sun. It takes one year for Earth to move in an orbit around the Sun.

(a) Holly Kuchera/Getty Images; (b) Dr. John Brackenbury/Science Source; (c) NASA/NOAA/GSFC/Suomi NPP/VIIRS/Norman Kuring

Figure 10-1 Rotational motion Three examples of rotating extended objects.

have to move together rigidly, as an **extended object**. (The skater could swing her arms around more quickly than her torso is rotating, for instance.) If we look instead at the Ferris wheel in **Figure 10-1a,** we see a system where the objects are rigidly connected. It is common to call such a system a **rigid body**. Motion of extended objects (including the special case of rigid bodies) about an axis is called **rotation**. The figure skater rotates around an axis that extends upward from the skate that touches the ground. The Ferris wheel rotates about a horizontal axis through its center. Both are examples of rotational motion. (**Figures 10-1b** and (**10-1c** show two more examples.) In this chapter, we'll investigate this type of motion.

In some situations, an extended object undergoes rotation but not translation. An example is the Ferris wheel shown in Figure 10-1a: The Ferris wheel's center of mass remains at the same location in the amusement park, but the wheel rotates continuously to entertain its riders. Other extended objects that rotate without translation are a spinning ceiling fan and the rotating platter in a microwave oven. In other situations, like a rolling caterpillar (Figure 10-1b) or the wheels of a fast-moving bicycle, an extended object undergoes rotation and translation at the same time. You've spent your entire life on a planet that experiences both rotation and translation (Figure 10-1c). In this chapter, we'll look at rotation both without and with translation.

NEED TO REVIEW?
Turn to Chapter 9, Section 9-7, for a quick review of center of mass.

We saw in Chapters 7 and 8 that it can be easier to describe motion in terms of work and energy than in terms of forces. With this in mind, let's first ask the following question: How can we find the kinetic energy due to an extended object's rotation—that is, its *rotational kinetic energy*? In Chapter 11, we'll use conservation of energy to solve many problems involving rotational motion. In this chapter, we'll begin with developing an understanding of the similarities between translational and rotational motion. So we will first approach these problems using dynamics. We will find that *torque* and *angular acceleration* bear the same relationship to rotational motion that force and acceleration bear to translational motion.

THE TAKEAWAY for Section 10-1

✔ Translation refers to the motion of the center of mass of a system (or an object) from one point to another through space.

✔ Rotation refers to the spinning motion of a system of connected objects (an extended object).

✔ An extended object can rotate around a fixed axis (rotation without translation), or its axis can move as the extended object rotates (rotation with translation).

Prep for the AP Exam

AP® Building Blocks

1.

The figure shows five views of a wheel as it rolls at a constant rate. There is a rock stuck in the wheel. Respond to each question using complete sentences that include your reasoning.

(a) In what direction is the hub at the center of the wheel translating?

(b) If the radius of the wheel is 1 m, what is the angular displacement of the rock along the circular path between the leftmost and rightmost views? Use the same convention usually used in math class: The counterclockwise direction is positive.

(c) If the time between one view of the wheel and the next is 0.125 seconds, what is the time for one complete revolution?

(d) If the wheel does not slip as it rolls, what is the translational speed of the hub, assuming the rotation given in part (c)?

(e) Is the translational direction of the rock at the top of the wheel different or the same as the translational direction of the rock at the bottom of the wheel?

(f) How does the translational velocity of the rock at the top of the wheel compare with the translational velocity of the hub?

AP® Skill Builder

2.

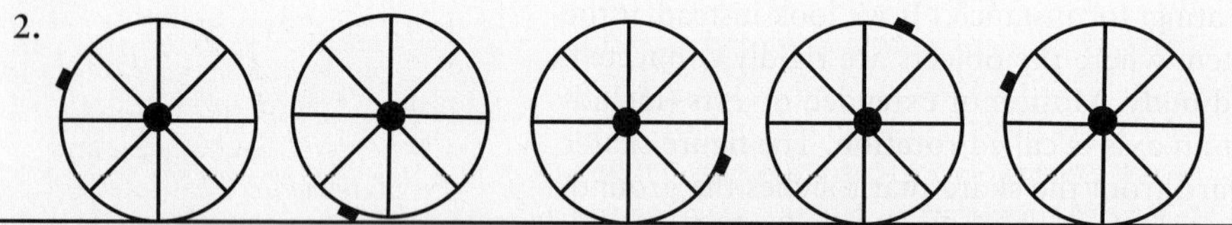

Explain why the sequence of views of a rolling wheel with a rock stuck on the rim as shown, is not possible if the wheel does not slip as it rolls.

10-2 An extended object's rotational kinetic energy is related to its angular velocity and how its mass is distributed

To simplify things, we'll begin our study of rotational motion by considering rotation without translation. We'll also limit ourselves to rigid bodies—that is, extended objects whose shape doesn't change as they rotate. While the general ideas apply to any extended object, the equations we will come up with assume the extended object keeps the same shape as it rotates, such as the spinning wind turbines shown in **Figure 10-2.** (By contrast, the figure skater in the photograph that opens this chapter can change her shape as she spins.) We'll also assume that the rotation axis is *fixed*—that is, it keeps the same orientation in space. The wind turbines in Figure 10-2 rotate around fixed axes.

Angular Velocity and Angular Speed

NEED TO REVIEW? Turn to the Math Tutorial, Section M8, in the back of the book for more information on trigonometry.

Let's see how to determine the rotational kinetic energy of the turbine shown in Figure 10-2. We'll begin by considering the motion of one of its blades, shown in **Figure 10-3.**

Recall from Section 7-3 that an object's kinetic energy depends on its mass m and its speed v: $K = \frac{1}{2}mv^2$. However, we *cannot* use this equation directly to find the

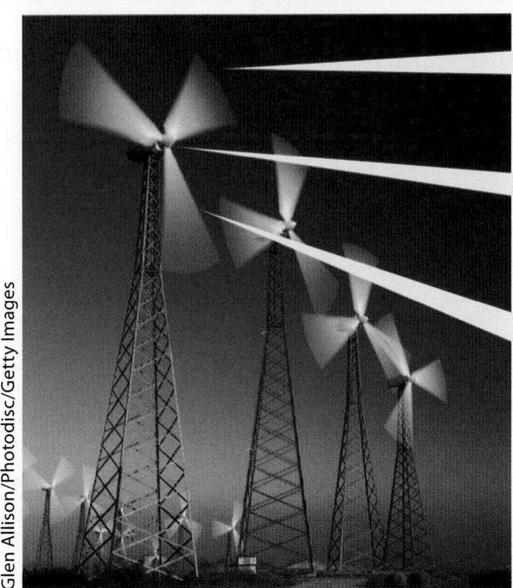

Glen Allison/Photodisc/Getty Images
Each blade of this rotating wind turbine is rigid: It maintains the same shape at all times.
The wind turbine rotates around a fixed axis: The axis remains in the same place and keeps the same orientation.
While all parts of the blade rotate together, different parts move at different speeds: The blurring shows that the outer parts (at a greater distance r from the axis) move more quickly than the inner parts of the blade.

Figure 10-2 A rotating rigid body The blades of a wind turbine undergo rotation but do not undergo translation (the wind turbine's center of mass doesn't go anywhere).

rotational kinetic energy of a turbine blade because that kinetic energy is associated with the motion of the center of mass of a system (or the motion of an object). The center of mass of the turbine is not moving, and the points on the blade move at different speeds around it! We need to somehow add up the motion of all of the pieces of the rigid body. During a time interval Δt, the entire blade rotates through an angle $\Delta\theta$ (pronounced "delta-theta"), the **angular displacement** of the blade. During this time interval, if the time interval is short, a point near the rotation axis travels only a short distance, while a point farther from the axis travels a greater distance. Hence a point farther out along the blade moves faster and has a greater speed than a point closer to the axis. You can see this in Figure 10-2: The blade is more blurred the farther it is from the axis. So the speed v is *different* for different parts of the blade.

To see how to account for different speeds at different locations, let's look in more detail at how the blade moves during a time interval Δt (Figure 10-3). Let's choose angles to increase in the counterclockwise direction; the blade in Figure 10-3 rotates counterclockwise, so with this choice the angular displacement $\Delta\theta$ in Figure 10-3 is

NEED TO REVIEW?
All of the terms discussed in this section, such as *average*, *constant*, and *instantaneous*, refer in the same way to rotational quantities as they do to translational ones. Turn to Chapter 2, Sections 2-2 and 2-3, to review.

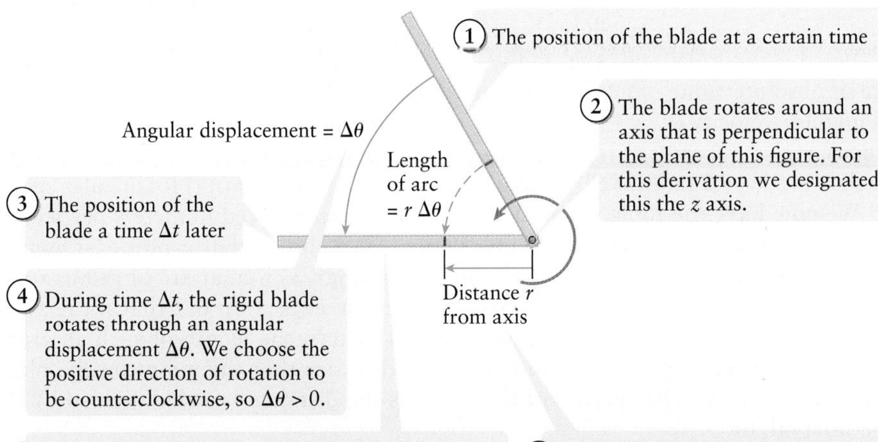

(1) The position of the blade at a certain time
(2) The blade rotates around an axis that is perpendicular to the plane of this figure. For this derivation we designated this the z axis.
Angular displacement = $\Delta\theta$
Length of arc = $r\,\Delta\theta$
(3) The position of the blade a time Δt later
Distance r from axis
(4) During time Δt, the rigid blade rotates through an angular displacement $\Delta\theta$. We choose the positive direction of rotation to be counterclockwise, so $\Delta\theta > 0$.
(5) The **angular velocity** ω_z of the blade is the angular displacement about the z axis $\Delta\theta$ divided by the elapsed time Δt:
$\omega_z = \frac{\Delta\theta}{\Delta t}$
The **angular speed** ω is the absolute value of ω_z.
(6) In time Δt, a point on the blade a distance r from the axis moves through an arc of length $r\,\Delta\theta$. The speed of this point is $v = r\omega$.

Figure 10-3 Angular velocity We define angular velocity by analogy to the way that we defined velocity for linear motion in Chapter 2.

positive. By analogy to how we defined average velocity for linear motion in Section 2-2, we define the average **angular velocity** of the blade to be

EQUATION IN WORDS
Average angular velocity

Average angular velocity of a rigid body over a time interval (can be + or −)

Angular displacement of the rigid body during the time interval (can be + or −)

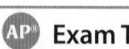

(10-1) $$\omega_{\text{average},z} = \frac{\Delta\theta}{\Delta t}$$

Duration of the time interval

AP® **Exam Tip**

The AP® Physics 1 equation sheet does not have average angular velocity (Equation 10-1). Rather than treating this as an equation to memorize, it is best to understand how average and instantaneous angular velocity are defined, and how they are related to each other.

Here ω is the Greek letter omega, and the z in the subscript tells us that the blade is rotating around an axis that we designated the z axis (Figure 10-3). For any rotating rigid body, the value of $\omega_{\text{average},z}$ is the same for all points on the rigid body. The average angular velocity can be positive or negative, depending on the direction in which the rigid body rotates. A common choice is to take counterclockwise rotation to be positive and clockwise rotation to be negative (this is based on some vector definitions we will learn in Section 10-5), but most of the time the choice is up to you, as long as you are consistent.

WATCH OUT !

Where did the *z* come from?

If you consider a tabletop to be an x–y plane, then to get a pencil to spin on the tabletop, you would have to make it rotate about an axis perpendicular to the tabletop, which would be the z direction. The axis of rotation will always be perpendicular to the plane in which the motion takes place, so our use of z implies the motion is taking place in the x–y plane.

Just as we did for ordinary velocity in Section 2-3, we'll define the *instantaneous* angular velocity ω_z, or just angular velocity for short, as the rate at which the rigid body is rotating at a given instant. That is, the instantaneous angular velocity is the average angular velocity for a very short time interval.

The preferred units of angular velocity are radians per second, or rad/s. As we discussed in Section 6-2, radians are a measure of angle; there are 2π radians in a circle. We can also describe angle in terms of the number of revolutions (rev) that a rigid body makes, so another common set of units for angular velocity is revolutions per second (rev/s). Since one revolution equals 2π radians, 1 rev/s = 2π rad/s.

 **Exam Tip**

Although rotational quantities are vectors just like their translational counterparts, you will never be required to do rotation about more than one axis at a time on the AP® Physics 1 exam, so you will need to think only in terms of clockwise or counterclockwise to define the vector direction. Imagine you are looking straight down the axis of rotation toward the plane the motion is in. If the rotation is going about the axis the same way the hands of a clock proceed around the clock's face, the motion is clockwise. If the rotation is instead in the opposite direction (going left at the 12 on the clock face), the motion is counterclockwise.

We also use rad/s and rev/s as the units for **angular speed**, which is the magnitude or absolute value of angular velocity. A rigid body that rotates 5.0 radians in 1 s has angular velocity $\omega_z = +5.0$ rad/s if it rotates in the positive direction and angular velocity $\omega_z = -5.0$ rad/s if it rotates in the negative direction, but in either case its angular speed is 5.0 rad/s. We'll use the symbol ω (with no subscript) for angular speed.

We now have the tools we need to find the *ordinary* speed in meters per second (m/s) of a point on the rotating blade. Figure 10-3 shows that a point (shown in red) a distance r from the rotation axis moves through a circular arc of radius r and angle $\Delta\theta$, which we've taken to be positive. We saw in Section 6-2 that if the angle $\Delta\theta$ is measured in radians and r is measured in meters, the length of such an arc in meters is $r\,\Delta\theta$ (see Figure 6-3). So $r\,\Delta\theta$ is the distance that this point on the blade travels in a time Δt. The speed v of this point is the distance traveled divided by the time interval, or

(10-2) $$v = \frac{r\Delta\theta}{\Delta t}$$

But $\Delta\theta/\Delta t$ is just the angular speed ω of the rotating blade. (Since the blade is rotating in the positive direction, its angular velocity $\omega_z = \Delta\theta/\Delta t$ is positive and so the angular speed ω—which is always positive—is the same as ω_z.) So we can rewrite Equation 10-2 for the speed of a point on the blade as

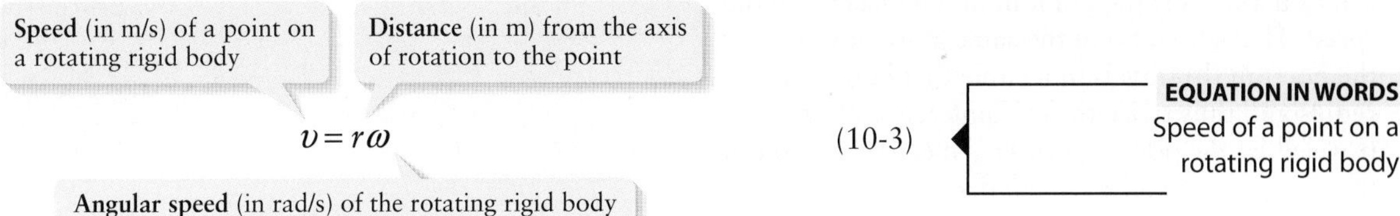

$$v = r\omega \quad (10\text{-}3)$$

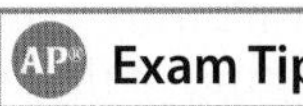

EQUATION IN WORDS
Speed of a point on a rotating rigid body

WATCH OUT!

Use correct units when relating speed and angular speed.

We derived Equation 10-3 from the statement that the length of the arc in Figure 10-3 is $r\,\Delta\theta$. This is true *only* if the angle $\Delta\theta$ is measured in radians (see Section 6-2). Hence you can safely use Equation 10-3 only if the angular speed ω is measured in radians per second (rad/s). If you are given the angular speed in other units, such as revolutions per second (rev/s), revolutions per minute (rev/min), or degrees per second, you must convert it to rad/s before you can use Equation 10-3.

AP® Exam Tip

Equation 10-3 is not on the AP® Physics 1 equation sheet. It is an important equation to understand in order to determine the relationship between the rotational and translational kinematic quantities.

We derived Equation 10-3 assuming that the rigid body is rotating in the positive direction. However, it's equally valid if the rigid body rotates in the negative direction, since the speed v and angular speed ω are always positive, no matter what the direction of motion.

Equation 10-3 agrees with our observations about the wind turbine shown in Figure 10-2. All points on each turbine blade have the same angular speed ω, but points that are farther from the axis are at a greater distance r and so are moving at a greater speed $v = r\omega$.

EXAMPLE 10-1 Speed Versus Angular Speed

A turbine blade is rotating at 25.0 rev/min. How fast (in m/s) is a point on the blade moving that is (a) 0.500 m from the rotation axis and (b) 1.00 m from the rotation axis?

Set Up

We are given the angular speed ω of the blade and the distance r from the rotation axis for each point. We'll use Equation 10-3 to find the speed v of these points on the blade.

Speed of a point on a rotating rigid body:

$$v = r\omega \quad (10\text{-}3)$$

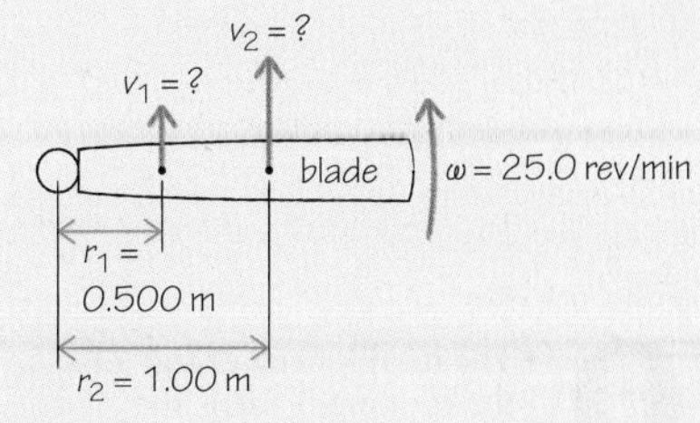

Solve

We are given the angular speed in revolutions per minute. We first need to convert this to radians per second so we can safely use Equation 10-3.

$\omega = 25.0$ rev/min

Convert revolutions to radians and minutes to seconds:

$$\omega = \left(25.0\frac{\text{rev}}{\text{min}}\right)\left(\frac{2\pi\ \text{rad}}{1\ \text{rev}}\right)\left(\frac{1\ \text{min}}{60\ \text{s}}\right)$$

$$= \frac{(25.0)(2\pi)}{60}\frac{\text{rad}}{\text{s}} = 2.62\ \text{rad/s}$$

For each case substitute $\omega = 2.62$ rad/s and the value of the distance r of the point from the axis.

(a) $v_1 = r_1\omega = (0.500\ \text{m})(2.62\ \text{rad/s}) = 1.31\ \text{m/s}$

(b) $v_2 = r_2\omega = (1.00\ \text{m})(2.62\ \text{rad/s}) = 2.62\ \text{m/s}$

Reflect

Note that the point in (b) is twice as far from the rotation axis as is the point in (a) and has twice the speed. That's because in the same amount of time, the point in (b) travels in a circle that has twice the radius and thus twice the circumference of the circle traveled by the point in (a). This means the point in (b) must cover twice as much distance as the point in (a) in the same amount of time.

Note that the units of radians disappeared when we calculated the speed of each point. We did this because a radian isn't truly a unit but is the measure of an angle as the ratio of the arc length through which a point has rotated about an axis to the radius of that point from the axis. Because it is a ratio of distances, it is actually a dimensionless number.

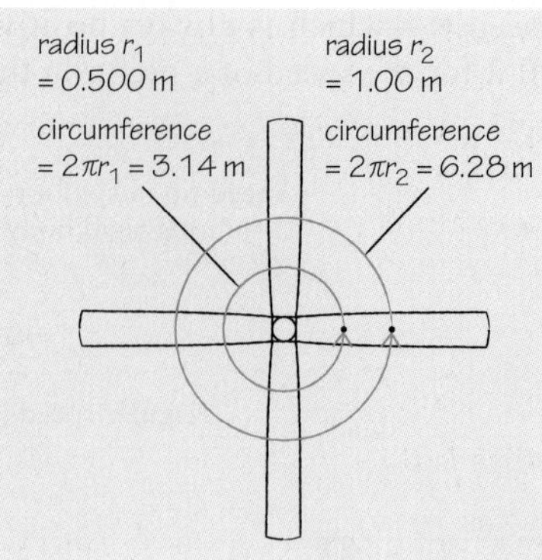

NOW WORK Problems 1–4 from The Takeaway 10-2.

Rotational Inertia and Rotational Kinetic Energy

Let's see how to use Equation 10-3 to find the kinetic energy of a rotating rigid body such as a wind turbine blade. Since it is rigid, all parts of a rigid body rotate at the same angular speed ω. As we saw, the speed v is different at different points along the blade, however. To account for this, let's imagine that the rigid blade is divided up into many small pieces (**Figure 10-4**). There are N such small pieces, where N is a large number. We label each piece using the subscript i, where i can equal any integer from 1 to N: $i = 1, 2, 3, \ldots, N$. The ith piece has mass m_i and is a distance r_i from the rotation axis. The reason that we have to use so many pieces is that we want each piece to be very small, so that there is only one value of the distance r_i for each piece.

From Equation 10-3 the speed of the ith small piece of the blade is $v_i = r_i\omega$. If we measure ω in radians per second and r_i in meters, v_i will be in meters per second (m/s). The kinetic energy of the ith piece of the blade is therefore

$$K_i = \frac{1}{2}m_i v_i^2 = \frac{1}{2}m_i(r_i\omega)^2 = \frac{1}{2}m_i r_i^2\omega^2$$

We can find the total kinetic energy of the entire rotating turbine blade by adding together the kinetic energies of all pieces of the blade:

$$K_{\text{rotational}} = K_1 + K_2 + K_3 + \ldots + K_N = \sum_{i=1}^{N} K_i = \sum_{i=1}^{N}\frac{1}{2}m_i r_i^2\omega^2 \quad (10\text{-}4)$$

- The total kinetic energy of the rotating blade is the sum of the kinetic energies of the N pieces:
  $K_{\text{rotational}} = K_1 + K_2 + K_3 + \ldots + K_N$
  $= \sum_{i=1}^{N} K_i = \sum_{i=1}^{N}\frac{1}{2}m_i r_i^2\omega^2$
- We can write this in terms of a quantity we define below, rotational inertia of the blade:
  $K_{\text{rotational}} = \frac{1}{2}I\omega^2$ where $I = \sum_{i=1}^{N} m_i r_i^2$

① Divide the blade into N small pieces.

② The ith piece has mass m_i and is a distance r_i from the rotation axis.

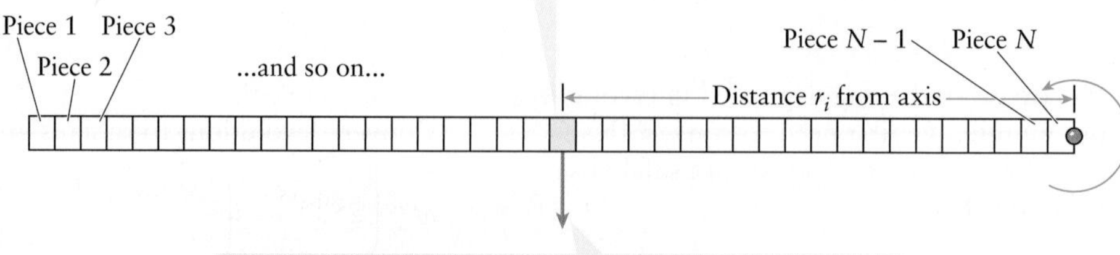

③ The blade is rotating with angular speed ω, so the speed of the ith piece is $v_i = r_i\omega$. The kinetic energy of this piece is
$K_i = \frac{1}{2}m_i v_i^2 = \frac{1}{2}m_i(r_i\omega)^2 = \frac{1}{2}m_i r_i^2\omega^2$

Figure 10-4 Calculating rotational kinetic energy and rotational inertia The kinetic energy of a rotating extended object is the sum of the kinetic energies of all of its component pieces.

The subscript "rotational" in Equation 10-4 reminds us that this is the kinetic energy of the turbine blade due to its rotation. The quantities $\frac{1}{2}$ and ω^2 have the same values for each term in the sum, so we can factor them out of the sum. (If you're not sure whether it's correct to factor constant terms out of a sum, try it with a few numbers; for example, $2(3) + 2(4) + 2(5) = 2(3 + 4 + 5)$.) The result is

$$K_{\text{rotational}} = \frac{1}{2}\left(\sum_{i=1}^{N} m_i r_i^2\right)\omega^2 \tag{10-5}$$

What is the quantity in parentheses in Equation 10-5, $\sum_{i=1}^{N} m_i r_i^2$? Although it involves a sum over all the little pieces into which we've divided the turbine blade, it is *not* simply the total mass M of the blade. That sum would be $M = \sum_{i=1}^{N} m_i$, without the factor of r_i^2. Instead the sum $\sum_{i=1}^{N} m_i r_i^2$ is a new quantity that also tells us how the mass of the blade is *distributed*: It depends on both the mass of each small piece (m_i) and how far away from the rotation axis that piece is (r_i). This quantity is called the **rotational inertia** of the turbine blade. We represent it by the symbol I:

To find the **rotational inertia** I of a rigid body...

...we imagine dividing the rigid body into **N small pieces,** and calculate the sum...

$$I = \sum_{i=1}^{N} m_i r_i^2 \tag{10-6}$$

...of the product of the **mass m_i of the ith piece**...

...and the square of the **distance** r_i from the ith piece to the rotation axis.

EQUATION IN WORDS
Rotational inertia of a rigid body

The SI units of rotational inertia are kilograms multiplied by meters squared ($\text{kg} \cdot \text{m}^2$).

WATCH OUT !

Rotational inertia is only constant for a rigid body.

While you can use Equation 10-6 to calculate the rotational inertia for any system, the rotational inertia will change if the configuration of that system changes. So, unless the system is a rigid body, rotational inertia is not constant.

WATCH OUT !

Take a moment to understand the meaning of *moment*.

Rotational inertia is also commonly called the *moment of inertia*. Despite this name, the term *moment of inertia* has nothing to do with a particular moment of time. Rather, *moment* is a mathematical term for a quantity that tells you how things are distributed. For example, when your class takes an exam, the average score on that exam (which tells you something about how the scores were distributed between zero and perfect) is an example of a moment.

We can use Equation 10-6 to write the expression for the **rotational kinetic energy** of a rigid body, Equation 10-5, in terms of the rotational inertia:

Rotational kinetic energy of a rigid body spinning around an axis

Rotational inertia of the rigid body for that rotation axis

$$K_{\text{rotational}} = \frac{1}{2} I\omega^2 \tag{10-7}$$

Angular speed of the rigid body

EQUATION IN WORDS
Rotational kinetic energy of a rigid body

We've derived this equation for a rotating turbine blade, but it applies to *any* rotating rigid body. The rotational kinetic energy of a rigid body is equal to one-half the rigid body's rotational inertia multiplied by the square of the rigid body's angular speed. We can use it for extended objects that are not always rigid, such as the ice skater, only for times when they do hold their shape as they rotate. Note that in Equation 10-7 the value of the rotational inertia I is for a specific rotation axis.

As we'll see in the next section, the value of I for a given rigid body can be different, depending on the axis around which it rotates.

WATCH OUT !

Use correct units in calculating rotational kinetic energy.

In deriving Equation 10-7 we used Equation 10-3, $v = r\omega$, which is valid only if angles are measured in radians. To have the units of rotational kinetic energy work out properly with Equation 10-7, you *must* express angular speed ω in radians per second (rad/s). The following example illustrates the need to use correct units.

EXAMPLE 10-2 Turbine Kinetic Energy

The blades of a wind turbine have a combined rotational inertia of $2.00 \times 10^3\ \text{kg} \cdot \text{m}^2$ for rotation around the turbine axis. If the blades make 4.00 complete revolutions every minute, what is the rotational kinetic energy of the blades?

Set Up

We use the definition of rotational kinetic energy from Equation 10-7. We are given the value of the rotational inertia I, but we are not told the value of the angular speed ω. Instead we are told that the turbine makes 4.00 complete revolutions (4.00 rev for short) in 1 min. We can use this value to calculate the angular velocity from its definition (Equation 10-1). The angular speed is the magnitude of angular velocity $|\omega_{\text{average},z}|$.

Rotational kinetic energy:

$$K_{\text{rotational}} = \frac{1}{2}I\omega^2 \quad (10\text{-}7)$$

Angular velocity:

$$\omega_{\text{average},z} = \frac{\Delta\theta}{\Delta t} \quad (10\text{-}1)$$

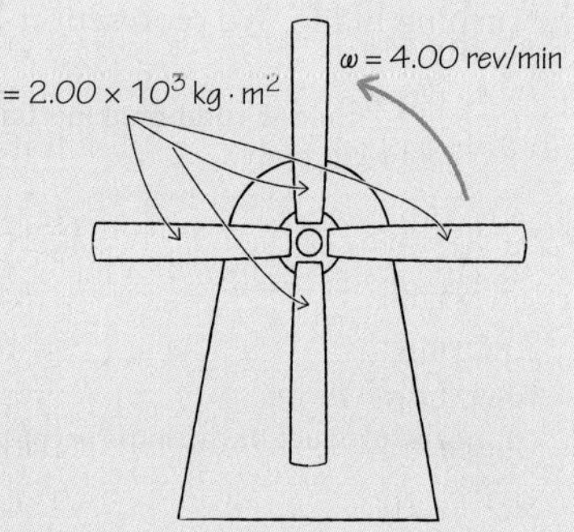

Solve

First calculate the angular velocity and angular speed. Be careful to convert revolutions per minute to radians per second.

$$\omega_{\text{average},z} = \frac{\Delta\theta}{\Delta t} = \left(\frac{4.00\ \text{rev}}{1\ \text{min}}\right)\left(\frac{2\pi\ \text{rad}}{1\ \text{rev}}\right)\left(\frac{1\ \text{min}}{60\ \text{s}}\right)$$

$$= \frac{(4.00)(2\pi)}{60}\frac{\text{rad}}{\text{s}} = 0.419\ \text{rad/s}$$

The angular speed is the magnitude of $\omega_{\text{average},z}$, which in this case is equal to $\omega_{\text{average},z}$ because its value is positive: $\omega = 0.419$ rad/s.

Substitute this value of ω and the given value of I into the expression for rotational kinetic energy. We can drop the "rad" from the units of the final answer because radians are a dimensionless ratio.

$$K_{\text{rotational}} = \frac{1}{2}I\omega^2 = \frac{1}{2}(2.00 \times 10^3\ \text{kg} \cdot \text{m}^2)(0.419\ \text{rad/s})^2$$

$$= 175\ \text{kg} \cdot \text{m}^2/\text{s}^2 = 175\ \text{J}$$

Reflect

According to the work-energy theorem, this answer means that 175 J of work must be done on the turbine blades to bring them up to this angular speed. The blades are shaped so that the wind pushes on them, doing work on them. To make the blades spin this quickly on a day with no wind, you would have to push on the blades yourself to do 175 J of work. Remember that work is the product of the external force and displacement of the point of application of the force in the direction of the force, so if you pushed the tip of one blade a distance of 1.00 m to start it moving, you would have to exert a force of 175 N in the direction of its motion over that distance. (You'd actually have to exert even more force than that because you also have to overcome friction in the bearings of the turbine.)

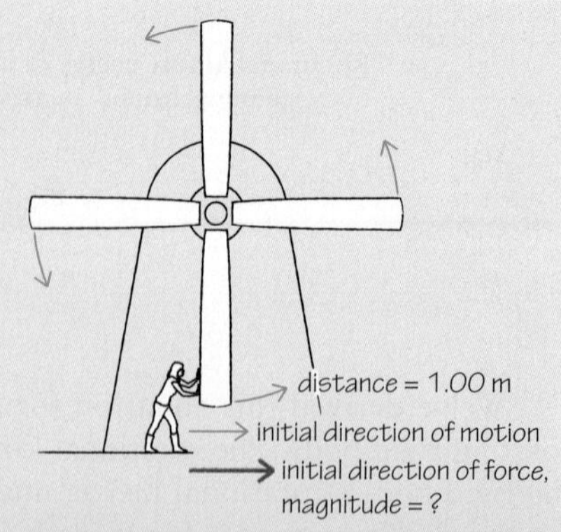

NOW WORK Problems 5 and 6 from The Takeaway 10-2.

Comparison of Rotational Kinetic Energy and Translational Kinetic Energy

It's useful to compare the equation for *rotational* kinetic energy from Equation 10-7 with the equation for *translational* kinetic energy that we learned in Section 7-3:

Translational kinetic energy	Rotational kinetic energy
$K_{\text{translational}} = \frac{1}{2}mv^2$	$K_{\text{rotational}} = \frac{1}{2}I\omega^2$

Notice that mass m in the translational kinetic energy equation corresponds to rotational inertia I in the rotational kinetic energy equation. Mass is a property of matter that represents the resistance of an object to a change in its velocity $\vec{v}$; by analogy, an extended object's rotational inertia represents the resistance of the extended object to a change in its *angular* velocity ω_z.

Both translational and rotational kinetic energy depend on mass. However, the rotational inertia (Equation 10-6), and therefore the rotational kinetic energy, both depend on how the mass is *distributed* with respect to the axis of rotation. A bit of mass far from the rotation axis has a larger effect on the value of the rotational inertia than the same amount of mass close to the axis. The following example illustrates this effect. We will also see that a single object that is somehow constrained to go around an axis can undergo rotational motion and have a rotational inertia.

EXAMPLE 10-3 Whirling a Ball

The physicist in **Figure 10-5** whirls a small red ball of mass 0.200 kg in a nearly horizontal circle at the end of a lightweight string. The ball starts from rest and increases in speed until it is rotating at 5.00 revolutions per second. How much energy must the physicist supply to cause this motion to start if the string is (a) 0.300 m long or (b) 0.600 m long?

Figure 10-5 An object moving in a circle What is the rotational kinetic energy of the small red ball moving in a horizontal circle?

Set Up

The statement that the string is "lightweight" means that we can ignore its mass and kinetic energy, so we need to calculate only the kinetic energy of the ball. We are given its final rotational speed, so we can calculate its final rotational kinetic energy. Figure 10-5 shows that the ball is small compared to the length of the string, so it's reasonable to treat the ball as though all its mass were concentrated at its center of mass (so we may model it as an object). We'll use Equation 10-6 to find the rotational inertia I of the ball in each case, then substitute its value into Equation 10-7 to calculate the rotational kinetic energy $K_{\text{rotational}}$. We're told that the angular speed is $\omega = 5.00$ rev/s.

Rotational inertia:

$$I = \sum_{i=1}^{N} m_i r_i^2 \qquad (10\text{-}6)$$

Rotational kinetic energy:

$$K_{\text{rotational}} = \frac{1}{2}I\omega^2 \qquad (10\text{-}7)$$

case (a)
$K_{\text{rotational}} = ?$
$r = 0.300$ m
$m = 0.200$ kg
$\omega = 5.00$ rev/s

case (b)
$K_{\text{rotational}} = ?$
$r = 0.600$ m
$m = 0.200$ kg
$\omega = 5.00$ rev/s

Solve

Because we can treat the red ball as an object, there's only one term in the sum for rotational inertia (that is, $N = 1$). We then calculate the rotational inertia for each case.

$$I = \sum_{i=1}^{N} m_i r_i^2 = m_{\text{object}}\, r_{\text{object}}^2$$

(a) String 0.300 m long: $I = (0.200 \text{ kg})(0.300 \text{ m})^2 = 0.0180 \text{ kg} \cdot \text{m}^2$

(b) String 0.600 m long: $I = (0.200 \text{ kg})(0.600 \text{ m})^2 = 0.0720 \text{ kg} \cdot \text{m}^2$

We are given the angular speed in revolutions per second, so we must convert it to radians per second.

$$\omega = \left(5.00\frac{\text{rev}}{\text{s}}\right)\left(\frac{2\pi\ \text{rad}}{1\ \text{rev}}\right) = (5.00)(2\pi)\frac{\text{rad}}{\text{s}} = 31.4\ \text{rad/s}$$

Finally, insert the values of I and ω into the equation for rotational kinetic energy.

(a) String 0.300 m long:

$$K_{\text{rotational}} = \frac{1}{2}I\omega^2 = \frac{1}{2}(0.0180\ \text{kg}\cdot\text{m}^2)(31.4\ \text{rad/s})^2$$
$$= 8.88\ \text{kg}\cdot\text{m}^2/\text{s}^2 = 8.88\ \text{J}$$

(b) String 0.600 m long:

$$K_{\text{rotational}} = \frac{1}{2}I\omega^2 = \frac{1}{2}(0.0720\ \text{kg}\cdot\text{m}^2)(31.4\ \text{rad/s})^2$$
$$= 35.5\ \text{kg}\cdot\text{m}^2/\text{s}^2 = 35.5\ \text{J}$$

Reflect

The angular velocity ω is the same in both situations. However, the physicist must supply four times as much kinetic energy to the rotating ball when he uses the longer string: $4 \times 8.88\ \text{J} = 35.5\ \text{J}$. This makes sense because the ball has to have twice the linear speed to move through the same angle in the same time, and kinetic energy is proportional to speed squared.

NOW WORK Problem 7 from The Takeaway 10-2.

Because the ball in the preceding example can be treated as an object, its rotational kinetic energy must be its total kinetic energy. We could actually calculate its kinetic energy without using the rotational inertia approach. Since all of the mass of an object moving in a circle is at the radius of the motion, and the angular speed is the translational speed divided by the radius of the motion, the radius terms cancel and you get $K = \frac{1}{2}mr^2(\omega)^2 = \frac{1}{2}mr^2(v/r)^2 = \frac{1}{2}mr^2v^2(1/r^2) = \frac{1}{2}mv^2$.

THE TAKEAWAY for Section 10-2

✔ Angular velocity is a measure of how rapidly and in what direction an extended object rotates. Angular speed is the magnitude of angular velocity.

✔ The speed of a point on a rotating extended object depends on the point's distance from the rotation axis and the extended object's angular speed.

✔ The rotational inertia of a rigid body depends on the rigid body's mass and how that mass is distributed relative to the rigid body's rotation axis.

✔ The rotational kinetic energy of a rigid body spinning around an axis depends on its angular speed and its rotational inertia for that axis.

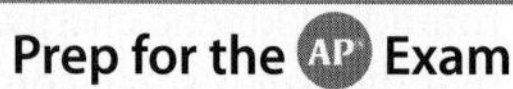

AP Building Blocks

1. Convert the following:
   (a) 45.0 rev/min = _____ rad/s
   (b) $33\frac{1}{3}$ rpm = _____ rad/s
   (c) 2π rev/s = _____ rad/s

EX 10-1
2. Calculate the average angular speed, in rad/s, of
   (a) the second hand of a clock.
   (b) the minute hand of a clock.

EX 10-1
3. An airplane flying at a constant speed of 380 m/s changes course with a circular path with a radius of 7600 m. Calculate the angular speed in rad/s. Remember that in the relationship between linear speed and angular speed, Equation 10-3, ω must be expressed in rad/s.

EX 10-1
4. The Moon completes one orbit about Earth in 27.4 days and the Earth–Moon distance is 3.84×10^8 m.
   (a) Calculate the average angular speed of the Moon, ω, as it orbits Earth.
   (b) Calculate the average orbital speed of the Moon, v, from ω.

EX 10-2
5. A rigid body rotates about the x axis. It has a rotational inertia of $2.00\ \text{kg}\cdot\text{m}^2$ and an angular speed of 3.00 rad/s. Calculate the rotational kinetic energy of this rigid body.

EX 10-2
6. A disk rotating about the y axis with a rotational inertia of $0.330\ \text{kg}\cdot\text{m}^2$ has a rotational kinetic energy of 6.00 J. Calculate the angular speed of the disk.

EX 10-3
7. A 1.00-g ball rotates at 300.0 rev/s about an axis that is 12.0 cm away from the center of the ball. Approximate the ball as an object.
   (a) Calculate the rotational inertia of the object about the axis of rotation.
   (b) Calculate the kinetic energy of the object.

AP® Skill Builders

8. You use a felt-tip pen to draw the diameter line on a dinner plate and then spin the dinner plate on the table around the plate's center. Imagine the appearance of this line as the plate is spinning. Use the evidence provided by your observation of the line to justify the claim that as a rigid disk spins about its center all points in the disk move with the same angular speed.
9. As we saw in Chapter 6, the centripetal acceleration of a satellite that has an orbital speed v and orbital radius r is v^2/r.
   (a) Express the centripetal acceleration for a point on a rotating rigid body in terms of the angular velocity ω and r. (Make sure to define your r.)
   (b) Explain why the result in part (a) shows that for all points on a rotating disk the centripetal acceleration increases linearly with distance from the center of the disk.
10. Justify the claim that points on the rim at the top and bottom of a vertical wheel spinning about a stationary axis have linear velocities with the same magnitude and opposite signs.

AP® Skills in Action

11. Calculate the angular speed of a rotating wheel that has a rotational inertia of $0.330\ \text{kg}\cdot\text{m}^2$ and a rotational kinetic energy of 2.75 J. Give your answer in both rad/s and rev/min.
12. A 0.25-kg object rotates at 3.0 rev/s about an axis that is 0.50 m away from the object. Calculate the kinetic energy of the object.
13. Evaluate the angular displacement (in radians and degrees) of Earth in one day of its orbit around the Sun.

10-3 An extended object's rotational inertia depends on its mass distribution and the choice of rotation axis

We saw in Section 10-2 that the rotational inertia I of an extended object plays the same role for rotational motion that the object's mass m does for translational motion. The rotational inertia of an extended object depends on how its mass is *distributed* so its shape has to be fixed for this value to be constant. This means the equations we develop will always hold for rigid bodies but can be used exactly only in more limited contexts with extended objects (when for a period of time they are holding their shape). These ideas are useful in considering what will happen as an extended object does change shape, as we will see in the next chapter.

An example of how the distribution of the mass is as important as the mass itself in rotation is the door of a bank vault (**Figure 10-6**). The door takes effort to move in part because it's so massive. But an additional factor is that a substantial fraction of the door's mass is in its locking mechanism, and this mechanism is located a good distance away from the hinges around which the door rotates. The greater the fraction of the door's mass that's located far from the hinges, the more difficult it is to start the door rotating if it's at rest or to stop it moving if it's already in motion.

Imagestate Media Partners Limited-Impact Photos/Alamy

Figure 10-6 Moving a bank vault door How easy it is to start this door moving depends on its rotational inertia around the hinges. The rotational inertia is determined by the mass of the door and how that mass is distributed.

Because the rotational inertia of a rigid object is such an important property, we'll devote this section to some examples of calculating its value. We'll see how the rotational inertia depends on the way in which the extended object's mass is distributed and how the rotational inertia of a given rigid body can change depending on the particular axis around which the rigid body rotates.

Rotational Inertia of a Collection of Small Pieces

In Section 10-2, we found if we imagine dividing a rigid body into a large number of small objects of masses $m_1, m_2, m_3, \ldots, m_N$, the rotational inertia I of the rigid body is

$$I = \sum_{i=1}^{N} m_i r_i^2 \tag{10-6}$$

In this expression, r_i is the distance from the axis around which the rigid body rotates to the position of the ith small piece.

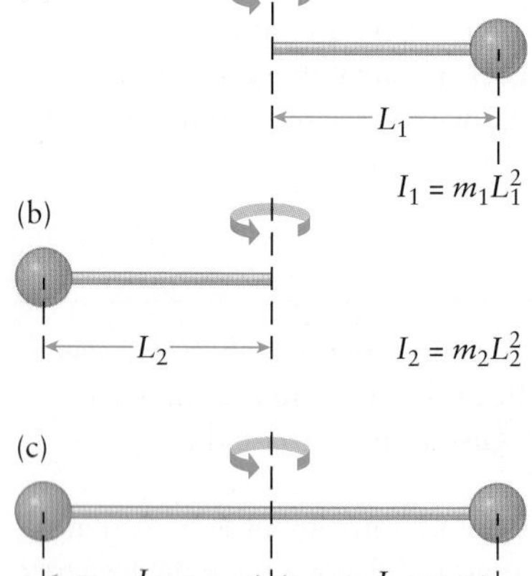

Figure 10-7 Calculating rotational inertia The sum of the rotational inertias of the objects in (a) and (b) about the axis of rotation equals the rotational inertia of the composite extended object in (c).

Equation 10-6 reveals three important properties of the rotational inertia:

1. *The rotational inertia is additive.* Each term $m_ir_i^2$ in Equation 10-6 represents the rotational inertia of the ith piece of the rigid body, and the total rotational inertia is the sum of the terms for all pieces. It follows that if you know the rotational inertias of two rigid bodies or objects around some rotation axis and you attach them to form a single rigid body, the rotational inertia of the new rigid body—around the same axis—is the sum of the rotational inertias of the two separate rigid bodies or objects (**Figure 10-7**).
2. *The farther away from the rotation axis a rigid body's mass lies, the greater the rigid body's rotational inertia.* In Equation 10-6, the farther the small pieces of the rigid body are from the rotation axis, the greater the values of r_i and the greater the rotational inertia. Example 10-4 demonstrates this idea.
3. *The rotational inertia depends on the rotation axis.* If a different rotation axis is used, the quantities r_i in Equation 10-6 change and so the rotational inertia changes as well. In general, the rotational inertia for a given rigid body is different for each different rotation axis. We'll see how this works in Example 10-5.

Let's illustrate these statements by looking at rigid bodies that we can describe as being composed of just a few objects.

EXAMPLE 10-4 Rotational Inertia for Barbells I

A barbell is made up of two small, massive spheres connected by a lightweight rod with a length much greater than the radii of the spheres. For a barbell made with two identical 50.0-kg spheres, find the rotational inertia for rotation around the midpoint of the barbell if the spheres are separated by (a) 1.20 m or (b) 2.40 m.

Set Up

The problem describes the rod connecting the spheres as "lightweight," so we can ignore the mass of the rod. The two spheres are also described as "small" and much smaller than the length of the rod separating them, so we can treat each sphere as an object. Because the barbell is made up of only two objects, the sum in Equation 10-6 has only two terms, one for each sphere. That is, the rotational inertia of the barbell is the sum of the rotational inertias of the two spheres.

Rotational inertia:

$$I = \sum_{i=1}^{N} m_ir_i^2 = m_1r_1^2 + m_2r_2^2 \qquad (10\text{-}6)$$

rotation axis
1.20 m
50.0 kg 50.0 kg
2.40 m
50.0 kg 50.0 kg

Solve

Each sphere has a mass of 50.0 kg, so $m_1 = m_2 = 50.0$ kg in each case. For the first barbell each sphere is one-half of 1.20 m from the rotation axis, so $r_1 = r_2 = 0.600$ m; similarly, in the second case each sphere is one-half of 2.40 m from the rotation axis, so $r_1 = r_2 = 1.20$ m. Substitute these values into the equation for rotational inertia I.

For both barbells:

$$I = m_1r_1^2 + m_2r_2^2 = (50.0\text{ kg})r_1^2 + (50.0\text{ kg})r_2^2$$

(a) For the first barbell:

$$I = (50.0\text{ kg})(0.600\text{ m})^2 + (50.0\text{ kg})(0.600\text{ m})^2 = 36.0\text{ kg}\cdot\text{m}^2$$

(b) For the second barbell:

$$I = (50.0\text{ kg})(1.20\text{ m})^2 + (50.0\text{ kg})(1.20\text{ m})^2 = 144\text{ kg}\cdot\text{m}^2$$

Reflect

Although both barbells have the same total mass $m_1 + m_2 = 50.0\text{ kg} + 50.0\text{ kg} = 100.0\text{ kg}$, the mass is distributed in different ways. Hence, they have very different values for the rotational inertia. The farther the massive spheres are from the rotation axis, the greater the value of the rotational inertia. For mass twice as far away from the rotation axis, we expect the rotational inertia to increase by 2 squared, or 4, as we found.

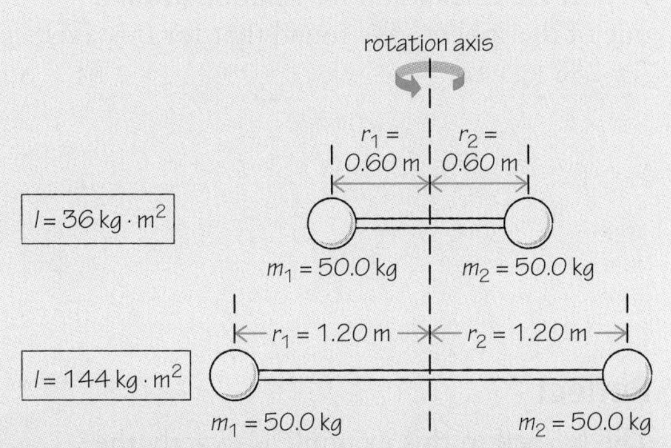

NOW WORK **Problem 2 from The Takeaway 10-3.**

EXAMPLE 10-5 Rotational Inertia for Barbells II

Consider again the second barbell from Example 10-4, which has two 50.0-kg spheres separated by 2.40 m. You may assume the spheres are very small compared to the separation. (a) Calculate the rotational inertia of this same barbell if it rotates around an axis through the center of one of the spheres, perpendicular to the length of the rod. (b) Determine the kinetic energy of this barbell if it rotates at 1.00 rad/s around its midpoint as in the preceding example, and if it rotates at 1.00 rad/s around the axis through the center of one of the spheres, as shown.

Set Up

As in Example 10-4 the rotational inertia I of the barbell is the sum of two terms, one for each sphere. We expect a different answer for I in part (a) because the values of r_1 and r_2 are different from those in Example 10-4. Once we know the value of I for each choice of axis, we can calculate the rotational kinetic energy using Equation 10-7.

Rotational inertia:

$$I = \sum_{i=1}^{N} m_i r_i^2 = m_1 r_1^2 + m_2 r_2^2 \quad (10\text{-}6)$$

Rotational kinetic energy:

$$K_{\text{rotational}} = \frac{1}{2} I\omega^2 \quad (10\text{-}7)$$

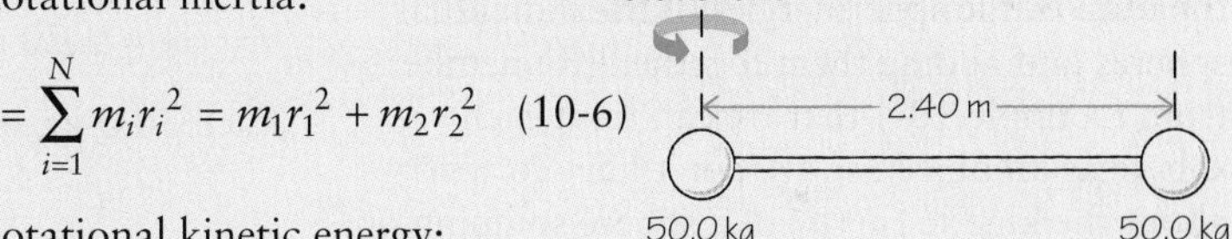

Solve

(a) Again each sphere has mass 50.0 kg, so $m_1 = m_2 = 50.0$ kg in each case. Now one sphere is *on* the rotation axis, so $r_1 = 0$; the other sphere is 2.40 m from the rotation axis, so $r_2 = 2.40$ m. Substitute these values into the equation for I.

$$I = m_1 r_1^2 + m_2 r_2^2 = (50.0\text{ kg})(0)^2 + (50.0\text{ kg})(2.40\text{ m})^2$$
$$= 288\text{ kg}\cdot\text{m}^2$$

(b) In Example 10-4 we found that the rotational inertia of this barbell for rotation around its midpoint was $I = 144\text{ kg}\cdot\text{m}^2$. Use this in Equation 10-7 to find the rotational kinetic energy when the barbell rotates around this axis at 1.00 rad/s.

For rotation around the midpoint of the barbell:

$$I = 144\text{ kg}\cdot\text{m}^2$$

Calculate the rotational kinetic energy for $\omega = 1.00$ rad/s:

$$K_{\text{rotational}} = \frac{1}{2} I\omega^2 = \frac{1}{2}(144\text{ kg}\cdot\text{m}^2)(1.00\text{ rad/s})^2 = 72.0\text{ J}$$

Repeat the calculation for rotation around one of the spheres. We found that for this axis, $I = 288\ \text{kg} \cdot \text{m}^2$.

For rotation around one of the spheres of the barbell:

$I = 288\ \text{kg} \cdot \text{m}^2$

Calculate the rotational kinetic energy for $\omega = 1.00$ rad/s:

$$K_{\text{rotational}} = \frac{1}{2}I\omega^2 = \frac{1}{2}(288\ \text{kg} \cdot \text{m}^2)(1.00\ \text{rad/s})^2 = 144\ \text{J}$$

Reflect

The barbell in this example is exactly the same rigid body as the second barbell in Example 10-4, yet its rotational inertia is different. This shows that the rotational inertia of a rigid body depends not only on its mass, but how that mass is distributed relative to the particular axis around which it rotates. The distance from the axis is twice as far, but the rotational inertia does not go up by a factor of 4, because there is only half as much mass that is located off the axis. So 4/2 is a factor of 2, and that is what we found.

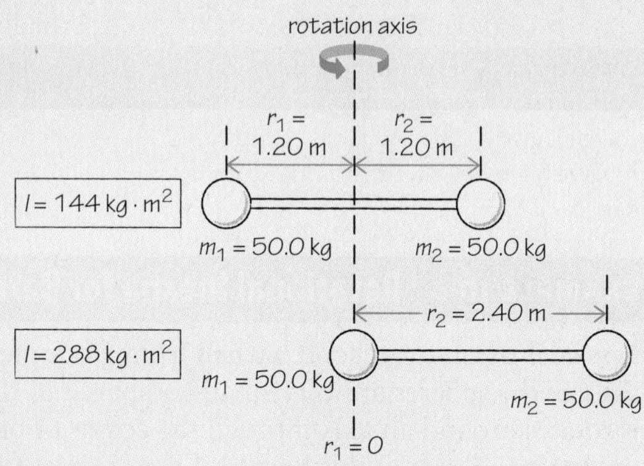

Extend: Just as we discussed at the end of Section 10-2, we can check our results for rotational energy in the two cases by finding the *translational* kinetic energies of the individual spheres and adding them together. (Remember from Example 10-4 that we can treat each sphere as an object, so we don't have to worry about the kinetic energy of a sphere spinning on its axis.) Our calculations show that changing the rotation axis while keeping the angular speed ω the same caused the speed v of the two spheres to change and hence changed the total kinetic energy—with numerical values that exactly match the ones we calculated previously.

For rotation around the midpoint of the barbell at angular speed $\omega = 1.00$ m/s: Each sphere is a distance $r = 1.20$ m from the rotation axis, so the speed of each sphere is

$v = r\omega = (1.20\ \text{m})(1.00\ \text{rad/s}) = 1.20\ \text{m/s}.$

Each 50.0-kg sphere has translational kinetic energy

$$K_{\text{sphere}} = \frac{1}{2}mv^2 = \frac{1}{2}(50.0\ \text{kg})(1.20\ \text{m/s})^2 = 36.0\ \text{J}$$

So the total kinetic energy of the system of two spheres rotating around its midpoint is 36.0 J + 36.0 J = 72.0 J.

For rotation around one of the spheres at angular speed $\omega = 1.00$ rad/s: One sphere is at the rotation axis and its center of mass is at rest, so it has zero translational kinetic energy. The other sphere is a distance $r = 2.40$ m from the rotation axis, so the speed of this sphere is

$v = r\omega = (2.40\ \text{m})(1.00\ \text{rad/s}) = 2.40\ \text{m/s}.$

The translational kinetic energy of this 50.0-kg sphere is

$$K_{\text{sphere}} = \frac{1}{2}mv^2 = \frac{1}{2}(50.0\ \text{kg})(2.40\ \text{m/s})^2 = 144\ \text{J}$$

So the total kinetic energy of the system of two spheres rotating around one of the spheres is 0 J + 144 J = 144 J.

NOW WORK Problem 3 from The Takeaway 10-3.

WATCH OUT!

A rigid body has more than one rotational inertia.

If your friend points to a DVD and asks, "What is the rotational inertia of that DVD?" it could be a trick question! Example 10-5 shows that rigid bodies do *not* have a single rotational inertia; rather, the value of the rotational inertia depends on the specific rotation axis. Any rigid body, even one like a DVD that commonly rotates around one particular axis (**Figure 10-8a**), can be made to rotate around any number of axes (**Figure 10-8b**). The axis does not even have to pass through the extended body—for example, imagine tying a string to the edge of a DVD and swinging it around in a circle like the physicist did the ball in Example 10-3 (**Figure 10-8c**). The DVD would be rotating around an axis that lies completely outside the DVD itself, just like the two objects that made up the barbell, since we were ignoring the rod. The rotational inertia of the DVD will be different for each of the axes shown in Figure 10-8. *Always* make sure that you identify the axis of rotation before determining the rotational inertia of a rigid body!

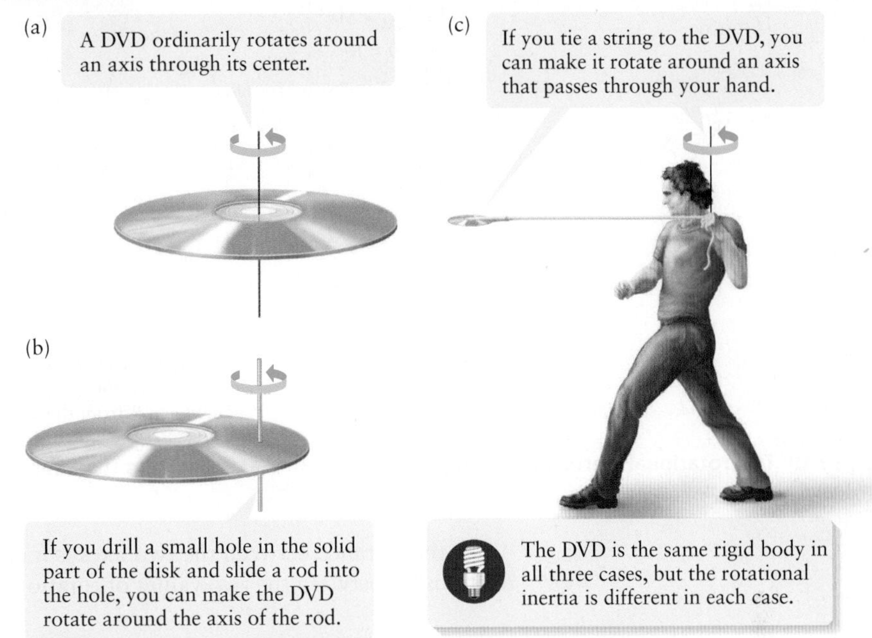

(a)
A DVD ordinarily rotates around an axis through its center.
(b)
If you drill a small hole in the solid part of the disk and slide a rod into the hole, you can make the DVD rotate around the axis of the rod.
(c)
If you tie a string to the DVD, you can make it rotate around an axis that passes through your hand.
The DVD is the same rigid body in all three cases, but the rotational inertia is different in each case.

Figure 10-8 Three rotation axes for the same extended object Changing the rotation axis of a rigid body changes the value of its rotational inertia.

The Parallel-Axis Theorem

Finding the rotational inertia of a rigid body can be challenging for certain choices of rotation axis. An example is a gymnastics hoop of mass M and radius R. A gymnast rotates the hoop around an axis that is perpendicular to the plane of the hoop and is near the hoop's rim (**Figure 10-9a**). If we imagine dividing the hoop into a large number of small pieces labeled $i = 1, 2, 3, \ldots, N$, every piece is a different distance r_i from the rotation axis, so it's very difficult to calculate the rotational inertia using Equation 10-6.

However, a remarkable relationship exists between the rotational inertia of a rigid body for an axis through its center of mass, and the rotational inertia when the rigid body rotates around any other parallel axis. This relationship, called the **parallel-axis theorem**, is useful because it's often relatively easy to determine a rigid body's rotational inertia around its center of mass.

Here's the statement of the parallel-axis theorem: Suppose the rotational inertia of a rigid body for a certain rotation axis is I. The rotational inertia of the same rigid body for a second axis that's parallel to the first one but passes through the rigid body's center of mass is I_{CM} (**Figure 10-10**).

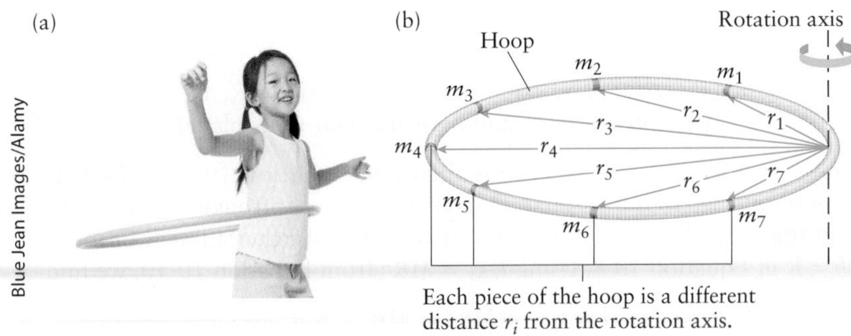

Figure 10-9 A gymnastics hoop (a) A hoop rotating around an axis that passes through its rim. (b) Calculating the hoop's rotational inertia for this axis.

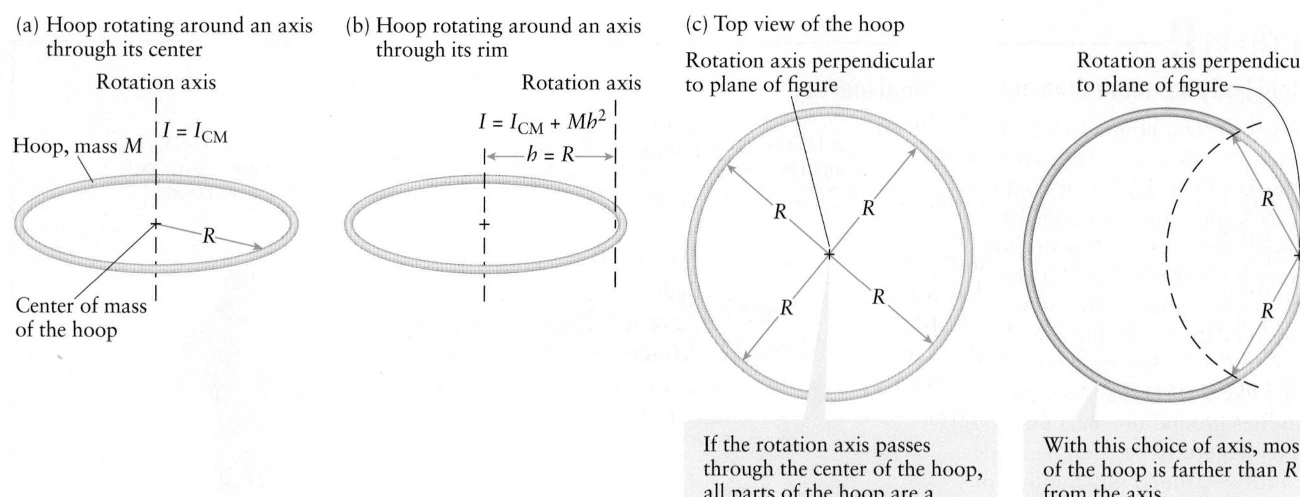

Figure 10-10 Two rotational inertias for a hoop and the parallel-axis theorem The parallel-axis theorem relates the hoop's rotational inertia for two different axes.

If the distance between the two axes is h and the mass of the rigid body is M, the relationship between the two values of rotational inertia is

EQUATION IN WORDS
The parallel-axis theorem

$$I = I_{CM} + Mh^2 \quad (10\text{-}8)$$

Rotational inertia of a rigid body for a **certain rotation axis** (I)

Rotational inertia of the same rigid body for a second, **parallel axis through its center of mass** (I_{CM})

Mass of the rigid body (M)

Distance between the two parallel axes (h)

To see the parallel-axis theorem in action, let's return to the gymnast's hoop of mass M and radius R shown in **Figure 10-9b.** The theorem tells us that to determine the hoop's rotational inertia for an axis at its rim, we should first calculate its rotational inertia for a parallel axis that passes through its center of mass (**Figure 10-10a**). The center of mass of a symmetrical rigid body such as a hoop is at its geometrical center. (Recall from Section 9-7 that there doesn't actually have to be any mass *at* an extended object's center of mass.) If we divide the hoop into many small pieces, each of those pieces is the *same* distance R from this rotation axis. So in Equation 10-6 for the rotational inertia we can replace r_i for each value of i by R. This equation then becomes

$$I_{CM} = \sum_{i=1}^{N} m_i r_i^2 = \sum_{i=1}^{N} m_i R^2 = \left(\sum_{i=1}^{N} m_i\right) R^2 \quad (10\text{-}9)$$

The quantity $\sum_{i=1}^{N} m_i$ in Equation 10-9 is just the sum of the masses of all of the individual pieces that make up the hoop. This sum is just the mass M of the hoop as a whole, so we can rewrite Equation 10-9 as

$$I_{CM} = MR^2 \quad (10\text{-}10)$$

(rotational inertia of a hoop of mass M and radius R, axis through its center of mass and perpendicular to its plane)

Now we can use the parallel-axis theorem, Equation 10-8, to find the rotational inertia of this hoop for a parallel axis that passes through the rim of the hoop (**Figure 10-10b**). The distance between the two axes is just the radius R of the hoop, so we let $h = R$ in Equation 10-8. Using $I_{CM} = MR^2$ from Equation 10-10, we find

$$I = I_{CM} + MR^2 = MR^2 + MR^2$$

or

$$I = 2MR^2 \qquad (10\text{-}11)$$

(rotational inertia of a hoop of mass M and radius R, axis through its rim and perpendicular to its plane)

This result would have been very difficult to calculate without using the parallel-axis theorem.

Why is the rotational inertia of the hoop of radius R for an axis through its rim (Equation 10-11) greater than that for an axis through its center (Equation 10-10)? The explanation is that with the rotation axis at the rim, the average distance from the axis to the pieces that make up the hoop is greater than R (**Figure 10-10c**). (A portion of the hoop is closer than R, but a larger portion is farther away than R.) As we have seen before, the farther a rigid body's mass lies from the rotation axis, the greater its rotational inertia.

EXAMPLE 10-6 Using the Parallel-Axis Theorem

Use the parallel-axis theorem and the results of Example 10-4 to find the rotational inertia of the barbell in Example 10-5 rotated about one end of the barbell.

Set Up

The two spheres at the ends of the barbell have the same mass, so the center of mass of the barbell is at its midpoint. In Example 10-4 we found the rotational inertia of this barbell for an axis that passes through the midpoint, so this expression is I_{CM}. The barbell in Example 10-5 is identical but has a different, parallel rotation axis, so we can find its rotational inertia using the parallel-axis theorem, Equation 10-8.

Parallel-axis theorem:

$$I = I_{CM} + Mh^2 \qquad (10\text{-}8)$$

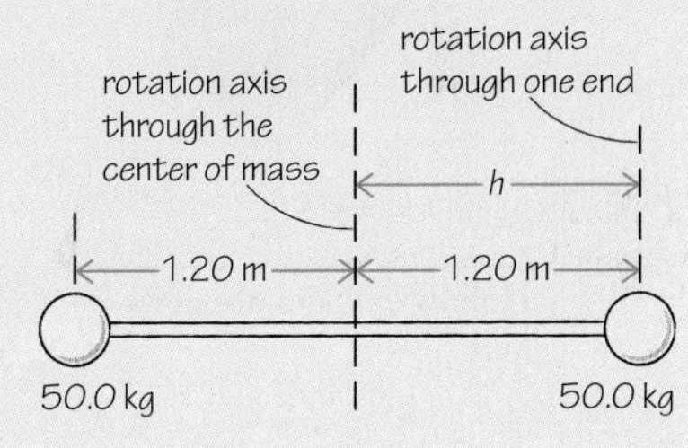

Solve

From Example 10-4, the rotational inertia of the barbell through its center of mass is $I_{CM} = 144\ \text{kg}\cdot\text{m}^2$. The axis of rotation for the barbell in Example 10-5 is a distance $h = 1.20$ m from the center of mass, and the total mass M of the barbell is the sum of the masses of the two 50.0-kg spheres. Substitute these values into the parallel-axis theorem.

$$I_{CM} = 144\ \text{kg}\cdot\text{m}^2$$
$$h = 1.20\ \text{m}$$
$$M = 50.0\ \text{kg} + 50.0\ \text{kg} = 100.0\ \text{kg}$$
$$I = I_{CM} + Mh^2 = 144\ \text{kg}\cdot\text{m}^2 + (100.0\ \text{kg})(1.20\ \text{m})^2$$
$$= 288\ \text{kg}\cdot\text{m}^2$$

Reflect

We get the same answer as in Example 10-5, which is a nice check that the parallel-axis theorem works.

NOW WORK Problem 4 from The Takeaway 10-3.

Rotational Inertia for Common Shapes

It was straightforward to calculate the rotational inertia for the hoop shown in Figure 10-10 because all parts of the hoop are the same distance from its rotation axis. For solid extended objects with other shapes, calculating the rotational inertia is more difficult and requires the use of calculus, which is beyond our scope. Instead, we'll just present the results for some common shapes that show up in various situations involving rotation (**Table 10-1**). We assume that these rigid bodies are *uniform*; that is, each has the same density throughout its volume.

TABLE 10-1 Rotational Inertias of Uniform Rigid Bodies of Various Shapes

Thin cylindrical shell about its axis	Thin cylindrical shell about diameter through center	Thin rod about perpendicular line through center	Thin spherical shell about diameter
$I = MR^2$	$I = \frac{1}{2}MR^2 + \frac{1}{12}ML^2$	$I = \frac{1}{12}ML^2$	$I = \frac{2}{3}MR^2$
Solid cylinder about its axis	**Solid cylinder about diameter through center**	**Thin rod about perpendicular line through one end**	**Solid sphere about diameter**
$I = \frac{1}{2}MR^2$	$I = \frac{1}{4}MR^2+\frac{1}{12}ML^2$	$I = \frac{1}{3}ML^2$	$I = \frac{2}{5}MR^2$
Hollow cylinder about its axis	**Hollow cylinder about diameter through center**		**Rectangular solid about axis through center perpendicular to face**
$I = \frac{1}{2}M(R_1^2 + R_2^2)$	$I = \frac{1}{4}M(R_1^2 + R_2^2) + \frac{1}{12}ML^2$		$I = \frac{1}{12}M(a^2 + b^2)$

A disk is a solid cylinder whose length L is negligible. By setting $L = 0$, the above equations for solid cylinders hold for disks. A hoop is a thin cylindrical shell whose length L is negligible. By setting $L = 0$, the above equations for thin cylindrical shells hold for hoops. Similarly, a ring is a hollow cylinder whose length L is negligible. By setting $L = 0$, the above equations for hollow cylinders hold for rings.

You can see from Table 10-1 that the rotational inertia of a thin cylindrical shell of mass M and radius R around its central axis ($I = MR^2$) is twice that of a solid cylinder of the same mass and radius ($I = \frac{1}{2}MR^2$). You should expect the cylindrical shell of the same mass to have a larger rotational inertia because rotational inertia is strongly influenced by how far the mass is from the rotation axis. In the case of the thin cylindrical shell, all of the mass is located a distance R from the axis, while for the solid cylinder only a tiny fraction of the mass is that far from the axis. The rest of the mass of the solid cylinder is a distance less than R from the axis. Hence the rotational inertia of the solid cylinder must be less than the rotational inertia of the cylindrical shell of the same mass and radius.

We can use the parallel-axis theorem to verify some of the results shown in Table 10-1. For example, consider a thin rod that has mass M and length L. The center of mass of this rod is at its geometric center. From Table 10-1, the rotational inertia of such a thin uniform rod rotating around an axis perpendicular to the rod and through its center is

$$I_{CM} = \frac{ML^2}{12}$$

What is the rotational inertia of this same rod for an axis through one end and perpendicular to the rod? This axis is parallel to an axis through the center of mass of the rod and a distance $h = L/2$ from that axis. From Equation 10-8, the rotational inertia for an axis through the end is

$$I = I_{CM} + Mh^2 = \frac{ML^2}{12} + M\left(\frac{L}{2}\right)^2$$
$$= \left(\frac{1}{12} + \frac{1}{4}\right)ML^2 = \left(\frac{1}{12} + \frac{3}{12}\right)ML^2 = \frac{4ML^2}{12}$$

or, simplifying,

$$I = \frac{ML^2}{3}$$

This expression is just the result shown in Table 10-1 for the thin uniform rod, axis through one end, perpendicular to rod.

If one of the shapes shown in Table 10-1 rotates around an axis that is different than but parallel to the axis shown in the table, we can use the parallel-axis theorem to find the new rotational inertia. Because the rotational inertia is additive, you can also use the results shown in Table 10-1 to find the rotational inertia of a more complex rigid body made up of two or more of the simple rigid bodies shown in the table.

EXAMPLE 10-7 Earring Rotational Inertia

A woman's earring is a thin, uniform disk that has a mass M and a radius R. The earring hangs from the earring post by a small hole near the edge of the disk and is free to rotate. Find the rotational inertia of the disk around this rotation axis.

Set Up

A solid disk is an example of a solid cylinder that has a negligible length. Table 10-1 shows that the rotational inertia of such a disk for an axis perpendicular to the plane of the disk and passing through its center is $I = MR^2/2$. Length has no effect on the value of I for this axis. (For a disk, the length L is much less than R (in other words, negligible), and we can set $L = 0$ in the equations in Table 10-1.)

Because the disk is uniform, its geometrical center is its center of mass, so $MR^2/2$ equals I_{CM}. The axis we want is parallel to the axis through the disk's center, so we can use the parallel-axis theorem, Equation 10-8, to find the rotational inertia around the axis at the rim.

Parallel-axis theorem:

$$I = I_{CM} + Mh^2 \qquad (10\text{-}8)$$

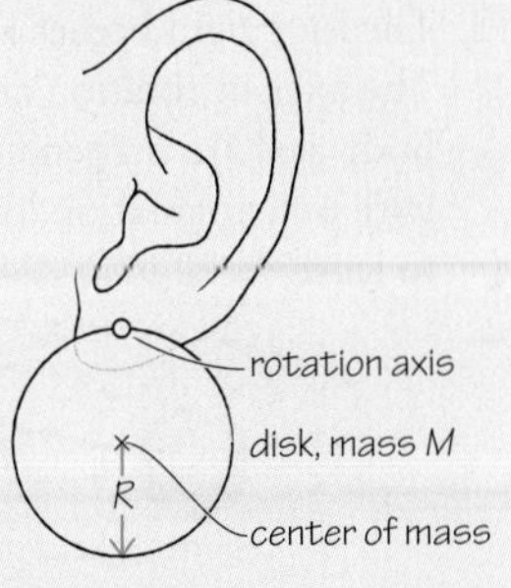

Solve

The edge of the disk is a distance $h = R$ from the center of mass of the disk. Substitute this value and $I_{CM} = MR^2/2$ into the parallel-axis theorem.

$$I = \frac{MR^2}{2} + MR^2$$

or

$$I = \frac{3MR^2}{2}$$

Reflect

If the wearer of these earrings nods her head up and down while dancing at a club, she can make the disks rotate back and forth around the posts in an eye-catching way. The more massive the disks and the larger their radius, the greater their rotational inertia and the more effort will be required to get them moving (recall that rotational kinetic energy is given by $K_{\text{rotational}} = \frac{1}{2}I\omega^2$). Making the earrings small and lightweight will require less effort from the wearer.

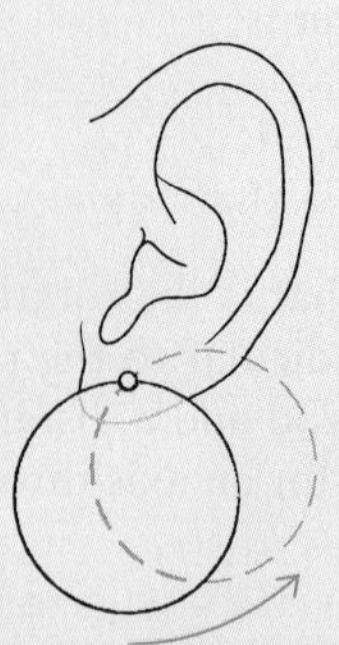

NOW WORK Problem 5 from The Takeaway 10-3.

THE TAKEAWAY for Section 10-3

✔ Since rotational inertia depends on shape, we can find exact equations of rotational inertia only for extended objects that do not change shape (rigid bodies).

✔ For other extended objects these equations are good only while shape is not changing, or to qualitatively determine how change of shape will affect rotational inertia.

✔ The rotational inertia of an extended object is additive and increases as the extended object's mass is moved farther from the rotation axis.

✔ An extended object's rotational inertia depends on the axis around which the extended object rotates.

✔ The parallel-axis theorem relates the rotational inertia for an axis through an extended object's center of mass to the rotational inertia for a second, parallel axis.

Prep for the AP® Exam

AP® Building Blocks

1.

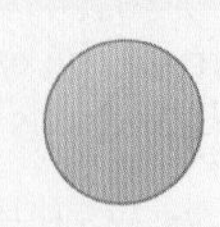

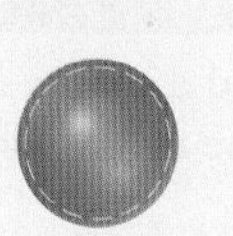

Hoop Solid cylinder Solid sphere Hollow sphere

The four rigid bodies shown have equal masses and radii. The axes of rotation are located at the center of each rigid body and are perpendicular to the plane of the paper. Rank, with an explanation, the rotational inertias from greatest to least.

EX 10-4 2.

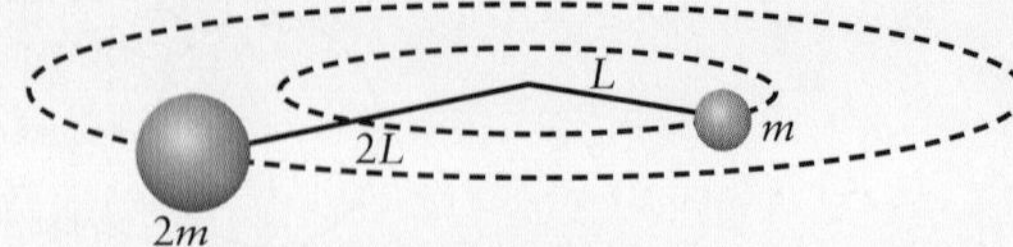

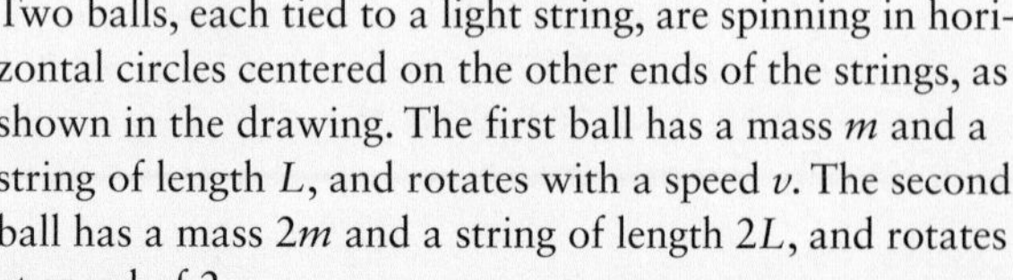

Two balls, each tied to a light string, are spinning in horizontal circles centered on the other ends of the strings, as shown in the drawing. The first ball has a mass m and a string of length L, and rotates with a speed v. The second ball has a mass $2m$ and a string of length $2L$, and rotates at speed of $2v$.

(a) Express the rotational inertia of the two-ball system in terms of L and m, approximating the balls as objects.

(b) Express the rotational kinetic energy of the system in terms of L, m, and v.

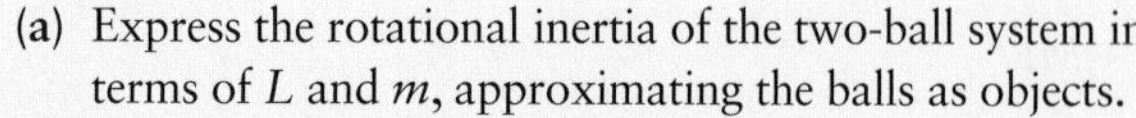

(c) Find the ratio of the rotational kinetic energies, K_{2L}/K_L.

EX 10-5 3.

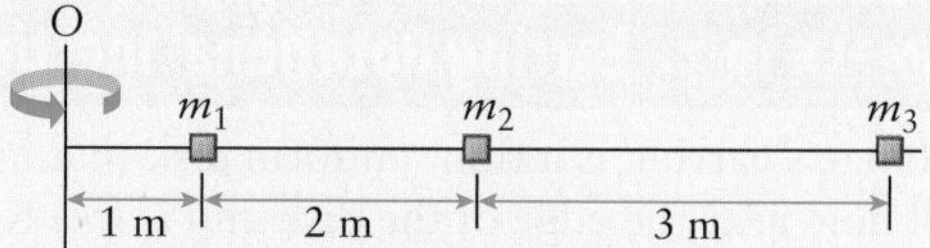

Determine the rotational inertia of three small beads (m_1 = 10 g, m_2 = 15 g, m_3 = 20 g) that can be approximated as objects, which are threaded onto a string with negligible mass. The beads are fixed in the positions shown above. The system is rotating about the axis O, as shown.

EX 10-6 4.

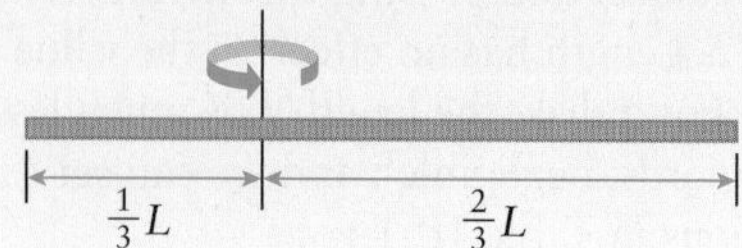

A thin uniform rod rotates about a point one-third of the way from the left end, as shown. Compared to the rotational inertia for rotation about the center of the rod, predict—using qualitative reasoning (not equations)—whether the rotational inertia for the rotation shown is greater or smaller.

EX 10-7 5.

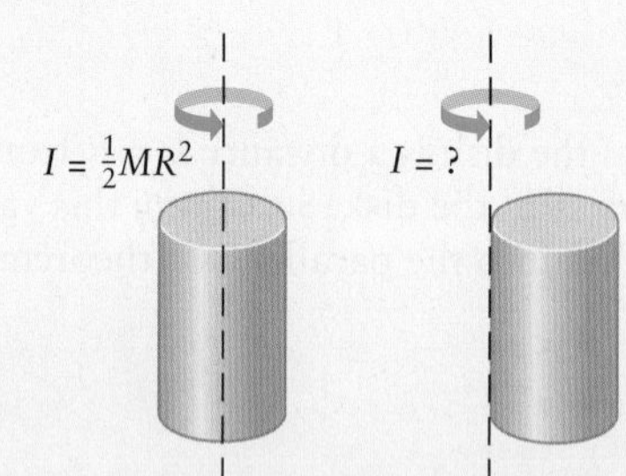

Express the rotational inertia for a uniform, solid cylinder in terms of the mass M and radius R with the axis of rotation tangent to the side of the cylinder, as shown. Recall the parallel-axis theorem, $I = I_{CM} + Mh^2$.

AP® Skill Builders

6. 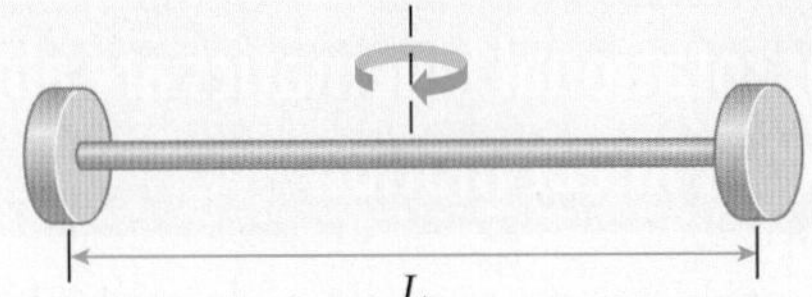

A baton twirler in a marching band complains that her baton is defective. The manufacturer specifies that the baton has an overall length of $L = 60.0$ cm and a rod diameter of 0.95 cm; that the small disks on each end have the same mass of 350 g; that the total mass of the baton is between 940 and 950 g; and that the rotational inertia of the baton about the central axis passing through the baton, as shown, should fall between 0.075 and 0.080 kg · m^2. The twirler claims this is impossible.

(a) Evaluate the validity of the twirler's claim by using the manufacturer's data to calculate the rotational inertia. Use geometric models: Mr^2 for a small disk with mass M treated as an object a distance r from the axis of rotation, and $mL^2/12$ for a thin rod with mass m and length L rotating about its center.

(b) The manufacturer refuses to refund the twirler's money. The small disks at the end of the rod can be removed by unscrewing them. Select the data that should be collected before again asking for a refund.

(c) Although the method was undisclosed, the manufacturer may have evaluated the rotational inertia experimentally, and the twirler is using geometric models for the components of the baton. Identify any approximations that the twirler may be making and look for bias in the evaluation of the rotational inertia.

(d) An improved geometric model for the rotational inertia of a rod with a diameter d is found in a reference.

$$I = \frac{mL^2}{12} + \frac{md^2}{16}$$

Evaluate the possibility that the thin-rod approximation could lead to the observed discrepancy between the calculated and manufacturer's value.

7. A model fish is cut out of a piece of cardboard. Two small holes are cut in the cardboard fish at points A and B, and a light string is attached through A. A thumb tack is then used to hang the fish from the wall, as shown in the figure.

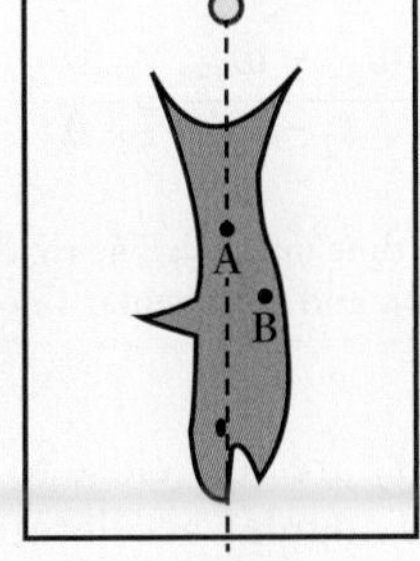

(a) Design a procedure that can be used to determine the location of the center of mass of the cardboard fish.

After the experimental determination of the center of mass, the mass of the fish is determined with a pan balance. Measurements are to be made to determine the rotational inertia about a rotation axis that runs through the center of mass, as shown in the figure below.

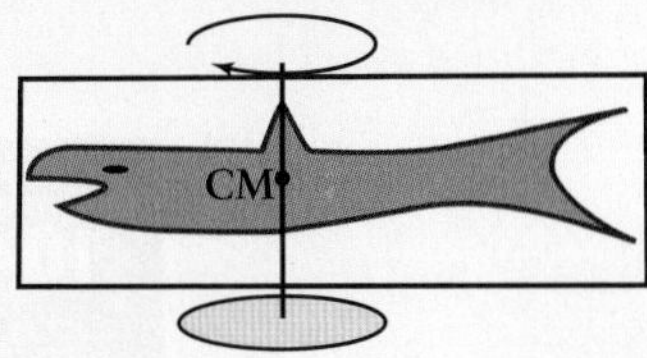

(b) Design a procedure using a sheet of paper with equally spaced lines to estimate the rotational inertia of the cardboard fish about the axis of rotation shown. Assume the paper can be cut, but the cardboard fish cannot, and that the paper is larger than the fish.

8. Recall that the rotational inertia of a hoop is MR^2 and the rotational inertia of a disk is $MR^2/2$, where M is the mass and R is the radius. So a hoop with the same mass and radius as a disk has a larger rotational inertia. Possible fan blades with the following shapes all have the same mass. Each is circumscribed by a circle with the same radius, shown in the figure as a dashed line.

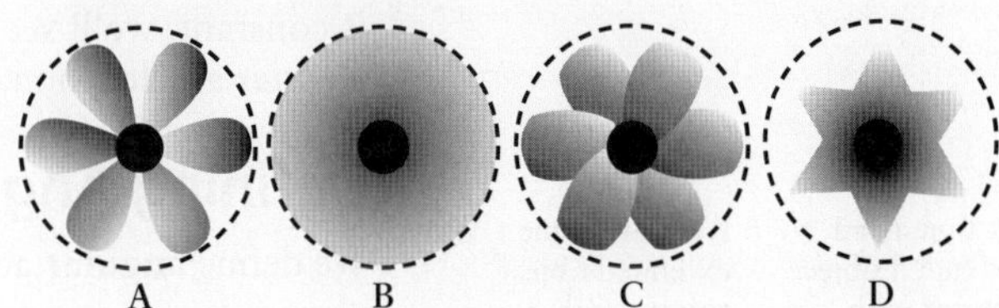

(a) Construct and justify a claim comparing the magnitudes of the rotational inertias of these four shapes. Keep in mind that each configuration has the same total mass.

The purpose of a fan is to increase the average velocity of the air as it passes through the fan.

(b) Using complete sentences, qualitatively apply the work-energy theorem to describe the work done by the fan blades to a small volume of air as it moves through the fan in terms of the change in energy of the air.

(c) Explain why the conventional fan design (A in the figure) is the best of these designs in terms of the mass of air that moves through the fan and work done to that air, and explain why the design of wind-powered water pumps used on farms that use the wind to turn the blades of the fan are more similar to C. (You will not need to discuss rotational inertia to explain this!)

AP® Skills in Action

9. Three beads with masses m_1, m_2, and m_3, with $m_1 < m_2 < m_3$, are each to be placed on a rod at distances of L_A, L_B, and L_C, with $L_A < L_B < L_C$, from the axis of rotation of the rod with rotational inertia I_{rod}. Express the smallest rotational inertia of the system in terms of I_{rod}, and the masses and locations of the beads.

10. A sphere with mass m and radius r is rigidly attached to one end of a rod of length L, and the rod–sphere system pivots about the other end of the rod. It is a useful approximation to represent the rotational inertia of the sphere as an object with no volume. In a paragraph-length response, predict whether the object–rod approximation overestimates or underestimates the true rotational inertia of the system. Your response should justify your prediction and discuss how the approximation could be modified to improve the prediction.

10-4 The equations for rotational kinematics are almost identical to those for linear motion

In the next chapter, we will look at problems more easily addressed by conservation principles. We will start our study just as we did for translational motion. In some situations we need to use **rotational kinematics**: a description not just of how rapidly an extended object rotates but also how rapidly its angular velocity is changing, that is, its *angular acceleration*.

Angular acceleration is an important factor in the design of rotating machinery. When you turn on an electric fan on a hot day, you don't just want the fan to start spinning but to come up to speed right away. If you're riding on a carnival merry-go-round, you want it to gain speed gradually and smoothly at the beginning of the ride (and lose speed in the same way at the end of the ride) rather than in a sudden jerk. In this section we'll see how angular acceleration is defined and how to analyze motion using this definition. Just as motion in a line is easiest to analyze when the acceleration is constant, we'll see that the simplest kind of rotational motion is that with constant *angular* acceleration.

Defining Angular Acceleration

At time t_1 an extended object rotates with angular velocity ω_{1z}.

At a later time t_2 the extended object rotates with a different angular velocity ω_{2z}.

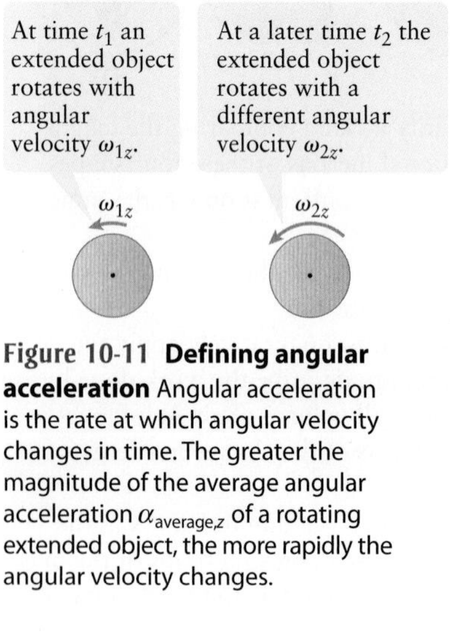

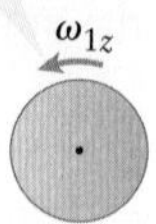

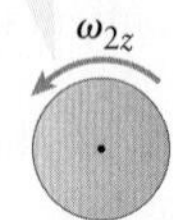

Figure 10-11 Defining angular acceleration Angular acceleration is the rate at which angular velocity changes in time. The greater the magnitude of the average angular acceleration $\alpha_{\text{average},z}$ of a rotating extended object, the more rapidly the angular velocity changes.

We define **angular acceleration**, the rate at which the angular velocity is changing with respect to time, in much the same way that we defined the acceleration of an object moving along a line in Section 2-4. In that section we described the average acceleration $a_{\text{average},x}$ of an object moving along the x axis as the change Δv_x in its x velocity during a time interval divided by the time interval, Δt:

$$a_{\text{average},x} = \frac{\Delta v_x}{\Delta t} \text{(average } x \text{ acceleration)}$$

If the time interval becomes very short, the ratio $\Delta v_x/\Delta t$ becomes the *instantaneous* acceleration a_x or, for short, simply the acceleration.

To derive an equation, we follow the same steps for rotational motion. Suppose an extended object is rotating around the z axis with angular velocity ω_{1z} at time t_1 and with angular velocity ω_{2z} at a later time t_2 (**Figure 10-11**). The *average angular acceleration* $\alpha_{\text{average},z}$ (the Greek letter α, or "alpha") for the time interval between t_1 and t_2 is the change in angular velocity, $\Delta\omega_z = \omega_{2z} - \omega_{1z}$, divided by the time interval, $\Delta t = t_2 - t_1$:

EQUATION IN WORDS
Average angular acceleration

Average angular acceleration of a rotating rigid body

Change in angular velocity of the rigid body over a certain time interval: The angular velocity changes from ω_{1z} to ω_{2z}.

$$\alpha_{\text{average},z} = \frac{\omega_{2z} - \omega_{1z}}{t_2 - t_1} = \frac{\Delta\omega_z}{\Delta t} \tag{10-12}$$

Elapsed time for the time interval: The rigid body has angular velocity ω_{1z} at time t_1, and has angular velocity ω_{2z} at time t_2.

If the two times t_1 and t_2 are very close to each other, so that Δt is very short, this becomes the *instantaneous* angular acceleration α_z, or just angular acceleration for short. At any instant angular acceleration is equal to the rate of change of angular velocity at that instant. Since angular velocity is measured in rad/s, angular acceleration is measured in rad/s^2.

Just as all pieces of a rigid body rotate with the same angular velocity ω_z at any time t, all pieces of a rigid body have the same angular acceleration. The value of α_z is positive if the angular velocity ω_z is becoming more positive and negative if ω_z is becoming more negative.

WATCH OUT !

The sign of angular acceleration can be misleading.

Just as for the acceleration of an object moving along a line, a positive value of angular acceleration α_z does *not* necessarily mean the extended object's rotation is speeding up, and a negative value of α_z does not necessarily correspond to slowing down. An extended object's rotation speeds up if ω_z and α_z have the same algebraic sign—if both are positive, the extended object is rotating in the positive direction and speeding up, while if both are negative, the extended object is rotating in the negative direction and speeding up. If ω_z and α_z have opposite signs (one positive and one negative), the extended object is rotating in the direction of the angular velocity and that rotation is slowing down (**Figure 10-12**).

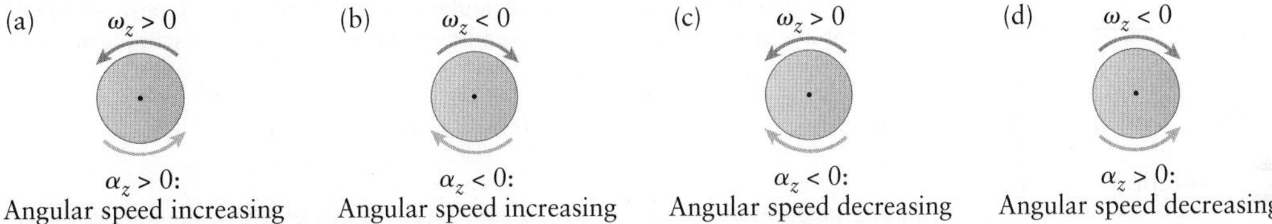

Figure 10-12 The sign of angular acceleration A rotating extended object is speeding up if its angular velocity ω_z and angular acceleration α_z are either (a) both positive or (b) both negative. It is slowing down if (c) ω_z is positive and α_z is negative or (d) ω_z is negative and α_z is positive.

Motion with Constant Angular Acceleration

We saw in Chapter 2 that an important case of motion in a line is when the acceleration a_x is constant, so that the velocity v_x changes at a steady rate. Let's look at the analogous situation of rotational motion with constant *angular* acceleration. A ball or wheel rolling downhill from rest moves with nearly constant angular acceleration, as do many types of rotating machinery as they start up or slow down.

We can write the equations for rotational motion with constant angular acceleration by looking at the equations for linear velocity v_x and angular velocity ω_z, and the equations for linear acceleration a_x and angular acceleration α_z:

Linear motion	Rotational motion
$v_x = \dfrac{\Delta x}{\Delta t}$ (linear velocity)	$\omega_z = \dfrac{\Delta \theta}{\Delta t}$ (angular velocity)
$a_x = \dfrac{\Delta v_x}{\Delta t}$ (linear acceleration)	$\alpha_z = \dfrac{\Delta \omega_z}{\Delta t}$ (angular acceleration)

Comparing these equations shows that the rotational quantities θ, ω_z, and α_z are related to each other in exactly the same way that x, v_x, and a_x are related to each other. We choose a single point to describe the motion of an object, its center of mass. For a rotating rigid body, we choose a reference line to describe exactly we mean by angular position (Figure 10-13). Because θ, ω_z, and α_z are related in the same way as x, v_x, and a_x, graphs for these rotational quantities have the same relationships as the linear quantities that we practiced in Chapter 2.

To see what we mean by **angular position**, imagine a line drawn on the extended object outward from the rotation axis (**Figure 10-13**). The angle of this line from a reference direction changes as the extended object rotates, and it's this angle that we call the angular position θ. The angular displacement from time 0 to time t is $\theta - \theta_0$.

So we can take the equations for constant linear acceleration and convert them to the equations for constant angular acceleration by replacing x with θ, v_x with ω_z,

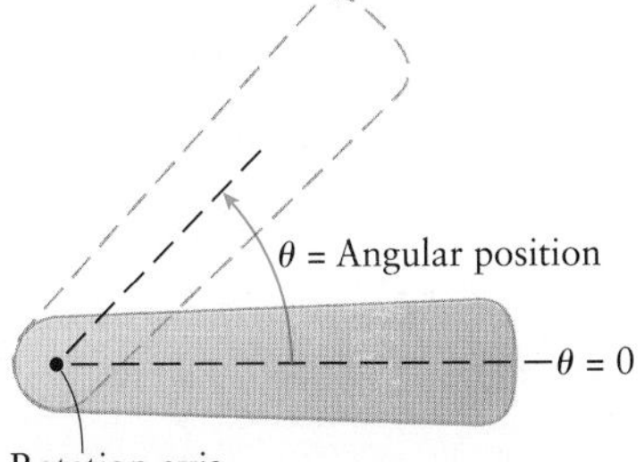

Figure 10-13 Angular position Imagine a line drawn on this rigid body from its rotation axis outward. The angle of this line at a given time is the extended object's angular position θ at that time.

and a_x with α_z. The equations for linear motion in the x direction for constant acceleration are

(2-5) $$v_x = v_{0x} + a_x t$$

(2-9) $$x = x_0 + v_{0x}t + \frac{1}{2}a_x t^2$$

(2-11) $$v_x^2 = v_{0x}^2 + 2a_x(x - x_0)$$

Hence the equations for constant angular acceleration for a rigid body are

EQUATION IN WORDS
Angular velocity, angular acceleration, and time for constant angular acceleration

(10-13) $$\omega_z = \omega_{0z} + \alpha_z t$$

- ω_z: **Angular velocity** at time t of a rotating rigid body with constant angular acceleration
- ω_{0z}: **Angular velocity** of the rigid body at time $t = 0$
- α_z: **Constant angular acceleration** of the rigid body
- t: **Time** at which the rigid body has angular velocity ω_z

EQUATION IN WORDS
Angular position, angular acceleration, and time for constant angular acceleration

(10-14) $$\theta = \theta_0 + \omega_{0z}t + \frac{1}{2}\alpha_z t^2$$

- θ: **Angular position** at time t of a rotating rigid body with constant angular acceleration
- ω_{0z}: **Angular velocity** of the rigid body at time $t = 0$
- α_z: **Constant angular acceleration** of the rigid body
- θ_0: **Angular position** at time $t = 0$ of a line on the rigid body
- t: **Time** at which the line on the rigid body is at angular position θ

EQUATION IN WORDS
Angular velocity, angular acceleration, and angular position for constant angular acceleration

(10-15) $$\omega_z^2 = \omega_{0z}^2 + 2\alpha_z(\theta - \theta_0)$$

- ω_z: **Angular velocity** at angular position θ of a rotating rigid body with constant angular acceleration
- ω_{0z}: **Angular velocity** of the rigid body when a line on it is at angular position θ_0
- α_z: **Constant angular acceleration** of the rigid body
- θ, θ_0: Two **angular positions** of the line on the rigid body

Exam Tip

While Equations 10-13 and 10-14 are on the AP® Physics 1 equation sheet, Equation 10-15 is not. The linear equivalent (Equation 2-11) is on the equation sheet, so you can reproduce Equation 10-15 by understanding the relationships between the linear and angular kinematic equations.

We can use Equations 10-13, 10-14, and 10-15 to solve a wide range of problems in rotational motion, just as we used Equations 2-5, 2-9, and 2-11 for linear motion problems in Chapter 2. Just as we saw in Chapter 2 for Equation 2-11, Equation 10-15 is derived from Equations 10-13 and 10-14, so they are not all three independent equations and we can solve for only two unknowns. Here's a representative example.

WATCH OUT !

Equations 10-13, 10-14, and 10-15 are for rigid bodies only.

The three equations we've just presented can be used only if *all parts* of a rotating extended object have the same angular velocity and angular acceleration at any given time—in other words, only if the extended object is a *rigid body*. They *cannot* be used for a rotating extended object that isn't rigid, such as water in a bathtub as it swirls down the drain or a cake mix being stirred in a bowl. In Example 10-8, and in the problems at the end of the chapter, always assume that the rotating extended objects are rigid bodies unless we specifically tell you otherwise.

EXAMPLE 10-8 A Stopping Top

A top spinning at 4.00 rev/s comes to a complete stop in 64.0 s. Assuming the top slows down at a constant rate, how many revolutions does it make before coming to a stop?

Set Up

The statement that the top slows down at a constant rate means that its angular acceleration is constant, so we can use Equations 10-13 and 10-14. (We won't use Equation 10-15 because that equation doesn't involve time, which is one of the quantities we're given.) These two equations involve six quantities (t, θ_0, θ, ω_{0z}, ω_z, and α_z).

To find constant angular acceleration α_z from ω_{0z}, ω_z, and t:

$$\omega_z = \omega_{0z} + \alpha_z t \qquad (10\text{-}13)$$

In this case we are given the initial angular velocity $\omega_{0z} = 4.00$ rev/s, the final angular velocity $\omega_z = 0$, and the time interval $\Delta t = 64.0$ s. If we consider the time the top was at its initial position $t = 0$, then this is just our final t. If we let the top start at an angular position of $\theta_0 = 0$, we want to find the top's final angular position θ at the end of this time interval but don't know the value of the angular acceleration α_z. We'll use Equation 10-13 to determine α_z from the known information, then substitute this value into Equation 10-14 to find the value of θ.

To find the final angular position θ from θ_0, ω_{0z}, α_z, and t:

$$\theta = \theta_0 + \omega_{0z}t + \frac{1}{2}\alpha_z t^2 \qquad (10\text{-}14)$$

Solve

Rewrite Equation 10-13 to solve for α_z. Then substitute the values of ω_z, ω_{0z}, and t. At some point we need to convert revolutions to radians: Because these are in the expected units for angular velocity, position, and acceleration, they allow us to smoothly transition between angular and linear quantities. One full revolution is 2π radians, so we just multiply by 2π.

Rewrite Equation 10-13: $\omega_z - \omega_{0z} = \alpha_z t$

$$\alpha_z = \frac{\omega_z - \omega_{0z}}{t}$$

The resulting value of α_z is negative, whereas ω_{0z} is positive, which correctly says that the rotation is slowing down. Note that by writing $\omega_{0z} = 2\pi \times 4.00$ rad/s, we made the assumption that the top is initially rotating in the positive direction. If we had made the opposite choice for angular velocity, we would have gotten a positive acceleration, so only the sign, not the magnitude of our calculations, would change.

Substitute values: $\alpha_z = \dfrac{0 - 4.00 \text{ rev/s} \times (2\pi \text{ rad/rev})}{64.0 \text{ s}} = -0.393 \text{ rad/s}^2$

Solve for θ by plugging this value of α_z into Equation 10-14 along with the given values of ω_{0z} and t.

Equation 10-14: $\theta = \theta_0 + \omega_{0z}t + \dfrac{1}{2}\alpha_z t^2$

$$= 0 \text{ rad} + (25.13 \text{ rad/s})(64.0 \text{ s}) + \frac{1}{2}(-0.393 \text{ rad/s}^2)(64.0 \text{ s})^2$$

$$= 1608.5 \text{ rad} - 804.2 \text{ rad}$$

$$= 804 \text{ rad} \Rightarrow 804 \text{ rad}/(2\pi \text{ rad/rev}) = 128 \text{ rev}$$

Reflect

We can check our answer by substituting the values for ω_z, ω_{0z}, and α_z into Equation 10-15 (which we did not use earlier) and solving for θ. This gives us the same answer for θ as above.

In translational motion, an increasing positive displacement would move an object farther in the positive direction from its initial position. In rotational motion, each full trip around, each 2π, brings the rigid body back to its initial position. If we want to see how much different the final position is, we can subtract off factors of 2π. Here, because the angular velocity was a whole number of revolutions per second, and the angular acceleration was divided and then multiplied by the same factor, we know that there was a whole number of revolutions, so, even though the angular position has changed a lot, the rigid body is in the same orientation in which it started!

Equation 10-15: $\omega_z^2 = \omega_{0z}^2 + 2\alpha_z(\theta - \theta_0)$

Rewrite to solve for θ: $\omega_z^2 - \omega_{0z}^2 = 2\alpha_z(\theta - \theta_0)$

$$\theta = \frac{\omega_z^2 - \omega_{0z}^2}{2\alpha_z} + \theta_0 = \frac{(0\ \text{rev/s})^2 - (4.00\ \text{rev/s})^2}{2(-0.0625\ \text{rev/s}^2)} + 0\ \text{rev}$$
$$= 128\ \text{rev}$$

NOW WORK Problems 1–3 from The Takeaway 10-4.

THE TAKEAWAY FOR Section 10-4

✔ Just as acceleration in a line is the rate at which velocity changes with respect to time, angular acceleration is the rate at which angular velocity changes with respect to time.

✔ If the angular acceleration is constant for a rigid body, there are three equations that relate time, angular position, angular velocity, and angular acceleration that are of the same structure as the relationships among the corresponding linear quantities.

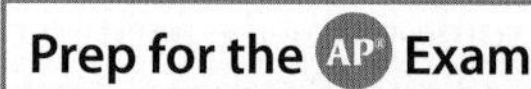

AP® Building Blocks

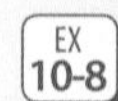

1. A wheel is supported by an axle through its center and is rotating at 1.0 rev/s.
   (a) The wheel has an angular acceleration of −0.020 rev/s². Calculate the interval of time before the wheel comes to rest.
   (b) Calculate the interval of time for the wheel to come to rest if the angular acceleration is −0.020 rad/s².
   (c) Calculate the number of complete rotations for the situation in part (b). Record the result in both revs and rads.

2. A spinning top completes 6.00×10^3 rotations before it starts to topple over. The average angular speed of the rotations is 8.00×10^2 rpm. Calculate how long the top spins before it begins to topple.

3. A communication satellite circles Earth in a geosynchronous orbit, which means that the satellite remains directly above the same point on the surface of Earth.
   (a) What angular displacement (in radians) does the satellite undergo in 1 hour of its orbit?
   (b) Calculate the angular speed of the satellite in rev/min and rad/s.
4. A disk is initially spinning clockwise at 60 rpm when it is given a constant angular acceleration of 30 rev/min².
   (a) Describe the motion in 1-min intervals up to 4 min.
   (b) Construct a graph of ω versus t in units of rad/s and s.
   (c) Add annotations to the graph showing the slope and its numerical value, the time at which the disk instantaneously stops, and the range of ω where the rotation is clockwise and counterclockwise.

AP® Skill Builders

5. A disk rotates in a horizontal plane about its center. At $t = 0$ the angular position of the disk is 0. The table shows the signs of the initial angular velocity and constant angular acceleration under six different conditions. Remember, angular displacement is by how much the angular position of the disk has changed. To reduce confusion, we will consider the angular position of the disk as it passes through its starting position when it has a positive angular displacement as $n2\pi$ (so $+2\pi$ the first time around, $+4\pi$ the second, and so on).

Condition	Initial velocity	Constant acceleration
A	+	+
B	+	−
C	−	+
D	−	−
E	+	0
F	−	0

(a) On a sheet of paper sketch representations of the angular displacement, θ, versus time, t, and angular velocity, ω, versus time for each condition.
(b) Using your sketches, construct claims that identify conditions that will
 i. cause the disk to momentarily stop.
 ii. bring the disk to $\theta = 0$, perhaps given enough time.
 iii. definitely not bring the disk to $\theta = 0$.
 iv. result in a displacement of $-2n\pi$, where n is an integer not equal to zero, given enough time.
 v. result in a displacement of $2n\pi$, where n is an integer not equal to zero, given enough time.

6.

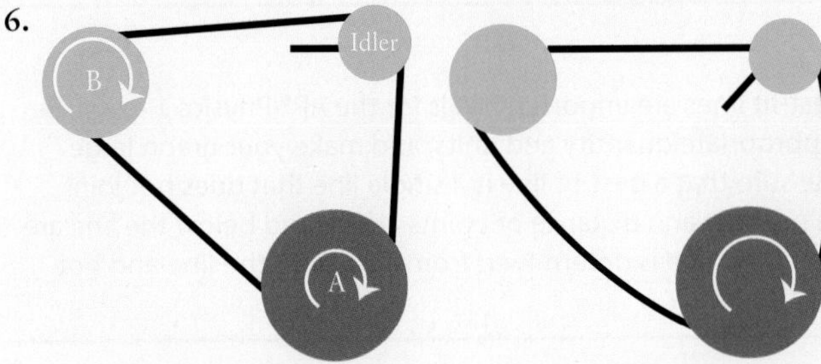

Machines driven by pulleys often have a moveable idler pulley whose position can be adjusted to remove slack in the belt. Pulley A in the figure is turned by a shaft. When the pulley spins, a rubber belt is pulled without slipping, so that the idler pulley makes the belt taut. The rotation of pulley A then drives the rotation of pulley B. When the idler pulley is lowered so that the belt is not taut the driving pulley A continues to spin, but now the belt slips and the driven pulley B stops rotating.

(a) Pulley A has a radius R_A and pulley B has a smaller radius R_B. Express the linear speed, v, of the belt in terms of R_A and ω_A, when the belt is taut and the driving pulley is rotating at a constant speed.
(b) Express the rotational speed of B in terms of R_A, R_B, and ω_A, when the belt is taut and the driving pulley is rotating at a constant speed.

The idler pulley can act as a safety switch. When the idler pulley is in the down position a brake presses against pulley B. If there is a sudden tug on the belt the idler pulley changes position, the belt becomes slack, and the brake brings pulley B to a halt.

A circular saw blade is bolted onto pulley B. The blade has a radius $5R_B = 3R_A$. The blade rotates at 7000 rpm.

(c) Calculate the angular speeds of pulleys A and B.
(d) The blade comes to a complete stop 10.0 ms after the brake is activated. Calculate the angular acceleration of the saw blade caused by the brake and the number of rotations of the saw blade after the brake is activated.

7. Optical disk drives, such as CDs and DVDs, have information stored in small depressions "burned" onto the disk surface. A laser is used to detect changes in the depth on the surface as the disk spins. A change in depth is interpreted as a 0 and no change in depth is interpreted as a 1. There are millions of these burns on the disk's surface. The zeros and ones are interpreted by a small computer within the player to retrieve the stored information. Most optical disk drive technologies are designed to maintain a constant rate of information retrieval. If the laser samples depth on the surface at constant intervals of time, independent of where the data are on the disk, explain why the angular velocity of the disk must vary for the sampling rate to be constant.

8. The reduction of gravitational force exerted on the human body during long times in space affects health. A substitute for Earth's gravitational force is to simulate gravity by exerting a contact force between the astronaut's body and the surface of a wheel-shaped space habitat.
(a) Construct a diagram of an object (an astronaut) standing on the inside of the outer surface of a toroidal (doughnut-shaped) space station, showing the direction of the contact force exerted on the astronaut by the space station.
(b) Express the relationship between this contact force and the centripetal acceleration of the astronaut in terms of the angular speed of rotation of the habitat and the outer radius of the wheel.
(c) Calculate the rotational speed that a toroidal space station would need to produce a centripetal acceleration that has the same magnitude as the acceleration caused by gravity near the surface of Earth, if the outer surface of the toroidal space station has a diameter of 2 km and the station is spinning about its center.
(d) Calculate the time for one revolution of the space station.

AP® Skills in Action

9. A child pushes a merry-go-round that has a diameter of 4.00 m and goes from rest to an angular speed of 18.0 rpm in a time of 43.0 s, and then hops on the merry-go-round.
(a) Calculate the average angular acceleration (in rad/s^2) of the merry-go-round.
(b) Calculate the angular displacement (in rad) of the merry-go-round while the child is pushing it.
(c) What is the maximum linear speed of the child as she rides on the edge of the platform?

10. A disk has an angular velocity of 60 rpm in the clockwise direction. At $t = 0$ the angular position of a point on the rim of the disk is $\theta = -30°$ and the disk is given an angular acceleration of π rad/s^2.
(a) Express the angular displacement of the point as a function of time, where the units of each term are rads and seconds.
(b) Calculate the angular displacement of the disk when it has an angular velocity of zero.
(c) Construct a θ–t graph from 0 to 5 s with angular displacement measured in radians.

11.

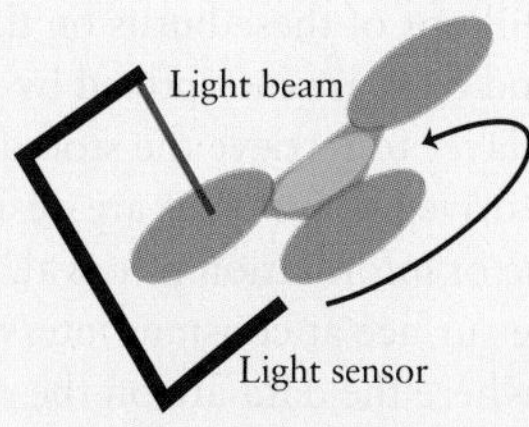
Light beam
Light sensor

A fidget spinner is positioned so that the arms of the spinner can block the beam of a photogate timer. The data collection software is set up to calculate angular speed from the time interval an arm blocks the light every 20 seconds.

The data shown here were collected from this apparatus.

t (s)	0	20	40	60	80	100	120
ω (rad/s)	219	192.6	166.2	139.8	113.4	87	60.6

(a) Construct an $\omega - t$ graph from these data.
(b) Using your graph, calculate the angular acceleration of the arms of the spinner.

Exam Tip

Graphing and identifying best-fit lines are important skills for the AP® Physics 1 exam. Be sure to label axes with the appropriate quantity and units, and make your graph large enough to be understood. Be sure that a best-fit line is a single line that does not join data points. Be sure that the number and distance of points above and below the line are balanced, and that the slope of the line is determined from points on the line and not data points on the graph.

10-5 Torque is to rotation as force is to translation

WATCH OUT !

New terminology alert!

A force is a way of describing an interaction, and is always exerted by something on something else. It is common, especially in rotation, to see forces referred to as "applied forces" since we will be worried about exactly where the force is exerted. This is often referred to as the point of application. "Applied" is really just another way to say "exerted on the extended object," so don't let it bother you.

Where do you push or pull on a door to open it most easily? In what direction should you exert that push or pull? These may seem like odd questions to ask in a chapter about rotational motion. But keep in mind that when you open a door you are giving it an *angular acceleration*: The door starts at rest and begins rotating around its hinges, so you're changing the door's angular velocity. Let's rephrase our questions like this: Where, and in what direction, should we push or pull on a door to give it an angular acceleration around its hinges?

Experience tells you that it's nearly impossible to open a door by pushing or pulling near its hinges, as with force $\vec{F}_1$ in **Figure 10-14.** A much better place to exert a force on the door is on the opposite end from its hinges, which is why the doorknob is located there. Even then, it's ineffective to push or pull on the doorknob in a direction parallel to the plane of the door (as with force $\vec{F}_2$ in Figure 10-14). The easiest way to open the door is to exert a force far from the hinges (that is, at the doorknob) in a direction perpendicular to the plane of the door, as with force $\vec{F}_3$ in Figure 10-14.

This example shows that to give an extended object an angular acceleration, what matters is not just how hard you push or pull on the extended object, but also *where* and *in what direction* that push or pull is applied. To simplify our discussion, if it is

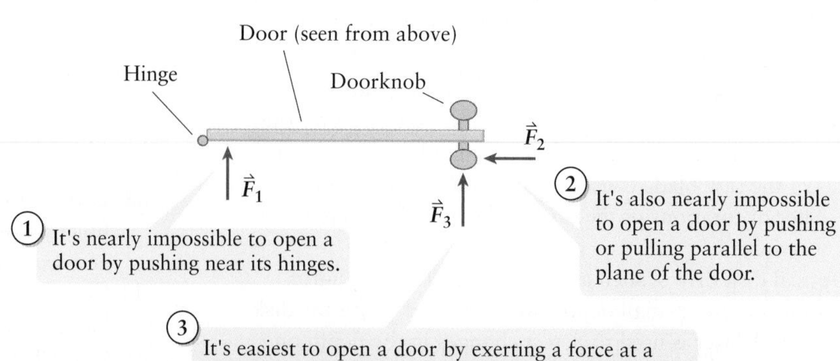
Door (seen from above)
Hinge
Doorknob
$\vec{F}_1$
$\vec{F}_2$
$\vec{F}_3$
1 It's nearly impossible to open a door by pushing near its hinges.
2 It's also nearly impossible to open a door by pushing or pulling parallel to the plane of the door.
3 It's easiest to open a door by exerting a force at a point far from the hinges (where the doorknob normally is) in a direction perpendicular to the plane of the door.

Figure 10-14 Opening a door When you push or pull on a door to open it, it matters where and in what direction you exert the force.

clear what we are trying to accelerate, we usually just say "applied force" instead of "force exerted on the extended object." Then the point where the force is "exerted on the extended object" is just where the force is "applied." A physical quantity that relates all these aspects of an applied force is the *torque* associated with that force. In this section we'll find that just as the net force exerted on an object determines its translational or linear acceleration (Newton's second law), the net *torque* exerted on an extended object determines its *angular* acceleration.

Defining the Magnitude and Direction of Torque

Suppose that you exert a force $\vec{F}$ on an extended object, as shown in **Figure 10-15a.** We use the symbol $\vec{r}$ to denote the displacement from the rotation axis to the point where the force is applied, and we use the symbol ϕ (the Greek letter "phi") for the angle between the directions of $\vec{r}$ and $\vec{F}$. The component of $\vec{F}$ that points straight out from the rotation axis, $F\cos\phi$, doesn't have any tendency to make the extended object rotate. But the perpendicular component of $\vec{F}$, $F\sin\phi$, *does* tend to make the extended object rotate—in this case in a clockwise direction. We describe the rotational effect of the force $\vec{F}$ by a quantity called the **torque** τ (the Greek letter "tau") associated with the force. This is given by

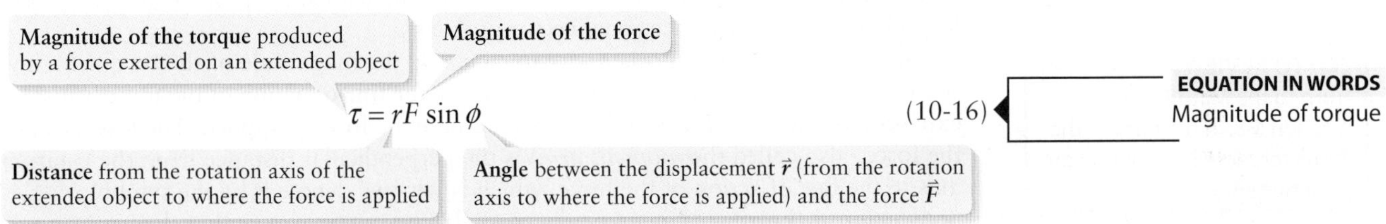

$$\tau = rF\sin\phi \qquad (10\text{-}16)$$

EQUATION IN WORDS
Magnitude of torque

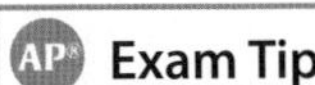

Exam Tip

Be sure to identify your angles in a problem; do not just use any angle given without thinking about whether it is the correct angle for what you are calculating: ϕ is the angle between the force and displacement, which may need to be determined from other angles given. Note that on the AP® Physics 1 equation sheet, the angle between the force and displacement $\vec{r}$ is given as θ rather than ϕ, and this is a different angle from the θ given in the rotational kinematics equation for angular position.

Views of extended object from along its rotation axis

(a) A force that tends to cause clockwise rotation

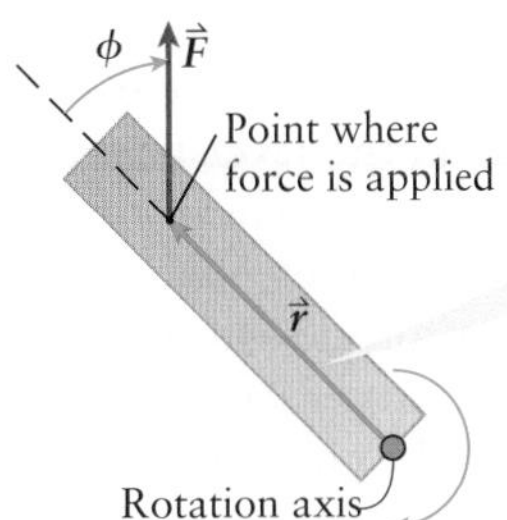

The displacement $\vec{r}$ points from an extended object's rotation axis to where a force $\vec{F}$ is exerted on the extended object.

(b) A force that tends to cause counterclockwise rotation

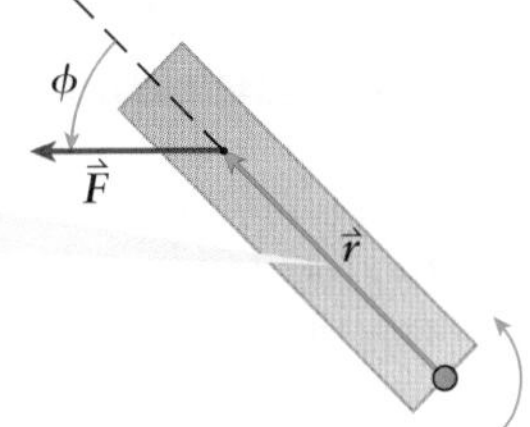

(c) Determining the line of action of a force and its lever arm

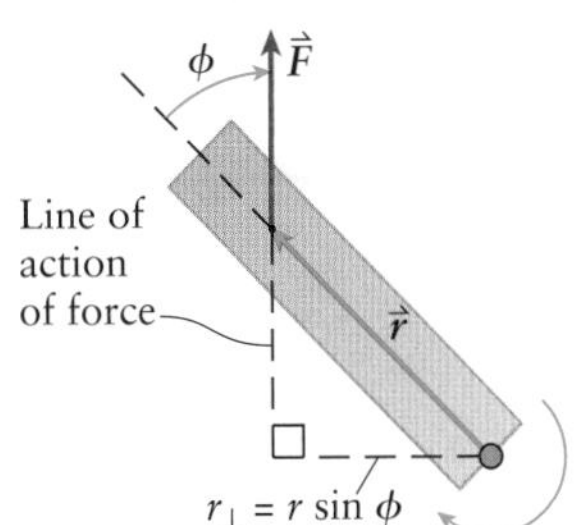

The lever arm $r_\perp$ is perpendicular to the line of action of the force, and extends from the rotation axis to the line of action of the force.

Figure 10-15 Torque (a) The displacement $\vec{r}$ points from an extended object's rotation axis to where a force $\vec{F}$ is applied (exerted on the extended object). The torque produced by the force has magnitude $\tau = rF\sin\phi$. In this case the torque is clockwise because it tends to make the extended object rotate clockwise. (b) This torque is counterclockwise because it tends to make the extended object rotate counterclockwise. (c) Finding the torque in terms of the lever arm $r_\perp$.

Equation 10-16 tells us the *magnitude* of the torque, but torque also has *direction*. For the force shown in Figure 10-15a, we say that the associated torque is clockwise because it tends to make the extended object rotate clockwise. By contrast, the force depicted in **Figure 10-15b** tends to make the extended object rotate counterclockwise, so we say that the torque associated with this force is counterclockwise. Although shortly, we'll learn more about how to represent torque as a vector, for rotation around a single axis with fixed direction, like a door rotating around its hinges, the vector nature of torque is adequately communicated by clockwise or counterclockwise.

In Section 10-2 we denoted the rotation axis as the z axis and used the symbol ω_z to denote angular velocity, a quantity that can be positive or negative depending on the direction of rotation. We did this to distinguish ω_z from the angular speed ω, which is the magnitude of angular velocity. In the same way we'll use τ_z as the symbol for torque and τ (with no subscript) to denote the magnitude of torque, as in Equation 10-16. If we choose the positive rotation direction to be counterclockwise, then in Figure 10-15a the torque is negative (it tends to cause clockwise rotation) and $\tau_z < 0$. In Figure 10-15b the torque has the same magnitude τ (the distance r, force magnitude F, and angle ϕ are the same as in Figure 10-15a), but the torque tends to cause counterclockwise rotation and so $\tau_z > 0$. We don't have to assign signs, and could just label them clockwise or counterclockwise, but some find it easier to use signs when we have to keep track of more than one torque, which might be in different directions, in a problem, just as we did for right or left in one-dimensional vector problems.

NEED TO REVIEW?
Turn to the Math Tutorial, Section M8, in the back of the book for more information on trigonometry.

Figure 10-15c shows another way to calculate the magnitude of the torque for the situation in Figure 10-15a. As this figure shows, the **line of action** of the force is just an extension of the force $\vec{F}$ through the point where the force is applied. The **lever arm** of the force (also called the *moment arm*) is the perpendicular distance from the rotation axis to the line of action of the force, which is why we denote it by the symbol $r_\perp$ ($\perp$ is mathematical shorthand for perpendicular). Trigonometry shows that the lever arm $r_\perp$ equals $r \sin \phi$, the same quantity that appears in Equation 10-16. Using these definitions, we can rewrite that equation as

EQUATION IN WORDS
Magnitude of torque in terms of lever arm

Magnitude of the torque produced by a force exerted on an extended object

Magnitude of the force

$$\tau = r_\perp F \tag{10-17}$$

Lever arm = perpendicular distance from the rotation axis of the extended object to the line of action of the force

Equation 10-17 is mathematically equivalent to Equation 10-16 but defines torque in a slightly different way: It says that the magnitude of the torque produced by a force equals the lever arm of the force multiplied by the force magnitude. The longer the lever arm for a given force magnitude, the greater the torque.

Figure 10-16 illustrates these ideas about torque using the door that we discussed previously (see Figure 10-14). For a given force magnitude F, the lever arm $r_\perp$ and hence the torque magnitude τ will be as large as possible if r is as large as possible and if $\sin \phi$ has its maximum value of 1, which occurs if $\phi = 90°$. In other words, a given force exerted on an extended object produces the maximum torque—and hence the maximum rotating effect—if it is applied as far from the rotation axis as possible and in a direction perpendicular to a line from the rotation axis to the point of application

(a) Pulling on a door far from the hinge (view as seen looking down on the door from above): large lever arm $r_\perp$

(b) Pulling on a door close to the hinge: small lever arm $r_\perp$

(c) Pushing on a door directly toward the hinge: zero lever arm $r_\perp$

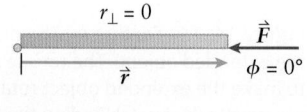

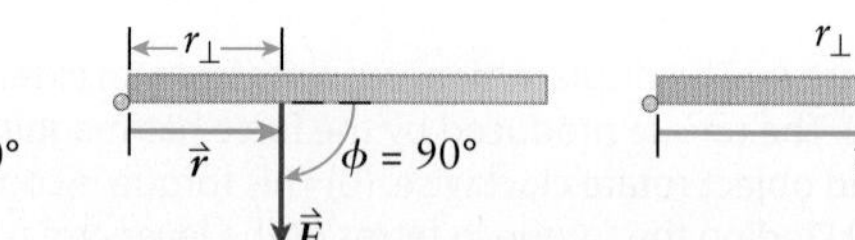

Figure 10-16 Torques on a door (a) For a force $\vec{F}$ of a given magnitude, the torque is greatest when r is large and $\phi = 90°$. The torque is small if (b) r is small or (c) $\sin \phi$ is small.

(see **Figure 10-16a**). That's just what we said in our discussion of the door. By contrast if the force is applied close to the rotation axis so the distance r is small—like pushing on a door at a point close to the hinges—the lever arm $r_\perp = r \sin \phi$ is short, so the torque magnitude and rotating effect are small (**Figure 10-16b**). The torque is also small if the force is applied in nearly the same direction as $\vec{r}$, as in **Figure 10-16c.** In that case the angle ϕ is nearly zero, $\sin \phi$ has a small value, and the lever arm $r_\perp = r \sin \phi$ is again short. This is like pulling on a door in a direction nearly parallel to the door's plane. (The same is true if the force is directed nearly opposite to $\vec{r}$, so ϕ is close to 180° and again $\sin \phi$ is close to zero. Imagine you are leaning on the wall near the hinges and pulling the door handle straight toward yourself—it's hard to open the door!)

From Equation 10-17, the SI units of torque are newtons multiplied by meters, or newton-meters (abbreviated N · m).

WATCH OUT !

Torque and work have the same units but are very different quantities.

Both torque and work have units of newton-meters. But you can't set one equal to the other. Work involves the product of the component of a force parallel to a displacement and the magnitude of the displacement. Torque is the product of the component of a force perpendicular to a lever arm and the magnitude of the lever arm. Also, work is a scalar quantity and torque is a vector quantity. Because physically they represent two very different things, it makes no sense to set one equal to the other.

Lever Arms in People

As Equation 10-17 shows, even a small force F can generate a large torque τ if the lever arm $r_\perp$ is long enough. Humans and other animals take advantage of this principle through the arrangement of their muscles and bones. As an example, **Figure 10-17a** shows that the point at which the masseter muscle is attached to the lower jawbone is far from the joint around which the jaw rotates. Hence the force exerted by the masseter muscle has a long lever arm, and so produces a large torque on the lower jawbone. This large torque enables you to crack a nut between your back teeth or use your front teeth to tear into a raw carrot.

A very different arrangement is used in your forearm. One end of the biceps muscle attaches to the bone of the upper arm and the other to the lower arm just below the elbow (**Figure 10-17b**). Even though the biceps muscle can exert a large force, this muscle–joint arrangement doesn't generate a large torque because the lever arm is relatively short. In chimpanzees the bicep is attached farther from the elbow joint, giving it the ability to generate more forearm torque than a human despite the chimpanzee's smaller size and smaller muscles.

Torque, Angular Acceleration, and Newton's Laws for Rotation: Rotational Dynamics

We have suggested that a net torque exerted on an extended object (like the torque produced by pushing or pulling on a door) affects its rotation. What is the precise

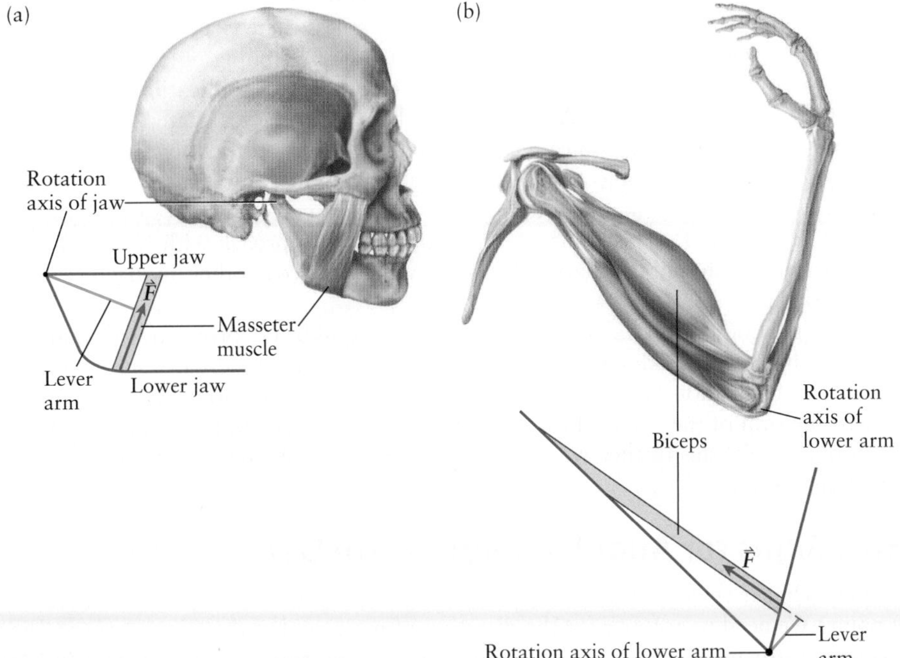

Figure 10-17 Lever arms in the human jaw and arm (a) The masseter muscle exerts a force on the lower jaw with a relatively long lever arm, so the torque on the lower jaw is quite large. (b) The lever arm for the force of the biceps muscle on the forearm is relatively short, so the torque is relatively small.

relationship between net torque and the rotational effect that it causes? To find out let's recall Newton's second law for *linear* motion along the x axis:

(4-3a) $$\sum F_{\text{ext},x} = ma_x$$

This equation states that an object accelerates in response to the sum of all the external forces that are exerted on the object: This sum $\sum F_{\text{ext},x}$ equals the object's mass m multiplied by its acceleration a_x. To find the rotational equivalent of this equation, remember from Section 10-2 that rotational inertia I plays the same role for rotational motion that mass m does for linear motion and recall from Section 10-4 that angular acceleration α is the rotational equivalent of linear acceleration a_x. If torque plays the same role for rotational motion as force plays for translational motion, then we should be able to replace F_x with τ_z, m with I, and a_x with α_z in Equation 4-3a to get the rotational version of Newton's second law. Since this depends on the extended object keeping a fixed shape, we again derive the equations for a rigid body:

EQUATION IN WORDS
Newton's second law for rotational motion

The **net torque (the sum of all external torques)** exerted on a rigid body for the axis about which it can rotate

The **angular acceleration** produced by the net torque

(10-18) $$\sum \tau_{\text{ext},z} = I\alpha_z$$

The rigid body's **rotational inertia** for the axis about which it can rotate

Experiment supports Equation 10-18. An extended object acquires an angular acceleration α_z in response to the sum $\sum \tau_{\text{ext},z}$ of all the external torques that are exerted on the extended object. This angular acceleration depends on the mass and shape of the extended object (in other words, it depends on the rotational inertia). For a rigid body, angular acceleration is directly proportional to the net torque on the rigid body, and inversely proportional to the rigid body's rotational inertia I around the rotation axis. The basic physics is the same as the original form of Newton's second law, Equation 4-3a.

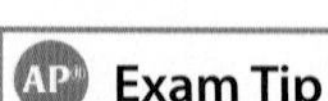

Equilibrium does not mean at rest in rotational motion either! Translational equilibrium means that the system has zero linear acceleration, so it could have a constant velocity. Rotational equilibrium means the system has zero rotational acceleration, so it could have a constant angular velocity.

Because Newton's second law holds for rotational motion, we can also apply Newton's first law in rotational motion. We know from Newton's first law in linear motion that the center of mass of an extended object that is at rest will remain at rest unless there is a net force exerted on the extended object. When we were only considering translational motion, this was sufficient to describe a state of equilibrium for a system. Now we consider rotation about the center of mass. We saw that it took a torque to cause a rotational acceleration of a rigid body, so we know that a rigid body that is spinning will not change how it is spinning unless there is a net external torque exerted on it and one that is not spinning will not start spinning unless there is a net external torque exerted on it. **Rotational equilibrium** is the application of Newton's first law in rotational motion. As we saw for the case of the ice skater and as is true for other nonrigid extended objects, application of this concept involves the following subtlety: If the extended object can change the distribution of its mass about its rotational axis, the rate at which it rotates about that axis can change even if there is no external net torque. This last concept is very difficult to approach with Newton's laws, so discussion is deferred to Chapter 11, when we introduce another important conservation principle.

For rigid bodies, we can approach a rotational equilibrium problem with Equation 10-18, with angular acceleration necessarily zero. This is achieved when the sum of the external torques about a rotational axis in the clockwise direction is equal in magnitude to the sum of the external torques about that same axis in the counterclockwise direction. To take this further, we must consider more deeply the vector nature of rotational quantities.

How Signs for Angular Velocity Are Determined

Similar to linear velocity, angular velocity is a *vector, denoted* $\vec{\omega}$. In what direction does $\vec{\omega}$ point? We need to account for the direction of rotation, but there is no way to align a single vector in the plane of the rotation to indicate this direction (**Figure 10-18a**).

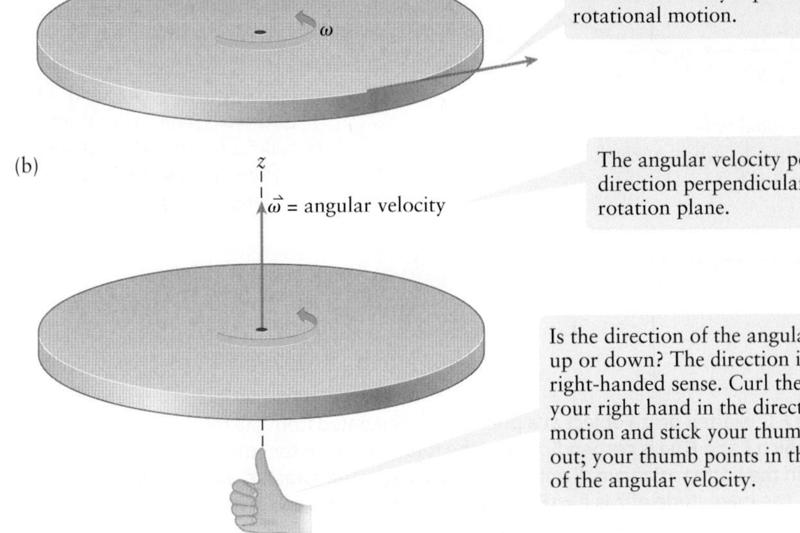
(a)
ω
No single vector in the plane of rotation correctly represents the rotational motion.
(b)
z
$\vec{\omega}$ = angular velocity
The angular velocity points in a direction perpendicular to the rotation plane.
Is the direction of the angular velocity up or down? The direction is up in a right-handed sense. Curl the fingers on your right hand in the direction of motion and stick your thumb straight out; your thumb points in the direction of the angular velocity.

Figure 10-18 Angular velocity Rather than lying in (a) the plane of rotation, the angular velocity of a rotating system is (b) oriented along the rotation axis. The direction of $\vec{\omega}$ is given by a right-hand rule.

Instead, we take the angular velocity to point along the axis of rotation, which we have been denoting the z axis. By convention the specific direction of $\vec{\omega}$ is given by a **right-hand rule**. Curl the fingers of your right hand in the direction of motion and stick your thumb straight out. By this right-hand rule, your thumb points in the direction of the angular velocity (**Figure 10-18b**). If the rotation is in the direction shown in Figure 10-18b, $\vec{\omega}$ points in the positive z direction so the component ω_z is positive; if the rotation is in the opposite direction, $\vec{\omega}$ points in the negative z direction and the component ω_z is negative. That's the origin of the positive and negative signs for ω_z that we introduced in Section 10-2: If you wrap your fingers in a counterclockwise direction in the x–y plane, your thumb points in the positive z direction.

Torque as a Vector

Like angular velocity $\vec{\omega}$, torque is a vector $\vec{\tau}$ that has both a magnitude and a direction. Equation 10-16 tells us the magnitude of the torque: $\tau = rF \sin \phi$.

In Equation 10-16, r is the displacement measured from the rotation axis to the point where a force of magnitude F is applied. Both the displacement $\vec{r}$ and the force $\vec{F}$ are vectors, and ϕ is the angle between $\vec{r}$ and $\vec{F}$. The direction of $\vec{\tau}$ is given by a right-hand rule that's similar to the one we described to find the direction of the angular velocity vector: Curl the fingers of your right hand in the direction that the torque would tend to make the object rotate, and your right thumb will point in the direction of $\vec{\tau}$ (**Figure 10-19**). Another right-hand rule is to point the fingers of your right hand along the direction of $\vec{r}$ so that your palm faces $\vec{F}$, then curl your fingers from the direction of $\vec{r}$ to the direction of $\vec{F}$ along the shortest path. Your extended right thumb then points in the direction of $\vec{\tau}$. (Practice this with the two situations shown in Figure 10-19.) As Equation 10-18 shows, and as we would expect from Newton's laws in linear systems, the direction of the angular acceleration will always be in the direction of the net torque. All right-hand rules will give the same result.

Note that $\vec{\tau}$ is perpendicular to the plane defined by $\vec{r}$ and $\vec{F}$. Mathematically, a combination of $\vec{r}$ and $\vec{F}$ that has these properties is called the **cross product** (or **vector product**) of $\vec{r}$ and $\vec{F}$. You can learn more about the properties of the cross product in the Math Tutorial. In this text, though, we will rely on Equation 10-16 to calculate the magnitude of the torque and on the right-hand rule to determine its direction.

NEED TO REVIEW?

Turn to the Math Tutorial, Section M10, in the back of the book for more information on the cross product.

WATCH OUT!

There is no motion along the direction of torque.

It's important to recognize that no motion occurs in the direction of the torque. For example, the torques shown in Figure 10-19 are directed either straight up or straight down, but they don't cause the bar on which the torque is exerted to *move* either up or down: Instead, the bar rotates around the rotation axis. The direction of the torque is a mathematical tool.

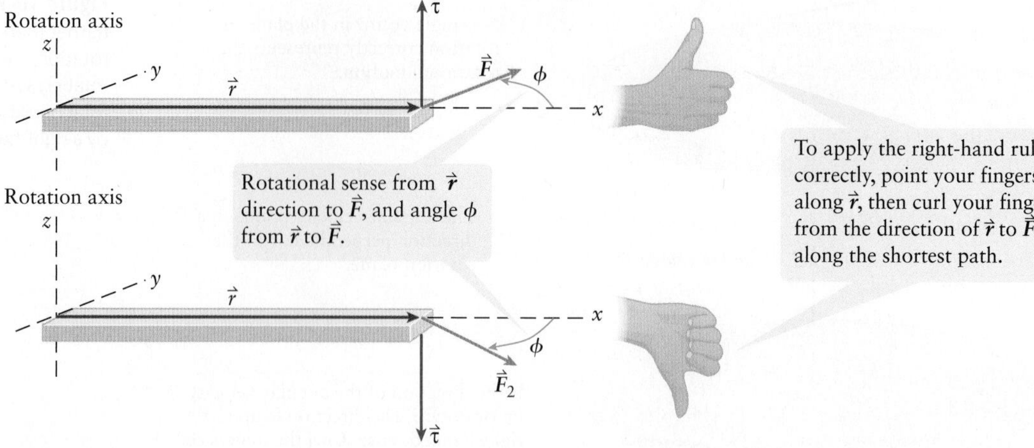

Figure 10-19 Torque A force $\vec{F}$ is exerted on an object at a point that is separated from the rotation axis, z, by a displacement $\vec{r}$. The plane in which $\vec{r}$ and $\vec{F}$ lie is perpendicular to the rotation axis. In the top diagram, $\vec{F}$ is directed into the page of the textbook. In the bottom diagram, $\vec{F}$ points out of the page. The torque $\vec{\tau}$ produced by this force is perpendicular to both $\vec{r}$ and $\vec{F}$. The magnitude of $\vec{\tau}$ is $\tau = rF \sin\phi$ and the direction of $\vec{\tau}$ is given by a right-hand rule.

If the *vector* sum $\sum\vec{\tau}$ of all external torques exerted on the system is zero, its angular acceleration is zero. For equilibrium, the net force on the system also needs to be zero to keep the center of mass from having a translational acceleration!

EXAMPLE 10-9 Balancing on a Seesaw

A seesaw is a uniform plank supported at its midpoint. Akeelah and her little sister Bree sit on opposite sides of the seesaw. Akeelah weighs 1.50 times as much as Bree, and Akeelah sits 80.0 cm from the midpoint of the seesaw. If the seesaw is tilted by an angle θ from the horizontal, where should Bree sit so that they just balance?

Set Up

When the system made up of Akeelah, Bree, and the seesaw is balanced, the angular acceleration of the system is constant, which in this case is zero: It has no tendency to start rotating one way or the other. Hence, the net torque on the system must be zero.

We are told that Akeelah's distance from the pivot is $r_A = 80.0 \text{ cm} = 0.800 \text{ m}$ and that Akeelah's weight is 1.50 times that of Bree; that is, $F_{gA} = 1.50\ F_{gB}$. Given that the system is not accelerating, the seesaw must be exerting forces upward on Akeelah and Bree to exactly balance the downward force of gravity on them. By Newton's third law, the forces they exert on the seesaw, $F_{A,seesaw}$ and $F_{B,seesaw}$, are the same as their weights. The upward force of the pivot on the seesaw balances all of the downward forces on the seesaw, so the seesaw does not accelerate up or down. Our goal is to find the distance r_B from the pivot to Bree that will satisfy the condition of zero torque. We'll write this condition in terms of the torque, using Equation 10-16 to calculate the magnitudes of the torques, and the right-hand rule to determine their directions.

Condition that the net torque is zero:

$$\sum\vec{\tau} = 0$$

Magnitude of the torque:

$$\tau = rF \sin\phi \qquad (10\text{-}16)$$

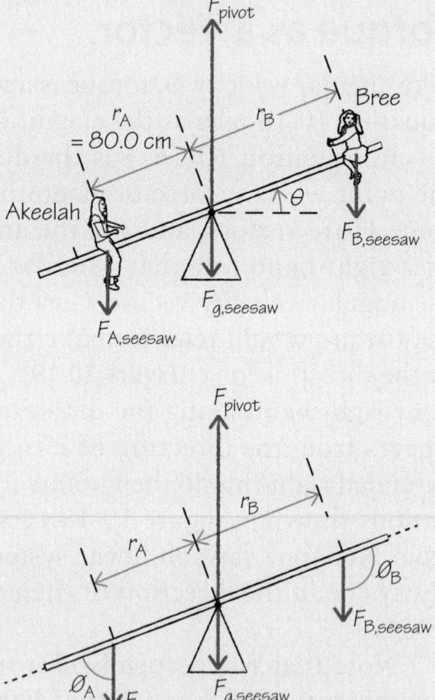

Solve

Neither the weight of the seesaw nor the support force exerts any torque on the system around the pivot point, because both of these forces are exerted at the pivot and have $\vec{r} = 0$. The only torques are due to forces exerted by Akeelah, $\vec{F}_{A,seesaw}$, and Bree, $\vec{F}_{B,seesaw}$. The right-hand rule tells us that the corresponding torques, $\vec{\tau}_A$ and $\vec{\tau}_B$, are in opposite directions, into and out of the page, respectively, so their vector sum can be zero (as it must be to keep the seesaw in balance). We can also see this just from thinking about the system. If you pull downward to the left of the pivot, where Akeelah is, you would cause the seesaw to rotate counterclockwise. If you pull down where Bree is, you would cause it to rotate clockwise. It is always important to check your signs by thinking this way.

Net torque on the system of Akeelah, Bree, and the seesaw:

$$\sum \vec{\tau} = \vec{\tau}_A + \vec{\tau}_B = 0$$

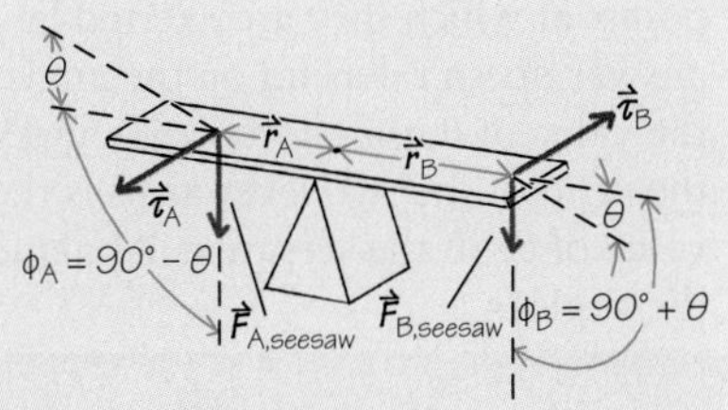

For the net torque to be zero, the opposite torques $\vec{\tau}_A$ and $\vec{\tau}_B$ must have the same magnitude. The angle between $\vec{r}_A$ and $\vec{F}_{A,seesaw}$ is $\phi_A = 90° - \theta$, and the angle between $\vec{r}_B$ and $\vec{F}_{B,seesaw}$ is $\phi_B = 90° + \theta$. From trigonometry, $\sin(90° - \theta) = \sin(90° + \theta) = \cos\theta$.

Opposite torques $\vec{\tau}_A$ and $\vec{\tau}_B$ have the same magnitude:

$$\tau_A = \tau_B$$

$$r_A F_{A,seesaw} \sin\phi_A = r_B F_{B,seesaw} \sin\phi_B$$

Substitute $\phi_A = 90° - \theta$, $\phi_B = 90° + \theta$:

$$r_A F_{A,seesaw} \sin(90° - \theta) = r_B F_{B,seesaw} \sin(90° + \theta)$$

$$r_A F_{A,seesaw} \cos\theta = r_B F_{B,seesaw} \cos\theta$$

$$r_A F_{A,seesaw} = r_B F_{B,seesaw}$$

Solve for Bree's distance from the pivot, r_B.

From $r_A F_{A,seesaw} = r_B F_{B,seesaw}$, $r_B = r_A \dfrac{F_{A,seesaw}}{F_{B,seesaw}}$

Substitute $r_A = 0.800$ m, $F_{A,seesaw} = 1.50 F_{B,seesaw}$:

$$r_B = r_A \frac{F_{A,seesaw}}{F_{B,seesaw}} = (0.800 \text{ m})\left(\frac{1.50 F_{B,seesaw}}{F_{B,seesaw}}\right) = (0.800 \text{ m})(1.50)$$

$$= 1.20 \text{ m}$$

Bree must sit 1.20 m from the pivot in order to balance Akeelah.

Reflect

Notice the significance of the lever arm in this problem. Bree, who is lighter than Akeelah, can create a balance by sitting farther from the center. A small force can give rise to a large torque when the lever arm is large. Bree has 1/1.50 the weight of Akeelah, so Bree's lever arm must be 1.50 times greater than Akeelah's—she must sit 1.50 times as far from the pivot as Akeelah. This equilibrium is only when they are not accelerating, however. Newton's third law means the forces that the children exert on the seesaw and the seesaw exerts on the children will always be the same size and oppositely directed. Imagine that they were in motion and Bree were accelerating upward: The force the seesaw exerted on her would have to be greater than gravity to cause this acceleration. At the same time, the force that the seesaw exerted on Akeelah would decrease, because she would be accelerating downward. The forces they exert on the seesaw would thus change in size. The torques caused by these forces would cause the rotation to slow even before Akeelah's feet hit the ground. Can you show that the torques as Bree starts coming back down, before they reach equilibrium, cause the rotation to speed up?

Lever arm for Akeelah:

$$r_{A\perp} = r_A \cos\theta$$

Lever arm for Bree:

$$r_{B\perp} = r_B \cos\theta$$

Ratio of the two lever arms:

$$\frac{r_{B\perp}}{r_{A\perp}} = \frac{r_B \cos\theta}{r_A \cos\theta} = \frac{r_B}{r_A} = 1.50$$

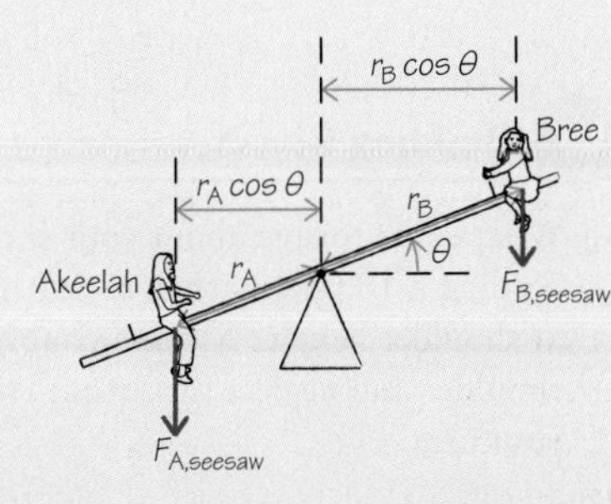

Because of the geometry (parallel forces intersecting a line), this is the same for any value of θ.

Note that because the forces are parallel and the points at which they are exerted lie along a line, the answer doesn't depend on the angle θ. The greater the value of θ, the shorter each girl's lever arm, but the *ratio* of the two lever arms is the same for any value of θ. So the seesaw will balance no matter what the angle θ.

NOW WORK Problems 5, 6, and 8 from The Takeaway 10-5.

WATCH OUT!

Equilibrium has two parts, and where the pivot is matters!

In Chapter 4 we introduced the concept that an object is in equilibrium when the net external force on it is zero. An important word here is *object*. For the object model to be valid for a system, every point in that system must move in the same way as the center of mass. Even if the object model is not valid, if the vector sum of the external forces exerted on a system is zero, the center of mass of that system will not accelerate. However, when rotation gets considered, even if the external forces add to zero, if the torques about an axis they produce do not, the system will rotate about that axis. To truly achieve equilibrium for any system, it is necessary to have both the sum of the external forces and the sum of the torques be zero.

THE TAKEAWAY for Section 10-5

✔ If a force exerted on an extended object generates a torque, that torque changes that extended object's rotation. The magnitude of the torque depends on where the force is applied relative to the rotation axis and on the orientation of the force.

✔ The magnitude of a torque is equal to the magnitude of the force that causes the torque multiplied by the lever arm of the force (the perpendicular distance from the rotation axis to the line of action of the force).

✔ Newton's second law for rotation says that if a net external torque $\sum \tau_{\text{ext},z}$ is exerted on a rigid body, the extended object acquires an angular acceleration α_z. These are related by the rigid body's rotational inertia I: $\sum \tau_{\text{ext},z} = I\alpha_z$.

✔ The angular velocity of a rotating extended object, or an object moving about a rotation axis, is a vector that lies along the rotation axis. The direction of this vector is determined by a right-hand rule.

✔ Torque is also a vector whose direction is determined by a right-hand rule. If a system is free to rotate around an axis, the torque around that axis points along the axis.

Prep for the AP® Exam

AP® Building Blocks

1. What is the torque about your shoulder axis produced by holding a 10.0-kg barbell in one hand at arm's length and at shoulder height? Assume your hand is 75 cm from your shoulder and neglect the torque caused by the weight of your arm.

2. 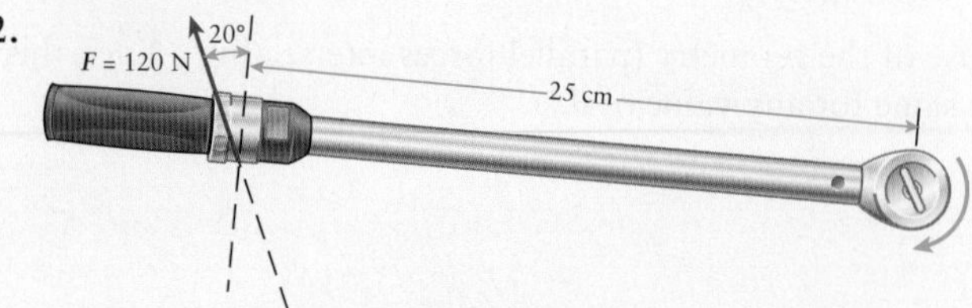

A torque wrench is used to tighten a nut on a bolt. The wrench is 25 cm long, and a force of 120 N is applied at the end of the wrench, as shown. Calculate the torque about the axis that passes through the center of the bolt.

3. 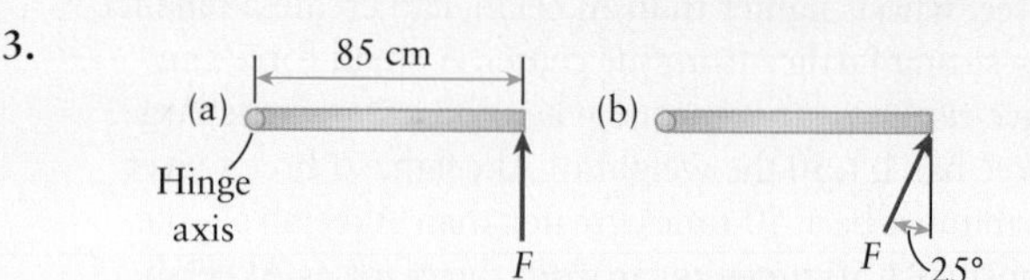

An 85.0-cm-wide door is pushed open with a force of $F = 75.0$ N. The force is applied at two different angles, as shown. Calculate the torque about an axis that passes through the hinges in each of the cases in the figure.

4. 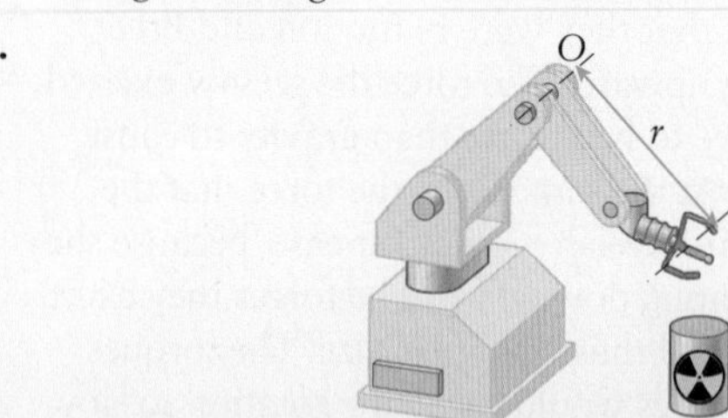

A robotic arm lifts a barrel of radioactive waste. If the maximum torque that can be exerted by the arm about the axis O is 3.00×10^3 Nm and the distance r in the diagram is 3.00 m, what is the maximum mass of the barrel? Assume that the mass of the arm is negligible.

EX 10-9 **5.** A 0.05-kg meter stick is balanced at its midpoint (50.0 cm). Then 0.1- and 0.2-kg objects are hung with light string from the 10- and 70-cm points, respectively.

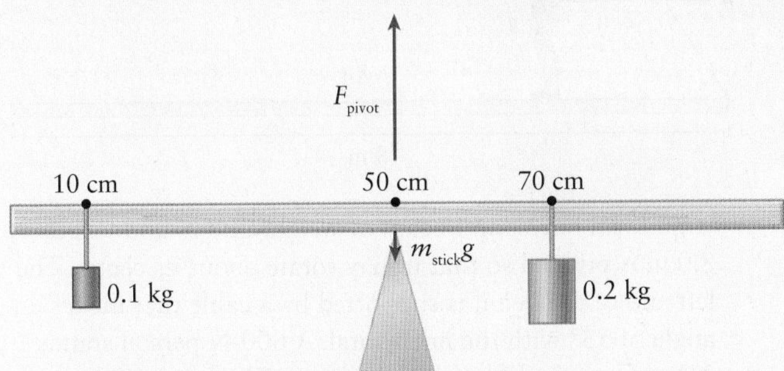

(a) Identify and calculate the clockwise and counterclockwise torques exerted on this system.
(b) Calculate the angular acceleration of the system.
(c) The fulcrum is moved to a point below the 40-cm mark on the meter stick. Identify and calculate the clockwise and counterclockwise torques exerted on this system.
(d) Calculate the angular acceleration of the system when the fulcrum is at the 40-cm mark. Recall the parallel-axis theorem, $I = I_0 + MD^2$, where $I_0 = ML^2/12$ is the rotational inertia about the center of mass of the meter stick, M is the mass of the meter stick with length L, and D is the magnitude of the displacement from the center of mass to the point about which the rotation axis is.
(e) Identify the direction of the angular acceleration.

EX 10-9 **6.** A 325-kg merry-go-round with a radius of 1.40 m is spinning clockwise as viewed from above at 4.70 rad/s. A 36.0-kg child is hanging on tightly 1.25 m from the rotation axis of the merry-go-round. Her father applies friction to the outer rim and the merry-go-round comes to a stop in 5.00 s. Model the merry-go-round as a solid disk and the child as an object. The rotational inertia of a solid disk with mass M and radius R is $MR^2/2$.
(a) Calculate the angular acceleration of the merry-go-round.
(b) Calculate the torque exerted by the father.
(c) Describe the directions of the initial angular velocity, torque, and angular acceleration.

7. A typical adult can deliver approximately 10 Nm of torque when attempting to open a twist-off cap on a bottle. In reality, the person's fingers have multiple points of contact with the cap. To simplify, assume the person exerts equal and opposite forces of magnitude F on the cap tangentially at the ends of a diameter, and the net force on the cap is thus zero. What is the value of the maximum tangential force F that the average person can exert with his or her fingers if most bottle caps are about 2 cm in diameter?

EX 10-9 **8.** A tightrope walker is walking between two buildings using a 15.0-m-long, 18.0-kg pole for balance. He grips the pole with each hand 0.60 m from the center of the pole. A 0.54-kg bird lands on the very end of the left-hand side of the pole. The rotational inertia of a rod about the midpoint is $mL^2/12$.

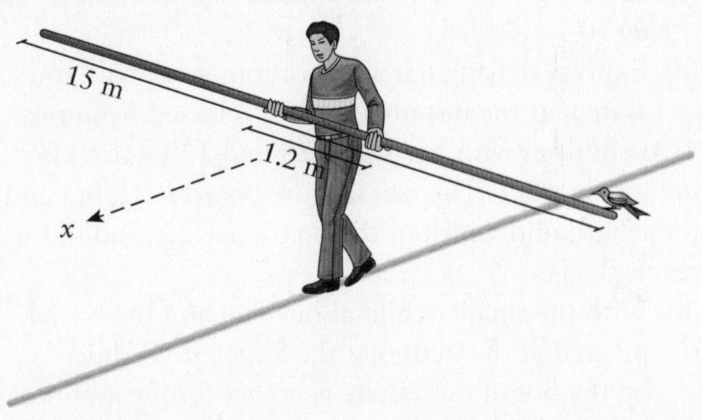

(a) Before the bird lands on the pole the tightrope walker grips the pole with palms up and the hands exert upward forces with equal magnitude perpendicular to the pole. From the perspective of the tightrope walker, looking forward in the direction of motion, identify the clockwise (cw) and counterclockwise (ccw) contributions from each hand to the net torque on the pole.
(b) The tightrope walker maintains his balance when the bird lands. Describe the direction of the torque due to the bird on the pole. How must the net torque due to the two hands change to keep the pole balanced? How might this be accomplished?
(c) Calculate the torque exerted by the bird about the center of the pole when the pole is horizontal.
(d) Assuming the tightrope walker still exerts forces with each hand in a direction perpendicular to the pole and does not move his hands from their original positions, how much force must each hand exert to counteract the torque of the bird? The total upward force exerted by the hands must support the pole and the bird against the force of gravity.

AP® Skill Builders

9.

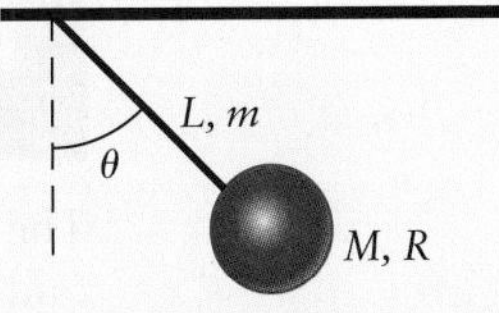

A ball with mass M and radius R is attached to the ceiling by a rod of mass m and length L. The rod can pivot about its point of attachment to the ceiling. The rod–ball system is initially held at rest in the position shown where the rod makes an angle θ with the vertical. The rotational inertia for the components of the system are: a rod about one end ($ML^2/3$), and a sphere about its center ($2MR^2/5$). Recall the parallel-axis theorem, $I = I_{CM} + Mh^2$.

(a) Derive an expression for the rotational inertia of the ball and rod in terms of m, M, L, and R.

(b) Derive an expression for the net torque exerted on the ball and rod in terms of m, M, L, R, and θ.

(c) Predict the value of the angular acceleration when $\theta = 0$.

(d) Express the angular acceleration of the ball–rod system at the instant after it is released from rest.

10. Two children with masses of 23 and 12 kg are playing on a seesaw. The mass of the board is 18 kg, and the length and width of the board are 2.5 and 0.3 m, respectively.

(a) With the smaller child at one end and the board pivoted at the midpoint the larger child slides on the board until there is no net torque and the board–children system is in rotational equilibrium. Predict the distance from the larger child to the midpoint of the board when the seesaw is in rotational equilibrium.

(b) Calculate the rotational inertia of the system of two children and the board. The rotational inertia for a board about its center is approximately mass × length^2/12.

(c) When the two children are balanced on the board the smaller child suddenly jumps off. Predict the angular acceleration of the board and linear acceleration of the larger child just after the smaller child jumps.

(d) Bored and bruised, the children decide to play a new game. The board can be taken off the fulcrum. They want to place the board on the fulcrum so that both children can sit at the ends and the system will be at rotational equilibrium. Predict the location of the fulcrum (relative to the midpoint of the board) that will allow for the board–children system to be at rotational equilibrium with both children at the ends of the board.

AP® Skills in Action

11.

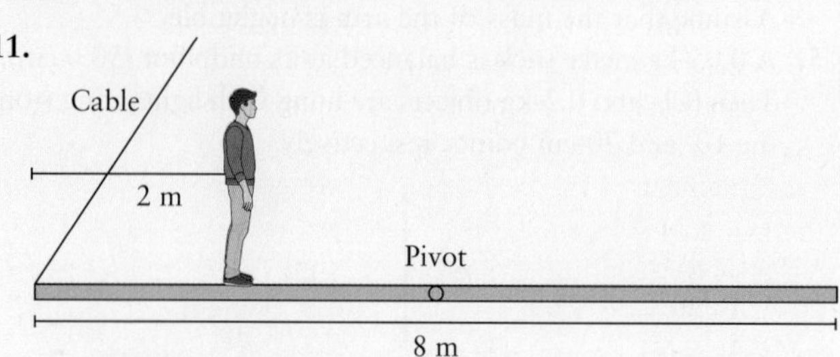

A uniform horizontal beam of weight 200 N and length 8.0 m is pivoted so that it may rotate about its center. The left end of the beam is supported by a cable that makes an angle of 53° with the horizontal. A 600-N person stands 2.0 m from the left end of the beam. The beam is in equilibrium (rotational and center of mass).

(a) Draw the extended free-body diagram for the beam for this situation.

(b) Assuming the cable can provide as large a tension force as necessary, how far in each direction could the person move and still allow the system to remain in rotational equilibrium?

12.

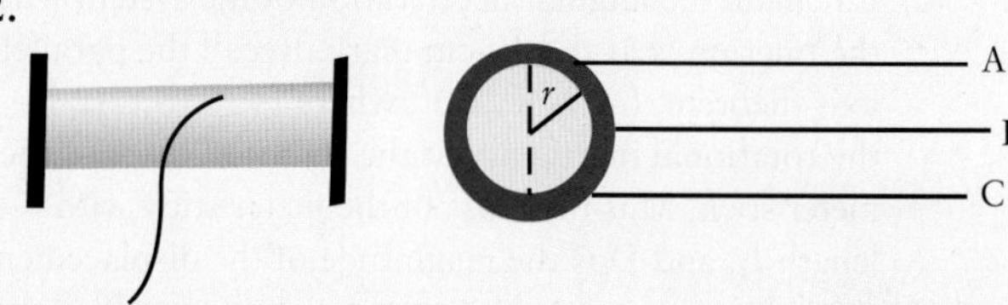

A string is attached to the cylindrical surface between the disks on either end of a spool, as shown in the figure. For three different points of attachment of the string (as shown from the end of the spool), A, B, and C, predict the direction of the rotation of the spool when the string is pulled horizontally along the line shown for each point of attachment.

10-6 The techniques used for solving problems with Newton's second law also apply to rotation problems

The examples in this section show how to use the rotational form of Newton's second law. Just as for the translational form of this law, in each example we'll begin our solution by drawing a free-body diagram. Unlike translational motion, rotation requires the free-body diagram to be extended to show the lever arm and line of action of any forces exerted on the extended object. In problems involving both translation and rotation, you should still also draw the simple free-body diagram. It makes it easier for you to visualize the net translational acceleration (the acceleration of the center of mass).

In many problems it's necessary to use *both* the translational and rotational forms of Newton's second law. Example 10-11 shows how to tackle problems like these.

WATCH OUT!

In rotational problems, it matters *where* the force is exerted.

In problems that involve Newton's second law for translational motion, we draw all the forces on an extended object as if they were exerted at a single point because we are modeling the system as if every part of it moves in the same way as the center of mass. But in problems that involve rotation, it's crucial to indicate where on the extended object each force is exerted. That's because torque exerted by a force depends on where the force is applied relative to the rotation axis. We call this an **extended free-body diagram** or simply a **force diagram**. In the examples that follow, pay attention to how the forces are depicted in extended free-body (force) diagrams and how the corresponding torques are calculated.

AP® Exam Tip

On the AP® Physics 1 exam, if you are asked to draw a free-body diagram for a translational motion problem, you will be given a dot on which you are to draw all the forces, so all the forces emanate from a point (the center of mass). If you are asked to draw a force diagram for an extended object that will rotate, you will be given or will need to draw a representation of the extended object, and you will then draw all the forces where they are applied, not all emanating from the same point. In a force diagram of a rigid body, the force of gravity should point downward, and is always drawn at the center of mass, which is where this force is always exerted. Gravity is the one force that is always the same in either representation.

WATCH OUT!

For circular rotating objects, similar equations can mean something different!

When a circular rigid body of radius R is rotating about a fixed pivot at its center, the speed v of any point on its rim is $v = \omega R$. Unlike pure rolling motion, this is not the speed of the center of mass, which in this case is not moving! In the case of something like the pulley in Example 10-10, this will also be the speed of anything attached to the rope moving with the pulley as long as the rope does not stretch or slip. The same holds for the tangential acceleration, what we referred to as a_{along} in Figure 6-2. The linear acceleration of the rope, as long as it is not slipping or stretching, is equal to the linear acceleration of a point on the rim, $a = \alpha R$. This will also be the acceleration of something moving with the string as the pulley moves, as in Example 10-11.

EXAMPLE 10-10 A Simple Pulley I

In this example the pulley can be modeled as a solid uniform cylinder of mass M_{pulley} and radius R that is free to rotate around an axis through its center. A lightweight rope is wound around the pulley. You exert a constant force of magnitude F on the rope, which makes the rope unwind and rotates the pulley (**Figure 10-20**). The rope does not stretch and does not slip on the pulley. What is the angular acceleration of the pulley about its axle?

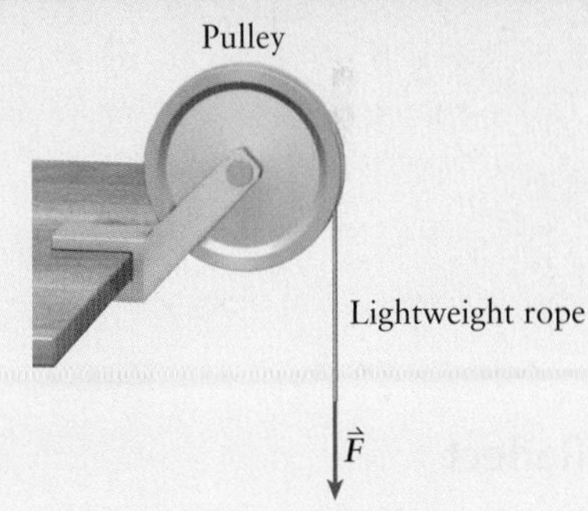

Figure 10-20 Exerting a torque on a pulley What is the angular acceleration of the pulley as the force $\vec{F}$ makes the rope unwind?

Set Up

Begin by drawing a force diagram for the pulley, taking care to draw each force at the point where it is exerted. We are told that the rope is lightweight (that is, it has much less mass than the pulley), so the force that you exert on the free end of the rope has the same magnitude F as the force that the rope exerts on the pulley. This force exerts a torque on the pulley and causes its angular acceleration. We'll use the rotational form of Newton's second law to determine this angular acceleration.

Newton's second law for rotational motion:

$$\sum \tau_{ext,z} = I_{pulley}\alpha_{pulley,z} \quad (10\text{-}18)$$

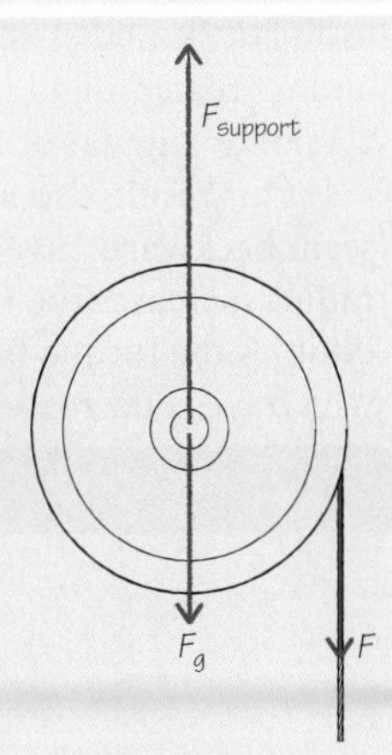

We aren't concerned with any translational motion in this problem, so we need only the extended free-body (force) diagram. However, we know that the center of mass is not accelerating and the net external force on the pulley must be zero, so the support force must balance both the pulling force and the force due to gravity. Even though the sum of the external forces on the pulley is zero, because the torques they produce about the center of mass do not cancel, there is a net torque and the pulley is not in rotational equilibrium.

Solve

The force diagram shows that only the tension force (which has the same magnitude as the pulling force) F exerts a torque on the pulley, causing it to rotate around its axle. (The support force $F_{support}$ and the weight of the pulley F_g are both exerted at its rotation axis, so the lever arm is zero for both these forces.) The lever arm for the tension force is R, so the tension torque is R multiplied by F.

Torque due to tension force:

$$\tau_z = r_\perp F = RF$$

(This torque makes the pulley rotate in the clockwise direction, so let's choose clockwise to be the positive rotation direction.)

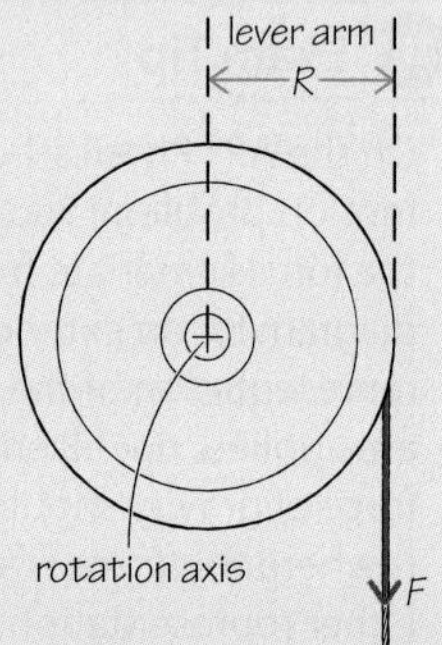

This is the only torque exerted on the pulley, so

$$\sum \tau_{ext,z} = RF$$

From Table 10-1 in Section 10-3, the rotational inertia of a solid cylinder around its central axis is $I = MR^2/2$. Insert this into Newton's second law for rotational motion and solve for $\alpha_{pulley,z}$.

$$\sum \tau_{ext,z} = I_{pulley}\alpha_{pulley,z}$$

Substitute values of $\sum \tau_{ext,z}$ and $I_{pulley} = M_{pulley}R^2/2$:

$$RF = \frac{1}{2}M_{pulley}R^2\alpha_{pulley,z}$$

Solve for angular acceleration α_z: $\alpha_{pulley,z} = \dfrac{2RF}{M_{pulley}R^2} = \dfrac{2F}{M_{pulley}R}$

Reflect

Our result for the angular acceleration $\alpha_{pulley,z}$ depends on the pulling force F, the pulley mass M_{pulley}, and the pulley radius R. It makes sense that our result is proportional to the ratio F/M_{pulley}: A greater force F means a stronger pull and a greater angular acceleration, whereas a greater mass M_{pulley} means the pulley is more difficult to rotate and gives a smaller angular acceleration.

Our result also shows that the larger the pulley radius R, the smaller the angular acceleration $\alpha_{pulley,z}$. This may seem backward, given that a larger radius means that the force F causes a larger torque $\tau = RF$. However, increasing the radius increases the rotational inertia $I = M_{pulley}R^2/2$ by a larger factor than it increases the torque. (Doubling the radius doubles the torque but quadruples the rotational inertia.) So the rotational inertia plays a more important role, which is why $\alpha_{pulley,z}$ decreases with increasing pulley radius.

NOW WORK Problems 3 and 4 from The Takeaway 10-6.

WATCH OUT !

The athlete on the cover of this textbook demonstrates several challenging concepts that you need to master for the AP® Physics 1 Exam and is also a great example that practice improves performance in challenging activities.

Lizzie Armato is the first female skater to complete the Loop, a daunting 360° ramp! (Related concepts are discussed in Chapter 6.) She is a recognized expert in navigating *bowls* (modeled on empty swimming pools) and *vert ramps* (vertical surfaces). We get a glimpse of the necessary skills behind one such maneuver in the cover image of this textbook, a frontside invert, reproduced in **Figure 10-21.** In addition to the image of Lizzie, some important markings that help us explore applicable physics concepts have been added. Lizzie began skateboarding in a park with a similar bowl. Early on, she fell the entire depth of the bowl. In an interview she commented, "It was kind of gnarly." She also recommends working your way up to a frontside invert, practicing moves that require fewer actions all performed together, just like you should start with the AP Building Blocks and Skill Builders!

To understand Lizzie's motion at the top of the bowl, we combine information about her center of mass with some of the techniques about extended objects that we have just learned about. Her approximate center of mass is marked with the usual red dot of free-body diagrams. Earth exerts its force of gravity on Lizzie at that point. She is at the apex of her projectile motion, instantaneously at rest in the air. The other external force on her at this instant is the normal force exerted by the edge of the bowl (the coping) on her hand. The normal force is denoted by the upward-pointing red arrow.

Although Lizzie cannot control the force of gravity, she can control the angle and magnitude of this normal force by adjusting the angle of her arm with respect to her center of mass and how hard she pushes on the surface. For a frontside invert she has to push much harder, so the coping pushes her harder by Newton's third law. (Backside inverts let you experience being inverted and planting your hand, but do not require the same push to get rotated.) The vertical component of the normal force the coping exerts decreases the downward acceleration of her center of mass, and the horizontal component accelerates her center of mass away from the edge of the bowl. The way she positions her body allows her to use these forces to generate a torque, accelerating her rotation and permitting Lizzie to land on her board in the orientation she wishes. The light blue line connects the point of application of the two forces. Either way we analyze the motion, the point of application of one of these forces will be our rotation axis and the other will be the force causing the torque, so this blue line will be the radius of the rotation.

Because a free extended object rotates about its center of mass, we could treat the center of mass as the axis of rotation—in which case we can ignore gravity—and just consider the rotation caused by the normal force.

Figure 10-21 Lizzie Amato and the physics of skateboarding The downward-pointing red arrow represents the force of gravity on Lizzie. The upward-pointing red arrow represents the normal force exerted by the edge of the bowl on her hand. The light blue line connects the point of application of the two forces.

The component of the normal force perpendicular to the radius of rotation (the blue line) is directed upward and to the right, thus providing Lizzie a rotational "kick" in the counterclockwise direction of rotation. If instead we treat the point at which she grasps the edge as the axis of rotation, the normal force would have no lever arm and we would look at the rotation about that point due to the force of gravity. The component of the gravitational force perpendicular to the blue line is down and to the left, which would also provide a counterclockwise kick. Either way of analyzing the motion yields the same result, as it must. As she pulls her legs down to land the board she is lowering where her center of mass is with respect to her head and she seems to float a bit longer (since our eyes are drawn to her face rather than to her center of mass), making the move even more impressive!

Although she has not practiced working such problems for this exam—and there is certainly no time to think about it when in action—Lizzie's efforts to hone the skills demanded by her sport have made these calculations part of her muscle memory.

EXAMPLE 10-11 A Simple Pulley II

You attach a block of mass M_{block} to the free end of the rope in Example 10-10 (**Figure 10-22**). When the block is released from rest and falls downward, the rope unwinds, and the pulley rotates around its central axis. As in Example 10-10 the rope neither stretches nor slips on the pulley. Derive expressions for the acceleration of the block, the angular acceleration of the pulley, and the tension in the string.

Figure 10-22 Block and pulley What is the downward acceleration of the block as it falls, and what is the angular acceleration of the pulley as the rope unwinds?

Set Up

As in Example 10-10, the pulley has an angular acceleration because the tension of the rope exerts a torque on it. The difference here is that the tension is caused by the block attached to the rope's free end.

Another difference from Example 10-10 is that we have *two* motions to consider: the pulley (a rigid body that rotates but does not translate) and the block (an object that moves linearly). So we have to write two Newton's second law equations: a rotational equation for the pulley, as in Example 10-10, and a translational equation for the block.

These two equations won't be enough to solve the problem, because we're trying to find *three* quantities: the block's acceleration $a_{block,x}$, the pulley's angular acceleration $\alpha_{pulley,z}$, and the string tension F_T. Happily, we can get a third equation because the string doesn't stretch. This tells us that the speed of the block equals the speed of a point on the pulley's rim. We'll use this to find a relationship between $a_{block,x}$ and $\alpha_{pulley,z}$.

Newton's second law for rotational motion applies to the pulley:

$$\sum \tau_{ext,z} = I_{pulley}\alpha_{pulley,z} \quad (10\text{-}18)$$

Newton's second law for translational motion applies to the block:

$$\sum F_{ext,x} = M_{block}a_{block,x} \quad (4\text{-}3a)$$

Statement that rope doesn't stretch:

$$v_{block} = v_{rim\ of\ pulley}$$

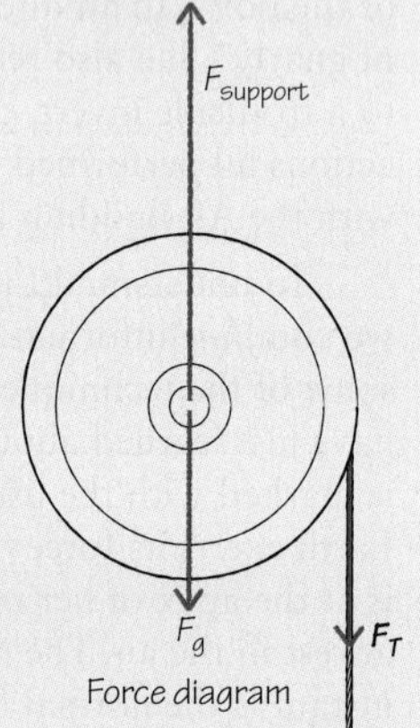

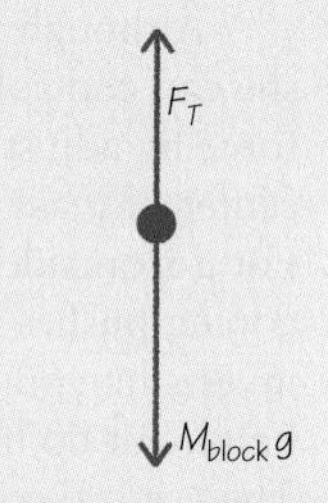

Solve

As in Example 10-10, the only force that exerts a torque on the pulley is the rope tension, which we call F_T. The lever arm of this force is R, so the net torque is $\sum \tau_{ext,z} = RF_T$. (As in Example 10-10, this torque makes the pulley rotate in the clockwise direction. So we take clockwise to be the positive rotation direction.) Insert this and $I_{pulley} = M_{pulley}R^2/2$ into the rotational Newton's second law equation and solve for the tension.

For the pulley,

$$\sum \tau_{ext,z} = I_{pulley}\alpha_{pulley,z}$$

Substitute values of $\sum \tau_{ext,z}$ and I_{pulley}:

$$RF_T = \frac{1}{2}M_{pulley}R^2\alpha_{pulley,z}$$

Divide through by R:

$$F_T = \frac{1}{2}M_{pulley}R\alpha_{pulley,z}$$

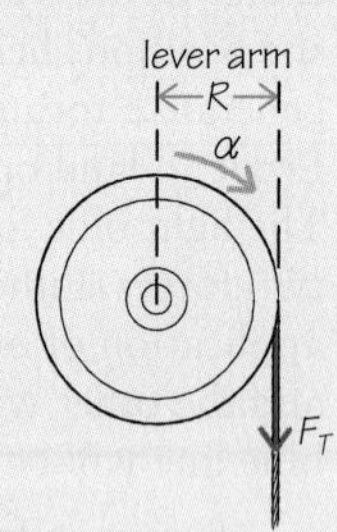

We take the positive x direction to point downward, which is the direction in which the block will move when the pulley rotates clockwise. With this choice the acceleration $a_{\text{block},x}$ of the block and the angular acceleration $\alpha_{\text{pulley},z}$ of the pulley are both positive, the gravitational force $M_{\text{block}}g$ is in the positive x direction, and the tension force F_{T} on the block is in the negative x direction. (The rope has negligible mass, so the tension is the same throughout its length.) Use this to write the Newton's second law equation for the block.

For the block:

$$\sum F_{\text{ext},x} = M_{\text{block}}a_{\text{block},x}$$

Substitute the individual forces:

$$M_{\text{block}}g - F_{\text{T}} = M_{\text{block}}a_{\text{block},x}$$

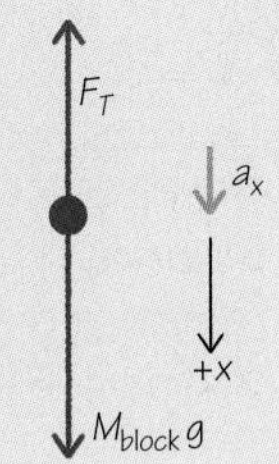

Because the rope doesn't stretch, at any instant the speed of the block v_{block} equals the speed of a point on the pulley rim, which is equal to $R\,\omega$ (the product of the pulley radius and the pulley angular speed given by Equation 10-3). As the block falls, it gains speed and the pulley gains angular speed, but the speed of the block remains equal to the speed of the pulley's rim. Use this to relate the linear acceleration of the block to the angular acceleration of the pulley.

At a certain time,

$v_{\text{block}} = v_{\text{rim of pulley}}$ or

$v_{\text{block},x} = R\omega_{\text{pulley}} = R\omega_{\text{pulley},z}$

(The block is moving downward in the positive x direction, so its x velocity $v_{\text{block},x}$ is positive and equal to its speed v_{block}. The pulley is rotating in the positive direction, so its angular velocity $\omega_{\text{pulley},z}$ is positive and equal to the pulley's angular speed ω_{pulley}.)

A time Δt later,

$$v_{\text{block},x} + \Delta v_{\text{block},x} = R(\omega_{\text{pulley},z} + \Delta\omega_{\text{pulley},z})$$

so

$$v_{\text{block},x} + \Delta v_{\text{block},x} = R\omega_{\text{pulley},z} + R\Delta\omega_{\text{pulley},z}$$

Because $v_{\text{block},x} = R\omega_{\text{pulley},z}$ from above, it follows that

$$\Delta v_{\text{block},x} = R\Delta\omega_{\text{pulley},z}$$

Divide both sides by Δt:

$$\frac{\Delta v_{\text{block},x}}{\Delta t} = R\frac{\Delta\omega_{\text{pulley},z}}{\Delta t}$$

The ratios on the two sides are just the average acceleration of the block and the average angular acceleration of the pulley. We can take the time Δt to be as small as we like, so these can be instantaneous instead of average. So

$$a_{\text{block},x} = R\alpha_{\text{pulley},z} \text{ or } \alpha_{\text{pulley},z} = \frac{a_{\text{block},x}}{R}$$

Combine the three equations to solve for the three unknowns.

Substitute $\alpha_{\text{pulley},z} = a_{\text{block},x}/R$ into the pulley equation:

$$F_{\text{T}} = \frac{1}{2}M_{\text{pulley}}R\alpha_{\text{pulley},z} = \frac{1}{2}M_{\text{pulley}}R\left(\frac{a_{\text{block},x}}{R}\right) = \frac{1}{2}M_{\text{pulley}}a_{\text{block},x}$$

Substitute this expression for F_{T} into the block equation:

$M_{\text{block}}g - F_{\text{T}} = M_{\text{block}}a_{\text{block},x}$ so

$$M_{\text{block}}g - \frac{1}{2}M_{\text{pulley}}a_{\text{block},x} = M_{\text{block}}a_{\text{block},x}$$

Solve for $a_{\text{block},x}$:

$$M_{\text{block}}g = M_{\text{block}}a_{\text{block},x} + \frac{1}{2}M_{\text{pulley}}a_{\text{block},x}$$

$$= \left(M_{\text{block}} + \frac{1}{2}M_{\text{pulley}}\right)a_{\text{block},x}$$

$$a_{\text{block},x} = \left(\frac{M_{\text{block}}}{M_{\text{block}} + M_{\text{pulley}}/2}\right)g$$

Now find $\alpha_{\text{pulley},z}$ by substituting this into $\alpha_{\text{pulley},z} = a_{\text{block},x}/R$:

$$\alpha_{\text{pulley},z} = \frac{a_{\text{block},x}}{R} = \left(\frac{M_{\text{block}}}{M_{\text{block}} + M_{\text{pulley}}/2}\right)\frac{g}{R}$$

Finally, solve for F_{T} from the pulley equation:

$$F_{\text{T}} = \frac{1}{2}M_{\text{pulley}}R\alpha_{\text{pulley},z} = \frac{1}{2}M_{\text{pulley}}R\left(\frac{M_{\text{block}}}{M_{\text{block}} + M_{\text{pulley}}/2}\right)\frac{g}{R}$$

$$= M_{\text{block}}g\left(\frac{M_{\text{pulley}}}{2M_{\text{block}} + M_{\text{pulley}}}\right)$$

Reflect

The downward acceleration of the block is less than g because the tension of the rope exerts an upward pull on the block that partially cancels the downward pull of gravity. The expression for tension F_{T} confirms this: F_{T} is less than $M_{\text{block}}g$, which says that the tension force doesn't completely balance out the gravitational force. If it did, the block would just hang there when released and wouldn't move at all!

Here's a way to check our results: If the pulley is very much lighter than the block, so M_{pulley} is nearly zero, it's as though the block is connected to nothing at all. In this case the block's downward acceleration would be g (it would be in free fall), and there would be no tension in the rope. Can you use our results for $a_{\text{block},x}$ and F_{T} to show that this is the case?

$$a_{\text{block},x} = \left(\frac{M_{\text{block}}}{M_{\text{block}} + M_{\text{pulley}}/2}\right)g$$

Inside the parentheses the numerator M_{block} is less than the denominator $M_{\text{block}} + M_{\text{pulley}}/2$, so the ratio inside the parentheses is less than 1 and $a_{\text{block},x} < g$.

$$F_{\text{T}} = M_{\text{block}}\,g\left(\frac{M_{\text{pulley}}}{2M_{\text{block}} + M_{\text{pulley}}}\right)$$

Inside the parentheses the numerator M_{pulley} is less than the denominator $2M_{\text{block}} + M_{\text{pulley}}$, so the ratio inside the parentheses is less than 1 and $F_{\text{T}} < M_{\text{block}}g$.

$F_T < M_{\text{block}}\,g$

$a_x < g$

$M_{\text{block}}\,g$

NOW WORK Problems 1, 2, and 5 from The Takeaway 10-6.

THE TAKEAWAY for Section 10-6

✔ Always draw an extended free-body (force) diagram as part of the solution of any problem that involves torques. To be able to calculate torques correctly, draw each force at the point where it is applied.

✔ In problems of this kind apply Newton's second law for translation to any object that changes position and Newton's second law for rotation to any rigid body that rotates. Both laws must be applied to any rigid body that undergoes both translation and rotation.

✔ In problems that involve both translation and rotation, you will need to relate the translational acceleration to the rotational acceleration.

Prep for the AP Exam

Building Blocks

1. A solid cylindrical pulley with a mass of 1.00 kg and a radius of 0.25 m is free to rotate about its axis. An object of mass 0.250 kg is attached to the pulley with a light string. Assuming the string does not stretch or slip over the pulley, calculate the tension in the string and the angular acceleration of the pulley. The rotational inertia of the pulley modeled as a solid disk rotating about its center with mass m and radius r is $mr^2/2$.

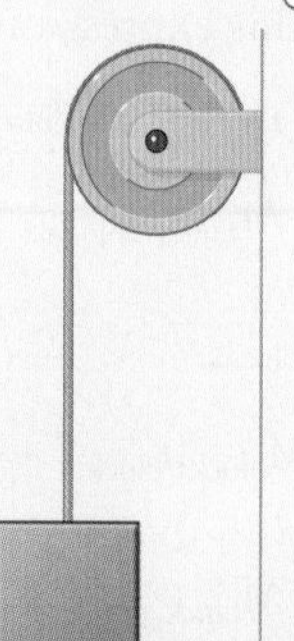

2.

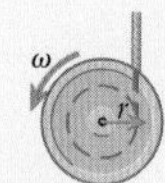

A yo-yo with a mass of 0.0750 kg and a rolling radius of 2.50 cm (the distance from the axis of the pulley to where its string comes off the spool) rolls down a string with a linear acceleration of 6.50 m/s^2. Approximate the rotational inertia of the yo-yo as that of a solid disk with mass m and radius r rotating about its center.

(a) Calculate the tension in the string and the angular acceleration of the yo-yo.

(b) Use Newton's second law for rotation to find the rotational inertia of this yo-yo. If this varies from your disk approximation, construct an explanation of the direction of the difference (larger or smaller).

X -10

3. Some 2.5-in.-diameter (6.35-cm-diameter) computer hard disks spin at a constant 7200-rpm operating speed. The disks have a mass of about 7.50 g and are essentially uniform throughout with a very small hole at the center. The rotational inertia of a thin disk is $MR^2/2$.
   (a) If the disk reaches its operating speed 2.50 s after the drive is turned on, what average torque does the disk drive supply to the disk during the acceleration?
   (b) Calculate the change in kinetic energy of the disk and the average power delivered to the drive as it accelerates to operating speed.

EX -10

4. A crane winch is lifting a 2300-kg mass. The winch can be modeled as a cable of negligible mass wound around a solid uniform cylinder with a 12-cm radius and a mass of 320 kg that rotates around its central axis. What is the torque that must be applied to the winch by its motor in order to accelerate a 2300-kg object straight upward at 0.35 m/s^2? Assume there is no energy dissipated by friction in the system, that the cable does not slip or stretch, and that the winch radius remains constant. The rotational inertia of a cylinder is $MR^2/2$.

EX 0-11

5. 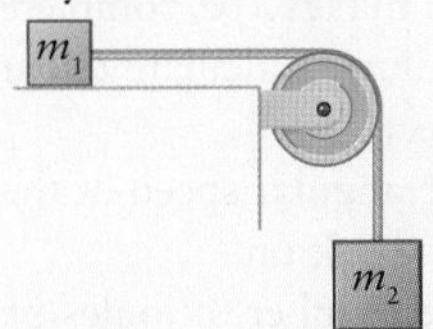

   A block with mass m_1 = 2.00 kg rests on a surface that exerts a negligible friction force on the block. The block is connected with a light string over a pulley to a hanging block of mass m_2 = 4.00 kg. The pulley is a uniform disk with a radius of 4.00 cm and a mass of 0.500 kg. The pulley rotates with negligible friction and the string does not slip or stretch. The rotational inertia of a uniform disk is $MR^2/2$.
   (a) Calculate the acceleration of each block and the tension in each segment of the string.
   (b) Calculate how long it takes the blocks to move a distance of 2.25 m.
   (c) Calculate the angular speed of the pulley at the instant that the blocks have moved 2.25 m.

AP® Skill Builders

6. The data in the table were collected on the motion of a rotating disk. The experimental data are to be used to estimate the rotational inertia of the disk. While the data were taken in 5-s intervals, the observation was made that the rotation of the disk did not change direction throughout the time data were being taken.

t (s)	θ (rad)
0	0.00
5	1.75
10	3.50
15	3.51
20	11.0
25	31.0

   (a) Construct a θ–t graph from these data.
   (b) Interpret the data by describing the motion in terms of the directions and relative magnitudes of angular velocity, angular acceleration, and torque during these time intervals: $0 \le t \le 10$ s, $10 \text{ s} \le t \le 15$ s, and $15 \text{ s} \le t \le 25$ s.
   (c) Estimate the numerical value of the angular velocity in the intervals $0 \le t \le 10$ s, $15 \text{ s} \le t \le 20$ s, and $20 \text{ s} \le t \le 25$ s.
   (d) Estimate the numerical value of the angular acceleration in the interval $15 \text{ s} \le t \le 25$ s.
   (e) Using your estimate of the angular acceleration, the estimate of the angular speed at t = 15 s obtained from angular displacements at 15 and 20 s, and the value of the angular displacement at t = 15 s, predict the displacements at t = 20 s and t = 25 s.
   (f) Using your estimate of the angular acceleration, the estimate of the angular speed at t = 15 s obtained from angular displacements at 10 and 15 s, and the value of the angular displacement at t = 15 s, predict the displacements at t = 20 s and t = 25 s.
   (g) Construct and support a claim based on data regarding biases in the estimates of the angular displacements made in parts (e) and (f).
   (h) In the time interval between 15 and 25 s a constant torque of 112 Nm was applied to the disk. Using these data, predict the rotational inertia of the disk about the rotational axis, including an estimate of uncertainty.

7. A cylinder rolls down an inclined plane. Explain why, if the axis of rotation of the cylinder is taken to be its central axis, the gravitational force does not exert a torque on the cylinder.

AP® Skills in Action

8. Flywheels are mechanical energy storage devices with ancient origins. It is thought they may be used in the next generation of power grids, storing renewable energy that depends on variable solar or wind sources.
   (a) The power delivered by a constant force exerted on an object moving with a constant velocity is $P = Fv$. Multiply the right-hand side by $r/r = 1$, where r is the radius of a flywheel, to obtain an expression for the power delivered by a torque exerted on a flywheel rotating at angular speed ω.
   (b) A flywheel is mounted on powerful magnets that make friction forces negligible. When the flywheel is rotating at 2000 rpm, what torque must be applied to the flywheel to store energy at a rate of 1 MW?

9. 
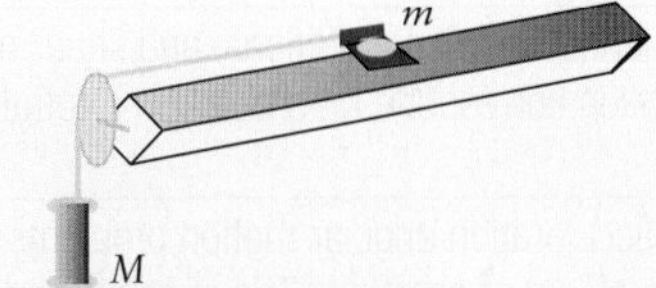

   Using the apparatus shown here, an experiment can be conducted to confirm Newton's second laws (translational and rotational). A cart with mass m glides with negligible friction over the air track. A thread is attached to the cart and to a falling object with mass M, which pulls the cart as the thread runs over a pulley with rotational inertia I and radius R. Linear acceleration of the falling object is measured with a motion sensor positioned below the falling object.

A student begins by investigating the dependence of linear acceleration of the system on the inertial mass of the system $(m + M)$, neglecting the rotational inertia of the pulley. The student graphs the measured values of acceleration versus $M/(m + M)$ and obtains 8.8 m/s^2 as the value of g.

In her data collection, the student used a falling object of mass $M = 200$ g. The glider mass was 340 g. Several 10-g disks were then added to increase the glider mass.

Dissatisfied with her results, the student measured the mass of the pulley to be $m_{\text{pulley}} = 30.0$ g. The pulley is not a simple disk; it has a complex shape with cutouts and a nonuniform thickness. She decides to simply assume that the rotational inertia is equal to $m_{\text{pulley}} R^2$.

(a) Justify the claim that in the new data collection, acceleration should be graphed versus a new variable: $x = M/(m + M + I/R^2)$.

m (kg)	a (m/s^2)	x
0.35	3.38	
0.37	3.31	
0.39	3.19	
0.41	3.06	
0.43	2.98	
0.45	2.92	
0.47	2.80	
0.49	2.76	

(b) The data given in the above table were collected, where a is the acceleration and m is the mass of the glider. Complete the table.

(c) Analyze the data graphically to obtain the slope as a measurement of g.

(d) Looking for a bias that could account for the error in her measured value of g, the student claims that the slope of the graph will increase if she assumes that $I = cm_{\text{pulley}} R^2$, where c is positive and less than 1. Evaluate her claim.

(e) To look for a bias in the student's experimental design, identify scientific questions that she should pose and propose strategies that she can use to answer these questions.

10. A flywheel of mass 35.0 kg and diameter 60.0 cm spins at 400 rpm when it experiences a sudden power loss. The flywheel slows due to friction in its bearings during the 20.0 s the power is off.

(a) Express the angular displacement following the power failure in terms of the angular acceleration of the flywheel and the time, Δt, while the power is off.

(b) The flywheel makes 102 complete revolutions during the power failure. Calculate the angular acceleration of the flywheel.

(c) Calculate the angular speed of the flywheel when the power comes back on.

(d) Calculate the number of joules of kinetic energy converted to internal energy while the power was off. The rotational inertia of a solid disk rotating about its central axis is $mr^2/2$.

WHAT DID YOU LEARN?

Prep for the AP Exam

Chapter learning goals	Section(s)	Related example(s)	Relevant section review exercises
Define the rotational motion of a system and describe when and how it can be simplified to an object model.	10-2 10-3	10-3 10-4	7 2
Define and explain the relationships among angular displacement, angular velocity, and angular acceleration and use these quantities to describe the rotation of an object or a system of rigidly connected objects, which is a rigid body.	10-2 10-4	10-1 10-8	2, 4 1, 2, 3
Explain what is meant by the rotational inertia of a system and how to use it to calculate the rotational kinetic energy of that system.	10-2 10-3	10-2 10-5	5, 6, 7 2
Describe the techniques for finding the rotational inertia of a rigid body, including use of the parallel-axis theorem.	10-2	10-3	7
Explain the similarities between rotational and linear motion, and describe how the two forms of motion correspond for a point on a rotating rigid system.	10-2 10-4 10-6	10-1 10-8 10-10, 10-11	3, 10 1, 2, 3, 10 4, 5, 10
Solve constant-acceleration angular motion problems when given a description of the motion of an object or a system. The description could be in the form of words, graphs, or experimental data.	10-4 10-6	10-8 10-10, 10-11,	1, 2, 3 1, 2, 3, 4, 5,
Define the concept of lever arm and know how to use it to calculate the torque caused by a force exerted on a rigid body.	10-5		2, 3
Define rotational equilibrium and calculate it for a rigid body.	10-5	10-9	8, 9
Explain how to find the direction of angular velocity and torque.	10-5	10-9	5, 6, 8

Chapter 10 Review

Key Terms

All the Key Terms can be found in the Glossary/Glosario on page G1 in the back of the book.

Chapter Summary

Topic	Equation or Figure

Angular velocity and angular speed: The angular velocity ω_z of a rigid body is the rate at which it rotates. (The z axis is the rotation axis of the extended object.) Angular velocity can be positive or negative, depending on which direction of rotation you choose to be positive. Angular speed ω is the magnitude of angular velocity. A point a distance r from the rotation axis has speed $v = r\omega$.

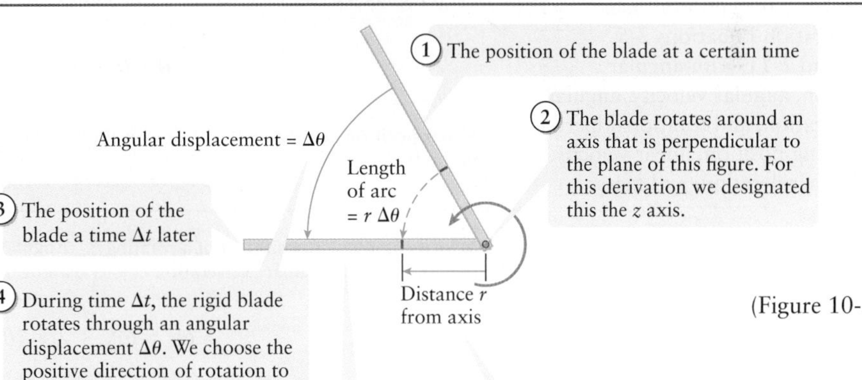

① The position of the blade at a certain time
② The blade rotates around an axis that is perpendicular to the plane of this figure. For this derivation we designated this the z axis.
Angular displacement = $\Delta\theta$
Length of arc = $r\,\Delta\theta$
③ The position of the blade a time Δt later
④ During time Δt, the rigid blade rotates through an angular displacement $\Delta\theta$. We choose the positive direction of rotation to be counterclockwise, so $\Delta\theta > 0$.
Distance r from axis

(Figure 10-3)

⑤ The **angular velocity** ω_z of the blade is the angular displacement about the z axis $\Delta\theta$ divided by the elapsed time Δt:

$$\omega_z = \frac{\Delta\theta}{\Delta t}$$

The **angular speed** ω is the absolute value of ω_z.

⑥ In time Δt, a point on the blade a distance r from the axis moves through an arc of length $r\,\Delta\theta$. The speed of this point is $v = r\omega$.

Rotational inertia and rotational kinetic energy: The quantity that plays the role of mass for rotational motion is an extended object's rotational inertia. This depends on the mass of the extended object and on how the components of the extended object are positioned relative to the rotation axis, so is only constant for a rigid body. (Table 10-1 lists the rotational inertias for various common shapes.) The rotational kinetic energy of a rigid body is related to its rotational inertia and angular speed.

To find the **rotational inertia** I of a rigid body...

...we imagine dividing the rigid body into **N small pieces,** and calculate the sum...

$$I = \sum_{i=1}^{N} m_i r_i^2 \tag{10-6}$$

...of the product of the **mass m_i of the ith piece**...

...and the square of the **distance r_i** from the ith piece to the rotation axis.

Rotational kinetic energy of a rigid body spinning around an axis

Rotational inertia of the rigid body for that rotation axis

$$K_{\text{rotational}} = \frac{1}{2} I\omega^2 \tag{10-7}$$

Angular speed of the rigid body

Parallel-axis theorem: If a rigid body's rotational inertia for an axis through its center of mass is I_{CM}, its rotational inertia for a second axis parallel to the center-of-mass axis and separated by a distance h is $I = I_{CM} + Mh^2$. Here M is the mass of the rigid body.

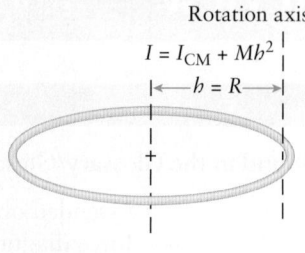

(a) Hoop rotating around an axis through its center

Rotation axis

$I = I_{CM}$

Hoop, mass M

R

Center of mass of the hoop

(b) Hoop rotating around an axis through its rim

Rotation axis

$I = I_{CM} + Mh^2$

$h = R$

(Figure 10-10 a and b)

Rotational kinematics: Angular acceleration is to angular velocity as linear acceleration is to linear velocity. If the angular acceleration of a rigid body is constant, three basic equations relate the time, angular position, angular velocity, and angular acceleration of the rigid body. These three equations are directly parallel to the constant linear acceleration Equations 2-5, 2-9, and 2-11, with angular position, angular velocity, angular acceleration, and rotational inertia replacing position, linear velocity, linear acceleration, and mass.

Angular velocity at time t of a rotating rigid body with constant angular acceleration

Angular velocity of the rigid body at time $t = 0$

$$\omega_z = \omega_{0z} + \alpha_z t \quad (10\text{-}13)$$

Constant angular acceleration of the rigid body

Time at which the rigid body has angular velocity ω_z

Angular position at time t of a rotating rigid body with constant angular acceleration

Angular velocity of the rigid body at time $t = 0$

Constant angular acceleration of the rigid body

$$\theta = \theta_0 + \omega_{0z}t + \frac{1}{2}\alpha_z t^2 \quad (10\text{-}14)$$

Angular position at time $t = 0$ of a line on the rigid body

Time at which the line on the rigid body is at angular position θ

Angular velocity at angular position θ of a rotating rigid body with constant angular acceleration

Angular velocity of the rigid body when a line on it is at angular position θ_0

$$\omega_z^2 = \omega_{0z}^2 + 2\alpha_z(\theta - \theta_0) \quad (10\text{-}15)$$

Constant angular acceleration of the rigid body

Two **angular positions** of the line on the rigid body

Views of extended object from along its rotation axis

(a) A force that tends to cause clockwise rotation

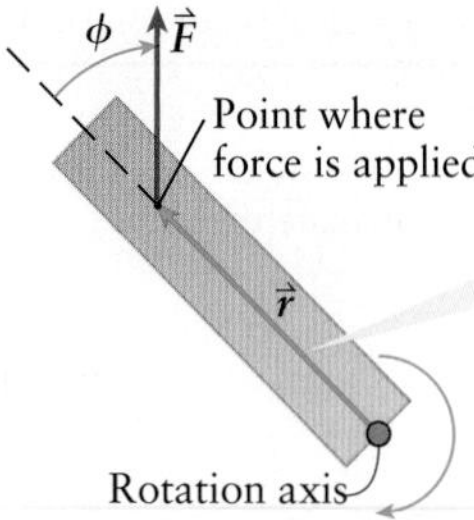

The displacement $\vec{r}$ points from an extended object's rotation axis to where a force $\vec{F}$ is exerted on the extended object.

(b) A force that tends to cause counterclockwise rotation

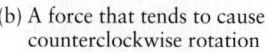

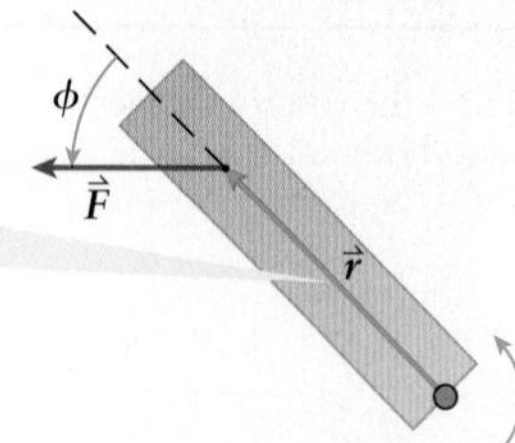

(c) Determining the line of action of a force and its lever arm

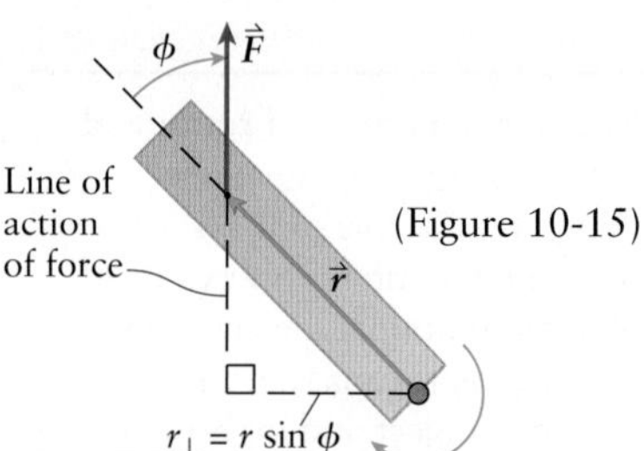

(Figure 10-15)

The lever arm $r_\perp$ is perpendicular to the line of action of the force, and extends from the rotation axis to the line of action of the force.

Torque: If a force $\vec{F}$ is exerted on an extended object (applied) at a displacement $\vec{r}$ from the extended object's rotation axis, it can change the rotation of the extended object. The torque associated with the force has magnitude $\tau = rF \sin\varphi = r_\perp F$, where $r_\perp$ is the lever arm of the force. The torque τ_z can be positive or negative.

Rotational form of Newton's second law: A net external torque on a rigid body gives it an angular acceleration. To analyze a rigid body that is able to translate as well as rotate (such as one that rolls without slipping), both Newton's second law in its original form $\left(\sum \vec{F}_{\text{ext}} = m\vec{a}\right)$ and this in its rotational form must be used.

The net torque (the sum of all external torques) exerted on a rigid body for the axis about which it can rotate

The angular acceleration produced by the net torque

$$\sum \tau_{\text{ext},z} = I\alpha_z \tag{10-18}$$

The rigid body's rotational inertia for the axis about which it can rotate

Rotational quantities as vectors: The angular velocity $\vec{\omega}$ and torque $\vec{\tau}$ are vector quantities that lie along the rotation axis. Their directions are given by a right-hand rule.

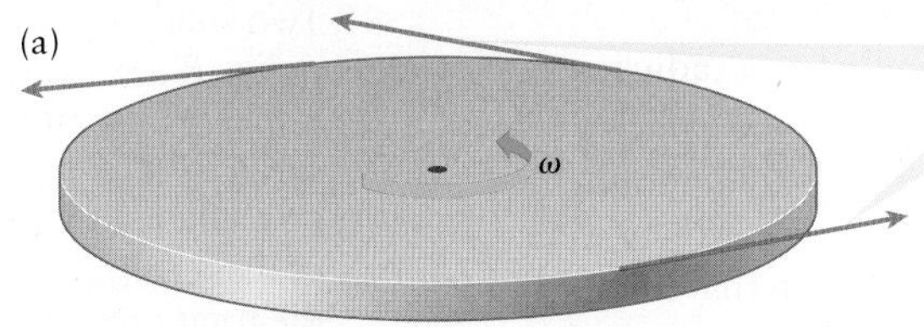

No single vector in the plane of rotation correctly represents the rotational motion.

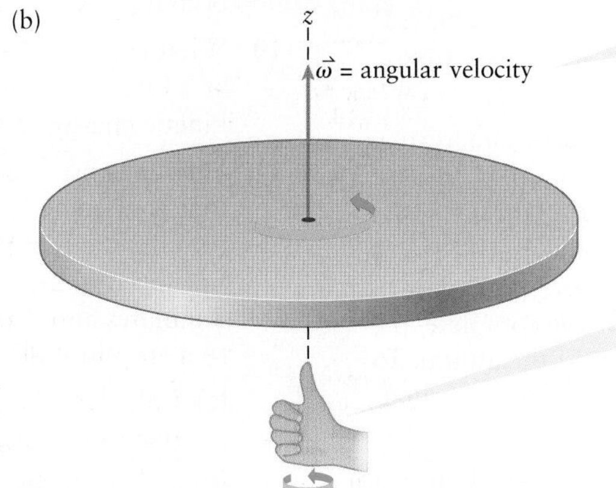

The angular velocity points in a direction perpendicular to the rotation plane.

Is the direction of the angular velocity up or down? The direction is up in a right-handed sense. Curl the fingers on your right hand in the direction of motion and stick your thumb straight out; your thumb points in the direction of the angular velocity.

(Figure 10-18)

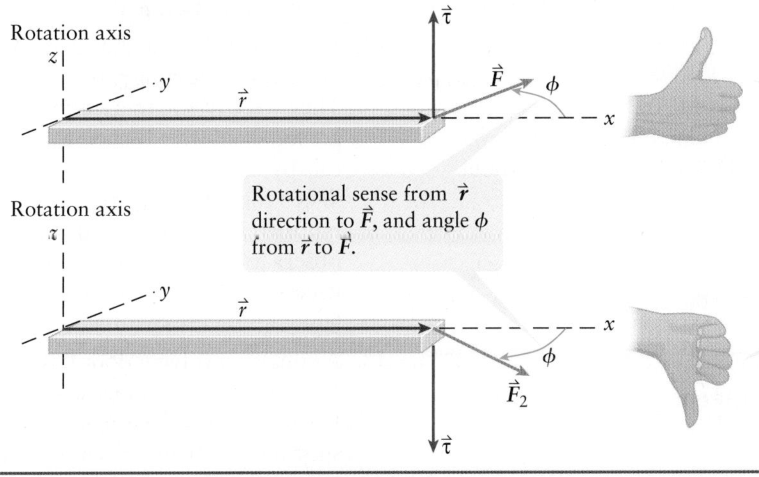

Rotational sense from $\vec{r}$ direction to $\vec{F}$, and angle ϕ from $\vec{r}$ to $\vec{F}$.

To apply the right-hand rule correctly, point your fingers along $\vec{r}$, then curl your fingers from the direction of $\vec{r}$ to $\vec{F}$ along the shortest path.

(Figure 10-19)

Chapter 10 Review Problems

1. How would the kinetic energy of a flywheel (a spinning disk) change if its rotational inertia was five times smaller but its angular speed was five times larger?
   (A) 0.1 times as large as before
   (B) 0.2 times as large as before
   (C) Same as before
   (D) 5 times as large as before

2. A DVD disk rotates so that a laser maintains a constant linear speed relative to the disk as it tracks over the playing area. Predict whether the rotational speed of the disk increases or decreases as the laser moves nearer to the edge of the disk.

3. You have two solid steel spheres; sphere 2 has twice the radius of sphere 1. Both rotational

inertias are measured about an axis that passes through the center of the spheres for which the rotational inertia is $2MR^2/5$. What is the ratio of the rotational inertias I_2/I_1?

(A) 32 (B) 16 (C) 8 (D) 4

4. Allison twirls an open umbrella around its central axis so that it completes 24.0 rotations in 30.0 s.
   (a) If the umbrella starts from rest, calculate the angular acceleration (in rad/s^2) of a point on the outer edge.
   (b) What is the maximum tangential speed of a point on the edge if the umbrella has a radius of 55.0 cm?

5. Prior to the music CD, stereo systems had a phonographic turntable on which vinyl disk recordings were played. The information was encoded so that it could be played at a constant angular speed. A particular phonographic turntable starts from rest and achieves a final constant angular speed of $33\frac{1}{3}$ rpm in a time of 4.5 s.
   (a) How many rotations did the turntable undergo during that time?
   (b) The classic Beatles album *Abbey Road* is 47 min and 7 s in duration. If the turntable requires 8 s to come to rest once the album is over, calculate the total number of rotations for the complete start-up, playing, and slow-down of the album. To simplify the problem, assume the entire recording is on one side of the disk.

6. A CD player varies its speed as it changes circular tracks on the CD. A CD player is rotating at 300 rpm. To read another track, the angular speed is increased to 450 rpm in a time of 0.75 s. Calculate the average angular acceleration in rad/s^2 during the change.

7. A meter stick that has a mass of 0.103 kg is rotating about a fixed axis with an angular speed of 3.00 rev/s. The rotational inertia for such a rod in rotation about its center of mass is $\frac{1}{12}ML^2$.
   (a) Calculate the rotational energy of this meter stick if the axis of rotation is at the center of mass of the meter stick.
   (b) Calculate the rotational energy of this meter stick if the axis of rotation is at one end of the meter stick.

8.

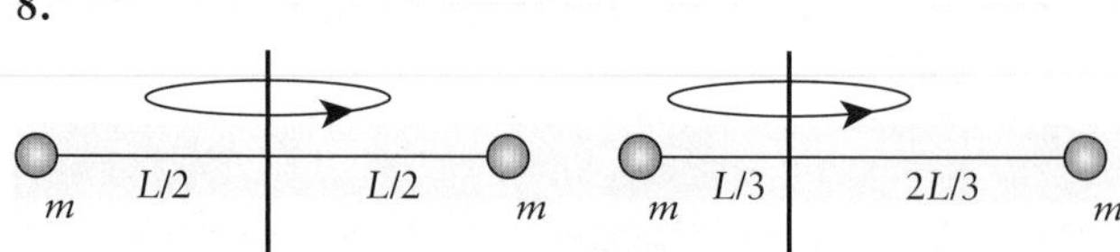

Two small balls attached on opposite ends of a rigid rod with length L and negligible mass rotate about an axis at the midpoint of the rod with a rotational inertia $I_{L/2}$. The rotation axis is shifted so that the rod rotates about a point a distance $L/3$ from one end with a rotational inertia $I_{L/3}$. Calculate the ratio $I_{L/2}/I_{L/3}$.

9.

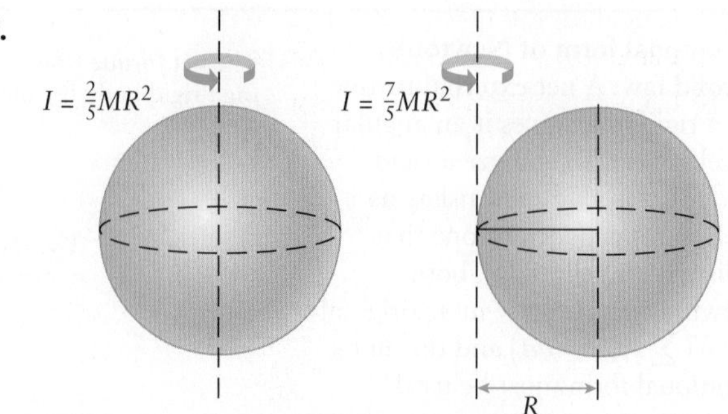

Two solid, uniform spheres with the same mass M and radius R are rotating, as shown. The rotational inertia for a sphere in rotation about an axis that is tangent to its surface is $I_{\text{tangential axis}} = 7MR^2/5$. The rotational inertia of a sphere in rotation about an axis through the center of mass is $I_{\text{central axis}} = 2MR^2/5$. Calculate the ratio of the rotational kinetic energies $K_{\text{central axis}}/K_{\text{tangential axis}}$ if both objects have the same angular speed.

10. What is the rotational inertia of an object that rotates at 13.0 rev/min about an axis and has a rotational kinetic energy of 18.0 J?

11. Hit songs used to be released as singles on records that rotated at 45.0 rpm while the song was playing. Assume the record starts from rest and increases to full speed in 2.50 seconds. The song then plays for 3 minutes and 40 seconds, and then the record slows to a stop in 4.80 seconds.
   (a) Calculate the initial angular acceleration of the record in rad/s^2.
   (b) Calculate the total number of rotations the record makes during the time from when the rotation begins until it ends.

12. A certain TV show that debunks scientific misconceptions got something wrong in one of their episodes. They were trying to ascertain if a person would really get knocked over by being shot by a bullet. They tested this by hanging an extended object from the ceiling. When the bullet hit the extended object, it rotated out, but then it swung back down, so they said from this observation that someone hit by a bullet would not necessarily tend to fall over. What is the biggest problem with their experimental setup?

13. A circular revolving door has a radius of $R = 1.00$ m and total rotational inertia of $I = 119\ \text{kg}\cdot\text{m}^2$, and is free to rotate on bearings with negligible friction, as shown in the figure. Each of the door "arms" makes a tight seal with the frame, so that $F_k = 45.0$ N of friction force is exerted by the frame on each arm when the door rotates. An 85.0-kg adult pushes perpendicularly on one of the doors at a distance $r = 0.60$ m from the rotation axis. Assume all four arms are in contact with the frame.

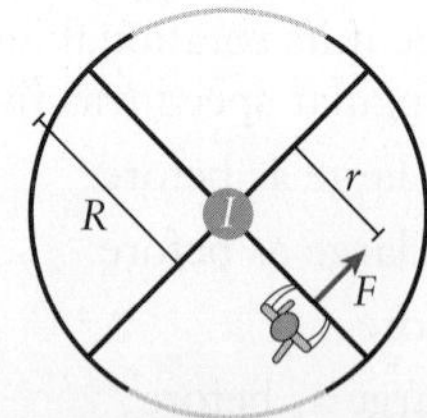

(a) Calculate the force that must be exerted by the person to make the door spin at a constant rate.

(b) Calculate the force that must be exerted by the person to accelerate the door at 1 rad/s^2.

14.

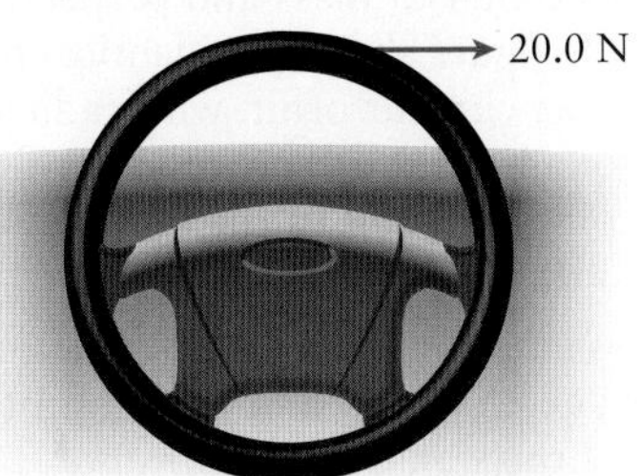

A driver applies a horizontal force of 20.0 N tangentially to the right at the top of a steering wheel, as shown in the figure. The steering wheel has a radius of 18 cm and a rotational inertia of 0.097 kg · m^2. Calculate the angular acceleration of the steering wheel about the central axis due only to this force. (Of course, the steering column is also exerting an opposing torque on the steering wheel, so your answer will be much greater than the actual acceleration of a steering wheel.)

15. A carnival game, sometimes called the strength tester, is played by striking the end of a board, mounted on a fulcrum, with a sledgehammer. The rotation of the board causes a puck, which was initially at rest on the other end of the board, to rise along a wire track. If the puck strikes the bell, you win.

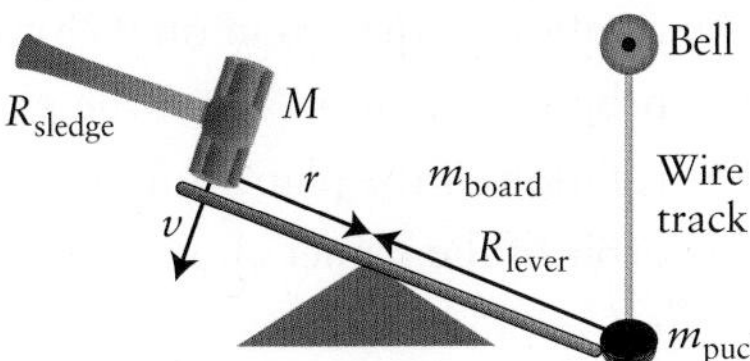

Use these parameter values to analyze the game: the mass of the sledge, $M = 5.0$ kg; the mass of the puck, $m_{puck} = 1.0$ kg; the distance from the point of impact of the sledge to the fulcrum, $r = 0.50$ m; the distance from the puck to the fulcrum, $R_{lever} = 1.0$ m; the length of the handle of the sledgehammer, $R_{sledge} = 0.80$ m; and the mass of the board that is struck whose length is $r + R_{lever}$, $m_{board} = 5.0$ kg. Assume that the board is rigid.

(a) The velocity the sledge gives to the point on which it strikes the board is v. If V/R_{lever} is the angular speed of the puck immediately after the board is struck, express the relationship between v and V.

(b) The bell to be rung by the puck is 6.0 m above the point at which the puck leaves the lever. Neglecting any friction forces, calculate the minimum speed of the puck as it leaves the lever that will allow the puck to strike the bell.

(c) An unenthusiastic contestant could just hold the head of the hammer above the board and let it fall. Estimate the minimum height from which the head of the sledgehammer must fall to ring the bell. (To make this estimation, you may assume the angular speed given to the lever does not change between the lever being struck and the puck leaving the lever, and that the puck leaves the lever directly upward along the track but dissipates no energy on the track.)

(d) A more enthusiastic contestant can increase the speed of the sledge head by exerting a torque that rotates the sledgehammer, as shown. The rotational axis is approximately a fixed position where one hand stays in place at the end of the sledgehammer handle. As shown in the figure, for this contestant's technique, the distance from the rotational axis to the point of application of the contact force on the handle decreases as the head rises. Describe all of the torques exerted on the sledgehammer as it goes through this rotation, including any changes in their sizes or directions.

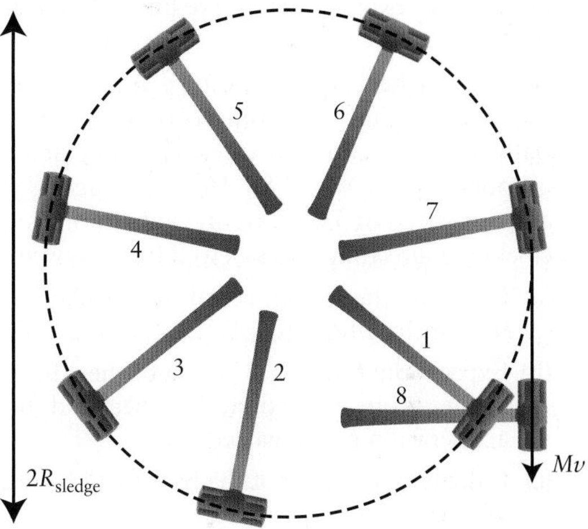

16. A rigid, uniform, 4.00-m-long, 40.0-kg beam rests on a pivot that is placed 3.00 m from one end of the beam. An 80.0-kg man walks slowly up the beam from the end resting on the ground.

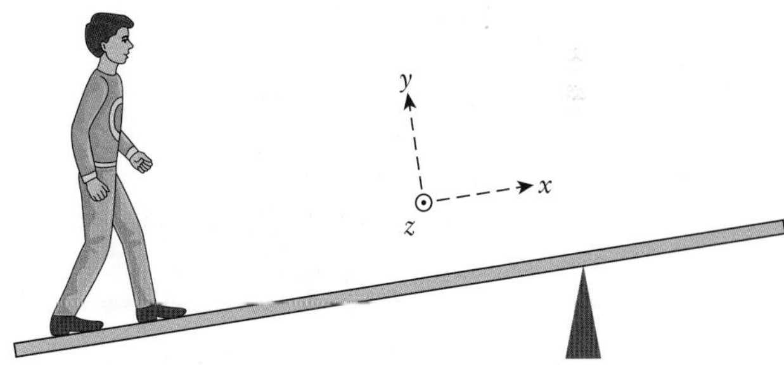

(a) Describe the net torque on the board as the man moves from his initial position to a position such that the lower end of the board just starts to lose contact with the ground, but the board is not yet rotating.

(b) Qualitatively, describe where the man is relative to the fulcrum when the situation described in part (a) occurs. Justify your answer in terms of net torque on the board.

17. A 45-kg high-diver launches himself from a springboard that is 3.0 m above the water's surface, so that he ends up with a 5-m fall from rest before he reaches the surface of the water. Calculate the time it takes for the diver to descend the 5.0 m to the surface of the water and the average angular speed of the diver that will allow him to complete a "two and a half" (2.5 full rotations) while in the air.

18. 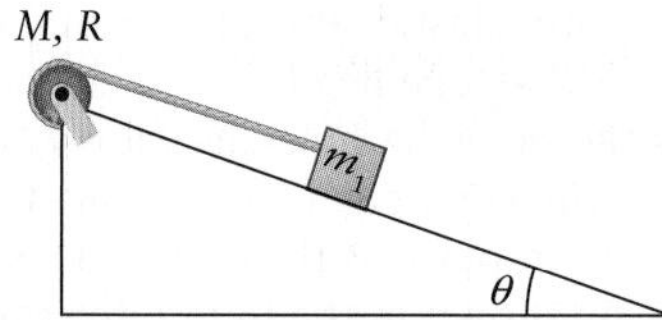

A block of mass m_1 is initially at rest on a surface that makes an angle θ with the horizontal, as shown. Assume there is negligible friction between the surface and the block. The block is attached to a string of negligible mass that is wrapped around a pulley of mass M and radius R. Assume you can model the pulley as a disk. Derive expressions for the linear acceleration of the block on the surface, and the tension in the string.

19. A potter's wheel is mounted on a shaft with bearings that exert negligible friction force. The wheel is initially at rest. A constant external torque of 75.0 Nm is applied to the wheel for 15.0 s, giving the wheel an angular speed of 5.00×10^2 rev/min in the counterclockwise direction when viewed from above the wheel.

(a) Calculate the magnitude of the angular acceleration of the wheel in rad/s^2. What is its direction?

(b) Express the rotational inertia of the wheel in terms of the torque exerted on the wheel and the angular acceleration of the wheel.

(c) Calculate the value of the rotational inertia of the wheel.

(d) The external torque is then removed, and a brake is applied. If it takes the wheel 2.00×10^2 s to come to rest after the brake is applied, what is the magnitude and direction of the torque exerted by the brake?

20. 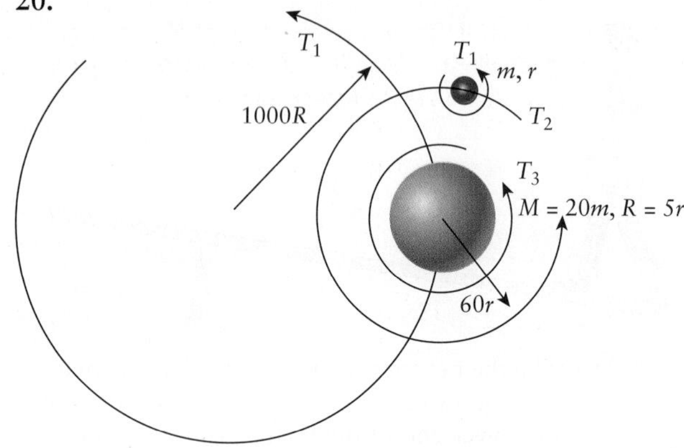

A moon with mass m and radius r orbits a planet with mass $M = 20m$ and radius $R = 5r$. The orbital radius of the moon about the planet is $60r$. The time, T_1, for one complete rotation of the moon about an axis through its center of mass and perpendicular to the plane of the figure shown, is identical to the time for one complete circular orbit, with radius $1000R$, of the planet around a star, as shown in the figure. The time for one complete orbit of the moon about the planet is $T_2 = T_1/10$.

(a) Express the angular speeds of the rotation of the moon, $\omega_{\text{moon rotation}}$, and the orbit of the planet around the star, $\omega_{\text{planet orbit}}$.

(b) Compare the linear speed of a point on the surface of the moon due to the moon's rotation on its axis with the linear speed of the center of mass of the planet.

(c) In time $T_3 = T_1/500$ the planet completes one rotation about an axis that is through its center of mass and is perpendicular to the plane of the figure. A point on the surface of the moon can be seen from the surface of the planet. Explain why this point is visible during each rotation of the planet. The planet has no atmosphere.

(d) The rotational inertia of a solid sphere with mass m and radius r rotating about a central axis is $2mr^2/5$. Express the rotational energies of each motion and rank them by

i. The rotation of the moon on its axis

ii. The orbit of the moon about the planet

iii. The rotation of the planet on its axis

iv. The orbit of the planet about the star

21. A 50.0-kg uniform horizontal beam of length 3.0 m is attached to a brick wall by a bracket at the wall and a cable attached to the far end that makes an angle of 60° with the horizontal. A 75.0-kg person stands 1.0 meter from the wall.

(a) Draw a free-body diagram and extended free-body diagram, to the same scale, for the beam.

(b) Find the tension in the cable and the force exerted by the wall on the beam.

Prep for the AP Exam

AP® Group Work

Directions: The following problem is designed to be done as group work in class.

Knife throwers, high-divers, and gymnasts must practice to "stick the landing" (land with the appropriate orientation). When a knife is thrown the center of mass follows a parabolic motion whose shape depends on the ratio of the components of the launch velocities, v_{0y}/v_{0x}. Balanced throwing knives have a center of mass at the midpoint of the knife.

The American Knife Throwers Alliance has competitions in which knives with lengths of either 9 or 16 inches are thrown at targets, using either a half-spin (thrown holding the blade) or a full spin (thrown holding the hilt). The distances in these competitions are half-spin: 7 and 13 ft; full spin: 10 and 16 ft. In the figure, a full-spin throw with three spins and a half-spin throw with 2½ spins are shown.

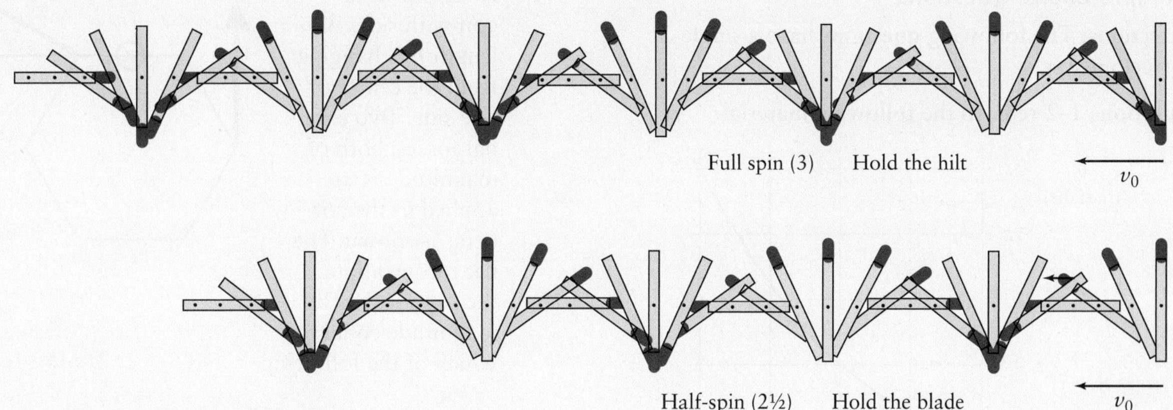

In the language used in knife throwing, "half-spin" and "full spin" refer to the number of rotations relative to an orientation of the knife when pointed at the target as the starting point. In fact, the knife is released when it is in the vertical position shown at the rightmost orientation in these diagrams.

(a) Justify the claim that for a wheel rolling without slipping the center of mass velocity, v_{CM}, is $R\omega$, where R is the radius of the wheel and ω is the angular speed. Explain why this no-slip condition that is used to describe rolling does not apply to a knife while it tumbles through the air.

(b) Justify the claim that the speed of the tip of a balanced knife with length $2R$ leaving the hand with a speed v_{CM} has a speed of $v_{CM} + R\omega$.

(c) When the knife is thrown it leaves the hand in a vertical orientation, as shown in the rightmost orientation of the knife in the diagram. Interpret the diagram to evaluate the angular displacement of the knife in radians for the half-spin throw in the diagram.

Shown in the figure is a trajectory for a full-spin throw with a target distance of 16 ft (4.88 m) using a knife with a length of 16 in. (0.41 m). The initial velocity of the center of mass of the knife is horizontal.

(d) Explain why the horizontal distance that the knife tip travels is $9\pi R/2 + v_{CM}t_{flight}$, where R is half of the length of the knife and t_{flight} is the time of flight.

(e) The knife leaves the hand just above the ear and strikes the target at chest level. The vertical displacement is h in the figure. Calculate the time of flight if h is equal to the length of the 16-inch knife.

(f) Calculate the distance traveled by each term in the expression for distance in part (d).

(g) Calculate the initial velocity of the throw, $v_{CM,16ft}$, and predict the throwing speed when the target is at the other competition distance for a full-spin throw, $v_{CM,10ft}$.

(h) If the target distance is between 10 and 16 ft, predict how the thrower can still strike the target with the tip of the knife. *Hint*: Competitions sponsored by the World Axe Throwing League are less prescriptive about distance to target for tomahawks, whose handle length is roughly 0.5 m.

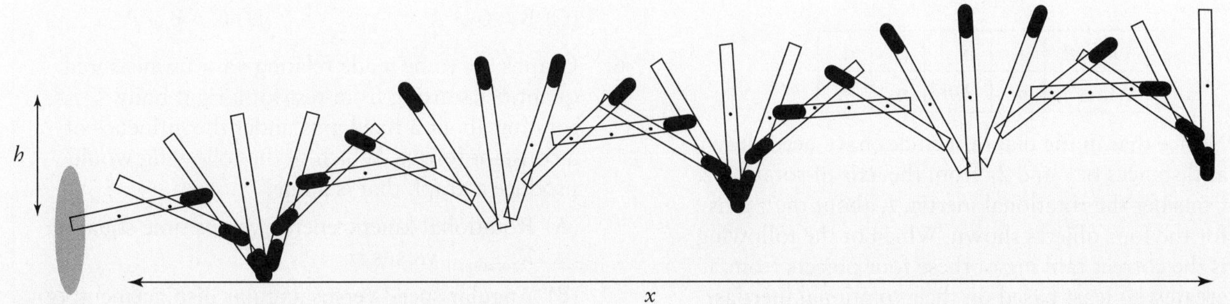

AP® PRACTICE PROBLEMS

Prep for the AP® Exam

Multiple-Choice Questions

Directions: The following questions have a single correct answer.

Questions 1–2 refer to the following material.

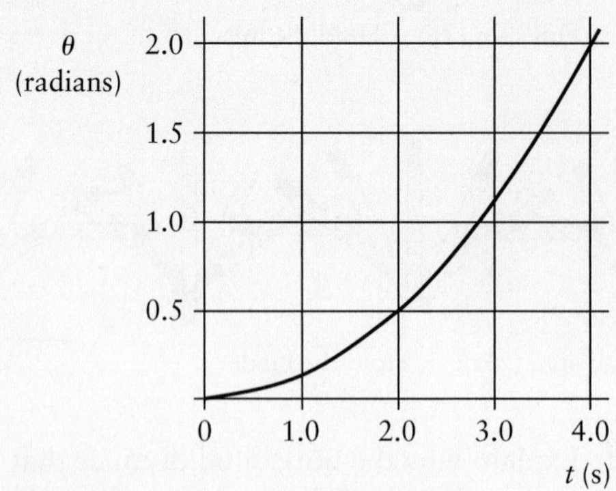

The graph shows the rotational displacement of a wheel as a function of time. The wheel started from rest at $t = 0$ and accelerated at a constant rate over the 4.0-s period shown.

1. What would be the rotational speed of the wheel at $t = 4.0$ s?
   (A) 0.25 rad/s (B) 0.50 rad/s
   (C) 0.75 rad/s (D) 1.0 rad/s
2. What is the rate of rotational acceleration during the 4-s period?
   (A) 0.25 rad/s^2 (B) 0.50 rad/s^2
   (C) 0.75 rad/s^2 (D) 1.0 rad/s^2

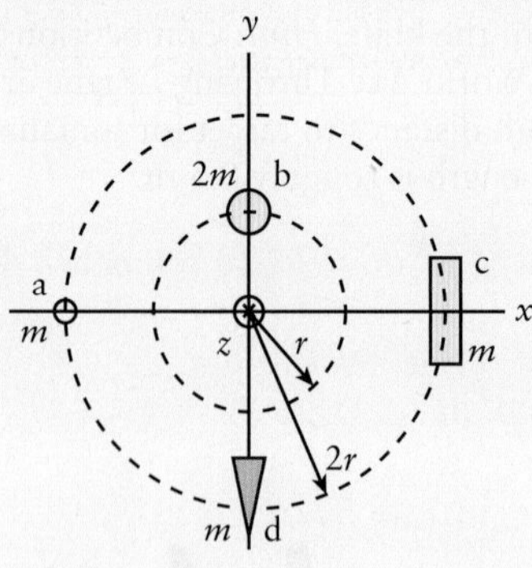

Object	a	b	c	d
Mass	m	$2m$	m	m

3. Notice that in the diagram circles have been drawn at distances of r and $2r$ from the axis of rotation z. Consider the rotational inertia, I, about the z axis for the four objects shown. Which of the following is the correct ranking of these four objects from greatest to least based on their rotational inertias?
   (A) c > a > d > b (B) a > d > c > b
   (C) b > a = c > d (D) c = a > d > b
4. A regular hexagon with sides of length L is hanging from the center of one side. Two external forces, both of magnitude F, are applied to the corners, as shown. The net torque about the support has a magnitude equal to which of the following?

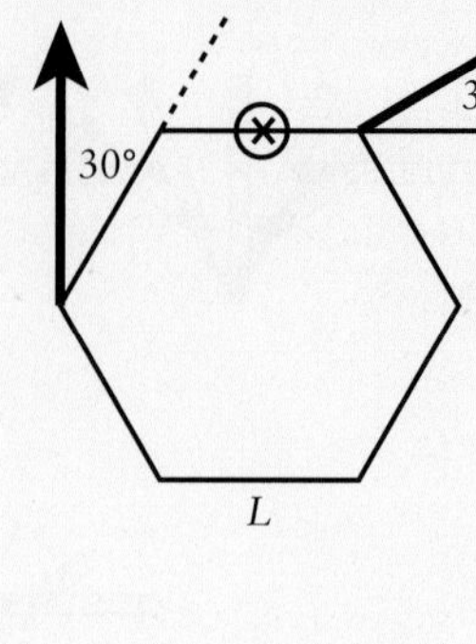

   (A) $\frac{5}{4}LF$ (B) $\frac{3}{4}LF$
   (C) LF (D) $\frac{1}{4}LF$

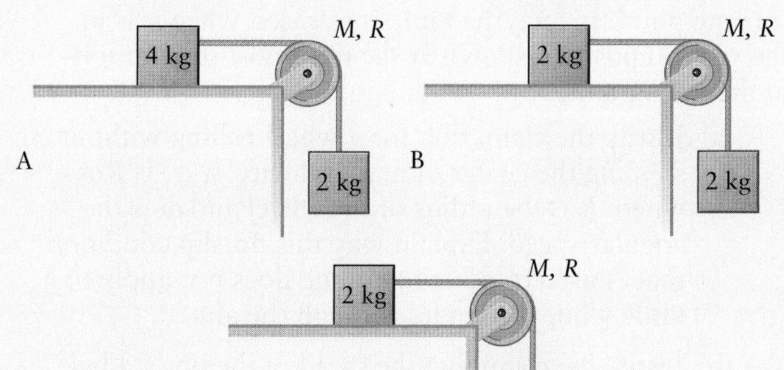

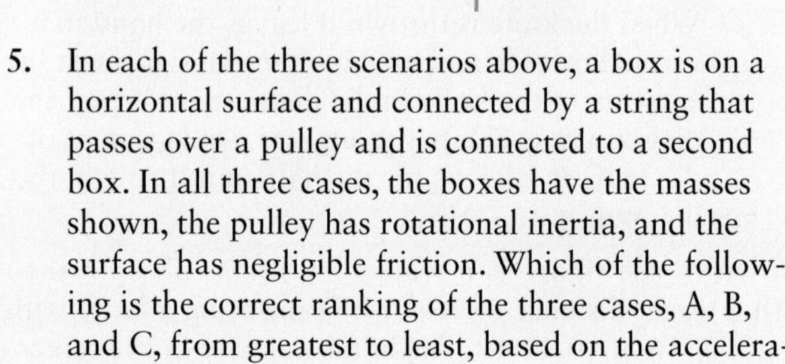

5. In each of the three scenarios above, a box is on a horizontal surface and connected by a string that passes over a pulley and is connected to a second box. In all three cases, the boxes have the masses shown, the pulley has rotational inertia, and the surface has negligible friction. Which of the following is the correct ranking of the three cases, A, B, and C, from greatest to least, based on the acceleration of the box on the table?
   (A) A > B > C (B) C > A = B
   (C) B = C > A (D) C > B > A
6. Graphs are to be made relating various measured quantities starting from rest for a rigid body rotating about a fixed axis under the influence of a constant torque. Which of the following would produce a graph that is a line?
   (A) Rotational kinetic energy versus time squared: $K_{\text{rotational}}$ versus t^2
   (B) Angular speed versus angular displacement: ω versus $\Delta\theta$

(*Continued*)

(C) Rotational kinetic energy versus angular displacement: $K_{rotational}$ versus $\Delta\theta$

(D) Angular displacement versus time: $\Delta\theta$ versus t

7. A long thin uniform rod originally has a mass M, a cross-sectional area A, and a length L. It is in rotation about its center of mass with an angular speed ω, and has rotational kinetic energy K. The rotational inertia for such a rod in rotation about its center of mass is $\frac{1}{12}ML^2$.

Without changing the material used or the angular speed, it is desired to reduce the kinetic energy for this rate of rotation to one-quarter of the original value. Which of the following changes would achieve that goal? Make the rod

(A) one-half of the original length and twice the cross-sectional area.

(B) one-half as long.

(C) one-half of the original cross-sectional area.

(D) one-half of the original cross-sectional area and one-half of the length.

Free-Response Questions

1. Two objects with masses $3m$ and m are connected by a thread of negligible mass. The thread runs over a pulley of radius r. The pulley rotates with negligible friction and the thread does not slip or stretch.

(a) i. Assume the pulley has negligible mass. The dots below represent the two blocks. Draw free-body diagrams showing and labeling the forces (not components) exerted on each block. Draw the relative lengths of all vectors to reflect the relative magnitude of the forces.

Block 1 Block 2

ii. The pulley has mass M, and rotational inertia I. Draw new free-body diagrams for each of the blocks for the case of a pulley with mass and also draw a force diagram for the pulley in the space indicated below. Draw the relative lengths of all vectors to reflect the relative magnitude of the forces. Specify the positive directions of translational and rotational motion on your diagrams.

Block 1 Block 2 Pulley

(b) In your calculations use the subscripts $3m$ and m to refer to the tensions in the two ends of the thread and the accelerations of the objects.

i. Apply Newton's second laws for translation and rotation to express the translational acceleration for each hanging object and rotational acceleration for the pulley.

ii. For $a_{3m} = a_m = r\alpha = a$, use your equations from (i) to show that

$$a = \frac{2g}{(4 + I/mr^2)}$$

(c) Explain what conditions are necessary for the translational accelerations of the hanging objects and the rotational acceleration to satisfy the expression given in (b):

$$a_{3m} = a_m = r\alpha = a$$

2. A seesaw is a uniform plank supported at its midpoint. Akeelah and her little sister Bree sit on opposite sides of the seesaw. Akeelah weighs 1.50 times as much as Bree. The girls sit spaced on the plank such that it is in equilibrium. Then one of the girls leans forward so that the plank starts accelerating.

(a) Describe how the sizes of the forces the plank exerts on the girls when Bree was accelerating downward differ from the forces the plank exerts on the girls in equilibrium. Justify your claim with reasoning based on physics principles.

(b) In terms of Bree's mass m_B, acceleration a, and any necessary constants, derive equations for the forces the plank exerts on the two girls at the instant Bree begins accelerating downward.

(c) Does your equation in (b) agree with your claim in (a)?

Energy and Momentum of Rotating Systems

Chapter 11 Torque and Rotation II: Work, Energy, and Angular Momentum

left: Al Bello/Getty Images; right: REUTERS/Michael Dalder/Newscom

Case Study: How does a diver control her rotation rate during a dive?

U.S. diver Jessica Parratto leaped gently from the 10-meter board during the 2016 Summer Olympics. As she plunged downward, she began to rotate more rapidly. Just before she hit the water her rotation stopped, and she entered the water vertically. To understand this motion, we need to further explore the physics of rotation.

The rotational motion of an extended object is related to its angular velocity and how its mass is distributed around the rotation axis. This is true for the kinetic energy associated with rotation and also for angular momentum, the rotational equivalent of linear momentum. We just saw that applying Newton's laws in rotation required the rotating system to be rigid, unlike this diver. Energy and angular momentum will be key to understanding motion of extended objects like this diver.

In this chapter, we will learn that the angular momentum of an extended object decreases if the angular speed remains constant but more of its mass is moved closer to the rotation axis. The diver's rotation axis remains the same, passing through her center of mass, but her angular speed about that axis changes as her mass distribution changes.

To understand how a diver controls her angular speed throughout the entire dive, we must recognize that *total angular momentum is always conserved*, and *it is constant when there is zero net torque exerted on a system*. When Jessica folded her body, she brought more of her mass close to her rotational axis, and her angular speed had to *increase* in order to keep angular momentum the same. Then Jessica straightened out, shifting her mass away from her rotation axis, *reducing* her angular speed so she could enter the water vertically.

The physics that underlies the way a falling cat manages to land on its feet is not unlike the physics of an Olympic diver. Can you describe, in general terms, what a cat must do to accomplish this feat? Imagine that the cat can control its front and hind legs independently.

By the end of this chapter, you should be able to define and explain the relationships among angular speed, angular momentum, and rotational inertia. Additionally, you should be able to use these quantities to describe the motion of a system, explain the circumstances under which angular momentum is constant, and apply the principles of the conservation of angular momentum and conservation of energy to rotating systems.

Torque and Rotation II: Work, Energy, and Angular Momentum

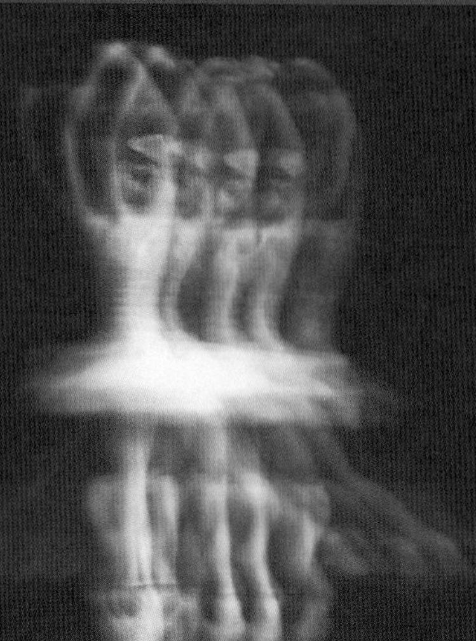

Image by cuppyuppycake/Getty Images

YOU WILL LEARN TO:

- Apply the conservation of mechanical energy to rotating systems, including in the case of rolling without slipping.
- Describe what is meant by the angular momentum of a rotating system and of a moving object, and explain the circumstances under which angular momentum is constant.
- Predict the behavior of rotational collision situations by the same processes that are used to analyze linear collision situations.
- Apply the law of universal gravitation and the expression for gravitational potential energy, with the concept of conservation of angular momentum, to analyze the orbits of satellites and planets.

11-1 Angular momentum and the next conservation law: conservation of angular momentum

When a spinning ballerina pulls her arms and legs in close to her body, the rate at which she spins will increase (see chapter opening photo). You can demonstrate this same effect by sitting in an office chair that can spin. Sit with your arms outstretched and hold a weight like a brick or a full water bottle in each hand. Now use your feet to start your body and the chair rotating, lift your feet off the ground, and then pull your arms inward. Your rotation will speed up quite noticeably!

Why does this happen? For both cases, the ballerina and you in the office chair, there's no net torque exerted on the system to make it rotate more quickly. Instead, both of these situations are examples of the *conservation of angular momentum*. This principle is the rotational analog of the conservation of *linear* momentum that we introduced in Chapter 9. Linear momentum and angular momentum are different quantities and have different units, as we will see in the next section. We saw that an object with linear momentum could transfer that linear momentum to another object during an interaction between them: to set another object in linear motion, or to change or stop its linear motion. **Angular momentum** similarly represents the ability of an object or system to set another object or system in some sort of rotational motion.

NEED TO REVIEW? Turn to the **Glossary** in the back of the book for definitions of bolded terms.

In Chapter 9, we saw that linear momentum depends only on an object's mass and velocity. In Chapter 10, we saw that rotational quantities, in addition, depend on shape and rotation axis. We will see that an object or system does not need to be rotating to have angular momentum; it just needs to have motion perpendicular to a rotation axis! Since a rotating system can have internal energy, its kinetic energy can change even as its angular momentum stays the same, because it can convert internal energy to kinetic energy, such as pulling water bottles closer to your body to spin more quickly in the office chair. This is similar to the explosive collisions we studied in Chapter 9.

Angular momentum and its conservation provide the additional concept we need to predict the motion of objects and systems in space. It explains the arrangement of solar systems and galaxies. The conservation of angular momentum on the human scale helps explain the rotation of ocean currents; the changing spin of a dancer, diver, or skater; or the speed and direction of winds in hurricanes or cyclones. On the atomic and subatomic scales, the language of angular momentum is used to describe the spin of atoms and spins and orbits of their constituents. If you study further, you will find that "spin" is actually a property of fundamental particles, not really a description of motion; but it still behaves as an angular momentum (truly tiny things have weird properties). How these particles interact is governed by conservation of angular momentum.

THE TAKEAWAY for Section 11-1

✔ Linear momentum and angular momentum are different quantities with different units.

✔ An object's linear momentum is the product of its mass and velocity, but an object's angular momentum also depends on its motion perpendicular to a rotation axis. For a rotating extended object, angular momentum also depends on its shape with respect a rotation axis.

✔ Angular momentum is always conserved.

Prep for the AP® Exam

AP® Building Blocks

1. Just as linear momentum is a measure of an object's ability to set other objects in linear motion, a quantity called angular momentum, L, is a measure of an object's ability to set other objects in rotational motion. What properties of an object determine its momentum, and what additional property must be involved to introduce angular momentum?
2. In Chapter 9, we saw that force is change in linear momentum with respect to time. In Chapter 10, we saw that torque is to rotational motion as force is to linear motion. Given this relationship, what do you predict is the relationship between torque and angular momentum?
3. If you sit on a stool that rotates easily with your arms outstretched and hold a weight in each hand, what property of the system of you and the weights are you changing when you pull your arms inward?

11-2 Conservation of mechanical energy also applies to rotating extended objects

Every year thousands of girls and boys race in the All-American Soap Box Derby. They race in homemade cars powered only by the pull of gravity (**Figure 11-1**). In building their cars, competitors have the flexibility to be creative, but they can be disqualified for using unapproved wheels—specifically, wheels for which too much of the mass is concentrated toward the center of the wheel. How would such wheels affect the outcome of the race? We can answer this and many related questions by considering the conservation of energy.

Joe Raedle/Getty Images News/Getty Images

Figure 11-1 Soap box derby racers These unpowered race cars roll downhill, propelled only by gravity. How do the properties of a car's spinning wheels affect the car's speed?

We learned in Section 8-1 that the mechanical energy of a system—the sum of kinetic energy and potential energy—is *constant* if no nonconservative interactions exist within that system and no external forces do work on that system. In the last chapter we introduced the *rotational* kinetic energy of a rigid body with rotational inertia I rotating with angular speed ω: $K_{\text{rotational}} = \frac{1}{2}I\omega^2$. Does conservation of energy still hold true if we include rotational kinetic energy? And in particular does it hold true for systems that are *both* translating (moving through space) *and* rotating, like a ball rolling downhill or the spinning wheels of a soap box derby car?

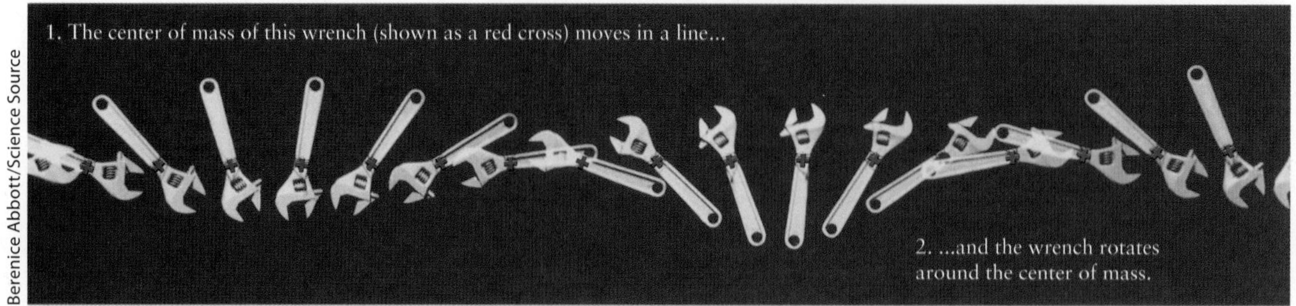

1. The center of mass of this wrench (shown as a red cross) moves in a line...
2. ...and the wrench rotates around the center of mass.
Berenice Abbott/Science Source

Figure 11-2 Combined translation and rotation In this time-lapse image, you are looking down on a spinning wrench as it slides across a table. The wrench naturally rotates around its center of mass (shown by the red "cross"), so we can think of its motion as the translation of the center of mass plus the rotation of the wrench around the center of mass.

Happily, the answer to both questions is "yes" because rotational kinetic energy isn't really a new kind of energy. As we saw in the last chapter, the rotational kinetic energy of a system is just the combined kinetic energy of all of its components due to the motion of those pieces around the rotation axis. So the principle of conservation of energy also holds if we include rotational kinetic energy.

We can also use the principle of conservation of energy for a system that's rotating as its center of mass is moving through space. In such a situation it turns out that we can write the system's *total* kinetic energy as the sum of two terms: the *translational* kinetic energy associated with the motion of the system's center of mass (see Section 7-5), and the *rotational* kinetic energy associated with the system's rotation around its center of mass (**Figure 11-2**). If the system has mass M and its center of mass is moving with speed v_{CM}, its translational kinetic energy is $K_{\text{translational}} = \frac{1}{2}Mv_{CM}^2$; this is always true. If the system is a rigid body, then we can write equations for its rotational inertia, so we will assume this special case to develop our equations. If the rigid body's rotational inertia for an axis through its center of mass is I_{CM} and it rotates with angular speed ω, its rotational kinetic energy is $K_{\text{rotational}} = \frac{1}{2}I_{CM}\omega^2$. The total kinetic energy (translational plus rotational) of the rigid body is then

(1) Total kinetic energy of a system that is both translating and rotating around its center of mass

(2) The **translational kinetic energy** is the same as if all of the mass of the system were moving at the speed of the center of mass.

EQUATION IN WORDS
Total kinetic energy for a rigid body undergoing both translation and rotation

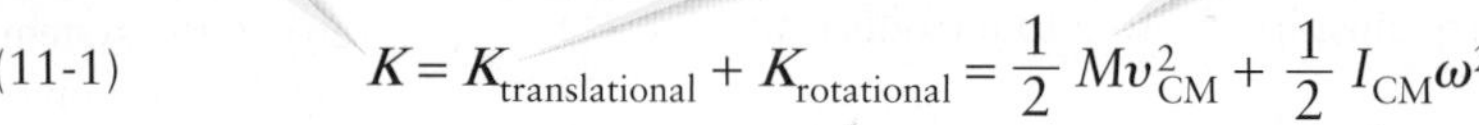

$$K = K_{\text{translational}} + K_{\text{rotational}} = \frac{1}{2}Mv_{CM}^2 + \frac{1}{2}I_{CM}\omega^2 \quad (11\text{-}1)$$

(3) The **rotational kinetic energy** is the same as if the system were not translating, and all of the mass were rotating around the center of mass.

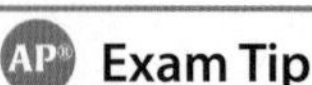

Exam Tip

For problems in which a rigid body is spinning or rolling and the center of mass is translating, be sure to sum *both* translational and rotational kinetic energy as the total kinetic energy.

Many rigid bodies that undergo both translation and rotation will be symmetrical, like a rolling bowling ball or bicycle tire. In such a case the center of mass is at the rigid body's geometrical center, and Table 10-1 will help you determine the value of I_{CM} for that rigid body. (The wrench shown in Figure 11-2 is not symmetric: It has more mass at one end than the other, so its center of mass is *not* at its geometrical center.)

Whenever we have a situation involving a rigid body that undergoes both translation and rotation, we can make use of the conservation of energy, provided that we use Equation 11-1 as the expression for kinetic energy so that we correctly include all of the kinetic energy. If no work is done on the system, and there are only conservative interactions in the system, then mechanical energy (the sum of kinetic energy K and potential energy U) is constant, so the value of $K + U$ is the same at any two times during the motion:

$$K_i + U_i = K_f + U_f \quad (8\text{-}3)$$

If work is done on the system, then it is necessary to use the full work-energy theorem:

$$W = \Delta K + \Delta U + \Delta E_{\text{internal}} \quad (7\text{-}12)$$

For any equation involving energy conservation, the kinetic energy K is now given by Equation 11-1 and the potential energy U includes a term for each conservative interaction within the system (for example, gravitational or spring interactions). For gravitational potential energy, the location of the "object" in the Earth–rigid body system is the center of mass of the rigid body.

Example 11-1 illustrates how to use these ideas to solve a problem that involves both translational and rotational motion.

EXAMPLE 11-1 Flying Disk Energy

A flying disk with a mass of 0.175 kg and a diameter of 0.266 m is used in the team sport called Ultimate. A player takes a disk at rest and does 1.00 J of work on it, causing the disk to fly off in a horizontal direction. When the disk leaves the player's hand, 9/10 of its kinetic energy is translational and 1/10 is rotational. Find (a) the speed of the flying disk's center of mass and (b) the angular speed of the rotating disk. Treat the disk as uniform.

Set Up

Work is done by the player, so the player is outside of the system. Let's define our system as the Earth–disk system. Then we can use the work-energy theorem to describe how the disk's energy changes, given that any changes in Earth's energy will be negligible. We use Equation 11-1 to describe the kinetic energy of the disk, which involves its rotational inertia I_{CM} for an axis through its center of mass (the same as its geometrical center). This rotational inertia is $I_{CM} = MR^2/2$, where $M = 0.175$ kg and R is one-half of the disk's diameter (see the first entry in the second row of Table 10-1). Our goal is to find the values of v_{CM} and ω just after the disk leaves the player's hand.

Work-energy theorem:

$$W = \Delta K + \Delta U + \Delta E_{internal} \qquad (7\text{-}12)$$

Total kinetic energy for a rigid body undergoing both translation and rotation:

$$K = K_{translational} + K_{rotational}$$
$$= \frac{1}{2}Mv_{CM}^2 + \frac{1}{2}I_{CM}\omega^2 \qquad (11\text{-}1)$$

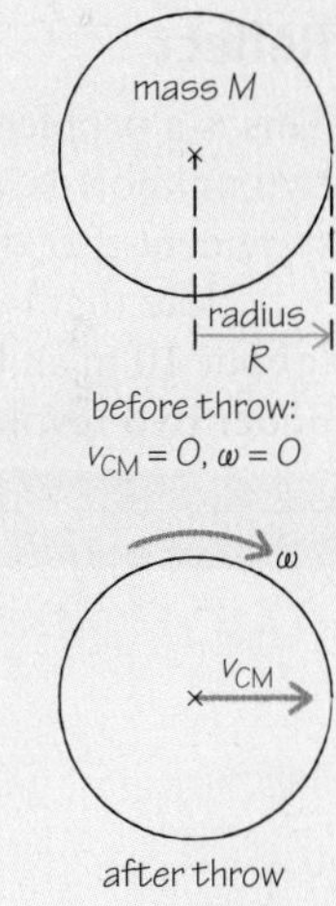

Solve

During the throw the disk moves horizontally, so its height stays the same and its gravitational potential energy is constant; that is, $U_f = U_i$. The disk starts at rest, so $K_i = 0$ (it has zero initial kinetic energy). The player is outside the system, so changes in the player's internal energy are not in the system. The disk does not have internal energy. So the work done on the disk by the player is transformed into the disk's final kinetic energy K_f.

$$W = \Delta K + \Delta U + \Delta E_{internal}$$

Because $\Delta E_{internal} = 0$, $U_f = U_i$, $K_i = 0$, and $W = 1.00$ J, this becomes

$$1.00\text{ J} = K_f$$

The disk's 1.00 J of kinetic energy is divided so that 9/10 is translational and 1/10 is rotational.

$$K_{translational} = \frac{1}{2}Mv_{CM}^2 = (9/10)(1.00\text{ J}) = 0.900\text{ J}$$

$$K_{rotational} = \frac{1}{2}I_{CM}\omega^2 = (1/10)(1.00\text{ J}) = 0.100\text{ J}$$

(a) Find the speed of the disk's center of mass.

From the expression for translational kinetic energy,

$$v_{CM}^2 = \frac{2K_{translational}}{M}$$

$$v_{CM} = \sqrt{\frac{2K_{translational}}{M}} = \sqrt{\frac{2(0.900\text{ J})}{0.175\text{ kg}}} = \sqrt{\frac{2(0.900\text{ kg}\cdot\text{m}^2/\text{s}^2)}{0.175\text{ kg}}}$$
$$= 3.21\text{ m/s}$$

(b) Calculate the disk's rotational inertia for the axis of rotation through its center, then solve for its angular speed ω. Note that if $K_{rotational}$ is in joules and I_{CM} is in $kg \cdot m^2$, ω is in radians per second.

Calculate the disk's rotational inertia through its center of mass:

$$I_{CM} = \frac{1}{2}MR^2 = \frac{1}{2}(0.175 \text{ kg})\left(\frac{0.266 \text{ m}}{2}\right)^2$$
$$= 1.55 \times 10^{-3} \text{ kg} \cdot \text{m}^2$$

From the expression for rotational kinetic energy,

$$\omega^2 = \frac{2K_{rotational}}{I_{CM}}$$
$$\omega = \sqrt{\frac{2K_{rotational}}{I_{CM}}} = \sqrt{\frac{2(0.100 \text{ J})}{1.55 \times 10^{-3} \text{ kg} \cdot \text{m}^2}}$$
$$= \sqrt{\frac{2(0.100 \text{ kg} \cdot \text{m}^2/\text{s}^2)}{1.55 \times 10^{-3} \text{ kg} \cdot \text{m}^2}} = 11.4 \text{ rad/s}$$

Reflect

This is a problem for which the conservation of energy equation is essential but not sufficient. In order to solve for the two unknowns, v_{CM} and ω, we needed a second relationship telling us how the kinetic energy was distributed—the statement that the disk's kinetic energy was 9/10 translational and 1/10 rotational.

Note that v_{CM} is greater than a typical walking speed (1 to 2 m/s) but less than the speed of a world-class sprinter (about 10 m/s). Because 2π (just over 6) radians represents a full revolution, the value of ω tells us that the disk spins just under two revolutions per second. These values seem reasonable for a hand-thrown flying disk.

NOW WORK Problems 1, 3, and 4 from The Takeaway 11-2.

Rolling Without Slipping

Let's return to one of the soap box derby cars that we thought about at the beginning of this section (Figure 11-1) and ask how we can use energy concepts to analyze the motion of one. For simplicity let's think about a single wheel rolling downhill (**Figure 11-3a**) rather than the entire car. After the wheel has traveled a certain distance down the hill, how fast is its center of mass moving, and how rapidly is it spinning?

Before the start of the race, the wheel–Earth system has gravitational potential energy but the wheel has no kinetic energy. The wheel picks up speed as it rolls down the race course, conserving energy as the system's potential energy decreases. Unlike the

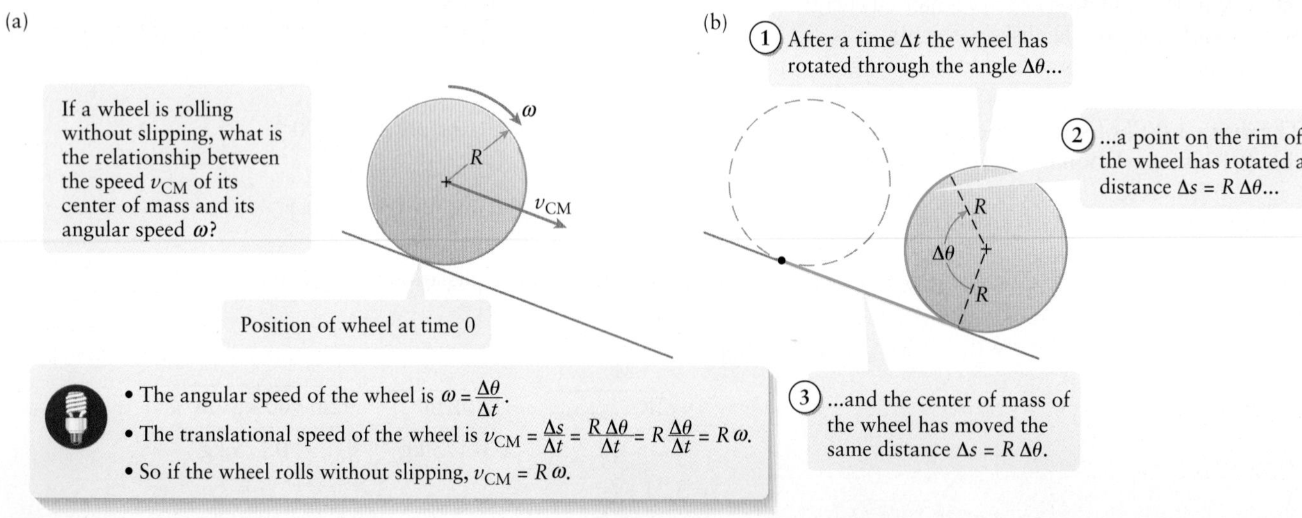

(a)
If a wheel is rolling without slipping, what is the relationship between the speed v_{CM} of its center of mass and its angular speed ω?
ω
R
v_{CM}
Position of wheel at time 0
• The angular speed of the wheel is $\omega = \frac{\Delta\theta}{\Delta t}$.
• The translational speed of the wheel is $v_{CM} = \frac{\Delta s}{\Delta t} = \frac{R\,\Delta\theta}{\Delta t} = R\frac{\Delta\theta}{\Delta t} = R\,\omega$.
• So if the wheel rolls without slipping, $v_{CM} = R\,\omega$.
(b)
1 After a time Δt the wheel has rotated through the angle $\Delta\theta$...
2 ...a point on the rim of the wheel has rotated a distance $\Delta s = R\,\Delta\theta$...
R
$\Delta\theta$
R
3 ...and the center of mass of the wheel has moved the same distance $\Delta s = R\,\Delta\theta$.

Figure 11-3 A wheel rolling downhill If a wheel rolls without slipping, there is a definite relationship between its angular speed and the speed of its center of mass.

flying disk in Example 11-1, however, we are not given the relationship between the amounts of rotational and translational kinetic energy. We need to use a physical principle to figure out how much of that kinetic energy is translational energy and how much is rotational. Furthermore, there must be friction between the wheel and the road in order for the wheel to roll. (If the road exerted no friction on the wheel, the wheel would simply slide downhill.) Therefore, we also need to know how much mechanical energy is lost due to friction as the wheel rolls.

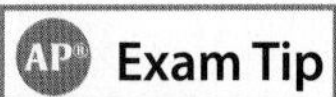

Exam Tip

Work is the product of an external force exerted on an object and the distance the point at which the force is exerted moves in the direction of the force, $W = Fd$. We can also look at this in terms of a torque exerted at a point on a rigid body. For a rigid body, the distance a point on the rigid body moves as it rotates about its axis due to that rotation is Δs. When a constant torque is used to angularly accelerate a rigid body without slipping, we can use the definitions of torque and angular displacement to find a new expression for work: $W = (1)Fd = (1)F\Delta s = (R/R)F\Delta s = (RF)(\Delta s/R) = \tau\Delta\theta$. Particularly in problems where there is no translational motion, it is often easier to use this expression for work.

To make further progress, we use the idea that the wheel is **rolling without slipping**—that is, the wheel does not skid or slide over the road. In this situation there is a specific relationship between the speed of the center of the wheel, v_{CM}, and the angular speed ω of the wheel's rotation. To see what this relationship is, consider what happens when the wheel rolls through a small angle $\Delta\theta$ over a time interval Δt (**Figure 11-3b**). Imagine that a thread has been wrapped around the circumference of the wheel and unwinds as the wheel rolls, marking the distance traveled. During the interval Δt a point on the rim of the wheel of radius R rotates a distance $\Delta s = R\Delta\theta$. Since the wheel doesn't slip on the road, $\Delta s = R\Delta\theta$ is also the length of thread that the wheel unwinds onto the road and hence the distance that the center of mass of the wheel has moved down the road. Hence v_{CM} is equal to the distance Δs divided by the duration Δt of the time interval:

$$v_{CM} = \frac{\Delta s}{\Delta t} = \frac{R\Delta\theta}{\Delta t} = R\frac{\Delta\theta}{\Delta t}$$

But $\Delta\theta/\Delta t$ is just equal to ω, the angular speed of the wheel, so

Speed of the center of mass of a rolling rigid body

$$v_{CM} = R\omega \qquad (11\text{-}2)$$

Radius of the rigid body

Angular speed of the rigid body's rotation

EQUATION IN WORDS
Condition for rolling without slipping

Equation 11-2 says that there is a direct proportionality between the linear speed v_{CM} and angular speed ω of a circular rigid body that rolls without slipping. As the wheel in Figure 11-3 rolls downhill it gains speed, v_{CM} increases, it rotates more quickly, and ω increases, but these increases are always proportional to each other so that v_{CM} and ω are always related by Equation 11-2. This has another very interesting effect. Picture the wheel simply rotating about an axis through its center of mass. As the wheel is rotating about its center of mass, the top of the wheel is moving with a velocity relative to the center of mass of $R\omega$ in the direction of motion. The bottom of the wheel is moving with a velocity relative to the center of mass of $R\omega$ opposite the direction of motion. When we add the center of mass velocity with which the whole wheel is moving forward, we find that at any instant the top of the wheel is moving twice as quickly as its center, and the bottom of the wheel is not moving. That is why this is *rolling without slipping*: The bottom of the wheel is at rest with respect to the ground.

How much mechanical energy is lost due to friction if a rigid body rolls without slipping, like the wheel in Figure 11-3? If the surface is also perfectly rigid, the answer is simple: The force of friction does *no* work since there is no relative motion: *No* energy is lost. In rolling with slipping, a rigid body *slides* over a surface, and, as we saw in Section 7-2, the kinetic friction force dissipates mechanical energy. But there is no sliding if the rigid body rolls without slipping, so friction does no work. The static friction force plays an important role: It's what makes the rigid body roll. Remember that even though an external force that is exerted on a point on a system that does not move cannot do work, it can cause the types of energy in the system to change. So as the wheel in Figure 11-3 rolls downhill and gravitational potential energy is converted to kinetic energy, the friction force is what is responsible for converting a portion of the kinetic energy to go into the rotational form.

Exam Tip

Rolling with slipping, the effects of nonrigid extended objects, and rolling friction will be only qualitatively covered on the AP® exam. When a rotating system is slipping relative to a surface, the point at which the kinetic friction force is exerted on the system moves in the opposite direction of the force, and thus the force of kinetic friction dissipates energy from the system. When slipping, $v_{cm} \neq R\omega$!

Nonrigid Extended Objects and Rolling Friction

Although friction does no work on a rigid body that rolls without slipping, if the extended object or surface is not perfectly rigid—that is, if it can *deform*—there *is* a related force that does work on the extended object as it rolls. An example is a bicycle

NEED TO REVIEW?
Turn to Chapter 5, Section 5-3, to review rolling friction.

tire or automobile tire, which flexes and flattens on its bottom where it contacts the road. As the tire rolls, different parts of the tire successively get compressed as they touch the ground and relax as they lose contact with the ground. This flexing causes the tire to get warmer, which dissipates mechanical energy. The energy comes from the mechanical energy of the rolling tire, which means that left to itself on a horizontal surface the tire will lose kinetic energy and slow down. The net effect is the same as if a force were exerted to oppose the rolling tire's motion. This is the force that we called *rolling friction*. Rolling friction is the force resisting the motion of any rolling extended object. It is generally increased if the surface on which the extended object rolls is deformable, like a wrestling mat or the green baize that covers a billiards table. This type of friction is typically a combination of several friction forces at the point of contact between the rolling extended object and the ground or other surface.

You can minimize rolling friction on your bicycle or automobile by keeping tires properly inflated. Underinflated tires flex more, which means that they dissipate more energy and reduce the car's gas mileage. That's also why transporting goods via railroad is more efficient than using trucks: Railroad cars have very rigid steel wheels, which flex much less than even properly inflated truck tires. The coefficient of *rolling friction* for steel on steel is typically $\mu_r = 0.002$. For rubber tires on pavement, according to the type and inflation of the tire, μ_r can be anywhere from 0.006 to 0.035. As the rigidity of the wheel and the surface decrease it also makes it harder to steer, so trains must ride on rails, which not only support the wheels but help guide them.

There are multiple types of friction involved in controlling your car. The wheels are purposefully not rigid. Only perfectly rigid bodies experience rolling without slipping. We saw that for rolling without slipping to occur, the single point or line of contact at the bottom of the rolling rigid body must be momentarily at rest relative to the surface on which it is rolling. This would make $\mu_r = 0$ and is rarely what actually happens. As the surface of an extended object flattens out, the friction force does not just cause the rolling motion but also allows more control of the motion, although flattening increases rolling friction. The coefficient of static friction is much larger than the coefficient of rolling or sliding friction, so you need to use static friction to control your car's acceleration, either to go more quickly, to go more slowly, or to change direction. The centripetal force that the road must exert on the car when the car drives around a curve ultimately comes primarily from the interaction of the tires and the ground. If only a single line on the bottom of the tire is in contact with the ground, you have no rolling friction, but you have much less surface to provide the static friction you need for these velocity changes. If you are trying to turn and to speed up or slow down, it is not hard to exceed the amount of static friction force that the road can provide on your tires (sometimes this is referred to as traction). This leaves you in a situation of sliding, where the friction force is even less and you have lost control of your car. The situation gets worse in cases where the coefficients of friction are reduced, such as wet or icy roads.

This scenario builds on Example 8-3 and provides a good real-world life lesson: Slamming on the brakes when the road is slippery is very dangerous because, without static friction, you lose control of the car; the wheels stop spinning, but the car doesn't stop. If you are approaching a stretch of road that you worry might be slippery, make sure your speed is reasonable and that you do not need to accelerate (change speed or direction). In such a case, conservation of momentum will keep you going in the right direction even if you lose traction. Even though we count on tire deformation to give us control of our car or bicycle, the amount of deformation required is actually very small, so it is still a great approximation to treat the wheels as rigid bodies in most sorts of problems.

AP® Exam Tip

Rolling doesn't have to imply rolling friction. An extended object rolling on a surface experiences rolling friction *only* if the extended object, the surface, or both are deformable. For problems on the AP® exam, assume that all rolling extended objects are rigid bodies and the surfaces on which they roll are perfectly rigid. Then there's no rolling friction. Static friction is what makes an extended object roll rather than slide. The static friction force does no work on the rigid body, so in this case mechanical energy is constant.

We now see how to use the energy approach to solve problems involving rigid bodies that roll without slipping. Since no work is done by the force of friction, if we can neglect all other external interactions, we can conserve mechanical energy as the rigid body rolls. The kinetic energy of the rolling rigid body is given by Equation 11-1, $K = \frac{1}{2}Mv_{CM}^2 + \frac{1}{2}I_{CM}\omega^2$, and the speed v_{CM} and angular velocity ω are related by Equation 11-2, $v_{CM} = R\omega$. The following example makes use of these equations.

EXAMPLE 11-2 Downhill Race: Disk Versus Hoop

A uniform disk and a hoop are both allowed to roll without slipping down a ramp of height H (**Figure 11-4**). Both are rigid bodies and have the same radius R and the same mass M. If the disk and hoop start from rest at the same time, which one reaches the bottom of the ramp first?

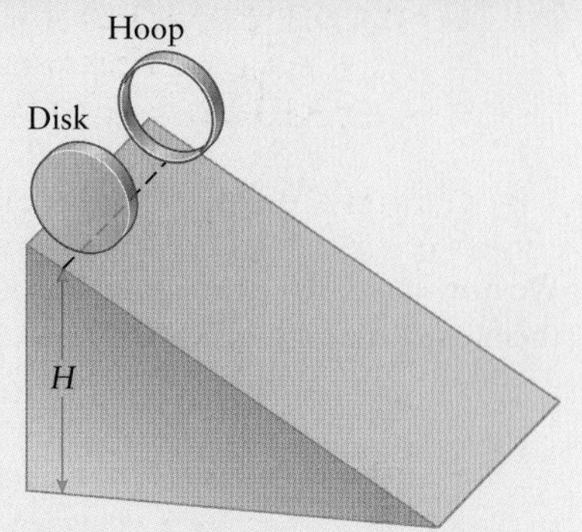

Figure 11-4 Which one wins the race? A competition between two rigid bodies rolling down a ramp.

Set Up

Both rigid bodies roll without slipping, so we can use Equations 8-3, 11-1, and 11-2 to describe their motion. The only real difference between the two is their rotational inertia: From Table 10-1, $I_{CM} = MR^2/2$ for the uniform disk, whereas $I_{CM} = MR^2$ for the hoop. To decide which one wins the race, we'll calculate v_{CM} at the bottom of the ramp for each. The one with the faster translational speed will get to the bottom first and be declared the winner. Because acceleration is constant, whichever rigid body is going more quickly at the end will have been going more quickly all along, so we'll solve for $v_{CM,f}$ (the speed at the bottom of the ramp). We will do this for a general rigid body that rolls down the ramp starting from rest: Only at the end will we plug in the value of I_{CM} for each rigid body.

Constant mechanical energy:

$$K_i + U_i = K_f + U_f \qquad (8\text{-}3)$$

Kinetic energy for a rigid body that undergoes both translation and rotation:

$$K = \frac{1}{2}Mv_{CM}^2 + \frac{1}{2}I_{CM}\omega^2 \qquad (11\text{-}1)$$

Condition for rolling without slipping:

$$v_{CM} = R\omega \qquad (11\text{-}2)$$

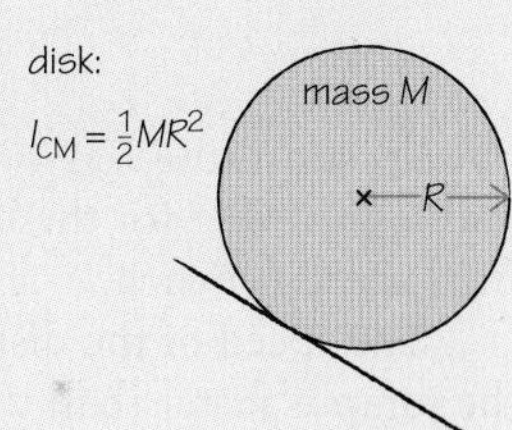

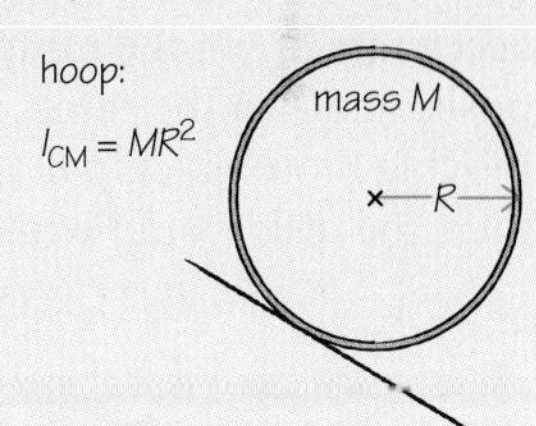

Solve

Initially the rigid body is at rest with zero kinetic energy, so $K_i = 0$. The initial gravitational potential energy of the Earth–rigid body system is $U_i = Mgy_i$, and the final gravitational potential energy is $U_f = Mgy_f$. The figure shows that $y_i - y_f = H$, so we can solve for the rigid body's final kinetic energy, K_f, at the bottom of the ramp.

Constant mechanical energy equation:

$$0 + Mgy_i = K_f + Mgy_f, \text{ or}$$

$$K_f = Mgy_i - Mgy_f = Mg(y_i - y_f)$$

Because $y_i - y_f = H$,

$$K_f = MgH$$

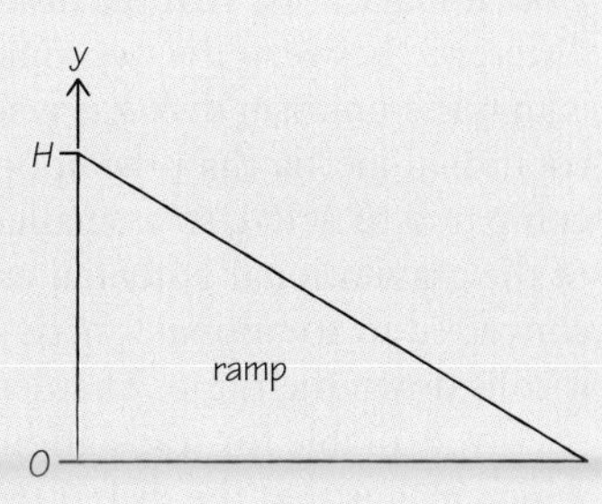

The kinetic energy is part translational and part rotational. We can use Equation 11-2 to write ω_f (the angular velocity at the bottom of the ramp) in terms of $v_{CM,f}$ (the linear speed at the bottom of the ramp), and so can express the final kinetic energy in terms of the linear speed $v_{CM,f}$ only.

Kinetic energy at the bottom of the ramp:

$$K_f = \frac{1}{2}Mv_{CM,f}^2 + \frac{1}{2}I_{CM}\omega_f^2$$

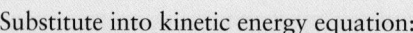

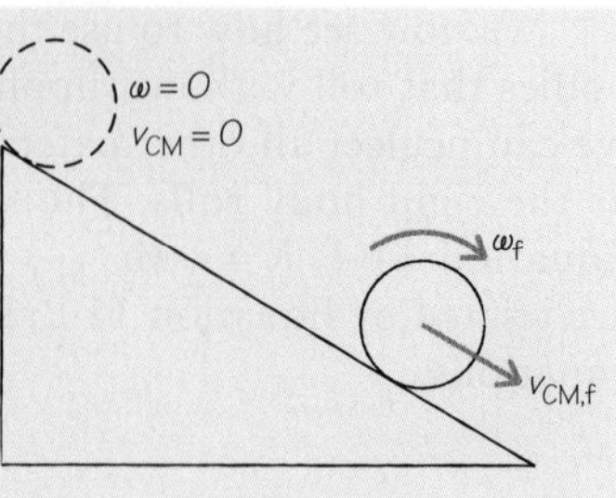

Rolling without slipping:

$v_{CM,f} = R\omega_f$, so

$$\omega_f = \frac{v_{CM,f}}{R}$$

Substitute into kinetic energy equation:

$$K_f = \frac{1}{2}Mv_{CM,f}^2 + \frac{1}{2}I_{CM}\left(\frac{v_{CM,f}}{R}\right)^2 = \frac{1}{2}Mv_{CM,f}^2 + \frac{1}{2}\left(\frac{I_{CM}}{R^2}\right)v_{CM,f}^2$$

$$= \frac{1}{2}\left(M + \frac{I_{CM}}{R^2}\right)v_{CM,f}^2$$

We now have two expressions for K_f. Set them equal to each other and solve for $v_{CM,f}$.

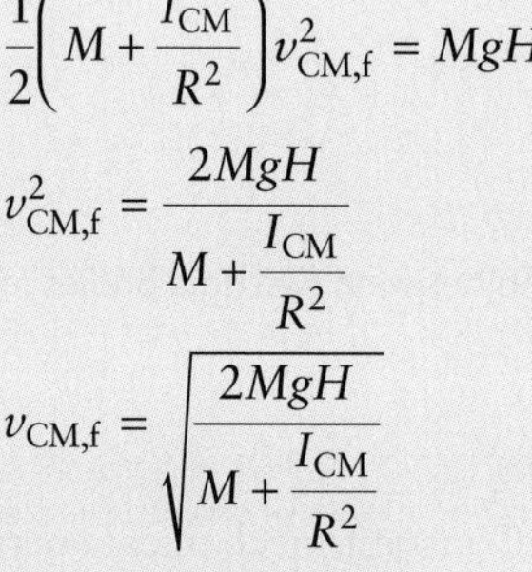

$$\frac{1}{2}\left(M + \frac{I_{CM}}{R^2}\right)v_{CM,f}^2 = MgH$$

$$v_{CM,f}^2 = \frac{2MgH}{M + \dfrac{I_{CM}}{R^2}}$$

$$v_{CM,f} = \sqrt{\frac{2MgH}{M + \dfrac{I_{CM}}{R^2}}}$$

Finally, substitute $I_{CM} = MR^2/2$ for the disk and $I_{CM} = MR^2$ for the hoop and find the final speed for each.

The speed of the disk at the bottom of the ramp is faster than that of the hoop: $\sqrt{(4/3)\,gH}$ compared to $\sqrt{gH}$. Both will accelerate down the ramp, but at any given position down the ramp, the disk will be moving more quickly than the hoop. In a race, the disk would win.

For the uniform disk:

$$I_{CM} = \frac{MR^2}{2}$$

$$M + \frac{I_{CM}}{R^2} = M + \frac{M}{2} = \frac{3M}{2}$$

$$v_{CM,f} = \sqrt{\frac{2MgH}{3M/2}} = \sqrt{\left(\frac{2}{3M}\right)(2MgH)} = \sqrt{\frac{4gH}{3}}$$

For the hoop:

$$I_{CM} = MR^2$$

$$M + \frac{I_{CM}}{R^2} = M + M = 2M$$

$$v_{CM,f} = \sqrt{\frac{2MgH}{2M}} = \sqrt{gH}$$

Reflect

Does it make sense that the disk wins? The difference between the two rigid bodies is that the uniform disk has a smaller rotational inertia than the hoop ($MR^2/2$ compared to MR^2), so a smaller fraction of the gravitational potential energy is converted to rotational kinetic energy as it rolls down the ramp. That leaves more energy available for translational motion, which means that the disk travels more quickly down the ramp and wins the race.

Ratio of rotational kinetic energy to translational kinetic energy:

$$\frac{K_{rotational}}{K_{translational}} = \frac{\frac{1}{2}I_{CM}\omega^2}{\frac{1}{2}Mv_{CM}^2} = \frac{I_{CM}\omega^2}{Mv_{CM}^2}$$

We can verify this by calculating the ratio of rotational kinetic energy $K_{\text{rotational}}$ to translational kinetic energy $K_{\text{translational}}$ for a rolling extended object. This ratio is I_{CM}/MR^2, so the smaller the rotational inertia I_{CM} for a given mass M and radius R—like the uniform disk compared to the hoop—the smaller the rotational kinetic energy and the more energy is available for translation. Note that this result does not depend on the mass of the rigid body; because mass is also a factor in the rotational inertia, it cancels out of the ratio. This also makes sense because the change in potential energy will also be proportional to mass, so what is important is how much of that energy becomes translational kinetic energy.

If a rigid body is rolling without slipping,

$$v_{CM} = R\omega$$

$$\omega = \frac{v_{CM}}{R}, \text{ so}$$

$$\frac{K_{\text{rotational}}}{K_{\text{translational}}} = \frac{I_{CM}\omega^2}{Mv_{CM}^2} = \frac{I_{CM}}{Mv_{CM}^2}\left(\frac{v_{CM}}{R}\right)^2 = \frac{I_{CM}}{MR^2}$$

NOW WORK Problems 2 and 5 from The Takeaway 11-2.

At the beginning of this section we mentioned that soap box derby cars are not allowed to have wheels with too much of their mass concentrated toward their centers. We can now explain this using Example 11-2, which shows that a wheel in the form of a solid disk rolls down an incline more rapidly than a wheel shaped like a hoop (like a bicycle tire). The reason for this difference is that the solid disk has more of its mass located near the rotation axis and so has a smaller rotational inertia around its rotation axis than does the hoop ($I_{CM} = MR^2/2$ for the solid disk versus $I_{CM} = MR^2$ for the hoop). A smaller rotational inertia I_{CM} means less energy is required to rotate the wheel, so more energy is available for translational kinetic energy and the center-of-mass velocity of the car is greater. Having wheels with too much mass toward the central rotation axis, and hence too small a rotational inertia, would give a soap box derby car an unfair advantage!

EXAMPLE 11-3 Rolling Without Slipping Revisited

In Example 11-2, we considered a uniform disk of mass M and radius R that rolls down a ramp without slipping. Use Newton's second law for rotational motion and for linear motion to determine the downhill acceleration of the disk's center of mass. The ramp is inclined at an angle θ to the horizontal.

Set Up

Unlike Example 10-10 in Section 10-6, in which we had one rigid body (the pulley) undergoing purely rotational motion and an object (the block) undergoing purely translational motion, here we have *one* rigid body that undergoes *both* rotational and translational motion. The principles are the same, however. We'll write equations for both the rotational and translational forms of Newton's second law for the disk.

We know that because the disk rolls without slipping, there's a relationship between the speed of the disk's center of mass v_{CM} and its angular speed ω. In a manner similar to what we did in Example 10-10, we'll relate the disk's center-of-mass acceleration and its angular acceleration.

Newton's second law for rotational motion of the disk around its central axis (the z axis):

$$\sum \tau_{\text{ext},z} = I\alpha_z \qquad (10\text{-}18)$$

Newton's second law for linear motion of the disk—note that we include both x and y equations:

$$\sum F_{\text{ext},x} = Ma_{CM,x}$$
$$\sum F_{\text{ext},y} = Ma_{CM,y} \qquad (4\text{-}3)$$

Disk rolls without slipping:

$$v_{CM} = R\omega \qquad (11\text{-}2)$$

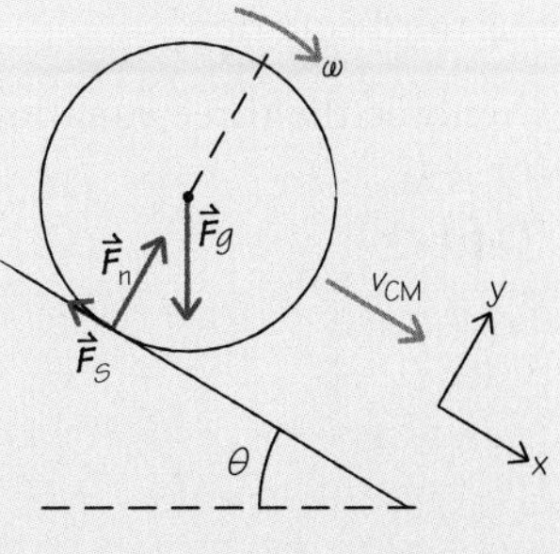

Solve

Three forces are exerted on the disk as it rolls: the downward force of gravity $F_g = Mg$, the normal force F_n that is exerted perpendicular to the surface of the ramp, and the uphill force of friction F_S that keeps the disk from sliding downhill.

The force of gravity exerts no torque around the center of mass because it is exerted at the disk's center of mass, so its lever arm is $r_\perp = 0$. The normal force is exerted along a line that passes through the center of mass, so its lever arm is also zero and it exerts no torque. For the friction force, $r_\perp = R$ and $\tau_{\text{friction},z} = RF_S$. Substitute this and $I = MR^2/2$ into Newton's second law for rotation.

Newton's second law for rotation:

$$\sum \tau_{\text{ext},z} = I\alpha_z$$

Substitute $\sum \tau_{\text{ext},z} = \tau_{\text{friction},z} = RF_S$ and $I = MR^2/2$:

$$RF_S = \frac{1}{2}MR^2\alpha_z$$

Divide both sides of this equation by R:

$$F_S = \frac{1}{2}MR\alpha_z$$

For translational motion, write the equations for the net force in the x and y directions. There is no acceleration in the y direction (perpendicular to the ramp) so $a_{\text{CM},y} = 0$. This conclusion results from the normal force and the component of the gravitational force in this direction exerted on the disk. The acceleration downhill in the x direction, $a_{\text{CM},x}$, results from the component of the force of gravity in this direction and the force of friction exerted on the disk.

Net force in each direction:

$$\sum F_x = Mg\sin\theta - F_S$$

$$\sum F_y = F_n - Mg\cos\theta$$

Substitute into Newton's second law for translational motion:

$$Mg\sin\theta - F_S = Ma_{\text{CM},x}$$

$$F_n - Mg\cos\theta = 0$$

(because $a_{\text{CM},y} = 0$)

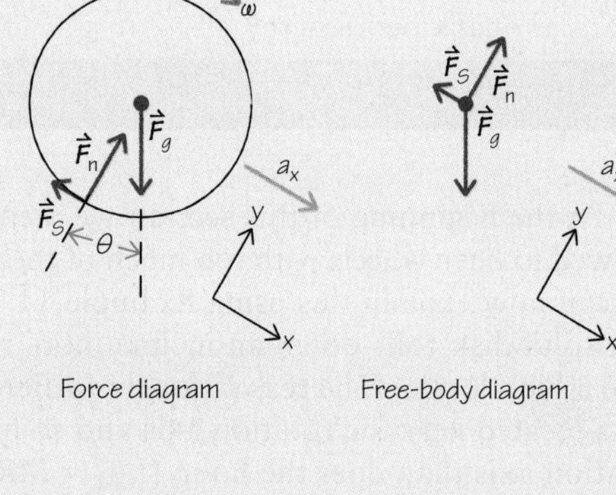

The y equation for linear motion isn't useful (it would be of interest only if we wanted to calculate the normal force F_n). The other two equations—one for rotational motion and one for linear motion along the x axis—involve three unknowns: F_S, α_z, and $a_{\text{CM},x}$. So we need a third equation. This comes from the condition for rolling without slipping, $v_{\text{CM}} = R\omega$.

Rotation: $F_S = \frac{1}{2}MR\alpha_z$

Translational motion along x axis: $Mg\sin\theta - F_S = Ma_{\text{CM},x}$

Rolling without slipping: $v_{\text{CM}} = R\omega$

In Example 10-10 the similar equation $v_{\text{block},x} = R\omega_{\text{pulley},z}$ led to $a_{\text{block},x} = R\alpha_{\text{pulley},z}$. The same mathematical steps (which we won't repeat here) lead us to the same relationship between $a_{\text{CM},x}$ and α_z for the disk, which is our third equation.

Hence $a_{\text{CM},x} = R\alpha_z$, or $\alpha_z = \dfrac{a_{\text{CM},x}}{R}$

Combine the three equations to solve for $a_{\text{CM},x}$.

Substitute $\alpha_z = a_{\text{CM},x}/R$ into the rotational equation:

$$F_S = \frac{1}{2}MR\alpha_z = \frac{1}{2}MR\left(\frac{a_{\text{CM},x}}{R}\right) = \frac{1}{2}Ma_{\text{CM},x}$$

Substitute this expression for F_S into the linear motion equation and solve for $a_{\text{CM},x}$:

$$Mg\sin\theta - F_S = Ma_{\text{CM},x}, \text{ so } Mg\sin\theta - \frac{1}{2}Ma_{\text{CM},x} = Ma_{\text{CM},x}$$

$$Mg\sin\theta = Ma_{\text{CM},x} + \frac{1}{2}Ma_{\text{CM},x} = \frac{3}{2}Ma_{\text{CM},x}$$

$$g\sin\theta = \frac{3}{2}a_{\text{CM},x}$$

$$a_{\text{CM},x} = \frac{2}{3}g\sin\theta$$

Reflect

If there were no friction, the disk would slide downhill rather than roll (all of its kinetic energy would be translational) and its acceleration would be $g \sin \theta$ (the component along the ramp of the acceleration due to gravity), as in Example 4-5. Friction slows the disk's translational motion, which is why its acceleration is only two-thirds of $g \sin \theta$. This means that the magnitude of the friction force must be $(Mg/3) \sin \theta$. It would be good practice to prove that! This is pure rolling motion, so this friction force does not dissipate any of the energy (it cannot because the point of contact of the force is not moving!) but it does allow some of the gravitational potential energy of the disk–Earth system that would normally go into translational kinetic energy to go into rotational kinetic energy instead.

We can check our result for acceleration by asking how quickly the disk will be moving at the bottom of the ramp if it starts from rest at a height H. The answer we get is $\sqrt{(4/3)gH}$. This is the same result as we found in Example 11-2 using energy conservation—as it must be!

Motion with constant x acceleration:

$$v_{CM,x}^2 = v_{CM,0x}^2 + 2a_{CM,x}(x - x_0)$$

Disk starts at rest: $v_{CM,0x} = 0$

Displacement down the ramp:

$$x - x_0 = \frac{H}{\sin \theta}$$

We calculated the acceleration earlier:

$$a_{CM,x} = \frac{2}{3} g \sin \theta, \text{ so}$$

$$v_{CM,x}^2 = 0 + 2\left(\frac{2}{3} g \sin \theta\right)\left(\frac{H}{\sin \theta}\right) = \frac{4gH}{3}$$

$$v_{CM,x} = \sqrt{\frac{4gH}{3}}$$

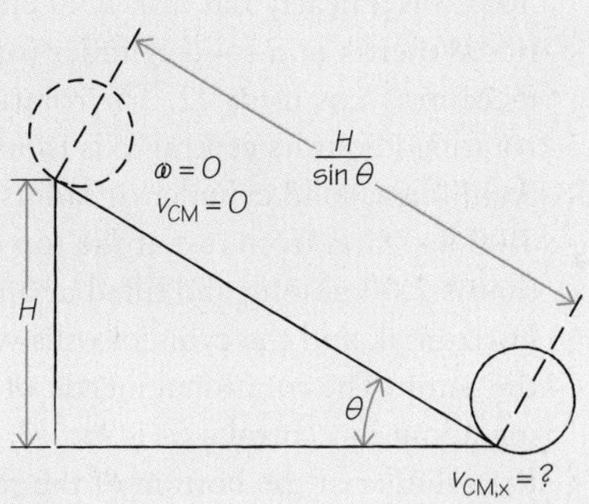

NOW WORK Problem 6 from The Takeaway 11-2.

THE TAKEAWAY for Section 11-2

✔ The total kinetic energy K of an extended object or a system of mass M is the sum of the translational kinetic energy of its center of mass, $K_{translational} = \frac{1}{2}Mv_{CM}^2$, and the rotational kinetic energy due to its rotation around its center of mass. For a rigid body this rotational kinetic energy can be written $K_{rotational} = \frac{1}{2}I_{CM}\omega^2$.

✔ The same energy conservation equations we learned in Chapters 7 and 8 also apply when there is rotational motion. The only difference is that the kinetic energy K now includes both translational and rotational kinetic energies.

✔ If a rigid body rolls without slipping on a surface, the speed of its center of mass is directly proportional to its angular speed. If the surface over which the rigid body rolls is also rigid, a static friction force prevents slipping but does no work on the rigid body as it rolls.

Prep for the AP Exam

AP Building Blocks

EX 11-1 1. Earth is approximately a solid sphere. It has a mass of 5.97×10^{24} kg, a radius of 6.38×10^6 m, and completes one rotation about its central axis each day. The orbit of Earth is approximately a circle with a radius of 1.5×10^{11} m, and Earth completes one orbit around the Sun each year. Calculate the ratio of the rotational and translational kinetic energies of Earth in its orbit. Even though Earth is orbiting the Sun in an almost circular path, for this motion we can neglect its size and treat it like an object, so its rotational kinetic energy for the orbit is the same as its translational kinetic energy. The rotational inertia of a solid sphere with mass m and radius r about a rotation axis through the center of the sphere is $2mr^2/5$.

EX 11-2 2. A bowling ball that has a radius of 11.0 cm and a mass of 5.00 kg rolls without slipping on a level lane at 2.00 rad/s. Calculate the ratio of the rotational kinetic energy to the translational kinetic energy of the bowling ball. Approximate the bowling ball as a uniform solid sphere. The rotational inertia of a solid sphere about its center is $2mr^2/5$.

3. A potter's flywheel is made of a 3.50-cm-thick, circular slab of concrete that has a mass of 60.0 kg and a diameter of 35.0 cm. This disk rotates about an axis that passes through its center, perpendicular to its circular surface. Calculate the angular speed of the slab about its center if the rotational kinetic energy is 15.0 J. Express your answer in both rad/s and rev/min. The rotational inertia of a solid cylinder rotating about its central axis is $mr^2/2$.

4. A flying disk (160 g, 25.0 cm in diameter) spins at a rate of 3.00×10^2 rpm when horizontal with its center balanced on a fingertip. What is the rotational kinetic energy of the disk if 70.0% of its mass is on the outer edge (basically a thin ring 25.0 cm in diameter) and the remaining

30.0% is a nearly flat disk 25.0 cm in diameter? The rotational inertia of a solid cylinder (or disk) rotating about its central axis is $mr^2/2$. The rotational inertia of a hoop rotating about its central axis is mr^2.

5. A uniform, solid cylinder of radius 5.00 cm and mass 3.00 kg starts from rest at the top of an inclined plane that is 2.00 m long and tilted at an angle of 25.0° with the horizontal, and the cylinder rolls without slipping down the ramp. The rotational inertia of a solid cylinder rotating about its central axis is $mr^2/2$. Calculate the speed of the cylinder at the bottom of the ramp.

6. A uniform disk that has mass $M = 0.300$ kg and radius $R = 0.270$ m rolls up a ramp of angle $\theta = 55.0°$ with an initial speed $v = 4.8$ m/s. If the disk rolls without slipping, how far up the ramp does it go? The rotational inertia of the disk is $MR^2/2$.

7.

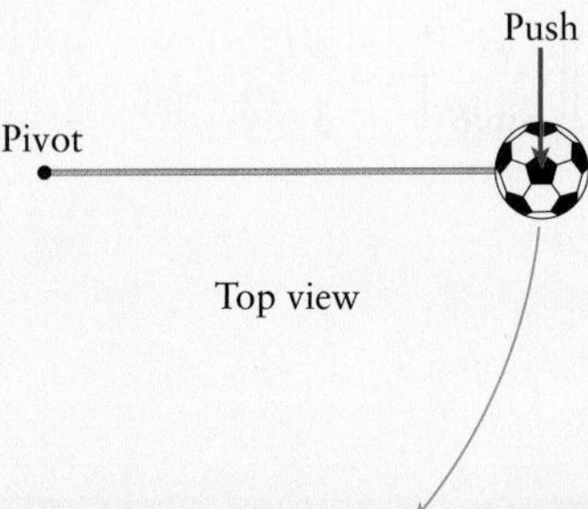

You give a quick push to a ball at the end of a rigid rod that has negligible mass, causing the ball to rotate clockwise in a horizontal circle over the top surface of a table that exerts negligible friction forces on the ball. Friction forces at the rod's pivot and air resistance are also negligible. After the push has ended, the ball's angular velocity will

(A) steadily decrease because mechanical energy is conserved.
(B) remain constant because there is no torque exerted on the ball.
(C) increase for a while, then remain constant as the effect of the push wears off.
(D) increase for a while, then remain constant because mechanical energy is conserved.

Skill Builders

8. A cart with mass M rolls on four wheels—a pair of wheels on the front and a pair on the back, each pair connected by an axle with negligible mass and radius. Each wheel has a radius r and turns with negligible friction on the axle. Each of these two wheel–axle combinations has a mass m and a rotational inertia $I_{\text{wheel–axle}}$. The cart is released from rest at a height h above the base of an inclined ramp with length L and angle of inclination θ. The rotational inertia of a solid disk rotating about its central axis is $mr^2/2$.

(a) Apply the conservation of mechanical energy to find the speed of the center of mass of the cart at the bottom of the ramp, $v_{\text{CM,f}}$, in terms of M, m, r, g, h, and $I_{\text{wheel–axle}}$.

(b) Justify the following model for $I_{\text{wheel–axle}}$:

$$I_{\text{wheel-axle}} = \frac{1}{2}mr^2$$

9. A spherical marble that has a mass of 50.0 g and a radius of 0.500 cm rolls without slipping down a loop-the-loop track that has a radius of 20.0 cm. The marble starts from rest and *just barely* makes it completely around the loop to emerge on the other side of the track. Apply the conservation of mechanical energy to predict the minimum height that the marble must start from to make it around the loop.

Skills in Action

10.

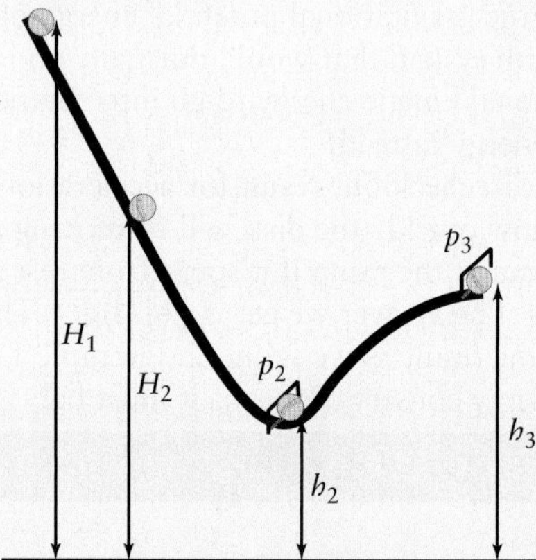

A spherical marble with an internal structure that makes the distribution of mass nonsymmetrical has a mass of 50.0 g and a radius of 0.500 cm. A track with photogate timers at P_2 and P_3 at track heights of h_2 and h_3, respectively, is to be used to determine the translational speeds of the marble at those heights. Starting points for runs from which the marble starts at rest are at heights $H_1 = 1.0$ m and $H_2 = 0.70$ m. Speeds were measured at heights $h_2 = 0.15$ m and $h_3 = 0.35$ m. The marble rolls without slipping.

The data shown here were collected.

Release height (m)	Speed measurement height (m)	Speed (m/s)
0.7	0.35	2.10
0.7	0.15	2.64
1.0	0.35	2.87
1.0	0.15	3.28

(a) Predict the dependence of the squared speed of the center of mass on the change in height by applying the conservation of mechanical energy.
(b) Justify the claim that the rotational inertia of the marble can be obtained from these data.
(c) Construct a data table of values of the squared speed and change in height.
(d) Analyze the data in the table from part (c) to obtain the rotational inertia of the marble.

(e) The rotational inertia for a spherical shell is $2MR^2/3$ and the rotational inertia of a solid sphere is $2MR^2/5$. Justify the claim that the internal structure of the marble might include air or other low-density material.

11. A solid ball, a solid disk, and a hoop, all with the same mass and the same radius, are set rolling without slipping up an incline. Each has the same initial linear speed. The rotational inertia for rotation about the center of the objects can be expressed as

$$I = cmr^2$$

with solid sphere ($c = 2/5$), disk ($c = 1/2$), and hoop ($c = 1$). Rank these three objects from greatest to least based on the distance that each object ascends along the incline before momentarily coming to rest. Justify your ranking.

12. A record is a vinyl disk with a mass of 125 g and a radius of 30 cm, which is placed on a turntable. When the disk is used to play music, it spins up from rest to $33\frac{1}{3}$ rpm in 0.80 s. What is the average power delivered by the turntable as it starts the disk?

11-3 Angular momentum is always conserved; it is constant when there is zero net torque (or angular impulse) exerted on a system

To see what we mean by angular momentum, consider a system rotating around a fixed z axis that doesn't change its orientation. The system's rotational inertia around this axis is I and its angular velocity is ω_z. We've seen that these quantities are analogous to the mass m and velocity v_x of an object in linear motion along the x axis. Applying conservation of linear momentum in just the x direction, the linear momentum for such motion is

$$p_x = mv_x \qquad (11\text{-}3)$$

By analogy to Equation 11-3 we'll mathematically define a the quantity called the *angular momentum* of a system rotating around the z axis. We'll give this the symbol L_z:

The **angular momentum** of a system rotating around a fixed axis...

...is proportional to the **angular velocity** at which the system rotates around that axis.

$$L_z = I\omega_z \qquad (11\text{-}4)$$

The angular momentum is also proportional to the system's **rotational inertia** for that rotation axis.

EQUATION IN WORDS
Angular momentum for rotation around a fixed axis

The greater the rotational inertia and the faster the angular velocity, the greater the value of angular momentum. Note that, like angular velocity, angular momentum can be positive or negative, depending on which way the system is rotating and which direction of rotation you choose to be positive.

WATCH OUT !

Angular momentum and linear momentum are different quantities.

Although linear momentum $\vec{p}$ and angular momentum $\vec{L}$ are analogous to each other, they are *not* the same quantity and do not have the same units. Linear momentum has units of mass times velocity, or kg · m/s. As we suspected, angular momentum must depend on a rotation axis, so the units of angular momentum are the units of rotational inertia (kg · m^2, which depends on a rotation axis) multiplied by the units of angular velocity (rad/s), which we can write as kg · m^2/s. (As usual, we eliminate the "rad" since radians are truly dimensionless.)

Exam Tip

Since they are different quantities, angular momentum should never be added to or set equal to linear momentum in a calculation or derivation.

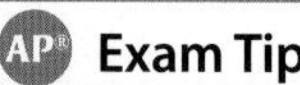

Exam Tip

When answering a conceptual question about angular momentum, it is important to carefully define your system and its rotation axis. For the discussion to the right, you could imagine a figure skater standing on the ice. For the gravitational force to not cause a torque, her center of mass must be directly above the point at which the skates contact the ice. She could be standing upright or lean her head in one direction and reach out with her arms in the opposite direction. To represent this visually, we draw the rotation axis as a line connecting the point of contact of her skates with the ice, and her center of mass. You can see a version of this in Example 11-4.

Why is angular momentum important? We saw in Section 9-3 that the total *linear* momentum of a system of two or more objects is *constant* if there is no net external force on the system. The analogous statement about angular momentum is

If there is no net external torque on a system, the angular momentum of the system is constant.

This is the principle of **conservation of angular momentum.** It explains what happens to a spinning figure skater when she pulls her arms in. As she does so her rotational inertia I decreases because more of her body's mass is closer to her rotation axis. There are no external torques on her: The force of gravity pulls straight down on her center of mass, causing no torque about the axis passing upward from her skates through her center of mass. The normal force also doesn't have any effect on her rotation, since it points upward from the ice along her rotation axis. The friction force on her skates is slightly away from the rotation axis, so it could cause a torque, but the force is very small and very close to the rotation axis, so it provides negligible torque. Hence, her angular momentum L_z must stay constant, so as I decreases her angular velocity ω_z has to increase so that the product $L_z = I\omega_z$ doesn't change. The same thing happens in the office chair experiment we described at the beginning of this chapter.

We can see this same effect in a child's game, tetherball, where a rope attached to the top of a tall pole is tied to a ball. Players hit the ball in opposite directions to attempt to wrap the rope around the pole. Of course, each of the hits (a force exerted on the ball) changes the motion of the ball, but even in between the hits, as the rope wraps around the pole, the angular speed of the ball increases. We can explain this in terms of angular momentum. Let's treat the rope as having negligible mass and neglect any resistive forces (such as air resistance and friction). The force of gravity is downward on the ball, so it doesn't have any component perpendicular to the rotation axis (the ball is constrained to move around the pole in a circle by the rope, and the rotation axis is the pole, so the radius of the rotation is the horizontal distance from the pole to the ball). As the rope gets shorter because it is wrapping around the pole, the radius of the ball's rotation around the pole gets smaller. For angular momentum to remain constant, the ball must move more quickly as the radius of its rotation becomes smaller.

EXAMPLE 11-4 Spinning Figure Skater

A figure skater executing a "scratch spin" gradually pulls her arms in toward her body while spinning on one skate. During the spin her angular speed increases from 1.50 rad/s [approximately one revolution (a complete circle of 2π radians) every 4 s] to 15.0 rad/s (approximately 2.5 rev/s). (a) By what factor does her rotational inertia around her central axis change as she pulls in her arms? (b) By what factor does her rotational kinetic energy change?

Set Up

There is very little friction between the skater and the ice, so the net torque on her is essentially zero and her angular momentum can be considered constant. We'll use this to determine her rotational inertia after she pulls her arms in, I_{after}, in terms of her rotational inertia before pulling them in, I_{before}. (Note that we are given the values $\omega_{before} = 1.50$ rad/s and $\omega_{after} = 15.0$ rad/s.) Then we'll compare her rotational kinetic energy before and after pulling her arms in by using Equation 10-7.

Angular momentum of the skater:

$$L_z = I\omega_z \qquad (11\text{-}4)$$

Rotational kinetic energy of the skater:

$$K_{rotational} = \frac{1}{2}I\omega^2 \qquad (10\text{-}7)$$

Solve

(a) Angular momentum is always conserved, and because there is negligible external torque, this means angular momentum is constant. So we write an equation which says that the angular momentum after she pulls her arms in equals the angular momentum before she pulls them in. Then we solve for I_{after} in terms of I_{before}.

Conservation of angular momentum:

$$L_{after,z} = L_{before,z}$$

From Equation 11-4, this says

$$I_{after}\omega_{after,z} = I_{before}\omega_{before,z}$$

Let the direction of the skater's rotation be positive. Then her angular velocity ω_z at any instant is positive and equal to her angular speed ω at that instant.

$$I_{after}\omega_{after} = I_{before}\omega_{before}$$

Rearrange to find the ratio of I_{after} to I_{before}:

$$\frac{I_{after}}{I_{before}} = \frac{\omega_{before}}{\omega_{after}} = \frac{1.50\text{ rad/s}}{15.0\text{ rad/s}} = 0.100$$

The rotational inertia after she pulls her arms in is only 0.100 times (one-tenth) as large as the value before she pulls her arms in.

(b) Now we can use Equation 10-7 to compare the skater's rotational kinetic energy before and after she pulls her arms in. Because the ice skater system is a rotating extended object, we could not calculate her rotational inertia from one of our equations for a rigid body, but we are given the values we need. We would be able to tell that her rotational inertia decreased by examining the distance of her various parts from the rotational axis as her shape changed.

$$K_{rotational,before} = \frac{1}{2}I_{before}\omega_{before}^2$$

$$K_{rotational,after} = \frac{1}{2}I_{after}\omega_{after}^2$$

Calculate the ratio of her rotational kinetic energy before and after:

$$\frac{K_{rotational,after}}{K_{rotational,before}} = \frac{(1/2)I_{after}\omega_{after}^2}{(1/2)I_{before}\omega_{before}^2} \text{ so}$$

$$\frac{K_{rotational,after}}{K_{rotational,before}} = \left(\frac{I_{after}}{I_{before}}\right)\left(\frac{\omega_{after}}{\omega_{before}}\right)^2 = (0.100)\left(\frac{15.0\text{ rad/s}}{1.50\text{ rad/s}}\right)^2$$

$$= (0.100)(10.0)^2 = 10.0$$

Her rotational kinetic energy after she pulls her arms in is 10.0 times as great as the value before she pulls her arms in.

Reflect

The skater's rotational inertia changes by a factor of 0.100 = 1/10.0. This might seem large, because her arms are a relatively small fraction of her total mass. However, the rotational inertia depends on the square of distance of the mass from the rotational axis (see Equation 10-6), and most of her body is close to that axis so holding her arms extended rather than close to her body has a significant effect on the skater's rotational inertia.

Although the skater's angular momentum maintains the same value as she pulls her arms in and perhaps also curls in her shoulders, her rotational kinetic energy does not. You can see why by rewriting the expression for rotational kinetic energy in terms of the angular momentum magnitude $L = I\omega$ and the angular speed ω. Because L is unchanged and ω increases when the skater pulls her arms in, her rotational kinetic energy must increase. There is no external source of energy. The extra kinetic energy must come from the skater herself: Chemical energy from food she ate is used to flex muscles and bring her arms in.

Rotational inertia of the skater:

$$I = \sum_{i=1}^{N} m_i r_i^2 \qquad (10\text{-}6)$$

Rotational kinetic energy:

$$K_{rotational} = \frac{1}{2}I\omega^2 = \frac{1}{2}(I\omega)\omega = \frac{1}{2}L\omega$$

The skater's angular speed ω increases while the magnitude L of her angular momentum stays the same, so $K_{rotational}$ increases.

NOW WORK Problems 1 and 3 in The Takeaway 11-3.

Angular Momentum of an Object

Example 11-4 demonstrates how the angular momentum concept helps us solve problems about rotating extended objects (systems that are connected but can change shape). It can also be useful to think about the angular momentum of a single object. We've already seen the utility of this concept for the example of a tetherball moving around a pole. Let's look at the general case of an object of mass m moving with velocity $\vec{v}$, so that its linear momentum is $\vec{p} = m\vec{v}$. As an example, let's consider a girl (who for this example we regard as an object, since we will describe her motion by that of her center of mass) running at constant speed toward a playground merry-go-round (**Figure 11-5**). We take the rotation axis to be the axis of the merry-go-round. At a given instant the displacement of the girl from the rotation axis, $\vec{r}$, points directly away from the rotation axis, and there is an angle ϕ between the vectors $\vec{r}$ and $\vec{p}$. What is the girl's angular momentum around this rotation axis?

In Equation 10-17 we defined the magnitude of the torque due to a force $\vec{F}$ as $\tau = r_\perp F$, where $r_\perp$ is the perpendicular distance from the rotation axis to the point where the force is applied. For example, in Figure 11-5 the perpendicular distance $r_\perp$ equals the radius R of the merry-go-round. We define the angular momentum of an object with linear momentum $\vec{p}$ in an analogous way:

EQUATION IN WORDS
Magnitude of the angular momentum of an object

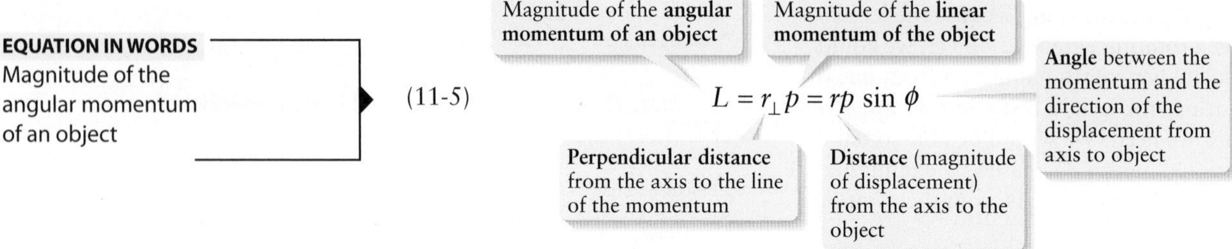

Does Equation 11-5 agree with our earlier definition $L_z = I\omega_z$ (Equation 11-4)? To find out, note that $p \sin \phi = mv \sin \phi = mv_\perp$, where $v_\perp$ is the component of the object's velocity that's perpendicular to $\vec{r}$ in Figure 11-5. We can also write $v_\perp = r\omega$, where ω is the angular speed of the object around the rotation axis. If we substitute these into Equation 11-5 for the magnitude of the object's angular momentum, we get

$$L = rp \sin \phi = rmv \sin \phi = rmv_\perp = rm(r\omega) = mr^2\omega \tag{11-6}$$

Equation 10-6 tells us that if there's only a single object of mass m a distance r from the rotation axis, its rotational inertia is $I = mr^2$. This means Equation 11-6 for the magnitude of the object's angular momentum becomes $L = I\omega$. But this is just the magnitude of the angular momentum $L_z = I\omega_z$ as given by Equation 11-4. So Equation 11-5 for the angular momentum of an object is completely consistent with our earlier work.

Figure 11-5 A girl running with linear momentum and angular momentum The angular momentum of the girl around the rotation axis is $L = r_\perp p = rp \sin \phi$ (see Equation 11-5).

(4) The perpendicular distance from the axis to the line of the momentum is $r_\perp = R$ (the radius of the merry-go-round).

Rotation axis of merry-go-round

(3) The distance r from the girl to the rotation axis and the angle ϕ are both continually changing as she runs.

C

R

(2) The girl's linear momentum has constant magnitude $p = mv$.

M

$\vec{r}$

ϕ

$\vec{p}$

G

(5) The radius R is one leg of the right triangle GCM, which has hypotenuse r. So $r_\perp = R = r \sin \phi$.

(1) A girl of mass m runs at constant speed v along this line toward the rim of the merry-go-round. This is the line of her linear momentum.

(6) So the girl's angular momentum around the rotation axis is $L = r_\perp p = Rp = rp \sin \phi$, as in Equation 11-4.

WATCH OUT !

Angular momentum doesn't require the motion to be in a circle.

Note that an object doesn't have to move in a circle, or even along a curved path, to have angular momentum. Consider the girl in Figure 11-5, who moves along a line at constant velocity. You can think of her as rotating around a fixed point (the axis of the merry-go-round) with a radius equal to the perpendicular component of her displacement. This component is measured from the axis perpendicular to her velocity, as long as her velocity is not directed straight at the axis. Although the girl is not attached to the center of rotation, if her velocity is perpendicular to her displacement from the axis, when she jumps on the merry-go-round she sets it into rotational motion. Because an object in linear motion can set another object in rotational motion, it has angular momentum.

The girl in Figure 11-5 has a constant linear momentum because her velocity is constant. She also has a constant *angular* momentum: Figure 11-5 shows that her angular momentum is $L = r_{\perp}p = Rp$, where R is the radius of the merry-go-round. Since R and her linear momentum p are both constant, her angular momentum L is constant. That happens because there is also zero net *torque* exerted on her, because there is no force exerted on her perpendicular to her radius from the axis of rotation.

If there *is* a net force that is exerted on an object, but if that force is always directed toward the rotation axis, that force exerts zero torque (there is a 180° angle between the directions of the displacement $\vec{r}$ from axis to object and the force $\vec{F}$ that points from object to axis, so the torque is $\tau = rF \sin \phi = rF \sin 180° = 0$). In that case the angular momentum is again constant. One example of this is the motion of water as it circles the drain of a bathtub. The water is pulled directly toward the drain, so the force on each drop of water exerts no torque and the angular momentum of the water is constant. As the water approaches the drain, r decreases and so $p = mv$ increases to keep the angular momentum $L = rp_{\perp}$ constant. The same thing happens on a much more dramatic scale as air is drawn into the low pressure at the center of a hurricane. The circulating air gains tremendous speed, which gives the hurricane great destructive power.

Angular Momentum and a Rotational "Collision"

Here's another application of the angular momentum of an object. The girl shown in Figure 11-5 runs across the playground and finally jumps onto the merry-go-round so that both end up rotating together. This seems very much like the inelastic collisions we discussed in Chapter 9, in which a moving object collided with an initially stationary one and the two stuck together afterward. We approached those inelastic collisions by demanding that linear momentum remain constant if the system is isolated. In Figure 11-5, however, the girl has linear momentum before the collision, and the system of girl and merry-go-round has *zero* total linear momentum after the collision (the system rotates but its center of mass does not move from one place to another). So *linear* momentum is not constant in this collision because the merry-go-round is attached to the ground, so the system is not isolated.

However, the *angular* momentum of the system of girl and merry-go-round *is* constant because there are no external torques exerted on the system around the merry-go-round's rotation axis. (The vertical force of gravity and normal force don't affect the rotation, and there is very little friction in the bearings of the merry-go-round.) During the collision the girl and the merry-go-round exert torques on each other, but these torques are *internal* to the system of girl plus merry-go-round and so don't affect the system's *total* angular momentum. All that happens in the collision is that angular momentum is transferred between the girl and the merry-go-round. Example 11-5 shows how to analyze what happens.

Exam Tip

If angular momentum of a system is constant, it does not mean that angular momentum of each object in the system is constant. If a system's angular momentum is constant, and two objects in the system exchange angular momentum, then each must exert a torque on the other. This means that each object exerts a force on the other that has a component perpendicular to the direction of that object's displacement with respect to the rotation axis.

EXAMPLE 11-5 A Girl on a Merry-Go-Round

The girl in Figure 11-5 has a mass of 30.0 kg. She runs toward the merry-go-round at 3.0 m/s, then jumps on the edge. The merry-go-round is initially at rest, has a mass of 100.0 kg and a radius of 2.0 m, and can be treated as a uniform disk. Find the rotation speed of the merry-go-round after the girl jumps on.

Set Up

As we described above, angular momentum of the girl–merry-go-round system is constant in this process. Before she jumps on, the girl has all of the angular momentum, with a magnitude given by Equation 11-5. After she jumps on, she and the merry-go-round rotate together with the same angular speed ω. Together they behave like a single extended object, which can be modeled as a rigid body made up of a rotating disk with an extra mass (the girl) at its rim.

Angular momentum of the girl before jumping on:

$$L = rp_{\perp} = rp \sin \phi \qquad (11\text{-}5)$$

Angular momentum of the merry-go-round with the girl riding on it:

$$L_z = I\omega_z \qquad (11\text{-}4)$$

Rotational inertia of the merry-go-round plus girl:

$$I = I_{\text{merry-go-round}} + I_{\text{girl}} \qquad (10\text{-}6)$$

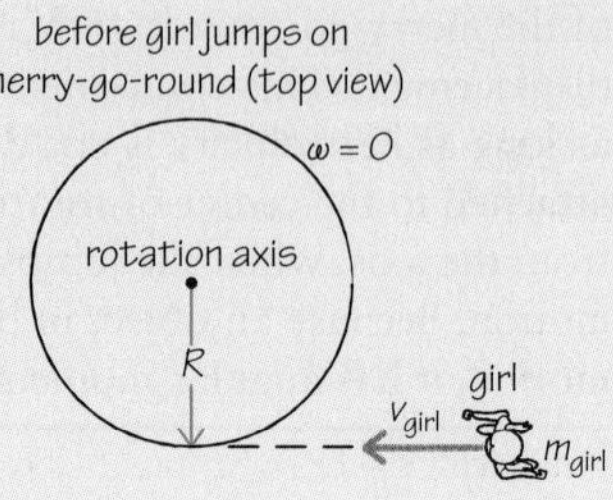

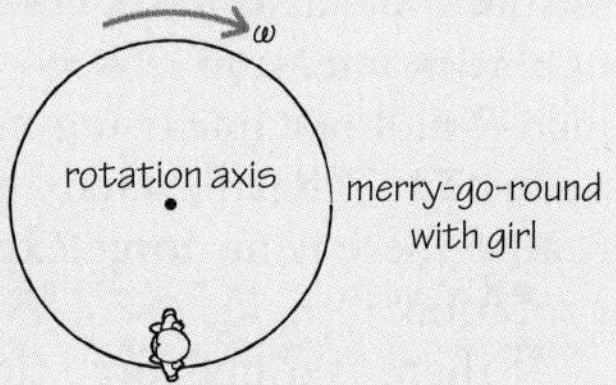

Solve

Figure 11-5 shows that the distance $r \sin \phi$ in the case $\phi = 90°$ is just equal to the radius of the merry-go-round, $R = 2.0$ m. Use this to find the magnitude of the angular momentum of the girl–merry-go-round system.

Before the girl jumps on the merry-go-round, her linear momentum has magnitude

$$p = m_{\text{girl}}v_{\text{girl}} = (30.0 \text{ kg})(3.0 \text{ m/s}) = 90 \text{ kg} \cdot \text{m/s}$$

Her angular momentum is

$$\begin{aligned} L &= rp \sin \phi = (r \sin \phi)p \\ &= Rp = (2.0 \text{ m})(90 \text{ kg} \cdot \text{m/s}) = 1.8 \times 10^2 \text{ kg} \cdot \text{m}^2/\text{s} \end{aligned}$$

After she jumps on, the magnitude of the angular momentum of the girl–merry-go-round system is $L = I\omega$, where ω is what we are trying to find. Calculate the value of I for the system of merry-go-round plus girl.

Rotational inertia of the merry-go-round alone, considered as a uniform disk: from Table 10-1,

$$\begin{aligned} I_{\text{merry-go-round}} &= \frac{1}{2}M_{\text{merry-go-round}}R^2 = \frac{1}{2}(100.0 \text{ kg})(2.0 \text{ m})^2 \\ &= 2.0 \times 10^2 \text{ kg} \cdot \text{m}^2 \end{aligned}$$

Rotational inertia of the girl alone, considered as an object of mass m_{girl} rotating a distance R from the axis:

$$I_{\text{girl}} = m_{\text{girl}}R^2 = (30.0 \text{ kg})(2.0 \text{ m})^2 = 1.2 \times 10^2 \text{ kg} \cdot \text{m}^2$$

Rotational inertia of the merry-go-round plus girl:

$$\begin{aligned} I = I_{\text{merry-go-round}} + I_{\text{girl}} &= 2.0 \times 10^2 \text{ kg} \cdot \text{m}^2 + 1.2 \times 10^2 \text{ kg} \cdot \text{m}^2 \\ &= 3.2 \times 10^2 \text{ kg} \cdot \text{m}^2 \end{aligned}$$

Angular momentum is constant, so set the magnitude of angular momentum before the girl jumps on equal to the magnitude after she jumps on. Then solve for the final angular speed ω. Note that if L is in $\text{kg} \cdot \text{m}^2/\text{s}$ and I is in $\text{kg} \cdot \text{m}^2$, then ω is in rad/s.

Before the girl jumps, $L = 1.8 \times 10^2 \text{ kg} \cdot \text{m}^2/\text{s}$.

After she jumps the angular momentum is $L_z = I\omega_z$ with magnitude $L = I\omega = (3.2 \times 10^2 \text{ kg} \cdot \text{m}^2)\omega$.

Angular momentum of the girl—merry-go-round system is constant, so the magnitude L is the same before and after:

$$1.8 \times 10^2 \text{ kg} \cdot \text{m}^2/\text{s} = (3.2 \times 10^2 \text{ kg} \cdot \text{m}^2)\omega$$

$$\omega = \frac{1.8 \times 10^2 \text{ kg} \cdot \text{m}^2/\text{s}}{3.2 \times 10^2 \text{ kg} \cdot \text{m}^2} = 0.56 \text{ rad/s}$$

Reflect

To get a better sense of our result, we convert it to rev/min. Our answer says that the merry-go-round spins 5.4 times per minute. In other words, it makes one revolution every (1/5.4) minute, or about once every 11 s—a reasonable pace for a merry-go-round.

Convert ω to revolutions per minute:

$$\omega = (0.56 \text{ rad/s})\left(\frac{1 \text{ rev}}{2\pi \text{ rad}}\right)\left(\frac{60 \text{ s}}{1 \text{ min}}\right) = \frac{(0.56)(60)}{2\pi}\frac{\text{rev}}{\text{min}}$$
$$= 5.4 \text{ rev/min}$$

NOW WORK Problems 2, 4, and 5 from The Takeaway 11-3.

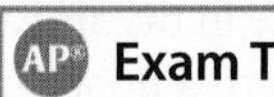

Exam Tip

In Example 11-5, you were asked to "find" the rotation speed. On the AP® Physics 1 exam free-response questions, when you are asked for a quantitative response (an answer given in terms of numbers or given quantities), you will only be asked to "derive," "calculate," or "determine," and each of these has a particular meaning. If asked to derive, you must start from a fundamental principle, such as angular momentum conservation in the case of Example 11-5; in other words, you should make sure to start from an explicitly stated fundamental principle and/or an equation on the equation sheet, and be sure to show every step needed to get to the final answer. If asked to calculate, you do not necessarily need to start from a fundamental principle, but you must show some logical flow, show some steps in the calculation. You need to show steps in derive and calculate problems even if you think they are obvious! If asked to determine, while steps can help for partial credit, you do not necessarily have to show any steps to the final answer.

Another way we could think about this angular collision would be if, after we solved for the final angular velocity with the girl in the system, we considered what effect she would have if we placed the girl as external to the system. This case is more closely related to the idea of impulse in collisions that we discussed in Chapter 9. Just like the calculation of impulse as the change in momentum in Equation 9-24 ($\vec{F}_{\text{collision}}\Delta t = \vec{p}_\text{f} - \vec{p}_\text{i} = \Delta\vec{p}$), we can think of the change in angular momentum as angular impulse: $\tau\Delta t = \Delta L$. In the same way as impulse depends on the force exerted on an object or a system over a fixed amount of time, the angular impulse depends on the torque exerted on an object or a system over a fixed amount of time. In the previous example, the merry-go-round's initial angular momentum was zero; it was at rest. After the collision, the angular momentum of the merry-go-round is the angular velocity of the merry-go-round, 0.56 rad/s, times its rotational inertia, 200 kg · m²: $L_z = I\omega_z = 112 \text{ kg} \cdot \text{m}^2/\text{s}$. So the change in the angular momentum of the merry-go-round is $\Delta L = L_\text{f} = L_\text{i} = 112 \text{ kg} \cdot \text{m}^2/\text{s} - 0 = 112 \text{ kg} \cdot \text{m}^2/\text{s}$. Because of the parallel to the impulse-momentum theorem, we can use this to estimate the torque the girl exerted on the merry-go-round! We know the change in angular momentum, and we can estimate how long it took the girl to jump on. Let's guess 0.5 s from when she first touched the merry-go-round until she was standing on it. So using $\tau\Delta t = \Delta L$, we can rearrange to solve for $\tau = \Delta L/\Delta t$. Substituting our value for change in angular momentum and our guess for the amount of time in which that change occurred, we get an estimate of the torque that the girl exerted on the merry-go-round of $224 \text{ kg} \cdot \text{m}^2/\text{s}^2$. Given that she is 2 m away from the axis of rotation, this implies that she exerted a horizontal force of 112 N on the merry-go-round. This is a reasonable force, about what it would take to lift a 25-lb object against Earth's gravity. If our estimate for time was too long or too short, then the force would have to be greater or less, but not by enough to make the answer unreasonable, since it is very doubtful that we are off by more than a factor of 5!

Exam Tip

Just as with momentum and impulse, students will be expected to relate angular momentum and angular impulse graphically as well as numerically on the AP® Physics 1 Exam. The mathematical relationships have the same form, so just as the slope of a momentum-time graph for a system gives the average force exerted on the system, the slope of an angular momentum-time graph gives the average torque exerted on the system.

Angular Momentum as a Vector

As we saw with angular velocity, and paralleling the relationship of linear velocity and linear momentum, angular momentum is a vector, denoted $\vec{L}$. Just like angular velocity, we take the direction of angular momentum to point along the axis of rotation, which we call the z axis, with the specific direction of $\vec{L}$ found with the right hand rule we introduced in Section 10-5. If we look at Figure 10-18, where this rule was introduced,

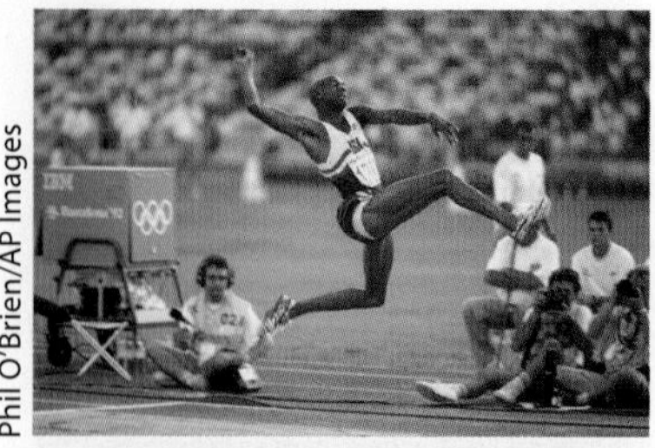
Phil O'Brien/AP Images

Figure 11-6 Angular momentum during a long jump This long jumper causes his legs to rotate counterclockwise (upward) by swinging his arms clockwise. This works because the total angular momentum of his body about his center of mass is equal to zero and is constant.

since $\vec{\omega}$ and $\vec{L}$ must point in the same direction, we can see how this rule applied to $\vec{L}$. If the component ω_z is positive (or negative), the component L_z will have the same sign.

Figure 11-6 shows an athlete doing the long jump. Although his body is essentially in an upright running position at the beginning of the jump, while in midair his legs swing forward. By the time he lands, his legs are out in front of his torso. This significantly increases the distance of his jump. How does the athlete manage this in midair, with nothing to push against? What physics underlies his motion?

The answer is that there was no net torque exerted on the athlete during his leap, so his angular momentum was constant. (The force of gravity is exerted on him by Earth, but the force of gravity is always exerted at an object or system's center of mass and so exerts no torque around his center of mass.) As the athlete left the ground, the net angular momentum of his body around its center of mass was zero, or nearly so. Once in the air he rotated his legs up and forward (counterclockwise in Figure 11-6), giving them a nonzero angular momentum. In order for the athlete's angular momentum to be constant, some other part of his body had to rotate in such a way that the two contributions to his angular momentum canceled. He rotated his arms rapidly clockwise while in the air, which helped rotate his upper torso clockwise in Figure 11-6. The angular momentum of his arms in one direction canceled the angular momentum of his legs in the opposite direction.

How does this apply to the long jumper? If more than one element of (an object within) a system is rotating, the net angular momentum $\vec{L}$ of the system is the vector sum of the angular momenta of each individual element. Therefore, the angular momentum of the jumper's arms and upper torso cancels out that of his legs. This stabilizes his motion so that both arms and legs are forward at the end of the jump, resulting in a longer flight.

Alligator Angular Momentum

Let's explore this physics more deeply by considering the feeding behavior of alligators, which involves rotations of their head, body, and tail. The structure of an alligator's jaws and teeth makes it impossible for these reptiles to cut large prey into chunks small enough to swallow. To make matters worse, alligators have no leverage when they clamp onto their prey while swimming—their legs are too short to hold the food, and they can't push against the river bottom. However, by tightly biting the prey and then executing a spinning maneuver, they can exert forces large enough to remove bite-sized pieces of food. This "death roll" arises from the conservation of angular momentum.

After the alligator has securely grabbed hold of its prey, it bends its tail and head to one side, forming a C shape (**Figure 11-7**). These motions have angular momentum

① The combined angular momentum about the center of mass of the free alligator is initially zero, so must stay zero.
$\vec{L}_{\text{Alligator}} = 0$
$\vec{L}_{\text{body}}$
$\vec{L}_{\text{head}}$
$\vec{L}_{\text{tail}}$
$\vec{L}_{\text{tail}}$
$\vec{L}_{\text{head}}$
Center of mass
② Using one of the right-hand rules, we can wrap our fingers in the direction of the tail's motion and find that on its own the motion would result in an angular momentum mostly out of the page (causing the alligator to rotate counterclockwise on the page). The twist to its tail gives it a component slightly up the page along the axis through the center of mass and parallel to the long axis of the alligator. This component is shown in the diagram.
③ Wrap your fingers in the direction of the head's motion and you'll find that on its own the motion would result in an angular momentum mostly into the page (causing the alligator to rotate clockwise on the page). It too has a twist that gives it a component slightly up the axis through the center of mass parallel to the long axis of the alligator. This component is shown in the diagram.
④ Because the total angular momentum of the whole free alligator is zero about its center of mass, the alligator's body will rotate about the center of mass to cancel the component of the angular momentum of the head and tail that does not already cancel. The clockwise–counterclockwise component cancels, leaving just the component shown in the diagram.

Figure 11-7 The alligator "death roll" Alligators use the conservation of angular momentum to rotate in a death roll that helps them consume their prey. In this simplified version we see how the alligator causes its body to rotate by moving its head and tail. Note that the green arrow points to the center of mass at its actual location. It is redrawn displaced to the right so the angular momentum vectors can be shown.

about the center of mass of the system. The total angular momentum of the system must be equal to zero if the alligator originally had no angular momentum and the alligator and its prey are free to rotate about their combined center of mass. The axis of rotation passes through this center of mass of the system of alligator + prey if the prey is floating freely in the water when the alligator starts the motion. This is a three-dimensional rotation problem with angular momentum about different axes, so you would not be asked to do this. The revolution of the alligator's head and tail have an angular momentum about the death roll axis that scientists have to use models of the shapes of the various body parts and the parallel axis theorem to calculate. We are going to look at a simplified version where we can draw diagrams for rotation around an axis and use a right-hand rule as we have learned to understand the basics of how this works, a case where the alligator is just causing itself to roll, so we do not have to worry about the location of the center of mass of the combined system or the motion of the prey, and we can narrow things down to two vector components, one of which cancels right away, so we have something on which we can use the techniques we are learning. This simplified version is illustrated in Figure 11-7.

In the case of the isolated alligator, the *total* angular momentum of the alligator remains constant because it experiences no net torque. As a result of rotating its head and tail, pulling into a C shape, it causes its body to roll as well since the angular momenta of its head and tail do not quite cancel and its total angular momentum about its center of mass must remain zero. As a reaction to the angular momentum the alligator gives its head and tail, the alligator's body acquires a rotation in the opposite direction. This angular momentum has the same magnitude as its head and tail about this axis but points in the opposite direction, so these vectors cancel and the *total* angular momentum of the alligator is zero. An alligator cannot initiate a death roll if the alligator's tail is restrained. When an alligator actually carries out this maneuver, the "death-roll axis" will fall along the center of mass of the system of alligator and prey. Although there is no net torque on the isolated system, this rotation results in internal torques, and therefore forces, on the alligator and its prey. The alligator's snout is very sturdy so these forces do not hurt the alligator, but the force on the less sturdy prey in the alligator's mouth is large enough to detach bite-sized pieces of meat. The force generated by this rolling maneuver on the alligator's prey is estimated to be well over 130 N even for smaller alligators and increases with the length of the alligator's tail, as you might imagine from even this simplified analysis.

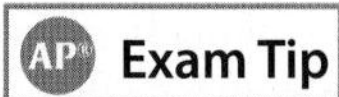

Torque and angular momentum are vectors; hence, they have a direction or sign, but are not shown as such on the AP® Physics 1 equation sheet. This is because in AP® Physics 1, you will only be given problems with one axis of rotation so it is adequate to treat them as quantities that are only clockwise or counterclockwise—and you assign a plus sign to one of those directions, and a minus sign to the other. This is just like using signs to indicate the direction of a single component of a force or linear momentum.

THE TAKEAWAY for Section 11-3

✔ Parallel to how an object's linear momentum is the product of its mass and velocity, a rotating extended object's angular momentum is the product of its rotational inertia around its rotation axis and its angular velocity.

✔ If there is no net external torque on a system about an axis of rotation, the angular momentum of the system is constant about that axis. This holds true even if the system includes an extended object whose mass distribution and rotational inertia change.

✔ An object moving past a point in space with a perpendicular distance between the velocity of the moving object and the point has angular momentum relative to that point, even if the object is not following a curved path.

✔ The angular momentum of a rotating extended object, or an object moving about a rotation axis, is a vector that lies along the rotation axis, just like angular velocity. The direction of this vector is determined by a right-hand rule.

Prep for the AP® Exam

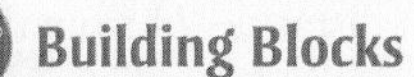

1. In what direction is the angular momentum of the tire of a bicycle that is moving forward, away from you as the observer?

2. A rod with a circular cross section has a length $l = 5.0$ cm, a radius $r = 5.0$ cm, and a mass $M = 2.50$ kg.
   (a) Calculate the angular momentum of the rod as it rotates at a constant angular speed of 1.25 rad/s about the central axis (along its length). The rotational inertia of a rod with mass M and radius R relative to its central axis is $MR^2/2$.
   (b) Calculate the angular momentum of the rod as it rotates at a constant angular speed of 1.25 rad/s about the midpoint of the rod. The rotational inertia of a rod with mass M, length ℓ, and radius R relative to its central axis (perpendicular to the rod at the midpoint) is $MR^2/4 + M\ell^2/12$.

EX 11-4

3. Earth is approximately a uniform, solid sphere that has a mass $m = 5.98 \times 10^{24}$ kg and a radius $R = 6.38 \times 10^6$ m

in a nearly circular orbit with a radius $r = 1.5 \times 10^{11}$ m about the Sun. The rotational inertia of a solid sphere with mass M and radius R is $2MR^2/5$.

(a) Calculate the angular momentum of Earth as it spins on its central axis once each day.

(b) Calculate the angular momentum of Earth as it orbits the Sun in 365.3 days, treating Earth as an object a distance r from the axis of rotation.

EX 11-5 4. A giant space station in the shape of a wheel is being evacuated. The space station is rotating about an axis through its center of mass that is perpendicular to the wheel. Each wave of evacuees consists of one thousand 4.000×10^3-kg escape pods (where the pod mass includes the passengers, originally included in the mass of the system). Assume the launchers are constructed such that the pods leave the station with a velocity directed straight outward from the rotation axis. Model the pods–station system before the release of the first wave of evacuees as a cylindrical hoop of radius 1.800 km and mass 1.000×10^8 kg rotating at a rate of 2π rad/min. The rotational inertia of a hoop with mass M and radius R is MR^2.

(a) Explain why the ejection of the pods changes the rotational speed and angular momentum of the station itself but does not change the angular momentum of the pods–station system.

(b) Calculate the change in angular speed of the station after the first wave of evacuees has left the station. Express your answer in rad/s.

(c) Now, assume the pods had been released from rest relative to the surface of the rotating station, not launched. Provide a paragraph-length explanation of the motion of the pods, and whether or not the angular speed and/or angular momentum of the station changes in this case.

EX 11-5 5. A 2.150-kg, 16.00-cm-radius, high-end turntable is rotating freely at 33.33 rpm when a naughty child drops 11.00 g of chewing gum straight down onto the turntable, 10.00 cm from the rotation axis. Assume that the gum sticks where it lands; the turntable can be modeled as a solid, uniform disk; and friction in the turntable axis can be ignored. The rotational inertia of a disk with mass M and radius R about its central axis (perpendicular to the plane of the disk) is $MR^2/2$.

(a) Explain why the gum changes the rotational speed of the turntable, which is rotating freely horizontally about its central axis without power supplied by a motor, but the angular momentum of the gum–turntable system is constant.

(b) Calculate the new angular speed of the turntable in rpm.

6. A chef tosses 0.500 kg of pizza crust dough upward and then catches it. During each toss, the dough starts out uniformly distributed and expands. On one toss the dough has an initial angular speed of 5.20 rad/s, starts as a 20.0-cm-diameter disk, and expands to 22.0 cm in diameter when it is caught. Assume the mass remains uniformly distributed.

(a) Explain why the change in shape of the dough does change the rotational speed of the dough but does not change the angular momentum of the dough.

(b) Calculate the angular speed of the dough when the chef catches it.

7. Provide a paragraph-length explanation (not just a definition) of what it means for an object moving in a line to have nonzero angular momentum. Your explanation should include an example.

AP® Skill Builders

8. On a snowy winter day, two small children with the same mass take a long narrow board whose mass is the same as one of the children and place it on a sheet of smooth ice, that exerts negligible friction on the board. There is a hole in the middle of the board and a pole upward through the ice, so they can make the board pivot in a circle about its midpoint. They sit on opposite ends of the board. Dad gives them a push to get the system of the two children and the board turning. Mom looks down at them out the window and sees them turning clockwise. Then the children crawl along the board toward each other at the same constant speed while the board is still turning. Use the variables board length ℓ, each child's mass M, and initial angular velocity ω_i to build a model of the situation. The rotational inertia of a board of mass M and length ℓ rotating about the midpoint is $M\ell^2/12$.

(a) Express the rotational inertia of the system when each child is a distance x from the midpoint of the board in terms of M, ℓ, and x.

(b) Construct a claim regarding the direction of the angular momentum of the system and the change in that vector as the children crawl toward the center.

(c) Describe the change in rotational kinetic energy as the children crawl toward each other and express the change from the initial value and the value when they are both at $x = \ell/4$.

(d) Justify the claim that no work was done on the system. Then explain the change in kinetic energy.

9.

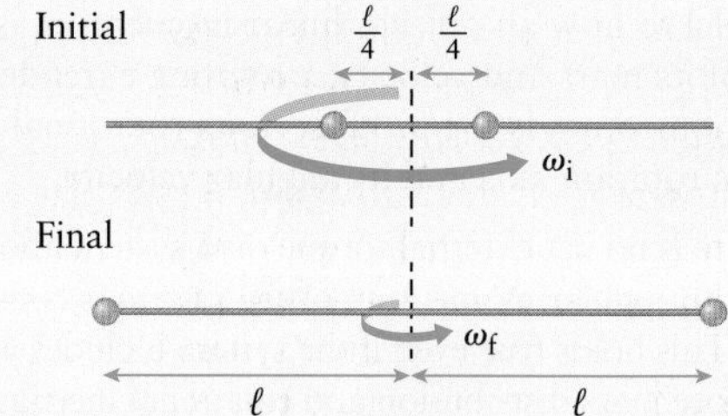

Two beads each have mass M and negligible radius. The beads are attached to a thin rod that has length 2ℓ and mass $M/8$. Each bead is initially a distance $\ell/4$ from the center of the rod. The rotational inertia for a thin rod with mass m and length l rotating about its midpoint is $ml^2/12$.

(a) Express the rotational inertia of the rod–beads system in terms of M and ℓ. The whole system is set into uniform rotation about the center of the rod, with initial angular velocity ω_i.

(b) Express the kinetic energy of the rod–beads system in terms of M and ℓ.

The beads are initially attached to the rod by a very light substance that evaporates, leaving them free to move. The beads then slide to the ends of the rod, where they stick.

(c) Express the final angular velocity of the system in terms of M, ℓ, and ω_i.

(d) Find the change in kinetic energy of the system when the positions of the beads changed.

10. Abbott and Costello were a comedy team in black-and-white films of the 1940s. Their most famous stand-up routine was *Who's on First?* In another routine, *Ice Capade!*, Abbott is attending an ice-skating party, and the skaters are holding onto a horizontal pole as they skate in a circle about the system's (pole plus skaters) center of mass. Abbott is at one end of the pole. When Costello crashes the party he clumsily grabs the other end of the pole and all the other skaters flee. Now Abbott, Costello, and the pole are the entire system, rotating at 1 rad/s. They are at the far ends of the 3-m-long pole.

(a) Costello has a mass of 100 kg and Abbott has a mass of 50 kg. Calculate the center of mass about which the pole is rotating. The mass of the pole is negligible.

(b) Costello is frightened. He can barely stand on his skates. He begins to pull himself toward Abbott. Explain why Abbott waves at him to come no closer. What will happen to their motion if Costello continues moving closer? Justify your answer.

(c) Nonetheless, Costello persists in moving closer to Abbott. Calculate their angular speed when the two are separated by just 1 m.

Note, because the friction interaction with the ice was negligible, they would be spinning still if two women hadn't ducked under the pole and skated up to save them.

11. Re-express the following statement, correcting any physical inconsistencies: If a ball is rotated on the end of a string of length ℓ, the angular momentum remains constant as long as the length and angular speed are fixed. When the ball is pulled inward and the length of the string is shortened, the rotational kinetic energy will remain constant due to conservation of energy, but the angular momentum will not remain constant because there is an external force exerted on the ball to pull it inward, in the same direction as its motion.

AP® Skills in Action

12. A physics teacher sits on a rotating stool that spins at 10.0 rpm while she holds weights in each of her hands. Her outstretched hands are 0.750 m from the axis of rotation, which passes through her head into the center of the stool. When she draws the weights in toward her body, her angular speed increases to 20.0 rpm. Neglecting the rotational inertia of her body about the axis of rotation and the mass of her arms, estimate the distance of her hands from the rotational axis at the faster angular speed.

11-4 Newton's law of universal gravitation along with gravitational potential energy and angular momentum explains Kepler's laws for the orbits of planets and satellites

In Chapter 6 we learned about circular motion and the properties of the gravitational force, including how Newton deduced that this force must grow weaker as objects move farther apart. From this, we started to understand circular orbits and in Chapter 8 we saw how to treat the gravitational potential energy of a system of two objects interacting with each other, such as a satellite and Earth. Now, we're ready to use these ideas about gravitational force and gravitational potential energy in conjunction with what we have just learned about conservation of angular momentum to understand the nature of orbits. We'll find that the same ideas that apply to a satellite orbiting Earth also apply to planets orbiting the Sun. Newton's idea of universal gravitation allowed him to develop a mathematical framework for the three laws of planetary motion that had been discovered—but not understood—decades before. We will use Newton's universal gravitation in conjunction with potential energy and angular momentum to develop a conceptual understanding of Kepler's laws.

Exam Tip

Note that the speed of a satellite in a circular orbit is not given on the AP® Physics 1 equation sheet. Equation 6-9 comes from using Newton's second law: setting the gravitational force F_g equal to ma, where the acceleration is centripetal, $a = v^2/r$.

Another Look at Circular Orbits, Now Considering Energy

Before placing a satellite into a circular orbit, an important question to ask is "How much energy will it take to put the satellite in the desired orbit?" The answer determines how powerful a rocket will be needed for the task. We now have all the tools we need to answer this question: Equation 6-9 tells us the speed the satellite must have in its circular orbit, which allows us to determine the required kinetic energy, and Equation 8-8 tells us the gravitational potential energy. Using the speed from Equation 6-9, the kinetic energy for a satellite of mass m in a circular orbit of radius r is

$$K = \frac{1}{2}mv^2 = \frac{1}{2}m\left(\sqrt{\frac{Gm_{\text{Earth}}}{r}}\right)^2 = \frac{Gm_{\text{Earth}}m}{2r}$$

Making the orbital radius r larger means the satellite moves more slowly, and the kinetic energy decreases. The gravitational potential energy is

$$U_{\text{grav}} = -\frac{Gm_{\text{Earth}}m}{r}$$

As we discussed in Section 8-5, the gravitational potential energy is negative. Making the orbital radius r larger makes the gravitational potential energy less negative—that is, closer to zero—which means that the gravitational potential energy increases. The total mechanical energy is the sum of K and U_{grav}:

$$E = K + U_{\text{grav}} = \frac{1}{2}mv^2 + \left(-\frac{Gm_{\text{Earth}}m}{r}\right) = \frac{Gm_{\text{Earth}}m}{2r} + \left(-\frac{Gm_{\text{Earth}}m}{r}\right)$$

or

EQUATION IN WORDS
Total mechanical energy for a circular orbit

(11-7)

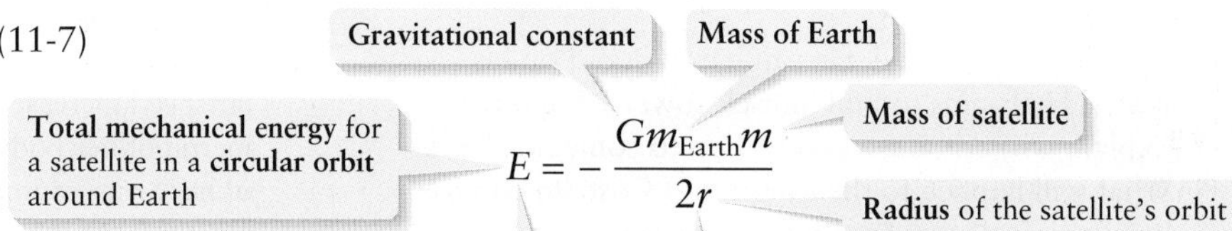

The total mechanical energy E for any orbit is always negative because of the way we have defined gravitational potential energy. A zero value of E corresponds to an orbit with an infinitely large radius r. In this limiting case the gravitational potential energy is zero, and the satellite would have zero speed and zero kinetic energy. If an object still had kinetic energy at infinity, its total energy would be positive, but it would not be held in an orbit.

As an example of a stable orbit, for a satellite in low Earth orbit (so that we can approximate $r = R_{\text{Earth}}$) the total mechanical energy is $E = -Gm_{\text{Earth}}m/2R_{\text{Earth}}$. Before the spacecraft was launched and was sitting on Earth's surface, its kinetic energy was zero, and E was equal to the gravitational potential energy: $E = U_{\text{grav}} = -Gm_{\text{Earth}}m/R_{\text{Earth}}$. So the amount of energy that had to be imparted to the satellite to put it into low Earth orbit was

$$\Delta E = E_{\text{in orbit}} - E_{\text{on surface}} = \left(-\frac{Gm_{\text{Earth}}m}{2R_{\text{Earth}}}\right) - \left(-\frac{Gm_{\text{Earth}}m}{R_{\text{Earth}}}\right) = +\frac{Gm_{\text{Earth}}m}{2R_{\text{Earth}}}$$

For a 1-kg satellite ($m = 1.00$ kg), $\Delta E = 3.13 \times 10^7$ J. So 3.13×10^7 J of energy has to be imparted to each kilogram of a satellite placed into low Earth orbit. In practice the energy requirements are much greater because the rocket and its fuel (which are much more massive than the satellite itself) also have to be given kinetic energy. One way to *reduce* the energy requirements is to take advantage of Earth's rotation. Every point on Earth's surface is rotating to the east (which is why we see the Sun rise in the east), and the speed in meters per second is greatest near the equator. That's why NASA and the European Space Agency launch satellites into orbit from locations relatively close to the equator (Florida and French Guiana, respectively) and launch them toward the east (**Figure 11-8**).

NASA

Figure 11-8 Launching toward the east This aerial photograph shows a space shuttle launch. Having climbed through the clouds, the spacecraft climbs upward and to the east (to the left in the photograph) from Florida. In this way the spacecraft takes advantage of Earth's eastward rotation, which gives the spacecraft 406 m/s of speed even before it lifts off.

EXAMPLE 11-6 The Energy Needed for a Satellite for Satellite Television

In Example 6-11 we calculated the speed and radius for a geostationary broadcasting satellite. (Geostationary means the satellite orbits Earth's equator with a period exactly the same as Earth's rotation period, so it always remains over the same spot on the equator, and appears to be in the same position in the sky as seen from anywhere on Earth's surface.) We found (a) the distance from Earth's center at which the television satellite orbits, 4.22×10^7 m, and (b) the satellite's orbital speed, 3.07×10^3 m/s. If you cannot remember how to do this, please review Example 6-11. In this example we will find (c) if a television satellite has a mass of 3.50×10^3 kg, how much energy must it be given to place it in orbit?

Set Up

Equation 11-7 tells us the total mechanical energy of the spacecraft in orbit, and Equation 8-8 tells us the gravitational potential energy of the spacecraft when it is at rest ($K = 0$, so $U_{grav} = E$) on Earth's surface before being launched. The difference between these two values is the energy that must be given to the satellite.

Total mechanical energy for a circular orbit:

$$E = -\frac{Gm_{Earth}m}{2r} \qquad (11\text{-}7)$$

Gravitational potential energy:

$$U_{grav} = -\frac{Gm_1 m_2}{r} \qquad (8\text{-}8)$$

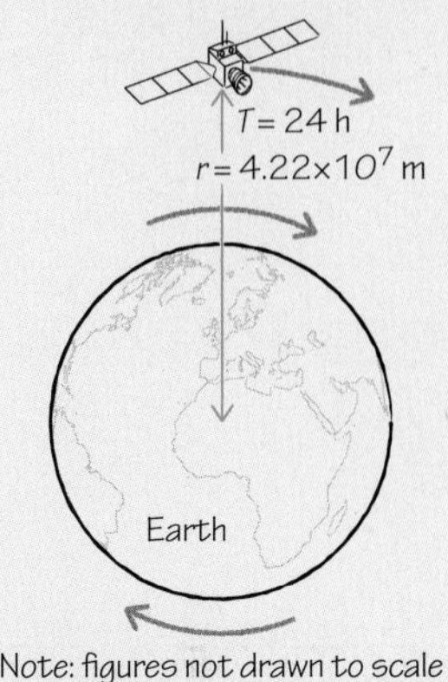

Solve

(c) Find the total mechanical energy in orbit, the total mechanical energy on Earth's surface, and the difference between these values for the satellite of mass $m = 3.50 \times 10^3$ kg.

When the satellite is in orbit, the total mechanical energy is

$$E_{in\ orbit} = -\frac{Gm_{Earth}m}{2r}$$

$$= -\frac{(6.67 \times 10^{-11}\ \text{Nm}^2/\text{kg}^2)(5.97 \times 10^{24}\ \text{kg})\ (3.50 \times 10^3\ \text{kg})}{2(4.22 \times 10^7\ \text{m})}$$

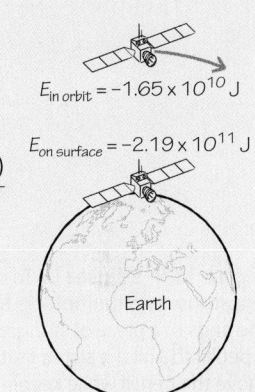

$$= -1.65 \times 10^{10}\ \text{Nm}$$

$$= -1.65 \times 10^{10}\ \text{J}$$

With the spacecraft at rest on Earth's surface, the total mechanical energy is just the gravitational potential energy:

$$E_{on\ surface} = U_{grav,on\ surface} = -\frac{Gm_{Earth}m}{r}$$

$$= -\frac{(6.67 \times 10^{-11}\ \text{Nm}^2/\text{kg}^2)(5.97 \times 10^{24}\ \text{kg})(3.50 \times 10^3\ \text{kg})}{6.38 \times 10^6\ \text{m}}$$

$$= -2.18 \times 10^{11}\ \text{Nm} = -2.18 \times 10^{11}\ \text{J}$$

The amount of energy that must be given to the satellite to put it into orbit is

$$E_{in\ orbit} - E_{on\ surface} = (-1.65 \times 10^{10}\ \text{J}) - (-2.18 \times 10^{11}\ \text{J})$$

$$= 2.02 \times 10^{11}\ \text{J}$$

Reflect

This amount of energy is considerably greater than the number of kilograms of the satellite times the energy per kilogram we found would be necessary to put it into low Earth orbit, which makes sense because we had to put it almost seven times farther from Earth.

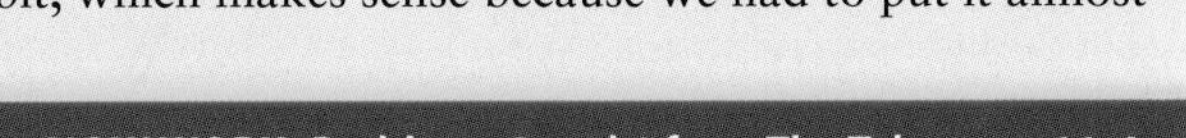

NOW WORK Problems 3 and 4 from The Takeaway 11-4.

Kepler's Laws of Planetary Motion

Decades before Newton published his law of universal gravitation, astronomers had carefully observed and recorded the positions of the planets as they traced out their orbits. By analyzing these observations, the German mathematician and astronomer Johannes Kepler discovered three laws that summarize the motions of the planets.

Kepler's laws describe the shape of planetary orbits, the speed at which a planet moves along its orbit, and the time it takes a planet to complete an orbit. One of Newton's great accomplishments was to show that his law of universal gravitation, in conjunction with his laws of motion, explained *why* the planets move according to Kepler's laws. As we found with solving many problems starting in Chapter 7, sometimes it is easier to use conservation laws, and conservation of angular momentum helps make the explanation of planetary motion simpler. Let's look at Kepler's three laws in turn.

The first of Kepler's laws is the **law of orbits**:

The orbit of each planet is an ellipse with the Sun located at one focus of the ellipse.

You can draw an ellipse by using a loop of string, two thumbtacks, and a pencil, as **Figure 11-9a** shows. Each thumbtack in the figure is at a *focus* (plural *foci*) of the ellipse; an ellipse has two foci. The longest diameter of an ellipse, called the *major axis*, passes through both foci. Half of that distance is called the **semimajor axis** and is usually designated by the symbol a. (Unfortunately, this is also the symbol for acceleration. We promise never to use semimajor axis and acceleration in the same equation!) A circle is a special case of an ellipse in which the two foci are at the same point; this corresponds to using only a single thumbtack in Figure 11-9a. The semimajor axis of a circle is equal to its radius.

The **eccentricity** (symbol e) of an ellipse describes how elongated the ellipse is. The value of e can range from 0 (a circle) to just under 1 (nearly a line). The greater the eccentricity, the more elongated the ellipse (**Figure 11-9b**). All the objects that orbit the Sun have orbits that are at least slightly elliptical, with e greater than zero. The most circular of any planetary orbit is that of Venus, with an eccentricity e of just 0.007; Mercury's orbit has $e = 0.206$. The minor planet Pluto has an orbit with $e = 0.249$, and several small bodies called comets move in very elongated orbits with eccentricities just less than 1 (**Figure 11-10**). For any elliptical orbit, the Sun is at one focus; there is nothing at the other focus.

(a) The geometry of an ellipse

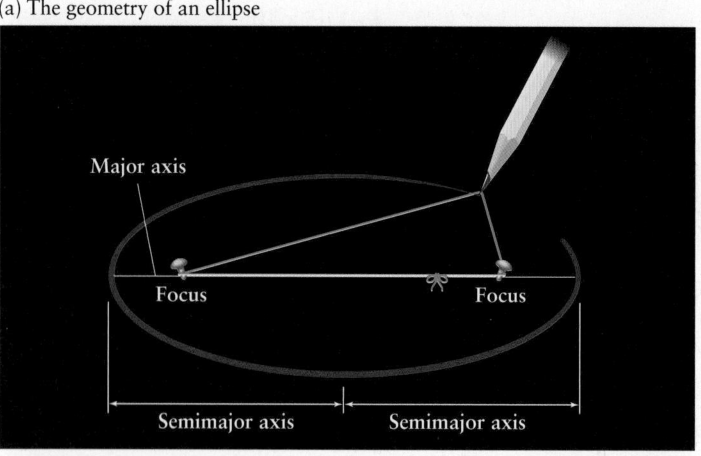

(b) Ellipses with different eccentricities

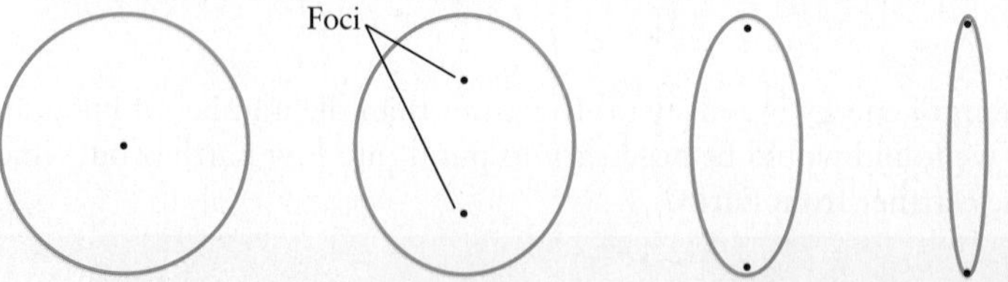

Figure 11-9 Ellipses (a) To draw an ellipse, use two thumbtacks to secure the ends of a piece of string, then use a pencil to pull the string taut. If you move the pencil while keeping the string taut, the pencil traces out an ellipse. The thumbtacks are located at the two foci of the ellipse. The major axis is the greatest distance across the ellipse; the semimajor axis is half of this distance. (b) These ellipses have the same major axis but different eccentricities. An ellipse can have any eccentricity from $e = 0$ (a circle) to just under $e = 1$ (virtually a line).

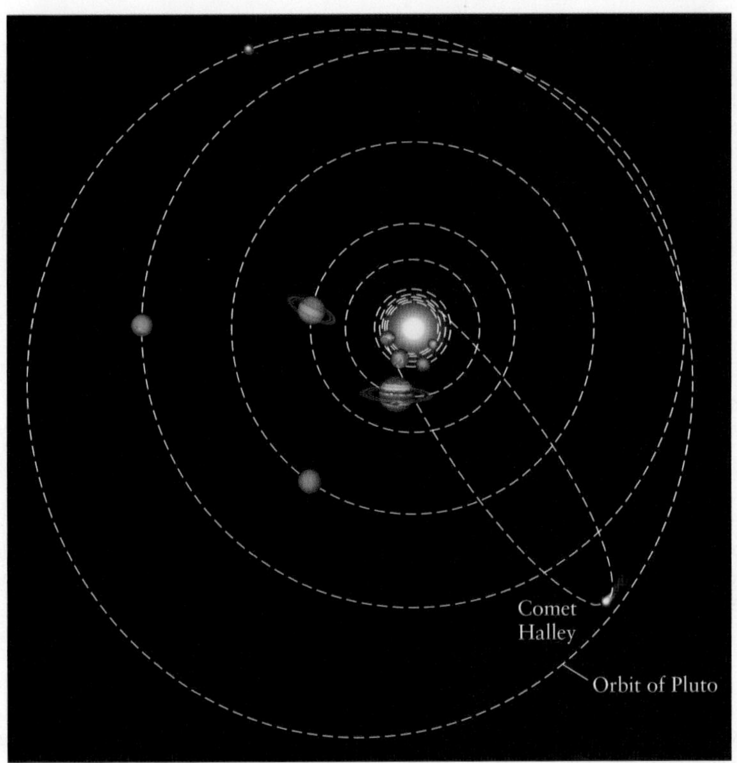

Comet
Halley
Orbit of Pluto

Figure 11-10 Orbits in the solar system The orbits of the planets and the minor planet Pluto are ellipses with relatively small eccentricities. Comet Halley, which moves around its elliptical orbit once every 76 years, has an eccentricity of 0.967. The planets, Sun, and comet are all depicted much larger than their actual size relative to orbits.

One of Newton's triumphs was to show that the planets can have elliptical orbits only if the force F attracting them toward the Sun is in inverse proportion to the square of the distance r from the Sun to planet. In other words, the elliptical shape of the planets' orbits is a verification of the law of universal gravitation, which states that $F = Gm_{\text{Sun}}m_{\text{planet}}/r^2$. (The proof is mathematically complex and beyond our scope.)

Kepler's second law states that, unlike the case of a circular orbit, a planet does *not* move at a constant speed on an elliptical orbit. His **law of areas** describes how the speed changes around the orbit:

A line joining the Sun and a planet sweeps out equal areas in equal intervals of time, regardless of the position of the planet in the orbit.

Figure 11-11 illustrates this law. Suppose that it takes 30 days for a planet to go from point A to point B. During that time an imaginary line joining the Sun and the

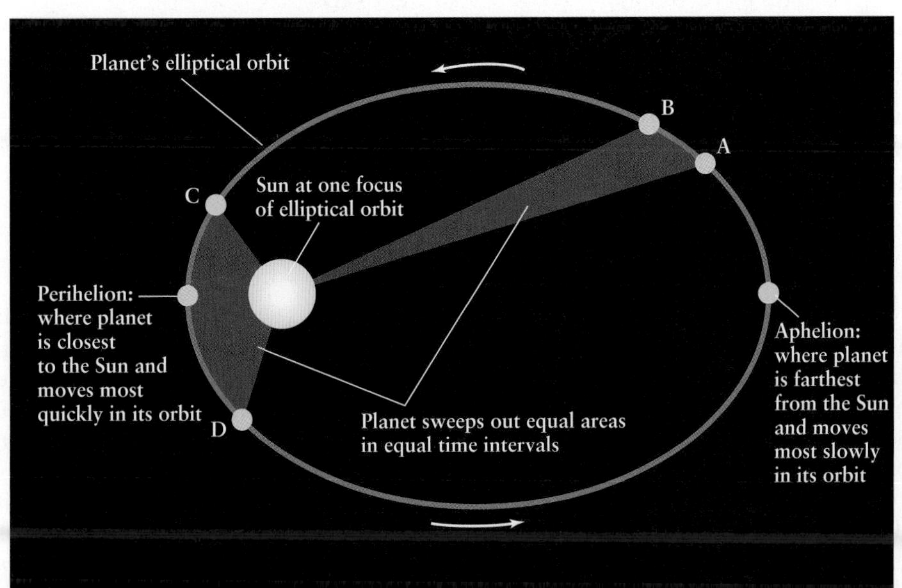

Planet's elliptical orbit
B
A
C
Sun at one focus of elliptical orbit
Perihelion: where planet is closest to the Sun and moves most quickly in its orbit
D
Planet sweeps out equal areas in equal time intervals
Aphelion: where planet is farthest from the Sun and moves most slowly in its orbit

Figure 11-11 Kepler's law of orbits and law of areas According to Kepler's law of orbits, a planet travels around the Sun along an elliptical orbit with the Sun at one focus. (There is nothing at the other focus.) According to his law of areas, as the planet moves, an imaginary line joining the planet and the Sun sweeps out equal areas in equal intervals of time (from A to B or from C to D). By using these laws in his calculations, Kepler found a perfect fit to the apparent motions of the planets.

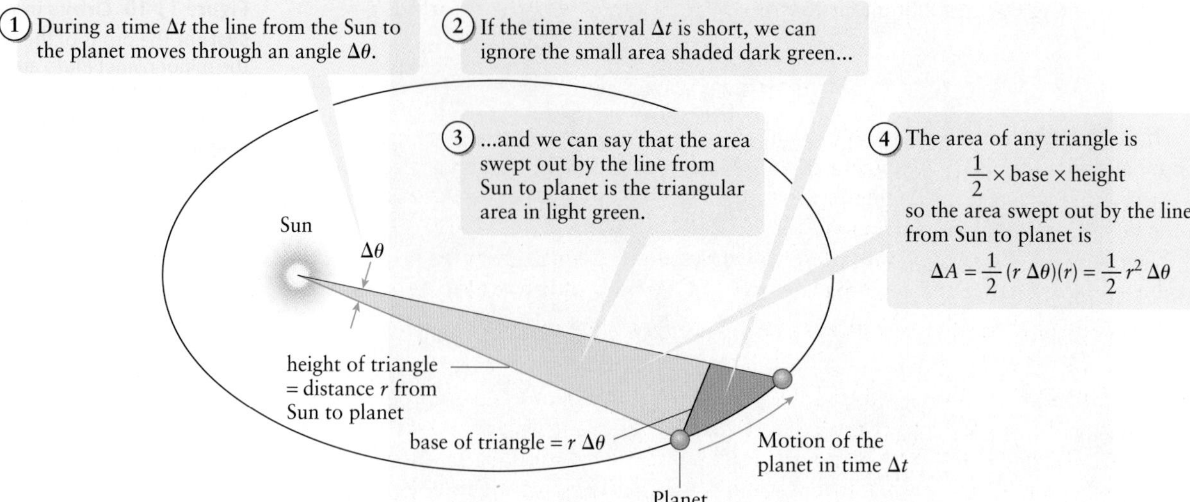

(1) During a time Δt the line from the Sun to the planet moves through an angle $\Delta\theta$.
(2) If the time interval Δt is short, we can ignore the small area shaded dark green...
(3) ...and we can say that the area swept out by the line from Sun to planet is the triangular area in light green.
(4) The area of any triangle is $\frac{1}{2}$ × base × height so the area swept out by the line from Sun to planet is $\Delta A = \frac{1}{2}(r\,\Delta\theta)(r) = \frac{1}{2}r^2\,\Delta\theta$
Sun
$\Delta\theta$
height of triangle = distance r from Sun to planet
base of triangle = $r\,\Delta\theta$
Motion of the planet in time Δt
Planet

Figure 11-12 Explaining Kepler's law of areas As a planet moves around its elliptical orbit, a line from the Sun to the planet sweeps out the same area $\Delta A = \frac{1}{2}r^2\Delta\theta$ in any time interval Δt. This turns out to be equivalent to the statement that the planet's angular momentum is constant. This means that its angular velocity must change as the radius changes.

planet sweeps out a nearly triangular area. Kepler discovered that a line joining the Sun and the planet also sweeps out exactly the same area during any other 30-day interval. In other words, if the planet also takes 30 days to go from point C to point D, then the two shaded segments in Figure 11-11 are equal in area.

The law of areas tells us that the planet moves most quickly at *perihelion* (the point in its orbit closest to the Sun) and most slowly at *aphelion* (the point when the planet is farthest from the Sun). We would expect this from the law of universal gravitation: A planet should speed up as it moves toward the Sun and approaches perihelion, and it should slow down as it moves away from the Sun and toward aphelion. But why does the speed vary in the particular manner described by the law of areas?

To answer this question, let's consider how a planet moves during a very short time interval Δt. **Figure 11-12** shows that if Δt is short, an imaginary line from the Sun to the planet moves through an angle $\Delta\theta$ and sweeps out an area $\Delta A = \frac{1}{2}r^2\Delta\theta$, where r is the distance from the Sun to planet. Hence, the *rate* at which this line sweeps out the area is ΔA divided by Δt, or

$$\text{rate at which area is swept out} = \frac{\Delta A}{\Delta t} = \frac{1}{2}r^2\frac{\Delta\theta}{\Delta t} = \frac{1}{2}r^2\omega$$

In this expression, $\omega = \Delta\theta/\Delta t$ is the *angular speed* of the planet around the Sun (see Section 10-2). Kepler's law of areas says that $\Delta A/\Delta t$ has the same value at all points along the orbit, so

$$\frac{1}{2}r^2\omega = \text{constant (Kepler's law of areas)} \tag{11-8}$$

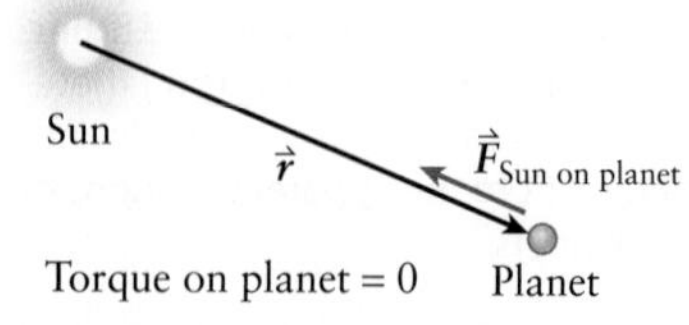

Figure 11-13 The Sun's gravity produces zero torque The gravitational force of the Sun on a planet is directed toward the Sun. Hence this force produces no torque around the Sun, so the planet's angular momentum, but not angular velocity, remains constant.

Now that we have the law of areas in equation form, we can understand how it arises. **Figure 11-13** shows that the gravitational force $\vec{F}$ exerted on a planet by the Sun is always directed opposite to the displacement vector that points from the Sun to the planet. This means that the gravitational force exerts zero *torque* on the planet (see Section 10-5, where we learned that a force that is exerted either in the direction of $\vec{r}$ or opposite to $\vec{r}$—as is the case in Figure 11-13—produces zero torque).

We learned in Section 11-3 that the *angular momentum* of a system is constant if there is no net torque on it. Hence the angular momentum of a planet is constant as it orbits the Sun. For a rotating system with rotational inertia I and angular speed ω, the magnitude of the angular momentum is $L = I\omega$. We also know that the rotational inertia of an object of mass m at a distance r from the rotation axis is $I = mr^2$

(see Section 10-2). So the statement that the angular momentum of an orbiting planet is constant is

$$L = mr^2\omega = \text{constant} \tag{11-9}$$

If we divide Equation 11-9 by the planet's mass m (which does not change as it orbits) and also divide by 2, we get Equation 11-8. So the law of areas is a consequence of the gravitational force on a planet being directed toward the Sun, precisely as stated in the law of universal gravitation (see Figure 6-13a).

The first two of Kepler's laws describe how a given planet moves in its orbit. His third law, the **law of periods**, compares the orbital periods of orbits of different sizes:

The square of the period of a planet's orbit is proportional to the cube of the semimajor axis of the orbit.

This is precisely the relationship that we found for *circular* orbits in Equation 6-10. Newton was able to show that the law of periods follows from the law of universal gravitation even for elliptical orbits, if we replace the orbital radius r in Equation 6-10 with the semimajor axis a. For objects orbiting the Sun we must also replace Earth's mass m_{Earth} with the mass of the Sun:

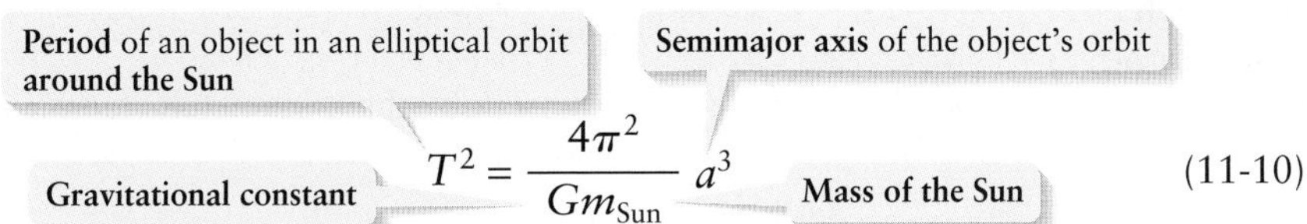

$$T^2 = \frac{4\pi^2}{Gm_{Sun}}a^3 \tag{11-10}$$

EQUATION IN WORDS
Newton's form of Kepler's law of periods

We've described Kepler's laws for the orbits of the planets. But the same laws apply to the orbits of satellites around Earth, to moons orbiting Saturn (see Figure 6-1b), and in general to any system where one object orbits another.

Newton's explanation of Kepler's three laws was perhaps his ultimate achievement. By extrapolating from the laws of nature that he saw here on Earth, he became the first human to understand the behavior of objects in the heavens.

WATCH OUT !

As the planets orbit, the Sun moves as well.

In this discussion we've ignored the forces that the planets exert on the Sun, and we have assumed that the Sun remains at rest. In fact these forces cause the Sun to "wobble" around the center of mass of the solar system in the same way the Moon causes Earth to rotate around the center of mass of the Earth–Moon system. Although this wobble is larger in amplitude than the radius of the Sun (6.96×10^5 km), we've neglected it because the amplitude is small compared to the semimajor axes of planetary orbits. However, astronomers have made use of this effect to search for planets orbiting other stars. By detecting the wobble of a star, astronomers have been able to detect the presence of that star's planets even though the planets themselves are too faint to be seen with even the most powerful telescopes. Hundreds of *extrasolar planets* have been discovered in this way.

The following example demonstrates one application of Equation 11-10.

EXAMPLE 11-7 A Comet's Orbit

Distances in the solar system are typically measured not in meters or kilometers but in *astronomical units* (au), where 1 au is the average distance between the Sun and Earth (1 au = 1.50×10^8 km). Earth's orbital period around the Sun is 1 year. Find the orbital period of a comet that is 0.50 au from the Sun at perihelion and 17.50 au from the Sun at aphelion.

Set Up

We'll use the data about the comet to find its semimajor axis a. From this we'll use Newton's form of Kepler's law of periods to determine the period T.

Newton's form of Kepler's law of periods:

$$T^2 = \frac{4\pi^2}{Gm_{Sun}}a^3 \tag{11-10}$$

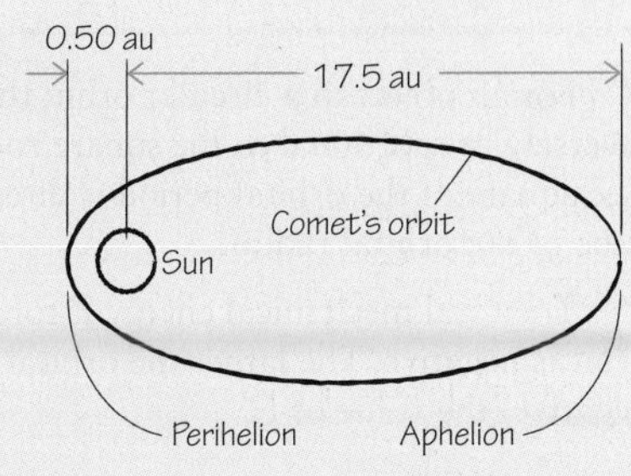

Solve

The drawing shows that the length of the major axis of the comet's orbit is the sum of the perihelion and aphelion distances. The semimajor axis a is one-half the length of the major axis.

Length of the major axis: (distance from the Sun to comet at perihelion) + (distance from the Sun to comet at aphelion) $= 0.50 \text{ au} + 17.50 \text{ au} = 18.0 \text{ au}$

major axis = 0.50 au + 17.5 au = 18.0 au

comet's orbit

Sun

semimajor axis a = 9.0 au

Semimajor axis:

$$a_{\text{comet}} = \frac{1}{2} \times (\text{length of the major axis}) = 9.0 \text{ au}$$

We could calculate T directly from Equation 11-10 by substituting the values of Gm_{Sun} and the semimajor axis of the comet's orbit (which we would have to convert to meters). But a much simpler approach is to compare the comet's orbit to that of Earth, for which the semimajor axis is 1 au and the period is 1 year.

For Earth: $T_{\text{Earth}}^2 = \dfrac{4\pi^2}{Gm_{\text{Sun}}} a_{\text{Earth}}^3$

For the comet: $T_{\text{comet}}^2 = \dfrac{4\pi^2}{Gm_{\text{Sun}}} a_{\text{comet}}^3$

If we divide the second equation by the first one, the factor of $4\pi^2/Gm_{\text{Sun}}$ cancels out, leaving

$$\frac{T_{\text{comet}}^2}{T_{\text{Earth}}^2} = \frac{a_{\text{comet}}^3}{a_{\text{Earth}}^3}$$

Substitute values:

$$\frac{T_{\text{comet}}^2}{(1 \text{ y})^2} = \frac{(9.0 \text{ au})^3}{(1 \text{ au})^3} = 729$$

$$T_{\text{comet}} = \sqrt{729} \times 1 \text{ y} = 27 \text{ y}$$

Reflect

Our technique for solving this problem really used Kepler's original form of the law of periods: The square of the period is directly proportional to the cube of the semimajor axis. The semimajor axis for the comet's orbit is 9.0 times larger than that of Earth's orbit, so the square of the orbital period for the comet must be $(9.0)^3 = 729$ times larger than the square of the 1-year orbital period of Earth. Therefore, the comet's orbital period is $\sqrt{729}$ years, or 27 years.

NOW WORK Problem 5 from The Takeaway 11-4.

Exam Tip

While Kepler's laws are not part of the AP® Physics 1 curriculum, the underlying theories of applying Newton's second law, energy conservation, and angular momentum conservation to an object moving in a circle are part of the required knowledge. Understanding how angular speed of an object changes as its distance from the rotation axis changes when angular momentum is constant is also a topic. Deriving Kepler's laws serves as good examples to practice for the AP® Physics 1 exam.

THE TAKEAWAY for Section 11-4

✔ For an object in a circular orbit, the orbital speed is inversely proportional to the square root of the orbital radius. The square of the orbital period is directly proportional to the cube of the orbital radius.

✔ The total mechanical energy E for an object in a circular orbit is negative. The larger the orbital radius, the greater (less negative) the value of E.

✔ Kepler's three laws of planetary motion—the law of orbits, the law of areas, and the law of periods—were deduced from observations of how the planets move, but can all be explained by Newton's law of universal gravitation and are made mathematically easier to understand using conservation of angular momentum and energy.

AP® Building Blocks

1. From the graph of energies of an object in a circular orbit as shown below, match the graphs A, B, and C with the following energies.
   - Total mechanical energy
   - Gravitational potential energy
   - Kinetic energy

   Justify your answers.

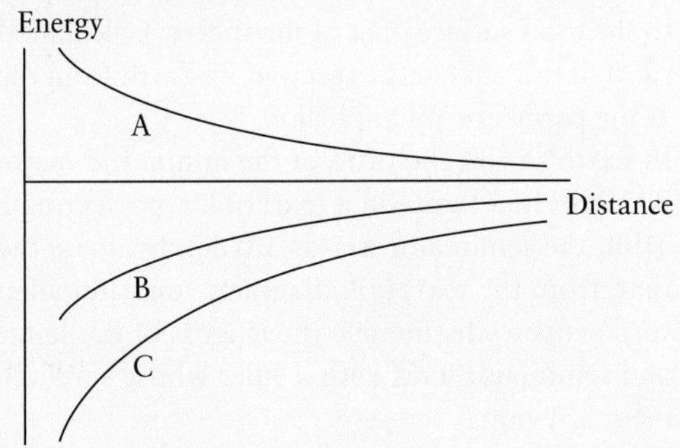

2. Express the centripetal acceleration of a circular orbit in terms of (a) r and v, the radius and velocity; (b) T, the period, and r; and (c) r and ω, the angular speed of the orbit.

3. Questions about Earth orbits are often stated in terms of altitude, the height above the surface of Earth. The space shuttle usually orbited Earth at altitudes of around 3.00×10^5 m.
   (a) Determine the time for one orbit of the shuttle about Earth. Take the mass of Earth as 5.97×10^{24} kg and the radius of Earth to be 6.371×10^6 m.
   (b) How many sunrises per day did the astronauts witness?
   (c) What was the tangential speed of this orbit of the space shuttle?

EX 11-6

4. Given that Earth orbits the Sun in a nearly circular orbit at an average distance of 1.000 au and an approximate orbital period T of 365.25 days, determine the mass of the Sun (1.000 au is 1.496×10^{11} m). Given how close to circular Earth's orbit is, the radius of its orbit is approximately the same as the semimajor axis.

EX 11-7

5. Measurements of the asteroid Apophis have shown that its aphelion (farthest distance from the Sun) is 1.099 au, its perihelion (closest distance from the Sun) is 0.746 au, and its mass is 2.7×10^{10} kg(1 au = 1.496×10^{11} m).
   (a) Sketch an ellipse and on the ellipse, show the foci as large dots; the semiminor axis, labeled b; and the semimajor axis, labeled a. Label the semimajor axis, the aphelion, and the perihelion relative to a focus that is identified by the distance to the center of the ellipse, labeled c.
   (b) From your sketch, explain why the sum of the aphelion and perihelion distances is equal to twice the length of the semimajor axis.
   (c) Apply Kepler's third law to calculate the period of the orbit of Apophis in days.

Prep for the AP® Exam

   (d) Apply Kepler's second law to show that the ratio of speeds at the aphelion and perihelion is inversely proportional to the ratio of these distances and explain why the object in orbit speeds up as it approaches the perihelion.

AP® Skill Builders

6. Planets (including Pluto, whose classification as a planet is disputed) orbit the Sun with periods (T) and semimajor axes (a) as shown in the table. 1 au = 1.496×10^{11} m.

Object	T (days)	a (au)
Mercury	87.97	0.3871
Venus	224.7	0.7233
Earth	365.2	1.000
Mars	687.0	1.5237
Jupiter	4332	5.203
Saturn	10,832	9.537
Uranus	30,799	19.20
Neptune	60,190	30.10
Pluto	90,560	39.48

   (a) Analyze the data to determine the mass of the Sun by first re-expressing the relationship between period and semimajor axis (Equation 11-10) and then by defining a new variable x, based on Kepler's third law. This variable should be the semimajor axis length to the appropriate power to allow you to write your new expression as $T = cx$, where c is a constant for the solar system.
   (b) Construct a graph of T (in days) versus x and obtain an equation for your line of best fit for your data.
   (c) Calculate the mass of the Sun based on the line of your best fit.
   (d) The orbit of an asteroid has a semimajor axis of 2.7675 au. Use your best-fit line equation to predict the period for this asteroid.
   (e) Comet Halley has an elliptical orbit with a period of 27,507 days. Use your best-fit line equation to predict the semimajor axis for Comet Halley.

7. A spacecraft with mass m is in a uniform circular motion of radius r_i, shown with a solid line in the figure. The spacecraft will use a thruster at point P to change its orbital radius and shape. A thruster effectively delivers an impulse, $I = F\Delta t$, on the spacecraft by pushing exhaust mass away from the spacecraft in the direction opposite to F. Conservation of momentum is responsible for this effect. The thruster converts internal energy of the fuel to cause a change in kinetic energy of the spacecraft. The spacecraft's center of mass is displaced a distance Δr in the direction of its motion (tangent to the circle) or opposite its direction of motion while this force, F, due

to the impulse delivered by the thruster, is exerted on the spacecraft:

$$W = F\Delta r \cos \theta_{rF}$$

If the thruster displaces the craft in the direction of the current motion (it causes a forward force on the spacecraft by pushing the exhaust mass backward), then $\cos \theta_{rF} = 1$. If the thruster displaces the craft in the direction opposite to the current motion (a backward force), then $\cos \theta_{rF} = -1$.

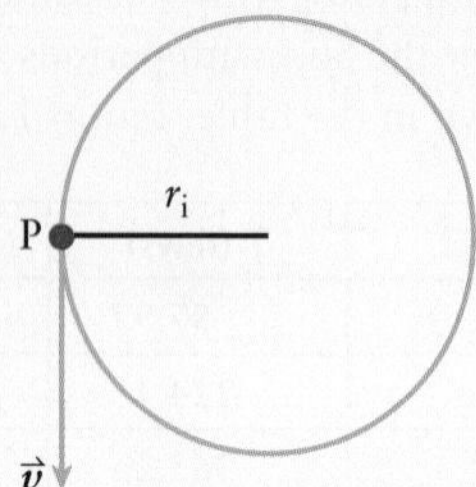

(a) Use conservation of angular momentum and the requirements for circular motion to justify the claim that a backward force decreases the orbital radius and a forward force increases the orbital radius. Summarize, in complete sentences, how the change in the spacecraft's kinetic energy due to the force is represented in your argument and how the mechanical energy is related to the orbital radius. *Hint*: The new orbit will include the point P but the semimajor axis will be either larger or smaller than r_i.

(b) Construct a diagram to qualitatively predict the shapes of orbits for the cases of the thruster causing a forward or a backward force on the spacecraft at point P on the initial circular orbit.

(c) Discuss what the astronaut would need to do to make the orbit circular again.

8. There is a documentary film called *Private Universe*, in which Harvard graduates and professors are asked to explain why the Earth has seasons. Most of them get it wrong. They say that Earth is closer to the Sun during the summer in the northern hemisphere. A physics graduate is interviewed and lists the advanced physics courses he has taken before giving the incorrect explanation.

The film claims that the reason this misconception is so prevalent is that textbooks often show Earth's orbit in perspective as a wildly exaggerated ellipse. Earth's orbital distance at perihelion (closest to the Sun) is 1.47×10^8 km and the aphelion (farthest from the Sun) distance is 1.52×10^8 km.

(a) Radiant energy (light and warmth) emitted by a light bulb shines in nearly all directions. Imagine surrounding the bulb with a spherical surface, which has the bulb at the center. The same amount of radiant energy would be measured per square meter anywhere on the surface. We can model the radiant energy reaching Earth from the Sun in exactly the same way. If the radius of the spherical surface increases, the same total amount of radiant energy still falls on the entire surface; but the total surface area of the sphere with this larger radius is larger, so the amount of radiant energy received by each square meter decreases. Thus, the amount of radiant energy received by each square meter is inversely proportional to the total surface area of the sphere. Calculate the ratio of radiant energy received by Earth from the Sun at the perihelion and aphelion.

(b) In Earth's orbit the ratio of the minor and major axes is 0.994. In a figure in a textbook representing Earth's orbit, the semimajor axis is 5.0 cm. Evaluate the claim that, from the textbook diagram, you can make measurements to distinguish the lengths of the semiminor and semimajor axes with a ruler whose smallest increment is 1 mm.

(c) Earth's angular momentum for its rotation about its own axis is tilted at 23° from the angular momentum for its orbit around the Sun. Assume this angle is constant as Earth orbits the Sun, and that at the distance Earth is from the Sun, the radiant energy striking Earth from the Sun is uniform over its cross-sectional area. In your own words summarize how this 23° tilt of the angular momentum explains the seasons. Drawing a picture showing the angle and indicating the energy per area on Earth's surface could be helpful.

AP® Skills in Action

9. The speed of the orbit of the Moon about Earth can be determined by observations of the radius of the orbit and the period. The centripetal acceleration of the Moon can be calculated and the mass of Earth can be measured.

(a) Explain why these measurements are sufficient to measure the mass of Earth but cannot be used to measure the mass of the Moon.

(b) In 1969 *Apollo 11* orbited the Moon, whose mean radius is 1737 km, at an altitude of 111 km and recorded the orbital period as 1.98 hours. Until then the Moon's mass had been known with low accuracy. Newton's measurement was based on tidal data and gave $m_{\text{Earth}}/m_{\text{Moon}} = 39.788$. Calculate the Moon's mass from these observational data and calculate the percent relative error of Newton's measurement.

10. Analyze the data below to determine the mass of Uranus from the properties of some of its satellites. *Hint*: Rearrange the expression for T and r to get a linear equation $T = cx$, with x defined as $x = r^{3/2}$, then graph or fit a line to T versus x, and use the slope of the best-fit line to determine the mass of Uranus.

Satellite	Distance from center of Uranus to center of satellite (m)	Orbital period (s)
Miranda	1.294×10^8	1.221×10^5
Ariel	1.910×10^8	2.178×10^5
Umbriel	2.663×10^8	3.580×10^5
Titania	4.359×10^8	7.522×10^5
Oberon	5.835×10^8	1.163×10^6

11. The most valuable and useful of all of the possible orbits around Earth is the geostationary orbit, a geosynchronous orbit over the equator. Geosynchronous satellites have orbital periods of 24 hours and are always at the same place above Earth, so that antennae that receive satellite TV signals never need to be moved.

Satellites are susceptible to magnetic storms caused by solar flares, which can interrupt transmissions and even damage the satellite. Transmission from a particular satellite is turned off by a magnetic storm. Imagine a situation where, when the satellite comes back online, it has moved 2° east of the assigned parking spot, and is moving faster.

For the described situation, predict whether or not the satellite has its original altitude and, if it has changed, whether the orbit is at a lower altitude or a higher altitude.

WHAT DID YOU LEARN?

Prep for the AP Exam

Chapter learning goals	Section(s)	Related example(s)	Relevant section review exercises
Apply the conservation of mechanical energy to rotating systems, including in the case of rolling without slipping.	11-2	11-1, 11-2, 11-3	1, 2, 3, 4, 5, 6
Describe what is meant by the angular momentum of a rotating system and of a moving object, and explain the circumstances under which angular momentum is constant.	11-3	11-4, 11-5	4, 5, 7
Predict the behavior of rotational collision situations by the same processes that are used to analyze linear collision situations.	11-3	11-5	4, 5
Apply the law of universal gravitation and the expression for gravitational potential energy, with the concept of conservation of angular momentum, to analyze the orbits of satellites and planets.	11-4	11-6	3, 4, 5

Chapter 11 Review

Key Terms

All the Key Terms can be found in the Glossary/Glosario on page G1 in the back of the book.

Chapter Summary

Topic	Equation or Figure
Total kinetic energy and conservation of energy: The total kinetic energy K of an extended object is the sum of its translational kinetic energy and rotational kinetic energy. We can write an expression for K for a rigid body. Using that expression, the same equations for conservation of energy that we learned in Chapter 7 apply for a rigid body that is rotating, translating, or both, with the inclusion of rotational work = $\tau\Delta\theta$ and rotational kinetic energy.	(1) **Total kinetic energy** of a system that is both translating and rotating around its center of mass (2) The **translational kinetic energy** is the same as if all of the mass of the system were moving at the speed of the center of mass. $K = K_{\text{translational}} + K_{\text{rotational}} = \frac{1}{2}Mv_{\text{CM}}^2 + \frac{1}{2}I_{\text{CM}}\omega^2$ (11-1) (3) The **rotational kinetic energy** is the same as if the system were not translating, and all of the mass were rotating around the center of mass.
Rolling without slipping: If a rigid body rolls on a surface without slipping, the speed of the center of mass of the rigid body must be related to its angular speed. A friction force is exerted on this rigid body to maintain this relationship. Because the point of contact of the force does not move, this friction force does no work on the rigid body.	**Speed** of the center of mass of a rolling rigid body $v_{\text{CM}} = R\omega$ (11-2) **Radius** of the rigid body **Angular speed** of the rigid body's rotation
Angular momentum and the conservation of angular momentum: A rotating system has an angular momentum L_z whose sign depends on the direction of rotation. An object moving relative to an axis also has angular momentum around that axis. If there is no net external torque on an object or system of objects, its angular momentum remains constant. Just as an impulse changes the momentum of an object, an angular impulse changes the angular momentum of an extended object, $\tau\Delta t = \Delta L$. Angular momentum is conserved: It is transferred into or out of a system by an external torque τ exerted on the system while the system rotates through an angle $\Delta\theta$ and constant when there is no torque.	The **angular momentum** of a system rotating around a fixed axis... ...is proportional to the **angular velocity** at which the system rotates around that axis. $L_z = I\omega_z$ (11-4) The angular momentum is also proportional to the system's **rotational inertia** for that rotation axis. Magnitude of the **angular momentum of an object** Magnitude of the **linear momentum of the object** $L = r_\perp p = rp\sin\phi$ (11-5) **Angle** between the momentum and the direction of the displacement from axis to object **Perpendicular distance** from the axis to the line of the momentum **Distance** (magnitude of displacement) from the axis to the object

Rotational quantities as vectors: The angular velocity $\vec{\omega}$, angular momentum $\vec{L}$, and torque $\vec{\tau}$ are actually all vector quantities that lie along the rotation axis. Angular momentum is always in the same direction as angular velocity. Their directions are given by a right-hand rule.

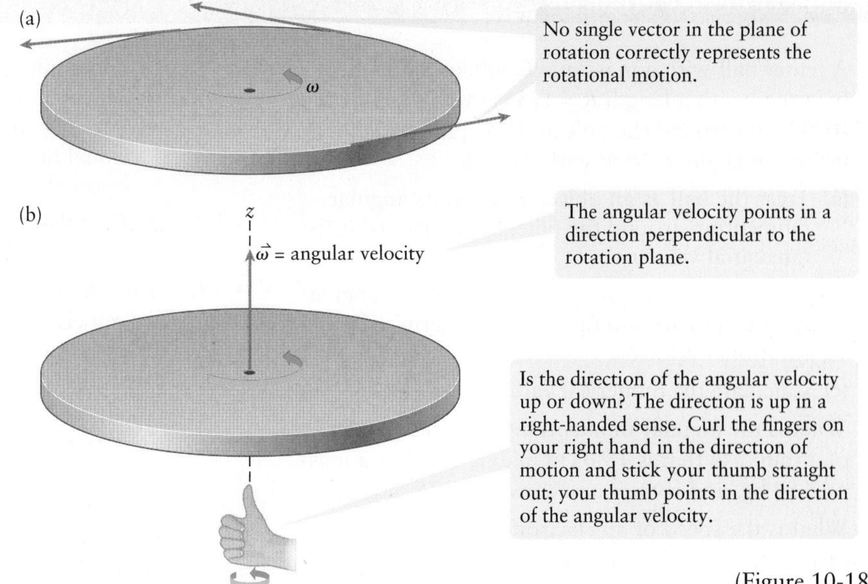

(a)
ω
No single vector in the plane of rotation correctly represents the rotational motion.
(b)
z
$\vec{\omega}$ = angular velocity
The angular velocity points in a direction perpendicular to the rotation plane.
Is the direction of the angular velocity up or down? The direction is up in a right-handed sense. Curl the fingers on your right hand in the direction of motion and stick your thumb straight out; your thumb points in the direction of the angular velocity.

(Figure 10-18)

Circular orbits: A satellite in a circular orbit around Earth moves with a constant speed v. The orbital period T and the total mechanical energy E in a circular orbit both increase with increasing orbital radius r.

Period of an Earth satellite in a circular orbit

Radius of the satellite's orbit

$$T^2 = \frac{4\pi^2}{Gm_{\text{Earth}}} r^3 \tag{6-10}$$

Gravitational constant

Mass of Earth

Gravitational constant

Mass of Earth

Mass of satellite

Total mechanical energy for a satellite in a **circular orbit** around Earth

$$E = -\frac{Gm_{\text{Earth}}m}{2r} \tag{11-7}$$

Radius of the satellite's orbit

The greater the radius r of the circular orbit, the greater (less negative) the total mechanical energy.

Elliptical orbits: A circular orbit is actually a special case of an elliptical orbit. The planets move in elliptical orbits with the Sun at one focus. The speed varies in accordance with Kepler's law of areas, which is equivalent to the statement that the angular momentum of a planet is constant. The orbital period is given by the same expression as for circular orbits around Earth, with Earth's mass replaced by the mass of the Sun and the orbital radius replaced by the semimajor axis (one-half the length of the long axis of the ellipse).

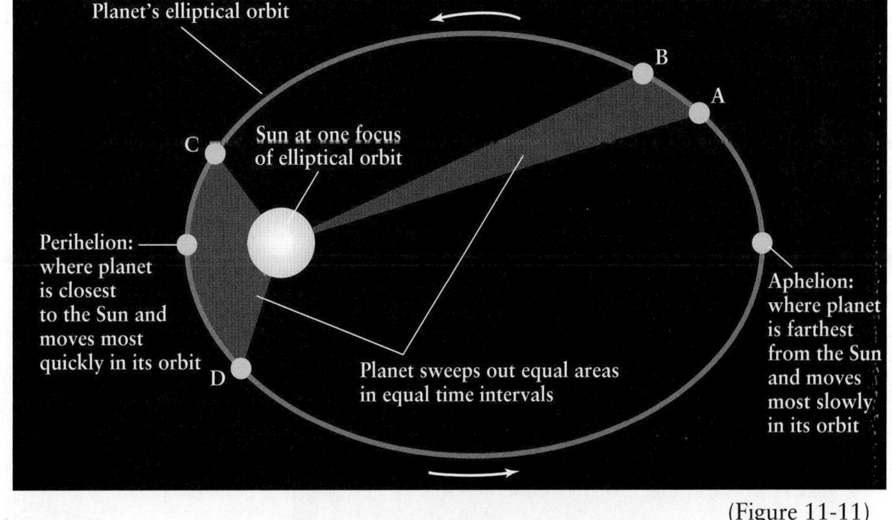

Planet's elliptical orbit
B
A
C
Sun at one focus of elliptical orbit
Perihelion: where planet is closest to the Sun and moves most quickly in its orbit
D
Planet sweeps out equal areas in equal time intervals
Aphelion: where planet is farthest from the Sun and moves most slowly in its orbit

(Figure 11-11)

Period of an object in an elliptical orbit **around the Sun**

Semimajor axis of the object's orbit

$$T^2 = \frac{4\pi^2}{Gm_{\text{Sun}}} a^3 \tag{11-10}$$

Gravitational constant

Mass of the Sun

Chapter 11 Review Problems

1. A tether ball with a mass $m = 0.300$ kg is attached by a rope with a length $R = 125$ cm to a pole. As the ball whirls around the pole at 41.5 rpm, the taut rope makes an angle of 28.0° with the pole.
   (a) Treat the ball as an object. Express its angular momentum around the pole, L, in terms of v, its tangential velocity.
   (b) Treat the ball as a rigid body. Express its angular momentum around the pole, L, in terms of ω, its rotational velocity.
   (c) Calculate the numerical value of L.
2. What is the angular momentum about the central axis of a thin disk that is 18.0 cm in diameter, has a mass of 2.50 kg, and rotates at a constant 1.25 rad/s?
3. What is the speed of an electron in the lowest energy orbital of hydrogen, of radius equal to 5.29×10^{-11} m? The mass of an electron is 9.11×10^{-31} kg, and its angular momentum in this orbital is 1.055×10^{-34} J · s.
4. The original Ferris wheel, shown in the figure, was the star of the 1893 Columbian Exposition in Chicago. It was built like the wheel of a bicycle, with a radius of 76.2 m rotating on a horizontal hollow hub 13.9 m in length and 0.84 m in diameter, with a mass of 40 Mg. Connecting the rim to the hub were spokes anchored by two disks perpendicular to the hub, each with a mass of 24 Mg and a radius of 2.5 m. There were cabins for 2160 passengers. Treat the mass of the spokes as negligible. The total mass of the wheel when filled to capacity with riders was approximately 1 Gg. It completed one rotation in 9 minutes.

   (a) Calculate the angular momentum of the Ferris wheel–rider system, modeling the system as a hollow cylinder (the hub), the two disks, and a hoop containing the rest of the mass of the system at the outer radius of the wheel. The rotational inertia of a disk about its central axis is $mr^2/2$. The rotational inertia of a hollow cylinder and also of a hoop about its central axis is mr^2.
   (b) Calculate the angular momentum of a 60.0-kg passenger riding on the Ferris wheel.
5. A freely rotating turntable moves at a steady angular velocity. A glob of cookie dough falls straight down and attaches to the very edge of the turntable. Consider the turntable–glob system and describe which quantities (angular velocity, angular acceleration, torque, rotational kinetic energy, internal energy, rotational inertia, or angular momentum) are constant during the process. If the property changes, predict whether the property increases or decreases. Each claim should be supported by reasoning.
6. Explain which physical quantities change when an ice skater moves her arms toward her axis of rotation as she rotates in a pirouette. If her angular speed changes, explain what causes the change. Assume that the ice exerts a negligible friction force on the skates.
7. While watching two people balanced on a seesaw you notice that they are sitting equidistant from the pivot point of the seesaw.
   (a) What, if anything, can you say about the relative masses of the two riders?

   As the two people "seesaw," you notice that the person on the way up always leans forward, while the person on the way down always leans backward.

   (b) How does this movement of the riders alter the angular momentum of the seesaw?
8. A yo-yo, with an inner cylinder radius of r and outer disk radius of $R = 3r$, has a center of mass that is at rest on a 30.0° incline. It is held in place by friction on the outer disk and the tension in the string that is wrapped around its inner cylinder, as shown in the figure.

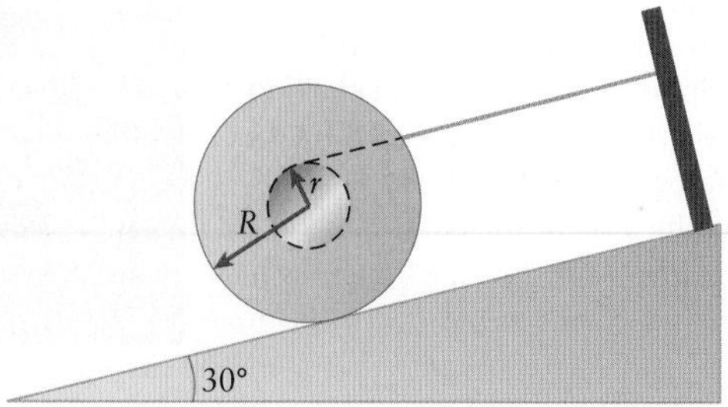

   (a) Predict the direction of the friction force.
   (b) Construct an extended free-body (force) diagram of the situation.
   (c) What assumption must you make to ignore a torque due to the force of gravity?

9. It is easier to balance a bicycle when the bicycle is in motion than when the bicycle is not in motion.
   (a) Describe the direction of the angular momentum of a spinning wheel.
   (b) Express the angular acceleration α of a spinning wheel in terms of the change in angular speed $\Delta\omega$ that occurs over a time interval Δt.
   (c) Angular acceleration and angular velocity are vectors. If the angular velocity of the wheel is decreasing to zero, is the direction of the angular acceleration in the same direction as the angular velocity, or in the opposite direction? In which direction is the torque necessary to cause this angular acceleration?
   (d) A bicycle placed so that it is standing on its wheels (with no rider to try to keep it balanced) will topple over. Describe the torque that causes this to happen: the force that causes it, where it is exerted, and the axis of rotation about which the bicycle rotates.

10. A student stands on a platform that can rotate on a shaft with bearings that make the torque required to rotate the platform negligible. The student is holding a shaft that extends through the hub of a rotating wheel. The wheel spins in a counterclockwise direction when viewed from the right in the figure. The angular momentum for the spinning wheel is shown.

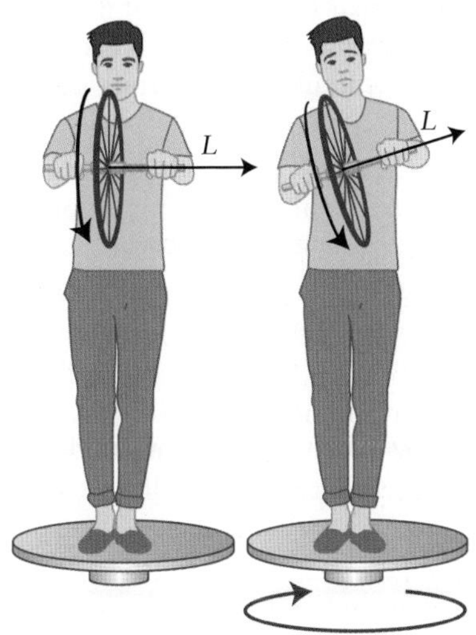

   (a) Evaluate the claim that no external torques are exerted on the student–wheel–platform system and that the angular momentum cannot change.
   (b) The student tilts the wheel as shown in the figure. Construct a representation of the horizontal and vertical components of the angular momentum of the wheel before and after the wheel is tilted.
   (c) Use representations of the vertical components contributing to the total angular momentum of the student–wheel–platform system before and after the wheel is tilted to explain why the platform begins to rotate in the direction shown, clockwise when viewed from above the student, when the student tilts the wheel.
   (d) Explain why the student–wheel–platform system does not rotate about a horizontal axis when the wheel is tipped and predict the direction of rotation of the student–wheel–platform system if the effects you identified in your explanation were absent.

11. Fusion in a main sequence star like the Sun produces helium from hydrogen. The heavier helium migrates to the core of the star. The forces exerted on atoms within the star are the gravitational force pulling inward and the outward pressure of expanding hot gases fueled by fusion reactions of hydrogen. (There is some other important stuff going on, but these will do for a simplified model!) If these are balanced the star is stable. When the amount of hydrogen fuel is insufficient to provide the compensatory outward force, the star will collapse under the gravitational force. If it is a small star, it will become a *dwarf* star and begin to cool. If the material from which the star formed had an initial net angular momentum, then the star will continue to rotate.
   (a) Select a mathematical relationship that will allow you to determine the rotational speed of the star as it collapses. Justify your selection in terms of physical principles.
   (b) Predict, in terms of the initial angular speed, the angular speed of a star whose radius shrinks to one-tenth of its original size. Assume the distribution of the star's mass remains uniform.

12. A thin rod of mass M and length ℓ is pivoted about its center. Attached to the beam at the ends are additional small cubes of mass M and $2M$, as shown in the figure. The figure is a top view of the system, and gravitational forces are into the page and do not cause a torque. The system of rod and cubes is initially at rest. A force F is exerted on the cube of mass $2M$ for a period of time Δt. The force F remains constant and tangential to the path of the cube. The rotational inertia of a rod about its center is $\frac{1}{12}M\ell^2$.
   (a) Derive an expression for the angular momentum of the system after this time interval in terms of the given quantities.
   (b) Derive an expression for the angular speed of the system after this time interval in terms of the given quantities.

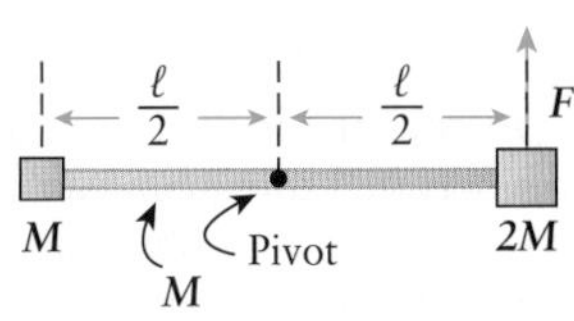

Top view

13. Fidget spinners, small toys designed for entertainment, have rotational inertias on the order of 10^{-3} kg · m^2 and can spin with speeds of about 3000 rpm. Astronauts took a fidget spinner to the International Space Station. They made a video distributed on YouTube in which it appeared that by grabbing the rotating spinner so that the direction of its angular momentum aligned with a rotational axis of their own bodies, they could execute spins and rolls of their bodies as they "floated" in mid-air. What really happened was they grabbed the spinner while they were holding themselves stationary, and then let themselves float and flipped the spinner over along its axis of rotation, reversing its angular momentum. This meant their own angular momentum had to change to keep the angular momentum of the system of spinner and astronaut constant.

(a) Describe how a fidget, modeled as a flat disk around a stationary axis of rotation, should be held and moved for an astronaut to execute rotations about an axis through the center of mass of his body (along his length) in terms of the orientations of the angular momentum of the fidget and the astronaut.

(b) Estimate the angular speed of an initially stationary astronaut if he grabs a fidget spinner rotating at 3000 rpm about the axis you described in part (a). Treat the astronaut as a 60-kg rod with a radius of 0.50 m that will rotate about the long axis of the rod. The rotational inertia of a rod of mass M about the center axis of the rod of radius R is $MR^2/2$.

(c) Estimate the angular speed of a lengthwise spin about an axis through the center of mass of the astronaut's body, perpendicular to his length, caused by flipping the spinner with its axis in a different orientation. Treat the astronaut as a 60-kg cylinder, with a radius of 0.50 m and a length of 1.8 m spinning about an axis perpendicular to the long axis of his body. The rotational inertia of a cylinder of radius R, length ℓ, and mass M rotating about an axis at the midpoint and perpendicular to the cylinder is $MR^2/4 + M\ell^2/12$.

14. A solid sphere of mass M and radius R is rolling without slipping along a horizontal surface. The speed of the center of mass is v_0 when it reaches the bottom of an incline made of the same material as the floor.

(a) Derive an expression for the maximum height, h, of the point on the incline reached by the sphere. The rotational inertia of a sphere about its center of mass is $\frac{2}{5}MR^2$.

(b) Now assume that the incline is an air table and the friction between the incline and the sphere is negligible. Assume the sphere again reaches the bottom of the incline with a speed v_0. Is the height to which the sphere rises on this new incline greater than, less than, or the same as the height, h, from part (a)? Justify your choice.

15. A planet orbits a star. The planet's orbital radius is 1.00 au. If the star has a mass that is 1.75 times our own Sun's mass, determine the period, in days, for one revolution of the planet around the star.

16. The four largest of Jupiter's moons are listed in the table. Use the data already given in the table to complete the table and determine the mass of Jupiter.

Moon	Semimajor axis (km)	Orbital period (days)
Io	421,700	1.769
Europa	671,034	
Ganymede		7.155
Callisto		16.689

17. In a new model of a machine, a spinning solid spherical part of radius R must be replaced by a ring of the same mass, which is to have the same rotational kinetic energy. Both parts need to spin at the same rate, the sphere about an axis through its center and the ring about an axis perpendicular to the plane parallel to its center.

(a) What should the radius of the ring be in terms of R?

(b) Will both parts have the same angular momentum? If not, which one will have more?

18. It is estimated that 60,000 tons of meteors and other space debris accumulate on Earth each year. Assume the debris is accumulated uniformly across the surface of Earth.

(a) How much does Earth's rotation rate change per year as a result of this accumulation? (That is, find the change in angular velocity.)

(b) How long would it take the accumulation of debris to change the rotation period by 1 s? Neglect any change in radius of Earth due to this additional material.

19. In a little over 5 billion years, our Sun will collapse to a *white dwarf* approximately 16,000 km in diameter. (Ignore the fact that the Sun will lose mass as it ages.)

(a) What will our Sun's angular momentum and rotation rate be as a white dwarf? (Express your answers as multiples of its present-day values.)

(b) Compared to its present value, will the Sun's rotational kinetic energy increase, decrease, or stay the same when it becomes a white dwarf? If it does change, by what factor will it change? The radius of the Sun is presently 6.96×10^8 m.

20. If all the people in the world (~7 billion) lined up along the equator, would Earth's rotation rate increase or decrease?

(a) Justify your answer in a paragraph-length explanation.

(b) How would the rotation rate change if all people were no longer on Earth? Assume the average mass of a human is 70.0 kg.

21.

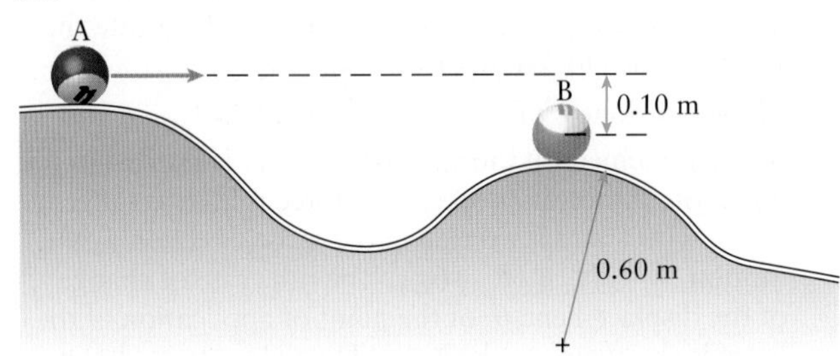

A billiard ball of mass 160 g and radius 2.50 cm starts with a translational speed of 2.0 m/s at point A on the track, as shown. If point B is at the top of a hill that has a radius of curvature of 0.60 m, what is the normal force exerted on the ball at point B? Assume the billiard ball rolls without slipping on the track.

22. A uniform, solid sphere of radius $r = 5.00$ cm and mass $m = 3.00$ kg starts with a translational speed of $v_i = 2.00$ m/s at the top of an inclined plane that is $L = 2.00$ m long and tilted at an angle of $\theta = 25.0°$ with the horizontal and rolls without slipping down the ramp.

(a) Construct a representation of the motion on the ramp, identifying initial and final positions and the location of the reference from which potential energy of the Earth–sphere system is measured.

(b) Express the change in potential energy as the sphere moves from the initial to the final position in terms of L, θ, m, and g.

(c) Express the change in kinetic energy as the sphere moves from the initial position to the final position in terms of v_i, v_f, I, r, and m.

(d) The rotational inertia of a sphere rotating about an axis through its center is $2mr^2/5$. Express the final speed of the sphere in terms of L, θ, g, and v_i.

(e) Calculate the sphere's speed at the bottom of the ramp.

23.

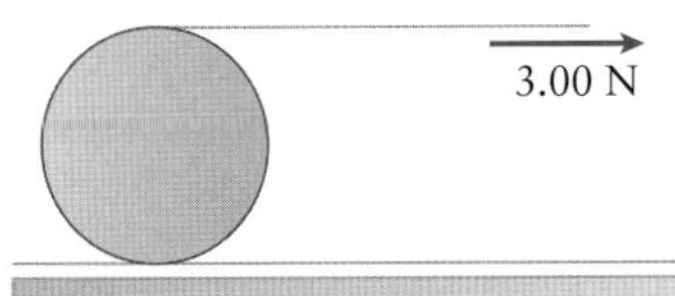

The figure shows a solid, uniform cylinder of mass 7.00 kg and radius 0.450 m with a light string wrapped around it. A 3.00-N tension force is applied as shown to the string, causing the cylinder to roll without slipping across a level surface that exerts a static friction force, F_S, on the cylinder. The rotational inertia of a solid cylinder rotating about its center is $MR^2/2$.

(a) Calculate the linear acceleration of the center of mass of the cylinder.

(b) Calculate the magnitude and direction of the angular acceleration of the cylinder.

(c) Calculate the magnitude and direction of the friction force that is exerted on the cylinder.

24. A flywheel of mass 35.0 kg and diameter 60.0 cm spins at 400 rpm when it experiences a sudden power loss. The flywheel slows due to friction in its bearings during the 20.0 s the power is off.

(a) Express the angular displacement following the power failure in terms of the angular acceleration of the flywheel and the time, Δt, while the power is off.

(b) The flywheel makes 102 complete revolutions during the power failure. Calculate the angular acceleration of the flywheel.

(c) Calculate the angular speed of the flywheel when the power comes back on.

(d) Calculate the number of joules of kinetic energy converted to internal energy while the power was off. The rotational inertia of a solid disk rotating about its central axis is $mr^2/2$.

25. A solid disk of mass M and radius R is initially at rest just above another solid disk that is rotating with an angular speed ω_{lower} about an axis that passes through the centers of mass of both disks, as shown in the figure. The lower disk has mass $2M$ and radius $2R$. The upper disk is released and collides with the lower disk, and the two eventually rotate with the same final angular speed, ω_{both}. What fraction of the kinetic energy of the lower disk is dissipated in this collision?

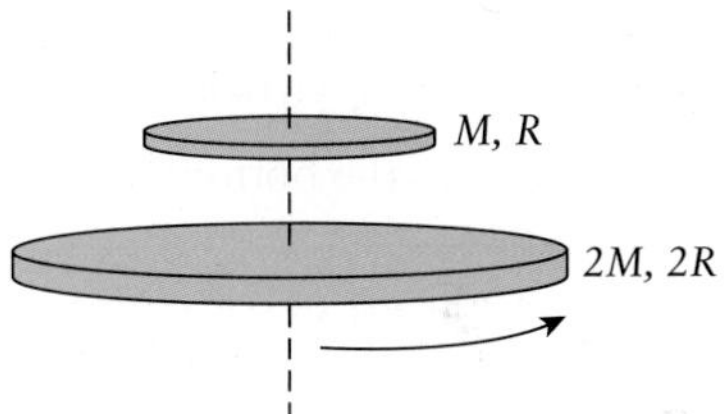

26. You are asked to do an experiment to study the motion of a disk of mass M and radius R, which spins about the axle at its center, as shown. Your measurements show that, when given an initial positive angular velocity, ω_0, the disk comes to rest at a steady rate. It comes to rest at a time t_f.

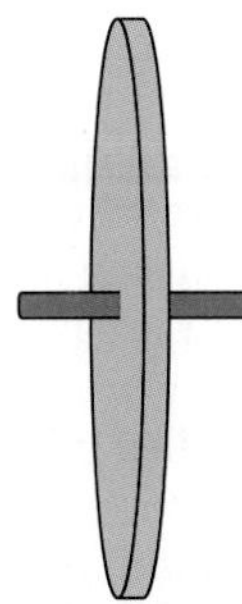

(a) Sketch graphs of the disk's angular velocity and angular acceleration as functions of time, from $t = 0$ to t_f.

(b) Calculate the average torque exerted by the axle on the disk. Describe how this compares to the torque exerted by the axle on the disk as a function of time.

(c) Describe the size and direction of the force of friction responsible for the torque in terms of the average torque and the radius of the axle.

You decide to do the experiment over again, first covering the axle with graphite to reduce the friction. You again collect data to graph the angular velocity and acceleration of the disk. The graphite does make friction negligible until a time t_1, and then the friction increases with time. The disk comes to rest at a time t_2.

(d) Sketch graphs of the disk's angular velocity and angular acceleration as functions of time, from $t = 0$ to t_2. Make sure t_1 is clearly marked on your graphs.

27. A 1.00×10^3-kg merry-go-round (a flat, solid cylinder) is spinning, while supporting 10 acrobats, each with a mass of 50.0 kg. Initially the acrobats support each other to form a column located very close to the axis of rotation (thus you may assume the acrobats have no angular momentum at that location). Describe a plan to move the acrobats such that the angular velocity of the merry-go-round decreases to one-half its initial value.

28. Three uniform disks are released from rest at the top of an incline. Disk A has mass M and radius R. Disk B has mass M and radius $2R$. Disk C has mass $2M$ and radius R. All three disks roll without slipping to the bottom of the incline. The rotational inertia of a disk about its center of mass is $\frac{1}{2}MR^2$.

(a) Which disk reaches the bottom of the incline first? Justify your choice.

(b) Which disk has the greatest rotational energy at the bottom of the incline? Justify your choice.

29.

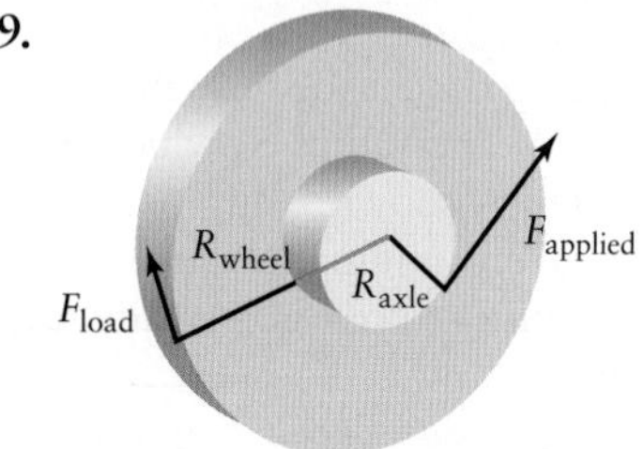

Electric power is supplied to the grid almost entirely by turbines. A turbine is a device used to turn a magnet within a coil of wire when hot water vapor strikes the turbine's blades, causing it to rotate. Hydroelectric power and wind power are produced by the conversion of streams of water or air into the rotation of a wheel. Power is supplied by internal combustion engines with the conversion of the up-and-down translation of a piston into the rotation of a crankshaft by the tension in the rod connecting the piston to the crank. In each of these devices the design begins with the ratio of torque due to an external force and the torque due to the force exerted by the load of the wheel that is driven.

The torque due to the force exerted by the load (F_{load}) opposes the torque due to the external force ($F_{applied}$). In the figure, the torque due to $F_{applied}$ is positive (it would produce counterclockwise rotation) and the torque due to F_{load} is negative (it would rotate the wheel in the clockwise direction).

The work done by $F_{applied}$ is the product of $F_{applied}$ and the displacement of the point at which it is exerted in the direction of motion. Because the force is always tangent to the wheel, and is exerted in the same direction as the motion, $W = F_{applied} R_{axle} \theta_{axle}$. The time rate of change of the displacement, θ, of the point of application of the force is the angular velocity, ω. The time rate of change of energy of the wheel due to the work done by $F_{applied}$ is the power supplied, P_{input}.

(a) Using these definitions, justify the claim that

$$P_{input} = \tau_{applied}\omega_{axle}$$

When energy is transferred from the wheel to the load, an angular velocity on the load side results. Then

$$P_{output} = \tau_{load}\omega_{wheel}$$

A fraction, ε, of the input power converted to output power is

$$P_{output} = -\varepsilon P_{input}$$

(b) Justify the claim that, when the wheel and axle in the figure can be treated as a rigid body in equilibrium, the transfer is perfect and $\varepsilon = 1$.

30.

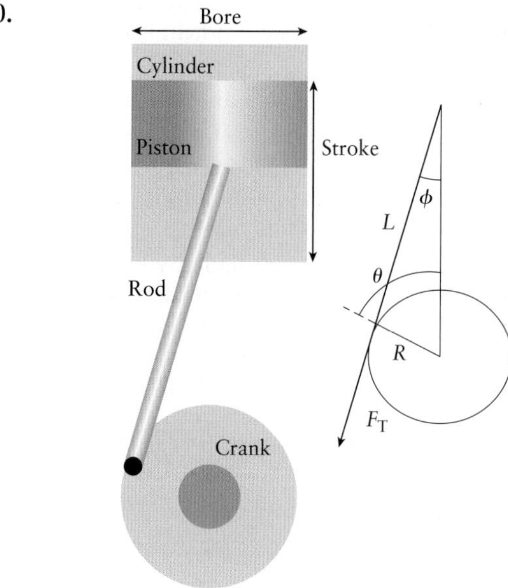

The diagram illustrates the operation of the piston-crank that converts the internal energy of gasoline into the rotational energy of the crank. Through a series of linkages, the rotational motion of the crank and crankshaft on which the crank is fixed is converted into the rotational motion of the wheel of a car.

The cylinder can be closed by a valve above the piston. In the configuration shown, a mixture of fuel and air is injected into a closed cylinder above the piston's upper face. The mixture is ignited by a spark, expanding the gas in the cylinder and exerting a downward force on the face of the piston. The piston exerts a force on the rod and the rod exerts a torque on the crank.

(a) Express the angle that the displacement $\vec{R}$ from the rotational axis to the point at which the force is exerted makes with the force exerted by the rod in terms of ϕ and θ. *Hint*: The interior angles of a triangle sum to 180°, or π in radians.

(b) Express the torque exerted on the crank in terms of the tension, F_T; the radius of the crank, R; and the angles ϕ and θ.

(c) What is called the power stroke of the engine begins with the ignition of the fuel and ends when the torque is zero. Explain why the power stroke ends when the angle $\phi = 0°$.

(d) Describe in terms of work why F_T causes the largest change in energy of the system when F_T and R from the crank axis to the rod are perpendicular.

Prep for the AP® Exam

AP® Group Work

Directions: The problem below is designed to be done as group work in class.

A wheel supported by a rope tied to the ceiling is attached to a hub, which extends through the center of the wheel, as shown on the left in the figure below. The wheel is then held so that it is upright, as shown on the right in the figure below, and spun rapidly. The wheel is released and remains upright.

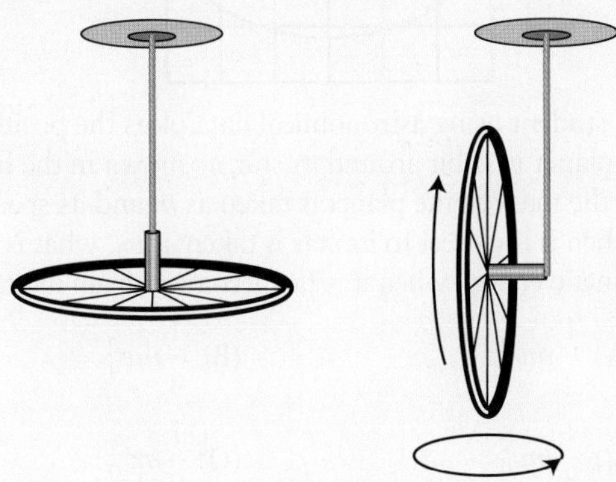

(a) Make a diagram of the wheel when it is upright as you view it from the front, with the hub and rope as shown in the figure above. The tangential velocity at the top of the wheel is into the page and the tangential velocity at the bottom of the wheel is out of the page. Add a vector representing the angular momentum of the wheel to your diagram.

(b) The gravitational force is exerted on the wheel at the center of mass of the wheel. The rotational axis for a torque caused by the gravitational force is the connection of the hub and the rope. The lever arm for this torque then lies along the hub shaft connecting the center of mass of the wheel and the rope. Construct a representation to show the direction of the torque caused by gravity.

When the wheel is spinning and upright, the hub will rotate in a horizontal plane around the point of contact between the rope and the hub shaft. As it rotates, the angular momentum of the wheel changes direction. The angular momentum is constantly turning in the direction of the torque. This motion is called precession.

(c) Although precession is a complicated motion, we can use an analogy to begin to make sense of this fascinating gyroscopic effect (and it's helpful to enter this term into a search engine to see many examples). Summarize how the precession of the spinning wheel caused by a torque exerted about an axis of rotation is similar to the uniform circular motion of an object moving in a circular path caused by a central force exerted on the object.

AP® PRACTICE PROBLEMS

Prep for the AP® Exam

Multiple-Choice Questions

Directions: The following questions have a single correct answer.

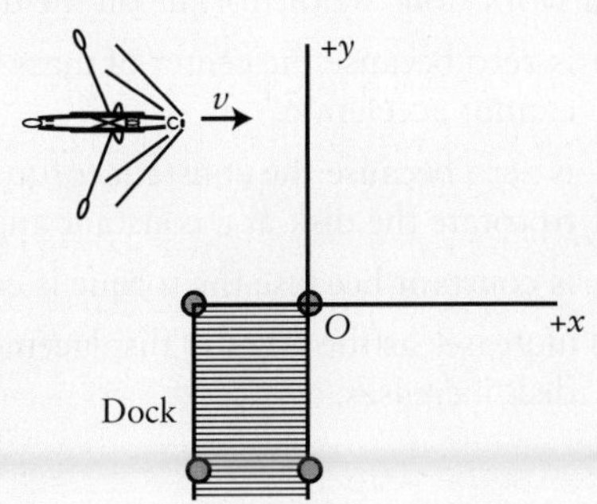

1. A person rows a boat in a line, keeping it moving at a constant speed v. Identify the correct graph of the angular momentum L for the boat relative to the origin O at a corner of the dock as shown to the left.

(A) L vs. $+x$ (0,0)

(B) L vs. $+x$ (0,0)

(C) L vs. $+x$ (0,0)

(D) L vs. $+x$ (0,0)

(Continued)

Questions 2–3 refer to the following material.

A thin ring of mass $m = 0.20$ kg and radius $r = 0.050$ m is initially at rest on a smooth horizontal tabletop that exerts negligible friction on the ring, as shown in the figure above. A string, wound around the ring, is pulled so that it exerts a constant 0.40-N force F tangent to the ring and is directed to the left of the picture as the string unwinds for 2.0 s. You may neglect the mass of the string.

2. What will be the translational momentum of the center of mass of the ring at $t = 2.0$ s?
   (A) 0.80 kg · m/s
   (B) 0.40 kg · m/s
   (C) 0.20 kg · m/s
   (D) 0.00 kg · m/s

3. What will be the angular momentum of the ring about its center at 2.0 s?
   (A) 0.080 kg · m^2/s
   (B) 0.040 kg · m^2/s
   (C) 0.010 kg · m^2/s
   (D) 0 kg · m^2/s

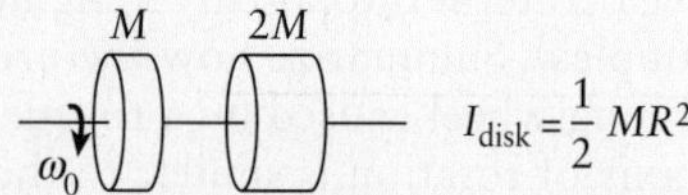

4. Two disks with the same radius R are free to rotate about a common axle with negligible friction, as shown in the figure above. One disk with a mass M is initially spinning freely with an angular speed ω_0. The other with a mass $2M$ is at rest. The rotating disk is then moved against the stationary disk and friction gradually causes the disks to rotate together. What will be the final angular momentum of the disk of mass $2M$?
   (A) $\frac{MR^2\omega_0}{3}$ (B) $\frac{MR^2\omega_0}{2}$
   (C) $\frac{2MR^2\omega_0}{3}$ (D) $\frac{3MR^2\omega_0}{2}$

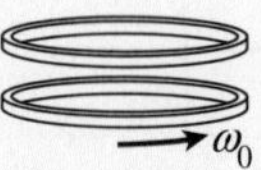

5. Two identical thin rings are shown above. The lower ring, supported by very thin spokes of negligible mass (not shown), spins freely about its vertical axis with an initial angular speed ω_0. The other ring, initially stationary, is then lowered on top of the first and friction forces cause the two rings to spin together. Which of the following is a correct expression for the kinetic energy of the system of two rings after they are spinning together?
   (A) $I\omega_0^2$ (B) $\frac{I\omega_0^2}{2}$
   (C) $\frac{I\omega_0^2}{4}$ (D) $\frac{I\omega_0^2}{8}$

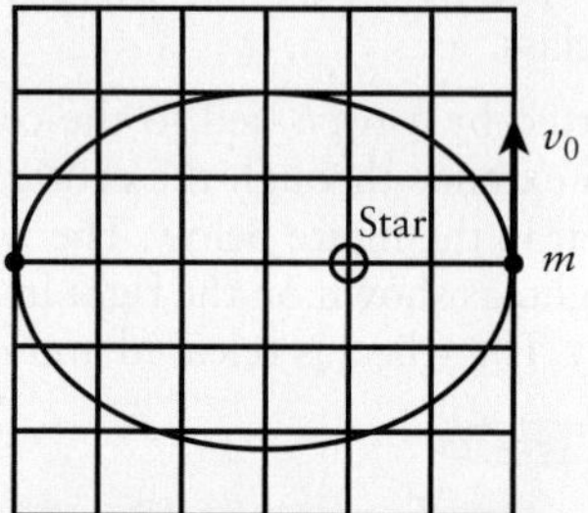

6. A student using astronomical data plots the position of a planet in orbit around its star, as shown in the figure. If the mass of the planet is taken as m and its speed when it is closest to its star is taken as v_0, what is its kinetic energy when it is farthest away from its star?
   (A) $\frac{1}{2}mv_0^2$ (B) $\frac{1}{3}mv_0^2$
   (C) $\frac{1}{4}mv_0^2$ (D) $\frac{1}{8}mv_0^2$

7. Three disks are rolling without slipping along a horizontal surface with a center-of-mass speed v_0. They all reach the bottom of an incline and roll up the incline. Disk A has mass M and radius R. Disk B has mass M and radius $2R$. Disk C has mass $2M$ and radius R. Which of the following is the correct ranking of the disks based on the maximum height they reach on the incline?
   (A) A = B = C (B) B > (A = C)
   (C) (A = C) > B (D) (A = B) > C

8. A constant external torque τ is exerted on a solid disk of mass M and radius R. The disk is free to rotate about a fixed axle with negligible friction (it can rotate but its center of mass cannot accelerate). The work done by the torque on the disk
   (A) is zero because the center of mass of the disk cannot accelerate.
   (B) is zero because the constant torque is necessary to rotate the disk at a constant angular speed.
   (C) is constant because the torque is constant.
   (D) increases as the angular displacement of the disk increases.

Free-Response Questions

1. A disk of mass M and radius R has a rocket motor attached to its edge. Assume the rocket motor has negligible mass compared to the disk. The disk is free to rotate with negligible friction about an axis through its center of mass perpendicular to the disk. The rocket motor fires, causing the disk to begin to rotate about this axis. The rocket, while it is firing, provides a constant force of F_0, tangent to the edge of the disk, for a time Δt. The rotational inertia of a disk of mass M and radius R is $\frac{1}{2}MR^2$.

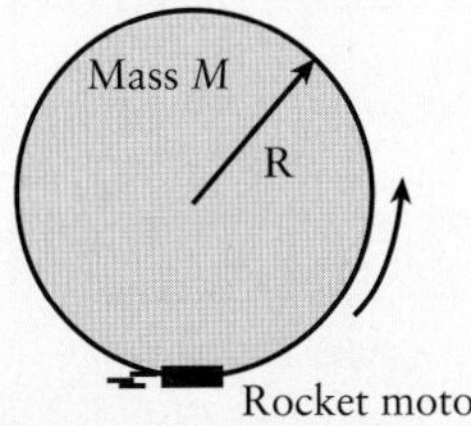

(a) Derive expressions, in terms of the given quantities and physical constants, for (i) the change in angular momentum of the disk, and (ii) the angular speed of the disk after the rocket has fired.

The rocket motor is now adjusted so that it is at an angle to the edge of the disk, pointed inward. Assume the rocket fires for the same amount of time and applies a force of the same magnitude.

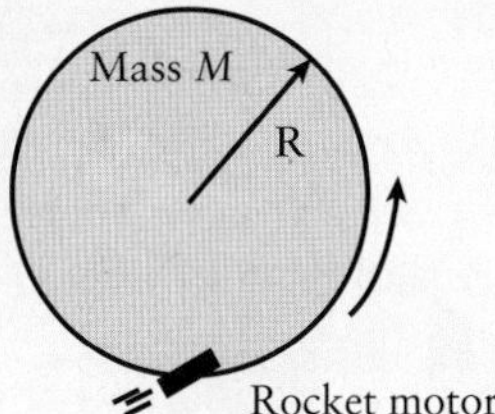

(b) On the axes below, sketch graphs of the angular speed as a function of time for the two systems described. Clearly label them so they may be distinguished.

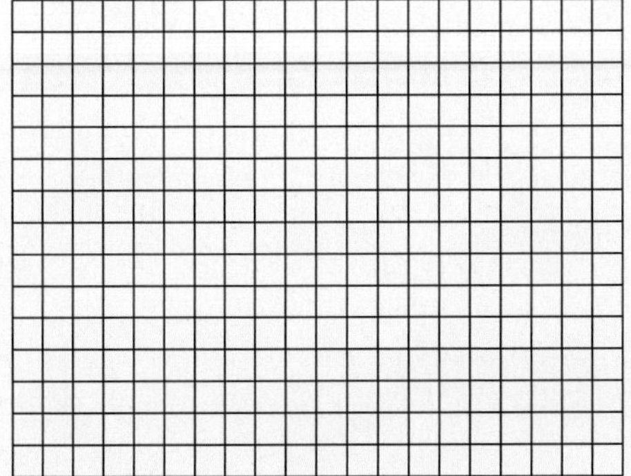

(c) How do the angular speeds for the disk for the two systems shown in your graph in (b) relate to your answer in (a)? Justify your answer.

2.

Case 1 Case 2

v_0 Clay $\frac{\ell}{4}$ Axis — Top view

v_0 Ball $\frac{\ell}{4}$ Axis — Top view

A long, thin rod of mass $3m$ and length ℓ is pivoted around a vertical axis and rotates in a horizontal plane. A top view of the rod is shown above. In case 1, a lump of clay of mass m is launched toward the rod with a speed v_0. It strikes the rod and sticks to it a distance $\ell/4$ from the axis of rotation. The rotational inertia of a long, thin rod of mass M and length l is $\frac{1}{12}Ml^2$.

(a) i. Derive an expression for the angular velocity of the clay–rod system after the collision, in terms of the given quantities and physical constants, as appropriate.

ii. What fraction of the initial kinetic energy of the clay is dissipated in the collision with the rod?

In case 2, a rubber ball of the same mass is launched at the rod with the same speed as the clay. It strikes the rod and bounces back in the opposite direction with a speed $v_f < v_0$.

(b) On the axes below, sketch graphs of torque on the rod as a function of time for each of the two cases. Clearly label your graphs so they may be distinguished.

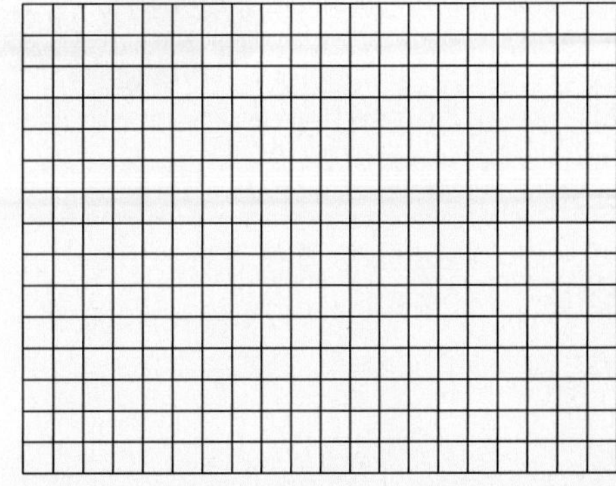

(c) Is the angular momentum of the rod greater in case 1, greater in case 2, or the same in both cases?

_____ $L_2 > L_1$ _____ $L_2 < L_1$ _____ $L_2 = L_1$

Justify your selection.

Oscillations

Chapter 12 Oscillations Including Simple Harmonic Motion

Dmitry Feoktistov/Getty Images

Case Study: At what point during a bungee jump will you reach your greatest speed?

If Chloe were just hanging on the bungee cord, she would remain stationary at the point where the forces exerted on her by the stretched cord and gravity exactly balance. This position of zero net force is her equilibrium position. But when she falls toward this point, inertia causes her to overshoot it. The more the bungee cord is stretched the greater the force it exerts. When Chloe is *below* the equilibrium position, the cord exerts a force greater than the gravitational force: She is pulled upward. She slows and the moves back toward equilibrium, but inertia again causes her to overshoot. When Chloe is *above* equilibrium, the cord is less stretched, and exerts a force less than the gravitational force. The net force on Chloe is downward. In each case, the net force is opposite the direction from Chloe's displacement from equilibrium. We call a force that is always directed toward the equilibrium position a *restoring force*. Chloe's oscillating motion, up and down around the equilibrium, is the result of this *interplay between a restoring force and inertia*.

Chloe is moving most quickly as she passes through the equilibrium. Her kinetic energy is therefore at a maximum there, and we define this as the zero of potential energy for the Earth–Chloe–cord system. She comes to a stop at the end points of her motion as she reverses direction. At those points, then, all energy is gravitational and spring potential. In this way *energy is transformed back and forth from kinetic energy to potential energy* in systems that oscillate. In this chapter you'll discover that the period of systems like this depends only on the stretchiness of the cord and Chloe's mass, as long as the cord is not stretched too much. The harder the cord is to stretch, the shorter the period. Chloe makes another jump, now while holding her heavy backpack. Do you think the period of her motion will increase, decrease, or remain the same?

By the end of this chapter, you should be able to explain key properties of a common form of oscillating motion called simple harmonic motion and what conditions are necessary for simple harmonic motion to occur. You should be able to predict both qualitatively and quantitatively how kinetic energy and potential energy vary during an oscillation of a system. You should be able to explain what properties determine the period, frequency, and angular frequency for simple harmonic motion and how these characteristics of the motion depend on those properties.

12 Oscillations Including Simple Harmonic Motion

Heri Mardinal/Getty Images

YOU WILL LEARN TO:

- Define oscillation and give everyday examples of systems that oscillate.
- Explain, qualitatively and quantitatively, characteristics of oscillations, including the key properties of simple harmonic motion, and what is required for simple harmonic motion to occur.
- Qualitatively describe a system using tensile stress and compressive stress and describe the relationship to Hooke's law.
- Explain the connection between Hooke's law and simple harmonic motion, and be able to calculate velocity and acceleration for any given displacement from equilibrium of an object oscillating on a spring.
- Explain what properties determine the period, frequency, and angular frequency of an object oscillating on a spring and how these characteristics of the motion depend on those properties.
- Predict, qualitatively and quantitatively, how kinetic energy and potential energy vary during an oscillation of a system and relate these changes to changes in the internal structure of the system.
- Explain what properties determine the period, frequency, and angular frequency of a simple pendulum and how these characteristics of the motion depend on those properties.
- Explain what properties determine the period, frequency, and angular frequency of a physical pendulum and how these characteristics of the motion depend on those properties.

12-1 We live in a world of oscillations

NEED TO REVIEW?
Turn to the **Glossary** in the back of the book for definitions of bolded Key Terms.

Try sitting by yourself in your room and keeping absolutely still. If no one else is in the room, it may seem as though there is no motion anywhere around you. Not so! Your heart is in continuous motion, pulsing at a rate of about one cycle per second. Your eardrums are also vibrating softly in response to the sound of your breathing. And in the electrical wires within the walls of your room, electrons move back and forth about 60 times per second as part of the processes that supply energy to the room lights, your computer, and other appliances. These are just three examples of a kind of motion called **oscillation**, in which an object moves back and forth around a position of *equilibrium*—that is, a point at which it experiences zero net force. (We introduced the idea of equilibrium in Section 4-3.) An object swinging back and forth on the end

(a)

The wings of common hummingbirds oscillate up and down about 50 times per second.

(b)

The timekeeper at the heart of most wristwatches is a tiny piece of quartz that oscillates 32,768 times per second.

(c)

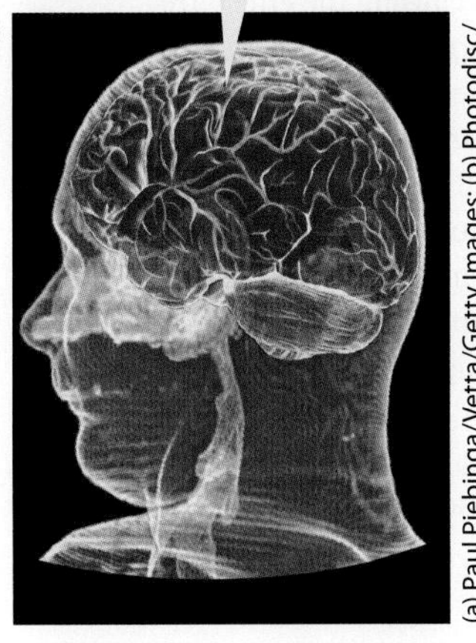

This MRI image is made by mapping how protons oscillate within molecules in the brain in response to radio waves.

(a) Paul Piebinga/Vetta/Getty Images; (b) Photodisc/Thinkstock/Getty Images; (c) Zephyr/Science Source

Figure 12-1 Oscillations The kind of motion that we call oscillation is commonplace (a) in nature, (b) in technology, and (c) in medicine.

of a string is in equilibrium when it is hanging straight down (the force of gravity exerted on the object is exactly balanced by the tension force exerted by the string). **Figure 12-1** shows three examples of oscillations in nature and technology. Oscillations have a **cycle**, that is, they occur as a repeating pattern. The time for one complete cycle of an oscillation is called the **period** of the oscillation.

We'll begin this chapter by studying the causes of oscillation. We'll find that oscillation is possible only when an object experiences a *restoring force*—a force that always pushes or pulls the object toward the equilibrium position. Oscillations are particularly simple in nature if the restoring force is directly proportional to how far the object is displaced from equilibrium. This proves to be an important special case that's a very good approximation to many kinds of oscillation.

As for other kinds of motion, we'll find that it's useful to describe oscillations in terms of kinetic and potential energy. The energy approach will also help us to understand what happens to make the oscillations of a swinging pendulum die out so that the pendulum eventually ends up at rest. It is the energy in an oscillation that is responsible for the destruction due to the oscillatory motion of the ground during an earthquake, as shown in the photograph that opens this chapter.

THE TAKEAWAY for Section 12-1

✔ Oscillation is a kind of motion in which an object moves back and forth around an equilibrium position.

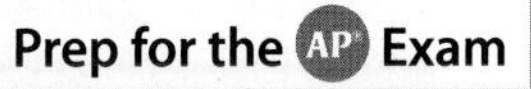

AP® Building Blocks

1. The rise and fall of your lungs and the beating of your heart are oscillations. Compare the frequency of breaths (number of breaths per second) with the frequency of heartbeats while sitting quietly for one minute. Does there appear to be a relationship? Pose scientific questions whose pursuit could reveal a cause of the relationship if one exists or reveal the isolation of these periodic behaviors if no relationship exists.
2. Give an everyday example of an oscillating system, and explain why you described that system as oscillating and what interaction makes it oscillate.
3. The ENSO cycle (El Niño–Southern Oscillation) describes how the subsurface temperature of the Pacific Ocean changes. The following figure indicates cold (La Niña) episodes with blue shading and warm (El Niño) episodes with red shading for one region of the Pacific Ocean.

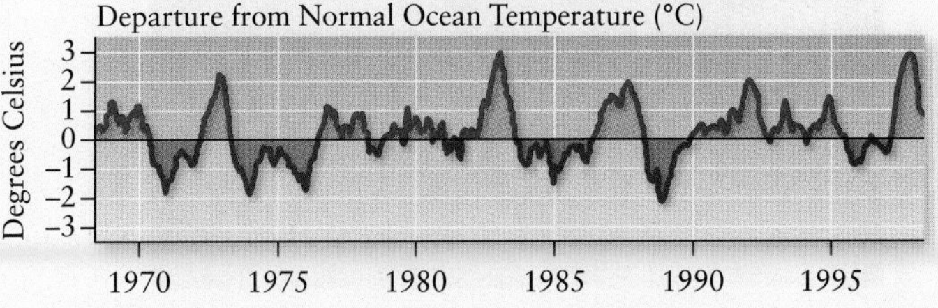

(a) What is "equilibrium" in the graph shown, and what does it represent?
(b) The cold and warm phases alternate but the period is not regular. The mechanisms causing the effect are not understood. On Earth, periods of day follow periods of night with regularity. The cycle of the seasons periodically repeats. The mechanisms that cause these effects are understood. Why is it easier to understand the mechanisms that cause periodic events than it is to understand the causes of seemingly random events, such as the period of a warm phase in the ENSO cycle?
(c) Earthquakes are presently thought to be random and unpredictable. Recently, data for earthquakes with magnitudes greater than 5 near Earth's surface were analyzed and nonrandomness was observed in earthquake frequency (see graph).

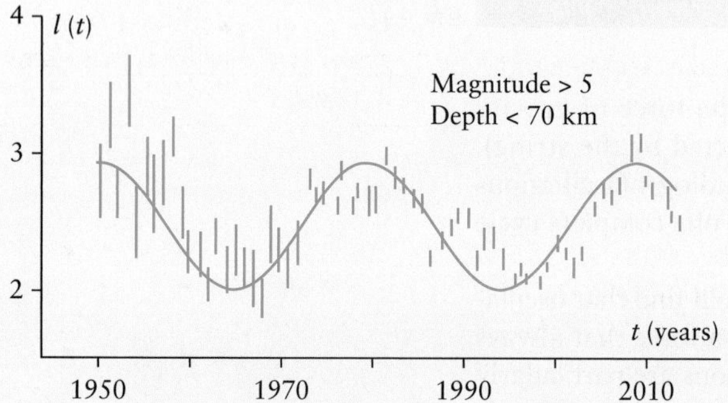

Researchers search for other phenomena with occurrences repeating between 30.98 and 42.20 years (the periods of whole Earth and deep Earth data fits), including motion of the mantle in the interior of Earth and variations in Earth's rotational cycle, the Moon's orbit, and Earth's precession. What is the question that these researchers are pursuing?

4. One of the greatest achievements of astrophysics was the collection of data that could be used to test Copernicus's claim of a Sun-centered solar system. These data were provided by Galileo's observation of the cyclic appearance of Venus, which can be seen in the recent photographs, as shown.

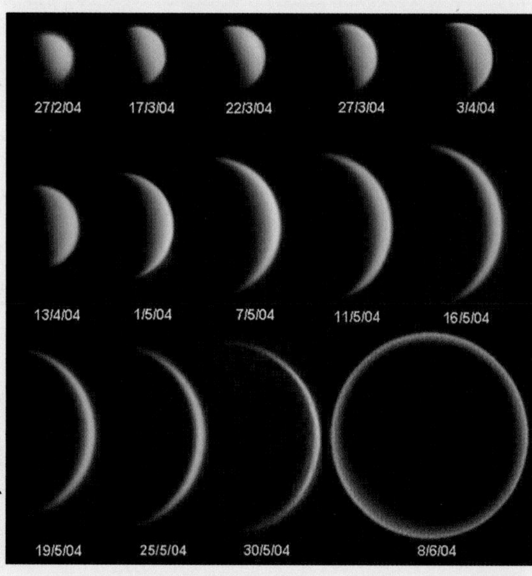

(a) In the photographs, the third-quarter phase of Venus occurred near 3/17/2004 (77th day of the leap year) and the new phase occurred near 6/8/2004 (the 160th day). (Instead of our month/day/year convention, Europeans record time as day/month/year, which is used in the photographs.) Estimate the period of the phases of Venus as observed from Earth.

The orbital period of Venus from a fixed point in the solar system is 225 days, much smaller than the period of phases of Venus you probably estimated in part (a). This is because the observation point on Earth is also moving as Venus moves. It is also much smaller than the full observed cycle, 584 days, because the cycle also depends on the time it takes Venus to overtake Earth in its orbit. Venus is at full phase when it is on the opposite side of the Sun from Earth. Venus is a thin crescent as viewed from Earth as it comes around between Earth and the Sun. Its new phase is when it is directly between Earth and the Sun, and its atmosphere appears in a telescope as a halo of light.

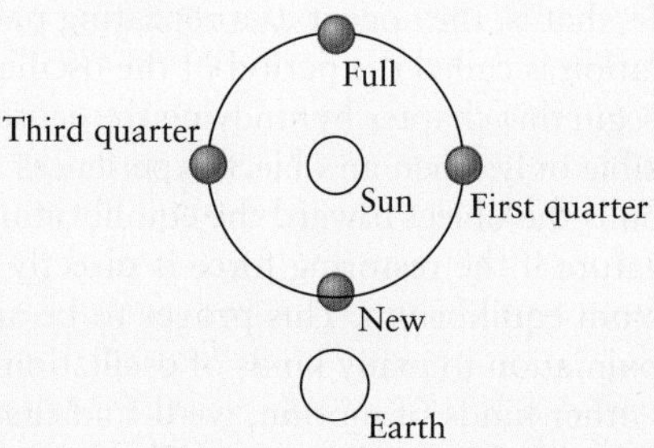

(b) Venus is an inferior planet, which means that its orbital radius is less than the orbital radius of Earth. Use the photographic data to justify the representation above of the orbit of Venus as observed from Earth and use the diagram to explain why Galileo's observations provided evidence of a Sun-centered solar system, as proposed by Copernicus, rather than an Earth-centered system in which the Sun orbited Earth with a period of 24 hours.

12-2 Oscillations are caused by the interplay between a restoring force and inertia

Figure 12-2 shows one of the simplest ways to make an object oscillate. A block is attached to a horizontal spring and is free to slide back and forth. (For simplicity we'll neglect friction between the block and the surface over which it slides.) We'll call the horizontal direction along which the block can move the x axis and let $x = 0$ be the point at which the spring is relaxed (**Figure 12-2a**). Then the quantity x represents the *position* of the block as well as the *displacement* of the block from equilibrium.

When the spring is stretched, corresponding to $x > 0$, it exerts a force on the block that points in the negative x direction, so $F_x < 0$ (**Figure 12-2b**). If instead the spring is compressed, corresponding to $x < 0$, it exerts a force on the block that points in the positive x direction, so $F_x > 0$ (**Figure 12-2c**). In other words, no matter which way the block is displaced from equilibrium, the spring exerts a force that tends to pull or push the block back toward the equilibrium position. A force of this kind is called a **restoring force**.

To start the block in Figure 12-2 oscillating, we move the block so as to stretch the spring by a distance A and then release it (**Figure 12-3a**). The restoring force of the spring pulls the block back toward the equilibrium position at $x = 0$, so the block accelerates and continues to gain speed as it moves back toward equilibrium (**Figure 12-3b**). The block reaches its maximum speed as it passes through equilibrium (**Figure 12-3c**). Even though there is zero net force on the block at the equilibrium position, the block's inertia causes it to overshoot this point. As a result the block compresses the spring, so the restoring force is now directed opposite to the block's motion and slows the block down (**Figure 12-3d**). Thus, the block eventually comes momentarily to rest with the spring compressed (**Figure 12-3e**). The restoring force of the spring is still pushing toward the equilibrium position, so the block again starts moving and gains speed as it heads back toward the equilibrium position (**Figure 12-3f**). The object reaches equilibrium a second time and again overshoots the equilibrium position (**Figure 12-3g**). The block stretches the spring as it overshoots, and the restoring force causes the block to slow down (**Figure 12-3h**). The block ends up once again at its initial position with zero speed (**Figure 12-3i**). The cycle then repeats. If there were no friction or air drag, the oscillation would go on forever!

Not every oscillation involves a spring, but *every* oscillation follows the same general sequence of steps shown in Figure 12-3. This figure shows that the two key factors for oscillation are a restoring force that always draws the oscillating object back toward the equilibrium position and inertia that causes the object to overshoot equilibrium.

Oscillation Period and Frequency

We'll use the symbol T for the period of an oscillation. It makes sense that a stiffer spring (which will exert a stronger restoring force) will push and pull the block more rapidly through the steps shown in Figure 12-3 and so will cause the period to be shorter. Similarly, a block with more inertia (more resistance to acceleration) will go through the steps more slowly, causing the period to be longer. We'll verify that these ideas are correct later in this section.

WATCH OUT!

When working in one dimension, do not forget that force and position are vectors.

The motion we will discuss in oscillations will be constrained to one dimension, so one vector component is enough to describe vector quantities: A sign is sufficient to designate the direction of the vector in one-dimensional motion. Do not interpret the lack of arrows to imply that these are not vectors.

(a)

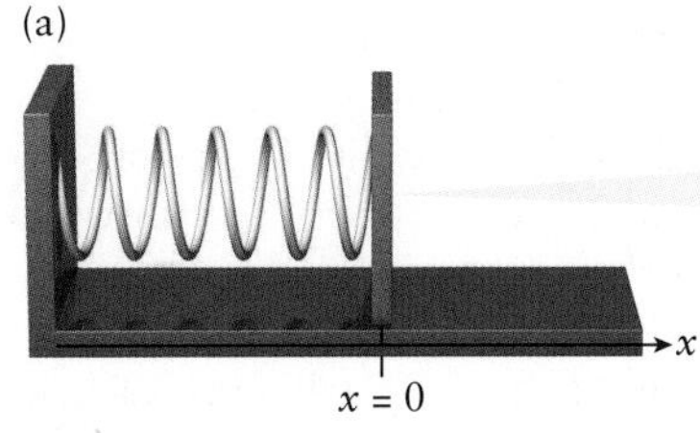

(1) The object at the free end of the spring is in equilibrium when the spring is neither stretched nor compressed. We call this point $x = 0$.

(b)

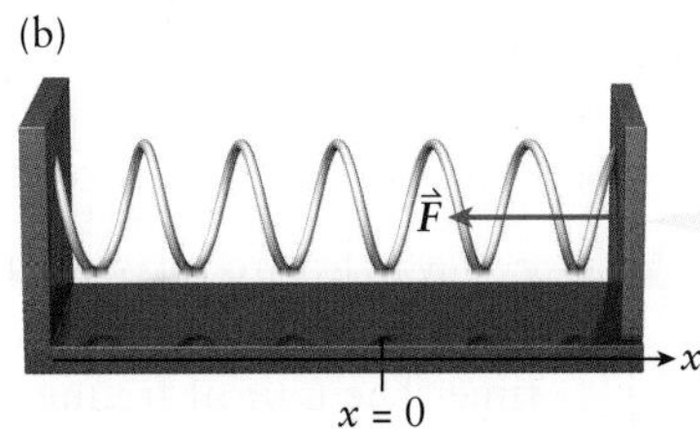

(2) When the spring is stretched the object is at $x > 0$. The spring force pulls the object back toward equilibrium ($F_x < 0$).

(c)

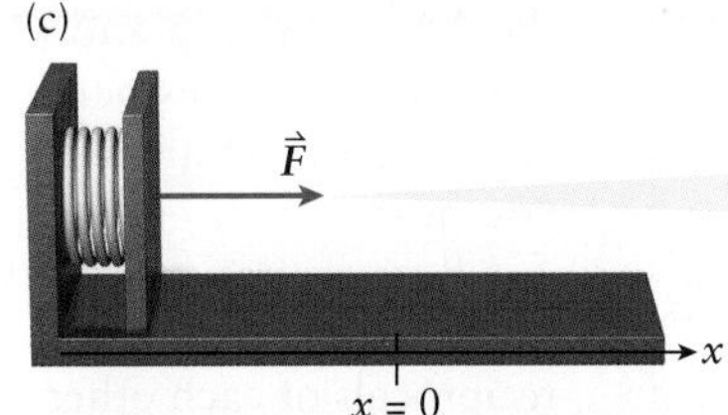

(3) When the spring is compressed the object is at $x < 0$. The spring force pushes the object back toward equilibrium ($F_x > 0$).

The force exerted by the spring is a restoring force: No matter which way the object is displaced from equilibrium, the force exerted by the spring on the object is always in the direction that returns the object to equilibrium.

Figure 12-2 A restoring force In order for oscillations to take place, the force must always be exerted on the oscillating object in the direction toward equilibrium.

Figure 12-3 A cycle of oscillation Nine stages in the oscillation of a block attached to a spring.

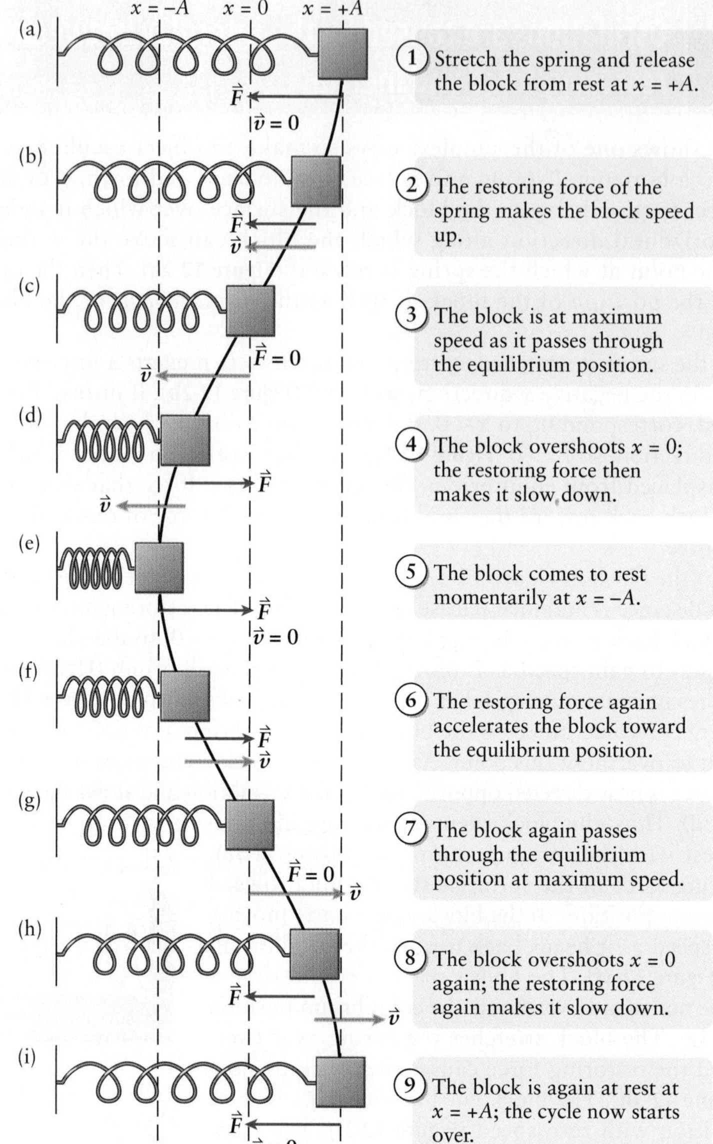

x = −A
x = 0
x = +A
(a)
(b)
(c)
(d)
(e)
(f)
(g)
(h)
(i)
1 Stretch the spring and release the block from rest at x = +A.
2 The restoring force of the spring makes the block speed up.
3 The block is at maximum speed as it passes through the equilibrium position.
4 The block overshoots x = 0; the restoring force then makes it slow down.
5 The block comes to rest momentarily at x = −A.
6 The restoring force again accelerates the block toward the equilibrium position.
7 The block again passes through the equilibrium position at maximum speed.
8 The block overshoots x = 0 again; the restoring force again makes it slow down.
9 The block is again at rest at x = +A; the cycle now starts over.

AP® Exam Tip

Note that here and on the AP® Physics 1 equation sheet, T is used for period. Sometimes you might see T used for tension, torque, or other physics quantities. Because oscillations are often talked about in the context of vibrating strings under tension, it is easy to confuse the tension with period in an equation that is not familiar. To ensure that you do not use an incorrect relationship to describe a problem you are solving, make sure to understand what each of the equations on the equation sheet represents.

A quantity related to the oscillation period is the **frequency** f of the oscillation. The frequency is equal to the number of complete cycles of the oscillation per unit time. The unit of frequency is the **hertz** (abbreviated Hz): 1 Hz = 1 cycle per second. A pendulum that oscillates once per second has a frequency $f = 1$ Hz. If you play the A4 key on a piano (A above middle C, also called concert A), the piano string vibrates 440 times per second, so the frequency is $f = 440$ Hz. The vibrating string compresses the air at this same frequency, which is the frequency of sound that you hear.

Because period T is the number of seconds that elapse per cycle and frequency f is the number of cycles that happen per second, it follows that these quantities are the reciprocals of each other:

EQUATION IN WORDS
Frequency and period

Period T of an oscillation

$$f = \frac{1}{T} \text{ and } T = \frac{1}{f} \quad (12\text{-}1)$$

Frequency f of the oscillation

EXAMPLE 12-1 Frequency and Period

(a) What is the oscillation period of your eardrum when you are listening to the A4 note on a piano (frequency 440 Hz)? (b) A bottle floating in the ocean bobs up and down once every 2.00 min. What is the frequency of this oscillation?

Set Up

We'll use Equation 12-1 to find the period T from the frequency f and vice versa.

Frequency and period:

$$f = \frac{1}{T} \quad \text{and} \quad T = \frac{1}{f} \qquad (12\text{-}1)$$

Solve

(a) The period of this oscillation is the reciprocal of the frequency. Note that a cycle is not a true unit but simply a way of counting, so we can remove it from the final answer as needed.

$$T = \frac{1}{f} = \frac{1}{440\ \text{Hz}} = \frac{1}{440\ \text{cycles/s}} = \frac{1}{440}\frac{\text{s}}{\text{cycle}}$$
$$= 2.3 \times 10^{-3}\ \text{s}$$

(b) The frequency of this oscillation is the reciprocal of the period. To get a result in hertz, we must first convert the period to seconds. Note that the period is the number of seconds per cycle, so we can use *cycle* to help us think about what we are calculating, but we remove it from the final answer.

$$T = 2.00\ \text{min} \times \frac{60\ \text{s}}{1\ \text{min}} = 120\ \text{s}$$

$$f = \frac{1}{T} = \frac{1}{120\ \text{s}} = \frac{1}{120\ \text{s/cycle}} = \frac{1}{120}\frac{\text{cycle}}{\text{s}}$$
$$= 8.33 \times 10^{-3}\ \text{Hz}$$

Reflect

Because frequency and period are the reciprocals of each other, a large value of f (high frequency) corresponds to a small value of T (short period) as in (a) and a small value of f (low frequency) corresponds to a large value of T (long period) as in (b), which is what we found.

NOW WORK Problem 2 in The Takeaway 12-2.

Oscillation Amplitude

If the spring in Figures 12-2 and 12-3 exerts the same magnitude of force on the block when it is stretched or compressed the same distance, the block in Figure 12-3 will move as far to the left of equilibrium ($x < 0$) as it does to the right of equilibrium ($x > 0$). In this chapter we'll consider only restoring forces that have this symmetric property. In this case the block will oscillate between $x = +A$ and $x = -A$, as shown in Figure 12-3. The distance A is called the **amplitude** of the oscillation. It's equal to the magnitude of the maximum displacement of the object from equilibrium. The amplitude is *always* a positive number, never a negative one.

WATCH OUT !

The amplitude of oscillation is the maximum distance from equilibrium.

A common mistake is to think of amplitude as the distance from where the spring is stretched the most (Figure 12-3a) to where the spring is compressed the most (Figure 12-3e). That's not correct! The amplitude A equals the distance from the *equilibrium* position to the point of maximum stretch, and it also equals the distance from the equilibrium position to the point of maximum compression. Keep this principle in mind to avoid giving answers that are wrong by a factor of 2.

Period, frequency, and amplitude are just part of a complete description of an oscillation. In the following section we'll see how to get such a complete description for an important special kind of oscillation.

THE TAKEAWAY for Section 12-2

✔ Oscillation is caused by the interplay between (i) a restoring force that returns an object to its equilibrium position (where it experiences zero net force) and (ii) inertia that makes the object overshoot equilibrium.

✔ The period of an oscillation is the time for one complete cycle. The frequency, which is the reciprocal of the period, is the number of cycles per second.

✔ The amplitude of an oscillation equals the magnitude of the maximum displacement from equilibrium.

Prep for the AP® Exam

AP® Building Blocks

1. One condition for oscillatory motion is that a force is exerted in a direction that restores the system to the equilibrium position. The oscillation depends upon the inertia of the system carrying it past equilibrium, even though the restoring force goes to zero at that point. Can you state a second condition that is necessary in real-world systems? *Hint*: Consider a case in which the system is restored to equilibrium very gradually.

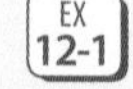

2. An event occurs at the same time every day. What is the frequency in hour^{-1}? What is the frequency in second^{-1}? What is the frequency in hertz?

AP® Skill Builder

3. The inventor of the vacuum tube, J. A. Fleming, illustrated the oscillatory electric circuits that are the basis of modern telecommunication with a mechanical example involving liquid mercury in a U-shaped glass tube. As shown in the figure, starting with a bead of mercury in equilibrium at the bottom of the U-tube, the U-tube is turned until the bead is lying in one arm of the U-tube. Fleming described two situations: In the upper sequence, you move the U-tube slowly from sideways to upright and then hold the U-tube steady. In the lower sequence, you instead quickly move the U-tube to the upright position, and then keep the U-tube stationary. In the two sequences, Δt represents the same amount of time.

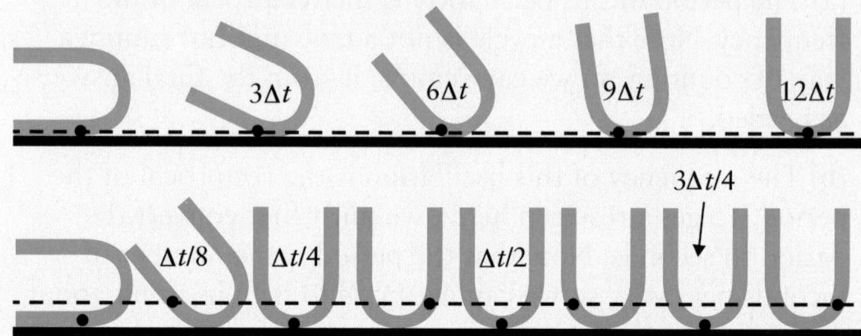

The bead of mercury moves as the U-tube rotates into an upright position in the upper and lower sequences in the figure. Explain, in terms of the restoring force and overshoot, the motion of the bead in the two sequences.

AP® Skills in Action

4. The period of an oscillatory motion is 0.125 s. What is its frequency in hertz?
5. An oscillatory motion completes 1250 cycles in 20.0 min. Calculate
(a) the frequency of the motion, and
(b) the period, in seconds.

12-3 An object changes length when under tensile or compressive stress; Hooke's law is a special case

We say that the forces exerted on each of the gymnast's legs in **Figure 12-4a** apply a stress on the gymnast's leg that makes it deform, and we call the resulting deformation the strain. We'll see that there are three distinct kinds of stress, which differ in the manner in which they are applied (**Figure 12-5**). One of our principal goals will be to see how much something deforms when we apply a particular amount of stress—for example, how much a tendon stretches when you flex a muscle, or how much a vertical spring stretches when an object with mass is suspended from its end.

Some materials, such as an ideal spring, a billiard ball (for small deformations), your Achilles tendon, your earlobes, or the tip of your nose, are elastic: If you apply a stress to stretch or squeeze them, they snap back to their original shape after the stress is removed. Other materials, such as a well-chewed piece of bubble gum, are plastic: They exhibit a permanent change of shape when stretched. Even an elastic material like rubber can remain irreversibly deformed if large enough forces are exerted on it. A spring can similarly be stretched too far, changing how it behaves. The material might even break apart and fracture, as with a plastic fork that's bent too far.

(a)

This gymnast's upper legs are under **tension:** The rings pull up on her legs and the weight of her body pulls down on her legs. For clarity, the forces are drawn for only one leg.

(b)

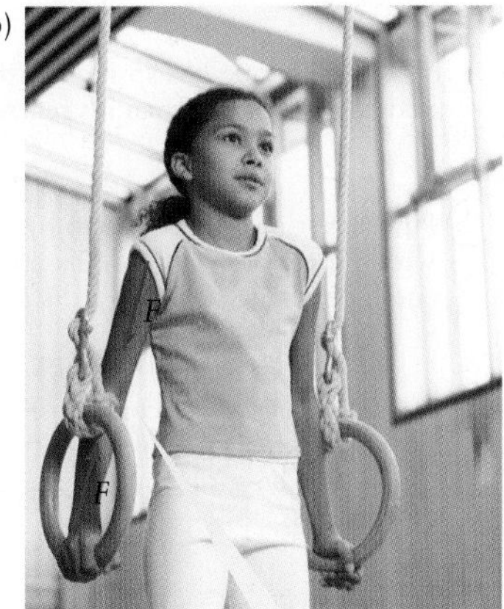

This gymnast's arms are under compression: Her shoulders push down on her arms and the rings in her hands push up on her arms. For clarity, the forces are drawn for only one arm.

(a) Muriel de Seze/Getty Images; (b) Christoph Wilhelm/Getty Images

Figure 12-4 Tension and compression (a) A system under tension tends to stretch along its length; (b) a system under compression tends to shorten along its length.

The upper legs of the gymnast shown in Figure 12-4a and the rope shown in Figure 12-5a are under a kind of stress called tension or tensile stress: A force is exerted on each end that makes the tendon or rope stretch. (Don't confuse this with the emotion of feeling "under tension," such as may happen before taking a big exam!) A closely related kind of stress is compression or compressive stress, which occurs when a force is exerted on each end of a system that tends to squeeze the system. The gymnast's arms in **Figure 12-4b** are under compression. If you grab each end of a pencil and try to pull the ends apart, you're putting the pencil under tension; if you push the ends toward each other, you're putting the pencil under compression. We will focus on just tensile stress and compressive stress, which will allow us to better understand the most common physical systems.

WATCH OUT !

Stress and strain are different quantities.

In everyday language the words *stress* and *strain* can mean pretty much the same thing. You might say, "I'm under a lot of stress" or "I'm under a lot of strain." In physics, however, *stress* refers to forces that make a system deform, and *strain* refers to how much deformation those forces cause.

In Chapter 5, we saw how one kind of system, an ideal spring (**Figure 12-6a**) behaves under compression or tension. The fact that the spring as a whole doesn't accelerate from rest tells us that the net external force exerted on the spring is zero. When we stretch a spring along its length, exerting a force of magnitude F on one end, whatever the other end is attached to also exerts a force of magnitude F on that other end (**Figure 12-6b**).

(a)

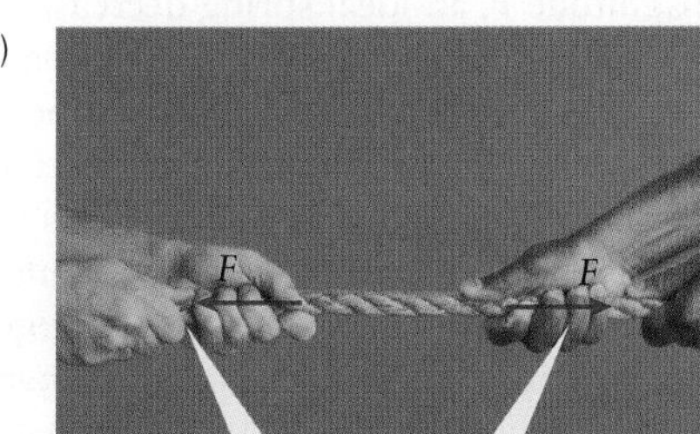

This rope is under **tensile stress**: Forces are exerted along its length to stretch it.

(b)

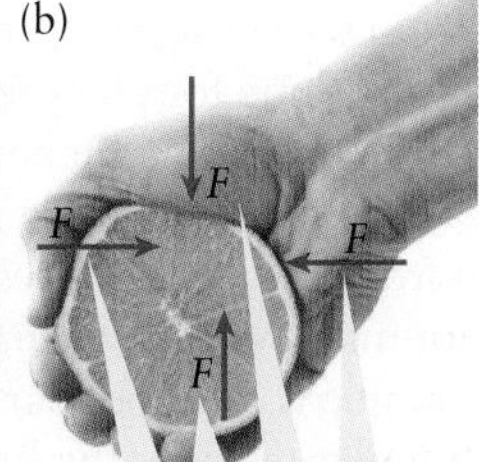

This orange is under **volume stress**: Forces are exerted on all sides to squeeze it.

(c)

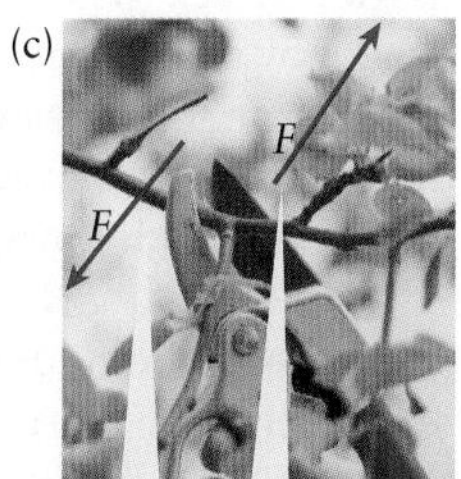

This branch is under **shear stress:** Offset forces are exerted to deform it.

(a) Tetra Images, LLC/Alamy; (b) pbombaert/Shutterstock; (c) Nobilior/Fotolia

Figure 12-5 Three kinds of stress You apply (a) tensile stress when you stretch a system along its length, (b) volume stress when you squeeze it on all sides, and (c) shear stress when offset forces are exerted (as by a pair of shears or scissors). We will focus on (a).

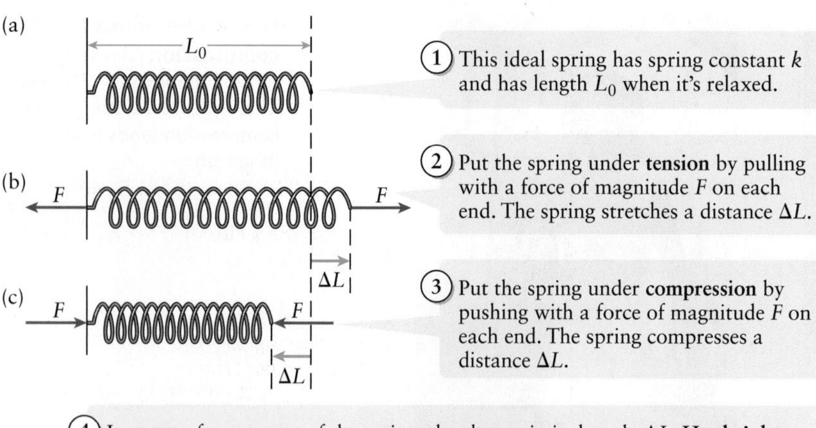

(a)
L_0
1 This ideal spring has spring constant k and has length L_0 when it's relaxed.
(b)
F
F
2 Put the spring under **tension** by pulling with a force of magnitude F on each end. The spring stretches a distance ΔL.
ΔL
(c)
F
F
3 Put the spring under **compression** by pushing with a force of magnitude F on each end. The spring compresses a distance ΔL.
ΔL
4 In terms of a property of the spring, the change in its length, ΔL, **Hooke's law** relates the force magnitude F to the distance ΔL that the spring compresses or stretches in response:
$F = k\Delta L$

Figure 12-6 Hooke's law for an ideal spring When an ideal spring is under tension or compression, the change in the spring's length is directly proportional to the forces exerted on its ends.

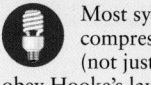

Most systems under compression or tension (not just ideal springs) obey Hooke's law, provided F and ΔL aren't too large.

According to Hooke's law, which we first encountered in Section 5-6, the force magnitude F is proportional to the distance ΔL that the spring stretches, provided that ΔL is small compared to the relaxed length of the spring, L_0:

EQUATION IN WORDS
Hooke's law, magnitude only in terms of spring length

$$F = k\Delta L \quad (12\text{-}2)$$

Magnitude of the **force exerted on each end** of the spring

Spring constant k tells you how stiff the spring is: Larger k means stiffer.

Magnitude of the **change in length of the spring from its relaxed length** as a result of the force exerted

[In Chapter 5 and Figure 12-3 we used x to describe the position of the end of the spring from its equilibrium length that we defined as $x = 0$. Then, x referred to the position of the end of the spring at a given time. To begin to develop an understanding of how systems behave, it is useful to think in terms of properties of the system instead of a position in space. So, we are going to use ΔL to emphasize that this distance is the change in the overall length L of the spring, and L_0 to represent the equilibrium length of the spring (or other system).]

The quantity k in Equation 12-2, the spring constant, is a measure of the stiffness of the spring. To see this, note that we can rewrite Equation 12-2 as $k = F/\Delta L$. The stiffer the spring, the shorter the amount of stretch ΔL for a given force magnitude F and so the greater the value of $k = F/\Delta L$. If the spring is very flexible, there is more stretch for a given force magnitude and $k = F/\Delta L$ has a smaller value. The SI unit of k is newtons per meter (N/m).

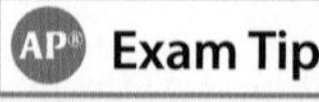

Exam Tip

The spring constant k is always a positive quantity, as it is a measure of how far an elastic system will stretch or compress for a given magnitude of force applied.

Remember, Hooke's law also applies to an ideal spring under compression (**Figure 12-6c**): When compressed by a force of magnitude F, an ideal spring decreases in length by a distance ΔL from the spring's relaxed length, given by Equation 12-2. The value of k for an ideal spring is the same for compression as for tension. Hence, if stretching an ideal spring with forces of magnitude F makes its length increase by 1.00 mm, squeezing that spring with forces of the same magnitude F will make its length decrease by 1.00 mm.

Hooke's law is important because it doesn't apply to just ideal springs. Experiment shows that it holds true for most systems when they are stretched, provided that the amount of stretch is small. An example is a rubber bar made by cutting apart a thick rubber band (**Figure 12-7a**). When put under tension, the bar stretches by a distance ΔL that's directly proportional to the force magnitude F (**Figure 12-7b**). Likewise, when the bar is compressed, it shrinks by a distance ΔL that's directly proportional to F, although you have to be a lot more careful to compress the bar without just causing it to bend in order to see this effect.

Experiment shows that the spring constant k for a given system depends on three things. It depends on the reciprocal of its relaxed length L_0, and it depends directly on

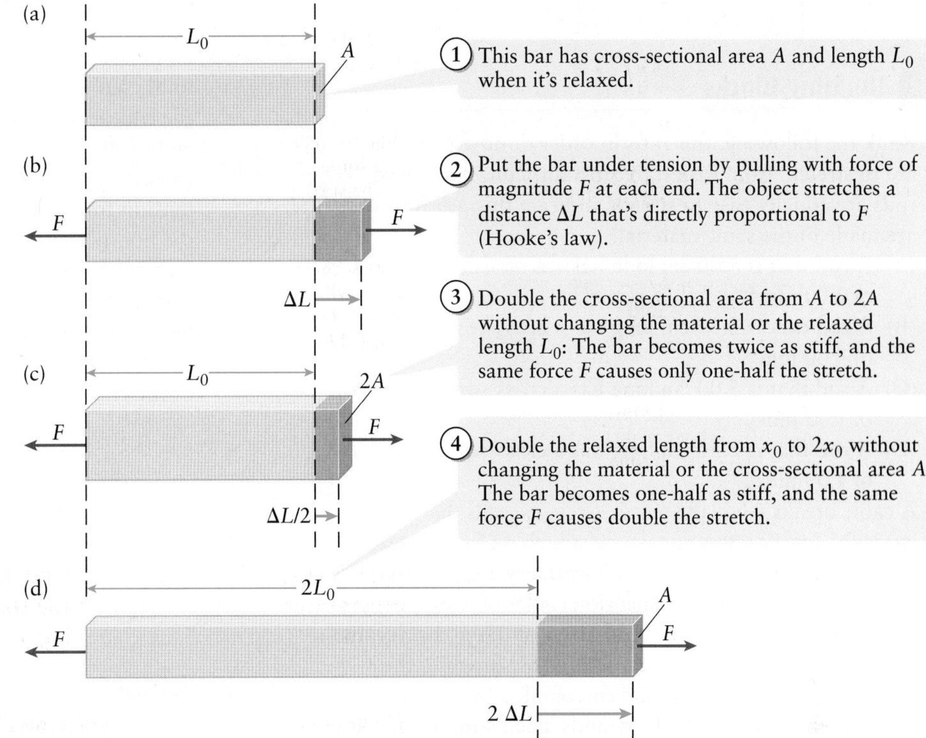

(a)
L_0
A
1 This bar has cross-sectional area A and length L_0 when it's relaxed.
(b)
F
F
2 Put the bar under tension by pulling with force of magnitude F at each end. The object stretches a distance ΔL that's directly proportional to F (Hooke's law).
ΔL
3 Double the cross-sectional area from A to $2A$ without changing the material or the relaxed length L_0: The bar becomes twice as stiff, and the same force F causes only one-half the stretch.
(c)
L_0
$2A$
F
F
4 Double the relaxed length from x_0 to $2x_0$ without changing the material or the cross-sectional area A: The bar becomes one-half as stiff, and the same force F causes double the stretch.
$\Delta L/2$
(d)
$2L_0$
A
F
F
$2\,\Delta L$

Figure 12-7 Hooke's law for tension and compression Almost anything placed under tension or compression obeys Hooke's law if the stress is not too great.

The bar's spring constant is proportional to the stiffness of the material from which it is made and its cross-sectional area but inversely proportional to its length.

its cross-sectional area A and the material of which it is made (for example, the particular kind of rubber or steel). This expression says that a short, thick piece of material (small L_0, large A) is stiffer than a long, thin (large L_0, small A) piece of the same material (**Figures 12-7c** and **12-7d**). That's why a rubber eraser is harder to stretch than a rubber band. It also says that the stiffer the material used to make something the stiffer that thing will be. This also relates to our discussions of normal force in Chapter 4 and elastic collisions in Chapter 9. To support an object by exerting a normal force on it, a stiffer surface deforms less to cause the same normal force. The higher the spring constant of something in a collision, the less it has to deform in the process of changing velocity, and the greater the percentage of the initial kinetic energy will get restored from potential energy. In both these cases, we want something that is stiff, but with some elasticity. Something too brittle will break instead of deforming. Materials science is an active field of research, ensuring we have materials with the right properties for many sorts of applications, from cushioning shoe soles to shielding nuclear reactors.

WATCH OUT !

Hooke's law has important limitations.

This simple form of Hooke's law applies only when the strain is small, that is, when the change in length is small relative to the overall initial length of the spring or an elastic material. For Hooke's law to apply to a spring or elastic material, both the stress and strain have to be able to be positive or negative. Hooke's law also applies only to elastic materials that are of a uniform composition, such as a solid bar of iron. For many purposes, however, it's a reasonable approximation to apply Hooke's law even to biological materials like we did with the tendon in Chapter 5.

Exam Tip

Stress and strain are not topics on the AP® Physics 1 exam, but the results for Hooke's law and the object–spring system are topics that are covered on the test.

THE TAKEAWAY for Section 12-3

✔ Elastic materials return to their original shape and size when the stress is removed, while plastic materials exhibit a permanent change in shape.

✔ The spring constant, k, of an elastic system is larger for stiffer materials of the same shape; shorter lengths of the same cross-sectional area and the same material; or a larger cross-sectional area, if length and material are the same.

✔ If a system obeys Hooke's law, the force exerted by or on the system and the resulting change in length of the system are directly proportional.

Prep for the AP Exam

AP Building Blocks

1. Rank the following four rods in order of how easy they are to stretch, from easiest to most difficult. If any two rods are equally easy to stretch, indicate this. All four rods are made of the same material.
   (A) A rod that is 5.00 cm long has a cross-sectional area of 2.00 mm^2.
   (B) A rod that is 2.50 cm long has a cross-sectional area of 2.00 mm^2.
   (C) A rod that is 5.00 cm long has a cross-sectional area of 1.00 mm^2.
   (D) A rod that is 2.50 cm long has a cross-sectional area of 1.00 mm^2.
2. A cable breaks when the tensile stress exceeds the tensile strength. Tensile stress equals the stretching force exerted on each end of a system, divided by the system's cross-sectional area. The four suspension cables that support the Brooklyn Bridge each have a diameter of 0.4 m. Each cable is composed of 19 strands, configured as shown in the figure, with one central strand surrounded by 6 strands, which are surrounded by 12 strands. Each strand is made of 286 8-gauge (3.2-mm or 1/8-in.-diameter) wires. Each wire can support up to 15.1 kN. Calculate the tensile strength of each of the four suspension cables of the Brooklyn Bridge.

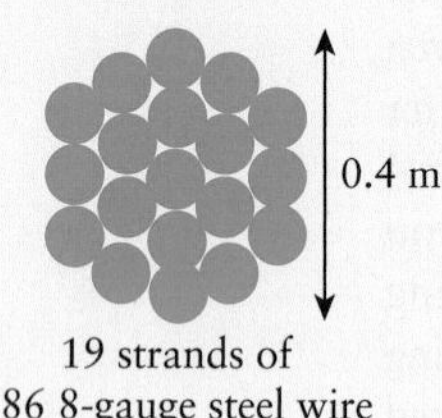

AP Skill Builders

3. Human skin is under tension like a rubber glove that has had air blown into it. Often as humans age their skin wrinkles. Pose a scientific question whose pursuit could contribute to an understanding of changes in a material property of skin as people get older.
4. When an object with mass is hung from the free end of a board it causes the board to bend. A 2.5-cm by 9-cm (1-in. by 4-in.) pine board, with its 9-cm side oriented horizontally, is securely clamped at one end to an immovable object. The length of the board is 2.5 m. The fibers of the board (the grain) are parallel to the long axis.

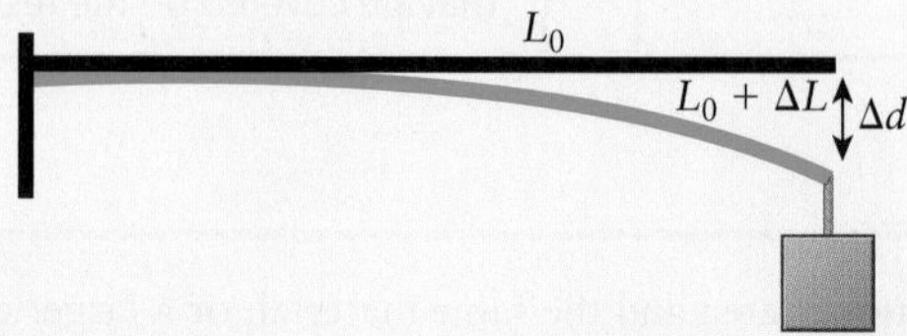

   (a) In this case, ΔL is not the displacement of the end of the board, but how much the board stretches as it bends under the load. (The blue line represents the final configuration of the board.) The deflection of the end of the board, Δd, is the displacement of the end of the board in the direction of the force. It is caused by the combination of positive and negative tension forces. The geometry of the deformation of the board makes the relationship between stretch ΔL and displacement Δd complicated, as is the relationship between the direction of the force and the stretch, but an approximate representation of the relationship between the stretch of the board and the displacement of the end of the board under a load mg is

$$mg = k\Delta L = k\frac{h^2}{L_0 w}\Delta d$$

   where L_0 is the length of the board when it carries no load, and w is the width and h the thickness of the board. Using this approximation, justify the claim that

$$\frac{\Delta d}{\Delta L} = \frac{L_0 w}{h^2}$$

   (b) The data in the table were collected for a board with width w = 9.0 cm, thickness h = 2.5 cm, and length L_0 = 2.5 m. Analyze the data graphically, plotting Δd versus m to determine the spring constant in units of kg · s^{-2}.

Mass m (kg)	Deflection Δd (m)
25	1.7×10^{-3}
50	3.9×10^{-3}
75	5.6×10^{-3}
100	7.5×10^{-3}

   (c) Evaluate the magnitude of the spring constant for the board in part (b) by comparing it with the stretch of a typical spring in your physics lab resulting from a hanging mass of 100 g.
5. The stiffness of a uniform material, such as brass, is independent of shape. However, a material can have an internal structure that causes the stiffness of the material to vary depending on the direction in which a force is exerted on the material. For example, the internal structure of a leg bone is different in the transverse (perpendicular to the long axis of the leg bone) direction than it is in the longitudinal (parallel to the long axis of the leg bone) direction. Forensic medical analyses and diagnoses of athletic injuries make use of this difference. Do not worry about what causes this difference (fibers in the bone), but just answer the following question in terms of the properties of elastic materials.

   The data in the graph show how the stiffness, denoted by Y, depends on the orientation of the applied stress for a cow's leg bone. When θ = 0°, the stress is applied perpendicular to the long axis of the bone. A bone fractures when the displacement of the bone caused by the force exerted on it exceeds how far the bone can "stretch" elastically. Use these data to explain why a bone is harder to break (a transverse fracture) than to split (a longitudinal fracture). Fortunately, in our daily lives we are less likely to hit our bones as hard on the ends!

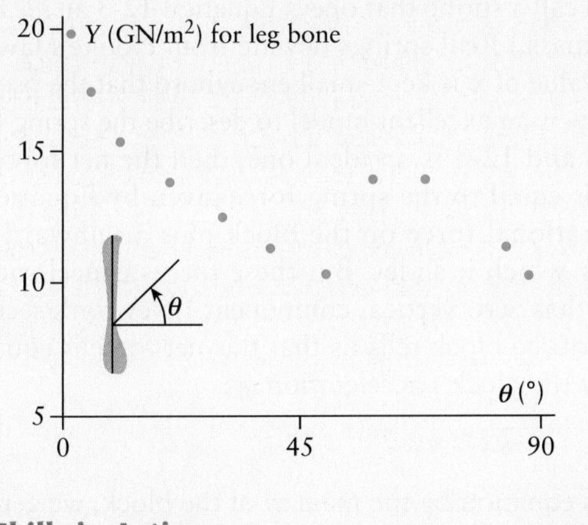

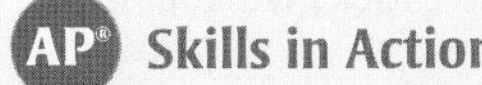

6.

L

m_2 m_1

Two objects with masses m_1 and m_2 are attached to opposite ends of a spring with spring constant k, negligible mass, and rest length L, as shown above. The system rests on a horizontal surface that exerts negligible friction force on the blocks or spring. The system is stretched by exerting equal magnitude forces in opposite directions and then released from rest.

(a) With respect to a coordinate origin in the horizontal surface, the two objects are at positions x_1 and x_2 with $x_1 > x_2$. When the objects are separated by a displacement $(x_1 - x_2) > L$, the force exerted by the spring on object 1 is $-k((x_1 - x_2) - L)$. Predict the magnitude and direction of the force exerted by the spring on object 2. Draw free-body diagrams for m_1 and m_2. Referring to those diagrams and fundamental laws, justify your prediction, showing why this form of the force makes sense in terms of the behavior of springs.

(b) Describe the location of the center of mass of the system and explain why the center-of-mass location is unchanging. Remember that the spring has negligible mass.

12-4 The simplest form of oscillation occurs when the restoring force obeys Hooke's law

To have a complete description of what happens to an object that is part of an oscillating system, we need to know the position, velocity, and acceleration of the object at all times during its motion. We can actually find these for the case in which the restoring force exerted on the object is *directly proportional* to the distance that the oscillating object is displaced from equilibrium. **Hooke's law** provides such a relationship. In the last section, to develop an understanding of how materials act, we called the relaxed length of a spring L_0, and $\Delta L = L - L_0$ the change in length of the spring (the difference in the position of the end of the spring when it was stretched or compressed from when it was relaxed). To use Hooke's law to mathematically describe oscillatory motion we don't have to worry about the length of the spring so we will go back to the expression we developed in Chapter 5, Equation 5-10, $F_{s,x} = -k\Delta x$. When $x = 0$ is defined as the position of the end of the spring when it is relaxed, $\Delta x = x$, and we can rewrite this as Equation 12-3. This equation applies equally well to an ideal spring whether it is compressed or stretched.

The **force the spring exerts on you** is equal and opposite to the force you exert on the spring. We choose our axes so that this is one component of a vector: The force is exerted along the axis in the direction the spring can stretch.

Spring constant k (measure of stiffness)

$$F_{s,x} = -kx \tag{12-3}$$

The force exerted by the spring is always in the opposite direction to the displacement of the end of the spring.

Change in the spring's length x (measured from the spring's relaxed equilibrium length) as a result of the force you exert, when the location of the equilibrium length is defined as $x = 0$.

EQUATION IN WORDS
Hooke's law for cases in which the equilibrium length of the spring is defined as $x = 0$

Like the minus sign in Equation 5-10, the minus sign in Equation 12-3 tells us that if the spring is stretched or compressed it exerts a force in the opposite direction. That's exactly the relationship between displacement x and the x component of the spring

Exam Tip

On the AP® Physics 1 equation sheet, Equation 12-3 is shown without a minus sign, using absolute values (magnitudes). You do need to understand the vector nature–sign relationship in Hooke's law for the exam.

Exam Tip

The equations in Chapter 2 describe the relationships between the kinematic variables for constant acceleration. Constant acceleration is a special kind of motion. There are many other kinds of motion, such as oscillatory motion, that do not demonstrate constant acceleration. It is easy to forget that the kinematic relationships we spent so much time with earlier in the book do not apply to all situations; Equation 12-4 shows an acceleration that is not constant because it depends on another kinematic variable.

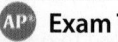

force, F_x shown in Figure 12-2. Making this change to where we define $x = 0$ makes no difference in how we describe the behavior of the spring force, but greatly simplifies our work as we seek to mathematically model oscillations.

As we did in Section 5-6, we'll call a spring that obeys Equation 12-3 an *ideal* spring. (We also ignore an ideal spring's mass.) Real springs deviate from Hooke's law a little. But experiment shows that if the value of x is kept small enough, so that the oscillations have small amplitudes, Hooke's law is an excellent model to describe the spring force.

If the spring in Figures 12-2 and 12-3 is an ideal one, then the net force on the block in Figures 12-2 and 12-3 is equal to the spring force given by Equation 12-3. (There's also a downward gravitational force on the block plus an upward normal force exerted by the surface over which it slides. But these forces cancel each other so that the net force on the block has zero vertical component.) Newton's second law for the motion in the x direction of the block tells us that this net force is equal to the mass m of the block multiplied by the block's acceleration a_x:

$$-kx = ma_x$$

If we divide both sides of this equation by the mass m of the block, we get

$$a_x = -\left(\frac{k}{m}\right)x \tag{12-4}$$

Equation 12-4 tells us that *if a block oscillates because it is attached to an ideal spring, the acceleration of the block is proportional to the negative of the block's displacement.* The acceleration is negative when the displacement x is positive, positive when the displacement is negative, and zero when the displacement is zero (that is, when the block is at equilibrium).

Because the acceleration of the block is not constant, we *cannot* describe the motion of the block in Figure 12-3 using the equations from Chapter 2 for linear motion with constant acceleration. Instead, we'll have to solve Equation 12-4 to find x, v_x, and a_x as functions of time t. This sounds like a complicated mathematical problem, but it turns out that we can solve it by using what we know about another, seemingly unrelated, kind of motion: an object moving around a circle at constant speed.

Uniform Circular Motion and Hooke's Law

Figure 12-8a shows an object of mass m moving at constant speed v around a circular path of radius A. (We'll see shortly why we choose this symbol for radius.) The object travels once around the circle, at an angle of 2π radians, in a time T, so the object's constant *angular* speed is $\omega = 2\pi/T$.

How does this uniform circular motion relate to oscillation? Imagine that you view the circular path edge-on so that all you can see is the x component of the object's motion (**Figure 12-8b**). From this vantage point you'll see the object oscillating back and forth along a linear path, moving a distance A to the left of center and the same distance A to the right of center (Figure 12-8b). The object will take a time T for one back-and-forth cycle. So what you'll see is the object moving in the same way as the block on a spring in Figure 12-3, with period T and amplitude A.

Let's show that the x component of the uniform circular motion obeys Equation 12-4 so that the x component of the object's acceleration is proportional to the object's x coordinate. (Remember that coordinates are the location of a position with respect to the origin of the coordinate system.) To do this

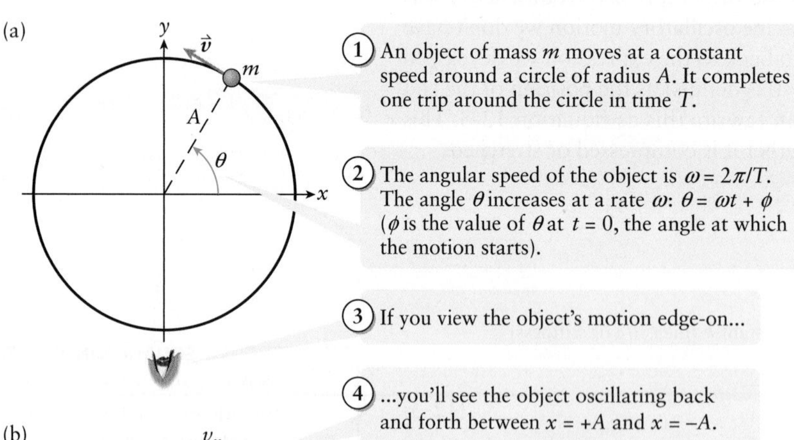
(a)
y
$\vec{v}$
m
A
θ
x
1 An object of mass m moves at a constant speed around a circle of radius A. It completes one trip around the circle in time T.
2 The angular speed of the object is ω = 2π/T. The angle θ increases at a rate ω: θ = ωt + ϕ (ϕ is the value of θ at t = 0, the angle at which the motion starts).
3 If you view the object's motion edge-on...
4 ...you'll see the object oscillating back and forth between x = +A and x = –A.
(b)
v_x
m
x = –A
x = 0
x = +A

The x component of uniform circular motion looks like straight-line oscillation around an equilibrium position at $x = 0$.

Figure 12-8 Oscillation and uniform circular motion If you look at one component of uniform circular motion, what you see is an oscillation.

we'll use three important facts about uniform circular motion. From Section 10-2 the linear speed v of an object in uniform circular motion is equal to the angular speed ω multiplied by the radius of the circle. The radius of the circle in Figure 12-8a is A, so the speed of the object's circular motion is

$$v = \omega A \qquad (12\text{-}5)$$

We also know from Section 6-2 that an object in uniform circular motion has a centripetal acceleration (that is, an acceleration directed toward the center of the circle). This acceleration has a constant magnitude a_{cent} equal to the square of the speed v divided by the radius of the circle, which in this case is equal to A. Using Equation 12-5 we can write the magnitude of the centripetal acceleration as

$$a_{\text{cent}} = \frac{v^2}{A} = \frac{(\omega A)^2}{A} = \frac{\omega^2 A^2}{A} = \omega^2 A \qquad (12\text{-}6)$$

Finally, from Equation 10-14 we have the following equation for the angle θ from the positive x axis for the angular position of an object at time t, assuming that the object has constant angular acceleration:

$$\theta = \theta_0 + \omega_{0z} t + \frac{1}{2}\alpha_z t^2 \qquad (10\text{-}14)$$

In Equation 10-14, α_z is the angular acceleration of the object as it moves around the circle, which is zero in this case because the motion is uniform and the angular velocity is constant. The quantity ω_{0z} is the initial z component of the angular velocity of the object, which is equal to the object's constant angular speed ω. (Figure 12-8a shows that the object moves around the circle in the positive, counterclockwise direction, so the angular velocity is positive and equal to the angular speed.) The angle θ_0 is the value of θ at time $t = 0$; we'll use the symbol ϕ (the Greek letter "phi") for this quantity. Then we can write Equation 10-14 as $\theta = \phi + \omega t$, or

$$\theta = \omega t + \phi \qquad (12\text{-}7)$$

Now let's put all of these pieces together. **Figure 12-9a** shows that, for a circular path of radius A, with its center located at (0, 0) the x coordinate of the object as it moves around the circle is $x = A\cos\theta$. Because a coordinate is always measured from the origin, it is the same as the object's displacement from the origin. From Equation 12-7 we can write this as

$$x = A\cos(\omega t + \phi) \qquad (12\text{-}8)$$

(a)

(1) The vector from the center of the circle to the object has length A and is at an angle θ from the $+x$ axis.

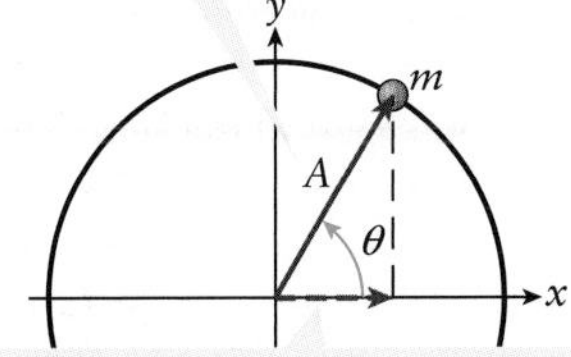

(2) The x component of the object's displacement from the center of the circle is $x = A\cos\theta = A\cos(\omega t + \phi)$.

(b)

(3) The object's velocity has magnitude $v = \omega A$ and is at an angle $90° + \theta$ from the $+x$ axis.

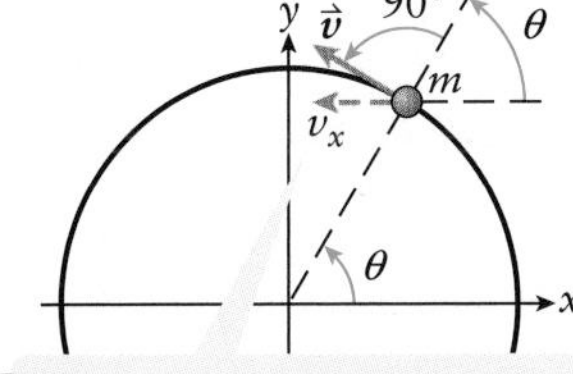

(4) The x component of the object's velocity is $v_x = v\cos(90° + \theta) = -v\sin\theta$ or $v_x = -\omega A\sin(\omega t + \phi)$.

(c)

(5) The object's acceleration has magnitude $a_{\text{cent}} = \omega^2 A$ and is at an angle $180° + \theta$ from the $+x$ axis.

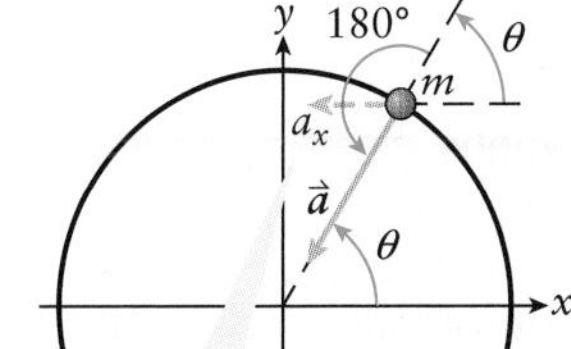

(6) The x component of the object's acceleration is $a_x = a_{\text{cent}}\cos(180° + \theta) = -a_{\text{cent}}\cos\theta$ or $a_x = -\omega^2 A\cos(\omega t + \phi)$.

Compare the equations for x and a_x: You'll see that $a_x = -\omega^2 x$. That's the same relationship as for Hooke's law, $a_x = -(k/m)x$, with $\omega^2 = k/m$. So the x component of uniform circular motion about the origin is the same as the simple harmonic motion of an object oscillating on the end of an ideal, Hooke's law spring.

Figure 12-9 Oscillation, uniform circular motion, and Hooke's law Finding the x components of the displacement (position measured from an origin set at the equilibrium for a spring on the center of the circle for uniform circular motion), velocity, and acceleration in uniform circular motion reveals an important connection to simple harmonic motion.

Figure 12-9b shows that the velocity in the x direction of the object is $v_x = -v \sin \theta$. From Equation 12-5 and Equation 12-7, we can write this as

$$v_x = -\omega A \sin(\omega t + \phi) \tag{12-9}$$

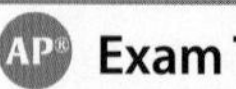

NEED TO REVIEW?
Turn to the Math Appendix, Section M8, in the back of the book for more information on trigonometry.

From **Figure 12-9c** the acceleration in the x direction of the object is $a_x = -a_{cent} \cos \theta$. Equation 12-6 and Equation 12-7 together tell us that we can write the x acceleration as

$$a_x = -\omega^2 A \cos(\omega t + \phi) \tag{12-10}$$

When we compare Equation 12-8 and Equation 12-10, we see that the x acceleration of an object in uniform circular motion is directly proportional to the object's x coordinate:

$$a_x = -\omega^2 x \tag{12-11}$$

AP® Exam Tip

For the AP® Physics 1 exam, you need to be able to manipulate Hooke's law and Equation 12-4 (note that the latter is derived from combining Hooke's law with Newton's second law). You should also know where the maxima and minima of acceleration and velocity are, and you should be able to sketch the oscillatory graphs of acceleration, velocity, and position with respect to equilibrium versus time. Although you will need to be able to interpret Equations 12-8, 12-9, and 12-10, you will not need to manipulate them on the exam.

That's *precisely* the relationship between acceleration a_x and displacement x that we found for an object of mass m attached to a spring with spring constant k that obeys Hooke's law: $a_x = -(k/m)x$ (Equation 12-4). The two equations are identical if we set the square of the angular speed ω in Figure 12-8 and Figure 12-9 equal to the ratio of the quantities k/m:

$$\omega^2 = \frac{k}{m} \quad \text{or} \quad \omega = \sqrt{\frac{k}{m}} \tag{12-12}$$

In other words, Equation 12-11 and Equation 12-12 tell us something quite profound and unexpected: *A block of mass m attached to an ideal spring of spring constant k oscillates in precisely the same way as the x coordinate of an object in uniform circular motion with angular speed ω, provided that* $\omega = \sqrt{k/m}$. We made the detour into analyzing uniform circular motion so that we could arrive at this insight.

With this observation in mind, we can now go back and write a complete description of how an object oscillates when a restoring force that obeys Hooke's law is exerted on it.

Simple Harmonic Motion: Angular Frequency, Period, and Frequency

We've seen that the period T of the oscillation in Figure 12-8b is the same as the time T that it takes the object in uniform circular motion in Figure 12-8a to travel once around the circle. The angular speed ω equals $2\pi/T$, so T equals $2\pi/\omega$. Using Equation 12-12 for the value of ω along with Equation 12-1 for the relationship between period T and frequency f, we get the following results for the oscillations of a block of mass m attached to an ideal spring of spring constant k:

EQUATION IN WORDS
Angular frequency, period, and frequency for a block attached to an ideal spring

(12-13)

Angular frequency: $$\omega = \sqrt{\frac{k}{m}}$$

Period: $$T = \frac{2\pi}{\omega} = 2\pi\sqrt{\frac{m}{k}}$$

Frequency: $$f = \frac{1}{T} = \frac{1}{2\pi}\sqrt{\frac{k}{m}} = \frac{\omega}{2\pi}$$

k = spring constant of ideal spring

m = mass of the object connected to the spring

In the first of Equations 12-13, we've introduced a new quantity called the **angular frequency**. This has the same symbol and the same value as the angular *speed* of the circular motion in Figure 12-8. That's because our analysis of Figure 12-8 showed that the back-and-forth motion of a block attached to an ideal spring is equivalent to the x component of uniform circular motion. If you find this confusing to remember, just keep in mind that the ordinary frequency f equals $\omega/2\pi$, or equivalently $\omega = 2\pi f$ (see the third of Equations 12-13). As we'll see, the quantity ω appears naturally in the equations for oscillation, which is why we use it. Like angular speed, angular frequency has units of radians per second (rad/s).

The second of Equations 12-13 tells us that the period of oscillation T is proportional to the square root of the mass m and inversely proportional to the square root of the spring constant k. This means that increasing the mass m makes the period longer, while increasing the spring constant k makes the period shorter. That's exactly what we predicted in Section 12-2 for the oscillations of a block attached to a spring. The last of Equations 12-13 is just the statement that the frequency f is the reciprocal of the period T. This equation says that increasing the mass causes the frequency to decrease, while increasing the spring constant makes the frequency increase.

Note that the amplitude A does *not* appear in Equations 12-13, which means that the angular frequency, period, and frequency of an oscillation are *independent* of the amplitude if the restoring force obeys Hooke's law. This is called the **harmonic property**. This may come as a surprise, because doubling the amplitude means the oscillating object has to cover twice as much distance during an oscillation and so it seems like it should take a longer time to complete an oscillation. But doubling the amplitude also means that the object reaches larger displacements x and so experiences a stronger Hooke's law restoring force $F_{s,x} = -kx$. This effect makes the object move faster during the oscillation. The net result is that the period T, the time for one oscillation, is unaffected by changing the amplitude. That means the frequency $f = 1/T$ and the angular frequency $\omega = 2\pi f = 2\pi/T$ are unaffected as well. This is the case *only* if the restoring force obeys Hooke's law. If the restoring force depends on displacement in a different way, the oscillations do not have the harmonic property.

The word *harmonic* sounds like a musical term, and indeed the harmonic property is important in musical instruments. If you continue your study of physics, you will learn that the pitch of a musical sound is determined primarily by the frequency of oscillation of the musical instrument that makes the sound, while the loudness of the sound is determined primarily by the amplitude. If changing the amplitude caused a change in frequency, playing the same key on a piano would make a different pitch if you pressed the key softly or pushed hard on the key. That would render a piano almost completely useless. But because the strings of a piano have the harmonic property, they vibrate at the same frequency and produce sounds of the same pitch whether they are played softly or loudly.

As we've described, the harmonic property is the result of the restoring force being directly proportional to the displacement from equilibrium (Hooke's law, Equation 12-3). For this reason the kind of oscillation that results from a Hooke's law restoring force is called **simple harmonic motion (SHM)**. Later in this chapter we'll encounter kinds of oscillations that are not simple harmonic motion; the oscillations of a pendulum are one example. We'll see that in many such cases, however, these kinds of oscillations are *approximately* SHM if the amplitude of oscillation is sufficiently small.

WATCH OUT !

Simple harmonic motion is not the same as circular motion.

Even though we describe simple harmonic motion in terms of an *angular* frequency, the oscillating object is *not* actually moving in a circle! As we've seen we can *visualize* simple harmonic motion as one component of the motion of an object moving in a circle with an angular speed ω, but there doesn't need to be an object that's actually moving in a circle.

EXAMPLE 12-2 SHM I: Angular Frequency, Period, and Frequency

An object of mass 0.80 kg is attached to a horizontal ideal spring that has a spring constant 1.8×10^2 N/m and is set into oscillation as in Figure 12-3. Friction between the block and surface is negligible. (a) Calculate the angular frequency, period, and frequency of the object. (b) Calculate the angular frequency, period, and frequency if the mass of the object is quadrupled to 4×0.80 kg = 3.2 kg.

Set Up

An ideal spring obeys Hooke's law, given that the force exerted by the spring on the object is proportional to the negative of the object's displacement. Therefore, the oscillations are simple harmonic motion. This means that in part (a) we can use Equations 12-13 to calculate the angular frequency ω, period T, and frequency f from the spring constant $k = 1.8 \times 10^2$ N/m and the mass $m = 0.80$ kg. In part (b) we'll use these equations to determine how the values of ω, T, and f change when we change the value of the mass m.

Angular frequency, period, and frequency:

$$\omega = \sqrt{\frac{k}{m}}$$

$$T = \frac{2\pi}{\omega} = 2\pi\sqrt{\frac{m}{k}}$$

$$f = \frac{1}{T} = \frac{1}{2\pi}\sqrt{\frac{k}{m}} = \frac{\omega}{2\pi} \quad (12\text{-}13)$$

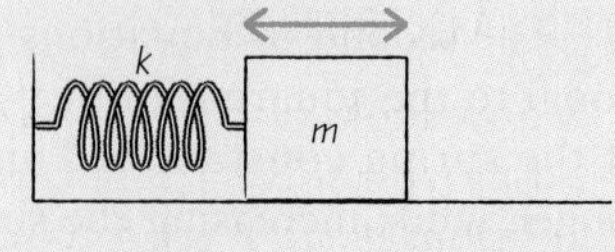

Solve

(a) Substitute the given values for the mass m and spring constant k into the first of Equations 12-13 to calculate the angular frequency.

$$\omega = \sqrt{\frac{k}{m}} = \sqrt{\frac{1.8 \times 10^2 \text{ N/m}}{0.80 \text{ kg}}}$$

Recall that $1 \text{ N} = 1 \text{ kg} \cdot \text{m/s}^2$, so

$$\frac{1 \text{ N/m}}{\text{kg}} = \frac{1 \text{ kg/s}^2}{\text{kg}} = \frac{1}{\text{s}^2} = \frac{1 \text{ rad}^2}{\text{s}^2}$$

(A radian is a way of counting, not a true unit, so we can insert it or remove it as needed.) Then

$$\omega = \sqrt{\frac{k}{m}} = \sqrt{\frac{1.8 \times 10^2}{0.80}\frac{\text{rad}^2}{\text{s}^2}} = 15 \text{ rad/s}$$

Then use the second and third of Equations 12-13 to find the period and frequency.

If the angular frequency ω is in radians per second, the period T will be in seconds:

$$T = \frac{2\pi}{\omega} = \frac{2\pi}{15 \text{ rad/s}} = 0.42 \text{ s}$$

The reciprocal of the period in seconds is the frequency in hertz:

$$f = \frac{1}{T} = \frac{1}{0.42 \text{ s}} = 2.4 \text{ Hz}$$

(b) We could substitute the new value of the mass m into Equations 12-13 to find the new values of ω, T, and f. Instead, we use proportional reasoning to find the new values from the old ones.

The relationships $\omega = \sqrt{\frac{k}{m}}$ and $f = \frac{1}{2\pi}\sqrt{\frac{k}{m}}$ tell us that the angular frequency and the frequency are inversely proportional to the square root of the mass. The new mass is four times greater than the old mass, so the new values of ω and f are $1/\sqrt{4} = 1/2$ the old values:

$$\omega_{\text{new}} = \frac{\omega_{\text{old}}}{\sqrt{4}} = \frac{\omega_{\text{old}}}{2} = \frac{15 \text{ rad/s}}{2} = 7.5 \text{ rad/s}$$

$$f_{\text{new}} = \frac{f_{\text{old}}}{2} = \frac{2.4 \text{ Hz}}{2} = 1.2 \text{ Hz}$$

In the same way, the relationship $T = 2\pi\sqrt{\frac{m}{k}}$ tells us that the period T is directly proportional to the square root of the mass. Again the mass has increased by a factor of 4, so the new period is

$$T_{\text{new}} = \sqrt{4}T_{\text{old}} = 2T_{\text{old}} = 2(0.42 \text{ s}) = 0.84 \text{ s}$$

Reflect

Our results show that when the mass of the oscillating object increases, the frequency decreases and the period increases, which makes sense. To check your results, you could confirm that f_{new} is equal to $\omega_{\text{new}}/2\pi$ and that T_{new} is equal to $1/f_{\text{new}}$.

NOW WORK Problems 2 and 3 from The Takeaway 12-4.

Simple Harmonic Motion: Position, Velocity, and Acceleration

Let's now look in more detail at the position x, velocity v_x, and acceleration a_x of an object moving in simple harmonic motion. Because this position is measured from the equilibrium position, defined as $x = 0$, position and displacement are the same. From Equation 12-8, the position is given by

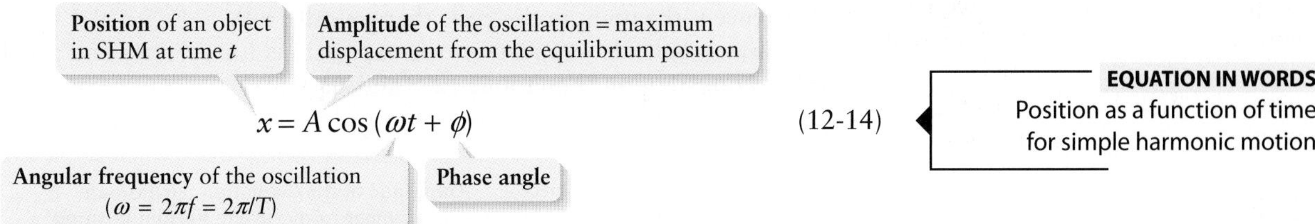

$$x = A\cos(\omega t + \phi) \qquad (12\text{-}14)$$

EQUATION IN WORDS
Position as a function of time for simple harmonic motion

The cosine function has values from +1 to −1. So Equation 12-14 says that the value of x varies from $x = A$ (when the oscillating object is at its most positive displacement from equilibrium) through $x = 0$ (the equilibrium position) to $x = -A$ (when the object is at its most negative displacement from equilibrium). The cosine is one of the simplest of all oscillating functions, which helps justify the name *simple* harmonic motion. The cosine and sine functions are called **sinusoidal functions** (from the Latin word *sinus*, meaning "bent" or "curved") because of their shape.

The angular frequency ω is measured in radians per second and time t is measured in seconds, so the product ωt in Equation 12-14 is in radians. The argument of a sinusoidal function must always be an angle, either in radians or in degrees. One cycle of oscillation lasts a time T, over which time the product ωt varies from 0 at $t = 0$ to ωT at $t = T$. But because $T = 2\pi/\omega$ from the second of Equations 12-13, it follows that $\omega T = \omega(2\pi/\omega) = 2\pi$. So over the course of one cycle the value of ωt varies from 0 to $\omega t = 2\pi$. The cosine function goes through one complete cycle when its value increases by 2π, so this tells us that the position x goes through one complete cycle when the time increases by T. That's just what we mean by saying that T is the period of the oscillation.

WATCH OUT!

Be careful your calculator is set in the correct mode.

If you use your calculator for sinusoidal functions, always make sure it is in the mode that is the same as you are using for the angle you enter, radians or degrees.

The one quantity in Equation 12-14 that seems a bit mysterious is the **phase angle** ϕ, which is measured in radians. To explain its significance recall from our comparison between uniform circular motion and oscillation that ϕ represents the angular position at $t = 0$ of an object in uniform circular motion. For oscillation the value of ϕ tells us where in the oscillation cycle the object is at $t = 0$. As an example, if $\phi = 0$ Equation 12-14 becomes

$$x = A\cos\omega t \text{ (simple harmonic motion, phase angle } \phi = 0) \qquad (12\text{-}15)$$

Figure 12-10a is the x–t graph that corresponds to Equation 12-15. This graph of x versus t is an ordinary cosine function, which has its most positive value ($x = +A$) when the argument of the function is zero (at $t = 0$) and its most negative value ($x = -A$) one-half cycle later (at $t = T/2$). You can see that the slope of this x–t graph is zero at $t = 0$, which means that the object has zero velocity at $t = 0$. In other words, the value $\phi = 0$ corresponds to starting the object at rest at $t = 0$ at the position $x = A$. That's just the motion that we depicted in Figure 12-3.

By comparison **Figure 12-10b** is the graph that corresponds to Equation 12-14, $x = A\cos(\omega t + \phi)$, for a case in which ϕ is *not* zero but has a positive value. As the figure shows, you can think of the phase angle

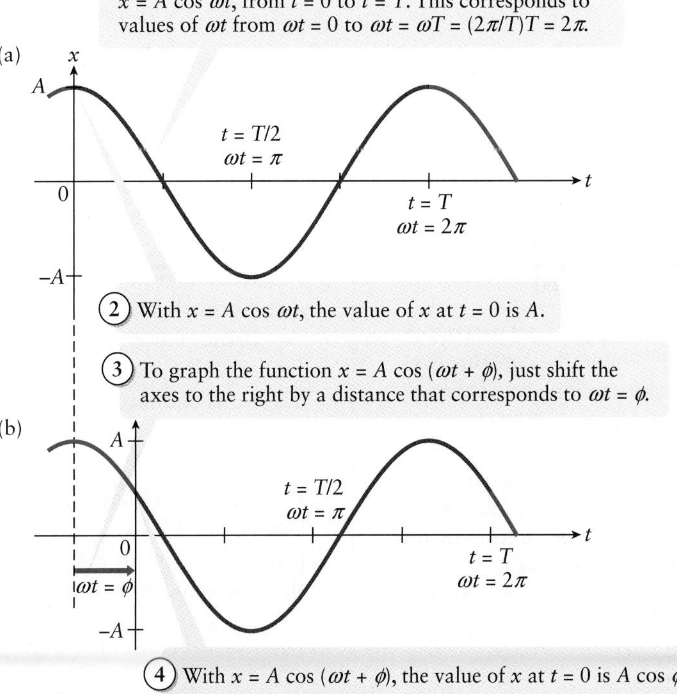

1 The blue curve graphs the first full cycle of the function $x = A\cos\omega t$, from $t = 0$ to $t = T$. This corresponds to values of ωt from $\omega t = 0$ to $\omega t = \omega T = (2\pi/T)T = 2\pi$.
(a)
x
A
$t = T/2$
$\omega t = \pi$
0
$t = T$
$\omega t = 2\pi$
t
−A
2 With $x = A\cos\omega t$, the value of x at $t = 0$ is A.
3 To graph the function $x = A\cos(\omega t + \phi)$, just shift the axes to the right by a distance that corresponds to $\omega t = \phi$.
(b)
A
$t = T/2$
$\omega t = \pi$
0
$\omega t = \phi$
$t = T$
$\omega t = 2\pi$
t
−A
4 With $x = A\cos(\omega t + \phi)$, the value of x at $t = 0$ is $A\cos\phi$.

Figure 12-10 Phase angle The two cycles shown in blue differ by a phase angle ϕ.

as shifting the axes to the right if ϕ is positive, as in Figure 12-10b, or to the left when ϕ is negative. Note that in this x–t graph the value of x at $t = 0$ is *not* $+A$; rather, from Equation 12-14 it is equal to

$$x(0) = A \cos (0 + \phi) = A \cos \phi$$

So for an object in simple harmonic motion the value of the phase angle ϕ tells you the point in the oscillation cycle that corresponds to $t = 0$. Note that Figures 12-10a and 12-10b really depict the *same* oscillation, with the same amplitude A and period T; the only difference is the point in the cycle that we call $t = 0$.

The phase angle also appears in the expressions for velocity and acceleration in simple harmonic motion (Equations 12-9 and 12-10):

EQUATION IN WORDS
Velocity as a function of time for simple harmonic motion

Velocity of an object in SHM at time t

Amplitude of the oscillation = maximum displacement from the equilibrium position

$$v_x = -\omega A \sin (\omega t + \phi) \tag{12-16}$$

Angular frequency of the oscillation ($\omega = 2\pi f = 2\pi/T$)

Phase angle

EQUATION IN WORDS
Acceleration as a function of time for simple harmonic motion

Acceleration of an object in SHM at time t

Amplitude of the oscillation = maximum displacement from the equilibrium position

$$a_x = -\omega^2 A \cos (\omega t + \phi) \tag{12-17}$$

Angular frequency of the oscillation ($\omega = 2\pi f = 2\pi/T$)

Phase angle

In the case $\phi = 0$, Equations 12-16 and 12-17 become

$$\begin{aligned} v_x &= -\omega A \sin \omega t \\ a_x &= -\omega^2 A \cos \omega t \end{aligned} \tag{12-18}$$

(simple harmonic motion, phase angle $\phi = 0$)

Figure 12-11 graphs x versus t, v_x versus t, and a_x versus t for the case $\phi = 0$. In this case the x–t graph is a cosine function, the v_x–t graph is a negative sine function, and the a_x–t graph is a negative cosine function. You can confirm that these graphs are consistent with each other by recalling from Section 2-4 that the slope of the x–t graph equals the velocity v_x and the slope of the v_x–t graph equals the acceleration a_x. Note also that the a_x–t graph looks like the negative of the x–t graph. That's what we expect from Hooke's law, $a_x = -(k/m)x$ (Equation 12-4), which tells us that the acceleration in simple harmonic motion is proportional to the negative of the displacement x.

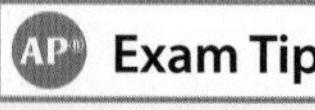

Phase angle is not part of the AP® Physics 1 exam. However, if you continue your study of physics, you will learn about two waves being "in phase" or "out of phase," which relies on phase angle, and you will be tested on that concept on the AP® Physics 2 exam.

If the phase angle ϕ is not equal to zero, the vertical axes of all of the graphs in Figure 12-11 are shifted to the left or right as in Figure 12-10. In this case the velocity at $t = 0$ is equal to

$$v_x(0) = -\omega A \sin (0 + \phi) = -\omega A \sin \phi$$

If $\phi = 0$, the velocity at $t = 0$ is zero and the oscillating object is at rest at that instant (see Figure 12-11). If ϕ is not zero, at $t = 0$ the velocity may not be zero. Figure 12-10b illustrates this: In this case the x–t graph has a negative slope at $t = 0$, which shows that the velocity at that instant is not zero.

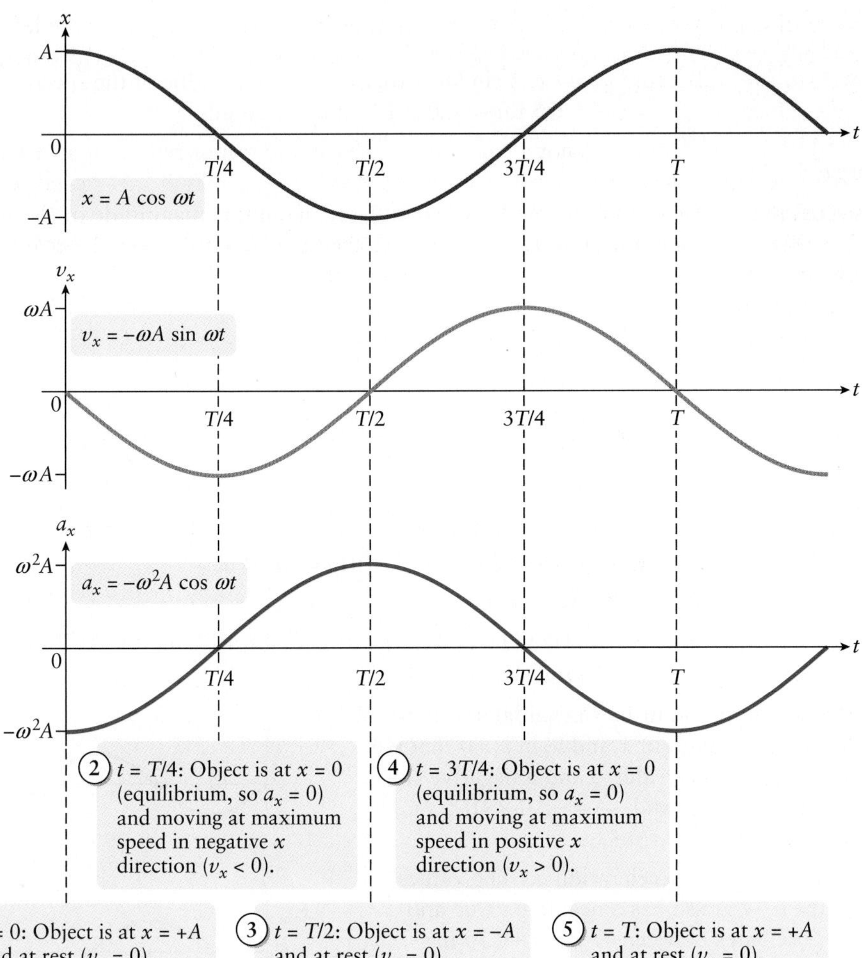

x
A
0
−A
T/4
T/2
3T/4
T
t
x = A cos ωt
v_x
ωA
−ωA
v_x = −ωA sin ωt
a_x
ω²A
−ω²A
a_x = −ω²A cos ωt
1 t = 0: Object is at x = +A and at rest (v_x = 0). Acceleration is negative (toward x = 0).
2 t = T/4: Object is at x = 0 (equilibrium, so a_x = 0) and moving at maximum speed in negative x direction (v_x < 0).
3 t = T/2: Object is at x = −A and at rest (v_x = 0). Acceleration is positive (toward x = 0).
4 t = 3T/4: Object is at x = 0 (equilibrium, so a_x = 0) and moving at maximum speed in positive x direction (v_x > 0).
5 t = T: Object is at x = +A and at rest (v_x = 0). Acceleration is negative (toward x = 0).

Figure 12-11 Displacement, velocity, and acceleration in SHM These graphs show displacement x (Equation 12-15) as well as velocity v_x and acceleration a_x (Equations 12-18) for simple harmonic motion for the case $\phi = 0$. Compare to Figure 12-3, which shows the same motion.

EXAMPLE 12-3 SHM II: Position, Velocity, and Acceleration

As in Example 12-2, an object of mass 0.80 kg is attached to a horizontal ideal spring of spring constant 1.8×10^2 N/m and set into oscillation (see Figure 12-3) on a surface with negligible friction. The amplitude of the motion is 2.0×10^{-2} m, and the phase angle is $\phi = \pi/2$. (a) Find the maximum speed of the object and the maximum magnitude of its acceleration during the oscillation. (b) Find the position, velocity, and acceleration of the object at $t = 0$. (c) Sketch graphs of the position, velocity, and acceleration of the object as functions of time.

Set Up

We can use the simple harmonic motion equations because the spring is ideal (it obeys Hooke's law). From Example 12-2 we know that the angular frequency is $\omega = \sqrt{k/m} = 15$ rad/s, and we are given the values of the amplitude $A = 2.0 \times 10^{-2}$ m and the phase angle $\phi = \pi/2$. Notice that a_x from Equation 12-17 is equal to $-\omega^2$ multiplied by x from Equation 12-14.

Position as a function of time in SHM:

$$x = A \cos(\omega t + \phi) \qquad (12\text{-}14)$$

Velocity as a function of time in SHM:

$$v_x = -\omega A \sin(\omega t + \phi) \qquad (12\text{-}16)$$

Acceleration as a function of time in SHM:

$$a_x = -\omega^2 A \cos(\omega t + \phi) \qquad (12\text{-}17)$$

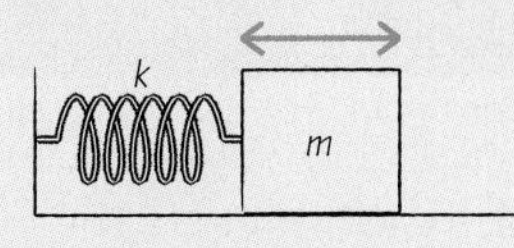

Solve

(a) Equation 12-16 describes the velocity v_x as a function of time. The sine function has values from +1 to −1, so v_x has values from $+\omega A$ to $-\omega A$. The maximum speed is therefore ωA.

Because $v_x = -\omega A \sin(\omega t + \phi)$, the maximum value of the speed is $v_{max} = \omega A = (15 \text{ rad/s})(2.0 \times 10^{-2} \text{ m}) = 0.30 \text{ m/s}$

(A radian is not a true unit, so we can insert it or remove it as needed.)

In the same way, Equation 12-17 describes the acceleration a_x as a function of time. The cosine varies from +1 to −1, so a_x has values from $+\omega^2 A$ to $-\omega^2 A$ and the maximum magnitude of a_x is $\omega^2 A$.

Because $a_x = -\omega^2 A \cos(\omega t + \phi)$, the maximum magnitude of the acceleration is $a_{max} = \omega^2 A = (15 \text{ rad/s})^2 (2.0 \times 10^{-2} \text{ m}) = 4.5 \text{ m/s}^2$

(b) To find the values of x, v_x, and a_x at $t = 0$, substitute the values of A, ω, and ϕ as well as $t = 0$ into Equations 12-14, 12-16, and 12-17. Note that $\cos(\pi/2) = 0$ and $\sin(\pi/2) = 1$.

Position at $t = 0$:

$$x(0) = A\cos(0 + \phi) = (2.0 \times 10^{-2} \text{ m}) \cos(\pi/2) = (2.0 \times 10^{-2} \text{ m})(0) = 0$$

Velocity at $t = 0$:

$$v_x(0) = -\omega A \sin(0 + \phi) = -(15 \text{ rad/s})(2.0 \times 10^{-2} \text{ m}) \sin(\pi/2) = -(15 \text{ rad/s})(2.0 \times 10^{-2} \text{ m})(1) = -0.30 \text{ m/s}$$

Acceleration at $t = 0$:

$$a_x(0) = -\omega^2 A \cos(0 + \phi) = -(15 \text{ rad/s})^2 (2.0 \times 10^{-2} \text{ m}) \cos(\pi/2) = -(15 \text{ rad/s})^2 (2.0 \times 10^{-2} \text{ m})(0) = 0$$

(c) To draw the x–t graph, recognize that this graph must be sinusoidal and that the slope of the x–t graph is the velocity v_x. The values of x and v_x at $t = 0$ that we found in part (b) tell us that at $t = 0$ the x–t graph has value 0 and a negative slope. The value of x oscillates between $+A = +2.0 \times 10^{-2}$ m and $-A = -2.0 \times 10^{-2}$ m. The graph repeats itself after a time t equal to the period $T = 0.42$ s.

The v_x–t graph is also sinusoidal, and its slope is the acceleration a_x. The values of v_x and a_x at $t = 0$ tell us that at $t = 0$ the v_x–t graph has a negative value and zero slope. From part (a), the value of v_x oscillates between $+v_{max} = +0.30$ m/s and $-v_{max} = -0.30$ m/s. The graph repeats itself after a time t equal to the period $T = 0.42$ s.

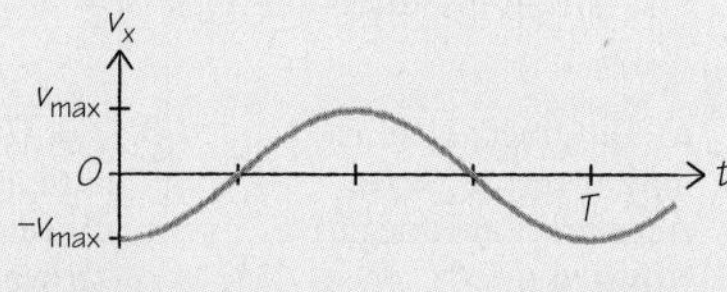

Finally, comparing Equations 12-14 and 12-17 shows that $a_x = -\omega^2 x$, so the a_x–t graph looks like the negative of the x–t graph (compare Figure 12-11). From part (a), the value of a_x oscillates between $+a_{max} = +4.5 \text{ m/s}^2$ and $-a_{max} = -4.5 \text{ m/s}^2$. The graph repeats itself after a time t equal to the period $T = 0.42$ s. The shape makes sense, given that the acceleration would be zero when the velocity is maximum, and is always oppositely directed to the displacment of the object from the relaxed equilibrium of the spring.

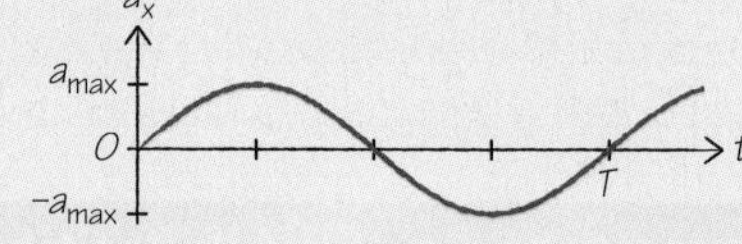

Reflect

To start an oscillation going so that the graphs look like what we have drawn, begin with the block at rest at equilibrium ($x = 0$) and give it a sharp hit (impulse) with your hand in the negative x direction so that it starts with a negative x velocity v_x at $t = 0$. Note that the x–t, v_x–t, and a_x–t graphs look like those in Figure 12-11, but with the vertical axis shifted to the right by a quarter-cycle. That's what we would expect: Figure 12-10 shows that the effect of a phase angle ϕ is to shift the vertical axis horizontally by ϕ, and $\pi/2$ is one-quarter of 2π (the number of radians in a complete cycle).

NOW WORK Problems 4 and 5 from The Takeaway 12-4.

Our discussion of simple harmonic motion has concentrated on blocks attached to springs. But the same basic equations, and the sinusoidal graph of the oscillation as a function of time, apply to many different oscillating systems. **Figure 12-12a** shows an example from cell biology. In some cells the intracellular concentration of calcium ions (Ca^{2+}) oscillates sinusoidally. (These oscillations can regulate enzyme activity, mitochondrial metabolism, or gene expression.) The red sinusoidal curve drawn on top of the data matches these oscillations well. **Figure 12-12b** shows another biological example, the volume of air in a person's lungs as a function of time. Again we've drawn a red sinusoidal curve on top of the data; the close match between the curves shows that the oscillations are well-modeled as simple harmonic motion.

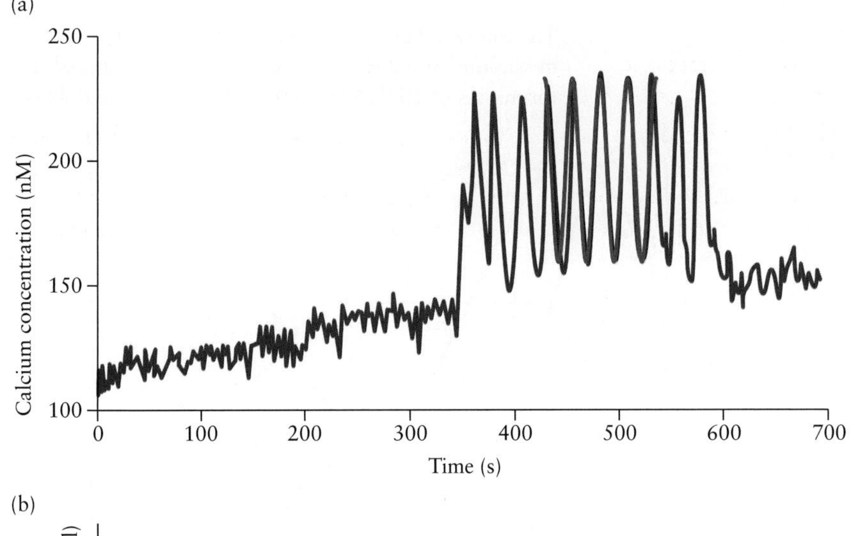

Figure 12-12 Simple harmonic motion in physiology (a) The concentration of calcium ions (Ca^{2+}) within a cell can oscillate under certain conditions. A sinusoidal curve (red) is a good match to these oscillations. (b) This graph shows the volume of air in a person's lungs as a function of time. The oscillations in lung volume are also well matched to a sinusoidal curve (red).

THE TAKEAWAY for Section 12-4

✔ The amplitude of an oscillation is the maximum positive displacement from equilibrium.

✔ The motion of an object oscillating under the influence of a Hooke's law restoring force is called simple harmonic motion (SHM). The angular frequency, period, and frequency in SHM are unaffected by changes in the oscillation amplitude.

✔ SHM is identical to the projection onto the x axis of the motion of an object in uniform circular motion in a plane including the x axis.

✔ Graphs of the position, velocity, and acceleration in SHM are sinusoidal curves. In each of these graphs the phase angle shifts the location of the vertical axis (where time equals zero) to the right or left.

Prep for the AP® Exam

AP® Building Blocks

1. Answer these questions with complete sentences.
   (a) What are the units of frequency f?
   (b) What is the relationship between ω and f?
   (c) How is the period T related to f?
   (d) How is the period T related to ω?
2. An oscillatory motion is represented by $x = (0.06 \text{ m}) \cos (3\pi t + \pi/2)$, where t is in seconds. What is the period of the motion? What is the amplitude? What is the maximum displacement of the motion? Express the motion in terms of the sine function.
3. A force $-kx$ is exerted by a spring on an object with mass m, causing it to oscillate as $x = A \cos (\omega t)$ with an acceleration $a = -\omega^2 A \cos (\omega t)$.
   (a) Justify the claim that the square of the angular frequency ω^2 is equal to k/m.
   (b) If $k = 2.00 \times 10^2$ N/m and 14 oscillations occur every 16 s, calculate the mass of the object that is attached to the spring.

EX 12-3
4. A block is attached to one end of a horizontal spring. When the spring is relaxed, the block is at rest at $x = 0$ on a table that exerts negligible friction force on the block. The other end of the spring is attached to a stationary support. The spring is then stretched so the block is located at $x = A$, and the block is released. At what point(s) in the resulting simple harmonic motion is the speed of the block at its maximum? Choose the correct answer.
   (A) $x = A$ and $x = -A$ because the displacement of the block relative to the equilibrium position is a maximum at these points.
   (B) $x = A$ because the displacement and the velocity are in phase.
   (C) $x = 0$ because at the equilibrium position the spring–block system potential energy is zero.
   (D) $x = A/2$ because at the equilibrium position the magnitude of the acceleration is a maximum.

EX 12-3 **5.** A system consisting of a 1.2-kg object attached to a spring oscillates on a horizontal surface that exerts a negligible friction force on the object. The mass of the spring is negligible. A graph of the position x of the object as a function of time is shown in the figure.

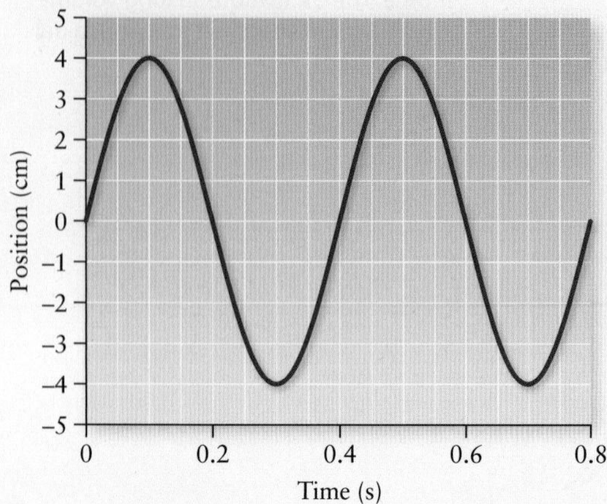

(a) Express the position of the object as a function of time.
(b) What are the frequency and angular frequency of the motion?
(c) What is the spring constant of the spring?
(d) What is the maximum speed of the object?
(e) What is the maximum acceleration of the object?

AP® Skill Builders

6. The functions $x_1(t)$ and $x_2(t)$ represent the SHM of two blocks on a surface that exerts negligible friction on them. The blocks have the same mass and are attached by horizontal ideal springs to the same stationary wall.

$$x_1(t) = 0.1 \text{ m} \cos(\pi t) \text{ and } x_2(t) = 0.1 \text{ m} \cos(2\pi t)$$

(a) Evaluate the ratio of spring constants, k_1 and k_2, for the two functions. Recall that when $x = A\cos(\omega t)$, $a = -A\omega^2\cos(\omega t)$, and $a = kx/m$, so $\omega^2 = k/m$.
(b) Construct a graph of $x_1(t)$ and $x_2(t)$ for $0 \le t \le 2$ s.
(c) The two blocks are moved so that the springs are initially stretched to $x(t = 0) = 0.1$ m and are released simultaneously. Predict the frequency of instances where the blocks are passing each other: $x_1(t) = x_2(t)$.

7. A new material is being developed. The material is assumed to expand according to Hooke's law for small displacements from equilibrium and behave harmonically. However, for larger displacements the material deforms plastically (it does not recover its original shape). An experiment is conducted by measuring the time dependence of the recovery of a sheet of the material following small displacements. These are produced by impacts of small blocks with differing masses dropped from a fixed height. As they strike the surface of the material, it depresses and then it springs back to equilibrium before rising higher and then going back down to equilibrium. This is like starting an object on a spring with an initial velocity at equilibrium at $t = 0$.

The following graphical data (amplitude, in terms of the maximum amplitude, A_{max}, for each block) were collected for masses of 10.0 g (•), 30.0 g (•), 60.0 g (•), and 90.0 g (•).

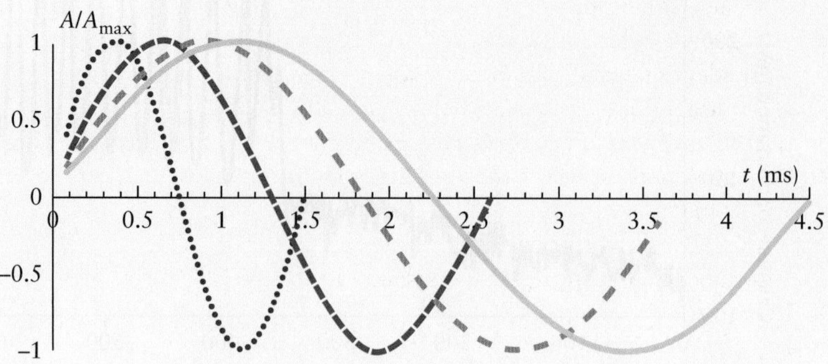

(a) Interpret the graphical data and complete the following data table.

Mass (g)	Period (ms)	Angular frequency (s^{-1})
10.0		
30.0		
60.0		
90.0		

(b) Analyze these data using the relationship among mass, angular frequency, and the spring constant to obtain the value of the spring constant k for this material.

$$\omega^2 = \frac{k}{m}$$

Skills in Action

8. The green curve below, C, is the graph of the position of an oscillatory motion during an interval of 12 s. The other two curves, A and B, represent sketches of the time dependence of the velocity function and the acceleration function. (For them, the vertical axis would be in appropriate units; these are just sketches of time dependence with appropriate increasing or decreasing behavior.)

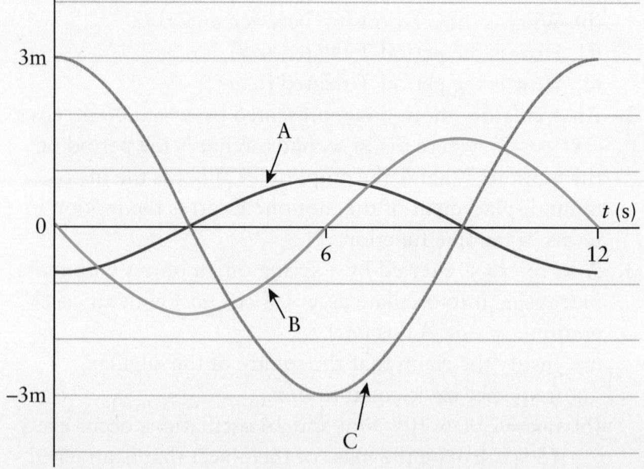

(a) Interpret the data to determine the numerical values of the following properties of the motion.
  i. Period
  ii. Amplitude
  iii. Angular frequency

(b) Identify A and B as either velocity or acceleration and justify your identification.

(c) Identify the times between $t = 0$ and $t = 12$ s of the maximum speed of the oscillatory motion of the object.

9. A 0.200-kg object is attached to the end of a 55.0-N/m ideal horizontal spring. It is displaced 10.0 cm to the right of equilibrium and released on a horizontal surface that exerts negligible friction force on the object. Predict the period of the motion.

10. The acceleration of an object that has a mass of 0.025 kg and exhibits simple harmonic motion is given by $a(t) = (10 \text{ m/s}^2) \cos(\pi t + \pi/2)$.
   (a) Calculate the period of the oscillation.
   (b) Predict the times when the magnitude of the velocity is largest.
   (c) Calculate the spring constant for the oscillator.

12-5 Mechanical energy is constant in simple harmonic motion

We've been discussing how a spring can provide the restoring force necessary for oscillation to take place. But as we learned in Section 7-6, a spring can also be used to store *potential energy*. One creature that applies this principle to oscillation is the kangaroo (**Figure 12-13**). A kangaroo's tendons act like springs to store spring potential energy, which is transformed into kinetic energy as the animal hops; and then the tendons get stretched again as the kangaroo lands. This increases the efficiency of movement when the hopping is repeated. It doesn't work just like an object on a spring, but there are similarities! Dolphins use a similar mechanism in swimming. Even when a dolphin increases its speed by beating its tail more quickly, there is hardly any change in the rate at which it consumes oxygen (an indication of the dolphin's power output). The explanation is that the tissue in a dolphin's tail acts much like a spring, storing potential energy as the tail flips either up or down and then transforming the potential energy into kinetic energy. As a result, the dolphin can swim efficiently at high speeds. There is some increase in power output because the water does resist the motion. The shape of the tail and extremely smooth skin help minimize this, however.

These observations suggest that it's worthwhile to look at oscillations again from an energy perspective. To do this we'll study the interplay between kinetic energy and potential energy for an object–spring system, assuming the spring is ideal (obeys Hooke's law, has negligible mass), and oriented horizontally as in Figures 12-2 and 12-3.

The tendons on the backs of this kangaroo's legs are fully stretched. As such they store spring potential energy...

...which is released during a hop, as the kangaroo straightens its legs and lets its tendons relax...

...and the released spring potential energy is converted into kinetic energy of the kangaroo.

David Tauck

Figure 12-13 Spring potential energy When stretched, tendons in a kangaroo's legs store potential energy in much the same way as a spring.

Our starting point is the expression from Section 7-6 for the potential energy stored in an ideal spring:

EQUATION IN WORDS
Spring potential energy

Spring potential energy of a stretched or compressed spring

Spring constant of the spring

(7-16) $$U_s = \frac{1}{2}kx^2$$

Extension of the spring, when the equilibrium position of the end of the spring when it is relaxed is defined as $x = 0$ ($x > 0$ if spring is stretched, $x < 0$ if spring is compressed)

This equation tells us that the potential energy stored in a spring varies during a cycle of oscillation. When the object is at equilibrium ($x = 0$), the spring is neither stretched nor compressed, and the spring potential energy is zero: $U_s = 0$. Whenever the spring is stretched ($x > 0$) or compressed ($x < 0$), the spring potential energy is greater than zero. The potential energy is greatest when the spring is either at its maximum extension (when $x = A$) or at its maximum compression ($x = -A$). At either of those points $U_s = \frac{1}{2}kA^2$.

The *kinetic* energy of the oscillating object $K = \frac{1}{2}mv^2$ also varies during an oscillation cycle. The object has its maximum speed v_{max} when passing through the equilibrium position, so the kinetic energy has its maximum value $K = \frac{1}{2}mv_{max}^2$ there as well. At any other point in the oscillation the object is moving more slowly than when it passes through equilibrium, so the kinetic energy is less than its maximum value. The kinetic energy has its minimum value $K = 0$ when the oscillating object is momentarily at rest; this happens when the spring is at maximum extension ($x = A$) or maximum compression ($x = -A$).

This analysis shows that for an object oscillating on an ideal spring, the object's kinetic energy is maximum where the spring potential energy is minimum (at $x = 0$), and the object's kinetic energy is minimum where the spring potential energy is maximum (at $x = A$ and $x = -A$). This is just what we would expect! If there are no nonconservative forces such as friction exerted on the system, the total mechanical energy E (the sum of the kinetic energy of the object and the potential energy of the spring) should remain constant, and energy will be transferred back and forth between kinetic energy of the object and potential energy stored in the spring as the object oscillates. Because we are modeling the spring as ideal, it has negligible mass, and therefore only the object has kinetic energy. Thus, the kinetic energy of the system and of the object are the same.

To analyze this in more detail, let's look at the expression for the total mechanical energy E of the system of object and spring:

(12-19) $$E = K + U_s = \frac{1}{2}mv^2 + \frac{1}{2}kx^2$$

Let's use our results from Section 12-4 to see how E, K, and U_s change during the course of an oscillation. From Equation 12-14 the position as a function of time is $x = A\cos(\omega t + \phi)$, so the spring potential energy as a function of time is

(12-20) $$U_s = \frac{1}{2}kx^2 = \frac{1}{2}kA^2\cos^2(\omega t + \phi)$$

Because the value of $\cos(\omega t + \phi)$ ranges from +1 through zero to −1, you can see that the value of $\cos^2(\omega t + \phi)$ ranges from 0 to 1. So the value of U_s ranges from 0 to a maximum value $\frac{1}{2}kA^2$, just as we reasoned earlier.

Equation 12-16 tells us that the velocity as a function of time is $v_x = -\omega A\sin(\omega t + \phi)$. The velocity can be negative, but the square of the velocity is positive and equal to the square of the speed. Substituting this speed, the kinetic energy of the object as a function of time is

$$K = \frac{1}{2}mv^2 = \frac{1}{2}m\omega^2A^2\sin^2(\omega t + \phi)$$

From the first of Equations 12-13, we know that the angular frequency for an object oscillating on the end of an ideal spring is $\omega = \sqrt{k/m}$. So $m\omega^2 = m(k/m) = k$, and we can rewrite our expression for the kinetic energy as a function of time as

$$K = \frac{1}{2}mv^2 = \frac{1}{2}kA^2 \sin^2 (\omega t + \phi) \qquad (12\text{-}21)$$

Like the cosine function the value of $\sin(\omega t + \phi)$ ranges from +1 through zero to −1, and the value of $\sin^2 (\omega t + \phi)$ ranges from 0 to 1. So just like the spring potential energy, the kinetic energy of the object ranges in value from 0 to a maximum value $\frac{1}{2}kA^2$.

We can now insert our expressions for K from Equation 12-21 and U_s from Equation 12-20 into Equation 12-19 for the total mechanical energy E. We get

$$E = K + U_s = \frac{1}{2}kA^2 \sin^2 (\omega t + \phi) + \frac{1}{2}kA^2 \cos^2 (\omega t + \phi)$$

$$= \frac{1}{2}kA^2 \left[\sin^2 (\omega t + \phi) + \cos^2 (\omega t + \phi)\right] \qquad (12\text{-}22)$$

Equation 12-20 looks like a complicated function of time. But we can simplify it thanks to an important result from trigonometry:

$$\sin^2 \theta + \cos^2 \theta = 1 \text{ for any value of } \theta$$

Then Equation 12-22 becomes

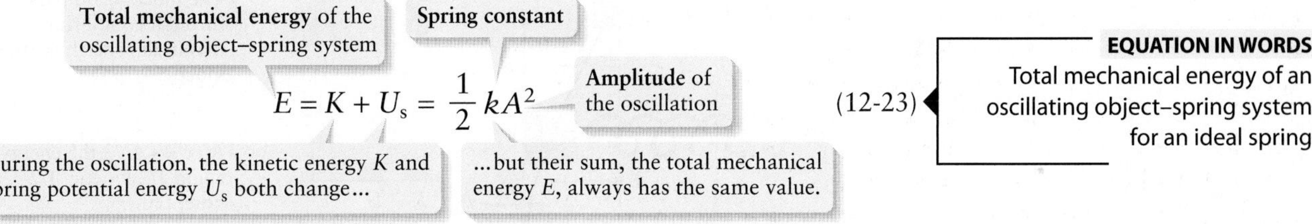

Because both k and A are constant for a specific motion of a given oscillator, the total energy of the oscillating object–spring system is constant. This is a statement of the conservation of energy: The energy of the system is transferred between kinetic and potential, but the total mechanical energy remains constant. This is shown in **Figure 12-14,** where for the object we have chosen a block.

Equations 12-20 and 12-21 give the spring potential energy and kinetic energy of the block as functions of time. It's also helpful to write these relationships as functions of the displacement x. We know from Equation 7-16 that the spring potential energy is $U_s = \frac{1}{2}kx^2$; the graph of this as a function of x is a parabola with its minimum value at $x = 0$ (**Figure 12-15**). From Equation 12-23 the kinetic energy K of the block is equal to the total mechanical energy $E = \frac{1}{2}kA^2$ (a constant) minus the potential energy U_s:

$$K = E - U_s = \frac{1}{2}kA^2 - \frac{1}{2}kx^2 \qquad (12\text{-}24)$$

This is an "upside-down" parabola that has its *maximum* value at $x = 0$, as Figure 12-15 shows.

Note that U_s is greatest at the two extremes of displacement ($x = +A$ and $x = -A$) and zero at equilibrium ($x = 0$). The kinetic energy of the object K is zero at the two extremes, where the object momentarily comes to a stop, and greatest as the object passes through $x = 0$. So at $x = 0$ the energy is 100% kinetic and 0% potential, while at $x = +A$ and $x = -A$ the energy is 0% kinetic and 100% potential.

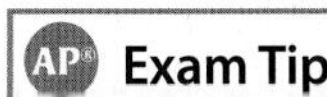

Exam Tip

When an object–spring system is oriented vertically instead of horizontally and gravitational force is the only additional force that is exerted along the spring force direction, the equilibrium length of the spring changes. If x is measured from the *new* equilibrium position, then Equation 12-19 still holds! Gravitational potential energy is completely included by redefining the equilibrium position. The proof of this is in problem 5 in The Takeaway 12-5; the equations of motion for a vertical ideal spring are the same as for a horizontal one.

Figure 12-14 Energy during a cycle of oscillation The bar graphs show the kinetic energy K of the block, spring potential energy U_s, and total mechanical energy $E = K + U_s$ at nine stages in the oscillation of a block–spring system (compare with Figure 12-3).

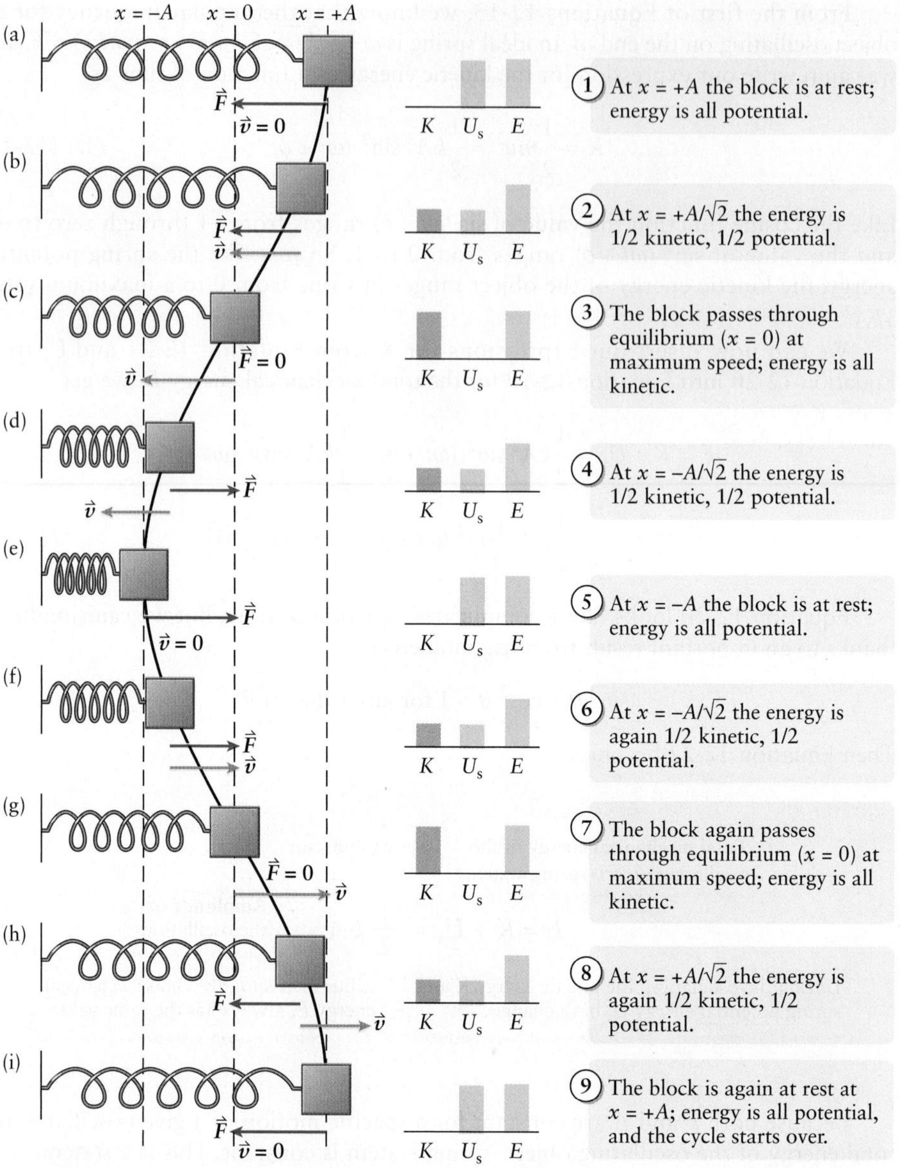

x = –A
x = 0
x = +A
(a)
(b)
(c)
(d)
(e)
(f)
(g)
(h)
(i)
F
v = 0
F = 0
v
K
U_s
E
1 At x = +A the block is at rest; energy is all potential.
2 At x = +A/√2 the energy is 1/2 kinetic, 1/2 potential.
3 The block passes through equilibrium (x = 0) at maximum speed; energy is all kinetic.
4 At x = –A/√2 the energy is 1/2 kinetic, 1/2 potential.
5 At x = –A the block is at rest; energy is all potential.
6 At x = –A/√2 the energy is again 1/2 kinetic, 1/2 potential.
7 The block again passes through equilibrium (x = 0) at maximum speed; energy is all kinetic.
8 At x = +A/√2 the energy is again 1/2 kinetic, 1/2 potential.
9 The block is again at rest at x = +A; energy is all potential, and the cycle starts over.

The kinetic energy of the block K and potential energy U_s change during a cycle of oscillation, but the total mechanical energy $E = K + U_s$ remains constant.

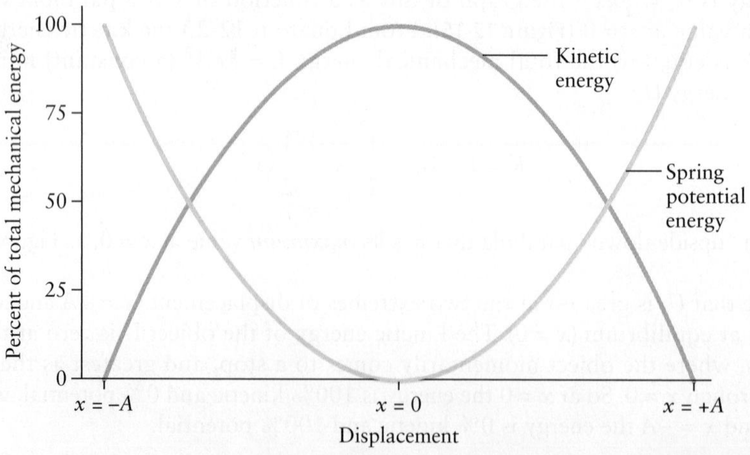

Figure 12-15 Kinetic energy and potential energy versus displacement The total mechanical energy—the sum of the kinetic energy of the object and spring potential energy—associated with an object–spring system in simple harmonic motion is constant. The percentage of the total mechanical energy that is kinetic and potential depends on the object's displacement from equilibrium.

100
75
50
25
0
Percent of total mechanical energy
Kinetic energy
Spring potential energy
x = –A
x = 0
x = +A
Displacement

EXAMPLE 12-4 SHM III: Kinetic and Potential Energy

As in Example 12-3 an object of mass 0.80 kg is attached to a horizontal ideal spring of spring constant 1.8×10^2 N/m. The object is initially at rest at equilibrium. You then start the object oscillating with amplitude 2.0×10^{-2} m on a horizontal surface of negligible friction. (a) How much work do you have to do on the object to set it into oscillation? (b) What is the speed of the object when the spring is compressed by 1.0×10^{-2} m? (c) How far is the object from equilibrium when the kinetic energy of the object equals the spring potential energy?

Set Up

We'll use energy ideas to answer these questions. Equation 12-23 will let us find the (constant) total mechanical energy E of the system from the given spring constant and amplitude. Equation 7-16 will let us find the spring potential energy for any displacement x; we can then find the kinetic energy K of the object using Equation 12-23, and from that find the speed for that value of x.

Total mechanical energy of an oscillating object–spring system for an ideal spring:

$$E = K + U_s = \frac{1}{2}kA^2 \qquad (12\text{-}23)$$

Spring potential energy:

$$U_s = \frac{1}{2}kx^2 \qquad (7\text{-}16)$$

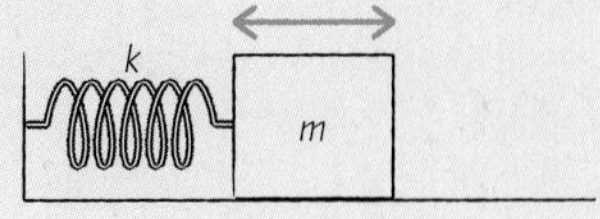

Solve

(a) Initially the system has zero kinetic energy and zero potential energy. You can start the oscillation by pulling the object so that you stretch the spring by a distance $A = 2.0 \times 10^{-2}$ m from equilibrium. The work you do goes into the potential energy of the spring.

(work you do) = (change of potential energy of the spring as you stretch it from $x = 0$ to $x = A$)

Initial potential energy

$$= \frac{1}{2}k(0)^2 = 0$$

Final potential energy $= \frac{1}{2}kA^2$

So the work you do is

$$W = \frac{1}{2}kA^2 - 0$$

$$= \frac{1}{2}(1.8 \times 10^2 \text{ N/m})(2.0 \times 10^{-2} \text{ m})^2$$

$$= 3.6 \times 10^{-2} \text{ Nm} = 3.6 \times 10^{-2} \text{ J}$$

(b) When the spring is compressed by 1.0×10^{-2} m, the displacement is $x = -1.0 \times 10^{-2}$ m. The first step in determining the object's speed at this value of x is to determine its kinetic energy.

From (a) the total mechanical energy is

$$E = \frac{1}{2}kA^2 = 3.6 \times 10^{-2} \text{ J}$$

From Equation 7-16 the spring potential energy for $x = -1.0 \times 10^{-2}$ m is

$$U_s = \frac{1}{2}kx^2$$

$$= \frac{1}{2}(1.8 \times 10^2 \text{ N/m})(-1.0 \times 10^{-2} \text{ m})^2$$

$$= 9.0 \times 10^{-3} \text{ J}$$

So from Equation 12-23 the kinetic energy at this value of x is

$$K = E - U_s = 3.6 \times 10^{-2} \text{ J} - 9.0 \times 10^{-3} \text{ J}$$

$$= 2.7 \times 10^{-2} \text{ J}$$

Given the kinetic energy K of the object and its mass $m = 0.80$ kg, calculate the speed of the object.

Solve for the speed v:

Kinetic energy is $K = \frac{1}{2}mv^2$, so

$$v^2 = \frac{2K}{m}$$

$$v = \sqrt{\frac{2K}{m}} = \sqrt{\frac{2(2.7 \times 10^{-2}\ \text{J})}{0.80\ \text{kg}}} = 0.26\ \text{m/s}$$

(Recall from Chapter 7 that if the kinetic energy is in joules and the mass is in kilograms, the speed is in meters per second.)
The *velocity* of the object could be +0.26 m/s or –0.26 m/s, depending on what direction it is moving as it passes through this point.

(c) Use the same ideas as in part (b) to solve for the value of x at which the kinetic energy K equals the spring potential energy U_s.

From earlier, the total mechanical energy is

$$E = \frac{1}{2}kA^2 = 3.6 \times 10^{-2}\ \text{J}$$

The spring potential energy is

$$U_s = \frac{1}{2}kx^2$$

and the kinetic energy is

$$K = E - U_s$$

At the value of x for which $K = U_s$,

$$\frac{1}{2}kA^2 - \frac{1}{2}kx^2 = \frac{1}{2}kx^2$$

Multiply both sides by $2/k$ and solve for x:

$$A^2 - x^2 = x^2 \quad \text{so} \quad 2x^2 = A^2 \quad \text{and} \quad x^2 = \frac{A^2}{2}$$

$$x = \pm\sqrt{\frac{A^2}{2}} = \pm\frac{A}{\sqrt{2}}$$

$$= \pm\frac{2.0 \times 10^{-2}\ \text{m}}{\sqrt{2}} = \pm 1.4 \times 10^{-2}\ \text{m}$$

Note that x can be either positive (the spring is stretched by 1.4×10^{-2} m) or negative (the spring is compressed by 1.4×10^{-2} m).

Reflect

The points $x = \pm A/\sqrt{2} = \pm 0.71A$, where the kinetic energy of the object and potential energy stored in the spring are equal, are as shown in Figure 12-14 [see parts (b), (d), (f), and (h) of that figure]. These points are *not* halfway between the equilibrium position ($x = 0$) and the extremes of the motion ($x = \pm A$); they are actually closer to the extremes. This makes sense as potential energy depends on x^2.

Notice in the previous example, that the result for part (b) would have been very difficult to find without using the energy approach. You would have needed to use the equations from Section 12-4 to solve for the time t at which the object passes through $x = -1.0 \times 10^{-2}$ m, then use this value of t to find the velocity of the object at that time. The energy approach makes this much easier.

NOW WORK Problems 1–3 from The Takeaway 12-5.

THE TAKEAWAY for Section 12-5

✔ Energy is transformed back and forth from kinetic energy to potential energy in systems that contain mechanical or biological springs.

✔ When an object attached to a spring is at its maximum displacement, the energy of the object–spring system is entirely potential energy stored in the spring.

✔ When the object passes through equilibrium, the energy of the object–spring system is entirely kinetic energy of the object.

✔ The total energy of an object–spring system, the maximum potential energy, and the maximum kinetic energy are all equal to $\frac{1}{2}kA^2$.

Prep for the AP® Exam

AP® Building Blocks

EX 2-4 1. An object–spring system undergoes simple harmonic motion. If the amplitude increases but the mass of the object is not changed, which of the following answer choices is correct? Support your acceptance or rejection of each of the claims using physical principles.
(A) The total energy of the system increases.
(B) The total energy of the system decreases.
(C) The total energy of the system doesn't change.
(D) The total energy of the system undergoes a sinusoidal change.

EX 2-4 2. A 0.250-kg object attached to an ideal spring oscillates on a horizontal table that exerts negligible friction on the object. The object oscillates with a frequency of 4.00 Hz and an amplitude of 20.0 cm.
(a) Calculate the maximum potential energy U_{max} of the system.
(b) Calculate the displacement of the object and the first time when the potential energy of the system is equal to $U_{max}/2$ if the object is at its maximum amplitude at $t = 0$.
(c) Calculate the kinetic energy when the displacement is 10.0 cm.

EX 12-4 3. The potential energy of an object–spring system is 2.4 J at a location where the kinetic energy of the object is 1.6 J. The amplitude of the simple harmonic motion is 20.0 cm.
(a) Calculate the spring constant.
(b) Calculate the maximum value of the force exerted on the object by the spring.

AP® Skill Builders

4. Galileo was one of the first scientists to observe that the period of a simple harmonic oscillator is independent of its amplitude.

One end of a spring is attached to a rigid wall. The other end of the spring is attached to an object with mass. The object is initially lying at rest at the relaxed equilibrium length of the spring, perpendicular to the rigid wall on a horizontal surface that exerts a negligible friction force on the object. Define this initial position as the origin of the coordinate system. The object is pulled a distance R from this equilibrium and released from rest. After time T, the object again passes through the position R. Use the fact that the period of the oscillation is independent of the amplitude, even though the energy of the system, E, depends on the amplitude, to find a ratio in terms of the lowest powers of E and R that gives a constant. What physical property of the system is related to this constant?

5. In this section, we said that we could ignore gravitational potential energy in a vertical spring–object system. In this problem, we are going to convince ourselves that there is a reason it works, and that what we know about conservation of energy makes sense. In the figure, the spring on the left represents the spring when no object is on it, which is attached to a rigid support (this is then the relaxed equilibrium length of the spring; we will call the position of the end of the spring $x = 0$). On the right, an object with mass m hangs at rest suspended from the same spring. The spring constant is k. Take upward to be positive throughout the problem.

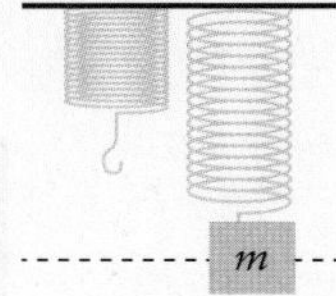

(a) When the object was attached to the spring, and the system reached a new equilibrium, the length of the spring increased by the amount Δx_{eq}. Show that this increase in length is mg/k.
(b) For the object–spring system, let's define a new coordinate to represent the oscillation, so that our equilibrium is this new position of the object, $y = 0$, where $y = x - mg/k$. What is the value we are setting for the gravitational potential energy at $y = 0$ if we are saying that the potential energy of the Earth–object–spring system is zero when $y = 0$? Remember, you cannot redefine the zero of a spring's potential energy, but you can always set any height to be zero for gravitational potential energy of an Earth–object system. What this means is you can define any height to be a particular constant, as long as you carry that constant through.
(c) At the highest point of the oscillation, when $y = A$, the displacement of the spring from its relaxed equilibrium length $x = 0$ is then $x = A - mg/k$. At the bottom of the oscillation, when $y = -A$, the displacement of the spring from its relaxed equilibrium length $x = 0$ is then $x = -(A + mg/k)$. Show that at both of these points, the total potential energy for the system measured from $y = 0$, including gravitational potential energy, is $1/2\ kA^2$. The math will work out a lot more easily if

you calculate the two types of potential energy separately, and write the gravitational potential energy of the system in the form $U_{grav} = U_{grav}(y = 0) + mg\Delta y$. (This is why the shift of equilibrium point allows us to apparently ignore gravity!)

6. Consider a linear restoring force $F = -kx$, and a nonlinear force $F = -kx - kx^2/2$. Justify your acceptance or rejection of each of the following claims, A and B, about the dynamics of systems constrained by either of these forces after being released from an initial displacement of $x = 1$ m.
   (A) The system with the linear restoring force will oscillate about $x = 0$ m between $x = 1$ m and $x = -1$ m because the acceleration is zero at $x = 0$ m. The system with the nonlinear restoring force will oscillate about $x = 0$ m between $x = 1$ m and $x = -3$ m because the acceleration is zero at $x = 0$ m and the potential energy at these two points is the same.
   (B) The system with the linear restoring force will oscillate about $x = 0$ m between $x = 1$ m and $x = -1$ m because as the potential energy decreases between $x = 1$ m and $x = 0$ m, the kinetic energy increases such that the sum is constant. As the object moves between $x = 0$ m and $x = -1$ m, the kinetic energy decreases. The system with the nonlinear restoring force will not oscillate because the potential energy change is not converted to kinetic energy for $x < 0$ and there is no stopping point.

AP® Skills in Action

7. An object attached to the end of a spring slides on a horizontal surface with negligible friction. The motion is simple harmonic between A and $-A$. Predict the times at which the object's kinetic energy and the potential energy of the system have the same value.
8. The potential energy of a simple harmonic oscillator is given by $U = kx^2/2$, when $x = 0$ is the relaxed equilibrium position. The spring constant is $k = 10$ N/m.
   (a) If $x(t) = A\cos(\omega t)$, where $A = 0.25$ m and $\omega = 2\pi\ \text{s}^{-1}$, plot the potential energy versus time for three full periods of motion.
   (b) Explain why the period of the potential energy $U(t)$ is different than the period of $x(t)$.
   (c) Calculate the value of the mass of the oscillating object.
   (d) Write the expression for the kinetic energy, using the fact that $U(t) + K(t) =$ constant and the fact that $\cos^2(\theta) + \sin^2(\theta) = 1$, as we showed in deriving Equation 12-23.
   (e) Add the plot of the kinetic energy $K(t)$ to your graph.

12-6 The motion of a pendulum is approximately simple harmonic

You've probably had a physician strike the patellar tendon just below your kneecap with a hammer then watch your lower leg swing upward and back (**Figure 12-16a**). For a physician this is a test of your patellar reflex: Hitting that tendon sends a signal to your spinal cord, which in turn sends a signal to the quadriceps muscle on your upper thigh. This makes that muscle flex and makes the lower leg move. But to a physicist this is the same kind of motion as an object swinging at the end of a string (**Figure 12-16b**). When allowed to move freely, both your leg and the object hang straight down in their equilibrium positions thanks to the influence of gravity. When either your leg or the object is displaced from equilibrium, released, and allowed to move freely, the force of gravity pulls it back toward equilibrium—that is, gravity serves as a restoring force. In either case inertia causes the leg or object to overshoot the equilibrium position, resulting in an oscillation. Both a swinging leg and an object on the end of a string are examples of **pendula**—systems that oscillate back and forth due to the restoring force of gravity.

Because pendula actually *rotate* around a point, we'll describe them using the language of rotational motion that we developed in Chapter 10. So instead of relating a restoring force to the acceleration that it produces, as we did in Section 12-4 for a block attached to an ideal spring, we'll describe pendulum motion in terms of a restoring *torque* that produces an *angular* acceleration.

In this section we'll explore oscillations of a **simple pendulum**. This is one in which the mass of the system is concentrated at a single point. It's an idealized version of the cylinder–string system shown in Figure 12-16b. In the idealized version, the string of length L has zero mass and we can use the object model for whatever is tied to the end of the string such that the string length is also the pendulum length. For the cylinder, to model it as an object we treat its location as being its center of mass, and the length of the pendulum actually represents the length from the pivot to that center of mass, not just the length of the string. For the object model to be valid, the string must be much longer than whatever we swing from it. When we refer to L as the length of the pendulum, we are referring to this distance from the pivot to the center of mass representing the location of the object. (In Section 12-7 we'll look at other, less idealized pendula.)

Figure 12-17a shows a simple pendulum. In equilibrium the string hangs straight down; the figure shows the pendulum displaced by an angle θ from the vertical. The blue dashed curve shows the path that the object of mass m at the end of the pendulum will take to return to equilibrium. **Figure 12-17b** shows the free-body diagram for this object. The two forces exerted on this object are a tension force of magnitude F_T exerted by the string and a gravitational force of magnitude mg. The tension force points along the string, so the line of action of the tension force passes through the pivot point. Hence the tension force exerts no torque around the pivot (it does nothing to make the pendulum rotate around the pivot). The gravitational force does exert a torque, however. From Equation 10-16 in Section 10-5 we find the torque's magnitude is the perpendicular component of the gravitational force, which, if the pendulum is at an angle θ from the vertical, is $mg \sin \theta$, multiplied by distance from the pivot to where the force is applied, L. Figure 12-17b shows that we get the same answer if we think about torque in terms of the magnitude of the force, mg, and the lever arm (the perpendicular distance to the line of action of the force), which is $L \sin \theta$. Hence, we can write the torque as

$$\tau_z = -mgL \sin \theta \qquad (12\text{-}25)$$

The subscript z indicates that the pendulum tends to rotate around the z axis shown in Figure 12-17a. The minus sign in Equation 12-25 indicates that this is a *restoring* torque. If the pendulum swings to the right of vertical (counterclockwise) so that θ and $\sin \theta$ are positive, as in Figure 12-17a, the torque will be negative (clockwise): That means it slows the object down as it moves away from equilibrium and speeds it up as the pendulum returns to the equilibrium position, $\theta = 0$. If the pendulum swings instead to the left of vertical (clockwise) so that θ and $\sin \theta$ are negative, the torque will be positive (counterclockwise) and so will again restore the pendulum to equilibrium.

The torque in Equation 12-25 is the only torque that is exerted on the pendulum, so this is also the *net* torque. Newton's second law for rotational motion (see Section 10-5) says that the net torque on an object is equal to the rotational inertia multiplied by the angular acceleration. For a system made up of a single object of mass m a distance L from the rotation axis, the rotational inertia is $I = mL^2$. Using the torque from Equation 12-25, we have

$$-mgL \sin \theta = mL^2 \alpha_z$$

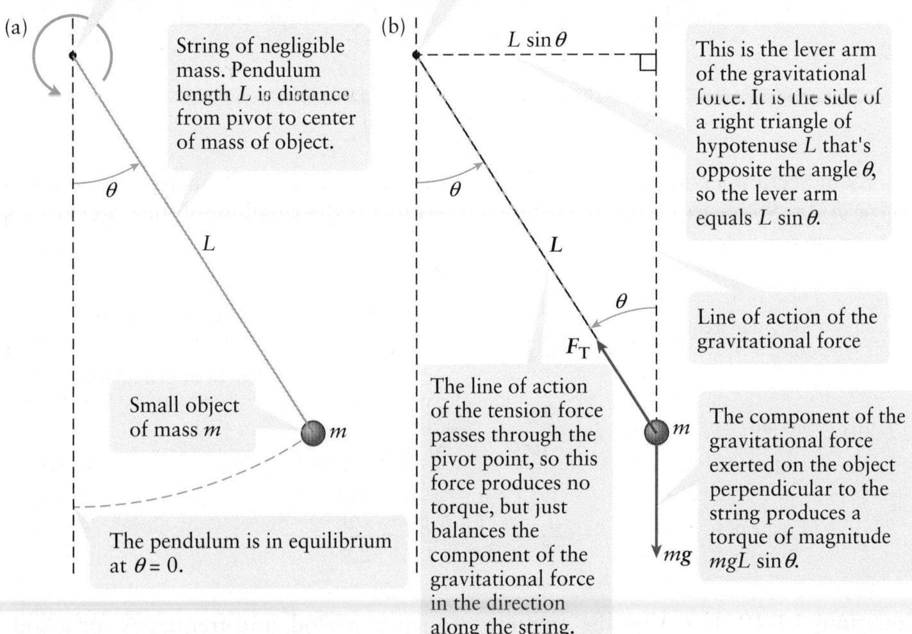
Pivot point: The positive z axis (the rotation axis of the pendulum) points outward from this point. We take positive rotation to be counterclockwise.
(a)
String of negligible mass. Pendulum length L is distance from pivot to center of mass of object.
θ
L
Small object of mass m
m
The pendulum is in equilibrium at $\theta = 0$.
(b)
$L \sin \theta$
This is the lever arm of the gravitational force. It is the side of a right triangle of hypotenuse L that's opposite the angle θ, so the lever arm equals $L \sin \theta$.
θ
L
θ
Line of action of the gravitational force
F_T
The line of action of the tension force passes through the pivot point, so this force produces no torque, but just balances the component of the gravitational force in the direction along the string.
m
The component of the gravitational force exerted on the object perpendicular to the string produces a torque of magnitude $mgL \sin \theta$.
mg

Figure 12-17 A simple pendulum (a) Layout of a simple pendulum. (b) The forces on a simple pendulum.

(a)

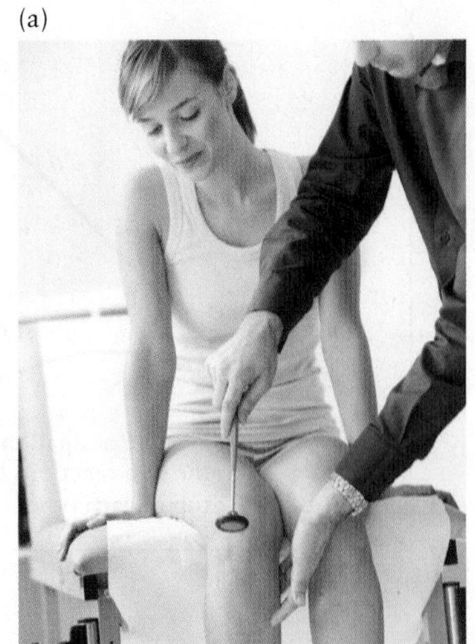

(b)

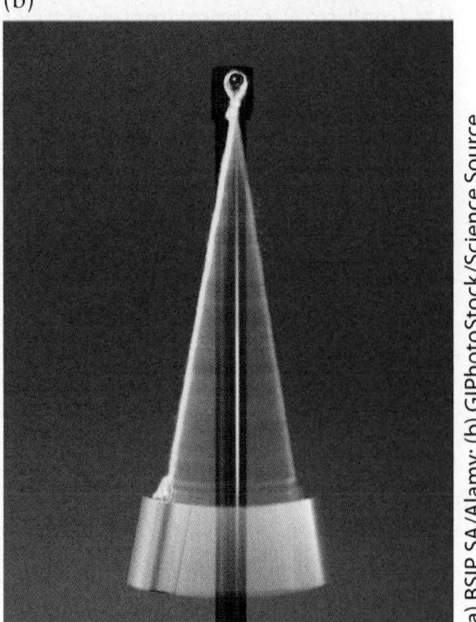

(a) BSIP SA/Alamy; (b) GIPhotoStock/Science Source

Figure 12-16 Swinging leg, swinging pendulum The motions of both (a) your lower leg in a physician's office and (b) an object hanging from a string have the ingredients for oscillation: a restoring force (gravity) that tends to pull the object toward equilibrium (hanging straight down), and inertia that causes the object to overshoot equilibrium.

AP® Exam Tip

As L is also used for angular momentum, on the AP® Physics 1 equation sheet, pendulum length is denoted with a script lowercase l. As always, be sure to understand what each variable or constant means in each equation so that you do not confuse one physical quantity for another when solving problems.

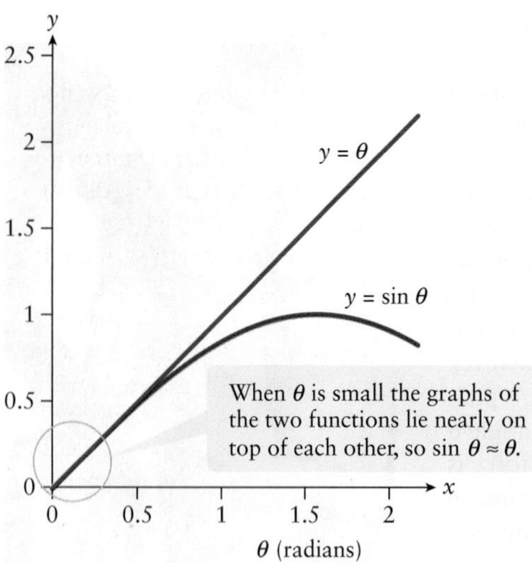

Figure 12-18 Approximating sin θ If the value of θ is small, sin θ is approximately equal to θ (measured in radians).

If we divide both sides of this equation by mL^2 and rearrange, we get

$$\alpha_z = -\frac{g}{L}\sin\theta \tag{12-26}$$

Let's see whether Equation 12-26 is equivalent to Hooke's law. If it is, that means the oscillations of simple pendula are simple harmonic motion and we can use our results from Section 12-4. We saw in Section 12-4 that we can write Hooke's law for linear oscillation as

$$a_x = -\omega^2 x$$

That is, the acceleration is directly proportional to the displacement, and the proportionality constant is the negative of the square of the angular frequency ω. From Section 10-4 the quantities x and a_x for linear motion correspond to the quantities θ and α_z, respectively, for rotational motion. So the rotational version of Hooke's law is

$$\alpha_z = -\omega^2\theta \tag{12-27}$$

The difference between Equation 12-26 for the pendulum and Equation 12-27, the rotational version of Hooke's law, is that Equation 12-26 involves sin θ rather than θ. So in general the oscillations of a simple pendulum do *not* obey Hooke's law, and so the motion of the pendulum is *not* simple harmonic motion.

If, however, the angle θ is relatively small and we measure θ in radians, it turns out that sin θ is *approximately* equal to θ (**Figure 12-18**):

$$\sin\theta \approx \theta \quad \text{when } \theta \text{ is small}$$

[You can confirm this with your calculator. You can switch your calculator from degree to radian mode (remember we warned you about checking degree mode) and calculate sin θ for $\theta = 1$ rad, 0.5 rad, 0.2 rad, 0.1 rad, and 0.01 rad. You'll find that as you try smaller values of θ, the value of sin θ gets closer and closer to θ.] With this approximation, Equation 12-26 for the simple pendulum becomes

$$\alpha_z \approx -\frac{g}{L}\theta \quad \text{if } \theta \text{ is small} \tag{12-28}$$

Compare this to Hooke's law for rotational motion, Equation 12-27, and you'll see that the oscillations of a pendulum *with small amplitude* (so that θ is always small) obey Hooke's law and that the angular frequency of the simple pendulum's oscillations is given by

$$\omega^2 = \frac{g}{L} \quad \text{or} \quad \omega = \sqrt{\frac{g}{L}}$$

As in Section 12-4, the period T is equal to $2\pi/\omega$, and the frequency f is equal to $1/T$ or $\omega/2\pi$. So we can write the following results for the small-amplitude oscillations of a simple pendulum:

WATCH OUT !

Large-amplitude oscillations of a simple pendulum are not simple harmonic motion.

Equations 12-29 apply only if the angle θ reached by the pendulum is always small enough that sin θ is approximately equal to θ. Figure 12-18 shows that this is not a good approximation if θ is greater than about 0.3 rad (~18°). If the amplitude is larger than this, the motion is *not* simple harmonic and the values of ω, T, and f do depend on the amplitude. As an example, if the amplitude is $\pi/2$ (90°), the period of oscillation is about 18% greater than the value given by the second of Equations 12-29.

EQUATION IN WORDS

Angular frequency, period, and frequency for a simple pendulum (small amplitude)

(12-29)

Angular frequency: $\omega = \sqrt{\frac{g}{L}}$

Period: $T = \frac{2\pi}{\omega} = 2\pi\sqrt{\frac{L}{g}}$

Frequency: $f = \frac{1}{T} = \frac{1}{2\pi}\sqrt{\frac{g}{L}} = \frac{\omega}{2\pi}$

g = acceleration due to gravity

L = length of pendulum

Equations 12-29 show that the angular frequency, period, and frequency for a simple pendulum do *not* depend on the mass of the object at the end of the pendulum. That's because the restoring torque is provided by the gravitational force. Doubling the mass doubles the rotational inertia, which by itself would make the oscillations happen more slowly, but it also doubles the restoring torque, which by itself would make the oscillations happen more rapidly. The two effects cancel each other out so that ω, T, and f are unaffected by changes in the pendulum mass.

Because the small-amplitude oscillations of a simple pendulum obey Hooke's law, it also follows that the angular frequency, period, and frequency do not depend on the *amplitude* of the oscillations. Here the amplitude is the maximum angle from the vertical that the pendulum attains. If you were to pull the simple pendulum in Figure 12-17a to an angle $\theta = 0.1$ rad—about 6°—and let it go, the pendulum would oscillate between $\theta = +0.1$ rad (to the right of vertical) and $\theta = -0.1$ rad (to the left of vertical), and the amplitude of the oscillation would be 0.1 rad.

Although the interaction resulting in the potential energy differs in different types of systems undergoing simple harmonic motion, the energy description for any motion that can be approximated as simple harmonic is fundamentally the same. Any motion that can be modeled as simple harmonic shares the constant transfer of energy between kinetic and potential, with the total energy being constant. If simple harmonic motion is a good model, it will always be possible to model the interaction responsible for that restoring force of torque in terms of potential energy. In any simple harmonic motion, the total energy of the motion is constant (mechanical energy is conserved). The kinetic energy is equal to this total mechanical energy as the oscillator passes through equilibrium, and the potential energy is zero at this point. At the maximum amplitude positions, the potential energy of the system is equal to the total mechanical energy and the kinetic energy is zero at these points.

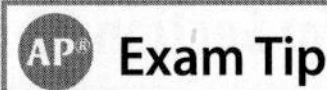

Exam Tip

You will be expected to describe the energy transfers involved in any simple harmonic motion, no matter the type of system in which that motion is occurring. As any oscillator passes through the equilibrium position, the kinetic energy is maximum and equal to the total energy, and the potential energy is zero. As with any oscillator is at the end points of its motion, the kinetic energy is zero and the potential energy equal to the total energy.

EXAMPLE 12-5 Changing a Pendulum

You make a simple pendulum by hanging a small 60-g marble from a light thread of negligible mass. The length L from the pivot to the center of the marble is 0.40 m. (a) Calculate the period when the marble is pulled a small angle to one side and released. (b) You then make a new pendulum with a small marble of mass 260 g and a pendulum length that is 10 cm longer than the initial pendulum. Calculate the period of the new pendulum.

Set Up

We'll use the second of Equations 12-29 to calculate the period of this simple pendulum.

Period of a simple pendulum:

$$T = 2\pi\sqrt{\frac{L}{g}} \qquad (12\text{-}29)$$

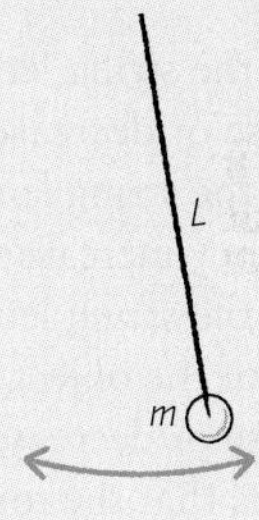

Solve

(a) Calculate the period of the initial pendulum with length $L = 0.40$ m.

Then the period is

$$T = 2\pi\sqrt{\frac{L}{g}} = 2\pi\sqrt{\frac{0.40\ \text{m}}{9.80\ \text{m/s}^2}} = 2\pi\sqrt{\frac{0.40\ \text{s}^2}{9.80}} = 1.3\ \text{s}$$

(b) Repeat the calculation with the new value of the length. Note that we know the value of g to three significant digits, but our answers for the period are given to only two significant digits. Can you see why? (*Hint*: Review Section 1-5 if you're not sure.)

The new length of the pendulum is

$$L_{\text{new}} = 0.40\ \text{m} + 0.10\ \text{m} = 0.50\ \text{m}$$

So the new period is

$$T_{\text{new}} = 2\pi\sqrt{\frac{L_{\text{new}}}{g}} = 2\pi\sqrt{\frac{0.50\ \text{m}}{9.80\ \text{m/s}^2}} = 1.4\ \text{s}$$

Reflect

Although the mass of the marble changed, we did *not* have to use its mass in the calculation. As we've seen, the period of a simple pendulum depends only on its length, not on its mass.

NOW WORK Problems 1–4 from The Takeaway 12-6.

THE TAKEAWAY for Section 12-6

✔ In a simple pendulum the mass of the pendulum is concentrated at a fixed distance from a rotation point.

✔ The angular frequency ω, the period T, and the frequency f of a simple pendulum depend on the length of the pendulum and the acceleration due to gravity. They do not depend on the pendulum mass, or the amplitude of the motion as long as the amplitude is small.

✔ The energy of a simple pendulum parallels that of any other simple harmonic oscillator.

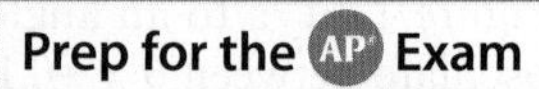

AP Building Blocks

1. Consider an object with mass m, hanging from the end of a string of negligible mass. The object is given an initial small angular displacement, θ_0, and is released, and its subsequent motion is recorded. Then the length of the string is increased and the experiment is repeated with the initial angular displacement and mass kept the same. These two cases are shown in the accompanying figure.

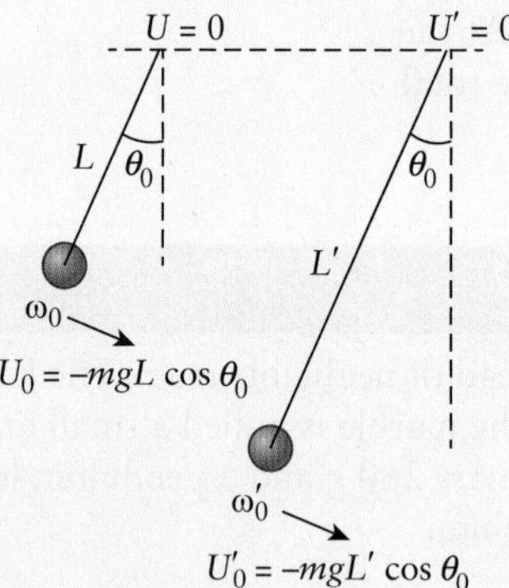

(a) When the string length is increased, does the period increase or decrease?
(b) When the string length is increased, does the angular frequency increase or decrease?
(c) When the string length is increased, does the linear speed of the object, when it is at its lowest point, increase or decrease?
(d) Assume that the total energy of each of the pendula shown in the drawing is constant. Starting from $\Delta(U + K) = \Delta(mgy + mv^2/2)$, express the angular speeds of the two pendula at $\theta = 0°$ (the lowest point in the oscillation) in terms of m, g, θ_0, and L or L'.
(e) What is the difference between angular frequency and angular speed, and why is the same symbol used to represent both?

2. If the period of a simple pendulum is T and we increase its length so that it's four times longer, what will the new period be?
(A) $T/2$ (B) T (C) $2T$ (D) $4T$

EX 12-5 3. A simple pendulum on the surface of Earth is 1.24 m long. What is the angular frequency of oscillation?

EX 12-5 4. In 1851 Jean Bernard Leon Foucault suspended a pendulum (later named the Foucault pendulum) from the dome of the Pantheon in Paris. The mass of the pendulum was 28.00 kg and the length of the rope was 67.00 m. The period was 16.42 s. Calculate the gravitational acceleration in Paris from these data, neglecting the mass of the rope.

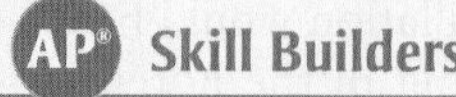

AP Skill Builders

5. A photogate timer was used to determine if:
i. the energy of a simple pendulum is constant, and
ii. the period is independent of the initial angular displacement of the pendulum.

The pendulum was constructed from a light string that has length L of 1.8 m attached to a 1.00-cm-diameter cylindrical object with mass m of 40 g. The pendulum was hung from a stationary support. The object was pulled to an angular displacement θ_0 from its natural equilibrium position (hanging straight down) where the photogate timer was positioned. The interval of time that elapsed as the object blocked the beam of the optical timer was reported. In addition, the total elapsed times for 15 complete oscillations of the pendulum were recorded.

Approximate angle from vertical, θ_0	Elapsed time of 15 cycles (seconds)	Elapsed time at photogate (milliseconds)
5°	40.41	27.3
10°	40.53	13.6
15°	40.72	9.0
20°	40.99	6.9
25°	41.35	5.5
30°	41.80	4.6

(a) Construct a diagram representing the apparatus. On your diagram denote the vertical coordinate as y and indicate your choice of the coordinate origin. On the diagram, also express the initial potential energy of the system, U_i, and the potential energy of the Earth–pendulum system as it passes through equilibrium, U_f, in terms of L, m, and θ_0, and the change in potential energy, ΔU.
(b) Apply the principle of energy conservation to express the change in potential energy of the system $\Delta U_{conservation}$ in terms of the kinetic energy of the pendulum at the lowest point in the arc, where the speed is measured.
(c) Explain why a comparison of ΔU and $\Delta U_{conservation}$ can be used to test the claim that the energy of a simple pendulum is constant for any initial angular displacement of the pendulum.
(d) Complete the following table by using measured values to calculate the kinetic energy of the object while it blocks the beam of the photogate timer.

Approximate angle from vertical	Period (s)	Elapsed time at photogate (s)	Speed of object at $\theta = 0°$ (m/s)	Kinetic energy of object at $\theta = 0°$ (J)
5°				
10°				
15°				
20°				
25°				
30°				

(e) Complete the following table by using the expressions from parts (a) and (b) to calculate the initial potential energy from the value of the initial angular displacement θ_0 and from the kinetic energy at $\theta = 0°$, assuming total energy remains constant.

Approximate angle from vertical	ΔU (J) from part (a)	$\Delta U_{\text{conservation}}$ (J) from part (b)
5°		
10°		
15°		
20°		
25°		
30°		

(f) Summarize the result of your comparison of potential energies in terms of the initial question about whether the energy of a simple pendulum is constant, independent of the initial angular displacement of the pendulum. In your comparison, consider the average relative percent difference of potential energies calculated using these two methods.

(g) A *simple* pendulum satisfies the condition that the period is independent of the pendulum's angular displacement for small initial displacements. Evaluate this condition for the system studied by graphing the percent relative error versus the initial angular displacement and state your conclusion based on these data.

6. The current best estimate of the effective gravitational field is based on variations on a spherical Earth with radius $r_{\text{average}} = 6367.88$ km. The local value of g depends on the local radius

$$g_{\text{local}} = g_{\text{spherical Earth}}\left(\frac{R_{\text{Earth average}}}{R_{\text{local}}}\right)^2$$

where

$$g_{\text{spherical Earth}} = \left(\frac{GM_E}{R^2_{\text{Earth average}}}\right) = 9.830 \text{ N/kg}$$

The value of the standard gravitational product GM_E is known to very high precision to be $3.986004418 \times 10^{14}$ m^3 s^{-2}. The spherical Earth approximation gives a value of g greater than most locations on Earth, because of Earth's actual nonspherical shape. It gives the correct value when the local radius is used.

The highest point on Earth's surface is Chimborazo in the Andes of central Ecuador. The peak is 6384.4 km from Earth's center. The distance from Earth's center for Mount Everest is 6382.3 km. Mount Everest has a greater elevation than Chimborazo but the peak of Chimborazo is farther from the center due to Earth's equatorial bulge.

(a) A grandfather clock, one whose mechanism is based on the simple harmonic oscillation of a pendulum, is calibrated to keep good time in Florence, Italy, whose distance from the center of Earth is 6367.9 km. The mechanism advances a wheel by one tooth, $\Delta\theta$, in one-half period. So each complete periodic swing of the pendulum advances the wheel by two teeth. Calculate the length of the pendulum.

(b) The clock described in part (a) is placed at the top of Chimborazo. Predict if the clock keeps good time, runs slow, or runs fast.

(c) Evaluate your prediction in part (b) by calculating the period in Florence and on Chimborazo, and explain why the calculated period tests the prediction.

AP® Skills in Action

7. Galileo investigated the interrupted pendulum shown in the figure. After the bob is released, the string strikes a peg that is situated directly below the point of attachment of the pendulum string to a support.

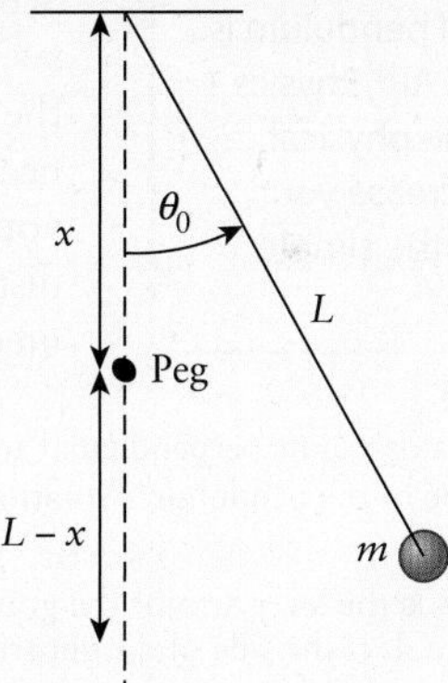

(a) Let the reference position at which the gravitational potential energy is set equal to zero be the lowest point in the arc of the bob where it would pass over the dashed line in the figure if the peg were absent. Express the initial total energy of the system in terms of g, m, L, and θ_0.

(b) Use a sketch of the system to qualitatively predict two possible paths of the pendulum bob after the string strikes the peg. In one, let the initial height of the bob lie below the peg and in the second, let the initial height of the bob lie above the height of the peg.

(c) Suppose that, after the string has struck the peg, the bob rises through an angular displacement of magnitude $0 < \theta < 180°$ before instantaneously stopping. Apply energy conservation to express the cosine of the angle θ, $\cos(\theta)$, in terms of g, m, L, and $\cos(\theta_0)$, and confirm that

$$\cos\theta = \frac{\cos\theta_0 - c}{(1 - c)}$$

where $c = x/L$.

(d) Locate the initial angle θ_c of the string where the bob just reaches a displacement of −180°. Predict what happens if the initial angular displacement is greater than θ_c.
(e) Predict the period of a cycle where the initial angular displacement is less than θ_c, expressed in terms of g, x, and L.

A light string, a small bob that can be attached to the string, a meter stick, a rigid support from which the pendulum is hung, and a stopwatch are provided. Below the point of attachment, a bench clamp supports a vertical rod, to which a wooden dowel with a small diameter can be attached.

(f) Design an experiment to test the prediction in part (e).

8. On the surface of Earth a simple pendulum of length $L = 0.500$ m and mass $m = 40.0$ g starts from rest at a maximum displacement of $\theta_0 = 20.0°$ from the equilibrium position. Assume that the small-angle approximation can be used.
(a) When will the pendulum return to its starting position?
(b) At what time will the pendulum be located at a displacement angle of 8.00°? *Hint*: The angular displacement is a periodic function of time and, for ideal pendula, has an amplitude of θ_0.
(c) Construct a diagram showing that the torque $Lmg \sin \theta$ exerted on a pendulum bob by gravitational force causes it to rotate about the point of attachment of the string to its support.
(d) Explain why $\theta = \theta_0 + \alpha t^2/2$, where $\alpha = \tau/I = g \sin \theta/L$ cannot be used to calculate the time from the displacement in part (b).

9. A small block is attached by a long light string to the ceiling. An identical block and string are set up in the same way in a science outpost on the Moon. Both blocks are displaced the same small angle from their equilibrium positions and released from rest. Predict which pendulum has a smaller period of oscillation. Justify your claim by describing the motion of the pendulum and the restoring force, without referencing any equations.

12-7 A physical pendulum has its mass distributed over its volume

AP® Exam Tip

While the physical pendulum is not a topic on the AP® Physics 1 exam, studying the physical pendulum can increase your understanding of the simple pendulum.

We began the previous section by comparing two examples of pendulum motion: the swing of your lower leg when a physician tests your patellar reflex (Figure 12-16a) and the motion of a mass on the end of a string (Figure 12-16b). The object on a string is nearly an ideal simple pendulum, as the mass is concentrated into a very small volume. But your lower leg is *not* a simple pendulum because its mass is distributed along the entire distance from the knee to the foot. A pendulum like your lower leg whose mass is distributed throughout its volume is called a **physical pendulum** or compound pendulum. Other examples of physical pendula include a swinging church bell, a chandelier swaying back and forth after an earth tremor, and the pendulum of an old-fashioned grandfather clock. Let's see how to find the angular frequency, period, and frequency for the oscillations of a physical pendulum.

Figure 12-19 shows an example of a physical pendulum of mass m. Two forces are exerted on the pendulum: a gravitational force of magnitude mg that is exerted at the center of mass of the pendulum, a distance h from the pivot point, and a support force that is exerted at the pivot. We haven't drawn the support force because it is exerted at the pivot point and so exerts no torque around that point. The torque exerted by the gravitational force is equal to the magnitude mg of the gravitational force multiplied by the lever arm of the gravitational force, which is the perpendicular distance from the line of action of the gravitational force to the pivot. Figure 12-19 shows that if the pendulum is displaced from the vertical by an angle θ, the lever arm equals $h \sin \theta$. Hence the torque on the pendulum is

$$\tau_z = -mgh \sin \theta \tag{12-30}$$

Just as for the torque on a simple pendulum, the minus sign in Equation 12-30 means that this is a restoring torque that always pulls the pendulum back toward its equilibrium position $\theta = 0$, where the center of mass is directly below the pivot.

The torque in Equation 12-30 is the only torque on the pendulum and so is equal to the net torque $\sum \tau_z$. From Newton's second law for rotation (Section 10-5), $\sum \tau_z = I\alpha_z$, we have

$$-mgh \sin \theta = I\alpha_z \tag{12-31}$$

The z axis points perpendicular to the plane of the pendulum's rotation.

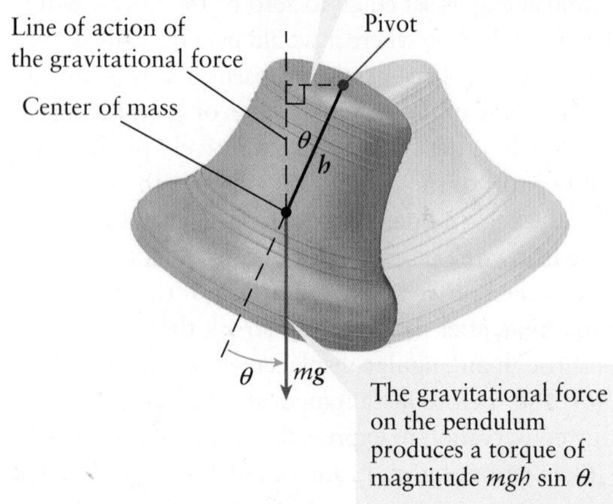
This is the lever arm of the gravitational force. It is the side of a right triangle of hypotenuse h that's opposite the angle θ, so the lever arm equals $h \sin \theta$.
Line of action of the gravitational force
Pivot
Center of mass
θ
h
θ
mg
The gravitational force on the pendulum produces a torque of magnitude $mgh \sin \theta$.

Figure 12-19 A physical pendulum The forces on a physical pendulum (one whose mass is not all concentrated in a single small blob, unlike the simple pendulum shown in Figure 12-17).

In Equation 12-31 the quantity I is the rotational inertia of the pendulum around the pivot point. Just as for the simple pendulum, if the angle θ is small then $\sin\theta$ is approximately equal to θ (measured in radians). With this approximation we can rewrite Equation 12-31 as

$$\alpha_z \approx -\left(\frac{mgh}{I}\right)\theta \quad \text{if } \theta \text{ is small} \tag{12-32}$$

Compare Equation 12-32 to the corresponding equation for the simple pendulum that we derived in Section 12-6, $\alpha_z \approx -(g/L)\theta$ (Equation 12-28): The equation is identical except that g/L has been replaced by mgh/I. So we conclude that just as for the simple pendulum, the oscillations of a physical pendulum are simple harmonic motion provided that the amplitude is relatively small. Because I always depends on m, the mass of the pendulum cancels and the properties of the physical pendulum do not truly depend on the mass, just the shape of the pendulum—how that mass is distributed. In practice, we cannot cancel the mass unless we know the functional form of the rotational inertia. To find the angular frequency, period, and frequency of a physical pendulum, we take Equations 12-29 for a simple pendulum and replace g/L with mgh/I:

Angular frequency $\omega = \sqrt{\dfrac{mgh}{I}}$

Period $T = \dfrac{2\pi}{\omega} = 2\pi\sqrt{\dfrac{I}{mgh}}$

Frequency $f = \dfrac{1}{T} = \dfrac{1}{2\pi}\sqrt{\dfrac{mgh}{I}} = \dfrac{\omega}{2\pi}$

(12-33)

m = **mass** of physical pendulum

g = **acceleration due to gravity**

h = **distance** from pivot point to center of mass of physical pendulum

I = **rotational inertia** of physical pendulum about pivot

EQUATION IN WORDS
Angular frequency, period, and frequency for a physical pendulum (small amplitude)

Here's a check on our results for a physical pendulum. If all of the mass of the physical pendulum is concentrated at the center of mass, then the rotational inertia of the physical pendulum around the pivot is $I = mh^2$. Then the quantity mgh/I in Equations 12-33 becomes

$$\frac{mgh}{I} = \frac{mgh}{mh^2} = \frac{g}{h} \tag{12-34}$$

If we change the symbol for the distance from the pivot to the center of mass from h to L, the quantity in Equation 12-34 becomes g/L. Then, from the first of Equations 12-33, the angular frequency for a physical pendulum with all of its mass concentrated at the center of mass becomes $\omega = \sqrt{g/L}$. That's exactly the result we found in Section 12-6 for a simple pendulum, which is just a physical pendulum with all of its mass concentrated a distance L from the pivot (see Figure 12-17 and the first of Equations 12-29). So our results for a *physical* pendulum give us the correct answers for the special case of a *simple* pendulum, just as they should.

EXAMPLE 12-6 An Oscillating Rod

A uniform rod has length L and mass m and is supported so that it can swing freely from one end. Derive an expression for the period of oscillation when the rod is pulled slightly from the vertical and released.

Set Up

Equations 12-33 tell us the period of a physical pendulum. The center of mass of a uniform rod is at its center, and its rotational inertia around one end is given in Table 10-1 (Section 10-3).

Period of a physical pendulum:

$$T = 2\pi\sqrt{\frac{I}{mgh}} \quad \text{(from Equations 12-33)}$$

Rotational inertia for a uniform rod of mass m and length L rotating about one end:

$$I = \frac{1}{3}mL^2$$

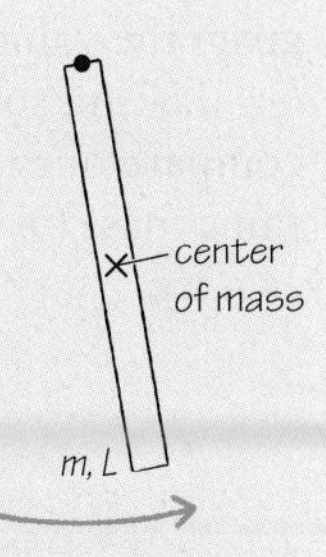

Solve

Given the rotational inertia I and the distance h, solve for the period. Given that we know the equation for the rotational inertia for a rod, the m's will cancel and we get a period that does not depend on mass.

The center of mass is a distance $L/2$ from the pivot, so

$$h = \frac{L}{2}$$

Substitute I and h into the expression for the period T from Equations 12-33:

$$T = 2\pi\sqrt{\frac{(1/3)mL^2}{mg(L/2)}} = 2\pi\sqrt{\frac{2mL^2}{3mgL}}$$

$$= 2\pi\sqrt{\frac{2L}{3g}}$$

Reflect

Compared to the period of a simple pendulum of length L, $T_{\text{simple pendulum}} = 2\pi\sqrt{L/g}$, the period of a uniform rod free to rotate about its end is smaller by a factor of $\sqrt{2/3} = 0.816$. In other words, the physical pendulum oscillates more quickly. That may seem surprising if you only compare the torques on the two objects. The center of mass of the rod is a distance $h = L/2$ from the pivot, so from Equation 12-30 the torque on the rod is $\tau_z = -mgh \sin\theta = -(mgL/2)\sin\theta$. By contrast, for a simple pendulum all of the mass m is a distance L from the pivot, so $h = L$ and the torque on the pendulum is $\tau_z = -mgh \sin\theta = -mgL \sin\theta$. So the rod experiences half as much torque pulling it back toward equilibrium as the simple pendulum. Why, then, does the rod oscillate more rapidly?

To see the explanation we also need to compare the rotational inertia of the rod ($I = mL^2/3$) to the rotational inertia of the simple pendulum ($I = mL^2$). So while the rod experiences half as much torque as the simple pendulum, the rod has only one-third as much rotational inertia as the simple pendulum. Hence the rod ends up oscillating faster. Notice that the period (and hence the frequency or angular frequency) does not actually depend on the mass, which cancels as soon as we know the form of the rotational inertia. This makes sense, because the force of gravity is still providing the restoring force, just as it does for simple pendula.

NOW WORK Problems 1 and 3 from The Takeaway 12-7.

EXAMPLE 12-7 Rotational Inertia of a Human Leg

How you walk is affected in part by how your legs swing around the rotation axis created by your hip joints. Because your leg has a complex shape and contains bone, muscle, skin, and other materials, it would be very difficult to calculate the rotational inertia I of a leg around the hip. Instead, I can be found experimentally by measuring the period of the leg when allowed to swing freely. In a clinical study, a person's leg of length 0.88 m is estimated to have a mass of 6.5 kg and a center of mass 0.37 m from the rotation axis through the hip. When allowed to swing freely, it oscillates with a period of 1.2 s. Determine the rotational inertia of the leg in rotation around the axis through the hip. Compare your answer to a uniform rod that has the same mass and length (see Example 12-6).

Set Up

The expression for the period T of a physical pendulum depends on the rotational inertia I for rotation around the pivot, the mass m, and the distance h from pivot to center of mass. We're given the values of $T = 1.2$ s, $m = 6.5$ kg, and $h = 0.37$ m, so we can solve for I. We'll then compare our result to a uniform rod with the same mass ($m = 6.5$ kg) and length ($L = 0.88$ m) as a leg.

Period of a physical pendulum:

$$T = 2\pi\sqrt{\frac{I}{mgh}} \quad \text{(from Equations 12-33)}$$

Rotational inertia for a uniform rod of mass m and length L rotating about one end:

$$I = \frac{1}{3}mL^2$$

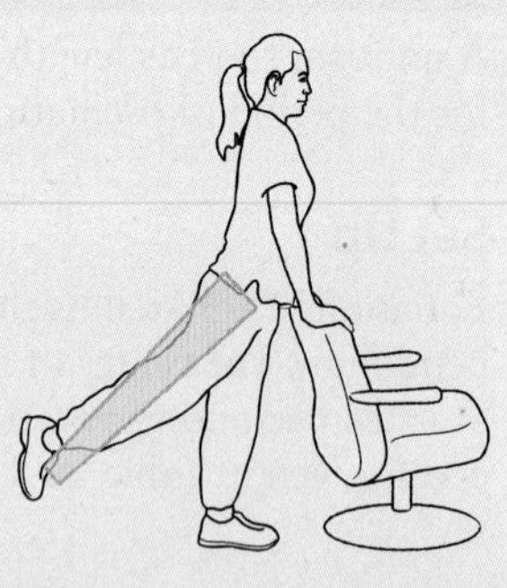

Solve

Rearrange the expression for period T to find a formula for the rotational inertia I, then substitute the known values.

Begin with the expression

$$T = 2\pi\sqrt{\frac{I}{mgh}}$$

Square both sides to get rid of the square root, then solve for I:

$$T^2 = \frac{4\pi^2 I}{mgh} \text{ so } I = \frac{mghT^2}{4\pi^2}$$

Substitute numerical values:

$$I = \frac{(6.5 \text{ kg})(9.80 \text{ m/s}^2)(0.37 \text{ m})(1.2 \text{ s})^2}{4\pi^2} = 0.86 \text{ kg} \cdot \text{m}^2$$

As a comparison, calculate the rotational inertia of a uniform rod with the same mass and length of the leg rotating about one end.

For a rod rotating around one end,

$$I = \frac{1}{3}mL^2 = \frac{1}{3}(6.5 \text{ kg})(0.88 \text{ m})^2 = 1.7 \text{ kg} \cdot \text{m}^2$$

Reflect

The rotational inertia of the uniform rod is *twice* as large as the rotational inertia of that leg. Is that reasonable? Consider that your upper leg (above the knee) is more massive than your lower leg, as suggested by the sketch, so more of the leg's mass is above the knee than below the knee. By contrast, the mass of a uniform rod is distributed uniformly along its length. We learned in Section 10-3 that the farther an object's mass lies from the rotation axis, the greater the rotational inertia for that axis. So it's not surprising that the uniform rod, which has more of its mass farther from the rotation axis, has a greater rotational inertia than the leg.

NOW WORK Problem 2 from The Takeaway 12-7.

THE TAKEAWAY for Section 12-7

✔ A physical pendulum has its mass distributed throughout its volume. Unlike a simple pendulum, it cannot be treated as an object suspended at the end of a string of negligible mass.

✔ The gravitational torque that arises when the pendulum is displaced from equilibrium provides the restoring torque that causes the pendulum to return to its equilibrium position.

✔ The angular frequency, period, and frequency of a physical pendulum depend on the way its mass is distributed, the distance from the pivot point to the pendulum's center of mass, and the pendulum's rotational inertia around that point. Just as for simple pendula, these quantities do not depend on the amplitude of the motion for small amplitudes.

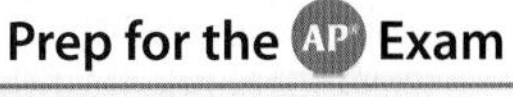

AP® Building Blocks

EX 12-6 1. A uniform rod of length L hangs from one end and oscillates with a small amplitude. The rotational inertia for a rod rotating about one end is $I = ML^2/3$. What is the period of the rod's oscillation?

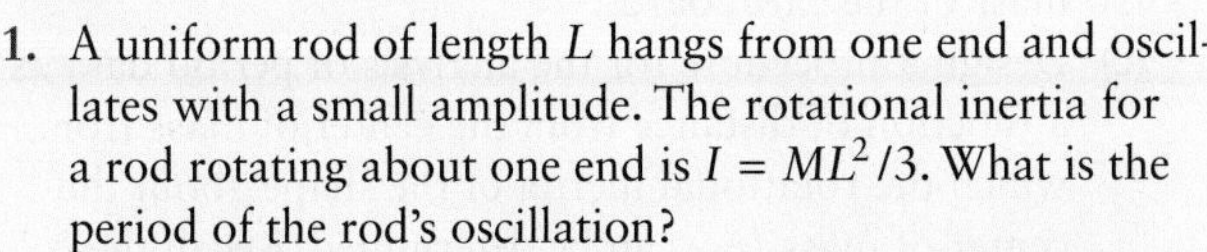

(A) $2\pi\sqrt{\frac{L}{g}}$ (B) $2\pi\sqrt{\frac{2L}{3g}}$ (C) $2\pi\sqrt{\frac{L}{2g}}$ (D) $2\pi\sqrt{\frac{L}{3g}}$

EX 12-7 2. A pendulum made of a uniform rod with a length of 30.0 cm is set into simple harmonic motion about one end, close to the surface of Earth.

(a) Using $I = mL^2/3$, calculate the period of its motion.

(b) Calculate the percent relative error in the period if the rod were treated as an object at a distance $2L/3$ from the pivot.

(c) Explain why the approximation you made to find the period in part (b) works for any length rod set into simple harmonic motion.

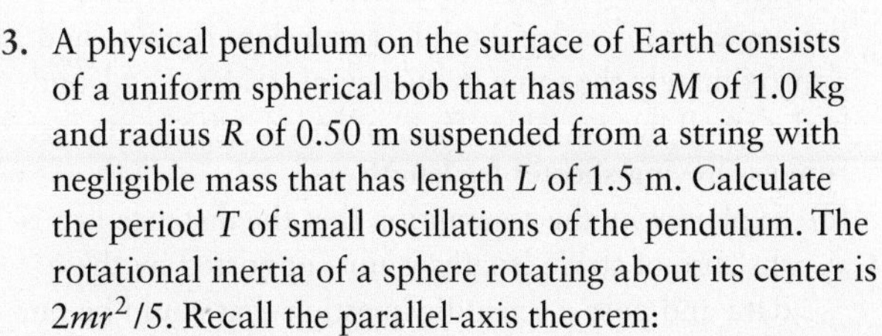

EX 12-6 3. A physical pendulum on the surface of Earth consists of a uniform spherical bob that has mass M of 1.0 kg and radius R of 0.50 m suspended from a string with negligible mass that has length L of 1.5 m. Calculate the period T of small oscillations of the pendulum. The rotational inertia of a sphere rotating about its center is $2mr^2/5$. Recall the parallel-axis theorem:

$$I = I_{CM} + md^2$$

where m is the mass and d is the distance separating the center of mass from the axis of rotation.

AP® Skill Builders

4. A position in the human running gait is traced in the figure, with the right leg drawn more darkly than the left leg. The upper (above the knee) and lower (below the knee) parts of the legs rotate differently. In the configuration shown, the foot of the left leg is about to make contact with the

ground. In one stride the two parts of the left leg will rotate into the current position of the right leg.

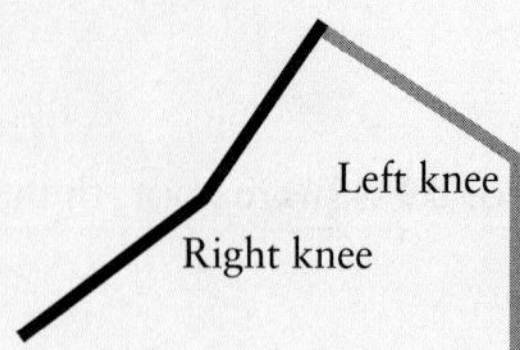

(a) Construct a diagram to describe the rotation of the upper left leg into the line of the upper right leg. Construct a diagram to describe the rotation of the lower left leg into the line of the lower right leg using the left knee as the rotation axis. Include the following in the diagram:
- The direction of rotation (clockwise or counterclockwise) around the pivot points
- The approximate angular displacement based on the drawing provided here during one-half period

(b) The rotations of the upper and lower legs occur in the same time interval, approximately one-half of a period of the full cycle of motion. Express the relative average angular speeds of the upper and lower legs over this time interval.

(c) The mass of a human leg is approximately 15% of body mass. The lengths of the upper and lower legs are similar and each is approximately 25% of total height. However, the mass of the upper leg is approximately a factor of 2 larger than the mass of the lower leg. Express the relative magnitudes and directions of the torques and angular momenta of the upper leg about the hip and the lower leg about the knee. Neglect any consideration of the overall translational motion, and the effect of the relative rotational motions of the two parts of the leg.

Average speed is distance divided by time. If it is assumed that the total distance is the sum of all the strides taken (remember frequency is the inverse of the period), Weyand and colleagues found experimentally that running speed was independent of the frequencies of the motion of the upper or lower leg. Both fast and slow runners had approximately the same frequency of strides. Leg length had a small but variable effect on speed, with some faster runners having shorter leg length.

(d) Explain why the assumption that the distance run is the sum of stride lengths is not supported by these data and pose a scientific question that could be pursued to explain why Usain Bolt is faster than you.

5. A physical pendulum can be constructed by cutting a random shape from a piece of cardboard and making a small hole with a sharp pencil near one end of the shape. When the shape is hung from a nail through this hole, it will remain motionless when the center of mass of the shape lies along a vertical line that includes the pivot point. When the shape is at rest when hanging from the nail and is given a small angular displacement, the shape oscillates with a measurable period. Let the vertical line from the hole to the center of mass be an x axis with an origin at the center of mass. Denote the location of the hole as x_7.

(a) Design a procedure to locate the center of mass of this shape experimentally, making measurements rather than using mathematical models. Denote this point as x_0. The cardboard shape can be taken off the nail, extra holes can be made, and lines can be drawn on the shape.

A series of six equidistant holes are made along the line between x_0 and x_7. A rotational motion sensor is used to collect numerical data on the angular displacement of the shape as a function of time while the shape oscillates. When the piece of cardboard is hung on the nail at each of these holes and given an angular displacement, the cardboard oscillates and the period is measured. Results for the period and the distance from the center of mass to the axis of rotation are tabulated in the table shown below. Neglect the mass of the small amount of cardboard removed in making the holes.

n	Period, T_n (s)	Pivot point, x_n (m)
1	0.805	0.11187
2	0.778	0.0992
3	0.775	0.0801
4	0.761	0.0658
5	0.791	0.0484
6	0.852	0.0341
7	1.16	0.0134

(b) Justify the claim that the rotational inertia about each rotation axis can be written as

$$I_n = \frac{mgx_n T_n^2}{4\pi^2}$$

where T_n is the period, x_n is the distance from the center of mass to the axis of rotation, and m is the mass of the cardboard.

(c) Design a procedure for the analysis of period data as a function of distance from the center of mass from which the rotational inertia of the shape about the center of mass I_{CM} can be determined. Recall the parallel-axis theorem for rotations a distance d from the center of mass:

$$I = I_{CM} + md^2$$

(d) Apply your method using the data displayed in the table to determine the rotational inertia of a cardboard shape. The mass of the cardboard shape used for these data was 24.0 g.

WHAT DID YOU LEARN?

Prep for the AP Exam

Chapter learning goals	Section(s)	Related example(s)	Relevant section review exercises
Define oscillation and give everyday examples of systems that oscillate.	12-1		2
Explain, qualitatively and quantitatively, characteristics of oscillations, including the key properties of simple harmonic motion, and what is required for simple harmonic motion to occur.	12-2	12-1	2
Qualitatively describe a system using tensile stress and compressive stress and describe the relationship to Hooke's law.	12-3		1
Explain the connection between Hooke's law and simple harmonic motion, and be able to calculate velocity and acceleration for any given displacement of an object oscillating on a spring.	12-4	12-3	4, 5
Explain what properties determine the period, frequency, and angular frequency of an object oscillating on a spring and how these characteristics of the motion depend on those properties.	12-4	12-2	3
Predict, qualitatively and quantitatively, how kinetic energy and potential energy vary during an oscillation of a system and relate these changes to changes in the internal structure of the system.	12-5	12-4	1, 2, 3
Explain what properties determine the period, frequency, and angular frequency of a simple pendulum and how these characteristics of the motion depend on those properties.	12-6	12-5	1, 2
Explain what properties determine the period, frequency, and angular frequency of a physical pendulum and how these characteristics of the motion depend on those properties.	12-7	12-6, 12-7	1, 2, 3

Chapter 12 Review

Key Terms

All the Key Terms can be found in the Glossary/Glosario on page G1 in the back of the book.

Chapter Summary

Topic	Equation or Figure
Oscillations: A system will oscillate if there is (a) a restoring force (or torque) that always pulls (or rotates) the system back toward equilibrium and (b) inertia that causes the system to overshoot equilibrium. The period T is the time for one oscillation cycle; the frequency $f = 1/T$ is the number of cycles per second (measured in hertz); the angular frequency ω equals $2\pi f$; and the amplitude A is the maximum displacement from equilibrium.	 (Figure 12-16b)
Simple harmonic motion: If the restoring force obeys Hooke's law $F_x = -kx$, the resulting oscillatory motion is called simple harmonic motion: The angular frequency, period, and frequency are independent of the oscillation amplitude.	$\omega = \sqrt{\frac{k}{m}}$ $T = \frac{2\pi}{\omega} = 2\pi\sqrt{\frac{m}{k}}$ (12-13) $f = \frac{1}{T} = \frac{1}{2\pi}\sqrt{\frac{k}{m}} = \frac{\omega}{2\pi}$
Equations of simple harmonic motion: In simple harmonic motion the position, velocity, and acceleration are all sinusoidal functions of time. The phase angle ϕ describes where in the oscillation cycle the system is at $t = 0$. We can understand these equations by comparing a single vector component for each of these quantities from uniform circular motion to the motion of a block attached to an ideal spring.	$x = A\cos(\omega t + \phi)$ (12-14) $v_x = -\omega A\sin(\omega t + \phi)$ (12-16) $a_x = -\omega^2 A\cos(\omega t + \phi)$ (12-17)
Energy in simple harmonic motion: If friction can be neglected, the total mechanical energy of a system in simple harmonic motion is constant. At equilibrium the kinetic energy is maximum, and the potential energy of the oscillating system zero; at the extremes of the motion, the kinetic energy is zero, and the potential energy is maximum. For a system of an object on a horizontal spring, the potential energy of the system is spring potential energy, as in the figure. For a pendulum it is gravitational potential energy.	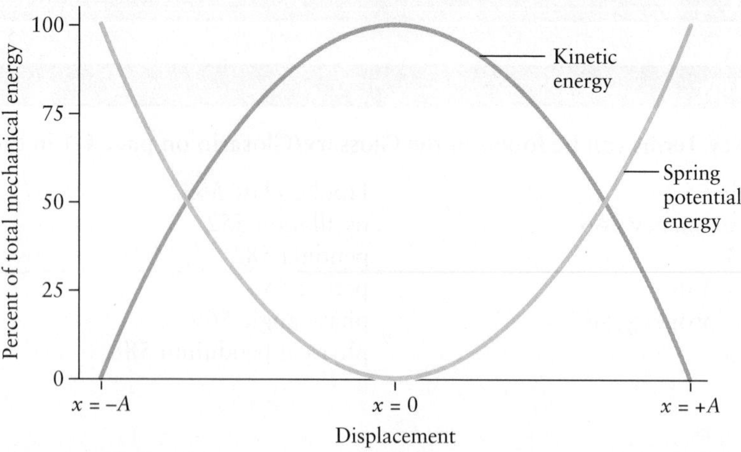 (Figure 12-15)

The simple pendulum: The oscillations of a simple pendulum are simple harmonic motion if the amplitude of oscillation is small. In this case the angular frequency of oscillation of a simple pendulum of length L is $\omega = \sqrt{g/L}$; this does not depend on the mass m.

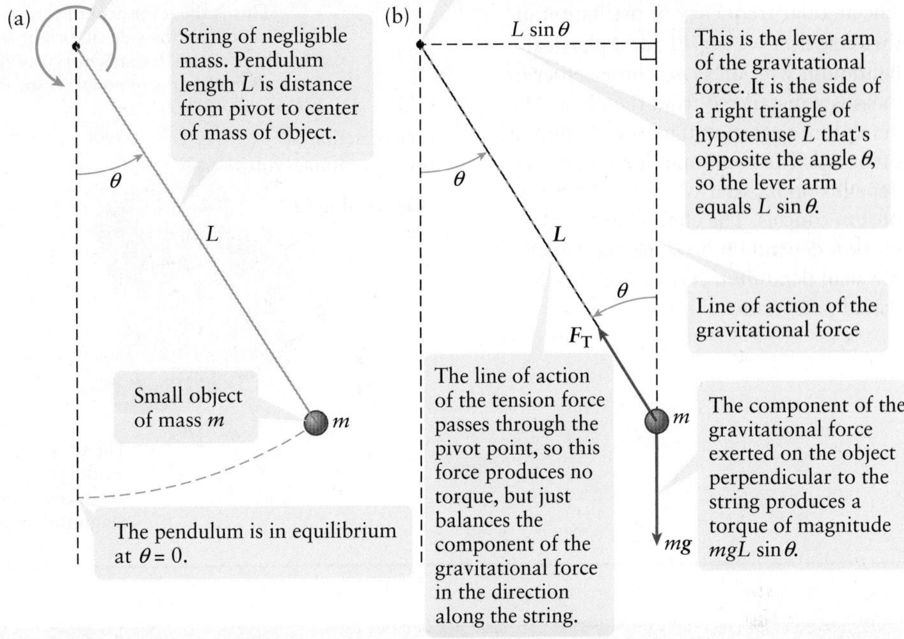

Pivot point: The positive z axis (the rotation axis of the pendulum) points outward from this point. We take positive rotation to be counterclockwise.
(a)
String of negligible mass. Pendulum length L is distance from pivot to center of mass of object.
θ
L
Small object of mass m
m
The pendulum is in equilibrium at $\theta = 0$.
(b)
$L \sin\theta$
This is the lever arm of the gravitational force. It is the side of a right triangle of hypotenuse L that's opposite the angle θ, so the lever arm equals $L \sin\theta$.
θ
L
θ
F_T
Line of action of the gravitational force
The line of action of the tension force passes through the pivot point, so this force produces no torque, but just balances the component of the gravitational force in the direction along the string.
m
The component of the gravitational force exerted on the object perpendicular to the string produces a torque of magnitude $mgL \sin\theta$.
mg

(Figure 12-17)

Small-angle approximation: For angles less than about 30° or about 0.5 rad, the length of the curved path the object has traveled from equilibrium is approximately equal to its horizontal displacement from equilibrium, which gives an equation of motion for a linear restoring force and means the system moved in simple harmonic motion. The smaller the angle, the better the approximation.

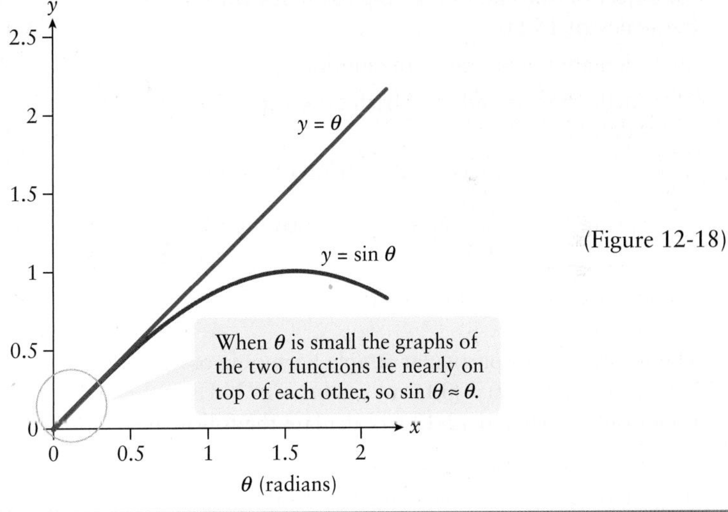

(Figure 12-18)

The physical pendulum: A pendulum of arbitrary shape also oscillates in simple harmonic motion if the amplitude is small. The angular frequency of oscillation in this case is $\omega = \sqrt{mgh/I}$ for a physical pendulum with mass m whose center of mass is a distance h from the pivot. The pendulum's rotational inertia around the pivot is I. The rotational inertia always depends on m, so if we know the form of I, m cancels. The characteristics of the motion depend on how the mass of the physical pendulum is distributed, but not actually its mass.

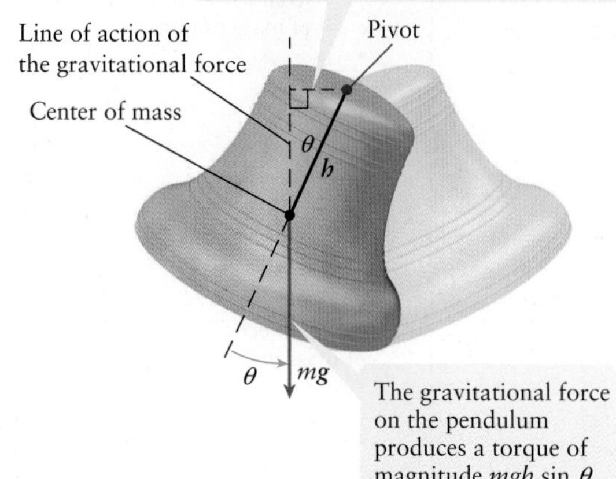

The z axis points perpendicular to the plane of the pendulum's rotation.
This is the lever arm of the gravitational force. It is the side of a right triangle of hypotenuse h that's opposite the angle θ, so the lever arm equals h sin θ.
Line of action of the gravitational force
Pivot
Center of mass
θ
h
θ
mg
The gravitational force on the pendulum produces a torque of magnitude mgh sin θ.

(Figure 12-19)

Chapter 12 Review Problems

1. An object on the end of a spring oscillates with a frequency of 15 Hz.
   (a) Calculate the period of the motion.
   (b) Calculate the number of oscillations that the object undergoes in 120 s.
2. Show that the formulas for the period of an object on a spring $T = 2\pi\sqrt{\dfrac{m}{k}}$ and a simple pendulum $T = 2\pi\sqrt{\dfrac{L}{g}}$ are dimensionally correct.
3. The position for a particular simple harmonic oscillator is given by $x(t) = (0.15\text{ m})\cos(\pi t + \pi/3)$. Use Equations 12-16 and 12-17 to calculate the following:
   (a) The velocity of the oscillator at $t = 1.0$ s
   (b) The oscillator's acceleration at $t = 2.0$ s
4. An ideal spring of unstretched length L and spring constant k is attached to a wall at its left end and an object of mass M at the right end. The object is at rest on a horizontal surface. The surface exerts a negligible friction force on the object. The object is pulled such that the spring is stretched a distance A to the right, then released. Let the $+x$ direction be to the right.
   (a) Express the position function $x(t)$ for the object. Take $x = 0$ to be the position of the object when the spring is relaxed; $t = 0$ is when the object is released.
   (b) Express the velocity of the object at $t = (7/6)T$, where T is the period.
   (c) Express the acceleration of the object at $t = T/4$.
5. A simple harmonic oscillator is observed to start its oscillations at its positive maximum amplitude when $t = 0$. Write a position function that is consistent with this initial condition, choosing either a sine or a cosine. Repeat when the oscillations start at the equilibrium position when $t = 0$. Are either of these functions incomplete?
6. Explain why each of following doubles the maximum speed of a simple harmonic oscillator.
   (a) Doubling the amplitude
   (b) Reducing the mass to one-quarter its original value
   (c) Increasing the spring constant to four times its original value
7. Which of the following best represents the change in frequency for an object–spring system when an object of mass M on the end of the spring is replaced with an object of mass $M/4$. Justify your choice.
   (A) 1/4 (B) 1/2 (C) 2 (D) 4
8. A block rests on a horizontal surface that exerts a negligible friction force on the block. A horizontal spring is attached to the block, and the other end of the spring is attached to a wall. The spring is stretched so that the block moves from $x = -A$ to $x = A$. The block is released at time $t = 0$, resulting in simple harmonic motion. Evaluate the validity of each of the following claims. The magnitude of the velocity of the block is at its maximum where the values of x and t are (T is the period):
   (a) $x = A$ and $x = -A$, when $t = 0$ and T
   (b) $x = 0$, when $t = T/4$ and $3T/4$
   (c) $x = A$ and $x = -A$, when $t = 0$ and $T/2$
   (d) $x = 0$, when $t = T/2$ and $3T/2$
9. An object with mass m is attached to a horizontal ideal spring with spring constant k, whose other end is attached to a wall. The object is pushed to position $x = -A$ on a horizontal surface with negligible friction

and released at $t = 0$. Describe each of the following during one full cycle of its motion.

(a) The total distance traveled by the object

(b) The displacement of the object

(c) The times at which the velocity of the object was most positive and the speed of the object was greatest

(d) The average velocity of the object

(e) The average speed of the object

(f) The maximum force on the object and at what times in the first period this occurs

10. A small object is attached to a horizontal spring and set in simple harmonic motion with amplitude A and period T. How long does it take for the object to travel a total distance of $6A$?

11. A 1.00-kg object is fixed to the end of a spring that has a spring constant of 16.0 N/m. The object is displaced 20.0 cm to the right and released from rest at $t = 0$ to slide on a horizontal table that exerts negligible friction force on the object. In terms of the period T:

(a) At what time, after release, does the object first pass the equilibrium position? Also find the second, third, and fourth times.

(b) Find the first two times after release that the object is 10.0 cm to the left of equilibrium.

(c) Find the first two times after it is released that the object is 10.00 cm to the right of equilibrium, moving toward the left.

12. A 0.200-kg object is attached to a horizontal spring that has a force constant of 75.0 N/m. The object is pulled 8.00 cm to the right of equilibrium and released from rest to slide on a horizontal table that exerts negligible friction on the object.

(a) Calculate the maximum kinetic energy of the object.

(b) Find the location of the object when the speed is equal to one-third of the maximum speed, the object is moving to the right, and the acceleration is positive.

(c) Calculate the value of the spring potential energy at this point.

13. Geoff counts the number of oscillations of a simple pendulum that he has released at a small angle from the vertical at a location where the acceleration due to gravity is 9.80 m/s^2, and finds that it takes 25.0 s for 14 complete cycles. Calculate the length of the pendulum.

14. What is the period of a 1.00-m-long simple pendulum on each of the planets in our solar system when it is released from a small angle? You will need to look up the acceleration due to gravity on each planet.

15. A simple pendulum oscillates between +8° and –8° (as measured from the vertical) near the surface of Earth. The length of the pendulum is 0.50 m.

(a) Calculate the period of this oscillation.

(b) Compare the time intervals between

i. +8° and –8°

ii. +4° and –4°

16. The period of a simple pendulum near the surface of Earth where $g = 9.80$ m/s^2 is 2.50 s. Calculate the length of this simple pendulum.

17. A simple pendulum has a period of 2.0 seconds on Earth. What is the period of the pendulum on Mars? The mass of Mars is 6.4×10^{23} kg and the radius of Mars is 3.4×10^6 m.

18. A meter stick is pivoted about the 20-cm mark and oscillates in simple harmonic motion. Calculate the period of oscillation of the meter stick.

19. Consider an object attached to an ideal spring.

(a) Using conservation of energy, derive an expression for the speed of an object that has a mass M, attached to a spring that has a force constant k, when the object–spring system is oscillating with an amplitude of A, as a function of position x (measured from the spring's equilibrium), $v(x)$.

(b) If M has a value of 250 g, the spring constant is 85 N/m, and the amplitude is 10.0 cm, use the expression you derived in part (a) to calculate the speed of the object at $x = 0$ cm, 2.0 cm, 5.0 cm, 8.0 cm, and 10.0 cm.

20. A glider of mass M on a horizontal air track is attached to a spring, which is connected to the end of the air track. The air track allows the glider to oscillate back and forth with negligible friction. The glider is initially oscillating with an amplitude A_1 and a period T_1.

(a) At a moment when the block is at its maximum displacement from equilibrium, a lump of clay, also of mass M, is dropped from a very small height onto the glider and sticks to it. Determine the new amplitude A_2 and period T_2 of the oscillations with the clay attached in terms of the original period and amplitude. Explain your reasoning.

(b) In a second experiment, the clay is removed, and the glider is set oscillating again. This time, the clay is dropped onto the glider as the glider passes through the equilibrium position. Determine the new amplitude A_3 and period T_3 of the oscillations in terms of the original period and amplitude. Explain your reasoning.

21. A block of mass M is attached to a horizontal spring of constant k, which is attached to a wall. The block oscillates with a frequency f_1 on a surface with negligible friction. A second block of mass $2M$ is attached to an identical spring and oscillates on an identical surface. The two block–spring systems have the same total energy.

(a) What is the frequency of motion of the second block in terms of f_1?

(b) How do the maximum speeds of the two blocks compare?

(c) How do the magnitudes of the maximum accelerations of the two blocks compare?

22. When a small ball of mass m swings at the end of a very light, uniform rod, the period of the pendulum is 2.00 s. What is the period of a thin uniform rod of mass m and the same length as the original pendulum?

23. Your space ship lands on a moon of a small planet that orbits a distant star. As you initially circled the moon, you measured its diameter to be 5480 km. After landing you observe that a simple pendulum that had a frequency of 3.50 Hz on Earth now has a frequency of 1.82 Hz.

(a) Predict the mass of the moon.

(b) Evaluate the claim that the oscillations of a spring–object system could also be used to determine the moon's mass.

24. A solid sphere, made of acrylic plastic with a density of 1.1 g/cm^3, has a radius of 5.0 cm. As shown below, a very small "eyelet" is screwed into the surface of the sphere and a horizontal support rod is passed through the eyelet, allowing the sphere to pivot around this fixed axis. If the sphere is displaced slightly from equilibrium near the surface of Earth, find the period of its harmonic motion when it is released. The rotational inertia of a sphere rotating about its center is $2mr^2/5$. Recall the parallel-axis theorem:

$$I = I_{CM} + md^2$$

where d is the distance from the center of mass to the axis of rotation.

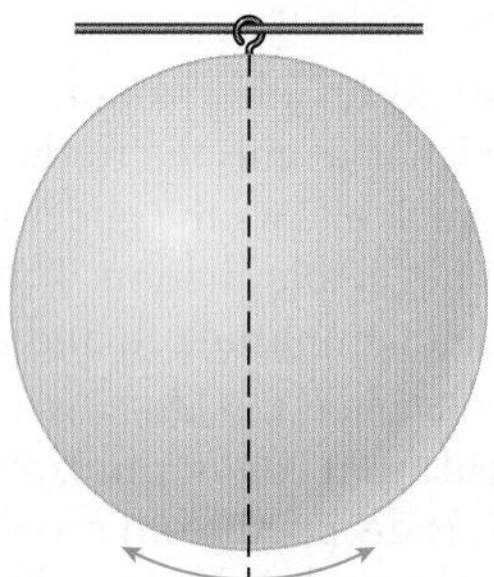

25. A block with a mass of 0.750 kg resting on a surface with negligible friction is attached to an unstretched spring with a length of 15.0 cm and a spring constant of $k = 9.80 \times 10^3$ N/m. The spring is attached to a wall at its other end. A 7.50-g, 9-mm-diameter bullet is fired into the block at a speed of 3.55×10^2 m/s and comes to a halt inside the block. Letting the initial position of the block be $x = 0$ and the positive direction be toward the wall, what is the position function $x(t)$ for the bullet–block system after the collision?

26. When an insect gets caught in a spider web, its struggles cause the web to vibrate. This alerts the spider to a potential meal. The frequency of vibration of the web gives the spider an indication of the mass of the insect. If we model the web as an ideal spring:

(a) Would a rapidly vibrating web indicate a large (massive) or a small insect? Explain your reasoning.

(b) Suppose that a 15-mg insect lands on a horizontal web and depresses it 4.5 mm. What would be the web's effective spring constant?

(c) At what rate would the web–insect system vibrate, assuming that the web's mass is negligible compared to that of the insect?

(d) Predict how the vibration rate would differ if the web were on a planet with a different gravitational constant. Explain how you made this prediction.

27. A 2.0-kg object is attached to a spring and undergoes simple harmonic motion. At $t = 0$ the object is released from rest, 10.0 cm from the equilibrium position. The spring constant, k, of the spring is 75 N/m. Calculate (a) the maximum speed and (b) the maximum acceleration of the object as the system oscillates. (c) Calculate the velocity of the object at $t = 5.0$ s.

28. A cable is to be chosen to hang a delicate electronic device, isolating it from mechanical vibrations with frequencies greater than 25 Hz. We discussed how the spring constant of a material is proportional to the stiffness of the material, the cross-sectional area of a shape made from the material, and inversely proportional to the length of that shape. To use stiffness in a calculation, we need a symbol, so we will denote the stiffness of the material as Y. Given Y, we can write

$$F_{\text{restoring}} = -Y\frac{A\Delta L}{L} = -k\Delta L$$

(a) Express the period of the vibration in terms of the length L, cross-sectional area A, and stiffness of the material used in the cable Y.

(b) The electronic device that will be hung has a mass of 475 kg. A 2.80-m-long nylon cable ($Y_{\text{nylon}} = 2.0$ GN/m^2) is considered. Predict the maximum diameter the cable can have.

(c) In addition, the design specifications require that the cable supporting the equipment cannot stretch by more than 0.1%. Calculate the minimum diameter of the cable that satisfies these design specifications.

Prep for the AP Exam

AP® Group Work

Directions: The problem below is designed to be done as group work in class.

When an object with mass is hung from the end of a cable, the cable stretches a little, ΔL, as shown in the figure. If the object was removed, the cable might behave elastically and return to its original length. Or the cable might behave plastically and remain stretched. Or the cable might break before the object is removed. Which of these responses that occurs depends on the material used to make the cable; the cross-sectional area, A, of the cable; and the tension in the cable, F. Thin cables stretch more than thick cables when the same force is exerted and the cables are made from the same material. Assume that the amount of stretch is proportional to the original length L. Assume that ΔL is measured from this unstretched original length.

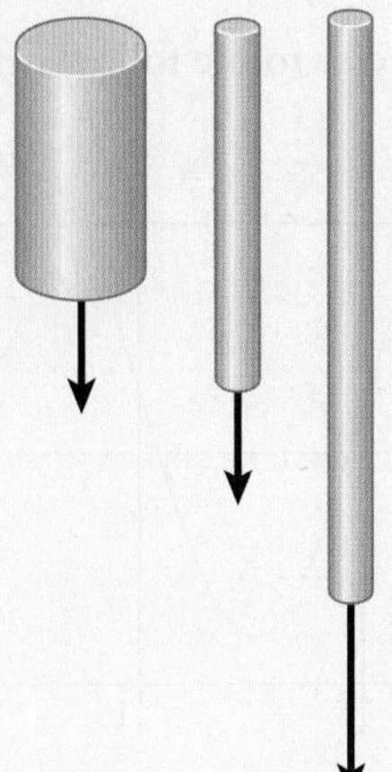

(a) Use mathematical reasoning to model ΔL in terms of the original length L, the area A, and the magnitude of the force exerted on the cable, F_0. *Hint*: Write a mathematical equation for ΔL that can account for the dependencies on A, L, and F_0, as described.

(b) Apply Newton's third law to express the elastic restoring force exerted by the cable, $F_{restoring}$, in terms of $\Delta L/L$ and A. As your constant of proportionality use the symbol Y. It is a property of the material used to make the cable that is independent of the cable's shape and related to its stiffness.

(c) Connect your model to the Hooke's law behavior of a spring stretched by a force of magnitude F_0 by expressing the relationship between the material constant Y and the spring constant k.

(d) Justify the claim that the spring constant k depends on the original shape of the spring before it is stretched or compressed but Y does not, so that the same material can be used to make springs with different values of k.

AP® PRACTICE PROBLEMS

Prep for the AP® Exam

Multiple-Choice Questions

Directions: The following questions have a single correct answer.

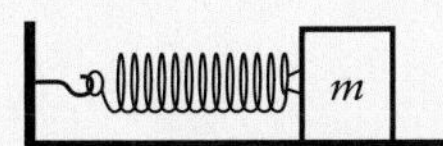

1. A block of mass m is attached to the end of a horizontal ideal spring and rests on a surface that exerts negligible friction on the block. The other end of the spring is attached to a wall. The block is pulled to the right and released from rest. Graphing which of the following sets of variables would produce a line?
   (A) Acceleration versus velocity of the block
   (B) Spring potential energy versus position of the block
   (C) Velocity of the block versus position of the block
   (D) Kinetic energy versus the square of the position of the block

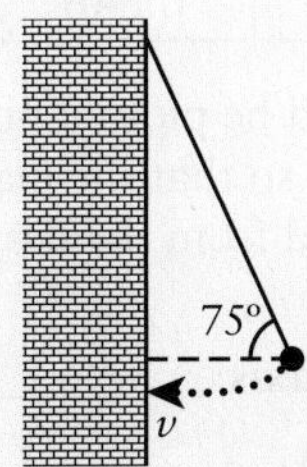

2. An object is attached to a wall by a long string that initially makes an angle of 75° with the horizontal. The object is released and strikes the vertical wall with speed v. What is the approximate time it takes for the object to reach the wall?
   (A) $\frac{\pi v}{2g}$
   (B) $\frac{\pi v}{g}$
   (C) $\frac{2\pi v}{g}$
   (D) $\frac{\sqrt{2}\pi v}{2g}$

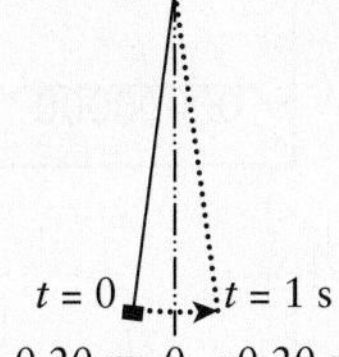

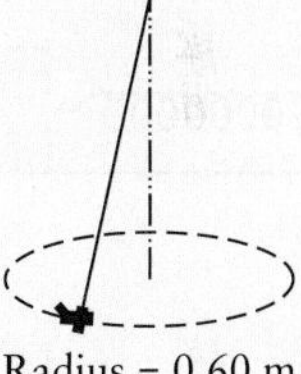

3. A toy battery-powered plane is hanging from the ceiling. With the motor turned off it is pulled to one side 0.20 m and released to swing as a pendulum. It takes 1 s to get to the far side of its swing. The plane now has the motor turned on and it flies in a circular path of radius 0.60 m. If the period of the circular motion is the same as the period of the simple harmonic motion, what is the approximate speed of the plane?
   (A) 0.6 m/s
   (B) 1.3 m/s
   (C) 1.9 m/s
   (D) 3.8 m/s

(*Continued*)

Questions 4 and 5 refer to the following material.

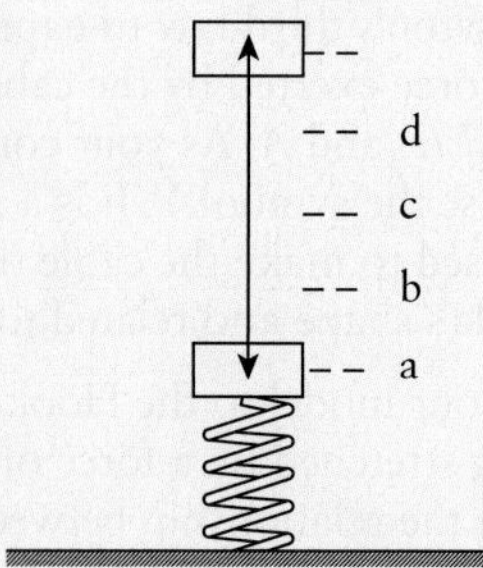

A block attached to a spring oscillates up and down as illustrated. The five levels indicated by the symbol—are equally spaced.

4. At which of the levels indicated is the potential energy of the spring–Earth–block system a minimum?
(A) a
(B) b
(C) c
(D) d

5. At which of the indicated levels is the magnitude of the acceleration of the block the greatest?
(A) a
(B) b
(C) c
(D) d

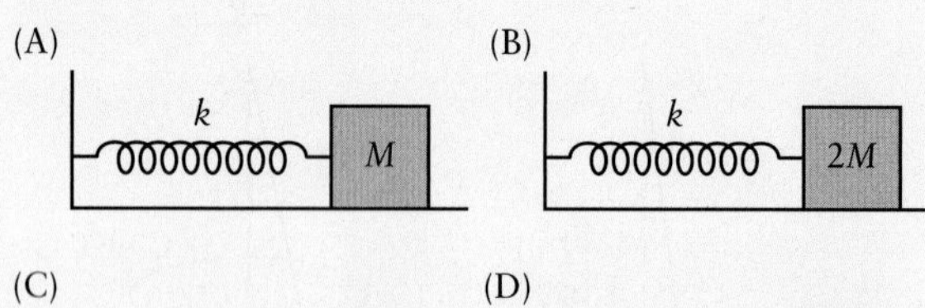

6. The four block–spring systems shown above have various masses and spring constants. They are all oscillating on surfaces of negligible friction. Which of the following is the correct ranking of the period of oscillation of the systems from greatest to least?
(A) (A = B) > (C = D)
(B) (C = D) > (A = B)
(C) D > (B = C) > A
(D) B > (A = D) > C

Questions 7 and 8 refer to the following material.

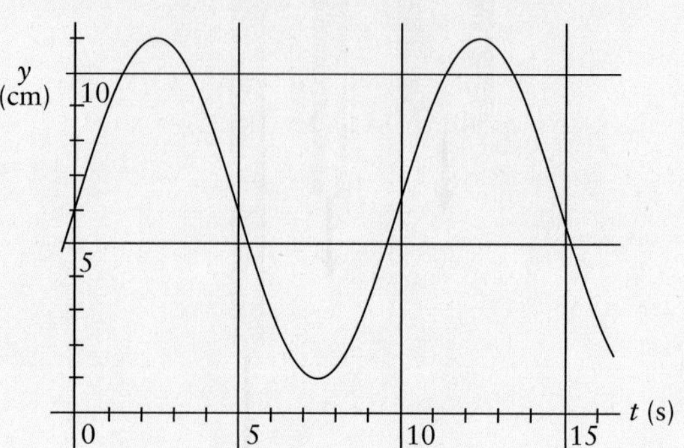

A 2.0-kg object hanging on the end of a spring moves up and down as indicated by the graph.

7. What is the amplitude of the motion?
(A) 5 cm
(B) 6 cm
(C) 10 cm
(D) 12 cm

8. What is the spring constant?
(A) 0.2 N/m
(B) 0.02 N/m
(C) 0.8 N/m
(D) 3 N/m

Free-Response Questions

1. Students are given five different springs with different spring constants k, a meter stick, and a stopwatch. They choose to measure the period T of oscillation of an object of unknown mass hanging from each of the springs. They obtain the data in the table below.

k (N/m)	T (s)	
25	0.902	
30	0.825	
35	0.771	
40	0.729	
50	0.646	

(a) What should be plotted on the vertical and horizontal axes so that the mass of the object can be calculated from the best-fit line of the linear data?

Horizontal axis ____________________

Vertical axis ____________________

Justify your choices.

(b) Use the empty space in the table to calculate any necessary quantities and plot the data you listed in part (a) on the axes below. Clearly scale and label all axes, including units if appropriate. Draw a line of best fit for the data.

(c) Using your best-fit line, calculate the experimental value of the unknown mass.

The teacher reveals that the actual mass of the object is 0.500 kg.

(d) Calculate the percent difference between the experimental and actual values of the mass.

(e) The uncertainty in the spring constant of each spring is 10% and the uncertainty in the period of each oscillation is 0.007 s. Based on this information, is the experimental value obtained for the mass of the object the same as the actual value to within the limits of precision of the measurements? Justify your response.

(f) Briefly describe another experiment students could do with the same equipment to determine the unknown mass of the object. Include a diagram of what students should measure and how students should analyze the data to obtain a value for the mass.

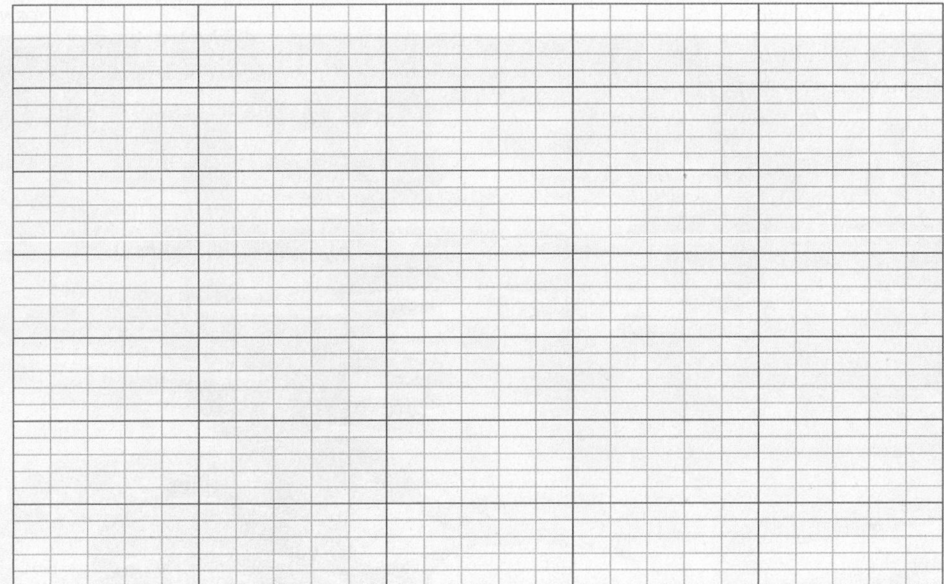

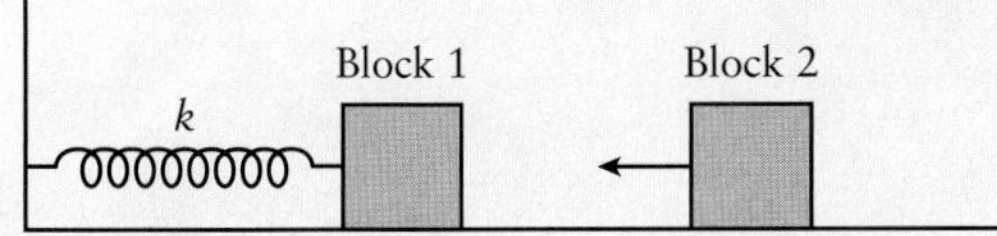

2. A spring with a spring constant k is initially at rest, unstretched, and is attached to block 1 of mass M, which is at rest on a horizontal surface, as shown in the figure. Block 2, also of mass M, is traveling toward the first block with an unknown speed. Friction between the blocks and the surface can be neglected. The blocks stick together after the collision and compress the spring a maximum distance D before coming momentarily to rest. Choose the positive x direction to be to the right. Give your answers to the following in terms of M, k, D, and physical constants, as appropriate.

(a) Derive an expression for the speed of block 2 before it strikes block 1 in terms of after the collision occurs.

(b) Derive an expression for the time it takes the two blocks to compress the spring a distance D and then return to the original equilibrium length of the spring.

(c) On the axes below, sketch a graph of the position of the blocks as a function of time after the collision for one complete cycle of the oscillation. Label the axes with appropriate values.

Fluids

Chapter 13 The Physics of Fluids

TORWAISTUDIO/Shutterstock

Case Study: How do we explain why only some things float?

We experience fluids every day. But they are not what people usually think they are. We saved fluids for last as their study incorporates aspects of everything we have learned so far, pulling together concepts from conservation of energy (the full work-energy theorem), mass, and momentum, and of forces. The interactions causing buoyancy, which will be defined in our study of fluids, explain why we "feel lighter" (as does the swimmer in the photo) when in water and why objects float. The study of fluids will also allow us to investigate in more detail the meaning of pressure and density.

To this point, we have focused on only one of the four fundamental states of matter, solids. Unlike systems that consist entirely of solid components—in which all constituents are connected in some way—no one point can be used to describe the motion of a fluid as a whole. In everyday language, the word *fluid* means something that flows (doesn't retain a fixed shape and moves easily around obstacles). There are two types of fluids—liquids and gases—which are two of the other fundamental states of matter. Their behaviors are similar in some ways but differ in others. Liquids tend to retain their volume; gases expand unless constrained.

Most of us are familiar with liquids. We breathe gases (air!) and we have some idea of how they behave (such as blowing up a balloon). But gases also exist in states much denser than those in a balloon. For example, the Sun and other stars are in part made of gases. These gases are so densely packed, so hot, and under so much pressure that electrons are stripped from atoms, creating the fourth fundamental state of matter, plasma. A plasma has similar properties to a gas, but is electrically charged, a topic in Physics 2. Familiar examples of plasma include lightning strikes—produced as electrons are stripped from air molecules!—and the Sun, in which most of the matter is plasma. Describing the behavior of plasmas requires a fundamental understanding of the other states of matter and a few other topics we haven't addressed yet. Because the Sun contains more than 99% of the mass in the solar system, our solar system is over 99% plasma!

Let's come back down to Earth. Each yellowback fusilier fish in the opening photo of Chapter 13 has an upward buoyant force exerted on it by the water around it. If you replaced one of these fish with a same-sized sculpture made of iron (which is denser than the fish's body), would the magnitude of the buoyant force increase, decrease, or remain the same?

By the end of this chapter, you should be able to describe the similarities and differences in the properties of liquids and gases, apply the definition of density to predict changes in density under different conditions for natural phenomena, explain the origin of fluid pressure, describe the source of the buoyant force, and use principles based on conservation laws to describe the behavior of moving fluids.

13 The Physics of Fluids

Jane Gould/Alamy

YOU WILL LEARN TO:

- Describe the similarities and differences in the properties of liquids and gases and how these relate to the interactions of their constituent molecules or atoms.
- Apply the definition of density to predict the densities, differences in densities, or changes in densities under different conditions for natural phenomena and design an investigation to verify the prediction.
- Explain the origin of fluid pressure in terms of molecular motion.
- Calculate the pressure at a given depth in a fluid in hydrostatic equilibrium.
- Explain the difference between absolute pressure and gauge pressure.
- Calculate the force on an object due to a difference in pressure on its sides.
- Explain how to apply Pascal's principle to a fluid at rest.
- Explain the source of the buoyant force on an object in a fluid, including how that determines its direction, and use Archimedes' principle to find that force.
- Use conservation of mass (the continuity equation) to analyze the flow of an incompressible fluid.
- Relate Bernoulli's principle to conservation of energy.
- Apply Bernoulli's equation to relate fluid pressure, height, and flow speed in an incompressible fluid.
- Use Bernoulli's equation and the continuity equation to describe the behavior of a moving ideal fluid.
- Describe the role of surface tension in the behavior of liquids.

13-1 Liquids and gases are both examples of fluids

Liquids and gases—collectively known as *fluids* for their ability to flow—are the most common states of matter in the universe. Indeed, part of our planet's interior, the entirety of our Sun, and most of your body are composed of fluids (**Figure 13-1**). Solid systems in our daily environment such as rocks and trees are exceptional cases in a universe dominated by fluid systems.

Solids are substances with individual molecules that cannot move freely but remain in essentially fixed positions relative to one another. That's why a solid system such as a rigid body maintains its shape, though it may bend or deform if you exert

Part of Earth's interior is a liquid substance, some of which occasionally rises to the surface as lava.
(a)

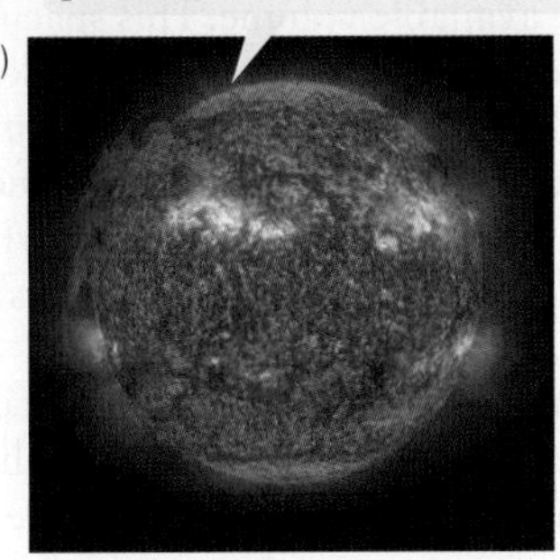

The Sun's material is in the form of a gas. In the high-temperature interior the gas is ionized, a state of matter called plasma.
(b)

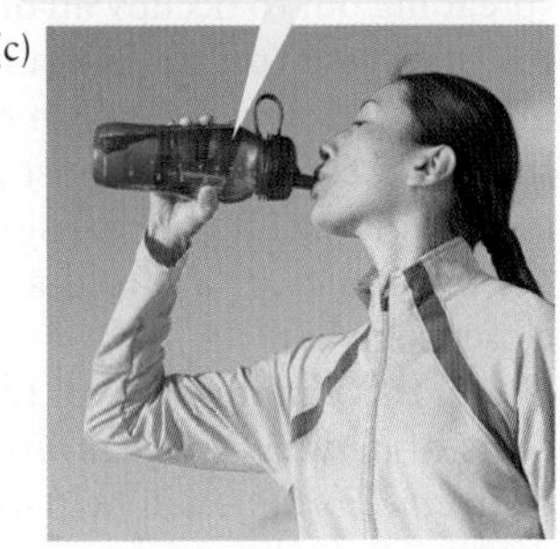

The body of an adult human is mostly liquid (by mass, about 50 to 60% water). Hydration keeps the liquid level constant.
(c)

Werner Van Steen/Getty Images; SOHO (ESA & NASA); Image Source Plus/Alamy

Figure 13-1 Fluids Materials that can flow make up (a) part of our Earth; (b) all of our star, the Sun (as well as all of the other stars visible in the night sky); and (c) most of our bodies. Note that the photo in (a) shows at least three different fluids: molten lava flowing down the rocks, liquid water in the ocean, and the air in our atmosphere.

large enough forces on it. In contrast, **fluids** are substances that can flow because their molecules can move with respect to each other and are not tied to fixed locations.

Fluids fall into two broad categories: liquids and gases. **Liquids** tend to maintain the same volume regardless of the shape and size of their container. For example, 1 cubic meter (1 m^3) of water taken from a storage tank will exactly fill a 1-m^3 container into which it is poured (**Figure 13-2a**). **Gases**, however, expand to fill whatever volume is available to them. If you place a cylinder containing 1 m^3 of compressed oxygen gas in a large room and open the valve, the oxygen gas spreads out over the entire room so that its volume is much greater than 1 m^3 (**Figure 13-2b**). It can also be compressed further to fill an even smaller container. (The gas in the interior of the Sun, shown in Figure 13-2b, is ionized: The gas atoms have lost one or more electrons. Such an ionized gas is called a *plasma*. Despite having a similar name, blood plasma is a liquid and is not ionized.)

What explains the difference between these two types of fluids? In a liquid, molecules are close to one another (almost touching). At close range the attractive forces between molecules in a liquid are strong, keeping the molecules together so that the

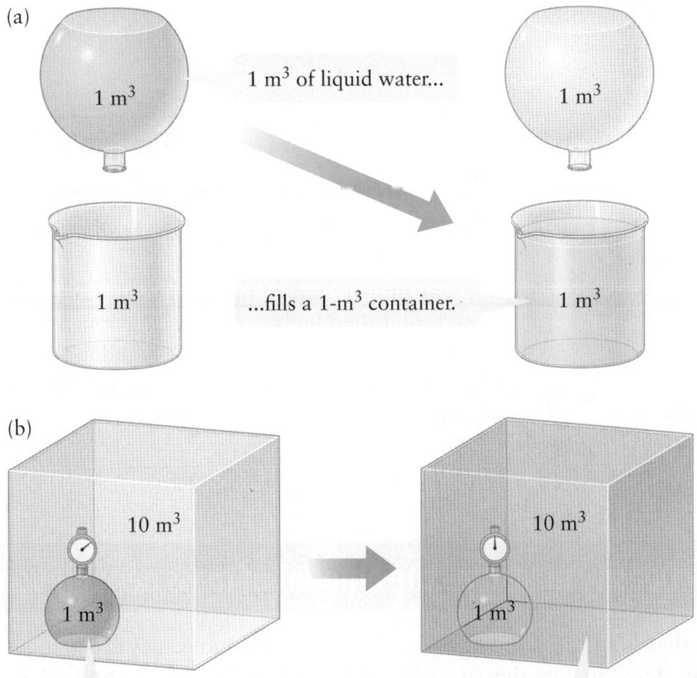

Figure 13-2 Liquids versus gases (a) A liquid maintains the same volume when you pour it from one container to another. (b) A gas expands to fill whatever volume is available.

volume of the liquid stays the same. In a gas, however, the molecules are much farther apart; in the air you're breathing now, the average distance between molecules is about 10 times the size of a single molecule. At these greater separations the molecules exert little or no attractive forces on each other, so nothing prevents the molecules from spreading out to occupy all of the available volume. This spacing also makes gases easier to compress. You can use a hand pump to put air into a bike tire or a basketball, for instance. Because gases can easily be compressed by squeezing, they are **compressible fluids.** By contrast, for most practical purposes, liquids are **incompressible fluids.** That is, the volume and the density of a liquid change very little when it is squeezed because the molecules of a liquid are almost touching and thus resist being squeezed much closer together.

In this chapter we will touch on many of the key aspects of fluids. The central concepts of *density* and *pressure*, coupled with the notion of what it means for a fluid to be in *equilibrium*, will help us understand why water pressure increases with increasing depth. They also explain the phenomenon of *buoyancy* that enables marine animals to float under water without sinking or rising. We will relate all of these factors to the conservation principles we have learned in previous chapters. Although fluids in *motion* are influenced not only by pressure and density but also by *viscous* forces, the frictional forces of one part of the fluid rubbing past another, we will adopt another ideal model, an ideal liquid, which neglects this viscosity. This is similar to neglecting air resistance or the mass of an ideal spring or string. As long as we understand when this model is valid it will allow us to examine some of the many ways in which fluids flow and understand important features of phenomena like the flight of airplanes and birds and the behavior of blood in the circulatory system. Finally, we will see how the mutual attraction of molecules in a liquid gives rise to *surface tension*, a phenomenon that impacts the function of your lungs.

THE TAKEAWAY for Section 13-1

✔ Fluids (which include liquids and gases) do not have a definite shape. They can flow to take on the shape of their container because their molecules are in constant motion.

✔ Liquids maintain essentially the same volume as they flow, but gases will expand or contract as necessary to fill the available space.

✔ The attractive forces between molecules are strong in liquids but are weak in gases because the molecules are farther apart in gases.

✔ Ideal fluids are incompressible and have negligible viscosity.

Prep for the AP® Exam

AP® Building Blocks

1. When we neglect air resistance (which is a fluid resistance to the motion of an object through the air) what features of the motion does that change for:
   (a) A ball thrown horizontally?
   (b) A ball thrown upward?
   (c) A car driving along a highway at constant speed?
2. When gas is compressed significantly, in some situations it can behave like a liquid. (The "some situations" will become clear if you take AP® Physics 2!) Why might this happen, based only on what we know about gases and liquids so far?
3. One way of determining the volume of an irregularly shaped solid rigid body is to submerge it in a liquid to measure the volume change of the liquid in the container. What assumption is being made about the rigid body when performing such experiments?

13-2 Density measures the amount of mass per unit volume

One quantity that helps describe how tightly packed the molecules are in a gas, liquid, or solid is the *density*. The **density** of a substance, denoted by the Greek letter ρ ("rho"), is its *mass per volume*: the mass of the substance divided by the volume that it occupies. The greater the density of a substance, the more kilograms of that substance are packed into a given number of cubic meters of volume:

The **density** ρ ("rho") of a certain substance... ...equals the **mass** m of a given quantity of that substance...

$$\rho = \frac{m}{V} \qquad (13\text{-}1)$$

...divided by the **volume** V of that quantity of the substance.

EQUATION IN WORDS
Definition of density

A given number of water molecules of combined mass m occupy a smaller volume V in liquid water than they do in water vapor (a gas). Hence the density of liquid water is greater than that of water vapor.

WATCH OUT!

Density alone doesn't tell you how closely packed the molecules are.

A certain number of oxygen gas molecules have 14% greater density than the same number of nitrogen gas molecules with the same spacing between molecules. That's because density is *mass* per unit volume, and the mass of an oxygen molecule is 14% greater than that of a nitrogen molecule. To find the spacing between molecules in a given substance, you need to know both the density of the substance *and* the mass per molecule.

Density is usually stated in kilograms per cubic meter (kg/m^3) or grams per cubic centimeter (g/cm^3; 1 g/cm^3 = 1000 kg/m^3). As an example, one cubic meter of liquid water at a temperature of 4°C has a mass of 1000 kg and therefore a density of $\rho = (1000\text{ kg})/(1\text{ m}^3) = 1000\text{ kg/m}^3 = 1\text{ g/cm}^3$. Twice the mass of liquid water (2000 kg) occupies twice the volume (2 m^3) and so its density is the same: $\rho = (2000\text{ kg})/(2\text{ m}^3) = 1000\text{ kg/m}^3$. This example shows that density depends only on the nature of the substance, *not* on how much of the substance is present. When we say that gold has a density of 1.93×10^4 kg/m^3, we mean that *any* quantity of gold has this same density. **Table 13-1** lists the densities of some common substances.

An alternative way to describe the density of a substance is to compare it to another substance whose density serves as a standard. One common standard is liquid water at 4°C ($\rho = 1000$ kg/m^3), and the **specific gravity** of a substance is equal to its density divided by that of 4°C liquid water. Human blood, for instance, has a specific gravity of 1.06, which means it is 1.06 times as dense as water: $\rho_{\text{blood}} = (1.06)(1000\text{ kg/m}^3) = 1.06 \times 10^3\text{ kg/m}^3$. The term *specific gravity* in this case means that a certain volume of blood has 1.06 times the mass and hence 1.06 times the weight as the same volume of water. On average, red blood cells constitute about 38% of the volume of blood in women and 42% in men; these cells are only slightly denser than water, which is why the specific gravity of blood is just a little greater than 1.

In general, the density of a liquid depends on temperature. If you take water from the refrigerator (temperature about 4°C) and let it warm to room temperature (about 20°C), the water expands slightly. Because the mass m of the water is the same but its volume V has increased, the density $\rho = m/V$ of 20°C water is slightly less (by about 0.2%) than that of 4°C water. Most liquids expand and become less dense when the temperature increases. Between 0°C and 4°C, water is a notable exception to this general rule: Within this temperature range, water *contracts* and becomes *more* dense as the temperature increases. That's why we give the densities of water and salt water in Table 13-1 at 4°C, the temperature at which these liquids are densest.

TABLE 13-1 Densities of Various Substances

Substance	Density (kg/m^3)
air at sea level at 15°C	1.23
dry timber (white pine)	370
fresh wood (American elm)	570
fresh wood (red oak)	740
gasoline	740
ethanol (ethyl alcohol) at 20°C	789
lipids (fats and oils)	915–945
ice at 0°C	917
fresh water at 4°C	1.000×10^3
fresh water at 20°C	998
seawater at 4°C	1.025×10^3
muscle	1.06×10^3
deer antler (low-density bone)	1.76×10^3
cow femur (typical bone)	2.06×10^3
tympanic bulla of a fin whale (densest bone)	2.47×10^3
aluminum (Al)	2.70×10^3
mollusk shell	2.7×10^3
calcite (mineral of shell)	2.71×10^3
tooth enamel, human	2.9×10^3
apatite (mineral of bone, tooth)	3.2×10^3
planet Earth (average density)	5.52×10^3
iron (Fe)	7.8×10^3
mercury (Hg) at 0°C	13.595×10^3
gold (Au)	19.3×10^3

The density of gases is sensitive to both temperature and pressure. Heated air expands, which lowers its density; applying pressure makes the air occupy a smaller volume, which raises its density. Although you can easily compress air to a higher density with a hand pump used for inflating bicycle tires, to compress water appreciably takes tremendous pressures. (At the deepest point in the Pacific Ocean, at a depth of 11 km, the pressure is about a thousand times greater than at the surface. Yet the density of seawater there is only about 5% greater than at the surface. In Section 13-3 we'll discuss why pressure increases with depth.)

While the main focus of this chapter is on fluids, the idea of density applies to solids as well. The following examples show some applications.

EXAMPLE 13-1 Density of Chicken

A typical whole chicken for sale at a supermarket has a mass of 2.3 kg and is approximately spherical in shape, with a radius of about 8.0 cm. What is the approximate density of such a chicken in kg/m^3?

Set Up

We are given the chicken's mass and its radius, and our goal is to calculate its density ρ. We use the definition of density in Equation 13-1, $\rho = m/V$.

The chicken is very nearly spherical, so to determine its volume we use the expression for the volume of a sphere of radius R (see the Math Tutorial).

Definition of density:

$$\rho = \frac{m}{V} \tag{13-1}$$

Volume of a sphere of radius R:

$$V = \frac{4\pi R^3}{3}$$

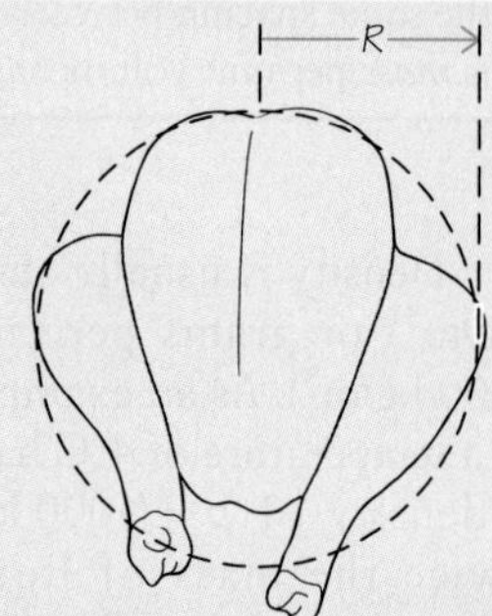

Solve

To calculate the density in the desired units, we need to find the chicken's volume in cubic meters (m^3). We first express its radius in meters.

$$R = (8.0\text{ cm})\left(\frac{1\text{ m}}{100\text{ cm}}\right) = 8.0 \times 10^{-2}\text{ m}$$

We then calculate the volume of the chicken in cubic meters.

$$V = \frac{4\pi R^3}{3} = \frac{4\pi(8.0 \times 10^{-2}\text{ m})^3}{3} = 2.1 \times 10^{-3}\text{ m}^3$$

The density is the mass of the chicken ($m = 2.3$ kg) divided by this volume.

$$\rho = \frac{m}{V} = \frac{2.3\text{ kg}}{2.1 \times 10^{-3}\text{ m}^3} = 1.1 \times 10^3\text{ kg/m}^3$$

Reflect

Our answer is close to some of the values for biological materials given in Table 13-1, but isn't a precise match for any of them. That's because a chicken's body is not a pure substance but a mixture of substances of different densities: fats, water, muscle, and bone. Our calculated density of 1.1×10^3 kg/m^3 is actually the *average* density of the chicken and falls between the density values for muscle and bone, so seems possible. We explore the idea of average density further in Example 13-2.

You may be wondering if it's reasonable to approximate a chicken as being spherical. If you look at the sketch, some parts of the chicken extend beyond the sphere, and some of the sphere is empty. But the volume of the chicken outside the sphere is nearly equal to the volume of the "empty" part of the sphere, so the volume of the chicken is actually pretty close to the volume of the sphere.

NOW WORK Problems 1–3 and 4(c) from The Takeaway 13-2.

EXAMPLE 13-2 Average Density

A submersible vessel used for deep-sea research has a metal hull with a large cavity in its interior for the crew and their equipment. Suppose that the hull is made of 1.125×10^3 m^3 of a particular steel alloy of density 7.910×10^3 kg/m^3 and that the cavity in the hull has volume 7.605×10^3 m^3 and is filled with air of density 1.230 kg/m^3. Find the *average* density of the submersible, that is, the density of the submersible as a whole.

Set Up

The submersible contains two materials—the steel of the hull and the air in the interior spaces—which have different densities and occupy different volumes. We want to find a third density, the *average* density of the entire submersible, equal to its total mass m_{sub} divided by its total volume V_{sub}. We again use the definition of density, Equation 13-1. We'll apply this definition three times, once for each density.

Definition of density:

$$\rho = \frac{m}{V} \quad (13\text{-}1)$$

Solve

The submersible's average density $\rho_{average}$ is its total mass m_{sub} divided by its total volume V_{sub}, so we need to calculate both m_{sub} and V_{sub}. The volume of the submersible (V_{sub}) is the sum of the volume of the hull (V_{hull}) and the volume of the cavity (V_{cavity}).

Average density of the sub as a whole:

$$\rho_{average} = \frac{m_{sub}}{V_{sub}}$$

Total volume of the sub:

$$V_{sub} = V_{hull} + V_{cavity} = 1.125 \times 10^3 \text{ m}^3 + 7.605 \times 10^3 \text{ m}^3$$
$$= 8.730 \times 10^3 \text{ m}^3$$

From Equation 13-1 the mass of steel in the hull (m_{hull}) and the mass of air in the cavity (m_{cavity}) are given by their respective densities multiplied by their respective volumes.

Rearrange Equation 13-1:

$$m = \rho V$$

Solve for m_{hull} and m_{cavity}:

$$m_{hull} = \rho_{steel} V_{hull} = (7.910 \times 10^3 \text{ kg/m}^3)(1.125 \times 10^3 \text{ m}^3)$$
$$= 8{,}898{,}750 \text{ kg}$$

$$m_{cavity} = \rho_{air} V_{cavity} = (1.230 \text{ kg/m}^3)(7.605 \times 10^3 \text{ m}^3)$$
$$= 9354 \text{ kg}$$

The total mass of the submersible (m_{sub}) is the sum of the mass of steel in the hull and the mass of air in the cavity.

$$m_{sub} = m_{hull} + m_{cavity} = 8{,}898{,}750 \text{ kg} + 9354 \text{ kg}$$
$$= 8.908 \times 10^6 \text{ kg}$$

To get the average density, substitute the total volume V_{sub} and the total mass m_{sub} into the expression for $\rho_{average}$.

$$\rho_{average} = \frac{m_{sub}}{V_{sub}}$$

$$= \frac{8.908 \times 10^6 \text{ kg}}{8.730 \times 10^3 \text{ m}^3} = 1.020 \times 10^3 \text{ kg/m}^3$$

Reflect

No one component of the submersible has a density of 1.020×10^3 kg/m^3; instead, this represents an average of the densities of the two components. The average density is intermediate between the density of air (1.230 kg/m^3) and the density of steel (7.910×10^3 kg/m^3), so our answer makes sense.

Extend: This particular value of average density is greater than the density of fresh water (1.000×10^3 kg/m^3) but is almost the *same* as that of seawater (1.025×10^3 kg/m^3). As we will see in Section 13-7, this means that the submersible will sink in fresh water but will float when submerged in seawater—which is a useful characteristic for a deep-sea research vessel to have. It has to have a slightly lower density so that it can still float when the crew and equipment are on board. Most such vessels have tanks they can fill with water to adjust the mass as needed so that the average density of the vessel and all its contents allows it to float, but also makes it possible for the submersible to go completely below the surface.

NOW WORK Problem 10 from The Takeaway 13-2.

EXAMPLE 13-3 How Much Space for a Molecule?

A single molecule of water (H_2O) has a mass of 2.99×10^{-26} kg. Find the average volume per water molecule (a) in liquid water at 4°C and (b) in water vapor (that is, water in gas form) at 120°C, which has a density of 0.559 kg/m^3. In each case, what is the length of each side of a cube with this volume? Compare these lengths to the diameter of a water molecule, which is about 2.0×10^{-10} m.

Set Up

The density ρ of a substance tells us the mass per volume. We want to find the volume per *molecule* in the substance. We'll do this calculation by combining the value of ρ for each substance (liquid water at 4°C and water vapor at 120°C) with the mass per water molecule. We'll imagine that each molecule is at the center of a cube of side L, and we'll determine L from the formula for the volume of a cube.

Definition of density:

$$\rho = \frac{m}{V} \quad (13\text{-}1)$$

Volume of a cube of side L:

$$V = L \times L \times L = L^3$$

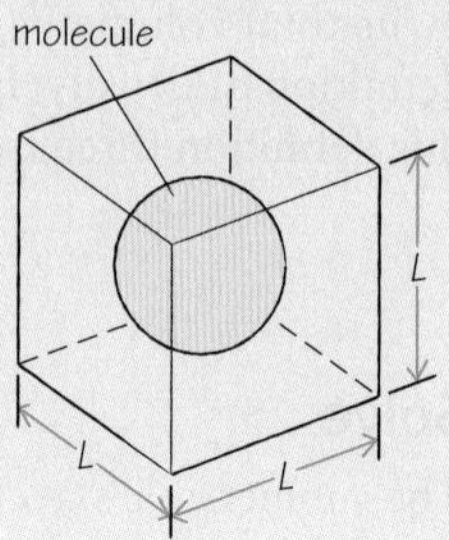

Solve

Rewrite Equation 13-1 to solve for the volume occupied by a quantity m of the substance.

From Equation 13-1:

$$V = \frac{m}{\rho}, \text{ or}$$

$$\text{volume occupied by a molecule} = \frac{\text{mass of a molecule}}{\text{density}}$$

(a) Table 13-1 tells us that water at 4°C has a density of 1.000×10^3 kg/m^3. Use this to find the volume per molecule and the length L of a cube with this volume.

For water at 4°C:

volume occupied by a molecule = V

$$= \frac{2.99 \times 10^{-26} \text{ kg}}{1.00 \times 10^3 \text{ kg/m}^3}$$

$$= 2.99 \times 10^{-29} \text{ m}^3$$

The volume V is equal to L^3; that is, V is the cube of L. So L is the cube *root* of V:

$$L = \sqrt[3]{V} = \sqrt[3]{2.99 \times 10^{-29} \text{ m}^3}$$

$$= 3.10 \times 10^{-10} \text{ m}$$

This is about 50% larger than the size of a water molecule we were given (roughly 2.0×10^{-10} m).

(b) Repeat the calculation for water vapor at 120°C. The molecule is the same and so has the same mass, but the density is far less than that for the liquid at 4°C.

For water vapor at 120°C:

volume occupied by a molecule = V

$$= \frac{2.99 \times 10^{-26} \text{ kg}}{0.559 \text{ kg/m}^3}$$

$$= 5.35 \times 10^{-26} \text{ m}^3$$

The length L of a cube with this volume is

$$L = \sqrt[3]{V} = \sqrt[3]{5.35 \times 10^{-26} \text{ m}^3}$$

$$= 3.77 \times 10^{-9} \text{ m}$$

This is larger than the size of a water molecule by a factor of $(3.77 \times 10^{-9} \text{ m})/(2.0 \times 10^{-10} \text{ m}) = 19$.

Reflect

Our results show that in the liquid state a water molecule almost fills the volume that it occupies on average. So there is very little extra room to move the molecules closer together, which means that it is very difficult to compress a liquid. In the gaseous state, however, on average each molecule is surrounded by lots of empty space. In this state it's much easier to push the molecules closer together, so a gas is relatively easy to compress, which matches what we expect for liquids and gases.

Extend: In both liquids and gases, the molecules are in constant motion. That's why the volumes we calculated in this example are the *average* volume per molecule.

NOW WORK Problems 5, 6, and 7 from The Takeaway 13-2.

THE TAKEAWAY for Section 13-2

✔ The density of a substance describes how much mass of a particular substance occupies a given volume.

✔ The specific gravity of a substance equals the density of that substance relative to the density of fresh water at 4°C.

✔ A system's average density equals its total mass divided by its overall volume. It is intermediate in value between the densities of the least dense and most dense constituents of the system.

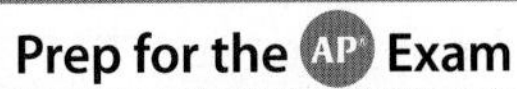

AP Building Blocks

EX 13-1 1. Earth has radius 6380 km and mass 5.98×10^{24} kg. Determine the average density of our planet.

EX 13-1 2. Determine the mass of a cube of iron that is 2.0 cm × 2.0 cm × 2.0 cm in size.

EX 13-1 3. What is the radius of a sphere made of aluminum, if its mass is 24.8 kg?

4. A cylinder is 0.20 m long and 1.0 cm in radius, and has a mass of 37 g.
(a) Determine the average density of the cylinder.
(b) Referring to Table 13-1, of what material might the cylinder be composed?
EX 13-1 (c) What is the specific gravity of the material?

EX 13-3 5. How long are the sides of an ice cube if its mass is 0.35 kg?

EX 13-3 6. Approximately 65% of a person's body weight is water. Find the volume of water in a 65-kg man.

AP Skill Builders

EX 13-3 7. When a massive star reaches the end of its life, it is possible for a supernova to occur. This may result in the formation of a very small, but very dense, neutron star, the density of which is about the same as a neutron. Determine the radius of a neutron star that has the mass of our Sun and the same density as a neutron. A neutron has a mass of 1.7×10^{-27} kg and an approximate radius of 1.2×10^{-15} m.

8. Object A has density ρ_1. Object B has the same shape and dimensions as object A, but it is three times as massive. Object B has density ρ_2 such that
(A) $\rho_2 = 3\rho_1$.
(B) $\rho_2 = \dfrac{\rho_1}{3}$.
(C) $\rho_2 = \rho_1$.
(D) $\rho_2 = 2\rho_1$.

AP Skills in Action

9. Which is heavier, a block of iron or a block of ice?
(A) The block of iron is heavier.
(B) The block of ice is heavier.
(C) They have the same weight.
(D) Not enough information is given to decide.

10. A 20 ft × 8.0 ft × 8.5 ft shipping container has a mass of 2350 kg when empty and has about 33 m^3 of cargo space. The container is 2/3 full of bags of cat litter that have a density of 0.54 g/cm^3.
(a) What is the average density of the container?
EX 13-2 (b) Will it float if it falls off a ship?

13-3 Pressure in a fluid is caused by the impact of molecules

Pressure is a commonplace idea. Everyone has had their blood pressure measured during a visit to the physician. An air-filled balloon bursts if the pressure within it is too great. And weather reports refer to low-pressure and high-pressure weather systems in the atmosphere. But what *is* pressure?

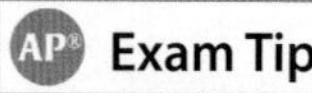

Exam Tip

As we stressed in Chapter 5, it is not uncommon to see the same symbol reused, or for different texts to use different symbols to represent the same quantity. That is why it is so important to understand what the symbols in any equation you use mean, so that you can ensure you are using the correct quantity.

Everything around you has forces exerted on it by the surrounding material. Air exerts forces on your body; water exerts forces on a submerged fish or the swimmer in the unit-opening photo; the surrounding soil exerts a force on a rock buried in the ground. The magnitude of the force per unit area on an object or system is called the **pressure** P. (Unfortunately, we used this same symbol for power in Chapter 8. We will not use them both in the same equation without at least adding a subscript to distinguish them!)

Pressure is always defined as the force perpendicular to a surface per unit area of that surface. **Figure 13-3** shows this definition of pressure for a fluid. If you place a small, thin, flat plate in a fluid—a gas or a liquid—you'll find that the fluid exerts forces on the plate that are perpendicular to either face of the plate. You can feel these forces if you stick your hand into a swimming pool or a sink full of water. The same kind of forces are exerted on your hand when it's surrounded by air, but are very feeble compared to the forces exerted by water. The net force that the fluid exerts on either face of the plate is perpendicular to the face. The pressure of the fluid exerted on the face is the magnitude of this perpendicular force divided by the area of the face:

EQUATION IN WORDS
Definition of pressure

The **pressure** P exerted by a fluid on an object in the fluid...

...equals the **force** $F_\perp$ that the fluid exerts **perpendicular to the face of the object...**

$$P = \frac{F_\perp}{A} \quad (13\text{-}2)$$

...divided by the **area** A of that face.

WATCH OUT!

The symbols for pressure and density are not the same.

We use an uppercase P to represent pressure and the lowercase Greek letter ρ ("rho") to represent density. Unfortunately, these two symbols look similar when handwritten. Be careful that you properly distinguish between the two.

To understand why a fluid exerts forces like those shown in Figure 13-3, keep in mind that the molecules of a liquid or gas are in constant motion. Any object (or system) placed in a fluid is struck repeatedly by these moving molecules. As molecules bounce off the object's surface, their momentum is changed. This requires the surface to have delivered an impulse to them, resulting in a force exerted by the surface on the molecules. By Newton's third law, the molecules must exert a force that is equal in magnitude and in the opposite direction on the surface. The force of each of these individual impacts is miniscule, but there are so many molecules and so many impacts that their combined effect is appreciable. It's that combined force that gives rise to the forces labeled $F_\perp$ in Figure 13-3. As an example, more than 10^{23} air molecules collide with a square centimeter of your skin every second, exerting a combined force of about 10 N on that square centimeter. In a fluid at rest (such as water in the kitchen sink that is not flowing or draining out, or the air in a room with no wind currents), the molecules of the fluid move in entirely random directions. At any point in the fluid there are as many molecules moving to the left as to the right, as many moving upward as downward, and so on. Hence a small, thin, flat plate placed in the fluid (Figure 13-3) experiences the same number of impacts, and hence the same magnitude of force, on

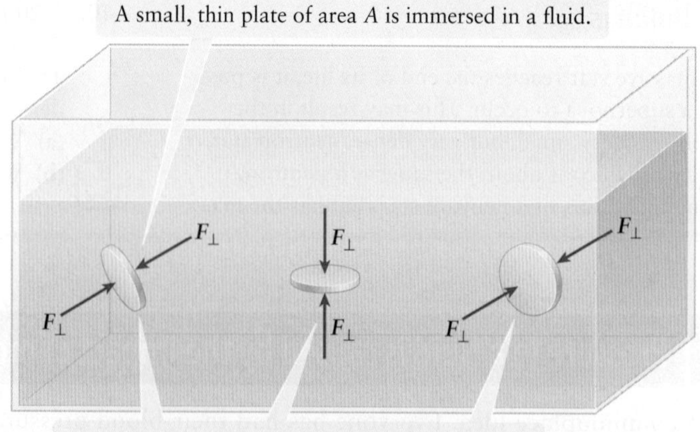
A small, thin plate of area A is immersed in a fluid.
F⊥
F⊥
F⊥
F⊥
F⊥
F⊥
No matter how the plate is oriented, the fluid exerts the same magnitude of force F⊥ and the same pressure P = F⊥/A on both sides of the plate.

Figure 13-3 Pressure on a thin plate in a fluid The fluid pressure on this plate does not depend on its orientation.

both of its sides no matter which way the plate is oriented. Here, because we specified a small plate, we are neglecting the difference in pressure as a function of depth in the fluid, which we will discuss in the next section.

If we double the surface area A exposed to the fluid, the force magnitude $F_{\perp}$ also doubles because there is twice as much area to be hit by the molecules of the fluid. However, the pressure $P = F_{\perp}/A$ remains the same. Pressure is a property of the fluid, not of the object that's immersed in the fluid. Furthermore, because the pressure is the same no matter how the disk in Figure 13-3 is oriented, pressure has no direction: Unlike force, pressure is *not* a vector.

The SI units of pressure are newtons per square meter (N/m^2). One newton per square meter is also called a **pascal** (abbreviated Pa): 1 Pa = 1 N/m^2.

To explore additional units, let us get a bit more precise. The air around you pushes in on you, exerting a considerable pressure from every direction. The air pressure at sea level varies somewhat due to the passage of weather systems, but on average is equal to 1.01325×10^5 Pa in SI units or about 14.7 pounds per square inch (lb/in^2) in English units. This average value of atmospheric pressure at sea level is defined to be one **atmosphere**, abbreviated atm:

$$1 \text{ atm} = 1.01325 \times 10^5 \text{ Pa} = 14.7 \text{ lb/in}^2$$

To three significant figures, 1 atm = 1.01×10^5 Pa. Because 1 Pa = 1 N/m^2, this means that if you paint a square 1 m on a side onto the ground at sea level, the weight of the column of air that sits atop that square and extends to the upper limit of Earth's atmosphere is 1.01×10^5 N. A metric ton has a mass $m = 1000$ kg and a weight $mg = (1000 \text{ kg})(9.80 \text{ m/s}^2) = 9.80 \times 10^3$ N; the weight of air above that 1-m square is 10.3 times greater (10.3 metric tons). If you prefer to think in English units, paint a 1-in square onto the ground at sea level: The weight of air above that square is 14.7 lb.

The atmosphere (atm) is a convenient unit for dealing with relatively large pressures. For example, the pressures found at the bottom of the ocean can exceed 10^3 atm, and high-pressure gas systems can operate at pressures of hundreds or even thousands of atmospheres. Very *low* pressures are often measured in pascals rather than atmospheres; for example, a high-quality vacuum pump can reduce the pressure inside a container to 10^{-5} Pa or lower.

WATCH OUT !

Don't confuse force and pressure.

Although force and pressure are related, they are *not* the same quantity. Here's an example. If you take off your shoes and stand up, your mass is distributed evenly over the surface area of your two feet. The pressure on the soles of your feet is your mass divided by the surface area of the bottom of your two feet and is probably not uncomfortable. But if you stood on a bed of nails, it would hurt! Your weight hasn't changed, so the force you exert on the surface supporting you hasn't changed. However, because the supporting surface—the tips of a few nails—has a much smaller surface area, the pressure on your feet is much greater (and much more painful) in this case. However, people can comfortably *lie* on a suitably prepared bed of nails. By lying flat, a person's weight is distributed over the area of the tips of many nails, resulting in a tolerable pressure at each tip (**Figure 13-4**).

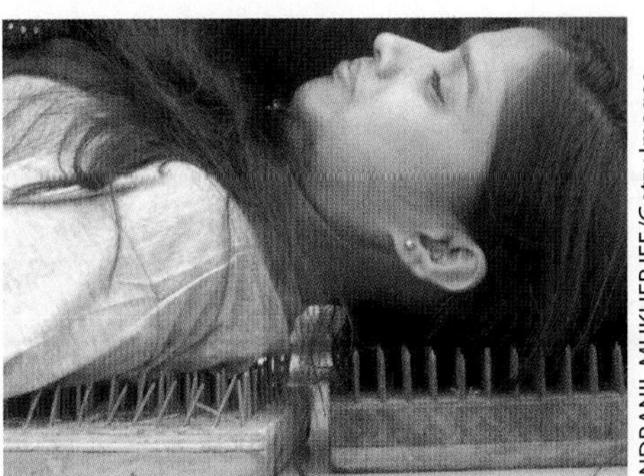

Figure 13-4 A bed of nails It doesn't hurt to lie on this bed of nails. That's because the force that the nails exert is distributed over the area of many nails, resulting in a tolerable pressure on the person's back.

While at sea level on Earth, the pressure of the atmosphere is about 1.01×10^5 N/m^2, or 1.01×10^5 Pa, the pressure at an altitude of 12,500 m (41,000 ft, about the highest altitude at which airliners fly) is only 1.78×10^4 Pa, or 17.6% of the pressure at sea level. One reason the pressure is lower at high altitude is that the air is less dense there (0.289 kg/m^3 at 12,500 m versus 1.23 kg/m^3 at sea level). Hence there are fewer air molecules to produce pressure through collisions. A second reason is that the molecules at high altitude move more slowly than those near sea level, so when collisions do take place they are relatively gentle. We'll learn in AP® Physics 2 that gas molecules move more slowly in a cold gas than a warm gas. The air at 12,500 m is indeed quite cold: The average temperature there is –56°C (–69°F) compared to 15°C (59°F) at sea level. The air pressure at 12,500 m is too low to sustain human life, which is why airliners have *pressurized* cabins: Air from the jet engine intakes is pumped into the cabin, increasing the pressure inside the cabin to what it would be at an altitude of 1800 to 2400 m (6000 to 8000 ft) above sea level.

The variation of pressure with elevation isn't unique to our atmosphere. In oceans, in lakes, and even in a glass of water there is greater pressure as you go deeper and lower pressure as you ascend. In the next section we'll learn more about how and why pressure increases with depth in a fluid.

EXAMPLE 13-4 Pressures and Forces on an Airliner Door

A typical passenger door on an airliner (**Figure 13-5**) is 183 cm high and 84 cm wide, and has an area (height times width) of 1.54 m^2. When the airliner is sitting on the ground at sea level, the air pressure on the outside of the door (in the surrounding air) and on the inside of the door (inside the cabin) are both equal to 1.01×10^5 Pa. When the airliner is cruising at an altitude of 12,500 m above sea level, the air pressure outside the cabin is 1.78×10^4 Pa and is 7.53×10^4 Pa inside the pressurized cabin. Find the forces that the air exerts on the inside and outside of the door, and the net force that the air exerts on the door (a) on the ground at sea level and (b) at an altitude of 12,500 m.

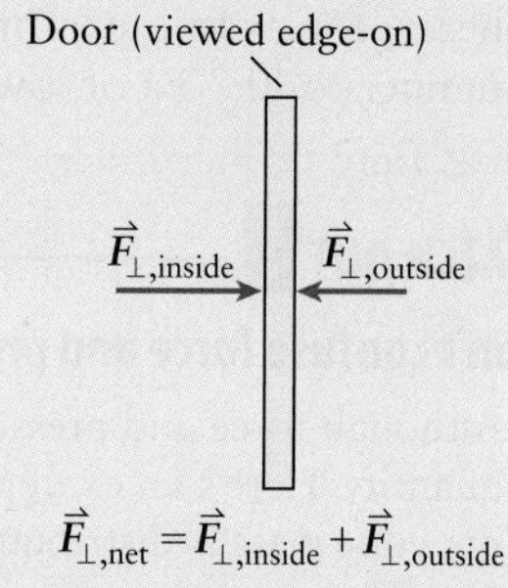

Figure 13-5 An airliner door How do the forces on this airliner door due to air pressure vary as the airliner climbs to its cruising altitude?

Set Up

Most airliner doors are curved to match the shape of the fuselage, but for simplicity we'll assume that the door is flat. We're given the pressure P on each side of the door as well as its area A, so we can use Equation 13-2 to solve for the force $F_\perp$ that the air exerts perpendicular to each side of the door. The forces on the two sides of the door are in opposite directions: Their vector sum is the *net* force that the air exerts on the door.

Definition of pressure:

$$P = \frac{F_\perp}{A} \quad (13\text{-}2)$$

Door (viewed edge-on)

$\vec{F}_{\perp,\text{inside}}$ $\vec{F}_{\perp,\text{outside}}$

$$\vec{F}_{\perp,\text{net}} = \vec{F}_{\perp,\text{inside}} + \vec{F}_{\perp,\text{outside}}$$

Solve

Rewrite Equation 13-2 to solve for the force exerted perpendicular to each side of the door.

From Equation 13-2:

$$F_\perp = PA$$

(a) At sea level the pressure is the same on both sides of the door, so the air exerts forces of equal magnitude on both sides of the door.

At sea level:

$$P_{\text{inside}} = P_{\text{outside}} = 1.01 \times 10^5 \text{ Pa} = 1.01 \times 10^5 \text{ N/m}^2$$

$$F_{\perp,\text{inside}} = P_{\text{inside}}A = (1.01 \times 10^5 \text{ N/m}^2)(1.54 \text{ m}^2) = 1.56 \times 10^5 \text{ N}$$

$$F_{\perp,\text{outside}} = P_{\text{outside}}A = 1.56 \times 10^5 \text{ N}$$

Because the forces on the inside and outside have the same magnitude but are in opposite directions, the *net* force exerted by the air on the door is zero:

$$F_{\perp,\text{net}} = 0$$

(b) Repeat the calculation for an altitude of 12,500 m.

At an altitude of 12,500 m:

$$P_{\text{inside}} = 7.53 \times 10^4 \text{ Pa} = 7.53 \times 10^4 \text{ N/m}^2$$

$$F_{\perp,\text{inside}} = P_{\text{inside}}A = (7.53 \times 10^4 \text{ N/m}^2)(1.54 \text{ m}^2) = 1.16 \times 10^5 \text{ N}$$

$$P_{\text{outside}} = 1.78 \times 10^4 \text{ Pa} = 1.78 \times 10^3 \text{ N/m}^2$$

$$F_{\perp,\text{outside}} = P_{\text{outside}}A = (1.78 \times 10^3 \text{ N/m}^2)(1.54 \text{ m}^2) = 2.74 \times 10^4 \text{ N}$$

The outward force on the inside of the door is greater in magnitude than the inward force on the outside of the door. Hence the net force exerted by the air on the door is outward. The magnitude of the net force is

$$F_{\perp,\text{net}} = F_{\perp,\text{inside}} - F_{\perp,\text{outside}} = (1.16 \times 10^5 \text{ N}) - (2.74 \times 10^4 \text{ N}) = 8.86 \times 10^4 \text{ N}$$

Reflect

The net force on the airliner door while cruising at 12,500 m is tremendous, equivalent to the weight of an object of mass 9040 kg! The door must be carefully engineered to withstand such a force.

Mass m of an object with weight $F_g = mg = 8.86 \times 10^4$ N:

$$m = \frac{F_g}{g} = \frac{8.86 \times 10^4 \text{ N}}{9.80 \text{ m/s}^2} = \frac{8.86 \times 10^4 \text{ kg} \cdot \text{m/s}^2}{9.80 \text{ m/s}^2} = 9.04 \times 10^3 \text{ kg}$$

NOW WORK Problems 1 and 6 from The Takeaway 13-3.

THE TAKEAWAY for Section 13-3

✔ A fluid exerts pressure on any object with which it is in contact.

✔ The pressure equals the force that the fluid exerts perpendicular to the object's face divided by the area of that face.

✔ The pressure exerted by air at sea level is approximately equal to 1 atm. This pressure is due to the weight of a column of air that extends to the top of our atmosphere.

Prep for the AP® Exam

AP® Building Blocks

1. An elephant that has a mass of 3.0×10^3 kg evenly distributes her weight on all four feet.
   (a) If her feet are approximately circular and each has a diameter of 0.50 m, estimate the pressure on each foot.

   (b) Compare the answer in part (a) with the pressure on each of your feet when you are standing up. Make some rough but reasonable assumptions about the area of your feet.
2. Convert the following pressures to the SI unit of pascals (Pa) (1 Pa = 1 N/m^2).
   (a) 1500 kPa
   (b) 35 psi
   (c) 2.85 atm

AP® Skill Builders

3. On an airplane, the door has to be reinforced against blowing outward. What sort of reinforcement would be needed on the door of a submarine? Briefly justify your answer in terms of physical principles.
4. Elaine wears her wide-brimmed hat at the beach. If the atmospheric pressure at the beach is exactly 1.0 atm, determine the weight of the imaginary column of air that "rests" on her hat if its diameter is 45 cm.

AP® Skills in Action

5. A tropical fish is at rest in the middle of an aquarium. Compared to the pressure on the left side of the fish, the pressure on the right side of the fish
   (A) has the same magnitude and the same direction.
   (B) has the same magnitude and opposite direction.
   (C) has a different magnitude and opposite direction.
   (D) has the same magnitude.

6. The head of a 16-penny nail is 0.32 cm in diameter. A carpenter hits it with a hammer with a force of 25 N.
   (a) Calculate the pressure on the head of the nail.

   (b) If the pointed end of the nail, opposite to the head, is 0.032 cm in diameter, find the pressure on that end due to the hammer blow.

(a)

Roger Freedman

(b)

Roger Freedman

Figure 13-6 Air pressure varies with elevation (a) This empty plastic bottle was opened to the air in a nonpressurized airplane at an altitude of 9000 ft (2740 m) then tightly capped. (b) Outside air pressure made the bottle collapse when it was returned to sea level.

13-4 In a fluid at rest pressure increases with increasing depth

Figure 13-6 shows the results of an experiment with air in our atmosphere. An empty plastic bottle is opened at an altitude of 9000 ft (2740 m) above sea level so that the bottle fills with air at that altitude. The bottle is then tightly capped (**Figure 13-6a**). As the bottle is brought back to sea level, it compresses as though it had been squeezed by an invisible hand (see **Figure 13-6b**).

Why does this happen? The answer is that the pressure in a fluid *increases* as you go *deeper* into the fluid. For example, air pressure in the atmosphere is about 40% greater at sea level than it is at an altitude of 2740 m. The air inside the bottle in Figure 13-6 was at the pressure found at 2740 m, so it compressed when it was brought to sea level and was surrounded by air at higher pressure. In this section we'll explore the relationship between pressure and depth in a fluid.

Hydrostatic Equilibrium

Let's consider a large tank filled with a fluid at rest (**Figure 13-7**). When a fluid is at rest (that is, not flowing), we say it is in **hydrostatic equilibrium**. The term *hydrostatic* specifically refers to water at rest, but it's common to use this term to refer to *any* kind of fluid in equilibrium.

Now imagine a box-shaped volume of fluid within the tank, as in Figure 13-7. The area of the top and bottom of the box is A, and the height of the box is d. We won't put an actual box in the fluid; instead we simply imagine a boundary that separates the box-shaped volume from the rest of the fluid.

The weight of the fluid above the box exerts a downward force of magnitude F_{down} on the box. Because fluids exert a force in all directions, the fluid below the box pushes upward on the box with a force of magnitude F_{up}. (We'll see in a moment that F_{up} is *not* equal to F_{down}.) The fluid also exerts forces on the sides of the box, but the *net* force on the sides is zero: There is as much force on the left side as on the right side, and these forces cancel. Finally, there is a downward force of gravity, $\vec{F}_g$ on the fluid inside the box—that is, the weight of the fluid in the box. Figure 13-7 shows these three forces on the box, $\vec{F}_{\text{up}}$, $\vec{F}_{\text{down}}$, and $\vec{F}_g$.

We started with the assumption that the fluid in the tank is at rest. This requires that the *net* force on the fluid volume is zero; that is, the vector sum of the three forces exerted on the box must be zero:

$$\Sigma\vec{F} = \vec{F}_{\text{up}} + \vec{F}_{\text{down}} + \vec{F}_g = 0 \tag{13-3}$$

All of these forces are vertically upward or downward, so we need only one component to describe these forces. Taking the positive y direction to be upward, Equation 13-3 becomes

$$\Sigma F_y = F_{\text{up}} - F_{\text{down}} - F_g = 0 \tag{13-4}$$

We can see that because the fluid in the box has a nonzero weight F_g, the forces F_{up} and F_{down} *cannot* be equal. From Equation 13-2, $P = F_\perp/A$, we can express the

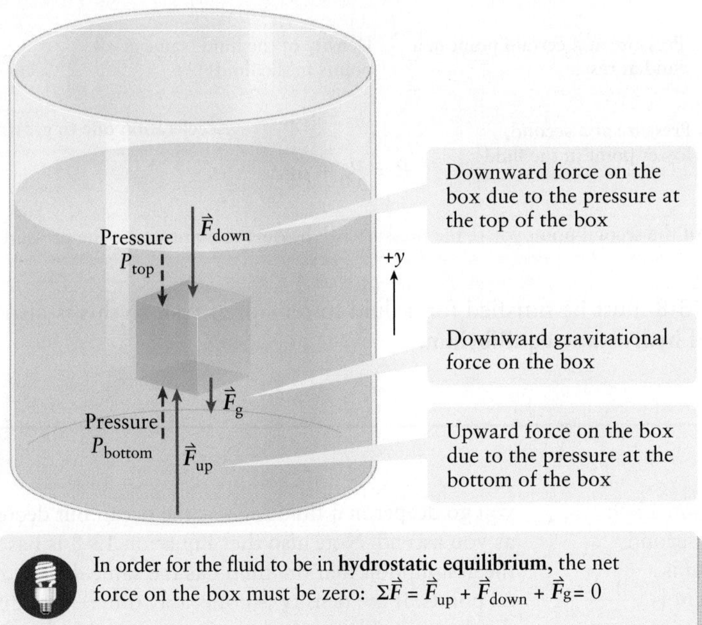
Consider a box-shaped volume of fluid that is part of a larger quantity of fluid at rest.
Pressure P_{top}
$\vec{F}_{down}$
+y
Downward force on the box due to the pressure at the top of the box
Downward gravitational force on the box
$\vec{F}_g$
Pressure P_{bottom}
$\vec{F}_{up}$
Upward force on the box due to the pressure at the bottom of the box
In order for the fluid to be in **hydrostatic equilibrium**, the net force on the box must be zero: $\Sigma\vec{F} = \vec{F}_{up} + \vec{F}_{down} + \vec{F}_g = 0$

Figure 13-7 Hydrostatic equilibrium We can derive the equation of hydrostatic equilibrium by considering a box-shaped portion of a fluid at rest. For the box to be in equilibrium, the pressure at its bottom must be greater than the pressure at its top.

magnitude F_{up} of the upward force (which acts perpendicular to the underside of the box) as the pressure P_{bottom} at the bottom of the box multiplied by the area A of the bottom of the box. Similarly, the magnitude F_{down} of the downward force equals the pressure P_{top} at the top of the box multiplied by the area A of the top of the box. Furthermore, the weight F_g of the fluid in the box equals the product of the mass m of fluid in the box and the acceleration due to gravity g. So Equation 13-4 becomes

$$P_{bottom}A - P_{top}A - mg = 0 \quad \text{or} \quad P_{bottom}A = P_{top}A + mg \tag{13-5}$$

Equation 13-5 tells us that because of the weight mg of the fluid in the box, the pressure P_{bottom} has to be greater than the pressure P_{top}. In other words the pressure of the fluid is greater at a point that's deep in the fluid (the bottom of the box) than at a point that's less deep (the top of the box) and so pressure increases with increasing depth in the fluid.

A Uniform-Density Fluid: The Equation of Hydrostatic Equilibrium

We can simplify Equation 13-5 even further if we assume that the fluid has **uniform density**; that is, the density of the fluid has the same value throughout the fluid. The mass m of the volume of fluid in the box is the product of the density of the fluid, ρ, and the volume of the box. The height of the rectangular box is d, so its volume is $V = Ad$, and the mass of the fluid in the box is $m = \rho V = \rho Ad$. So Equation 13-5, the statement that the fluid in the box remains at rest, becomes

$$P_{bottom}A = P_{top}A + (\rho Ad)g \tag{13-6}$$

Now divide each term in Equation 13-6 by A:

$$P_{bottom} = P_{top} + \rho gd \tag{13-7}$$

This says that the pressure at the bottom of the box is greater than the pressure at the top of the box. Furthermore, the box is just an imaginary construct, so the real meaning of Equation 13-7 is as follows: At *any* two points in the fluid with vertical separation d, the pressure at the lower point is greater than the pressure at the upper point by an amount $\rho g d$. If we let P_0 be the pressure at the upper point, the pressure P at a point a distance d below the upper point is

EQUATION IN WORDS
Variation of pressure with depth in a fluid with uniform density (equation of hydrostatic equilibrium)

$$P = P_0 + \rho g d \quad (13\text{-}8)$$

Pressure at a certain point in a fluid at rest

Pressure at a second, lower point in the fluid

Density of the fluid (same at all points in the fluid)

Acceleration due to gravity

Depth of the second point where the pressure is P below the point where the pressure is P_0

Equation 13-8 must be satisfied for a fluid to remain at rest, so this is also called the **equation of hydrostatic equilibrium.**

WATCH OUT !

Understand what Equation 13-8 tells you.

Be careful with the sign of the $\rho g d$ term in Equation 13-8: Remember that the symbol d means *depth*. If the second point is *below* the point where the pressure is P_0, d is positive and P is greater than P_0. If the second point is *above* the point where the pressure is P_0, d is *negative* and P is less than P_0. In other words, pressure increases as you go deeper in a fluid such as the ocean but decreases as you ascend. Note also that Equation 13-8 is based on the assumption that the fluid has the same density ρ at all points. If the density is noticeably different at different depths (as is the case for the air in our atmosphere), Equation 13-8 gives only approximate results.

EXAMPLE 13-5 Air Pressure Versus Water Pressure

A swimming pool at the YMCA is 2.00 m deep. What is the pressure at the bottom of the pool (a) before it is filled with water and (b) after it is filled with water? At the time that you make the measurements, the air pressure at ground level is 1.0100×10^5 Pa.

Set Up

We are given the pressure $P_0 = 1.0100 \times 10^5$ Pa at the top of the pool and want to find the pressure P at a depth $d = 2.00$ m below the top of the pool. We'll use Equation 13-8 for this purpose. The only difference between parts (a) and (b) is the density ρ of the fluid: In part (a) we use the density of air, while in part (b) we use the density of water.

Variation of pressure with depth:

$$P = P_0 + \rho g d \quad (13\text{-}8)$$

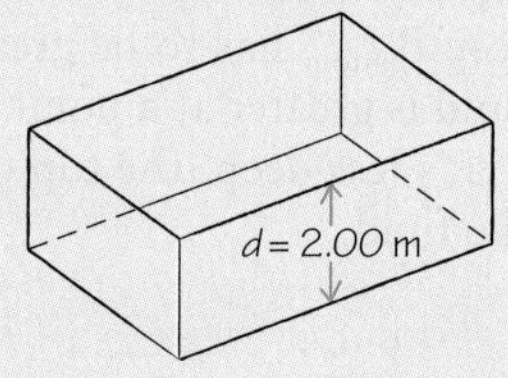

Solve

(a) From Table 13-1 the density of air at sea level is $\rho = 1.23$ kg/m^3. Substitute this into Equation 13-8.

Pressure at the bottom of the pool filled with air:

$$P = (1.0100 \times 10^5 \text{ Pa}) + (1.23 \text{ kg/m}^3)(9.80 \text{ m/s}^2)(2.00 \text{ m})$$
$$= 1.0100 \times 10^5 \text{ Pa} + 24.1 \text{ kg/(m} \cdot \text{s}^2)$$

Convert units: Recall that

$$1 \text{ N} = 1 \text{ kg} \cdot \text{m/s}^2 \text{ so}$$
$$1 \text{ Pa} = 1 \text{ N/m}^2 = 1 \text{ kg/(m} \cdot \text{s}^2)$$

So the pressure at the bottom of the air-filled pool is

$$P = 1.0100 \times 10^5 \text{ Pa} + 24.1 \text{ Pa}$$
$$= 1.0100 \times 10^5 \text{ Pa} + 0.000241 \times 10^5 \text{ Pa}$$
$$= 1.0102 \times 10^5 \text{ Pa}$$

(b) When the pool is filled with water, the calculation is the same as in part (a) except that we use the density of water from Table 13-1, $\rho = 1.000 \times 10^3$ kg/m^3.

Pressure at the bottom of the pool filled with water:

$$P = (1.0100 \times 10^5 \text{ Pa}) + (1.000 \times 10^3 \text{ kg/m}^3)(9.80 \text{ m/s}^2)(2.00 \text{ m})$$
$$= 1.0100 \times 10^5 \text{ Pa} + 1.96 \times 10^4 \text{ kg/(m} \cdot \text{s}^2)$$
$$= 1.0100 \times 10^5 \text{ Pa} + 0.196 \times 10^5 \text{ Pa}$$
$$= 1.206 \times 10^5 \text{ Pa}$$

Reflect

The pressure at the bottom of the air-filled pool is only 0.02% greater than at the top of the pool. That's such a tiny difference that we can ignore it, which makes sense. It takes a much greater variation in altitude than 2.00 m for the pressure difference in the air to be noticeable. (Your ears—which are sensitive pressure sensors—may "pop" when you drive up into the mountains, but won't "pop" when you climb a ladder or a single flight of stairs.)

As we saw in this example, there *is* a noticeable pressure difference between the top and bottom of the *water*-filled pool: The pressure is about 20% greater at the bottom than at the surface. The difference is that water is almost a thousand times denser than air. A swimming pool full of water weighs almost a thousand times more than a swimming pool full of air, so it produces an additional pressure almost a thousand times greater. We will explore the pressures in water further in the following example.

NOW WORK Problems 3, 4, 6, 7, and 11 from The Takeaway 13-4.

EXAMPLE 13-6 Diver's Rule of Thumb

To what depth in a freshwater lake would a diver have to descend for the pressure to be 1.00 atm greater than at the surface?

Set Up

Equation 13-8 tells us how the pressure P at a depth d below the lake's surface compares to the pressure P_0 at the surface. We can use this information to find the value of d such that P is greater than P_0 by 1.00 atm.

Variation of pressure with depth:

$$P = P_0 + \rho g d \quad (13\text{-}8)$$

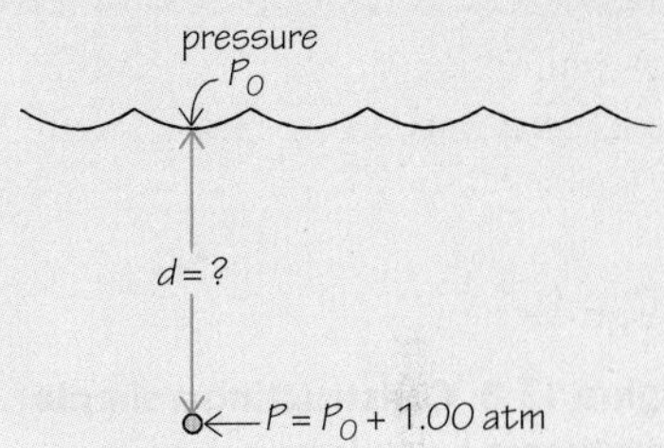

Solve

Rearrange Equation 13-8 to solve for the depth d at which the pressure has a certain value P.

From Equation 13-8:

$$P - P_0 = \rho g d$$
$$d = \frac{P - P_0}{\rho g}$$

We want the pressure P at depth d to be greater than the pressure P_0 at the surface by 1.00 atm, so $P - P_0 = 1.00$ atm. Substitute this value as well as the value of density ρ for fresh water from Table 13-1. Then use the conversion factors 1 atm $= 1.01 \times 10^5$ Pa $= 1.01 \times 10^5$ N/m^2 and 1 N $= 1$ kg $\cdot$ m/s^2.

Pressure difference:

$$P - P_0 = 1.00 \text{ atm}$$

Density of fresh water from Table 13-1:

$\rho = 1.000 \times 10^3$ kg/m^3 so

$$d = \frac{P - P_0}{\rho g}$$
$$= \frac{1.00 \text{ atm}}{(1.000 \times 10^3 \text{ kg/m}^3)(9.80 \text{ m/s}^2)} \times \left(\frac{1.01 \times 10^5 \text{ N/m}^2}{1 \text{ atm}}\right)$$
$$= 10.3 \, \frac{\text{m}^3}{\text{kg}} \, \frac{\text{s}^2}{\text{m}} \, \frac{\text{N}}{\text{m}^2} \times \left(\frac{1 \text{ kg} \cdot \text{m/s}^2}{1 \text{ N}}\right)$$
$$= 10.3 \text{ m}$$

Reflect

Our answer is a little bit more than 10 m, so it's a good approximation (a "rule of thumb" for divers) to say that our diver experiences a pressure change of 1 atm going from the surface to a depth of 10 m. You can see that descending an additional 10 m would increase the pressure by another atmosphere, and so on.

NOW WORK Problem 8 from The Takeaway 13-4.

Measuring Pressure

Blood pressure readings and atmospheric pressure values in weather reports are usually given in units of inches or millimeters of mercury. To understand these units of pressure, we'll examine how to construct a simple device called a *barometer* that's used to measure air pressure.

Fill a tube closed at one end with mercury (Hg), which has density ρ_{Hg}. While keeping the open end sealed, turn the tube upside down into a pan partially filled with mercury. (*Warning*: Mercury is poisonous, so do *not* try this experiment at home!) Now release the seal on the end of the tube, which is beneath the surface of the mercury in the pan. No air can get into the tube, so an equal volume of vacuum must remain in the place of whatever volume of fluid drains from the tube. **Figure 13-8** shows this process.

Let's consider what happens once the system has come to hydrostatic equilibrium and a column of mercury of height h (measured from the surface of the mercury of the pan) remains in the tube. The pressure at the top of this column is $P_0 = 0$ because the pressure in a vacuum is zero. The pressure at the exposed surface of the mercury in the pan equals atmospheric pressure P_{atm}. Because the mercury is in hydrostatic equilibrium, the pressure at a point inside the mercury column at the same elevation as the exposed mercury surface is *also* equal to P_{atm}. From the equation of hydrostatic equilibrium, Equation 13-8, the pressure P_{atm} is greater than P_0 by $\rho_{Hg}gh$:

$$P_{atm} = P_0 + \rho_{Hg}gh$$

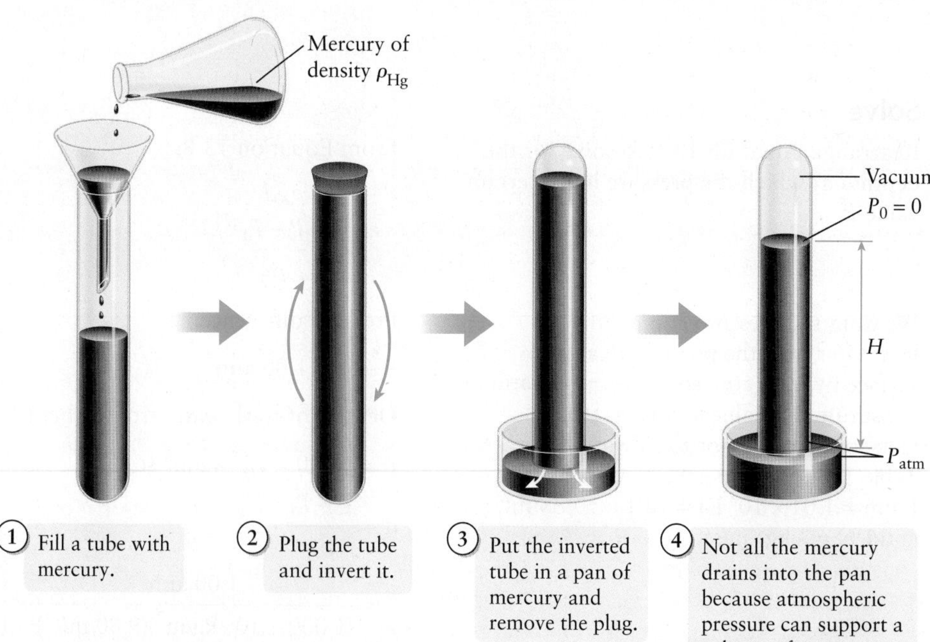

Mercury of density ρ_{Hg}
Vacuum
$P_0 = 0$
H
P_{atm}
1 Fill a tube with mercury.
2 Plug the tube and invert it.
3 Put the inverted tube in a pan of mercury and remove the plug.
4 Not all the mercury drains into the pan because atmospheric pressure can support a column of mercury about three-quarters of a meter tall.

Figure 13-8 Constructing a simple barometer A simple barometer to measure atmospheric pressure can be made by inverting a tube of fluid such as mercury in a pan. A height h of liquid remains in the tube, supported by the outside air pressure.

As $P_0 = 0$ in the vacuum at the top of the mercury column, this equation becomes

$$P_{atm} = \rho_{Hg} g h \quad (13\text{-}9)$$

(atmospheric pressure measured with a mercury barometer)

As the atmospheric pressure P_{atm} rises and falls, the height h of the mercury column rises and falls along with it. Hence the height of the mercury column is a direct measure of atmospheric pressure. For that reason, pressures are sometimes given in **millimeters of mercury**, or mmHg for short.

A mercury barometer like that shown in Figure 13-8 is not a perfect instrument for measuring pressure: The density of mercury changes with temperature, and the value of the acceleration due to gravity depends on location. At a temperature of 0°C so that the density of mercury is $\rho_{Hg} = 1.3595 \times 10^4$ kg/m^3 and at a location where $g = 9.80665$ m/s^2 (a "standard" value of g at the equator), the height of a column of mercury supported by exactly 1 atm of pressure is 760 mm, so 1 atm = 760 mmHg. Another unit you might see for pressure is the **torr**, defined as 1/760 atm. This unit is named for the Italian scientist and mathematician Evangelista Torricelli, who invented the mercury barometer in the 1640s. The torr is a more appropriate unit than mmHg because many modern barometers work on different principles that don't involve columns of mercury. For instance, the altimeter in a smartphone is an electronic barometer.

Why is toxic mercury used in barometers? It's a case of convenience prevailing over safety. Equation 13-9 shows that the greater the density of the fluid, the lower the height h of the fluid column in the barometer. Because mercury is so dense—13.6 times denser than water—a mercury barometer is relatively compact and convenient to use. If you wanted to make a barometer like that shown in Figure 13-8 but using water rather than mercury, it would have to be 13.6 times taller than a mercury barometer. Can you show that if $P_{atm} = 1$ atm, the height of the liquid column in a water barometer would be 10.3 m—far too tall to be convenient?

Gauge Pressure

What is the air pressure inside a flat automobile tire? It's not zero: Rather, it's equal to atmospheric pressure (P_{atm}, about 1 atm, 1.01×10^5 Pa, or 14.7 lb/in^2 at sea level). The same is true for a balloon that hasn't been inflated and for an empty, open bottle, because air at atmospheric pressure is inside these vessels. But if you use a tire pressure gauge to measure the air pressure in the deflated tire, it *will* read zero. This **gauge pressure** shows how much the pressure *exceeds* atmospheric pressure. The **absolute pressure**—that is to say, the true value of pressure P—inside a deflated tire is equal to atmospheric pressure. If a tire pressure gauge shows that the air inside an inflated tire is at a pressure of 30.0 lb/in^2, that's the gauge pressure P_{gauge}. *The absolute pressure equals the gauge pressure plus atmospheric pressure,* $P = P_{gauge} + P_{atm}$. So if the gauge pressure inside the tire is 30.0 lb/in^2 and atmospheric pressure is 14.7 lb/in^2, the absolute pressure inside the tire is $(30.0 \text{ lb/in}^2) + (14.7 \text{ lb/in}^2) = 44.7 \text{ lb/in}^2$.

Note that the tire and the balloon are both nonrigid containers. The way the pressure gauge you use to measure your tire's pressure works is that it lets out a little air and it compares the pressures inside and outside. If instead the container were rigid, and strong enough to keep the atmosphere from pressing on its contents, then the pressure in the fluid inside would not include atmospheric pressure. A device to measure the pressure inside would measure something different in either of these cases than a standard pressure gauge that compares the pressure inside and out. In that empty space above the mercury in the barometer, the pressure is very small, close to vacuum (not the true vacuum we approximated, perhaps). So, a tiny device of some sort in there would measure a pressure of zero. If we had done this in a column where we have a place to connect a pressure gauge at the top, we would measure approximately –1 atm of pressure, as the outside would be at this much higher a pressure than the vacuum, but quickly, with this large of a difference, the pressures would even out, and the mercury column would fall, because the pressure above it would be the same as that of the atmosphere!

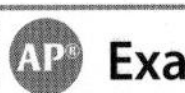

Exam Tip

Gauge pressure is an important concept on the AP® exam. Understanding what pressure you are measuring, and when you need to add an atmosphere of pressure to a measurement to determine behavior of a fluid, is critical to solving problems.

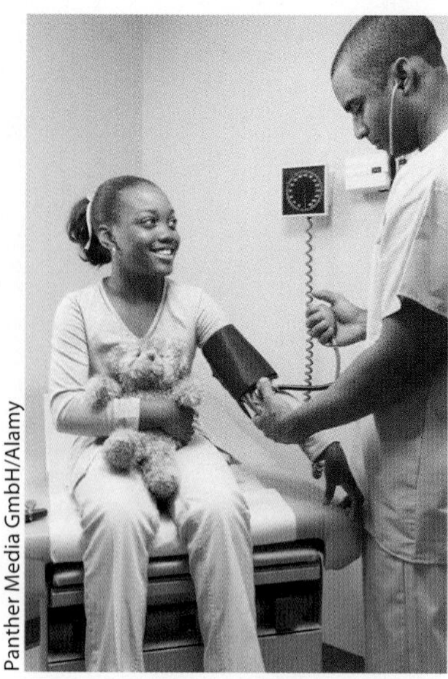

Panther Media GmbH/Alamy

Figure 13-9 Measuring blood pressure A blood pressure cuff is applied around the upper arm, at the same elevation as the patient's heart. If the pressure were measured at a different elevation, the values would be either lower or higher than the actual pressures produced by the heart.

Blood Pressure: An Example of Gauge Pressure and Its limitations

Many of us encounter a pressure gauge when we visit the doctor and have our blood pressure measured (**Figure 13-9**). Because the heart is a pump, the pressure in the system changes over the period of each heartbeat. Blood pressure is usually given as two numbers, for example, "120 over 80." These values are measured in units of mmHg and represent the high and low values of the *gauge* pressure in the arteries at the level of the heart.

Blood pressure is usually measured using a cuff around your upper arm, at the same elevation as your heart. We can understand why this is so by using the relationship between pressure and depth given by Equation 13-8. The blood within the body behaves rather like the fluid in the container shown in Figure 13-7: The farther down you go within the fluid, the greater the pressure. Because the heart is the principal organ of the circulatory system, let's have P_0 be the pressure at the position of the heart. Then Equation 13-8 tells us that the pressure of blood in the feet will be higher than the pressure in the heart, and the pressure in the head will be lower. To measure the actual pressure produced by the heart's pumping action, it's important to measure that pressure at the same elevation as the heart itself—which explains the placement of a blood pressure cuff.

The average blood pressure (gauge pressure) in an upright person's feet is about 100 mmHg higher than when it leaves the heart, or about twice as great as the gauge pressure measured in a person's arm. In contrast, the pressure in the head is about 50 mmHg less than that at the heart. If you could somehow stretch your neck upward by a meter or so, the blood pressure in your head would drop to zero and no blood would reach your brain. So it's all for the best that you can't stretch your neck that far! (Some animals do have necks that are even longer: The head of an adult male giraffe is about 2 m higher than his heart. To get blood to the brain, the giraffe heart has to generate much higher pressure than a human heart does—about 260 mmHg.)

These differences in pressure between different regions of the body affect the flow of blood in the veins returning to the heart from the head and neck. Blood pressure is always lower in veins than in arteries; if you can feel a pulse in your neck, you're feeling the carotid *artery* with its relatively high pressure. The gauge pressure of blood entering the heart is very close to zero. Because the pressure above the heart is even lower, the veins of the head and neck tend to collapse! As blood continues to enter the veins, pressure builds up, and the vessels reopen. But as the blood flows through them, the pressure will again drop, and the veins will collapse again. The result is intermittent venous blood flow in the head and neck.

You may be wondering why the gauge pressure of the blood entering the heart is so much lower than the gauge pressure of the blood leaving the heart, when the blood is at essentially the same height. Equation 13-8 seems to indicate that these two pressures should be the same. However, you need to remember that Equation 13-8 is valid only under the conditions of *hydrostatic equilibrium*. Because blood is pumped by the heart and flows through the circulatory system, the blood is certainly *not* in hydrostatic equilibrium and we cannot use Equation 13-8. The difference in arterial and venous pressure depends on viscous flow, which is beyond the scope of the AP® physics courses.

THE TAKEAWAY for Section 13-4

✔ In a fluid in hydrostatic equilibrium (at rest), pressure increases as you go deeper into the fluid.

✔ The pressure difference between any two points within a uniform fluid in hydrostatic equilibrium depends only on the difference in the vertical height of the two points and the density of the fluid.

✔ Gauge pressure is measured relative to some reference pressure. Usually this reference pressure is atmospheric pressure.

Prep for the AP® Exam

AP® Building Blocks

1. When you cut your finger badly, why might it be wise to hold it high above your head?
2. When you donate blood, is the collection bag held below or above your body? Why?
3. EX 3-5 Calculate the absolute pressure at the bottom of a 0.25-m-tall graduated cylinder that is half full of mercury and half full of water.
4. EX 3-5 A swimming pool is filled to a depth of 3.0 m. What is the pressure on the bottom of the pool when the temperature is 4°C?

AP® Skill Builders

5. If the gauge pressure is doubled, the absolute pressure will
   (A) be halved.
   (B) be doubled.
   (C) be unchanged.
   (D) be increased, but not necessarily doubled.
6. EX 3-5 At 15°C the density of ether is 713 kg/m^3 and the density of ethanol is 789 kg/m^3. A cylinder is filled with ethanol to a depth of 1.5 m. How tall would a cylinder filled with ether need to be so that the pressure at the bottom is the same as the pressure at the bottom of the cylinder filled with ethanol?
7. EX 13-5 The Hoover Dam is approximately 726 ft (221 m) tall and approximately 1244 ft (379 m) wide. Estimate the *hydrostatic force* (the force from the water) on the Hoover Dam. Assume that the top 50 m of the dam is above the surface of Lake Mead.
8. EX 13-6 A diver is 10.0 m below the surface of the ocean (seawater). The surface pressure is 1 atm. What are the absolute pressure and gauge pressure experienced by the diver?
9. Suppose that your pressure gauge for determining the blood pressure of a patient measured absolute pressure instead of gauge pressure. How would you write the normal value of systolic blood pressure, 120 mmHg, in such a case?

AP® Skills in Action

10. You pour water into a U-shaped tube, as shown in the figure. The left-hand leg of the tube is 2.00 cm in radius, while the right-hand leg is 1.00 cm in radius. When the water is in equilibrium, how will the height of the water in the left-hand leg and the pressure at the bottom of the left-hand leg compare to the height of water in the right-hand leg and the pressure at the bottom of the right-hand leg?

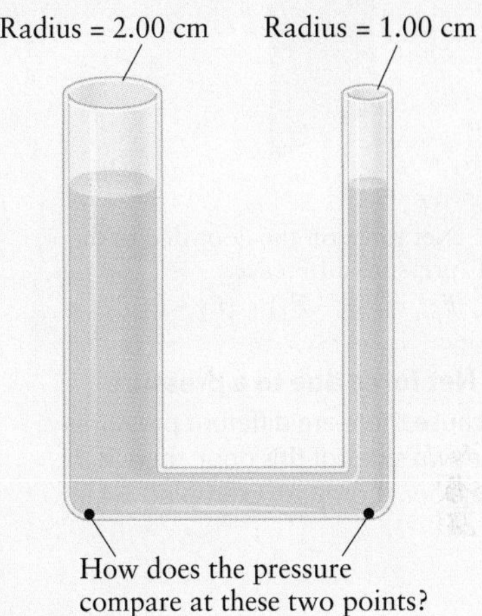

   (A) Greater height, greater pressure
   (B) Same height, same pressure
   (C) Lower height, lower pressure
   (D) Same height, greater pressure
11. EX 13-5 Find the difference in blood pressure (mmHg) between the top of the head and bottom of the feet of a 1.75-m-tall person standing vertically. The density of blood is $\rho_{blood} = 1.06 \times 10^3$ kg/m^3.
12. You have two balloons at sea level. One is deflated, and the other is inflated to 0.50 atm according to a pressure gauge connected to the mouth of the balloon. Rank the following from highest to lowest: (A) the gauge pressure inside the deflated balloon; (B) the absolute pressure inside the deflated balloon; (C) the gauge pressure inside the inflated balloon; (D) the absolute pressure inside the inflated balloon.

13-5 A difference in pressure on opposite sides of an object produces a net force on the object

We have seen that the pressure within a fluid is a result of the external pressure on the fluid and the internal interactions between the molecules that constitute the fluid. The many randomly directed collisions within the fluid, through the impulse-momentum theorem, produce the pressure within the fluid and the fluid's ability to exert a force on something with which it is in contact. Because almost everything we interact with is surrounded by some fluid, liquid or gas, pressure within a fluid alone cannot describe many common phenomena. It turns out it is the *pressure difference* that is

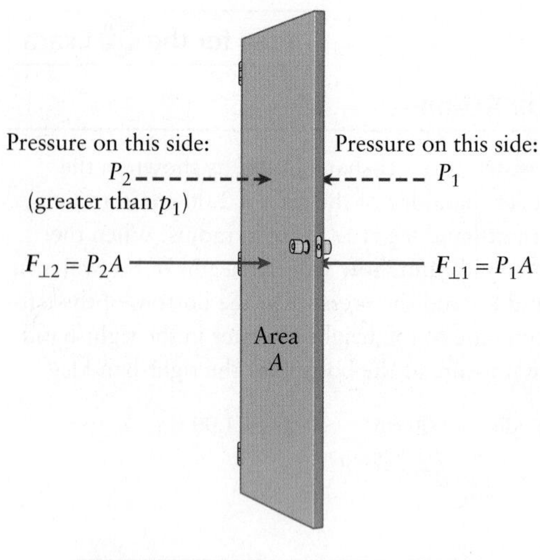

Figure 13-10 Net force due to a pressure difference Because there are different pressures P_1 and P_2 on the two sides of this door, there is a net force on the door. (Compare Example 13-4 in Section 13-3.)

responsible for many phenomena we observe. Examples include a beverage moving up a straw into your mouth as you sip, the pistons moving in the cylinders of an automobile engine, and a soda spraying liquid when you open the bottle. The soda sprays out because carbon dioxide gas inside the bottle is at a higher pressure than the air outside the bottle. Burning gasoline in a cylinder produces hot gas on one side of the piston that is at much higher pressure than the air on the other side, forcing the piston to move. When you suck on a straw, you reduce the pressure in your mouth to a lower value than in the outside air, and the pressure difference drives liquid up the straw.

As a simple example of the force produced by a pressure difference, consider an object like a door with fluid on each side. (This would be a good time to review Example 13-4 in Section 13-3.) The object has area A, there is pressure P_1 on one side of the object, and there is a greater pressure P_2 on the other side (**Figure 13-10**). By rearranging Equation 13-2, $P = F_\perp / A$, we can calculate the magnitude of the force that each fluid exerts perpendicular to the object's surface:

$$\text{Force on side 1: magnitude } F_{\perp 1} = P_1 A$$

$$\text{Force on side 2: magnitude } F_{\perp 2} = P_2 A$$

Because $P_2 > P_1$, the force $F_{\perp 2}$ on side 2 is greater than the force $F_{\perp 1}$ on side 1. These forces are in opposite directions, so the *net* force on the object has magnitude

$$F_{net} = F_{\perp 2} - F_{\perp 1} = P_2 A - P_1 A$$

EQUATION IN WORDS
Net force on an object due to a pressure difference

$$F_{net} = (P_2 - P_1)\,A \quad (13\text{-}10)$$

A **pressure difference** on two opposite sides of an object **produces a net force.**
Area of each side of the object
High pressure on one side of the object
Lower pressure on the other side of the object

That is, the net force on the object due to the pressures exerted on it is proportional to the difference in pressure between the object's two sides. Note that Equation 13-10 refers *only* to the net force on an object due to a difference in pressure on its two sides. It doesn't include other forces such as gravity that may be exerted on the object.

WATCH OUT!

We are still using the object model, even though surface area is important.

The "objects" we are submersing in fluids do have a size, for instance, their bottom surface and top surface might be at different depths, and they have a surface area. These objects are what we referred to in earlier chapters as rigid bodies. However, if the model we use for their motion is just a net force giving a single acceleration (they don't rotate, squeeze, or change shape), the object model is still the best to use. If their size were to change shape during the motion, then we would have to change our approach, but that will not be expected on the AP® Physics 1 Exam.

EXAMPLE 13-7 Submarine Hatch

The air in the crew compartment of a research submarine is maintained at a pressure of 1.00 atm so that the crew can breathe normally. The sub operates at a depth of 85.0 m below the surface of the ocean. At this depth what force would be required to push open a hatch of area 2.00 m^2?

Set Up

There is 1.00 atm of air pressure on the inside of the hatch (call this P_{air}) pushing the hatch outward, but this is much less than the outside water pressure (call this P_{water}) pushing the hatch inward. The net inward force F_{net} due to this pressure difference is given by Equation 13-10; $A = 2.00\ m^2$ is the surface area of the hatch. To open the hatch the crew would have to exert an outward force of the same magnitude F_{net}.

To calculate F_{net} we need to know the value of the water pressure P_{water}. We'll find this value using Equation 13-8 for the pressure at depth $d = 85.0$ m. In this equation P_0 equals the pressure at the surface of the ocean (that is, at zero depth), which we also take to be 1.00 atm.

Net force on the hatch due to the pressure difference:

$$F_{net} = (P_{water} - P_{air})A \qquad (13\text{-}10)$$

Variation of pressure with depth:

$$P_{water} = P_0 + \rho g d \qquad (13\text{-}8)$$

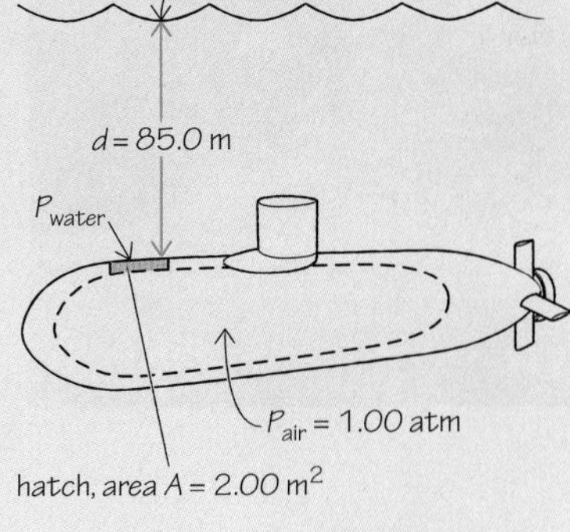

Solve

First find the pressure of the water at depth $d = 85.0$ m. Use the density of seawater from Table 13-1, $\rho = 1.025 \times 10^3\ kg/m^3$; convert atmospheres to newtons per square meter; and note that $1\ kg/(m \cdot s^2) = 1\ N/m^2$.

Water pressure on the outside of the hatch:

$$P_{water} = P_0 + \rho g d$$

$$P_0 = 1.00\ atm = 1.01 \times 10^5\ N/m^2, \text{ so}$$

$$\begin{aligned} P_{water} &= 1.01 \times 10^5\ N/m^2 \\ &\quad + (1.025 \times 10^3\ kg/m^3)(9.80\ m/s^2)(85.0\ m) \\ &= 1.01 \times 10^5\ N/m^2 + 8.54 \times 10^5\ kg/(m \cdot s^2) \\ &= 1.01 \times 10^5\ N/m^2 + 8.54 \times 10^5\ N/m^2 \\ &= 9.55 \times 10^5\ N/m^2 \end{aligned}$$

Now we can calculate the magnitude of the net force on the hatch due to the pressure difference. This is the same as the force that the crew must exert to open the hatch.

Air pressure on the inside of the hatch:

$$P_{air} = 1.00\ atm = 1.01 \times 10^5\ N/m^2, \text{ so}$$

$$\begin{aligned} F_{net} &= (P_{water} - P_{air})A \\ &= (9.55 \times 10^5\ N/m^2 - 1.01 \times 10^5\ N/m^2)(2.00\ m^2) \\ &= (8.54 \times 10^5\ N/m^2)(2.00\ m^2) \\ &= 1.71 \times 10^6\ N \end{aligned}$$

Reflect

This is an immense amount of force, equivalent to nearly 200 U.S. tons! It's more than 10 times the force on the airliner door in Example 13-4, as we would expect because of the difference in the density of water and air.

Extend: Clearly the crew would not be able to open the hatch. For a crew member in diving gear to exit the submerged vessel, a portion of the crew compartment would have to be flooded with water from the outside. This action will equalize the inside and outside pressures and make it possible to open the hatch.

NOW WORK Problems 1, 4, 5, and 6 from The Takeaway 13-5.

Pressure differences drive air into and out of the lungs. We humans suck air into our lungs by contracting the diaphragm and the muscles of the chest wall to increase the volume of our chest cavity. This pulls the lungs open, increasing their volume and therefore decreasing the pressure inside them. As this pressure becomes less than atmospheric pressure, the pressure difference pushes air through the airways and into the expanding lungs. Other animals create a pressure difference in different ways (**Figure 13-11**).

Michael Willis/Alamy

Figure 13-11 How frogs breathe Frogs do not have an expandable chest cavity like humans and so cannot breathe in the same way. Instead a frog closes its mouth and lowers the floor of its mouth, causing the mouth cavity to expand, then takes air into its mouth through its nostrils. It then closes its nostrils and raises the floor of its mouth, increasing the pressure there and forcing the trapped air into the lungs. This process is relatively inefficient, so frogs take in most of their oxygen by diffusion into blood vessels that lie just below the frog's permeable skin.

It takes energy for us to contract the muscles required for *inhalation*. However, normal *exhalation* does not require our muscles to do additional work on our chest wall. Relaxing the diaphragm and chest muscles allows the volume of the chest cavity to decrease. This process causes the pressure in the chest to rise above atmospheric pressure and forces air out of the lungs.

Breathing becomes more difficult if the chest is under pressure, such as if you're at the bottom of a football pile-on. The outside pressure on the chest makes it more difficult to expand the chest to inhale. Measurements show that if the pressure on a person's chest exceeds atmospheric pressure by 0.05 atm, the chest can't expand and inhalation is no longer possible. Example 13-8 shows that this places a limit on how deep underwater a person can be and still breathe through a hollow tube that opens above the surface.

EXAMPLE 13-8 Breathing Underwater

Secret agent Cassian Andor dives into a shallow freshwater pond to avoid capture by his nemesis. He intends to lie flat on his back on the bottom of the pond and breathe through a hollow reed. What is the length of the longest reed for which this could work?

Set Up

Cassian needs to ensure that the difference between the water pressure P on his chest and the pressure P_0 of the atmosphere at the surface (where the open end of the reed is) does not exceed 0.05 atm. Equation 13-8 tells us how the water pressure varies with depth d, so our goal is to find the depth such that $P - P_0$ equals 0.05 atm.

Variation of pressure with depth:

$$P = P_0 + \rho g d \qquad (13\text{-}8)$$

atmospheric pressure P_0
$d = ?$
water pressure P

Solve

Solve Equation 13-8 to find the depth d in terms of the water pressure on Cassian's chest (P) and atmospheric pressure (P_0). Then find the depth d for which $P - P_0 = 0.05$ atm. Use the density of fresh water given in Table 13-1 as well as the definitions 1 atm $= 1.01 \times 10^5$ N/m^2 and 1 N $= 1$ kg $\cdot$ m/s^2.

Rearrange Equation 13-8:

$$P - P_0 = \rho g d$$

Solve for d:

$$d = \frac{P - P_0}{\rho g}$$

$$= \frac{0.05 \text{ atm}}{(1.000 \times 10^3 \text{ kg/m}^3)(9.80 \text{ m/s}^2)} \times \frac{1.01 \times 10^5 \text{ N/m}^2}{1 \text{ atm}}$$

$$= 0.5 \frac{\text{N} \cdot \text{s}^2}{\text{kg}} \times \frac{1 \text{ kg} \cdot \text{m/s}^2}{1 \text{ N}} = 0.5 \text{ m}$$

Reflect

The longest hollow reed through which Cassian can breathe underwater is about half a meter—not that long, really. To check this result let's calculate the weight of water half a meter deep on the area of a person's chest, about 30 cm × 60 cm. The answer is about 900 N or 200 lb. Certainly if a 200-lb person were sitting on your chest, you'd find it hard to breathe!

Area of chest:

$$A = 30 \text{ cm} \times 60 \text{ cm}$$
$$= 0.30 \text{ m} \times 0.60 \text{ m}$$
$$= 0.18 \text{ m}^2$$

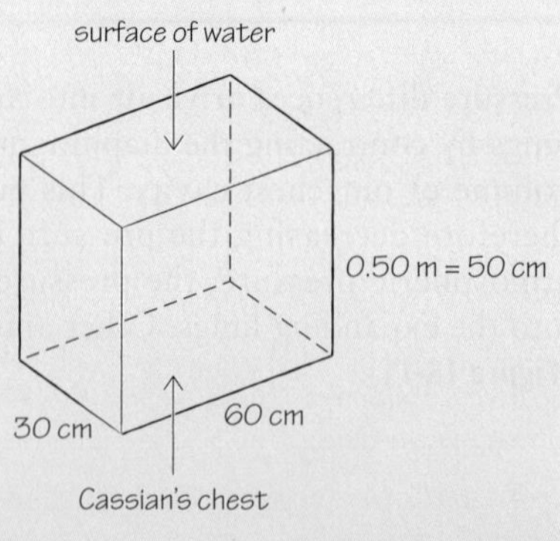

Scuba divers can descend to depths much greater than 0.5 m, but to do so they carry tanks of pressurized air. The regulator on a scuba tank delivers this air to the diver's mouth at the same pressure as the surrounding water, so $P - P_0 = 0$. This means it's just as easy for a scuba diver to breathe underwater as on dry land.

Mass of water in a rectangular volume with base 0.18 m^2 and height 0.50 m:

$$m = \rho V = (1.000 \times 10^3 \text{ kg/m}^3) \times (0.180 \text{ m}^2)(0.50 \text{ m}) = 90 \text{ kg}$$

Weight of that water:

$$mg = (90 \text{ kg})(9.8 \text{ m/s}^2) = 900 \text{ N} = 200 \text{ lb}$$

NOW WORK Problem 2 from The Takeaway 13-5.

THE TAKEAWAY for Section 13-5

✔ If the fluid pressure is different on two sides of an object, the result is a net force on the object. This net force is proportional to the pressure difference and to the cross-sectional area of the object.

Prep for the AP® Exam

AP® Building Blocks

EX 13-7 1. Calculate the net force on an airplane window of area 1000 cm^2 if the pressure inside the cabin is 0.95 atm and the pressure outside is 0.85 atm.

EX 13-8 2. An Andean condor with a wingspan of 300 cm and mass 12 kg soars along a horizontal path. Modeling its wings as a rectangle with a width of 30 cm, what must be the difference in pressure between the top and bottom surfaces of its wings, $p_{top} - p_{bottom}$?

AP® Skill Builders

3. Hold a piece of paper in your two hands so that the plane of the paper is horizontal. Considering the air pressure on the top of the paper and the air pressure on the bottom of the paper, is the force F_{net} on the paper described by Equation 13-10 upward, downward, or zero? Briefly support your answer using relevant physics principles.

EX 13-7 4. Suppose that the hatch on the side of a Mars lander is built and tested on Earth so that the internal pressure just balances the external pressure. The hatch is a disk 50.0 cm in diameter. When the lander goes to Mars, where the external pressure is 650 N/m^2, what will be the net force (in newtons and pounds) on the hatch, assuming that the internal pressure is the same in both cases? Will it be an inward or an outward force?

AP® Skills in Action

EX 13-7 5. What is the net force on the curved sides of a 55-gal drum that is standing upright on the bottom of the ocean with its center at a depth of 250 m? Assume the drum is a cylinder with a height of $34\frac{1}{2}$ in and a diameter of $21\frac{5}{8}$ in. (Remember to convert inches to meters!) The interior pressure of the drum is exactly 1 atm. Neglect the variation in pressure over the height of the drum.

EX 13-7 6. A rectangular swimming pool is 8.0 m × 35 m in area. The depth varies uniformly from 1.0 m in the shallow end to 2.0 m in the deep end.

(a) Find the pressure at the bottom of the deep end of the pool and at the shallow end.

(b) Calculate the net force on the bottom of the pool due to the water in the pool. (Ignore the effects of the air above the pool for this part.)

13-6 A pressure increase at one point in a fluid causes a pressure increase throughout the fluid

When you squeeze one end of a tube of toothpaste, the pressure you apply is transmitted throughout the tube, and toothpaste comes out the other end. This is a simple example of a more general principle first proposed by the seventeenth-century French philosopher, mathematician, and scientist Blaise Pascal, who did pioneering investigations of the nature of pressure in fluids:

Pascal's principle: *Pressure applied to a confined, static fluid is transmitted undiminished to every part of the fluid as well as to the walls of the container.*

We've already seen **Pascal's principle** in action. We learned that if a fluid is at rest—that is, in hydrostatic equilibrium—and the pressure at a certain point in the fluid is P_0, the pressure at a second point a distance d deeper in the fluid is $P = P_0 + \rho g d$ (Equation 13-8). If we were to increase the value of the pressure P_0 at the first point by, say, 50 Pa, the value of the pressure P at the second point would also have to increase by 50 Pa to maintain hydrostatic equilibrium.

A common practical application of Pascal's principle is a *hydraulic jack*. This device makes it possible to lift heavy objects using a force much less than the object's weight. You'll find these devices in operation wherever cars are being worked on: A hydraulic lift raises the car so that the mechanic can work on the car's underside. Smaller versions are used to raise and lower the chairs you sit in at the dentist and the barber. **Figure 13-12** shows the construction of a simplified hydraulic lift. Start with a tube bent into a U shape. On the side labeled 1 the tube is much narrower than on the side labeled 2; we'll call the cross-sectional areas of the two sides of the tube A_1 and A_2, respectively. We partially fill the tube with an incompressible liquid such as oil so that its density does not change when pressure is applied. The liquid rises to the same height on both sides of the tube because the atmospheric pressure pushing down on each side is the same, pressure is constant throughout a liquid in equilibrium, and pressure depends just on the height of the column. We'll put a moveable cap at each end of the U-tube to keep the fluid from leaking out.

Now apply a downward force of magnitude F_1 to the cap on side 1. This causes the pressure in the fluid under the cap to increase by an amount $\Delta P = F_1/A_1$. According to Pascal's principle, this change in pressure is transmitted throughout the tube. We will therefore see this same pressure increase ΔP below the cap on side 2, and that will cause an upward force $F_2 = \Delta P \times A_2$ on that cap. Substituting $\Delta P = F_1/A_1$ into this expression for F_2 gives

$$F_2 = \Delta P \times A_2 = \left(\frac{F_1}{A_1}\right) A_2$$

$$F_2 = F_1\left(\frac{A_2}{A_1}\right) \quad (13\text{-}11)$$

Because A_2 is greater than A_1, F_2 is greater than F_1, which means that a small force exerted on side 1 results in a larger force exerted on the cap on side 2. This is why a hydraulic jack is a useful piece of equipment for an auto mechanic: A relatively small force exerted on side 1 can exert enough force on side 2 to raise a heavy car off the ground.

NEED TO REVIEW? Turn to page M3 in the Math Tutorial in the back of the book for more information on direct and inverse proportions.

Equation 13-11 may make it seem like we get extra force "for free." However, there is a trade-off: The cap on side 1 must be pushed down a large distance to make the cap on side 2 move up a short distance. Here's why: When force F_1 is applied to

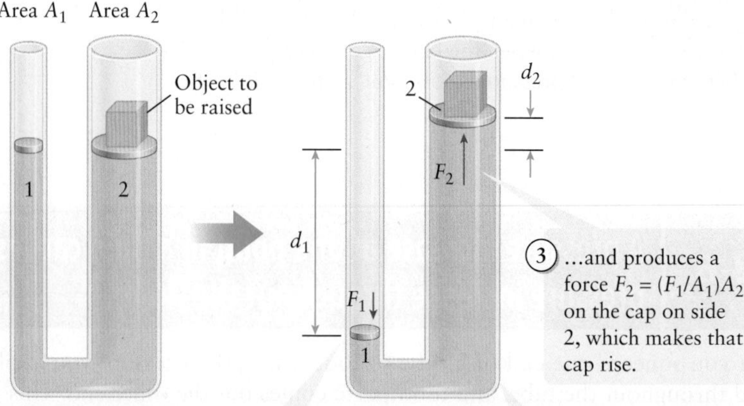

Area A_1
Area A_2
Object to be raised
1
2
d_1
d_2
F_1
F_2
1 When a downward force F_1 is applied to the cap on side 1, which has area A_1...
2 ...the increase in pressure $\Delta P = F_1/A_1$ is transmitted throughout the fluid...
3 ...and produces a force $F_2 = (F_1/A_1)A_2$ on the cap on side 2, which makes that cap rise.

Figure 13-12 A hydraulic jack Hydraulic jacks are used in workshops to elevate cars under repair and in dental offices to elevate the patient's chair. They operate using Pascal's principle.

the cap on side 1, that cap moves down a distance d_1 and displaces a volume of liquid $V_1 = d_1A_1$. The increase in pressure throughout the tube causes liquid to rise on side 2, pushing up the cap on that side by a distance d_2. The increase in liquid volume under cap 2 is $V_2 = d_2A_2$. Because the liquid used in the U-tube is assumed to be incompressible, these two volumes of liquid must be equal. Hence

$$d_2 A_2 = d_1 A_1$$

or

$$d_2 = d_1\left(\frac{A_1}{A_2}\right) \qquad (13\text{-}12)$$

The factor A_1/A_2 in Equation 13-12 is the reciprocal of the factor that relates F_2 to F_1 in Equation 13-11. Although the *force* that is exerted on side 2 is larger than the force on side 1, the *distance* the cap on side 2 moves is smaller than the distance the cap on side 1 was moved.

THE TAKEAWAY for Section 13-6

✔ Pascal's principle states that pressure applied to a confined, static fluid is transmitted undiminished to every part of the fluid as well as to the walls of the container.

✔ Pascal's principle explains the hydraulic lift, which allows a small force exerted on a fluid at one location to translate into a larger force exerted by the fluid at another location.

✔ This macroscopic behavior of the fluid is the result of the collisions between the fluid's constituent molecules and external forces exerted on the fluid.

Prep for the AP® Exam

AP® Building Blocks

1. Find the maximum weight that can be raised by the hydraulic lift shown in the figure.

$F_1 = 150$ N $\quad A_1 = 8.0$ cm^2 $\quad A_2 = 750$ cm^2 $\quad F_2 = ?$

2. A hydraulic lift is designed to raise a 9.00×10^2-kg car. If the "large" piston has a radius of 35.0 cm and the "small" piston has a radius of 2.00 cm, determine the minimum force exerted on the small piston to accomplish the task.

AP® Skill Builders

3. Consider a hydraulic lift like that shown in Figure 13-12. The cap on side 1 has a radius of 1.00 cm, and the cap on side 2 has a radius of 5.00 cm. As you push down on the cap on side 1, you do a certain amount of work W_1. How much work is done on the cap on side 2 as it rises?
(A) $25W_1$
(B) $5W_1$
(C) W_1
(D) $W_1/5$

4. A hydraulic lift has a leak so that it is only 75.0% efficient in raising its load. At this efficiency, if the large piston exerts a force of 150 N when the small piston is depressed with a force of 15.0 N and the radius of the small piston is 5.00×10^{-2} m, what is the radius of the large piston?

AP® Skills in Action

5. For the hydraulic lift shown in the figure:
(a) What force will the large piston provide if the small piston in a hydraulic lift is moved down as shown in the figure?
(b) If the small piston is depressed a distance of Δy_1, by how much will the large piston rise?
(c) How much work is done in pushing down the small piston compared to the work done in raising the large piston if $\Delta y_1 = 0.20$ m?

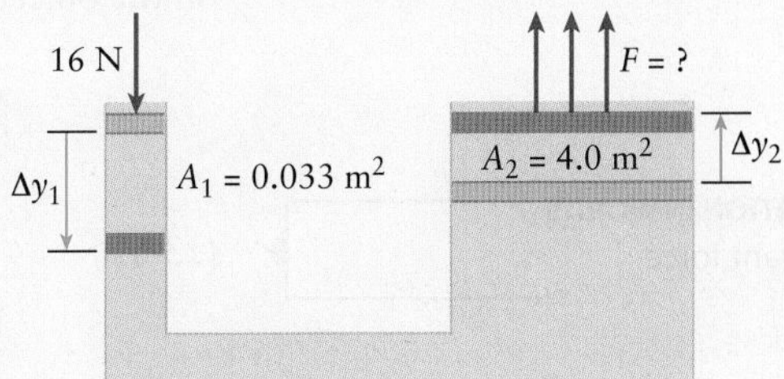

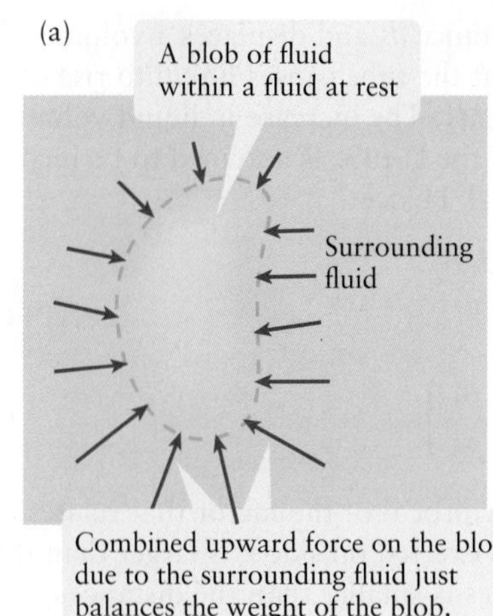
(a)
A blob of fluid within a fluid at rest
Surrounding fluid
Combined upward force on the blob due to the surrounding fluid just balances the weight of the blob.

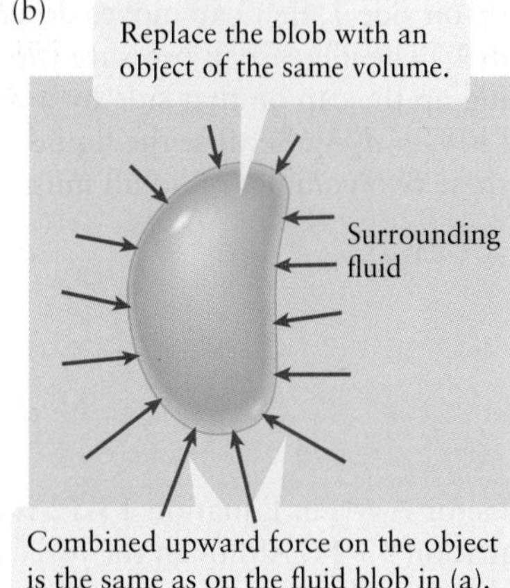
(b)
Replace the blob with an object of the same volume.
Surrounding fluid
Combined upward force on the object is the same as on the fluid blob in (a).

Figure 13-13 Archimedes' principle (a) The blob is a portion of a fluid at rest. The pressure from the surrounding fluid is greater at the bottom than at the top. (b) If the blob is replaced by some other object of the same volume, that object feels the same upward (buoyant) force as did the blob.

13-7 Archimedes' principle helps us understand buoyancy

Drop a plastic toy into a pond, and you're not surprised when it pops back up to the surface. A boat anchor won't come back up—which is also not surprising. But why does a boat, which weighs much more than its anchor, float on the surface rather than sink?

To answer these questions, think of a fluid (gas or liquid) at rest. To maintain hydrostatic equilibrium of any portion of the fluid, like the shaded blob shown in **Figure 13-13a**, the net force on that portion of fluid must be zero. Hence the combined force due to the pressure of the surrounding fluid—greater pressure on the bottom, less pressure on top—must be an upward force that exactly balances the weight of the shaded blob of fluid. We call this combined upward force the **buoyant force**; it "buoys up" the blob of fluid just enough to keep it from sinking.

Now we replace the blob of fluid in Figure 13-13a with an object of exactly the same dimensions (**Figure 13-13b**). We say that this object *displaces* a volume of fluid just equal to the volume of the object that's immersed in the fluid. For a boat floating in water, the displaced volume equals that portion of the boat's volume that's underwater. For a submarine under water, a balloon in air, or any other object that's totally immersed in a fluid, the displaced volume equals the total volume of the object.

AP® Exam Tip

An object model is acceptable in buoyancy problems unless the system being submerged changes shape. As long as the system stays rigid, we can describe its motion as if the forces exerted on it were exerted at a single point, even though in reality the buoyant force is due to the fluid pressure differences over the surface of the system. The transition in thinking from Figure 13-13 to **Figure 13-14** makes the object model work.

Here's the critical observation: At each point on the object's surface, the *pressure* of the surrounding fluid (caused by fluid molecules colliding with the object) is *exactly* the same as it was before we swapped the object for the blob of fluid. So the same buoyant force is exerted on the submerged object as on the blob of fluid, with a magnitude just equal to the weight of that fluid blob. This statement about buoyant force is called **Archimedes' principle:**

The buoyant force on an object immersed in a fluid is equal to the weight of the fluid that the object displaces.

To be specific, suppose the object displaces a volume $V_{\text{displaced}}$ of fluid. If the fluid has density ρ_{fluid}, the fluid that the object displaces has mass $m_{\text{fluid}} = \rho_{\text{fluid}} V_{\text{displaced}}$ and weight $m_{\text{fluid}} g = \rho_{\text{fluid}} V_{\text{displaced}}\, g$. Hence the magnitude F_b of the buoyant force exerted on the object is

EQUATION IN WORDS
Buoyant force

Magnitude of the buoyant force on an object in a fluid; Density of the fluid

$$F_b = \rho_{\text{fluid}} V_{\text{displaced}}\, g \qquad (13\text{-}13)$$

Volume of fluid that the object displaces; Acceleration due to gravity

If the object's weight is exactly the same as the magnitude of the buoyant force F_b—that is, exactly the same as the weight of the displaced fluid—the net force on the object is zero and the object neither sinks nor rises; instead, it floats, remaining at the same height within the fluid (**Figure 13-14a**) at which it is placed. If the object weighs more than the displaced fluid, the net force is downward and the object accelerates downward (sinks) (**Figure 13-14b**). If the object weighs less than the displaced fluid, the net force is upward and the object accelerates upward (rises) (**Figure 13-14c**).

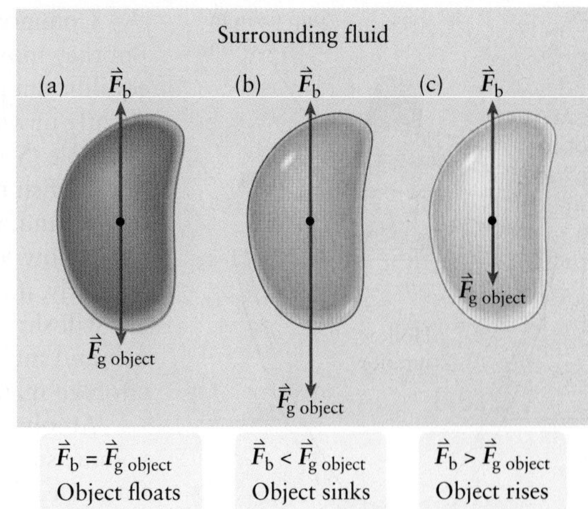

Figure 13-14 Buoyancy: Floating, sinking, and rising All three of these objects are immersed in the same fluid. All three have the same shape and size and so experience the same buoyant force F_b. Whether the object floats, sinks, or rises depends on its weight.

Floating: Submarines, Fish, Ships, and Balloons

Many types of fish, as well as submarines, are able to float while submerged. In this case the volume $V_{displaced}$ of displaced fluid in Equation 13-13 is the same as the volume of the object (fish or submarine). If the object has average density ρ, its mass is $m = \rho V_{displaced}$ and its weight is $\rho V_{displaced} g$. Then the condition that the buoyant force on the object has the same magnitude as the object's weight is

$$\rho_{fluid} V_{displaced} g = \rho V_{displaced} g$$

or, dividing through by $V_{displaced} g$,

$$\rho_{fluid} = \rho$$

(condition that an object of average density ρ floats while submerged in a fluid of density ρ_{fluid})

In words, a submerged object can float only if its density is the same as the fluid in which it is immersed. If it had a lower density it would keep popping to the surface.

Submarines and deep-sea research vessels are made of steel, which is far denser than water. They are nonetheless able to float underwater because they are hollow, with much of the internal volume filled with low-density air. The amount of air can be adjusted by filling ballast tanks with seawater or emptying the ballast tanks so that the submarine's *average* density (the mass of the submarine divided by its volume) is equal to that of water (see Example 13-2 in Section 13-2). For the same reason, many species of fish have a flexible gas-filled sac called a *swimbladder* (**Figure 13-15**). These fish are able to float while submerged because they can regulate the amount of gas in the swimbladder to make their average density equal to that of water. (In cod and other species this regulation is done by exchanging gas between the swimbladder and the blood.)

(a)

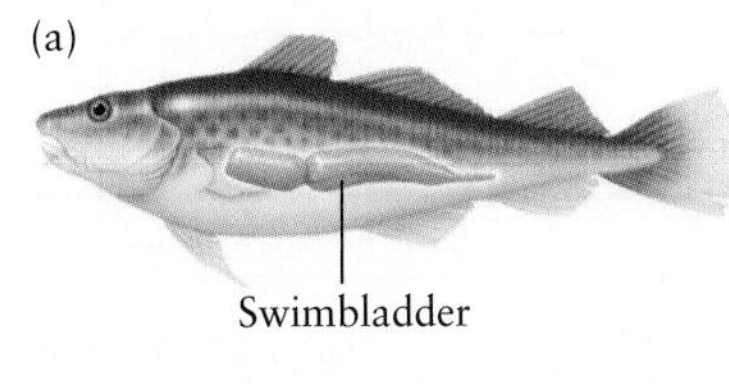

(b)

Image Source/Alamy

Figure 13-15 Marine animals with and without swimbladders (a) The gas-filled swimbladder gives this fish the same average density as the surrounding water so that it floats while submerged. Many fish species have a swimbladder, including salmon, herring, mackerel, cod, and the fish shown in the photo that opens this chapter. (b) Fish that live near the ocean bottom, like flounder and dogfish, lack swimbladders. They are denser than water and hence cannot float; they must swim continuously to avoid sinking to the bottom. The same is true for sharks like the great hammerhead (*Sphyrna mokarran*) shown here.

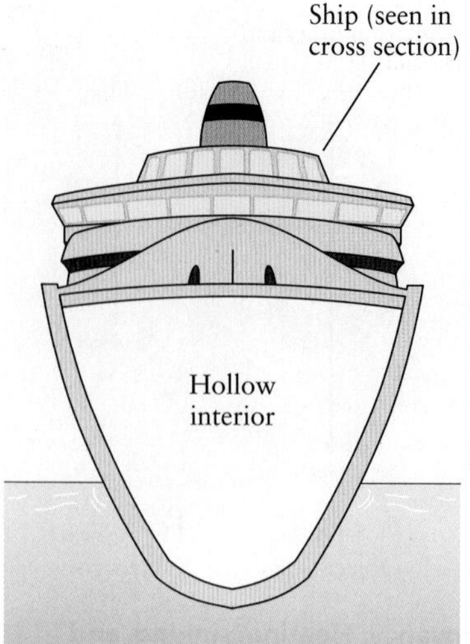
Ship (seen in cross section)
Hollow interior

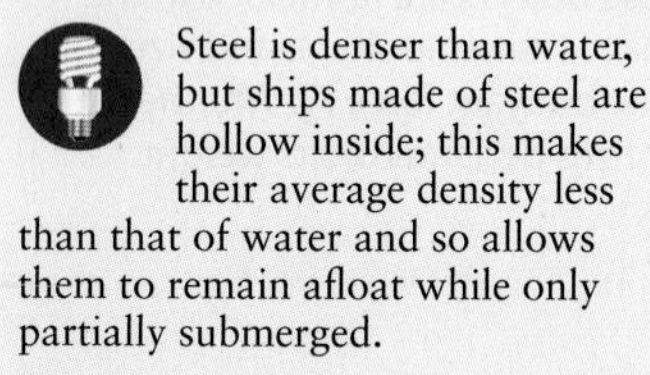
Steel is denser than water, but ships made of steel are hollow inside; this makes their average density less than that of water and so allows them to remain afloat while only partially submerged.

Figure 13-16 How steel ships float What determines whether an object floats is how its overall density compares to the density of water.

Common aquarium fish have swimbladders and are able to float while submerged. But they move their fins constantly even if they are standing still. The reason is that the equilibrium provided by a swimbladder is *unstable*. If a water current displaces the fish slightly upward, the pressure at its new position is slightly less than before, and Pascal's principle (Section 13-6) tells us that this reduced pressure will be transmitted throughout the fish to the swimbladder. The reduced pressure makes the gas in the swimbladder expand, which decreases the average density of the fish as a whole. As a result the fish is now less dense than the surrounding water and will continue to move upward. Similarly, if the fish should be displaced below its equilibrium position, the swimbladder will shrink in response to the increased pressure, thus increasing the density of the fish and making it sink even farther. The continuous motions of the fish's fins are an effort to maintain a constant depth.

An object that is *less* dense than water floats on the surface with only part of its volume submerged; the lower the object's density relative to that of water, the higher it floats (see Example 13-8). Steel is denser than water, but ships made of steel are hollow inside; this makes their average density less than that of water and so allows them to remain afloat while only partially submerged (**Figure 13-16**).

We've mostly been discussing objects in liquids such as water, but there is also a buoyant force on an object immersed in a gas such as the atmosphere. The buoyant force on a balloon filled with helium (which is less dense than air) is greater than the balloon's weight, which is why it rises if released. By contrast, a balloon filled with room-temperature air falls if released: The air inside the balloon is at higher pressure and has a higher density than the outside air, so the balloon's weight is more than the buoyant force exerted on it. Hot air is less dense than cold air, so an air-filled balloon can be made to float if it's equipped with a burner to warm the balloon's contents. That's the principle of a hot-air balloon.

EXAMPLE 13-9 What Lies Beneath the Surface?

A solid block of density ρ_{block} is placed in a liquid of density ρ_{liquid}. If the block is less dense than the liquid (that is, $\rho_{\text{block}} < \rho_{\text{liquid}}$), derive an expression for the fraction of the block's volume that is submerged.

Set Up

Archimedes' principle tells us that the buoyant force exerted on the block by the liquid equals the weight of the displaced liquid. In order for the block to float, the upward buoyant force must balance the weight of the block. We'll use this principle to compare the volume of displaced liquid $V_{\text{displaced}}$—equal to the volume of the block that's submerged in the liquid—to the total volume of the block.

Buoyant force:

$$F_b = \rho_{\text{liquid}} V_{\text{displaced}}\, g \qquad (13\text{-}13)$$

block (seen from the side), volume V_{block}

$V_{\text{displaced}}$

Solve

First write expressions for the block's mass m_{block} and weight $F_{g,\text{block}}$. When the block is in equilibrium, the upward buoyant force on the block equals the block's weight. Use this and the expression for $F_{g,\text{block}}$ to solve for the ratio of $V_{\text{displaced}}$ to V_{block}. Both forces have only a vertical component, so we will take $+y$ to be upward.

volume of block = V_{block}

mass of block = (density of block) × (volume of block):

$$m_{\text{block}} = \rho_{\text{block}} V_{\text{block}}$$

weight of block = (mass of block) × g:

$$F_{g,\text{block}} = m_{\text{block}}\, g = \rho_{\text{block}} V_{\text{block}}\, g$$

$\vec{F}_b$

$\vec{F}_{g\,\text{block}}$

Net force on the block in equilibrium is zero:

$$\sum F_y = F_b - F_{g,\text{block}} = 0 \text{ so}$$

$$F_b = F_{g,\text{block}}$$

From Equation 13-13:

$$F_b = \rho_{\text{liquid}} V_{\text{displaced}} g, \text{ so}$$

$$\rho_{\text{liquid}} V_{\text{displaced}} g = \rho_{\text{block}} V_{\text{block}} g$$

Divide through by g:

$$\rho_{\text{liquid}} V_{\text{displaced}} = \rho_{\text{block}} V_{\text{block}}$$

Solve for ratio of $V_{\text{displaced}}$ to V_{block}:

$$\text{fraction of block's volume that is submerged} = \frac{V_{\text{displaced}}}{V_{\text{block}}} = \frac{\rho_{\text{block}}}{\rho_{\text{liquid}}}$$

Reflect

Our answer makes sense because the fraction of the block's volume that is submerged, $V_{\text{displaced}}/V_{\text{block}}$, is equal to the ratio of the block's density ρ_{block} to the density of the liquid, ρ_{liquid}.

Extend: Let's try this for a couple of real-life examples: ice floating in fresh water and ice floating in salt water. We find that a cube of pure ice in a glass of water floats with 91.7% of its volume submerged, while an iceberg in salt water floats a bit higher with 89.5% of its volume below the surface.

Block of ice floating in fresh water:

$$\rho_{\text{block}} = \text{density of ice} = 917 \text{ kg/m}^3$$

$$\rho_{\text{liquid}} = \text{density of fresh water} = 1.000 \times 10^3 \text{ kg/m}^3$$

$$\text{fraction of ice that's submerged} = \frac{V_{\text{displaced}}}{V_{\text{block}}} = \frac{\rho_{\text{block}}}{\rho_{\text{liquid}}} = \frac{917 \text{ kg/m}^3}{1.000 \times 10^3 \text{ kg/m}^3} = 0.917$$

So 91.7% of the ice is submerged.

Block of ice floating in salt water:

$$\rho_{\text{block}} = \text{density of ice} = 917 \text{ kg/m}^3$$

$$\rho_{\text{liquid}} = \text{density of salt water} = 1.025 \times 10^3 \text{ kg/m}^3$$

$$\text{fraction of ice that's submerged} = \frac{V_{\text{displaced}}}{V_{\text{block}}} = \frac{\rho_{\text{block}}}{\rho_{\text{liquid}}} = \frac{917 \text{ kg/m}^3}{1.025 \times 10^3 \text{ kg/m}^3} = 0.895$$

So, 89.5% of the ice is submerged.

NOW WORK Problems 5, 6, 8, and 10 from The Takeaway 13-7.

EXAMPLE 13-10 Underwater Float

A solid plastic ball of density 6.00×10^2 kg/m^3 and radius 2.00 cm is attached by a lightweight string to the bottom of an aquarium filled with fresh water. What is the tension in the string?

Set Up

Three forces are exerted on the ball: the downward force of gravity, the upward buoyant force exerted by the water, and the downward tension force exerted by the string (which is what we want to find). We'll first use Equation 13-13 to determine the buoyant force on the ball and then use Newton's first law to solve for the tension force. All forces have only a vertical component, so we will take $+y$ to be upward.

Buoyant force:

$$F_b = \rho_{\text{water}} V_{\text{displaced}} g \qquad (13\text{-}13)$$

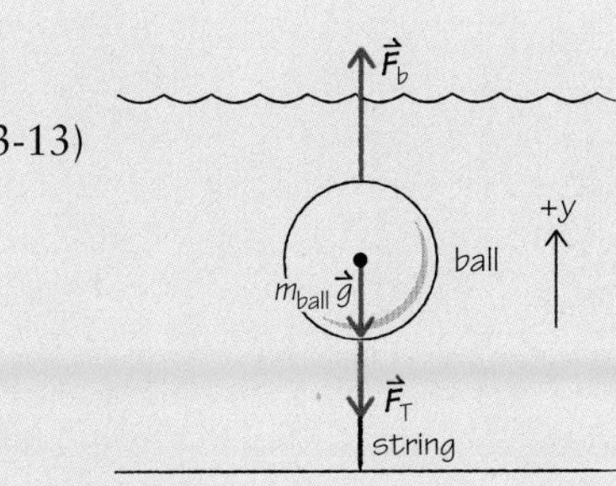

Solve

Newton's first law tells us that because the ball is at rest, the net force on the ball must be zero. Solve for the tension F_T in the string in terms of the buoyant force F_b and the weight of the ball $m_{ball}g$.

Newton's first law:

$$\sum F_y = F_b - F_T - m_{ball}g = 0 \text{ so}$$

$$F_T = F_b - m_{ball}g$$

The entire ball is submerged, so the volume of water that it displaces is equal to the volume of the ball of radius $R_{ball} = 2.00 \text{ cm} = 2.00 \times 10^{-2}$ m. Use this to calculate the buoyant force.

Buoyant force:

$$F_b = \rho_{water} V_{displaced} g$$

$$V_{displaced} = V_{ball}$$

$$= \frac{4}{3}\pi R_{ball}^3 = \frac{4}{3}\pi(2.00 \times 10^{-2} \text{ m})^3$$

$$= 3.35 \times 10^{-5} \text{ m}^3 \text{ so}$$

$$F_b = (1.000 \times 10^3 \text{ kg/m}^3)(3.35 \times 10^{-5} \text{ m}^3)(9.80 \text{ m/s}^2)$$

$$= 0.328 \text{ kg} \cdot \text{m/s}^2 = 0.328 \text{ N}$$

Use the density of the ball to calculate its weight.

$$\text{weight of ball} = m_{ball}g = \rho_{ball} V_{ball} g$$

$$= (6.00 \times 10^2 \text{ kg/m}^3)(3.35 \times 10^{-5} \text{ m}^3)(9.80 \text{ m/s}^2)$$

$$= 0.197 \text{ kg} \cdot \text{m/s}^2 = 0.197 \text{ N}$$

Finally, use the relationship that we derived from Newton's first law to solve for the tension F_T.

$$F_T = F_b - m_{ball}g = 0.328 \text{ N} - 0.197 \text{ N}$$

$$= 0.131 \text{ N}$$

Reflect

The ball is less dense than water, so the buoyant force ($F_b = 0.328$ N) is greater than the weight of the ball ($m_{ball}g = 0.197$ N). Hence the string must exert a downward tension force to keep the ball from rising. If the string were to break, the ball would rise to the surface of the water. This makes sense!

Extend: Can you see from Example 13-9 that the ball would end up floating with 60.0% of its volume submerged?

NOW WORK Problem 11 from The Takeaway 13-7.

Apparent Weight

In Example 13-10 the submerged plastic ball is less dense than water and so has to be tethered to keep it from floating upward. Let's now think about a submerged object that's denser than water. The problem now is how to keep the object from sinking, which we solve by suspending it from above. **Figure 13-17a** shows such a submerged object hanging from a spring balance, and **Figure 13-17b** shows an identical object that is surrounded by vacuum. Although both objects have the same mass M and the same volume V, the weights registered on the spring balances are *not* the same. Why? The difference is that for the submerged object in Figure 13-17a, the buoyant force due to the displaced water opposes the force of gravity and makes the object seem to weigh less. Thus, the scale reads the object's apparent weight (remember this is also sometimes referred to as an object's effective weight) rather than its true weight. In Chapter 6 we saw that an object in orbit has an apparent weight of zero due to its acceleration. The kind of apparent weight we're discussing here is for an object that is *not* accelerating but upon which a buoyant force is exerted. Apparent or effective weight can always be thought of as the magnitude of the normal force that would be exerted on an object to support it by a scale. The needed force can be changed by the effects of buoyancy or acceleration.

You can experience the difference between apparent weight and true weight in a swimming pool. If you swim below the surface of the water, you feel as though you weigh less than normal. Just as for the plastic ball in Example 13-10, the upward buoyant force

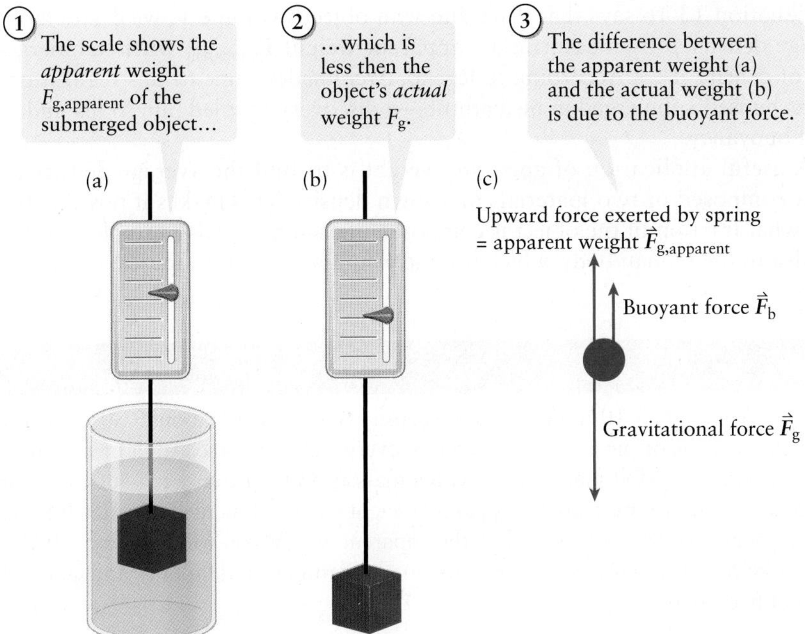

1 The scale shows the *apparent* weight $F_{g,apparent}$ of the submerged object...
2 ...which is less then the object's *actual* weight F_g.
3 The difference between the apparent weight (a) and the actual weight (b) is due to the buoyant force.
(a)
(b)
(c)
Upward force exerted by spring = apparent weight $\vec{F}_{g,apparent}$
Buoyant force $\vec{F}_b$
Gravitational force $\vec{F}_g$

Figure 13-17 Apparent weight (a) An object hung from a spring balance has an apparent weight that depends on the fluid in which it is submerged. The reading on the scale indicates the force exerted by the spring, which is equal in magnitude to the object's apparent weight. (b) The reading on the scale shows the object's actual weight only if the object is surrounded by vacuum, so no buoyant force acts. (c) A free-body diagram showing the three forces acting on the submerged object in (a).

on you partially cancels the force of gravity. Astronauts use this effect to train for working in low gravity: They practice while submerged in an immense pool of water.

How is apparent weight related to true weight? **Figure 13-17c** shows the free-body diagram for the submerged object on the left in Figure 13-17a. The three forces exerted on the object are the downward force of gravity (whose magnitude is the object's true weight F_g), the upward buoyant force F_b exerted by the water, and the upward force exerted by the spring balance. The scale on the spring balance measures how much force the balance exerts, so this force is just equal to the apparent weight $F_{g,apparent}$. The net force on the hanging object is zero (because the object is at rest), so

$$\sum F_y = F_{g,apparent} + F_b - F_g = 0$$

and

$$F_{g,apparent} = F_g - F_b \tag{13-14}$$

(apparent weight of a submerged object)

The apparent weight of a submerged object equals its true weight minus the buoyant force exerted on the object by the surrounding fluid.

If the object is completely submerged, the volume of fluid that it displaces equals the volume V of the object. Then we can replace $V_{displaced}$ in Equation 13-13 (the statement of Archimedes' principle) by V, and Equation 13-14 becomes

$$F_{g,apparent} = F_g - \rho_{fluid} V g \tag{13-15}$$

(apparent weight of an object of weight F_g and volume V completely submerged in a fluid of density ρ_{fluid})

So by measuring the apparent weight of an object while submerged ($F_{g,apparent}$) and its true weight when not submerged (F_g), you can determine the object's volume. This method can be quite handy when the object is irregularly shaped!

We can also write the apparent weight in terms of the average density ρ of the object. Because the object has volume V, its mass is $m = \rho V$ and its true weight is $F_g = mg = \rho V g$. Then Equation 13-15 becomes

$$F_{g,apparent} = \rho V g - \rho_{fluid} V g = (\rho - \rho_{fluid}) V g \tag{13-16}$$

(apparent weight of an object of average density ρ and volume V completely submerged in a fluid of density ρ_{fluid})

Equation 13-16 says that if we know an object's volume V, we can determine its average density ρ by measuring its apparent weight $F_{g,apparent}$ when submerged in a fluid of density ρ_{fluid}. According to legend, Archimedes came to this realization while he was himself submerged in his bathtub—a discovery that led him to his understanding of buoyancy.

A useful application of apparent weight is to find the average density ρ of an object composed of two materials of known density. This makes it possible to determine what fraction of the object is composed of each material. Example 13-11 applies this idea to the human body, which is largely composed of lean muscle and fat.

EXAMPLE 13-11 Measuring Body Fat

In the human body the density of lean muscle is about 1.06×10^3 kg/m^3, and the density of fat tissue is about 9.30×10^2 kg/m^3. Two-thirds of Americans have a percent body fat of 25% or more and are therefore overweight or obese. Adult men with between 18 and 24% body fat are considered healthy. An adult male patient with a mass of 85.0 kg comes to your clinic with the claim that he has 20.0% body fat. To test this claim, you measure his apparent weight when submerged in water. If his claim is correct, what are (a) the volume of his body, (b) his average density, and (c) the apparent weight you will measure? (d) What do you conclude if the apparent weight you measure is 19.7 N? Assume that he is made of muscle and fat only, and ignore the bones (which typically constitute about 15% of body mass).

Set Up

We're given the densities of fat and of muscle, the fluid density $\rho_{fluid} = 1000$ kg/m^3, and the person's mass $m = 85.0$ kg (from which we can find his actual weight $F_g = mg$). We're also given his claimed percentage of body fat. We'll use Equation 13-1 to determine the volume of his body that is fat and the volume that is muscle according to his claim, then use these together to calculate what his claim says about his overall volume V and his average density ρ. We can then use either Equation 13-15 or Equation 13-16 to calculate what his apparent weight should be if his claim is correct.

Definition of density:

$$\rho = \frac{m}{V} \qquad (13\text{-}1)$$

Two equations for the apparent weight $F_{g,apparent}$ of an object of weight F_g, volume V, and average density ρ submerged in fluid of density ρ_{fluid}:

$$F_{g,apparent} = F_g - \rho_{fluid} V g \qquad (13\text{-}15)$$

$$F_{g,apparent} = (\rho - \rho_{fluid}) V g \qquad (13\text{-}16)$$

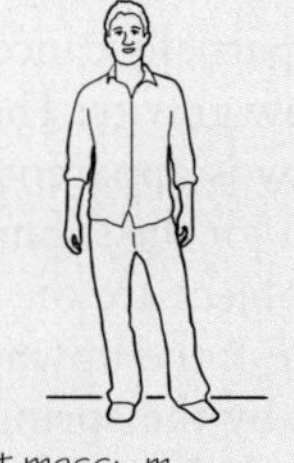

Solve

(a) Use the person's claim about his percentage of body fat to determine the mass of fat and mass of muscle in his body. Then use Equation 13-1 to determine the volume of the fat, V_{fat}, and the volume of the muscle, V_{muscle}. The total volume V of the person is the sum of V_{fat} and V_{muscle}.

The person claims that 20.0% = 0.200 of their body is fat, so 80.0% = 0.800 is muscle:

$$m_{fat} = 0.200m = (0.200)(85.0 \text{ kg}) = 17.0 \text{ kg}$$

$$m_{muscle} = 0.800m = (0.800)(85.0 \text{ kg}) = 68.0 \text{ kg}$$

Rearrange Equation 13-1 to solve for the volume of fat and volume of muscle in his body:

$$\rho = \frac{m}{V}, \text{ so } V = \frac{m}{\rho}$$

$$V_{fat} = \frac{m_{fat}}{\rho_{fat}} = \frac{17.0 \text{ kg}}{9.30 \times 10^2 \text{ kg/m}^3} = 0.0183 \text{ m}^3$$

$$V_{muscle} = \frac{m_{muscle}}{\rho_{muscle}} = \frac{68.0 \text{ kg}}{1.06 \times 10^3 \text{ kg/m}^3} = 0.0642 \text{ m}^3$$

The person's total volume V is

$$V = V_{fat} + V_{muscle} = 0.0183 \text{ m}^3 + 0.0642 \text{ m}^3 = 8.25 \times 10^{-2} \text{ m}^3$$

(b) If the person's body fat claim is correct, his average density from Equation 13-1 will be his mass divided by the volume V that we found in part (a).

Average density of person from Equation 13-1:

$$\rho = \frac{m}{V} = \frac{85.0\ \text{kg}}{0.0825\ \text{m}^3}$$
$$= 1.031 \times 10^3\ \text{kg/m}^3$$

or, rounded to three significant figures,

$$\rho = 1.03 \times 10^3\ \text{kg/m}^3$$

Given that the person is part fat and part muscle, the value of ρ is intermediate between the values of ρ_{fat} and ρ_{muscle}.

(c) Use Equation 13-15 to calculate the apparent weight of the person submerged in water, provided his body fat claim is correct.

From Equation 13-15:

$$F_{\text{g,apparent}} = F_{\text{g}} - \rho_{\text{fluid}} Vg$$

The person's true weight is

$$F_{\text{g}} = mg = (85.0\ \text{kg})(9.80\ \text{m/s}^2)$$
$$= 833\ \text{kg} \cdot \text{m/s}^2 = 833\ \text{N}$$

The fluid density $\rho_{\text{fluid}} = 1.00 \times 10^3\ \text{kg/m}^3$ is only a little less than the person's average density $\rho = 1.03 \times 10^3\ \text{kg/m}^3$, so the buoyant force of magnitude $\rho_{\text{fluid}} Vg$ is only a little less than the person's weight F_{g}. Hence the apparent weight (true weight minus buoyant force) is only a small fraction of the true weight.

The fluid in which he is immersed is water with density $\rho_{\text{fluid}} = 1.00 \times 10^3\ \text{kg/m}^3$. Using his body volume V from part (a)

$$F_{\text{g,apparent}} = 833\ \text{N} - (1.00 \times 10^3\ \text{kg/m}^3)(0.0825\ \text{m}^3)(9.80\ \text{m/s}^2)$$
$$= 833\ \text{N} - 808\ \text{N}$$
$$= 25\ \text{N}$$

(d) The person's measured apparent weight is 19.7 N, which is less than the 25 N that we calculated in part (c) based on the person's claim of 20.0% body fat. The *smaller* measured value of $F_{\text{g,apparent}}$ means that the person has a *greater* body fat percentage than his claim suggests.

From Equation 13-15, the apparent weight is

$$F_{\text{g,apparent}} = F_{\text{g}} - \rho_{\text{fluid}} Vg$$

The measured value of $F_{\text{g,apparent}}$ is 19.7 N, versus the 25-N value we calculated based on the person's claim. Because the person's weight $F_{\text{g}} = 833$ N is known, the only way to get a lower value of $F_{\text{g,apparent}}$ is for the buoyant force $\rho_{\text{fluid}} Vg$ to have a greater value. This means that the volume V of the person must be *greater* than we calculated in part (a), and so the person's average density $\rho = m/V$ must be *less* than we calculated in part (b). For this to be so, the person's body must have a greater percentage of low-density fat ($\rho_{\text{fat}} = 9.30 \times 10^2\ \text{kg/m}^3$) and a smaller percentage of high-density muscle ($\rho_{\text{muscle}} = 1.06 \times 10^3\ \text{kg/m}^3$) than he claims.

Reflect

We can check our calculation for the person's apparent weight (based on his optimistic body fat claim) by using Equation 13-16. Happily, we get the same result as we did using Equation 13-15.

$$F_{\text{g,apparent}} = (\rho - \rho_{\text{fluid}})Vg$$

Use the value of V from part (a) and the unrounded value $\rho = 1.031 \times 10^3\ \text{kg/m}^3$ from part (b):

$$F_{\text{g,apparent}} = (1.031 \times 10^3\ \text{kg/m}^3 - 1.00 \times 10^3\ \text{kg/m}^3)(0.0825\ \text{m}^3)(9.80\ \text{m/s}^2)$$
$$= (31\ \text{kg/m}^3)(0.0825\ \text{m}^3)(9.80\ \text{m/s}^2)$$
$$= 25\ \text{N}$$

Extend: Can you show that the person's actual body fat percentage is 25.0%?

NOW WORK Problems 7 and 9 from The Takeaway 13-7.

THE TAKEAWAY for Section 13-7

✔ A buoyant force is a net upward force exerted on an object by a fluid, and is equal to the weight of the fluid displaced by the object.

✔ The buoyant force exerted on an object is the vector sum of all the forces exerted on the object by collisions with molecules making up the fluid.

✔ An object immersed in a fluid has an apparent weight equal to its true weight minus the magnitude of the buoyant force on the object.

Prep for the AP® Exam

AP® Building Blocks

1. A plastic cube with a coin taped to its top surface is floating partially submerged in water. You mark the level of the water on the cube then remove the coin and tape it to the *bottom* of the cube. When you put the cube back in the water, will the cube sit higher in the water, lower in the water, or at the same height in the water as it was when the coin was on the top of the cube? Briefly explain your prediction. *Hint*: The buoyant force on an object is the weight of the fluid that it displaces, and in this problem water is displaced by whatever is below the surface.
2. An ice cube floats in a glass of water so that the water level is exactly at the rim. After the ice cube melts, will all the water still be in the glass? Explain your answer using physics principles.
3. An open, empty glass soda bottle will float in a tub of water. First you place an empty, open glass bottle into a tub of water so that it floats and mark the level of the water on the wall of the tub. You then submerge the bottle so that it fills with water and sinks to the bottom. Is the level of the water in the tub higher, lower, or the same as it was when the bottle was floating?
4. An object floats in water with 5/8 of its volume submerged. The ratio of the density of the object to that of water is
   (A) 8/5.
   (B) 5/8.
   (C) 1/2.
   (D) 3/8.

 Skill Builders

5. A rectangular block of wood, 10.0 cm × 15.0 cm × 40.0 cm, has a specific gravity of 0.600.
   (a) Draw a free-body diagram labeling all of the forces on the block when it is placed in a pool of fresh water.
   (b) Determine the buoyant force exerted on the block.
   (c) What fraction of the block is submerged?
   EX 13-9 (d) Determine the weight of the water that is displaced by the block.
6. EX 13-9 A rectangular block of wood floats in fresh water with its lower 10.0 cm submerged. What distance will be submerged when it floats in seawater (specific gravity 1.025)?
7. The person in Example 13-11 has a weight of 833 N. When submerged in water (density 1.00×10^3 kg/m^3), his apparent weight is 19.7 N.
   (a) Calculate his average density.
   EX 13-11 (b) Find the percent of his body mass that is fat, assuming that the person is made of muscle and fat only.
8. EX 13-9 A log raft is 3.00 m × 4.00 m × 0.150 m and is made from trees that have an average density of 7.00×10^2 kg/m^3. How many people can stand on the raft and keep their feet dry, assuming an average person has a mass of 70.0 kg?

AP® Skills in Action

9. A crown that is supposed to be made of solid gold is under suspicion. When the crown is weighed in air, it has a weight of 5.15 N. When it is suspended from a digital balance and lowered into water, its apparent weight is measured to be 4.88 N. The specific gravity of gold is 19.3.
   (a) Calculate the apparent weight of the crown if it is gold.
   EX 13-11 (b) Identify the material from which the crown is made.
10. EX 13-9 A woman floats in a region of the Great Salt Lake where the water is about four times saltier than the ocean and has a density of about 1130 kg/m^3. The woman has a mass of 55 kg, and her density is 985 kg/m^3 after exhaling as much air as possible from her lungs. Determine the percentage of her volume that will be above the waterline.
11. A cube of side s is completely submerged in a pool of fresh water. Your expressions may include some or all of the following quantities: atmospheric pressure, P_0; the density of fresh water, ρ; the length of the side of the cube, s; the mass of the cube, m_{cube}; and the acceleration due to gravity, g.
    (a) Derive an expression for the pressure difference between the bottom and top of the cube.
    (b) After drawing a free-body diagram, derive an algebraic expression for the net force on the cube.
    EX 13-10 (c) Find the weight of the displaced water when the cube is submerged.

In a wind tunnel smoke trails display the smooth, steady pattern of laminar flow around this automobile.

(a)

The smoke rising from this incense stick changes from laminar flow to chaotic, turbulent flow.

(b)

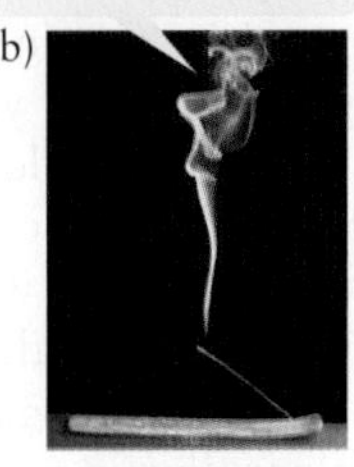

Although water rotates around the center of a whirlpool, the flow is largely irrotational (a physicist's way of saying that the flow velocity changes in a special and gradual way from one point to another).

(c)

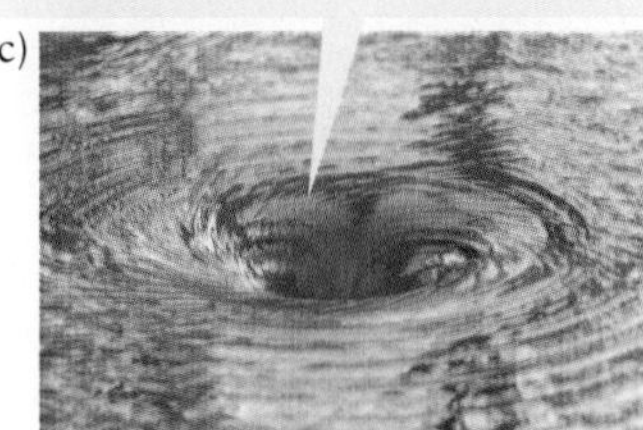

Thinkstock; Nick Stephens/Alamy

Figure 13-18 Some examples of fluid flow Three very different situations in which a fluid is in motion.

13-8 Fluids in motion: a more robust definition of an ideal fluid, and application of conservation of mass

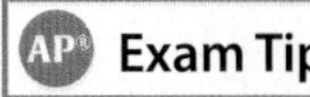

Exam Tip

The AP® Physics 1 Exam is limited to ideal fluids. A brief introduction to nonideal fluid flow is provided to help you understand when the ideal fluid approximation is appropriate.

In the preceding two sections we considered fluids at rest. All around us, however, we find fluids in *motion*. Masses of air shift through the atmosphere to bring today's weather, river water courses downhill, and the oceans move in and out with the tides. Within our bodies, moving fluids—blood, lymph, and the air used in respiration—play essential roles in sustaining life. To cause motion between the ends of the path of the motion there needs to be a difference in pressures (causing a force) or a difference in potential energy. In the next section, we will consider some of these causes of fluid motion. But just as we first discussed how to describe motion, and *then* introduced its causes, for fluids we will first focus on how to describe the motion of fluids.

Examples of fluid flow in both nature and technology are breathtaking in their diversity (**Figure 13-18**). We will continue our study of fluids with the ideal fluid approximation. To understand when the ideal fluid approximation is appropriate, we first begin our study of fluids in motion by briefly examining how physicists classify different types of fluid flow.

Steady Flow and Unsteady Flow

The simplest type of fluid flow is one in which the flow pattern doesn't change with time, like a stream in which water moves very smoothly with no variations. Such fluid motion is called **steady flow**. The direction and speed of the flow can be different from one point in the fluid to another (**Figure 13-19**). At any one point, however, the flow velocity remains constant from one moment to the next.

In **unsteady flow** the velocity at a given point can change with time. You experience the unsteady flow of air when you stand outside on a gusty day: The direction and speed of the wind at your position change erratically. A less erratic example of unsteady flow is the pulsing motion of blood as it exits your heart through the aorta. Ocean waves, too, involve an unsteady pulsing motion that carries water (and anyone bobbing in the water) alternately up and down.

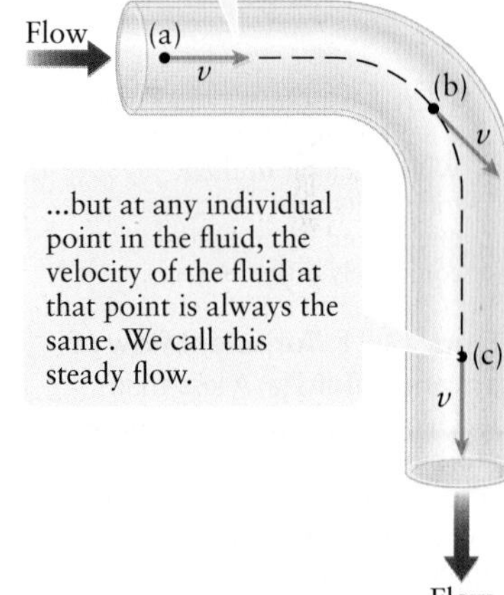

As a particle of fluid moves through this pipe from (a) to (b) to (c), its velocity changes in direction and magnitude...
Flow
(a)
v
(b)
v
...but at any individual point in the fluid, the velocity of the fluid at that point is always the same. We call this steady flow.
(c)
v
Flow

Figure 13-19 Steady flow of fluid in a pipe In steady fluid flow the velocity can be different at different points in the flow, but the velocity at any one point maintains the same value at all times.

WATCH OUT!

Steady flow doesn't mean that the fluid velocity is the same everywhere.

Note that even though the fluid velocity at any one position in steady flow remains constant, the velocity of any given quantity of fluid *can* change as it moves. For example, water flowing through a curved pipe such as the one shown in **Figure 13-20** changes its direction of motion as it follows the bends of the pipe. When we say that the flow is steady, we mean that when the next quantity of fluid passes through the same point in the pipe, its velocity at that given point will be the same as the previous quantity of fluid that passed through that same point.

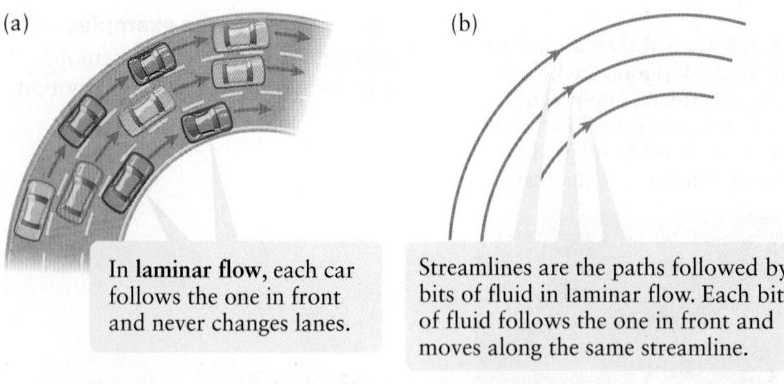

(a)
In **laminar flow**, each car follows the one in front and never changes lanes.
(b)
Streamlines are the paths followed by bits of fluid in laminar flow. Each bit of fluid follows the one in front and moves along the same streamline.

Figure 13-20 Laminar flow of cars and a fluid (a) An idealized highway in which cars move in laminar flow. (b) Streamlines in a fluid with laminar flow.

Laminar Flow and Turbulent Flow

Imagine a multilane highway filled with moving cars. If none of the cars ever changed lanes, you'd have an efficient, smooth-running transportation system in which each car would follow exactly the same trajectory as the one in front of it and the one behind it (**Figure 13-20a**). If you now replace the cars with bits of fluid, you would have a type of fluid motion called **laminar flow**. Each bit of fluid follows a path called a **streamline** that is the direct equivalent of one of the lanes of our idealized highway (**Figure 13-20b**). The smoke trails in Figure 13-18a show the streamlines of laminar flow in a wind tunnel.

Real highways are not as well organized as those in our imaginary example. Some cars change lanes, some cars move faster than others in the same lane, and occasionally there are collisions (**Figure 13-21a**). A fluid that behaves in this way is undergoing **turbulent flow** (**Figure 13-21b**). There are no streamlines in this case, because adjacent bits of fluid can follow very different paths. Figure 13-18b shows the turbulent flow of smoke.

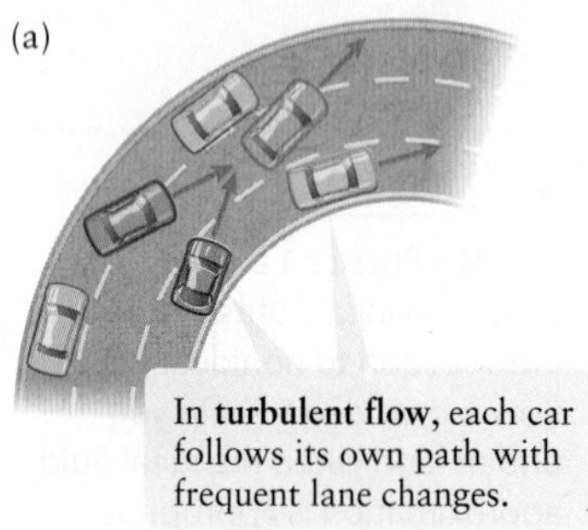

(a)
In **turbulent flow**, each car follows its own path with frequent lane changes.

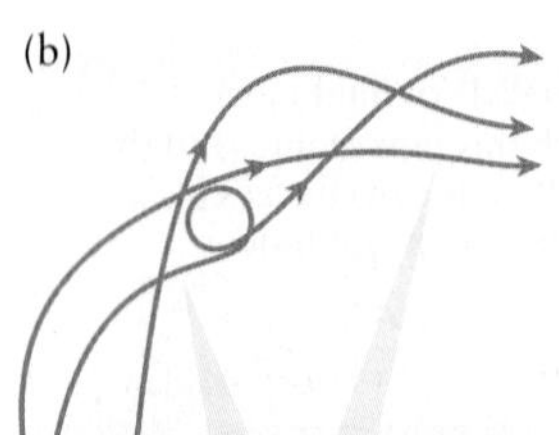

(b)
There are no streamlines for a fluid in turbulent flow. While each bit of fluid follows its own path, the one behind it can follow a completely different path.

Figure 13-21 Turbulent flow of cars and a fluid (a) A less-than-ideal highway in which cars move in turbulent flow. (b) In a fluid with turbulent flow, bits of fluid move in a haphazard and seemingly unpredictable way.

Generally speaking, a given type of flow changes from laminar to turbulent if the flow speed exceeds some critical value (which depends on the particular type of fluid). That's why airliners approaching Denver, which is downwind of the Rocky Mountains, can have a smooth ride when the wind is light and the airflow is laminar but a much bumpier ride when the wind is howling and the airflow turbulent.

Turbulence is much noisier than laminar flow. As an example, when you make a hissing or "S" sound, you blow air past your teeth in a way that produces turbulence. By contrast, if you form your lips into an "o" and blow with equal force, your teeth are out of the way, the airflow is more laminar, and the sound is much softer. Your diastolic blood pressure—the second of the two numbers in a blood pressure report such as 120/80 (see Section 13-5)—is measured by putting a high-pressure cuff around your upper arm, gradually lowering the pressure of the cuff, and listening for a change in sound. At pressures just above the diastolic value, the artery is partially compressed, making the flow turbulent and noisy; at the diastolic pressure and below, the noise disappears because the artery is fully expanded and the flow is once again in its normal laminar state.

Viscous Flow and Inviscid Flow

We already discussed the idea of **viscosity**, which is caused by adjacent parts of a fluid moving at different velocities. These parts exert frictional forces on each other as they rub past. Just as kinetic friction opposes the sliding motion of a block on a ramp, this intrinsic resistance to flow opposes the sliding of one bit of fluid past another. As an example, motor oil is more *viscous*—that is, it has a greater viscosity—than water, which in turn is more viscous than air. Many liquids are less viscous at high temperatures; for example, warm maple syrup flows more easily than cold. Most automobiles with gasoline engines use a *multiviscosity* oil designed to flow (and hence lubricate the engine's moving parts) equally well over a broad range of temperatures.

Fluids also experience friction when they flow past a solid surface. This friction is so great that it leads to what is called the **no-slip condition**: Right next to the solid surface, the velocity of the fluid is *zero* so the fluid does not "slip" over the surface. Instead, if fluid is flowing past the surface with a speed v, there is a **boundary layer** next to the surface within which the fluid speed increases from zero at the solid surface to the full speed v at the edge of the layer (**Figure 13-22**). A boundary layer of this kind develops around an automobile in motion, which is why driving even at freeway speeds doesn't blow dirt off the car. Any dirt particles lie well within the boundary layer, where the air is hardly moving at all. Even running water over the car won't dislodge all of the dirt particles, because there is also a boundary layer for water flow; only scrubbing with a sponge will complete the job.

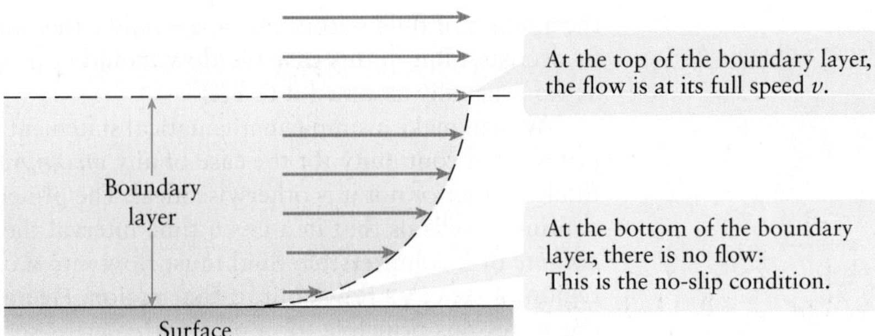
At the top of the boundary layer, the flow is at its full speed v.
Boundary layer
At the bottom of the boundary layer, there is no flow: This is the no-slip condition.
Surface

Figure 13-22 A boundary layer in a viscous fluid A boundary layer in a fluid moving past a flat surface.

Figure 13-22 shows that at different depths within the boundary layer, the fluid flows at different speeds. Hence frictional (viscous) forces are prevalent in the boundary layer, and these forces oppose the flow past the solid surface. This effect is called *viscous drag*, and it makes an important contribution to the force of air resistance felt by a moving car. If the flow in the boundary layer is turbulent, there is a greater difference in flow speeds between bits of fluid close to the car's skin, so the viscous forces are greater and the vehicle experiences more drag. Giving a car a "streamlined" shape is a way to make the flow of air around the car less turbulent and more laminar so that the air follows streamlines, as in Figure 13-18a. This reduces the drag so that less power has to be provided by the engine, and the car gets better mileage.

Although every fluid has some viscosity, in many physical situations the viscosity is relatively unimportant. (An example is the airflow around a bird in flight. The viscous forces are much less important than the forces due to the pressure of air on the bird.) In these situations, it is reasonable to simplify the problem by imagining that the moving fluid has *zero* viscosity, referred to as **inviscid flow**. It is also a good approximation when the flow is approximately laminar, and the surface boundary layer is small compared to the overall cross-sectional area of the container through which the fluid flows.

In real fluids, in which the flow is laminar, viscosity is sufficiently high that adjoining bits of fluid move at nearly the same speed. Then the speed varies gradually from one part of the fluid to another, with no abrupt jumps. Physicists call such a phenomenon **irrotational flow**. This does *not* mean that the fluid can't rotate—it certainly can, like water in a whirlpool (Figure 13-18c) or air in a hurricane. Rather, the curious term *irrotational* refers to what would happen to a small paddlewheel that was put in the fluid and allowed to move with it. If the fluid speed on one side of the paddlewheel were sharply different from that on the other side, the paddlewheel would start turning. In ideal fluids, there are no such differences, and the paddlewheel wouldn't rotate—which is why the flow is called irrotational.

As we discussed in our earlier consideration of static fluids, we'll examine only the special case of incompressible fluids—that is, fluids in which changes in pressure do not affect the density of the fluid. Flowing water, blood, and in many situations, air, usually behave as incompressible fluids. All of these fluids *can* be compressed by exerting pressure on them, but in many practical situations the pressures are low enough that the compression can be ignored. So, for the remainder of our study of fluids, we will focus on ideal fluids, which are incompressible and have steady, inviscid, and irrotational flow. It is important to start with simpler models to develop understanding of the fundamentals of phenomena, but it is also important to know the limitations of those models!

Conservation of Mass Results in the Continuity Equation

No matter what other properties a moving fluid may have, it must obey the following restriction: Mass can be neither created nor destroyed as the fluid flows. Therefore, in a *steady* flow, if a certain mass of fluid flows *into* a given region of space (say, the interior of a certain segment of a pipe) in a given time interval, the same amount of mass must flow *out of* that region during that same time interval. This is called the *principle of continuity*. For example, if 1 kg of fluid flows into one end of a pipe each second, then 1 kg/s must flow out the other end. (If the principle of continuity were violated,

① During the time it takes this quantity of fluid to move a distance Δx_1...

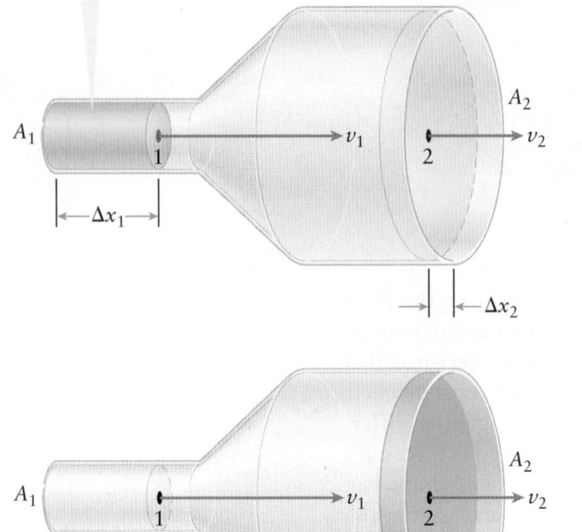

② ...this identical quantity of fluid moves a distance Δx_2. The amount of fluid between points 1 and 2 remains constant.

Figure 13-23 The continuity equation This illustration shows the flow of an incompressible fluid through a pipe of varying diameter. A slug of fluid of volume $A_1 \Delta x_1$ moving at speed v_1 enters the region between points 1 and 2, which causes a slug of fluid of volume $A_2 \Delta x_2$ moving at speed v_2 to exit the region. The continuity equation says that fluid enters the region at the same volume flow rate that it leaves the region.

the amount of fluid within the pipe would either increase or decrease. But in this case the flow wouldn't be steady as we originally assumed it to be.)

We can make a simple mathematical statement of the principle of continuity for the case of any *incompressible* fluid whether or not it is otherwise ideal. The principle of continuity tells us that in a given time interval the same *volume* of incompressible fluid must flow into a certain region of space as flows out of that region. **Figure 13-23** illustrates this principle for a pipe with a cross-sectional area that varies along its length. Let 1 and 2 be two different points along the pipe, and consider the segment of the pipe between these two points. While a quantity of fluid of volume $A_1 \Delta x_1$ enters the pipe segment at point 1, a quantity of volume $A_2 \Delta x_2$ exits the segment at point 2. The principle of continuity then tells us that

$$A_1 \Delta x_1 = A_2 \Delta x_2 \tag{13-17}$$

It takes some time interval Δt for the quantity of fluid to enter the region at point 1. And as the volume of fluid between the points must remain constant, the quantity of fluid at point 2 must exit the region in the same time interval. If we divide Equation 13-17 by the time interval Δt it takes for the quantities of fluid to enter or exit the region, we arrive at

$$A_1 \frac{\Delta x_1}{\Delta t} = A_2 \frac{\Delta x_2}{\Delta t} \tag{13-18}$$

To see why we divided through by Δt, note that the fluid at point 1 moves a distance Δx_1 during the time interval Δt and so has speed $v_1 = \Delta x_1 / \Delta t$. Similarly, the fluid at point 2 (which moves a distance Δx_2 during the same time interval) has speed $v_2 = \Delta x_2 / \Delta t$. So we can rewrite Equation 13-18 as the following relationship, called the **continuity equation:**

EQUATION IN WORDS
Continuity equation for steady flow of an incompressible fluid

Cross-sectional area of the flow at point 1 — Flow speed at point 1

$$A_1 v_1 = A_2 v_2 \tag{13-19}$$

Cross-sectional area of the flow at point 2 — Flow speed at point 2

In laminar flow, fluid enclosed by these streamlines remains in the enclosed volume.

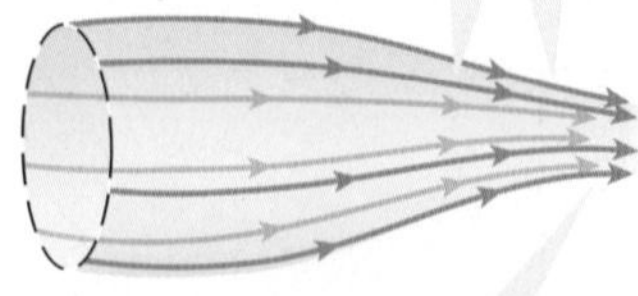

Fluid moves fastest where streamlines are closest together.

Fluid moves slowest where streamlines are far apart.

Figure 13-24 Streamline spacing and flow speed In laminar flow of an incompressible fluid, the spacing between streamlines tells you about the flow speed.

In other words, the product of the pipe's cross-sectional area A and the flow speed v has the same value at point 1 as at point 2. The choice of these points is quite arbitrary; they could be any two points along the pipe's length. So Equation 13-19 tells us that the product Av has the same value *everywhere* along the pipe. *Where a pipe is narrow, an incompressible fluid moves rapidly; where the pipe is broad, the fluid moves more slowly.* If the flow is laminar, the "pipe" doesn't have to be a solid object with walls; it can be a volume enclosed by a set of streamlines (**Figure 13-24**). It follows that *where streamlines are close together, an incompressible fluid moves rapidly; where streamlines are far apart, the fluid moves more slowly*. The airflow around the car shown in Figure 13-18a is mostly incompressible, so the flow is rapid over the roof of the car where streamlines are close together.

The continuity equation explains what happens when you put your thumb over the end of a garden hose. This action reduces the area through which water can flow out of the hose and so the water emerges at a faster speed. The same principle explains why there are often strong winds through mountain passes, where the air is forced into a "pipe" of narrow cross section.

Note that the product Av in Equation 13-19 has units of $(\text{m}^2)(\text{m/s}) = \text{m}^3/\text{s}$, or volume per unit time. This quantity, called the **volume flow rate**, tells you the number of cubic meters of fluid that pass a given point each second. For example, the average volume flow rate of blood through the aorta of a resting human is about 10^{-4} m^3/s, or 6 L/min. (This is an *average* volume flow rate because the flow is in pulses rather than steady.) So the continuity equation states that *the volume flow rate of an incompressible fluid moving through a pipe is the same at all points.*

This principle needs a little restatement if the pipe branches into a number of small pipes. An example is the human circulatory system, where the flow of oxygenated blood from the aorta is spread out into an enormous number of narrow capillaries (**Figure 13-25**). In this case the volume flow rate in the aorta is the same as the *combined* volume flow rate through all of the tiny capillaries. If the combined cross-sectional area of the capillaries, A_c, were the same as the cross-sectional area A_a of the aorta, blood would flow at the same speed throughout the system. In fact, in the human circulatory system A_c for the capillaries is greater than A_a for the aorta, so the flow speed in the capillaries is *slower* than in the aorta. This slower speed gives the blood in the capillaries more time for exchange of nutrients, gases, and waste products between the blood and the surrounding tissues.

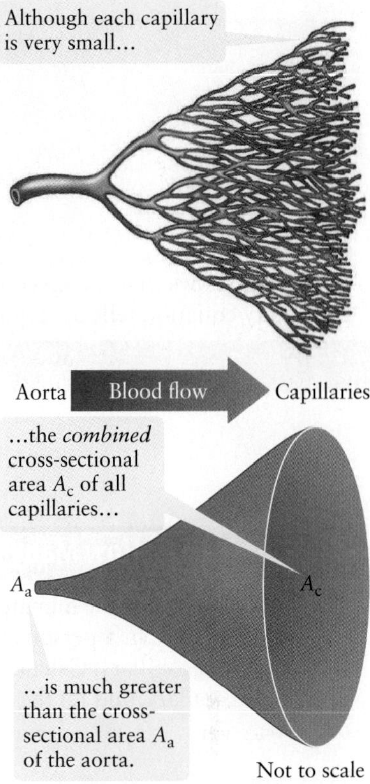
Although each capillary is very small...
Aorta
Blood flow
Capillaries
...the *combined* cross-sectional area A_c of all capillaries...
A_a
A_c
...is much greater than the cross-sectional area A_a of the aorta.
Not to scale
Hence the *speed* at which blood flows is much slower in the capillaries then in the aorta.

Figure 13-25 The human circulatory system Blood leaving the heart flows first through a single vessel, the aorta. Branches off the aorta eventually lead to billions of very small capillaries.

WATCH OUT !

The continuity equation is for incompressible flow only.

Note that Equation 13-19 applies only if the flow is *incompressible*, so its density is the same in all circumstances. In certain cases of fluid motion, however, the fluid does change density, and we say that the flow is *compressible*. Air flowing faster than sound behaves like a compressible fluid: When such fast-moving air enters a constriction such as a narrow pipe, it slows down and the molecules of the fluid get squeezed together so that the density of the fluid increases. In such a case Equation 13-19 doesn't apply. An analogy for a compressible fluid is how automobiles behave on a highway. You can think of the cars as "molecules" that make up a "fluid" that "flows" along the highway. In light traffic the cars in this "fluid" are far apart, but when this "fluid" passes through a narrower "pipe"—for example, a section of a two-lane highway where one lane is closed due to construction—the flow slows down, the cars get squeezed together, and a traffic jam results.

EXAMPLE 13-12 Flow in a Constriction

An incompressible fluid in a pipe of cross-sectional area 4.0×10^{-4} m^2 flows at a speed of 3.0 m/s. What is the flow speed in a part of the pipe that is constricted and has a cross-sectional area of 2.0×10^{-4} m^2?

Set Up

We are given the cross-sectional area and flow speed at one point in the pipe, and we wish to determine the flow speed at another point in the same pipe that has a different cross-sectional area. Because the fluid is incompressible, we can use the continuity equation, Equation 13-19.

Continuity equation for steady flow of an incompressible fluid:

$$A_1 v_1 = A_2 v_2 \quad (13\text{-}19)$$

point 1
area = 4.0×10^{-4} m^2
speed = 3.0 m/s

point 2
area = 2.0×10^{-4} m^2
speed = ?

Solve

Let point 1 be a location where the pipe has its full cross-sectional area. Then $A_1 = 4.0 \times 10^{-4}$ m^2 and $v_1 = 3.0$ m/s. Choose point 2 to be in the constriction, where $A_2 = 2.0 \times 10^{-4}$ m^2. Use the continuity equation to solve for v_2.

Rearrange Equation 13-19 to solve for v_2:

$$v_2 = \frac{A_1}{A_2} v_1$$

Substitute values:

$$v_2 = \frac{(4.0 \times 10^{-4}\ \text{m}^2)}{(2.0 \times 10^{-4}\ \text{m}^2)}(3.0\ \text{m/s})$$

$$= 6.0\ \text{m/s}$$

Reflect

Our result shows that the speed in the narrow constriction is faster than in the broad part of the pipe, just as the continuity equation tells us it must be.

NOW WORK Problems 3, 4, and 7–9 from The Takeaway 13-8.

EXAMPLE 13-13 How Many Capillaries?

The inner diameter of the human aorta is about 2.50 cm, while that of a typical capillary is about 6.00 μm = 6.00×10^{-6} m (see Figure 13-25). In a person at rest, the average flow speed of blood is about 20.0 cm/s in the aorta and about 1.00 mm/s in a capillary. Calculate (a) the volume flow rate (in m^3/s) of blood in the aorta, (b) the volume flow rate in a single capillary, and (c) the total number of open capillaries into which blood from the aorta is distributed at any one time.

Set Up

Figure 13-25 shows the situation. We are given the dimensions of the aorta and each capillary as well as the flow speed in each of these pipes. Our goal is to determine the volume flow rate in the aorta and in a capillary as well as the number of capillaries into which the aorta empties. The volume flow rate in a pipe equals its cross-sectional area times the speed of the fluid in the pipe. Like water, blood acts like an incompressible fluid. (It will compress appreciably only under pressures much higher than those found in the body.) So we can use the continuity equation: The volume flow rate through the aorta must be equal to the flow rate through all of the open capillaries combined.

Continuity equation for steady flow of an incompressible fluid:

$$A_1v_1 = A_2v_2 \tag{13-19}$$

1 = aorta

2 = all open capillaries combined

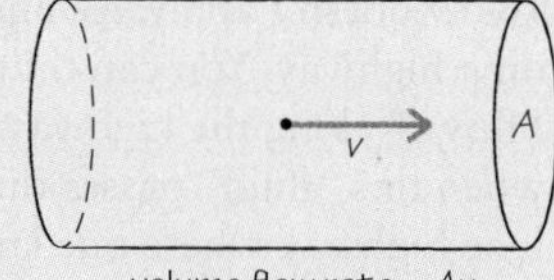

Solve

(a) The volume flow rate in the aorta is equal to the product of its cross-sectional area and the flow speed of aortal blood ($v_{\text{aorta}} = 20.0\ \text{cm/s} = 0.200\ \text{m/s}$).

Radius of aorta:

$$r_{\text{aorta}} = (1/2) \times (\text{diameter of aorta}) = (1/2) \times 2.50\ \text{cm}$$
$$= 1.25\ \text{cm} = 1.25 \times 10^{-2}\ \text{m}$$

Cross-sectional area of aorta:

$$A_{\text{aorta}} = \pi r_{\text{aorta}}^2 = \pi(1.25 \times 10^{-2}\ \text{m})^2$$
$$= 4.91 \times 10^{-4}\ \text{m}^2$$

Volume flow rate in aorta:

$$A_{\text{aorta}}v_{\text{aorta}} = (4.91 \times 10^{-4}\ \text{m}^2)(0.200\ \text{m/s})$$
$$= 9.82 \times 10^{-5}\ \text{m}^3/\text{s}$$

(b) Do the same calculations for a single capillary, in which the flow speed is $v_{capillary} = 1.00 \text{ mm/s} = 1.00 \times 10^{-3}$ m/s.

Radius of a capillary:

$$r_{capillary} = (1/2) \times (\text{diameter of capillary}) = (1/2) \times 6.00 \times 10^{-6} \text{ m} = 3.00 \times 10^{-6} \text{ m}$$

Cross-sectional area of a capillary:

$$A_{capillary} = \pi r_{capillary}^2 = \pi(3.00 \times 10^{-6} \text{ m})^2 = 2.83 \times 10^{-11} \text{ m}^2$$

Volume flow rate in a capillary:

$$A_{capillary} v_{capillary} = (2.83 \times 10^{-11} \text{ m}^2)(1.00 \times 10^{-3} \text{ m/s}) = 2.83 \times 10^{-14} \text{ m}^3\text{/s}$$

(c) Our results from (a) and (b) show that compared to the volume flow rate through a *single* capillary, the volume flow rate through the aorta is 3.47×10^9 times greater. The idea of continuity tells us that the combined volume flow rate through *all* the open capillaries must be equal to the volume flow rate through the aorta. We therefore learn the total number of open capillaries.

$$\frac{\text{volume flow rate in aorta}}{\text{volume flow rate in a capillary}} = \frac{9.82 \times 10^{-5} \text{ m}^3\text{/s}}{2.83 \times 10^{-14} \text{ m}^3\text{/s}} = 3.47 \times 10^9$$

(volume flow rate in aorta) = (total volume flow rate in all open capillaries combined) . . . so there must be 3.47×10^9 open capillaries.

Reflect

As a check on our results, note that the *combined* cross-sectional areas of all capillaries is 9.82×10^{-2} m^2, which is 200 times greater than the cross-sectional area of the aorta. By the continuity equation, the flow speed in the capillaries should therefore be *slower* than in the aorta by a factor of $1/(2.00 \times 10^2)$; that is, $v_{capillary} = v_{aorta}/(2.00 \times 10^2) = (0.200 \text{ m/s})/(2.00 \times 10^2) = 1.00 \times 10^{-3}$ m/s. This gives us back one of the numbers we started with, so our calculation is consistent.

Total cross-sectional area of all open capillaries combined:

$$A_{\text{all open capillaries}} = (3.47 \times 10^9) A_{capillary} = (3.47 \times 10^9)(2.83 \times 10^{-11} \text{ m}^2) = 9.82 \times 10^{-2} \text{ m}^2$$

$$\frac{\text{area of all open capillaries combined}}{\text{area of aorta}} = \frac{9.82 \times 10^{-2} \text{ m}^2}{4.91 \times 10^{-4} \text{ m}^2} = 2.00 \times 10^2$$

Extend: Our results show that the human circulatory system is truly extensive!

NOW WORK Problem 2 from The Takeaway 13-8.

EXAMPLE 13-14 From Capillaries to the Vena Cavae

Blood returns to the heart from the capillaries through two veins known as the *vena cavae*. The combined cross-sectional area of the vena cavae is 10.0 cm^2. At what average speed does blood move through these veins?

Set Up

We know the net volume flow rate of blood in the aorta from Example 13-13, and we're given the cross-sectional area $A_{vc} = 10.0$ cm^2 for the vena cavae. We'll use the continuity equation to find the flow speed of blood in the vena cavae, v_{vc}.

Continuity equation for steady flow of an incompressible fluid:

$$A_1 v_1 = A_2 v_2 \qquad (13\text{-}19)$$

1 = aorta

2 = vena cavae

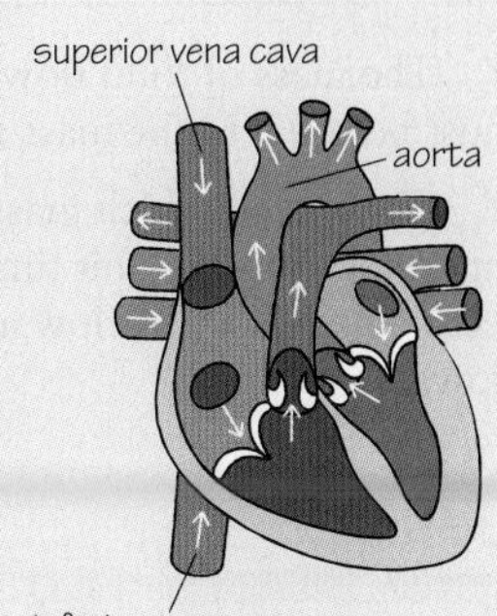

Solve

If we assume no blood volume is lost as it circulates through the body, the volume flow rate in the aorta is the same as in the vena cavae.

From Example 13-13:

$$\text{volume flow rate in aorta} = A_{\text{aorta}}v_{\text{aorta}}$$
$$= (4.91 \times 10^{-4}\ \text{m}^2)(0.200\ \text{m/s})$$
$$= 9.82 \times 10^{-5}\ \text{m}^3/\text{s}$$

volume flow rate in venae cavae = $A_{\text{vc}}v_{\text{vc}}$, where

$$A_{\text{vc}} = (10.0\ \text{cm}^2)\left(\frac{1\ \text{m}}{100\ \text{cm}}\right)^2$$
$$= 1.00 \times 10^{-3}\ \text{m}^2$$

The two volume flow rates are the same, so

$$A_{\text{aorta}}v_{\text{aorta}} = A_{\text{vc}}v_{\text{vc}}$$
$$v_{\text{vc}} = \frac{A_{\text{aorta}}v_{\text{aorta}}}{A_{\text{vc}}}$$
$$= \frac{9.82 \times 10^{-5}\ \text{m}^3/\text{s}}{1.00 \times 10^{-3}\ \text{m}^2}$$
$$= 0.0982\ \text{m/s} = 9.82\ \text{cm/s} = 98.2\ \text{mm/s}$$

Reflect

Compared to an average speed of 1.00 mm/s in the capillaries, blood moves 98.2 times faster in the vena cavae. The reason is that the cross-sectional area of the vena cavae is 1/98.2 as great as the combined cross-sectional area of the open capillaries (see Example 13-13). So the blood has to speed up as it returns to the heart to compensate for the decreased cross-sectional area.

Speed of blood in vena cavae compared to speed of blood in capillaries is

$$\frac{v_{\text{vc}}}{v_{\text{all open capillaries}}} = \frac{98.2\ \text{mm/s}}{1.00\ \text{mm/s}} = 98.2$$

Explanation: Cross-sectional area of vena cavae compared to cross-sectional area of all open capillaries combined is

$$\frac{A_{\text{vc}}}{A_{\text{all open capillaries}}} = \frac{1.00 \times 10^{-3}\ \text{m}^2}{9.82 \times 10^{-2}\ \text{m}^2} = \frac{1}{98.2}$$

Extend: It's a common misconception that the slow-moving blood in the capillaries continues to move slowly as it returns to the heart. This example shows that this isn't the case! Note that blood does flow more slowly in the vena cavae (0.0982 m/s) than in the aortae (0.200 m/s). That's because the venae cavae have a larger cross-sectional area than the aortae.

NOW WORK Problems 1 and 6 from The Takeaway 13-8.

THE TAKEAWAY for Section 13-8

✔ The mass of fluid flowing into a location per unit time must be equal to the mass flowing out of that location.

✔ The rate at which mass of an incompressible fluid flows into a location per unit time is proportional to the cross-sectional area of the flow and the speed at which the fluid flows.

✔ An ideal fluid model works best for fluids that are incompressible, with steady, irrotational flows and in situations in which viscosity can be neglected.

✔ The continuity equation for fluid flow describes the conservation of mass flow rate for incompressible, steady flow fluids: The product of the cross-sectional area of the flow and the speed of the flow is constant.

Prep for the AP® Exam

AP® Building Blocks

EX 13-14 1. A river runs through a wide valley and then through a narrow channel. How do the velocities of the flows of water compare between the wide valley and the narrow channel?

EX 13-13 2. A hose is connected to a faucet and used to fill a 5.0-L container in a time of 45 s.

(a) Determine the volume flow rate in m^3/s.

(b) Determine the speed of the water in the hose in part (a) if the hose has a radius of 1.0 cm.

(c) If instead of a hose you used drinking straws of radius 0.5 cm to fill the container, how many drinking straws would it take to fill the container in the same amount of time as the hose in (b)?

EX 13-12 3. Calculate the speed of the water leaving the 7.5-mm-diameter nozzle of a hose with a volume flow rate of 0.45 m^3/s.

EX 13-12 4. Determine the time required for a 50.0-L container to be filled with water when the speed of the incoming water is 25.0 cm/s and the cross-sectional area of the hose carrying the water is 3.00 cm^2.

AP® Skill Builders

5. While walking past a construction site, you notice a pipe sticking out of a second-floor window, with water rushing out. As the water flows to the ground, it must speed up due to the effect of gravity. How does the diameter of the flowing stream of water change as it descends? Briefly explain your prediction using physics principles.

EX 13-14 6. You inject your patient with 2.50 mL of medicine. If the inside diameter of the 31-gauge needle is 0.114 mm and the injection lasts 0.650 s, determine the average speed of the fluid as it leaves the needle.

EX 13-12 7. A cylindrical blood vessel is partially blocked by the buildup of plaque. At one point, the plaque decreases the diameter of the vessel by 60.0%. The blood approaching the blocked portion has speed v_0. Find the speed of the blood as it enters the blocked portion of the vessel, in terms of v_0.

EX 13-12 8. The return-air ventilation duct in a home has a cross-sectional area of 9.0×10^2 cm^2. The air in a room with dimensions 7.0 m × 10.0 m × 2.4 m is to be completely circulated in a 30-min cycle. What is the speed of the air in the duct?

AP® Skills in Action

EX 13-12 9. In July 1995 a spillway gate broke at the Folsom Dam in California. During the uncontrolled release the flow rate through the gate peaked at 40,000 ft^3/s, and about 1.35 billion gallons of water were lost (nearly 40% of the reservoir). Estimate the time that the gate was open.

10. The flow of water in a pipe of radius 0.500 mm = 5.00×10^{-4} m has an average flow speed of 0.800 m/s.

(a) If the radius is reduced to 0.125 mm = 1.25×10^{-4} m, calculate the new average flow speed.

(b) Given this radius of the second pipe, is there any reason to question using the ideal fluid model? Briefly justify your answer.

(c) Briefly explain if your answer to (b) affects the accuracy of your calculation in (a).

13-9 Bernoulli's equation, an expression of the work-energy theorem, helps us relate pressure and speed in fluid motion

Hold a piece of notebook paper by two corners so that the paper droops downward. Then blow on the top of the paper as shown in **Figure 13-26**. You might expect that the force of the air expelled from your mouth would push the paper downward. But remarkably, the paper actually lifts *up* where it is struck by the moving air.

What's happened is that by making air move over the top of the paper, you've lowered the pressure of that air. Hence there is greater air pressure on the underside of the paper than on the top, and the paper lifts up. This simple experiment illustrates that certain kinds of fluid flow have a property described by **Bernoulli's principle**: In a moving fluid, the pressure is low where the fluid is moving rapidly. This principle was first identified by the eighteenth-century mathematician Daniel Bernoulli.

Bernoulli's principle explains why an open door may swing closed on a windy day. The pressure in the moving air outside the house is lower than the pressure of the still air inside the house. The difference in air pressure on the two sides of the door pulls the door toward the outside, slamming it shut. An umbrella bulges upward in the wind for the same reason; there is low-pressure, fast-moving air on top of the umbrella, but high-pressure still air in the space underneath.

Figure 13-27 illustrates the origins of Bernoulli's principle. A pipe of varying diameter carries an incompressible fluid from left to right. According to the continuity

Blowing on the top of a sheet of paper lowers the pressure there.
Higher pressure below pushes the paper up.

Figure 13-26 Bernoulli's principle A piece of paper lifts up if you blow over the top of the paper.

equation that we discussed in Section 13-8, the fluid moves fastest in the narrow part of the pipe (point 2). Hence a parcel of fluid must speed up as it enters the narrow part. A net force must be exerted on the fluid to change its velocity (that is, cause it to accelerate). This net force must be to the right as a parcel of fluid enters the narrow part. If we assume that there is negligible viscosity (remember, ideal fluid model!), the only forces are those due to differences in pressure on the left and right sides of the parcel. As Figure 13-27 shows, to produce the required acceleration, the pressure must be lower in the narrow part of the pipe (where the fluid moves rapidly) than in the wide part (where the fluid moves slowly). This is just what Bernoulli's principle says: The pressure is lowest where the fluid moves the fastest.

WATCH OUT!

Pressure differences cause velocity changes, not the other way around.

It's *not* correct to say that the pressure differences in Figure 13-27 are *caused* by the changing velocity of the fluid. In fact, just the reverse is true: The changes in fluid velocity are caused by the pressure differences! Recall the meaning of Newton's second law: An object accelerates (that is, changes its velocity) in response to a net force exerted on it. In other words, a change in velocity is a result of a net force, not the other way around. In the same way, the meaning of Bernoulli's principle is that a fluid undergoes a change in velocity as a result of pressure differences, not the reverse. As we begin to look at changes in height, we will see it is easier to think of this force in terms of work, and use conservation of energy. But just as with our study of kinematics, a force or energy approach can be used.

Fluid must move faster at point 2 than at point 1 to satisfy the continuity equation.

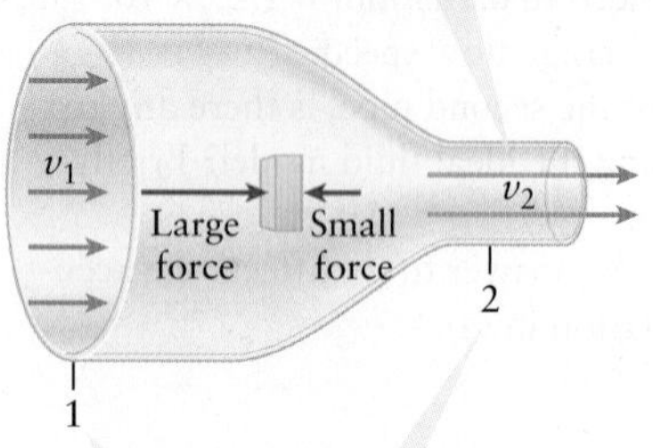

To make the fluid speed up between point 1 and point 2, there must be a higher pressure at 1 than at 2.

Figure 13-27 Interpreting Bernoulli's principle Pressure differences drive a fluid through a constriction. The larger pressure on the left means that a parcel of fluid entering the constriction has a stronger force pushing it to the right than to the left, so the parcel speeds up as it approaches the constriction.

Bernoulli's Equation

Let's expand on these ideas and see how to express Bernoulli's principle as a rather simple and useful equation. By seeing where this equation comes from, we'll be able to understand why Bernoulli's principle holds only under certain special conditions. Many kinds of flow satisfy these conditions, but many others—including, for example, the flow of blood in the circulatory system—do not.

The shaded volume in **Figure 13-28a** shows a quantity of a fluid in motion. If we assume that the flow is *laminar*, we can think of this quantity of fluid as being enclosed within streamlines just as though it were flowing inside a pipe. As time goes by this quantity of fluid moves along its "pipe," displacing other fluid at its front end and being displaced by fluid at its back end. **Figure 13-28b** shows our quantity of fluid at a brief time interval Δt after the instant shown in Figure 13-28a. The fluid has vacated a volume $A_1 \Delta x_1$ and has moved into a volume $A_2 \Delta x_2$.

Let's further assume that the flow is *steady* so that the fluid velocity at any fixed position in the fluid remains the same. In this case the motion of the shaded volume of

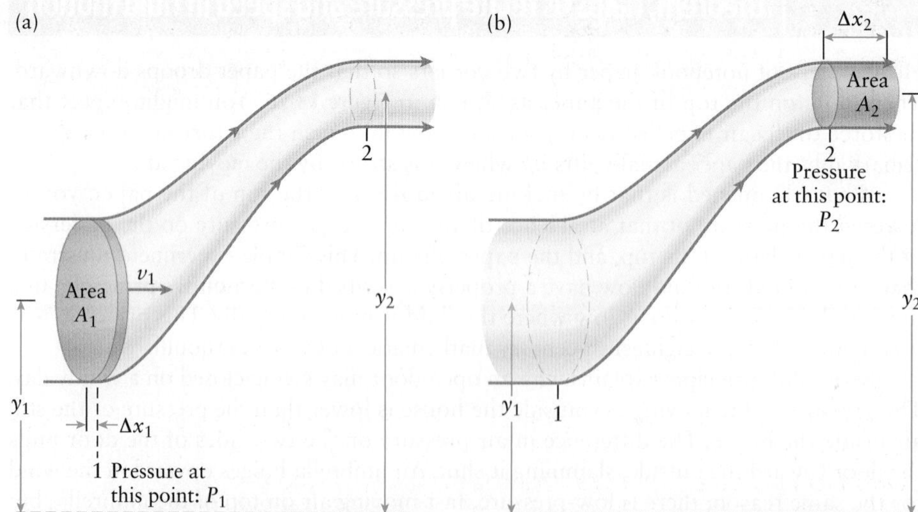

Figure 13-28 Deriving Bernoulli's equation (a) A slug of incompressible fluid of volume $A_1 \Delta x_1$ moving at speed v_1 takes a time Δt to enter the region between points 1 and 2. (b) During this same time Δt, a slug of fluid of volume $A_2 \Delta x_2$ moving at speed v_2 exits the region. The continuity equation says that fluid enters the region at the same rate that it exits. (Compare with Figure 13-27.)

fluid from the situation in Figure 13-28a to that in Figure 13-28b is the same as if the fluid had "lost" the volume $A_1\Delta x_1$ at its back end and "gained" the volume $A_2\Delta x_2$ at its front end, with no other changes. If we make an additional assumption that the fluid is *incompressible*—so that its density ρ is the same at all points in the fluid—then the volume "lost" at the back end must be the same as the volume "gained" at the front end:

$$\text{volume "lost"} = A_1\Delta x_1 = A_2\Delta x_2 = \text{volume "gained"} \tag{13-20}$$

While the volume of the moving incompressible fluid remains the same, its *energy* can change. Over the time interval Δt the shaded volume of fluid has lost the kinetic energy and gravitational potential energy associated with the "lost" volume at point 1 but has gained the kinetic and potential energies associated with the "gained" volume at point 2. The total energy change associated with the shaded volume of fluid is

$$\Delta E = \Delta K + \Delta U \tag{13-21}$$

We can write the mass of this volume of fluid as $\rho A_1\Delta x_1$ (when it is at point 1) or $\rho A_2\Delta x_2$ (when it is at point 2). We'll use the symbols v_1 and v_2 to denote the fluid speed at points 1 and 2, respectively, and use the symbols y_1 and y_2 for the heights at points 1 and 2, respectively. Then, using the familiar formulas $K = \frac{1}{2}mv^2$ for kinetic energy and $U = mgy$ for gravitational potential energy of the fluid–Earth system, we can write ΔK and ΔU in Equation 13-21 as

$$\Delta K = \text{change in kinetic energy} = +\frac{1}{2}(\rho A_2\Delta x_2)v_2^2 - \frac{1}{2}(\rho A_1\Delta x_1)v_1^2 \tag{13-22}$$

$$\Delta U = \text{change in gravitational potential energy} = +(\rho A_2\Delta x_2)gy_2 - (\rho A_1\Delta x_1)gy_1 \tag{13-23}$$

The law of conservation of energy informs us that the total energy of a system (in this case, Earth and the shaded volume of fluid) changes only when work is done on it by external forces. The total mechanical energy of a closed system changes only when there is a nonconservative interaction within the system, so we'll assume that there's no friction of any kind and that the fluid is *inviscid*. Then the only forces that are exerted on the shaded volume of fluid are forces from the pressure of the surrounding fluid. There is pressure on all sides of the shaded volume of fluid, but work is done only by those forces that are exerted on the *moving* parts of the fluid. Just like in our study of buoyancy, the other forces all tend to cancel. Unlike our study of buoyancy, we will no longer need to consider the vertical difference in forces, because that will be taken care of by the gravitational potential energy term, because we put Earth in the system. From Figure 13-28 we can determine that the forces (pressure times area) on the back and front ends of the shaded volume of fluid are P_1A_1 and P_2A_2, respectively. The force on the back end pushes in the same direction as the displacement Δx_1 and so does positive work; the force on the front end does negative work because it pushes opposite to the displacement Δx_2. Hence, the total work done on the shaded volume of fluid in Figure 13-28 is

$$W = +(P_1A_1)\Delta x_1 - (P_2A_2)\Delta x_2 \tag{13-24}$$

Now we can put all the pieces together.

$$E_{\text{total}} = \Delta K + \Delta U = W$$

or

$$\begin{aligned} &+\frac{1}{2}(\rho A_2\Delta x_2)v_2^2 - \frac{1}{2}(\rho A_1\Delta x_1)v_1^2 + (\rho A_2\Delta x_2)gy_2 - (\rho A_1\Delta x_1)gy_1 \\ &= +(P_1A_1)\Delta x_1 - (P_2A_2)\Delta x_2 \end{aligned} \tag{13-25}$$

Equation 13-25 looks like a horrible mess. But because we assumed the fluid is incompressible, $A_1\Delta x_1$ is equal to $A_2\Delta x_2$ (see Equation 13-20). Hence, we can divide out all the factors of $A_1\Delta x_1$ and $A_2\Delta x_2$, leaving the simpler expression

$$+\frac{1}{2}\rho v_2^2 - \frac{1}{2}\rho v_1^2 + \rho g y_2 - \rho g y_1 = P_1 - P_2$$

Rearranging this expression, we derive **Bernoulli's equation**.

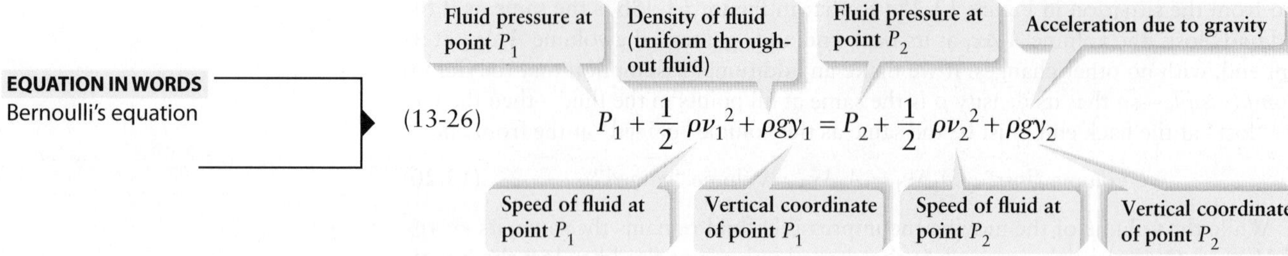

$$P_1 + \frac{1}{2}\rho v_1^2 + \rho g y_1 = P_2 + \frac{1}{2}\rho v_2^2 + \rho g y_2 \tag{13-26}$$

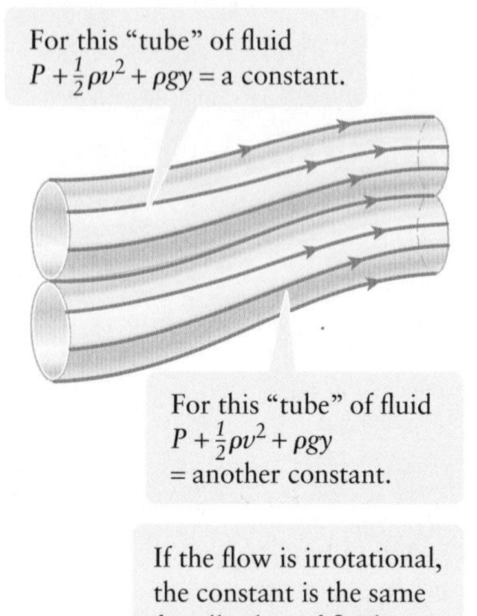

For this "tube" of fluid $P + \frac{1}{2}\rho v^2 + \rho g y$ = a constant.
For this "tube" of fluid $P + \frac{1}{2}\rho v^2 + \rho g y$ = another constant.
If the flow is irrotational, the constant is the same for all tubes of fluid.

Figure 13-29 Irrotational flow and Bernoulli's equation Two adjacent volumes of fluid, drawn as tubes. If the flow is irrotational, the quantity $p + \frac{1}{2}\rho v^2 + \rho g y$ $p + \frac{1}{2}\rho v^2 + \rho g y$ has the same value for both tubes (adjacent volumes) and for *all* parts of the fluid.

Bernoulli's equation states that the quantity $P + \frac{1}{2}\rho v^2 + \rho g y$ has the same value at point 1 (the back end of the shaded volume of fluid) as at point 2 (the front end). But we could have chosen the back and front ends to be *anywhere* along the length of the shaded volume of fluid in Figure 13-28. So, the quantity $P + \frac{1}{2}\rho v^2 + \rho g y$ must have the same value at *any* point within the streamlines that enclose this shaded volume. Because we are assuming an ideal fluid, the fluid flow shown in Figure 13-28 is irrotational. Remember for ideal fluids our approximation is that the fluid has some viscosity but not so much that we cannot ignore it. *Some* viscosity must be present to make the flow irrotational. If we assume irrotational flow, which, as shown in **Figure 13-29**, implies the speed of the water does not change in adjacent volumes, the quantity $P + \frac{1}{2}\rho v^2 + \rho g y$ has the same value in the shaded volume in Figure 13-28 as it does in any adjacent volume. It also has the same value in the volume adjacent to the other side of that adjacent volume, and so on. So, our ideal fluid has the same value of $P + \frac{1}{2}\rho v^2 + \rho g y$ at *all* points in the fluid. Put another way, points 1 and 2 in Bernoulli's equation could be any two points in the fluid, without consideration of streamlines.

Note that if the fluid is at rest, so that $v = 0$ at all points, Equation 13-26 becomes $P_1 + \rho g y_1 = P_2 + \rho g y_2$; that is, as y decreases and you go deeper in a static fluid, the quantity $\rho g y$ decreases and the pressure P increases. This is just the relationship for pressure at various depths in a static fluid that we found in Section 13-4. Note also that if we compare two points at the same height y, Equation 13-26 tells us that the pressure P is high where the fluid speed v is low—which is just the statement of Bernoulli's principle that we made at the beginning of this section.

To derive Bernoulli's equation, we had to make several approximations: The fluid flow had to be *laminar*, *steady*, *incompressible*, *inviscid*, and *irrotational*. In many real-life situations these assumptions aren't valid. For example, the flow of water down a waterfall is turbulent, not laminar; viscosity is important for blood flow in capillaries; and the air flowing around a supersonic airplane undergoes substantial compression. But there are a number of situations where Bernoulli's equation gives very good results. In the examples that follow we'll examine how to use Bernoulli's equation in some of these situations.

EXAMPLE 13-15 Lift on a Wing

Figure 13-30 shows a computer simulation of air flowing around an airplane wing (seen end-on). As the figure shows, air flows faster over the top of the wing so that there is lower pressure on the top of the wing than on the bottom. Suppose the air (density 1.20 kg/m^3) moves at 75 m/s past the lower surface of a small airplane's wing and at 85 m/s past the upper surface. If the area of the wing as seen from above is 10.0 m^2 and the top-to-bottom thickness of the wing is 7.0 cm, what is the overall upward force (the *lift*) that the air exerts on the wing?

Figure 13-30 A computer simulation of airflow around a wing A vertical column of parcels of air (shown by colored dots) starts at the left and moves to the right, flowing around a wing. The parcels passing over the top of the wing go faster than those passing below the bottom of the wing (they're spaced farther apart horizontally), so there must be lower pressure on the top. (The dots are colored red in regions of low pressure.)

Set Up

The pressure on the bottom of the wing exerts an upward force, and the pressure on the top exerts a downward force. Our goal is to calculate the *combined* vertical force that these pressures exert on the wing. We use Bernoulli's equation to relate the pressure on the two surfaces of the wing. Pressure is force per area, so the force on either surface of the wing is the pressure multiplied by the wing area. All of the forces we are interested in are vertical, so we can just use a single component with ± to describe the direction of vectors.

Bernoulli's equation:

$$P_1 + \frac{1}{2}\rho v_1^2 + \rho g y_1 = P_2 + \frac{1}{2}\rho v_2^2 + \rho g y_2 \qquad (13\text{-}26)$$

Definition of pressure:

$$P = \frac{F_\perp}{A} \qquad (13\text{-}2)$$

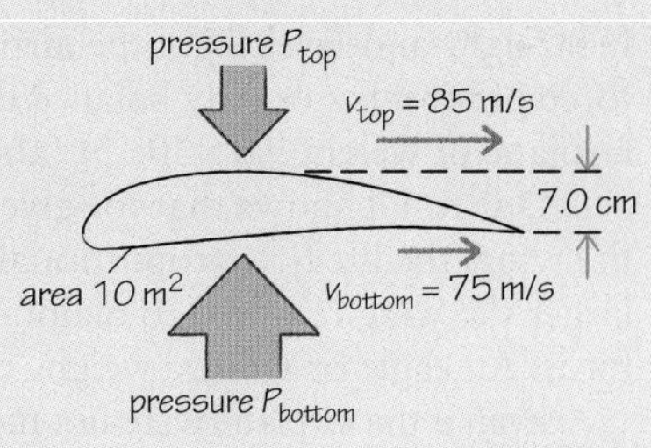

Solve

We want to calculate the net upward force on the wing, or the lift L. Let P_{bottom} and P_{top} be the pressures on the bottom and top of the wing, respectively. Express L in terms of these pressures and the wing area A.

From Equation 13-2 the upward force on the bottom of the wing is

$$F_{bottom} = P_{bottom}A$$

and the downward force on the top of the wing is

$$F_{top} = -P_{top}A$$

The vector sum of the forces on the wing (the lift):

$$\begin{aligned} F_L &= F_{bottom} + F_{top} \\ &= P_{bottom}A - P_{top}A \\ &= (P_{bottom} - P_{top})A \end{aligned}$$

Use Bernoulli's equation to write an expression for the pressure difference $P_{bottom} - P_{top}$:

$$P_{bottom} + \frac{1}{2}\rho v_{bottom}^2 + \rho g y_{bottom} = P_{top} + \frac{1}{2}\rho v_{top}^2 + \rho g y_{top}$$

Solve for $P_{bottom} - P_{top}$:

$$P_{bottom} - P_{top} = \frac{1}{2}\rho v_{top}^2 - \frac{1}{2}\rho v_{bottom}^2 + \rho g y_{top} - \rho g y_{bottom}$$

Calculate the pressure difference using $\rho = 1.20\ \text{kg/m}^3$, $v_{top} = 85$ m/s, $v_{bottom} = 75$ m/s, and $y_{top} - y_{bottom} = 7.0\ \text{cm} = 7.0 \times 10^{-2}$ m (the top-to-bottom thickness of the wing).

$$\begin{aligned} \frac{1}{2}\rho v_{top}^2 - \frac{1}{2}\rho v_{bottom}^2 &= \frac{1}{2}\rho(v_{top}^2 - v_{bottom}^2) \\ &= \frac{1}{2}(1.20\ \text{kg/m}^3)[(85\ \text{m/s})^2 - (75\ \text{m/s})^2] \\ &= 9.6 \times 10^2\ \text{kg/(m}\cdot\text{s}^2) = 9.6 \times 10^2\ \text{N/m}^2 \end{aligned}$$

$$\begin{aligned} \rho g y_{top} - \rho g y_{bottom} &= \rho g(y_{top} - y_{bottom}) \\ &= (1.20\ \text{kg/m}^3)(9.80\ \text{m/s}^2)(7.0 \times 10^{-2}\ \text{m}) \\ &= 0.82\ \text{kg/(m}\cdot\text{s}^2) = 0.82\ \text{N/m}^2 \end{aligned}$$

so

$$\begin{aligned} P_{bottom} - P_{top} &= 9.6 \times 10^2\ \text{N/m}^2 + 0.82\ \text{N/m}^2 \\ &= 9.6 \times 10^2\ \text{N/m}^2 \text{ to two significant figures} \end{aligned}$$

Calculate the lift F_L by multiplying the pressure difference by the wing area.

$$\begin{aligned} F_L &= (P_{bottom} - P_{top})\,A = (9.6 \times 10^2\ \text{N/m}^2)(10.0\ \text{m}^2) \\ &= 9.6 \times 10^3\ \text{N} \end{aligned}$$

Reflect

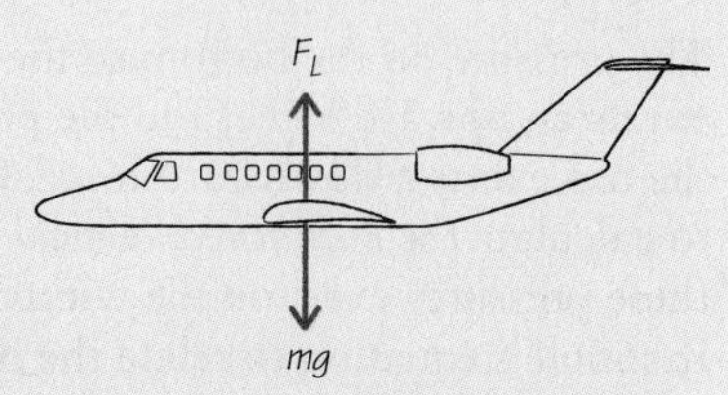

In straight-and-level flight the airplane is not accelerating vertically, so the upward force of lift must exactly balance the weight of the airplane. So our wing can keep an airplane of weight 9.6×10^3 N (about 2200 lb) in the air.

Our result shows that for given speeds of airflow along the top and bottom of the wing, the lift F_L is proportional to the wing area A. The heavier the airplane, the larger the wing required to maintain flight. The same principle applies to gliding birds: An eagle or vulture weighs more than a hawk and so has a larger wing.

Even if the airplane were not moving, there would still be a small pressure difference of $\rho g y_{top} - \rho g y_{bottom} = 0.82\ \text{N/m}^2$ between the upper and lower surfaces of the wing. This pressure difference means that the air exerts a small buoyant force on the wing. Our calculations show that this small pressure difference is totally negligible compared to the pressure difference $\frac{1}{2}\rho v_{top}^2 - \frac{1}{2}\rho v_{bottom}^2 = 9.6 \times 10^3\ \text{N/m}^2$ due to air traveling faster past the top of the wing than past the bottom of the wing.

NOW WORK Problems 5 and 7 from The Takeaway 13-9.

WATCH OUT !

Air molecules don't meet at the trailing edge of a wing.

Why does air travel at different speeds over the two surfaces of the wing? A common misconception is that air traveling along the top of the wing takes just as much time to go from the wing's leading edge to its trailing edge as does the air that travels along the bottom. According to this misconception, air has to move faster along the top of the wing because the upper surface is curved more than the lower, and molecules that parted company at the leading edge must somehow meet up at the trailing edge. But the computer simulation in Figure 13-30 shows that this isn't the case at all. Air travels over the upper surface of the wing *much* faster than the common misconception would have us believe, and the molecules *don't* meet up at the trailing edge. A better explanation is that air follows the curvature of the wing due to its (slight) viscosity. This curvature is such that the wing pushes air downward. By Newton's third law the air must push the wing *upward* equally hard. To produce this upward force, or lift, there has to be a pressure difference between the two surfaces of the wing, and this pressure difference is what causes the air to flow at different speeds over the top and bottom of the wing.

EXAMPLE 13-16 A Venturi Meter

Figure 13-31 shows a simple device called a *Venturi meter* for measuring fluid velocity in a gas such as air. When gas passes from left to right through the horizontal pipe, it speeds up as it passes through the constriction at point 2. Bernoulli's principle tells us that the gas pressure must be lower at point 2 than at point 1, and the pressure difference causes the liquid in the U-tube to drop on the left-hand side and rise on the right-hand side. Suppose the gas is air (density $1.20\ \text{kg/m}^3$) that enters the left-hand side of the Venturi meter at 25.0 m/s. The horizontal tube has cross-sectional area $2.00\ \text{cm}^2$ at point 1 and cross-sectional area $1.00\ \text{cm}^2$ at point 2. If the liquid in the U-tube is water, what is the difference in height between the water columns on the left-hand and right-hand sides of the tube?

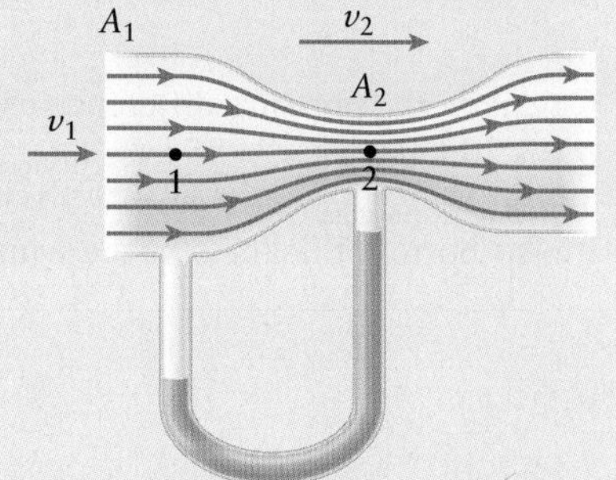

Figure 13-31 A Venturi meter A Venturi meter, or flow meter, can be used to measure the flow speed of a gas.

Set Up

In this problem there are *two* fluids, the air that flows through the horizontal pipe and the water in the U-tube. Hence, we'll use Bernoulli's equation twice: once to relate the moving air at point 1 to the moving air at point 2, and once to relate the heights of the water on the two sides of the U-tube (which is what we're trying to find). We'll also use the continuity equation to relate the speeds of the air at points 1 and 2.

Bernoulli's equation:

$$P_1 + \frac{1}{2}\rho v_1^2 + \rho g y_1 = P_2 + \frac{1}{2}\rho v_2^2 + \rho g y_2 \qquad (13\text{-}26)$$

Continuity equation for steady flow of an incompressible fluid:

$$A_1 v_1 = A_2 v_2 \qquad (13\text{-}19)$$

relate properties of air here and here

1

2

relate properties of water here and here

Solve

We need to find the difference in air pressure at points 1 and 2, as this is what causes the difference in height of the water on the two sides of the U-tube. Find this difference using Bernoulli's equation and the continuity equation applied to the air in the horizontal pipe, keeping in mind that we know the values of v_1, A_1, and A_2.

Bernoulli's equation for the air at points 1 and 2:

$$P_1 + \frac{1}{2}\rho_{\text{air}}v_1^2 + \rho_{\text{air}}gy_1 = P_2 + \frac{1}{2}\rho_{\text{air}}v_2^2 + \rho_{\text{air}}gy_2$$

At the center of the horizontal pipe, $y_1 = y_2$, so

$$P_1 + \frac{1}{2}\rho_{\text{air}}v_1^2 = P_2 + \frac{1}{2}\rho_{\text{air}}v_2^2$$

$$P_1 - P_2 = \frac{1}{2}\rho_{\text{air}}v_2^2 - \frac{1}{2}\rho_{\text{air}}v_1^2$$

From the continuity equation:

$$v_2 = \frac{A_1}{A_2}v_1 \text{ so}$$

$$P_1 - P_2 = \frac{1}{2}\rho_{\text{air}}(v_2^2 - v_1^2) = \frac{1}{2}\rho_{\text{air}}\left[\left(\frac{A_1}{A_2}\right)^2 v_1^2 - v_1^2\right]$$

$$= \frac{1}{2}\rho_{\text{air}}v_1^2\left[\left(\frac{A_1}{A_2}\right)^2 - 1\right]$$

$$= \frac{1}{2}(1.20\ \text{kg/m}^3)(25.0\ \text{m/s})^2\left[\left(\frac{2.00\ \text{cm}^2}{1.00\ \text{cm}^2}\right)^2 - 1\right]$$

$$= 1.13 \times 10^3\ \text{kg/(m} \cdot \text{s}^2) = 1.13 \times 10^3\ \text{N/m}^2$$

The pressure difference between points 1 and 2 is also the pressure difference between the water at two points: the top of the water column on the left-hand side of the U-tube and the water column on the right-hand side. The water is at rest ($v = 0$) at both points. Use this technique to find the height difference between the two water columns.

Bernoulli's principle for the water at the tops of the two columns:

$$P_1 + \frac{1}{2}\rho_{\text{water}}v_{\text{water},1}^2 + \rho_{\text{water}}gy_{\text{water},1}$$

$$= P_2 + \frac{1}{2}\rho_{\text{water}}v_{\text{water},2}^2 + \rho_{\text{water}}gy_{\text{water},2}$$

Water is at rest: $v_{\text{water},1} = v_{\text{water},2} = 0$

Solve for the height difference of the two water columns:

$$P_1 + \rho_{\text{water}}gy_{\text{water},1} = P_2 + \rho_{\text{water}}gy_{\text{water},2}$$

$$\rho_{\text{water}}gy_{\text{water},2} - \rho_{\text{water}}gy_{\text{water},1} = P_1 - P_2$$

$$y_{\text{water},2} - y_{\text{water},1} = \frac{P_1 - P_2}{\rho_{\text{water}}g} = \frac{1.13 \times 10^3\ \text{kg/(m} \cdot \text{s}^2)}{(1.000 \times 10^3\ \text{kg/m}^3)(9.80\ \text{m/s}^2)}$$

$$= 0.115\ \text{m} = 11.5\ \text{cm}$$

Reflect

This height difference is large enough to easily measure, so the Venturi meter is a practical device.

We made an approximation in our solution that the pressure difference between points 1 and 2 within the horizontal pipe is the same as the pressure difference between the two water columns. This isn't exactly correct because there is a greater weight of air above the column on the left-hand side. However, the resulting additional pressure difference is so small that we can ignore it.

Additional pressure difference between the tops of the two water columns due to the extra weight of air on the left-hand side:

$$\rho_{\text{air}}g(y_{\text{water},2} - y_{\text{water},1}) = (1.20\ \text{kg/m}^3)(9.80\ \text{m/s}^2)(0.115\ \text{m})$$

$$= 1.35\ \text{kg/(m} \cdot \text{s}^2) = 1.35\ \text{N/m}^2$$

This is 0.12% of the pressure difference calculated above ($P_1 - P_2 = 1.13 \times 10^3\ \text{N/m}^2$), so we can neglect it.

NOW WORK Problems 4 and 6 from The Takeaway 13-9.

Applications of Bernoulli's Principle

Can a thrown baseball be a "curveball"—that is, can it be made to follow a path that curves left or right? In the early days of baseball, most people thought a curveball was just an optical illusion. It wasn't until 1941, when *Life* magazine published photographs of a curveball taken with a strobe light, that baseball fans (and everyone else) had proof that a properly thrown baseball can be made to curve. Bernoulli's principle helps to explain why.

Imagine a baseball thrown without any spin, what ball players call a knuckleball. As the ball flies, the air rushes past the left and right sides of the ball at the same speed. Because the speed of the air is the same on both sides, according to Bernoulli's equation the air pressure in the air is also the same on both sides. Hence there is no tendency for the ball's trajectory to curve either right or left.

Now imagine the ball spinning as it moves (**Figure 13-32**). In this figure we're looking at the ball from above, and it is rotating counterclockwise as seen from this vantage point. The ball is moving in the direction shown by the arrow, and air rushes past it in the opposite direction. Because the surface of the ball is rough, and because the baseball has raised seams that hold the leather cover of the ball together, viscosity drags a layer of air around in the same direction as the spin of the ball (compare Figure 13-22). As a result, the net speed of air relative to the ball is slower on the right-hand side of the ball than on the left-hand side. Bernoulli's principle tells us that a speed difference corresponds to a pressure difference, and the slower speed to the right of the ball means that there is higher pressure there. This pressure difference means that as viewed from above, this baseball feels a force to the left and will curve to the left!

As the ball rotates counterclockwise (in this picture) a layer of air close to the ball is also dragged around counterclockwise. This boundary layer moves in the same direction as the air rushing past the ball on the left but opposes the airflow on the right.

Motion of ball from the pitcher toward the batter

Motion of air relative to the ball

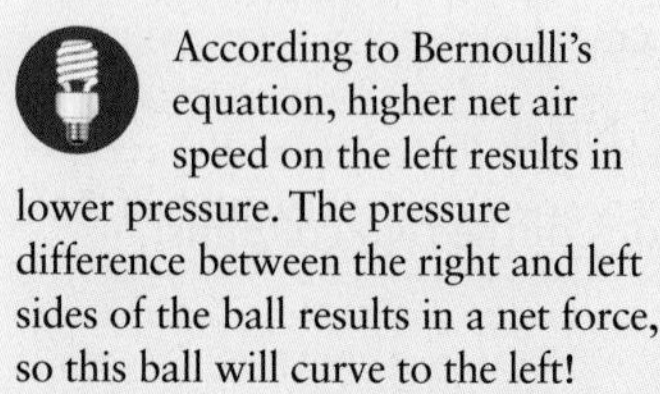

According to Bernoulli's equation, higher net air speed on the left results in lower pressure. The pressure difference between the right and left sides of the ball results in a net force, so this ball will curve to the left!

Figure 13-32 What makes a curveball curve If a baseball rotates, viscosity drags a layer of air around the ball in the same direction as the ball's rotation.

Bernoulli's principle can help baseball pitchers win games; it also helps fish such as mackerel stay alive. A mackerel swims with its mouth open, allowing water to enter through the mouth, pass through the gills where oxygen is extracted, and exit through an aperture called the *operculum*. The streamlines of water flow around the mackerel to follow the contours of its body. **Figure 13-33** shows that these streamlines are close together near the operculum. As we learned in Section 13-8, flow in an incompressible fluid such as water is rapid where streamlines are close together but slow where the streamlines are far apart. Hence the water flow is *faster* at the operculum than at the mouth, and so according to Bernoulli's principle the pressure at the operculum is *lower* than at the mouth. This pressure difference helps to force water from mouth to operculum by way of the mackerel's gills. This effect, called *ram ventilation*, is essential to the mackerel's ability to extract sufficient oxygen from the water. Other species of fish use ram ventilation only at high swimming speeds; at lower speeds, where ram ventilation is less effective, they use muscular action to pump water through their gills.

Streamlines close together: Water is fast-moving and pressure is low.

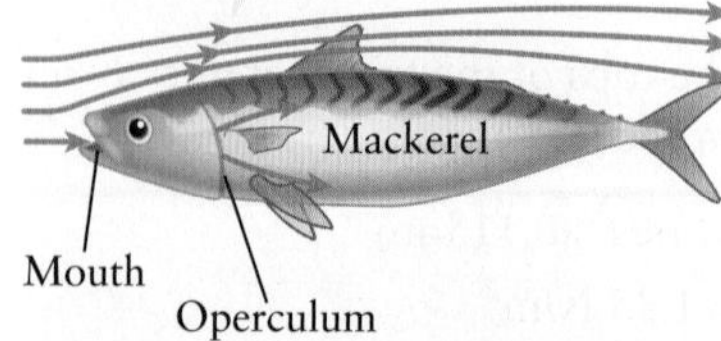

Figure 13-33 Bernoulli's principle and fish respiration Mackerel breathe by letting water flow from the mouth through the gills (where oxygen is extracted from the water) and out the operculum. Bernoulli's principle tells us that the water pressure at the operculum is lower when the fish is in motion, thus facilitating flow through the gills.

A special case of Bernoulli's equation, called Torricelli's theorem, can be used to determine the speed of a liquid exiting a small opening in the side of a container. It works for the special case of an opening much smaller than the horizontal cross-section of the container. For such a scenario, for an ideal fluid the speed of the fluid is the same as that of an object that had dropped the same distance that the opening is below the surface of the fluid.

We can show that this result is valid by applying Bernoulli's equation with the first point taken at the liquid's surface, and the second point just outside the opening in the side of the container. When the area of the opening is very small relative to the horizontal cross-sectional area of the container (the area of the surface of the liquid) and we apply the continuity equation, we see that the velocity of the surface, $v_{\text{surface}} = v_{\text{opening}}\left(\frac{A_{\text{opening}}}{A_{\text{surface}}}\right)$, can be assumed to be negligible. The pressure on the top of the water and the water just outside the opening is atmospheric pressure. Because the changes in atmospheric pressure are not very large unless we make the container exceptionally tall, the pressure at both points can be assumed to be equal, so substituting this same pressure into Bernoulli's theorem, we get:

$$\frac{1}{2}\rho v_{\text{opening}}^2 + \rho g y_{\text{opening}} = \frac{1}{2}\rho v_{\text{surface}}^2 + \rho g y_{\text{surface}} \tag{13-27}$$

Torricelli's Theorem

Limitations of Bernoulli's equation

Bernoulli's equation holds true only with certain very special assumptions. For the equation we derived to be correct, we need the fluid to be ideal. Neglecting viscosity turns out to be critical to the model, as we did not include a nonconservative work term. However, it turns out fluids must have a little bit of viscosity to make the laminar flow be irrotational. But we couldn't allow the fluid to have too much viscosity, as that would mean that *viscous forces*—that is, forces on fluid due to friction—would play an important role in determining the acceleration of a bit of moving fluid.

For many situations involving fluid flow, there is "too much" viscosity, and Bernoulli's equation is *not* valid. One example is water flowing in a garden hose (**Figure 13-34**). The hose has the same cross-sectional area A throughout its length, so to satisfy the continuity equation (which states that the product Av is a constant), the flow speed v must be the same at all points along the hose's length. Furthermore, if the hose is horizontal, then all points along the hose are at the same height y. If Bernoulli's equation applied to this situation, then the water pressure P would also have to have the same value everywhere inside the hose. But in fact the pressure at the open end of the hose is *lower* than at the end attached to the faucet. The reason is that frictional forces oppose the flow of water through the hose, so there has to be additional pressure at the faucet to sustain the flow.

The human circulatory system is another situation where Bernoulli's equation doesn't hold. The speed of blood flow in capillaries (about 10^{-3} m/s) is hundreds of times slower than in the aorta (about 0.2 m/s), so the term $\frac{1}{2}\rho v^2$ in Equation 13-26 has a much smaller value in the capillaries. Bernoulli's equation would therefore predict that blood pressure is much higher in the capillaries than in the aorta. In fact, the pressure is higher in the *aorta*; this is necessary to push against the frictional forces that the blood encounters on its way to the capillaries.

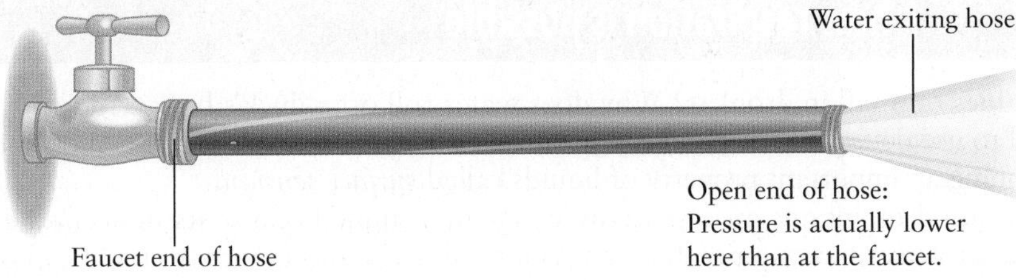

Figure 13-34 When Bernoulli's equation doesn't work A horizontal garden hose is attached to a faucet. If Bernoulli's equation held true, the water pressure would be the same everywhere along the length of the hose. Bernoulli's equation does not apply in such a case because viscous forces are important in this situation. Although we often choose to ignore viscous forces for systems like hoses, we cannot neglect viscosity at other times, such as when modeling the human circulatory system.

THE TAKEAWAY for Section 13-9

✔ Bernoulli's equation relates the pressure, speed, and height of an ideal fluid.

✔ Bernoulli's equation is the result of applying the work-energy theorem to an ideal fluid that is flowing with negligible resistance.

✔ Bernoulli's equation becomes an approximation when fluid resistance increases, in the same way neglecting friction or air resistance impacts kinematic calculations.

AP Building Blocks

1. Imagine holding two pieces of paper vertically, with a small gap between them, and then blowing gently into the gap. Would you expect the pieces of paper to (A) be drawn together, (B) be pushed apart, or (C) be unaffected? (Try it!) Justify your answer in terms of physics principles.
2. Use Bernoulli's equation to explain why a house roof is easily blown off during a tornado or hurricane.
3. A cylindrical container is filled with water. If a hole is cut on the side of the container so that the water shoots out, what is the direction of the water flow the instant it leaves the container?

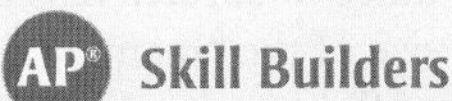

4. At one point, Hurricane Katrina had maximum sustained winds of 175 mi/h (282 km/h) and a low pressure in the eye of 676.52 mmHg (0.890 atm). **(a)** Using the given air speed, determine the pressure predicted by Bernoulli's equation and calculate the percentage of the predicted value that the measured value is. **(b)** Explain why the value in (a) does not equal 100%. Assume the pressure of the air is normally 1.00 atm, 1.013 kPa.

EX 13-15 5. When the atmospheric pressure is 1.00 atm, a water fountain ejects a stream of water that rises to a height of 5.00 m. There is a 1.00-cm-radius pipe that leads from a pressurized tank to the opening that ejects the water. Calculate the height of the fountain if it were operational while the eye of a hurricane passed by. Assume that the atmospheric pressure in the eye is 0.877 atm and the tank's pressure remains the same.

Skills in Action

EX 13-16 6. A cylinder that is 20 cm tall and open on top is filled with water as shown in the figure. If a hole is made in the side of the cylinder, 5.0 cm below the top level, how far from the base of the cylinder will the stream land? Assume that the cylinder is large enough so that the level of the water in the cylinder does not drop significantly.

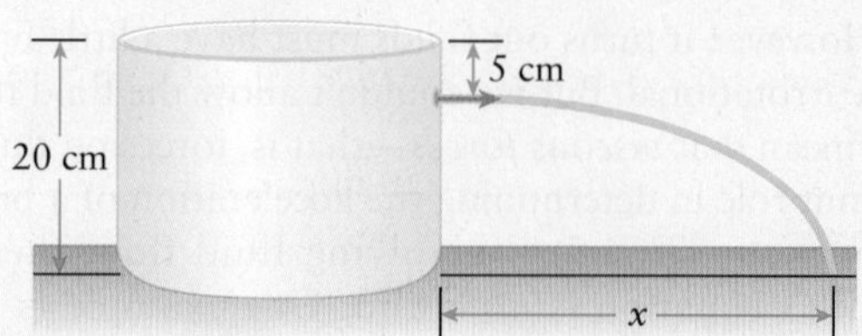

EX 13-15 7. Water flows from a fire truck through a hose that is 11.7 cm in diameter and has a nozzle that is 2.00 cm in diameter. The firefighters stand on a hill 5.00 m above the level of the truck. When the water leaves the nozzle, it has a speed of 20.0 m/s. Determine the minimum gauge pressure in the truck's water tank. Assume the water has no kinetic energy in the tank, and that the hose is large enough to ignore viscous forces.

AP® Exam Tip

This information will not be directly tested on the AP® Physics 1 Exam. It is content that many find interesting, and it does give an additional insight into how energy is important in fluid behavior.

13-10 Surface tension explains the shape of raindrops and how respiration is possible

Why does rain fall in droplets? Why does water roll off a duck's back? And why is it useful to use detergent when washing clothes? We can answer all of these questions by examining an important property of liquids called *surface tension*.

A liquid maintains an essentially constant volume because its molecules exert strong attractive forces on each other. As a result the molecules try to huddle close to each other. The potential energy associated with these forces is lowest if as many of the molecules as possible are completely surrounded by other molecules, and if as few as possible are on the surface (where they are surrounded by other molecules on one side only). Therefore, to minimize this potential energy, the molecules arrange themselves so as to minimize the surface area of the liquid (**Figure 13-35**). A sphere has the smallest surface area for a given volume, which is why raindrops are spherical. The drop behaves as though its surface were a membrane that resists being stretched, which is why this behavior is called **surface tension**.

WATCH OUT !

Raindrops aren't shaped like "teardrops."

It's a common belief that raindrops are teardrop-shaped, with a rounded lower half and a pointed upper half. But the aerodynamic forces that would force a raindrop into this shape are quite weak compared to the attractive forces between molecules. In fact, due to surface tension, raindrops are nearly spherical in shape.

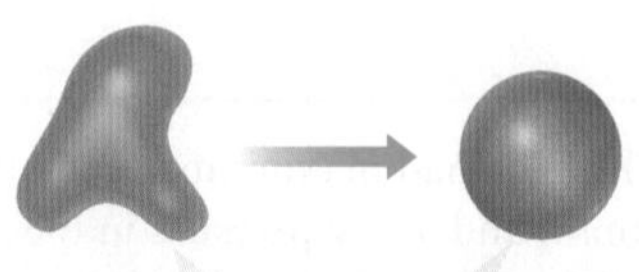

An irregular blob of liquid tends to become spherical due to surface tension.

Figure 13-35 Surface tension The effect of surface tension on a liquid blob or drop.

A spherical drop of radius R has volume $(4\pi/3)R^3$ and surface area $4\pi R^2$. The importance of surface tension is proportional to the ratio of the drop's surface area (which is the size of the "membrane" enclosing the drop) to its volume (which tells you the total amount of fluid in the drop). This ratio is $4\pi R^2/[(4\pi/3)R^3] = 3/R$, which *increases* as the radius R *decreases*. Hence surface tension is more important for small drops than for large ones. If you pour enough liquid into a container, surface tension is overwhelmed by gravitational forces and the liquid assumes a shape that fills the container and has a horizontal upper surface. In zero gravity even large quantities of liquid form spherical drops (**Figure 13-36**).

The surface tension of water enables ducks to remain warm and dry even after submerging themselves in search of a meal. A duck's feather has a crisscross grid of tiny barbules that form a fine mesh (**Figure 13-37a**). To push through this mesh the water surface must stretch as shown in **Figure 13-37b**. Surface tension resists such stretching of the surface "membrane," so water does not penetrate the mesh, and the duck's skin remains dry.

The same effect makes it difficult for water to penetrate between the closely spaced fibers of clothing. That's why clothing is best washed using a detergent, which is a substance that reduces the surface tension of water.

Surface tension also plays an important role in human respiration. In the lungs oxygen exchange takes place across the walls of tiny, balloon-like structures called *alveoli* (**Figure 13-38**). The inside surface of the alveoli is wet, and the surface tension of this fluid generates a pressure that tends to pull the walls of the alveoli inward. The magnitude of this pressure is proportional to the surface tension and inversely proportional to the radius of the alveolus:

$$\text{pressure} = (2 \times \text{surface tension})/(\text{alveolar radius})$$

The problem is that the alveoli do not all have the same radius, so smaller alveoli experience a greater pressure than larger ones. As a result, small alveoli would tend to collapse into larger ones if not for the presence of a *surfactant*, which is a substance that coats the inside surface of each alveolus. This remarkable substance decreases surface tension as a function of its concentration in the fluid lining the inside of the alveoli. As the radius of the alveolus decreases during exhalation, the concentration of surfactant increases on the surface of the fluid lining the alveolus and therefore decreases surface tension and the resulting pressure in the alveolus. This prevents smaller alveoli from collapsing into larger ones. Also, during inhalation the reduced surface tension makes it easier to inflate the lungs with the relatively small pressure difference between inhaled air and the low-pressure air in the thoracic cavity—about 130 Pa, or 0.13% of atmospheric pressure. A condition called respiratory distress syndrome occurs in premature infants whose lungs are deficient in surfactant; the increased surface tension pressure can make their alveoli collapse. Administering a replacement surfactant helps these newborn babies survive.

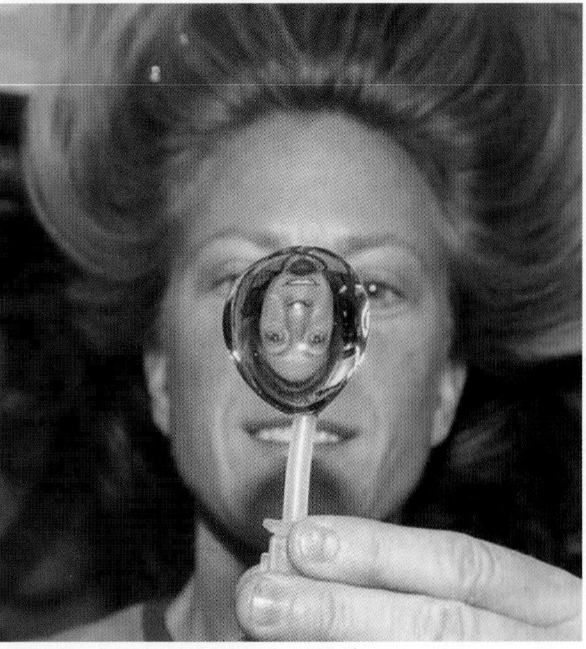
NASA

Figure 13-36 Surface tension on an apparently weightless water drop In the apparent weightlessness of the orbiting International Space Station, surface tension allows this astronaut to squeeze water out of her beverage container and form a spherical drop about 3 cm in diameter.

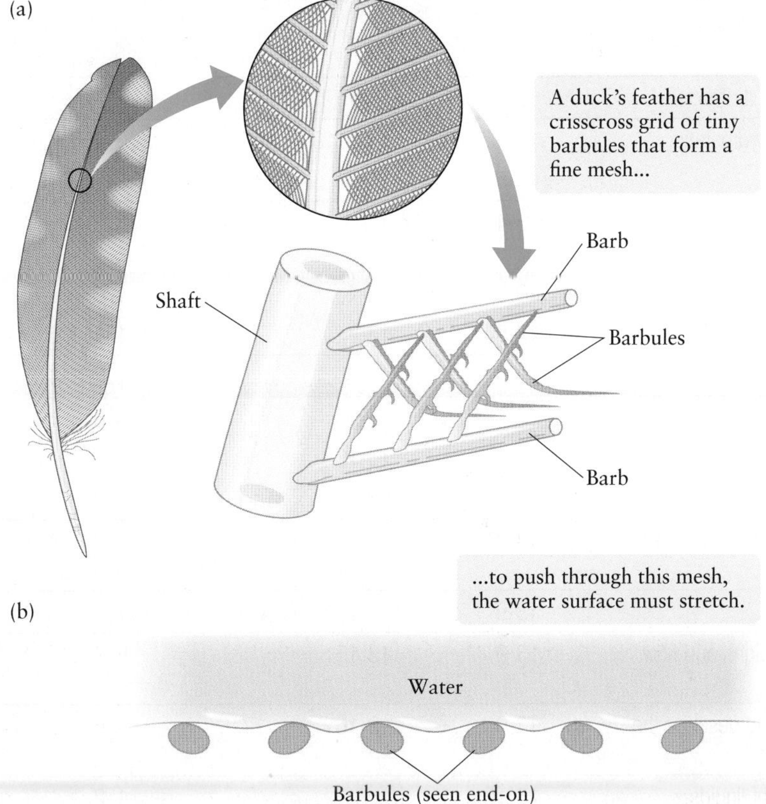
(a)
A duck's feather has a crisscross grid of tiny barbules that form a fine mesh...
Barb
Shaft
Barbules
Barb
...to push through this mesh, the water surface must stretch.
(b)
Water
Barbules (seen end-on)

Figure 13-37 Surface tension and bird feathers (a) This drawing of a bird feather shows the tiny interlocking barbules. (b) Due to surface tension, water cannot easily pass through the tiny spaces between barbules.

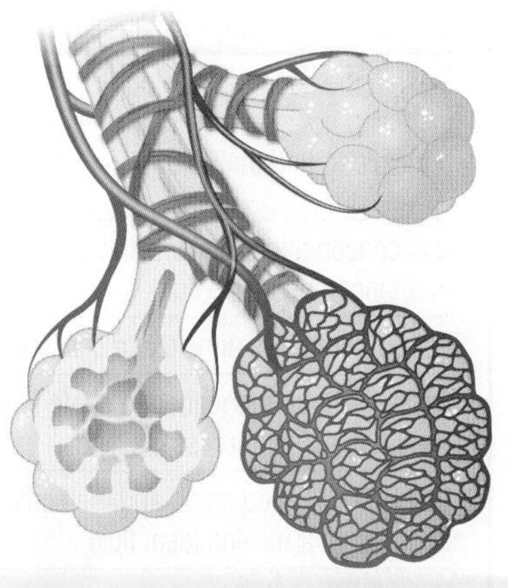

Figure 13-38 Alveoli This illustration shows a cluster of the tiny balloon-like sacs called alveoli. A human lung contains about 3×10^8 alveoli.

THE TAKEAWAY for Section 13-10

✔ The molecules of a liquid minimize their energy by surrounding themselves with their fellow molecules. As a result, liquids tend to form spherical drops.

✔ The surface of a liquid drop resists being stretched. This behavior is called surface tension.

Skill Builders

1. We did not include Earth in the system when discussing minimizing potential energy in this section. Briefly explain why we can discuss just the potential energy of a raindrop when talking about its shape, but need to include Earth in the system to use potential energy to calculate the change in velocity of a raindrop as it falls.

2. Calculate the ratio of surface area to volume for (a) a sphere of radius R, (b) a cube with side R, and (c) a rectangular box of height $2R$ with square ends of side R. (d) Using your answers from (a), (b), and (c), briefly discuss the statement that a spherical shape minimizes surface area for a given volume.

WHAT DID YOU LEARN?

Chapter learning goals	Section(s)	Related example(s)	Prep for the AP Exam: Relevant section review exercises
Describe the similarities and differences in the properties of liquids and gases and how these relate to the interactions of their constituent molecules or atoms.	13-1		2
Apply the definition of density to predict the densities, differences in densities, or changes in densities under different conditions for natural phenomena and design an investigation to verify the prediction.	13-2	13-1, 13-2, 13-3	1, 3, 4, 5, 6, 7, 10
Explain the origin of fluid pressure in terms of molecular motion.	13-3	Figure 13-3	15
Calculate the pressure at a given depth in a fluid in hydrostatic equilibrium.	13-4	13-5, 13-6	3, 4, 6, 7, 8, 11
	13-5	13-8	2
Explain the difference between absolute pressure and gauge pressure.	13-4		5, 9, 12
Calculate the force on an object due to a difference in pressure on its sides.	13-3	13-4	1, 6
	13-5	13-7	1, 4, 5, 6
Explain how to apply Pascal's principle to a fluid at rest.	13-6		1, 2, 3, 4, 5
Explain the source of the buoyant force on an object in a fluid, including how that determines its direction, and use Archimedes' principle to find that force.	13-7	13-9	5, 6, 8, 10
		13-10	11
		13-11	7, 9
Use conservation of mass (the continuity equation) to analyze the flow of an incompressible fluid.	13-8	13-12, 13-13, 13-14	1, 2, 3, 4, 5, 6, 7, 8, 9, 10
Relate Bernoulli's principle to conservation of energy.	13-9		1
Apply Bernoulli's equation to relate fluid pressure, height, and flow speed in an incompressible fluid.	13-9	13-15	5, 7
Use Bernoulli's equation and the continuity equation to describe the behavior of a moving ideal fluid.	13-9	13-16	4, 6
Describe the role of surface tension in the behavior of liquids.	13-10		1, 2

Chapter 13 Review

Key Terms

All the Key Terms can be found in the Glossary/Glosario on page G1 in the back of the book.

Chapter Summary

Topic	Equation or Figure
Fluids: The molecules in a fluid are not arranged in any organized structure. A fluid does not have any shape of its own, and flows to conform to the shape of the container in which it is placed. A quantity of liquid maintains nearly the same volume, while a quantity of gas expands to fill the volume available.	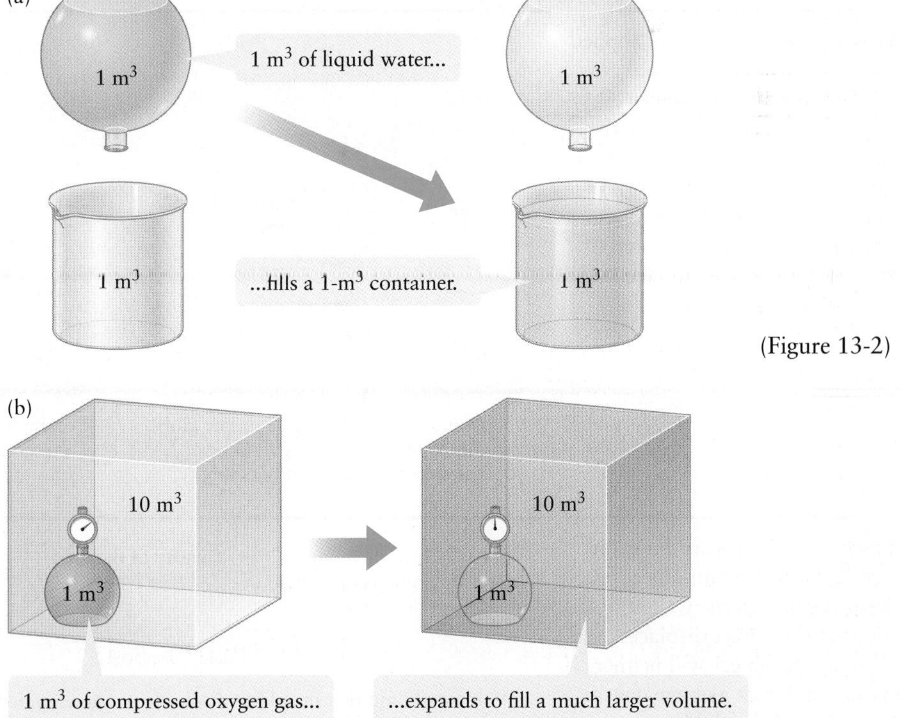 (Figure 13-2)

Density: The density ρ of a substance is its mass per volume. Density depends on how closely packed the molecules of the substance are, as well as on the mass of the individual molecules.

The **density** ρ ("rho") of a certain substance... ...equals the **mass** m of a given quantity of that substance... ...divided by the **volume** V of that quantity of the substance.

$$\rho = \frac{m}{V} \tag{13-1}$$

Pressure: The pressure P exerted by a fluid is the perpendicular force it exerts on a given area in the fluid divided by the area. Unlike force, pressure is not a vector. Pressure measured relative to atmospheric pressure is called gauge pressure.

The pressure P exerted by a fluid on an object in the fluid... ...equals the **force** $F_\perp$ that the fluid exerts **perpendicular to the face of the object...** ...divided by the **area** A of that face.

$$P = \frac{F_\perp}{A} \tag{13-2}$$

Fluids at rest: A stationary fluid must be in hydrostatic equilibrium so that the sum of the forces on any parcel of fluid is zero. In a fluid of uniform density, the pressure at a given point increases with depth to support the weight of the fluid above that point.

Pressure at a certain point in a **fluid at rest**

Density of the fluid (same at all points in the fluid)

Pressure at a second, lower point in the fluid

Acceleration due to gravity

$$P = P_0 + \rho g d \tag{13-8}$$

Depth of the second point where the pressure is P below the point where the pressure is P_0

Pressure difference: If there are different pressures on the two sides of an object, the different pressures exert a net force on the object.

A **pressure difference** on two opposite sides of an object **produces a net force.**

Area of each side of the object

$$F_{\text{net}} = (P_2 - P_1)A \tag{13-10}$$

High pressure on one side of the object

Lower pressure on the other side of the object

Pascal's principle: Pressure applied to a confined, static fluid is transmitted undiminished to every part of the fluid as well as to the walls of the container. A hydraulic jack, which converts the work of a small force exerted over a large distance into an equal amount of work done by a large force exerted over a small distance, relies on Pascal's principle.

(a)

A blob of fluid within a fluid at rest
Surrounding fluid
Combined upward force on the blob due to the surrounding fluid just balances the weight of the blob.

(b)

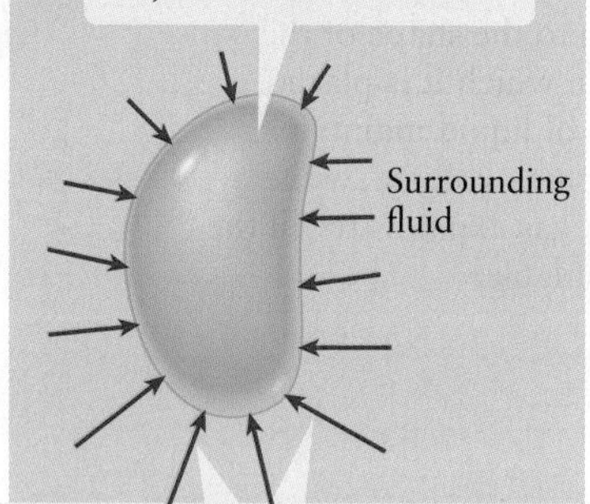
Replace the blob with an object of the same volume.
Surrounding fluid
Combined upward force on the object is the same as on the fluid blob in (a).

(Figure 13-13)

Buoyancy: The upward buoyant force F_b exerted on an object by a fluid is equal to the weight of fluid that the object displaces. A submerged object will neither sink nor rise if its average density equals that of the fluid.

Magnitude of the buoyant force on an object in a fluid

Density of the fluid

$$F_b = \rho_{\text{fluid}} V_{\text{displaced}}\, g \tag{13-13}$$

Volume of fluid that the object displaces

Acceleration due to gravity

Fluid flow: The motion of a fluid is classified as being steady or unsteady; as laminar or turbulent; and as viscous (with internal friction) or inviscid (without internal friction). An ideal fluid flow approximation is an incompressible fluid in a steady, laminar flow with a negligible amount of viscosity, just enough to make the fluid irrotational (no abrupt jumps in flow without obstacles).	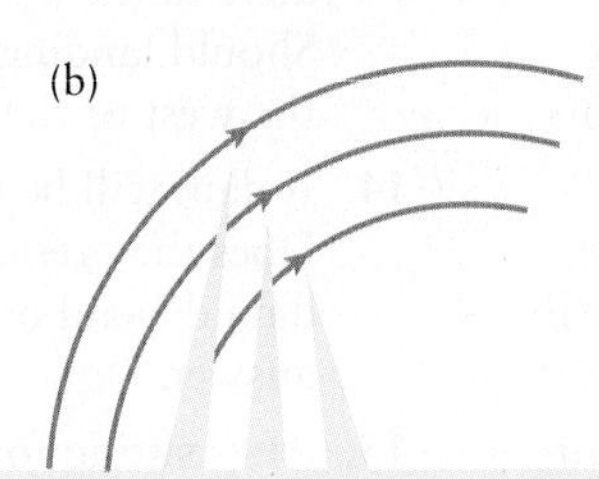 Streamlines are the paths followed by bits of fluid in laminar flow. Each bit of fluid follows the one in front and moves along the same streamline. 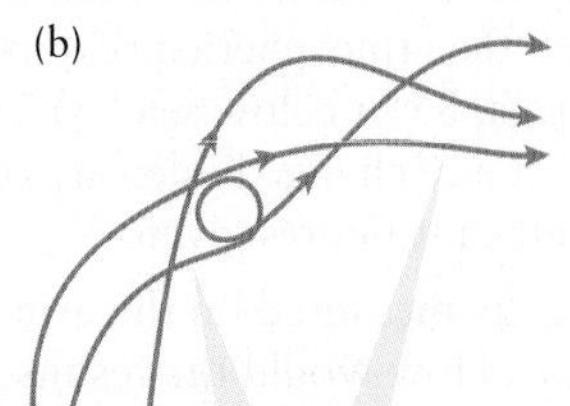 There are no streamlines for a fluid in turbulent flow. While each bit of fluid follows its own path, the one behind it can follow a completely different path.	(Figure 13-20b) (Figure 13-21b)
Continuity equation: If an incompressible (constant-density) fluid flows through a pipe, the size of the pipe and the flow speed may change, but because mass is conserved the volume flow rate (the cross-sectional area of the flow multiplied by the flow speed) remains the same.	Cross-sectional area of the flow at point 1; Flow speed at point 1 $$A_1v_1 = A_2v_2$$ Cross-sectional area of the flow at point 2; Flow speed at point 2	(13-19)
Bernoulli's principle and Bernoulli's equation: The pressure in a moving fluid is reduced at locations where the fluid moves rapidly. The mathematical statement of this principle, called Bernoulli's equation, is accurate only for idealized laminar, inviscid flows. This mathematical relationship is a result of the conservation of energy, applying the work-energy theorem.	Fluid pressure at point P_1; Density of fluid (uniform throughout fluid); Fluid pressure at point P_2; Acceleration due to gravity $$P_1 + \frac{1}{2}\rho v_1^2 + \rho g y_1 = P_2 + \frac{1}{2}\rho v_2^2 + \rho g y_2$$ Speed of fluid at point P_1; Vertical coordinate of point P_1; Speed of fluid at point P_2; Vertical coordinate of point P_2	(13-26)
Surface tension: Attractive interactions between the molecules of a liquid make it favorable for the liquid to minimize its surface area, reducing the potential energy stored in the configuration of the liquid. This effect causes water to form droplets and makes it difficult to force water through small apertures.	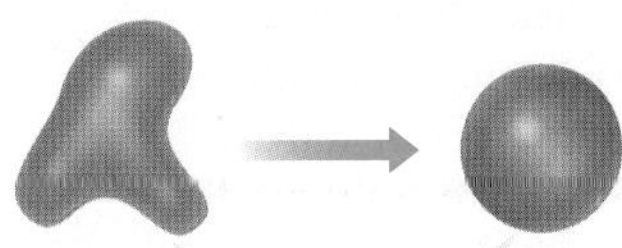 An irregular blob of liquid tends to become spherical due to surface tension.	(Figure 13-35)

Chapter 13 Review Problems

1. Aluminum is more dense than plastic. You are given two identical, closed and opaque boxes, which are designed to also muffle sound. One contains a piece of aluminum, and the other contains a piece of plastic. Without opening the boxes is it possible to tell which box contains the plastic and which box contains the aluminum? Explain your answer.

2. Blood flows through an artery that is partially blocked. As the blood moves from the wider region into the narrow region, the blood speed
(A) increases.
(B) decreases.
(C) stays the same.
(D) drops to zero.

3. Assuming the atmospheric pressure is exactly 1 atm at sea level, determine the atmospheric pressure in Death Valley, California, 85 m below sea level. Air is light enough to neglect any change in density over this depth within the significant figures given.

4. Blood pressure is usually measured on the arm at the same level as the heart. How would the results differ if the measurement were made on the leg instead?

5. What is the *absolute* pressure in pascals (Pa) of the air inside a bicycle tire that is inflated to 65 psi?

6. Using physics principles, explain why a stream of water ejected from a fountain becomes wide as it rises.

7. A hybrid car travels about 50.0 mi/gal of gasoline. The density of gasoline is 737 kg/m^3 and 1 gal equals 3.788 L. Express the car's mileage in miles per kilogram (mi/kg) of gas.

8. In 2009 a person in Italy won a lottery prize of 146.9 million euros, which at the time was worth 211.8 million U.S. dollars. Gold at the time was worth $953 per troy ounce, and silver was worth $14.16 per troy ounce. A troy ounce is 31.1035 g, the density of gold is 19.3 g/cm^3, and the density of silver is 10.5 g/cm^3.
   (a) If the lucky lottery winner opted to be paid in a single cube of pure gold, how high would the cube be?
   (b) Find the height of a silver cube of the same value.
   (c) Could a person carry the cube of gold you found in (a)? Why or why not?

9. A regulation men's basketball has an inflated circumference of 75 cm and an uninflated mass of 623.69 g. What is the average density of the basketball when it is inflated with air at an absolute pressure of 1.544 atm? Air at this pressure has a density of 1.89 kg/m^3. You may assume the vinyl shell has negligible thickness.

10. The salinity of the Great Salt Lake varies from place to place ranging from about two to as much as eight times the salinity of ocean water. Is it easier or harder for a person to float in the Great Salt Lake compared to floating in ocean water? Why?

11. A wooden boat floats in a small pond, and the level of the water at the edge of the pond is marked. Will the level of the water rise, fall, or stay the same when the boat is removed from the pond?

12. You are given two objects of identical size, one made of aluminum and the other of lead. You hang each object separately from a spring balance. Because lead is denser than aluminum, the lead object weighs more. Now you weigh each object while it is submerged in water. Will the difference between the measured weights of the aluminum object and the lead object be greater than, less than, or the same as it was when the objects were weighed in air? Explain your answer.

13. When landing, airplanes maintain a certain speed relative to the air. The wind is blowing from west to east. Should landing airplanes approach the runway from the west or the east? Briefly justify your answer.

14. A dam will be built across a river to create a reservoir. Does the pressure in the reservoir at the base of the dam depend on the shape of the dam? Explain your answer.

15. After sitting for many hours during a trans-Pacific flight, a passenger jumps up quickly as soon as the "fasten seat belt" light is turned off and immediately falls to the floor, unconscious! Briefly explain why, within seconds of being in the horizontal position, the passenger's consciousness is restored.

16. An ice cube floats in a glass of water. As the ice melts what happens to the water level?
   (A) It rises.
   (B) It remains the same.
   (C) It falls by an amount that cannot be determined from the information given.
   (D) It falls by an amount proportional to the volume of the ice cube that was initially above the water line.

17. Water flows through a 0.5-cm-diameter pipe connected to a 1-cm-diameter pipe. Compared to the speed of the water in the 0.5-cm pipe, the speed in the 1-cm pipe is
   (A) one-quarter the speed in the 0.5-cm pipe.
   (B) one-half the speed in the 0.5-cm pipe.
   (C) double the speed in the 0.5-cm pipe.
   (D) quadruple the speed in the 0.5-cm pipe.

18. Blood pressure is normally expressed as the ratio of the *systolic* pressure (when the heart just ejects blood) to the *diastolic* pressure (when the heart is relaxed). The measurement is made at the level of the heart (usually at the middle of the upper arm), and the pressures are given in millimeters of mercury, although the units are not usually written. Normal blood pressure is typically 120/80. How would you write normal blood pressure in **(a)** pascals, **(b)** atmospheres, and **(c)** pounds per square inch (lb/in^2, psi)? **(d)** Is the blood pressure, as typically stated, the absolute pressure or the gauge pressure? Explain your answer.

19. Oceans as deep as 0.50 km once may have existed on Mars. The acceleration due to gravity on Mars is $0.379g$.
   (a) If there were any organisms in the Martian ocean in the distant past, what pressure (absolute and gauge) would they have experienced at the bottom, assuming the surface pressure was the same as it is on present-day Earth? Assume that the salinity of Martian oceans was the same as oceans on Earth (sea water).

(b) If the bottom-dwelling organisms in part (a) were brought from Mars to Earth, how deep could they go in our ocean without exceeding the maximum pressure they experienced on Mars?

20. Blood pressure is normally taken on the upper arm at the level of the heart. Suppose, however, that a patient has both arms in casts so that you cannot take his blood pressure in the usual way. If you have him stand up and take the blood pressure at his calf, which is 95.0 cm below his heart, what would normal blood pressure be? The density of blood is 1060 kg/m^3.

21. A syringe that has an inner diameter of 0.60 mm is attached to a needle with an inner diameter of 0.25 mm. A nurse uses the syringe to inject an ideal fluid into a patient's artery in which the blood pressure is 140/100.

(a) Calculate the minimum force the nurse must exert on the syringe, neglecting any effects due to the speed of the fluid.

(b) Assume the fluid has the same density as water. If the fluid has a speed of 5.0 mm/s in the needle, calculate the change in pressure as the fluid flows into the needle.

(c) Given your calculation in (b), is the assumption we made to ignore the change in pressure due to the change in cross-sectional area of the flow valid for this low speed in the needle?

22. Two identically shaped containers in the shape of a truncated cone are placed on a table, but one is inverted such that the small end is resting on the table. The containers are filled with the same height of water. The pressure at the bottom of each container is the same. However, the weight of the water in each container is different. Using physics principles, briefly explain why the pressures are the same.

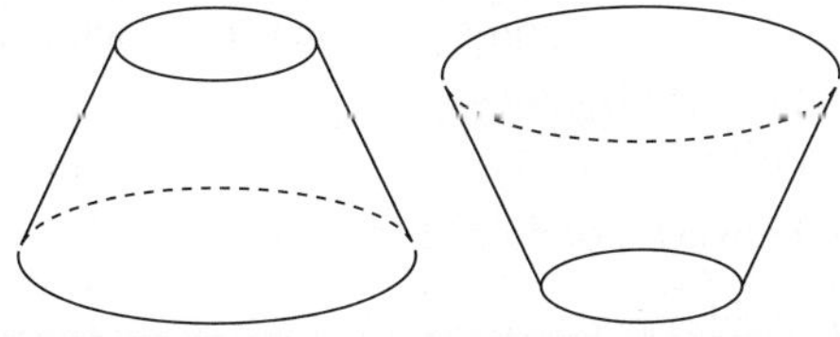

23. A large water tank is 18.0 m above the ground, as shown in the figure. Suppose a pipe with a diameter of 8.0 cm is connected to the base of the tank and leads down to the ground. How fast does the water rush out of the pipe at ground level? Assume that the tank is open to the atmosphere.

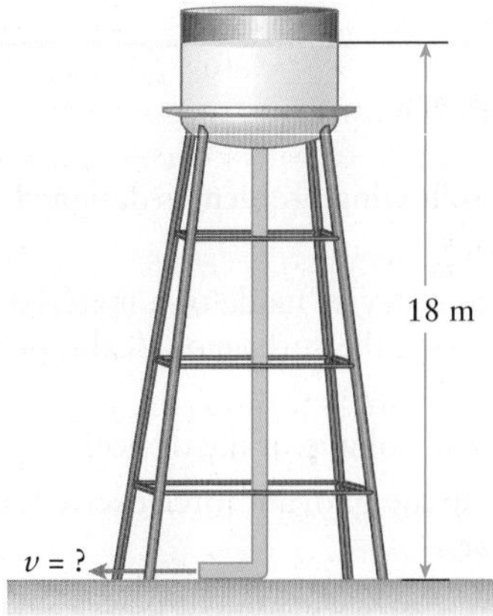

24. The aorta is approximately 25 mm in diameter. The mean pressure there is about 100 mmHg, and the blood flows through the aorta at approximately 60 cm/s. Suppose that at a certain point a portion of the aorta is blocked so that the cross-sectional area is reduced to 3/4 of its original area. The density of blood is 1060 kg/m^3.

(a) Find the speed of the blood moving just as it enters the blocked portion of the aorta.

(b) Calculate the gauge pressure (in mmHg) of the blood just as it enters the blocked portion.

25. Two wooden boxes of equal mass but different density are held beneath the surface of a large container full of water. Box A has smaller average density than box B. When the boxes are released, they accelerate upward to the surface. Which box has the greater acceleration?

(A) Box A

(B) Box B

(C) They are the same.

(D) We need to know the actual densities of the boxes to answer the question.

26. The human body contains about 5.0 L of blood that has a density of 1060 kg/m^3. Approximately 45% (by mass) of the blood is cells and the rest is plasma. The density of blood cells is approximately 1125 kg/m^3, and about 1% of the cells are white blood cells, with the rest being red blood cells. The red blood cells are about 7.5 μm across (modeled as spheres).

(a) Calculate the mass of the blood in a typical human.

(b) Find approximately how many blood cells (of both types) are in the blood.

Prep for the AP® Exam

AP® Group Work

Directions: The following problem is designed to be done as group work in class.

A 10.0-kg flotation device made of a material with specific gravity 0.05 floats on the surface of a calm pond. (The density of water is 1000 kg/m^3.)

(a) What is the volume of the device?

(b) Determine the buoyant force exerted on the device as it floats.

To test the device a student places small iron (density = 7.8×10^3 kg/m^3) weights on the device one at a time until the moment when the device becomes completely submerged.

(c) Determine the new buoyant force on the device.

(d) To effectively serve as a flotation device, the device must be designed so as to ensure that 20% of the volume of each person using it remains above water when the device is completely submerged. Assuming each person has a density of 1.1×10^3 kg/m^3, what is the maximum mass of people that can be supported by the device? On the grid, draw free-body diagrams for the device and the people. Forces that are the same magnitude should be equal in length.

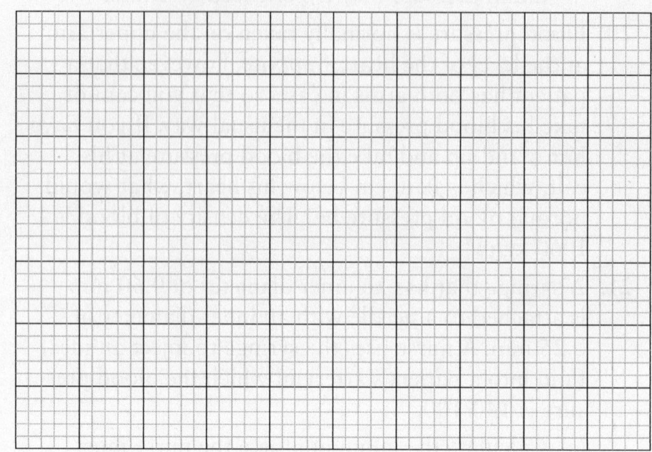

(e) If instead, the 10.0-kg raft were in sea water (density of sea water is 1025 kg/m^3), how would the answer to part (d) change? Choose one answer and justify it briefly.

____ More mass could be supported.

____ Less mass could be supported.

____ The same amount of mass could be supported.

AP® PRACTICE PROBLEMS

Prep for the AP® Exam

Multiple-Choice Questions

Directions: The following questions have a single correct answer.

1. A toy floats in a swimming pool. The buoyant force exerted on the toy depends on the volume of

(A) water in the pool.

(B) the toy under water.

(C) the toy above water.

(D) none of the above choices.

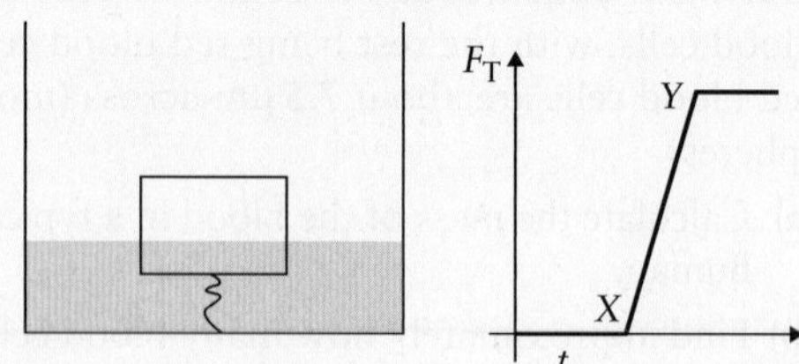

2. A block of wood is floating in water while attached to the bottom of a container by a light thread. A cross section of this setup is shown above. The container is slowly filled with water at a constant rate until it is full. The length of the thread is approximately one-half the depth of the container. A graph of the tension in the string as a function of time is shown. X and Y are the points at which the slope of the graph changes. Where on the graph is the point at which the block becomes completely submerged?

(A) Point X

(B) Between points X and Y

(C) Point Y

(D) After point Y

3. A container is in the shape of a cube, with each side of the cube of length 10.0 cm. If it needs to be filled with water in 1.5 minutes, the minimum volume flow rate of the water is approximately which of the following?

(A) 6.7×10^{-2} m^3/s (B) 1.1×10^{-3} m^3/s

(C) 6.7×10^{-4} m^3/s (D) 1.1×10^{-5} m^3/s

(*Continued*)

4. Which of the following graphs could be used to represent the pressure on the bottom of the container as a function of time from time $t = 0$, when the container is empty, until the container is completely full?

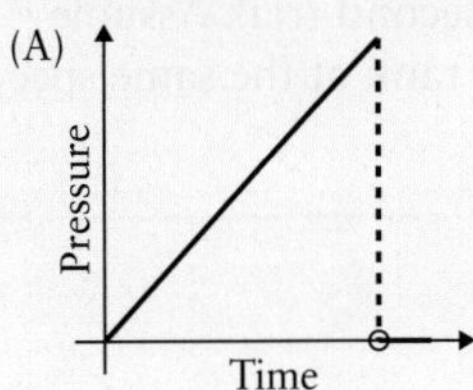

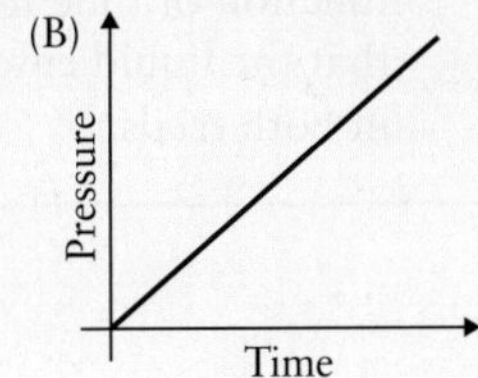

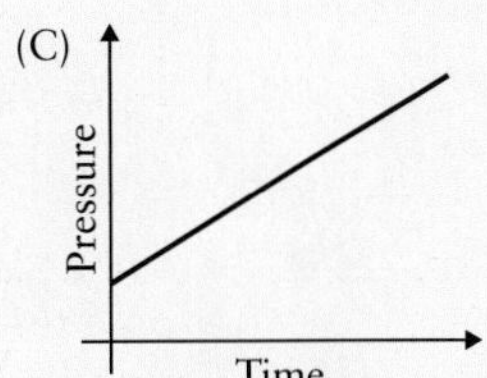

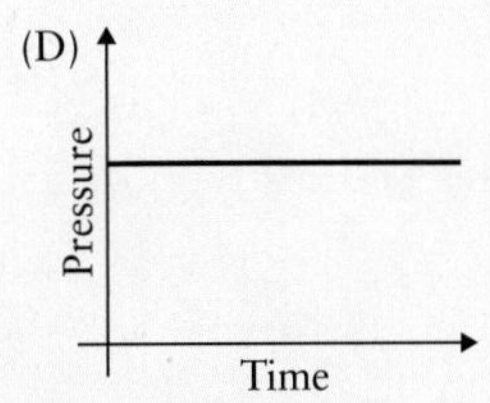

5. The water supply for a water ride that is a set of parallel streams is transported through a pipe with a cross-sectional area of 3.0 m^2. The speed of the flowing water in this pipe is 3.0 m/s. If the parallel streams are modeled as a number of pipes whose combined cross-sectional area is 300 m^2, what is the speed of the water in these parallel streams? The density of water is 1000 kg/m^3.

(A) 0.030 m/s

(B) 3.0 m/s

(C) 300 m/s

(D) The speed cannot be determined without knowing the number of streams.

Free-Response Questions

1. On a field trip to an aquarium, students find a large tank with a clear wall to allow observation of large sea mammals. A student notices an air bubble rising through the water.

(a) The bubble is an approximate sphere of air (density ρ_{air}), surrounded by water. The bubble has a volume V_{bubble}. The density of the water is ρ_{water}. The bubble starts at rest and has a speed v_f when it has risen a distance Δy. Assuming that the change in the bubble's volume is negligible, derive an expression for the mechanical energy dissipated by drag forces as the bubble rises this distance. Express your answer in terms of the given quantities and fundamental constants, as appropriate.

(b) At a particular instant, the bubble is 2.5 m below the water's surface. The surface of the water is at sea level, and the density of the water is $\rho_{\text{water}} = 1025$ kg/m^3. Determine the absolute pressure in the bubble at this location.

(c) The bubble has a volume V_{bubble} when it is 2.5 m below the water's surface. As it rises the size of the bubble changes. Using physics principles, briefly and qualitatively explain

i. if the bubble gets larger or smaller, and why.

ii. what potential effects this change in volume has on your derivation in (a).

2.

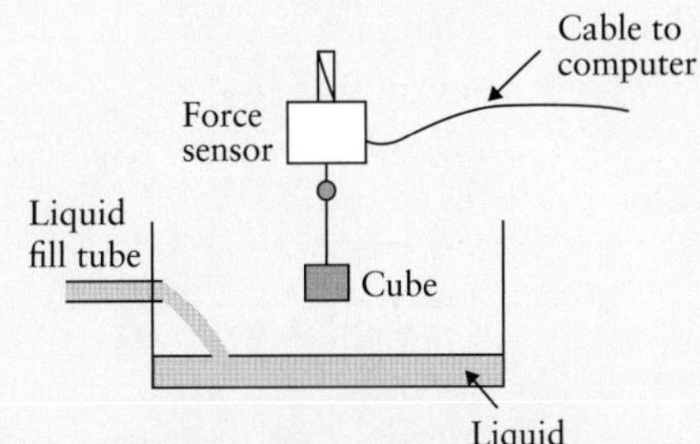

A solid metal cube of side $S = 0.040$ m is suspended from a light string, which is connected to a force sensor, as shown in the diagram. The force sensor is connected to a computer that graphs the force exerted by the string as a function of time. The block is initially suspended above the bottom of a glass tank. The tank has a horizontal cross-sectional area of $A_{\text{tank}} = 0.030$ m^2. Liquid is added to fill the tank slowly, which completely submerges the cube. The graph below shows the reading on the force sensor as a function of time.

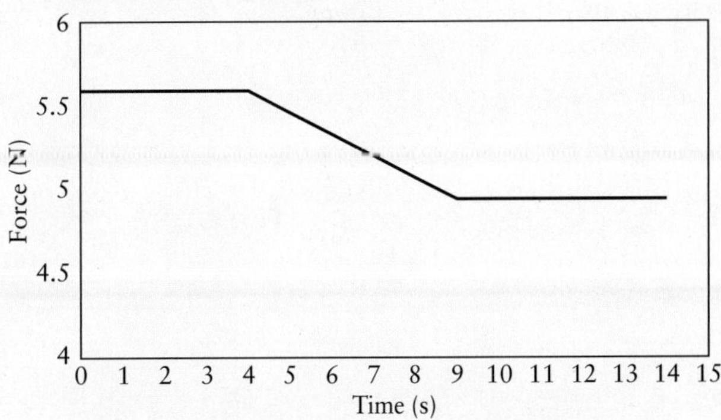

(a) Explain the shape of the graph based on physics principles.

(b) i. Use the data in the graph and the given information to calculate the density of the metal cube.

(Continued)

ii. Use the data in the graph and the given information to calculate the density of the liquid added to the tank.

iii. The liquid flows into the tank through a fill tube that has a diameter of 0.010 m. Calculate the speed of the liquid as it flows through the fill tube.

(c) The original liquid is emptied from the tank and a second trial of the experiment is conducted with a different liquid that is less dense than the original liquid. On the graph above add a sketch of the reading on the force sensor as a function of time for the second trial. Assume that the liquid enters the tank at the same speed in both trials.

AP® Physics 1 Practice Exam 2

Prep for the AP® Exam

Multiple-Choice Questions

Directions: In the following questions, select the best answer choice from the four options listed.

Questions 1–2 refer to the following diagram.

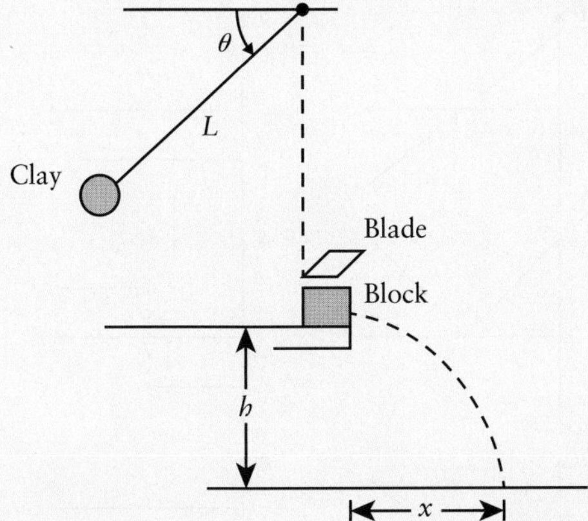

A lump of sticky clay of mass m is fixed to the end of a light string, and the string is tied to a fixed hook to form a pendulum of length L. The string is pulled back to an angle θ below the horizontal, as shown. A block, also of mass m, is placed on a table of height h above the floor, so that when the pendulum is released at the bottom of its swing the clay strikes the block and sticks to it. At the instant the clay sticks to the block, a blade cuts the pendulum string. The block and clay system then becomes a projectile, landing on the floor below at a horizontal distance x.

1. In terms of the given quantities and fundamental constants, the speed of the clay and block system immediately after the collision is

(A) $\dfrac{\sqrt{2mgL(1-\cos\theta)}}{2m}$.

(B) $\dfrac{\sqrt{mgL(1-\sin\theta)}}{m}$.

(C) $\dfrac{\sqrt{2gL(1-\sin\theta)}}{2m}$.

(D) $\dfrac{\sqrt{2gL(1-\sin\theta)}}{2}$.

2. Which of the following statements is true regarding the clay and block?

(A) The potential energy of the clay–Earth system before the clay is released is equal to the kinetic energy of the clay–block system immediately after the collision.

(B) The potential energy of the clay–Earth system before the clay is released is equal to the kinetic energy of the clay–block system immediately before it strikes the floor.

(C) The momentum of the clay immediately before it strikes the block is equal to the total momentum of the clay–block system just before it strikes the floor.

(D) The momentum of the clay–block system immediately after the clay strikes the block is less than the total momentum of the clay–block system just before it strikes the floor.

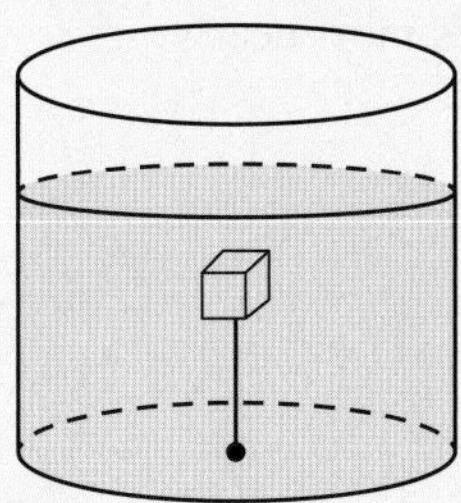

3. A block of wood is submerged in a container of liquid. A string is attached to the block and connected to the bottom of the container such that the block remains completely submerged, as shown above. The block has a mass M and a density that is $\frac{1}{3}$ the density of the liquid in which it is submerged. Which of the following is a correct expression for the tension in the string?

(A) Mg

(B) $2\ Mg$

(C) $3\ Mg$

(D) $4\ Mg$

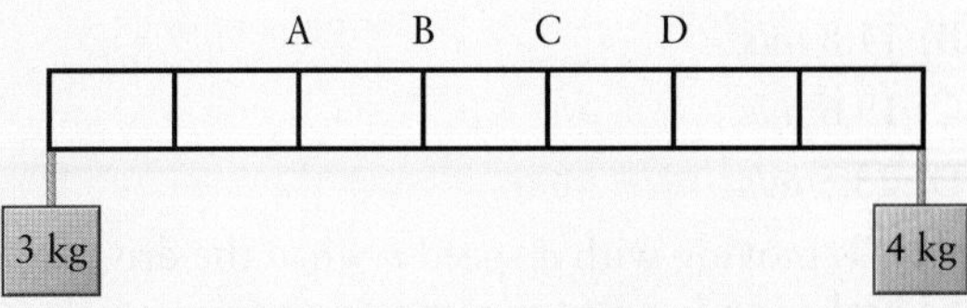

4. Two blocks of mass 3 and 4 kg hang from the ends of a rod of negligible mass, which is marked in seven equal parts, as shown. A string is to be attached to the rod at one of the points and the rod is to be suspended from the string. To which point should the string be attached so the rod remains horizontal?

(A) A

(B) B

(C) C

(D) D

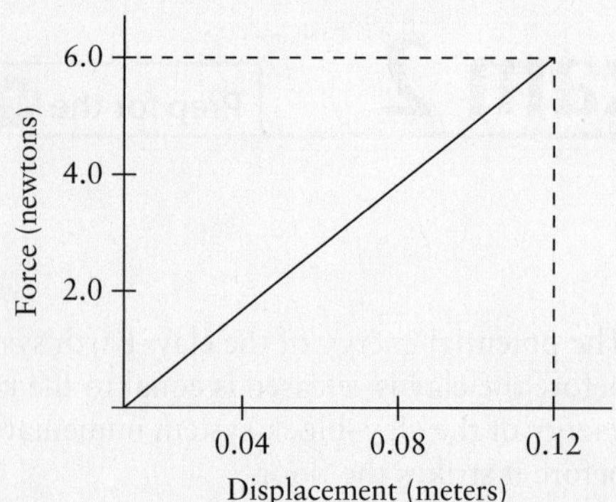

5. An object with a mass of 0.5 kg oscillates on the end of a spring on a horizontal surface with negligible friction according to the equation $x = A\cos(2\pi ft)$. The graph of the magnitude of the force F exerted by the spring on the object versus displacement of the object x for this motion is shown. The last data point corresponds to the maximum displacement of the object. The displacement from equilibrium position ($x = 0$) at a time of 2.0 s is most nearly

(A) 0.01 m
(B) 0.02 m
(C) 0.05 m
(D) 0.12 m

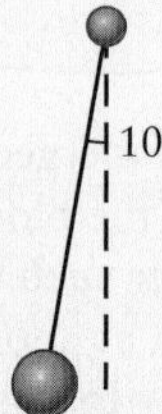

6. A pendulum with a bob of mass 0.40 kg is pulled back and released from an angle of 10° with the vertical. The period of the pendulum on Earth is 1.55 s. This same pendulum is taken to another planet, where its period is 1.00 s. What is the acceleration due to gravity, g_P, on the second planet?

(A) 4.08 m/s^2
(B) 11.8 m/s^2
(C) 19.6 m/s^2
(D) 23.5 m/s^2

7. A car is moving with a speed v when the driver applies the brakes and comes to rest at a uniform acceleration in a distance D. A second car of the same mass is moving with a speed $2v$ and also comes to rest with the same uniform acceleration. What is the stopping distance for the second car?

(A) D
(B) $2D$
(C) $4D$
(D) $8D$

8. A baseball is thrown vertically upward, rises to a maximum height, and returns to the height from which it was thrown. Consider the upward direction to be positive. Which of the following pairs of graphs represents the velocity and acceleration of the ball during its motion?

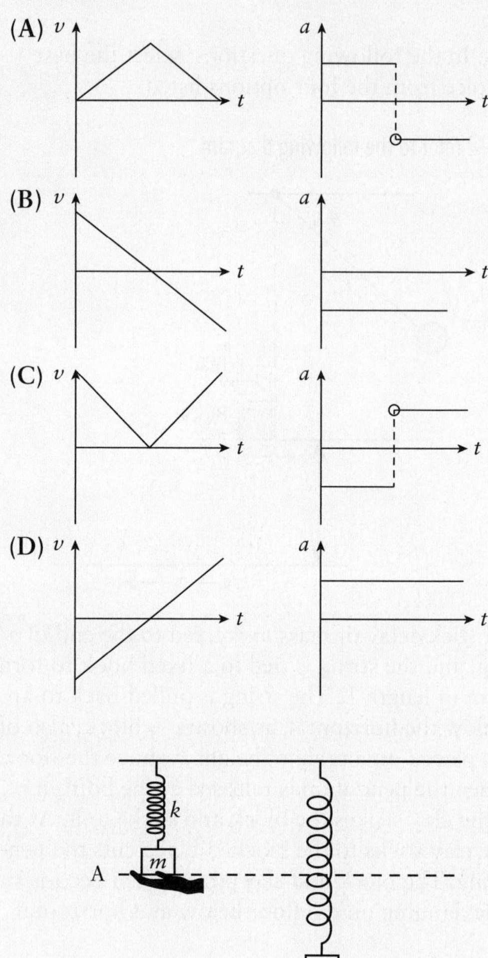

9. A student holds a block of mass m attached to a spring of constant k at point A, so that the spring is not stretched. The student slowly lowers the block until it remains at rest on the end of the stretched spring at point B, when her hand is removed. Which of the following statements is true?

(A) The energy of the block–spring–Earth system at point B is less than the energy of the block–spring–Earth system at point A.
(B) The energy of the block–spring–Earth system at point B is greater than the energy of the block–spring–Earth system at point A.
(C) The energy of the block–spring–Earth system at point B is equal to the energy of the block–spring–Earth system at point A.
(D) The energy of the block–spring–Earth system at point B is equal to the energy stored in the spring at point A.

10. Scientists would like to place a satellite in a circular orbit around a planet so that the satellite orbits with a particular constant speed v. Which quantities determine the speed required for a satellite to maintain a stable circular orbit?

(A) The radius of the planet only

(B) The mass of the satellite only

(C) The altitude of the satellite above the planet's surface, the radius of the planet, and the mass of the planet

(D) The mass of the satellite, the altitude of the satellite above the planet's surface, the radius of the planet, and the mass of the planet

Questions 11–12 refer to the following diagram.

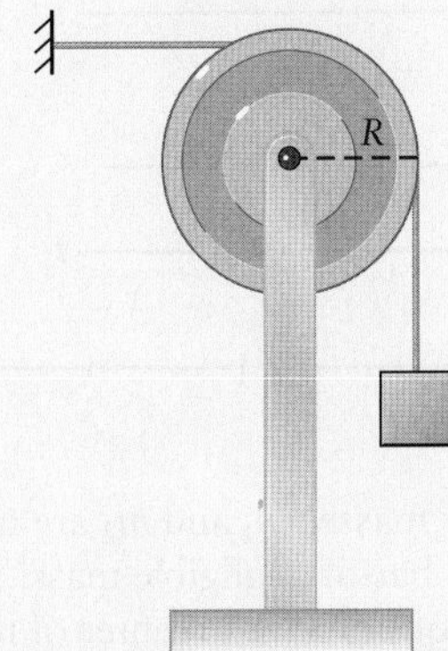

The apparatus shown consists of a pulley of mass m and radius R mounted to a pole on a stand. The stand rests on a table. A block is attached to a string which is wrapped around the pulley, and another string is attached to the top of the pulley and connected horizontally to a wall, preventing the pulley from turning. The pole, stand, and block each also each have mass m, making the total mass of the apparatus $4m$.

11. In terms of the given quantities, the magnitude of the torque provided by the horizontal string is

(A) $4mgR$

(B) $2mgR$

(C) mgR

(D) $\frac{1}{2}mgR$

12. If the horizontal string is cut, the normal force exerted by the table on the apparatus is

(A) equal to mg.

(B) less than $4mg$ but greater than mg.

(C) equal to $4mg$.

(D) greater than $4mg$.

Questions 13–14 refer to the following diagram.

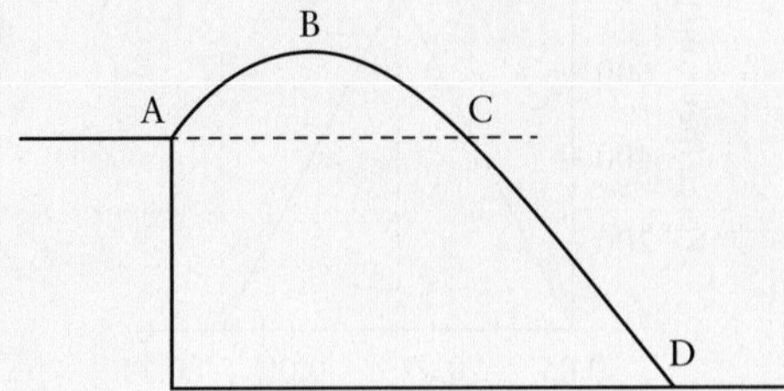

A ball is thrown with a speed v at an angle θ above the horizontal at point A in the figure above. It travels along the parabolic path shown to point D just before it strikes the ground. Point B is the highest point in the trajectory.

13. Which of the following is the direction of the acceleration of the ball at point B?

(A) $\longrightarrow$

(B) $\downarrow$

(C) $\longleftarrow$

(D) There is no direction: The acceleration is zero.

14. Which of the following is the correct ranking of the speeds of the ball at each of the points?

(A) $v_D > v_C > v_B > v_A$

(B) $v_B > (v_A = v_C) > v_D$

(C) $(v_A = v_C) > v_B > v_D$

(D) $v_D > (v_C = v_A) > v_B$

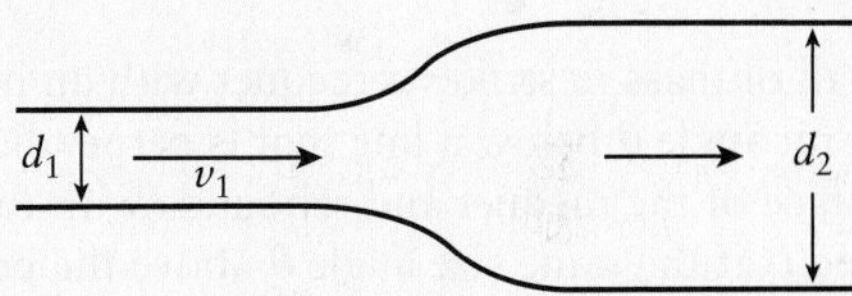

15. Water flows through a pipe as shown above. At the smaller end, the pipe has a diameter d_1 and at the larger end, the pipe has a diameter $d_2 > d_1$. The speed of the water at the smaller end is v_1. Which of the following is a correct expression for the speed of the water at the larger end?

(A) $\frac{d_1^2}{d_2^2}v_1$

(B) $\frac{d_2^2}{d_1^2}v_1$

(C) $\frac{d_1}{d_2}v_1$

(D) $\frac{d_2}{d_1}v_1$

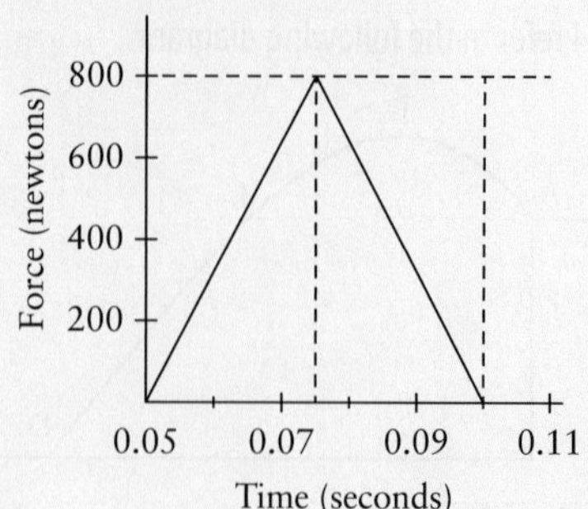

16. A 2-kg block slides along a floor of negligible friction with a speed of 20 m/s when it collides with a 3-kg block, which is initially at rest. The graph represents the force exerted on the 3-kg block by the 2-kg block as a function of time. The momentum of the 2-kg block after the collision is most nearly

(A) –20 kg · m/s
(B) 20 kg · m/s
(C) 40 kg · m/s
(D) 60 kg · m/s

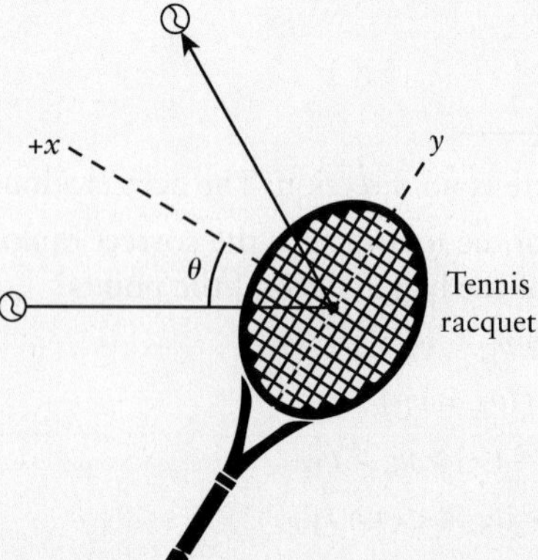

17. A ball of mass m strikes a racquet with an initial speed v at an angle θ below a line that is perpendicular to the face of the racquet and rebounds with the same speed v at the same size angle θ above the perpendicular line, as shown. The ball is in contact with the strings of the racquet for a time interval Δt. Which of the following is a correct expression for the magnitude of the average force exerted on the ball by the racket?

(A) $mv \sin\theta$
(B) $2mv \cos\theta$
(C) $\dfrac{mv \sin\theta}{\Delta t}$
(D) $\dfrac{2mv \cos\theta}{\Delta t}$

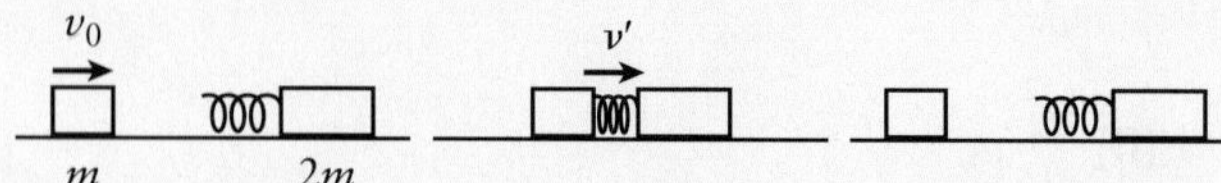

18. A block of mass m is moving on a horizontal surface with a speed v_0 as it approaches a block of mass $2m$, which is at rest on the surface but free to move. The friction force exerted by the surface on the blocks is negligible. A spring is attached to the block of mass $2m$. When the two blocks collide, the spring compresses and the two blocks momentarily move at the same speed. The smaller block is then pushed away by the spring, separating them again, with each continuing to move. Which of the following *kinetic energy* versus *time* graphs represents the motion of the blocks from the time block m approaches block $2m$ until the two blocks are separated after the collision?

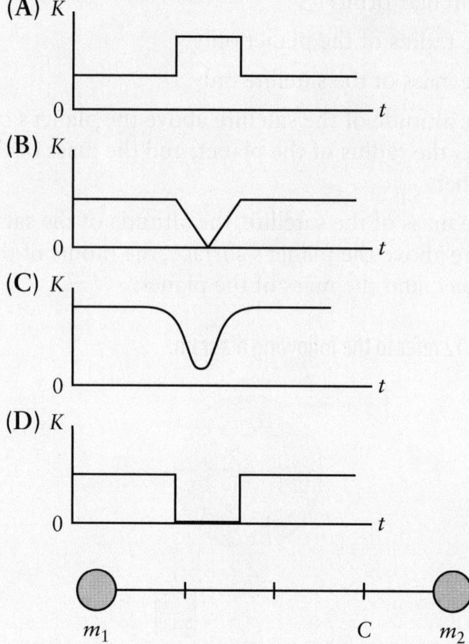

19. Two objects of masses m_1 and m_2 are mounted on either end of a bar of negligible mass. The bar is marked off in quarters. The center of mass of the system C is labeled as shown. Which of the following is the ratio of m_2 to m_1?

(A) $\dfrac{4}{1}$
(B) $\dfrac{3}{1}$
(C) $\dfrac{2}{1}$
(D) $\dfrac{1}{3}$

Questions 20–21 refer to the diagram below.

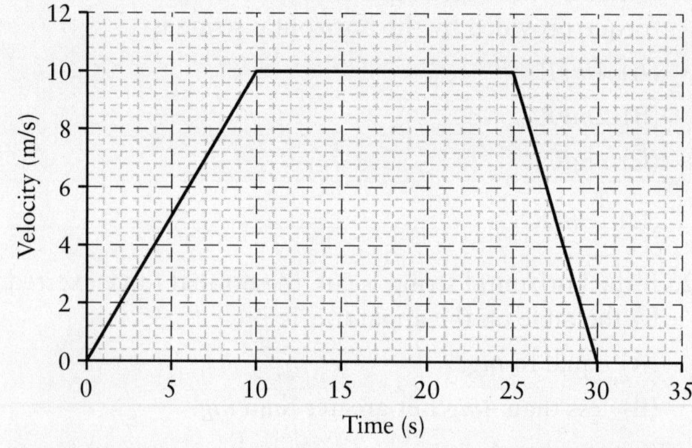

Consider an object moving according to the velocity versus time graph above.

20. What is the average acceleration of the car over the first 25 s of motion?

(A) 0.4 m/s^2
(B) 1.0 m/s^2
(C) 2.5 m/s^2
(D) Zero

21. What is the distance traveled by the object during the 30-s time interval shown?

(A) 150 m (B) 200 m

(C) 225 m (D) 300 m

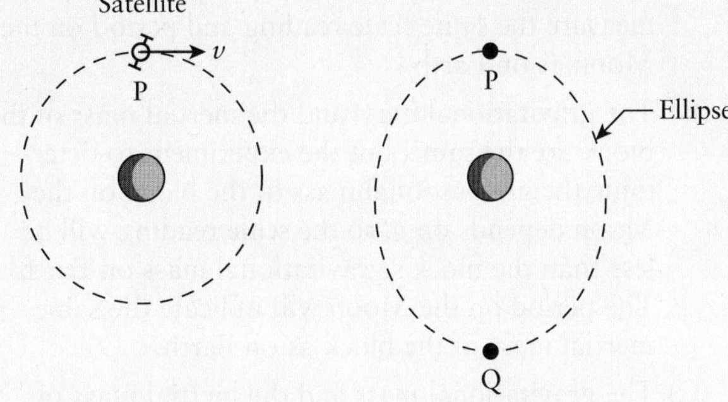

Figure 1 Figure 2

22. An engineer intends to place a satellite in a circular orbit around Earth at a speed v, as shown in Figure 1. The satellite is initially located at point P in the figure. The engineer has made a mistake and the orbit is instead an elliptical orbit, as shown in Figure 2. Which of the following is true about the elliptical orbit?

(A) Because the speed of the satellite at point P is v, and angular momentum is conserved because the force of gravity exerted by the planet is always directed toward the planet, the speed of the satellite will remain v as the satellite passes through point Q.

(B) Because the speed of the satellite at point P is v, and angular momentum is conserved because the force of gravity exerted by the planet is always directed toward the planet, the speed of the satellite will be less than v as the satellite passes through point Q.

(C) Angular momentum is not conserved in this case, because the orbit of the satellite is not circular.

(D) Although angular momentum is conserved, it is not possible to tell what will happen to the speed of the satellite as it passes through point Q, because the orbit of the satellite is not circular.

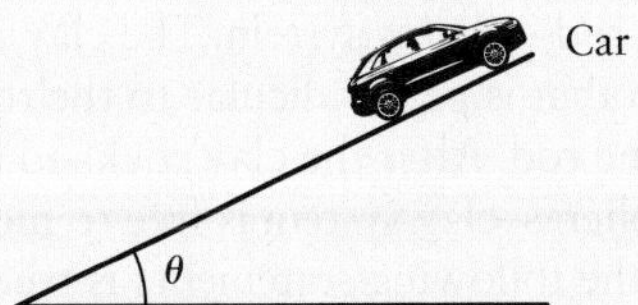

23. A toy car consists of a block of mass m and four wheels, each with mass m and radius R. The car is placed at the top of a ramp having a surface with a coefficient of static friction μ. The car is released from rest at the top of the ramp, and the wheels roll without slipping on the surface. Take the zero for potential energy of the car–Earth system to be the bottom of the ramp. Which of the following statements is true?

(A) The torque causing the wheels to turn is provided by the gravitational force Earth exerts on the car, and the translational kinetic energy of the car at the bottom of the ramp is equal to the potential energy of the car–Earth system at the top of the ramp.

(B) The torque causing the wheels to turn is provided by the friction force the ramp exerts on the car, and the translational kinetic energy of the car at the bottom of the ramp is less than the potential energy of the car–Earth system at the top of the ramp because energy is dissipated by the friction force.

(C) The torque causing the wheels to turn is provided by the friction force the ramp exerts on the wheels, and the translational kinetic energy of the car at the bottom of the ramp is equal to the potential energy of the car–Earth system at the top of the ramp.

(D) The torque causing the wheels to turn is provided by the friction force the ramp exerts on the wheels, and the translational kinetic energy of the car at the bottom of the ramp is less than the potential energy of the car–Earth system at the top of the ramp because the wheels have gained rotational kinetic energy.

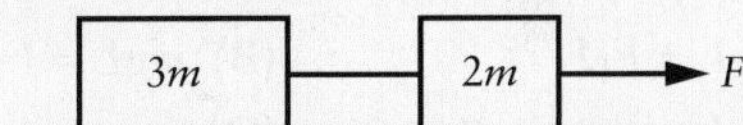

24. A block of mass $3m$ and a block of mass $2m$ are connected by a string. A force of magnitude F is exerted on the block of mass $2m$ to accelerate the system to the right along a horizontal surface of negligible friction. Which of the following is a correct expression for the tension in the string between the blocks?

(A) $\frac{2}{5}F$ (B) $\frac{3}{5}F$ (C) $\frac{3}{2}F$ (D) $3F$

25. A 1.00×10^2 N block is being pulled at a constant speed along a horizontal table by a force of 30.0 N that is exerted parallel to the table. What is the coefficient of kinetic friction between the block and the table?

(A) 0.03 (B) 0.30 (C) 0.70 (D) 3.3

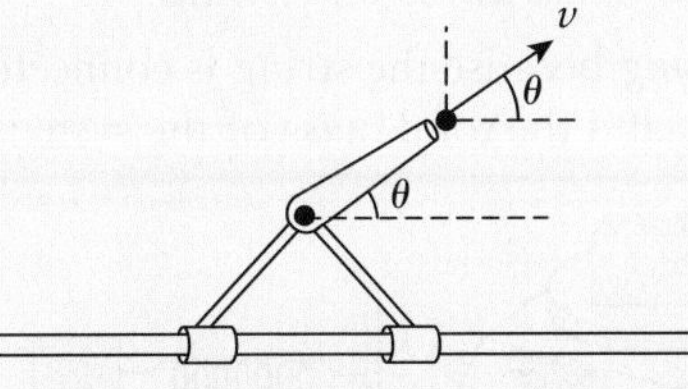

26. A toy cannon is mounted on two horizontal rails on which it can slide with negligible friction. The cannon fires a ball at speed v at angle θ above the horizontal as shown in the figure. Which of the following statements is true as the cannon fires the ball?

(A) Only the horizontal component of the total momentum of the cannon and ball system is constant, given that the cannon is restricted to move only on the horizontal rails.

(B) Only the vertical component of the total momentum of the cannon and ball system is constant, given that the cannon is restricted to move only on the horizontal rails.

(C) The total momentum of the cannon and ball system is constant, given that there are no external forces exerted on the cannon or ball.

(D) The momentum of the cannon and ball system is not constant in any direction, given that energy is added to the system to fire the cannon.

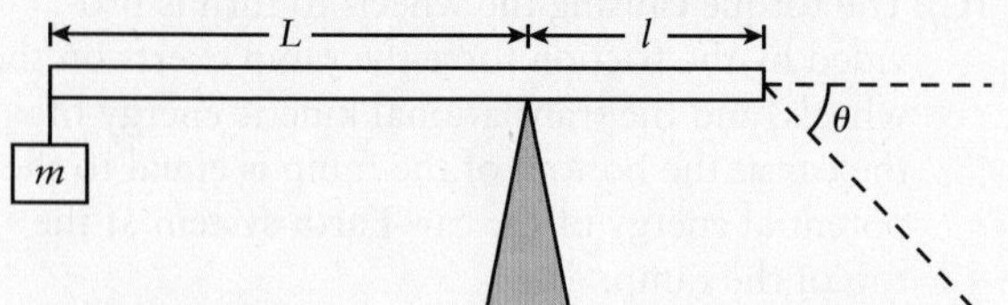

27. The figure shows an object of mass m hanging from the end of a rod of negligible mass, which is pivoted on a fulcrum as shown. A string is tied at an angle θ from the horizontal at the right end of the rod to keep the rod from rotating. Which of the following is a correct equation that could be used to calculate the tension F_T in the string?

(A) $mgL = F_T l$ (B) $mgL = F_T l \cos\theta$

(C) $mgL = F_T l \sin\theta$ (D) $mg = F_T$

28. A block of mass m hangs from one end of a string that passes over a pulley. The other end of the string is connected to another hanging block of mass $2m$. Assume the pulley and string have negligible mass and the pulley rotates without friction. Which of the following must be true about the tension, F_T, in the string?

(A) $F_T = mg$ because the system of two blocks is at rest, so the upward force on the block of mass m must equal the downward force on the block.

(B) $F_T = 2mg$ because the block of mass $2m$ is pulling the block of mass m upward as it moves downward.

(C) $mg < F_T < 2mg$ because the net force on the block of mass m is upward while the net force on the block of mass $2m$ is downward.

(D) $F_T = 3mg$ because the string is connected to both blocks and the total mass of the system is $3m$.

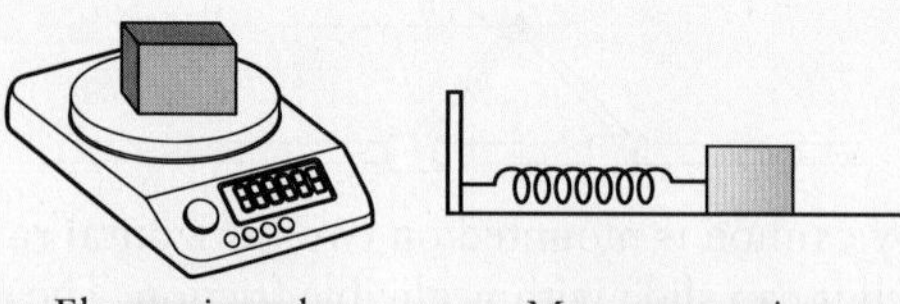

29. A student determines the gravitational mass of a block by placing it on an electronic scale on Earth and recording the number of grams shown on the screen of the scale. The student then determines the inertial mass of the block by attaching the block to a spring of known spring constant, placing the block and spring system on a horizontal surface of negligible friction, allowing it to oscillate, and measuring the period of the oscillation. The student then gives all of the same equipment to an astronaut, who takes the equipment to the Moon and repeats the student's experiment. Which of the following statements is true?

(A) The gravitational mass and the inertial mass of the block are the same, and the experiments will measure the same scale reading and period on the Moon as on Earth.

(B) The gravitational mass and the inertial mass of the block are the same, but the experiment to determine the gravitational mass of the block on the Moon depends on g, so the scale reading will be less than the block's gravitational mass on Earth. The period on the Moon will indicate the same inertial mass of the block as on Earth.

(C) The gravitational mass and the inertial mass of the block are the same. The experiment using the period to calculate the inertial mass of the block on the Moon depends on g, so it will measure less than its inertial mass on Earth, but the gravitational mass of the block will measure the same as its gravitational mass on Earth.

(D) The gravitational mass and the inertial mass of the block are only the same when the block is on Earth. The gravitational mass of the block is less than the inertial mass of the block when the block is on the Moon. The experiments will correctly measure this difference.

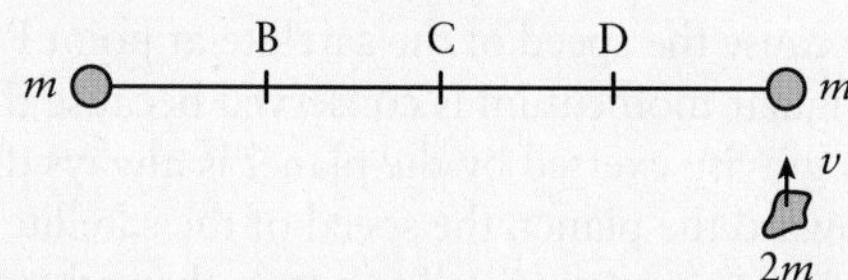

30. An astronaut floating in space attaches two spheres, each of mass m, to either end of a rod of negligible mass, as shown. In the figure, three equally spaced points on the rod are labeled B, C, and D, with point C being the center of the rod. The astronaut places the rod and sphere system at rest in space, and then launches a lump of sticky clay of mass $2m$ toward one end of the rod–spheres system. The clay is moving in a direction that is perpendicular to the rod just before it strikes the rod. After the clay sticks to one sphere, the rod–spheres–clay system is free to move in space. Which of the following statements is true of the rod–spheres–clay system immediately after the collision?

(A) Linear momentum of the system is constant, kinetic energy of the system is restored, and the system will rotate about point C.

(B) Linear momentum and angular momentum of the system are constant, and the system will rotate about point D.

(C) Linear momentum and angular momentum of the system are constant, and the system will rotate about point B.

(D) Linear momentum is constant, and the system will not rotate.

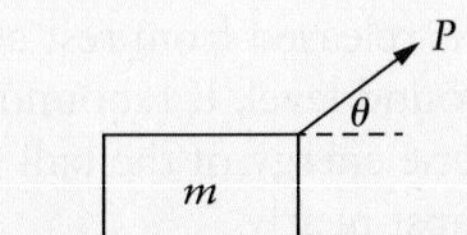

31. A block of mass m is pulled along a horizontal surface at a constant speed by a pulling force P directed at an angle θ above the horizontal. The coefficient of friction between the surface and the block is μ_k. Which of the following is the correct expression for the friction force exerted on the block by the surface?

(A) $\mu_k mg$

(B) $\mu_k (mg + P \sin \theta)$

(C) $\mu_k (mg + P \cos \theta)$

(D) $\mu_k (mg - P \sin \theta)$

32. A block of mass 5.00 kg is initially placed at rest on an incline that makes an angle of 30.0° with the horizontal. The coefficient of static friction between the block and the incline is 0.750 and the coefficient of kinetic friction is 0.320. The magnitude of the friction force exerted by the incline on the block is closest to

(A) 14 N

(B) 25 N

(C) 32 N

(D) 50 N

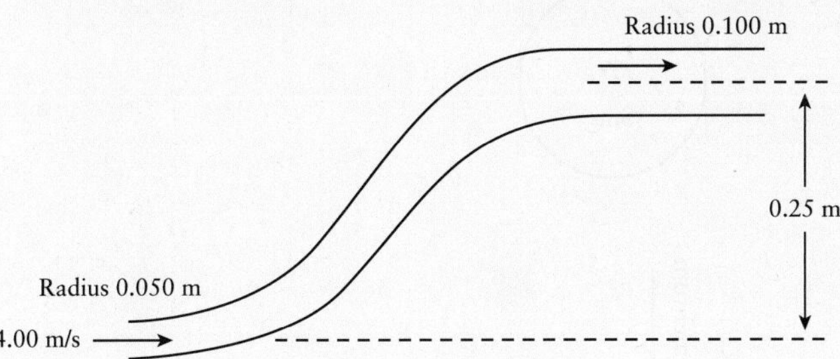

33. Water of density 1.00×10^3 kg/m^3 flows at a speed of 4.00 m/s through a section of pipe of radius 0.050 m. The pipe bends upward to a height of 0.25 m above its original height and increases to a radius of 0.100 m, as shown in the figure above. If the water pressure is 1.80×10^5 Pa in the lower part of the pipe, which of the following is most nearly the pressure in the upper part of the pipe?

(A) 0.90×10^5 Pa

(B) 1.77×10^5 Pa

(C) 1.83×10^5 Pa

(D) 1.85×10^5 Pa

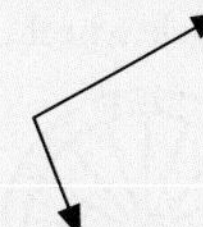

34. The vector diagram above represents the momenta of two objects after they collide. One of the objects is initially at rest. Which of the following vectors may represent the initial momentum of the other object before the collision?

(A) 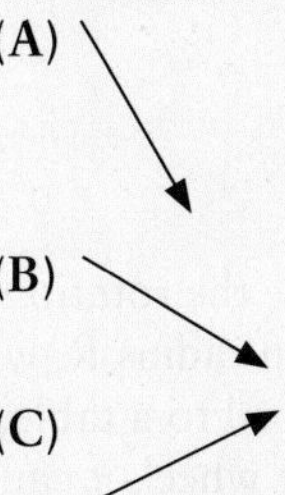

(B)

(C)

(D)

35. In which of the following cases is the kinetic energy of an object increasing?

(A) A baseball is moving downward from the top of its path toward the fielder's glove.

(B) The Moon is orbiting Earth in a circular orbit.

(C) A crane is lifting a beam at constant speed.

(D) A car is approaching a red light and applies its brakes.

36. A student is asked to calculate the work done by friction on a metal block as it is pulled across the lab table. The student is given a stopwatch, a meter stick, and a force sensor. The student makes a graph of this motion and uses the graph to calculate the work. Which of the following will give the correct work done by friction?

(A) The student pulls the block at a constant speed and finds the area under the force versus time graph.

(B) The student pulls the block at a constant speed and finds the area under the force versus distance graph.

(C) The student pulls the block at a steadily increasing speed and finds the area under a force versus distance graph.

(D) The student pulls the block at a steadily increasing speed and finds the area under the force versus time graph.

37. A planet is discovered that has a radius equal to four times Earth's radius, and has four times the mass of Earth. If g is the acceleration due to gravity near the surface of Earth, the gravitational field near the surface of the discovered planet is

(A) $\frac{g}{4}$ (B) $\frac{g}{2}$ (C) g (D) $4g$

38. A group of students want to calculate the rotational inertia, I, of a bicycle wheel of known radius R, which can turn on an axle on a stand clamped to a table, as shown. Given the construction of the wheel, it cannot be modeled as a uniform disk. They begin by wrapping a string of negligible mass around the wheel and attaching an object of known mass, m, to the end of the string. They then allow the object to descend, unwinding the string and rotating the wheel. Which additional quantities could best be directly measured and, together with m and R, are sufficient to calculate the rotational inertia I of the wheel?

(A) The angular acceleration of the wheel, the time for the mass to descend, and the mass of the wheel

(B) The mass of the wheel and the time for the block to descend

(C) The distance the block descends and the time for the mass to descend

(D) The distance the block descends and the speed of the mass just before striking the floor

39. A lump of clay of mass 0.500 kg is thrown at a speed of 9.00 m/s at a block of mass 1.00 kg that is initially at rest. The clay sticks to the block and the two move off together. What is the kinetic energy of the clay–block system just after the collision?

(A) 60.8 J (B) 20.3 J

(C) 6.75 J (D) 4.50 J

40. A construction crew slowly raises a bucket from rest at ground level to rest at the top of a building using a rope. Which of the following correctly describes the change in total mechanical energy of the bucket–Earth system?

(A) The total mechanical energy increases because the rope does positive work on the bucket.

(B) The total mechanical energy of the system remains the same because there is no change in kinetic energy of the bucket.

(C) The total mechanical energy of the system decreases because the tension in the rope is upward and the gravitational force is downward.

(D) The total mechanical energy of the system increases because the work done by the rope on the bucket is negative.

41. A 0.15-kg ball is released from rest at a height of 3.0 m above ground level. It rebounds to a height of 2.2 m. The kinetic energy of the ball just after it leaves the ground is most nearly

(A) 1.2 J

(B) 3.3 J

(C) 4.5 J

(D) 7.8 J

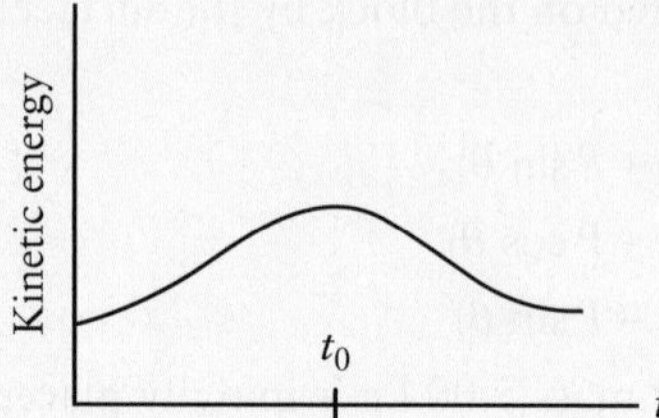

42. A comet orbits the Sun. The graph of the kinetic energy of the comet versus time for one-half of the orbital period is shown. The position of the comet indicated in each orbit diagram represents the position of the comet at the marked time, t_0, in the kinetic energy and angular momentum graphs. Which of the following diagrams of the orbit and graphs of the angular momentum of the comet versus time for one-half of the orbital period best represents the orbit of the comet around the Sun?

(A)

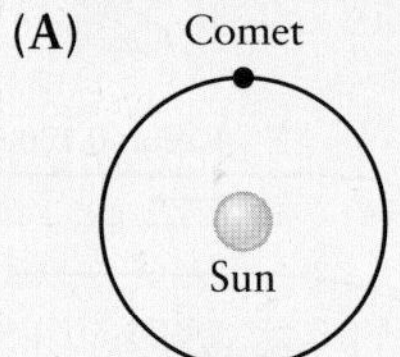

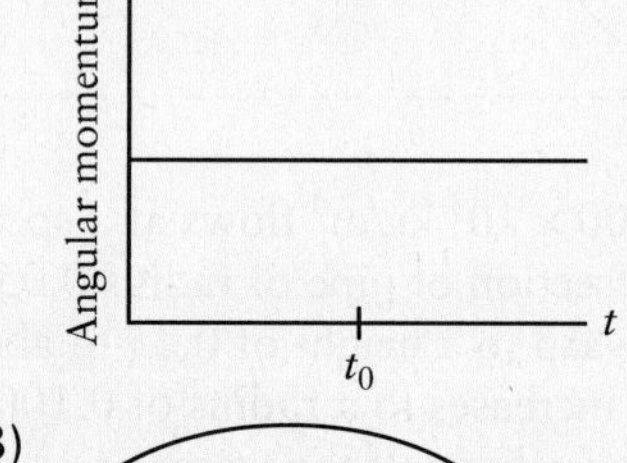

(B)

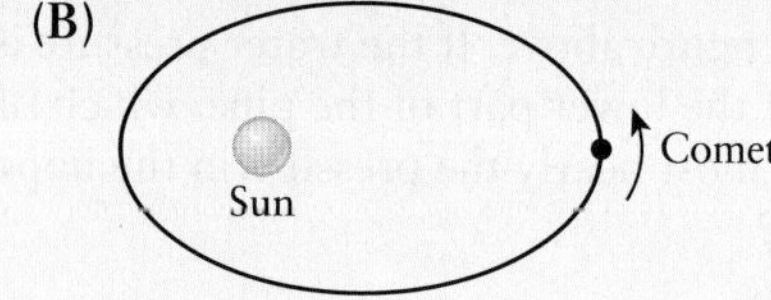

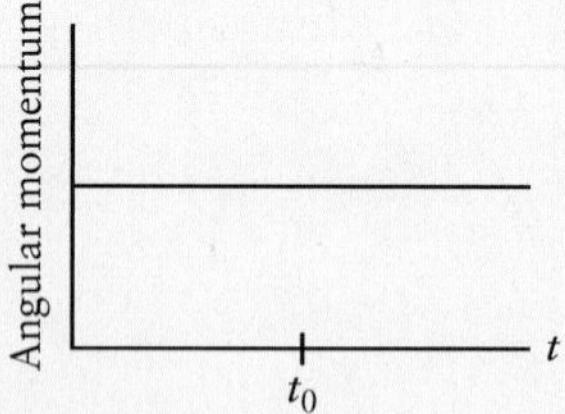

(C)

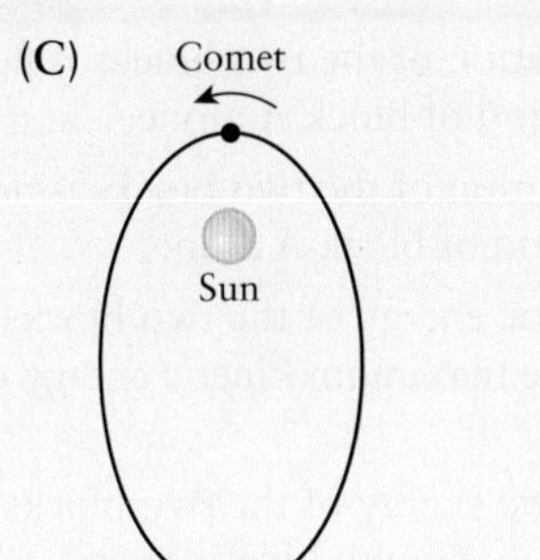

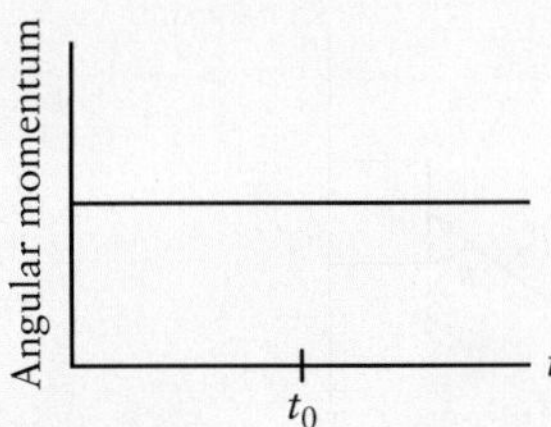

(D)

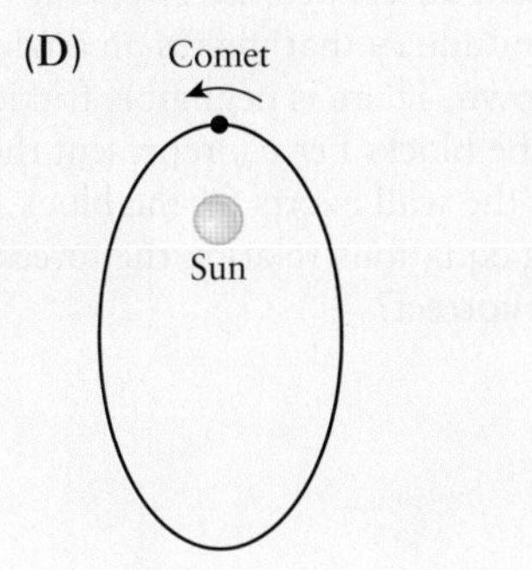

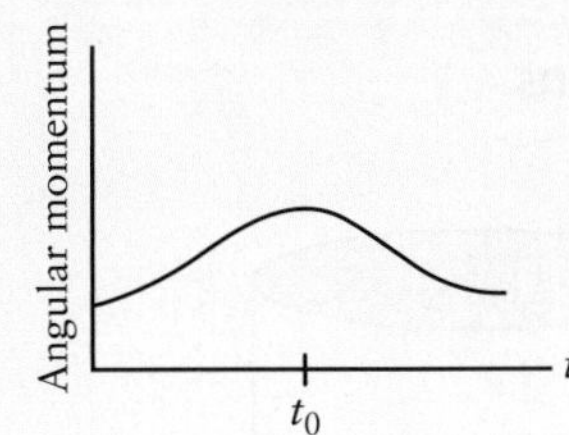

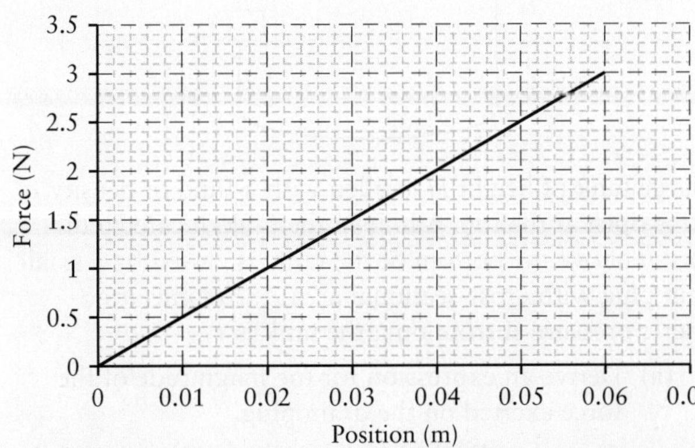

43. The figure above shows the net force exerted on an object as a function of the position of the object. The object starts from rest at position $x = 0$ m and acquires a speed of 3.0 m/s after traveling a distance of 0.060 m. Calculate the mass of the object.

(A) 0.010 kg (B) 0.020 kg
(C) 0.040 kg (D) 0.080 kg

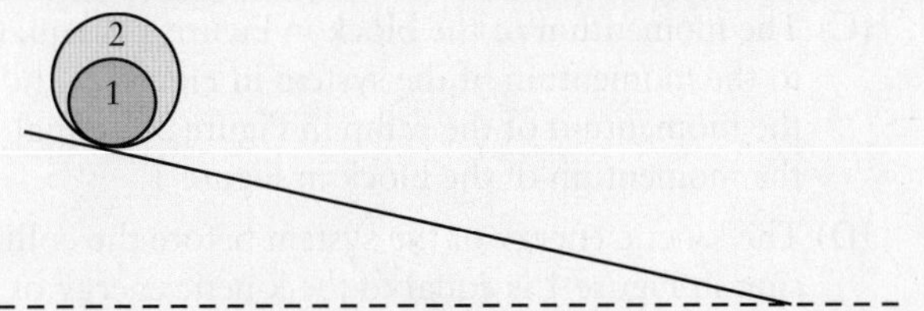

44. Two solid disks are released from the top of a ramp of height h. Both disks are released from rest at the same time and roll without slipping down the incline. Disk 1 has mass M and radius R. Disk 2 has mass $2M$ and radius $2R$. Let K_1 be the translational kinetic energy of disk 1 at the bottom of the incline, and K_2 be the translational kinetic energy of disk 2 at the bottom of the incline. What is the ratio K_2/K_1?

(A) 4/1
(B) 2/1
(C) 1/1
(D) 1/2

45. The escape velocity from the surface of Earth is v_E. What is the escape velocity from the surface of a planet that has the same mass and twice the radius of Earth?

(A) $\frac{v_E}{2}$ (B) $\frac{v_E}{\sqrt{2}}$
(C) v_E (D) $2v_E$

46. You are riding in an elevator. Under what conditions would the normal force exerted on you by the floor be less than your weight?

(A) You are moving upward at increasing speed.
(B) You are moving upward at decreasing speed.
(C) You are moving downward at constant speed.
(D) You are moving downward at decreasing speed.

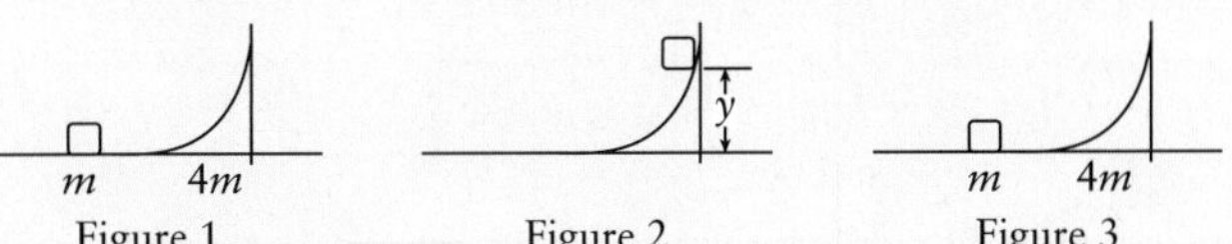

Figure 1 Figure 2 Figure 3

47. A small block of mass m is sliding to the right on a horizontal surface with speed v when it collides with a ramp of mass $4m$, which is free to move, as shown in Figure 1. The block slides up the ramp to a height y, and the block and ramp momentarily slide together, as shown in Figure 2. The small block then slides back down the ramp, as shown in Figure 3. Assume the friction between all surfaces is negligible. Which of the following statements is true?

(A) The momentum of the block in Figure 1 is equal to the momentum of the system in Figure 3, and the speed of the block is zero in Figure 3.
(B) The kinetic energy of the system before the collision in Figure 1 is equal to the potential energy of the block–ramp–Earth system in Figure 2, and the speed of the block is momentarily zero in Figure 2.

(C) The momentum of the block in Figure 1 is equal to the momentum of the system in Figure 2, and the momentum of the ramp in Figure 3 is equal to the momentum of the block in Figure 1.

(D) The kinetic energy of the system before the collision in Figure 1 is equal to the kinetic energy of the system in Figure 3, and the speed of the block–ramp system in Figure 2 is $\frac{v}{5}$.

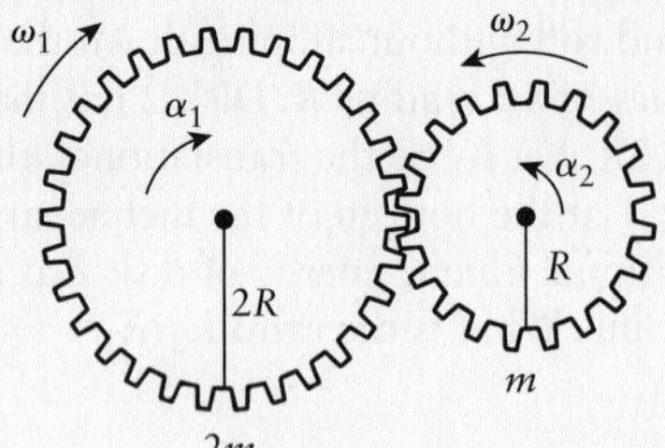

48. A large gear of mass $2m$ and radius $2R$ is interlocked with a smaller gear of mass m and radius R and rotated with a constant angular acceleration, causing both gears to rotate. The tangential velocity of gear 1 at any time is v_1, and the tangential velocity of gear 2 at any time is v_2. The angular velocity of gear 1 at any time is ω_1, and the angular velocity of gear 2 is ω_2. The angular acceleration of gear 1 at any time is α_1, and the angular acceleration of gear 2 is α_2. Which of the following is true at any time for the two gears as they are rotating?

(A) $v_1 > v_2$ and $\omega_1 = \omega_2$

(B) $v_1 = v_2$ and $\omega_1 = \omega_2$

(C) $\omega_1 = \omega_2$ and $\alpha_1 = \alpha_2$

(D) $\omega_1 < \omega_2$ and $\alpha_1 < \alpha_2$

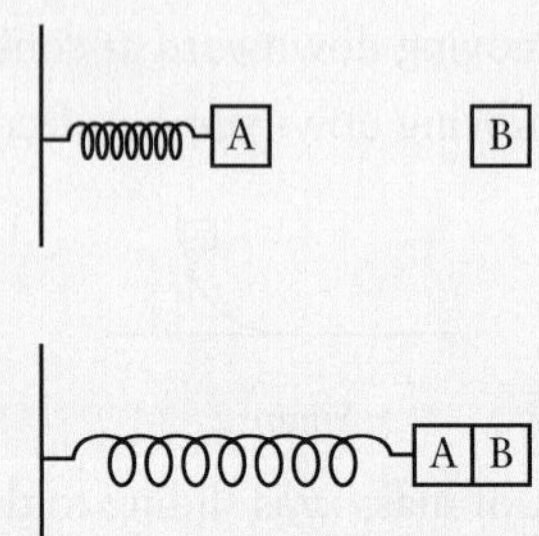

49. Block A is attached to the end of a spring and oscillates on a horizontal surface with negligible friction. Block B is then placed so that exactly when block A reaches maximum displacement during an oscillation, block B touches and sticks to block A without interrupting the oscillation. Which of the following statements is true of the quantities related to both blocks after they stick together, compared to block A alone before it sticks to block B?

(A) The period of oscillation of the two blocks is now shorter than the period of block A alone.

(B) The period of oscillation of the two blocks is now longer than the period of block A alone.

(C) The maximum kinetic energy of the two blocks is now greater than the maximum kinetic energy of block A alone.

(D) The maximum kinetic energy of the two blocks is now less than the maximum kinetic energy of block A alone.

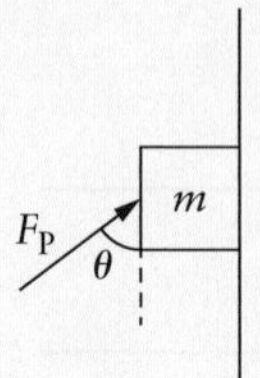

50. A block of mass m is held at rest against a vertical wall by a force of magnitude F_P that makes an angle θ with the vertical, as shown. There is negligible friction between the wall and the block. Let F_W represent the magnitude of the force the wall exerts on the block. Which of the following equations relating the forces exerted on the block is correct?

(A) $F_P \cos\theta = mg$

(B) $F_W = F_P$

(C) $F_P \sin\theta = mg$

(D) $F_W = F_P \cos\theta$

Free-Response Questions

Question 1

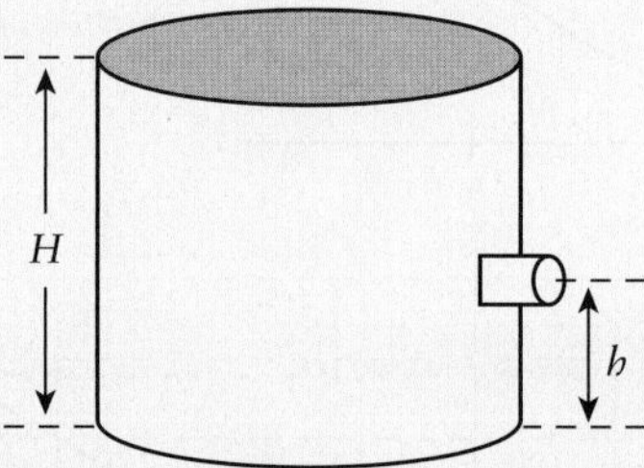

A large tank of height H is filled with a fluid of density ρ and is open at the top. A hole in the tank, located a distance h above the bottom of the tank, is closed by a small drain plug with cross-sectional area A. Assume atmospheric pressure surrounding the tank is P_0.

(a) Derive an expression for the magnitude of the force exerted on the drain plug.

At some time, the plug breaks loose from the tank and the fluid flows from the drain.

(b) Considering conservation of energy, sketch a graph of the speed of the water as it leaves the open drain near the bottom of the tank as a function of $(H - h)$.

(c) If the drain's cross-sectional area A were increased, but the diameter of the opening were still small compared to the height h, would the speed with which the water leaves the drain be increased, decreased, or remain the same? Justify your answer.

Question 2

A disk of known radius, R, and unknown mass is mounted on a fixed axle of negligible mass and rotates in a horizontal plane, as shown in the diagram below.

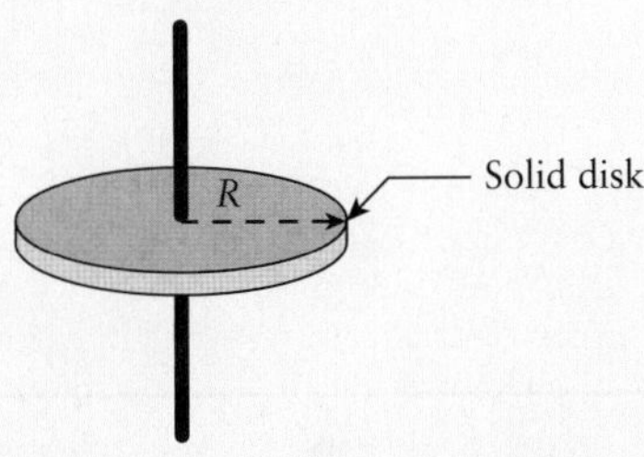

Students are given the task of applying a torque to the disk to cause it to rotate and making measurements to calculate the mass of the disk. The students have access to most of the equipment usually found in a physics classroom, including a set of known masses, but not a balance.

(a) Develop an experimental procedure that the students could use to collect the data needed to calculate the mass of the disk, indicating the quantities to be measured and including any steps that the students take to reduce the uncertainty in their measurements.

(b) Explain how the data from the measurements you listed in (a) could be analyzed to determine the mass of the disk.

(c) Another lab group takes the measurements shown in the table below and measures $R = 0.2$ m. For each trial, they start the disk rotating from rest and exert the constant force for 0.25 s.

$F_{\text{spring scale}}$ (N)	$\Delta\theta$ (rad)		
0.2	88		
0.4	183		
0.6	273		
0.8	350		
1.0	455		

i. What should the students graph to find the mass of the disk from the slope of a straight line? Label and fill in the blank columns in the table for any calculations needed based on these measured values.

ii. On the axes below, graph the data you identified in part i. Clearly scale and label all axes, including units.

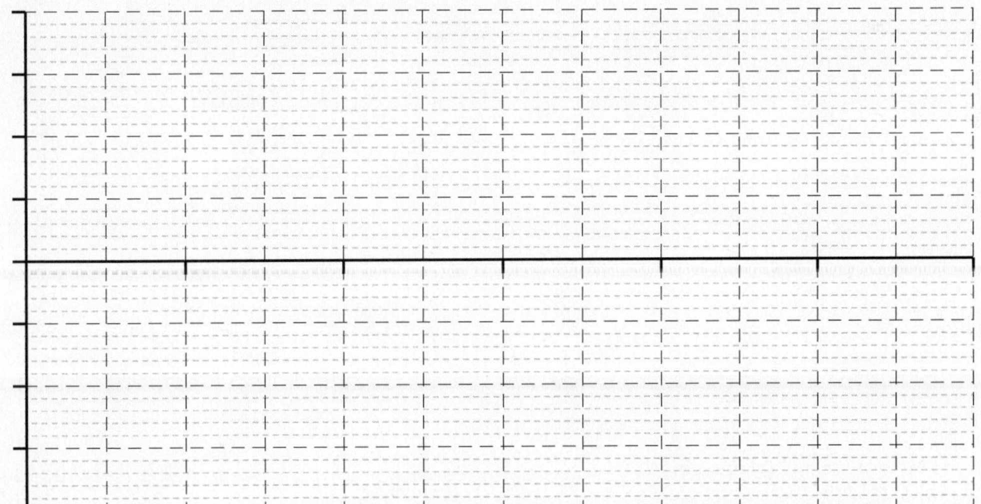

(d) Use the graph you created in (c) to determine the mass of the disk.

The teacher removes the disk from the axle and uses a digital balance to measure the mass of the disk. The measured mass is less than the mass the students calculated from their experiments.

(e) Cite one experimental reason why the calculated mass is greater than the actual mass of the disk. Explain your choice.

Question 3

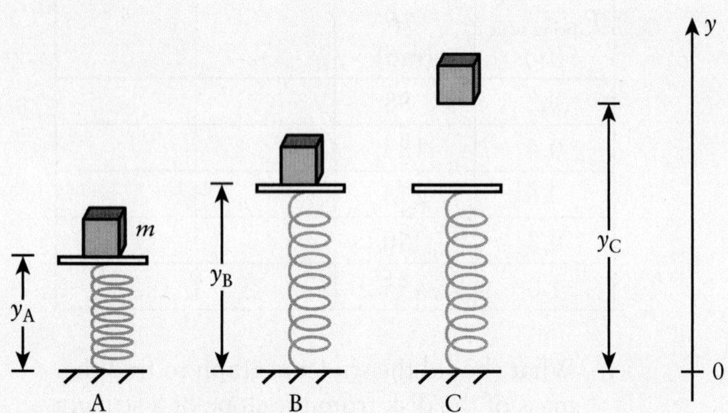

A block of mass m is placed on a lightweight platform at the top of a spring of constant k, and the block is pushed down, compressing the spring until the block is at a height y_A above the ground, as shown in Figure A above. The block is released from rest and is launched from the spring and reaches a maximum height, y_C, as shown in Figure C. Figure B represents the position of the block when the spring is at its uncompressed, equilibrium length.

(a) On the axes below, sketch graphs of the energy of the block–spring–Earth system as a function of the vertical position of the block. Draw and label lines for the kinetic energy, the elastic potential energy, and the gravitational potential energy. Take $y = 0$ to be ground level at the bottom of the spring.

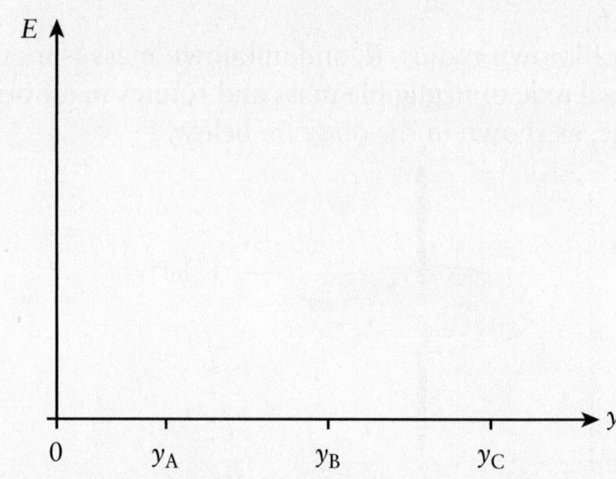

(b) Derive an expression for the maximum height of the block, y_C, in terms of m, k, y_A, and y_B.

(c) At some point in its path between y_A and y_C, the block has a maximum speed. Is that location at y_B, above y_B, or below y_B? Justify your choice referring to your graph and your derivation as needed.

Question 4

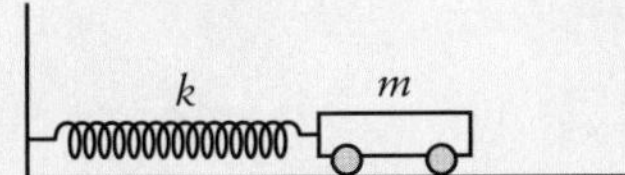

A cart of unknown mass, m, oscillates on the end of a horizontal spring of spring constant $k = 15$ N/m, as shown above. Assume the cart moves with negligible friction along the horizontal surface. The velocity of the cart as a function of time is measured by a motion sensor (not shown in diagram) and shown in the graph below. Take the positive direction to the right.

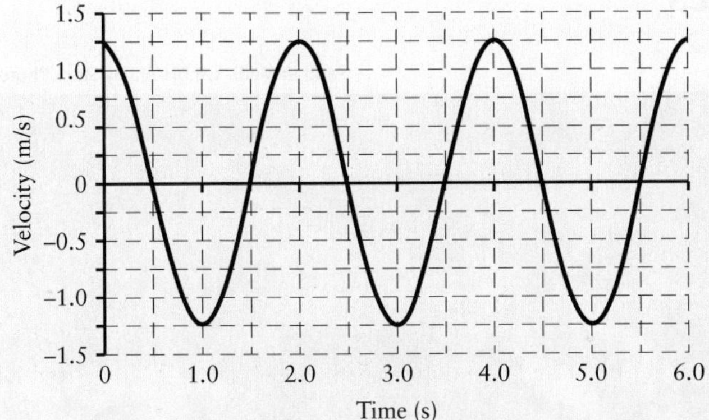

(a) Draw a free-body diagram for the cart at time $t = 1.5$ s. Each force must be represented by a distinct arrow starting on and pointing away from the dot below, which represents the cart. The lengths of the arrows should reflect the relative magnitudes of the forces at this instant for forces where this can be determined from examining the experimental set up.

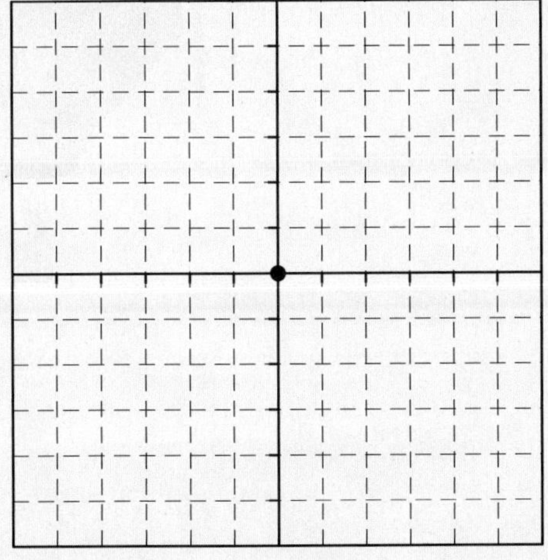

(b) Calculate the mass of the cart and the amplitude of its motion.

(c) On the axes provided below, graph the position as a function of time for this cart.

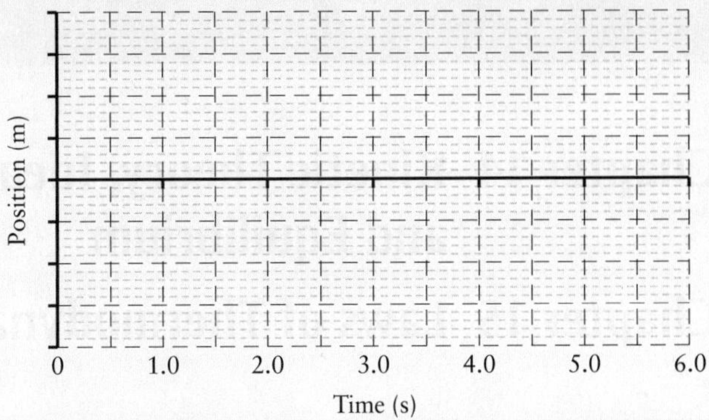

(d) i. Briefly describe how your graph in (c) agrees with your free-body diagram in (a).

ii. Briefly describe how your graph in (c) agrees with your calculation in (b).

Thermodynamics

Panther Media GmbH/Alamy Stock Photo

Case Study: Why do hot-air balloons rise?

The air pressure inside a hot-air balloon remains relatively constant as it is inflated. To help us understand why, in this chapter we will learn that under some circumstances pressure (P), volume (V), and temperature (T) are related by the ideal gas law, $PV = nRT$ (n and R are constants). Let us apply this law to the case of the balloon. To start, a burner is ignited, warming air in the balloon. That is, when energy is added to the air in the balloon, air temperature increases. But because the interior of the balloon is open to the atmosphere, the pressure of the air in the balloon remains relatively constant. By inspection of the ideal gas law, the volume of the air inside the balloon must increase. That is, the balloon inflates.

As we learned in Chapter 13, an object floats in a fluid when it is less dense than the fluid that surrounds it. For a hot-air balloon, as the air inside the balloon expands it occupies a larger volume, but the *amount* of air remains constant. Thus, the density of the hot air inside the balloon is lower than the density of the cooler air outside. Why then does the balloon float? Even considering the (relatively small) contribution of the mass of the envelope of the balloon to the overall density, the balloon is of lower density than the air that surrounds it—a balloon filled with hot air rises!

Consider another example. The bubbles of air that form as a scuba diver exhales are buoyant because the air inside a bubble is significantly less dense than the water surrounding the bubbles. So, these air bubbles rise to the surface. These air bubbles encounter changing pressure as they rise. The pressure at any depth in a fluid—let us consider water for this example—depends on the amount of overlying water (that is, water depth). This means that the pressure exerted by the water is lower closer to the surface. An air bubble, therefore, encounters increasingly lower pressure as it rises to the water's surface. Treating the trapped air as an ideal gas, we conclude, from $PV = nRT$, that when the air in a rising bubble remains at a relatively constant temperature, the volume of the bubble increases. For example, for a bubble that rises from a depth of 10 m up to the surface, pressure decreases by one-half (from 2 to 1 atm); therefore, the volume of an air bubble doubles as it rises through that distance. Watch a video of a scuba diver, and you'll see the bubbles of their exhaled breath increase in volume as they rise to the surface.

By the end of this unit, you should be able to describe the causes of pressure and temperature within a system; understand the thermal processes by which energy is transferred and the effects of such transfers; and describe the behavior of an ideal gas.

Kinetic Theory, Ideal Gases, Energy Transfer, and Equilibrium

University of Illinois at Urbana-Champaign Physics Department

YOU WILL LEARN TO:

- Define thermodynamics.
- Explain the meaning of temperature and thermal equilibrium.
- Describe the origin of pressure in an ideal gas.
- Explain the relationship between molecular kinetic energy and temperature of an ideal gas.
- Examine the relationship between the quantity of energy that enters or leaves a system through thermal processes and the temperature change of that system and calculate the temperature change of a system for a given energy transfer to or from the system.
- Explain the transfer of energy by thermal processes necessary to cause a substance to change between the solid, liquid, and gas phases.
- Describe the key properties of the thermal processes through which energy is transferred: radiation, convection, and conduction.

14-1 A knowledge of thermodynamics is essential for understanding almost everything around you—including your own body

Thermodynamics, broadly defined, is the branch of physics that allows us to describe the behavior of measurable macroscopic physical quantities of a system that depend on the microscopic objects making up the system. We know how to calculate kinetic energy for an object or for a system made up of a few objects. When the center of mass of a system is not moving, then that system, if modeled as an object, does not have translational kinetic energy, yet that doesn't mean the objects that make up the system are not moving. The temperature of a system, we will see, is a measure of the kinetic energy associated with the random motion of the microscopic objects (usually molecules) making up that system. When the systems contain so many objects, we cannot do the same sort of calculations we have done in the previous chapters. Thermodynamics deals with the relationships among measurable macroscopic properties of systems such as temperature, pressure, and volume, as well as the energy and transfer of energy associated with these properties and the processes by which those energy transfers take place, for instance, why one of the identical ice cubes in the

A liquid-in-glass thermometer works on the principle that substances like the liquid expand as the temperature increases.
(a)
iStockphoto/Thinkstock

This ice cube melts because energy is transferred into the ice, but its temperature remains at a constant 0°C until it has completely melted. The energy transferred is called heat.
(b)
Sebastian Duda/Shutterstock

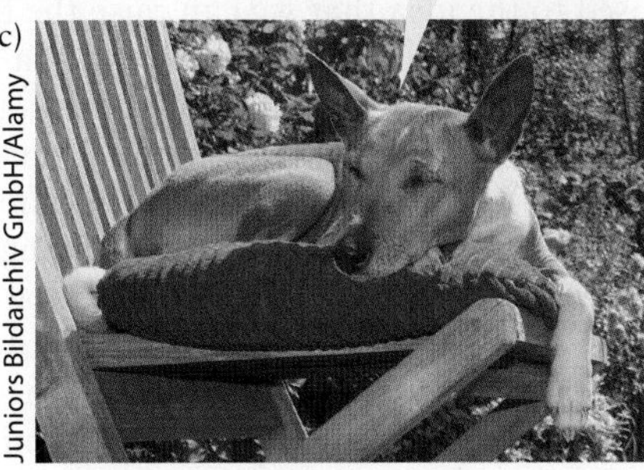

Energy is transferred to the system of the sleeping dog by radiation from the Sun and out of the system through convection of the air and conduction into the cooler pillow below.
(c)
Juniors Bildarchiv GmbH/Alamy

Figure 14-1 Thermodynamics The subject matter of thermodynamics includes (a) the meaning of temperature and how it is measured, (b) what heat is and what happens when energy is transferred into or out of a system by thermal processes, and (c) the thermal processes by which energy is transferred from one system to another.

chapter opening photograph is melting so much more rapidly because of the plate on which it rests. It is based on a statistical treatment of the microscopic objects (atoms and molecules) making up a system or interacting between systems.

Thermodynamics applies to almost *everything* in your surroundings. Thermodynamics explains how the ice cubes slowly melting in your beverage keep the beverage cold and how the clothing you wear keeps you warm. Your body is an example of a complex thermodynamic system: Some of the chemical energy released when you digest food goes into maintaining the temperature of your body at a healthy value, and some of the energy is transferred to your surroundings and warms the air around you. Indeed, to understand any aspect of nature or technology that involves the idea of *temperature*, we have to invoke thermodynamic concepts.

Thermodynamics is such a broad subject that it will take us this chapter and the next to introduce and apply its key concepts. We'll begin by defining what we mean by the temperature of a system, and we'll see that the temperature of a gas has a particularly simple interpretation. Almost all substances undergo physical changes when their temperature changes, which allows us to design thermometers, devices to measure temperature (**Figure 14-1a**). We'll define *heat* as the amount of energy that is transferred by a thermal process, just like work is defined as the amount of energy transferred by a mechanical process, and we'll see how to analyze the energy transfer that takes place when a substance changes phase, such as from solid to liquid (**Figure 14-1b**). We'll conclude this chapter by looking at the ways in which energy is transferred from one system to another by thermal processes (**Figure 14-1c**). When we focused on mechanical energy, we saw that energy was often dissipated, which usually meant that some of the energy was converted into the sort of random motion of molecules associated with temperature!

THE TAKEAWAY for Section 14-1

✔ Thermodynamics is the study of the relationships among measurable macroscopic properties of a system and transfers of energy associated with these properties that result from internal interactions between the system's microscopic constituents. These properties include temperature, pressure, and volume.

✔ The macroscopic thermodynamic properties of a system depend on the behavior of the incredibly large number of microscopic objects (atoms or molecules) that make up the system.

Prep for the AP® Exam

AP® Building Blocks

1. In Chapter 13, we introduced the concept of pressure in a fluid. How did that relate a microscopic interaction with a macroscopic measurable property?
2. The phrase "temperature measures kinetic energy" is somewhat difficult to understand and can be seen as contradictory given that a system does not need to be moving to have temperature. Give a reason why the explanation makes sense and some reasons why this is not the kinetic energy we have previously learned to calculate for a system.

14-2 Temperature is a measure of the energy within a system

A fundamental quantity in thermodynamics is the *temperature* of a system. We are used to the idea that we can raise the temperature of a system like a slice of eggplant or a cup of milk by heating it, say in a pan. We also know that we can lower the temperature of a system by cooling it, as in a refrigerator or freezer. But these everyday experiences fail to answer an important question: What *is* temperature?

Often, for the purposes of work and energy we discussed in previous chapters, we could think of what we are considering a system now as an object. The center of mass of the system can be used to describe the system's translational kinetic energy. The reason we use system here is that we cannot neglect the internal interactions of the system's many constituent molecules when we discuss thermodynamic properties. Every system that is not at absolute zero temperature has atoms or molecules moving within it. They move in random directions, though, so their individual motions do not cause translational motion of the system.

In our study of thermodynamics it is often necessary to identify substances, which will have given properties based on the individual molecules that make up the substances. In *monatomic* substances, such as helium, the molecules are actually individual atoms. In *polyatomic* substances the molecules are made up of combinations of atoms; for example, in water each molecule is made up of two hydrogen atoms and an oxygen atom. We'll use the term *molecule* to refer to the constituents of both monatomic and polyatomic substances. To define what we mean by the temperature of a quantity of a substance (a quantity of a substance is a system, or a system can be made up of quantities of more than one substance) we have to examine what happens on the *microscopic* level—that is, how the individual molecules that make up the substance(s) behave.

Molecules in any substance are always in motion. In a gas or liquid, molecules move in every direction with a range of speeds. A molecule in a solid has an equilibrium position, but it constantly wiggles back and forth around that equilibrium position. We can define **temperature** as a measure of this random-motion kinetic energy associated with molecular motion. As the temperature of a given substance increases, so does the average kinetic energy of a molecule in that substance.

This microscopic definition of temperature isn't very precise: The relationship between the average kinetic energy of a molecule and the temperature is different for gases, liquids, and solids, and depends on what type of molecule makes up the substance. What's more, this definition doesn't tell us how to *measure* temperature, since it's not very practical to analyze individual molecules in a substance to determine the temperature of that substance.

A practical approach to defining temperature is to say that *temperature is the property that you measure with a* **thermometer**. There are many kinds of thermometers that work in different ways, but they're all based on the principle that certain substances change their properties when the temperature changes. For example, nearly all liquids increase in volume as temperature increases. A liquid thermometer uses a column of alcohol or mercury inside a glass tube, and the height of that column goes up as the temperature increases and the liquid expands (see Figure 14-1a).

Strictly speaking, a thermometer measures its *own* temperature. The reason you can use a thermometer to measure the temperature of a system is the phenomenon of *thermal equilibrium*. Experiment shows that when two systems that have different temperatures are in **thermal contact**, which means that energy can be transferred from one to the other by thermal processes we will shortly learn about, energy moves between the two systems until both systems reach the same temperature. This final temperature lies between the two original temperatures of the systems. For example, when you pour cold water into a room-temperature glass, energy is transferred from the glass into the water. As a result, the temperature of the glass decreases and the temperature of the water increases. The energy transfer stops (an equilibrium is established) when the temperatures of the two systems are the *same*: Two systems in **thermal equilibrium** are at the same temperature. If one of those systems is a thermometer, the reading on the thermometer indicates the temperature of both the thermometer *and* the system with which it is in contact (**Figure 14-2**).

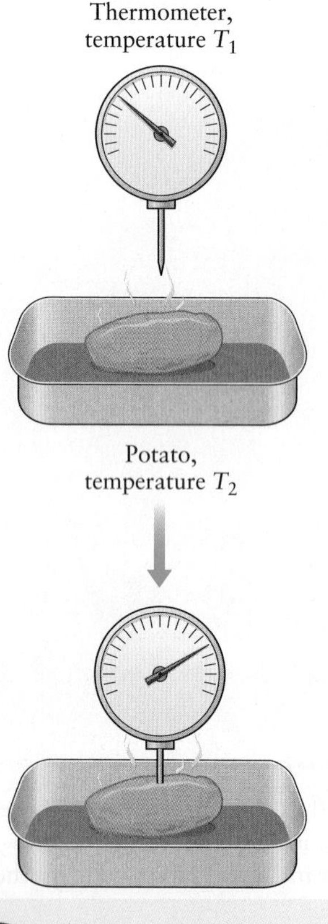

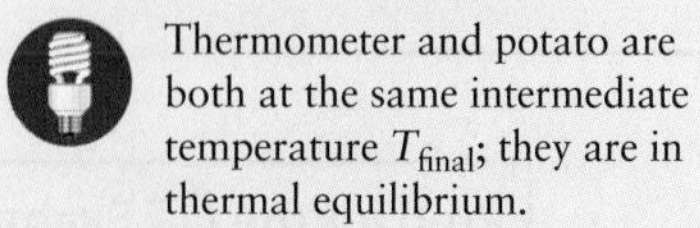

Thermometer and potato are both at the same intermediate temperature T_{final}; they are in thermal equilibrium.

Figure 14-2 Using a thermometer When a thermometer is placed in contact with a system (such as a baking potato) and allowed to come to thermal equilibrium, the thermometer and the system end up at the same temperature.

(a)

A and C are in thermal equilibrium: The reading on the thermometer tells us the temperature of A.

(b)

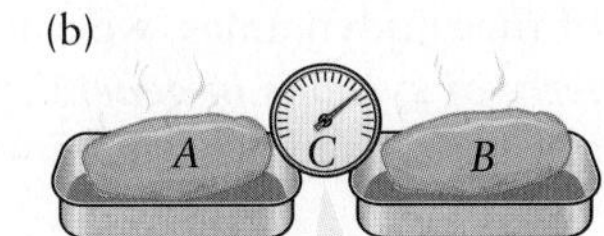

A and C are in thermal equilibrium, as are B and C. The reading on the thermometer tells us the temperature of both A and B.

(c)

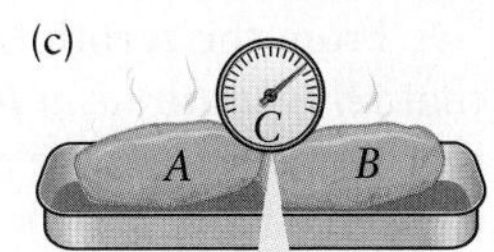

A and B are placed in contact; the thermometer reading does not change. Hence A and B are in thermal equilibrium.

The **zeroth law of thermodynamics:** If two systems A and B are in thermal equilibrium with a third system C, then A and B are also in thermal equilibrium with each other.

Figure 14-3 The zeroth law of thermodynamics If systems A and B are both in thermal equilibrium with a third system (at the same temperature), they are in thermal equilibrium with each other.

A good measurement is one that doesn't change the property being measured. Unfortunately, any thermometer causes a change in the temperature of the system being measured. That's because some energy transfer has to take place between the system and the thermometer for them to come to equilibrium, and this will cause the system's temperature to either increase or decrease. A good thermometer minimizes this transfer so that the temperature of the system hardly changes before the thermometer and system are in thermal equilibrium. Then, the thermometer reading is very close to the temperature that the system had before you made it interact with the thermometer.

Now imagine we put a thermometer C in thermal contact with system A (**Figure 14-3a**). Energy is transferred until they reach thermal equilibrium and the reading on the thermometer tells us the temperature of system A. We now also put the thermometer C in thermal contact with system B, allowing energy transfer between the thermometer and system B until they come to thermal equilibrium (**Figure 14-3b**). The reading on the thermometer then tells us the temperature of *both* A and B, so at this point the two systems have the same temperature. Finally, we put A and B directly into thermal contact with each other (**Figure 14-3c**). When we do this last step, we find that nothing changes: There is no energy transfer between A and B, which means that these two systems at the same temperature must have been in thermal equilibrium *before* we put them in direct contact with each other. In other words, *two systems at the same temperature must be in thermal equilibrium, even if they do not interact.* If we combine this observation with our previous observations about thermal equilibrium, we end up with a single statement called the **zeroth law of thermodynamics:**

If two systems are each in thermal equilibrium with a third system, they are also in thermal equilibrium with each other.

WATCH OUT !

Living organisms avoid thermal equilibrium.

Although we've emphasized the idea of thermal equilibrium, note that we humans are in general *not* in thermal equilibrium with our surroundings. Instead, our bodies regulate our internal temperature to keep it very constant (**Figure 14-4**). This is a characteristic of living organisms in general. Our bodies come to thermal equilibrium with their surroundings only after our life functions cease.

(a)

Jaren Wicklund/Alamy

(b)

Michael Honegger/Alamy

Figure 14-4 Maintaining a constant body temperature Whether in (a) a frigid snowstorm or (b) a sweltering desert, your internal body temperature remains nearly constant at about 37°C.

From the zeroth law of thermodynamics, we can draw the important conclusion that *temperature is a property of systems in general*, not just of special devices called thermometers. (There are also a *first* and a *second* law of thermodynamics; we'll study these in Chapter 15.)

To measure temperature, we need a scale such as the numbers marked on the tube of a liquid thermometer (Figure 14-1a). The numerical values associated with the markings are arbitrary, and throughout history a variety of schemes have been used to determine what the numbers on a thermometer should be.

The most common temperature scale around the world is the **Celsius scale**, based on the work of the eighteenth-century Swedish astronomer Anders Celsius. In this scale the freezing point of water is approximately 0°C and the boiling point is approximately 100°C (both values are at a standard pressure of 1 atm). In a handful of countries (including the United States) the official temperature scale is the **Fahrenheit scale**, originated by the German scientist Daniel Fahrenheit (who also invented the alcohol thermometer and mercury thermometer). On this scale water freezes at 32°F and boils at 212°F at a pressure of 1 atm.

The range of Fahrenheit temperatures between freezing and boiling (212°F − 32°F = 180°F) corresponds almost exactly to a difference of 100°C on the Celsius scale. So a change of 1°C is nearly equivalent to (100/180)°F, or (5/9)°F. If we combine this with the observation that 0°C corresponds to 32°F, we get an approximate conversion between Fahrenheit and Celsius scales:

$$T_C = \frac{5}{9}(T_F - 32) \quad \text{and} \quad T_F = \frac{9}{5}T_C + 32 \tag{14-1}$$

You should use Equations 14-1 to verify that 0°C corresponds to 32°F, and 100°C corresponds to 212°F.

The Celsius and Fahrenheit scales were based on the properties of a particular substance (water) under particular circumstances (normal atmospheric pressure on Earth). A scale that has its basis in much more fundamental physics is the **Kelvin scale**, first proposed by the nineteenth-century Scottish physicist William Thomson (1st Baron Kelvin). The Kelvin scale is based on the relationship between the pressure and temperature of low-density gases. (In Section 14-3 we'll see why it's important that the density be low.) Experiment shows that the pressure in a sealed volume of gas decreases as temperature decreases and that a graph of pressure versus temperature is a straight line (**Figure 14-5**). For a given volume both the slope of this line and the pressure at a given temperature depend on the quantity of gas in the volume. But no matter what kind of gas or what quantity of gas is used, if we extrapolate the lines in Figure 14-5 to low temperatures and pressures, we find that the pressure goes to zero at the *same* temperature. (We have to *extrapolate* each line because at sufficiently low temperatures

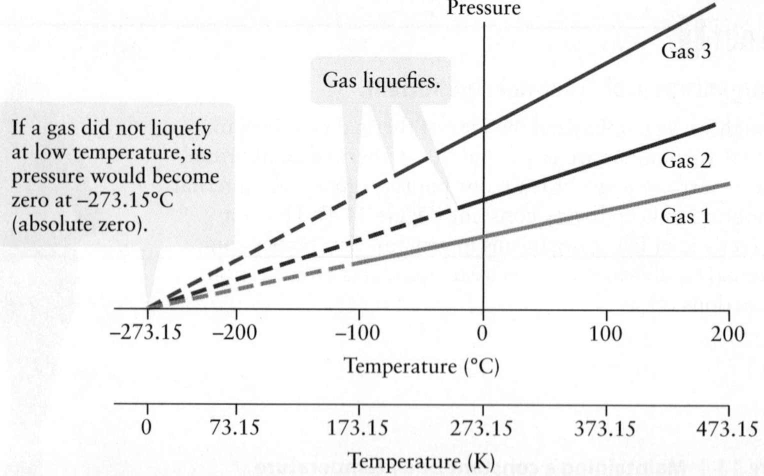

Figure 14-5 Ideal gases and absolute zero The rate at which pressure in a sealed volume of gas decreases as temperature decreases depends on the amount of gas present. However, extrapolating the graphs shows that pressure would become zero for *all* gases at the same temperature. That temperature, −273.15°C, is termed absolute zero.

any gas becomes a liquid.) This temperature is called **absolute zero** because a lower temperature is not physically possible. Zero temperature in the Kelvin scale is absolute zero, so temperatures in the Kelvin scale are never negative.

The unit of temperature used in the Kelvin scale is the **kelvin**, abbreviated K. The value of the kelvin is based on an easily reproduced temperature that can be precisely measured, which is the temperature of the **triple point** of water. The triple point of a substance is the pressure and temperature of that substance at which the solid, liquid, and **vapor** (gas) phases of the substance all coexist. (We'll learn more about the phases of matter and the triple point in Section 14-5.) For water the triple-point pressure is 0.00603 atm, and the triple point temperature of water is defined to be 273.16 K. With this choice of the value of the kelvin, a change of 1 degree Celsius is *exactly* equal to a change of 1 kelvin. The triple point of water turns out to be at 0.01°C, so 0°C corresponds to 273.15 K, and 0 K (absolute zero) corresponds to –273.15°C. To convert from a temperature in kelvins to degrees Celsius, you simply *subtract* 273.15, and to convert a Celsius temperature to a Kelvin temperature, you *add* 273.15:

$$T_C = T_K - 273.15 \text{ and } T_K = T_C + 273.15 \tag{14-2}$$

Water freezes at approximately 0°C = 273.15 K and boils at approximately 100°C = 373.15 K. (The precise values of these temperatures depend on the atmospheric pressure.)

As we will see in the following section, the Kelvin scale is a natural one to use for many purposes in physics. Here's an example of how to convert among temperature values in Fahrenheit, Celsius, and Kelvin.

WATCH OUT !

Temperature is in kelvins, not "degrees kelvin."

It's correct to say that a system has a temperature of a certain number of degrees Celsius or of degrees Fahrenheit, as in "20 degrees Celsius." But in the Kelvin scale we say a system has a temperature of a certain number of kelvins, as in "293 kelvins." We do *not* say, "293 degrees kelvin." Note also that the name of the scale (Kelvin) is capitalized, but the unit (kelvin) is not.

EXAMPLE 14-1 Cold, Hot, and In Between

On the Fahrenheit scale, a cold day might be 5°F. A comfortable temperature is 68°F, which is widely accepted as "room temperature," and a hot day might be 95°F. Express all three temperatures in degrees Celsius and kelvins.

Set Up

We can use the first of Equations 14-1 to convert from degrees Fahrenheit to degrees Celsius and the second of Equations 14-2 to convert from degrees Celsius to temperatures on the Kelvin scale.

Conversion between Celsius and Fahrenheit:

$$T_C = \frac{5}{9}(T_F - 32) \text{ and } T_F = \frac{9}{5}T_C + 32 \tag{14-1}$$

Conversion between Celsius and Kelvin:

$$T_C = T_K - 273.15 \text{ and } T_K = T_C + 273.15 \tag{14-2}$$

Solve

Convert the cold temperature of 5°F to degrees Celsius and to kelvins.

$$T_{C,\text{cold}} = \frac{5}{9}(T_{F,\text{cold}} - 32) = \frac{5}{9}(5 - 32)$$
$$= -15°\text{C}$$
$$T_{K,\text{cold}} = -15 + 273.15$$
$$= 258 \text{ K}$$

Do the conversions for the comfortable room temperature of 68°F.

$$T_{C,\text{room}} = \frac{5}{9}(T_{F,\text{room}} - 32) = \frac{5}{9}(68 - 32)$$
$$= 20°\text{C}$$
$$T_{K,\text{room}} = 20 + 273.15$$
$$= 293 \text{ K}$$

Finally, convert the hot temperature of 95°F to the Celsius and Kelvin scales.

$$T_{C,\text{hot}} = \frac{5}{9}(T_{F,\text{hot}} - 32) = \frac{5}{9}(95 - 32)$$
$$= 35°\text{C}$$
$$T_{K,\text{hot}} = 35 + 273.15$$
$$= 308 \text{ K}$$

Reflect

Our calculations show that typical temperatures on Earth's surface are close to 300 K. That's a good reference number to keep in mind. Another more precise equivalence to remember is that a typical room temperature is 68°F = 20°C = 293 K.

NOW WORK Problems 3 and 4 from The Takeaway 14-2.

Most Substances Expand When the Temperature Increases

This topic is not covered in AP® Physics 2, but is something many people have observed, so it seemed like we should mention it! Nearly all systems expand when heated and contract when cooled. For example, the lid on a jar of pickles may be too tight to unscrew when you first take the jar from the refrigerator, but when the jar warms up the lid expands and is easier to remove. This is known as **thermal expansion**. Thermal expansion is the basis of many thermometers, including those that use alcohol or mercury (see Figure 14-1a).

Liquid water is a conspicuous exception to this rule at temperatures below 4°C. Below that temperature liquid water actually *expands* as the temperature decreases further (**Figure 14-6**). So water is less dense at 0°C (the freezing point) than at slightly warmer temperatures. As a result, as the water in rivers and lakes gets cold during the winter, the coldest water (closest to the freezing point) floats to the surface, while denser, warmer water sinks to the bottom. This means that the surface water freezes first and that rivers and lakes freeze from the surface down. Were water like most substances, the coldest water would be denser than warmer water and would sink to the bottom, and bodies of water would freeze from the bottom up. In this case rivers and lakes could freeze throughout their volume during extended periods of cold temperatures, possibly destroying aquatic life. Life on Earth might not have been able to survive if water did not exhibit the unusual property seen in Figure 14-6.

Figure 14-6 shows another aspect of water that affects life on Earth: Above 4°C an increase in temperature causes liquid water to expand. The average global temperature on Earth has been increasing for decades due to the burning of fossil fuels such as coal and gasoline, which releases carbon dioxide (CO_2) in the atmosphere and causes a temperature increase. (We'll discuss the physics of this process in Section 14-6.) As a result, the water in the oceans is expanding, which causes the sea level to rise. At present the global average rate of rise is about 3 mm/y (millimeters per year), about half of which is due to thermal expansion of seawater; the rest is caused by the melting of land-based ice such as glaciers, causing water to run off into the ocean.

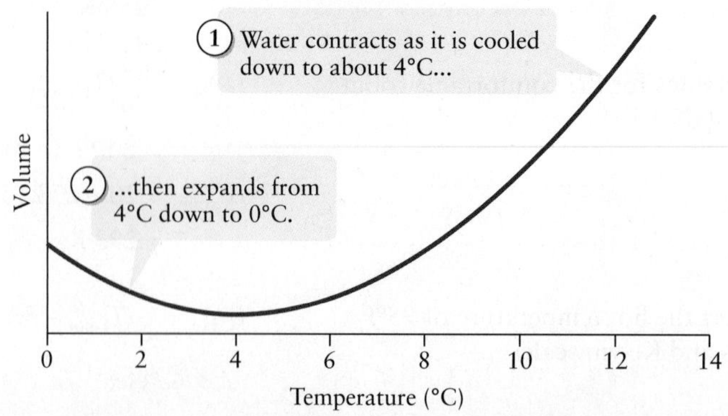

Figure 14-6 The strange case of cold water Although most substances expand when heated and contract when cooled, water is different. As water is cooled below 4°C, it stops contracting and actually expands as the temperature decreases from 4°C to the freezing point.

THE TAKEAWAY for Section 14-2

✔ The temperature of a system is a measure of the kinetic energy of the molecules that are the constituents of that system.

✔ If two systems with different temperatures are in thermal contact (energy can transfer from one to the other by thermal processes), energy will transfer from the hotter system to the cooler system until the two systems are at the same temperature. This final state is called thermal equilibrium.

✔ Two systems are in thermal equilibrium if, and only if, they are at the same temperature.

Prep for the AP Exam

AP® Building Blocks

1. Explain the physical significance of the value −273.15°C.
2. Search the Internet for *temperature scales*. How many different temperature scales can you find, and what are they?
3. (EX 14-1) Convert the following temperatures:
   (a) 28°C = __________ °F
   (b) 58°F = __________ °C
   (c) 128 K = __________ °F
   (d) 78°F = __________ K
   (e) 37°C = __________ °F
4. (EX 14-1) The highest temperature ever recorded on Earth before 2021 was 56.7°C, in Death Valley, California, in 1913. The lowest temperature on record is −89.2°C, measured at Vostok Research Station in Antarctica in 1983. Convert these extreme temperatures to °F and K.

AP® Skill Builders

5. Describe some possible uses for the following thermometers whose temperature ranges are provided in kelvins.
   Thermometer A: 200–270 K
   Thermometer B: 230–370 K
   Thermometer C: 300–550 K
   Thermometer D: 300–315 K
6. A physics student has decided that reading the textbook is very time-consuming and has further decided that she will simply attempt the problems in the book without any prior background reading. During the completion of one problem that involves a temperature conversion, the student concludes that the answer is −508°F. Discuss the validity of the answer.
7. Rank these temperatures from coldest to warmest: (a) 280 K; (b) −13°C; (c) −13°F.
8. "Temperature is the physical quantity that is measured with a thermometer." Discuss the limitations of this working definition.

AP® Skills in Action

9. A very old mercury thermometer is discovered in a physics lab. All the markings on the glass have worn away. How could you recalibrate the thermometer?
10. Starting from $T_C = 5/9(T_F - 32)$ (the first of Equations 14-1), derive a formula for converting from °C to °F (the second of Equations 14-1).

14-3 In a gas, temperature and molecular kinetic energy are directly related

We said in Section 14-2 that the temperature of a system is related to the kinetic energy of its individual molecules. Let's take a look at this relationship in more detail for the case of *gases*, which are in many ways the simplest form of matter. In a liquid or solid, adjacent molecules nearly touch each other and so the forces between molecules are strong. But, as we saw in Example 13-3, in a low-density gas, the average distance between molecules is large compared to the size of a molecule. Because gas molecules are very far apart on average, the forces they exert on each other are so weak that we can ignore them. As we'll see, that leads to tremendous simplifications in the relationship between the temperature of a gas and the kinetic energy of its molecules.

The Ideal Gas Law

Figure 14-5 shows that for a low-density gas at a constant volume, the graph of pressure versus temperature is a straight line, and the pressure P is zero when the Kelvin temperature T is zero. This means that the pressure is directly proportional to the Kelvin temperature. We can write this in equation form as

$$P = (\text{constant}) \times T \quad (14\text{-}3)$$

(low-density gas at constant volume)

NEED TO REVIEW?
Turn to page M3 in the Math Tutorial in the back of the book for more information on direct and inverse proportions.

The pressure of a given quantity of gas also depends on the volume V that the gas occupies. For example, think of a quantity of gas in a flexible container whose volume we can change. Experiment shows that if the temperature of the gas is held constant and the volume is decreased, the pressure increases. In fact, the pressure turns out to be *inversely proportional* to the volume, or in equation form

$$P = \frac{(\text{constant})}{V} \quad (14\text{-}4)$$

(low-density gas at constant temperature)

We also find that for a given kind of gas, if the volume and temperature are held fixed but the *mass* of gas in a container is increased, the pressure increases in direct proportion to the mass. Increasing the mass means increasing the number N of molecules present in the gas, so we can write this relationship as

$$P = (\text{constant}) \times N \quad (14\text{-}5)$$

(low-density gas at constant volume and temperature)

We can combine all three equations, 14-3, 14-4, and 14-5, together into a single relationship:

$$P = (\text{constant}) \times \frac{NT}{V}$$

This relationship is most commonly written in the form

EQUATION IN WORDS
Ideal gas law in terms of number of molecules

$$PV = NkT \quad (14\text{-}6)$$

Volume occupied by the gas
Number of molecules present in the gas
Pressure of an ideal gas
Temperature of the gas on the Kelvin scale
Boltzmann constant

The constant k in Equation 14-6, called the **Boltzmann constant**, turns out to have the same value for *all* gases. To four significant digits,

$$k = 1.381 \times 10^{-23} \text{ J/K}$$

Equation 14-6 is called the **ideal gas law**, and a gas that would obey this equation exactly is called an **ideal gas**. Real gases are *not* ideal: They do not obey this equation exactly, especially at high pressures or at temperatures close to the point at which the gases become liquids. At relatively low gas densities, however, the ideal gas law does a good job of representing the relationship between pressure, volume, and temperature for real gases. That's why an ideal gas is a useful idealization.

Given the very large number of molecules in most samples of gas, it's often more convenient to write Equation 14-6 in terms of the number of *moles* of a gas. One **mole** of a substance (abbreviated mol) is defined to contain exactly $6.02214076 \times 10^{23}$ molecules. This value is called Avogadro's number N_A. To four significant digits,

$$N_A = 6.022 \times 10^{23} \text{ molecules/mol}$$

The number N of molecules in a substance is therefore equal to the number of moles n multiplied by Avogadro's number: $N = nN_A$. Then Equation 14-6 becomes

$$pV = nN_A kT$$

The product $N_A k$ has a special name: We call it the **ideal gas constant** and give it the symbol R. To four significant digits,

$$R = N_A k = 8.314 \text{ J/(mol} \cdot \text{K)}$$

With this definition the ideal gas law (Equation 14-6) becomes

Volume occupied by the gas

Number of moles present in the gas

Pressure of an ideal gas

$$PV = nRT \qquad (14\text{-}7)$$

Temperature of the gas on the Kelvin scale

Ideal gas constant

EQUATION IN WORDS
Ideal gas law in terms of number of moles

The pressure P, volume V, and temperature T characterize the physical state of a system. A relationship among these quantities, like Equation 14-6 or Equation 14-7 for an ideal gas, is called an **equation of state**. The equation of state for a real gas, a liquid, or a solid can be substantially more complicated than that for an ideal gas. Earlier, we discussed the origin of the absolute zero of temperature as the point at which a line fitting the pressure versus temperature of a fixed volume of an idea gas would intercept $P = 0$. This made sense because for pressure to go to zero, the motion of the molecules in the volume of gas must go to zero. (Remember, this is an extrapolation of the line, because the gas will become a liquid before reaching this temperature.) Because of the relationship we see in Equations 14-6 and 14-7, we see it must also be true that if we plot volume versus temperature for a sample of gas at a fixed pressure, the absolute zero of temperature must be the point at which V goes to zero. We can see this makes sense by considering that to get the volume to go to zero and not have the pressure increase, the molecules must stop moving, so again $T = 0$.

EXAMPLE 14-2 An Ideal Gas at Room Temperature

At ordinary room temperature (20°C) and a pressure of 1 atm (1.013×10^5 Pa $= 1.013 \times 10^5$ N/m^2), how many molecules are there in 1.00 m^3 of an ideal gas? How many moles are there?

Set Up

We can use the two forms of the ideal gas law to determine the number of molecules N and number of moles n. Note that the temperature in the ideal gas law is the *Kelvin* temperature, but we're given the value of the *Celsius* temperature. We'll have to convert the temperature using the second of Equations 14-2.

Ideal gas law in terms of number of molecules:

$$PV = NkT \qquad (14\text{-}6)$$

Ideal gas law in terms of number of moles:

$$PV = nRT \qquad (14\text{-}7)$$

Conversion between Celsius and Kelvin:

$$T_C = T_K - 273.15 \text{ and}$$
$$T_K - T_C + 273.15 \qquad (14\text{-}2)$$

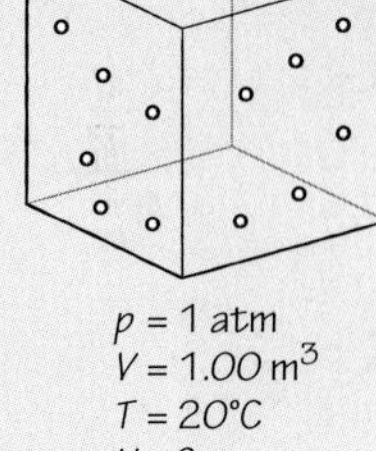

Solve

First convert the temperature to the Kelvin scale.

From the second of Equations 14-2, a temperature of 20°C corresponds to a Kelvin temperature of

$$T = 20 + 273.15 = 293 \text{ K}$$

Solve Equation 14-6 for the number of molecules N and substitute numerical values.

From Equation 14-6:

$PV = NkT$, so

$$N = \frac{PV}{kT} = \frac{(1.013 \times 10^5 \text{ N/m}^2)(1.00 \text{ m}^3)}{(1.381 \times 10^{-23} \text{ J/K})(293 \text{ K})}$$

$$= 2.50 \times 10^{25} \frac{\text{Nm}}{\text{J}}$$

Since 1 Nm = 1 J, N has *no* units: It is a pure number (in this case, the number of molecules in 1.00 m^3). So

$$N = 2.50 \times 10^{25} \text{ molecules}$$

In a similar way, find the number of moles using Equation 14-7.

From Equation 14-7:

$PV = nRT$, so

$$n = \frac{PV}{RT} = \frac{(1.013 \times 10^5 \text{ N/m}^2)(1.00 \text{ m}^3)}{(8.314 \text{ J/(mol} \cdot \text{K)) (293 K)}}$$

$$= 41.6 \frac{\text{N} \cdot \text{m} \cdot \text{mol}}{\text{J}} = 41.6 \text{ mol}$$

Reflect

We can check our results by confirming that the number of molecules equals the number of moles multiplied by Avogadro's number (the number of molecules per mole).

We can also calculate N from the number of moles n and Avogadro's number N_A:

$$N = nN_A = (41.6 \text{ mol})(6.022 \times 10^{23} \text{ molecules/mol})$$

$$= 2.50 \times 10^{25} \text{ molecules}$$

Extend: To put our result into perspective, the number of *stars* in the observable universe is estimated to be about 10^{24}. The number of molecules in a cubic meter of gas is truly astronomical!

NOW WORK Problems 1, 4, and 6 from The Takeaway 14-3.

Temperature and Translational Kinetic Energy and Absolute Zero

Remember from Chapter 13 that a gas is a fluid, so we know that the pressure arises from the forces that individual molecules exert when they collide with a surface. We will use this concept and the ideal gas law to understand the relationship between pressure, volume, and the *average translational kinetic energy* of molecules in a gas in a container.

Figure 14-7 shows a cubical box that has sides of length L and contains N molecules of an ideal gas. Each molecule has mass m. When the gas is in equilibrium, the gas as a whole exhibits no net motion. Although every molecule is moving and has momentum, the *total* momentum of all molecules in the gas averages to zero since the velocities of the individual molecules are randomly directed. In Figure 14-7 one representative molecule has just collided with and bounced off the left wall. We'll follow this molecule for the time Δt that it takes for the single molecule to return to this same position after colliding with the wall on the right side of the container. We'll assume that the molecule doesn't collide with any other molecules during this round trip. (Later in this section we'll see why this is a reasonable assumption to make.)

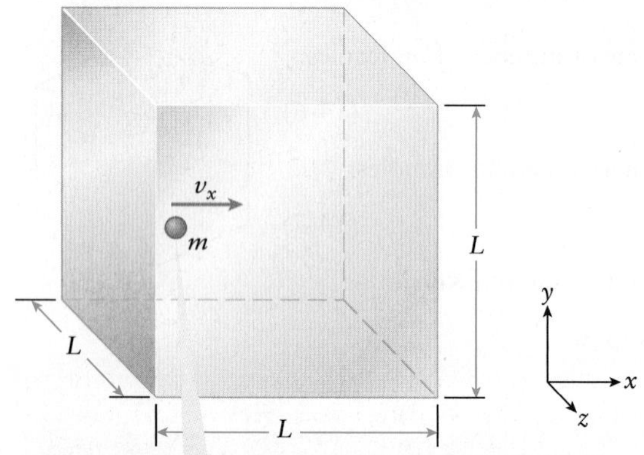
The gas molecule shown has just collided with and been reflected from the wall on the left.
v_x
m
L
L
L
y
x
z
The molecule, moving in the positive x direction, travels a distance L, collides with the right wall, and travels a distance L again before returning to this same position. This round trip takes a time Δt.

Figure 14-7 The origin of gas pressure A cubical container contains a gas. This figure shows one representative gas molecule that has just bounced off the left-hand wall. The combined effect of all molecules bouncing off the walls gives rise to the pressure of the gas.

For simplicity we'll initially consider the molecule to be moving in the x direction only. We'll also make the simplifying assumptions that the walls of the container are rigid and the gas molecules behave like hard spheres. Then the collisions are *elastic*, and no energy is lost when a molecule collides with a wall (we introduced elastic collisions in Chapter 9).

Since an individual molecule has negligible mass compared to the container, when the molecule in Figure 14-7 hits the right-hand wall of the box, it reverses direction but loses no kinetic energy. Its velocity changes from $+v_x$ (to the right) to $-v_x$ (to the left) and its momentum changes from $+mv_x$ to $-mv_x$. So the *change* in the molecule's momentum is $(-mv_x) - (+mv_x) = -2mv_x$. This momentum change, the impulse delivered to the molecule by the wall, allows us to calculate the force that the wall exerts on the molecule during the hit. The hit lasts just a short time, and there is one such hit by the right-hand wall on the molecule per time Δt (the time needed for the molecule to make a round trip and return to the right-hand wall). If we call $F_{\text{wall on molecule},x}$ the *average* force that the right-hand wall exerts on the molecule over the time Δt, the impulse-momentum theorem states that this force multiplied by Δt equals the change in the molecule's momentum due to its interaction with the right-hand wall:

$$F_{\text{wall on molecule},x}\Delta t = -2mv_x \quad \text{so} \quad F_{\text{wall on molecule},x} = -\frac{2mv_x}{\Delta t}$$

This is negative because the right-hand wall pushes the molecule to the left, in the negative x direction. By Newton's third law, the average force that the *molecule* exerts on the *wall* is the opposite of this and is *positive* (the molecule exerts a force on the wall in the positive x direction):

$$F_{\text{molecule on wall},x} = -F_{\text{wall on molecule},x} = \frac{2mv_x}{\Delta t}$$

We emphasize that this is the *average* force that this molecule exerts on the wall: During an actual hit the force is large, and the rest of the time (when the molecule is not in contact with the wall) the force is zero. Since Δt is the time it takes the molecule moving at speed v_x to move a distance $2L$ (back and forth across the length L of the box), and time equals distance divided by speed, it follows that $\Delta t = 2L/v_x$. Substituting this into our expression for $F_{\text{molecule on wall},x}$ we get

$$F_{\text{molecule on wall},x} = \frac{2mv_x}{(2L/v_x)} = \frac{mv_x^2}{L} \tag{14-8}$$

Equation 14-8 tells us the average force on the right-hand wall from *one* molecule. There are N molecules of mass m in the box shown in Figure 14-7, and each of them collides periodically with the right-hand wall. If every molecule had the same value of v_x, the total force from all gas molecules combined would just be N times the force in Equation 14-8. This is extremely unlikely to be the case, however. So we get the total force that the gas as a whole exerts on the wall by replacing v_x^2 in Equation 14-8 with the *average value* of v_x^2 (that is, averaged over its values for all of the molecules in the box) and then multiplying by N:

$$F_{\text{gas on wall},x} = \frac{Nm(v_x^2)_{\text{average}}}{L} \tag{14-9}$$

WATCH OUT!

The average of the square is not the square of the average.

The quantity $(v_x^2)_{\text{average}}$ in Equation 14-9 is the *average value* of the *square* of the quantity v_x. Since the square of a quantity is always positive, it follows that $(v_x^2)_{\text{average}}$ has a positive value. Suppose instead that you calculated the *square* of the *average value* of v_x. You might think this would give you the same answer, but it doesn't: On average there are as many molecules moving to the left in Figure 14-7 as there are moving to the right, so there are as many with a positive value of v_x as with a negative value. So the average value of v_x is zero, and the square of that is also zero. When we say to take the average value of the square, we mean square first, *then* average!

What we actually measure is the *pressure* that the gas exerts on the walls of the box. The right-hand wall is a square of side L and area L^2, so the pressure (force divided by area) on the right-hand wall is

$$P = \frac{F_{\text{gas on wall},x}}{L^2} = \frac{Nm(v_x^2)_{\text{average}}}{L^3} = \frac{Nm(v_x^2)_{\text{average}}}{V} \tag{14-10}$$

In Equation 14-10 we've used the symbol V for the volume L^3 of the cubical container of side L.

In general the gas molecules are moving in the y and z directions as well as the x direction. Since there's no preferred direction inside the box, we expect that the average values of v_y^2 and v_z^2 are the same as the average value of v_x^2: $(v_x^2)_{\text{average}} = (v_y^2)_{\text{average}} = (v_z^2)_{\text{average}}$. Furthermore, the square of the *speed* v of a given

WATCH OUT !

This is the pressure everywhere in the sample of gas, not just at the walls.

Because all the molecules are in motion throughout the volume, this is the same pressure you would measure at any point in the gas, just as we found for fluids in general in Chapter 13. This is *not* just the pressure at the walls.

molecule is the sum of the squares of its velocity components: $v^2 = v_x^2 + v_y^2 + v_z^2$. If we put these observations together, we can write an expression for the average value of the square of the speed of gas molecules:

$$(v^2)_{\text{average}} = (v_x^2)_{\text{average}} + (v_y^2)_{\text{average}} + (v_z^2)_{\text{average}} = 3(v_x^2)_{\text{average}} \tag{14-11}$$

Equation 14-11 tells us that the quantity $(v_x^2)_{\text{average}}$ in Equation 14-10 is equal to one-third of the average value of the square of the speed: $(v_x^2)_{\text{average}} = (v^2)_{\text{average}}/3$. Then we can rewrite Equation 14-10 for the gas pressure as

$$P = \frac{Nm(v^2)_{\text{average}}}{3V} \tag{14-12}$$

Now $K_{\text{translational}} = (1/2)mv^2$ is the translational kinetic energy of a single molecule. (We emphasize *translational* since the molecule could also be rotating around its axis and have rotational kinetic energy. Its constituent atoms could also be vibrating within the molecule, so there could be vibrational kinetic energy as well. We will see why this is important shortly.) So $K_{\text{translational,average}} = (1/2)m(v^2)_{\text{average}}$ is the *average* translational kinetic energy of a single molecule in the gas. In terms of this we can rewrite Equation 14-12 as

$$P = \frac{Nm(v^2)_{\text{average}}}{3V} = \frac{2N}{3V}\left[\frac{1}{2}m(v^2)_{\text{average}}\right] = \frac{2N}{3V}K_{\text{translational,average}} \tag{14-12}$$

The pressure of a given quantity of ideal gas is directly proportional to the average translational kinetic energy of a molecule of the gas. The more kinetic energy that the molecules have on average, the greater the pressure.

We can learn something equally important by comparing Equation 14-12 with the ideal gas law in terms of number of molecules, Equation 14-6: $PV = NkT$, or $P = (N/V)kT$. For both Equations 14-6 and 14-12 to be correct, it must be true that $(2/3)K_{\text{translational,average}} = kT$, or

EQUATION IN WORDS
Temperature and average translational kinetic energy of an ideal gas molecule

Average translational kinetic energy of a molecule in an ideal gas

Average value of the square of a gas molecule's speed

Temperature of the gas on the Kelvin scale

$$K_{\text{translational,average}} = \frac{1}{2}m(v^2)_{\text{average}} = \frac{3}{2}kT \tag{14-13}$$

Mass of a single gas molecule

Boltzmann constant

In words, Equation 14-13 says that *the average translational kinetic energy of a molecule in an ideal gas is directly proportional to the Kelvin temperature of the gas.* This justifies the statement we made in Section 14-2 about the physical meaning of temperature. Equation 14-13 relates a *microscopic* property of an ideal gas (the average kinetic energy of an individual gas molecule) to a *macroscopic* property (the temperature of the gas that you might measure with a thermometer).

Equation 14-13 also gives us a microscopic interpretation of absolute zero. If the temperature is 0 K, the average translational kinetic energy of a gas molecule must also be zero and the speed of the molecule must be zero, too. So absolute zero is the temperature at which all molecular motion would cease.

Note from Equation 14-13 that the average translational kinetic energy of a gas molecule does *not* depend on the *mass* of the molecule, just on the temperature. For example, the air around you is mostly nitrogen (N_2) and oxygen (O_2). A mole of N_2 molecules has a mass of 28.0 g, and a mole of O_2 molecules has a mass of 32.0 g, so the mass m of an individual N_2 molecule is (28.0/32.0) = 0.875 as much as the mass of an individual O_2 molecule. Nonetheless, on average each kind of molecule in the air has the *same* translational kinetic energy: An average nitrogen molecule must therefore be moving faster than an average oxygen molecule.

A measure of how fast gas molecules move on average is the **root-mean-square speed** or **rms speed.** To see how this is defined, first imagine finding the average value of v^2, the square of the speed v, for all the molecules in a gas that have a given mass m. This is the quantity that we've called $(v^2)_{\text{average}}$. Another word for average is *mean*, which is why $(v^2)_{\text{average}}$ is called the *mean-square* of the speed. The *root*-mean-square speed is the square root of the mean-square:

$$v_{\text{rms}} = \sqrt{(v^2)_{\text{average}}} \tag{14-14}$$

Note that v_{rms} has units of m/s. From Equation 14-13, $(1/2)m(v^2)_{\text{average}} = (3/2)kT$, so $(v^2)_{\text{average}} = 3kT/m$. If we substitute this into Equation 14-14, we get

$$v_{\text{rms}} = \sqrt{\frac{3kT}{m}} \tag{14-15}$$

Root-mean-square speed of a molecule in an ideal gas

Boltzmann constant

Temperature of the gas on the Kelvin scale

Mass of a single gas molecule

EQUATION IN WORDS
Root-mean-square speed of molecules in an ideal gas

Note that while v_{rms} is a typical speed of molecules in a gas, there is in fact a distribution of speeds. Many molecules move more slowly than v_{rms}, and many move more rapidly (**Figure 14-8**).

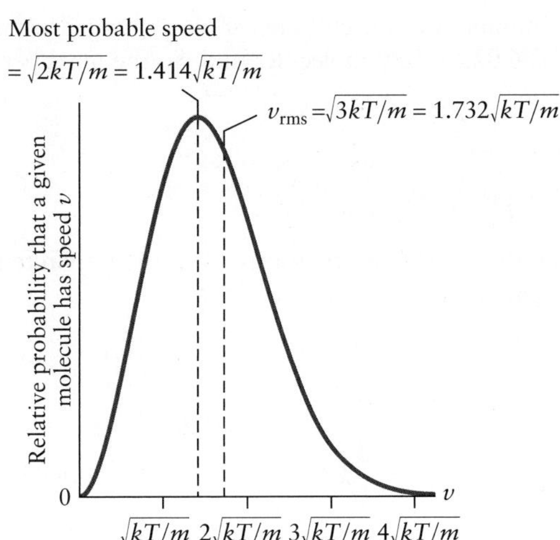

Figure 14-8 Molecular speeds in a gas The molecules in an ideal gas have a distribution of speeds. This graph shows the relative probability that any one molecule has a particular speed v. The maximum of the curve is at the most probable speed, which turns out to be $\sqrt{2kT/m} = 1.414\sqrt{kT/m}$. The rms speed $v_{\text{rms}} = \sqrt{3kT/m} = 1.732\sqrt{kT/m}$ is a bit faster than this. This type of graph is called a probability density function. For ideal gases, this function is referred to as the Maxwell–Boltzman distribution.

WATCH OUT!

The root-mean-square value is not the same as the average value.

We use the root-mean-square speed v_{rms} as a measure of the typical molecular speed in a gas because it follows naturally from Equation 14-13, which relates temperature to translational kinetic energy. It is *not*, however, the same as the *average* molecular speed. To see the difference consider a gas with just five molecules that have speeds 1.00, 2.00, 3.00, 4.00, and 5.00 m/s. The average of these speeds is

$$v_{\text{average}} = \frac{\left[\begin{array}{l}1.00\text{ m/s} + 2.00\text{ m/s} + 3.00\text{ m/s}\\ + 4.00\text{ m/s} + 5.00\text{ m/s}\end{array}\right]}{5} = 3.00\text{ m/s}$$

but the rms speed is

$$v_{\text{rms}} = \sqrt{(v^2)_{\text{average}}} = \sqrt{\frac{\left[\begin{array}{l}(1.00\text{ m/s})^2 + (2.00\text{ m/s})^2 + (3.00\text{ m/s})^2\\ + (4.00\text{ m/s})^2 + (5.00\text{ m/s})^2\end{array}\right]}{5}} = 3.32\text{ m/s}$$

It turns out that the average value of any collection of numbers is *always* less than the rms value. For the special case of an ideal gas, a detailed analysis shows that the average speed is $\sqrt{8kT/\pi m} = 1.596\sqrt{kT/m}$, which is less than $v_{\text{rms}} = \sqrt{3kT/m} = 1.732\sqrt{kT/m}$ but greater than the most probable speed shown in Figure 14-8.

EXAMPLE 14-3 Oxygen at Room Temperature

Calculate (a) the average translational kinetic energy and (b) the root-mean-square speed of an oxygen molecule in air at room temperature (20°C). One mole of oxygen molecules has a mass of 32.0 g = 32.0×10^{-3} kg.

Set Up

Equation 14-13 tells us the average translational kinetic energy, and Equation 14-15 tells us the rms speed of molecules. Note that we're given the mass per mole of O_2, but we'll need to convert this to m, the mass per molecule. We'll also need to convert the temperature from Celsius to Kelvin.

Temperature and average translational kinetic energy of an ideal gas molecule:

$$K_{\text{translational,average}} = \frac{1}{2}m(v^2)_{\text{average}} = \frac{3}{2}kT \quad (14\text{-}13)$$

Root-mean-square speed of molecules in an ideal gas:

$$v_{\text{rms}} = \sqrt{\frac{3kT}{m}} \quad (14\text{-}15)$$

Conversion between Celsius and Kelvin:

$$T_C = T_K - 273.15 \text{ and } T_K = T_C + 273.15 \quad (14\text{-}2)$$

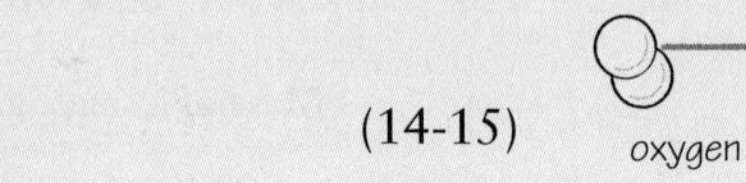

Solve

(a) First find the mass per molecule m and the Kelvin temperature T.

The number of molecules per mole is Avogadro's number, $N_A = 6.022 \times 10^{23}$ molecules/mol. So the mass per O_2 molecule is

$$m = \frac{(32.0 \times 10^{-3} \text{ kg/mol})}{(6.022 \times 10^{23} \text{ molecules/mol})}$$
$$= 5.31 \times 10^{-26} \text{ kg/molecule}$$

From the second of Equations 14-2, the Kelvin temperature that corresponds to 20°C is

$$T = 20 + 273.15 = 293 \text{ K}$$

Use Equation 14-13 to find the average kinetic energy per molecule.

From Equation 14-13:

$$K_{\text{translational,average}} = \frac{3}{2}kT = \frac{3}{2}(1.381 \times 10^{-23} \text{ J/K})(293 \text{ K})$$
$$= 6.07 \times 10^{-21} \text{ J}$$

(b) Use Equation 14-15 to find the rms speed of an O_2 molecule.

From Equation 14-15:

$$v_{\text{rms}} = \sqrt{\frac{3kT}{m}} = \sqrt{\frac{3(1.381 \times 10^{-23} \text{ J/K})(293 \text{ K})}{5.31 \times 10^{-26} \text{ kg}}}$$
$$= 478 \text{ m/s}$$

Reflect

The rms speed of O_2 molecules in air is tremendous, 478 m/s = 1720 km/h = 1070 mi/h. Very few O_2 molecules travel at precisely that speed: Their speeds range from nearly zero to many times faster than 478 m/s (see Figure 14-8).

Note that *nitrogen* (N_2) molecules in air at 20°C have the *same* translational kinetic energy (which does not depend on the mass m of the molecules) but a *different* rms speed (which is proportional to the reciprocal of $\sqrt{m}$). As we discussed above, an N_2 molecule has 0.875 the mass of an O_2 molecule, so the rms speed for N_2 is $1/\sqrt{0.875} = 1.07$ times faster than the rms speed for O_2.

NOW WORK Problems 2, 3, and 5 from The Takeaway 14-3.

Degrees of Freedom

The factor of three in Equation 14-13 for the average translational kinetic energy of a molecule, $K_{\text{translational,average}} = (1/2)m(v^2)_{\text{average}} = (3/2)kT$, is significant. It arises because $(v^2)_{\text{average}}$ is the sum of three terms, one for each component of molecular motion (Equation 14-11): $(v^2)_{\text{average}} = (v_x^2)_{\text{average}} + (v_y^2)_{\text{average}} + (v_z^2)_{\text{average}}$. As we discussed above, each of these terms has the same value. That means we can write the average kinetic energy of a molecule as the sum of three equal terms:

$$\frac{1}{2}m(v^2)_{\text{average}} = \frac{1}{2}m(v_x^2)_{\text{average}} + \frac{1}{2}m(v_y^2)_{\text{average}} + \frac{1}{2}m(v_z^2)_{\text{average}} = \frac{3}{2}kT$$

In other words, on average the translational kinetic energy is shared equally between energies associated with the motions in each of the three possible directions. Each direction of motion available to the molecule contributes an average translational kinetic energy of one-third of $(3/2)kT$, or $(1/2)kT$. We refer to each of these possible directions of motion as a **degree of freedom** of the system. (An object that could move only along a line would have one degree of freedom, and one that could move only in a plane would have two degrees of freedom. An object that can move in all three dimensions of space has three degrees of freedom.)

The number of degrees of freedom of a gas can be more than the number of directions in which gas molecules can translate. If the molecules contain more than one atom, the molecule can have energy of motion even if its translational velocity is zero. An example is the ammonia molecule, NH_3 (**Figure 14-9a**). The entire ammonia molecule can rotate (**Figure 14-9b**), and it can vibrate in two ways: The nitrogen (N) and hydrogen (H) atoms can oscillate along the length of the bond (**Figure 14-9c**), and the hydrogen atoms can be disturbed so that the angle of the bonds oscillates (**Figure 14-9d**). Each of these also represents a possible degree of freedom of the gas molecule, and each can contribute an additional $(1/2)kT$ to the average energy per molecule. Thus, the average energy per molecule is likely to be more than $(3/2)kT$ for a gas that is composed of polyatomic molecules.

Two other examples in the air around you are O_2 and N_2 at room temperature. For both, the total energy per molecule is actually close to $(5/2)kT$. That's because the molecules have five degrees of freedom: They can translate, for which there are three degrees of freedom, and they can rotate, for which there are two degrees of freedom (if the long axis of the O_2 or N_2 molecule is the z axis, one degree of freedom corresponds to rotation around the x axis and the other to rotation around the y axis). Unlike the ammonia molecule depicted in Figure 14-9, however, O_2 and N_2 molecules in the air around you do not oscillate. Quantum mechanics tells us that there's not enough energy at room temperature for bond length oscillations to contribute significantly to the energy of these diatomic molecules.

The fact that the energy of a molecule is shared equally among each degree of freedom is called the **equipartition theorem**.

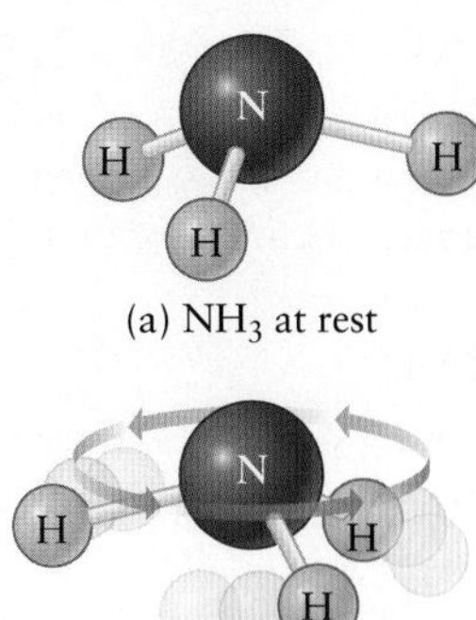

(a) NH_3 at rest

(b) Rotation of NH_3

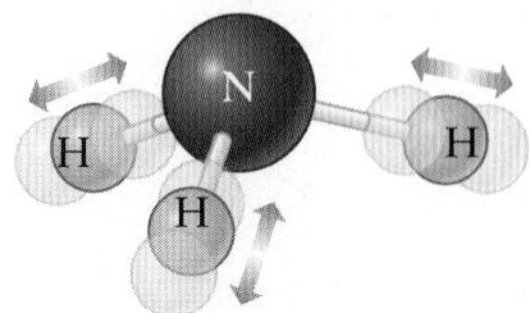

(c) Bond length oscillations of NH_3

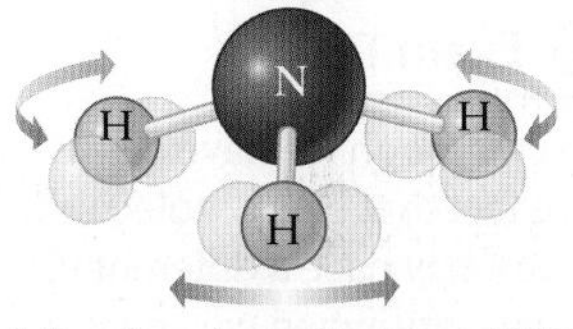

(d) Bond angle oscillations of NH_3

Figure 14-9 Motions of an ammonia molecule If a gas is made of molecules with two or more atoms, a molecule can have energy of motion even if the translational velocity of the molecule is zero. For an ammonia (NH_3) molecule, the molecule can rotate, and both the bond length and the bond angle can oscillate. This means that NH_3 has more degrees of freedom than the three translational degrees of freedom.

Mean Free Path

In our analysis of what happens in an ideal gas at the microscopic level, we considered what happens when a gas molecule hits one of the container walls, but we ignored the effects of collisions between gas molecules. How frequently do such collisions between molecules occur? One way to answer this question is in terms of the **mean free path** of a gas molecule. This is the average distance that a molecule travels from the time at which it collides with one molecule to when it collides with another molecule. If this distance is large compared to the size of the container that holds the gas, we can conclude that a molecule is unlikely to have any collisions with other molecules as it bounces back and forth between the container walls. But if the mean free path is short, a molecule may undergo many collisions with other molecules during a single trip across the container.

The mean free path depends on how densely packed the gas molecules are. If there are very few gas molecules in a container, collisions between molecules are unlikely and the mean free path is long. If, however, there are many molecules in the container, the likelihood of a collision is greater and the mean free path is short. The mean free path also depends on the *size* of the molecules: The larger the gas molecules are, the bigger "targets" they are for other molecules, the more likely they are to undergo collisions, and the shorter the mean free path.

If there are N gas molecules in a volume V, and if we model the molecules as spheres of radius r, the mean free path (to which we give the symbol λ, Greek letter "lambda") turns out to be

$$\lambda = \frac{1}{4\sqrt{2}\pi r^2 (N/V)} \tag{14-16}$$

Just as we expected, the mean free path is short if r is large (the molecules are large) or if N/V is large (there are many molecules present per volume, so the molecules are close together). Equation 14-6 tells us that $PV = NkT$, which we can rearrange to read $N/V = P/kT$, so Equation 14-16 becomes

$$\lambda = \frac{kT}{4\sqrt{2}\pi r^2 P} \tag{14-17}$$

AP® Exam Tip

Even though we have discussed the fact that every molecule in a gas may have a different value of velocity, when we analyze the properties associated with thermodynamic behavior of any ideal gas, we treat the gas as if every molecule in the gas has the average value of whatever property we are analyzing for that gas.

Test out using Equation 14-17 by calculating the mean free path for oxygen (O_2) molecules, which are about 2.0×10^{-10} m in radius, at ordinary room temperature (20°C or 293 K) and a pressure of 1 atm $= 1.013 \times 10^5$ N/m^2. You'll find that $\lambda = 5.6 \times 10^{-8}$ m, a distance that is large compared to the size of a molecule but hundreds of millions of times smaller than the size of a typical container for air. An O_2 molecule trying to travel from one side to another of a room 2 m wide would undergo tens of millions of collisions along the way!

How, then, can we justify the assumption we made earlier in this section that we can *ignore* collisions between molecules? The explanation is that a given molecule collides with other molecules that are moving in *all* directions. Some collisions make the molecule speed up, and others make it slow down; some deflect it to the left, and others deflect it to the right. The result is that the effects of all these collisions largely cancel out, so on average we can treat molecules as if they were unaffected by collisions. Note that collisions also tend to *randomize* the velocities of the molecules that make up the gas, which justifies our assumption that the average values of v_x^2, v_y^2, and v_z^2 are all the same.

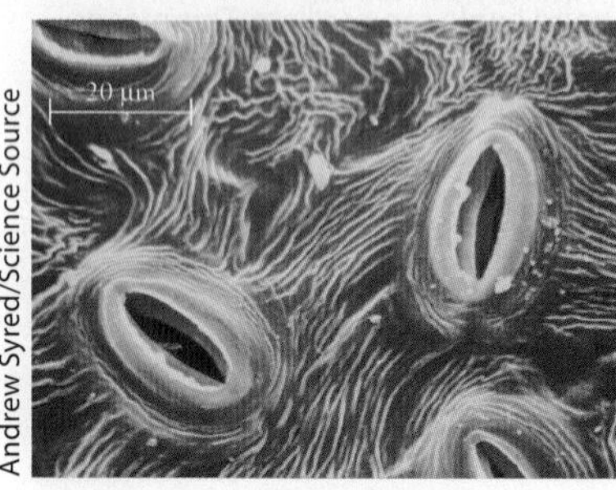

Andrew Syred/Science Source

Figure 14-10 Stomata on the epidermis of a leaf This photomicrograph shows fine details on the leaf of an elder tree (*Sambucus nigra*). The size of the stomata, through which the leaf exchanges gases with the atmosphere, is related to the mean free path of gas molecules.

The concept of mean free path plays an important role in botany. The leaves of plants "breathe" through their pores or *stomata*, tiny apertures through which they exchange gas molecules with the surrounding air (**Figure 14-10**). In photosynthesis, carbon dioxide (CO_2) is taken in through the pores and oxygen (O_2) is released; in respiration, O_2 is taken in and CO_2 is released; and in transpiration, water vapor (H_2O in its gas phase) is released. Each pore is like a narrow tube, and gas molecules passing through the tube experience collisions with the walls of the tube as well as with each other. When the diameter of the tube is large compared to the mean free path of the gas molecules, gas molecules collide mostly with each other and only infrequently with the tube walls. As a result, gas can flow relatively easily through the tube. But if the tube diameter is about the same as the mean free path, collisions with the walls become more frequent and the rate of gas flow slows dramatically. Stomata take advantage of this: They are able to expand when conditions require increased gas flow (for example, when intense sunlight shines on the leaf and the rate of photosynthesis goes up).

THE TAKEAWAY for Section 14-3

✔ The ideal gas law relates the pressure, volume, temperature, and number of molecules (or number of moles) for a low-density gas.

✔ The pressure that a gas exerts on the walls of its container is due to collisions that the gas molecules make with the walls and exists throughout the gas.

✔ The temperature of a gas is a measure of the average translational kinetic energy per molecule. In a gas at Kelvin temperature T, this energy is $(3/2)kT$. This does not depend on the mass of the molecule. Molecules with more than one atom can have additional energy associated with rotation or vibration.

✔ The average distance that a molecule travels between collisions with other molecules is called the mean free path. The larger the molecule and the more densely molecules are packed, the shorter the mean free path.

Prep for the AP® Exam

AP® Building Blocks

EX 4-2 1. The boiling point of water at 1.00 atm is 373 K. What is the volume occupied by water vapor due to evaporation of 10.0 g of liquid water at 1.00 atm and 373 K?

EX 4-3 2. An ideal monatomic gas is confined to a container at a temperature of 300 K. What is the average translational kinetic energy of an atom of the gas?

EX 4-3 3. Calculate v_{rms} for a helium atom if 1.00 mol of the gas is confined to a 1.00-L container at a pressure of 10.0 atm.

EX 4-2 4. One mole of an ideal gas is at a pressure of 1.00 atm and occupies a volume of 1.00 L. (a) What is the temperature of the gas? (b) Convert the temperature to °C and °F.

AP® Skill Builders

EX 4-3 5. Calculate the energy of a sample of 1.00 mol of ideal oxygen (O_2) gas molecules at a temperature of 300 K. Assume that the molecules are free to rotate and move in three dimensions, but also note that this temperature is too low to allow any other motions.

EX 14-2 6. A 55.0-g sample of a certain gas occupies 4.13 L at 20.0°C and 10.0 atm pressure. What is the gas?

AP® Skills in Action

7. If you halve the value of the root-mean-square speed, or v_{rms}, of an ideal gas, the absolute temperature must be
(A) reduced to one-half its original value.
(B) reduced to one-quarter its original value.
(C) increased to twice its original value.
(D) increased to four times its original value.

8. The mean free path for O_2 molecules at a temperature of 300 K and at 1.00 atm pressure is 7.10×10^{-8} m. Use these data to estimate the size of an O_2 molecule.

9. In Section 14-3 we show that $K_{translational,average} = (3/2)kT$. The result is valid for a three-dimensional collection of atoms. (a) Discuss any changes in the formula that would correspond to a one-dimensional system. (b) What change would be necessary for a ten-dimensional space?

10. A small fraction of the molecules in air are atoms of helium. A mole of helium has a mass of 4.00 g, and a helium atom has a radius of about 3×10^{-11} m. State whether each of the following quantities will be greater, less, or the same for helium than for oxygen (O_2, with a mass of 32.0 g/mol and a radius of about 2.0×10^{-10} m) in a given quantity of air: (a) the average translational kinetic energy per molecule; (b) the total energy of all kinds per molecule; (c) the rms speed; (d) the mean free path. Justify your answers in sentences referring to equations and physics principles.

14-4 Heat is the amount of energy that is transferred in a thermal process

We've seen that the temperature of a system is a measure of the average kinetic energy of its molecules. We've also seen that if two systems at different temperatures are placed in contact, their temperatures eventually come to the same value. So there must be a transfer of energy between systems at different temperatures. While we will explore this first, we will also see that there is a similar transfer of energy whenever a system undergoes a *phase change* such as from solid to liquid, liquid to solid, liquid to gas, or gas to liquid even though these occur at a constant temperature. We will define and discuss what happens in a phase change in the next section. The mechanism that *drives* a thermal process of energy transfer is, however, always a temperature difference.

In Chapter 7, we introduced the concept of work. We use the term *work* and the symbol W for the energy that is transferred into or out of a system as a result of a mechanical process (described in terms of a force). Since work is an amount of energy, it has units of joules (J). We called such a process "doing work." When there is work done on a system, if that work is positive (the force and the displacement of the point of contact of the force on the system are in the same direction) then energy is transferred

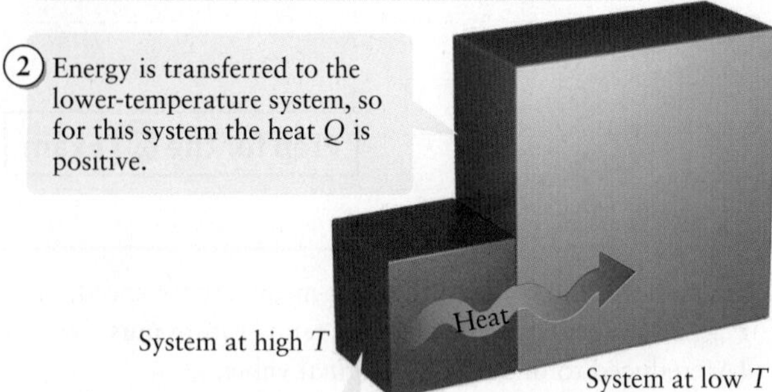

(1) If two systems at different temperatures T are in contact, energy is transferred from one to the other due to the temperature difference. The amount of energy transferred is called heat (symbol Q).
(2) Energy is transferred to the lower-temperature system, so for this system the heat Q is positive.
Heat
System at high T
System at low T
(3) Energy is transferred from the higher-temperature system, so for this system the heat Q is negative.

Figure 14-11 Energy transfers for systems in thermal contact Energy is always transferred from a higher-temperature system to a lower-temperature system. The amount of energy transferred is called heat (Q). If undisturbed, the process will continue until the two systems are in thermal equilibrium (at the same temperature).

into the system, and the value of W for that transfer is positive. When the work done is negative (the force and the displacement of the point of contact of the force on the system are in opposite directions) then energy is transferred out of the system, and the value of W for that transfer is negative. While doing work was the only process for mechanical energy transfer, in the final section of this chapter we will discuss in more detail the three thermal processes (described in terms of a temperature difference) by which energy is transferred into or out of a system.

We use the term **heat** and the symbol Q for the energy that is transferred into or out of a system as a result of a thermal process. Since heat is an amount of energy, it has units of joules (J). If energy is transferred into a system through a thermal process (we call a thermal process that transfers energy into a system **heating**), the value of Q for that transfer is positive. If energy is transferred out of the system (we call a thermal process that transfers energy out of a system **cooling**), the value of Q for that transfer is negative (**Figure 14-11**). In the next chapter we will revisit the work-energy theorem to include energy transferred by thermal processes, just as we expanded the work-energy theorem for an object to its more general form by including potential and internal energy. With the addition of including the contribution of thermal processes to understanding the energy changes of a system, the work-energy theorem becomes the first law of thermodynamics.

Internal energy, *a totally different physical quantity from heat*, is the energy within a system due to the kinetic and potential energies associated with the individual molecules that comprise the system. Note also that *temperature* (a measure of the kinetic energy per molecule) and *heat* (the amount of energy transferred by a thermal process) are also *not* the same quantity. The word *hot* doesn't mean "high heat" but rather "high temperature."

WATCH OUT !

Systems do not contain heat.

We emphasize that *heat* is the amount of energy that is transferred in a thermal process.

You can understand heating and cooling by considering what happens on the molecular level for two systems. Where the two systems touch, molecules on the surfaces of the two systems can collide with each other. The molecules in the higher-temperature system have more kinetic energy than those in the lower-temperature system. As a result, collisions between surface molecules in the two systems end up transferring kinetic energy from the molecules in the higher-temperature system to those in the lower-temperature system. Collisions between neighboring molecules in each system share these energy transfers among all of the system's molecules. The net result is that energy is transferred from the higher-temperature system to the lower-temperature one.

When there is no phase change, heating or cooling a system changes the system's temperature. If the amount of energy Q transferred into a system is relatively small, the resulting temperature change ΔT turns out to be *directly* proportional to Q and *inversely* proportional to the mass m of the system. In other words, the greater the amount of energy Q that transferred into a system, the more its temperature changes; the more massive the system and so the more material that makes up the system, the smaller the temperature change for a given amount of energy Q. (The same amount of energy that will cook a single meatball will cause hardly any temperature change in a pot roast.) We can write this relationship as

$$\Delta T = (\text{constant}) \times \frac{Q}{m}$$

The constant in this equation depends on the substance of which the system is made. It's conventional to express this equation as

EQUATION IN WORDS
Heat and the resulting temperature change

$$Q = mc\Delta T \quad (14\text{-}18)$$

Amount of energy that is transferred (heat) into (if $Q > 0$) or out of ($Q < 0$) an system

Specific heat of the substance of which the system is made

Temperature change of the system that results from the energy transfer

Mass of the system

The quantity c is called the **specific heat** of the substance that makes up the system. Its units are joules per kilogram per Kelvin [J/(kg · K) or $J \cdot kg^{-1} \cdot K^{-1}$]. For example, the value of c for aluminum is 910 J/(kg · K); this means that 910 J of energy must be transferred into a 1-kg block of aluminum to raise its temperature by 1 K (or, equivalently, 1°C). **Table 14-1** lists the values of specific heats for a range of substances. The value of the specific heat for any substance varies somewhat with the temperature; however, over the range of temperatures we typically experience, these variations are small enough that we'll ignore them.

Equation 14-18 states that if energy is transferred *into* a system, so $Q > 0$, it will cause the temperature to increase so $\Delta T > 0$. That's what happens when you warm a saucepan on a hot stove: Energy is transferred from the stove into the saucepan (so $Q > 0$ for the saucepan), and the saucepan's temperature increases (so $\Delta T > 0$ for the saucepan). If energy is transferred *out* of a system, so $Q < 0$, it will cause the temperature to decrease, so $\Delta T < 0$. This happens to the saucepan when you take it off the stove and allow it to cool: Energy is transferred from the saucepan to its surroundings (so $Q < 0$ for the saucepan), and the saucepan's temperature decreases (so $\Delta T < 0$ for the saucepan). We know that energy is conserved, so the surroundings warm up, although it might not be noticeable in a big room! Example 14-4 shows how to use Equation 14-18 to find the temperature change of a system due to a certain amount of heating or cooling.

A slightly more complicated application of Equation 14-18 is to find the final temperature of two systems that begin at different temperatures and are placed in contact (for example, hot coffee poured into a cold container). Example 14-5 shows how to solve this sort of problem. In problems of this kind we'll make the simplifying assumption that the two systems are *thermally isolated* from their environment; that is, they can exchange energy with each other but don't exchange energy with anything in their environment. So, the amount of energy transferred out of the higher-temperature system is equal to the amount of energy transferred into the lower-temperature system.

TABLE 14-1 Specific Heats

Substance	Specific heat ($J \cdot kg^{-1} \cdot K^{-1}$)
Air (50°C)	1046
Aluminum	910
Benzene	1750
Copper	387
Glass	840
Gold	130
Ice (−10°C to 0°C)	2093
Iron/steel	452
Lead	128
Marble	858
Mercury	138
Methyl alcohol	2549
Silver	236
Steam (100°C)	2009
Water (0°C to 100°C)	4186
Wood	1700

EXAMPLE 14-4 Camping Thermodynamics

A certain camping stove releases 5.00×10^4 J of energy per minute from burning propane. (This requires that it use up the propane at a rate of about 1 g/min.) If half of the released energy is transferred to 2.00 kg of water in a pot above the flame, how much does the temperature of the water change in 1.00 min? (The other half of the energy released by the stove goes into warming the surrounding air and the material of the pot.)

Set Up

Energy is transferred from the stove into the water because of the temperature difference between them, so this is energy transferred by a thermal process. In 1 min an amount of energy Q equal to one-half of 5.00×10^4 J is transferred into the water. Our goal is to find the resulting temperature change ΔT of the water.

Amount of energy transferred and the resulting temperature change:

$$Q = mc\,\Delta T \tag{14-18}$$

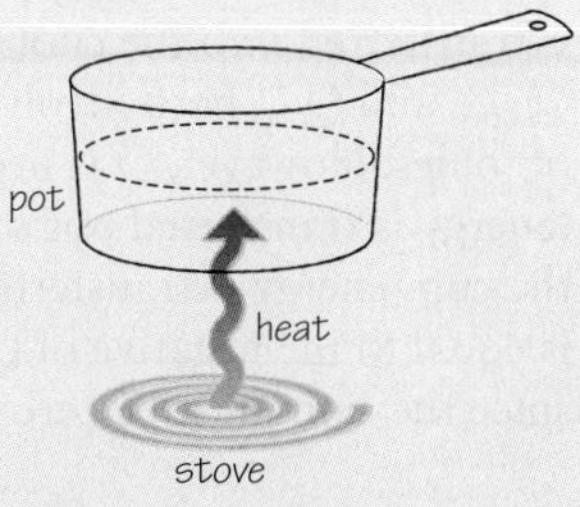

Solve

We are given the values of Q and m, and Table 14-1 lists the value of c for water. Use this and Equation 14-18 to find the temperature change of water in 1.00 min.

Solve Equation 14-18 for the temperature change ΔT:

$$\Delta T = \frac{1}{mc}Q$$

The mass is $m = 2.00$ kg, $c = 4186\ \text{J}\cdot\text{kg}^{-1}\cdot\text{K}^{-1}$ for water from Table 14-1, and $Q = (1/2) \times (5.00 \times 10^4\ \text{J})$. So

$$\Delta T = \frac{(1/2)(5.00 \times 10^4\ \text{J})}{(2.00\ \text{kg})(4186\ \text{J}\cdot\text{kg}^{-1}\cdot\text{K}^{-1})} = 2.99\ \text{K} = 2.99°\text{C}$$

If the water starts at a temperature of 20.0°C, after 1.00 min its temperature will be 20.0°C + 2.99°C = 23.0°C. After another 1.00 min its temperature will be 23.0°C + 2.99°C = 26.0°C, and so on.

Reflect

If you've ever used a camping stove, you know that it takes quite a while to bring water to a boil. Using the numbers in this problem, we see it would take roughly 27 min to increase the temperature of 2.00 kg of water (corresponding to a volume of 2.00 L) from 20°C to 100°C. That's because water has a very large specific heat, so it takes a lot of energy to raise its temperature by a given amount.

NOW WORK Problems 1, 4, 5, and 9 from The Takeaway 14-4.

EXAMPLE 14-5 Cooling Coffee

A coffee maker produces coffee at a temperature of 95.0°C (203°F), which you find is a bit too warm to drink comfortably. To cool the coffee you pour 0.350 kg of brewed coffee (mostly water) at 95.0°C into a 0.250-kg aluminum cup that is initially at room temperature (20.0°C). What is the final temperature of the coffee and cup? Assume that the coffee and cup are thermally isolated from their environment.

Set Up

There are three unknown quantities in this problem: the amount of energy transferred to the cup by thermal processes, Q_{cup}; the amount of energy transferred to the coffee by thermal processes, Q_{coffee} (which will be negative since energy is transferred *out* of the coffee); and the final equilibrium temperature T_f of the two systems (which is what we want to find). So we need three equations that relate these quantities.

We can get two equations by writing Equation 14-18 twice, once for the cup and once for the coffee. To get a third equation we'll use the idea that the two systems are thermally isolated, so the energy that is transferred *out* of the hot coffee must equal the energy that is transferred *into* the cool aluminum cup. So Q_{coffee} and Q_{cup} both involve the same number of joules. However, Q is negative for the coffee (energy is transferred out of it) and positive for the cup (energy is transferred into it), so Q_{coffee} is equal to the negative of Q_{cup}, as they must be since the sum must be zero to conserve energy.

Amount of energy transferred and the resulting temperature change:

$$Q = mc\,\Delta T \qquad (14\text{-}18)$$

Energy is conserved:

$$Q_{coffee} = -Q_{cup}$$

Solve

We are given the masses and initial temperatures of both the cup and the coffee, and we can get the specific heats from Table 14-1. Write equations for the three unknowns Q_{cup}, Q_{coffee}, and T_f. Note that for each system, ΔT is the difference between the final and initial temperatures for that system.

The cup has mass $m_{cup} = 0.250$ kg and initial temperature $T_{cup,i} = 20.0°\text{C}$, is made of aluminum with $c_{Al} = 910\ \text{J}\cdot\text{kg}^{-1}\cdot\text{K}^{-1}$, and ends up at temperature T_f. Equation 14-18 for the cup says

$$Q_{cup} = m_{cup}c_{Al}\Delta T_{cup} = m_{cup}c_{Al}(T_f - T_{cup,i})$$

The coffee has mass $m_{\text{coffee}} = 0.350$ kg and initial temperature $T_{\text{coffee,i}} = 95.0°\text{C}$, is made almost completely of water with $c_{\text{water}} = 4186\ \text{J} \cdot \text{kg}^{-1} \cdot \text{K}^{-1}$, and ends up at the same final temperature T_f as the cup. Equation 14-18 for the coffee says

$$Q_{\text{coffee}} = m_{\text{coffee}} c_{\text{water}} \Delta T_{\text{coffee}} = m_{\text{coffee}} c_{\text{water}} (T_f - T_{\text{coffee,i}})$$

The equation of energy conservation is

$$Q_{\text{coffee}} = -Q_{\text{cup}}$$

Combine these three equations to get a single equation for T_f, the quantity we are trying to find.

Substitute Q_{cup} and Q_{coffee} from the first two equations into the third equation for energy conservation:

$$Q_{\text{coffee}} = -Q_{\text{cup}}$$

$$m_{\text{coffee}} c_{\text{water}} (T_f - T_{\text{coffee,i}}) = -m_{\text{cup}} c_{\text{Al}} (T_f - T_{\text{cup,i}})$$

The only unknown quantity in this equation is the final temperature T_f.

Solve for the final temperature T_f. Recall that a temperature change of 1°C is equivalent to a temperature change of 1 K, so we can use kelvins and degrees Celsius interchangeably in our calculation.

Multiply out both sides of the equation:

$$m_{\text{coffee}} c_{\text{water}} T_f - m_{\text{coffee}} c_{\text{water}} T_{\text{coffee,i}} = -m_{\text{cup}} c_{\text{Al}} T_f + m_{\text{cup}} c_{\text{Al}} T_{\text{cup,i}}$$

Rearrange so that all the terms with T_f are on the same side of the equation:

$$m_{\text{coffee}} c_{\text{water}} T_f + m_{\text{cup}} c_{\text{Al}} T_f = m_{\text{coffee}} c_{\text{water}} T_{\text{coffee,i}} + m_{\text{cup}} c_{\text{Al}} T_{\text{cup,i}}$$

or

$$(m_{\text{coffee}} c_{\text{water}} + m_{\text{cup}} c_{\text{Al}}) T_f = m_{\text{coffee}} c_{\text{water}} T_{\text{coffee,i}} + m_{\text{cup}} c_{\text{Al}} T_{\text{cup,i}}$$

Solve for T_f:

$$T_f = \frac{m_{\text{coffee}} c_{\text{water}} T_{\text{coffee,i}} + m_{\text{cup}} c_{\text{Al}} T_{\text{cup,i}}}{m_{\text{coffee}} c_{\text{water}} + m_{\text{cup}} c_{\text{Al}}}$$

$$= \frac{\begin{bmatrix} (0.350\ \text{kg})(4186\ \text{J} \cdot \text{kg}^{-1} \cdot \text{K}^{-1})(95.0°\text{C}) \\ + (0.250\ \text{kg})(910\ \text{J} \cdot \text{kg}^{-1} \cdot \text{K}^{-1})(20.0°\text{C}) \end{bmatrix}}{(0.350\ \text{kg})(4186\ \text{J} \cdot \text{kg}^{-1} \cdot \text{K}^{-1}) + (0.250\ \text{kg})(910\ \text{J} \cdot \text{kg}^{-1} \cdot \text{K}^{-1})}$$

$$= 84.9°\text{C}$$

The coffee cools from 95.0°C to 84.9°C, and the aluminum cup warms from 20.0°C to 84.9°C.

Reflect

The temperature of the coffee decreases and the temperature of the cup increases, just as we expected. We can check our results by calculating Q_{cup} and Q_{coffee}. This calculation shows that $Q_{\text{coffee}} = -Q_{\text{cup}}$, which must be true for energy to be conserved.

Note that the final temperature of 84.9°C is closer to the initial temperature of the coffee (95.0°C) than to the initial temperature of the cup (20.0°C). That's because the mass and the specific heat are both greater for the coffee than for the cup, so a given amount of energy transferred produces a smaller temperature change in the coffee than in the cup.

The temperature change for the cup is

$$\Delta T_{\text{cup}} = T_f - T_{\text{cup,i}} = 84.9°\text{C} - 20.0°\text{C} = +64.9°\text{C} = +64.9\ \text{K}$$

(Recall that 1°C and 1 K represent the same temperature change.) The energy that is transferred into the cup is

$$Q_{\text{cup}} = m_{\text{cup}} c_{\text{Al}} (T_f - T_{\text{cup,i}})$$

$$= (0.250\ \text{kg})(910\ \text{J} \cdot \text{kg}^{-1} \cdot \text{K}^{-1})(+64.9\ \text{K})$$

$$= 1.48 \times 10^4\ \text{J}$$

The temperature change for the coffee is

$$\Delta T_{\text{coffee}} = T_f - T_{\text{coffee,i}} = 84.9°\text{C} - 95.0°\text{C} = -10.1°\text{C} = -10.1\ \text{K}$$

The energy that is transferred into the coffee is

$$\begin{aligned} Q_{\text{coffee}} &= m_{\text{coffee}} c_{\text{water}} \Delta T_{\text{coffee}} \\ &= (0.350\ \text{kg})\ (4186\ \text{J} \cdot \text{kg}^{-1} \cdot \text{K}^{-1})(-10.1\ \text{K}) \\ &= -1.48 \times 10^{4}\ \text{J} \end{aligned}$$

This is negative because energy is transferred *out* of the coffee.

NOW WORK Problems 2, 3, 6, 7, and 8 from The Takeaway 14-4.

In Examples 14-4 and 14-5 we used the SI unit of heat or energy, the joule. However, other units are also commonly used. The **calorie** (cal), equal to 4.186 J, is defined as the amount of energy that must be transferred to increase the temperature of 1 gram (1 g) of pure water from 14.5°C to 15.5°C. So in terms of calories, the specific heat of water at 14.5°C is $c = 1\ \text{cal} \cdot \text{g}^{-1} \cdot \text{K}^{-1}$. The energy content of foods is given in *food* calories, often denoted by a capital C, equal to 1000 calories or 1 kilocalorie: $1\ \text{C} = 1\ \text{kcal} = 4186\ \text{J} = 4.186\ \text{kJ}$. In countries other than the United States, the energy content on food labels is given in units of kilojoules (**Figure 14-12**).

The unit of heat in the English system is the **British thermal unit** (BTU), defined as the amount of energy that must be transferred to increase the temperature of 1 lb of pure water from 63°F to 64°F. The energy transfer rate of air conditioners and heaters is often given in BTUs, but really references how many BTUs per hour are produced by the device. For example, a 12,000-BTU burner on a stove could transfer up to 12,000 BTU to a pot set on top of it (and the surrounding air) in an hour. One BTU is equal to 1055 J, 252 cal, or 0.252 kcal.

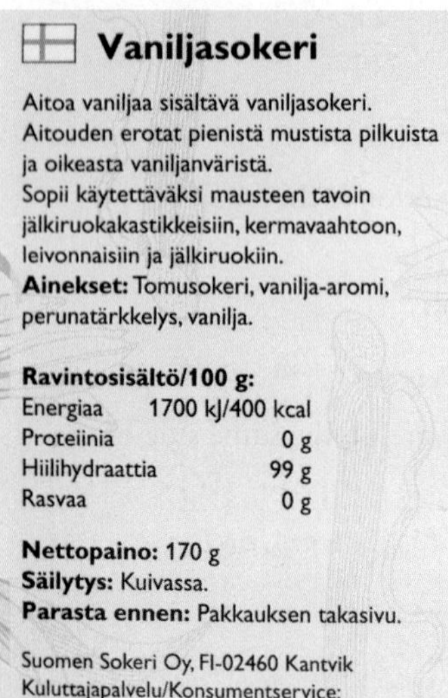

Vaniljasokeri
Aitoa vaniljaa sisältävä vaniljasokeri.
Aitouden erotat pienistä mustista pilkuista ja oikeasta vaniljanväristä.
Sopii käytettäväksi mausteen tavoin jälkiruokakastikkeisiin, kermavaahtoon, leivonnaisiin ja jälkiruokiin.
Ainekset: Tomusokeri, vanilja-aromi, perunatärkkelys, vanilja.
Ravintosisältö/100 g:
Energiaa 1700 kJ/400 kcal
Proteiinia 0 g
Hiilihydraattia 99 g
Rasvaa 0 g
Nettopaino: 170 g
Säilytys: Kuivassa.
Parasta ennen: Pakkauksen takasivu.
Suomen Sokeri Oy, FI-02460 Kantvik
Kuluttajapalvelu/Konsumentservice:
Puh/Tel. 0800-0-4400, klo/kl 12–15
kuluttajapalvelu@dansukker.fi
www.dansukker.fi
David Tauck

Figure 14-12 Counting calories and kilojoules The label on a package of flavored sugar from Finland lists the energy content in both kilojoules (kJ) and food calories (1 food calorie = 1 kilocalorie = 1 kcal).

THE TAKEAWAY for Section 14-4

✔ Heat is the amount of energy that is transferred from one system to another through a thermal process. Heat is denoted by the symbol Q. Heat is positive if energy is transferred into a system by thermal processes (heating) and negative if energy is transferred out of a system by thermal processes (cooling).

✔ The specific heat c of a substance is the amount of energy (in J) required to raise the temperature of 1 kg of the substance by 1 K.

Prep for the AP® Exam

AP® Building Blocks

1. You wish to use 0.250 kg of water to make a hot cup of coffee. If the water starts at 20.0°C and you want your coffee to be 95.0°C, calculate the minimum amount of heat required.

2. Provide a brief, simplified explanation of how specific heat differences between soil and water may contribute to the global demographic factoid that 90% of the world's population lives within 100 km of a coastline, even though the average temperature at the coast and inland (California and Kansas, for instance) are very similar.

3. Two systems that are not initially in thermal equilibrium are placed in close contact. After a while
   (A) the specific heats of both systems will be equal.
   (B) the thermal conductivity of each system will be the same.
   (C) the increase in temperature of the cooler system will be as great as the temperature decrease of the hotter one.
   (D) the temperature of each system will be the same.

4. What is the specific heat of a 0.500-kg metal sample for which the temperature rises 4.80°C when it is heated by 307 J?

AP® Skill Builders

5. Two systems that are made of the same substance are placed in thermal contact. System A is far more massive than system B.
   (a) The final temperature of the two systems will be (choose one)
   close to the initial temperature of system A.
   close to the initial temperature of system B.
   about midway between the two initial temperatures.
   (b) Briefly justify your prediction referring to physics principles.
6. In a thermodynamically sealed container, 20.0 g of 15.0°C water is mixed with 40.0 g of 60.0°C water. Calculate the final equilibrium temperature of the water.

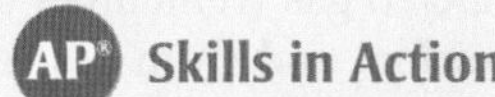

AP® Skills in Action

7. A 0.250-kg sample of copper is heated to 100.0°C and placed into a cup containing 0.300 kg of water initially at 30.0°C. Ignoring the container holding the water, calculate the final equilibrium temperature of the copper and water, assuming no energy is transferred to or from the environment.
8. EX 14-5 A 50.0-g calorimeter cup made from aluminum contains 0.100 kg of water. Both the aluminum and the water are at 25.0°C. A 0.320-kg cube of some unknown metal is heated to 98.0°C and placed into the calorimeter; the final equilibrium temperature for the water, aluminum, and metal sample is 41.0°C. Calculate the specific heat of the unknown metal and, based on your calculation, use Table 14-1 to identify its most likely composition.
9. EX 14-4 A hacksaw is used to cut a 20.0-g steel bolt. Each stroke of the saw supplies 30.0 J of energy. How many strokes of the saw will it take to raise the temperature of the bolt from 20.0°C to 80.0°C? Assume none of the energy goes into heating the surroundings. Of course, it will take significantly more energy than this to also cut the metal!
10. Calculate the temperature increase in a 1.00-kg sample of water that results from the heating of the water due to conversion of gravitational potential energy in the world's tallest waterfall, the 807-m-tall Salto Angel in Canaima National Park, Venezuela. Assume no energy transfers to or from the surroundings and that the translational velocity of the river is not affected.

14-5 Heating and cooling do not always result in a temperature change

While the temperature of a system that is being heated will not decrease, it does not always increase. Suppose you heat a cup of ice at –20°C (**Figure 14-13a**) at a constant rate. The temperature of the ice will indeed increase up to its melting point of 0°C (**Figure 14-13b**). But as you continue heating it, the temperature of the ice will *remain* at 0°C as the ice melts (**Figure 14-13c**). Once all the ice has melted so that only liquid water remains (**Figure 14-13d**), the heating again causes the temperature to increase (**Figure 14-13e**).

AP® Exam Tip

Phase changes are not emphasized on the AP® Physics 2 Exam. However, they are included here because they are part of developing a more complete conceptual understanding of heating, cooling, and equilibrium. The topic of phase change is covered more thoroughly on the AP® Chemistry Exam.

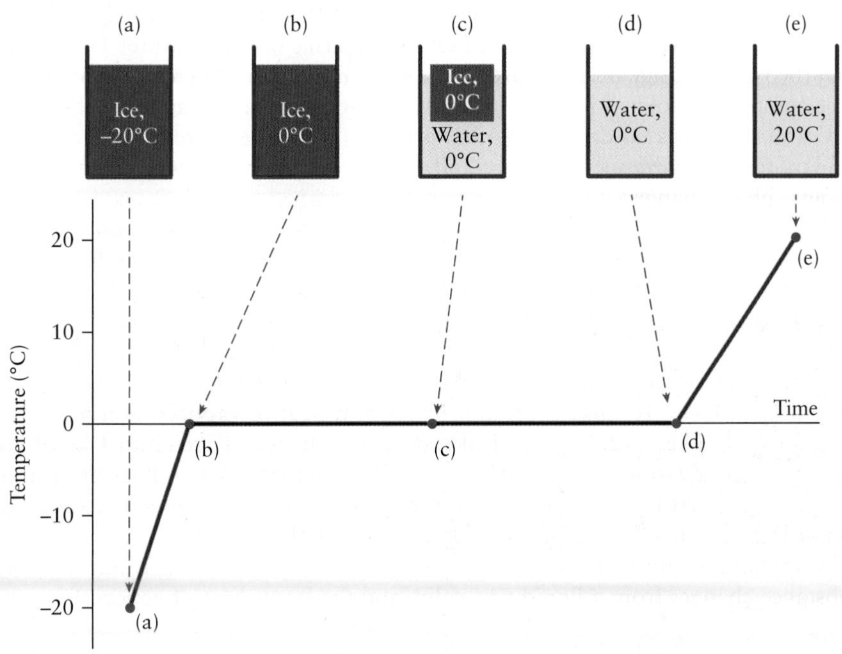

Figure 14-13 A change in phase Ice initially at a temperature below 0°C is heated at a constant rate. The temperature increases until the melting point of 0°C, then remains constant until the solid ice has all undergone a phase change to liquid water. Only then will the temperature increase above 0°C. If you reverse the process and cool liquid water at a constant rate, it will go through the same stages in reverse order.

The solid, liquid, and gaseous states of water are called its **phases**, and when ice melts (as happens between the stages in Figure 14-13b and Figure 14-13d) the water undergoes a **phase change**. Energy must be either absorbed or released for a substance to change from one phase to another, but the temperature of the substance stays the same during a phase change. For example, in boiling water at 100°C the energy that the water absorbs goes entirely into rearranging the organization of the water molecules to effect the phase change, with none left over to raise the temperature. You can verify this by putting a thermometer in a pot of water and heating the pot on a stove. The water temperature increases until it reaches 100°C and the water begins to boil. As the water boils away, the temperature will remain at 100°C; at this temperature all the energy transferred into the water goes into changing the phase of the water. (Pressure cookers work because they increase the pressure in the water, which increases the temperature at which it boils. Adding salt or sugar to water also changes the boiling temperature since it modifies the substance.)

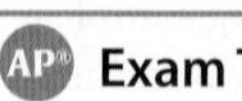
Exam Tip

The names of phase changes may help you in reading other scientific information, but are not required on the AP® Physics 2 Exam.

You're familiar with several kinds of phase change. When a liquid becomes a solid, the process is called **freezing**; when a solid becomes a liquid, the process is called melting or **fusion**. Other common phase changes include from liquid to gas (remember a substance in its gas phase is also referred to as vapor), **vaporization**, and from gas to liquid, **condensation**. Some substances can change from solid to gas without an intermediate liquid phase. A common example is carbon dioxide (CO_2); at room temperature atmospheric pressure CO_2 goes from a solid form, commonly known as *dry ice*, directly to CO_2 gas. A phase change from solid directly to gas is called **sublimation**; the reverse process, from gas directly to solid, is called **deposition**.

For a given substance at a given pressure, there is a specific temperature at which a phase change occurs. For example, at 1 atm of pressure water can change between its solid and liquid forms at 0°C. This is the temperature at which these two phases can *coexist*. The stages between Figure 14-13b and Figure 14-13d show solid ice and liquid water coexisting at 0°C as *energy is transferred* into the ice–water system.

Latent Heat

Suppose a quantity of a substance is at the temperature at which a phase change can occur. The amount of energy per unit mass that must be added to or removed from the substance to cause the phase change is called the **latent heat**. For solid ice at 0°C the latent heat is the amount of energy per kilogram that must be added to the ice to melt it; for liquid water at 0°C this same latent heat is the amount of energy per kilogram that must be removed from the water to freeze it. The units of latent heat are joules per kilogram (J/kg). *Latent* literally means "hidden." The energy required to change the phase is not associated with a temperature change, so it is considered hidden from observation, since we initially discussed energy in terms of temperature! The system in which the phase change occurs must have an external source of energy, that is, it cannot be completely isolated from its environment. If a system of water and the ice were at 0°C and there was no driving thermal process (no temperature difference across a boundary to the environment), the ice and water would remain in thermal equilibrium without a phase change.

The value of the latent heat depends on the substance as well as on the kind of phase change. For example, 2.47×10^5 J of energy must flow into a 1-kg block of iron to melt it at its melting temperature of 1811 K, but even more energy (3.34×10^5 J) must flow into a 1-kg block of ice to melt it at its melting temperature of 273 K. We say that the **latent heat of fusion** (so named since the latent heat for fusion, freezing, is the same as for melting, which is just the reverse process) is $L_F = 2.47 \times 10^5$ J/kg for iron and $L_F = 3.34 \times 10^5$ J/kg for water. The **latent heat of vaporization** for water is $L_V = 2.26 \times 10^6$ J/kg, so 2.26×10^6 J of energy must be transferred into 1 kg of liquid water at its vaporization temperature of 100°C to vaporize it. The values of L_F and L_V for water show that it takes about seven times more energy to vaporize a kilogram of water at 100°C than it does to melt a kilogram of ice at 0°C.

Note that the latent heat of fusion is also the energy *released* per unit mass when a substance changes from a liquid to a solid, and the latent heat of vaporization is the

TABLE 14-2 Melting and Boiling Temperatures (T_F and T_V) and Latent Heats of Fusion and Vaporization (L_F and L_V) for Various Substances at 1 atm of Pressure

Substance	T_F (K)	T_V (K)	L_F(J/kg) × 10^3	L_V(J/kg) × 10^3
Alcohol (ethyl)	159	351	104	830
Aluminum	933	2792	397	10,900
Copper	1357	2835	209	4730
Gold	1337	3129	63.7	1645
Hydrogen (H_2)	14	20	59.5	445
Iron	1811	3134	247	6090
Lead	600	2022	24.5	866
Nitrogen (N_2)	63	77	25.3	199
Oxygen (O_2)	54	90	13.7	213
Water	273	373	334	2260

energy *released* per unit mass when a substance changes from gas to liquid. **Table 14-2** lists the latent heat of fusion and the latent heat of vaporization for various substances.

The amount of energy needed to cause a phase change to occur is proportional to the mass m of the substance. The more massive a block of ice or iron, for example, the more energy is required to melt it. We can summarize these relationships in a single equation:

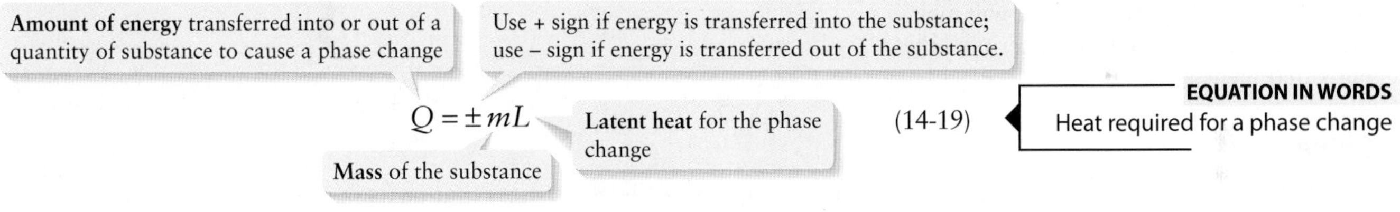

Amount of energy transferred into or out of a quantity of substance to cause a phase change
Use + sign if energy is transferred into the substance; use – sign if energy is transferred out of the substance.
Latent heat for the phase change
Mass of the substance

$$Q = \pm mL \quad (14\text{-}19)$$

EQUATION IN WORDS
Heat required for a phase change

Equation 14-19 applies to all phase changes. For example, if the phase change is fusion (a solid becomes a liquid), then L is the latent heat of fusion L_F and we use the plus sign because Q is positive (energy is transferred into the substance to make it melt). If the phase change is condensation (a gas becomes a liquid), L is the latent heat of vaporization L_V and we use the minus sign because Q is negative (energy is transferred out of the substance as it condenses).

WATCH OUT !

Equation 14-19 describes what happens in the phase change, not what it takes to get there.

Remember that a phase change can happen only if the substance is at the appropriate temperature (for example, it must be at the vaporization temperature for vaporization or condensation to take place). Equation 14-19 does *not* include the heating or cooling of a substance to get the substance to that temperature. To calculate that, you must use Equation 14-18, $Q = mc\Delta T$. So if you want to know how much energy would need to be transferred into the system to take a quantity of ice at –20°C and melt it to liquid water at 20°C (Figure 14-13), you would (1) use Equation 14-18 to find the heat needed to raise the ice temperature from –20°C to 0°C, (2) use Equation 14-19 to find the heat needed to melt that ice at 0°C, and (3) use Equation 14-18 again to find the heat to raise the water temperature from 0°C to 20°C.

Equation 14-19 states that energy must be transferred into a substance to cause a phase change from solid to liquid or from liquid to gas. Often the source of that energy is a second system in thermal contact with the first. For example, when an ice cube (one system) is dropped into a glass of water (a second system), energy is transferred from the water (so $Q_{water} < 0$) into the ice (so $Q_{ice} > 0$), warming the ice to 0°C and then melting it: The number of joules of energy that leaves the water equals the number of joules that enters the ice, so $Q_{water} = -Q_{ice}$. The temperature of the water decreases, and if the water loses enough energy, some or all of it could even freeze (which would require the quantity of ice to be large and the initial temperature of the ice to be below 0°C).

The following examples illustrate how to solve some typical problems that involve phase changes.

EXAMPLE 14-6 Melting Ice I

A 1.00-kg block of ice initially at a temperature of 0°C is placed inside an experimental apparatus that precisely controls heating, and 250 kJ of energy is transferred into the ice. What mass of ice melts as a result? What is the final temperature of the melted ice?

Set Up

The ice starts at the melting temperature, so no heating is required to the ice to get it to that temperature. Energy must be added to melt the ice, so $Q_{ice} > 0$. From Table 14-2 the latent heat of fusion for water is $L_F = 334$ kJ/kg $= 334 \times 10^3$ J/kg, so it would take 334 kJ of heat to completely melt 1.00 kg of ice. The energy transferred is less than that, so only part of the ice will melt. We'll use Equation 14-19 to determine the mass that melts.

Heat required to melt a mass m of ice:

$$Q_{ice} = +mL_F \quad (14\text{-}19)$$

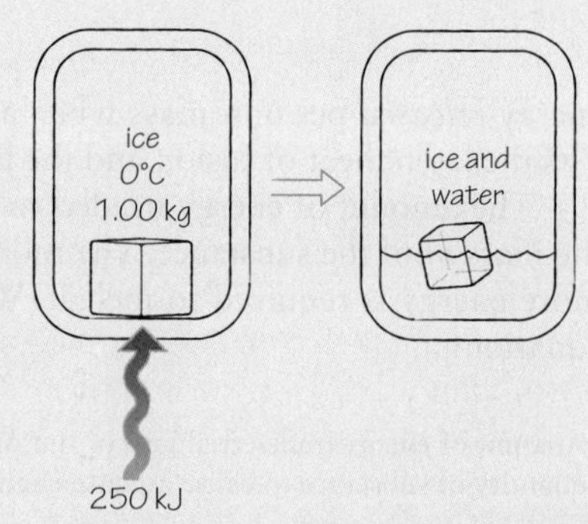

Solve

Find the mass of ice that melts.

From Equation 14-19 the mass that melts is

$$m = \frac{Q_{ice}}{L_F} = \frac{250 \text{ kJ}}{334 \text{ kJ/kg}} = 0.75 \text{ kg}$$

Three-quarters of the 1.00 kg of ice melts. Because all the heat goes into melting the ice, we end up with solid ice and liquid water that are both at the initial temperature of 0°C.

Reflect

If 334 kJ of energy were transferred to the block of ice, it would completely melt. Any more heating would go into raising the temperature of the resulting liquid water to a value greater than 0°C.

NOW WORK Problems 1, 2, 3, and 6 from The Takeaway 14-5.

EXAMPLE 14-7 Melting Ice II

A chunk of ice at −10.0°C is placed in an insulated container that holds 1.00 kg of water at 20.0°C. When the system comes to thermal equilibrium, all of the ice has melted and the entire system is at a temperature of 0.00°C. What was the initial mass of the ice?

Set Up

Three processes occur as the system reaches thermal equilibrium: (a) The temperature of the water decreases from 20.0°C to 0.0°C, so energy is transferred out of the water; (b) the temperature of the ice increases from −10.0°C to 0.0°C, so energy is transferred into the ice; and (c) the ice melts at 0°C, so additional energy is transferred into the ice. Equation 14-18 describes the energy transfer associated with the temperature changes, and Equation 14-19 describes the energy transfer associated with melting the ice. Energy conservation says that the energy that is transferred into the ice equals the energy that is transferred out of the water. Our goal is to find the initial mass of ice, m_{ice}.

Amount of energy transferred and the resulting temperature change:

$$Q = mc\,\Delta T \qquad (14\text{-}18)$$

Heat required to melt a mass m of ice:

$$Q_{ice} = +mL_F \qquad (14\text{-}19)$$

Energy is conserved:

$$Q_{ice} = -Q_{water}$$

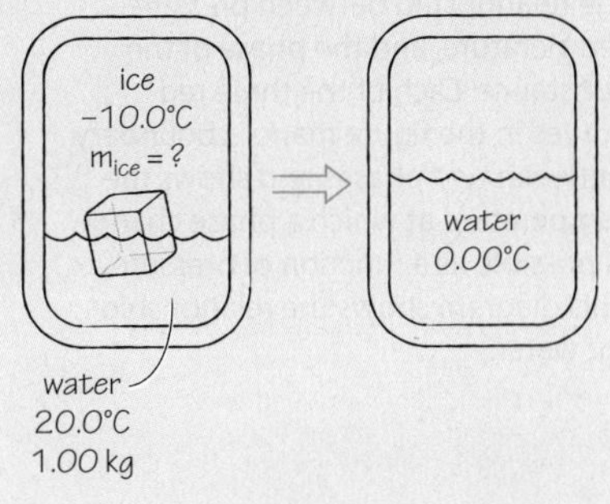

Solve

Write Equation 14-18 for the water and for the ice, and write Equation 14-19 for the ice. The specific heats for water and ice are given in Table 14-1, and the latent heat of fusion is given in Table 14-2.

Q_{water} is negative: Energy is transferred *out* of the water to make its temperature decrease.

Q_{ice} is positive: Both processes require that energy be transferred *into* the ice. Note that the term that involves melting the ice is about 16 times larger than the term that involves raising the temperature of the ice from −10.0°C to 0.0°C.

(We again use the idea that a temperature change in degrees Celsius is equivalent to a temperature change in kelvins.)

The energy transfer into the water that is required to lower its temperature from 20.0°C to 0.0°C:

$$\begin{aligned} Q_{water} &= m_{water}c_{water}\Delta T_{water} \\ &= (1.00\ \text{kg})(4186\ \text{J}\cdot\text{kg}^{-1}\cdot\text{K}^{-1})(0.0°\text{C} - 20.0°\text{C}) \\ &= (1.00\ \text{kg})(4186\ \text{J}\cdot\text{kg}^{-1}\cdot\text{K}^{-1})(-20.0\ \text{K}) \\ &= -8.37\times10^{4}\ \text{J} \end{aligned}$$

The energy transfer into the ice is the sum of that required to raise its temperature from −10.0°C to 0.0°C and that required to melt it at 0.0°C:

$$\begin{aligned} Q_{ice} &= m_{ice}c_{ice}\Delta T_{ice} + m_{ice}L_F = m_{ice}(c_{ice}\Delta T_{ice} + L_F) \\ &= m_{ice}[(2093\ \text{J}\cdot\text{kg}^{-1}\cdot\text{K}^{-1})[0.0°\text{C} - (-10.0°\text{C})] \\ &\quad + 3.34\times10^{5}\ \text{J/kg}] \\ &= m_{ice}(2.093\times10^{4}\ \text{J/kg} + 3.34\times10^{5}\ \text{J/kg}) \\ &= m_{ice}(3.55\times10^{5}\ \text{J/kg}) \end{aligned}$$

Use energy conservation to relate Q_{water} and Q_{ice}, then solve for m_{ice}.

The energy transfer into the ice equals the negative of the (negative) energy transfer into the water:

$$Q_{ice} = -Q_{water}$$

$$m_{ice}(3.55\times10^{5}\ \text{J/kg}) = -(-8.37\times10^{4}\ \text{J})$$

Solve for m_{ice}:

$$m_{ice} = \frac{8.37\times10^{4}\ \text{J}}{3.55\times10^{5}\ \text{J/kg}} = 0.236\ \text{kg}$$

Reflect

The amount of ice required has a mass of 0.236 kg. In more familiar terms, 1 kg is the mass of 1 L of water, and an ice cube from a typical ice cube tray has a mass of about 0.04 kg (40 g). So it would take about six ice cubes to reproduce the process described in this problem, which doesn't seem unreasonable.

Extend: If the container that holds the ice and water were *not* insulated, we would also have to worry about energy transfers between the water–ice mixture and the container and ultimately the surrounding environment. Using an insulated container allows us to avoid this complication.

NOW WORK Problems 8, 9, and 10 from The Takeaway 14-5.

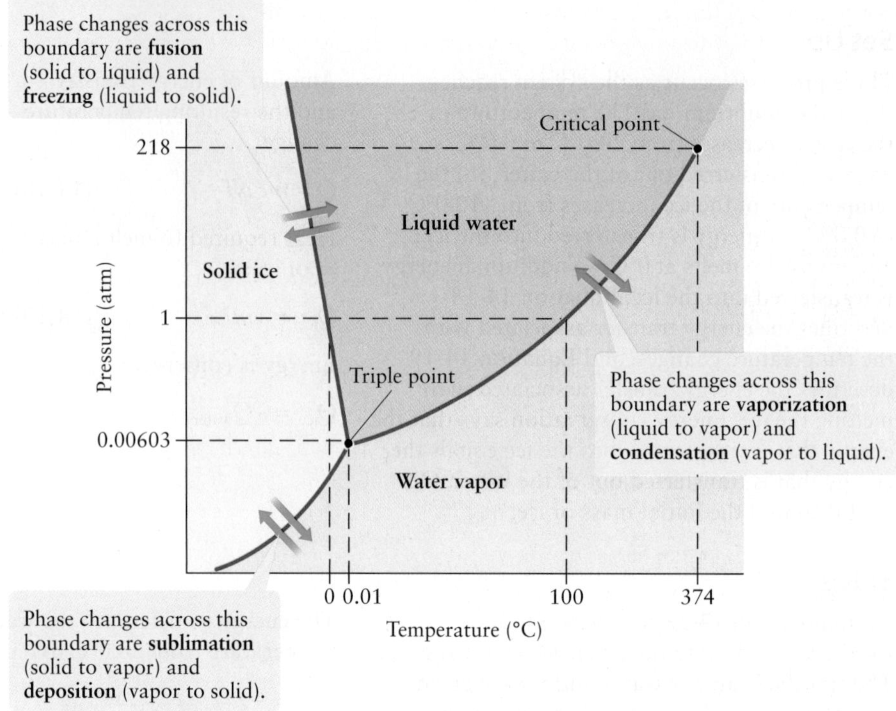
Phase changes across this boundary are **fusion** (solid to liquid) and **freezing** (liquid to solid).
Critical point
218
Liquid water
Solid ice
Pressure (atm)
1
Triple point
Phase changes across this boundary are **vaporization** (liquid to vapor) and **condensation** (vapor to liquid).
0.00603
Water vapor
0 0.01
100
374
Temperature (°C)
Phase changes across this boundary are **sublimation** (solid to vapor) and **deposition** (vapor to solid).

Figure 14-14 The phase diagram for water A phase diagram shows the relationship between pressure, temperature, and the phase of the substance. Each of the three red curves in the figure marks a boundary between two phases and shows the temperature at which a phase change is possible as a function of pressure. This diagram shows the relationship for water.

Phase Diagrams

A useful tool for visualizing the possible phase changes for a particular substance is a **phase diagram.** This is a graph of pressure versus temperature for the substance. For example, **Figure 14-14** is the phase diagram for water. The three red curves in the figure show the boundaries in terms of pressure and temperature between phases. Each point on one of the red curves represents a combination of pressure and temperature at which two phases can coexist and a phase change can take place. For example, at a pressure of 1 atm, solid ice and liquid water can coexist at 0°C; this is the melting temperature of ice at that pressure. Similarly, at 1 atm pressure, liquid water and water vapor can coexist at 100°C, the temperature at which water boils and vaporizes. The phase diagram shows that at different pressures the phase changes occur at different temperatures. For example, if the pressure is less than 1 atm, the boundary between liquid water and water vapor is at a temperature less than 100°C. That's why water boils at only 95°C at the elevation of Denver or Albuquerque (1610 m or 5280 ft), where air pressure is only about 0.82 atm. Due to the reduced boiling temperature, foods prepared in boiling water at such elevations have to be cooked longer than they would be at sea level.

In Section 14-2 we introduced the idea of the *triple point* of a substance, which is the particular combination of pressure and temperature at which the solid, liquid, and gas phases can all coexist. In Figure 14-14 the triple point is where the three red curves meet. If the pressure is greater than the triple-point pressure, the substance can exist in the solid, liquid, or gas (vapor) phase depending on the temperature. That's the case for water at sea level on Earth: The triple point of water is 0.00603 atm and 0.01°C, so at normal atmospheric pressure (1 atm) water can exist in any of the three phases. By contrast, the triple point of carbon dioxide (CO_2) is 5.10 atm and –56.6°C. Since atmospheric pressure is less than its triple-point pressure, CO_2 can exist in only the solid and gas phases. That's why solid CO_2 sublimates rather than melting at atmospheric pressure and room temperature (**Figure 14-15**).

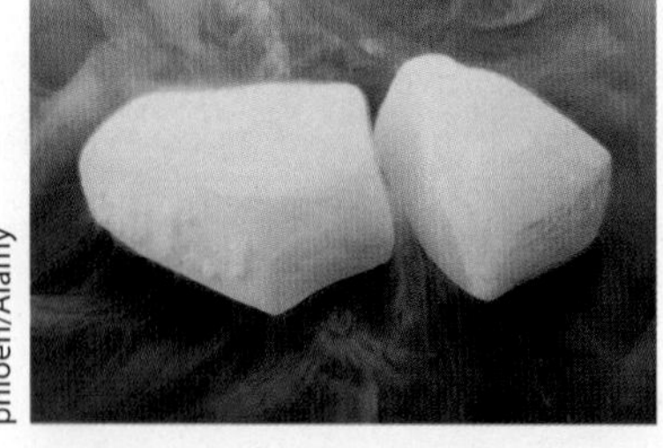
phloen/Alamy

Figure 14-15 When air pressure is less than the triple-point pressure Carbon dioxide (CO_2) is a solid at atmospheric pressure (1 atm) and temperatures below –78.5°C. The solid is also called dry ice because it does not melt at room temperature; instead, because atmospheric pressure is less than its triple-point pressure of 5.10 atm, CO_2 sublimates and goes directly to the vapor phase. The escaping cold CO_2 vapor is invisible, but due to its low temperature it causes a fog of water droplets to form in the air, which can be seen.

Figure 14-14 shows that the red curve that forms the boundary between the liquid and gas (vapor) phases comes to an end at a certain point, which for water occurs at 218 atm and 374°C. This special combination of pressure and temperature is called the **critical point.** If the pressure is greater than the critical-point pressure, the temperature is greater than the critical-point temperature, or both, there is no sharp dividing line between gas and liquid: The substance exists in a single phase called a *supercritical fluid*. Above the critical temperature, unlike below it, no increase in pressure will cause a gas to condense into liquid form.

THE TAKEAWAY for Section 14-5

✔ Energy must either be absorbed or released by a volume of a substance for that substance to change from one phase to another. The amount of energy transferred in such a thermal process is called latent heat.

✔ The latent heat of fusion is the amount of energy per unit mass required to cause a substance to undergo a phase change between the solid and liquid phases. Similarly, the latent heat of vaporization is the amount of energy per unit mass required to cause a substance to undergo a phase change between the liquid and gaseous phases.

✔ The triple point is the particular combination of pressure and temperature at which the solid, liquid, and gas phases of a substance can coexist.

Prep for the AP Exam

AP® Building Blocks

EX 14-6 1. Describe the sequence of thermodynamic steps that a very cold block of ice (below 0°C) will undergo as it transforms into steam at a temperature above 100°C.

EX 14-6 2. A sealed container (with negligible heat capacity) holds 30.0 g of 120°C steam. Describe the final state of the system in the container if this system is cooled by 100,000 J.

EX 14-6 3. Calculate the amount of energy required to change 25.0 g of ice at –40.0°C to 25.0 g of steam at 140°C.

4. Briefly explain the term *latent* as it applies to phase changes.

AP® Skill Builders

5. Assume steam is at the lowest temperature possible for the corresponding pressure.
   (a) Which has a higher temperature: a kilogram of boiling water or a kilogram of steam?
   (b) Which one can cause a more severe burn? Why?

EX 14-6 6. A 60.0-kg ice hockey player is moving at 8.00 m/s when he skids to a stop. If 40% of his kinetic energy goes into melting ice, how much water is created as he comes to a stop? Assume that the surface layer of the ice in the hockey rink has a temperature of 0°C.

AP® Skills in Action

7. Use the information in Tables 14-1 and 14-2 to rank the following from largest to smallest: (A) The energy needed to raise the temperature of 1 kg of ice from –5°C to –4°C; (B) the energy needed to melt 1 kg of ice at 0°C; (C) the energy needed to raise the temperature of 1 kg of liquid water from 20°C to 21°C; (D) the energy needed to boil 1 kg of water at 100°C; and (E) the energy needed to raise the temperature of 1 kg of steam from 110°C to 111°C.

EX 14-7 8. Suppose 20.0 g of ice at –10.0°C is placed into 0.300 kg of water in a 0.200-kg copper calorimeter. If the final temperature of the water and copper calorimeter is 18.0°C, what was the initial common temperature of the water and copper?

EX 14-7 9. Rina is a physics student who enjoys hot tea. She wants to determine how much ice is needed to cool 0.225 kg of tea to an optimal drinking temperature of 57.8°C. The method Rina uses to prepare her tea results in an initial temperature of 80.0°C. The average mass of an ice cube from the ice tray she uses is 20.0 g and has a starting temperature of –5.0°C. (a) What mass of ice in grams is needed to reach the optimal temperature of 57.8°C? (b) How many whole ice cubes should Rina use to cool her tea as close as possible to 57.8°C? Assume that the heat capacity of tea is the same as pure water, and that no energy is transferred to the surroundings.

EX 14-7 10. A 3.5-kg block of –10.0°C ice is placed in a thermally isolated chamber with 2.5 kg of 100.0°C steam. When the system in the chamber reaches equilibrium, what are the final state and temperature of the H_2O?

14-6 Thermal processes of energy transfer are radiation, convection, and conduction

Animals have evolved to live in a broad range of climates. The jackrabbit in **Figure 14-16a** has a scrawny body, long skinny legs, enormous ears, and thin fur, all of which promote energy transfer and enable the animal to survive in the desert. In contrast, the arctic hare's compact body, stubby legs, relatively small ears, and thick insulating fur help prevent energy transfer (**Figure 14-16b**).

Figure 14-16 Controlling energy transfer Both (a) jackrabbits and (b) arctic hares belong to the same genus (*Lepus*). However, jackrabbits evolved to radiate energy in the hot desert environment, while arctic hares evolved to not radiate energy in the cold conditions of the Arctic.

Figure 14-17 Three thermal energy transfer processes These cold-blooded iguanas lose energy by convection but take energy in by radiation from the Sun and by conduction from the hot rock beneath them.

The iguanas shown in **Figure 14-17** use different strategies for controlling their body temperature: After a cold swim in the Pacific Ocean, these reptiles warm themselves by lying in the tropical sun atop rocks that have already been heated by the Sun. In this section we'll look at how energy is transferred from one place to another by thermal processes, such as from a jackrabbit to its environment or from a hot rock to a cold iguana.

Energy can be transferred from one system to another through three thermal processes: *radiation*, *convection*, and *conduction*.

Radiation is energy transfer by the emission (or absorption) of electromagnetic waves. Some examples of electromagnetic waves are visible light, infrared radiation, and microwaves. (We'll see in Chapter 22 that all of these are different manifestations of the same physical phenomenon.) You use radiation when you lie out on a sunny day or warm yourself under a heat lamp. The iguanas in Figure 14-17 are using radiation to gather energy directly from the Sun.

Convection is energy transfer by the motion of a liquid or gas (such as air). Convection is what happens in a pot of water boiling on the stove: Warm water from the bottom of the pot rises, carrying energy to the cooler water at the top of the pot. Convection carries energy away from the arctic hare in Figure 14-16b and the iguanas in Figure 14-17, as the air warmed by the animal's body rises away from it.

Conduction is energy transfer by the collision of molecules in one system with the molecules in another. It requires physical contact between the two systems. If you use an electric blanket to keep warm on a cold winter's night, the energy is transferred from the blanket into your body by conduction. The iguanas in Figure 14-17 also use conduction to take in energy from the hot rock on which they are lying.

Let's look at each of these thermal processes of energy transfer individually.

Radiation

Both convection and conduction rely on molecules bumping against one another. But because radiation carries energy in the form of electromagnetic waves, it can travel through a vacuum. The energy of radiation emitted by the Sun reaches Earth after passing through 150 million kilometers of almost completely empty space.

Experiment shows that *any* system emits energy in the form of radiation. The rate at which radiation is emitted by a system—that is, the radiated *power* P in joules per second or watts—is proportional to the system's surface area A and to the fourth power of the Kelvin temperature T of the system:

EQUATION IN WORDS
Rate of energy transfer by radiation

$$P = e\sigma AT^4 \qquad (14\text{-}20)$$

Rate at which an system emits energy in the form of radiation (P)

Emissivity of the system (a number between 0 and 1) (e)

Stefan–Boltzmann constant $= 5.6704 \times 10^{-8}\ \mathrm{W \cdot m^{-2} \cdot K^{-4}}$ (σ)

Surface area of the system (A)

Temperature of the system on the Kelvin scale (T)

The **Stefan–Boltzmann constant** σ (Greek letter "sigma") has the same value for all systems. The quantity e is the **emissivity** of the surface; its value indicates how well or how poorly a surface radiates. A surface with a value of e close to 1 is a good radiator of internal energy.

The factor of T^4 in Equation 14-20 means that the radiated power can change substantially with a change in temperature. For example, a kiln for making ceramic pots can easily reach temperatures of 600 K; a pot at this temperature will radiate energy at $2^4 = 16$ times the rate when the pot is at room temperature of about 300 K. The red glow from the heating element in an electric toaster is an example of radiation of this kind. The factor A in Equation 14-20 shows that for a given surface temperature T, a system will radiate energy at a greater rate if it has a larger surface area. That's why it's useful for the jackrabbit in Figure 14-16a to have a lanky frame and big ears; this maximizes its surface area and makes it easier for the jackrabbit to get rid of excess energy in the hot desert. The more compact arctic hare in Figure 14-16b has much less surface area for its volume, so it radiates away less of its internal energy to its frigid surroundings.

Just as for jackrabbits and arctic hares, radiation plays a significant role in cooling the human body. Under typical conditions approximately half of the energy transferred from the body to the environment is in the form of radiation. This radiation is emitted almost entirely in the form of infrared light to which your eyes are not sensitive, so you can't see yourself glowing in the dark. But specialized devices can detect this radiation (**Figure 14-18**). One such device now in common use in medicine is the *temporal artery thermometer*. The nurse runs the thermometer over your head in the region of the temporal artery, and the circuitry in the thermometer detects the power radiated by the blood in the artery. This gives a very accurate measurement of body temperature while being noninvasive.

Radiation and Climate

Radiation is also the mechanism that keeps Earth from freezing. Our planet's interior releases very little energy, so what keeps the surface warm is energy that reaches us from the Sun in the form of electromagnetic radiation. In equilibrium the rate at which Earth *absorbs* solar energy (about 1.21×10^{17} W) must be equal to the rate at which it *emits* energy into space as given by Equation 14-20. Since the emission rate depends on the surface temperature, this is what determines our planet's average surface temperature T. If Earth had no atmosphere, T would be about 254 K (= −19°C = −2°F), so cold that oceans and lakes around the world should be frozen over. In fact, Earth's actual average surface temperature is a much more livable 287 K (= 14°C = 57°F). The explanation for this discrepancy is called the **greenhouse effect:** Our atmosphere prevents some of the radiation emitted by Earth's surface from escaping into space. Certain gases in our atmosphere called **greenhouse gases**, among them water vapor and carbon dioxide, are transparent to visible light but not to infrared radiation. Consequently,

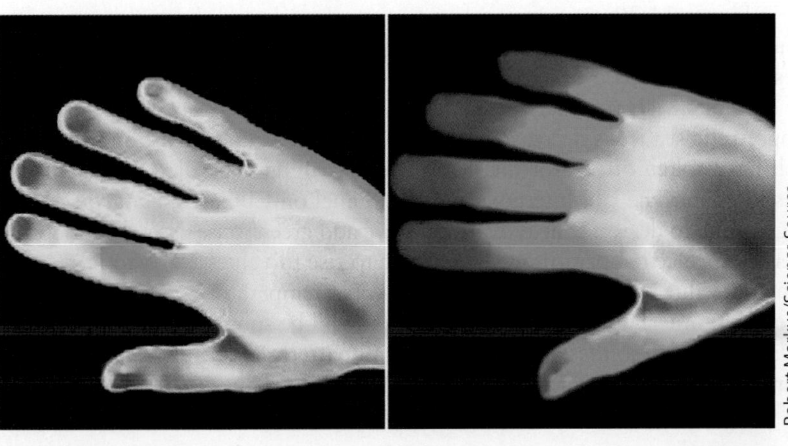

Robert Markus/Science Source

Figure 14-18 Radiation from a human hand A special infrared camera was used to record these images of a person's right hand before (left) and after (right) smoking a cigarette. The intensity of the radiation from different parts of the hand, and hence the temperature of those parts, is indicated by colors from red (warmest) through yellow, green, and blue (coldest). The temperature of the fingers drops after smoking because the nicotine in the tobacco causes blood vessels to contract and reduces blood circulation to the extremities.

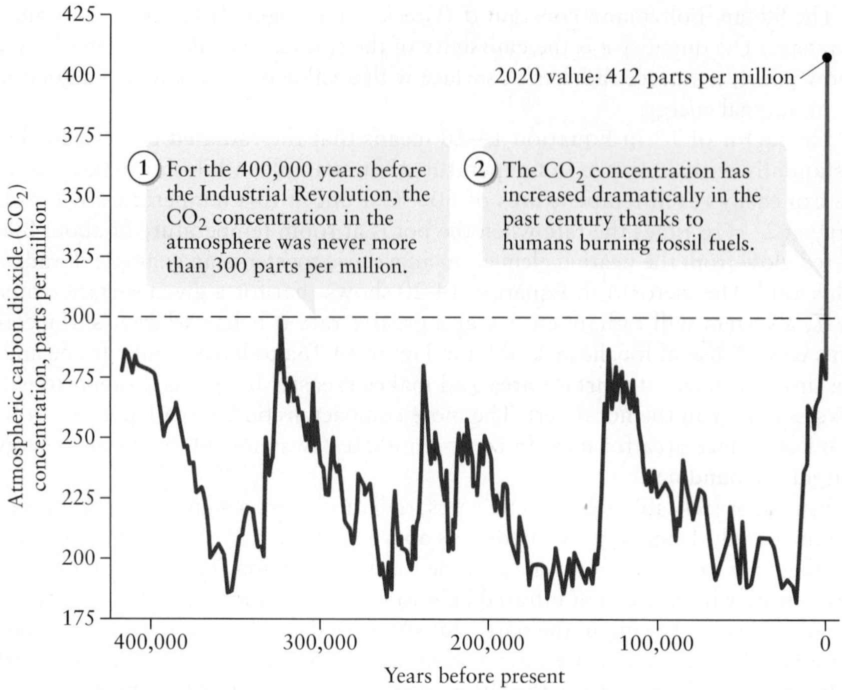

2020 value: 412 parts per million
① For the 400,000 years before the Industrial Revolution, the CO_2 concentration in the atmosphere was never more than 300 parts per million.
② The CO_2 concentration has increased dramatically in the past century thanks to humans burning fossil fuels.
Atmospheric carbon dioxide (CO_2) concentration, parts per million
425
400
375
350
325
300
275
250
225
200
175
400,000
300,000
200,000
100,000
0
Years before present

Figure 14-19 Carbon dioxide in the atmosphere Because carbon dioxide (CO_2) absorbs infrared radiation emitted by Earth's surface, its presence in our atmosphere decreases Earth's emissivity and causes the average surface temperature to increase. Thanks to the burning of fossil fuels, the level of CO_2 in our atmosphere is now greater than it has been at any time in the past 400,000 years.

visible sunlight has no trouble entering our atmosphere and warming the surface. But the infrared radiation coming from the heated surface is partially trapped by the atmosphere and cannot escape into space. This lowers Earth's net emissivity e in Equation 14-20. To have the power radiated into space equal to the power received from the Sun, the factor of T^4 in Equation 14-20 must be greater to compensate for the reduced value of e. The net effect is that Earth's surface is some 33°C (59°F) warmer than it would be without the greenhouse effect, and water remains unfrozen over most of the planet.

The warming caused by the greenhouse effect gives our planet the moderate temperatures needed for the existence of life. For more than a century, however, our technological civilization has been adding greenhouse gases to the atmosphere at an unprecedented rate. **Figure 14-19** shows how the concentration of carbon dioxide (CO_2) has varied in our atmosphere over the past 400,000 years. (Data from past centuries come from analyzing bubbles of air trapped in ancient ice in the Antarctic.) While there is natural variation in the CO_2 concentration, the value has skyrocketed since the beginning of the Industrial Revolution thanks to our burning of fossil fuels such as coal and petroleum. The result has been an amplification of the greenhouse effect and an increase in the average surface temperature, an effect known as **global warming** (**Figure 14-20**). Other explanations for global warming have been proposed, such as changes in the Sun's brightness, but these do not stand up to close scrutiny. Only greenhouse gases produced by human activity can explain the steep temperature increase shown in Figure 14-20.

The effects of global warming can be seen around the world. Nine out of 10 of the warmest years on record have occurred since 2005, and we have seen increasing numbers of droughts, water shortages, and unprecedented heat waves. Glaciers worldwide are receding, Arctic sea ice is decreasing by 13% per decade, and a portion of the Antarctic ice shelf has broken off. Unfortunately, global warming is predicted to intensify in the decades to come. The UN Intergovernmental Panel on Climate Change predicts that if nothing is done to decrease the rate at which we add greenhouse gases to our atmosphere, the average surface temperature will continue to rise to values 2.5°C to 7.8°C above the preindustrial value by the year 2100. What is worse, temperatures will rise in some regions and decline in others and the patterns of rainfall will be substantially altered. Agriculture depends on rainfall, so these changes in rainfall patterns can cause major disruptions in the world food supply. Studies suggest that the climate changes caused by a 3°C increase

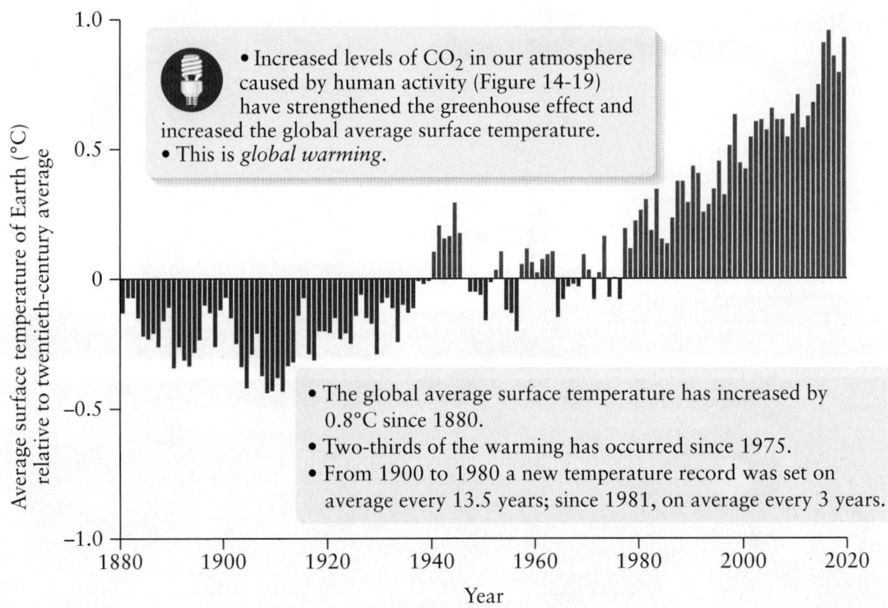

• From 1900 to 1980 a new temperature record was set on average every 13.5 years; since 1981, on average every 3 years.

Figure 14-20 Earth's average surface temperature is increasing The increase in greenhouse gases shown in Figure 14-19 has led to elevated temperatures averaged over our planet's surface. The rate of increase has accelerated over recent decades.

in the average surface temperature would cause a worldwide drop in cereal crops of 20 to 400 million tons, putting 400 million more people at risk of hunger.

The solution to global warming will require concerted and thoughtful action. Global warming cannot be stopped completely: Even if we were to immediately halt all production of greenhouse gases, the average surface temperature would increase an additional 2°C by 2100 thanks to the natural inertia of Earth's climate system. Instead, our goal must be to minimize the effects of global warming by changing how we produce energy, making choices about how to decrease our requirements for energy, and searching for ways to remove CO_2 from the atmosphere and trap it in the oceans or beneath our planet's surface. Confronting global warming is perhaps the greatest challenge to face our civilization in the twenty-first century.

Convection

You probably learned long ago that "hot air rises." It does, and the rising air carries energy with it from one region to another in a process called convection. **Figure 14-21** shows a *convection current* in a room: a continuously circulating flow of air that forms

(1) In convection a quantity of heated air rises from the radiator...
(2) ...transfers energy to the surrounding air, cooling off in the process...
28°C
26°C
25°C
23°C
35°C
21°C
19°C
17°C
16°C
(3) ...and sinks to the floor, where it returns to the radiator to start the process all over again.

Figure 14-21 A convection current A continuous circulating flow of air forms when warm air rises, forcing the air above it to move out of the way and then downward somewhere else.

Figure 14-22 Seaside convection Because the specific heat of land is much lower than the specific heat of water, the land heats up and cools down more quickly than the water. This creates onshore sea breezes in late afternoon and offshore land breezes during early morning.

when rising air forces the air above it to move out of the way and then downward somewhere else.

Convection is possible because fluids (gases and liquids) expand when heated. As a quantity of fluid expands, it becomes less dense than its surroundings and so is buoyed up according to Archimedes' principle (Chapter 13). The hot air "floats" in cooler air, and the cooler air sinks.

Convection plays an important role in driving motions in the atmosphere. People who live near the coasts of large bodies of water often experience a breeze that blows toward the shore in late afternoon and toward the water in early morning (**Figure 14-22**). Because the specific heat of land is much lower than the specific heat of water, the land heats up and cools down more quickly than the water. Especially when it is hot and the sky is clear, the temperature of the land becomes higher than the temperature of the water as the day progresses; the air above the land gets warmer and less dense. As this warm air rises, it is replaced by cooler air from above the water, resulting in an onshore, or sea, breeze. Thermal processes then automatically work to move energy from the warmer land to the cooler water. In the early morning, after the land and the air above it have cooled to a temperature below the temperature of the water, air rises above the relatively warmer water, and cooler air from above the land flows to replace it. The result is an offshore, or land, breeze. Thermal processes then automatically work to move energy from the warmer water to the cooler land.

Conduction

Perhaps you've made the mistake of touching the handle of a metal spoon that has been resting in a pot of soup on the stove. Even though the handle is touching neither the stove nor the soup, it can get hot enough to hurt your fingers if you touch it. The explanation is *conduction*: Energy absorbed from the soup is transmitted along the length of the spoon by collisions between adjacent atoms within the spoon. Through this process, the end of the spoon farthest from the soup eventually will also be hot.

Figure 14-23 shows an idealized situation in which conduction takes place. The cylinder of length L and cross-sectional area A is in thermal contact at one end with a hot system at temperature T_H and in thermal contact at the other end with a cold system at temperature T_C. Experiment shows that the rate of energy transfer is proportional

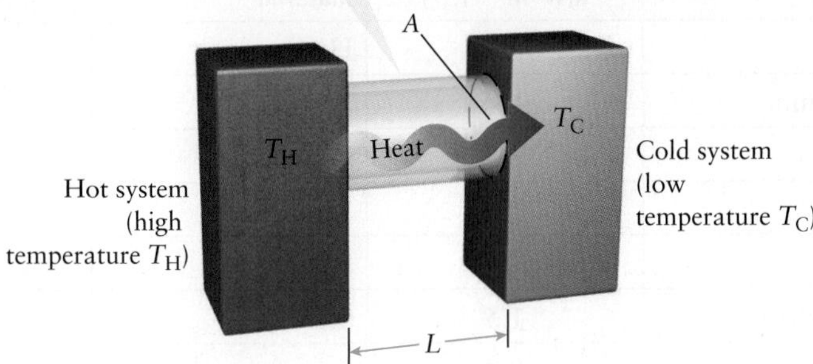

(1) The rate H of energy transfer (heating or cooling) through the cylinder is proportional to the temperature difference $T_H - T_C$.
A
T_H
Heat
T_C
Cold system (low temperature T_C)
Hot system (high temperature T_H)
L
(2) The rate of heating through the cylinder is greater when its length L is small and its cross-sectional area A is large.

Figure 14-23 Conduction The rate of energy transfer by conduction between two systems depends on the temperature difference as well as on the cross-sectional area and length of the contact region between them.

to the temperature difference $T_H - T_C$ (the greater the temperature difference, the more rapidly energy is transferred). The rate of energy transfer is also greater if the cylinder is short (L is small) and wide (A is large). We can put these observations together into a single equation:

Rate of energy transfer (heating or cooling) H in conduction = amount of energy transferred Q divided by the time Δt the transfer takes.

Temperature difference of the two systems between which energy is transferred

$$H = \frac{Q}{\Delta t} = k\frac{A}{L}(T_H - T_C) \qquad (14\text{-}21)$$

Thermal conductivity of the material through which the energy is transferred

Cross-sectional area and length of the material through which the energy is transferred

EQUATION IN WORDS
Rate of energy transfer by conduction

The quantity k, called the **thermal conductivity**, is a constant that depends on the material of which the cylinder is made. The units of $H = Q/\Delta t$ are watts (1 W = 1 J/s), so the units of k are watts per meter per kelvin ($W \cdot m^{-1} \cdot K^{-1}$). **Table 14-3** lists the thermal conductivity of various substances. A good thermal *conductor* has a high value of k; the thermal conductivity of aluminum, for example, is 235 $W \cdot m^{-1} \cdot K^{-1}$. Materials that are poor thermal conductors—those that make good thermal *insulators*—have k values less than 1 $W \cdot m^{-1} \cdot K^{-1}$. You can actually feel the difference between thermal conductors and thermal insulators. If you pick up a room-temperature aluminum rod ($k = 235\ W \cdot m^{-1} \cdot K^{-1}$) with one hand and a room-temperature wooden stick ($k = 0.12\ W \cdot m^{-1} \cdot K^{-1}$) of the same size with the other hand, the aluminum rod will feel colder. Both systems are at the same temperature (which is lower than the temperature of your hand), but the much higher thermal conductivity of the aluminum means that energy can flow into it from your hand at a faster rate.

Table 14-3 shows that the thermal conductivity of air is relatively low. This small value makes air, in particular air trapped between other materials, a good insulator. Most animals rely on air trapped between fur or feathers to slow the rate of energy loss in cold conditions. This phenomenon is even more developed in polar bears. The hair that makes up their fur is hollow; the air trapped within the hair shafts makes it a good thermal insulator. Clothing, particularly that made from cotton, wool, and other woven cloth, keeps us warm primarily because of the air trapped between the fibers. Animal fat is also a good insulator. A typical value of thermal conductivity for human fat is 0.2 $W \cdot m^{-1} \cdot K^{-1}$, which is about the same as whale blubber and nearly a third smaller than the value of water ($k = 0.58\ W \cdot m^{-1} \cdot K^{-1}$). That's one reason blubber is so important to marine mammals in frigid polar waters.

TABLE 14-3 Thermal Conductivities

Material	Thermal conductivity $k(\mathrm{W \cdot m^{-1} \cdot K^{-1}})$	Material	Thermal conductivity $k(\mathrm{W \cdot m^{-1} \cdot K^{-1}})$
Air	0.024	Hydrogen	0.168
Aluminum	235	Nitrogen	0.024
Brick	0.9	Plywood	0.13
Copper	401	Sand (dry)	0.35
Cotton	0.03	Silver	429
Earth (dry)	1.5	Steel	46
Human fat	0.2	Styrofoam	0.033
Fiberglass	0.04	Water	0.58
Glass (window)	0.96	Wood (white pine)	0.12
Granite	1.7–4.0	Wool	0.04
Gypsum (plaster) board	0.17		

EXAMPLE 14-8 Energy Loss Through a Window

Windows are a major source of energy loss from a house. That's because of all the materials typically used in construction, glass has one of the highest thermal conductivities. Determine the rate of energy loss for a house on an evening when the temperatures of the outer and inner surfaces of the windows are 14.0°C and 15.0°C, respectively. Take the total window area of the house to be 28.0 m^2 and the thickness of the windows to be 3.80 mm.

Set Up

The rate of energy transfer H through all of the windows combined is the same as if there were a single window with area $A = 28.0\ \mathrm{m}^2$ and thickness $L = 3.80\ \mathrm{mm} = 3.80 \times 10^{-3}\ \mathrm{m}$. We use Equation 14-21 to calculate the value of H.

Rate of energy transfer in conduction:

$$H = k\frac{A}{L}(T_\mathrm{H} - T_\mathrm{C}) \qquad (14\text{-}21)$$

Solve

Substitute values into Equation 14-21.

From Table 14-3 the thermal conductivity of window glass is $k = 0.96\ \mathrm{W \cdot m^{-1} \cdot K^{-1}}$. The rate of energy transfer through the windows is

$$H = k\frac{A}{L}(T_\mathrm{H} - T_\mathrm{C})$$

$$= (0.96\ \mathrm{W \cdot m^{-1} \cdot K^{-1}})\left(\frac{28.0\ \mathrm{m}^2}{3.80 \times 10^{-3}\ \mathrm{m}}\right)(15.0^\circ\mathrm{C} - 14.0^\circ\mathrm{C})$$

$$= 7.1 \times 10^3\ \mathrm{W} = 7.1\ \mathrm{kW}$$

(Recall that a temperature difference of 1.0°C is the same as a temperature difference of 1.0 K.)

Reflect

To maintain the interior of the house at the same temperature, energy has to be provided to the interior by the heating system at a rate of 7.1×10^3 W or 7.1 kW. To put that into perspective, a typical microwave oven or portable hair dryer requires about 1 kW of power to operate. So keeping this house warm with electric heat will require the same amount of electric power as seven microwave ovens or seven hair dryers running continuously. This seems to be a reasonably sized number.

Extend: One way to keep heating costs down is to replace the windows in this house with *dual-pane* or *triple-pane* windows. These windows have two or three sheets of glass separated by a gap of 1 to 2 cm, with gas filling the gap. The greater thickness of glass increases the value of L in Equation 14-21, and the gas between the panes reduces the overall thermal conductivity k, just like the air in the polar bear's hair.

NOW WORK Problems 5, 6, and 9 from The Takeaway 14-6.

THE TAKEAWAY for Section 14-6

✔ Energy transfer from one place or system to another can occur through three thermal processes: radiation, convection, and conduction.

✔ In radiation, energy is transferred from one system to a cooler one in the form of electromagnetic waves. The rate at which a system emits energy in the form of radiation is proportional to its surface area and to the fourth power of its Kelvin temperature.

✔ In convection, energy is transferred through a liquid or gas by the motion of that liquid or gas.

✔ In conduction, energy is transferred through a system by collisions between the molecules that make up that system. The rate of energy transfer through a system by conduction depends on the dimensions of the system and on the temperature difference between its ends.

Prep for the AP® Exam

AP® Building Blocks

1. The photograph that opens the chapter was taken 30 s after identical ice cubes were placed on each of two black plates. An ice cube placed on the left-hand plate melts more rapidly than one placed on the right-hand plate. The plates are the same size and were initially at the same temperature. Which of the following is the most likely composition of the two plates?
(A) Left-hand plate is wood, right-hand plate is aluminum
(B) Left-hand plate is aluminum, right-hand plate is wood
(C) Left-hand plate is brick, right-hand plate is glass
(D) Left-hand plate is glass, right-hand plate is brick
2. A heated bar of gold radiates at a temperature of 300°C.
(a) By what factor does the radiated power increase if the temperature is increased to 600°C?
(b) By what factor does the radiated power increase if the temperature is increased to 900°C?
(c) If the surface area of the gold bar is doubled, how will the answers to parts (a) and (b) be affected?
3. An astrophysicist determines the surface temperature of a distant star is 12,000 K. The surface temperature of the Sun is about 5800 K. If the surface temperature of the Sun were to suddenly increase to 12,000 K, by how much would the radiated power increase?

AP® Skill Builders

4. Write a brief description for each of the three different thermal processes of energy transfer: radiation, convection, and conduction.
5. EX 14-8 Calculate the rate at which energy is transferred by heating through a glass window that is 30.0 cm × 150 cm in area and 1.20 mm thick. Assume the temperature on the inside of the window is 25.0°C while the outside temperature is 8.00°C.
6. EX 14-8 What is the rate at which energy is transferred by heating due to conduction through a single-pane glass window that is 3.0 m^2 and 3.175 mm thick, when the exterior temperature is 37°C and the interior temperature is 24°C?

AP® Skills in Action

7. A clay pot at room temperature is placed in a kiln (an oven for baking clay), and the pot's temperature (in kelvins) doubles. How much more energy per second is the pot radiating when hot compared to when cool?
(A) 2 times
(B) 4 times
(C) 8 times
(D) 16 times
8. Which thermal process(es) is/are important in the transfer of energy from the Sun to Earth?
(A) Radiation
(B) Convection
(C) Conduction
(D) Radiation, convection, and conduction
9. EX 14-8 The surface area of the human body is about 1.8 m^2. If an average 37.0°C human is surfing in 10.0°C seawater in a 3.0-mm-thick neoprene wetsuit with a thermal conductivity of 0.050 W/(m · K), how much energy does the surfer transfer through heating to the ocean in 1 hour?

10. The skin temperature of a person is 34.0°C, and the surroundings are at 20.0°C. The emissivity of skin is 0.900, and the surface area of the person is 1.50 m^2. Neglect any insulating effects of clothing.

(a) What is the rate at which energy radiates from the person?

(b) What is the net energy loss from the body in 1 min by radiation?

WHAT DID YOU LEARN?

Prep for the AP® Exam

Chapter learning goals	Section(s)	Related example(s)	Relevant section review exercises
Define thermodynamics.	14-1		
Explain the meaning of temperature and thermal equilibrium.	14-2	14-1	3
Describe the origin of pressure in an ideal gas.	14-3	14-2	1, 4, 6
Explain the relationship between molecular kinetic energy and temperature of an ideal gas.	14-3	14-3	2, 3, 5, 7, 10
Examine the relationship between the quantity of energy that enters or leaves a system through thermal processes and the temperature change of that system and calculate the temperature change of a system for a given energy transfer to or from the system.	14-4	14-4	1, 4, 5, 9
		14-5	2, 3, 6, 7, 8
Explain the transfer of energy by thermal processes necessary to cause a substance to change between the solid, liquid, and gas phases.	14-5	14-6	1, 2, 3, 6
		14-7	8, 9, 10
Describe the key properties of the thermal processes through which energy is transferred: radiation, convection, and conduction.	14-6	14-8	5, 6, 9
			4, 8

Chapter 14 Review

Key Terms

All the Key Terms can be found in the Glossary/Glosario on page G1 in the back of the book.

absolute zero 687
Boltzmann constant 690
British thermal unit 704
calorie 704
Celsius scale 686
condensation 706
conduction 712
convection 712
cooling 700
critical point 710
degree of freedom 697
deposition 706
emissivity 713
equation of state 691
equipartition theorem 697
Fahrenheit scale 686
freezing 706
fusion 706
global warming 714
greenhouse effect 713
greenhouse gas 713
heat 700
heating 700
ideal gas 690
ideal gas constant 690
ideal gas law 690
internal energy 700
kelvin 687
Kelvin scale 686
latent heat (of fusion or vaporization) 706
mean free path 697
mole 690
phase 706
phase change 706
phase diagram 710
radiation 712
root-mean-square speed (rms speed) 694
specific heat 701
Stefan–Boltzmann constant 713
sublimation 706
temperature 684
thermal conductivity 717
thermal contact 684
thermal equilibrium 684
thermal expansion 688
thermodynamics 682
thermometer 684
triple point 687
vapor 687
vaporization 706
zeroth law of thermodynamics 685

Chapter Summary

Topic	Equation or Figure
Temperature and thermal equilibrium: The temperature of a system is a measure of the kinetic energy of its molecules. Two systems are in thermal equilibrium (no energy is transferred from one to the other) if and only if they are at the same temperature. We use both the Celsius and Kelvin temperature scales; on the Kelvin scale zero temperature is absolute zero.	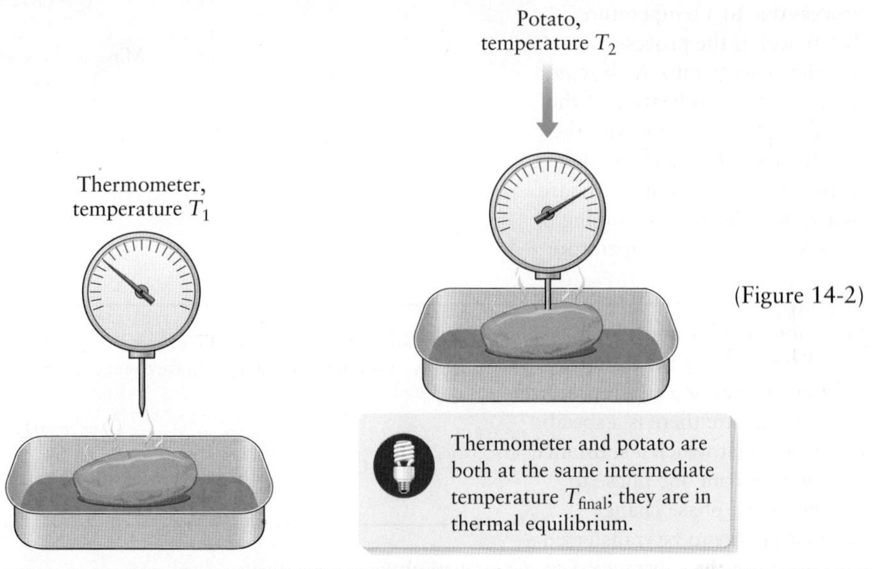 (Figure 14-2)
Ideal gases: In an ideal gas there are no interactions between molecules of the gas. The ideal gas law relates the pressure, volume, and Kelvin temperature of the gas; it can be expressed in terms of either the number of molecules or the number of moles.	$PV = NkT$ (14-6) **Volume** occupied by the gas; **Number of molecules** present in the gas; **Pressure** of an ideal gas; **Temperature** of the gas on the Kelvin scale; **Boltzmann constant** $PV = nRT$ (14-7) **Volume** occupied by the gas; **Number of moles** present in the gas; **Pressure** of an ideal gas; **Temperature** of the gas on the Kelvin scale; **Ideal gas constant**
Molecular motion in an ideal gas: For a gas system, the average translational kinetic energy of a molecule of the gas is directly proportional to the Kelvin temperature T of that system. The root-mean-square molecular speed is proportional to the square root of T. If the gas molecules contain more than one atom, there can also be energy in the vibration or rotation of the molecules.	$K_{translational,average} = \frac{1}{2}m(v^2)_{average} = \frac{3}{2}kT$ (14-13) **Average translational kinetic energy** of a molecule in an ideal gas; Average value of the square of a gas molecule's speed; **Temperature** of the gas on the Kelvin scale; **Mass** of a single gas molecule; **Boltzmann constant** $v_{rms} = \sqrt{\frac{3kT}{m}}$ (14-15) **Root-mean-square speed** of a molecule in an ideal gas; **Boltzmann constant**; **Temperature** of the gas on the Kelvin scale; **Mass** of a single gas molecule

Heat and temperature change: Heat is the amount of energy transferred into or out of a system through a thermal process (a process due to a temperature difference). If the process transfers energy into the system, it is referred to as heating. If the process transfers energy out the system, it is referred to as cooling. If there is no change in the phase of a system, heating or cooling of the system causes a temperature change, ΔT.

$$Q = mc\Delta T \tag{14-18}$$

Amount of energy that is transferred (heat) into (if $Q > 0$) or out of ($Q < 0$) an system

Specific heat of the substance of which the system is made

Mass of the system

Temperature change of the system that results from the energy transfer

Heat and phase change: Solid, liquid, and gas are the different possible phases of a substance. At a given pressure there is a specific temperature at which a substance can change from one phase to another. Each phase change requires energy to be transferred into or out of the substance. For one particular combination of pressure and temperature, called the triple point, the three phases can coexist.

$$Q = \pm mL \tag{14-19}$$

Amount of energy transferred into or out of a quantity of substance to cause a phase change

Use + sign if energy is transferred into the substance; use – sign if energy is transferred out of the substance.

Mass of the substance

Latent heat for the phase change

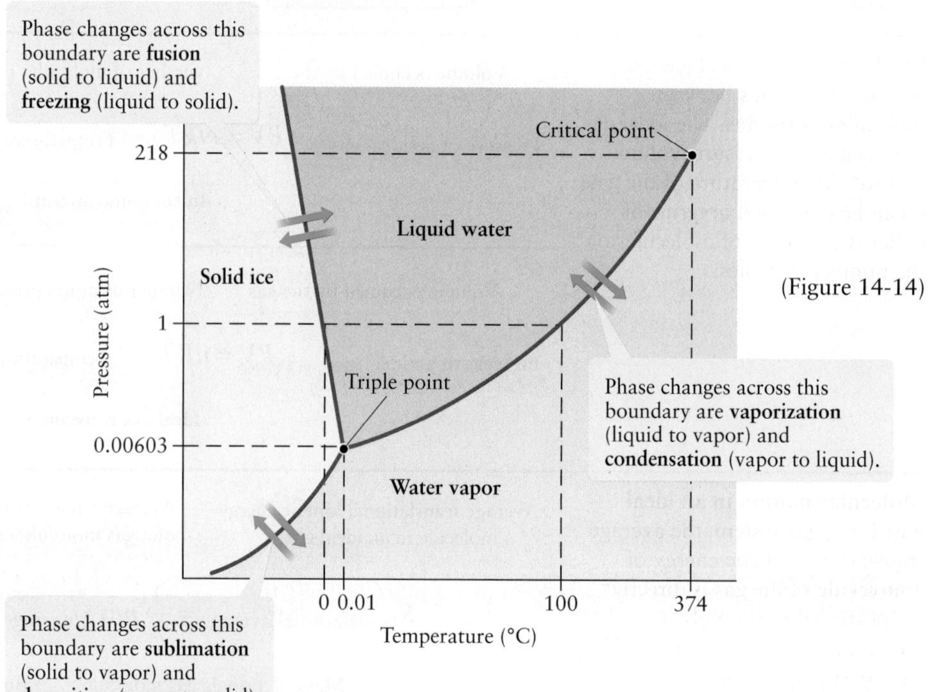
Phase changes across this boundary are **fusion** (solid to liquid) and **freezing** (liquid to solid).
Critical point
218
Liquid water
Solid ice
1
Pressure (atm)
Triple point
0.00603
Water vapor
Phase changes across this boundary are **vaporization** (liquid to vapor) and **condensation** (vapor to liquid).
0
0.01
100
374
Temperature (°C)
Phase changes across this boundary are **sublimation** (solid to vapor) and **deposition** (vapor to solid).

(Figure 14-14)

Radiation, convection, and conduction: Energy can be transferred from place to place by three thermal processes: radiation (electromagnetic waves), convection (circulation of a liquid or gas), or conduction (direct contact between solid systems). The rate at which a system emits energy by radiation is proportional to the fourth power of the system's Kelvin temperature. The rate at which energy is transferred through a system by conduction depends on the temperature difference between the ends of the system, the dimensions of the system, and the system's thermal conductivity. Convection relies on buoyancy and the property of fluids that they become less dense as they warm, so that a warmer gas or liquid floats upward.

Rate at which an system emits energy in the form of radiation

Emissivity of the system (a number between 0 and 1)

$$P = e\sigma A T^4 \qquad (14\text{-}20)$$

Temperature of the system on the Kelvin scale

Stefan–Boltzmann constant $= 5.6704 \times 10^{-8}\ \text{W}\cdot\text{m}^{-2}\cdot\text{K}^{-4}$

Surface area of the system

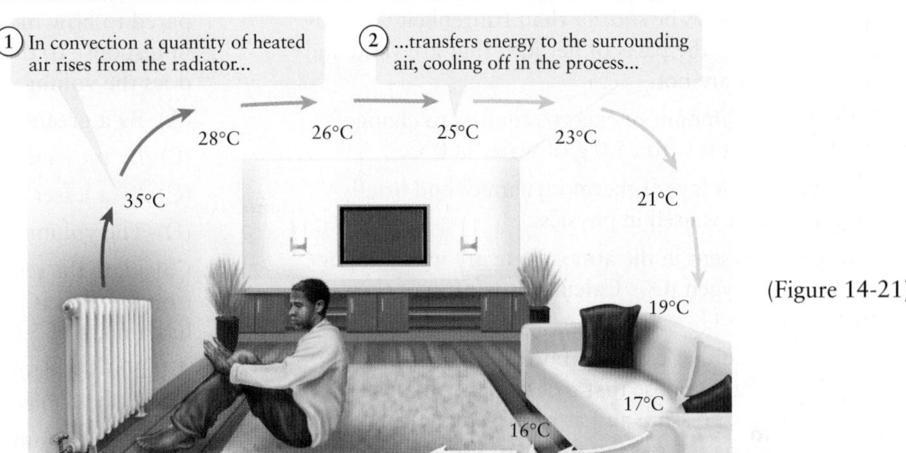

① In convection a quantity of heated air rises from the radiator...
② ...transfers energy to the surrounding air, cooling off in the process...
28°C
26°C
25°C
23°C
35°C
21°C
19°C
17°C
16°C
③ ...and sinks to the floor, where it returns to the radiator to start the process all over again.

(Figure 14-21)

Rate of energy transfer (heating or cooling) H in conduction = amount of energy transferred Q divided by the time Δt the transfer takes.

Temperature difference of the two systems between which energy is transferred

$$H = \frac{Q}{\Delta t} = k\frac{A}{L}(T_H - T_C) \qquad (14\text{-}21)$$

Thermal conductivity of the material through which the energy is transferred

Cross-sectional area and length of the material through which the energy is transferred

Chapter 14 Review Problems

1. In adults the normal range for oral (under the tongue) temperature is approximately 36.7°C to 37.0°C.
   (a) Calculate this temperature range in °F and in K.
   (b) For which temperature scale is the difference between the temperatures the same as the difference in °C?

2. Let's call T_{eq} the temperature at which the numerical value on the Celsius and Fahrenheit scales are the same.
   (a) Write an equation for the temperature T_{eq}.
   (b) Calculate T_{eq}.

3. A thermally isolated system has a temperature of T_A. The temperature of a second isolated system is T_B. When the two systems are placed in thermal contact with each other, they come to an equilibrium temperature of T_C. Describe all the possibilities regarding the relative magnitudes of T_A and T_B when (a) $T_C < T_A$, and (b) $T_C > T_A$.

4. Convert the following temperatures and comment on any physical significance of each:
   (a) 0°C = _________ °F
   (b) 212°F = _________ °C
   (c) 273 K = _________ °F
   (d) 68°F = _________ K

5. A careful physics student is reading about the concept of temperature and has what she thinks is a bright idea. While reading the text, she discovers that the Celsius temperature scale and the Kelvin temperature scale both have the *same* increments for equal temperature differences. However, the Fahrenheit scale was described as having *larger* increments for the same temperature differences. From this information the student determines that Celsius (or Kelvin) thermometers will always be *shorter* than Fahrenheit thermometers. Explain the parts of her idea that are valid and the parts that are not.

6. Calculate the amount of energy required to change 25.0 g of ice at 0°C to 25.0 g of water at 0°C.

7. State the zeroth law of thermodynamics and briefly explain how it is used in physics.

8. Two gases present in the atmosphere are water vapor (H_2O) and oxygen (O_2). Calculate the ratio of their rms speeds.

9. A vacuum chamber can attain pressures as low as 7.0×10^{-11} Pa. Suppose that a chamber contains helium at that pressure and at room temperature (300 K). Estimate the mean free path for helium in the chamber. Assume the diameter of a helium atom is 1.0×10^{-10} m.

10. A star radiates 3.75 times less power than our own Sun. What is the ratio of the temperature of the star to the temperature of our Sun? Assume the star and our Sun have the same radius.

11. What is the average kinetic energy per molecule for an ideal gas made up of diatomic molecules that can move in two dimensions when the temperature is 75.0°F?

12. Now that you know a bit more about the relationship of average molecular kinetic energy to temperature, let's revisit a question, and extend it a bit: The phrase "temperature measures kinetic energy" can be seen as contradictory. Give a few reasons why the explanation actually makes sense and a few reasons why it seems contradictory.

13. Describe briefly, using physics principles, how convection causes hot air to rise in a room.

14. Explain how a material made of light filaments, such as fiberglass insulation, makes an effective barrier to help keep a home warm in the winter and cool in the summer.

15. How much energy is transferred to the environment through heating when the temperature of 2.00 kg of water drops from 88.0°C to 42.0°C?

16. A copper pot has a mass of 1.00 kg and is at 100.00°C. How much energy must be removed from it by cooling to decrease its temperature to precisely 0.00°C?

17. A lake has a specific heat of 4186 J/(kg · K). Calculate the mass of the water in the lake if warming the lake's water from 10.0°C to 15.0°C took 1.70×10^{14} J of energy. Neglect energy transferred to the surroundings.

18. Fire beetles (*Melanophila acuminata*) converge on burning forests from distances of up to 10 km to lay their eggs in trees weakened by fires. These insects can detect a fire from so far away because they are equipped with special chambers that serve as sensitive thermometers. When heated even by a small amount, the fluid inside the chambers expands and causes the pressure within to increase; sensory neurons in the chamber walls respond to the pressure change. Compared to how much the volume of the fluid increases in response to the temperature increase, by how much does the volume of the chambers increase?

(A) By a greater amount

(B) By the same amount

(C) By a lesser but nonzero amount

(D) The volume actually decreases.

19. Calculate the temperature increase of a superheated iron bar that is heated in a blacksmith's fire. The bar has a mass of 5.00 kg and $Q = 2.50 \times 10^6$ J.

20. A 0.200-kg block of ice is at −10.0°C. How much must it be cooled to lower its temperature to −40.0°C?

21. Calculate the amount of energy required to melt a 0.400-kg sample of copper that starts at 20.0°C.

22. A skunk is threatened by a great horned owl and emits its foul-smelling fluid to get away. Will the odor be more easily detected on a day when it is cooler or when it is warmer? Briefly explain your answer using physics principles.

23. If ideal gas A is thermally in contact with ideal gas B for a significant time, what can you say (if anything) about the following variables: P_A versus P_B, V_A versus V_B, and T_A versus T_B? (P, V, and T stand for pressure, volume, and temperature, respectively.)

24. A 37°C chef needs to handle a 165°C cast iron skillet. With a cotton hot pad, the chef can handle the skillet safely, but if the hot pad is wet, the chef can get burned. If the hot pad is 1.0 mm thick, what is the ratio of heating through 1.0 cm^2 of the pad if it is dry versus if it is wet? Use the thermal conductivity of water for the wet hot pad.

25. The ideal gas law can be written as $PV = NkT$ or $PV = nRT$. (**a**) Explain the different contexts in which you might use one or the other and (**b**) define all the variables and constants.

26. Helium condenses at −268.93°C and has a latent heat of vaporization of 21,000 J/kg. If you start with 5.00 g of helium gas at 30.0°C, calculate the amount of energy that must be transferred from the gas to change the sample to liquid helium. The specific heat of helium is 5193 J/(kg·K) at 300 K.

27. You have 50.0 g of iron at 120°C, 60.0 g of copper at 150°C, and 30.0 g of water at 40.0°C. If the ice's initial temperature is −5.00°C, how much ice does each melt?

28. A distant star radiates 1000 times more power than our own Sun, even though the temperature of the star is only 70% of the Sun's. If both stars have an

emissivity of 1, estimate the radius of the distant star. Recall that the radius of the Sun is 6.96×10^8 m.

29. A 1.0-g sample of metal is slowly heated. The temperature increase for each increment of energy added is tabulated in the following table. Plot a graph for the data and predict the specific heat for this common metal. If the measurements are ±10%, make a guess as to the type of metal it might be.

Heat (J)	Temperature change (°C)
0	0
3.9	10
7.9	20
20	50
40	100

30. The table below provides data from a lake as it is freezing.

(**a**) Plot water density as a function of temperature for the data listed in the table.

(**b**) Use the graph to make a conclusion about the way that lakes freeze.

Density (g/cm^3)	Temperature (°C)
0.99990	8.0
0.99996	6.0
1.00000	4.0
0.99996	2.0
0.99988	0.0

31. Susan ordinarily eats 2000 kcal of food per day. If her mass is 60 kg and her height is 1.7 m, her surface area is probably close to 1.7 m^2. Although the body is only about two-thirds water, we will model it as being all water. (**a**) Typically 80% of the calories we consume are converted to internal energy. If Susan's body has no way of transferring this energy by thermal processes, by how many degrees Celsius would her body temperature rise in a day? (**b**) Would this be a noticeable increase? (**c**) How does her body prevent the increase from happening?

32. Suppose a person who lives in a house next to a busy urban freeway attempts to "harness" the sound energy from the nonstop traffic to warm the water in his home. He places a transducer on his roof, "catches" the sound waves, and converts the sound waves into an electrical signal that warms a cistern of water. However, after running the system for 7 days, the 5 kg of water increases in temperature by only 0.01°C! Assuming 100% transfer efficiency, calculate the acoustic power "caught" by the transducer.

33. Calculate the total power radiated by our Sun. Assume it is a perfect emitter of radiation ($e = 1$) with a radius of 6.96×10^8 m and a surface temperature of 5800 K.

34. Estimate the amount of energy required to vaporize Lake Superior.

35. Jane's surface area is approximately 1.5 m^2. Calculate the rate of heating by conduction through her skin when the temperature difference across her skin is 1.00°C. Assume the average thickness of the skin is 1.00 mm.

36. You may have noticed that small mammals (such as mice) seem to be constantly eating, whereas some large mammals (such as lions) eat much less frequently. Let us investigate the phenomenon. For simplicity, we can model an animal as a sphere.

(**a**) Show that the internal energy stored by an animal is proportional to the cube of its radius, but the rate at which the animal radiates energy away is proportional to the square of its radius.

(**b**) Show that the fraction of the animal's stored energy that it radiates away per second is inversely proportional to the animal's radius.

(**c**) Use the result in part (b) to explain why small animals must eat much more per gram of body weight than very large animals.

37. A person can generate about 300 W of power on a treadmill. If the treadmill is inclined at 3.00° and a 70.0-kg man runs at 3.00 m/s for 45.0 min, (**a**) calculate the percentage of the power output that goes into heating up his body and the percentage that keeps him moving on the treadmill. (**b**) How much water would need to be evaporated to keep his temperature from increasing, assuming no other thermal processes of energy transfer?

38. A 1.88-m (6 ft 2 in.) man has a mass of 80.0 kg, a body surface area of 2.1 m^2, and a skin temperature of 30.0°C. Normally 80% of the food calories he consumes go to internal energy; the rest goes to mechanical energy. To keep his body's temperature constant, how many food calories should he eat per day if he is in a room at 20.0°C and he loses internal energy only through radiation? Does the answer seem reasonable? His emissivity e is 1 because his body radiates almost entirely nonvisible infrared energy, which is not affected by skin pigment. (*Careful!* His body at 30.0°C radiates into the air at 20.0°C, but the air also radiates back into his body. The *net* rate of radiation is $P_{net} = P_{body} - P_{air}$.)

39. About 65 million years ago an asteroid struck Earth in the area of the Yucatán Peninsula and wiped out the dinosaurs and many other life forms. Judging from the size of the crater and the effects on Earth, the asteroid was about 10.0 km in diameter (assumed spherical) and probably had a density of 2.0 g/cm^3 (typical of asteroids). Its speed was at least 11 km/s.

(**a**) What is the maximum amount of ocean water (originally at 20.0°C) that the asteroid could have evaporated if all of its kinetic energy were transferred to the water? Express your answer in kilograms and treat the ocean as though it were fresh water.

(**b**) If this amount of water were formed into a cube, how high would it be?

40. The Arctic perennial sea ice does not melt during the summer and thus lasts all year. NASA found that the perennial sea ice decreased by 14% between 2004 and 2005. The melted ice covered an area of 720,000 km^2 (the size of Texas!) and was 3.00 m thick on average. The ice is pure water (not salt water) and is 92% as dense as liquid water. Assume that the ice was initially at –10.0°C.
    (a) How much heat was required to melt the ice?
    (b) Given that 1.0 gal of gasoline releases 1.3×10^8 J of energy when burned, how many gallons of gasoline contain as much energy as in part (a)?
    (c) A ton of coal releases around 21.5 GJ = 21.5×10^9 J of energy when burned. How many tons of coal would need to be burned to produce the energy to melt the ice?

41. What mass of ice at –20.0°C must be added to 50.0 g of steam at 120.0°C to end up with water at 40.0°C?

42. Titan, a satellite of Saturn, has a nitrogen atmosphere with a surface temperature of –179°C and pressure of 1.5 atm. The mass of a nitrogen molecule is 4.7×10^{-26} kg, and we can model it as a sphere of diameter 2.4×10^{-10} m. The average surface temperature of Earth's atmosphere is about 10°C.
    (a) What is the density of atmospheric molecules near the surface of Titan?
    (b) Which has a denser atmosphere, Titan or Earth? Justify your answer by calculating the ratio of the molecule density on Titan to the molecule density on Earth.
    (c) What is the average distance that a nitrogen molecule travels between collisions on Titan? How does this result compare with the distance for oxygen calculated in Section 14-3? Is your result reasonable?

Prep for the AP® Exam

AP® Group Work

Directions: The problem below is designed to be done as group work in class.

The Sun has mass 1.99×10^{30} kg and radius 6.96×10^8 m. The mass of a hydrogen atom is 1.68×10^{-27} kg.

(a) Calculate the escape speed for hydrogen atoms from the surface (the *photosphere*) of our Sun.

(b) If the root-mean-square speed of the hydrogen atoms were equal to the speed you found in part (a), what would be the temperature of the Sun's photosphere?

(c) Given that the actual temperature of the photosphere is 5800 K, is the Sun likely to lose its atomic hydrogen?

AP® PRACTICE PROBLEMS

Prep for the AP® Exam

Multiple-Choice Questions

Directions: The following questions have a single correct answer.

1. Two gases each have the same number of molecules, same volume, and same atomic radius, but the atomic mass of gas B is twice that of gas A. Compare the mean free paths.
    (A) The mean free path of gas A is four times larger than that of gas B.
    (B) The mean free path of gas A is two times larger than that of gas B.
    (C) The mean free path of gas A is the same as that of gas B.
    (D) The mean free path of gas A is two times smaller than that of gas B.

2. Two systems that have the same mass are placed in thermal contact. System A is made of a substance that has a much larger specific heat than the substance from which system B is made. The final temperature of the two systems will be
    (A) close to the initial temperature of system A.
    (B) close to the initial temperature of system B.
    (C) about midway between the two initial temperatures.
    (D) exactly midway between the two initial temperatures.

3. Two systems that have different sizes, masses, and temperatures are placed in thermal contact with each other. Energy is transferred
    (A) from the larger system to the smaller system.
    (B) from the system that has more mass to the one that has less mass.
    (C) from the system that has the higher temperature to the system that has the lower temperature.
    (D) from the system that has the lower temperature to the system that has the higher temperature.

(*Continued*)

4. If the thickness of a uniform wall is doubled, the rate of conduction through the wall is
(A) quadrupled.
(B) doubled.
(C) unchanged.
(D) halved.

Free-Response Question

1. Three samples of a monatomic ideal gas are prepared. Each sample contains the same number of molecules of the gas, but the samples are at different temperatures. Sample Y has temperature T_Y, sample X has a lower temperature, and sample Z has a higher temperature $(T_Z > T_Y > T_X)$.
(a) The graph below shows the distribution of the kinetic energies of the molecules in sample Y. On the graph, sketch and label the distributions for samples X and Z.

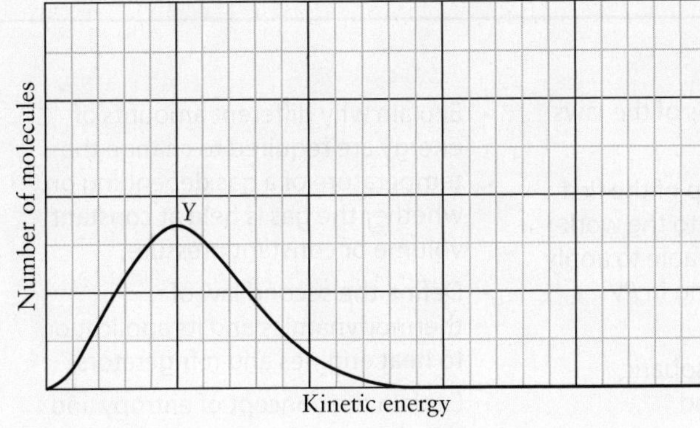

The three samples are placed in thermal contact, with sample X in the middle as shown below, and the samples are insulated from their surroundings. The sample can exchange heat but not gas molecules. The samples eventually reach equilibrium.

Z	X	Y

(b) Describe the net energy transfers between the three samples from initial contact until they reach equilibrium.
(c) For each of the samples, indicate whether the pressure of the sample increases, decreases, or remains the same from the initial state to the equilibrium state.
X ______ Increases ______ Decreases ______ Remains the same
Z ______ Increases ______ Decreases ______ Remains the same

Justify your answer.

(d) Starting with the same initial temperatures, the gas in sample X was replaced with a sample that has the same number of molecules, but the molecules have a higher mass. The experiment was repeated. Briefly describe any changes and how these changes would impact the results of the experiment. Justify your answer by referring to physics principles.

15 Laws of Thermodynamics

David J. Mitchell/Alamy

YOU WILL LEARN TO:

- Explain the general ideas of the laws of thermodynamics.
- Describe the relationship of the first law of thermodynamics to the work-energy theorem and be able to apply it quantitatively, including in *PV* diagrams.
- Describe the nature of isobaric, isothermal, adiabatic, and isovolumetric processes.
- Explain why different amounts of energy are required to change the temperature of a gas depending on whether the gas is held at constant volume or constant pressure.
- Define the second law of thermodynamics and its application to heat engines and refrigerators.
- Explain the concept of entropy and the circumstances under which entropy changes.

15-1 The laws of thermodynamics involve energy and entropy

In Chapter 14 we learned about the zeroth law of thermodynamics, which basically says that the concept of temperature is a useful one. (You should review Section 14-2 about the meaning of temperature and the zeroth law.) But what are the other laws of thermodynamics, and what do they tell us?

In this chapter we'll begin by learning about the *first law of thermodynamics*, which is a generalized statement about the conservation of energy. This law tells us about the interplay between the internal energy of a system, the work done on that the system by its surroundings (the work-energy theorem), and the heating or cooling of the system (**Figure 15-1a**). We'll use the first law to analyze a variety of *thermodynamic processes* in which the state of a system—as measured by its pressure, volume, and temperature—changes due to external influences. The photograph that opens this chapter shows an example of one of these processes, thunderstorm clouds formed when moist air rises rapidly to high altitudes. The temperature of the rising air decreases so much that the moisture condenses into droplets of liquid water or pieces of solid ice, which fall to the ground as rain or hail, respectively.

According to the first law, some very remarkable things could happen: Room-temperature water could spontaneously lower its temperature and freeze at the same

(a)

A cat demonstrates the first law of thermodynamics: The energy it takes in as food is either stored (as fat), converted into work, or transferred to its surroundings through heating.
Trevor Hirst/iStockphoto

(b)

An aircraft engine demonstrates the second law of thermodynamics: The burning of fuel releases energy, but it is impossible to convert 100% of this energy transferred through thermal processes into useful work.
Roger A. Freedman

Figure 15-1 The laws of thermodynamics The laws of thermodynamics are universal: They apply to (a) natural systems such as living organisms and (b) manufactured devices such as engines.

time that the room-temperature glass holding the water spontaneously raises its temperature and melts. The *second law of thermodynamics* describes why such remarkable things are never observed in the real world. It also places a firm limit on the efficiency of *heat engines*, devices that convert the energy transfer due to a temperature difference between systems into work: It tells us that no matter how carefully the engine is designed or built, only part of that heat can be converted to work. Most vehicles in our technological society use a heat engine, so this aspect of the second law is of tremendous importance (**Figure 15-1b**).

We'll finish the chapter with a discussion of *entropy*, a physical quantity that measures the amount of disorder in a system. We'll find that as a result of the second law of thermodynamics, entropy is not conserved: The total amount of entropy in the universe is increasing.

THE TAKEAWAY for Section 15-1

✔ The first law of thermodynamics relates changes in a system's internal energy to the work that the system does and the heating or cooling of the system.

✔ The second law of thermodynamics describes what thermodynamic processes are possible. It also limits how efffcient a heat engine can be.

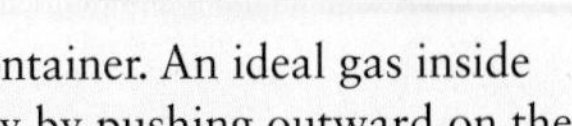

AP® Building Blocks

1. Can energy be transferred to a system through a thermal process without increasing the system's internal energy? Explain.
2. A cylindrical container that is thermally isolated is designed so that its volume can change by moving a piston in one end of the container. An ideal gas inside the cylinder expands slowly by pushing outward on the piston. The temperature of the gas
   (A) increases.
   (B) decreases.
   (C) remains the same.
   (D) can increase or decrease.

15-2 The first law of thermodynamics applies conservation of energy to thermal processes

Suppose you take some unpopped kernels of popcorn at room temperature, put them in a pot with cooking oil, and put a lid on the pot. You then put the pot on the stove and warm it. In a few minutes the popcorn has popped and expanded so much that

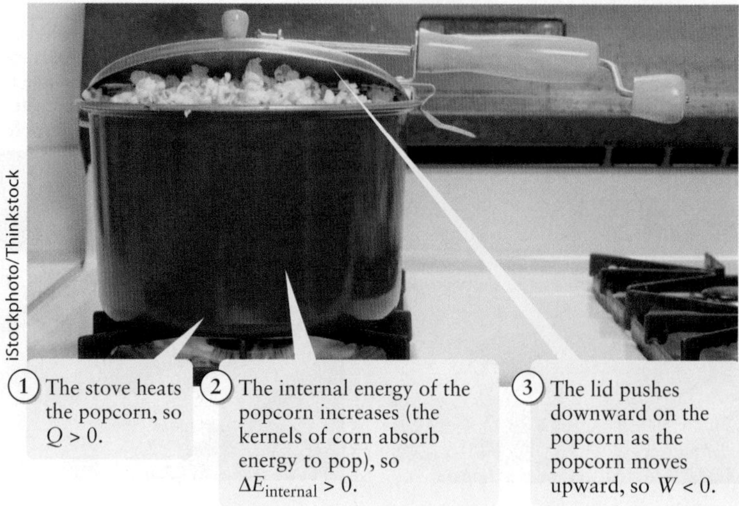

iStockphoto/Thinkstock
1 The stove heats the popcorn, so $Q > 0$.
2 The internal energy of the popcorn increases (the kernels of corn absorb energy to pop), so $\Delta E_{internal} > 0$.
3 The lid pushes downward on the popcorn as the popcorn moves upward, so $W < 0$.

Figure 15-2 A thermodynamic process In a general thermodynamic process for a closed system the internal energy of the system may change as the system is heated or cooled and/or work is done on the system.

AP® Exam Tip

The systems we will discuss in thermodynamics are always going to be closed (no transfer of mass across the boundary) unless you are specifically told otherwise. They are not isolated, because energy will be transferred into the system by both thermal and mechanical processes.

it has lifted the lid off the pot (**Figure 15-2**). By popping the popcorn you've changed its volume (each kernel has expanded) and made its temperature increase. You've also increased its **internal energy**. That's because the chemical reaction involved in popping the kernels requires that each kernel absorb energy. Quantities such as volume, temperature, and pressure are called **state variables** because they depend on the state or condition of the popcorn. These state variables all depend on the microscopic interactions within the system. Any process that changes the state of a closed system (remember that closed means no mass is transferred into or out of the system) related to these variables is called a **thermodynamic process**. While the velocity of the system (as opposed to its constituent molecules) is also a state variable, it is related to the translational kinetic energy of a system and does not depend on the internal structure of the system. Velocity is related to a mechanical process so will not be included in our discussion here, which focuses on thermodynamic processes.

The values of the state variables of a system do *not* depend on the system's history—that is, the details of how the system got to that state. For example, the state of a glass of water at room temperature is the same whether the water came out of the tap at room temperature, was heated in a pot and allowed to cool, or was originally a collection of ice cubes that gradually warmed and melted. In other words the state variables of a system in a given state do not depend on the particular thermodynamic processes that led to that state.

Figure 15-2 shows that in popping the popcorn, the popcorn is heated by the stove; the internal energy of the popcorn increases; and the popcorn does work on its surroundings as it lifts the lid of the pot. These are each statements about energy. We will look at each more closely, then we will see how to relate them.

We'll use the same convention as in Chapter 14 that heat Q is positive if energy is transferred *into* a system and negative if it is transferred *out of* a system (see Section 14-4). Since the popcorn in Figure 15-2 is heated by the stove, $Q > 0$ for the popcorn.

Internal Energy Means Something Very Specific in Thermodynamics

You may remember we first introduced the term *internal energy*, denoted $\Delta E_{internal}$, in Chapter 7, when we introduced the full work-energy theorem, $W = \Delta E = \Delta K + \Delta U + \Delta E_{internal}$. There we were using the term very generically, to refer to all the forms of energy that were internal to a system. We have since unpacked two forms of potential energy that could be due to reversible changes in the configuration of a system, U_{grav} and U_{spring}. We defined $\Delta E_{internal}$ as the change in all other types of energy inside the system due to its configuration or the internal motion of its constituent parts or something like chemical energy. This is very close to the internal energy of a system we introduced in Section 14-4, and will be specifically discussed in thermodynamics, the kinetic and potential energies of the individual molecules making up the system. This definition of internal energy depends on temperature and phase. When we used $\Delta E_{internal}$ we did not rule out ordered motion relative to the center of mass of objects within a system, irreversible changes in configuration, or chemical energy, none of which depends on temperature or phase, so for thermodynamics we want to make it very clear that we are using the definition from Section 14-4, and will use a different symbol for internal energy to remind you.

It is common to use the symbol U (without a subscript) for the internal energy of a system, as we defined it in Section 14-4, and we shall do so in our study of thermodynamics. This internal energy, U, depends only on the kinetic and potential energies of the molecules that make up the system, and directly depends on temperature, a state variable. Because it depends only on a state variable and constants, U is referred to as a state function.When the internal energy of a closed system increases, its change is positive and $\Delta U > 0$. Generally speaking, if the temperature of a system increases in a thermodynamic process, the internal energy also increases and so $\Delta U > 0$. If the system temperature decreases, the internal energy generally decreases and so $\Delta U < 0$. Because the shape of the popcorn permanently changes, the change in its total internal energy, $\Delta E_{internal}$, is different than ΔU, although we can still get the sign right!

WATCH OUT!

In some textbooks *W* is defined differently.

Some textbooks when discussing thermodynamics will use the symbol W to represent the work done *by* a system on its surroundings, so it is positive when energy is taken out of the system by a mechanical process. We will continue to use the same convention we introduced in Chapter 7, that W is the work done on the system, so it is positive when energy is added to the system by a mechanical process.

For the process shown in Figure 15-2, the popcorn does positive work on the lid: The popcorn exerts an upward force on the lid, and the lid's displacement is also upward. If a system increases in volume and pushes against its surroundings, as the popcorn does in Figure 15-2, then the work the system does is positive. Because the force the popcorn exerts on the lid is upward, by Newton's third law, the force the lid exerts on the popcorn must be downward. So, the work done on the system of the popcorn is negative. If the system decreases in volume, like air being compressed in a bicycle pump, positive work is done on the system since it is being pushed on and it is compressing in the direction of that force. In the work-energy theorem, W represents the energy being transferred to a system by mechanical processes. We now know Q, the energy transferred to a system by thermal processes. We can now expand our conservation of energy equation to include both of these terms as inputs, $W + Q = \Delta E$. **Figure 15-3** shows the conventions that we use for the sign of Q and the sign of W. This expanded conservation of energy equation is applied in Example 15-1.

WATCH OUT!

Internal energy isn't the same as potential energy.

In Chapter 7 we used the symbol U to represent the *potential energy* associated with a particular interaction within a system. In specific cases, we added a subscript for clarity. It can be confusing that the same symbol U is used in thermodynamics to represent the *internal energy* of a system, because these two kinds of energy are *not* the same. In thermodynamics the value of U for a system includes the kinetic energy of its molecules and the potential energy of their interactions with each other. U does *not* include the gravitational potential energy of the system. For example, if you carry the popcorn in Figure 15-2 to your neighbor's apartment above you, the gravitational potential energy associated with the popcorn increases (the popcorn is at a greater height) but the popcorn's internal energy stays the same. Keep the new definition of the quantity U in mind throughout your study of thermodynamics.

WATCH OUT!

In chemistry, work *W* is definitely different than in physics.

Note that in most discussions of thermodynamics in chemistry textbooks, W is defined to be the work done *on* a system, which is the exact opposite of how it's defined in physics. Remember this if you're taking physics and chemistry at the same time!

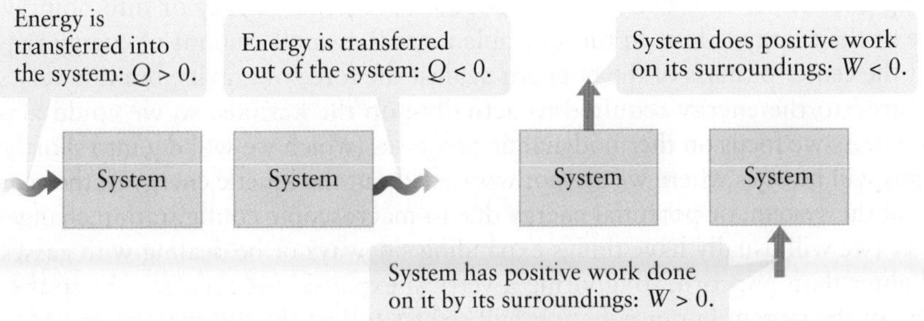
Energy is transferred into the system: $Q > 0$.
Energy is transferred out of the system: $Q < 0$.
System does positive work on its surroundings: $W < 0$.
System
System
System
System
System has positive work done on it by its surroundings: $W > 0$.

Figure 15-3 The signs of work and heat Heat Q is considered positive if energy is transferred into a system, negative if it is transferred out through thermal processes. Work W for a system follows the same convention, for mechanical processes.

EXAMPLE 15-1 Cycling It Off

At the end of a 15-min session on a stationary bicycle at the gym, the bike's display indicates that you have delivered 165 kJ of energy to the bike in the form of work. Your heart rate monitor shows that you have "burned off" 155 food calories of energy. Although we know that this is a mix of chemical and thermal energy (look back at our expanded example of tossing a pencil in Chapter 8 if this doesn't sound familiar), assume this change in energy is the change in your total energy. Determine the amount of energy your body has transferred to your surroundings through heating.

Set Up

In this problem your body is the system. You do work on your surroundings (the bicycle), so the work done on you, W, is negative and equal to 165 kJ. The 155 food calories (155 kcal) that you have "burned off" represent a decrease in your body's energy, so ΔE is negative and equal to 155 kcal. We'll use conservation of energy to determine the heat Q.

Conservation of energy:

$$Q + W = \Delta E$$

Your body does work on the bicycle: $W < 0$.
system = your body
Your body heats the environment: $Q < 0$.

Solve

First convert ΔE from kilocalories to kilojoules.

1 kcal = 4186 J = 4.186 kJ, so

$$\Delta E = -155 \text{ kcal}\left(\frac{4.186 \text{ kJ}}{1 \text{ kcal}}\right) = -649 \text{ kJ}$$

Rearrange our expression for conservation of energy to solve for Q. A positive value of Q means energy is transferred *into* your body. A negative value of Q means that energy is transferred *out of* your body.

$$\begin{aligned} Q &= \Delta E - W \\ &= -649 \text{ kJ} - (-165 \text{ kJ}) \\ &= -484 \text{ kJ} \end{aligned}$$

This is negative, so your body transfers 484 kJ of energy to the environment through heating as you exercise.

Reflect

During this exercise session your body expended 649 kJ of internal energy, of which 165 kJ went into doing work and 484 kJ went into heating your surroundings. Only (165 kJ)/(649 kJ) = 0.254, or 25.4%, of the energy that you expended went into doing work. So your body has an *efficiency* of 25.4% in this situation.

Note that 3500 kcal is the energy content of one pound (0.45 kg) of fat. If all of the 155 kcal that you expended came from stored fat, the net result of your exercise is that you will have lost an amount of fat equal to (155 kcal)/(3500 kcal/lb) = 0.044 lb (0.020 kg, or 0.71 ounce). Losing weight requires a *lot* of exercise!

NOW WORK Problems 1, 2, 5, and 6 from the Takeaway 15-2.

For the system (popcorn) shown in Figure 15-2, the popcorn is heated ($Q > 0$), the internal energy of the popcorn increases ($\Delta U > 0$), and the popcorn does positive work on its surroundings ($W < 0$).

Careful measurement confirms that the energy that enters the popcorn through heating goes either into changing the popcorn's internal energy or into doing work. None of the energy is lost. In this example there is a small amount of energy required to lift the center of mass of the popcorn against the force of gravity, but it is very small compared to the energy required to actually pop the kernals, so we could actually ignore it. As we focus on thermodynamic processes (which we will dig into shortly) our systems will be ones where we are not worried about the kinetic energy of the center of mass of the system, or potential energy due to macroscopic configuration changes like gravity (we will usually have things expanding sideways or be dealing with gasses that are lighter than popcorn, so ignoring a vertical expansion is even easier), so the total energy of the system in our equation will reduce to just the internal energy, $\Delta E = \Delta U$.

With this limitation, our expanded conservation of energy equation becomes the first law of thermodynamics, which is written as a mathematical expression as

During a thermodynamic process the **change in the internal energy of a system...**

$$W + Q = \Delta U \quad (15\text{-}1)$$

...plus the **work done on the system** during the process.

...equals **the energy transferred to or from the system through heating or cooling** during the process...

EQUATION IN WORDS
The first law of thermodynamics

The **first law of thermodynamics** is a generalization of the law of conservation of energy for thermodynamic processes. It says that the internal energy of a system can change *only* if the system gains energy from its surroundings (if the surroundings heat the system, or its surroundings do positive work on the system) or if the system loses energy to its surroundings (if the surroundings cool the system, which means the system heats the surroundings, or the surroundings do negative work on the system). No exception to this rule has ever been found.

Internal Energy Change and Thermodynamic Processes

As we mentioned above, the internal energy U is a state function that depends only on the current state of the system, not on how it got into that state. To see what this implies let's consider two *different* thermodynamic processes that take a system from the *same* initial state to the *same* final state.

In **Figure 15-4a** a cylinder with a moveable piston encloses a thermodynamic system: a quantity of gas. The gas has an initial pressure, volume, and temperature. We use a candle to transfer energy slowly into the gas (so $Q > 0$) and allow the piston to move so that the gas expands. The expanding gas does positive work on the piston (the gas pushes the piston to the right as the piston moves to the right), so $W < 0$. If we regulate how fast the piston moves, we can arrange it so that the rate at which the gas does work on the piston is exactly the same as the rate at which the gas is heated. Then $Q = -W$, and from the first law of thermodynamics (Equation 15-1), the internal energy of the gas will remain unchanged: $\Delta U = Q + W = 0$. For an ideal gas the average energy per molecule is proportional to the Kelvin temperature T of the gas (see Section 14-3), so the internal energy U—the average energy per molecule, multiplied by the number of molecules—is also proportional to T. Since U remains constant for the process shown in Figure 15-4a, the temperature of the gas also remains constant.

Now imagine a second thermodynamic process that starts with the system (the gas) in the same initial state. In **Figure 15-4b** we use the same quantity of gas as in Figure 15-4a, and the gas starts at the same pressure, volume, and temperature. The difference is that now the cylinder is equipped with a thin barrier of lightweight material instead of a piston, and there is a vacuum between the barrier and the other end of the cylinder. If we puncture the barrier, the gas expands freely and fills the entire cylinder. Since nothing pushes against the gas as it expands (it expands into a vacuum), $W = 0$. If we insulate the walls of the cylinder so that there is no heating or cooling of the gas, $Q = 0$ for this process. So the first law of thermodynamics tells us that in this process again the internal energy of the gas will not change: $\Delta U = Q + W = 0 + 0 = 0$. Because U is proportional to Kelvin temperature T for an ideal gas, it follows that the temperature of the gas remains constant for this process. In other words *both* the processes shown in Figure 15-4 take the same quantity of gas from the same initial volume to the same final volume while leaving the temperature unchanged. (The ideal gas law $PV = nRT$, Equation 14-7, tells us that the final pressure P will also be the same. That's because the final volume V, final temperature T, and number of moles n are the same in both processes.)

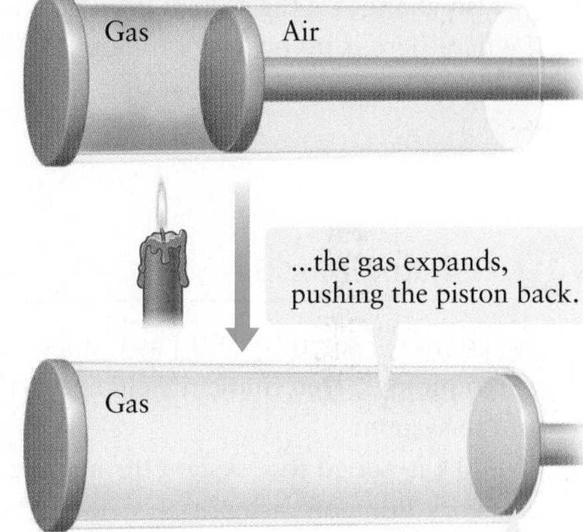
(a)
When energy is added slowly to gas trapped in a cylinder with a moveable piston...
Gas
Air
...the gas expands, pushing the piston back.
Gas
The gas has done work on the piston.

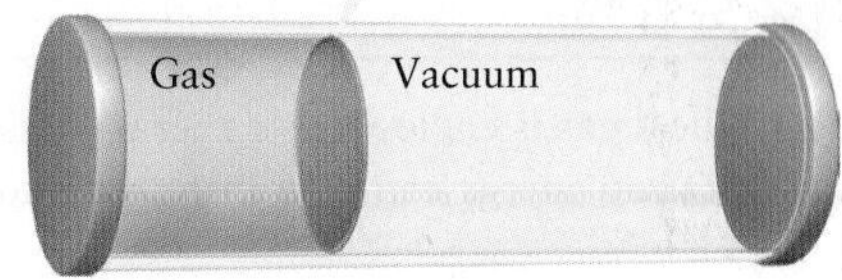
(b)
A thin barrier of lightweight material separates a gas from a vacuum.
Gas
Vacuum
When the barrier ruptures, the gas expands to fill the entire cylinder. No energy was added to the gas in this case, but the final pressure and volume are the same as in part (a).

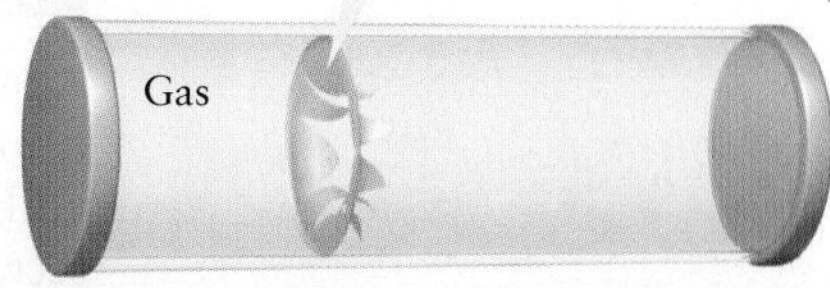
Gas

Figure 15-4 Two thermodynamic processes (a) If an expanding gas is heated at just the right rate, the gas temperature remains constant. (b) The gas temperature is also unchanged in a free expansion. An initial state may be changed to the same final state by different thermodynamic processes.

The two very different processes in Figure 15-4 take the system from the same initial state to the same final state. In the first process the system is heated, and the system does positive work on its surroundings. The second process requires neither of these things. This is why we *cannot* say a particular state of a system *contains* work or heat. The heat or work required depends on the process. For any thermodynamic process that takes a system between the same two states, the change in internal energy is the *same*, independent of the process, so internal energy is a state function.

THE TAKEAWAY for Section 15-2

✔ The internal energy of a system (the sum of the kinetic energy and potential energy of every atom and molecule in the system) describes the state of the system in a way that does not depend on the specific thermodynamic processes that led to it.

✔ The internal energy of a system changes if the system is heated or cooled but no work is done on the system. Internal energy also changes if the system does positive or negative work when there is no heating or cooling. When there are energy transfers by both thermal and mechanical processes, the internal energy will not change if those transfers are of equal magnitude and opposite sign.

✔ The first law of thermodynamics relates the change in internal energy of a thermodynamic system to energy transfers into the system by thermal or mechanical processes, when the macroscopic velocity of the system can be ignored.

Prep for the AP Exam

AP Building Blocks

1. A system is heated by 500 J as it does 200 J of work on its surroundings. Determine the change in the internal energy of the system.

2. If 800 J is added to a system through heating and there is no work done on or by the system, by how much does the internal energy of the system increase?
3. In a slow, steady expansion of an ideal gas against a piston, if the temperature doesn't change the work done is equal to the heating. Is this consistent with the first law of thermodynamics?

AP Skill Builders

4. An ideal gas is enclosed in a thermally isolated cylinder so that no energy can be transferred into or out of the gas by thermal processes. One end of the cylinder is sealed by a moveable piston like that shown in Figure 15-4a. If you allow the gas to expand slowly by pulling back the piston, what will happen to the temperature T of the gas? Explain your answer.

EX 15-1

5. Why is it possible for the temperature of a system to remain constant even when the system is heated or cooled?

EX 15-1

6. An ideal gas is contained in a closed cylinder of fixed length and diameter. Eighty joules of energy is added to the gas by heating. The work done by the gas on the walls of the cylinder is
(A) 80 J.
(B) 0 J.
(C) less than 80 J.
(D) more than 80 J.

15-3 A graph of pressure versus volume helps to describe what occurs in a thermodynamic process

A thermodynamic process involves a change in the state of a system. In most cases this involves a change in the pressure of the system, the volume that the system occupies, or both. That's the case for the gas that expands against a piston in Figure 15-4a. In this section we'll use the first law of thermodynamics to help us analyze four special but important kinds of thermodynamic processes:

- **Isobaric process:** The pressure of the system remains constant.
- **Isothermal process:** The temperature of the system remains constant.
- **Adiabatic process:** There are no transfers of energy by thermal processes into or out of the system.
- **Isovolumetric process:** The volume of the system remains constant.

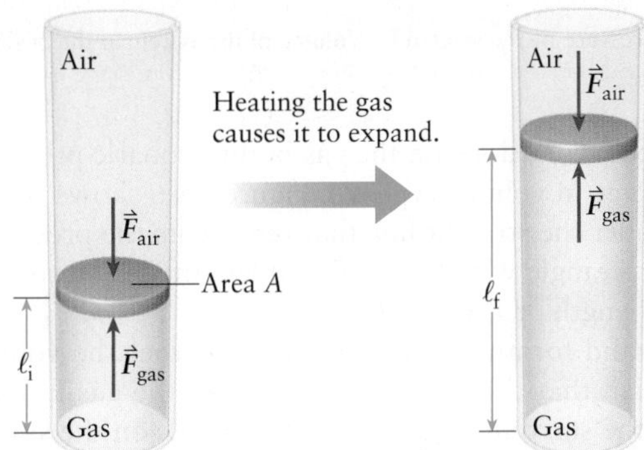
The piston reaches its equilibrium position when the pressure of the gas in the cylinder equals atmospheric pressure.
Air
Heating the gas causes it to expand.
$\vec{F}_{air}$
Area A
ℓ_i
$\vec{F}_{gas}$
Gas
Air
$\vec{F}_{air}$
$\vec{F}_{gas}$
ℓ_f
Gas
Even after the gas expands, the gas pressure must equal atmospheric pressure. Because neither gas pressure nor atmospheric pressure has changed, the expansion is isobaric.

Figure 15-5 An isobaric expansion The piston exerts a constant pressure on the gas as it is heated, so the pressure of the gas itself remains constant.

AP® Exam Tip

Thermodynamic processes and thermal process sound similar, but they are different. On the AP® exam, you are responsible for being able to describe the thermal processes we learned in Chapter 14, which are ways energy is transferred due to a temperature difference and these four special cases of thermodynamic processes.

Many thermodynamic processes do not fit into any of these simple categories. But by analyzing these four special cases, we'll learn general rules that can be applied to *any* thermodynamic process.

Isobaric Processes

Isobaric processes occur with no change in pressure. An example is when you cook food in a frying pan without a lid: The food is heated and its temperature increases, but the pressure that the atmosphere exerts on the food remains constant.

Let's consider an isobaric process in which work is done. Suppose our system is a quantity of gas inside a cylinder that's sealed with a moveable piston of negligible mass, as in **Figure 15-5**. At equilibrium the upward force on the piston due to the pressure of the gas equals the downward force due to atmospheric pressure. These forces also balance if we heat the gas gradually and allow it to expand slowly so that the piston moves up at constant velocity: The net force on the piston remains zero (it is not accelerating). Since atmospheric pressure doesn't change, the pressure in the gas must remain constant as well and so the process is isobaric.

To make it easier to visualize and understand what happens in this isobaric process, it's useful to make a ***PV* diagram.** This is a graph that plots the pressure of a system on the vertical axis versus the volume of the system on the horizontal axis. On a PV diagram an isobaric process is represented by a horizontal line (**Figure 15-6**).

In Figure 15-5 the expanding gas does work W on the piston to lift it. To calculate how much work is done, let the constant pressure of the gas be P and the surface area of the piston be A. Figure 15-5 shows that the original length of the cylinder containing the gas is ℓ_i, and the final length of the cylinder after the gas has expanded is ℓ_f. Recall that pressure is force per unit area, so force is pressure multiplied by area. This means that the force that the gas exerts on the piston has magnitude $F_{gas} = PA$, exerted upward, and by Newton's third law, the force the piston exerts on the gas is the same magnitude and downward. Because the force exerted on the gas is in the opposite direction to the one in which the piston moves, and the piston moves a distance $\ell_f - \ell_i$, the total work done on the gas as it expands is

$$W = -F_{piston}(\ell_f - \ell_i) = -PA(\ell_f - \ell_i) = -P(A\ell_f - A\ell_i) \quad (15\text{-}2)$$

The volume of a cylinder is equal to the area of its base multiplied by its height. So $A\ell_i$ and $A\ell_f$ are, respectively, the initial volume V_i and final volume V_f occupied by the gas, and we can rewrite Equation 15-2 as

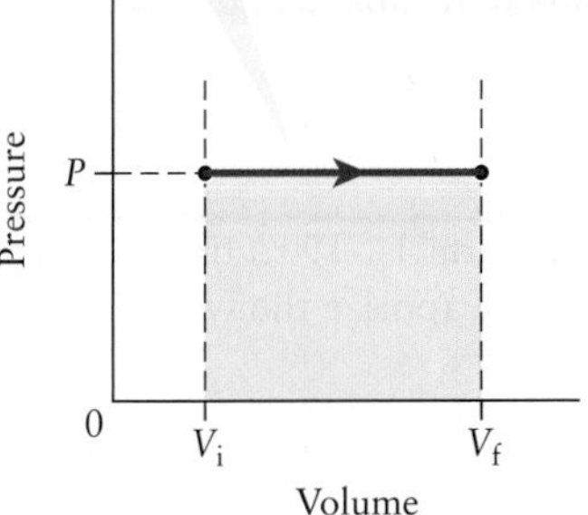
The area under a PV graph is the magnitude of the work done on a system in an isobaric process
Pressure
P
0
V_i
V_f
Volume

Figure 15-6 Work done in an isobaric expansion On a *PV* diagram (a graph of pressure versus volume) for a thermodynamic process, the area under the curve represents the magnitude of the work done on the system. If the volume decreases from the initial to the final state, the work done is positive. If the volume is increasing as shown here, the work done is negative.

EQUATION IN WORDS
Work done in an isobaric (constant-pressure) process

Work done on a system in an isobaric process | Volume of the system at the end of the process

(15-3) $$W = -P(V_f - V_i)$$

Constant pressure of the system | Volume of the system at the beginning of the process

The magnitude of the work done on the gas in this isobaric process is the pressure P multiplied by the change in volume $(V_f - V_i)$. Figure 15-6 shows that this is the *area* of the shaded region under the straight line that represents the process on the PV diagram. This region is a rectangle with height P and length $V_f - V_i$, so its area equals its height multiplied by its length or $P(V_f - V_i)$.

Equation 15-3 is valid for an isobaric process whether the volume increases or decreases. If $V_f > V_i$, such that the system expands like the gas in Figure 15-5, then $V_f - V_i$ is negative and the surroundings do negative work on the system. Such a process is called an *isobaric expansion*. If instead, $V_f < V_i$, which would be like the process in Figure 15-5 in reverse, then $V_f - V_i$ is negative and the surroundings do positive work on the system. This is called an *isobaric compression*.

Equation 15-3 is valid *only* for the case of constant pressure. But it turns out to be true in general that in *any* process the work done on the system is the area under the curve that represents the process on a PV diagram, positive when the volume of the system decreases and negative when the volume of the system increases. We'll use this idea later in this section.

EXAMPLE 15-2 Boiling Water Isobarically

Boiling 1.00 g (1.00 cm^3) of water at 1.00 atm results in 1671 cm^3 of water vapor. A cylinder like that shown in Figure 15-5 contains 1.00 g of water at 100°C. The pressure on the outside of the piston is 1.00 atm. Determine the change in internal energy of the water when it is heated just enough to boil all of the liquid.

Set Up

In this process the water is heated to change its phase from liquid to vapor, and the vapor does work on the piston as it expands. Because the pressure on the outside of the piston is constant, the pressure that the piston exerts on the water (liquid plus vapor) is also constant and this is an isobaric process. We'll apply the first law of thermodynamics to determine the change in internal energy.

First law of thermodynamics:

$$\Delta U = Q + W \quad (15\text{-}1)$$

Work done in an isobaric (constant-pressure) process:

$$W = -P(V_f - V_i) \quad (15\text{-}3)$$

Heat required for a phase change:

$$Q = \pm mL \quad (14\text{-}19)$$

The water vapor pushes upward, so the piston does negative work on the water+vapor system: $W < 0$.
system = water
The water is heated: $Q > 0$.

Solve

Use Equation 14-19 to find the heat Q needed to vaporize the water.

The water is heated, so Q in Equation 14-19 is positive. The mass of water is m = 1.00 g, and the latent heat of vaporization is $L_V = 2260 \times 10^3$ J/kg $= 2.26 \times 10^6$ J/kg (from Table 14-2). So

$$Q = mL_V$$
$$= (1.00\text{ g})\left(\frac{1\text{ kg}}{10^3\text{ g}}\right)(2.26 \times 10^6\text{ J/kg})$$
$$= 2.26 \times 10^3\text{ J}$$

Use Equation 15-3 to find the work W that the piston does on the expanding water vapor.

The pressure of the water vapor is P = 1.00 atm $= 1.01 \times 10^5$ Pa $= 1.01 \times 10^5$ N/m^2 and remains constant in this isobaric expansion.

The volume occupied by the water increases from V_i = 1.00 cm^3 in the liquid state to V_f = 1671 cm^3 in the vapor state. In calculating the work done on the expanding vapor, we must convert cm^3 to m^3 using the relationship 1 m = 10^2 cm:

$$W = -P(V_f - V_i)$$

$$= -(1.01 \times 10^5 \text{ N/m}^2)(1671 \text{ cm}^3 - 1.00 \text{ cm}^3)\left(\frac{1 \text{ m}}{10^2 \text{ cm}}\right)^3$$

$$= -(1.01 \times 10^5 \text{ N/m}^2)(1670 \times 10^{-6} \text{ m}^3)$$

$$= -169 \text{ N} \cdot \text{m} = -169 \text{ J}$$

Substitute the values for Q and W into Equation 15-1, the first law of thermodynamics. This tells us the change in internal energy of the water in this process.

$$\Delta U = Q + W$$

$$= 2.26 \times 10^3 \text{ J} - 169 \text{ J}$$

$$= 2.09 \times 10^3 \text{ J}$$

Reflect

We have to add 2.26×10^3 J of energy to boil the water. Of this energy, 2.09×10^3 J goes into increasing the internal energy of the system. The rest of the added energy (169 J) leaves the system in the form of work done on the piston.

NOW WORK Problem 2 from The Takeaway 15-3.

Isothermal Processes

In an isothermal process temperature remains constant. Although the process in **Example 15-2** is isobaric (constant pressure), it is also isothermal: When a substance freezes or boils, the phase transition occurs with no change in temperature.

Gases can also expand or be compressed in such a way as to maintain a constant temperature, even though no phase transition occurs. As we discussed in Section 15-2, the internal energy U of an ideal gas is directly proportional to the Kelvin temperature of the gas. So if the temperature of an ideal gas remains constant during a thermodynamic process, the internal energy U remains constant as well and there is zero change in the internal energy: $\Delta U = 0$. The first law of thermodynamics, Equation 15-2, tells us that $\Delta U = Q + W$, so for an isothermal process that involves an ideal gas, we have

$$0 = Q + W$$

or

$$Q = -W \qquad (15\text{-}4)$$

(isothermal process for an ideal gas)

So *when an ideal gas undergoes an isothermal process, the amount of energy added to the gas by heating Q balances the work W done on the gas.* If the gas is heated so that $Q > 0$, it must be that $W < 0$: The surroundings do negativetive work on the gas, which means it must expand. This is called an *isothermal expansion.* If the gas is cooled so that $Q < 0$, then $W > 0$ and the surroundings do positive work on the gas, so the volume of the gas decreases. This is called an *isothermal compression.*

How much work is done on a gas as it undergoes an isothermal expansion or compression? To get insight into the answer, let's use a PV diagram to depict such a process. The ideal gas law (which we studied in Section 14-3) says that the pressure P, volume V, and temperature T of n moles of an ideal gas are related by

$$PV = nRT \qquad (14\text{-}7)$$

In an isothermal process T is constant, so the product PV is likewise constant. If the gas expands and the volume V increases, the pressure P must decrease; if the gas compresses and the volume V decreases, the pressure P must increase. **Figure 15-7a** shows three **isotherms**, or curves of constant temperature, on a graph of pressure versus volume.

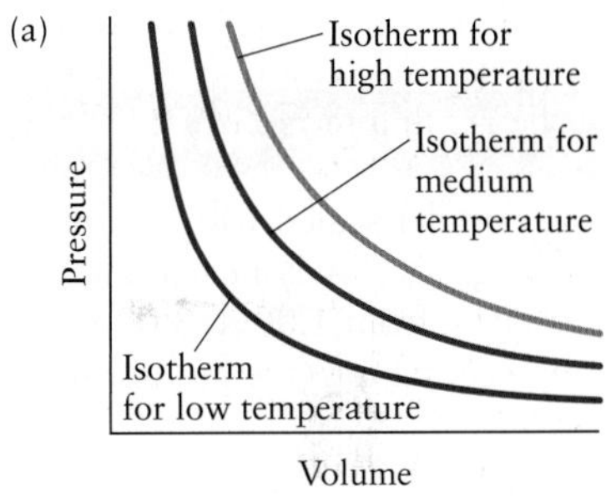

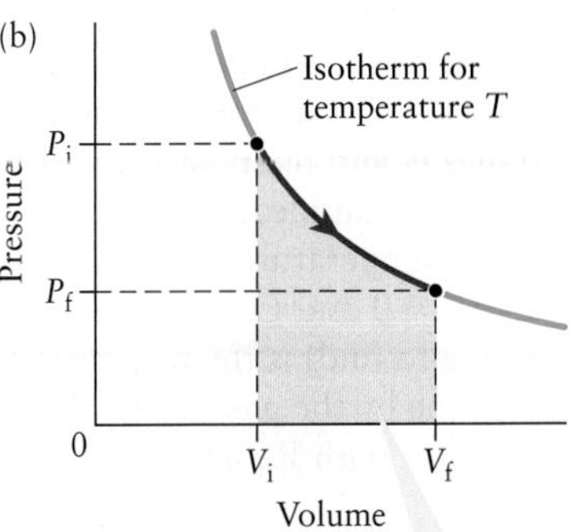

Figure 15-7 Isothermal processes in an ideal gas (a) This PV diagram shows isotherms (curves of constant temperature) for an ideal gas. (b) The work done on an ideal gas (the system) in an isothermal expansion. (If the volume increases as shown here, the work done on the gas is negative.)

On a PV diagram, the area under the curve again represents the magnitude of the work W done on the gas; **Figure 15-7b** shows this area for an isothermal expansion. Unlike the isobaric case shown in Figure 15-6, the area under the curve is *not* a rectangle, so we cannot find W simply by multiplying the pressure and the change in volume. An analysis using calculus (which is beyond our scope) shows that the work that the gas does as the volume changes is

EQUATION IN WORDS
Work done on an ideal gas in an isothermal (constant-temperature) process

$$W = -nRT \ln \frac{V_f}{V_i} \quad (15\text{-}5)$$

Work done on an ideal gas (system) in an isothermal process

Number of moles of gas

Ideal gas constant

Kelvin temperature of the gas

Volume of the gas at the end of the process

Volume of the gas at the beginning of the process

NEED TO REVIEW?
Turn to page M8 in the Math Tutorial in the back of the AP® Physics 1 book for more information on logarithms.

In Equation 15-5 "ln" is the natural logarithm function. It has the properties that $\ln x$ is positive if x is greater than 1 and $\ln x$ is negative if x is less than 1; if $x = 1$, then $\ln x = \ln 1 = 0$. So Equation 15-5 says that negative work is done on the gas ($W < 0$) if the gas expands, so V_f is greater than V_i and V_f/V_i is greater than 1. Positive work is done *on* the gas by its surroundings if the gas is compressed, so V_f is less than V_i and V_f/V_i is less than 1. An alternative version of Equation 15-5 in terms of the number of molecules N present in the gas is

$$W = -NkT \ln \frac{V_f}{V_i} \quad (15\text{-}6)$$

EXAMPLE 15-3 An Isothermal Expansion

A cylinder sealed with a moveable piston as in Figure 15-5 contains 0.100 mol of an ideal gas at a temperature of 295 K. The gas is slowly heated and allowed to expand isothermally from an initial volume of 1.00×10^{-3} m^3 to 3.00×10^{-3} m^3 (that is, from 1.00 to 3.00 L). (a) What are the initial and final pressures of the gas? (b) How much energy is transferred by heating into the gas during this process?

Set Up

Since this is an ideal gas, we can use the ideal gas equation to determine the pressure P from the volume V, number of moles n, and temperature T. The internal energy of an ideal gas does not change if the temperature remains constant, so $\Delta U = 0$ and Q (which is what we're trying to find) is the negative of the work W done by the gas. We'll find W and hence Q with Equation 15-5.

Ideal gas law:

$$PV = nRT \quad (14\text{-}7)$$

Isothermal process for an ideal gas:

$$Q = -W \quad (15\text{-}4)$$

Work done by an ideal gas in an isothermal (constant-temperature) process:

$$W = -nRT \ln \frac{V_f}{V_i} \quad (15\text{-}5)$$

The gas pushes upward and so the piston does negative work on the gas: $W < 0$.

system = gas

The gas is heated to make it expand: $Q > 0$.

Solve

(a) Use Equation 14-7 to find the initial and final pressures.

Rewrite the ideal gas law $PV = nRT$ to solve for the pressure:

$$P = \frac{nRT}{V}$$

The pressure P_i when the gas occupies the initial volume $V_i = 1.00 \times 10^{-3}$ m^3 is

$$P_i = \frac{nRT}{V_i} = \frac{(0.100\ \text{mol})[8.314\ \text{J/(mol}\cdot\text{K)}](295\ \text{K})}{1.00 \times 10^{-3}\ \text{m}^3}$$
$$= 2.45 \times 10^5 \frac{\text{J}}{\text{m}^3}\left(\frac{1\ \text{N}\cdot\text{m}}{1\ \text{J}}\right)$$
$$= 2.45 \times 10^5\ \text{N/m}^2 = 2.45 \times 10^5\ \text{Pa}$$

(We used the relationship that $1\ \text{J} = 1\ \text{N}\cdot\text{m}$.) The pressure when the gas occupies the final volume $V_f = 3.00 \times 10^{-3}\ \text{m}^3$ is

$$P_f = \frac{nRT}{V_f} = \frac{(0.100\ \text{mol})[8.314\ \text{J/(mol}\cdot\text{K)}](295\ \text{K})}{3.00 \times 10^{-3}\ \text{m}^3}$$
$$= 8.18 \times 10^4\ \text{Pa}$$

(b) From the first law of thermodynamics, the sum of energy that enters the gas through heating in this process and the work done on the gas must be zero.

For an isothermal process for an ideal gas, $\Delta U = Q + W = 0$ and $Q = -W$.

Use Equation 15-5 to calculate the work done on the gas, the negative of which is the energy that is added to the gas through heating.

From Equation 15-5 the work done on the gas is

$$W = -nRT \ln \frac{V_f}{V_i}$$
$$= -(0.100\ \text{mol})[8.314\ \text{J/(mol}\cdot\text{K)}](295\ \text{K}) \times \ln\left(\frac{3.00 \times 10^{-3}\ \text{m}^3}{1.00 \times 10^{-3}\ \text{m}^3}\right)$$
$$= -(245\ \text{J}) \ln 3.00 = -(245\ \text{J})(1.10)$$
$$= -269\ \text{J}$$

Since $Q = -W$, $Q = 269$ J.

Reflect

For an ideal gas that expands isothermally, *all* of the energy that is transferred into the gas through heating (in this case 269 J) is converted to work done by the gas on its surroundings (the surroundings do negative work on the gas).

Extend: Note that to keep the temperature constant the gas pressure must decrease during the expansion. To make this happen it's not enough to just slowly heat the gas: The difference in pressure between the two sides of the piston must be such that the volume increases to match the desired pressure of the gas (so that the piston moves slowly and doesn't accelerate). This tells us that an isothermal expansion or compression of an ideal gas requires rather special circumstances.

NOW WORK Problems 1, 4, 5, and 8 from The Takeaway 15-3.

Adiabatic Processes

In an adiabatic process there is no heating or cooling of a system. An important system that undergoes adiabatic processes is our atmosphere. Air is a very poor thermal conductor (see Section 14-6), so there is little energy transferred into or out of a mass of air by thermal processes as its pressure and volume change. This helps explain why clouds often appear near mountain summits (**Figure 15-8**).

When winds carry air up and over mountains, the lower pressure at higher altitudes causes the air to expand and do work on the surrounding air, so the surrounding air does negative work on the air that is our system ($W < 0$). This expansion occurs nearly adiabatically thanks to the low thermal conductivity of air, so Q is effectively zero. From the first law of thermodynamics, Equation 15-2, the change in the internal energy of the air is

$$\Delta U = Q + W = 0 + W \quad \text{or} \quad \Delta U = W \quad \text{(adiabatic process)} \qquad (15\text{-}7)$$

Since $W < 0$, it follows that $\Delta U < 0$: The internal energy U of the air decreases as it is pushed up the flanks of the mountains. For an ideal gas U is proportional to the

Figure 15-8 Adiabatic cooling and mountain clouds The prevailing winds push moist air up and over the mountains that form the Bavarian Alps. The temperature of the uplifted air drops and airborne moisture condenses into droplets, forming cumulus clouds like this one.

Kelvin temperature of the gas, so the air temperature drops by as much as 10°C for every 1 km of elevation gained. As the temperature decreases, water vapor in the air condenses into clouds like the one shown in Figure 15-8.

The air we have described expands as it is pushed up the slopes of the mountains, so this is an *adiabatic expansion*. In an *adiabatic compression* the volume of the gas decreases, so $W > 0$ (the surroundings do work on the gas) and $\Delta U > 0$ (the internal energy and temperature of the gas increase).

WATCH OUT !

Remember that heat and temperature are not the same thing.

Even though no energy is transferred through thermal processes into or out of air pushed up a range of mountains, the temperature of the air changes. If this statement seems contradictory, it's probably because you're still thinking of *heat* as meaning approximately the same thing as *temperature*. Remember that these two concepts are different: Heat is the energy that is transferred into or out of a system due to a temperature difference, whereas temperature is a measure of the kinetic energy of a system's molecules. The temperature can change even if there is *no* heating or cooling, provided work is done on (or done by) the system.

Isovolumetric Processes

An isovolumetric process is one for which the volume remains constant. As an example, consider a quantity of gas sealed inside a rigid container. If the gas is heated, the pressure and temperature of the gas will increase but the volume of the gas will remain the same because the container cannot expand.

The pressure of the gas exerts forces on the walls of its container, but since the walls do not move (there is no moveable piston), there is zero displacement. Work is force multiplied by the displacement of the point of application of the force, so the surroundings do *zero* work on the gas in an isovolumetric process: $W = 0$. From the first law of thermodynamics, Equation 15-1,

$$\Delta U = Q + W = Q + 0 \quad \text{or} \quad \Delta U = Q \quad \text{(isovolumetric process)} \tag{15-8}$$

This says that for an isovolumetric process any thermal transfer of energy into or out of a system goes entirely into changing the internal energy of the system. If Q is positive so that energy is transferred into the system, the process is called *isovolumetric heating*; if Q is negative and energy is transferred out of the system, the process is called *isovolumetric cooling*.

EXAMPLE 15-4 Two Thermodynamic Processes: Isovolumetric and Isobaric

The PV diagram in **Figure 15-9** shows two thermodynamic processes that occur in an ideal diatomic gas, for which the internal energy is $U = (5/2)nRT$. The gas is in a cylinder with a moveable piston (see Figure 15-5) and is initially in the state labeled A on the diagram, at pressure $P_i = 3.03 \times 10^5$ Pa and volume $V_i = 1.20 \times 10^{-3}$ m^3. The piston is first locked in place so that the volume of the gas cannot change, and the cylinder is cooled so that the temperature and pressure of the gas both decrease. When the gas is in state B, the pressure of the gas is $P_f = 1.01 \times 10^5$ Pa = 1.00 atm, the same as the air pressure

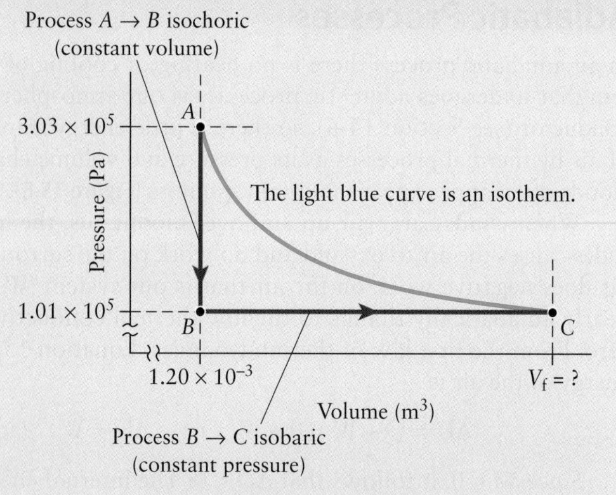

Figure 15-9 Isovolumetric cooling and isobaric expansion In this two-step process for an ideal gas, how much work is done on the gas, how much is the gas heated or cooled, and by how much does the internal energy change?

outside the cylinder. The piston is then unlocked so that it is free to move, and the cylinder is slowly heated so that the gas expands at constant pressure. The temperature of the gas increases until in the final state (C in Figure 15-9), when the gas is at the same temperature as in the initial state A. (a) What is the final volume of the gas? (b) Find the values of W, ΔU, and Q for the isovolumetric process $A \to B$. (c) Find the values of W, ΔU, and Q for the isobaric process $B \to C$. (d) Find the values of W, ΔU, and Q for the net process $A \to B \to C$.

Set Up

We'll use the ideal gas law to determine the final volume of the gas. For the isovolumetric (constant-volume) process $A \to B$, the gas does no work, so the quantity of energy Q that enters the gas by heating is equal to the internal energy change ΔU of the gas. (The gas is *cooled* in this process, so $Q < 0$ and $\Delta U < 0$.) We'll determine ΔU from the temperature change. For the isobaric (constant-pressure) process $B \to C$, the work W done on the gas is given by Equation 15-3; this is negative because the gas expands. The internal energy change ΔU is again given by the temperature change. We'll then determine Q for this process from the first law of thermodynamics.

Ideal diatomic gas:

$$U = \frac{5}{2}nRT$$

Ideal gas law:

$$PV = nRT \qquad (14\text{-}7)$$

First law of thermodynamics:

$$\Delta U = Q + W \qquad (15\text{-}1)$$

Work done in an isobaric (constant-pressure) process:

$$W = -P(V_f - V_i) \qquad (15\text{-}3)$$

Isovolumetric (constant-volume) process:

$$\Delta U = Q \qquad (15\text{-}8)$$

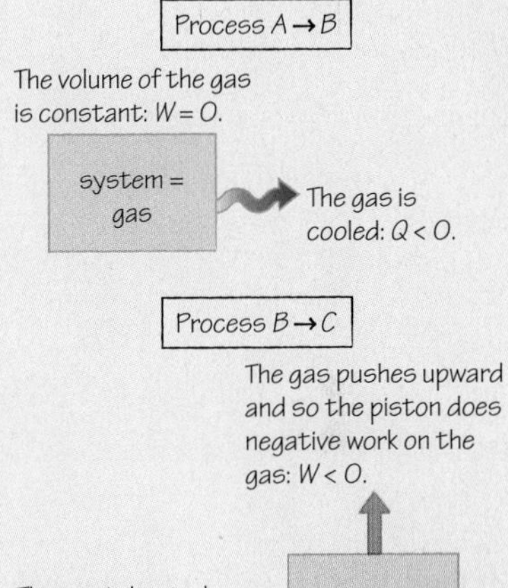
Process A→B
The volume of the gas is constant: W = 0.
system = gas
The gas is cooled: Q < 0.
Process B→C
The gas pushes upward and so the piston does negative work on the gas: W < 0.
The gas is heated to make it expand: Q > 0.
system = gas

Solve

(a) Use the ideal gas law to find the final volume V_f.

The initial pressure and volume of the gas are

$$P_i = 3.03 \times 10^5 \text{ Pa}$$
$$V_i = 1.20 \times 10^{-3} \text{ m}^3$$

From the ideal gas law, Equation 14-7,

$$P_iV_i = nRT_i$$

where T_i is the initial temperature of the gas in state A. The final pressure of the gas is

$$P_f = 1.01 \times 10^5 \text{ Pa}$$

The ideal gas law tells us that

$$P_fV_f = nRT_f$$

where T_f is the final temperature of the gas in state C. The initial and final temperatures are the same ($T_i = T_f$), so

$$P_iV_i = P_fV_f$$

$$V_f = \frac{P_iV_i}{P_f} = \frac{(3.03 \times 10^5 \text{ Pa})(1.20 \times 10^{-3} \text{ m}^3)}{1.01 \times 10^5 \text{ Pa}} = 3.60 \times 10^{-3} \text{ m}^3$$

The final pressure is one-third of the initial pressure, and the final volume is three times the initial volume.

(b) For the isovolumetric process $A \to B$, the work done is $W_{A\to B} = 0$. Find the change in internal energy ΔU and the heat Q for this process.

The change in internal energy U is the difference between the value of U in state B and the value of U in state A:

$$\Delta U_{A\to B} = U_B - U_A = \frac{5}{2}nRT_\text{B} - \frac{5}{2}nRT_\text{i}$$

We don't know the number of moles of gas n, nor do we know the temperatures T_i (in the initial state A, for which the pressure is P_i and the volume is V_i) and T_B (in state B, for which the pressure is P_f and the volume is V_i). But from the ideal gas law $PV = nRT$, so

$$P_\text{i}V_\text{i} = nRT_\text{i} \text{ and } P_\text{f}V_\text{i} = nRT_\text{B}$$

So the change in internal energy is

$$\begin{aligned}\Delta U_{A\to B} &= \frac{5}{2}P_\text{f}V_\text{i} - \frac{5}{2}P_\text{i}V_\text{i} = \frac{5}{2}(P_\text{f} - P_\text{i})V_\text{i}\\ &= \frac{5}{2}(1.01\times10^5\text{ Pa} - 3.03\times10^5\text{ Pa})(1.20\times10^{-3}\text{ m}^3)\\ &= -606\text{ J}\end{aligned}$$

(Remember that $1\text{ Pa} = 1\text{ N}\cdot\text{m}^2$ and $1\text{ N}\cdot\text{m} = 1\text{ J}$.) From Equation 15-8 for an isovolumetric process,

$$Q_{A\to B} = \Delta U_{A\to B} = -606\text{ J}$$

Energy is transferred out of the gas through cooling, and the internal energy of the gas decreases.

(c) Find W, ΔU, and Q for the isobaric process $B \to C$.

The work done on the system in an isobaric process is given by Equation 15-3. The constant pressure is $P_\text{f} = 1.01\times10^5$ Pa, and the volume increases from $V_\text{i} = 1.20\times10^{-3}\text{ m}^3$ to $V_\text{f} = 3.60\times10^{-3}\text{ m}^3$:

$$\begin{aligned}W_{B\to C} &= -P_\text{f}(V_\text{f} - V_\text{i})\\ &= -(1.01\times10^5\text{ Pa})(3.60\times10^{-3}\text{ m}^3 - 1.20\times10^{-3}\text{ m}^3)\\ &= -242\text{ J}\end{aligned}$$

The surroundings do negative work on the gas as it expands. As in part (b) we calculate the internal energy change with the aid of the ideal gas law:

$$\begin{aligned}\Delta U_{B\to C} &= U_C - U_B = \frac{5}{2}nRT_\text{f} - \frac{5}{2}nRT_B\\ &= \frac{5}{2}P_\text{f}V_\text{f} - \frac{5}{2}P_\text{f}V_\text{i} = \frac{5}{2}P_\text{f}(V_\text{f} - V_\text{i})\end{aligned}$$

This is just $-5/2$ times the above expression for $W_{B\to C}$, so

$$\Delta U_{B\to C} = -\frac{5}{2}W_{B\to C} = -\frac{5}{2}(-242\text{ J}) = +606\text{ J}$$

From the first law of thermodynamics, the energy that is transferred into the gas through heating in this process is

$$\begin{aligned}Q_{B\to C} &= \Delta U_{B\to C} - W_{B\to C} = (+606\text{ J}) - (-242\text{ J})\\ &= +848\text{ J}\end{aligned}$$

(d) Find the values of W, ΔU, and Q for the combined process $A \to B \to C$.

The total amount of work done on the gas is

$$W_{A\to B} + W_{B\to C} = 0 + (-242\text{ J}) = -242\text{ J}$$

The total change in internal energy of the gas is

$$\Delta U_{A\to B} + \Delta U_{B\to C} = (-606 \text{ J}) + (+606 \text{ J}) = 0$$

The total quantity of energy transferred to the gas by heating is

$$Q_{A\to B} + Q_{B\to C} = -606 \text{ J} + (+848 \text{ J}) = +242 \text{ J}$$

Reflect

For the combined process $A \to B \to C$, there is *zero* net change in the internal energy U of the ideal gas. That's because U depends only on the number of moles of gas and the temperature, which have the same values in the final state C as in the initial state A. In the combined process the gas is heated by 242 J, and the gas uses this energy to do 242 J of work on the environment (the environment does the negative of this work on the gas).

NOW WORK Problems 3, 6, and 7 from The Takeaway 15-3.

THE TAKEAWAY for Section 15-3

✔ A PV diagram can be used to describe thermodynamic processes that take a system from one state to another.

✔ Isobaric processes occur with no change in pressure.

✔ Temperature remains constant in isothermal processes.

✔ There is no heating or cooling of a system during an adiabatic process.

✔ Isovolumetric processes occur with no change in volume.

Prep for the AP® Exam

AP® Building Blocks

EX 15-3 1. An ideal gas expands isothermally, and does 8.80 kJ of work on its surroundings in the process. Calculate the heat absorbed during the expansion.

EX 15-2 2. A gas contained in a cylinder that has a piston is kept at a constant pressure of 2.80×10^5 Pa. The gas expands from 0.500 to 1.50 m^3 when 300 kJ of heat is added to the cylinder. Calculate the change in internal energy of the gas.

EX 15-4 3. An ideal gas is held in a container with rigid walls. The pressure in the gas is slowly reduced to $\frac{1}{4}$ its initial value. In the process the gas transfers 800 kJ of energy to the environment through heating. Determine the change in internal energy of the gas during this process.

EX 15-3 4. In an isothermal expansion process, 8.00 m^3 of helium gas in an expandable chamber is heated such that its volume increases by 2.00 m^3 and the gas does 2.00 kJ of work on its surroundings. Determine the initial pressure of the gas.

AP® Skill Builders

EX 15-3 5. A cylinder that has a piston contains 2.00 mol of an ideal gas. The gas undergoes a reversible isothermal expansion at 400 K from an initial pressure of 12.0 atm down to 3.00 atm. Determine the amount of work done on the gas.

EX 15-4 6. A gas is heated and is allowed to expand such that a PV diagram from its initial state (1.0×10^5 Pa, 1.0 m^3) to its final state (1.0×10^5 Pa, 2.0 m^3) is a horizontal line. Calculate the work done on the gas by its surroundings.

EX 15-4 7. A gas is heated such that a PV diagram from its initial state (1.0×10^5 Pa, 3.0 m^3) to its final state (2.0×10^5 Pa, 3.0 m^3) is a vertical line. Calculate the work done on the gas by its surroundings.

EX 15-3 8. A sealed cylinder has a piston and contains 8.00×10^3 cm^3 of an ideal gas at a pressure of 8.00 atm. The gas is slowly heated, and isothermally expands to 1.60×10^4 cm^3. Calculate the work the piston does on the gas.

AP® Skills in Action

Problems 9 and 10 refer to **Figure 15-10**.

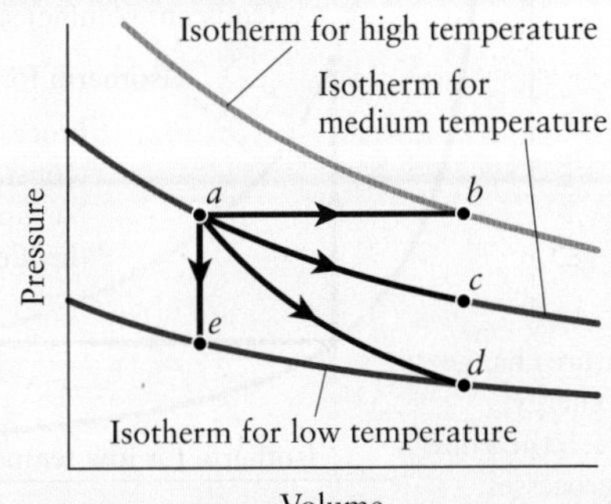

Figure 15-10 Four thermodynamic processes A PV diagram shows four thermodynamic processes for a certain quantity of an ideal gas: (i) $a \to b$, (ii) $a \to c$, (iii) $a \to d$, and (iv) $a \to e$. All four processes start in the same initial state a. The volume is the same for states b, c, and d. Process $a \to c$ lies along an isotherm.

9. (a) Rank the four processes illustrated in Figure 15-10 according to the internal energy change ΔU of the gas, from most positive to most negative.
(b) Rank these four processes according to the work W done on the gas, from most positive to most negative.

10. Which one of the processes shown in Figure 15-10 could be adiabatic?
(A) $a \to b$
(B) $a \to c$
(C) $a \to d$
(D) $a \to e$

15-4 The concept of molar specific heat helps us understand isobaric, isovolumetric, and adiabatic processes for ideal gases

Our discussion in Section 15-3 of various types of thermodynamic processes was fairly general. Let's now look more closely at some of these processes for the special case in which the thermodynamic system is an *ideal gas*. This special case has a tremendous number of applications: It will help us understand why a bicycle pump gets warm when you use it to inflate a tire and how a diesel engine can make a fuel–air mixture ignite even though there are no spark plugs in such an engine.

Specific Heats of an Ideal Gas

Suppose you have a quantity of ideal gas. How much energy does it take to raise the temperature of the gas to a new, higher temperature? The best answer is "It depends on how you do it." The *PV* diagram in **Figure 15-11** helps us understand why this is so.

This figure shows two possible ways that an ideal gas in a cylinder with a moveable piston can be slowly heated from a lower temperature to a higher one. Process $A \to B$ is *isovolumetric:* It takes place at constant volume (the moveable piston is locked in position), so the pressure of the gas increases as you heat the gas. Process $A \to C$ is *isobaric:* It takes place at constant pressure. The piston is allowed to move so that the gas expands as it is heated, and the pressure inside the cylinder remains equal to the pressure on the other side of the piston. The final temperature is the same for both processes because both state B and state C are on the same isotherm in Figure 15-11.

The internal energy U of an ideal gas depends only on the number of moles and the temperature, so the internal energy change ΔU is the same for both processes. But the amount of work W done on the gas is *different* for the two processes: $W = 0$ for process $A \to B$ (no work is done on the gas if the volume it occupies doesn't change), while $W < 0$ for process $A \to C$ (the piston exerts a force on the gas opposite its direction of motion). From the first law of thermodynamics, Equation 15-1,

$$\Delta U = Q + W \quad \text{so} \quad Q = \Delta U - W$$

Because ΔU is the same for both processes, *more* energy is required for process $A \to C$ (in which negative work is done) than for process $A \to B$ (for which $W = 0$). So the heating required to increase the temperature of an ideal gas is *greater* if the temperature is increased at constant *pressure* than if it is increased at constant *volume*.

Let's be more quantitative about the heat required to change the temperature of a certain quantity of ideal gas by a given amount. In Section 14-4 we wrote an equation for the heat required to change the temperature of a mass m of substance by an amount ΔT:

(14-18) $$Q = mc\,\Delta T$$

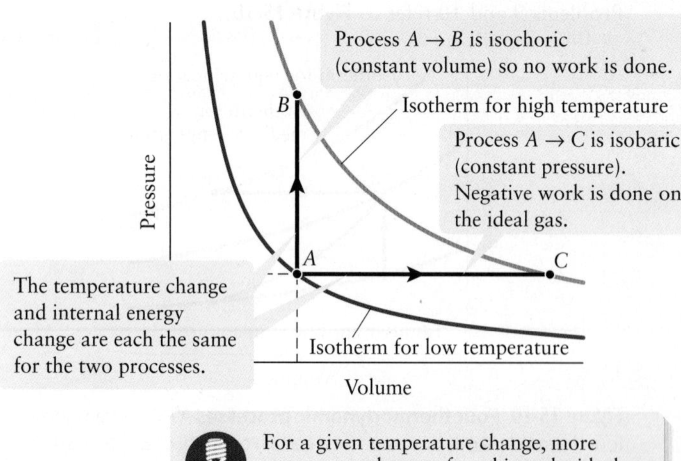

Process $A \to B$ is isochoric (constant volume) so no work is done.
Isotherm for high temperature
Process $A \to C$ is isobaric (constant pressure). Negative work is done on the ideal gas.
Pressure
B
A
C
The temperature change and internal energy change are each the same for the two processes.
Isotherm for low temperature
Volume
For a given temperature change, more energy must be transferred into the ideal gas in process $A \to C$ than in process $A \to B$.

Figure 15-11 Two ways to increase temperature Processes $A \to B$ and $A \to C$ for an ideal gas lead to different final states but the same final temperature.

In Equation 14-18 the quantity c is the *specific heat* of the substance. For an ideal gas we usually write the internal energy U in terms of the number of *moles* of gas present rather than the amount of mass. We follow the same approach for the heat Q, so we write

Amount of energy that is transferred (heat) into (if $Q > 0$) or out of (if $Q < 0$) an system

Molar specific heat of the substance of which the system is made

$$Q = nC\Delta T \quad (15\text{-}9)$$

Number of moles present of the substance

Temperature change of the system that results from the energy transfer

EQUATION IN WORDS
Heat and the resulting temperature change in terms of molar specific heat

The quantity C in Equation 15-9 is called the **molar specific heat** of the substance. It has units of joules per mole per kelvin, or J/(mol · K).

We've seen that the heat Q required to make an ideal gas undergo a temperature change ΔT is different if the pressure P is held constant than if the volume V is held constant. To keep track of this difference, we use the symbol C_p for the **molar specific heat at constant pressure** and the symbol C_V for the **molar specific heat at constant volume.** Then we can write Equation 15-9 for these two special cases as

$$Q = nC_p\Delta T \quad \text{(constant pressure)}$$

$$Q = nC_V\Delta T \quad \text{(constant volume)} \quad (15\text{-}10)$$

Our discussion of Figure 15-11 showed that for a given temperature increase ΔT, the value of Q was greater for a constant-pressure process than for a constant-volume process. So we conclude that C_p, the molar specific heat at constant pressure, must be greater than the molar specific heat at constant volume C_V. **Table 15-1**, which lists experimental values of C_p and C_V for various gases, shows that this is indeed the case.

C_p, C_V, and Degrees of Freedom

Notice the following from Table 15-1:

(a) The *difference* $C_p - C_V$ between the two different molar specific heats has almost the same value for *all* gases.

(b) The value of C_V is nearly the same for all monatomic gases, nearly the same for all diatomic gases, and nearly the same for all triatomic gases. The value increases as the number of atoms increases.

WATCH OUT !

Don't confuse specific heat and molar specific heat.

It's important to recognize the difference between the specific heat c that appears in Equation 14-18 and the molar specific heat C in Equation 15-9. Specific heat c (lowercase) tells you the number of joules of heat required to raise the temperature of one *kilogram* of a substance by one kelvin. Molar specific heat C (uppercase) tells you the number of joules of heat required to raise the temperature of one *mole* of the substance by one kelvin. If the substance has a molar mass (mass per mole) of M, the two quantities are related by $C = Mc$.

TABLE 15-1 Molar Specific Heats of Gases [in J/(mol • K) for a gas at $T = 300$ K and $P = 1$ atm, except where otherwise indicated]

	Gas	C_p	C_V	$C_p - C_V$	$\gamma = C_p/C_V$
Monatomic	Ar	20.8	12.5	8.28	1.66
	He	20.8	12.5	8.28	1.66
	Ne	20.8	12.7	8.08	1.63
Diatomic	H_2	28.8	20.4	8.37	1.41
	N_2	29.1	20.8	8.33	1.40
	O_2	29.4	21.1	8.33	1.39
	CO	29.3	21.0	8.28	1.39
Triatomic	CO_2	37.0	28.5	8.49	1.30
	SO_2	40.4	31.4	9.00	1.29
	H_2O (373 K)	34.3	25.9	8.37	1.32

We can understand the first of these observations by using Equation 15-9, the first law of thermodynamics, and the ideal gas law. For a constant-volume process $W = 0$, so from the second of Equations 15-10, the internal energy change for n moles of an ideal gas that undergoes a given temperature change ΔT is

$$\Delta U = Q + W = nC_V\,\Delta T + 0 = nC_V\,\Delta T \tag{15-11}$$

(ideal gas, constant volume)

For an isobaric process that starts at the same temperature, the gas changes its volume by ΔV and an amount of work is done on the gas

$$W = -P\Delta V = -P(V_f - V_i) = -(PV_f - PV_i)$$

where V_f and V_i are the final and initial volumes of the gas. But from the ideal gas law,

$$PV_f = nRT_f \quad \text{and} \quad PV_i = nRT_i$$

where T_f and T_i are the final and initial temperatures of the gas, respectively. The temperature change of the gas is $\Delta T = T_f - T_i$, so we can write the work done by the gas in a constant-pressure process as

$$W = -(PV_f - PV_i) = -(nRT_f - nRT_i) = -nR(T_f - T_i) = -nR\Delta T \tag{15-12}$$

(ideal gas, constant pressure)

Using Equation 15-12 and the first of Equations 15-10, we can now write the internal energy change for this constant-pressure process as

$$\Delta U = Q + W = nC_p\Delta T - nR\Delta T = n(C_p - R)\Delta T \tag{15-13}$$

(ideal gas, constant pressure)

Equation 15-11 for a constant-volume process and Equation 15-13 for a constant-pressure process look rather different from each other. But we know that the internal energy change ΔU for an ideal gas is the *same* in both cases, as long as the number of moles n and the temperature change ΔT are the same for both processes. So ΔU in Equation 15-11 is equal to ΔU in Equation 15-13:

$$nC_V\,\Delta T = n(C_p - R)\Delta T$$

For this to be true it must be that $C_V = C_p - R$, or

EQUATION IN WORDS
Molar specific heats of an ideal gas

Molar specific heat of an ideal gas at **constant pressure**

$$C_p - C_V = R \tag{15-14}$$

Ideal gas constant

Molar specific heat of an ideal gas at **constant volume**

Equation 15-14 indicates that for an ideal gas C_p is greater than C_V by $R = 8.314$ J/(mol · K), the ideal gas constant. The experimentally determined values of $C_p - C_V$ given in Table 15-1 for *real* gases are very close to this theoretical prediction.

How can we understand our second observation from Table 15-1, that the value of C_V is essentially the same for all gases with the same number of atoms? To see the answer, note from Equation 15-11 that for an ideal gas with constant volume, $\Delta U = nC_V\Delta T$. In other words, C_V tells us about the change in internal energy of a gas that results from a temperature change. We learned something important about the internal energy of an ideal gas in Section 14-3: On average each molecule in an ideal gas at temperature T has energy $(1/2)kT$ for each of its *degrees of freedom*. (You'll find it helpful to review the discussion of degrees of freedom in Section 14-3. Recall that k is the Boltzmann constant, equal to the ideal gas constant R divided by Avogadro's number N_A.) A monatomic gas has three degrees of freedom, corresponding to its ability to move in the x, y, and z directions. A diatomic gas at room temperature has two additional degrees of freedom that correspond to its ability to rotate around either of two axes perpendicular to the long axis of the molecule, and a triatomic gas (which at room temperature can both rotate and vibrate) has even more degrees of freedom.

We'll use the symbol D to represent the number of degrees of freedom. Then the average energy per gas molecule is $(D/2)kT$. One mole of the gas contains Avogadro's number of molecules, so the energy per mole is $N_A(D/2)kT = (D/2)(N_Ak)T = (D/2)RT$. (Again recall that $k = R/N_A$, so $R = N_Ak$.) Then the internal energy of n moles of ideal gas is

$$U = n\left(\frac{D}{2}R\right)T \qquad (15\text{-}15)$$

(ideal gas, D degrees of freedom)

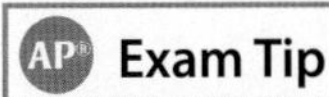

Exam Tip

For a monatomic gas $D = 3$, so Equation 15-15 becomes $U = 3/2nRT$. While this is the form you will use most often, it is important to remember that the 3/2 comes from $D = 3$, and be able to explain that internal energy at a given temperature increases with increasing degrees of freedom.

If the temperature of the gas changes by ΔT, it follows from Equation 15-15 that the internal energy change is

$$\Delta U = n\left(\frac{D}{2}R\right)\Delta T \qquad (15\text{-}16)$$

(ideal gas, D degrees of freedom)

But we saw above that for an ideal gas, $\Delta U = nC_V\Delta T$. Comparing this to Equation 15-16 we see that

$$C_V = \left(\frac{D}{2}\right)R \qquad (15\text{-}17)$$

Number of degrees of freedom for a molecule of the gas

Molar specific heat of an ideal gas at **constant volume**

Ideal gas constant

EQUATION IN WORDS
Molar specific heat at constant volume for an ideal gas

Let's see how well Equation 15-17 works for the real gases listed in Table 15-1. For a monatomic gas there are three degrees of freedom per molecule, so $D = 3$ and Equation 15-17 becomes

$$C_V = \frac{3}{2}R = \frac{3}{2}[8.314 \text{ J/(mol}\cdot\text{K)}] = 12.47 \text{ J/(mol}\cdot\text{K)} \quad \text{(monatomic gas)}$$

Table 15-1 shows that this is a very close fit to the experimental values of C_V for argon, helium, and neon. For a diatomic gas with five degrees of freedom ($D = 5$), we expect

$$C_V = \frac{5}{2}R = \frac{5}{2}[8.314 \text{ J/(mol}\cdot\text{K)}] = 20.79 \text{ J/(mol}\cdot\text{K)} \quad \text{(diatomic gas)}$$

This is also a very close fit to the experimental values for the diatomic atoms listed in Table 15-1. Indeed, this is how we know that such molecules have five degrees of freedom! Experiments show that at temperatures well above room temperature, the value of C_V for diatomic molecules increases and becomes closer to $7R/2 = 29.10$ J/(mol · K). What's happening is that at higher temperatures it becomes possible for these molecules to vibrate. Different configurations of molecules can have different numbers of possible vibration modes.The simplest to imagine is a molecule with two atoms that can start oscillating toward and away from each other. Each vibrational mode will give each molecule two additional degrees of freedom (one for the kinetic energy of atoms moving toward and away from each other, the other for the potential energy of the spring-like chemical bonds that provide the restoring force for the vibration).

We encourage you to use Equation 15-17 to determine the number of degrees of freedom for the triatomic gases listed in Table 15-1.

The molar specific heat at constant pressure C_p also depends on D, the number of degrees of freedom. Since $C_p - C_V = R$ (Equation 15-14) and $C_V = (D/2)R$ (Equation 15-17), you can see that

$$C_p = C_V + R = \left(\frac{D}{2}\right)R + R = \left(\frac{D+2}{2}\right)R \qquad (15\text{-}18)$$

We can get an expression for the number of degrees of freedom D by taking the *ratio* of C_p to C_V. This **ratio of specific heats** is denoted by the Greek letter γ ("gamma"):

$$\gamma = \frac{C_p}{C_V} \quad \text{(ratio of specific heats)} \tag{15-19}$$

If we substitute Equation 15-17 for C_V and Equation 15-18 for C_p into Equation 15-19, we get

$$\gamma = \frac{C_p}{C_V} = \frac{\left(\frac{D+2}{2}\right)R}{\left(\frac{D}{2}\right)R} = \frac{D+2}{D} \quad \text{(ideal gas)} \tag{15-20}$$

You can check Equation 15-20 against the experimental values for γ given in Table 15-1. For example, for a monatomic ideal gas with three degrees of freedom ($D = 3$), Equation 15-20 predicts

$$\gamma = \frac{3+2}{3} = 1.67$$

which is a very good match to the experimental values for argon, helium, and neon. Try Equation 15-20 for a diatomic gas with $D = 5$. How well does the predicted value of γ compare to the experimental values in Table 15-1?

Adiabatic Processes for an Ideal Gas

The ratio of specific heats $\gamma = C_p/C_V$ for an ideal gas plays an important role in *adiabatic* processes in which there is no heating or cooling of the gas. If we have a quantity of gas in a cylinder with a moveable piston and then allow the piston to move in or out, the gas will compress or expand adiabatically if the cylinder and piston are made of an insulating material that does not allow heating or cooling. It can be shown using calculus (which is beyond our scope) that if an ideal gas expands or contracts adiabatically, the pressure P and volume V of the gas are related by

EQUATION IN WORDS
Pressure and volume for an ideal gas, adiabatic process

$$PV^{\gamma} = \text{constant} \tag{15-21}$$

Pressure of an ideal gas

Volume of an ideal gas

Ratio of specific heats of the gas = C_p/C_V

The quantity PV^{γ} has the same value at all times during an **adiabatic process** for an ideal gas.

As long as the thermodynamic changes that take place in an ideal gas are adiabatic, the product of pressure and volume to the power γ remains the same. As volume increases in an adiabatic expansion, pressure decreases; as volume decreases in an adiabatic compression, pressure increases. Note that in an adiabatic process, the pressure change is *greater* than in an isothermal expansion or compression between the same initial and final volumes. You can see this in **Figure 15-12**. The path in black is that followed in an adiabatic expansion $A \rightarrow B$ of an ideal gas. If the expansion were isothermal instead, but the initial and final volumes were the same, the path followed would be along the green isotherm in Figure 15-12, which leads to the point labeled C. The pressure at B is less than at C, so there is a greater pressure decrease in an adiabatic expansion than in an isothermal one.

Figure 15-12 also shows that the final state of the ideal gas after an adiabatic expansion lies on an isotherm for a lower temperature than for the initial state. In other words, the temperature of an ideal gas decreases in an adiabatic expansion. This

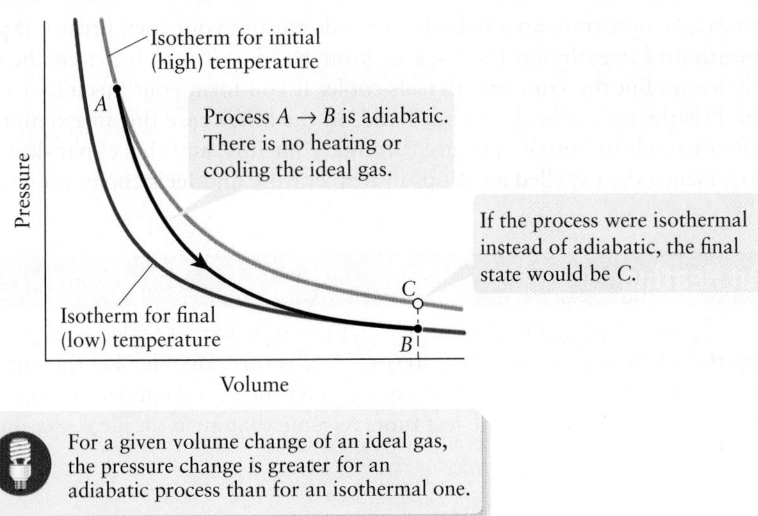
Isotherm for initial (high) temperature
A
Process $A \rightarrow B$ is adiabatic. There is no heating or cooling the ideal gas.
Pressure
If the process were isothermal instead of adiabatic, the final state would be C.
C
Isotherm for final (low) temperature
B
Volume
For a given volume change of an ideal gas, the pressure change is greater for an adiabatic process than for an isothermal one.

Figure 15-12 Adiabatic expansion of an ideal gas When an ideal gas expands adiabatically from an initial volume to a greater final volume, its pressure drops more than in an isothermal expansion between the same two volumes. Hence the gas temperature decreases. In the reverse process, an adiabatic compression, the gas temperature increases and the pressure increases more than in an isothermal compression between the same two volumes.

makes sense based on what we learned about adiabatic processes in Section 15-3: In an adiabatic expansion, $Q = 0$ and $W < 0$ (the surroundings do negative work on the gas as it expands), so $\Delta U < 0$: The internal energy of the gas decreases. The internal energy of an ideal gas is proportional to its absolute temperature T, so it follows that T decreases in an adiabatic expansion of an ideal gas. In the same way, the temperature of an ideal gas increases in an adiabatic compression.

We can find a relationship between temperature and volume in an adiabatic process for an ideal gas. To do this we'll use the ideal gas law $PV = nRT$, which we can rewrite as $P = nRT/V$. If we substitute this expression for the pressure P into Equation 15-21, we get

$$\frac{nRT}{V}V^{\gamma} = nRTV^{\gamma-1} = \text{constant}$$

Since the number of moles n does not change (no gas enters or leaves the system) and R is itself a constant, we can rewrite this expression as

$$TV^{\gamma-1} = \text{constant} \tag{15-22}$$

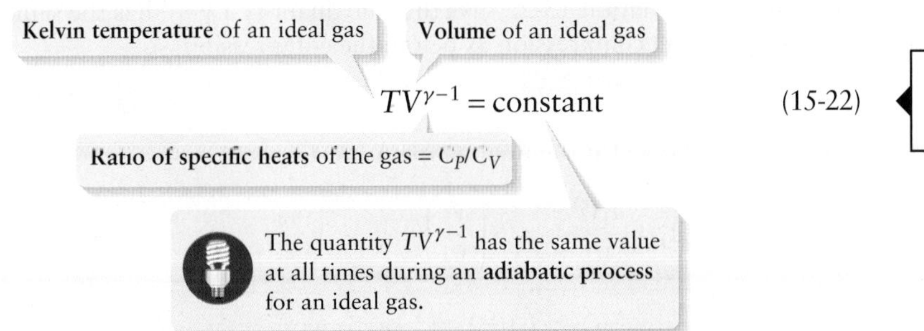
Kelvin temperature of an ideal gas
Volume of an ideal gas
Ratio of specific heats of the gas = C_P/C_V
The quantity $TV^{\gamma-1}$ has the same value at all times during an **adiabatic process** for an ideal gas.

EQUATION IN WORDS
Temperature and volume for an ideal gas, adiabatic process

Note that the value of γ is greater than 1 for all gases (see Table 15-1), so $\gamma - 1$ is greater than zero. Hence as the volume V of an ideal gas increases in an adiabatic expansion, the quantity $V^{\gamma-1}$ increases as well. For the quantity $TV^{\gamma-1}$ to maintain the same value, the temperature T must decrease in an adiabatic expansion, just as we described above. In the same way, as volume decreases in an adiabatic compression of an ideal gas, temperature increases.

Equation 15-22 explains what happens when you use a bicycle pump to inflate a tire: The pump rapidly becomes warm to the touch. Although the pump is not made of an insulating material, the air in the pump cylinder is compressed so rapidly that there's no time for much heating or cooling of the air during the compression. As a result, the compression is adiabatic, so the temperature of the gas (and the pump that holds it) increases as the air is compressed.

You can demonstrate an adiabatic expansion using your own breath. If you open your mouth and breathe on the back of your hand, you can feel that the expelled breath is warm. But the same breath feels cooler if you form your lips into a small "O" and then breathe on the back of your hand. In the latter case the air expands rapidly as it exits through the small aperture through your lips, and this expansion is nearly adiabatic. Hence the expelled air drops in temperature and feels cooler on your hand.

EXAMPLE 15-5 Burning Paper Without Heating

In a classroom physics demonstration a piece of paper is placed in the bottom of a test tube with a sealed plunger. When the plunger is quickly depressed, compressing the air inside, the paper combusts. How is this possible? Let the initial volume of the gas in the test tube be 10.0 cm^3 and the final volume be 1.00 cm^3, and let the initial temperature be room temperature (20.0°C). Calculate the final temperature of the gas in the test tube. Assume that air is an ideal, diatomic gas.

Set Up

Because the compression happens quickly, there is no time for any heating or cooling of the air in the tube. So we can consider this to be an adiabatic process. We use Equation 15-22 to relate volume and temperature for this process, and we use Equation 15-20 to find the ratio of specific heats γ for air. For diatomic gases, there are five degrees of freedom.

Ideal gas, adiabatic process:

$$TV^{\gamma-1} = \text{constant} \qquad (15\text{-}22)$$

Ratio of specific heats for an ideal gas with D degrees of freedom:

$$\gamma = \frac{C_p}{C_V} = \frac{D+2}{D} \qquad (15\text{-}20)$$

Solve

First find the value of γ for the gas and convert the temperature from Celsius to Kelvin.

From Equation 15-20, with $D = 5$:

$$\gamma = \frac{C_p}{C_V} = \frac{5+2}{5} = \frac{7}{5} = 1.40$$

The initial temperature of the air is 20.0°C or

$$T_i = (20.0 + 273.15)\text{ K} = 293\text{ K}$$

We know the initial values of the temperature (T_i = 293 K) and volume (V_i = 10.0 cm^3) and the final value of the volume (V_f = 1.00 cm^3). Use these in Equation 15-22 to solve for the final value of temperature.

Equation 15-22 says that the quantity $TV^{\gamma-1}$ has the same value after the compression as before the compression, so

$$T_f V_f^{\gamma-1} = T_i V_i^{\gamma-1}$$

Solve for the final temperature T_f:

$$T_f = \frac{T_i V_i^{\gamma-1}}{V_f^{\gamma-1}} = T_i\left(\frac{V_i}{V_f}\right)^{\gamma-1}$$

$$= (293\text{ K})\left(\frac{10.0\text{ cm}^3}{1.00\text{ cm}^3}\right)^{\frac{7}{5}-1}$$

$$= (293\text{ K})(10.0)^{\frac{2}{5}}$$
$$= 736\text{ K or }463°\text{C}$$

Reflect

The final temperature is above the flash point of paper (the temperature at which paper catches fire—around 230°C), so the paper will burn until the supply of paper or oxygen is used up. This same process is used in diesel engines: A mixture of fuel and air is rapidly compressed in the cylinders of the engine, raising the temperature to such a high value that the mixture spontaneously ignites without a spark plug (as is used in a conventional gasoline engine).

NOW WORK Problems 2–5 and 7 from The Takeaway 15-4.

THE TAKEAWAY for Section 15-4

✔ More heat is required at constant pressure than at constant volume to produce the same change in temperature of a gas.

✔ Thermodynamic processes that are carried out quickly can often be treated as adiabatic because the amount of energy transferred through thermal processes is relatively slow.

✔ Specific heat depends on the interactions of the atoms that constitute the substance, as well as the arrangement of the atoms.

Prep for the AP® Exam

AP® Building Blocks

1. A container holds 32.0 g of oxygen gas at a pressure of 8.0 atm. How much heat is required to increase the temperature by 100°C at constant pressure?
2. EX 5-5 A container holds 32.0 g of oxygen gas at a pressure of 8.0 atm. How much heat is required to increase the temperature by 100°C at constant volume?
3. EX 5-5 What ratio of initial volume to final volume V_i/V_f will raise the temperature of air from 27.0°C to 857°C in an adiabatic process? The molar specific heat ratio $\gamma = C_p/C_V$ for air is 1.4.
4. EX 5-5 A monatomic ideal gas at a pressure of 1.00 atm expands adiabatically from an initial volume of 1.50 m^3 to a final volume of 3.00 m^3. What is the new pressure?

AP® Skill Builders

5. EX 15-5 Two moles of an ideal monatomic gas expand adiabatically, performing 8.00 kJ of work in the process. Determine the change in temperature of the gas during the expansion.
6. One mole of an ideal monatomic gas ($\gamma = 1.66$), initially at a temperature of 0.00°C, undergoes an adiabatic expansion from a pressure of 10.0 atm to a pressure of 2.00 atm. Determine the work done on the gas.
7. EX 15-5 The temperature of 4.00 g of helium gas is increased at constant volume by 1.00°C. For the same heat, the temperature of what mass of oxygen gas will increase at constant volume by 1.00°C?
8. A temperature increase of 100°C results when 1.00 mol of air is heated at constant pressure. If the same amount of energy is instead added through heating when the gas is kept at constant volume, what is the temperature increase? The molar specific heat ratio $\gamma = C_p/C_V$ for the air is 1.4.
9. The volume of a gas is halved during an adiabatic compression that increases the pressure by a factor of 2.6. Determine the molar specific heat ratio $\gamma = C_p/C_V$.

AP® Skills in Action

10. A quantity of helium gas at room temperature is placed in a cylinder with a moveable piston. If the cylinder is locked in place so that the volume cannot change and a quantity of energy Q_0 is added to the gas, the temperature of the gas increases by ΔT_0. Now the gas is cooled to its original temperature, and the piston is unlocked so that the gas is free to expand. If the pressure exerted on the gas by the piston is held constant and the same quantity of energy Q_0 is added to the gas, will the temperature change of the gas be ΔT_0, the same as before, less than ΔT_0, or more than ΔT_0? Justify your choice.
11. The volume of a gas is halved during an adiabatic compression that increases the pressure by a factor of 2.5. Determine the factor by which the temperature increases.
12. Two cylinders, each equipped with a moveable piston, contain equal numbers of moles of a gas. One cylinder contains helium (He), while the other contains nitrogen (N_2). If each gas is heated the same amount, Q, and each is allowed to expand at constant pressure, what can you say about the temperature change ΔT and the magnitude of work done on the gas $|W|$ for the two gases?
(A) ΔT and $|W|$ are the same for both gases.
(B) ΔT and $|W|$ are both less for He than for N_2.
(C) ΔT and $|W|$ are both greater for He than for N_2.
(D) ΔT is greater for He than for N_2, but $|W|$ is less for He than for N_2.

15-5 The second law of thermodynamics describes why some processes are impossible

Figure 15-13 shows two frames from a video in which a person inflates a balloon until it pops. How do you know that the left frame happens before the right one? None of the physical laws that we have studied so far precludes a set of balloon fragments from flying toward each other and reassembling as an intact balloon.

For example, neither conservation of energy nor momentum would be violated were the balloon fragments to reassemble. What about the pictures of a dropped

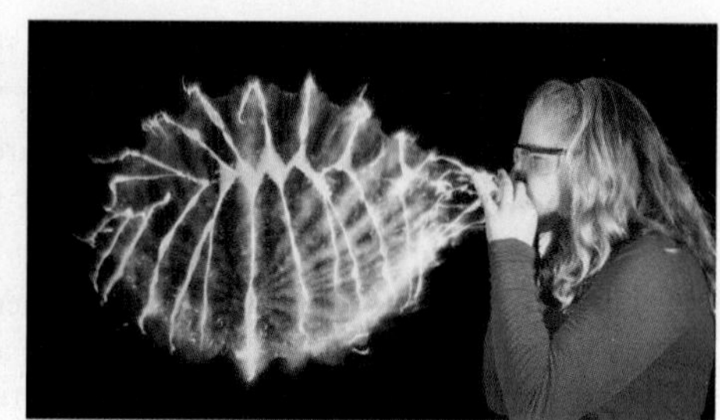

Ted Kinsman/Science Source

Figure 15-13 Bursting your balloon Both images were taken from a video of a person blowing up a balloon until it pops. How do you know that the left frame happens before the right one?

popsicle melting on a carpet in **Figure 15-14**? Each image was taken before the one to its right, but would it be possible for this sequence to run in reverse? Here again the reverse process would be permitted by the fundamental laws we know. The total energy of the system would be conserved if the carpet drew enough energy from the melted popsicle (mostly water) to cause it to freeze: The amount of energy required to melt a chunk of ice is exactly the same as the energy needed to freeze the resulting water. Yet we have declared that thermal transfer of energy is always from a hotter to a cooler system. Why?

For both the balloon in Figure 15-13 and the popsicle in Figure 15-14, events happen in a sequence that tends to decrease the level of *organization* or *order*. In Figure 15-13 the air that was confined within the balloon in an orderly way spreads out across the room when the balloon is popped, and the balloon itself breaks into a collection of random fragments. In Figure 15-14 the water molecules begin in an ordered state in which they are locked into the crystalline structure of the popsicle, and they end up free to move in a disordered way into the fibers of the carpet. The reverse processes, in which air and balloon fragments spontaneously coalesce into a filled balloon or water spontaneously freezes, never occur in nature. That is, natural events always happen in a direction from ordered to disordered, or toward increasing *randomness*. We can generalize this observation to the **second law of thermodynamics**:

The amount of disorder in an isolated system either always increases or, if the system is in equilibrium, stays the same.

We emphasize that a system described by the second law of thermodynamics must be *isolated*. This is to ensure that nothing outside the system can cause its state to change in an unnatural way. For example, we could manually sort out the fragments of the balloon in Figure 15-13, carefully reassemble them, and then inflate the reassembled balloon. The balloon system would experience increasing order, but only as a result of our external intervention; the balloon would not be isolated.

Let's see how to use the second law of thermodynamics to analyze systems that convert heating to work or work to heating. We'll look in particular at mechanical systems such as the internal combustion engine in a car (which takes the heating that results from burning gasoline and converts it to work) and a kitchen refrigerator (which takes work done by the refrigerator motor and uses it to transfer energy out of the refrigerator's contents). We'll see that the second law imposes strict limits on how efficient an engine can be and on how much performance you can get from a refrigerator.

Joe Belanger/Alamy

Figure 15-14 Melting a popsicle A dropped popsicle melts atop a piece of carpet. Each image in the sequence was taken after the one to its left. Could this sequence run in reverse?

In the following section we'll extend the second law to include the concept of entropy, which is a measure of the amount of randomness in a system. We'll use the entropy concept to gain further insight into the second law and to see what this law tells us about living systems.

Heat Engines and the Second Law

A system or device that converts heat to work is a **heat engine** or simply an *engine*. Heat engines are cyclic: Some part of the system absorbs energy, work is done, and the system returns to its original state for the cycle to begin again. Although the term *engine* might call to mind the complex device that powers an automobile, in thermodynamics a heat engine can be as simple as gas in a piston that expands as the gas is heated and contracts as the gas cools.

There are actually tiny molecular engines in living cells. In the mitochondria of all human cells, hydrogen ions flow through ATP synthase, an enzyme that has a structure not unlike a waterwheel (**Figure 15-15**). ATP synthase converts the kinetic energy of hydrogen ions into rotational kinetic energy of the enzyme, which in turn is used to convert adenosine diphosphate (ADP) into adenosine triphosphate (ATP). The chemical bond energy in the resulting ATP is in a form that can be used by all cells.

Figure 15-16 shows a simple model for any heat engine. The engine is thermally connected to a *reservoir* of higher (hotter) temperature (T_H) and to a reservoir of lower (cooler) temperature (T_C). A **reservoir** is a part of a system large enough to either absorb or supply energy without a change in temperature. In an old-time steam engine, for example, the furnace serves as the hot reservoir and the surrounding atmosphere acts as the cold reservoir. Energy Q_H is transferred from the hot reservoir into the engine; during this process some of the energy goes into work W, and the remainder $|Q_C|$ is transferred out of the engine and into the cold reservoir.

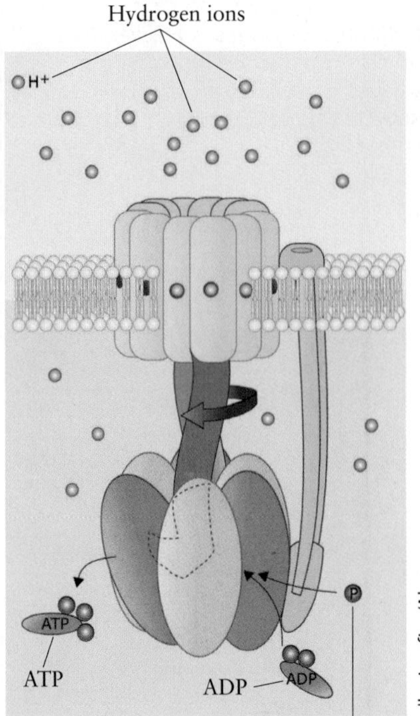
Hydrogen ions
H+
ATP
ATP
ADP
ADP
P
Phosphate ion
ellepigrafica/Alamy

Figure 15-15 A molecular heat engine This illustration shows an ATP synthase enzyme attached to the inner membrane (yellow) of a mitochondrion. The enzyme extracts energy from hydrogen ions flowing through its stationary part (blue), and uses this energy to do work on the stalk of the enzyme (purple and orange) and make it rotate. This rotation pushes phosphate ions and adenosine diphosphate molecules (ADP) together to form adenosine triphosphate (ATP). An ATP synthase molecule is about 10 nm in diameter and 20 nm tall.

WATCH OUT !

The quantity Q_C is negative.

Remember the convention that Q is positive if it is transferred into the system in question and negative if it is transferred out of the system. So for our engine Q_H is positive (energy is transferred into the engine from the hot reservoir) and Q_C is negative (energy is transferred out of the engine into the cold reservoir). However, in analyzing the engine in Figure 15-16 we'll be interested in how much energy is transferred *into* the *cold reservoir*. So we've labeled Figure 15-16 with $|Q_C|$ (the absolute value of Q_C, which is positive) to indicate the positive quantity of energy that is transferred into the cold reservoir through heating.

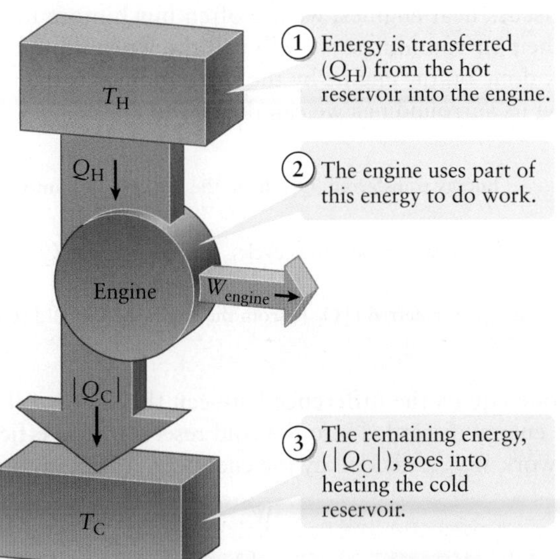
T_H
Q_H
Engine
W_{engine}
$|Q_C|$
T_C
1 Energy is transferred (Q_H) from the hot reservoir into the engine.
2 The engine uses part of this energy to do work.
3 The remaining energy, ($|Q_C|$), goes into heating the cold reservoir.

Figure 15-16 What a heat engine does A generic heat engine (in purple) is thermally connected to a reservoir of higher temperature T_H (in red) and to a reservoir of lower temperature T_C (in blue). Energy Q_H is transferred from the hot reservoir into the engine. The engine uses some of the energy to do work W; the remaining energy is transferred to the cold reservoir. (Note that Q_C is negative because it is energy that *leaves* the engine, so the energy that enters the cold reservoir is $|Q_C|$).

In a perfect engine *all* of the energy Q_H taken in from the hot reservoir would be converted to work, with no energy at all going into the cold reservoir (that is, $Q_C = 0$ in Figure 15-16). This would be a 100% efficient engine: None of the energy from the hot reservoir would be "thrown away" to the cold reservoir. Unfortunately the second law of thermodynamics says that such a perfect engine is *impossible*. To see why, note that there is an increase in order of the hot reservoir: Because energy is transferred out of this reservoir, some of the molecules of the reservoir must lose their random motion and so become more orderly. Over one cycle of operation the engine itself returns to its original state, so it has zero net change in order, and there is no change at all in the cold reservoir (because no energy is transferred into it). So the net result would be that over one cycle the system of engine plus reservoirs undergoes an increase in order. But this violates the second law of thermodynamics, which says that the order of an isolated system can decrease or stay the same but cannot increase! We conclude that you simply cannot build a perfect engine that's 100% efficient. The *Kelvin–Planck statement* is a rewording of the second law of thermodynamics and describes the inherent inefficiency of heat engines as follows:

No process is possible in which energy transferred from a reservoir through heating is converted completely into work.

For a heat engine to satisfy the second law of thermodynamics, some energy *must* be transferred into the cold reservoir to make the cold reservoir less ordered. (This happens because the energy added to the cold reservoir increases the random motion of the reservoir's molecules.) The decrease in order of the cold reservoir can then compensate for the increase in order of the hot reservoir. So not all of the energy that is transferred from the hot reservoir can be converted to work; some of this energy must go into heating the cold reservoir and cannot be recovered. That's what happens in a car's gasoline engine: Only about a quarter of the energy released by burning gasoline gets converted into work to propel the car. The remainder goes into heating up the engine and the exhaust gases. Similarly, in the human body only about a quarter of the energy in food is actually used by your cells to do work. The rest of the energy heats your body.

Let's apply the *first* law of thermodynamics, Equation 15-1, to the generic engine shown in Figure 15-16. In one cycle the engine takes in energy Q_H from the hot reservoir and sends energy $|Q_C|$ to the cold reservoir, so the *net* quantity of energy that is transferred into the engine is $Q = Q_H - |Q_C|$. In that same cycle the engine does work W. Because the cycle returns the engine to its initial state at the beginning of the cycle, there is zero net change in its internal energy: $\Delta U = 0$. So the first law of thermodynamics, $\Delta U = Q + W$, becomes

$$0 = Q_H - |Q_C| + W$$

This W is the work done on the engine (system) by its surroundings, as is our convention. When we discuss heat engines, we are often much more interested in the work they can do on their surroundings; we will call the work done by the engine W_{engine}. Because the work done on the engine by the surroundings is the negative of the work the engine does on its surroundings we can define the

EQUATION IN WORDS
Work done by a heat engine

Energy transferred (Q_H) from the hot reservoir into the engine in one cycle

(15-23) Work done by a heat engine in one cycle $$W_{\text{engine}} = Q_H - |Q_C|$$

Energy transferred ($|Q_C|$) from the engine to the cold reservoir in one cycle

The work done equals the difference between the energy taken in from the hot reservoir and the energy discarded into the cold reservoir. The **efficiency** of the engine is defined as the work W_{engine} divided by the energy Q_H taken in to do the work:

(15-24) $$e = \frac{W_{\text{engine}}}{Q_H}$$

Efficiency is essentially what you get out of an engine divided by what you put in. If all of the input energy were converted to work, then W would equal Q_H and the efficiency would be 100%, or $e = 1$. But the Kelvin–Planck form of the second law of thermodynamics tells us this is impossible. The best combustion engines, such as the ones found in automobiles, operate at an efficiency of $e \approx 0.30$ (about 30%). Similarly, when human muscles contract, only about 25% of the input energy does work, with the rest going into heating, so $e \approx 0.25$. Using Equation 15-23 we can express the efficiency of a heat engine (Equation 15-24) in terms of the energy taken in from the hot reservoir and the energy rejected to the cold reservoir:

$$e = \frac{W_{\text{engine}}}{Q_H} = \frac{Q_H - |Q_C|}{Q_H} = 1 - \frac{|Q_C|}{Q_H} \qquad (15\text{-}25)$$

(efficiency of a heat engine)

EXAMPLE 15-6 Efficiency of an Engine

The combustion of gasoline in a lawnmower engine releases 44.0 J of energy per cycle. Of that, 31.4 J are lost to warming the body of the engine and the surrounding air. (a) How much work does the lawnmower engine do per cycle? (b) What is the efficiency of the engine?

Set Up

We use Equation 15-23 to determine the work done by the engine and Equation 15-25 to find the efficiency.

Work done by a heat engine:

$$W_{\text{engine}} = Q_H - |Q_C| \qquad (15\text{-}23)$$

Efficiency of a heat engine:

$$e = \frac{W_{\text{engine}}}{Q_H} = \frac{Q_H - |Q_C|}{Q_H} = 1 - \frac{|Q_C|}{Q_H} \qquad (15\text{-}25)$$

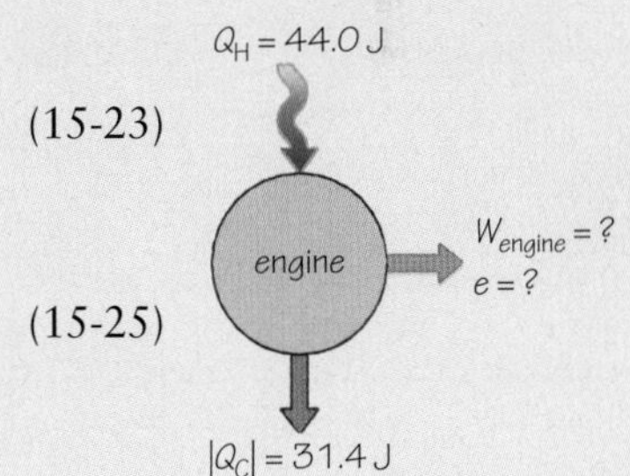

Solve

(a) The work done by the engine equals the difference between the energy Q_H taken in by burning gasoline and the energy $|Q_C|$ that is transferred from the engine to its surroundings.

We are given that in one cycle

$$Q_H = 44.0 \text{ J}$$
$$|Q_C| = 31.4 \text{ J}$$

From Equation 15-23 the work that the engine does in one cycle is

$$W_{\text{engine}} = Q_H - |Q_C| = 44.0 \text{ J} - 31.4 \text{ J} = 12.6 \text{ J}$$

(b) The efficiency of the engine equals the work done divided by the energy taken in. It's also equal to 1 minus the ratio of energy transferred to the surroundings that cannot be recovered to energy taken in.

From Equation 15-25 the efficiency is

$$e = \frac{W_{\text{engine}}}{Q_H} = \frac{12.6 \text{ J}}{44.0 \text{ J}} = 0.286$$

Alternatively,

$$e = 1 - \frac{|Q_C|}{Q_H} = 1 - \frac{31.4 \text{ J}}{44.0 \text{ J}} = 1 - 0.714 = 0.286$$

Reflect

An efficiency of around 0.25 to 0.30 (25% to 30%) is typical for a gasoline engine.

NOW WORK Problem 1 from The Takeaway 15-5.

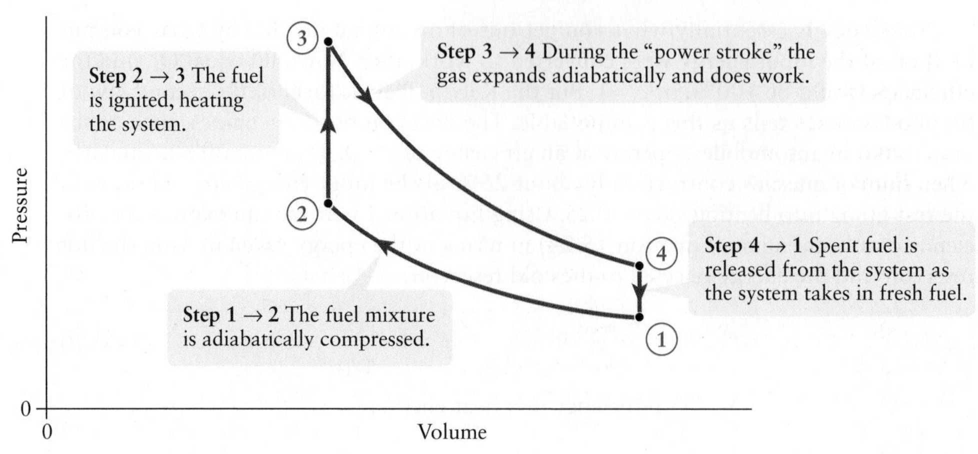

Step 2 → 3 The fuel is ignited, heating the system.
Step 3 → 4 During the "power stroke" the gas expands adiabatically and does work.
Step 4 → 1 Spent fuel is released from the system as the system takes in fresh fuel.
Step 1 → 2 The fuel mixture is adiabatically compressed.
Pressure
Volume
0
0
1
2
3
4

Figure 15-17 A simplified Otto cycle This *PV* diagram shows a simplified version of the cycle used in most automobile engines. (The aircraft engine shown in Figure 15-1b uses this same cycle.) It is named for the nineteenth-century German engineer Nicolaus Otto, who was the first to build an engine that worked on this cycle.

Figure 15-17 shows the PV diagram for the cyclic process (called the *Otto cycle*) used in a typical automobile engine. (The process is somewhat simplified in this figure.) Beginning at the state marked 1, a mixture of fuel and air is rapidly compressed in a cylinder with a moveable piston. Because the compression is rapid, step $1 \to 2$ is adiabatic and there is no heating of the fuel–air mixture. However, work is done *on* the mixture by the piston, so for this step the work done *by* the fuel–air mixture (which in this case is our engine) is negative: $W_{\text{engine},1\to 2} < 0$. At state 2 a spark ignites the fuel, resulting in energy Q_H transferred into the system. The associated increase in pressure and temperature from state 2 to state 3 occurs so rapidly that the volume of the gas in the piston remains nearly constant, so no work is done. At state 3, the high pressure causes the gas to expand rapidly and so adiabatically. Because $Q = 0$ in an adiabatic process, the first law of thermodynamics ($\Delta U = Q + W$) tells us that all of the change in internal energy in step $3 \to 4$ results in positive work done by the gas on the piston: $W_{\text{engine},3\to 4} > 0$. As the system returns to its initial state (step $4 \to 1$, for which the volume again remains constant so no work is done) exhaust is expelled and new fuel is injected into the cylinder.

The net work output in the cycle is the sum of the positive work in step $3 \to 4$ (in which the gas expands) and the negative work in step $1 \to 2$ (in which the gas compresses). Remember that the work done equals the area under the curve in a PV diagram; step $3 \to 4$ happens at higher pressure than step $1 \to 2$, so there is more area under the curve for step $3 \to 4$ than for step $1 \to 2$ and there is more positive work done by the engine than negative work done by the engine. So the net work output of the engine over one cycle is positive. This work is used to turn the driveshaft of the automobile and make the wheels turn.

The Carnot Cycle

The cycle shown in Figure 15-17 is not the only cycle used in heat engines, nor does it have the highest efficiency. It turns out that the most efficient of all possible cyclic processes in a heat engine is the **Carnot cycle** (pronounced "car-noe"), first proposed in the early 1800s by French engineer Sadi Carnot. Carnot based his cycle on a requirement that all of the processes that comprise the cycle be **reversible processes**—that is, the reverse process is physically possible. Let's explore what this means.

Most thermodynamic processes are **irreversible processes**: They can proceed in only one direction. An example is what happens when ice at 0°C is placed in a large metal pot at 20°C. Energy is transferred from the pot into the ice through heating, so the ice begins to melt, and the temperature of the pot begins to drop. When the ice and pot come to equilibrium, the ice has completely melted, and the pot and melted water both reach the same final temperature. The reverse of this process—in which liquid water in a pot spontaneously freezes, and the pot warms up—never happens in nature.

An example of a reversible process is ice at 0°C placed in a large metal pot that is also at 0°C. If a small amount of energy is transferred from the pot into the ice, the ice will partially melt and the pot will contain ice and liquid water at 0°C. Because the energy transferred is so small, the temperature of the pot changes very little and is still 0°C. If the same small amount of energy is now transferred from the water–ice mixture back into the pot, the water will freeze back to ice and the ice and pot will both be at 0°C.

In general, a process that involves heating will be reversible *only* if the systems between which energy is transferred are at the same temperature. If they are at different temperatures, the energy transfer by any thermal process will always be from the higher-temperature system to the lower-temperature system and the process will be irreversible.

A process can also be reversible if it involves *no* heat ($Q = 0$). An example is a *slow* adiabatic expansion or compression, such as would happen with gas in an insulated cylinder. If the gas is allowed to expand slightly, it does a small amount of positive work on the piston ($W_{engine} > 0$) and the internal energy of the gas decreases slightly (from the first law of thermodynamics, where we have replaced $+W$ with $-W_{engine}$, so $\Delta U = Q - W_{engine}$ is negative because Q is zero, and W_{engine} is positive). In the reverse process the gas is compressed slightly and does negative work on the piston ($W_{engine} < 0$, meaning that the piston does work on the gas), and the internal energy of the gas increases. The system then returns to its initial state. By contrast, a free expansion of a gas, in which a gas under pressure suddenly expands to a larger volume (as in Figure 15-4b), is *irreversible*. It is not possible to return the system to the initial pressure and volume without adding energy by doing work on the gas.

Carnot understood that energy is conserved in both reversible and irreversible processes. But he also understood that in an irreversible process much of the energy becomes *unavailable*. For example, when the 20°C pot heats the 0°C ice to melt it, the energy that went into the ice cannot be returned to the pot. In order not to "lose" energy in this way, Carnot realized that the most efficient heat engine cycle possible must therefore involve only reversible processes. As we've seen, this means either slow processes that involve energy transfer with zero temperature difference or slow adiabatic processes.

Figure 15-18 shows the *PV* diagram for the Carnot cycle. The system is an ideal gas in a cylinder with a moveable piston. In state 1 the cylinder is in contact and in equilibrium with the cold reservoir at temperature T_C, so the gas is also at temperature T_C. The four *reversible* steps that constitute the cycle:

- **Step 1 → 2:** Adiabatic compression. You take the cylinder filled with gas and wrap it with a perfect insulator so that there can be no energy transfer by thermal processes into or out of the gas, and you do work on the moveable piston to slowly push it inward (so that the gas does negative work). As the gas is compressed, its temperature increases from T_C to T_H.
- **Step 2 → 3:** Isothermal expansion. You remove the insulation and put the cylinder in contact with the hot reservoir at temperature T_H (which is the same

The Carnot cycle consists of:
- Two adiabatic processes, for which no energy is transferred in or out by thermal processes of the system.
- Two isothermal processes, one at the temperature of the hot reservoir and one at the temperature of the cold reservoir.

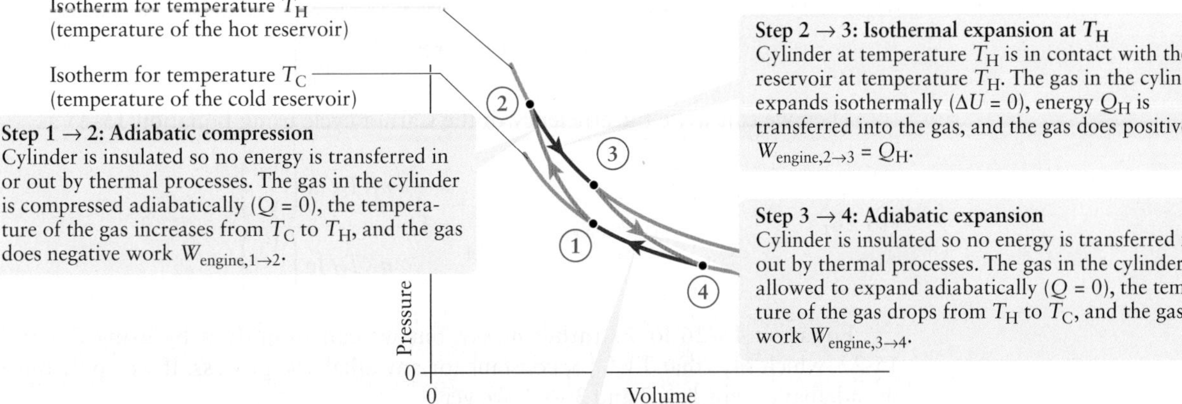

Isotherm for temperature T_H
(temperature of the hot reservoir)
Isotherm for temperature T_C
(temperature of the cold reservoir)
Step 1 → 2: Adiabatic compression
Cylinder is insulated so no energy is transferred in or out by thermal processes. The gas in the cylinder is compressed adiabatically ($Q = 0$), the temperature of the gas increases from T_C to T_H, and the gas does negative work $W_{engine,1\to2}$.
Step 2 → 3: Isothermal expansion at T_H
Cylinder at temperature T_H is in contact with the hot reservoir at temperature T_H. The gas in the cylinder expands isothermally ($\Delta U = 0$), energy Q_H is transferred into the gas, and the gas does positive work $W_{engine,2\to3} = Q_H$.
Step 3 → 4: Adiabatic expansion
Cylinder is insulated so no energy is transferred in or out by thermal processes. The gas in the cylinder is allowed to expand adiabatically ($Q = 0$), the temperature of the gas drops from T_H to T_C, and the gas does work $W_{engine,3\to4}$.
1
2
3
4
Pressure
0
0
Volume
Step 4 → 1: Isothermal compression at T_C
Cylinder at temperature T_C is in contact with the cold reservoir at temperature T_C. The gas in the cylinder is compressed isothermally ($\Delta U = 0$), energy $|Q_C|$ is transferred into the gas, and the gas does negative work $W_{engine,4\to1} = -|Q_C|$.

Figure 15-18 The Carnot cycle The most efficient cycle possible for a heat engine is made up of four reversible steps.

as the current temperature of the gas). You now allow the gas to expand so that it does positive work on the piston. The temperature remains equal to T_H because the gas and hot reservoir are in equilibrium.

- **Step 3 → 4:** Adiabatic expansion. You again wrap the cylinder with a perfect insulator and allow the gas to expand. The expanding gas does more positive work on the piston, and the temperature of the gas decreases from T_H back down to T_C.
- **Step 4 → 1:** Isothermal compression. You again remove the insulation, but now you put the cylinder in contact with the cold reservoir at temperature T_C (which is the same as the current temperature of the gas). You do work on the piston to slowly push it inward, so the gas does negative work. The temperature remains equal to T_C because the gas and cold reservoir are in equilibrium. You stop when the gas is once again in state 1, with the same values of pressure, volume, and temperature as it had initially.

You can see from the description that the Carnot cycle is an idealization (like a massless rope or a frictionless ramp) and is *not* a practical cycle for a real engine. The Carnot cycle is nonetheless tremendously important. Because all four of its processes are reversible, the Carnot cycle is the *most efficient* heat engine cycle possible. It therefore represents the standard against which all other cycles should be compared. Let's see how to determine the efficiency of a Carnot engine.

From Equation 15-25 the efficiency of any heat engine is $e = 1 - |Q_C|/Q_H$. For the Carnot cycle depicted in Figure 15-18, Q_H is the energy that is transferred into the ideal gas during step 2 → 3, when the gas cylinder is in contact with the hot reservoir at temperature T_H, and Q_C is the (negative) energy that is transferred into the gas during step 4 → 1, when the gas cylinder is in contact with the cold reservoir at temperature T_C. Both of these processes are isothermal, so $\Delta U = 0$ and thus $Q = W_{engine}$. Furthermore, we can see from Equation 15-5 that the work done by n moles of an ideal gas in an isothermal process at temperature T is $W_{engine} = -W = nRT \ln (V_f/V_i)$, where V_i and V_f are the volume of the gas at the beginning and end of the process, respectively. So for the two isothermal steps in the Carnot cycle, we have

$$Q_H = W_{2\to3} = nRT_H \ln\left(\frac{V_3}{V_2}\right)$$

$$Q_C = W_{4\to1} = nRT_C \ln\left(\frac{V_1}{V_4}\right)$$

Note that Q_C is negative for step 4 → 1. To see this, notice that V_1, the volume in state 1, is less than V_4, the volume in state 4 (see Figure 15-18). So the ratio V_1/V_4 is less than 1, and the natural logarithm of a number less than 1 is negative. Since $\ln (1/x) = -\ln x$, we can write the absolute value of Q_C as

$$|Q_C| = -Q_C = nRT_C \ln\left(\frac{V_4}{V_1}\right)$$

Then we can write the efficiency of the Carnot cycle using Equation 15-25 as

$$e_{Carnot} = 1 - \frac{|Q_C|}{Q_H} = 1 - \frac{nRT_C \ln\left(\frac{V_4}{V_1}\right)}{nRT_H \ln\left(\frac{V_3}{V_2}\right)} \tag{15-26}$$

Equation 15-26 looks rather messy, but we can simplify it by using Equation 15-22, which says that $TV^{\gamma-1}$ = constant for any adiabatic process. If we apply this to the adiabatic steps 1 → 2 and 3 → 4, we get

$$\text{Step } 1 \to 2\text{: } T_C V_1^{\gamma-1} = T_H V_2^{\gamma-1} \quad \text{or} \quad \frac{T_C}{T_H} = \frac{V_2^{\gamma-1}}{V_1^{\gamma-1}}$$

$$\text{Step } 3 \to 4\text{: } T_C V_4^{\gamma-1} = T_H V_3^{\gamma-1} \quad \text{or} \quad \frac{T_C}{T_H} = \frac{V_3^{\gamma-1}}{V_4^{\gamma-1}}$$

If we set the two expressions for T_C/T_H equal to each other and rearrange, we get

$$\frac{V_2^{\gamma-1}}{V_1^{\gamma-1}} = \frac{V_3^{\gamma-1}}{V_4^{\gamma-1}} \quad \text{so} \quad \frac{V_2}{V_1} = \frac{V_3}{V_4} \quad \text{and} \quad \frac{V_4}{V_1} = \frac{V_3}{V_2}$$

The last of these tells us that the quantities ln (V_4/V_1) and ln (V_3/V_2) in Equation 15-26 are equal and so cancel out, as do the factors of nR. So we're left with a very simple expression for the efficiency of a Carnot engine:

Efficiency of a Carnot engine — **Kelvin temperature of the cold reservoir** — **Kelvin temperature of the hot reservoir**

$$e_{\text{Carnot}} = 1 - \frac{T_C}{T_H} \qquad (15\text{-}27)$$

EQUATION IN WORDS
Efficiency of a Carnot engine

The efficiency of an ideal Carnot heat engine depends *only* on the hot and cold temperatures between which the engine operates. Equation 15-27 is the maximum theoretical efficiency of *any* heat engine operating between temperatures T_H and T_C. It is impossible for any engine to do better than this.

The efficiency of an ideal Carnot engine can be made closer and closer to $e_{\text{Carnot}} = 1$ (100% efficiency) by decreasing T_C and increasing T_H. For the efficiency to be exactly one, T_C must equal 0 K. It is not possible, however, to attain a temperature of absolute zero. This statement is referred to as the **third law of thermodynamics:**

It is possible for the temperature of a system to be arbitrarily close to absolute zero, but it can never reach absolute zero.

A consequence of the third law of thermodynamics is that it is impossible to make a heat engine that is 100% efficient.

EXAMPLE 15-7 Carnot Efficiency and Actual Efficiency

An automotive engine is designed to operate between 290 and 450 K. The engine produces 1.50×10^2 J of mechanical energy for every 6.00×10^2 J of the energy absorbed from the combustion of fuel. Calculate (a) the actual efficiency of this engine and (b) the theoretical maximum efficiency of a Carnot engine operating between these temperatures.

Set Up

We'll use Equation 15-25 to determine the actual efficiency and Equation 15-27 to find the efficiency of a Carnot engine operating between 290 and 450 K.

Efficiency of a heat engine:

$$e = \frac{W_{\text{engine}}}{Q_H} = \frac{Q_H - |Q_C|}{Q_H} = 1 - \frac{|Q_C|}{Q_H} \qquad (15\text{-}25)$$

Efficiency of a Carnot engine:

$$e_{\text{Carnot}} = 1 - \frac{T_C}{T_H} \qquad (15\text{-}27)$$

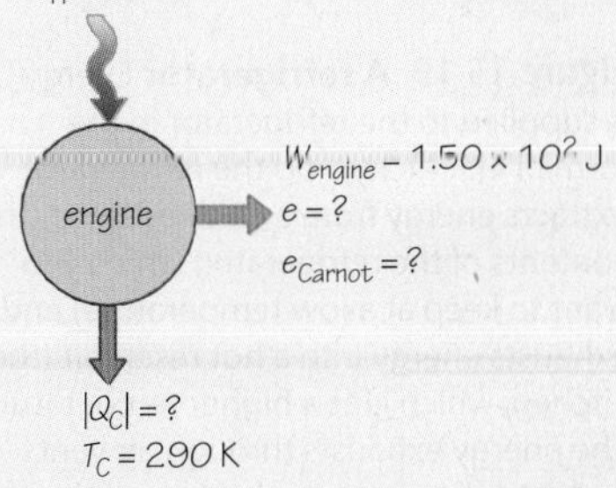

Solve

(a) We are given $W_{\text{engine}} = 1.50 \times 10^2$ J and $Q_H = 6.00 \times 10^2$ J. Use this and the first expression in Equation 15-25 to find the actual efficiency of the engine.

Actual efficiency:

$$e = \frac{W_{\text{engine}}}{Q_H} = \frac{1.50 \times 10^2 \text{ J}}{6.00 \times 10^2 \text{ J}} = 0.250$$

(b) The efficiency of the engine in (a) is less than the theoretical maximum efficiency of a Carnot engine with $T_C = 290$ K and $T_H = 450$ K.

Efficiency for a theoretical Carnot engine operating between these temperatures:

$$e_{\text{Carnot}} = 1 - \frac{T_C}{T_H} = 1 - \left(\frac{290 \text{ K}}{450 \text{ K}}\right) = 0.356$$

Reflect

Note that even a Carnot engine—the most efficient heat engine that could possibly operate between these temperatures—is not terribly efficient: $e_{\text{Carnot}} = 0.356$ means that only 35.6% of the energy released by combustion can be used to do work. The other 64.4% warms up the engine and its surroundings. This is one reason electric cars, which are powered by electric motors that are not subject to the limitations of heat engines, are much more energy efficient than are gasoline-powered cars.

NOW WORK Problems 2, 3, 6, and 7 from The Takeaway 15-5.

Refrigerators

A **refrigerator** is a device that takes in energy and uses it to transfer energy from a system at low temperature to a system at high temperature. An ordinary kitchen refrigerator (**Figure 15-19**) works this way. In many ways a refrigerator is simply an engine running backward.

Figure 15-19 A refrigerator Energy is supplied to the refrigerator in the form of electricity, and the refrigerator extracts energy from a cold reservoir (the contents of the refrigerator, which you want to keep at a low temperature) and exhausts energy into a hot reservoir (the kitchen, which is at a higher temperature). The energy exhausts through air vents in the back or near the bottom of the refrigerator, which is why it's warm there.

Like heat engines, refrigerators are cyclic, and like heat engines, refrigerators are constrained by the second law of thermodynamics. The *Clausius form* of the second law of thermodynamics, named after the nineteenth-century German physicist Robert Clausius, says that:

No process is possible in which energy absorbed from a cold reservoir is transferred completely to a hot reservoir.

In other words it isn't possible to make a perfect refrigerator.

Figure 15-20 is a diagram of a generic refrigerator. This looks similar to the generic heat engine shown in Figure 15-16, but the heat and work arrows are in the opposite direction. Work is an *input* to this device; as a result, heat Q_C is made to go from the colder region to the hotter one. This transfer is, of course, exactly what you'd want to cool down a container in which to store food (a refrigerator) or a building (an air conditioner). A heat pump, a device used to warm up a building, is identical to a refrigerator; the difference is that a heat pump transfers energy from the cold exterior of the building into the warm interior to keep it warm.

Figure 15-21 shows the cycle used in a typical kitchen refrigerator like that in Figure 15-19. This cycle uses the inert compound R134a (chemical formula CH_2FCF_3). R134a in the liquid state is forced through a tube (the evaporator coil) that passes through the freezer and refrigerator compartments. The R134a absorbs energy from the air in these compartments to cool them, which raises the temperature of the R134a and turns it into a gas. A compressor pump then pressurizes the R134a gas, which drives the gas around the system and also increases its temperature even more. As the hot gas passes through the long, narrow condenser coil tube (usually in the

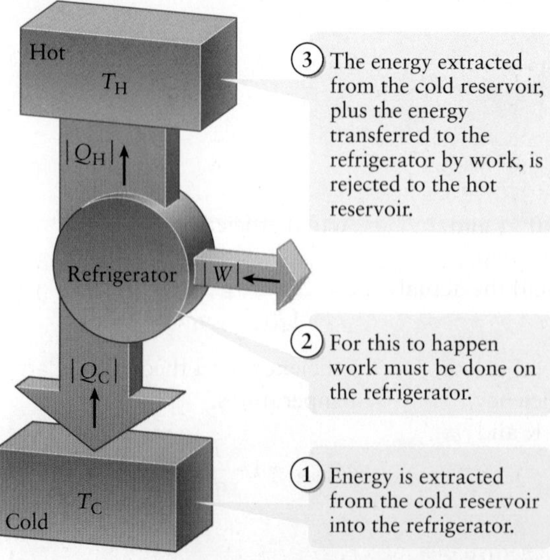

Hot
T_H
$|Q_H|$
Refrigerator
$|W|$
$|Q_C|$
Cold
T_C
3 The energy extracted from the cold reservoir, plus the energy transferred to the refrigerator by work, is rejected to the hot reservoir.
2 For this to happen work must be done on the refrigerator.
1 Energy is extracted from the cold reservoir into the refrigerator.

Figure 15-20 What a refrigerator does A generic refrigerator (in purple) extracts heat Q_C from a cold reservoir (in blue) at temperature T_C. To do this, work must be done on the refrigerator (so W_{engine}, the work done *by* the refrigerator, is negative, and the work done *on* the refrigerator is $W > 0$). All of the energy added to the engine by heating and work is rejected to the hot reservoir (in red) at temperature T_H. Here Q_H is negative (it represents energy that *leaves* the refrigerator), so the energy that is rejected *into* the hot reservoir is $|Q_H|$.

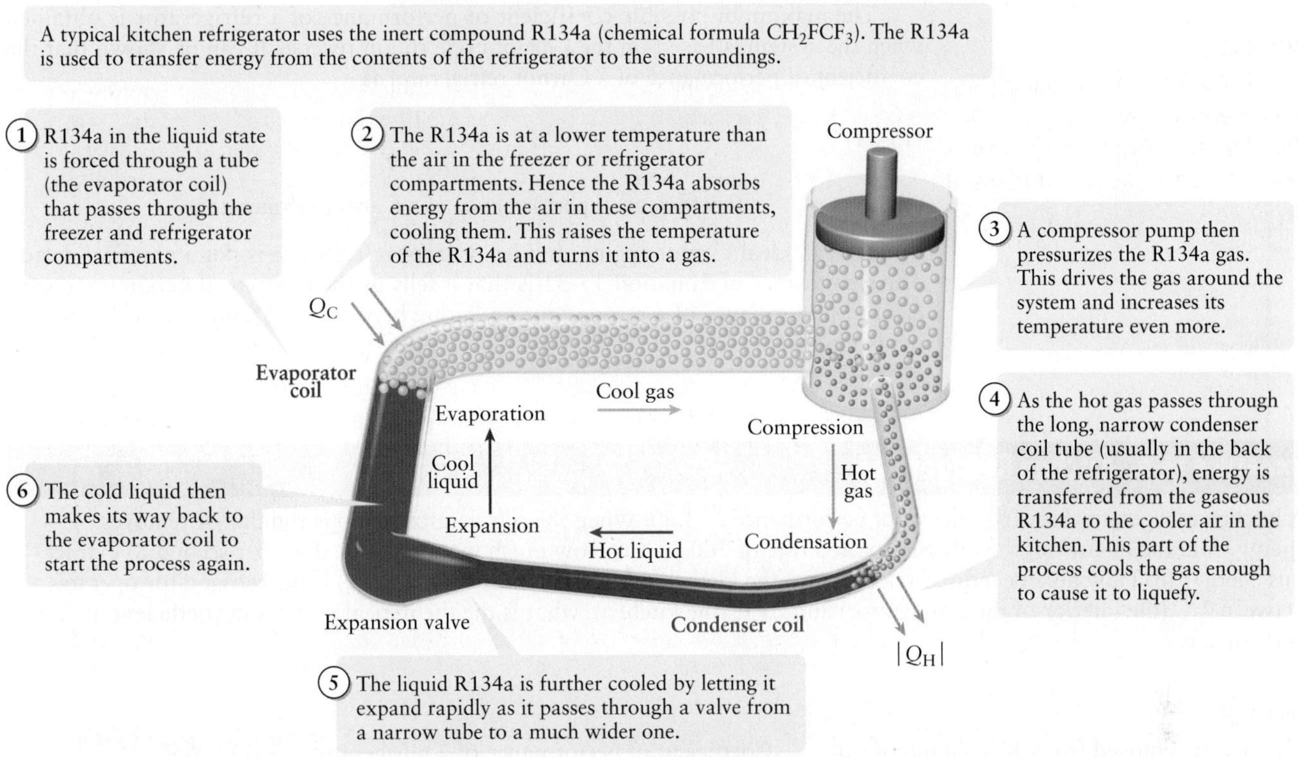

A typical kitchen refrigerator uses the inert compound R134a (chemical formula CH_2FCF_3). The R134a is used to transfer energy from the contents of the refrigerator to the surroundings.

(1) R134a in the liquid state is forced through a tube (the evaporator coil) that passes through the freezer and refrigerator compartments.

(2) The R134a is at a lower temperature than the air in the freezer or refrigerator compartments. Hence the R134a absorbs energy from the air in these compartments, cooling them. This raises the temperature of the R134a and turns it into a gas.

Compressor

(3) A compressor pump then pressurizes the R134a gas. This drives the gas around the system and increases its temperature even more.

Q_C

Evaporator coil

Evaporation

Cool gas

Compression

Cool liquid

Hot gas

(4) As the hot gas passes through the long, narrow condenser coil tube (usually in the back of the refrigerator), energy is transferred from the gaseous R134a to the cooler air in the kitchen. This part of the process cools the gas enough to cause it to liquefy.

(6) The cold liquid then makes its way back to the evaporator coil to start the process again.

Expansion

Hot liquid

Condensation

Expansion valve

Condenser coil

$|Q_H|$

(5) The liquid R134a is further cooled by letting it expand rapidly as it passes through a valve from a narrow tube to a much wider one.

Figure 15-21 Inside a refrigerator This simplified diagram shows the cyclic process that cools a kitchen refrigerator. This process involves compressing and expanding a refrigerant that transitions between being a liquid and a gas.

back of the refrigerator), energy is transferred from the gaseous R134a to the cooler air in the kitchen. This part of the process cools the gas enough to cause it to liquefy, and the liquid R134a is further cooled by letting it expand rapidly as it passes through a valve from a narrow tube to a much wider one. The cold liquid then makes its way back to the evaporator coil to start the process again. In this way energy is transferred from the interior of the refrigerator to the surroundings.

The less work required to extract energy Q_C from the cold reservoir, the more efficient the refrigerator. A perfect refrigerator would require no work to be done in order to cause energy to be transferred from the cold to the hot reservoir, but this process isn't possible according to the second law of thermodynamics. The efficiency of a refrigerator is given by the **coefficient of performance**, for which we use the symbol *CP*:

$$CP = \frac{|Q_C|}{|W_{\text{engine}}|} \tag{15-28}$$

As with efficiency, *CP* is essentially what you get out divided by what you put in. The smaller the amount of work *W* for a given quantity of energy Q_C extracted from the cold reservoir, the larger the coefficient of performance *CP* and the better the performance of the refrigerator. A typical value for *CP* for a kitchen refrigerator or a home air conditioner is 3: This means that for every 1 J of work input to the refrigerator, 3 J of energy is extracted from the cold interior of the refrigerator or home and 1 J + 3 J = 4 J is rejected to the warmer surroundings.

The first law of thermodynamics requires that in one cycle, the total energy that goes into the refrigerator (in the form of work *W* done on it and energy Q_C extracted from the cold reservoir) equals the amount of energy $|Q_H|$ delivered to the hot reservoir). So

$$Q_C + W = |Q_H| \quad \text{or} \quad W = |Q_H| - Q_C \quad \text{(refrigerator)} \tag{15-29}$$

Equation 15-28 then becomes

$$CP = \frac{Q_C}{W} = \frac{Q_C}{|Q_H| - Q_C} \tag{15-30}$$

(coefficient of performance of a refrigerator)

The maximum possible coefficient of performance of a refrigerator is obtained when the system is based on the Carnot cycle run in reverse. It can be shown that the coefficient of performance of a Carnot refrigerator is

(15-31) $$CP_{\text{Carnot}} = \frac{T_C}{T_H - T_C}$$

(coefficient of performance of a Carnot refrigerator)

Like an ideal Carnot engine, an ideal Carnot refrigerator is not a practical device. The significance of Equation 15-31 is that it tells us the theoretical maximum coefficient of performance of a refrigerator operating between cold temperature T_C and hot temperature T_H.

EXAMPLE 15-8 Kitchen Refrigerator

A kitchen refrigerator has a coefficient of performance of 3.30. When the refrigerator pump is running, it removes energy from the interior of the refrigerator at a rate of 760 J/s. (a) How much work must be done per second to extract this energy? (b) How much energy is rejected into the kitchen per second in this process? (c) If the refrigerator operates between 2°C (the interior of the refrigerator) and 25°C (the kitchen), what is the theoretical maximum coefficient of performance?

Set Up

The energy removed from the cold interior of the refrigerator is represented by Q_C, and the work done to remove that energy is W. We can relate these to the coefficient of performance by using Equation 15-30, and to the energy $|Q_H|$ rejected into the warm kitchen by using Equation 15-29. The theoretical maximum value of CP is the Carnot value, given by Equation 15-31.

Coefficient of performance of a refrigerator:

$$CP = \frac{Q_C}{W} = \frac{Q_C}{|Q_H| - Q_C} \quad (15\text{-}30)$$

Energy relationships for a refrigerator:

$$Q_C + W = |Q_H| \text{ or } W = |Q_H| - Q_C \quad (15\text{-}29)$$

Coefficient of performance of a Carnot refrigerator:

$$CP_{\text{Carnot}} = \frac{T_C}{T_H - T_C} \quad (15\text{-}31)$$

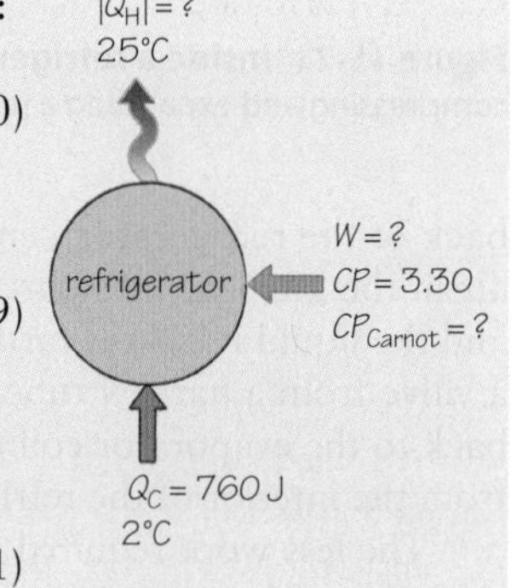

Solve

(a) Use Equation 15-30 to determine the amount of work that must be done per second.

In 1 s an amount of energy $Q_C = 760$ J is removed from the interior of the refrigerator. The amount of work W that must be done to make this happen depends on Q_C and the coefficient of performance:

$$CP = \frac{Q_C}{W}$$

$$W = \frac{Q_C}{CP} = \frac{760 \text{ J}}{3.30} = 230 \text{ J}$$

So the rate at which work must be done is 230 J/s = 230 W.

(b) Find the rate at which energy is rejected into the kitchen.

From Equation 15-30 the amount of heat $|Q_H|$ rejected into the kitchen in 1 s is

$$|Q_H| = Q_C + W = 760 \text{ J} + 230 \text{ J} = 990 \text{ J}$$

So the rate at which the refrigerator rejects heat is 990 J/s = 990 W.

(c) Calculate the coefficient of performance of a Carnot refrigerator operating between the same hot and cold temperatures.

The temperatures between which the refrigerator operates are

2°C or $T_C = (273.15 + 2)$ K = 275 K and

25°C or $T_H = (273.15 + 25)$ K = 298 K

The coefficient of performance of a Carnot refrigerator operating between these temperatures is

$$CP_{\text{Carnot}} = \frac{T_C}{T_H - T_C} = \frac{275\text{ K}}{298\text{ K} - 275\text{ K}} = \frac{275\text{ K}}{23\text{ K}}$$
$$= 12$$

Reflect

The coefficient of performance *CP* is substantially less than what could be achieved by an ideal Carnot refrigerator. Notice that the higher the actual value of *CP*, the less work that must be done to achieve the same amount of refrigeration. Less work means less waste, and as a result, less heating of the kitchen. This actual refrigerator heats the kitchen at a rate of 990 J/s or 990 W, about the same as the power output of an electric space heater.

NOW WORK Problems 4, 5, 9, and 10 from The Takeaway 15-5.

THE TAKEAWAY for Section 15-5

✔ One statement of the second law of thermodynamics is that randomness in systems tends to increase over time.

✔ A system or device that converts heat to work is a heat engine. It is not possible to create an engine or a process in which energy absorbed from a reservoir is converted completely into work.

✔ A refrigerator is a heat engine in reverse. It uses work to move energy from a cold reservoir to a hot reservoir.

✔ The third law of thermodynamics states that the temperature of a system can never attain absolute zero.

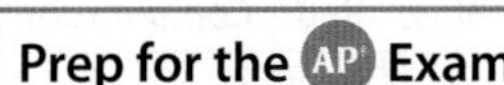

1. An engine doing work takes in 10.0 kJ of energy and exhausts 6.00 kJ of energy to its surroundings through heating. What is the efficiency of the engine?

2. What is the theoretical maximum efficiency of a heat engine operating between 100°C and 500°C?

3. A heat engine operating between 473 and 373 K runs at 70.0% of its theoretical maximum efficiency. What is its efficiency?

4. A kitchen refrigerator extracts 75.0 kJ per second of energy from a cool chamber while exhausting 1.00×10^2 kJ per second to the room. What is its coefficient of performance?

5. What is the coefficient of performance of a Carnot refrigerator operating between 0.00°C and 80.0°C?

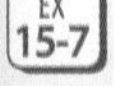

6. An engine operates between 10.0°C and 200°C. What is the least possible amount of energy that needs to be supplied from the high-temperature reservoir to output 1.00×10^3 J of work?

7. A furnace supplies 28.0 kW of power at 300°C to an engine that exhausts waste energy at 20.0°C. Determine the maximum work that could be expected from the system per second.

8. A fellow student announces that he has invented a gasoline engine that is 90% efficient. He claims that he gets such efficiency by running the engine at very high temperatures. The cylinders of his engine are made of titanium, which has a melting temperature of 1941 K. If he asks you to invest in this engine, should you agree?

9. A certain refrigerator requires 35.0 J of work to remove 190 J of heat from its interior.
(a) What is its coefficient of performance?
(b) How much energy is transferred to the surroundings at 22.0°C?
(c) If the refrigerator cycle is reversible, what is the temperature inside the refrigerator?

EX 15-8
10. An electric refrigerator removes 13.0 MJ of heat from its interior for each kilowatt-hour of electric energy used. What is its coefficient of performance?

15-6 The entropy of a system is a measure of its disorder

The second law of thermodynamics uses the concept of the *order* of a system. How can we quantify just how ordered or disordered a system is? Searching for the answer to this question will lead us to introduce a new physical quantity called *entropy* that plays an essential role in thermodynamics.

To get the best understanding of how ordered or disordered a thermodynamic system is, you would have to note the position and behavior of every one of its molecules. That's a challenging task for a system even as small as a single drop of water, which contains about 10^{22} H_2O molecules. (That's roughly 1000 times more molecules than there are grains of sand on all of Earth's beaches.) Instead let's look at a system with a much smaller number of constituents. Imagine that your sock drawer is an unorganized jumble of an equal number of blue socks and red socks. If you grab four socks without looking at their colors, what distribution of colors within the four-sock "thermodynamic system" are you most likely to have?

There is only one way to end up with four red socks: Each individual selection must be a red sock. There are four ways to have three red socks and one blue one, however. One way is to get a red sock during each of the first three selections and then get one blue one. Another way is for the first sock chosen to be blue and the last three red. We can summarize the possible ways to get three red and one blue sock this way:

(• • • •), (• • • •), (• • • •), and (• • • •)

TABLE 15-2 Possible Combinations of Red and Blue Socks

Blue	Red	Combinations	Number
0	4	(• • • •)	1
1	3	(• • • •), (• • • •), (• • • •), (• • • •)	4
2	2	(• • • •), (• • • •), (• • • •), (• • • •), (• • • •), (• • • •)	6
3	1	(• • • •), (• • • •), (• • • •), (• • • •)	4
4	0	(• • • •)	1

Table 15-2 summarizes all 16 possible four-sock combinations. The most probable grouping is two red socks and two blue socks (6 of the 16 possible combinations). This is also the most randomly distributed, least ordered four-sock state. The most ordered state would be one that was made of either four red socks or four blue socks. Table 15-2 shows that such states are also the least likely ones.

These conclusions about socks taken from a drawer turn out to be true in general: *The most probable final or equilibrium state of a system is the one that is the least ordered.* For example, a hot system and a cold system put into thermal contact come to the same final temperature not because no other possibility is allowed but because this state is the least ordered and therefore the most probable. Energy would still be conserved if, say, the hot system got hotter while the cold one got colder. But this would be highly improbable because this is a highly ordered state (the energy of the system ends up concentrated in the hot system rather than being spread between the two systems).

To represent order, we must take the list of variables that describe the state of a system—its temperature T, pressure p, and volume V, and the state function internal energy U—and add to it a new variable that we call **entropy**. Entropy, which for historical reasons we give the rather unexpected symbol S, is defined so that the larger its value the *less order* and the *more disorder* are present in a system. So we can think of entropy S as a measure of *disorder*. The second law of thermodynamics therefore states that *the entropy of an isolated system must either remain constant or increase.*

Entropy Change in a Reversible Process

The second law of thermodynamics suggests that what matters most is not the *amount* of order in a system but rather how much *change* in order there is in a thermodynamic process. The first law of thermodynamics says the same thing about internal energy. The first law tells us that the change in internal energy U must be equal to $Q + W$. Let's see how to express the entropy change for the special case of a system that undergoes a process that happens at a constant temperature T. To be specific we'll look at the reversible isothermal expansion of an ideal gas.

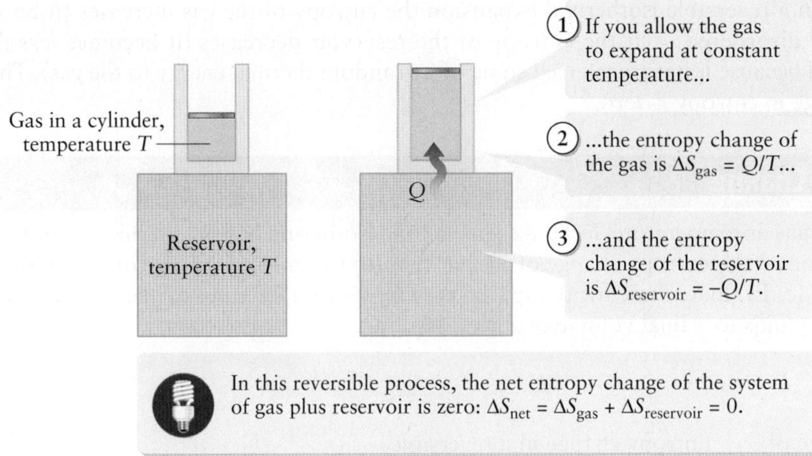

Gas in a cylinder, temperature T
Reservoir, temperature T
Q
1 If you allow the gas to expand at constant temperature...
2 ...the entropy change of the gas is $\Delta S_{gas} = Q/T$...
3 ...and the entropy change of the reservoir is $\Delta S_{reservoir} = -Q/T$.
In this reversible process, the net entropy change of the system of gas plus reservoir is zero: $\Delta S_{net} = \Delta S_{gas} + \Delta S_{reservoir} = 0$.

Figure 15-22 Entropy changes in an isothermal expansion In a reversible isothermal expansion of an ideal gas, the entropy of the gas increases; the entropy of the reservoir with which the gas is in contact decreases by an equal amount.

Figure 15-22 shows a quantity of ideal gas at temperature T confined within a cylinder with a moveable piston. We put the cylinder in thermal contact with a large reservoir at the same temperature T then allow the gas to expand. Because the gas is in thermal contact with the reservoir, its temperature remains constant and the internal energy of the gas does not change: $U = 0$. The piston does work on the gas as the gas expands, so $W < 0$. From the first law of thermodynamics, $U = Q + W$ and $Q = U - W$, or in this case (since U is zero) $Q = -W$. All of the energy that is transferred into the gas from the reservoir by heating, Q, is used to do work on the piston.

As the gas expands, the volume V that the gas occupies increases, and the molecules of the gas are now more spread out. This makes the gas more disordered, just as taking socks from a drawer and strewing them around the larger volume of your bedroom makes the socks more disordered. The greater the quantity of energy that enters the gas, the more the gas expands, and the greater the increase in disorder. So in this isothermal expansion the entropy change ΔS of the gas is *proportional* to the energy Q that flows into the gas.

The entropy change ΔS in an isothermal expansion also depends on the temperature T. For a given quantity of energy Q, the increase in disorder is greater if the system is at a low temperature (so that the molecules are moving slowly and are relatively well ordered to start with) than if the system is at a high temperature (so that the molecules are moving rapidly in a relatively disordered state). We conclude that the entropy change ΔS in this reversible isothermal process is *inversely proportional* to the Kelvin temperature T.

Putting these ideas together, we *define* the entropy change ΔS in a reversible isothermal process as

Energy transferred to ($+Q$) or from ($-Q$) **the system** by thermal processes

Entropy change of a system in **a reversible isothermal process**

$$\Delta S = \frac{Q}{T} \qquad (15\text{-}32)$$

Kelvin temperature of the system

EQUATION IN WORDS
Entropy change in a reversible isothermal process

Equation 15-32 shows that the units of entropy are the units of heat (that is, energy) divided by the units of temperature, or joules per kelvin (J/K).

We can also apply Equation 15-32 to the *reservoir* with which the gas cylinder is in contact. If a positive quantity of energy $Q > 0$ is transferred *into* the gas *from* the reservoir, the energy that is transferred into the reservoir from the gas is $-Q$. (This means that the quantity of energy that enters the gas equals the quantity of energy that leaves the reservoir.) The gas and reservoir are both at the same temperature T, so the *net* entropy change of the system of gas plus reservoir is

$$\Delta S_{net} = \Delta S_{gas} + \Delta S_{reservoir} = \frac{Q}{T} + \frac{(-Q)}{T} = 0$$

In a reversible isothermal expansion the entropy of the gas increases (it becomes more disordered), but the entropy of the reservoir decreases (it becomes *less* disordered because it has transferred some of its random thermal energy to the gas). The net change in entropy is zero.

EXAMPLE 15-9 Calculating Entropy Change

A cylinder containing n moles of an ideal gas at temperature T changes its volume isothermally from an initial volume V_i to a final volume V_f. Find expressions for (a) the entropy change of the gas and (b) the entropy change of the thermal reservoir with which the gas is in contact. (c) Evaluate these for the special case in which 0.050 mol of gas at 20.0°C has an initial volume of 1.00×10^{-3} m^3 and expands to a final volume of 2.00×10^{-3} m^3.

Set Up

Equation 15-32 tells us the entropy change of the gas in terms of its temperature T and the energy Q that is transferred into the gas during the volume change. From Equation 15-4 the quantity of energy added to the gas by heating, Q, equals the work W that the gas does. Equation 15-5 tells us the work done in this process. We'll put all of these pieces together to find the entropy changes of the gas and the thermal reservoir.

Entropy change in a reversible isothermal process:

$$\Delta S = \frac{Q}{T} \qquad (15\text{-}32)$$

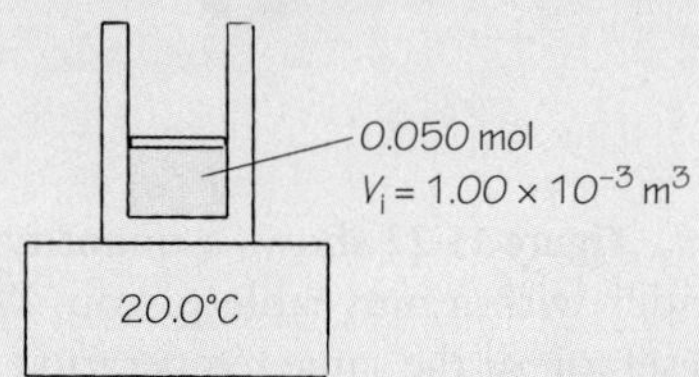

Isothermal process for an ideal gas:

$$Q = -W \qquad (15\text{-}4)$$

Work done by an ideal gas in an isothermal process:

$$W = -nRT \ln \frac{V_f}{V_i} \qquad (15\text{-}5)$$

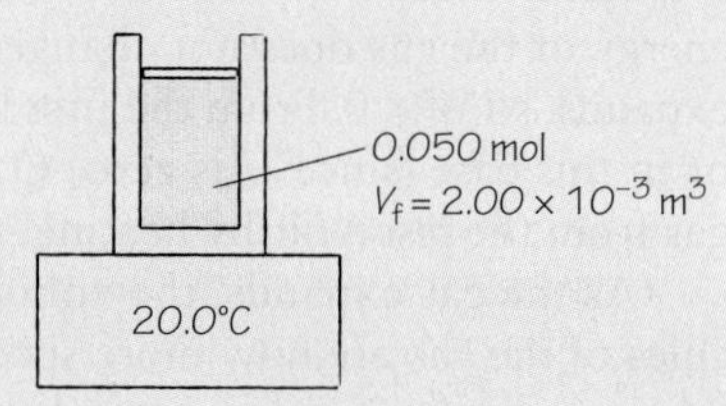

Solve

(a) Find an expression for the entropy change of the gas.

Combine Equation 15-4 and Equation 15-5 to get an expression for the heat that enters the gas:

$$Q_{gas} = -W = nRT \ln \frac{V_f}{V_i}$$

From Equation 15-32 the entropy change of the gas in this process is

$$\Delta S_{gas} = \frac{Q_{gas}}{T} = \frac{nRT}{T} \ln \frac{V_f}{V_i} = nR \ln \frac{V_f}{V_i}$$

If the gas expands so that the final volume is greater than the initial volume, the ratio V_f/V_i is greater than 1, ln (V_f/V_i) is positive, and ΔS_{gas} is positive (the entropy of the gas increases). If the gas is compressed so that the final volume is less than the initial volume, the ratio V_f/V_i is less than 1, ln (V_f/V_i) is negative, and ΔS_{gas} is negative (the entropy of the gas decreases).

(b) Find an expression for the entropy change of the reservoir.

The energy that enters the gas comes from the reservoir, so the value of Q for the reservoir equals the negative of the value of Q for the gas:

$$Q_{reservoir} = -Q_{gas} = -nRT \ln \frac{V_f}{V_i}$$

The entropy change of the reservoir is then

$$\Delta S_{reservoir} = \frac{Q_{reservoir}}{T} = -\frac{nRT}{T} \ln \frac{V_f}{V_i} = -nR \ln \frac{V_f}{V_i}$$

If we compare this to the result from part (a), we see that

$$\Delta S_{reservoir} = -\Delta S_{gas}$$

(c) Calculate the numerical values of ΔS_{gas} and $\Delta S_{reservoir}$.

With $n = 0.050$ mol, $V_i = 1.00 \times 10^{-3}$ m^3, and $V_f = 2.00 \times 10^{-3}$ m^3, we get

$$\Delta S_{gas} = nR \ln \frac{V_f}{V_i}$$
$$= (0.050\ \text{mol})\left(8.314\ \frac{\text{J}}{\text{mol} \cdot \text{K}}\right) \ln\left(\frac{2.00 \times 10^{-3}\ \text{m}^3}{1.00 \times 10^{-3}\ \text{m}^3}\right)$$
$$= (0.416\ \text{J/K}) \ln 2.00$$
$$= 0.288\ \text{J/K}$$

$$\Delta S_{reservoir} = -\Delta S_{gas} = -0.288\ \text{J/K}$$

The net entropy change of the system of gas and reservoir is

$$\Delta S_{net} = \Delta S_{gas} + \Delta S_{reservoir}$$
$$= 0.288\ \text{J/K} + (-0.288\ \text{J/K}) = 0$$

Reflect

Note that the entropy change of the gas, $\Delta S_{gas} = nR \ln (V_f/V_i)$, depends only on the number of moles n and the ratio of the final and initial volumes occupied by the gas. The temperature doesn't matter. This is consistent with the idea that a change in entropy corresponds to a change in disorder. The greater the increase in volume, the more disordered the system becomes; the greater the number of moles (and hence the greater the number of molecules) that expand into the new volume, the greater the increase in disorder. To see why the number of moles matters, think about taking socks from your sock drawer and strewing them around your bedroom. If there are only two socks involved, the amount of disorder isn't great; if there are a hundred socks, it's a hugely disordered mess.

NOW WORK Problems 1–3, 6, and 9–12 from The Takeaway 15-6.

In **Example 15-9** the gas and reservoir together make up an *isolated* system: The gas and reservoir are in thermal contact with each other but are not in thermal contact with anything else. In this case, in which the thermodynamic process is *reversible*, the entropy of that isolated system remains constant.

Entropy Change in an Irreversible Process

Let's now consider a different isothermal process that's *irreversible*. In **Figure 15-23** we use a metal bar to put a hot reservoir at temperature T_H in thermal contact with a cold reservoir

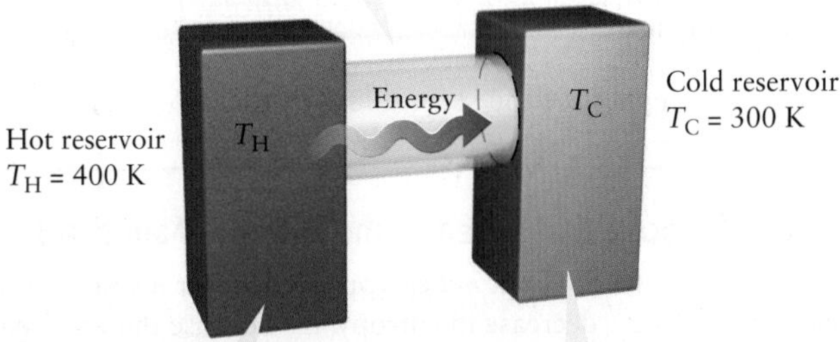

① A quantity of energy $Q = 1.20 \times 10^3$ J is transferred through the metal rod from the hot reservoir to the cold reservoir.
Energy
T_H
T_C
Hot reservoir
$T_H = 400$ K
Cold reservoir
$T_C = 300$ K
② Entropy change of the hot reservoir:
$\Delta S_H = \frac{(-Q)}{T_H} = \frac{(-1.20 \times 10^3\ \text{J})}{400\ \text{K}} = -3\ \text{J/K}$
③ Entropy change of the cold reservoir:
$\Delta S_C = \frac{(+Q)}{T_C} = \frac{(+1.20 \times 10^3\ \text{J})}{300\ \text{K}} = +4\ \text{J/K}$

The net entropy change in this irreversible process is positive: $\Delta S_{net} = \Delta S_H + \Delta S_C = (-3\ \text{J/K}) + (+4\ \text{J/K}) = +1$ J/K. The net entropy always increases in an irreversible process.

Figure 15-23 Entropy changes in irreversible energy transfers In an irreversible transfer of energy by thermal processes from a hot system to a cold one, the *net* entropy of the two systems together increases.

at temperature T_C. Energy is transferred from the hot reservoir to the cold one, but the reservoirs are so large that their temperatures remain essentially unchanged. Just as we did for the reservoir in Example 15-9, we can calculate the entropy change of each reservoir using Equation 15-32, $\Delta S = Q/T$. If a quantity of energy Q is transferred between the reservoirs, then $Q_C = +Q$ (energy is transferred *into* the cold reservoir from the hot one) and $Q_H = -Q$ (an equal quantity of energy is transferred *out* of the hot reservoir into the cold one). The entropy of the metal bar doesn't change: As much energy enters one end of the bar as leaves the other. The entropy changes of the two reservoirs are then

$$\text{Cold reservoir: } \Delta S_C = \frac{(+Q)}{T_C} \text{ (positive)}$$

$$\text{Hot reservoir: } \Delta S_H = \frac{(-Q)}{T_H} \text{ (negative)}$$

$$\text{Net entropy change: } \Delta S_{net} = \Delta S_C + \Delta S_H = \frac{(+Q)}{T_C} + \frac{(-Q)}{T_H}$$

As in Example 15-9 the entropy of one system (the cold reservoir) increases while the entropy of the other system (the hot reservoir) decreases. The difference is that the process shown in Figure 15-23 is *irreversible* because the two systems are at *different* temperatures T_C and T_H: By itself energy will be transferred only from the high-temperature system at T_H to the low-temperature one at T_C, never the other way, so the process cannot run in reverse. And because the two temperatures are different, the entropy changes ΔS_C and ΔS_H in this irreversible process do *not* cancel. To be specific, suppose a quantity of energy $Q = 1.20 \times 10^3$ J flows from a hot reservoir at $T_H = 400$ K to a cold reservoir at $T_C = 300$ K, as shown in Figure 15-23. In this case

$$\text{Cold reservoir: } \Delta S_C = \frac{(+Q)}{T_C} = \frac{1.20 \times 10^3 \text{ J}}{300 \text{ K}} = +4 \text{ J/K}$$

$$\text{Hot reservoir: } \Delta S_H = \frac{(-Q)}{T_H} = \frac{-1.20 \times 10^3 \text{ J}}{400 \text{ K}} = -3 \text{ J/K}$$

$$\text{Net entropy change: } \Delta S_{net} = \Delta S_C + \Delta S_H = 4 \text{ J/K} + (-3 \text{ J/K}) = +1 \text{ J/K}$$

For this irreversible process the cold reservoir gains more entropy than the hot reservoir loses, so there is a *net increase* in the entropy of the system of two reservoirs. So the system becomes increasingly disordered. While energy is conserved, the energy becomes less available to do useful work (for instance, to run a heat engine between the temperatures of the high and low reservoirs).

Entropy and the Second Law of Thermodynamics

We can summarize our observations about reversible and irreversible processes as follows:

> *In a reversible process there is no net change in entropy. In an irreversible process the net entropy increases.*

This is the *entropy statement* of the second law of thermodynamics.

WATCH OUT !

The entropy of a system can decrease, but only through a process that increases the entropy of another system.

For both the reversible isothermal expansion of an ideal gas (Figure 15-22) and the irreversible energy transfer from a hot reservoir to a cold one (Figure 15-23), one of the systems undergoes a *decrease* in entropy. This may seem to contradict the statement that entropy stays the same in a reversible process and increases in an irreversible process. In fact there's no contradiction, because this statement is about *net* entropy. Whenever a system undergoes a decrease in entropy, it's because there's a second system that undergoes an *increase* in entropy (the gas in the example of the reversible isothermal expansion and the cold reservoir in the example of irreversible thermal energy transfer). When all of the interacting systems are taken into account, the *net* entropy never decreases.

The entropy statement of the second law of thermodynamics gives us additional insight into the Carnot cycle, which is the most efficient heat engine theoretically possible. Each step in the Carnot cycle—an isothermal compression, an adiabatic compression, an isothermal expansion, and an adiabatic expansion—is *reversible*. So there is no net entropy change of the system of a Carnot engine, a hot reservoir, and a cold reservoir. (In Example 15-9 we saw that $\Delta S_{net} = 0$ for an isothermal expansion; the same is true for an isothermal compression, in which the gas loses entropy and the reservoir with which it is in contact gains an equal amount of entropy. There is no heating or cooling at all in a reversible adiabatic compression or expansion, so the entropy does not change in either of these steps.) At the end of each cycle, the system is just as ordered as it was at the beginning of the cycle. By contrast, if we use a heat engine that utilizes irreversible processes (like the idealized automotive engine cycle shown in Figure 15-17), there is a net increase in entropy—and hence in disorder—per cycle. So the *most efficient* engine is also the one that gives rise to the *least* increase in net entropy.

We've considered only situations in which the entropy of systems changes due to the transfer of energy by thermal processes between systems. One situation in which there is *no* transfer of energy is the free expansion of a gas shown in Figure 15-4b. This process is irreversible: After the barrier has ruptured and the expansion has taken place, it would be impossible for all of the gas molecules to spontaneously reassemble on the left-hand part of the cylinder in Figure 15-4b. So we would expect the entropy of the gas to increase in the irreversible free expansion. However, we *cannot* use Equation 15-32 to calculate the entropy change; there is no heating of the gas (the cylinder that encloses the gas is insulated), so the equation $\Delta S = Q/T$ predicts *incorrectly* that there would be *zero* entropy change.

The reason Equation 15-32 doesn't work is that it assumes that the process occurs slowly enough that the system is never far from equilibrium, which is definitely not the case for the sudden rush of gas in a free expansion. Instead we find the entropy change by noting that the final state—pressure, volume, and temperature—of the gas after a free expansion is the same as if it had undergone a reversible isothermal expansion to the same final volume [compare parts (a) and (b) of Figure 15-4]. We saw in Example 15-9 that the entropy of the gas increases in such a process, and we saw how to calculate ΔS_{gas}. Because entropy is a state function whose value does not depend on the history of the system, the change in entropy ΔS_{gas} must be the same for a free expansion. In a reversible isothermal expansion the gas is in thermal contact with a reservoir at the same temperature, and the entropy of the reservoir decreases by as much as the entropy of the gas increases so that the net entropy change is zero. But for a free expansion the cylinder is isolated, so the only entropy change is the increase in entropy ΔS_{gas} of the gas. So there is indeed a net entropy increase, just as we expect for an irreversible process.

It's instructive to ask what the entropy statement of the second law of thermodynamics says about the existence and evolution of life. Living organisms such as the cat shown in Figure 15-1a undergo a decrease in entropy as they develop. For example, humans start out as a single cell and grow into a highly ordered network of cells by rearranging raw materials (food) from the environment. Complex organisms are more efficient, require less energy, and are thereby better able to survive in an environment of limited resources. The processes, like evolution, by which complex organisms arise from simpler ones necessarily introduce more order and therefore lower entropy. Does this mean that they violate the second law of thermodynamics?

The answer to this question is no: Life most definitely does *not* violate the second law of thermodynamics. While the entropy of a living organism decreases as it develops, the entropy of its surroundings *increases*. That's because when an organism consumes and metabolizes food, it uses some of the energy from the food to do work (for example, to grow) but also wastes some of that energy. It is this inefficiency that causes the *net* effect to be an increase in entropy. Another way to say this is that the entropy of an organism decreases, but the entropy of its waste products—for example, the carbon dioxide that you exhale, the perspiration that you release from your skin, and the liquid and solid material that you excrete—increases more than the organism's entropy decreases.

THE TAKEAWAY for Section 15-6

✔ The entropy of a system is a measure of the disorder of the system. Systems with a low value of entropy are more ordered.

✔ The most probable final state of a system is the one that is the most disordered and has the highest entropy.

✔ In a process in which two systems interact, it's possible for the entropy of one system to decrease. However, the *net* entropy remains constant in a reversible process and increases in an irreversible process.

Prep for the AP® Exam

AP® Building Blocks

EX 15-9 1. A room is at a constant 295 K maintained by an air conditioner. How much does the air conditioner change the entropy of the room for each 5.00 kJ of energy it removes through cooling?

EX 15-9 2. A reservoir at a temperature of 400 K gains 100 J through heating from another reservoir. What is its entropy change?

EX 15-9 3. What is the minimum change of entropy that occurs in 0.200 kg of ice at 273 K when it is heated by 6.68×10^4 J so that it melts to water?

4. Which is greater, the entropy of 1 kg of liquid iron or of 1 kg of solid iron? Justify your answer in terms of physics principles.

5. A pot full of hot water is placed in a cold room, and the water gradually cools. Briefly describe how the entropy of the water changes.

AP® Skill Builders

EX 15-9 6. If, in a reversible process, heating changes a 500-g block of ice to water at a temperature of 273 K, what is the change in the entropy of the ice–water system?

7. For water the latent heat of fusion (melting) is $L_F = 3.34 \times 10^5$ J/kg, and the latent heat of vaporization is $L_V = 2.26 \times 10^6$ J/kg. Which involves a greater change in entropy?

(A) Melting 1.00 kg of ice at 0°C
(B) Vaporizing 1.00 kg of liquid water at 100°C
(C) The entropy change is the same for both.

8. One mole of ideal gas expands isothermally from 1.00 to 2.00 m^3. What is the entropy change for the gas?

AP® Skills in Action

EX 15-9 9. A 1.80×10^3-kg car traveling at 80.0 km/h crashes into a concrete wall. If the temperature of the air is 27.0°C, what is the entropy change of the universe as a result of the crash? Assume all of the car's kinetic energy is transferred to its surroundings by thermal processes.

EX 15-9 10. A 1.00×10^3-kg rock at 20.0°C falls 1.00×10^2 m into a large lake, also at 20.0°C. Assuming that all of the rock's kinetic energy on entering the lake heats the lake, what is the change in entropy of the lake?

EX 15-9 11. The surface of the Sun is about 5700 K, and the temperature of Earth's surface is about 293 K. What is the net change in entropy of the Sun–Earth system when the Sun heats the Earth by 8000 J?

EX 15-9 12. A 0.750-L cup of coffee at 70°C is left outside where the temperature is 4°C. When the coffee reaches thermal equilibrium with the atmosphere, by how much has the entropy of the atmosphere changed? Assume the properties of coffee are identical to those of water.

WHAT DID YOU LEARN?

Chapter learning goals	Sections	Related example(s)	Prep for the AP® Exam Relevant section review exercise(s)
Explain the general ideas of the laws of thermodynamics.	15-1		1, 2
Describe the relationship of the first law of thermodynamics to the work-energy theorem and be able to apply it quantitatively, including in *PV* diagrams.	15-2	15-1	1, 2, 5, 6
	15-2		3, 4
Describe the nature of isobaric, isothermal, adiabatic, and isovolumetric processes.	15-3	15-2	2
		15-3	1, 4, 5, 8
		15-4	3, 6, 7
Explain why different amounts of energy are required to change the temperature of a gas depending on whether the gas is held at constant volume or constant pressure.	15-4	15-5	3, 4, 5, 6, 8
	15-4		10, 12
Define the second law of thermodynamics and its application to heat engines and refrigerators.	15-5	15-6	1
		15-7	2, 3, 6, 7, 8
		15-8	4, 5, 9, 10
Explain the concept of entropy and the circumstances under which entropy changes.	15-6	15-9	1, 2, 3, 6, 9, 10, 11, 12
	15-6		5, 7

Chapter 15 Review

Key Terms

adiabatic process 734
Carnot cycle 756
coefficient of performance 761
efficiency 754
entropy 764
first law of thermodynamics 733
heat engine 753
internal energy 730
irreversible process 756
isobaric process 734
isovolumetric process 734
isotherm 737
isothermal process 734
molar specific heat 745
molar specific heat at constant pressure 745
molar specific heat at constant volume 745
PV diagram 735
ratio of specific heats 748
refrigerator 760
reservoir 753
reversible process 756
second law of thermodynamics 752
state variable 730
thermodynamic process 730
third law of thermodynamics 759

Chapter Summary

Topic	Equation or Figure
The first law of thermodynamics: In any thermodynamic process the change in internal energy of the system is related to the heating Q of the system and the work W done on the system during the process. When energy is transferred into the system by heating or work Q and W are positive, and they are negative when energy is transferred out of the system.	During a thermodynamic process the **change in the internal energy of a system...** $$W + Q = \Delta U \quad (15\text{-}1)$$ ...plus the **work done on the system** during the process. ...equals **the energy transferred to or from the system through heating or cooling** during the process...
	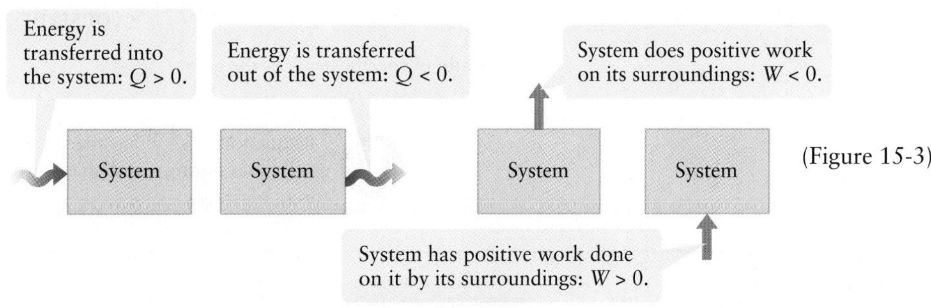 (Figure 15-3)
Thermodynamic processes: Four important types of thermodynamic processes are isobaric (constant pressure), isothermal (constant temperature), adiabatic (no heating or cooling), and isovolumetric (constant volume). In *any* process, the work done is represented by the area under the curve of the *PV* diagram for the process. There are general expressions for the work done in an isobaric process and for the work done by an ideal gas in an isothermal process. In an isovolumetric process zero work is done.	**Work** done on a system in an isobaric process **Volume** of the system **at the end** of the process $$W = -P(V_f - V_i) \quad (15\text{-}3)$$ **Constant pressure** of the system **Volume** of the system **at the beginning** of the process
	Number of moles of gas **Ideal gas constant** **Work done on an ideal gas (system) in an isothermal process** $$W = -nRT \ln \frac{V_f}{V_i} \quad (15\text{-}5)$$ **Volume** of the gas **at the end** of the process **Volume** of the gas **at the beginning** of the process **Kelvin temperature** of the gas

Molar specific heats of a gas: The molar specific heat is the energy required to raise the temperature of one mole of substance by one kelvin. For an ideal gas, the molar specific heats at constant pressure (C_p) and constant volume (C_V) differ by a fixed amount; the value of C_V depends on the number of degrees of freedom of a gas molecule.

Amount of energy that is transferred (heat) into (if $Q > 0$) or out of (if $Q < 0$) an system

Molar specific heat of the substance of which the system is made

$$Q = nC\Delta T \tag{15-9}$$

Number of moles present of the substance

Temperature change of the system that results from the energy transfer

Molar specific heat of an ideal gas at constant pressure

$$C_p - C_V = R \tag{15-14}$$

Ideal gas constant

Molar specific heat of an ideal gas at **constant volume**

Number of degrees of freedom for a molecule of the gas

Molar specific heat of an ideal gas at **constant volume**

$$C_V = \left(\frac{D}{2}\right)R \tag{15-17}$$

Ideal gas constant

Adiabatic processes for an ideal gas: How pressure, volume, and temperature change in an adiabatic process for an ideal gas depends on the value of the ratio of specific heats for the gas $\gamma = C_p/C_V$.

Pressure of an ideal gas

Volume of an ideal gas

$$PV^{\gamma} = \text{constant} \tag{15-21}$$

Ratio of specific heats of the gas = C_p/C_V

The quantity PV^{γ} has the same value at all times during an **adiabatic process** for an ideal gas.

Kelvin temperature of an ideal gas

Volume of an ideal gas

$$TV^{\gamma-1} = \text{constant} \tag{15-22}$$

Ratio of specific heats of the gas = C_p/C_V

The quantity $TV^{\gamma-1}$ has the same value at all times during an **adiabatic process** for an ideal gas.

The second law of thermodynamics: The second law states that the amount of order in an isolated system cannot increase. This implies that a heat engine—a device that converts heat into work—cannot be 100% efficient but can convert only part of its energy intake Q_H (from a high-temperature source) into work. The rest is rejected through heating $|Q_C|$ the engine's low-temperature surroundings. The efficiency is $e = W/Q_H$.

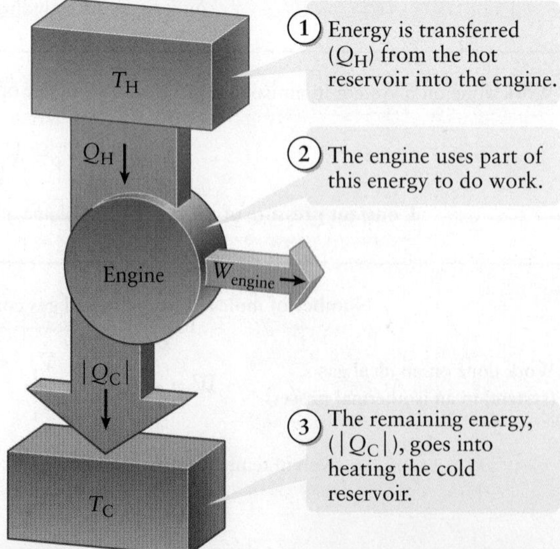

T_H
Q_H
Engine
W_{engine}
$|Q_C|$
T_C
(1) Energy is transferred (Q_H) from the hot reservoir into the engine.
(2) The engine uses part of this energy to do work.
(3) The remaining energy, ($|Q_C|$), goes into heating the cold reservoir.

(Figure 15-16)

The Carnot cycle: The most efficient heat engine theoretically possible is a Carnot engine. It uses the Carnot cycle, which is based on two reversible adiabatic processes and two reversible isothermal processes. No other engine can have a greater efficiency. A Carnot engine run in reverse is also the best-performing refrigerator theoretically possible.

Efficiency of a Carnot engine; Kelvin temperature of the cold reservoir; Kelvin temperature of the hot reservoir

$$e_{\text{Carnot}} = 1 - \frac{T_C}{T_H} \quad (15\text{-}27)$$

Entropy: The entropy of a system is a measure of the amount of disorder present in the system. In a reversible process in which two systems interact with each other, the entropy of one system increases while the entropy of the other system decreases by the same amount. In an irreversible process, the entropy changes do not cancel and there is a net increase in entropy.

Energy transferred to (+Q) or from (–Q) the system by thermal processes; Entropy change of a system in a reversible isothermal process; Kelvin temperature of the system

$$\Delta S = \frac{Q}{T} \quad (15\text{-}32)$$

Chapter 15 Review Problems

1. In an isothermal process there is no
   (A) change in pressure.
   (B) change in temperature.
   (C) change in volume.
   (D) heating or cooling.
2. In an isobaric process there is no
   (A) change in pressure.
   (B) change in temperature.
   (C) change in volume.
   (D) heating or cooling.
3. In an isovolumetric process there is no change in
   (A) change in pressure.
   (B) change in temperature.
   (C) change in volume.
   (D) heating or cooling.
4. A gas quickly expands in an isolated environment. During the process there is no heating or cooling of the gas. The process is
   (A) isothermal.
   (B) isobaric.
   (C) isovolumetric.
   (D) adiabatic.
5. If you drop a glass cup on the floor, it will shatter into fragments. If you then drop the fragments on the floor, why will they not become a glass cup? Justify your claim in terms of numbers of states.
6. Clearly define and give an example of each of the following thermodynamic processes: (**a**) isothermal, (**b**) adiabatic, (**c**) isobaric, and (**d**) isovolumetric.
7. Briefly explain why the temperature of a gas increases when it is quickly compressed.
8. When we say, "engine," we think of something mechanical with moving parts. In such an engine, friction always reduces the engine's efficiency. Briefly justify this using physics principles.
9. Why do engineers designing a steam-electric generating plant always try to design for as high a feed-steam temperature as possible?
10. Is the operation of an automobile engine reversible? Briefly justify your answer.
11. Is a process necessarily reversible if there is no exchange of energy between a system in which the process takes place and its surroundings? Why or why not?

12. Conduction across a temperature difference is an irreversible process, but the system that lost energy can always be rewarmed, and the one that gained energy can be recooled. A ball sliding across the grass slows down and warms up as mechanical energy dissipates. This process is irreversible, but the ball can be cooled and set moving again at its original speed. So in just what sense are these processes "irreversible"?

13. There are people who try to keep cool on a hot summer day by leaving the refrigerator door open, but you can't cool your kitchen this way! Why not?

14. If the coefficient of performance is greater than 1, do we get more energy out than we put in, violating conservation of energy? Why or why not?

15. How does the time required to freeze water vary with each of the following parameters: mass of water, power of the refrigerator, and temperature of the outside air?

16. How is the entropy of the universe changed when energy is transferred from a hotter system to a colder one? In what sense does this correspond to energy becoming unavailable for doing work?

17. If a gas expands freely into a larger volume in an insulated container that allows no heating or cooling of the gas from its surroundings, its entropy increases. Explain this using the idea that this process is irreversible.

18. In discussing the Carnot cycle, we say that extracting energy from a reservoir isothermally does not change the entropy of the universe. In a real process, this is a limiting situation that can never quite be reached. Why not? What is the effect on the entropy of the universe?

19. A gas is taken through the cyclic process shown in the accompanying PV diagram. At point 1, the pressure is 2.50×10^3 Pa and the volume 3.50 m^3. At point 2, the volume is 6.00 m^3. At point 3, the pressure is 8.50 is $\times 10^3$ Pa. Over one complete cycle, how much energy is transferred to the system by thermal processes?

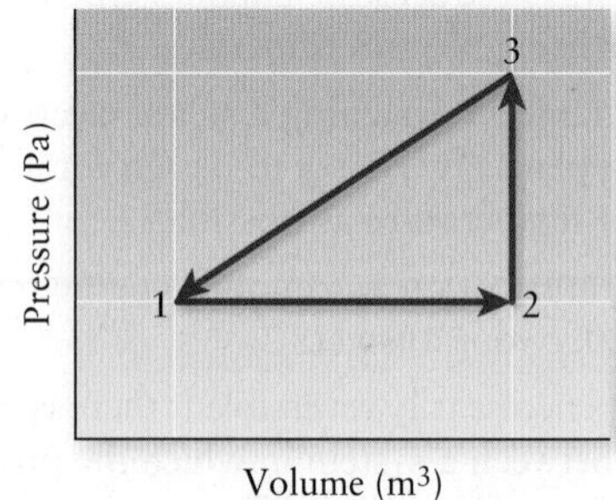

20. Calculate the amount of work done on a gas that undergoes a change of state described by the PV diagram shown in the accompanying figure.

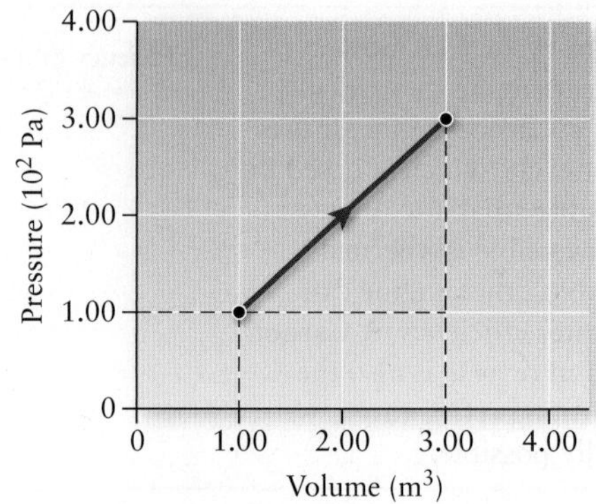

21. Calculate the amount of work done on a gas that undergoes a change of state described by the PV diagram shown in the accompanying figure.

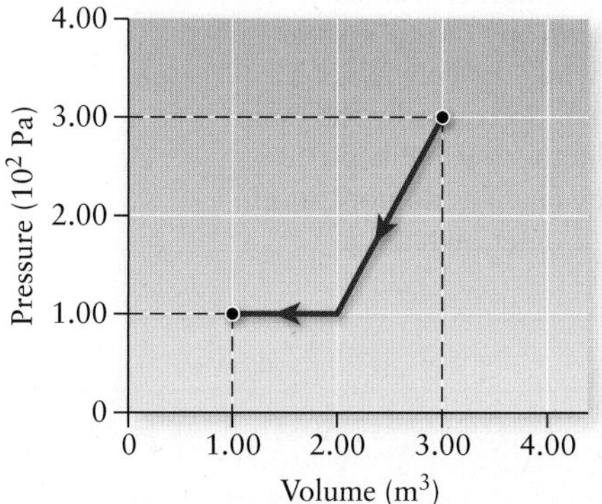

22. An ideal gas is compressed adiabatically to half its volume. In doing so 1888 J of work is done on the gas. What is the change in internal energy of the gas?

23. In an international diving competition divers fall from a platform 10.0 m above the surface of the water into a very large pool. A diver leaves the platform with negligible initial speed. Assuming that all of a diver's kinetic energy is transferred to the water by thermal processes, what is the maximum change in the entropy of the water in the pool at 25.0°C when a 75.0-kg diver executes this dive? Does the pool's entropy increase or decrease?

24. A refrigerator is rated at 370 W. Its interior is at 0°C, and its surroundings are at 20°C. If the refrigerator operates at 66% of its theoretical maximum efficiency, how much energy can it transfer from its interior to its surroundings in 1 min?

25. Liquid nitrogen, with a latent heat of fusion of $L_F = 25.3 \times 10^3$ J/kg, solidifies at a temperature of 63 K. Calculate the change in entropy of a 1.5-kg sample of liquid nitrogen as it transitions from a liquid to a solid.

26. A Carnot engine takes in 1.20×10^3 J of energy from a high-temperature source and transfers 6.00×10^2 J of that energy to the atmosphere at 20.0°C.

(a) What is the efficiency of the engine?

(b) What is the temperature of the hot reservoir?

27. A balloon containing 0.50 mol of helium in a 20.0°C chamber undergoes an isothermal expansion as the pressure in the chamber is slowly reduced. If the pressure drops to half its initial value during this process, what is the entropy change of the helium?

28. The volume of 20°C air taken in during a typical breath is 0.5 L. The inhaled air is heated to 37°C (the internal body temperature) as it enters the lungs. Because air is about 80% nitrogen N_2, we can model it as an ideal gas. Assume that the pressure does not change during the process.
   (a) How many joules of energy does it take to warm the air in a single breath?
   (b) Suppose that you take two breaths every 3.0 s. How many food calories (kcal) are used up per day in heating the air you breathe? Is this a significant amount of typical daily caloric intake?

29. A heat engine works in a cycle between reservoirs at 273 and 490 K. In each cycle, the engine absorbs 1250 J of energy from the high-temperature reservoir and does 475 J of work.
   (a) What is its efficiency?
   (b) By how much is the entropy of the universe changed when the engine goes through one full cycle?
   (c) How much energy becomes unavailable for doing work when the engine goes through one full cycle?

30. The thermodynamic cycle of a heat engine using ideal helium gas is shown in the *PV* diagram. The ratio of heat capacities γ (also called the ratio of specific heats) is the ratio of the heat capacity at constant pressure C_P to the heat capacity at constant volume C_V. For ideal helium gas $\gamma = 1.67$. A confused physics student determines that the cycle comprises the following processes: $1 \rightarrow 2$ adiabatic; $2 \rightarrow 3$ isovolumetric; $3 \rightarrow 4$ isothermal; $4 \rightarrow 1$ isobaric. Which process did the student identify correctly?

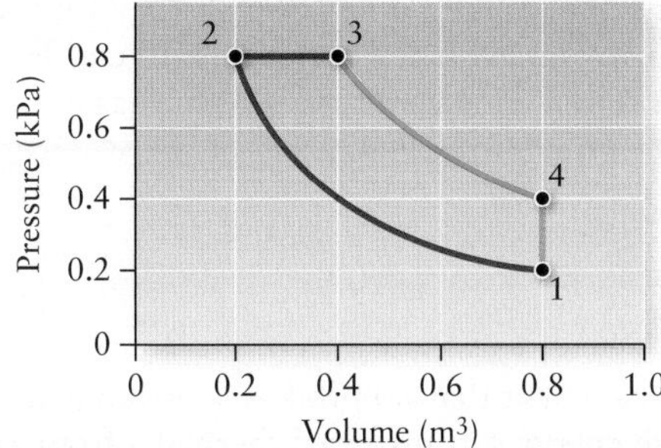

31. During a high fever a 60.0-kg patient's normal metabolism is increased by 10.0%. This results in an increase of 10.0% in the person heating their surroundings. When the person slowly walks up five flights of stairs (20.0 m), she normally transfers 1.00×10^5 J of energy to her surroundings through heating. Compare her efficiency when she has a fever to when her temperature is normal.

32. A rigid 5.50-L pressure cooker contains steam initially at 100°C under a pressure of 1.00 atm. The density of steam at this temperature and pressure is 0.6 kg/m^3. Consult Table 15-1 as needed and assume that the values given there remain constant. The mass of a water molecule is 2.99×10^{-26} kg.
   (a) Calculate the temperature (in °C) that the steam must have when its pressure is 1.25 atm.
   (b) Calculate the amount of energy you would need to transfer to the water in part (a).
   (c) Calculate the specific heat of the steam in part (a) in units of J/(kg · K).

33. A certain engine operates at 85% of its theoretical maximum efficiency. During each cycle it absorbs 4.80×10^2 J of energy from a reservoir at 300°C and transfers 3.00×10^2 J of energy to a cold-temperature reservoir.
   (a) Determine the temperature of the cold reservoir.
   (b) How much more work could be done by a Carnot engine working between the same two reservoirs and extracting the same 4.80×10^2 J of heat in each cycle?

34. A certain electric generating plant produces electricity by using steam that enters its turbine at a temperature of 320°C and leaves it at 40°C. Over the course of a year, 4.40×10^{16} J of energy is transferred to the plant by thermal processes and the plant produces an average electric power output of 600 MW. What is the efficiency of this generating plant compared to a Carnot engine operating between the same temperatures?

35. As we drill down into the rocks of Earth's crust, the temperature typically increases by 3.0°C for every 100 m of depth. Oil wells can be drilled to depths of 1830 m. If water is pumped into the shaft of the well, it will be heated by the hot rock at the bottom, and the resulting steam can be used as a heat engine. Assume that the surface temperature is 20°C.
   (a) Calculate the maximum possible efficiency of such a 1830-m well used as a heat engine.
   (b) If a combination of such wells is to produce a 2.5-MW power plant, determine the energy that must be transferred to the engine from the interior of Earth each day.

36. The energy efficiency ratio (or rating)—the EER—for air conditioners, refrigerators, and freezers is defined as the ratio of the input transfer rate of energy $\left(\dfrac{Q_C}{t}, \text{ in } \dfrac{\text{BTU}}{\text{h}}\right)$ to the output rate of work (W/t, in W):

$$\text{EER} = \frac{Q_C/t(\text{BTU/hr})}{W/t(\text{W})}.$$

An energy-efficient home freezer has an EER of 6.50. In preparation for a picnic you put 1.50 L of water at 20.0°C into the freezer to make ice at 0.00°C for your ice chest. (See Tables 14-1 and 14-2 as needed.)

(a) How much electric energy (which runs the freezer) is required to make the ice? Express your answer in J and kWh.

(b) How much energy does the refrigerator transfer into your kitchen, which is at 22.0°C, through thermal processes in freezing the ice?

(c) How much does making the ice change the entropy of your kitchen?

37. A 68.0-kg person typically eats about 2250 kcal per day, 20.0% of which goes to mechanical energy; the rest is radiated to her surroundings. If she spends most of her time in her apartment at 22.0°C, how much does the entropy of her apartment change in one day? Does the entropy of the apartment increase or decrease?

Prep for the AP® Exam

AP® Group Work

Directions: The following problem is designed to be done as group work in class.

Consider an engine in which the working substance is 1.23 mol of an ideal gas for which $\gamma = 1.41$. The engine runs reversibly in the cycle shown on the *PV* diagram. The cycle consists of an isobaric (constant-pressure) expansion *a* at a pressure of 15.0 atm, during which the temperature of the gas increases from 300 to 600 K, followed by an isothermal expansion *b* until its pressure becomes 3.00 atm. Next is an isobaric compression *c* at a pressure of 3.00 atm, during which the temperature decreases from 600 to 300 K, followed by an isothermal compression *d* until its pressure returns to 15 atm. Find the work done by the gas, the heat absorbed by the gas, and the internal energy change of the gas, first for each part of the cycle and then for the complete cycle. Also, find the total entropy change for a complete cycle.

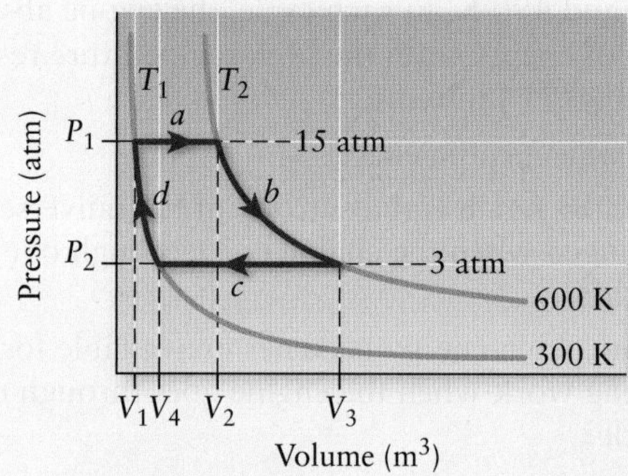

AP® PRACTICE PROBLEMS

Prep for the AP® Exam

Multiple-Choice Questions

Directions: The following questions have a single correct answer.

1. A student suggests two methods to improve the theoretical efficiency of a heat engine: lower T_C by 10 K or raise T_H by 10 K. Select the best answer.
   (A) Lowering T_C by 10 K will improve the theoretical efficiency the most.
   (B) Raising T_H by 10 K will improve the theoretical efficiency the most.
   (C) Both changes will improve the theoretical efficiency by the same amount.
   (D) There is nothing you can do to improve the theoretical efficiency of a heat engine.

2. The temperature of rapidly rising air decreases, which can cause precipitation. Why does this temperature decrease happen?
   (A) The air loses internal energy to match the low temperature of the atmosphere at high altitude.
   (B) The air's temperature naturally drops as its gravitational potential energy increases.
   (C) The air expands rapidly as it moves into the low pressure of the upper atmosphere.
   (D) The air molecules radiate most of their energy into space.

3. A gas is compressed adiabatically by a force of 800 N exerted over a distance of 5.0 cm. The net change in the internal energy of the gas is
   (A) −800 J.
   (B) −40 J.
   (C) +40 J.
   (D) +800 J.

4. The statement that no process is possible in which energy is transferred thermally from a cold reservoir completely to a hot reservoir is
   (A) not always true.
   (B) only true for isothermal processes.
   (C) the first law of thermodynamics.
   (D) the second law of thermodynamics.

Free-Response Question

1. A sample of 0.219 mol of a monatomic gas undergoes the cyclic process, where $P_1 = 400$ kPa, $P_2 = 640$ kPa, $V_1 = V_2 = 1.25 \times 10^3$ cm^3, and $V_3 = 2.53 \times 10^3$ cm^3. Between stage 3 and stage 1, the gas is at a constant temperature, and between stage 2 and stage 3, no energy is transferred by thermal processes. The temperature of the gas at stage 2 is 440 K.

   (a) Using a grid like the one at right, sketch a *PV* diagram of the cycle.

   (b) i. Calculate the temperature between stages 3 and 1.

   ii. Determine the work done on this system during one complete cycle.

   (c) Identify the process involved in each step of the cycle, supporting your claims in terms of physical principles.

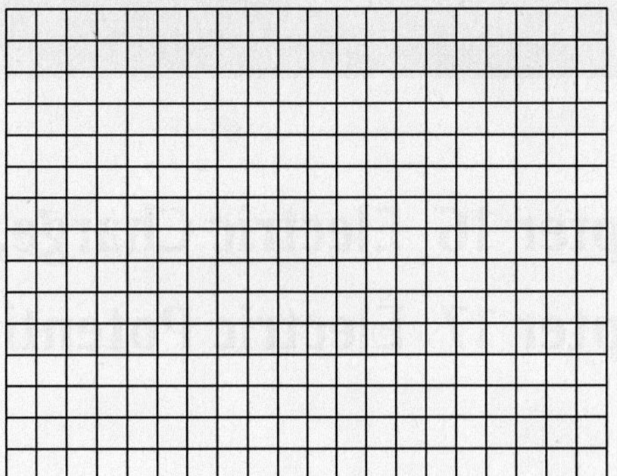

Electric Force, Field, and Potential

TED M. KINSMAN/Science Source

Case Study: How does electric charge make your hair stand on end?

All ordinary matter is made of atoms. Atoms, in turn, are composed of electrons, protons, and (except for most hydrogen atoms) neutrons. The first two of these are electrically charged—electrons carry negative charge and protons carry positive charge. Neutrons, as the name implies, are electrically neutral, meaning they have a total charge of zero. Under most conditions, atoms are also electrically neutral; that is, they carry the same number of electrons and protons, and since electrons and protons have the same size charge, this gives atoms a combined charge of zero.

Electrically charged objects exert forces on each other. Oppositely charged objects attract, whereas objects that carry like charge, such as electrons, repel. In the photo above, the hair on Meredith's head is electrically charged. If all of her hair is, for example, negatively charged, each strand of hair has excess (extra) electrons, which repel other excess electrons. The result? Each hair will position itself as far from all the other hairs as possible.

But wait! If the atoms in Meredith's hair are electrically neutral, how does a repulsive electric force arise? The key to the answer lies in the fact that *charge can be transferred between objects*. Meredith is touching the top of a Van de Graaff generator, a device used to accumulate electric charge on a hollow, metallic sphere. By touching the sphere, Meredith is in electrical contact with it, which allows electrons to move between her and the sphere. Meredith's hair accumulates excess electrons, and the repulsive force between these excess electrons drives her hair apart. The result . . . her hair stands on end.

How close do the excess electrons on Meredith's hair have to be for there to be an appreciable force between them? In this chapter we will encounter Coulomb's law, which describes the electrostatic force between two charged objects. Coulomb's law is similar to Newton's law of universal gravitation: The electrostatic force is inversely proportional to the square of the distance between charged objects, and proportional to the product of the magnitude of the excess charge on them. Just as in Newton's law, there is a constant in Coulomb's law. The constant in Coulomb's law is large, so even for small amounts of excess charge the inter-charge (and inter-hair!) forces are large enough to keep her hairs apart even over a distance of centimeters.

Rubbing certain materials with certain other materials can cause charge to be transferred from one to the other. Why do you think a balloon sticks to your clothes after you rub it on them?

By the end of this chapter, you should be able to explain the contact forces between objects as arising from interatomic electric forces, compare the similarities and differences between electric and gravitational forces, use Coulomb's law qualitatively and quantitatively to make predictions about the interactions between two point charges, model the force between charged objects or systems using the concept of electric field, and describe the electric permittivity of a material or medium.

16 Electric Charge, Force, and Field

VisionDive/Shutterstock

YOU WILL LEARN TO:

- Define open and closed systems for everyday situations involving charge transfer and apply conservation of electric charge to those systems, including a description of the smallest unit of electric charge.
- Describe the behavior of a system using the law of conservation of electric charge, including sign and relative quantity of net charge after various charging processes.
- Construct an explanation of the two-charge model based on evidence produced through scientific practices.
- Recognize the differences between insulators and conductors.
- Describe the point charge model.
- Make claims about contact forces based on the microscopic cause of those forces, and explain them as arising from interatomic electric forces.
- Use Coulomb's law qualitatively and quantitatively to make predictions about the interactions between two point charges.
- Compare the similarities and differences between electric and gravitational forces.
- Describe the electric field produced by a charged object or a configuration of point charges.

16-1 Electric forces and electric charge are all around you—and within you

You may think of *electricity* as the shock you get when you walk across a carpet and then touch a metal doorknob, or a commodity that you purchase from the power company (although what you are really purchasing from them is energy). The reality is that electric phenomena are all around you, from the behavior of grains of pollen to the drama of a thunderstorm (**Figure 16-1**). As you read this sentence, your eye casts an image of the text onto your retina, and this image is transmitted to your brain along the optic nerve as a stream of electrical impulses. As was introduced in AP® Physics 1, and will be discussed further in Section 16-5, electric forces are responsible for all the contact forces we use to describe interactions between objects and systems.

(a) Grains of pollen wafting through the air carry a small amount of electric charge. This attracts them to the oppositely charged stigma of a flower and so aids in pollination.

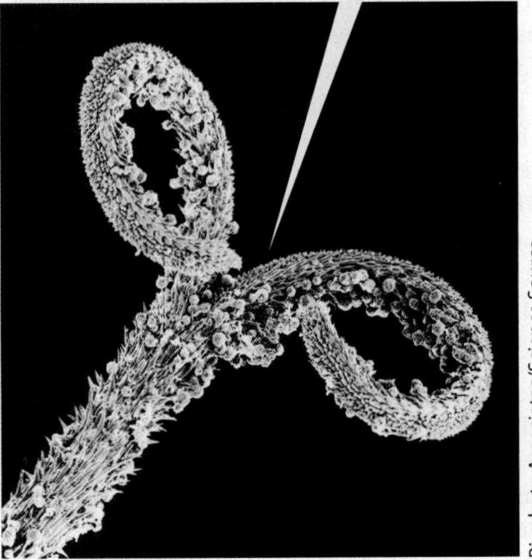

Biophoto Associates/Science Source

(b) The motion of air in a thundercloud causes positive and negative charges to separate, with the bottom of the cloud becoming negatively charged. This attracts positive charge on the ground. Electrons are removed from air molecules by the strong electric field generated by this separated positive and negative charge, and the motion of large amounts of charge carriers through the air is what we see as lightning.

KingWu/iStockphoto

Figure 16-1 Electric charge and force Electrostatics—the physics that describes how oppositely charged objects attract and like charged objects repel—plays an important role in (a) biology and (b) the weather.

The molecules on the surface of the floor exert electric forces that repel the molecules on the underside of your shoes, and these forces are strong enough to balance Earth's downward gravitational force on you, so you do not fall through the floor!

Electric phenomena are important because *all* ordinary matter is made of electrically charged objects. The simplest place to begin our study of electricity is with *electrostatics*, the branch of physics that deals with electrically charged objects at rest and how they interact with each other. We'll first learn about the kinds of electric charge in nature and describe the forces that two electrically charged objects exert on each other. We'll see how these forces manifest themselves in two important classes of materials called conductors and insulators. We'll take another look at *fundamental forces* and their relationship to contact forces and conclude this chapter with the concept of an *electric field*, a powerful way of understanding how charged objects can exert forces on each other over a distance that is analogous to the gravitational field introduced in Section 6-4.

THE TAKEAWAY for Section 16-1

✔ Electric phenomena are important because all ordinary matter is made of electrically charged objects.

✔ Electrostatics is the study of electrically charged objects at rest and the forces that they exert on each other.

Prep for the AP® Exam

Building Blocks

1. Gravitational forces exerted between objects with mass are always attractive. When you push an object over a surface or pull on a string you exert a force that has a direction. As discussed in Chapter 4, interactions between charged objects cause the normal, tension, and friction forces. You may not have any direct experimental evidence that supports this claim of the role of charge. But if you accept the claim that interactions between charged objects cause these forces, do you have any evidence from your experience to support the claim that interactions between charged objects can be both attractive and repulsive?

16-2 Matter contains objects with positive and negative electric charge

AP Exam Tip

A neutral object contains no net charge, which typically means that it has the same amount of positive charge and negative charge, but there are some neutral fundamental particles that carry no electric charge.

Figure 16-2 shows some simple experiments that demonstrate the nature of electric charge. If you hold two rubber rods next to each other, they neither attract nor repel each other. The same is true for two glass rods held next to each other, or for a rubber rod and a glass rod (**Figure 16-2a**). But if you rub the ends of each rubber rod with fur and each glass rod with silk, you will find that the ends of the rubber rods repel each other and the ends of the glass rods repel each other, but the ends of the rubber and glass rods attract each other (**Figure 16-2b**). What's more, the end of the rubber rod attracts the piece of fur used to rub it, and the end of the glass rod attracts the piece of silk used to rub it (**Figure 16-2c**). In each case where there is a repulsion or an attraction, the repulsive or attractive force is stronger the closer the objects are held to each other.

Here's how we explain these experiments and others like them:

1. *Matter has electric charge.* In addition to having mass, matter also possesses a property called **electric charge**. This charge comes in two forms, positive and negative. An object that has a net electric charge (more positive than negative, or more negative than positive) is **charged**. Most matter, however, contains equal amounts of positive and negative charge, which makes it electrically **neutral**. In Figure 16-2a the rubber and glass rods are electrically neutral, as are the pieces of fur and silk.
2. *Electric charge is a property of the constituents of atoms.* All ordinary matter is made up of atoms. Atoms are in turn composed of electrons, protons, and (except for the most common type of hydrogen atom) neutrons. Electrons are fundamental particles (they don't have internal structure, they really are objects!) and carry a negative charge, while protons carry an equally large positive charge; neutrons are neutral. Protons and neutrons have about the same mass and make up the dense nucleus at the center of an atom. Protons and neutrons are made up of smaller fundamental particles we will learn about in

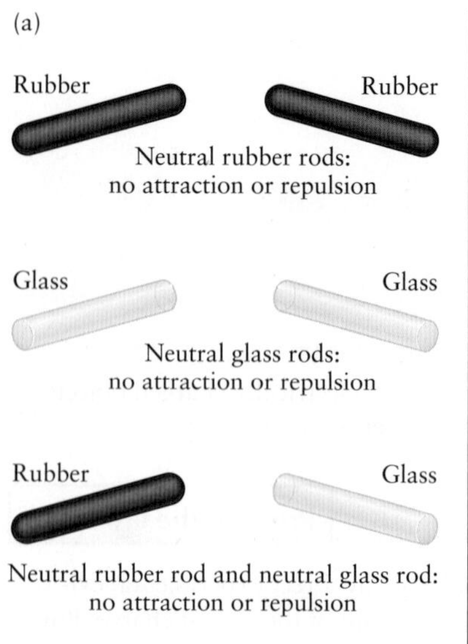

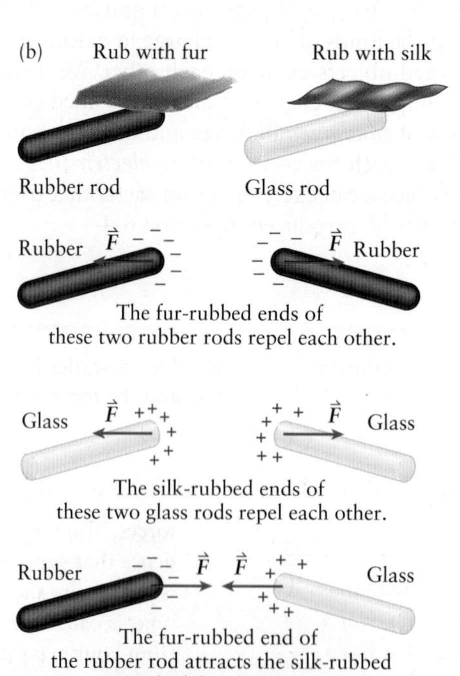

(b)
Rub with fur
Rub with silk
Rubber rod
Glass rod
Rubber
Rubber
The fur-rubbed ends of these two rubber rods repel each other.
Glass
Glass
The silk-rubbed ends of these two glass rods repel each other.
Rubber
Glass
The fur-rubbed end of the rubber rod attracts the silk-rubbed end of the glass rod.

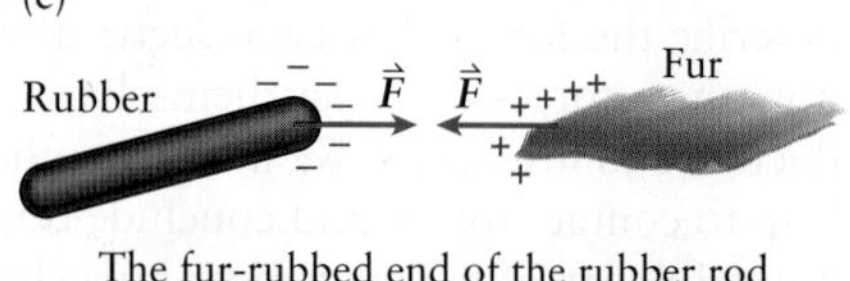

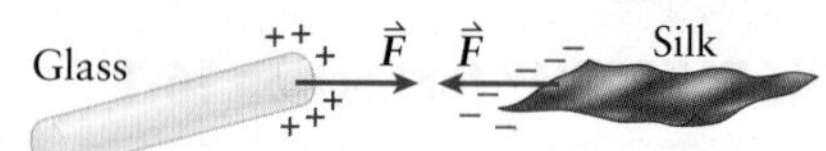

Figure 16-2 Experiments in electrostatics Electric charge can be transferred from one object to another by rubbing. Electrically charged objects exert forces on each other. Plus and minus signs indicate the sign of the net charge on each object (+ for positive, – for negative).

Chapter 25. A negatively charged electron has only about 0.05% the mass of a proton, and the electrons are mostly found outside the massive, positively charged nucleus (**Figure 16-3**). A normal atom contains as many electrons as it does protons and so is electrically neutral. Although we assign positive and negative signs to describe charge, doing so does not imply direction. Thus, charge, like temperature, is a scalar quantity.

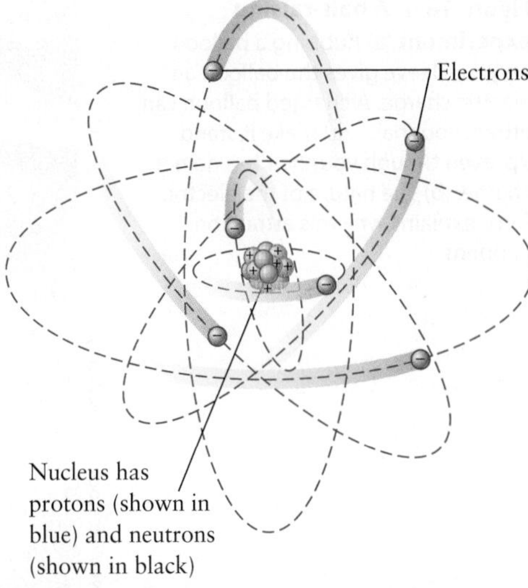

Figure 16-3 A very simplified model of an atom Negatively charged electrons orbit the atom's nucleus, which contains most of the atom's mass. The nucleus contains two types of particles, positively charged protons and neutral neutrons.

3. *Charge can be transferred between objects.* When dissimilar materials are rubbed together, negatively charged electrons can be transferred from one material to the other. As an example, consider the rubber rod and fur shown in Figure 16-2b. Initially the rod and fur are electrically neutral, so each has equal amounts of positive and negative charge. But when they are rubbed together, electrons are transferred from the fur to the rubber. This leaves the fur with a net positive charge (more positive charge than negative charge) and the rubber with a net negative charge. By contrast, when the glass rod and silk in Figure 16-2b (both of which are initially neutral) are rubbed together, electrons are transferred from the glass to the silk. So the silk ends up with a net negative charge and the glass rod with a net positive charge.
4. *Electrically charged objects exert forces on each other.* Objects with electric charge of the same sign (both positive or both negative) repel each other, while objects with electric charge of opposite sign (one positive and the other negative) attract each other (Figure 16-2c). The force between charged objects is called the **electric force.** The magnitude of the force depends on the amount of charge on each object and on the distance between the objects. The force increases for a greater amount of charge, and it increases if the charged objects are brought closer to each other.
5. *Charge is never created or destroyed.* In the experiments shown in Figure 16-2, charge is *transferred* between objects. Although charge moves between the rubber rod and the fur and between the glass rod and the silk, in each case the *total* charge of the two objects remains the same as before they were rubbed together. That is, charge is conserved; it is never created or destroyed. No one has ever observed a process in which charge is not conserved, and many very careful experiments have been made to test this, so to the best of our knowledge the conservation of electric charge is an absolute law of nature, just like the conservation of momentum or energy. If the two objects rubbed together are both in the system you are observing, then the conservation of charge tells you the total charge is constant in that system. If you were looking at just the silk as your system, conservation of charge would tell you that the amount of charge on the silk would have to equal the amount of charge removed from the glass.

Just as with energy and momentum, to use charge conservation we must first define a system. The type of system we need is a closed system, one where there is no exchange of matter (electrons). What the conservation law tells us is that the total amount of charge in a closed system is constant. Because charge is a property of objects with mass, if mass cannot transfer across the system boundary (definition of a closed system), charge cannot transfer. Electrons are tiny, so this mass transfer might be too small for us to see. If the amount of charge in a system we are observing changes, there must be a way for charge to enter or leave the system, so the system would not be a closed system. We will discuss several common ways to move charge. The character of the electric force explains the remarkable experiment shown in **Figure 16-4a**. A balloon is first rubbed against the girl's sleeve, causing electrons to transfer from her sleeve onto the rubber surface of the balloon (just like the rubber rod and fur in Figure 16-2b). When the negatively charged balloon is held above the girl's head, electric forces attract the hair to the balloon—even though the hair itself is *neutral.* What happens is that the negatively charged electrons on the balloon repel the electrons in the atoms that comprise the hair, but also attract the positively charged nuclei

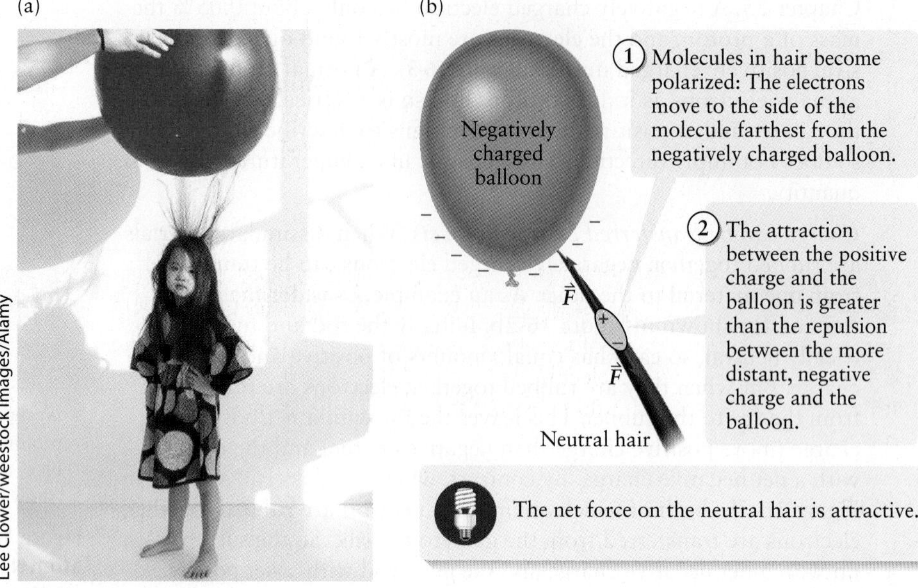

(a)
(b)
Lee Clower/weestock Images/Alamy
Negatively charged balloon
1 Molecules in hair become polarized: The electrons move to the side of the molecule farthest from the negatively charged balloon.
2 The attraction between the positive charge and the balloon is greater than the repulsion between the more distant, negative charge and the balloon.
Neutral hair
The net force on the neutral hair is attractive.

Figure 16-4 A hair-raising experiment (a) Rubbing a balloon on your sleeve gives the balloon an electric charge. A charged balloon can attract your hair and make it stand up, even though your hair has no net charge. (b) The nature of the electric force explains why this attraction happens.

of these atoms. The electrons and nuclei remain within their atoms but end up slightly displaced from each other (**Figure 16-4b**). We say that the hair becomes *polarized*. As we described earlier, the electric force is greater the closer charged objects are to each other. Figure 16-4b shows that the nuclei in the polarized hair are pulled slightly closer to the negatively charged balloon than the electrons in the hair. So the attractive force between the hair nuclei and the balloon is slightly greater than the repulsive force between the hair electrons and the balloon. The net force thus causes an attraction of the hair toward the balloon.

The unit of electric charge is the **coulomb** (C), named after the eighteenth-century French physicist Charles-Augustin de Coulomb who uncovered the fundamental law that governs the interaction of charged objects. (We'll discuss this law in Section 16-4.) As we mentioned earlier, electrons and protons have the same magnitude of electric charge; we call this magnitude e. Precise measurements show that

$$e = 1.60217662 \times 10^{-19}\ \text{C}$$

We use q or Q as the symbol for the charge of an object, so the charge on a proton is (to four significant digits) $q_{\text{proton}} = +e = +1.602 \times 10^{-19}$ C, and the charge on an electron is $q_{\text{electron}} = -e = -1.602 \times 10^{-19}$ C. No free particle has ever been detected with a charge smaller in magnitude than e. For that matter, no object has ever been found whose charge was not a multiple of e, such as $+2e, -3e$, and so on. That is, charge is *quantized*: The charge of an object is always increased or decreased by an amount equal to an integer multiple of the fundamental charge, e, also known as the fundamental unit of charge. It is impossible to add a fraction of the charge of a proton or an electron to an object or system. Even though we can and will often treat protons and neutrons as if they were fundamental particles, we will see that there is significant evidence that protons and neutrons are not fundamental, being composed of fundamental particles called *quarks* that have charges of $+2e/3$ and $-e/3$. However, no isolated quarks have ever been observed because the strong force that binds the quarks together into systems like protons and neutrons ensures no quark can be isolated.

Exam Tip

The electron is a fundamental particle—in other words, it cannot be broken down into smaller pieces—while the proton and neutron are made up of smaller particles called quarks. It is the quark composition that determines the charge of the proton and neutron.

Before scientists could perform experiments that made it possible to identify electrons and protons, they already believed there were just two types of charge. If two objects repel each other, they will each also repel any object the other repels. In other words, no charge has ever been found that attracts or repels both of the known existing types of charge. (Neutral doesn't count, since it is just an equal sum of the two kinds!)

While electrons and protons have opposite signs, it's completely arbitrary whether we choose electrons to have negative charge and protons to have positive charge or the other way around. The choice of sign is due to the American scientist and statesman Benjamin Franklin, who decided in the mid-eighteenth century that the sign of charge on a glass rod rubbed with silk (see Figure 16-2b) is positive. The electron and proton were not discovered until much later (1897 and 1919, respectively).

EXAMPLE 16-1 Electrons in a Raindrop

A water molecule is made up of two hydrogen atoms, each of which has one electron, and an oxygen atom, which has eight electrons. One water molecule has a mass of 2.99×10^{-26} kg. How many electrons are there in a single raindrop with a radius of $1.00 \text{ mm} = 1.00 \times 10^{-3}$ m, which has a mass of 4.19×10^{-6} kg? What is the total charge of all these electrons?

Set Up

Each water molecule has 10 electrons (2 from the hydrogen atoms and 8 from the oxygen atom). So we need to determine the number of water molecules in the raindrop and multiply by 10 to find the number of electrons.

Charge of an electron:

$$q_{\text{electron}} = -e = -1.602 \times 10^{-19}\ \text{C}$$

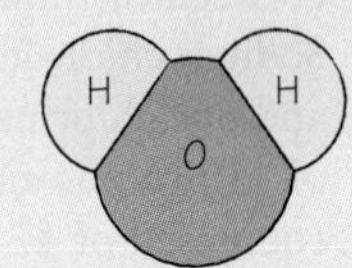

Solve

Determine the number of molecules in the raindrop, and from that determine the number of electrons.

The total number of molecules in the raindrop is the mass of the raindrop divided by the mass of a single water molecule:

$$\frac{\text{mass of raindrop}}{\text{mass of one water molecule}} = \frac{4.19 \times 10^{-6}\ \text{kg}}{2.99 \times 10^{-26}\ \text{kg/molecule}}$$
$$= 1.40 \times 10^{20}\ \text{molecules}$$

The number of electrons in the raindrop is

$$N = (1.40 \times 10^{20}\ \text{molecules})(10\ \text{electrons/molecule})$$
$$= 1.40 \times 10^{21}\ \text{electrons}$$

To find the charge of this number of electrons, multiply by the charge per electron.

Total charge of this number of electrons:

$$q = N \cdot q_{\text{electron}} = (1.40 \times 10^{21}\ \text{electrons})(-1.602 \times 10^{-19}\ \text{C/electron})$$
$$= -224\ \text{C}$$

Reflect

How large a number is 1.40×10^{21}? The human body contains about 10^{14} cells. Even if you include all the cells in all the approximately 9 million (9×10^{6}) people in New York City, that's only about 9×10^{20} cells—still fewer than the number of electrons in a single raindrop.

The charge of –224 C is quite substantial, but remember that for every electron in a water molecule, there is also one proton (one in each hydrogen atom and eight in each oxygen atom). Since a proton carries exactly as much positive charge as an electron carries negative charge, our raindrop has zero *net* charge. In Section 16-4 we'll see how difficult it would be to separate the positive and negative charges of this raindrop.

Finally, it is important to note that because charge is a property of electrons and protons, when we talk about something like a balloon being charged, we should technically be thinking of it as a system. However, because for most situations we will not be interested in describing the motion inside that system, just the initial and final states, we will find the object model is usually sufficient. So, we will typically talk about "charging an object," unless we need to worry about a change in its shape.

NOW WORK Problems 2–5 from The Takeaway 16-2.

THE TAKEAWAY for Section 16-2

✔ Electric charge is quantized. The smallest amount of charge that can be added to or removed from an object is equal to the fundamental charge $e = 1.602 \times 10^{-19}$ C.

✔ There are two kinds of charge. All ordinary matter contains positive charge in the form of protons, which have a charge of $+e$, and negative charge in the form of electrons, which have a charge of $-e$. An object is electrically neutral if it contains equal amounts of positive and negative charge. If these amounts are not equal, the object has a net charge.

✔ Charge can be neither created nor destroyed. However, charge can be transferred from one object to another, for example, by moving electrons between objects.

✔ Objects with a net charge exert electric forces on each other. These forces become weaker with increasing distance. If the objects have the same sign of charge (both positive or both negative), the forces are repulsive. If the objects have opposite signs of charge (one positive and one negative), the forces are attractive.

Prep for the AP Exam

AP Building Blocks

1. An object has a charge of -1 μC. Calculate the difference between the number of protons and electrons.

EX 16-1 2. Sodium chloride is composed of one sodium ion, Na^+, and one chlorine ion, Cl^-. The sodium ion is composed of 11 protons and 10 electrons. The chlorine ion is composed of 17 protons and 18 electrons. The mass of each sodium atom is 3.82×10^{-26} kg. The mass of each chlorine atom is 5.81×10^{-26} kg. For a paper packet of salt containing 1.00 g of sodium chloride calculate:
   (a) The number of electrons and protons
   (b) The number of coulombs of negative charge and positive charge
   (c) The net charge

EX 16-1 3. The charge per unit length on a glass rod that has been rubbed by a silk cloth is +0.000050 C/m.
   (a) If the charged segment of the rod is 10 cm long, how many electrons have been removed from the glass rod?
   (b) The glass rod is not a closed system with respect to charge transfer because electrons can be removed from the rod. Assuming the glass rod–silk cloth system is a closed system, how many electrons have been added to the silk cloth?

AP Skills in Action

EX 16-1 4. How many coulombs of negative charge are there in 0.500 kg of water?

EX 16-1 5. A particular ion of oxygen is composed of 8 protons, 10 neutrons, and 6 electrons. In terms of the fundamental charge e, what is the total charge of this ion?

6. Consider the demonstration shown in Figure 16-2.
   (a) Imagine you are the first investigator to label charge with signs. Explain how your choice of the sign of the charge on the rubber rod after it has been rubbed with fur determines the sign of the glass rod after it has been rubbed with silk.
   (b) Justify the claim that the identification of the sign of the charge of a third object by the direction of force exerted by the approach of the rubber rod would be consistent with an identification of the sign of the charge of the same object by the approach of the glass rod. Identify any problem you might encounter in determining the charge of the third object if you could use only one of the rods.
   (c) Support the claim of the two-charge model using evidence produced through scientific practices.

7. Two light rods, each hanging by a thread that does not affect the distribution of charge, are charged as shown in the figure, with dark shading indicating one sign of charge and light shading indicating the other sign of charge. They are hanging far enough apart that they do not affect each other. The hanging rods are each approached simultaneously by two other rods, charged as shown in cases (a) and (b). Predict the direction of rotation of the hanging rod in each case.

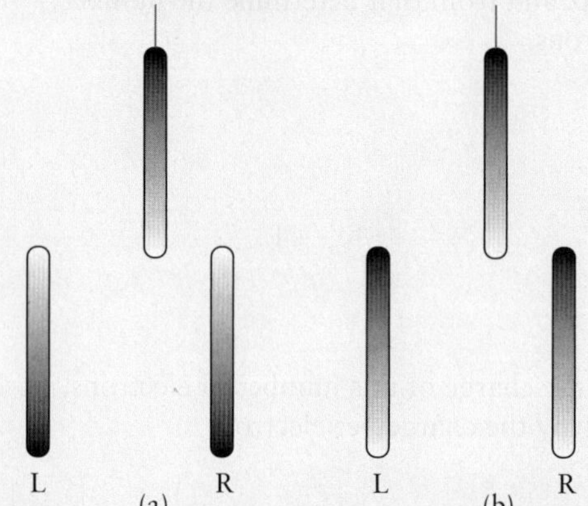

8. Consider the rubber rod and glass rod in Figure 16-2b. Compared to their masses before they are rubbed with fur and silk, respectively, what are their masses after being rubbed?
   (A) Both rods have less mass.
   (B) The rubber rod has more mass and the glass rod has less mass.
   (C) The rubber rod has less mass and the glass rod has more mass.
   (D) The masses of both rods are unchanged.

9. Describe a set of experiments that might be used to determine if you have discovered a third type of charge other than positive and negative.

10. A very light object is suspended by a thread.
   (a) A positively charged glass rod attracts the object. Does it follow that the object is negatively charged?
   (b) If, instead, the rod repels it, does it follow that the suspended object is positively charged?

16-3 Charge moves freely in a conductor but not in an insulator

All substances contain positive and negative charge. But how *mobile* that charge is depends on the specific material. The rubber, glass, silk, and fur shown in Figure 16-2 are all examples of **insulators**, substances in which charge is not able to move freely. All the electrons in an insulator are bound tightly to the nuclei of atoms, and any excess charge added to an insulator tends to stay wherever it is placed. This means that when electrons are placed on one end of a rubber rod by rubbing it with fur, the excess electrons cannot redistribute themselves along the rod (**Figure 16-5a**). So there is no excess charge on the opposite end of the rod to attract the positively charged fur. (Due to a polarization effect like that shown in Figure 16-4, there is still a very weak attraction between the positively charged fur and the neutral, unrubbed end of the rubber rod.) If you put a charged rubber rod in contact with an uncharged rubber rod, charge does not move between them. Most nonmetals are insulators.

By contrast, a metal such as copper is an example of a **conductor**, a substance in which charge *can* move freely. In a copper atom the outermost, or valence, electron can easily be dislodged. (The valence electron is relatively far from the positively charged nucleus, and the many electrons closer to the nucleus tend to shield the valence electron from the charge of the nucleus.) As a result, electrons—including excess electrons added to the copper—can move between copper atoms relatively freely. This explains what happens when you rub one end of a copper rod with nylon (**Figure 16-5b**). Electrons are transferred from the nylon to the copper, giving the copper rod a net negative charge and the nylon a net positive charge. But unlike what we saw in Figure 16-5a, after one end of the copper rod has been rubbed, the nylon attracts *both* ends of the copper rod equally. That's because the excess electrons deposited on one end of the copper rod can easily move along the rod. When the positively charged nylon is brought close to *either* end of the copper rod, the excess electrons on the copper rush to that end. As a result, that end has a net negative charge and is attracted to the nylon. Since charge can move easily in a conductor, and like charge repels, the charge will move around until it reaches equilibrium: There is no net force on any charge in a direction it can move. Thus, the charge placed on a conductor will always eventually come to rest on the surface of the conductor, spread out as much as possible so that the only force exerted on it is perpendicular to the surface.

If you put two conductors in contact, the charge spreads out over both of them. If they are identical, they end up with equal amounts of charge on their surfaces. If enough charge is placed on a conductor, it can actually strip electrons out of the air to move some charge off of its surface, sometimes transporting charge from another object. This is the spark you sometimes get when you reach for a doorknob, after dragging your feet on a carpet. You are a pretty good conductor, and can build up a lot of charge with the right carpet and shoes. Lightning is a similar, much larger, effect.

Conductors and insulators are an essential part of all electric circuits, which are systems in which there is an ongoing current (a movement of charge) around a closed path. Electric circuits are at the heart of any device that uses a battery (such as a mobile

Exam Tip

You will be expected to describe the transfer of charge in a variety of charging processes. It is important to understand the differences in the behavior of conductors and insulators.

(a)

The fur-rubbed end of the rubber rod attracts the fur with which it was rubbed...

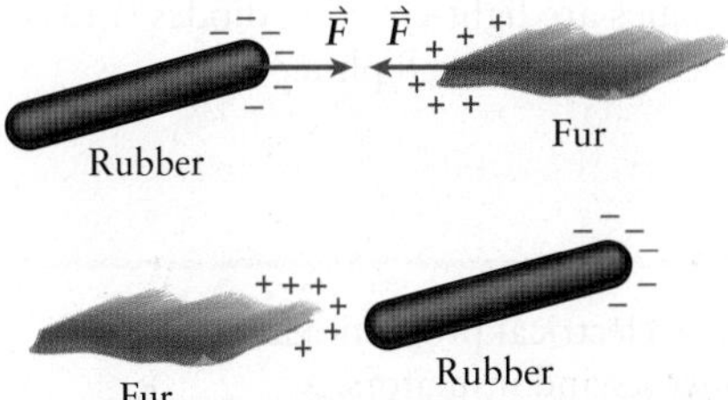

...but the end of the rubber rod that was not rubbed does not feel a strong attraction.

(b)

The nylon-rubbed end of the copper rod attracts the nylon with which it was rubbed...

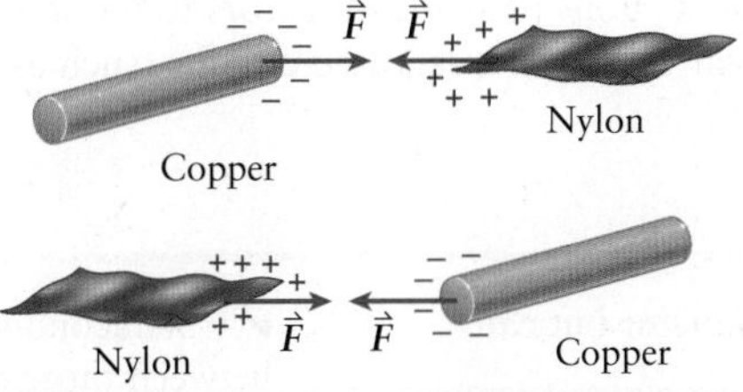

...and the end of the copper rod that was not rubbed also feels a strong attraction.

Figure 16-5 Insulators versus conductors (a) If you place excess charge at one location on an insulator, it remains at that location. (b) If you place excess charge at one location on a conductor, the excess charge can move freely through the conductor.

phone, a flashlight, or an electric vehicle) or that you plug into a wall socket (such as a desktop computer, a toaster, or an electric fan). An electric circuit uses moving charge (typically electrons) to transfer energy along a conductor from one point in the circuit to another. In a flashlight, for example, electrons moving through a copper wire carry energy from the battery (which we will find in the next chapter is a repository of *electric potential energy*) to the light bulb, where the energy is converted into visible light. The electrons then return to the battery through a second copper wire to pick up more energy and repeat the process. Insulators play a crucial role in this process: Each conducting wire is clad in a sheath made of an insulator, which helps ensure that the electrons only move along the length of the wires. (The visible part of an ordinary extension cord or power cord is actually the insulating sheath. The copper wires are contained within the sheath.) We'll explore these ideas in Chapter 18.

AP® Exam Tip

It is important to note that electrons do not get used up in a light bulb or in any device using electricity, and that a battery (or any power supply) is not a source of electrons. The electrons are already in the device's wires, and the power supply is there to move the electrons around the circuit. The current in the wires before and after a device is the same, and the power supply is there to give the electrons energy to make them move.

Most metals are good electrical conductors *and* also good conductors of energy through thermal processes, with large values of thermal conductivity. That's because the motion of electric charge and the transfer of energy by thermal processes within a material both require that objects that make up the material be free to move. In Chapter 14 we learned that the temperature of a material is related to the kinetic energy of the objects that make up that material. So the presence of free electrons makes metals good thermal conductors as well as good conductors of electricity. Materials that are electrical insulators, in which the electrons are generally not free to move between atoms, tend to be thermal insulators with low thermal conductivity.

Not all electrical conductors are metals, however. In nearly all biological systems, electric charge is carried by *ions*—atoms with an excess or deficit of electrons. These ions are suspended in water, that is, in aqueous solution. Water is particularly good at holding ions in solution because the water molecule (H_2O), while electrically neutral, has slightly more negative charge at the oxygen atom of the molecule and slightly more positive charge at the hydrogen atoms. So when an ionic compound such as sodium chloride (NaCl) is dissolved in water, the positive ions (Na^+) are attracted to the negative (oxygen) end of H_2O molecules while the negative ions (Cl^-) are attracted to the positive (hydrogen) end. Because the H_2O molecules to which the ions are attracted are free to move, the aqueous solutions within living things are conductors. It doesn't matter whether charge is carried by an electron or by an ion. As we'll see in Chapter 18, it also doesn't matter whether the moving charge is positive or negative: In either case there is a motion of charge.

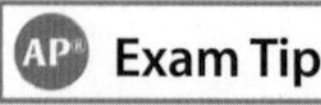

Exam Tip

Semiconductors and diodes are not part of the AP® Physics 2 curriculum, but they are important reasons why your computer-based devices work.

There's an important third class of substances called **semiconductors.** These substances have electrical properties that are intermediate between those of insulators and conductors. A common example of a semiconductor is silicon. A silicon atom has four outer electrons, compared to just one for copper, which might suggest that it would be a good conductor. However, in pure silicon each of those electrons forms part of a chemical bond with a neighboring silicon atom, so the outer electrons have limited mobility. As a result, pure silicon conducts electricity far worse than a conductor such as copper, though still far better than an insulator such as rubber.

Semiconductors are of tremendous practical use in electric circuits. That's because it's possible to adjust their electrical properties by *doping*—that is, by adding a small amount of a second substance. Here's an example: If we take pieces of silicon that have been doped in different ways, we can arrange them so that charge will move through the combination in one direction but will *not* move in the opposite direction! Such a combination, called a *diode*, plays the same role in an electric circuit as the valves in the human heart, which allow blood to flow through the heart in one direction only. Some of the applications of semiconductors to modern technologies are light-emitting diodes (LEDs), solar cells, and integrated circuits (such as in your computer or cell phone).

THE TAKEAWAY for Section 16-3

✔ Charges are free to move within a conductor but can move very little in an insulator.

✔ In a metal conductor such as copper, the charge carriers are typically electrons. In biological systems, ions in aqueous solution are the charge carriers.

✔ Semiconductors have electrical properties intermediate between those of conductors and insulators.

✔ When excess charge added to a conductor is allowed to come to rest in equilibrium, the excess charge resides only on the surfaces of the conductor.

AP® Building Blocks

1. Suppose 2.00 μC of positive charge is distributed evenly throughout a solid sphere of 1.27-cm radius.
   (a) Calculate the charge per unit volume ρ using the fact that volume is $4\pi/3$ times the cube of the radius.
   (b) Support the claim that the material from which the sphere is constructed is an insulator.
2. Why is it so difficult to charge a conductor by rubbing, compared to charging an insulator by rubbing?

AP® Skill Builders

3. A static charge of −5.00 μC is transferred to a flat sheet of steel (a conductor) with a negligible thickness. The area of one side of the sheet is 16.0 cm^2.
   (a) Calculate the surface charge per unit area σ, keeping in mind that the sheet has two sides, and assuming the charge is distributed uniformly.
   (b) It is claimed that all excess charge in a metal resides only on the surface of the metal. It turns out that the charge does not distribute uniformly on something that has points or sharp edges, so you must use spheres for this sort of experiment. Suppose that you had a method of measuring surface charge density. Identify the dependent and independent variables and method of data analysis in an experimental approach to test this claim by making use of the measurement of surface charge density and a way of fabricating metal spheres of arbitrary radius. You may not simply measure the charge and confirm it is the right amount. You must make use of a graph to analyze your data. (While this may seem artificial, there are actually calculations you could do that would let you set up an experiment that is quite similar, given that you probably won't have a way to measure surface charge density directly, but you don't need to think about those right now.)
4. A positively charged rod is brought near one end of an uncharged metal bar. The end of the metal bar farthest from the charged rod will be charged
   (A) positively.
   (B) negatively.
   (C) neutral.
   (D) twice as much as the end nearest the rod.
5. A balloon can be charged by rubbing it with your sleeve while holding it in your hand. You can conclude from this that the balloon is a(n)
   (A) conductor.
   (B) insulator.
   (C) neutral object.
   (D) semiconductor.

Prep for the AP® Exam

6. The triboelectric series is a description of the observed direction of transfer of electrons during charge by friction. In contact between a pair of materials, the material on the right in the following list is observed to receive negative charges from the material on the left.

   rabbit fur / glass / quartz / wool / cat fur / silk / cotton / wood / amber / rubber / metals / silicon / Teflon

   When the rubber rod in Figure 16-2 is frictionally charged with rabbit fur, electrons are transferred from the rabbit fur to the rubber rod, and the rod acquires a net negative charge. When the glass rod is rubbed with silk, electrons are transferred to the silk, and the glass rod acquires a net positive charge.

   Now, use this idea to design a mechanical system for use in space, where the absence of an atmosphere makes it impossible to use typical liquid lubricants. For example, rotational maneuvers use gyroscopes, in which the orientation and speed of a spinning cylinder controls torques. Design a method of using charging by friction to eliminate friction forces between the disk and axle, both labeled A in the figure. *Hint*: Surfaces between cylinders labeled A and B can be in relative motion.

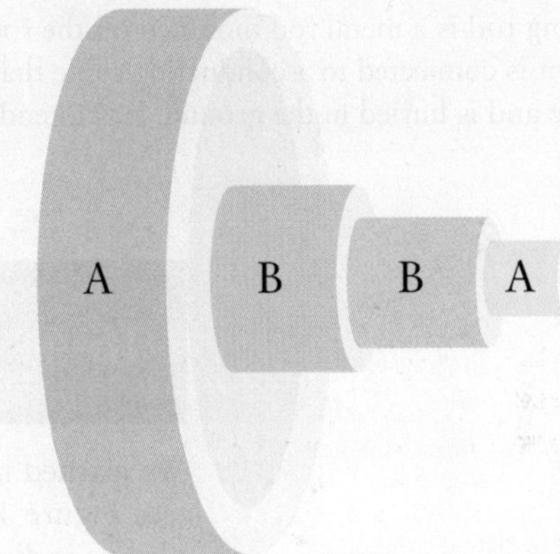

7. Small transfers of charge between a human body and an object with an excess charge may be undetectable. A study was reported of the threshold amount of charge transferred in a small time period that was perceptible. Tests were made on 70 (28 female, 42 male) adult personnel of the Electric Testing Laboratories in New York, New York. Different methods of contact were studied, as shown in the table. The subject could tap a charged plate with the index finger (tap), pinch a charged plate between the index finger and the thumb (pinch), or grab a charged rod in the closed palm (grip). The self-reported amount of charge per unit time that was perceptible is shown. Maxima and minima are the largest and smallest thresholds of perception among the male and female subjects. The amount of time over which charge was transferred was much less than one second, so the amounts of charge transferred were much smaller than would be transferred in a second.

	28 women (W)		42 men (M)			
Method of contact	Minimum charge per second (mC/s)		Average charge per second (mC/s)		Maximum charge per second (mC/s)	
	W	M	W	M	W	M
Tap	0.2	0.2	0.27	0.80	0.4	0.8
Pinch	0.2	0.3	0.59	0.87	1.2	2.4
Grip	0.5	0.3	0.84	1.19	1.4	3.0

(a) Based on these data, it was reported that there is a slight difference in sensitivity between women and men. Evaluate the validity of this claim based on the data.

(b) The three methods of contact involve different surface areas of contact. Specialized neurons, called nociceptors, are embedded in the skin. The neurons signal the central nervous system and then the brain when cells are damaged by charge transfer. Evaluate the validity of the claim that as the surface charge density increases, sensitivity increases.

Skills in Action

8. A lightning rod is a metal rod mounted on the roof of a house that is connected to a conducting cable that leaves the house and is buried in the ground. It is intended to protect the house from a lightning strike. If lightning, a sudden transfer of charge, hits the house, the charge will strike the metal rod and go through the conducting cable, instead of passing through the house, where it could start a fire or shock someone.
   (a) Explain the purpose of this chain of conducting materials that terminates in the ground in terms of the transfer of electrons and the risk of static charge.
   (b) Explain why doorknobs and car doors are common surfaces where shocks occur.

9. After combing your hair with a plastic comb, you find that when you bring the comb near an empty aluminum soft-drink can that is lying on its side on a nonconducting tabletop, the can rolls toward the comb. After being touched by the comb, the can is still attracted by the comb. Explain these observations.

10. A normal atom of sodium (chemical symbol Na) has 11 electrons and is electrically neutral. But in the compound sodium chloride (NaCl) the sodium atom is actually an ion of charge $+e$. A sodium atom acquires this charge
    (A) by destroying one of its electrons.
    (B) by transferring an electron to another atom or molecule.
    (C) by acquiring an electron from another atom or molecule.
    (D) by acquiring a proton from another atom or molecule.

16-4 Coulomb's law describes the force between charged objects

We learned in Section 16-2 that electrically charged objects exert forces on each other (see Figure 16-2). Careful measurements reveal that the force between two charged objects is directly proportional to the magnitude of each charge. The greater the magnitudes of the charges on the objects, the larger the force that is exerted on each. The force also depends on the distance between the charged objects; the closer the objects, the larger the force.

Coulomb's law summarizes the results of these measurements. Specifically, this law tells us about the force between two **point charges**, which are charged objects, often very small, with sizes much smaller than the separation between them (**Figure 16-6a**). The point -charge model is an idealization, just like a rope of negligible mass or an incline with negligible friction, but it's a good description in many situations where charged objects interact. Coulomb's law tells us the magnitude of the electric force that two point charges, q_1 and q_2, separated by a distance r exert on each other:

EQUATION IN WORDS
Coulomb's law

Any two point charges q_1 and q_2 exert electric forces of equal magnitude on each other.

Coulomb constant

Absolute values of the charge on each of the **point charges**

$$F_{q_1 \text{ on } q_2} = F_{q_2 \text{ on } q_1} = \frac{k|q_1||q_2|}{r^2} \quad (16\text{-}1)$$

Distance between the point charges

The value of the **Coulomb constant** k in Equation 16-1 is, to three significant digits,

$$k = 8.99 \times 10^9 \text{ Nm}^2/\text{C}^2$$

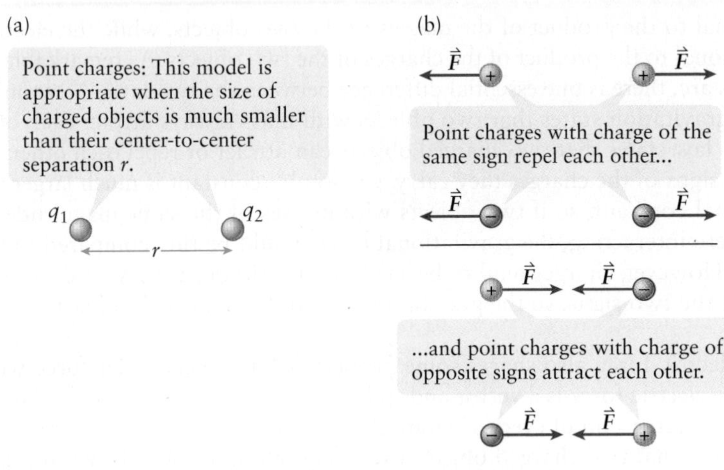

(a)
Point charges: This model is appropriate when the size of charged objects is much smaller than their center-to-center separation r.
q1
q2
r
(b)
Point charges with charge of the same sign repel each other...
...and point charges with charge of opposite signs attract each other.

Figure 16-6 Coulomb's law (a) Coulomb's law describes the electric force that two point charges (not drawn to scale) exert on each other. (b) The direction of the electric force between point charges depends on the signs (positive or negative) of the charges.

Note that Equation 16-1 tells you just the *magnitude* of the electric force between two point charges. As **Figure 16-6b** shows, the *direction* of the electric force is such that objects with charge of the same sign (both positive or both negative) repel each other, whereas objects with charge of opposite sign (one positive and one negative) attract each other. **Figure 16-7** shows an application of this principle.

You should notice the similarity between Coulomb's law and Newton's law of universal gravitation:

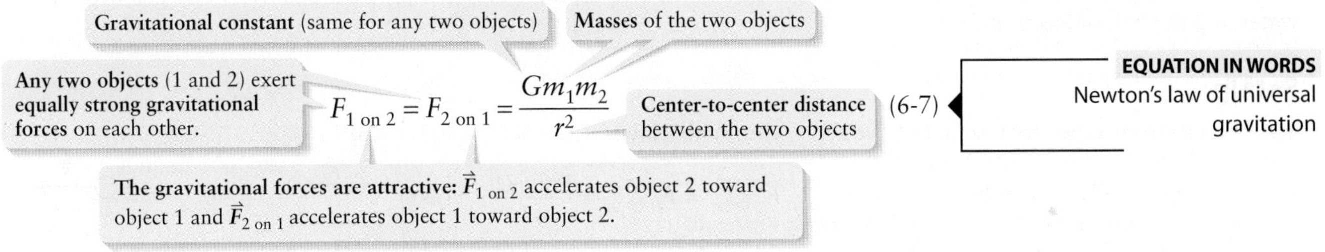

$$F_{1 \text{ on } 2} = F_{2 \text{ on } 1} = \frac{Gm_1m_2}{r^2} \quad (6\text{-}7)$$

EQUATION IN WORDS
Newton's law of universal gravitation

In both Coulomb's law (Equation 16-1) and the law of universal gravitation (Equation 6-7), the force that one object exerts on the other is inversely proportional to the square of the distance between them. In addition, the gravitational force is

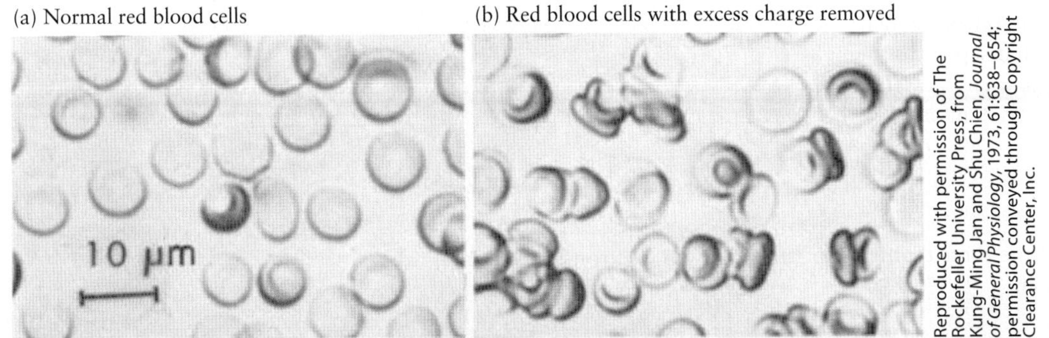

(a) Normal red blood cells
(b) Red blood cells with excess charge removed
10 μm

Figure 16-7 Coulomb's law and red blood cells Red blood cells carry oxygen from your lungs to other parts of your body through the circulatory system. (a) Each red blood cell has a slight excess of electrons that gives it a net negative charge. Since all cells are negatively charged, they exert repulsive electric forces on each other that help keep the cells apart. (b) The red blood cells in this photo have been treated with an enzyme that removes the excess electrons. Without electric forces to keep them apart, the red blood cells tend to clump. If red blood cells in your body behaved like this, their flow through your circulatory system would be impeded and your body would be starved for oxygen.

proportional to the product of the masses of the two objects, while the electric force is proportional to the product of the charges of the two objects. As remarkable as these similarities are, there is one essential difference between the two laws: Newton's law of universal gravitation states that two objects with mass always attract each other, but Coulomb's law states that two charged objects can attract or repel each other, depending on the signs of the charges they carry. Coulomb's constant is much larger than the gravitational constant, so if two objects with masses of the same magnitude as their charges were interacting, the gravitational force would be tiny compared to the electric force. However, charges tend to be small in magnitude, as they tend to cancel out because of the two signs, so the gravitational field dominates all other forces at long distances.

The electric force also shares some properties with every other force we've discussed. The electric force is a vector and so has a direction. The net electric force on an object is the vector sum of every separate electric force exerted on it. Furthermore, the electric forces that two charged objects exert on each other obey Newton's third law. When objects with charges q_1 and q_2 interact, the force that object 1 exerts on object 2 is equal in magnitude to the force that object 2 exerts on object 1, but the two forces are in opposite directions (see Figure 16-6b).

EXAMPLE 16-2 Calculating Electric Force

(a) What is the electric force (magnitude and direction) between two electrons separated by a distance of 10.0 cm = 0.100 m? (b) Suppose you could remove all the electrons from a drop of water 1.00 mm in radius (see Example 16-1 in Section 16-2) and clump them into a ball 1.00 mm in radius. If this ball of electrons is 10.0 cm from the drop of water from which they were removed, what is the magnitude of the electric force between the now positively charged drop of water and the ball of electrons?

Set Up

The two electrons repel because both have a negative charge $q = -e = -1.602 \times 10^{-19}$ C. From Example 16-1, the combined charge of all of the electrons in a water drop of this size is −224 C; the water drop was initially neutral, so the charge of the water drop after all of the electrons have been removed is +224 C. Since the ball of electrons and the water drop (with electrons removed) have opposite signs of charge, they attract each other.

We can use Coulomb's law (Equation 16-1) in both parts of this problem. That's because in both cases the charged objects are much smaller than the distance that separates them, so we can treat them as point charges. (Indeed, electrons are very small, even compared to the dimensions of an atom.)

Coulomb's law:

$$F_{q_1 \text{ on } q_2} = F_{q_2 \text{ on } q_1} = \frac{k|q_1||q_2|}{r^2} \quad (16\text{-}1)$$

Solve

(a) Find the magnitude of the force that each electron exerts on the other.

For the two electrons, the charges are

$$q_1 = q_2 = -e = -1.602 \times 10^{-19}\ \text{C}$$

The distance between the electrons is $r = 0.100$ m. From Equation 16-1 the magnitude of the force that each electron exerts on the other is

$$F = \frac{(8.99 \times 10^9\ \text{Nm}^2/\text{C}^2)\left|-1.602 \times 10^{-19}\ \text{C}\right|\left|-1.602 \times 10^{-19}\ \text{C}\right|}{(0.100\ \text{m})^2}$$

$$= 2.31 \times 10^{-26}\ \text{N}$$

(b) Find the magnitude of the force between the ball of electrons and the water drop from which they were extracted.

For the ball of electrons and the water drop with electrons removed, the charges are

$$q_1 = -224\ \text{C},\ q_2 = +224\ \text{C}$$

The distance between the two objects is $r = 0.100$ m. From Equation 16-1 the two objects exert forces on each other of magnitude

$$F = \frac{(8.99 \times 10^9\ \text{Nm}^2/\text{C}^2)|-224\ \text{C}||+224\ \text{C}|}{(0.100\ \text{m})^2}$$
$$= 4.51 \times 10^{16}\ \text{N}$$

Reflect

The repulsive force between the two electrons is tiny because the electron charge is tiny. By contrast, the attractive force between the ball of electrons and the electron-free water drop is immense. To put this force into perspective, a solid cube of lead with a weight of 4.51×10^{14} N would be 7.4 *kilometers* on a side! This weight is 100 times smaller than the magnitude of force that you would have to exert to keep the electrons from flying back into the water drop. There is no known way to produce a force of this magnitude, which is why you'll never see an object with all of its electrons removed. It's relatively easy to remove a small fraction of an object's electrons, as for the fur and the glass rod in Figure 16-2b. But removing *all* of the electrons from a piece of fur or a glass rod is not something we can do.

NOW WORK Problems 2 and 3 from The Takeaway 16-4.

EXAMPLE 16-3 Three Point Charges in a Line

An object with negative charge q is placed halfway between two identical objects, each of which carries the same positive charge: $Q_1 = Q_2 = +Q$. The distance d between adjacent objects is much larger than their sizes. If each of the three objects experiences a net electric force of zero, what is the magnitude of charge q in terms of Q?

Set Up

We want the net force on each point charge—that is, the *vector* sum of the forces on that point charge due to the other two point charges—to be equal to zero. We'll use Coulomb's law, Equation 16-1, to solve for the magnitude of the force of one point charge on another. We'll also use the idea that objects with charge of the same sign repel while objects with charge of opposite sign attract.

Coulomb's law:

$$F_{q_1 \text{ on } q_2} = F_{q_2 \text{ on } q_1} = \frac{k|q_1||q_2|}{r^2} \quad (16\text{-}1)$$

$Q_1 = +Q > 0$ $q < 0$ $Q_2 = +Q > 0$

d d

Solve

The objects are small compared to the distance between them, so we can use the point charge model. Let's start by considering the forces on positive point charge Q_2. The other positive point charge, Q_1, exerts a repulsive force $\vec{F}_{Q_1 \text{ on } Q_2}$ that pushes Q_2 away from Q_1, that is, to the right. The negative point charge q exerts an attractive force $\vec{F}_{q \text{ on } Q_2}$ that pulls Q_2 toward q, that is, to the left.

The net force on charge Q_2 must be zero:

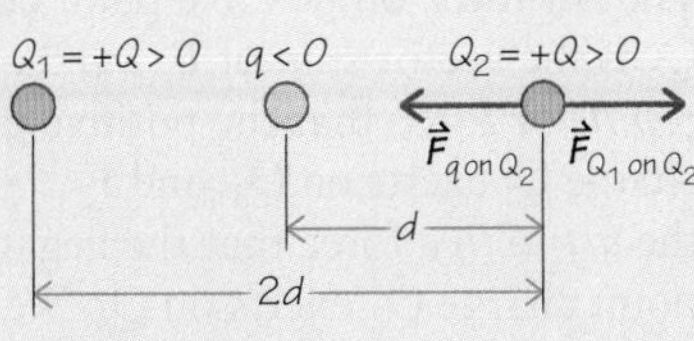

$$\vec{F}_{Q_1 \text{ on } Q_2} + \vec{F}_{q \text{ on } Q_2} = 0$$

For this to be true, $\vec{F}_{Q_1 \text{ on } Q_2}$ and $\vec{F}_{q \text{ on } Q_2}$ must have the same magnitude:

$$F_{Q_1 \text{ on } Q_2} = F_{q \text{ on } Q_2}$$

Use Coulomb's law (Equation 16-1) to find the magnitudes $F_{Q_1 \text{ on } Q_2}$ and $F_{q \text{ on } Q_2}$. Set these equal to each other and solve for the magnitude (absolute value) of q.

The distance between Q_1 and Q_2, each of which has a charge of magnitude Q, is $2d$. From Equation 16-1, the force exerted by Q_1 on Q_2 has magnitude

$$F_{Q_1 \text{ on } Q_2} = \frac{k|Q_1||Q_2|}{(2d)^2} = \frac{kQ^2}{4d^2}$$

d is the distance between Q_1 and q, and also between Q_2 and q. Notice that Q_1 and Q_2 are separated by distance $2d$. Remember that q is negative, so we need to keep its absolute value.

The distance between q and Q_2 is d, so the magnitude of the force exerted by q on Q_2 is

$$F_{q \text{ on } Q_2} = \frac{k|q||Q_2|}{d^2} = \frac{k|q|Q}{d^2}$$

In order for the forces to have the same magnitude,

$$\frac{kQ^2}{4d^2} = \frac{k|q|Q}{d^2}$$

Solve for the absolute value of q:

$$|q| = \frac{Q}{4}$$

Reflect

The value $|q| = Q/4$ satisfies the condition that there is zero net force exerted on Q_2. Because Q_1 is twice as far from Q_2 as q, and because the electric force is inversely proportional to the square of the distance, the charge Q_1 must be $(2)^2 = 4$ times greater than the magnitude of q in order for the forces these charges exert on Q_2 to have the same magnitude.

You can see that since Q_1 and Q_2 have the same charge, the forces on Q_1 are the mirror images of those on Q_2 (a repulsive force to the *left* exerted by Q_2, which is a distance $2d$ from Q_1, and an attractive force to the *right* exerted by q, which is a distance d from Q_1). So the net force on Q_1 will be zero, too.

You can see that the net force on the negative point charge q is guaranteed to be zero because it is the same distance d from the two equal positive point charges Q_1 and Q_2, so the force from Q_1 that pulls q to the left is just as great as the force from Q_2 that pulls q to the right.

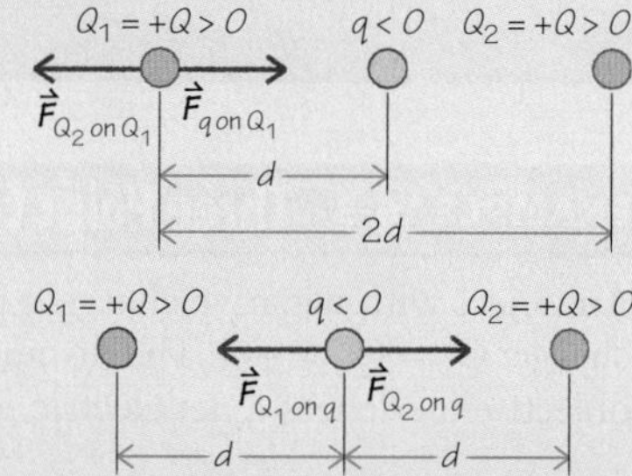

NOW WORK Problem 4 from The Takeaway 16-4.

EXAMPLE 16-4 Three Point Charges in a Plane

Two objects that can be modeled as point charges, Q_1 and Q_3, are both positive and carry a charge equal to 1.50×10^{-9} C; point charge Q_2 is negative and equal to -1.50×10^{-9} C. Point charges Q_1 and Q_2 are placed at a fixed position a distance $D = 6.00$ cm apart, and Q_3 is placed at a fixed position a distance $H = 4.00$ cm above the midpoint of the line that connects Q_1 and Q_2. Calculate the magnitude and direction of the force on Q_3 due to the other two point charges.

Set Up

The net force on positive point charge Q_3 is the vector sum of $\vec{F}_{1 \text{ on } 3}$, the *repulsive* force that the positive point charge Q_1 exerts on Q_3, and $\vec{F}_{2 \text{ on } 3}$, the *attractive* force that the negative point charge Q_2 exerts on Q_3. We'll use Coulomb's law to find the magnitude of each of these forces. We'll then add the two vectors using components.

Coulomb's law:

$$F_{q_1 \text{ on } q_2} = F_{q_2 \text{ on } q_1} = \frac{k|q_1||q_2|}{r^2} \quad (16\text{-}1)$$

Solve

Use Equation 16-1 to find the magnitudes of the forces $\vec{F}_{1\text{ on }3}$ and $\vec{F}_{2\text{ on }3}$.

The figure shows that the distance from Q_1 to Q_3 is the same as the distance from Q_2 to Q_3:

$$r_{13} = r_{23} = \sqrt{\left(\frac{D}{2}\right)^2 + H^2} = \sqrt{\left(\frac{6.00\text{ cm}}{2}\right)^2 + (4.00\text{ cm})^2}$$

$$= 5.00\text{ cm} = 5.00 \times 10^{-2}\text{ m}$$

All three point charges carry a charge of the same magnitude:

$$|Q_1| = |Q_2| = |Q_3| = 1.50 \times 10^{-9}\text{ C}$$

So, from Equation 16-1, $\vec{F}_{1\text{ on }3}$ and $\vec{F}_{2\text{ on }3}$ have the same magnitude:

$$F_{1\text{ on }3} = \frac{k|Q_1||Q_3|}{r_{13}^2}$$

$$= \frac{(8.99 \times 10^9\text{ Nm}^2/\text{C}^2)(1.50 \times 10^{-9}\text{ C})^2}{(5.00 \times 10^{-2}\text{ m})^2}$$

$$= 8.09 \times 10^{-6}\text{ N}$$

$$F_{2\text{ on }3} = \frac{k|Q_2||Q_3|}{r_{23}^2} = F_{1\text{ on }3} = 8.09 \times 10^{-6}\text{ N}$$

Choose the positive x direction to be to the right and the positive y direction to be upward. Then find the x and y components of $\vec{F}_{1\text{ on }3}$ and $\vec{F}_{2\text{ on }3}$, and use these to calculate the components of the net force exerted on Q_3.

The components of $\vec{F}_{1\text{ on }3}$ and $\vec{F}_{2\text{ on }3}$ are

$$F_{1\text{ on }3,x} = F_{1\text{ on }3}\cos\theta$$

$$F_{1\text{ on }3,y} = F_{1\text{ on }3}\sin\theta$$

$$F_{2\text{ on }3,x} = F_{2\text{ on }3}\cos\theta$$

$$F_{2\text{ on }3,y} = -F_{2\text{ on }3}\sin\theta$$

Because the magnitudes $F_{1\text{ on }3}$ and $F_{2\text{ on }3}$ are equal, the components of the net force on Q_3 are

$$F_{\text{net on }3,x} = F_{1\text{ on }3,x} + F_{2\text{ on }3,x}$$

$$= F_{1\text{ on }3}\cos\theta + F_{2\text{ on }3}\cos\theta$$

$$= 2F_{1\text{ on }3}\cos\theta$$

$$F_{\text{net on }3,y} = F_{1\text{ on }3}\sin\theta + (-F_{2\text{ on }3}\sin\theta)$$

$$= F_{1\text{ on }3}\sin\theta - F_{1\text{ on }3}\sin\theta = 0$$

From the figure,

$$\cos\theta = \frac{(D/2)}{r_{13}} = \frac{3.00\text{ cm}}{5.00\text{ cm}} = 0.600$$

So

$$F_{\text{net on }3,x} = 2(8.09 \times 10^{-6}\text{ N})(0.600) = 9.71 \times 10^{-6}\text{ N}$$

$$F_{\text{net on }3,y} = 0$$

So the net force on Q_3 is to the right and has magnitude 9.71×10^{-6} N.

Reflect

The two individual forces on Q_3 add to a net force that is neither directly away from Q_1 nor directly toward Q_2. Since one point charge is pushing Q_3 and one point charge is pulling Q_3 and the two forces are the same size, it makes sense that the net force is horizontal when Q_3 is on the line directly between the other two point charges.

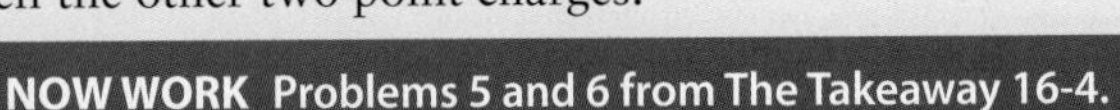
NOW WORK Problems 5 and 6 from The Takeaway 16-4.

THE TAKEAWAY for Section 16-4

✔ Coulomb's law tells us the magnitude of the electric force that two point charges exert on each other. This magnitude is proportional to the product of the magnitudes of the two charges and inversely proportional to the square of the distance between them.

✔ The electric force between two point charges is repulsive if the two charges have the same sign (both positive or both negative) and attractive if the two charges have opposite signs (one positive and one negative).

✔ The electric force and the gravitational force are similar in form, but whereas the electric force can be attractive or repulsive, the gravitational force is only attractive.

✔ The Coulomb constant is much larger than the universal gravitational constant, so over short distance scales, the electric force dominates.

✔ Because the electric force can be attractive or repulsive, over large distances the electric forces between systems will tend to cancel out. In contrast, the gravitational force is only attractive, so the gravitational force dominates on large scales.

Prep for the AP® Exam

AP® Building Blocks

1. Two insulated spherical objects, each with a uniform charge distribution, a radius R, and equal charges q, are separated by a distance r.
   (a) The nearest surfaces of the two spheres are separated by a distance $r - 2R$. In terms of R, calculate the separation r at which $(r - 2R)/r > 0.999$.
   (b) Calculate the percent relative error in the distance-dependence of Coulomb's law where the distance of nearest separation $(r - 2R)$ is used rather than the center–center distance r.
   (c) Explain why the concept of a point charge is not a limitation on the size of the objects but is a limitation on the distance between the objects when applying Coulomb's law.

EX 16-2

2. A point charge $q_1 = 10.0\ \mu C$ is separated from a second point charge $q_2 = -10.0\ \mu C$, at a distance of 1.0 m. Recall that the constant k in Coulomb's law is $8.99 \times 10^9\ Nm^2/C^2$.
   (a) Calculate the magnitude of the force exerted on the 10.0-μC point charge.
   (b) Describe the directions of the forces exerted on each point charge.

EX 16-2

3. The mass of an electron is 9.11×10^{-31} kg and the charge is 1.602×10^{-19} C.
   (a) Calculate the separation between two electrons near the surface of Earth such that the magnitude of the gravitational force exerted by Earth on q_1 is equal to the electric force exerted on q_1 by q_2.
   (b) For two electrons, can the magnitude of the gravitational force exerted by q_2 on q_1 ever be equal to the electric force exerted by q_2 on q_1? If so, calculate the separation when this is true. If not, support your answer with mathematical reasoning.

EX 16-3

4. Point charge A with charge $q_A = 11.00$ nC is located at the origin. Point charge B with charge $q_B = -11.00$ nC is on the x axis at $x = 3.00$ m. Point charge C with charge $q_C = 22.00$ nC is on the x axis at $x = -6.00$ m. And point charge D with charge $q_D = 22.00$ nC is on the x axis at $x = 8.00$ m.
   (a) Construct a representation of the array of point charges. On the representation, label the positions of the point charges and their magnitudes.
   (b) Calculate the magnitude of each force exerted on point charge A.
   (c) Below each point charge, B through D, add vectors to your representation with relative lengths that are proportional to the relative magnitudes of the forces exerted on A. Draw these vectors to indicate the direction in which the force on point charge A is exerted.
   (d) Calculate the net electric force on point charge A due to the other three charges.
   (e) Add a vector with the correct length and direction to your representation to show the net force on point charge A.

EX 16-4

5. A point charge $q_1 = 3.00$ nC is located at the origin, and a second point charge of $q_2 = -6.00$ nC is located on the x–y plane at the point (3.0 m, 2.5 m).
   (a) Determine the magnitude of the electric force exerted by the point charge q_2 on the point charge q_1.
   (b) Draw a diagram showing the direction of the force on point charge q_1 and calculate the angle that the vector makes with the positive x axis.

EX 16-4

6. A point charge $Q_1 = 4.00$ nC is located on the y axis at $y = 4.00$ m. A second point charge $Q_2 = -4.00$ nC is located on the x axis at $x = -4.00$ m. And a third point charge $Q_3 = 4.00$ nC is located on the x axis at $x = 4.00$ m. A 1.00 nC point charge q_0 is located at the origin.
   (a) Construct a representation of the x–y plane with the positions and charges of the four point charges.
   (b) Calculate the x and y components of the net electric force exerted on the 1.00-nC point charge located at the origin.

(c) Draw vectors showing the forces exerted by each of the three point charges on the point charge at the origin and the net force on the point charge q_0.

AP® Skill Builder

7. A helium-4 atom has a nucleus of two protons and two neutrons and has four electrons. The diameter of its nucleus is approximately 4×10^{-15} m. At low temperatures, like those on the surface of Earth, the electrons are tightly bound to the nucleus. At very high temperatures, like those in the interior of the Sun, the atom is ionized, electrons are stripped away, and the nucleus is bare.
   (a) Support the claim that at low temperatures a helium-4 atom exerts no net electric force on a point charge q in the vicinity of the atom.
   (b) Support the claim that at very high temperatures a helium-4 nucleus exerts a nonzero electric force on a point charge q in the vicinity of the nucleus.

AP® Skills in Action

8. Point charge A, which is 15.00 nC, is positioned at the origin of a coordinate system. Point charge B, which is 23.00 nC, is fixed on the x axis at $x = 3.00$ m.
   (a) Calculate the magnitude and describe the direction of the force that point charge B exerts on point charge A.
   (b) Apply Newton's third law to the forces exerted by the pair of point charges and explain why Coulomb's law and Newton's third law must be consistent and how that consistency is expressed mathematically in the electric force law.
9. Two point charges are separated by a distance of 12.0 cm. The numerical value of one charge is twice that of the other. Each charge exerts a repulsive force of magnitude 4.5×10^{-5} N on the other.
   (a) Support the claim that there is a unique pair of values of the charge on the point charges that is consistent with this condition.
   (b) Describe a strategy to determine the charges.
   (c) Apply the strategy in part (b) to calculate the magnitudes of the charges.
   (d) Explain why the signs of the charges can be either positive or negative but both must have the same sign.
10. In part (b) of Example 16-2 we imagined removing all of the electrons from a drop of water and moving them a given distance from the water drop. We then calculated the electric force between the drop and the electrons. Suppose instead we removed only one-half of the electrons from the water drop and then moved them to the same distance as in Example 16-2. Compared to the force we calculated in part (b) of Example 16-2, what would be the force between the electrons and the water drop in the scenario outlined here?
    (A) $1/\sqrt{2}$ as great
    (B) 1/2 as great
    (C) 1/4 as great
    (D) 1/16 as great.

16-5 Electric forces are the true cause of many other forces you experience, and electric fields can help model electric forces

Most forces in our daily experience arise only when one object is in direct contact with another object, like the normal force that is exerted on your body when you sit in a chair or when you push directly on an object. In Chapter 4, we introduced the idea of the fundamental forces, those that do not require contact (action-at-a-distance forces). We have discussed two of the three fundamental forces, the electroweak and the gravitational force. The other force is the strong force, which is exerted over a much smaller distance, and as its name implies is very strong.

The strong force, which we will learn about in Chapter 25, is what holds atomic nuclei together, despite the large repulsive electric forces between the positively charged protons we will learn about in this chapter. The electroweak force (which is a related group of interactions we can separate into electric, magnetic, and weak forces) is the fundamental force responsible for contact forces. The strong force and the weak force primarily deal with interactions within atoms, while as we saw gravity dominates on very large scales. The electric and magnetic forces are involved in the various interactions between electrically charged objects, including more than just the electric force we will introduce in this chapter. (The electric and magnetic forces apply to all charged objects. For interactions between stationary charged objects, such as those we will work with, these are sometimes called the electrostatic and magnetostatic forces.) The interaction between macroscopic objects can be roughly described as resulting from the electromagnetic interactions between their atomic constituents. Everyday objects do not *actually* touch; rather, contact forces are in large part due to the interactions of the electrons at or near the surfaces of the objects. Normal and friction forces are used to describe these surface interactions. Tension forces are the result of these interactions holding an object together when you try to stretch it. Even the buoyant force, which we describe in terms of pressure, is caused by these interactions, since the molecules actually interact through them. In Chapter 19 the magnetic force will be discussed.

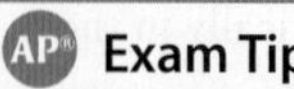

Exam Tip

You should be able to discuss the circumstances in which each of the fundamental forces is dominant and when the other forces can be ignored.

We can describe action-at-a-distance forces by using the concept of a *field*. In Chapter 4 we found that we could express the gravitational force $\vec{F}_g$ on an object of mass m as $\vec{F}_g = m\vec{g}$, where $\vec{g}$ is the acceleration due to gravity, but that we could also think of $\vec{g}$ as the *gravitational field* that Earth sets up in the space around it; $\vec{g}$ represents the value of that field vector (magnitude and direction) at a given point. An object of mass m placed at the point then experiences a gravitational force $\vec{F}_g = m\vec{g}$. The concept of a gravitational field helps us visualize how two objects with mass (Earth and the object of mass m) can interact without touching each other. In this same way we can visualize that a charged object modifies all of space by producing an electric field, which is strongest closest to the object but extends infinitely far away. A second charged object interacts with this electric field and experiences an electric force (**Figure 16-8**).

If we place a point charge q in an electric field $\vec{E}$, the force exerted on the point charge can be written in exactly the same form as the gravitational force and field:

EQUATION IN WORDS
Electric field and electric force

If an object with charge q is placed at a position where the **electric field** due to other charges is $\vec{E}$...

$$\vec{F}_E = q\vec{E} \tag{16-2}$$

...then the **electric force** on the object is $\vec{F}_E = q\vec{E}$.

An equivalent way to write Equation 16-2 is

$$\vec{E} = \frac{\vec{F}_E}{q} \tag{16-3}$$

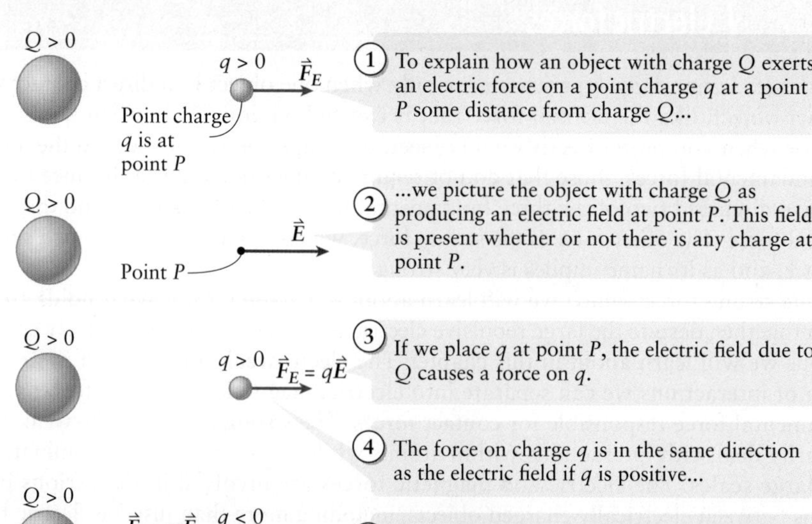

Q > 0
q > 0
$\vec{F}_E$
Point charge q is at point P
(1) To explain how an object with charge Q exerts an electric force on a point charge q at a point P some distance from charge Q...
Q > 0
$\vec{E}$
Point P
(2) ...we picture the object with charge Q as producing an electric field at point P. This field is present whether or not there is any charge at point P.
Q > 0
q > 0
$\vec{F}_E = q\vec{E}$
(3) If we place q at point P, the electric field due to Q causes a force on q.
(4) The force on charge q is in the same direction as the electric field if q is positive...
Q > 0
$\vec{F}_E = q\vec{E}$
q < 0
(5) ...and the force on charge q is in the direction opposite to the electric field if q is negative.

Figure 16-8 Electric field and electric force The electric field concept helps us visualize how an object with charge Q exerts an electric force on a second object with charge q.

Equation 16-3 tells us that we can interpret the electric field $\vec{E}$ at a certain point as the *electric force per charge* that is exerted on a charged object placed at that point. If an object with double the charge is placed at that point, it will experience double the force.

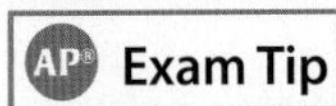

Exam Tip

In Section 16-3 we saw that because there are charged objects in a conductor that move easily, the charged objects always rearrange themselves until the net force on any mobile charged object is zero. This means the electric field inside any conductor in equilibrium is zero. For insulators, there is not an easily mobile charge, so the electric field in an insulator does not have to be zero. We will explore this distinction more in Chapter 17.

WATCH OUT !

Be aware of the direction of *force* versus the direction of *field*.

Although we can think of the electric field as the electric force per unit charge, that is determined by assuming a positive point charge. So the direction of the electric force exerted on a charged object at some point depends on the *sign* of the charge as well as the *direction* of the electric field at that point. Equation 16-2 tells us that the force exerted on a positive charge ($q > 0$) is in the same direction as the electric field $\vec{E}$, but the force exerted on a negative charge ($q < 0$) is in the direction opposite to $\vec{E}$ (Figure 16-8). So the electric field at a certain point is in the direction of the electric force that would be exerted on an object with *positive* charge placed at that point.

From Equation 16-3 we see that the SI units of electric field are newtons per coulomb, or N/C. If a 1-C charge experiences a 1-N force at a certain point in space, then at that point in space there is an electric field with a magnitude of 1 N/C.

Note that Equations 16-2 and 16-3 are strictly valid *only* if the charged object experiencing the force is a point charge q (one with a very small size). That's because the value of $\vec{E}$ can be different at different places. In that case it wouldn't be clear which value of $\vec{E}$ to use in Equation 16-2 or 16-3. But if the object with charge q is a point charge, it's clear what value of $\vec{E}$ to use: the value at the point where that point charge is located. We can arrange charge to create electric fields that are fairly uniform over some region of space. In this sort of field, we can do the exact same uniform acceleration problems we can do with the gravitational field near the surface of Earth. The source of the force does not affect how physics works!

EXAMPLE 16-5 Determining Charge-to-Mass Ratio

When released from rest in a uniform electric field of magnitude 1.00×10^4 N/C, a certain object that can be modeled as a point charge travels 2.00 cm in 2.88×10^{-7} s in the direction of the field. You can ignore any nonelectric forces exerted on the object. (a) What is the *charge-to-mass ratio* of this object (that is, the ratio of its charge q to its mass m)? (b) Other experiments show that the mass of this object is 6.64×10^{-27} kg. What is the charge of this object?

Set Up

The only force exerted on the object is the electric force given by Equation 16-2. Since the object accelerates in the direction of the electric field $\vec{E}$, the force on the object must also be in the direction of $\vec{E}$. So the charge on the object must be positive. Since $\vec{E}$ is uniform (it has the same value at all points), the force on the object, and so its acceleration a_x, will be constant. Thus, we can use the constant-acceleration equations from Chapter 2 to determine a_x. We'll use this with Newton's second law and Equation 16-2 to learn what we can about this object.

Electric field and electric force:

$$\vec{F}_E = q\vec{E} \quad (16\text{-}2)$$

Straight-line motion with constant acceleration:

$$x = x_0 + v_{0x}t + \frac{1}{2}a_x t^2 \quad (2\text{-}9)$$

Newton's second law:

$$\sum \vec{F}_{\text{ext}} = m\vec{a} \quad (4\text{-}2)$$

$q > 0$, $\vec{a}$, $\vec{E}$, x

Solve

(a) We are given that the object travels 2.00 cm = 2.00×10^{-2} m in 2.88×10^{-7} s. Use this to determine the object's constant acceleration.

Take the positive x direction to be the direction in which the object moves. The object begins at rest, so $v_{0x} = 0$. If we take the initial position of the object to be $x_0 = 0$, then Equation 2-9 becomes

$$x = \frac{1}{2}a_x t^2$$

Solve for the acceleration:

$$a_x = \frac{2x}{t^2} = \frac{2(2.00 \times 10^{-2}\ \text{m})}{(2.88 \times 10^{-7}\ \text{s})^2} = 4.82 \times 10^{11}\ \text{m/s}^2$$

Relate the acceleration to the net external (electric) force exerted on the object and solve for the charge-to-mass ratio.

The net force on the object of charge q is the electric force in the x direction, which from Newton's second law is equal to the mass m of the object multiplied by the acceleration a_x:

$$qE_x = ma_x$$

This shows that the acceleration of an object in an electric field depends on the object's charge-to-mass ratio:

$$a_x = \frac{q}{m}E_x$$

In this example we know both a_x and E_x, so the charge-to-mass ratio is

$$\frac{q}{m} = \frac{a_x}{E_x} = \frac{4.82 \times 10^{11}\ \text{m/s}^2}{1.00 \times 10^4\ \text{N/C}}$$

Since $1\ \text{N} = 1\ \text{kg} \cdot \text{m/s}^2$, the charge-to-mass ratio is

$$\frac{q}{m} = 4.82 \times 10^7\ \text{C/kg}$$

(b) Given the charge-to-mass ratio and the mass of the object, determine the charge q.

The charge of the object is

$$q = m\left(\frac{q}{m}\right)(6.64 \times 10^{-27}\ \text{kg})(4.82 \times 10^7\ \text{C/kg})$$

$$= 3.20 \times 10^{-19}\ \text{C}$$

The charge on a proton is $e = 1.60 \times 10^{-19}$ C; the charge on this object is $2e$.

Reflect

The mass of this object is similar to constituents of atoms, so a charge close to that of an electron is not unexpected. The object is probably missing two electrons! In fact, the real object in this example has about four times the mass of a proton but only double the charge of a proton. For historical reasons it's known as an *alpha particle*; in fact it's the nucleus of a helium atom, which contains two protons (each with charge e) and two neutrons (each with nearly the same mass as a proton but with zero charge). A neutral helium atom would have two electrons, one for each proton.

NOW WORK Problem 2 from The Takeaway 16-5.

An important application of the electric force that is most easily described in terms of a uniform electric field is *electrophoresis*. Chemists use this technique to separate molecules of different kinds according to their charge and mass. In the simplest kind of electrophoresis, a small amount of a sample containing molecules of different kinds is placed on a strip of filter paper, and the paper is soaked with a solution that conducts electricity (**Figure 16-9**). A uniform electric field of magnitude E is then applied along the length of the strip. Each molecule accelerates in response to the field, and that acceleration (taking into account fluid resistance on the molecules) is proportional to the charge-to-mass ratio q/m of the molecule. As a result, molecules with larger values of q/m move farther along the paper than do those with small values of q/m. The many applications of this specific technique, called *paper electrophoresis*, include analyzing currency to determine whether it is counterfeit (a forger's ink may have a different chemical composition than the ink used in legal currency) and looking for the presence of cancer antibodies or human immunodeficiency virus (HIV) in blood.

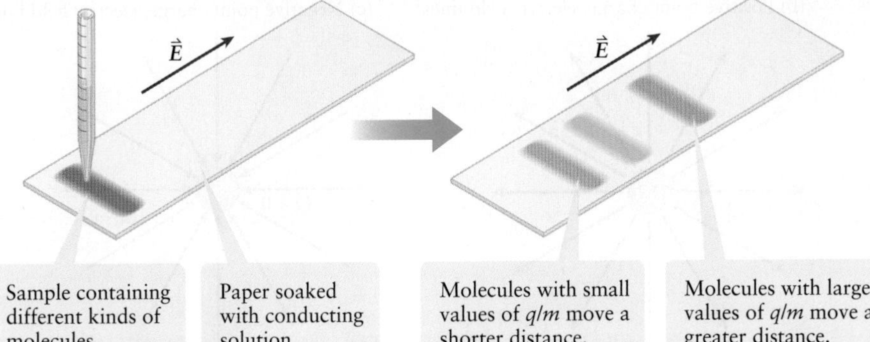

Figure 16-9 Paper electrophoresis This analytical technique used by chemists makes use of the electric field concept.

A different sort of electrophoresis is used for DNA profiling (also called genetic fingerprinting), which is an essential part of modern forensic science. A sample of human DNA is treated with an enzyme that breaks the long DNA strand into shorter segments. The sizes of these segments are characteristic of the person's genetic code, so measuring the segment sizes is a powerful technique for forensic identification. Unfortunately, the ratio q/m for a segment of DNA is nearly the same for segments of any size, so paper electrophoresis isn't useful. Instead, a sample containing the DNA segments is placed in a special gel that is permeated by many microscopic pores. When a uniform electric field is applied, all the segments move with about the same initial acceleration, but the smaller segments move through the gel pores more easily than the larger segments do. The result is that the DNA segments are spread out according to their size, allowing a genetic "fingerprint" to be made. Similar techniques are used in medical research for studying both DNA and proteins.

Electric Field of a Point Charge

Equation 16-2 explains how a point charge with charge q responds to a given electric field $\vec{E}$. It also tells us how to determine the value of $\vec{E}$ at any point. As an example, **Figure 16-10** shows how we might determine the electric field around a positive point charge of charge Q. We place a small positive point charge with charge q (which we call a *test charge*) at various locations around the point charge and measure the force $\vec{F}$ on that test charge. The electric field at each location is given by Equation 16-3, $\vec{E} = \vec{F}/q$; since q is positive, the electric field is in the same direction as the force on the test charge. The Coulomb force exerted by Q repels a positive test charge, q; thus, the direction of force at each location—and so the direction of the electric field at each location, shown by the blue vectors—is radially away from Q. As the lengths of the vectors show, the electric field magnitude decreases with increasing distance from Q. That's because the force between Q and q decreases with increasing distance in accordance with Coulomb's law (Section 16-4).

There are two ways we can depict the entire electric field around a positive charge Q. In **Figure 16-11a** we draw vectors to represent the electric field at a large

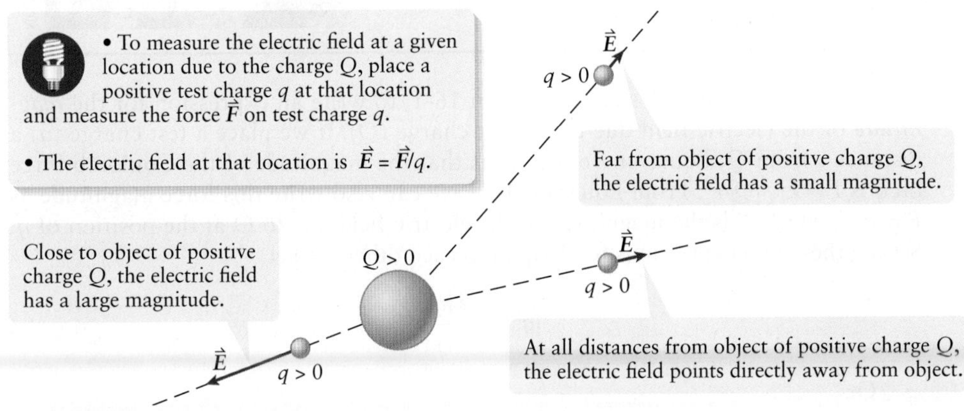

• To measure the electric field at a given location due to the charge Q, place a positive test charge q at that location and measure the force $\vec{F}$ on test charge q.
• The electric field at that location is $\vec{E} = \vec{F}/q$.
$\vec{E}$
$q > 0$
Far from object of positive charge Q, the electric field has a small magnitude.
Close to object of positive charge Q, the electric field has a large magnitude.
$Q > 0$
$\vec{E}$
$q > 0$
$\vec{E}$
$q > 0$
At all distances from object of positive charge Q, the electric field points directly away from object.

Figure 16-10 Mapping the electric field We can map out the electric field surrounding a point charge of charge Q—in this case a positive charge—by placing a positive test charge ($+q$) at various locations around Q.

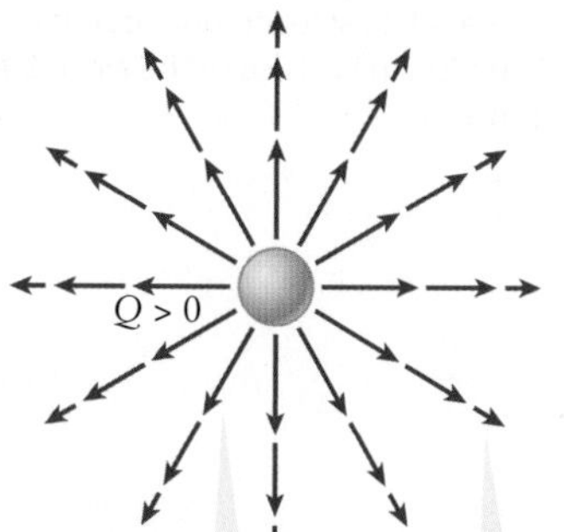
(a) Positive point charge: electric field vectors
$Q > 0$
Field due to a *positive* charge points *away* from the charge.
Field strength decreases with increasing distance from the charge.

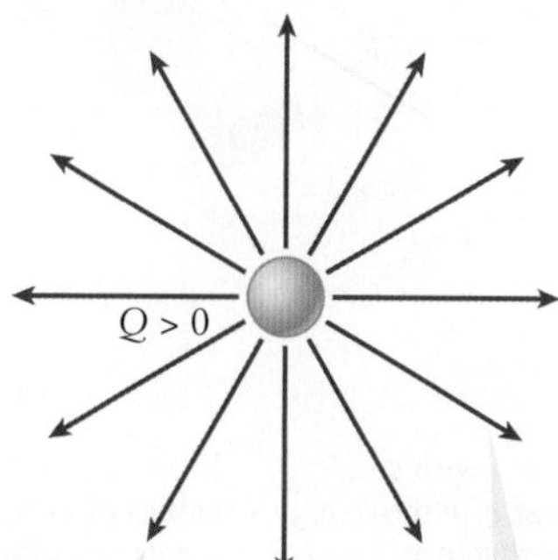
(b) Positive point charge: electric field lines
$Q > 0$
Field due to a *positive* charge points *away* from the charge.
Farther from the charge, where the field strength is weaker, field lines are farther apart.

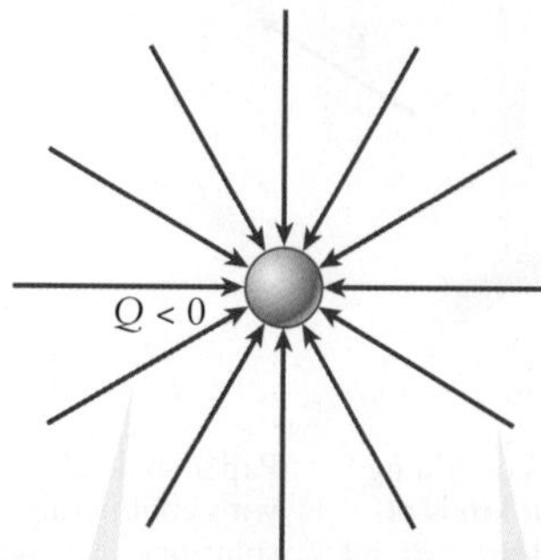
(c) Negative point charge: electric field lines
$Q < 0$
Field lines due to a *negative* charge point *toward* the charge.
Farther from the charge, where the field strength is weaker, field lines are farther apart.

Figure 16-11 Electric field and electric field lines The electric field around a charged object can be represented by (a) electric field vectors or (b), (c) electric field lines.

number of points around Q. **Figure 16-11b** shows an approach that's easier to draw but only an approximation: We connect adjacent vectors to form lines, called **electric field lines.** The direction of the field line passing through any point represents the direction of the field at that point. The magnitude of the electric field is related to the density of the field lines—that is, how close they are to each other. Close to Q, where the field lines are close together, the field has a large magnitude. Far from Q, where the field lines are far apart, the magnitude of the electric field is smaller. If the charge Q is negative, a positive test charge q experiences an attractive force $\vec{F}$ toward Q, so the electric field $\vec{E} = \vec{F}/q$ due to charge Q is directed *toward* that charge (**Figure 16-11c**). Although this can help predict behavior, it is important to realize that it is just an approximation for spherical or point charges as the number of lines represents a three-dimensional quantity and we lay it out in just two dimensions.

WATCH OUT!

Electric fields are three-dimensional.

Figure 16-11 may give you the incorrect impression that the electric field and electric field lines of a point charge lie only on the plane of the page. Not so! The electric field completely surrounds the charge; it is three-dimensional. The field lines are arranged around the charge rather like the spines of a sea urchin (**Figure 16-12**). More importantly, because we draw this three-dimensional representation in just two dimensions, we don't get the density correct, so the spacing is not exactly proportional to the field strength, just qualitatively connected.

NOAA

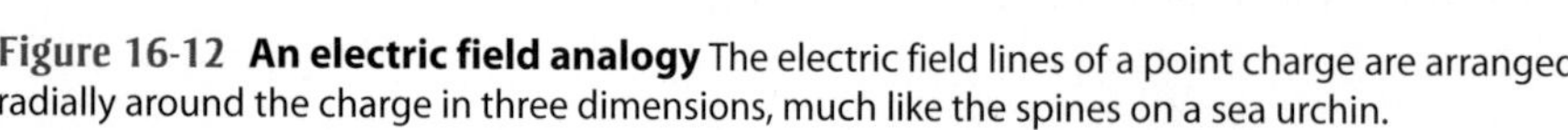

Figure 16-12 An electric field analogy The electric field lines of a point charge are arranged radially around the charge in three dimensions, much like the spines on a sea urchin.

We can use Coulomb's law, Equation 16-1, to write an expression for the *magnitude* of the electric field due to a point charge (Q). If we place a test charge (q) a distance r from Q, Equation 16-1 tells us that the magnitude of the Coulomb force on q is $F = k|q||Q|/r^2$. From Equation 16-2 we can also write this force magnitude as $F = |q|E$, where E is the magnitude of the electric field due to Q at the position of q. Setting these two expressions for F equal to each other, we get

$$|q|E = \frac{k|q||Q|}{r^2}$$

Or

Coulomb constant — Absolute value of the charge Q on the point charge

Magnitude of the electric field due to a point charge

$$E = \frac{k|Q|}{r^2} \quad (16\text{-}4)$$

Distance between point charge and the location where the field is measured

EQUATION IN WORDS
Magnitude of the electric field due to a point charge

The *magnitude* of the electric field due to a point charge Q decreases with increasing distance r from the charge. As Figure 16-11 shows, the *direction* of the electric field depends on the sign of the charge Q. The field points radially outward from a positive point charge and radially inward toward a negative point charge.

Electric Field of an Arrangement of Charges

The results in Figure 16-11 and Equation 16-4 describe the electric field due to a single point charge (positive or negative). If several point charges $Q_1, Q_2, Q_3, \ldots$ are present, experiment shows that the net electric force $\vec{F}$ that these charges exert on a test charge at any location is just the vector sum of the forces $\vec{F}_1, \vec{F}_2, \vec{F}_3, \ldots$ that these charges *individually* exert on that test charge. If we use the symbol $\vec{E}$ for the net electric field produced at a given location by point charges of magnitude $Q_1, Q_2, Q_3, \ldots$ together, and the symbols $\vec{E}_1, \vec{E}_2, \vec{E}_3, \ldots$ for the electric fields that these point charges produce individually, we can use Equation 16-2 to express this experimental result as

$$\vec{F} = \vec{F}_1 + \vec{F}_2 + \vec{F}_3 + \ldots \quad \text{or} \quad q\vec{E} = q\vec{E}_1 + q\vec{E}_2 + q\vec{E}_3 + \ldots$$

If we divide both sides of the second of these equations by q, we get

$$\vec{E} = \vec{E}_1 + \vec{E}_2 + \vec{E}_3 + \ldots \quad (16\text{-}5)$$

In other words, *when there are two or more point charges present, the electric field at any point in space is the vector sum of the fields due to each charge separately*. The following examples illustrate how to use this principle.

WATCH OUT !

Charge is a property of an object, not an object.

It is not unusual to see people talk about, for example, "charge Q_1," instead of saying "an object with charge Q_1." Although we try to avoid using this shortcut, we, when working with multiple charged objects, use the symbols for the amounts of charge to represent the objects to make it easier to keep track of which object we are talking about.

EXAMPLE 16-6 Where *Is* the Electric Field Zero?

A point charge with charge $Q_1 = +4.00$ nC (1 nC = 1 nanocoulomb = 10^{-9} C) is placed 0.500 m to the left of a point charge with charge $Q_2 = +9.00$ nC. Find the position between the two point charges where the net electric field is zero.

Set Up

At any point between the two charges, the electric field $\vec{E}_1$ due to Q_1 points to the right (away from this positive charge), and the electric field $\vec{E}_2$ due to Q_2 points to the left (away from this positive charge). We want to find the point P where the total field $\vec{E} = \vec{E}_1 + \vec{E}_2$ equals zero. We'll use the symbol D for the 0.500-m distance between the two charges and x for the distance from Q_1 to point P: Then the distance from Q_2 to point P is $D - x$. Our goal is to find the value of x for which $\vec{E} = 0$.

Magnitude of the electric field due to a point charge:

$$E = \frac{k|Q|}{r^2} \quad (16\text{-}4)$$

Total electric field:

$$\vec{E} = \vec{E}_1 + \vec{E}_2 \quad (16\text{-}5)$$

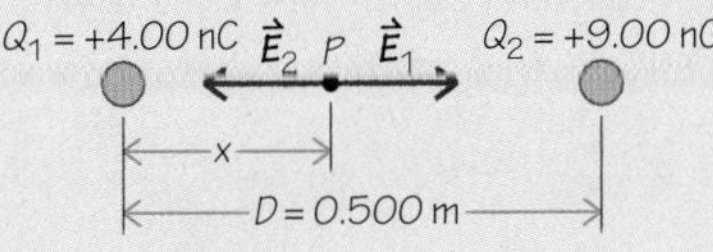

Solve

If the net electric field at P is zero, then $\vec{E}_1$ (to the right) and $\vec{E}_2$ (to the left) must have equal magnitudes so that these two vectors cancel. Use Equation 16-4 to write this statement in equation form.

For the net electric field at P to be zero,

$$E_1 = E_2$$

The distance from Q_1 to P is x, and the distance from Q_2 to P is $D - x$, so from Equation 16-4

$$E_1 = \frac{k|Q_1|}{x^2} \quad \text{and} \quad E_2 = \frac{k|Q_2|}{(D-x)^2}$$

If these are equal to each other,

$$\frac{k|Q_1|}{x^2} = \frac{k|Q_2|}{(D-x)^2}$$

Solve this equation for x.

The factors of k cancel in the above equation, so

$$\frac{|Q_1|}{x^2} = \frac{|Q_2|}{(D-x)^2} \quad \text{or}$$

$$|Q_1|(D-x)^2 = |Q_2|x^2$$

Multiply out the quantity $(D - x)^2$ and rearrange:

$$|Q_1|(D^2 - 2Dx + x^2) = |Q_2|x^2$$

$$(|Q_2| - |Q_1|)x^2 + (2D|Q_1|)x + (-D^2|Q_1|) = 0$$

We can simplify this if we divide through by $|Q_1|$:

$$\left(\frac{|Q_2|}{|Q_1|} - 1\right)x^2 + 2Dx + (-D^2) = 0$$

This is a quadratic equation of the form $ax^2 + bx + c = 0$, with $a = |Q_2|/|Q_1| - 1 = |9.00 \text{ nC}|/|4.00 \text{ nC}| - 1 = 1.25$, $b = 2D = 2(0.500 \text{ m}) = 1.00 \text{ m}$, and $c = -D^2 = -(0.500 \text{ m})^2 = -0.250 \text{ m}^2$. The solutions are

$$\begin{aligned} x &= \frac{-b \pm \sqrt{b^2 - 4ac}}{2a} \\ &= \frac{-(1.00 \text{ m}) \pm \sqrt{(1.00 \text{ m})^2 - 4(1.25)(-0.250 \text{ m}^2)}}{2(1.25)} \\ &= \frac{-(1.00 \text{ m}) \pm \sqrt{2.25 \text{ m}^2}}{2.50} \\ &= \frac{-(1.00 \text{ m}) \pm (1.50 \text{ m})}{2.50} \\ &= +0.200 \text{ m or } -1.00 \text{ m} \end{aligned}$$

We want a positive value of x to correspond to a point to the right of Q_1, so the solution we want is $x = +0.200$ m. We conclude that point P is a distance $x = 0.200$ m to the right of charge Q_1 and a distance $D - x = 0.500 \text{ m} - 0.200 \text{ m} = 0.300$ m to the left of charge Q_2.

Reflect

Because the amount of charge Q_1 is smaller than that of Q_2, the location of the point where the net electric field is zero must be closer to the point charge of charge Q_1 than to the point charge of charge Q_2. That's just what we found. You can check the result $x = 0.200$ m by substituting this value into the above expressions for E_1 and E_2 and confirming that $E_1 = E_2$ for this value of x.

But what's the significance of the second solution, $x = -1.00$ m? This refers to a point 1.00 m to the *left* of Q_1 and 1.00 m + 0.500 m = 1.50 m to the left of Q_2. Our calculation shows that at this point E_1 is equal to E_2. However, at this point $\vec{E}_1$ and $\vec{E}_2$ *both* point to the *left* (both fields point away from the positive charges that produce them). So, at this point the electric fields $\vec{E}_1$ and $\vec{E}_2$ do *not* cancel, and the total field is not zero.

NOW WORK Problem 8 from The Takeaway 16-5.

EXAMPLE 16-7 Field of an Electric Dipole

A combination of two point charges of the same magnitude but opposite signs is called an **electric dipole. Figure 16-13** shows an electric dipole made up of a point charge with charge $+q$ and a point charge with charge $-q$ separated by a distance $2d$. (This is a simple model for an ionic molecule like NaCl, which has a charge $+q$ on the Na^+ ion and a charge $-q$ on the Cl^- ion.)

Derive expressions for the magnitude and direction of the net electric field due to these two point charges at a point P a distance y along the midline of the dipole.

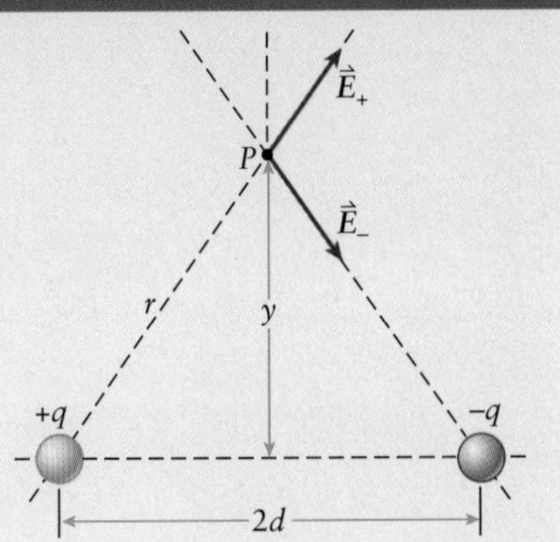

Figure 16-13 An electric dipole The field produced by an electric dipole at any point is the vector sum of the fields $\vec{E}_+$ and $\vec{E}_-$ caused by the point charge of charge $+q$ and the point charge of charge $-q$, respectively.

Set Up

The net field is the vector sum of the field $\vec{E}_+$ due to the positive point charge ($+q$) (which points away from $+q$) and the field $\vec{E}_-$ due to the negative point charge ($-q$) (which points toward $-q$). In Example 16-4 in Section 16-4 we used vector addition to find the net electric *force* exerted by two point charges (one positive and one negative) on a third charge; here we use vector addition to find the net electric *field* due to the two point charges. As in Example 16-4, we'll choose the positive x direction to be to the right and the positive y axis to be upward and add the two vectors using components.

Magnitude of the electric field due to a point charge:

$$E = \frac{k|Q|}{r^2} \quad (16\text{-}4)$$

Total electric field:

$$\vec{E} = \vec{E}_+ + \vec{E}_- \quad (16\text{-}5)$$

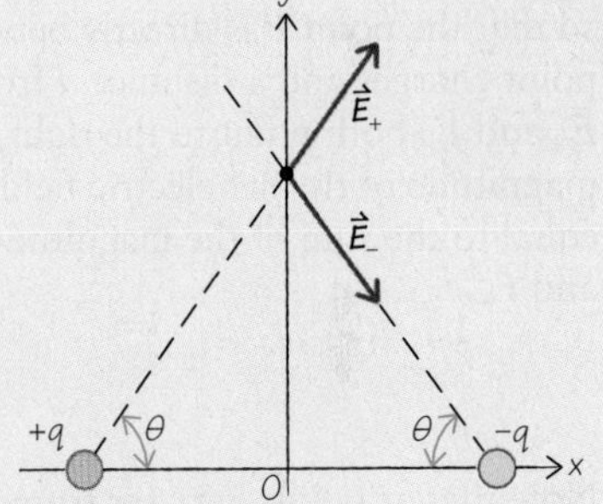

Solve

Use Equation 16-4 to find the magnitudes of the fields $\vec{E}_+$ and $\vec{E}_-$ at P.

The distance from $+q$ to point P is the same as the distance from $-q$ to P. Call this distance r:

$$r = \sqrt{y^2 + d^2}$$

Since $+q$ and $-q$ have the same magnitude (q, which is positive) and are the same distance from P, Equation 16-4 tells us that the fields that the two charges produce at P have the same magnitude:

$$E_+ = E_- = \frac{kq}{r^2} = \frac{kq}{y^2 + d^2}$$

Find the x and y components of $\vec{E}_+$ and $\vec{E}_-$, and use these to calculate the components of the net field at P.

The components of $\vec{E}_+$ and $\vec{E}_-$ are

$$E_{+,x} = E_+ \cos\theta$$
$$E_{+,y} = E_+ \sin\theta$$
$$E_{-,x} = E_- \cos\theta$$
$$E_{-,y} = -E_- \sin\theta$$

Since the magnitudes E_+ and E_- are equal, the components of the net field at P are

$$E_x = E_{+,x} + E_{-,x} = E_+ \cos\theta + E_- \cos\theta$$
$$= \frac{2kq}{y^2 + d^2}\cos\theta$$
$$E_y = E_{+,y} + E_{-,y} = E_+ \sin\theta + (-E_- \sin\theta)$$
$$= \frac{kq}{y^2 + d^2}\sin\theta - \frac{kq}{y^2 + d^2}\sin\theta = 0$$

From the figure:

$$\cos\theta = \frac{d}{r} = \frac{d}{\sqrt{y^2 + d^2}}$$

So the components of the net electric field are

$$E_x = \frac{2kq}{(y^2 + d^2)} \frac{d}{\sqrt{y^2 + d^2}} = \frac{2kqd}{(y^2 + d^2)^{3/2}}$$

$$E_y = 0$$

The net electric field at point P is to the right and has magnitude $E = 2kqd/(y^2 + d^2)^{3/2}$.

Reflect

We can check our result by substituting $y = 0$ so that the point P is directly between the two point charges and a distance d from each. Then $\vec{E}_+$ and $\vec{E}_-$ both point to the right, and the magnitude of the net electric field should be equal to the sum of the magnitudes of $\vec{E}_+$ and $\vec{E}_-$.

At $y = 0$, the net electric field has magnitude

$$E = \frac{2kqd}{(0 + d^2)^{3/2}} = \frac{2kqd}{d^3} = \frac{2kq}{d^2} = 2\left(\frac{kq}{d^2}\right)$$

This is just twice the magnitude of the field due to each individual point charge:

$$E_+ = E_- = \frac{kq}{d^2}$$

Note that at points very far from the dipole, so that y is much greater than d, the magnitude of the field is inversely proportional to the *cube* of y: At double the distance, the field of a dipole is $(1/2)^3 = 1/8$ as great. This is a much more rapid decrease with distance than the field of a single point charge, for which E is inversely proportional to the *square* of the distance: At double the distance, the field of a point charge is $(1/2)^2 = 1/4$ as great. The dipole field decreases much more rapidly because the fields of $+q$ and $-q$ partially cancel each other.

If y is much greater than d, $y^2 + d^2$ is approximately equal to y^2. Then the magnitude of the net electric field due to the dipole is approximately

$$E_{\text{net}} = \frac{2kqd}{(y^2)^{3/2}} = \frac{2kqd}{y^3}$$

NOW WORK **Problem 9 from The Takeaway 16-5.**

By using techniques like those we employed in Example 16-7, it's possible to calculate and map out the electric field at all points around an electric dipole. **Figure 16-14** shows the field lines. Note that as you move away from the dipole along its midline, the field lines become farther apart. This is a graphical way of showing that the magnitude of the field decreases with increasing distance, just as we found in Example 16-7.

Another way of representing the Coulomb constant k, introduced in the equations in this chapter, is to replace the k that appears in these equations with $1/4\pi\varepsilon_0$. Here ε_0, called the **permittivity of free space**, is equal to $1/(4\pi k) = 8.85 \times 10^{-12}\ \text{C}^2/(\text{N} \cdot \text{m}^2)$ to three significant digits. This is an important physical constant that we will see again in both the next chapter, when we study capacitance, and in Chapter 22, where we will discover that it is actually a determining factor in the speed of light. This constant is related to the capability of a vacuum to permit electric fields to exist. To this point we have only thought about electric fields in air. In air, the permittivity is very close to that of vacuum—the same value to three significant digits. In other materials, the permittivity is different. In an insulating material, the permittivity is a measure of how much the material polarizes in response to an applied electric field. If you think back to our

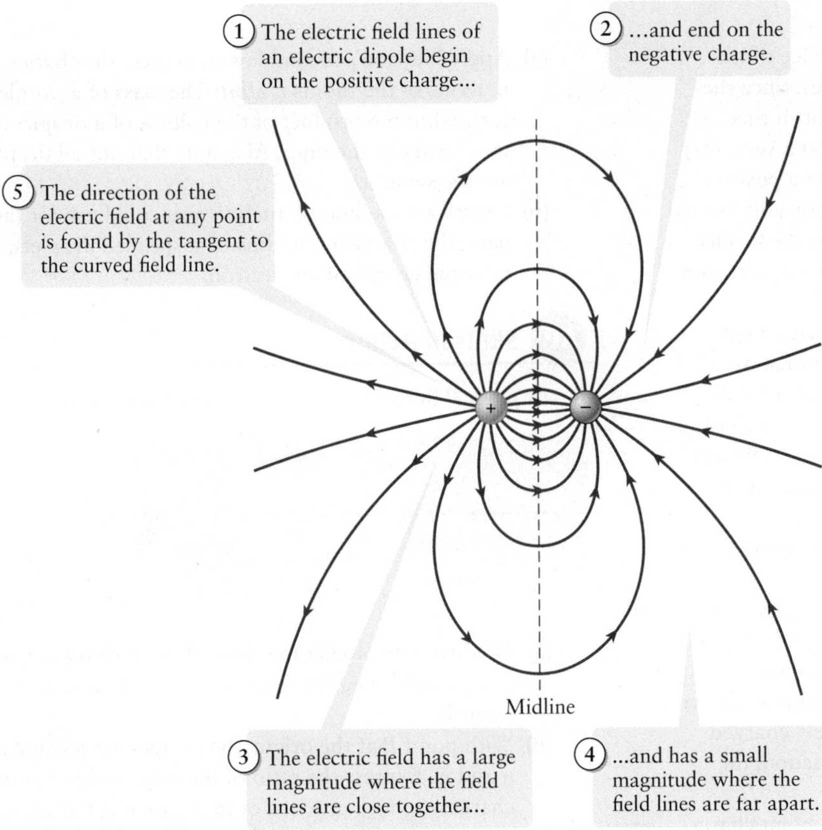

(1) The electric field lines of an electric dipole begin on the positive charge...
(2) ...and end on the negative charge.
(5) The direction of the electric field at any point is found by the tangent to the curved field line.
+
−
Midline
(3) The electric field has a large magnitude where the field lines are close together...
(4) ...and has a small magnitude where the field lines are far apart.

Figure 16-14 Field lines of an electric dipole At any point the electric field due to an electric dipole is the vector sum of the field due to the positive charge and the field due to the negative charge. An electric field is a vector, and so is always in a single direction at any point in space. When field lines are curved, as in this example, the electric field is in the direction tangent to the field line at the point of interest.

discussion of attracting a neutral object, we can see that a larger permittivity would mean a neutral object made of that material would be more strongly attracted to a charged object, since the charge separation would be of greater magnitude.

THE TAKEAWAY for Section 16-5

✔ Contact forces we experience every day are simplified models of the many interatomic electric forces that cause them. By understanding the underlying force (such as repulsion for normal forces and attraction for tension forces), we can predict the direction in which these contact forces will be exerted.

✔ Fundamental forces are action-at-a-distance forces, and can be modeled by fields.

✔ Any charged object produces an electric field in the space around it. A second charged object interacts with this electric field; this interaction is the origin of the electric force that the first object exerts on the second.

✔ The electric field of a positively charged object points directly away from that object; the electric field of a negatively charged object points directly toward that object.

✔ The net electric field due to two or more point charges is the vector sum of the electric fields due to the individual point charges.

Prep for the AP Exam

AP® Building Blocks

1. A force is exerted on a positive point charge (q) when it is placed near a negative point charge, ($-Q$).
   (a) Explain the direction of the electric field of the point charge of charge $-Q$ in terms of the sign of the point charge of charge q and the direction of the force exerted on the positive point charge.
   (b) Construct a representation of the electric force on q due to $-Q$ for any position at the same distance from $-Q$.
   (c) Compare the direction of this force to that exerted by Earth on the Moon.
   (d) In terms of your descriptions in part (c), compare Earth's gravitational field caused by Earth's mass M_E and the electric field caused by a charge of $-Q$ How are these fields similar? How are the electric field and the gravitational field different in general?

(e) Describe the electric field near the surface of the "point charge" of charge $-Q$ (in quotes, since the point charge model would no longer fit) if this amount of charge is now distributed on a very, very large sphere. Think about it in terms of a positive point charge of charge q that is very tiny and located very close to the surface of $-Q$, so that the surface appears flat, just as Earth's surface does to a person standing on it.

EX 16-5 2. At a point near the surface of Earth an electric field points radially downward and has a magnitude of approximately 1.00×10^2 N/C. At that point a penny, with a charge q and a mass of 3.11 g, has an acceleration of (i) 0.00 m/s^2 or, for another value of q (ii) 0.190 m/s^2 upward. (This is not a very realistic problem, but this once, let's pretend!)

(a) Calculate the magnitude and sign of the charge q for each case.

(b) Calculate the excess ($N > 0$) or deficiency ($N < 0$) from charge neutrality in the number N of electrons on the penny for an acceleration of 0.00 m/s^2.

3. The positively charged point charge in the dipole shown in Figure 16-14 is attracted to the negatively charged point charge. To find the force of this attraction, the electric field $\vec{E}$ to use in Equation 16-2 is

(A) the field due to the positively charged point charge.
(B) the field due to the negatively charged point charge.
(C) the net field due to both point charges.
(D) none of these.

4. Suppose both of the point charges in Figure 16-13 were negatively charged and had charges of the same magnitude. At point P in that figure, the net electric field due to these point charges would

(A) point to the left.
(B) point to the right.
(C) point straight up.
(D) point straight down.

AP® Skill Builders

5. Of the fundamental forces, the text claims that "gravity dominates on very large scales." However, both the gravitational force and the electric force decrease in proportion to the inverse squared distance. Explain why the electric force isn't as strong as the gravitational force at very large distances.

6. Apply the conservation of charge and the form of the electric force law to explain why the normal force and friction forces appear to be contact forces. That is, explain why these forces are so short ranged.

7. The charge of an electron was first measured by Robert Millikan and Harvey Fletcher (1910). The motions of small droplets of oil with static charges were observed. A charged oil droplet with a charge q and a radius r is suspended, motionless, at the point where the electric field is E.

(a) Apply Newton's second law to express the charge q in terms of the radius r. *Hint*: The mass of a droplet is equal to the product of the volume of a droplet and the density of the oil ρ. Also note that not all droplets are the same size.

(b) Describe a method of analysis you could use for the data, the charge and radius of suspended droplets, to infer the charge of an electron.

AP® Skills in Action

EX 16-6 8. At point P in the diagram the electric field is zero.

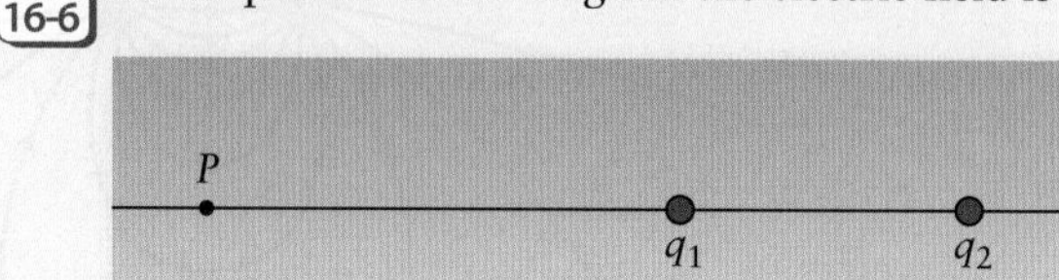

(a) Qualitatively predict the signs of point charges q_1 and q_2 by considering the force on a point charge q at point P.

(b) With point P at the origin, the charges are located at r_1 and r_2. Express the ratio of the magnitude of point charges $|q_1|/|q_2|$, in terms of the ratio of distances r_1/r_2.

EX 16-7 9. In the figure there is a -12-nC point charge Q at the origin. No charges are initially present other than the charge at the origin.

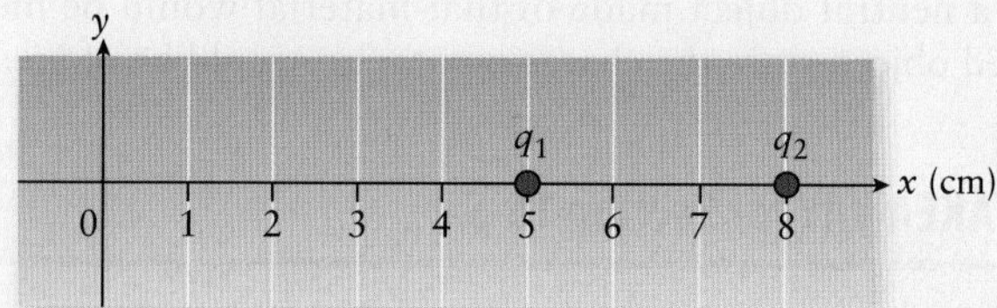

(a) Calculate the electric field at $x = 5.00$ cm due to the point charge at the origin.

(b) Point charge q_1, with a charge of 1.00×10^{-7} C, is placed at $x = 5.00$ cm. Calculate the magnitude and direction of the electric force exerted on the point charge q_1.

(c) Point charge q_1 is removed and point charge q_2 is placed at $x = 8.00$ cm. The electric force exerted on point charge q_2 has the same magnitude as the electric force that was exerted on q_1. Calculate the magnitude and sign of q_2.

10. A helium nucleus, also known as an alpha particle, is composed of two neutrons and two protons and has a mass of 6.64×10^{-27} kg. Calculate the magnitude and direction, relative to the direction of the electric field, of the acceleration of the alpha particle in an electric field with a magnitude of 2.4 kN/C. Neglect gravitational forces exerted on the nucleus.

WHAT DID YOU LEARN?

Prep for the AP Exam

Chapter learning goals	Section(s)	Related example(s)	Relevant section review exercise(s)
Define open and closed systems for everyday situations involving charge transfer and apply conservation of electric charge to those systems, including a description of the smallest unit of electric charge.	16-2	16-1	2
Describe the behavior of a system using the law of conservation of electric charge, including sign and relative quantity of net charge after various charging processes.	16-2	16-1	3
Construct an explanation of the two-charge model based on evidence produced through scientific practices.	16-2		4
Recognize the differences between insulators and conductors.	16-3		2, 3
Describe the point charge model.	16-4	16-2	1
Make claims about contact forces based on the microscopic cause of those forces, and explain them as arising from interatomic electric forces.	16-1 16-5		1, 6 6
Use Coulomb's law qualitatively and quantitatively to make predictions about the interactions between two point charges.	16-4	16-2, 16-3, 16-4	2, 4, 5, 6
Compare the similarities and differences between electric and gravitational forces.	16-4 16-5		3 1, 2, 5
Describe the electric field produced by a charged object or a configuration of point charges.	16-5	16-6, 16-7	3, 4, 8, 9

Chapter 16 Review

Key Terms

All the Key Terms can be found in the Glossary/Glosario on page G1 in the back of the book.

charged 782
conductors 787
coulomb 784
Coulomb constant 790
Coulomb's law 790
electric charge 782
electric dipole 805
electric field lines 802
electric force 783
insulators 787
neutral 782
permittivity of free space 806
point charge 790
semiconductors 788

Chapter Summary

Topic	Equation or Figure
Electric charge: Ordinary matter contains equal amounts of positive and negative charge. Charge can be transferred from one object to another, for example, by rubbing. In this and all other processes, charge is conserved: It can be moved from place to place but can be neither created nor destroyed.	(b) 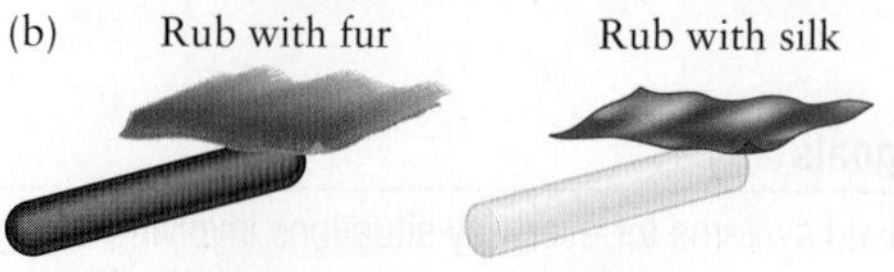 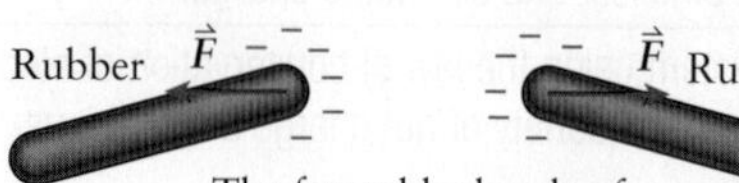 (Figure 16-2b)  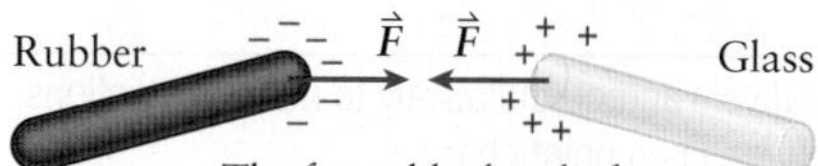
Conductors, insulators, and semiconductors: In a conductor, charge is free to move with ease; in an insulator, charge can move very little. Semiconductors have properties that are intermediate between those of conductors and insulators.	(b) The nylon-rubbed end of the copper rod attracts the nylon with which it was rubbed... 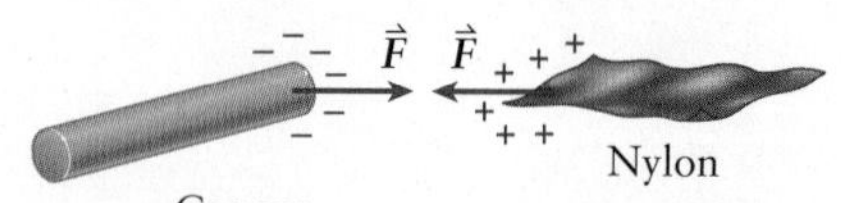 (Figure 16-5b) 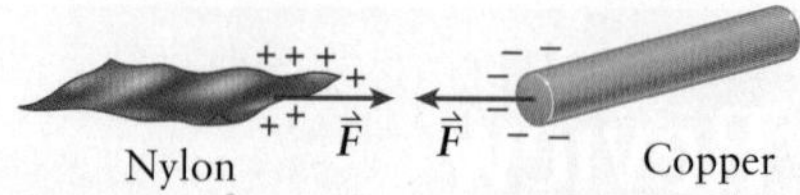 ...and the end of the copper rod that was not rubbed also feels a strong attraction.
Coulomb's law: The electric forces that two point charges exert on each other are proportional to the magnitudes of their charges and inversely proportional to the square of the distance between them. These forces are attractive if the two point charges have opposite signs, and repulsive if the two point charges have the same sign.	$F_{q_1 \text{ on } q_2} = F_{q_2 \text{ on } q_1} = \dfrac{k\lvert q_1\rvert\lvert q_2\rvert}{r^2}$ (16-1) Any two point charges q_1 and q_2 exert electric forces of equal magnitude on each other. Coulomb constant **Absolute values** of the charge on each of the **point charges** **Distance** between the point charges

Electric field: We can regard the interaction between point charges as a two-step process: One point charge sets up an electric field, and the other point charge responds to that field. The electric field points away from a positive charge and toward a negative charge. This same idea can be applied to interactions with more complicated systems of charge. The electric field due to a system of point charges is the vector sum of the fields due to the individual point charges.

If an object with charge q is placed at a position where the **electric field** due to other charges is $\vec{E}$...

$$\vec{F}_E = q\vec{E} \quad (16\text{-}2)$$

...then the **electric force** on the object is $\vec{F}_E = q\vec{E}$.

Coulomb constant — **Absolute value of the charge Q on the point charge**

Magnitude of the electric field due to a point charge

$$E = \frac{k|Q|}{r^2} \quad (16\text{-}4)$$

Distance between point charge and the location where the field is measured

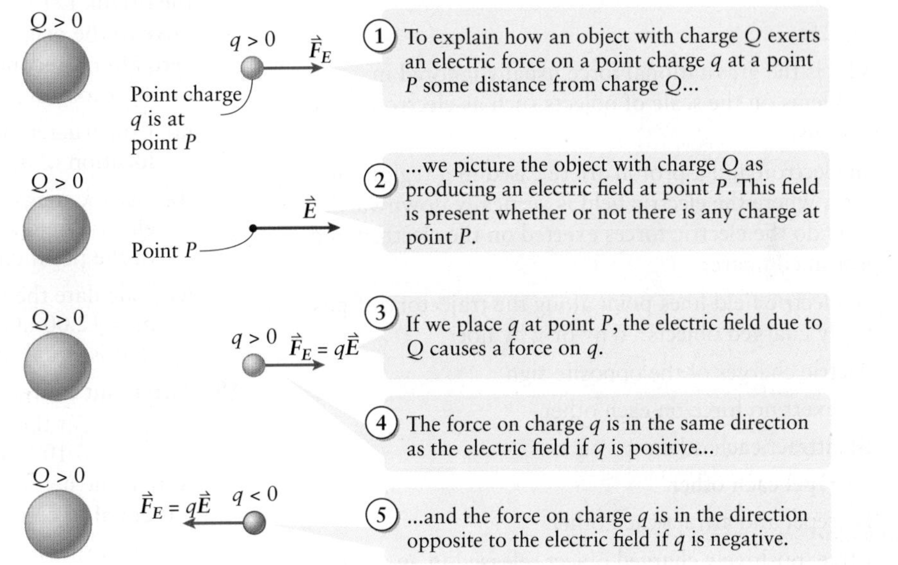

Q > 0
q > 0
$\vec{F}_E$
Point charge q is at point P
① To explain how an object with charge Q exerts an electric force on a point charge q at a point P some distance from charge Q...
Q > 0
$\vec{E}$
Point P
② ...we picture the object with charge Q as producing an electric field at point P. This field is present whether or not there is any charge at point P.
Q > 0
q > 0
$\vec{F}_E = q\vec{E}$
③ If we place q at point P, the electric field due to Q causes a force on q.
④ The force on charge q is in the same direction as the electric field if q is positive...
Q > 0
$\vec{F}_E = q\vec{E}$
q < 0
⑤ ...and the force on charge q is in the direction opposite to the electric field if q is negative.

(Figure 16-8)

Chapter 16 Review Problems

1. The nucleus of a copper atom has 29 protons and 35 neutrons. What is the total charge of the nucleus?
2. How many electrons must be transferred from an object to produce a charge of –1.61 C?
3. After combing your hair with a plastic comb, you find that when you bring the comb near a small bit of paper, the bit of paper moves toward the comb. Then, shortly after the paper touches the comb, it moves away from the comb. Explain these observations.
4. When you remove socks from a hot dryer, they tend to cling to everything. Two identical socks, however, usually repel. Why?
5. When the demonstrations in Figure 16-2 are done in the spring they may be less successful than when they are done in the winter. Evaluate the validity of each of the following explanations. If the claim is invalid, or if it is valid but with insufficient evidence supporting the claim, please modify the claim.
   (a) In the spring, the electric force between two point charges may be smaller.
   (b) In the spring, the amount of water vapor in the air is higher, and the friction force exerted between the fur and the rubber rod is smaller so that fewer protons in the rod are converted to electrons.
   (c) In the winter, there is more electricity in the air to be transferred to the rod when the friction force is exerted between the rod and the fur.
   (d) In the winter, the amount of water vapor in the air is lower, so that there is less water on the surface of the rods that distributes electrons by conduction through the water.
   (e) In the spring, the amount of water vapor in the air is higher, so that there is more water insulating the rods.

6. If two uncharged objects are rubbed together and one of them acquires a negative charge, then the other one
   (A) remains uncharged.
   (B) also acquires a negative charge.
   (C) acquires a positive charge.
   (D) acquires a positive charge equal to twice the negative charge.
7. Metal sphere *A* has a charge of $-Q$. An identical metal sphere *B* has a charge of $+2Q$. The magnitude of the electric force on *B* due to *A* is *F*. The magnitude of the electric force on *A* due to *B* is
   (A) $F/4$.
   (B) $F/2$.
   (C) F.
   (D) $2F$.
8. Why is the gravitational force usually ignored in problems on the scale of objects such as electrons and protons?
9. An electron and a proton are released in a region of space where the electric field is vertically downward. How do the electric forces exerted on the electron and proton compare?
10. Do electric field lines point along the trajectory of positively charged objects? Why or why not?
11. Electric charges of the opposite sign
    (A) exert no force on each other.
    (B) attract each other.
    (C) repel each other.
    (D) repel and attract each other.
12. A free, positively charged object released in an electric field will
    (A) remain at rest.
    (B) accelerate in the direction opposite to the electric field.
    (C) accelerate in the direction perpendicular to the electric field.
    (D) accelerate in the same direction as the electric field.
13. A red blood cell may carry an excess charge, *Q*, of about -2.5×10^{-12} C distributed uniformly over its surface. The cells, modeled as spheres, are approximately 7.5 μm in diameter and have a mass of 9.0×10^{-14} kg.
    (a) How many excess electrons does a typical red blood cell carry?
    (b) Do these extra electrons appreciably affect the mass of the cell?
14. Point charge A with a charge of 3.0 nC is located at the origin. Point charge B with a charge of −2.0 nC is located on the *x* axis at $x = 7.00$ cm. Calculate the force exerted on point charge A by point charge B (magnitude and direction).
15. A point charge q_1 equal to 0.600 nC is at the origin, and a second point charge q_2 equal to 0.800 nC is on the *x* axis at $x = 5.00$ cm.
    (a) Calculate the force (magnitude and direction) that each point charge exerts on the other.
    (b) Predict the forces on q_1 when the second charge is changed: (i) $q_2 = -0.800$ nC and (ii) $q_2 = -3.200$ nC.
16. A 13.00-nC point charge is located at the origin, and a 26.00-nC point charge is located on the *x*–*y* plane at (6.00 m, 4.00 m). Determine the electric force exerted by the 26.00-nC point charge on the 13.00-nC point charge.
17. A point charge with charge $q_1 = -3.00$ nC is fixed at the origin. Let x_q be the displacement of a point relative to the origin where the electric field is equal to zero after a second positive point charge q_2 is placed on the *x* axis at x_2.
    (a) Construct a diagram of the situation showing the location of q_1, q_2, and x_q, labeling all variables.
    (b) Use the diagram to support the claim that if the electric field at x_q is equal to zero, the position x_2 of the point charge q_2 must satisfy $x_2 > x_q$.
    (c) Calculate the position of the second point charge $q_2 = 12.00$ nC that results in an electric field at x_q that is equal to zero.
18. Two point charges lie on the *x* axis: a 24.00-nC point charge, q_1, at the origin and a −6.00-nC point charge, q_2, at $x = 0.100$ m. A third point charge will be placed so that the net force on the third positive point charge *q* is equal to zero.
    (a) Construct a diagram of the situation showing the location of q_1 and q_2, labeling both variables and the distance between them.
    (b) Use the diagram to support the claim that the third point charge cannot be placed at any point between the origin and 0.100 m.
    (c) Explain why the magnitude of the third point charge does not appear in your reasoning in part (b).
    (d) Support the claim that the field is zero at the point $x = 0.200$ m.
19. When a 15.00-nC point charge is placed at a certain point, the force exerted on it has a magnitude of 8.000×10^{-5} N.
    (a) Calculate the electric field at that point and describe the direction of the force in terms of the direction of the electric field.
    (b) If the charge were −22.00 nC instead, calculate the magnitude and direction of the force that would be exerted on the point charge.
20. Two small spheres each have a mass *m* of 0.100 g and are suspended as pendulums by light insulating strings from a common point, as shown in the figure. The spheres are given the same electric charge, and the two

come to equilibrium when each string is at an angle of $\theta = 3.00°$ with the vertical. If each string is 1.00 m long, what is the magnitude of the charge on each sphere?

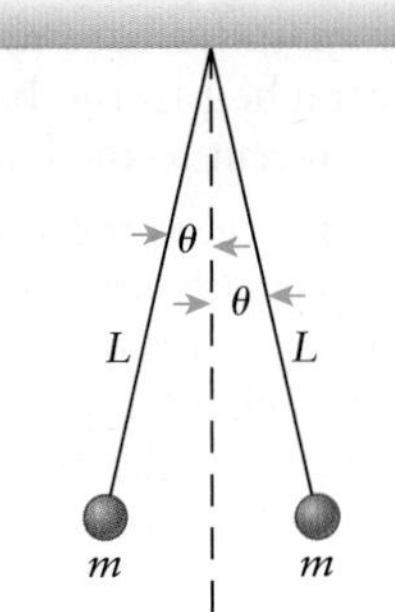

21. Three point charges are placed on the x–y plane: $q_1 = +50.0 \times 10^{-9}$ C at the origin, $q_2 = -50.0 \times 10^{-9}$ C on the x axis at 10.0 cm, and $q_3 = +150 \times 10^{-9}$ C at the point (10.0 cm, 8.00 cm).
    (a) Find the components of the total electric force on the 150-nC point charge due to the other two point charges.
    (b) What are the components of the electric field at the location of the 150-nC point charge due to the presence of the other two point charges?
22. In the Bohr model, the hydrogen atom consists of an electron in a circular orbit of radius $a_0 = 5.29 \times 10^{-11}$ m around the nucleus. Using this model, and ignoring relativistic effects, what is the speed of the electron?
23. An electron, released in a region where the electric field is uniform, is observed to have an acceleration of 3.00×10^{14} m/s^2 in the positive x direction.
    (a) Determine the electric field producing the acceleration.
    (b) Assuming the electron is released from rest, determine the time required for it to reach a speed of 11,200 m/s, the escape speed from Earth's surface.
24. Most workers in nanotechnology are actively monitored for excess static charge buildup. The human body acts like an insulator as one walks across a carpet, collecting −50 nC per step.
    (a) What charge buildup will a worker in a manufacturing plant accumulate if she walks 25 steps?
    (b) How many electrons are present in that amount of charge?
    (c) If a delicate manufacturing process can be damaged by an electrical discharge of greater than 10^{12} electrons, what is the maximum number of steps that any worker should be allowed to take before touching the components?
25. An electron with an initial speed of 5.00×10^5 m/s enters a region in which there is an electric field directed along its direction of motion. If the electron travels 5.00 cm in the field before being stopped, what are the magnitude and direction of the electric field?
26. Point charge A with a charge of +3.00 μC is located at the origin. Point charge B with a charge of +6.00 μC is located on the x axis at $x = 7.00$ cm. Point charge C with a charge of +2.00 μC is located on the y axis at $y = 6.00$ cm. What is the net force (magnitude and direction) exerted on each point charge by the others?
27. In the famous Millikan oil-drop experiment, tiny spherical droplets of oil are sprayed into a uniform vertical electric field. The drops get a very small charge (just a few electrons) due to friction with the atomizer as they are sprayed. The field is adjusted until one drop (which is viewed through a small telescope) is just balanced against gravity and therefore remains stationary. Using the measured value of the electric field, we can calculate the charge on the drop and from this calculate the charge e of the electron. In one apparatus the drops are 1.10 μm in diameter, and the oil has a density of 0.850 g/cm^3.
    (a) If the drops are negatively charged, which way should the electric field point to hold them stationary (up or down)? Support your answer using physics principles.
    (b) If a certain drop contains four excess electrons, what magnitude electric field is needed to hold it stationary?
    (c) You measure a balancing field of 5183 N/C for another drop. How many excess electrons are on this drop?
28. Three point charges (q_A, q_B, and q_C) are placed at the vertices of the equilateral triangle that has sides of length s in the figure below.

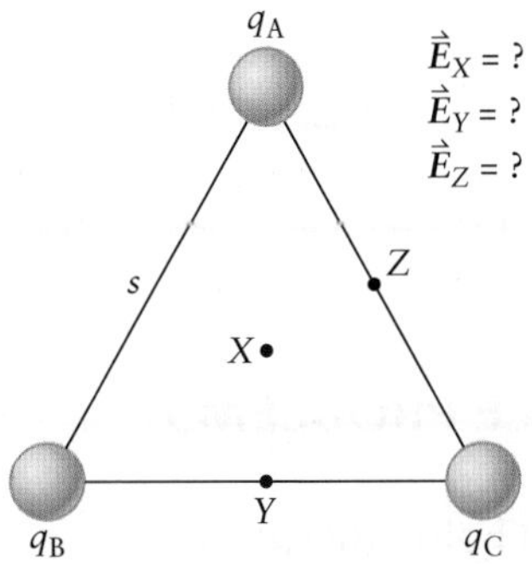

    (a) Derive expressions for the electric field at (i) X (at the center of the triangle), (ii) Y (at the midpoint of the side between q_B and q_C), and (iii) Z (at the midpoint of the side between q_A and q_C).
    (b) Now use the following numerical values and calculate the electric field at those same points: $s = 10.0$ cm, $q_A = +20.0$ nC, $q_B = -8.00$ nC, and $q_C = -10.0$ nC.

Prep for the AP® Exam

AP® Group Work

Directions: This problem is designed to be done as group work in class.

A simple Van de Graaff generator is based on charging by contact of dissimilar materials, causing the transfer of some electrons from one material to the other. One version consists of a belt of rubber moving over two rollers, one of which is surrounded by a hollow metal sphere. An electrode, a comb-shaped row of sharp metal points, is connected to the sphere over the upper roller, which is made of acrylic. The rubber of the belt will become negatively charged while the acrylic glass of the upper roller will become positively charged. The belt carries away negative charge on its inner surface while the upper roller accumulates positive charge. The strong electric field surrounding the positive upper roller induces a very high electric field near the points of the nearby electrode. At the points, the field becomes strong enough to ionize air molecules, and the electrons are attracted to the outside of the belt while positive ions go to the electrode, where they are neutralized by electrons that were on the electrode, thus leaving the comb and the attached outer shell with fewer net electrons. Because the shell is a conductor, the excess positive charge is accumulated on the outer surface of the outer shell. Near the bottom, the lower roller is also a conductor and connected by a wire to the ground, so it removes some of the excess negative charge from the belt. As the belt rotates, the conducting shell becomes increasingly depleted of electrons.

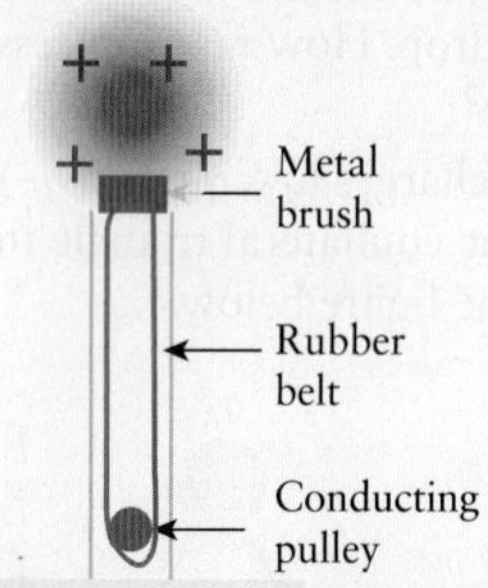

A second, smaller, light metal shell is supported by a vertical, flexible conducting rod. The conducting rod is inserted into a heavy metallic base and a wire from the conducting pulley is attached to the base. This supported second shell is placed adjacent to the larger shell.

(a) As the charging proceeds the small shell begins to oscillate: It is repeatedly attracted, contacted, and repelled by the large shell of the Van de Graaff generator. Predict what happens in terms of the distance between the shells, the signs of the charges on each shell, and the path of electrons through the system.

Sometimes as a dramatic demonstration, a second metallic cylindrical shell supported by a rigid insulating column is connected by the wire to the conducting pulley.

(b) Predict what will happen by describing the distribution of electrons and comparing the outcome with what happens when you touch a metallic doorknob after walking over a carpet.

Another dramatic classroom demonstration involves two (or more) students standing on an insulating material. One student places one hand on the uncharged shell of the Van de Graaff. The first student's hand clasps the hand of the second student.

(c) Predict what would happen if, when the shell was fully charged, (i) the first student released contact with the shell while the two students maintained contact, and (ii) both students maintained contact with each other while the second student stepped from the insulating material to the floor.

AP® PRACTICE PROBLEMS

Prep for the AP® Exam

Multiple-Choice Questions

Directions: The following questions have a single correct answer.

1. Two point charges, with charges Q and q, are placed a distance d apart and exert a repulsive force, F, on each other. Of the following changes, which would keep the force the same?
   (A) The first charge becomes $-Q$ and the other $-2q$ and the separation is doubled to $2d$.
   (B) Both charges are doubled and the separation is doubled.
   (C) The first charge is made $4Q$ and the other remains q but the separation is increased by a factor of 4.
   (D) Both charges are halved and the separation is doubled.

(Continued)

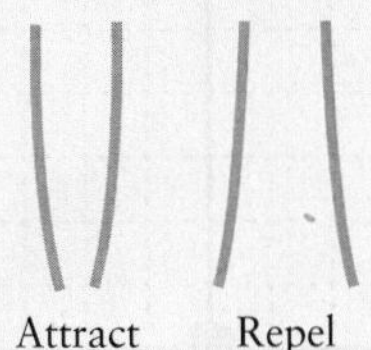

2. Students prepare two pairs of tape "sandwiches" as follows. A strip of clear plastic tape is put sticky side down on the surface of a table with one end folded over to form a small tab. A second strip is then placed sticky side down on top of the first, again leaving a small tab folded over at the end so the strips can be separated. Each of the two pairs is then peeled off the table and then the strips are labeled and separated into "Top" and "Bottom" strips as shown in the figure above. It is noticed that the strips seem to be electrically charged, so the students experiment to see what is going on. They hang each of the strips from one end and watch each strip's behavior with other strips and with an uncharged piece of cardboard. Which of the following is likely to describe the results of the experiments?
   (A) A "Top" strip and a "Bottom" strip will repel each other, but each will be attracted to an uncharged piece of cardboard.
   (B) A "Top" strip and a "Bottom" strip will attract each other, and neither will be attracted to an uncharged piece of cardboard.
   (C) Two "Top" strips will attract each other and will not be attracted to the uncharged piece of cardboard.
   (D) A "Top" strip and a "Bottom" strip will attract each other, and both will be attracted to the uncharged cardboard.

3. Suppose in a different system the unit of charge is the tb. A team of scientists measures the charge on five oil drops and finds them to be 3.0 μ_{tb}, 4.2 μ_{tb}, 5.4 μ_{tb}, and 6.6 μ_{tb}.
   From their data the scientists would have an upper limit on the size of an elementary charge as
   (A) 0.3 μ_{tb}.
   (B) 0.6 μ_{tb}.
   (C) 3.0 μ_{tb}.
   (D) 1.2 μ_{tb}.

Questions 4–5 refer to the following figure.

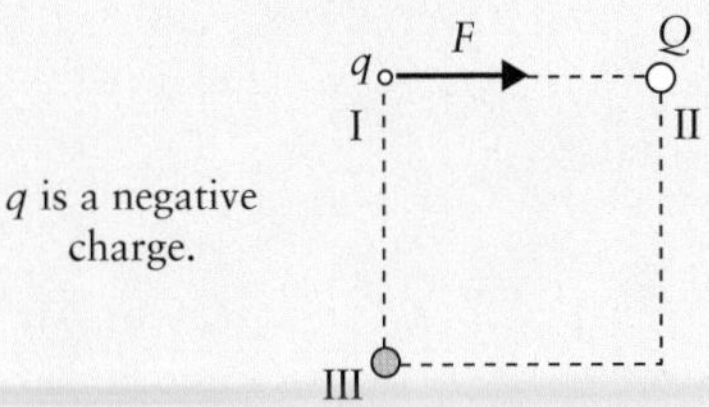

4. Three small conducting objects are placed at the corners of a very large square, as shown. The object at corner I has charge $q < 0$, and the one at II has charge Q. The object at III is initially neutral. At the start the charges at I and II attract with a force F. The neutral conducting object at III is then brought into contact with the one at II and returned to its original position. The result is that the force between the charges at I and II is still attractive but is reduced to $\frac{F}{4}$. What is now the charge on III?
   (A) $\frac{Q}{8}$ (B) $\frac{Q}{4}$ (C) $\frac{Q}{2}$ (D) $\frac{3Q}{4}$

5. Consider three charged conducting objects placed at the corners of a very large square, as shown. The charge on the object at I is $q < 0$. The force between the objects at I and II is attractive and of magnitude F. The object at III is identical in size to that at II but has a different charge. It is now brought into contact with the object at II and then returned to its original position. The force F between the objects at I and II is still the same size but is now repulsive. What was the original charge on the object at III?
   (A) $-Q$ (B) $-2Q$ (C) $-3Q$ (D) $-4Q$

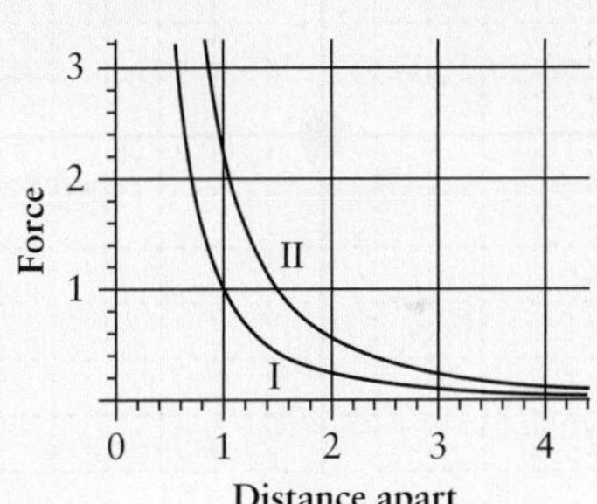

6. The force between two identical charged objects, which are small compared to their separation, is measured for different distances of separation. A graph using arbitrary units of both force and distance is plotted and labeled I as shown above. Then both charged objects have the same additional amount of charge added to each and the force and distance are again measured and plotted as graph II. Of the following which best describes the new charge on each?
   (A) 1.25 times greater
   (B) 1.50 times greater
   (C) 2.0 times greater
   (D) 2.25 times greater

(Continued)

Free-Response Question

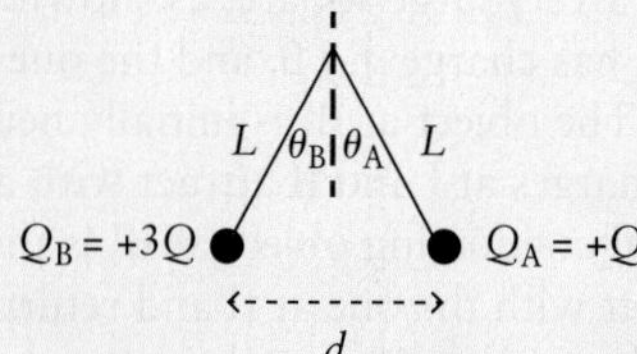

Two identical small conducting spheres of mass M have charges $Q_A = +Q$ and $Q_B = +3Q$. They hang in equilibrium by insulating threads from the same point. The angles that the threads make with the vertical are labeled θ_A and θ_B, respectively, as shown. Each thread has length L and the spheres are separated by distance d.

(a) On the dots that represent the spheres on the figure below, draw all of the forces on each sphere. The lengths of the arrows should be to scale and reflect the motion of each sphere.

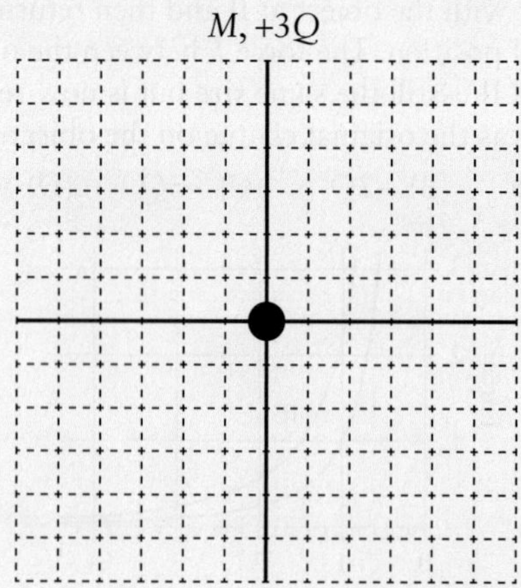

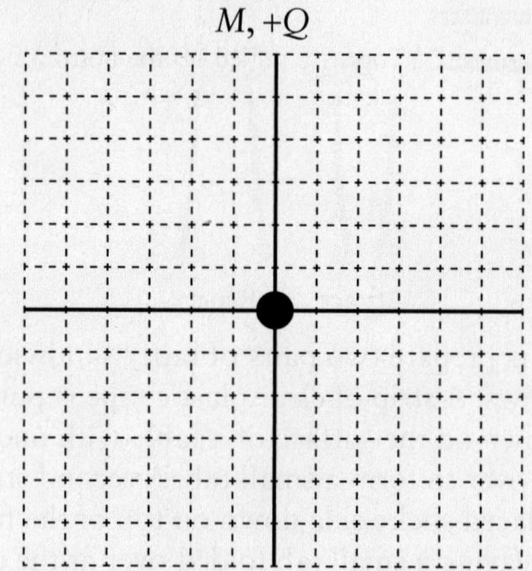

(b) (i) Calculate the magnitude of the force that sphere A initially exerts on sphere B.

(ii) The two spheres are then brought into contact briefly and then separated and the strings are allowed to come to equilibrium again. Find the new charge on each sphere.

(iii) Assume the length of the strings is adjusted so the spheres have the same separation as before. Calculate the ratio of the magnitude of the force that sphere A now exerts on sphere B to the magnitude of the force sphere A exerted on sphere B before they were brought into contact.

(c) When the spheres are in equilibrium, how do the two angles that the strings make compare to each other?

(i) $\theta_A < \theta_B$ ______

(ii) $\theta_A > \theta_B$ ______

(iii) $\theta_A = \theta_B$ ______

Support your claim, using physical principles.

17 Electric Potential and Electric Potential Energy

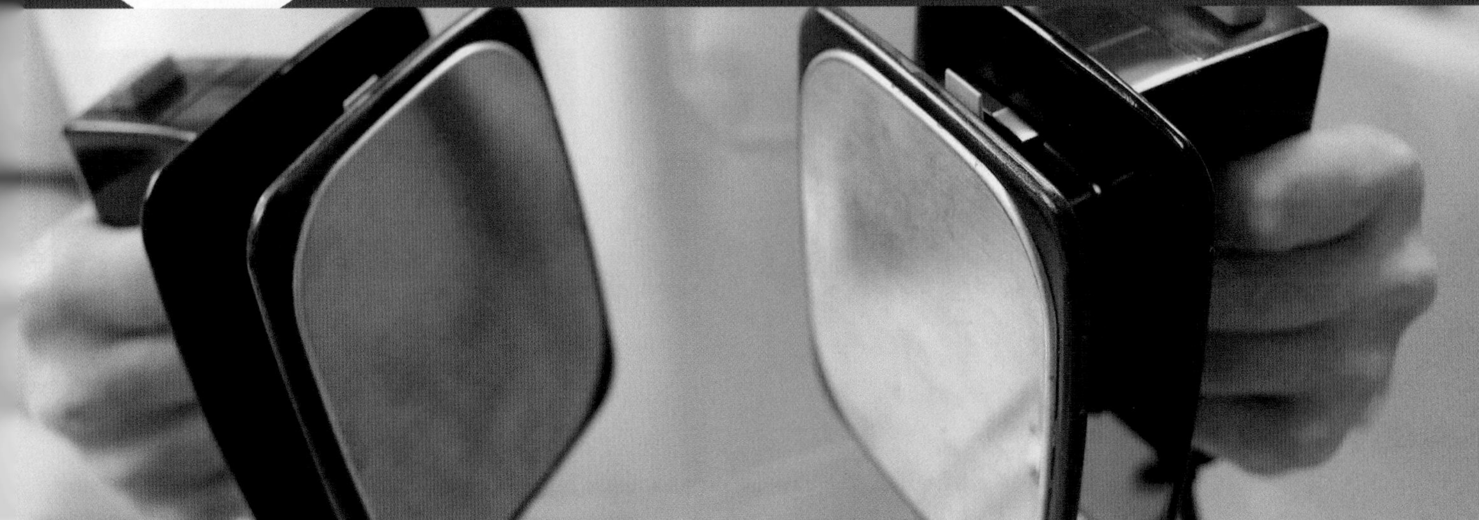

Racha Phuangpoo/Shutterstock

YOU WILL LEARN TO:

- Explain the significance of energy in electrostatics.
- Describe electric potential difference in terms of charge and energy.
- Discuss how the work done on a charged object by the electric field relates to changes in electric potential energy.
- Explain the difference between electric potential and electric potential energy.
- Use electric potential energy in conservation of energy relationships.
- Recognize why equipotential surfaces are perpendicular to electric field lines.
- Explain what is meant by capacitance and describe how the capacitance of a parallel-plate capacitor depends on the size of the plates and their separation.
- Calculate the electric energy stored in a capacitor.
- Describe how the capacitance of a capacitor increases when an insulating material other than a vacuum is placed between the plates of the capacitor.
- Describe the equivalent capacitance of multiple capacitors connected in series and/or parallel.

17-1 Electric energy is important in nature and for technology; electric and gravitational potential energy have similar forms

Many people are familiar with the idea that electricity implies energy. The energy to run your computer, lighting, and television is delivered to your home by electric circuits (which we will study in the next chapter). A bolt of lightning releases so much energy that it heats air to a temperature of approximately 30,000°C and strips electrons from atoms, causing the air to glow with the characteristic light that we call a lightning flash (see Figure 16-1b). A defibrillator saves lives by delivering a sharp punch of electric energy to a malfunctioning heart (see the photo that opens this chapter). When you purchase an ordinary electric battery, you are really paying for the electric energy that the battery can deliver to whatever electric circuit you plug it into (**Figure 17-1a**). Some specialized species of fish such as electric rays (**Figure 17-1b**) are equipped with organic

Figure 17-1 Electric energy Reservoirs of electric energy in (a) technology and (b) nature.

(a) An ordinary battery is a source of electric energy. A battery is made up of one or more electrochemical cells. Devices that use multiple batteries or larger batteries have greater energy requirements.

(b) An electric ray (order *Torpediniformes*) is equipped with a large number of organic batteries that work together to deliver intense electric shocks.

[a] iStockphoto/Thinkstock; [b] NHPA/Superstock

batteries; they can deliver an intense burst of electric energy to stun their prey or ward off predators.

There are similarities between electric and gravitational potential energy. Just as there is a change in gravitational potential energy when an object with mass moves up or down in the presence of Earth's gravitational field, there is a change in electric potential energy when a charged object moves along with, or opposite to, an electric field. In this chapter, we will introduce the useful concept of electric potential, or electric potential energy per unit charge, and electric potential difference, which is the difference in the value of electric potential at two positions. We'll also learn about *equipotential surfaces*, which are surfaces on which the electric potential has the same value at every point.

An important device for storing electric energy is a *capacitor*. In its simplest form a capacitor is just two pieces of metal, called capacitor plates, placed close to each other but not in contact. We'll examine the key properties of capacitors, including how to combine two or more of them for even greater energy storage. We'll conclude the chapter by seeing how capacitors can be made even more effective by inserting an insulating material between the plates. In Chapter 18 we'll learn how batteries like those shown in Figure 17-1a provide electric energy to the components of an electric circuit.

THE TAKEAWAY for Section 17-1

✔ Electric potential energy associated with a point charge is analogous to the gravitational potential energy associated with an object with mass.

✔ Electric potential is electric potential energy per charge.

✔ A capacitor is a device for storing electric energy.

Prep for the AP® Exam

AP® Building Blocks

1. A ball dropped near Earth's surface always falls down toward the surface due to gravity.
   (a) What happens to the gravitational potential energy of the object–Earth system as the object falls toward Earth?
   (b) If instead of gravity, the force pulling objects toward Earth were electric, what sign of charge would be needed on Earth if the objects dropped were positively charged?
   (c) What must happen to the potential energy of a system consisting of oppositely charged objects as they attract each other? Support your claim by comparing this system to the gravitational interaction in (a).

17-2 Electric potential energy of a system changes when a charged object moves in an electric field

We learned in Section 7-6 that Earth's gravitational force is a conservative force: As an object moves from one position to another, the work done on the object by the gravitational force depends only on where the object starts and where it ends up, not on how it gets there. We learned that if we put Earth in the system we could instead talk about the change in potential energy of the system of Earth and the object.

Like gravity, the electric force due to the interaction between charged objects is a conservative force. This means we can express the work done by the electric force in terms of a change in electric potential energy if we put the source of the force (other charged objects) in the system.

To be specific, suppose an object of mass m moves near Earth's surface, so the acceleration $\vec{g}$ due to Earth's gravity and the gravitational force $\vec{F} = m\vec{g}$ on the object are the same at all positions. The object's displacement is $\vec{d}$, and that displacement is at an angle θ to the direction of the force $\vec{F} = m\vec{g}$ and so at the same angle θ to the direction of $\vec{g}$ (**Figure 17-2a**). Then the work done by gravity is $W = Fd\cos\theta = mgd\cos\theta$ when Earth is not in the system. Instead, if we put Earth in the system, instead of talking about work, which is energy added by something external to the system, we can describe the change in gravitational potential energy for the same change in position as

$$\Delta U_{\text{grav}} = -mgd\cos\theta \tag{17-1}$$

Figure 17-2b shows three special cases. If the object moves a distance d straight upward, $\theta = 180°$ and $\cos\theta = \cos 180° = -1$. In this case we can describe the changes in energy as either the downward gravitational force does negative work on the object, or the gravitational potential energy increases: $\Delta U_{\text{grav}} = -mgd(-1) = +mgd$. If the object moves a distance d straight downward, in the direction of the gravitational force, $\theta = 0$ and $\cos\theta = \cos 0° = 1$. Then gravity does positive work on the object or the gravitational potential energy decreases: $\Delta U_{\text{grav}} = -mgd(+1) = -mgd$. And if the object moves horizontally, gravity does zero work and there is no change in gravitational potential energy: $\theta = 90°$ and $\Delta U_{\text{grav}} = -mgd\cos 90° = 0$.

Like gravity, the electric force due to the interaction between charged objects is a conservative force. You can see this easily for the special case in which an object moves in a region of space where there is a *uniform* electric field $\vec{E}$—that is, where the

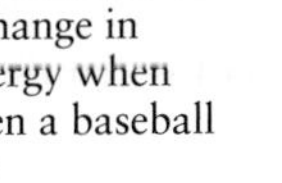
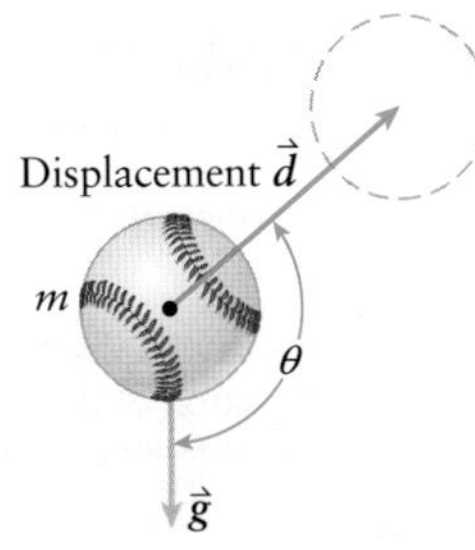
(a) Work done by gravity when Earth is not in the system and change in gravitational potential energy when Earth is in the system when a baseball undergoes a displacement
Displacement $\vec{d}$
m
θ
$\vec{g}$
Work done by gravity:
$W_{\text{grav}} = mgd\cos\theta$
Change in gravitational potential energy:
$\Delta U_{\text{grav}} = -mgd\cos\theta$

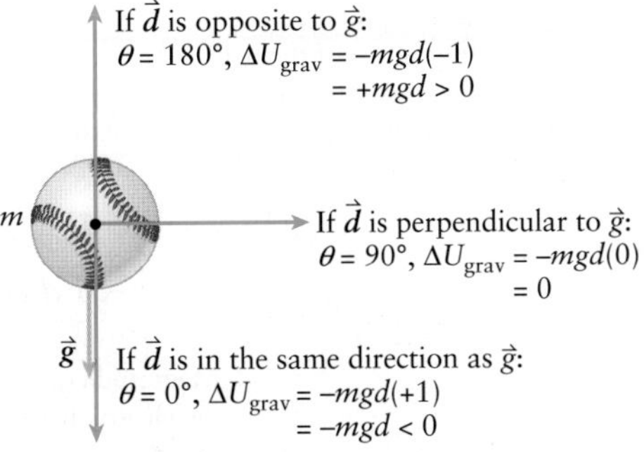
(b) Three special cases
If $\vec{d}$ is opposite to $\vec{g}$:
$\theta = 180°$, $\Delta U_{\text{grav}} = -mgd(-1) = +mgd > 0$
m
If $\vec{d}$ is perpendicular to $\vec{g}$:
$\theta = 90°$, $\Delta U_{\text{grav}} = -mgd(0) = 0$
$\vec{g}$
If $\vec{d}$ is in the same direction as $\vec{g}$:
$\theta = 0°$, $\Delta U_{\text{grav}} = -mgd(+1) = -mgd < 0$

Figure 17-2 Change in gravitational potential energy When an object of mass m moves near Earth, the change in energy can be described by either the gravitational force doing work on it (Earth not in the system) or a change in gravitational potential energy (Earth in the system).

(a) Work done by electric force when source is not in the system and change in electric potential energy when source is in the system when a charged object undergoes a displacement

(b) Three special cases for a positively charged object

(c) Three special cases for a negatively charged object

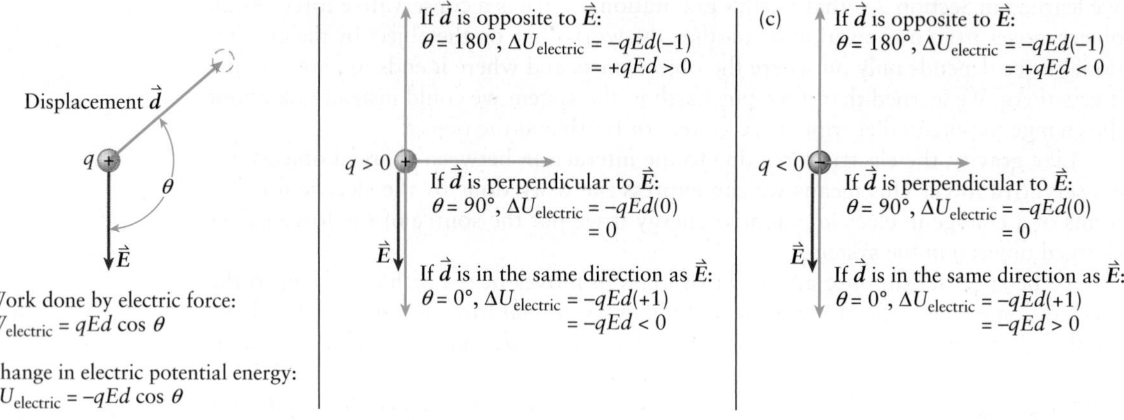

Displacement $\vec{d}$
q
θ
$\vec{E}$
Work done by electric force:
$W_{\text{electric}} = qEd\cos\theta$
Change in electric potential energy:
$\Delta U_{\text{electric}} = -qEd\cos\theta$
If $\vec{d}$ is opposite to $\vec{E}$:
$\theta = 180°$, $\Delta U_{\text{electric}} = -qEd(-1) = +qEd > 0$
$q > 0$
If $\vec{d}$ is perpendicular to $\vec{E}$:
$\theta = 90°$, $\Delta U_{\text{electric}} = -qEd(0) = 0$
If $\vec{d}$ is in the same direction as $\vec{E}$:
$\theta = 0°$, $\Delta U_{\text{electric}} = -qEd(+1) = -qEd < 0$
(c)
If $\vec{d}$ is opposite to $\vec{E}$:
$\theta = 180°$, $\Delta U_{\text{electric}} = -qEd(-1) = +qEd < 0$
$q < 0$
If $\vec{d}$ is perpendicular to $\vec{E}$:
$\theta = 90°$, $\Delta U_{\text{electric}} = -qEd(0) = 0$
If $\vec{d}$ is in the same direction as $\vec{E}$:
$\theta = 0°$, $\Delta U_{\text{electric}} = -qEd(+1) = -qEd > 0$

Figure 17-3 Change in electric potential energy When an object of charge q moves in a uniform electric field $\vec{E}$, the change in energy of the system can be expressed as the electric force doing work on the system of just the charged object or as a change in electric potential energy if the source of the field is also in the system (compare Figure 17-2).

value of $\vec{E}$ is the same at all positions (**Figure 17-3**). The electric force on an object of charge q is the same everywhere in this region and equal to $\vec{F} = q\vec{E}$. That's exactly like the situation in Figure 17-2: In a region where $\vec{g}$ is the same at all positions, the gravitational force on an object of mass m is the same everywhere and equal to $\vec{F} = m\vec{g}$. This uniform gravitational force is conservative, so the force $\vec{F} = q\vec{E}$ due to a uniform electric field must be conservative as well. Hence, we can express the work done by the electric force in terms of a change in **electric potential energy.**

Electric Potential Energy in a Uniform Field

This analogy between gravitational force and electric force tells us how to write the change in electric potential energy for an object of charge q that undergoes a displacement $\vec{d}$ in the presence of a uniform electric field $\vec{E}$ (**Figure 17-3a**). Following the same steps that we used to find Equation 17-1 above, you can see that if θ is the angle between the directions of $\vec{d}$ and $\vec{E}$, then the work done on the charged object by the electric force $\vec{F} = q\vec{E}$ is $W_{\text{electric}} = qEd\cos\theta$. The change in electric potential energy of the system, including the charged object and the source of the field, equals the negative of the work done on the charged object by the field when the system is just the charged object:

EQUATION IN WORDS
Electric potential energy for a charged object in a uniform electric field

The **change in electric potential energy** for a system of the source of the field and an object of charge q that moves in a uniform electric field $\vec{E}$...

...equals the **negative of the work done** on the object by the electric force when the source of the field is external to the system.

$$\Delta U_{\text{electric}} = -W_{\text{electric}} = -qEd\cos\theta \qquad (17\text{-}2)$$

Charge of the object

Angle between the displacement and the direction of the electric field

Magnitude of the electric field

Displacement of the object

Just as for the gravitational case, Equation 17-2 tells us how much the electric potential energy *changes* when the charged object is displaced as shown in Figure 17-3a. Note that to use electric potential energy the system must include the charged object q and the charged objects that produce the electric field $\vec{E}$. If you know the value of the field, you do not need to worry about the objects causing it, except that you must know that they do not move or change their charges in any way (otherwise, the field would change).

Suppose q is positive, so the force $\vec{F} = q\vec{E}$ on the charged object is in the *same* direction as the uniform electric field $\vec{E}$ (**Figure 17-3b**). If this positive charge moves a distance d opposite to $\vec{E}$, then $\theta = 180°$ and $\cos\theta = -1$. In this case the electric force does negative work on the charge, or $\Delta U_{\text{electric}} = (-qEd)(-1) = +qEd$ is positive, and the electric potential energy of the system of charged object and the source of the field increases. That's exactly what happens to *gravitational* potential energy when an object with mass moves upward from Earth's surface, in the direction opposite to the gravitational force (Figure 17-2b). If the positive charge q instead moves in the same direction as the electric field, and so in the same direction as the electric force, then $\theta = 0$ and $\cos\theta = +1$. Then the electric force does positive work on the object, or $\Delta U_{\text{electric}} = (-qEd)(+1) = -qEd$ is negative and the electric potential energy of the system of charged object and sources of the field decreases. That's the same thing that happens when an object with mass falls toward Earth's surface, in the same direction as the gravitational force on the object: Gravitational potential energy decreases (Figure 17-2b).

For an object with *negative* charge ($q < 0$), the force $\vec{F} = q\vec{E}$ on the object is in the direction *opposite* to the electric field $\vec{E}$ (**Figure 17-3c**). (The electric field is caused by objects not included in our system.) If such a negatively charged object moves a distance d opposite to the direction of $\vec{E}$ so that $\theta = 180°$ and $\cos\theta = -1$ in Equation 17-2, the electric force does *positive* work on the charge because the force and displacement are in the same direction. For this same motion of a negative charge, if we instead put the sources of the field into the system we are considering, $\Delta U_{\text{electric}} = +qEd$. When the sources are in the system, we talk about potential energy instead of work. Then, because $q < 0$ is *negative*, the electric potential energy of the system decreases. In the same way, if a negative charge moves a distance d in the direction of $\vec{E}$ so that $\theta = 0$ and $\cos\theta = +1$ in Equation 17-2, the electric force does *negative* work on the object (the electric force is opposite to E and so opposite the displacement). The electric potential energy change of the system including the sources for the case of motion in the direction of the field is $\Delta U_{\text{electric}} = (-qEd)(+1) = -qEd$, which is positive because q is negative. So the electric potential energy of the system increases. These results are exactly opposite to what happens for a positively charged object that moves a distance d in the direction of the electric field.

EXAMPLE 17-1 Electric Potential Energy Difference in a Uniform Field

An electron (charge $q = -e = -1.60 \times 10^{-19}$ C, mass $m = 9.11 \times 10^{-31}$ kg) is released from rest in a uniform electric field. The field points in the positive x direction and has magnitude 2.00×10^{2} N/C. Find the speed of the electron after it has moved 0.300 m.

Set Up

The only force exerted on the electron of charge q is the conservative electric force. If we consider the source of the field as part of the system, we can use conservation of energy to solve this problem. We'll find the change in electric potential energy using Equation 17-2 and from this find the change in kinetic energy of the electron. The electron's initial kinetic energy is zero because it starts at rest; once we know the final kinetic energy, we can determine the electron's final speed.

The electron has a negative charge, so the force $\vec{F} = q\vec{E}$ is directed opposite to the electric field $\vec{E}$, and the electron will move in the direction opposite to $\vec{E}$ when released from rest. So the angle in Equation 17-2 between electric field and displacement will be $\theta = 180°$.

Conservation of mechanical energy:

$$K_i + U_{\text{electric,i}} = K_f + U_{\text{electric,f}} \quad (8\text{-}3)$$

Electric potential energy change for a charge in a uniform electric field:

$$\Delta U_{\text{electric}} = -qEd\cos\theta \quad (17\text{-}2)$$

Kinetic energy:

$$K = \frac{1}{2}mv^2 \quad (7\text{-}8)$$

Solve

Use Equation 17-2 to solve for the change in electric potential energy.

The change in electric potential energy equals the difference between the final and initial electric potential energy:

$$\Delta U_{\text{electric}} = U_{\text{electric,f}} - U_{\text{electric,i}}$$

From Equation 17-2,

$$\begin{aligned}\Delta U_{\text{electric}} &= U_{\text{electric,f}} - U_{\text{electric,i}} \\ &= -qEd\cos\theta \\ &= -(-1.60\times10^{-19}\text{ C})(2.00\times10^{2}\text{ N/C})(0.300\text{ m})\cos 180^\circ \\ &= -(-9.60\times10^{-18}\text{N}\cdot\text{m})(-1) \\ &= -9.60\times10^{-18}\text{ N}\cdot\text{m} = -9.60\times10^{-18}\text{ J}\end{aligned}$$

Find the final kinetic energy and the final speed of the electron.

The initial kinetic energy of the electron is $K_i = 0$. Solve the energy conservation equation for the final kinetic energy of the electron:

$$\begin{aligned}K_f &= K_i + U_{\text{electric,i}} - U_{\text{electric,f}} = 0 - (U_{\text{electric,f}} - U_{\text{electric,i}}) \\ &= 0 - (-9.60\times10^{-18}\text{ J}) = 9.60\times10^{-18}\text{ J}\end{aligned}$$

Finally, solve for the final speed of the electron:

$$K_f = \frac{1}{2}mv_f^2 \text{ so}$$

$$v_f = \sqrt{\frac{2K_f}{m}} = \sqrt{\frac{2(9.60\times10^{-18}\text{ J})}{9.11\times10^{-31}\text{ kg}}} = 4.59\times10^{6}\text{ m/s}$$

Reflect

The electric potential energy decreases when we release the electron in the electric field, in the same way that the gravitational potential energy decreases when you release a ball in Earth's gravitational field. In both situations, the kinetic energy of the system increases by the same amount that the potential energy of the system decreases. If the distribution of charged objects causing the field is held in place, only the charged object in the field can have a change in kinetic energy. Although Earth is not held in place, the ball's kinetic energy is the only kinetic energy that changes, because Earth is so much larger and momentum is conserved.

You should verify that the force on the electron is very small, only 3.2×10^{-17} N in magnitude. The electron has only a tiny mass, however, and our results show that it acquires a ferocious speed (about 1.5% of the speed of light) after moving just 0.300 m.

NOW WORK Problem 4 from The Takeaway 17-2.

Electric Potential Energy of Point Charges

Electric potential energy is particularly important for the case of two *point charges* interacting with each other. An example from chemistry is the dissociation energy of an ionic compound such as sodium chloride (NaCl)—that is, the energy that must be given to the molecule to break it into its component atoms. The dissociation energy is determined in large part by the change in electric potential energy required to separate the positive sodium ion (Na^+) and the negative chloride ion (Cl^-), both of which behave much like point charges, as well as by the change in electric potential energy required to move an electron from the Cl^- ion to the Na^+ ion to make the atoms neutral.

Just as for a charge moving in a uniform electric field, we can use our knowledge of gravitation to find an expression for the electric potential energy of two point charges, q_1 and q_2, separated by a distance r. In Section 6-4 we saw the following expression for the attractive *gravitational* force between two objects with *masses* m_1 and m_2 separated by a distance r:

$$F_{1 \text{ on } 2} = F_{2 \text{ on } 1} = \frac{Gm_1m_2}{r^2} \quad (6\text{-}7)$$

Gravitational constant (same for any two objects)

Masses of the two objects

Any two objects (1 and 2) exert **equally strong gravitational forces** on each other.

Center-to-center distance between the two objects

The gravitational forces are attractive: $\vec{F}_{1 \text{ on } 2}$ accelerates object 2 toward object 1 and $\vec{F}_{2 \text{ on } 1}$ accelerates object 1 toward object 2.

EQUATION IN WORDS
Newton's law of universal gravitation

The gravitational potential energy of these two objects is given by Equation 8-8 in Section 8-5:

$$U_{\text{grav}} = -\frac{Gm_1m_2}{r} \quad (8\text{-}8)$$

Gravitational constant (same for any two objects)

Masses of the two objects

Gravitational potential energy of a system of two objects (1 and 2)

Center-to-center distance between the two objects

The gravitational potential energy of a system of two objects is zero when those objects are infinitely far apart. If the objects are brought closer together (so r is made smaller), U_{grav} decreases (it becomes more negative).

EQUATION IN WORDS
Gravitational potential energy

Now compare Equation 6-7 to Coulomb's law for the electric force between two point charges, Equation 16-1:

$$F_{q_1 \text{ on } q_2} = F_{q_2 \text{ on } q_1} = \frac{k|q_1||q_2|}{r^2} \quad (16\text{-}1)$$

Any two point charges q_1 and q_2 exert equally strong **electric forces** on each other.

Coulomb constant

Absolute values of the **point charges**

Distance between the point charges

EQUATION IN WORDS
Coulomb's law

If we replace m and G in Equation 6-7 with q and k, we see that Equation 16-1 is identical to Equation 6-7 for the gravitational force, but with an important difference: The electric force is attractive (like the gravitational force) if the two point charges, q_1 and q_2, have opposite signs (one negative and the other positive) but repulsive if q_1 and q_2 have the same sign (either both positive or both negative). Because of that difference, we need to remove the minus sign as well as replace m and G in Equation 8-8 with q and k to develop an equation for electric potential energy of a system of two point charges:

$$U_{\text{electric}} = \frac{kq_1q_2}{r} \quad (17\text{-}3)$$

Coulomb constant

Values of the **two charges**

Electric potential energy of a system of two point charges

Distance between the point charges

EQUATION IN WORDS
Electric potential energy of a system of two point charges

This expression will give us a negative value, similar to Equation 8-8 for gravitational potential energy, when the point charges have opposite signs and the interaction is attractive. **Figure 17-4a** graphs the gravitational potential energy given by Equation 8-8. Remember, we chose the point at which potential energy U_{grav} is zero to be where the two objects are infinitely far apart, $r \to \infty$. We can define the zero of a potential energy of a system however is most convenient (once per system, only!) and by choosing infinity to be zero, we don't have to carry some constant value around in

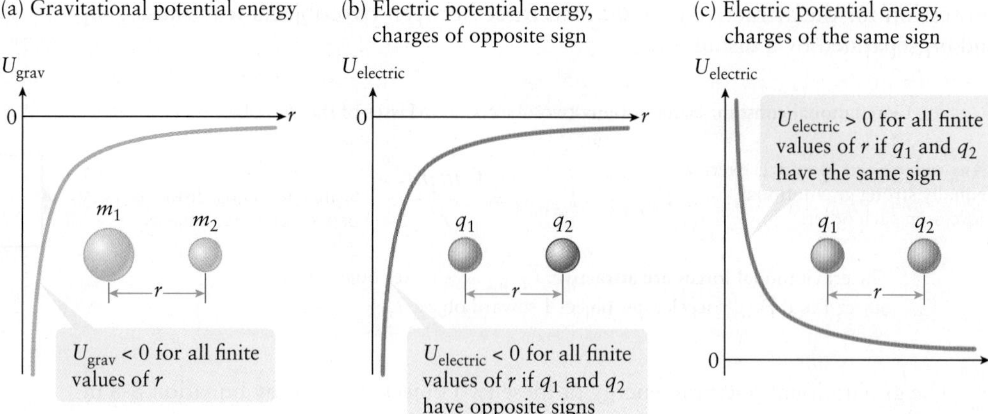

Figure 17-4 Potential energy for two objects with mass and for two objects with charge The electric potential energy $U_{electric}$ for two point charges is similar to the gravitational potential energy for two objects with mass. The sign of $U_{electric}$ depends on the signs of the charges of the two objects. (a) Gravitational potential energy. (b) Electric potential energy; charges have opposite signs. (c) Electric potential energy; charges have the same sign.

all of our equations. With this choice, the gravitational potential energy is negative for any finite value of r. For this or any other choice of zero, the gravitational potential energy of the system increases—that is, becomes less negative—as the objects in the system move farther apart. That's because we would have to pull on one of the objects (do positive work on it) to get it to move away from the other object.

Equation 17-3 shows that $U_{electric}$ of the two point charge system is inversely proportional to the distance between the point charges (r), so the electric potential energy is zero when the two charges are infinitely far apart ($r \to \infty$). But unlike gravitational potential energy, $U_{electric}$ can be either negative or positive depending on the signs of the two point charges. If q_1 and q_2 have different signs (one positive and one negative), so that the two point charges attract each other, then $U_{electric} < 0$ for any finite distance r between the point charges (**Figure 17-4b**).

Here's a useful way to interpret $U_{electric}$, as given by Equation 17-3:

The electric potential energy of a system of two point charges equals the amount of work you would have to do to bring the point charges to their current positions from infinitely far away.

If the two point charges have opposite signs as in Figure 17-4b, you would have to do negative work to oppose the attractive electric force that pulls the two point charges together. (If you didn't exert a force opposite their direction of motion to do this work, the electric attraction would make them crash into each other instead of stopping a distance r apart.) So in this case the electric potential energy of the system of two point charges is negative, as Figure 17-4b shows. By contrast, if the two point charges have the same sign as in **Figure 17-4c**, you would have to do positive work to push the two point charges together to overcome the repulsive electric force that pushes the two point charges apart. So, the electric potential energy of this system of two point charges is positive, as depicted in Figure 17-4c.

The same idea holds for a system of three or more point charges. To put point charges q_1, q_2, and q_3 into proximity to each other, you would first have to do work to bring q_1 and q_2 from infinity to a distance r_{12} from each other. If you then brought in the third charge q_3 while keeping the other two charges stationary, you would have to do additional work against the electric force that q_1 exerts on q_3 *and* against the electric force that q_2 exerts on q_3. If q_3 ends up a distance r_{13} from q_1 and a distance r_{23} from q_2, the electric potential energy of the system of three point charges is the total amount of work that you did to bring them to that configuration:

$$U_{electric} = \frac{kq_1q_2}{r_{12}} + \frac{kq_1q_3}{r_{13}} + \frac{kq_2q_3}{r_{23}} \tag{17-4}$$

(electric potential energy of three charges)

The total electric potential energy of a system of three point charges is the sum of three terms, one for each pair of charges in the system (q_1 and q_2, q_1 and q_3, and q_2 and q_3). Each such term is the same as Equation 17-3. You can easily extend this idea to a system made up of any number of point charges.

The following examples show how to do calculations using Equations 17-3 and 17-4.

EXAMPLE 17-2 Electric Potential Energy and Nuclear Fission

When a nucleus of uranium-235 (92 protons and 143 neutrons) absorbs an additional neutron, it undergoes a process called *nuclear fission* in which it breaks into two smaller nuclei. One possible fission is for the uranium nucleus to divide into two palladium nuclei, each of which has 46 protons and is 5.9×10^{-15} m in radius. The palladium nuclei then fly apart due to their electric repulsion. If we assume that the two palladium nuclei begin at rest and are just touching each other, what is their combined kinetic energy when they are very far apart?

Set Up

We can treat the two spherical nuclei as though they were point charges of $q = +46e$ located at the centers of the two nuclei. This is just like when we calculate the gravitational force between two objects. Equation 17-3 then gives the electric potential energy of the two palladium nuclei when they begin at rest. We'll then use energy conservation to find the combined kinetic energy when the palladium nuclei are very far apart.

Electric potential energy of two point charges:

$$U_{\text{electric}} = \frac{kq_1q_2}{r} \quad (17\text{-}3)$$

Conservation of mechanical energy:

$$K_{\text{i}} + U_{\text{electric,i}} = K_{\text{f}} + U_{\text{electric,f}} \quad (8\text{-}3)$$

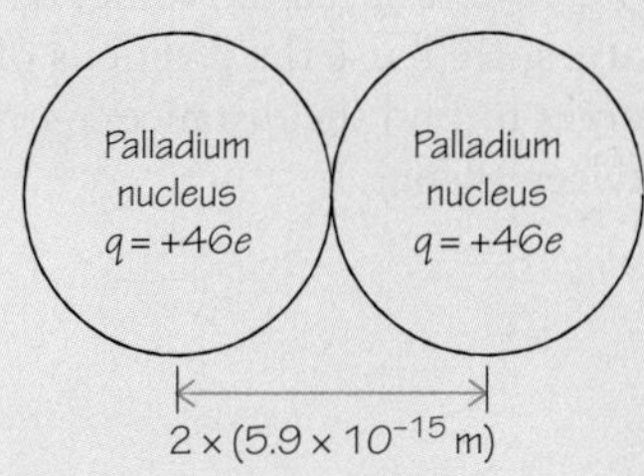

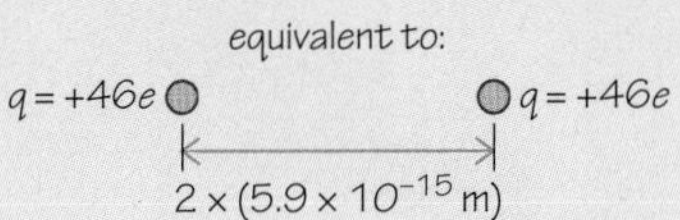

Solve

Use Equation 17-3 to solve for the initial electric potential energy when the two nuclei are just touching.

Each nucleus has charge $+46e$:

$$q_1 = q_2 = +46(1.60 \times 10^{-19}\ \text{C}) = +7.36 \times 10^{-18}\ \text{C}$$

The separation between the point charges is twice the radius of either nucleus:

$$r = 2(5.9 \times 10^{-15}\ \text{m})$$

The initial electric potential energy of the two-point-charge system is

$$U_{\text{electric,i}} = \frac{kq_1q_2}{r} = \frac{(8.99 \times 10^9\ \text{N} \cdot \text{m}^2/\text{C}^2)(7.36 \times 10^{-18}\ \text{C})^2}{2(5.9 \times 10^{-15}\ \text{m})}$$

$$= 4.1 \times 10^{-11}\ \text{N} \cdot \text{m} = 4.1 \times 10^{-11}\ \text{J}$$

Use energy conservation to find the combined final kinetic energy of the two palladium nuclei.

The palladium nuclei begin at rest, so the initial kinetic energy is zero:

$$K_{\text{i}} = 0$$

The palladium nuclei end up very far apart, so their separation is essentially infinite ($r \to \infty$). The final electric potential energy for the system is therefore zero:

$$U_{\text{electric,f}} = 0$$

From the energy conservation equation, the final kinetic energy is

$$K_{\text{f}} = K_{\text{i}} + U_{\text{electric,i}} - U_{\text{electric,f}}$$

$$= 0 + 4.1 \times 10^{-11}\ \text{J} - 0 = 4.1 \times 10^{-11}\ \text{J}$$

Reflect

All of the initial electric potential energy is converted into kinetic energy of the palladium nuclei.

Extend: The energy released by the fission of a single uranium-235 nucleus, 4.1×10^{-11} J, is very small. But imagine that you could get 1.0 kg of uranium-235, which contains about 2.6×10^{24} uranium atoms, to undergo fission at once. (This is the principle of a fission bomb.) The released energy would be $(2.6 \times 10^{24}) \times (4.1 \times 10^{-11}\ \text{J}) = 1.1 \times 10^{14}$ J, equivalent to the energy given off by exploding 26,000 tons of TNT. This suggests the terrifying amount of energy that can be released by a fission weapon.

NOW WORK Problems 1, 2, 7, and 9 from The Takeaway 17-2.

EXAMPLE 17-3 Electric Potential Energy of Three Charges

A point charge $q_1 = +4.30$ nC is located at $x = 0$, $y = 0$. A second point charge $q_2 = -9.80$ nC is located at $x = 0$, $y = 4.00$ cm, and a third point charge $q_3 = +5.00$ nC is located at position $x = 3.00$ cm, $y = 0$. (Note that 1 nC = 1 nanocoulomb = 10^{-9} C.) What is the total electric potential energy of these three point charges?

Set Up

We'll use Equation 17-4 to find the value of U_{electric}. We're given the values of the three charges; we'll use the positions of the point charges to find the distances r_{12}, r_{13}, and r_{23} between them.

Electric potential energy of three charges:

$$U_{\text{electric}} = \frac{kq_1q_2}{r_{12}} + \frac{kq_1q_3}{r_{13}} + \frac{kq_2q_3}{r_{23}} \quad (17\text{-}4)$$

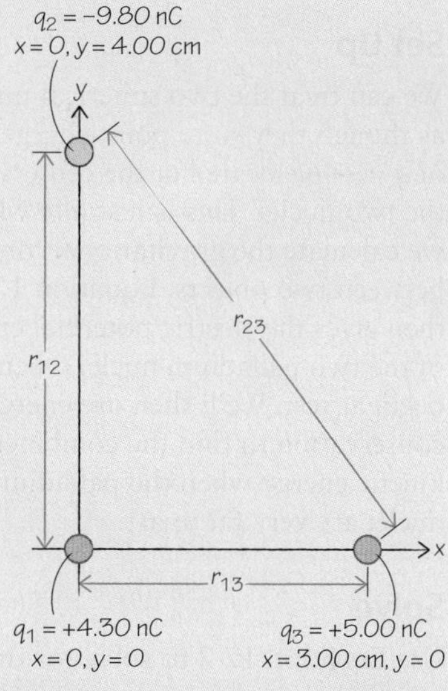

Solve

The figure above shows that charges 1 and 2 are $r_{12} = 4.00$ cm apart, and that charges 1 and 3 are $r_{13} = 3.00$ cm apart. The distance r_{23} between charges 2 and 3 is the hypotenuse of a right triangle of sides r_{12} and r_{13}.

From the figure above:

$$r_{12} = 4.00 \text{ cm} = 4.00 \times 10^{-2} \text{ m}$$

$$r_{13} = 3.00 \text{ cm} = 3.00 \times 10^{-2} \text{ m}$$

From the Pythagorean theorem:

$$r_{23} = \sqrt{r_{12}^2 + r_{13}^2} = \sqrt{(4.00 \times 10^{-2} \text{ m})^2 + (3.00 \times 10^{-2} \text{ m})^2} = 5.00 \times 10^{-2} \text{ m}$$

Calculate each term in the expression for electric potential energy, Equation 17-4.

We can write Equation 17-4 as

$$U_{\text{electric}} = U_{12} + U_{13} + U_{23}$$

Each term on the right-hand side of this equation represents the contribution to the electric potential energy due to a specific pair of point charges. The contribution due to the interaction between point charges q_1 and q_2 is

$$U_{12} = \frac{kq_1q_2}{r_{12}} = \frac{(8.99 \times 10^9 \text{ N} \cdot \text{m}^2/\text{C}^2)(+4.30 \times 10^{-9} \text{ C})(-9.80 \times 10^{-9} \text{ C})}{4.00 \times 10^{-2} \text{ m}} = -9.47 \times 10^{-6} \text{ J}$$

The contribution due to the interaction between point charges q_1 and q_3 is

$$U_{13} = \frac{kq_1q_3}{r_{13}}$$

$$= \frac{(8.99 \times 10^9\ \text{N}\cdot\text{m}^2/\text{C}^2)(+4.30 \times 10^{-9}\ \text{C})(+5.00 \times 10^{-9}\ \text{C})}{3.00 \times 10^{-2}\ \text{m}}$$

$$= +6.44 \times 10^{-6}\ \text{J}$$

The contribution due to the interaction between charges q_2 and q_3 is

$$U_{23} = \frac{kq_2q_3}{r_{23}}$$

$$= \frac{(8.99 \times 10^9\ \text{N}\cdot\text{m}^2/\text{C}^2)(-9.80 \times 10^{-9}\ \text{C})(+5.00 \times 10^{-9}\ \text{C})}{5.00 \times 10^{-2}\ \text{m}}$$

$$= -8.81 \times 10^{-6}\ \text{J}$$

Finally, calculate the total electric potential energy.

The total electric potential energy is the sum of the three terms calculated above:

$$U_{\text{electric}} = U_{12} + U_{13} + U_{23}$$

$$= (-9.47 \times 10^{-6}\ \text{J}) + (+6.44 \times 10^{-6}\ \text{J}) + (-8.81 \times 10^{-6}\ \text{J})$$

$$= -1.184 \times 10^{-5}\ \text{J} \approx -1.18 \times 10^{-5}\ \text{J}$$

Reflect

The contributions U_{12} and U_{23} are negative because these pairs of point charges (q_1 and q_2 for U_{12}, q_2 and q_3 for U_{23}) have opposite signs. The point charges in these pairs attract each other, so you must do negative work to move the point charges from infinity to the positions shown in the figure (hold them back to keep them from flying toward each other). The contribution U_{13}, however, is positive because point charges q_1 and q_3 have the same sign (both positive). These two point charges repel, so you must do positive work to move these point charges from infinity to their positions. The total potential energy is negative, which shows that the total amount of work you would do to move all three point charges from infinity is negative.

NOW WORK Problems 3, 5, and 6 from The Takeaway 17-2.

THE TAKEAWAY for Section 17-2

✔ Electric potential energy associated with a charged object is analogous to the gravitational potential energy associated with an object with mass.

✔ The electric potential energy associated with a point charge can change if the point charge changes position in an electric field.

✔ The electric potential energy of a pair of point charges equals the amount of work you would have to do to move those point charges from infinity to their present positions.

✔ The electric potential energy of a system of three or more point charges is the sum of the electric potential energies for each pair of point charges in the system.

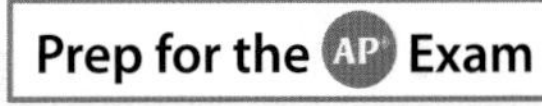

1. A point charge $q_0 = 0.500$ nC is at the origin of a coordinate system.
   (a) An object with a charge q of 1.00 nC and a mass of 0.0800 kg is placed at $x = 0.500$ m. Calculate the electric potential energy of this system of charged objects.
   (b) The object at $x = 0.500$ m with charge q is released from rest and the point charge at the origin is held fixed in place. Calculate the speed of the object when it reaches $x = 2.00$ m.

EX 17-2

2. A point charge $q_0 = 6.00$ nC is held fixed at rest at the origin of a coordinate system.
   (a) An external force is exerted on a 12.00-nC point charge to move it from point A at $x = 230.0$ cm to point B at $x = 150.0$ cm while the point charge at the origin remains at rest. Apply energy conservation to calculate the work done on the system of two point charges if the 12.00-nC point charge begins and ends at rest.

(b) Justify the sign of the work you calculated in part (a).

(c) The same point charge is released from rest at point B while the point charge at the origin again remains fixed in position. What is its kinetic energy when it passes through point A?

(d) Describe the work that must be done through exerting an external force on a point charge of −12.00 nC to displace it from point A to point B with no change in its kinetic energy. Again, the point charge at the origin remains at rest.

EX 17-3 3. Three point charges lie on the x axis. Point charge $q_1 =$ +2.20 nC is at $x = -30.0$ cm, point charge $q_2 = -3.10$ nC is at the origin, and point charge $q_3 = +1.70$ nC is at $x =$ 25.0 cm. Calculate the potential energy of the system of point charges.

EX 17-1 4. A uniform electric field of 2.00 kN/C points in the $+x$ direction.

(a) What is the change in potential energy $U_{\text{electric,B}} - U_{\text{electric,A}}$ of a +2.00-nC test charge as it is moved from point A at $x = -30.0$ cm to point B at $x = +50.0$ cm?

(b) The same test charge is released from rest at point A. What is its kinetic energy when it passes through point B?

(c) If a negative test charge instead of a positive test charge were used in this problem, how would your answers change qualitatively?

EX 17-3 5. Four point charges are arranged in a rectangle of length x and width y, as shown in the figure below. Two point charges are positive and two are negative, but all have the same magnitude Q. Derive an expression for the amount of energy required to disassemble the configuration of charges so that each charge is far away from all the others. *Hint*: This is the change in potential energy of the system.

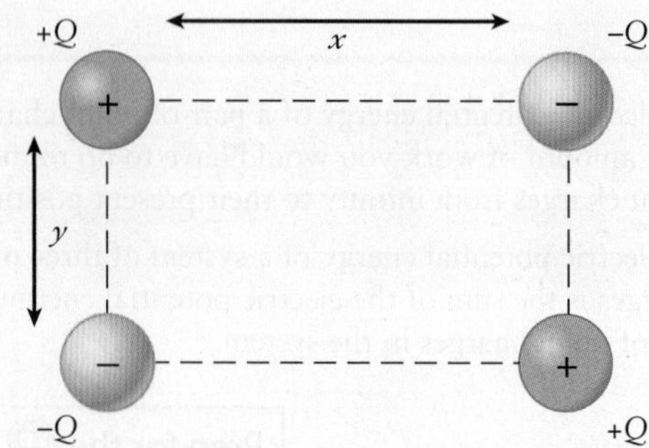

AP® Skill Builders

EX 17-3 6. The potential energy change of a system of point charges, relative to the potential energy when all of the point charges are at separations large enough for the potential energy of each pair to be negligible, is the work required to assemble the point charges.

(a) Evaluate the work that must be done to place two point charges with opposite signs, such that one is located at the origin of an x axis and one is at $x = 0.1$ m. Use two steps, moving only one point charge at a time.

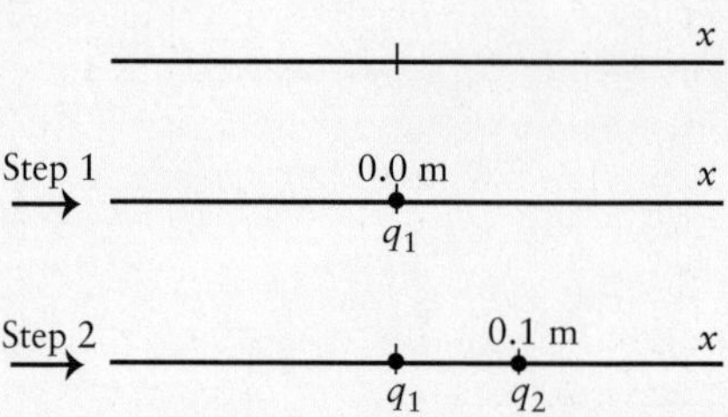

(b) Evaluate the work that must be done to place a point charge q at the origin among the other point charges present in the following figure, which is a circle of radius 1 m.

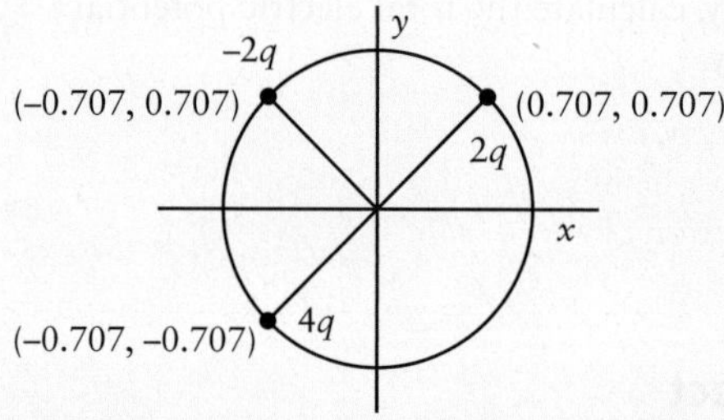

AP® Skills in Action

EX 17-2 7. Two red blood cells each have a mass of 9.0×10^{-14} kg and carry a negative charge spread uniformly over their surfaces. The repulsion arising from the excess charge prevents the cells from clumping together. One cell carries −2.50 pC of charge and the other −3.10 pC, and each cell can be modeled as a sphere 7.5 μm in diameter.

(a) What speed would they need when very far away from each other to get close enough to just touch? Ignore viscous drag from the surrounding liquid.

(b) What is the magnitude of the maximum acceleration of each cell in part (a)?

8. The gravitational potential energy U_{grav} of an object with mass near Earth's surface depends on the distance of the object from Earth's surface (y). For larger separations, the more general form of gravitational potential energy depends on the distance of the object from Earth's center (r).

In situations involving the orbital motion of planets and satellites, as described in Section 8-5, the universal law of gravitation is used with gravitational potential energy, as shown by the graph in Figure 8-10d. In situations involving the gravitational potential energy near the surface of Earth, the gravitational acceleration is treated as constant, and potential energy is more usefully described by the graph in Figure 8-10b.

Justify the claim that U_{electric} has the same form as U_{grav} for the electric potential energy function for a positive point charge as a function of position r from a negative point charge.

9. Nuclear fission reactors with fuel rods maintained in water emit neutrons with low kinetic energy, "slow neutrons" that are captured by the strong nuclear force and increase the atomic mass of a uranium-235 (U-235) nucleus beyond 235 nucleons (protons and neutrons). One isotope produced in this way is U-236. Half of the U-236 produced will still be present in a spent fuel rod in 27 million years.

   The principal decay process of U-236 is alpha decay, leaving behind thorium-232, which has a lifetime comparable to the age of the universe (14 billion years). Thorium has 90 protons and helium has 2.

   (a) A model of nuclear radii predicts that the thorium-232 nucleus has a radius of $r_{Th} = 7.4\text{ fm} = 7.4 \times 10^{-15}$ m, and that of the helium-4 nucleus is $r_{He} = 1.8 \times 10^{-15}$ m. Calculate the work done against the repulsive electric force to displace a helium-4 nucleus from an infinite distance to a distance equal to the sum of the helium and thorium nuclei.

   (b) A helium-4 nucleus is released from rest at a separation distance from a thorium-232 nucleus equal to the sum of the radii. Predict the velocity of the helium-4 nucleus when the separation is equal to $1000r_{He}$. Assume that the position of the thorium-232 nucleus remains fixed. The mass of the helium nucleus is approximately 6.6×10^{-27} kg.

   (c) Explain what conservation principle would be violated if the thorium-232 nucleus did not move in this decay. Estimate the ratio of the velocity of the thorium-232 nucleus to that of the helium-4 nucleus. What does this tell you about the error due to your assumption in part (b)? The mass of the thorium-232 nucleus is approximately 3.9×10^{-25} kg.

10. Suppose you reversed the signs of the three point charges in Example 17-3, so that $q_1 = -4.30$ nC, $q_2 = +9.80$ nC, and $q_3 = -5.00$ nC. If the positions of the point charges remain unchanged, the total electric potential energy of this system would be

    (A) more negative than -1.18×10^{-5} J.
    (B) -1.18×10^{-5} J.
    (C) between -1.18×10^{-5} J and $+1.18 \times 10^{-5}$ J.
    (D) +11.84 J.

17-3 Electric potential difference relates to the change in electric potential energy

Our discussion in Section 17-2 shows that if a point charge q changes position relative to a second point charge, the potential energy change $\Delta U_{electric}$ depends on both the magnitude and the sign (positive or negative) of q (see Figure 17-3). We can add other charged objects to the system by calculating the additional electric potential energy between each new object and those already in place. Often, we will consider situations in which many of the charged objects in the system are fixed in place. Such problems can be simplified by considering the potential energy per unit charge—that is, the electric potential energy of a system of charged objects including a charged object at a given position divided by the value of the charge on that object. We call this quantity the **electric potential** V:

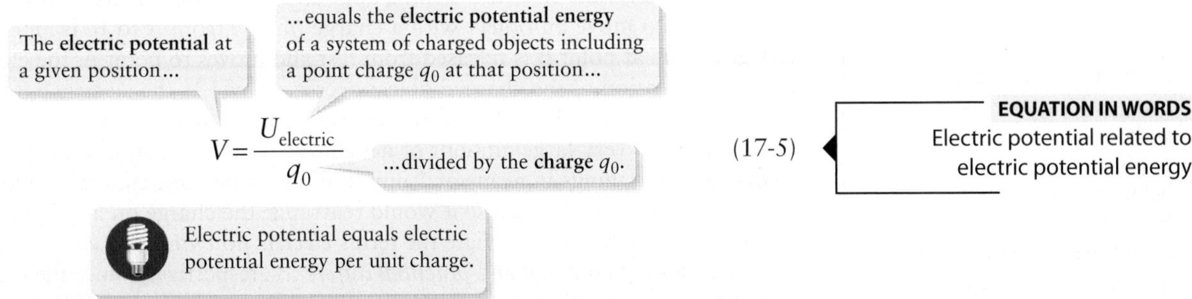

The electric potential at a given position...
...equals the electric potential energy of a system of charged objects including a point charge q_0 at that position...
...divided by the charge q_0.
Electric potential equals electric potential energy per unit charge.

$$V = \frac{U_{electric}}{q_0} \qquad (17\text{-}5)$$

EQUATION IN WORDS
Electric potential related to electric potential energy

We call the point charge q_0 the **test charge**: Its charge has such a small magnitude that it doesn't affect the other charged objects that create the electric field in which q_0 moves. Because we divide out the value of q_0, the value of the potential V at a given position does not depend on the value of the point charge q_0 that we place there. Instead, V is determined by the arrangement and charge of the other charged objects that cause the electric force that the test charge would experience at that position.

Remember that potential energy always depends on the configuration of a system. Although we are discussing the electric potential energy when a charged object is at a point in space, that potential energy depends on the other charged objects in the system and their arrangement.

Note that electric potential V has the same relationship to electric potential energy U_{electric} as electric field $\vec{E}$ has to electric force $\vec{F}$: V is electric potential energy per charge, just as $\vec{E}$ is electric force per charge. Like electric potential energy, but unlike electric field, potential V is a *scalar* quantity.

As when solving problems in mechanics, often the value of U_{electric} for the system is not as important as the change in electric potential energy $\Delta U_{\text{electric}}$ when one charged object moves from a point A to a different point B (assuming the rest of the charged objects in the system do not change configuration). Since this is useful in many situations, it is convenient to define electric potential difference. From Equation 17-5, the **electric potential difference** ΔV between two points is

EQUATION IN WORDS
Electric potential difference related to electric potential energy difference

The **difference in electric potential** between two positions...

...equals the **change in electric potential energy** of the system of charged objects when q_0 is moved between these two positions...

$$\Delta V = \frac{\Delta U_{\text{electric}}}{q_0} \tag{17-6}$$

...divided by the **charge** q_0.

Electric potential difference equals electric potential energy difference per unit charge.

Understanding how electric potential energy and electric potential are determined for systems of charged objects helps us understand these concepts and apply them correctly. Electric potential can also be caused by a chemical process that causes charge to separate, such as within a battery. The SI unit of electric potential and electric potential difference is the **volt** (V), named after the Italian scientist Alessandro Volta. For example, a common AA or AAA flashlight battery has "1.5 volts" written on its side. This means that the electric potential at the positive terminal of the battery (labeled +) is 1.5 V greater than the electric potential at the negative terminal of the battery (labeled −). In other words, 1.5 V is the electric potential difference between the terminals of the battery. In the next section, we'll see how this electric potential difference causes electric charge to move when a battery is included in an electric circuit. (Volta is credited with inventing the electric battery in 1800.) Equations 17-2 and 17-3 show that 1 volt is equal to 1 joule per coulomb, and we know that 1 joule is equal to 1 newton multiplied by 1 meter. So

$$1 \text{ V} = 1 \text{ J/C} = 1 \text{ Nm/C}$$

This means that if the electric potential at point B is 1 V higher than at point A, it takes 1 J of work to move an object with a charge of +1C from A to B. If an object with a charge of +1C at point B is released from rest and moves to point A, the electric potential at the location of the charged object decreases by 1 V, the electric potential energy associated decreases by 1 J, and the charged object acquires 1 J of kinetic energy. A coulomb is a very large amount of net charge to have on an object. It is convenient to discuss these things in terms of 1 unit, but an actual "test charge" could never have this much charge on it, because it would rearrange the charge on anything near it!

People sometimes abbreviate the terms *electric potential* and *electric potential difference* as simply *potential* and *potential difference*, respectively. Since the unit of electric potential is the volt, it's also common to refer to electric potential difference as voltage.

WATCH OUT !

Electric potential and electric potential energy are not the same.

Remember that electric potential V and electric potential energy U_{electric} are related but different scalar quantities. Electric potential energy U_{electric} refers to the energy of a system associated with an object with a particular amount of charge being at a given position. Electric potential V is the energy associated with a unit of charge at that position.

Electric Potential in a Uniform Electric Field

We can use Equation 17-6 to find the electric potential difference between two points A and B in a uniform electric field (**Figure 17-5**). From Equation 17-2 the change or difference in electric potential energy when a charge q_0 is moved from A to B is $\Delta U_{electric} = -q_0 Ed \cos\theta$. Equation 17-6 tells us that the electric potential difference between A and B is $\Delta U_{electric}$ divided by q_0:

$$\Delta V = \frac{\Delta U_{electric}}{q_0} = \frac{-q_0 Ed \cos\theta}{q_0} = -Ed \cos\theta \quad (17\text{-}7)$$

Equation 17-7 shows that ΔV is positive if $\cos\theta$ is negative, which is the case if the angle θ is greater than 90° as depicted in Figure 17-5. In other words, *if you move in a direction opposite to the electric field, the electric potential increases.* If the angle θ is less than 90°, $\cos\theta$ is positive and ΔV is negative; *if you move in the direction of the electric field, the electric potential decreases.*

We know that the electric force on a positive charge is in the direction of the electric field $\vec{E}$, and the electric force on a negative charge is in the direction opposite to $\vec{E}$. So our observations about how electric potential changes with position tell us that

If an object has positive charge, the electric force on that object pushes it toward a region of lower electric potential. If an object has negative charge, the electric force on that object pushes it toward a region of higher electric potential.

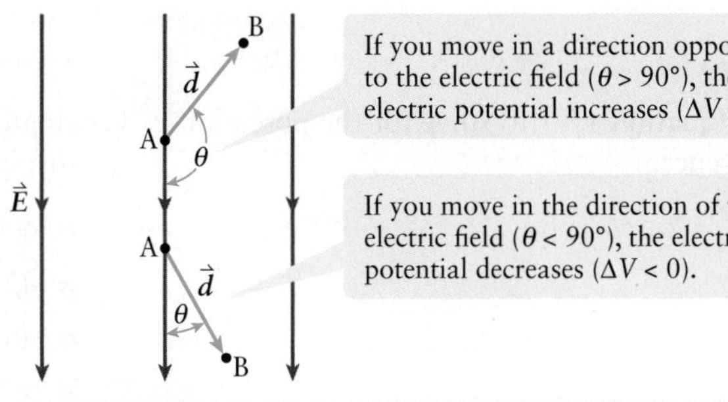
B
$\vec{d}$
A
θ
$\vec{E}$
A
$\vec{d}$
θ
B
If you move in a direction opposite to the electric field ($\theta > 90°$), the electric potential increases ($\Delta V > 0$).
If you move in the direction of the electric field ($\theta < 90°$), the electric potential decreases ($\Delta V < 0$).

If you move from point A to point B in a uniform electric field $\vec{E}$, the change in electric potential is $\Delta V = -Ed \cos\theta$.

Figure 17-5 Change in electric potential Calculating the difference in electric potential between two points in a uniform electric field $\vec{E}$.

As we'll see below, these observations hold true whether or not the electric field is uniform.

Comparing Equations 17-2 and 17-7 shows that the right-hand side of Equation 17-7 equals the negative of the work done by the electric field ($q_0 Ed \cos\theta$) divided by the charge q_0. So, another way to think of the potential difference between two points is as the work that *you* must do per unit charge against the electric force caused by the other charged objects in the system to move a charged object between those points. That is, if you move an object with charge +1 C (not a test charge–sized charge for sure) that is at rest at point A to point B where it is again at rest, and the potential at B is 1 V higher than the potential at A, you must do +1 J of work on that object to move it through that field. If we treat the field as external and the object as the system, then the field would do −1 J of work on the object as it was moved between those points.

EXAMPLE 17-4 Electric Potential Difference in a Uniform Field I

A uniform electric field points in the positive x direction and has magnitude 2.00×10^2 V/m. Points A and B are both in this field: Point B is a distance 0.300 m from A in the negative x direction. Determine the electric potential difference $V_B - V_A$ between points A and B.

Set Up

We apply Equation 17-7 to the path shown in the figure. Note that the magnitude of the displacement is $d = 0.300$ m, and the angle between the electric field and the displacement is $\theta = 180°$. Note also that the potential difference ΔV equals the potential at the *end* of the displacement $\vec{d}$ (that is, at point B) minus the potential at the *beginning* of the displacement (that is, at point A).

Potential difference between two points in a uniform electric field:

$$\Delta V = V_B - V_A = -Ed \cos\theta \quad (17\text{-}7)$$

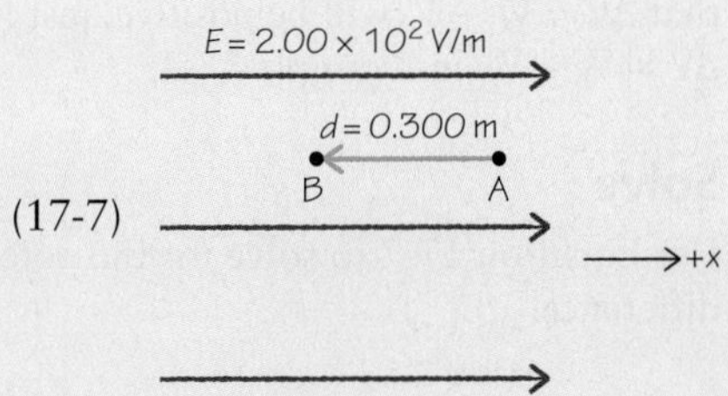
E = 2.00 × 10² V/m
d = 0.300 m
B
A
+x

Solve

Use Equation 17-7 to solve for the potential difference.

Calculate the potential difference from the electric field magnitude E, the displacement d, and the angle θ:

$$\Delta V = V_B - V_A = -Ed\cos\theta$$
$$= -(2.00 \times 10^2 \text{ V/m})(0.300 \text{ m}) \cos 180°$$
$$= -(60.0 \text{ V})(-1)$$
$$= +60.0 \text{ V}$$

Reflect

We can check our result by comparing with Example 17-1 in Section 17-2, where we considered the change in electric potential *energy* $\Delta U_{\text{electric}}$ for an electron that undergoes the same displacement in this same electric field. Using Equation 17-6 we find the same value of $\Delta U_{\text{electric}}$ as in Example 17-1.

The positive value of $\Delta V = V_B - V_A$ means that point B is at a higher potential than point A. This agrees with our observation above that if you travel opposite to the direction of the electric field, the electric potential increases. The value $E = 2.00 \times 10^2$ V/m means that the electric potential increases by 2.00×10^2 V for every meter that you travel opposite to the direction of $\vec{E}$.

If a charge $q_0 = -1.60 \times 10^{-19}$ C travels from A to B, the change in electric potential energy is given by Equation 17-6:

$$\Delta V = \frac{\Delta U_{\text{electric}}}{q_0}$$

so

$$\Delta U_{\text{electric}} = q_0 \Delta V = (-1.60 \times 10^{-19} \text{ C})(+60.0 \text{ V})$$
$$= -9.60 \times 10^{-18} \text{ V} \cdot \text{C} = -9.60 \times 10^{-18} \text{ J}$$

(Recall from above that 1 V = 1 J/C.)

NOW WORK Problem 5 from The Takeaway 17-3.

EXAMPLE 17-5 Electric Potential Difference in a Uniform Field II

Determine the electric potential difference $\Delta V = V_C - V_A$ between points A and C in the uniform electric field of Example 17-4. Point C is a distance 0.500 m from point A, and the linear path from A to C makes an angle of 126.9° with respect to the electric field.

Set Up

Again we'll use Equation 17-7 to calculate the potential difference between the two points. Because the displacement from point A to point C points generally opposite to the direction of the electric field, we expect that $\Delta V = V_C - V_A$ will be positive, just like $\Delta V = V_B - V_A$ in Example 17-4.

Potential difference between two points in a uniform electric field:

$$\Delta V = V_C - V_A = -Ed\cos\theta \quad (17\text{-}7)$$

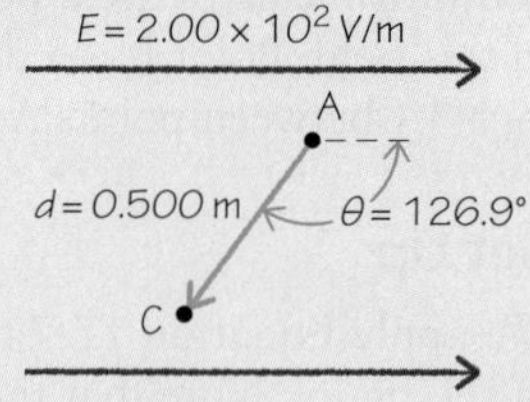

Solve

Use Equation 17-7 to solve for the potential difference.

Calculate the potential difference from the electric field magnitude E, the displacement d, and the angle θ:

$$\Delta V = V_C - V_A = -Ed\cos\theta$$
$$= -(2.00 \times 10^2 \text{ V/m})(0.500 \text{ m}) \cos 126.9°$$
$$= -(2.00 \times 10^2 \text{ V/m})(0.500 \text{ m})(-0.600)$$
$$= +60.0 \text{ V}$$

Reflect

The potential difference between points A and C is the *same* as the potential difference between points A and B in Example 17-4. Equation 17-7 tells us why this should be: The potential difference $\Delta V = -Ed\cos\theta$ involves the magnitude E of the electric field multiplied by $d\cos\theta$, which is the component of the displacement $\vec{d}$ in the direction of $\vec{E}$. In both examples the electric field magnitude has the same value (2.00×10^2 V/m), as does the component of displacement in the direction of the electric field (–0.300 m).

Another way to come to this same conclusion is to recognize that the displacement from A to C can be broken down into two parts: a displacement in the negative x direction from A to B, followed by a displacement from B to C in the y direction. The potential difference for the displacement from A to B is $V_B - V_A = +60.0$ V as we calculated in Example 17-4; the potential difference for the displacement from B to C is $V_C - V_B = -Ed\cos 90° = 0$, because that displacement is perpendicular to the electric field. So the net potential difference between A and C is $V_C - V_A = (V_C - V_B) + (V_B - V_A) = 0\text{ V} + 60.0\text{ V} = +60.0\text{ V}$.

NOW WORK Problem 4 from The Takeaway 17-3.

EXAMPLE 17-6 Transmission Electron Microscope

A *transmission electron microscope* forms an image by sending a beam of fast-moving electrons—rather than a beam of light—through a thin sample. As we will see later, such fast-moving electrons behave very much like light. If the electrons have sufficiently high energy, the image that they form can show much finer detail than the best optical microscope. The electrons are emitted from a heated metal filament and are then accelerated toward a second piece of metal called the *anode* that is at a potential 2.50 kV (1 kV = 1 kilovolt = 10^3 V) higher than that of the filament. If the electrons leave the filament initially at rest, how fast are the electrons traveling when they pass the anode?

Set Up

As the electrons move through the potential difference between the filament and the anode, the electric potential energy will change in accordance with Equation 17-6. Each electron has a negative charge $q_0 = -e$, so an *increase* in electric potential ($\Delta V > 0$) means a *decrease* in electric potential energy ($\Delta U_{electric} < 0$). The total mechanical energy (kinetic energy plus potential energy) is conserved as the electron moves because there are no forces acting on it other than the conservative electric force, so the electron kinetic energy will increase as the potential energy decreases.

Electric potential difference related to electric potential energy difference:

$$\Delta V = \frac{\Delta U_{electric}}{q_0} \quad (17\text{-}6)$$

Conservation of mechanical energy:

$$E = K + U_{electric} = \text{constant} \quad (8\text{-}3)$$

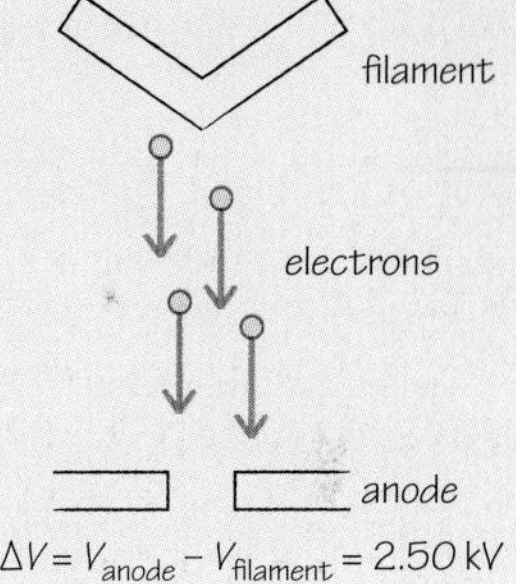

Solve

Solve Equation 17-6 for the change in electric potential energy as the electron moves from the filament to the anode, which is at a potential 2.50 kV higher than the filament.

From Equation 17-6:

$$\Delta U_{electric} = q_0\Delta V$$

The electron has charge $q_0 = -e$, so

$$\begin{aligned}\Delta U_{electric} &= -e\Delta V \\ &= -(1.60 \times 10^{-19}\text{ C})(+2.50 \times 10^3\text{ V}) \\ &= -4.00 \times 10^{-16}\text{ J}\end{aligned}$$

(Recall that 1 V = 1 J/C, so 1 J = 1 C · V.)

The conservation of mechanical energy tells us that the change in the kinetic energy of the electron is equal to the negative of the change in the electric potential energy of the system.

Conservation of mechanical energy:

$$E = K + U_{electric} = \text{constant}$$

So the *change* in mechanical energy is zero:

$$\Delta E = \Delta K + \Delta U_{electric} = 0$$

The change in the kinetic energy of the electron is

$$\Delta K = -\Delta U_{\text{electric}} = -(-4.00 \times 10^{-16}\ \text{J})$$
$$= +4.00 \times 10^{-16}\ \text{J}$$

Each electron begins with zero kinetic energy, so the change in its kinetic energy is equal to its final kinetic energy as it reaches the anode.

Because the electron begins with zero kinetic energy at the filament, the change in its kinetic energy is

$$\Delta K = +4.00 \times 10^{-16}\ \text{J} = K_{\text{anode}} - K_{\text{filament}} = K_{\text{anode}}$$

Use this to find the speed of the electron at the anode:

$$K_{\text{anode}} = \frac{1}{2} m_{\text{electron}} v_{\text{anode}}^2, \text{ so}$$

$$v_{\text{anode}} = \sqrt{\frac{2K_{\text{anode}}}{m_{\text{electron}}}} = \sqrt{\frac{2(4.00 \times 10^{-16}\ \text{J})}{9.11 \times 10^{-31}\ \text{kg}}}$$
$$= 2.96 \times 10^7\ \text{m/s}$$

Reflect

The speed of light is $c = 3.00 \times 10^8$ m/s; the electrons are accelerated to nearly one-tenth of that speed.

NOW WORK Problem 3 from The Takeaway 17-3.

Example 17-6 is just one of many situations in which an object with a charge of $-e$ (such as an electron) or $+e$ (such as a proton) moves through a potential difference. A common unit for the potential energy change in such situations is the **electron volt** (eV), which is equal to the magnitude e of the charge on the electron multiplied by 1 volt. Since $e = 1.60 \times 10^{-19}$ C, it follows that

$$1\ \text{eV} = (1.60 \times 10^{-19}\ \text{C})(1\ \text{V}) = 1.60 \times 10^{-19}\ \text{C} \cdot \text{V} = 1.60 \times 10^{-19}\ \text{J}$$

In Example 17-6, the electron of charge $-e$ moves through a potential difference of $+2.50 \times 10^3$ V, the change in electric potential energy is -2.50×10^3 eV, and the kinetic energy that the electron acquires is $+2.50 \times 10^3$ eV. We also use the abbreviations 1 keV = 10^3 eV, 1 MeV = 10^6 eV, and 1 GeV = 10^9 eV. [The largest particle accelerator in the world, the Large Hadron Collider at the European Organization for Nuclear Research (CERN) near Geneva, Switzerland, accelerates protons to a kinetic energy of 7×10^{12} eV = 7 TeV. This is equivalent to making the protons pass through a potential difference of 7×10^{12}, or 7 trillion, volts.] In later chapters we'll see that the electron volt is a useful unit for expressing energies on the atomic or nuclear scale.

Electric Potential Due to a Point Charge

Note that Equation 17-7 is useful for calculating potential differences only in the case of a uniform electric field. Another important case is the electric potential due to a *point charge* Q. From Equation 17-3, if we place a test charge q_0 a distance r from a point charge Q, the electric potential energy of the system of two charges is

$$U_{\text{electric}} = \frac{kq_0Q}{r}$$

Equation 17-5 tells us that to find the electric potential due to the point charge Q, we must divide U_{electric} by the value of the test charge q_0:

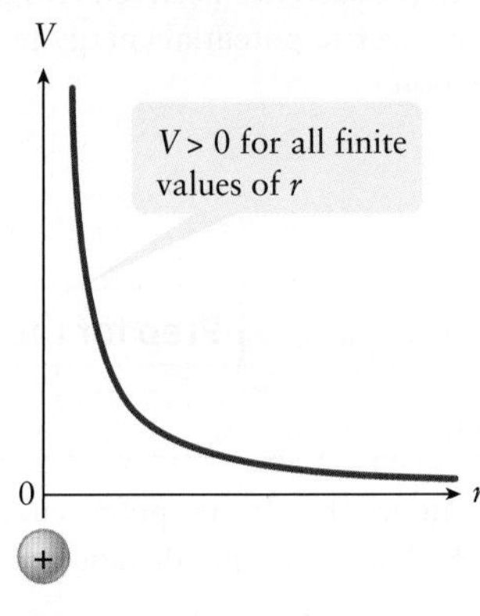

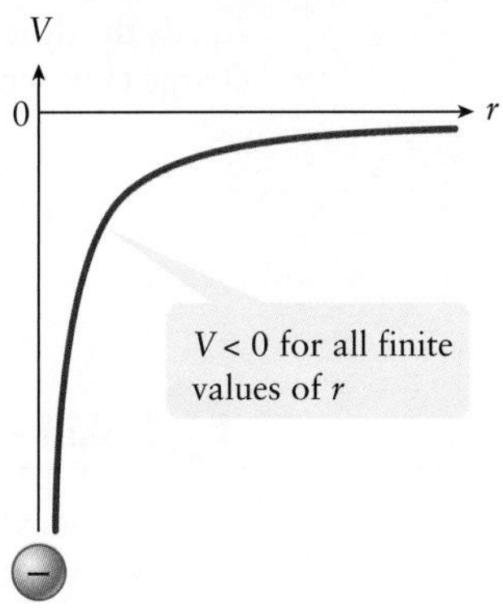

Figure 17-6 Electric potential due to a point charge The electric potential V due to a point charge is (a) positive if its charge is positive and (b) negative if its charge is negative.

Electric potential due to a point charge Q — **Coulomb constant**

$$V = \frac{kQ}{r} \tag{17-8}$$

Value of the point **charge**

Distance from the point charge Q to the location where the potential is measured

EQUATION IN WORDS
Electric potential due to a point charge

Equation 17-8 states that all points that are the same distance r from a point charge have the same electric potential due to that point charge. It also says that if $Q > 0$, the electric potential due to the point charge is positive and decreases (becomes less positive) as you move farther away from it so that r increases (**Figure 17-6a**). If $Q < 0$, the electric potential due to the point charge is negative and decreases (becomes more negative) as you move closer to it (**Figure 17-6b**). For either sign of Q, the electric potential goes to zero at an infinite distance from the point charge.

These observations about Equation 17-8 are consistent with our previous statements about electric potential and electric force. A positive test charge q_0 placed near a positive point charge Q feels an electric force that pushes it farther away from Q, toward regions where the potential V due to Q is lower (less positive). If instead that positive test charge is placed near a negative point charge Q, the test charge feels an electric force that pulls it toward Q—again toward regions where the potential V is lower (in this case more negative).

If there is not a single point charge but a system of point charges, the total electric potential at a given position due to these point charges is the sum of the individual potentials. For example, for the case of three point charges Q_1, Q_2, and Q_3, the potential at a point that is a distance r_1 from the first charge, a distance r_2 from the second charge, and a distance r_3 from the third charge is

$$V = \frac{kQ_1}{r_1} + \frac{kQ_2}{r_2} + \frac{kQ_3}{r_3} \tag{17-9}$$

(electric potential of three charges)

WATCH OUT !

Don't confuse the formulas for electric potential and electric field due to a point charge.

Be sure that you recognize the differences between Equation 17-8, $V = kQ/r$, and Equation 16-4 for the magnitude of the electric field due to a point charge Q, $E = k|Q|/r^2$. Equation 17-8 says that the potential V due to a point charge is inversely proportional to r and can be positive or negative, depending on the sign of Q. By contrast, Equation 16-4 tells us that the magnitude E of the field due to a point charge is inversely proportional to the *square* of r. Furthermore, because E is the magnitude of a vector, it is always positive (it is proportional to the absolute value of Q).

THE TAKEAWAY for Section 17-3

✔ The electric potential at a certain position equals the electric potential energy per unit charge of a system of charged objects including a test charge at that position. Electric potential is a scalar, not a vector, quantity. Electric potential decreases as you move in the direction of the electric field.

✔ The electric potential due to a point charge is positive if the charge is positive and negative if the charge is negative.

✔ As you move farther from an isolated point charge, the electric potential becomes closer to zero.

✔ The electric potential due to a collection of charges is the sum of the potentials due to the individual charges.

✔ The electric potential difference between two points equals the difference in electric potential energy per unit charge between those points.

Prep for the AP® Exam

AP® Building Blocks

1. Consider the system of point charges described by problem 6(b) in the Takeaway 17-2.
   (a) Express the change in electric potential V_q experienced by a point charge moved from infinity to the origin if the other three point charges are already in place.
   (b) Consider a modified figure for Takeaway 17-2 problem 6(b) in which the charges are rotated, so that the point charge $2q$ lies along the y axis. Justify the claim that the electric potential V_q is not changed by the rotation of the coordinate system, but the direction of the electric field at the origin is changed.
2. Explain why an electron will accelerate toward a region of higher electric potential and not lower electric potential.
3. (EX 17-6) At a certain point in space, x_1, there is a potential $V = 800 \text{ V} = 800 \text{ J/C}$ relative to the electric potential, 0 V, at an infinite distance. Calculate the change in electric potential energy when an 11.0-nC charge is moved from an infinite distance to the point x_1.
4. (EX 17-5) A uniform electric field of magnitude 28.0 V/m makes an angle of 30.0° with the x axis. If a charged particle moves along the x axis from the origin to $x = 10.0$ m, what is the potential difference of its final position relative to its initial position?
5. (EX 17-4) As shown in the figure below, two large parallel plates, which are aligned along the y axis, are separated by a distance $d = 30.0$ cm and are at different electric potentials. The center of each plate has a small opening that lies on the x axis. A proton, traveling on the x axis, passes through the first plate with a speed of 2.50×10^5 m/s, and then leaves through the second plate with a speed of 7.80×10^5 m/s. Calculate the potential difference $V_2 - V_1$ between the two plates. Note that a positive potential difference indicates the second plate is at a higher potential than the first plate.

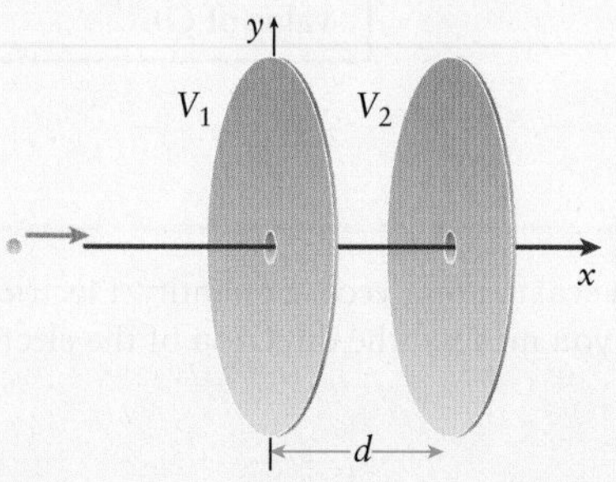

AP® Skill Builders

6. At point P in the figure the electric potential, relative to the electric potential at an infinite distance, is zero.

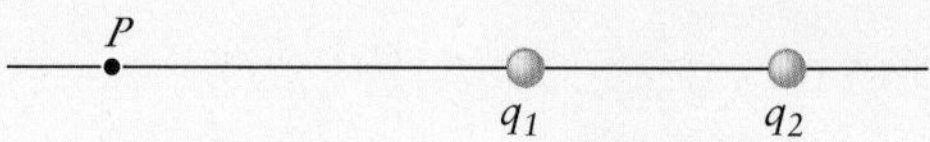

   (a) Make and support claims regarding the sign and magnitude of the point charges q_1 and q_2.
   (b) Make and support a claim regarding the possibility of a point on the line between the two charges where the value of the electric potential difference is zero.
7. A positive point charge that has a charge of 12.00 nC is at the origin, and a positive point charge that has a charge of 23.00 nC is on the y axis at $y = 40.0$ cm.

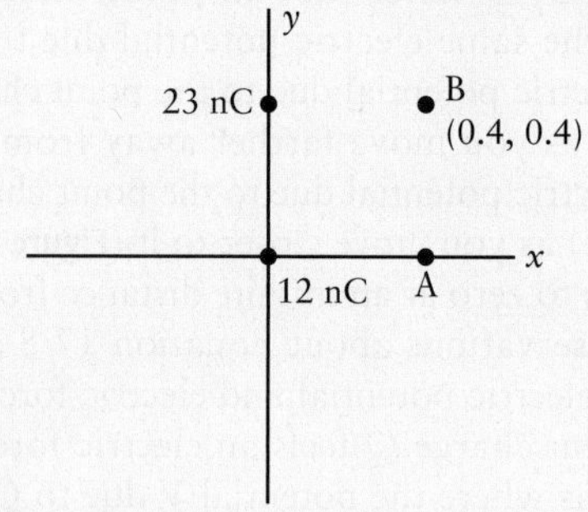

   (a) Calculate the electric potential at point A, which is on the x axis at $x = 40.0$ cm.
   (b) Calculate the potential difference $V_B - V_A$ when point B is at (40.0 cm, 40.0 cm). Describe what this means in terms of energy and charge.
   (c) Calculate the work required to move an electron at rest from point A to rest at point B.
   (d) Suppose that the charges of the point charge at the origin and the point charge at $y = 40.0$ cm had the same magnitude. Explain why no work is done when an electron is displaced from A to B in this situation.

AP® Skills in Action

8. For a positive point charge moving in the direction of the electric field, its potential energy
   (A) increases and its electric potential increases.
   (B) increases and its electric potential decreases.
   (C) decreases and its electric potential increases.
   (D) decreases and its electric potential decreases.

9. If a negative point charge is released in a uniform electric field, it will move
   (A) in the direction of the electric field.
   (B) from high potential to low potential.
   (C) from low potential to high potential.
   (D) in a direction perpendicular to the electric field.
10. An electric potential difference can be used to accelerate a charged particle.
   (a) Apply the conservation of mechanical energy to express the speed of an electron, initially at rest, when it is accelerated through an electric potential difference ΔV.
   (b) A sharply focused beam of electrons is used to cut paths in silicon crystals to make integrated circuits. The electron is accelerated from rest through an electric potential difference of 1.0 kV. Calculate the speed of the electron.
   (c) Explain how a static charge can be avoided when using the instrument described in part (b) in terms of an electric potential difference and a sink for excess electrons. It may help to start your discussion by thinking about which direction a positively charged object would "fall" in an electric potential difference.

17-4 The electric potential has the same value everywhere on an equipotential surface

In Examples 17-4 and 17-5 in the previous section, we looked at the potential differences between two pairs of points in an electric field, A and B in Example 17-4 and A and C in Example 17-5. Although the electric field is the same in both examples, the distance from A to C in Example 17-5 is clearly longer than the distance from A to B in Example 17-4. Nonetheless, the potential differences that we found in the two examples were equal, which tells us that the potential is the *same* at points B and C. You can see why from Equation 17-7, $\Delta V = -Ed\cos\theta$. A displacement $\vec{d}$ from point B to point C is perpendicular to the electric field $\vec{E}$, so the angle in Equation 17-7 is $\theta = 90°$, and $\cos\theta = \cos 90° = 0$. Thus, the potential difference ΔV between points B and C must be zero.

In general, the electric potential will be the same at any two points that lie along a curve perpendicular to electric field lines. Such a curve is called an *equipotential curve* or simply an **equipotential**. For the case of a uniform electric field, the electric potential has the same value anywhere on a plane that's perpendicular to the electric field. Such a plane is an example of an **equipotential surface**, one on which the electric potential has the same value at all points (**Figure 17-7**). No work is required to move a point charge from one point to another along any path on an equipotential surface. As we described in Section 17-3, the electric potential decreases as you move in the direction of the electric field $\vec{E}$, so the value of the potential V is lower for equipotential surfaces that are "downstream" in the electric field than on surfaces that are "upstream."

Figure 17-8 shows both electric field lines and equipotentials for the case of a *nonuniform* electric field. The equipotential surfaces are perpendicular at *all* points to the

A plane perpendicular to a uniform electric field $\vec{E}$ is an **equipotential surface:** Potential V has the same value at all points on this surface.

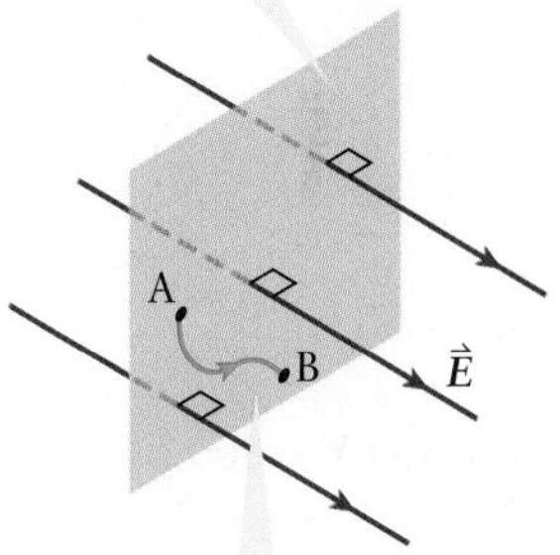

Any curve that lies on an equipotential surface, such as this curve from A to B, is an equipotential curve.

Figure 17-7 Equipotential surfaces I In a uniform electric field equipotential surfaces are planes perpendicular to the electric field lines.

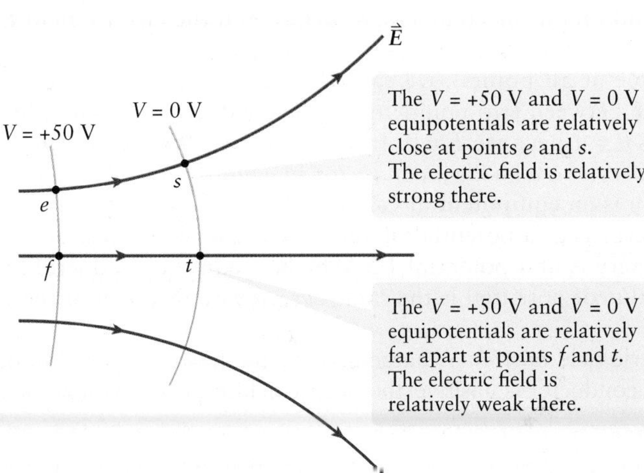

The $V = +50$ V and $V = 0$ V equipotentials are relatively close at points e and s. The electric field is relatively strong there.

The $V = +50$ V and $V = 0$ V equipotentials are relatively far apart at points f and t. The electric field is relatively weak there.

Figure 17-8 Equipotential surfaces II In a nonuniform electric field equipotential surfaces are curved surfaces (seen here from the side) that are everywhere perpendicular to the electric field lines.

field lines, just as for the case of a uniform field in Figure 17-7. But since the electric field lines are not parallel lines, the equipotential surfaces are not flat planes. In general, *any* surface that is everywhere perpendicular to the electric field is an equipotential surface, and the value of the potential V on an equipotential surface is lower the farther "downstream" in the electric field that surface is.

In Figure 17-8 points e and f are both on the equipotential for which $V = +50$ V, and points s and t are both on the equipotential for which $V = 0$ V. So the potential difference between points e and s is the same as the potential difference between points f and t:

$$V_e - V_s = V_f - V_t = (+50 \text{ V}) - (0 \text{ V}) = +50 \text{ V}$$

However, the distance between points e and s is less than that between points f and t. Equation 17-7 tells us that the potential difference between two points is proportional to the distance d between the points and the electric field magnitude E. (This equation is strictly valid only for a uniform field, but is approximately true for a nonuniform field.) So the electric field must be greater between the points e and s that are closer together, and less between the points f and t that are farther apart. This is an example of a general rule:

The magnitude of the electric field is related to the spacing of equipotential surfaces in that field. Where two equipotential surfaces that have a certain potential difference between them are closer together, the electric field is stronger. Where equipotential surfaces with the same potential differences are farther apart, the electric field is weaker.

Figure 17-9 shows an application of this idea. For a positive point charge (**Figure 17-9a**) or a negative point charge (**Figure 17-9b**), the electric field points radially outward or inward. The equipotential surfaces are everywhere perpendicular to the field lines, so they are spheres centered on the point charge. The radial distance from one spherical equipotential surface to the next is the same no matter where you are around the sphere, so the electric field magnitude is the same at all points a given distance from the point charge. This agrees with Equation 16-4 for the field magnitude E due to a point charge, $E = k|Q|/r^2$; the value of E depends only on the distance from the charge, not on where you are around the charge. For an electric dipole, however, the situation is different (**Figure 17-9c**). The field lines point from the positive point charge to the negative one, and the equipotential surfaces are neither spherical nor centered on the point charges. (They are, however, everywhere perpendicular to the field lines.) Equipotential surfaces for the same potential difference between surfaces (this is sometimes referred to as *adjacent equipotential surfaces*) are close together between the two point charges because the electric field is strong there. To the left of the left-hand point charge or to the right of the right-hand point charge, the electric field is relatively weak and adjacent equipotential surfaces are farther apart.

The equipotential concept is helpful for understanding the electric field around a charged conductor. We learned in Section 16-3 that if we put excess charge on a conductor when the charge carriers are in equilibrium positions, all of the excess charge will end up on the conductor's surface and the electric field $\vec{E}$ will be zero everywhere inside the conductor. Since $\vec{E}$ is uniform inside the conductor (it has the same value at all points), we can use Equation 17-7 to calculate the potential difference between two points inside the conductor separated by a distance d: $\Delta V = -Ed\cos\theta = 0$ because $E = 0$. In other words, *the electric potential has the same value everywhere inside a conductor in equilibrium*. We say that a conductor in equilibrium is an **equipotential volume**. That's why we can make statements like "This conductor is at a potential of +20 V" or "The conductor at the positive end of an AA battery is at a potential 1.5 V higher than the conductor at the negative end"—the value of potential is the same *everywhere* throughout the volume of the conductor.

The electric field *outside* a charged conductor, however, is *not* zero. If the excess charge on the conductor is positive, the electric field $\vec{E}$ outside will point away from the

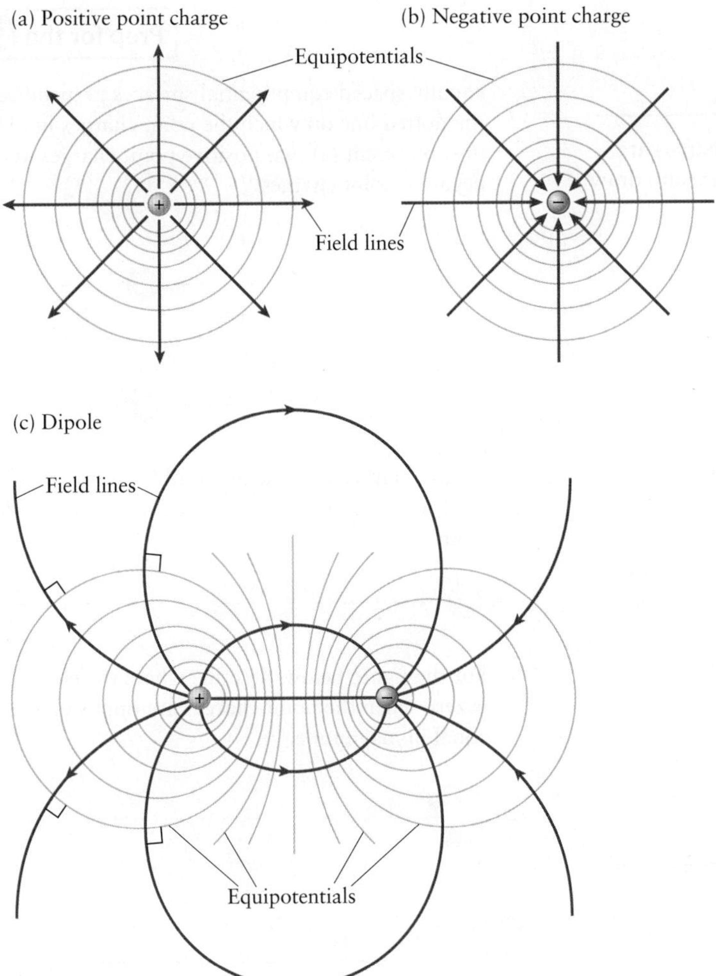

Figure 17-9 Equipotential surfaces III (a), (b) The equipotential surfaces for a positive or negative point charge are spheres centered on the point charge. (You can see this from Equation 17-8, $V = kQ/r$. This says that the potential V has the same value at all points that are the same distance from a point charge Q—that is, at all points on a sphere of radius r centered on the charge.) (c) For a dipole (point charges $+Q$ and $-Q$) the equipotentials are more complicated.

surface of the conductor; if the excess charge is negative, $\vec{E}$ outside will point toward the surface. Because the surface of the conductor is part of the equipotential volume, it is itself an equipotential surface. Because field lines and equipotential surfaces are always perpendicular, we conclude that *the electric field just outside a conductor in equilibrium must be perpendicular to the surface of the conductor*. Since the electric field always points toward lower electric potential, we can also conclude that *a positively charged conductor is at a higher electric potential than an adjacent negatively charged conductor*. We'll use these ideas about conductors in the following section to help us understand an important device called a *capacitor*.

THE TAKEAWAY for Section 17-4

✔ The potential is the same everywhere along an equipotential curve or equipotential surface.

✔ Any curve or surface that is perpendicular to the electric field lines is an equipotential.

✔ The spacing of equipotential surfaces with equal potential differences between them is related to the magnitude of the electric field. The electric field is stronger when they are closer together.

✔ A conductor in equilibrium has the same potential throughout its volume. The electric field at the surface of a conductor is perpendicular to the surface.

Prep for the AP® Exam

AP® Building Blocks

1. Electric field lines for a system of two point charges are shown in the figure below. Reproduce the figure and draw on it some equipotential lines for the system.

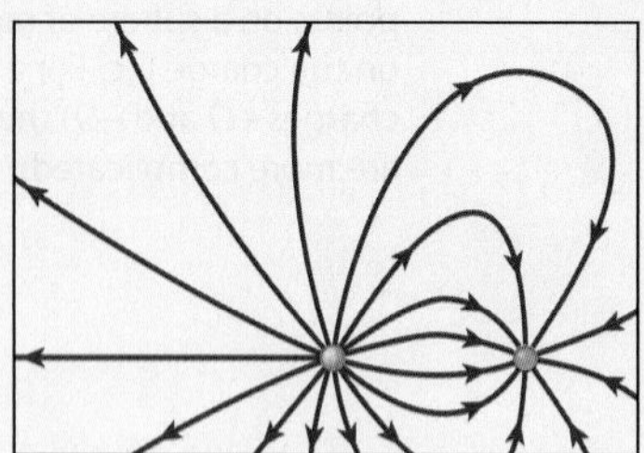

2. In the figure below, equipotential lines, whose values are given on the figure, are shown at 1-m intervals. What is the average magnitude and the direction of the electric field at (a) point A and (b) point B?

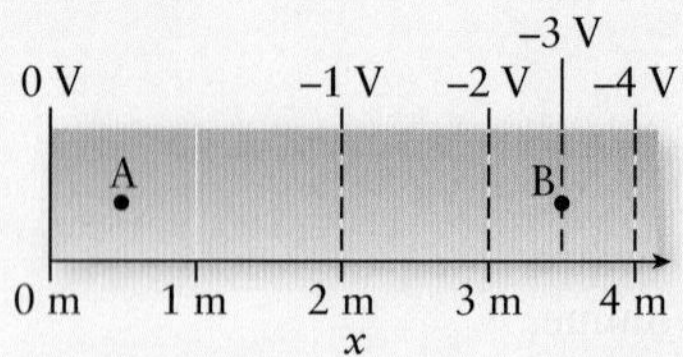

3. Equipotential lines for some region of space are given in the figure below. What is the approximate electric field (magnitude and direction) at (a) point A and (b) point B?

0 V −1 V −2 V −3 V −4 V
A B
0 m 1 m 2 m 3 m 4 m
x

4. A positive charge is moved from one point to another point along an equipotential surface. The work required to move the charge
(A) is positive.
(B) is negative.
(C) is zero.
(D) depends on the magnitude of the potential.

5. How much work is required to move a charged object from one end of an equipotential path to the other? Explain your answer.

AP® Skill Builders

6. On the figures below, draw vertical lines to the outside of the two point charges that approximately represent equally spaced equipotential surfaces perpendicular to the dotted line on which the point charges lie. These figures represent (a) two positive point charges and (b) two negative point charges.

(a)

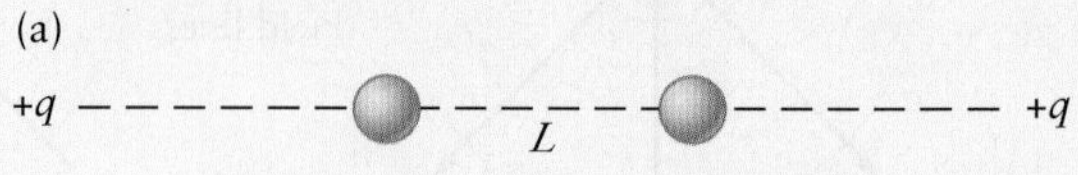

(b)

−q L −q

7. For the figure below, identify all locations where the electric potential is zero for the dipole ($+q$ is a distance L from $-q$).

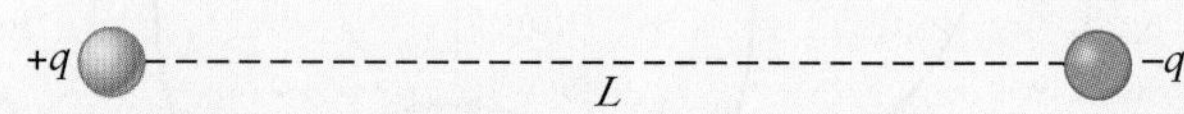

8. For the figure below, describe where the electric potential is zero when the point charges are opposite in sign but not equal in magnitude.

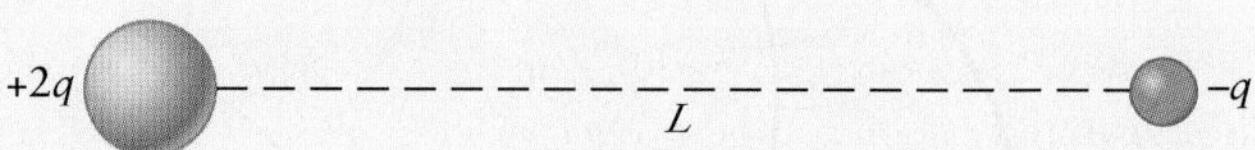

AP® Skills in Action

9. The orange vertical lines in the figure below represent equipotential surfaces in some region of space. Which statement is correct about the electric field $\vec{E}$ at point A compared to the electric field at point B?

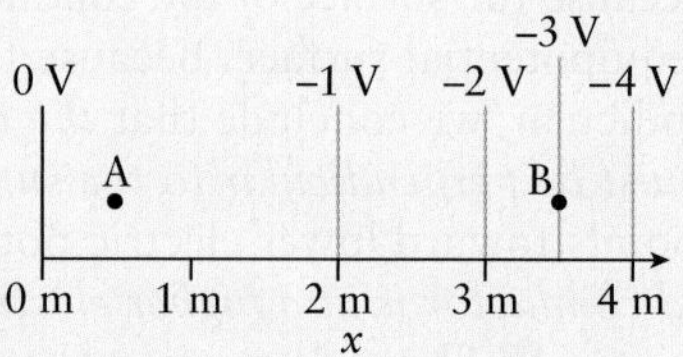

(A) At A the field $\vec{E}$ points to the right and has a greater magnitude than at B.
(B) At A the field $\vec{E}$ points to the right and has a smaller magnitude than at B.
(C) At A the field $\vec{E}$ points to the left and has a greater magnitude than at B.
(D) At A the field $\vec{E}$ points to the left and has a smaller magnitude than at B.

10. Support a claim that a topographical map showing various elevations around a mountain is analogous to the equipotential lines surrounding a charged object.

17-5 A capacitor stores equal amounts of positive and negative charge

The surface of every cell in your body is a *membrane* composed of a phospholipid bilayer that separates the fluid inside the cell and the fluid outside the cell. Negative charge accumulates on the membrane's interior surface, and this attracts positive charge onto the exterior surface. The result is a potential difference between the inner and outer surfaces of the membrane, and an electric field within the membrane that points from the outside in (**Figure 17-10**). This field helps drive essential ions through apertures in the membrane. They are also the source of the electrical signal used by the specialized cells called neurons to code, process, and transmit information.

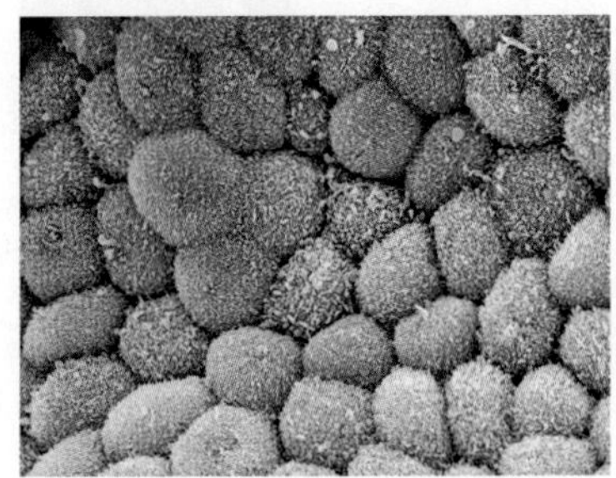

Figure 17-10 Human cells and potential difference Each of the cells in your body, like these cells from the internal lining of the gall bladder, is enclosed by a cell membrane. Typically, the electric potential of the membrane's outer surface is about 0.1 V higher than the electric potential of the inner surface.

A system or device that can store positive and negative charge like a cell membrane is called a **capacitor**. In technological applications a capacitor uses two pieces of metal called **plates**. One plate holds an amount of positive charge q, and the other a negative charge $-q$. (Note that the *net* charge on the capacitor is zero.) It takes work to separate the positive and negative charge against the electric forces that attract them to each other, and this work goes into increasing the electric potential energy of the system. So, a capacitor is a device for storing electric potential energy.

We often need to draw capacitors in diagrams that represent electric circuits. The standard symbol for a capacitor in the diagram of a circuit is

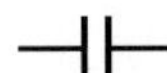

The two closely spaced parallel lines represent the two plates of the capacitor, and the horizontal lines represent wires that can connect the capacitor to an electric circuit.

The simplest geometry for a capacitor is two large parallel plates, one with charge $+q$ and the other with charge $-q$. We can get a good approximation of what the field looks like close to a large, charged plate and far from its edges if we consider such a plate as a series of very many point charges, uniformly spaced in a horizontal plane. **Figure 17-11a** shows this approximation. In **Figures 17-11b** and **17-11c** we pick one line of point charges from the plane and find the electric field for points near the center and near the edge. For such a grid of point charges, as long as we are not close to the edge, the horizontal components of the fields due to the point charges to the left and the right (and behind and in front) of any given point charge cancel, leaving only a field perpendicular to the plane. The electric field due to such a plate, far from its edges, is uniform and perpendicular to the plane of the plate. The magnitude of this electric field is $E = \sigma/2\varepsilon_0$. In this expression, σ is the charge per unit area, equal to the magnitude of the total charge on the plate Q divided by the area A of the plate: $\sigma = Q/A$.

For a capacitor, we place the plates very close together, so this is an excellent approximation for the electric field over most of the area of the plates. **Figure 17-12a** shows these two plates separated from each other. Because each plate has the same magnitude of charge, the field $\vec{E}_+$ due to the charge on the positive plate has the same magnitude as the field $\vec{E}_-$ due to the charge on the negative plate. [Recall from Section 16-5 that the *permittivity of free space* ε_0 is a constant equal to $1/(4\pi k)$, where is k is the constant in Coulomb's law: $\varepsilon_0 = 8.85 \times 10^{-12} \mathrm{C}^2/(\mathrm{N \cdot m}^2)$.] The electric field of the positive plate points away from the plate, and that of the negative plate points toward the negative plate, so the two vectors will point in the same direction. This means that for a capacitor the *net* electric field, the vector sum of $\vec{E}_+$ and $\vec{E}_-$, is $\vec{E} = \vec{E}_+ + \vec{E}_-$. **Figure 17-12b** shows the two plates moved into position to form a **parallel-plate capacitor.** Because the magnitude of each field does not depend on the distance from the plate that generates the field as long as the plates are large compared to their separation, $\vec{E}_+$ cancels $\vec{E}_-$ in the region above the upper plate and below the lower plate. Between the plates the fields $\vec{E}_+$ and $\vec{E}_-$ have the same magnitude and point in the same direction, so

(a)

Top view of a large uniformly charged plane of total charge Q

Imagine dividing the plane into n small squares of equal area. The charge on each square is $q = Q/n$. The larger n, the better the point charge approximation. We have chosen $n = 49$ for ease of illustration.

Point charges used to model the electric field of the plane

(b)

Side view of the center row of point charges from those modeling the plane, with the electric field vectors at the center of the plane for point P, due to each point charge, $q = Q/49$. For each q, a vector is drawn to show the relative magnitude and the direction of the electric field due to that point charge. The vectors the field due to q_1 and q_2 are too small to see on this scale. Since the horizontal components cancel, E_4 dominates.

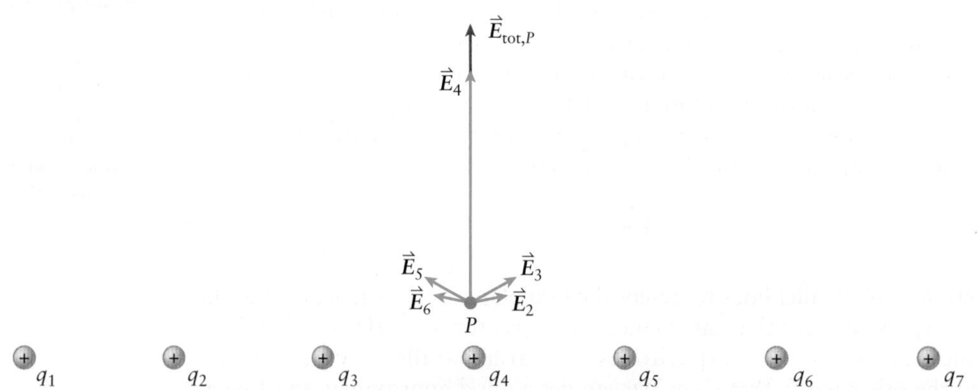

(c)

Side view of the center row of point charges from those modeling the plane, with the electric field vectors near the edge of the plane for point R, due to each point charge, $q = Q/49$. For each q, a vector is drawn to show the relative magnitude and the direction of the electric field due to that point charge. The vectors for the field due to q_1, q_2, q_3, and q_4 are too small to see on this scale. Since there are not additional point charges beyond q_7 the horizontal components do not cancel, so although E_7 dominates, the field points outward with a small horizontal component near the edge of the plate.

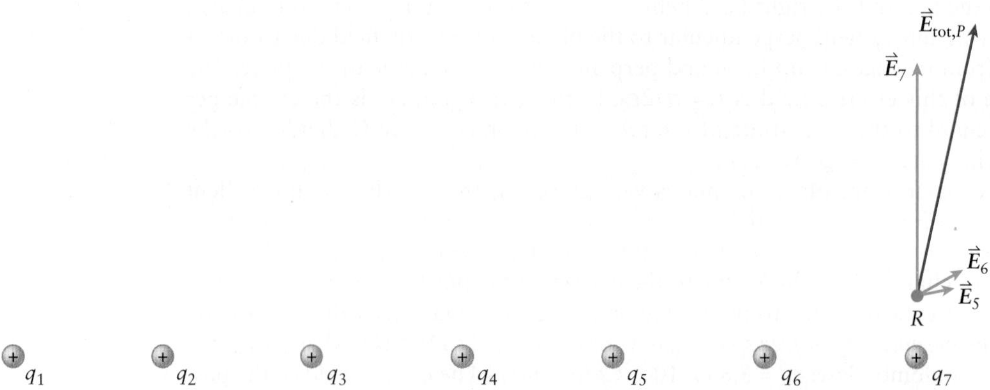

When we chose this row of point charges we could have chosen it along either direction in the plane, so the components of the electric field not pointing directly away from the plane in either direction will cancel for points not near the edge. Also, if we were to use many more point charges in our model, it would be easier to see that as you get farther from the plane, although the field due to the closest point charge decreases, the field due to the nearby charge becomes more vertical, so cancels less and contributes more to the field away from the plane so the magnitude of the field stays fairly constant for points close to the plane compared to its size.

Figure 17-11 The electric field of a large, charged plate approximated as the electric field of many uniformly spaced point charges (a) The arrangement of point charges for this approximation. The approximate electric field for several distances from the plane (b) near the center and (c) closer to the edge along one line of point charges from this arrangement.

(a) Two large plates (viewed from the side) carry charge $+q$ and $-q$. Close to the plates the electric fields are nearly constant, so we represent them by straight, parallel field lines.

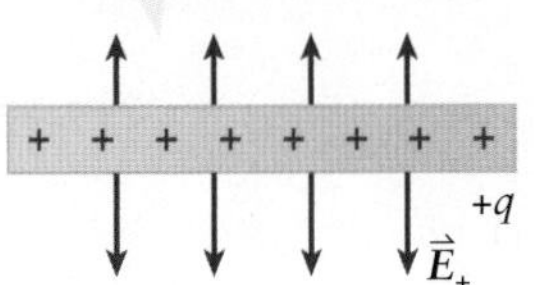

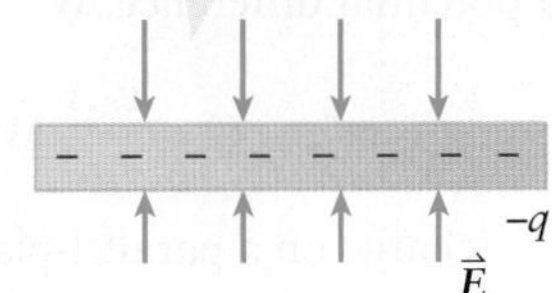

(b) Because the fields are nearly constant, the magnitude of the field due to each plate is the same at any point near the plates...

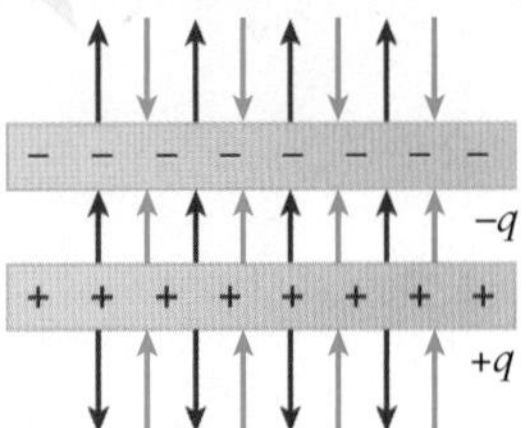

(c) ...so when the plates are placed close together, the fields cancel in the region outside the plates. The fields add in between the plates.

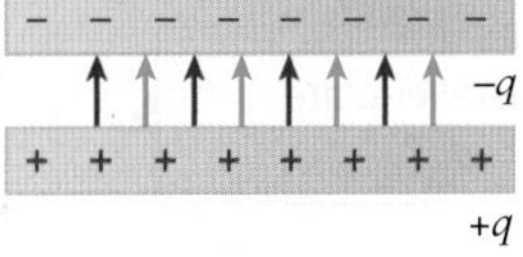

Figure 17-12 Electric field of a parallel-plate capacitor The field due to two oppositely charged plates is the vector sum of the fields due to each plate. Each plate in part (c) has area A, and the distance between the plates is d.

the net field between the plates has twice the magnitude of the field due to either plate by itself:

$$E = 2\left(\frac{\sigma}{2\varepsilon_0}\right) = \frac{\sigma}{\varepsilon_0} = \frac{Q}{\varepsilon_0 A} \tag{17-10}$$

(electric field in a parallel-plate capacitor)

Note that the electric field points from the positive plate to the negative plate, just as we described in our discussion of conductors in Section 17-4.

(Equation 17-10 and the claim that $\vec{E}_+$ and $\vec{E}_-$ cancel outside the capacitor are strictly true only if the plates are very large and very close together, and when we consider points in space far from the edges of the capacitor. This idealization is physically reasonable for the kinds of problems we will encounter. Many commercial capacitors are designed to meet these requirements by rolling two thin conducting sheets with paper in between them as a spacer.)

We can substitute Equation 17-10 into Equation 17-7 to calculate the *potential difference* across (between the plates of) a parallel-plate capacitor. The electric field points from the positive plate to the negative plate. If we let d be the distance between the plates, and we travel from the negative plate to the positive plate, the angle between the electric field and the displacement is $\theta = 180°$. The potential difference is therefore

$$\begin{aligned}\Delta V &= V_+ - V_- = -Ed\cos\theta = -Ed(-1)\\ &= Ed = \left(\frac{Q}{\varepsilon_0 A}\right)d \\ &= \frac{Qd}{\varepsilon_0 A}\end{aligned} \tag{17-11}$$

For a capacitor of plate area A and plate separation d, the potential difference ΔV between the plates is proportional to the magnitude of the charge Q on each plate. The total charge on the capacitor is zero, as there are equal and opposite amounts of charge on the two plates.

We can rewrite Equation 17-11 as an expression for the amount of charge on each plate. To charge an initially uncharged capacitor, we apply a potential difference ΔV between its plates as in **Figure 17-13**. Electrons are driven to one of the plates, giving it

When this battery is connected to the plates of a capacitor, it creates a potential difference ΔV between the plates...

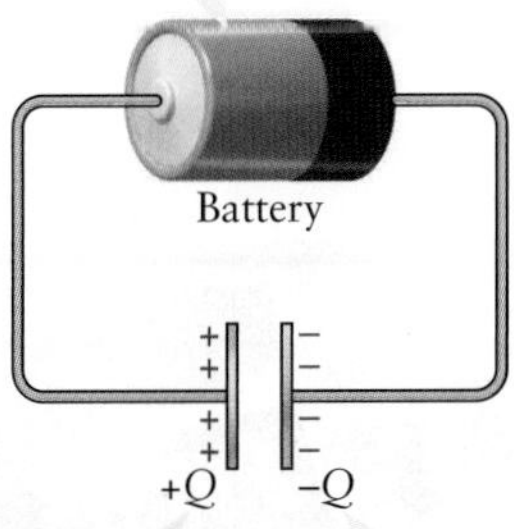

...which drives electrons from one plate, through the battery, and onto the other plate. The plate that lost electrons ends up with charge $+Q$, and the other plate ends up with charge $-Q$.

Figure 17-13 Charging a capacitor The charges that appear on the two capacitor plates have the same magnitude but opposite signs.

a charge $-Q$; the plate from which the electrons were taken is left with a charge $+Q$. (As we mentioned previously the net charge on the capacitor remains zero.) Equation 17-11 tells us that the magnitude Q of the charge that each plate acquires is proportional to the applied potential difference ΔV:

$$Q = \frac{\varepsilon_0 A}{d}\Delta V \quad (17\text{-}12)$$

(charge on a parallel-plate capacitor)

The quantity $\varepsilon_0 A/d$ in Equation 17-12 relates charge the capacitor can hold to the energy per unit charge being applied to the properties of the capacitor (its capacity to hold charge) and is therefore called the **capacitance** of the capacitor. We use the symbol C for capacitance:

EQUATION IN WORDS
Capacitance of a parallel-plate capacitor

$$C = \frac{\varepsilon_0 A}{d} \quad (17\text{-}13)$$

Capacitance of a parallel-plate capacitor

Permittivity of free space = $1/(4\pi k)$

Area of each capacitor plate

Distance between the capacitor plates

For a parallel-plate capacitor, C depends only on the area A of the plates, the distance d between them, and the material between them. We've assumed that the plates are separated by vacuum; in Section 17-8 we'll explore what happens if the space between the capacitor plates is filled with a different material.

Capacitance tells us the amount of charge that can be stored on a capacitor held at a given potential difference. From Equations 17-12 and 17-13,

EQUATION IN WORDS
Charge, potential difference, and capacitance for a capacitor

$$Q = C\Delta V \quad (17\text{-}14)$$

A capacitor carries a charge $+Q$ on its positive plate and a charge $-Q$ on its negative plate.

The magnitude of Q is directly proportional to ΔV, the potential difference between the plates.

The constant of proportionality between charge Q and potential difference ΔV is the **capacitance** C of the capacitor.

The greater the capacitance, the more charge is present for a given potential difference. Note that while Equations 17-12 and 17-13 are valid for a parallel-plate capacitor only, Equation 17-14 is valid for capacitors of *any* geometry. In the problems at the end of this chapter, you'll analyze some other simple types of capacitors.

Equation 17-14 shows that the unit of capacitance is the coulomb per volt, also known as the **farad** (symbol F): $1\text{ F} = 1\text{ C/V}$. The farad is named for the nineteenth-century English physicist Michael Faraday. The capacitors used in consumer electronics are typically in the range of 10 pF (10×10^{-12} F) to 10,000 μF ($10{,}000 \times 10^{-6}$ F); 1 F is an extremely large capacitance. Using $1\text{ F} = 1\text{ C/V}$, $1\text{ V} = 1\text{ J/C}$, and $1\text{ J} = 1\text{ N}\cdot\text{m}$, we see that $1\text{ F} = (1\text{ C})/(1\text{ J/C}) = 1\text{ C}^2/\text{J} = 1\text{ C}^2/(\text{N}\cdot\text{m})$. We can express the value of the constant ε_0 as

$$\varepsilon_0 = 8.85 \times 10^{-12}\frac{\text{C}^2}{\text{N}\cdot\text{m}^2} = 8.85 \times 10^{-12}\frac{\text{F}}{\text{m}}$$

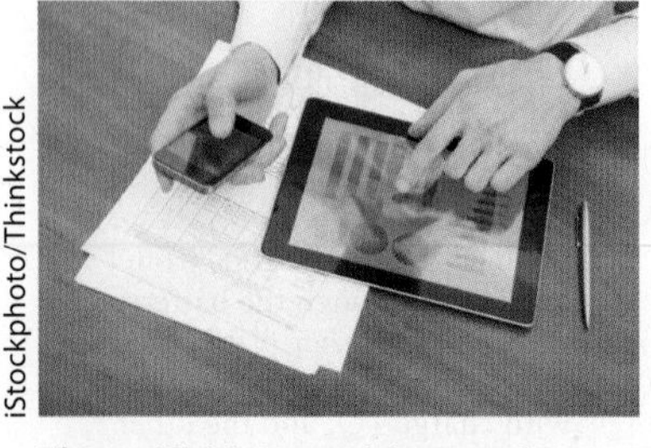
iStockphoto/Thinkstock

Figure 17-14 Capacitive touchscreens Mobile devices such as these use the physics of capacitors to determine where your finger touches the screen.

These units are useful in calculations of capacitance, as we'll see below.

An everyday application of Equation 17-13 is the touchscreen on a mobile device such as a smartphone or tablet (**Figure 17-14**). Behind the device's glass screen is a layer of a special transparent conductor called indium tin oxide (ITO), which is actually a solid mixture of indium oxide, In_2O_3, and tin oxide, SnO_2. When you touch your finger to the screen, the conducting ITO layer acts as one plate of a capacitor and your finger—which is also a conductor—acts as the other plate. Sensor circuits in the mobile device detect where on the screen a capacitance appears due to this capacitor, which is how the device "knows" where on the screen it has been touched. (To verify that the object touching the screen has to be a conductor, try using the rubber eraser on a pencil to touch the screen of a mobile device. Because rubber is an insulator, not a conductor, the screen won't respond.) The sensor circuits are adjusted so that they will register only if the capacitance C is above a certain minimum value. That explains why you have to physically touch the screen: According to Equation 17-13, C increases as the distance d between the plates decreases, so the capacitance is largest when your finger is touching the screen and so is closest to the ITO layer. Your finger still acts as a capacitor plate if you hold it a slight distance away from the screen, but the capacitance is now too low to trigger the sensor circuits.

EXAMPLE 17-7 A Parallel-Plate Capacitor

Two square, parallel conducting plates each have dimensions 5.00 cm by 5.00 cm and are placed 0.100 mm apart. Determine the capacitance of this configuration.

Set Up

We'll use Equation 17-13 to determine the value of C for this capacitor.

Capacitance of a parallel-plate capacitor:

$$C = \frac{\varepsilon_0 A}{d} \qquad (17\text{-}13)$$

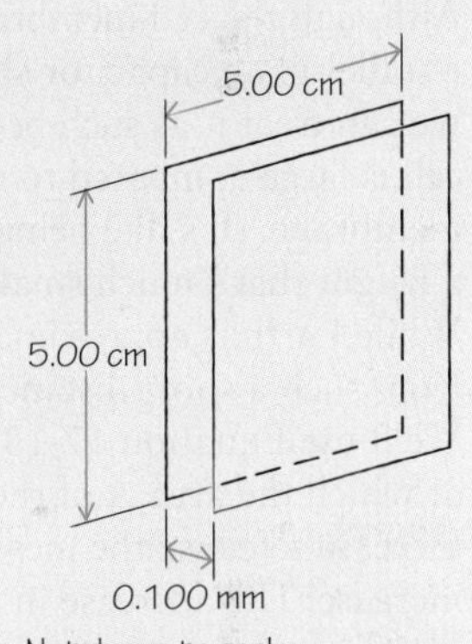

Solve

The area of each plate is the product of the length of the two sides, and the plate separation is given. We need to convert all dimensions into SI units.

Area $A = (5.00 \times 10^{-2}\ \text{m})(5.00 \times 10^{-2}\ \text{m}) = 2.50 \times 10^{-3}\ \text{m}^2$

Plate separation $d = 1.00 \times 10^{-4}$ m

Using Equation 17-13:

$$C = \frac{(8.85 \times 10^{-12}\ \text{F/m})(2.50 \times 10^{-3}\ \text{m}^2)}{1.00 \times 10^{-4}\ \text{m}} = 2.21 \times 10^{-10}\ \text{F}$$

$$= 221\ \text{pF}$$

(Recall that the prefix p, or *pico-*, represents 10^{-12}.)

Reflect

Capacitors with capacitances of a few hundred picofarads are found in calculators and mobile phones.

To make a 1-F capacitor with the same separation $d = 0.100$ mm between plates, we would have to increase the plate area by a factor of $(1\ \text{F})/(2.21 \times 10^{-10}\ \text{F}) = 4.52 \times 10^9$. The length of each side would have to be increased by a factor of the square root of 4.52×10^9, or 6.73×10^4. The sides of the plates would be $(6.73 \times 10^4)(5.00 \times 10^{-2}\ \text{m}) = 3.36 \times 10^3$ m, or 3.36 *kilometers* (about 2 miles). One farad is a *very* large capacitance.

NOW WORK Problems 3, 4, 5, 6, and 9 from The Takeaway 17-5.

EXAMPLE 17-8 Insulin Release

The hormone insulin minimizes variations in blood glucose levels. Pancreatic beta cells (β cells) synthesize insulin and store it in *vesicles*, bubble-like organelles approximately 150 nm in radius within the cytoplasm of the cells. To release insulin, vesicles fuse with the membrane of the β cell. This increases the surface area of the β cell by the surface area of the fused vesicles (**Figure 17-15**). The thickness of the cell membrane does not change, so the increase in surface area increases the capacitance of the β cell membrane. Experiment shows that the membrane capacitance increases at 1.6×10^{-13} F/s during insulin release. If the membrane capacitance is approximately 1 μF per square centimeter of surface area, estimate the number of vesicles that fuse with the cell membrane per second during insulin release.

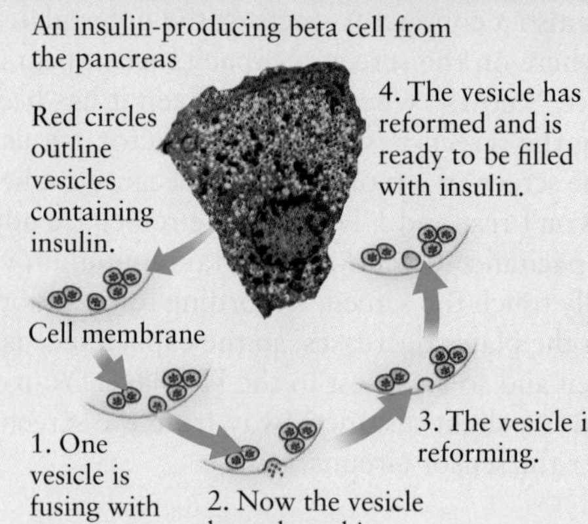

An insulin-producing beta cell from the pancreas
Red circles outline vesicles containing insulin.
4. The vesicle has reformed and is ready to be filled with insulin.
Cell membrane
1. One vesicle is fusing with the cell membrane.
2. Now the vesicle has released its insulin to the outside of the cell. The vesicle membrane is part of the cell membrane.
3. The vesicle is reforming.
NASA/Tim McClanahan

Figure 17-15 Changing membrane, changing capacitance To study this important process in a pancreatic β cell, scientists monitor changes in the capacitance of the cell membrane.

Set Up

Although the cell membrane is not flat like the parallel-plate capacitor shown in Figure 17-12, we can treat it as such because the size of the cell is large compared to the thickness of the cell membrane. (It's like being above Earth's surface at a height that's much smaller than Earth's radius. While Earth is approximately spherical, when seen from such a short distance, it appears to be flat.) We'll use Equation 17-13 to calculate the rate at which the area A of the cell membrane must increase to cause the measured rate of capacitance increase. This increase in area comes from fusing vesicles.

Capacitance of a parallel-plate capacitor:

$$C = \frac{\varepsilon_0 A}{d} \qquad (17\text{-}13)$$

Surface area of a sphere of radius r:

$$A_{\text{sphere}} = 4\pi r^2$$

Solve

Find the rate at which the membrane area must increase to cause this rate of capacitance increase.

Rate of change of the capacitance of the membrane:

$$\frac{\Delta C_{\text{m}}}{\Delta t} = 1.6 \times 10^{-13}\ \text{F/s}$$

Equation 17-13 tells us that the capacitance C_{m} is directly proportional to the membrane surface area A_{m}. We are told that the capacitance per unit area is approximately 1 μF/cm^2, so the change in capacitance ΔC_{m} that corresponds to a given change in area ΔA_{m} is

$$\Delta C_m = \left(1\,\frac{\mu\text{F}}{\text{cm}^2}\right)\Delta A_m$$

Since the capacitance increases by 1.6×10^{-13} F in 1 s, the increase in area of the membrane in 1 s must be

$$\Delta A_{\text{m}} = \frac{\Delta C_{\text{m}}}{1\ \mu\text{F/cm}^2}$$

$$= (1.6 \times 10^{-13}\ \text{F})\left(1\,\frac{\text{cm}^2}{\mu\text{F}}\right)\left(\frac{1\ \mu\text{F}}{10^{-6}\ \text{F}}\right)\left(\frac{1\ \text{m}}{100\ \text{cm}}\right)^2$$

$$= 1.6 \times 10^{-11}\ \text{m}^2$$

(We used 1 μF = 1 microfarad = 10^{-6} F and 1 m = 100 cm.)

Now we can calculate how many vesicles must fuse with the membrane per second to cause this increase in area.

The surface area of a single vesicle is

$$A_{\text{vesicle}} = 4\pi r_{\text{vesicle}}^2 = 4\pi(150 \text{ nm})^2 = 4\pi(150 \times 10^{-9} \text{ m})^2$$
$$= 2.8 \times 10^{-13} \text{ m}^2$$

The number of vesicles that must add their area to the membrane in 1 s to cause an area increase $\Delta A_m = 1.6 \times 10^{-11}$ m² is

$$\frac{\Delta A_m}{A_{\text{vesicle}}} = \frac{1.6 \times 10^{-11} \text{ m}^2}{2.8 \times 10^{-13} \text{ m}^2/\text{vesicle}} = 57 \text{ vesicles}$$

Our final answer should have just one significant digit, since we were given the capacitance per unit area 1 μF/cm² to just one significant digit. So our final answer is that about 60 vesicles fuse with the membrane wall per second.

Reflect

The ability to detect small changes in capacitance makes it possible to study the molecular mechanisms of hormone release from single cells and to verify previous biochemical measurements of insulin release from β cells.

Extend: After insulin release, new vesicles form by pinching off from the cell membrane and are refilled with insulin. By recycling the vesicle membrane, β cells maintain their size over the long term.

NOW WORK Problem 10 from The Takeaway 17-5.

THE TAKEAWAY for Section 17-5

✔ A parallel-plate capacitor has two plates, one with charge $+Q$ and the other with charge $-Q$.

✔ For a given capacitor the potential difference between the plates is proportional to the magnitude of the charge Q on each plate. The proportionality constant, called the capacitance, depends on the geometry of the plates and on the material in the space between the plates.

AP® Building Blocks

1. Using a single 10.0-V battery as a source of potential difference across a capacitor, what capacitance do you need to store 10.0 μC of charge?
2. A 2.00-μF capacitor is connected to a 12.0-V battery. What is the magnitude of the charge on each plate of the capacitor?
3. (EX 17-7) A parallel-plate capacitor has a plate separation of 1.00 mm. If the material between the plates is air, what plate area is required to provide a capacitance of 2.00 pF?
4. (EX 17-7) A parallel-plate capacitor has square plates that have edge lengths equal to 1.00×10^2 cm and are separated by 1.00 mm. What is the capacitance of this device?

AP® Skill Builders

5. An air-filled parallel-plate capacitor has plates measuring 10.0 cm × 10.0 cm and a plate separation of 1.00 mm. If you want to construct a parallel-plate capacitor of the same capacitance but with plates measuring 5.00 cm × 5.00 cm, what plate separation do you need?
6. (EX 17-7) A parallel-plate capacitor has square plates that have edge length equal to 1.00 m. If the material between the plates is air, calculate the separation distance required to provide a capacitance of 8850 pF.
7. Describe three methods by which you might increase the capacitance of a parallel-plate capacitor.
8. Explain why capacitance depends neither on the stored charge Q nor on the potential difference ΔV across the plates of a capacitor.

AP® Skills in Action

9. (EX 17-7) A parallel-plate capacitor has a potential difference of 1V across its plates. If the potential difference is increased to 2V, what effect does this have on the capacitance C?
   (A) C increases by a factor of 4.
   (B) C increases by a factor of 2.
   (C) C becomes $\frac{1}{2}$ as great.
   (D) C is unchanged.

EX 17-8 10. A parallel-plate, air-filled capacitor has a charge of 20.0 μC and a gap width of 0.100 mm. The potential difference across the plates is 200 V.
(a) Calculate the magnitude of the electric field between the plates.
(b) Calculate the surface charge density on the positive plate.
(c) If the plates are moved closer together while the charge remains constant, how do the electric field magnitude, surface charge density, and potential difference change, if at all? Explain your answers.

17-6 A capacitor is a storehouse of electric potential energy

In electric circuits, capacitors are most useful because of their ability to store electric potential energy. An applied potential difference ΔV across the plates, such as that supplied by the battery in Figure 17-13, charges a capacitor by effectively pulling electrons (negatively charged) from one of the plates and depositing them on the other. To move negative charge away from positive charge requires work, and it is this work that results in electric potential energy being stored in the capacitor. At a later time the potential energy can be transferred by charge leaving the capacitor and passed on to other parts of the circuit, which we will learn about in the next chapter. (That's what happens in the electronic flash unit in a camera or mobile phone. The device's battery charges a capacitor, and the energy stored in the charged capacitor is used to produce a short, intense burst of light.)

Let's see how to calculate the amount of electric potential energy stored in a capacitor that has charge $+Q$ and $-Q$ on its positive and negative plates, respectively. This is equal to the amount of electric potential energy that's added to the capacitor if we start with both plates uncharged (which we can regard as a state of zero potential energy) and move charge $+Q$ from the first plate to the second one, leaving charge $-Q$ on the first plate. If the potential difference between the plates had a constant value ΔV, Equation 17-6 tells us that the change in electric potential energy $\Delta U_{\text{electric}}$ in this process would be

$$\Delta V = \frac{\Delta U_{\text{electric}}}{q} \quad \text{so} \quad \Delta U_{\text{electric}} = q\,\Delta V$$

However, the potential difference between the plates (that is, across the capacitor) does *not* stay constant as we transfer charge from one plate to the other! Equation 17-14 tells us that the potential difference between the plates is proportional to the amount of charge on the positive plate. So as we transfer more charge from one plate to the other, the potential difference across which the charge must move increases from zero (its starting value) to a final value ΔV given by Equation 17-14: $Q = C\Delta V$, so $\Delta V = Q/C$. To correctly calculate the amount of potential energy stored in the capacitor when it is charged, we have to replace ΔV in Equation 17-6 by the *average* value of the potential difference during the charging process. Because potential difference increases in direct proportion to the charge, this average value is just the average of the starting potential difference and the final potential difference V:

$$\begin{aligned}\Delta U_{\text{electric}} &= Q\,\Delta V_{\text{average}} \\ &= Q\left[\frac{(\text{starting potential difference}) + (\text{final potential difference})}{2}\right] \\ &= Q\left(\frac{0+\Delta V}{2}\right) = \frac{1}{2}Q\Delta V \qquad (17\text{-}15)\end{aligned}$$

Equation 17-14 tells us that $Q = CV$ and $\Delta V = Q/C$, so we can also write Equation 17-15 as

$$\Delta U_{\text{electric}} = \frac{1}{2}(C\Delta V)\Delta V = \frac{1}{2}C\Delta V^2 \quad \text{or} \quad \Delta U_{\text{electric}} = \frac{1}{2}Q\left(\frac{Q}{C}\right) = \frac{Q^2}{2C} \qquad (7\text{-}16)$$

If we say that the electric potential energy of the initial uncharged capacitor was zero, the final potential energy U_{electric} is just equal to the increase in potential energy $\Delta U_{\text{electric}}$ given by Equation 17-15 or Equation 17-16. So, we can write the potential energy stored in the capacitor in three ways:

The **electric potential energy** stored in a charged capacitor...

$$U_{\text{electric}} = \frac{1}{2}Q\Delta V = \frac{1}{2}C\Delta V^2 = \frac{Q^2}{2C} \qquad (17\text{-}17)$$

...can be expressed in three ways in terms of the **charge** Q, **potential difference** ΔV, and **capacitance** C.

EQUATION IN WORDS
Electric potential energy stored in a capacitor

Equation 17-17 says that the energy stored in a charged capacitor is proportional to the *square* of the charge q or, equivalently, to the square of the potential difference ΔV across the capacitor. These results are true for capacitors of all kinds, not just the simple parallel-plate capacitor that we discussed in Section 17-5.

The last of the three expressions for electric potential energy in Equation 17-17, $U_{\text{electric}} = Q^2/(2C)$, is very similar to the equation for spring potential energy that we learned in Section 7-6:

Spring potential energy of a stretched or compressed spring

Spring constant of the spring

$$U_s = \frac{1}{2}kx^2 \qquad (7\text{-}16)$$

Extension of the spring, when the equilibrium position of the end of the spring when it is relaxed is defined as $x = 0$ ($x > 0$ if spring is stretched, $x < 0$ if spring is compressed)

EQUATION IN WORDS
Spring potential energy

The only differences between $U_{\text{electric}} = Q^2/(2C)$ from Equation 17-17 and the expression in Equation 7-16 is that the spring displacement x is replaced by the charge Q, and the spring constant k is replaced by the reciprocal of the capacitance, $1/C$. This similarity isn't surprising. To add potential energy to a spring by stretching it, you have to pull against the force of magnitude $F = kx$ that the spring exerts on you. The greater the distance the spring is already stretched, the more force it exerts and the harder it is to stretch it farther. In exactly the same way, to add potential energy to a capacitor by increasing the magnitude of charge on the two plates, you have to transfer charge against a potential difference $\Delta V = (1/C)Q$. The greater the charge already on the plates, the greater the potential difference and the harder it is to increase Q.

EXAMPLE 17-9 A Defibrillator Capacitor

A defibrillator, like the one shown in the photograph that opens this chapter, is essentially a capacitor that is charged by a high-potential difference source and then delivers the stored energy to a patient's heart. (a) How much charge does the 80.0-μF capacitor in a certain defibrillator store when it is fully charged by applying 2.50 kV? (b) How much energy can this defibrillator deliver?

Set Up

We're given the capacitance $C = 80.0\ \mu\text{F}$ (recall $1\ \mu\text{F} = 1$ microfarad $= 10^{-6}$ F) and the potential difference $V = 2.50$ kV (recall 1 kV = 1 kilovolt = 10^3 V) between the capacitor plates. We'll use Equation 17-14 to determine the magnitude q of the charge on each capacitor plate and Equation 17-17 to find the potential energy U_{electric} stored in the charged capacitor. If we assume that no energy is lost in the process of being transferred to the patient, this is equal to the energy that the defibrillator delivers.

Charge, potential difference, and capacitance for a capacitor:

$$Q = C\Delta V \tag{17-14}$$

Electric potential energy stored in a capacitor:

$$U_{\text{electric}} = \frac{1}{2}Q\Delta V = \frac{1}{2}C\Delta V^2 = \frac{Q^2}{2C} \tag{17-17}$$

Solve

(a) Substitute the given values of C and ΔV into Equation 17-14 to solve for the charge q.

Charge on the capacitor plates:

$$\begin{aligned} Q = C\Delta V &= (80.0\ \mu\text{F})(2.50\ \text{kV}) \\ &= (80.0 \times 10^{-6}\ \text{F})(2.50 \times 10^3\ \text{V}) \\ &= 0.200\ \text{F}\cdot\text{V} = 0.200\ \text{C} \end{aligned}$$

(Recall that 1 F = 1 C/V, so 1 F · V = 1 C.)

(b) Since the values of C and ΔV are given, let's use the second of the three relationships in Equation 17-17 to find the stored electric potential energy.

Energy stored in the capacitor:

$$\begin{aligned} U_{\text{electric}} &= \frac{1}{2}C\Delta V^2 \\ &= \frac{1}{2}(80.0 \times 10^{-6}\ \text{F})(2.50 \times 10^3\ \text{V})^2 \\ &= 2.50 \times 10^2\ \text{F}\cdot\text{V}^2 \\ &= 2.50 \times 10^2\ \text{J} \end{aligned}$$

(Recall that 1 V = 1 J/C and 1 F = 1 C/V, so $1\ \text{F}\cdot\text{V}^2 = 1\,[\text{C/V}]\text{V}^2 = 1\ \text{C}\cdot\text{V} = 1\ \text{C}\cdot[\text{J/C}] = 1\ \text{J}$.)

Reflect

The American Heart Association recommends that a defibrillator shock should deliver between 40 and 360 J to be effective, so our numerical result is in the recommended range.

We can double-check our answers by using the other two relationships in Equation 17-17. Happily, by using the value of Q that we calculated in part (a), we get the same result for U_{electric} in part (b) as we found above.

Remember that in solving any problem, you should *always* take advantage of alternative ways to find the answer in order to check your results.

One alternative way to calculate the energy stored in the capacitor:

$$\begin{aligned} U_{\text{electric}} &= \frac{1}{2}Q\Delta V \\ &= \frac{1}{2}(0.200\ \text{C})(2.50 \times 10^3\ \text{V}) \\ &= 2.50 \times 10^2\ \text{C}\cdot\text{V} \\ &= 2.50 \times 10^2\ \text{J} \end{aligned}$$

Another alternative way to calculate the energy stored in the capacitor:

$$\begin{aligned} U_{\text{electric}} &= \frac{Q^2}{2C} \\ &= \frac{(0.200\ \text{C})^2}{2(80.0 \times 10^{-6}\ \text{F})} \\ &= 2.50 \times 10^2\ \frac{\text{C}^2}{\text{F}} \\ &= 2.50 \times 10^2\ \text{J} \end{aligned}$$

(Recall that 1 V = 1 J/C and 1 F = 1 C/V. You should verify that $1\ \text{C}^2/\text{F} = 1$ J.)

NOW WORK Problems 1–6 from The Takeaway 17-6.

THE TAKEAWAY for Section 17-6

✔ A charged capacitor stores electric potential energy. The amount of energy stored is proportional to the square of the magnitude of the charge on each plate and also proportional to the square of the potential difference between the plates.

Prep for the AP® Exam

AP® Building Blocks

EX 7-9 1. Using a single 10.0-V battery, what capacitance do you need to store 1.00×10^{-4} J of electric potential energy?

EX 7-9 2. A parallel-plate capacitor has square plates that have edge length equal to 1.00×10^{2} cm and are separated by 1.00 mm. It is connected to a battery and is charged to 12.0 V. How much energy is stored in the capacitor?

EX 7-9 3. You charge a 2.00-μF capacitor to 50.0 V. How much additional energy must you add to charge it to 100 V?

EX 7-9 4. A capacitor has a capacitance of 80.0 μF. If you want to store 160 J of electric energy in this capacitor, what potential difference do you need to apply to the plates?

AP® Skill Builders

EX 17-9 5. For each case below, find the value of the necessary capacitance.

(a) You want to store 1.00×10^{-5} C of charge on a capacitor, and you have a 100-V potential difference source with which to charge it.

(b) You want to store 1.00×10^{-3} J of energy on a capacitor, and you have a 100-V potential difference source with which to charge it.

EX 17-9 6. A defibrillator containing a 20.0-μF capacitor is used to shock the heart of a patient by holding it to the patient's chest. Just prior to discharging, the capacitor has a potential difference of 10.0 kV across its plates. How much energy is transferred to the patient, assuming no energy losses?

7. You charge a capacitor and then remove it from the battery. The capacitor consists of large movable plates with air between them. You pull the plates a bit farther apart. What happens to the stored energy?

8. A parallel-plate capacitor is connected to a battery that maintains a constant potential difference across the plates. If the separation between the plates is doubled, the electric energy stored in the capacitor will be

(A) halved.
(B) doubled.
(C) quadrupled.
(D) quartered.

AP® Skills in Action

9. Suppose you increase the distance between the plates of a charged parallel-plate capacitor without changing the amount of charge stored on the plates. What will happen to the energy stored in the capacitor?

(A) It will decrease.
(B) It will remain the same.
(C) It will increase.
(D) There is not enough information given to decide.

10. Suppose you increase the distance between the plates of a parallel-plate capacitor while holding constant the potential difference between the plates. (You could do this by keeping the plates connected to a battery, as in Figure 17-13.) What will happen to the energy stored in the capacitor?

(A) It will decrease.
(B) It will remain the same.
(C) It will increase.
(D) There is not enough information given to decide.

17-7 Capacitors can be combined in series or in parallel

In both biological systems and electric circuits, it is not uncommon for more than one capacitor to be connected in some way. The net result is that the capacitor combination behaves as though it were a *single* capacitor, with an **equivalent capacitance** that depends on the properties of the individual capacitors present. (An analogy is lifting a heavy conference table to move it across a room. Four people of normal strength could do the job, or it could be done by a single weight lifter. We would say that the four people together have an "equivalent strength" comparable to that of the weight lifter.) Our goal in this section is to find the equivalent capacitance in different situations.

Capacitors in Series

Figure 17-16 shows three initially uncharged capacitors with capacitances C_1, C_2, and C_3 that are connected end to end. The capacitors become charged when the combination

Figure 17-16 Capacitors in series
What is the equivalent capacitance of this series combination?

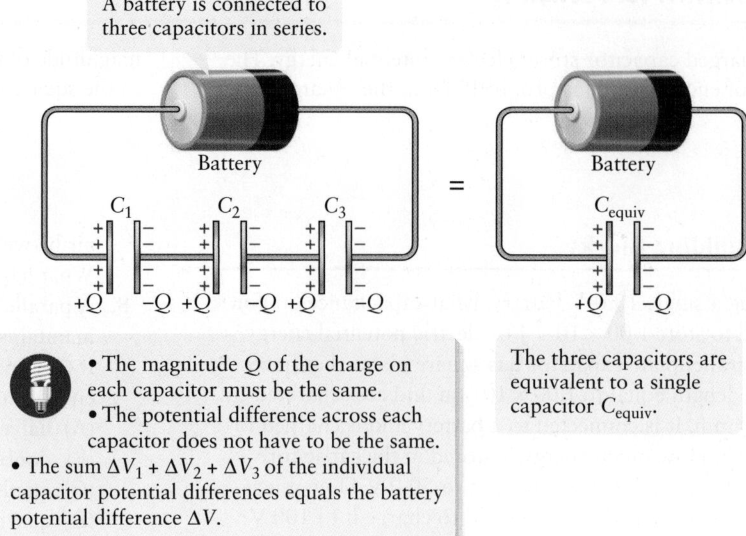

is connected to a battery of potential difference ΔV as shown in the figure. Note that the negative plate of one capacitor is connected to the positive plate of the next capacitor in the combination. Capacitors connected in this way are said to be in **series**.

A battery is connected to three capacitors in series.
Battery
C_1
C_2
C_3
+Q
−Q
=
Battery
C_{equiv}
• The magnitude Q of the charge on each capacitor must be the same.
• The potential difference across each capacitor does not have to be the same.
• The sum $\Delta V_1 + \Delta V_2 + \Delta V_3$ of the individual capacitor potential differences equals the battery potential difference ΔV.
The three capacitors are equivalent to a single capacitor C_{equiv}.

What is the equivalent capacitance of this series combination? To answer this question let's first determine the charges Q_1, Q_2, and Q_3 on the individual capacitors and the potential differences ΔV_1, ΔV_2, and ΔV_3 across the individual capacitors. For each capacitor the magnitudes of the charge on the positive and negative plates must be equal. So as the battery draws negative charge from the left-hand plate of C_1, whatever positive charge $+Q$ that plate acquires must be balanced by charge $-Q$ on the right plate. The charging of the right plate of C_1 occurs as negatively charged electrons are drawn from the left plate of C_2, and because this whole section is initially uncharged, the left plate of C_2 acquires charge $+Q$. If we apply the same reasoning to C_3, we see that all three capacitors acquire the *same* charge:

(17-18)
$$Q_1 = Q_2 = Q_3 = Q$$
(capacitors in series)

Thus, the series combination of three capacitors is equivalent to a single capacitor with charge Q. You can think of the charge on the negative plate of C_1 as canceling the charge on the positive plate of C_2 and likewise for the charges on the negative plate of C_2 and the positive plate of C_3.

The charges and potential differences for each capacitor are also given by Equation 17-14, $Q = C\Delta V$. If we substitute this into Equation 17-18, we get

$$Q = C_1\Delta V_1 = C_2\Delta V_2 = C_3\Delta V_3$$
(capacitors in series)

which we can rewrite as expressions for the potential differences across the individual capacitors:

(17-19)
$$\Delta V_1 = \frac{Q}{C_1} \quad \Delta V_2 = \frac{Q}{C_2} \quad \Delta V_3 = \frac{Q}{C_3}$$
(capacitors in series)

Now, the potential difference ΔV of the battery equals the potential difference ΔV across the combination of three capacitors. This is just the sum of the potential differences across the individual capacitors:

(17-20)
$$\Delta V = \Delta V_1 + \Delta V_2 + \Delta V_3$$
(capacitors in series)

If we substitute Equations 17-19 into Equation 17-20, we get a relationship between the charge Q on each capacitor and the potential difference ΔV across the combination—that is, between the charge Q on the equivalent capacitor and the potential difference ΔV across the equivalent capacitor:

$$\Delta V = \frac{Q}{C_1} + \frac{Q}{C_2} + \frac{Q}{C_3} = Q\left(\frac{1}{C_1} + \frac{1}{C_2} + \frac{1}{C_3}\right) \quad (17\text{-}21)$$

(capacitors in series)

For the equivalent capacitor alone, of capacitance C_{equiv}, Equation 17-14 says that $Q = C_{equiv}\Delta V$ or $\Delta V = Q/C_{equiv}$. If we compare this to Equation 17-21, we see that

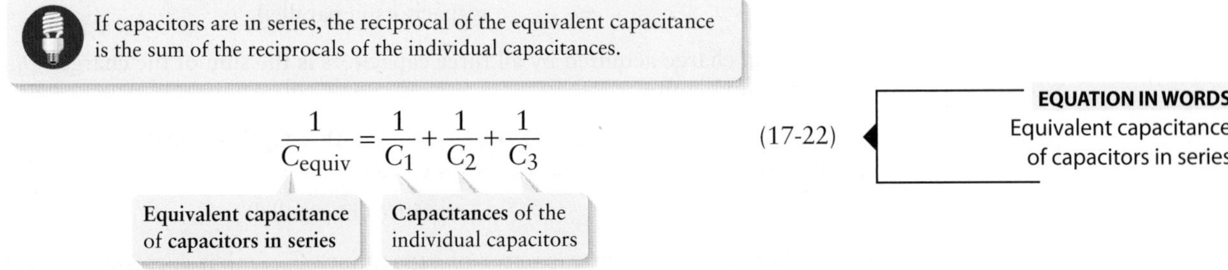
If capacitors are in series, the reciprocal of the equivalent capacitance is the sum of the reciprocals of the individual capacitances.
Equivalent capacitance of capacitors in series
Capacitances of the individual capacitors
EQUATION IN WORDS
Equivalent capacitance of capacitors in series

$$\frac{1}{C_{equiv}} = \frac{1}{C_1} + \frac{1}{C_2} + \frac{1}{C_3} \quad (17\text{-}22)$$

If there are just two or more than three capacitors in series, the same rule given in Equation 17-22 applies. The equivalent capacitance of a series combination is always *less* than the smallest capacitance of any of the individual capacitors.

Capacitors in Parallel

Figure 17-17 shows an alternative way to connect the three initially uncharged capacitors C_1, C_2, and C_3 to a battery with potential difference ΔV. Capacitors connected in this way are said to be in **parallel**. Notice the difference in the arrangement of the capacitors in parallel (Figure 17-17) compared to capacitors in series (Figure 17-16). In a series arrangement, the right-hand plate of one capacitor is connected to the left-hand plate of the capacitor to its right. In a parallel arrangement, all of the right-hand plates are connected, and all of the left-hand plates are connected.

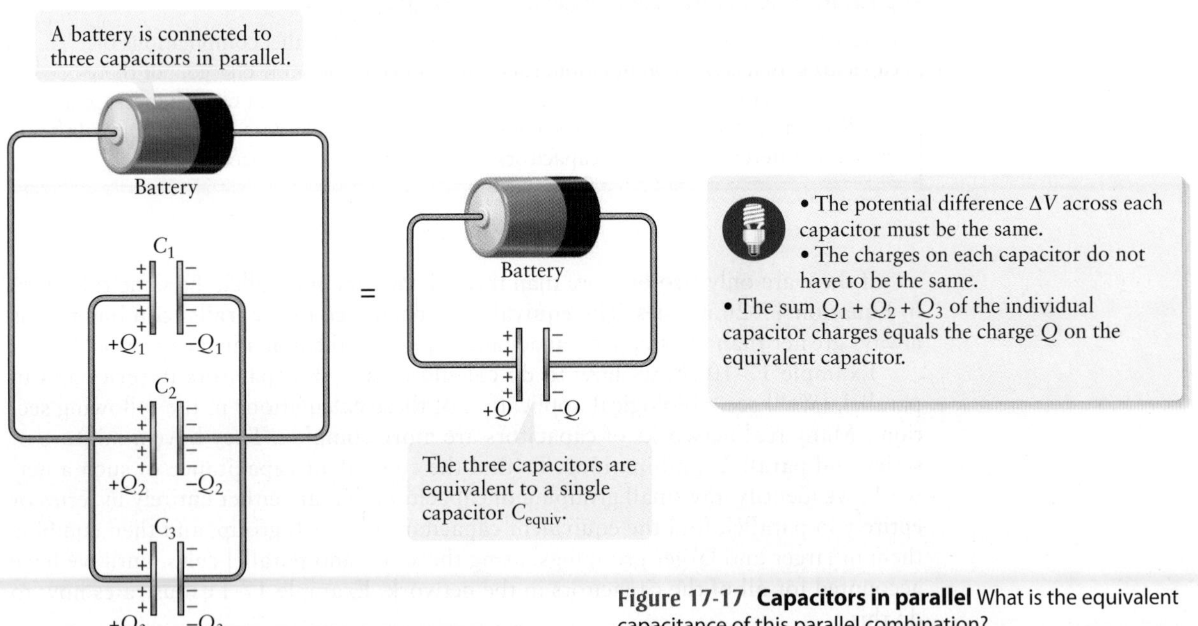
A battery is connected to three capacitors in parallel.
Battery
C_1
$+Q_1$
$-Q_1$
C_2
$+Q_2$
$-Q_2$
C_3
$+Q_3$
$-Q_3$
=
Battery
$+Q$
$-Q$
The three capacitors are equivalent to a single capacitor C_{equiv}.
• The potential difference ΔV across each capacitor must be the same.
• The charges on each capacitor do not have to be the same.
• The sum $Q_1 + Q_2 + Q_3$ of the individual capacitor charges equals the charge Q on the equivalent capacitor.

Figure 17-17 Capacitors in parallel What is the equivalent capacitance of this parallel combination?

To find the equivalent capacitance of capacitors in parallel, first note that all of the right-hand capacitor plates are connected to one terminal of the battery and all of the left-hand plates are connected to the other terminal. So each of the potential differences ΔV_1, ΔV_2, and ΔV_3 across the individual capacitors is equal to the potential difference ΔV across the battery:

$$\Delta V = \Delta V_1 = \Delta V_2 = \Delta V_3 \quad \text{(capacitors in parallel)} \tag{17-23}$$

Equation 17-23 coupled with Equation 17-14, $Q = C\Delta V$, then tells us the charges Q_1, Q_2, and Q_3 on the individual capacitors:

$$Q_1 = C_1\Delta V \quad Q_2 = C_2\Delta V \quad Q_3 = C_3\Delta V \quad \text{(capacitors in parallel)} \tag{17-24}$$

The *total* charge acquired by all three capacitors is the sum of the charges Q_1, Q_2, and Q_3. From Equation 17-24,

$$\begin{aligned} Q &= Q_1 + Q_2 + Q_3 = C_1\Delta V + C_2\Delta V + C_3\Delta V \\ &= (C_1 + C_2 + C_3)\Delta V \end{aligned} \quad \text{(capacitors in parallel)} \tag{17-25}$$

For the equivalent capacitor of capacitance C_{equiv} that corresponds to the three capacitors in parallel, Equation 17-14 says that $Q = C_{equiv}\Delta V$. Comparing this to Equation 17-25 shows that

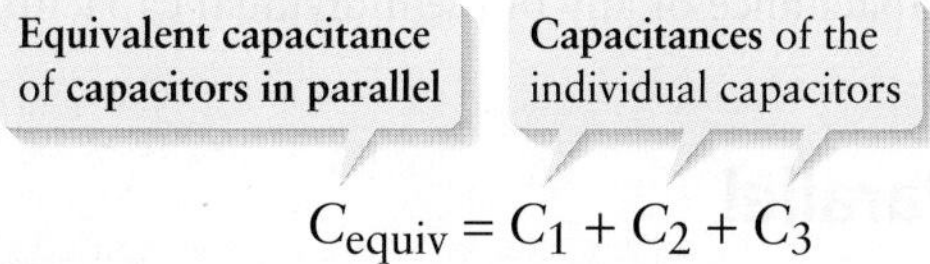

EQUATION IN WORDS
Equivalent capacitance of capacitors in parallel

$$C_{equiv} = C_1 + C_2 + C_3 \tag{17-26}$$

If capacitors are in parallel, the equivalent capacitance is the sum of the individual capacitances.

WATCH OUT!

Capacitors in series and parallel have different properties.

Here's a summary of the differences between series and parallel combinations of capacitors. In a *series* combination, each capacitor has the same charge, but there are different potential differences across capacitors with different capacitances. In a *parallel* combination, there is the same potential difference across each capacitor, but there are different charges on capacitors with different capacitances.

If there are only two or more than three capacitors in parallel, the same rule given in Equation 17-26 applies. The equivalent capacitance of a parallel combination is always greater than the smallest capacitance of any of the individual capacitors.

Example 17-10 shows how to do calculations with capacitors in series and in parallel. (We'll see a biological application of these calculations in the following section.) Many real networks of capacitors are more complex: They have a mixture of series and parallel combinations. To find the equivalent capacitance of such a network, we identify any small grouping of capacitors that are either entirely in series or entirely in parallel, find the equivalent capacitance of each group, and then combine them in larger and larger groupings, using the series and parallel rules, until we have accounted for all of the capacitors in the network. Example 17-11 illustrates how to do this.

EXAMPLE 17-10 **Two Capacitors in Series or in Parallel**

(a) If two capacitors are connected in series, find their equivalent capacitance for the case when the capacitors have different capacitances $C_1 = 2.00\ \mu F$ and $C_2 = 4.00\ \mu F$ and for the case where both have the same capacitance $C_1 = C_2 = 2.00\ \mu F$. (b) Repeat part (a) for the two capacitors connected in parallel.

Set Up

We'll use Equations 17-22 and 17-26 to find the equivalent capacitance C_{equiv} in each case. Because there are only two capacitors in each combination, we drop the C_3 term from these equations.

Two capacitors in series:

$$\frac{1}{C_{equiv}} = \frac{1}{C_1} + \frac{1}{C_2} \quad (17\text{-}22)$$

Two capacitors in parallel:

$$C_{equiv} = C_1 + C_2 \quad (17\text{-}26)$$

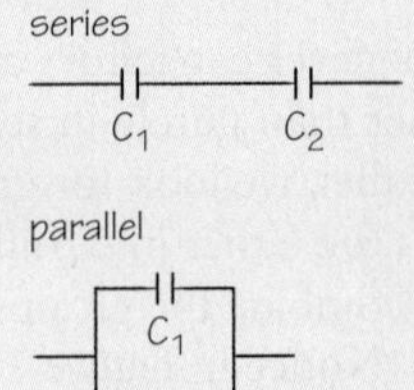

Solve

(a) We first apply Equation 17-22 to the case where $C_1 = 2.00\ \mu F$ and $C_2 = 4.00\ \mu F$.

With capacitors $C_1 = 2.00\ \mu F$ and $C_2 = 4.00\ \mu F$ in series, Equation 17-22 becomes

$$\frac{1}{C_{equiv}} = \frac{1}{2.00\ \mu F} + \frac{1}{4.00\ \mu F}$$

$$= 0.500\ \mu F^{-1} + 0.250\ \mu F^{-1} = 0.750\ \mu F^{-1}$$

To find C_{equiv} take the reciprocal of $1/C_{equiv}$:

$$C_{equiv} = \frac{1}{0.750\ \mu F^{-1}} = 1.33\ \mu F$$

Note that C_{equiv} is less than either C_1 or C_2.

Now repeat the calculation for the case where $C_1 = C_2 = 2.00\ \mu F$.

Follow the same steps but with both capacitances equal to $2.00\ \mu F$:

$$\frac{1}{C_{equiv}} = \frac{1}{2.00\ \mu F} + \frac{1}{2.00\ \mu F}$$

$$= 0.500\ \mu F^{-1} + 0.500\ \mu F^{-1} = 1.00\ \mu F^{-1}$$

$$C_{equiv} = \frac{1}{1.00\ \mu F^{-1}} = 1.00\ \mu F$$

For this case of identical capacitors in series, C_{equiv} is exactly one-half of $C_1 = C_2 = 2.00\ \mu F$.

(b) Apply Equation 17-26 to the case where $C_1 = 2.00\ \mu F$ and $C_2 = 4.00\ \mu F$.

With capacitors $C_1 = 2.00\ \mu F$ and $C_2 = 4.00\ \mu F$ in parallel, Equation 17-26 becomes

$$C_{equiv} = 2.00\ \mu F + 4.00\ \mu F = 6.00\ \mu F$$

Note that C_{equiv} is greater than either C_1 or C_2.

Now repeat the calculation for the case where $C_1 = C_2 = 2.00\ \mu F$.

Follow the same steps but with both capacitances equal to $2.00\ \mu F$:

$$C_{equiv} = 2.00\ \mu F + 2.00\ \mu F = 4.00\ \mu F$$

For this case of identical capacitors in parallel, C_{equiv} is exactly twice as great as $C_1 = C_2 = 2.00\ \mu F$.

Reflect

Our results illustrate the following general results: Connecting capacitors in series reduces the capacitance, whereas connecting them in parallel increases the capacitance.

NOW WORK Problems 2, 3, 6, and 7 from The Takeaway 17-7.

EXAMPLE 17-11 Multiple Capacitors

Find the equivalent capacitance of the three capacitors shown in **Figure 17-18.** The individual capacitances are $C_1 = 1.00\ \mu\text{F}$, $C_2 = 2.00\ \mu\text{F}$, and $C_3 = 6.00\ \mu\text{F}$.

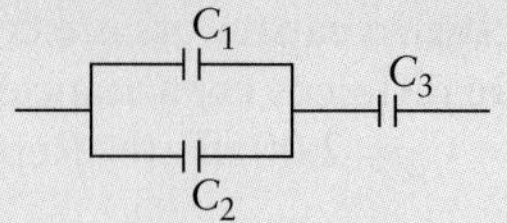

Figure 17-18 A capacitor network These three capacitors are neither all in series nor all in parallel.

Set Up

Whenever capacitors are combined in ways other than purely in series or purely in parallel, we look for groupings of capacitors that *are* either in parallel or in series, and then we combine the groups one at a time.

Notice in Figure 17-18 that C_1 and C_2 are in parallel because their two right plates are directly connected, as are their two left plates. We can therefore use Equation 17-26 to find C_{12} (the equivalent capacitance of the combination of C_1 and C_2) by using our relationship for capacitors in parallel.

Capacitor C_3 is in series with C_{12}, so we can find their combined capacitance by using Equation 17-22. This result is C_{123}, the equivalent capacitance of all three capacitors.

Two capacitors in series:

$$\frac{1}{C_{\text{equiv}}} = \frac{1}{C_1} + \frac{1}{C_2} \qquad (17\text{-}22)$$

Two capacitors in parallel:

$$C_{\text{equiv}} = C_1 + C_2 \qquad (17\text{-}26)$$

Replace the parallel capacitors C_1 and C_2 by their equivalent capacitor C_{12}:

C_{12} C_3

Then find the equivalent capacitance of C_{12} and C_3 in series.

Solve

First find the equivalent capacitance of the parallel capacitors C_1 and C_2.

For $C_1 = 1.00\ \mu\text{F}$ and $C_2 = 2.00\ \mu\text{F}$ in parallel, the equivalent capacitance C_{12} is given by Equation 17-26:

$$C_{12} = C_1 + C_2 = 1.00\ \mu\text{F} + 2.00\ \mu\text{F} = 3.00\ \mu\text{F}$$

Then find C_{123}, the equivalent capacitance of the series capacitors C_{12} and C_3. This is the equivalent capacitance of the entire network of C_1, C_2, and C_3.

Given $C_{12} = 3.00\ \mu\text{F}$ and $C_3 = 6.00\ \mu\text{F}$ are in series, their equivalent capacitance C_{123} is given by Equation 17-22:

$$\frac{1}{C_{123}} = \frac{1}{C_{12}} + \frac{1}{C_3} = \frac{1}{3.00\ \mu\text{F}} + \frac{1}{6.00\ \mu\text{F}}$$

$$= 0.333\ \mu\text{F}^{-1} + 0.167\ \mu\text{F}^{-1} = 0.500\ \mu\text{F}^{-1}$$

Take the reciprocal of this to find C_{123}:

$$C_{123} = \frac{1}{0.500\ \mu\text{F}^{-1}} = 2.00\ \mu\text{F}$$

Reflect

As we described previously, when capacitors are connected in parallel, the equivalent capacitance is always greater than the greatest individual capacitor. That's why C_{12} is greater than either C_1 or C_2. When capacitors are connected in series, the equivalent capacitance is always less than the least individual capacitor, which is why C_{123} is less than either C_{12} or C_3.

NOW WORK Problems 1, 4, and 8 from The Takeaway 17-7.

THE TAKEAWAY for Section 17-7

✔ Whenever two or more capacitors are connected, the equivalent capacitance is the capacitance of a single capacitor that is the equivalent of the combined capacitors.

✔ Capacitors are connected in series when they are connected one after another. In a series combination, all capacitors have the same magnitude of charge on each plate, and the equivalent capacitance is less than that of any of the individual capacitors.

✔ Capacitors are connected in parallel when all of their right-hand plates are directly connected to each other and all of their left-hand plates are directly connected to each other. In a parallel combination, the potential difference is the same across all capacitors, and the equivalent capacitance is greater than that of any of the individual capacitors.

Prep for the AP® Exam

AP® Building Blocks

1. What is the equivalent capacitance of the network of three capacitors shown in the figure below?

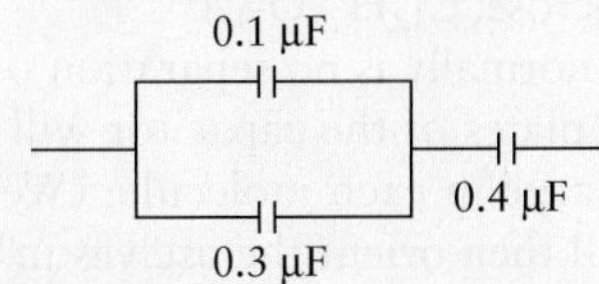

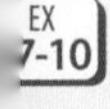

2. A 0.0500-μF capacitor and a 0.100-μF capacitor are connected in parallel across a 220-V battery. Determine the charge on each of the capacitors.

3. A 10.0-μF capacitor, a 40.0-μF capacitor, and a 100.0-μF capacitor are connected in parallel across a 12.0-V battery.
   (a) Find the equivalent capacitance of the combination.
   (b) What is the charge on each capacitor?
   (c) Calculate the potential difference across each capacitor.

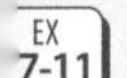

4. Calculate the equivalent capacitance between A and B for the combination of capacitors shown in the figure below.

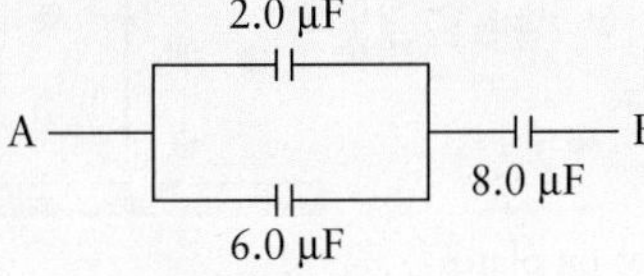

5. Capacitors connected in series have the same
   (A) charge.
   (B) potential difference.
   (C) dielectric.
   (D) plate separation.

AP® Skill Builders

6. A 10.0-μF capacitor, a 40.0-μF capacitor, and a 100.0-μF capacitor are connected in series across a 12.0-V battery.
   (a) Calculate the equivalent capacitance of the combination.
   (b) Calculate the charge on each capacitor.
   (c) Find the potential difference across each capacitor.

EX 17-10

7. Three capacitors have capacitances 10.0 μF, 15.0 μF, and 30.0 μF. Find their equivalent capacitance if the three are connected (a) in parallel and (b) in series.

EX 17-11

8. How should four 1.0-pF capacitors be connected to have a total capacitance of 0.25 pF?

AP® Skills in Action

9. An isolated parallel-plate capacitor carries a charge Q. If the separation between the plates is doubled, the electric energy stored in the capacitor will be
   (A) halved.
   (B) doubled.
   (C) quadrupled.
   (D) quartered.

10. Capacitor 1 has capacitance $C_1 = 1.0$ μF and capacitor 2 has capacitance $C_2 = 2.0$ μF. You connect the initially uncharged capacitors to each other then connect the capacitor combination to a battery. Which capacitor stores the greater amount of electric potential energy if the two capacitors are connected in series? If they are connected in parallel?
    (A) Capacitor 1 for both the series and parallel cases
    (B) Capacitor 2 for both the series and parallel cases
    (C) Capacitor 1 for the series case, capacitor 2 for the parallel case
    (D) Capacitor 2 for the series case, capacitor 1 for the parallel case

17-8 Placing a dielectric between the plates of a capacitor increases the capacitance

So far in our discussion of capacitors, we've assumed that there is only vacuum (or air, which is close to the same) between the capacitor plates. In most situations, however, the two plates are separated by a layer of a **dielectric**, a material that is both an insulator and *polarizable*—that is, in which there is a separation of positive and negative charge within the material when it's exposed to an electric field. Dielectrics used in commercial capacitors include glass, ceramics, and plastics such as polystyrene. These dielectrics not only help to keep the positive and negative plates from touching each other so they can be placed closer together, but also directly increase the capacitance of the capacitor. In this section we'll see how dielectrics make this possible.

Let's consider an isolated parallel-plate capacitor with charges $+Q$ and $-Q$ on its plates and with vacuum in the space between its plates (**Figure 17-19a**). The charge creates a uniform electric field $\vec{E}_0$ between the plates. We now insert a dielectric material that fills the space between the plates of this capacitor. What happens then depends on the kind of molecules that make up the dielectric.

If the molecules are *polar*—that is, if one end of the molecule has a positive charge and the other end has a negative charge of the same magnitude—the molecules will orient themselves so that their positive ends are pointed toward the negatively charged plate and their negative ends toward the positively charged plate, as in **Figure 17-19b**. (The most common molecule in your body, the water molecule H_2O, is a polar molecule. Others include ammonia, NH_3, and sucrose, $C_{12}H_{22}O_{11}$.)

If the molecules are not polar, so there normally is no separation of charge within the molecule, the electric field between the plates of the capacitor will *induce* a slight separation of the positive and negative charge in each molecule. (We described this process in Section 16-2.) The molecules will then orient themselves in the same manner as polar molecules would (**Figure 17-19c**). In either case we say that the dielectric becomes *polarized* once it has been inserted between the plates of the capacitor.

In the interior of the polarized dielectric in Figures 17-17b and 17-17c, there are as many positive ends of molecules as there are negative ends, so the interior of the dielectric is neutral. But there *is* a net charge at the surfaces of the dielectric: There is a layer

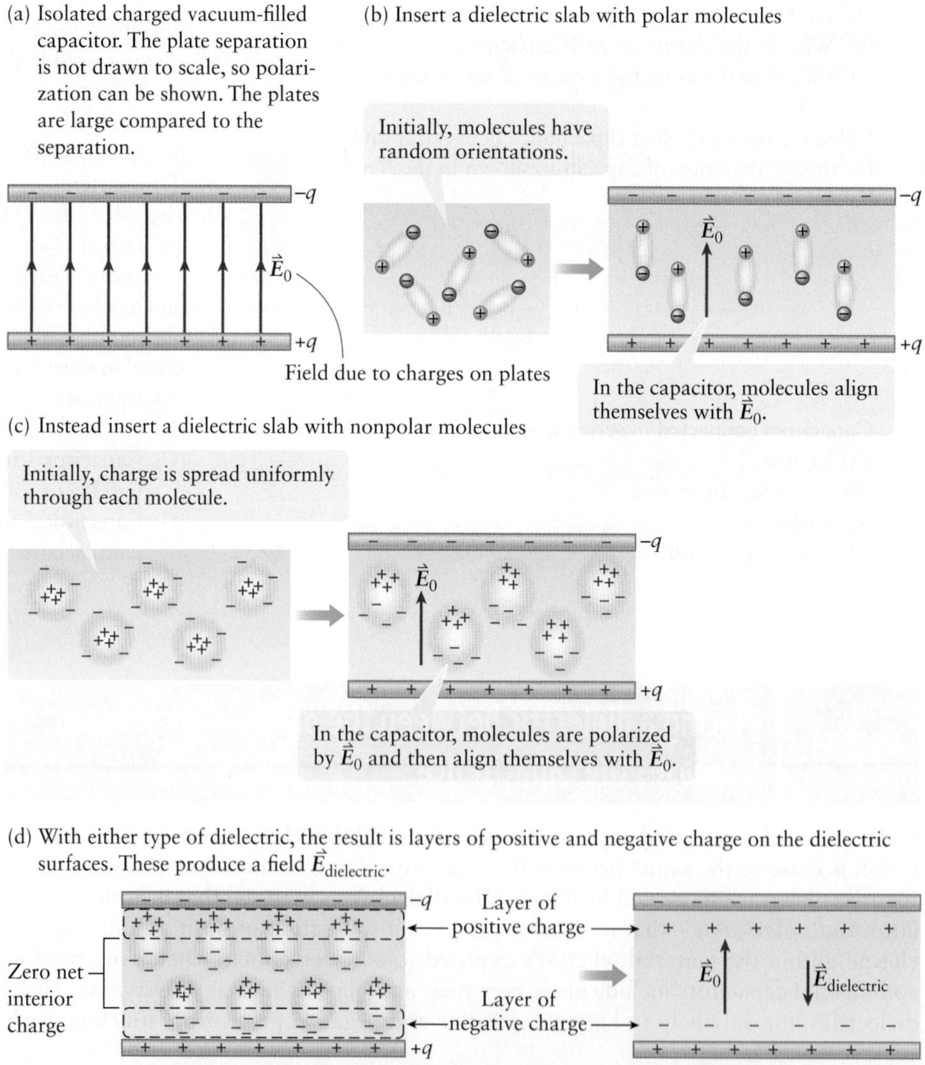

(a) Isolated charged vacuum-filled capacitor. The plate separation is not drawn to scale, so polarization can be shown. The plates are large compared to the separation.

Field due to charges on plates

(b) Insert a dielectric slab with polar molecules

Initially, molecules have random orientations.

In the capacitor, molecules align themselves with $\vec{E}_0$.

(c) Instead insert a dielectric slab with nonpolar molecules

Initially, charge is spread uniformly through each molecule.

In the capacitor, molecules are polarized by $\vec{E}_0$ and then align themselves with $\vec{E}_0$.

(d) With either type of dielectric, the result is layers of positive and negative charge on the dielectric surfaces. These produce a field $\vec{E}_{\text{dielectric}}$.

Zero net interior charge

Layer of positive charge

Layer of negative charge

Figure 17-19 A dielectric in a parallel-plate capacitor A dielectric slab inserted between the plates of an isolated capacitor reduces the electric field between the plates.

of positive charge on the surface next to the negatively charged plate, and there is a layer of negative charge on the surface next to the positively charged plate. (The dielectric as a whole is neutral, so these two layers have the same magnitude of charge.) These two layers of charge create an electric field $\vec{E}_{\text{dielectric}}$ that points opposite to the electric field $\vec{E}_0$ created by the charges on the plates of the capacitor (**Figure 17-19d**). Hence $\vec{E}_{\text{dielectric}}$ partially cancels $\vec{E}_0$, and the *net* electric field $\vec{E} = \vec{E}_0 + \vec{E}_{\text{dielectric}}$ between the plates is less than $\vec{E}_0$. So the effect of the dielectric is to *reduce* the electric field between the plates. We can express this as

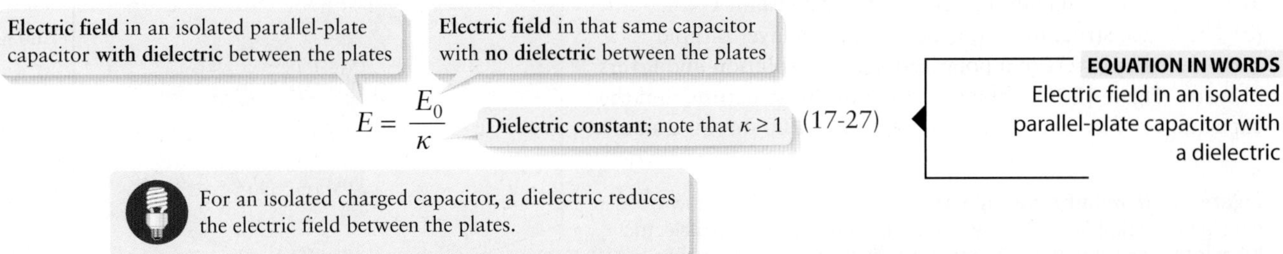

$$E = \frac{E_0}{\kappa} \qquad (17\text{-}27)$$

The quantity κ (the Greek letter "kappa") is the **dielectric constant** of the material. The greater the value of κ, the more the dielectric is polarized when it is placed between the plates of a charged capacitor, so the greater the magnitude of the field $\vec{E}_{\text{dielectric}}$ due to the dielectric and the smaller the magnitude E of the net field. **Table 17-1** lists values of the dielectric constant for a variety of materials.

TABLE 17-1 Dielectric Constants (at 20°C and 1 atm)

Material	κ
vacuum	1
air	1.00058
lipid	2.2
paraffin	2.2
paper	2.7
ceramic (porcelain)	5.8
water	80

What does Equation 17-27 tell us about how the dielectric affects the capacitance of the capacitor? We saw in Section 17-5 that the potential difference ΔV between the positive and negative plates is proportional to the magnitude E of the electric field between the plates (see Equation 17-11, which states that $\Delta V = Ed$). Equation 17-27 says that the dielectric reduces the value of the electric field by a factor $1/\kappa$, so ΔV is also reduced by a factor $1/\kappa$. Now, Equation 17-14 tells us that the capacitance C is related to potential difference V and the magnitude q of the charges on the plates by $q = C\Delta V$, or $C = Q/\Delta V$. While the dielectric reduces ΔV by a factor $1/\kappa$, it has no effect on the value of Q. (Since we specified that the capacitor is isolated, no charge can leave either plate.) Because the capacitance $C = Q/\Delta V$ is inversely proportional to the potential difference ΔV, which is reduced by a factor $1/\kappa$, it follows that C *increases* by a factor of $1/(1/\kappa) = \kappa$. If C_0 is the capacitance without the dielectric, the capacitance with the dielectric is

Capacitance of a parallel-plate capacitor **with dielectric** between the plates

Capacitance of that same capacitor with **no dielectric** between the plates

$$C = \kappa C_0 \qquad (17\text{-}28)$$

A dielectric increases the capacitance of a capacitor.

Dielectric constant; note that $\kappa \geq 1$

EQUATION IN WORDS
Capacitance of a parallel-plate capacitor with a dielectric

As an example, Table 17-1 tells us that the dielectric constant of porcelain is $\kappa = 5.8$. From Equation 17-28 the capacitance of a capacitor with porcelain filling the space between its plates is 5.8 times greater than an identical capacitor with vacuum between its plates. That's fundamentally because the electric field in the porcelain-filled capacitor is only $1/5.8 = 0.17$ as great as for an identical vacuum capacitor carrying the same charge (Equation 17-27).

Example 17-12 shows how to apply these ideas to the membrane of a cell, which we can regard as a capacitor with multiple layers of dielectric between its plates.

EXAMPLE 17-12 Cell Membrane Capacitance

In all cells, it is easier for positive potassium ions (K^+) to flow out of the cell than it is for negative ions. As a result, there is negative charge on the inside of the cell membrane and positive charge on the outside of the membrane, much like a capacitor. Consider a typical cell with a membrane of thickness $7.60\text{ nm} = 7.60 \times 10^{-9}\text{ m}$. (a) Find the capacitance of a square patch of membrane $1.00\ \mu\text{m} = 1.00 \times 10^{-6}\text{ m}$ on a side, assuming that the membrane has dielectric constant $\kappa = 1$. (b) The actual structure of the membrane is a layer of lipid surrounded by layers of a polarized aqueous solution and layers of water (**Figure 17-20**). Taking account of this structure, find the capacitance of a square patch of membrane 1.00 μm on a side.

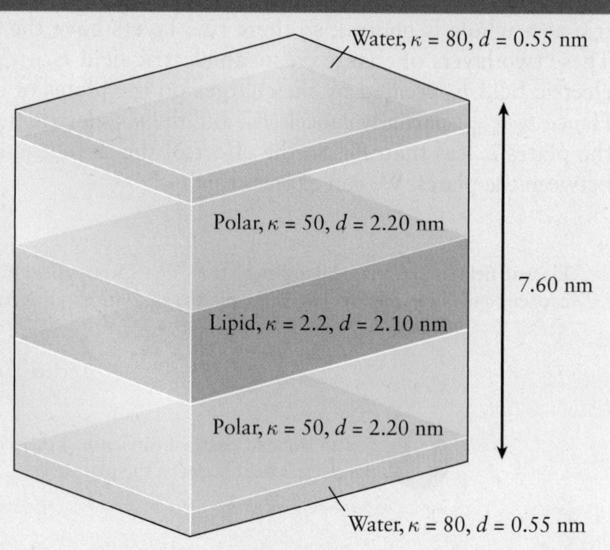

Figure 17-20 Membrane capacitance A cell membrane (shown here in highly simplified form) acts as a capacitor filled with a dielectric. What is the capacitance of a membrane that consists of a layer of lipid surrounded by layers of a polarized aqueous solution and water?

Set Up

A dielectric constant $\kappa = 1$ corresponds to vacuum, so in part (a) we can find the capacitance C_0 of the membrane under the assumption $\kappa = 1$ by using Equation 17-13 from Section 17-5.

In part (b) we must deal with the more complicated situation shown in Figure 17-20. We can't simply use Equation 17-28 to find the capacitance because there's not a single value of κ for the arrangement of layers that make up the membrane. Instead we'll imagine that there's a very thin conducting sheet separating each layer from the next. If there are charges $+Q$ and $-Q$ on the outer and inner surfaces of the multilayer membrane, there will also be charges $+Q$ and $-Q$ on the surfaces of these conducting sheets. This is exactly like the situation with capacitors in series (Section 17-7). So we'll treat each of the five layers in Figure 17-20 as an individual capacitor, then use Equation 17-22 to find the capacitance of the combination.

Parallel-plate capacitor with vacuum between the plates:

$$C_0 = \frac{\varepsilon_0 A}{d} \qquad (17\text{-}13)$$

Parallel-plate capacitor with dielectric:

$$C = \kappa C_0 \qquad (17\text{-}28)$$

Five capacitors in series:

layer 1, layer 2, layer 3, layer 4, layer 5 = C_1, C_2, C_3, C_4, C_5

$$\frac{1}{C_{\text{equiv}}} = \frac{1}{C_1} + \frac{1}{C_2} + \frac{1}{C_3} + \frac{1}{C_4} + \frac{1}{C_5} \qquad (17\text{-}22)$$

Solve

(a) First use Equation 17-13 to find the capacitance, assuming that the membrane has $\kappa = 1$ (equivalent to vacuum).

The area A for the capacitor of interest is a square 1.00 μm on a side:

$$A = (1.00\ \mu\text{m})^2 = (1.00 \times 10^{-6}\text{ m})^2 = 1.00 \times 10^{-12}\text{ m}^2$$

A parallel-plate capacitor with this area and with $d = 7.60 \times 10^{-9}$ m has capacitance

$$C_0 = \frac{\varepsilon_0 A}{d} = \frac{(8.85 \times 10^{-12}\text{ F/m})(1.00 \times 10^{-12}\text{ m}^2)}{7.60 \times 10^{-9}\text{ m}} = 1.16 \times 10^{-15}\text{ F}$$

(b) Calculate the capacitance of each of the five layers separately using Equations 17-13 and 17-28.

The capacitance of the first layer of water is $\kappa_1 = 80$ times the capacitance of a vacuum capacitor of thickness $d_1 = 0.55$ nm:

$$C_1 = \kappa_1\left(\frac{\varepsilon_0 A}{d_1}\right) = (80)\frac{(8.85 \times 10^{-12}\ \text{F/m})(1.00 \times 10^{-12}\ \text{m}^2)}{0.55 \times 10^{-9}\ \text{m}}$$

$$= 1.3 \times 10^{-12}\ \text{F}$$

Similarly, for the second layer of polarized aqueous solution with $\kappa_2 = 50$ and thickness $d_2 = 2.20$ nm, the capacitance is

$$C_2 = \kappa_2\left(\frac{\varepsilon_0 A}{d_2}\right)$$

$$= (50)\frac{(8.85 \times 10^{-12}\ \text{F/m})(1.00 \times 10^{-12}\ \text{m}^2)}{2.20 \times 10^{-9}\ \text{m}}$$

$$= 2.0 \times 10^{-13}\ \text{F}$$

The capacitance of the third lipid layer with $\kappa_3 = 2.2$ and thickness $d_3 = 2.10$ nm is

$$C_3 = \kappa_3\left(\frac{\varepsilon_0 A}{d_3}\right)$$

$$= (2.2)\frac{(8.85 \times 10^{-12}\ \text{F/m})(1.00 \times 10^{-12}\ \text{m}^2)}{2.10 \times 10^{-9}\ \text{m}}$$

$$= 9.3 \times 10^{-15}\ \text{F}$$

The fourth layer of polarized aqueous solution is identical to the second layer, and the fifth water layer is identical to the first layer. So

$$C_4 = C_2 = 2.0 \times 10^{-13}\ \text{F}$$

$$C_5 = C_1 = 1.3 \times 10^{-12}\ \text{F}$$

Calculate the net capacitance of the five layers in series using Equation 17-22.

The equivalent capacitance C_{equiv} of the five-layer stack is given by

$$\frac{1}{C_{\text{equiv}}} = \frac{1}{C_1} + \frac{1}{C_2} + \frac{1}{C_3} + \frac{1}{C_4} + \frac{1}{C_5}$$

Because $C_4 = C_2$ and $C_5 = C_1$, this is

$$\frac{1}{C_{\text{equiv}}} = \frac{1}{C_1} + \frac{1}{C_2} + \frac{1}{C_3} + \frac{1}{C_2} + \frac{1}{C_1} = \frac{2}{C_1} + \frac{2}{C_2} + \frac{1}{C_3}$$

$$= \frac{2}{1.3 \times 10^{-12}\ \text{F}} + \frac{2}{2.0 \times 10^{-13}\ \text{F}} + \frac{1}{9.3 \times 10^{-15}\ \text{F}}$$

$$= 1.2 \times 10^{14}\ \text{F}^{-1}$$

The reciprocal of this is C_{equiv}:

$$C_{\text{equiv}} = \frac{1}{1.2 \times 10^{14}\ \text{F}^{-1}} = 8.4 \times 10^{-15}\ \text{F}$$

Reflect

The equivalent capacitance of the stack of five layers is less than that of any individual layer. This is just what we saw in Section 17-7 for capacitors in series.

Compared to the capacitance $C_0 = 1.16 \times 10^{-15}$ F calculated assuming $\kappa = 1$, the capacitance $C_{\text{equiv}} = 8.4 \times 10^{-15}$ F is greater by a factor $C_{\text{equiv}}/C_0 = (8.4 \times 10^{-15}\ \text{F})/(1.16 \times 10^{-15}\ \text{F}) = 7.2$. So the effective dielectric constant of the membrane is 7.2, intermediate between the largest ($\kappa_1 = 80$) and smallest ($\kappa_3 = 2.2$) values of dielectric constant for the individual layers.

NOW WORK Problems 1, 3, 5, 7, 8, and 10 from The Takeaway 17-8.

THE TAKEAWAY for Section 17-8

✔ When a dielectric material is placed in an electric field $\vec{E}_0$, it becomes polarized. This produces an additional electric field that partially cancels $\vec{E}_0$ and so reduces the net field in the dielectric. The dielectric constant is a measure of how much the field is reduced.

✔ If a dielectric material fills the space between the plates of a capacitor, the capacitance is greater than if there is vacuum or air between the plates.

Prep for the AP Exam

Building Blocks

1. A parallel-plate capacitor has plates of 1.00 cm by 2.00 cm. The plates are separated by a 1.00-mm-thick piece of paper. What is the capacitance of this capacitor? The dielectric constant for paper is 2.7.
2. Does inserting a dielectric into a capacitor increase or decrease the energy stored in the capacitor? Explain your answer.

3. What is the dielectric constant of the material that fills the gap between a parallel-plate capacitor with plate area of $20.0\ \text{cm}^2$ and plate separation of 1.00 mm if the capacitance is measured to be 0.0142 μF?
4. What are the benefits, if any, of filling a capacitor with a dielectric other than air?

Skill Builders

5. A parallel-plate capacitor has square plates 1.00×10^2 cm on a side that are separated by 1.00 mm. It is connected to a battery and charged to 12.0 V. How much energy is stored in the capacitor if a ceramic dielectric material (κ is 5.8) fills the space between the plates?
6. Capacitors A and B are identical except that the region between the plates of capacitor A is filled with a dielectric. As shown in the figure below, the plates of these capacitors are maintained at the same potential difference by a battery. Is the electric field magnitude in the region between the plates of capacitor A smaller than, the same as, or larger than the field in the region between the plates of capacitor B? Explain your answer.

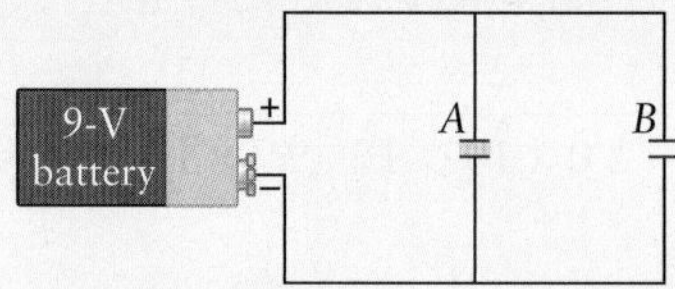

7. A parallel-plate capacitor that has a plate separation of 0.50 cm is filled halfway with a slab of dielectric material (κ is 5.0), as shown in the following figure. If the plates are 1.25 cm by 1.25 cm in area, what is the capacitance of this capacitor?

8. A 2800-pF air-filled capacitor is connected to a 16-V battery. If you now insert a ceramic dielectric material ($\kappa = 5.8$) that fills the space between the plates, calculate the magnitude of charge the battery will then put on each plate.

Skills in Action

9. An isolated parallel-plate capacitor (one that is not connected to anything else) has a vacuum between its plates and has charges $+Q$ and $-Q$ on its plates. If you insert a slab of dielectric with dielectric constant κ that fills the space between the plates, the capacitance increases. What happens to the energy stored in the capacitor as you insert the dielectric slab? Will you have to push the slab into the capacitor, or will you feel the capacitor pulling the slab in?
   (A) Stored energy increases; you will have to push the slab in.
   (B) Stored energy increases; the slab will be pulled in.
   (C) Stored energy decreases; you will have to push the slab in.
   (D) Stored energy decreases; the slab will be pulled in.
10. A dielectric with constant κ_1 fills up one-quarter of the area, but the full separation of the plates in the parallel-plate capacitor is shown in the figure below.
    (a) Determine the capacitance of the parallel-plate capacitor when materials with constants κ_2 and κ_3 fill the other three-quarters of the area and divide the separation of the plates in half.
    (b) What happens to the capacitance if the material with constant κ_3 is replaced by air?

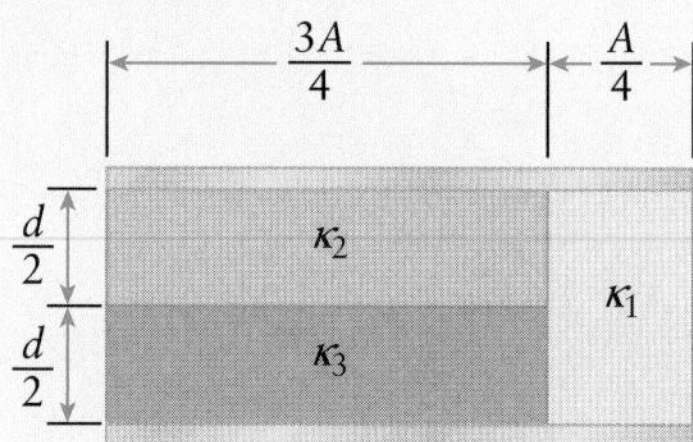

WHAT DID YOU LEARN?

Prep for the AP® Exam

Chapter learning goals	Section(s)	Related example(s)	Relevant section review exercise(s)
Discuss how the work done on a charged object by the electric field relates to changes in electric potential energy.	17-2	17-1 17-3	4 6
Explain the difference between electric potential and electric potential energy and describe electric potential difference in terms of charge and energy.	17-3	 17-4 17-5	2, 6, 7 5 4
Use electric potential energy in conservation of energy relationships.	17-2 17-3	17-1 17-2 17-3 17-6	4 1, 2, 7, 9 3, 5, 6, 10 3
Recognize why equipotential surfaces are perpendicular to electric field lines, use that relationship to identify equipotentials for a charge distribution, and find average electric field from equipotential values.	17-4		1, 6, 9, 10 2, 3
Explain what is meant by capacitance and describe how the capacitance of a parallel-plate capacitor depends on the size of the plates and their separation.	17-5	 17-7 17-8	7, 8 3, 4, 5, 6, 9 10
Describe how the electric energy stored in a capacitor depends on the properties of the capacitor and calculate the electric energy stored in a capacitor.	17-6	17-9	1–10
Describe how the capacitance of a capacitor increases when an insulating material other than a vacuum is placed between the plates of the capacitor.	17-8	 17-12	2, 4, 6 1, 3, 5, 7, 8, 10
Describe the equivalent capacitance of multiple capacitors connected in series and/or parallel.	17-7	17-10 17-11	2, 3, 6 1, 4, 8

Chapter 17 Review

Key Terms

All the Key Terms can be found in the Glossary/Glosario on page G1 in the back of the book.

capacitance 844
capacitor 841
dielectric 857
dielectric constant 859
electric potential 829
electric potential difference 830
electric potential energy 820
electron volt 834
equipotential 837
equipotential surface 837
equipotential volume 838
equivalent capacitance 851
farad 844
parallel (capacitors) 853
parallel-plate capacitor 841
plate 841
series (capacitors) 852
test charge 829
volt 830

Chapter Summary

Topic	Equation or Figure
Electric potential energy: Like the gravitational force, the electric force is a conservative force and has an associated potential energy. The change in electric potential energy of a system, when a charged object moves from one point to another in an electric field created by the rest of the system, depends on the sign of the object's charge. The sign of the electric potential energy of a system of two point charges depends on whether the two charges have the same sign or different signs.	The **change in electric potential energy** for a system of the source of the field and an object of charge q that moves in a uniform electric field $\vec{E}$... ...equals the **negative of the work done** on the object by the electric force when the source of the field is external to the system. $\Delta U_{\text{electric}} = -W_{\text{electric}} = -qEd\cos\theta$ **Charge** of the object; **Magnitude** of the **electric field**; **Displacement** of the object; **Angle** between the displacement and the direction of the electric field (17-2)
	Electric potential energy of a system of two point charges $U_{\text{electric}} = \frac{kq_1q_2}{r}$ **Coulomb constant**; Values of the **two charges**; **Distance** between the point charges (17-3)
Electric potential: The electric potential at a given position equals the electric potential energy for a system including a point charge q_0 at that position and the charged objects causing the field, divided by the value of q_0. If you move in a direction opposite to the electric field, the electric potential increases; if you move in the direction of the electric field, the electric potential decreases. The electric potential due to a point charge is inversely proportional to the distance from the point charge. Electric potential difference is the difference in electric potential between two locations.	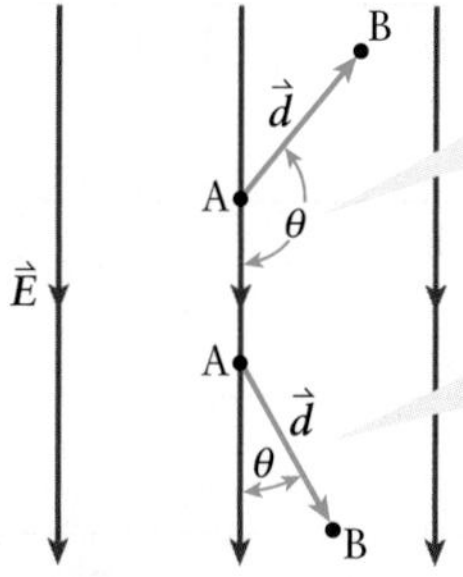 If you move in a direction opposite to the electric field ($\theta > 90°$), the electric potential increases ($\Delta V > 0$). If you move in the direction of the electric field ($\theta < 90°$), the electric potential decreases ($\Delta V < 0$). (Figure 17-5)  If you move from point A to point B in a uniform electric field $\vec{E}$, the change in electric potential is $\Delta V = -Ed\cos\theta$.
	Electric potential due to a point charge Q $V = \frac{kQ}{r}$ **Coulomb constant**; Value of the point **charge**; **Distance** from the point charge Q to the location where the potential is measured (17-8)

Equipotentials: The electric potential has the same value everywhere on an equipotential surface. An equipotential surface is everywhere perpendicular to the electric field. The electric potential has the same value throughout the volume of a conductor in equilibrium.

A plane perpendicular to a uniform electric field $\vec{E}$ is an **equipotential surface:** Potential V has the same value at all points on this surface.

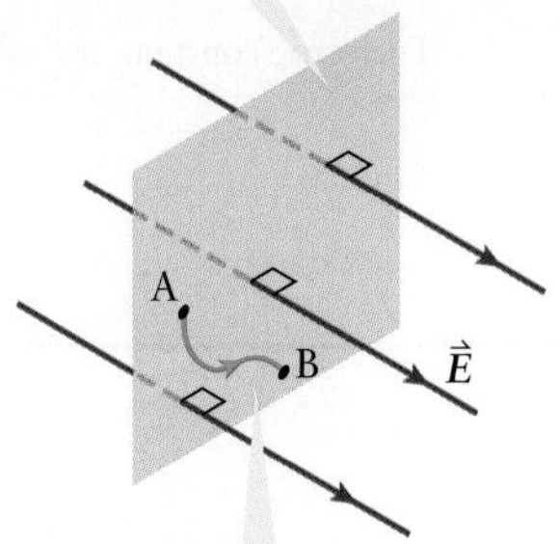

Any curve that lies on an equipotential surface, such as this curve from A to B, is an equipotential curve.

(Figure 17-7)

Capacitors: A parallel-plate capacitor has two conducting plates that store charge $+Q$ and $-Q$, of the same magnitude but opposite sign. The potential difference ΔV between the plates is proportional to the charge Q; the proportionality constant, the capacitance C, depends only on the geometry of the capacitor and the substance between the plates.

A capacitor carries a charge $+Q$ on its positive plate and a charge $-Q$ on its negative plate.

The magnitude of Q is directly proportional to ΔV, the potential difference between the plates.

$$Q = C\Delta V \quad (17\text{-}14)$$

The constant of proportionality between charge Q and potential difference ΔV is the **capacitance** C of the capacitor.

Energy stored in a capacitor: A charged capacitor stores electric potential energy. The stored energy can be expressed by any two of the three quantities: capacitance C, charge Q, and potential difference ΔV.

The **electric potential energy** stored in a charged capacitor...

$$U_{\text{electric}} = \frac{1}{2}Q\Delta V = \frac{1}{2}C\Delta V^2 = \frac{Q^2}{2C} \quad (17\text{-}17)$$

...can be expressed in three ways in terms of the **charge** Q, **potential difference** ΔV, and **capacitance** C.

Capacitors in series and parallel: A collection of capacitors in a circuit behaves as though it were a single capacitor, with an equivalent capacitance C_{equiv}. The equivalent capacitance is different for capacitors in series (with the positive plate of one capacitor connected to the negative plate of another) than for capacitors in parallel (with positive plates connected to positive plates and negative plates to negative plates).

If capacitors are in series, the reciprocal of the equivalent capacitance is the sum of the reciprocals of the individual capacitances.

$$\frac{1}{C_{\text{equiv}}} = \frac{1}{C_1} + \frac{1}{C_2} + \frac{1}{C_3} \quad (17\text{-}22)$$

Equivalent capacitance of capacitors in series

Capacitances of the individual capacitors

Equivalent capacitance of capacitors in parallel

Capacitances of the individual capacitors

$$C_{\text{equiv}} = C_1 + C_2 + C_3 \quad (17\text{-}26)$$

If capacitors are in parallel, the equivalent capacitance is the sum of the individual capacitances.

Dielectrics: A dielectric is an insulator that becomes polarized when placed in an electric field. When the space between the plates of a capacitor is filled with a dielectric, the electric field between the plates decreases (for a given charge Q on the plates) and the capacitance increases.

Capacitance of a parallel-plate capacitor **with dielectric** between the plates

Capacitance of that same capacitor **with no dielectric** between the plates

$$C = \kappa C_0 \quad (17\text{-}28)$$

Dielectric constant; note that $\kappa \geq 1$

A dielectric increases the capacitance of a capacitor.

Chapter 17 Review Problems

1. What is the difference between electric potential and electric field?
2. What is the difference between electric potential and electric potential energy?
3. Explain why electric potential requires the existence of only one charged object, but a finite electric potential energy requires the existence of two charged objects.
4. If the potential difference across a capacitor is doubled, by how much does the stored energy change?
5. Does it make sense to say that the potential difference at some point in space is 10.3 V? Explain your answer.
6. Explain why an electron will accelerate toward a region of lower electric potential energy but higher electric potential.
7. Support your claim for each of the following questions.
   (a) If the electric potential throughout some region of space is zero, does it necessarily follow that the electric field is zero?
   (b) If the electric field throughout a region is zero, does it necessarily follow that the electric potential is zero?
8. The capacitance of several capacitors in series is less than any of the individual capacitances. What, then, is the advantage of having several capacitors in series?
9. What is the advantage to arranging several capacitors in parallel?
10. Which way of connecting (series or parallel) three identical capacitors to a battery would store more energy?
11. Qualitatively explain why the equivalent capacitance of a parallel combination of identical capacitors is larger than the individual capacitances.
12. A pollen grain with a diameter of 55 μm has a maximum electric charge limit it can carry before the electric field it generates exceeds the dielectric limit for air of 3×10^6 N/C. Estimate the maximum potential at the surface of a pollen grain that is carrying its maximum electric charge, assuming $V = 0$ at infinity.
13. A potential difference exists between the inner and outer surfaces of the membrane of a cell. The inner surface is negative relative to the outer surface. If 1.5×10^{-20} J of work is required to eject a positive sodium ion (Na^+) from the interior of the cell, what is the potential difference between the inner and outer surfaces of the cell?
14. What is the electric potential due to the nucleus of hydrogen at a distance of 5.00×10^{-11} m? Assume the potential is equal to zero as $r \to \infty$.
15. A point charge is located at the origin. Assume the potential is equal to zero as $r \to \infty$.
    (a) Calculate the electric potential at $x = 0.500$ cm, if the point charge has a value of +2.00 nC.
    (b) How will the answer change if the charge is, instead, −2.00 nC?
16. The electric potential has a value of −200 V at a distance of 1.25 m from a point charge. What is the value of that charge? Assume the potential is equal to zero as $r \to \infty$.
17. Two point charges are placed on the x axis: +0.500 nC at $x = 0$ and −0.200 nC at $x = 10.0$ cm. At what point(s), if any, on the x axis is the electric potential equal to zero?
18. At point P in the following figure, the electric potential is zero. (As usual, we take the potential to be zero at infinite distance.) **(a)** What can you say about the two

point charges? (b) Are there any other points of zero potential on the line connecting P and the two point charges?

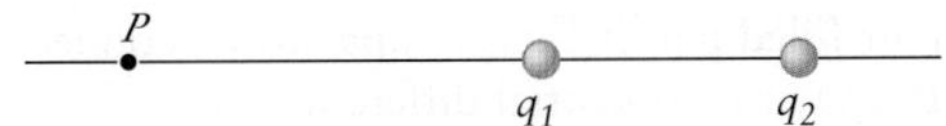

19. A point charge of +2.00 nC is at the origin and a charge of −3.00 nC is on the y axis at $y = 40.0$ cm. (a) What is the potential at point A, which is on the x axis at $x = 40.0$ cm? (b) What is the potential difference $V_B - V_A$ when point B is at (40.0 cm, 30.0 cm)? (c) How much work is required to move an electron at rest from point A to rest at point B?

20. In 2004, physicists at the SLAC National Accelerator Laboratory fired electrons toward each other at very high speeds so that they came within 1.0×10^{-15} m of each other (approximately the diameter of a proton). (a) What was the electric force on each electron at closest approach? (b) Would you be able to feel a force of this magnitude exerted on you? (c) To be able to get this close, what kinetic energy must each electron have had when all of them were far apart?

21. Potassium ions (K^+) move across an 8.0-nm-thick cell membrane from the inside to the outside. The potential inside the cell is −70.0 mV, and the potential outside is zero.

    (a) What is the change in the electric potential energy of the potassium ions as they move across the membrane?

    (b) Does their potential energy increase or decrease?

22. Calculate the equivalent capacitance of the combination shown in the figure below.

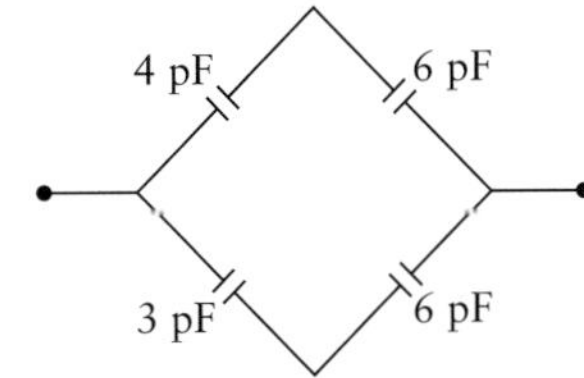

23. In the Bohr model of the hydrogen atom, an electron in the lowest energy state moves around the nucleus at a speed of 2.19×10^6 m/s at a distance of 0.529×10^{-10} m from the nucleus.

    (a) What is the electric potential due to the hydrogen nucleus at this distance?

    (b) How much energy is required to ionize a hydrogen atom, whose electron is in this lowest energy state? Assume the electric potential goes to zero as $r \to \infty$.

24. Calculate the electric potential at the origin O due to the point charges shown in the figure below.

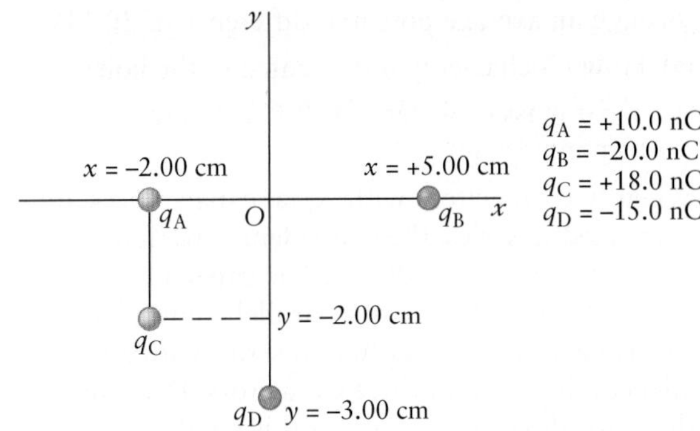

25. A 0.50-μF capacitor and a 0.10-μF capacitor are connected in series with a 220-V battery. Determine the charge on each of the capacitors.

26. Two capacitors provide an equivalent capacitance of 8.00 μF when connected in parallel and 2.00 μF when connected in series. Find the capacitance of each capacitor.

27. A 2.00-μF capacitor is first charged by being connected across a 6.00-V battery. It is then disconnected from the battery and connected across an uncharged 4.00-μF capacitor. Calculate the final charge on each of the capacitors.

28. Two point charges equal in magnitude but opposite in sign are placed on a plane. Three points, A, B, and C, are defined, as shown in the figure below.

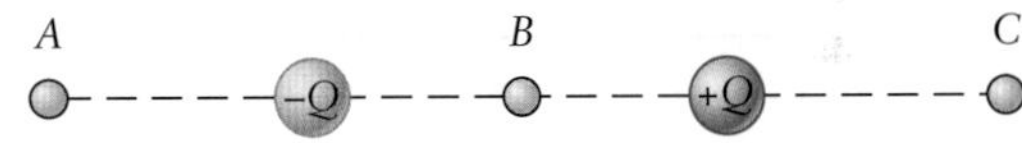

    (a) Order the values V_A, V_B, and V_C of the electric potential at points A, B, and C, respectively, from highest to lowest.

    (b) A single, negative test charge is moved among the three points. Order the potential energies U_A, U_B, and U_C of the system when the test charge is at each point, from highest to lowest.

29. For the capacitor network shown in the figure below, the potential difference between A and B is 75.0 V. How much charge and how much energy are stored in this system?

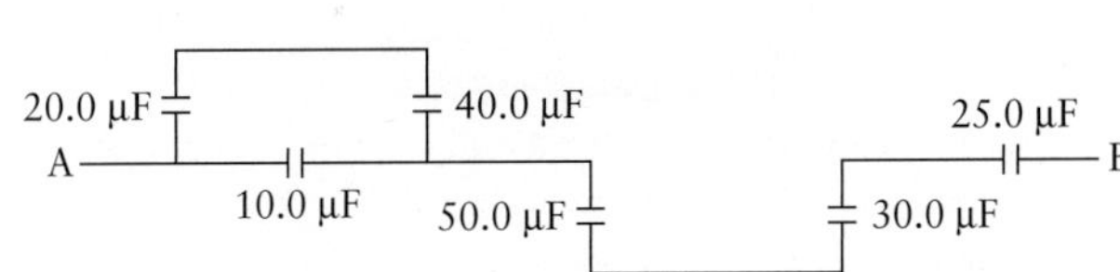

30. A parallel-plate capacitor has a plate separation of 1.5 mm and is charged to 600 V. If an electron leaves the negative plate, starting from rest, how fast is it going when it hits the positive plate?

31. A lightning bolt transfers 20 C of charge to Earth through an average potential difference of 30 MV.

(a) How much energy is dissipated in the bolt?

(b) What mass of water at 100°C could this energy turn into steam?

32. When five capacitors with equal capacitances are connected in series, the equivalent capacitance of the combination is 6.00 mF. The capacitors are then reconnected so that a parallel combination of two capacitors is connected in series with a parallel combination of three capacitors. Determine the equivalent capacitance of this combination in millifarads.

33. The arrangement of four capacitors shown in the figure below has an equivalent capacitance of 8.00 μF. Calculate the value of C_x.

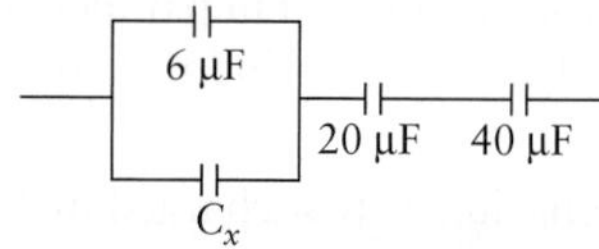

34. Calculate the magnitude of the charge stored on each plate on the capacitor in the circuit shown in the figure below.

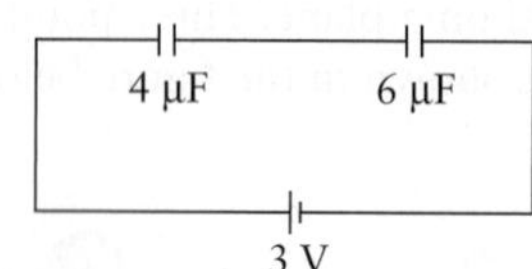

35. A parallel-plate capacitor is made by sandwiching 0.100-mm sheets of paper (dielectric constant 2.7) between three sheets of aluminum foil (A, B, and C as shown in the figure below) and rolling the layers into a cylinder. A capacitor that has an area of 10 m^2 is fabricated this way. What is the capacitance of this capacitor?

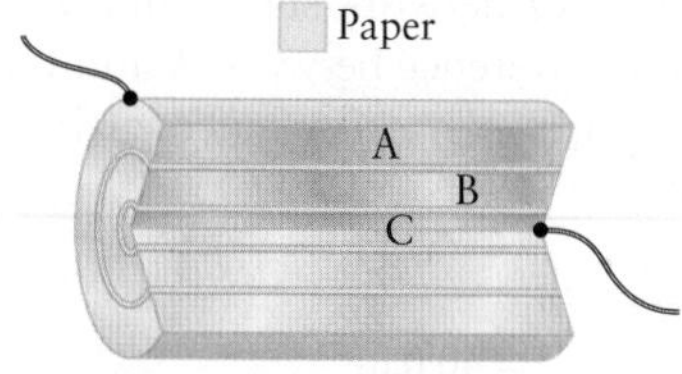

36. A parallel-plate capacitor with a capacitance of 5.00 μF is fully charged with a 12.0-V battery. The battery is then removed. How much work is required to triple the separation between the plates?

37. An air-filled parallel-plate capacitor is connected to a battery with a potential difference ΔV.

(a) The plates are pulled apart, doubling the gap width, while they remain connected to the battery. By what factor does the potential energy of the capacitor change?

(b) If the capacitor is first removed from the battery, what happens to the stored potential energy when the gap width is doubled? Support your claim.

38. An air-filled parallel-plate capacitor is attached to a battery with a potential difference ΔV. While attached to the battery, the area of the plates is doubled and the separation of the plates is halved. During this process, by what factor do (a) the capacitance, (b) the charge on the positive plate, (c) the potential difference across the plates, and (d) the potential energy stored in the capacitor change? (e) By what factor do all the listed quantities change if, once the capacitor is charged, it is first disconnected from the battery before the area and separation are changed as described?

39. A honey bee of mass 130 mg has accumulated a static charge of +1.8 pC. The bee is returning to her hive by following the path shown in the figure below. Because Earth has a naturally occurring electric field near ground level of around 100 V/m pointing vertically downward, the bee experiences an electric force as she flies.

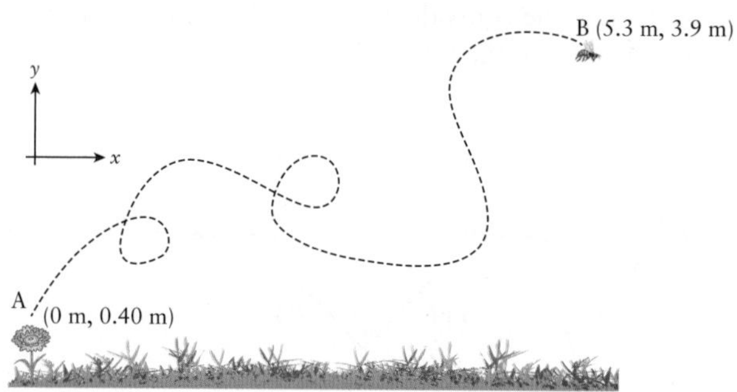

(a) What is the change in the bee's electric potential energy, $\Delta U_{\text{electric}}$, as she flies from point A to point B?

(b) Compute the ratio of the bee's change in electric potential energy to her change in gravitational potential energy, $\Delta U_{\text{electric}}/\Delta U_{\text{grav}}$.

40. A parallel-plate capacitor has area A and separation d.
(a) What is its new capacitance if a *conducting* slab of thickness $d' < d$ is inserted between, and parallel to, the plates as shown below?

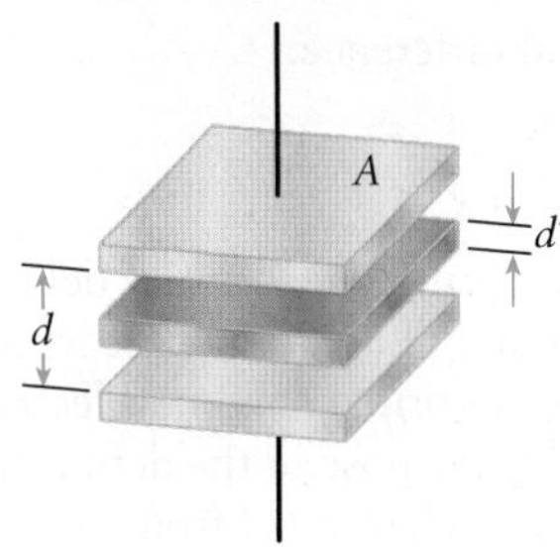

(b) Does your answer depend on where the slab is positioned vertically between the plates?

41. The figure below shows equipotential curves for 30 V, 10 V, and −10 V. A proton follows the path shown in the figure. If the proton's speed at point A is 80 km/s, what is its speed at point B?

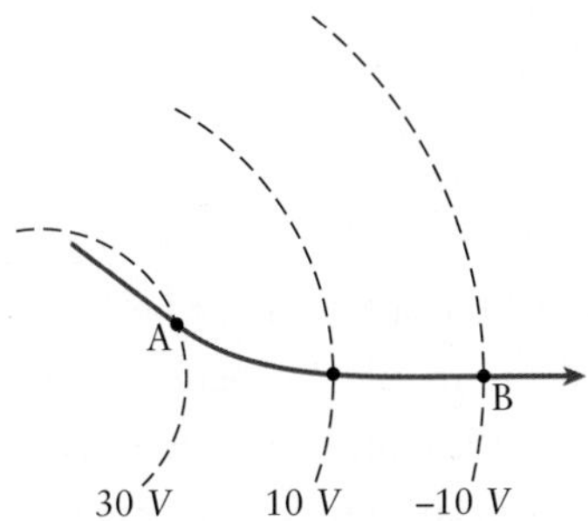

42. In the region shown in the figure below, there is a uniform electric field of magnitude 40.0 N/C that points in the positive y direction. Points 2, 3, and 4 are all 0.650 m away from point 1, and the angle $\varphi = 35.0°$. Calculate the following potential differences: (a) $V_2 - V_1$, (b) $V_3 - V_1$, (c) $V_4 - V_1$, and (d) $V_2 - V_4$.

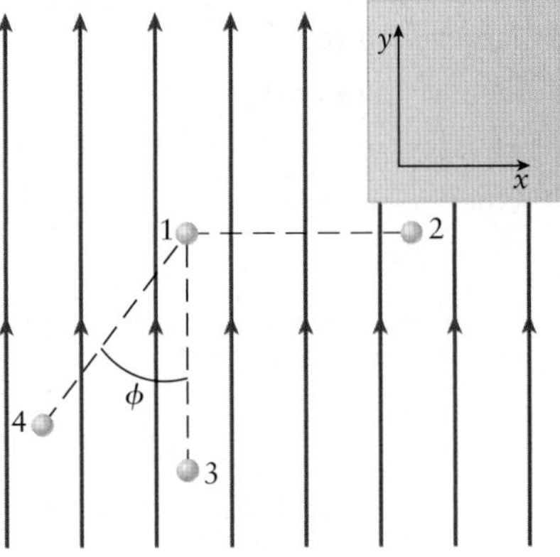

Prep for the AP Exam

AP® Group Work

Directions: The following problem is designed to be done as group work in class.

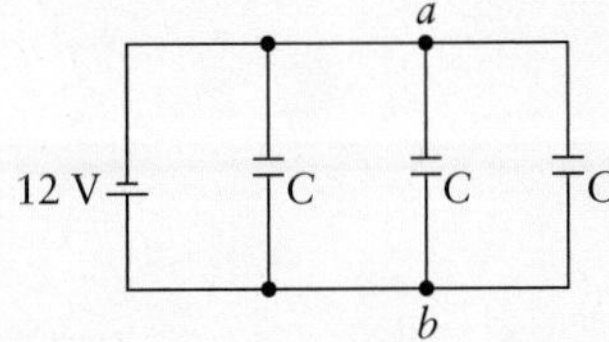

Three 0.18-μF capacitors are connected in parallel across a 12-V battery, as shown in the figure above. The battery is then disconnected. Next, one capacitor is carefully disconnected so that it doesn't lose any charge and is reconnected with its positively charged and negatively charged sides reversed.

(a) What is the potential difference across the capacitors now?

(b) What is the stored energy of the combination of capacitors after they have been rearranged?

(c) Then the space between the plates of one of the capacitors is completely filled with a conducting material without changing the arrangement. What happens to the energy stored in the combination of capacitors after the conducting material is added?

AP® PRACTICE PROBLEMS

Prep for the AP® Exam

Multiple-Choice Questions

Directions: The following questions have a single correct answer.

1. The electric potential at a point equidistant from two point charges that have charges $+Q$ and $-Q$ is
   (A) larger than zero.
   (B) smaller than zero.
   (C) equal to zero.
   (D) equal to the average of the two distances times the charges.

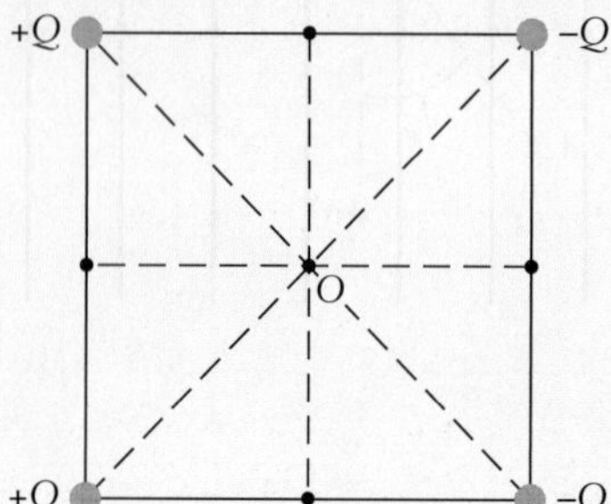

2. Four point charges of equal magnitude but differing signs are arranged at the corners of a square, as shown in the figure above. The electric field E and the potential V at the center of the square are
   (A) $E = 0$ and $V \neq 0$, respectively.
   (B) $E = 0$ and $V = 0$, respectively.
   (C) $E \neq 0$ and $V \neq 0$, respectively.
   (D) $E \neq 0$ and $V = 0$, respectively.

3. Capacitors connected in parallel have the same
   (A) charge.
   (B) potential difference.
   (C) dielectric.
   (D) surface area.

4. A cardiac defibrillator works by delivering an electric shock to a malfunctioning heart to jump-start it back into its normal rhythm. Electric energy is stored for this purpose in the defibrillator by separating positive charge $+Q$ from an equal amount of negative charge $-Q$. To transfer this energy to the heart, conducting paddles are placed on the patient's chest, and charge $-Q$ in the form of electrons is allowed to flow through the paddles and patient until it reaches the stored charge $+Q$ within the defibrillator. If the value Q of the charge stored in a defibrillator is doubled, by what factor will the stored energy increase?
   (A) $\sqrt{2}$
   (B) 2
   (C) 4
   (D) 8

5. An equipotential surface must be
   (A) parallel to the electric field at every point.
   (B) equal to the electric field at every point.
   (C) perpendicular to the electric field at every point.
   (D) tangent to the electric field at every point.

(Continued)

Free-Response Question

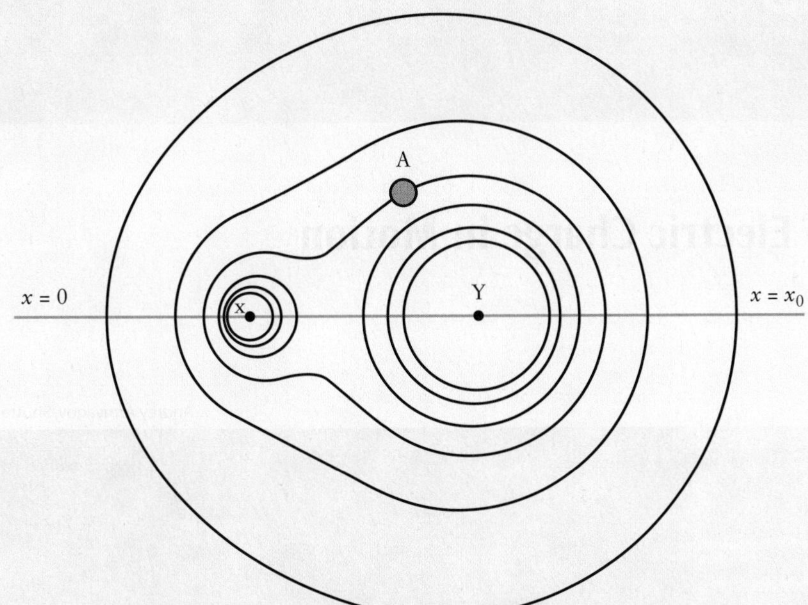

In this figure, the two dots represent the locations of two charged objects, X and Y, that are fixed in place. The lines are isolines of electric potential, with a potential difference of 10 V between each set of adjacent lines. The absolute value of the electric potential of the outermost line is 50 V, and the center-to-center distance between X and Y is 50 cm.

(a) (i) Indicate the value of the potential for each line on the figure, including the signs, if the charge on object Y is positive.

(ii) Assume a small positively charged conducting disk with a diameter of 1 cm and a charge of 3 nC, lying in the plane of the page, is placed at point A. Sketch it on the diagram above and indicate the distribution of charge on the object before it is released.

(b) Indicate the initial direction of motion of the disk if it is released at rest from point A. If its mass is 2.0 μg, determine the disk's speed when it has crossed two isolines of electric potential.

(c) On the axes below, sketch a qualitative graph of the electric force exerted on a positive test charge at points along the x axis.

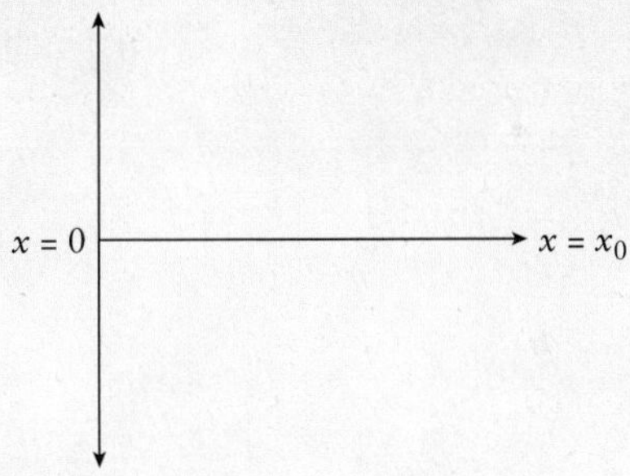

(d) How do the magnitudes and the signs of the charges of the objects compare? Justify your prediction in terms of isolines of electric potential, and relate this to the electric force graph you sketched in part (c).

DC Circuits

Chapter 18 DC Circuits: Electric Charge in Motion

Andrey Armyagov/Shutterstock

Case Study: How does a flashlight work?

Holden knows from experience that light radiates from his flashlight when he pushes the on button. He also knows that fresh batteries make the light shine brightly. What's really going on? How do the battery, button, and bulb produce light?

At the heart of Holden's flashlight is electric charge in motion. The movement of charge, or electric *current*, occurs in many situations in nature, and is also embedded in every manner of today's technology. In many electronic devices, the moving charge carriers are electrons, and the device is designed so that this movement of charge follows a closed path or *circuit*. One of the simplest electric circuits is found inside a flashlight.

An electric circuit includes a battery or similar device that provides the energy required for the electrons to move. A battery is analogous to a person providing the external energy necessary to store gravitational potential energy in an Earth–object system by raising the object. Instead of separating the object and Earth, the battery separates positive and negative charge at its two terminals. Electrons will move through the circuit from the negatively charged terminal of the battery to the positively charged terminal, causing the bulb to light. Chemical energy in the battery is used to move electrons to the negatively charged terminal so that the process continues. We will learn that we can describe this property of a battery to provide electric potential energy to a circuit as electric potential difference.

Another element in this flashlight's circuit is an incandescent bulb. When an electric current passes through the bulb's filament (a thin thread of the metal tungsten), the energy it carries is converted into warmth and light. *Incandescent* comes from the Latin word meaning "to glow"—when electrons move through the filament, the bulb glows.

Electrons need to follow a conducting path to move from one terminal of the battery to the other. The flashlight's "on" button is a switch that either completes or breaks the circuit. When the circuit is broken, electrons do not move and the bulb is not lit.

A characteristic of the filament is its resistance; as it resists the motion of the electrons, some of their energy is transferred to it, heating it up until it glows. The faster energy transfers to the filament, the brighter it glows. Energy transferred per second is power; the power to the bulb decreases with increasing resistance of the filament and increases with the magnitude of the electric potential difference supplied by the battery. Holden replaces the battery in his flashlight with one with a greater potential difference. What would you expect to happen to the brightness of the flashlight? Can you think of a reason that it might not be a good idea for Holden to do this?

By the end of this chapter, you should be able to describe the relationships among electric potential difference, current, resistance, and resistivity for electric charge carriers moving in a wire. You should also be able to calculate the resistance of a resistor and the current that a given electric potential difference produces in that resistor, and the power into or out of a circuit element. You should also be able to calculate these quantities for elements in more complicated circuits containing multiple resistors and batteries.

18 DC Circuits: Electric Charge in Motion

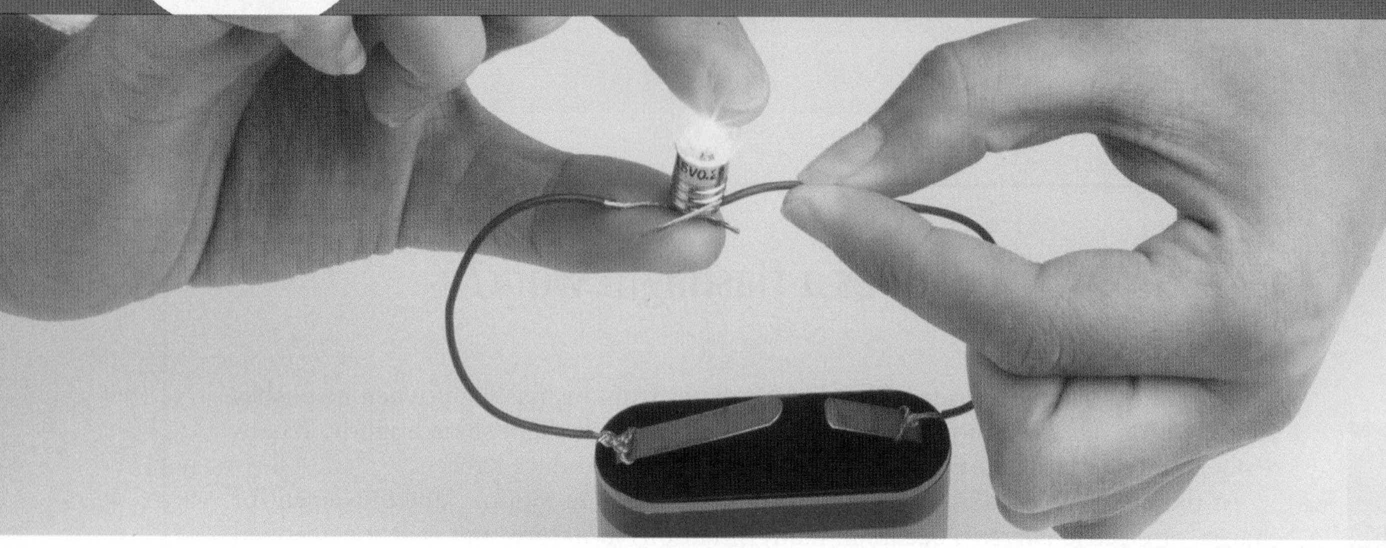

Dave King/Dorling Kindersley/Science Source

YOU WILL LEARN TO:

- Define current and describe the relationships among electric potential difference, current, resistance, and resistivity for charge carriers moving in a wire.
- Calculate the resistance of a resistor and the current that a given electric potential difference produces in that resistor.
- Describe the equivalent resistance of multiple resistors connected in series and/or parallel.
- Discuss Kirchhoff's rules in terms of conservation of energy and conservation of charge, and apply them to simple and compound circuits containing batteries and resistors, including nonideal batteries and resistive wires.
- Create a schematic diagram of a circuit from a description (or a description from a diagram) and use it to calculate unknown values of current in or potential difference across various segments or branches of the circuit or to predict how those values would change if the configuration of the circuit changed.
- Describe how to measure current and potential difference within a circuit.
- Describe the transfer of energy into, out of, or within a circuit in terms of power.
- Describe the behavior of a circuit containing combinations of resistors and capacitors.

18-1 Life on Earth and our technological society are possible only because of charge in motion

In the previous chapter we investigated electric charge and force. In our investigations we considered charged objects that were fixed in place. But in many important situations in nature and technology electric charge is in motion. You are able to read these words thanks to electric charge carriers that travel along the optic nerve from your eye to your brain, transmitting the image of the words in the form of a coded electrical signal. If you are reading these words after sunset, you are either looking at a printed page illuminated by a light bulb that's powered by moving electric charge or else reading them on a tablet or other electronic device that operates using complex electric circuitry. The simplest of all electric circuits is a battery connected to a light bulb, such

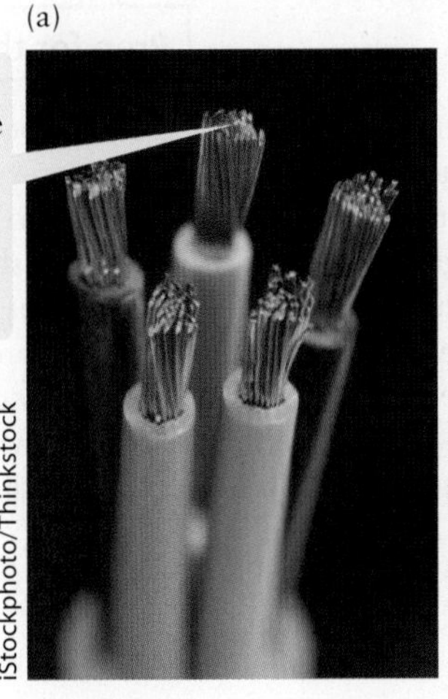

(a)
Wires and cables are made of conductors that allow charge carriers to move along their length.
iStockphoto/Thinkstock

(b)
The current in a mobile device is provided by a battery (a source of emf).
Daniel M Ernst/Shutterstock

Figure 18-1 Charge in motion
(a) Conducting wires are an essential part of any electric circuit. (b) A smartphone is a common device that uses currents (moving electric charge).

as the one you find inside an ordinary flashlight. The battery causes electrons to move through the circuit. As electrons pass through the light bulb, they transfer energy to the light bulb to make it shine. We will learn what type of energy this is in this chapter.

In this chapter we'll look at the energy associated with configurations of electric charge, and the basic physics of electric charge in motion. We'll introduce the idea of *current*, which measures the rate at which charge moves through a conductor (**Figure 18-1a**). Ordinary conductors have *resistance* to the motion of charge, so it's necessary to set up a change in potential energy between the ends of a conductor to produce this motion. You can think of it just as if you were trying to get a box to slide down an incline when there is friction between the incline and the box. The decrease in gravitational potential energy of the box–Earth system as the box moves down the incline would become the kinetic energy of the box. The friction, if it is the right amount, keeps the box from speeding up by dissipating this energy (transferring this mechanical energy into warmth and damage of the surface of the incline and the box), as we saw in Chapter 8. In circuits, providing the increase in potential energy is the role of a **source of emf** (pronounced *ee-em-eff*), the most common example of which is an ordinary battery (**Figure 18-1b**). We'll analyze simple circuits that include a source of emf and one or more *resistors* (circuit elements that have resistance), and see how to treat combinations of resistors. Although we will primarily use batteries as our source of emf, we will also consider capacitors. There are many other ways to provide an increase in electric potential energy. For instance, in Chapter 20, we will learn about generators, which commonly convert energy from coal or nuclear sources (such as power plants), or can use other sources of energy, such as wind, water, or a hand crank.

Fundamentally, an electric circuit is used to transfer energy from one place to another (for instance, from a battery to a light bulb, as in the photograph that opens this chapter). We will see how to describe electric potential energy, and how conservation of charge and energy determine the behavior of all circuits. We'll see how to describe the *power*, or rate of energy transfer, associated with any circuit element. We will learn the limiting cases on circuits that involve capacitors as well as resistors. During their functioning they are mathematically complicated to describe, but their behavior at short and long times is more straightforward.

THE TAKEAWAY for Section 18-1

✔ Technological devices and biological systems depend on electric charge in motion.

✔ An electric circuit is fundamentally a means of transferring energy.

Prep for the AP® Exam

AP® Building Blocks

1. In this section, a box sliding down an incline with friction was mentioned as an analogy for moving charged objects through a circuit, such as that in a flashlight. The box stops when it hits the floor, but the flashlight light bulb keeps glowing. You could stand next to the incline and return the box to the top, so that the box keeps sliding down the incline. Describe the energy transfers involved in this process, where you would serve as a human "battery" for this mechanical analogy to a circuit. The process eventually would slow down when you got tired. What would cause this?
2. When a waterfall is used to turn a wheel to grind grain or a turbine to generate electricity, the water slows down a tiny bit as it turns the wheel or turbine, but all the water keeps going (water is "conserved" in this process). In complete sentences, briefly compare the energy conversions and transfers of the waterfall turning the wheel and the box sliding down an incline with friction that was discussed in this section.

18-2 Electric current equals the rate at which charge moves

Figure 18-2 shows a common situation in which electric charge carriers are in motion. The source of emf that sets them into motion is the **battery**, which contains one or more *electrochemical cells*. Inside the cell are two different substances that, due to their different chemical properties, undergo a chemical reaction, so that each substance ends up with an excess or deficit of electrons. In an ordinary alkaline battery, like an AA or a D cell, the two substances are zinc and manganese dioxide. The zinc ends up with an electron excess, while the manganese dioxide ends up with an electron deficit.

The two terminals of the battery are each connected to one of these substances. The terminals are conductors, so they are each equipotential surfaces at the same potential as the substances with which they are in contact. Thus, the terminal attached to the substance with an electron excess is at lower potential and is called the *negative* terminal. The other terminal, attached to the substance with an electron deficit, is at a higher potential and is called the *positive* terminal. These terminals are marked on the battery (and in Figure 18-2) by a minus sign and a plus sign, respectively. The electric potential difference between the terminals depends on the two substances within the battery. For zinc and manganese dioxide, the electric potential difference is 1.5 V, a number that you will see written on the case of many alkaline batteries. A 9.0-V alkaline battery is made up of six 1.5-V cells. A 1.5-V battery is a single cell.

We know from Chapter 17 that charged objects tend to move from one point to another when there is a potential difference between those points: The electric force

(1) A simple electric circuit is made up of a battery connected to a wire and a device (such as a light bulb), forming a closed circuit.

(3) The electric field causes charged objects within the wire that are free to move, called charge carriers or mobile charges, to move along the wire. The current—the rate at which charge moves past a given point in the wire—is I.

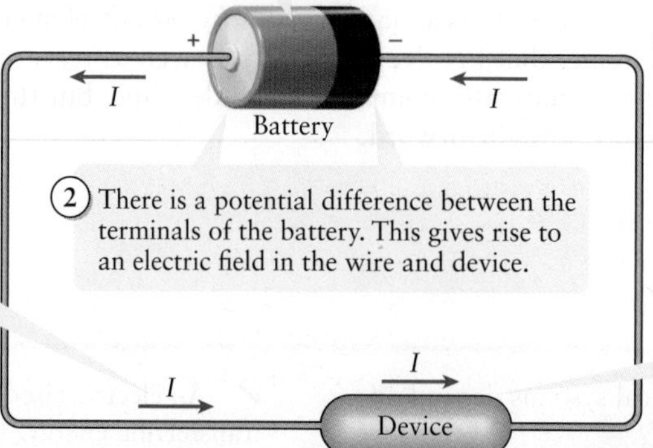

(2) There is a potential difference between the terminals of the battery. This gives rise to an electric field in the wire and device.

(4) Except for at the terminals of the battery where the chemical reaction keeps the configuration of charge relatively constant, charge does not pile up at any point in the circuit. So the current has the same value I at all points in the circuit, including inside the battery, in the wire, and inside the device. Charge passing through a "point" in a circuit refers to the charge passing through a cross-sectional area of the circuit element located at that point.

Figure 18-2 Current in a circuit Charge moves in a wire loop due to the electric potential difference supplied by a battery.

pushes positively charged objects from high to low potential and negatively charged objects from low to high potential. (The direction of the force changes with the wire, at each point directed so that it would push a positive point charge along the length of the wire from the positive to the negative terminal.) If a battery is isolated and not connected to anything, there is no way for charged objects to be pushed outside the battery from one terminal to another. Because the positive terminal is at a higher electric potential than the negative terminal when a conducting path is established between them, charge carriers can move to try to reduce the excess charge produced by the separation in the substances, but the chemical reaction continues the separation, replenishing the excess charge. Such a conducting path is a metallic wire used to connect the two terminals to each other, as depicted in Figure 18-2. As we saw in Section 16-3, metals are especially good electrical conductors because one or more of the electrons associated with a metal atom are only weakly bound to that atom. As a result, these loosely bound electrons can move with relative freedom within the metal. In most circuits it is electrons that move freely, but in some, such as fluorescent lights, it is charged ions, which can be positive or negative. It is common just to call whatever is moving freely in a circuit charge. When the battery is connected to the wire, so that the wire forms a path for charge to move from one terminal of the battery to another, it is called a complete loop or closed **circuit**; an incomplete circuit is sometimes referred to as an open circuit. For a closed circuit, charge moves continuously through the circuit from one terminal of the battery to the other, as if it were an object sliding at constant speed down an incline with friction near Earth's surface. Inside the battery, the chemical reaction "lifts" the charge back up to the top of the incline.

The electric force on the charge is necessary because the moving charged objects (usually electrons) within the wire collide very frequently with the atoms that make up the wire. The net effect of the collisions is to slow or retard the motion of charge through the wire, rather like the motion of an algal spore through water is retarded by fluid resistance (see Section 5-5). This similarity to fluid resistance is one of the reasons why you will often see this motion of charge referred to as a flow of charge. The speed at which charge moves is determined by the balance between the retarding force due to collisions and the electric force in the direction of motion.

Later in this chapter we'll see how the motion of charge delivers energy to a device (such as a light bulb) that's part of the circuit. First, let's develop a mathematical model for the motion of charge in a circuit.

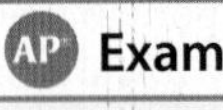
Exam Tip

A common misconception is that the current is used up in a device (a light bulb, for example), and then the current would be smaller after the device than before the device. This is incorrect. The correct analysis is that the current is the same everywhere in a closed loop regardless of the devices and where they are being powered in that loop. It is energy, not charge, that is transferred to the bulb!

Current

The **current** in a circuit, such as that shown in Figure 18-2, equals the rate at which charge moves past any point in the circuit. In particular, if an amount of charge Δq moves past a certain point in a time Δt, the current I is

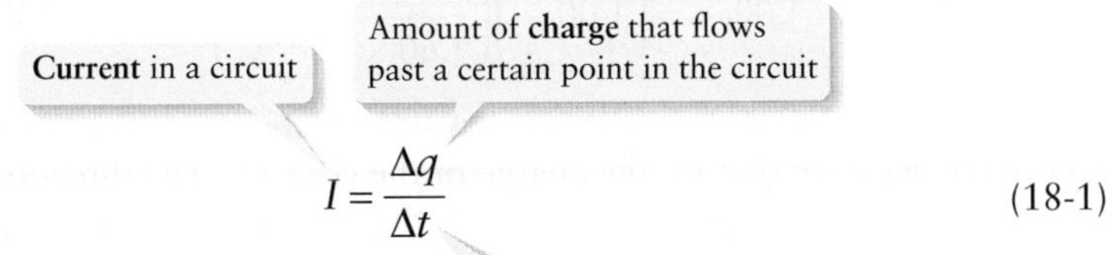

$$I = \frac{\Delta q}{\Delta t} \quad (18\text{-}1)$$

EQUATION IN WORDS
Definition of current

The SI unit of current is the **ampere**, named after the French scientist and mathematician André-Marie Ampère. One ampere (abbreviated amp or A) is equivalent to one coulomb of charge passing a given point per second: 1 A = 1 C/s.

We will initially limit our discussion of currents to situations that result in a *steady* movement of charge. Then the current I has the same value at all times. In addition, the current has the same value at all points in a simple circuit, such as that in Figure 18-2, in which charges move around a single loop. The charge carriers cannot "pile up" or accumulate at any point in the circuit except at the terminals of the battery (if they did, their mutual electrostatic repulsion would make them spread apart again). Since charge is conserved, there is also no way for more moving charged objects to join the current or for charged objects to leave the current: Whatever amount of charge moves into a point in the circuit, the same amount of charge must move out of that point in the circuit. This means the value of the current is the same everywhere around a closed loop.

Note that current is *not* a vector quantity, even though it does have a direction. That direction is associated with which way positive charge would flow, but not with some coordinate system fixed in space. For example, in Figure 18-2 the current is to the left in the upper part of the circuit, downward in the left-hand part of the circuit, to the right in the lower part of the circuit, and upward in the right-hand part of the circuit. So, there is no single vector that describes the direction of current in every part of the circuit. Instead, we simply say that the current in Figure 18-2 is counterclockwise around the circuit. If we reversed the battery, so that the positive terminal were on the right rather than on the left, the current would instead be clockwise. Circular motion, torque, angular momentum, and—as we will discuss in the next chapter—magnetism include vector quantities where the direction of the vector is defined by a right-hand rule, and thus terms such as *clockwise* are useful. For those cases, the vectors have directions in space that are well defined, with that definition depending upon if the related property of the system is oriented clockwise or counterclockwise.

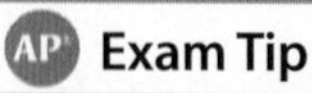

The AP® Physics 2 equation sheet reminds you of the convention that the direction of current is defined as the direction positive charge would move, regardless of what actually moves.

WATCH OUT !

The direction of current is chosen to be the direction in which *positive* charge would move.

In ordinary wires and in common electric devices such as light bulbs and toasters, the charge carriers are negatively charged electrons. In an electric circuit, such as that shown in Figure 18-2, electrons move through the wire from low potential to high potential and so from the negative terminal of the battery toward the positive terminal. But in Figure 18-2 we show the current directed through the wire from the battery's *positive* terminal to its *negative* terminal. That's because the convention is that current is in the direction that positively charged objects would move, regardless of whether the moving objects are positively or negatively charged. We'll use this convention throughout this book. (This convention is often attributed to the eighteenth-century American scientist and statesman Benjamin Franklin. The discovery that the charge carriers in wires are negatively charged came decades after Franklin's death.) Note that in most biological systems, such as neurons and muscle cells, the charge carriers actually *are* positive (they are positive ions, atoms that have lost one or more electrons each).

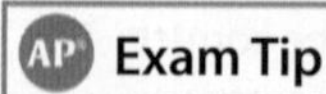

Note that the current is a rate at which charge passes through a cross section of a wire; it is not a *speed*. It does not make sense to talk about current as a speed, and it also is incorrect to talk about the speed of current—current is already a rate. The correct terminology is to talk about or calculate the *amount* of current.

Because an electron has a very small charge ($q = -e = -1.60 \times 10^{-19}$ C), typical currents involve the movement of a very large number of electrons. For example, suppose the current I in the circuit shown in Figure 18-2 is 1.00 A, or 1.00 C/s. (That's roughly the current in a large flashlight.) Then the rate at which electrons move past any specific point on the wire is

$$1.00 \text{ A} = 1.00 \frac{\text{C}}{\text{s}} = \left(1.00 \frac{\text{C}}{\text{s}}\right)\left(\frac{1 \text{ electron}}{1.60 \times 10^{-19} \text{ C}}\right) = 6.25 \times 10^{18} \frac{\text{electron}}{\text{s}}$$

Since a positive current is defined as the direction positive charge would move, we ignore the negative sign on the charge of the electron. The direction of current would be in the direction opposite that which the electrons are moving. This result states that if we could count the number of electrons passing through the cross section of the wire, we would count about 6×10^{18} per second. We don't get a huge shock from walking near this wire because for every electron moving through the wire, there is a positively charged atom missing an electron, so the wire is neutral.

EXAMPLE 18-1 Charging a Sphere

A large, hollow metal sphere is electrically isolated from its surroundings, except for a wire that can carry a current to charge the sphere. Initially the sphere is uncharged and is at electric potential zero. What must be true of the potential at the other end of the wire, away from the sphere, for charge to move? If the sphere has a radius of 0.150 m and the current is a steady 5.00 μA = 5.00×10^{-6} A, how long does it take for the sphere to attain a net charge of −6.67 μC? Which direction is the current going in the wire?

Set Up

If the electric potential at the surface of the sphere is different from the electric potential of the device being used to charge it, charge can move from the device to the sphere through the wire. To get negative charge to move onto the sphere, the other end of the wire must be at a negative electric potential. We are told the current is constant, so we can use the equation for current to find the time required for this charge to reach the sphere.

Definition of current:

$$I = \frac{\Delta q}{\Delta t} \quad (18\text{-}1)$$

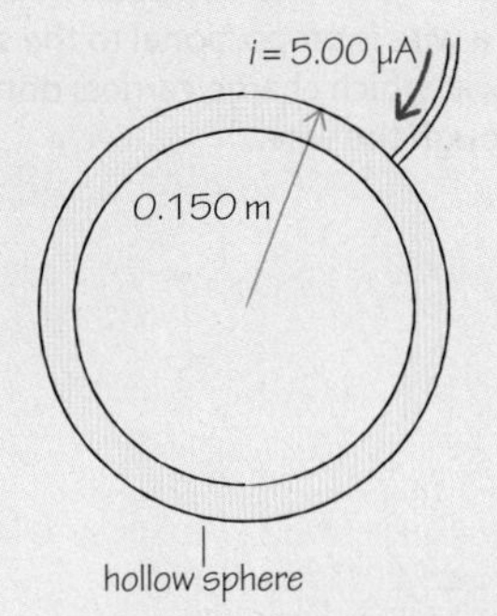

Solve

We are told charge $Q = -6.67\ \mu\text{C}$ is the amount of charge Δq that must move onto the sphere in a time Δt. The minus sign means negative charge is accumulating on the sphere, so electrons are moving onto the sphere. To get electrons to move onto the sphere, the sphere must be at a higher potential than the other end of the wire. Use Equation 18-1 to solve for Δt. Recall that $1\ \text{A} = 1\ \text{C/s}$. Because the direction of current is defined as the direction in which positive charge moves, the direction of the current is from the sphere toward the other end of the wire. We can use the givens and the definition of current to find the time for this magnitude of charge to move. Unlike when there is a closed path, for the current to remain constant as negative charge collects on the sphere, the potential at the opposite end of the wire must be continually increasing.

From Equation 18-1:

$$\Delta t = \frac{\Delta q}{I} = \frac{6.67 \times 10^{-6}\ \text{C}}{5.00 \times 10^{-6}\ \text{A}} = \frac{6.67 \times 10^{-6}\ \text{C}}{5.00 \times 10^{-6}\ \text{C/s}}$$
$$= 1.33\ \text{s}$$

Reflect

A charged sphere like this is used in a Van de Graaff generator, which you may have seen demonstrated in your physics class. If you have, you know that when the generator is turned on to charge the sphere, the sphere can begin to throw off sparks within seconds. So, a result for Δt on the order of 1 s is reasonable.

NOW WORK Problems 1–3, 7, and 9 from The Takeaway 18-2.

Drift Speed

The value of the current I tells us what quantity of charge passes a given point in a circuit per second. Let's see how to relate this to the **drift speed** v_{drift}, which is the average speed at which the charge carriers that are free to move throughout the conducting material in the circuit move (or "drift") through the circuit.

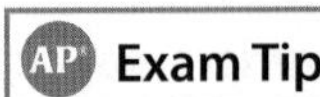

AP Exam Tip

Drift speed is not covered on the AP® Physics 2 Exam, but helps to deepen your qualitative understanding of current, which is tested.

Figure 18-3 shows a wire that has a cross-sectional area A and carries a current I. Charge carriers are moving ("drifting") through the green region at an average speed of v_{drift}. At any time the total charge of the charge carriers in that region is Δq. Note that Δq is the product of n (the number of charge carriers per volume), the volume $A\Delta x$ of the green region (a cylinder of area A and length Δx), and the amount of charge e on each charge carrier:

$$\Delta q = n(\text{volume})e = n(A\Delta x)e \quad (18\text{-}2)$$

If the mobile charges are electrons, each has a charge $-e$ rather than e. But since we take the direction of current to be the direction in which positive charge carriers would move, we'll use the convenient fiction that this charge is positive (remembering

Figure 18-3 Drift speed The current I in a wire is proportional to the speed v_{drift} at which charge carriers drift through the wire.

① Current I is present in the wire of cross-sectional area A.

② Within the wire there are n charge carriers per unit volume. Each has a charge e.

③ The charge carriers drift through the wire with speed v_{drift}, and travel a distance Δx in a time $\Delta t = \Delta x / v_{drift}$.

④ A length Δx of the wire (shown in green) has volume $A\Delta x$. This length contains $n(A\Delta x)$ charge carriers, with total charge $\Delta q = n(A\Delta x)e$.

⑤ The current in the wire equals the amount of charge Δq that passes this point divided by the time Δt it takes them to pass this point:

$$I = \frac{\Delta q}{\Delta t} = \frac{n(A\Delta x)e}{\Delta x / v_{drift}} = nAev_{drift}$$

that if the charge carriers are electrons, they are moving in the opposite direction). The time required for this volume of charge to drift the distance Δx along the length of the wire is $\Delta t = \Delta x / v_{drift}$ (time equals distance divided by speed). So from Equations 18-1 and 18-2, the current in the wire is

$$I = \frac{\Delta q}{\Delta t} = \frac{n(A\Delta x)e}{\Delta x / v_{drift}}$$

or, simplifying,

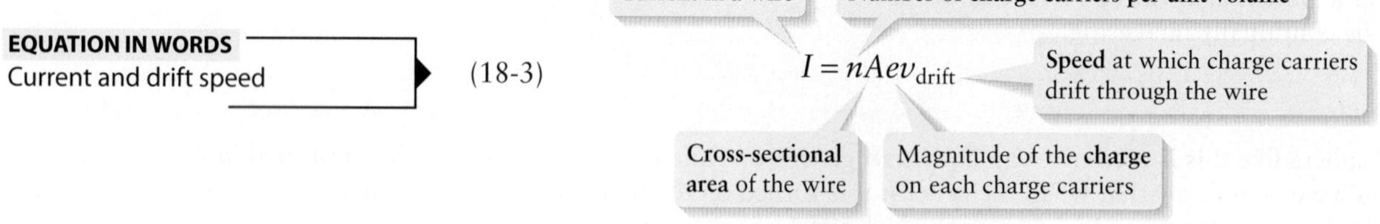

The number of moving charges per volume (n) is different for different materials. Equation 18-3 tells us that for a wire made of a given material and with a given cross-sectional area A, the current i is directly proportional to the drift speed.

WATCH OUT !

Individual charge carriers moving does not necessarily mean there is a current.

Even if there is no current through a wire, the charge carriers within that wire are still in motion. However, their motions are in random directions, so there is no *net* motion of charge. It's rather like a swarm of bees flying around a hive: Individual bees move in different directions, but the swarm as a whole stays in the same place. If the swarm moves to a different hive, the motions of individual bees will be different but the swarm as a whole will "drift" together to the new hive. In the same way, if a current is present in the wire in Figure 18-2, the individual charge carriers can be moving at different speeds and in different directions, but this "swarm" of charge drifts along the wire at speed v_{drift}.

EXAMPLE 18-2 Electron Drift Speed in a Flashlight

In a large flashlight the distance from the on–off switch to the light bulb is 10.0 cm. How long does it take electrons to drift this distance if the flashlight wires are made of copper, are 0.512 mm in radius, and carry a current of 1.00 A? There are 8.49×10^{28} atoms in 1 m^3 of copper, and one electron per copper atom can move freely through the metal.

Set Up

We'll use Equation 18-3 to determine the drift speed from the information given about the number of electrons serving as charge carriers, the radius of the wire, and the current. From this we'll be able to find the drift time by using the familiar relationship between speed, distance, and time.

Current and drift speed:

$$I = nAev_{\text{drift}} \qquad (18\text{-}3)$$

Solve

Calculate the drift speed of the electrons in the wire.

Solve Equation 18-3 for the drift speed:

$$v_{\text{drift}} = \frac{I}{nAe}$$

There is one moving electron per atom and 8.49×10^{28} atoms/m^3, so the value of n is 8.49×10^{28} electrons/m^3. The wire has a circular cross section with radius $r = 0.512$ mm $= 0.512 \times 10^{-3}$ m, so its cross-sectional area is

$$A = \pi r^2 = \pi(0.512 \times 10^{-3}\ \text{m})^2 = 8.24 \times 10^{-7}\ \text{m}^2$$

The magnitude of the charge per electron is $e = 1.602 \times 10^{-19}$ C/electron, and the current is $I = 1.00$ A $= 1.00$ C/s. So the drift speed is

$$v_{\text{drift}} = \frac{I}{nAe}$$

$$= \frac{1.00\ \text{C/s}}{(8.49 \times 10^{28}\ \text{electrons/m}^3)(8.24 \times 10^{-7}\ \text{m}^2)(1.602 \times 10^{-19}\ \text{C/electron})}$$

$$= 8.928 \times 10^{-5}\ \text{m/s} = 0.0893\ \text{mm/s}$$

Then calculate how long it takes electrons to drift the distance from switch to light bulb.

At this drift speed the time it takes electrons to travel a distance $d = 10.0$ cm $= 0.100$ m from the flashlight on–off switch to the light bulb is

$$t = \frac{d}{v_{\text{drift}}} = \frac{0.100\ \text{m}}{8.93 \times 10^{-5}\ \text{m/s}} = 1.12 \times 10^3\ \text{s}$$

$$= (1.12 \times 10^3\ \text{s})\left(\frac{1\ \text{min}}{60\ \text{s}}\right) = 18.7\ \text{min}$$

Reflect

The phrase "a snail's pace" refers to something that moves very slowly. But an ordinary snail moves at about *twice* the drift speed of electrons in this wire. The drift speed is so slow because electrons in the wire are continually colliding with the copper atoms, which slows their progress tremendously. (More sophisticated physics shows that an electron in copper moves in *random* motion at an average speed of about 10^6 m/s, about 10^{10} times faster than the drift speed. In the analogy we made earlier between electrons and a swarm of bees, you should think of the electrons in this wire as *very* fast-moving bees within a swarm that's drifting very, very slowly because of collisions with many obstacles.)

Extend: At their slower-than-a-snail pace, it takes electrons more than a quarter of an hour to travel from the on–off switch to the light bulb. Why, then, does the flashlight turn on immediately when you move the switch to the on position?

The explanation is that the wire is completely made of copper, and each atom has an electron that can move, and these electrons drift in response to the electric field in the wire. With the switch in the off position, there is no electric field and so no drift. An electric field is set up only when the switch is put in the on position, making a complete circuit like that shown in Figure 18-2. Changes in the electric field propagate through the wire at close to the speed of light (3×10^8 m/s), so the field is set up throughout the circuit, and the electrons begin to drift, in a tiny fraction of a second. That's why the light comes on nearly instantaneously.

NOW WORK Problems 4, 8, and 10 from The Takeaway 18-2.

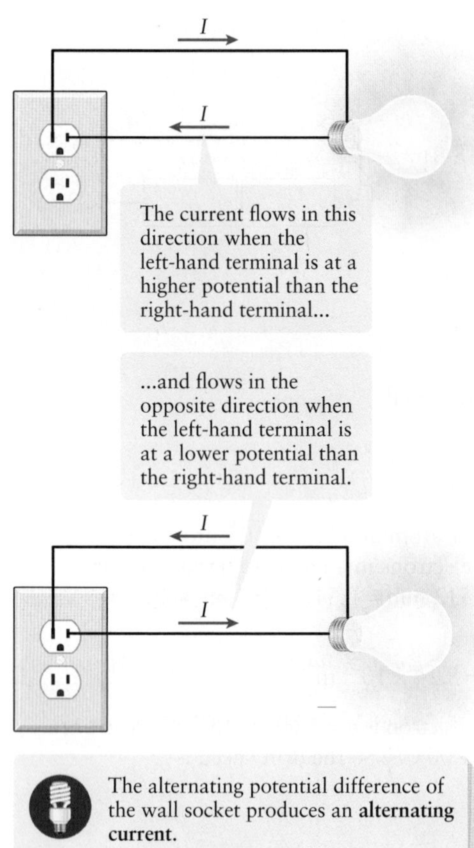

I
I
The current flows in this direction when the left-hand terminal is at a higher potential than the right-hand terminal...
...and flows in the opposite direction when the left-hand terminal is at a lower potential than the right-hand terminal.
I
I
The alternating potential difference of the wall socket produces an **alternating current.**

Figure 18-4 Alternating current The potential difference between the terminals of a wall socket varies sinusoidally. As a result, the current in a circuit that contains this socket alternates direction.

Direct Current and Alternating Current

In the circuit shown in Figure 18-2 the current around the circuit is always in the same direction, as shown by the arrows labeled I. Current of this kind is called **direct current**, or **DC** for short. You'll find direct current in any device that's powered by a battery, such as a flashlight, a television remote control, or a mobile phone. That's because the potential difference between the two terminals of the battery that powers the circuit always has the same sign: The positive terminal is always at a higher potential than the negative terminal.

Although only DC circuits are discussed in AP® Physics, AC circuits are very common in modern society—and in your life! When you plug a toaster or a table lamp into a wall socket, that is an AC circuit. The potential difference between the two terminals in a wall socket is not constant, but instead oscillates or *alternates*. At one instant the left-hand terminal is at a higher potential than the right-hand terminal; a short time later the left-hand terminal is the one at the lower potential, and a short time after that the left-hand terminal is again at the higher potential. As a result, the current in a device plugged into a wall socket alternates direction (**Figure 18-4**). This is called **alternating current**, or **AC** for short. (The third terminal found in most wall sockets, called the *ground*, remains at a constant potential and is not directly involved in providing the current. The ground terminal only comes into play if there is a failure in the wiring in the appliance or its power cord, in which case the resulting current could be dangerously high. In this case the unwanted current is diverted into the ground terminal. As the name suggests, the ground terminal directly connects the circuit to the ground—Earth—which has a nearly infinite capacity to accept excess electrons from such a source.) The potential difference between the two terminals of a wall socket varies with time in a sinusoidal way, just as the position of an object in simple harmonic motion does. So, in an AC circuit, electrons just oscillate back and forth around an equilibrium position. That's very different from a DC circuit, in which electrons plod slowly in the same direction around the circuit.

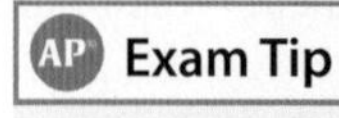
Exam Tip

AC currents are not part of the AP® Physics 2 curriculum.

THE TAKEAWAY for Section 18-2

✔ An electric potential difference applied across a wire produces an electric current (a net motion of charge in the wire).

✔ The SI unit of current is the ampere (A): 1 A = 1 C/s.

✔ Current has a magnitude and a direction, but it is not a vector.

✔ By convention, the direction assigned to a current is the direction in which positive charge carriers would move. In typical metals such as those used in wires, it is the negative electrons that are actually the charge carriers.

✔ Direct current (DC) always travels the same direction around a circuit.

✔ The drift speed is the speed at which actual charge carriers in a current move, but current is almost instantaneous because charge carriers exist throughout the circuit and no individual charge carrier has to travel the complete circuit.

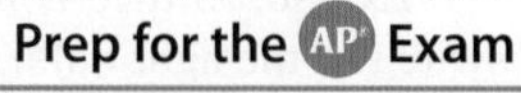
Prep for the AP® Exam

AP® Building Blocks

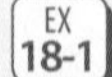
EX 18-1 1. A steady current of 35 mA exists in a wire.
   (a) Calculate the number of electrons that pass any given point in the wire per second.
   (b) Identify the direction of the current relative to the motion of the electrons.

EX 18-1 2. Biological membranes contain channels that allow ions to pass through. A particular Na^+ channel carries a current of 1.9 pA.
   (a) Calculate the number of Na^+ ions that pass through it in 1.0 ms.
   (b) Identify the direction of the current relative to the motion of the ions.

EX 8-1 **3.** A synchrotron storage ring maintains a circulation of charged elementary particles (typically electrons or protons) at a constant speed. Particles can be diverted from the storage ring for experiments involving a collision of the particles with a target and provide an intense source of synchrotron radiation. At the Argonne Advanced Photon Source (APS) electrons are diverted along beamlines tangent to the 1104-m-circumference circular storage ring. A current of 300 mA can be diverted into a beamline.

(a) Calculate the number of electrons that can be diverted into this beamline in 1 min.

(b) Electrons in a storage ring have a speed of 0.9999 c, where the speed of light $c = 2.998 \times 10^8$ m/s. Calculate the time for one complete circular path of an electron at the APS (neglect relativistic effects).

EX 8-2 **4.** Aluminum wiring can be used as an alternative to copper wiring. Suppose a 40-ft run of 12-gauge aluminum wire (2.05232 mm in diameter) carrying a 7.0-A current is used in a circuit. How long would it take electrons to travel the length of this run and back? There are 6.03×10^{28} atoms in 1 m^3 of aluminum, with each atom contributing three free electrons to the metal.

AP Skill Builders

5. We justified a number of electrostatic phenomena by the argument that there can be no electric field in a conductor. Now we say that the current in a conductor is driven by a potential difference and thus there is an electric field in the conductor. Is this statement a contradiction?

6. We show electric current in a wire by an arrow to indicate the direction of motion of positive charge. Explain why current, unlike velocity, is a scalar and not a vector quantity.

7. 2.00×10^{-3} mol of potassium ions pass through a cell membrane in 4.00×10^{-2} s.

(a) Calculate the electric current.

(b) Describe its direction relative to the motion of the potassium ions.

8. A copper wire that has a diameter of 2.00 mm carries a current of 10.0 A. Assuming that each copper atom contributes one free electron to the metal, find the drift speed of the electrons in the wire. The molar mass of copper is 63.5 g/mol, and the density of copper is 8.95 g/cm^3.

AP Skills in Action

9. If a current-carrying wire has a cross-sectional area that gradually becomes smaller along the length of the wire, the drift speed

(A) increases along the length of the wire.

(B) decreases along the length of the wire.

(C) increases along the length of the wire only if the resistivity increases too.

(D) decreases along the length of the wire only if the resistivity decreases too.

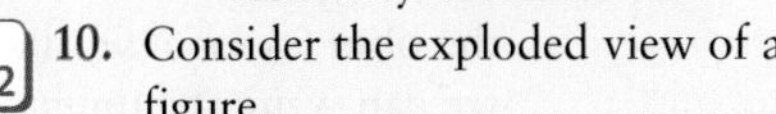

10. Consider the exploded view of a light bulb and base in the figure.

(a) Explain why the tungsten filaments in incandescent light bulbs don't have "+" and "–" printed on them, but batteries do, in terms of how one might connect batteries or light bulbs in a circuit.

(b) Describe the circuit that lights the bulb in terms of a sequence of lettered connections between the conducting elements shown in the figure.

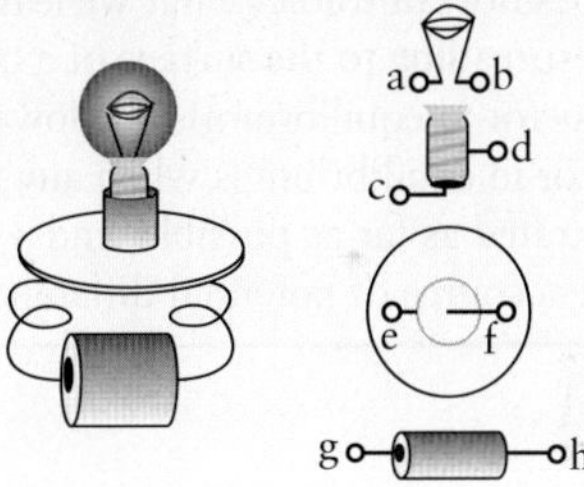

18-3 The resistance to current through an object depends on the object's resistivity and dimensions

We saw in Section 18-2 that current exists in a wire only if there is a potential difference between the ends of the wire. We discussed in Section 18-1 that we can think of it like trying to get a box to slide down an incline when there is friction between the incline and the box. The decrease in gravitational potential energy of the box–Earth system as the box moves down the incline would become the kinetic energy of the box. The friction, if it is the right amount, keeps the box from speeding up by dissipating this energy. To better understand current, we need to know: Just how is the resulting current related to the potential difference between the ends of the wire?

To answer this question, consider a straight wire of uniform cross-sectional area A and length L (**Figure 18-5**). If there is a potential difference between the ends of the wire, a current will exist. The direction of current, which we choose to be the direction in which positive charge would move, has to be in the direction of decreasing potential energy, and so the direction of the current is from high potential to low potential.

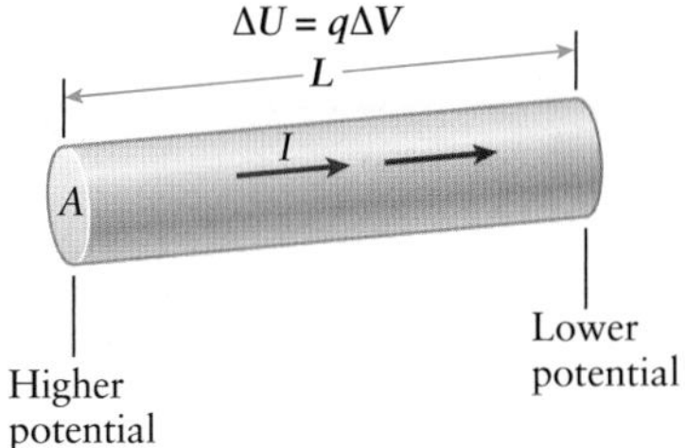

Figure 18-5 Inside a current-carrying wire By conservation of energy, decreasing potential energy provides increasing energy for motion or is dissipated by friction, so positive charge "falls" without speeding up from the high to the low value of the electric potential difference across the ends of a current-carrying wire. The current is in the direction of decreasing potential energy for positive charge.

AP Exam Tip

The resistance of wires in a common circuit is very, very small compared to the resistance of other elements of the circuit. In almost all cases, the resistance of the wires is negligible in determining the behavior of any circuit containing more than a wire and a battery. Therefore, the model used on the AP® Physics 2 Exam (and for most standard circuits in any problem) is to neglect the resistance of the wires used to connect a circuit.

For many materials (including common conductors such as copper), experiments on current-carrying wires show that the current I, which arises in a wire when a potential difference is placed across the wire, is directly proportional to both the magnitude of the electric potential difference and the cross-sectional area A of the wire. The current decreases with the length of the wire L, and depends upon the material. We can write this relationship as

$$I = \frac{\Delta V A}{\rho L} \tag{18-4}$$

In Equation 18-5 the quantity ρ (the Greek letter "rho") is called the **resistivity.** The value of ρ depends on the material of which the wire is made and tells us how well or poorly this material inhibits the flow of electric charge. Equation 18-4 states that for a given cross-sectional area A, a length L, and a given electric potential difference, a *greater* value of the resistivity means a *smaller* amount of current I. (It's unfortunate that the same Greek letter that we use as the symbol for density is also used as the symbol for resistivity. We promise never to use density and resistivity in the same equation. If you get to a point in your studies where you do need both, you can put the subscript m on the mass density!)

The units of resistivity are volt-meters per ampere, or $\text{V} \cdot \text{m/A}$. For reasons that we will see, the unit V/A (volts per ampere) appears very often and is given its own name, the **ohm** (symbol Ω, the uppercase Greek letter "omega"): $1\ \Omega = 1\ \text{V/A}$. In terms of this we can write the units of resistivity as ohm-meters, or $\Omega \cdot \text{m}$.

WATCH OUT !

Notice the electric field is not zero inside this conductor.

We learned in Chapter 17 that an electric field points in the direction of decreasing electric potential, so it looks like there must be an electric field in a wire if there is a current. This realization does not contradict what we learned in Chapter 17 about charge spreading to the surface of a conductor (that is, for a conductor in equilibrium). The lowest energy state for a conductor in equilibrium is when any like-signed excess charge is separated as far as possible, and a conductor that does not have a source of potential difference across it will reach equilibrium very quickly. When we are discussing current—that is, charge in motion—there is a source of potential difference constantly putting energy into the circuit to keep the charge moving against the resistance it encounters in the wires. There isn't any excess charge, just the charge carriers that are free to move and the oppositely signed atoms from which they are separated (opposite sign because they are missing the charge carriers). So, a wire in a circuit in which there is a current is not a conductor in equilibrium!

For copper, which is a good conductor of electricity, the resistivity is very low: $\rho = 1.725 \times 10^{-8}\ \Omega \cdot \text{m}$ at 21°C. For hard rubber, which is a very poor conductor (and a good insulator), $\rho = 10^{13}\ \Omega \cdot \text{m}$ at 21°C. The explanation for the huge ratio between these two values of resistivity is that copper has many mobile electrons per cubic meter, while hard rubber has hardly any. The value of resistivity also depends on temperature. In metals, atoms move more rapidly and are likely to be less well organized at higher temperatures compared to lower temperatures. As a result, the charge carriers (usually electrons) collide more frequently with the atoms in a metal when the temperature is higher. Consequently, the resistivity of most metals increases as temperature increases. In some other materials, such as ceramics, resistivity *decreases* with increasing temperature. **Table 18-1** lists the values of resistivity for a variety of substances at 21°C.

If we rearrange Equation 18-4, we can now answer our question:

$$\Delta V = I\left(\frac{\rho L}{A}\right) \tag{18-5}$$

We define the **resistance** R of a wire as

EQUATION IN WORDS
Definition of resistance

$$R = \frac{\rho L}{A} \tag{18-6}$$

Resistance of a wire

Resistivity of the material of which the wire is made

Length of the wire

Cross-sectional area of the wire

In terms of this new quantity, we can rewrite Equation 18-5 as

Potential difference between the ends of a wire

$$\Delta V = IR \quad (18\text{-}7)$$

Current in the wire | **Resistance** of the wire

EQUATION IN WORDS
Relationship among potential difference, current, and resistance

Equation 18-7 states that the electric potential difference ΔV required to produce a current I in a wire is proportional to I and to the resistance of the wire. Equation 18-6 further tells us that the resistance of a wire depends on the resistivity of the material of which the wire is made, the length of the wire, and the cross-sectional area of the wire. For a given material, the resistance is greater for a wire that is long (large L) and thin (small A) than for a wire that is short (small L) and thick (large A). You can see an example by looking inside the slots of an ordinary kitchen toaster. The wire that makes up the toaster's heating coils is very thin and if uncoiled would be very long. Hence, the coils have a high resistance. The resistance is made even higher by making the wire out of nichrome, which has a resistivity about 10^2 times greater than that of copper (see Table 18-1).

Since resistivity ρ has units of ohm-meters ($\Omega \cdot$ m), length L has units of meters, and area A has units of meters squared, Equation 18-7 states that the unit of resistance R is the ohm. This agrees with Equation 18-7, because $1\ \Omega = 1$ V/A: To produce a 1-A current in a wire with a resistance of $1\ \Omega = 1$ V/A, an electric potential difference of 1 V is required.

An ohm is a relatively small resistance. The heating coils of a toaster have a resistance R of about 10 to 20 Ω. A typical **resistor**—a circuit component intended to add resistance to the flow of charge—such as you can find for sale in an electronics store will likely have a resistance in the range from 1 kΩ (1 kΩ = 1 kilohm = $10^3\ \Omega$) to 10 MΩ (1 MΩ = 1 megohm = $10^6\ \Omega$). The resistance of a human body measured on the skin can be more than 0.5 MΩ.

Equation 18-7 is often referred to as Ohm's law, after the nineteenth-century physicist Georg Ohm, whose pioneering experiments increased our understanding of electric current (and for whom the ohm is named). By itself it suggests that electric potential difference and current are directly proportional to each other. This proportionality holds true if the resistance R remains constant as the electric potential difference is changed, and this is in fact the case for many conducting materials over a wide range of electric potential differences. This means for Ohmic materials, the resistance is also constant for all values of current. Materials that have this property are referred to as *ohmic*. However, for many materials (including those used in a variety of electronic devices) the value of the resistance changes as the potential difference across the material changes. We can still use Equation 18-7 for such *nonohmic* materials, provided we keep in mind that R is not a constant. We will see that a current moving through a resistor causes the resistor to warm up, which we just saw changes its resistivity. This can also cause nonohmic behavior.

TABLE 18-1 Resistivity of Some Conductors and Insulators at 21°C

Conductor	ρ ($\Omega \cdot$ m)
Aluminum	2.733×10^{-8}
Copper	1.725×10^{-8}
Gold	2.271×10^{-8}
Iron	9.98×10^{-8}
Nichrome	150×10^{-8}
Nickel	7.2×10^{-8}
Silver	1.629×10^{-8}
Titanium	43.1×10^{-8}
Tungsten	5.4×10^{-8}
Insulator	**ρ ($\Omega \cdot$ m)**
Glass	10^{12}
Hard rubber	10^{13}
Fused quartz	7.5×10^{17}

EXAMPLE 18-3 Stretching a Wire

A 10.0-m-long wire has a radius of 2.00 mm and a resistance of 50.0 Ω. If the wire is stretched to 10.0 times its original length, what will be its new resistance?

Set Up

Equation 18-6 states that resistance of the wire depends on its resistivity (which is a property of the material of which the wire is made, and doesn't change if the dimensions change). It also depends on the length and cross-sectional area, both of which change when the wire is stretched. To find the new cross-sectional area, we'll use the idea that

Definition of resistance:

$$R = \frac{\rho L}{A} \quad (18\text{-}6)$$

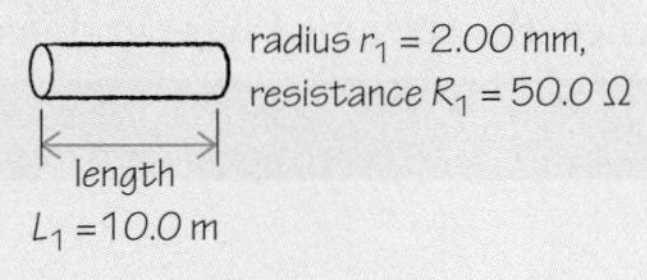

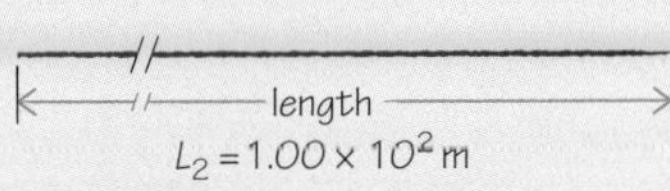

the volume of the wire (the product of its length and cross-sectional area) does not change as it's stretched.

Solve

Find the cross-sectional area of the wire after it has been stretched.

The original cross-sectional area A_1 of the wire of radius $r_1 = 2.00 \text{ mm} = 2.00 \times 10^{-3}$ m is

$$A_1 = \pi r_1^2 = \pi(2.00 \times 10^{-3} \text{ m})^2 = 1.26 \times 10^{-5} \text{ m}^2$$

The volume of the wire is A_1L_1, where $L_1 = 10.0$ m is the wire's initial length. If we stretch the wire to a new length $L_2 = 10.0\ L_1 = 1.00 \times 10^2$ m, the cross-sectional area will change to a new value A_2, but the volume will be the same:

$$A_2L_2 = A_1L_1$$

$$A_2 = \frac{A_1L_1}{L_2} = \frac{(1.26 \times 10^{-5} \text{ m}^2)(10.0 \text{ m})}{1.00 \times 10^2 \text{ m}} = 1.26 \times 10^{-6} \text{ m}^2$$

The new cross-sectional area is 1/10.0 of the initial area.

Calculate the new resistance of the wire.

The initial resistance of the wire (before being stretched) is

$$R_1 = \frac{\rho L_1}{A_1} = 50.0\ \Omega$$

The resistance of the wire after being stretched is

$$R_2 = \frac{\rho L_2}{A_2}$$

If we take the ratio of the two resistances, the unknown value of resistivity will cancel out:

$$\frac{R_2}{R_1} = \frac{\rho L_2/A_2}{\rho L_1/A_1} = \frac{L_2}{L_1}\frac{A_1}{A_2}$$

$$= \left(\frac{1.00 \times 10^2 \text{ m}}{10.0 \text{ m}}\right)\left(\frac{1.26 \times 10^{-5} \text{ m}^2}{1.26 \times 10^{-6} \text{ m}^2}\right)$$

$$= (10.0)(10.0) = 1.00 \times 10^2$$

The length has increased by a factor of 10.0, and the cross-sectional area has decreased by a factor of 10.0, so the new value of resistance is greater than the old value by a factor of $(10.0)^2 = 1.00 \times 10^2$:

$$R_2 = 1.00 \times 10^2\ R_1 = (1.00 \times 10^2)(50.0\ \Omega)$$

$$= 5.00 \times 10^3\ \Omega = 5.00 \text{ k}\Omega$$

Reflect

Stretching the wire makes it longer and thinner, and both of these changes make the resistance of the wire increase. Hence, the stretched wire in this problem has a much higher resistance than the wire had initially.

NOW WORK Problems 1 and 2 from The Takeaway 18-3.

EXAMPLE 18-4 Calculating Current

If a 12.0-V potential difference is set up between the ends of the wire in Example 18-3 both before and after it is stretched, what is the current in the wire at each time?

Set Up

We know the resistance of the wire in each case from Example 18-3, so we can use Equation 18-7 to solve for the current I in the wire in each case.

Relationship among potential difference, current, and resistance:

$$\Delta V = IR \quad (18\text{-}7)$$

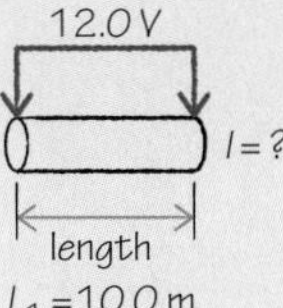

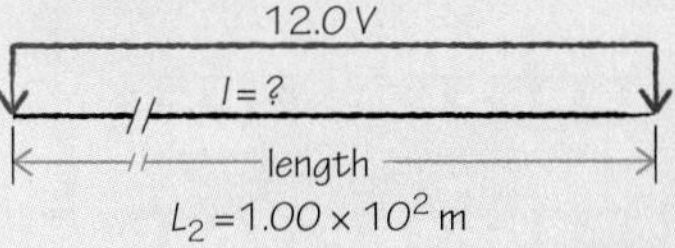

Solve

Rewrite Equation 18-7 as an expression for the current in terms of the electric potential difference and resistance. Then solve for the current in each case.

From Equation 18-7:

$$I = \frac{\Delta V}{R}$$

For the wire before it is stretched, $R = R_1 = 50.0\ \Omega$. The current that results from a 12.0-V potential difference between the ends of this wire is

$$I_1 = \frac{\Delta V}{R_1} = \frac{12.0\text{ V}}{50.0\ \Omega} = 0.240 \frac{\text{V}}{\Omega}$$

Recall that $1\ \Omega = 1$ V/A, so $1\text{ A} = 1\text{ V}/\Omega$. Therefore, the current in the wire before it is stretched is

$$I_1 = 0.240\text{ A}$$

After the wire is stretched the resistance is $R = R_2 = 5.00 \times 10^3\ \Omega$. A 12.0-V potential difference between the ends of this wire produces a current

$$I_2 = \frac{\Delta V}{R_2} = \frac{12.0\text{ V}}{5.00 \times 10^3\ \Omega} = 2.40 \times 10^{-3}\text{ A} = 2.40\text{ mA}$$

Reflect

The stretched wire has a greater resistance than the original wire, so the same potential difference produces a smaller current. Note that the milliampere ($1\text{ mA} = 10^{-3}$ A) is a commonly used unit of current, as are the microampere ($1\ \mu\text{A} = 10^{-6}$ A), the nanoampere ($1\text{ nA} = 10^{-9}$ A), and the picoampere ($1\text{ pA} = 10^{-12}$ A).

NOW WORK Problems 3 and 4 from The Takeaway 18-3.

Electrical resistance can be found in every technological device, from the wires in an automobile ignition system to the resistive elements (resistors) in the circuits of a computer or mobile phone. In electronic devices, resistors are often small and cylindrical with colored bands (**Figure 18-6**). Other resistors look more like tiny, black cubes with the number of ohms etched on one face.

What purpose does a resistor serve? As we have seen, a potential difference ΔV between two points in a conducting material causes a current I. But according to Equation 18-8, $\Delta V = IR$, the amount of current that the potential difference produces depends on the resistance of the conducting material. The greater the resistance R, the smaller the current I. Stated another way, the resistance allows us to control the current due to any particular applied electric potential difference.

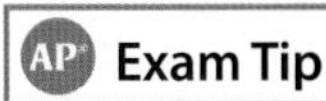

AP® Exam Tip

On the AP® Physics 2 equation sheet, Equation 18-7 may be presented as it is rearranged in the previous example:

$$I = \frac{\Delta V}{R}$$

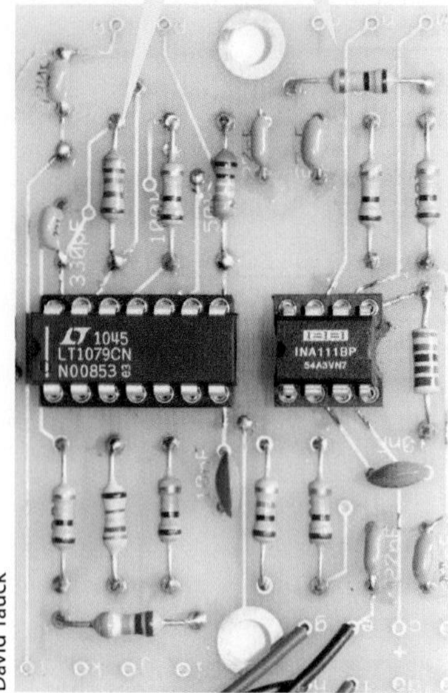

The circuit elements with colored bands are resistors. The particular colors on each resistor indicate the value of its resistance.
LT1079CN
INA118P

David Tauck

Figure 18-6 Resistors in a circuit A typical electronic circuit is likely to contain many resistors.

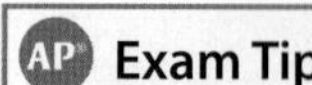

Exam Tip

For a constant potential difference, such as a battery, across a resistor in a closed loop, the current will be the same everywhere in the loop (the same before, through, and after the resistor). If you change the value of the resistance (by changing the resistor or by adding or taking away a resistor), then the amount of current will change compared to the original circuit, but the current will still be the same everywhere in the loop. A common misconception that arises from this is that because *changing* a resistor will result in a change in the current in a circuit, a resistor changes a current as the charge is moving through the circuit, with the incorrect conclusion that the current is different before and after the resistor.

An important application of this idea takes place in every one of the millions of cells in your body. For a cell to live, there must be a higher concentration of positively charged potassium ions (K^+) inside the cell than in the surrounding fluid. This difference in concentration means that K^+ ions tend to leak out of the cell through pathways called *potassium channels* (**Figure 18-7**). The motion of K^+ ions constitutes a current, and each potassium channel acts like a resistor. For a given potential difference between the interior and exterior of the cell, the amount of charge that moves through these channels depends on their resistance. So the movement of potassium through the membranes of your cells is determined by the electrical resistance of the membrane channels.

All biological cells are surrounded by a membrane that serves as both an insulator and a barrier to the movement of ions. Embedded in the membrane are proteins that are electrically equivalent to batteries as well as the membrane channels that are equivalent to resistors. These create a potential difference between the two sides of the membrane, called the *membrane potential*. Figure 18-7 shows a simplified model of how the membrane potential impacts the motion of ions.

As the following example shows, we can use Equation 18-7 to determine the resistance of the membrane to potassium ion flow.

(1) Potassium ions (shown as green circles) cannot diffuse through a membrane until proteins in the membrane cause a membrane potential.

(2) Potassium channels embedded in the membrane allow only potassium ions to diffuse out of the cell, driven by the membrane potential.

Cell membrane

Ions

(3) Positively charged potassium ions leave the cell, but negative charges (orange circles) are unable to follow.

(4) The result is a potential difference across the membrane.

(5) When the potential difference due to the separated charge is the same size as the membrane potential, the cell is in equilibrium and it prevents any additional net movement of positively charged potassium ions out of the cell.

Figure 18-7 Membrane potential The concentration of positively charged potassium ions (K^+) is always higher inside cells than outside, which causes K^+ ions to leak out. This motion of K^+ ions is a current; the channels that allow K^+ ions to leak out of the cell behave like resistors.

EXAMPLE 18-5 Resistance of a Potassium Channel

Using an instrument called a patch clamp, scientists control the electric potential difference across a tiny patch of cell membrane and measure the current through individual potassium channels. In one experiment, an electric potential difference of 0.120 V was applied across a patch of membrane, and as a result K^+ ions carried 6.60 pA of current (1 pA = 1 picoampere = 10^{-12} A). What is the resistance of this K^+ channel?

Set Up

Whether the charge carriers are negative electrons or positive ions, we can use Ohm's law to relate potential difference, current, and resistance.

Relationship among electric potential difference, current, and resistance:

$$\Delta V = IR \qquad (18\text{-}7)$$

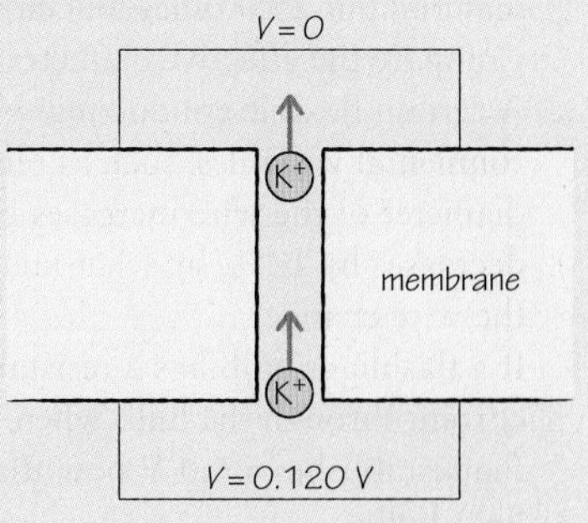

Solve

Rewrite Ohm's law to solve for the resistance.

From Equation 18-7:

$$R = \frac{\Delta V}{I}$$

The current is $I = 6.60 \text{ pA} = 6.60 \times 10^{-12}$ A, so

$$R = \frac{0.120 \text{ V}}{6.60 \times 10^{-12} \text{ A}} = 1.82 \times 10^{10}\ \Omega$$

Reflect

This is an immensely high resistance compared to the values found in circuits such as that shown in Figure 18-6. Yet it is very characteristic of the channels in cell membranes, which typically have resistances in the range from 10^9 to $10^{11}\ \Omega$. A potassium channel is only about 10^{-9} m wide—so small that K^+ ions must pass through it in single file—so it's not surprising that the resistance is so high.

NOW WORK Problems 5 and 6 from The Takeaway 18-3.

WATCH OUT !

The larger the current through a light bulb, the brighter the light bulb glows and the hotter it gets!

Light bulbs have a resistance that depends on their temperature, which depends on the current through the bulb. In general, we will treat bulbs as resistors and give you the resistance or ask you to determine their resistance. Be prepared to explain that this is an approximation, and that light bulbs are in general non-Ohmic.

THE TAKEAWAY for Section 18-3

✔ Resistivity ρ is a measure of how well or poorly a particular material inhibits an electric current. The SI units of resistivity are volt-meters per ampere or ohm-meters.

✔ The value of resistivity differs from material to material, and by temperature for any given material. The resistivity is low for conductors and high for insulators.

✔ The resistance R of an object is a measure of the current through the object for a given potential difference between its ends. The value of R depends on the object's shape and on the resistivity of the material from which it is made. A long, narrow wire has a much higher resistance than a short, thick one.

✔ For a given applied electric potential difference, the greater the resistance in an electrical system, the smaller the current.

Prep for the AP Exam

AP Building Blocks

1. There is a current of 112 pA when a certain potential difference is set up between the ends of a certain cylindrical resistor. When that same potential difference is set up across another cylindrical resistor made of the identical material but 25.0 times longer, the current is 0.0440 pA. Compare the effective diameters of the two resistors.

2. A certain flexible conducting wire changes shape as environmental variables, such as temperature, change. If the diameter of the wire increases by 25% while the length decreases by 12%, by what factor does the resistance of the wire change?

3. If a flashlight bulb has a resistance of 12.0 Ω, what is the current through the bulb when it is connected to a circuit that establishes a 6.0-V potential difference across the light bulb?

4. A light bulb has a resistance of 8.0 Ω and a current of 0.5 A. At what potential difference is it operating?

5. Cell membranes contain channels that allow K^+ ions to transport out of the cell. Consider a channel that has a diameter of 1.0 nm and a length of 10.0 nm. If the channel has a resistance of 18 GΩ, what is the resistivity of the solution in the channel?

6. A single ion channel is selectively permeable to K^+ ions and has a resistance of 1.0 GΩ. During an experiment, the channel is open for approximately 1.0 ms while the potential difference across the channel is maintained at 180.0 mV. Calculate the number of K^+ ions transported through the channel.

AP Skill Builders

7. Fertilizers not taken up by growing plants in agricultural fields run off the field into ground water. Freshwater availability and ocean biodiversity are both threatened, and the United Nations 2030 Agenda for Sustainable Growth includes the goal of minimizing this form of pollution. Fertilizer applied at a rate that supports more productive areas in a field will end up in ground water when applied to less productive areas. Measurements of soil concentrations of nutrients, such as potassium, nitrogen, and phosphorus, are too expensive to apply to many samples from small areas within a field. Measurement of soil resistivity is a possible proxy for soil nutrient measurements. Salinity, clay fraction, and porosity are often correlated with available nutrients and thus affect resistivity. A research project is aimed at the development of technology that continuously measures resistivity to control the amount of fertilizer applied and avoid excess. To calibrate field measurements of resistivity, cores (cylindrical soil samples with a diameter of 4 cm and length of 20 cm) are taken from the field and brought into the lab. A core is shown in the figure. The current through the core and the electric potential difference between any two points, such as a and b, can be measured.

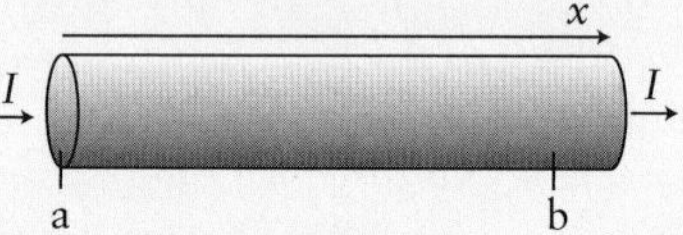

Point a in the figure is at the soil–air surface. Because the composition of material in the column of soil varies with depth, resistivity can depend on depth.
(a) Design a method of data collection that can be used to determine the resistivity to a depth d below the soil–air surface.
(b) Justify the claim that measurement of the current can be made at either end of the soil core.

AP Skills in Action

8. Explain why reference materials concerned with electric circuits provide resistivity data, such as Table 18-1, rather than resistance data.

EX 18-4
9. When a thin wire is connected across a potential difference of 1 V, the current is 1 A. If we connect the same wire across a potential difference of 2 V, the current is
(A) (1/4) A.
(B) (1/2) A.
(C) 1 A.
(D) 2 A.

10. Power lines transport electric energy. Birds commonly perch on these lines with both feet. A line worker repairing a broken line grabs both ends and is electrocuted. Explain why the birds are not harmed but the line worker is.

18-4 Conservation of energy and conservation of charge make it possible to analyze electric circuits

Many simple electric circuits are made up of a source of electric potential difference and one or more resistors. An example is a flashlight such as that shown in the photograph that opens this chapter: The batteries provide the potential difference, and the filament of the light bulb acts as a resistor. We'll refer to batteries and resistors collectively as **circuit elements**.

In this section we'll see how to analyze circuits. We will start with those made up of a single battery and one or more resistors and then apply what we learn to more complicated arrangements. In particular, we'll see how to determine the electric potential difference across each circuit element as well as the current through each circuit element.

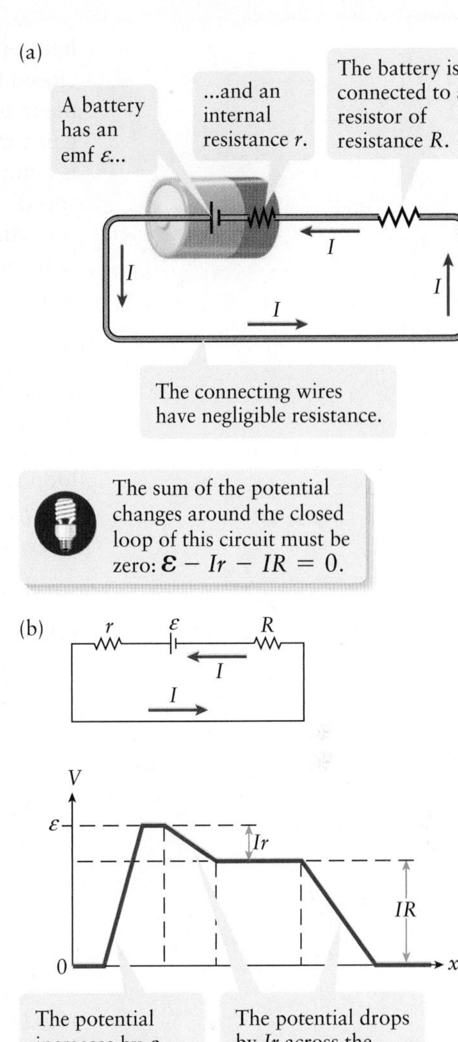

(a)
A battery has an emf $\mathcal{E}$...
...and an internal resistance r.
The battery is connected to a resistor of resistance R.
I
I
I
I
The connecting wires have negligible resistance.
The sum of the potential changes around the closed loop of this circuit must be zero: $\mathcal{E} - Ir - IR = 0$.
(b)
r
$\mathcal{E}$
R
I
I
V
$\mathcal{E}$
Ir
IR
0
x
The potential increases by $\mathcal{E}$ from the negative terminal to the positive terminal of the emf.
The potential drops by Ir across the internal resistance and drops by IR across the resistor.

Figure 18-8 Kirchhoff's loop rule (a) A circuit made up of a battery (which has an emf $\mathcal{E}$ and an internal resistance r) and a resistor R. (b) The changes in electric potential around the circuit.

A Single-Loop Circuit and Kirchhoff's Loop Rule

Figure 18-8a shows the standard symbols used in drawing circuit diagrams. This figure represents a battery connected to a single resistor of resistance R. In fact, there are *two* resistors in this circuit; the other one, which we label r, is the **internal resistance** of the battery itself. Internal resistance is what we neglect in the ideal battery model. The ideal battery model is fairly good for fresh batteries. Understanding that real batteries have internal resistance explains why testing for potential difference may not tell you if your battery is dead. We can think of the battery as having two components: its emf—the electric potential difference across the battery due to its makeup—and its internal resistance, which impedes the motion of charge carriers. We just add the internal resistance into the diagram as a second resistor. When we say that a D or an AA cell is a "1.5-volt battery" or that the battery in an automobile is a "12-volt battery," we're actually stating the value of the battery's emf. We use the symbol $\mathcal{E}$, a Greek epsilon, for emf. So $\mathcal{E} = 1.5$ V for a D or an AA cell, and $\mathcal{E} = 12$ V for a standard automotive battery. (Note that the term *emf* comes from the older term *electromotive force*. Although an emf is what causes electric charge to move through the circuit, it is a change in potential energy per unit charge, not a force, so it has units of volts, not newtons.) The symbol in Figure 18-8a for a source of constant emf is two parallel lines of unequal length: —|⊢. Another common symbol for a battery is just two of these same lines, —||⊢. We will use both. The symbol for a resistor is a jagged line: —\/\/\/—.

The longer of the two parallel lines represents the positive terminal of the source. Since we regard the direction of the current as being in the direction that positive charge moves, such a source of emf causes current in the direction out of its positive terminal and into its negative terminal (see Figure 18-4).

We call the circuit in Figure 18-8a a **single-loop circuit** because there's only a single path along which charge can move around the circuit. Think about what happens to a positive point charge as it travels through this circuit in the direction of the current I. (We'll continue to use the convenient fiction that current is the motion of positive charge.) As the point charge passes through the source of emf (the battery) from the negative terminal to the positive terminal, it experiences an *increase* in potential energy because it goes through a positive electric potential difference of $\mathcal{E}$. When it moves through the internal resistance, however, the point charge experiences a *decrease* in electric potential (a negative electric potential difference) of Ir. (Recall that the direction of the current is the same as the direction of the electric force on a positively charged object and points in the direction from high potential to low potential. So the potential decreases as you move with the current through a resistor.) The point charge experiences an additional decrease in electric potential of IR when it moves through the resistor of resistance R.

There's also resistance in the wires that connect the battery and the resistor. However, the resistance of the connecting wires is generally quite small compared to the values of r and R. So we'll ignore the wire resistance and assume that a moving charge experiences no change in potential as it traverses the wires.

The current in Figure 18-8a does not change with time. So when the point charge finishes a trip around the circuit and returns to where it started, the value of the electric potential at its starting point must have the same value as when the point charge

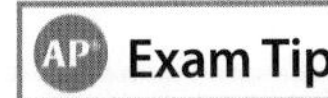

Unless otherwise stated, on the AP® Physics 2 Exam, you may always assume wires are ideal—that they have no resistance—in a circuit. This means that the length of the wire or the position of a circuit element along a wire in a closed loop has no impact on the current anywhere or the potential difference across any circuit element. As mentioned, this is an excellent approximation in any circuit that contains other elements.

left that point. In other words, the *net change* in potential for a round trip around the closed loop must be *zero* (**Figure 18-8b**). This idea is not new to us. It is really a statement of the conservation of energy. The potential energy our positive point charge gains in going through the battery from the negative terminal to the positive terminal is comparable to the point charge being "picked up"; it is then "thrown with an initial speed" as it leaves the battery to "fall" the rest of the way around the circuit. Electrical resistance R is directly related to resistance to the motion of the point charge, keeping it from accelerating around the loop. This means the energy dissipated as the point charge travels through the resistances will exactly balance the energy the battery gave it, and it will have the same energy it had when it gets back to where it started. Once the current is established, the changes in energy of a point charge as it moves around a closed path must add to zero. As we learned, electric potential difference is change in energy per unit charge. This idea was first proposed as applied to circuits by the Prussian physicist Gustav Kirchhoff in the mid-nineteenth century and is called **Kirchhoff's loop rule:**

The sum of the changes in potential around a closed loop in a circuit must equal zero.

In equation form, we can write Kirchhoff's loop rule for the circuit in Figure 18-8a as

$$\sum \Delta V = 0 \qquad (18\text{-}8)$$

We can write Equation 18-8 to describe exactly the example of the battery with two resistors we just gave:

$$\mathcal{E} - Ir - IR = 0$$

We can then rearrange it to solve for the current in the circuit:

$$\mathcal{E} = Ir + IR = I(r + R)$$

$$I = \frac{\mathcal{E}}{r + R}$$

To give a specific numerical example, suppose $\mathcal{E} = 12.0$ V, $r = 1.00\ \Omega$, and $R = 19.0\ \Omega$. Then the current is

$$I = \frac{12.0\text{ V}}{1.00\ \Omega + 19.0\ \Omega} = \frac{12.0\text{ V}}{20.0\ \Omega} = 0.600\text{ A}$$

(Recall that 1 A = 1 V/Ω.) You can see that if the battery had a greater internal resistance r, the current would be smaller. As an example, the internal resistance r of a disposable battery in a flashlight or television remote control increases as the battery is used. Eventually r becomes so great that the current becomes too small to make the device work, which means that it's time to replace the battery. The emf $\mathcal{E}$ of the used battery is almost the same as when it was new; it's the internal resistance that makes the battery no longer useful. This fact can be important in your everyday life: If you measure the potential difference across the battery with no current, you would measure about 12 V even when it is almost dead. The potential difference you would measure would decrease significantly if you did it while the battery was connected to a circuit so there is a current through the battery.

Note that for this numerical example, the negative electric potential difference across the internal resistance is $Ir = (0.600\text{ A})(1.00\ \Omega) = 0.600$ V, and the electric potential difference across the 19.0-Ω resistor is $IR = (0.600\text{ A})(19.0\ \Omega) = 11.4$ V. The sum of the potential changes is zero, just as Kirchhoff's loop rule states it must be:

$$\mathcal{E} - Ir - IR = 12.0\text{ V} - 0.600\text{ V} - 11.4\text{ V} = 0$$

For any single-loop circuit, the *current* $I = 0.600$ A is the same through every element in the circuit. That's because no charge can appear or disappear at any place in the circuit (see Section 18-2). However, the *electric potential differences* are different for r and R because the values of resistance are different.

AP® Exam Tip

For ideal batteries, the emf is always equal to the potential difference across the battery's terminals. In AP® Physics 2, it is normally safe to consider a battery ideal unless told otherwise, or if you are asked about experimental uncertainties.

WATCH OUT!

The electric potential difference across a battery in a circuit is less than the emf.

The example we have just given shows that when a battery is in a circuit, the potential difference between the battery's terminals is *not* equal to the emf $\mathcal{E}$. Rather, it is equal to the emf *minus* the decrease in electric potential Ir across the internal resistance of the battery. However, as we are learning about simple circuits and developing models in this course, we are going to model all batteries as **ideal batteries**; that is, batteries for which the internal resistance can be neglected. For an ideal battery the electric potential difference across the battery is always the emf of the battery.

Resistors in Series

An important application of Kirchhoff's loop rule is to resistors in **series**. That is, the resistors are connected end to end, like the two in our last example. As we saw from Kirchhoff's loop rule, the total energy given to a point charge as it moves through the battery must be exactly equal to the energy it loses as it moves through the resistors. This is true whether there is one resistor, or there are two, three, or a thousand resistors. **Figure 18-9a** shows a circuit that has three resistors R_1, R_2, and R_3 in series, connected to a source of emf such as a battery. This potential difference across the battery, ΔV_{bat}, includes the emf and internal resistance of the battery, but we are just going to treat it as if it is ideal, and not worry about internal resistance. Imagine we replace the three separate resistors with an **equivalent resistance**—that is, a single resistor of a resistance R_{equiv} that gives the same current as the series combination, as shown in **Figure 18-9b**. What is the equivalent resistance in terms of R_1, R_2, and R_3?

For R_{equiv} to have the same effect in the circuit as R_1, R_2, and R_3 together, it must give rise to the same current I and the same electric potential difference. As we discussed earlier, the electric potential difference across a resistance R that carries current I is IR. The same current is present through each of the three resistors shown in Figure 18-9a, and the total electric potential difference across the three resistors in series is the sum of the individual electric potential differences: $IR_1 + IR_2 + IR_3$. The electric potential difference through the equivalent resistance is IR_{equiv}, so

$$IR_{equiv} = IR_1 + IR_2 + IR_3 = I(R_1 + R_2 + R_3)$$
$$R_{equiv} = R_1 + R_2 + R_3$$
(resistors in series)

(a)

The current through each resistor in series is the same.

(b)

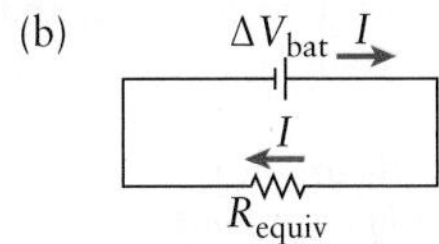

• The current I through R_{equiv} is the same as through each of the resistors in series.
• The potential difference ΔV_{bat} across R_{equiv} is the same as that across the combination of resistors in series.

Figure 18-9 Resistors in series (a) A circuit contains three resistors connected in series to a battery. (b) The three resistors have been replaced by a single, equivalent resistor.

For example, if the three resistors have resistances $R_1 = 25.0\ \Omega$, $R_2 = 12.0\ \Omega$, and $R_3 = 36.0\ \Omega$, they are equivalent to a single resistor with resistance $R_{equiv} = 25.0\ \Omega + 12.0\ \Omega + 36.0\ \Omega = 73.0\ \Omega$. This means that if we replace the three resistors with a single $73.0\ \Omega$ resistor, the current in the circuit will be exactly the same. The net electric potential difference is also the same. If the current in our example is $I = 1.00$ A, the electric potential differences across the individual resistors are $\Delta V_1 = IR_1 = (1.00\text{ A})(25.0\ \Omega) = 25.0$ V, $\Delta V_2 = IR_2 = (1.00\text{ A})(12.0\ \Omega) = 12.0$ V, and $\Delta V_3 = IR_3 = (1.00\text{ A})(36.0\ \Omega) = 36.0$ V, and the net electric potential difference is $25.0\text{ V} + 12.0\text{ V} + 36.0\text{ V} = 73.0$ V. The electric potential difference across the equivalent resistance is $IR_{equiv} = (1.00\text{ A})(73.0\ \Omega) = 73.0$ V, the same as for the three resistors in series.

In general, if there are N resistors arranged in series the equivalent resistance is

Equivalent resistance of N resistors in series

Resistances of the individual resistors

$$R_{equiv} = R_1 + R_2 + R_3 + \ldots + R_N \qquad (18\text{-}9)$$

EQUATION IN WORDS
Equivalent resistance of resistors in series

The equivalent resistance of N resistors in series is the sum of the individual resistances.

Equation 18-9 tells us that by combining resistors in series we create a circuit with a higher equivalent resistance than that of any of the individual resistors. Looking back to calculating the resistance of a single resistor based on its shape, this makes sense. Essentially, we made the resistor longer, by placing the resistors end to end. The numerical example presented earlier illustrates this: with $R_1 = 25.0\ \Omega$, $R_2 = 12.0\ \Omega$, and $R_3 = 36.0\ \Omega$ in series, $R_{\text{equiv}} = 73.0\ \Omega$ is greater than R_1, R_2, or R_3.

Our analysis tells us that for resistors in series, the *current* is the same through each resistor, but the electric potential difference is different for different resistors. As we will see, this is not the case if the resistors are in an arrangement other than series.

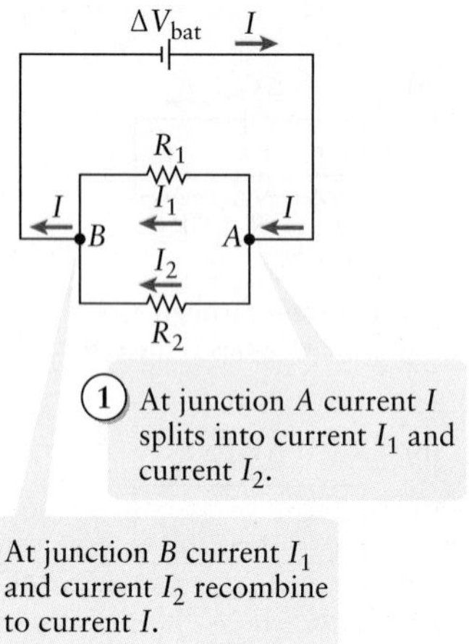

Kirchhoff's Junction Rule and Resistors in Parallel

(1) At junction A current I splits into current I_1 and current I_2.

(2) At junction B current I_1 and current I_2 recombine to current I.

Figure 18-10 Kirchhoff's junction rule At the circuit junctions *A* and *B*, the net current into the junction must equal the net current out of that junction.

Figure 18-10 shows a circuit with two resistors, R_1 and R_2, connected in **parallel** to an ideal battery of electric potential difference ΔV_{bat}. Unlike the single-loop circuits shown in Figures 18-8 and 18-9, this is a **multiloop circuit**: There is more than one pathway that the moving charge can take from the positive terminal of the battery through the circuit to the negative terminal. In particular, the circuit in Figure 18-10 has two **junctions** at A and B where the current either breaks into two currents (as at A) or comes together (as at B). These pathways into which the current breaks are called **circuit branches**. What is the relationship among the current I that passes through the battery, the current I_1 that passes through resistor R_1, and the current I_2 that passes through resistor R_2?

One condition on I, I_1, and I_2 is that charge is conserved, so since current is charge per unit time, then current must also be conserved. Since this means charge can be neither created nor destroyed, nor pile up anywhere in the circuit, the rate at which charge *arrives* at a junction must be equal to the rate at which charge *leaves* that junction. This is our second rule of electric circuits. Like the loop rule, it grows out of a general conservation principle. This was also proposed for electric circuits by Kirchhoff and is called **Kirchhoff's junction rule:**

The sum of the currents into a junction equals the sum of the currents out of it.

Let's apply this rule to the junctions shown in Figure 18-10. Current I represents the flow of charge into junction A, and currents I_1 and I_2 represent the charge flow out of it. So the junction rule tells us that I (the sole current into junction A) must equal $I_1 + I_2$ (the sum of the currents out of A). At junction B the sum of the currents in is $I_1 + I_2$ and the sole current out is I. So by analyzing either junction we can conclude that $I = I_1 + I_2$. The current divides itself into I_1 and I_2 when it reaches junction A, with no extra current being added and no current being lost. These currents rejoin at junction B. We can express this relationship more generally as the current going into a junction (I_{in}) must equal the current coming out of that same junction (I_{out}): $I_{\text{in}} = I_{\text{out}}$. Either of these currents would be properly expressed as the sum of the currents either directed in or out of the junction in the branches at that junction.

By itself, this relationship doesn't tell us how much of current I takes the branch through resistor R_1 (as current I_1) and how much takes the branch through resistor R_2 (as current I_2). To determine this, let's apply the *loop* rule to two different loops through the circuit in Figure 18-10. First, consider a loop that starts at the negative terminal of the battery, passes through the battery to the positive terminal (electric potential increases by ΔV_{bat}), then follows the path of current I_1 through resistor R_1 (electric potential decreases by I_1R_1), and returns to the negative terminal of the battery. (As before, we'll ignore the resistance of the wires, and treat the battery as ideal.) From the loop rule, the net change in electric potential for this loop is zero, so:

$$\Delta V_{\text{bat}} = I_1R_1$$

The second loop we'll consider also starts at the negative terminal of the battery and passes through the battery to the positive terminal (electric potential increases by ΔV_{bat}), but then follows the path of current I_2 through resistor R_2 (electric potential decreases by I_2R_2) before returning to the battery's negative terminal. The loop rule states that the net change in electric potential is also zero for this loop:

$$\Delta V_{\text{bat}} = I_2R_2$$

If you compare the equations we got for the two loops, you'll see that these equations can both be true only if

$$I_1R_1 = I_2R_2$$

In other words, for resistors in parallel the electric potential difference must be the same for each resistor. However, the *currents* are different for different resistors in parallel: If R_1 is less than R_2, the current will be greater in R_1 (with the smaller resistance) and smaller in R_2 (with the greater resistance). Compare this to resistors in series, for which the current is the same but the electric potential differences are different for different resistors.

As an illustration, suppose $\Delta V_{bat} = 12.0\text{ V}$, $R_1 = 3.00\ \Omega$, and $R_2 = 6.00\ \Omega$. Applying our loop rule to the loop with resistor R_1, we find the current I_1 through R_1 is given by

$$\Delta V_{bat} - I_1R_1 = 0 \quad \text{so} \quad I_1R_1 = \Delta V_{bat} \quad \text{and} \quad I_1 = \frac{\Delta V_{bat}}{R_1} = \frac{12.0\text{ V}}{3.00\ \Omega} = 4.00\text{ A}$$

We can then find the current I_2 in the same way:

$$I_1R_1 = I_2R_2 \quad \text{and} \quad I_2 = \frac{I_1R_1}{R_2} = \frac{(4.00\text{ A})(3.00\ \Omega)}{6.00\ \Omega} = 2.00\text{ A}$$

Current $I_1 = 4.00$ A is twice as great as current $I_2 = 2.00$ A because resistance $R_1 = 3.00\ \Omega$ is half as great as resistance $R_2 = 6.00\ \Omega$. Note that the total current I that passes through the battery is $I = I_1 + I_2 = 4.00\text{ A} + 2.00\text{ A} = 6.00\text{ A}$.

We can now find the equivalent resistance of a set of resistors in parallel. **Figure 18-11a** shows three resistors, R_1, R_2, and R_3, connected in parallel to an ideal battery with potential difference ΔV_{bat}. The currents through these three resistors are I_1, I_2, and I_3, respectively. **Figure 18-11b** shows the three resistors replaced by an equivalent resistance R_{equiv}. As we did for resistors in series, we'll determine R_{equiv} by demanding that the net current I and the electric potential difference be the same for the actual set of resistors in Figure 18-11a and the equivalent resistance in Figure 18-11b. From our earlier discussion we see that the electric potential difference through each of the resistors in Figure 18-11a is equal to ΔV_{bat}:

$$\Delta V_{bat} = I_1R_1, \quad \Delta V_{bat} = I_2R_2, \quad \Delta V_{bat} = I_3R_3$$

If we divide each of these equations by R_1, R_2, and R_3, respectively, we get expressions for the current through each resistor:

$$I_1 = \frac{\Delta V_{bat}}{R_1}, \quad I_2 = \frac{\Delta V_{bat}}{R_2}, \quad I_3 = \frac{\Delta V_{bat}}{R_3}$$

The junction rule tells us that the total current that passes through the battery is the sum of the currents through the individual resistors: $I = I_1 + I_2 + I_3$. From the equations we have seen in this chapter, we can write this as

$$I = \frac{\Delta V_{bat}}{R_1} + \frac{\Delta V_{bat}}{R_2} + \frac{\Delta V_{bat}}{R_3} = \Delta V_{bat}\left(\frac{1}{R_1} + \frac{1}{R_2} + \frac{1}{R_3}\right)$$

The total current must be the same for the circuit in Figure 18-11b with the equivalent resistance. If we apply the loop rule to this circuit, we get $\Delta V_{bat} - IR_{equiv} = 0$, so $IR_{equiv} = \Delta V_{bat}$ or

$$I = \frac{\Delta V_{bat}}{R_{equiv}} = \Delta V_{bat}\left(\frac{1}{R_{equiv}}\right)$$

Since the two preceding equations are both expressions for the total current I, they can both be valid only if the right-hand sides of these equations are equal to each other:

$$\frac{1}{R_{equiv}} = \frac{1}{R_1} + \frac{1}{R_2} + \frac{1}{R_3} \qquad (18\text{-}10)$$

(resistors in parallel)

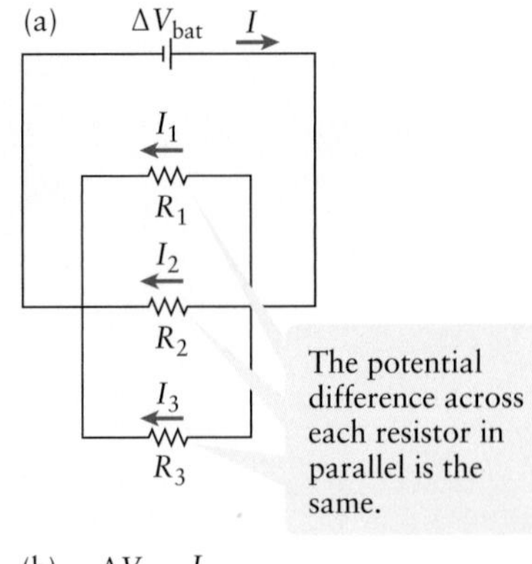

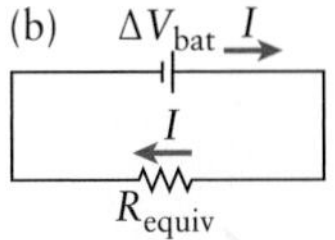

• The current I through R_{equiv} is the same as through the combination of resistors in parallel.
• The potential difference ΔV_{bat} across R_{equiv} is the same as that across each of the resistors in parallel.

Figure 18-11 Resistors in parallel
(a) A circuit contains three resistors connected in parallel to a source of emf. (b) The three resistors have been replaced by a single, equivalent resistor.

AP® Exam Tip

Note that electrically parallel does not mean that the circuit elements need to be geometrically parallel. It just means that both ends of each element are connected either directly or by nothing but wire.

As an illustration, suppose the three resistors in Figure 18-11a are $R_1 = 25.0\ \Omega$, $R_2 = 12.0\ \Omega$, and $R_3 = 36.0\ \Omega$. According to Equation 18-10 they are equivalent to a single resistor with resistance R_{equiv} given by

$$\frac{1}{R_{\text{equiv}}} = \frac{1}{25.0\ \Omega} + \frac{1}{12.0\ \Omega} + \frac{1}{36.0\ \Omega} = 0.151\ \Omega^{-1}$$

$$R_{\text{equiv}} = \frac{1}{0.151\ \Omega^{-1}} = 6.62\ \Omega$$

If the net current through the battery is 1.00 A, the electric potential difference across the equivalent resistance is $IR_{\text{equiv}} = (1.00\ \text{A})(6.62\ \Omega) = 6.62\ \text{V}$. The electric potential difference across each of the individual resistors in parallel must be the same, so $6.62\ \text{V} = I_1R_1 = I_2R_2 = I_3R_3$. You can use these relationships to show that $I_1 = 0.265$ A, $I_2 = 0.551$ A, and $I_3 = 0.184$ A (note that the current is greatest through resistor R_2, which has the smallest of the three resistances). The net current through the battery is $I = I_1 + I_2 + I_3 = 1.00$ A, just as for the battery connected to the equivalent resistance.

If we have N resistors in parallel, Equation 18-10 becomes

The reciprocal of the equivalent resistance of N resistors in parallel is the sum of the reciprocals of the individual resistances.

EQUATION IN WORDS
Equivalent resistance of resistors in parallel

$$\frac{1}{R_{\text{equiv}}} = \frac{1}{R_1} + \frac{1}{R_2} + \frac{1}{R_3} + \ldots + \frac{1}{R_N} \quad (18\text{-}11)$$

Equivalent resistance of N resistors in parallel

Resistances of the individual resistors

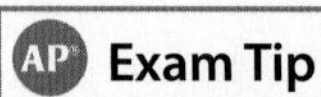

Exam Tip

Equations 18-9 (equivalent resistance of resistors in series) and 18-11 (equivalent resistance of resistors in parallel) are on the AP® Physics 2 equation sheet using slightly different notation; be sure to understand that the equations on the equation sheet say the same thing as the equations here in the text.

Equation 18-11 tells us that by combining resistors in parallel, we create a circuit with a smaller equivalent resistance than any of the individual resistors. The numerical example earlier in this chapter illustrates this: With $R_1 = 25.0\ \Omega$, $R_2 = 12.0\ \Omega$, and $R_3 = 36.0\ \Omega$ in parallel, the equivalent resistance $R_{\text{equiv}} = 6.62\ \Omega$ is less than R_1, R_2, or R_3. Again, we can make sense of this by thinking back to calculating the resistance of a single resistor based on its dimensions. When we put the resistors side by side (in parallel) we are essentially increasing the cross-sectional area of the equivalent resistor.

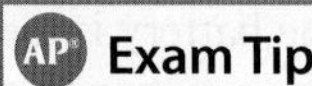

Exam Tip

The analysis of multiple resistors in a parallel or series circuit using Kirchhoff's rules shows that if you have multiple resistors, the order of the resistors in the circuit does not matter and the relative distance of any resistor from the battery does not matter when you are determining the current through or potential difference across any circuit element. The current through or potential difference across any circuit element depends only on the value of the emf, the value of each resistor, and whether the resistors are in parallel or series.

The following examples illustrate some applications of the relationships we've developed so far in this section. In the second example, we'll see how to analyze resistors arranged in a combination that is neither purely series nor purely parallel.

EXAMPLE 18-6 Giant Axons in Squid

In a squid, axons of approximately 30,000 nerve cells fuse together to form giant axons, which are nerve fibers that transmit information in the form of electrical signals. A typical axon has a diameter of 15.0 μm, is 10.0 cm long, and has a resistivity of about 3100 Ω · m. Find the resistance of a giant squid axon by considering it as 30,000 separate axons in parallel.

Set Up

We can find the resistance of a typical axon by using Equation 18-6. Each of these individual axons acts as a separate conducting path that mobile charges can follow between a point of high potential and a point of low potential. That's just like the three resistors in parallel shown in Figure 18-11a, so we can use Equation 18-11 to determine the equivalent resistance of the giant axon as a whole in terms of the resistances of the individual axons.

Definition of resistance:

$$R = \frac{\rho L}{A} \quad (18\text{-}6)$$

Equivalent resistance of resistors in parallel:

$$\frac{1}{R_{\text{equiv}}} = \frac{1}{R_1} + \frac{1}{R_2} + \frac{1}{R_3} + \cdots + \frac{1}{R_N} \quad (18\text{-}11)$$

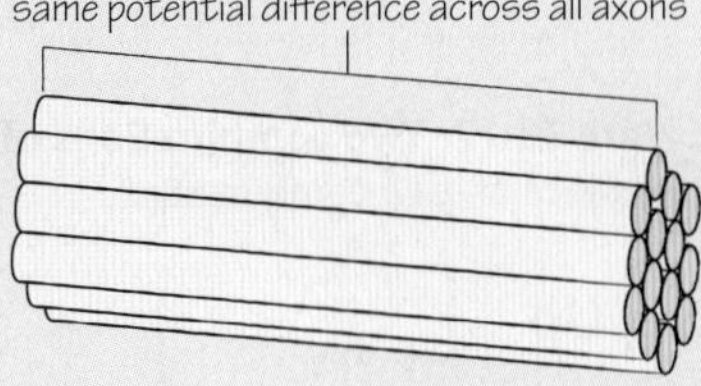

Solve

Calculate the resistance of an individual axon.

Each individual axon has resistivity $\rho = 3100\ \Omega \cdot \text{m}$, length $L = 10.0\ \text{cm} = 0.100\ \text{m}$, and diameter $15.0\ \mu\text{m} = 15.0 \times 10^{-6}\ \text{m}$. The radius r is one-half of the diameter:

$$r = \frac{1}{2}(15.0 \times 10^{-6}\ \text{m}) = 7.50 \times 10^{-6}\ \text{m}$$

If the axon has a circular cross section, its cross-sectional area is $A = \pi r^2$. So from Equation 18-7 the resistance is

$$R = \frac{\rho L}{A} = \frac{\rho L}{\pi r^2} = \frac{(3100\ \Omega \cdot \text{m})\,(0.100\ \text{m})}{\pi (7.50 \times 10^{-6}\ \text{m})^2} = 1.75 \times 10^{12}\ \Omega$$

This resistance is very large because the axon is long and thin, and because the axon resistivity is much higher than that of metallic conductors (see Table 18-1).

Calculate the resistance of the giant axon, as if it were 30,000 individual axons in parallel.

If N axons are in parallel and each has the same resistance R, there are N identical terms on the right-hand side of Equation 18-11. So

$$\frac{1}{R_{\text{equiv}}} = \frac{1}{R_1} + \frac{1}{R_2} + \frac{1}{R_3} + \cdots + \frac{1}{R_N} = \frac{N}{R}$$

Take the reciprocal of both sides:

$$R_{\text{equiv}} = \frac{R}{N}$$

There are $N = 30{,}000$ individual axons, each of which has resistance $R = 1.75 \times 10^{12}\ \Omega$, so the equivalent resistance of the giant axon is

$$R_{\text{equiv}} = \frac{1.75 \times 10^{12}\ \Omega}{30{,}000} = 5.85 \times 10^{7}\ \Omega$$

Reflect

In Example 18-5 we saw that the channels in cell membranes typically have resistances in the range from 10^9 to $10^{11}\ \Omega$. Our individual axons are very tiny, and longer, so the high resistance makes sense. Then we got a much smaller resistance as we added them together, as we expect for a parallel connection (effectively increasing the cross-sectional area). So, as best we can tell, this answer seems reasonable.

NOW WORK Problems 3 and 5 from The Takeaway 18-4.

EXAMPLE 18-7 Resistors in Combination

Figure 18-12 shows two different combinations of three identical resistors, each with resistance R. Find the equivalent resistance of the combination in (a) **Figure 18-12a** and (b) **Figure 18-12b**.

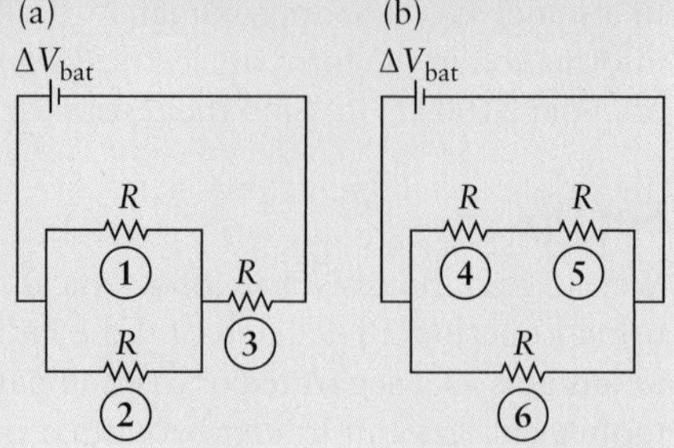

Figure 18-12 Two combinations of three identical resistors What is the equivalent resistance of each combination?

Set Up

Neither combination in Figure 18-12 is a simple series or parallel arrangement of resistors. But in Figure 18-12a resistors 1 and 2 are in parallel with each other, and that combination is in series with resistor 3. Similarly, in Figure 18-12b resistors 4 and 5 are in series with each other, and that combination is in parallel with resistor 6. So we can use Equations 18-9 and 18-11 together to find the equivalent resistances of both arrangements of resistors.

Equivalent resistance of resistors in series:

$$R_{\text{equiv}} = R_1 + R_2 + R_3 + \cdots + R_N \qquad (18\text{-}9)$$

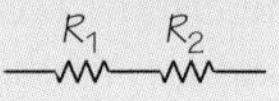

Equivalent resistance of resistors in parallel:

$$\frac{1}{R_{\text{equiv}}} = \frac{1}{R_1} + \frac{1}{R_2} + \frac{1}{R_3} + \cdots + \frac{1}{R_N} \qquad (18\text{-}11)$$

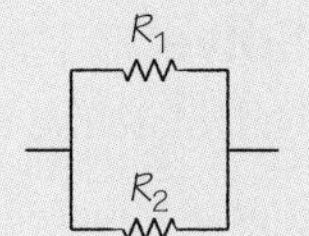

Solve

(a) For the arrangement in Figure 18-12a, first find the equivalent resistance of resistors 1 and 2.

Resistors 1 and 2 in Figure 18-12a are in parallel, so their equivalent resistance R_{12} is given by Equation 18-11:

$$\frac{1}{R_{12}} = \frac{1}{R} + \frac{1}{R} = \frac{2}{R}$$

$$R_{12} = \frac{R}{2}$$

The combination of resistors 1 and 2 is in series with resistor 3. This tells us the overall equivalent resistance.

Equivalent resistor R_{12} is in series with resistor 3. The equivalent resistance R_{123} of the entire combination is given by Equation 18-9:

$$R_{123} = R_{12} + R = \frac{R}{2} + R = \frac{3R}{2}$$

(b) For the arrangement in Figure 18-12b, first find the equivalent resistance of resistors 4 and 5.

Resistors 4 and 5 in Figure 18-12b are in series, so their equivalent resistance R_{45} is given by Equation 18-9:

$$R_{45} = R + R = 2R$$

The combination of resistors 4 and 5 is in parallel with resistor 6. This tells us the overall equivalent resistance.

Equivalent resistor R_{45} is in parallel with resistor 6. The equivalent resistance R_{456} of the entire combination is given by Equation 18-11:

$$\frac{1}{R_{456}} = \frac{1}{R_{45}} + \frac{1}{R} = \frac{1}{2R} + \frac{1}{R} = \frac{3}{2R}$$

$$R_{456} = \frac{2R}{3}$$

Reflect

Notice that the equivalent resistance for the arrangement in Figure 18-12a equals the series combination of the last two resistances, and so is greater than any of the individual resistance values. Similarly, the equivalent resistance for the arrangement in Figure 18-12b is a combination of resistors in parallel, and so is smaller than any of the individual resistances. So these values behave as we would expect.

NOW WORK Problems 2, 4, and 6 from The Takeaway 18-4.

Analysis of More Complex Circuits: The Same Approach

Thus far, we have approached the analysis of circuits that fall under certain guidelines: multiple resistors in series and parallel, and a single constant source of potential difference (a single battery). We have been able to analyze these circuits by reducing the network of resistors to a single equivalent resistance, then expanding the network back to its original configuration. We will now consider the task of how to analyze circuits with multiple batteries. The complication that this task presents is that sometimes a battery can occur *between* resistors so that the simple series or parallel relation no longer holds. Kirchhoff's loop and junction rules are still our tools in this new situation (energy and charge are still, as always, conserved). We just need to be careful and also, usually, solve more than one equation simultaneously. This also means we usually have ways to check our answers to make sure they are consistent, which is important because algebra mistakes are easy to make.

As we have already practiced with resistors to solve a multiloop circuit with more than one battery, we need to identify the junctions and branches of the circuit. In each branch, the direction of the current needs to be specified. (The direction of current can be a guess. There are only two possible directions and if you choose the wrong one, the mathematical analysis will signal this incorrect guess at the problem's end.) A loop is just a series of branches that form a closed path. For any circuit you can always draw one more loop than is mathematically independent, since you can always add other loops together to get the last loop. This means when you count how many independent equations you need to solve a problem, which is the same as the number of unknowns, you can only use one less than the total number of loop equations. You get the right number if you just write as many loop equations as you need to include all the circuit elements. The same is true for junctions (except for those, you need equations that include each current). The junction and loop equations are independent, so between the two types of equations we will have what we need.

To see how to solve such problems, first, we will briefly describe how to identify circuit loops and the sign conventions for the potential differences across elements in those loops. We need these to be able to apply Kirchhoff's laws. Then we will work through an example in detail to explicitly show the steps in solving any multiloop circuit problem with more than one battery.

 **Exam Tip**

It is important to be able to identify the correct number of independent loops in a Kirchhoff's law problem. There will always be one extra loop, which is a sum of the others, and only independent loops can be used to solve for unknowns.

A circuit loop can be identified as the path from a point on a circuit along a series of branches, back to the initial point. First, you identify any junctions, remembering that branches are paths between junctions. In the circuit in **Figure 18-13a**, the junctions are b and e and there are three distinct branches between them: b–c–d–e, b–e, and b–a–f–e. In **Figure 18-13b**, these three branches of the circuit are shown in separated views for clarity: b–c–d–e, b–a–f–e, and b–c–d–e–f–a. In **Figure 18-13c**, the three loops of the circuit are shown in separated views: b–c–d–e–b, b–a–f–e–b, and b–c–d–e–f–a–b. As

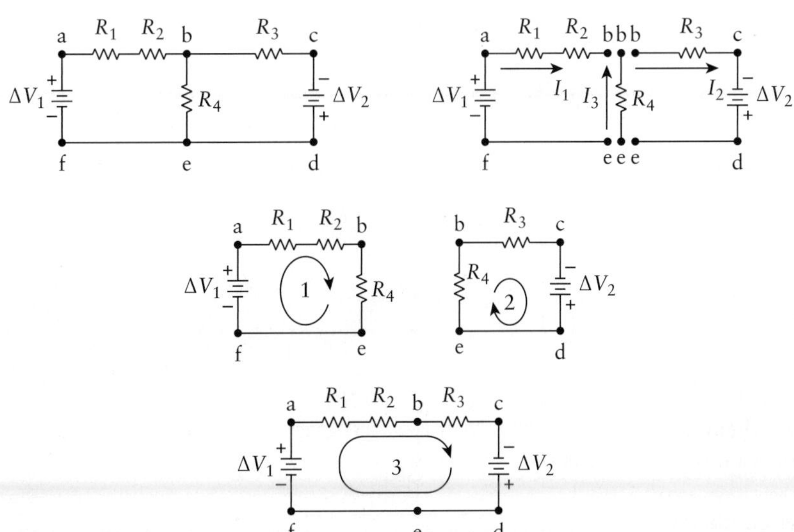

Figure 18-13 Process for identifying the loop in a multiloop circuit. (a) Label any junctions. In the circuit shown, those would be b and e. We have also labeled all the corners to clarify descriptions of the loops. (b) The branches are shown separated for clarity. (c) The three possible loops that can be drawn for the circuit shown. The arrows 1, 2, and 3 each indicate one of the two potential choices for the direction of current in each loop.

TABLE 18-2 Sign Conventions for Potential Differences in Circuits

Circuit element and direction	Magnitude and sign of potential difference
Through battery from negative terminal to positive terminal	Add $+\Delta V_{\text{battery}}$ to the loop equation, potential increase
Through battery from positive terminal to negative terminal	Add $-\Delta V_{\text{battery}}$ to the loop equation, potential drop
Through resistor in the direction of the stated current	Add $-IR$ to the loop equation, potential drop
Through resistor in the opposite direction of the stated current	Add $+IR$ to the loop equation, potential increase

Exam Tip

While the choice of direction of current in any branch and the direction to go around any loop to create your equations in Kirchhoff's law problems are always up to you, you must stick with your choices, since they affect the equations you create. It is useful to draw arrows indicating the direction of the current in each branch, and to always go around the loops in the same direction (clockwise or counterclockwise) so you don't make a mistake and switch mid-loop! Paying attention to these details and the signs of the potential differences based on current and direction will ensure that your equations are consistent.

we mentioned before, there are only *two independent* loops for this circuit and adding any two of the loop equations together will always yield the third. Also, remember that choosing the direction of the current and how you go around the loop is completely arbitrary. Once you choose it, however, you must consistently use that chosen direction as you write your equations!

When applying Kirchhoff's loop rule, it is important to identify the sign convention for potential differences. There is a distinction between a negative potential difference (we will refer to this as a *potential drop* for short) and a positive potential difference (a *potential increase*), which depends on the type of circuit element and the direction you have chosen for your current. As long as you consistently use your chosen direction for each current, you will get a negative sign on that current at the end to show you it was wrong, if you guessed incorrectly. When traversing the loop in the direction chosen (see previous step) the sign of the potential difference will either be negative (a potential drop) or positive (a potential increase) according to the conventions shown in **Table 18-2**.

EXAMPLE 18-8 Solving Multiloop Kirchhoff's Laws Problems

The solution to this example is produced as an outline to solving any Kirchhoff's circuit problem.

Consider the circuit in Figure 18-13 with the following values for the circuit elements:

$R_1 = 2.0\ \Omega$, $R_2 = 4.0\ \Omega$, $R_3 = 3.0\ \Omega$, $R_4 = 8.0\ \Omega$, $\Delta V_1 = 6.0$ V, and $\Delta V_2 = 9.0$ V. Determine the current through each resistor and the potential difference across each resistor.

Set Up

We need to use Kirchhoff's loop and junction rules to produce a sufficient set of independent equations to find the currents. Then we can use the currents to find the potential differences.

We will use the additional symbols below:

R_s: resistance of series combination

I_i: current through resistor i

ΔV_i: magnitude of potential difference across resistor i

(a) Draw the circuit: Draw a circuit diagram for the circuit to be reduced as it is given.

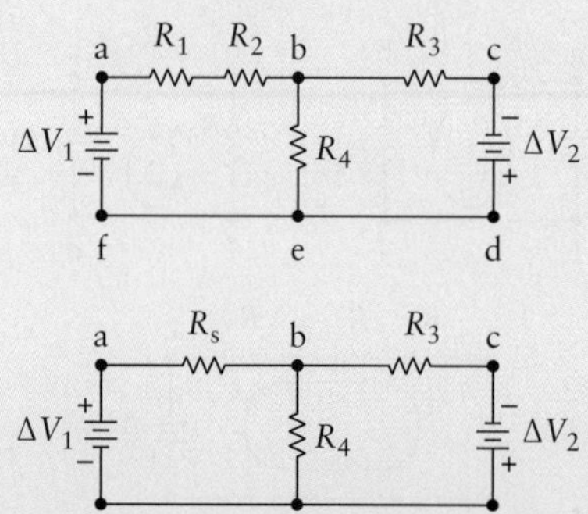

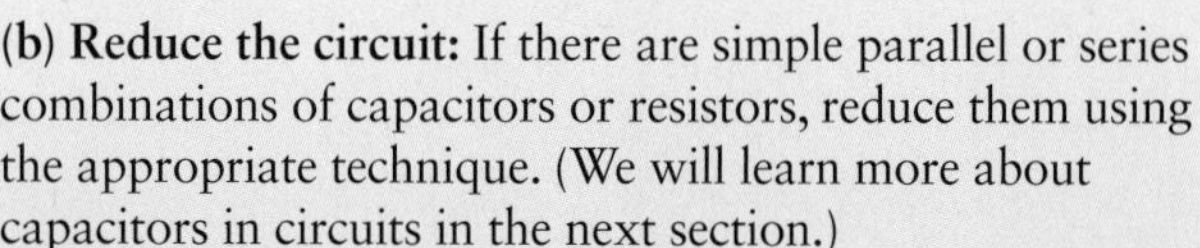

(b) Reduce the circuit: If there are simple parallel or series combinations of capacitors or resistors, reduce them using the appropriate technique. (We will learn more about capacitors in circuits in the next section.)

For this example, R_1 and R_2 form a series combination: $R_s = R_1 + R_2 = 6.0\ \Omega$.

(c) Draw the currents: Assign a current, I_i, to each branch of the circuit. Draw a labeled arrow to represent each on the diagram drawn in (b). Be careful not to introduce currents that are redundant, as they complicate the math. The directions for each of the currents is a guess; if any of the numerical values for current turn out to be negative, then we know that the direction represented here is opposite to the actual current direction.

For this example, there are three branches, and each gets a unique current, I_1, I_2, and I_3.

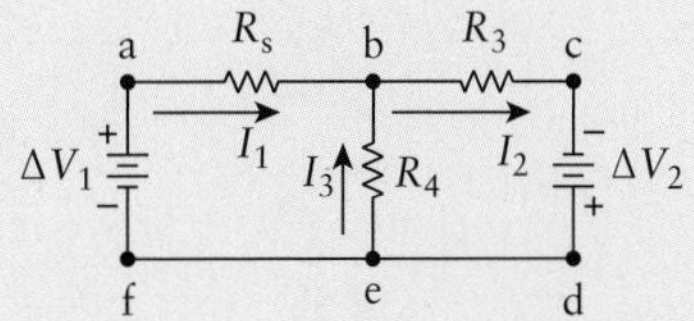

(d) Write junction equations: A junction is any place in the circuit where more than two wires are connected.

Write a junction equation for each junction until you have a junction equation containing each current. For this example, there are two junctions, b and e.

The junction equations $\sum I_{in} = \sum I_{out}$ for both junctions b and e give equivalent results, including all three currents. Using junction b, the junction equation is $I_1 + I_3 = I_2$.

(e) Draw the loops: Draw circuit loops on the circuit diagram. A loop is a path that returns to its starting point.

Draw enough loops so that a loop goes through each circuit element. For this example, two loops are sufficient to go through each circuit element. Loop 1 is a–b–e–f–a, and loop 2 is d–e–b–c–d.

(f) Write the loop equations:

For each loop, write a loop equation. Make sure each circuit element appears in at least one loop equation. Further, recall the sign convention for potential differences. Kirchhoff's rule for circuit loops is $\Sigma\Delta V = 0$.
For this example, we chose loops 1 and 2, but we could have picked any two of the three loops to include all the circuit elements.

The loop equation for loop 1 is

$$-I_1R_s + I_3R_4 + \Delta V_1 = 0$$

The loop equation for loop 2 is

$$-I_3R_4 - I_2R_3 + \Delta V_2 = 0$$

(g) Count unknowns: Your unknowns are the currents. Identify the same number of independent loop or junction equations as you have unknown currents.

We have three unknowns so we need three independent equations: one junction equation and two loop equations.

(h) Solve the independent equations: Some folks use linear algebra, but we will just substitute equations into each other to reduce the number of unknowns in an equation to one, so we can solve for it, and then substitute it back into other equations to get the rest of the unknowns.

Solve the three independent equations:

$$I_1 + I_3 = I_2 \quad (1)$$

$$-I_1R_s + I_3R_4 + \Delta V_1 = 0 \quad (2)$$

$$-I_3R_4 - I_2R_3 + \Delta V_2 = 0 \quad (3)$$

Choose two equations to use to eliminate a variable.

Substitute (1) into (3) to eliminate I_2, giving (4):

$$-I_3R_4 - (I_1 + I_3)R_3 + \Delta V_2 = 0 \quad (4)$$

Regroup the equation so that the unknowns are in separate terms.

$$-I_1R_3 - (R_3 + R_4)I_3 + \Delta V_2 = 0 \quad \text{(4, regrouped)}$$

Solve the equation you did not previously use, to get one of your two remaining unknowns in terms of the other.

Solve equation (2) for I_1 in terms of I_3 to give (5):

$$I_1 = \frac{I_3R_4 + \Delta V_1}{R_s} \quad (5)$$

Substitute this new expression into your regrouped equation, to reduce your equation to a single unknown.

Substitute (5) into (4, regrouped) giving (6):

$$-\left(\frac{I_3R_4 + \Delta V_1}{R_s}\right)R_3 - (R_3 + R_4)\,I_3 + \Delta V_2 = 0 \quad (6)$$

Solve for the single unknown.

Solve (6) for I_3:

$$I_3 = \frac{\Delta V_2 - \left(\frac{\Delta V_1 R_3}{R_S}\right)}{\left(\frac{R_3 R_4}{R_S} + R_3 + R_4\right)} = \frac{9.0\text{ V} - \left(\frac{6.0\text{ V}\cdot 3.0\ \Omega}{6.0\ \Omega}\right)}{\left(\frac{8.0\ \Omega \cdot 3.0\ \Omega}{6.0\ \Omega} + 3.0\ \Omega + 8.0\ \Omega\right)}$$

$$= \frac{2}{5}\text{ A}$$

Substitute this value back into an equation that only contained one other unknown (so it will now only have one unknown) and solve. It usually simplifies things to leave the answers as fractions.

Solve (2) for I_1 (this was equation 5):

$$I_1 = \frac{I_3 R_4 + \Delta V_1}{R_s}$$

$$= \frac{\left(\frac{2}{5}\text{ A}\right)(8.0\ \Omega) + 6.0\text{ V}}{6.0\ \Omega} = \frac{16.0\text{ V} + 30.0\text{ V}}{30.0\ \Omega}$$

$$= \frac{23}{15}\text{ A}$$

Once you have solved for two of the unknowns, you can use any of the equations including the third unknown to solve for the third unknown. If the third unknown is a current, the junction equation is the most straightforward equation to use.

Use (1) to compute I_2:

$$I_2 = I_1 + I_3 = \frac{23}{15}\text{ A} + \frac{2}{5}\text{ A}$$

$$= \frac{29}{15}\text{ A}$$

(i) Calculate Potential Differences Across Resistors: Use Ohm's law, $\Delta V = IR$, to calculate the potential difference across each resistor.

Using Ohm's law:

$$\Delta VR_3 = I_2 R_3 = \frac{29}{5}\text{ V}$$

$$\Delta VR_4 = I_3 R_4 = \frac{16}{5}\text{ V}$$

$$\Delta V_s = I_1 R_s = \frac{46}{5}\text{ V}$$

In this example, the potential difference across R_s is actually the potential difference across two resistors, R_1 and R_2.

Because they are in series, we know that the current is the same through both, I_1.

Use Ohm's law for these two resistors:

$$\Delta VR_1 = I_1 R_1 = \frac{46}{15}\text{ V}$$

$$\Delta VR_2 = I_1 R_2 = \frac{92}{15}\text{ V}$$

Reflect:

(j) Check for consistency: Check that the junction equation and the loop equations work with the numeric answers. I often make mistakes in Kirchhoff's law problems, and checking the loops allows me to find the errors and fix them. You should check all three equations, but always check at least one.

Our loop equations should always add to zero, and these do, so this checks!

We used the junction equation for the last calculation in solving for current, so we don't need to check that one.

Check symbolic equations to check numerical answers:

$$-I_3 R_4 - I_2 R_3 + \Delta V_2 = 0 \qquad (3)$$

$$-\left(\frac{2}{5}\text{ A}\cdot 8.0\ \Omega\right) - \left(\frac{29}{15}\text{ A}\cdot 3.0\ \Omega\right) + 9.0\text{ V} =$$

$$-\frac{16}{5}\text{ V} - \frac{29}{5}\text{ V} + 9.0\text{ V} = -\frac{45}{5}\text{ V} + 9.0\text{ V} = 0$$

$$-I_1 R_s + I_3 R_4 + \Delta V_1 = 0 \qquad (2)$$

$$-\left(\frac{23}{15}\text{ A}\cdot 6.0\ \Omega\right) + \left(\frac{2}{5}\text{ A}\cdot 8.0\ \Omega\right) + 6.0\text{ V} =$$

$$-\frac{46}{5}\text{ V} + \frac{16}{5}\text{ V} + 6.0\text{ V} = -\frac{30}{5}\text{ V} + 6.0\text{ V} = 0$$

NOW WORK Problems 1 and 7–9 from The Takeaway 18-4.

Measuring Properties of Circuit Elements

When building and testing circuits, there are tools to help you measure the sorts of quantities you just calculated in Example 18-8. The names of the most common come from the units they measure. A voltmeter is designed to measure the potential difference across an element in a circuit. This means a voltmeter must actually be attached across (in parallel with) the circuit element for which you are trying to measure the potential difference, since elements connected in parallel have the same potential difference across them. An ammeter is designed to measure the current at a point in a circuit, so is connected in series in the branch you wish to measure the current in, since things in series have the same current. The way they have to be connected informs their design. If something of very low resistance were to be connected in parallel with another circuit element, the majority of the current would pass through the low-resistance element, changing the circuit. So, a voltmeter must have a very high resistance, to keep the current going through the voltmeter as close to zero as possible, so that it does not affect the circuit. On the other hand, the ammeter must be connected in series, so to not increase the resistance of the branch in which it is placed, it should have as low a resistance as possible. We usually model ammeters as having a zero resistance, and voltmeters as having an infinite resistance. While this is not quite true, it is adequate for most measurements we will make in AP® Physics. Understanding the model and the reality will help you understand if you do find experimental error. The other thing it can help you understand is how to use a multimeter. A multimeter is a device that on one setting can be used as a voltmeter, and on another can be used as an ammeter. So, if you place a multimeter in a circuit intending to measure current, but you use the voltmeter setting, the huge resistance will reduce the current to near zero in the branch you place it in. The more problematic mistake is to connect a multimeter as a voltmeter when it is on the ammeter setting. In that case, the majority of the current diverts through the multimeter, typically blowing out a fuse, sometimes permanently damaging the device. The symbol in a circuit diagram for either of these devices is the unit with a circle around it. Common symbols you should recognize and be able to produce in a circuit diagram for these tools and other common circuit elements are illustrated in **Figure 18-14.**

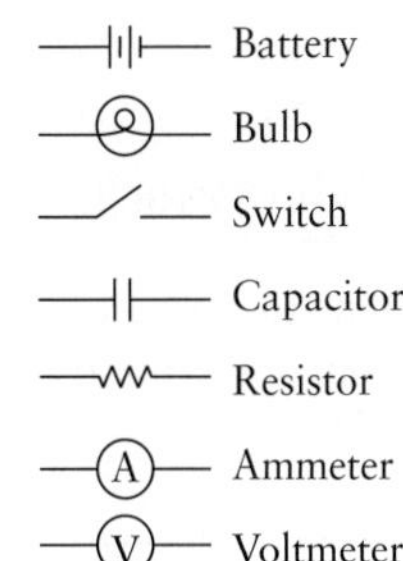

Figure 18-14 Seven common symbols used in circuit drawings There are seven common symbols you are responsible for recognizing and being able to produce when interpreting or drawing circuit diagrams on the AP® Physics exam. Batteries may be drawn as one or more sets of a long and a short parallel line (two sets are shown), and light bulbs may be represented either as a resistor or as a curly line inside a circle as shown. The other elements are consistently represented as shown.

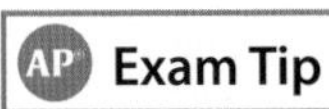

Exam Tip

You are responsible for understanding the basics of how an ammeter and a voltmeter work. You must understand how to connect each device (and which one) in a circuit to measure current or potential difference. There are other potential tools, but these are the only ones you must know. Voltmeters must be connected in parallel with the element across which potential difference is to be measured, and are modeled as having infinite resistance. Ammeters must be connected in series with the element in which current is to be measured, and are modeled as having zero resistance.

THE TAKEAWAY for Section 18-4

✔ From the negative terminal of a battery to its positive terminal, the electric potential increases by an amount that depends on the battery. Across a resistor of resistance R in the direction of the current, the electric potential decreases by IR.

✔ Kirchhoff's loop rule states that the sum of electric potential differences around a closed loop in a circuit is zero. This is an application of the conservation of energy.

✔ Kirchhoff's junction rule states that the net current into a circuit junction equals the net current out of that junction. This is an application of the conservation of charge.

✔ If resistors are connected in series, the current is the same in each resistor but the electric potential differences are different across different resistors. The equivalent resistance of the resistors connected in series equals the sum of the individual resistances.

✔ If resistors are connected in parallel, the electric potential difference is the same for each resistor but the currents are different for different resistors. The reciprocal of the equivalent resistance of the resistors connected in parallel equals the sum of the reciprocals of the individual resistances.

✔ Multiloop circuits with just one battery can be solved by simplifying combinations of resistors. When more than one battery is included, Kirchhoff's laws can still be applied to solve for unknown currents or potential differences.

Prep for the AP® Exam

AP® Building Blocks

EX 18-8 **1.** Calculate the potential difference $V_A - V_B$ in each of the situations shown in the figure. Resistance in the wire is negligible.

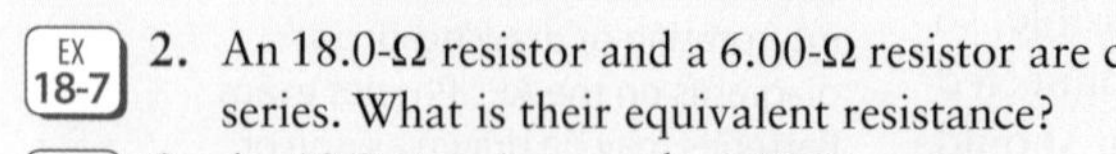

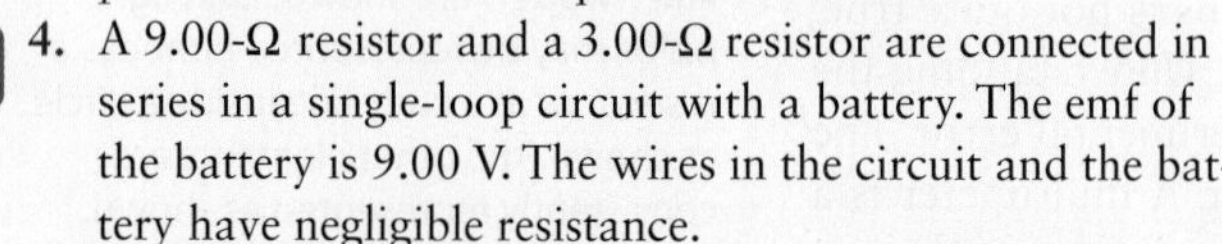

EX 18-7 **2.** An 18.0-Ω resistor and a 6.00-Ω resistor are connected in series. What is their equivalent resistance?

EX 18-6 **3.** An 18.0-Ω resistor and a 6.00-Ω resistor are connected in parallel. What is their equivalent resistance?

EX 18-7 **4.** A 9.00-Ω resistor and a 3.00-Ω resistor are connected in series in a single-loop circuit with a battery. The emf of the battery is 9.00 V. The wires in the circuit and the battery have negligible resistance.

(a) Draw a circuit diagram and calculate the current through the battery.

(b) Calculate the current through each resistor.

(c) Calculate the potential difference across each resistor. Justify the reasonableness of your answer using conservation of energy.

(d) If the two resistors are cylindrical resistors made of the same material, describe how the geometry of the 9-Ω resistor would be different from that of the 3-Ω resistor, and why that change in geometry would affect the current at a given potential difference in the way that it does.

AP® Skill Builders

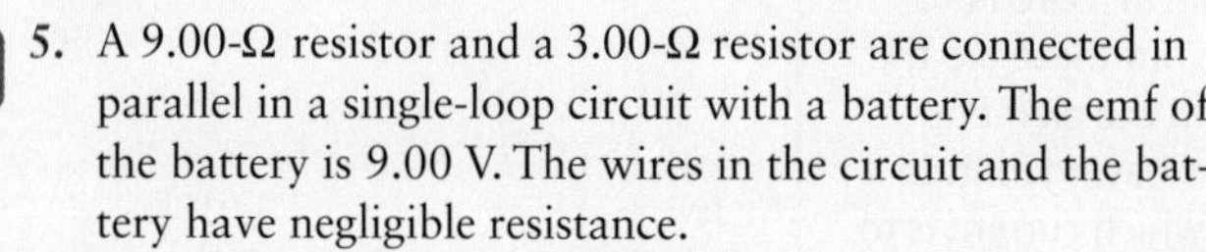

EX 18-6 **5.** A 9.00-Ω resistor and a 3.00-Ω resistor are connected in parallel in a single-loop circuit with a battery. The emf of the battery is 9.00 V. The wires in the circuit and the battery have negligible resistance.

(a) Draw a circuit diagram.

(b) Calculate the equivalent resistance of the two resistors.

(c) Calculate the current through the battery.

(d) Calculate the potential difference across each resistor.

(e) Calculate the current through each resistor.

(f) How would your answers to (c)–(e) change if the 3.00-Ω resistor were removed from the circuit? Justify your answers in terms of properties of the circuit elements.

EX 18-7 **6.** A segment of an electric circuit is represented in the figure. Take the resistance in the wires between resistors to be zero.

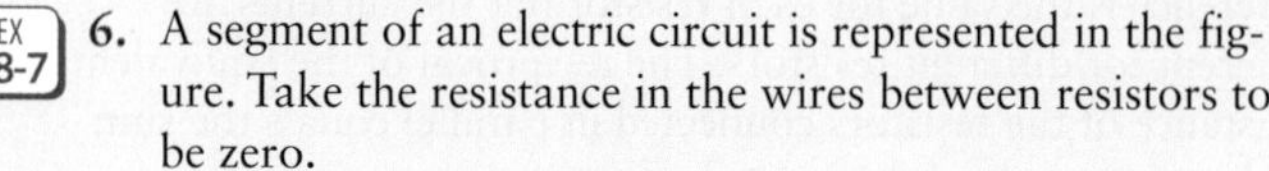

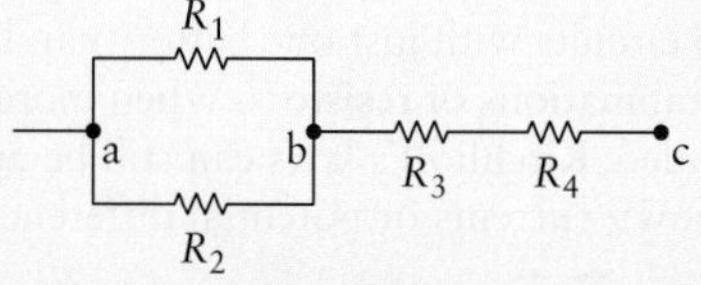

(a) Apply both charge conservation and energy conservation to explain why $V_b - V_c = I_c(R_3 + R_4) = I_b(R_3 + R_4)$, where I_c is the current entering the junction labeled c and I_b is the current leaving the junction labeled b.

(b) Apply energy conservation to explain why $V_a - V_b = I_1R_1 = I_2R_2$, where I_1 is the current in the resistor labeled R_1, and I_2 is the current in the resistor labelled R_2.

(c) Apply charge conservation to explain why the current leaving the junction labeled b, I_b, can be expressed as

$$I_b = (V_a - V_b)\left[\frac{1}{R_1} + \frac{1}{R_2}\right]$$

(d) How would you connect an (i) ammeter to measure I_1, and (ii) a voltmeter to measure potential difference $V_a - V_b$?

AP® Skills in Action

EX 18-8 **7.** In the diagrams, assumptions about the direction of currents in wires 2 and 3 are indicated.

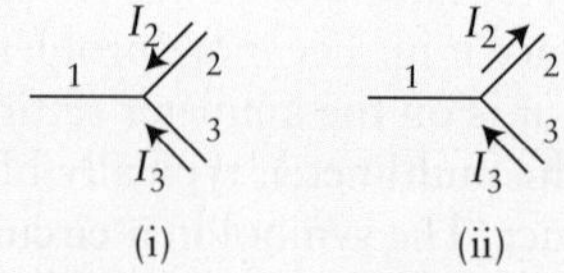

(a) For each possible assumption regarding the direction of current in the wire labeled 1 (either into or out of the junction), express I_1 in terms of the currents I_2 and I_3.

(b) The currents in branches 2 and 3 of the circuit are measured using the assumed directions of current shown in (i) and found to be $I_3 = -0.1$ A and $I_2 = 0.2$ A. Describe what this means about the directions of the currents, and predict the value of the current I_1 and its direction into or out of the junction that would be measured if these measurements for 2 and 3 are correct.

EX 18-8 **8.** Considering the diagram shown, label your assumption for the directions of current in each branch.

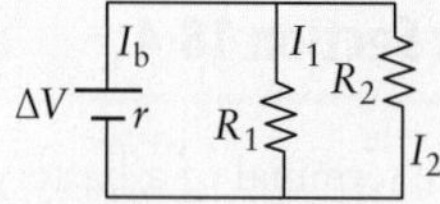

(a) Construct an expression of charge conservation.

(b) Construct three expressions of energy conservation.

(c) Confirm the claim that only two of the energy conservation expressions are independent.

(d) Explain why the claim in part (c) is necessary if the analysis of this circuit has a solution for currents I_b, I_1, and I_2, given the potential difference ΔV and resistances R_1 and R_2.

EX 18-8 **9.** A circuit is constructed using two batteries and three resistors as shown in the figure. The batteries have potential differences $\Delta V_A = 3.0$ V and $\Delta V_B = 1.5$ V. The resistors

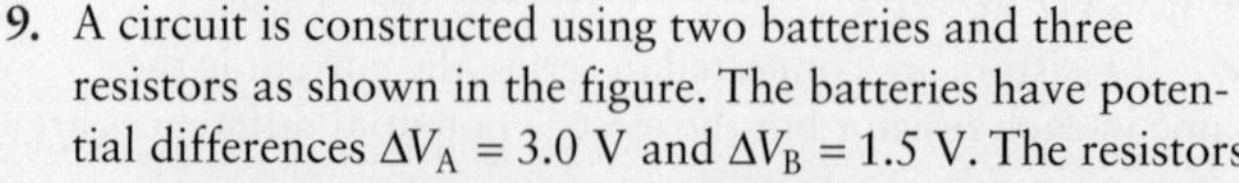

have resistances $R_1 = 25\ \Omega$, $R_2 = 250\ \Omega$, and $R_3 = 50\ \Omega$. Find the currents through each resistor. Carry out your calculations to four significant digits.

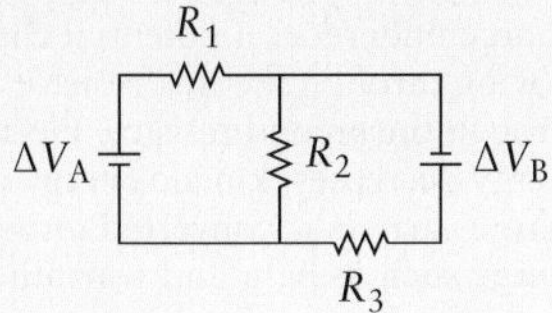

10. For the circuit drawn in problem 10 in The Takeaway for Section 18-2, explain why connecting these pairs of points with two ends of a piece of wire turns out the light: (g and h), (e and f), (c and d), and (a and b).

18-5 The rate at which energy is transferred by a circuit element depends on the current through the element and the electric potential difference across it

The aspects of electric circuits that we've concentrated on so far are electric potential difference, current, and resistance. While we have framed these discussions by thinking about energy, we have not yet discussed how a circuit transfers energy from one place to another. Yet, this transfer of energy, such as from a battery to a flashlight bulb (where the energy is converted into visible light) or from a wall socket to a toaster (where the energy is used to heat your bread or bagel) is the most fundamental use of a circuit. In most applications what's of interest is the *rate* at which energy is transferred into or out of a circuit element. For example, in order for a toaster to be useful, it must heat the bread rapidly enough that it becomes toast in a minute or so, not an hour, and the brightness of a flashlight's bulb increases with the rate of energy transfer.

WATCH OUT !

Transfer or Convert: It all depends on your system.

Rate of energy transfer is always the rate at which energy is converted between different types of energy, which allows it to cross a system boundary. For instance, in Example 8-5, the rowers' internal energy decreased as they increased their and the water's thermal energy and the kinetic energy of the boat. Instead of talking about the rate at which they were converting their internal energy into the kinetic energy of the boat, we could talk about the rate they were transferring energy to the boat. We would mean exactly the same thing. The only difference is how we define the system. The rower converts internal energy into kinetic energy of the boat (same system), or the rower transfers energy to the boat (rower and boat considered different systems). In this chapter, we are talking about electric potential energy, so the transfers of energy we will be interested in are the conversions of electric potential energy into or from other types of energy.

As we learned in Section 8-4, **power** is the rate at which energy is transferred into or out of a system. The unit of power is the joule per second, or watt (abbreviated W): 1 W = 1 J/s. You can see the importance of power in electric circuits from the numbers that are used to describe various electric devices: An amplifier for a home audio system is rated by its power output (perhaps 75 to 100 W), and every light bulb is stamped with the power that must be supplied to it for normal operation (say, 13 or 60 W).

In this section, we'll see how to calculate the power *output* of a source of emf such as a battery, which is fundamentally a source of electric potential energy. We'll also see how to calculate the power *input* of a resistor, which converts electric potential energy into other types of energy, such as warmth or light.

Power in a Circuit Element

The key to understanding energy and power in electric circuits is that there is a potential difference across each circuit element, which means that a change takes place in electric potential energy as charge traverses a circuit element. Recall from Chapter 17

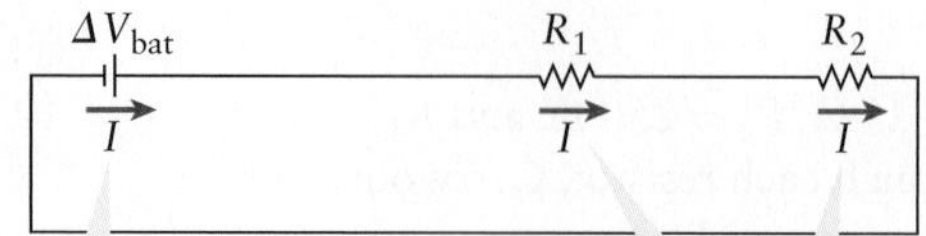
ΔV_{bat}
R_1
R_2
I
I
I
Charge moves through an electric potential difference of $+\Delta V_{bat}$ as it travels through the battery, so electric potential energy increases; chemical energy in the battery is converted to electric potential energy of the circuit. The battery transfers energy into the circuit.
Charge undergoes a potential change of $-IR_1$ moving through the first resistor and $-IR_2$ moving through the second resistor. Electric potential energy decreases; kinetic energy of the moving charge carriers is converted into other types of energy such as light and warmth by the resistors. The resistors transfer this energy out of the circuit.

Figure 18-15 Energy and power in a single-loop circuit Chemical energy is converted to electric potential energy in the battery (the battery transfers energy to the circuit). This keeps the charge carriers moving as their kinetic energy is converted into other types of energy in the resistors (the resistors transfer warmth and light out of the circuit). For any loop in a circuit, the rate at which energy is transferred into the circuit by circuit elements in the loop (in this case the battery) must exactly equal the rate at which energy is transferred out by elements in the loop (resistors).

that the relationship between electric potential difference and electric potential energy difference:

$$\Delta V = \frac{\Delta U_{\text{electric}}}{q_0}$$

WATCH OUT !

Brightness is only a measure of power for the same kind of light bulbs.

Because energy goes into warmth as well as light, light bulbs that produce less warmth are brighter at lower power. This is why a 10 W LED bulb is as bright as a 60 W incandescent.

Let's rewrite this equation for the case in which a small amount of charge (Δq) moves from one end of a circuit element to the other. If the potential difference between the ends of the element is ΔV, then Equation 17-6 becomes

$$\Delta V = \frac{\Delta U_{\text{electric}}}{\Delta q} \quad \text{or} \quad \Delta U_{\text{electric}} = (\Delta q)\Delta V \tag{18-12}$$

We'll continue to use the idea that current is in the direction in which positively charged objects would move, so the amount of charge that moves through the circuit element is positive: $\Delta q > 0$. Then Equation 18-12 tells us that there is an *increase* in electric potential energy ($\Delta U_{\text{electric}} > 0$) of the circuit if Δq traverses a circuit element from low potential to high potential, so that $\Delta V > 0$. That's the case for charge that travels through the source of emf in **Figure 18-15** from the negative terminal to the positive terminal. (We get the same amount of energy if we send negative charge the other way, but this just keeps makes things easier to write down!) Since energy is conserved, it must be that the amount of energy extracted from the source is equal to the increase in electric potential energy of the circuit.

There is a *decrease* in electric potential energy ($\Delta U_{\text{electric}} < 0$) if Δq traverses a circuit element from high potential to low potential, so that $\Delta V < 0$. That's what happens when Δq travels through the resistor R in Figure 18-15 in the direction of the current. The electric potential energy decreases as the kinetic energy of the moving charge carriers is converted into other types of energy, causing an increase in the resistor's temperature. If the resistor is the filament of an incandescent flashlight bulb, the increased temperature causes the filament to **radiate** energy, some of which is in the form of visible light as well as warmth, so the greater the power, the brighter the bulb.

WATCH OUT !

Potential energy may change as charges move around a circuit, but current does not.

It's a common misconception that current is "used up" when it passes through a resistor and that current is "added to" when it passes through a source of emf. In fact, there is *no* change in the current as it passes through either a resistor or a source: The charge carriers leave any circuit element at the same rate as they enter the element. The signs of the energy change show which way electric potential energy gets converted, or which way we would say energy was transferred: In the resistor, electric potential energy is converted to warmth and light (energy is transferred out of the system); in the battery, chemical energy is converted to electric potential energy (energy is transferred into the system).

With a current I in a circuit, the power P for each circuit element is just the rate at which electric potential energy is converted by that element. This equals the change in electric potential energy ($\Delta U_{\text{electric}}$) for an amount of charge (Δq) that travels through the element, as given by Equation 18-12, divided by the time Δt that it takes each new bit of charge Δq to travel through the element:

$$P = \frac{\Delta U_{electric}}{\Delta t} = \frac{\Delta q \Delta V}{\Delta t} = \left(\frac{\Delta q}{\Delta t}\right)\Delta V = (I)\Delta V$$

In the last part of this equation, we've used the idea that the ratio $\Delta q/\Delta t$ equals the rate at which charge moves through the circuit element—in other words, this ratio equals the current I through the element (see Equation 18-1). Let's rewrite this relationship:

Power produced by or transferred into or out of a circuit element

$$P = I\Delta V \tag{18-13}$$

Current through the circuit element

Electric potential difference across the circuit element

EQUATION IN WORDS
Power for a circuit element

Source of emf: Power is the rate at which other types of energy are converted into electric potential energy, a transfer of energy into the circuit.
Resistor: Power is the rate at which electric potential energy is converted into other types of energy, a transfer of energy out of the circuit.

Equation 18-13 states that the power that flows into or out of any circuit element is equal to the product of the current through the element multiplied by the potential difference across the element. The power can be positive or negative, depending on the sign of the potential difference ΔV between the ends of the circuit element. For a given potential difference ΔV, each small amount of charge Δq that traverses the circuit element transfers the same amount of energy into or out of the element. The more charge that traverses the element per unit time, the greater the current I, and the greater the *rate* of energy transfer P.

Note that the power company charges its customers based not on how much *power* they use at a given time but on the total amount of *energy* that they use. Since power is energy per time, the units of energy are the units of power multiplied by the units of time. That's why the power company bills in terms of kilowatt-hours (kWh): 1 kWh = 1000 watt-hours, or the amount of energy it takes to run a device that uses 1000 W of power (typical for a microwave oven) for 1 hour. Since 1 h = 3600 s, 1 kWh equals (1000 W)(3600 s) = 3.6×10^6 W · s = 3.6×10^6 J.

If the circuit element is a resistor with resistance R, Equation 18-7 tells us that the potential difference ΔV across the resistor is equal to the product of the current and the resistance: $\Delta V = IR$, or equivalently $I = \Delta V/R$. We can use these two expressions to rewrite Equation 18-13 in two equivalent forms for the special case of a resistor:

$$P = I(IR) \quad \text{or} \quad P = \left(\frac{\Delta V}{R}\right)\Delta V$$

We can simplify these to

Power for a **resistor**

Current through the resistor

$$P = I^2R = \frac{\Delta V^2}{R} \tag{18-14}$$

Electric potential difference across the resistor

Resistance of the resistor

EQUATION IN WORDS
Power for a resistor

The expression $P = I^2R$ is useful if we know the current through a resistor of known resistance, while $P = \Delta V^2/R$ is useful if we know the potential difference across that resistor. The following examples show how to use Equations 18-13 and 18-14.

WATCH OUT !

The units of power take time into account.

The units of power sometimes cause confusion because it seems that watts require an additional time measurement. For example, to find the power in watts of a flashlight bulb, a student might ask, "Find the power for what amount of time?" That's not a sensible question: Time is already included in the units of power, since 1 watt is equal to 1 joule *per second*. When a light bulb is rated at 120 W, that means it requires 120 J of energy every second to operate. Power is a rate of energy transfer with respect to time.

EXAMPLE 18-9 Power in a Single-Loop Circuit

A battery of electric potential difference $\Delta V_{bat} = 12.0$ V is connected in series to a resistor with resistance 1.00 Ω and another with resistance 19.0 Ω. Find (a) the rate at which energy is supplied by the battery, (b) the rate at which energy flows into the 1.00-Ω resistor, and (c) the rate at which energy flows into the 19.0-Ω resistor.

Set Up

We can use Kirchhoff's loop rule (the sum of the changes in potential around a closed loop in a circuit must equal zero) to find the current in the circuit. Equation 18-7 then tells us the potential difference across either resistance. We'll use Equation 18-13 to determine the power for each element of the circuit, and check our results for the two resistances using Equations 18-14.

Relationship among potential difference, current, and resistance:

$$\Delta V = IR \quad (18\text{-}7)$$

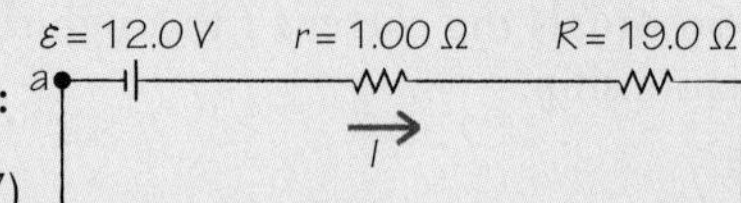

Power for a circuit element:

$$P = I\Delta V \quad (18\text{-}13)$$

Power for a resistor:

$$P = I^2R = \frac{\Delta V^2}{R} \quad (18\text{-}14)$$

Solve

(a) First apply Kirchhoff's loop rule to determine the current I. Start at the point a in the circuit and go around in the direction of the current.

There is an electric potential increase ($\Delta V_{bat} = \mathcal{E}$) of 12.0 V going across the battery, an electric potential decrease of Ir going across the resistor with resistance $r = 1.00$ Ω, and an electric potential decrease of IR going across the resistor with resistance $R = 19.0$ Ω. The sum of the potential changes around the circuit is zero:

$$+\mathcal{E} + (-Ir) + (-IR) = 0$$

Rearrange this to solve for the current I:

$$\mathcal{E} - I(r + R) = 0, \text{ so } I(r + R) = \mathcal{E} \text{ and}$$

$$I = \frac{\mathcal{E}}{r + R} = \frac{12.0 \text{ V}}{1.00\ \Omega + 19.0\ \Omega} = \frac{12.0 \text{ V}}{20.0\ \Omega} = 0.600 \text{ A}$$

(Recall that 1 A = 1 V/Ω.)

Use Equation 18-13 to find the power for the battery.

The potential difference across the battery is 12.0 V. The rate at which the battery transfers energy to the circuit is

$$P_{bat} = I\Delta V_{bat} = (0.600 \text{ A})(12.0 \text{ V}) = 7.20 \text{ A} \cdot \text{V}$$

Because 1 A = 1 C/s, 1 V = 1 J/C, and 1 W = 1 J/s,

$$P_{bat} = 7.20 \frac{\text{C}}{\text{s}} \cdot \frac{\text{J}}{\text{C}} = 7.20 \frac{\text{J}}{\text{s}} = 7.20 \text{ W}$$

(b) First use Equation 18-7 to find the potential difference across the resistor $r = 1.00$ Ω. Then find the power for the resistor using Equation 18-13.

From Equation 18-7, the electric potential difference across the resistor r is

$$\Delta V_r = Ir = (0.600 \text{ A})(1.00\ \Omega) = 0.600 \text{ V}$$

From Equation 18-13, the rate at which the resistor transfers energy out of the circuit is

$$P_r = I\Delta V_r = (0.600 \text{ A})(0.600 \text{ V}) = 0.360 \text{ W}$$

(c) Repeat part (b) for the resistor of resistance $R = 19.0$ Ω.

The electric potential difference across the resistor R is

$$\Delta V_R = IR = (0.600 \text{ A})(19.0\ \Omega) = 11.4 \text{ V}$$

The rate at which energy is converted in the resistor is

$$P_R = I\Delta V_R = (0.600 \text{ A})(11.4 \text{ V}) = 6.84 \text{ W}$$

Reflect

Note that the rate at which electric potential energy is transferred into the circuit by the battery is equal to the *net* rate at which energy is converted into other types by the resistors. This is equivalent to saying that energy is conserved in the circuit.

The net rate of energy flow into the two resistors is

$$P_r + P_R = 0.360\ \text{W} + 6.84\ \text{W} = 7.20\ \text{W}$$

This is the same as the rate at which energy is transferred into the circuit by the battery, $P_{bat} = 7.20$ W.

We can check our result $P_R = 6.84$ W for the power into the resistor by showing that we get the same result using Equations 18-14. Can you use the same approach to check the result $P_r = 0.360$ W for the resistor r?

From the first of Equations 18-14, the power into the 19.0-Ω resistor is

$$P_R = I^2R = (0.600\ \text{A})^2\ (19.0\ \Omega) = 6.84\ \text{W}$$

(Note that $1\ \text{A}^2 \cdot \Omega = 1\ \text{A} \cdot \text{V} = 1\ \text{W}$.)

To use the second of Equations 18-14, use the potential difference, $\Delta V_R = 11.4$ V across the resistor:

$$P_R = \frac{\Delta V_R^2}{R} = \frac{(11.4\ \text{V})^2}{19.0\ \Omega} = 6.84\ \text{W}$$

(Note that $1\ \text{V}^2/\Omega = 1\ \text{V} \cdot (\text{V}/\Omega) = 1\ \text{V} \cdot \text{A} = 1\ \text{W}$.)

This agrees with the result for P_R found earlier.

NOW WORK Problems 2, 3, and 8 from The Takeaway 18-5.

EXAMPLE 18-10 Power in Series and Parallel Circuits

Two resistors, one with $R_1 = 2.00\ \Omega$ and one with $R_2 = 3.00\ \Omega$, are both connected to a battery with $\Delta V_{bat} = 12.0$ V. Find the power delivered by the battery and the power transferred out of the circuit by each resistor if the resistors are connected to the battery (a) in series and (b) in parallel.

Set Up

We'll use the same tools as in the previous example, plus Equations 18-9 and 18-11 for the equivalent resistance of resistors in series and in parallel. In each case we'll use the equivalent resistance to find the current through the battery, then use Equation 18-13 to determine the power provided by the battery. If the resistors are in series, the current through each is the same as the current through the battery, so we'll use the first of Equations 18-14 ($P = I^2R$) to determine the power into each resistor. For resistors in parallel, the potential difference is the same across each resistor, so in that case we'll find the power into each resistor using the second of Equations 18-14 ($P = \Delta V^2/R$).

Relationship among potential difference, current, and resistance:

$$\Delta V = IR \qquad (18\text{-}7)$$

Equivalent resistance of two resistors in series:

$$R_{equiv} = R_1 + R_2 \qquad (18\text{-}9)$$

Equivalent resistance of two resistors in parallel:

$$\frac{1}{R_{equiv}} = \frac{1}{R_1} + \frac{1}{R_2} \qquad (18\text{-}11)$$

Power for a circuit element:

$$P = I\Delta V \qquad (18\text{-}13)$$

Power for a resistor:

$$P = I^2R = \frac{\Delta V^2}{R} \qquad (18\text{-}14)$$

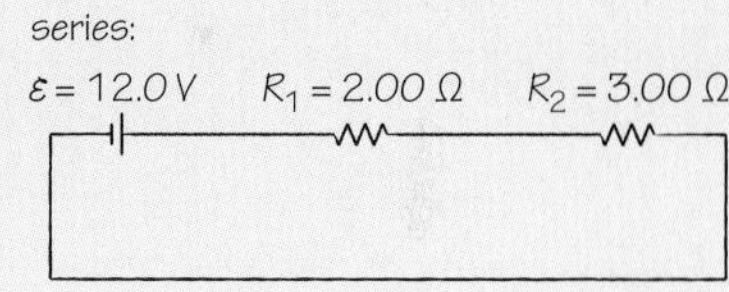

parallel: ℰ = 12.0 V, $R_1 = 2.00\ \Omega$, $R_2 = 3.00\ \Omega$

Solve

(a) The two resistors in series are equivalent to a single resistor R_{equiv}. This is connected directly to the terminals of the battery, across which

From Equation 18-9, the equivalent resistance of the two resistors in series is

$$R_{equiv} = R_1 + R_2 = 2.00\ \Omega + 3.00\ \Omega = 5.00\ \Omega$$

the potential difference is ΔV_{bat}. So the electric potential difference across R_{equiv} is also equal to ΔV_{bat}, which tells us the current through the circuit. Equation 18-13 then tells us the power transferred into the circuit by the battery.

The electric potential difference across this equivalent resistance is $\Delta V = IR_{equiv}$ from Equation 18-7, which is also equal to the emf $\mathcal{E} = 12.0$ V of the battery. So the current through the equivalent resistance is given by

$$\mathcal{E} = IR_{equiv} \quad \text{or} \quad I = \frac{\mathcal{E}}{R_{equiv}} = \frac{12.0 \text{ V}}{5.00\ \Omega} = 2.40 \text{ A}$$

This is also the current through the battery. Because the potential difference across the battery is equal to $\mathcal{E} = 12.0$ V, the power delivered by the battery is

$$P_{bat} = I\mathcal{E} = (2.40 \text{ A})(12.0 \text{ V}) = 28.8 \text{ W}$$

The current $I = 2.40$ A is the same through both resistors in this series circuit. Use this to calculate the power that is transferred out of the circuit by each resistor.

Using the first of Equations 18-14, the power into resistor $R_1 = 2.00\ \Omega$ is

$$P_1 = I^2R_1 = (2.40 \text{ A})^2\ (2.00\ \Omega) = 11.5 \text{ W}$$

The power into resistor $R_2 = 3.00\ \Omega$ is

$$P_2 = I^2R_2 = (2.40 \text{ A})^2\ (3.00\ \Omega) = 17.3 \text{ W}$$

The net power transferred out of the circuit by the two resistors is equal to the power transferred into the circuit by the battery, as it should be:

$$P_1 + P_2 = 11.5 \text{ W} + 17.3 \text{ W} = 28.8 \text{ W} = P_{bat}$$

(b) Follow the same steps as in part (a) to find the equivalent resistance of the two resistors in parallel, the current through the battery connected to that parallel arrangement, and the power delivered by the battery in this situation.

From Equation 18-11 the equivalent resistance of the two resistors in parallel is given by

$$\frac{1}{R_{equiv}} = \frac{1}{R_1} + \frac{1}{R_2} = \frac{1}{2.00\ \Omega} + \frac{1}{3.00\ \Omega} = 0.833\ \Omega^{-1}$$

$$R_{equiv} = \frac{1}{0.833\ \Omega^{-1}} = 1.20\ \Omega$$

Equation 18-7 states that the electric potential difference across this equivalent resistance is $\Delta V = IR_{equiv}$; because this equivalent resistance is connected to the terminals of the battery, the electric potential difference is also equal to the battery $\Delta V_{bat} = 12.0$ V. So the current through the equivalent resistance and through the battery is given by

$$\mathcal{E} = IR_{equiv} \text{ or } I = \frac{\mathcal{E}}{R_{equiv}} = \frac{12.0 \text{ V}}{1.20\ \Omega} = 10.0 \text{ A}$$

From Equation 18-13 the power delivered by the battery is

$$P_{bat} = I\Delta V_{bat} = (10.0 \text{ A})(12.0 \text{ V}) = 1.20 \times 10^2 \text{ W}$$

Note that this is more than four times as much power as the same battery transfers into the circuit for the same resistors in series.

The electric potential difference is the same across resistors in parallel. Each resistor is effectively connected directly to the terminals of the battery, so the electric potential difference across each resistor is $\Delta V = \Delta V_{bat} = 12.0$ V. Use this to calculate the power that is transferred out of the circuit by each resistor.

Using the second of Equations 18-14, we find that the power transferred out of the circuit by resistor $R_1 = 2.00\ \Omega$ is

$$P_1 = \frac{\Delta V^2}{R_1} = \frac{\mathcal{E}^2}{R_1} = \frac{(12.0 \text{ V})^2}{2.00\ \Omega} = 72.0 \text{ W}$$

The power transferred out of the circuit by resistor $R_2 = 3.00\ \Omega$ is

$$P_2 = \frac{\Delta V^2}{R_2} = \frac{\mathcal{E}^2}{R_2} = \frac{(12.0 \text{ V})^2}{3.00\ \Omega} = 48.0 \text{ W}$$

As for the series case, the net power into the two resistors is equal to the power supplied by the battery:

$$P_1 + P_2 = 72.0 \text{ W} + 48.0 \text{ W} = 1.20 \times 10^2 \text{ W} = P_{bat}$$

Reflect

Although the same battery and same resistors are used in both circuits, the power transferred into the circuit by the battery is *much* different in the two circuits. That's because the current through the battery is different for the two circuits: $I = 2.40$ A for the series circuit, $I = 10.0$ A for the parallel circuit.

What's more, the power transferred out of the circuit by each resistor is very different in the two circuits, because the electric potential difference and current for each resistor are greater for the parallel circuit than for the series circuit. If the resistors are light bulbs that convert electric potential energy to produce light, the lights will glow brighter in the parallel circuit than in the series circuit.

Use Equation 18-7 to find the electric potential difference across each resistor in the series circuit:

$$\Delta V_1 = IR_1 = (2.40 \text{ A})(2.00 \ \Omega) = 4.80 \text{ V}$$

$$\Delta V_2 = IR_2 = (2.40 \text{ A})(3.00 \ \Omega) = 7.20 \text{ V}$$

(compared to $\Delta V = \mathcal{E} = 12.0$ V for each resistor in the parallel circuit)

Use Equation 18-7 to find the current through each resistor in the parallel circuit:

$$I_1 = \frac{\mathcal{E}}{R_1} = \frac{12.0 \text{ V}}{2.00 \ \Omega} = 6.00 \text{ A}$$

$$I_2 = \frac{\mathcal{E}}{R_2} = \frac{12.0 \text{ V}}{3.00 \ \Omega} = 4.00 \text{ A}$$

(compared to $I = 2.40$ A for each resistor in the series circuit)

Extend: Note that the current is the same for both resistors in the series circuit, so more power is transferred out of the circuit by the resistor with the greater resistance ($P_2 = 17.3$ W by $R_2 = 3.00 \ \Omega$ versus $P_1 = 11.5$ W by $R_1 = 2.00 \ \Omega$). That follows from the relationship $P = I^2R$ for resistors. By contrast, the electric potential difference is the same for the two resistors in the parallel circuit, so more power is transferred out of the circuit by the resistor with the smaller resistance ($P_1 = 72.0$ W into $R_1 = 2.00 \ \Omega$ versus $P_2 = 48.0$ W into $R_2 = 3.00 \ \Omega$). That agrees with the relationship $P = \Delta V^2/R$ for resistors. So if the resistors are light bulbs, the light bulb with $R_2 = 3.00 \ \Omega$ is the brighter one in the series circuit, but the bulb with $R_1 = 2.00 \ \Omega$ is the brighter one in the parallel circuit! When comparing the power in resistors, choose wisely among the relationships $P = I\Delta V$, $P = I^2R$, and $P = \Delta V^2/R$.

NOW WORK Problems 4, 5, 7, 9, and 10 from The Takeaway 18-5.

We've seen how to apply Equation 18-13, $P = I\Delta V$, and Equations 18-14, $P = I^2R = \Delta V^2/R$, to DC circuits in which there is a steady current that does not vary with time. But these same equations also apply to circuits in which the current is *not* constant. We will address these latter types of circuits in the next section.

AP® Exam Tip

The power transferred into a circuit by a battery or other source of emf is not a fixed quantity dependent only on the battery; it depends also on what is hooked up to the battery and how it is hooked up. It is a common misconception to believe that the more elements hooked up to a battery, the more power is used. This is not necessarily true, depending on how they are hooked up. For instance, the same resistors connected in parallel will always transfer more energy out of the circuit in the same time than if they were connected in series.

THE TAKEAWAY for Section 18-5

✔ Power is the rate at which energy is transferred into or out of a system or converted from one type to another within a system. The unit of power is the watt (1 W = 1 J/s).

✔ The power into, out of, or converted by any circuit element equals the current through the element multiplied by the electric potential difference across the element.

✔ The power into a resistor is proportional to the square of the current through the resistor or, equivalently, the square of the electric potential difference across the resistor.

✔ For identical light bulbs, the brightness of a bulb depends on the rate electrical energy is being converted to light by the bulb, so indicates the power.

Prep for the AP® Exam

AP® Building Blocks

1. Confirm that each of the following representations of power P in terms of current I, electric potential difference ΔV, and resistance R are dimensionally consistent. ($1\ \Omega = 1\ \text{J} \cdot \text{s/C}^2$.)

$$P = I\Delta V \qquad P = I^2R \qquad P = \frac{\Delta V^2}{R}$$

EX 18-9 2. A 4.0-Ω resistor is connected to a 12-V potential difference source. Calculate the rate at which energy is transferred out of the circuit by the resistor.

EX 18-9 3. Assume the current in a 40.0-Ω high-voltage, DC transmission line is 1200 A. Calculate the rate at which electric potential energy would be converted to thermal energy due to resistance.

EX 18-10 4. Two resistors are connected in parallel in a circuit with a single emf. The emf is 9.0 V and the resistances are 1.5 and 2.0 Ω.

(a) Draw a diagram of the circuit, and calculate the equivalent resistance of the circuit.
(b) Calculate the current through the battery.
(c) Calculate the current in each resistor.
(d) Calculate the rate of energy dissipation through each of the resistors and the total power provided by the emf to the circuit. Explain how a comparison of these values could provide an experimental test of the conservation of energy.
(e) How could some combination of measurements of the values you calculated above serve to test the law of conservation of charge?

AP® Skill Builders

EX 18-10 5. Three resistors are connected to a battery with a potential difference of $\Delta V_{bat} = 9.00$ V, as shown. The rate of energy dissipation in each resistor is 1.50 W. The internal resistance in the battery is negligible.

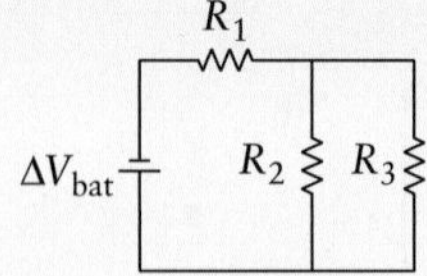

(a) Justify the claims that (i) the potential differences across the two parallel resistors must be equal and (ii) the resistances are equal, $R_2 = R_3$.
(b) Calculate the current through the battery from the total rate of energy dissipation by the circuit.
(c) Apply charge conservation to calculate the current in each resistor.
(d) Calculate the resistance of each resistor.
(e) Evaluate the equivalent resistance of the circuit to test the preceding results.

6. A voltmeter is designed to measure the potential difference across an element in a circuit. An ammeter is designed to measure the current at a point in a circuit. These measurement tools should change the measured property of the circuit as little as possible.

(a) Justify the claim that the voltmeter should be connected in parallel with a resistor to measure the potential difference across the resistor and that the internal resistance of the voltmeter should be very large.
(b) Justify the claim that the ammeter should be connected in series with a resistor to measure the current through the resistor and that the internal resistance of the ammeter should be very small.

AP® Skills in Action

EX 18-10 7. A 10.0-V battery is connected in series with two resistors, $R_1 = 10.0\ \Omega$ and $R_2 = 40.0\ \Omega$, as shown.

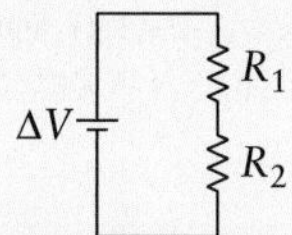

(a) Calculate the equivalent resistance of the circuit.
(b) Predict the current in each of the resistors.
(c) Calculate the potential difference across each resistor.
(d) Calculate the rate energy is transferred out of the circuit (converted to other types of energy such as warming of the environment and light) by each resistor.
(e) Test your calculation of the power for each resistor by calculating the total power transferred to the circuit.

EX 18-9 8. A stereo speaker has a resistance of 8.00 Ω. The power output is 40.0 W. Calculate the current passing through the speaker wires.

EX 18-10 9. When connected in parallel across a 120-V source, two light bulbs convert energy at a rate of 60 and 120 W, respectively. What powers do the light bulbs convert if instead they are connected in series across the same source? Assume the resistance of each light bulb is constant.

EX 18-10 10. A battery with potential difference $V_b = 12.0$ V is connected to resistors $R_1 = 8.00\ \Omega$, $R_2 = 10.0\ \Omega$, and $R_3 = 12.0\ \Omega$, as shown in the figure below. Calculate the total power provided by the battery and the power dissipated by each of the three resistors.

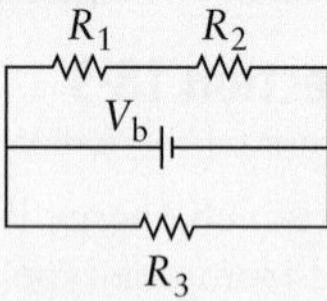

WATCH OUT !

When currents are changing, there are effects we can ignore, for now.

For now, we are ignoring magnetic effects, and usually it is safe to do so. As we go on, we will find that the magnetic flux linking a circuit will cause a potential difference and any changing current generates a changing magnetic field. When a capacitor is charged, a changing magnetic field exists which changes the potential difference in that region of the circuit. For circuits containing only resistors and batteries at steady state the magnetic field is constant, so the change in electric potential around any closed path is zero. The quick change in current if you quickly unplug a lit lamp from the wall can cause a visible spark across the plug prongs, due to this effect.

18-6 A circuit containing a resistor and a capacitor has a current that varies with time

In an ordinary DC circuit, the battery provides a steady current and delivers energy to the other circuit elements at a steady rate. In some circuits, however, what's required is a short burst of energy. That's the case for the electronic flash in a camera or mobile phone (Figure 18-1c): Energy has to be delivered to the flash lamp in a very brief time interval to produce a short-duration, high-intensity flash.

The simplest way to deliver a short burst of electric potential energy to a circuit element is by using a *capacitor* as an energy source. (We discussed capacitors in Chapter 17.) We'll first examine a circuit in which a battery is used to charge a capacitor, then see what happens when the charged capacitor is discharged and its stored energy is delivered to a resistor.

A Series *RC* Circuit: Charging the Capacitor

Figure 18-16a shows a **series *RC* circuit**, with a resistor of resistance R and a capacitor of capacitance C connected in series to a battery of emf $\mathcal{E}$. (We will assume that the internal resistance of the battery is so small compared to R that it can be ignored.)

Initially the capacitor in Figure 18-16a is uncharged. If the switch in this circuit is moved to the up position (**Figure 18-16b**), the circuit is completed and the battery will cause charge to begin to flow. We'll call $t = 0$ the time when the switch is thrown. As time passes, positive charge $+Q$ builds up on the upper capacitor plate, and an equal amount of negative charge $-Q$ builds up on the lower capacitor plate. (The charge on the two plates must be of equal magnitude since at any instant the current is the same throughout the circuit. Hence positive charge leaves the initially uncharged lower plate at the same rate that it arrives at the initially uncharged upper plate, leaving as much negative charge on the lower plate as there is positive charge on the upper plate.) So the circuit in Figure 18-16b is a *charging* series RC circuit.

Let's apply Kirchhoff's loop rule to the circuit shown in Figure 18-16b. In Chapter 17 we found that the magnitude of the charge q on the capacitor plates is proportional to the voltage V across the capacitor, $Q = C\Delta V$, Equation 17-14.

We can rewrite Equation 17-14 as $\Delta V = Q/C$. If we start at point p in Figure 18-16b and move clockwise around the circuit, we encounter a potential increase of $+\mathcal{E}$ as we cross the source of emf, a potential drop $-IR$ as we cross the resistor, and a potential drop $-Q/C$ as we cross the capacitor from the positive plate to the negative plate (which is at lower potential). Kirchhoff's loop rule tells us that the sum of these potential differences must be zero:

$$\mathcal{E} + (-IR) + \left(-\frac{Q}{C}\right) = 0 \tag{18-15}$$

If we solve Equation 18-15 for the current i in the circuit, we get

$$I = \frac{\mathcal{E}}{R} - \frac{Q}{RC} \tag{18-16}$$

Equation 18-16 tells us that the current in the circuit of Figure 18-16b *cannot* be constant. As charge builds up on the capacitor, the value of Q increases. Hence the

(a) The position of this switch determines whether the source of emf $\mathcal{E}$ is part of the circuit.

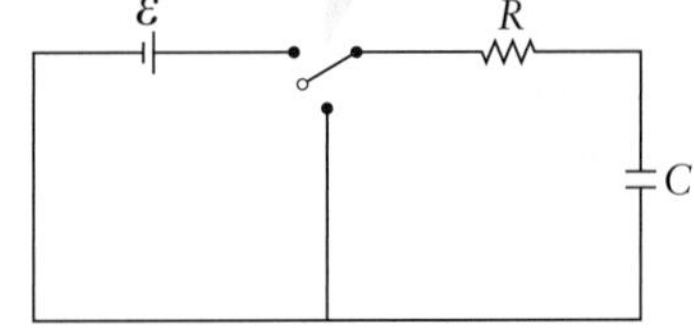

(b) With the switch in the up position, the emf causes a current. This current charges the capacitor, so Q increases.

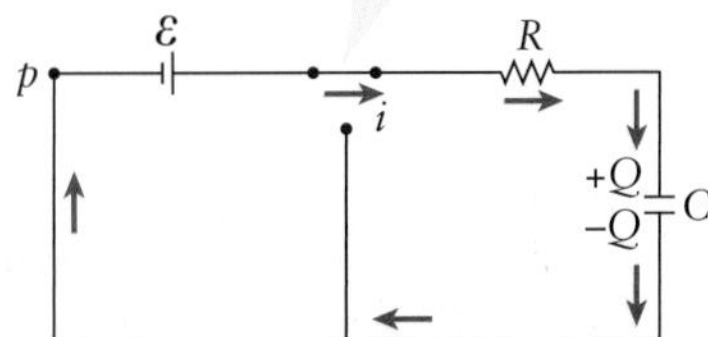

Figure 18-16 Charging a series *RC* circuit (a) In this circuit a resistor and an initially uncharged capacitor are connected in series to a source of emf. (b) The switch is moved to the up position at $t = 0$, beginning the charging process.

right-hand side of Equation 18-16 decreases with time, so the current I must decrease with time as well. What's happening is that the potential difference across the capacitor increases as the charge increases. This potential difference opposes that of the source of emf, so the current decreases. The current is what's causing the charge to increase, so Q increases at an ever-slower rate as time goes by.

We'd like to know the capacitor charge Q and the current I as functions of time. Solving Equation 18-16 for these functions is a problem in calculus that's beyond our scope. Instead, we'll present the solutions and see that they make sense:

EQUATION IN WORDS
Capacitor charge and current in a charging series *RC* circuit

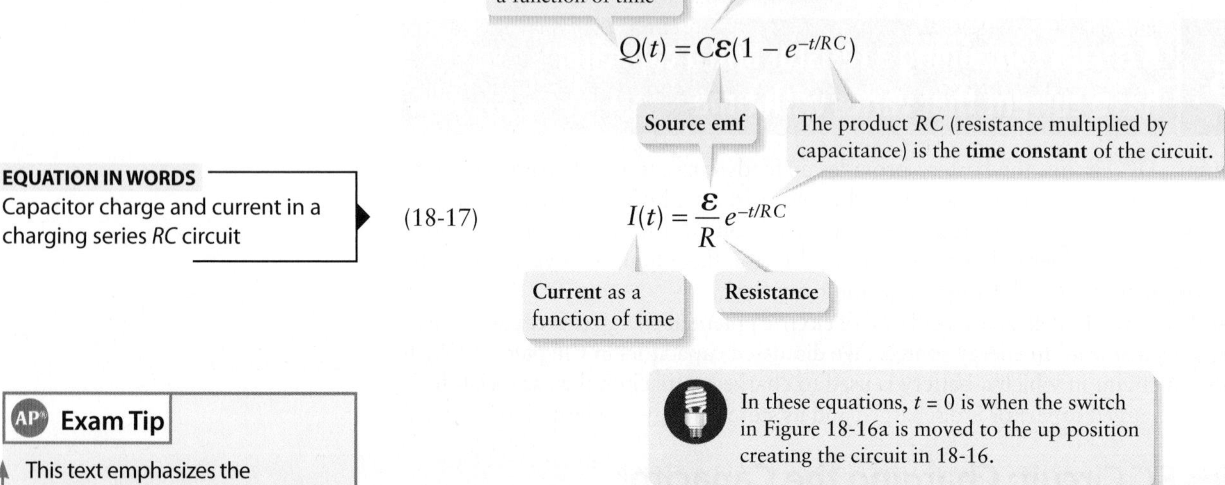
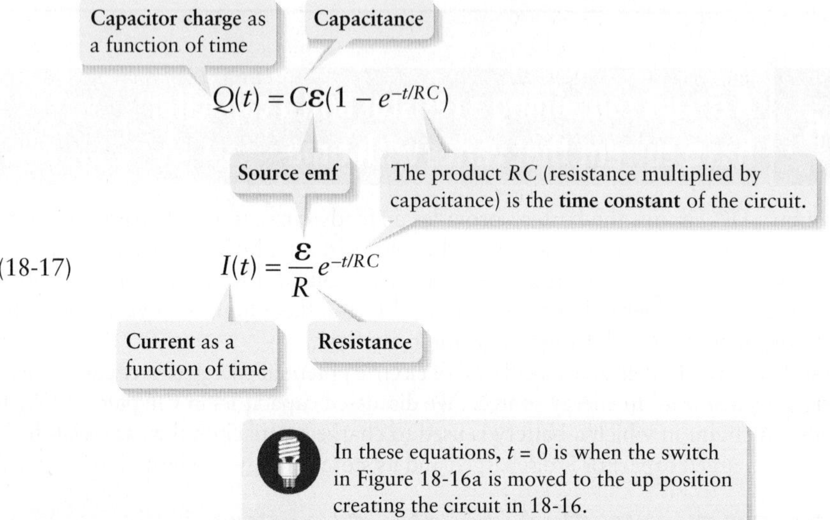

Capacitor charge as a function of time
Capacitance
Source emf
The product RC (resistance multiplied by capacitance) is the time constant of the circuit.
Current as a function of time
Resistance

$$Q(t) = C\mathcal{E}(1 - e^{-t/RC})$$

$$I(t) = \frac{\mathcal{E}}{R}e^{-t/RC} \quad (18\text{-}17)$$

In these equations, $t = 0$ is when the switch in Figure 18-16a is moved to the up position creating the circuit in 18-16.

Exam Tip

This text emphasizes the dependence of the time constant on the properties of the circuit, so the time constant is always written *RC*. The AP® Physics 2 equation sheet uses τ to represent the time constant, $\tau = RC$.

Figure 18-17 graphs the charge Q and current I as given by Equations 18-17. These equations involve the **exponential function**, the irrational number $e = 2.71828...$ raised to a power. (Note that e is not the same as the magnitude of the charge on the electron, for which we unfortunately use the same symbol.) The number e has special properties: If a population has N_0 members and grows at a rate r per year (for example, if the population grows by

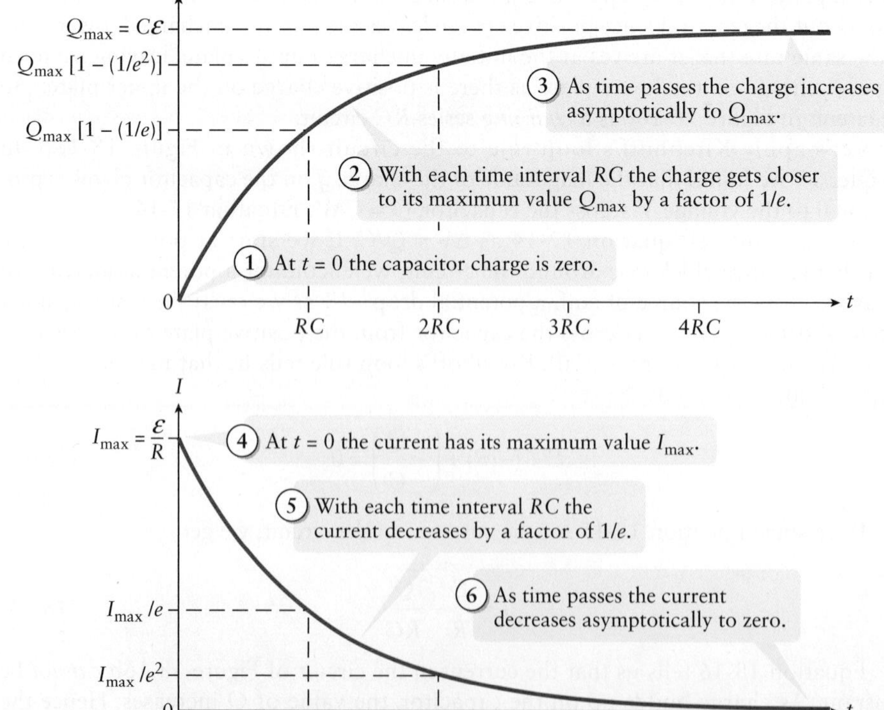

Q
Qmax = Cℰ
Qmax [1 – (1/e²)]
Qmax [1 – (1/e)]
3 As time passes the charge increases asymptotically to Qmax.
2 With each time interval RC the charge gets closer to its maximum value Qmax by a factor of 1/e.
1 At t = 0 the capacitor charge is zero.
0
RC
2RC
3RC
4RC
t
I
Imax = ℰ/R
4 At t = 0 the current has its maximum value Imax.
5 With each time interval RC the current decreases by a factor of 1/e.
6 As time passes the current decreases asymptotically to zero.
Imax /e
Imax /e²
0
RC
2RC
3RC
4RC
t

Figure 18-17 Charge and current in a charging series *RC* circuit Capacitor charge Q starts at zero and approaches a maximum value; current I starts at a maximum value and approaches zero.

2% per year, $r = 0.02$), then after t years the population will be $N(t) = N_0e^{rt}$. In Equations 18-17 the exponential function has a negative exponent, which means that this function decreases with time. In particular, $e^{-t/RC}$ decreases by a factor of $1/e = 0.36787...$ every time the quantity t/RC increases by 1—that is, whenever time t increases by RC. At $t = 0$ (when the switch in Figure 18-16 is closed), $e^{-t/RC} = e^0 = 1$; at $t = RC, e^{-t/RC} = e^{-1} = 1/e = 0.368$ to three significant digits; at $t = 2RC, e^{-t/RC} = e^{-2} = 1/e^2 = 0.135$; and so on.

The quantity RC is called the **time constant** of a series RC circuit. [Note that the product RC has units of ohms times farads. Since $1\ \Omega = 1$ V/A, 1 F $= 1$ C/V, and 1 A $= 1$ C/s, it follows that $1\ \Omega \cdot \text{F} = 1\ (\text{V/A})(\text{C/V}) = 1\ \text{C/A} = 1$ s. So the quantity RC does indeed have units of time.] The smaller the time constant, the more rapidly the charge Q approaches its maximum value Q_{max} and the more rapidly the current decreases to its final value of zero.

NEED TO REVIEW?

Turn to page M-6 in the Math Tutorial in the back of the book for more information on exponents and logarithms.

At $t = 0$ the capacitor is uncharged, so there is zero potential difference across the capacitor. As a result, at $t = 0$ the potential difference IR across the resistor equals the potential difference $\mathcal{E}$ across the source of emf, so $\mathcal{E} = IR$ and $I = I_{max} = \mathcal{E}/R$. After a very long time the current has dropped to zero, so the potential difference IR across the resistor is zero. Then the potential difference Q/C across the capacitor equals the potential difference $\mathcal{E}$ across the source, so $\mathcal{E} = Q/C$ and $Q = Q_{max} = C\mathcal{E}$.

A Series *RC* Circuit: Discharging the Capacitor

Some time after the switch in Figure 18-16b was thrown to the up position, the capacitor is charged to a charge Q_{max}. (This may be less than $C\mathcal{E}$, depending on how long the source has had to charge the capacitor.) We now throw the switch to the down position, as in **Figure 18-18**, and restart our clock so that $t = 0$ is the time when the switch is moved to the new position. The emf is no longer part of the circuit, so the positive charge on the upper plate of the capacitor is free to move counterclockwise around the circuit to cancel the negative charge on the lower plate. So the value of the capacitor charge Q decreases, and the electric potential energy stored in the capacitor is transferred into the resistor. We call this a *discharging* series RC circuit.

With the switch in the down position, the potential difference across the capacitor causes a current. This current discharges the capacitor, so q decreases.

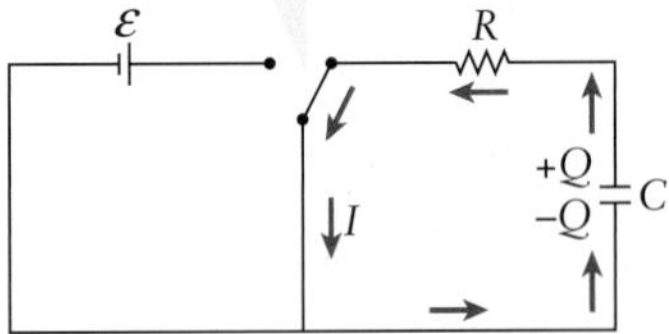

Figure 18-18 Discharging a series *RC* circuit When the switch in the circuit of Figure 18-16 is moved to the down position, the discharging process begins.

Kirchhoff's loop rule for the discharging circuit is now the same as Equation 18-15, but with the emf $\mathcal{E}$ removed:

$$(-IR) + \left(-\frac{Q}{C}\right) = 0 \tag{18-18}$$

Solving Equation 18-28 for the current i in the circuit gives

$$I = -\frac{Q}{RC} \tag{18-19}$$

The minus sign in Equation 18-19 means that the current is in the direction opposite to that in the charging RC circuit of Figure 18-16b. (You can see this in Figure 18-18.) As the capacitor discharges and the charge Q decreases, the current I will decrease in magnitude. The charge and current as functions of time are

Capacitor charge as a function of time

$$Q(t) = Q_{max}\,e^{-t/RC}$$

Capacitor charge at $t = 0$

The product RC (resistance multiplied by capacitance) is the **time constant** of the circuit.

$$I(t) = -\frac{Q_{max}}{RC}\,e^{-t/RC} \tag{18-20}$$

Current as a function of time

EQUATION IN WORDS

Capacitor charge and current in a discharging series *RC* circuit

In these equations, $t = 0$ is when the switch in Figure 18-18 is moved to the down position.

Figure 18-19 Charge and current in a discharging series *RC* circuit Capacitor charge *Q* starts at a maximum value and approaches zero; current *I* starts at a maximum (negative) value and approaches zero.

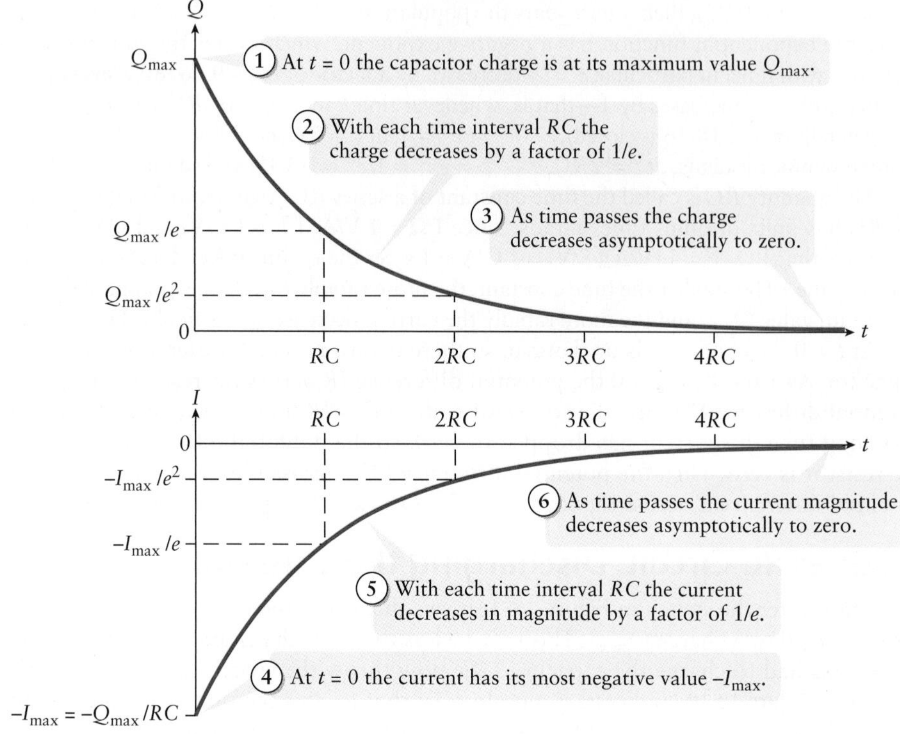

Q
Qmax
1 At t = 0 the capacitor charge is at its maximum value Qmax.
2 With each time interval RC the charge decreases by a factor of 1/e.
Qmax /e
3 As time passes the charge decreases asymptotically to zero.
Qmax /e2
0
t
RC
2RC
3RC
4RC
I
RC
2RC
3RC
4RC
0
t
–Imax /e2
6 As time passes the current magnitude decreases asymptotically to zero.
–Imax /e
5 With each time interval RC the current decreases in magnitude by a factor of 1/e.
4 At t = 0 the current has its most negative value –Imax.
–Imax = –Qmax /RC

The graphs in **Figure 18-19** show the charge Q and current I as given by Equations 18-20. Both functions are proportional to $e^{-t/RC}$, so both decrease by a factor of $1/e$ whenever time t increases by one time constant RC. So the value of RC determines how rapidly the capacitor charges *and* how rapidly it discharges.

EXAMPLE 18-11 A Charging Series *RC* Circuit

A 10.0-MΩ resistor is connected in series with a 5.00-μF capacitor. When a switch is thrown these circuit elements are connected to a 24.0-V battery of negligible internal resistance. The capacitor is initially uncharged. (a) What is the current in the circuit immediately after the switch is closed so that charging begins? (b) What is the charge on the capacitor once it is fully charged? (c) Find the capacitor charge, current, power provided by the battery, power taken in by the resistor, and power taken in by the capacitor at $t = 50.0$ s. (d) When the capacitor is fully charged, find the total energy that has been delivered by the battery and the total energy that has been delivered to the capacitor.

Set Up

We are given $R = 10.0\ \text{M}\Omega = 10.0 \times 10^6\ \Omega$, $C = 5.00\ \mu\text{F} = 5.00 \times 10^{-6}$ F, and $\mathcal{E} = 24.0$ V. Equations 18-17 tell us the capacitor charge and current at any time, including at $t = 0$ (when the switch is first closed) and $t \to \infty$ (long after the switch is closed, so the capacitor is fully charged). We'll use Equations 18-13 and 18-14 to find the power out of the battery and into the resistor and capacitor. (Equation 17-14 will help us in this.) To charge the capacitor to its maximum charge Q_{max}, the total charge that must pass through the battery is Q_{max}; we'll use this and Equation 17-6 to find the total energy delivered by the battery. Equation 17-17 allows us to find the total energy that is stored in the charged capacitor.

$\mathcal{E}$ = 24.0 V R = 10.0 MΩ switch closed at t = 0 C = 5.00 μF

Capacitor charge and current in a charging series RC circuit:

$$Q(t) = C\mathcal{E}(1 - e^{-t/RC})$$

$$I(t) = \frac{\mathcal{E}}{R}e^{-t/RC} \qquad (18\text{-}17)$$

Power for a circuit element:

$$P = I\Delta V \qquad (18\text{-}13)$$

Power for a resistor:

$$P = I^2R = \frac{\Delta V^2}{R} \qquad (18\text{-}14)$$

Charge, voltage, and capacitance for a capacitor:

$$Q = C\Delta V \qquad (17\text{-}14)$$

Electric potential difference related to electric potential energy difference:

$$\Delta V = \frac{\Delta U_{\text{electric}}}{q_0} \quad (17\text{-}6)$$

Electric potential energy stored in a capacitor:

$$U_{\text{electric}} = \frac{1}{2}Q\Delta V = \frac{1}{2}C\Delta V^2 = \frac{Q^2}{2C} \quad (17\text{-}17)$$

Solve

(a) Find the current at $t = 0$.

From the second of Equations 18-17, the current when the switch is first closed at $t = 0$ is

$$I(0) = \frac{\mathcal{E}}{R}e^{-(0)/RC} = \frac{\mathcal{E}}{R}e^0$$

Since any number raised to the power 0 equals 1, we have $e^0 = 1$ and

$$I(0) = I_{\text{max}} = \frac{\mathcal{E}}{R} = \frac{24.0 \text{ V}}{10.0 \times 10^6 \ \Omega}$$
$$= 2.40 \times 10^{-6} \text{ A} = 2.40 \ \mu\text{A}$$

(b) Find the capacitor charge long after the switch is closed ($t \to \infty$).

The first of Equations 18-17 tells us the capacitor charge $Q(t)$. As $t \to \infty$, the exponent $-t/RC \to -\infty$. Any number raised to the power $-\infty$ is zero, so

$$e^{-t/RC} \to 0$$
$$Q(t) = C\mathcal{E}(1 - e^{-t/RC}) \to Q_{\text{max}} = C\mathcal{E}(1-0) = C\mathcal{E}$$
$$= (5.00 \times 10^{-6} \text{ F})(24.0 \text{ V})$$
$$= 1.20 \times 10^{-4} \text{ C} = 0.120 \text{ mC}$$

(c) The time constant for this circuit is $RC = 50.0$ s, so we are actually being asked about the behavior of the circuit one time constant after the switch is closed. Use this to find charge Q, current I, and the power out of or into each circuit element.

The time constant for this circuit is

$$RC = (10.0 \times 10^6 \ \Omega)(5.00 \times 10^{-6} \text{ F}) = 50.0 \text{ s}$$

so at $t = 50.0$ s, $t/RC = (50.0 \text{ s})/(50.0 \text{ s}) = 1.00$

From Equations 18-17,

$$Q = C\mathcal{E}(1 - e^{-1.00}) = (5.00 \times 10^{-6} \text{ F})(24.0 \text{ V})(1 - 0.368)$$
$$= 7.59 \times 10^{-5} \text{ C} = 0.632Q_{\text{max}}$$

$$I = \frac{\mathcal{E}}{R}e^{-1.00} = \frac{24.0 \text{ V}}{10.0 \times 10^6 \ \Omega}(0.368)$$
$$= 8.83 \times 10^{-7} \text{ A} = 0.368I_{\text{max}}$$

The potential difference across the battery is $\mathcal{E} = 24.0$ V, so from Equation 18-13 the power the battery is delivering is

$$P_{\text{bat}} = I\mathcal{E} = (8.83 \times 10^{-7} \text{ A})(24.0 \text{ V})$$
$$= 2.12 \times 10^{-5} \text{ W} = 21.2 \ \mu\text{W}$$

From the first of Equations 18-14, the power into the resistor (the rate it is converting energy) is

$$P_R = I^2R = (8.83 \times 10^{-7} \text{ A})^2 (10.0 \times 10^6 \ \Omega)$$
$$= 7.80 \times 10^{-6} \text{ W} = 7.80 \ \mu\text{W}$$

Equation 17-14, $Q = C\Delta V$, tells us that the potential difference across the capacitor is $\Delta V = Q/C$. Combining this with

Equation 18-13 gives the power into the capacitor (the rate it is storing energy):

$$P_C = I\left(\frac{Q}{C}\right) = (8.83 \times 10^{-7}\ \text{A})\left(\frac{7.59 \times 10^{-5}\ \text{C}}{5.00 \times 10^{-6}\ \text{F}}\right)$$
$$= 1.34 \times 10^{-5}\ \text{W} = 13.4\ \mu\text{W}$$

Note that the net power into the resistor and capacitor combined equals the power out of the battery:

$$P_R + P_C = 7.80\ \mu\text{W} + 13.4\ \mu\text{W} = 21.2\ \mu\text{W} = P_{bat}$$

(d) Use the maximum charge stored by the capacitor, which is the total charge moved across the battery, and Equation 17-6 to calculate the change in electric potential energy imparted by the battery. Use Equations 17-7 to calculate the electric potential energy stored in the capacitor.

Long after the switch is closed, the total amount of charge that has passed through the battery and to the positive capacitor plate is $Q_{max} = 1.20 \times 10^{-4}$ C. From Equation 17-6 the potential energy change that was imparted by moving this charge across the 24.0-V emf of the battery is

$$\Delta U_{bat} = Q_{max}\mathcal{E} = (1.20 \times 10^{-4}\ \text{C})(24.0\ \text{V})$$
$$= 2.88 \times 10^{-3}\ \text{J} = 2.88\ \text{mJ}$$

The last of Equations 17-17 tells us the amount of energy that went into the capacitor to store charge Q_{max} there:

$$U_C = \frac{Q_{max}^2}{2C} = \frac{(1.20 \times 10^{-4}\ \text{C})^2}{2(5.00 \times 10^{-6}\ \text{F})}$$
$$= 1.44 \times 10^{-3}\ \text{J} = 1.44\ \text{mJ}$$

So exactly one-half of the energy provided by the battery is stored in the capacitor: $U_C = (1/2)\Delta U_{bat}$.

Reflect

Our results for charge Q and current I in part (c) agree with Figure 18-17: After one time constant the capacitor charge has reached $[1 - (1/e)] = 0.632 = 63.2\%$ of its fully charged value Q_{max}, and the current has decreased to $1/e = 0.368 = 36.8\%$ of its initial value I_{max}.

The power calculations in part (c) show that all of the energy delivered by the battery is accounted for: Part of the energy from the battery goes into the resistor, and the rest goes into adding to the electric potential energy stored in the capacitor. We've shown this for a specific instant, but it's true at *all* times during the charging process.

These results from part (c) also help us understand our calculations in part (d): Only one-half of the energy provided by the battery goes into the capacitor, so the other half must have gone into the resistor. This is a general result for charging *any* series RC circuit.

NOW WORK Problems 1, 2, 4, 5, and 7–9 from The Takeaway 18-6.

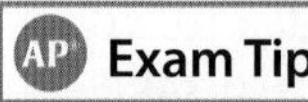

Exam Tip

Only very short and very long times are required for calculations with *RC* circuits. The examples in this section also include calculation of properties as one time constant. While you will not be asked to calculate properties at one time constant on the exam, it will help deepen your understanding of the behavior of the circuits for short and long times.

EXAMPLE 18-12 A Discharging Series *RC* Circuit

After the 5.00-μF capacitor in Example 18-11 has attained its full charge of 1.20×10^{-4} C, you disconnect it from the charging circuit and place it in the circuit shown below, in series with a 0.300-kΩ resistor and a switch. When the switch is thrown, the capacitor discharges and produces a current in the 0.300-kΩ resistor. (a) What is the current in the circuit immediately after the switch is closed, so that discharging begins? (b) Find the capacitor charge, the current, the rate at which electric potential energy is being converted to other types of energy by the resistor, and the rate at which the capacitor is providing electric potential energy at $t = 1.50 \times 10^{-3}\ \text{s} = 1.50$ ms.

Set Up

This situation is similar to Example 18-11, except that the capacitor is now discharging rather than charging and we are using a smaller resistance. We are given $R = 0.300\ \text{k}\Omega = 0.300 \times 10^3\ \Omega$, $C = 5.00\ \mu\text{F} = 5.00 \times 10^{-6}$ F, and the initial charge on the capacitor $Q_{max} = 1.20 \times 10^{-4}$ C. Equations 18-20 tell us the capacitor charge and current at any time, including at $t = 0$ (when the switch is first closed). As in Example 18-11, we'll use Equations 18-13, 18-14, and 17-14 to find the power for the capacitor and resistor.

Capacitor charge and current in a discharging series RC circuit:

$$Q(t) = Q_{max}e^{-t/RC}$$

$$I(t) = -\frac{Q_{max}}{RC}e^{-t/RC} \qquad (18\text{-}20)$$

Power for a circuit element:

$$P = I\Delta V \qquad (18\text{-}13)$$

Power for a resistor:

$$P = I^2R = \frac{\Delta V^2}{R} \qquad (18\text{-}14)$$

Charge, voltage, and capacitance for a capacitor:

$$Q = C\Delta V \qquad (17\text{-}14)$$

Solve

(a) Find the current at $t = 0$.

The second of Equations 18-20 gives us the current at $t = 0$ when the switch is first closed:

$$I(0) = -\frac{Q_{max}}{RC}e^{-(0)/RC} = -\frac{Q_{max}}{RC}e^{0}$$

We saw in Example 18-11 that $e^0 = 1$, so

$$I(0) = -\frac{Q_{max}}{RC} = -\frac{1.20 \times 10^{-4}\ \text{C}}{(0.300 \times 10^3\ \Omega)(5.00 \times 10^{-6}\ \text{F})}$$
$$= -8.00 \times 10^{-2}\ \text{A} = -0.0800\ \text{A}$$

The minus sign means that the current in this *discharging RC* circuit is directed as in Figure 18-18, which is opposite to the direction in a *charging RC* circuit.

(b) The time $t = 1.50$ ms of interest is equal to the time constant for this circuit, $RC = 1.50$ ms. So, we are being asked about the properties of the circuit (charge Q, current I, and power out of or into each circuit element) one time constant after the switch is closed.

The time constant for this circuit is

$$RC = (0.300 \times 10^3\ \Omega)(5.00 \times 10^{-6}\ \text{F}) = 1.50 \times 10^{-3}\ \text{s} = 1.50\ \text{ms}$$

so at $t = 1.50$ ms, $t/RC = (1.50\ \text{ms})/(1.50\ \text{ms}) = 1.00$. From Equations 18-20,

$$Q = Q_{max}e^{-1.00} = (1.20 \times 10^{-4}\ \text{C})(0.368)$$
$$= 4.41 \times 10^{-5}\ \text{C} = 0.368Q_{max}$$

$$I = -\frac{Q_{max}}{RC}e^{-1.00} = -\frac{1.20 \times 10^{-4}\ \text{C}}{(0.300 \times 10^3\ \Omega)(5.00 \times 10^{-6}\ \text{F})}(0.368)$$
$$= -2.94 \times 10^{-2}\ \text{A} = 0.368I_{max}$$

The first of Equations 18-14 tells us the rate at which the resistor is converting energy (the power into the resistor):

$$P_R = I^2R = (-2.94 \times 10^{-2}\ \text{A})^2\,(0.300 \times 10^3\ \Omega)$$
$$= 0.260\ \text{W}$$

From Equation 17-14, the potential difference across the capacitor is $\Delta V = Q/C$. Combine this with Equation 18-13 to find the rate at which the capacitor is providing energy (the power out of the capacitor):

$$P_C = I\left(\frac{Q}{C}\right) = (-2.94 \times 10^{-2}\ \text{A})\left(\frac{4.41 \times 10^{-5}\ \text{C}}{5.00 \times 10^{-6}\ \text{F}}\right)$$
$$= -0.260\ \text{W}$$

The minus sign means that energy is leaving the capacitor at a rate of 0.260 W = 0.260 J/s.

Reflect

In agreement with Figure 18-18, after one time constant the current and capacitor charge have both decreased to $1/e = 0.368 = 36.8\%$ of their initial values.

Note that as in Example 18-11, all of the energy in the circuit is accounted for: At $t = 1.50$ ms, the rate at which energy is being provided by the capacitor has the same magnitude (0.260 W) as the rate at which energy is being converted to other types by the resistor. The same will be true at any time after the switch is closed.

If you compare this example with Example 18-11, you'll see that the initial current $I(0)$ is *much* greater in magnitude for the discharging RC circuit (0.0800 A) than for the charging circuit (2.40 μA). Likewise the power into the resistor after one time constant for the discharging RC circuit (0.260 W) is much greater than for the charging RC circuit (7.80×10^{-6} W). That's because we used a much smaller resistance for the discharging circuit (0.300 kΩ) than for the charging circuit (10.0×10^{6} Ω), which made the time constant RC much smaller (1.50×10^{-3} s for the discharging circuit compared to 50.0 s for the charging circuit) and so allowed the capacitor to discharge much more rapidly than it charged. Rapid discharge is important in many capacitor applications where a powerful surge of energy is required, such as in an electronic flash unit for photography (see Figure 18-1c) or a cardiac defibrillator (see the photo that opens Chapter 17).

NOW WORK Problems 6 and 10 from The Takeaway 18-6.

THE TAKEAWAY for Section 18-6

✔ A circuit containing a resistor and a capacitor connected in series is a series RC circuit.

✔ Connecting the resistor and an uncharged capacitor in series with a source of emf causes a current that charges the capacitor. Switching the circuit to remove the source of emf, leaving the resistor in series with the capacitor in a complete circuit causes the capacitor to discharge through the resistor. In either case an exponential function describes how the current and the capacitor charge vary with time.

✔ The rate of charge or discharge in a series RC circuit depends on the time constant, equal to the product RC. In a time interval RC the current decreases by a factor of $1/e \approx 37\%$. In a charging circuit, this is the time in which the capacitor reaches $\approx (100 - 37)\% = 63\%$ of its final charge. In a discharging circuit, RC is the time in which the capacitor reaches $\approx 37\%$ of its initial charge.

✔ After a time much greater than the time constant, a charging capacitor and circuit branch in which it is located approach a steady-state condition: The capacitor approaches a state of being fully charged, the potential difference across the capacitor is at a maximum value and there is zero current in the circuit branch in which the capacitor is located.

✔ After a time much greater than the time constant, a discharging capacitor and circuit branch in which it is located approach a steady-state condition: The capacitor approaches a state of having zero charge, the potential difference across the capacitor approaches zero, and there is zero current in the circuit branch in which the capacitor is located.

Prep for the AP Exam

AP Building Blocks

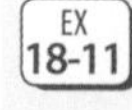
1. A 4.00-MΩ resistor and a 3.00-μF capacitor are connected in series with a power supply. What is the time constant for the circuit?

2. An ideal battery is connected in series with a resistor and a capacitor.
   (a) Does the time required to fully charge the capacitor through the given resistor with the battery depend on the potential difference of the battery?
   (b) Does it depend on the total amount of charge to be placed on the capacitor? Explain your answers.

3. Give a simple physical explanation for why the charge on a capacitor in an RC circuit can't be changed instantaneously.

4. A capacitor of 20.0 μF and a resistor of 1.00×10^{2} Ω are quickly connected in series to a battery of 6.00 V. What is the charge on the capacitor 0.00100 s after the connection is made?

5. A capacitor has a value of 160 μF. What is the resistance of the charging circuit if it takes 10.0 s to charge the capacitor to 80.0% of its maximum charge?

AP Skill Builders

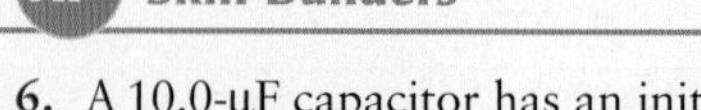
6. A 10.0-μF capacitor has an initial charge of 80.0 μC. A 25.0-Ω resistor is connected across it.
   (a) What is the initial current in the resistor?
   (b) What is the time constant of the circuit?

7. The current through a resistor R of 25 Ω is measured at time intervals of 500 μs after closing the switch that connects it to a 5-V battery through a capacitor C with an unknown value, as shown in the figure below. The magnitude of the current through the resistor is shown in the table on the next page. Graph the current as a function of time and determine the capacitance.

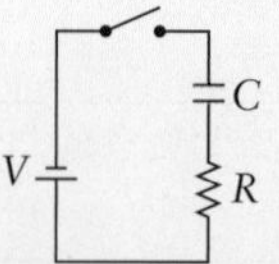

Time (s)	Current (A)
0	0.200
5.00×10^{-4}	0.109
1.00×10^{-3}	0.060
1.50×10^{-3}	0.032
2.00×10^{-3}	0.018
2.50×10^{-3}	0.010
3.00×10^{-3}	0.005
3.50×10^{-3}	0.003

EX 18-11 8. Early heart pacemakers used an RC circuit to stimulate the heart at regular intervals. A particular pacemaker is designed to operate at 70 beats per minute, and sends a pulse to the heart every time the potential difference across the capacitor reaches 8.0 V. If the RC circuit is powered by a 12-V battery and incorporates a 30-μF capacitor, what resistance is required for the RC circuit? Assume that the capacitor discharges completely, and instantly, when it sends the pulse to the heart.

AP® Skills in Action

EX 18-11 9. If we use a 2-V battery instead of a 1-V battery to charge the capacitor shown in the figure, the time constant will

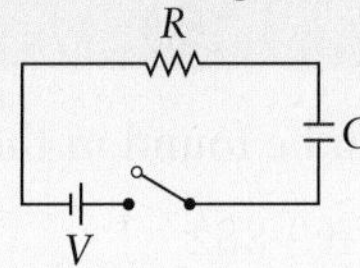

(A) be four times greater.
(B) double.
(C) remain the same.
(D) be half as much.

EX 18-12 10. A 3.0-μF capacitor is put across a 12-V battery. After it is fully charged the capacitor is disconnected and placed in series through an open switch with a 2.00×10^2-Ω resistor. (a) Determine the charge on the capacitor before it is discharged. (b) What is the initial current through the resistor when the switch is closed? (c) At what time will the current reach 37% of its initial value?

WHAT DID YOU LEARN?

Prep for the AP Exam

Chapter learning goals	Section(s)	Related example(s)	Relevant section review exercises
Define current and describe the relationships among electric potential difference, current, resistance, and resistivity for charge carriers moving in a wire.	18-2	18-1	1, 2, 3, 7, 9
	18-3	18-2	4, 8, 10
		18-3	1, 2
Calculate the resistance of a resistor and the current that a given electric potential difference produces in that resistor.	18-3	18-4	3, 4
	18-4	18-5	5, 6
Describe the equivalent resistance of multiple resistors connected in series and/or parallel.	18-4	18-6	3, 5
		18-7	2, 4
Discuss Kirchhoff's rules in terms of conservation of energy and conservation of charge, and apply them to simple and compound circuits containing batteries and resistors, including nonideal batteries and resistive wires.	18-1	18-7	1, 2
	18-4	18-8	4, 6
			8
Create a schematic diagram of a circuit from a description (or a description from a diagram) and use it to calculate unknown values of current in or potential difference across various segments or branches of the circuit or to predict how those values would change if the configuration of the circuit changed.	18-4	18-6	5
		18-7	4
		18-8	1, 7, 9
Describe how to measure current and potential difference within a circuit.	18-4		6
	18-5		6
Describe the transfer of energy into, out of, or within a circuit in terms of power.	18-5	18-9	2, 3, 8
		18-10	4, 5, 7, 9, 10
Describe the behavior of a circuit containing combinations of resistors and capacitors.	18-6	18-11	1, 2, 4, 5, 7, 8, 9
		18-12	6, 10

Chapter 18 Review

Key Terms

All the Key Terms can be found in the Glossary/Glosario on page G1 in the back of the book.

alternating current (AC) 882
ampere 877
battery 876
circuit 877
circuit branch 894
circuit element 890
current 877
direct current (DC) 882
drift speed 879
equivalent resistance 893
exponential function 914
ideal battery 893
internal resistance 891
junction 894
Kirchhoff's junction rule 894
Kirchhoff's loop rule 892
multiloop circuit 894
ohm 884
parallel (resistors) 894
power 905
radiate 906
resistance 884
resistivity 884
resistor 885
series RC circuit 913
series (resistors) 893
single-loop circuit 891
source of emf 875
time constant 915

Chapter Summary

Topic	Equation or Figure
Electric potential energy: Like the gravitational force, the electric force is a conservative force and has an associated potential energy.	(see figure below) (Figure 18-3)
Current: Electric charge moves around a circuit in response to an electric potential difference, such as that provided by a battery (a source of emf). Current is the rate at which charge moves through the circuit. In ordinary metals the mobile charges are negatively charged electrons, but it's conventional to take the direction of the current to be the direction in which positively charged objects would move.	$I = \frac{\Delta q}{\Delta t}$ (18-1)

① Current I is present in the wire of cross-sectional area A.

② Within the wire there are n charge carriers per unit volume. Each has a charge e.

③ The charge carriers drift through the wire with speed v_{drift}, and travel a distance Δx in a time $\Delta t = \Delta x/v_{drift}$.

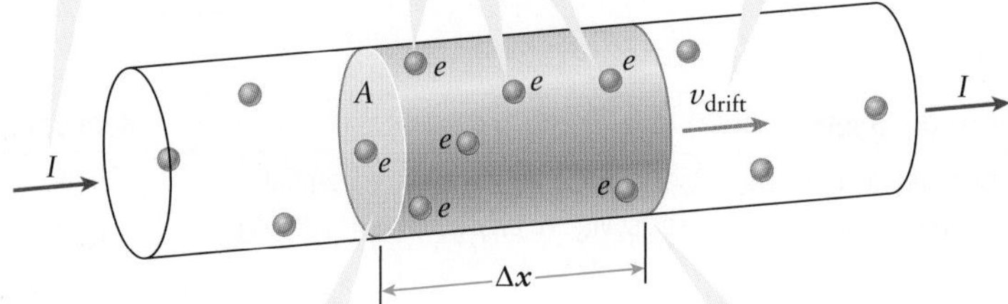

④ A length Δx of the wire (shown in green) has volume $A\Delta x$. This length contains $n(A\Delta x)$ charge carriers, with total charge $\Delta q = n(A\Delta x)e$.

⑤ The current in the wire equals the amount of charge Δq that passes this point divided by the time Δt it takes them to pass this point:

$$I = \frac{\Delta q}{\Delta t} = \frac{n(A\Delta x)e}{\Delta x/v_{drift}} = nAev_{drift}$$

(Figure 18-3)

Current in a circuit

Amount of **charge** that flows past a certain point in the circuit

$$I = \frac{\Delta q}{\Delta t} \qquad (18\text{-}1)$$

Time required for that amount of charge to flow past that point

Resistance: Resistivity is a measure of how difficult it is for charge to move through a material. The resistance of a wire or circuit element made from a uniform material depends on the dimensions of the wire or element as well as the resistivity of the material of which it is made. Resistance and resistivity are not constants but vary with temperature. A resistor is a circuit element whose most important property is its resistance.

Resistance of a wire

Resistivity of the material of which the wire is made

$$R = \frac{\rho L}{A} \quad (18\text{-}6)$$

Length of the wire

Cross-sectional area of the wire

Potential difference, current, and resistance: For current to exist in a conductor with resistance, there must be a potential difference (sometimes referred to as voltage) between the ends of the conductor. For a given resistance, a greater potential difference produces a greater current.

Potential difference between the ends of a wire

$$\Delta V = IR \quad (18\text{-}7)$$

Current in the wire

Resistance of the wire

Rules for circuits: Kirchhoff's loop rule is based on conservation of energy and states that the sum of the changes in electric potential around a closed loop in a circuit must equal zero. Kirchhoff's junction rule is based on conservation of charge and states that the sum of the currents into a junction equals the sum of the currents out of it. These rules allow us to analyze the currents and potential differences in electric circuits.

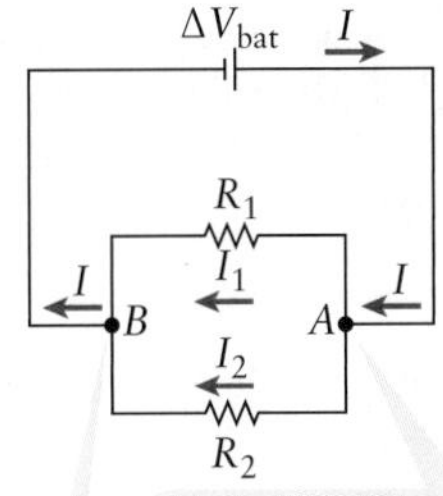

(1) At junction A current I splits into current I_1 and current I_2.

(2) At junction B current I_1 and current I_2 recombine to current I.

(Figure 18-10)

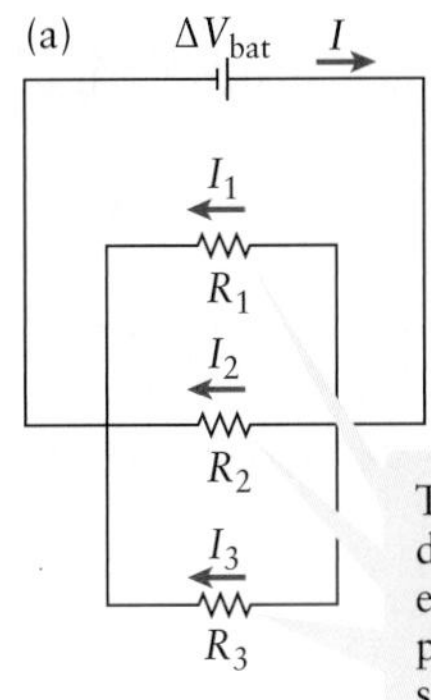

The potential difference across each resistor in parallel is the same.

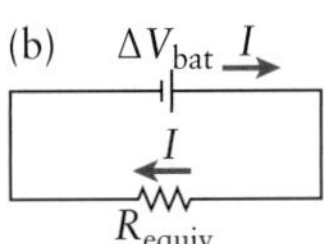

- The current I through R_{equiv} is the same as through the combination of resistors in parallel.
- The potential difference ΔV_{bat} across R_{equiv} is the same as that across each of the resistors in parallel.

(Figure 18-11)

Resistors in series and parallel: A collection of resistors connected in a circuit behaves as though it were a single resistor, with an equivalent resistance R_{equiv}. The equivalent resistance is different for resistors in series (all of which carry the same current) than for resistors in parallel (all of which have the same potential difference across them).

Equivalent resistance of N resistors in series

Resistances of the individual resistors

$$R_{equiv} = R_1 + R_2 + R_3 + \ldots + R_N \quad (18\text{-}9)$$

The equivalent resistance of N resistors in series is the sum of the individual resistances.

The reciprocal of the equivalent resistance of N resistors in parallel is the sum of the reciprocals of the individual resistances.

$$\frac{1}{R_{\text{equiv}}} = \frac{1}{R_1} + \frac{1}{R_2} + \frac{1}{R_3} + \ldots + \frac{1}{R_N} \qquad (18\text{-}11)$$

Equivalent resistance of N resistors in parallel

Resistances of the individual resistors

(resistors in parallel)

Power in circuits: Power is the rate of energy transfer or conversion. In an electric circuit, electric potential energy is transferred into the circuit by the source of emf and out of the circuit (converted into other types) by resistors. The power transferred out of the circuit by a resistor can be expressed in terms of the resistance and either the current through the resistor or the potential difference across the resistor.

Power produced by or transferred into or out of a circuit element

$$P = I\Delta V \qquad (18\text{-}13)$$

Current through the circuit element

Electric potential difference across the circuit element

Source of emf: Power is the rate at which other types of energy are converted into electric potential energy, a transfer of energy into the circuit.
Resistor: Power is the rate at which electric potential energy is converted into other types of energy, a transfer of energy out of the circuit.

Power for a resistor

Current through the resistor

$$P = I^2R = \frac{\Delta V^2}{R} \qquad (18\text{-}14)$$

Electric potential difference across the resistor

Resistance of the resistor

Series *RC* circuits: If a resistor R and capacitor C are connected in series to a source of emf, the capacitor charge increases toward a maximum value. The current in the circuit begins with a large value and gradually decreases to zero as the potential difference across the capacitor increases to the value of the emf. If the source of emf is taken out of the circuit and the circuit is reconnected, the capacitor discharges through the resistor. The current in the circuit now is caused by the excess charge separated on the capacitor redistributing until the potential difference across the capacitor reaches zero. So the current again starts out at its maximum and decreases to zero. The rate of charging or discharging depends on the time constant of the circuit, equal to the product RC.

(b) With the switch in the up position, the emf causes a current. This current charges the capacitor, so Q increases.

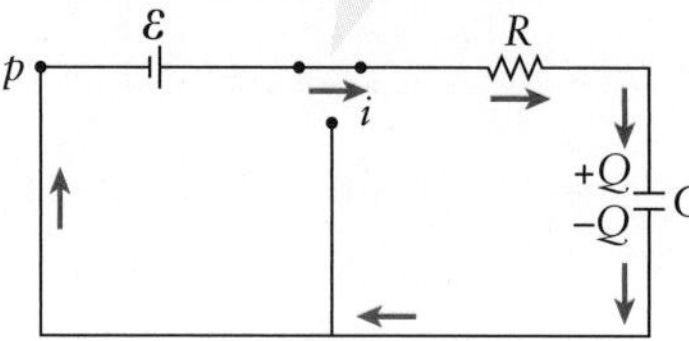

(Figure 18-16b)

With the switch in the down position, the potential difference across the capacitor causes a current. This current discharges the capacitor, so q decreases.

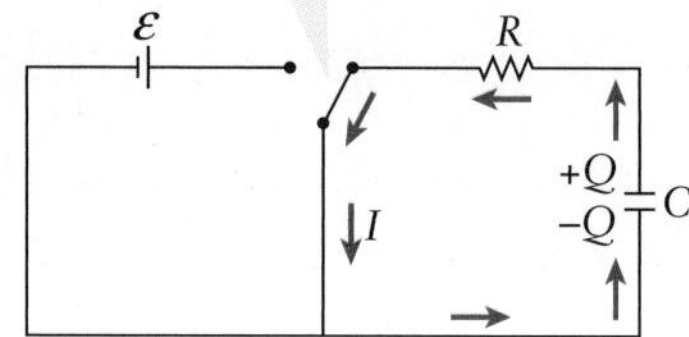

(Figure 18-18)

Chapter 18 Review Problems

1. The average drift speed of electrons in a wire carrying a steady current is constant even though the electric field within the wire is doing work on the electrons. What happens to this energy?

2. Under ordinary conditions the drift speed of electrons in a metal is around 10^{-4} m/s or less. Why doesn't it take a long time for a light bulb to come on when you flip the wall switch that is several meters away?

3. A particular light bulb requires a current of 0.50 A to emit a normal amount of light. If the light is left on for 1.0 h, how many electrons pass through the bulb?

4. An ammeter measures the current through a particular circuit element.
   (a) How should it be connected with that element, in parallel or in series?
   (b) Should an ammeter have a very large or a very small resistance? Why?

5. How much chemical energy must be converted to electric potential energy to move 2.0 C of positive charge from the negative terminal of a 9.0-V battery to the positive terminal (how much energy is transferred from the battery to the circuit)?

6. Find the equivalent resistance of the combination of resistors shown in the figure.

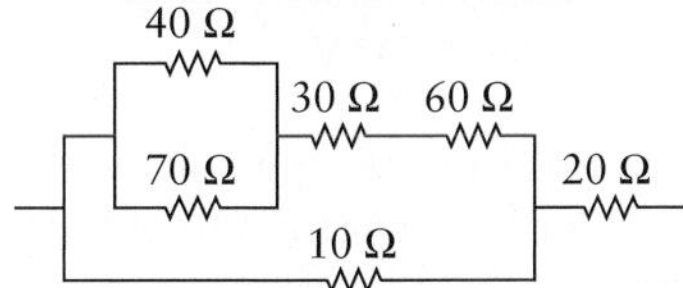

7. If 1.2×10^8 potassium ions, K^+, pass through a cell membrane in 4.00×10^{-2} s:
   (a) Calculate the electric current.
   (b) Identify the direction of the current relative to the motion of the potassium ions.
   (c) Calculate the electric current if the same number of Ca^{2+} ions were pumped over a membrane in the same time interval.

8. If a copper wire has a resistance of 2.00 Ω and a diameter of 1.00 mm, how long is it?

9. The resistance ratio of two conducting wires that have equal cross-sectional areas and equal lengths is 1:3. What is the ratio of the resistivities of the materials from which they are made?

10. An 8.00-m-long wire has a resistance of 4.00 Ω. The wire is uniformly stretched to a length of 16.0 m. Find the resistance of the wire after it has been stretched.

11. In the figure, a potential difference of 5.00 V is applied between points a and b. Analyze this circuit segment, including the total equivalent resistance, the total current, the current in each resistor, and the power for each resistor.

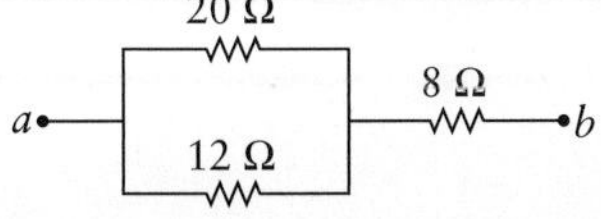

12. Lightning bolts can carry as much as 30 C of charge and can travel between a cloud and the ground in around 100 ms. Potential differences have been measured as high as 400 million V.
    (a) Calculate the average current in such a lightning strike.
    (b) Calculate the resistance of the air during such a lightning strike.
    (c) Calculate the total energy delivered during such a strike.
    (d) Approximately 100 strikes from cloud to ground occur on Earth every second. Suppose each lightning strike is as described in the introduction to this problem and calculate the maximum power transferred to Earth by lightning.
    (e) Electric energy end use in the United States in the year 2017 was approximately 1.38 billion kWh. Calculate the number of minutes required per year for lightning strikes with maximum power to deliver the equivalent of the 2017 U.S. electric energy end use.

13. Two resistors, A and B, are connected in series to a 6.0-V battery; the potential difference across resistor A is 4.0 V. When A and B are connected in parallel across a 6.0-V battery, the current through B is 2.0 A. The battery has negligible internal resistance. Calculate the resistances of A and B.

14. A metal wire of resistance 48 Ω is cut into four equal pieces that are then connected side by side to form a new wire, which is one-quarter of the original length. Calculate the resistance of the new wire.

15. A potential difference of 3.6 V is applied between points a and b in the figure. Analyze the circuit segment, including the current through each resistor and the potential difference across each resistor.

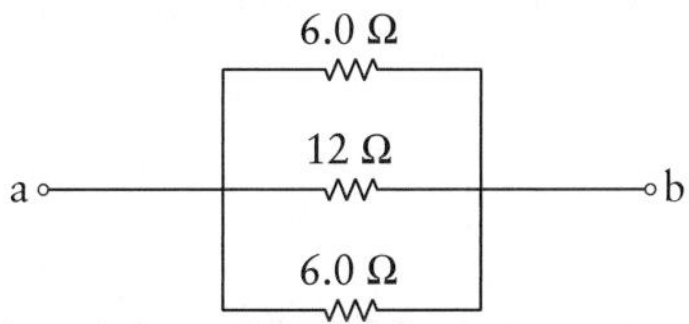

16. A potential difference of 7.50 V is applied between points a and c in the figure. Analyze the circuit segment, including the current through each resistor and the potential difference across each resistor.

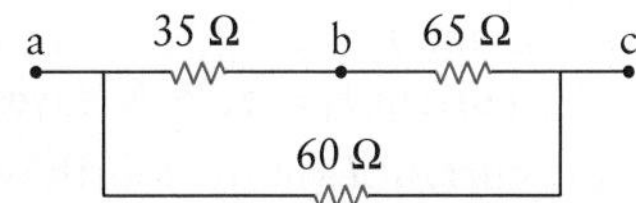

17. When a 120-V potential difference is placed across the filament of a 75-W light bulb, the current through the filament is 0.63 A. When a potential difference across the same filament is changed to 3.0 V, the current is 0.086 A. Is the filament made of an ohmic material? Explain your answer.

18. Three resistors are connected to a battery with an emf of $\Delta V_{bat} = 3.00$ V, as shown in the figure. The power transferred out of the circuit by each resistor is 0.50 W. What is the resistance of each resistor?

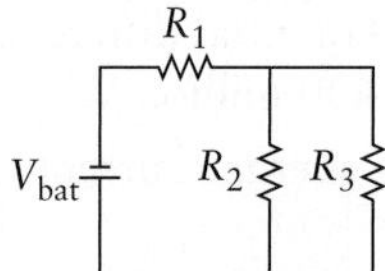

19. How long is a tungsten wire if it has a diameter of 0.150×10^{-3} m and a resistance of 3.00 Ω? The resistivity of tungsten is $5.62 \times 10^{-8}\ \Omega \cdot \text{m}$.

20. A battery with potential difference $V_{bat} = 12.0$ V is connected to resistors $R_1 = 8.00\ \Omega$, $R_2 = 10.0\ \Omega$, and $R_3 = 12.0\ \Omega$, as shown in the figure. Calculate the total power transferred into the circuit by the battery (P_b) and the power transferred out of the circuit by each of the three resistors.

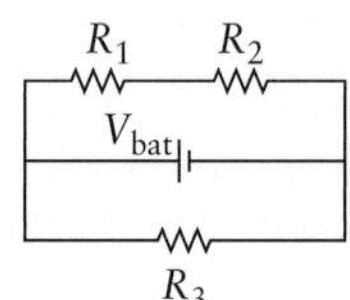

21. If your local power company charges \$0.11 per kW · h, what would it cost to run a 1500-W heater continuously during an 8.0-h night?

22. You need a 75-Ω resistor, but you have only a box of 50-Ω resistors on hand.
   (a) How can you make a 75-Ω resistor using the resistors you have on hand?
   (b) How could you use your 50-Ω resistors to make a 60-Ω resistor?

23. What is the resistance of a 50.0-m-long aluminum wire that has a diameter of 8.00 mm? The resistivity of aluminum is $2.83 \times 10^{-8}\ \Omega \cdot \text{m}$.

24. A power transmission line is made of copper that is 1.80 cm in diameter. What is the resistance of 1.00 mi of the line?

25. If a light bulb draws a current of 1.0 A when connected to a 12-V circuit, what is the resistance of its filament?

26. If a flashlight bulb has a resistance of 12.0 Ω, how much current will the bulb draw when it is connected to a 6.0-V circuit?

27. A light bulb has a resistance of 8.0 Ω and a current of 0.50 A. At what potential difference is it operating?

28. Four wires meet at a junction. In two of the wires, currents $I_1 = 1.25$ A and $I_2 = 2.50$ A enter the junction. In one of the wires, current $I_3 = 5.75$ A leaves the junction.
   (a) What is the current I_4 in the fourth wire?
   (b) Does I_4 enter or leave the junction?

29. The four resistors shown in the figure have an equivalent resistance of 8 Ω. Calculate the value of R_x.

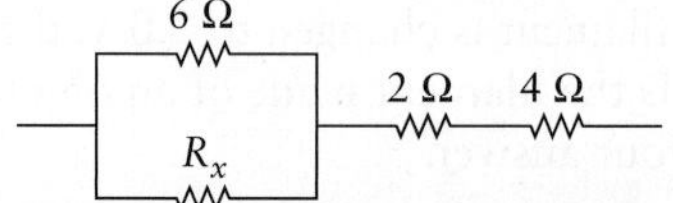

30. There is a current of 112 pA when a certain potential difference is applied across a certain resistor. When that same potential difference is applied across a resistor made of the identical material but 25 times longer, the current is 0.044 pA. Calculate the ratio of the diameter of the first resistor to that of the second.

31. Most of the resistance of the human body comes from the skin, as the interior of the body contains aqueous solutions that are good electrical conductors. For dry skin, the resistance between a person's hands is typically 500 kΩ. The skin is on average about 2.0 mm thick. We can model the body between the hands as a cylinder 1.6 m long and 14 cm in diameter with the skin wrapped around it. What is the resistivity of the skin?

32. A house is heated by a 24-kW electric furnace. The local power company charges \$0.10 per kW · h and the heating bill for January is \$218. How long must the furnace have been running on an average January day?

33. How much power is dissipated in each resistor shown in the figure?

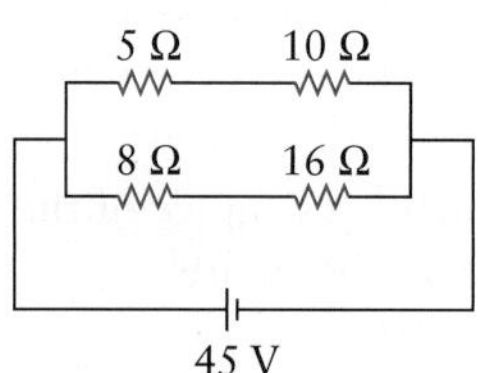

34. Giant electric eels can deliver a voltage shock of 500 V and up to 1.0 A of current for a brief time. A snorkeler in salt water has a body resistance of about 600 Ω. A current of about 500 mA can cause heart fibrillation and death if it lasts too long. **(a)** What is the maximum power a giant electric eel can deliver to its prey? **(b)** If the snorkeler is struck by the eel, what current will pass through her body? Is this large enough to be dangerous? **(c)** What power does the snorkeler receive from the eel?

35. Batteries release their energy rather slowly and are very damaging environmentally. Capacitors would be much cleaner for the environment and can be quickly recharged. Unfortunately they don't store much energy. **(a)** A new 1.5-V AAA battery has a "capacity" (*not* capacitance) of 1250 mA · h. What does this "capacity" actually represent? Express it in standard SI units. **(b)** How many joules of energy can be stored in the AAA battery? **(c)** At a steady current of 400 mA, how many hours will the AAA battery last? **(d)** How much energy can be stored in a typical 10-μF capacitor charged to a potential of 1.5 V? How does that compare to the energy stored in the AAA battery?

Prep for the AP Exam

AP® Group Work

Directions: The problem below is designed to be done as group work in class.

A space heater is plugged into a wall socket where the potential difference across the socket is 120 V. The rate at which the heater transfers energy out of the circuit is 1500 W.

(a) Calculate the current through the resistors converting electric potential energy into thermal energy in the heater and calculate the equivalent resistance of the heater.

As a precaution, when the current in a circuit exceeds a threshold, the resistor in a circuit breaker or a fuse can break the circuit. Whereas circuits in residential rooms other than the kitchen have threshold currents of 15 A, a kitchen or a workshop is often designed to allow a 20-A current.

(b) Justify the claim that the appliances and lights in a residence must be connected in parallel.

(c) A coffeemaker with a resistance of 15 Ω is in parallel with the space heater. Complete the following circuit diagram connecting the circuit elements so that when both switches are closed both appliances operate and when only one switch is closed only one of the appliances operates.

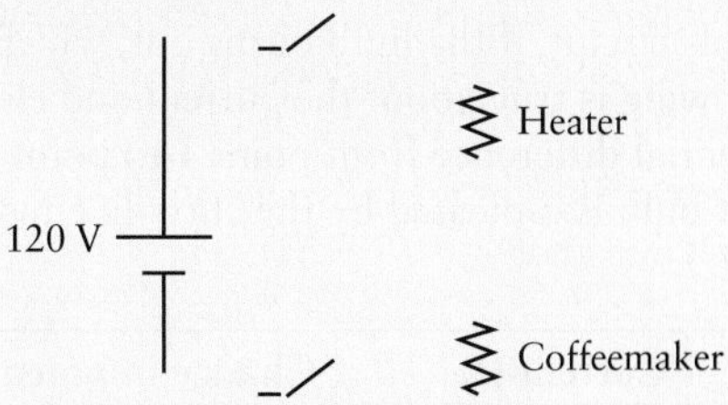

(d) Calculate the equivalent resistance and total current through the 120-V source of emf when both space heater and coffeemaker are switched on.

AP® PRACTICE PROBLEMS

Prep for the AP Exam

Multiple-Choice Questions

Directions: The following questions have a single correct answer.

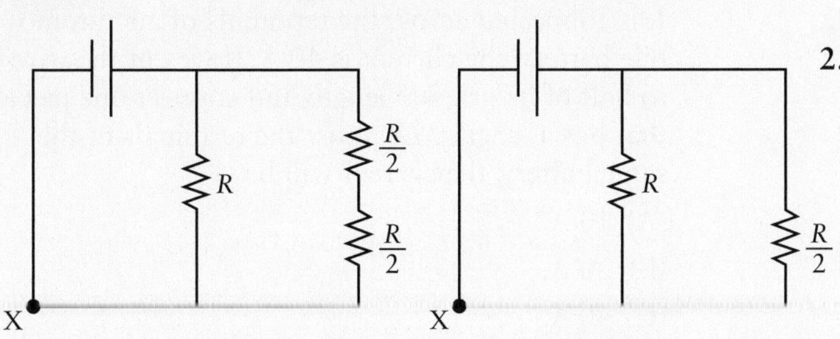

1. Consider the two circuits shown. The electric potential difference supplied by the battery is the same for both circuits. In the second circuit, one of the $R/2$ resistors has been removed. If the current at the point X in the first diagram is I_1 the current at the same point in the second circuit, I_2, will equal

(A) $\frac{I_1}{2}$

(B) $\frac{2I_1}{3}$

(C) $\frac{3I_1}{2}$

(D) $\frac{4I_1}{3}$.

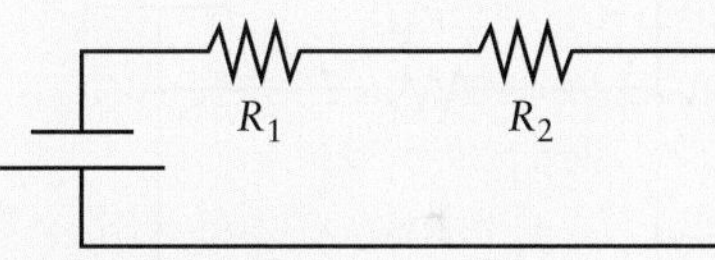

2. Two resistors, R_1 and R_2, are connected as shown. The resistance of R_1 is twice that of R_2. Compare the amount of energy dissipated by R_2 to that by R_1 during the same period of time.

(A) R_2 dissipates twice as much as R_1.

(B) R_2 dissipates one-half as much as R_1.

(C) R_2 dissipates four times as much as R_1.

(D) R_2 dissipates one-fourth as much as R_1.

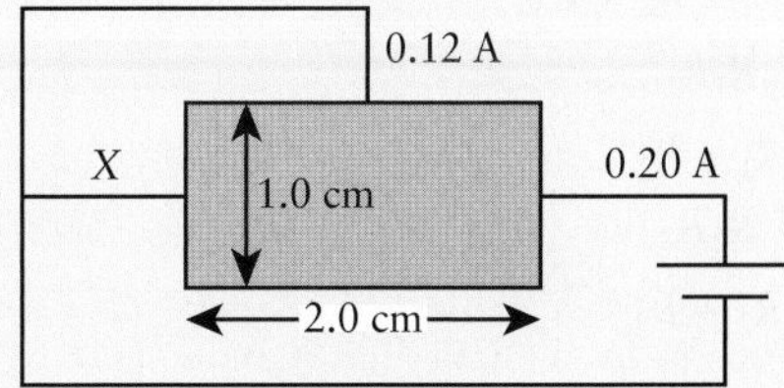

3. A rectangular flat plate that is 1.0 cm wide and 2.0 cm long is connected to a battery, as shown. What would be the current in the wire labeled X?

(A) 0.06 A

(B) 0.08 A

(C) 0.14 A

(D) 0.32 A

(Continued)

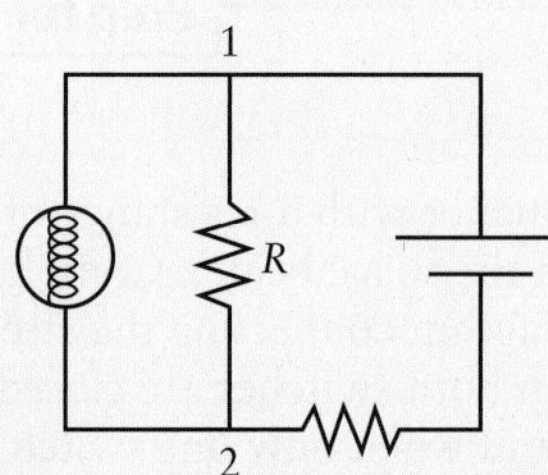

4. In this circuit, if the bulb burns out, which of the following is true about the current and electric potential difference from point 1 to point 2? The light bulb is indicated by the curly line inside a circle.

	Current (I)	Change in potential (ΔV)
(A)	Reduced	Reduced
(B)	Stays the same	Stays the same
(C)	Increased	Stays the same
(D)	Increased	Increased

Questions 5 and 6 refer to the figure shown.

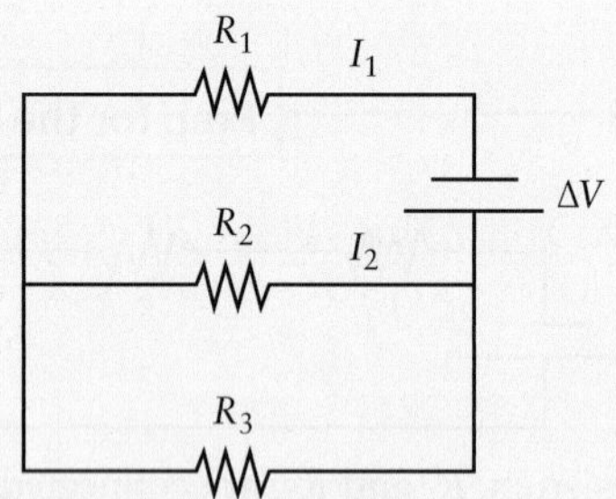

Assume the source of the electric potential difference in this circuit has no internal resistance.

5. Of the following expressions, which one best represents the resistance for the circuit?
(A) $R_1 + R_2 + R_3$
(B) $\dfrac{R_1R_2R_3}{R_1 + R_2 + R_3}$
(C) $\dfrac{R_1R_2}{R_1 + R_2} + R_3$
(D) $\dfrac{R_2R_3}{R_2 + R_3} + R_1$

6. Given a loop rule equation $\Delta V - I_2R_2 - I_1R_1 = 0$, which of the equations could be used with it for correct solutions?
(A) $\Delta V - I_1R_3 - I_1R_1 = 0$
(B) $(I_1 - I_2)R_3 - I_2R_2 = 0$
(C) $I_1R_1 + \Delta V + (I_1 + I_2)R_3 = 0$
(D) $I_2\left(\dfrac{R_2R_3}{R_2 + R_3}\right) - \Delta V = 0$

7. The switch in an initially open circuit containing a charged capacitor and a resistor is closed, causing the capacitor to discharge. Suppose that the resistor is a light bulb. Compared to the brightness of the bulb when the switch is closed at $t = 0$, the brightness of the bulb at $t = RC$ is
(A) $1 - (1/e^2) = 0.865$ as great.
(B) $1 - 1/e = 0.632$ as great.
(C) $1/e = 0.368$ as great.
(D) $1/e^2 = 0.135$ as great.

8. Two copper wires have the same length, but one has twice the diameter of the other. Compared to the one that has the smaller diameter, the one that has the larger diameter has a resistance that is
(A) larger by a factor of 2.
(B) larger by a factor of 4.
(C) smaller by a factor of 1/2.
(D) smaller by a factor of 1/4.

9. When a wire that has a large diameter and a length L is connected across the terminals of an automobile battery, the current is 40 A. If we cut the wire to half of its original length and connect one piece that has a length $L/2$ across the terminals of the same battery, the current will be
(A) 10 A.
(B) 20 A.
(C) 40 A.
(D) 80 A.

10. When a second light bulb is added in series to a circuit with a single light bulb, the resistance of the circuit
(A) increases.
(B) decreases.
(C) remains the same.
(D) doubles.

(Continued)

Free-Response Questions

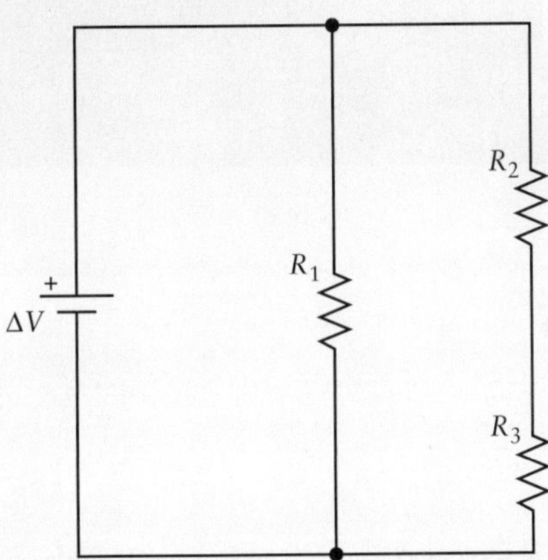

1. Consider the circuit shown: Three resistors are connected to a battery with negligible internal resistance. R_1 and R_2 are identical ($R_1 = R_2 = R$) and R_3 has twice the resistance of R_1 and R_2 ($R_3 = 2R$).
   (a) Let I_1, I_2, and I_3 be the currents through resistors R_1, R_2, and R_3, respectively. Let I_{bat} be the current through the battery. Let ΔV_1, ΔV_2, and ΔV_3 be the potential differences across resistors R_1, R_2, and R_3, respectively. Express all of the currents in terms of I_1 and all the potential differences in terms of ΔV_1.
   (b) On the circuit diagram, add another wire that will cause the current through resistor R_3 to decrease, while allowing the current through resistor R_1 to remain unchanged. Explain your reasoning.
   (c) Through which of the resistors shown is electric energy being converted to other types at the greatest rate? Explain your choice.

2. You wish to design a circuit using wires, a switch, an ideal battery, a light bulb, and an initially uncharged capacitor. In one position of the switch, (i) you want the bulb to be brightest when the switch is first closed, getting dimmer with time, and going out completely when the switch has been closed for a long time. Then you want to be able to change the position of the switch (ii) without touching any of the other circuit elements and have the bulb behave in the same way.
   (a) Draw the circuit diagram showing the arrangement of the circuit elements given. Clearly label which position of the switch satisfies conditions (i) and (ii).
   (b) You are given a circuit where a resistor of resistance R, a capacitor of capacitance C, and a battery with potential difference $\mathcal{E}$ are in series. Assume the capacitor is uncharged when the circuit is closed. In terms of given quantities and any necessary constants, calculate the current in the resistor, the potential difference across the capacitor, and the energy stored in the capacitor as soon as the switch is closed and after the switch has been closed for a very long time.
   (c) On the grids below, sketch the potential difference across the capacitor and the current in the resistor as functions of time for the process in part (b).

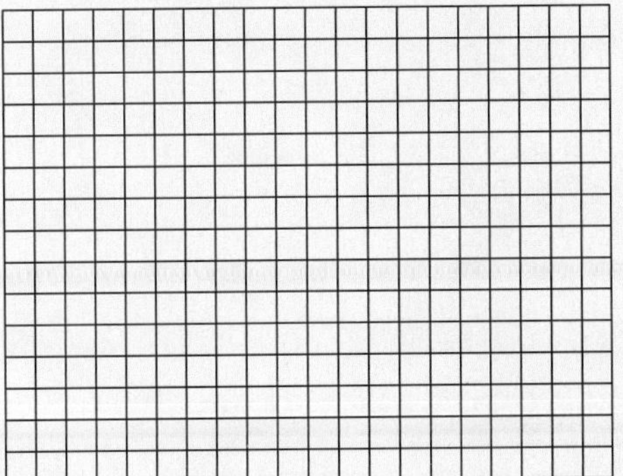

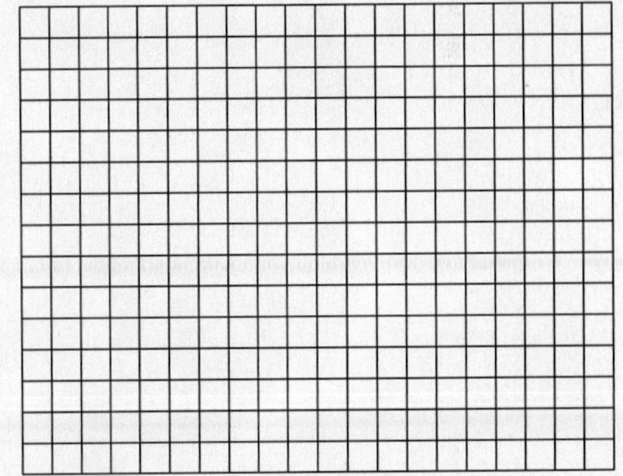

   (d) Describe how the graphs you sketched in part (c) relate to one of the positions of the switch you drew in part (a).

Magnetism and Electromagnetism

NASA, ESA, and J. Nichols (University of Leicester); NASA and the Hubble Heritage Team (STScI/AURA) Acknowledgment: NASA/ESA, John Clarke (University of Michigan)

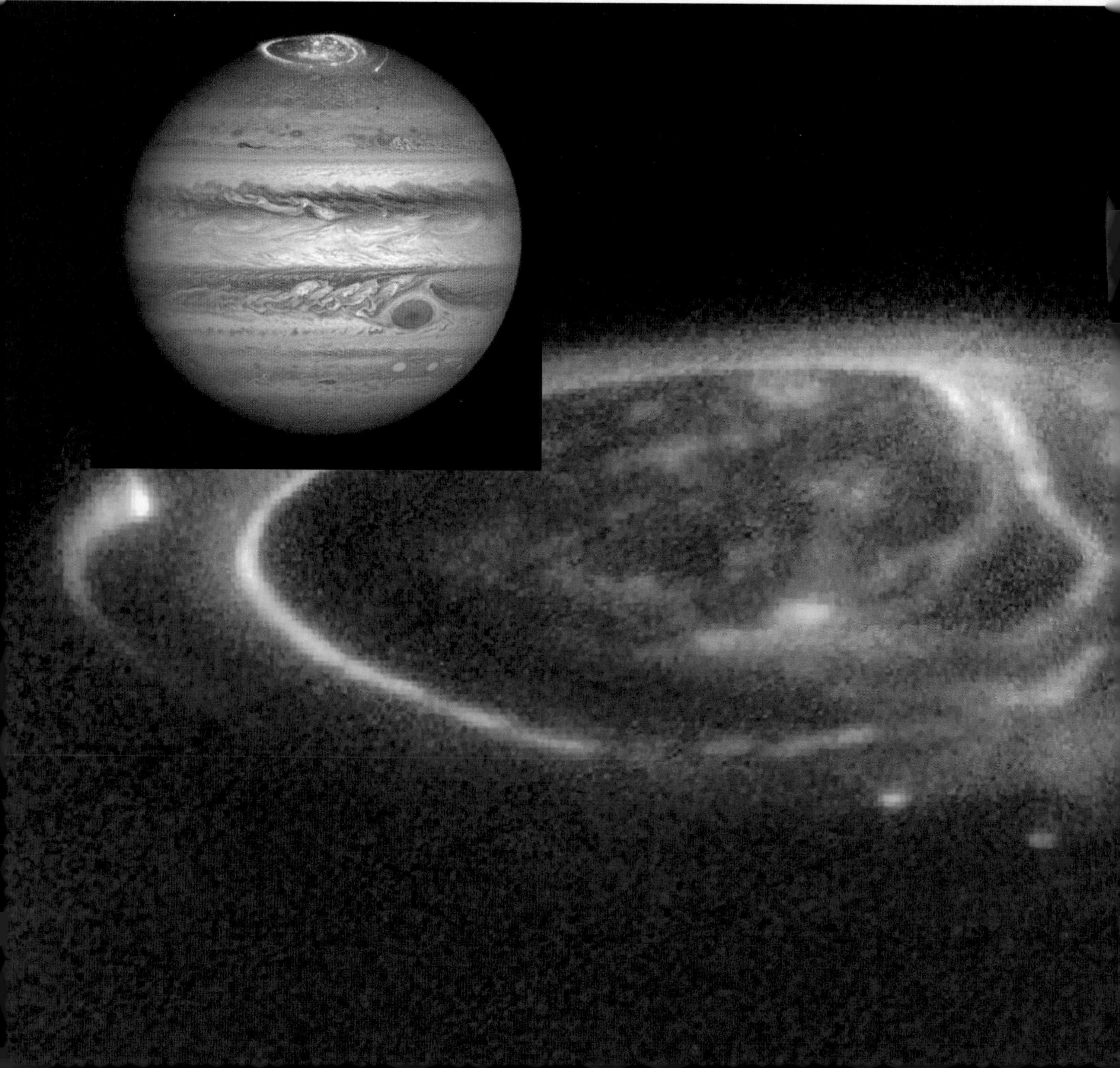

Case Study: How does a planet's magnetic field cause lights in its sky?

In your kitchen, you may have a few refrigerator magnets with humorous or pithy quotes. How do these rubbery strips stick to the outside of the refrigerator? A paper clip sticks to a bar magnet. Both phenomena result from the magnetic field surrounding the magnetic material causing electrons in iron atoms to align in a way that creates a second, magnetic field around the atoms. That field interacts with the magnetic field in the refrigerator magnet or bar magnet, generating an attractive force between them (if your refrigerator is made of a metal alloy that contains iron).

The swirling of electrically conducting iron and nickel atoms in Earth's molten outer core also produces a magnetic field. The force between that field and the magnetic needle of a compass causes the needle to align with the north–south direction on Earth.

Earth's magnetic field also exerts a force on the billions of charged particles that stream out from the Sun every second. Those particles travel in a "solar wind" that passes near Earth and are drawn in—accelerated!—toward the magnetic poles of our planet. They gain energy as they spiral in, and when those energetic particles collide with gas molecules in the atmosphere, energy is transferred to the gas, which glows.

Most of the gas in our atmosphere is nitrogen, which generally emits a bluish light when it glows. The atmosphere also contains a significant amount of oxygen, which glows either greenish-yellow or crimson, and a small amount of argon, which emits light blue light. Neon gas constitutes a small fraction of our atmosphere. When excited by energetic particles from the Sun, neon can glow in an array of colors similar to how it glows in neon lights! The variety of colors given off as charged particles from the Sun stream down toward Earth's magnetic poles result in the otherworldly light show known as the northern lights (the aurora borealis) and the southern lights (the aurora australis) as shown in the Chapter 19 opening photo. Can you guess why the northern and southern lights appear to shimmer and dance?

A magnetic field exists on other planets in our solar system. Some of these are also enveloped in an atmosphere. Can an aurora be seen on them? Yes! Scientists have observed this phenomenon on Jupiter, Saturn, Uranus, and Neptune. However, the aurora seen on Jupiter, shown here, is not as colorful as Earth's. Astronomers at the National Radio Astronomy Observatory have detected the faint glow of an aurora associated with the star LSR J1835 + 3259, more than 30,000 times farther from Earth than the outer reaches of our solar system.

By the end of these chapters, you should be able to describe the properties of a magnetic field; the conditions required for magnetic fields to be produced and to exert forces on electrically charged objects, current carrying wires, and other magnets; the magnetic properties of matter; and how changing magnetic fields create electric potential difference, generating the electricity we use every day.

19 Magnetism: Forces and Fields

Rowan Romeyn/Alamy

YOU WILL LEARN TO:

- Describe the properties of a magnetic field.
- Describe the magnetic behavior of material in terms of the configuration of magnetic dipoles within the material.
- Describe the magnetic permeability of free space and matter.
- Describe the magnetic field produced by a current, and from this, the approximation of the magnetic field for a charged object that is moving at much less than the speed of light.
- Describe the magnetic force exerted on charged objects moving in a magnetic field.
- Describe the magnetic field produced by a long, straight current-carrying wire.
- Describe the force exerted on a current-carrying wire placed in a magnetic field.
- Explain why a current loop in a uniform magnetic field experiences a net torque but zero net force.

19-1 The magnetic force, like the electric force, is a long-range force

If you rub a balloon on your head, then move the balloon a few centimeters away, the hairs on your head will stand up (see Figure 16-4). This is a result of *electric* forces. The rubbing transfers electrons between your hair and the balloon, giving one a net negative charge and the other a net positive charge; these attract each other, even over a distance, and this makes your hair stand up.

Figure 19-1a shows another long-range force exerted at a distance like the electric force. The fundamental force that describes both electric and magnetic forces is the electromagnetic force. In Chapter 26 we will introduce the modern physics topic relativity. It turns out that magnetism is related to relativistic effects; it is the force that electrically charged objects exert on each other because of relative motion. Since two iron bars will each contain equal quantities of positive and negative charge, the electric force between them will be zero. If even some of the charged objects within them move in a coordinated way, we will see this causes a magnetic force that can be *much* stronger than that between the balloon and your hair. Certain objects called **magnets**,

(a)
The north pole (red) of one bar magnet attracts the south pole (blue) of another magnet.

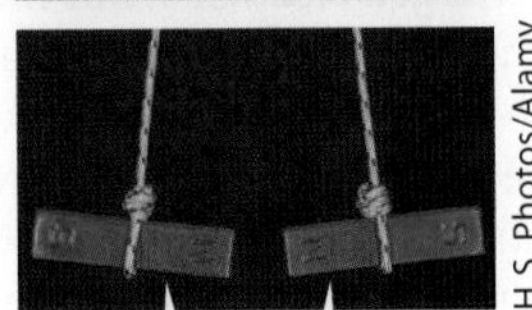

The north poles (red) of the two magnets repel each other.
H.S. Photos/Alamy

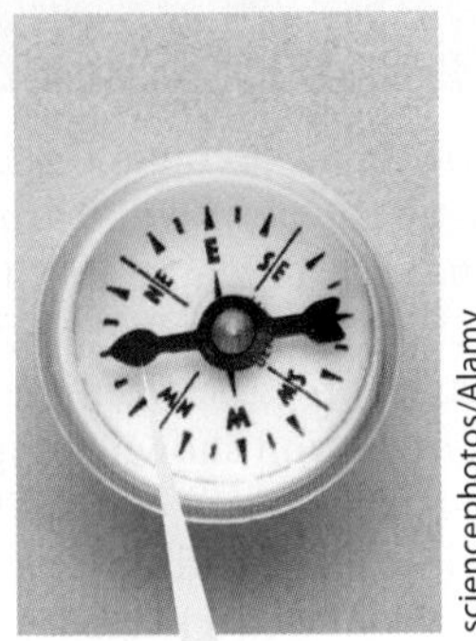

(b)
The magnetized compass needle aligns with Earth's magnetic field.
sciencephotos/Alamy

(c)
The magnetic field required to pick up pieces of iron at a scrap yard comes from an electromagnet—a device that generates a magnetic field when there is current through its wires.
Cordelia Molloy/Science Source

Figure 19-1 Magnetic forces (a) The force between two bar magnets. (b) A compass needle interacting with Earth's magnetic field. (c) A large electromagnet.

made of one of a handful of special materials such as iron, cobalt, and nickel, can exert strong **magnetic forces** on other magnets. (The name magnet comes from the region of ancient Greece known as Magnesia, where these objects were discovered more than 2500 years ago.) Just as we use the umbrella term *electricity* to refer to interactions between electrically charged objects, we use the term **magnetism** to describe the interactions between magnets. One application of magnetic interactions is the compass (**Figure 19-1b**). Earth's core acts like a giant magnet and produces a *magnetic field* in the surrounding space. (We'll see that magnetic fields are analogous to electric fields, but with important differences.) The magnetic needle of a compass aligns with Earth's field and points north. Earth's magnetic field is relatively weak; the field produced by the electromagnet in **Figure 19-1c** is thousands of times stronger.

Our goal in this chapter is to understand magnetism. We begin in the next section by investigating the properties of magnets. While we will not need to understand the connection to relativity to describe magnets, we will find that magnets are created by moving electrically charged objects. From there we develop a full understanding of magnetic forces, which requires us to understand two things: (1) the nature of the magnetic field that moving charged objects *produce* and (2) how moving charged objects *respond* to the magnetic fields produced by other moving charged objects. It turns out to be easiest to look at the second of these first. We'll analyze the magnetic force that is exerted on a moving charge placed in a magnetic field, as well as the magnetic force on a wire that carries a current (a collection of charged objects moving in the wire) and is placed in a magnetic field. Later in the chapter we'll see how to calculate the magnetic field produced by a given collection of charged objects in motion, and from this, how to approximate the field of a single moving charged object.

THE TAKEAWAY for Section 19-1

✔ Magnetic forces, like electric forces, are long-range forces.

Prep for the AP® Exam

AP® Building Blocks

1. Your refrigerator is not moving, and yet a magnet might stick to it (meaning the magnet exerts a force on it). What must be true for this to happen?

2. You are given three iron rods. Two of them are magnets but the third one is not. We know that magnets can attract or repel each other and that they can also attract magnetic materials that aren't magnetized. How could you use the two magnets to find that the third rod is not magnetized?

19-2 Magnetism is an interaction between moving charged objects

Like the electric force, magnetic forces become stronger as the objects are moved closer together. However, the magnetic force we will learn to calculate is different from the electric force. The attracting magnets in Figure 19-1a are *not* electrically charged: Their atoms are made of positively charged nuclei and negatively charged electrons, but their net charges are zero. Any electric forces between these magnets are very weak and are not responsible for the strong attraction shown in Figure 19-1a.

As we did for the electric force, we can explain how the magnetic force is exerted at a distance by invoking the idea of a field. Just as an electrically charged object sets up an electric field in the space around it, a magnet sets up a magnetic field in the space around it (**Figure 19-2a**). A second magnet placed in this field experiences a magnetic force that depends on the magnitude and direction of the field. We use the symbol $\vec{B}$ for magnetic field. Unlike the electric field $\vec{E}$ of a point charge, which points either directly away from or directly toward the charge, the magnetic field $\vec{B}$ of a magnet points away from one end of a magnet and toward the other end. These two ends are called the **magnetic poles** of the magnet, one of which is called the *north pole* and the other the *south pole*. The meaning of the names is that Earth itself acts like a giant magnet. If allowed to swing freely, a magnet will orient itself so that its north pole is pointing approximately toward Earth's geographic north pole and its south pole toward Earth's geographic south pole (**Figure 19-2b**). Since we saw in Figure 19-1a that like poles repel and opposite poles attract, this tells us that Earth's geographic north pole must be close to its magnetic south pole, and its geographic south pole must be near its magnetic north pole! We see the field lines point away from geographic south and toward geographic north, as we just described for a magnet.

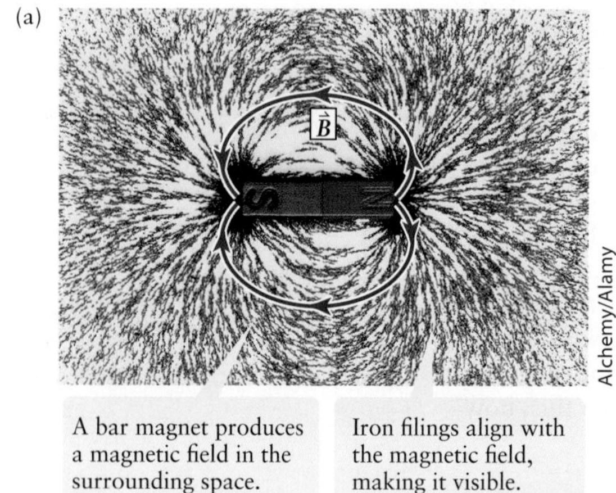

(a)
$\vec{B}$
S
N
Alchemy/Alamy
A bar magnet produces a magnetic field in the surrounding space.
Iron filings align with the magnetic field, making it visible.

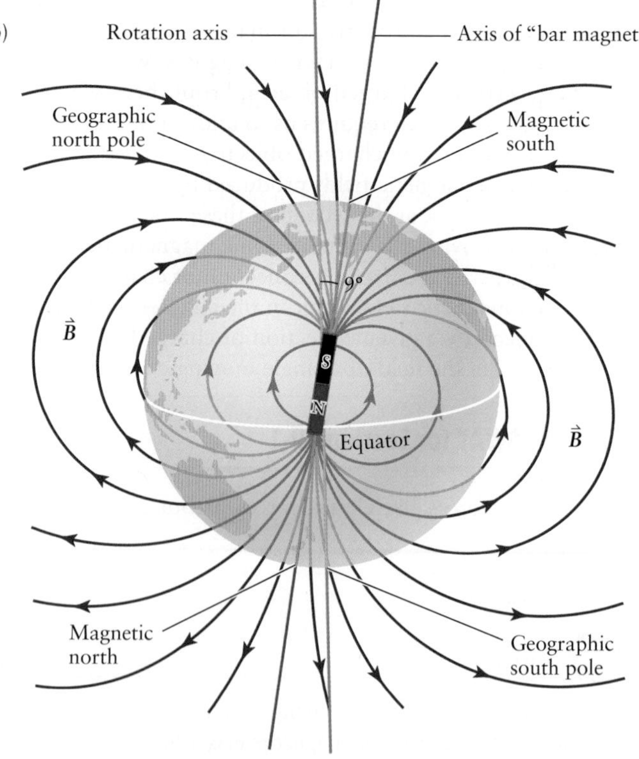

Figure 19-2 The magnetic fields of a bar magnet and Earth (a) The magnetic field of a bar magnet points out of its north pole (colored red) and into its south pole (colored black). (b) Earth's magnetic field has a similar pattern. Although Earth's field is produced in a different way—by electric currents in the liquid portion of our planet's interior—the field is much the same as if there were a giant bar magnet inside Earth. As of this writing (2022), this "bar magnet" is tipped by about 9° from Earth's rotation axis, which is why Earth's magnetic poles are not at the same locations as the true, or geographic, poles. A compass needle points toward the magnetic south pole, not the true north pole. Most of the time, these poles move slowly enough with time that compasses remain useful for most purposes, but randomly (intervals ranging from 100,000 to 50 million years) the poles flip. The last time this happened was about 780,000 years ago. Scientists can tell by looking at the direction magnetic sediment lines up in core samples.

By convention we choose the direction of the magnetic field $\vec{B}$ produced by a magnet to be such that $\vec{B}$ points away from the magnet's north pole and toward the magnet's south pole. This should remind you of an electric dipole (see Section 16-5) made up of an object with positive charge $+q$ and one with negative charge $-q$, for which the electric field $\vec{E}$ points away from the positively charged object and toward the negatively charged object. Indeed, a magnet like that shown in Figure 19-2a is often called a **magnetic dipole.**

The magnetic forces between magnets can be either attractive or repulsive. If the north pole of one magnet is close to the south pole of a second magnet, the two magnets attract each other (**Figure 19-3a**); if the north poles are close together or the south poles are close together, the two magnets repel each other (**Figure 19-3b**). This is very different from the behavior of electrically charged objects but analogous to the way in which electric dipoles interact.

(a) These magnets attract.

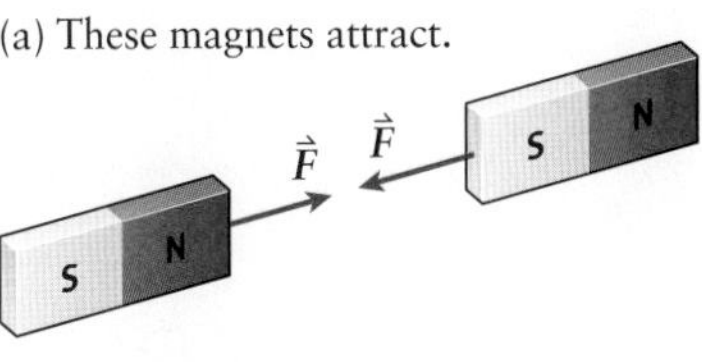

(b) These magnets repel.

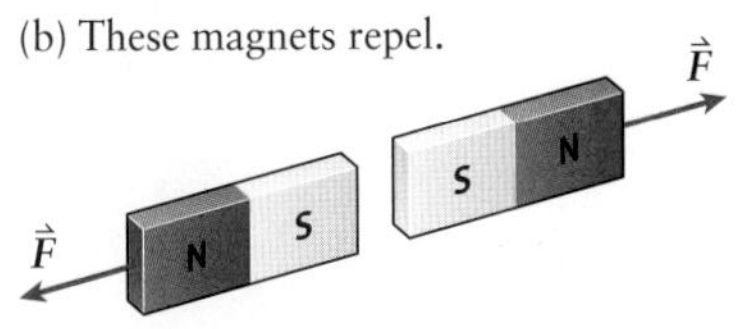

(c) If we cut a magnet in half, we always find each half has a north and south pole.

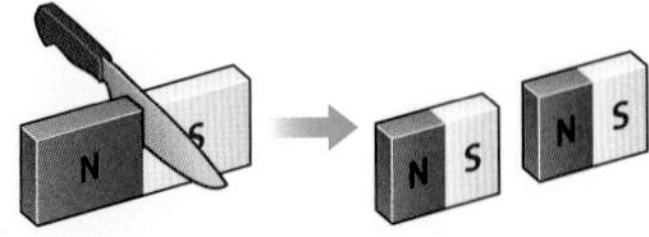

Figure 19-3 Bar magnets and magnetic poles (a) The opposite poles of these magnets attract. (b) Like poles of these magnets repel. (c) Every magnet has both a north and south pole; they cannot be separated.

WATCH OUT!

Magnetic poles are not like electric charge.

Note that there is an important distinction between electric and magnetic dipoles. You can take an electric dipole apart by separating the component positive and negative charged objects. But you *cannot* separate the north and south poles of a magnet. If you cut a magnet in half, each half has a north pole and a south pole (**Figure 19-3c**). The same is true no matter how many small pieces you cut a magnet into; a very small piece produces only a weak magnetic field, but is still a magnetic dipole with both a north pole and a south pole. (Some physicists have speculated about the existence of *magnetic monopoles*, objects that have the properties of an isolated north or south magnetic pole. Many experiments have been performed to look for evidence of a magnetic monopole. No evidence has yet been found.)

What is it about magnets that causes them to produce and respond to magnetic fields? And why is it impossible to separate their north and south poles? The answers to these questions were revealed by a set of crucial experiments in the nineteenth century. **Figure 19-4a** shows a version of one of these experiments. A coil of copper wire loops through a flat piece of clear plastic, and iron filings are spread over the plastic. Since copper is not magnetic, the filings lie wherever they were placed. But when there is an electric current through the coil, the filings line up just as they do around the magnet in Figure 19-2. This demonstrates that the moving charged objects that make up the current in the coil *produce* a magnetic field. The field points out of the coil at

(a) The magnetic field created by current in a coil.

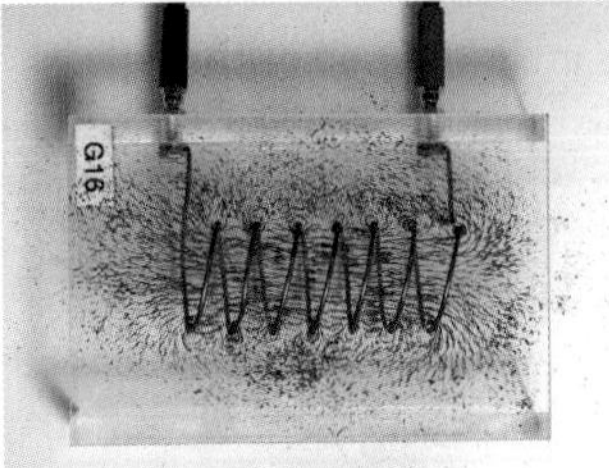

Eli Sidman, Technical Services Group, MIT

(b) This coil and magnet attract.

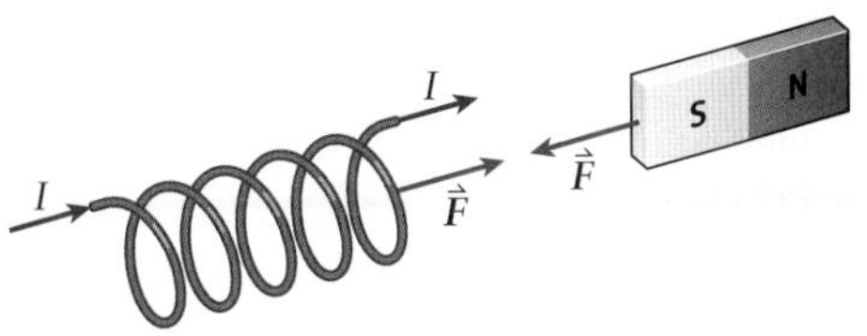

(c) This coil and magnet repel.

(d) These coils attract, just like two magnets.

Figure 19-4 A current-carrying coil acts like a magnet (a) The field of a current-carrying coil is very similar to that of a bar magnet (Figure 19-1a). (b), (c) Such a coil interacts with a bar magnet just as though it were a magnet itself. (d) Two current-carrying coils interact like two bar magnets.

one end, which is the "north pole" of the coil, and points into the other end, which is the coil's "south pole."

We can verify that the coil produces a magnetic field by carrying out the experiment shown in **Figure 19-4b**, in which we've replaced one of the two magnets from Figure 19-3a with a coil. When the current in the coil is turned on, the coil and the magnet are attracted to each other. This can happen only if the current-carrying coil is indeed producing a magnetic field, and the magnet is responding to that field. If we flip the coil over, as shown in **Figure 19-4c**, the coil and the magnet now repel each other. This is just what would happen if the coil were an iron magnet that we flipped over to interchange the north and south poles (see Figures 19-3a and 19-3b).

For the coil in Figures 19-4b and 19-4c to be attracted to or repelled from the magnet, it must also be true that the moving charged objects that make up the current in the coil respond to the magnetic field of the magnet. In other words, a current-carrying coil and a magnet are fundamentally the same, in that both objects *produce* as well as *respond to* magnetic fields. **Figure 19-4d** shows an experiment that verifies this: Two current-carrying coils with the same orientation attract each other, just as do two magnets with the same orientation.

The series of experiments shown in Figure 19-4 suggests that *magnetism is an interaction between charged objects in motion*. For two systems such as wires or magnets to exert magnetic forces on each other, there must be moving charged objects in both systems. We use the umbrella term **electromagnetism** to include both electricity and magnetism, since both involve interactions between charged objects. The distinction between electricity and magnetism is that two charged objects exert *electric* forces on each other whether or not they are moving, while two charged objects exert *magnetic* forces on each other only if both are in motion. (The wires and magnets in the experiments shown in Figure 19-4 are all electrically neutral. So there are no electric forces in those experiments, only magnetic forces.)

You may be wondering if magnetism must always involve charged objects in motion, since the ordinary magnets shown in Figures 19-1, 19-2, and 19-3 are not attached to sources of emf and so do not carry currents. The answer is that there *are* charged objects in motion inside an ordinary iron magnet: The moving charged objects are the electrons within the atoms of the magnet, and their motions within the atom are like those of electrons in the circular coils of Figure 19-4. Each individual iron atom in a magnet produces only a tiny magnetic field. But because many of the atoms that make up the magnet are oriented so that their electron motions are in the same direction as in the surrounding atoms, their magnetic fields add together to make a substantial total field. This explains why cutting a magnet into pieces doesn't leave you with a separate north pole and south pole: Each piece is simply a smaller version of the original magnet.

THE TAKEAWAY for Section 19-2

✔ Magnetism is an interaction between moving charged objects. A moving charged object may produce a magnetic field, and a moving charged object in a magnetic field may experience a force. A stationary charged object neither produces nor responds to a magnetic field.

✔ In a magnet, charged objects are in motion at the atomic level to produce a magnetic field.

Prep for the AP Exam

AP Building Blocks

1. Two identical objects each have a positive charge Q. One of the objects is held in place, while the other object is in motion a short distance from the first object. Which of the following describes the forces the two objects exert on each other?
   (A) Both electric and magnetic forces
   (B) Electric forces but not magnetic forces
   (C) Magnetic forces but not electric forces
   (D) Neither electric forces nor magnetic forces.
2. Explain why no magnetic north pole is ever found in isolation from a magnetic south pole.
3. You are standing next to a large electromagnet that creates a magnetic field pointing to geographic east, that is much stronger than Earth's magnetic field. If you were to look at a compass, in which direction would it point?

19-3 A moving point charge can experience a magnetic force

We'll begin our study of magnetic forces by considering the force on a single charged object moving in a magnetic field $\vec{B}$. For example, this could be an electron moving in a current-carrying wire in the vicinity of a magnet, or a moving proton in Earth's upper atmosphere on which a force is exerted by our planet's magnetic field. Here's what experiments tell us about the magnetic force on such a moving charged object:

1. A charged object can experience a force when placed in a magnetic field, but *only when it is moving*.
2. The magnetic force depends on the direction of the charged object's velocity relative to the direction of the magnetic field. The charged object does not experience a magnetic force when its velocity is parallel to or opposite to the magnetic field.
3. The magnitude of the force is proportional to the charge on the object, to the magnitude of the magnetic field, and to the speed of the object.
4. The direction of the force is perpendicular to both the direction of motion of the charged object and the direction of the magnetic field.
5. The direction of the force depends on the sign of the charge.

Figure 19-5a shows the magnetic force on an object with positive charge q moving with velocity $\vec{v}$ in the presence of a magnetic field $\vec{B}$. If θ is the angle between the velocity of the charged object and the magnetic field, the magnitude of the magnetic force $\vec{F}_M$ is

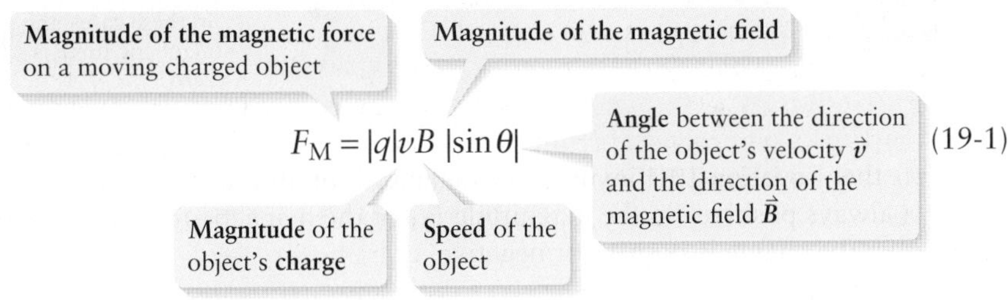

$$F_M = |q|vB\,|\sin\theta| \quad (19\text{-}1)$$

EQUATION IN WORDS
Magnitude of magnetic force on a moving charged object

The SI unit of magnetic field is the **tesla** (T). In Equation 19-1, if charge q is in coulombs (C), speed v is in meters per second (m/s), and magnetic field magnitude B is in tesla (T), the force F is in newtons (N). So 1 T is the same as $1\ \text{N}\cdot\text{s}/(\text{C}\cdot\text{m})$ or $1\ \text{N/A}\cdot\text{m}$. (Recall that one ampere equals one coulomb per second: $1\ \text{A} = 1\ \text{C/s}$.) The strongest magnetic field most of us will experience directly is the roughly 2-T field inside a magnetic resonance imaging (MRI) scanner. Other magnetic fields in our environment are much weaker: The strength of Earth's magnetic field is about 5×10^{-5} T, the electrical impulses that drive the contraction and relaxation of your heart produce magnetic fields of about 5×10^{-11} T, and the magnetic fields in your brain are on the order of 10^{-13} T. An alternative unit for magnetic field is the *gauss* (G): $1\ \text{G} = 10^{-4}\ \text{T}$.

AP® Exam Tip

On the AP® Physics 2 Exam, you will only be asked to calculate the magnetic force on a charged object when the field and velocity are at angles of 0, 90, or 180 degrees. You can be asked for qualitative descriptions for the force or subsequent behavior of the object for other angles.

Here's a **right-hand rule** for determining the direction of the magnetic force on a moving charge. First extend the fingers of your right hand along the direction of the velocity $\vec{v}$ and orient your hand so that your palm is facing in the direction that the magnetic field $\vec{B}$ points. Then extend your right thumb so that it is perpendicular to the four fingers of your right hand and swivel your right hand so that the fingers now point in the direction of $\vec{B}$. If the object has a positive charge, so $q > 0$, your thumb points in the direction of the force $\vec{F}_M$. If the object has a negative charge, so $q < 0$ (**Figure 19-5b**), the force points in the direction opposite to that given by the right-hand rule. In either case, the force is perpendicular to the velocity $\vec{v}$ *and* perpendicular to the magnetic field $\vec{B}$. If we draw the vectors $\vec{v}$ and $\vec{B}$ with their tails together, they form a plane; the force $\vec{F}_M$ is perpendicular to that plane, with a direction given by the right-hand rule. (We will encounter a number of other right-hand rules for magnetism in this chapter.)

(a) Magnetic force on a positively charged object

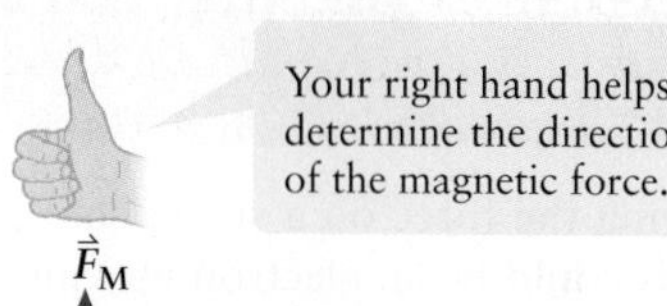

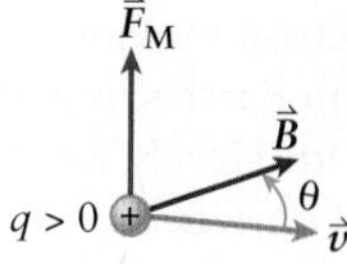

The force $\vec{F}_M$ is perpendicular to both the velocity $\vec{v}$ and the magnetic field $\vec{B}$.

(b) Magnetic force on a negatively charged object

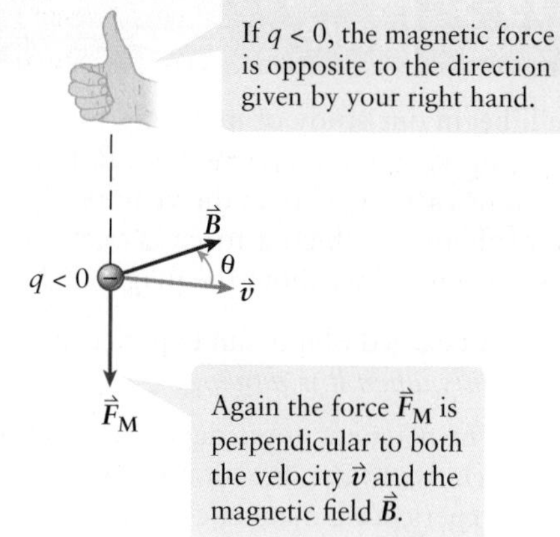

Again the force $\vec{F}_M$ is perpendicular to both the velocity $\vec{v}$ and the magnetic field $\vec{B}$.

(c) Magnetic force on a positively charged object, $\theta = 90°$

For a given speed v, the magnetic force is greatest if the velocity $\vec{v}$ is perpendicular to the magnetic field $\vec{B}$.

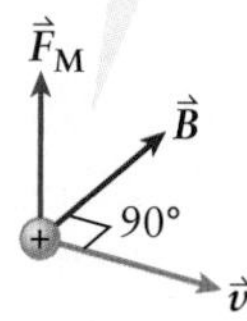

(d) Magnetic force on a positively charged object, $\theta = 0$ or 180°

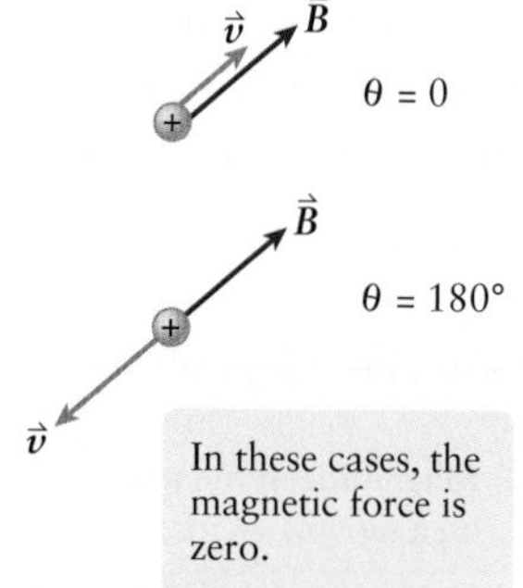

In these cases, the magnetic force is zero.

Figure 19-5 Magnetic force on a moving charged object The direction of the magnetic force on an object with charge q depends on whether (a) q is positive or (b) q is negative. The magnitude of the force $F_M = |q|vB \sin\theta$ goes from (c) a maximum value if $\theta = 90°$ to (d) zero if $\theta = 0$ or $\theta = 180°$.

AP® Exam Tip

A special case of the magnetic force on moving charges is that of charge carriers in a wire. Because positive or negative charge carriers move in opposite directions to cause a given current, the direction of the magnetic force will be the same for either sign charge carrier. This deflection of charge in the wire causes a potential difference across the diameter of the wire that can be used to determine the sign of the charge carriers. This is known as the **Hall effect**, and you may be asked to describe it on the exam.

Note that Equation 19-1 involves the magnitude or absolute value of the charge, $|q|$, which is always positive. So the magnitude F_M of the magnetic force does *not* depend on whether the charge is positive or negative.

WATCH OUT!

The magnetic force on a moving charged object is never in the same direction as the magnetic field.

From Chapter 16 you're used to the idea that the electric force on an object with charge q placed in an electric field $\vec{E}$ is $\vec{F}_{electric} = q\vec{E}$. This force is in the same direction as $\vec{E}$ if q is positive; if q is negative, it is in the direction opposite to $\vec{E}$. Figures 19-5a and 19-5b show that the magnetic force on a moving charged object is very different: This force is *perpendicular* to the direction of the magnetic field. Electric and magnetic forces are quite different from each other!

Equation 19-1 tells us that the magnitude of the magnetic force on a moving charged object depends on the angle θ between the directions of $\vec{v}$ and $\vec{B}$. For a given speed v, the magnetic force has its greatest magnitude if the velocity $\vec{v}$ is perpendicular to the magnetic field $\vec{B}$; then $\theta = 90°$, $\sin\theta = 1$, and $F_M = |q|vB$ (**Figure 19-5c**). If $\theta = 0$ or $\theta = 180°$, so the object is moving either in the same direction as $\vec{B}$ or in the direction opposite to $\vec{B}$, $\sin\theta = 0$ and the force on the object is *zero*. Thus a charged object moving in the direction of the magnetic field, or in the direction opposite to the magnetic field, experiences no magnetic force (**Figure 19-5d**).

We've shown the vectors $\vec{v}$, $\vec{B}$, and $\vec{F}$ in perspective in Figure 19-5. Drawing in perspective isn't easy to do, so we'll often use a simple convention for drawing vectors that are pointed either into or out of the pages of this book (**Figure 19-6**). Think of a vector as an arrow, with a sharp point at its head and feathers at its tail. If a vector is directed perpendicular to the plane of the page and pointed toward

(a)

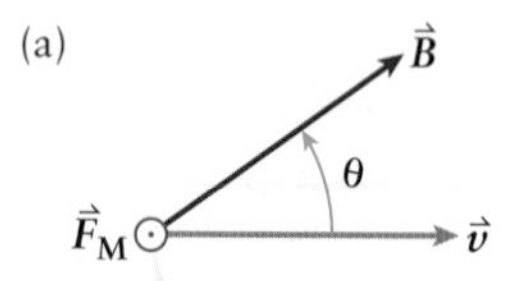

If $q > 0$, the force vector in this situation points out of the page (as indicated by the circle with a dot).

(b)

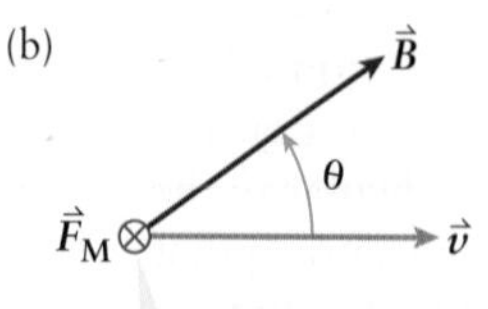

If $q < 0$, the force vector in this situation points into the page (as indicated by the circle with an X).

Figure 19-6 Vectors out of the page and into the page We use the symbols ⊙ and ⊗ to denote vectors that point out of the page and into the page, respectively.

you—that is, *out of* the page—we'll depict it with the symbol ⊙. The dot in the center of this symbol represents the sharp point of the arrowhead. If a vector is directed perpendicular to the plane of the page and pointed away from you—that is, *into* the page—we'll depict it with the symbol ⊗. The "X" in the center of this symbol represents the feathers on the tail of the arrow as you would see if the arrow were moving away from you.

EXAMPLE 19-1 Magnetic Forces on a Proton and an Electron

At a location near our planet's equator, the direction of Earth's magnetic field is horizontal (that is, parallel to the ground) and due north, and the magnitude of the field is 2.5×10^{-5} T. Find the direction and magnitude of the magnetic force on an object moving at 1.0×10^{4} m/s if the object is (a) a proton moving horizontally and due east, (b) an electron moving horizontally and due east, and (c) a proton moving horizontally in a direction 25° east of north. Recall that a proton has charge $e = 1.60 \times 10^{-19}$ C and an electron has charge $-e$.

Set Up

In each case we'll use Equation 19-1 to find the magnitude of the force on the moving proton or electron. The right-hand rule (Figure 19-5) will tell us the direction of the magnetic force.

Magnetic force on a moving charged object:

$$F_M = |q|vB|\sin\theta| \quad (19\text{-}1)$$

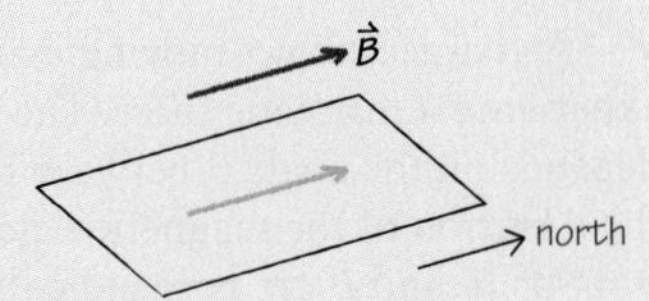

Solve

(a) The velocity $\vec{v}$ of the proton is perpendicular to the magnetic field $\vec{B}$, so in Equation 19-1 the angle $\theta = 90°$ and $\sin\theta = 1$. Both $\vec{v}$ and $\vec{B}$ lie in a horizontal plane; the magnetic force $\vec{F}$ must be perpendicular to this plane, so $\vec{F}$ is vertical. Applying the right-hand rule shows that the force points vertically upward.

Charge on the proton:

$$q = e = 1.60 \times 10^{-19}\ \text{C}$$

The proton velocity is perpendicular to the magnetic field ($\theta = 90°$), so the magnitude of the magnetic force on the proton is

$$F_M = |q|vB|\sin\theta| = evB|\sin\theta|$$
$$= (1.60 \times 10^{-19}\ \text{C})(1.0 \times 10^{4}\ \text{m/s})(2.5 \times 10^{-5}\ \text{T})\sin 90°$$
$$= 4.0 \times 10^{-20}\ \text{N}$$

(b) The electron has the same magnitude of charge as the proton and the same velocity, so it experiences the same magnitude of magnetic force. But its charge is negative, so the direction of the magnetic force $\vec{F}$ on the electron is *opposite* to the direction given by the right-hand rule for the proton. Hence the force F points vertically downward.

Charge on the electron:

$$q = -e = -1.60 \times 10^{-19}\ \text{C}$$

The electron velocity is perpendicular to the magnetic field ($\theta = 90°$), so the magnitude of the magnetic force on the electron is

$$F_M = |q|vB|\sin\theta| = evB|\sin\theta|$$
$$= (1.60 \times 10^{-19}\ \text{C})(1.0 \times 10^{4}\ \text{m/s})(2.5 \times 10^{-5}\ \text{T})\sin 90°$$
$$= 4.0 \times 10^{-20}\ \text{N}$$

(c) The proton velocity $\vec{v}$ and the magnetic field $\vec{B}$ again both lie in a horizontal plane, but now the angle between these vectors is $\theta = 25°$. So the magnitude of the magnetic force is less than in part (a). The direction of the force is the same as in part (a), however.

The proton velocity is at an angle $\theta = 25°$ to the magnetic field, so the magnitude of the magnetic force on the proton is

$$F_M = |q|vB|\sin\theta| = evB|\sin\theta|$$
$$= (1.60 \times 10^{-19}\ \text{C})(1.0 \times 10^{4}\ \text{m/s})$$
$$\times (2.5 \times 10^{-5}\ \text{T})\sin 25°$$
$$= 1.7 \times 10^{-20}\ \text{N}$$

This is smaller than in part (a) because $\sin 25° = 0.42$ compared to $\sin 90° = 1$.

Reflect

This example illustrates how the magnetic force on a moving charged object depends on the direction in which the object is moving. Note that the force magnitudes in parts (a), (b), and (c) are very small because a single electron or proton carries very little charge. These objects also have very small mass, however (1.67×10^{-27} kg for the proton and 9.11×10^{-31} kg for the electron), so the resulting *accelerations* are tremendous. Can you use Newton's second law to show that the proton in part (a) and the electron in part (b) have accelerations of 2.4×10^7 m/s^2 and 4.4×10^{10} m/s^2, respectively?

NOW WORK Problems 1–3 and 6–8 from The Takeaway 19-3.

THE TAKEAWAY for Section 19-3

✔ A charged object moving in a magnetic field can experience a magnetic force. The magnitude of this force depends on the angle θ between the object's velocity $\vec{v}$ and the direction of the magnetic field $\vec{B}$. The force is maximum if $\theta = 90°$ ($\vec{v}$ and $\vec{B}$ are perpendicular) and zero if $\theta = 0$ or 180° ($\vec{v}$ and $\vec{B}$ are either in the same direction or opposite directions).

✔ The magnitude of the magnetic force on a charged object is proportional to the amount of charge, to the speed of the charged object, and to the magnitude of the magnetic field.

✔ The direction of the magnetic force is perpendicular to both the object's velocity $\vec{v}$ and the direction of the magnetic field $\vec{B}$ and depends on the sign of the charge. A right-hand rule helps tell us the force direction.

✔ The SI unit of magnetic field is the tesla (T).

Prep for the AP® Exam

AP® Building Blocks

1. A +1-nC charge moving at 1 m/s makes an angle of 30° with a uniform, 1-T magnetic field. What is the magnitude of the magnetic force that the charge experiences?

2. Determine the directions of the magnetic forces that are exerted on positive charged objects moving in the magnetic fields as shown in the figure below.

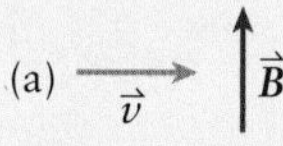

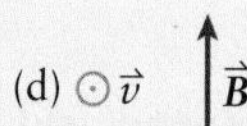

(b) $\vec{v}$ ⊙$\vec{B}$

(c) $\vec{B}$ ⊗$\vec{v}$

(f) ⊙$\vec{v}$ $\vec{B}$

3. An electron is moving with a speed of 18 m/s in a direction parallel to a uniform magnetic field of 2.0 T. What are the magnitude and direction of the magnetic force on the electron?

AP® Skill Builders

4. If a magnetic field exerts a force on moving charged objects, is it capable of doing work on the objects? Explain your answer.

5. In a lightning strike there is a negative charge moving rapidly from a cloud to the ground. In what direction is a lightning strike deflected by Earth's magnetic field?

EX 19-1

6. Determine the direction of the missing vector, $\vec{v}$, $\vec{B}$, or $\vec{F}$, in the scenarios shown in the figure below. All moving charged objects are positive.

(a) $\vec{v}$ $\vec{F}_M$ (b) $\vec{v}$ ⊗$\vec{F}_M$ (c) $\vec{F}_M$ ⊗$\vec{v}$

(d) ⊙$\vec{v}$ ⊗$\vec{F}_M$ (e) $\vec{B}$ $\vec{F}_M$ (f) ⊙$\vec{B}$ $\vec{F}_M$

EX 19-1

7. A proton P travels with a speed of 18 m/s toward the top of the page through a uniform magnetic field of 2.0 T directed into the page, as shown in the figure below. What are the magnitude and direction of the magnetic force on the proton?

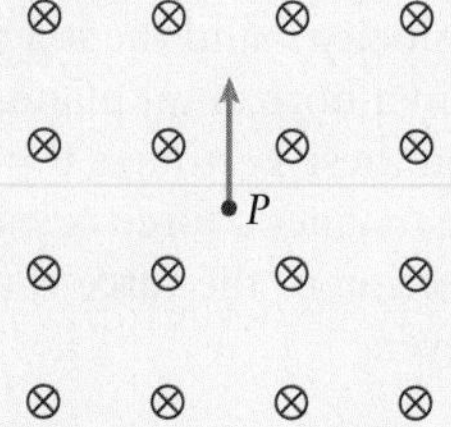

EX 19-1

8. A proton is propelled at 2×10^6 m/s perpendicular to a uniform magnetic field. If it experiences a magnetic force of 5.8×10^{-13} N, what is the magnitude of the magnetic field?

AP® Skills in Action

Questions 9–10 refer to the following figure.

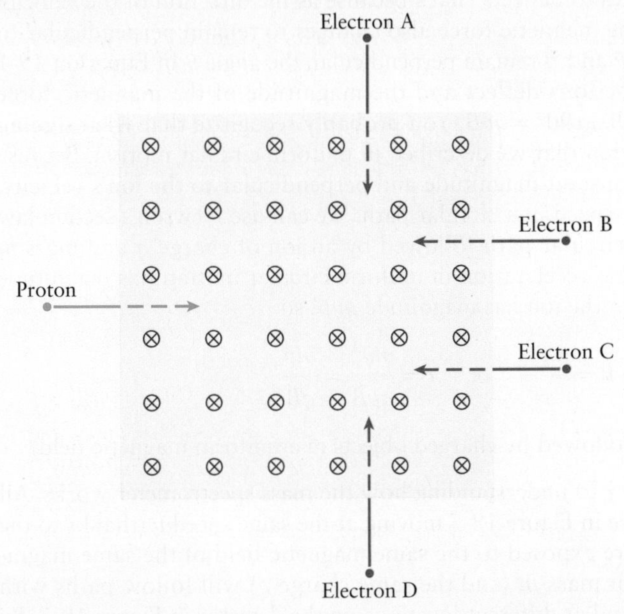

9. A proton is fired into a region of uniform magnetic field pointing into the page as shown in the figure at the left. The proton's initial velocity is shown by the blue vector, which lies in the plane of the page. In which direction does the proton's trajectory bend?
(A) Toward the top of the figure
(B) Toward the bottom of the figure
(C) Out of the figure
(D) Into the figure

10. Four electrons, A, B, C, and D, are fired into a region of uniform magnetic field pointing into the page. The initial velocities of the electrons are shown by the red vectors that lie in the plane of the page in the figure at the left. Each electron will follow a trajectory that is bent by the magnetic force. For which electron(s) could that trajectory lead to the point on the left-hand side where the proton enters the field region?

19-4 A mass spectrometer uses magnetic forces to differentiate atoms of different masses

An important application of the magnetic force on a moving charged object is the **mass spectrometer**, a device used to determine the masses of individual atoms and molecules. **Figure 19-7** shows a simplified version of one type of mass spectrometer. A sample to be analyzed is first vaporized, and its atoms and molecules are ionized by removing one or more electrons so all of the ions have a positive charge. A beam of the moving ions then passes through a *velocity selector*, a region where there is *both* an electric field $\vec{E}$ and a magnetic field $\vec{B}$. The fields $\vec{E}$ and $\vec{B}$ are oriented perpendicular to the ion beam as well as perpendicular to each other. You should be able to show in Figure 19-7 that in this region the electric force on a positive ion is exerted to the left while the magnetic force is exerted to the right. An ion will pass through this region without deflection only if these two forces cancel so that the *net* force on the ion is zero. For an ion of positive charge q, the electric force has magnitude $F_{\text{electric}} = qE$, and the magnetic force has magnitude $F_{\text{magnetic}} = qvB \sin 90° = qvB$ (the ion velocity and the magnetic field are perpendicular, so $\theta = 90°$ in Equation 19-1). So the condition for an ion to continue through this region without being deflected is

$$qE = qvB \quad \text{or} \quad v = \frac{E}{B} \tag{19-2}$$

(speed of ions emerging from a velocity selector)

If an ion has a speed different from $v = E/B$, it will be deflected out of the beam. This is why we call this arrangement of fields a velocity selector: After passing through the

③ In this region ions follow a circular path under the influence of a magnetic field. For ions with the same charge, the trajectory depends on the ion mass.

$\odot\vec{B}$ $\odot\vec{B}$ $\odot\vec{B}$ $\odot\vec{B}$

Aperture

Low-mass ion

High-mass ion

Detector

$\vec{v}$ $\vec{E}$ $\vec{E}$ $\vec{v}$ $\vec{B}\odot$ $\odot\vec{B}$ $\vec{v}$

② Only ions with speed $v = E/B$ pass through the aperture: Slow ions are deflected to the left of the aperture, fast ions to the right.

① Positive ions move upward through the velocity selector. A uniform magnetic field points out of the plane of the figure.

Figure 19-7 A mass spectrometer Electric and magnetic forces are used in this device to separate different ions according to their mass.

Exam Tip

A mass spectrometer or velocity selector is an important example of the forces on a charged object sometimes seen on the AP® Physics 2 Exam. It is not considered a separate topic in the AP® Physics curriculum framework, but we discuss it separately because of its importance as an example.

$\vec{E}$ and $\vec{B}$ fields, the only ions that remain in the beam are those with a speed given by Equation 19-2.

As Figure 19-7 shows, the ion beam then passes into a second region with a magnetic field as before but no electric field. The beam deflects to the right as a result of the magnetic force and continues to deflect. That's because as the direction of the velocity changes, the direction of the magnetic force also changes to remain perpendicular to the ion velocity $\vec{v}$. Because $\vec{v}$ and $\vec{B}$ remain perpendicular, the angle θ in Equation 19-1 remains equal to 90° as the ions deflect and the magnitude of the magnetic force remains the same, $F_M = qvB \sin 90° = qvB$. You probably recognize that what's going on here is exactly the situation that we described in uniform circular motion. Because the force on the ion is of constant magnitude and perpendicular to the ion's velocity, the ion moves at a constant speed in a circular path. We can use Newton's second law to find the radius r of the circular path followed by an ion of charge q and mass m that moves with speed v. The acceleration in uniform circular motion has magnitude $a = v^2/r$, and the net force on the ion has magnitude qvB, so

(19-3) $$qvB = \frac{mv^2}{r} \quad \text{or} \quad r = \frac{mv^2}{qvB} = \frac{mv}{qB}$$

(radius of circular path followed by charged objects in a uniform magnetic field)

Equation 19-3 is the key to understanding how the mass spectrometer works. All of the ions enter the aperture in Figure 19-7 moving at the same speed v (thanks to the velocity selector), and all are exposed to the same magnetic field of the same magnitude B. But ions of different mass m (and the same charge q) will follow paths with different radii and so will land at different locations on the detector in Figure 19-7. By counting how many ions land at each location, we can learn about the composition of the sample being analyzed. Example 19-2 illustrates how this works.

EXAMPLE 19-2 Measuring Isotopes with a Mass Spectrometer

Most oxygen atoms are the isotope ^{16}O ("oxygen-16"), which contains eight electrons, eight protons, and eight neutrons. The mass of this atom is 16.0 u, where 1 u = 1 atomic mass unit = 1.66×10^{-27} kg. The second most common isotope is ^{18}O ("oxygen-18"), which has two additional neutrons and so is more massive than an atom of ^{16}O: Each atom has a mass of 18.0 u. To determine the relative abundances of ^{16}O and ^{18}O, you send a beam of singly ionized oxygen atoms ($^{16}O^+$and $^{18}O^+$, each with a charge $+e = +1.602 \times 10^{-19}$ C) through a mass spectrometer like that shown in Figure 19-7. The magnetic field strength in both parts of the spectrometer is 0.0800 T, and the electric field strength in the velocity selector is 4.00×10^3 N/C. (a) What is the speed of the beam that emerges from the velocity selector? (b) How far apart are the points where the ^{16}O and ^{18}O ions land?

Set Up

We use Equation 19-2 to find the speed v of ions that pass undeflected through the velocity selector and Equation 19-3 to find the radius of the circular path that each isotope follows. The distance from where each ion enters the region of uniform magnetic field to where it strikes the detector is the diameter (twice the radius) of the circular path.

Speed of ions emerging from a velocity selector:

$$v = \frac{E}{B} \quad (19\text{-}2)$$

Radius of circular path followed by charged objects in a uniform magnetic field:

$$r = \frac{mv}{qB} \quad (19\text{-}3)$$

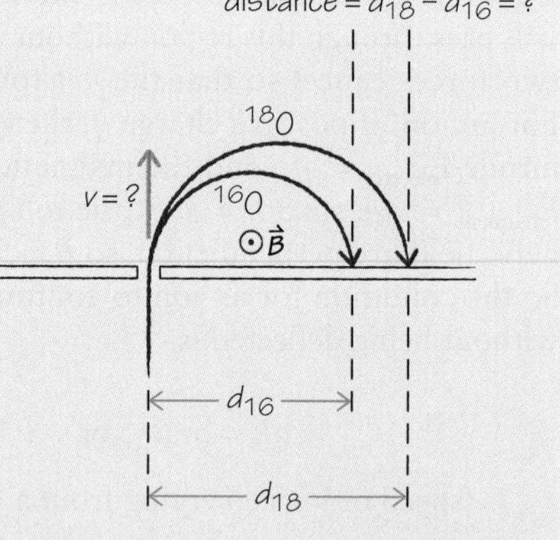

Solve

(a) Calculate the speed of ions that emerge from the velocity selector. Note that this speed does not depend on the charge or mass of the ion, so the $^{16}O^+$ and $^{18}O^+$ ions both emerge with this speed.

From Equation 19-2:

$$v = \frac{E}{B} = \frac{4.00 \times 10^3 \text{ N/C}}{0.0800 \text{ T}} = 5.00 \times 10^4 \ \frac{\text{N}}{\text{T} \cdot \text{C}}$$

Because force can be found from the product of charge, velocity, and magnetic field, the SI unit of magnetic field, T, must be equivalent to $\frac{\text{N}}{\text{C}\frac{\text{m}}{\text{s}}}$, so 1 T · m = 1 N · s/C. Thus, we can write the speed as

$$v = 5.00 \times 10^4 \ \frac{\text{N}}{\text{C} \cdot \text{T}} = 5.00 \times 10^4 \left(\frac{\text{N}}{\text{C}}\right)\left(\frac{\text{C}\frac{\text{m}}{\text{s}}}{\text{N}}\right)$$

$$= 5.00 \times 10^4 \text{ m/s}$$

(b) Calculate the radius r and diameter d of the circular path followed by each isotope.

The masses of the two atoms are

For ^{16}O: $m_{16} = (16.0 \text{ u})(1.66 \times 10^{-27} \text{ kg/u}) = 2.66 \times 10^{-26}$ kg

For ^{18}O: $m_{18} = (18.0 \text{ u})(1.66 \times 10^{-27} \text{ kg/u}) = 2.99 \times 10^{-26}$ kg.

The mass of each positive ion ($^{16}O^+$and $^{18}O^+$) is slightly less than the mass of the neutral atom; the difference is the mass of one electron, which is 9.11×10^{-31} kg = $0.0000911 \times 10^{-26}$ kg. This difference is so small that we can ignore it.

For ^{16}O:

$$r_{16} = \frac{m_{16}v}{qB} = \frac{(2.66 \times 10^{-26} \text{ kg})(5.00 \times 10^4 \text{ m/s})}{(1.602 \times 10^{-19} \text{ C})(0.0800 \text{ T})}$$

$$= 0.104 \ \frac{\text{kg} \cdot \text{m}}{\text{T} \cdot \text{C} \cdot \text{s}}$$

Since 1 T = 1 N · s/(C · m) and 1 N = 1 kg ·m/s^2, it follows that 1 T = 1 kg/(C · s) and 1 T · C · s = 1 kg. So for ^{16}O we have

$$r_{16} = 0.104 \ \frac{\text{kg} \cdot \text{m}}{\text{kg}} = 0.104 \text{ m}$$

$$d_{16} = 2r_{16} = 0.207 \text{ m} = 20.7 \text{ cm}$$

For ^{18}O:

$$r_{18} = \frac{m_{18}v}{qB} = \frac{(2.99 \times 10^{-26} \text{ kg})(5.00 \times 10^4 \text{ m/s})}{(1.602 \times 10^{-19} \text{ C})(0.0800 \text{ T})}$$

$$= 0.1166 \text{ m}$$

$$d_{18} = 2r_{18} = 0.233 \text{ m} = 23.3 \text{ cm}$$

The distance between the positions where the ^{16}O and ^{18}O ions land is the difference between the two diameters.

The distance between where the ^{16}O and ^{18}O ions land is

$$d_{18} - d_{16} = 23.3 \text{ cm} - 20.7 \text{ cm} = 2.6 \text{ cm}$$

This is a substantial distance, so the mass spectrometer does a good job of separating the two isotopes.

Reflect

Experiments of this kind show that, on average, 99.8% of oxygen atoms are ^{16}O and 0.2% are ^{18}O. (Making up a small fraction of a percent is a third isotope, ^{17}O.)

Extend: Measurements of the ratio of ^{18}O to ^{16}O are important to the science of *paleoclimatology*, the study of Earth's ancient climate. One way to determine the average temperature of the planet in the distant past is to examine ancient ice deposits in Greenland and Antarctica. These deposits endure for hundreds of thousands of years; deposits near the surface are more recent, while deeper deposits are older. The ice comes from ocean water that evaporated closer to the equator and then fell as snow in the far north or far south. Each molecule of water is made up of two hydrogen atoms and one oxygen atom, which could be ^{16}O or ^{18}O. A water molecule can more easily evaporate if it contains lighter ^{16}O than if it contains heavier ^{18}O, so the water that evaporated and fell on Greenland and Antarctica as snow contains an even smaller percentage of ^{18}O than ocean water does. This deficiency becomes even more pronounced in colder climates. It has been shown that a decrease of one part per million of ^{18}O in ice indicates a 1.5°C drop in sea-level air temperature at the time it originally evaporated from the oceans.

Using mass spectrometers to analyze ancient ice from Greenland and Antarctica, paleoclimatologists have been able to determine the variation in Earth's average temperature over the past 160,000 years. They have also analyzed the amount of atmospheric carbon dioxide (CO_2) that was trapped in the ice as it froze. An important result of these studies is that higher levels of atmospheric CO_2 have been correlated with elevated temperatures for the past 160,000 years, which is just what we would expect from our discussion of global warming. The tremendous increase in CO_2 levels in the past century due to burning fossil fuels has correlated with recent dramatic increases in our planet's average temperature.

NOW WORK Problems 1–4 from The Takeaway 19-4.

THE TAKEAWAY for Section 19-4

✔ In a mass spectrometer, a magnetic field causes ions of the same speed and charge but different masses to follow different paths.

✔ In a velocity selector, the only charged objects that pass through undeflected are those traveling at the speed for which the electric and magnetic forces have equal magnitudes but opposite directions.

Prep for the AP® Exam

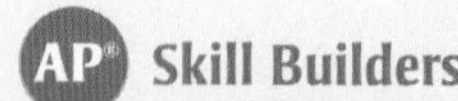

AP® Skill Builders

1. A beam of protons is directed in a straight line along the $+z$ direction through a region of space in which there are crossed electric and magnetic fields. If the electric field is 500 N/C in the $-y$ direction and the protons move at a constant speed of 10^5 m/s, what must be the magnitude and direction of the magnetic field such that the beam of protons continues along its straight-line trajectory?

EX 19-2

2. A beam of ions (each ion has a charge $q = -2e$ and a kinetic energy of 4.00×10^{-13} J) is deflected by the magnetic field of a bending magnet as shown in the figure below. The radius of curvature of the beam is 20.0 cm, and the strength of the magnetic field is 1.50 T.
(a) What is the mass of one of the ions in the beam?
(b) Sketch the path for the given ions in the given magnetic field as a reference path. Then sketch a path for a more massive doubly ionized negative ion and a less massive doubly ionized positive ion, both with the same speed as the given ions, for comparison.

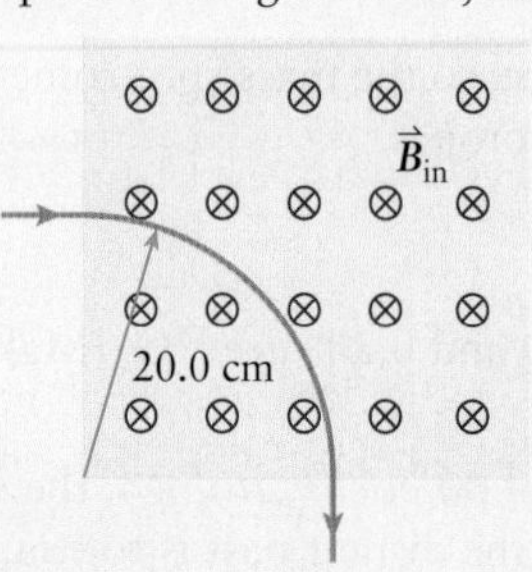

EX 19-2

3. An electron is in a region of space containing a uniform 2.00×10^{-5}-T magnetic field. The electron's speed is 150 m/s and it travels perpendicularly to the field. Under these conditions, the electron undergoes circular motion. (a) Find the radius of the electron's path, and (b) the frequency of its motion.

AP® Skills in Action

EX 19-2

4. When passed through the mass spectrometer of Example 19-2, which of the following ions would follow nearly the same path as an $^{16}O^+$ ion? Assume that all ions are moving at the speed given by Equation 19-2.
(A) A doubly charged oxygen-16 ion ($^{16}O^{2+}$) of mass 16.0 u
(B) A singly charged sulfur-32 ion ($^{32}S^{2+}$) of mass 32.0 u
(C) A doubly charged sulfur-32 ion ($^{32}S^{2+}$) of mass 32.0 u
(D) None of these

19-5 Magnetic fields exert forces on current-carrying wires

The magnetic force that we've described may seem to apply only in certain very special circumstances. But in fact, you use magnetic forces whenever you listen to recorded music through earbuds (**Figure 19-8a**). Within each earbud is a small but powerful magnet adjacent to a flexible plastic cone with a coil of wire attached to it (**Figure 19-8b**). This coil is connected through the earbud wires to your music player, which sends the musical signal to the coil in the form of a varying electric current. Charge carriers within the coil are thus set into motion, and these charge carriers experience magnetic forces exerted by the magnet. These forces pull on the coil that contains the charge carriers and on the plastic "cone" to which the coil is attached. In large speakers, the cone really is cone-shaped, but in earbuds it is almost flat, although it is still called a cone, since it serves the same purpose. We learned in Section 19-3 that the direction of magnetic force depends on the direction in which charged objects move. So the force on the charge carriers, coil, and plastic cone reverses whenever the current in the coil changes direction. The result is that the plastic cone oscillates back and forth in response to the signal coming from your player. This oscillation pushes on the surrounding air, producing a sound wave—the sound of music—that travels to your eardrum. Wireless earbuds use a different technology to get the signal to the coil, based on concepts we will learn in the next chapter.

Let's see how to find the magnitude of the magnetic force on a current-carrying wire, such as a segment of the coil in Figure 19-8b. **Figure 19-9** shows a straight wire of length ℓ that carries a current I. A magnetic field $\vec{B}$ points at an angle θ to the direction of the current and has the same value along the entire length of the wire. We learned that the current in the wire is related to the speed at which individual charged objects drift through the wire:

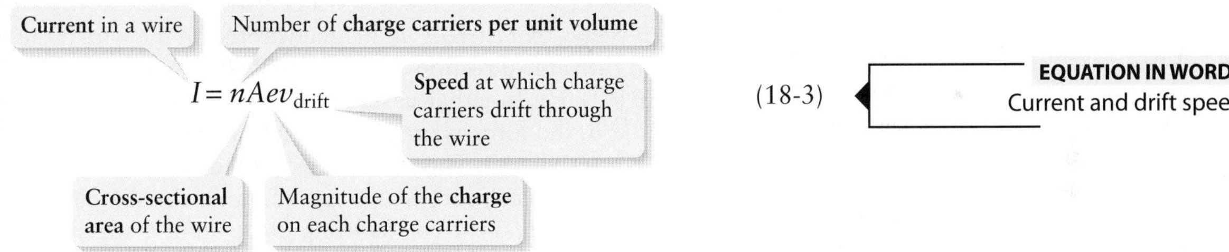

$$I = nAev_{\text{drift}} \quad (18\text{-}3)$$

EQUATION IN WORDS
Current and drift speed

From Equation 19-1, the magnetic force on an individual charge e moving through the wire at speed v_{drift} has magnitude $ev_{\text{drift}}B \sin\theta$. Each such moving charge feels a

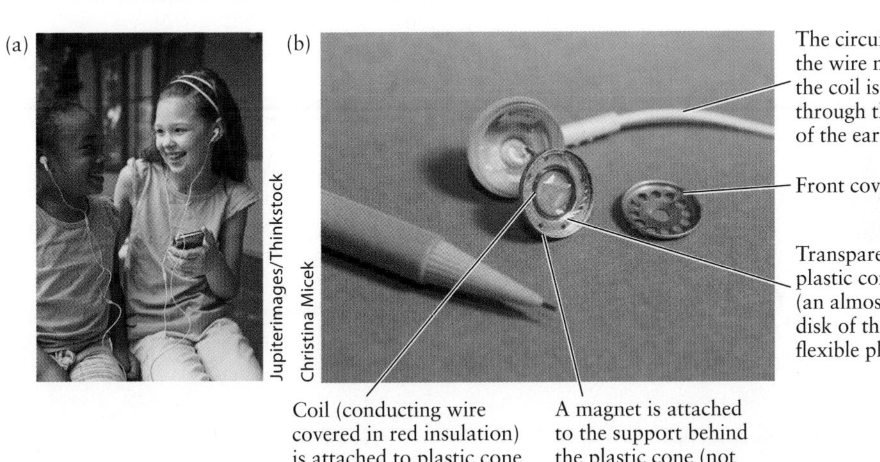

Figure 19-8 Earbud physics (a) Earbuds and other speakers use magnetic forces to produce sound. (b) Internal construction of an earbud.

The force $\vec{F}_M$ on a current-carrying wire in a magnetic field $\vec{B}$ is perpendicular to both $\vec{B}$ and the length of the wire. The direction is given by a right-hand rule.

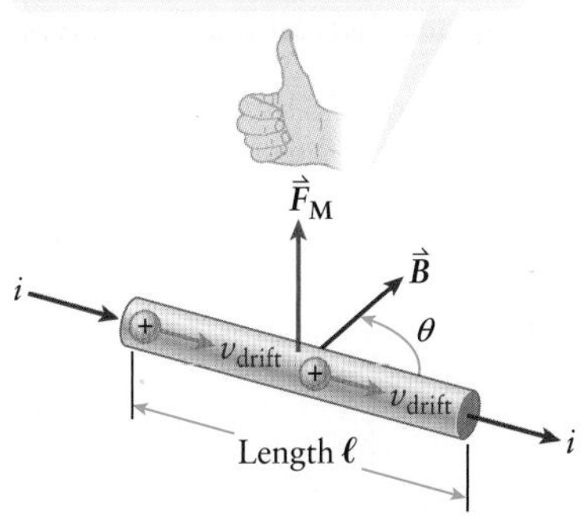

Figure 19-9 Magnetic force on a current-carrying wire The force that a magnetic field exerts on this wire is the sum of the forces on all the charge carriers within the wire.

force of the same magnitude and in the same direction. The total number of such charge carriers in the wire shown in Figure 19-9 is n (the number of charge carriers per unit volume) multiplied by the volume V of the wire, where $V = A\ell$ (the cross-sectional area of the wire multiplied by its length). So the magnitude of the *net* magnetic force on all of the charge carriers in the wire is

$$F_M = (nA\ell)\,(ev_{drift}B\,|\sin\theta|)$$

If we rearrange the terms in this equation, we get

$$F_M = (nAev_{drift})\,\ell\,|B\sin\theta| \qquad (19\text{-}4)$$

The quantity in parentheses in Equation 19-4 is just the current I as given by Equation 18-3. So the magnitude of the magnetic force on a current-carrying wire is

EQUATION IN WORDS
Magnitude of magnetic force on a current-carrying wire

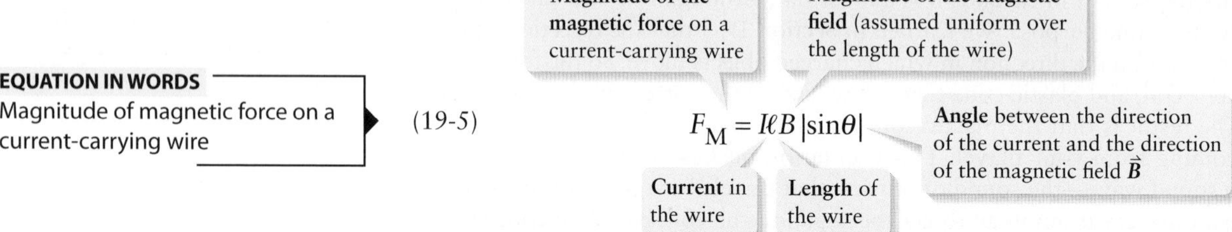

Just as for the magnetic force on a moving charged object, a right-hand rule tells you the direction of the magnetic force on a current-carrying wire. Extend the fingers of your right hand in the direction of the current and orient your hand so that the palm is facing the direction that the magnetic field $\vec{B}$ points. With your right thumb extended, swivel your right hand so that the fingers now point in the direction of $\vec{B}$. Your outstretched thumb points in the direction of the magnetic force exerted on the wire (Figure 19-10).

Equation 19-5 says that there is *no* magnetic force on the wire if the magnetic field $\vec{B}$ points either in the same direction as the current ($\theta = 0$) or in the direction opposite to the current ($\theta = 180°$). In either case, $\sin\theta = 0$ and $F_M = 0$. For a given magnetic field of magnitude B, the force is greatest if $\vec{B}$ points perpendicular to the current so $\theta = 90°$ and $\sin\theta = \sin 90° = 1$. This idea is used in the design of the earbuds shown in Figure 19-8. The wires in each earbud are curved into a circular coil, not straight as in Figure 19-9. But we can treat the coil as being made up of many short segments, each of which is effectively a straight piece of wire. As **Figure 19-10a** shows, the magnetic

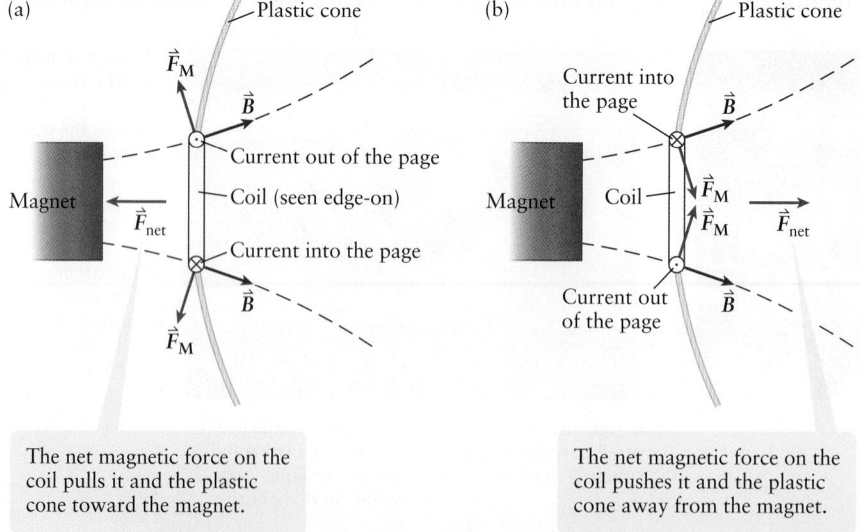

Figure 19-10 Magnetic forces in an earbud (a) The magnetic field produced by an earbud magnet (see Figure 19-8) exerts a net force on the coil (seen here edge-on) and the plastic cone to which it is attached. (b) Reversing the direction of the current reverses the direction of the net force.

field from the magnet in each earbud is perpendicular to each such segment. The forces on different segments are in different directions, but the vector sum of these forces is toward the magnet. If the current is in the reverse direction, as in **Figure 19-10b**, all of the individual forces also reverse direction and the vector sum of the forces is away from the magnet. The current from the music player continually reverses direction, so the coil and attached diaphragm oscillate back and forth, producing a sound wave.

In the following section we'll see how to use Equation 19-5 to help us understand how electric motors work. For now, let's see how to use a magnetic field to levitate a current-carrying wire.

EXAMPLE 19-3 Magnetic Levitation

You set up a uniform horizontal magnetic field that points from south to north and has magnitude 2.00×10^{-2} T. (This is about 400 times stronger than Earth's magnetic field, but is achievable with common magnets.) You want to place a straight copper wire of diameter 0.812 mm in this field, then run enough current through the wire so that the magnetic force will make the wire "float" in midair. This is called *magnetic levitation*. What minimum current is required to make this happen? The density of copper is 8.96×10^3 kg/m^3.

Set Up

To make the wire "float," there must be an upward magnetic force on the wire that just balances the downward gravitational force. We know from Equation 19-5 that to maximize the magnetic force the current direction should be perpendicular to the magnetic field $\vec{B}$. The right-hand rule then shows that the current should be from west to east so that the magnetic force is directed upward. We're not given the mass of the wire, but we can express the mass (and hence the gravitational force on the wire) in terms of its density. We're also not given the length of the wire; as we'll see, this will cancel out of the calculation.

Magnetic force on a current-carrying wire:

$$F_M = I\ell B\,|\sin\theta| \quad (19\text{-}5)$$

Definition of density:

$$\rho = \frac{m}{V} \quad (11\text{-}1)$$

Volume of a cylindrical wire of cross-sectional area A and length ℓ:

$$V = A\ell$$

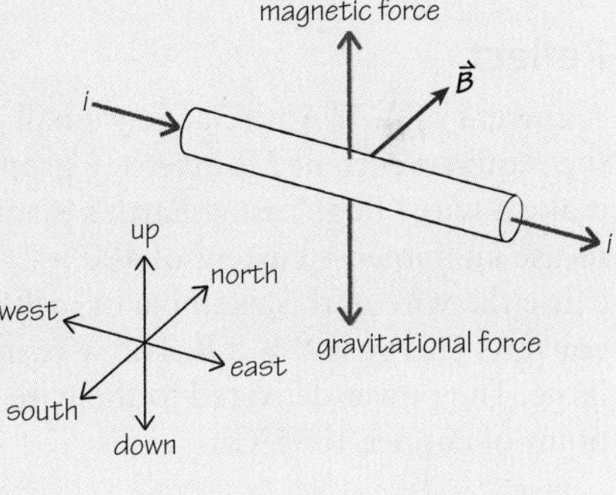

Solve

Write expressions for the two forces (magnetic and gravitational) exerted on the wire.

Assume the wire has length ℓ. Since the current is in a direction perpendicular to the magnetic field, the angle θ in Equation 19-5 is 90°. So the magnitude of the upward magnetic force is

$$F_M = I\ell B\,|\sin 90°| = I\ell B$$

The magnitude of the downward gravitational force on the wire of mass m is

$$F_g = mg$$

The mass of the wire equals the density of copper multiplied by the volume of the wire:

$$m = \rho V = \rho A\ell$$

The cross-sectional area A of the wire is that of a circle of radius r:

$$A = \pi r^2 \quad \text{so} \quad m = \rho A\ell = \rho(\pi r^2)\ell$$

So the magnitude of the gravitational force is

$$F_g = mg = \rho(\pi r^2)\ell g$$

If the wire is floating in equilibrium, the upward magnetic force must balance the downward gravitational force. Use this to solve for the required current I.

In equilibrium the net vertical force on the wire is zero:

$$F_M - F_g = 0, \text{ so } F_M = F_g \text{ and } I\ell B = \rho(\pi r^2)\ell g$$

Both the magnetic force and the gravitational force are proportional to the length ℓ of the wire, so ℓ cancels out of the equation:

$$IB = \rho(\pi r^2)g$$

Solve for the current I:

$$I = \frac{\rho(\pi r^2)g}{B}$$
$$= \frac{(8.96 \times 10^3 \text{ kg/m}^3)(\pi)(0.406 \times 10^{-3} \text{ m})^2 (9.80 \text{ m/s}^2)}{2.00 \times 10^{-2} \text{ T}}$$
$$= 2.27 \frac{\text{kg}}{\text{T} \cdot \text{s}^2} = 2.27 \text{ A}$$

[Check on units: We know from Section 19-3 that $1 \text{ T} = 1 \text{ N/(A} \cdot \text{m)}$, and we also know that $1 \text{ N} = 1 \text{ kg} \cdot \text{m/s}^2$. Therefore, $1 \text{ T} = 1 \text{ kg/(A} \cdot \text{s}^2\text{)}$, so $1 \text{ kg/(T} \cdot \text{s}^2\text{)} = 1 \text{ A}$.]

Reflect

A current of 2.27 A is relatively small, so this experiment in magnetic levitation is not too difficult to perform. Note that the required current I is inversely proportional to the magnitude B of the magnetic field. You can see that if you tried to make a wire "float" using Earth's magnetic field, which is about 1/400 as strong as the field used here, you would need to use an immense current of $400 \times 2.27 \text{ A} = 909 \text{ A}$. That's not practical because a current of that magnitude would cause the wire in this example to melt! (Recall that the rate at which a resistor with resistance R that carries current I converts energy is $P = I^2R$. The wire in this example has a small cross-sectional area, so its resistance R will be fairly large. The power delivered to the wire by a 909-A current will quickly increase its temperature to above the melting point of copper, 1085°C.)

Extend: A practical application of magnetic levitation is train design. By using magnetic forces to make a train float just above the track, the rolling friction between the wheels and the track can be completely eliminated and very high speeds achieved. (Magnetic forces are also used to propel the train forward.) A train of this type in commercial operation in Shanghai, China, reaches a top speed of 431 km/h (268 mi/h). Such train lines require special magnets and wires capable of sustaining very high currents.

NOW WORK Problems 1, 3, and 5–8 from the Takeaway 19-5.

THE TAKEAWAY for Section 19-5

✔ The magnetic force exerted on a current-carrying wire in a magnetic field is perpendicular to both the current direction and the magnetic field direction.

✔ The magnitude of this magnetic force is maximum if the current and magnetic field directions are perpendicular, and zero if the wire axis is along the magnetic field.

AP® Building Blocks

1. A straight wire is positioned in a uniform magnetic field so that the maximum magnetic force on it is 4.0 N. If the wire is 80 cm long and carries a current that is 2.0 A, what is the magnitude of the magnetic field?

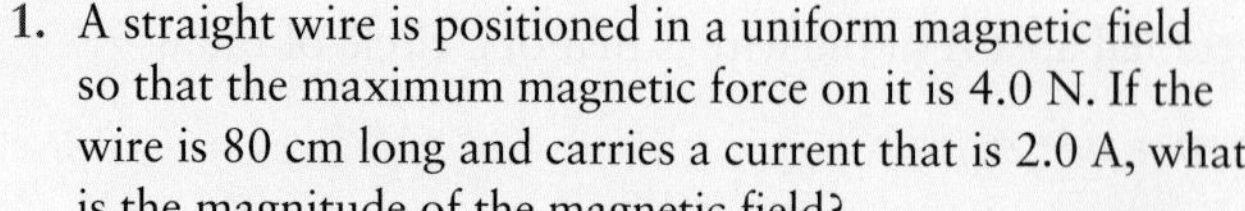

2. A long, straight current-carrying wire is placed in a cubic region that has a uniform magnetic field as shown in the figure below. Does the force on the wire depend on the width of the magnetic field? Explain your answer.

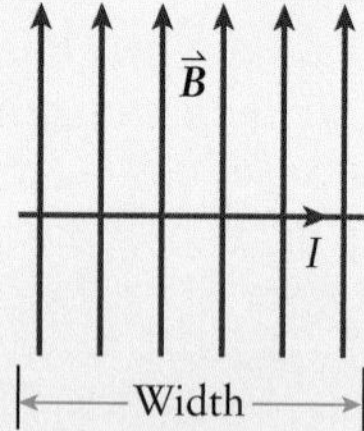

3. A 1.5-m length of straight wire experiences a maximum force of 2.0 N when in a uniform magnetic field that is 1.8 T. What current must be passing through it?

AP® Skill Builders

4. A current-carrying wire is in a region where there is a magnetic field, but there is no magnetic force exerted on the wire. How can this be?

EX 19-3 5. A straight wire of length 0.50 m is conducting a current of 2.0 A and makes an angle of 30° with a 3.0-T uniform magnetic field. What is the magnitude of the force exerted on the wire?

EX 19-3 6. A wire of length 0.50 m is conducting a current of 8.0 A in the $+x$ direction through a 4.0-T uniform magnetic field directed parallel to the wire. What are the magnitude and direction of the magnetic force on the wire?

EX 19-3 7. A wire of length 0.50 m is conducting a current of 8.0 A toward the top of the page and through a 4.0-T uniform magnetic field directed into the page, as shown in the figure below. What are the magnitude and direction of the magnetic force on the wire?

$I = 8.0$ A

8. A straight segment of wire 35.0 cm long carrying a current of 1.40 A is in a uniform magnetic field. The segment makes an angle of 53.0° with the direction of the magnetic field. If the force on the segment is 0.200 N, what is the magnitude of the magnetic field?

AP® Skills in Action

9. A stream of charged objects moving in the same direction can be thought of as a current. Imagine a stream of positively charged objects is moving along the $+x$ axis in a region where there is a uniform magnetic field $\vec{B}$ in the $+y$ direction. You want to balance the magnetic force with an electric field so that the force on the charged objects will be zero. The electric field should be in the
(A) $+x$ direction.
(B) $-x$ direction.
(C) $+z$ direction.
(D) $-z$ direction.

10. A long, straight wire carrying a current can be placed in various orientations with respect to a constant magnetic field as shown in the figure below. What is the direction of the force in each case? Give your answers in terms of the positive and negative x, y, and z directions. If the force is zero, say so. (In each case, the positive z direction is out of the plane of the figures.)

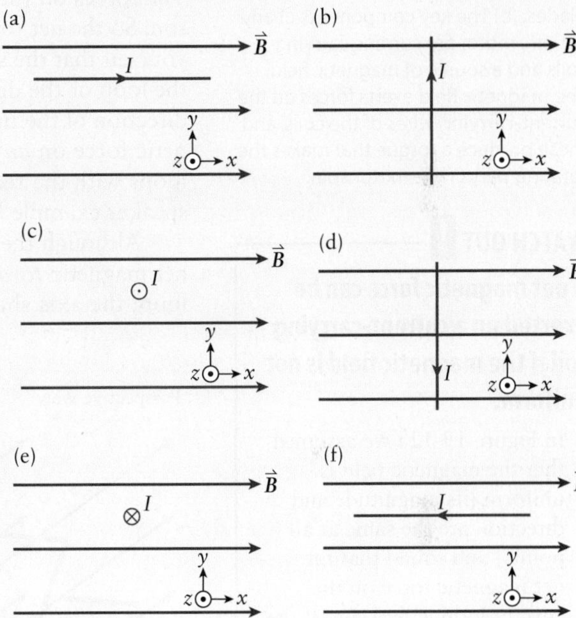

19-6 A magnetic field can exert a torque on a current loop

A common everyday application of magnetic forces on current-carrying wires is an electric motor. An electric motor is at the heart of a kitchen blender, a vacuum cleaner, the starter for an internal-combustion automobile, and many other devices (**Figure 19-11a**). Inside any electric motor you'll find a rotating portion (called the *rotor*) that's wrapped with coils of wire (**Figure 19-11b**), as well as magnets in the stationary part of the motor. When you turn the motor on, causing a current through the coils, the magnets exert forces on the current-carrying wire of the coils. The net effect is that there is a magnetic *torque* that makes the rotor spin. Let's look at a simplified version of this process to see how such a magnetic torque arises.

(a)

iStockphoto/Thinkstock

(b)

iStockphoto/Thinkstock

Figure 19-11 Electric motors and magnetic torque (a) This fan uses an electric motor to rotate the fan blades. (b) The key components of any electric motor are current-carrying coils and a source of magnetic field. The magnetic field exerts forces on the current-carrying wires of the coils, and these produce a torque that makes the rotating part of the motor spin.

Figure 19-12a shows a straight wire bent into a single rectangular loop of wire with sides of length L and W. The loop carries a current i provided by a source of emf (not shown), so we call it a **current loop**. This loop is immersed in a uniform magnetic field of magnitude B and is free to rotate around an axis (shown in green). You should apply the right-hand rule for magnetic forces on a current-carrying wire (see Section 19-5) to each side of this loop. You'll see that the left-hand side of the loop experiences a force to the left, the right-hand side feels a force to the right, the top of the loop feels an upward force, and the bottom of the loop feels a downward force, as Figure 19-12a shows. Each segment of the loop carries the same current and is in the same magnetic field. Since the left- and right-hand sides are the same length and at the same angle to the magnetic field, the forces on these two sides have the same magnitude but opposite directions. Thus these forces cancel, and there is zero net force to the left or the right. The forces on the top and bottom segments of the wire also cancel for the same reason. So the net *force* on this rectangular loop is zero. (You should be able to convince yourself that the same would be true if we reversed the direction of the current around the loop or the direction of the magnetic field, or if we changed the angle between the direction of the field and the plane of the loop.) It turns out that there is zero net magnetic force on *any* closed current-carrying loop in a uniform magnetic field, not just loops with the rectangular shape shown in Figure 19-12a. (The magnetic field in our speaker example before was not uniform!)

Although the net magnetic force on the current loop in Figure 19-12a is zero, the net magnetic *torque* around the axis is not. **Figure 19-12b** is a side view of the current loop; the axis shown in green in Figure 19-12a is perpendicular to the plane of this

WATCH OUT !

A net magnetic force can be exerted on a current-carrying coil if the magnetic field is not uniform.

In Figure 19-12a we assumed that the magnetic field is uniform (its magnitude and direction are the same at all points) and found that the net magnetic force on the current loop is zero. But if the magnetic field magnitude were greater at the left-hand side of the loop than at the right-hand side, then a greater magnetic force would be exerted on the left-hand side. This would result in a net force on the loop to the left. In this chapter we'll restrict our discussion to the simplified case in which the field is uniform.

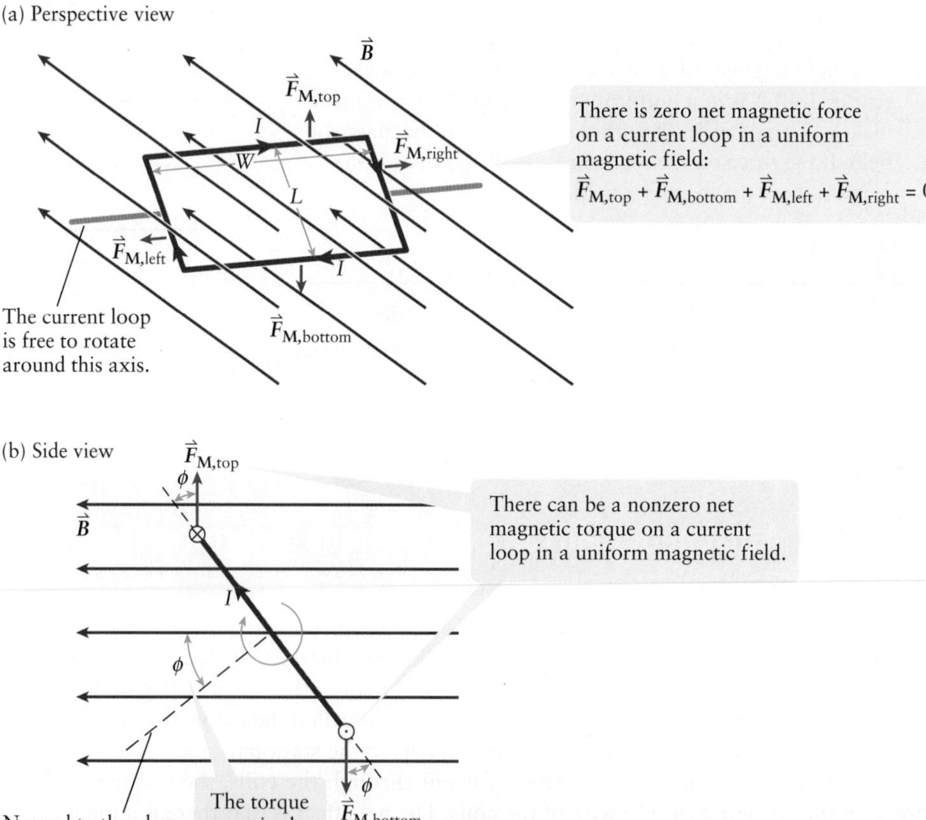
(a) Perspective view
$\vec{B}$
$\vec{F}_{M,top}$
I
W
L
$\vec{F}_{M,right}$
$\vec{F}_{M,left}$
$\vec{F}_{M,bottom}$
There is zero net magnetic force on a current loop in a uniform magnetic field:
$\vec{F}_{M,top} + \vec{F}_{M,bottom} + \vec{F}_{M,left} + \vec{F}_{M,right} = 0$
The current loop is free to rotate around this axis.
(b) Side view
$\vec{F}_{M,top}$
ϕ
$\vec{B}$
There can be a nonzero net magnetic torque on a current loop in a uniform magnetic field.
$\vec{F}_{M,bottom}$
Normal to the plane of the current loop
The torque magnitude depends on the angle ϕ.

Figure 19-12 A current loop in a uniform magnetic field (a) The magnetic forces on the sides of the current loop. (b) These forces can give rise to a net magnetic torque.

figure and passes through the center of the loop. The forces exerted on the near and far sides of the loop, which point out of and into the page, respectively, are directed parallel to this axis and so produce no torque. However, the forces exerted on the top and bottom sides of the loop both give rise to a torque. For the situation shown in Figure 19-12b, both of these forces tend to make the loop rotate clockwise around the axis, so the net torque is clockwise.

Let's calculate the magnitude of the magnetic torque on the loop in Figure 19-12b. First note that the magnetic field is perpendicular to the current in both the top side of the loop and the bottom side of the loop. So in Equation 19-5 for the magnetic force on a current-carrying wire, the angle θ equals 90°. Both the top and bottom sides of the loop have length W (see Figure 19-12a), so the magnitude of the magnetic force on each of these sides is

$$F_M = IWB \sin 90° = IWB \tag{19-6}$$

NEED TO REVIEW?
Turn to page M12 in the Math Tutorial in the back of the book for more information on trigonometry.

To find the torque around the axis that each of these forces produces, we use Equation 9-10:

Magnitude of the torque produced by a force exerted on an extended object

Magnitude of the force

$$\tau = rF_M \sin \phi \tag{10-16}$$

Distance from the rotation axis of the extended object to where the force is applied

Angle between the displacement $\vec{r}$ (from the rotation axis to where the force is applied) and the force $\vec{F}$

EQUATION IN WORDS
Magnitude of torque

In Figure 19-12b the distance r is one-half of the length L of the near or far side of the loop, and ϕ is the angle between the direction of the near or far side of the current loop and the force exerted on either the top or bottom of the loop. This angle is also equal to the angle between the direction of the magnetic field and an imaginary line that's perpendicular to the plane of the current loop. We call this imaginary line the **normal** to the plane of the loop. From Equations 10-16 and 19-6, the magnitude of the torque due to the force on either the top or bottom side of the loop is

$$\tau_{\text{one side}} = \left(\frac{L}{2}\right)F_M \sin \phi = \left(\frac{L}{2}\right)(IWB) \sin \phi = \frac{1}{2} I(LW)B \sin \phi \tag{19-7}$$

The product LW (length times width) in Equation 19-7 is just the area of the current loop: $A = LW$. As we mentioned above, the torque from the top of the loop and the torque from the bottom of the loop are both in the same direction, so the *net* torque on the loop is just double the torque on one side given by Equation 19-7. Using $A = LW$, we have

$$\tau = 2\tau_{\text{one side}} = 2\left(\frac{1}{2}\right)I(LW)B \sin \phi = 2\left(\frac{1}{2}\right)IAB \sin \phi$$

or

Magnitude of the magnetic torque on a current loop

Magnitude of the magnetic field

$$\tau = IAB \sin \phi \tag{19-8}$$

Angle between the normal to the plane of the loop and the direction of the magnetic field $\vec{B}$

Current in the loop

Area of the loop

EQUATION IN WORDS
Magnitude of magnetic torque on a current loop

Although we've assumed a rectangular loop, Equation 19-8 applies to a current loop of area A of *any* shape.

Typically the rotating coil in an electric motor is not just a single loop of wire but has many *turns* (equivalent to many single coils stacked on top of each other). If a coil has N turns, the magnetic torque on it is N times greater than the value given by Equation 19-8.

EXAMPLE 19-4 Angular Acceleration of a Current Loop

A length of copper wire is formed into a square loop with 50 turns. The loop is free to turn with negligible friction about an axis that lies in the plane of the loop and passes through its center. Each side of the loop is 2.00 cm long, and the rotational inertia of the loop about the axis of rotation is 4.00×10^{-6} kg · m^2. The loop lies in a region where there is a uniform magnetic field of magnitude 1.50×10^{-2} T that is perpendicular to the rotation axis of the loop. The current in the loop is 0.500 A. Find the angular acceleration of the loop (magnitude and direction) (a) when the loop is released from rest from the orientation shown in **Figure 19-13**, (b) after the loop has rotated 90°, and (c) after the loop has rotated 180°.

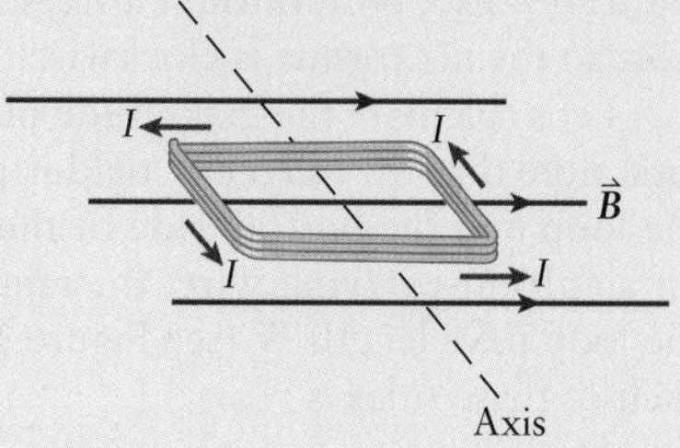

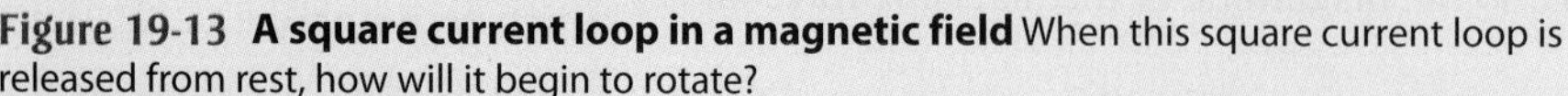

Figure 19-13 A square current loop in a magnetic field When this square current loop is released from rest, how will it begin to rotate?

Set Up

For each orientation we'll use Equation 19-8 to find the magnitude of the magnetic torque on the loop. There is no other torque on the loop (we were told to ignore friction, and gravity exerts zero torque, given the center of mass of the loop lies on the rotation axis). So the magnetic torque is also the net torque, and we can use Equation 10-18 to find the magnitude of the angular acceleration. We'll find the direction of the angular acceleration by looking at the directions of the magnetic forces on the individual sides of the loop. Figure 19-12 shows that forces on the sides parallel to the rotation axis tend to affect rotation. The forces on the other sides cause no torque (compare Figure 19-12a).

Magnitude of magnetic torque on a current loop:

$$\tau = IAB\,|\sin\phi| \qquad (19\text{-}8)$$

Newton's second law for rotational motion (note we have added a subscript to the symbol for rotational inertia to alleviate any confusion in the equations below):

$$\sum \tau_{\text{ext},z} = I_{\text{rot}}\alpha_z \qquad (10\text{-}18)$$

Solve

(a) First find the direction of the magnetic torque and hence the direction of the angular acceleration.

The right-hand rule for the magnetic force on a current-carrying wire shows that the forces on the wire tend to cause a clockwise rotation. So the angular acceleration is clockwise, and when the loop is released it will begin to rotate in the clockwise direction.

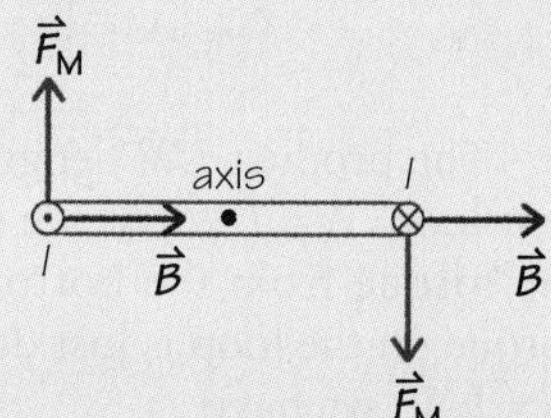

Find the magnitude of the angular acceleration.

The normal to the loop is perpendicular to the magnetic field, so in Equation 19-8 $\sin\phi = \sin 90° = 1$. The loop is square, so the cross-sectional area of the loop is just the square of the length of one side (2.00 cm = 0.0200 m). Because there are 50 turns, the total torque is $N = 50$ times greater than that given by Equation 19-8:

$$\begin{aligned}\tau &= NIAB\,|\sin\phi| \\ &= (50)(0.500\ \text{A})(0.0200\ \text{m})^2(1.50 \times 10^{-2}\ \text{T})(1) \\ &= 1.50 \times 10^{-4}\ \text{T} \cdot \text{A} \cdot \text{m}^2 = 1.50 \times 10^{-4}\ \text{N} \cdot \text{m}\end{aligned}$$

[Recall that $1\ \text{T} = 1\ \text{N/(A} \cdot \text{m)}$, so $1\ \text{T} \cdot \text{A} \cdot \text{m}^2 = 1\ \text{N} \cdot \text{m}$.]

This is the net torque on the loop, so from Equation 10-18 the magnitude of the loop's angular acceleration is

$$\begin{aligned}\alpha &= \frac{\tau}{I_{\text{rot}}} = \frac{1.50 \times 10^{-4}\ \text{N} \cdot \text{m}}{4.00 \times 10^{-6}\ \text{kg} \cdot \text{m}^2} \\ &= 37.5\ \text{rad/s}^2\end{aligned}$$

(If the torque is in N · m and the rotational inertia is in kg · m^2, the angular acceleration in Equation 10-18 is in rad/s^2.)

When released, the loop will begin to rotate in the clockwise direction, as shown above. Note that as the loop rotates, the angle ϕ and the value of sin ϕ will decrease, so the torque and angular acceleration will decrease: This is *not* a situation with constant angular acceleration.

(b) Find the angular acceleration when the loop has rotated 90° from its initial orientation.

When the loop has rotated 90°, the normal to the loop is in the same direction as the magnetic field, so $\phi = 0$ and sin $\phi = 0$. From Equation 19-8 it follows that there is zero torque on the loop at this point in its rotation. We can also see this using the right-hand rule for the magnetic forces on the wires of the loop: These forces do not exert any torque around the axis. Therefore, this orientation represents an equilibrium position for the loop, and the angular acceleration is zero. Note that the loop will be in motion as it passes through this position (there has been an angular acceleration ever since the loop was released). As a result, the loop doesn't stop at this position but keeps on rotating.

(c) Find the angular acceleration when the loop has rotated 180° from its initial orientation.

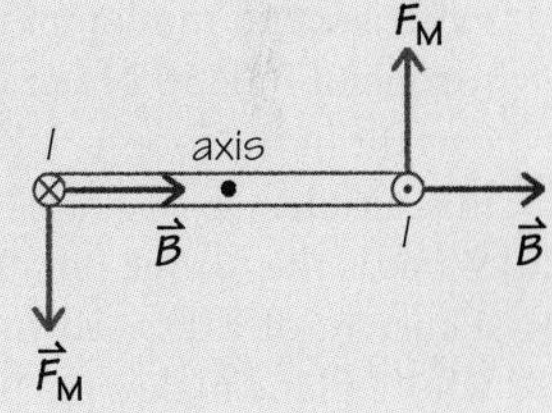

At the 180° position, the normal to the loop is again perpendicular to the direction of the magnetic field, just as in part (a). So again the angle $\phi = 90°$, and again the magnitude of the angular acceleration is $\alpha = 37.5$ rad/s^2. However, because the loop has been flipped over relative to its original orientation, the directions of the forces are reversed. So the torque and angular acceleration are now *counterclockwise*. In fact, the angular acceleration has been increasingly counterclockwise ever since the loop moved past the position in (b). Since passing that position, the loop has been rotating in a clockwise direction but has been slowing down due to the counterclockwise angular acceleration.

Reflect

The motion of the loop in this example should remind you of the motion of a pendulum. When displaced from equilibrium and released, the pendulum will swing toward its equilibrium orientation (hanging straight down) but overshoot that equilibrium and swing to the other side of equilibrium. If friction were truly negligible, the pendulum would keep swinging back and forth indefinitely. The same is true for this current loop: If the axis on which it rotates has negligible friction, it will oscillate back and forth between the orientation in part (a) and the orientation in part (c).

NOW WORK **Problems 1–4, 8, and 9 from the Takeaway 19-6.**

In an electric motor we want the coil to continue rotating in the same direction, not oscillate back and forth like the coil in Example 19-4. To make this happen, most commonly the connection between the coil and the source of emf is arranged so that when the coil is at its equilibrium position [as in part (b) of Example 19-4], the direction of the current *reverses*. As a result, when the coil moves past this equilibrium position the torque is in the same direction as the rotation, and the rotation continues to speed up. The current reverses again after another half-rotation, so the torque is always in the same direction. Another way to keep the direction of rotation the same is to block the current for half the rotation, instead of reversing it. Then the rotational inertia keeps the motor spinning until the next part of the cycle where there is again a torque in the same direction.

With this arrangement the coil would continue to gain rotational speed without limit if there were no other torques exerted on it. In practice there are other torques that oppose the rotation, and the rotational speed reaches an upper limit. That's the case for the electric fan in Figure 19-11a: Air resistance on the fan blades increases as the fan turns faster, and the fan speed stabilizes when the torque due to air resistance just balances the torque of the electric motor (and there is *some* friction in the rotation).

THE TAKEAWAY for Section 19-6

✔ There is zero net force on a current loop in a uniform magnetic field.

✔ A uniform magnetic field can exert a torque on a current loop. The torque is maximum if the normal to the plane of the loop is perpendicular to the magnetic field direction.

Prep for the AP® Exam

AP® Building Blocks

1. A square loop 10.0 cm on a side with 100 turns of wire experiences a minimum torque of zero and a maximum torque of 0.0450 N · m in a uniform magnetic field. If the current in the loop is 2.82 A, calculate the magnetic field magnitude.

2. What is the torque on a round loop of wire that carries a current of 1.00×10^{-2} A, has a radius of 10.0 cm, and whose plane makes an angle of 30.0° with a uniform magnetic field of 0.244 T? How does the answer change if the angle decreases to 10.0°? Increases to 50.0°?

3. A rectangular loop of wire with width 10.0 cm and length 20.0 cm carries 100 mA of current and sits in the y–z plane with current directed clockwise when viewed from $+x$. A magnetic field $B = 0.10$ T in the $+x$ direction fills all space. Compute the torque on the loop.

AP® Skill Builders

4. A circular loop of wire has radius 10.0 cm and carries 0.50 A of current. The normal of the loop makes an angle of 25° with a 0.25-T uniform magnetic field.
   (a) Compute the torque the loop experiences in the magnetic field.
   (b) Sketch the loop and field and indicate the direction of rotation on your drawing.
5. A circular flat coil that has N turns, encloses an area A, and carries a current I has its central axis of rotation parallel to a uniform magnetic field $\vec{B}$. The net force on the coil is
   (A) zero.
   (B) $NIAB$.
   (C) IBA.
   (D) NIA.
6. How is it possible for an object that experiences no net magnetic force to experience a net magnetic torque?

7. A circular flat coil that has N turns, encloses an area A, and carries a current I has its central axis of rotation parallel to a uniform magnetic field $\vec{B}$. The net torque on the coil is
   (A) zero.
   (B) $NIAB$.
   (C) IBA.
   (D) NIA.

AP® Skills in Action

EX 19-4

8. A wire loop with 50 turns is formed into a square with sides of length s. The loop is completely within a 1.5-T uniform magnetic field that points in the negative y direction. The plane of the loop is tilted off the x axis by 15°, as shown in the figure below. When the current in the loop is 2.0 A, the torque on the loop has magnitude 0.035 N · m. Calculate the lengths of the sides s of the square loop.

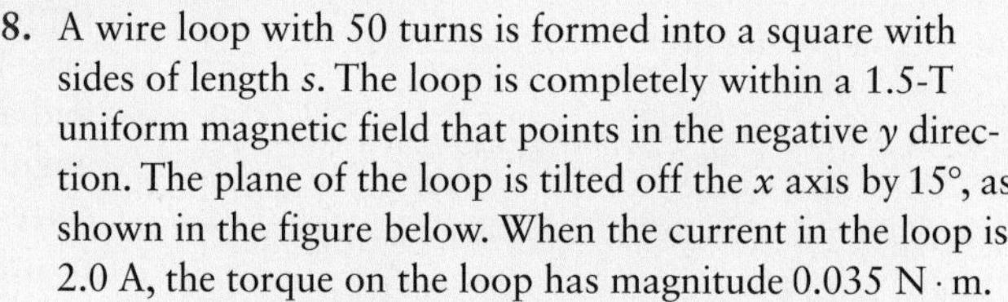

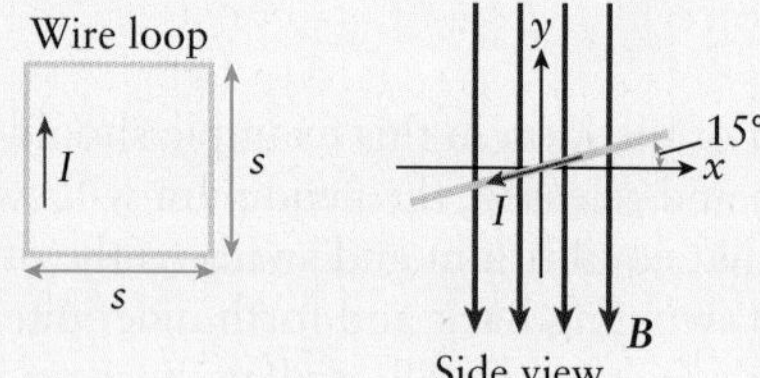

9. MRI (magnetic resonance imaging) scans are often prohibited for patients with an implanted device such as a pacemaker. There are multiple reasons for this, including electromagnetic interference that confuses the device. Here we consider another potential issue: the physical torque on the device due to the magnetic field. Suppose a patient's pacemaker implant requires 2.0 mA of current to function. The leads of the device travel close together until they reach the heart and then split off to touch the top and bottom of the heart. So we can approximate the size of the current loop as half the cross-sectional area of the heart, and treat the heart as a box that is 12 cm long and

8.0 cm wide. Most MRI machines use a magnetic field of 1.5 T. Calculate the maximum torque on the current loop.

10. Suppose the number of turns in a current loop in a uniform magnetic field were increased from 50 to 100. The maximum angular acceleration of the loop would then be
(A) twice as great.
(B) 1/2 as great.
(C) 1/4 as great.
(D) the same as before.

19-7 Current-carrying wires create magnetic fields

So far in our discussion of magnetism, we've looked at the forces and torques that a magnetic field exerts on a current-carrying wire or the force on a moving charged object. As we described in Section 19-2, all magnetic fields are produced by electric charged objects in motion, whether it's a current in the coils of the electromagnet shown in Figure 19-1c, electrons in motion within the atoms of an iron bar magnet, or electric currents in the human brain as shown in **Figure 19-14**. So current-carrying wires also *produce* magnetic fields (see Figure 19-1c and Figure 19-4). To complete our understanding of magnetic forces, we need to be able to calculate the magnetic field produced by an arrangement of charged objects in motion. This is much like Chapter 16, where we needed to learn how to calculate the electric field due to an arrangement of charged objects to complete our understanding of electric forces.

We saw that the electric field due to a charge distribution is the vector sum of the electric fields due to all the charged objects in the distribution. These calculations can be rather challenging unless the charge distribution is very simple. When we look at the analogous problem of finding the *magnetic* field due to a distribution of moving charged objects, we find that the problem is even more complicated. That's because the magnetic field due to even a single moving charged object is itself rather complex: The field does not point directly away from or toward the moving charge, but in a direction perpendicular to the velocity and the field. Rather than looking at how to do calculations of this kind, we'll look at the result for one important situation, the magnetic field due to a long, straight current-carrying wire. We'll then see an alternative approach for calculating a magnetic field using *Ampère's law*. This law is a very powerful one, but it can be used for field calculations only in certain simple situations. We will then see how this lets us approximate the magnetic field for a charged object that is not moving at close to the speed of light (most things we encounter)! We'll also look at the magnetic field produced by a current loop, which helps us understand the dipole nature of the magnetic field, and dig a little deeper into the property of a material that determines how well it is attracted to a permanent magnet, or can become a magnet.

Electric currents in the human brain generate weak magnetic fields. The colors in this magnetoencephalogram represent the strength of the magnetic field produced in different regions.

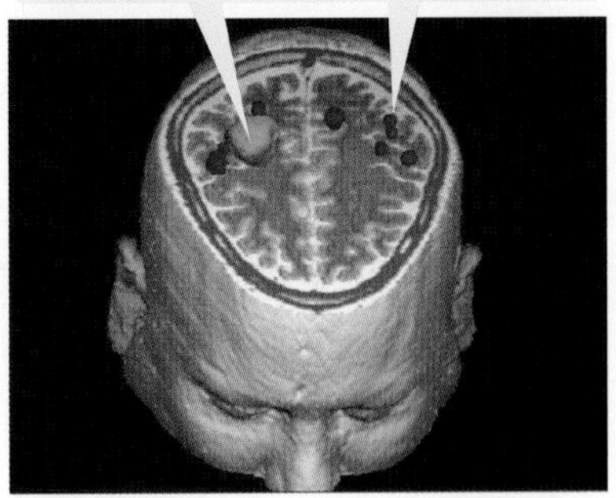

Figure 19-14 Currents as sources of magnetic field Electric currents in biological systems produce magnetic fields.

A Long, Straight Wire and Ampère's Law

Consider a very long, straight wire that carries a constant current I. Experiment and calculation both show that the magnetic field due to this current has the properties shown in **Figure 19-15**. Note that the wire and the magnetic field that it produces have *cylindrical symmetry*: The wire and the field pattern look exactly the same if you rotate the wire around its length. The magnitude of the field is given by

Magnitude of the magnetic field due to a long, straight wire

Permeability of free space

$$B = \frac{\mu_0 I}{2\pi r} \quad (19\text{-}9)$$

Current in the wire

Distance from the wire to the location where the field is measured

EQUATION IN WORDS
Magnitude of magnetic field due to a long, straight wire

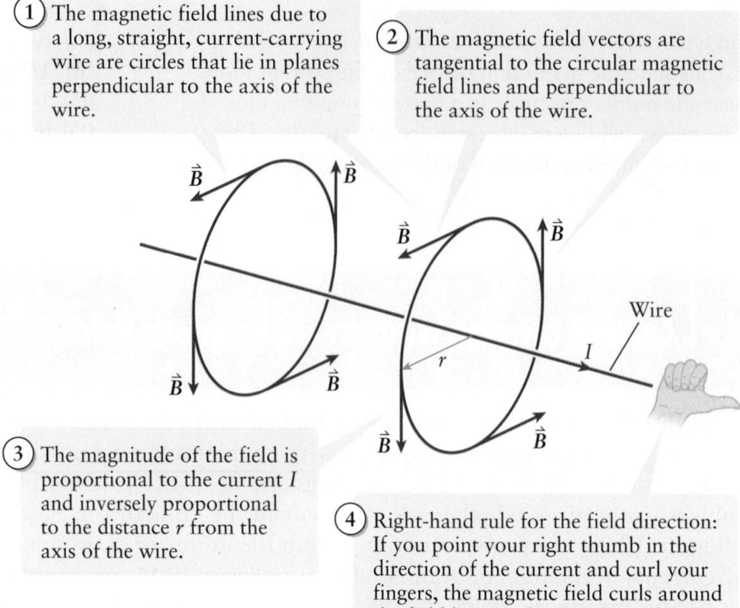

1 The magnetic field lines due to a long, straight, current-carrying wire are circles that lie in planes perpendicular to the axis of the wire.
2 The magnetic field vectors are tangential to the circular magnetic field lines and perpendicular to the axis of the wire.
B
Wire
I
r
3 The magnitude of the field is proportional to the current I and inversely proportional to the distance r from the axis of the wire.
4 Right-hand rule for the field direction: If you point your right thumb in the direction of the current and curl your fingers, the magnetic field curls around the field lines in the direction of the curled fingers of your right hand.

Figure 19-15 Magnetic field of a long, straight current-carrying wire The magnitude of this field is given by Equation 19-9.

Equation 19-9 and the field properties shown in Figure 19-15 are strictly correct only if the wire is infinitely long. But they are very good approximations if the distance r is small compared to the length of the wire.

The constant μ_0 in Equation 19-9, called the **permeability of free space** for historic reasons, plays a role in magnetism that's comparable to the role of the constant ε_0, the *permittivity* of free space, in electricity. Its value μ_0 determined experimentally is $\mu_0 = 1.25663706 \times 10^{-7}\ \text{T} \cdot \text{m/A}$. Remarkably, this is equal to $4\pi \times 10^{-7}\ \text{T} \cdot \text{m/A}$ to nine significant figures. This is not a coincidence, but a result of the way that the ampere was defined prior to 2018. In the old definition of the ampere, at a distance of 1 m from a long, straight wire carrying a current of 1 A, the magnetic field strength was exactly

$$B = \frac{\mu_0 I}{2\pi r} = \frac{(4\pi \times 10^{-7}\ \text{T}\cdot\text{m/A})(1\ \text{A})}{2\pi(1\ \text{m})} = 2 \times 10^{-7}\ \text{T}$$

The current definition of 1 ampere is 1 coulomb per second, with the coulomb defined in terms of the magnitude of charge e on a proton or electron. With this definition μ_0 is not exactly equal to $4\pi \times 10^{-7}\ \text{T} \cdot \text{m/A}$, but is close enough that we can safely use $\mu_0 = 4\pi \times 10^{-7}\ \text{T} \cdot \text{m/A}$ in calculations.

WATCH OUT !

Remember that vectors are straight, never curved.

When drawing the magnetic field around a long, straight wire, it may be tempting to draw curved arrows to represent how the magnetic field lines curl around the wire. But a vector always denotes a *single* direction and so *cannot* be curved. At any point along a magnetic field line, the direction of the field is always along the *tangent* to the field line, as shown in Figure 19-15.

We can use Equation 19-9 to illustrate a useful principle about magnetic fields and their sources. Imagine that we draw a circle of radius r around a long, straight wire as shown in **Figure 19-16**. Imagine further that we break the circle into a number of segments of length $\Delta\ell$. If $\Delta\ell$ is sufficiently small, we can treat each segment as being straight. Then for each segment take the component $B_{\parallel}$ of the magnetic field parallel to that segment and multiply it by the segment length $\Delta\ell$. If we add up the values of these products for every segment in the circle, the result is a quantity called the **circulation** of the magnetic field around the circle:

(19-10) $$\text{circulation} = \sum B_{\parallel}\Delta\ell$$

For the circle shown in Figure 19-16, Equation 19-9 tells us that $B_{\parallel}$ is equal to $\mu_0 I/(2\pi r)$ at every point around the circle. That's because the magnetic field is everywhere tangent to the circle and so parallel to a short segment of length on the circle. Therefore, the circulation of the magnetic field as defined by Equation 19-10 is

(19-11) $$\text{circulation} = \sum\left(\frac{\mu_0 I}{2\pi r}\right)\Delta\ell = \frac{\mu_0 I}{2\pi r}\sum \Delta\ell$$

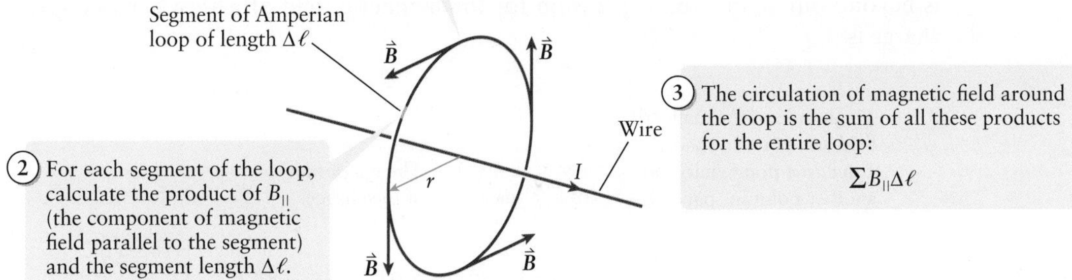

① We draw an Amperian loop that encircles the current-carrying wire. (This particular loop is a circle, and so coincides with a field line.)
Segment of Amperian loop of length $\Delta\ell$
$\vec{B}$
$\vec{B}$
Wire
I
r
② For each segment of the loop, calculate the product of $B_{\parallel}$ (the component of magnetic field parallel to the segment) and the segment length $\Delta\ell$.
$\vec{B}$
$\vec{B}$
③ The circulation of magnetic field around the loop is the sum of all these products for the entire loop:
$\Sigma B_{\parallel}\Delta\ell$

Figure 19-16 Ampère's law and circulation Ampère's law states that the circulation of the magnetic field around a closed path (called an *Amperian loop*) is proportional to the current that passes through that path.

In Equation 19-11 we've taken the quantity $\mu_0 I/(2\pi r)$ outside the sum because it has the same value everywhere around the circle and so in all terms of the sum. The quantity $\Sigma\,\Delta\ell$ is the sum of the lengths of all of the segments that make up the circle—that is, the circumference of the circle, which is equal to $2\pi r$. So we can rewrite Equation 19-11 as follows:

$$\text{circulation} = \frac{\mu_0 I}{2\pi r}(2\pi r) = \mu_0 I \qquad (19\text{-}12)$$

The circulation as given by Equation 19-12 does *not* depend on the radius r of the circle. (If the circle is larger, the magnetic field has a smaller magnitude, but it takes more segments of length $\Delta\ell$ to go all the way around the circle.) Remarkably, the same result holds true even if we draw a noncircular path around the wire: The circulation around *any* path that encloses the wire is equal to $\mu_0 I$. This is an example of **Ampère's law**, which was discovered by the French physicist André-Marie Ampère (pronounced "ahm-pair") in 1826:

Circulation of magnetic field around an Amperian loop

Permeability of free space

$$\sum B_{\parallel}\Delta\ell = \mu_0 I_{\text{through}} \qquad (19\text{-}13)$$

Current through the Amperian loop

EQUATION IN WORDS
Ampère's law

An **Amperian loop** is simply a closed path in space; the circle in Figure 19-16 is an example. The subscript "through" reminds us that the right-hand side of Equation 19-13 should include only current that passes through the interior of the Amperian loop.

It can be shown that Ampère's law is true for *any* magnetic field and *any* Amperian loop, no matter what the geometry of the current that produces the magnetic field. (The proof is beyond our scope.) However, it's only in certain very symmetric situations that Ampère's law helps us calculate the value of magnetic field.

This does not tell us about the magnetic field of a single moving charged object, but it does let us think about what that field should look like. We can think of the current passing through any point in the wire in a very small interval of time as the motion of a charged object. So, if we look at the direction of current as the direction of motion of a charged object, the magnetic field around it should be perpendicular to its velocity. Let's assume we want to know the magnetic field of a point charge—that field will be the same shape as the magnetic field in Figure 19-16, if we imagine the location of the point charge as being at the center of the circle of radius r drawn in that figure. Because there is not a constant line of charge carriers in front and behind the single point charge, it seems reasonable that the field would change in size and direction as you moved either in front of or behind the location of the test charge. While the proof is beyond our scope, the expression for the magnetic field of a slowly moving test charge is

EQUATION IN WORDS
Magnitude of the magnetic field due to a slowly moving point charge

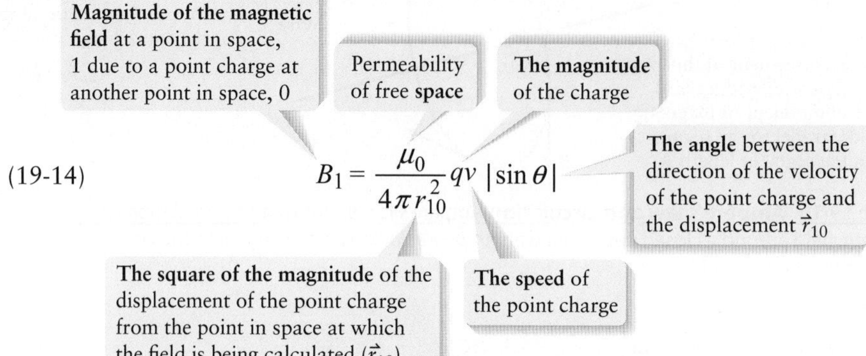

$$B_1 = \frac{\mu_0}{4\pi r_{10}^2} qv\,|\sin\theta| \qquad (19\text{-}14)$$

> **AP Exam Tip**
>
> The only locations at which the calculation of the magnetic field of a moving point charge will be asked are along the circle instantaneously centered on its location and direction of motion. Thus, the r_{10} is the radius of the circle, and $\theta = 90°$.

We keep reminding ourselves that this is for a slowly moving test charge (compared to the speed of light). While we won't get into the details until Chapter 26, remember that we mentioned magnetism has something to do with relativity. We didn't worry about this when we were discussing current. It turns out we don't have to! Remember that the speed of the charge carriers in a current, the drift velocity, is very small. In fact the average drift velocity is more than 12 orders of magnitude smaller than the speed of light so the "slowly moving" approximation is great. Don't get too concerned about the direction of the vectors in Equation 19-14. For any location on the circle around the direction of motion of a positive point charge, the magnetic field can be found with the same right-hand rule as the magnetic field of a current-carrying wire. If it is a negative point charge, the magnetic field can be found by assuming the current is in the opposite direction to the motion of the point charge.

In the following example we use Ampère's law to find the magnetic field due to a rather different distribution of current.

EXAMPLE 19-5 Magnetic Field Due to a Coaxial Cable

A coaxial cable consists of a solid conductor of radius R_1 surrounded by insulation, which in turn is surrounded by a thin conducting shell of radius R_2 made of either fine wire mesh or a thin metallic foil (**Figure 19-17**). The combination is enclosed in an outer layer of insulation. The inner conductor carries current in one direction, and the outer conductor carries it back in the opposite direction. For a coaxial cable that carries a constant current I, find expressions for the magnetic field (a) inside the inner conductor at a distance $r < R_1$ from its central axis, (b) in the space between the two conductors, and (c) outside the coaxial cable. Assume that the moving charge in the inner conductor is distributed uniformly over the volume of the conductor.

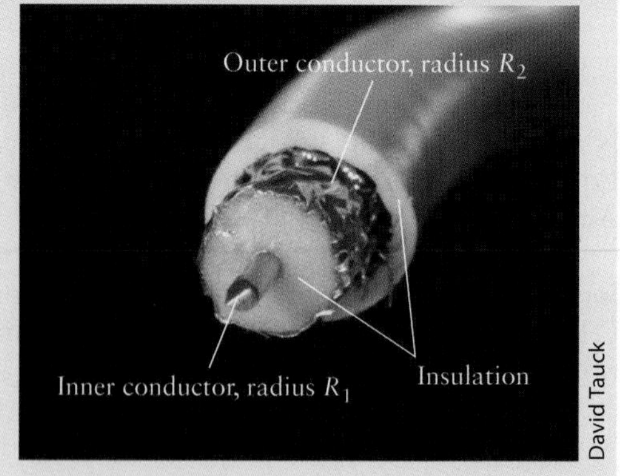

Figure 19-17 A coaxial cable There are equal amounts of current in opposite directions in the inner and outer conductors of this cable.

Set Up

Both the inner conductor separately and the coaxial cable as a whole have the same cylindrical symmetry as a long, straight wire (Figure 19-15). So we expect that the field lines are circles concentric with the axis of the cable. Just as for the long, straight wire, this means that it's natural to choose circular paths concentric with the cable axis as the Amperian loops. To find the field in the three regions, we'll choose the radius r of the Amperian loop to be less than R_1 in part (a), between R_1 and R_2 in part (b), and greater than R_2 in part (c).

Ampère's law:

$$\sum B_{\|}\Delta\ell = \mu_0 I_{\text{through}} \quad (19\text{-}13)$$

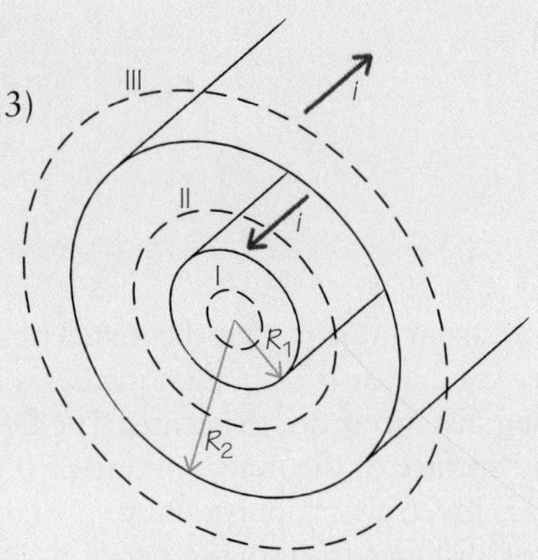

Dashed circles labeled I, II, and III: Amperian loops for parts (a), (b), and (c), respectively

Solve

(a) Find the field inside the inner conductor by using an Amperian loop of radius $r < R_1$.

Inside the inner conductor the magnetic field has magnitude B_{inner} and points tangent to the Amperian loop, so $B_{\|} = B_{\text{inner}}$. The left-hand side of the Ampère's law equation is

$$\sum B_{\|}\Delta\ell = B_{\text{inner}} \sum \Delta\ell = B_{\text{inner}}(2\pi r)$$

The Amperian loop encloses area πr^2, which is less than the cross-sectional area πR_1^2 of the inner conductor. The current through the loop is therefore a fraction $(\pi r^2)/(\pi R_1^2)$ of the total current I in the inner conductor:

$$I_{\text{through}} = I\left(\frac{\pi r^2}{\pi R_1^2}\right) = I\frac{r^2}{R_1^2}$$

Insert these into Equation 19-13 and solve for B_{inner}:

$$B_{\text{inner}}(2\pi r) = \mu_0 I \frac{r^2}{R_1^2}$$

$$B_{\text{inner}} = \mu_0 I \frac{r^2}{2\pi r R_1^2} = \frac{\mu_0 I r}{2\pi R_1^2}$$

(b) Find the field between the conductors by using an Amperian loop of radius r, where $R_1 < r < R_2$.

Between the two conductors the magnetic field of magnitude B_{between} also points tangent to the Amperian loop, so as in part (a) the left-hand side of Equation 19-13 is

$$\sum B_{\|}\Delta\ell = B_{\text{between}}(2\pi r)$$

The Amperian loop encloses the entire inner conductor, so $I_{\text{through}} = I$. Insert these into Equation 19-13 and solve for B_{between}:

$$B_{\text{between}}(2\pi r) = \mu_0 I$$

$$B_{\text{between}} = \frac{\mu_0 I}{2\pi r}$$

(c) Find the field outside the cable by using an Amperian loop of radius $r > R_2$.

Just as in parts (a) and (b), outside the outer conductor the magnetic field of magnitude B_{outer} points tangent to the Amperian loop, so

$$\sum B_{\|}\Delta\ell = B_{\text{outer}}(2\pi r)$$

The Amperian loop encloses both conductors, each of which carries current I. Because the currents are in opposite directions, the *net* current through the loop is $I_{\text{through}} = 0$. So Equation 19-13 gives

$$B_{\text{outer}}(2\pi r) = \mu_0(0)$$
$$B_{\text{outer}} = 0$$

Reflect

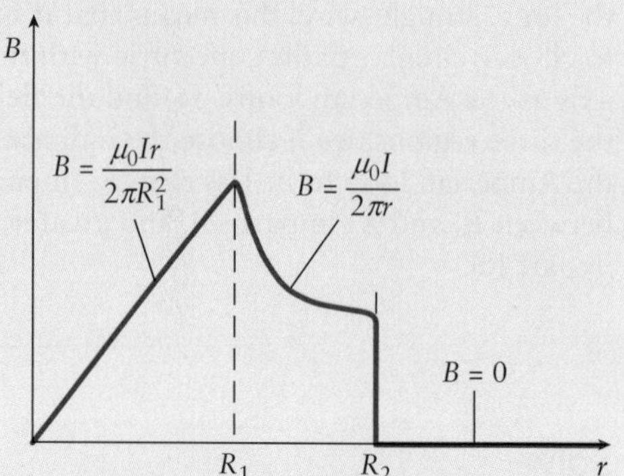

Our result from (a) says that the magnetic field is zero at the center of the inner conductor ($r = 0$), then increases in direct proportion to r with increasing distance from the center. The field reaches its maximum value at the outer surface of the inner conductor ($r = R_1$). Between the conductors the field is inversely proportional to r, so the magnitude decreases with increasing distance from the center of the cable. Outside the outer conductor there is *zero* magnetic field.

Coaxial cables are often referred to as "shielded" cables. The arrangement of the two conductors eliminates the presence of stray magnetic fields outside the cable. The shielding also serves to isolate the inner conductor from external electromagnetic signals. You'll find a coaxial cable connected to the back of most television sets (it's the "cable" in the term *cable TV*); the signal carried by this cable involves an alternating current and hence a varying magnetic field rather than a steady one, but the shielding principle is the same.

NOW WORK Problems 2–4, 7, and 8 from The Takeaway 19-7.

Magnetic Field of a Current Loop

An important special case for which Ampère's law is *not* helpful is the magnetic field produced by a current loop (a current-carrying wire bent into a circle). **Figure 19-18** shows some of the magnetic field lines for such a loop.

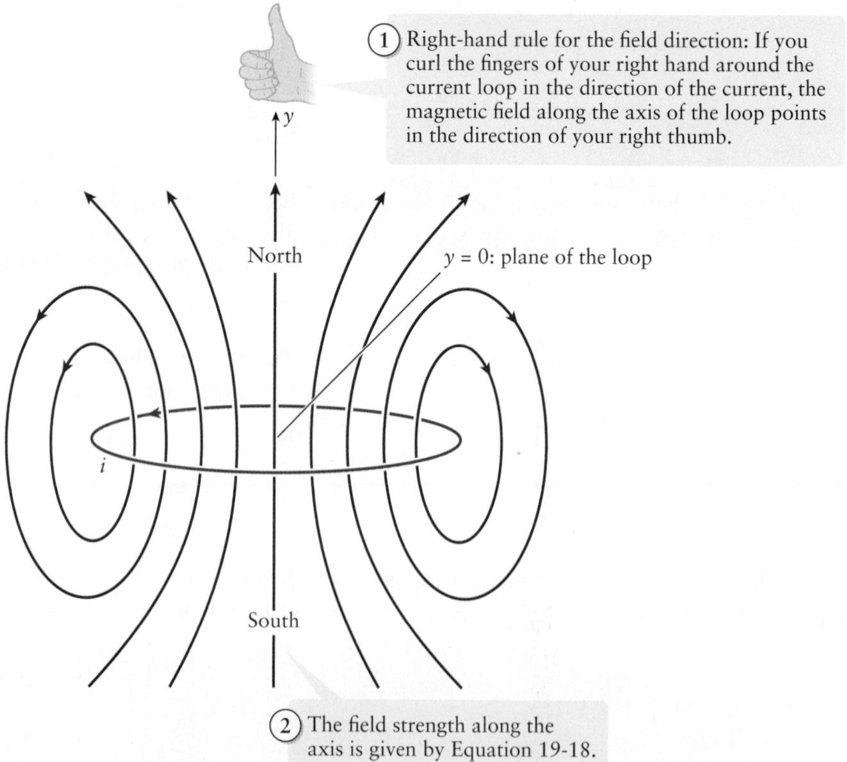

1 Right-hand rule for the field direction: If you curl the fingers of your right hand around the current loop in the direction of the current, the magnetic field along the axis of the loop points in the direction of your right thumb.
y
North
y = 0: plane of the loop
i
South
2 The field strength along the axis is given by Equation 19-18.

Figure 19-18 Magnetic field due to a current loop This magnetic field pattern is similar to the electric field pattern of an electric dipole (Figure 16-14).

Unlike the case for a long, straight wire (Figure 19-15) or for a coaxial cable (Example 19-5), the magnetic field does *not* have the same magnitude at all points on a field line: The magnitude B is greater where the lines are closer together. So choosing an Amperian loop that coincides with a field line will not give us a simple equation for the magnitude B, as was the case in Example 19-5. To calculate the magnetic field at a given point in this situation, it's necessary to find the contribution to the field due to each short segment of the loop, then add those contributions using vector arithmetic. Such a calculation is beyond our scope. Here's the result for the magnitude of the magnetic field along the axis of the current loop, labeled y in Figure 19-18:

$$B = \frac{\mu_0 I}{2} \frac{R^2}{(R^2 + y^2)^{3/2}} \tag{19-15}$$

In Equation 19-15 I is the current in the loop, R is the radius of the loop, and y is the coordinate along the y axis, where $y = 0$ represents the plane of the loop. If we substitute $y = 0$ into Equation 19-15, we get the field magnitude at the very center of the loop. At points far from the loop, so y is much greater than R, we can replace $R^2 + y^2$ with y^2 to good approximation. Equation 19-15 then becomes

$$B = \frac{\mu_0 i}{2} \frac{R^2}{(y^2)^{3/2}} = \frac{\mu_0 i\, R^2}{2|y|^3} \tag{19-16}$$

(We've added the absolute value signs because y can be positive or negative, but the field magnitude B must be positive.) So at large distances from a current loop, the magnetic field is inversely proportional to the *cube* of the distance from the loop.

This result is reminiscent of the *electric* field of an electric dipole (a combination of a positive charge q and a negative charge $-q$): We found in Example 16-7 (Section 16-5) that at large distances from an electric dipole, the electric field due to that dipole is inversely proportional to the cube of the distance. The overall magnetic field pattern of a current loop also has some similarities to the electric field pattern of an electric dipole. (Compare Figure 19-18 and Figure 16-14; if you rotate Figure 16-14 clockwise 90°, the similarities will be more evident.)

In light of these similarities, we use the term *magnetic dipole* to refer to a current loop. The two "poles" of a magnetic dipole are the points just above and just below the loop, as Figure 19-18 shows. We call these poles north and south by analogy to the poles of a bar magnet: The magnetic field points away from the current loop at its north pole and points toward the current loop at its south pole, just as for a bar magnet (see Section 19-2).

Note that unlike the two charged objects that make up an electric dipole, the north and south poles of a current loop can never be separated: The current loop must always have two sides! As we discussed in Section 19-2, a permanent magnet such as a bar magnet acts as a magnetic dipole. That's because a permanent magnet is really just a collection of *atomic* current loops, each the result of electron motions within the atom. Their combined effect is the same as electrons moving around the circular loop of wire in Figure 19-18. The poles of such a magnet can no more be separated than can the two sides of a current loop.

Earth's magnetic field is nearly that of a dipole, with the axis of the field tilted slightly from Earth's rotation axis (Figure 19-2). The magnetic field is produced because molten material in the outer regions of Earth's core is in a state of continuous motion, and this motion gives rise to electric currents that generate the field. The photo that opens this chapter illustrates one dynamic consequence of our planet having a magnetic field.

Anyone who uses a compass to navigate makes use of Earth's magnetic field, which points generally from south to north. Other living organisms also take advantage of Earth's field to guide them from location to location. Sea turtles, for example, have been observed to travel hundreds of kilometers and still find their way back to their nesting sites along relatively direct paths. Yet when the turtles are transported away from their nests after a magnet has been attached to their heads, they take wildly circuitous routes back to the nesting site. The field of the attached magnet clearly disrupts the turtles' ability to determine their position using Earth's magnetic field.

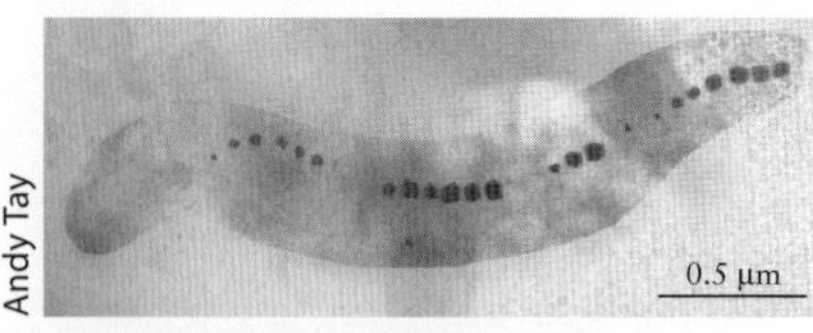

Figure 19-19 Ferromagnetic materials in bacteria Several species of bacteria, called *magnetotactic bacteria,* contain tiny pieces of ferromagnetic material (either iron oxide or iron sulfide). These appear as dark dots in this microscope image. Magnetotactic bacteria actively swim along geomagnetic field lines and use these membrane-encased bits of ferromagnetic material to orient themselves in the swampy environments in which they are found.

Magnetic Materials

While there are circulating electrons within every kind of atom, not all materials have the same magnetic properties. In some materials there is a net rotation of electrons within the atom, so each atom behaves like a current loop. In most cases these atomic current loops are randomly oriented, so their effects cancel out. But if the material is placed in a strong magnetic field, the atomic current loops experience a torque and align themselves with the magnetic field (see Figure 19-12b). As a result, the material behaves like a much larger current loop. If the magnetic field is turned off, random thermal motion will cause the atomic current loops to return to their original, non-aligned orientations. Materials that display this behavior are called **paramagnetic**. Everyday paramagnetic materials include aluminum and sodium. The net magnetic effect in a paramagnetic material is generally quite small; while an empty can made of (paramagnetic) aluminum acts like a current loop when brought next to a magnet, the magnetic force on the aluminum can is so small that a magnet can't pick it up.

In a handful of materials the interactions between adjacent atomic current loops are very strong. As a result, once the material is placed in a magnetic field the atomic current loops not only align with the field but can *remain* aligned after the field is turned off, leaving the material permanently magnetized. Iron is the most common of these materials, which are called **ferromagnetic** (**Figure 19-19**). (*Ferro* derives from the Latin word for iron.) Any permanent magnet is made of a ferromagnetic material. A permanent magnet can pick up objects made of a ferromagnetic material, such as a steel paper clip. The field of the permanent magnet causes the atomic current loops in the paper clip to align, making the paper clip a magnet itself. The magnetized paper clip is then attracted to the permanent magnet.

WATCH OUT !

Earth is not a permanent magnet.

Our planet's core is made primarily of iron and nickel, both of which are ferromagnetic materials. It's common to conclude from this that the core is magnetized like a permanent magnet and that this gives rise to our planet's magnetic field. However, this cannot be true. Any ferromagnetic material loses its magnetism if it is heated above a certain temperature specific to that material: This critical temperature is 773°C for iron and 354°C for nickel. The temperature in Earth's core is in excess of 4400°C, so the iron and nickel in the core do *not* act like ferromagnetic materials. Instead, Earth's magnetic field is caused by electric currents in the molten material that makes up the outer regions of the core.

Most materials are neither paramagnetic nor ferromagnetic because their atoms have zero net electron current. When placed in a magnetic field, a small amount of atomic current appears, but the current loops end up aligned in the direction *opposite* to what happens for paramagnetic or ferromagnetic materials. (This is a consequence of *electromagnetic induction*, which we'll discuss in Chapter 20.) As a result, these materials, called **diamagnetic**, are slightly repelled by magnets rather than being attracted. In most cases, however, the repulsion is very weak.

In each of these cases, the magnetic properties of the material can be described by magnetic permeability, which is proportional to the magnetic permeability of free space. It is a measure of the strength of the magnetic field that will exist inside the material when it is placed in an external magnetic field. A diamagnetic material has a constant permeability relative to μ_0 slightly less than 1. For a paramagnetic material this ratio is constant and slightly more than 1. A ferromagnetic material, such as iron, does not have a constant magnetic permeability. As the magnetizing field increases, the magnetic permeability increases, reaches a maximum, and then decreases. Purified iron and many magnetic alloys can have permeabilities 100,000 or more times μ_0.

The magnetic field around current-carrying wires with other geometries is much more complicated. As an example, **Figure 19-20** shows some of the magnetic field lines for a straight helical coil of wire. Such a coil is called a **solenoid**. Close to an individual wire, the field lines resemble those around the long, straight wire shown in Figure 19-15. In the space outside the solenoid, the magnetic field is very weak, as you can see from the large spacing between adjacent field lines. (Recall from Section 16-5

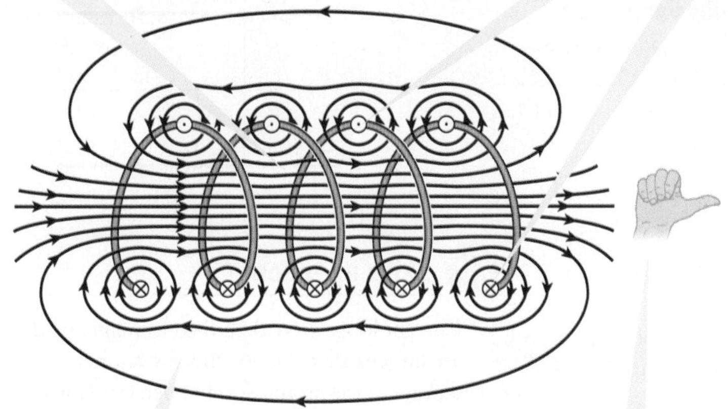

(1) The magnetic field in the interior of a long, straight solenoid is essentially uniform (the field lines are very nearly evenly spaced).
(2) The direction of current is represented as into the page at the bottom, and out of the page at the top, of each loop.
(3) The magnetic field outside the solenoid is very weak (the field lines are far apart).
(4) Right-hand rule for the field direction: If you curl the fingers of your right hand around the solenoid in the direction of the current, the magnetic field inside the solenoid points in the direction of your right thumb.

Figure 19-20 Magnetic field of a solenoid Compare this illustration to the photograph of a solenoid and its field in Figure 19-4a.

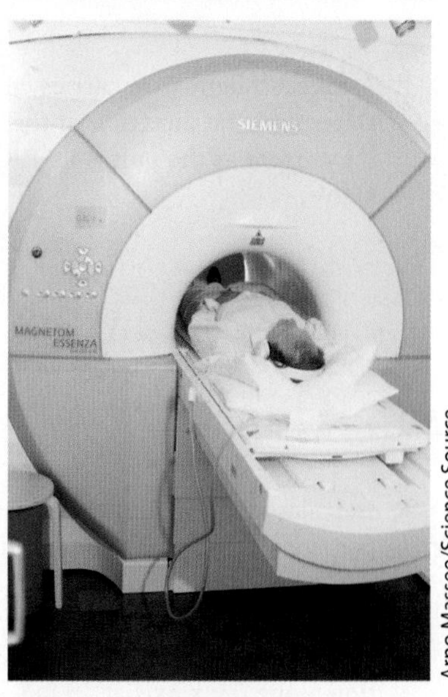

Arno Massee/Science Source

Figure 19-21 Magnetic resonance imaging (MRI) The medical imaging technique known as MRI requires that the patient be immersed in a strong, uniform magnetic field. In this MRI device this is done by having the patient lie inside a solenoid.

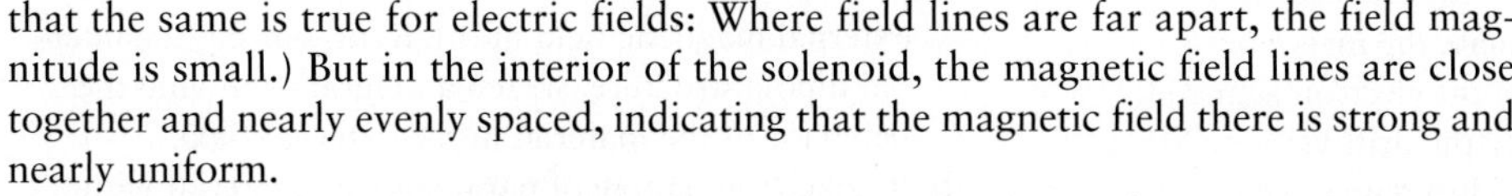

that the same is true for electric fields: Where field lines are far apart, the field magnitude is small.) But in the interior of the solenoid, the magnetic field lines are close together and nearly evenly spaced, indicating that the magnetic field there is strong and nearly uniform.

This property of solenoids explains why a conventional magnetic resonance imaging (MRI) scanner is in the form of a long tube inside which the patient lies (**Figure 19-21**). This tube is actually the interior of a solenoid, like that shown in Figure 19-20, so the patient is bathed in a strong, uniform magnetic field—which is just what MRI requires.

THE TAKEAWAY for Section 19-7

✔ Ampère's law relates the current through a wire to the magnetic field it generates.

✔ Current through a long, straight wire produces a magnetic field with circular field lines centered on the wire.

✔ A moving test charge produces a magnetic field that is directed the same way as a magnetic field around a current-carrying wire, proportional to its charge and that falls off as $1/r^2$.

✔ A current loop produces a more complicated magnetic field. A magnetic material can be thought of as a collection of atomic current loops.

✔ The magnetic properties of a material depend on the magnetic permeability of the material.

AP Building Blocks

1. A very long, straight wire carries a constant current. The magnetic field a distance d from the wire and far from its ends varies with distance d according to
   (A) d^{-3}.
   (B) d^{-2}.
   (C) d^{-1}.
   (D) d.

EX 19-5

2. A long, straight wire carries current in the $+z$ direction (out of the page). Determine the direction of the magnetic field due to the current at the points O, P, Q, and R in the figure below.

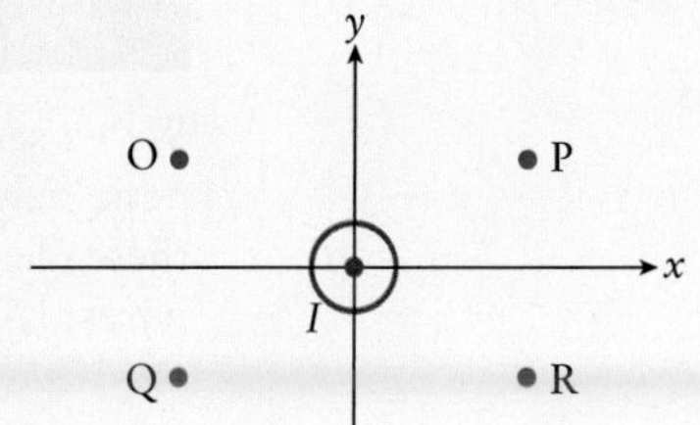

EX 19-5 3. A long, straight wire carries current in the $+x$ direction. Determine the direction of the magnetic field due to the current at the points O, P, Q, and R in the figure below.

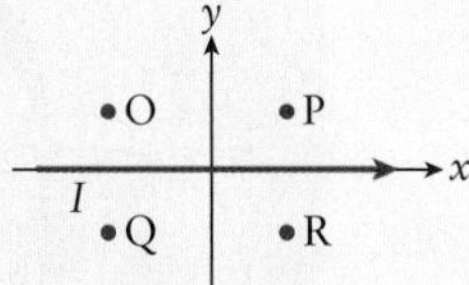

4. Calculate the magnitude of the magnetic field at a perpendicular distance of 2.20 m from a long copper pipe that has a diameter of 2.00 cm and carries a current of 20.0 A.

5. A point charge with charge 1 μC is shot in the $+x$ direction along the x axis at 10 m/s. Calculate the magnitude of the magnetic field at $y = 1$ cm when the point charge is passing through the origin.

AP® Skill Builders

6. An electron moves along the x axis in the $+x$ direction with a speed of 3.5×10^{-5} m/s. Calculate the magnetic field at the position $y = 2.0$ cm when the electron is at $x = 0$. (This velocity is approximately the drift velocity for electrons in a current-carrying wire. The comparatively large fields of wires are due to the number of electrons moving.)

7. Jerry wants to predict the magnetic field that a power line (the wire that brings electricity from the electric power plant to your home) creates in his apartment. A current of 100 A passes through a wire that is 5 m from his window. Calculate the magnitude of the magnetic field. How does the field compare to the magnitude of Earth's magnetic field of about 5×10^{-5} T in New York City?

EX 19-5 8. Two long, straight wires parallel to the x axis are at $y = \pm 2.5$ cm as shown in the figure. Each wire carries a current of 16 A in the $+x$ direction. Calculate the magnetic field on the y axis at (a) $y = 0$, (b) $y = 1.0$ cm, and (c) $y = 4.0$ cm.

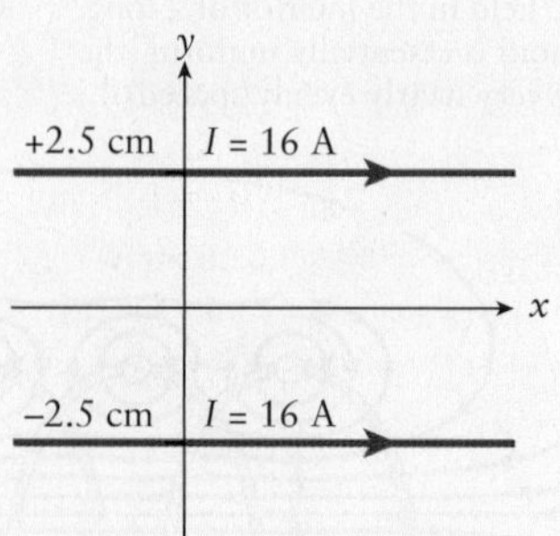

AP® Skills in Action

9. Current loops were described as magnetic dipoles. These are large-scale magnetic dipoles but help to make magnetic dipoles at the atomic level understandable. Imagine a loop of current in the plane of the page. The direction of the current is clockwise.
(a) Sketch a diagram showing the loop and the magnetic dipole of the current in the loop.
(b) The magnetic dipole you drew in (a) is placed in an external magnetic field oriented upward on the page. In what direction will your dipole align?
(c) Instead of the dipole you drew in part (a), imagine placing a sample of a paramagnetic material in the external magnetic field in (b). Representing the internal dipole structure, sketch a diagram showing the behavior of the material in the external field.
(d) Replace the sample of paramagnetic material with an identically shaped one of ferromagnetic material. Sketch the same sort of representation as you did in part (c), emphasizing any differences.
(e) The ferromagnetic sample and the paramagnetic sample are left in the external magnetic field for the same significant amount of time. Referencing your diagrams in (c) and (d), describe any changes in the representation you would see after they are removed from the field, and how the two materials' behaviors would compare.

10. In telephone lines, two wires carrying currents in opposite directions are twisted together. Using physics principles, explain how this reduces the magnetic fields surrounding the wires.

19-8 Two current-carrying wires exert magnetic forces on each other

We've seen that a current-carrying wire experiences a force when placed in a magnetic field, and also that a current-carrying wire generates a magnetic field. Let's put these ideas together and look at the magnetic interaction between *two* current-carrying wires. (Note that these two wires do not exert *electric* forces on each other. That's because each wire has as much positive charge as negative charge and so is electrically neutral.)

Experiment shows that two parallel, straight wires carrying current in the same direction attract each other, and that two parallel, straight wires carrying current in opposite directions repel. Let's see why this is the case.

Figure 19-22 shows the situation. Wires 1 and 2 are long, straight, and parallel to each other and are separated by a distance d. The current I_1 in wire 1 sets up a magnetic field $\vec{B}_1$ at the position of wire 2, which carries current I_2. From Equation 19-9 the magnitude B_1 of this field is

$$B_1 = \frac{\mu_0 I_1}{2\pi d} \tag{19-17}$$

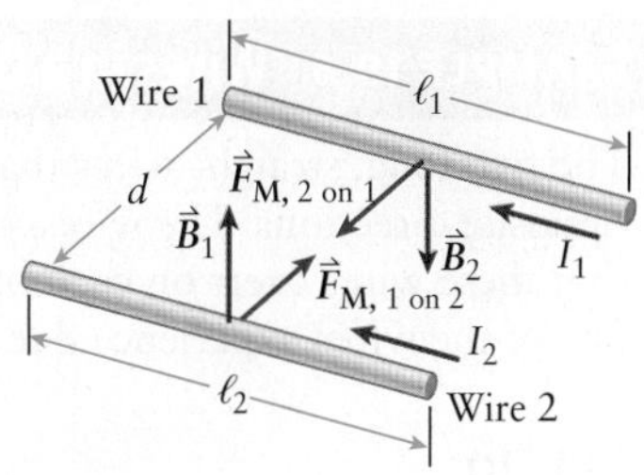

$\vec{B}_1$ = Magnetic field due to wire 1 at the position of wire 2

$\vec{B}_2$ = Magnetic field due to wire 2 at the position of wire 1

Figure 19-22 Magnetic forces between two current-carrying wires These two parallel wires carry current in the same direction and exert attractive magnetic forces on each other. If we reverse the direction of one of the currents, the forces become repulsive.

The right-hand rule for the field produced by a long, straight wire (Section 19-7) tells us that at the position of wire 2, $\vec{B}_1$ points upward and perpendicular to wire 2. To find the direction of the force $\vec{F}_{1\text{ on }2}$ that this field exerts on wire 2, use the right-hand rule for the direction of the magnetic force on a current-carrying wire (Section 19-5): This tells us that $\vec{F}_{1\text{ on }2}$ points toward wire 1, so the force attracts wire 2 to wire 1. The magnitude of the force on wire 2, of length ℓ_2, is given by Equation 19-5 with $\theta = 90°$ (since the direction of the current in wire 2 is perpendicular to the direction of $\vec{B}_1$):

$$F_{1\text{ on }2} = I_2 \ell_2 B_1 \sin 90° = I_2 \ell_2 B_1 \tag{19-18}$$

If we substitute B_1 from Equation 19-17 into Equation 19-18, we get

$$F_{1\text{ on }2} = I_2 \ell_2 \left(\frac{\mu_0 I_1}{2\pi d}\right) = \frac{\mu_0 I_1 I_2 \ell_2}{2\pi d} \tag{19-19}$$

The force per unit length on wire 2 is $F_{1\text{ on }2}$ (given by Equation 19-19) divided by the length ℓ_2 of wire 2:

$$\text{Magnetic force per unit length exerted by wire 1 on wire 2} = F_{1\text{ on }2}/\ell_2 = \frac{\mu_0 I_1 I_2}{2\pi d} \tag{19-20}$$

We can use the same procedure to find the magnetic force per unit length that wire 2 exerts on wire 1. The field $\vec{B}_2$ that the current I_2 in wire 2 produces at the position of wire 1 has magnitude $B_2 = \mu_0 I_2/(2\pi d)$ and points *downward* in Figure 19-22. From the right-hand rule for the force on a current-carrying wire, the force $\vec{F}_{2\text{ on }1}$ on wire 1 points toward wire 2 (the force is attractive); its magnitude is $F_{2\text{ on }1} = I_1 \ell_1 B_2 = I_1 \ell_1 [\mu_0 I_2/(2\pi d)] = \mu_0 I_1 I_2 \ell_1/(2\pi d)$, where ℓ_1 is the length of wire 1. The force per unit length on wire 1 is then $F_{1\text{ on }2}$ divided by ℓ_1:

$$\text{Magnetic force per unit length exerted by wire 2 on wire 1} = F_{2\text{ on }1}/\ell_1 = \frac{\mu_0 I_1 I_2}{2\pi d} \tag{19-21}$$

The force magnitudes per unit length in Equations 19-20 and 19-21 are equal, and the forces $\vec{F}_{1\text{ on }2}$ and $\vec{F}_{2\text{ on }1}$ are opposite in direction. That's just what we would expect from Newton's third law.

What changes if we reverse the direction of the current I_1 in wire 1? The force *magnitudes* given by Equations 19-20 and 19-21 won't be affected, but the force *directions* will be. This will reverse the direction of the magnetic field $\vec{B}_1$ that wire 1 produces at the position of wire 2, and so will reverse the direction of the force $\vec{F}_{1\text{ on }2}$ on wire 2. So in this case wire 2 will be pushed away from wire 1 (it will be repelled). Reversing the direction of I_1 will also reverse the direction of the force $\vec{F}_{2\text{ on }1}$ that wire 2 exerts on wire 1, so this force will push wire 1 away from 2 (again, it will be repelled). So we conclude that

Two parallel current-carrying wires attract each other if they carry current in the same direction, and repel each other if they carry current in opposite directions.

EXAMPLE 19-6 Wires in a Computer

The two long, straight wires that run along the back of a computer case to power the cooling fan carry 0.110 A in opposite directions. The wires are separated by 5.00 mm. (a) Find the force per unit length (magnitude and direction) that these wires exert on each other. (b) The mass per unit length of the wire is 5.00×10^{-3} kg/m. What acceleration does one of the wires experience due to this force?

Set Up

We'll use Equation 19-20 to find the force per unit length that one wire exerts on the other. (As we saw with Equation 19-21, the force per unit length has the same magnitude for either wire.) If we assume that this force equals the net external force on the wire, we can use Newton's second law to calculate the acceleration of the wire.

Magnetic force per unit length exerted by wire 1 on wire 2:

$$F_{1 \text{ on } 2}/\ell_2 = \frac{\mu_0 I_1 I_2}{2\pi d} \quad (19\text{-}20)$$

Newton's second law:

$$\sum \vec{F}_{\text{ext}} = m\vec{a} \quad (4\text{-}2)$$

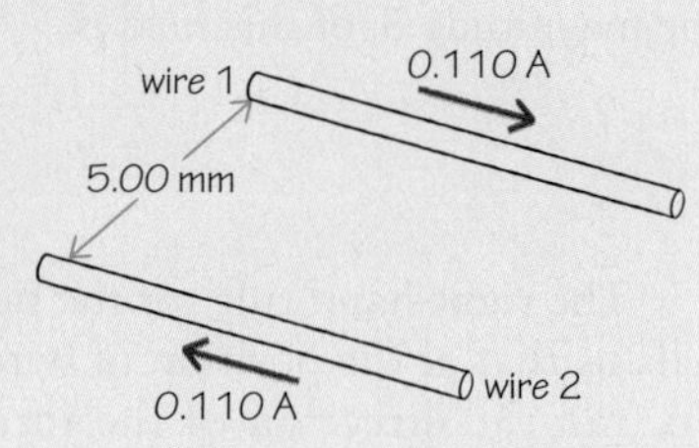

Solve

(a) The currents are in opposite directions, so the force is repulsive (it pushes the two wires apart). Use Equation 19-20 to find the magnitude of the force per unit length.

The two wires are separated by $d = 5.00 \text{ mm} = 5.00 \times 10^{-3}$ m and carry currents with the same magnitude: $I_1 = I_2 = 0.110$ A. The force per unit length on either wire is

$$= \frac{(4\pi \times 10^{-7}\ \text{T} \cdot \text{m/A})(0.110\ \text{A})(0.110\ \text{A})}{2\pi(5.00 \times 10^{-3}\ \text{m})}$$
$$= 4.84 \times 10^{-7}\ \text{T} \cdot \text{A} = 4.84 \times 10^{-7}\ \text{N/m}$$

[Recall that $1\ \text{T} = 1\ \text{N/(A} \cdot \text{m)}$.]

(b) Use Newton's second law to find the acceleration of the wire.

Our result for part (a) says that a 1-m length of wire would experience a force of magnitude 4.84×10^{-7} N. Given that this wire has mass per unit length 5.00×10^{-3} kg/m, a 1-m length would have mass 5.00×10^{-3} kg. The acceleration is

$$a = |\vec{a}| = \frac{\left|\sum \vec{F}_{\text{ext}}\right|}{m} = \frac{4.84 \times 10^{-7}\ \text{N}}{5.00 \times 10^{-3}\ \text{kg}}$$
$$= 9.68 \times 10^{-5}\ \text{m/s}^2$$

Reflect

The force and acceleration are both very gentle, so the effect on these wires will be almost imperceptible. In applications with very large currents, however, the magnetic forces between conductors can be substantial.

NOW WORK Problems 1, 3, and 6–8 from The Takeaway 19-8.

THE TAKEAWAY for Section 19-8

✔ Two wires attract each other when carrying current in the same direction and repel each other when carrying current in opposite directions.

✔ The forces per unit length on the wires are equal in magnitude and opposite in direction, exactly as required by Newton's third law.

Prep for the AP® Exam

AP® Building Blocks

EX 9-6 1. If wire 1 carries 2.00 A of current north, wire 2 carries 3.60 A of current south, and the two wires are separated by 1.40 m, calculate the force (magnitude and direction) exerted on a 1.00-cm section of wire 1 due to wire 2.

2. A power cord for an electronic device consists of two parallel straight wires carrying currents in opposite directions. Do they exert any forces on each other? Explain your answer.

3. The fasteners on overhead power lines are 50.0 cm long. What force must they be able to withstand if two power lines are 2.00 m apart, each carrying 2500 A in the same direction?

 Skill Builders

4. Parallel wires exert magnetic forces on each other. What about perpendicular wires? Explain your answer.

5. Two parallel wires carry currents in opposite directions, as shown in the figure below. Which of the following statements is correct?

(A) The force on the I_2 wire is upward, and the force on the I_1 wire is upward.

(B) The force on the I_2 wire is downward, and the force on the I_1 wire is upward.

(C) The force on the I_2 wire is upward, and the force on the I_1 wire is downward.

(D) The force on the I_2 wire is downward, and the force on the I_1 wire is downward.

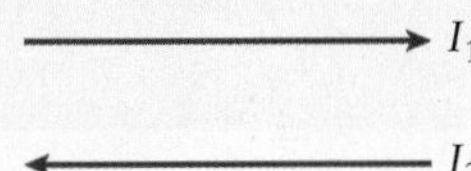

6. A long, straight wire carries a current of 1.2 A toward the south. A second, parallel wire carries a current of 3.8 A toward the north and is 2.8 cm from the first wire. What is the magnitude of the magnetic force per unit length each wire exerts on the other?

 Skills in Action

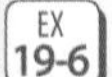

7. What is the net force (magnitude and direction) on a rectangular loop of wire that is 2.00 cm wide, 6.00 cm long, and located 2.00 cm from a long, straight wire that carries $I = 40.0$ A of current, as shown in the figure below? Assume a current of 20.0 A in the loop.

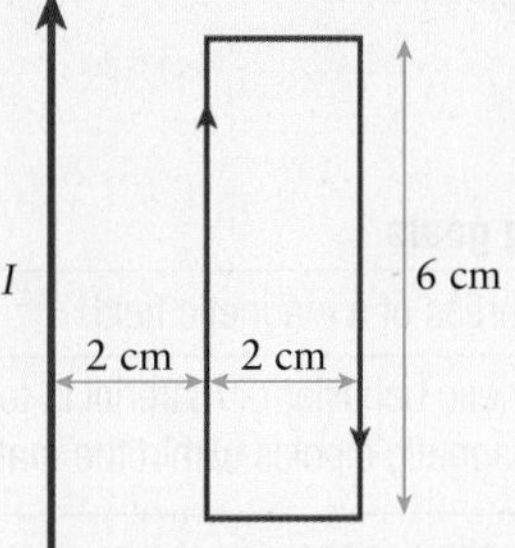

EX 19-6 8. A 2.0-m lamp cord leads from a 110-V outlet to a lamp having a 75-W lightbulb. The cord consists of two insulated parallel wires 4.0 mm apart that are held together by the insulation. One wire carries the current into the bulb, and the other carries it out. What is the magnitude of the magnetic field the cord produces (**a**) midway between the two wires and (**b**) outside the wires; 2.0 mm from one of the wires in the same plane in which the two wires lie? (**c**) Compare each of the fields in parts (a) and (b) with Earth's magnetic field (5×10^{-5} T). (**d**) What magnetic force (magnitude and direction) do the two wires exert on one another?

9. Two current-carrying wires are perpendicular to each other. One wire lies horizontally with the current directed toward the east. The other wire is vertical with the current directed upward. What is the direction of the net magnetic force on the horizontal wire due to the vertical wire?

(A) West

(B) South

(C) North

(D) Zero force

10. A very flexible helical coil is suspended as shown in the figure. What will happen when a sizable current I is sent through the coil?

(A) The coils will be pulled together.

(B) The coils will be pushed apart.

(C) Some of the coils will be pulled together, while others will be pulled apart

(D) There will be no net effect on the coils.

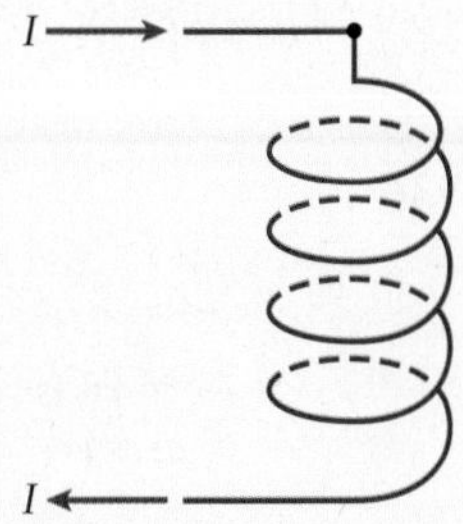

WHAT DID YOU LEARN?

Prep for the AP® Exam

Chapter learning goals	Section(s)	Related example(s)	Relevant section review exercises
Describe the properties of a magnetic field.	19-2		2
Describe the magnetic behavior of material in terms of the configuration of magnetic dipoles within the material.	19-2		
	19-7		9
Describe the magnetic permeability of free space and matter.	19-7		
Describe the magnetic field produced by a current, and from this, the approximation of the magnetic field for a charged object that is moving at much less than the speed of light.	19-7		5, 6
Describe the magnetic force exerted on charged objects moving in a magnetic field.	19-3	19-1	1-10
	19-4	19-2	1-4
Describe the magnetic field produced by a long, straight current-carrying wire.	19-7	19-5	1-4, 7, 8, 10
	19-8		8
Describe the force exerted on a current-carrying wire placed in a magnetic field.	19-5	19-3	1-8, 10
	19-8	19-6	1-7
Explain why a current loop in a uniform magnetic field experiences a net torque but zero net force.	19-6	19-4	1-10

Chapter 19 Review

Key Terms

All the Key Terms can be found in the Glossary/Glosario on page G1 in the back of the book.

Ampère's law 957
Amperian loop 957
circulation 956
current loop 950
diamagnetic 962
electromagnetism 936
ferromagnetic 962
Hall effect 938
magnet 932
magnetic dipole 935
magnetic force 933
magnetic poles 934
magnetism 933
mass spectrometer 941
normal 951
paramagnetic 962
permeability of free space 956
right-hand rule 937
solenoid 962
tesla 937

Chapter Summary

Topic	Equation or Figure

Magnetism and magnetic forces: Magnetic forces are interactions between moving charged objects. (By comparison, electric forces are interactions between charged objects, whether moving or not.) A magnet contains charged objects in continuous motion. A magnet sets up a magnetic field in the space around it; a second magnet responds to that field and can be attracted or repelled, depending on its orientation. No matter how a magnet is divided, it will still have two poles, north and south.

(a) These magnets attract.

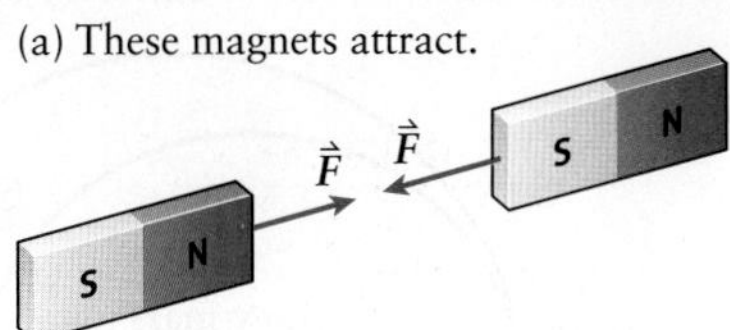

(b) These magnets repel.

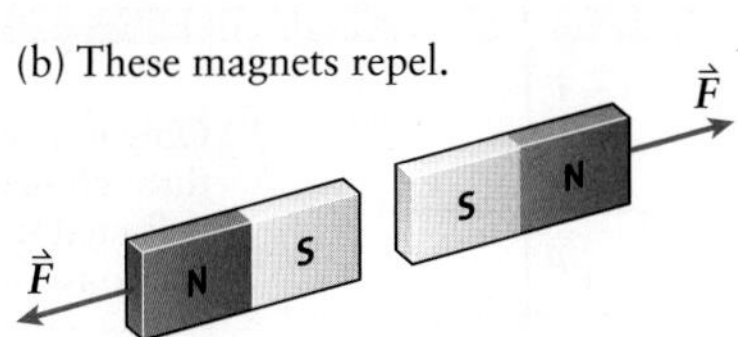

(c) If we cut a magnet in half, we always find each half has a north and south pole.

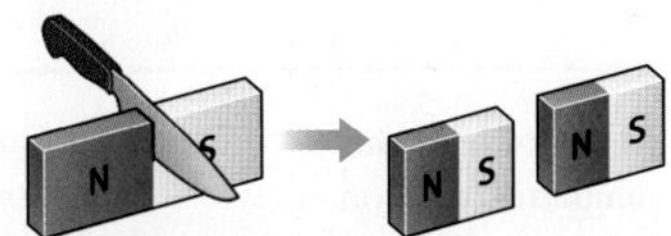

(Figure 19-3 a/b/c)

Magnetic force on a moving charged object: A magnetic force can be exerted on a single charged object moving in a magnetic field $\vec{B}$. The magnitude of the force depends on both the speed of the object and the direction of the object's velocity $\vec{v}$ relative to the magnetic field. The direction of the force is perpendicular to both $\vec{v}$ and $\vec{B}$, and is given by a right-hand rule.

$$F_M = |q|vB\,|\sin\theta| \quad (19\text{-}1)$$

Magnitude of the magnetic force on a moving charged object

Magnitude of the magnetic field

Angle between the direction of the object's velocity $\vec{v}$ and the direction of the magnetic field $\vec{B}$

Magnitude of the object's **charge**

Speed of the object

(a) Magnetic force on a positively charged object

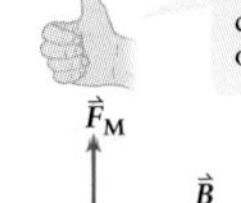

Your right hand helps determine the direction of the magnetic force.

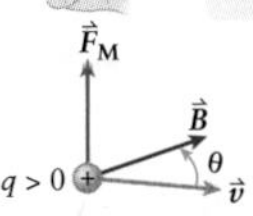

The force $\vec{F}_M$ is perpendicular to both the velocity $\vec{v}$ and the magnetic field $\vec{B}$.

(b) Magnetic force on a negatively charged object

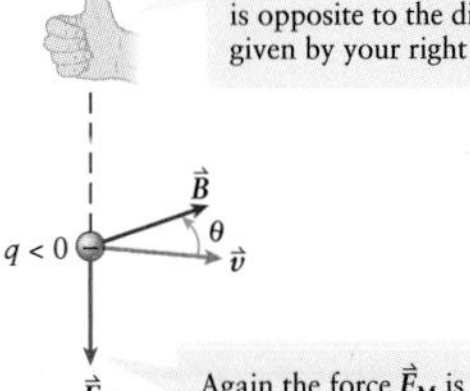

If $q < 0$, the magnetic force is opposite to the direction given by your right hand.

Again the force $\vec{F}_M$ is perpendicular to both the velocity $\vec{v}$ and the magnetic field $\vec{B}$.

(c) Magnetic force on a positively charged object, $\theta = 90°$

For a given speed v, the magnetic force is greatest if the velocity $\vec{v}$ is perpendicular to the magnetic field $\vec{B}$.

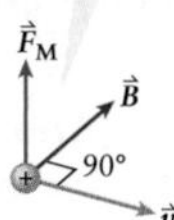

(d) Magnetic force on a positively charged object, $\theta = 0$ or 180°

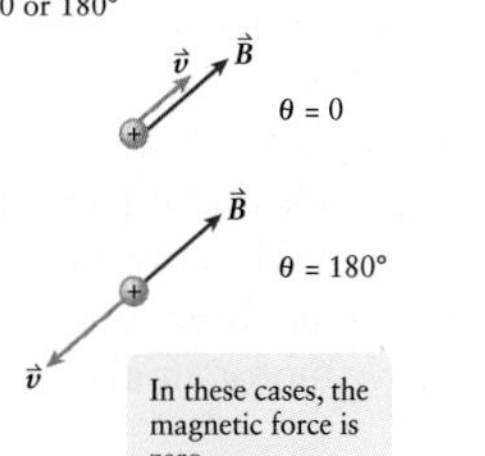

In these cases, the magnetic force is zero.

(Figure 19-5)

Object trajectories in a magnetic field: A charged object moving in a magnetic field and subject to no other forces can move in a circular trajectory whose radius depends on its speed, mass, and charge as well as the magnetic field magnitude. This is the principle of the mass spectrometer.

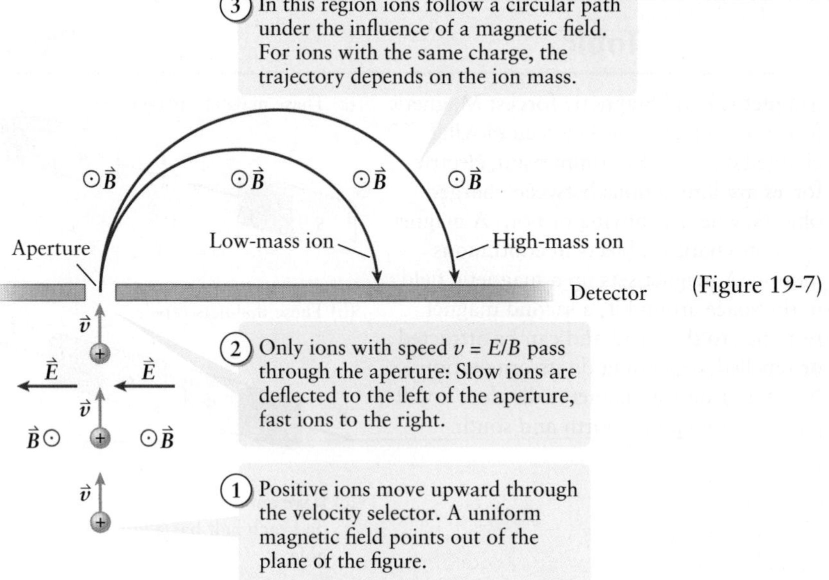

(Figure 19-7)

Magnetic forces on current-carrying wires: A magnetic force is exerted on a wire that carries a current and is placed in a magnetic field. This force is the sum of the magnetic forces exerted on the individual charge carriers within the wire. The force magnitude depends on the amount of current and the orientation of the wire relative to the magnetic field; its direction is given by a right-hand rule.

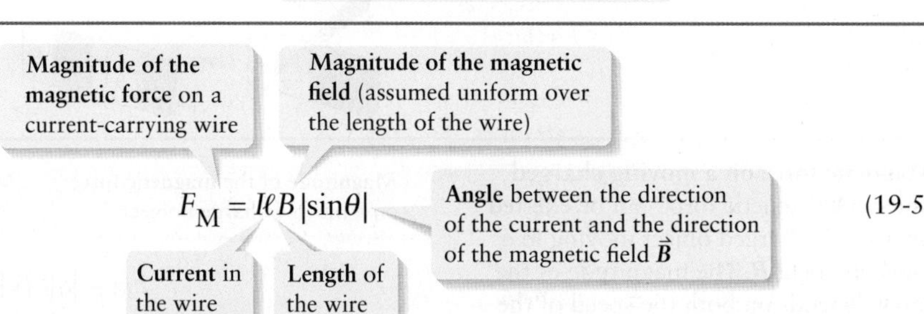

$$F_M = I\ell B\,|\sin\theta| \tag{19-5}$$

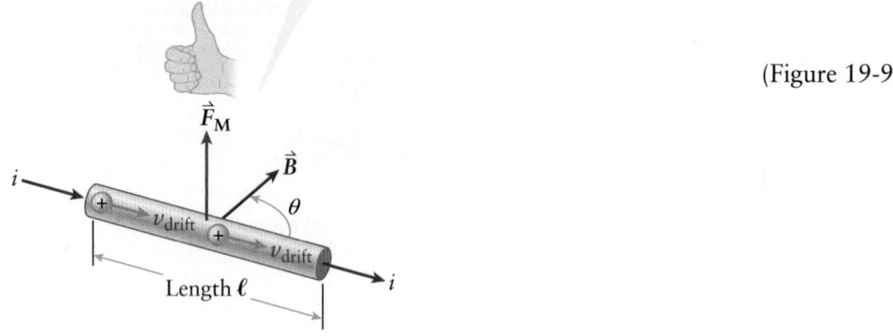

The force $\vec{F}_M$ on a current-carrying wire in a magnetic field $\vec{B}$ is perpendicular to both $\vec{B}$ and the length of the wire. The direction is given by a right-hand rule.
Length ℓ

(Figure 19-9)

Magnetic torque on a current loop: A current-carrying loop of wire can experience a torque when placed in a magnetic field. The magnitude and direction of the torque depend on how the current loop is oriented relative to the direction of the magnetic field. Electric motors make use of this principle.

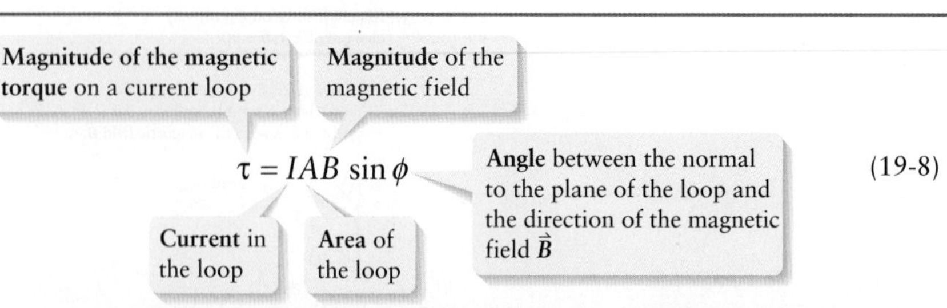

$$\tau = IAB\sin\phi \tag{19-8}$$

(b) Side view

$\vec{F}_{M,top}$

ϕ

$\vec{B}$

I

ϕ

ϕ

$\vec{F}_{M,bottom}$

There can be a nonzero net magnetic torque on a current loop in a uniform magnetic field.

Normal to the plane of the current loop

The torque magnitude depends on the angle ϕ.

(Figure 19-12b)

Magnetic field produced by electric currents and moving point charges: A long, straight current-carrying wire produces a relatively simple magnetic field in the space around it. Ampère's law—which relates the circulation of magnetic field around a closed loop to the amount of current through that loop—can be used to find the magnetic field produced by currents with other simple geometries. A single moving point charge creates a magnetic field at a point perpendicular to its path with the same orientation as the magnetic field centered on a long current carrying wire.

Magnitude of the magnetic field due to a long, straight wire

Permeability of free space

$$B = \frac{\mu_0 I}{2\pi r} \qquad (19\text{-}9)$$

Current in the wire

Distance from the wire to the location where the field is measured

Circulation of magnetic field around an Amperian loop

Permeability of free space

$$\sum B_{||}\Delta\ell = \mu_0 I_{\text{through}} \qquad (19\text{-}13)$$

Current through the Amperian loop

Magnitude of the magnetic field at a point in space, 1 due to a point charge at another point in space, 0

Permeability of free **space**

The magnitude of the charge

$$B_1 = \frac{\mu_0}{4\pi r_{10}^2} qv\,|\sin\theta| \qquad (19\text{-}14)$$

The angle between the direction of the velocity of the point charge and the displacement $\vec{r}_{10}$

The square of the magnitude of the displacement of the point charge from the point in space at which the field is being calculated ($\vec{r}_{10}$)

The speed of the point charge

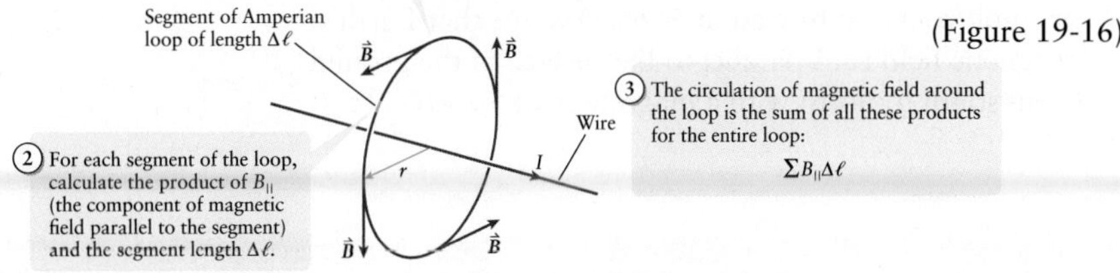

1 We draw an Amperian loop that encircles the current-carrying wire. (This particular loop is a circle, and so coincides with a field line.)
Segment of Amperian loop of length $\Delta\ell$
$\vec{B}$
$\vec{B}$
r
I
Wire
$\vec{B}$
$\vec{B}$
2 For each segment of the loop, calculate the product of $B_{||}$ (the component of magnetic field parallel to the segment) and the segment length $\Delta\ell$.
3 The circulation of magnetic field around the loop is the sum of all these products for the entire loop:
$\Sigma B_{||}\Delta\ell$

(Figure 19-16)

Current loops and magnetic materials: A current loop is called a magnetic dipole because the magnetic field that it produces is similar to the electric field produced by an electric dipole. A permanent magnet is a material in which the atoms behave like individual current loops, many of which are oriented in the same direction so that their individual magnetic fields add to make a strong field. The magnetic permeability of the material is determined by the alignment of these individual magnetic fields in the presence of an external field.

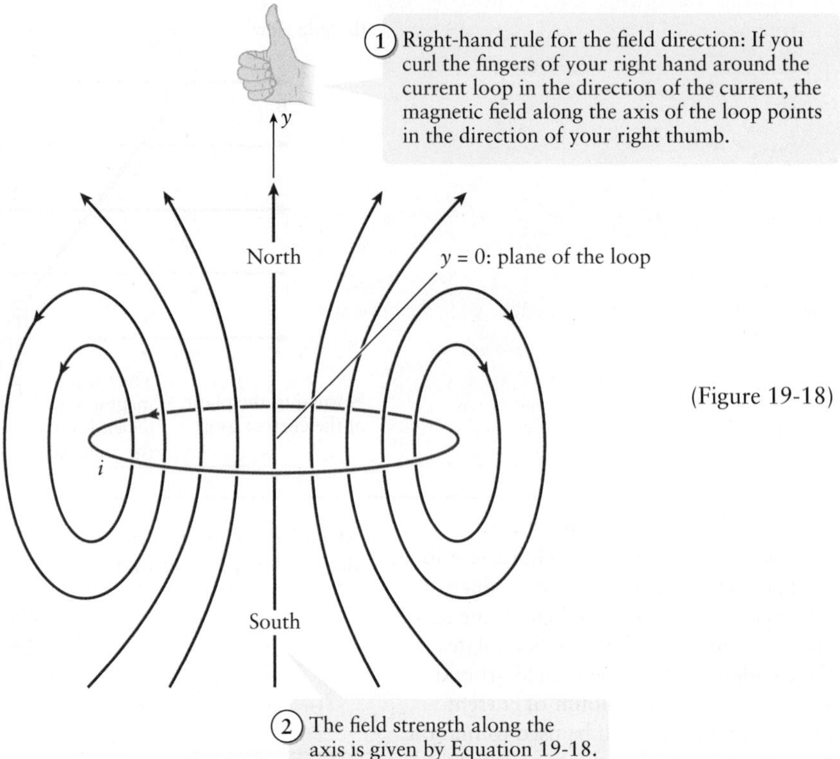

(1) Right-hand rule for the field direction: If you curl the fingers of your right hand around the current loop in the direction of the current, the magnetic field along the axis of the loop points in the direction of your right thumb.
y
North
y = 0: plane of the loop
i
South
(2) The field strength along the axis is given by Equation 19-18.

(Figure 19-18)

Force between current-carrying wires: Two parallel current-carrying wires exert magnetic forces on each other: The current in one wire produces a magnetic field at the location of the second wire, and the current in the second wire responds to that field. The two wires attract if the currents are in the same direction and repel if the currents are in opposite directions.

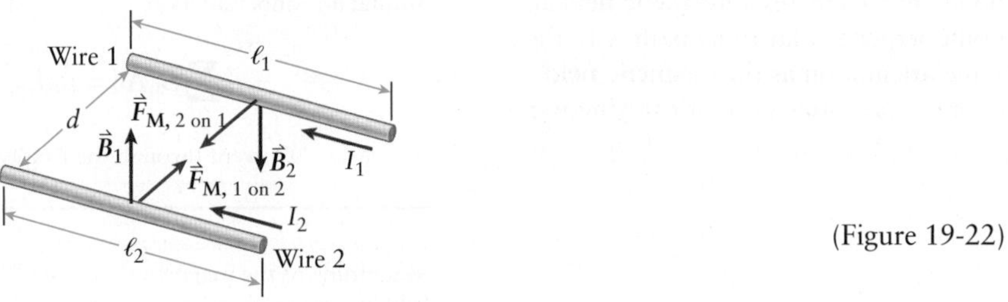

(Figure 19-22)

Chapter 19 Review Problems

1. Honeybees can acquire a small net charge on the order of 1 pC as they fly through the air and interact with plants. Estimate the magnetic force on a honeybee due to Earth's magnetic field as the bee flies near the ground from east to west at 6 m/s. Assume that Earth's magnetic field runs parallel to the surface of the ground from south to north with a magnitude of 5×10^{-5} T.

2. Physicists refer to crossed electric and magnetic fields as a *velocity selector*. In the same sense, the deflection of charged objects in a strong magnetic field perpendicular to their motion can be thought of as a *momentum selector*. Why?

3. Convert the units for the following expressions for magnetic fields as directed (recall 1 G = 10^{-4} T):

(a) 5.00 T = ________ G

(b) 25,000 G = ________ T

(c) 7.43 mG = ________ μT

(d) 1.88 mT = ________ G

4. Horizontal electric power lines supported by vertical poles can carry large currents. Assume that Earth's magnetic field runs parallel to the surface of the ground from south to north with a magnitude of 0.50×10^{-4} T and that the supporting poles are 32 m apart. Find the magnitude and direction of the force that Earth's magnetic field exerts on a 32-m segment of power line (a wire) carrying 95 A if the current runs (a) from north to south, (b) from east to west, or (c) toward the northeast making an angle of 30.0° north of east. (d) Are any of the above forces large enough to have an appreciable effect on the power lines?

5. A levitating train is three cars long (180 m) and has a mass of 100 metric tons (1 metric ton = 1000 kg). The current in the superconducting wires is about 500 kA, and even though the traditional design calls for many small coils of wire, assume for this problem that there is a 180-m-long wire carrying the current. Find the magnitude of the magnetic field needed to levitate the train.

6. A small 20-turn current loop with a 4.00-cm diameter is suspended in a region with a magnetic field of 1.00×10^3 G, with the plane of the loop parallel with the magnetic field direction.

(a) What is the current in the loop if the torque exerted by the magnetic field on the loop is 4.00×10^{-5} N · m?

(b) Describe the subsequent motion of the loop if it is allowed to rotate.

7. A straight wire carries a current of 8.00 A toward the top of the page. What are the magnitude and direction of the magnetic field at point P, which is 8.00 cm to the right of the wire, as shown in the figure below?

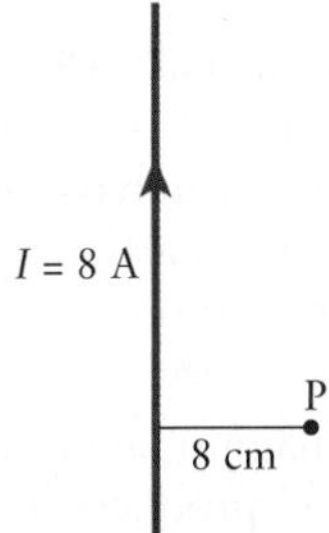

8. A velocity selector consists of crossed electric and magnetic fields, with the magnetic field directed toward the top of the page. A beam of positively charged objects passing through the velocity selector from left to right is undeflected by the fields.

(a) In what direction is the electric field? (left, right, toward the top of the page, toward the bottom of the page, into the page, out of the page)

(b) The direction of the beam is reversed so that it travels from right to left. Is it then deflected? If so, in what direction?

(c) A beam of electrons (negatively charged) moving with the same speed is passed through from left to right. Is it deflected? If so, in what direction?

9. The National High Magnetic Field Laboratory holds the world record for creating the largest magnetic field—100 T. To see if such a strong magnetic field could pose health risks for nearby workers, calculate the maximum acceleration the field could produce on Na^+ ions (of mass 3.8×10^{-26} kg) in blood traveling through the aorta. The speed of blood is highly variable, but 50 cm/s is reasonable in the aorta. Does your result indicate that it would be dangerous to expose workers to such a large magnetic field?

10. The magnetic field due to a current-carrying cylinder of radius 1 cm is measured at various points ($r < 1$ cm and $r > 1$ cm).

(a) Graph the magnitude of the magnetic field B as a function of r.

(b) Use this graph to find the functional relationship between the magnetic field and the radial distance. *Hint:* The magnetic field will be described by two different functions—one for inside the cylinder and one for outside the cylinder. After making an initial graph, you may want to graph the data for the inside and outside separately.

r (m)	B (T)	r (m)	B (T)
0.001	0.00050	0.015	0.00353
0.002	0.00100	0.020	0.00250
0.003	0.00152	0.025	0.00200
0.004	0.00200	0.030	0.00180
0.005	0.00252	0.035	0.00143
0.006	0.00300	0.040	0.00125
0.007	0.00350	0.045	0.00110
0.008	0.00401	0.050	0.00103
0.009	0.00453	0.100	0.000502
0.010	0.00500		

11. An electron and a proton have the same kinetic energy upon entering a region of constant magnetic field, and their velocities are perpendicular to the magnetic field. Suppose the magnetic field is strong enough to allow these particles to circle in the field. What is the ratio $r_{proton}/r_{electron}$ of the radii of their circular paths?

12. During electrical storms, a bolt of lightning can transfer 10 C of charge in 2.0 μs (the amount of charge and time can vary considerably). We can model such a bolt as a very long current-carrying wire.

(a) What is the magnetic field 1.0 m from such a bolt? What is the field 1.0 km away? How do these fields compare with Earth's magnetic field?

(b) Compare the fields in part (a) with the magnetic field produced by a typical household current of 10 A in a very long wire at the same distances from the wire as in (a).

(c) How close would you have to get to the wire in part (b) for its magnetic field to be the same as the field produced by the lightning bolt at 1.0 km from the bolt?

13. A long, straight wire carries a current as shown in the figure below. A point charge moving parallel to the wire experiences a force of 0.80 N at point P. Assuming the same charge and same velocity, what would be the magnitude of the magnetic force on the point charge at point S?

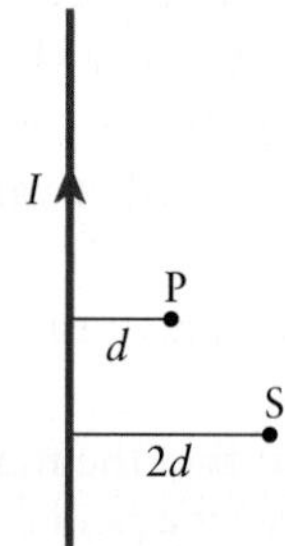

14. A wire of mass 40 g slides with negligible friction on two horizontal conducting rails spaced 0.8 m apart. A steady current of 100 A is in the circuit formed by the wire and the rails. A uniform magnetic field of 1.2 T, directed into the plane of the drawing, exerts a force on the wire.

(a) In which direction in the figure below will the wire accelerate?

(b) What is the magnetic force on the wire?

(c) How long must the rails be if the wire, starting from rest, is to reach a speed of 200 m/s? How would your answers differ if the magnetic field were (d) directed out of the page or (e) in the plane of the drawing, directed toward the top of the drawing?

I
Wire
B_{in}
0.8 m

15. Transcranial magnetic stimulation (TMS) is a noninvasive method to stimulate the brain using magnetic fields. It is used in treating strokes, Parkinson's disease, depression, and other physical conditions. In the procedure, a circular coil is placed on the side of the forehead to generate a magnetic field inside the brain. Although values can vary, a typical coil would be about 15 cm in diameter and contain 250 thin circular windings. The magnetic field in the cortex (3.0 cm from the coil measured along a line perpendicular to the coil at its center) is typically 0.50 T.

(a) What current in the coil is needed to produce the desired magnetic field inside the brain?

(b) What is the magnetic field at the center of the coil at the forehead?

(c) If the current needed in part (a) seems too large, how could you easily achieve the same magnetic field with a smaller current?

16. When operated on a household 110-V line, typical hair dryers draw about 1650 W of power. We can model the current as a long, straight wire in the handle. During use, the current is about 3.0 cm from the user's head.

(a) What is the current in the dryer?

(b) What is the resistance of the dryer?

(c) What magnetic field does the dryer produce at the user's head? Compare your answer with Earth's magnetic field (5×10^{-5} T) to decide if we should have health concerns about the magnetic field created when using a hair dryer.

17. Three very long, straight wires lie at the corners of a square of side *d*, as shown in the figure below. The magnitudes of the currents in the three wires are the same, but the two diagonally opposite currents are directed into the page while the other one is directed outward. Derive an expression for the magnetic field (magnitude and direction) at the fourth corner of the square.

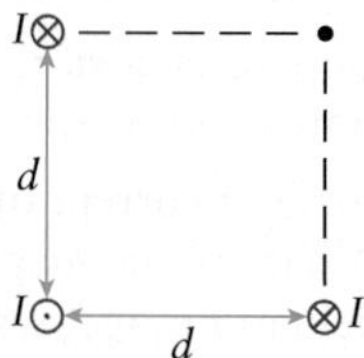

18. Some people have raised concerns about the magnetic fields produced by current-carrying power lines in residential neighborhoods. Currents in such lines can be up to 100 A. Suppose you have such a line near your house. Assume the wires are supported horizontally 5.0 m above the ground on vertical poles and your living room is 12 m from the base of the poles.

(a) Calculate the magnitude of the magnetic field strength the wire produces in your living room if it carries 100 A.

(b) Express your answer as a multiple of Earth's magnetic field (5×10^{-5} T). Does the magnetic field from such wires seem strong enough to cause health concerns?

19. In the mass spectrometer shown in the figure on the next page, an object with charge $-e = -1.60 \times 10^{-19}$ C enters a region of magnetic field that has a magnitude of 0.00242 T and points into the page. The velocity of the object is confirmed with a velocity selector. The electric field is 9.00×10^4 N/C (down) and the magnetic field is 0.00530 T (into page) in the velocity selector. If the radius of curvature of the object's path is 4.00 cm, calculate the mass of the object.

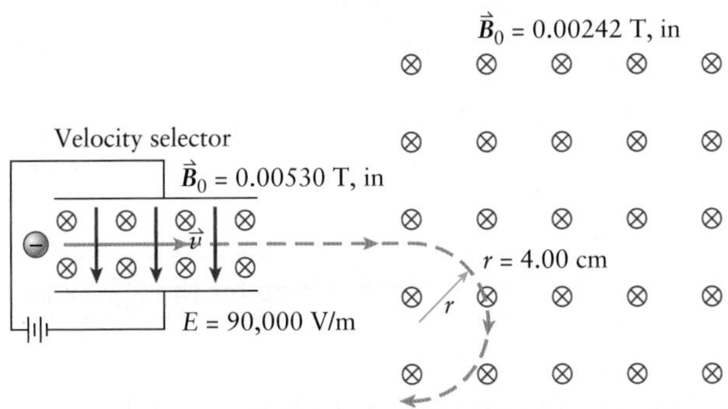

20. Medical magnetoencephalography (MEG) is a technique for measuring changes in the magnetic field of the brain caused by external stimuli such as being touched or viewing images of food. Such a change in the field occurs due to electrical activity (current) in the brain. During the process, magnetic sensors are placed on the skin to measure the magnetic field at that location. Typical field strengths are a few femtoteslas (1 femtotesla = 1 fT = 10^{-15} T). An adult brain is about 140 mm wide, divided into two sections (called hemispheres, although the brain is not truly spherical) each about 70 mm wide. We can model the current in one hemisphere as a circular loop, 65 mm in diameter, just inside the brain. The sensor is placed so that it is along the axis of the loop 2.0 cm from the center. A reasonable magnetic field is 5.0 fT at the sensor. According to this model, (a) what is the current in the brain and (b) what is the magnetic field at the center of the hemisphere of the brain?

21. Helmholtz coils are composed of two coils of wire that have their centers on the same axis, separated by a distance that is equal to the radius of the coils, as shown in the figure. The coils have N turns of wire that carry a current I in the same direction. If one coil is centered at the origin, and the other at $x = R$, derive expressions for the net magnetic field due to the coils at the points (a) $x = R/2$ and (b) $x = 2R$.

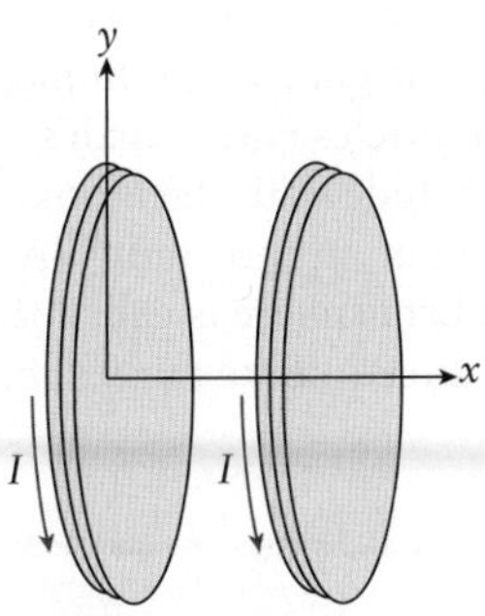

22. Geophysicists often measure magnetic field in gauss ($1\text{ G} = 10^{-4}$ T). Earth's magnetic field at the equator can be taken as 0.7 G directed north. At the center of a flat circular coil that has 10 turns of wire and is 1.4 m in diameter, the coil's magnetic field exactly cancels Earth's field at this location.
(a) Calculate the current in the coil.
(b) How should the coil be oriented?

23. A square loop of wire lies on a horizontal table, with one side of the loop constrained to stay on the table, as shown in the figure below. The loop is in the presence of a magnetic field that is parallel to the surface of the table. When there is a current of $I_1 = 0.350$ A in the loop, the loop lifts off the table to an angle $\theta_1 = 15.0°$ relative to the surface of the table. When there is a different current of $I_2 = 1.18$ A in the loop, the angle the loop makes relative to the table increases to $\theta_2 = 42.1°$. Calculate the ratio of the magnetic torque on the loop when the current in the loop is I_1 to the magnetic torque when the current in the loop is I_2.

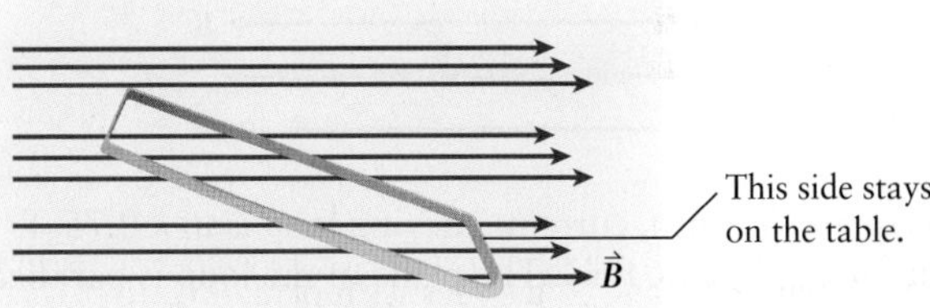

24. Migratory birds use Earth's magnetic field to guide them. Some people are concerned that human-caused magnetic fields could interfere with bird navigation. Suppose that a pair of parallel power lines, each carrying 100 A, are 3.00 m apart and lie in the same horizontal plane. Find the magnitude and direction of the magnetic field the lines produce at a point 15.0 m above them equidistant from both lines in each of the following cases.
(a) Both lines run in the north–south direction, and both currents run from north to south.
(b) Both lines run in the north–south direction, and the current in the eastern line runs northward while the current in the western line runs southward.
(c) Both lines run in the east–west direction, and both currents run from west to east.
(d) Is it reasonable to think that the fields caused by the wires are likely to interfere with bird migration?

25. A coaxial cable consists of a solid inner conductor of radius R_i, surrounded by a concentric outer conducting shell of radius R_o. Insulating material fills the space between the conductors. The inner conductor carries current to the right, and the outer conductor carries the same current to the left down the outer surface of the cable, as shown in the figure.

(a) Draw a diagram representing the system of two wires if you were looking at it from a position to the left in the diagram, along the axis of the wires. On your diagram show the directions of current and indicate in what regions there is magnetic field, and the direction of the field in each region where one exists.

(b) Using Ampère's law, derive an expression for the magnitude of the magnetic field in three separate regions of space: inside the inner conductor, between the two conductors, and outside of the outer conductor.

(c) If the magnitude of the current in the inner wire is 1.5 A, $R_i = 0.15$ cm, and $R_o = 0.35$ cm calculate the magnitude of the magnetic field at (i) $R_1 = 0.10$ cm, and (ii) $R_2 = 0.30$ cm.

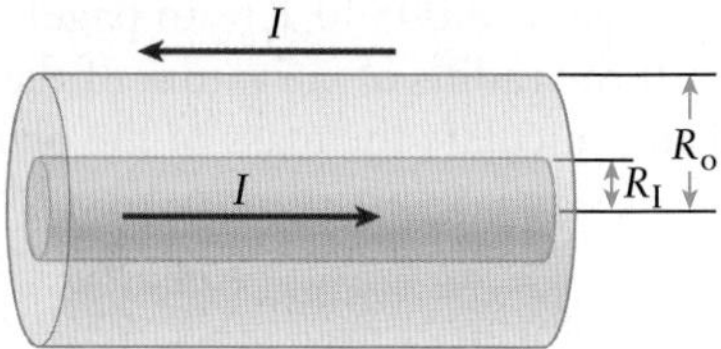

Prep for the AP® Exam

AP® Group Work

Directions: The following problem is designed to be done as group work in class.

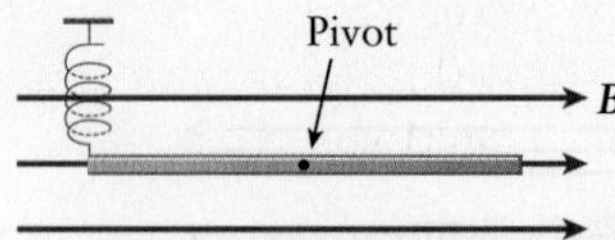

A square, 35-turn, current-carrying loop is in a 0.75-T uniform magnetic field. The plane of the loop is parallel to the direction of the magnetic field, as shown in the figure above (an edge-on view of the loop). The loop is kept from rotating by a stretched spring that is attached to one side of the loop. The spring constant is 450 N/m, and the spring is stretched 5.6 cm.

(a) If the length of each side of the square loop is 42 cm, what is the magnitude of the current in the loop?

(b) When viewed from above, is the current in the loop clockwise or counterclockwise?

AP® PRACTICE PROBLEMS

Prep for the AP® Exam

Multiple-Choice Questions

Directions: The following questions have a single correct answer.

1. The magnetic force on a moving charged object
   (A) depends on the magnetic field at the object's instantaneous position.
   (B) is in the direction that is mutually perpendicular to the direction of motion of the charged object and the direction of the magnetic field.
   (C) is proportional both to the charge and to the magnitude of the magnetic field.
   (D) is described by all of the above.

2. A proton traveling to the right enters a region of uniform magnetic field that points into the page. When the proton enters this region, it will be
   (A) deflected out of the plane of the page.
   (B) deflected into the plane of the page.
   (C) deflected toward the top of the page.
   (D) deflected toward the bottom of the page.

3. An electron is moving northward in a magnetic field. The magnetic force on the electron is toward the northeast. What is the direction of the magnetic field?
   (A) Up
   (B) Down
   (C) West
   (D) This situation cannot exist because of the orientation of the velocity and force.

4. The aurora borealis ("northern lights") is caused by fast-moving, electrically charged subatomic particles ejected from the Sun. Earth's magnetic field exerts a magnetic force on these particles that steers them toward our planet's north magnetic pole. When these particles enter Earth's upper atmosphere, they collide with the atoms there and cause the atoms to emit an eerie glow. (A similar effect in the southern hemisphere is called the aurora australis.) In what direction are these subatomic particles

(*Continued*)

moving when the magnetic force on them is strongest, assuming their speeds are equal?

(A) The same direction as the magnetic field

(B) Opposite to the magnetic field

(C) Perpendicular to the magnetic field

(D) Either (A) or (B)

5. A proton with a velocity along the $+x$ axis enters a region where there is a uniform magnetic field $\vec{B}$ in the $+y$ direction. You want to balance the magnetic force with an electric field so that the proton will continue along a straight line. The electric field should be in the

(A) $+z$ direction.

(B) $-z$ direction.

(C) $+x$ direction.

(D) $-x$ direction.

6. A wire is oriented such that the current in the wire is directed into the page. A magnetic field points downward in the region occupied by the wire. For positive charge carriers, the electric potential across the wire due to current in the wire in the presence of the field

(A) decreases into the page, because current always goes in the direction of decreasing potential.

(B) is greater on the right edge of the wire, because negative charge in the wire will be pushed to the left.

(C) is greater on the right edge of the wire because the positive charge carriers will be pushed to the right.

(D) is greater on the left edge of the wire because the positive charge carriers will be pushed to the left.

Free-Response Question

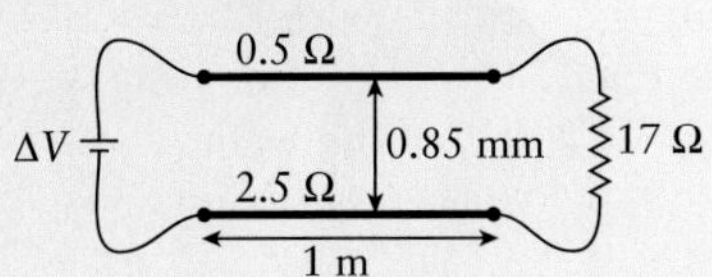

1. Two straight conducting rods, which are 1.0 m long, exactly parallel, and separated by 0.85 mm, are connected by an ideal battery and a 17-Ω resistor, as shown in the figure above. The 0.5-Ω rod "floats" above the 2.5-Ω rod, in equilibrium.

(a) On the grid below, draw a free-body diagram for the floating rod, when it is in equilibrium.

(b) If the mass of each rod is 25 g, calculate the potential difference ΔV provided by the battery.

(c) You remove the resistor and use additional wires of negligible resistance to connect the two rods directly to the battery. How would this change the separation between the rods in equilibrium? Refer to your diagram in (a), your calculations in (b), and physics principles as needed to support your claim.

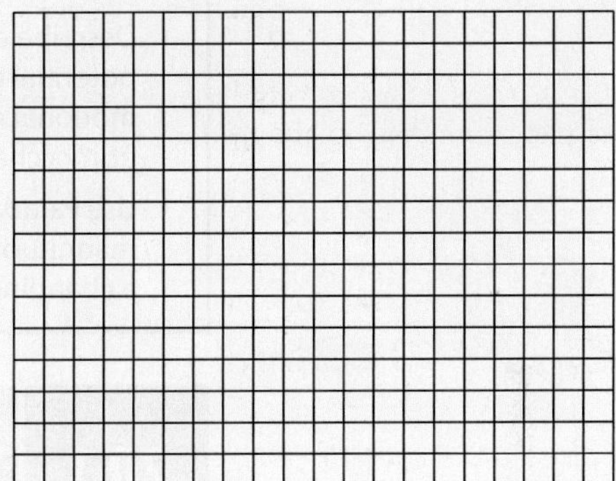

20 Electromagnetic Induction

pick-uppath/Getty Images

YOU WILL LEARN TO:

- Describe and calculate magnetic flux.
- Describe an induced electric potential difference (often called motional or induced emf) resulting from a change in magnetic flux.
- Use Faraday's law to determine the magnitude of an emf in a circuit with a changing magnetic flux.
- Use Lenz's law to determine the direction of an emf in a circuit with a changing magnetic flux.
- Describe the functioning of a generator.

20-1 The world runs on electromagnetic induction

In Chapter 18 we discussed *direct-current* electric circuits in which the current is always in the same direction. In these circuits what makes the charge carriers move is the emf provided by a battery. This electric potential difference between the battery terminals is caused by chemical processes inside the battery (see Section 18-2).

But many of the electric circuits around you are *alternating-current* circuits in which the current constantly changes direction. That includes the current in light fixtures, toasters, electric fans, and other devices plugged into wall sockets. (Alternating current also indirectly powers mobile devices like cell phones and laptop computers. These devices have batteries, but the batteries are recharged by plugging them into a wall socket.) What kind of emf produces an alternating current?

The answer to this question comes from a remarkable discovery made by physicists around 1830: *If the magnetic field in a region of space changes, the change gives rise to an electric field.* This electric field, called an *induced* field, is very different in character from the electric field produced by point charges that we described in Chapter 16: An induced electric field does not point away from positive charge and toward negative charge but instead has field lines that form closed loops like magnetic field lines. This induced electric field can push charge carriers around a loop of wire and generate an electric current. As we'll see later in this chapter, it's easy to make this induced electric field flip its direction back and forth, which makes a current that flips

An electric generator produces current by electromagnetic induction: Coils of wire move relative to a magnetic field, which generates an emf in the coils. The motion can be powered by the wind, as in these wind turbines.

(a)

GregC/iStock/Getty Images

Creating a changing magnetic field in the brain induces and electric current in the brain, causing electric currents. Areas in red are where the currents are strongest.

(b)

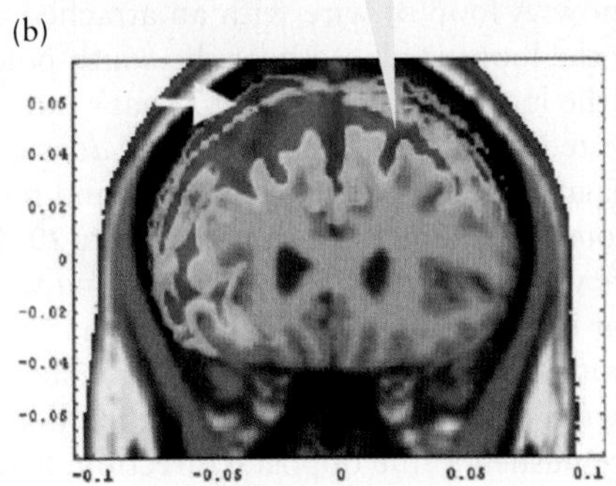

Republished with permission of Elsevier Science and Technology Journals, from Post, A. and Keck, M.E., "Transcranial Magnetic Stimulation as a Therapeutic Tool in Psychiatry: What Do We Know about the Neurobiological Mechanisms?" *Journal of Psychiatric Research* 35:4, page 193–215, (2001); permission conveyed through Copyright Clearance Center, Inc.

Figure 20-1 Electromagnetic induction Two examples of the phenomenon of electromagnetic induction, in which electric currents are induced by the presence of a changing magnetic flux.

back and forth—in other words, an alternating current. The vast amount of electric current used by our technological civilization is produced in this way (**Figure 20-1a**).

We use the term **electromagnetic induction** for the process whereby a changing magnetic field induces an electric field. (The word *electromagnetic* shows that this process involves both electric and magnetic fields.) Electromagnetic induction has many applications beyond producing an alternating current to be delivered to wall sockets. It's how a credit card reader decodes the information on the card's magnetized strip (see the photo that opens this chapter). It's also at the heart of a relatively new medical technique called *transcranial magnetic stimulation* (TMS), which allows physicians to stimulate electrical activity in the brain without sticking electrodes to the scalp or inserting them through the skull. In TMS, a time-varying magnetic field is produced inside the brain by current-carrying coils around the head. This causes an induced electric field, which in turn causes currents within the brain (**Figure 20-1b**). TMS has been used with some success to treat cases of depression that have not responded to more conventional therapy.

In this chapter we'll begin by describing the relationship between a changing magnetic flux and the electric field that it induces. We'll introduce two important laws that describe electromagnetic induction. The first of these, Faraday's law, will tell us how the emf that appears in a closed loop (such as an electric circuit) due to an induced electric field is related to the rate of change of the magnetic flux through the loop. The second, Lenz's law, will tell us the direction of this induced emf. We'll see how induced emf makes possible the important device called a *generator*, which converts mechanical energy into electric energy and creates an alternating emf. (Each of the wind turbines shown in Figure 20-1 uses its spinning blades to run a generator.) We will conclude the chapter with a discussion of *Maxwell's equations*, the summary of electromagnetism and the development that led to the prediction of electromagnetic waves, which we will learn about in the next unit.

THE TAKEAWAY for Section 20-1

✔ In electromagnetic induction, a time-varying magnetic flux in a certain region gives rise to an electric field in that same region.

✔ Electromagnetic induction is used to produce an alternating current.

Prep for the AP Exam

Building Blocks

1. In hospitals with magnetic resonance imaging facilities and at other locations where large magnetic fields are present, there are usually signs warning people with pacemakers and other electronic medical devices not to enter. Why?

20-2 A changing magnetic flux creates an electric field

Figure 20-2 shows an experiment that we can understand with the physics we already know. A loop of wire with an attached ammeter (a device for measuring the current in the loop) is placed near the south pole of a stationary magnet. There is no current if the loop is held stationary. That's not surprising, since there's no source of emf connected to the loop. But a current *does* exist in the loop when it is moved toward the magnet's south pole (**Figure 20-2a**) and reverses direction when the loop is moved away from the magnet's south pole (**Figure 20-2b**). What's happening is that the charge carriers within the loop are moving along with the loop through the magnetic field of the bar magnet and so a magnetic force is exerted on them that pushes the charge carriers around the loop (**Figure 20-2c**). Reversing the direction in which the loop and its charge carriers move also reverses the direction of the magnetic force, so the charge carriers are pushed in the opposite direction, and the current direction reverses (**Figure 20-2d**). In either case we say that the magnetic force on the charge carriers is equivalent to an emf that drives the current. There is no magnetic force, and hence no emf, if the loop and its charge carriers are at rest. (Recall from Section 19-2 that magnetic forces are exerted only on *moving* charged objects.) Because the loop must be in motion for the emf to appear, it is often called **motional emf**.

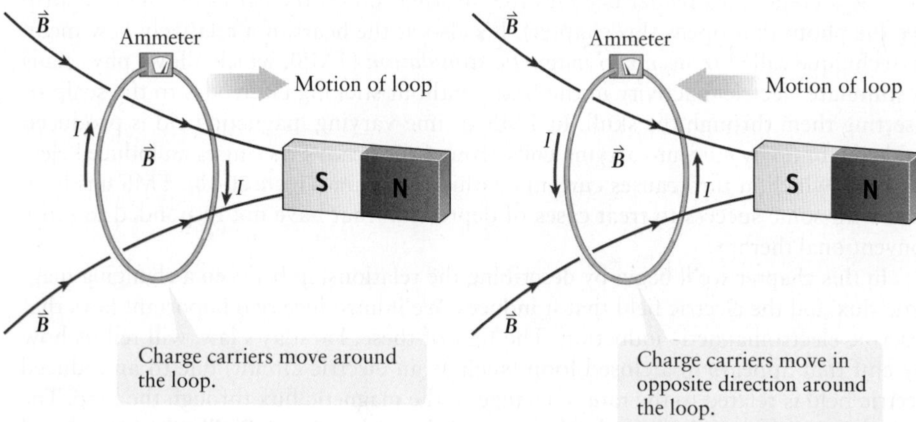
(a) Moving the loop toward a stationary magnet
$\vec{B}$
Ammeter
Motion of loop
I
$\vec{B}$
I
S
N
$\vec{B}$
Charge carriers move around the loop.
(b) Moving the loop away from a stationary magnet
$\vec{B}$
Ammeter
Motion of loop
I
$\vec{B}$
I
S
N
$\vec{B}$
Charge carriers move in opposite direction around the loop.

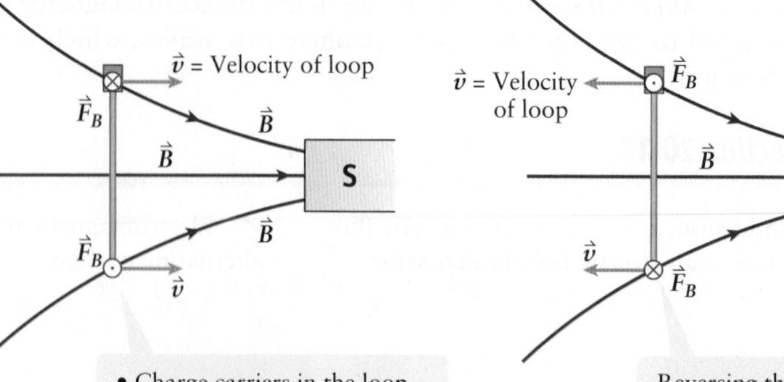
(c) Side view of loop moving toward a stationary magnet
$\vec{v}$ = Velocity of loop
$\vec{F}_B$
$\vec{B}$
$\vec{B}$
S
$\vec{B}$
$\vec{F}_B$
$\vec{v}$
• Charge carriers in the loop move along with the loop in the magnetic field.
• Magnetic forces push charge carriers around the loop, causing a current.
(d) Side view of loop moving away from a stationary magnet
$\vec{v}$ = Velocity of loop
$\vec{F}_B$
$\vec{B}$
$\vec{B}$
S
$\vec{B}$
$\vec{v}$
$\vec{F}_B$
Reversing the velocity of the loop reverses the directions of the magnetic forces, so charge carriers move in the opposite direction.

Figure 20-2 A loop of wire moving with respect to a magnet If a loop of wire moves toward or away from a magnet, a current appears in the loop. The current is caused by magnetic forces.

Figure 20-3 shows an experiment that looks similar but, in our stationary reference frame, involves entirely different physics. Now we hold the loop stationary and move the south pole of the magnet either toward the loop (**Figure 20-3a**) or away from the loop (**Figure 20-3b**). In this case there can be no magnetic force on the charge carriers within the loop because those charge carriers are at rest in the stationary loop. Nonetheless, there is an emf in the loop and a current around the loop in response, but only when the magnet is moving relative to the loop. Since there is no magnetic force in this situation, it must be that the emf is due to an *electric* force on the charge carriers (**Figures 20-3c** and **20-3d**).

What's happening is that when the magnet is moving, the magnetic field at the location of the loop is changing: Its magnitude increases when the magnet's south pole moves toward the loop (Figure 20-3a) and decreases when the magnet's south pole moves away from the loop (Figure 20-3b). So this experiment shows that an electric field is *induced* by the changing magnetic field. For this reason we call the emf in the experiment of Figure 20-3 an **induced emf**. We use the term *electromagnetic induction* for situations in which a changing magnetic field causes, or induces, an electric field.

If we were to consider a different inertial frame of reference, one moving at the velocity of the magnet, then the magnet would appear to be at rest, and the loop would again be moving. Either the stationary observer or the one moving with the magnet should measure the same current for the same magnet and coil, so motional emf and induced emf are the same. You could even imagine a reference frame moving as some intermediate velocity, so both the magnet and the loop appear to be moving. The results would be the same!

So, although the experiments in Figures 20-2 and 20-3 are different, they can be thought of as the same situation viewed from different reference frames and they must

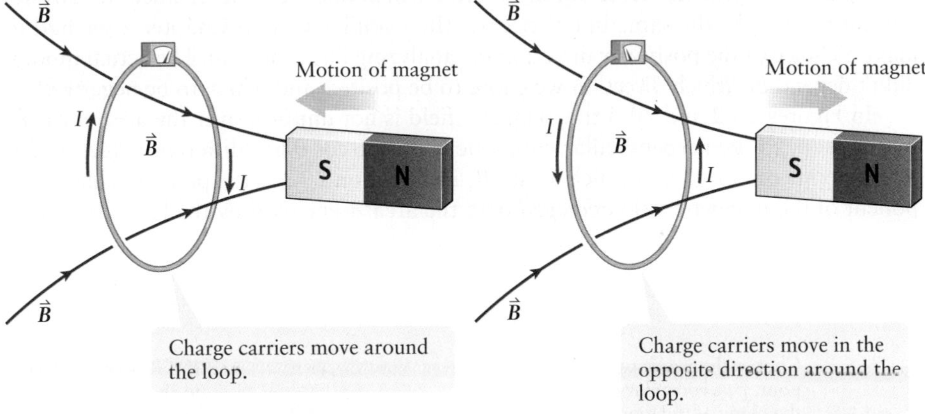
(a) Moving the magnet toward a stationary loop
(b) Moving the magnet away from a stationary loop
$\vec{B}$
Motion of magnet
I
S
N
Charge carriers move around the loop.
Charge carriers move in the opposite direction around the loop.

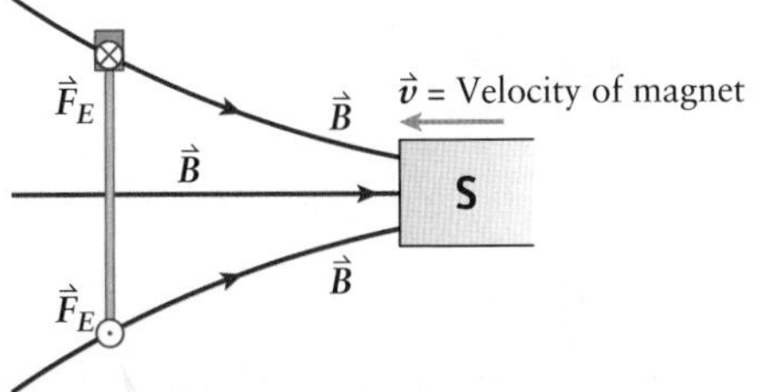
(c) Side view of magnet moving toward a stationary loop
$\vec{F}_E$
$\vec{B}$
$\vec{v}$ = Velocity of magnet
S
• Charge carriers in the loop experience electric forces.
• These forces push charge carriers around the loop, causing a current.

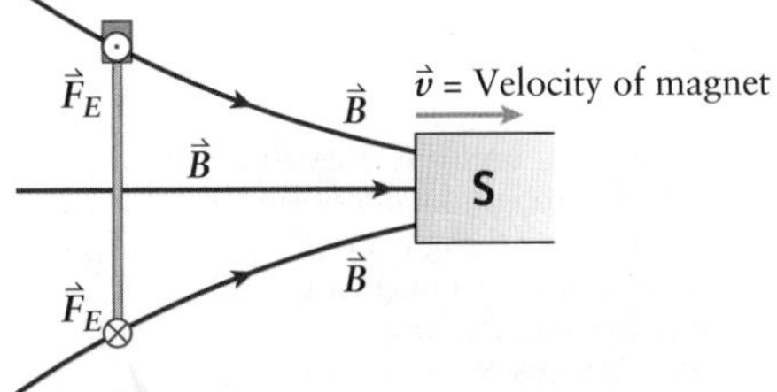
(d) Side view of magnet moving away from a stationary loop
$\vec{F}_E$
$\vec{B}$
$\vec{v}$ = Velocity of magnet
S
Reversing the velocity of the magnet reverses the directions of the electric forces, so charge carriers move in the opposite direction.

Figure 20-3 A magnet moving with respect to a loop of wire If a magnet moves toward or away from a loop of wire, a current appears in the loop. Magnetic forces cannot explain why this happens, so electric forces must be present to produce the current.

have the *same* result: Whether the loop moves toward the stationary magnet at 1 m/s, as in Figure 20-2a, or the magnet moves toward the stationary loop at 1 m/s, as in Figure 20-3a, the same emf appears in the loop. We find the same emf as long as the magnet and loop approach each other at a relative speed of 1 m/s. Since the result is the same in each of these cases, we should be able to describe all of these effects in terms of a single equation. But what equation is that?

It turns out that we can describe the emf in any of these situations in terms of the change in *magnetic flux* through the loop in Figures 20-2 and 20-3. We define flux as the area A of the surface outlined by the loop, multiplied by $B\cos\theta$, the component of the magnetic field that's perpendicular to that surface (see **Figure 20-4a**). In equation form, the **magnetic flux Φ_B** ("phi-sub-B") through the loop is

EQUATION IN WORDS
Magnetic flux

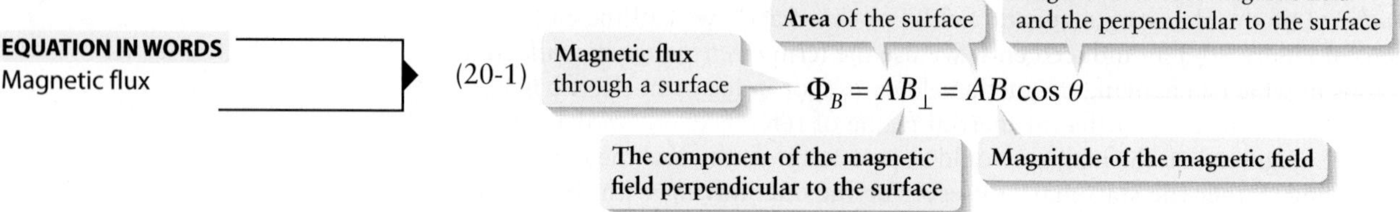

$$\Phi_B = AB_\perp = AB\cos\theta \tag{20-1}$$

NEED TO REVIEW? See the Math Tutorial, Section M-8, in the back of the book for more information on trigonometry.

The subscript B on the symbol Φ_B in Equation 20-1 reminds us that this is the flux of the magnetic field $\vec{B}$. We will later see an example of when it is useful to calculate the flux of an electric field. As parts (b) and (c) of Figure 20-4 show, the flux Φ_B can be positive or negative. Note that the choice of the positive x direction is arbitrary; in Figure 20-4 we chose the positive x direction to be up, so Φ_B is positive for the case shown in **Figure 20-4b** and negative for the case shown in **Figure 20-4c**. Had we chosen the positive x direction to be downward, we would have had $\Phi_B < 0$ in Figure 20-4b and $\Phi_B > 0$ in Figure 20-4c. It doesn't matter which one we choose, since the physics will turn out to be the same in either case. (In a similar way, in Chapter 2 we had to make a choice of the positive x direction for analyzing linear motion. The actual motion didn't depend on which direction we chose to be positive and which to be negative.)

In Figures 20-2 and 20-3 the magnetic field is not uniform over the area enclosed by the loop, so the perpendicular component $B_\perp = B\cos\theta$ has different values at different points on this area. In such a case $B_\perp$ in Equation 20-1 is the perpendicular component of the magnetic field *averaged* over the area A enclosed by the loop. Note that

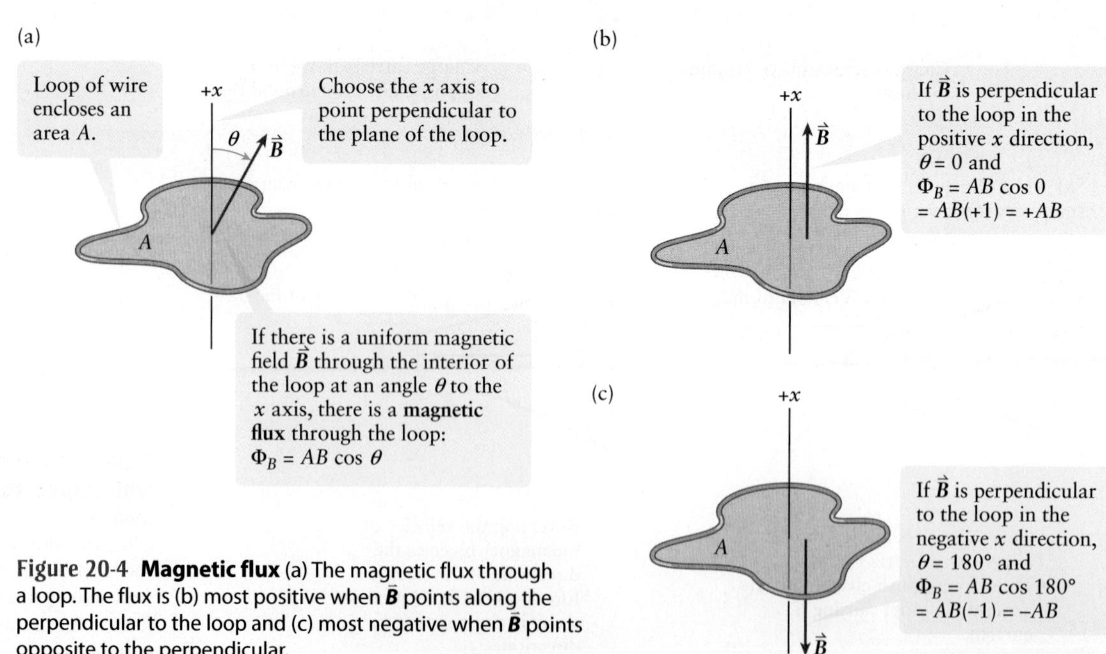

(a)
Loop of wire encloses an area A.
+x
θ
B
Choose the x axis to point perpendicular to the plane of the loop.
A
If there is a uniform magnetic field B through the interior of the loop at an angle θ to the x axis, there is a magnetic flux through the loop:
Φ_B = AB cos θ
(b)
+x
B
A
If B is perpendicular to the loop in the positive x direction, θ = 0 and Φ_B = AB cos 0 = AB(+1) = +AB
(c)
+x
A
B
If B is perpendicular to the loop in the negative x direction, θ = 180° and Φ_B = AB cos 180° = AB(−1) = −AB

Figure 20-4 Magnetic flux (a) The magnetic flux through a loop. The flux is (b) most positive when $\vec{B}$ points along the perpendicular to the loop and (c) most negative when $\vec{B}$ points opposite to the perpendicular.

if the loop is actually a coil with N turns of wire, the net magnetic flux through the coil is N multiplied by the flux through one turn of the coil.

If the magnet and loop in Figures 20-2 and 20-3 are not moving with respect to each other, the magnetic flux through the loop remains the same. In this case there is no emf and no current in the loop. The flux changes, however, when either the loop moves relative to the magnet (Figure 20-2) or the magnet moves relative to the loop (Figure 20-3). In these cases there *is* an emf in the loop and the current. This suggests that *an emf appears in a loop when the magnetic flux through that loop changes*. This observation is known as **Faraday's law of induction**, named for the nineteenth-century English physicist Michael Faraday:

Magnitude of the induced emf in a loop

Change in the magnetic flux through the surface outlined by the loop

$$|\mathcal{E}| = \left|\frac{\Delta\Phi_B}{\Delta t}\right| \qquad (20\text{-}2)$$

Time interval over which the change in magnetic flux takes place

EQUATION IN WORDS
Faraday's law of induction

This law states that the magnitude of the emf that appears in a loop is equal to the magnitude of the *rate of change* of the magnetic flux through the loop. If a large change in flux $\Delta\Phi_B$ happens in a short time interval Δt, the resulting emf has a large magnitude; if the change in flux is relatively small and happens over a long time interval, the resulting emf has a small magnitude.

Note that Equation 20-2 tells us only the *magnitude* of the emf, not its direction. In the following section we'll see how the direction is determined.

WATCH OUT !

It's not the magnetic flux that causes an emf, but the rate at which the flux changes.

The mere presence of magnetic flux through a loop does not cause an emf to appear in the loop. If a flux is present but does not change, such as what happens when a magnet and loop are held stationary with respect to each other, there is *no* resulting emf. An emf appears only when the flux *changes*, such as when the magnet and loop in Figures 20-2 and 20-3 move either toward or away from each other.

In the following example we'll check Faraday's law. We'll do this by considering a situation in which we can use our knowledge of magnetic forces to calculate the emf, then compare this to the emf calculated using Equation 20-2.

EXAMPLE 20-1 Changing Magnetic Flux I: A Sliding Bar in a Magnetic Field

A copper bar of length L slides at a constant speed v along stationary, U-shaped copper rails (**Figure 20-5**). A uniform magnetic field of magnitude B is directed perpendicular to the plane of the bar and rails. The moving bar and stationary rails form a closed circuit, and an emf is produced in this circuit because the bar is moving in a magnetic field. Determine the emf in the circuit (a) by using the expression for the magnetic force on a charge in the moving wire and (b) by using Faraday's law of induction, Equation 20-2.

Copper bar slides on the rails at constant velocity $\vec{v}$.
Copper rails
$\vec{v}$
L
A uniform magnetic field $\vec{B}$ points perpendicular to the plane of the rails.

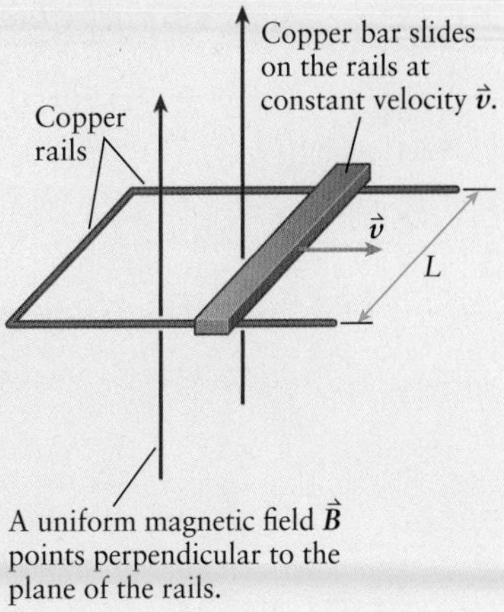

Figure 20-5 A sliding copper bar What emf is generated in the bar as it slides in the presence of a magnetic field $\vec{B}$?

Set Up

For a battery the magnitude of the emf equals the change in electric potential (which is potential energy per unit charge); that is, it's equal to the *work per unit charge* that the battery does on charge carriers that travel from one terminal to the other. We'll use the same idea in part (a) to calculate the emf in terms of the work done by the magnetic force on a charged object that travels the length of the moving bar. In part (b) we'll find the emf by instead using Equation 20-2. The magnetic field doesn't change, but the area of the loop outlined by the moving bar and the rails *does* change, and so the magnetic flux through this loop changes.

Magnetic force on a moving charged object:

$$F_M = |q|\, vB \sin\theta \quad (19\text{-}1)$$

Work done by a constant force that points in the same direction as the displacement:

$$W = Fd \quad (6\text{-}1)$$

Magnetic flux:

$$\Phi_B = AB_\perp = AB\cos\theta \quad (20\text{-}1)$$

Faraday's law of induction:

$$|\mathcal{E}| = \left|\frac{\Delta\Phi_B}{\Delta t}\right| \quad (20\text{-}2)$$

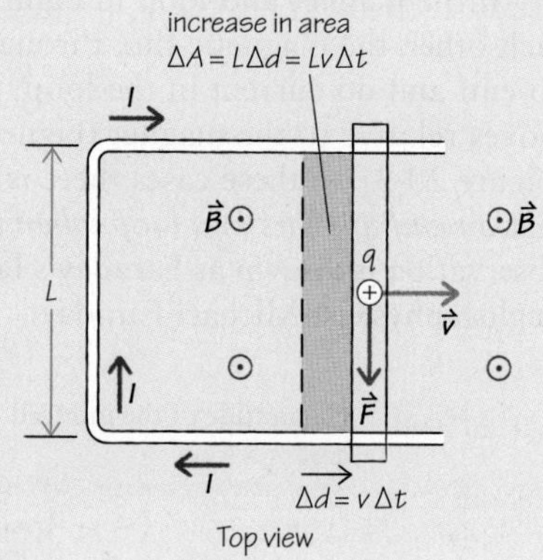

Solve

(a) Find the magnetic force on a charged object moving along with the copper bar.

For a positive point charge q moving with the bar, the velocity $\vec{v}$ is perpendicular to the magnetic field $\vec{B}$. So $\theta = 90°$ in Equation 19-1, and the magnetic force $\vec{F}_M$ on such a point charge has magnitude

$$F_B = qvB\sin 90° = qvB(1) = qvB$$

The force $\vec{F}_M$ is perpendicular to both $\vec{v}$ and $\vec{B}$, so it is directed along the length of the moving bar.

Use the magnetic force on a charged object to find the emf produced in the bar.

The magnetic force $\vec{F}_M$ on a point charge q causes it to move along the length L of the bar. Because $\vec{F}_M$ is in the same direction as the displacement of the point charge, the work done on the point charge as it travels this length is

$$W = F_M L = qvBL$$

The magnitude of the emf in the bar equals the work done per unit charge:

$$|\mathcal{E}| = \frac{W}{q} = \frac{qvBL}{q} = vBL$$

(b) Use Faraday's law of induction to find the emf.

The magnetic field $\vec{B}$ points perpendicular to the plane of the loop outlined by the moving copper bar and the copper rails. If the area of this loop is A and we take the positive x direction to point out of the plane of the above figure (in the same direction as $\vec{B}$), then $\theta = 0$ in Equation 20-1. The magnetic flux through the loop is then

$$\Phi_B = AB\cos 0 = AB(1) = AB$$

The magnetic field is constant, but the area A changes with time because the bar moves. The speed v of the bar is just the distance Δd that the bar moves divided by the time Δt that it takes to move that distance, so

$$v = \frac{\Delta d}{\Delta t} \quad \text{and} \quad \Delta d = v\Delta t$$

During time Δt the area A of the loop outlined by the moving bar and rails increases by an amount $\Delta A = L\Delta d = Lv\Delta t$. Therefore, the change in magnetic flux through the loop during this time is

$$\Delta\Phi_B = (\Delta A)B = (Lv\Delta t)B = vBL\Delta t$$

From Equation 20-2 the magnitude of the emf in the loop is

$$|\mathcal{E}| = \left|\frac{\Delta\Phi_B}{\Delta t}\right| = \left|\frac{vBL\,\Delta t}{\Delta t}\right| = vBL$$

Reflect

We find the same expression for the emf in both parts (a) and (b), as we must. This gives us added confidence that Equation 20-2 is valid, and a host of experiments back up its validity.

NOW WORK Problem 5 from The Takeaway 20-2.

EXAMPLE 20-2 Changing Magnetic Flux II: A Varying Magnetic Field

A uniform magnetic field of magnitude $B = 1.50$ T is directed at an angle of 60.0° to the plane of a circular loop of copper wire. The loop is 3.50 cm in diameter. (a) What is the magnetic flux through the loop? What is the induced emf in the loop if the magnetic field decreases to zero (b) in 10.0 s or (c) in 0.100 s?

Set Up

The magnetic flux is given by Equation 20-1. Note that θ in this equation is the angle between the direction of the magnetic field $\vec{B}$ and the *perpendicular* to the loop, so $\theta = 90.0° - 60.0° = 30.0°$. The magnetic flux through the loop changes when the field magnitude changes, so an emf will be induced in the loop. We'll use Equation 20-2 to calculate the magnitude of this induced emf.

Magnetic flux:

$$\Phi_B = AB_\perp = AB\cos\theta \qquad (20\text{-}1)$$

Area of a circle of radius r:

$$A = \pi r^2$$

Faraday's law of induction:

$$|\mathcal{E}| = \left|\frac{\Delta\Phi_B}{\Delta t}\right| \qquad (20\text{-}2)$$

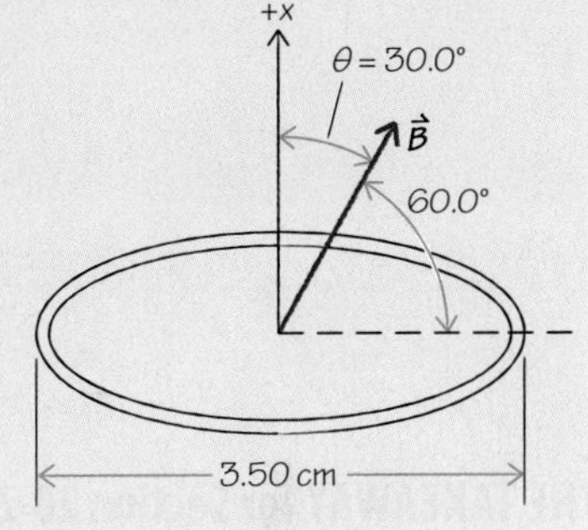

Solve

(a) Find the area of the loop, then use Equation 20-1 to calculate the magnetic flux through the loop.

The radius r of the loop is one-half of the diameter:

$$r = \frac{1}{2}(3.50\text{ cm}) = 1.75\text{ cm} = 1.75\times10^{-2}\text{ m}$$

The area of the loop is

$$A = \pi r^2 = \pi(1.75\times10^{-2}\text{ m})^2 = 9.62\times10^{-4}\text{ m}^2$$

From Equation 20-1 the magnetic flux through the loop is

$$\Phi_B = AB\cos\theta = (9.62\times10^{-4}\text{ m}^2)(1.50\text{ T})\cos 30.0° = 1.25\times10^{-3}\text{ T}\cdot\text{m}^2$$

(b) The change in magnetic flux is the final value (zero) minus the initial value that we found in (a). Equation 20-2 tells us that to find the magnitude of the induced emf, we divide this change by the time, $\Delta t = 10.0$ s, over which the flux change takes place.

The change in magnetic flux is

$$\Delta\Phi_B = (\text{final flux}) - (\text{initial flux}) = 0 - 1.25\times10^{-3}\text{ T}\cdot\text{m}^2 = -1.25\times10^{-3}\text{ T}\cdot\text{m}^2$$

If the flux decreases to zero in $\Delta t = 10.0$ s, the magnitude of the induced emf is

$$|\mathcal{E}| = \left|\frac{\Delta\Phi_B}{\Delta t}\right| = \left|\frac{-1.25\times10^{-3}\text{ T}\cdot\text{m}^2}{10.0\text{ s}}\right| = 1.25\times10^{-4}\text{ T}\cdot\text{m}^2/\text{s} = 1.25\times10^{-4}\text{ V}$$

(c) Repeat part (b) with $\Delta t = 0.100$ s.

If the flux decreases to zero in just $\Delta t = 0.100$ s, the magnitude of the induced emf is

$$|\mathcal{E}| = \left|\frac{\Delta\Phi_B}{\Delta t}\right| = \left|\frac{-1.25\times10^{-3}\ \mathrm{T\cdot m^2}}{0.100\ \mathrm{s}}\right|$$

$$= 1.25\times10^{-2}\ \mathrm{T\cdot m^2/s} = 1.25\times10^{-2}\ \mathrm{V}$$

Reflect

The induced emf is 100 times greater in part (c) than in part (b) because the same flux change takes place in 1/100 as much time. The faster the flux change, the greater the induced emf that results. To create this electric potential energy, some other type of energy must be used, to provide a push for a magnet or a conductor. The faster the flux is changing the more energy that must be used to cause that push. Note that the emf is induced *only* during the time when the magnetic flux is changing. There is zero emf when the magnetic field is at its original value of 1.50 T, and there is zero emf when the magnetic field has stabilized at its final value of zero.

Extend: If we replace the loop by a coil of the same diameter with 500 turns of wire, the induced emf is 500 times greater: $500\times1.25\times10^{-4}$ V = 0.0625 V in part (b), $500\times1.25\times10^{-2}$ V = 6.25 V in part (c). The key to generating a large induced emf is to have many turns of wire and a rapid change in magnetic flux.

NOW WORK Problems 1–4 from The Takeaway 20-2.

 **Exam Tip**

You will never be responsible for calculating an instantaneous value of emf for a changing rate of flux. The emf depends on the rate of change of flux. If the rate is itself changing, then the value we calculate for the emf for a finite Δt would not be exactly what we would measure at any instant, but gives us an average value over the interval, just like we find average velocity from $\Delta x/\Delta t$, which is the value of the velocity for all times during the interval, unless the velocity is constant. On the AP® exam, you will not be asked to make this distinction.

THE TAKEAWAY for Section 20-2

✔ A motional emf appears in a conductor that moves in a magnetic field. The force that produces the emf is a magnetic one.

✔ An induced emf appears in any loop subjected to a changing magnetic field. The force that produces the emf is an electric one.

✔ Motional emfs are a form of induced emf. Induced emfs can be described by Faraday's law of induction: The magnitude of the emf in a loop is equal to the absolute value of the change in magnetic flux through the loop divided by the time over which the change takes place.

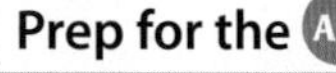

Prep for the AP Exam

Building Blocks

1. A single-turn circular loop of wire that has a radius of 5.0 cm lies in the plane perpendicular to a spatially uniform magnetic field. During a 0.12-s time interval, the magnitude of the field increases uniformly from 0.20 to 0.40 T. Determine the magnitude of the emf induced in the loop during the time interval.

2. In Example 20-2, we chose the positive x direction to be upward so that the angle θ between the magnetic field and the perpendicular to the loop was 30.0°. Show that the results for magnitude of the emf would be the same had we chosen the positive x direction to be downward so that $\theta = 150.0°$. (We will find the direction of the emf is the same in the next section.)

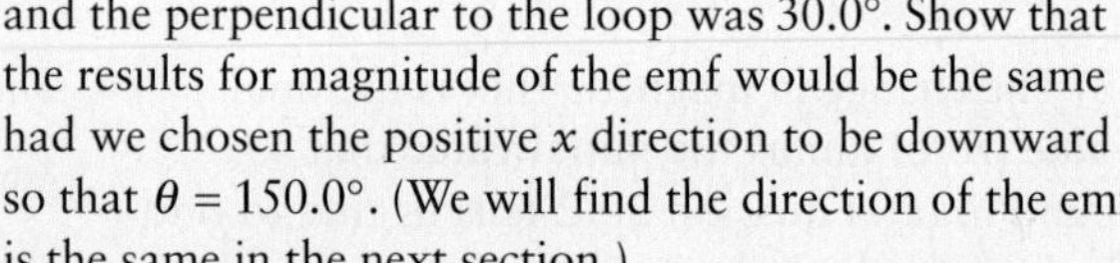

Skill Builders

3. A circular coil that has 100 turns and a radius of 10.0 cm lies in a magnetic field that has a magnitude of 0.0650 T

directed perpendicular to the coil. (a) What is the magnetic flux through the coil? (b) The magnetic field through the coil is increased steadily to 0.100 T over a time interval of 0.500 s. What is the magnitude of the emf induced in the coil during the time interval?

AP® Skills in Action

EX 20-2 4. A 30-turn coil with a diameter of 6.00 cm is placed in a constant, uniform magnetic field of 1.00 T directed perpendicular to the plane of the coil. Beginning at time $t = 0$ s, the field is increased at a uniform rate until it reaches 1.30 T at $t = 10.0$ s. The field remains constant thereafter. What is the magnitude of the induced emf in the coil at (a) $t < 0$ s, (b) $t = 5.00$ s, and (c) $t > 10.0$ s? (d) Plot the magnetic field and the induced emf as functions of time for the range $-5.00\text{ s} < t < 15.0\text{ s}$.

EX 20-1 5. A student takes a strand of copper wire and forms it into a circular loop of circumference 0.350 m. The student then places the loop in a uniform, constant magnetic field of magnitude 0.00250 T that is oriented perpendicular to the face of the loop. Pulling on the ends of the wire, the student reduces the circumference of the loop to 0.125 m in a time interval of 0.800 s. Assuming that the loop remains circular as it shrinks, what is the magnitude of the emf induced in the loop during this time interval?

6. Suppose the loop in Example 20-2 were made out of wood rather than copper wire, but the magnetic field changes in the same manner as in part (b) of Example 20-2. Compared to the emf calculated in part (b) of Example 20-2, the emf induced in the wooden loop would be
(A) zero.
(B) much smaller but not zero.
(C) slightly less.
(D) the same.

20-3 Lenz's law describes the direction of the induced emf

Equation 20-2 tells us the *magnitude* of the emf that appears in a loop when there is a change in magnetic flux through that loop: $|\mathcal{E}| = |\Delta\Phi_B/\Delta t|$. It does not, however, tell us the *direction* in which the emf drives current around that loop. As we'll see, there's a simple rule for determining this direction for any induced emf.

To learn about this rule let's think again about the loop of wire in Figure 20-2. In addition to the field $\vec{B}_{\text{magnet}}$, there's also a magnetic field produced by the current within the loop itself. As we learned in the last chapter, a current-carrying loop produces a magnetic field of its own. The direction of the field $\vec{B}_{\text{loop}}$ due to the loop depends on the direction of the current around the loop and is given by a right-hand rule: Curl the fingers of your right hand around the loop in the direction of the current, and the extended thumb of your right hand will point in the direction of $\vec{B}_{\text{loop}}$ in the interior of the loop (see **Figure 20-6a**). So whenever an induced emf appears in a loop, the current produced by that emf generates a magnetic field $\vec{B}_{\text{loop}}$ whose direction depends on the direction of the current and emf. We call $\vec{B}_{\text{loop}}$ an **induced magnetic field.**

The field $\vec{B}_{\text{loop}}$ itself produces a magnetic flux through the loop, and it's the sense of this flux that will tell us the direction of the emf in the loop. Let's choose the positive x direction for the loop in Figure 20-6 to point to the right, perpendicular to the plane of the loop. If the loop is close to the south pole of a bar magnet, as in **Figure 20-6b**, the field of the magnet causes a positive magnetic flux through the loop (the field $\vec{B}_{\text{magnet}}$ points generally to the right, in the positive x direction). If the loop moves toward the magnet as in the left-hand side of Figure 20-6b, the field $\vec{B}_{\text{magnet}}$ inside the loop increases and the positive flux increases. Experiment shows that the current induced in the loop gives rise to an induced magnetic field $\vec{B}_{\text{loop}}$ within the loop, which points in the *opposite* direction to $\vec{B}_{\text{magnet}}$. So while the flux due to $\vec{B}_{\text{magnet}}$ becomes more positive, $\vec{B}_{\text{loop}}$ gives rise to a negative flux that opposes the change in the flux of $\vec{B}_{\text{magnet}}$.

If instead the loop moves away from the magnet as in the right-hand side of Figure 20-6b, the field $\vec{B}_{\text{magnet}}$ inside the loop decreases and the positive flux decreases. In this case the direction of the induced current in the loop is reversed, as is the direction of the induced magnetic field $\vec{B}_{\text{loop}}$: Now $\vec{B}_{\text{loop}}$ inside the loop points in the *same* direction as $\vec{B}_{\text{magnet}}$. The magnetic flux due to $\vec{B}_{\text{loop}}$ is now positive, which opposes the negative change (decrease) in the flux of $\vec{B}_{\text{magnet}}$.

In both cases shown in Figure 20-6b, the induced magnetic field is in a direction opposite to the *change* in flux of the external magnetic field (in this case the field due to

Figure 20-6 Lenz's law The current induced in a loop by a change in flux always opposes the change in the flux.

(a)

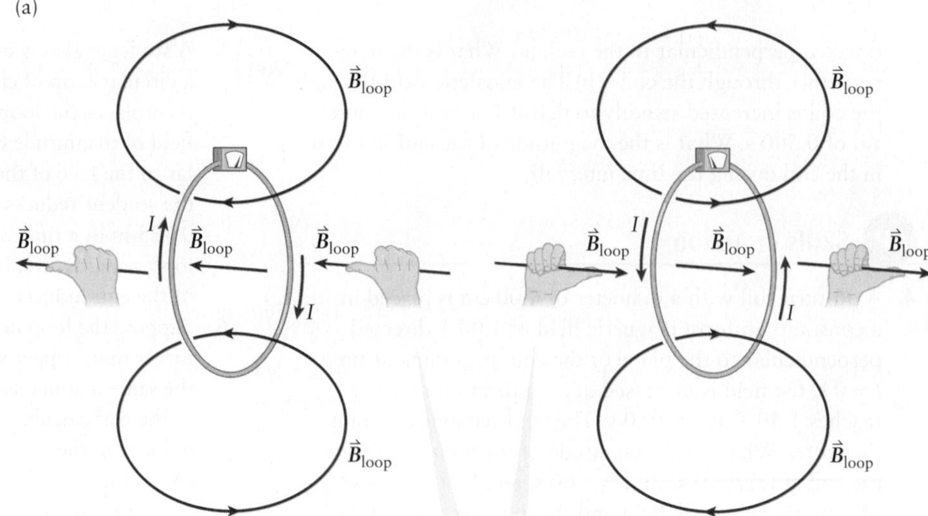

$\vec{B}_{loop}$
I
• A current-carrying loop generates a magnetic field $\vec{B}_{loop}$.
• To find the direction of $\vec{B}_{loop}$, curl the fingers of your right hand around the loop in the direction of the current I. Your extended right thumb points in the direction of $\vec{B}_{loop}$ in the interior of the loop.
• $\vec{B}_{loop}$ itself causes a magnetic flux through the loop.

(b)

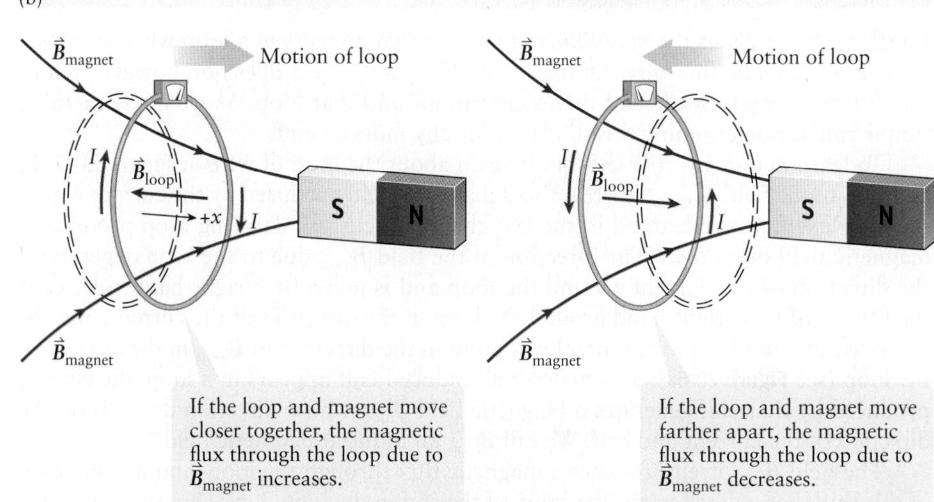

$\vec{B}_{magnet}$
Motion of loop
$\vec{B}_{loop}$
I
$+x$
S
N
If the loop and magnet move closer together, the magnetic flux through the loop due to $\vec{B}_{magnet}$ increases.
If the loop and magnet move farther apart, the magnetic flux through the loop due to $\vec{B}_{magnet}$ decreases.

the bar magnet). Many experiments show that this is always the case, no matter whether the magnet or the loop or both are moving. The nineteenth-century Russian physicist Heinrich Lenz summarized these observations in a principle that we now call **Lenz's law:**

The direction of the magnetic field induced within a conducting loop creates a magnetic flux that opposes the change in magnetic flux that created it.

It's common to combine Faraday's law and Lenz's law into a single equation:

EQUATION IN WORDS
Faraday's law and Lenz's law for induction

$$\mathcal{E} = -\frac{\Delta \Phi_B}{\Delta t} \quad (20\text{-}3)$$

Induced emf in a loop

Change in the magnetic flux through the surface outlined by the loop

Time interval over which the change in magnetic flux takes place

The minus sign indicates that the current caused by the emf induces a magnetic field that opposes the change in flux.

We can check Lenz's law by revisiting the sliding copper bar from Example 20-1 (Section 20-2). The upward external magnetic field of magnitude B causes an upward magnetic flux through the loop formed by the sliding bar and the rails on which it slides (**Figure 20-7**). The area enclosed by this loop increases as the bar slides to the right, so the upward flux increases as well. By Lenz's law an induced current in the loop will generate an induced magnetic field that opposes this change in flux. So this induced magnetic field must point downward, and to produce that induced field the current in the loop must be clockwise as seen from above the loop. That's just the direction of the current that we depicted in the figure that accompanies Example 20-1. So this situation is consistent with Lenz's law.

The sliding copper bar in Figure 20-7 illustrates another aspect of Lenz's law. Once the induced current occurs in the bar, the external magnetic field exerts a force on that current. Using the right-hand rule for this force (see Section 19-5), we see that this force points opposite to the direction in which the bar is moving. In other words, this force *opposes* the motion that gives rise to the change in flux through the loop made up of the bar and rails. That's always the case when a conductor moves through a magnetic field: A current is induced in the conductor, and the magnetic field exerts a force on the current that opposes the motion of the conductor. We can summarize this in an alternative statement of Lenz's law:

When the magnetic flux through a loop changes, the induced current will always be in the direction to cause a magnetic force that opposes that change.

Since a magnetic force opposes the motion of the bar in Figure 20-7, we need to apply an external force to keep the bar in motion. If we make the bar move faster, the magnetic flux through the loop changes more rapidly, the emf and resulting current in the loop are greater, and the magnetic force opposing the motion of the bar is greater. (The magnetic force on the bar is proportional to the speed of the bar, just like the drag force on a microscopic object moving through a fluid we discussed in Chapter 5.) So we must apply a greater force to make the bar slide at a faster speed.

This same effect explains the phenomenon of *magnetic braking*. If you try to make a magnet move past a conductor or a conductor move past a magnet, currents appear in the conductor. (These are called *eddy currents*, since their pattern resembles that of eddies in a body of water. The conductor does *not* need to be in the form of a loop for these currents to appear.) The magnetic force that the magnet exerts on the eddy currents opposes the motion of the conductor relative to the magnet, so by Newton's third law there is a force that opposes the motion of the magnet relative to the conductor. One application of this is to roller coasters. When a roller coaster car enters the part of the ride where it's supposed to slow down, a copper fin on the car passes through powerful permanent magnets mounted on the track. Eddy currents arise in the fin, and the interaction between the eddy currents and the field of the permanent magnets causes a force that smoothly brings the car to a slow speed. The car is then stopped by conventional mechanical braking. This same braking technique is used to help large trucks. Conventional mechanical brakes would often overheat and become less effective, resulting in runaway trucks that could cause accidents. Eddy current braking greatly increased safety by allowing trucks to brake without the friction that heated up the mechanical brakes. Then, at low speeds, when the relative motion is not enough to cause a large eddy current braking force, the mechanical brakes can be used safely.

Eddy currents are also used in an *electromagnetic flowmeter*, a device that can measure blood flow in an artery. Blood is an electrical conductor; eddy currents are induced in the blood as it flows past magnets in the flowmeter. The device records the small but measurable magnetic fields due to these currents and uses them to determine the rate of flow. The advantage of an electromagnetic flowmeter is that it is noninvasive: No component of the device need be surgically introduced into the body.

WATCH OUT!

Like Faraday's law, Lenz's law is about the *change* in flux.

Notice that Lenz's law refers to the direction of the *change* in magnetic flux ("Is the flux increasing or decreasing?"), not to the direction of the field that causes the flux. The field $\vec{B}_{magnet}$ points in the same direction in both of the situations shown in Figure 20-6b, but the flux change is different in the two situations, so the emf is in different directions as well. The minus sign in Equation 20-3 is a reminder about the direction of the induced emf. As we discuss below, this direction will always be such that the resulting current produces a magnetic field which opposes the change in flux.

(1) As the bar slides to the right, the upward magnetic flux through the loop increases.

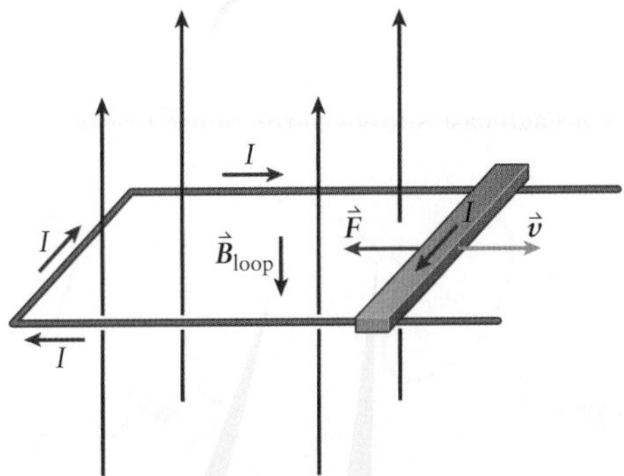

(2) The induced current I produces an induced magnetic field $\vec{B}_{loop}$ that opposes the change in flux.

(3) The magnetic force on the current in the moving bar opposes the bar's motion.

Figure 20-7 A sliding copper bar revisited Lenz's law helps explain the direction of induced current in this situation.

Another application of eddy currents is *magnetic induction tomography*, a relatively new experimental technique for medical imaging. In this technique changing magnetic fields created by coils placed near a part of the body induce eddy currents. Observing the fields produced by these eddy currents is a way to monitor brain swelling. Eddy currents can also be used for the controlled, repeated delivery of medication. A capsule containing the drug is implanted in the body; the capsule is made from a gel that heats up slightly when there are eddy currents, opening pores through which the medication is released. As for an electromagnetic flowmeter, no implanted electronics are required.

THE TAKEAWAY for Section 20-3

✔ An emf induced by a changing magnetic flux in a conductor causes a current. This current generates a magnetic field of its own, called the induced magnetic field.

✔ The induced magnetic field is in a direction that opposes the change in magnetic flux that created the emf that gave rise to the induced field.

✔ Eddy currents arise whenever a conducting material, even a nonmagnetic one, moves relative to a magnetic field.

Prep for the AP® Exam

AP® Building Blocks

1. Each situation shown in the figure below shows the initial and final magnitude and direction of a changing magnetic field. A conducting wire loop is placed perpendicular to the magnetic field as shown. For each situation, determine if the induced current in the loop is clockwise, counterclockwise, or zero as viewed from above while the magnetic field changes from its initial value to its final value.

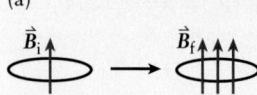

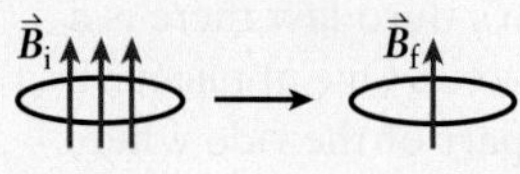

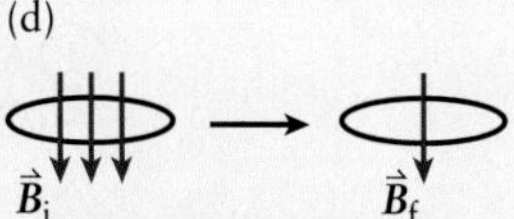

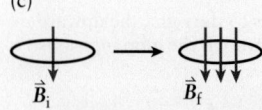

2. Determine the direction of the induced current in the loop for each case shown in the figure below.

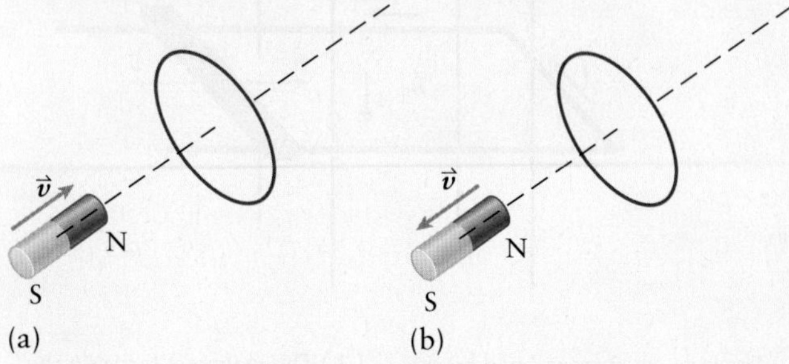

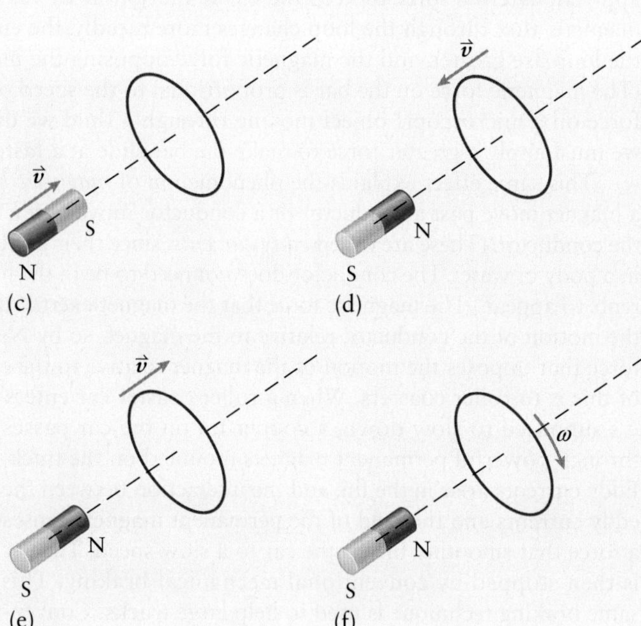

AP® Skill Builders

3. A bar magnet is moved steadily through a wire loop, as shown in the figure below. Make a qualitative sketch of the induced emf in the loop as a function of time (be sure to include the times t_1, t_2, and t_3). Consider the direction of positive emf to be as indicated in the figure.

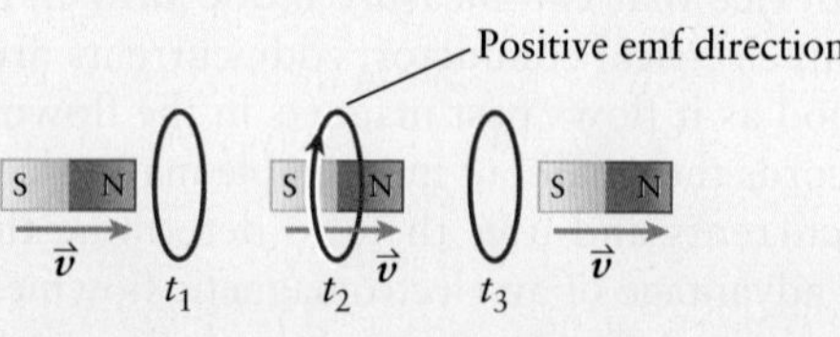

4. A bar magnet is moved steadily through a wire loop, as shown in the figure in problem 3, except that the leading edge of the magnet is the south pole instead of the north pole. Make a qualitative sketch of the induced emf in the loop as a function of time (be sure to include the times t_1, t_2, and t_3). Consider the direction of positive emf to be as indicated in the figure.

AP® Skills in Action

5. A square, 30-turn coil 10.0 cm on a side with a resistance of 0.820 Ω is placed between the poles of a large electromagnet. The electromagnet produces a constant, uniform magnetic field of 0.600 T directed into the page. As suggested by the figure below, the field drops sharply to zero at the edges of the magnet. The coil moves to the right at a constant velocity of 2.00 cm/s.

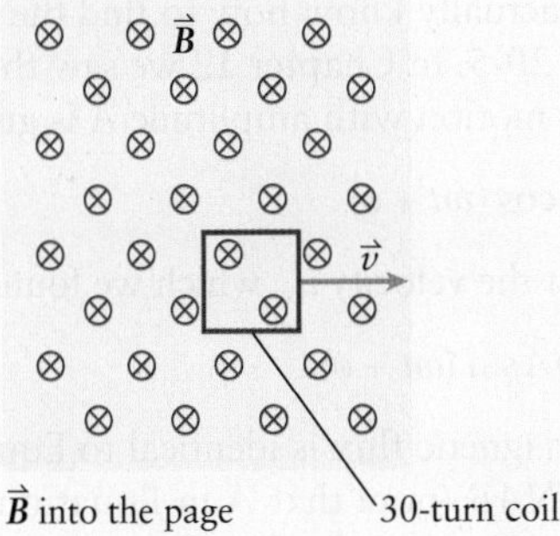

(a) What is the current through the wire coil
  (i) before the coil reaches the edge of the field,
  (ii) while the coil is leaving the field, and
  (iii) after the coil leaves the field?
(b) What is the total charge that flows past a given point in the coil as it leaves the field?
(c) Plot the induced current in the loop as a function of the horizontal position of the right side of the current loop. Let the right-hand edge of the magnetic field region be $x = 0$. Your plot should cover the range of $-5.00\text{ cm} < x < 20.0\text{ cm}$.

6. For the coil in problem 5:
(a) Determine the magnitude and direction of the force on each side of the coil for situations (a) through (c).
(b) As the loop enters the field region from the left, what is the direction of the induced current and the resulting force on each segment of the coil?

7. Two metal rings with a common axis are placed near each other, as shown in the figure below. If current I_a is suddenly set up and is increasing in ring a as shown, the current in ring b is
(A) zero.
(B) parallel to I_a.
(C) antiparallel to I_a.
(D) alternatively parallel and antiparallel to I_a.

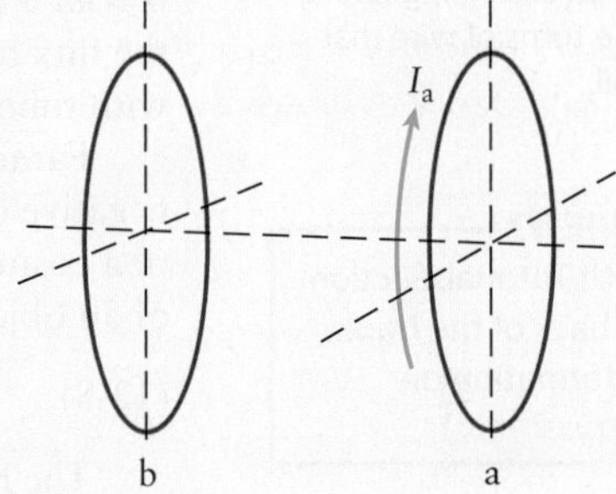

8. The figure below shows two coils wound around an iron ring, which directs the magnetic field of each coil around the ring. Current appears in the coil on the right
(A) the moment the battery is connected by closing the switch.
(B) the entire time the battery is connected with the switch closed.
(C) the moment the battery is disconnected by opening the switch.
(D) the moment the battery is connected by closing the switch and the moment the battery is disconnected by opening the switch.

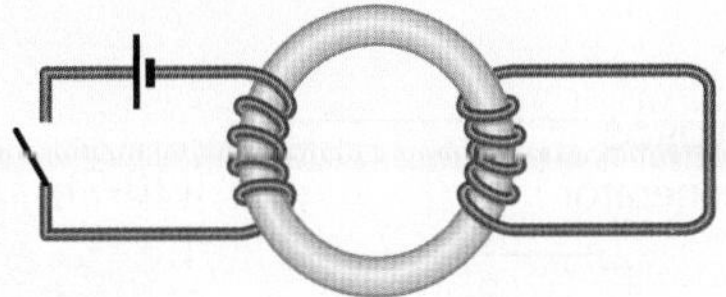

20-4 Faraday's law explains how alternating currents are generated

> **AP® Exam Tip**
> Note this topic is not specifically on the AP® exam, but it will deepen your understanding of Faraday's law. The questions asked in The Takeaway will help you prepare for the AP® Physics 2 Exam.

Alternating current is important in technology: If you're reading these words in a room lit by electric light, the light bulbs are powered by alternating current. If you're reading on a mobile device such as a tablet, the device's battery was charged by plugging it into a wall socket and using the alternating current delivered by that socket. We now have the physics we need to understand how alternating current is produced.

Let's look at a coil of wire with N turns, each of which has area A. As **Figure 20-8** shows, this coil is free to rotate around an axis that lies along a diameter of the coil. We place the coil in a region of uniform magnetic field $\vec{B}$, then rotate the coil at a constant angular speed ω. As the coil rotates, the magnetic flux through each turn of the

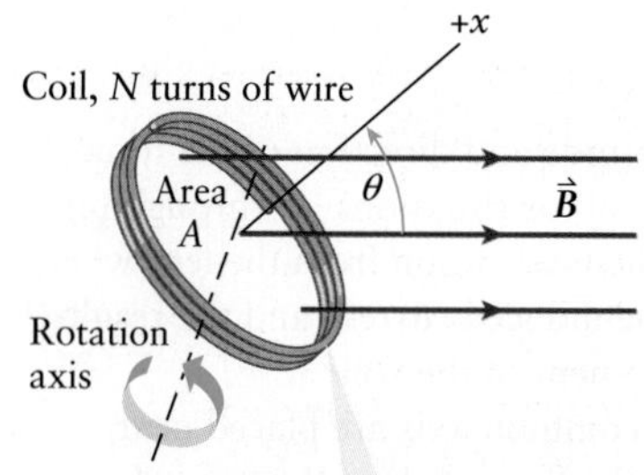

As the coil rotates with angular speed ω, the angle θ between the magnetic field $\vec{B}$ direction and the perpendicular to the coil changes: $\theta = \omega t + \phi$.

Figure 20-8 An ac generator As the coil rotates, an oscillating emf is generated in the turns of wire that make up the coil.

NEED TO REVIEW?

See the Math Tutorial, Section M-8, in the back of the book for more information on trigonometry.

coil changes, so an emf is generated. As we will see, this is an alternating emf of just the sort required to generate an alternating current. That's why a rotating coil of the sort shown in Figure 20-8 is called an **ac generator.**

We begin by writing an equation for the magnetic flux through the rotating coil. The angle θ between the magnetic field $\vec{B}$ and the perpendicular to the coil changes as the coil rotates:

(20-4) $$\theta = \omega t + \phi$$

In Equation 20-4 ϕ is the value of the angle at $t = 0$. From Equation 20-1 the magnetic flux through the N turns of the coil is N times that through one turn:

(20-5) $$\Phi_B = NAB \cos \theta = NAB \cos (\omega t + \phi)$$

When $\theta = \omega t + \phi = 0$ so $\cos \theta = 1$, the perpendicular to the coil is in the same direction as $\vec{B}$ and the flux has its most positive value $\Phi_B = NAB$; when $\theta = \omega t + \phi = \pi/2$ so $\cos \theta = 0$, the coil is edge-on to the magnetic field and the flux is zero; when $\theta = \omega t + \phi = \pi$ and $\cos \theta = -1$, the perpendicular to the coil points opposite to $\vec{B}$ and the flux has its most negative value $\Phi_B = -NAB$; and so on. So the magnetic flux varies with time, and it follows that there will be an emf in the coil.

Faraday's law and Lenz's law (Equation 20-3) tell us that the emf is equal to the negative of the rate of change of Φ_B. We actually know how to find the rate of change of a cosine function like that in Equation 20-5. In Chapter 12 we saw that the position of an object undergoing simple harmonic motion with amplitude A is given by

(12-8) $$x = A \cos (\omega t + \phi)$$

The rate of change of position x is just the velocity v_x, which we found was equal to

(12-9) $$v_x = -\omega A \sin (\omega t + \phi)$$

You can see that Equation 20-5 for magnetic flux is identical to Equation 12-8 for position, with amplitude A replaced by NAB (note that A in Equation 20-5 denotes area, not amplitude). Making the same replacement in Equation 12-9 tells us that the rate of change of magnetic flux through the rotating coil is

(20-6) $$\frac{\Delta \Phi_B}{\Delta t} = -\omega NAB \sin (\omega t + \phi)$$

Substituting Equation 20-6 into Equation 20-3 then gives us the emf in the rotating coil:

EQUATION IN WORDS
Emf in an ac generator

(20-7) $$\mathcal{E} = \omega NAB \sin (\omega t + \phi)$$

Angular speed of the rotating coil; **Time**; **Angle** of the coil at $t = 0$; **Emf produced by an ac generator**; **Number of turns** in the coil; **Area of** the coil; **Magnitude of the magnetic field** to which the coil is exposed

The emf alternates with angular frequency ω, which is the same as the angular speed of the rotating coil. The maximum value of the emf is $\mathcal{E}_{max} = \omega NAB$, which shows that we can increase the maximum emf by increasing the angular speed ω, the number of turns N, the coil area A, the magnetic field magnitude B, or a combination of these.

While the emf produced by an ac generator changes from positive to negative, the power delivered by the generator does not. As an example, suppose an ac generator is connected to a circuit device (such as a light bulb or a toaster) that we can represent as a resistor with resistance R. If we ignore the internal resistance of the coil, the emf is equal to the potential difference across the resistor: $\mathcal{E} = IR$. The current in the resistor is therefore

(20-8) $$I = \frac{\mathcal{E}}{R} = \frac{\omega NAB}{R} \sin (\omega t + \phi)$$

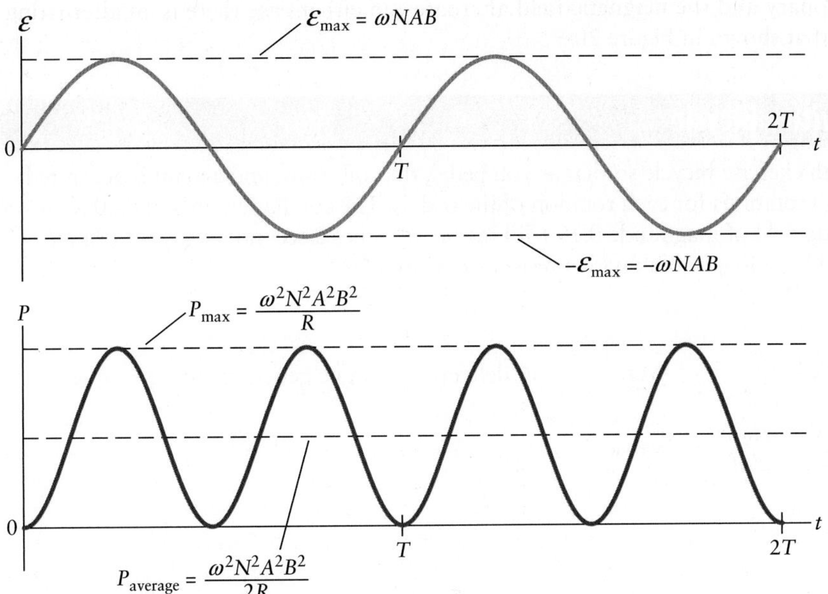

Figure 20-9 An ac generator: emf and power These graphs show the emf $\mathcal{E}$ generated in the coil shown in Figure 20-8 and the power P that this emf delivers to a resistor R connected to the coil. We assume $\phi = 0$ in Equations 20-7 and 20-9. Note that T is the time it takes for the coil to complete one rotation (the period of the rotation).

The current in the resistor alternates with the same angular frequency ω as the emf. From Section 18-5 the power into such a resistor is

$$P = I^2 R \tag{18-14}$$

If we substitute Equation 20-8 into Equation 18-24, we get

$$P = \left(\frac{\omega NAB}{R}\sin(\omega t + \phi)\right)^2 R = \frac{\omega^2 N^2 A^2 B^2}{R}\sin^2(\omega t + \phi) \tag{20-9}$$

Figure 20-9 shows graphs of the emf $\mathcal{E}$ (Equation 20-7) and resistor power P (Equation 20-9) as functions of time. The power P is *never* negative, which means that energy is always transferred from the ac generator into the resistor, never the other way. The average value of the function $\sin^2(\omega t + \phi)$ is $\frac{1}{2}$, so the average power into the resistor is

$$P_{average} = \frac{\omega^2 N^2 A^2 B^2}{2R} \tag{20-10}$$

It may seem like the power given by Equation 20-10 comes "for free": You let the coil rotate, and an emf is generated that makes energy that can be converted into other types by the resistor. However, as we discussed in Section 20-3, whenever a conductor (such as the coil of an ac generator) moves in the presence of a magnetic field, the conductor experiences a magnetic force that opposes its motion. So left to itself, the coil would quickly slow to a halt. To keep the coil in motion, you must apply a torque that just balances the effects of this magnetic force. At an electric generating station, this torque is applied to the blades of a turbine that is connected to the coil. The blades can be turned by the force of the wind (Figure 20-1a), by the force of flowing water at a hydroelectric plant, or by the force of fast-moving steam at a coal-fired or nuclear power plant (where burning fossil fuels or radioactive decay heats water to produce steam). Part of the mechanical power used to make the turbine spin goes into the electric power provided by the generator; the rest is lost due to friction in the turbine and generator. Note that in some generators, like the ones shown in **Figure 20-10**, the coil is

Lee Roberts/US Army Corps of Engineers

Figure 20-10 An ac hydroelectric generator At the Center Hill Dam in Tennessee, water flowing past a turbine turns this rotor, which is equipped with a series of electromagnets (the tall red rectangles around the rotor's rim). These are arranged with alternating orientations: One electromagnet has its north pole at the top, the next one has its north pole at the bottom, the next has its north pole at the top, and so on. (The red rotor shown here has been removed for maintenance.) As the rotor spins, the magnetic field through the stationary coils around the rotor alternates, which produces an alternating emf as depicted in Figure 20-9. Each of the three generators at the Center Hill Dam can produce an average power of 45 megawatts (4.5×10^7 W).

stationary and the magnetic field alternates; in either case there is an alternating emf like that shown in Figure 20-9.

EXAMPLE 20-3 Lighting the Gym with a Bicycle

You attach the coil of an ac generator to an exercise bicycle so that as you pedal, the coil turns, and an emf is generated. The generator is geared so that it makes 10 rotations for each rotation of the pedals. The coil has an area of $6.40 \times 10^{-3}\ \text{m}^2$, has 2000 turns of wire, and is in a magnetic field of magnitude 0.100 T. How many times a second must you turn the pedals to deliver an average power of 60.0 W to a light bulb with a resistance of 80.0 Ω?

Set Up

The situation is as shown in Figure 20-8. We'll use Equation 20-10 to solve for the angular speed ω of the generator coil. Because the coil makes 10 rotations for every rotation of the pedals, the angular speed of the pedals is equal to ω divided by 10.

Average power delivered by an ac generator to a resistor:

$$P_{\text{average}} = \frac{\omega^2 N^2 A^2 B^2}{2R} \qquad \text{(20-10)}$$

Solve

Rewrite Equation 20-10 to solve for ω.

Find the angular speed of the generator coil from Equation 20-10:

$$\omega^2 = \frac{2RP_{\text{average}}}{N^2A^2B^2}$$

$$\omega = \frac{\sqrt{2RP_{\text{average}}}}{NAB} = \frac{\sqrt{2(80.0\ \Omega)(60.0\ \text{W})}}{(2000)(6.40 \times 10^{-3}\ \text{m}^2)(0.100\ \text{T})}$$

$$= 76.5\ \text{rad/s}$$

Convert this from radians per second to revolutions per second:

$$\omega = \left(76.5 \frac{\text{rad}}{\text{s}}\right)\left(\frac{1\ \text{rev}}{2\pi\ \text{rad}}\right) = 12.2\ \text{rev/s}$$

The generator turns 10 times faster than the pedals, so the pedals must turn at a rate of

$$(12.2\ \text{rev/s})/10 = 1.22\ \text{rev/s} = 73.1\ \text{rev/min}$$

Reflect

A cycling cadence of 73.1 rev/min is just a bit faster than one turn per second, so you can power a light bulb in this way. Exercise bicycles with generators of this kind are commercially available and are used to return power to the electrical grid in the same manner as residential solar panels.

NOW Work Problem 3 from The Takeaway 20-4.

THE TAKEAWAY for Section 20-4

✔ An ac generator consists of a coil that rotates in a magnetic field.

✔ The emf produced by an ac generator oscillates at an angular frequency that equals the angular speed of the rotating coil.

Prep for the AP Exam

AP Building Blocks

1. A rectangular coil with sides 0.10 m by 0.25 m has 500 turns of wire. It is rotated about its long axis in a magnetic field of 0.58 T directed perpendicular to the rotation axis. At what frequency must the coil be rotated for it to generate a maximum potential difference of 110 V?
2. An electromagnetic generator consists of a coil that has 100 turns of wire, has an area of 400 cm^2, and rotates at

60 rev/s in a magnetic field of 0.25 T directed perpendicular to the rotation axis. What is the magnitude of the emf induced in the coil?

AP® Skill Builders

EX 20-3 3. You decide to build a small generator by rotating a coiled wire inside a static magnetic field of 0.30 T. You construct the apparatus by coiling wire into three loops of radius 0.16 m. If the coils rotate at 3.0 revolutions per second and are connected to a device with 1.0×10^2 Ω resistance, calculate the average power supplied to that device.

4. Perhaps it has occurred to you that we could tap Earth's magnetic field to generate energy. One way to do this would be to spin a metal loop about an axis perpendicular to Earth's magnetic field. Suppose that the metal loop is a square that is 45.0 cm on each side and that we want to generate in the loop an electric potential difference of amplitude 120 V at a place where Earth's magnetic field is 5×10^{-5} T. At what angular speed (in rev/s) would we have to spin the coil? Does this appear to be a feasible method for converting energy using Earth's magnetic field?

AP® Skills in Action

5. Certain types of fluorescent lamps flicker rapidly. That's because these lamps emit a pulse of light every time a burst of electric power is provided to the lamp. If such a lamp is powered by a source of emf that oscillates at 60 Hz, what is the frequency at which the lamp will flicker?
(A) 30 Hz
(B) 60 Hz
(C) 120 Hz
(D) 3600 Hz

20-5 Maxwell's equations tie electricity and magnetism together

By the middle of the nineteenth century, scientists understood a great deal about electricity and magnetism. The work of Gauss (discussed in this section), Ampère, and Faraday had established fundamental relationships that describe electric and magnetic phenomena; for example, it was understood that electric charge gives rise to electric fields and that electric currents give rise to magnetic fields. It was the Scottish physicist James Clerk Maxwell who added to these fundamental relationships and forged our understanding of electricity and magnetism into a unified theory. In this section we'll look at four basic equations, known as **Maxwell's equations**, which describe *all* electromagnetic phenomena. These equations help us understand the nature of electromagnetic waves that we will discuss in the next chapter.

AP® Exam Tip

The material in this section is not directly tested on the AP® exam. It may help provide a deeper understanding of the laws we have already introduced that are covered. We will introduce electromagnetic waves in the next unit. Maxwell's equations predict the existence of electromagnetic waves, and their speed, and help us understand their properties, tying concepts together across units.

Gauss's Laws for Electricity and Magnetism

The first of Maxwell's equations is **Gauss's law for the electric field.** We did not introduce this in Chapter 16, because first we needed to develop a stronger understanding of flux. Gauss's law for the electric field states that the net electric flux through a closed surface (called a *Gaussian surface*) is proportional to the total amount of electric charge enclosed within that surface:

Electric flux through a closed surface

Net amount of **charge enclosed** within the surface

$$\Phi_E = \frac{q_{\text{encl}}}{\varepsilon_0} \quad (20\text{-}11)$$

Permittivity of free space = $1/(4\pi k)$

EQUATION IN WORDS
Gauss's law for the electric field

Figure 20-11a illustrates Gauss's law for the electric field. If a surface encloses a net positive charge, as for Gaussian surface 1 in Figure 20-11a, there is a net outward (positive) electric flux, and electric field lines point out of the surface. If instead there is a net negative charge inside the surface, as for Gaussian surface 2 in Figure 20-11a, the net electric flux is inward (negative), and electric field lines point into the surface. If there is no charge at all inside the surface, as for Gaussian surface 3 in Figure 20-11a,

(a) Gauss's law for the electric field: $\Phi_E = q_{encl}/\mathcal{E}_0$

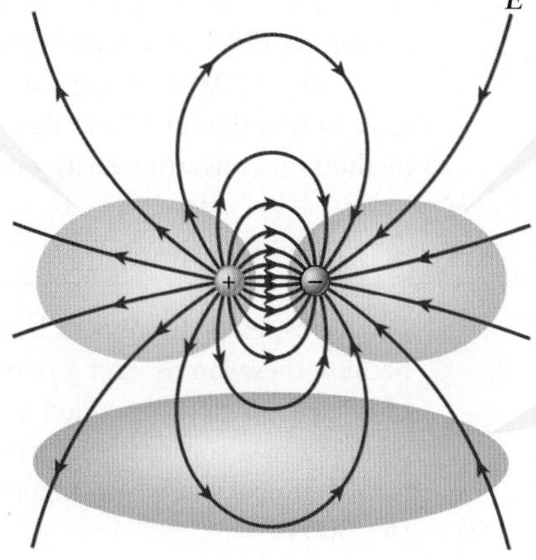

Gaussian surface 1 encloses positive charge. Field lines point out of this surface, and there is a net outward (positive) electric flux.
$\vec{E}$
Gaussian surface 2 encloses negative charge. Field lines point into this surface, and there is a net inward (negative) electric flux.
Gaussian surface 3 encloses zero charge. Each field line that points into this surface at one place points out at another place. There is zero net electric flux.

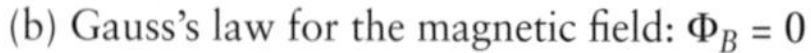

(b) Gauss's law for the magnetic field: $\Phi_B = 0$

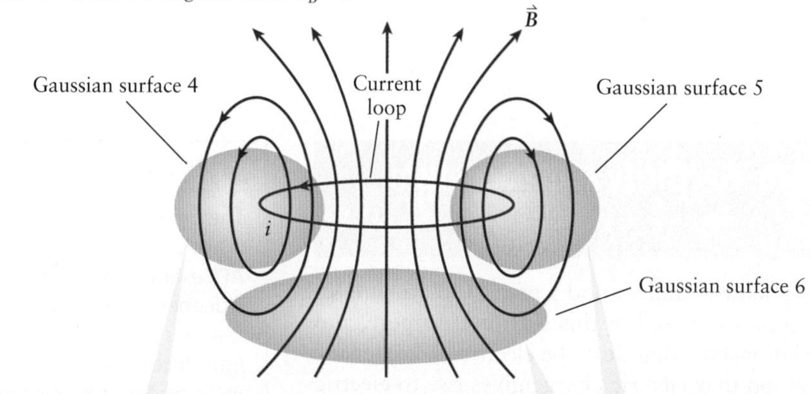

Figure 20-11 Gauss's laws for the electric and magnetic fields (a) The electric flux through a closed surface depends on the charge enclosed by that surface. (b) The magnetic flux through any closed surface is zero.

- In a region of space where there are no electric charges, there is zero net electric flux through any Gaussian surface in that region.
- In any region of space, there is zero net magnetic flux through any Gaussian surface.

there is zero net electric flux: Each field line that enters the surface at one point exits it at another point.

In addition to Gauss's law for the electric field, there is **Gauss's law for the magnetic field.** This states that for *any* closed Gaussian surface, there is *zero* net flux of the magnetic field $\vec{B}$ through that surface:

EQUATION IN WORDS
Gauss's law for the magnetic field

$$\Phi_B = 0 \qquad (20\text{-}12)$$

Magnetic flux through a closed surface

There is **zero magnetic flux through any closed surface,** no matter what the size or shape of the surface or what it contains.

As an example, **Figure 20-11b** shows three Gaussian surfaces in the magnetic field of a current loop. For all three surfaces, each field line that enters the surface at one point exits it at another point. This is true whether there is current enclosed within the Gaussian surface (as for surfaces 4 and 5) or if there is no enclosed current (as for surface 6). Just as for the electric flux for Gaussian surface 3 in Figure 20-11a, this implies that there is zero net flux of magnetic field for all three surfaces. (There *would* be a nonzero net magnetic flux through a Gaussian surface if the surface enclosed an isolated north magnetic pole that had no associated south pole, or an isolated south pole that had no associated north pole. These would be the magnetic analogs of an isolated positive or negative electric charge. Although some physicists have searched for decades for such isolated poles, referred to as *magnetic monopoles*, none have ever been observed.)

Faraday's Law: A Changing Magnetic Field Generates an Electric Field

The third of Maxwell's equations is *Faraday's law*, which we just introduced in Sections 20-2 and 20-3. It states that an emf is induced in a loop if the magnetic flux through that loop changes:

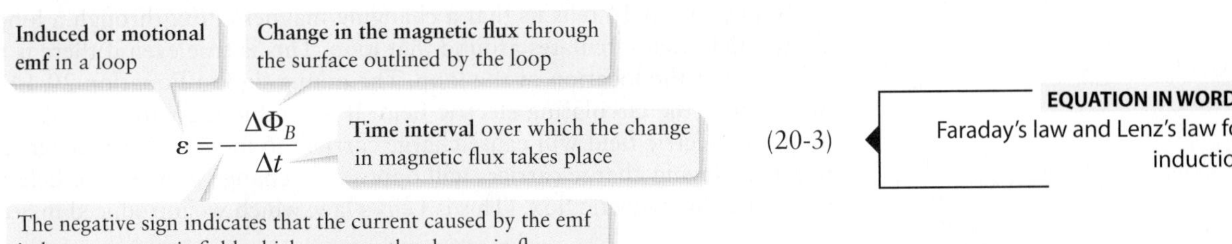

Induced or motional emf in a loop
Change in the magnetic flux through the surface outlined by the loop
$\varepsilon = -\frac{\Delta\Phi_B}{\Delta t}$
Time interval over which the change in magnetic flux takes place
(20-3)
The negative sign indicates that the current caused by the emf induces a magnetic field which opposes the change in flux.

EQUATION IN WORDS
Faraday's law and Lenz's law for induction

As an example, **Figure 20-12a** shows a wire loop placed between the poles of an electromagnet. As the current in the electromagnet increases, the magnetic field increases and the increasing magnetic flux through the wire loop induces an emf in the loop. It's important to note that the emf is present whether the loop is made of a conductor like copper wire, a semiconductor like silicon, or an insulator like wood; the only difference is the amount of current that's established in response to the emf. The emf is even present if there is no material substance there at all!

Fundamentally, what happens is that the changing magnetic field generates a circulating electric field $\vec{E}$ as **Figure 20-12b** shows. The emf around a loop is just the *circulation* of the electric field around the loop. We define this in exactly the same way that we defined the circulation of magnetic field in Section 19-7. First imagine breaking the loop into a number of small segments of length $\Delta\ell$; then find the component $E_{\|}$ of the electric field that's tangent to each segment; then sum the products $E_{\|}\Delta\ell$:

$$\text{circulation of the electric field} = \sum E_{\|}\Delta\ell \qquad (20\text{-}13)$$

(One simple case is the circulating electric field shown in Figure 20-12b. Here $\vec{E}$ is tangential to the circular loop of radius r, so the circulation is just the magnitude E multiplied by the total distance around the loop—that is, its circumference $2\pi r$.)

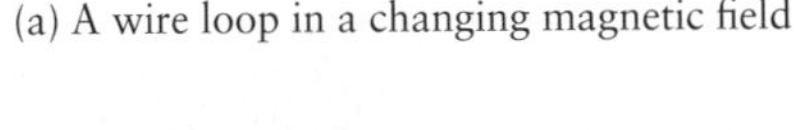

(a) A wire loop in a changing magnetic field

(b) A changing magnetic field produces a circulating electric field.

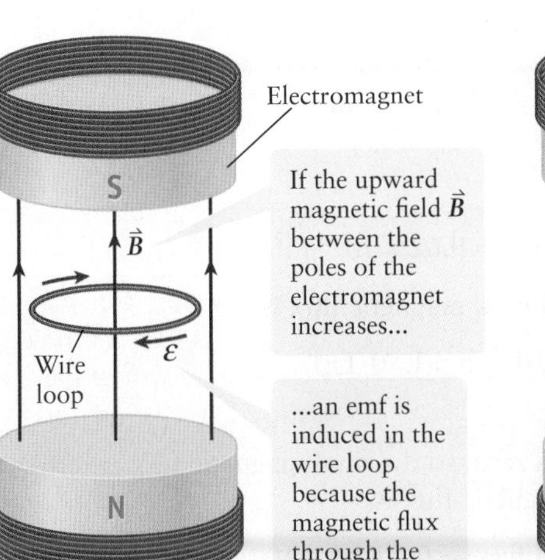

Electromagnet
S
$\vec{B}$
If the upward magnetic field $\vec{B}$ between the poles of the electromagnet increases...
ε
Wire loop
...an emf is induced in the wire loop because the magnetic flux through the loop changes.
N

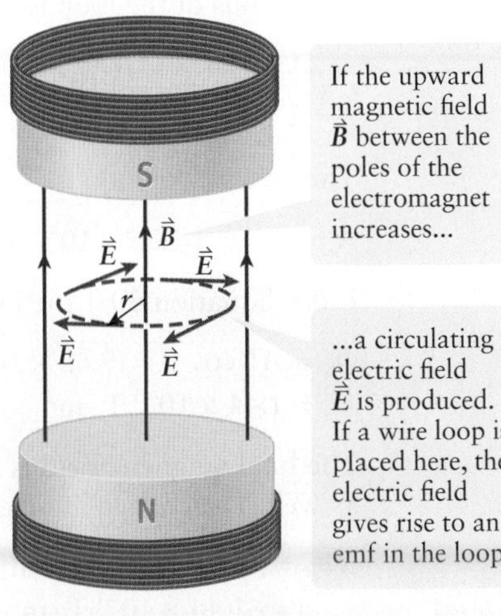

S
If the upward magnetic field $\vec{B}$ between the poles of the electromagnet increases...
$\vec{B}$
$\vec{E}$
$\vec{E}$
r
$\vec{E}$
$\vec{E}$
...a circulating electric field $\vec{E}$ is produced. If a wire loop is placed here, the electric field gives rise to an emf in the loop.
N

Figure 20-12 A changing $\vec{B}$ produces a circulating $\vec{E}$ (a) Faraday's law says that an emf is induced in the loop when the magnetic field changes. (b) Fundamentally what happens when the magnetic field changes is that a circulating electric field is produced.

We can now rewrite Faraday's law by replacing the induced emf $\mathcal{E}$ in Equation 20-3 with the circulation of the electric field, as given by Equation 20-13:

EQUATION IN WORDS
Faraday's law in terms of circulation

$$\sum E_{\parallel}\Delta\ell = -\frac{\Delta\Phi_B}{\Delta t} \qquad (20\text{-}14)$$

Circulation of electric field around a loop

Change in the magnetic flux through the surface outlined by the loop

Time interval over which the change in magnetic flux takes place

Equation 20-14 tells us that a changing magnetic flux through a loop causes an electric field that circulates around that loop. This is true even if there is no material substance at the location of the loop. The minus sign in Equation 20-14 tells us the direction of the circulating electric field: If a conducting wire is present around the loop, the electric field will cause charge carriers to move, and the magnetic field due to those moving charge carriers will oppose the change in magnetic field that caused the change in magnetic flux. (This is Lenz's law, which we introduced in Section 20-3.)

EXAMPLE 20-4 An Electric Field Due to a Changing Magnetic Field

The uniform magnetic field shown in Figure 20-12b decreases in magnitude from 1.50 T to zero in a time t, inducing an electric field. What is the magnitude of this electric field around a loop 3.50 cm in diameter if (a) $t = 10.0$ s and (b) $t = 0.100$ s?

Set Up

Equation 20-1 tells us the magnetic flux through the loop of diameter 3.50 cm. The magnetic field $\vec{B}$ is perpendicular to the plane of the loop, so θ (the angle between the direction of $\vec{B}$ and the perpendicular to the loop) is $\theta = 0$.

The electric field in this case has the same magnitude E all the way around the circle and is tangent to the circle, so $E_{\parallel}$ in Equation 20-14 is equal to E.

Magnetic flux:

$$\Phi_B = AB_{\perp} = AB\cos\theta \qquad (20\text{-}1)$$

Faraday's law in terms of circulation:

$$\sum E_{\parallel}\Delta\ell = -\frac{\Delta\Phi_B}{\Delta t} \qquad (20\text{-}14)$$

Area of a circle of radius r:

$$A = \pi r^2$$

Circumference of a circle of radius r:

$$C = 2\pi r$$

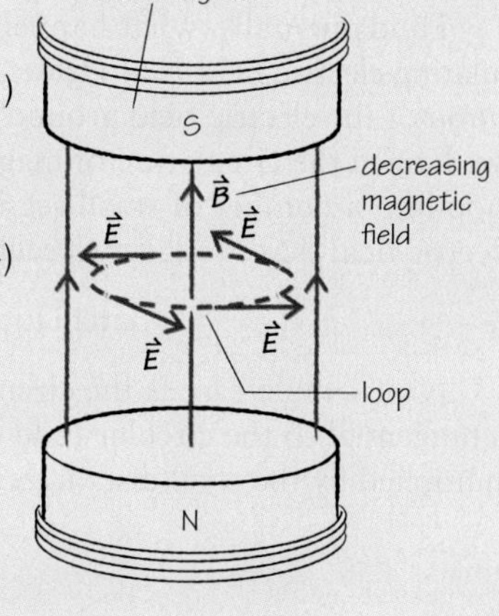

Solve

(a) First determine the initial and final values of magnetic flux and the change in magnetic flux for a circle of diameter 3.50 cm.

The radius of the loop is half the diameter:

$$r = \frac{1}{2}(3.50 \text{ cm}) = 1.75 \text{ cm} = 1.75 \times 10^{-2} \text{ m}$$

The area of the loop is

$$A = \pi r^2 = \pi(1.75 \times 10^{-2} \text{ m})^2 = 9.62 \times 10^{-4} \text{ m}^2$$

From Equation 20-1 the initial magnetic flux is

$$\Phi_B = AB\cos 0 = (9.62 \times 10^{-4} \text{ m}^2)(1.50 \text{ T})(1) = 1.44 \times 10^{-3} \text{ T}\cdot\text{m}^2$$

The final magnetic field is zero, so the final magnetic flux is zero as well. The change in magnetic flux is

$$\Delta\Phi_B = (\text{final flux}) - (\text{initial flux}) = 0 - 1.44 \times 10^{-3} \text{ T}\cdot\text{m}^2 = -1.44 \times 10^{-3} \text{ T}\cdot\text{m}^2$$

Use Equation 20-14 to calculate the circulation of the electric field in the case where $\Delta t = 10.0$ s. Because the field magnitude E has the same value around the circle and $E_{\|} = E$, we can use this to calculate E.

The circulation of the electric field is equal to $-\Delta\Phi_B/\Delta t$:

$$\text{circulation of the electric field} = \sum E_{\|}\Delta\ell = -\frac{\Delta\Phi_B}{\Delta t}$$

$$= -\frac{-1.44 \times 10^{-3}\ \text{T}\cdot\text{m}^2}{10.0\ \text{s}} = 1.44 \times 10^{-4}\frac{\text{T}\cdot\text{m}^2}{\text{s}}$$

$1\ \text{T} = 1\ \dfrac{\text{V}\cdot\text{s}}{\text{m}^2}$, so the magnitude of the circulation is

$$1.44 \times 10^{-4}\frac{\text{T}\cdot\text{m}^2}{\text{s}} = 1.44 \times 10^{-4}\left(\frac{\text{V}\cdot\text{s}}{\text{m}^2}\right)\left(\frac{\text{m}^2}{\text{s}}\right) = 1.44 \times 10^{-4}\ \text{V}$$

Because $E_{\|} = E$ and E has the same value at all points around the circle, we can write the circulation as

$$\sum E_{\|}\Delta\ell = \sum E\Delta\ell = E\sum\Delta\ell$$

The sum $\sum\Delta\ell$ is the total distance around the loop, equal to the loop circumference $2\pi r$. So

$$E(2\pi r) = 1.44 \times 10^{-4}\ \text{V}$$

$$E = \frac{1.44 \times 10^{-4}\ \text{V}}{2\pi r} = \frac{1.44 \times 10^{-4}\ \text{V}}{2\pi(1.75 \times 10^{-2}\ \text{m})} = 1.31 \times 10^{-3}\ \text{V/m}$$

(b) Repeat the calculation for the case where $\Delta t = 0.100$ s.

Equation 20-14 tells us that the circulation of the electric field, and hence the electric field itself, is inversely proportional to the time Δt over which the magnetic flux changes. If the magnetic field drops to zero in $\Delta t = 0.100$ s rather than $t = 10.0$ s, the elapsed time is smaller by a factor of

$$\frac{0.100\ \text{s}}{10.0\ \text{s}} = 1.00 \times 10^{-2}$$

and so the induced field is larger by a factor of

$$\frac{1}{1.00 \times 10^{-2}} = 1.00 \times 10^{2}$$

Therefore, the electric field in the case where $\Delta t = 0.100$ s is $E = (1.00 \times 10^2)(1.31 \times 10^{-3}\ \text{V/m}) = 0.131\ \text{V/m}$

Reflect

Our results show that the more rapid the change in magnetic field, the greater the magnitude of the electric field that is induced.

If a circular loop of conducting wire were placed along the circular path of diameter 3.50 cm, a current would be generated so as to produce a magnetic field that would oppose the change in magnetic flux. The upward magnetic field in the figure decreases, so the magnetic field produced in this way would have to be upward. The induced current that produces this magnetic field is in the same direction as the circulating electric field. So the electric field and current must both have the direction shown.

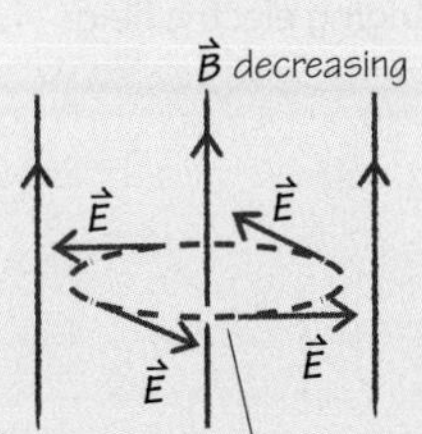
$\vec{B}$ decreasing
$\vec{E}$
$\vec{E}$
$\vec{E}$
$\vec{E}$
If a wire loop were placed here, current would be induced to produce an upward $\vec{B}$.

NOW WORK Problem 4 from The Takeaway 20-5.

The Maxwell–Ampère Law: A Changing Electric Field Generates a Magnetic Field

Equation 20-14 tells us that a circulating electric field is produced by a *magnetic* field that changes over time. In Chapter 19 we learned that a circulating *magnetic* field is produced by electric charge carriers in motion, that is, by a current. The mathematical expression of this statement is *Ampère's* law:

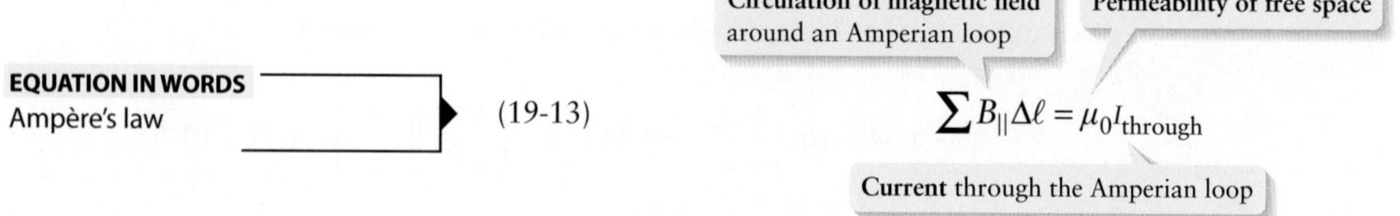

$$\sum B_{\|}\Delta\ell = \mu_0 I_{\text{through}} \qquad (19\text{-}13)$$

Figure 20-13a shows an application of Ampère's law that we introduced in Section 19-7: the magnetic field due to a long, straight, current-carrying wire. The current through each loop in the figure is equal to the current in the wire, so for each loop $I_{\text{through}} = I$. As a result, there is a magnetic field that circulates around each loop, and the circulation $\sum B_{\|}\Delta\ell$ of the magnetic field is equal to $\mu_0 I$.

Now suppose we break the wire in Figure 20-13a and insert two metal disks to form a parallel-plate capacitor (**Figure 20-13b**). If the same steady current exists as in Figure 20-13a, positive charge will build up on the lower plate, negative charge will build up on the upper plate, and the electric field between the two plates will increase in magnitude. There is no current from the lower plate to the upper plate, so $I_{\text{through}} = 0$ for the loop between the plates in Figure 20-13b. So Equation 19-13 predicts that there should be no circulating magnetic field between the plates. Yet experiment shows that there *is* a circulating magnetic field between the plates! How can this be?

James Clerk Maxwell's great insight was to propose that if a changing magnetic field could produce a circulating *electric* field, then a changing *electric* field could produce a circulating *magnetic* field. That's what is happening between the capacitor plates shown in Figure 20-13b: The electric field is increasing in magnitude as charge builds on the plates, and a circulating magnetic field results. To explain this effect, Maxwell

(a) A current-carrying wire

Wire
i
The current produces a circulating magnetic field $\vec{B}$.
$\vec{B}$
Current i
The magnetic field direction is given by a right-hand rule: Point your right thumb in the direction of the current, and the $\vec{B}$ field circulates in the direction of your fingers.

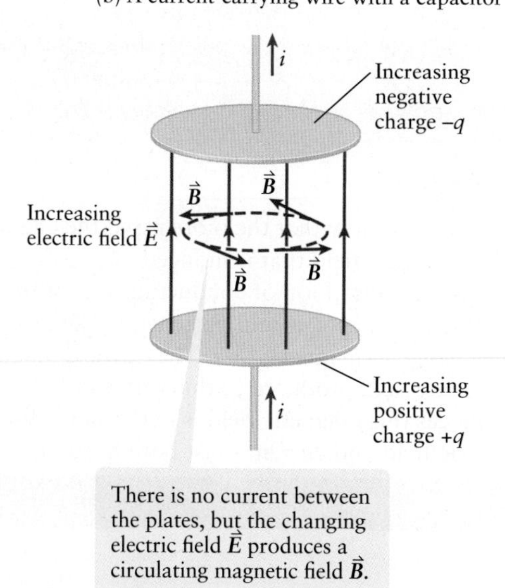

Figure 20-13 A changing $\vec{E}$ produces a circulating $\vec{B}$
(a) Ampère's law says that a magnetic field circulates around a current-carrying wire. (b) A circulating magnetic field can also be produced by a changing electric field.

expanded on Ampère's law as given by Equation 19-13 to include an additional term on the right-hand side. The result is called the **Maxwell–Ampère law:**

Circulation of magnetic field around an Amperian loop

Permittivity of free space

Change in the electric flux through the surface outlined by the loop

$$\sum B_{\parallel}\Delta\ell = \mu_0\left(I_{\text{through}} + \varepsilon_0\frac{\Delta\Phi_E}{\Delta t}\right) \qquad (20\text{-}15)$$

Permeability of free space

Current through the Amperian loop

Time interval over which the change in electric flux takes place

EQUATION IN WORDS
Maxwell–Ampère law

The electric flux Φ_E through a loop is defined in precisely the same manner as the magnetic flux Φ_B through a loop. The quantity $\varepsilon_0(\Delta\Phi_E/\Delta t)$ has units of amperes and is called the **displacement current**. Experiment confirms Equation 20-15: A circulating magnetic field can be produced by an electric current, by a changing electric field, or by a combination.

We saw in Section 17-5 that we can write the permittivity of free space as $\varepsilon_0 = 8.85 \times 10^{-12}$ F/m. Since 1 F = 1 C/V and 1 A = 1 C/s, we can write this as

$$\varepsilon_0 = 8.85 \times 10^{-12}\,\frac{\text{F}}{\text{m}} = 8.85 \times 10^{-12}\,\frac{\text{C}}{\text{V}\cdot\text{m}} = 8.85 \times 10^{-12}\,\frac{\text{A}\cdot\text{s}}{\text{V}\cdot\text{m}} \qquad (20\text{-}16)$$

We'll make use of this equation in the following example.

EXAMPLE 20-5 A Magnetic Field Due to a Changing Electric Field

A parallel-plate capacitor like that shown in Figure 20-13b has circular plates 5.00 cm in diameter. The electric field between the plates increases by 8.00×10^5 V/m in 1.00 s, inducing a magnetic field. What is the magnitude of that magnetic field around a loop 3.50 cm in diameter in the space between the capacitor plates?

Set Up

No charge actually moves through the loop in question, so $I_{\text{through}} = 0$ in Equation 20-15. However, the electric flux through the loop changes as the electric field changes, so the term $\Delta\Phi_E/\Delta t$ in Equation 20-15 is not zero, and a circulating magnetic field will result. This example is very similar to Example 20-4, except that now we need to calculate the change in *electric* flux to determine the magnitude of the circulating *magnetic* field.

By using the same approach as magnetic flux, we find the electric flux through the 3.50-cm-diameter loop; the electric field $\vec{E}$ is perpendicular to the plane of the loop, so $\theta = 0$. The magnetic field has the same magnitude B all the way around the loop and is tangent to the loop, so $B_{\parallel} = B$ in Equation 20-15.

Electric flux:

$$\Phi_E = AE_{\perp} = AE\cos\theta$$

Maxwell–Ampère law:

$$\sum B_{\parallel}\Delta\ell = \mu_0\left(I_{\text{through}} + \varepsilon_0\frac{\Delta\Phi_E}{\Delta t}\right) \qquad (20\text{-}15)$$

Permittivity of free space:

$$\varepsilon_0 = 8.85 \times 10^{-12}\,\frac{\text{A}\cdot\text{s}}{\text{V}\cdot\text{m}} \qquad (20\text{-}16)$$

Permeability of free space:

$$\mu_0 = 4\pi \times 10^{-7}\ \text{T}\cdot\text{m/A}$$

Area of a circle of radius r:

$$A = \pi r^2$$

Circumference of a circle of radius r:

$$C = 2\pi r$$

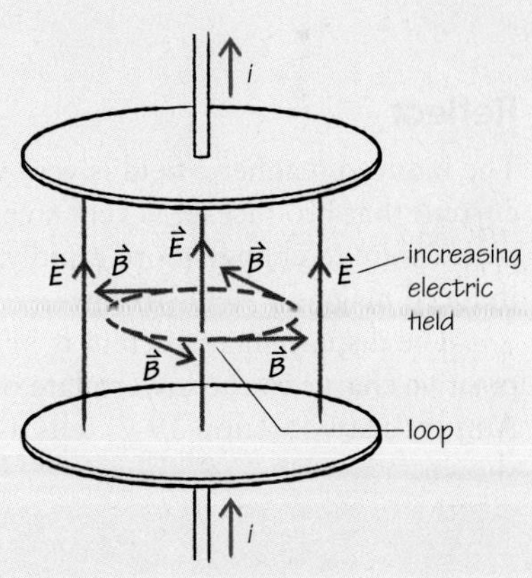

Solve

Find the displacement current $\varepsilon_0(\Delta\Phi_E/\Delta t)$ associated with the change in electric flux through the loop.

The change in electric flux in a time $\Delta t = 1.00$ s is equal to the area of the loop multiplied by the change in electric field in that time (recall that $\theta = 0$). The loop has radius $r = (1/2) \times (3.50 \text{ cm}) = 1.75 \text{ cm} = 1.75 \times 10^{-2}$ m, so

$$\Delta\Phi_E = A(\Delta E)\cos 0 = A(\Delta E)(1)$$
$$= (\pi)(1.75 \times 10^{-2} \text{ m})^2(8.00 \times 10^5 \text{ V/m})(1)$$
$$= 7.70 \times 10^2 \text{ V} \cdot \text{m}$$

The displacement current through the loop is

$$\varepsilon_0 \frac{\Delta\Phi_E}{\Delta t} = \left(8.85 \times 10^{-12} \frac{\text{A} \cdot \text{s}}{\text{V} \cdot \text{m}}\right)\left(\frac{7.70 \times 10^2 \text{ V} \cdot \text{m}}{1.00 \text{ s}}\right)$$
$$= 6.81 \times 10^{-9} \text{ A}$$

Use Equation 20-15 to determine the magnitude of the circulating magnetic field.

Because $I_{\text{through}} = 0$, the circulation of the magnetic field is equal to μ_0 times the displacement current:

$$\text{circulation of the magnetic field} = \sum B_{\parallel}\Delta\ell = \mu_0\left(\varepsilon_0 \frac{\Delta\Phi_E}{\Delta t}\right)$$
$$= \left(4\pi \times 10^{-7} \frac{\text{T} \cdot \text{m}}{\text{A}}\right)(6.81 \times 10^{-9} \text{ A}) = 8.56 \times 10^{-15} \text{ T} \cdot \text{m}$$

Because $B_{\parallel} = B$ and B has the same value at all points around the circle, we can write the circulation as

$$\sum B_{\parallel}\Delta\ell = \sum B\Delta\ell = B\sum\Delta\ell$$

The sum $\sum\Delta\ell$ is the total distance around the loop and is equal to the loop circumference $2\pi r$. So

$$B(2\pi r) = 8.56 \times 10^{-15} \text{ T} \cdot \text{m}$$
$$B = \frac{8.56 \times 10^{-15} \text{ T} \cdot \text{m}}{2\pi(1.75 \times 10^{-2} \text{ m})} = 7.78 \times 10^{-14} \text{ T}$$

Reflect

The induced magnetic field is very weak (7.78×10^{-14} T) because the displacement current that produces it is very small (6.81×10^{-9} A). If the electric field between the plates were to change more rapidly, the displacement current would be greater and the induced magnetic field stronger.

The displacement current is in the same direction as the current that brings positive charge to the lower plate of the capacitor. The right-hand rule for using Ampère's law (Section 19-7) tells us that the induced magnetic field is in the direction shown.

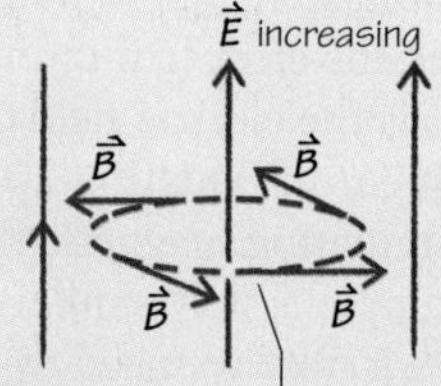

The magnetic field due to an increasing $\vec{E}$ circulates in the same direction as the electric field due to a decreasing $\vec{B}$ (see Example 20-4).

NOW WORK Problem 5 from The Takeaway 20-5.

In a vacuum, where no electric currents are present, we can write Faraday's law and the Maxwell–Ampère law as

$$\sum E_{\parallel}\Delta\ell = -\frac{\Delta\Phi_B}{\Delta t}$$

(20-17)
$$\sum B_{\parallel}\Delta\ell = +\mu_0\varepsilon_0\frac{\Delta\Phi_E}{\Delta t}$$

Notice that the first of Equations 20-17 (Faraday's law) has a minus sign, while the second of these equations (the Maxwell–Ampère law) does not. This says that the circulating electric field produced by a *decreasing* magnetic flux (so $\Delta\Phi_B/\Delta t < 0$) is in the same direction as the circulating magnetic field produced by an *increasing* electric flux (so $\Delta\Phi_E/\Delta t > 0$). That's why the electric field that we found in Example 20-4 due to a decreasing magnetic field circulates in the same direction as the magnetic field that we found in Example 20-5 due to an increasing electric field.

Faraday's law says that a varying magnetic field gives rise to a circulating electric field, and the Maxwell–Ampère law says that a varying electric field gives rise to a circulating magnetic field. Taken together, these two laws tell us how electromagnetic waves are possible and why they involve both electric and magnetic fields. Maxwell's equations successfully predict the value of the speed of light in terms of the permittivity of free space ε_0 and the permeability of free space μ_0. We will learn about electromagnetic waves in the next unit.

THE TAKEAWAY for Section 20-5

✔ Faraday's law says that a circulating electric field is produced around a loop by a time-varying magnetic flux through the loop. The Maxwell–Ampère law states that a circulating magnetic field is produced around a loop by a current through the loop or a time-varying electric flux through the loop.

✔ Faraday's law and the Maxwell–Ampère law explain how an electromagnetic wave sustains itself and why the electric and magnetic fields are mutually perpendicular.

1. In an RC circuit the capacitor begins to discharge. In the region of space between the plates of the capacitor, there
   (A) is an electric field but no magnetic field.
   (B) is a magnetic field but no electric field.
   (C) are both electric and magnetic fields.
   (D) are no electric and magnetic fields.
2. James Clerk Maxwell is credited with compiling the three laws of electricity and magnetism (Gauss's law, Ampère's law, and Faraday's law); adding his own law (also called Gauss's law for magnetism); modifying Ampère's law; and understanding the connections between electricity, magnetism, and optics. Were Maxwell's efforts more important than the individual discoveries of Gauss, Ampère, and Faraday? Explain your answer.
3. Maxwell's equations apply
   (A) to both electric fields and magnetic fields that are constant over time.
   (B) only to electric fields that are time-dependent.
   (C) to both electric fields and magnetic fields that are time-dependent.
   (D) to both time-independent and time-dependent electric and magnetic fields.

EX 20-4

4. A ring is in a region of space that contains a uniform magnetic field directed perpendicular to the plane of the ring. The diameter of the ring is 1.5 cm. The following graph shows the magnetic field as a function of time. What is the magnitude of the electric field along the perimeter of the ring for the time intervals $t = 0$ to 4.0 ms, 4.0 to 10.0 ms, and 10.0 to 14.0 ms?

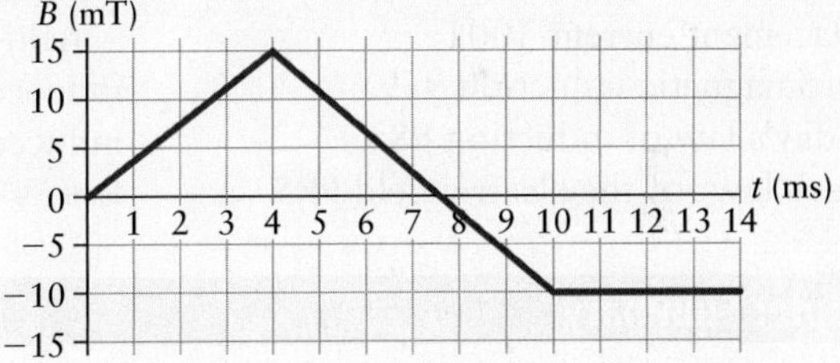

FX 20-5

5. Charge flows onto the positive plate of a 6.0-cm-diameter parallel-plate capacitor at the rate $I = \Delta q/\Delta t = 1.5$ A. What is the magnetic field between the plates at a distance of 3.0 cm from the axis of the plates?
6. According to Maxwell's equations, which of the following situations is possible?
   (A) A closed surface that has a net outward magnetic flux through its surface
   (B) A closed surface that has a net outward electric flux through its surface
   (C) A loop that has zero electric flux through the interior of the loop and has a magnetic field that circulates around the loop
   (D) (B) and (C)

WHAT DID YOU LEARN?

Prep for the AP® Exam

Chapter learning goals	Section(s)	Related example(s)	Relevant section review exercise(s)
Describe and calculate magnetic flux.	20-2	20-1 20-2	5 1–4
Describe an induced electric potential difference (often called motional or induced emf) resulting from a change in magnetic flux (Faraday's law).	20-2 20-3 20-5	20-1 20-2 20-4	5 1–4 6 5–8 4
Use Lenz's law to determine the direction of an emf in a circuit with a changing magnetic flux.	20-3		1–7
Describe the functioning of a generator.	20-4	20-3	1–5

Chapter 20 Review

Key Terms

All the Key Terms can be found in the Glossary/Glosario on page G1 in the back of the book.

ac generator 992
displacement current 1001
electromagnetic induction 979
Faraday's law of induction 983
Gauss's law for the electric field 995
Gauss's law for the magnetic field 996
induced emf 981
induced magnetic field 987
Lenz's law 988
magnetic flux 982
motional emf 980
Maxwell–Ampère law 1001
Maxwell's equations 995

Chapter Summary

Topic	Equation or Figure
Motional emf: An emf appears in a loop when that loop moves in the presence of a magnetic field. To an observer in this reference frame, the emf is a result of magnetic forces on the charge carriers within the loop. Motional emf is just one type of induced emf.	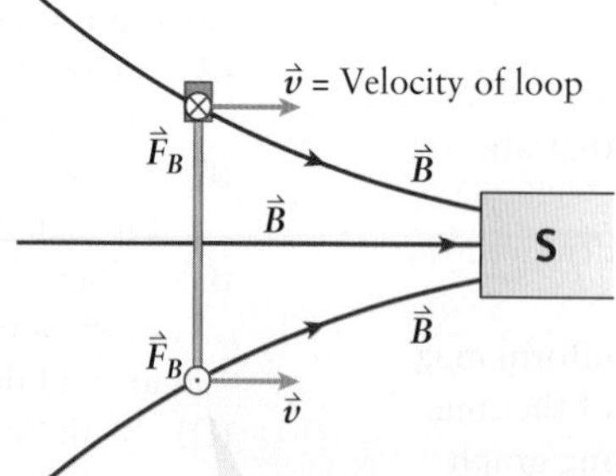 (Figure 20-2c)

Induced emf: An emf also appears in a loop when the magnetic field within the loop changes. To an observer in this reference frame the forces that create the emf are not magnetic, but electric; an electric field is produced by the changing magnetic field.

(c) Side view of magnet moving toward a stationary loop

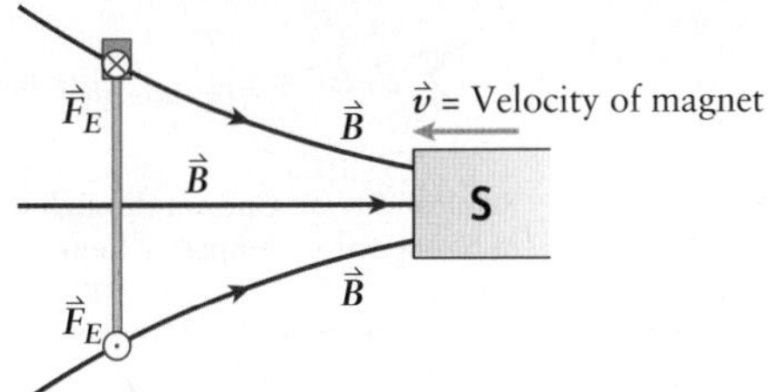
$\vec{v}$ = Velocity of magnet
$\vec{F}_E$
$\vec{B}$
$\vec{B}$
S
$\vec{B}$
$\vec{F}_E$
• Charge carriers in the loop experience electric forces.
• These forces push charge carriers around the loop, causing a current.

(Figure 20-3c)

Faraday's law and Lenz's law: Faraday's law states that if the magnetic flux through a loop changes, an emf is induced around that loop. The magnitude of the induced emf equals the magnitude of the rate of change of the magnetic flux through the loop.

Lenz's law states that the direction of the induced emf is such as to oppose the flux change: The induced current causes its own flux (in the opposite direction of the change). A conductor and magnet that move relative to each other experience magnetic forces that oppose this relative motion.

Induced emf in a loop

Change in the magnetic flux through the surface outlined by the loop

$$\mathcal{E} = -\frac{\Delta\Phi_B}{\Delta t} \qquad (20\text{-}3)$$

Time interval over which the change in magnetic flux takes place

The minus sign indicates that the current caused by the emf induces a magnetic field that opposes the change in flux.

Maxwell's equations: Four equations govern the behavior of electric and magnetic fields in all situations. Gauss's law for electric fields relates electric fields to the existence of net charge within a region of space. Gauss's law for magnetism shows that there are no magnetic monopoles. Faraday's law and the Maxwell–Ampère law explain how the oscillations of the magnetic field produce the electric field and how the oscillations of the electric field produce the magnetic field.

Electric flux through a closed surface

Net amount of **charge enclosed** within the surface

$$\Phi_E = \frac{q_{\text{encl}}}{\varepsilon_0} \qquad (20\text{-}11)$$

Permittivity of free space = $1/(4\pi k)$

Magnetic flux through a closed surface

There is **zero magnetic flux through any closed surface,** no matter what the size or shape of the surface or what it contains.

$$\Phi_B = 0 \qquad (20\text{-}12)$$

Circulation of electric field around a loop

Change in the magnetic flux through the surface outlined by the loop

$$\sum E_{||}\Delta\ell = -\frac{\Delta\Phi_B}{\Delta t} \qquad (20\text{-}14)$$

Time interval over which the change in magnetic flux takes place

Circulation of magnetic field around an Amperian loop; Permittivity of free space; Change in the electric flux through the surface outlined by the loop

$$\sum B_{\parallel}\Delta \ell = \mu_0\left(I_{\text{through}} + \varepsilon_0 \frac{\Delta \Phi_E}{\Delta t}\right) \tag{20-15}$$

Permeability of free space; Current through the Amperian loop; Time interval over which the change in electric flux takes place

Chapter 20 Review Problems

1. The stripe on the back of a credit card such as in the chapter opening photograph is magnetized in a pattern that encodes your account information. A credit card reader contains a loop of wire, and when you swipe the card through the reader, the magnetized card's motion generates an electric current in the wire that sends a signal to the credit card company. To generate this current, what kind of force must be exerted on electrons in the wires of the card reader?
   (A) A magnetic force
   (B) An electric force
   (C) A combination of electric and magnetic forces
   (D) A mechanical force

2. Two conducting loops with a common axis are placed near each other, as shown in the figure below. Initially the currents in both loops are zero. If a current is suddenly set up in loop a in the direction shown, is there also a current in loop b? If so, in which direction? What is the direction of the force, if any, that loop a exerts on loop b? Explain your answer.

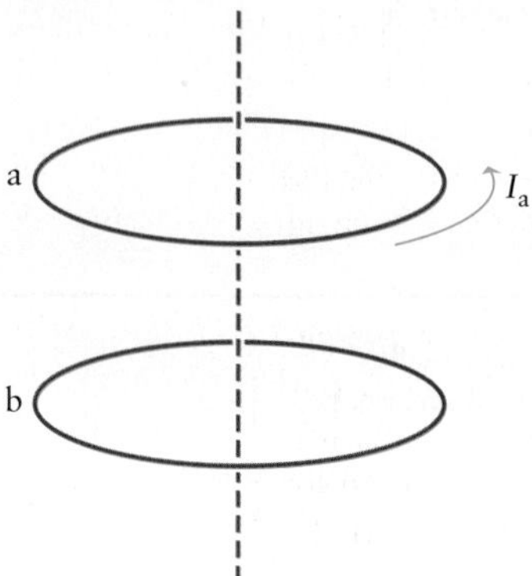

3. A conducting rod slides with negligible friction on conducting rails in a magnetic field, as shown in the figure below. The rod is given an initial velocity $\vec{v}$ to the right. Describe its subsequent motion and justify your answer.

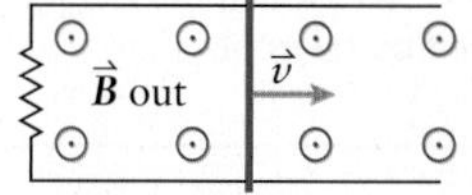

4. In a popular demonstration of electromagnetic induction, a metal plate is suspended in midair above a large electromagnetic coil, as shown in the following figure.
   (a) How does this work?
   (b) If your professor does the demonstration, one thing you'll notice is that the plate gets quite hot. (In fact, you can end the demonstration by frying an egg on the plate!) Why does the plate become hot?
   (c) Would the trick work if the plate were made of an insulating material?

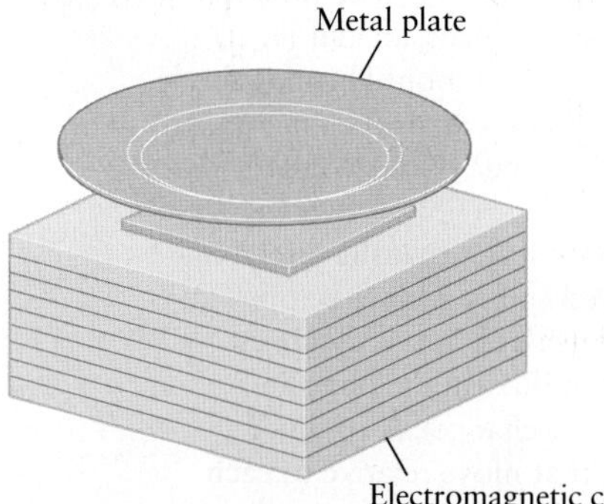

5. Activity on the Sun, such as solar flares and coronal mass ejections, hurls large numbers of charged objects into space. When the objects reach Earth, they can interfere with communications and the power grid by causing electromagnetic induction. As one example, a current of millions of amps (known as the *auroral electrojet*) that runs about 100 km above Earth's surface can be perturbed. The change in the current causes a change in the magnetic field it produces at Earth's surface, which induces an emf along Earth's surface and in the power grid (which is grounded). Induced electric fields as large as 6.0 V/km have been measured. We can model the circuit at Earth's surface as a rectangular loop made up of the power lines completed by a path through the ground beneath. We can treat the magnetic field created by the electrojet as being uniform (but not constant). Consider a 1.0-km-long stretch of power line that is 5.0 m above the surface of Earth. If the induced emf in the Earth–power line loop is 6.0 V, at what rate must the magnetic field through the loop be changing? Assume that the plane of the loop is perpendicular to Earth's surface.

6. A common physics demonstration is to drop a small magnet down a long, vertical aluminum pipe. Describe the motion of the magnet and the physical explanation for the motion.

7. During transcranial magnetic stimulation (TMS) treatment, a magnetic field typically of magnitude 0.50 T is produced in the brain using external coils. During the treatment the current in the coils (and hence the

magnetic field in the brain) rises from zero to its peak in about 75 μs. Assume that the magnetic field is uniform over a circular area of diameter 2.0 cm inside the brain. What is the magnitude of the induced emf around this area in the brain during the treatment?

8. The figure below depicts an electron at rest between the poles of an electromagnet. Explain how the electron is accelerated if the magnetic field is gradually being increased.

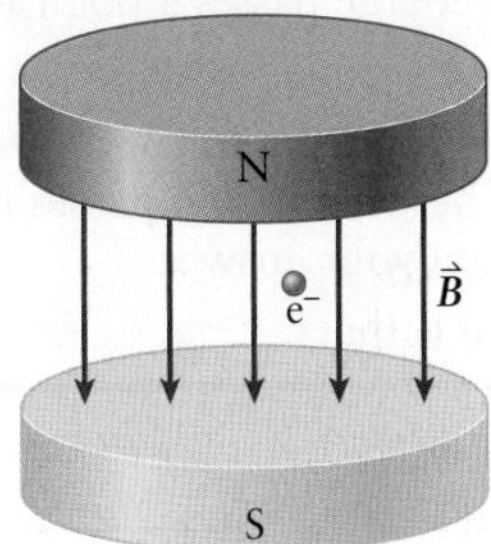

9. A permanent bar magnet with the north pole pointing downward is dropped into a solenoid.
   (a) Determine the direction of the induced current that would be measured in the ammeter shown in the figure below.
   (b) If the magnet is suddenly pulled upward through the solenoid, what is the direction of the induced current that would be measured in the ammeter?

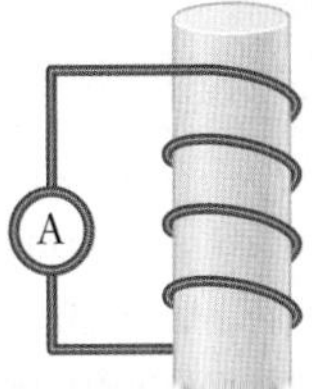

10. The induced emf versus time for a coil that has 100 circular turns of wire with radii 25 cm is given in the table. Plot $V(t)$ and using this graph, sketch the graph of the magnetic field as a function of time, $B(t)$, that is passing through the loop (assume the magnetic field is perpendicular to the plane of the loop and that the average value of $B(t)$ is 0).

t (s)	V (V)	t (s)	V (V)
0	0	9	2
1	2	10	4
2	4	11	2
3	2	12	0
4	0	13	−2
5	−2	14	−4
6	−4	15	−2
7	−2	16	0
8	0		

11. An ordinary gold wedding ring is tossed end over end into a running MRI machine with a 4.0-T field. The ring spends 1.3 s in the field, during which it completes 60 full rotations. The ring has a resistance of 6.0 Ω and a diameter of 18 mm. Calculate the average power dissipated by the ring.

12. A pair of parallel conducting rails that are 12 cm apart lie at right angles to a uniform magnetic field of 0.8 T directed into the page, as shown in the figure below. A 15-Ω resistor is connected across the rails. A conducting bar is moved to the right at 2 m/s across the rails.
    (a) What is the current in the resistor?
    (b) What direction is the current in the bar (up or down)?
    (c) What is the magnetic force on the bar?

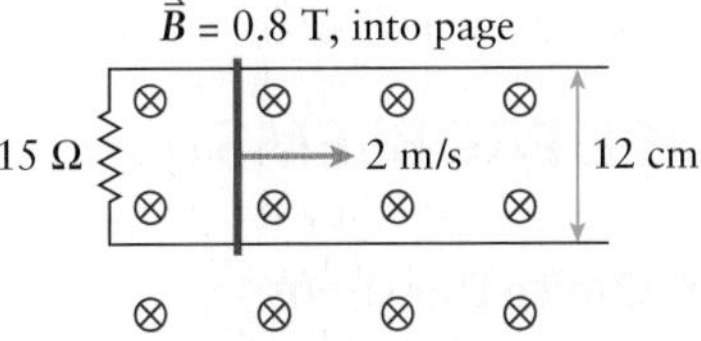

13. Marisol is designing a system to detect when the door to her room is opened. She has observed that the magnetic field in the vicinity of the door is perpendicular to the door when it is closed and has a magnitude of about $B = 5.50 \times 10^{-5}$ T. Knowing Faraday's law, Marisol uses electromagnetic induction as the basis of her alarm. She wraps a single loop of wire around the perimeter of her door and connects it to a voltmeter that can detect an emf as small as $V_{min} = 1.00 \times 10^{-6}$ V. The voltmeter is configured to take a time-averaged reading and will trigger the alarm if the emf over a certain period of time exceeds the threshold potential difference ΔV_{min}. What is the minimum time interval required for someone to open Marisol's door $\theta = 25.0°$, starting from the closed position, without the induced emf over that time interval exceeding the threshold? The dimensions of the door are 0.900 m by 2.00 m.

14. A long, rectangular loop of width w, mass m, and resistance R is being pushed into a magnetic field by a constant force $\vec{F}$, as shown in the figure below. Derive an expression for the speed of the loop while it is entering the magnetic field.

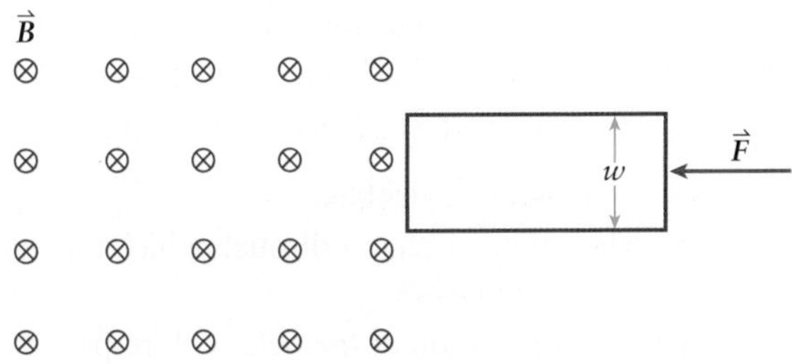

Prep for the 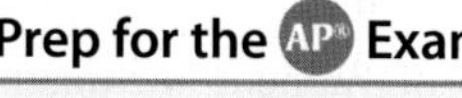 Exam

Group Work

Directions: The problem below is designed to be done as group work in class.

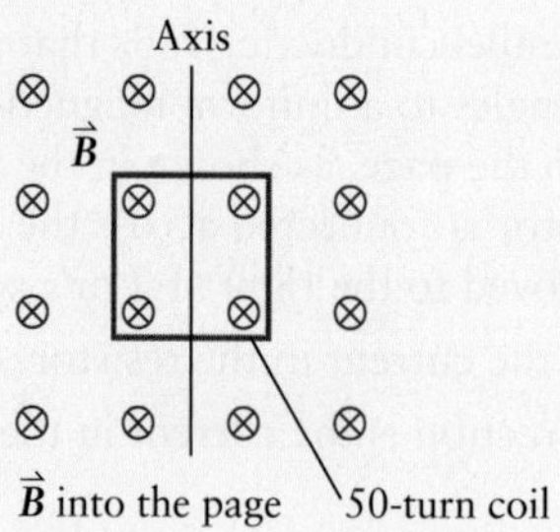

A 50-turn square coil with a cross-sectional area of 5.00 cm^2 has a resistance of 20.0 Ω. The plane of the coil is perpendicular to a uniform magnetic field of 1.00 T. The coil is suddenly rotated about the axis shown in the figure through an angle of 60.0° over a period of 0.200 s.

(a) How much charge passes a point in the coil during that time?

(b) If the loop is rotated a full 360° around the axis, what is the net charge that passes the point in the loop? Explain your answer.

AP® PRACTICE PROBLEMS

Prep for the AP® Exam

Multiple-Choice Questions

Directions: The following questions have a single correct answer.

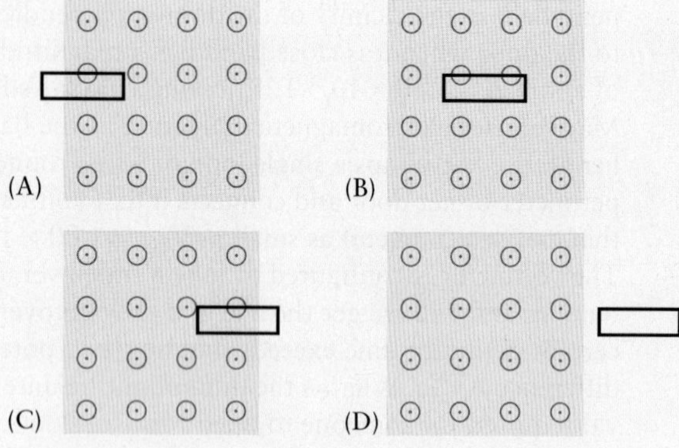

1. The figure above shows a sequence of sketches depicting a rectangular loop passing from left to right through a region of constant magnetic field. The field points out of the page and perpendicular to the plane of the loop. In which sketch is the magnetic flux through the loop decreasing?

(A) Entering the magnetic field

(B) Inside the magnetic field

(C) Leaving the magnetic field

(D) Moving away from the magnetic field

2. On which variable does the magnetic flux depend?

(A) The magnetic field

(B) The area of a region through which the magnetic field passes

(C) The orientation of the field with respect to the region through which it passes

(D) All of the above

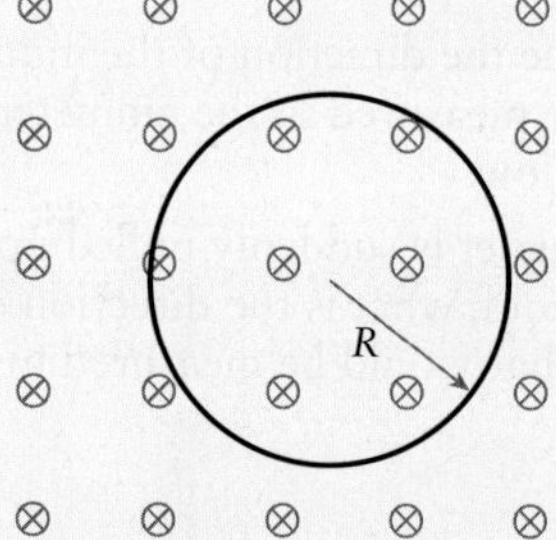

3. The copper ring of radius R in the figure above lies in a magnetic field pointed into the page. The field is uniformly decreasing in magnitude. The induced current in the ring is

(A) clockwise and constant.

(B) clockwise and changing.

(C) counterclockwise and constant.

(D) counterclockwise and changing.

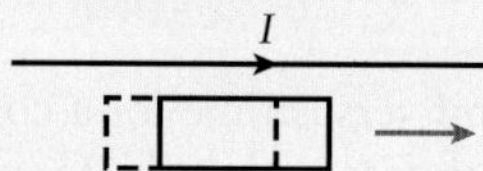

4. A conducting loop moves at a constant speed parallel to a long, straight wire carrying a constant current, as shown in the figure above.

(A) The induced current in the loop will be clockwise.

(B) There will be no induced current in the loop.

(C) The induced current in the loop will be counterclockwise.

(D) The induced current in the loop will be alternately clockwise and then counterclockwise.

(Continued)

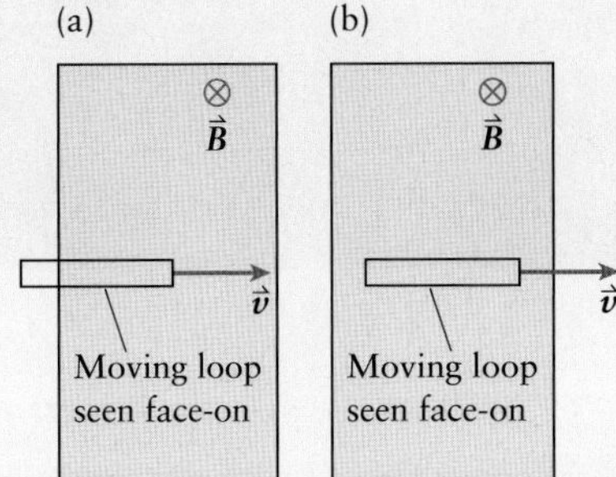

The following two questions refer to a rectangular loop of wire that moves to the right into a region of constant, uniform magnetic field, as shown in the figure above. The field points into the plane of the figure, in a direction perpendicular to the plane of the loop.

5. When the loop is entering the field region, as in figure (a), what is the direction of the current around the loop?
   (A) Clockwise
   (B) Counterclockwise
   (C) Current is zero
   (D) Not enough information given to decide

6. When the loop is moving and completely inside the field region, as in figure (b), what is the direction of the current around the loop?
   (A) Clockwise
   (B) Counterclockwise
   (C) Current is zero
   (D) Not enough information given to decide

Free-Response Question

1. A magnetic field of 0.45 G is directed straight down, perpendicular to the plane of a circular coil of wire that is made up of 250 turns and has a radius of 20 cm. The coil is designed so that its size may be changed without affecting any other property. The coil's size is increased.
   (a) Sketch the coil and the direction of the induced current in the coil as viewed from above as its size is increased.
   (b) The coil is stretched in a time of 15 ms, to a radius of 30 cm.
   (i) Calculate the magnitude of the emf induced in the coil during the process.
   (ii) Assuming the resistance of the coil is a constant 25 Ω, calculate the induced current in the coil during the process and indicate its direction.
   (c) What is the direction of the magnetic field produced by the induced current? Justify your claim in terms of your answers to (a) and (b) and physics principles as needed.

Waves, Sound, and Physical Optics

fotografixx/Getty Images

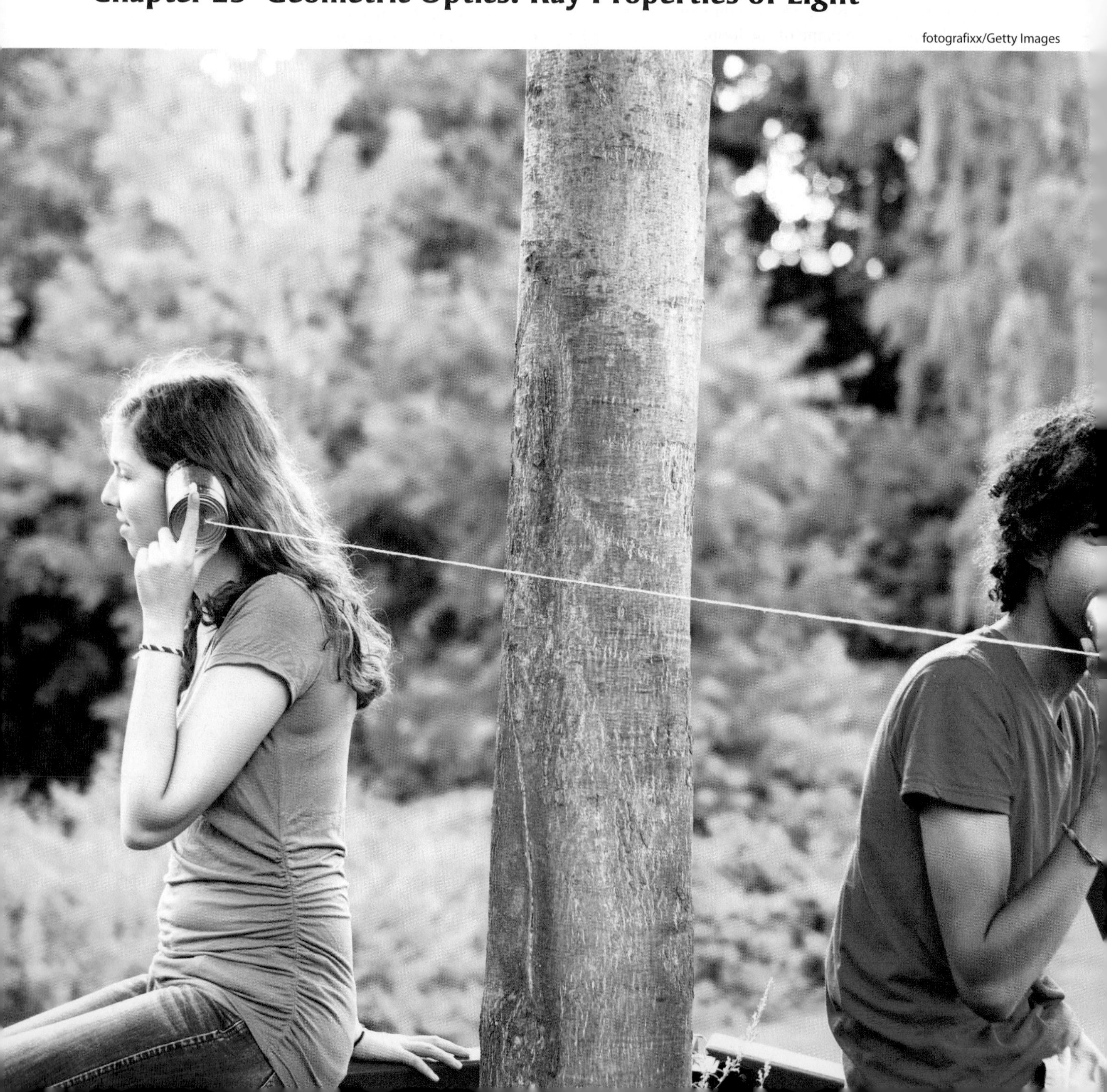

Case Study: How do you hear on a *can phone*?

Carolina is listening to Austin on a *can phone*. If you've never experimented with a can phone—make one! Punch a hole in two cans that are open on one end or paper cups, push a string through the holes, and tie knots to keep the string in place. When the string is taut, you'll easily be able to hear someone talking on the other end. How?

When Austin speaks, he's generating sound. Sound is a mechanical wave, a disturbance that propagates through a substance, which we refer to as the medium for the wave. In this case the mediums are air and string. All waves transfer energy and momentum as they propagate. However, unlike processes we've encountered before, this transfer does not result from the net motion of objects. Mechanical waves set each point in a medium oscillating back and forth about its equilibrium position. Oscillations are along the direction the sound is traveling, but there is no overall flow of the medium. Oscillating air molecules transfer energy to the bottom of Austin's can, which sets up an oscillation in the string. This propagates to Angelina's can where the original oscillation is recreated in the air. Because the original oscillation causes the oscillations of each successive medium, air–can–string–can–air, the frequency of the oscillations remains the same and the sound produced in the second can matches the original sound. Carolina can hear what Austin is saying!

When the oscillations are parallel to the direction of the propagation, the wave is referred to as *longitudinal*. All the waves for the parts of our can phone are longitudinal. If you grasp the end of a rope and move it from side to side, you see a wave wiggle its way down the rope: The oscillation of the rope is perpendicular to the direction of propagation. Such waves are referred to as *transverse*. Mechanical waves can combine these two modes. For example, waves on the surface of water appear to be transverse, but the water molecules move up and down *and* back and forth in a rolling motion, thus the expression *rolling seas*.

All waves maintain their frequency as they travel from one medium into another, but the speed with which they travel depends on the properties of the medium. For example, with the can phone the greater the tension in the string, the greater the speed of the wave in the string. Higher speeds mean the waves travel greater distances in the same time. What factors do you think determine the speed of sound in air? Will different media and speeds determine how you perceive the sound?

By the end of this unit, you should be able to describe the properties of mechanical and electromagnetic waves; explain the properties of transverse and longitudinal waves, and the relationships between periodic wave speed, wavelength, and period; and describe the interaction between waves, between a wave and a boundary, and how the properties of a wave change when there is relative motion between the wave and the observer.

21 Mechanical Waves and Sound

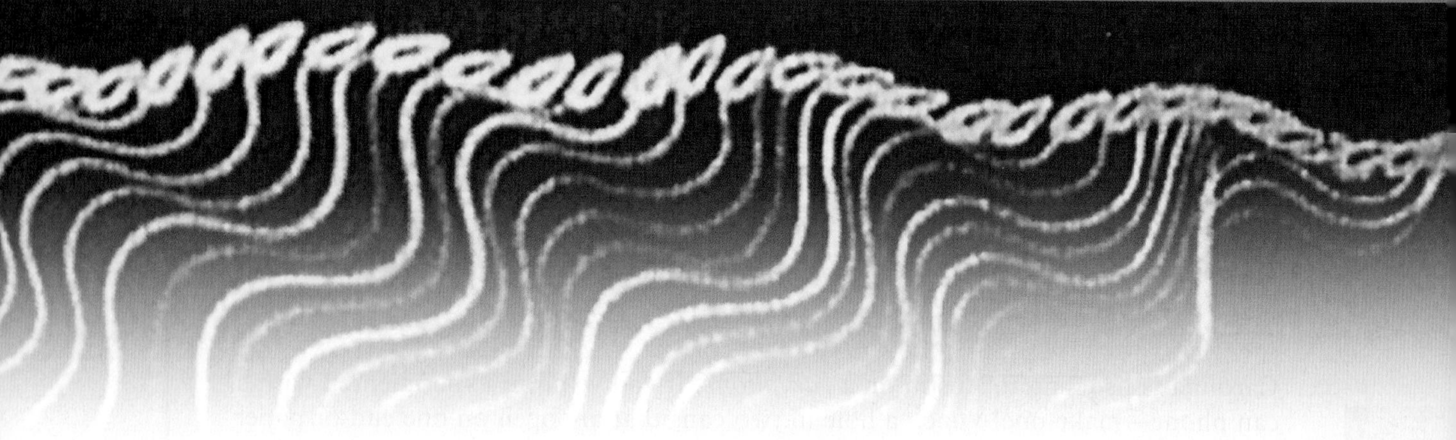

Republished with permission of John Wiley & Sons, from J. Cosson. "A Moving Image of Flagella: News and Views on the Mechanisms Involved in Axonemal Beating." *Cell Biology International*, 1996, 20(2), 83–94; permission conveyed through Copyright Clearance Center, Inc.

YOU WILL LEARN TO:

- Describe what a mechanical wave is.
- Explain the key properties of transverse, longitudinal, and surface waves, and create representations of transverse and longitudinal waves.
- Define the relationship between simple harmonic motion and what happens in a sinusoidal wave.
- Explain the relationship between periodic wave speed, wavelength, and frequency, and relate these to everyday examples.
- Describe what happens when two wave pulses interfere with each other.
- Describe sound in terms of transfer of energy and momentum in a medium and relate the concepts to everyday examples.
- Predict the properties of standing sound waves in open and closed pipes or on strings with fixed or unfixed ends.
- Describe how beats arise from combining two sound waves of slightly different frequencies.
- Describe how the observed frequency of a sound changes when the source and listener are moving relative to each other.

21-1 Waves transport energy and momentum from place to place without transporting matter

If you point your index finger downward and flex your wrist from side to side repeatedly, your finger oscillates. That's the kind of motion we discussed in Chapter 12: Your finger has an equilibrium position (straight down), and it oscillates on either side of that position. But if you dip that finger in a pool of water and repeat this motion, something new happens: A series of ripples spreads away from your finger. The disturbance that you create in this way is called a periodic **wave**. It's actually made up of countless miniature oscillations, because each part of the water's surface oscillates as the wave passes by. Let's first think about this for a simpler motion. Imagine quickly sticking your finger straight down into the water and immediately pulling it out. In this case you create a single **wave pulse**, a single circle in the water that spreads outward from your finger. If you have never done this before, try it in your sink. You can watch this

A disturbance at one place in the water, such as a falling drop...
...causes a wave to spread outward over the water's surface.
(a)
Don Farrall/Getty Images

The wave pulse created moves away from the source along the string.
(b)

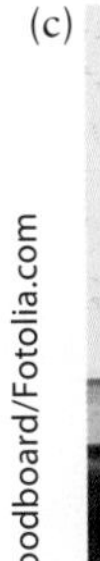
moodboard/Fotolia.com

Sound waves from the teacher travel through the classroom, but the air doesn't move along with them.
(c)

Figure 21-1 Waves traveling in a medium Three examples of waves that travel in a medium (called mechanical waves): (a) one in a liquid, (b) one on a string, and (c) one in a gas (air). Mechanical waves can also travel through solids that don't move easily like the string, such as walls or steel bars.

circle in the water expand. In **Figure 21-1a** a series of rings caused by several drops of water falling into a body of water one after the other moves outward from the source (the drops). We can also see an example of a single wave pulse on a string. If you tie one end of a string to a pole and hold the other end so that the string is fairly horizontal, and then snap the end of the string up and down once, a single wave pulse would move outward, constrained to move along the string (**Figure 21-1b**). A periodic wave can be thought of as a pattern of repeating wave pulses. The time over which a complete piece of the pattern that repeats to make the whole wave is a period.

NEED TO REVIEW? Turn to the **Glossary** in the back of the book for definitions of bolded Key Terms.

What's remarkable about water waves is that the disturbance travels a substantial distance across the water's surface, but the individual water molecules stay in pretty much the same place. This disturbance in the water is transferring energy and momentum outward from where the water was originally disturbed. The water at each point as the wave travels through moves up and down. If it is a single wave pulse, this up-and-down motion happens once. If it is a periodic wave, this happens over and over again.

The same thing is true for *sound* waves. When you sit in class, sound waves travel from the teacher to the back of the classroom, but there's no overall flow of air from the front to the back of the classroom (**Figure 21-1c**). If there were, by the end of the hour the air would be denser at the back of the classroom than at the front! What this disturbance does carry is momentum and energy. The air molecules are oscillating back and forth around their original positions, and as they strike your eardrum, you hear the sound. Your eardrum works because the wave collides with your eardrum, setting it in motion. In fact, this transfer of energy and momentum without transferring mass is true of *all* waves.

Each of these is an example of a **mechanical wave**, the subject of this chapter. Mechanical waves are disturbances that propagate only through a material (something that has mass) that we call the **medium** for the wave. The medium could be a liquid such as water (Figure 21-1a), a solid such as a string (Figure 21-1b), a gas such

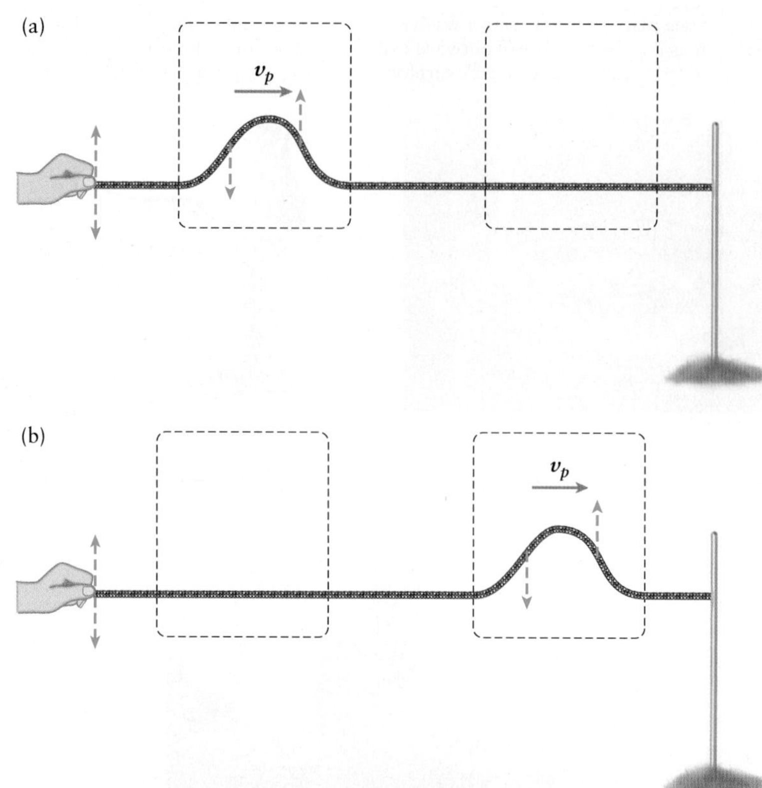

Figure 21-2 Energy and momentum in mechanical waves Energy and momentum transfer is shown for the wave pulse on a string. (a) As the pulse moves along the string, the up-and-down motion of the pulse has kinetic energy, and the resulting stretch stores potential energy in the configuration of the string. (b) At a later time, the location of the wave pulse (and therefore the energy) has moved down the string, although the actual bits of the string have not moved along with the wave pulse down the string. The ability of each bit of string to move the next shows transfer of momentum.

as air (Figure 21-1c), or a bulk solid such as a steel bar or plasterboard. (If you live in a residence hall or an apartment, you certainly know that sound waves can travel through walls, made up of solid materials.) Not all waves are mechanical: In the next chapter, we will cover *electromagnetic* waves, such as radio signals, light rays, and x-rays, which don't require a medium to propagate or carry energy and momentum. They actually propagate fastest in a vacuum!

Waves are actually a much different phenomenon than we have seen before. In mechanics, to transfer energy and momentum we had to move an object along with the energy and momentum, like throwing a ball from one person to another. The easiest wave in which to see the energy and momentum transfer is the wave pulse on the string from Figure 21-1b. In **Figure 21-2**, the wave pulse from Figure 21-1b is shown moving along the string with a velocity v_p. As the pulse moves along the string, the leading edge of the pulse is pulling the string in front of itself upward, and the string at the back edge of the pulse is headed back toward its equilibrium position. This vertical motion of the string means that these bits of the string have kinetic energy. Also, because the string is somewhat stretched, the displacement of these bits of the string stores potential energy in the configuration of the string. In **Figure 21-2a**, the wave pulse is still near the hand, and the undisturbed string is still flat (it has no kinetic energy or potential energy). In **Figure 21-2b**, showing a later time, the wave pulse is located where the string was previously undisturbed. If there is no energy loss in the string, the pulse looks the same. Since this means it is the same shape, it has the same kinetic and potential energies it had when it was near the hand. The string near the hand is now flat (it has no kinetic energy or potential energy). We cannot exactly see momentum but remember that momentum is the potential to set another object or system in motion. Each bit of string pulls on the next, so we can tell momentum is being transferred down the string because each bit moves in turn. We could also imagine setting something light on top of the string. As the wave pulse traveled under it, it would get bumped off, being set into motion by the wave pulse.

We'll begin this chapter by looking at the basic properties of mechanical waves. We'll see how to describe periodic waves using some mathematics, but primarily we will create visual representations of periodic waves and wave pulses, and we'll discover what determines the speed at which mechanical waves propagate. We'll go on to investigate what happens when two waves overlap and interfere with each other. In some

AP Exam Tip

AP® Physics 2 includes mechanical waves (waves that require a medium in which to propagate, such as waves on a string, water waves, and sound waves) and electromagnetic waves. We will first develop an understanding of waves in the context of mechanical waves, and in the next chapter, explore electromagnetic waves.

cases for periodic waves the result is a *standing* wave, in which the disturbance no longer propagates. In other cases we get a wave that propagates, but whose frequency varies with time in a curious way. We'll also see how the properties of sound change when the source, the listener, or both are in motion.

THE TAKEAWAY for Section 21-1

✔ A mechanical wave travels through and displaces a material medium (gas, liquid, or solid). No net flow of matter travels with the wave, which only transfers energy and momentum.

Prep for the AP Exam

AP® Building Blocks

1. When you talk to your friend, are the air molecules that reach her ear the same ones that were in your lungs? Explain your answer, making sure to use the concepts of collision and transfer of momentum to describe the communication.

2. Imagine you have a large, deep bowl of clear broth on the table in front of you with some crackers floating on the broth's surface. If you were to drip a large drop of a red soft drink into the center of the bowl (you don't want to stick your finger in because the broth is very warm), what would you expect to see the broth and the crackers doing, and why?

21-2 Mechanical waves can be transverse, longitudinal, or a combination of these; their speed depends on the properties of the medium

Most mechanical waves are of one of the two types shown in **Figure 21-3**. The wave shown in **Figure 21-3a** is called a **transverse wave** because the individual parts of the wave medium (the rope) move in a direction *perpendicular* to the direction in which

(a)

① If the hand holding the rope moves up and down, a **transverse wave** travels along the rope.

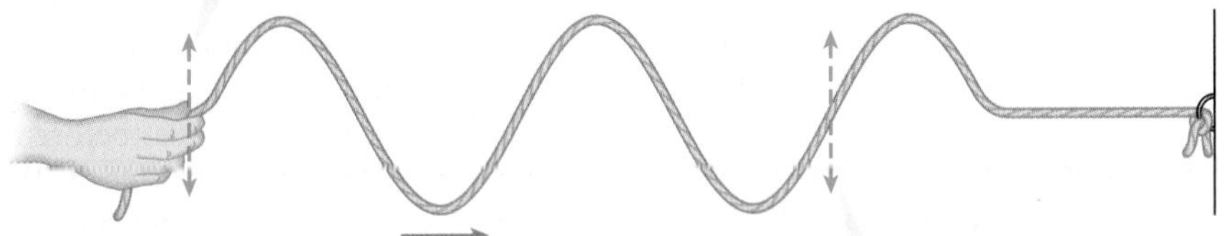

② The wave disturbance propagates (moves along the rope) horizontally...

③ ...and individual parts of the rope (the wave medium) move up and down, perpendicular (transverse) to the propagation direction.

(b)

① If the hand holding the spring moves back and forth, a **longitudinal wave** travels along the spring.

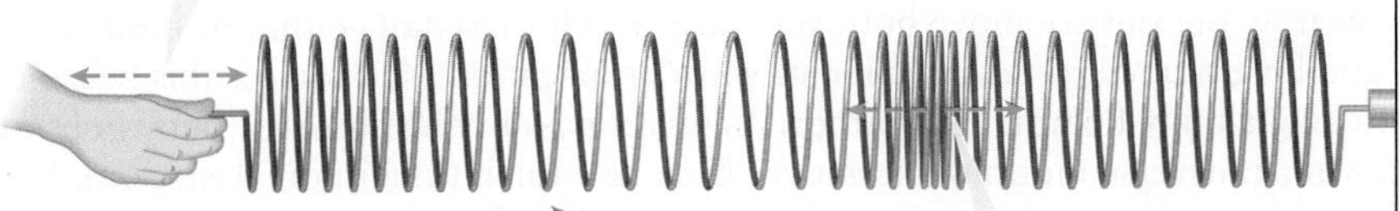

② The wave disturbance propagates (moves along the spring) horizontally...

③ ...and individual parts of the spring (the wave medium) move back and forth, parallel to the propagation direction (longitudinal).

Figure 21-3 Transverse and longitudinal waves These two types of waves differ in how the parts of the wave medium (the material carrying the wave) move relative to the direction that the wave propagates.

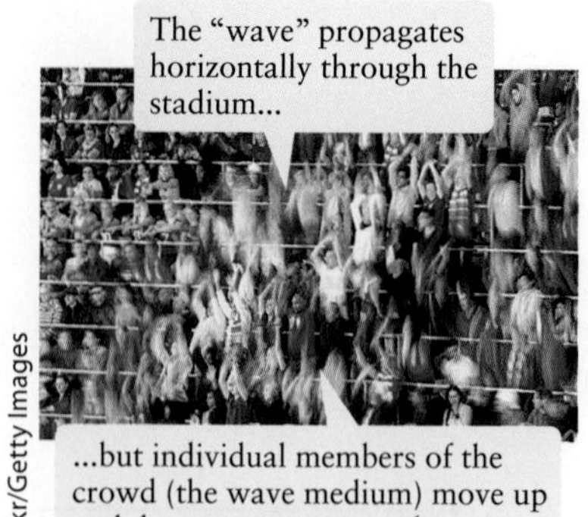

Figure 21-4 A transverse "wave" Sports fans call this "doing the wave," but a more descriptive name would be "doing the transverse wave."

the wave propagates. By contrast, in the **longitudinal wave** shown in **Figure 21-3b**, the individual parts of the wave medium (the spring) move in the direction *parallel* to the direction of wave propagation.

You've participated in a kind of transverse wave if you've been to a crowded sports stadium and done "the wave," in which fans jump up section by section and then sit back down (**Figure 21-4**). The disturbance moves horizontally through the crowd, but individual parts of the wave medium—that is, the individual fans—move up and down, perpendicular to the direction that the "wave" propagates. (Note that this "wave" has the key property of waves that we described in Section 21-1: The disturbance moves through the crowd, but the members of the crowd don't move along with it.)

You can easily make a longitudinal wave just by clapping your hands (**Figure 21-5**). If you clap your hands together sharply at the opening of a tube, your hands squeeze on the air between them. As a result, the air pressure between your hands increases. This higher-pressure air pushes on the air around it, causing the surrounding air to compress and undergo a pressure increase. The air behind the newly compressed air then relaxes back to its normal pressure. The result is that a pulse of increased pressure travels down the tube. (There's no net flow of air down the tube, just a disturbance in pressure.) If you clap your hands rhythmically, a series of pressure pulses propagates down the tube. We call this propagating disturbance a sound wave. These pulses spread out in all directions if you don't focus them with the tube. Sound waves are the most common form of **pressure waves**, waves which consist of regions of higher and lower pressure. (You've probably noticed that your ears, which you use to detect sound waves, are sensitive to changes in air pressure.)

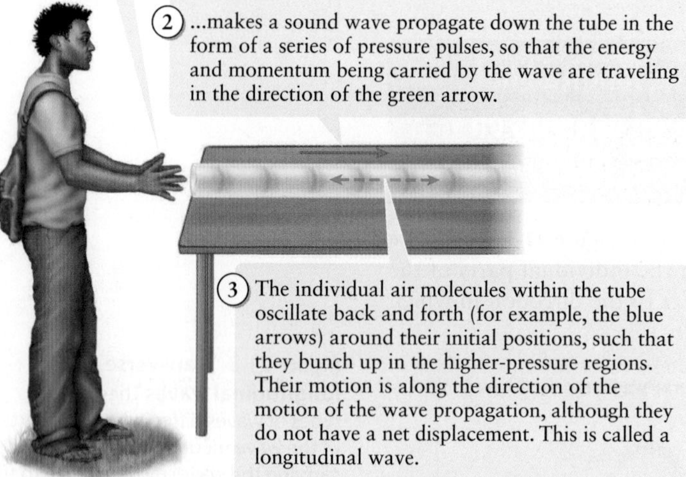

1 Clapping your hands rhythmically at the opening of a tube filled with air...
2 ...makes a sound wave propagate down the tube in the form of a series of pressure pulses, so that the energy and momentum being carried by the wave are traveling in the direction of the green arrow.
3 The individual air molecules within the tube oscillate back and forth (for example, the blue arrows) around their initial positions, such that they bunch up in the higher-pressure regions. Their motion is along the direction of the motion of the wave propagation, although they do not have a net displacement. This is called a longitudinal wave.

Figure 21-5 A longitudinal wave A sound wave is a longitudinal wave.

Here's how you can see that a sound wave is a longitudinal wave. Within each pressure pulse of the sound wave, air molecules are squeezed together; as a pulse passes, the molecules move apart again. So within the tube air molecules slosh back and forth parallel to the axis of the tube and so parallel to the direction of wave propagation. That's just our definition of a longitudinal wave (see Figure 21-3b).

An earthquake produces *both* transverse *and* longitudinal waves that propagate through Earth. The longitudinal waves travel at about twice the speed of the transverse waves, so a seismic monitoring station some distance away from the earthquake site (the *epicenter*) will receive the longitudinal waves before the transverse waves. The more distant the location from the epicenter, the greater the time delay, so by measuring the time delay at a given location scientists can determine how far from that location the earthquake took place. By correlating such measurements made at three or more locations, they can triangulate the position of the epicenter.

A third variety of wave is a **surface wave**, which is a wave that propagates on the surface of a medium. Two common examples are waves on the surface of the ocean and waves that spread away from a disturbance in a pond (Figure 21-1a). A surface wave is actually a *combination* of a transverse wave and a longitudinal wave, which is possible at a surface that is a place where one medium comes into contact with another medium. As a surface wave propagates parallel to the (horizontal) surface of the water, particles near the surface move both vertically and horizontally—that is, both transverse and longitudinal to the direction of wave propagation. You can see this behavior by watching a buoy floating in a harbor: As waves pass by the buoy, the buoy bobs up and down (transverse motion) and moves back and forth (longitudinal motion). Surface waves can occur on the surface of solids as well as liquids. As an example, earthquakes produce not only transverse and longitudinal waves that travel through Earth but also waves that travel along Earth's surface. The surface waves are often the most destructive (see the photo that opens Chapter 12, which shows the devastation caused by the surface waves of a powerful earthquake).

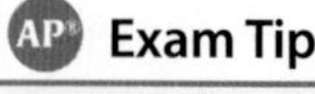

Exam Tip

In AP® Physics 2, we model waves as being either transverse or longitudinal and do not include a treatment of surface waves. Examples of surface waves are often treated as transverse only, but it is important to realize that liquids and gasses can only support transverse waves at surface boundaries.

While there are essential differences among transverse, longitudinal, and surface waves, they all have this characteristic in common: A mechanical wave propagates through its medium because of *restoring forces* that tend to return the medium to its undisturbed state. As an example, for the transverse wave on a rope shown in Figure 21-3a, the restoring force is the tension in the rope. To see the importance of tension, imagine detaching the rope from the post on the right-hand side of Figure 21-3a and placing the rope on top of a table. As the rope is now slack, there is zero tension. If you now wiggle one end of the rope, that end will move in response but *no* wave will travel along the length of the rope. With no restoring force due to tension, the medium (the rope) doesn't resist being disturbed and there is no wave propagation.

The restoring force is different for different types of mechanical waves. For the longitudinal waves on the spring shown in Figure 21-3b, it's the Hooke's law force that opposes the spring being either stretched or compressed. In the case of sound waves in air (Figure 21-5), the pressure of the air itself provides the restoring force. A region of higher pressure pushes against the neighboring regions, thus expanding and lowering its pressure back to the equilibrium value; a region of lower pressure contracts due to pressure from its neighbors, thus raising its pressure toward equilibrium. And for surface waves in water the restoring force is the gravitational force, which tends to make the water surface smooth and level and so opposes the kind of disturbance shown in Figure 21-1a. Because liquids and gases cannot exert tension (the molecules can bump into each other but not pull on each other) liquids and gases can support only longitudinal waves except at a surface.

For many common kinds of mechanical waves the propagation speed depends on the restoring force because it returns the medium to equilibrium when the medium is disturbed. The greater the magnitude of this restoring force, the more rapidly the medium goes back to equilibrium and the faster the wave will propagate. At the same time, the *inertia* of the medium slows down the wave propagation. If the same restoring force is applied to a medium with more inertia (think of a rope compared to a thin string), the medium will return to equilibrium more slowly and the wave will propagate at a lower speed.

Speed of a Transverse Wave on a Rope

We have discussed that for a string or rope, the restoring force is provided by the *tension* in the string or rope (let's just call it rope). If you pluck a rope, the transverse wave that you create will move more rapidly if you pull on the end of the rope to increase the tension in the rope.

The inertia of the rope depends on its *mass per unit length* or **linear mass density** (SI units kg/m). If the rope is uniform, its mass per unit length is just equal to the mass of the rope divided by its length. A thick rope has more mass per unit length than does a piece of ordinary string and so has more inertia. As the mass per unit length of a rope increases, the wave propagation speed decreases.

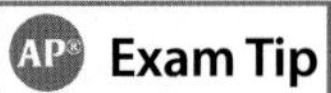 **Exam Tip**

Wave speed for a transverse wave on a rope may be an equation given on the exam (although it is not on the equation sheet), as well as something you should understand qualitatively. In our discussion of the speed of a transverse wave on a rope, if we introduce the symbol μ for linear mass density, the quantitative relationship that supports our qualitative discussion is $v_p = \sqrt{(F_T/\mu)}$. We give examples of applying this mathematical model in Example 21-1. While this equation can be derived by applying Newton's second law to a segment of a rope, the derivation is beyond the scope of this course. We let you use dimensional analysis to come up with the form of this equation for yourself in problem 4 in the Takeaway for Section 21-2.

The Speed of Longitudinal Waves

Just as for a transverse wave on a rope, the speed of a *longitudinal* wave in a medium depends on the restoring force and the inertia of the medium. For a longitudinal wave in a fluid (a gas or a liquid), for example a *sound* wave, the disturbance associated with the wave corresponds to changes in the pressure of the fluid. We saw in Section 12-3 that bulk materials behave somewhat like springs and this, in slightly different forms, is also true of liquids and gases. A stiffer material is more difficult to compress, which means that it has a greater tendency to return to its original volume when the pressure is released. In other words, the stiffness tells us about the restoring force exerted within a material when it is disturbed by the pressure changes associated with a longitudinal wave. A measure of the inertia of the material is its density ρ, equal to its mass per volume. The stiffer the material, the faster the propagation speed; the greater the density ρ, the slower the propagation speed.

Sound waves in *air* are particularly important. The "stiffness" and density ρ of air both depend on temperature and humidity, so the speed of sound in air depends on both of these quantities. The speed of sound in dry air at 20°C is $v_{sound} = 343$ m/s.

The speed of sound is slower in colder air, which is denser. For example, at the altitude where most jetliners fly (about 11,000 m, or 36,000 ft, above sea level), the average temperature is about −55°C and the speed of sound is only about 295 m/s. Even though water is almost a thousand times denser than air ($\rho_{water} = 1.0 \times 10^3$ kg/m^3 versus $\rho_{air} = 1.2$ kg/m^3), sound travels more than four times faster in water (1.5×10^3 m/s versus 343 m/s in air). The explanation is that water is much more difficult to compress than air and overwhelms the density factor so that the wave speed is greater in water than in air. In the same way, steel is *very* difficult to compress, and there is a tremendous restoring force when it is disturbed by a wave. This more than makes up for steel being eight times denser (and so having eight times more inertia) than water. So sound waves travel even faster along a steel rail than they do in water.

We've just stated that the propagation speed of waves is independent of the properties of the wave in most cases! One notable exception to this general rule is the speed of *surface* waves in water, where the propagation speed turns out to be greater for waves that are longer.

We saw in Section 12-2 that the presence of a restoring force was also an essential ingredient for oscillation. This suggests that there are deep connections between the physics of oscillation that we studied in Chapter 12 and the physics of waves. We'll explore these connections in the next several sections.

THE TAKEAWAY for Section 21-2

✔ In a transverse wave the elements of the medium are disturbed in the direction perpendicular to the direction in which the wave propagates.

✔ In a longitudinal wave the elements of the medium are disturbed along the direction in which the wave propagates.

✔ Surface waves propagate on the surface at the boundary of two media and are a combination of a transverse and a longitudinal wave.

✔ In mechanical waves of all types, a restoring force must be present to make the wave propagate.

✔ The propagation speed of a mechanical wave depends on the strength of the restoring force in the medium (a larger force makes the wave propagate faster) and on the inertia of the medium (a greater inertia makes the wave propagate more slowly).

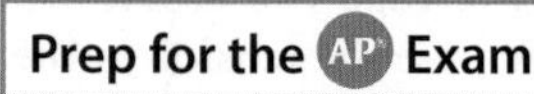

AP Building Blocks

1. A stone is dropped into still water. Answer the following questions in complete sentences describing your reasoning.
   (a) The geometry of the wave propagation on the surface is circular. What is the direction of displacement of the water? What is the restoring force that causes the wave propagation?
   (b) The geometry of the wave propagation below the surface is spherical. What is the direction of displacement of the water?
   (c) Identify the types of waves in parts (a) and (b) as primarily transverse or longitudinal.
2. As described in the text, when you clap your hands together you increase the pressure in the air between your hands slightly. This pressure wave propagates radially outward from the point of contact between your hands. Pressure has units of N/m^2. The inertia of the medium is proportional to the density.
   (a) Use dimensional analysis to determine the units of the ratio of pressure and density.
   (b) Express the speed of sound in air in terms of the ratio of pressure P and density ρ, P/ρ, and a dimensionless constant of proportionality C.
   (c) The pressure of the atmosphere at sea level is 1.013×10^5 N/m^2. The density of dry air at 15°C is 1.225 kg/m^3. The speed of sound under these conditions is 340 m/s. Calculate the constant of proportionality C.
3. We learned that the spring constant k for a given system depends on three things: the reciprocal of its relaxed length L_0, directly on its cross-sectional area A, and the stiffness of the material of which it is made (for example, the particular kind of rubber or metal).

   Two solid rods are made of different materials with the same stiffness, but one has greater density than the other. In which rod will the speed of longitudinal waves be greater? Explain your answer, referring to the variables that affect the wave speed.

AP® Skill Builders

4. A string with mass m and length L is hanging from the ceiling. An object with mass M is attached to the lower end of the string. If the string is given a horizontal displacement (perpendicular to the line of the hanging string, like plucking a guitar string) at a point, the displacement will propagate away from that point as a wave pulse. The restoring force is the tension in the string F, with units of newtons. A measure of the inertial resistance to propagation of the disturbance is the linear mass density of the string $\mu = m/L$.
   (a) Use dimensional analysis to determine the units of the ratio of tension and linear mass density.
   (b) Express the speed of wave propagation in the string in terms of the ratio F/μ and a dimensionless constant of proportionality C.
   (c) A video camera can be used to observe the transverse wave pulse on the string. Design a data collection procedure and a method of data analysis that can be used to measure the constant C.

AP® Skills in Action

5. A frog jumps into a pond, producing a wave on the pond surface. A leaf floats near the frog. When the surface wave reaches the leaf, the leaf is
   (A) displaced horizontally in the direction in which the wave is propagating at that instant.
   (B) displaced vertically in the direction in which the wave is propagating at that instant.
   (C) displaced horizontally in a direction perpendicular to the direction in which the wave is propagating while it travels.
   (D) displaced vertically in a direction perpendicular to the direction in which the wave is propagating while it travels.
6. As shown in Section 21-2, oscillations are described by the interplay between restoring force and inertia. The restoring force in a liquid is the resistance to compression. The density of liquid mercury is about 13,600 kg/m^3 and the density of water is about 1000 kg/m^3. The speed of sound in both liquid mercury and water at a temperature of 20°C is about 1500 m/s. Predict which liquid—water or mercury—has the greater resistance to compression.

21-3 Sinusoidal waves are related to simple harmonic motion

Waves can be very complicated. For example, when you're having a conversation with a friend, the sound waves coming from your mouth spread out in all directions, and the character of the wave changes from one moment to another as you vary the pitch and volume of your voice and pronounce different vowels and consonants. Instead of beginning our description of waves with complicated cases such as these, we'll start by considering a simple but important kind of wave.

To explore periodic waves, we will restrict ourselves to **sinusoidal waves**, in which the wave pattern at any instant is a sinusoidal function (a sine or cosine). This restriction may seem very limiting, but in fact, *any* periodic wave pattern can be formed by combining sinusoidal waves. That's the principle behind some types of musical synthesizers. By combining sinusoidal sound waves with different characteristics, a synthesizer can simulate the sound of other musical instruments or create wholly new sounds.

Because we will more often be interested in periodic waves, going forward we will just refer to them as waves, as is common. When we mean just a single wave pulse, we will refer to it as a wave pulse. In addition to using only sinusoidal waves as a model, we are also going to restrict ourselves to waves that travel along a straight line, such as the transverse and longitudinal waves shown in Figure 21-3 and the sound wave in a tube in Figure 21-5. We call these **one-dimensional waves** because they propagate along a single dimension of space.

To be specific let's consider a sinusoidal, one-dimensional wave on a rope. This is what is drawn in Figure 21-3a. If the hand holding the left-hand end of the rope in Figure 21-3a moves so that the end of the rope moves up and down in simple harmonic motion with frequency f and period T, the position of the hand as a function of time is a sinusoidal function (see Section 12-4). The wave that propagates down the

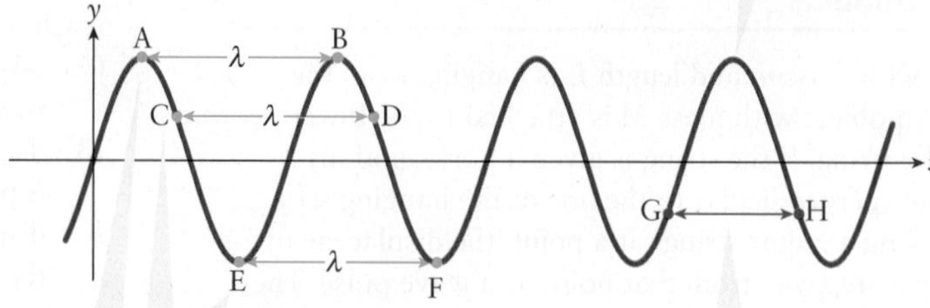

① The "snapshot" of the wave at a given instant is a graph at time t of the displacement y of the medium versus the coordinate x along the length of the medium.
② This is a sinusoidal wave: The graph of displacement y versus coordinate x is a sinusoidal function (sine or cosine).
y
x
A
B
C
D
E
F
G
H
λ
③ Points A and B are one wavelength λ apart: The distance between them corresponds to one complete cycle of the wave. The same is true for points C and D and for points E and F.
④ The displacement is the same at points G and H, but these points are not one wavelength apart: The distance between them does not correspond to a complete cycle.

Figure 21-6 A wave "snapshot": Wavelength The wavelength of a wave is the distance over which the wave repeats itself.

AP® Exam Tip

Although waves are a different phenomenon from others we have studied before, they are similar to simple harmonic oscillators. A wave can be thought of as a complex system of connected harmonic oscillators. While you will not be required to discuss this relationship, you should understand that for any wave at a given time, its amplitude is related to the energy carried in it, in the same way the amplitude is related to the energy in a simple harmonic oscillator (see Chapter 12). This means that the amplitude of a sound wave is related to the loudness of the sound, or that amplitude of a wave on a string is related to how much energy the wave is transporting from one end to the other. This makes sense as you would need to burn more internal energy to wiggle the string further up and down in the same amount of time! On the AP® Physics 2 Exam, you should be comfortable discussing this relationship of energy and amplitude.

rope will then also be a sinusoidal function. Let's look at the properties of such a sinusoidal, one-dimensional wave.

Periodic Waves: Wavelength, Amplitude, Period, Frequency, and Propagation Speed

There are two ways to visualize a wave. One is to imagine taking a "snapshot" or "freeze-frame" of the wave to see what the whole wave looks like at a given time. The other is to concentrate on a given piece of the medium and see how that piece moves as a function of time. (For the wave on a rope shown in Figure 21-3a, that would mean focusing on a single small piece of the rope and watching its motion.)

In **Figure 21-6** we show a snapshot of a wave on a rope at a given instant of time. This snapshot is a graph of the displacement y of the medium (the rope) from equilibrium versus position x along the length of the medium. For a transverse wave on the rope, the displacement y is indeed perpendicular to x, as shown in Figure 21-6. Figure 21-6 could also represent a snapshot of a longitudinal wave like that in Figure 21-3b; in that case y represents the displacement *in the x direction* of pieces of the medium from their undisturbed, equilibrium positions. We will explore that more at the end of this section.

Figure 21-6 shows that the **wavelength** λ (the Greek letter "lambda") is the distance the disturbance travels over one full cycle of the wave. You can measure λ from one *crest* (high point) of the wave to the next crest, one *trough* (low point) to the next trough, or any specific point in the cycle to the analogous point in the next cycle. A wavelength is a distance, so its units in the SI system are meters.

Figure 21-7, another snapshot of a wave, shows the **amplitude** A of the wave. Just like the amplitude of an oscillation, A is the maximum displacement from equilibrium that occurs as the wave moves through its medium. It is *not* the difference between the maximum positive and maximum negative displacements. For a single wave pulse, the displacement may be in only one direction; the maximum is still called the amplitude.

To help us understand how a wave propagates over time, imagine a series of snapshots of a sinusoidal wave equally spaced in time like the frames of a

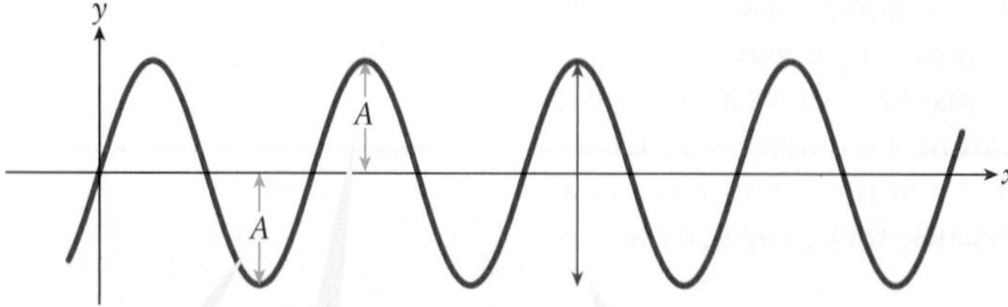

Figure 21-7 A wave "snapshot": Amplitude The amplitude of a wave is the maximum displacement from equilibrium.

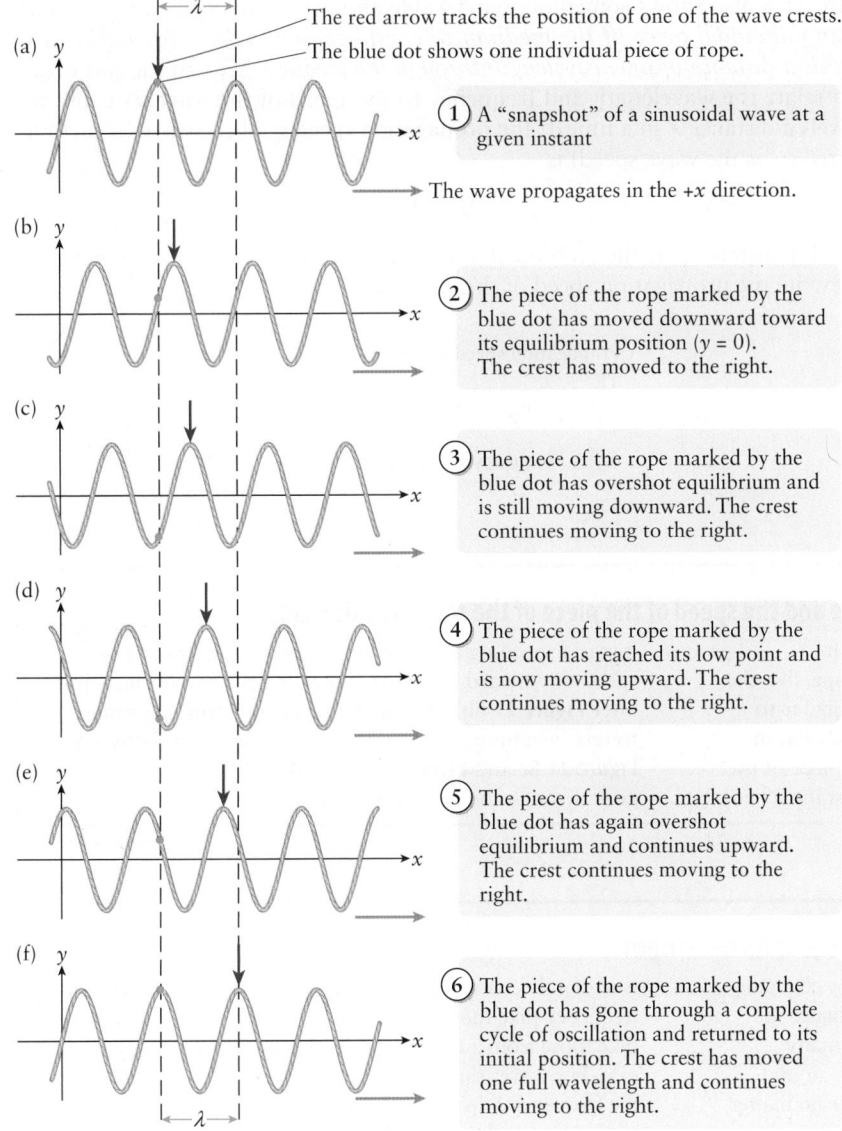

(a) The red arrow tracks the position of one of the wave crests. The blue dot shows one individual piece of rope.

1 A "snapshot" of a sinusoidal wave at a given instant

The wave propagates in the +x direction.

(b) 2 The piece of the rope marked by the blue dot has moved downward toward its equilibrium position (y = 0). The crest has moved to the right.

(c) 3 The piece of the rope marked by the blue dot has overshot equilibrium and is still moving downward. The crest continues moving to the right.

(d) 4 The piece of the rope marked by the blue dot has reached its low point and is now moving upward. The crest continues moving to the right.

(e) 5 The piece of the rope marked by the blue dot has again overshot equilibrium and continues upward. The crest continues moving to the right.

(f) 6 The piece of the rope marked by the blue dot has gone through a complete cycle of oscillation and returned to its initial position. The crest has moved one full wavelength and continues moving to the right.

Figure 21-8 Scenes from a wave "movie" These successive "snapshots" of a wave in motion reveal the connection between the period of a wave and the wavelength. The elapsed time from (a) to (f) is one period T.

During one period of oscillation, T, of the wave medium, the wave travels a distance of one wavelength, λ.

movie. **Figure 21-8** shows such a series. As the wave moves through the medium, an individual piece of the medium (such as the blue dot in Figure 21-8) oscillates up and down. Just as in Section 12-2, we use the term **period** and the symbol T for the time required for each piece of the medium to go through a complete oscillation cycle. Furthermore, just as in Sections 12-2 and 12-4, the **frequency** f (in hertz, or Hz) is the number of cycles that a piece of the medium goes through per second, and the **angular frequency** ω (in rad/s) equals the frequency multiplied by 2π $\left(\omega = 2\pi f = \frac{2\pi}{T}\right)$. So we simply restate Equations 12-1 for waves instead of oscillations.

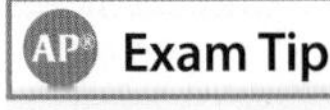

Exam Tip

The energy carried by a wave depends on and increases with the amplitude of the wave.

$$f = \frac{1}{T} \quad \text{and} \quad T = \frac{1}{f} \tag{21-1}$$

Period of a wave

Frequency of the wave

EQUATION IN WORDS
Frequency and period of a wave

The values of T, f, and ω are the same *everywhere* in the medium where the wave is present, so these quantities are properties of the wave as a whole. The frequency of a sound wave helps determine its pitch: The higher the frequency, the higher the pitch.

Figure 21-8 also shows something remarkable about waves: *During the time that it takes an individual piece of the medium to complete one cycle of oscillation, the wave travels a distance of one wavelength through the medium.* We can use this observation to relate the wavelength and frequency to the *speed* of the wave. Because the wave travels a distance λ in a time T, the **propagation speed** v_p of the wave (sometimes just referred to as the wave speed) is

$$v_p = \lambda/T = \lambda f$$

From Equations 12-1, the reciprocal $1/T$ of the period is just the frequency f. So we can rewrite the propagation speed of the wave as

EQUATION IN WORDS
Propagation speed, frequency, and wavelength of wave

(21-2)

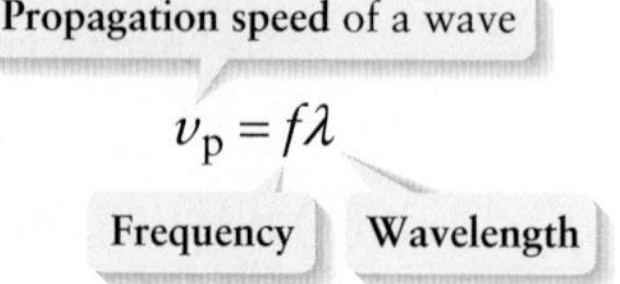

WATCH OUT!

The propagation speed of the wave on the rope and the speed of the piece of the rope are different.

The propagation speed of the transverse wave in Figure 21-8 is constant, but each piece of the rope, like the one marked by the blue dot, is moving perpendicular to the wave in simple harmonic motion. In **Figure 21-8a**, the blue dot is at the maximum amplitude, so that piece of the rope has a transverse velocity of zero at that instant. Then it starts speeding up, reaching its maximum transverse speed downward as it passes through the equilibrium, just after **Figure 21-8b**. Then it starts slowing down, coming to rest instantaneously and changing directions between **Figure 21-8c** and **Figure 21-8d**. As described in the figure, it goes through the complete cycle.

WATCH OUT!

Be careful relating propagation speed, frequency, and wavelength.

Equation 21-2 may give you the impression that by changing the frequency or wavelength of a wave, you can change the propagation speed. In general this is *not* true! For many common types of waves, including sound waves in air and waves on a rope, the propagation speed is the *same* no matter what the frequency. If sound waves of different frequencies traveled at different speeds, a concertgoer sitting in the back row of a theater would hear a time delay between notes of low frequency (like those from a tuba) and notes of high frequency (like those from a flute); experience tells us that this isn't the case. For waves that propagate at a given speed v_p, increasing the frequency f decreases the wavelength λ and vice versa, so that the product $f\lambda$ keeps the same value. On the AP® Physics 2 equation sheet, this formula is presented as $\lambda = v/f$.

WATCH OUT!

Propagation speed is not the same as the speed of pieces of the medium for any type of medium.

Be careful not to confuse v_p, the speed of propagation of the *wave*, with the speed of individual pieces of the medium. As Figure 21-8 shows, such pieces oscillate up and down in a transverse wave (or back and forth in a longitudinal wave), so their speed is continuously changing in magnitude. By contrast, the speed of a wave through a uniform medium (one that has the same properties throughout) stays the same. The magnitude of the propagation speed can also be very different from the speed of pieces of the wave medium. In longitudinal waves it is not as easy to visualize, but each piece of the medium is undergoing simple harmonic motion about its equilibrium along the direction of propagation of the wave, reaching its biggest speed as it passes through equilibrium and coming momentarily to rest when farthest from equilibrium. As an example, while a typical sound wave produced by a person speaking travels through dry air at a temperature of 20°C at the speed of sound for these conditions, v_p = 343 m/s (about 1230 km/h, or 767 mi/h), the maximum speed of the air itself due to the wave passing through is less than 10^{-4} m/s, or one-tenth of a millimeter per second.

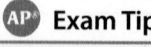

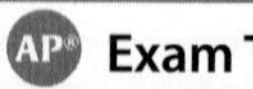

Exam Tip

On the AP® Physics 2 equation sheet, wave propagation speed is just given as speed *v* without the subscript. This is the speed that the disturbance travels along the medium. Be careful not to confuse this speed with other speeds in a given problem.

EXAMPLE 21-1 Wave Speed on a Sperm's Flagellum

The photograph that opens this chapter shows a sea urchin spermatozoon (sperm cell) in motion. These cells move by beating a long, tail-like flagellum against the surrounding fluid. Some aspects of the motor that drives the flagellum are distributed along the length of the flagellum and so it can be treated as if there is tension in the flagellum. This means we can approximate the flagellum as behaving like a uniform rope with a transverse wave moving along it. The flagellum is 40 μm long, the period of oscillation of the flagellum is 0.030 s, and there are approximately two complete cycles of the wave in the length of the flagellum. (a) Estimate the speed of the transverse wave as it propagates along the flagellum. (b) The material of which the flagellum is made has a linear mass density of 2.2×10^{-10} kg/m. Estimate the tension in the flagellum from the equation provided in Section 21-2.

Set Up

We are given the oscillation period T and information about the wavelength λ. We use Equations 21-1 to determine the frequency of oscillation of the flagellum. Given this, we use Equation 21-2 to find the wave speed. Once this speed is determined, we can use the relationship given in Section 21-2 to estimate the tension in the flagellum.

Period and frequency:

$$f = \frac{1}{T} \qquad (21\text{-}1)$$

Propagation speed, frequency, and wavelength of a wave:

$$v_\mathrm{p} = f\lambda \qquad (21\text{-}2)$$

Relationship between propagation speed, tension and linear mass density:

$$v_\mathrm{p} = \sqrt{(F_\mathrm{T}/\mu)}$$

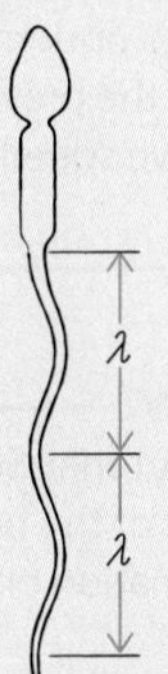

Solve

Find the frequency from the period.

From $T = 0.030$ s, the frequency is

$$f = \frac{1}{T} = \frac{1}{0.030\ \mathrm{s}} = 33\ \mathrm{Hz}$$

Two wavelengths fit into the length of the flagellum, so the wavelength is one-half of the 40-μm length. Use this to determine v_p.

Wavelength of the wave:

$$\lambda = \frac{1}{2} \times 40\ \mu\mathrm{m} \times \frac{10^{-6}\ \mathrm{m}}{1\ \mu\mathrm{m}} = 2 \times 10^{-5}\ \mathrm{m}$$

Then the wave speed is

$$v_\mathrm{p} = f\lambda = (33\ \mathrm{Hz})(2 \times 10^{-5}\ \mathrm{m}) = 7 \times 10^{-4}\ \mathrm{m/s}$$

to one significant digit.

Find the tension from the wave speed, square both sides, and solve for F_T.

$$F_\mathrm{T} = (v_\mathrm{p}^2)(\mu) = (7 \times 10^{-4}\ \mathrm{m/s})^2 \times (2.2 \times 10^{-10}\ \mathrm{kg/m})$$
$$= 1 \times 10^{-16}\ \mathrm{kg \cdot m/s^2}$$

Reflect

A wave moving at this propagation speed would travel about 2 m per hour. To put this in perspective, the world record holder in the World Snail Racing Championships (held annually in Congham, England) covered the 13-in. course in 2 min, which is just under 10 m/h. The propagation speed on the flagellum is only a factor of 5 slower—not bad for something 1000 times smaller than a snail. So this seems reasonable. The tension is truly tiny, beyond our normal experience, and in fact is almost impossible to measure directly. Wave behavior does give us a way to infer this property.

Convert v_p to meters per hour:

$$v_\mathrm{p} = 7 \times 10^{-4}\ \mathrm{m/s} \times \frac{3600\ \mathrm{s}}{1\ \mathrm{h}} = 2\ \mathrm{m/h}$$

NOW WORK Problems 3 and 4 from The Takeaway 21-3.

Exam Tip

On the AP® Physics 2 Exam, you may be required to explain wave phenomena conceptually and create visual representations of waves, and the mathematical treatment in this section helps build that conceptual understanding. You will not be asked to mathematically model waves using sines or cosines on the AP® Physics 2 Exam, but you should be able to understand and manipulate the previous equation for wave speed propagation.

NEED TO REVIEW?

See the Math Tutorial, Section M8, in the back of the book for more information on trigonometry.

Sinusoidal Waves: Displacement as a Function of Position and Time

For a complete description of a wave, we need to know the displacement y of the medium at every position x and every time t. A function that provides this description is called the **wave function** $y(x, t)$. As its symbol suggests, y is a function of *both* x and t.

To obtain this function for a sinusoidal wave, let's start with a "snapshot" of such a wave with amplitude A and wavelength λ. **Figure 21-9a** shows such a snapshot at a given instant of time, which we'll call $t = 0$. Because the wave is sinusoidal and repeats itself over the distance from $x = 0$ to $x = \lambda$, we can write the wave function at $t = 0$ as

(21-3) $$y(x,0) = A\cos\left(\frac{2\pi x}{\lambda} + \phi\right)$$

The factor $2\pi x/\lambda$ varies from 0 at $x = 0$ to 2π at $x = \lambda$, so the cosine function goes through a complete cycle over a distance of one wavelength. The **phase angle** ϕ tells us what point in the cycle corresponds to $x = 0$. (We used a phase angle ϕ in a similar way in our description of simple harmonic motion in Section 12-4.) In Figure 21-9a $\phi = 0$, so the function is a cosine that has its maximum value at $x = 0$. In Figure 21-8a, which shows a different sinusoidal wave at $t = 0$, $\phi = -\pi/2$ (the function is shifted by $\pi/2$ radians, or one-quarter cycle, compared to the wave function in Figure 21-9a).

How can we get the wave function for *any* time t from Equation 21-3? Let's suppose that the wave is propagating in the positive x direction at speed v_p. As **Figure 21-9b** shows we can get the wave function at time t by replacing x in the wave function at $t = 0$ with $x - v_pt$:

$$y(x,t) = A\cos\left[\frac{2\pi}{\lambda}(x - v_pt) + \phi\right]$$

and then multiplying through:

$$= A\cos\left(\frac{2\pi x}{\lambda} - \frac{2\pi v_pt}{\lambda} + \phi\right)$$

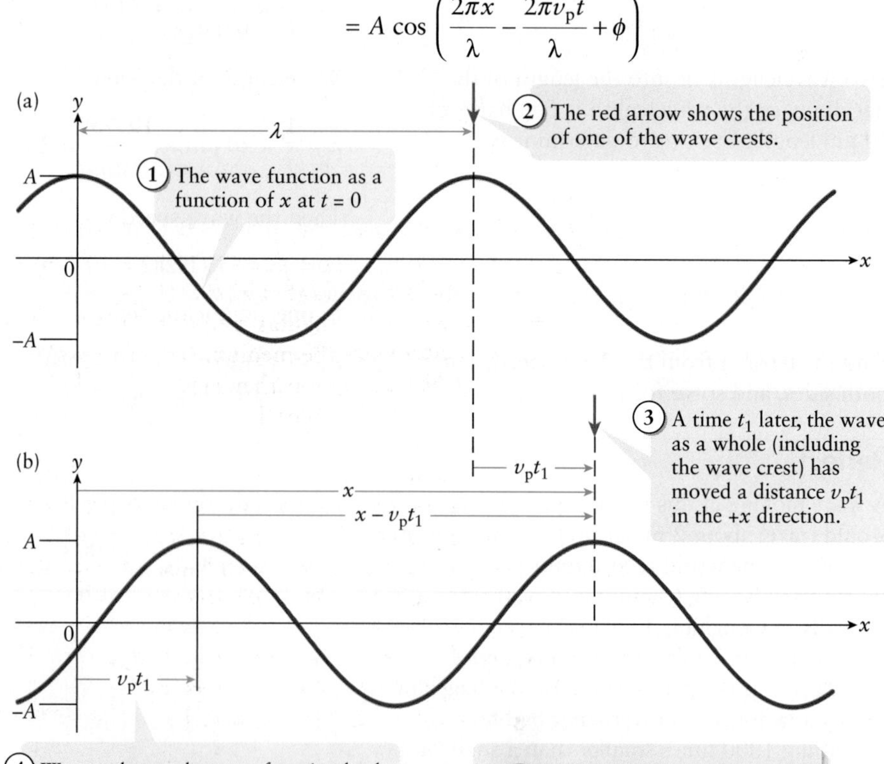
(a)
y
λ
A
0
–A
x
1 The wave function as a function of x at t = 0
2 The red arrow shows the position of one of the wave crests.
(b)
y
A
0
–A
x
x – v_pt_1
v_pt_1
3 A time t_1 later, the wave as a whole (including the wave crest) has moved a distance v_pt_1 in the +x direction.
4 We can change the wave function back to the way it was at t = 0 by shifting the vertical axis a distance v_pt_1 to the right, so coordinate x becomes x – v_pt_1.
The wave function at time t_1 is the same as the wave function at t = 0, but with x replaced by x – v_pt_1.

Figure 21-9 A sinusoidal wave function The wave function describes the displacement y as a function of position x and time t. If all parts of the wave move at the same propagation speed v_p, we can relate (a) the wave function at time $t = 0$ and (b) the wave function at a later time t_1.

We can simplify this expression by using Equation 21-2 for the propagation speed, $v_p = f\lambda$. Dividing both sides of Equation 21-2 by λ tells us that $v_p/\lambda = f$. Then the wave function becomes

$$y(x,t) = A\cos\left(\frac{2\pi x}{\lambda} - 2\pi ft + \phi\right) \tag{21-4}$$

The quantity $2\pi f$ is the angular frequency ω of the wave, and the combination $2\pi/\lambda$ in Equation 21-4 is called the **angular wave number**. We use the symbol k for this quantity:

$$k = \frac{2\pi}{\lambda} \tag{21-5}$$

Because 2π represents the number of radians in one cycle and wavelength λ is in meters, the angular wave number k is measured in radians per meter (rad/m) (2π radians in each cycle multiplied by the number of cycles per unit distance). We use the adjective *angular* because the term *wave number* is typically used for $1/\lambda$, the reciprocal of the wavelength. This quantity multiplied by 2π is the angular wave number $k = 2\pi/\lambda$, in the same fashion that frequency f multiplied by 2π is the angular frequency $\omega = 2\pi f$.

In terms of angular wave number and angular frequency, we can rewrite Equation 21-4 as

$$y(x,t) = A\cos(kx - \omega t + \phi) \tag{21-6}$$

Note that the angular frequency $\omega = 2\pi f$ and the angular wave number $k = 2\pi/\lambda$ are related by the propagation speed, which is determined by the properties of the medium. To see this relationship note that $f = \omega/2\pi$ and $\lambda = 2\pi/k$. Then, from Equation 21-2,

$$v_p = f\lambda = \left(\frac{\omega}{2\pi}\right)\left(\frac{2\pi}{k}\right)$$

We can use Equation 21-6 to verify a statement that we made earlier: If the wave is sinusoidal, each part of the medium oscillates in simple harmonic motion. To see that this is true, think of x as a constant (so that we are considering a single piece of the wave medium). Then Equation 21-6 becomes

$$\begin{aligned} y(x,t) &= A\cos(kx - \omega t + \phi) = A\cos[-\omega t - (-kx - \phi)] \\ &= A\cos[\omega t + (-kx - \phi)] \end{aligned}$$

In the last step we used the trigonometric identity $\cos(-\theta) = \cos\theta$. This equation looks *exactly* like Equation 12-14 for simple harmonic motion, $x = A\cos(\omega t + \phi)$, if we interpret the quantity $-kx - \phi$ as the phase angle of the simple harmonic oscillation of the wave medium at position x. This equation shows that all parts of the medium oscillate with the same amplitude A and the same angular frequency ω, but with a phase angle that depends on the position x within the medium. Two pieces of the medium oscillate together—that is, they are *in phase*—if they are a whole number (1, 2, 3,...) of wavelengths apart and oscillate opposite to each other—that is, they are *out of phase*—if they are $\frac{1}{2}$, $1\frac{1}{2}$, $2\frac{1}{2}$,... wavelengths apart.

We developed Equation 21-6 for a wave propagating in the positive x direction. If the wave propagates instead in the *negative* x direction, in time t it moves a distance $v_p t$ to the *left* in **Figure 21-9**, and the wave function at time t is the same as the wave function at time $t = 0$, with x replaced by $x + v_p t$. The result is that the wave function for a sinusoidal wave propagating in the negative x direction is the same as Equation 21-6, but with $kx - \omega t$ replaced by $kx + \omega t$.

EXAMPLE 21-2 A Wave on a Rope

A wave travels down a stretched rope. The wave has amplitude 2.00 cm and wavelength 15.0 cm, and each part of the rope goes through a complete cycle once every 0.400 s. Find the frequency, angular frequency, angular wave number, and propagation speed for this wave.

Set Up

We are given the amplitude $A = 2.00$ cm, the oscillation period $T = 0.400$ s, and the wavelength $\lambda = 15.0$ cm. We use Equation 21-1 to determine the frequency and Equation 21-2 to find the propagation speed. We then use Equations 12-13 and 21-5 to calculate the angular frequency and angular wave number, respectively. We could go on to find the displacement at any position x and any time t using Equation 21-6 because we are given the amplitude, but that is not required. To find frequency and speed, we do not need the amplitude.

Period and frequency:

$$f = \frac{1}{T} \qquad (21\text{-}1)$$

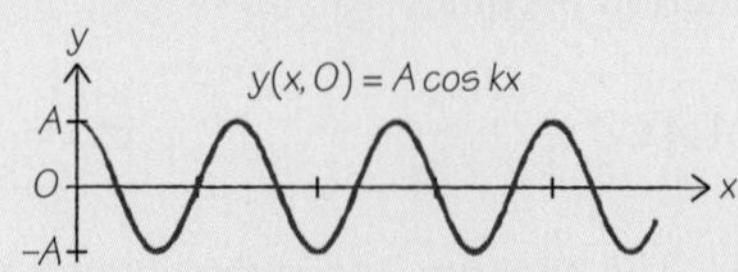

Propagation speed of a wave:

$$v_\mathrm{p} = f\lambda \qquad (21\text{-}2)$$

Angular frequency:

$$\omega = 2\pi f \qquad (12\text{-}13)$$

Angular wave number:

$$k = \frac{2\pi}{\lambda} \qquad (21\text{-}5)$$

Solve

Calculate the frequency f and propagation speed v_p.

We know $T = 0.400$ s, so the frequency is

$$f = \frac{1}{T} = \frac{1}{0.400\ \mathrm{s}} = 2.50\ \mathrm{Hz}$$

From Equation 21-2, the propagation speed is

$$v_\mathrm{p} = f\lambda = (2.50\ \mathrm{Hz})(15.0\ \mathrm{cm})\left(\frac{1\ \mathrm{m}}{100\ \mathrm{cm}}\right) = 0.375\ \mathrm{m/s}$$

Now calculate angular frequency ω and angular wave number k.

$$\omega = 2\pi f = 2\pi(2.50\ \mathrm{Hz}) = 15.7\ \mathrm{rad/s}$$

$$k = \frac{2\pi}{\lambda} = \frac{2\pi}{(0.150\ \mathrm{m})} = 41.9\ \mathrm{rad/m}$$

Reflect

The period of the wave is a little less than one-half of a second, so the frequency is a bit more than 2 Hz. This makes sense given that frequency is the number of cycles per second. Given this frequency and noting that the wavelength is only about one-sixth of a meter, it makes sense that the speed is less than half of a meter per second.

NOW WORK Problems 5 and 6 from The Takeaway 21-3.

Sinusoidal Sound Waves

We've described sinusoidal waves in terms of their displacement y as a function of position x and time t. For sound waves in air it's more convenient to describe them in terms of pressure variations as a function of x and t, which we can picture in terms of the longitudinal oscillations of the air molecules. Humans hear by detecting very small variations of pressure on the eardrum. The ear can detect pressure changes as small as 2×10^{-5} Pa, equivalent to a change in atmospheric pressure of 1 part in *5 billion* (5×10^9). If you've ever had difficulty getting your ears to "pop" when driving from the mountains down to sea level or on a descending airplane, you know how sensitive to pressure differences your ears can be.

Because a sound wave is a longitudinal wave, a positive value of $y(x, t)$ means that the medium is displaced in the positive x direction, while a negative value of $y(x, t)$ means that the medium is displaced in the negative x direction (**Figure 21-10**). As a result, as a sound wave propagates it squeezes together air molecules at some places along the wave, resulting in regions of air that have higher density and higher pressure. Adjacent regions of air are slightly depleted of air molecules and therefore have lower density and lower pressure. As Figure 21-10 shows, the pressure is greatest or least

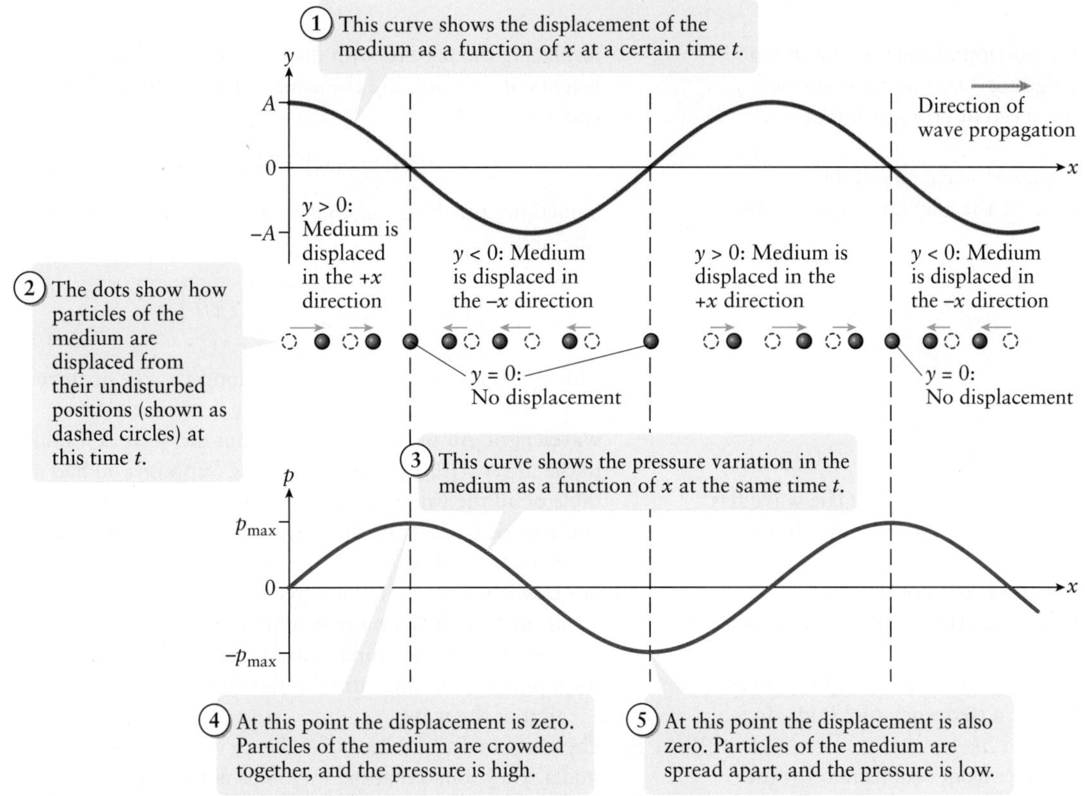
1 This curve shows the displacement of the medium as a function of x at a certain time t.
y
A
0
−A
Direction of wave propagation
x
y > 0: Medium is displaced in the +x direction
y < 0: Medium is displaced in the −x direction
y > 0: Medium is displaced in the +x direction
y < 0: Medium is displaced in the −x direction
2 The dots show how particles of the medium are displaced from their undisturbed positions (shown as dashed circles) at this time t.
y = 0: No displacement
y = 0: No displacement
3 This curve shows the pressure variation in the medium as a function of x at the same time t.
p
p_max
0
−p_max
x
4 At this point the displacement is zero. Particles of the medium are crowded together, and the pressure is high.
5 At this point the displacement is also zero. Particles of the medium are spread apart, and the pressure is low.

Figure 21-10 Pressure and displacement in a sinusoidal sound wave The pressure variation from normal in a sound wave is greatest where the displacement is zero and is zero where the displacement is greatest.

at points where the displacement is *zero*. So while the pressure is also described by a sinusoidal function, it is shifted by one quarter-cycle from the wave function for displacement, $y(x, t)$. That's why we've drawn the displacement graph in Figure 21-10 as a *cosine* curve but the pressure variation graph as a *sine* curve.

THE TAKEAWAY for Section 21-3

✔ In a sinusoidal wave each part of the wave medium goes through simple harmonic motion.

✔ The wavelength λ is the distance the disturbance travels over one full cycle of the wave. In one period T the wave travels (propagates) a distance of one wavelength.

✔ Because the motion of each part of the wave medium can be described by simple harmonic motion, the energy carried by the wave is related to the wave's amplitude.

✔ For many common types of waves, the propagation speed is the same no matter what the frequency. Increasing the wave frequency decreases the wavelength and vice versa.

Prep for the AP Exam

AP Building Blocks

1. Sitting on your porch on a stormy night you see a flash of light in the sky and 3.5 s later you hear thunder. If light travels much, much faster than sound and the speed of sound is 340 m/s, how far away was the lightning that generated the light and sound wave?
2. Two buoys floating on calm water in the harbor bob up and down in phase with a frequency of 0.2 s^{-1}. The buoys are separated by 15 m.
   (a) Could the line between the buoys be parallel to the direction of propagation of a wave and, if so, what is the maximum wave speed?
   (b) Could the line between the buoys be perpendicular to the direction of propagation of a wave and, if so, can the wave speed be estimated based solely on the motion of the buoys?
3. EX 21-1 The period of a sound wave is 0.0100 s. Calculate the frequency f and the angular frequency ω.

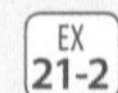

4. A disturbance on a taut, horizontal rope travels down the rope with a wavelength of 3.0 m and a frequency of 1.5 Hz. Calculate the speed of propagation of the wave.

EX 21-2

5. A wave on a string propagates at 22 m/s. If the frequency is 24 Hz, calculate the angular frequency and the wavelength.

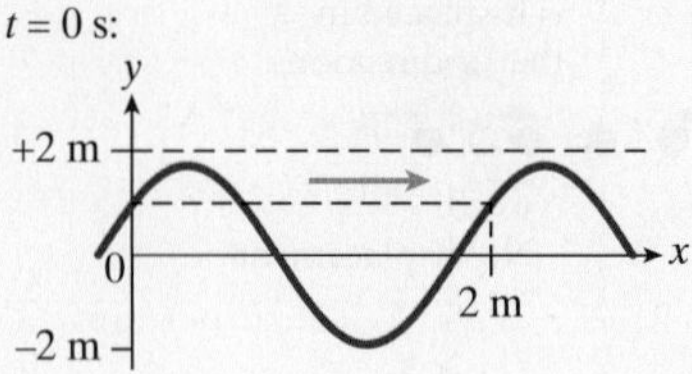

6. The graph of a wave is shown in the figure. The wave has a period of 4 s and the wave moves to the right (in the positive x direction).
(a) Calculate the propagation speed of the wave.
(b) Write the mathematical description of the wave using Equation 21-4.

7. A transverse wave on a string has an amplitude of 30 cm, a wavelength of 70 cm, and a frequency of 2.0 Hz. In parts (a)–(d) apply Equation 21-4.
(a) Express the displacement from equilibrium for the wave propagating in the positive direction if at $t = 0$, $x = 0$ and $y = 0$.
(b) Express the displacement from equilibrium for the wave propagating in the positive direction if at $t = 0$, $x = 0$ and $y = 0.30$ m.
(c) Express the displacement from equilibrium for the wave propagating in the negative direction if at $t = 0$, $x = 0$ and $y = 0$.
(d) Express the displacement from equilibrium for the wave propagating in the negative direction if at $t = 0$, $x = 0$ and $y = 0.30$ m.

8. A wave on a string is described by the equation

$$y(x,t) = 21.0 \text{ m} \cos\left(\frac{2.0 \text{ rad}}{\text{m}} x + \frac{20 \text{ rad}}{\text{s}} t\right)$$

What are the (a) frequency, (b) wavelength, (c) speed, and (d) direction of propagation of the wave?

AP® Skill Builders

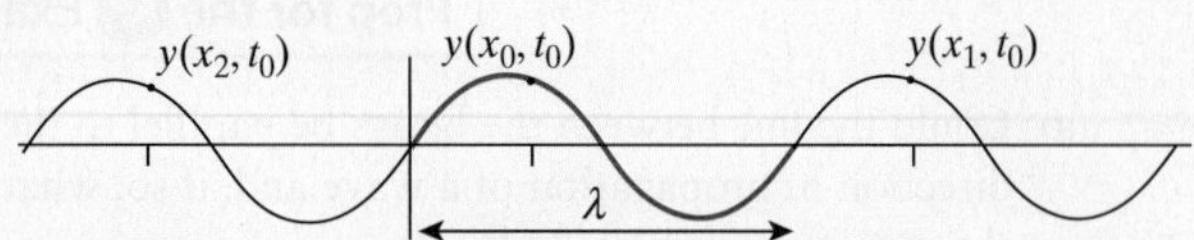

9. As shown in the diagram, the height y of a point on the wave at position x_0 and time t_0 is $y(x_0, t_0)$. This point is replicated at multiples of the wavelength $\lambda = v_p(t - t_0)$, where v_p is the speed of propagation of the wave. A periodic wave with a single wavelength propagates by displacing the pattern with displacement $v_p(t - t_0)$. The heights of the wave are the same as the height at x_0 at x_1 and x_2:

$$y(x_0) = y(x_1) = y(x_2)$$

Sometimes the minus sign in the argument of cosine (or sine) in Equation 21-4 is confusing:

$$y(x,t) = A\cos\left(\frac{2\pi x}{\lambda} - 2\pi f t\right)$$

This problem views the wave equation as a way of storing and retrieving information about the wave in the first wavelength. All the information about the wave is found in the interval $[0, \lambda]$ that contains x_0. Suppose you had a table of all the values of the wave heights in that interval and needed to describe the wave at another point, x, not in the interval $[0, \lambda]$.
(a) Describe how you would recover the value of $y(x_0)$ from x. Include as a first step in the description of your method a comparison of x and x_0.
(b) Connect your algorithm to the direction of propagation of the wave.

10. Physicists analyzed videos of audience waves in football stadia. They found that the wave was not initiated by an individual but by a critical cluster of 25 to 30 people in the stands who were independently displaying behaviors that excited others in their vicinity. They found that events on the field did not cause the initiation of the wave and that, in fact, initiation occurred during lulls between exciting events on the field. They found that, when viewed from the field, clockwise propagation was three times more common than counterclockwise propagation. They also found that the speed of propagation, 22 ± 3 seats/s, and the wavelength, 15 ± 5 seats, of audience waves showed little variation.
(a) Predict the frequency from these data. Test the prediction by measuring the elapsed time when you stand from a sitting position, raise your hands into the air, and sit down again to estimate the maximum frequency. Repeat the test with an unhurried motion that might be typical of an audience wave participant.
(b) Speed of mechanical wave propagation increases with increasing restoring force and decreases with increasing inertia. The narrow distribution of propagation speeds of an audience wave suggests that there are features of this phenomenon that should correspond to the restoring force and inertia of a medium. Inertia could be represented as density and so be the number of seats per meter. But there may be other ways of viewing inertia in this case (for example, more excitable audience members might be expected to propagate the wave more rapidly, and so have a smaller inertia). How can this behavior be interpreted as a greater restoring force or greater stiffness of the medium? Construct claims regarding analogies to the ratio of stiffness/density or tension/inertia for an audience wave.

(c) Pose questions whose pursuit might lead to an explanation as to why audience wave propagation is more common in the clockwise direction, so that a person is activated by a signal arriving from his right side.

Skills in Action

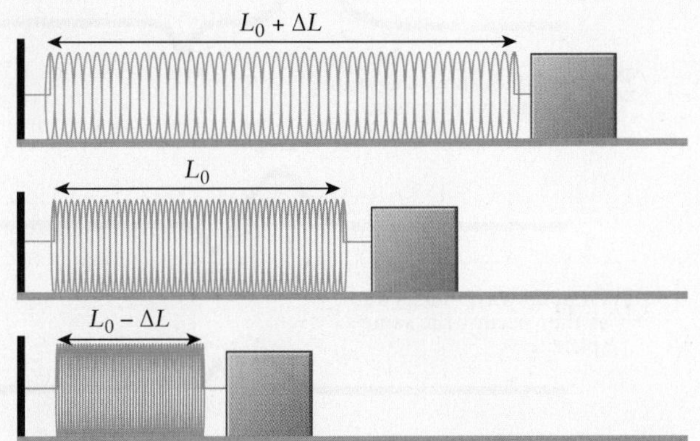

11. A spring with negligible mass is attached at one end to a fixed support clamped to a horizontal plate. The other end of the spring is attached to a block. The plate exerts a negligible friction force on the block. The block is initially displaced horizontally by $+\Delta L$ so that the spring is stretched to a length $\Delta L + L_0$, where L_0 is the equilibrium rest length, and released. The block oscillates along the y axis, as shown from three snapshots of the motion.
   (a) The mass of the block is 1.0 kg. The spring constant is 25 N/m. Predict the period of the harmonic oscillation of the block.
   (b) The initial displacement of the block $\Delta L = 0.10$ m. Construct a y versus t graph of the displacement of the block for four complete oscillations from the initial displacement. Use the period calculated in part (a) to add numerical values to the scale on the t axis. Add annotation to your sketch showing the period and the amplitude. Identify a crest and a trough.

12. A spring, with a small spring constant, rests on a table that exerts a negligible friction force on the spring. The spring is attached at one end to a rigid support clamped to the table. The free end of the spring is stretched gently, then quickly pushed forward toward the support and released.
   (a) Describe the resulting motion of the spring as either a transverse or a longitudinal wave pulse and justify your answer in a complete sentence.
   (b) A wave is set up in the spring as shown above. Copy the snapshot of the spring shown in the figure. Add annotation to your sketch, showing the wavelength, a crest, and a trough.
13. It was experimentally difficult for Newton to measure the speed of sound in air. In a corridor of known length at Trinity College (63.4 m) he measured the time interval of an echo of a loud noise. He reported the speed of sound as 305 ± 25 m/s.
   (a) Calculate the minimum and maximum time intervals recorded in his experiment.
   (b) In his measurement of the time interval, Newton used a pendulum. He adjusted the length of the pendulum so that its period was the same as the time that elapsed between the transmission and reception of the sound. How should the length of the pendulum be modified if the sound wave returned before the completion of a full swing of the pendulum?
   (c) Calculate the range of lengths of the pendulum, assuming simple harmonic motion.
   (d) The mean of Newton's measured speed of sound is smaller than the accepted value. Identify a possible bias in his measured time interval.

21-4 Waves pass through each other without changing shape; while they overlap, the net displacement is just the sum of the displacements of the individual waves

So far in our study of waves, we've considered only single waves in isolation. But in the real world we are bombarded by multiple waves simultaneously. If you are reading this book outside, you may hear sound waves from the conversations of others, from the wind blowing through leaves, and from vehicles such as buses or airplanes. How can we describe what happens when more than one wave is present in a given medium?

In most cases when more than one wave is present in a medium, nothing unusual happens! If you hear the sound of a singer's voice at the same time as the sound of a piano, each sound wave is unaffected by the other. What you hear is simply the *sum* of

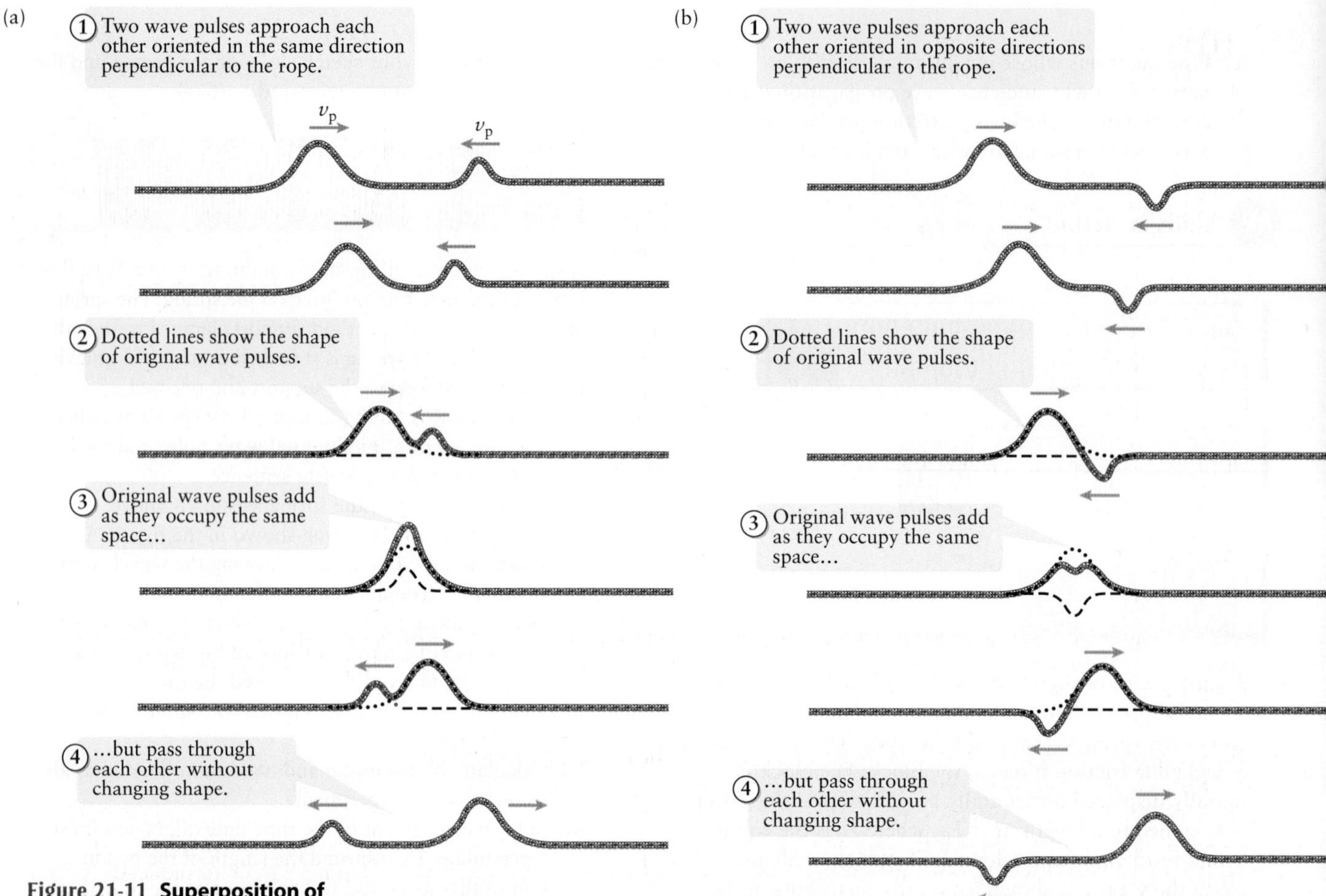

(a)
1 Two wave pulses approach each other oriented in the same direction perpendicular to the rope.
v_p
v_p
2 Dotted lines show the shape of original wave pulses.
3 Original wave pulses add as they occupy the same space...
4 ...but pass through each other without changing shape.
(b)
1 Two wave pulses approach each other oriented in opposite directions perpendicular to the rope.
2 Dotted lines show the shape of original wave pulses.
3 Original wave pulses add as they occupy the same space...
4 ...but pass through each other without changing shape.

Figure 21-11 Superposition of wave pulses (a) Wave pulses traveling along a rope in opposite directions interfere constructively when they displace the rope in the same direction perpendicular to the string. While they overlap, the net displacement is the algebraic sum of the displacements of the individual wave pulses. The wave pulses pass through each other with their shapes unchanged. (b) When wave pulses traveling on a rope in opposite directions displace the rope in opposite directions perpendicular to the string, they interfere destructively. The net displacement is the algebraic sum of the displacements of the individual wave pulses. The wave pulses still pass through each other with their shapes unchanged. If the waves were traveling in the same direction along the rope, they would still interfere in exactly the same ways, based on the direction of the transverse displacement (although because the waves travel with the same speed in the medium, it is difficult to think of how you might set up such a situation!).

the two sound waves. That observation is at the heart of an important physical principle called the *principle of superposition*:

> *When two waves are present simultaneously, the total wave is the sum of the two individual waves.*

Although we started this discussion thinking about sound waves, because that is a common example, single wave pulses also exhibit interference when they pass through the same point in space. In **Figure 21-11** we begin to explore this phenomenon by constructing a representation for the case of wave pulses on a rope. In **Figure 21-11a** the two wave pulses are both above the position of the undisturbed rope, so the net wave pulse as they pass through each other will also be above the position of the undisturbed rope. The amplitude of this resultant wave pulse is just the sum of the two individual pulses' amplitudes. If both wave pulses had been below the position of the undisturbed rope, they would have also added to give a bigger negative amplitude than either individual wave pulse. Because the wave pulses add to form a larger wave pulse in these cases, we say they interfere constructively. If one wave pulse is above the position of the undisturbed rope and one is below the position of the undisturbed rope (**Figure 21-11b**) when you add the wave pulses, you are really doing a subtraction, because one pulse is up and the other is down. In these cases (it doesn't matter which is up and which is down) because the resulting pulse is smaller, we say these wave pulses interfere destructively.

The principle of superposition can lead to surprising results if the two waves have the *same frequency*. We could use the mathematics presented in Figure 21-11, but now considering the displacement for waves instead of wave pulses. Two waves are said to be *in phase* when their crests, or high points, occur at the same time t and their troughs, or low points, also occur at the same time (**Figure 21-12a**). Because we know that waves simply add when they occupy the same space, if each wave has an amplitude A, the total wave has amplitude $2A$. In other words, the waves reinforce each other so that the amplitude of the total wave is twice that of either individual wave. We call this **constructive interference**. One way for two waves to be in phase is if

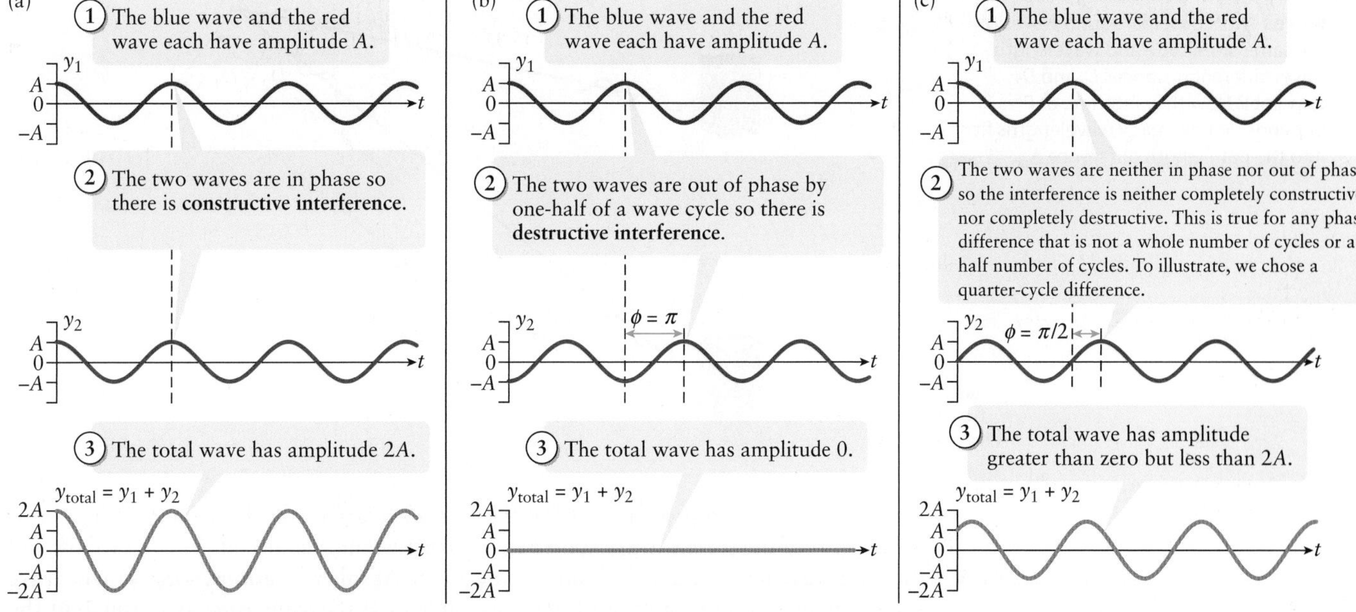

Figure 21-12 Superposition: Combining two waves with the same frequency When two waves arrive at the same location, the total wave is the sum of the two individual waves. The properties of the total wave depend on whether the individual waves are (a) in phase, (b) out of phase, or (c) something in between.

they start in phase and both travel the same distance to reach the point where they add. Because each wave repeats itself once per cycle, constructive interference also happens if either wave travels some whole number of wavelengths farther.

Two waves are *out of phase* if the two individual waves are one half-cycle out of step with each other (**Figure 21-12b**). When waves are *out of phase* a crest of the first wave occurs at the same time as a trough of the second wave, and vice versa. In this case the two waves cancel each other, so the amplitude of the total wave is 0. This could happen if the waves are produced a half cycle out of phase and travel the same distance. This situation is called **destructive interference.** Destructive interference also occurs if the waves are a half cycle plus any whole number of cycles out of phase. So they could be created out of phase, or the difference in the lengths of the paths the two waves travel, if they start in phase, is any multiple of wavelength plus one additional half-wavelength.

If the two waves are neither in phase nor out of phase (**Figure 21-12c**), we simply say that there is **interference** between the two waves: It is neither completely constructive nor completely destructive.

Figure 21-13 shows a real-life example of such interference. Two oscillating mechanical fingers dip in unison into a tank of water, producing two sets of surface waves like those in Figure 21-1a with the same amplitude, frequency, and wavelength. In certain locations, some of which are shown by the yellow dots in Figure 21-13, the two waves

AP Exam Tip

While the quantity phase angle (introduced in Section 21-3) is not part of the required curriculum, the concept of being in phase or out of phase is an important part of understanding wave interference.

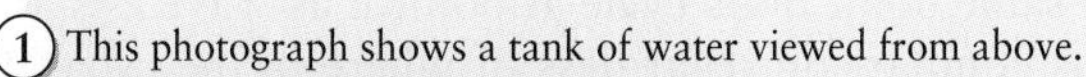

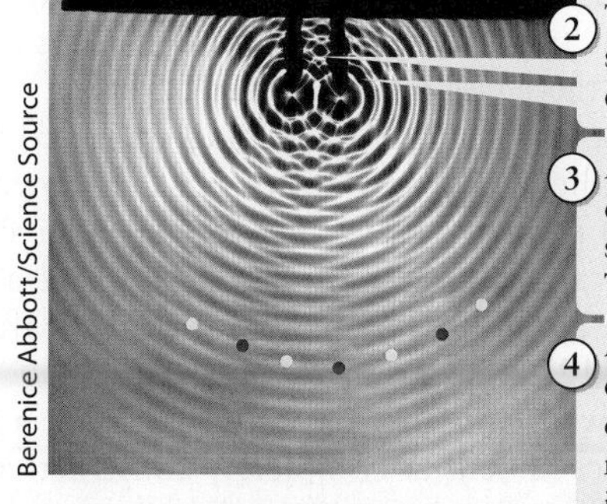

Berenice Abbott/Science Source
2 Two "fingers" oscillate up and down and dip into the suface of the water. This produces two sets of waves, each one spreading out in circles from each "finger."
3 At some positions in the water (shown by the yellow dots), a crest (or trough) of one wave arrives at the same time as a crest (or trough) of the other wave. This results in constructive interference.
4 At other positions (shown by the red dots •), a crest of one wave arrives at the same time as a trough from the other wave. There is destructive interference, and the result is a total wave of nearly zero amplitude, the water is undisturbed.

Figure 21-13 Combining water waves Circular surface waves in water spread outward from two sources that oscillate in unison.

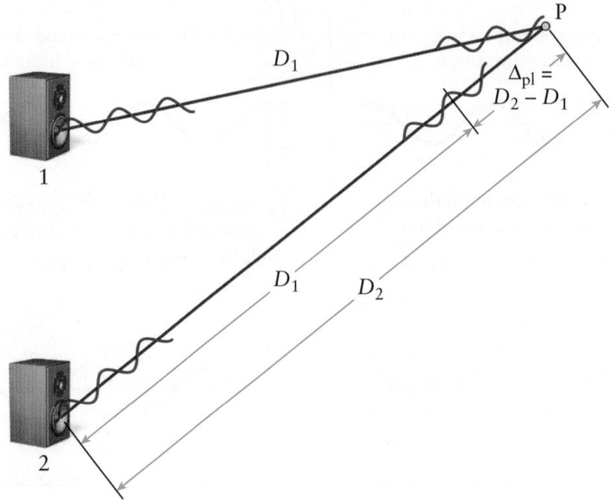

Figure 21-14 Combining sound waves Two sinusoidal sound waves of the same amplitude and wavelength travel different distances D_1 and D_2 to point P. How they interfere at P depends on how many wavelengths fit into the path length difference Δ_{pl}.

arrive in phase: A crest (or trough) of the first wave arrives at the same time as a crest (or trough) of the second wave. There is constructive interference at these locations, and the total wave has a large amplitude, up or down. At other locations, some of which are shown by the red dots, a crest of one wave arrives at the same time as a trough of the other wave. The two waves are out of phase, and there is destructive interference. At these locations the two waves essentially cancel each other, so that the total wave is nearly zero, and the water remains nearly flat. Because interference of the surface waves occurs at all points in Figure 21-13, we call the overall pattern in that figure an **interference pattern.**

In Figure 21-13, because the two wave *sources* are oscillating in phase, it is the combination of the initial phase of each wave and the *distances* from each source to the point where the waves meet and interfere that determine where there are constructive and destructive interferences. The same principle applies to waves of all kinds, so in **Figure 21-14** we've drawn two loudspeakers that emit sinusoidal sound waves of the same amplitude A, the same wavelength λ, and the same frequency f. We've drawn the sound waves as sinusoidal functions that indicate the displacement of air along the direction of propagation. The two loudspeakers oscillate *in phase*, so they both emit the exact same waves at any time t. A listener is at point P, a distance D_1 from speaker 1 and a distance D_2 from speaker 2. We call these distances the *path lengths* from either speaker to point P. The **path length difference** Δ_{pl} ("delta-sub-p-l") for the two waves is the difference between these two distances: $\Delta_{pl} = D_2 - D_1$, so $D_2 = D_1 + \Delta_{pl}$.

At the instant shown in Figure 21-14, wave 1 happens to reach a peak of a cycle at point P. Notice that wave 2 travels farther to reach P. The additional distance is the path length difference Δ_{pl}. Because the two waves are emitted in phase, each wave hits a peak at a distance D_1 from its source. So to determine what happens when the two waves overlap at P, we need only ask how many cycles of wave 2 fit into the extra distance Δ_{pl}.

If exactly one wavelength λ fits into Δ_{pl}, then wave 2 will be at a peak at point P, just as if it were at distance D_1 from the source. The same is true if 2λ, 3λ, 4λ, or *any* integer number of full wavelengths fit into Δ_{pl}. (It's also true if $\Delta_{pl} = 0$, which is the case if point P is equal distances from the two speakers.) And because we've already established that wave 1 is at the same point in its cycle at point P as wave 2 is at D_1 from source 2, we can make this comparison regardless of where in the cycles the waves are at P. If an integer number of wavelengths fit into Δ_{pl}, the two waves will rise and fall together as they arrive at point P. The two waves are therefore *in phase* at point P, and constructive interference takes place there (see Figure 21-12a).

Suppose instead that exactly one-half of a wavelength, or an odd number of half-wavelengths ($\lambda/2$, $3\lambda/2$, $5\lambda/2$, ...), fits into the path length difference Δ_{pl} shown in Figure 21-14. In this case, when wave 2 is at a peak at distance D_1 from the source, it will be at a trough in its cycle at point P. So waves 1 and 2 will be *out of phase* at point P, resulting in destructive interference. If the amplitudes of the two waves at P are equal, the total wave will be zero as in Figure 21-12b.

If the path length difference Δ_{pl} is neither a whole number of wavelengths nor an odd number of half-wavelengths, the interference is neither completely constructive nor completely destructive (Figure 21-12c).

Here's a summary of the conditions that must be met for constructive interference and destructive interference for two waves produced by sources that are in phase:

Path length difference from the two wave sources to the point where interference occurs

Wavelength

$$\text{Constructive interference:} \quad \Delta_{\text{pl}} = n\lambda \quad \text{where } n = 0, 1, 2, 3, \ldots$$

$$\text{Destructive interference:} \quad \Delta_{\text{pl}} = \left[n + \frac{1}{2}\right]\lambda \quad \text{where } n = 0, 1, 2, 3, \ldots \tag{21-7}$$

EQUATION IN WORDS
Conditions for constructive and destructive interference of waves from two sources that are in phase

Note that for constructive interference, $n = 0$ refers to the case where there is *zero* path length difference, so the two waves naturally arrive in phase. We draw the exact same representations for longitudinal waves; we just need to remember that what we are representing by displacement looks a little different in this case, as is shown in **Figure 21-15**.

WATCH OUT !

Interference effects require special conditions.

A sound system with two speakers resembles the setup in Figure 21-14. Based on our discussion of interference, you might expect that such a system would have "dead spots" where there is destructive interference between the sound waves coming from the two speakers (like the red dots in Figure 21-13) and "loud spots" where there is constructive interference (like the yellow dots in Figure 21-13). But even if you have such a system, you've probably never noticed any such dead spots or loud spots. One reason is that the positions where destructive interference occurs change if the wavelength changes (see Equations 21-7). Music contains sounds of many different frequencies and hence many different wavelengths, and places that are dead or loud spots for one of the wavelengths will not be dead or loud for other wavelengths. Another reason is that in a stereo system the signals coming from the left-hand speaker are not identical to those from the right-hand speaker. A final reason is that you also receive sound waves that come from the speakers and bounce off the walls or ceiling before reaching you. These have different path lengths than the waves that reach you directly from the speakers, and so they can "smooth out" any interference between the waves that reach you directly.

The locations of the air molecules causing the pressure pulses in the wave

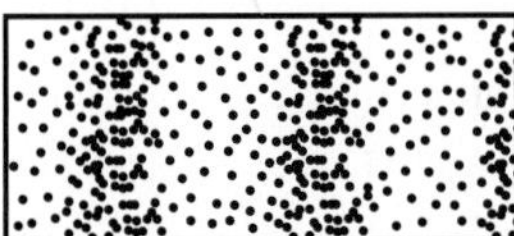

The first two sketches represent the location of the air molecules when the crest of one wave is located the same distance from the end of the tube as the crest of the other wave. The third sketch shows how those molecules would be distributed if you added the first two together: bigger amplitude!

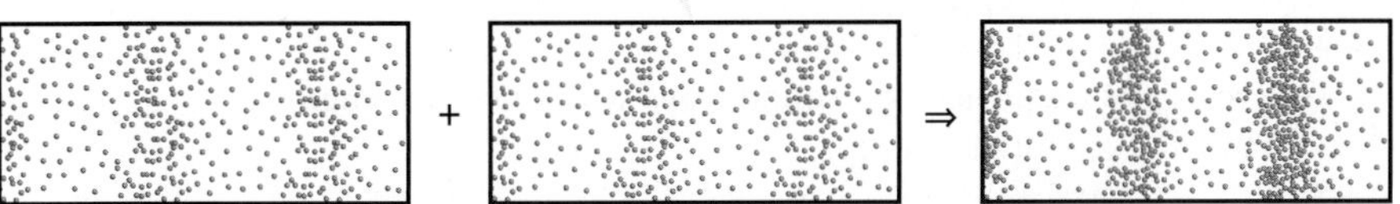

The first two sketches represent the location of the air molecules when the crest of one wave is located the same distance from the end of the tube as the trough of the other wave. The third sketch shows how those molecules would be distributed if you added the first two together: no amplitude!

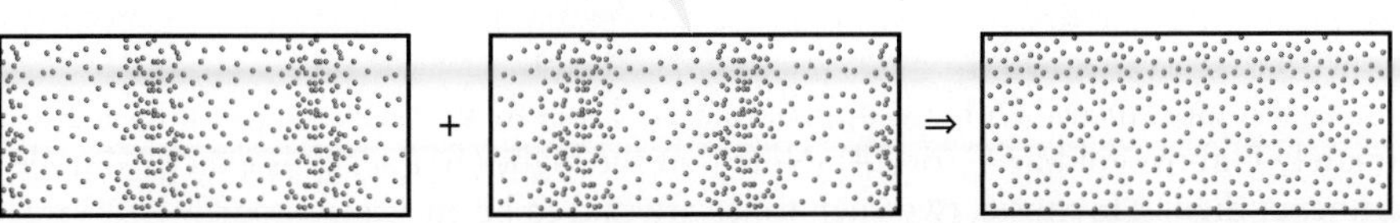

Figure 21-15 Interference in a longitudinal wave, such as a sound wave It is easier to draw a representation that looks more like the displacement graph in Figure 21-12. These figures show what this representation means about the distribution of air molecules for a sound wave for (a) constructive interference and (b) destructive interference.

EXAMPLE 21-3 Stereo Interference

Your sound system consists of two speakers 2.50 m apart. You sit 2.50 m from one of the speakers so that the two speakers are at the corners of a right triangle. As a test you have both speakers emit the same *pure tone* (that is, a sinusoidal sound wave). The speakers emit in phase. At your location what is the lowest frequency for which you will get (a) *destructive* interference? (b) *constructive* interference? (The speed of sound in dry air at 20°C is 343 m/s.)

Set Up

There is a path length difference for the two waves that reach your ear: One travels a distance $D_1 = 2.50$ m; the other travels a longer distance D_2 equal to the hypotenuse of the right triangle. Because the product of frequency and wavelength equals the constant speed of sound, the *lowest* frequency for each kind of interference corresponds to the *longest* wavelength.

Constructive interference:

$$\Delta_{pl} = n\lambda \text{ where } n = 0, 1, 2, 3, \ldots$$

Destructive interference:

$$\Delta_{pl} = \left(n + \frac{1}{2}\right)\lambda$$

where $n = 0, 1, 2, 3, \ldots$ (21-7)

Propagation speed of a sound wave:

$$v_{sound} = f\lambda \quad (21\text{-}2)$$

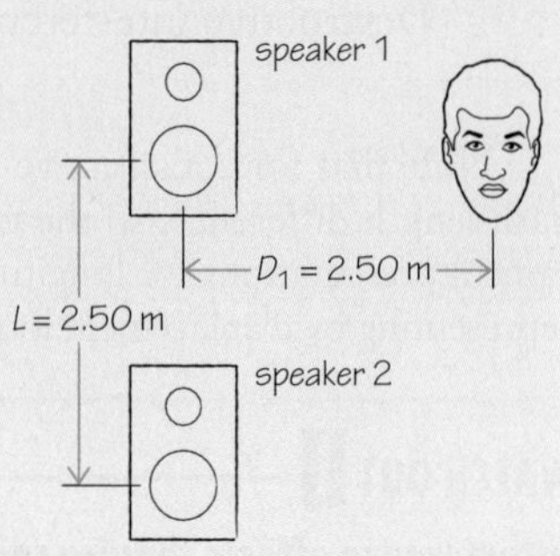

Solve

(a) The path length difference Δ_{pl} is equal to the difference between the two path lengths D_2 and D_1. Remember that we report our answers to the correct number of significant digits, but we use our intermediate numbers in further calculations.

Path length from speaker 2 to the listener:

$$D_2 = \sqrt{D_1^2 + L^2} = \sqrt{(2.50 \text{ m})^2 + (2.50 \text{ m})^2} = 3.54 \text{ m}$$

Path length difference:

$$\Delta_{pl} = 3.54 \text{ m} - 2.50 \text{ m} = 1.04 \text{ m}$$

The longest wavelength for which there is destructive interference is the one for which one half-wavelength fits into the distance Δ_{pl}. This corresponds to $n = 0$ in the destructive interference relation in Equations 21-7.

If one half-wavelength fits into the distance Δ_{pl},

$$\Delta_{pl} = \frac{1}{2}\lambda$$

$$\lambda = 2\Delta_{pl} = 2(1.0355 \text{ m}) = 2.07 \text{ m}$$

The frequency of a sound wave with this wavelength is

$$f = \frac{v_{sound}}{\lambda} = \frac{343 \text{ m/s}}{2.07 \text{ m}} = 166 \text{ s}^{-1} = 166 \text{ Hz}$$

If the sound has this frequency, there will be destructive interference at the listener's position and the sound level will be diminished.

(b) The longest wavelength for which there is constructive interference is the one for which one full wavelength fits into the distance Δ_{pl}. This corresponds to $n = 1$ in the constructive interference relation from Equations 21-7. Note that $n = 0$ is possible only if the path length difference is zero, which in this case it is not.

If one wavelength fits into the distance Δ_{pl},

$$\Delta_{pl} = \lambda = 1.04 \text{ m}$$

The frequency of a sound wave with this wavelength is

$$f = \frac{v_{sound}}{\lambda} = \frac{343 \text{ m/s}}{1.0355 \text{ m}} = 331 \text{ s}^{-1} = 331 \text{ Hz}$$

If the sound has this frequency, there will be constructive interference at the listener's position and the sound level will be elevated.

Reflect

The frequencies that we found are of musical importance: 166 Hz is approximately E below middle C (E3 in musical nomenclature), and 331 Hz is approximately E above middle C (or E4). Note that the higher frequency is exactly double that of the lower frequency (within significant digits). In music, two such tones are said to be an *octave* apart.

Notice that we didn't consider the number of wavelengths that fit into the distance between you and either speaker. All that matters is the *difference* between the two paths. Can you show that in the destructive case the listener is 1.20 wavelengths away from speaker 1 and 1.70 wavelengths away from speaker 2? Can you also show that in the constructive case the numbers are 2.40 and 3.40 wavelengths, respectively?

NOW WORK Problem 4 from The Takeaway 21-4.

THE TAKEAWAY for Section 21-4

✔ When two waves or wave pulses interfere with each other, they obey the principle of superposition. At every point where more than one wave passes simultaneously, the net disturbance of the medium equals the algebraic sum of the displacements that each wave would have caused individually.

✔ If two waves are in phase at some location, the two waves reinforce each other and there is constructive interference. If the two waves are out of phase at some location, they cancel each other and there is destructive interference.

Prep for the AP Exam

AP Building Blocks

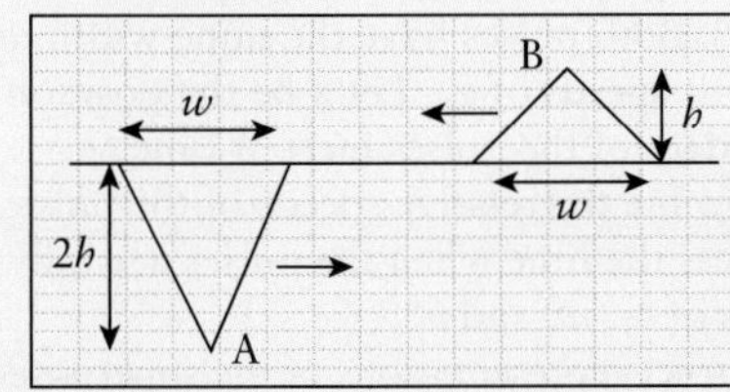

1. A triangular pulse (A) moving from left to right passes through a triangular pulse (B) moving from right to left. Using graph paper, draw the superposition of these pulses when (a) they begin to pass through each other, (b) when they are aligned, and (c) as they are close to completely passing through each other.

2. Two waves interfere at point x. The resultant wave is shown. Draw possible shapes for the two waves that interfere to produce this outcome. Explain how the wave height of the resultant wave would be calculated if the values of the wave heights of the interfering waves were known.

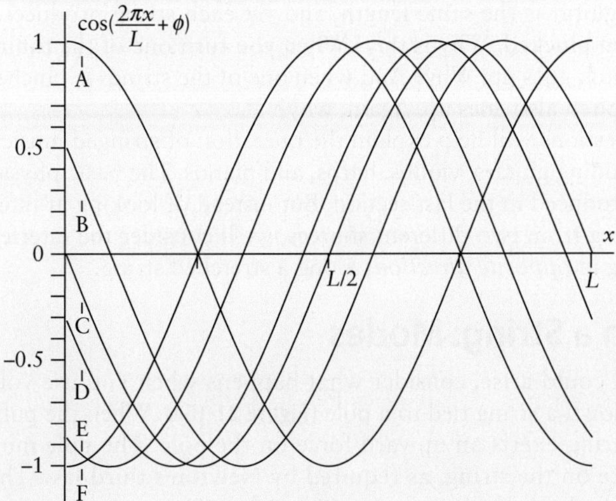

3. In the drawing, the wave labeled A has a phase of 0. Identify the phases for waves B through F among these possible values of the shift: $\pi/3$ radians, $\pi/2$ radians, $3\pi/5$ radians, $5\pi/6$ radians, and π radians.

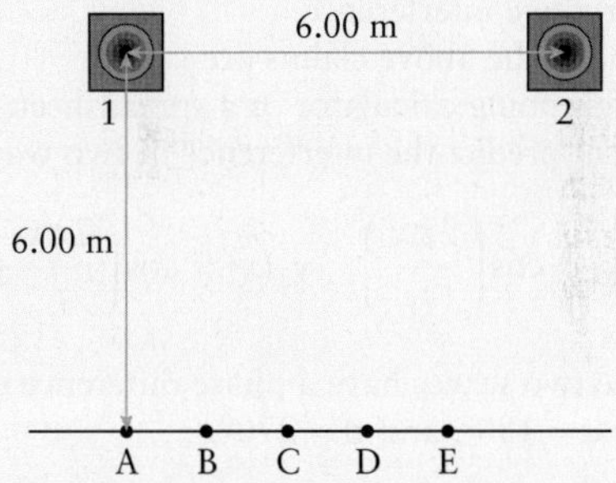

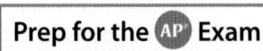

EX 21-3 4. Two identical speakers (1 and 2) are both playing the same tone that has a frequency of 193.45 Hz. The tones are in phase. The speakers are located 6.00 m apart, as shown. In parts (b)–(c) consider the path lengths from speaker 1, L_1, and from speaker 2, L_2, to each of the points A through E.

(a) Express the conditions, for points in the space (ignoring points A–E) in front of the speakers, at which there is (i) purely constructive interference, so that there is no phase difference and (ii) completely destructive interference, so that the phase difference is 180°. *Hint*: Consider differences in the path lengths from each speaker to the point in terms of the wavelength of the tone.

(b) Consider the points A, B, C, D, and E, all separated by 1.00 m along the line that is 6.00 m in front of the speakers. Point A is directly in front of speaker 1. Complete the table of path lengths L_1 and L_2 for each point shown.

Point	L_1	L_2	$\lvert L_1 - L_2 \rvert$
A			
B			
C			
D			
E			

(c) For each point predict when the interference of waves from the two speakers is completely constructive or completely destructive, or if the interference is neither completely constructive nor completely destructive. Use 343 m/s as the speed of sound.

AP® Skill Builders

5. Two point sources produce waves of the same wavelength that are in phase with each other. The sources are separated by a distance much larger than the wavelength. Evaluate each of the following claims about what you expect to observe as you move along the straight line between the sources.
   (a) At all points between the sources there should always be destructive interference.
   (b) At all points between the sources there should always be constructive interference.
   (c) At all points between the sources there will be interference with alternating points of constructive and destructive interference.
   (d) None of the above claims are valid.
6. Using a graphing calculator or a spreadsheet, sketch four waves that predict the interference of two waves

$$y_1(x) = \cos\left(\frac{2\pi x}{L}\right) \quad y_2(x) = \cos\left(\frac{2\pi x}{L} + \phi\right)$$

when the two waves have a phase difference of $\phi = 0$, $\phi = 90°$, $\phi = 180°$, and $\phi = 270°$.

AP® Skills in Action

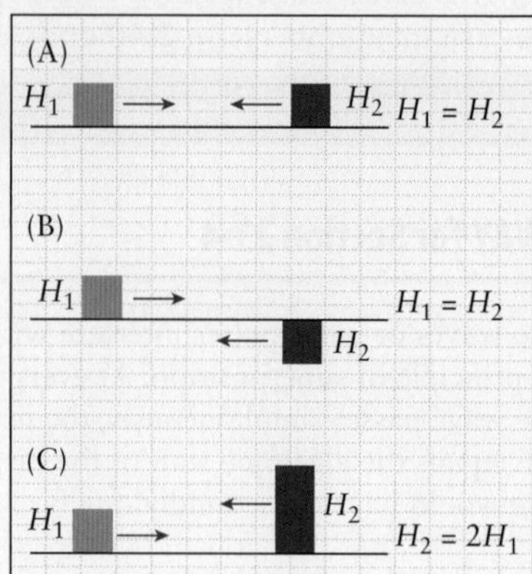

7. In the figure, pairs of rectangular pulses approach each other. The pulses have been drawn as solid rectangles to make it easier to draw the representations requested in part (c). Both pulses have the same width w and are traveling at speeds of v_p. Let x be the position where the leading edges of the pulses meet at time t_0.
   (a) Determine the relative speeds of the pulses.
   (b) Express the time when the pulses are fully aligned in terms of t_0, w, and v_p.
   (c) Construct representations of the superposed pulses graphically and in complete sentences at these instants:
   - The leading edges of the pulses first meet.
   - One-half of each pulse is yet to overlap.
   - The pulses are fully aligned.
   - One-half of each pulse still overlaps.
   - The trailing edges of the pulses are aligned.

21-5 A standing wave is caused by interference between waves traveling in opposite directions

Each of the six strings of a guitar is the same length, and yet each string produces a sound of different pitch when plucked. Why is this? When you turn one of the tuning keys to tighten a string, its pitch goes up. Why? And when one of the strings is pinched against the fingerboard, the pitch also goes up. Again, why?

The answers to these questions will help explain the operation of stringed musical instruments of all kinds, including guitars, violins, harps, and pianos. The basic physics is interference, which we introduced in the last section. But instead of looking at interference between waves coming from two different sources, we'll consider the interference between waves traveling in *opposite directions* along a stretched string.

Standing Waves on a String: Modes

To see how such a situation could arise, consider what happens when you use your hand to send a single pulse down a string tied to a pole (**Figure 21-16a**). When the pulse reaches the fixed end, the string exerts an upward force on the pole. The pole must then exert a downward force on the string, as required by Newton's third law. This action reflects an inverted pulse back along the string (**Figure 21-16b**).

If you were to wiggle the end of the string up and down periodically, instead of a pulse you would create a sinusoidal wave on the string. The reflection would be an inverted sinusoidal wave traveling back toward you with the same amplitude, wavelength, and frequency as the incoming wave.

Let's see what happens when we add together two sinusoidal waves traveling in opposite directions. We will talk through the outline of the mathematics in this paragraph, but then we are going to graphically show that the solution we get is the right sort of solution, and understand what it means. From Equation 21-6 the wave function for such a wave traveling in the positive x direction is $y(x,t) = A\cos(kx - \omega t + \phi)$. For simplicity we'll choose the phase angle ϕ to be zero. As we discussed in Section 21-3, a wave traveling in the *negative* x direction has the same wave function but with $kx - \omega t$ replaced by $kx + \omega t$. We'll also use $\phi = 0$ for this wave but we have to put a minus sign in front of the amplitude to indicate that the wave is inverted because of what happened when it was reflected. From the principle of superposition the total wave is the sum of these two sinusoidal waves with the same amplitude A, wavelength λ, and frequency f.

Using a trigonometric identity from Math Tutorial M-8, you can find that this becomes $y_{\text{total}}(x,t) = 2A(\sin kx)(\sin \omega t)$. This wave function describes a wave pattern unlike any we've seen yet. Note that the x and t terms do not appear in the same sinusoidal function. What does this equation tell us? The wave crests are at points where $\sin kx$ equals either +1 or −1, and at those points the value of y_{total} oscillates between $+2A$ and $-2A$ as the function $\sin \omega t$ varies between +1 and −1. At the points where $\sin kx = 0$, the wave is *always* zero! We call points on the wave that don't move **nodes**. This means the crests of the wave do *not* propagate along the string! Because the wave does not move but stays in the same place, it is called a **standing wave**. By contrast, a wave of the form $y(x,t) = A\cos(kx - \omega t)$ is referred to as a **traveling wave**. Note that we formed the standing wave by combining two traveling waves that propagate in opposite directions. When we add them together, the result is that, while the pieces of the string still move, the *pattern* created by the interference between two traveling sinusoidal waves—the standing wave—remains stationary.

(a) As the pulse arrives at the pole, the string exerts an upward force on the pole...

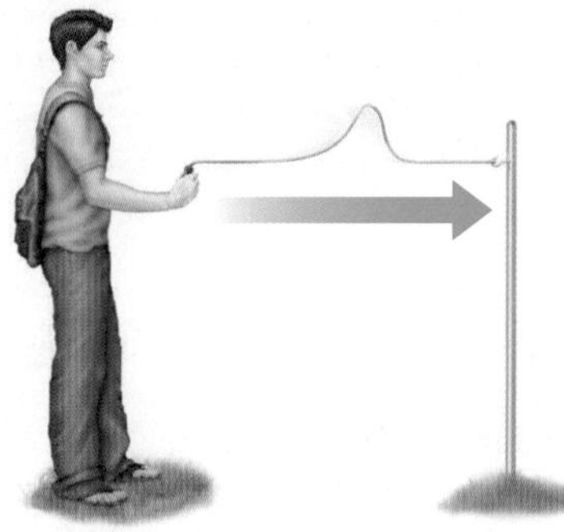

(b) ...so the pole exerts a downward force on the string. This causes the pulse to become inverted as it is reflected.

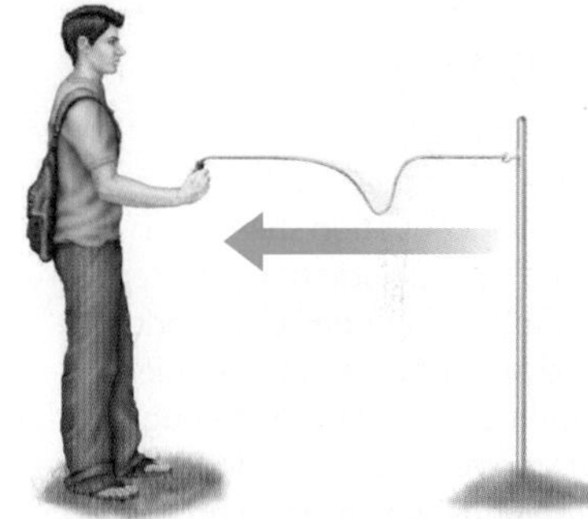

Figure 21-16 Reflecting a wave (a) A single pulse moves down a stretched string tied to a pole. (b) The pulse reflects back from the pole.

Now think again about the string in **Figure 21-16**. Let L be the length of the string, take $x = 0$ to be at the end of the string you hold in your hand, and let $x = L$ be at the end of the string attached to the pole. The wave cannot wiggle up and down there because that end is rigidly attached to the pole, so that has to be a node. The end that you hold in your hand moves very little compared to the amplitude of the wave on the string, so it's pretty accurate to say that the amplitude of the wave at your hand is zero, meaning this is also a node. So if we can put a wave on the string that has nodes on *both* ends of the string, a steady standing wave can be set up. Because a wavelength is the distance between any two points on a wave that are at exactly the same point in the motion of the wave, any number of whole wavelengths would work. But, it turns out that half a wavelength would work as well, because there is a node at every half-wavelength, as we can see in **Figure 21-17**. So the relationship of the wavelength λ (the wavelength of both the standing wave and also the original wave we created on the string) to the length of the string must be a whole number n of half-wavelengths.

Length of a string held at both ends — Wavelength of a standing wave on the string

$$L = \frac{n\lambda}{2} \quad \text{where } n = 1, 2, 3, \ldots \tag{21-8}$$

EQUATION IN WORDS
Wavelengths for a standing wave on a string

Figure 21-17 shows the patterns for $n = 1$ (for which half of a wavelength fits), $n = 2$ (for which two half-wavelengths, or one full wavelength, fits), and $n = 3$ (for which three half-wavelengths, or one and a half full wavelengths, fit). Each of these patterns is called a **standing wave mode**. If we wanted to see how that would fit back with our mathematical equation, because a sine function is zero for any multiple of π, we could multiply both sides of Equation 21-8 by 2π and divide both sides by the wavelength, and we get $\frac{2\pi L}{\lambda} = n\pi$. This is completely mathematically equivalent to

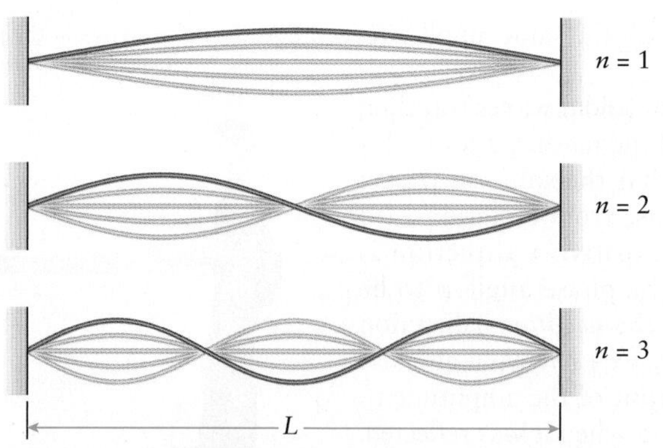

Figure 21-17 Standing waves on a string I These illustrations show the first three standing wave modes of a string connected to two supports. The mode number n counts the number of half-wavelengths that fit into the length of the string.

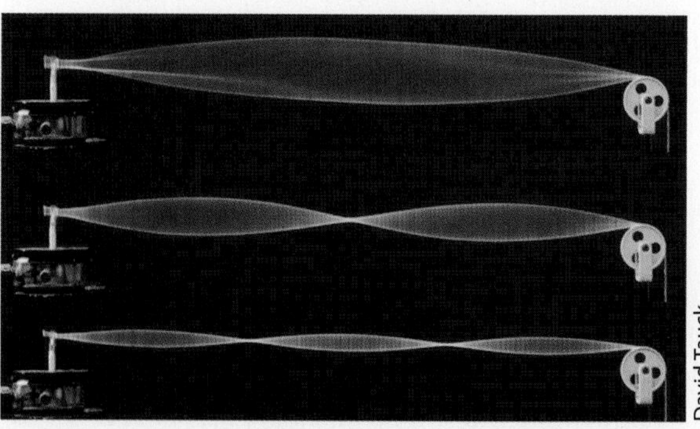

Figure 21-18 Standing waves on a string II These photographs show the first three standing wave modes on a string. The left-hand end of the string moves up and down (very slightly) to create each standing wave.

what we got by observing what had to happen, and it makes the math work out, so we must be on the right track!

AP® Exam Tip

Even though Equation 21-8 does not appear on the AP® Physics 2 equation sheet, you need to understand the concept exhibited, to understand nodes and antinodes, and to be able to determine possible wavelengths of standing waves in various situations, which are all on the exam.

Figure 21-18 shows photographs of the $n = 1$, 2, and 3 standing wave modes on a real string. In these photographs the movement of the end of the string that creates the standing waves is extremely slight. This justifies treating the end of this string as fixed, as we described earlier.

Figure 21-17 and Figure 21-18 show that for all modes except $n = 1$, there are other positions besides the ends which are nodes of the standing wave. For the $n = 1$ standing wave mode there are nodes at each end of the string; for the $n = 2$ mode there is an additional node at the center of the string, $x = L/2$; and for the $n = 3$ mode there are two nodes at $x = L/3$ and $x = 2L/3$.

Positions along the standing wave at which the oscillation of the string is maximal are called **antinodes.** We cannot call them crests, because the string moves very fast up and down at these points, between crest and trough positions. These points lie halfway between adjacent nodes. For the $n = 1$ mode in Figure 21-17 and Figure 21-18 there is one antinode at the center of the string ($x = L/2$); for the $n = 2$ mode there are two antinodes, at $x = L/4$ and $x = 3L/4$; and for the $n = 3$ mode there are three antinodes, at $x = L/6$, $x = L/2$, and $x = 5L/6$.

Standing Waves on a String: Frequencies and Musical Sound

We learned in Section 21-2 that the propagation speed of a *traveling* wave on a string goes up with tension and down with linear mass density. We also know that the frequency f and wavelength λ of a traveling wave are related by $v_p = f\lambda$ (Equation 21-2). Because a standing wave on a string is a superposition of two traveling waves, we can use these two relationships to understand how the *frequencies* associated with the standing wave modes on a string fixed at both ends behave.

By combining Equations 21-2 and 21-8, we get the following relationship between the frequency f and length of the string L of a standing wave on a string:

EQUATION IN WORDS
Frequencies for a standing wave on a string

(21-9)
$$f_n = \frac{n}{2L} v_p \text{ where } n = 1, 2, 3, \ldots$$

*n*th possible frequency of a standing wave on a string held at both ends

Speed that pulses are traveling on the string to produce the standing wave

Length of the string

Although the wave is not still a traveling wave, v_p represents the speed that pulses are traveling on the string to produce the standing wave due to its tension and mass density. The subscript n on the symbol f_n in Equation 21-9 denotes the frequency of the nth standing wave mode (see Figure 21-17 or Figure 21-18). Each of these frequencies represents a *natural* frequency of the string, at which it will oscillate if displaced from equilibrium and released. To make the string oscillate at the $n = 1$ frequency, displace it into the $n = 1$ shape shown in Figure 21-17 and let it go; to make it oscillate at the $n = 2$ frequency, displace it into the $n = 2$ shape (for instance, hold it still at the center and pluck it about a quarter of its length from one of the ends); and so on. They are also the *resonant* frequencies of the string. If one end of the string is set into oscillation at one of the frequencies given by Equation 21-9, the string will oscillate with large amplitude (see Figure 21-18). If the end of the string is forced to oscillate at a frequency other than those given by Equation 21-9, the result will be a jumble of small-amplitude wiggles rather than a well-behaved standing wave.

The frequency f_1, which is the lowest natural frequency of the string, is called the **fundamental frequency**. The corresponding ($n = 1$) standing wave mode is called the **fundamental mode**. Equation 21-9 shows that the fundamental frequency is equal to

$$f_1 = \frac{1}{2L} v_\text{p} \qquad (21\text{-}10)$$

For a fixed string length, the fundamental frequency increases with string tension and decreases with the string's linear mass density, because each of these increases the speed of propagation of the interfering waves. The fundamental frequency is inversely proportional to the length L of the string as this would increase the wavelength without changing the speed of propagation of the waves.

Equation 21-10 answers the questions we posed at the beginning of this section about the strings of a guitar. For stringed instruments such as a guitar, plucking a string normally causes it to vibrate primarily at its fundamental frequency. If it's an acoustic (not electric) guitar, the oscillation of the string actually causes the body of the guitar to oscillate at the same frequency. Those oscillations of the body in turn produce a sound wave of the same frequency in the surrounding air, which is what reaches the listener's ear. (If it's an electric guitar, sensors on the guitar detect the oscillation of the string and generate an electrical signal of the same frequency. This is then amplified in a loudspeaker to produce sound.) So Equation 21-10 is really an equation for the sound frequencies produced by a stringed musical instrument. It shows that for strings of a given length L, the frequency and hence the pitch of the sound are greater for strings with less mass per unit length. That's why the first string on a guitar, which has a fundamental frequency of 329.6 Hz, is much thinner than the sixth string, which has a fundamental frequency of 82.4 Hz. It also shows that if you increase the tension on a string, the frequency and pitch increase. And it explains what happens when you pinch one of the strings against the fingerboard: This shortens the length of string that's free to vibrate, thereby causing the frequency and pitch to go up.

Equation 21-9 shows that all of the other natural frequencies of a string held at both ends are integer multiples of f_1:

$$f_n = \frac{n}{2L} v_\text{p} = n\left(\frac{1}{2L} v_p\right) = nf_1$$

The $n = 2$ frequency f_2, called the *second harmonic* (or *first overtone*), is equal to $2f_1$; the $n = 3$ frequency f_3, called the *third harmonic* (or *second overtone*), is equal to $3f_1$; and so on.

The word *harmonic* should make you think of music, and indeed stringed musical instruments are an important application of the standing wave modes of a string. When you pluck a guitar string, the string actually vibrates in a *superposition* of standing waves with $n = 1$, $n = 2$, $n = 3$, and so on. (Just as two sinusoidal traveling waves can be present simultaneously on a string and combine to make a sinusoidal standing wave, more than one standing wave can be present on a string at the same time.) If it's an acoustic guitar, the guitar body is forced to oscillate simultaneously at the frequencies $f_1, f_2, f_3, \ldots$ and so produces a sound wave that is a superposition of sinusoidal waves at these frequencies. You might think that this would produce a hopeless jumble of sound. In fact, the result is a *nonsinusoidal* sound wave with the same frequency f_1

Exam Tip

As with most of the material on waves, the relationship exhibited by Equation 21-9 needs to be understood qualitatively for the AP® Physics 2 Exam. The equation does not need to be known, and for that reason is not on the equation sheet. However, you should understand that wave propagation speed increases with tension, and decreases with density.

Exam Tip

It is easy to get confused between *harmonic* and *overtone*. For the AP® Physics 2 Exam, you will need to know what *harmonic* means (you should understand the relationship between harmonic and frequency), but you will not be tested on *overtone*.

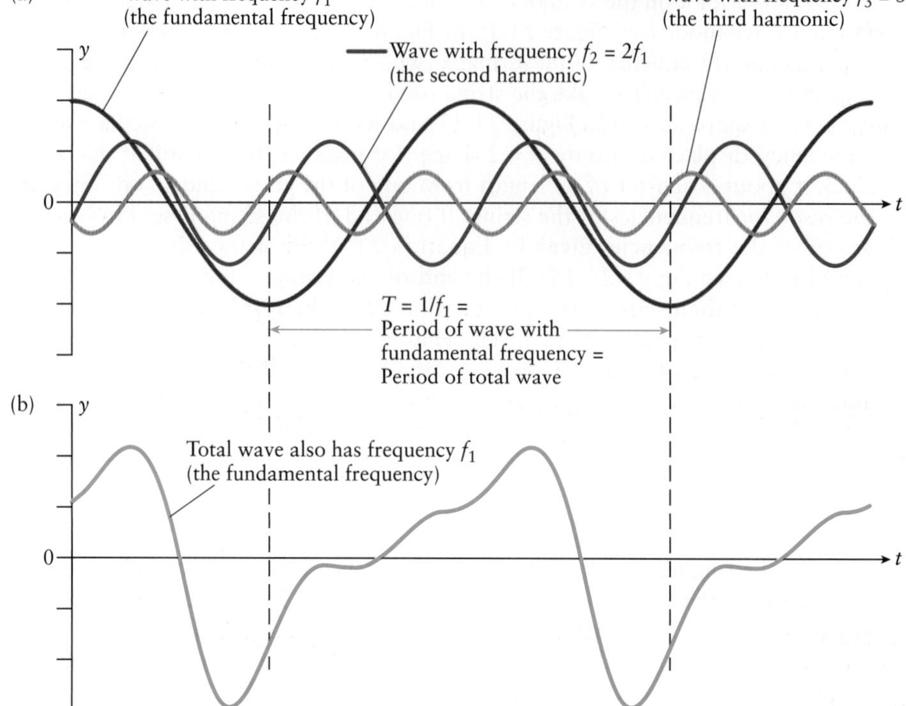

Figure 21-19 Adding harmonics (a) A sound wave is produced by a string oscillating simultaneously at its fundamental frequency f_1, its second harmonic f_2, and its third harmonic f_3. (b) The total sound wave is periodic with a frequency f_1 but is not a sinusoidal wave.

• Adding waves of frequencies $f_1, f_2, f_3,$... (the harmonics of f_1) produces a wave of frequency f_1 with a nonsinusoidal wave function.
• The shape of the sound wave produced by a musical instrument determines the tone quality of the sound.

as the fundamental frequency of the string (**Figure 21-19**). The shape of the wave function of this nonsinusoidal wave—which depends on the relative amplitudes and phases of the sinusoidal waves at frequencies $f_1, f_2, f_3, \ldots$—determines the *tone quality* or *timbre* of the sound. Different musical instruments playing the same note all generate sound waves of the same frequency, but with different amounts of the various harmonics. As a result, different instruments produce sound waves with differently shaped wave functions and hence different tone qualities, which is why they sound different.

EXAMPLE 21-4 Guitar String Physics

The third string on a guitar is tuned to play a note of frequency 196 Hz (to a musician, this note is G3 or G below middle C). The propagation speed of transverse waves on this string would be 251 m/s. The speed of sound in dry air at 20°C is 343 m/s. (a) What must be the length of the string for this to be its fundamental frequency? (b) What is the wavelength of the fundamental standing wave on the string? (c) What is the wavelength of the sound wave produced by standing waves on the string?

Set Up

We are given the fundamental frequency of the string $f_1 = 196$ Hz, and the speed a wave would propagate on the string, 251 m/s. We'll use Equation 21-8 to find the wavelength λ of the standing wave. The sound wave produced by the guitar has the same *frequency* as the standing wave on the string. We'll find the *wavelength* of the sound wave using Equation 21-2.

Frequencies for a standing wave on a string:

$$f_n = \frac{n}{2L} v_p \text{ where } n = 1, 2, 3, \ldots \quad (21\text{-}9)$$

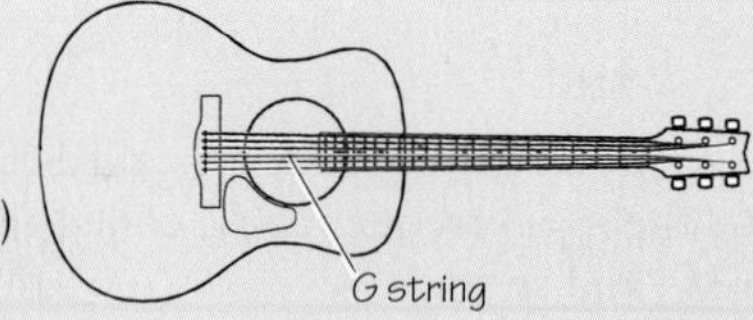

Wavelengths for a standing wave on a string:

$$L = \frac{n\lambda}{2} \text{ where } n = 1, 2, 3, \ldots \quad (21\text{-}8)$$

Propagation speed, frequency, and wavelength of a wave:

$$v_p = f\lambda \quad (21\text{-}2)$$

Solve

(a) The fundamental frequency is given by Equation 21-9 with $n = 1$. Solve this equation for the length of the string.

The fundamental frequency ($n = 1$) is

$$f_1 = \frac{1}{2L} v_p$$

To solve for the length of the string, multiply both sides by L and divide by the frequency:

$$L = v_p / 2f_1$$
$$= (251 \text{ m/s})/(2)(196 \text{ Hz}) = 0.640 \text{ m}$$

(Recall that $1 \text{ Hz} = 1 \text{ s}^{-1}$.)

(b) For the fundamental mode (the $n = 1$ standing wave) of a string held at both ends, the length of the string equals one half-wavelength.

From Equation 21-8:

$$L = \frac{\lambda}{2} \text{ for } n = 1$$

$$\lambda = 2L = 2(0.640 \text{ m}) = 1.28 \text{ m}$$

(wavelength of the standing wave on the string)

(c) Use Equation 21-2 to find the wavelength of the sound wave produced by the guitar.

For the sound wave:

$$v_{\text{sound}} = f\lambda$$

The frequency of the sound wave is the same as the frequency of the standing wave on the string, but the wavelength is different because the wave speed is different:

$$\lambda = \frac{v_{\text{sound}}}{f} = \frac{343 \text{ m/s}}{196 \text{ Hz}} = \frac{343 \text{ m/s}}{196 \text{ s}^{-1}}$$
$$= 1.75 \text{ m}$$

Reflect

There are six strings on the guitar, each of which is under about the same tension, or the guitar would bend and break! So the difference in the frequencies must be due to the difference in the mass density of the strings. Other than that, we probably don't have a good sense of the size of these numbers, so we really need to check the answer to (b) to see if (a) makes sense. Our answers for parts (b) and (c) are reasonable. A guitar is about this size, and the wavelength λ of the sound wave is greater than the wavelength of the standing wave on the string because, although the frequency f is the same for both waves, the propagation speed v_p is greater for the sound wave and $v_p = f\lambda$.

NOW WORK Problems 1 and 2 from The Takeaway for Section 21-5.

Exam Tip

While you should be able to graphically explain standing waves for different situations, including explaining the function of musical instruments, you will not need to create a mathematical representation of the phenomenon.

Exam Tip

For the AP® Physics 2 Exam, you should be ready to draw sinusoidal harmonics and identify nodes and antinodes on a graph. However, you will not need to create a mathematical representation of the phenomenon.

THE TAKEAWAY for Section 21-5

✔ Two identical waves traveling in opposite directions can interfere and form a standing wave. In a standing wave the points where there is zero displacement (the nodes) always remain at the same places in the medium.

✔ For a stretched string that is fixed at both ends, the only allowed standing waves are those that have an integer number of half-wavelengths in the length of the string.

✔ The length, linear mass density, and tension of a string determine the standing wave frequencies for that string, because the speed is the product of the wavelength and the frequency, and the speed depends on the linear mass density and tension of the string.

Prep for the AP® Exam

AP® Building Blocks

1. Each end of a string is fixed to a rigid support, and a standing wave is established in the string. The length of the string is 2.00 m, and it vibrates in the fundamental mode ($n = 1$).
   (a) What is the wavelength of this wave?
   (b) Construct a drawing of the wave in which a solid line is used to show the physical part of the wave and a dashed line is used to show the completion of a full wavelength of the wave beyond the physical part, if needed.
   (c) If the speed of the wave on the string is 60.0 m/s, calculate the frequency of the wave.

2. Each end of a 2.35-m-long string is tied to a rigid support. The string vibrates with a fundamental frequency of 24.0 Hz.
   (a) Calculate the speed of waves on the string.
   (b) Find the frequencies of the next four possible harmonics.
   (c) Make a sketch of these five standing waves showing both the forward and reflected wave with enough precision so that their similarities and differences are clearly distinguished.
   (d) Describe the motion of a piece of the string at an antinode.

AP® Skill Builders

3. Standing waves are set up on a string of length L that is fixed at both ends, so that the ends are nodes. Explain why the longest standing wave that can exist on this string has a wavelength of $2L$.
4. If two musical notes are an octave apart, the frequency of the higher note is twice that of the lower note.
   (a) Calculate the ratio of wavelengths for C_4 (middle C) and C_5 (one octave above middle C).

   In Western music, the octave is divided into 12 notes: C, $C^\#/D^b$, D, $D^\#/E^b$, E, F, $F^\#/G^b$, G, $G^\#/A^b$, A, $A^\#/B^b$, and B, followed by C one octave above the C beginning the series. Some of the notes have two designations, such as $C^\#$ and D^b. Each of the 12 notes is called a semitone. In the ideal tempered scale, the ratio of the frequency of any semitone to the frequency of the note below it is the same for all pairs of adjacent notes. So for example, $f_D/f_{C\#}$ is the same as $f_{A\#}/f_A$.
   (b) Express the ratio of the frequency of any semitone to the frequency of the note just below it as a power of 2. *Hint*: There are 12 semitones in an octave in which the frequency doubles, so the ratio multiplied by itself 12 times (raised to the power 12) has to equal the ratio of the frequency of one note from one octave to the next, 2.
   (c) There is a dispute among some people who enjoy music about whether it is best to tune an orchestra to an A_4 at 440 Hz or an A_4 at 432 Hz. Use your expression in part (b) to calculate the frequencies of $F_4^\#$ for these two tunings of the tempered scale.
   (d) Some stringed instruments, such as the guitar, are fretted. A fret is a small rigid bar that runs across the neck, perpendicular to the string. Explain how pressing a string down on the space between two frets changes the frequency of the plucked string. *Hint:* Pressing down on a fret changes the length of the string that is free to vibrate, but does not change the tension of the string or its mass density.
   (e) A Gibson acoustic Hummingbird guitar has a neck length of 62.9 cm and there are 20 frets (ridges) on the neck. Each fret separates two semitones. In standard tuning, the thinnest string is E_4, or 329.6 Hz, when the A_4 is tuned to 440 Hz. Calculate the highest frequency that can be played on the "E string."
5. Active noise reduction (ANR), or "noise canceling," is used in some headphones to remove sound waves produced in the environment from the sound waves produced by the speakers. A microphone on the outside of each headphone enclosure, or ear cup, gathers acoustic wave data from the environment. Even with headphones on, these sound waves could reach the eardrum. A microprocessor uses an algorithm to generate an electrical signal that is fed into the data stream from the music player before the data reach the speaker. Second-generation ANR headphones have multiple microphones on the outside of the ear cup and even a microphone on the interior of each ear cup.
   (a) Using the concept of wave interference, explain what the algorithm running in the microprocessor does in terms of the waveforms of the acoustic wave data gathered from the environment.

(b) Talking on a cell phone in public places lacks privacy and annoys others nearby. Design a technological solution using wave interference that could improve the quality of the voice signal received and could make your conversation less audible to others nearby.

AP® Skills in Action

6. Each end of a string of length L is tied to a rigid support. A harmonic mode is created within the string that has a frequency of $f_{n-1} = 40.0$ Hz. The next successive harmonic is at $f_n = 48.0$ Hz. The speed of transverse waves on the string is 56.0 m/s. Recall that the wavelength of allowed standing waves on a string of length L is $2L/n$, where n is an integer greater than 0.
   (a) Express the wavelengths λ_n and λ_{n-1} in terms of these successive frequencies and L.
   (b) Calculate the length of the string.
7. A wire of length L is fixed at each end and is under tension. The wire is made to vibrate so that a standing wave is formed in it. Evaluate each of the following claims about what you expect to observe at the midpoint of the wire, $L/2$. Recall that the wavelength of allowed standing waves on a string of length L fixed at both ends is $2L/n$, where n is an integer greater than 0. Choose the best answer below. At $L/2$
   (A) there should be a node caused by destructive interference in all harmonics.
   (B) there should be an antinode caused by constructive interference in all harmonics.
   (C) there should be alternating nodes and antinodes because the midpoint of the string at $L/2$ is a node when there is a whole number of wavelengths in L and is an antinode where there is not.
   (D) None of the above claims is valid.
8. A string is fixed on both ends. There is a standing wave vibrating at the fourth harmonic. Draw the shape of the standing wave and label the location of its antinodes (A) and its nodes (N).

21-6 Wind instruments, the human voice, and the human ear use standing sound waves

You've probably noticed that your singing voice sounds better in the shower than in the open. You may have also noticed that a flute (a musical instrument based on a long tube) produces lower notes than a piccolo (which uses a shorter tube). And you know that the sound of fingernails on a chalkboard is a particularly unpleasant one—so much so that just thinking about that sound may make you cringe.

We can help to explain all these effects by again invoking the idea of *standing waves*, which we introduced in the last section. The standing waves we need to consider are not waves on a string, but rather standing *sound* waves that result from traveling sound waves that reflect from the ends of an enclosure or tube.

Figure 21-20 shows an example of such a standing sound wave. Sound waves travel horizontally through the transparent tube and are reflected when they strike the ends,

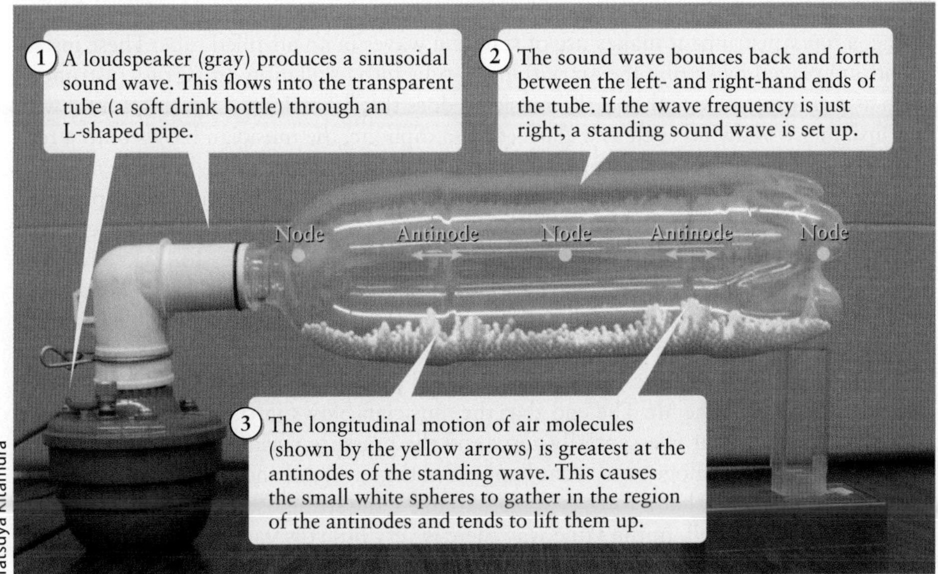

(1) A loudspeaker (gray) produces a sinusoidal sound wave. This flows into the transparent tube (a soft drink bottle) through an L-shaped pipe.
(2) The sound wave bounces back and forth between the left- and right-hand ends of the tube. If the wave frequency is just right, a standing sound wave is set up.
Node
Antinode
Node
Antinode
Node
(3) The longitudinal motion of air molecules (shown by the yellow arrows) is greatest at the antinodes of the standing wave. This causes the small white spheres to gather in the region of the antinodes and tends to lift them up.
Tatsuya Kitamura

Figure 21-20 Visualizing a standing sound wave The small white spheres inside this apparatus (known as *Kundt's tube* after the German physicist who devised it in the nineteenth century) help to identify the locations of the antinodes of a standing sound wave in air, which would otherwise be invisible.

just as transverse waves on a string are reflected at the fixed ends of the string. Sound is a longitudinal wave, so the motion of the air caused by the wave moving horizontally through the tube is also in the horizontal direction. The solid ends of the tube prevent this motion from taking place there, so there is zero displacement at the ends. This is just like the situation for a string with both ends fixed (see Figure 21-17). So just as for a standing wave on a string, a standing *sound* wave will be set up if a whole number of half-wavelengths of the wave fit into the length of the tube (Equation 21-8). In Figure 21-20 the frequency of sound waves provided by the loudspeaker has been adjusted so that two half-wavelengths (one full wavelength) fit into the tube's length, corresponding to $n = 2$ in Figure 21-17 and Equation 21-8.

Note that we can use Figure 21-17 for the displacement in a standing wave on a string to depict the displacement in a standing *sound* wave in a tube like the one shown in Figure 21-20. The difference is that because sound is a longitudinal wave, the displacement is now measured *along* the direction of wave propagation (see Figure 21-10).

Note also that the *pressure variation* in a traveling sound wave is a quarter-cycle out of phase with the displacement. Figure 21-10 shows this for a traveling sound wave, but the same is true for a standing sound wave. So where there is zero displacement in a standing sound wave (a displacement node), the pressure variation is maximum (a pressure antinode); where the displacement oscillates with maximum amplitude (a displacement antinode), the pressure variation is zero, so the pressure is always equal to the undisturbed air pressure (a pressure node).

Each standing wave sound frequency corresponds to a natural frequency of the tube. That means that if you produce a sound in the tube at one of those natural frequencies, a strong standing wave will be set up at that frequency. The net result will be that the sound wave you produce will be enhanced. That explains what happens when you sing in the shower. In the $n = 1$ standing wave mode, half a wavelength will fit between the walls of a shower stall. A typical shower stall is about 1 m wide, so the wavelength of the $n = 1$ mode is about 2 m. The frequency f_1 of this mode is related to the wavelength λ and the propagation speed v_{sound} for sound waves by $v_{\text{sound}} = f_1\lambda$ (Equation 21-2), so

$$f_1 = \frac{v_{\text{sound}}}{\lambda} = \frac{343 \text{ m/s}}{2 \text{ m}} \approx 170 \text{ Hz}$$

So if you sing at a frequency near 170 Hz (E or F below middle C), the shower stall will resonate and your voice will sound much fuller than it would outside the shower. The same will happen at frequencies that are multiples of 170 Hz (340 Hz, 510 Hz, 680 Hz, and so on).

Standing Waves and Wind Instruments: Closed Pipes

Just as a guitar or other stringed instrument produces tones by using standing waves on a string, a *wind* instrument makes use of standing waves in an air-filled tube. These instruments are played by setting the air within the tube into oscillation. In a brass instrument such as a trumpet or trombone, the musician does this by making a buzzing sound with her lips; in a woodwind such as a clarinet or saxophone, the musician blows onto a reed that vibrates the air in the instrument. For the sound to get out of the instrument, it has to be *open* at one or both ends. (The air outside of the tube is at atmospheric pressure, so when the low-pressure region of a sound wave traveling down the tube reaches the open end, air from the atmosphere rushes in and creates a compression wave heading back down the tube just as if the wave had been reflected, so the ingredients for making a standing wave are still present.) This means that a wind instrument has a different set of standing wave modes than those shown in Figure 21-17.

Most wind instruments behave like a **closed pipe**, in which one end is closed and the other is open to the air. The end that the musician blows into behaves like a closed end: The displacement is essentially zero, and the pressure variation has its maximum amplitude. (This is analogous to what happens at the left-hand end of the string in Figure 21-18. It's at that end that a force is applied to make the string oscillate, but the displacement there is very small. In the same way there is very little displacement of the air at the musician's end of a brass instrument or woodwind, but there is a large pressure variation due to the musician either buzzing her lips or making a reed vibrate.) The other end of the tube is open to the air. As a result the pressure there is essentially equal to the pressure of the surrounding air, so the pressure variation is close to zero and the displacement amplitude is large. In other words, a closed tube has a displacement node at the closed end but a pressure node (a displacement antinode) at the open end.

Figure 21-21 shows the first three standing wave modes in a closed pipe. We represent the magnitude of the displacement of the air from equilibrium by the separation of the blue curves from the dashed centerline. **Figure 21-21a** shows that in the fundamental mode of this closed pipe, only *one-quarter* of a wavelength fits in the length of the pipe. You can see this more clearly in **Figure 21-22a**, in which we've extended the sinusoidal representation of the fundamental mode out beyond the open end to show one full wavelength. For a closed pipe, then, the relationship between the length of the tube L and the wavelength λ of the sound wave that sets up the fundamental mode of the standing wave interference is

$$L = \frac{1}{4}\lambda$$

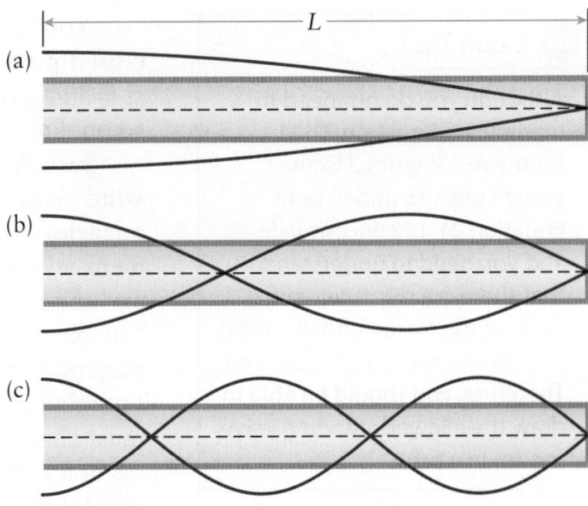

Figure 21-21 Standing sound waves in a closed pipe I (a), (b), and (c) show the displacement patterns for the first three standing sound wave modes of a pipe closed at its right-hand end and open at its left-hand end. There is a displacement node at the closed end and a pressure node at the open end.

WATCH OUT !

The standing sound wave in a closed pipe does not extend far beyond the pipe.

This extension of the wave pattern that we've drawn in **Figure 21-22a** does *not* represent the standing wave; the standing wave exists inside the pipe only. We've drawn this only to allow you to compare the wavelength of the sound with the length of the pipe.

Figures 21-21b and **21-21c** show the next two standing wave modes of a closed pipe, and **Figure 21-22** shows the displacement patterns extended outside the pipe to help visualize the dependence on wavelength. (The standing wave does not actually extend far beyond the end of the pipe.) You can see that three-quarters (3/4) and five-quarters (5/4) of a wavelength, respectively, fit into the pipe for these modes. You can also see that the general relationship between L and λ for a closed pipe is

Length of a closed pipe (open at one end, closed at the other)

Wavelength of a standing wave in the pipe

$$L = \frac{n\lambda}{4} \quad \text{where } n = 1, 3, 5, \ldots \tag{21-11}$$

EQUATION IN WORDS
Wavelengths for a standing sound wave in a closed pipe

Notice that only the *odd* harmonics ($n = 1, 3, 5, \ldots$) can exist in a pipe closed at one end. From $v_{\text{sound}} = f\lambda$ (Equation 21-2 for sound) the corresponding natural frequencies for a closed pipe are

$$f_n = \frac{v_{\text{sound}}}{\lambda} = \frac{v_{\text{sound}}}{(4L/n)} = n\left(\frac{v_{\text{sound}}}{4L}\right), \text{ where } n = 1,3,5,\ldots \tag{21-12}$$

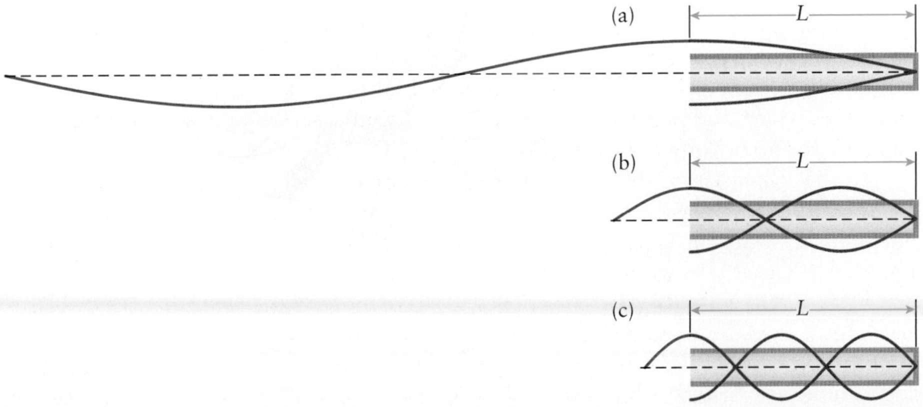

(a) L

(b) L

(c) L

Figure 21-22 Standing sound waves in a closed pipe II We've extended the sine curves that represent the displacement patterns for the first three standing sound wave modes in a closed pipe (compare Figure 21-21). This makes it easier to see what fractional number of wavelengths fit into the pipe in each mode.

AP® Exam Tip

Although you do not need to manipulate Equation 21-11 for the AP® Physics 2 Exam, you do need to understand Equation 21-11 conceptually, and you need to understand that there is a displacement node at the closed end of a tube, and an antinode at an open end. Therefore, you should be able to draw the allowed standing wave modes in a tube.

You can see from Equation 21-12 that the fundamental frequency of a closed pipe of length L is $f_1 = v_{\text{sound}}/4L$. The next harmonic is the *third* harmonic, $f_3 = 3(v_{\text{sound}}/4L) = 3f_1$; the next harmonic after that is the *fifth* harmonic, $f_5 = 5f_1$; and so on. For a closed pipe the even harmonics are absent.

Just like what happens when you pluck a guitar string, blowing into a closed-pipe wind instrument produces standing sound waves at more than one of the natural frequencies given by Equation 21-12. As a result, the sound that emanates from the instrument will be a combination of sinusoidal sound waves at frequencies $f_1, f_3, f_5, \ldots$, and the net result will be a nonsinusoidal wave with frequency f_1 (see Figure 21-19). The particular combination of harmonics is different for different wind instruments playing the same note, which gives each instrument its own distinctive tone quality. Equation 21-12 tells us that to change the fundamental frequency of a closed pipe, you must change the pipe length L. For a brass instrument this is done either with a slide (as in a trombone) or with valves that open and close air passages (as in a trumpet or tuba). For a woodwind such as a clarinet or oboe, this is done by opening holes in the pipe that effectively shorten the length of the vibrating air column.

You actually have closed pipes on either side of your head. These are the *auditory canals* of your left and right ears (**Figure 21-23**), which extend a distance of about 2.5 cm = 0.025 m from the opening (an open end) to the eardrum (a closed end). From Equation 21-12 the fundamental frequency ($n = 1$) of this short closed pipe is

$$f_1 = \frac{v_{\text{sound}}}{4L} = \frac{343 \text{ m/s}}{4(0.025 \text{ m})} = 3430 \text{ Hz}$$

This suggests that if a sound wave with a frequency of about 3430 Hz enters your ear, it will cause resonance to take place and the sound wave will be enhanced. In fact, a person with normal hearing is most sensitive to sounds around this frequency. And an important reason why the sound of fingernails scraping across a chalkboard is so unpleasant is that this is a sound in the range from 2000 to 4000 Hz—precisely where your hearing is the most sensitive.

Your vocal tract, which extends from the vocal folds in your throat to your lips, is also a closed pipe (closed at the vocal folds, which behave rather like the reed in a clarinet or oboe, and open at the lips). Unlike other musical instruments, you can reshape your vocal tract by the way that you hold your tongue and your lips. This changes the natural frequencies of the vocal tract (so that they no longer obey Equation 21-12) and makes each spoken sound distinct. To demonstrate this, try saying "a, e, i, o, u" and notice how your tongue and lips change position. Then use your fingers to hold your lips together and repeat the experiment; without the ability to reshape your vocal tract, all five vowels will now sound almost the same.

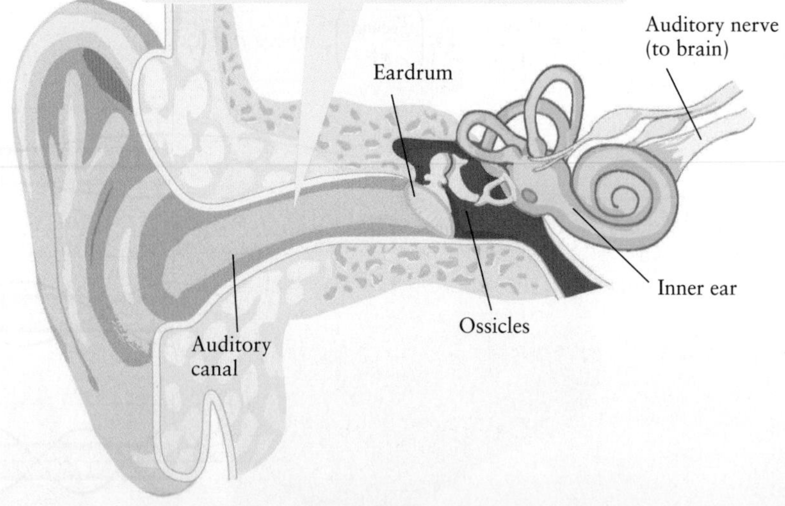
The auditory canal is a closed pipe (open to the outside air at one end and closed at the other end by the eardrum) that is 2.5 cm (0.025 m) in length.
Auditory nerve (to brain)
Eardrum
Inner ear
Ossicles
Auditory canal

Figure 21-23 The human ear Sound waves entering the human ear via the auditory canal cause the eardrum to vibrate, which in turn causes vibrations of the three ossicles (the smallest bones in the human body). These induce vibrations in the fluid that fills the inner ear. These vibrations are detected and converted through a chemical process into an electrical signal that is carried by the auditory nerve to the brain. The auditory canal acts like a closed pipe, which makes hearing especially sensitive for frequencies around 3430 Hz.

EXAMPLE 21-5 An Amorous Frog

The male tree-hole frog of Borneo (*Metaphrynella sundana*) attracts females by croaking out a simple call dominated by a tone of a single frequency. To amplify his call and enhance his attractiveness to potential mates, he finds a tree with a cavity in its trunk, sits inside the cavity, and adjusts the frequency of his call to match the fundamental frequency of the cavity. If the cavity is cylindrical and 11 cm deep, at what frequency should the frog make his mating call?

Set Up

The cavity in the tree trunk acts like a closed pipe (open on the outside of the tree, closed on the inside). Equation 21-12 gives the frequency of the fundamental ($n = 1$) mode of this cavity.

Frequencies for a standing sound wave in a closed pipe:

$$f_n = n\left(\frac{v_{\text{sound}}}{4L}\right), \text{ where } n = 1, 3, 5, \ldots \qquad (21\text{-}12)$$

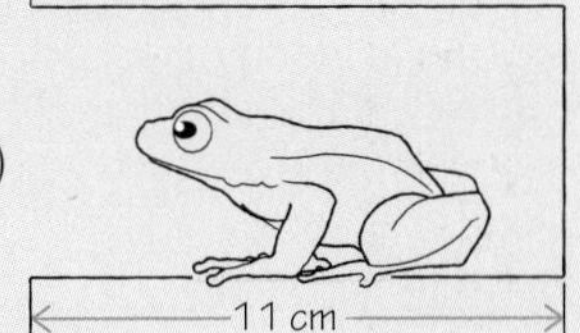

Solve

Calculate the fundamental frequency using $n = 1$ and $v_{\text{sound}} = 343$ m/s.

The fundamental frequency ($n = 1$) is

$$f_1 = \frac{v_{\text{sound}}}{4L} = \frac{343 \text{ m/s}}{4(11 \text{ cm})}\left(\frac{100 \text{ cm}}{1 \text{ m}}\right)$$
$$= 780 \text{ s}^{-1} = 780 \text{ Hz}$$

Reflect

If we look it up, we find our calculated result is close to the measured value of the frequency of a male tree-hole frog's mating call in a cavity 11 cm deep. These frogs have been observed to match the natural frequency of tree cavities from 10 to 15 cm in depth.

NOW WORK Problems 4 and 5 from The Takeaway for Section 21-6.

Standing Waves and Wind Instruments: Open Pipes

Unlike most other wind instruments, a flute is a tube that is open at both ends. This is called an **open pipe**. A flautist plays this instrument by blowing across the top of a hole in the pipe. This causes a pressure variation that causes the air inside the pipe to oscillate. Sounds made by striking solid rods can behave in a similar fashion because both ends are free to vibrate. Due to other vibrational modes (unlike air, solids can move transversely or twist) rods are more complicated, so we will stick with this wind instrument example.

Because the flute is open at both ends, there is a displacement antinode (pressure node) at each end. **Figure 21-24** shows the displacement patterns for the first three standing wave modes of such a pipe. (Just as at the open end of a closed pipe, a sound wave traveling through an open pipe is partially reflected at each open end. The traveling waves moving in opposite directions through the pipe give rise to a standing wave.) You can see in **Figure 21-24a** that one half-wavelength fits in the length L of the pipe. This represents the fundamental mode of the pipe, so the wavelength λ of the fundamental mode is given by

$$L = \frac{1}{2}\lambda$$

Similarly, the standing wave modes in **Figures 21-24b** and **21-24c** have two half-wavelengths (one full wavelength) and three half-wavelengths, respectively, in the length of the pipe. In general, the relationship between L and λ for an open pipe is

Length of an open pipe (open at both ends)

Wavelength of a standing wave in the pipe

$$L = \frac{n\lambda}{2} \quad \text{where } n = 1, 2, 3, \ldots \qquad (21\text{-}13)$$

EQUATION IN WORDS
Wavelengths for a standing sound wave in an open pipe

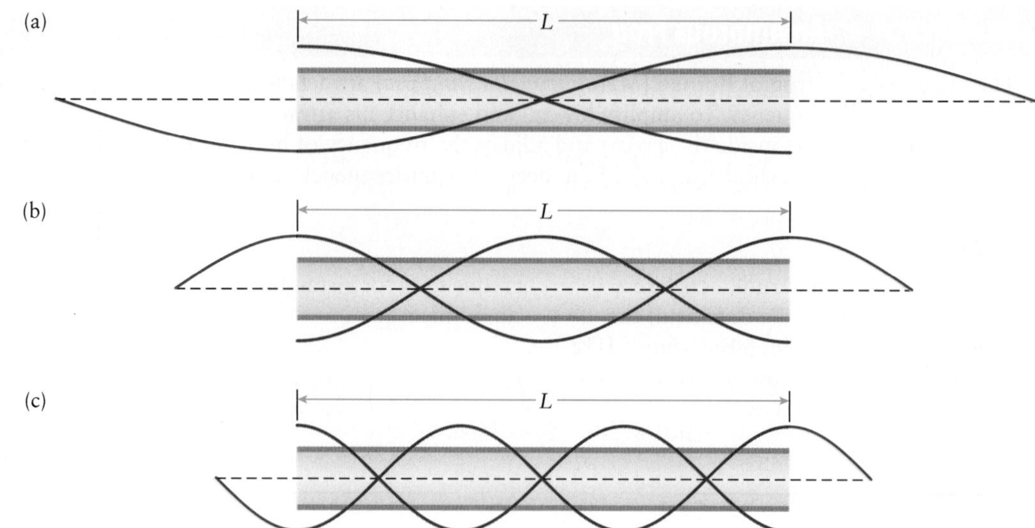

Figure 21-24 Standing sound waves in an open pipe (a), (b), and (c) show the displacement patterns for the first three standing sound wave modes of a pipe open at both ends. There is a displacement antinode at each open end.

Exam Tip

While you should be able to graphically explain standing waves for different situations, including explaining the function of musical instruments, you will not need to create a mathematical representation of the wave pattern. You will need to be able to calculate wavelength and frequency based on the boundary conditions (open or closed ends) and length of the pipe.

This is the same relationship as for standing waves on a string held at both ends (Equation 21-8) or for a pipe *closed* at both ends (see Figure 21-20). Although the wavelengths are the same as in those two cases, an open pipe has antinodes, not nodes, at its two ends.

The corresponding natural frequencies are

$$f_n = \frac{v_{\text{sound}}}{\lambda} = \frac{v_{\text{sound}}}{(2L/n)} = n\left(\frac{v_{\text{sound}}}{2L}\right), \text{ where } n = 1, 2, 3, \ldots \quad (21\text{-}14)$$

Unlike a closed pipe (Equation 21-12), an open pipe has *all* harmonics of its fundamental ($n = 1$) frequency. Note that for a given length L, the fundamental frequency of a closed pipe (from Equation 21-12, $f_1 = v_{\text{sound}}/4L$) is one-half of the fundamental frequency of an open pipe (from Equation 21-14, $f_1 = v_{\text{sound}}/2L$). If you cover one end of an open pipe while blowing into it, you change it into a closed pipe and the sound it produces will be lower in frequency and pitch.

THE TAKEAWAY for Section 21-6

✔ Traveling sound waves in a pipe can interfere to form a standing sound wave.

✔ A displacement node always appears at a closed end of the pipe, and a displacement antinode always appears at an open end.

✔ For a closed pipe (open at one end, closed at the other) the standing wave frequencies are odd multiples of the fundamental frequency. For an open pipe (open at both ends) the standing wave frequencies include both even and odd multiples of the fundamental frequency.

Prep for the AP Exam

Building Blocks

1. A pipe with a length L is open to the atmosphere on both ends and so the pressure of the air at the ends of the pipe equals the pressure of the air surrounding the pipe. Explain, supporting your explanation with a drawing, why displacement antinodes at the open ends of an open pipe support standing waves with wavelengths of $2L/n$, where n is an integer greater than zero.

2. A pipe that is closed at one end has length L. There is a displacement node at the closed end. Explain, supporting your explanation with a drawing, why a displacement antinode at the open end of a closed pipe restricts the allowed wavelengths of standing waves to $4L/n$, where n is an odd integer.

3. In a slide trombone the pipe increases in length as the musician extends the slide. When the slide is extended

(A) the fundamental frequency of the trombone remains the same because the fundamental is a fixed property of the instrument.
(B) the fundamental frequency of the trombone will increase because the wavelengths of all the standing waves in the tube increase.
(C) the fundamental frequency of the trombone will decrease because the wavelengths of all the standing waves in the tube decrease.
(D) the fundamental frequency of the trombone will decrease because the wavelengths of all the standing waves in the tube increase.

EX 1-5
4. An organ pipe of length L sounds its fundamental tone at 40.0 Hz. The pipe is open at both ends.
   (a) If the speed of sound in air is 343 m/s, calculate the length of the pipe.
   (b) Determine frequencies of the first four allowed harmonics.

EX 1-5
5. A narrow glass tube is 0.40 m long and sealed on the bottom end. It is held under a loudspeaker that sounds a tone at 220 Hz, causing the tube to radically resonate in its first harmonic. Calculate the speed of sound in the room.

AP® Skill Builders

6. Explain why for standing waves in a tube that is closed on one end and open on the other end, the closed end is a displacement node and a pressure antinode.
7. Explain why for standing waves in a tube that is closed on one end and open on the other end, the open end is a pressure node and a displacement antinode.
8. A recorder is a musical instrument that is a pipe with a hole that is closed at the end where the musician blows into the mouthpiece and is open on the other end of the pipe. There are holes bored along the length of the tube. Different notes are played when the musician covers holes with her fingers. Explain why for standing waves inside the recorder, each hole forms a pressure node and a displacement antinode, and describe what the effect of covering one of these holes with a finger produces.

AP® Skills in Action

9. A single organ pipe produces two successive tones at 228.6 and 274.3 Hz. Determine whether the pipe is open at both ends or open at one end and closed at the other. Use 343 m/s for the speed of sound in air. Recall that for a pipe that is closed at one end and open at the other the allowed wavelengths are $\lambda = 4L/n$, where n is an odd integer. For a pipe with two open ends the allowed wavelengths are $\lambda = 2L/n$, where n is an integer greater than 0.

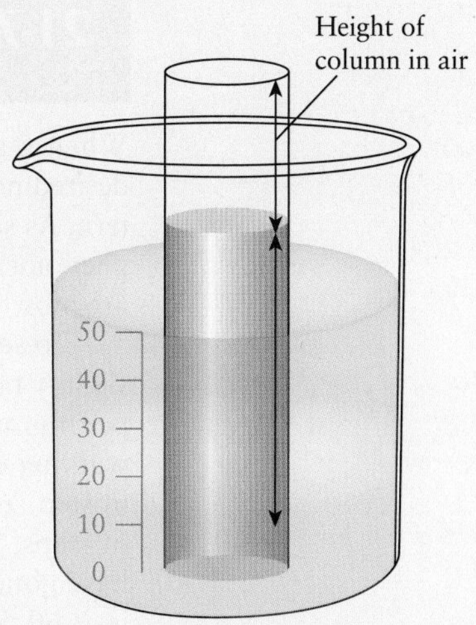

10. An apparatus used for the measurement of the speed of sound is shown in the figure. The bottom of the tube is open and embedded in a large beaker of water and the top of the tube is open. When the tube is raised and lowered, the distance between the top of the tube and the water level varies. When a tuning fork is struck near the top end of the tube a standing wave can be created, depending on the distance from the top of the tube to the water level. Tuning forks with several frequencies and a ruler are provided.
    (a) Design an experiment to collect data that can be used to measure the speed of sound with this apparatus.
    (b) Describe a method of graphical analysis that can be used to find the speed of sound in air using the relationship between the dependent and independent variables identified in part (a) that can be used to determine the speed of sound graphically.

    The dependence on temperature in dry air is represented with an error of less than 0.1% for temperatures T between 0°C and 30°C by

$$v_{\text{p,dry air}} = \sqrt{401.86 \text{ ms}^{-1}\text{K}^{-1}(T + 273.15)}$$

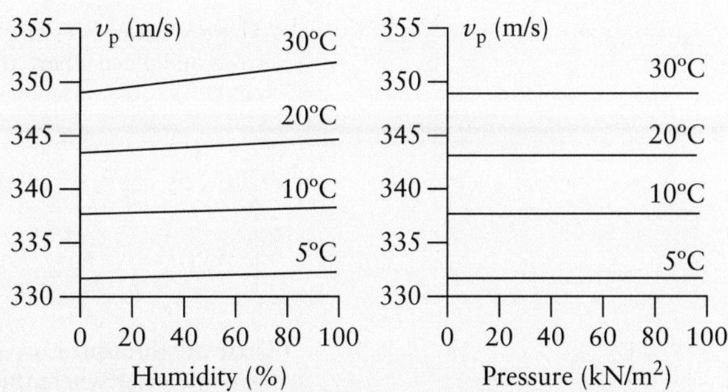

The dependence on pressure (relative to standard atmospheric pressure) and humidity over this same range of temperatures is summarized in the graphs above.
    (c) Describe the difference between accuracy and precision in experimental measurement and how the reference data above could be used to evaluate the accuracy of the data collection designed in part (a).

21-7 Two sound waves of slightly different frequencies produce beats

When a guitar player tunes her instrument by plucking a string while playing the desired note on a tuner you may hear the sound get louder and softer in a regular pattern. As she adjusts the tension in the string, this variation in loudness disappears. This phenomenon, known as **beats**, arises from the interference between sound waves. Beats are most pronounced when two waves of nearly identical frequencies interfere.

To see how beats arise, let's think our way through an analogy. Imagine you and a slightly taller friend are walking quickly side by side. Let's say you are walking 63 steps per minute (the frequency of your steps), and your friend, with slightly longer legs, is walking only 60 steps per minute. You start out exactly in step, but you do not stay in step. You will come back into step every time you both have gone a whole number of steps. So, when you have gone 21 steps, your friend will have gone 20, when you have gone 42, your friend will have gone 40, and when you have gone 63, your friend is at 60. So in each minute there are three times you are perfectly in step (you cannot count both the beginning and end, because the end is the beginning of the next minute!). This works out to be exactly the absolute value of the difference in your stepping frequencies. Somewhere in the middle of those times you are out of step. Now instead, let's imagine you increase your stride just a little and walk 62 steps per minute. Now, you and your friend will be in step only when you are at 31 and your friend is at 30, and when you are at 62 steps and your friend is at 60. So again, the number of times you are in step is the absolute value of the difference in your frequencies. If you really stretch yourself out, and go 60 steps per minute, you and your friend are always in step, and there is no variation (if your steps were waves, they would perfectly line up and just add constructively).

In **Figure 21-25**, we see how this would look for two sound waves traveling through air, using our same representation as in Figure 21-15. We can also see this by looking at what happens when we combine two sinusoidal waves with the same amplitude A but with slightly different frequencies f_1 and f_2. **Figure 21-26** shows the result: The total wave is also a sinusoidal wave whose frequency f is the average of f_1 and f_2, but with an amplitude that varies between 0 and $2A$. We use the term *beats* for this

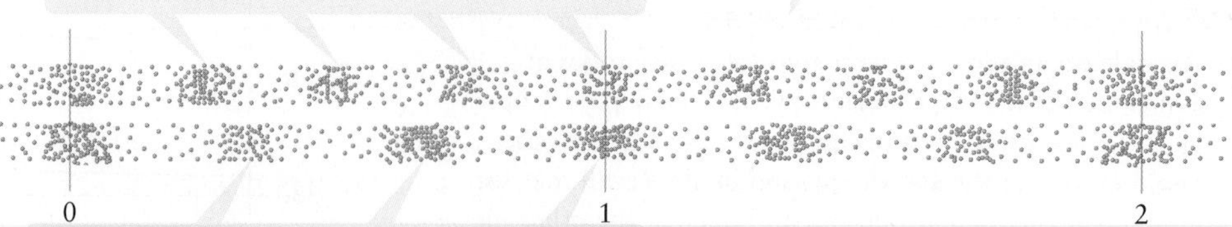

The waves are "in step" at the times indicated by the blue lines, and at those times add constructively. In between these points the waves still add, but because there is only ever a maximum for one, the sound in between the times denoted by the blue lines will be less.
There are four pressure pulses in the top wave from one place where they are "in step" until the next.
0
1
2
There are three pressure pulses in the bottom wave from one place where they are "in step" until the next.

Figure 21-25 Beats I Given the frequency of a typical sound wave, this number of pressure pulses would go by in only a small fraction of a second, but to keep our analogy simple, let's ignore that and pretend this is one second of traveling sound wave going by as you observe. The top wave has 8 complete cycles from the middle of the first pulse to the middle of the last pulse. The bottom wave has 6 complete cycles in the same time. Our reasoning says that we should have two places (8 – 6 = 2) where the waves are in step and we get maximum sound, not counting the first place they line up (labeled 0, which would be the last alignment from the previous second). We see that we indeed have two times when the waves are in step, labeled 1 and 2!

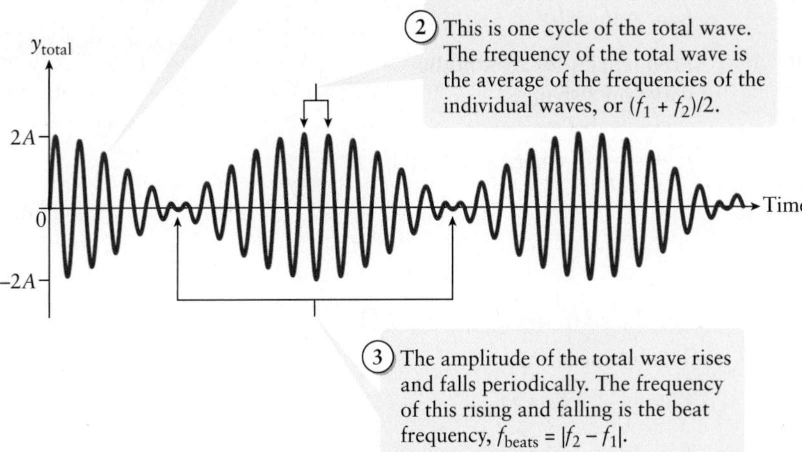

(1) This is the total wave that results when two individual sine waves of similar frequencies f_1 and f_2 are combined.
(2) This is one cycle of the total wave. The frequency of the total wave is the average of the frequencies of the individual waves, or $(f_1 + f_2)/2$.
y_{total}
2A
0
−2A
Time t
(3) The amplitude of the total wave rises and falls periodically. The frequency of this rising and falling is the beat frequency, $f_{beats} = |f_2 - f_1|$.

Figure 21-26 Beats II When two sinusoidal waves of nearly the same frequency are added together, the total wave has a rising and falling amplitude, or beats.

up-and-down variation in amplitude. The frequency of the beats, also called the **beat frequency**, is equal to the *absolute value* of the difference between the two frequencies:

Beat frequency heard when sound waves of two similar frequencies interfere

Frequencies of the individual sound waves

$$f_{beats} = |f_2 - f_1| \tag{21-15}$$

EQUATION IN WORDS
Beat frequency

(The absolute value ensures that we get a positive value for f_{beats}. We do this because a negative frequency has no meaning.) For example, if the two individual sound waves have frequencies $f_1 = 200$ Hz and $f_2 = 202$ Hz, what you will hear is a tone at a frequency of 201 Hz (the average of 200 and 202 Hz) with an amplitude that rises and falls at frequency $f_{beats} = |f_2 - f_1| = |202\text{ Hz} - 200\text{ Hz}| = 2$ Hz. That is, the amplitude will rise and fall twice per second, or once every 1/2 s.

The greater the difference between the frequencies f_1 and f_2, the greater the beat frequency and the more rapid the beats. If the frequency difference is large enough, the beats are no longer perceptible and you will hear two distinct frequencies.

AP® Exam Tip

For the AP® Physics 2 Exam, in addition to calculating the beat frequency, you should be able to qualitatively describe why beats happen, identify how to change the beat frequency, and identify how the corresponding sound will change for a particular situation.

WATCH OUT !

Don't confuse the beat frequency with the frequency of the sound.

When you hear two waves with slightly different frequencies f_1 and f_2, the beat frequency given by Equation 21-15 is the frequency at which the amplitude of the sound rises and falls. The frequency of the sound itself is different: Because of the rules for adding sine functions, the perceived frequency of the sound is the average of f_1 and f_2, or $(f_1 + f_2)/2$. If the frequencies are too different, you hear them as separate sounds.

EXAMPLE 21-6 Guitar Tuning

You have just replaced the first string on your guitar and wish to tune it to the correct frequency of 329.6 Hz (the note E_4). To do this you pluck both this string and the sixth (E_2) string, which you know is properly tuned to 82.4 Hz. You hear one beat every 2.50 s. How far out of tune (in hertz) is the first string?

Set Up

The frequencies 82.4 and 329.6 Hz are far apart, so it may seem surprising that you would get beats when both are sounded. But remember that a plucked string oscillates not only at its fundamental frequency but also at the higher *harmonics* of the string. So what you hear is beats between the sound made by the first string and one of the harmonics of the sixth string.

Beat frequency:

$$f_{\text{beats}} = |f_2 - f_1| \quad (21\text{-}15)$$

Frequencies for a standing wave on a string:

$$f_n = \frac{n}{2L} v_p \text{ where } n = 1, 2, 3, \ldots \quad (21\text{-}9)$$

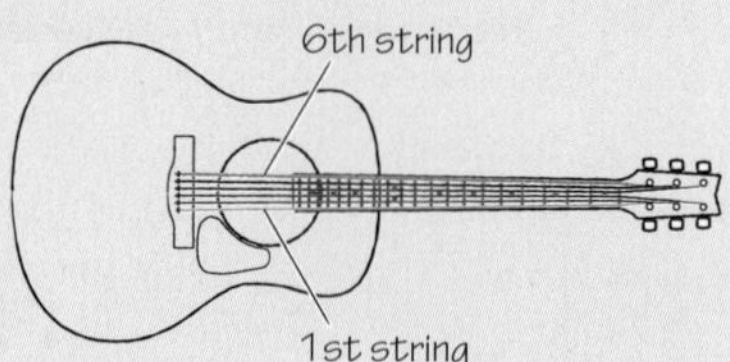

Solve

The harmonics of the sixth string are at integer multiples of its fundamental frequency.

For the sixth string, $f_1 = 82.4$ Hz.

The harmonics are

$$f_2 = 2f_1 = 164.8 \text{ Hz}$$

$$f_3 = 3f_1 = 247.2 \text{ Hz}$$

$$f_4 = 4f_1 = 329.6 \text{ Hz}$$

So if the guitar is properly tuned, the fundamental frequency of the first string is the same as the fourth harmonic frequency of the sixth string. If the first string is mistuned, the frequencies will be different and you will hear beats.

Find the beat frequency and how far out of tune the first string is.

There is one beat per 2.50 s, so the beat frequency (the number of beats per second) is

$$f_{\text{beats}} = \frac{1}{2.50 \text{ s}} = 0.400 \text{ s}^{-1} = 0.400 \text{ Hz}$$

This is equal to the difference between the fundamental frequency of the first string and the fourth harmonic frequency of the sixth string:

$$f_{\text{beats}} = 0.400 \text{ Hz} = |f_{1 \text{ for first string}} - f_{4 \text{ for sixth string}}|$$
$$= |f_{1 \text{ for first string}} - 329.6 \text{ Hz}|$$

So the first string is out of tune by 0.400 Hz.

Reflect

With just the information given, we don't know whether the first string is tuned 0.400 Hz too high (in which case its frequency is 330.0 Hz) or too low (in which case its frequency is 329.2 Hz). To find out, try increasing the tension on the first string to raise its fundamental frequency. If the beats slow down, the beat frequency is getting closer to zero as you bring the first string into tune, so the first string must have been tuned too low. Keep increasing the tension until the beats stop altogether, at which point the first string is properly tuned. If instead the beats become more rapid, the beat frequency is increasing, so the tuning of the first string is getting worse. This means the first string was tuned too high, so you should decrease the tension on the first string to lower its fundamental frequency and bring it into tune.

NOW WORK Problems 1–3 from The Takeaway for Section 21-7.

THE TAKEAWAY for Section 21-7

✔ When two waves of nearly identical frequencies and wavelengths interfere, the result is a total wave whose amplitude rises and falls periodically. This phenomenon is called beats.

✔ The frequency of the beats equals the absolute value of the difference in frequency between the two waves.

Prep for the AP Exam

AP Building Blocks

EX 21-6 1. The sound from a tuning fork of 440 Hz produces beats when combined with the unknown frequency of a vibrating string. If beats are heard at a frequency of 4 Hz, what is the vibrational frequency of the string?

EX 21-6 2. Two pianists sit down to play two identical pianos. However, a string is out of tune on Elizabeth's piano. The $G_3^{\#}$ key (208 Hz) appears to be the problem. When Greg plays the note on his piano and Elizabeth plays hers, a beat frequency of 6 Hz is heard. Luckily a piano tuner is present, and she is ready to correct the problem. However, she does not first check to ensure which piano is out of tune, and increases the tension in Greg's $G_3^{\#}$ string. Now both pianos are out of tune! Now when Elizabeth plays her $G_3^{\#}$ note and Greg plays his $G_3^{\#}$, there is no beat frequency. Explain what happened, including what frequency both piano's $G_3^{\#}$ keys are tuned to after the tuner increases the tension.

EX 21-6 3. A guitar string will be "in tune" at 440 Hz. When a standardized tuning fork rated at 440 Hz is simultaneously sounded with the guitar string, a beat frequency of 5 Hz is heard. The pegs on a guitar can be rotated to increase or decrease the tension. What steps should you take to tune the guitar? Your answer should be expressed in complete sentences.

21-8 The frequency of a sound depends on the motion of the source and the listener

Until now we have considered only waves generated by stationary emitters and observed by stationary listeners. Everyday experience, however, suggests that something curious happens when a moving object creates a sound. As a police car with sirens blaring zooms by you, for example, the frequency of the sound is higher as the car approaches and lower after it passes. Even without a siren, a fast-moving car generates a characteristic high-to-low frequency sound (something like "neee-urrrr") as it approaches and passes you. And although we don't often get the chance to experience this phenomenon in reverse, if you were to move at high speed toward a stationary police car with its sirens blaring, the sound of the siren would follow a similar high-to-low frequency shift as you approached and then passed it. This effect is known as the **Doppler effect**, named after the Austrian physicist Christian Doppler who in 1842 first proposed this phenomenon associated with waves.

The Doppler Effect and Frequency Shift

In **Figure 21-27a**, a police siren emits a periodic sound wave that has a fixed frequency and a fixed wavelength. Because the car is stationary, each wave crest—that is, each region of highest pressure along the spherical wave, shown as arcs of circles—is centered on the siren. On the left side of **Figure 21-27b** the car moves from left to right. The wave crest associated with the largest circle originated first, when the car was somewhere to the left, and the circle of the smallest radius represents the most recently generated wave crest. Because the car moves to the right as the sound wave propagates, the siren emits each new wave crest closer to the previous one on the right and farther from the previous one on the left. Because the distance between wave crests determines the wavelength heard by a stationary listener, someone standing in front of the car detects a shorter wavelength than the siren generates. The opposite is true on the right-hand side of Figure 21-27b, in which the car moves away from the listener.

The relationship $f = v_{\text{sound}}/\lambda$ (Equation 21-2 for sound) tells us that a shorter wavelength results in a higher frequency. So the listener in Figure 21-27b hears a higher frequency than the siren generates as the car approaches. Analogously, because the wave crests are spread farther apart behind the car, the observed wavelength lengthens, resulting in the listener hearing a lower frequency.

In **Figure 21-28** we expand the view of the moving police car to show the car at two instants in time separated by the period T of the tone created by the siren. A crest of the sound wave was created when the car was at the location marked by the red dashed line; in the time T the wave crest has propagated as indicated by the red semicircle. That distance is given by the product of the speed of sound v_{sound} and T, as

Figure 21-27 The Doppler effect: A moving source I A police car siren emits a periodic sound wave that has a fixed frequency and a fixed wavelength. (a) The police car is stationary, so each region of highest pressure along the wave, shown as arcs of circles, is centered on the siren. (b) The car moves from left to right. Because the source of the sound waves moves as the waves propagate, the wave crests are closer together as the car approaches a listener's ear and farther apart as the car recedes from the listener.

(a)

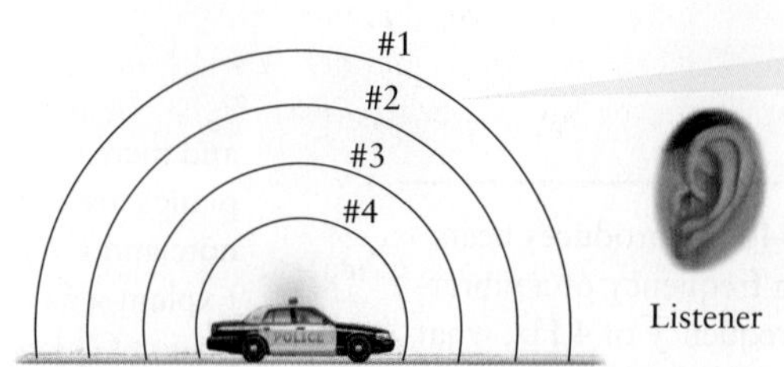

(b)

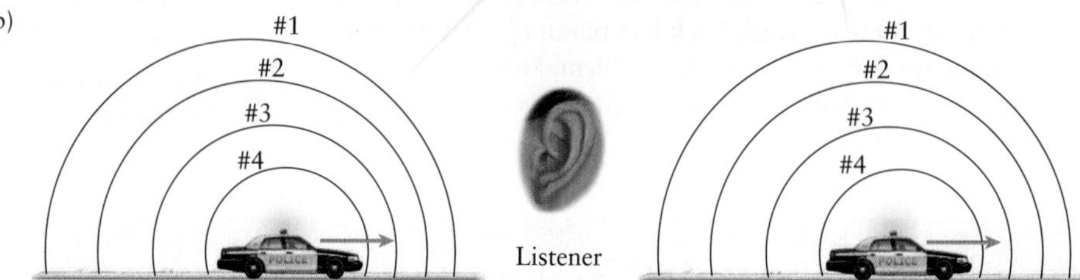

Exam Tip

Only qualitative understanding of the Doppler effect is required. A graphical representation and demonstrating whether the observed frequency will increase or decrease is sufficient for the AP® Physics 2 Exam. The mathematical description is provided as it helps to form a deeper understanding of the concept.

shown. The distance the car moves in that time is $v_{car}T$, where v_{car} is the speed of the car. This new location is where the car will emit the next wave crest. So a stationary listener in front of the car detects $\lambda_{listener}$ (the distance between two successive wave crests) as the wavelength of the siren's tone.

As seen in the figure these three distances—the distance the sound moves, the distance the car moves, and the distance between two successive wave crests at the place where the listener sits—are related by

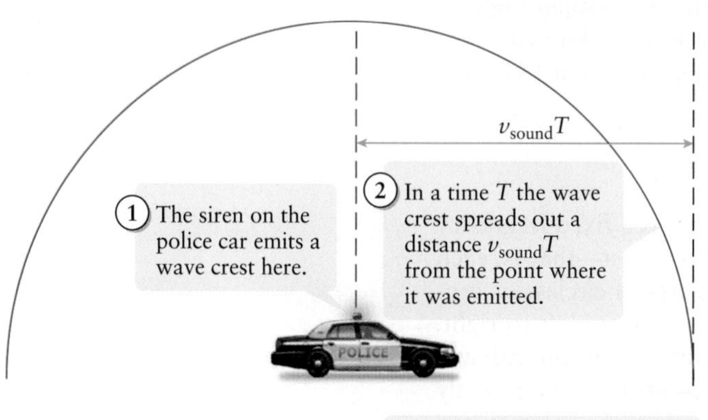

$v_{sound}T$
1 The siren on the police car emits a wave crest here.
2 In a time T the wave crest spreads out a distance $v_{sound}T$ from the point where it was emitted.
POLICE
3 In the same time T the police car travels a distance $v_{car}T$.
$v_{car}T$
$\lambda_{listener}$

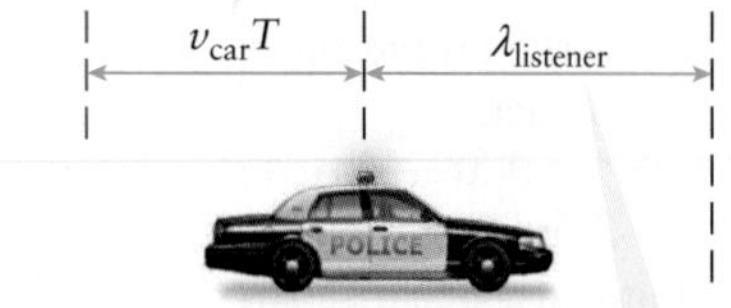

POLICE
Because of the car's motion, a listener in front of the police car hears a shortened wavelength.

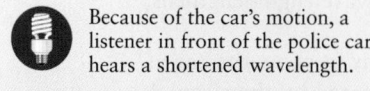

Figure 21-28 The Doppler effect: A moving source II The wavelength of the sound heard by a stationary listener as a source of sound approaches depends on the distance the wave travels in a certain time and the distance the source travels in that same time.

$$v_{car}T + \lambda_{listener} = v_{sound}T$$

or

$$\lambda_{listener} = v_{sound}T - v_{car}T \quad (21\text{-}16)$$

Although this equation mathematically describes the Doppler effect, by convention we write the relationship as one between the actual and the observed frequencies relative to the speed of the emitter rather than the speed of sound v. To write the expression we use $T = 1/f$ and Equation 21-2 for sound waves:

$$\lambda_{listener} = \frac{v_{sound}}{f_{listener}}$$

Combining these equations with Equation 21-16 gives

$$\frac{v_{sound}}{f_{listener}} = \frac{v_{sound}}{f} - \frac{v_{car}}{f}$$

or, rearranging,

$$f_{listener} = \left(\frac{v_{sound}}{v_{sound} - v_{car}}\right)f$$

Because v_{car} is a positive number, the fraction in parentheses must be greater than 1. So when the car is moving toward a stationary listener, $f_{listener} > f$ (the listener hears a higher frequency).

You can apply a similar approach to determine the observed frequency as the source moves away from a stationary listener, as well as for the cases in which the listener moves relative to a stationary source. Using *source* instead of *car* as a more general way to indicate the source of the sound, here's a single equation for the frequency heard by the listener, no matter how the source and listener are moving:

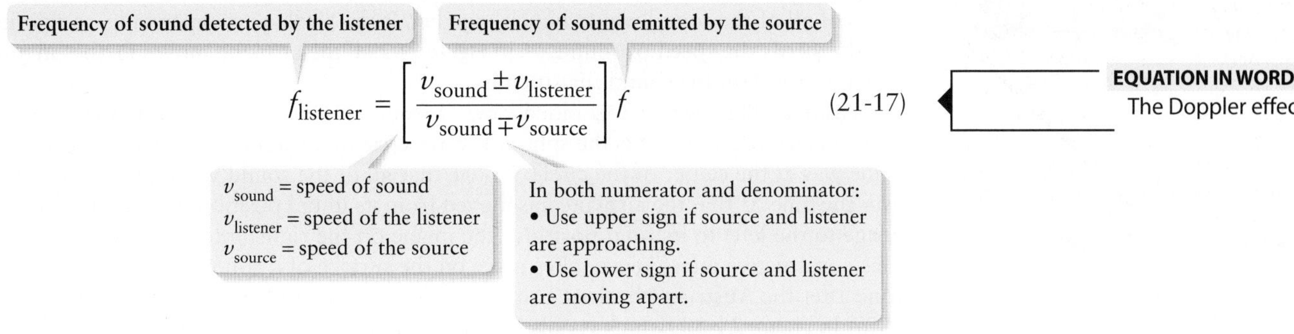

$$f_{listener} = \left[\frac{v_{sound} \pm v_{listener}}{v_{sound} \mp v_{source}}\right] f \qquad (21\text{-}17)$$

EQUATION IN WORDS
The Doppler effect

The speed of a moving object can be determined by measuring the shift in frequency associated with the Doppler effect. A common way to do this is to create a wave of known frequency and then bounce it off the moving object to be studied. A *radar speed gun* uses this principle: It sends out a *radio* wave of a known frequency and compares it to the frequency of the wave after it is reflected by a moving car. (Because this device uses radio waves rather than sound waves, v_{sound} in Equation 21-17 is replaced by the speed of light c.) Bats and dolphins use the same technique with sound waves to track the motion of prey.

Sound from a Supersonic Source

Something curious occurs when a source of sound moves faster than the speed of sound (that is, the source is *supersonic*). In **Figure 21-29a** a jet airplane sits on the ground before taking off; the concentric circles centered on the airplane represent the spherical wave crests of the sound it generates. The three drawings in parts (b), (c), and (d)

Crests of sound waves are expanding spheres, drawn as circles.

(a)

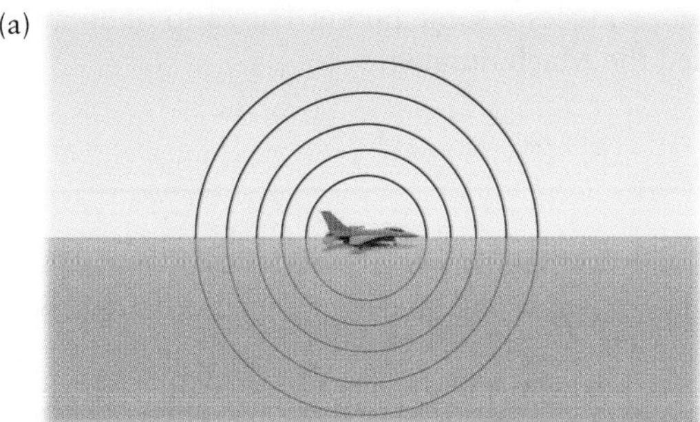

The airplane is stationary on the ground.

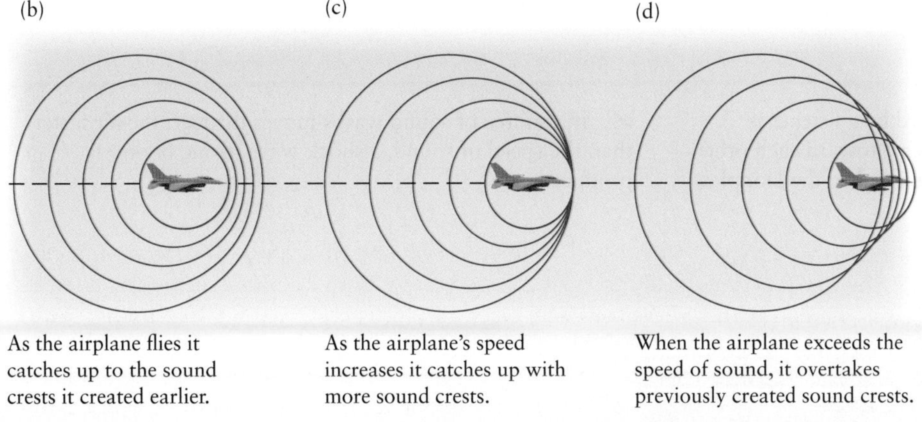

(b)
(c)
(d)
As the airplane flies it catches up to the sound crests it created earlier.
As the airplane's speed increases it catches up with more sound crests.
When the airplane exceeds the speed of sound, it overtakes previously created sound crests.

Figure 21-29 To the speed of sound—and beyond (a) Concentric circles centered on a stationary jet plane represent the spherical wave fronts of the sound it generates. As the speed of the plane increases after takeoff, the plane catches up to the sound wave fronts. (b) The plane is not moving as fast as the sound waves propagate. (c) When the plane moves at the speed of the sound waves, it just exactly catches up to the wave fronts. (d) When the speed of the plane exceeds the speed of sound, the plane moves farther in any time interval than the corresponding sound wave front.

of Figure 21-29 show the airplane flying at increasingly faster speeds. As the airplane increases its speed, it catches up to the sound it generated earlier in the motion, squeezing the wave crests closer together ahead of the airplane (**Figure 21-29b**). But notice that if the airplane moves at the same speed as the sound, as in **Figure 21-29c**, the airplane and the sound it generates travel the same distance in any time interval. In this case the crests of all of the sound waves bunch together. If the airplane is supersonic, so that its speed exceeds the speed of sound, as in **Figure 21-29d**, the airplane moves farther in any time interval than the sound that it emits.

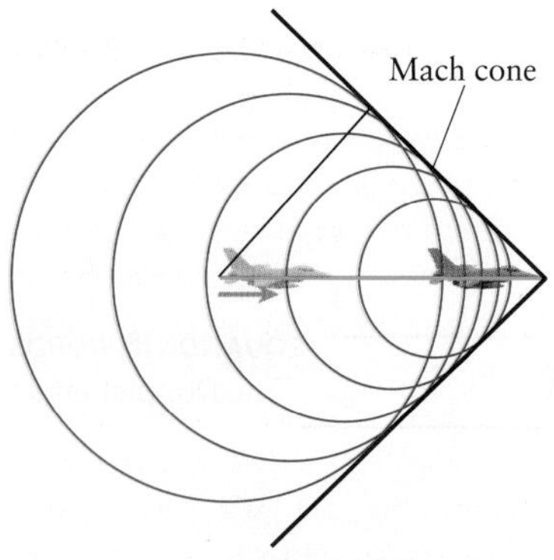

Figure 21-30 Faster than sound In an expanded view of a plane flying faster than the speed of sound, each circle represents the spherical wave crest of a sound generated when the plane was at the center of the circle. The crests of all of the sound waves generated during the time that the plane has moved from its initial position to its final position interfere constructively on the surface of the Mach cone.

Figure 21-30 shows an expanded view of what happens when the airplane is supersonic. Each circle represents the spherical wave crest of a sound generated when the airplane was at the center of the circle. Notice that all of the sound wave crests generated over the time Δt that the airplane has moved from its initial position (the lighter-colored image to the left) to its final position (the image on the right) interfere constructively. This constructive interference takes place on the surface of a cone, known as the **Mach cone** after the Austrian physicist Ernst Mach, who first explained this phenomenon in 1877. In Figure 21-30 (which is only two dimensional), we represent the cone by the two black lines. Note that there is *no* sound to the right of the Mach cone in Figure 21-30. An observer to the right of the cone will be able to see the airplane but will not be able to hear it: No sound from the airplane will have reached him yet.

Because all of the wave crests pile up at the Mach cone, this is a region of substantially higher pressure than the surrounding air. If you are standing on the ground when a supersonic airplane flies over, you experience this as a **shock wave** (so called because there is no advance warning that it is coming) or as a **sonic boom**. The pressure increase in a shock wave from an airplane is in the range of 5 to 50 Pa. The shock wave is relatively narrow in extent, as you can see in **Figure 21-31**. This is a photograph of an X-15 aircraft model in a wind tunnel with air moving at more than twice the speed of sound. In the figure a number of separate shock waves, each originating from various discontinuities on the airplane's surface, are made evident by a photographic technique that reveals regions of differing air density. Notice that each of these regions is only a fraction of the length of the airplane. For this reason you would hear a sharp, explosive sound from each shock wave as this aircraft passed overhead.

Figure 21-31 Shock waves This X-15 aircraft model shadowgraph shows airflow shock waves in a supersonic wind tunnel. Data taken in these wind tunnels (up to Mach 6.8) indicate shock waves that originate from various discontinuities on its surface. The rocket-powered X-15 was a true spacecraft, but it was designed to aid engineers and scientists in putting into practice what they had learned from such wind tunnels and small-scale models.

Note that the sonic boom is *not* the amplified sound of the aircraft engine. As Figure 21-31 shows, shock waves spread out from several parts of the airplane and are not associated with the engine at all. If an airplane is still flying supersonically after it runs out of fuel, it still produces a sonic boom. The ratio of an airplane's speed to the speed of sound is called the **Mach number**.

WATCH OUT !

The Mach cone moves along with the object that creates it.

Although the individual sound waves that go into making a sonic boom travel at the speed of sound, the Mach cone on which the crests pile up moves along with the supersonic airplane. For example, the sonic boom associated with a jet airplane flying horizontally at Mach 1.4 also moves horizontally at 1.4 times the speed of sound.

THE TAKEAWAY for Section 21-8

✔ The frequency of a sound wave heard by a listener is higher if the listener and source are moving toward each other and lower if they are moving apart. This is called the Doppler effect.

✔ If a source of sound waves moves through the air faster than the speed of sound, a shock wave (sonic boom) is produced.

Prep for the AP® Exam

AP® Building Blocks

1. A person sitting in a parked car hears an approaching ambulance siren at a frequency f, and as it passes him and moves away he hears a frequency f'. The actual frequency of the source is
   (A) greater than f' but less than f because the perceived period of the tone emitted by the ambulance is greater as the ambulance approaches and smaller as it moves away.
   (B) greater than f but less than f' because the perceived period of the tone emitted by the ambulance is smaller as the ambulance approaches and greater as it moves away.
   (C) greater than f' but less than f because the perceived period of the tone emitted by the ambulance is smaller as the ambulance approaches and greater as it moves away.
   (D) greater than f but less than f' because the perceived period of the tone emitted by the ambulance is greater as the ambulance approaches and smaller as it moves away.
2. The fundamental frequency of a whistle is f_1. A sound wave is produced by the whistle from a train in motion. Use complete sentences to predict (and justify the prediction of) the relationship of the frequency, f_{listener}, as it is perceived by a listener to f_1 from a
   (a) stationary coordinate frame at the side of the track as the train approaches.
   (b) stationary coordinate frame at the side of the track as the train recedes.
   (c) coordinate frame moving along the track behind the train at the same speed as the train.
   (d) coordinate frame fixed on the train.
3. A car sounding its horn (rated by the manufacturer at 600 Hz) is moving at 20 m/s toward the east. A stationary observer is standing due east of the oncoming car.
   (a) What frequency will he hear, assuming that the speed of sound is 343 m/s?
   (b) What if the observer is standing due west of the car as it drives away?
4. Sonar was used as a military technology aboard torpedo boat destroyers during World War II to locate torpedo boats (submarines) that were invisible under water. Sound waves are emitted by a device on the ship, and reflected by an object in the path of the wave. By measuring the time difference between sending the wave and receiving the reflected wave, and using the speed of sound in seawater, the distance from a submarine to the destroyer could be determined.
   (a) A sound wave transmitted from a stationary source and reflected by a stationary target returns to the source in a time Δt. If the wave travels only in a medium where the speed of sound is v_p, then the distance Δx between the stationary source and target can be calculated as $2\Delta x = v_p \Delta t$. If neither the source nor the target is stationary, describe how the distance could be calculated if the relative velocities of the source and target were known ($v_{\text{target}} - v_{\text{source}}$). For simplicity assume that $v_{\text{target}} > v_{\text{source}}$. Finally, describe why the information provided by the Doppler shift is needed when the relative velocities of the source and target are not known.
   (b) A sound wave with frequency f_{sent} is transmitted by the destroyer, and the frequency of the reflected sound wave is f_{received}. Suppose the submarine is moving away from the destroyer along the line joining the destroyer and the submarine with a speed v_S, and the destroyer is moving toward the submarine along the same line with a smaller speed v_D. The frequency of the sound wave arriving at the submarine does not change upon reflection from the submarine. Using Equation 21-17, express the frequency of the sound wave received by the destroyer in terms of the frequency of the sound wave sent by the destroyer and the speed of sound in seawater.

   Suppose that the submarine and destroyer are moving toward each other with speeds v_S and v_D.
   (c) Express the frequency received by the destroyer in terms of the speed of sound in seawater, the speed of the submarine, and the speed of the destroyer.
   (d) Calculate the percent change in the ratio $(f_{\text{received}} - f_{\text{sent}})/f_{\text{sent}}$ for the situations in parts (b) and (c) when the speed of the submarine is 40 m/s and the speed of the destroyer is 20 m/s. The speed of sound in seawater is 1500 m/s.
   (e) The total path length of the transmitted and received waves can be determined by the time interval Δt between an emitted wave and a received wave. But it is also important to know the distance from the surface, the depth, of the submarine. Use the following diagram to describe how a microphone towed beneath the water from the destroyer can be used to determine the depth of the submarine.

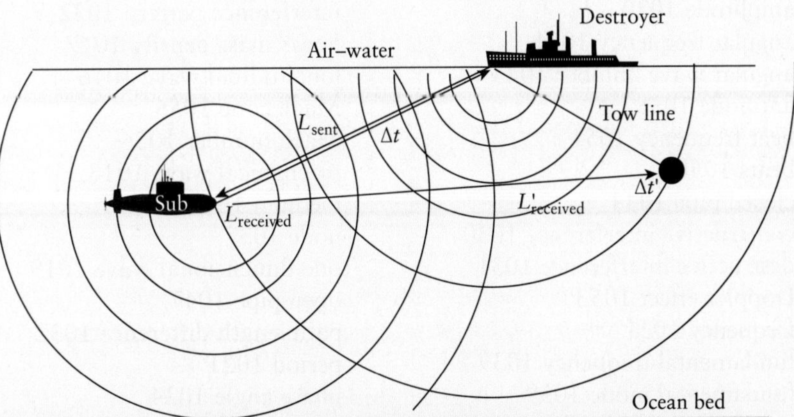

5. A bicyclist is moving toward a wall while holding a ringing tuning fork rated at 484.0 Hz. If the bicyclist detects a beat frequency of 6.0 Hz (between the waves coming directly from the tuning fork and the echo waves coming from the wall), calculate the speed of the bicycle (the speed of the source). Assume the speed of sound is 343.0 m/s. The frequency of the sound wave does not change when it is reflected by the wall.

WHAT DID YOU LEARN?

Prep for the AP Exam

Chapter learning goals	Section(s)	Related example(s)	Relevant section review exercise(s)
Describe what a mechanical wave is.	21-1		1
Explain the key properties of transverse, longitudinal, and surface waves, and create representations of transverse and longitudinal waves.	21-2		1
Define the relationship between simple harmonic motion and what happens in a sinusoidal wave.	21-3		1
Explain the relationship between periodic wave speed, wavelength, and frequency, and relate these to everyday examples.	21-3	21-1, 21-2	3, 4, 5, 6
Describe what happens when two wave pulses interfere with each other.	21-4	21-3	1, 2, 4
Describe sound in terms of transfer of energy and momentum in a medium and relate the concepts to everyday examples. Predict the properties of standing sound waves in open and closed pipes or on strings with fixed or unfixed ends.	21-1 21-5 21-6	21-4 21-5	1 1, 2 1, 2, 5
Describe how beats arise from combining two sound waves of slightly different frequencies.	21-7	21-6	1, 2, 3
Create a visual representation to demonstrate how the observed frequency of a sound changes if the source and listener are moving relative to each other.	21-8		1, 2

Chapter 21 Review

Key Terms

All the Key Terms can be found in the Glossary/Glosario on page G1 in the back of the book.

amplitude 1020
angular frequency 1021
angular wave number 1025
antinode 1038
beat frequency 1051
beats 1050
closed pipe 1044
constructive interference 1030
destructive interference 1031
Doppler effect 1053
frequency 1021
fundamental frequency 1039
fundamental mode 1039
interference 1031
interference pattern 1032
linear mass density 1017
longitudinal wave 1016
Mach cone 1056
Mach number 1056
mechanical wave 1013
medium 1013
node 1037
one-dimensional wave 1019
open pipe 1047
path length difference 1032
period 1021
phase angle 1024
pressure wave 1016
propagation speed 1022
shock wave 1056
sinusoidal wave 1019
sonic boom 1056
standing wave 1037
standing wave mode 1037
surface wave 1016
transverse wave 1015
traveling wave 1037
wave 1012
wave function 1024
wavelength 1020
wave pulse 1012

Chapter Summary

Topic	Equation or Figure

Mechanical waves: A mechanical wave is a disturbance that propagates through a material medium. There is no net flow of material through the medium. A wave transfers momentum and energy. Mechanical waves can be transverse, longitudinal, or a combination of the two.

(a)

① If the hand holding the rope moves up and down, a **transverse wave** travels along the rope.

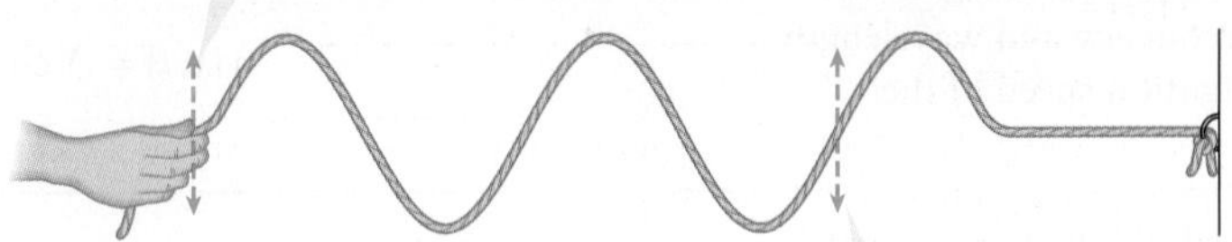

② The wave disturbance propagates (moves along the rope) horizontally...

③ ...and individual parts of the rope (the wave medium) move up and down, perpendicular (transverse) to the propagation direction.

(b)

① If the hand holding the spring moves back and forth, a **longitudinal wave** travels along the spring.

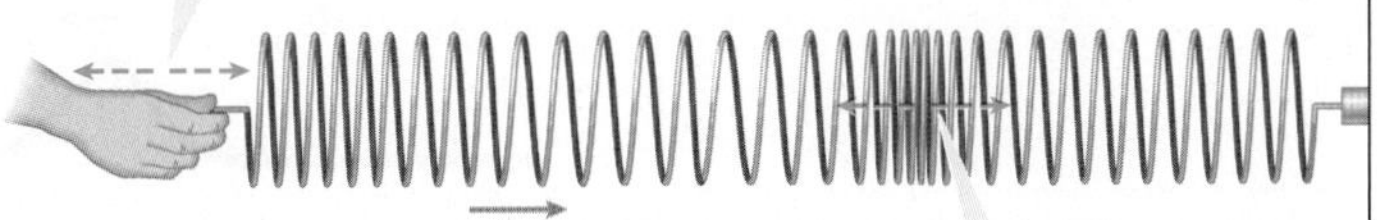

② The wave disturbance propagates (moves along the spring) horizontally...

③ ...and individual parts of the spring (the wave medium) move back and forth, parallel to the propagation direction (longitudinal).

(Figure 21-3)

(a)

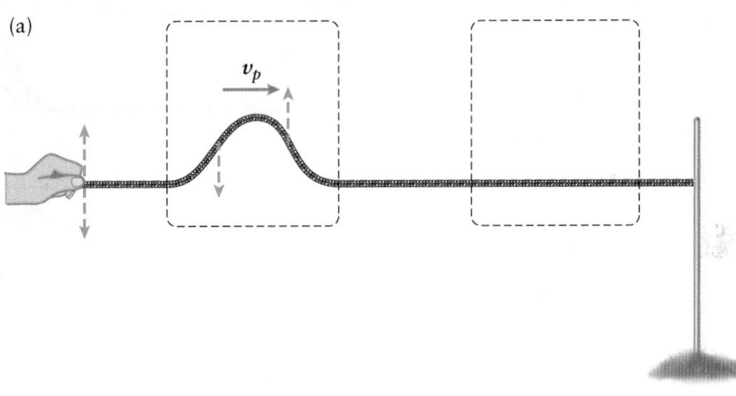

(b)

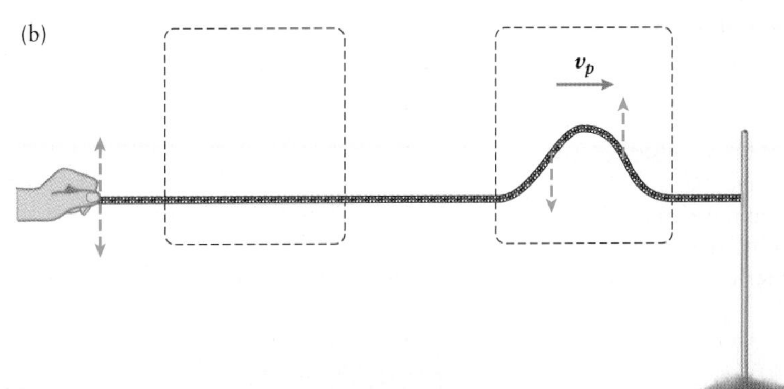

(Figure 21-2)

Sinusoidal waves: In a sinusoidal wave, each piece of the medium undergoes simple harmonic motion with the same amplitude and frequency. At any instant the wave pattern repeats itself over a distance known as the wavelength. The product of the frequency and wavelength equals the propagation speed of the wave.

Propagation speed of a wave

$$v_p = f\lambda \qquad (21\text{-}2)$$

Frequency **Wavelength**

$$y(x,t) = A\cos(kx - \omega t + \phi) \qquad (21\text{-}6)$$

Propagation speed of a wave: The speed at which a mechanical wave travels through a medium depends on the type of wave and the properties of the medium, increasing with restoring force and decreasing with mass density.

(a)

① Two wave pulses approach each other oriented in the same direction perpendicular to the rope.

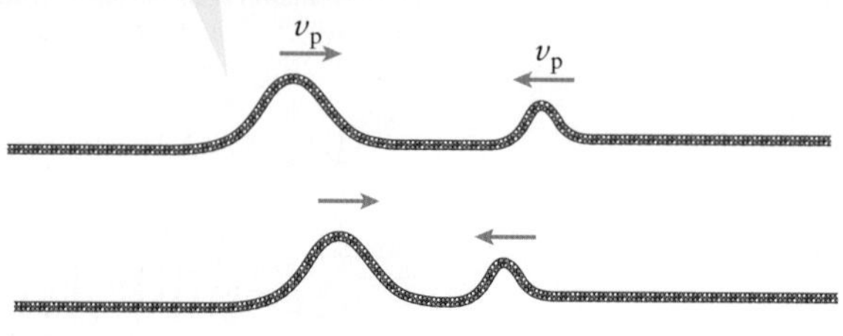

② Dotted lines show the shape of original wave pulses.

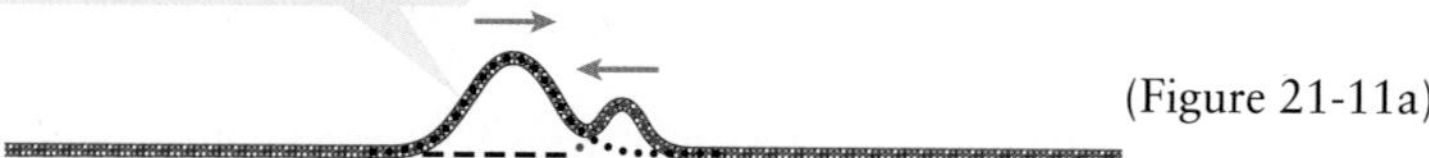

(Figure 21-11a)

③ Original wave pulses add as they occupy the same space...

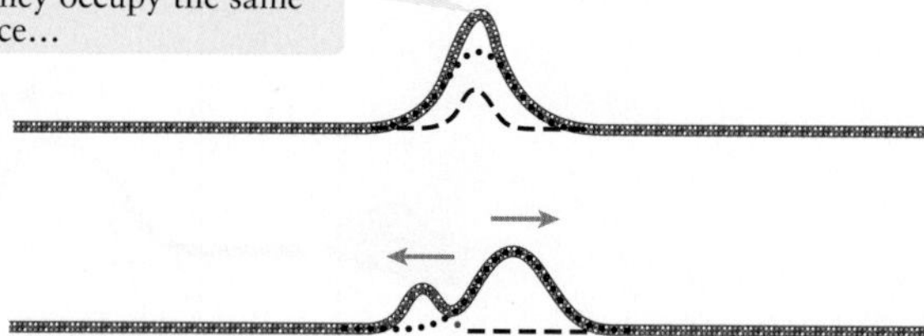

④ ...but pass through each other without changing shape.

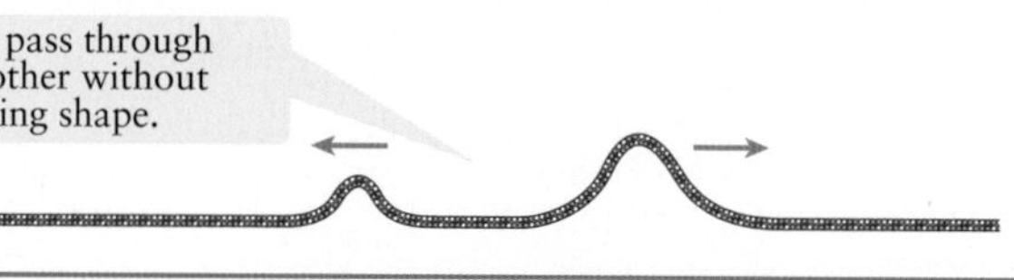

Superposition and interference: When two wave pulses or waves are present simultaneously in a medium, the total wave pulse or wave is the sum of the two individual wave pulses or waves. If two waves have the same frequency but emanate from different places, there will be positions where constructive and destructive interference occur.

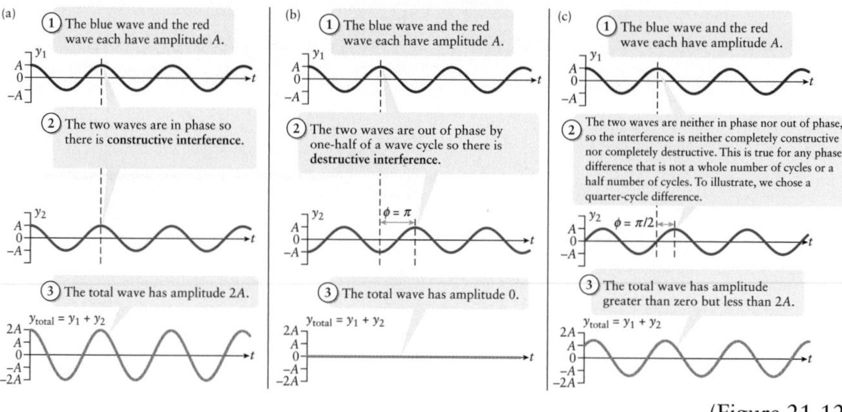

(Figure 21-12)

Standing waves on a string: When sinusoidal waves reflect back and forth from the ends of a string, the result can be a standing wave. Standing waves are possible only if a whole number of half-wavelengths fit into the length of the string. Each standing wave mode has its own natural frequency.

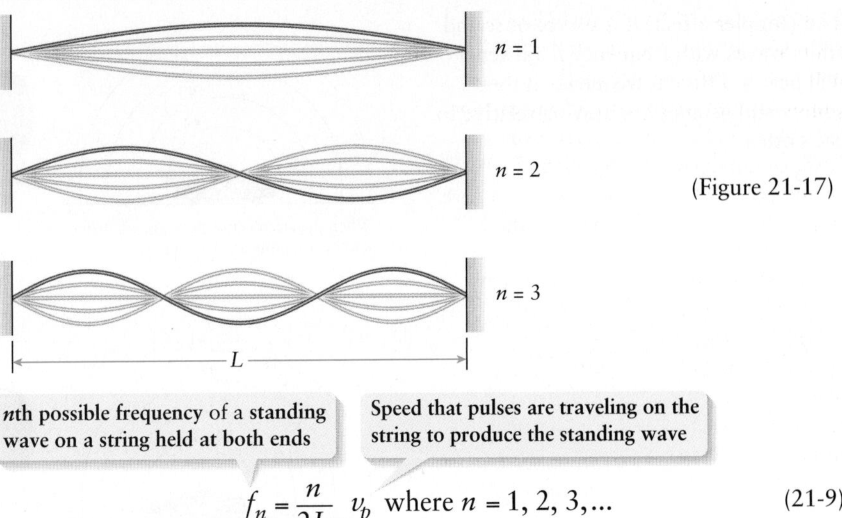

(Figure 21-17)

***n*th possible frequency** of a standing wave on a string held at both ends

Speed that pulses are traveling on the string to produce the standing wave

$$f_n = \frac{n}{2L} v_p \text{ where } n = 1, 2, 3, \ldots \tag{21-9}$$

Length of the string

Standing sound waves in a pipe: Standing waves can also occur when sinusoidal sound waves reflect back and forth from the ends of a pipe. The allowed wavelengths and frequencies depend on whether the pipe is closed (closed at one end and open at the other) or open (open at both ends).

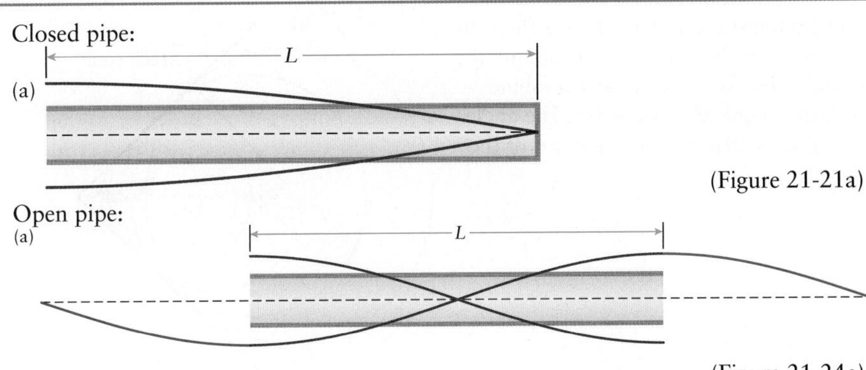

(Figure 21-21a)

(Figure 21-24a)

Beats: When two sound waves of similar frequencies f_1 and f_2 interfere, the result is a total wave with a frequency $(f_1 + f_2)/2$ and an amplitude that rises and falls at the beat frequency $|f_2 - f_1|$.

There are four pressure pulses in the top wave from one place where they are "in step" until the next.

The waves are "in step" at the times indicated by the blue lines, and at those times add constructively. In between these points the waves still add, but because there is only ever a maximum for one, the sound in between the times denoted by the blue lines will be less.

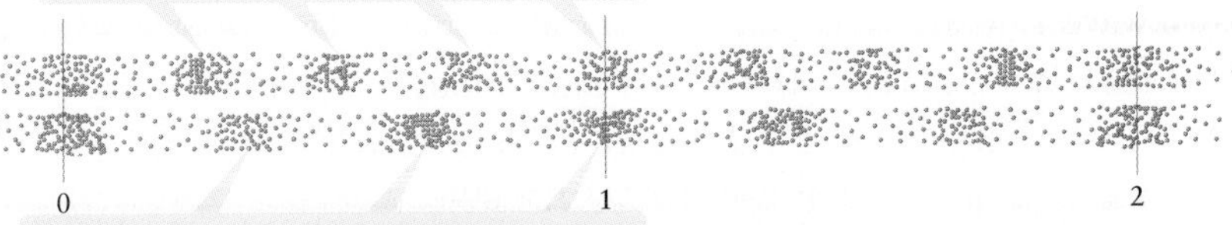

There are three pressure pulses in the bottom wave from one place where they are "in step" until the next.

(Figure 21-25)

Beat frequency heard when sound waves of two similar frequencies interfere

Frequencies of the individual sound waves

$$f_{\text{beats}} = |f_2 - f_1| \tag{21-15}$$

The Doppler effect: If a source of sound emits waves with frequency f, a listener will hear a different frequency if the source and listener are moving relative to each other.

(a)

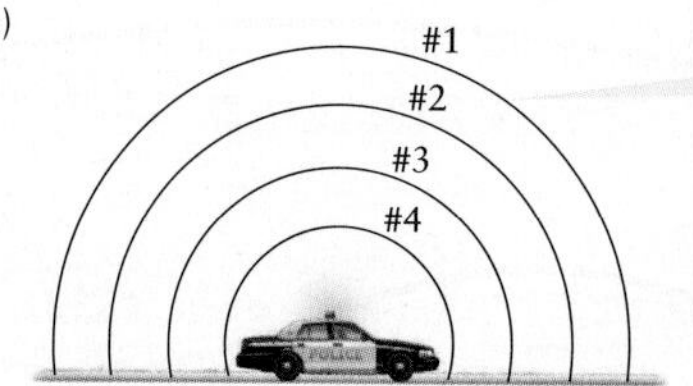

When the siren is not moving, a listener will hear a constant frequency.

Listener

Wave crests spread outward from the siren. The numbers on the crests show the sequence in which they were emitted: First #1, then #2, then #3, then #4, so the most recently emitted crest is closest to the siren.

(b)

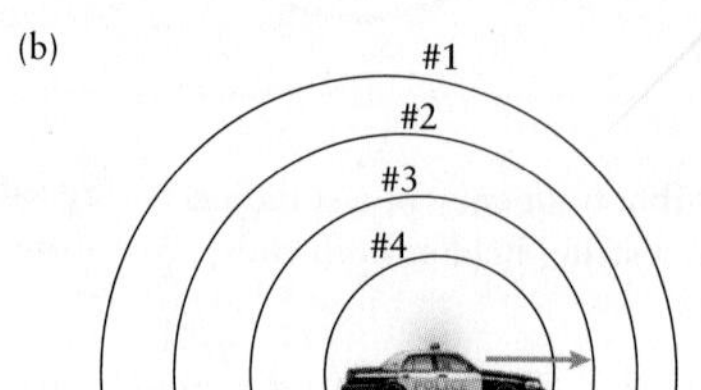

As the siren approaches the listener, the sound frequency is higher.

Listener

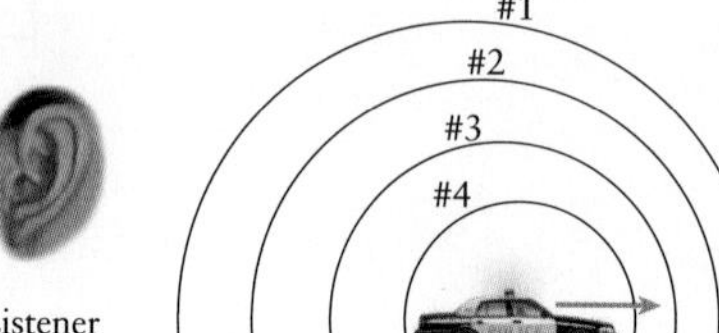

The frequency is lower as the siren moves away from the listener.

(Figure 21-27)

Sonic booms: If an object moves through the air faster than the speed of sound, the sound waves from the object combine to form a conical shock wave. The angle of this cone depends on the speed of the object.

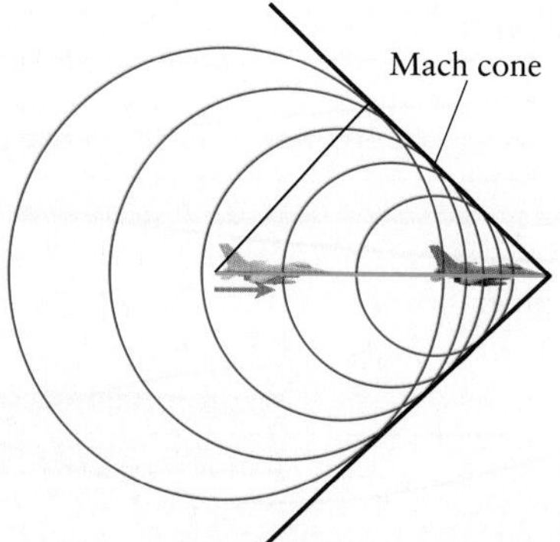

(Figure 21-30)

Chapter 21 Review Problems

1. Show that the dimensions of speed (distance/time) are consistent with the expression for the propagation speed of a wave: $v_p = \lambda f$.
2. Use the relationship $v_p = \lambda f$ for these calculations.
   (a) A wave is propagating with a speed of 10 m/s and a frequency of 2 Hz. Calculate the wavelength.
   (b) A wave is propagating with a frequency of 3 Hz and a wavelength of 0.2 m. Calculate the wave speed.
   (c) A wave is propagating with a wavelength of 0.2 m and a wave speed of 12 m/s. Calculate the frequency.
3. Describe the difference between a transverse and a longitudinal wave in terms of the displacement of the medium relative to the direction of propagation of the wave.
4. Write the wave equation, using Equation 21-4, for a periodic transverse wave traveling in the positive x direction at a speed of 20 m/s if it has a frequency of 10 Hz and an amplitude in the y direction of 0.10 m.

$t = 0$ s:

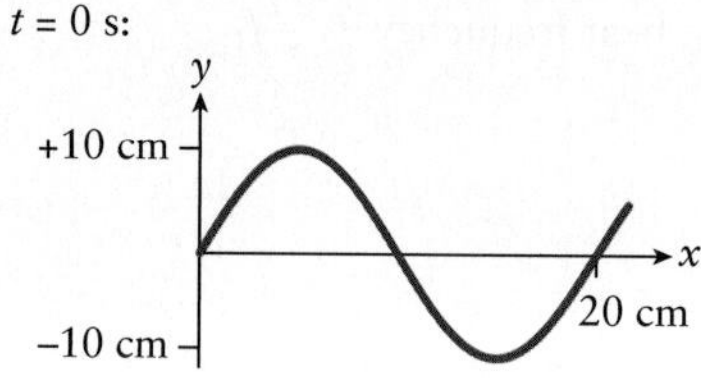

$x = 0$ m:

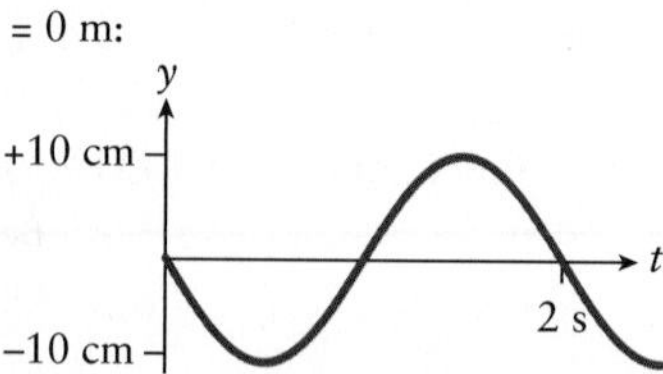

5. Using Equation 21-4, write a mathematical description of the wave represented by the graphs in the figure.
6. A transverse wave is propagating according to the following wave function:

$$y(x,t) = (1.25 \text{ m}) \cos [(5.00 \text{ rad/m})x - (4.00 \text{ rad/s})t]$$

(a) Plot a graph of y versus x when $t = 0$ s. Your graph should at least cover the range $-2\text{ m} \le x \le 2\text{ m}$.

(b) Repeat when $t = 1.00$ s.

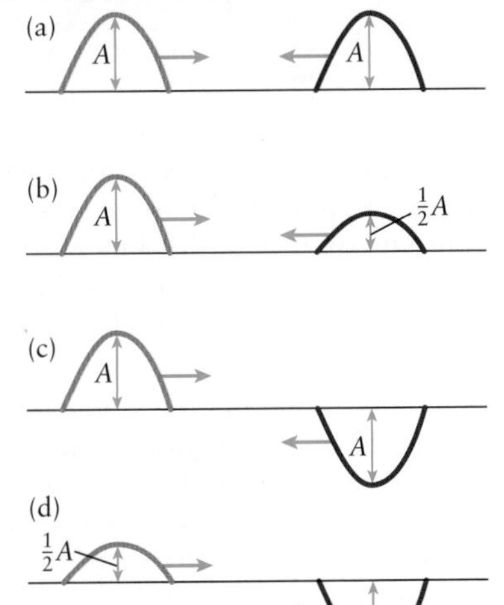

7. Describe the resultant pulse that is formed when the two pulses shown in each of the four cases occupy the same space and interfere. The pulses have the same width.

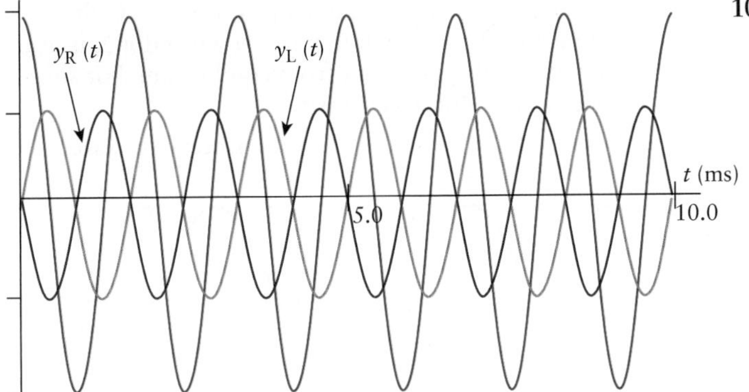

8. Two loudspeakers are placed facing each other. The speakers produce a single tone and oscillate in phase. A microphone is placed at the midpoint, $x = 0$, with the speaker on the right at $x = d$ and the speaker on the left at $x = -d$. The microphone receives the sound wave shown by the red line in the figure. The blue and green lines represent the individual sound waves that would be recorded when the microphone is moved slightly.

(a) What are the frequency and wavelength of the tone that the speakers emit? Use 343 m/s for the speed of sound.

(b) Explain why there is only a single wave received at the midpoint.

The microphone is then displaced to the right by Δx, so that $x_R = d - \Delta x$, and the distance to the left speaker is now $x_L = d + \Delta x$. The speakers are turned on one at a time. The microphone now receives the sound wave $y_R(t)$, shown in blue, from the speaker on the right when only it is on, and the wave $y_L(t)$, shown in green, from the speaker on the left when only it is on.

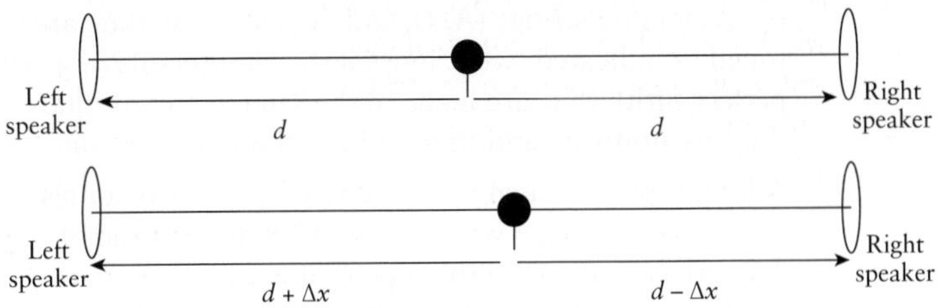

(c) In terms of the wavelength, what is the displacement Δx toward the speaker on the right?

9. The Wanamaker Pipe Organ in Macy's Center City department store in Philadelphia is the largest working pipe organ in the world. The organ occupies a seven-story hall, and has the longest pipe length of 9.75 m that occupies four floors and can be used as either an open or a closed pipe. Assume that the speed of sound is 345 m/s.

(a) Calculate the lowest fundamental frequency for the Wanamaker Pipe Organ.

(b) The highest note on an 88-key piano is 4186 Hz. What would be the length of a pipe open on one end that produced this as a fundamental frequency?

10. It has been proposed that alligators bellow to communicate their body size. A mating call is a complex sound wave but scientists analyzed alligator bellows to detect a dominant frequency of 120 Hz by collecting data from several individual alligators. Actual vocalizations contain sound waves with frequencies larger and smaller than the dominant frequency.

(a) Model the chest cavity as a pipe closed on only one end to estimate the range of lengths for the primary and third harmonic frequencies f_1 and f_3, based on a single effective wavelength corresponding to the dominant frequency. Although the air is warm and damp in an alligator's lungs, approximate the speed as that of sound in dry air, 345 m/s.

(b) Adult alligators are between 3 and 4.5 m in length and roughly half of that is the tail. Test your model by comparison with the length of adult alligators.

11. Two tuning forks of frequency 480 and 484 Hz are struck simultaneously. What is the beat frequency resulting from the two sound waves?

(A) 964 Hz because the beat frequency is the sum of the frequencies of each tone

(B) 482 Hz because the beat frequency is the average of the frequencies of each tone

(C) 4 Hz because the beat frequency is the difference of the frequencies of each tone

(D) 2 Hz because the beat frequency is the difference between the frequency of each tone and the average of the frequencies

12. The violin is a four-stringed instrument tuned so that the ratio of the frequencies of adjacent strings is 3 to 2. (This is the ratio when taken as high frequency to lower

frequency.) If the diameter of the E string (the highest frequency) on a violin is 0.25 mm, find the diameters of the remaining strings (A, D, and G), assuming they are tuned as indicated (what musicians call intervals of a perfect fifth), they are made of the same material, that they are uniform, and they all have the same tension.

13. A long rope is shaken up and down by a rodeo contestant. The transverse waves travel 12.8 m in a time of 2.1 s. If the tension in the rope is 80.0 N, calculate the mass per unit length for the rope.

14. Two tuning forks are both rated at 256 Hz, but when they are struck at the same time, a beat frequency of 4 Hz is created. If you know that one of the tuning forks is in tune (but you are not sure which one), what are the possible values of the "out-of-tune" fork?

15. A guitar string has a tension of 1.00×10^2 N and is supposed to have a frequency of 1.10×10^2 Hz. When a standard tone of that value is sounded while the string is plucked, a beat frequency of 2.00 Hz is heard. The peg holding the string is loosened (decreasing the tension), and the beat frequency increases. What should the tension in the string be in order to achieve perfect pitch?

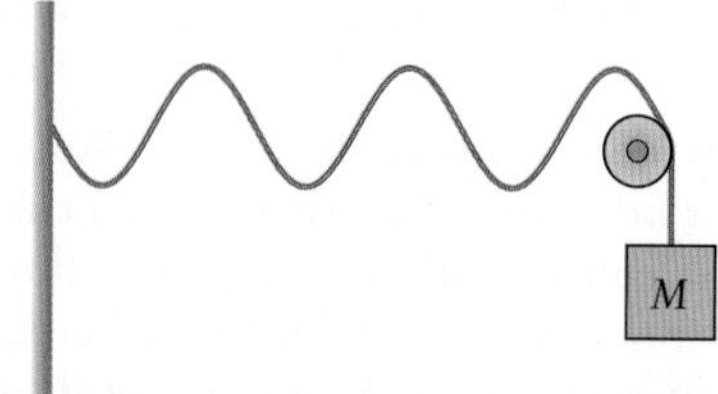

16. An object of mass M is used to provide tension in a 4.50-m-long string that has a mass of 0.252 kg, as shown in the figure. Assume that the string is uniform. A standing wave that has a wavelength equal to 1.50 m is produced by a source that vibrates at 30.0 Hz. Calculate the value of M.

17. A guitar string has a mass per unit length of 2.35 g/m. If the string is vibrating between points that are 60.0 cm apart, find the tension when the string is designed to play a note of 440 Hz (A_4).

18. An E string on a violin has a diameter of 0.25 mm and is made from steel (density of 7800 kg/m^3). The string is designed to sound a fundamental note of 660 Hz, and its unstretched length is 32.5 cm. Assume that the string is uniform. Calculate the tension in the string.

19. The trunk of a very large elephant may extend up to 3 m! It acts much like an organ pipe open only at one end when the elephant blows air through it.

(a) Calculate the fundamental frequency of the sound the elephant can create with its trunk.

(b) If the elephant blows even harder, the next harmonic may be sounded. What is the frequency of this first overtone?

20. Two dots have been painted on a light string at points labeled X and Y in the following figures. A transverse wave travels to the right along the string.

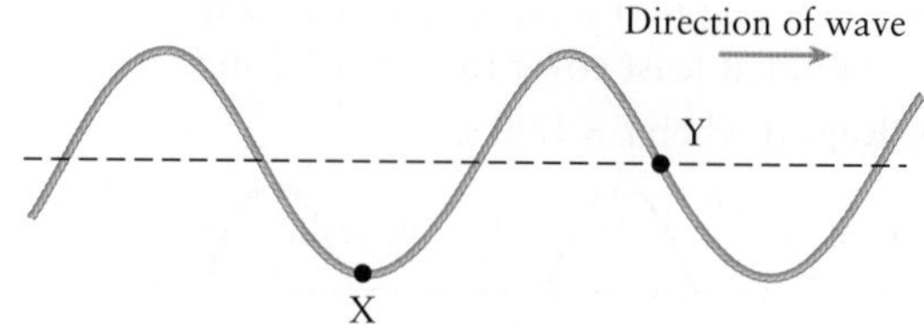

(a) At the instant in time shown, dot X has maximum displacement and dot Y has zero displacement from equilibrium. At each of the dots X and Y, draw an arrow indicating the direction of the instantaneous velocity of that dot. If either dot has zero velocity, write "$v = 0$" next to the dot.

(b) The figure below shows the string at the same instant as shown in part (a). At each of the dots X and Y, draw an arrow indicating the direction of the instantaneous acceleration of that dot. If either dot has zero acceleration, write "$a = 0$" next to the dot.

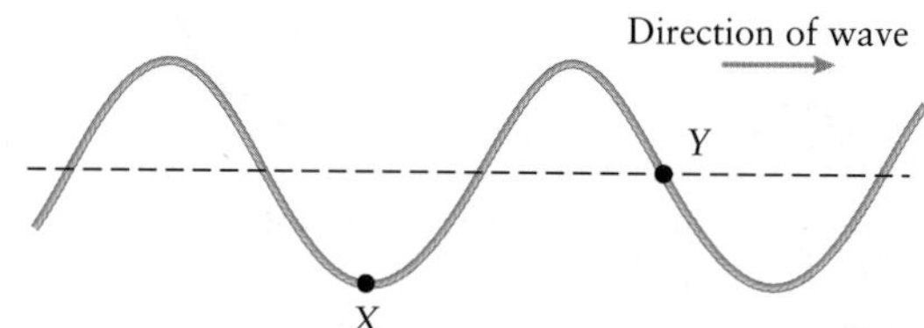

The graph represents the string at time $t = 0$, the same instant as shown in part (a) when dot X is at its maximum displacement from equilibrium. For simplicity, only dot X is shown.

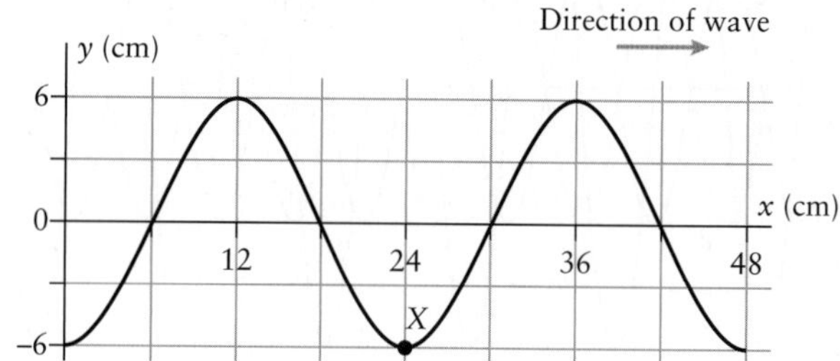

(c) Determine the distance traveled (not the displacement) by dot X during one full cycle of the wave (over a time interval of the period, T).

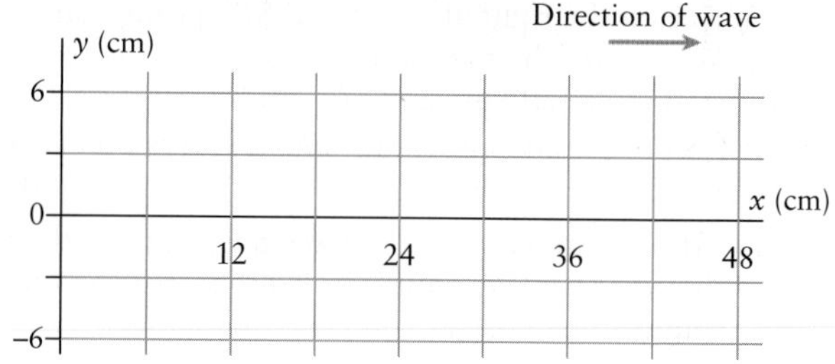

(d) Using the grid above, draw the string at a later time $t = T/4$, where T is the period of the wave. Draw a dot to indicate the position of dot X on the string at time $t = T/4$ and clearly label the dot with the letter X.

21. The third harmonic of an organ pipe that is open at both ends and is 2.25 m long excites the fourth harmonic in another organ pipe. Determine the length of the other pipe and whether it is open at both ends or open at one end and closed at the other. Assume the speed of sound is 343 m/s in air.

Prep for the AP® Exam

AP® Group Work

Directions: The problem below is designed to be done as group work in class.

The human ear canal has a typical length of about 2.2 cm and can be modeled as a tube open at one end and closed at the eardrum.

(a) Calculate the fundamental frequency that humans should be expected to hear best. Assume that the speed of sound is 345 m/s in air.

Looking deeply down your friend's throat, or perhaps looking at a photograph, you observe the vocal tract. At the bottom of the vocal tract, when you aren't swallowing, are membranous folds that close and open to modify the pressure at the vocal folds.

(b) If the vocal tract with closed vocal folds is modeled as a pipe closed only on one end, predict the length of the vocal tract corresponding to the fundamental frequency for hearing just calculated in part (a).

(c) Calculate the fundamental frequency of the human voice if $L = 17$ cm.

AP® PRACTICE PROBLEMS

Prep for the AP® Exam

Multiple-Choice Questions

Directions: The following questions have a single correct answer.

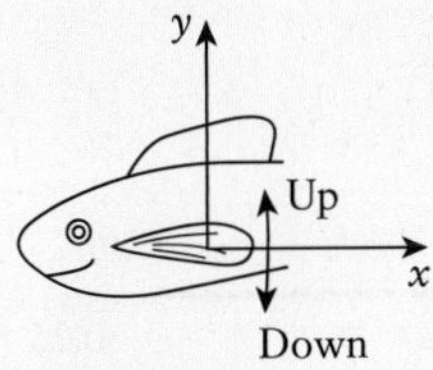

1. Deep under water, a fish moves the fins at its sides in an up-down circular motion, as shown in the figure above. Of the following choices, which one best describes the waves transmitted through the water at a significant distance from the fish?

	In the x direction	In the y direction
(A)	Transverse	Transverse
(B)	Longitudinal	Longitudinal
(C)	Transverse	Longitudinal
(D)	Longitudinal	Transverse

2. Resonance occurs in a tube closed at one end. If the speed of sound is taken as 300.0 m/s, which of the following could be resonance frequencies for a tube that is 0.50 m long?

(A) 75 Hz

(B) 300 Hz

(C) 600 Hz

(D) 750 Hz

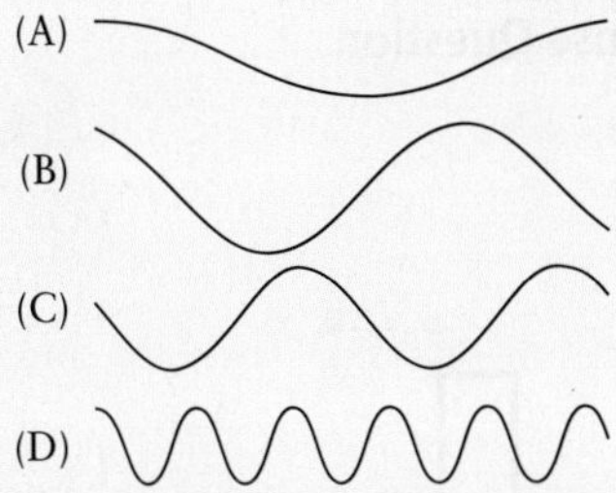

3. The figure above shows profiles indicating the periodic change in pressure for four sound waves. Which of these waves is transmitting the most energy?

(A) A

(B) B

(C) C

(D) D

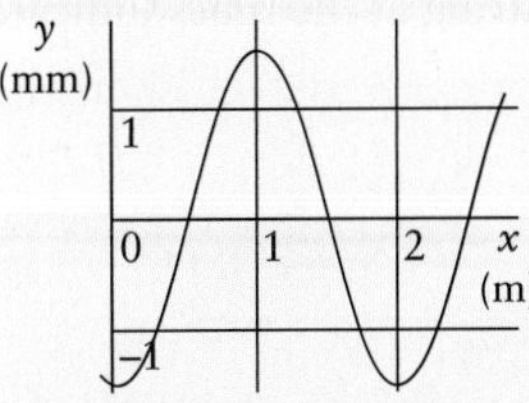

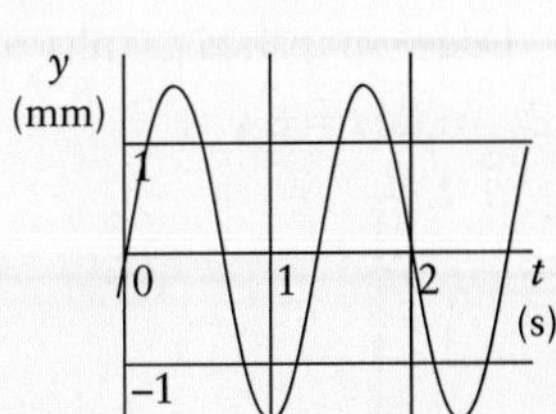

4. The two graphs above represent the same wave. Calculate the speed of the wave.

(A) 1.0 m/s

(B) 1.6 m/s

(C) 1.2 m/s

(D) 2.5 m/s

(Continued)

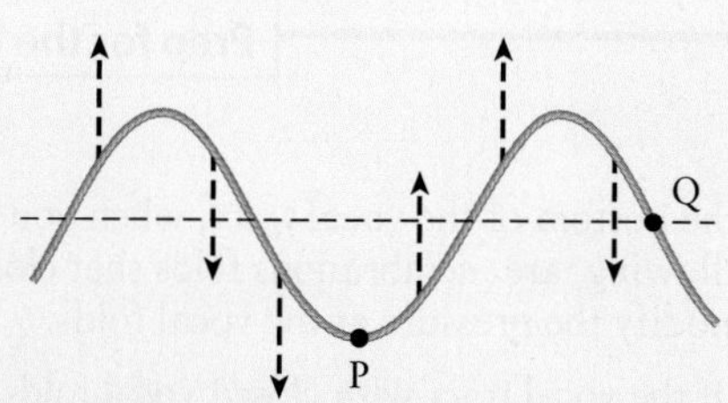

5. The figure above shows a portion of a periodic wave on a light string at a particular instant in time. The vertical arrows indicate the direction of the velocity of some of the points on the string. The wave is moving
   (A) to the left.
   (B) to the right.
   (C) in neither direction; the wave is a standing wave so it is not moving.
   (D) in either direction; the figure is consistent with wave motion to the right or to the left.

6. Which of the following best describes the situation when music is produced by a bowed string instrument, such as a violin? (In each case, the wave in the string actually causes the body of the violin to vibrate, which then causes the wave in the air; but for this question, just consider the waves on the string and in the air.)
   (A) A longitudinal wave is created on the string, which in turn produces a transverse wave in the air.
   (B) A longitudinal wave is created on the string, which produces a longitudinal wave in the air.
   (C) A transverse wave is created on the string, which in turn produces a longitudinal wave in the air.
   (D) A transverse wave is created on the string, which in turn produces a transverse wave in the air.

Free-Response Question

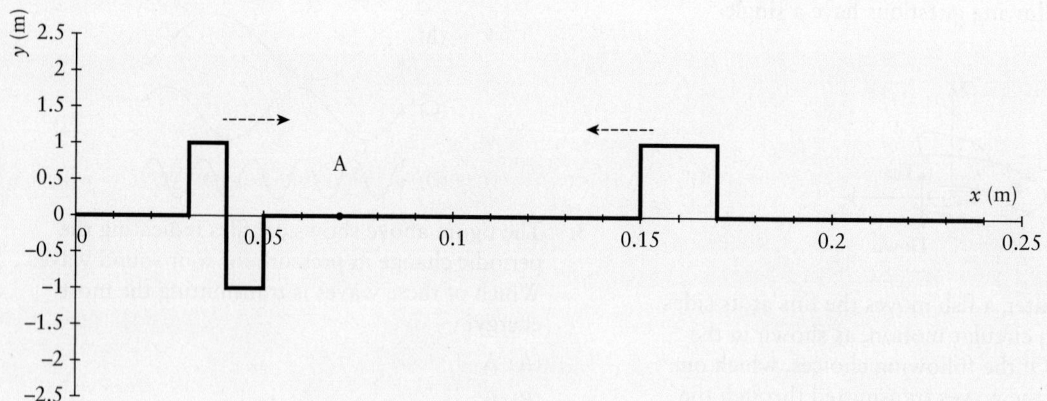

Two pulses on a string are approaching each other at 1 cm/s. At time $t = 0$ they are in the locations shown in the figure above.

Draw snapshot graphs of the string at the times indicated above each graph.

(a) $t = 6$ s

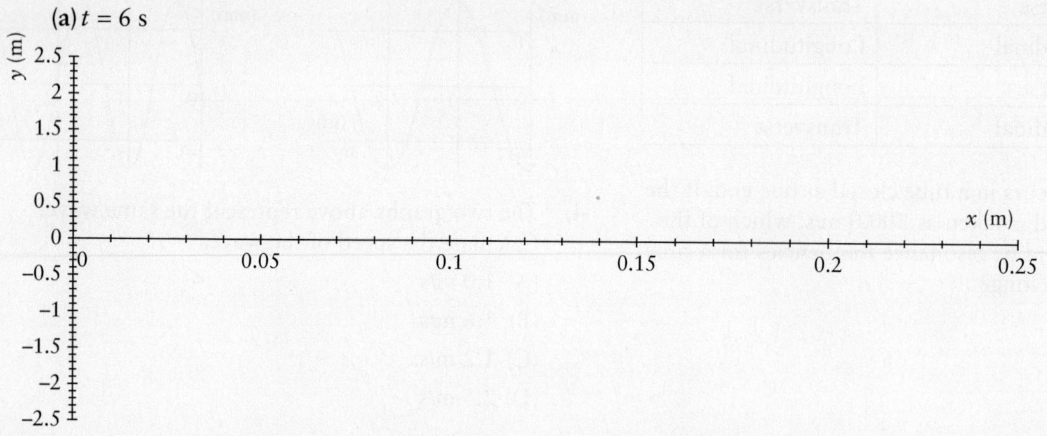

(Continued)

(b) $t = 10$ s

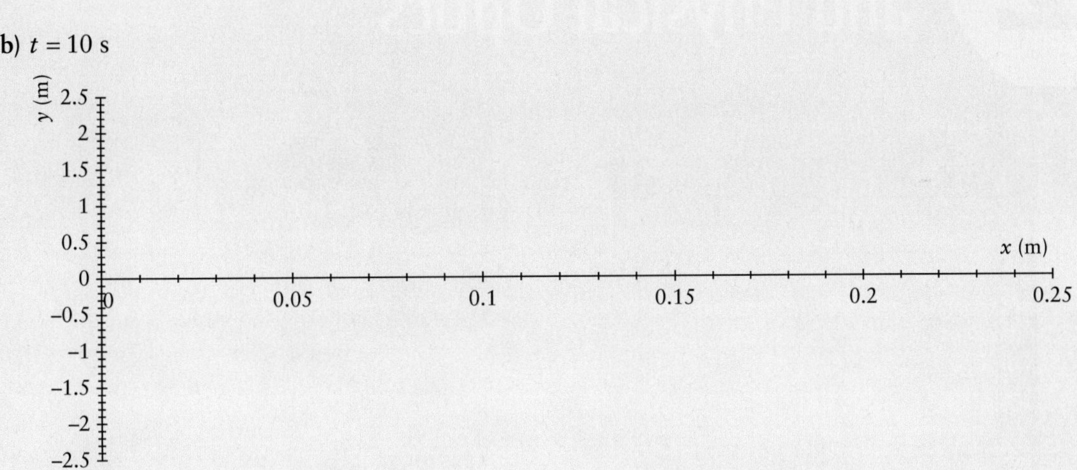

(c) Consider point A on the string as the two pulses pass point A. Sketch a graph of the position as a function of time for point A from $t = 0$ until $t = 15$ s.

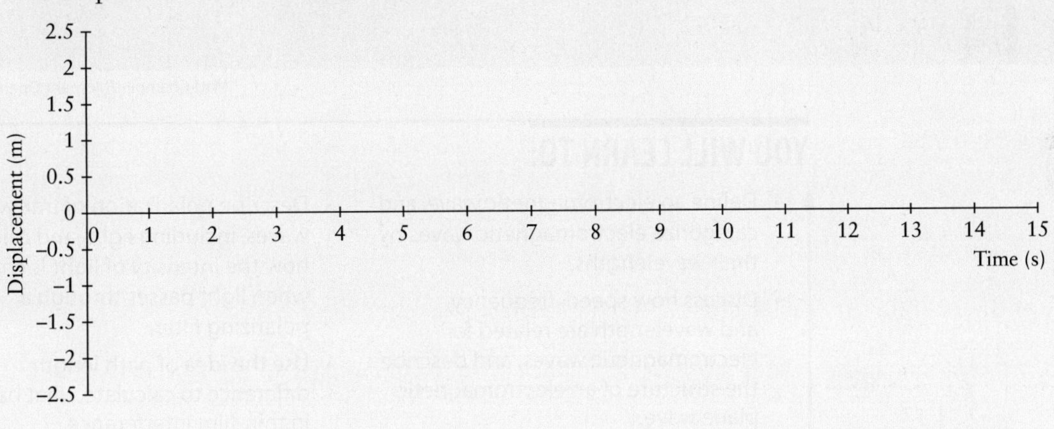

22 Electromagnetic Waves and Physical Optics

Matt Champlin/Moment Open/Getty Images

YOU WILL LEARN TO:

- Define an electromagnetic wave, and categorize electromagnetic waves by their wavelengths.
- Discuss how speed, frequency, and wavelength are related for electromagnetic waves, and describe the structure of an electromagnetic plane wave.
- Explain Huygens' principle and what it tells us about the laws of reflection and refraction.
- Recognize the special circumstances under which total internal reflection can take place and calculate the necessary angle of incidence.
- Describe dispersion and use it in calculations.
- Describe polarization of transverse waves, including light, and calculate how the intensity of light is changed when light passes through a polarizing filter.
- Use the idea of path length difference to calculate what happens in thin-film interference.
- Explain two-slit interference in terms of the wave properties of light and determine the properties of an interference pattern.
- Explain why light spreads out when it passes through a narrow opening and determine the pattern of light and dark fringes for a single slit.
- Calculate how the angular resolution of an optical device is limited by diffraction.

22-1 Light is one example of an electromagnetic wave, and its wave nature explains much about how light behaves

What is light? The answer to this question was not discovered until the nineteenth century, when the Scottish physicist James Clerk Maxwell realized that light is an **electromagnetic wave**—a traveling disturbance that, unlike a sound wave, does not require a physical material through which to propagate. Instead, an electromagnetic

We can see the light from this cluster of stars even though it is separated from us by 100,000 light years (about 10^{18} km) of nearly empty space. This is because electromagnetic waves, including visible light, can propagate in a vacuum.
(a)
S. Kafka and K. Honeycutt, Indiana University/ WIYN/NOAO/NSF

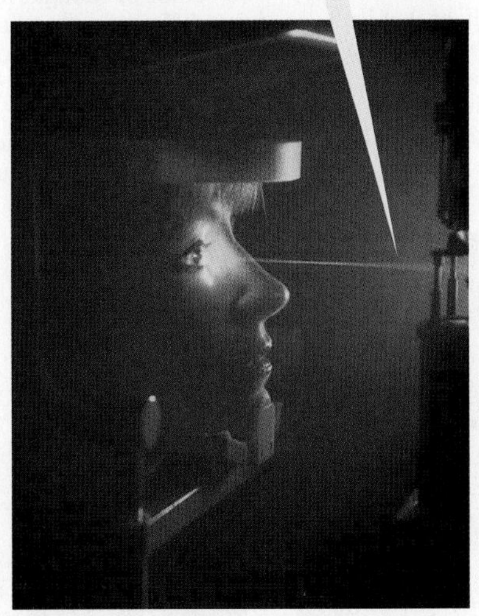

The intensity of an electromagnetic wave—including this laser beam used in ophthalmic surgery—depends on the amplitudes of the electric and magnetic fields that make up the wave.
(b)
Will & Deni McIntyre/Science Source

Figure 22-1 Electromagnetic waves Unlike sound waves or water waves, electromagnetic waves do not require the oscillation of any material substance. Instead, what oscillates are electric and magnetic fields. This explains (a) why these waves can propagate in a vacuum and (b) what determines their intensity.

wave involves oscillating electric and magnetic fields. These can exist even in the vacuum of space, which is why we can see the light from distant stars (**Figure 22-1a**).

In this chapter we'll study the properties of electromagnetic waves. We'll examine the broad variety of electromagnetic waves, which also includes x rays, microwaves, and radio waves. We'll see how wavelength, frequency, and speed are related for electromagnetic waves that propagate in a vacuum. We'll also look at the inner workings of a particularly simple kind of electromagnetic wave called a *sinusoidal plane wave*. We'll calculate the speed of light from the fundamental equations that govern electric and magnetic fields introduced in Chapter 20. We'll explore several of the consequences of the wave nature of light. For most of this exploration we won't need the details about electric and magnetic fields; what's important is simply that in many cases light can be treated as a wave. As a result, many of the properties of light waves that we'll encounter apply equally well to sound and other types of waves.

We'll begin by introducing *Huygens' principle*, a simplified model that describes how waves propagate through space. We'll use Huygens' principle to explain what happens when light is reflected (**Figure 22-2a**). Light waves travel at different speeds in different transparent materials, and we'll see how Huygens' principle explains *refraction*—the bending of light when it moves from one transparent material to another. We will see that this bending of light can affect its brightness, which we quantify as intensity (**Figure 22-1b**). We'll see that in certain circumstances light can be trapped inside a transparent material, just as if that material had mirrored surfaces. This effect, called *total internal reflection*, is essential for the medical technique of endoscopy. We'll see that the speed of light in a transparent material also depends on the wavelength of the light. This phenomenon, called *dispersion*, explains the vivid colors of a rainbow (**Figure 22-2b**).

One aspect of light that depends on its being a transverse wave is its *polarization*, which describes how the electric field vector of the wave is oriented. We'll see how light can become polarized by scattering or reflection, and we'll examine how polarizing filters work and why they're used in sunglasses.

We'll look at the phenomenon of *interference*, in which two light waves can add together constructively or destructively. Interference explains the colors seen in a thin film of soapy water (**Figure 22-2c**), as well as why cats and other animals have reflective eyes (see Figure 22-2a). We'll finish with a discussion of *diffraction*, an important effect in which light waves spread out when they pass through a small aperture.

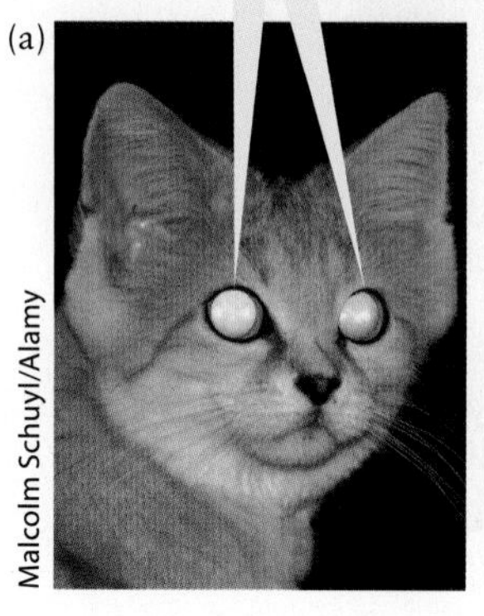

Cats, including this sand cat (*Felis margarita*), can see even in very low light levels thanks to reflection by a layer called the *tapetum lucidum* at the back of each eye. Incoming light that isn't absorbed by the retina reflects straight back, and some of the reflected light is detected on the second pass.
(a)
Malcolm Schuyl/Alamy

The colors of the rainbow are caused by dispersion: The speed of light in water depends on the frequency of the light. As a result, each color of sunlight follows a different path as it enters a raindrop and undergoes refraction, reflects off the back of the raindrop, and undergoes refraction again as it exits the raindrop.
(b)
iStockphoto/Thinkstock

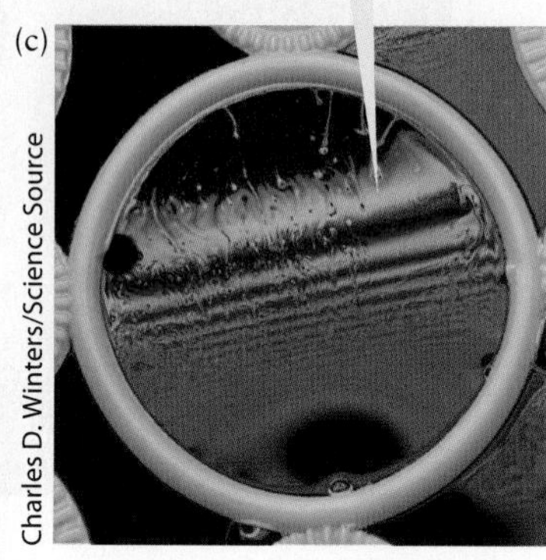

The colors of this soap film are caused by interference. Some light reflects from the front surface of the film, and some enters the film and reflects from the back surface. If the wavelength is just right, the two waves interfere constructively and produce a bright band.
(c)
Charles D. Winters/Science Source

Figure 22-2 Light waves in nature (a) Reflection, (b) dispersion, and (c) interference are among the many phenomena that light waves exhibit in the natural world.

THE TAKEAWAY for Section 22-1

✔ Visible light is an example of an electromagnetic wave. The properties of electromagnetic waves are explained by the equations of electricity and magnetism.

✔ Many of the key properties of light can be understood simply by using the idea that light can be treated as a wave.

Prep for the AP® Exam

Building Blocks

1. Which of the following does not require a physical medium through which to travel?
   (A) Transverse waves
   (B) Light
   (C) Sound
   (D) Longitudinal waves

22-2 In an electromagnetic plane wave, electric and magnetic fields both oscillate

Experiment shows that in a vacuum all electromagnetic waves—including radio waves, infrared radiation, visible light, ultraviolet radiation, x rays, gamma rays, and others—propagate at the same speed. This is the **speed of light**, to which we give the symbol c:

(22-1) $c = 2.99792458 \times 10^8$ m/s (= 3.00×10^8 m/s to three significant figures)

Different kinds of electromagnetic waves have different frequencies and wavelengths. In Section 21-3 we learned that for a mechanical wave the frequency f and wavelength λ are related to the propagation speed of the wave v_p by $v_p = f\lambda$

(Equation 21-2). The same relationship holds for electromagnetic waves in a vacuum, with v_p replaced by c:

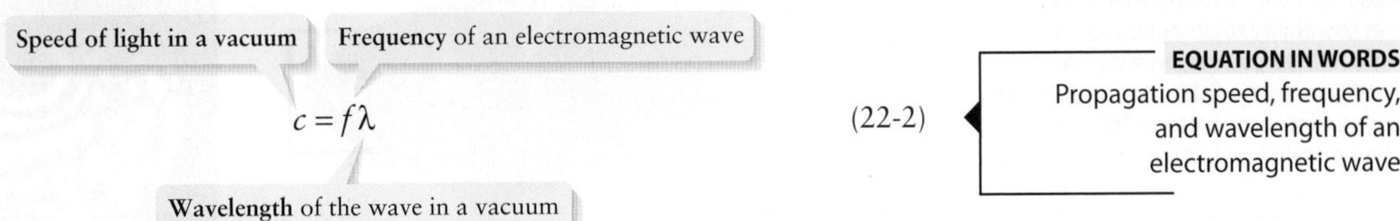

$$c = f\lambda \quad (22\text{-}2)$$

EQUATION IN WORDS
Propagation speed, frequency, and wavelength of an electromagnetic wave

Equation 22-2 tells us that the product of frequency f and wavelength λ has the same value, c, for *all* electromagnetic waves in a vacuum. The longer the wavelength, the lower the frequency; the shorter the wavelength, the higher the frequency.

Electromagnetic waves of any wavelength are possible. **Figure 22-3a** shows the names given to different wavelength ranges, which we refer to collectively as the **electromagnetic spectrum**. The human eye is sensitive to only a very narrow range of wavelengths known as **visible light** (**Figure 22-3b**). Visible light encompasses wavelengths from about $\lambda = 380$ nm to about $\lambda = 750$ nm (1 nm = 1 nanometer = 10^{-9} m). We perceive light of different wavelengths as having different colors; the shortest-wavelength light we can see is violet, and the longest-wavelength light we can see is red. At wavelengths shorter than visible light are ultraviolet light (UV), x rays, and gamma rays. At wavelengths longer than visible light are infrared light (IR), microwaves, and radio waves.

Other species can detect wavelengths longer or shorter than those visible to humans. Certain snakes (including pythons and rattlesnakes) have special pit organs on their heads that can sense infrared light. This enables these snakes to detect the radiation that both predators and prey emit due to their body temperature. Other species, such as damselfish, are able to sense ultraviolet light (**Figure 22-4**).

Note that in a medium other than a vacuum, electromagnetic waves propagate at speeds slower than c. For example, visible light travels at about 2.2×10^8 m/s ($0.73c$) in water and at about $0.9998c$ in air. The propagation speed in a medium other than vacuum also depends on the wave frequency; for example, in ordinary glass blue light travels slightly slower than does red light. We'll shortly see that this explains why a

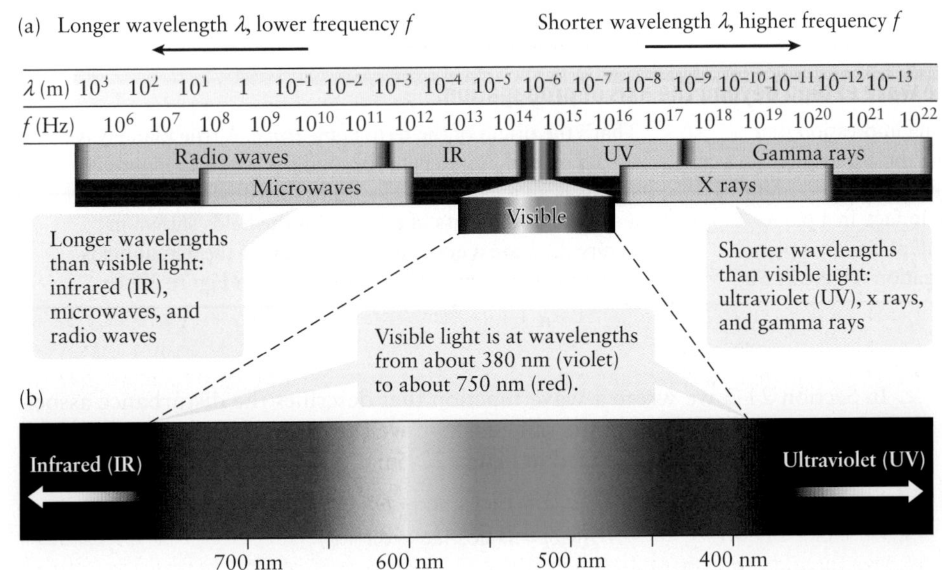

Figure 22-3 The electromagnetic spectrum (a) We classify electromagnetic waves according to their wavelength λ (or, equivalently, their frequency f). (b) Visible light makes up a tiny portion of the entire electromagnetic spectrum.

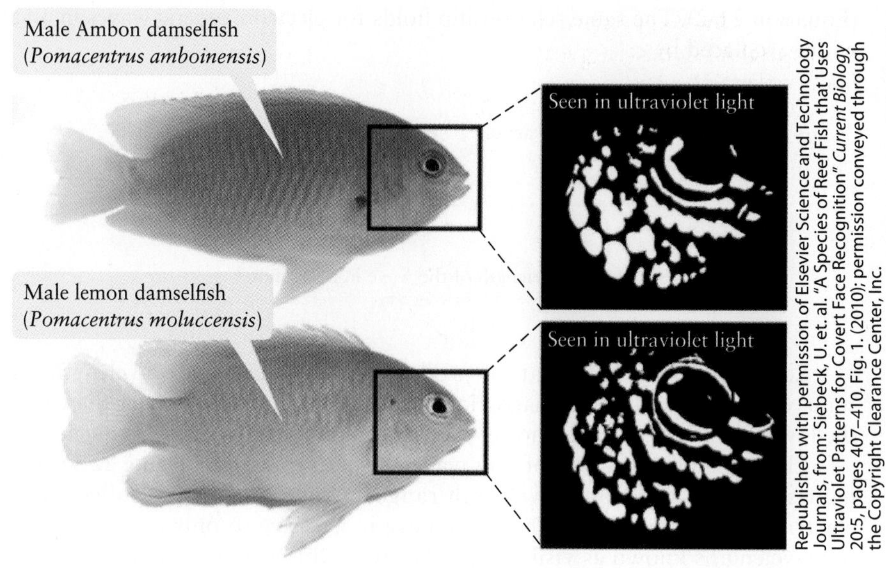

Figure 22-4 Using ultraviolet light for face recognition The Ambon damselfish (*Pomacentrus amboinensis*, a reef fish native to the western Pacific) has the ability to detect ultraviolet light. Where you see only dark and light bands on a fish's face, an Ambon sees an intricate pattern. This enables the territorial male Ambon damselfish (top) to identify and attack another Ambon to defend its territory but ignore a male lemon damselfish (*Pomacentrus moluccensis*, bottom) with its slightly different facial pattern.

prism is able to break white light into colors. In a vacuum, however, electromagnetic waves of all frequencies propagate at c.

The structure of electromagnetic waves can be quite complex. For example, the waves that make up the beam of laser light shown in Figure 22-1b are strong near the center of the beam, then taper off in strength toward the edge of the beam. A simpler, idealized kind of wave that has all of the key properties of a more general electromagnetic wave is a **sinusoidal plane wave** (**Figure 22-5a**). As the name suggests, the disturbance in such a wave oscillates in a sinusoidal fashion. There are actually two disturbances in this wave, an electric field $\vec{E}$ with amplitude E_0 and a magnetic field $\vec{B}$ with amplitude B_0. Both of these fields are *transverse*: They are perpendicular to the direction of propagation, just like the disturbance for waves propagating along a stretched rope or string (Sections 21-3 and 21-4). The fields are also perpendicular to each other. In a snapshot of the wave, as shown in Figure 22-5a, the $\vec{E}$ and $\vec{B}$ fields repeat over the same distance and so have the same wavelength λ. As the wave passes by a given point in space, the $\vec{E}$ and $\vec{B}$ fields both oscillate with the same frequency f. Equation 22-2 then tells us that both fields propagate at the same speed $c = f\lambda$.

WATCH OUT !

The fields of an electromagnetic plane wave extend beyond the axis of propagation.

Figure 22-5a may give you the misleading impression that the $\vec{E}$ and $\vec{B}$ fields are present only along the x axis (the axis along which the wave propagates). Such a wave would be like an infinitely narrow laser beam. In fact, in a plane wave the fields have the same value at *all* points on a plane perpendicular to the direction of propagation (**Figure 22-5b**). That's the origin of the term *plane wave*. A true plane wave would extend infinitely far beyond the x axis in Figure 22-5a, so this is an idealization like neglecting the friction of a surface or the mass of a rope. But the fields shown in Figure 22-5 are a good approximation of the actual fields found near the center of the laser beam in Figure 22-1b.

In Section 21-3 we wrote a wave function that describes the disturbance associated with a wave on a rope. In the same fashion we can write wave functions for the electric and magnetic fields depicted in Figure 22-5a:

$$E_y(x,t) = E_0\cos(kx - \omega t + \phi)$$
$$B_z(x,t) = B_0\cos(kx - \omega t + \phi) \qquad (22\text{-}3)$$

(sinusoidal electromagnetic plane wave)

Figure 22-5 The characteristics of a sinusoidal electromagnetic plane wave (a) This illustration shows two complete wavelengths of the wave. (b) How the electric field $\vec{E}$ and magnetic field $\vec{B}$ extend through space.

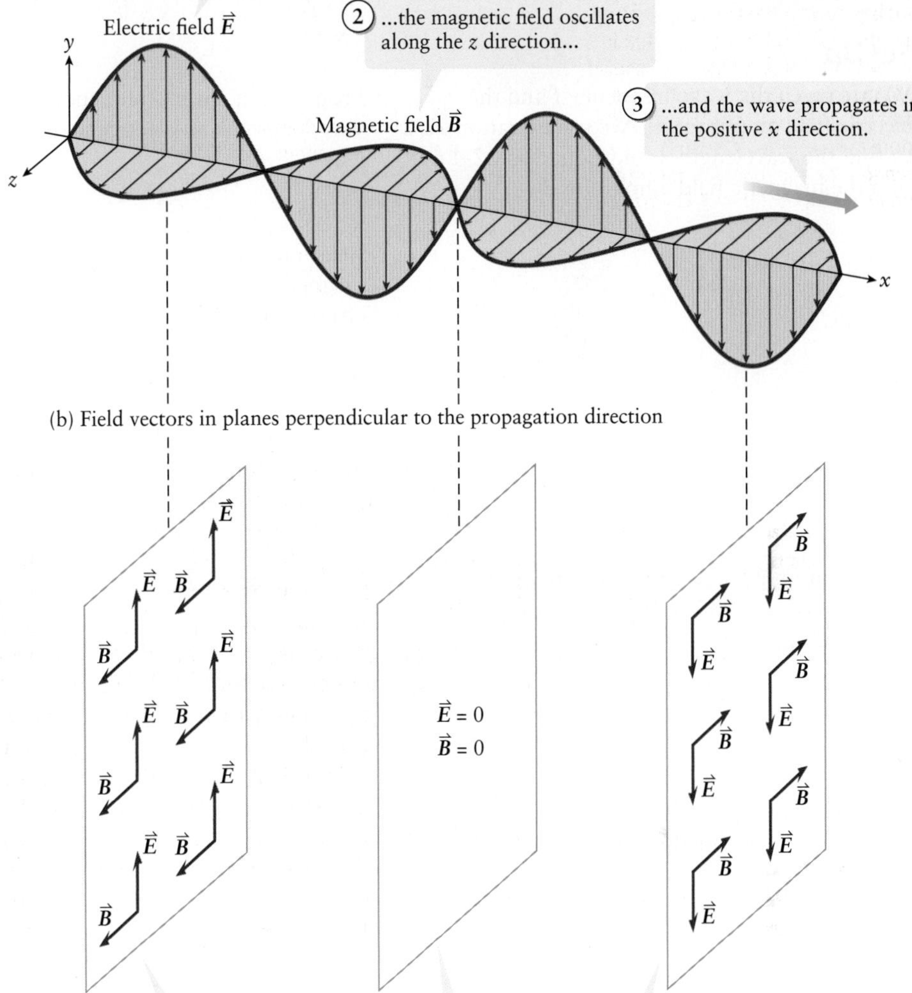

(a) An electromagnetic plane wave propagating in the positive x direction
1 In this sinusoidal electromagnetic plane wave, the electric field oscillates along the y direction...
Electric field $\vec{E}$
2 ...the magnetic field oscillates along the z direction...
Magnetic field $\vec{B}$
3 ...and the wave propagates in the positive x direction.
y
z
x
(b) Field vectors in planes perpendicular to the propagation direction
$\vec{E}$
$\vec{B}$
$\vec{E} = 0$
$\vec{B} = 0$
4 At any instant the electric field vector has the same value at all points on a plane perpendicular to the direction in which the wave propagates (for this wave, any plane perpendicular to the x axis). The same is true for the magnetic field vector.

- An electromagnetic wave is transverse: The electric and magnetic fields (the wave disturbances) are perpendicular to the propagation direction.
- In this sinusoidal plane wave, the electric and magnetic fields oscillate in phase.
- This is a plane wave because on any plane perpendicular to the direction in which the wave propagates, $\vec{E}$ has the same value at all points and $\vec{B}$ has the same value at all points.

In Equations 22-3 we use the same symbols that we used for waves on a rope in Section 21-3: $k = 2\pi/\lambda$ is the angular wave number, $\omega = 2\pi f$ is the angular frequency, and ϕ is the phase angle (which tells us what point in the oscillation cycle corresponds to $x = 0, t = 0$). Note that $E_y(x,t)$ and $B_z(x,t)$ both depend on position x and time t in the same way, which tells us that the electric and magnetic fields oscillate in phase with the same wavelength and frequency. In addition, the electric field amplitude E_0 and the magnetic field amplitude B_0 in a plane wave are directly proportional to each other:

$$B_0 = \frac{E_0}{c} \tag{22-4}$$

(sinusoidal electromagnetic plane wave)

EXAMPLE 22-1 A Radio Wave

A certain FM radio station broadcasts at a frequency of 98.7 MHz (1 MHz = 10^6 Hz). In the wave that reaches the radio in your car, the electric field amplitude is 6.00×10^{-2} V/m. Calculate the wavelength of the wave and the amplitude of the magnetic field.

Set Up

We are given the wave frequency f and the electric field amplitude E_0. We use Equation 22-2 to find the wavelength λ and Equation 22-4 to find the magnetic field amplitude B_0.

Propagation speed, frequency, and wavelength of an electromagnetic wave:

$$c = f\lambda \qquad (22\text{-}2)$$

Relation between the electric and magnetic field amplitudes in an electromagnetic wave:

$$B_0 = \frac{E_0}{c} \qquad (22\text{-}4)$$

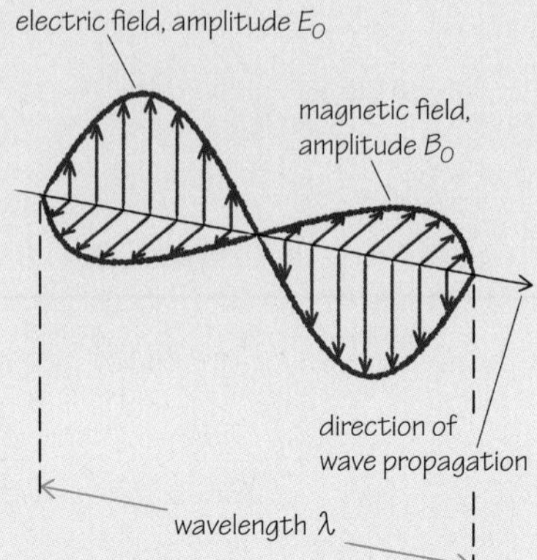

Solve

Use Equation 22-2 and the given frequency $f = 98.7$ MHz $= 98.7 \times 10^6$ Hz to solve for the wavelength.

From Equation 22-2:

$$\lambda = \frac{c}{f} = \frac{3.00 \times 10^8 \text{ m/s}}{98.7 \times 10^6 \text{ Hz}} = 3.04 \frac{\text{m}}{\text{s} \cdot \text{Hz}}$$

Recall that 1 Hz = 1 s^{-1}, so the units of s and Hz cancel:

$$\lambda = 3.04 \text{ m}$$

Use Equation 22-4 and the value $E_0 = 6.00 \times 10^{-2}$ V/m to solve for the magnetic field amplitude.

From Equation 22-4:

$$B_0 = \frac{E_0}{c} = \frac{6.00 \times 10^{-2} \text{ V/m}}{3.00 \times 10^8 \text{ m/s}} = 2.00 \times 10^{-10} \left(\frac{\text{V}}{\text{m}}\right)\left(\frac{\text{s}}{\text{m}}\right)$$

We learned in Section 17-3 that 1 V/m = 1 N/C, and in Section 19-3 we learned that 1 T = 1 (N · s)/(C · m). So

$$B_0 = 2.00 \times 10^{-10} \left(\frac{\text{N}}{\text{C}}\right)\left(\frac{\text{s}}{\text{m}}\right) = 2.00 \times 10^{-10} \frac{\text{N} \cdot \text{s}}{\text{C} \cdot \text{m}}$$
$$= 2.00 \times 10^{-10} \text{ T}$$

Reflect

Because the speed of light c has such a large value in m/s, the magnetic field amplitude B_0 in tesla (T) is much smaller than the electric field amplitude E_0 in volts per meter (V/m). As we'll see later in the chapter, however, the electric and magnetic fields prove to be equally important in an electromagnetic wave in a vacuum.

In this example we've seen how to relate the units of magnetic field (T) to those of electric field (V/m): 1 T = 1 (V/m) · (s/m) = 1 V · s/m^2. We'll make use of this result in later examples.

NOW WORK Problems 1–6 from The Takeaway 22-2.

Faraday's law and the Maxwell–Ampère law as a pair of equations have a solution that is a wave. When, from these equations, the speed at which such a wave should propagate through a vacuum was calculated, it was found to be c. Experiment showed that this was not a coincidence: Light is an electromagnetic wave. Maxwell's equations also show that an electromagnetic wave propagating in a vacuum must be transverse.

A longitudinal component of the wave, with an oscillating electric or magnetic field along the direction of propagation, would violate Gauss's law because it would involve a changing flux through a surface when there is no charge within that surface.

It turns out that the speed is related in a simple way to the values of the permittivity of free space ε_0 and the permeability of free space μ_0, both of which appear in Maxwell's equations:

$$c = \frac{1}{\sqrt{\mu_0 \varepsilon_0}}$$

Substitute $\varepsilon_0 = 8.85 \times 10^{-12}\ (\text{A} \cdot \text{s})/(\text{V} \cdot \text{m})$ and $\mu_0 = 4\pi \times 10^{-7}\ \text{T} \cdot \text{m/A}$:

$$c = \frac{1}{\sqrt{\left(4\pi \times 10^{-7}\ \dfrac{\text{T} \cdot \text{m}}{\text{A}}\right)\left(8.85 \times 10^{-12}\ \dfrac{\text{A} \cdot \text{s}}{\text{V} \cdot \text{m}}\right)}} = 3.00 \times 10^{8} \sqrt{\frac{\text{V}}{\text{T} \cdot \text{s}}}$$

These are very odd units! Note, however, that $1\ \text{V} = 1\ \text{N} \cdot \text{m/C}$ and $1\ \text{T} = 1\ \text{N} \cdot \text{s}/(\text{C} \cdot \text{m})$, so

$$1\sqrt{\frac{\text{V}}{\text{T} \cdot \text{s}}} = 1\sqrt{\left(\frac{\text{N} \cdot \text{m}}{\text{C}}\right)\left(\frac{\text{C} \cdot \text{m}}{\text{N} \cdot \text{s}}\right)\left(\frac{1}{\text{s}}\right)} = 1\sqrt{\frac{\text{m}^2}{\text{s}^2}} = 1\ \frac{\text{m}}{\text{s}}$$

The above expression for the speed of an electromagnetic wave then becomes

Speed of light in a vacuum

$$c = \frac{1}{\sqrt{\mu_0 \varepsilon_0}} = 3.00 \times 10^{8}\ \text{m/s} \qquad (22\text{-}5)$$

Permeability of free space — Permittivity of free space

EQUATION IN WORDS
Speed of light in a vacuum

This agrees with Equation 22-1 for the speed of light in a vacuum. The same mathematical analysis that leads to Equation 22-4 also relates the amplitudes E_0 and B_0 of the electric and magnetic fields in the wave: The result is that $B_0 = E_0/c$, the same result that we stated in Equation 22-4. Historically the value of ε_0 was determined experimentally and the value of μ_0 was determined by how the ampere was defined. Since 2018 the value of c has been *defined* to be exactly 2.99792458×10^8 m/s, which is now the basis of how the meter is defined. With this definition the values of both ε_0 and μ_0 are now determined experimentally, consistent with Equation 22-5.

WATCH OUT !

The speed of light in a medium other than a vacuum is less than *c*.

As we noted, electromagnetic waves propagate at speed c only in a vacuum. We determined the value of c using ε_0, the permittivity of free space, and μ_0, the permeability of free space; *free space* is equivalent to *in a vacuum*. Different materials have different values of permittivity and permeability, which is why the speed of light is different in them.

THE TAKEAWAY for Section 22-2

✔ Electromagnetic waves in a vacuum propagate at the speed of light.

✔ Different kinds of electromagnetic waves have different wavelengths and frequencies. The human eye can detect only a very narrow range of wavelengths.

✔ In a sinusoidal electromagnetic plane wave, the electric and magnetic fields both oscillate in phase with the same wavelength and frequency. The field directions are perpendicular to each other and perpendicular to the direction of propagation.

Prep for the AP® Exam

AP® Building Blocks

EX 22-1 **1.** Calculate the wavelengths of the electromagnetic waves with the following frequencies and classify the electromagnetic radiation of each (x ray, radio, and so on, including colors for visible light).
(a) $f = 4.14 \times 10^{15}$ Hz
(b) $f = 7.00 \times 10^{14}$ Hz
(c) $f = 8.00 \times 10^{16}$ Hz
(d) $f = 3.00 \times 10^{13}$ Hz
(e) $f = 9.00 \times 10^{12}$ Hz
(f) $f = 3.44 \times 10^{17}$ Hz
(g) $f = 8.23 \times 10^{15}$ Hz
(h) $f = 6.00 \times 10^{15}$ Hz

EX 22-1 **2.** Calculate the wavelengths of the electromagnetic waves with the following frequencies. Classify the electromagnetic radiation of each (x ray, radio, and so on, including colors for visible light).
(a) $f = 7.50 \times 10^{15}$ Hz
(b) $f = 6.00 \times 10^{14}$ Hz
(c) $f = 5.00 \times 10^{14}$ Hz
(d) $f = 4.29 \times 10^{14}$ Hz
(e) $f = 7.50 \times 10^{16}$ Hz
(f) $f = 2.66 \times 10^{16}$ Hz
(g) $f = 8.23 \times 10^{17}$ Hz
(h) $f = 6.00 \times 10^{18}$ Hz

EX 22-1 **3.** Calculate the frequencies of the electromagnetic waves that have the following wavelengths.
(a) $\lambda = 700$ nm
(b) $\lambda = 600$ nm
(c) $\lambda = 500$ nm
(d) $\lambda = 400$ nm
(e) $\lambda = 100$ nm
(f) $\lambda = 0.0333$ nm
(g) $\lambda = 500$ μm
(h) $\lambda = 63.3$ pm

EX 22-1 **4.** Calculate the frequencies of the electromagnetic waves that have the following wavelengths.
(a) $\lambda = 800$ nm
(b) $\lambda = 650$ nm
(c) $\lambda = 550$ nm
(d) $\lambda = 450$ nm
(e) $\lambda = 2.22$ nm
(f) $\lambda = 1.10 \times 10^{-8}$ m
(g) $\lambda = 50.0$ μm
(h) $\lambda = 33.4$ mm

AP® Skill Builders

EX 22-1 **5.** The FM radio band is from 88 to 108 MHz. Calculate the corresponding range of wavelengths.

EX 22-1 **6.** Describe how the frequency at which the changing electric and magnetic fields oscillate in electromagnetic waves is related to the speed of light.

7. The antenna for an AM radio station is a 75-m-high tower whose height is equivalent to one-quarter wavelength. At what frequency does the station transmit?

8. Rank the following electromagnetic waves from the shortest to the longest wavelength: (i) microwaves; (ii) red light; (iii) ultraviolet light; (iv) infrared light; (v) gamma rays.

AP® Skills in Action

9. Four electromagnetic waves in a vacuum have different frequencies. Which of these propagates at the fastest speed?
(A) $f = 3.95 \times 10^6$ Hz
(B) $f = 2.44 \times 10^7$ Hz
(C) $f = 1.26 \times 10^{11}$ Hz
(D) None is fastest. All have the same speed.

10. Our eyes are sensitive to the light from the setting sun, while the receivers in a cell phone tower are sensitive to radio waves coming from mobile phones. Both visible light and radio waves are kinds of electromagnetic waves. Compared to radio waves, visible light has
(A) much higher frequency and much faster speed.
(B) much lower frequency and much slower speed.
(C) much higher frequency and the same speed.
(D) much lower frequency and the same speed.

22-3 Huygens' principle explains the reflection and refraction of light

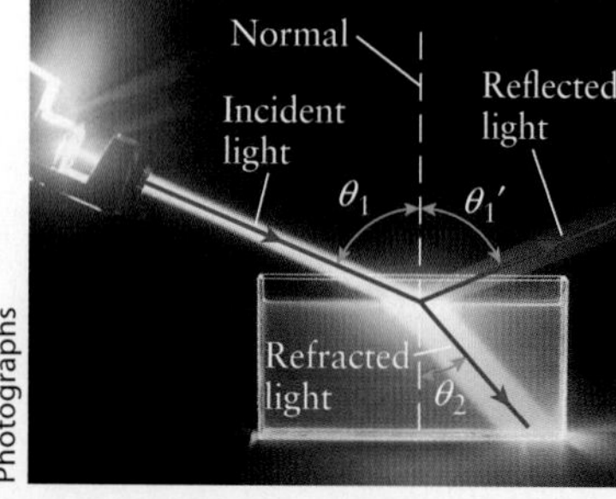

Light travels (propagates) in a straight line if the material through which the light travels—called the *medium* for the light—is uniform in its properties. But the direction in which light propagates changes when the light strikes a *boundary* between two different media, such as that between air and water (**Figure 22-6**). In general, some of the light reflects off the boundary, while the remainder travels into the second material at a different angle. The **law of reflection** states that the angle of the reflected light

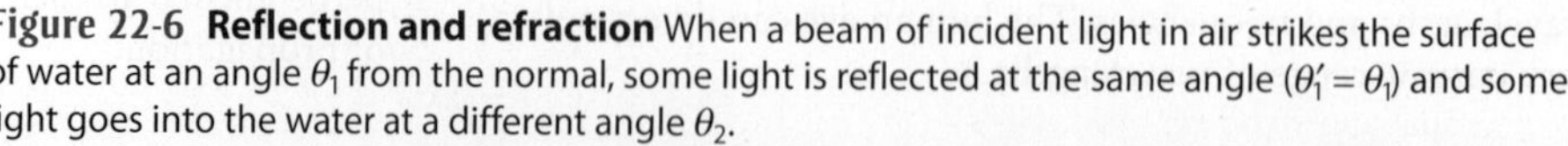
Figure 22-6 Reflection and refraction When a beam of incident light in air strikes the surface of water at an angle θ_1 from the normal, some light is reflected at the same angle ($\theta_1' = \theta_1$) and some light goes into the water at a different angle θ_2.

is the same as the angle of the incoming, or **incident light:** $\theta_1' = \theta_1$. **Refraction** is the change in direction of the light that travels into the second medium: The angle θ_2 for the refracted light is not equal to the angle θ_1 for the incident light.

WATCH OUT !

In reflection and refraction, angles are always measured from the normal to the boundary between two media.

Note that in Figure 22-6 the angles θ_1, θ_1', and θ_2 are all measured relative to the normal to the boundary, which is a line perpendicular to the boundary at the point where the incident light hits the boundary. A common mistake is to measure these angles relative to the boundary itself. If you make that mistake, you'll end up getting the wrong answer when you use the formulas that we'll derive in this section!

AP® Exam Tip

We will primarily focus on the properties of light that depend on its wave nature in this chapter, and study those properties that do not directly depend on its wave nature in Chapter 23, as we focus on geometric optics. However, these geometric properties of light build from an understanding of how a plane wave propagates, so we will lay the foundation for geometric optics in our study of waves in this and the next section.

Why is the law of reflection true? And what determines how the direction of the light changes when it travels from one medium into another? We'll answer both of these questions using a model of waves introduced by the seventeenth-century Dutch scientist Christiaan Huygens. (Huygens' model precedes by two centuries Maxwell's complete description of light as an electromagnetic wave, but is consistent with it.) We'll see that what determines the difference between the incident and refracted angles are the *speeds* at which light travels in the two media. Refraction isn't just for visible light but occurs for waves of all kinds: Radio waves, sound waves, and water waves may refract when crossing from one medium to another under the right circumstances.

Huygens' Principle and Reflection

Huygens considered waves that travel in two dimensions (like ripples on the surface of a pond) or in three dimensions (like light waves). He suggested that each point on a wave crest, or **front**, at time t can be treated as a source of tiny **wavelets** that themselves move at the speed of the wave (**Figure 22-7**). The wave front at a later time $t = t + \Delta t$ is then the superposition of all of these wavelets emitted at time t, and is tangent to the leading edges of the wavelets. This idea is called **Huygens' principle**. As Figure 22-7 shows, Huygens' principle helps explain how circular waves in water retain their shape as they propagate outward from a splash in a pond and how light waves spread out in spherical wave fronts from a light source.

To analyze what's happening to the light beams in Figure 22-6, let's apply Huygens' principle to a *plane* wave like the ones we introduced in the previous section. A plane wave propagates in a single direction, so it is a good description of the light beams shown in Figure 22-6. For each beam in Figure 22-6, the **ray** is an arrow that points in the direction of light propagation. Note that the ray is always perpendicular to the wave front.

(1) A drop of water falling into a pond produces waves on the pond's surface.
(2) The waves propagate at the same speed v in all directions. So each individual wave crest, or wave front, spreads outward in a circle from its point of origin.
iStockphoto/Thinkstock

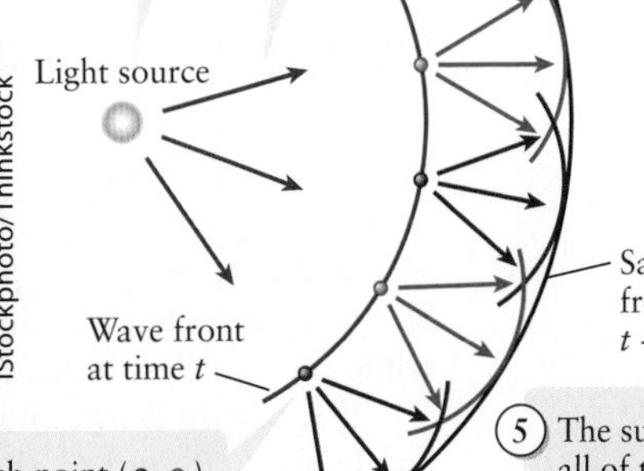

(3) In the same way, a light source produces light waves that spread out from the source. In a uniform medium the waves propagate at the same speed v in all directions.
Light source
Wave front at time t
Same wave front at time $t + \Delta t$
(4) We can treat each point (●, ●) on a wave front as the source of a wavelet that itself propagates at speed v.
(5) The superposition of all of the wavelets shows us the shape and size of the wave front at a later time.

Figure 22-7 Huygens' principle This principle provides a simple way to visualize wave propagation in terms of wavelets.

(a) Wave front ABC reflects from the boundary between medium 1 and medium 2. Note the right triangles $AA'B'$ and ABB'.

ABC: wave front at time t, moving toward the boundary

Reflected ray

Incident ray

A' B C C'

$v_1\Delta t$

Medium 1

Medium 2

A B'

$A'B'C'$: wave front at time $t + \Delta t$, moving away from boundary (reflected) between A' and B', still moving toward the boundary between B' and C'

(b) Comparing the right triangles $AA'B'$ and ABB'.

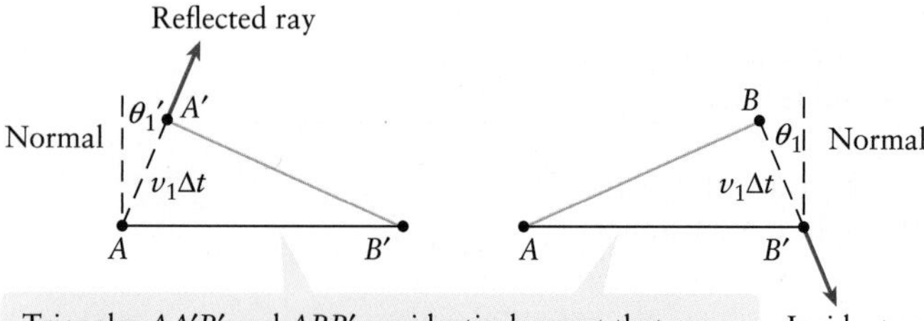

Triangles $AA'B'$ and ABB' are identical except that one is flipped left-to-right compared to the other. So $\theta_1' = \theta_1$.

Figure 22-8 Huygens' principle and reflection (a) A wave front reflected at a boundary between two media. (b) Finding the law of reflection, $\theta_1' = \theta_1$.

Figure 22-8a shows how to apply Huygens' principle to understand the law of reflection. A plane wave with wave front ABC is directed at an angle toward the boundary between medium 1 (say, air) and medium 2 (say, water or glass). At time t point A on the wave front has just arrived at the boundary. A time Δt later, the wavelets from points A, B, and C have each spread outward by a distance $v_1\Delta t$, where v_1 is the speed at which waves propagate in medium 1. The wavelets from points B and C propagate forward through medium 1, while the wavelet from point A is reflected at the boundary and so propagates *backward* from the boundary into medium 1. (We haven't drawn the wavelets that propagate into medium 2. We'll return to those a little later to help us understand refraction.)

If we draw a new wave front that's tangent to the leading edges of the wavelets that emanate from points A, B, and C, the result is $A'B'C'$ in Figure 22-8a. The wave front from B' to C' is still propagating toward the boundary; this represents the light still incident on the boundary. But the wave front from A' to B' is propagating away from the boundary and so represents the light reflected from the boundary.

The line BB' in Figure 22-8a is perpendicular to the incident wave front and so points in the direction of the incident ray. Likewise, the line AA' is perpendicular to the reflected wave front and so points in the direction of the reflected ray. To see how these directions are related to each other, notice that triangles ABB' and $AA'B'$ are both right triangles, both have the same hypotenuse of length AB', and both have one side of length $v_1\Delta t$ (**Figure 22-8b**). These two right triangles are identical, except that triangle $AA'B'$ has been flipped left-to-right compared to triangle ABB'. So the angle θ_1' of the line AA' measured from the vertical (that is, from the normal to the boundary) must be the same as the angle θ_1 of the line BB' measured from the vertical. We conclude that

NEED TO REVIEW?
Turn to page M12 in the Math Tutorial in the back of the book for more information on trigonometry.

EQUATION IN WORDS
The law of reflection for light waves at a boundary

When light reflects at the boundary between two media, the **angle of the reflected ray** from the normal...

$$\theta_1' = \theta_1 \qquad (22\text{-}6)$$

...is equal to the **angle of the incident ray** from the normal.

This is just the law of reflection that we mentioned above. We'll use Equation 22-6 extensively in Chapter 23 when we study the properties of mirrors.

Huygens' Principle and Refraction

Let's now use Huygens' principle to determine the direction of the refracted ray in Figure 22-6. **Figure 22-9a** is similar to Figure 22-8a, except that for point A we've drawn only the wavelet that emanates from that point and propagates into medium 2. We've assumed that the wave speed v_2 in medium 2 is slower than the speed v_1 in medium 1. So in a time Δt the wavelet that propagates into medium 2 travels a short distance $v_2\Delta t$ while the wavelets in medium 1 travel a longer distance $v_1\Delta t$.

If we again draw a new wave front that's tangent to the leading edges of the wavelets from A, B, and C, the result is $A''B'C'$. As in Figure 22-8a, the wave front from B' to C' represents incident light in medium 1 that is still propagating toward the boundary at speed v_1. The wave front from A'' to B' represents *refracted* light that is propagating in medium 2 at speed v_2. The angle of the wave front, and hence the angle of the ray that's perpendicular to the wave front, has changed because the wave speed has changed.

In **Figure 22-9b** we've redrawn the right triangles ABB' and $AA''B'$ from Figure 22-9a. Both triangles have the same hypotenuse of length AB', but the angles of the two triangles are different. Side BB' of triangle ABB' has length $v_1\Delta t$, points in the direction of the incident ray, and is at an angle θ_1 from the normal to the boundary. Because ABB' is a right triangle, you can see that the angle of side AB from the horizontal is also θ_1, the same as the angle of side BB' from the vertical (the normal to the boundary). If you look now at triangle $AA''B'$, you'll see that side AA'' has length $v_2\Delta t$, points in the direction of the refracted ray, and is at a different angle θ_2 from the normal to the boundary. And because $AA''B'$ is a right triangle, the angle of side $A''B'$ from the horizontal is θ_2, the same as the angle of side AA'' from the normal.

Recall that the sine of an angle in a right triangle equals the length of the side opposite to that angle divided by the length of the hypotenuse. So the sine of the angle θ_1 in triangle ABB' is

$$\sin\theta_1 = \frac{\text{length of side } BB'}{\text{length of side } AB'} = \frac{v_1\Delta t}{\text{length of side } AB'}$$

so

$$\frac{\sin\theta_1}{v_1} = \frac{\Delta t}{\text{length of side } AB'}$$

Similarly the sine of the angle θ_2 in triangle $AA''B'$ is

$$\sin\theta_2 = \frac{\text{length of side } AA''}{\text{length of side } AB'} = \frac{v_2\Delta t}{\text{length of side } AB'}$$

so

$$\frac{\sin\theta_2}{v_2} = \frac{\Delta t}{\text{length of side } AB'}$$

If you compare these two equations, you can see that

$$\frac{\sin\theta_1}{v_1} = \frac{\sin\theta_2}{v_2} \qquad (22\text{-}7)$$

Equation 22-7 states that the relationship between the angles θ_1 and θ_2 is determined by the speeds v_1 and v_2 of the wave in medium 1 and medium 2, respectively.

It's common to express the speed of light in a given medium in terms of a quantity called the **index of refraction:**

Index of refraction of a medium

Speed of light in a vacuum

$$n = \frac{c}{v} \qquad (22\text{-}8)$$

Speed of light in the medium

EQUATION IN WORDS
Index of refraction for light waves in a medium

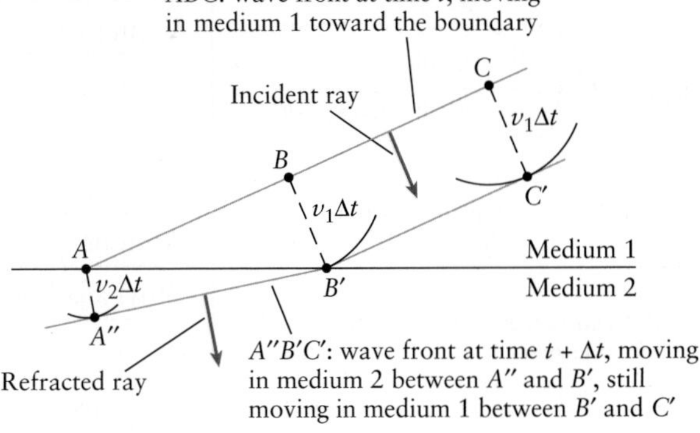

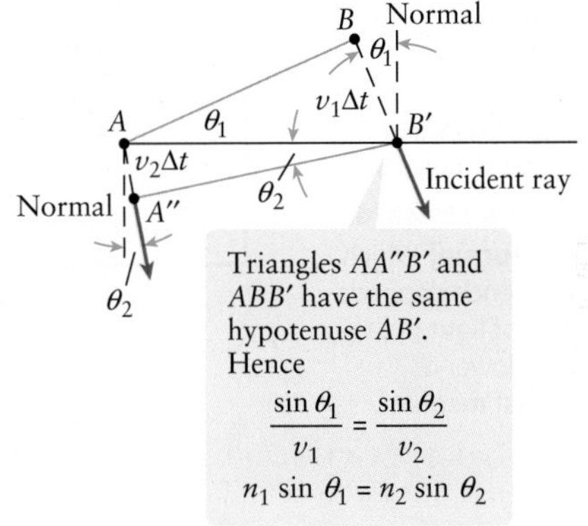

Figure 22-9 Huygens' principle and refraction (a) A wave front refracted at a boundary between two media. (b) Finding the law of refraction.

The index of refraction of vacuum is 1, because light travels at the speed of light c, so $v = c$ and $n = c/c = 1$. In any material medium light travels slower than c, so $v < c$ and $n > 1$. The greater the value of the index of refraction n in a given medium, the slower the speed v at which light propagates in that medium. **Table 22-1** lists the index of refraction of some common materials. Note that the index of refraction of air is equal to one to three significant digits, so we'll often take $n_{\text{air}} = 1$ in calculations.

Equation 22-8 states that $1/v = n/c$, so we can rewrite Equation 22-7 as

$$\left(\frac{n_1}{c}\right)\sin\theta_1 = \left(\frac{n_2}{c}\right)\sin\theta_2$$

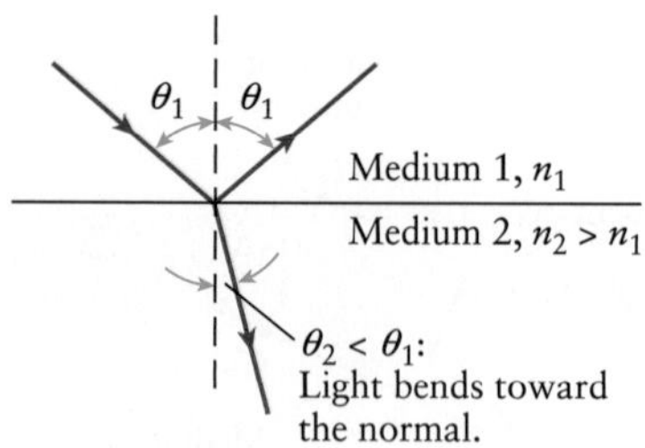

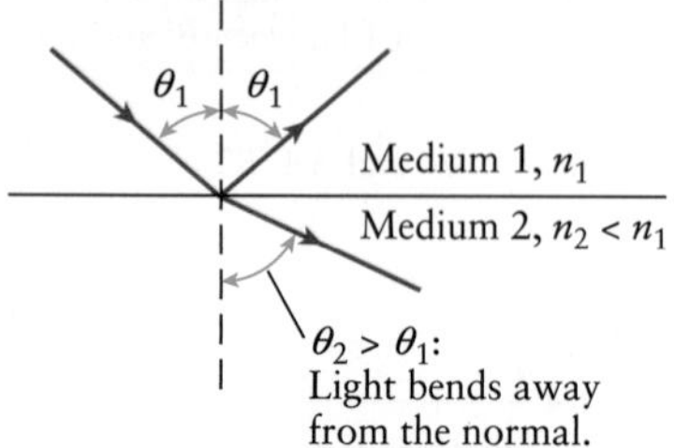

Figure 22-10 Refraction toward and away from the normal Which way the refracted ray bends depends on whether the speed of light in the second medium is (a) slower or (b) faster than in the first medium.

TABLE 22-1 Indices of Refraction

Material	Index of refraction
Vacuum	1 (exactly)
Air at 20°C, 1 atm pressure	1.00029
Ice	1.31
Water at 20°C	1.33
Acetone	1.36
Ethyl alcohol	1.36
Eye, cornea	1.38
Eye, lens	1.41
Sugar water (high concentration)	1.49
Plexiglas	1.49
Typical crown glass	1.52
Sodium chloride	1.54
Sapphire	1.77
Diamond	2.42

If we cancel the factors of c on both sides of this equation, we get

EQUATION IN WORDS
Snell's law of refraction for light waves at a boundary

$$n_1 \sin\theta_1 = n_2 \sin\theta_2 \quad (22\text{-}9)$$

Angle of the incident ray from the normal (θ_1); Angle of the refracted ray from the normal (θ_2); Index of refraction for the medium with the incident light (n_1); Index of refraction for the medium with the refracted light (n_2)

Equation 22-9 is known as **Snell's law of refraction.** (This law is named for the Dutch scientist Willebrord Snellius but was in fact first discovered by the Persian scientist Ibn Sahl in 984, more than 600 years before Snellius.)

Snell's law tells us that when a ray of light crosses from one medium to another, the product of the index of refraction and the sine of the angle the ray makes to the normal remains constant. When light passes into a material of higher index of refraction—for example, from air into glass—so that the speed of light is slower in the second medium and $n_2 > n_1$, the sine of the refracted angle and the angle itself both decrease. In this case $\theta_2 < \theta_1$, and the light bends closer to the normal (**Figure 22-10a**). When light instead passes into a material of lower index of refraction—for example, from glass into air—so that the speed of light is faster in the second medium and $n_2 < n_1$, the sine of the refracted angle and the angle itself both increase. In this situation $\theta_2 > \theta_1$, and the light bends away from the normal (**Figure 22-10b**).

The fraction of incident light that is reflected and the fraction that is refracted depend in part on the indices of refraction of the two media. (They also depend on the incident angle and on how the electric field vectors in the light wave are oriented relative to the boundary.) The index of refraction of a medium is often a function of the medium's density. One example is blood plasma; its density and index of refraction depend on the concentration of dissolved protein. Veterinarians can use this dependence to estimate protein levels in livestock at a clinic or on a farm by measuring how much light refracts as it passes through a sample of an animal's plasma. Winemakers determine the amount of sugar in their grapes by using the same technique.

EXAMPLE 22-2 Seeing Under Water

A surveyor (labeled S) looking at an aqueduct is just able to see the underwater edge at point F where the far wall meets the bottom (**Figure 22-11**). If the aqueduct is 4.2 m wide and her line of sight to the near, top edge at point N is 25° above the horizontal, find the actual depth of the aqueduct.

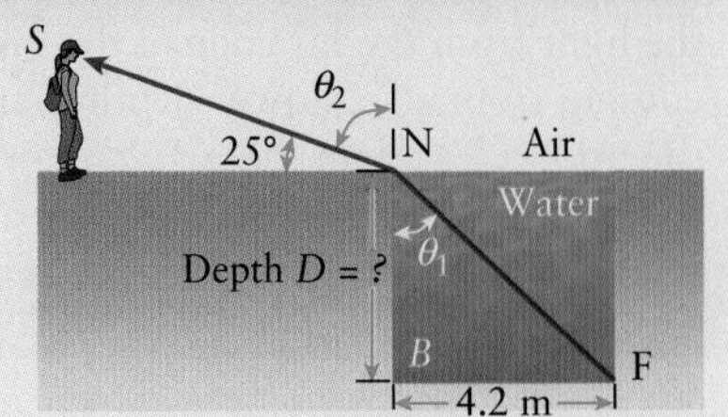

Figure 22-11 A refracted view If the surveyor just sees the bottom edge of the far wall of the aqueduct, how deep is the aqueduct?

Set Up

A light beam from point F refracts when it reaches the boundary between the water (medium 1) and the air (medium 2) at point N. Figure 22-11 shows that the angle of the refracted ray from the *normal* (shown as a vertical dashed line) is $\theta_2 = 90° - 25° = 65°$. We'll first use Snell's law to determine the angle θ_1 of the incident ray, and then use trigonometry and the given width of the aqueduct (4.2 m) to find the depth D.

Snell's law of refraction:

$$n_1 \sin \theta_1 = n_2 \sin \theta_2 \quad (22\text{-}9)$$

Solve

Find the angle θ_1 of the incident ray using Snell's law.

Solve Equation 22-9 for the sine of the incident angle θ_1:

$$\sin \theta_1 = \frac{n_2}{n_1} \sin \theta_2$$

Figure 22-11 shows that $\theta_2 = 90° - 25° = 65°$, and from Table 22-1 we see that $n_1 = n_{\text{water}} = 1.33$ and $n_2 = n_{\text{air}} = 1.00$. So

$$\sin \theta_1 = \frac{1.00}{1.33} \sin 65° = 0.68$$

$$\theta_1 = \sin^{-1} 0.68 = 43°$$

Use trigonometry to determine the depth D of the aqueduct.

The incident ray that travels from point F to point N is the hypotenuse of a right triangle NFB with vertical dimension D and horizontal dimension 4.2 m. The side of length 4.2 m is the side opposite the angle θ_1, and the side of length D is the side adjacent to this angle. The tangent of θ_1 equals the opposite side divided by the adjacent side:

$$\tan \theta_1 = \frac{4.2 \text{ m}}{D}$$

Solve for the distance D:

$$D = \frac{4.2 \text{ m}}{\tan \theta_1} = \frac{4.2 \text{ m}}{\tan 43°} = 4.5 \text{ m}$$

The aqueduct is 4.5 m deep.

Reflect

The refracted angle $\theta_2 = 65°$ is larger than the incident angle $\theta_1 = 43°$, just as in Figure 22-10b. This makes sense, because $n_2 < n_1$ (the index of refraction for air is smaller than the index for water).

Our brains are used to the idea that light travels in straight lines. If we extend the refracted ray backward, we see that to the surveyor's eye the light from the far edge of the bottom of the aqueduct appears to be coming from a shallower depth than $D = 4.5$ m. So the *apparent* depth of the pool is less than the *actual* depth D. You can easily see this effect in a swimming pool.

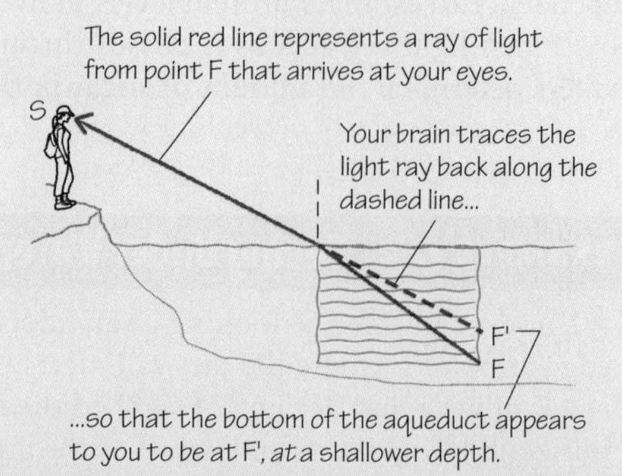

The solid red line represents a ray of light from point F that arrives at your eyes.
S
Your brain traces the light ray back along the dashed line...
F'
F
...so that the bottom of the aqueduct appears to you to be at F', at a shallower depth.

NOW WORK Problems 4, 5, and 8–10 from The Takeaway 22-3.

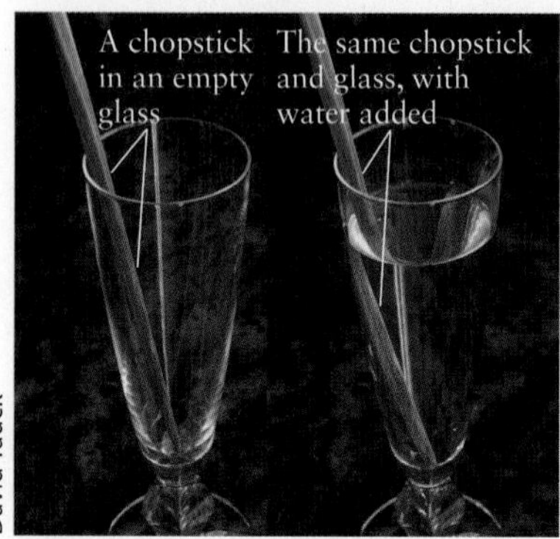

Figure 22-12 A refracted chopstick Due to refraction, the submerged part of the chopstick on the right appears displaced from its actual position.

The situation shown in **Figure 22-12** involves the same effect that makes the aqueduct in Example 22-2 appear shallower than it really is. When light from the submerged part of the chopstick passes from water to air at the boundary between the two, the light rays refract. As a result, it appears to our eyes that the submerged part is in a different position than its true location.

Frequency and Wavelength in Refraction

When a wave travels from one medium to another, the frequency of the wave remains the same. (In a given time interval, as many crests arrive at the boundary as leave the boundary. If this were not true, there would be a "traffic jam" of wave crests at the boundary.) However, because the wave speed is different in the two media, the wavelength must change. This follows from the relationship among the propagation speed v of the wave, the frequency f, and the wavelength λ. From Equation 21-2:

$$v = f\lambda \quad \text{so} \quad \lambda = \frac{v}{f}$$

In a vacuum light waves travel at speed $v = c$, so the wavelength is

$$\lambda_{\text{vacuum}} = \frac{c}{f} \qquad (22\text{-}10)$$

In a medium with index of refraction n, the wave speed from Equation 22-8 is $v = c/n$. Then we can write the wavelength of light in a medium as

EQUATION IN WORDS
Wavelength of light in a medium

$$\lambda = \frac{c}{nf} = \frac{\lambda_{\text{vacuum}}}{n} \qquad (22\text{-}11)$$

Wavelength of light in a medium (λ); Speed of light in a vacuum (c); Wavelength of the light in a vacuum (λ_{vacuum}); Frequency of the light (f); Index of refraction of the medium (n)

Equation 22-11 shows the wavelength is shorter in a medium with a higher index of refraction, where the propagation speed is slower. For example, red light that has a wavelength $\lambda_{\text{vacuum}} = 750$ nm in a vacuum has a wavelength in water ($n = 1.33$) equal to $\lambda = \lambda_{\text{vacuum}}/n = (750 \text{ nm})/(1.33) = 564$ nm. The frequency of this light is the same in both media: $f = c/\lambda_{\text{vacuum}} = (3.00 \times 10^8 \text{ m/s})/(750 \times 10^{-9} \text{ m}) = 4.00 \times 10^{14}$ Hz in a vacuum and $f = v/\lambda = c/(n\lambda) = (3.00 \times 10^8 \text{ m/s})/((1.33)(564 \times 10^{-9} \text{ m})) = 4.00 \times 10^{14}$ Hz in water.

THE TAKEAWAY for Section 22-3

✔ Huygens' principle says that each point on a wave front acts as a source of wavelets. The new wave front is the superposition of the individual wavelets.

✔ When waves encounter a boundary between two media in which the wave speed is different, the waves can bounce back into the first medium (reflect) or pass into the second medium at a different angle (refract).

✔ The angle of the reflected ray is the same as the angle of the incident ray (law of reflection).

✔ Snell's law describes the direction of the refracted ray. If the wave speed is lower in the second medium than in the first, the refracted ray bends toward the normal. If the wave speed is faster in the second medium, the refracted ray bends away from the normal.

✔ The index of refraction is a measure of the speed of light in a medium relative to the speed of light in a vacuum. The slower light travels in a medium, the larger its index of refraction.

Prep for the AP® Exam

AP® Building Blocks

1. The speed of light in a newly developed plastic is 1.97×10^8 m/s. Calculate the index of refraction.
2. Calculate the speed of light for each of the following materials.
   (a) Ice
   (b) Acetone
   (c) Plexiglas
   (d) Sodium chloride
   (e) Sapphire
   (f) Diamond
   (g) Water
   (h) Crown glass
3. The index of refraction for a vacuum is 1.00000. The index of refraction for air is 1.00029. Determine the ratio of time required for light to travel through 1000 m of air to the time required for light to travel through 1000 m of vacuum.

EX 22-2
4. Determine the unknown angle in each of the situations in the figure below.

(a)

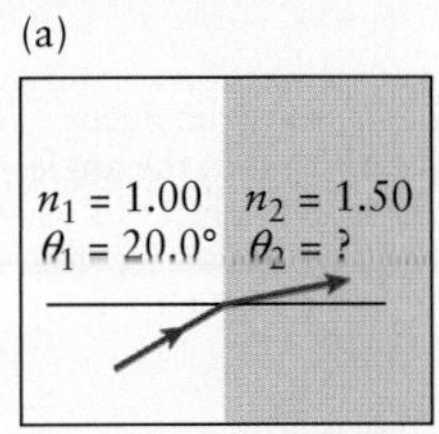

(b)

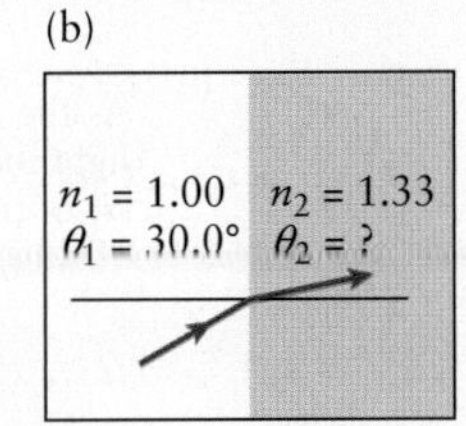

(c)

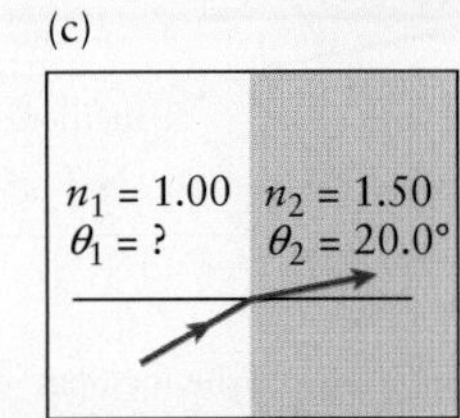

(d)

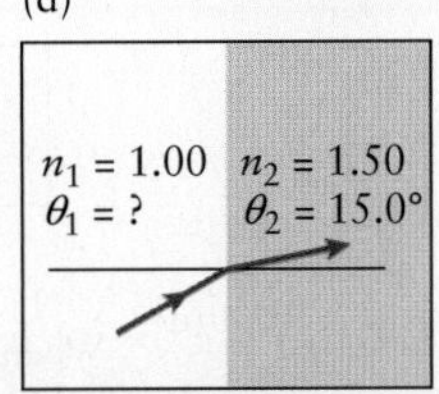

EX 22-2
5. Determine the unknown index of refraction in each of the four situations in the figure.

(a)

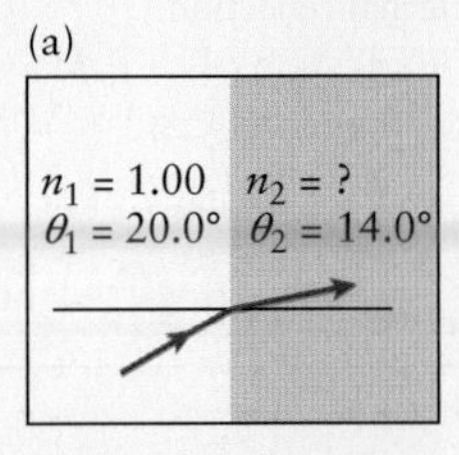

(b)

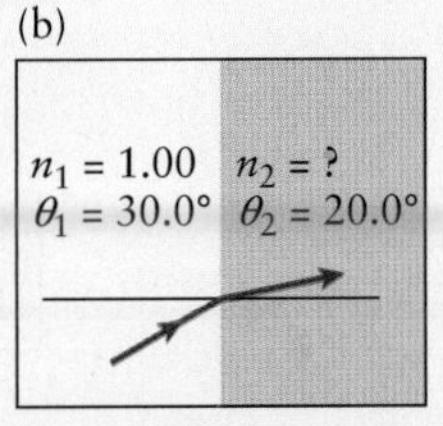

(c)

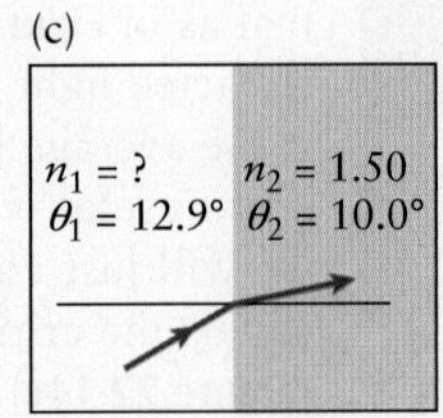

(d)

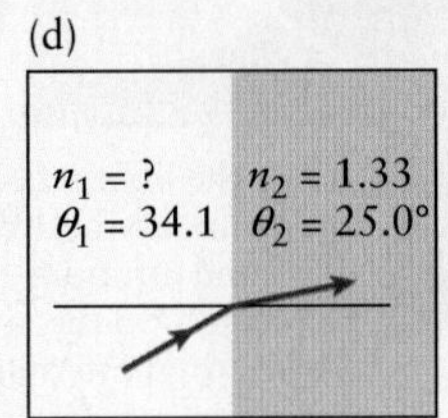

6. What is the wavelength of red light (700 nm in air) when it is inside a glass slab with $n = 1.55$?

AP® Skill Builders

7. The speed of light in methylene iodide is 1.72×10^8 m/s. The index of refraction of water is 1.33. Through what distance of methylene iodide must light travel such that the time to travel through the methylene iodide is the same as the time required for light to travel through 1000 km of water?

EX 22-2
8. Light travels from air toward water. If the angle that is formed by the light beam in air with respect to the normal line between the two media is 27.0°, calculate the angle of refraction of the light in the water.

AP® Skills in Action

EX 22-2
9. A beam of light in air travels into a transparent, flat-walled container made of Plexiglas. The incident angle is 5.00°. The light then travels from the Plexiglas into the fresh water inside the container. In each refraction, does the ray bend closer to the normal or farther away from it?
   (A) Closer in both refractions
   (B) Farther away in both refractions
   (C) Closer going from air to Plexiglas, farther away going from Plexiglas to water
   (D) Farther away going from air to Plexiglas, closer going from Plexiglas to water

EX 22-2
10. When light enters a piece of glass from air with an angle of θ with respect to the normal to the boundary surface
   (A) it bends with an angle larger than θ with respect to the normal to the boundary surface.
   (B) it bends with an angle smaller than θ with respect to the normal to the boundary surface.
   (C) it does not bend.
   (D) it bends with an angle equal to two times θ with respect to the normal to the boundary surface.

22-4 In some cases light undergoes total internal reflection at the boundary between media

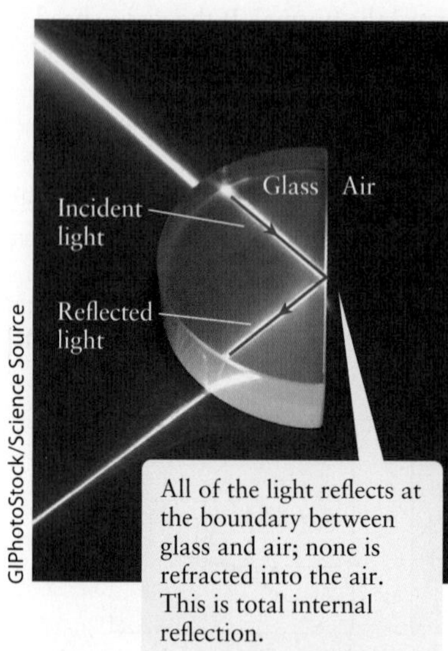

GIPhotoStock/Science Source

Figure 22-13 Total internal reflection There is nothing unusual about the place where the light beam strikes the glass, yet none of the light escapes into the air on the other side.

In the photograph shown in Figure 22-6, light in air encounters a boundary with water on the other side. Some of the incident light is reflected at the boundary, and some of it is refracted from the first medium (air) into the second medium (water). There are situations, however, in which *none* of the light is refracted into the second medium. **Figure 22-13** shows a light beam in glass that encounters a boundary with air on the other side. As the photograph shows, 100% of the light is reflected back into the glass. This effect is called **total internal reflection.**

Figure 22-14 shows how total internal reflection arises. Light travels more slowly in glass than air, so the index of refraction of glass is higher than the index of refraction of air. As light crosses the boundary from glass to air, it is bent away from the normal as in **Figure 22-14a**. As the incident angle of the light increases (**Figure 22-14b**), the refracted light gets farther from the normal and decreases in intensity. Wave **intensity** is the average power per unit area carried by a wave. In Chapter 24 we will see that it is a useful way to express the energy carried by an electromagnetic wave. For now, we will just consider intensity as a measure of brightness. When the incident angle equals the **critical angle** θ_c, the refracted light lies exactly in the plane of the surface (**Figure 22-14c**). At this angle the intensity of the refracted light is zero. If the incident angle is greater than the critical angle, as in **Figure 22-14d**, the light is completely reflected back into the glass. This is total internal reflection.

Total internal reflection is possible only when the first medium (in Figure 22-14, glass) has a higher index of refraction than the second medium (in Figure 22-14, air), so $n_1 > n_2$. Then the refracted angle θ_2 is greater than the incident angle θ_1, and the refracted angle can reach 90°, as in Figure 22-14c. If the first medium has a lower index of refraction than the second medium, so $n_1 < n_2$, the refracted angle θ_2 is less than the incident angle θ_1, and the refracted angle can never reach 90°.

We can calculate the critical angle θ_c using Snell's law of refraction, Equation 22-9. When the incident angle θ_1 equals the critical angle θ_c, the refracted angle θ_2 equals 90°. If we substitute these into Equation 22-9 and recall that sin 90° = 1, we get

$$n_1 \sin\theta_c = n_2 \sin 90° = n_2 \quad \text{so} \quad \sin\theta_c = \frac{n_2}{n_1} \tag{22-12}$$

(a) Light refracts as it passes from glass into air. The index of refraction of glass is greater than the index of refraction of air, so $\theta_2 > \theta_1$.

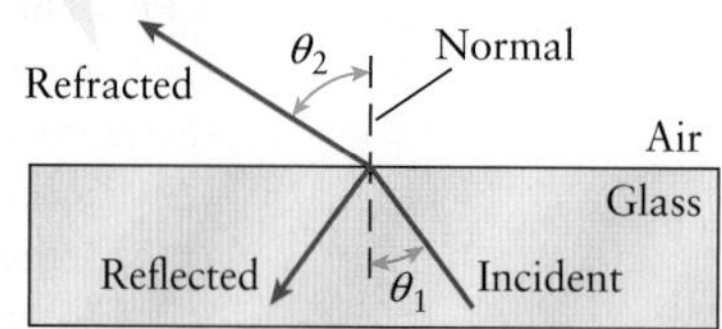

(b) As the angle θ_1 of the incident light increases, so does the angle θ_2 of the refracted light.

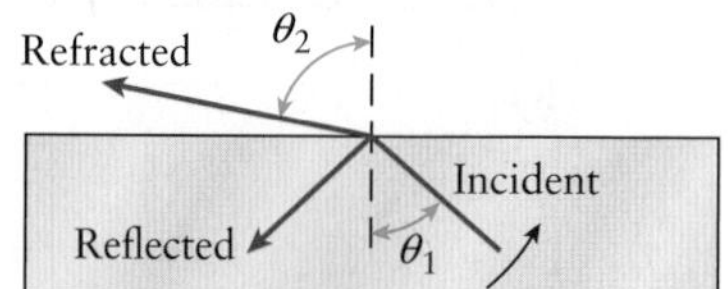

(c) When the angle θ_1 of the incident light equals the critical angle θ_c, the refracted angle is $\theta_2 = 90°$. The intensity of the refracted light becomes zero.

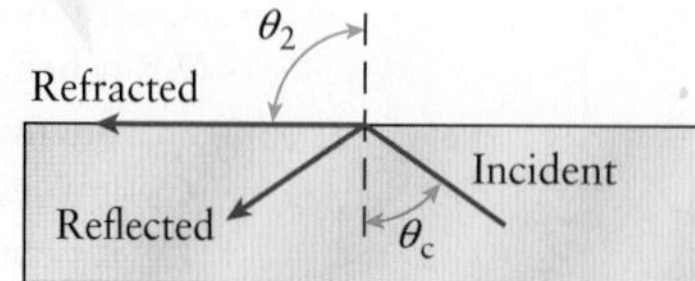

(d) When the angle θ_1 of the incident light is greater than the critical angle θ_c, the light is completely reflected back into the glass. This is total internal reflection.

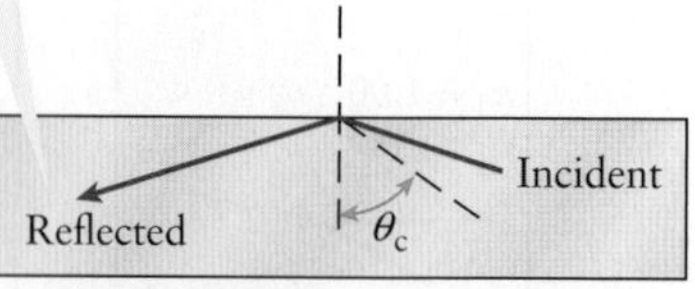

Figure 22-14 Approaching total internal reflection (a) and (b) Light approaches a boundary between glass and air at an incident angle θ_1 less than the critical angle θ_c. (c) When $\theta_1 = \theta_c$, the light ray is refracted along the surface. (d) Total internal reflection occurs for incident angles greater than the critical angle.

Note that the sine of an angle between 0 and 90° is between 0 and 1. If $n_1 > n_2$, the ratio n_2/n_1 is less than 1, and there will be some angle θ_c for which Equation 22-12 is satisfied. In this case total internal reflection is possible. But if $n_1 < n_2$, the ratio n_2/n_1 is greater than 1 and Equation 22-12 has no solution. This is another way of seeing that total internal reflection is possible only if $n_1 > n_2$.

If we solve Equation 22-12 for the critical angle θ_c, we get

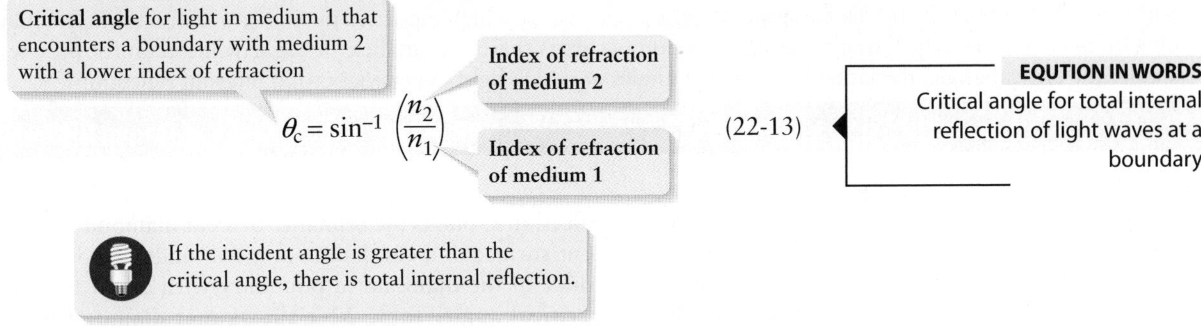

$$\theta_c = \sin^{-1}\left(\frac{n_2}{n_1}\right) \quad (22\text{-}13)$$

EQUTION IN WORDS
Critical angle for total internal reflection of light waves at a boundary

Notice that the critical angle depends on the indices of refraction of the media on *both* sides of a boundary.

EXAMPLE 22-3 Critical Angles

(a) A laser is aimed from under the water toward the surface, as in **Figure 22-15.** Find the critical angle of the light incident in the water beyond which total internal reflection occurs. (b) Find the critical angle if the liquid in the tank were replaced by water containing a high concentration of dissolved sugar.

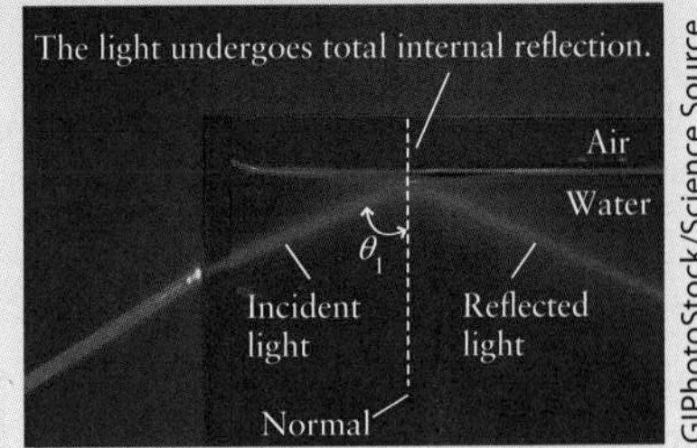

GIPhotoStock/Science Source

Figure 22-15 Total internal reflection in water Light from a laser aimed from under the water toward the surface is totally internally reflected. What is the minimum incident angle for which total internal reflection will occur?

Set Up

Before hitting the surface (the boundary between water and air), the light is propagating in water, so this is medium 1. In part (a) medium 1 is ordinary water with index of refraction $n_1 = 1.33$; in part (b) medium 1 is sugar water with $n_1 = 1.49$. In both parts, medium 2 on the other side of the surface is air with index of refraction $n_2 = 1.00$. We'll use Equation 22-13 to solve for the critical angle in each case.

Critical angle for total internal reflection:

$$\theta_c = \sin^{-1}\left(\frac{n_2}{n_1}\right) \quad (22\text{-}13)$$

Solve

(a) Find the critical angle if medium 1 is water.

With $n_1 = 1.33$ and $n_2 = 1.00$, the critical angle is

$$\theta_c = \sin^{-1}\left(\frac{1.00}{1.33}\right) = \sin^{-1} 0.752 = 48.8°$$

If the angle of incidence θ_1 is 48.8° or greater, the light will undergo total internal reflection and no light will go into the air. (Note that θ_1 in Figure 22-15 is approximately 60°, which is indeed greater than 48.8°.)

(b) Find the critical angle if medium 2 is sugar water.

With $n_1 = 1.49$ and $n_2 = 1.00$, the critical angle is

$$\theta_c = \sin^{-1}\left(\frac{1.00}{1.49}\right) = \sin^{-1} 0.671 = 42.2°$$

Reflect

Adding sugar to water decreases the speed of light in the water, which increases the index of refraction. The minimum angle of incidence for which total internal reflection occurs is therefore smaller for sugar water than for pure water. The smaller the critical angle, the larger the range of angles at which light experiences total internal reflection.

NOW WORK Problems 3–8 and 10 from The Takeaway 22-4.

Total internal reflection explains the brilliance of a cut diamond as in **Figure 22-16a**. If light enters the front surface of a cut diamond, it will undergo total internal reflection at the back surface of the diamond (a boundary with air) if the incident angle is greater than the critical angle. Table 22-1 shows that diamond has a very high index of refraction, 2.42, so from Equation 22-13 the critical angle is very small: $\theta_c = \sin^{-1}(1.00/2.42) = 24.4°$. A talented jeweler cuts a diamond so that two things happen: First, light entering the front of the diamond over a broad range of angles will strike the back at an incident angle greater than $\theta_c = 24.4°$ so that total internal reflection occurs there; and second, this reflected light strikes the *front* surface of the diamond at an incident angle *less* than 24.4° so that this light can escape and be seen by you. The result is a gem that sparkles with brilliant reflections.

Total internal reflection also explains the *mirage* that happens when you look down the highway on a hot day and see what appear to be puddles of water on the road, as in **Figure 22-16b**. The explanation is that air sits in layers above the road, each layer a bit warmer, less dense, and with a lower index of refraction than the one above it. Light from the sky is refracted as it encounters the boundary between one layer of air and the next (**Figure 22-17**). Because the index of refraction of the layer of air closer to the road is lower, the light is bent farther from the normal. The normal direction in these

These "puddles" are actually a mirage.
The Sydney Morning Herald/ Getty Images

Figure 22-16 Sparkles and mirages Total internal reflection (a) is responsible for the sparkle of a cut diamond and (b) causes the shimmering patches on the road that look like puddles.

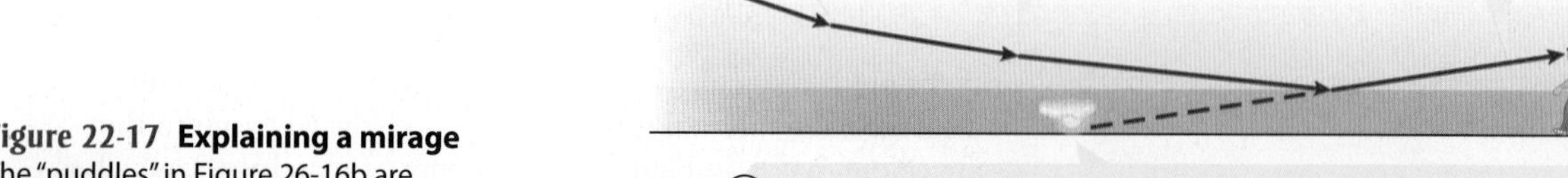

1 Light from a region of sky is refracted as it passes through layers of air above the road.
2 Light travels faster in the layers of warmer air closer to the road, so it is bent closer to horizontal.
3 Total internal reflection occurs for very shallow angles with respect to the air layer boundaries.
4 An image of the region of sky is formed close to the ground that gives the appearance of shimmery, blue puddles on the road.

Figure 22-17 Explaining a mirage The "puddles" in Figure 26-16b are an illusion caused by total internal reflection.

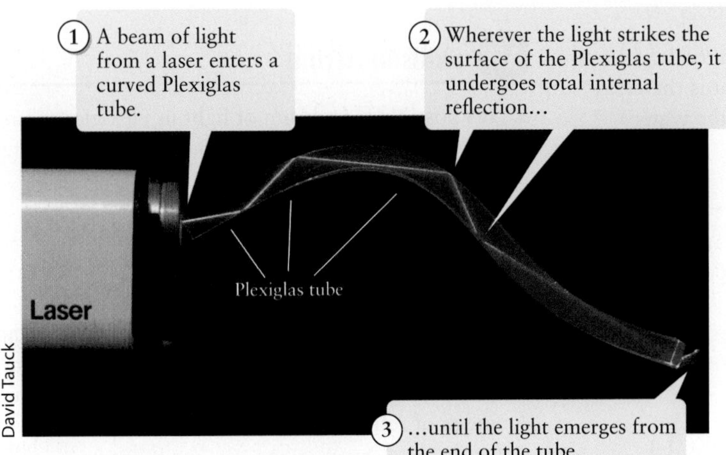

① A beam of light from a laser enters a curved Plexiglas tube.
② Wherever the light strikes the surface of the Plexiglas tube, it undergoes total internal reflection...
③ ...until the light emerges from the end of the tube.
Laser
Plexiglas tube
David Tauck

Figure 22-18 Light trapped by total internal reflection Light propagates through the interior of a curved bar of Plexiglas by a series of total internal reflections.

refractions is vertical, so the light is refracted closer to horizontal. Light that strikes the boundary between layers at a large angle with respect to the normal (a grazing angle as measured from the air layer boundary) can experience total internal reflection and reflect back up into the higher layer. Your eyes trace the light rays back along straight lines, so the light appears to be coming from a point close to the road. What looks like a puddle is actually a refracted image of the blue sky.

Endoscopy is a medical procedure used to see inside the body. It relies on a light fiber, an optical device used to carry light and sometimes data encoded in pulses of light, from one place to another. A beam of light sent down a light fiber experiences total internal reflection at the surface of the fiber, which results in multiple reflections that keep the beam inside the fiber. The bent bar of Plexiglas in **Figure 22-18** carries light in a similar way. Notice in Figure 22-18 that no light leaks out of the bar at the points where the beam hits the surface. This is because the angle of incidence at the surface is greater than the critical angle, so total internal reflection occurs. To maximize total internal reflection, most light fibers are made by surrounding a central core with one or two layers of a material of lower index of refraction than the core. Many endoscopes actually use bundles of several fibers. Light is sent down some fibers to illuminate the subject. Other fibers carry the image back to a camera.

THE TAKEAWAY for Section 22-4

✔ When light traveling in a medium with index of refraction n_1 reaches a boundary with a second medium of index of refraction n_2, total internal reflection can happen if $n_1 > n_2$.

✔ Total internal reflection takes place only if the incident angle measured from the normal at the boundary is greater than the critical angle θ_c given by Equation 22-13. In this case none of the light is refracted into the second medium.

Prep for the AP Exam

AP Building Blocks

1. Give two common uses of total internal reflection.
2. In your own words explain why the phenomenon of total internal reflection occurs only when light moves from a medium with a larger index of refraction toward a medium with a smaller index of refraction.
3. EX 22-3 Calculate the critical angle for the following.
   (a) Light travels from plastic (n = 1.50) to air (n = 1.00).
   (b) Light travels from water (n = 1.33) to air (n = 1.00).
   (c) Light travels from glass (n = 1.56) to water (n = 1.33).
   (d) Light travels from air (n = 1.00) to glass (n = 1.55).
4. EX 22-3 For each of the critical angles given, calculate the index of refraction for the optical materials that light travels out of toward air (n = 1.00).
   (a) $\theta_c = 48.5°$
   (b) $\theta_c = 47.0°$
   (c) $\theta_c = 42.6°$
   (d) $\theta_c = 35.0°$
   (e) $\theta_c = 55.7°$
   (f) $\theta_c = 38.5°$
   (g) $\theta_c = 22.2°$
   (h) $\theta_c = 75.0°$
5. EX 22-3 Calculate the critical angle for light traveling from sapphire to air.
6. EX 22-3 At what angle with respect to the vertical must a scuba diver look to see her friend standing on the very distant shore? Take the index of refraction of the water to be n = 1.33.

EX 22-3 7. A block of glass that has an index of refraction of 1.55 is completely immersed in water ($n = 1.33$). What is the critical angle for light traveling from the glass to the water?

AP® Skill Builders

EX 22-3 8. What is the largest angle θ_1 that will ensure that light is totally internally reflected in the fiber-optic pipe made of acrylic ($n = 1.50$) shown in the figure below?

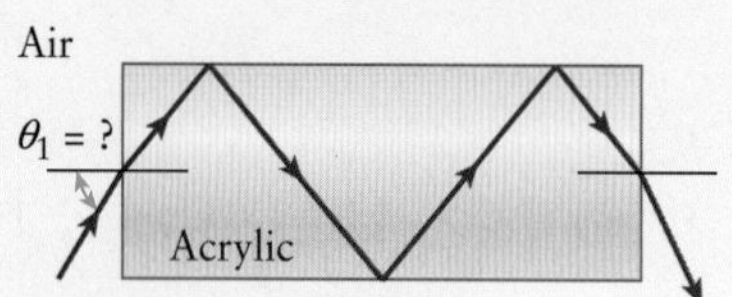

AP® Skills in Action

9. It's possible for a beam of light in Plexiglas to undergo total internal reflection at a boundary if the material on the other side of the boundary is
(A) ethyl alcohol.
(B) sapphire.
(C) diamond.
(D) more than one of these.

EX 22-3 10. A point source of light is 2.50 m below the surface of a pool. What is the diameter of the circle of light that a person above the water will see? Assume the water has an index of refraction of $n = 1.33$.

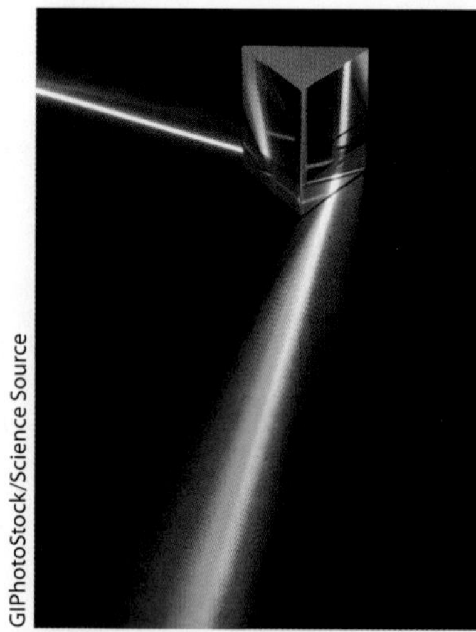

GIPhotoStock/Science Source

Figure 22-19 Dispersion Light of different wavelengths in a vacuum, and hence different colors, propagates at different speeds through the glass of which this prism is made. As a result, different colors refract along slightly different paths.

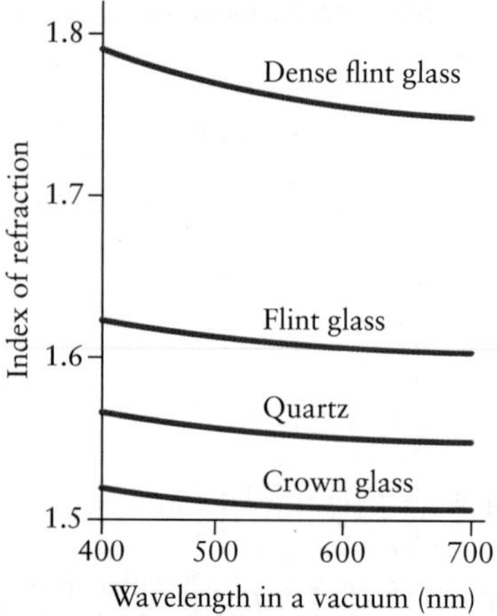

Figure 22-20 Dispersion in different materials The index of refraction in glass varies with the vacuum wavelength of light and with the type of glass.

22-5 The dispersion of light explains the colors from a prism or a rainbow

White light is a mixture of all the colors of the visible spectrum. You can see these colors by allowing sunlight to pass through a glass prism, as in **Figure 22-19**: The different colors emerge in different directions. This happens because the speed of light in a medium other than vacuum, such as the glass in a prism, is different for different frequencies of light. (In a vacuum the speed is equal to c for all frequencies.) This is a result of how a light wave interacts with the atoms of the medium. This variation of speed with frequency is called **dispersion**.

Because the speed v of light waves in a medium depends on the frequency, the index of refraction $n = c/v$ (Equation 22-8) depends on frequency as well. In most transparent materials the speed decreases with increasing frequency, from red to yellow to violet (see Figure 22-3): Red light travels fastest, and violet light travels slowest. This means that the index of refraction n increases with increasing frequency. Note that Equation 22-10 states that the wavelength in a vacuum is inversely proportional to frequency: $\lambda_{\text{vacuum}} = c/f$. So we can also say that the value of n decreases with increasing vacuum wavelength. (We specify *vacuum* wavelength, as the wavelength in the medium depends on the value of n; see Equation 22-11.) **Figure 22-20** shows how the index of refraction varies with vacuum wavelength for four different transparent materials. (The indices of refraction given in Table 22-1 are for yellow light, near the middle of the visible spectrum.)

Snell's law of refraction, Equation 22-9, tells us that the angle at which light refracts as it crosses the boundary between two transparent media depends on their indices of refraction. So it follows that different colors of light, with different vacuum wavelengths, refract at different angles. **Figure 22-21** shows this for light passing from vacuum into glass. The higher the index of refraction of the second medium, the more the refracted light is bent toward the normal. In common crown glass the index of refraction is about 1.51 for red light and about 1.53 for blue light. Hence the blue light bends more toward the normal than does the red light, and different colors of light are spread out or dispersed. (This is the origin of the term *dispersion*.)

The same effect explains the appearance of a rainbow (Figure 22-2b). When raindrops in midair are illuminated by the Sun, sunlight enters each raindrop, is partly reflected off the back of the drop, and then exits out the front of the drop. The index of refraction of water is different for different wavelengths, so each color of light emerges in a slightly different direction to form a rainbow.

The amount of dispersion is different for different transparent materials. In crown glass, for example, the index of refraction varies by 0.02 from red ($n = 1.51$) to blue

White light is composed of light across all wavelengths (colors) in the visible spectrum. The colors spread apart when light crosses the boundary between two light transmitting media.

When light passes into a region of higher index of refraction, it is bent toward the direction normal to the surface. The greater the index of refraction of the second medium, the more the light is bent.

Air

Glass

The index of refraction of blue light is higher than the index of refraction of red light. Therefore, blue light is bent more than red light.

Figure 22-21 Analyzing dispersion The difference in refracted angle between the red and blue light is greatly exaggerated for clarity.

($n = 1.53$), while for diamond the index of refraction varies by 0.04 from red ($n = 2.41$) to blue ($n = 2.45$). As a result, the colors of white light are spread out over a wider angle by a cut diamond than by a piece of glass cut to the same shape. This high value of dispersion contributes to the "sparkly" character of a cut diamond.

EXAMPLE 22-4 Dispersion in Dense Flint Glass

A narrow beam of white light enters a rectangular block of dense flint glass at 60.0° from the normal. The block is 0.500 m on a side. How far apart will the red and blue parts of the visible spectrum be when the light leaves the glass? The index of refraction of dense flint glass is 1.75 for red light and 1.79 for blue light.

Set Up

Both colors of light are incident on the block at the same angle: $\theta_1 = 60.0°$. However, the red and blue refract at different angles because the index of refraction is different for the two colors. We'll use Snell's law, Equation 22-9, to calculate the angles θ_{red} and θ_{blue}, and then use trigonometry to find the distances d_{red} and d_{blue} shown in the figure. The difference between these distances tells us how far apart the points are where the two colors emerge from the glass.

Snell's law of refraction:

$$n_1 \sin \theta_1 = n_2 \sin \theta_2 \qquad (22\text{-}9)$$

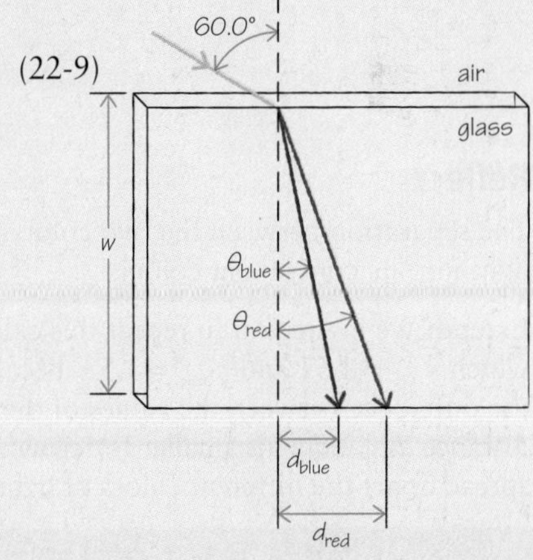

Solve

Find the refracted angles θ_{red} and θ_{blue} for the two colors.

For both colors $n_1 = n_{air} = 1.00$ and $\theta_1 = 60.0°$. For red light $n_2 = 1.75$, so Equation 22-9 becomes

$$1.00 \sin 60.0° = 1.75 \sin \theta_{red}$$

$$\sin \theta_{red} = \frac{1.00 \sin 60.0°}{1.75} = 0.4948$$

$$\theta_{red} = \sin^{-1} 0.4948 = 29.66°$$

(We've kept an extra significant digit in our result; we'll round off at the end of the calculation.) Do the same calculation for blue light, for which $n_2 = 1.79$:

$$1.00 \sin 60.0° = 1.79 \sin \theta_{blue}$$

$$\sin \theta_{blue} = \frac{1.00 \sin 60.0°}{1.79} = 0.4838$$

$$\theta_{blue} = \sin^{-1} 0.484 = 28.93°$$

Find the distances d_{red} and d_{blue} for the two colors of light, and from these find the separation between the two colors as they exit the glass.

For each color of light, the ray that extends from where the light enters the block of glass to where it exits the block forms the hypotenuse of a right triangle. The other two sides are the vertical dimension of the block, $w = 0.500$ m, and the distance d_{red} or d_{blue} that the light is displaced horizontally. In each case the side of length w is adjacent to the angle θ_{red} or θ_{blue}, and the side of length d_{red} or d_{blue} is opposite to that angle. In a right triangle the length of the opposite side divided by the length of the adjacent side equals the tangent of the angle, so

$$\tan \theta_{red} = \frac{d_{red}}{w}$$

$$\tan \theta_{blue} = \frac{d_{blue}}{w}$$

Solve for the distances d_{red} and d_{blue} using the angles that we calculated above:

$$d_{red} = w \tan \theta_{red} = (0.500 \text{ m}) \tan 29.66° = 0.2847 \text{ m}$$

$$d_{blue} = w \tan \theta_{blue} = (0.500 \text{ m}) \tan 28.93° = 0.2764 \text{ m}$$

The distance between where the red light exits the glass and where the blue light exits the glass is

$$d_{red} - d_{blue} = 0.2847 \text{ m} - 0.2764 \text{ m} = 0.0083 \text{ m} = 8.3 \text{ mm}$$

Reflect

The separation between the two colors is fairly substantial, so this block of flint glass does a good job of spreading white light into its constituent colors.

Extend: We invite you to repeat this calculation for crown glass, the kind of glass commonly used to make windows, for which $n_{red} = 1.51$ and $n_{blue} = 1.53$. Because crown glass has a smaller index of refraction than flint glass, and because the difference between the values of the two indices is smaller for crown glass than for flint glass, you'll find that the distance $d_{red} - d_{blue}$ is smaller for crown glass. Which type of glass would be a better choice for a prism intended to spread apart the different colors of light, as in Figure 22-19?

NOW WORK Problems 1, 3, 5, and 7 from The Takeaway 22-5.

THE TAKEAWAY for Section 22-5

✔ The speed of light in a material medium depends on the frequency of the light. This is called dispersion. In most materials the index of refraction for visible light increases from the red end of the spectrum (low frequency, long wavelength) to the blue end of the spectrum (high frequency, short wavelength).

✔ When light crosses a boundary into a medium of different index of refraction, different colors refract by different angles. This causes the colors to spread apart by an amount that depends on how strongly the index of refraction varies with wavelength.

Prep for the AP® Exam

AP® Building Blocks

EX 22-4 1. A light beam strikes a piece of glass with an incident angle of 45.0°. The beam contains two colors: 450 nm and an unknown wavelength. The index of refraction for the 450-nm light is 1.482. Determine the index of refraction for the unknown wavelength if the angle between the two refracted rays is 0.275°. Assume the glass is surrounded by air.

2. Sunlight striking a diamond throws rainbows of color in every direction. From where do the colors come?

AP® Skill Builders

EX 22-4 3. Blue light (500 nm) and yellow light (600 nm) are incident on a 12-cm-thick slab of glass as shown in the figure below. In the glass, the index of refraction for the blue light is 1.545, and for the yellow light it is 1.523. What distance along the glass slab (side AB) separates the points at which the two rays emerge back into air?

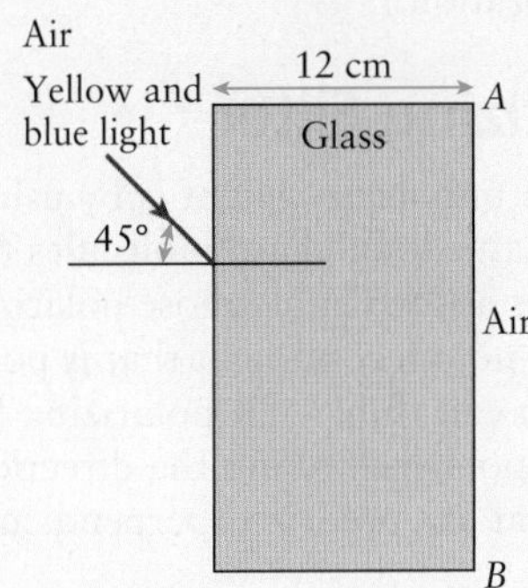

4. Explain why the Moon appears to change colors during a total lunar eclipse (when Earth's shadow completely blocks the light coming from the Sun).

EX 22-4 5. For a certain optical medium the speed of light varies from a low value of 1.90×10^8 m/s for violet light to a high value of 2.00×10^8 m/s for red light.
(a) Calculate the range of the index of refraction of the material for visible light.
(b) A white light is incident on the medium from air, making an angle of 30.0° with the normal. Compare the angles of refraction for violet light and red light.
(c) Repeat the previous part when the incident angle is 60.0°.

AP® Skills in Action

6. A flash of white light in air shines straight down on the surface of a pond of water. Which color of light from the flash reaches the bottom of the pond first? (Remember white light contains all of the colors of the visible spectrum.)
(A) The blue light
(B) The yellow light
(C) The red light
(D) All reach the bottom at the same time.

EX 22-4 7. A beam of light shines on an equilateral glass prism at an angle of 45.0° to one face as shown in figure (a) below.

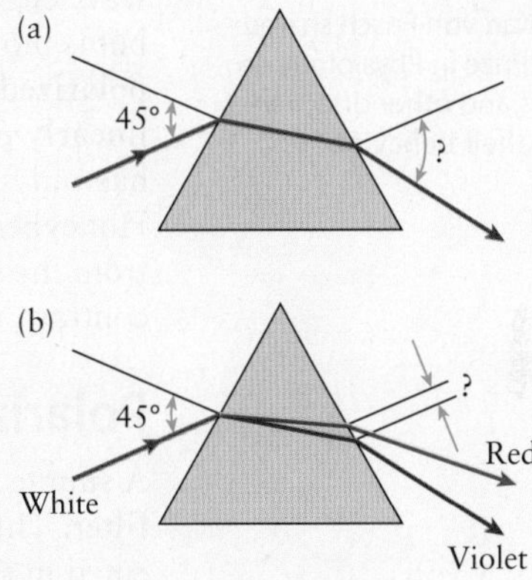

(a) What is the angle at which the light emerges from the opposite face, given that $n_{glass} = 1.57$?
(b) Now consider what happens when dispersion is involved, as shown in figure (b). Assume the incident ray of light spans the spectrum of visible light between 400 and 700 nm (violet to red, respectively). The index of refraction for violet light in the glass prism is 1.572, and it is 1.568 for red light in the glass prism. Find the distance along the right face of the prism between the points where the red light and violet light emerge back into air. Assume the prism is 10.0 cm on a side and the incident ray hits the midpoint of the left face.

8. Diamonds are renowned for how they reflect light and produce a rainbow of colors. Diamonds have these properties because the speed of light in diamond
(A) is faster than in air.
(B) is slower than in air.
(C) depends on the color of the light.
(D) both (B) and (C).

22-6 In a polarized light wave the electric field vector points in a specific direction

When a honeybee finds nectar, it communicates the location to other bees in the hive. In the 1940s Austrian ethologist Karl von Frisch established that if bees can see even a small patch of blue sky, they can use the position of the Sun to describe the path back to the food from the hive (**Figure 22-22**). How is it that bees can know the position of the Sun even if they can't see it directly?

Medioimages/Photodisc/Getty Images

Figure 22-22 Navigating by the light of the sky Honeybees have the ability to detect the polarization of light entering their eyes—that is, the orientation of the electric field $\vec{E}$ in the light wave. The polarization helps them determine the position of the Sun in the sky, even if the Sun isn't directly visible. Karl von Frisch shared the 1973 Nobel Prize in Physiology or Medicine for this and other discoveries about bees and their behavior.

The explanation is that light is a transverse electromagnetic wave, with an oscillating electric field $\vec{E}$ and magnetic field $\vec{B}$ that are perpendicular to each other and to the direction of wave propagation (see Figure 22-5). The orientation of the $\vec{E}$ field is called the **polarization** of the light wave. (We don't need to separately state the orientation of $\vec{B}$, because we know that it's perpendicular to both the direction of propagation and the orientation of $\vec{E}$.) Natural light such as that emitted by the Sun or an ordinary light bulb is **unpolarized:** The orientation of the electric field changes randomly from one moment to the next. For example, if the wave is propagating in the positive x direction, at one moment $\vec{E}$ may be oriented along the y axis, a short time later it may be oriented along the z axis, a short time after that it may be oriented at 23.7° to the y axis, and so on.

When sunlight scatters from molecules or small particles in the atmosphere, however, the scattered light that we see has its $\vec{E}$ field oriented predominantly in one direction (**Figure 22-23**). (Scattering is stronger for short-wavelength light than for long-wavelength light, which is why the color of the sky is dominated by short-wavelength blue light.) Light in which the orientation of the $\vec{E}$ field changes randomly, but is more likely to be in one orientation than in other orientations, is called **partially polarized.** Light for which the $\vec{E}$ field is oriented *completely* along one direction is called **linearly polarized.** For example, the $\vec{E}$ field for the light wave shown in Figure 22-5 has only a y component, so we say this light is linearly polarized along the y axis. Honeybees (Figure 22-22) have the ability to detect the polarization of light coming from the sky, and by using this they can infer the position of the Sun. (Human eyes, by contrast, are only weakly sensitive to polarization.)

Polarizing Light with a Polarizing Filter

A simple way to make polarized light from unpolarized light is by using a **polarizing filter.** This is a transparent sheet that contains long-chain molecules that are all oriented in the same direction. These molecules absorb light whose polarization direction is along the axis of the molecules, but have no effect on light that is polarized perpendicular to that direction. Equivalently, you can think of a polarizing filter as having slots that allow waves to pass if they are polarized along the direction of the slots, called the *polarization direction*. Waves that are polarized perpendicular to the slots

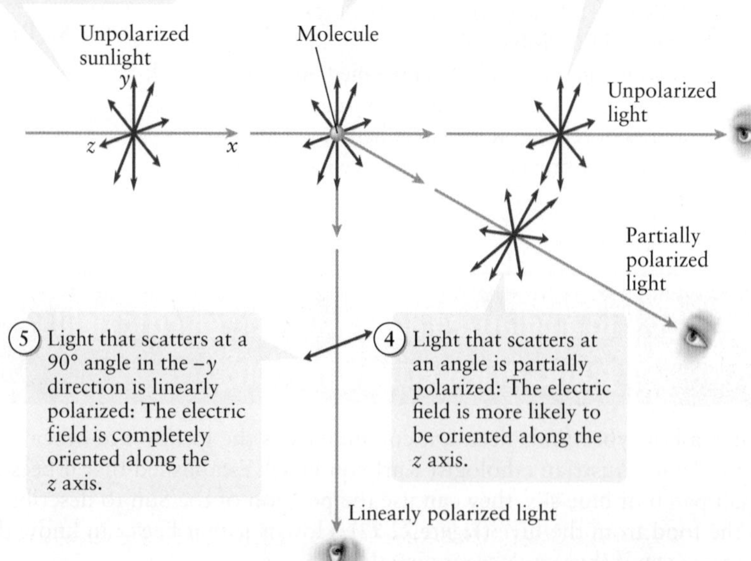

1 Sunlight propagates in the x direction. The light is unpolarized, so the orientation of the electric field (shown by the blue arrows) changes randomly in the yz plane.
2 When light strikes a molecule in the atmosphere, the polarization and direction of propagation can both change.
3 Light that scatters in the forward direction (that is, is not deflected) remains unpolarized.
Unpolarized sunlight
y
z
x
Molecule
Unpolarized light
Partially polarized light
5 Light that scatters at a 90° angle in the −y direction is linearly polarized: The electric field is completely oriented along the z axis.
4 Light that scatters at an angle is partially polarized: The electric field is more likely to be oriented along the z axis.
Linearly polarized light

Figure 22-23 Polarization by scattering The extent to which sunlight is polarized by scattering depends on the scattering angle.

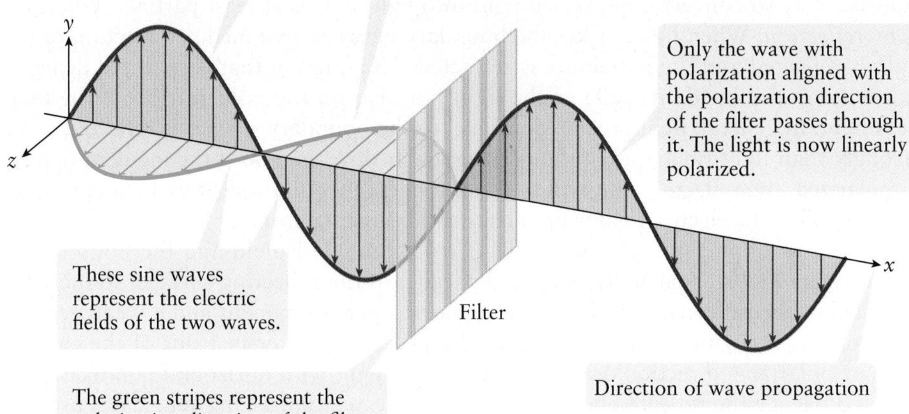
Two waves of different polarizations encounter a polarizing filter.
y
z
Only the wave with polarization aligned with the polarization direction of the filter passes through it. The light is now linearly polarized.
x
These sine waves represent the electric fields of the two waves.
Filter
The green stripes represent the polarization direction of the filter.
Direction of wave propagation

Figure 22-24 A linear polarizing filter I A filter of this kind allows only light with an electric field component aligned in a certain direction, called the polarization direction, to pass through it.

are blocked (**Figure 22-24**). If we send unpolarized light into the filter, at any instant the electric field $\vec{E}$ has a component along the polarization direction of the filter and a component perpendicular to that direction. Only the component of $\vec{E}$ along the polarization direction of the filter will pass through, so the light that emerges from the filter will be linearly polarized.

Because a polarizing filter absorbs some of the incident light that falls on it, the light exiting the filter is in general less intense than the incident light. Suppose the incident light has an oscillating electric field with amplitude E_0 that is oriented at an angle θ to the polarization direction of the filter (**Figure 22-25**). Only the component of this field that is aligned with the polarization direction is allowed to pass through the filter:

$$E_{\text{transmitted}} = E_0 \cos\theta \tag{22-14}$$

The intensity of the light is proportional to the square of the electric field amplitude. Equation 22-14 shows the amplitude $E_{\text{transmitted}}$ of the transmitted light equals $\cos\theta$ times the amplitude E_0 of the incident light. Hence the intensity I of the transmitted light equals $\cos^2\theta$ times the intensity I_0 of the incident light:

$$I = E_0^2 \cos^2\theta = I_0 \cos^2\theta \tag{22-15}$$

If the incident light is unpolarized, the value of the angle θ will vary randomly between 0 and 360°. The average value of $\cos^2\theta$ over this range is 1/2, so if unpolarized light incident on a polarizing filter has intensity I_0, the intensity of the light that emerges from the filter will be $I = I_0/2$.

(1) The polarizing filter allows light polarized along this direction to pass. It blocks light polarized in the perpendicular direction.

(2) As a result, this component of the electric field of a light wave is blocked by the filter...

- The transmitted light is linearly polarized along the polarization direction of the filter.
- The electric field amplitude of the transmitted light equals the incident amplitude E_0 multiplied by $\cos\theta$.
- The intensity of the transmitted light equals the incident intensity I_0 multiplied by $\cos^2\theta$.

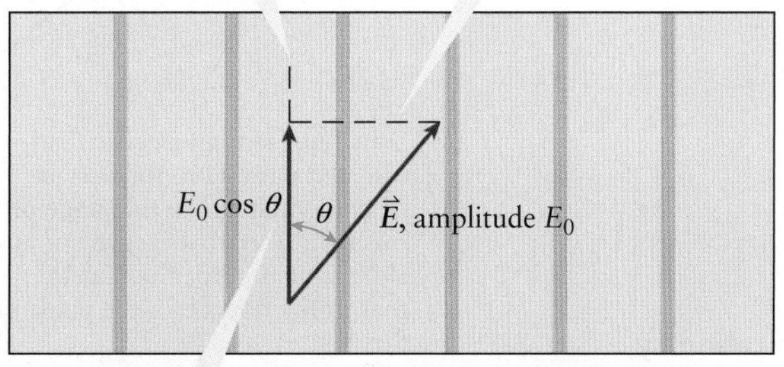
$E_0 \cos\theta$
θ
$\vec{E}$, amplitude E_0

(3) ...and only this component is allowed to pass through the filter.

Figure 22-25 A linear polarizing filter II What happens to light that enters a polarizing filter with its electric field $\vec{E}$ at an angle to the polarization direction?

(a) Photographed without a filter

David Tauck

(b) Photographed with a polarizing filter oriented vertically, to block light with horizontal polarization

David Tauck

Figure 22-26 Reducing reflections with a polarizing filter Reflected light is linearly polarized. The polarizing lenses often used for sunglasses can dramatically reduce glare from reflected light.

Polarizing Light by Reflection

Another way to convert unpolarized light into light that is at least partially polarized is by *reflection*. When light strikes the boundary between two media, a fraction of the light is reflected and the remainder is refracted. The fraction that is reflected depends not only on the incident angle of the light but also on the polarization of the incident light. In general, light polarized parallel to the boundary surface is reflected more strongly than light polarized in the perpendicular direction. So if the incident light is unpolarized, the reflected light will be at least partially polarized: It will contain more light in which the electric field is parallel to the boundary surface.

As an example, sunlight that reflects from the windshield and the hood of the car in **Figure 22-26a** is partially polarized parallel to the reflecting surface, so the electric field of the reflected light has a strong horizontal component and a weak vertical component. In **Figure 22-26b** we've placed a polarizing filter in front of the camera with its polarizing direction oriented vertically, so light with horizontal polarization is blocked. Hence the reflected light is greatly suppressed. Polarizing sunglasses use this same principle to minimize reflections from the road or the surface of a lake or ocean: Sunlight reflected from these horizontal surfaces is predominantly polarized in the horizontal direction, so to block this reflected light, the filters that make up the sunglass lenses have a vertical polarization direction. If you look at this reflected light through polarizing sunglasses and tilt your head to one side until one eye is directly above the other, you'll see the reflections reappear. That's because the polarization direction of the lenses is now horizontal, the same as the orientation of the electric field of the reflected light.

When unpolarized light is incident on the boundary between two media, there is one particular angle of incidence for which the reflected light is *completely* polarized parallel to the boundary. The angle of incidence θ_1 that results in this special condition is called **Brewster's angle θ_B**. When $\theta_1 = \theta_B$, it turns out that the sum of the incident angle θ_1 and the refracted angle θ_2 equals 90° (**Figure 22-27**):

(22-16) $$\theta_1 + \theta_2 = 90° \quad \text{when } \theta_1 = \theta_B$$

We can use this to solve for the value of Brewster's angle θ_B in terms of the indices of refraction n_1 and n_2 of the two media. Snell's law of refraction, Equation 22-9, gives us this relationship between θ_1 and θ_2:

(22-9) $$n_1 \sin \theta_1 = n_2 \sin \theta_2$$

From Equation 22-16, if θ_1 equals Brewster's angle θ_B,

$$\theta_2 = 90° - \theta_1 = 90° - \theta_B$$

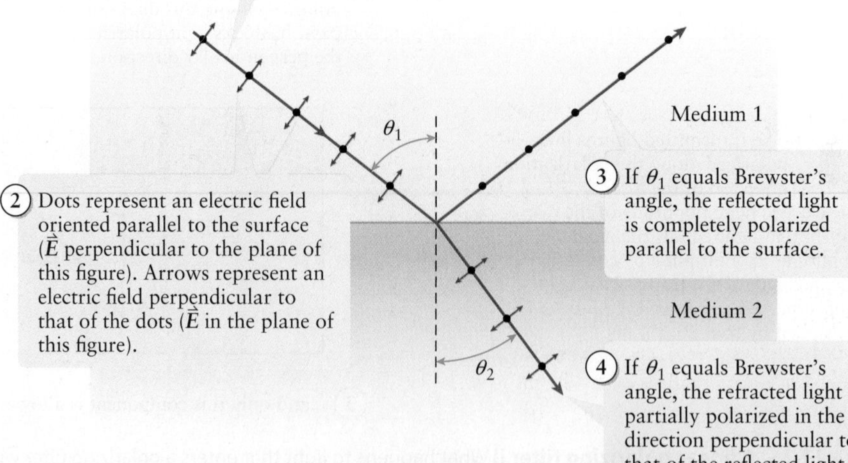

1 Unpolarized light propagating in medium 1 strikes the boundary with medium 2 at an incident angle θ_1.
θ_1
Medium 1
2 Dots represent an electric field oriented parallel to the surface ($\vec{E}$ perpendicular to the plane of this figure). Arrows represent an electric field perpendicular to that of the dots ($\vec{E}$ in the plane of this figure).
3 If θ_1 equals Brewster's angle, the reflected light is completely polarized parallel to the surface.
Medium 2
θ_2
4 If θ_1 equals Brewster's angle, the refracted light is partially polarized in the direction perpendicular to that of the reflected light.

Figure 22-27 Polarization by reflection Light reflected from a surface is partially polarized and becomes completely polarized when the incident angle equals Brewster's angle.

If we substitute this into Equation 22-9, we get

$$n_1 \sin \theta_B = n_2 \sin (90° - \theta_B)$$

We know from trigonometry that $\sin (90° - \theta_B) = \cos \theta_B$. So

$$n_1 \sin \theta_B = n_2 \cos \theta_B$$

Divide both sides of this equation by n_1, then divide both sides by $\cos \theta_B$. The result is

$$\frac{\sin \theta_B}{\cos \theta_B} = \frac{n_2}{n_1}$$

The sine of an angle divided by its cosine equals the tangent of the angle. So $\tan \theta_B = n_2/n_1$, or

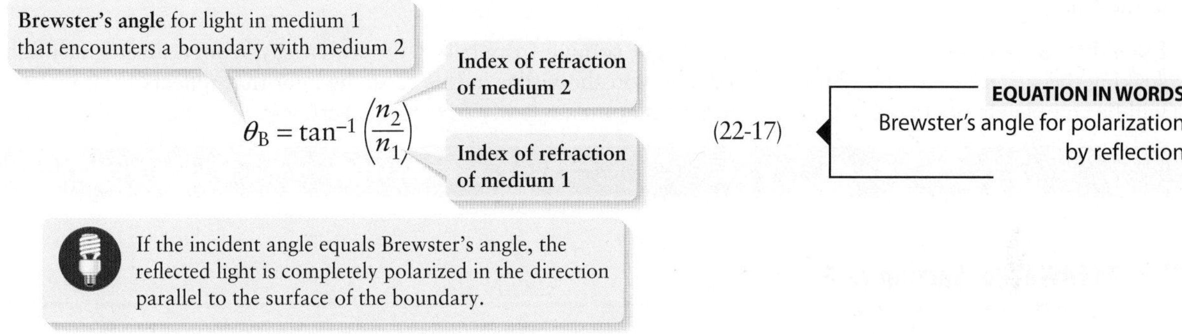

Brewster's angle for light in medium 1 that encounters a boundary with medium 2
Index of refraction of medium 2
Index of refraction of medium 1
EQUATION IN WORDS
Brewster's angle for polarization by reflection
If the incident angle equals Brewster's angle, the reflected light is completely polarized in the direction parallel to the surface of the boundary.

$$\theta_B = \tan^{-1}\left(\frac{n_2}{n_1}\right) \qquad (22\text{-}17)$$

If the angle of incidence for unpolarized light is something other than θ_B, the reflected ray is partly polarized. If unpolarized light strikes the boundary perpendicular to it, so $\theta_1 = 0$, there is no change in polarization of the reflected light.

WATCH OUT !

Brewster's angle is not the same as the critical angle.

Be careful not to confuse Equation 22-17 for Brewster's angle with the similar-looking Equation 22-13 for the critical angle. For a boundary between two transparent media, Brewster's angle (which involves an inverse tangent function) is the *one and only* angle of incidence for which the reflected light is completely polarized. This can happen whether the index of refraction n_1 for the first medium is larger or smaller than the index of refraction n_2 for the second medium. The critical angle (which involves an inverse sine function) is the minimum angle of incidence for which there is total internal reflection (Section 22-4); there is also total internal reflection if the angle of incidence is greater than the critical angle. Total internal reflection is possible only if the index of refraction n_1 for the first medium is greater than the index of refraction n_2 for the second medium.

EXAMPLE 22-5 Brewster's Angle for Air to Water

At what incident angle must light strike the surface of a pond so that the reflected light is completely polarized?

Set Up

This is the situation shown in Figure 22-27, with air as medium 1 and water as medium 2. The reflected light will be completely polarized if the incident angle θ_1 is equal to Brewster's angle θ_B given by Equation 22-17. Table 22-1 tells us the value of the indices of refraction: $n_1 = n_{air} = 1.00$ and $n_2 = n_{water} = 1.33$.

Brewster's angle for polarization by reflection:

$$\theta_B = \tan^{-1}\left(\frac{n_2}{n_1}\right) \qquad (22\text{-}17)$$

Solve

Use Equation 22-17 to find Brewster's angle for this situation.

With $n_1 = 1.00$ and $n_2 = 1.33$, Brewster's angle is

$$\theta_B = \tan^{-1}\left(\frac{1.33}{1.00}\right) = 53.1°$$

The reflected light will be completely polarized if the incident angle θ_1 is equal to $\theta_B = 53.1°$.

Reflect

The boundary between the two media (air and water) is the horizontal surface of the pond, so the normal to this surface is vertical. Our result tells us that if light from the sky strikes the surface at an angle of 53.1° from the vertical, the reflected light will be completely polarized. Light striking close to this angle will be strongly polarized but not 100% polarized.

Extend: You should repeat this calculation for light coming from below the water (say, from a diver's flashlight) and striking the surface of the pond from below. Can you show that in this case, the light that reflects back into the water will be completely polarized if the incident light is at an angle of 36.9° to the vertical?

NOW WORK Problems 3 and 8 from The Takeaway 22-6.

THE TAKEAWAY for Section 22-6

✔ The orientation of the oscillating electric field in a light wave tells you the polarization of that wave.

✔ In natural light the direction of the electric field changes randomly and is equally likely to be in any direction perpendicular to the propagation direction of the wave. Such light is called unpolarized.

✔ Light that has its electric field oriented in one specific direction is called linearly polarized. Unpolarized light can be polarized by scattering, by passing it through a polarizing filter, or by reflection.

1. Unpolarized light is passed through an optical filter that is oriented in the vertical direction. If the incident intensity of the light is 78 W/m^2, what are the polarization and intensity of the light that emerges from the filter?
2. Describe the physical interactions that take place when unpolarized light is passed through a polarizing filter. Be sure to describe the electric field of the light before and after the filter as well as the incident and transmitted intensities of the light source.
3.  What angle(s) does vertically polarized light make relative to a polarizing filter that diminishes the intensity of the light by 25% (so that it is at 75% of its initial intensity)?
4. What is Brewster's angle when light in water is reflected off a glass surface? Assume $n_{water} = 1.33$ and $n_{glass} = 1.55$.

AP Skill Builders

5. A pool has a light source underwater.
   (a) Calculate Brewster's angle for reflections off the surface of the pool from this underwater light source.
   (b) Calculate the angle for total internal reflection when light starts in water and reflects off air.
6. Light that passes through a series of three polarizing filters emerges from the third filter horizontally polarized with an intensity of 250 W/m^2. If the polarization angle between the filters increases by 25° from one filter to the next, find the intensity of the incident beam of light, assuming it is initially unpolarized.
7. Vertically polarized light that has an intensity of 4.0×10^2 W/m^2 is incident on two polarizing filters. The first filter is oriented 30.0° from the vertical, while the second filter is oriented 75.0° from the vertical. Predict the intensity and polarization of the light that emerges from the second filter.

AP Skills in Action

8.  The critical angle between two optical media is 60.0°. What is Brewster's angle at the same interface between the two media?

9. Two polarizing filters, A and B, are placed one behind the other with their polarization directions perpendicular to each other. A beam of unpolarized light is directed at filter A. What fraction of the original light intensity will remain after the light passes through both filters A and B?
(A) 1/2
(B) 1/4
(C) 1/8
(D) 0

10. Two polarizing filters, A and B, are placed one behind the other with their transmission directions perpendicular to each other. A third polarizing filter, C, is placed in between them. The transmission direction of C is halfway between those of filters A and B (that is, at 45° to that of filter A and at 45° to that of filter B). A beam of unpolarized light is directed at filter A. What fraction of the original light intensity will remain after the light passes through filters A, B, and C?
(A) 1/2
(B) 1/4
(C) 1/8
(D) 0

22-7 Light waves reflected from the surfaces of a thin film can interfere with each other, producing dazzling effects

The wings of the butterfly *Morpho menelaus* (**Figure 22-28**) show brilliant colors. Yet the material of which the wings are made is colorless! The explanation for this seeming contradiction is that the colors are produced by the interference of light, a process similar to the interference of sound waves we investigated in Chapter 21. In Section 21-5 we saw that sound waves interfere constructively or destructively depending on how the peaks and troughs of the two waves align. The same is true for light waves.

Figure 22-28 Butterfly interference The iridescent colors on the wings of this *Morpho menelaus* butterfly result from the interference of light waves that reflect from the wing surfaces.

To begin our investigation of light wave interference, let's consider what happens when light of a single wavelength, and hence a single color—called **monochromatic light**—strikes a thin layer of a transparent material. An example of such a transparent *thin film* is the *tapetum lucidum* ("shining carpet") that lines the back of some animals' eyes behind the retina (see Figure 22-2a). The material behind the tapetum lucidum is opaque (light cannot pass through it), so we can think of the tapetum lucidum as a **thin film** against an opaque backing.

As **Figure 22-29a** shows, some of the light waves that strike the front surface of the thin film are reflected. The remaining light enters the thin film (**Figure 22-29b**), strikes the opaque backing at the back surface of the film, and is reflected back up. (Some light is also absorbed by the backing.) As **Figure 22-29c** shows, the light waves reflected from the front surface and the light waves reflected from the back surface both end up above the thin film and traveling in the same direction.

The single light wave that was incident on the thin film has now been split into two reflected, outgoing waves that have traveled different paths. These two waves were in phase before they encountered the thin film because they were part of the same incident wave. But in general the two reflected waves are *not* in phase when they recombine above the front surface of the thin film. That's because the light that enters the film and reflects off the back surface travels farther than the light that reflects off the front surface.

Whether the interference that occurs between the two reflected waves is constructive or destructive depends on the number of wave cycles that fit into the extra distance traveled by the light that enters the film. If an integer number of cycles (1, 2, 3, . . . cycles) fit into that extra distance, then the two outgoing waves are in phase and constructively interfere. The surface of the film appears bright. If an odd number of half cycles (1/2, 3/2, 5/2, . . . cycles) fit into the extra distance, however, the two outgoing waves are 180° out of phase and destructively interfere. (We specify an *odd* number of half cycles because an even number of half cycles is the same as an integer number of full cycles; for example, 4/2 = 2. This case gives constructive, not destructive, interference.) When destructive interference occurs, the two outgoing waves cancel each other out and the surface of the film appears dark.

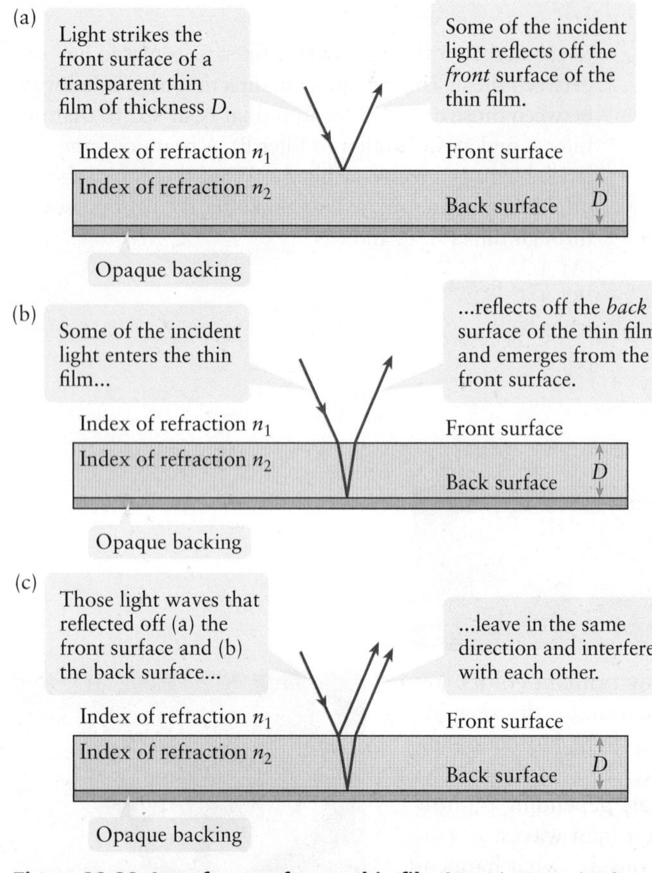
(a)
Light strikes the front surface of a transparent thin film of thickness D.
Some of the incident light reflects off the front surface of the thin film.
Index of refraction n1
Front surface
Index of refraction n2
Back surface
D
Opaque backing
(b)
Some of the incident light enters the thin film...
...reflects off the back surface of the thin film, and emerges from the front surface.
Index of refraction n1
Front surface
Index of refraction n2
Back surface
D
Opaque backing
(c)
Those light waves that reflected off (a) the front surface and (b) the back surface...
...leave in the same direction and interfere with each other.
Index of refraction n1
Front surface
Index of refraction n2
Back surface
D
Opaque backing

Figure 22-29 Interference from a thin film I For clarity in this figure we've drawn light rays that reach the surface of a thin film at an angle. In our calculations we'll assume that the light hits the surface face-on.

Let's consider the case in which the incident light is normal to the surface of the thin film (that is, the light strikes the film face-on). Then the path length difference Δ_{pl} for the two waves—one that reflects off the front of the film and the other that reflects off the back—is twice the thickness D of the film:

(22-18) $$\Delta_{pl} = 2D$$

Constructive interference occurs when an integer number of wavelengths of light fit into this path length difference of $2D$. Destructive interference occurs when an odd number of half-wavelengths fit into the distance $2D$. However, the wavelength of light inside the film is not the same as the wavelength outside the film. Recall from Section 22-3 that the wavelength of light in a medium with index of refraction n is

(22-11) $$\lambda = \frac{\lambda_{vacuum}}{n}$$

As in Figure 22-29, let's say that the material outside the film has index of refraction n_1, and the material of which the film is made has index of refraction n_2. Then from Equation 22-11, the wavelengths in the two materials are

$$\lambda_1 = \frac{\lambda_{vacuum}}{n_1} \text{ outside the film}$$

$$\lambda_2 = \frac{\lambda_{vacuum}}{n_2} \text{ inside the film}$$

Comparing these two, we see that

(22-19) $$\lambda_2 = \frac{n_1 \lambda_1}{n_2}$$

We now know everything we need to determine how the light reflected from the front surface of the thin film interferes with the light that enters the film and is reflected from the back surface. *Constructive* interference occurs when the path difference equals an integer number of wavelengths λ_2 (the wavelength of light inside the film):

(22-20) $$2D = m\lambda_2 = m\frac{n_1 \lambda_1}{n_2}, \quad m = 1, 2, 3, \ldots$$

(constructive interference, thin film with an opaque backing)

If we set $m = 1$ in Equation 22-20 and solve for D, we get the *minimum* thickness that a thin film must have to give constructive interference when light of wavelength λ_1 strikes the front surface face-on:

(22-21) $$D_{min} = \frac{n_1 \lambda_1}{2n_2}$$

(minimum thickness for constructive interference, thin film with an opaque backing)

Destructive interference occurs when the path difference equals an odd number of one-half the wavelength λ_2 inside the film:

(22-22) $$2D = (2m - 1)\frac{\lambda_2}{2} = (2m - 1)\frac{n_1 \lambda_1}{2n_2}, \quad m = 1, 2, 3, \ldots$$

(destructive interference, thin film with an opaque backing)

If we set $m = 1$ in Equation 22-22 and solve for D, we get the minimum thickness that a thin film must have to give destructive interference when light of wavelength λ_1 strikes the front surface face-on:

$$D_{\text{min}} = \frac{n_1 \lambda_1}{4n_2} \qquad (22\text{-}23)$$

(minimum thickness for destructive interference, thin film with an opaque backing)

This effect explains why the tapetum lucidum of an animal's eye improves the animal's ability to see in low light levels, and why this ability is greatest at certain wavelengths. The retina detects light by absorbing it, but some light passes through the retina without being absorbed. The tapetum lucidum behind the retina reflects some of this "lost" light back into the retina so it can have a second chance to be absorbed and detected. This reflection happens most strongly for certain wavelengths—to be specific, those that satisfy Equation 22-20, where n_2 is the index of refraction of the material of which the tapetum lucidum is made. Other colors with other wavelengths do not satisfy that relationship perfectly, so they are not reflected as strongly. Hence the animal's eye is less sensitive to light at these other wavelengths.

Even after a second pass through the retina, some of the light reflected from the tapetum lucidum escapes from the retina and exits the eye. This gives rise to the phenomenon of *eyeshine* that you can see in Figure 22-2a. The tapetum lucidum for the cat in that photo preferentially reflects green light because an integer number of wavelengths of that color fit into the path difference (twice the thickness of the tapetum lucidum). This explains why the cat's eye is most sensitive to green light.

The same effect explains the colors of the *Morpho* butterfly shown in Figure 22-28. The wings of *Morpho* are covered with microscopic scales that act like a thin film. The thickness of these scales is such that there is constructive interference for reflected light at wavelengths in the blue-green part of the spectrum. As a result, *Morpho* appears to glow at those wavelengths.

EXAMPLE 22-6 A Soapy Film

Monochromatic light that has wavelength 560 nm in air strikes a layer of soapy water, which has an index of refraction of 1.40 and rests on a bathroom tile. (a) If the layer of soapy water is 700 nm thick, does constructive interference, destructive interference, or neither occur when the light strikes the surface close to the normal? (b) What is the minimum thickness of soapy water that would result in no (or minimum) reflection from the surface?

Set Up

The situation is the same as in Figure 22-29. Interference occurs between (i) light that is reflected from the top surface of the soapy layer and (ii) light that enters the soapy layer and is eventually reflected back out. In part (a) we'll compare the given values of wavelength and film thickness to Equations 22-20 and 22-22 to decide whether the interference is constructive, destructive, or something in between. In part (b) we'll use Equation 22-23 to find the minimum thickness for destructive interference. In both parts, Equation 22-19 will help us relate the wavelength of the light in air to its wavelength in the soapy water.

Constructive interference:

$$2D = m\lambda_2 = m\frac{n_1 \lambda_1}{n_2}, \quad m = 1, 2, 3, \ldots \qquad (22\text{-}20)$$

Destructive interference:

$$2D = (2m - 1)\frac{\lambda_2}{2} = (2m - 1)\frac{n_1 \lambda_1}{2n_2}, \quad m = 1, 2, 3, \ldots \qquad (22\text{-}22)$$

Minimum thickness for destructive interference:

$$D_{\text{min}} = \frac{n_1 \lambda_1}{4n_2} \qquad (22\text{-}23)$$

Wavelength in two different media:

$$\lambda_2 = \frac{n_1 \lambda_1}{n_2} \qquad (22\text{-}19)$$

Solve

(a) In this situation medium 1 is air ($n_1 = 1.00$) and medium 2, of which the film is made, is soapy water ($n_2 = 1.40$). The wavelength in the soapy water is therefore shorter than in air.

Wavelength in air: $\lambda_1 = 560$ nm

Wavelength in soapy water:

$$\lambda_2 = \frac{n_1\lambda_1}{n_2} = \frac{(1.00)(560 \text{ nm})}{1.40} = 400 \text{ nm}$$

Find how many wavelengths of the wavelength λ_2 in the soapy water fit into the path length difference $\Delta_{pl} = 2D$, where $D = 700$ nm is the film thickness.

The path length difference between the waves that reflect off the top and bottom surfaces of the film is

$$\Delta_{pl} = 2D = 2(700 \text{ nm}) = 1400 \text{ nm}$$

The number of wavelengths that fit into this path length difference equals Δ_{pl} divided by λ_2:

$$\frac{\Delta_{pl}}{\lambda_2} = \frac{1400 \text{ nm}}{400 \text{ nm}} = 3.5 = \frac{7}{2}$$

The path length difference is an odd number of half-wavelengths. So there is destructive interference between the light that reflects from the top surface of the soapy water and the light that reflects from the bottom surface (where the soapy water touches the tile).

(b) The minimum thickness required for destructive interference is such that the path length difference $2D$ is one half-wavelength, so $2D = \lambda_2/2$ and $D = \lambda_2/4$.

From Equation 22-23, the minimum thickness for destructive interference is

$$D_{min} = \frac{n_1\lambda_1}{4n_2} = \frac{(1.00)(560 \text{ nm})}{4(1.40)} = 100 \text{ nm}$$

Alternatively, because $\lambda_2 = n_1\lambda_1/n_2$ from Equation 22-19,

$$D_{min} = \frac{\lambda_2}{4} = \frac{400 \text{ nm}}{4} = 100 \text{ nm}$$

Reflect

When light of wavelength 560 nm in air strikes the soapy layer close to the normal, destructive interference occurs both when the layer is 100 nm thick (so twice the thickness is 200 nm, or 1/2 the wavelength in the soapy water) and when the layer is 700 nm thick (so twice the thickness is 1400 nm, or 7/2 the wavelength in the soapy water). For these thicknesses, reflections from the surface would be minimized, and the surface would look dark.

Extend: Destructive interference always occurs when twice the thickness of the layer is an odd multiple of one-half of the wavelength. Can you see that this would also occur if the film of soapy water were either 300 or 500 nm thick?

NOW WORK Problems 3, 4, and 8 from The Takeaway 22-7.

Figure 22-2c shows another example of thin-film interference. This film was made by dipping an open ring into soapy water and then holding the ring vertically. Some of the light that strikes the soap film is reflected from the front surface, while some passes into the film before being reflected at the back surface. When these two light waves recombine, the wavelength of light (the color) that results in constructive interference appears bright. The thickness of the film increases from top to bottom, so the path difference for the two waves is different at different places on the film. That's why the brightest color you see (corresponding to the wavelength for which there is constructive interference) is different at different positions. The thickness of the film is relatively constant *across* the film, however, so the colors appear in bands.

Note that the very top of the soap film in Figure 22-2c appears dark. That may come as a surprise because the top of the film is very thin (far thinner than a wavelength of visible light), so the path difference between light that reflects from the front and back surfaces should be negligible. As a result, there should be constructive interference at the top of the film, and the top should appear bright rather than dark. Why is our prediction incorrect?

To see the explanation, we need to go back to our discussion in Chapter 21 of how waves are reflected from a boundary: A wave pulse on a rope is inverted as it reflects from a fixed boundary. This inversion is equivalent to a phase shift of one-half of a wavelength; the position along the wave that arrived at the boundary as a peak has been reflected as a trough. In general, a wave traveling in one medium is inverted when it reflects (either partially or completely) from a second medium in which the wave speed is lower. (For the rope, the wave speed is zero on the other side of the boundary, which is definitely lower than the speed along the rope.) An inversion (a one-half wavelength phase shift) happens to the light waves in **Figure 22-30a** when light waves reflect off the front surface of the thin film because light travels slower in the film of index of refraction $n_2 > n_1$. The same phase shift happens to the light waves that reflect off the opaque background at the back surface of the thin film. Because both waves undergo the same phase shift due to reflection, any phase difference between the two waves is due to the path difference, as we discussed above.

But as shown in **Figure 22-30b**, things are different for the soap film in Figure 22-2c. Again there is a one-half wavelength phase shift for waves that reflect off the front surface of the film. But there is *no* phase shift for waves that reflect off the back surface, because light waves travel *faster* in the medium on the other side of that boundary. So even if the film were extremely thin, so there were no path difference between the light waves that reflect off the front and back surfaces of the film, there would still be a one-half wavelength phase difference between these two waves. This would result in destructive interference. That's just what we see at the top of the soap film in Figure 22-2c, where the film is very thin compared to the wavelength of the light. This explains the dark band across the top of the film.

Thanks to the additional half-wavelength phase shift for a film like that in Figure 22-2c and Figure 22-30b, Equation 22-20 is no longer the condition for

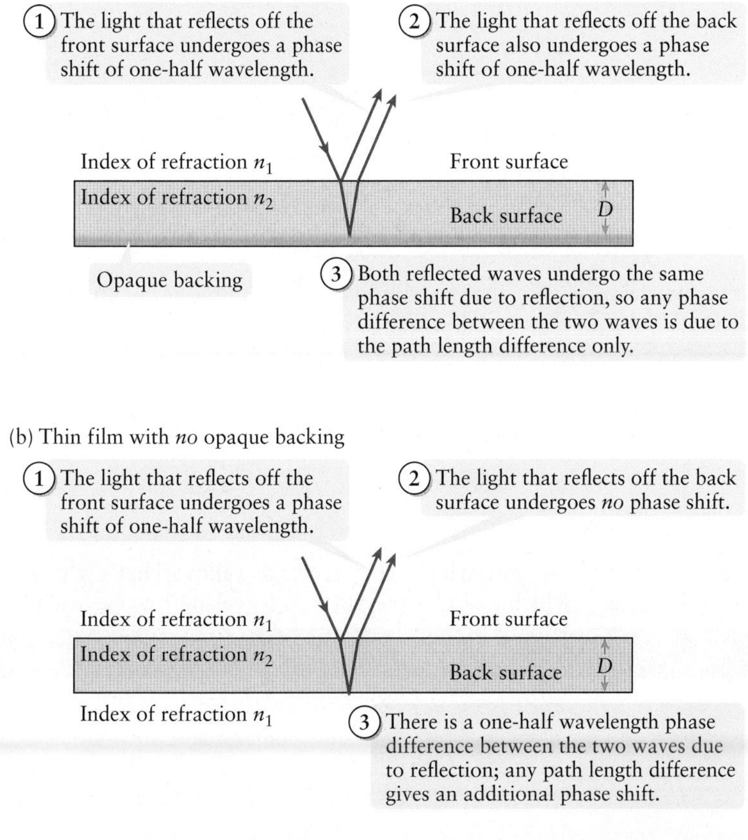

(a) Thin film with an opaque backing
1 The light that reflects off the front surface undergoes a phase shift of one-half wavelength.
2 The light that reflects off the back surface also undergoes a phase shift of one-half wavelength.
Index of refraction n_1
Front surface
Index of refraction n_2
Back surface
D
Opaque backing
3 Both reflected waves undergo the same phase shift due to reflection, so any phase difference between the two waves is due to the path length difference only.
(b) Thin film with *no* opaque backing
1 The light that reflects off the front surface undergoes a phase shift of one-half wavelength.
2 The light that reflects off the back surface undergoes *no* phase shift.
Index of refraction n_1
Front surface
Index of refraction n_2
Back surface
D
Index of refraction n_1
3 There is a one-half wavelength phase difference between the two waves due to reflection; any path length difference gives an additional phase shift.

Figure 22-30 Interference from a thin film II Phase shifts can occur when light reflects from the surfaces of a thin film.

constructive interference, and Equation 22-22 is no longer the condition for destructive interference. Instead the roles of these equations are reversed: Equation 22-20 is now the condition for *destructive* interference, and Equation 22-22 is now the condition for *constructive* interference.

EXAMPLE 22-7 Reducing the Reflection

The glass in an LCD display is sometimes coated with a transparent thin film to minimize glare. Interference from the front and back surfaces of such an *antireflective coating* minimizes the amount of light reflected from the display, particularly light that strikes normal to the display's surface. One material used in such coatings is zirconium acrylate, which has an index of refraction of 1.54. What is the minimum thickness of zirconium acrylate that will accomplish the desired reduction in reflection for light that has wavelength 560 nm in air? Note that the index of refraction of zirconium acrylate is higher than that of glass.

Set Up

We want the light reflected from the surface of the coating to interfere destructively with the light reflected from the zirconium acrylate-to-glass boundary. This requires that when these two light waves recombine, they are shifted by an odd number of half-wavelengths. One half-wavelength shift occurs because the light in air is inverted when it reflects off the surface of the zirconium acrylate, but the light in zirconium acrylate undergoes no shift when it reflects off the glass. (That's because light travels more slowly in zirconium acrylate than in air, but faster in glass than in zirconium acrylate.) Any additional shift arises from the path difference of $2D$ between the two waves. For the total shift to be equivalent to an odd number of half-wavelengths, the shift due to the path difference must be an *integer* number of wavelengths. The minimum thickness corresponds to $2D$ equal to one wavelength λ_2, the wavelength in the zirconium acrylate as given by Equation 22-19.

Wavelength in two different media:

$$\lambda_2 = \frac{n_1\lambda_1}{n_2} \quad (22\text{-}19)$$

Solve

First calculate the wavelength of the light in zirconium acrylate.

We are given the wavelength in air ($n_1 = 1.00$): $\lambda_1 = 560$ nm. The wavelength in zirconium acrylate ($n_2 = 1.54$) is given by Equation 22-16:

$$\lambda_2 = \frac{n_1\lambda_1}{n_2} = \frac{(1.00)(560 \text{ nm})}{1.54} = 364 \text{ nm}$$

The minimum film thickness D_{min} for destructive interference is such that $2D_{min}$ equals λ_2. Use this to solve for D_{min}.

The condition for the minimum thickness that leads to destructive interference is

$$2D_{min} = \lambda_2 = 364 \text{ nm}$$

$$D_{min} = \frac{364 \text{ nm}}{2} = 182 \text{ nm}$$

Reflect

For most people light sensitivity peaks in the range of 555 to 565 nm, which we perceive as yellow. That's why antireflective coatings are usually optimized for yellow light (which includes the 560-nm wavelength we've used here).

NOW WORK Problems 5–7 and 9 from The Takeaway 22-7.

Why have we been emphasizing interference due to light reflecting from a *thin* film? The explanation is that in this section we've assumed that a steady, continuous train of light waves is incident on the film. However, the light from ordinary sources such as light bulbs and the Sun is emitted in a sequence of short bursts, each of which is a segment of wave no more than a few micrometers to about a millimeter in length. This is called the *coherence length* of the light. The phase of the wave changes randomly from one burst to the next. If the thickness of the film is small compared to the coherence length, then the two waves that interfere—the light that reflects from the back of the film and the light that reflects from the front of the film—are part of the same burst. In this case the phase relationships we have developed in this section (which assumed a steady train of waves) are valid. But if the thickness of the film is large compared to the coherence length, the two waves are likely to be from different wave bursts and will differ in phase by a random and rapidly changing amount. As a result, the interference between the waves will be neither always constructive nor always destructive, and any interference effects will be wiped out. That's why you won't see interference effects like those we've described in this section from a thick film like an ordinary pane of glass, which is several millimeters deep. (However, you *can* see interference effects for a glass pane if the light source is a laser. A laser produces light in a very different way from an ordinary light bulb, and the coherence length can be several meters.)

THE TAKEAWAY for Section 22-7

✔ Light waves can interfere constructively or destructively depending on how the peaks and troughs of two waves align when they combine.

✔ When a single beam of light strikes a thin film of a transparent material, some of the light is reflected from the front surface, while some enters the film and is reflected off its back surface. These two reflected light waves may not be in phase when they recombine, resulting in interference. Whether there is destructive interference, constructive interference, or neither depends on the number of wave cycles that fit into the extra distance traveled by the light that enters the thin layer. It also depends on what material is in contact with the back surface of the film.

Prep for the AP® Exam

AP® Building Blocks

1. Give two or three examples of thin-film interference.
2. A thin layer of gasoline floating on water appears brightly colored in sunlight. What is the origin of these colors?
3. (EX 22-6) A ray of normal-incidence light is reflected from a thin film back into air. If the film is a coating on a slab of glass ($n_{film} < n_{glass}$), describe the phase changes that the reflected ray undergoes (a) as it reflects off the front surface of the film and (b) as it reflects off the back surface of the film (the film–glass interface).

AP® Skill Builders

4. (EX 22-6) When white light illuminates a thin film with normal incidence, it strongly reflects both indigo light (450 nm in air) and yellow light (600 nm in air) as shown in the figure. Calculate the minimum thickness of the film if it has an index of refraction of 1.28 and it sits atop a slab of glass that has $n = 1.50$.

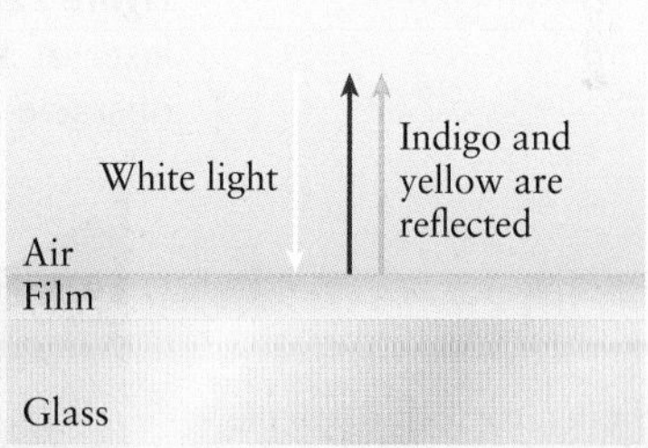

5. (EX 22-7) When white light illuminates a thin film normal to the surface, it strongly reflects both blue light (500 nm in air) and red light (700 nm in air) as shown in the figure. Calculate the minimum thickness of the film if it has an index of refraction of 1.35 and it "floats" on water with $n = 1.33$.

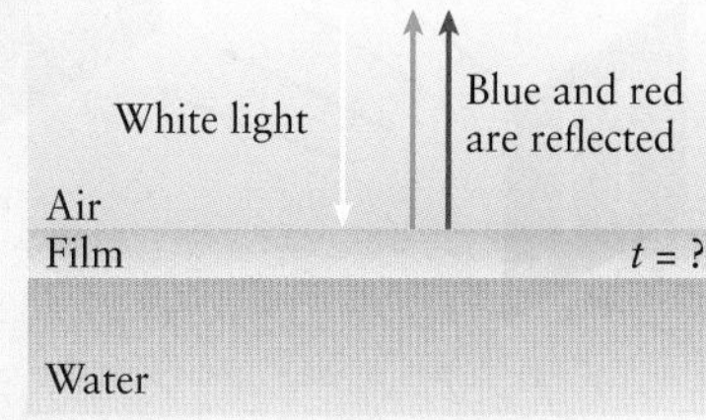

EX 22-7 6. A soap bubble is suspended in air. If the thickness of the soap is 625 nm and both blue light (500 nm in air) and red light (700 nm in air) are *not* observed to reflect from the soap film when viewed normal to the surface of the film, what is the index of refraction of the thin film?

EX 22-7 7. A thin film of cooking oil ($n = 1.38$) is spread on a puddle of water ($n = 1.33$). What are the minimum and the next three thicknesses of the oil that will strongly reflect blue light having a wavelength in air of 518 nm at normal incidence?

EX 22-6 8. What is the minimum thickness of a nonreflective coating of magnesium chloride ($n = 1.39$) so that no normal-incident light centered on 550 nm in air will reflect back off a glass lens ($n = 1.56$)?

Skills in Action

9. Water ($n = 1.33$) in a shallow pan is covered with a thin film of oil that is 450 nm thick and has an index of refraction of 1.45. What visible wavelengths will *not* be present in the reflected light when the pan is illuminated with white light and viewed from straight above?

10. The figure shows light shining on a thin layer of oil that has an index of refraction of 1.4. The oil layer is atop a piece of glass that has an index of refraction of 1.5. On the underside of the glass is air. At which boundary or boundaries does the reflected light undergo an inversion?
(A) The air–oil boundary
(B) The oil–glass boundary
(C) The glass–air boundary
(D) More than one of these

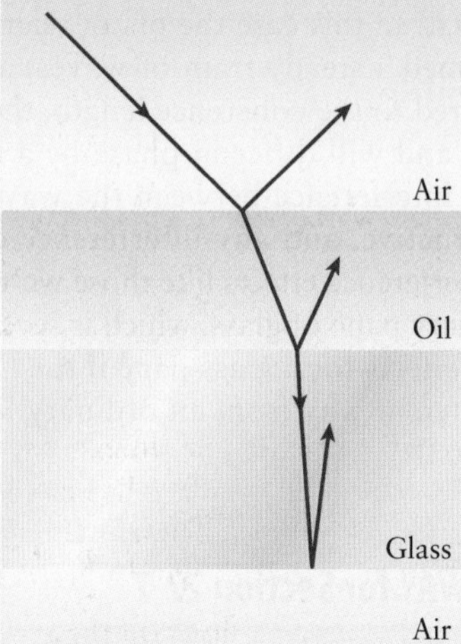

22-8 Interference can occur when light passes through two narrow, parallel slits

Figure 22-31 shows a remarkable experiment into the properties of light. Light from a laser of a single wavelength λ (that is, monochromatic light) is directed at an opaque obstacle that has two narrow slits, S_1 and S_2, separated by a distance d. The light that emerges from the slits falls on a screen a distance L away.

You might expect that there would be just *two* bright patches on the screen, one produced by the light that passed through slit S_1 and one produced by the light that passed through slit S_2. But in fact what we see is a *series* of bright patches with dark patches between them, collectively called **fringes**. We can explain this curious result using two key ideas about light waves: Huygens' principle and interference.

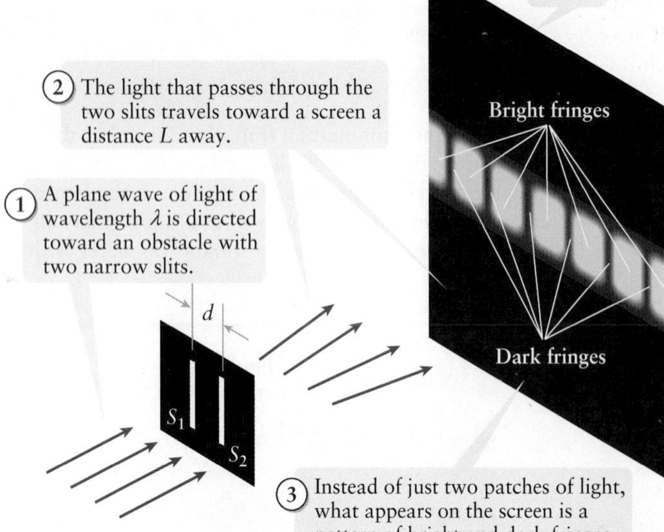

Screen
Bright fringes
Dark fringes
(2) The light that passes through the two slits travels toward a screen a distance L away.
(1) A plane wave of light of wavelength λ is directed toward an obstacle with two narrow slits.
d
S_1
S_2
(3) Instead of just two patches of light, what appears on the screen is a pattern of bright and dark fringes.

Figure 22-31 The double-slit experiment The light waves that pass through slits S_1 and S_2 produce a pattern of bright and dark fringes.

According to Huygens' principle, each point on the incident plane wave that strikes the slits in Figure 22-31 acts as a source of wavelets. Because the two slits S_1 and S_2 are very narrow, we can treat each slit as a *single* source of this kind. Each crest of the incident plane wave strikes both slits simultaneously, so the wavelets of light that emanate from each slit start off in phase with each other (**Figure 22-32**). But because the wavelets travel different distances to reach various locations on the screen, they may or may not arrive at the screen in phase. There are three possibilities:

1. At certain locations on the screen the two wavelets arrive in phase with each other, so crests from S_1

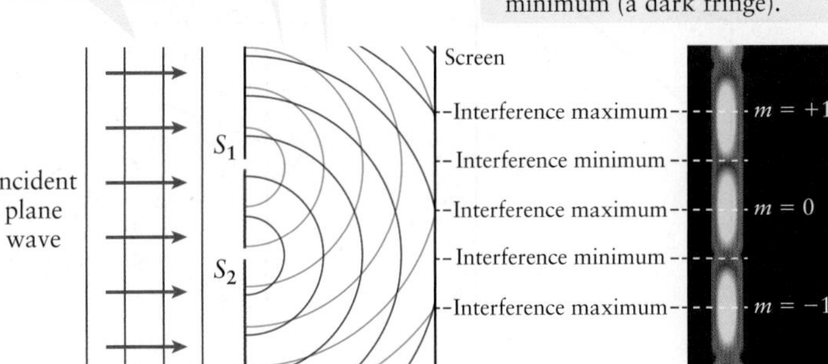

2 Wavelets emanate from slit S_1 (shown in blue) and from slit S_2 (shown in red).
3 Where wavelets from the two slits arrive at the screen in phase, there is an interference maximum (a bright fringe).
1 Successive crests of a plane wave of light are incident on the two slits S_1 and S_2.
4 Where wavelets from the two slits arrive at the screen out of phase, there is an interference minimum (a dark fringe).
Screen
Incident plane wave
S_1
S_2
Interference maximum
Interference minimum
Interference maximum
Interference minimum
Interference maximum
$m = +1$
$m = 0$
$m = -1$

Figure 22-32 Huygens' principle and double-slit interference The pattern of bright and dark fringes arises from interference between wavelets emanating from slit S_1 and from slit S_2.

and S_2 arrive at the same time. So at these locations there is constructive interference, the total wave has maximum amplitude, and the light reaching the screen is brightest. These **interference maxima** correspond to the centers of the bright fringes shown in Figures 22-31 and 22-32.

2. At certain other locations on the screen, the two wavelets arrive out of phase, so a crest from S_1 arrives at the same time as a trough from S_2 and vice versa. At these locations there is destructive interference, the total wave is zero, and no light reaches the screen. The centers of the dark fringes shown in Figure 22-31 and 22-32 correspond to these **interference minima.**

3. At all other locations on the screen, the interference between the wavelets from S_1 and S_2 is neither completely constructive nor completely destructive. At these locations a crest from S_1 does not arrive at the same time as either a crest or a trough from S_2, so the total wave has neither maximum amplitude nor zero amplitude. These are the locations in Figures 22-31 and 22-32 between the center of a bright fringe and the center of a dark fringe.

Because the wavelets that interfere with each other come from a pair of slits, the experiment shown in Figures 22-31 and 22-22 is called **double-slit interference** and the pattern of bright and dark fringes is called an interference pattern. Such interference patterns are conclusive evidence that light is a wave. (In Section 21-5 we discussed the same sort of interference that occurs when there are water waves or sound waves produced by two sources, analogous to the light waves emanating from the two slits in Figures 22-31 and 22-32.)

You may be wondering why it's important to use a laser to demonstrate double-slit interference. The reason is that the waves that enter slit S_1 and the waves that enter slit S_2 must have a constant phase relationship with each other. As we mentioned at the end of Section 22-7, laser light has a high degree of coherence, so the waves entering the two closely spaced slits are guaranteed to have such a constant relationship. (In Figure 22-32 the relationship is that the two waves are in phase.) If you were to use an ordinary light bulb with a color filter to make the light monochromatic before entering the slits, the waves entering slits S_1 and S_2 would be from different short bursts emitted by the light bulb. The phase relationship between these bursts would change from one moment to the next. Hence there would be no definite interference pattern, and on the screen in Figure 22-31 you would see two closely spaced bright spots, one caused by light from slit S_1 and one from slit S_2.

Locating the Interference Maxima and Minima

Let's see how to determine the locations of the double-slit interference maxima and minima on the screen in Figures 22-31 and 22-32. Just as for the thin-film interference

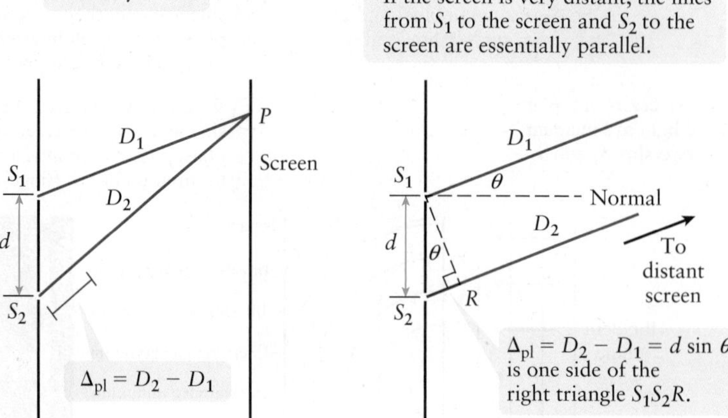

(a)
A nearby screen.
P
Screen
D_1
D_2
S_1
S_2
d
$\Delta_{pl} = D_2 - D_1$
(b)
If the screen is very distant, the lines from S_1 to the screen and S_2 to the screen are essentially parallel.
D_1
D_2
θ
Normal
To distant screen
S_1
S_2
d
R
$\Delta_{pl} = D_2 - D_1 = d \sin \theta$ is one side of the right triangle S_1S_2R.

Figure 22-33 Calculating the path difference for double-slit interference (a) The path difference Δ_{pl} for a point P on the screen. (b) If the distance to the screen is large compared to the distance d between the slits, Δ_{pl} is related to d and the angle θ to the point P.

that we studied in Section 22-7, what determines whether the interference is completely constructive, completely destructive, or intermediate between those two extremes is the *path difference* $\Delta_{pl} = D_2 - D_1$ (**Figure 22-33a**), the difference between the distance D_1 that a wave from slit S_1 travels to a point P on the screen and the distance D_2 that a wave from slit S_2 travels to that same point. The conditions for constructive and destructive interference are

(22-24)

Constructive interference:

$$\Delta_{pl} = \text{a whole number of wavelengths} = 0, \pm\lambda, \pm 2\lambda, \pm 3\lambda, \ldots$$

Destructive interference:

$$\Delta_{pl} = \text{an odd number of half-wavelengths} = \pm\lambda/2, \pm 3\lambda/2, \pm 5\lambda/2, \ldots$$

Note that the path difference Δ_{pl} is positive for locations in the upper half of the screen. These locations are farther from slit S_2 than from slit S_1, so D_2 is greater than D_1 (the case shown in Figure 22-32). Locations in the lower half of the screen are closer to slit S_2 than to slit S_1, so for these locations D_2 is less than D_1 and the path difference Δ_{pl} is negative.

In typical double-slit experiments the spacing d between the slits is about a millimeter, and the distance L from the slits to the screen is a meter or more. Because L is so large compared to d, it's a good approximation to treat the screen as being infinitely far away. Then the straight lines from slit S_1 to the point P on the screen and from slit S_2 to point P are parallel to each other, and are both at an angle θ from the normal to the opaque obstacle with the two slits (**Figure 22-33b**). The path difference Δ_{pl} is then one side of a right triangle S_1S_2R with hypotenuse d. The angle opposite the side of length Δ_{pl} is θ, and the sine of θ equals the length of this opposite side divided by the length d of the hypotenuse:

(22-25)
$$\sin\theta = \frac{\Delta_{pl}}{d} \text{ so } d\sin\theta = \Delta_{pl}$$

If we combine Equations 22-24 and 22-25, we get the following conditions for a bright fringe (constructive interference) and for a dark fringe (destructive interference):

EQUATION IN WORDS
Double-slit experiment: Condition for constructive interference

(22-26) Constructive interference (bright fringes):
$$d\sin\theta = m\lambda$$

Distance between the two slits

Wavelength of the light illuminating the two slits

Angle between the normal to the two slits and the location of a **bright fringe** on the screen

Number of the bright fringe: $m = 0, \pm1, \pm2, \pm3, \ldots$

Destructive interference (dark fringes): $$d \sin\theta = \left(m + \frac{1}{2}\right)\lambda \quad (22\text{-}27)$$

Distance between the two slits (d)

Wavelength of the light illuminating the two slits (λ)

Angle between the normal to the two slits and the location of a **dark fringe** on the screen (θ)

Number of the dark fringe: $m = 0, \pm1, \pm2, \pm3, \ldots$

EQUATION IN WORDS
Double-slit experiment: Condition for destructive interference

Equation 22-26 gives the angles θ at which the bright fringes are found. Each of the bright fringes in Figure 22-32 is labeled with the corresponding values of m. For example, for the $m = 0$ bright fringe it follows that $d \sin\theta = 0$, so $\sin\theta = 0$ and $\theta = 0$. Hence the $m = 0$ bright fringe is at the center of the pattern. For the other bright fringes, $\sin\theta = m\lambda/d = +\lambda/d$ (for $m = +1$), $-\lambda/d$ (for $m = -1$), $+2\lambda/d$ (for $m = +2$), $-2\lambda/d$ (for $m = -2$), and so on.

WATCH OUT!

Be careful with numbering the dark fringes.

Equation 22-27 gives the angles θ at which the dark fringes are found. Note that $m = 0$ in this equation does *not* correspond to the center of the interference pattern (where there is a bright fringe). Instead this is a dark fringe given by $d \sin\theta = (1/2)\lambda$, so $\sin\theta = \lambda/2d$. This dark fringe lies between the $m = 0$ bright fringe and the $m = +1$ bright fringe shown in Figure 22-29. For the $m = +1$ dark fringe, $\sin\theta = (1 + \frac{1}{2})\lambda/d = +3\lambda/2d$; this dark fringe lies between the $m = +1$ and $m = +2$ bright fringes. For the $m = -1$ dark fringe, $\sin\theta = (-1 + \frac{1}{2})\lambda/d = -\lambda/2d$; this dark fringe lies between the $m = 0$ and $m = -1$ bright fringes.

If we divide both sides of Equation 22-26 and both sides of Equation 22-27 by d, we see that that the value of $\sin\theta$ for either a given bright fringe or a given dark fringe is proportional to λ/d, the ratio of the wavelength to the slit spacing:

$$\text{Constructive interference (bright fringe): } \sin\theta = m\frac{\lambda}{d}$$

$$\text{Destructive interference (dark fringe): } \sin\theta = \left(m + \frac{1}{2}\right)\frac{\lambda}{d} \quad (22\text{-}28)$$

Equations 22-28 show that if we increase the wavelength λ, the value of $\sin\theta$ and hence of θ for each fringe increases. This means that the entire interference pattern becomes broader. If instead we increase the slit spacing d, the value of $\sin\theta$ and hence of θ for each fringe decreases. In this case the entire interference pattern becomes narrower.

EXAMPLE 22-8 Measuring Wavelength Using Double-Slit Interference

You send light from a laser through two narrow slits spaced 0.100 mm apart, producing an interference pattern on a screen 5.00 m away. The angle from the central bright fringe to the next bright fringe is 0.300°. (a) What is the wavelength of the light? (b) What is the distance on the screen from the center of the interference pattern to the first dark fringe?

Set Up

We are given the slit spacing $d = 0.100$ mm and the angle $\theta = 0.300°$ for the first bright fringe, which corresponds to $m = 1$ in Equation 22-26. In part (a) we'll use this equation to solve for the wavelength λ. Given the wavelength, in part (b) we'll find the angle of the first dark fringe using Equation 22-27 with $m = 0$. We'll then use trigonometry to find the desired distance.

Bright fringes: $d \sin\theta = m\lambda$ (22-26)

Dark fringes: $d \sin\theta = (m + \frac{1}{2})\lambda$ (22-27)

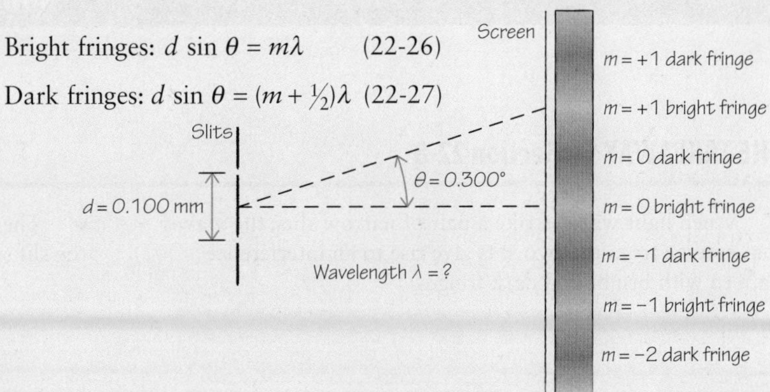

Solve

(a) Solve Equation 22-26 for the wavelength λ.

We have $d = 0.100$ mm $= 1.00 \times 10^{-4}$ m for the slit spacing, and $\theta = 0.300°$ for the angle of the first bright fringe away from the central bright fringe. From Equation 22-26 with $m = 1$,

$$\lambda = d \sin\theta = (1.00 \times 10^{-4}\text{ m}) \sin 0.300°$$
$$= (1.00 \times 10^{-4}\text{ m})(5.24 \times 10^{-3})$$
$$= 5.24 \times 10^{-7}\text{ m} = 524\text{ nm}$$

This is in the green part of the visible spectrum.

(b) Use Equation 22-27 to find the angle of the first dark fringe.

The first dark fringe corresponds to $m = 0$ in Equation 22-27. Solve this for θ:

$$d \sin\theta = (0 + \tfrac{1}{2})\lambda = \lambda/2$$

$$\sin\theta = \frac{\lambda}{2d} = \frac{5.24 \times 10^{-7}\text{ m}}{2(1.00 \times 10^{-4}\text{ m})}$$
$$= 2.62 \times 10^{-3}$$
$$\theta = \sin^{-1}(2.62 \times 10^{-3}) = 0.150°$$

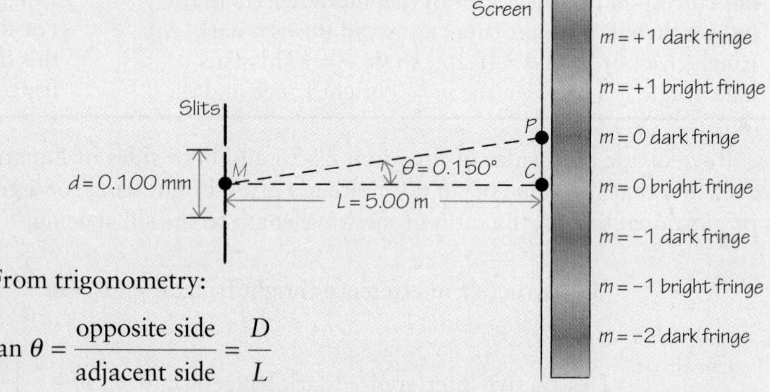

The point P on the screen where the first dark fringe lies, the center of the interference pattern C, and the point M midway between the two slits define a right triangle. The side of this triangle adjacent to the angle $\theta = 0.150°$ is the distance $L = 5.00$ m from the slits to the screen, and the side opposite this angle is the distance D that we want.

From trigonometry:

$$\tan\theta = \frac{\text{opposite side}}{\text{adjacent side}} = \frac{D}{L}$$

Solve for the distance D from the center of the interference pattern to the first dark fringe:

$$D = L \tan\theta = (5.00\text{ m}) \tan 0.150°$$
$$= (5.00\text{ m})(2.62 \times 10^{-3}) = 0.0131\text{ m} = 1.31\text{ cm}$$

Reflect

Part (a) shows how a double-slit interference experiment can be used to measure the wavelength of light. Note from part (b) that to three significant digits, the angle of the first dark fringe from the center of the interference pattern is one-half the corresponding angle for the first bright fringe. This agrees with Figure 22-31, which shows that the bright and dark fringes are equally spaced.

NOW WORK Problems 1 and 3–5 from The Takeaway 22-8.

THE TAKEAWAY for Section 22-8

✔ When light waves strike a pair of narrow slits, the waves that emerge from the two slits give rise to an interference pattern with bright and dark fringes.

✔ The positions of the fringes are determined by the ratio of the slit spacing to the wavelength.

Prep for the AP® Exam

AP® Building Blocks

1. Light that has a 650-nm wavelength is incident upon two narrow slits that are separated by 0.500 mm. An interference pattern from the slits is projected onto a screen that is 3.00 m away.
   (a) What is the separation distance on the screen of the first bright fringe from the central bright fringe?
   (b) What is the separation distance on the screen of the second dark fringe from the central bright fringe?
2. In a two-slit interference experiment, why must the incident light striking one slit have the same wavelength and phase as that striking the other slit?

3. An interference pattern from a double-slit experiment displays 12 bright fringes per centimeter on a screen that is 8.5 m away. The wavelength of light incident on the slits is 550 nm. What is the distance between the two slits?

AP® Skill Builders

4. An argon laser that has a wavelength of 455 nm shines on a double-slit apparatus, which produces an interference pattern on a screen that is 10.0 m away from the slits. The slit separation distance is 70.0 μm.
   (a) How many bright fringes are there on the screen within an angle of ±1° relative to the central axis (including the central maximum)?
   (b) How many dark fringes are there on the screen within an angle of ±2° relative to the central axis? Be careful to count *all* the fringes.

5. Conducting an experiment with a 532-nm-wavelength green laser, a researcher notices a slight shift in the image generated and suspects the laser is unstable and switching between two closely spaced wavelengths, a phenomenon known as mode-hopping. To determine if this is true, she decides to shine the laser on a double-slit apparatus and look for changes in the pattern. Measuring to the first bright fringe on a screen 0.50 m away, with a slit separation of 80 μm, she measures a distance of 3.325 mm. When the laser shifts, so does the pattern and she then measures the same fringe spacing to be 3.375 mm. What wavelength is the laser "hopping" to?

AP® Skills in Action

6. A red laser shines on a pair of slits separated by a distance *d*, producing an interference pattern on a screen. A green laser of the same intensity is then used to illuminate the same slit pair. Compared to the red laser, the green laser will produce an interference pattern with maxima that are
   (A) in the same locations.
   (B) brighter.
   (C) farther apart.
   (D) closer together.
7. In a certain double-slit experiment the $m = 3$ bright fringe is at an angle of 0.500° from the central bright fringe. Which of the following changes would not cause the $m = 2$ bright fringe to appear at this angle instead?
   (A) Increasing the slit spacing by a factor of 3/2 while leaving the wavelength unchanged
   (B) Decreasing the slit spacing by a factor of 2/3 while leaving the wavelength unchanged
   (C) Increasing the wavelength by a factor of 3/2 while leaving the slit spacing unchanged
   (D) Increasing the wavelength by a factor of 3 while doubling the slit spacing

22-9 Diffraction is the spreading of light when it passes through a narrow opening

In the last section we used Huygens' principle to understand the bright and dark interference fringes formed when light passes through two narrow slits. Huygens' principle will also allow us to understand **diffraction**, in which waves tend to spread out when they pass through a narrow opening or near the sharp edge of an object. We will see that diffraction can also give rise to a pattern of bright and dark fringes.

Consider a wave front of a plane wave that passes through an opening in an obstacle, as in **Figure 22-34a**. The opening is wide compared to the wavelength of the wave. Many of the wavelets pass through the opening, as in **Figure 22-34b**. The resulting wave on the other side of the obstacle is mostly a new plane wave, though the ends of the wave front are curved. This means that most of the wave energy continues straight through the opening, with only a small fraction "leaking" to the sides.

Something rather different happens when the opening is comparable in size to or smaller than the wavelength. That's the situation in **Figure 22-35a**, in which the opening is much narrower than in Figure 22-34a. Because the opening is narrow, only a very few of the wavelets that comprise each wave front pass through it. And because so few wavelets get through, they cannot reproduce the straight wave front that was incident

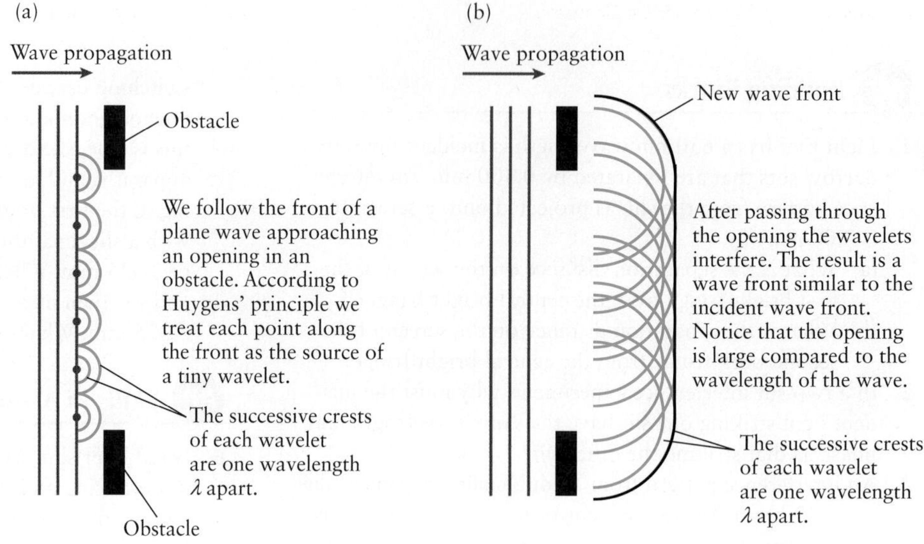

(a)
Wave propagation
Obstacle
We follow the front of a plane wave approaching an opening in an obstacle. According to Huygens' principle we treat each point along the front as the source of a tiny wavelet.
The successive crests of each wavelet are one wavelength λ apart.
Obstacle
(b)
Wave propagation
New wave front
After passing through the opening the wavelets interfere. The result is a wave front similar to the incident wave front. Notice that the opening is large compared to the wavelength of the wave.
The successive crests of each wavelet are one wavelength λ apart.

Figure 22-34 Diffraction I: A wide opening Huygens' principle predicts that if the width of the opening is large compared to the wavelength, most of the wave continues straight ahead through the opening. A slight amount diffracts to the sides.

on the opening. After passing through the narrow opening, the wave spreads out; it has undergone diffraction (**Figure 22-35b**). Diffraction is also present in Figure 22-34b, as shown by the waves that "leak" to the side of the much wider opening, but to a much smaller, greatly reduced extent.

Figure 22-36 shows these effects for water waves passing through an opening in an obstacle. If the width of the opening is large compared to the wavelength, as in **Figure 22-36a**, there are almost no effects of diffraction. But if the opening is comparable in size to the wavelength as in **Figure 22-36b**, the wave spreads out. This explains why you hear someone talking on the other side of an open door, even if you're not directly in front of it. The wavelengths of sound used in human speech are in the range of a few meters, comparable in size to the width of a typical door (about one meter). As a result sound waves emerging from a door spread out like the water waves in Figure 22-36b, making it easy to eavesdrop on conversations. However, you can't *see* through an open door if you stand to one side because the wavelengths of visible light (around 550 nm, or 5.5×10^{-7} m) are very small compared to the width of the door. So light waves passing through the door behave like the water waves in Figure 22-36a.

To consider diffraction in more detail, imagine that we send a beam of light of a single wavelength λ through a long, narrow slit of width w. After passing through

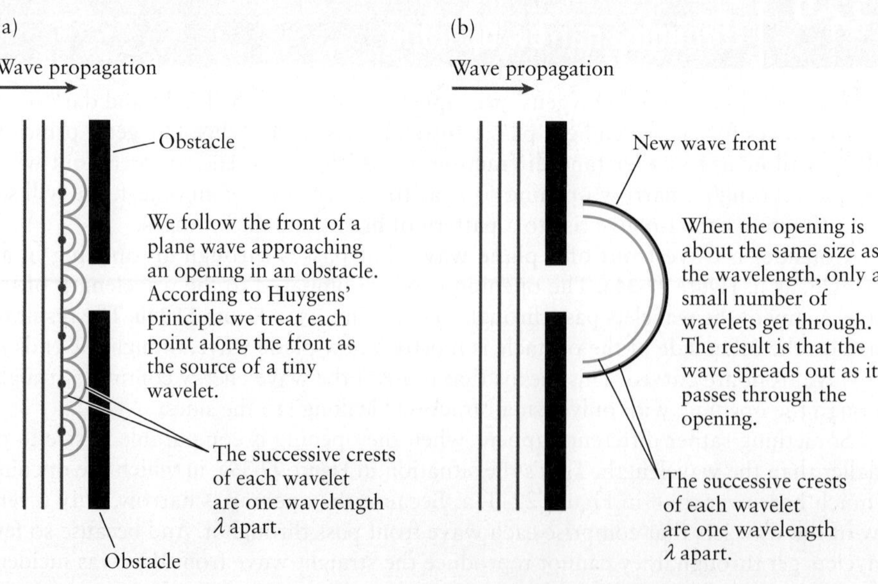

(a)
Wave propagation
Obstacle
We follow the front of a plane wave approaching an opening in an obstacle. According to Huygens' principle we treat each point along the front as the source of a tiny wavelet.
The successive crests of each wavelet are one wavelength λ apart.
Obstacle
(b)
Wave propagation
New wave front
When the opening is about the same size as the wavelength, only a small number of wavelets get through. The result is that the wave spreads out as it passes through the opening.
The successive crests of each wavelet are one wavelength λ apart.

Figure 22-35 Diffraction II: A narrow opening If the width of the opening is comparable to the wavelength, the wave spreads out substantially after exiting the opening. (Compare Figure 22-34.)

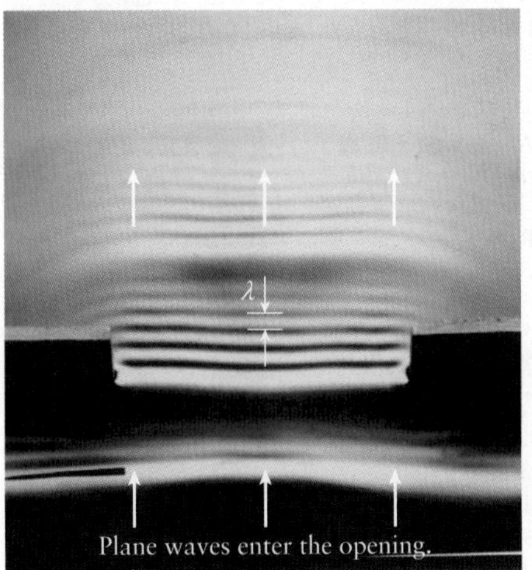

Plane waves enter the opening.

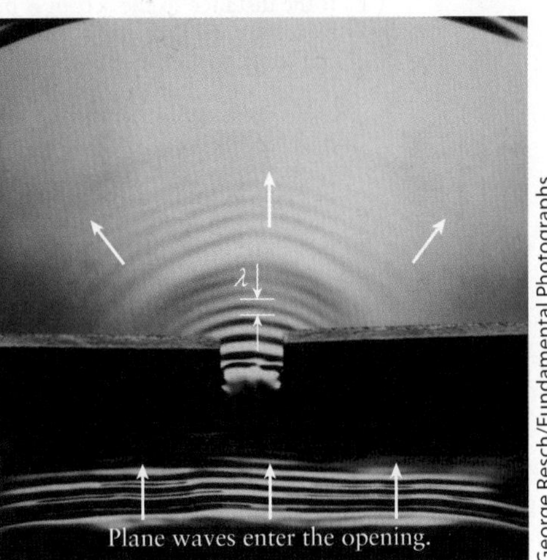

Plane waves enter the opening.

• The width of the opening is large compared to the wavelength λ.
• Most of the wave energy continues straight ahead, so there is very little diffraction.

Figure 22-36 Water wave diffraction Compare these photographs to Figures 22-34 and 22-35.

the slit the light falls on a screen a distance L away (**Figure 22-37**). You might expect that the pattern on the screen would be a single blob of light. But as the photo in Figure 22-37 shows, what actually appears is a series of bright and dark fringes called a **diffraction pattern.** In Section 22-8 we saw how such fringes arise in the *double*-slit experiment due to interference between Huygens wavelets emanating from the first slit and wavelets emanating from the second slit. In this **single-slit diffraction** experiment, the fringes arise due to interference between wavelets emanating from different parts of the *same* slit. Huygens wavelets emanate from each part of the slit. At some locations, these wavelets interfere destructively; these are the locations of the dark fringes, also called **diffraction minima.** At locations between the dark fringes, the wavelets interfere more or less constructively and we see bright fringes, also called **diffraction maxima.**

NEED TO REVIEW?
Turn to page M12 in the Math Tutorial in the back of the book for more information on trigonometry.

Let's see how to determine the positions of the dark fringes. The first dark fringe is found where light waves from the upper half of the slit, of width $w/2$, and the lower half, also of width $w/2$, interfere destructively. This means that the waves from one half of the slit must travel one half-wavelength farther from the slit to the screen than do the waves from the other half, so the path length difference Δ_{pl} equals $\lambda/2$. We can use this to easily determine the position of the first dark fringe if we assume that the distance L to the screen is much greater than the slit width w. With this assumption,

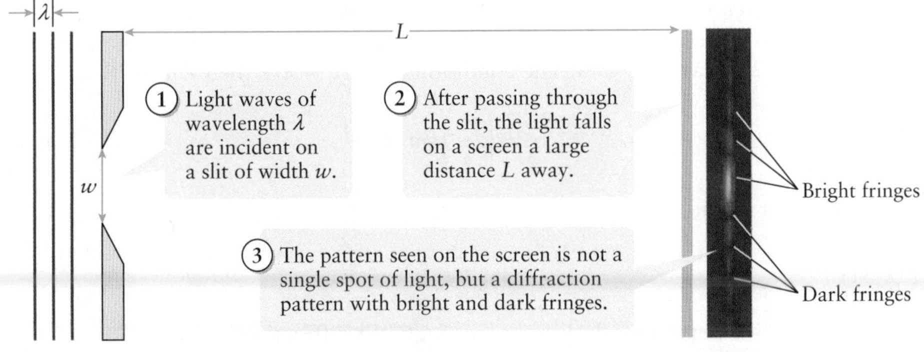

λ
L
w
1 Light waves of wavelength λ are incident on a slit of width w.
2 After passing through the slit, the light falls on a screen a large distance L away.
3 The pattern seen on the screen is not a single spot of light, but a diffraction pattern with bright and dark fringes.
Bright fringes
Dark fringes

Figure 22-37 Diffraction through a single slit A diffraction pattern arises when monochromatic light passes through a narrow slit.

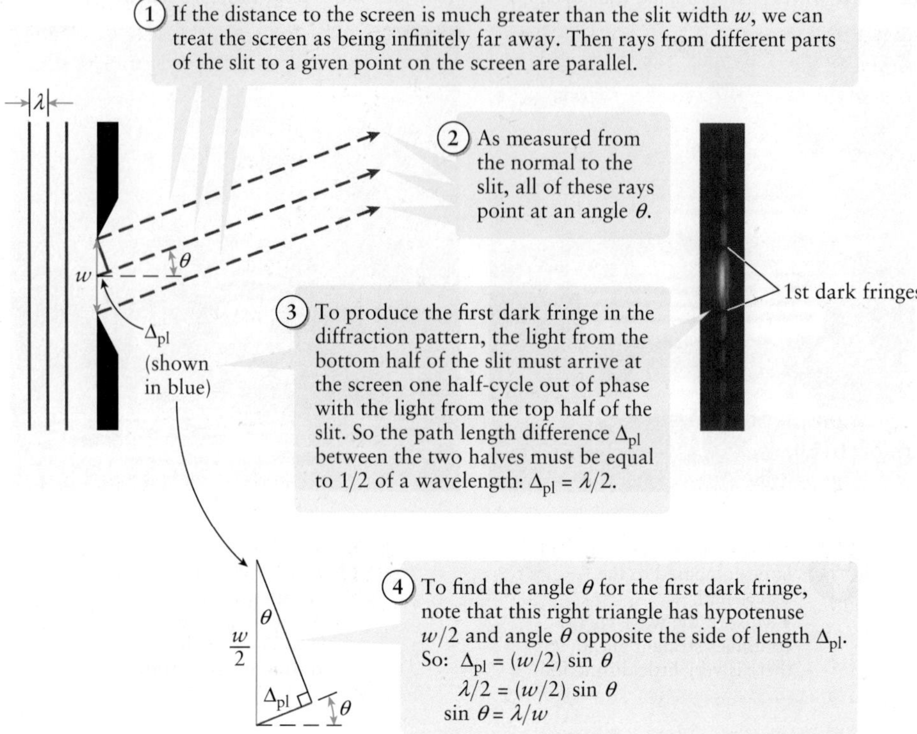

1 If the distance to the screen is much greater than the slit width w, we can treat the screen as being infinitely far away. Then rays from different parts of the slit to a given point on the screen are parallel.
λ
2 As measured from the normal to the slit, all of these rays point at an angle θ.
w
θ
Δ_{pl} (shown in blue)
3 To produce the first dark fringe in the diffraction pattern, the light from the bottom half of the slit must arrive at the screen one half-cycle out of phase with the light from the top half of the slit. So the path length difference Δ_{pl} between the two halves must be equal to 1/2 of a wavelength: $\Delta_{pl} = \lambda/2$.
1st dark fringes
w/2
Δ_{pl}
4 To find the angle θ for the first dark fringe, note that this right triangle has hypotenuse $w/2$ and angle θ opposite the side of length Δ_{pl}.
So: $\Delta_{pl} = (w/2)\sin\theta$
$\lambda/2 = (w/2)\sin\theta$
$\sin\theta = \lambda/w$

Figure 22-38 Calculating the single-slit diffraction pattern Dark fringes are found where light from each part of the slit interferes destructively with light from another part.

Figure 22-38 shows that for a given position on the screen, Δ_{pl} is related to the angle θ between the normal to the slit and a line to that position:

$$\Delta_{pl} = \left(\frac{w}{2}\right)\sin\theta \tag{22-29}$$

To find the angle for the first dark fringe, substitute $\Delta_{pl} = \lambda/2$ (the condition for destructive interference) into Equation 22-29:

$$\frac{\lambda}{2} = \left(\frac{w}{2}\right)\sin\theta$$

so

$$\sin\theta = \frac{\lambda}{w} \tag{22-30}$$

(first dark fringe in single-slit diffraction)

There are two of these first dark fringes, one on either side of the central bright fringe (see Figure 22-38).

To see how the second dark fringe arises, imagine breaking the slit into quarters, each of width $w/4$. Destructive interference occurs when light from the first quarter arrives at the screen out of phase with light from the second quarter, and light from the third quarter arrives at the screen out of phase with light from the fourth quarter. The path length difference Δ_{pl} between light from adjacent quarters is given by Equation 22-29, with $w/2$ replaced by $w/4$. The condition that $\Delta_{pl} = \lambda/2$ then tells us that

$$\frac{\lambda}{2} = \left(\frac{w}{4}\right)\sin\theta$$

If we solve this for $\sin\theta$, we get

$$\sin\theta = \frac{2\lambda}{w} \tag{22-31}$$

(second dark fringe in single-slit diffraction)

This gives a larger value for $\sin\theta$ and hence for θ, than for the first dark fringe given by Equation 22-30. For the third dark fringe the factor of 2 in Equation 22-31 is replaced by a 3, for the fourth dark fringe it is replaced by a 4, and so on. In general, the angle of the mth dark fringe is given by

Angle between the normal to the slit and the location of the mth **dark fringe**

Number of the dark fringe: $m = 1, 2, 3,...$

$$\sin\theta = \frac{m\lambda}{w} \quad (22\text{-}32)$$

Wavelength of the light

Width of the slit

EQUATION IN WORDS
Dark fringes in single-slit diffraction

Equations 22-30 and 22-31 are special cases of Equation 22-32, with $m = 1$ and $m = 2$, respectively.

The middle of the central *bright* fringe in Figures 22-37 and 22-38 is where all of the waves from the slit arrive in phase, because they all travel essentially the same distance to this point. This point, which corresponds to $\theta = 0$, is where the intensity is maximum. The next bright fringe, located between the first and second dark fringes, is fainter than the central bright fringe. Roughly speaking, at this bright fringe the light from the upper third of the slit interferes destructively with the light from the middle third of the slit, canceling it out. What remains is the light from the lower third of the slit. This wave has one-third the amplitude and hence $(1/3)^2 = 1/9$ the intensity of the wave that reaches the middle of the central bright fringe. (Recall from Section 22-6 that the intensity of an electromagnetic wave is proportional to the square of the amplitude of the electric field.) Successive bright fringes are even fainter. Note that the point of greatest intensity in the central bright fringe is at its center; for the other bright fringes the point of greatest intensity is close to (but slightly displaced from) a point halfway between the adjacent dark fringes.

WATCH OUT !

Keep in mind the difference between double-slit interference and single-slit diffraction.

It's common to get confused between double-slit interference and single-slit diffraction, because in both situations there is a pattern of bright and dark fringes. Keep in mind the essential difference between the two situations: In double-slit interference the pattern of fringes is caused by light waves from one slit interfering with light waves from the other slit, while in single-slit diffraction the pattern is caused by light waves from different parts of the same single slit interfering with each other.

EXAMPLE 22-9 Diffraction Through a Slit

A green laser pointer emits light at a wavelength of 532 nm. You aim the beam from this laser at a slit 1.50 μm wide. Find the angles of the first, second, and third dark fringes in the diffraction pattern.

Set Up

We are given the wavelength λ and the slit width w, and we want to find the value of the angle θ for the $m = 1$, 2, and 3 dark fringes. We'll use Equation 22-32 for this purpose.

Dark fringes in single-slit diffraction:

$$\sin\theta = \frac{m\lambda}{w} \quad (22\text{-}32)$$

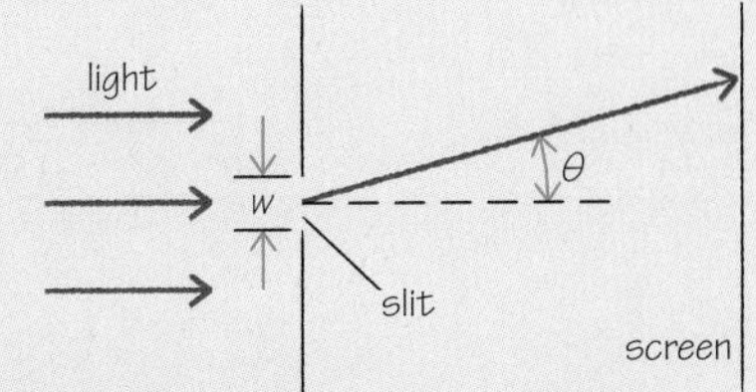

Solve

Find the angle of the first dark fringe ($m = 1$).

We have $\lambda = 532\text{ nm} = 5.32 \times 10^{-7}\text{ m}$ and $w = 1.50\ \mu\text{m} = 1.50 \times 10^{-6}\text{ m}$. From Equation 22-32, with $m = 1$:

$$\sin\theta_1 = \frac{\lambda}{w} = \frac{5.32 \times 10^{-7}\text{ m}}{1.50 \times 10^{-6}\text{ m}} = 0.355$$

$$\theta_1 = \sin^{-1} 0.355 = 20.8°$$

Find the angle of the second dark fringe ($m = 2$).

From Equation 22-32, with $m = 2$:

$$\sin\theta_2 = \frac{2\lambda}{w} = \frac{2(5.32 \times 10^{-7}\ \text{m})}{1.50 \times 10^{-6}\ \text{m}} = 0.709$$

$$\theta_2 = \sin^{-1} 0.709 = 45.2°$$

Find the angle of the third dark fringe ($m = 3$).

From Equation 22-32, with $m = 3$:

$$\sin\theta_3 = \frac{3\lambda}{w} = \frac{3(5.32 \times 10^{-7}\ \text{m})}{1.50 \times 10^{-6}\ \text{m}} = 1.06$$

This equation has *no* solution! The sine of an angle cannot be greater than 1, so there is no value of θ_3 that satisfies this equation. We are forced to conclude that the diffraction pattern in this situation has a first dark fringe at $\theta_1 = 20.8°$ and a second dark fringe at $\theta_2 = 45.2°$, but there is no third dark fringe before the end of the pattern at $\theta = 90°$.

Reflect

The number of dark fringes present in the diffraction pattern of a slit depends on the relative sizes of the wavelength λ and the slit width w. We explore this further below.

NOW WORK Problems 1–7 from The Takeaway 22-9.

Figure 22-39 shows the intensity in the diffraction pattern of a slit as a function of the angle θ. As the width of the slit is increased from $w = \lambda$ (**Figure 22-39a**) to $w = 4\lambda$ (**Figure 22-39b**) to $w = 8\lambda$ (**Figure 22-39c**), the pattern becomes narrower. Note that if $w = \lambda$, as in Figure 22-39a, the first dark fringe corresponds to $\sin\theta = \lambda/w = 1$, so $\theta = \sin^{-1} 1 = 90°$; there are no dark fringes at smaller angles. The light intensity is spread out very broadly over all angles, much like what happens to the water waves in Figure 22-36b. If the slit width is large compared to the wavelength, as in Figure 22-39c, the vast majority of the light emerging from the slit goes straight ahead (toward $\theta = 0$) or very nearly so. That's just like what happens to the water waves in Figure 22-36a, for which there is very little diffraction.

The intensity pattern of single-slit diffraction also plays a role in *double-slit* interference, as **Figure 22-40** shows. The closely spaced bright fringes in this image are due to

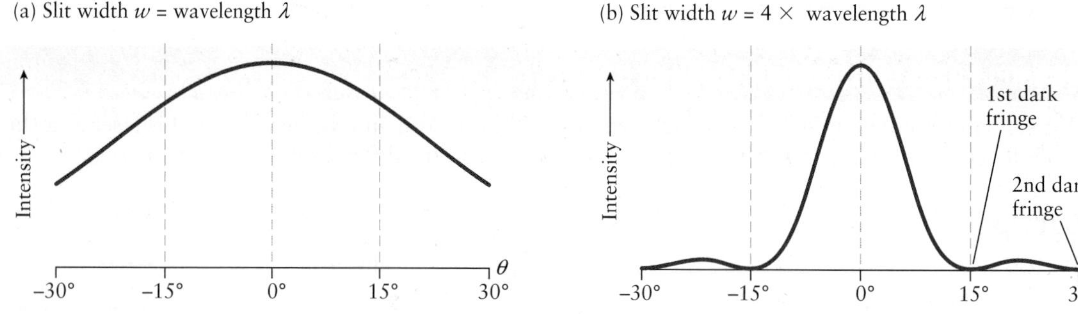

(c) Slit width $w = 8 \times$ wavelength λ

Intensity

1st dark fringe

3rd dark fringe

2nd dark fringe

4th dark fringe

θ

−30° −15° 0° 15° 30°

Figure 22-39 Intensity in single-slit diffraction The intensity in the diffraction pattern from a narrow slit depends on the relative size of the slit width w and the wavelength λ.

(1) This is the interference pattern in a double-slit experiment.
(2) The closely spaced bright and dark fringes are due to interference between light waves emanating from the two slits.
(3) The interference pattern disappears at certain locations. This is a result of diffraction of light through each slit (these are the locations of the dark *diffraction* fringes).
(4) The intensity of the bright interference fringes decreases with greater angle from the center of the pattern. This is also due to diffraction through the individual slits (see Figure 23-35).

Figure 22-40 Single-slit diffraction affects double-slit interference In a double-slit experiment there is both interference of the light waves emanating from the two slits and diffraction of light waves through each individual slit. The diffraction affects the brightness of the bright interference fringes.

interference between light coming from the two slits. However, not all of these fringes are equally bright. The explanation is that light also *diffracts* as it emanates from each of the slits. Each of the slits has a width w that is larger than the wavelength λ, so the effect of diffraction is that the intensity of light emanating from each slit varies with angle θ as shown in Figures 22-39b and 22-39c. Hence the brightness of the bright double-slit interference fringes in Figure 22-40 rises and falls with angle θ just as the intensity curves in Figures 22-39b and 22-39c rise and fall with θ.

Exam Tip

You are expected to know what a diffraction grating is and you should be able to explain what it does qualitatively. However, the AP® Exam will not ask you to do calculations or provide an explanation of how a diffraction grating functions beyond demonstrating an understanding of it as a collection of slits.

A diffraction grating is a collection of evenly spaced parallel slits or openings that produces an interference pattern that is the combination of numerous diffraction patterns superimposed on each other. This sort of optical element has many uses. Diffraction gratings are found in spectrometers and some cool toys, and used in making holograms. White light is made up of all the colors of visible light, and the location of the interference fringes for a diffraction grating are determined by the separation of the slits and the wavelength of the light, so when white light is incident on a diffraction grating, the center maximum is white, and the higher-order fringes disperse the light into the whole rainbow of colors with the longest wavelength light (red) appearing farthest from the central maximum.

In this section we've concentrated on the diffraction that takes place when waves pass through a narrow opening. But diffraction can also happen when waves encounter an obstacle. One example is a sound wave coming from one side of your head. As we mentioned above, sound waves used in speech have wavelengths of a meter or more, which is large compared to the diameter of a human head. As a result, these sound waves are able to diffract around your head, so you can hear the sound with both ears. Because the sound wave must travel a greater distance to one ear than to the other, there will be a phase difference between the waves that the two ears detect. Your brain detects and processes this information about phase and uses this to help determine the direction from which the sound is coming.

THE TAKEAWAY for Section 22-9

✔ Waves passing through a narrow opening tend to spread out, or diffract. The diffraction is more important the smaller the size of the opening.

✔ The diffraction pattern caused by waves passing through a narrow slit has bright and dark fringes. The positions of these depend on the relative size of the wavelength and the slit width.

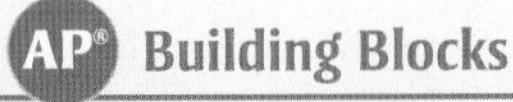

Building Blocks

EX 22-9 1. Light that has a wavelength of 550 nm is incident on a single slit that is 10.0 μm wide. Determine the angular location of the first three dark fringes that are formed on a screen behind the slit.

EX 22-9 2. Light that has a wavelength of 475 nm is incident on a single slit that is 800 nm wide. Calculate the angular location of the first three dark fringes that are formed on a screen behind the slit.

EX 22-9 3. What is the highest order dark fringe that is found in the diffraction pattern for light that has a wavelength of 633 nm and is incident on a single slit that is 1500 nm wide?

EX 22-9 4. The highest order dark fringe found in a diffraction pattern is 6. Determine the wavelength of light that is used with the single slit that has a width of 3500 nm.

AP® Skill Builders

5. When blue light ($\lambda = 500$ nm) is incident on a single slit, the central bright spot has a width of 8.75 cm. If the screen is 3.55 m distant from the slit, calculate the slit width.

6. A helium–neon laser illuminates a narrow single slit that is 1850 nm wide. The first dark fringe is found at an angle of 20.0° from the central peak. Determine the wavelength of the light from the laser.

7. Yellow light that has a wavelength of 625 nm produces a central maximum peak that is 24.0 cm wide on a screen that is 1.58 m from a single slit. Calculate the width of the slit.

AP® Skills in Action

8. A monochromatic light passes through a narrow slit and forms a diffraction pattern on a screen behind the slit. As the wavelength of the light decreases, the diffraction pattern
   (A) shrinks, with all the fringes getting narrower.
   (B) spreads out, with all the fringes getting wider.
   (C) remains unchanged.
   (D) spreads out, with all the fringes getting alternately wider and then narrower.

9. A monochromatic light passes through a narrow slit and forms a diffraction pattern on a screen behind the slit. As the slit width increases, the diffraction pattern
   (A) shrinks, with all the fringes getting narrower.
   (B) spreads out, with all the fringes getting wider.
   (C) remains unchanged.
   (D) spreads out, with all the fringes getting alternately wider and then narrower.

10. The photographs below show the diffraction pattern that is created when red laser light passes through a narrow slit. The wavelength of the light is the same in all three photographs, but the width of the slit is different. Order the photographs from the widest slit to the narrowest one.

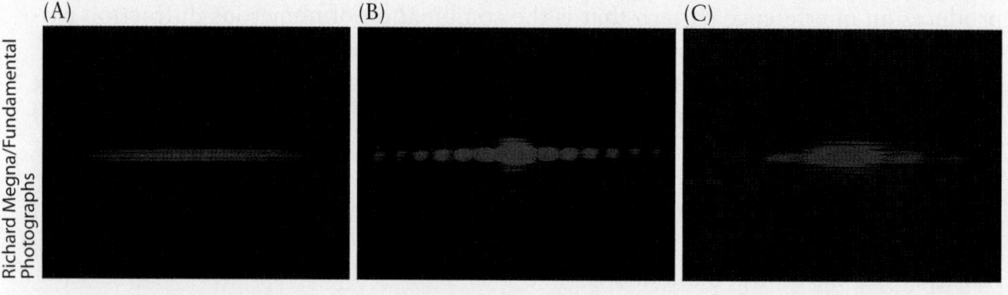

22-10 The diffraction of light through a circular aperture is important in optics

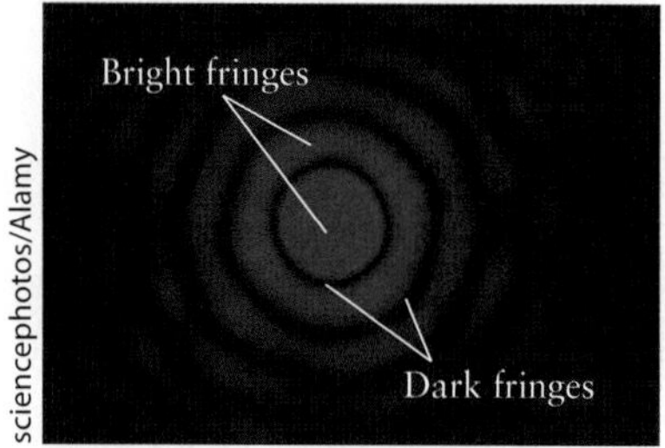

Figure 22-41 Diffraction by a circular aperture I Compare the diffraction pattern from a circular aperture with that for a narrow slit (Figure 22-37).

An important real-life application of diffraction is the case of light passing through a *circular* aperture. That's what happens whenever light enters the lens of a microscope, the circular mirror of an astronomical telescope, or the pupil of a human eye. **Figure 22-41** shows the diffraction pattern produced by red laser light passing through a circular aperture. Like the diffraction pattern of a slit (Section 22-9), there are bright and dark fringes. But because the aperture is circular, the fringes are circles.

Figure 22-42 shows how we define the angle θ of a point in the diffraction pattern of a circular aperture of diameter D through which passes light of wavelength λ. For a narrow slit we found that the diffraction pattern depends on the relative sizes of the wavelength λ and the slit width w; in the same way, the diffraction pattern of a circular aperture depends on the relative sizes of λ and D. In particular, the location of the center of the first dark fringe is given by

(22-33)
$$\sin\theta = 1.22\frac{\lambda}{D}$$
(circular aperture, first dark fringe)

This is similar to Equation 22-30 for the first dark fringe formed by light passing through a narrow slit, with the slit width w replaced by the diameter D. The factor of 1.22 results from the different geometry of a circular opening versus a rectangular one.

Figure 22-43 shows the diffraction pattern made by two truly point-like objects as the objects are moved closer and closer together. In **Figure 22-43a** the two objects are so far apart we see only one object and its associated diffraction pattern in the field of

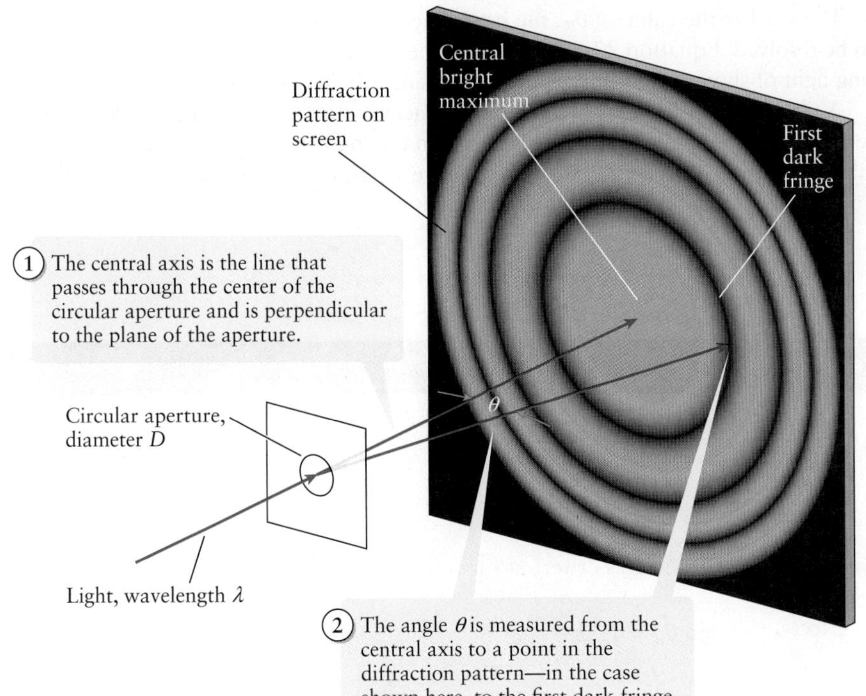

Central bright maximum
Diffraction pattern on screen
First dark fringe
1 The central axis is the line that passes through the center of the circular aperture and is perpendicular to the plane of the aperture.
θ
Circular aperture, diameter D
Light, wavelength λ
2 The angle θ is measured from the central axis to a point in the diffraction pattern—in the case shown here, to the first dark fringe.

Figure 22-42 Diffraction by a circular aperture II The size of the first dark fringe—and hence the size of the central bright maximum that it surrounds—depends on the ratio of wavelength λ to aperture diameter D.

view. In **Figure 22-43b** a second object has been brought close to the first; the diffraction patterns of the two objects overlap but are still distinct. In **Figure 22-43c**, however, the two objects are very close together. Their diffraction patterns overlap so much that it is barely possible to tell the two objects apart. We have run into the limit on our ability to **resolve**, or optically distinguish, the two objects.

The nineteenth-century English physicist John William Strutt, 3rd Baron Rayleigh, proposed that two point-like objects observed through a circular aperture can be resolved when the central maximum of one coincides with the center of the first dark fringe of the other. The angle θ_R that separates two point objects that are just barely resolved through a circular aperture, known as the **angular resolution** of the aperture, is then just the angle given by Equation 22-33:

Angle between two pointlike objects that can **barely be resolved** through an optical device

Wavelength of the light

$$\sin\theta_R = 1.22\,\frac{\lambda}{D} \qquad (22\text{-}34)$$

Diameter of the circular aperture of the device

EQUATION IN WORDS
Rayleigh's criterion for resolvability

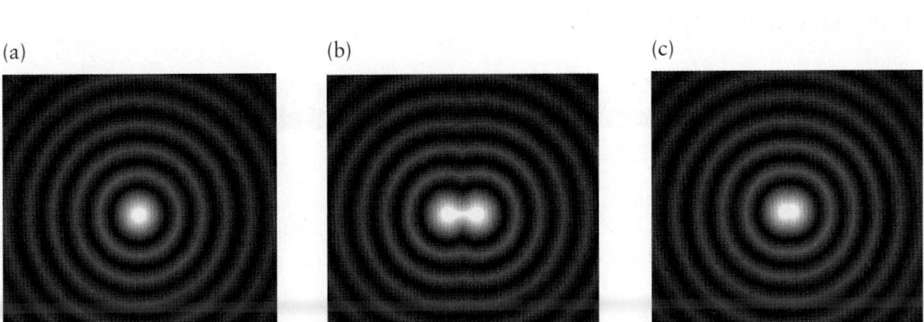

Figure 22-43 Angular resolution (a) Light from a single point source gives rise to a diffraction pattern when it passes through a circular aperture. (b) These diffraction patterns from two distant point sources partially overlap but are distinct. (c) At the limit of our ability to resolve two distant point sources, the diffraction patterns formed as their light passes through a circular aperture overlap and are barely distinguishable.

The smaller the value of θ_R, the better the resolution and the smaller the details that can be resolved. Equation 22-34 shows that better angular resolution can be obtained by using light of shorter wavelength λ and by using a larger circular aperture D.

A physician's eye chart is a device for measuring the value of θ_R for each of your eyes. An unaided eye with normal vision can distinguish objects (such as the lines that make up the letter E on an eye chart) that are separated by an angle of as small as 1/60 of a degree. So $\theta_R = (1/60)°$ for a person with normal vision.

EXAMPLE 22-10 The Hubble Space Telescope

The Hubble Space Telescope (HST) has a circular aperture 2.4 m in diameter. What is the theoretical angular resolution of the HST for light of wavelength 550 nm?

Set Up

Equation 22-34 gives the angular resolution, the smallest angular separation of two objects that can be resolved as a result of diffraction effects.

Rayleigh's criterion for resolvability:

$$\sin\theta_R = 1.22\frac{\lambda}{D} \tag{22-34}$$

Solve

We are given $\lambda = 550$ nm and $D = 2.4$ m. We calculate θ_R from these using Equation 22-34.

From Equation 22-34:

$$\sin\theta_R = 1.22\frac{(550 \times 10^{-9}\text{ m})}{(2.4\text{ m})} = 2.8 \times 10^{-7}$$

$$\theta_R = \sin^{-1}(2.8 \times 10^{-7}) = 1.6 \times 10^{-5}\text{ degree}$$

Reflect

Because the diameter of the HST mirror is so much greater than that of the human eye, θ_R for the HST is much smaller than the value $\theta_R = (1/60)° = 0.017°$ for the human eye, which means that the HST has far better angular resolution. The photograph shown here is an HST image of the minor planet Pluto and its largest moon, Charon. Pluto and Charon are separated by about 25×10^{-5} degrees in this image. They are easily resolved by the HST, for which $\theta_R = 1.6 \times 10^{-5}$ degrees.

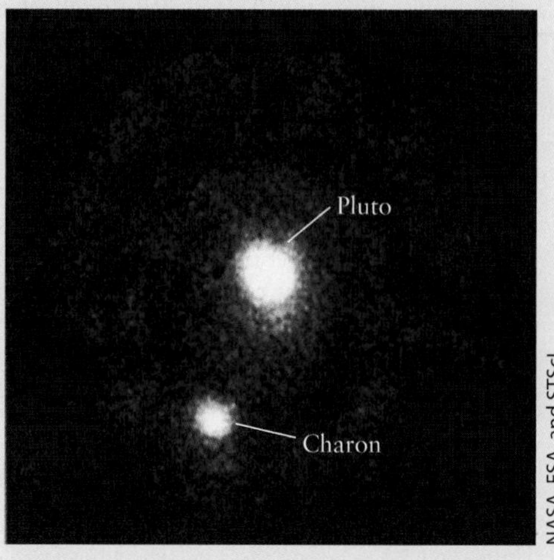

NASA, ESA, and STScI

Extend: Why is it important for a telescope like the HST to be in orbit? One key reason is that a telescope of the same diameter on Earth would have a much poorer resolution of about 0.03 degree, and would not be able to resolve Pluto and Charon. That's due to the blurring effects of our atmosphere, which is constantly in motion. By placing the HST in orbit high above the atmosphere, this blurring is eliminated. (A second important reason for placing telescopes in orbit is that it makes it possible to see at a variety of wavelengths, including infrared and ultraviolet, that do not penetrate through our atmosphere and so cannot be detected by telescopes on Earth.)

NOW WORK Problems 1–6 and 8–10 from The Takeaway 22-10.

THE TAKEAWAY for Section 22-10

✔ The angular resolution of an optical device is limited by diffraction, which blurs the images of even point-like objects.

✔ Rayleigh's criterion states that two point-like objects can just be resolved if the center of the bright maximum for one object coincides with the first dark fringe for the second object.

Prep for the AP® Exam

AP® Building Blocks

EX 22-10
1. In the human eye, the average pupil is 5.0 mm in diameter, and the average normal-sighted human eye is most sensitive at a wavelength of 555 nm. What is the eye's angular resolution in radians?

EX 22-10
2. The telescope at Mount Palomar has an objective mirror that has a diameter of 508 cm. What is the angular limit of resolution due to diffraction for 560-nm light in degrees and radians?

EX 22-10
3. The Hubble Space Telescope has a diameter of 2.4 m. What is the angular limit of resolution (in radians) due to diffraction when a wavelength of 540 nm is viewed?

AP® Skill Builders

EX 22-10
4. Light from a helium–neon laser with a wavelength of 633 nm passes through a 0.180-mm-diameter hole and forms a diffraction pattern on a screen 2.0 m behind the hole. Calculate the diameter of the central maximum.

EX 22-10
5. The distance from the center of a circular diffraction pattern to the first dark ring is 15,000 wavelengths on a screen that is 0.85 m away. What is the size of the aperture?

EX 22-10
6. Assume your eye has an aperture diameter of 3.00 mm at night when bright headlights are pointed at it. At what distance can you see two headlights separated by 1.5 m as distinct? Assume a wavelength of 550 nm, near the middle of the visible spectrum.

AP® Skills in Action

7. Rank the following telescopes from best to worst angular resolution. Assume that diffraction is the only limiting factor. (A) A radio telescope with an effective diameter of 27 km that observes waves at wavelength 21 cm; (B) a telescope in orbit that has a mirror of 0.85 m diameter and that observes infrared light of wavelength 4.5 μm; (C) an inexpensive telescope for amateur astronomers with a lens of 5.0 cm diameter that observes visible light of wavelength 550 nm.

EX 22-10
8. If you peek through a 0.75-mm-diameter hole at an eye chart, you will notice a decrease in visual acuity.
   (a) Calculate the angular limit of resolution if the wavelength is taken as 575 nm.
   (b) If the horizontal lines on the letter E are 2 mm apart, how close would you need to be to resolve them through the hole?

EX 22-10
9. The pupil of the human eye is the circular opening through which light enters. Its diameter can vary from about 2.0 to about 8.0 mm to control the intensity of the light reaching the retina.
   (a) Calculate the angular resolution, θ_R, of the eye for light that has a wavelength of 550 nm in both bright light and dim light. In which light can you see more sharply, dim or bright?
   (b) You probably have noticed that when you squint, objects that were a bit blurry suddenly become somewhat clearer. In light of your results in part (a), explain why squinting helps you see an object more clearly.

EX 22-10
10. Under bright light the pupil of the human eye is typically 2.0 mm in diameter. The diameter of the eye is about 25 mm. Suppose you are viewing something with light of wavelength 500 nm. Ignore the effect of the lens and the vitreous humor in the eye. The orders of minima for circular apertures are not simple integers; the first three correspond to $m = 1.220$, 2.233, and 3.238.
   (a) At what angles (in degrees) will the first three diffraction dark rings occur on either side of the central bright spot on the retina at the back of the eye?
   (b) Approximately how far (in millimeters) from the central bright spot would the dark rings in part (a) occur?
   (c) Explain why we do not actually observe such diffraction effects in our vision.

WHAT DID YOU LEARN?

Prep for the AP® Exam

Chapter learning goals	Section(s)	Related example(s)	Relevant section review exercise(s)
Define an electromagnetic wave, and categorize electromagnetic waves by their wavelengths.	22-1 22-2	Figure 22-3, Example 22-1	1 8–10
Discuss how speed, frequency, and wavelength are related for electromagnetic waves, and describe the structure of an electromagnetic plane wave.	22-2	22-1	1–6
Explain Huygens' principle and what it tells us about the laws of reflection and refraction.	22-3	22-2	4, 5, 7–9
Recognize the special circumstances under which total internal reflection can take place and calculate the necessary angle of incidence.	22-4	22-3	2–10
Describe dispersion and use it in calculations.	22-5	22-4	1, 3, 5, 7, 8
Describe polarization of transverse waves, including light, and calculate how the intensity of light is changed when light passes through a polarizing filter.	22-6	22-5	1-3, 6, 7, 9, 10
Use the idea of path length difference to calculate what happens in thin-film interference.	22-7	22-6, 22-7	3–9
Explain two-slit interference in terms of the wave properties of light and determine the properties of an interference pattern.	22-8	22-8	1–7
Explain why light spreads out when it passes through a narrow opening and determine the pattern of light and dark fringes for a single slit.	22-9	22-9	1–10
Calculate how the angular resolution of an optical device is limited by diffraction.	22-10	22-10	1–6, 8–10

Chapter 22 Review

Key Terms

All the Key Terms can be found in the Glossary/Glosario on page G1 in the back of the book.

angular resolution 1117
Brewster's angle 1094
critical angle 1084
diffraction 1109
diffraction maxima 1111
diffraction minima 1111
diffraction pattern 1111
dispersion 1088
double-slit interference 1105
electromagnetic spectrum 1071
electromagnetic wave 1068
fringe 1104
front 1077
Huygens' principle 1077
incident light 1077
index of refraction 1079
intensity 1084
interference maxima 1105
interference minima 1105
law of reflection 1076
linearly polarized (light) 1092
monochromatic light 1097
partially polarized (light) 1092
polarization 1092
polarizing filter 1092
ray 1077
refraction 1077
resolve 1117
single-slit diffraction 1111
sinusoidal plane wave 1072
Snell's law of refraction 1080
speed of light 1070
thin film 1097
total internal reflection 1084
unpolarized 1092
visible light 1071
wavelet 1077

Chapter Summary

Topic	Equation or Figure	
Speed of electromagnetic waves: In a vacuum all electromagnetic waves propagate at the same speed $c = 3.00 \times 10^8$ m/s. The shorter the wavelength, the higher the frequency of the wave. Our eyes are sensitive to only a narrow band of wavelengths known as the visible spectrum (visible light).	$$c = f\lambda$$ **Speed of light in a vacuum**; **Frequency** of an electromagnetic wave; **Wavelength** of the wave in a vacuum	(22-2)
Speed of light in a medium: In a medium (transparent material) other than vacuum, the speed of light is less than c. This is described in terms of the index of refraction n of the material. The value of n is different for different materials. The wavelength also changes when light enters a medium.	$$n = \frac{c}{v}$$ **Index of refraction** of a medium; **Speed of light in a vacuum**; **Speed of light in the medium**	(22-8)
	$$\lambda = \frac{c}{nf} = \frac{\lambda_{\text{vacuum}}}{n}$$ **Wavelength of light in a medium**; **Speed of light in a vacuum**; **Wavelength of the light in a vacuum**; **Frequency of the light**; **Index of refraction of the medium**	(22-11)
Huygens' principle, reflection, and refraction: Each point on a wave front (or wave crest) acts as a source of spherical waves called wavelets. This principle helps us explain the laws of reflection and refraction, which describe what happens when light encounters the boundary between two media. The angle of the reflected light is always equal to the angle θ_1 of the incident light, and the angle θ_2 of the refracted light is given by Snell's law.	(a) Refraction from one medium into a slower one 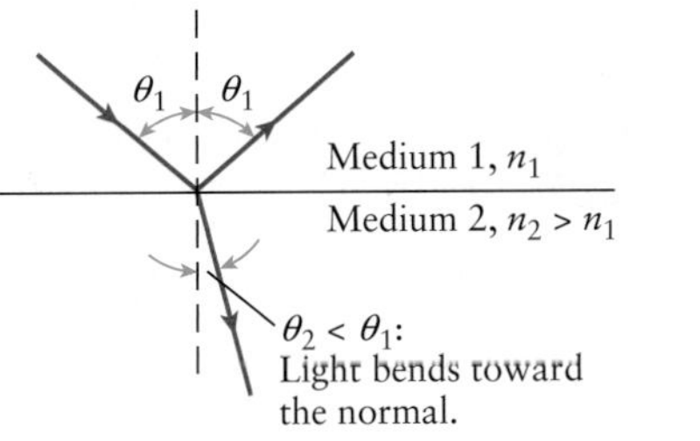	(Figure 22-10a)
	$$n_1 \sin\theta_1 = n_2 \sin\theta_2$$ **Angle of the incident ray** from the normal; **Angle of the refracted ray** from the normal; **Index of refraction for the medium with the incident light**; **Index of refraction for the medium with the refracted light**	(22-9)

Total internal refraction: If light in one medium reaches a boundary with a second medium of lower index of refraction ($n_1 > n_2$), total internal reflection is possible. It will take place only if the incident angle is greater than the critical angle for the two media.

Critical angle for light in medium 1 that encounters a boundary with medium 2 with a lower index of refraction

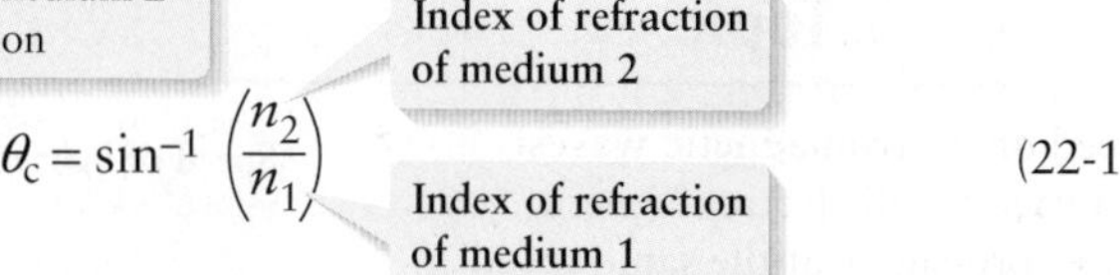
Index of refraction of medium 2
Index of refraction of medium 1

$$\theta_c = \sin^{-1}\left(\frac{n_2}{n_1}\right) \tag{22-13}$$

If the incident angle is greater than the critical angle, there is total internal reflection.

Dispersion: For a given material the index of refraction depends on the frequency of the light (or, equivalently, the wavelength in a vacuum). This is the reason a prism or water droplets can break white light into its constituent colors.

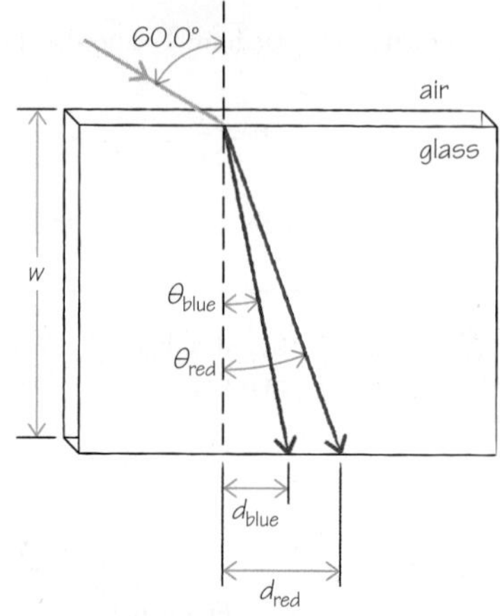
60.0°
air
glass
W
θblue
θred
dblue
dred

(Example 22-4)

Polarization: The polarization of a light wave is a description of the orientation of its electric field vector. In unpolarized light this orientation changes randomly. In linearly polarized light the orientation is in a fixed direction. Unpolarized light can become polarized by scattering from the atmosphere, by passing through a polarizing filter, or by reflection at Brewster's angle.

Brewster's angle for light in medium 1 that encounters a boundary with medium 2

Index of refraction of medium 2

$$\theta_B = \tan^{-1}\left(\frac{n_2}{n_1}\right) \tag{22-17}$$

Index of refraction of medium 1

If the incident angle equals Brewster's angle, the reflected light is completely polarized in the direction parallel to the surface of the boundary.

Thin-film interference: If a thin, transparent film is illuminated with monochromatic light, the light that reflects from the back surface of the film interferes with light that reflects from the front surface. If the two waves emerge in phase, the interference is constructive; if they are out of phase, it is destructive. The details of the interference depend on whether there is an inversion of the wave on each reflection.

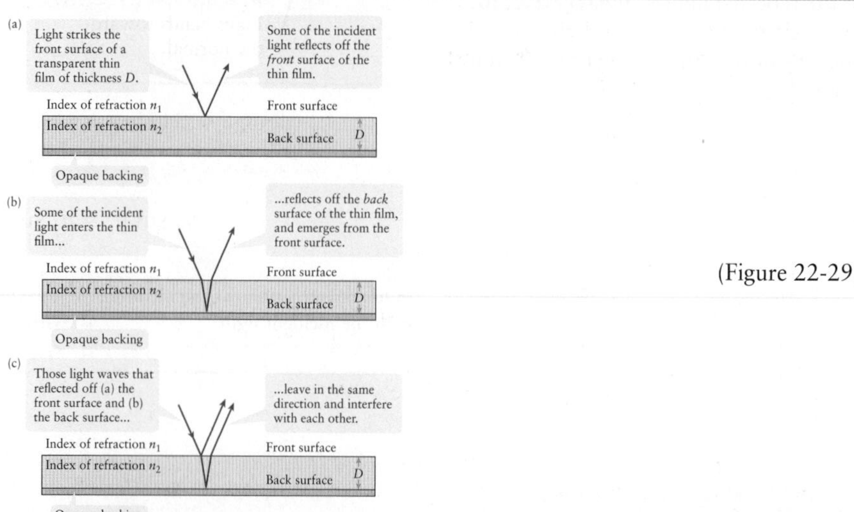
(a)
Light strikes the front surface of a transparent thin film of thickness D.
Some of the incident light reflects off the front surface of the thin film.
Index of refraction n1
Index of refraction n2
Front surface
Back surface
D
Opaque backing
(b)
Some of the incident light enters the thin film...
...reflects off the back surface of the thin film, and emerges from the front surface.
Index of refraction n1
Index of refraction n2
Front surface
Back surface
D
Opaque backing
(c)
Those light waves that reflected off (a) the front surface and (b) the back surface...
...leave in the same direction and interfere with each other.
Index of refraction n1
Index of refraction n2
Front surface
Back surface
D
Opaque backing

(Figure 22-29)

Double-slit interference: When monochromatic light illuminates a pair of closely spaced narrow slits, a pattern of bright and dark fringes results. The bright fringes appear where light waves from the two slits interfere constructively; the dark fringes appear where light waves from the two slits interfere destructively. The closer together the two slits, the broader the interference pattern.

Constructive interference (bright fringes):

$$d \sin\theta = m\lambda \qquad (22\text{-}26)$$

- **Distance between the two slits** (d)
- **Wavelength** of the light illuminating the two slits (λ)
- **Angle** between the normal to the two slits and the location of a **bright fringe** on the screen (θ)
- **Number of the bright fringe:** $m = 0, \pm1, \pm2, \pm3, \ldots$

Destructive interference (dark fringes):

$$d \sin\theta = \left(m + \frac{1}{2}\right)\lambda \qquad (22\text{-}27)$$

- **Distance between the two slits** (d)
- **Wavelength** of the light illuminating the two slits (λ)
- **Angle** between the normal to the two slits and the location of a **dark fringe** on the screen (θ)
- **Number of the dark fringe:** $m = 0, \pm1, \pm2, \pm3, \ldots$

Single-slit diffraction: A pattern of bright and dark fringes also results when monochromatic light illuminates a single narrow slit. When monochromatic light illuminates a narrow slit, a pattern of bright and dark fringes results. The dark fringes appear where light from any one part of the slit interferes destructively with light from some other part of the slit, so that zero net electromagnetic wave reaches the position of the dark fringe. The narrower the width of the slit, the broader the diffraction pattern.

$$\sin\theta = \frac{m\lambda}{w} \qquad (22\text{-}32)$$

- **Angle** between the normal to the slit and the location of the mth **dark fringe** (θ)
- **Number of the dark fringe:** $m = 1, 2, 3, \ldots$
- **Wavelength** of the light (λ)
- **Width of the slit** (w)

Diffraction by a circular aperture: Light also diffracts when it passes through a circular aperture. This sets a fundamental limit on the angular resolution of an optical device: If two objects are closer together than an angle θ_R, it will be impossible to tell with the device whether they are two objects or a single object.

$$\sin\theta_R = 1.22\,\frac{\lambda}{D} \qquad (22\text{-}34)$$

- **Angle** between two pointlike objects that can **barely be resolved** through an optical device (θ_R)
- **Wavelength** of the light (λ)
- **Diameter of the circular aperture** of the device (D)

Chapter 22 Review Problems

1. How far does light travel in a vacuum in 10 ns?
2. How long does it take light to travel 300 km in a vacuum?
3. How long does it take a radio signal from Earth to reach the Moon, which has an orbital radius of approximately 3.84×10^8 m?
4. Changing electric fields create changing magnetic fields. These oscillating fields propagate at the speed of light. Describe the directions of the fields and the velocity of the electromagnetic wave.
5. The frequency and wavelength of an electromagnetic wave are related to the speed of light by

$$c = f\lambda$$

Starting with this expression, derive a similar relationship between the speed of light, the angular frequency ($\omega = 2\pi f$), and the wave number ($k = 2\pi/\lambda$).

6. What is Huygens' principle, and why is it necessary to understand Snell's law of refraction?

7. Does the phenomenon of diffraction apply to wave sources other than light? Give an example if it does.
8. Does the refraction of light make a swimming pool seem deeper or shallower? Explain your answer.
9. Does the depth of a pool determine the critical angle that a light ray will have as it travels from the bottom of the pool and heads toward the air above the water? Explain your answer.
10. Linearly polarized light is incident at Brewster's angle on the surface of an optical medium. What can be said about the refracted and reflected beams if the incident beam is polarized (a) parallel to the plane of the surface and (b) perpendicular to the plane of the surface?
11. Describe how polarized sunglasses work. Why do such sunglasses have *vertically* polarized lenses (as opposed to *horizontally* polarized lenses)?
12. The sound waves used in speech have wavelengths of a few meters. Explain why some of the sound waves emerging from a person's mouth go to the sides so that they can be heard even if the listener is not in front of the person.
13. The closest star to our sun (Proxima Centauri) is 4.0×10^{16} m away. It has a planet (Proxima b) in orbit around it at about 1/20 of the radius of Earth's orbit around our Sun. Estimate the minimum diameter of the aperture of a space telescope that would be required to resolve Proxima b from its star, for light in the visible part of the spectrum.
14. A car radio is tuned to receive a signal from a particular radio station. While the car slows to a stop at a traffic signal, the reception of the radio seems to fade in and out. Use the concept of interference to explain the phenomenon. *Hint:* In broadcast technology, the phenomenon is known as multipathing.
15. One of the world's largest aquaria is the Monterey Bay Aquarium. The viewing wall is made of acrylic and is 0.33 m thick, and the tank holds 1.2 million gallons of water. A diagram of the tank is shown in the figure. If a ray of light is directed into the plastic from air at an angle of 40.0°, calculate the angle that the ray will make (a) when it enters the plastic and (b) when it enters the seawater. The indices of refraction for air, acrylic, and seawater are listed in the figure.

Air Acrylic Seawater

1.00 1.50 1.37

40.0°

16. What is the last color you expect to observe at sunset? Using physics principles, explain your answer.
17. At what angle θ above the horizontal, as shown in the figure, is the Sun when a person observing its rays reflected off water finds them linearly polarized along the horizontal?

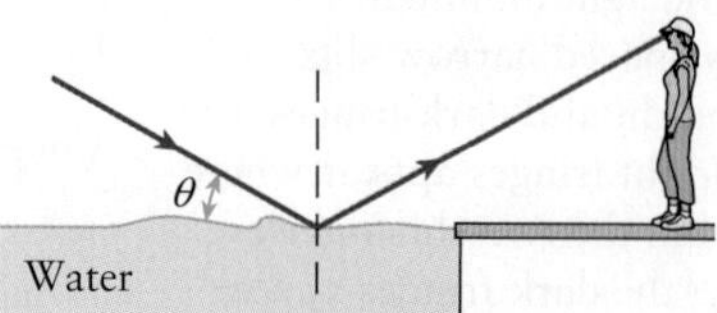

18. Suppose you are measuring double-slit interference patterns using an optics kit that contains the following options that you can mix and match: a red laser or a green laser; a slit width of 0.04 or 0.08 mm; a slit separation of 0.25 or 0.50 mm. Which elements would you select to produce an interference pattern with the most widely spaced fringes? Explain your answer using physics principles.
19. A beam of light travels from medium 1 to medium 2 in the x–y plane. Medium 1 is found in quadrants 2 and 3; medium 2 is in quadrants 1 and 4. The beam touches each of the points in the x–y plane given in the table below. Calculate the ratio of the index of refraction of medium 2 to medium 1.

x (cm)	y (cm)	x (cm)	y (cm)
−4.00	−2.00	+1.00	+0.296
−3.00	−1.52	+2.00	+0.595
−2.00	−1.02	+3.00	+0.901
−1.00	−0.514	+4.00	+1.20
0	0		

20. One way of describing the speed of light in an optical material is to specify the ratio of the time that is required for light to travel through a length of vacuum to the time required for light to travel through the same length of the optical material. For example, if light travels through a material in 150% of the time for light to travel through a vacuum, the speed of light in the material would be 2/3 = 1/1.5 that in a vacuum. Complete the table by giving the speed of light and the index of refraction for each of the following optical materials, listed with the corresponding percentage.

Optical material with percentage of time required for light to pass through compared to an equal length of vacuum	Speed of light	Index of refraction
100%		
125%		
150%		
200%		
500%		
1000%		

21. The distance between points B and C in the figure below is 0.75 cm.

(a) Determine the index of refraction for medium 2. Assume the index of refraction in medium 1 is 1.00.

(b) Suppose instead $n_2 = 1.55$. Calculate the distance between points B and C in this case.

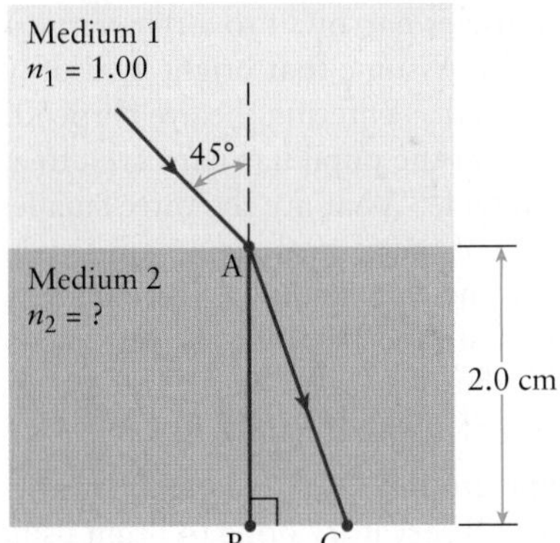

22. Suppose you are measuring double-slit interference patterns using an optics kit that contains the following options that you can mix and match: a red laser or a green laser; a slit width of 0.04 or 0.08 mm; a slit separation of 0.25 or 0.50 mm. Estimate how far you should place a screen from the double slit to give you an interference pattern on the screen that you can accurately measure using an ordinary ruler.

23. Prove that in the case where there are more than two optically different media sandwiched together, with air on the left and air on the right, the angle at which light returns to air is independent of the indices of refraction of the interior media, as shown in the figure. In other words, the refraction angle, θ_n, depends *only* on n_1, θ_1, and n_n (not $n_2, n_3, n_4, \ldots$).

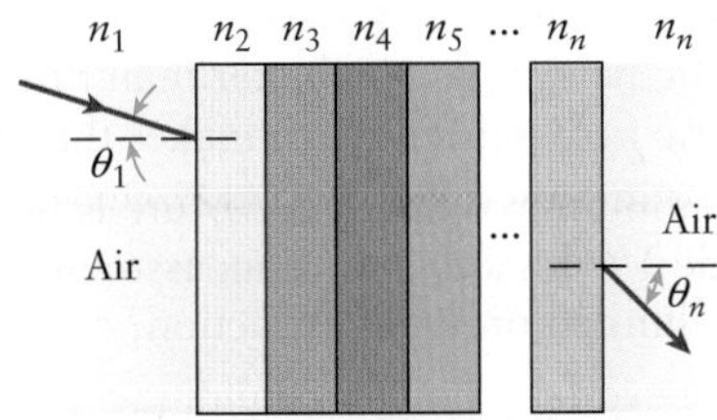

24. A flat glass table ($n = 1.54$) has a layer of water ($n = 1.33$) of uniform thickness directly above the glass. Calculate the minimum angle of incidence at which light in the glass must strike the glass–water interface for the light to be totally internally reflected at the water–air interface at the water layer's top surface.

25. Light rays in air fall normally on the vertical surface of a glass prism ($n = 1.55$), as shown in the figure below.

(a) Calculate the largest value of θ such that the ray is totally internally reflected at the slanted face.

(b) Repeat the calculation if the prism is immersed in water with $n = 1.33$.

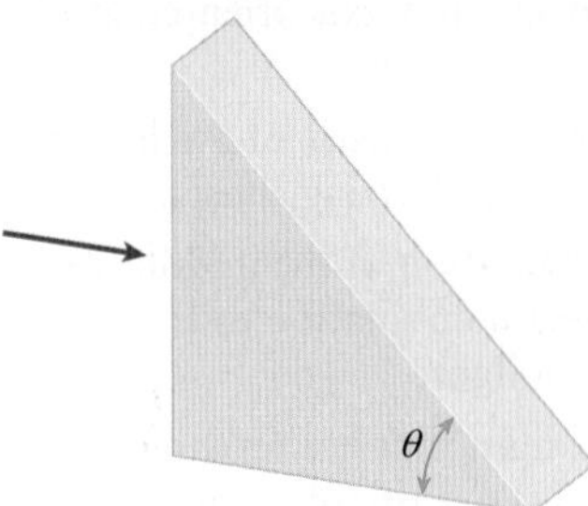

26. The object in the figure is at a depth $d = 0.85$ m below the surface of clear water.

(a) How far from the end of the dock, distance D in the figure, must the object be if it cannot be seen from any point on the end of the dock? The index of refraction of water is 1.33.

(b) If you could change the index of refraction of the water, how would you change it so that the object could be seen at any distance D beneath the dock? Assume that the dock is 2.0 m long.

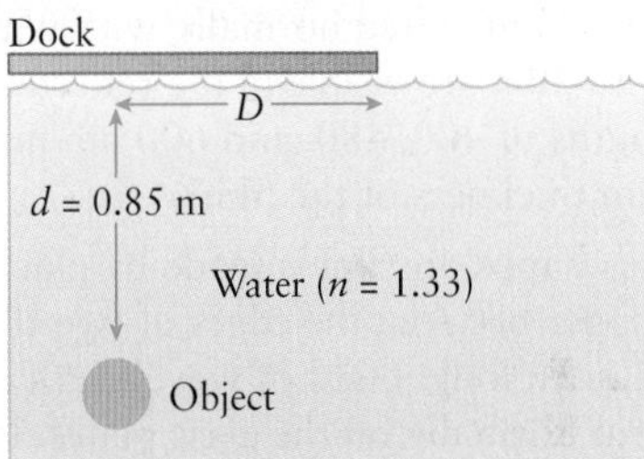

27. The polarizing angle for light that passes from water ($n = 1.33$) into a certain plastic is 61.4°. What is the critical angle for total internal reflection of the light passing from the plastic into air?

28. A thin layer of SiO, having an index of refraction of 1.45, is used as a coating on certain solar cells. The refractive index of the cell itself is 3.5.

(a) What is the minimum thickness of the coating needed to cancel visible light of wavelength 400 nm in the light reflected from the top of the coating in air? Are any other visible wavelengths also canceled?

(b) Suppose that technological limitations require you to make the coating 3.0 times as thick as in part (a). Which, if any, visible wavelengths in the reflected light will be canceled in air and which, if any, will be reinforced?

29. A glass sheet 1.40 μm thick is suspended in air. In reflected light, there are gaps in the visible spectrum at 560 and 640 nm. Calculate the minimum value of the

index of refraction of the glass sheet that produces this effect.

30. Unpolarized light of intensity 100 W/m^2 is incident on two ideal polarizing sheets that are placed with their transmission axes perpendicular to each other. An additional polarizing sheet is then placed between the two, with its transmission axis oriented at 30° to that of the first.

(a) Calculate the intensity of the light passing through the stack of polarizing sheets.

(b) What orientation of the middle sheet enables the three-sheet combination to transmit the greatest amount of light?

31. Unpolarized light that has an intensity of 850 W/m^2 is incident on a series of polarizing filters as shown in the figure. If the intensity of the light after the final filter is 75 W/m^2, what is the orientation of the second filter relative to the x axis? *Hint:* $\cos(90° - \theta) = \sin\theta$ and $2\sin\theta\cos\theta = \sin 2\theta$.

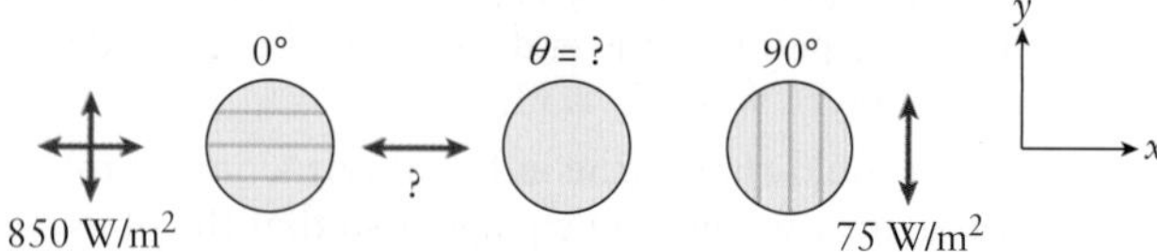

32. A thin film of soap solution ($n = 1.33$) has air on either side and is illuminated normally with white light. Interference minima are visible in the reflected light only at wavelengths of 400, 480, and 600 nm in air. What is the minimum thickness of the film?

33. A wedge-shaped air film is made by placing a small slip of paper between the edges of two thin plates of glass 12.5 cm long. Light of wavelength 600 nm in air is incident normally on the glass plates. If interference fringes with a spacing of 0.200 mm are observed along the plate, how thick is the paper? This form of interferometry is a very practical way of measuring small thicknesses.

34. You want to coat a pane of glass that has an index of refraction of 1.54 with a 155-nm-thick layer of material that has an index of refraction greater than that of the glass. The purpose of the coating is to cancel light reflected off the top of the film having a wavelength (in air) of 550 nm.

(a) What should be the index of refraction of the coating?

(b) If, due to technological difficulties, you cannot achieve a uniform coating at the desired thickness, what are the next three thicknesses of the coating you could use?

35. Two point sources of light, each of which has a wavelength of 500 nm, are photographed from a distance of 100 m using a camera with a 50.0-mm focal length lens. The camera aperture is 1.05 cm in diameter. What is the minimum separation of the two sources if they are to be resolved in the photograph, assuming the resolution is limited only by diffraction?

36. The pupil (the opening through which light enters the lens) of a house cat's eye is round under low light, but ciliary muscles narrow it to a thin vertical slit in very bright light. Assume that bright light of wavelength 550 nm in air is entering the eye perpendicular to the lens and that the pupil has narrowed to a slit that is 0.50 mm wide. What are the three smallest angles on either side of the central maximum at which no light will reach the cat's retina (a) if we imagine the eye is filled with air and (b) if we take into consideration that in reality the eye is filled with a fluid having index of refraction of approximately 1.4?

37. Sometime around 2025, astronomers at the European Southern Observatory hope to begin using the ELT (Extremely Large Telescope), which is planned to have a primary mirror 42 m in diameter. Let us assume that the light it focuses has a wavelength of 550 nm.

(a) What is the most distant Jupiter-sized planet the telescope could resolve, assuming its resolution is limited only by diffraction? Express your answer in meters and light years.

(b) The nearest known exoplanets (planets beyond the solar system) are around 20 light years away. What would have to be the minimum diameter of an optical telescope to resolve a Jupiter-sized planet at that distance using light of wavelength 550 nm? (1 light year = 9.461×10^{15} m.)

38. Under the best atmospheric conditions at the premium site for land-based observing (Mauna Kea, Hawaii, elevation ~4.27 km), an optical telescope can resolve celestial objects that are separated by one-fourth of a second of arc (1 second of arc = 1 arcsec = 1/3600 of a degree). The viewing never gets any better than this because of atmospheric turbulence, which makes the images jitter.

(a) What minimum diameter aperture is necessary to provide 1/4-arcsec resolution, assuming the resolution is limited only by diffraction?

(b) Is there ever any point in building a telescope much bigger than this? Explain your answer using physics principles.

39. The Herschel infrared telescope, launched into space in 2009, made observations from 2010 to 2013. Its primary mirror is 3.5 m in diameter, and the telescope focuses infrared light in the range of 55 to 672 μm. Because this telescope operated above Earth's atmosphere, its resolution was limited only by diffraction.

(a) What wavelength in its observing range will give the maximum angular resolution? What is that maximum resolution (in radians and seconds, $1° = 60'$ and $1' = 60''$)?

(b) To achieve the same resolution as in part (a) using visible light of wavelength 550 nm, what should be the mirror diameter of an optical telescope?

(c) What is the smallest infrared source that the Herschel infrared telescope can resolve at a distance of 150 light years? (A light year is the distance that light travels in one year—about 9.461×10^{15} m.)

40. The world's largest refracting telescope is at Yerkes Observatory in Williams Bay, Wisconsin. Its objective is 1.02 m in diameter. Suppose you could mount the telescope on a spy satellite 200 km above the ground.

(a) Assuming that the resolution is limited only by diffraction, what minimum separation of two objects on the ground could it resolve? Take 550 nm as a representative wavelength for visible light.

(b) Because of atmospheric turbulence, objects on the surface of Earth can be distinguished only if their angular separation is at least 1.00 arcsec (1 arcsec = 1/3600 of a degree). How far apart would two objects on Earth's surface be if they subtended an angle of 1.00 arcsec as measured from the satellite?

(c) Compare the separation you found in part (b) with your answer to part (a).

41. In a realistic two-slit experiment, you see two interference effects described in this chapter. There is the large-scale interference pattern from diffraction that depends on the width of each slit, and within that pattern there is the small-scale interference pattern that depends on the separation distance between the slits. For a slit width of 0.04 mm and a slit separation of 0.50 mm, how many of the bright two-slit interference fringes fit inside the central bright fringe of the diffraction pattern (from the first dark fringe on one side of the center to the first dark fringe on the other side) if the laser has a wavelength of (a) 650 nm or (b) 530 nm?

42. An equilateral triangular glass prism rests on a table as shown in the figure. A beam of white light is incident on one face of the prism, at an angle of 60° below the normal to the surface, as shown in the figure. If the index of refraction of the glass is 1.51 for red light and 1.53 for blue light, what is the angular separation of the red and blue portions of the spectrum that emerge from the prism?

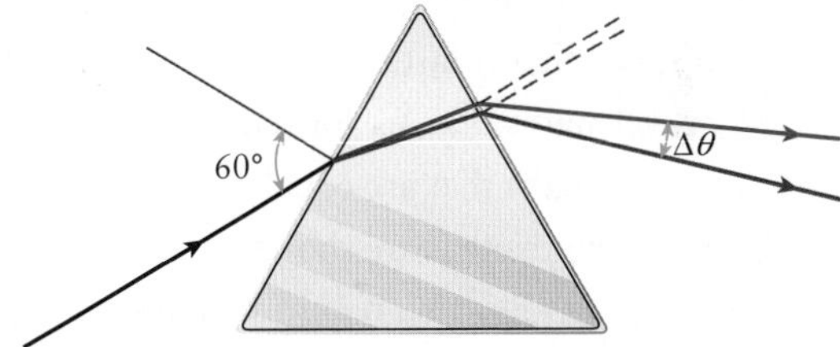

Prep for the AP Exam

AP® Group Work

Directions: The problem below is designed to be done as group work in class.

In 2009 researchers reported on evidence that a giant tsunami had hit the eastern coast of the Mediterranean Sea (present-day Lebanon and Israel) around 1600 B.C., causing huge damage to the civilizations located there. It is believed that the tsunami was caused by the eruption of the Thera volcano near the island of Crete. The waves would have passed through the 100-mi-wide opening between Crete and Rhodes, which would cause them to diffract and spread out. Satellite observations of tsunamis show that the waves measure about 250 mi from a crest to the adjacent trough, and the time between successive crests is typically 60 min.

(a) How fast do tsunami waves travel?

(b) How long after the eruption of Thera would the tsunami reach the eastern shore of the Mediterranean Sea, 600 mi from Thera?

(c) For these waves, could we apply the formula $w \sin\theta = m\lambda$ to find the angles at which the waves cancel after passing through the 100-mi "slit" between Crete and Rhodes? Explain why or why not.

AP® PRACTICE PROBLEMS

Prep for the AP Exam

Multiple-Choice Questions

Directions: The following questions have a single correct answer.

1. In comparison to x rays in a vacuum, visible light in a vacuum has
   (A) a speed that is faster.
   (B) wavelengths that are longer.
   (C) wavelengths that are equal.
   (D) wavelengths that are shorter.
2. X rays and gamma rays in a vacuum
   (A) have the same frequency.
   (B) have the same wavelength.
   (C) have the same speed.
   (D) have the same "color."
3. In comparison to radio waves in a vacuum, visible light in a vacuum has
   (A) a speed that is faster.
   (B) wavelengths that are longer.
   (C) wavelengths that are equal.
   (D) wavelengths that are shorter.
4. Which of the following requires a physical medium through which to travel?
   (A) Radio waves
   (B) Light
   (C) X rays
   (D) Sound
5. Which type of wave can refract when crossing from one medium to another if the wave has a different speed in each of the two media?
   (A) Electromagnetic waves
   (B) Sound waves
   (C) Water waves
   (D) Electromagnetic, sound, and water waves

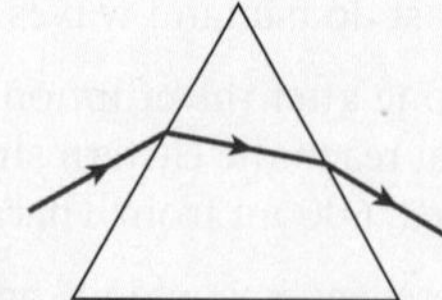

6. Which phenomenon would cause monochromatic light to enter the prism and follow along the path as shown in the figure above?
   (A) Reflection (B) Refraction
   (C) Interference (D) Diffraction
7. Two linear polarizing filters are placed one behind the other so that their transmission directions are parallel to one another. A beam of unpolarized light of intensity I_0 is directed at the two filters. What fraction of the light will pass through both filters?
   (A) I_0
   (B) $(1/2)I_0$
   (C) $(1/4)I_0$
   (D) 0
8. Two linear polarizing filters are placed one behind the other so that their transmission directions form an angle of 45°. A beam of unpolarized light of intensity I_0 is directed at the two filters. What fraction of the light will pass through both?
   (A) I_0
   (B) $(1/2)I_0$
   (C) $(1/4)I_0$
   (D) 0

9. The figure above shows two single-slit diffraction patterns created with the same source. The distance between the slit and the viewing screen is the same in both cases. Which of the following is true about the widths, w_a and w_b, of the two slits?
   (A) $w_a > w_b$
   (B) $w_a < w_b$
   (C) $w_a = w_b$
   (D) $w_a = (1/2)w_b$

(Continued)

Free-Response Question

Top view:

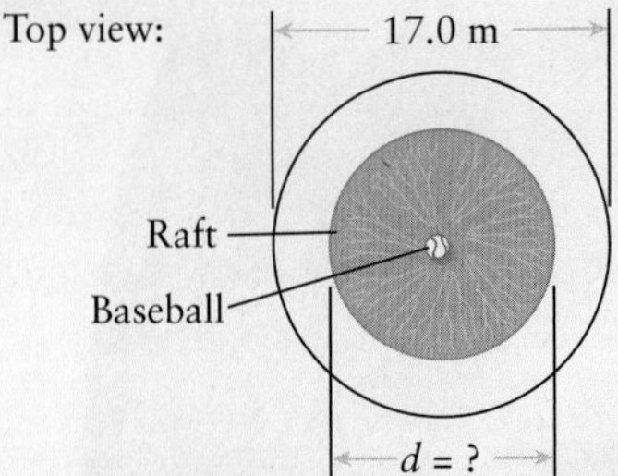

Side view:

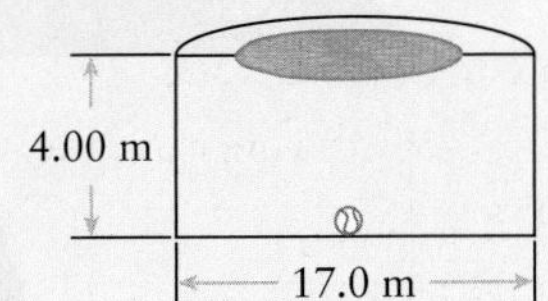

A baseball is on the bottom of a round pool of water that is 4.00 m deep and 17.0 m across as shown in the figure above. The ball is right in the center of the pool. A large round raft is floating in the pool, concentrically on top of the location of the ball. Assume that water has an index of refraction of 1.33.

(a) On the side view diagram, make a sketch showing the path of the light from the ball to an observer sitting beside the pool.

(b) Calculate the minimum diameter the raft must have to completely obscure the ball from sight.

(c) If chemicals added to the water increase the index of refraction of the water in the pool, how would that affect your answer to part (b)? Explain your answer in terms of physics principles.

23 Geometric Optics: Ray Properties of Light

moodboard/Alamy

YOU WILL LEARN TO:

- Describe how a plane mirror forms an image.
- Use ray diagrams to explain how the image formed by a concave mirror depends on the position of the object.
- Calculate the position and height of an image made by a concave mirror.
- Explain the differences between the images made by a convex mirror and a concave mirror.
- Calculate the position and height of an image made by a convex mirror.
- Describe how the curved surfaces of a lens make light rays converge or diverge.
- Calculate the focal length of a lens based on its composition and shape, and the position and magnification of an image made by a lens.
- Explain how the eye forms images on the retina, and find the needed focal length for corrective lenses used to compensate for deficiencies in vision.
- Calculate the angular magnification of a magnifying glass, a microscope, and a telescope.

23-1 Mirrors or lenses can be used to form images

We saw in Chapter 22 that the direction of a light ray changes when it either reflects from a surface (law of reflection) or refracts as it passes from one transparent medium to another (Snell's law). **Geometric optics** is the branch of physics that uses the laws of reflection and refraction to understand **optical devices**—instruments that change the direction of light rays in a regular way. The simplest optical devices are reflective objects, or *mirrors*, and pieces of transparent material with carefully shaped surfaces, or *lenses*. Mirrors or lenses whose surfaces are curved in just the right way can be used to change the apparent sizes of objects (**Figure 23-1a**). The lens of your eye is an essential part of vision and needs to be replaced if it becomes clouded with age (**Figure 23-1b**).

As the name geometric optics suggests, to study optical devices all we need besides the laws of reflection and refraction is a bit of geometry. We won't need to refer to the wave properties of light. (Much of the basic physics of mirrors and lenses was deduced before it was understood that light is a wave; although, in Chapter 22, we saw that the wave nature of light is crucial in other applications.)

We'll begin by considering a simple plane mirror. We'll see how the law of reflection explains the kind of image that it forms. We'll then use similar ideas to understand the kind of images formed by a mirror with a surface that's curved either inward or outward. In contrast to mirrors, lenses use the refraction of light to form an image: Light rays can change direction when they enter a lens and again when they exit the lens. Happily, we'll find that many of the same ideas that we'll develop by considering curved mirrors apply equally well to lenses. We'll discuss the human eye, how it forms images, how deficiencies in vision can be corrected, and how it compares to a camera. Finally, we'll look at three optical devices of particular importance: the magnifying glass, the microscope, and the telescope.

This dentist uses a curved mirror to make a magnified image of a patient's tooth. He can also get a magnified view by using the lenses attached to his eyeglasses.

(a)

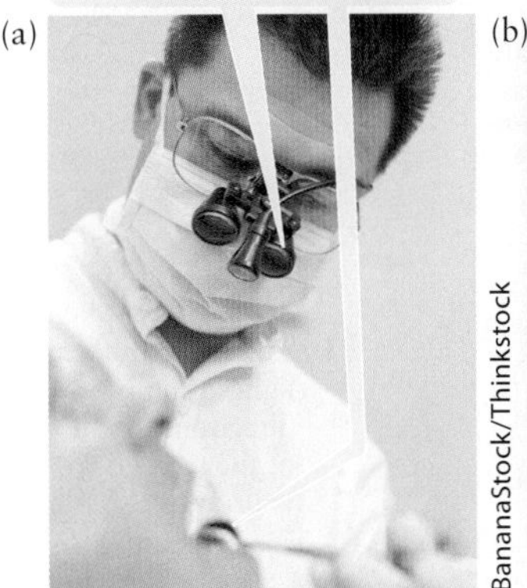

BananaStock/Thinkstock

The human eye contains a transparent lens that helps focus images onto the retina. Many people develop cataracts as the clear lens gradually becomes cloudy. The treatment is to surgically remove the lens and replace it with an artificial one.

(b)

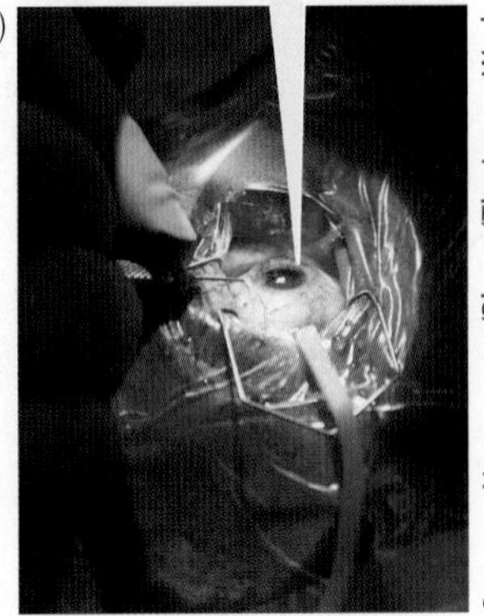

Syracuse Newspapers/Blume/The Image Works

Figure 23-1 Mirrors and lenses Mirrors and lenses are two different devices used to form images.

THE TAKEAWAY for Section 23-1

✔ A mirror forms images by the reflection of light from the mirror's surface.

✔ A transparent lens forms images by the refraction of light as it enters and exits the lens.

Prep for the AP® Exam

AP® Building Blocks

1. Which of the following does not depend on refraction to form an image?
   (A) A mirror
   (B) A lens
   (C) Both depend on refraction.
   (D) Neither depends on refraction.

AP® Exam Tip

In optics we use the word *object* in the same way as you might in normal conversation. Remember that in our modeling of motion, we used a very specific definition of object, one where we could neglect its size and anything internal, and describe its motion as if it were located at its center of mass. In geometric optics, we will use *object* to refer to something we might look at through an optical system. We will consider its size to be fixed, but not zero!

23-2 A plane mirror produces an image that is reversed back to front

Our visual system can detect only objects that emit or reflect light. (We can't see in the dark!) Although it's easy to find examples of luminous things, such as the Sun or the screen of a mobile device, most of what we see only reflects light. (In optics we will use the term *object* to refer to anything that acts as a source of light rays for an optical device.) As we learned in the last chapter, a ray of light that strikes a surface always reflects in such a way that the angle of incidence equals the angle of reflection. (This is the law of reflection.) However, because most objects have uneven surfaces, the light that they reflect goes off in many seemingly random directions. This is called **diffuse reflection** (**Figure 23-2a**).

(a)

Light that reflects from an object's surface in many random directions is called diffuse. Uneven surfaces produce diffuse light even when the incident rays come from the same direction. For each incident ray the light obeys the law of reflection: It is reflected at an angle from the normal equal to the incident angle, but the normals are not aligned due to the unevenness of the surface.

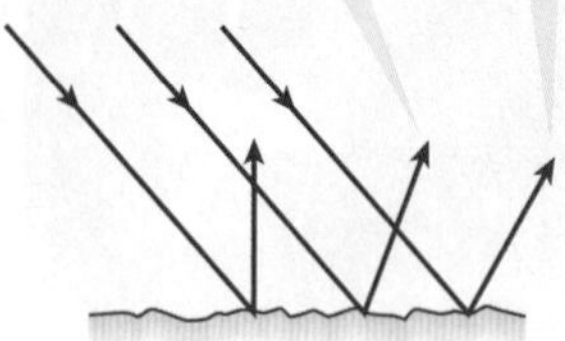

(b)

Light that reflects from a smooth surface is called specular. Light rays coming from the same general direction all reflect in the same general direction. The law of reflection is easier to see in this instance.

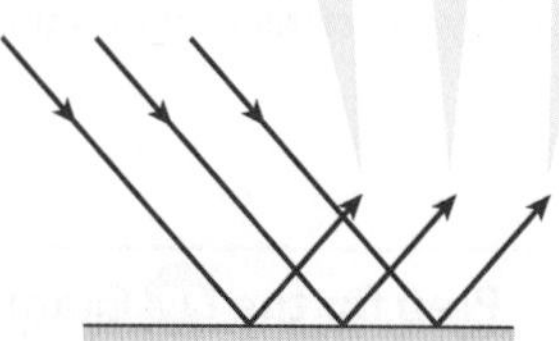

Figure 23-2 Diffuse and specular reflection Light reflecting from (a) an uneven surface and (b) a smooth surface.

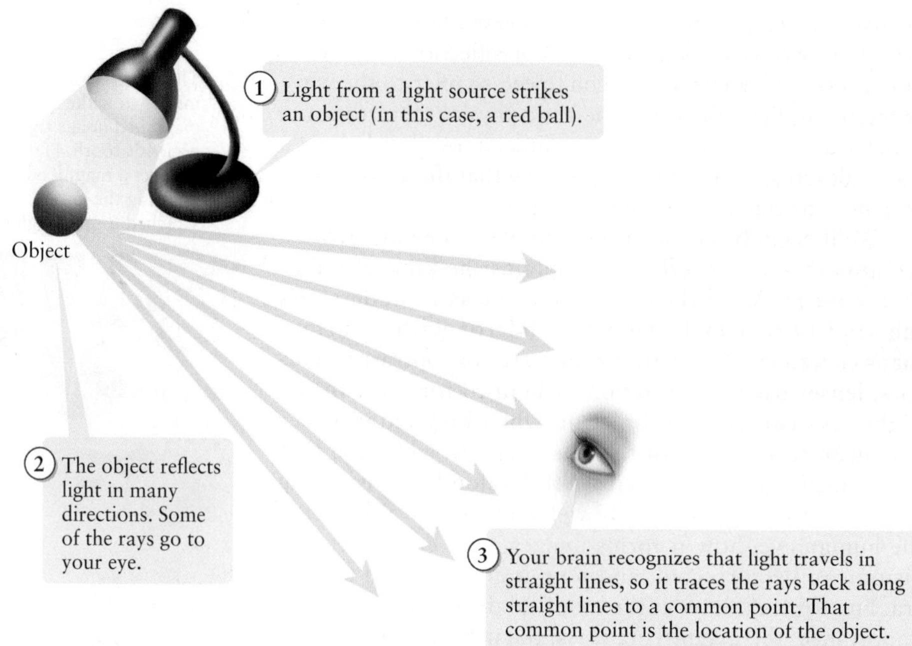

Figure 23-3 Locating an object When light strikes our eyes, we trace the rays of light back along straight lines to an apparent common source.

If an object has a flat surface, however, light rays that strike that surface are all reflected in the same general direction (**Figure 23-2b**). Such a flat, reflecting surface is called a **plane mirror**, and reflections from such a surface are called *specular* (from the Latin word for mirror, *speculum*). Your reflection in a bathroom mirror is an example of **specular reflection**. Let's look more closely at how a plane mirror creates an image.

Figure 23-3 shows how your eye and brain interpret light coming from an object. Some of the light rays coming from the object go to your eyes. Your brain traces those rays backward to a common origin and interprets that origin as the location of the object. **Figure 23-4** shows a similar situation, except that we have added a plane mirror. Some of the light from the object strikes the mirror and is reflected toward your eye. These rays appear to be coming from a point behind the mirror, so your brain interprets

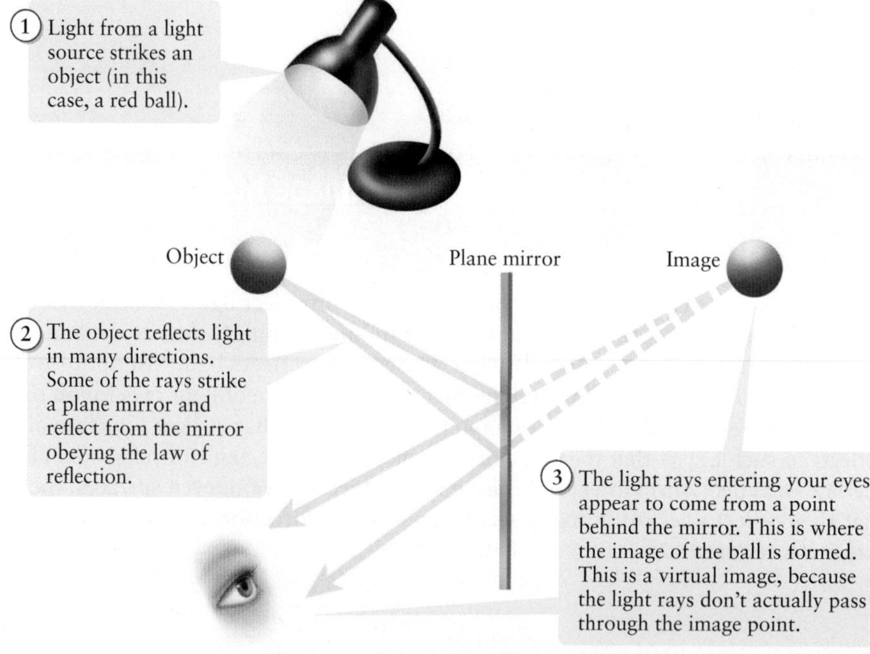

1 Light from a light source strikes an object (in this case, a red ball).
Object
Plane mirror
Image
2 The object reflects light in many directions. Some of the rays strike a plane mirror and reflect from the mirror obeying the law of reflection.
3 The light rays entering your eyes appear to come from a point behind the mirror. This is where the image of the ball is formed. This is a virtual image, because the light rays don't actually pass through the image point.

Figure 23-4 Locating an image When light from a mirror strikes our eyes, we use the same technique as in Figure 23-3 to determine the position of the image made by the mirror.

the light as coming from that point. We say that the mirror has formed an **image** of the object, and this image lies behind the mirror.

As you can see in Figure 23-4, no light rays actually pass through the location of the image. An image of this kind is called a **virtual image**. The image formed by a plane mirror is always a virtual image. We will shortly encounter some optical devices that cause light rays to bend toward each other so that the image forms where light rays do actually meet. An image formed by light rays coming together is called a **real image.**

We can determine the position of the image made by a plane mirror by using a **ray diagram** (**Figure 23-5**). In such a diagram, we draw a few light rays coming from the object and show how they reflect from the mirror. We represent the object by a red arrow of height h_o (the **object height**) located a distance s_o (the **object distance**) from the mirror. To determine the image position we must draw at least two rays from the tip of the arrow. (We've drawn three in Figure 23-5.) When each ray strikes the mirror, it obeys the law of reflection: The angle of the reflected ray equals the angle of the incident ray. Note that the horizontal ray strikes the mirror face-on, so it is reflected back horizontally toward the tip of the arrow.

The reflected rays diverge from each other and never actually meet. But if we trace these rays back to where they would meet, as shown by the dashed lines in Figure 23-5, we find the position of the tip of the image arrow. The geometry of the light rays requires that the image be as far behind the front of the mirror as the object is in front of the mirror. In other words, the **image distance** s_i has the same magnitude as the object distance s_o. We'll use the convention that a point on the reflective side of the mirror (to the left of the mirror in Figure 23-5) is at a positive distance, while a point on the back side of the mirror (to the right of the mirror in Figure 23-5) is at a negative distance. So in Figure 23-5 the object distance s_o is positive, but the image distance s_i is negative. We can write the relationship between s_i and s_o for a plane mirror as

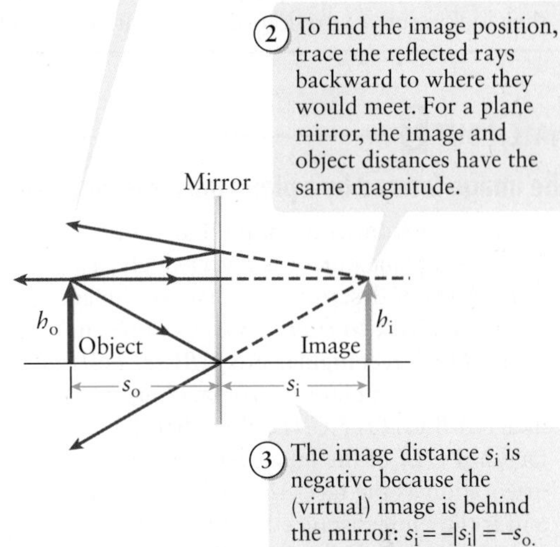

① Light rays from the object reflect from the plane mirror. (The ray that hits the mirror face-on reflects back along its initial path because the incident angle is zero.)
② To find the image position, trace the reflected rays backward to where they would meet. For a plane mirror, the image and object distances have the same magnitude.
Mirror
h_o
Object
s_o
h_i
Image
s_i
③ The image distance s_i is negative because the (virtual) image is behind the mirror: $s_i = -|s_i| = -s_o$.

Figure 23-5 A ray diagram for a plane mirror This diagram helps us determine the position, orientation, and size of the image.

The **image distance** is negative. The **object distance** is positive.

$$s_i = -|s_i| = -s_o \qquad (23\text{-}1)$$

The negative value of s_i indicates that the image is on the opposite side of the mirror from the object.

EQUATION IN WORDS
Image distance for a plane mirror

For example, if you stand a distance $s_o = 1.0$ m in front of a plane mirror, your image is at $s_i = -1.0$ m—that is, 1.0 m behind the mirror.

Figure 23-5 also shows that the **image height** h_i is the same as the object height h_o. We define the **lateral magnification** M as the ratio of the image height to the object height. (We will often refer to M simply as the *magnification*.)

Lateral magnification; Image height

$$M = \frac{h_i}{h_o} \qquad (23\text{-}2)$$

Object height

EQUATION IN WORDS
Lateral magnification

For a plane mirror $h_i = h_o$, so the magnification is $M = 1$. For an optical device that results in magnification greater than 1, the image is larger than the object. Such a device is commonly called a *magnifier*; it forms a magnified image of the object. When M is less than 1, the image formed by the optical device is smaller than the object. As we'll see in later sections, a *curved* mirror can form an image that is either larger or smaller than an object placed in front of it.

WATCH OUT !

The image formed by a plane mirror appears *behind* the mirror.

The image does not form "on" a plane mirror but rather behind it. This is evident from the ray diagrams in Figure 23-5. But if this diagram doesn't convince you, you can prove it to yourself by taping a bit of paper to a plane mirror, then looking at your reflection in the mirror from about 1 m away. You'll find it difficult, likely impossible, to focus your eyes on both your image in the mirror and the piece of paper at the same time. That's because your image in the mirror is twice as far from your eyes as the paper on the surface of the mirror. The image is behind the mirror, not on it.

WATCH OUT !

The image formed by a plane mirror is reversed back to front, *not* left to right.

It's a common misconception that your image in a plane mirror is reversed left to right. A better description is that your image is reversed *back to front*. As an example, in **Figure 23-6** a rectangular box *ABCDEFGH* sits in front of a plane mirror, making an image *A′B′C′D′E′F′G′H′*. Note that the face *A′B′C′D′* of the image has the same orientation as the face *ABCD* of the object and is the same distance from the mirror; the same is true of the face *E′F′G′H′* of the image and the corresponding face *EFGH* of the object. You can see that the net result is that the image is identical to the object but flipped from back to front.

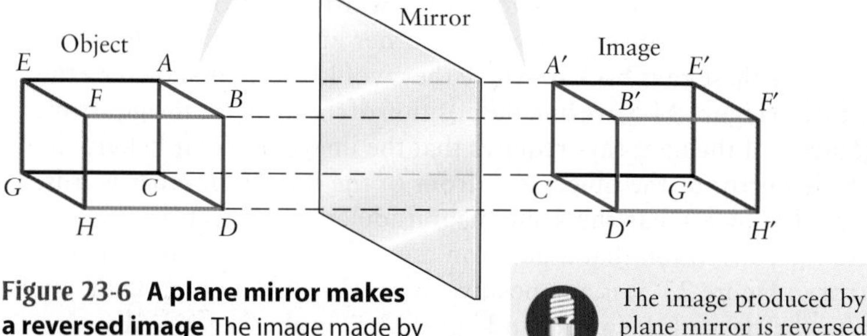

Each point on the image behind the mirror is directly opposite the corresponding point on the object in front of the mirror: A′ opposite A, B′ opposite B, and so on.
Mirror
Object
Image
E
A
F
B
G
C
H
D
A′
E′
B′
F′
C′
G′
D′
H′

Figure 23-6 A plane mirror makes a reversed image The image made by a plane mirror is actually reversed back to front, not left to right.

The image produced by a plane mirror is reversed from back to front.

THE TAKEAWAY for Section 23-2

✔ When light rays coming from the same general direction hit a plane mirror, they tend to be reflected in the same general direction.

✔ If an object is placed in front of a plane mirror, the image is as far behind the mirror as the object is in front. The image is the same size as the object, but reversed from back to front.

✔ A plane mirror makes a virtual image; the light rays coming from the image do not actually pass through the image position.

Prep for the AP® Exam

AP® Building Blocks

1. A plane mirror seems to invert your image left and right but not up and down. Is this what it really does?
2. Explain why reflected light is usually diffuse.
3. What is the difference between a real image and a virtual image?

AP® Skill Builders

4. Two flat mirrors are perpendicular to each other. An incoming beam of light makes an angle of $\theta = 30°$ with the first mirror, as shown in the figure. What angle will the outgoing beam make with respect to the normal of the second mirror?

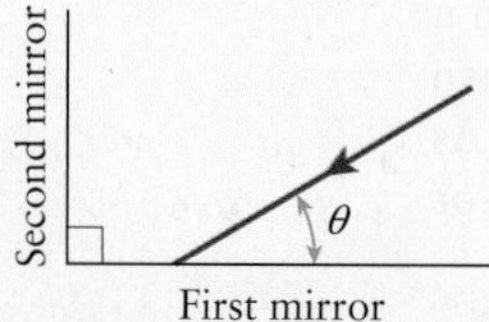

5. A 1.8-m-tall man stands 2.0 m in front of a vertical plane mirror. How tall will the image of the man be?
6. What must be the minimum height of a plane mirror for a 1.8-m-tall person to see a full image of himself?

AP® Skills in Action

7. You look into a mirror hanging near the corner of a hallway and see the eyes of someone standing around the corner. Is she able to see you?
   (A) Yes
   (B) No, because it is never possible to see around a corner with a mirror
   (C) No, because in this case it is not possible
   (D) Not enough information is given to decide.
8. A plane mirror is 10 m away from and parallel to a second plane mirror, as shown in the figure. Find the locations of the first five images formed by each mirror when

an object is positioned exactly in the middle between the two mirrors.

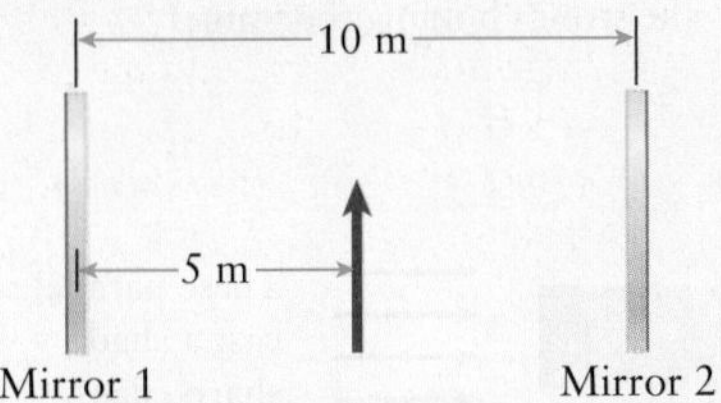

9. You look into a mirror hanging near the corner of a hallway. You see the right hand of someone standing around the corner, but not his eyes. Is he able to see you?
(A) Yes
(B) No, because it is never possible to see around a corner with a mirror
(C) No, because in this case it is not possible
(D) Not enough information is given to decide.

10. A plane mirror is 10 m away from and parallel to a second plane mirror as shown in the figure. Find the locations of the first five images formed by each mirror when an object is positioned 3 m from one of the mirrors.

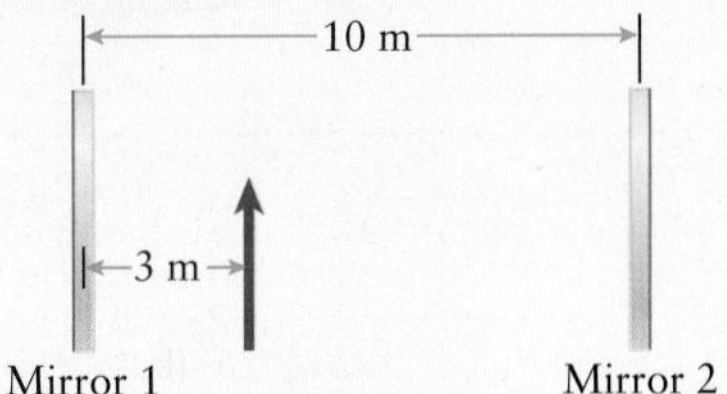

23-3 A concave mirror can produce an image of a different size than the object

Figure 23-7 shows a jalapeño pepper placed in front of two curved mirrors. The mirror in **Figure 23-7a** is **convex** (its reflective surface is curved outward) and produces an image that is smaller than the object. The mirror in **Figure 23-7b** is **concave** (its reflective surface is curved inward, or "caved in") and produces an image that is larger than the object. By contrast, the plane mirror that we studied in Section 23-2 always produces an image of the same size as the object. Let's now analyze what happens with curved mirrors.

Figure 23-8 shows ray diagrams for parallel light rays striking two *spherical* mirrors—that is, mirrors that are shaped like a section cut from a complete sphere. **Figure 23-8a** shows that parallel light rays converge after they strike a concave mirror, while **Figure 23-8b** shows that parallel light rays diverge after they strike a convex mirror. By comparison, parallel light rays that strike a plane mirror remain parallel after striking it (see Figure 23-2b). It's important to consider such light rays because light rays that come from distant objects are either parallel or nearly so. The Sun's rays, for example, are essentially parallel when they strike Earth. You can see this from the crisp shadows cast by an object placed in the Sun, such as the hanging frame in **Figure 23-9a.** In contrast the shadow cast by a light bulb in **Figure 23-9b** is fuzzy because the light rays from the light bulb are not parallel.

The concave mirror has many practical uses. For example, a bathroom mirror used for applying makeup or for shaving is concave so that it gives an enlarged image of your face (like the enlarged image of the pepper in Figure 23-7b). Telescopes used by professional and amateur astronomers have a large concave mirror that brings the light from distant objects to a focus, forming an image (we'll discuss this further in Section 23-10). Automobile headlights use the same principle in reverse: A light bulb is placed in front of a concave mirror, and the mirror reflects the light forward to illuminate the road ahead. In this section and the next, we'll look at the concave mirror in detail; we'll return to the convex mirror in Sections 23-5 and 23-6.

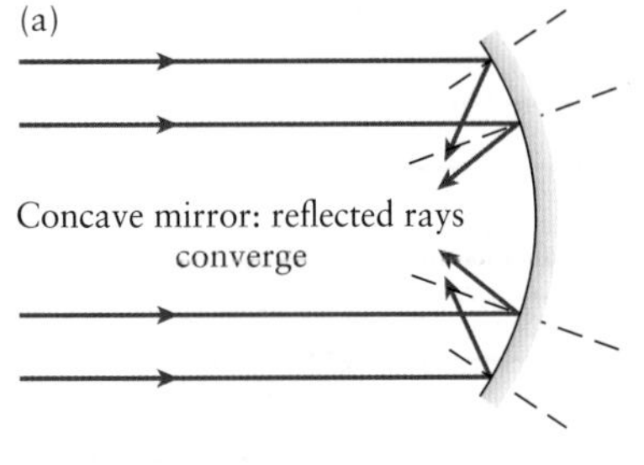

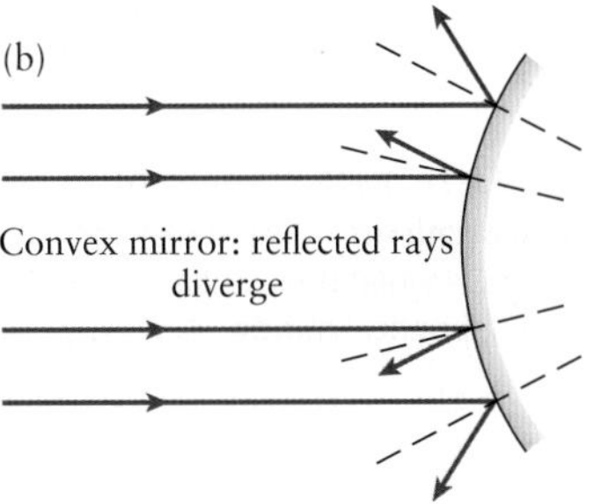

Figure 23-8 Concave and convex mirrors (a) Parallel light rays converge after reflecting from a concave mirror but (b) diverge after reflecting from a convex mirror.

(a) Convex mirror (b) Concave mirror

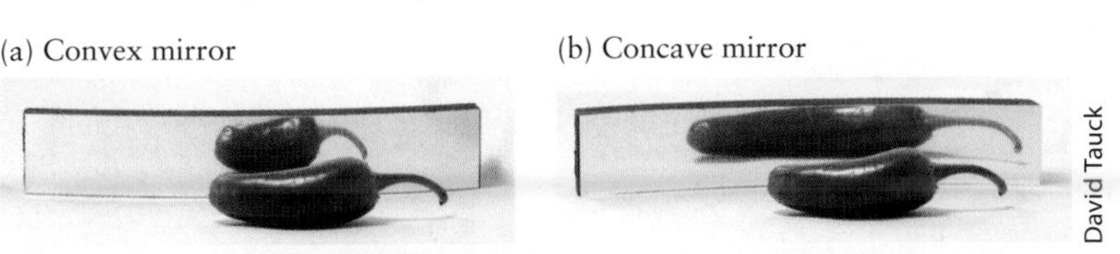

David Tauck

Figure 23-7 Images from curved mirrors A curved mirror can produce an image that is a different size than the object.

Figure 23-9 Parallel and nonparallel light rays These photographs show the difference between light rays that are (a) parallel and (b) nonparallel.

(a)

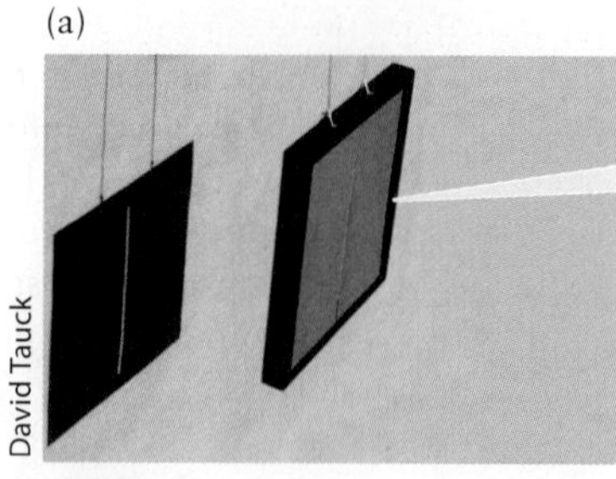
David Tauck

Rays of light from the distant Sun strike the hanging frame. Notice the sharp edges on all of the shadows, including those of the strings holding the frame.

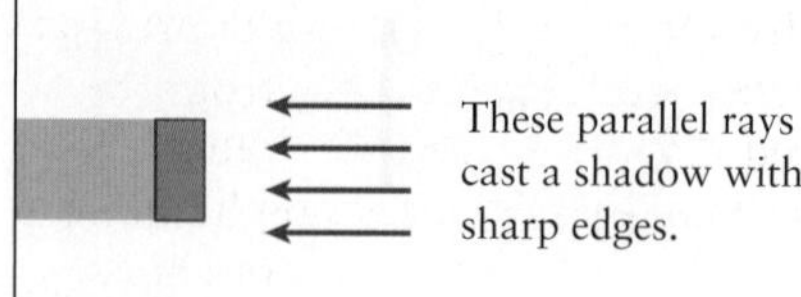
These parallel rays cast a shadow with sharp edges.

(b)

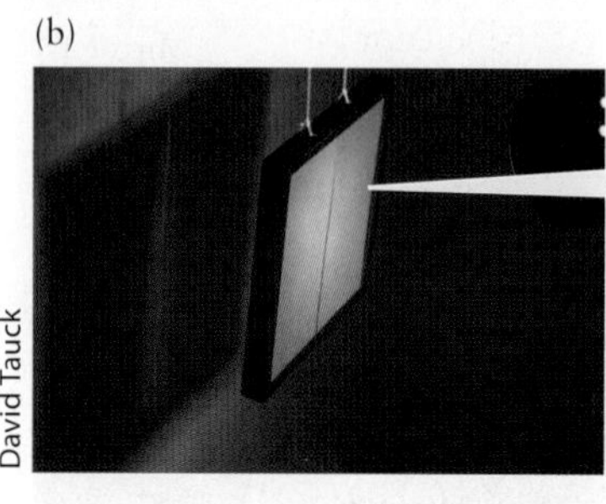
David Tauck

Light from a lamp strikes the frame. Because the light bulb is close to the frame, the central region of the shadow is darker than the outer regions.

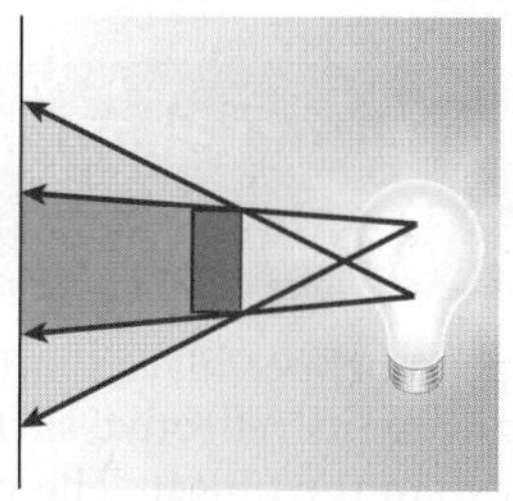
These nonparallel rays cast a shadow with fuzzy edges.

- A concave, spherical mirror causes parallel light rays to nearly converge along the principal axis of the mirror.
- If we consider only rays close to the principal axis, or if the mirror is only a small arc of the complete sphere, the light rays all converge to essentially a single point.

Figure 23-10 shows what happens when a series of parallel light rays strike a concave mirror formed from a fairly large section of a sphere. The incoming rays are parallel to each other and also parallel to the **principal axis** of the mirror, the axis that runs through the center of the sphere and also the center of the mirror. Notice that while the reflected rays generally converge along the principal axis of the mirror, they do not converge to the exact same point. This unfortunate characteristic of spherical mirrors is referred to as *spherical aberration.*

To avoid spherical aberration, we limit the reflective surface of the mirrors we consider to a relatively small section of a sphere. Here "relatively small" means that the size of the reflective surface is small compared to the radius of the sphere. There is no specific cut-off value; rather, the smaller the mirror compared to the radius, the more tightly focused the reflected rays will be. For the rest of this chapter, we will deal with spherical mirrors small enough that all rays parallel to the principal axis are focused to essentially a single point.

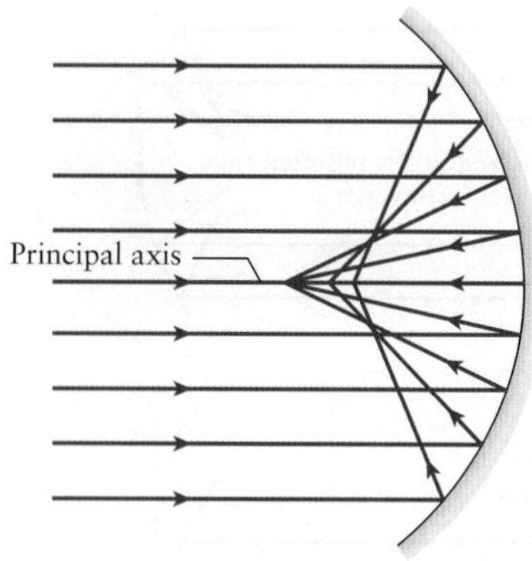
Principal axis

Figure 23-10 Reflection from a spherical mirror How parallel light rays behave when they strike a concave, spherical mirror.

Figure 23-11 defines many of the variables we use to describe a concave, spherical mirror. The **focal point** of the mirror is the point F along the principal axis at which incident rays parallel to the principal axis converge to a common focus when they reflect off the mirror. The distance from the focal point to the center of the mirror is f, the **focal length**. The mirror is a small section of a full sphere, and the point C—the **center of curvature** of the mirror—is the center of that sphere. The distance from C to any point on the mirror is the radius r of the sphere, also called the **radius of curvature** of the mirror.

For a concave mirror small enough that all rays parallel to the principal axis are focused at the focal point, the focal length is exactly half of the radius:

EQUATION IN WORDS
Focal length of a spherical mirror

Focal length of a spherical mirror | **Radius of curvature** of the mirror

$$f = \frac{r}{2} \qquad (23\text{-}3)$$

Figure 23-11 shows this relationship. Equation 23-3 says that the tighter the curve of a spherical mirror and hence the smaller the radius of curvature r, the shorter the focal length f. (We'll prove the relationship $f = r/2$ in Section 23-4.)

To see how a concave mirror forms an image, let's put our standard arrow at a point far from the mirror, as in **Figure 23-12a**. *Far* in this case means that the object distance s_o is greater than the radius of the mirror r; in other words, the base of the arrow is farther from the mirror, along the principal axis, than the center of curvature C. Now let's trace two rays of light from the tip of the arrow as they strike and then are reflected from the mirror. The image of the arrow's tip forms where those two rays intersect. Although any two light rays that originate at the tip of the arrow and that strike the mirror will work, we choose two that are particularly convenient. In **Figure 23-12b**, we trace a ray that starts parallel to the principal axis because all such rays are reflected through the focal point. In **Figure 23-12c** we add the trace of a ray that strikes the center of the mirror. The normal to the surface at this point lies along the principal axis, so it's easy to apply the law of reflection; the incident and reflected rays are symmetric around the principal axis.

Where does the image of the *base* of the arrow form? The base is on the principal axis, which is normal to the mirror's surface at the center of the mirror. So a light ray coming from the base of the arrow is reflected straight back from the center of the mirror (angles of incidence and reflection both zero), which means the image of the base of the arrow forms along the principal axis. **Figure 23-12d** shows the final image of the arrow. This is an **inverted image**: It's flipped upside down compared to the object. The image is also smaller than the object. Finally, the image is real; that is, it forms in front of the mirror where light rays reflected from different parts of the mirror's surface meet.

Figure 23-12d shows both the object distance s_o and the image distance s_i; both distances are positive, because the object and image are both on the reflective side of the mirror. It also shows that if s_o is large, the image is closer to the mirror than the object is, so $s_i < s_o$. Note that the image distance is greater than the focal length (the distance from the mirror to the focal point F). This is how a concave mirror is used in a telescope: The object is very far away, while the image is formed very close to the

r is the radius of curvature of the full sphere.

f is the focal length.

Principal axis

C is the center of curvature of the full sphere.

F is the focal point of the mirror.

Figure 23-11 Mirror nomenclature This drawing defines many of the variables we use to describe a concave, spherical mirror.

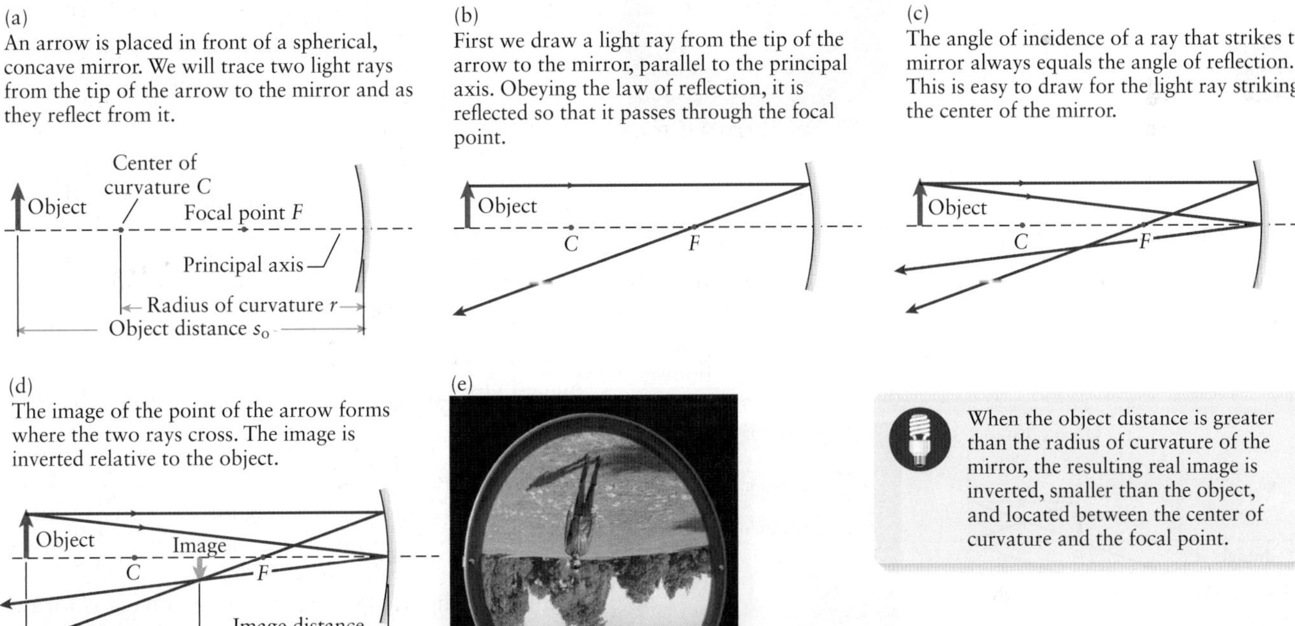

(a)
An arrow is placed in front of a spherical, concave mirror. We will trace two light rays from the tip of the arrow to the mirror and as they reflect from it.
Center of curvature C
Object
Focal point F
Principal axis
Radius of curvature r
Object distance s_o
(b)
First we draw a light ray from the tip of the arrow to the mirror, parallel to the principal axis. Obeying the law of reflection, it is reflected so that it passes through the focal point.
Object
C
F
(c)
The angle of incidence of a ray that strikes the mirror always equals the angle of reflection. This is easy to draw for the light ray striking the center of the mirror.
Object
C
F
(d)
The image of the point of the arrow forms where the two rays cross. The image is inverted relative to the object.
Object
Image
C
F
Image distance s_i
Object distance s_o
(e)
David Tauck

When the object distance is greater than the radius of curvature of the mirror, the resulting real image is inverted, smaller than the object, and located between the center of curvature and the focal point.

Figure 23-12 Ray diagram for a concave mirror I (a) A distant object in front of a concave mirror. (b), (c), (d) Locating the image of this object. (e) The image of a person standing far from a concave mirror.

mirror and is much smaller than the object. (For example, the Moon is 3.84×10^5 km distant and 1738 km in radius, but the Moon's image made by an amateur astronomer's telescope is formed only a meter or so from the mirror and is less than a centimeter in radius.) **Figure 23-12e** shows the inverted image of one of the authors of this book standing far away from a concave mirror.

Let's see what happens if we move the object closer to the mirror. **Figure 23-13a** shows the arrow placed at the center of curvature *C*, so the object distance equals the radius of curvature *r*. From Equation 23-3 the focal length $f = r/2$, so for the case shown in Figure 23-13, $s_o = r = 2f$. We trace the same two light rays as in Figure 23-12; where the two rays meet and the image forms is now farther from the mirror. This image is the same size as the object and, as in the previous case, is both real (the light rays pass through the image) and inverted. With the object placed at the center of curvature, the image forms at the center of curvature, so $s_i = s_o$. In **Figure 23-13b** one of the authors is standing at the center of curvature of a concave mirror; note that his inverted image is larger than that shown in Figure 23-12e.

In **Figure 23-14a** we've moved the arrow closer still to the mirror so that it sits between the center of curvature *C* and the focal point *F*. In this case the object distance is between $2f$ and f: $2f > s_o > f$. Again we've traced the same two light rays that we considered in the previous two cases. The image is still both real and inverted but has moved farther still from the mirror, so it is now farther from the mirror than the object (so $s_i > s_o$) and larger than the object. **Figure 23-14b** shows such an image of one of the authors; compare Figures 23-12e and 23-13b.

Comparing Figures 23-12, 23-13, and 23-14 shows that the image gets larger and moves farther from the mirror as we move a distant object closer to the focal point. In **Figure 23-15a** we place the object *at* the focal point so that the object distance equals the focal length, or $s_o = f$. After reflection, light rays from the tip of the arrow are parallel. (This is how a concave mirror is used in an automobile headlight: The lamp itself is placed close to the focal point of the curved mirror behind it, and the reflected light forms a beam of nearly parallel light rays.) With the object at the focal point, no image is formed because the parallel reflected rays never meet. The author in **Figure 23-15b** is standing close to the focal point of the concave mirror; there is no sharp image. If the object is slightly outside the focal point (s_o is slightly larger than f), the image is formed very far from the mirror and is much larger than the object.

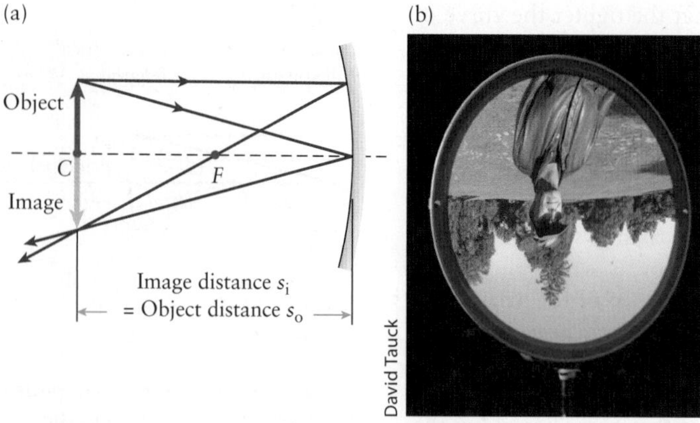
(a)
(b)
Object
C
F
Image
Image distance s_i = Object distance s_o
David Tauck

When the object is located at the radius of curvature of the mirror, the resulting real image is inverted, the same size as the object, and located at the radius of curvature.

Figure 23-13 Ray diagram for a concave mirror II (a) We move the object from Figure 23-12 to the center of curvature of the concave mirror. (b) The image of a person standing at the center of curvature of a concave mirror.

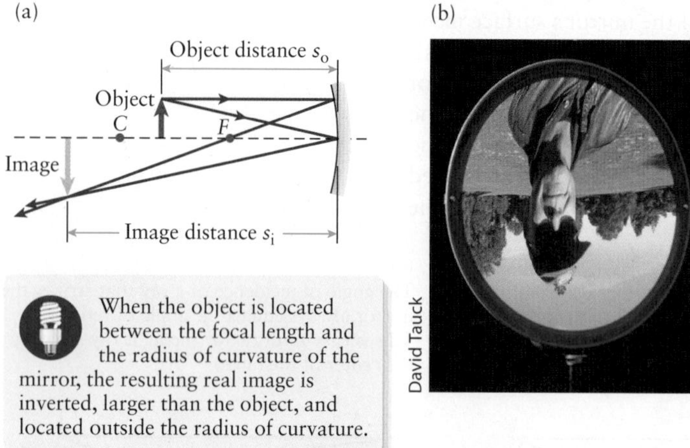
(a)
(b)
Object distance s_o
Object
C
F
Image
Image distance s_i
When the object is located between the focal length and the radius of curvature of the mirror, the resulting real image is inverted, larger than the object, and located outside the radius of curvature.
David Tauck

Figure 23-14 Ray diagram for a concave mirror III (a) We move the object from Figures 23-12 and 23-13 to a point between the center of curvature and the focal point of the concave mirror. (b) The image of a person standing at such a point.

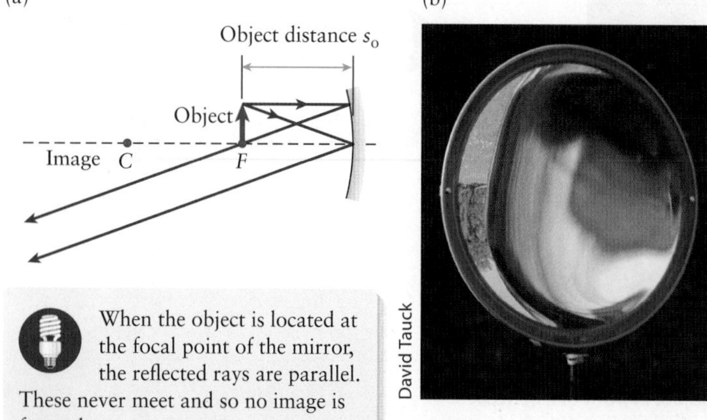
(a)
(b)
Object distance s_o
Object
Image
C
F
When the object is located at the focal point of the mirror, the reflected rays are parallel. These never meet and so no image is formed.
David Tauck

Figure 23-15 Ray diagram for a concave mirror IV (a) We move the object from Figures 23-12, 23-13, and 23-14 to the focal point of the concave mirror. (b) The image of a person standing near the focal point.

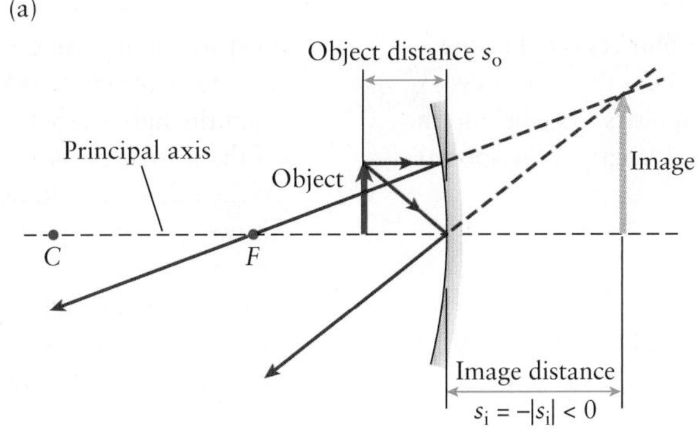

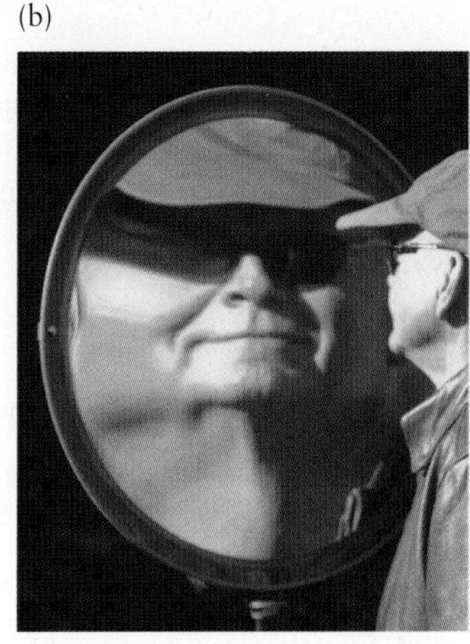

When the object is closer to the mirror than the focal point, the resulting virtual image is upright, larger than the object, and located behind the mirror.

Figure 23-16 Ray diagram for a concave mirror V (a) We move the object from Figures 23-12, 23-13, 23-14, and 23-15 to a point inside the focal point of the concave mirror. (b) The image of a person standing at such a point.

What happens when the object is placed inside the focal point, so the object distance is less than the focal length ($s_o < f$)? In this case the reflected rays never actually meet but rather appear to meet at a point behind the mirror (**Figure 23-16a**). Hence the image is virtual, like the image formed by the plane mirror of Section 23-2, and the image distance s_i is negative: $s_i = -|s_i| < 0$. The image is larger than the object and is an **upright image** (it has the same orientation as the object). That's the kind of enlarged image you see when you look in a curved bathroom mirror for shaving or applying makeup. **Figure 23-16b** shows such an enlarged image of one of the authors (compare Figures 23-12e, 23-13b, 23-14b, and 23-15b). The photograph in Figure 23-7b shows another such enlarged image.

So far our results for image formation by concave mirrors have been qualitative only. In the following section we'll take a more *quantitative* look at how the position of an object placed in front of a concave mirror determines the position and size of the resulting image.

THE TAKEAWAY for Section 23-3

✔ A concave spherical mirror is in the shape of the inner surface of a section of a sphere.

✔ When an object is outside the focal point of a concave mirror, the image is outside the focal point, real and inverted.

✔ As the object is moved closer to the focal point, the image becomes larger and forms farther from the mirror.

✔ When the object is at the focal point, the reflected rays are parallel. Effectively, the image is at infinity and infinitely large.

✔ If the object is placed inside the focal point, the image is virtual and upright.

Prep for the AP® Exam

AP® Building Blocks

1. Describe the image you see of yourself if your head is at the center of curvature of a spherical concave mirror.
2. Describe the difference between the images seen in a spherical, concave mirror when the object is "up close" (closer to the image than the focal length) compared to "far away" (outside the focal length).
3. Are there any situations where a real image is formed in a spherical, concave mirror? If so, describe the location of the object relative to the focal point for such situations.

AP® Skill Builders

4. What is the radius of curvature of a plane mirror? Explain your answer using physics principles.

AP® Skills in Action

5. A shiny spoon is not so different in shape than a spherical mirror: It's concave on one side and convex on the other. Your reflection from the concave side of the spoon, when held at arm's length, is upside down and appears to float in front of the spoon. When you hold the spoon about

6 cm from your eye, you see only a blur reflected in the spoon, but when you hold it about 4 cm from your eye, your reflection is right side up and appears to be behind the spoon. What is the approximate focal length of the spoon?
(A) > 6 cm
(B) 6 cm
(C) 4 cm
(D) < 4 cm

6. You place a light bulb oriented vertically (with its base on the bottom) just outside the focal point of a concave mirror. The resulting real image of the light bulb falls on a wall far from the mirror. This image is inverted and larger than the light bulb. If you were to paint the bottom half of the mirror black so that light rays that strike this half of the mirror would not be reflected, the image would show
(A) only the top of the light bulb.
(B) only the bottom of the light bulb.
(C) only one side of the light bulb.
(D) the entire light bulb.

23-4 Simple equations give the position and magnification of the image made by a concave mirror

Let's return to the concave spherical mirror that we considered in the previous section. We'll see that given the radius of the mirror and the object distance, we can determine exactly how far from the mirror the image forms and how much magnification the mirror provides. We just need some geometry and the law of reflection.

The Mirror Equation for a Concave Mirror

To find the mathematical relationship among the object distance s_o, image distance s_i, and the radius of curvature r, we'll imagine that the object is a small sphere located on the principal axis of the mirror as in **Figure 23-17**. Then the image will also lie along the principal axis. (The explanation for the middle of the sphere is the same one we used in Section 23-3 to show why the image of the base of the arrow in Figure 23-12 must lie on the principal axis. You can also sketch a few rays to see that the bottom of the sphere, an equal distance from the optical axis, will always be symmetrically placed about the principal axis from the top of the sphere. So we will just use a ray from the center of the object in our diagram.) Figure 23-17 shows an arbitrarily chosen light ray coming from the object at point O and reflecting off the mirror at point P. The image forms where this reflected ray intercepts the principal axis at point I.

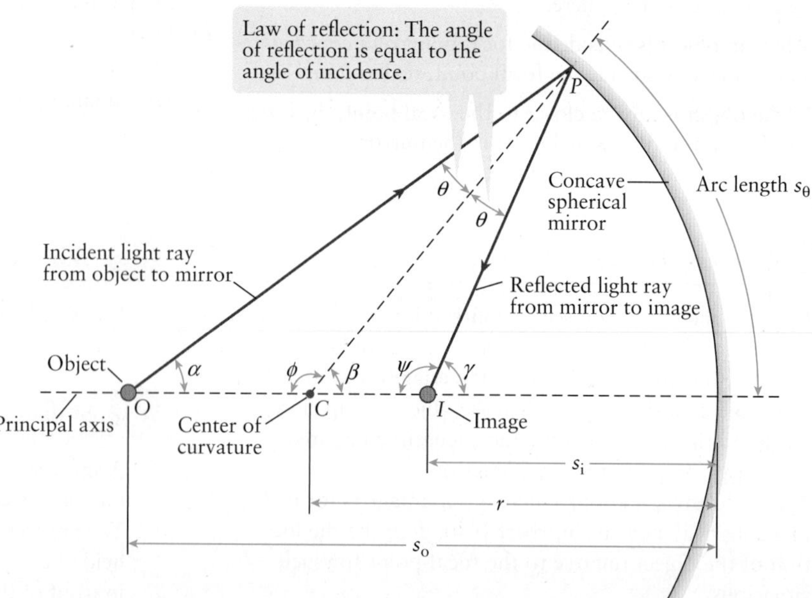

Figure 23-17 Analyzing a concave spherical mirror A ray diagram for an object on the principal axis of a concave spherical mirror.

WATCH OUT!

The path from object to mirror to image isn't just for one special light ray.

There's nothing special about the light ray we've drawn in Figure 23-17. *Any* ray that comes from the object point O will reflect off the mirror and pass through the *same* image point *I*, provided the ray is at a shallow enough angle α to the principal axis of the mirror. So every part of the mirror contributes to forming the image. (Note that in Figure 23-17 we've drawn the angle α as fairly large, even though this angle is actually quite small. We've done this just to make the angles in the figure easier to see.)

We can find a relationship among s_o, s_i, and r by first finding a relationship among the angles in Figure 23-17 labeled α (the angle between the principal axis and the light ray from O to P), β (the angle between the principal axis and a line from the center of curvature C to P), and γ (the angle between the principal axis and the light ray from P to I). Because we're considering α to be a small angle, then necessarily β and γ are small angles as well. For that reason we can treat each of the three regions formed by one of these angles and the arc length s_θ of the mirror as a sector (a pie-like slice) of a circle. This arc length is common to all three sectors. The length of the arc of a circle of radius r subtended by an angle θ equals r_θ (see Section 3-7), provided the angle θ is in radians. So in Figure 23-17

NEED TO REVIEW? See page M10 of the Math Tutorial for more information on geometry.

$$s_\theta = s_o\alpha; \quad s_\theta = r\beta; \quad s_\theta = s_i\gamma \tag{23-4}$$

Note also that the sum of the angles in any triangle equals 180° or π radians. For the triangle OPC we have

$$\alpha + \theta + \phi = \pi \tag{23-5}$$

However, the angles ϕ and β in Figure 23-17 must add to 180°, or π radians (together they make up half a circle). So $\phi + \beta = \pi$ and $\phi = \pi - \beta$, and Equation 23-5 becomes

$$\alpha + \theta + \pi - \beta = \pi \quad \text{or} \quad \theta = \beta - \alpha \tag{23-6}$$

Similarly, for the triangle OPI the sum of the angles is π:

$$\alpha + 2\theta + \psi = \pi \tag{23-7}$$

Figure 23-17 shows that the angles ψ and γ also make up half a circle, so $\psi + \gamma = \pi$ and $\psi = \pi - \gamma$. Triangle CPI shows that $\beta + \theta + \psi = \pi$. Comparing with Equation 23-7, it follows that $\alpha + 2\theta = \beta + \theta$. Equation 23-7 can then be written as

$$\beta + \theta + \pi - \gamma = \pi \quad \text{or} \quad \theta = \gamma - \beta \tag{23-8}$$

Equations 23-6 and 23-8 are two different expressions for the angle θ in Figure 23-17. If we set these equal to each other, we get

$$\beta - \alpha = \gamma - \beta \quad \text{or} \quad \alpha + \gamma = 2\beta \tag{23-9}$$

From Equations 23-4 we have $\alpha = s_\theta/s_o$, $\beta = s_\theta/r$, and $\gamma = s_\theta/d_I$. Substituting these into Equation 23-9 gives

$$\frac{s_\theta}{s_o} + \frac{s_\theta}{s_i} = 2\frac{s_\theta}{r}$$

To simplify we divide through by the arc length s_θ, giving us the relationship we've been seeking among the object distance, image distance, and radius of curvature of the mirror:

$$\frac{1}{s_o} + \frac{1}{s_i} = \frac{2}{r} \tag{23-10}$$

To help interpret Equation 23-10, recall from Figure 23-11 that if the incident light rays are parallel, they come to a focus at the focal point a distance f from the mirror. So in this case $s_i = f$. The incident rays will be parallel if the object is infinitely far away, so $s_o \to \infty$ and $1/s_o \to 0$. Then Equation 23-10 becomes

$$0 + \frac{1}{f} = \frac{2}{r} \quad \text{or} \quad f = \frac{r}{2}$$

This justifies Equation 23-3: The focal length of a concave spherical mirror is one-half of the radius of curvature.

If we replace $2/r$ with $1/f$ in Equation 23-10, we get the final form of the **mirror equation:**

EQUATION IN WORDS
Mirror equation and lens equation relating object distance, image distance, and focal length

(23-11)
$$\frac{1}{s_o} + \frac{1}{s_i} = \frac{1}{f}$$

Focal length (f); Object distance (s_o); Image distance (s_i)

We also call Equation 23-11 the **lens equation** because, as we shall see in Section 23-8, it's also applicable to the image formed by a lens.

We take the focal length f of a concave mirror to be positive because the center of curvature C of the mirror is in front of the mirror, that is, on its reflective side (to the left in Figure 23-17). Likewise, because the object point O is in front of the mirror, the object distance s_o is positive. The image distance s_i, however, can be positive if the image is real (light rays really converge, in front of the mirror) or negative if the image is virtual (light rays do not really converge at that point, behind the mirror). To see when the image distance is positive and when it is negative, let's rewrite Equation 23-11 to solve for s_i:

(23-12)
$$\frac{1}{s_i} = \frac{1}{f} - \frac{1}{s_o} = \frac{s_o}{s_o f} - \frac{f}{s_o f} = \frac{s_o - f}{s_o f} \quad \text{or} \quad s_i = \frac{s_o f}{s_o - f}$$

In Equation 23-12, s_o and f are both positive for a concave mirror, so the numerator $s_o f$ is positive. However, the denominator $s_o - f$ can be positive or negative depending on whether s_o is larger or smaller than f.

If the object is outside the focal point so that $s_o > f$, then the denominator $s_o - f$ in Equation 23-12 is positive and the image distance is positive. Therefore, in this case the image made by the mirror is real. As the object moves closer to the focal point, the difference between s_o and f gets smaller and so s_i given by Equation 23-12 gets larger. Hence the image moves farther away from the mirror. If the object is inside the focal point, however, then $s_o < f$ and the denominator $s_o - f$ is negative. Then s_i is negative as well, and the image is virtual. That's exactly the behavior that we deduced in Section 23-3 by analyzing ray diagrams.

As a further check on Equations 23-11 and 23-12, note that if the object is exactly two focal lengths from the mirror so $s_o = 2f$, the image distance is

$$s_i = \frac{s_o f}{s_o - f} = \frac{(2f)f}{2f - f} = \frac{2f^2}{f} = 2f$$

So when the object is a distance $2f$ from the mirror—which, because $r = 2f$, is at the center of curvature—the image is at the same position. That's the same conclusion we came to by using the ray diagram in Figure 23-13.

Magnification for a Concave Mirror

We can also use simple geometry to find an expression for the lateral magnification of the image produced by a concave spherical mirror. In **Figure 23-18** we represent the object at point O with an upright arrow of height h_o that extends from O (on the principal axis) to O′. Just as in Figure 23-12 we draw two rays from the tip of the arrow at O′ to determine the position I' of the tip of the image arrow.

Because the object in Figure 23-18 is outside the focal point F, the image is real and inverted and so the height of the image is negative: $h_i = -|h_i|$. By the law of reflection the ray from O′ to the point P' at the center of the mirror makes the same angle θ with the mirror's principal axis (shown as a dashed line in Figure 23-18) as does the reflected ray from P' to I'. If you look at the right triangle $OO'P'$, you'll see that the tangent of θ (the opposite side divided by the adjacent side) is $\tan\theta = h_o/s_o$; if you do the same for the right triangle $II'P'$, you'll see that $\tan\theta = |h_i|/s_i$. Setting these two expressions equal to each other, we see that

$$\frac{h_o}{s_o} = \frac{|h_i|}{s_i} \quad \text{or} \quad \frac{|h_i|}{h_o} = \frac{s_i}{s_o}$$

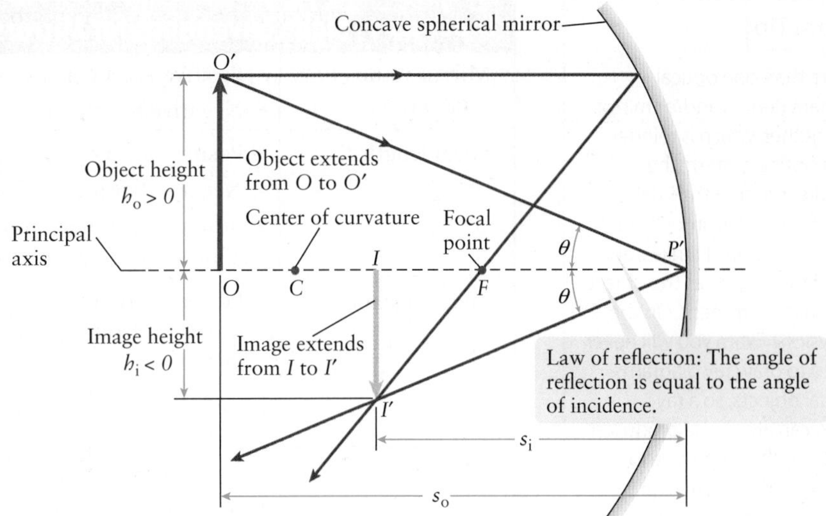

Figure 23-18 Magnification for a concave spherical mirror We replace the small sphere in Figure 23-17 with an arrow.

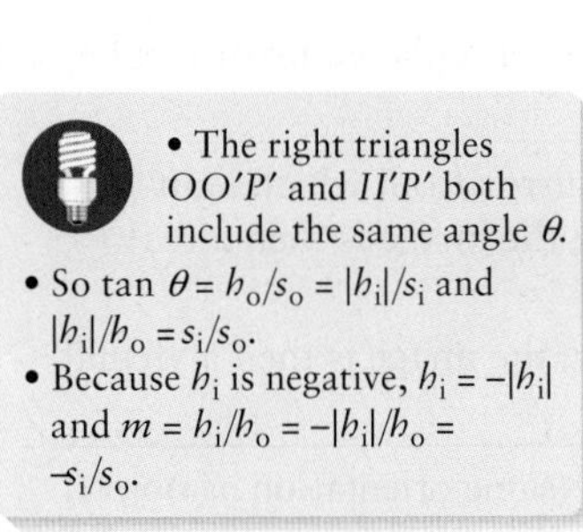

• The right triangles $OO'P'$ and $II'P'$ both include the same angle θ.
• So $\tan\theta = h_o/s_o = |h_i|/s_i$ and $|h_i|/h_o = s_i/s_o$.
• Because h_i is negative, $h_i = -|h_i|$ and $m = h_i/h_o = -|h_i|/h_o = -s_i/s_o$.

Because $h_i = -|h_i|$, we can rewrite this as

$$\frac{h_i}{h_o} = -\frac{s_i}{s_o} \tag{23-13}$$

From Equation 23-2 the lateral magnification is $M = h_i/h_o$. So Equation 23-13 becomes

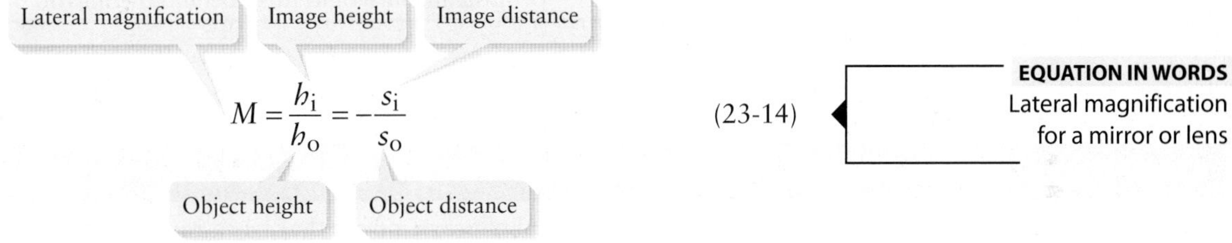

$$M = \frac{h_i}{h_o} = -\frac{s_i}{s_o} \tag{23-14}$$

EQUATION IN WORDS
Lateral magnification for a mirror or lens

A negative value of the magnification M means that the image is inverted, as in Figure 23-18. If M is positive, the image is upright. We've derived Equation 23-14 for the case in which the image is real (in front of the mirror), but it's also true when the image is virtual and behind the mirror, as in Figure 23-16. (As we'll see in Section 23-8, the same equation also applies to lenses.)

When the object is far from the mirror and the image is close to the focal point, s_i is positive and small compared to s_o, so from Equation 23-14 M is small and negative; the mirror produces a reduced, inverted image (see Figure 23-12). As the object distance decreases, the image distance increases and the ratio $M = -s_i/s_o$ increases in magnitude. As we saw above, when $s_o = 2f$ the image distance s_i is also equal to $2f$; then $M = -1$ and the inverted image is as large as the object (see Figure 23-13). If we move the object even closer to the focal point, but still outside it, the image distance is greater than the object distance and the absolute value of the magnification is greater than 1. Hence the inverted image is larger than the object (see Figure 23-14). If we move the object inside the focal point so that $s_o < f$, the image distance is negative (the image is behind the mirror) and so the image is virtual. In this case Equation 23-14 tells us that M is positive, so the virtual image is upright (see Figure 23-16). We see that Equation 23-14 gives us the same results as we deduced from the ray diagrams in Section 23-3, as it must.

Table 23-1 summarizes when the radius of curvature r, focal length f, image distance s_i, image height h_i, and magnification M are positive and when they are negative.

Exam Tip

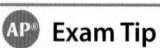

For more than one optical instrument (lenses and/or mirrors) used together, which is referred to as an optical system, the procedure is to find the image due to the first instrument that light from the object strikes, and then use that image as the object for the next instrument. On the AP® Physics 2 Exam you will never be asked to draw ray diagrams for virtual objects, so a ray diagram can always be produced by applying the technique for one optical instrument over again for each instrument. To mathematically solve a multiple instrument problem, it is necessary to find the image due to the first instrument the light from the object strikes, and locate the object distance to the second instrument from that image. You then repeat this for each subsequent instrument. On the AP® Physics 2 Exam, you will never be asked to solve for more than two instruments.

TABLE 23-1 Sign Conventions for Mirrors

Mirror radius of curvature r	• Positive for a concave mirror • Negative for a convex mirror
Focal length f	• Positive for a concave mirror • Negative for a convex mirror • Infinity for a plane mirror, as r for a plane mirror can be approximated as infinity
Image distance s_i	• Positive if on the side of the mirror from which light leaves (in front of the mirror); the image is then a real image • Negative if behind the mirror; the image is then a virtual image
Image height h_i	• Positive if the image is upright (same orientation as object) • Negative if the image is inverted (upside down relative to object)
Lateral magnification M	• Positive if the image is upright (same orientation as object) • Negative if the image is inverted (upside down relative to object)
Object distance s_o	• Positive when the object being observed is real, as in any single-instrument optical system • If the object being observed is the image of another optical instrument, then it has a positive object distance if the image of the first instrument is in front of the mirror or a negative object distance if the image of the first instrument would form behind the mirror.

EXAMPLE 23-1 Images Made by a Concave Mirror

An object is 1.50 cm tall and is placed in front of a spherical concave mirror that has a radius of curvature equal to 20.0 cm. Find the image distance and height if the object is (a) 14.0 cm from the mirror; (b) 6.00 cm from the mirror. In each case, draw a ray diagram as part of the solution.

Set Up

Our mathematical tools are the expression for the focal length of the mirror, the mirror equation, and the equation for lateral magnification. We're given the radius of curvature $r = 20.0$ cm and the object height $h_o = 1.50$ cm; our goal is to find the values of the image distance s_i and image height h_i for the cases $s_o = 14.0$ cm and $s_o = 6.00$ cm.

Focal length of a spherical mirror:

$$f = \frac{r}{2} \qquad (23\text{-}3)$$

Mirror equation:

$$\frac{1}{s_o} + \frac{1}{s_i} = \frac{1}{f} \qquad (23\text{-}11)$$

Lateral magnification of a mirror:

$$M = \frac{h_i}{h_o} = -\frac{s_i}{s_o} \qquad (23\text{-}14)$$

Solve

First calculate the focal length of the mirror.

From Equation 23-3:

$$f = \frac{r}{2} = \frac{20.0\text{ cm}}{2} = 10.0\text{ cm}$$

(a) The object distance $s_o = 14.0$ cm is greater than the focal length $f = 10.0$ cm, but less than $2f = 20.0$ cm. That is, the object is inside the center of curvature but outside the focal point. We've drawn a ray diagram that's similar to Figure 23-14. This tells us to expect that the image will be real, inverted, farther from the mirror than the object is, and larger than the object.

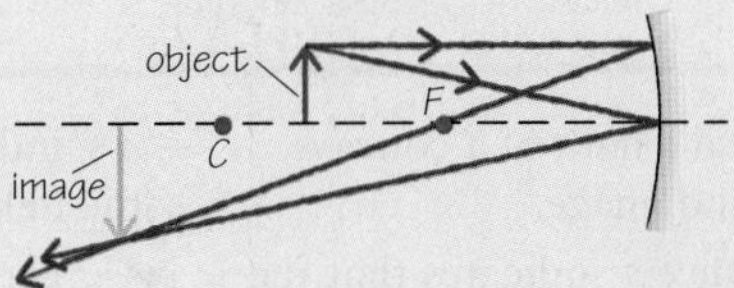

Calculate the image distance using the mirror equation.

From Equation 23-11:

$$\frac{1}{s_i} = \frac{1}{f} - \frac{1}{s_o} = \frac{1}{10.0 \text{ cm}} - \frac{1}{14.0 \text{ cm}}$$
$$= 0.1000 \text{ cm}^{-1} - 0.0714 \text{ cm}^{-1} = 0.0286 \text{ cm}^{-1} \text{ so}$$
$$s_i = \frac{1}{0.0286 \text{ cm}^{-1}} = 35.0 \text{ cm}$$

The image distance is positive, so the image is 35.0 cm in front of the mirror. An image that forms in front of the mirror is a real image.

Calculate the image height using the magnification equation.

From Equation 23-14:

$$M = \frac{h_i}{h_o} = -\frac{s_i}{s_o} = -\frac{35.0 \text{ cm}}{14.0 \text{ cm}} = -2.50$$

The image is 2.50 times larger than the object and, as the minus sign shows, inverted. Solve for the image height h_i:

$$h_i = Mh_o = (-2.50)(1.50 \text{ cm}) = -3.75 \text{ cm}$$

The inverted image is 3.75 cm high.

(b) Now the object distance $s_o = 6.00$ cm is less than the focal length $f = 10.0$ cm, so the object is inside the focal point. We've drawn a ray diagram that's similar to Figure 23-16. This tells us to expect that the image will be virtual, upright, farther from the mirror, and larger than the object.

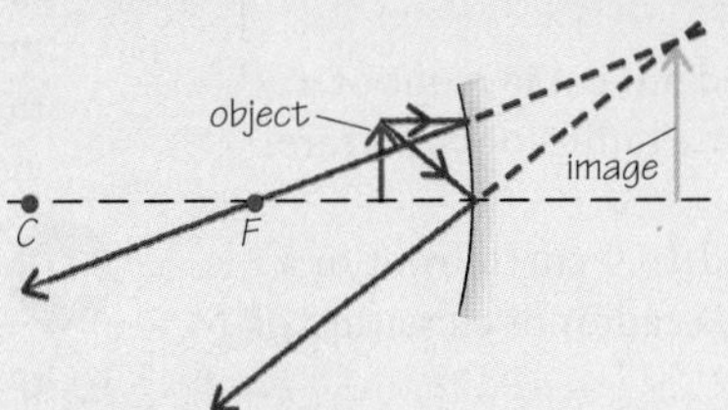

Calculate the image distance using the mirror equation.

From Equation 23-11:

$$\frac{1}{s_i} = \frac{1}{f} - \frac{1}{s_o} = \frac{1}{10.0 \text{ cm}} - \frac{1}{6.00 \text{ cm}}$$
$$= 0.1000 \text{ cm}^{-1} - 0.1667 \text{ cm}^{-1} = -0.0667 \text{ cm}^{-1} \text{ so}$$
$$s_i = \frac{1}{-0.0667 \text{ cm}^{-1}} = -15.0 \text{ cm}$$

The image distance is negative, so the image is 15.0 cm behind the mirror. An image that forms behind the mirror is a virtual image.

Calculate the image height using the magnification equation.

From Equation 23-14:

$$M = -\frac{s_i}{s_o} = -\frac{(-15.0 \text{ cm})}{6.00 \text{ cm}} = +2.50$$

Again the image is 2.50 times larger than the object but is now upright, as shown by the plus sign. Solve for the image height h_i:

$$h_i = Mh_o = (+2.50)(1.50 \text{ cm}) = +3.75 \text{ cm}$$

The upright image is 3.75 cm high.

Reflect

In both parts the position and size of the image are consistent with our ray diagrams. It's always a good idea to draw such diagrams as a check on your calculations using the mirror equation and magnification equation.

NOW WORK Problems 1–8 and 10 from The Takeaway for Section 23-4.

THE TAKEAWAY for Section 23-4

✔ The mirror equation relates the focal length of a concave mirror and the positions of the object and image.

✔ A positive value of the image distance s_i indicates that the image is in front of the mirror and is real. A negative value of s_i indicates that the image is behind the mirror and virtual (the light rays never actually go there).

✔ The magnification of the image is positive if the image is upright compared to the object and negative if the image is inverted.

Prep for the AP® Exam

AP® Building Blocks

EX 23-1 **1.** An object is placed 8.0 cm in front of a concave mirror with a 10.0-cm radius of curvature. Calculate the image distance and the magnification of the image. Determine whether the image is real or virtual and whether it is inverted or upright by using (a) a ray diagram and (b) the mirror equation.

EX 23-1 **2.** An object 1.0 cm tall is placed 3.0 cm in front of a spherical concave mirror with a radius of curvature equal to 10.0 cm. Calculate the image distance and height by using (a) a ray diagram and (b) the mirror equation.

EX 23-1 **3.** An object 1.0 cm tall is placed 6.0 cm in front of a spherical concave mirror with a radius of curvature equal to 10.0 cm. Calculate the image distance and height by using (a) a ray diagram and (b) the mirror equation.

AP® Skill Builders

EX 23-1 **4.** Construct ray diagrams to locate the images in the following cases.

(a) A 10.0-cm-tall object is located 5.0 cm in front of a spherical concave mirror with a radius of curvature of 20.0 cm.

(b) A 10.0-cm-tall object is located 10.0 cm in front of a spherical concave mirror with a radius of curvature of 20.0 cm.

(c) A 10.0-cm-tall object is located 20.0 cm in front of a spherical concave mirror with a radius of curvature of 20.0 cm.

EX 23-1 **5.** The radius of curvature of a spherical concave mirror is 20.0 cm. Describe the image formed when a 10.0-cm-tall object is (a) 5.0 cm from the mirror, (b) 20.0 cm from the mirror, (c) 50.0 cm from the mirror, and (d) 100.0 cm from the mirror. For each case give the image distance, the image height, the type of image (real or virtual), and the orientation of the image (upright or inverted).

EX 23-1 **6.** The radius of curvature of a spherical concave mirror is 15.0 cm. Describe the image formed when a 20.0-cm-tall object is (a) 10.0 cm from the mirror, (b) 20.0 cm from the mirror, and (c) 100.0 cm from the mirror. For each case give the image distance, the image height, the type of image (real or virtual), and the orientation of the image (upright or inverted).

EX 23-1 **7.** An object is 24.0 cm from a spherical concave mirror of unknown focal length. The image that is formed is 30.0 cm from the mirror and on the same side of the mirror as the object.

(a) Calculate the focal length.

(b) Is the image real or virtual?

(c) If the object is 10.0 cm tall, determine the height of the image(s).

EX 23-1 **8.** An object is 40.0 cm from a concave spherical mirror whose radius of curvature is 32.0 cm. Locate and describe the type and magnification of the image formed by the mirror (a) by calculating the image distance and lateral magnification and (b) by drawing a ray diagram. On the ray diagram draw an eye in a position from which it can view the image.

AP® Skills in Action

9. If an object is placed 12.0 cm from a concave mirror, the resulting real image is 2.00 times as large as the object. What is the focal length of the mirror?

(A) 6.00 cm

(B) 8.00 cm

(C) 16.0 cm

(D) 24.0 cm

EX 23-1 **10.** Derive a relationship between the radius of curvature of a spherical, concave mirror and the object distance that gives an upright image that is four times as tall as the object.

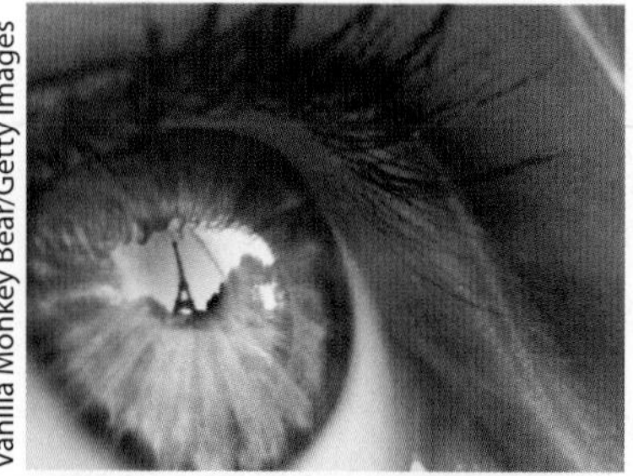

Vanilla Monkey Bear/Getty Images

Figure 23-19 Parisian reflections The Eiffel Tower can be seen reflected from the surface of this person's eye. The eye is relatively spherical, so its outer surface acts like a convex mirror.

23-5 A convex mirror always produces an image that is smaller than the object

Let's now turn our attention to images produced by a *convex* mirror. As Figure 23-7a shows, if an object is held next to a convex mirror, the resulting image is smaller than the object. The same is true if an object is far away from a convex mirror. **Figure 23-19** shows the reflection of the Eiffel Tower in the convex surface of a person's eye. Although the Eiffel Tower is 324 m tall, its image is only about a centimeter in height—smaller than the iris of the eye. In this section we'll use ray diagrams to understand the nature of the images formed by a convex mirror.

Recall from Figure 23-8 the key difference between concave and convex mirrors: Parallel light rays that reflect from a concave mirror converge toward a point in front

Figure 23-20 Ray diagram for a convex mirror How the image is formed for an object in front of a convex mirror.

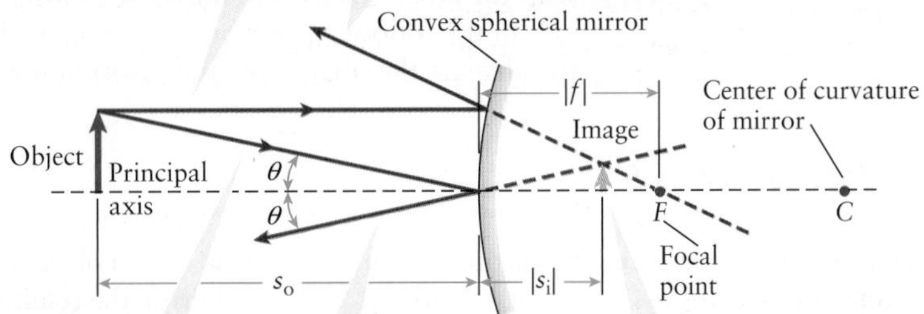

1 A light ray that starts off parallel to the principal axis is reflected so that when we trace the ray back behind the mirror, it appears to pass through the virtual focal point F.
Convex spherical mirror
|f|
Image
Center of curvature of mirror
Object
Principal axis
θ
θ
F
C
Focal point
s_o
$|s_i|$
2 A light ray that strikes the center of the mirror at an angle θ from the principal axis reflects at the same angle.
3 Trace the two reflected rays to where they appear to cross behind the mirror. This is where the virtual image forms.

- A convex mirror has a negative focal length: $f = -|f|$.
- For any object distance s_o a convex mirror produces a virtual image behind the mirror.
- The image distance s_i is negative: $s_i = -|s_i|$.
- The virtual image is upright and smaller than the object.

of the mirror, while parallel light rays that reflect from a convex mirror diverge from a point on the back side of the mirror. (If the mirrors are spherical, the rays don't truly converge on or diverge from a single point. But this is a good description of what happens if the rays are all close to the principal axis of the mirror. We'll make this assumption—the same that we made in Sections 23-3 and 23-4 for concave mirrors—throughout this section.)

Because the focal point of a convex mirror is behind the mirror, we say that the focal length f is negative. For a convex mirror, we call the focal point *virtual* because parallel light rays that reflect from the mirror seem to come from that point but don't actually pass through it.

Figure 23-20 shows an object placed a distance s_o in front of a convex mirror. The focal point F lies a distance $|f|$ behind the mirror. (Because the focal length f is negative, the distance is the absolute value of f.) The center of curvature C is also behind the mirror, so the radius of curvature r is also negative. It turns out that the focal length f is equal to one-half of the radius of curvature r, just as for a concave mirror:

$$f = \frac{r}{2} \tag{23-3}$$

(We'll justify this statement in Section 23-6.) For a concave mirror f and r are both positive; for a convex mirror f and r are both negative.

To find the location of the image in Figure 23-20, we draw two rays from the tip of the arrow, just as we did for the ray diagrams in Section 23-3 for a concave mirror. The image of the arrow tip forms where the two reflected rays appear to meet. The image of the base of the arrow must lie along the principal axis. (A light ray coming from the base of the arrow and traveling along the principal axis strikes the mirror normal to the surface and so is reflected straight back. The image of the base of the arrow must therefore form on the principal axis.) As Figure 23-20 shows, the image is virtual because it forms behind the mirror, just like the image formed by a plane mirror (Section 23-2): The reflected rays don't actually cross there. Hence the image distance s_i is negative. We saw in Section 23-3 that a concave mirror produces a virtual image only if the object is inside the focal point; a convex mirror produces a virtual image for *any* position of the object.

Figure 23-20 shows that the image formed by a convex mirror is smaller than the object, no matter what the object distance is. Figure 23-20 also shows that the image formed by the convex mirror will always be upright and will always be closer to the mirror than the virtual focal point. **Figure 23-21** shows that as the object distance s_o decreases, the image becomes larger and closer to the convex mirror, but remains virtual and upright. If the object were moved all the way to the surface of the mirror, the image would be exactly the same size as the object.

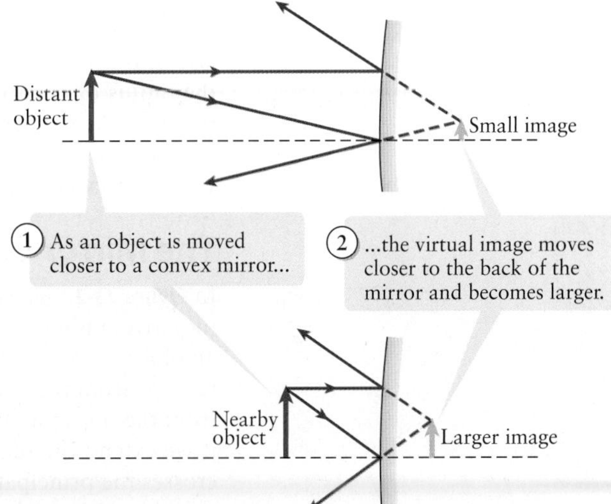

Distant object
Small image
1 As an object is moved closer to a convex mirror...
2 ...the virtual image moves closer to the back of the mirror and becomes larger.
Nearby object
Larger image

Figure 23-21 Moving closer to a convex mirror As the object far from a convex mirror is moved closer to the mirror, the image increases in size.

Because a convex mirror makes objects appear smaller, they provide a wide-angle view. Rearview mirrors on automobiles and trucks often include a convex mirror to allow the driver to see as much of the area behind the vehicle as possible. (Any mirror with the label "Objects in mirror are closer than they appear" is a convex mirror.)

In the following section we'll look at these ideas more quantitatively and see how to calculate the image size and position for a convex mirror.

THE TAKEAWAY for Section 23-5

✔ Parallel light rays that strike a convex mirror diverge and appear to come from a virtual focal point behind the mirror.

✔ If an object is placed anywhere in front of a convex mirror, the resulting image is virtual, upright, closer to the mirror than the object is, and smaller than the object.

Prep for the AP® Exam

AP® Building Blocks

1. Describe the difference between the images seen in a spherical, convex mirror when the object is "up close" (a shorter distance from the mirror than the focal distance of the mirror) compared to "far away" (a longer distance from the mirror than the focal distance of the mirror).
2. Are there any situations where a real image is formed in a spherical, convex mirror? Why or why not?
3. How can you remember the difference between the shapes of a spherical *concave* mirror and a spherical *convex* mirror?

AP® Skill Builders

4. Explain the meaning of the phrase etched on the passenger-side mirror of most cars, "Objects in mirror are closer than they appear," using physics principles.

AP® Skills in Action

5. The Sun is so far from Earth that rays of sunlight are effectively parallel. Sunlight reflected from a mirror can be focused onto a pot, raising its temperature enough to pasteurize water in the pot or cook food. What kind of mirror would be most effective for this purpose?
(A) A convex mirror
(B) A concave mirror
(C) Either a convex or concave mirror
(D) A plane mirror
6. A 10-cm-tall object is located in front of a spherical, convex mirror with a radius of curvature of 20 cm. Construct ray diagrams to locate the images and estimate the image height in each of the following cases.
(a) The object is located 5 cm in front of the mirror.
(b) The object is located 10 cm in front of the mirror.
(c) The object is located 20 cm in front of the mirror.

23-6 The same equations used for concave mirrors also work for convex mirrors

Just as we did for the concave mirror in Section 23-4, we can use geometry and the law of reflection to find an equation that relates the object and image distances and the radius of curvature for a convex mirror. We'll also find an equation that relates the sizes of the image and object for a convex mirror. Remarkably we'll see that these equations are exactly the same as those for a concave mirror, provided we're careful with the signs of the radius of curvature and focal length.

The Mirror Equation for a Convex Mirror

In **Figure 23-22** we've placed an object on the principal axis of a convex mirror with (negative) radius of curvature r. This is analogous to Figure 23-17, in which we placed an object on the principal axis of a concave mirror. The object is at position O, a distance s_o from the mirror. We've drawn an arbitrarily chosen light ray that travels away from the object at an angle α from the principal axis and reflects off the mirror at P; if we extend the reflected ray backward, the extension (shown as a dashed blue line) crosses the principal axis at I. A second light ray (not shown) that travels away from the object along the principal axis will be reflected straight back along that axis. If we extend this reflected ray backward, it will meet the extension of the first reflected

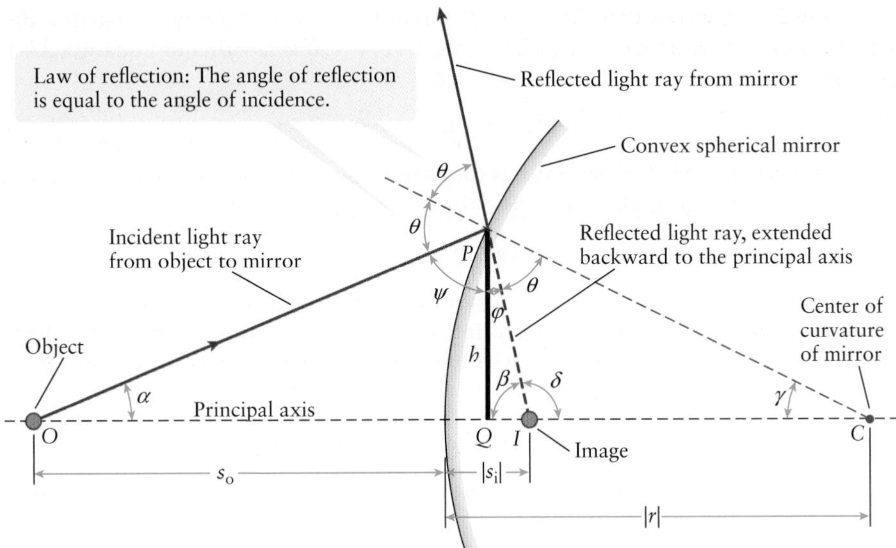

Figure 23-22 Analyzing a convex spherical mirror A ray diagram for an object on the principal axis of a convex spherical mirror.

ray at I. Hence I is the position of the image made by the mirror. The image is behind the mirror, so the image distance is negative. The center of curvature C of the mirror also lies behind the mirror, so the radius of curvature is also negative. Mathematically, $s_i = -|s_i| < 0$ and $r = -|r| < 0$.

Note that the red dashed line CP in Figure 23-22 is normal to the mirror at P, so by the law of reflection the incident ray and the reflected ray are both at the same angle θ relative to this normal. It follows that θ is also the angle between the line CP and the backward extension PI of the reflected ray.

To relate the object distance s_o, image distance s_i, and radius of curvature r, we'll first find a relationship among α, the angle between the principal axis and the light ray incident on the mirror; β, the angle between the principal axis and the reflected ray; and γ, the angle between the principal axis and the red dashed line CP that defines the normal at the point P where the light ray is reflected. Notice that each of these three angles is part of a right triangle for which one side is the thick black line PQ in Figure 23-22: triangle OPQ for angle α, triangle IPQ for angle β, and triangle CPQ for angle γ. In each case the line PQ, of length h, is the side opposite the angle. Because we're considering only light rays that are very close to the principal axis, the reflection point P is also very close to the principal axis. Hence point Q (directly below point P) is essentially at the point where the principal axis touches the mirror surface. As a result, we can treat s_o as the base of triangle OPQ, $|s_i|$ as the base of triangle IPQ, and $|r|$ as the base of triangle CPQ. For each triangle the tangent of the angle equals the length h of the side opposite the angle, divided by the length of the triangle's base. So

NEED TO REVIEW?
See page M12 of the Math Tutorial for more information on trigonometry.

$$\tan\alpha = \frac{h}{s_o}, \quad \tan\beta = \frac{h}{|s_i|}, \quad \tan\gamma = \frac{h}{|r|} \tag{23-15}$$

If the incident ray OP is at a small angle to the principal axis, then α, β, and γ are all small angles. (We've drawn the angle α fairly large in Figure 23-22 to make the geometry easier to visualize. In reality, it must be quite small to conform to the approximation that the rays are nearly parallel to the principal axis.) The tangent of a small angle is approximately equal to the angle in radians, so Equation 23-15 becomes

$$\alpha = \frac{h}{s_o}, \quad \beta = \frac{h}{|s_i|}, \quad \gamma = \frac{h}{|r|} \tag{23-16}$$

To relate the angles α, β, and γ, we'll do as in Section 23-4 and use the result that the sum of the angles of any triangle must be 180°, or π radians. For the triangle IPC that connects the image point I, reflection point P, and center of curvature C, the three angles are γ, θ, and δ. So

$$\gamma + \theta + \delta = \pi \tag{23-17}$$

Figure 23-22 shows that the angles β and δ must add to π radians (together they make up half a circle around the point I). So $\beta + \delta = \pi, \delta = \pi - \beta$, and Equation 23-17 becomes

$$\gamma + \theta + \pi - \beta = \pi \quad \text{or} \quad \theta = \beta - \gamma \tag{23-18}$$

For the triangle OPC that connects the object point O, reflection point P, and center of curvature C, the angles are α, $(\psi + \theta + \phi)$, and γ, so we have

$$\alpha = (\psi + \theta + \phi) + \gamma = \pi \tag{23-19}$$

We can simplify Equation 23-19 by noting from Figure 23-22 that the angles θ, ψ, ϕ, and θ form a half-circle around the point P, so their sum is π radians: $\theta + \psi + \phi + \theta = \pi$, or $\psi + \theta + \phi = \pi - \theta$. If we substitute this expression for $(\psi + \theta + \phi)$ into Equation 23-19, we get

$$\alpha + \pi - \theta + \gamma = \pi \quad \text{or} \quad \theta = \alpha + \gamma \tag{23-20}$$

Equations 23-18 and 23-20 are both expressions for the angle θ. If we set these equal to each other, we get

$$\beta - \gamma = \alpha + \gamma \quad \text{or} \quad \alpha - \beta = -2\gamma \tag{23-21}$$

Equation 23-21 is the relationship among the angles α, β, and γ we've been looking for. We can now get a relationship among the object distance s_o, image distance s_i, and radius of curvature r by substituting the expressions for α, β, and γ from Equation 23-16 into Equation 23-21:

$$\frac{h}{s_o} - \frac{h}{|s_i|} = -\frac{2h}{|r|}$$

If we divide through by the height h of the black line in Figure 23-22 and recall that the image distance and radius of curvature are both negative, so that $s_i = -|s_i|$ and $r = -|r|$, this becomes

$$\frac{1}{s_o} + \frac{1}{s_i} = \frac{2}{r} \tag{23-22}$$

Equation 23-22 is *exactly* the same as Equation 23-10, which we derived for a *concave* mirror. It's reassuring that we get the same expression for both kinds of mirrors.

Note that if the object is infinitely far away, so that $s_o \to \infty$ and $1/s_o \to 0$, the rays from the object to the mirror will all be parallel to the axis and the virtual image will be formed at the focal point F, a distance $|f|$ behind the mirror. Then $s_i = -|f| = f$ (recall that the focal length f is negative). For this situation, Equation 23-22 becomes

$$\frac{1}{f} = \frac{2}{r} \quad \text{or} \quad f = \frac{r}{2}$$

This result is exactly the same as that for a concave spherical mirror: The focal length f is equal to one-half of the radius of curvature r (Equation 23-3). The only difference is that for a convex mirror f and r are both negative.

If we substitute $1/f = 2/r$ into Equation 23-22, we get the mirror equation for a convex mirror:

$$\frac{1}{s_o} + \frac{1}{s_i} = \frac{1}{f} \tag{23-11}$$

This is Equation 23-11, the *same* mirror equation that we derived for a *concave* mirror in Section 23-4. This equation works equally well whether the focal length is positive (for a concave mirror) or negative (for a convex mirror).

We can use Equation 23-11 to explore the properties of the image made by a convex mirror. We saw in Section 23-4 that this equation can be rewritten as

$$s_i = \frac{s_o f}{s_o - f} \tag{23-12}$$

Using $f = -|f|$ for the negative focal length of a convex mirror, Equation 23-12 becomes

$$s_i = \frac{s_o(-|f|)}{s_o - (-|f|)} = -\left(\frac{s_o}{s_o + |f|}\right)|f| \quad (23\text{-}23)$$

The right-hand side of Equation 23-23 is always negative, so the image distance s_i for a convex mirror will always be negative (the image will always be behind the mirror). Note that the fraction in parentheses is always less than or equal to 1, so the image distance is always between 0 and $-f$. That is, the image always forms somewhere between the mirror and the focal point.

Magnification for a Convex Mirror

Figure 23-23 again shows the image made by a convex spherical mirror, but now the object is represented by an upright arrow of height h_o a distance s_o in front of the mirror. Just as for a concave mirror (see Figure 23-18 in Section 23-4), we draw two rays from the tip of the arrow at O'. One ray is parallel to the principal axis of the mirror; after reflection this travels away from the mirror as though it had come from the focal point F. The other ray strikes the center of the mirror at P', and the reflected ray makes the same angle θ with the principal axis as the incident ray from O' to P'. If we extend this reflected ray backward, the extension is at the same angle θ to the principal axis. The extensions of the two reflected rays meet at I', the position of the tip of the image arrow. The base of the image arrow is at I, directly underneath I' on the principal axis. The image is behind the mirror, so the image distance s_i is negative. This image is also upright, so the image height h_i is positive.

To relate the image and object heights, note that the right triangles $OO'P'$ and $II'P'$ both include the same angle θ. For $OO'P'$ the tangent of θ (the opposite side divided by the adjacent side) equals h_o/s_o; for $II'P'$ the tangent of θ equals $h_i/|s_i|$. These two expressions for tan θ must be equal, so

$$\frac{h_o}{s_o} = \frac{h_i}{|s_i|} \quad \text{or} \quad \frac{h_i}{h_o} = \frac{|s_i|}{s_o} = -\frac{s_i}{s_o} \quad (\text{because } s_i = -|s_i|)$$

The lateral magnification M equals h_i/h_o, so it follows that for a convex mirror

$$M = \frac{h_i}{h_o} = -\frac{s_i}{s_o} \quad (23\text{-}14)$$

This is the *same* expression for magnification as for a concave mirror. Because the image distance s_i is negative for a convex mirror, Equation 23-14 tells us that the magnification

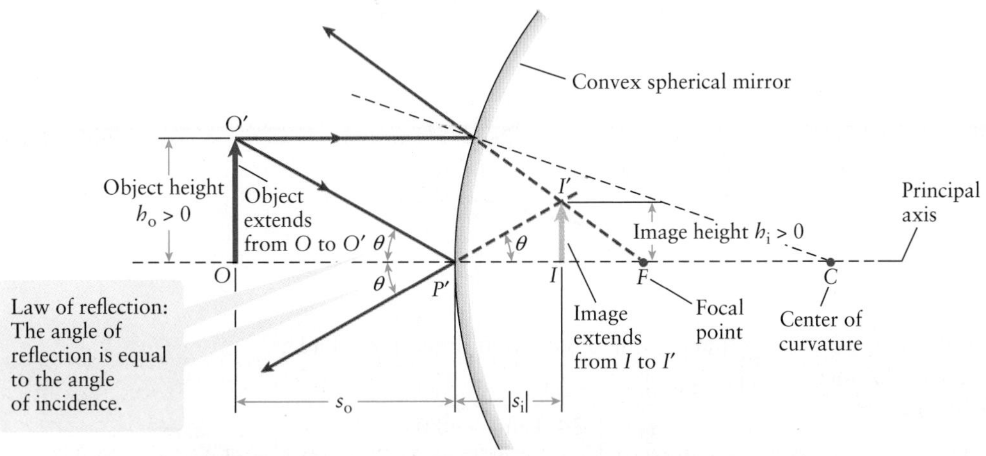

- The right triangles $OO'P'$ and $II'P'$ both include the same angle θ.
- So $\tan\theta = h_o/s_o = h_i/|s_i|$ and $h_i/h_o = |s_i|/s_o$.
- Because s_i is negative, $s_i = -|s_i|$ and $m = h_i/h_o = |s_i|/s_o = -s_i/s_o$.

Figure 23-23 Magnification for a convex spherical mirror We replace the small sphere in Figure 23-22 with an arrow.

is positive and so the image is upright. In addition, for a convex mirror the image is closer to the mirror than is the object, so $|s_i| < s_o$, and the value of M is less than 1. That is, the image made by a convex mirror is always smaller than the object, just as we saw in Section 23-5.

WATCH OUT !

Positive magnification indicates that an image is upright.

Remember that the *sign* of the magnification of a mirror doesn't indicate whether the image is larger or smaller than the object, only whether the image is upright or inverted. An upright image, for example, is always associated with a positive magnification, regardless of whether the image is larger than or smaller than the object.

EXAMPLE 23-2 An Image Made by a Convex Mirror

An object is 1.50 cm high and is placed in front of a spherical convex mirror with a radius of curvature of magnitude 48.0 cm. Find the image distance and height if the object is (a) 68.0 cm from the mirror; (b) 3.00 cm from the mirror.

Set Up

This is similar to Example 23-1 in Section 23-4. The key difference is that the mirror is now convex, so the radius of curvature is negative: $r = -48.0$ cm. We're given the object height $h_o = 1.50$ cm, and want to find the values of the image distance s_i and image height h_i for the cases $s_o = 68.0$ cm and $s_o = 3.00$ cm. Our tools are Equations 23-3, 23-11, and 23-14, which apply to convex mirrors as well as concave ones.

Focal length of a spherical mirror:

$$f = \frac{r}{2} \quad (23\text{-}3)$$

Mirror equation:

$$\frac{1}{s_o} + \frac{1}{s_i} = \frac{1}{f} \quad (23\text{-}11)$$

Lateral magnification of a mirror:

$$M = \frac{h_i}{h_o} = -\frac{s_i}{s_o} \quad (23\text{-}14)$$

Solve

Use Equation 23-3 to calculate the focal length of the mirror.

From Equation 23-3:

$$f = \frac{r}{2} = \frac{-48.0 \text{ cm}}{2} = -24.0 \text{ cm}$$

The negative value of focal length means that the focal point is behind the mirror.

In both cases we expect that the image will be virtual (behind the mirror), smaller than the object, and closer to the mirror than the object is. The specific position and size of the image will be different in the two cases, however.

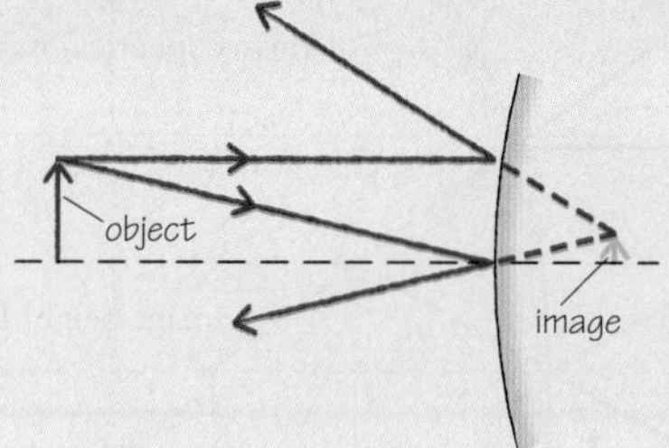

(a) With an object distance $s_o = 68.0$ cm, we use Equation 23-11 to find the image distance and Equation 23-14 to find the image size.

From the mirror equation (Equation 23-11):

$$\frac{1}{s_i} = \frac{1}{f} - \frac{1}{s_o} = \frac{1}{(-24.0 \text{ cm})} - \frac{1}{68.0 \text{ cm}}$$

$$= -0.0417 \text{ cm}^{-1} - 0.0147 \text{ cm}^{-1} = -0.0564 \text{ cm}^{-1}, \text{ so}$$

$$s_i = \frac{1}{(-0.0564 \text{ cm}^{-1})} = -17.7 \text{ cm}$$

The image distance is negative, so the image is 17.7 cm behind the mirror. An image that forms behind the mirror is a virtual image.

The magnification of the image is, from Equation 23-14,

$$M = -\frac{s_i}{s_o} = -\frac{(-17.7\text{ cm})}{68.0\text{ cm}} = +0.261$$

Because $M > 0$, the image is upright. The image is 0.261 as tall as the object, so its height is

$$h_i = Mh_o = (+0.261)(1.50\text{ cm}) = 0.391\text{ cm}$$

(b) Repeat the calculations of part (a) with $s_o = 3.00$ cm.

Calculate the image distance:

$$\frac{1}{s_i} = \frac{1}{f} - \frac{1}{s_o} = \frac{1}{(-24.0\text{ cm})} - \frac{1}{3.00\text{ cm}}$$
$$= -0.0417\text{ cm}^{-1} - 0.333\text{ cm}^{-1} = -0.375\text{ cm}^{-1}\text{, so}$$
$$s_i = \frac{1}{(-0.375\text{ cm}^{-1})} = -2.67\text{ cm}$$

Again the image distance is negative. Note that with a smaller object distance than in (a), the image distance is also smaller.

The magnification of the image is

$$M = -\frac{s_i}{s_o} = -\frac{(-2.67\text{ cm})}{3.00\text{ cm}} = +0.889$$

Again $M > 0$, and the image is upright. The image is 0.889 as tall as the object. Its height is

$$h_i = Mh_o = (+0.889)(1.50\text{ cm}) = +1.33\text{ cm}$$

Reflect

As the object is moved closer to the convex mirror, the image moves closer to the mirror and increases in size.

NOW WORK Problems 1–8 and 10 from The Takeaway for Section 23-6.

THE TAKEAWAY for Section 23-6

✔ The mirror equation that relates the image distance, object distance, and focal length of a convex mirror is the same as for a concave mirror. The same is true for the equation that relates the heights of the image and the object to the image and object distances, and the equation that relates the focal length and radius of curvature.

✔ The fundamental difference between concave and convex mirrors is that a convex mirror has a negative radius of curvature and so a negative focal length.

Prep for the AP® Exam

AP® Building Blocks

EX 23-2 1. A spherical convex mirror is placed at the end of a driveway on a corner with a limited view of oncoming traffic. The mirror has a radius of curvature of 1.85 m. An oncoming car is 12.6 m from the mirror. How far from the mirror does the image of the car appear to be?

EX 23-2 2. A girl sees her image in a shiny glass spherical tree ornament that has a diameter of 10.0 cm. The image is upright and is located 1.5 cm behind the surface of the ornament. How far from the ornament is the child located?

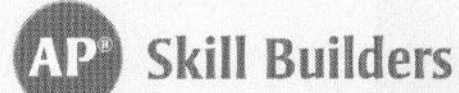

AP® Skill Builders

EX 23-2 3. A car's convex rearview mirror has a radius of curvature equal to 15.0 m. What are the magnification, type, and location of the image that is formed by an object that is 10.0 m from the mirror?

EX 23-2 4. An 18.0-cm-long pencil is placed beside a convex spherical mirror, and its image is 10.5 cm in length. If the radius of curvature of the mirror is 88.4 cm, find the image distance, the object distance, and the magnification of the pencil.

EX 23-2 5. A 1-cm-long horse fly hovers 1.0 cm from a shiny sphere with a radius of 25.0 cm. Calculate the location of the image of the fly, its type (real or virtual), and its length.

EX 23-2 6. Using the mirror equation, prove that all images of real objects seen in spherical convex mirrors are virtual.

EX 23-2 7. A shiny sphere, 30.0 cm in diameter, is placed in a garden for aesthetic purposes. Determine the type, location, and height of the image of a 6.00-cm-tall squirrel located 40.0 cm in front of the sphere.

Skills in Action

EX 23-2 8. The radius of curvature of a spherical convex mirror is 15.0 cm. Describe the image formed when a 20.0-cm-tall object is positioned **(a)** 5.0 cm from the mirror, **(b)** 20.0 cm from the mirror, and **(c)** 100.0 cm from the mirror. For each case, provide the image distance, the image height, the type of image (real or virtual), and the orientation of the image (upright or inverted).

9. If the image made by a convex mirror of focal length f is 1/2 the height of the object, the distance from the object to the mirror must be equal to
(A) $4|f|$.
(B) $2|f|$.
(C) $|f|$.
(D) $|f|/2$.

EX 23-2 10. The radius of curvature of a spherical convex mirror is 20.0 cm. Describe the image formed when a 10.0-cm-tall object is positioned **(a)** 20.0 cm from the mirror, **(b)** 50.0 cm from the mirror, and **(c)** 100.0 cm from the mirror. For each case, provide the image distance, the image height, the type of image (real or virtual), and the orientation of the image (upright or inverted).

23-7 Convex lenses form images like concave mirrors and vice versa

A curved mirror forms images by reflection. A **lens**—a piece of glass or other transparent material with a curved surface on its front side, back side, or both sides—forms images by the *refraction* of light as the light enters and leaves the lens. Just as for a curved mirror, the images made by a lens can be larger or smaller than the object (**Figure 23-24**). In this section we'll explore how lenses form images.

The key idea that we need to understand lenses is Snell's law of refraction. As we learned in Section 23-2, a light ray changes direction when it moves from one transparent medium to a second medium in which light travels at a different speed. The ray bends toward the normal if the speed of light is slower in the second medium (**Figure 23-25a**) and bends away from the normal if the speed of light is faster in the second medium (**Figure 23-25b**).

Figure 23-26 shows how parallel light rays refract when they enter or exit a glass sphere (in which the speed of light is relatively slow) surrounded by air (in which the speed of light is almost as fast as in vacuum). Both the front and back surfaces of the sphere are convex (they bulge outward). Parallel light rays converge when they enter the glass through the left-hand convex surface, and also converge when they

This rodent appears larger when viewed through a magnifying glass, a lens that is convex on both sides.

(a)

Monika Graff/The Image Works

The ruler appears smaller than actual size when viewed through this lens, which is concave on both sides.

(b)

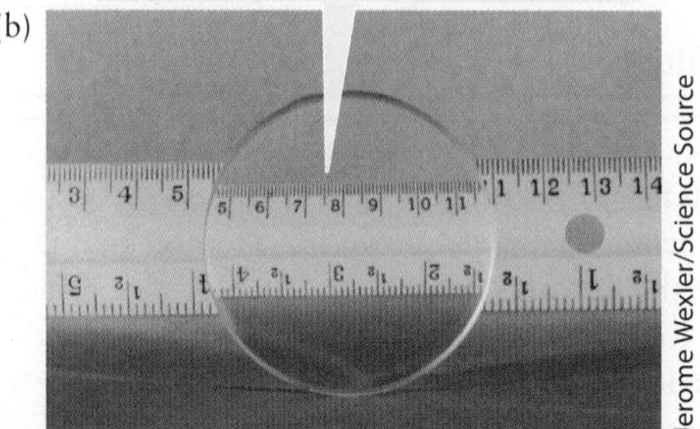

Jerome Wexler/Science Source

Figure 23-24 Lenses can magnify or shrink Different types of lenses can make (a) large or (b) small images.

(a) Refraction from one medium into a slower one

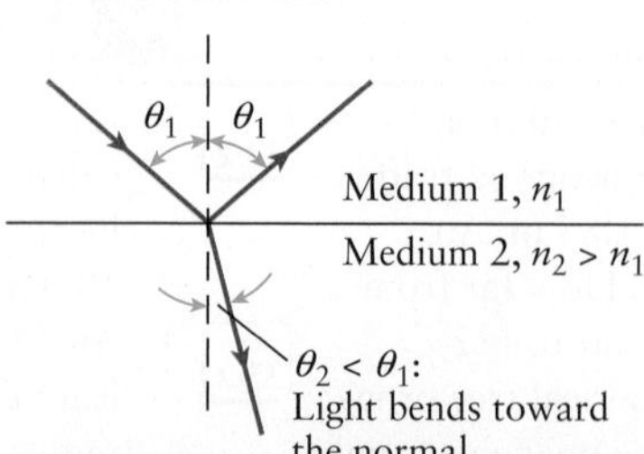

(b) Refraction from one medium into a faster one

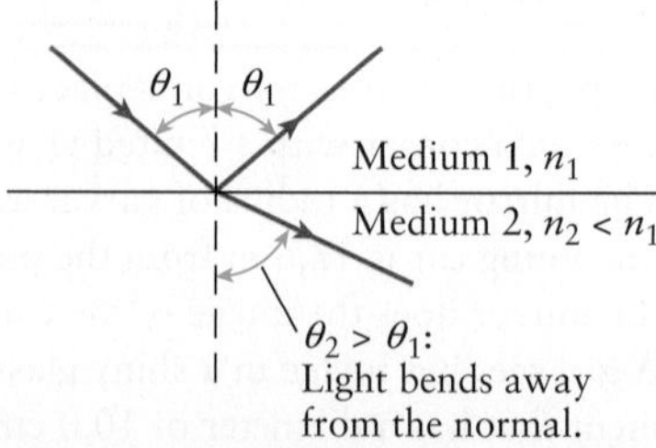

Figure 23-25 Snell's law of refraction A light ray crossing the boundary between two transparent media can refract either (a) toward or (b) away from the normal, depending on how the speed of light compares in the two media.

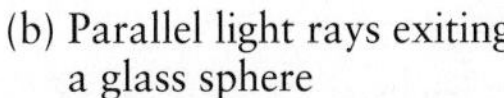

(a) Parallel light rays entering a glass sphere
Each dashed line represents the normal to the surface at the point where a light ray enters the sphere.
The incident angle θ_1 and the refracted angle θ_2 obey the law of refraction.
θ_2 is less than θ_1 for light crossing from air to glass, so the rays are bent toward the principal axis of the sphere. Hence the light rays converge as they enter the sphere.
θ_1
θ_2
Principal axis
C
Air
Glass
(b) Parallel light rays exiting a glass sphere
θ_2 is greater than θ_1 for light crossing from glass to air, so the rays are bent toward the principal axis of the sphere. Hence the light rays converge as they exit the sphere.

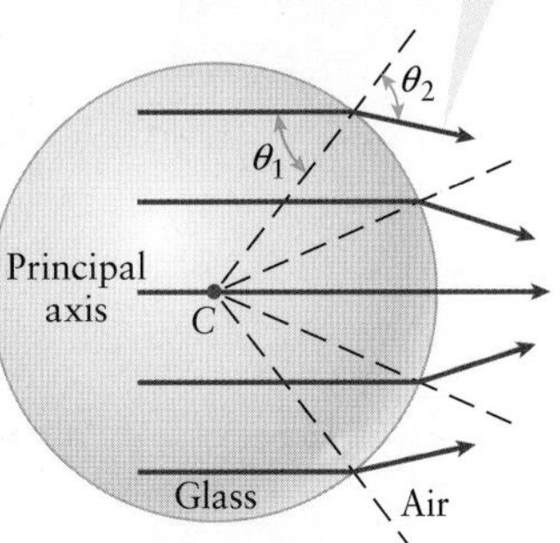

Figure 23-26 **Refraction by a glass sphere** If a glass sphere is surrounded by air, parallel light rays converge whether they (a) enter the sphere or (b) exit the sphere.

exit the glass through the right-hand convex surface. **Figure 23-27** shows what happens when parallel light rays enter or exit a piece of glass with a concave surface (one that bulges inward). In this case parallel rays diverge as they cross the concave surface into the glass, and diverge as they exit the glass through the concave surface.

The light rays in Figures 23-26 and 23-27 undergo only a single refraction when they enter or exit a glass object with a curved surface. In a lens there are *two* refractions: once when the light enters the front surface of the lens, and once when it exits the back surface. Figure 23-26 shows that the refractions will make the rays converge if each surface is convex, and Figure 23-27 shows that the refractions will make the rays diverge if each surface is concave. So a lens with two convex surfaces—called a *convex* lens—will be a **converging lens** that takes incoming parallel light rays and makes them converge toward the principal axis. The same is true for a lens with one convex surface and one flat surface, called a *plano-convex* lens. The refraction at the flat surface by itself causes neither convergence nor divergence. Similarly, a lens with two concave surfaces—called a *concave* lens—will be a **diverging lens** that takes incoming parallel light rays and makes them diverge away from the principal axis. A lens with one concave surface and one flat surface, called a *plano-concave* lens, will also be a diverging

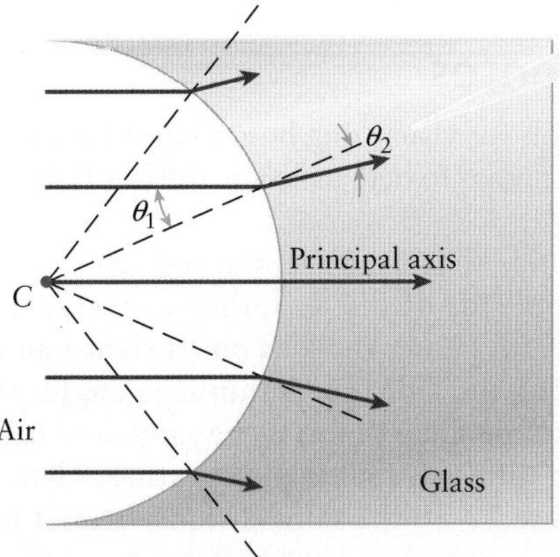

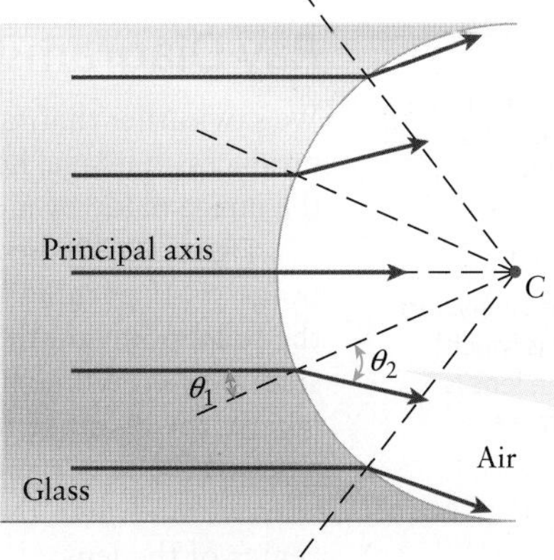

Figure 23-27 **Refraction by glass with a spherical cutout** If a piece of glass with a spherical cutout is surrounded by air, parallel light rays diverge whether they (a) enter the glass through the cutout or (b) exit the glass through the cutout.

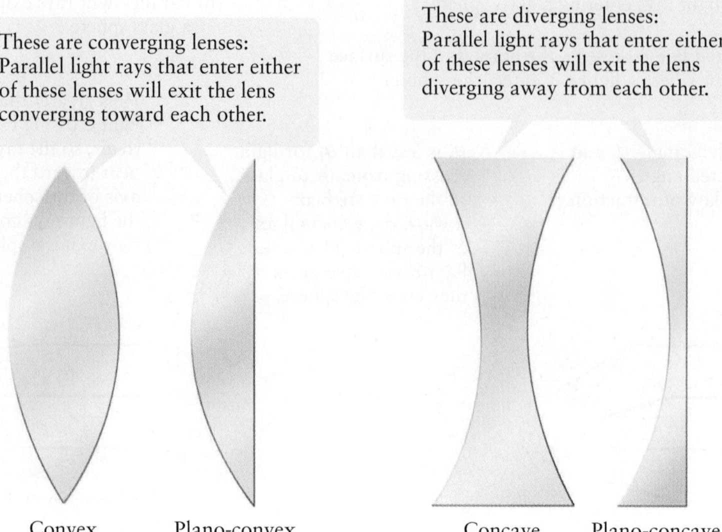

These are converging lenses: Parallel light rays that enter either of these lenses will exit the lens converging toward each other.
These are diverging lenses: Parallel light rays that enter either of these lenses will exit the lens diverging away from each other.
Convex
Plano-convex
Concave
Plano-concave

Figure 23-28 Converging and diverging lenses Two examples of converging lenses and two examples of diverging lenses.

lens (**Figure 23-28**). Here's a general rule: A lens that is thicker in the middle than at the edges will be a converging lens, while a lens that is thicker at the edges than at the middle will be a diverging lens, even if both sides of the lens are curved.

WATCH OUT !

The curvature of a lens has the opposite effect to the same curvature in a mirror.

We saw in Sections 23-3 and 23-5 that a concave mirror causes light rays to converge on reflection, while a convex mirror causes light rays to diverge on reflection. By contrast, a concave lens causes light rays to *diverge* as they pass through, and a convex lens causes light rays to *converge* as they pass through. Mirrors and lenses are different!

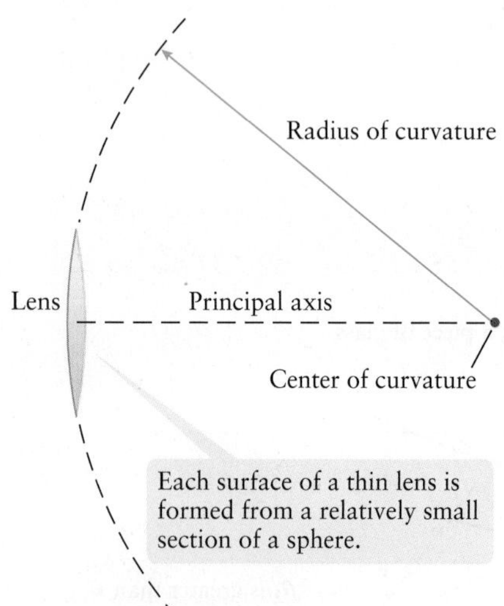

Figure 23-29 A thin lens Analyzing image formation by a lens is much easier if we assume that the lens is thin and has spherical surfaces.

A lens with spherical surfaces suffers from the same spherical aberration as a spherical mirror (Section 23-3): Parallel light rays do not all focus to the same point for a converging lens or appear to originate from a single point for a diverging lens. However, we can neglect this effect if the rays of light are all close to the principal axis. We can ensure this by using only a small section of a large spherical surface to form each surface of the lens (**Figure 23-29**). We'll also assume that there is very little thickness of material between the front and back surfaces of the lens. The result is a model called the **thin lens**. We'll consider only thin lenses for the rest of this chapter so that we can neglect spherical aberration and treat each lens as if parallel rays are focused to a single point. Eyeglasses and contact lenses are everyday examples of thin lenses.

Ray Diagrams for Converging Lenses

We saw earlier that ray diagrams are powerful tools for visualizing how a curved mirror produces an image. Let's see how to draw a ray diagram to help us locate the position of the image made by a converging lens.

Figure 23-30 shows a thin convex lens. Notice that we have marked *two* focal points F. If parallel light rays enter the left-hand side of the lens, they converge at the focal point on the right-hand side; if parallel rays enter the lens on its right-hand side, they converge at the focal point on the left-hand side. For a thin lens the focal length, the distance from the center of the lens to the focal point, is the same on both sides of the lens even if the two surfaces have a different radius of curvature. That's why we've drawn the two focal points F in Figure 23-30 the same distance from the center of the lens.

In Figure 23-30 we've drawn a red arrow to represent our object on the left-hand side of the lens and placed this arrow outside the left-hand focal point. We've also drawn three representative light rays from the tip of the arrow. While each light ray actually refracts twice, once on entering the lens and once on exiting, for simplicity

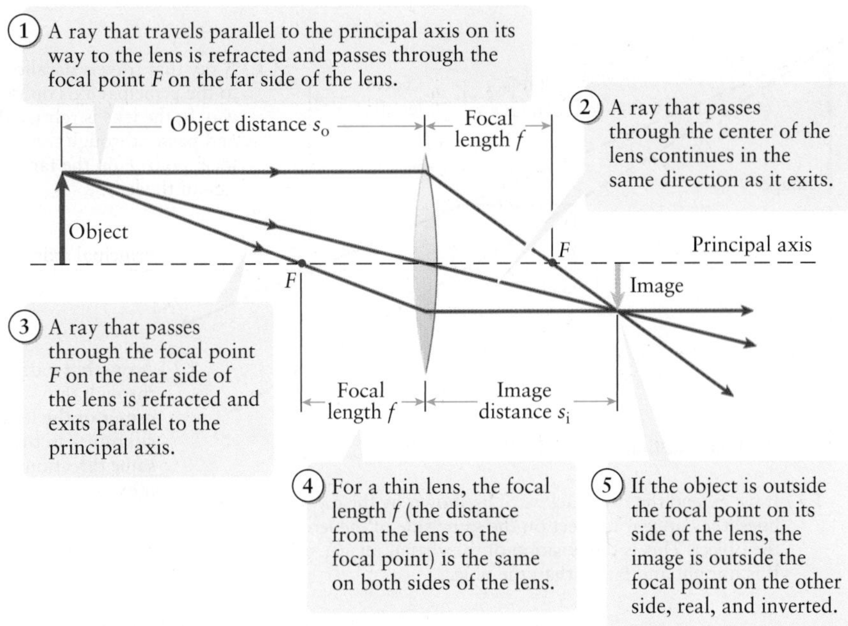

(1) A ray that travels parallel to the principal axis on its way to the lens is refracted and passes through the focal point F on the far side of the lens.
Object distance s_o
Focal length f
(2) A ray that passes through the center of the lens continues in the same direction as it exits.
Object
F
Principal axis
F
Image
(3) A ray that passes through the focal point F on the near side of the lens is refracted and exits parallel to the principal axis.
Focal length f
Image distance s_i
(4) For a thin lens, the focal length f (the distance from the lens to the focal point) is the same on both sides of the lens.
(5) If the object is outside the focal point on its side of the lens, the image is outside the focal point on the other side, real, and inverted.

Figure 23-30 Ray diagram for a converging lens I How the image is formed for an object placed outside the focal point of a converging lens.

we've drawn the rays as if they refract only once, along the centerline of the lens (the vertical line that runs through the center of the lens). The three rays are:

1. A ray that arrives at the lens traveling parallel to the principal axis is refracted so that it passes through the far (right-hand) focal point *F*.
2. A ray that enters the lens directly at its center continues in a straight line as it exits. This is only an approximation because it assumes that the ray undergoes no refraction on entering or exiting the lens. But this is a good approximation for a thin lens, which is nearly flat at its very center.
3. A ray that passes through the near (left-hand) focal point *F* on its way to the lens exits the lens traveling parallel to the principal axis. To see why this is the case, imagine that we could record a video of light traveling along this path, then run the video backward. Then we would see a light ray coming from the far right in Figure 23-30 and traveling parallel to the principal axis. After this ray passes through the lens from right to left, it would naturally pass through the left-hand focal point *F*. If we now run that same video forward, we see light from the object following the path shown in Figure 23-30.

The image of the tip of the arrow in Figure 23-30 forms where the rays coming from the tip cross. The image of the base of the arrow must form along the principal axis because a ray that comes from the base and travels along the axis strikes the center of the lens and therefore continues in the same direction. You can see that the image of the arrow is real (the light rays actually cross at the image point) and is inverted.

Note that in Figure 23-30 the object is relatively far from the lens (the object distance s_o is greater than twice the focal length f), and the image is real, inverted, *and* smaller than the object. That's the same result we found for a concave mirror in Section 23-3 (see Figure 23-12). This reinforces the idea that a convex lens behaves similarly to a concave mirror. **Figure 23-31** shows the small, inverted image formed when light from a distant object passes through a convex lens. You can also see this in Figure 13-36, in which a spherical drop of water acts as a convex lens.

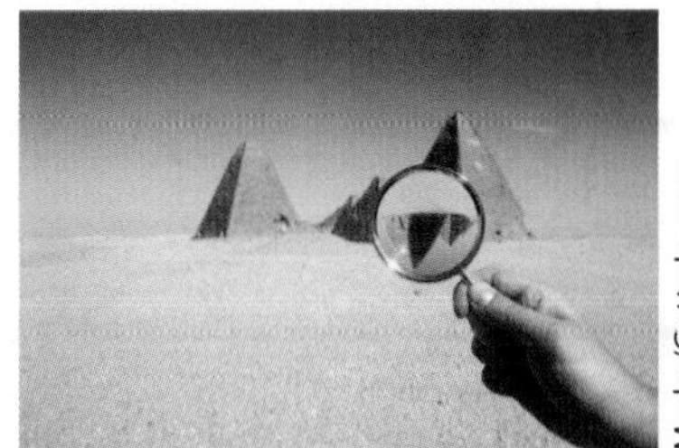
Marka/Getty Images

Figure 23-31 A real, inverted image made by a converging lens The lens in the person's hand makes a real, inverted, and small image of these pyramids in Sudan.

WATCH OUT !

A real image formed by a lens is on the side of the lens opposite to the object.

We've defined a real image as one that forms where reflected or refracted light rays converge. A concave mirror can cause light rays to converge only on the reflective side of the mirror, which is where we would put an object, so this is the only side on which a real image can form. A lens, however, can only cause light rays to converge on the side of the lens *opposite* to the object, as in Figure 23-30. As we'll see below, only a converging lens can produce a real image, but only if the object is outside the focal point as in Figure 23-30.

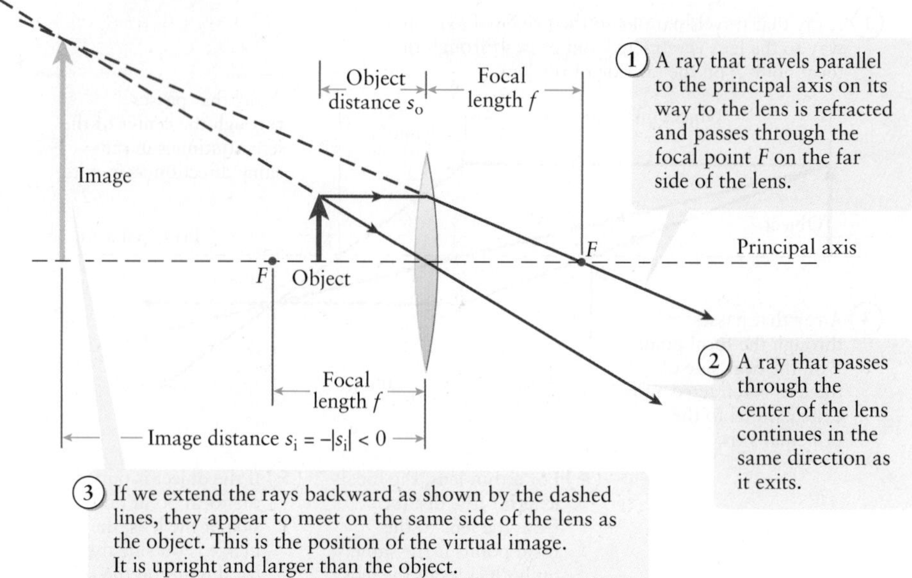
Object distance s_o
Focal length f
1 A ray that travels parallel to the principal axis on its way to the lens is refracted and passes through the focal point F on the far side of the lens.
Image
F
Principal axis
F
Object
2 A ray that passes through the center of the lens continues in the same direction as it exits.
Focal length f
Image distance $s_i = -|s_i| < 0$
3 If we extend the rays backward as shown by the dashed lines, they appear to meet on the same side of the lens as the object. This is the position of the virtual image. It is upright and larger than the object.

Figure 23-32 Ray diagram for a converging lens II How the image is formed for an object placed inside the focal point of a converging lens.

Recall that for mirrors our convention was that the image distance is positive if the image is real and so is on the same side of the mirror as the object. For a lens we'll also say that the image distance is positive if the image is real. This means that for a lens, a positive image distance s_i implies that the image is on the *opposite* side of the lens from the object. For example, the image distance is positive in Figure 23-30.

In **Figure 23-32** we've moved the object to a point between the lens and the focal point, so the object distance is less than the focal length: $s_o < f$. To locate the image, we've drawn just two light rays from the tip of the arrow. One ray travels parallel to the principal axis of the lens and therefore passes through the focal point on the far side after being refracted by the lens. The other ray we have drawn enters the center of the lens and so continues straight. (In Figure 23-30 we drew a third ray through the focal point on the same side of the lens as the object. We don't draw that ray here, because it leads away from the lens rather than toward it.) Notice that these two rays do not meet. But if we extend the rays backward, as shown by the dashed lines, they appear to meet on the same side of the lens as the object. Where they meet is the location of the image. The rays do not actually cross there, so this is a *virtual* image. Because the image is on the same side of the lens as the object, the image distance is negative: $s_i < 0$. We only ever actually need two rays to find the image, but a third ray lets us check our results. There is a third ray we could have drawn for this object. If we draw a ray from the object to the lens at an angle as if it was coming from the focal point on the object side of the lens, it would then leave the lens parallel to the principal axis. Its backward extension would intersect the other two rays. This would actually cross the plane of the lens much higher than we drew the lens, but you can always extend a line representing the center of the lens up or down. Try it for yourself!

As in Figure 23-30, the base of the image arrow must lie on the principal axis of the lens. It follows that the image is upright. Note that the image is also larger than the object. This is the kind of image shown in Figure 23-24a; when we hold a convex lens so that the rodent is inside the focal point of the lens, we see an enlarged, upright image of the rodent. This is very similar to the image made by a concave mirror when an object is placed inside the focal point of the mirror (see Figure 23-16).

Ray Diagrams for Diverging Lenses

We can also use ray diagrams to learn about the image made by a diverging lens. We've placed an object in **Figure 23-33**, represented by an arrow, in front of a thin lens with two concave surfaces. Because this lens causes parallel light rays to diverge, we say that the focal length is negative: $f = -|f| < 0$. (The focal length of a convex mirror, which also causes parallel light rays to diverge, is also negative.)

We've drawn two light rays to determine where the image forms, and as we did for the thin, convex lens, we've drawn the rays as if they refract only once along the

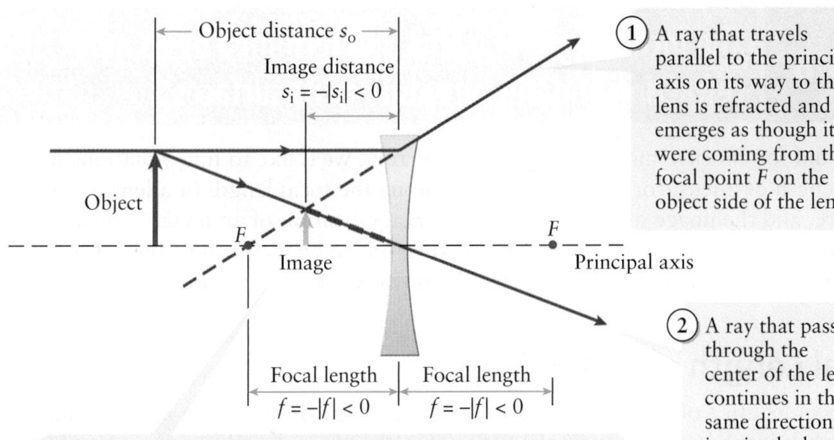
Object distance s_o
Image distance
$s_i = -|s_i| < 0$
Object
Image
Principal axis
F
F
Focal length
$f = -|f| < 0$
Focal length
$f = -|f| < 0$
① A ray that travels parallel to the principal axis on its way to the lens is refracted and emerges as though it were coming from the focal point F on the object side of the lens.
② A ray that passes through the center of the lens continues in the same direction as it exits the lens.
③ If we extend the rays backward as shown by the dashed lines, they appear to meet on the same side of the lens as the object. This is the position of the virtual image. It is upright and smaller than the object.

Figure 23-33 Ray diagram for a diverging lens How the image is formed for an object placed in front of a diverging lens. We could draw a third ray. In this case it would be straight from the tip of the arrow to the focal point on the far side of the lens. Such a ray would leave the lens parallel to the principal axis. While it would clutter up this figure, you should try it yourself and see that the backward extension does cross the others.

centerline of the lens. The refracted rays don't actually meet, but if we extend these rays backward, we find the location where the extensions meet. The image of the tip of the arrow forms at this point; the image is virtual (because the rays don't actually meet there), upright, and smaller than the object. This is the same behavior we saw in Section 23-5 for convex mirrors. As for a convex mirror, the image is virtual, upright, and smaller no matter what the object distance.

Figure 23-24b shows an image of a ruler made by a concave lens. Just like the image in Figure 23-33, this image is upright and smaller than the object.

THE TAKEAWAY for Section 23-7

✔ When parallel light rays enter one side of a convex lens, they exit the lens converging toward the focal point on the other side of the lens.

✔ When parallel light rays enter a concave lens, they exit the lens diverging away from the focal point on the same side of the lens that the rays entered.

✔ A convex lens produces a real image if the object is outside the focal point and produces a virtual image if the object is inside the focal point.

✔ A concave lens produces a virtual image no matter what the position of the object, for any real object.

Prep for the AP Exam

AP Building Blocks

1. A real image that is created due to reflection in a spherical mirror appears in front of the mirrored surface. Is this also the case for a real image that is created due to refraction in a lens?
2. For a certain glass lens in air, both radii of curvature are positive. Considering Figures 23-26 and 23-27, can you tell if it is a converging lens or a diverging lens, or do you need additional information to tell? Explain.
3. Under what circumstances will the images of a real object formed by converging or diverging lenses be designated as "real"? Indicate the type or types of lenses and the required position of the object.

AP Skill Builders

4. A convex lens made of clear ice ($n = 1.31$) acts as a *converging* lens when it is in air, but as a *diverging* lens when it is surrounded by acetone ($n = 1.36$). Referring to Figures 23-26 and 23-27, explain why this is true.
5. What *minimum* number of rays are required to locate the image that is formed by a lens in a ray diagram? Explain.
6. Where does the bending of light physically take place in a typical concave or convex lens? Is this how we draw ray diagrams? Why or why not?

AP Skills in Action

7. You make a lens with two convex surfaces out of ice ($n = 1.31$). You then submerge the lens in a large tank of concentrated sugar water ($n = 1.49$). If you place an object in the tank and in front of the lens, an image is formed at a position inside the tank. What kind of image is this?
(A) A real image
(B) A virtual image that is larger than the object
(C) A virtual image that is smaller than the object
(D) Any of (A), (B), or (C), depending on the distance from the object to the lens

23-8 The focal length of a lens is determined by its index of refraction and the curvature of its surfaces

As we did for concave mirrors and convex mirrors, we'd like to find equations for the focal length of a lens; for the relationship among the focal length of a lens, the object distance, and the image distance; and for the magnification of an image produced by a lens. As we'll see, the latter two equations turn out to be identical to those for curved mirrors. The expression for the focal length, however, is a little more complicated.

Focal Length of a Thin Lens

The focal length f of a thin lens depends on the index of refraction n of the material of which it is made. The greater the value of n, the more sharply a light ray is bent as it passes either from the surrounding air (for which the index of refraction is essentially 1) into the lens or from the lens into the air. The value of f also depends on how the front and back surfaces of the lens are curved. The mathematical expression of these relationships for a lens in air is called the **lensmaker's equation:**

EQUATION IN WORDS
Lensmaker's equation for the focal length of a thin lens

Focal length of a lens surrounded by air

Index of refraction of the lens material

$$\frac{1}{f} = (n-1)\left(\frac{1}{R_1} - \frac{1}{R_2}\right) \quad (23\text{-}24)$$

Radius of curvature of lens surface 1 (the surface closer to the object)

Radius of curvature of lens surface 2 (the surface farther from the object)

Equation 23-24 can be derived by applying Snell's law of refraction (Equation 23-6) to a ray of light refracted by both surfaces of a lens. The derivation is beyond our scope, however.

The values of the radii of curvature R_1 and R_2 depend on how sharply and in what direction the two surfaces of the lens are curved. As **Figure 23-34** shows, we take

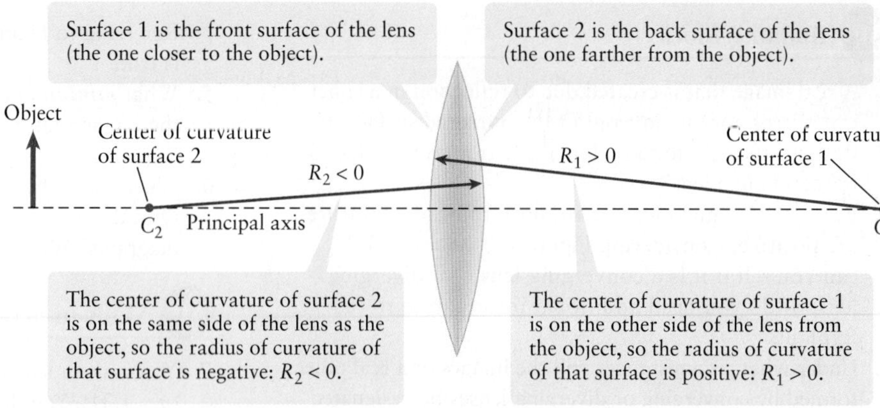

Surface 1 is the front surface of the lens (the one closer to the object).

Surface 2 is the back surface of the lens (the one farther from the object).

Object

Center of curvature of surface 2

$R_2 < 0$

$R_1 > 0$

Center of curvature of surface 1

C_2 Principal axis

C_1

The center of curvature of surface 2 is on the same side of the lens as the object, so the radius of curvature of that surface is negative: $R_2 < 0$.

The center of curvature of surface 1 is on the other side of the lens from the object, so the radius of curvature of that surface is positive: $R_1 > 0$.

- If the center of curvature of a lens surface is on the other side of the lens from the object, the radius of curvature R of that surface is positive.
- If the center of curvature of a lens surface is on the same side of the lens as the object, the radius of curvature R of that surface is negative.
- A flat surface has a radius of curvature that is defined to be infinite R, so $1/R = 0$ for that surface.

Figure 23-34 Interpreting the lensmaker's equation For each surface of a thin lens, the sign of the radius of curvature depends on where the center of curvature is located.

a radius to be positive if the center of curvature is on the other side of the lens from the object but negative if the center of curvature is on the same side as the object. For example, in Figure 23-34 the radius R_1 of surface 1 is positive, but the radius R_2 of surface 2 is negative.

Equation 23-24 tells us that the more tightly curved the surfaces of a lens are, the *smaller* the magnitudes of the radii R_1 and R_2 and the *smaller* the magnitude of the focal length f. (We saw a similar result for the focal length of a spherical mirror in Section 23-3.) The smaller the magnitude of f, the more sharply light rays are refracted by passing through the lens.

The lensmaker's equation is expressed in terms of the reciprocal of the focal length f. For that reason it's common, especially among optometrists, to characterize lenses (and also mirrors) in terms of the reciprocal of the focal length:

$$P = \frac{1}{f} \tag{23-25}$$

The quantity P is called the **power** of the lens or mirror. The units of P are m^{-1}; $1\ m^{-1}$ is known as 1 **diopter**. The larger the power of a lens, the greater the amount of refraction it causes (that is, the more "powerfully" it causes light rays to bend). If parallel light rays enter a thin, convex lens of power 2 diopters, they come to a focus 1/2 m (0.5 m) behind the lens; if instead these parallel light rays enter a 5-diopter converging lens, they are bent more sharply and come to a focus just 1/5 m (0.2 m) behind the lens. A thin, concave lens has a negative focal length and so has a negative power P: When parallel light rays enter a diverging lens with a power of –5 diopters, they emerge from the lens as though they were coming from a point 1/5 m (0.2 m) in front of the lens (see Figure 23-33). In general, the smaller the magnitude of the focal length of a lens, the greater the magnitude of its power.

EXAMPLE 23-3 Calculating Focal Length for a Convex Lens

In Figure 23-34 the front surface (surface 1) of the lens has a radius of curvature of magnitude 15.0 cm, and the back surface (surface 2) has a radius of curvature of magnitude 25.0 cm. The lens is made of crown glass with an index of refraction of 1.520. (a) Calculate the focal length of the lens. (b) Now flip the orientation of the lens so that the front surface is now the one with a radius of curvature of magnitude 25.0 cm and the back surface is now the one with a radius of curvature of magnitude 15.0 cm. Calculate the focal length of the lens in this case.

Set Up

For each situation we'll draw the lens and determine on which side of the lens the centers of curvature lie. That will tell us whether R_1 and R_2 are positive or negative. We'll then use Equation 23-24 to calculate the focal length.

Lensmaker's equation for the focal length of a thin lens:

$$\frac{1}{f} = (n-1)\left(\frac{1}{R_1} - \frac{1}{R_2}\right) \tag{23-24}$$

Solve

(a) Both surfaces are convex. Hence the center of curvature of each surface is on the other side of the lens from that surface. The center of curvature C_1 of the front surface (on the left in the figure) is on the far side of the lens, so R_1 is positive: $R_1 = +15.0$ cm. The center of curvature C_2 of the back surface (on the right in the figure) is on the near side of the lens, so R_2 is negative: $R_2 = -25.0$ cm.

Calculate the focal length.

From Equation 23-24:

$$\frac{1}{f} = (1.520 - 1)\left[\frac{1}{(+15.0\text{ cm})} - \frac{1}{(-25.0\text{ cm})}\right]$$
$$= 0.520\,[(0.0667\text{ cm}^{-1}) - (-0.0400\text{ cm}^{-1})]$$
$$= 0.0555\text{ cm}^{-1}$$

$$f = \frac{1}{0.0555\text{ cm}^{-1}} = 18.0\text{ cm} = 0.180\text{ m}$$

Although R_2 is negative, we subtract $1/R_2$ in Equation 23-24, so both the $1/R_1$ term and the $1/R_2$ term—that is, both surface 1 and surface 2—contribute to giving $1/f$ a positive value. As a result, the focal length is positive, as we expect for a convex lens.

(b) Because we have flipped the lens around, we have interchanged surfaces 1 and 2. The object is still to the left of the lens, however. As in part (a), R_1 is positive and R_2 is negative, but now $R_1 = +25.0$ cm and $R_2 = -15.0$ cm.

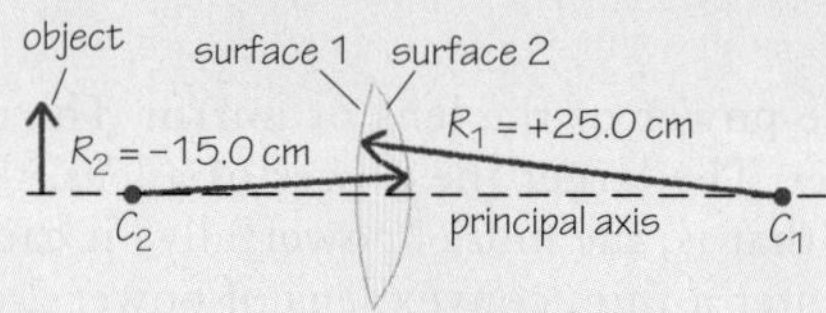

Calculate the focal length.

From Equation 23-24:

$$\frac{1}{f} = (1.520 - 1)\left[\frac{1}{(+25.0\text{ cm})} - \frac{1}{(-15.0\text{ cm})}\right]$$
$$= 0.520\,[(0.0400\text{ cm}^{-1}) - (-0.0667\text{ cm}^{-1})]$$
$$= 0.0555\text{ cm}^{-1}$$

$$f = \frac{1}{0.0555\text{ cm}^{-1}} = 18.0\text{ cm} = 0.180\text{ m}$$

Again both the $1/R_1$ term and the $1/R_2$ term contribute to giving $1/f$ a positive value. The focal length has the same positive value as in part (a).

Reflect

The focal length of this thin lens stays the same after we flip it back to front. As we stated in the previous section, the focal points on either side of the lens are both the same distance (that is, the same focal length) from the lens. Note that the power of this lens is $P = 1/f = 1/(0.180\text{ m}) = 5.55\text{ m}^{-1}$, or 5.55 diopters.

NOW WORK Problems 3 and 8 from The Takeaway for Section 23-8.

EXAMPLE 23-4 Calculating Focal Length for a Concave Lens

A certain lens made of crown glass ($n = 1.520$) has concave front and back surfaces. The front surface has a radius of curvature of magnitude 15.0 cm, and the back surface has a radius of curvature of magnitude 25.0 cm. Calculate the focal length of the lens.

Set Up

As in the previous example, we'll first draw the lens and use our drawing to decide whether R_1 and R_2 are positive or negative. Equation 23-24 will then allow us to calculate the focal length.

Lensmaker's equation for the focal length of a thin lens:

$$\frac{1}{f} = (n - 1)\left(\frac{1}{R_1} - \frac{1}{R_2}\right) \qquad (23\text{-}24)$$

Solve

Both surfaces are concave. Hence the center of curvature of each surface is on the same side of the lens as that surface. The center of curvature C_1 of the front surface (on the left in the figure) is on the near side of the lens, so R_1 is negative: $R_1 = -15.0$ cm. The center of curvature C_2 of the back surface (on the right in the figure) is on the far side of the lens, so R_2 is positive: $R_2 = +25.0$ cm.

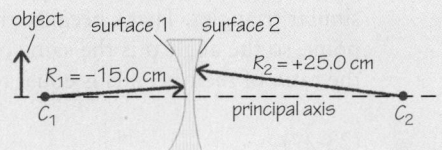

Calculate the focal length.

From Equation 23-24:

$$\frac{1}{f} = (1.520 - 1)\left[\frac{1}{(-15.0\ \text{cm})} - \frac{1}{(+25.0\ \text{cm})}\right]$$
$$= 0.520\left[(-0.0667\ \text{cm}^{-1}) - (0.0400\ \text{cm}^{-1})\right]$$
$$= -0.0555\ \text{cm}^{-1}$$

$$f = \frac{1}{-0.0555\ \text{cm}^{-1}} = -18.0\ \text{cm} = -0.180\ \text{m}$$

The $1/R_1$ term is negative, and we subtract from it the positive $1/R_2$ term. So both terms contribute to giving $1/f$ a negative value, and the focal length is negative.

Reflect

The focal length is negative, just as we expect for a concave lens. The power of such a lens is negative: $P = 1/f = 1/(-0.180\ \text{m}) = -5.55\ \text{m}^{-1}$, or -5.55 diopters. You should repeat the calculation with the lens reversed back to front, as in part (b) of Example 23-3. You should get the same result for the focal length. Do you?

NOW WORK Problem 5 from The Takeaway for Section 23-8.

Image Position and Magnification for a Thin Lens

To see how the image distance is related to the object distance and the focal length for a thin, convex lens, let's look again at a ray diagram like Figure 23-30. We've drawn such a diagram in **Figure 23-35**. Let's see how to use trigonometry to find relationships between the distances of interest.

The shaded regions in **Figure 23-35a** are similar triangles. That's because the ray from the tip of the object at O' to the tip of the image at I' passes through the center of the lens at C without deflection, so the angle θ is the same in both triangle $O'C'C$ and triangle ICI'. The ratio of the heights of the two triangles is therefore equal to the ratio of their bases; that is,

$$\frac{|h_i|}{h_o} = \frac{s_i}{s_o} \qquad (23\text{-}26)$$

(The height h_i of the image in Figure 23-35a is negative because the image is

(a)
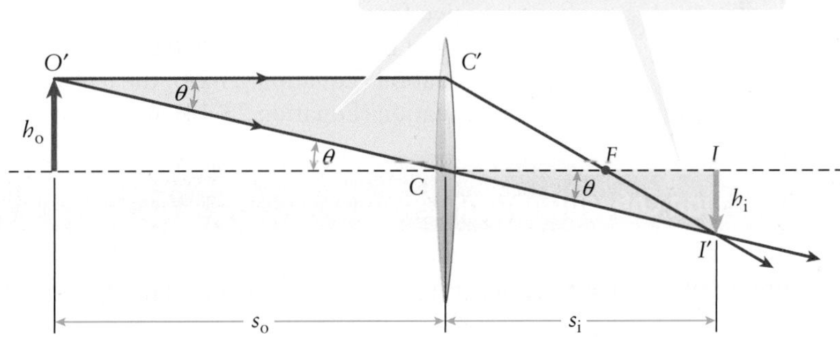

(b)
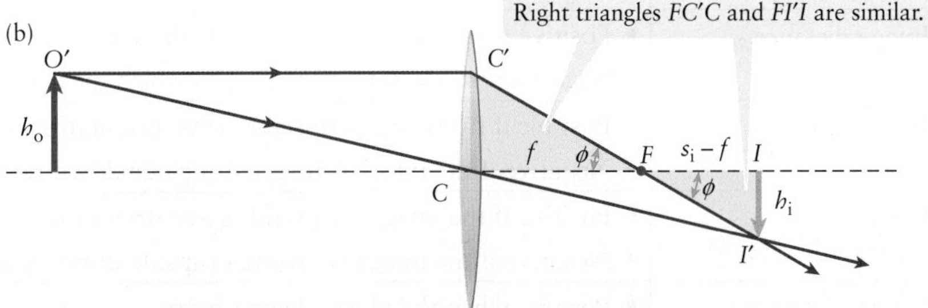

Figure 23-35 Analyzing a converging lens Ray diagrams for an object on the principal axis of a converging lens. The similar triangles in (a) and (b) help us determine the position and magnification of the image.

inverted. That's why we've used $|h_i|$ in Equation 23-26 for the distance from I to I'.) The two shaded regions in **Figure 23-35b**—the right triangles $FC'C$ and $FI'I$—are also similar triangles. That's because the straight ray from C' to I' passes through the focal point, so the angle ϕ is the same on either side of F. So for these two triangles as well, the ratio of their heights is equal to the ratio of their bases:

$$\frac{|h_i|}{h_o} = \frac{s_i - f}{f} \tag{23-27}$$

If we combine Equation 23-26 and Equation 23-27, we find a relationship among the image distance s_i, the object distance s_o, and the focal length f:

$$\frac{s_i}{s_o} = \frac{s_i - f}{f} \quad \text{or} \quad \frac{s_i}{s_o} = \frac{s_i}{f} - 1 \quad \text{or} \quad \frac{s_i}{s_o} + 1 = \frac{s_i}{f}$$

Divide both sides by s_i:

$$\frac{1}{s_o} + \frac{1}{s_i} = \frac{1}{f} \quad \text{(lens equation)} \tag{23-11}$$

This is the same equation that we deduced for spherical mirrors in Section 23-4. It applies just as well to thin, spherical convex lenses. Now we see why we were justified in calling Equation 23-11 the mirror and lens equation—it works for both.

Equation 23-26 above also tells us the magnification of the image. Because the image height h_i is negative in Figure 23-35a (the image is inverted), h_i is equal to $-|h_i|$ and $|h_i| = -h_i$. If we substitute this into Equation 23-26, we get

$$\frac{(-h_i)}{h_o} = \frac{s_i}{s_o} \quad \text{or} \quad \frac{h_i}{h_o} = -\frac{s_i}{s_o}$$

Lateral magnification is equal to the ratio of image height to object height: $M = h_i/h_o$ (Equation 23-2). So the magnification of the image produced by a thin lens is

$$M = \frac{h_i}{h_o} = -\frac{s_i}{s_o} \quad \text{(lateral magnification)} \tag{23-14}$$

That's the same Equation 23-14 that we derived in Section 23-4 for a curved mirror.

We've derived Equations 23-11 and 23-14 for a thin convex lens with a positive focal length, but they turn out to be equally valid for a thin concave lens with a negative focal length. As for mirrors, it's important to keep track of the signs of quantities in these equations. **Table 23-2** summarizes when the quantities in the lensmaker's equation (Equation 23-24), the lens equation (Equation 23-11), and the magnification equation (Equation 23-14) are positive and when they are negative.

TABLE 23-2 Sign Conventions for Lenses

Quantity	Sign convention
Lens surface radius of curvature R_1 or R_2	• Positive if the center of curvature is on the side of the lens from which the light leaves • Negative if the center of curvature is on the side of the lens on which the light is incident
Focal length f	• Positive for a converging lens • Negative for a diverging lens
Image distance s_i	• Positive if on the side of the lens from which the light leaves; the image is then a real image • Negative if on the side of the lens on which the light is incident; the image is then a virtual image
Image height h_i	• Positive if the image is upright (same orientation as object) • Negative if the image is inverted (upside down relative to object)
Lateral magnification M	• Positive if the image is upright (same orientation as object) • Negative if the image is inverted (upside down relative to object)
Object distance s_o	• Positive when the object being observed is real, as in any single-instrument optical system • If the object of a lens is the image of another optical instrument, then it has a positive object distance if the image of the first instrument is on the side from which light is incident on the lens or a negative object distance if the image of the first instrument would form on the side from which light exits the lens

Because the equations for mirrors and thin lenses are effectively identical, many of the same conclusions that we came to for images made by a curved mirror also apply to images made by a thin lens:

- If the lens has a positive focal length (a converging lens) and an object is placed outside the focal point, the image is real and inverted. Depending on the object distance, the image can be smaller, larger, or the same size as the object.
- If the lens has a positive focal length (a converging lens) and an object is placed inside the focal point of a converging lens, the image is virtual, upright, and larger than the object.
- If the lens has a negative focal length (a diverging lens), the image is virtual, upright, and smaller than the object. This is true for any object distance.

One key difference between mirrors and lenses is in the position of the image. For a mirror a real image is on the same side of the mirror as the object, and a virtual image is on the other side of the mirror. For a lens a real image is on the opposite side of the lens from the object, and a virtual image is on the same side of the lens as the object. This difference is because a real image is formed where light rays actually cross, and light rays leave a mirror to its front, but pass through a lens.

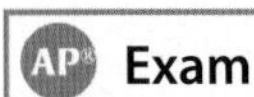

Exam Tip

Here is a quick strategy for remembering signs for optics. The object distance is always positive if the object is on the side from which the light is incident (the original real object). All other quantities are positive if they are on the transmitted side of the instrument (the side light leaves the instrument). For a mirror, the light reflects and leaves on the same side as it came in. The light transmits through the lens and leaves on the opposite side. For lenses, the focal point used to determine the sign is the one used to determine how an incoming ray parallel to the principal axis will bend. This is sometimes referred to as the primary focal point.

EXAMPLE 23-5 An Image Made by a Convex Lens

A thin convex lens has a focal length of 15.0 cm. How far from the lens does the image form when an object is placed 9.00 cm from the center of the lens? Is the image virtual or real? Is the image inverted or upright? By what factor is the image magnified relative to the object?

Set Up

We'll begin by drawing a ray diagram to help us visualize the kind of image that will be produced. We'll use Equation 23-11 to calculate the position of this image, and Equation 23-14 to calculate the image height compared to the object height.

Lens equation:

$$\frac{1}{s_o} + \frac{1}{s_i} = \frac{1}{f} \qquad (23\text{-}11)$$

Magnification:

$$M = \frac{h_i}{h_o} = -\frac{s_i}{s_o} \qquad (23\text{-}14)$$

Solve

We begin by drawing a ray diagram. We've drawn one ray from the object that enters the lens parallel to the principal axis, exits the lens, and passes through the focal point F on the other side of the lens. We've also drawn a ray that passes through the center of the lens. The rays never meet on the other side of the lens, but their extensions do meet on the same side of the lens as the object. So the image will be virtual. As the diagram shows, the image is also upright and larger than the object.

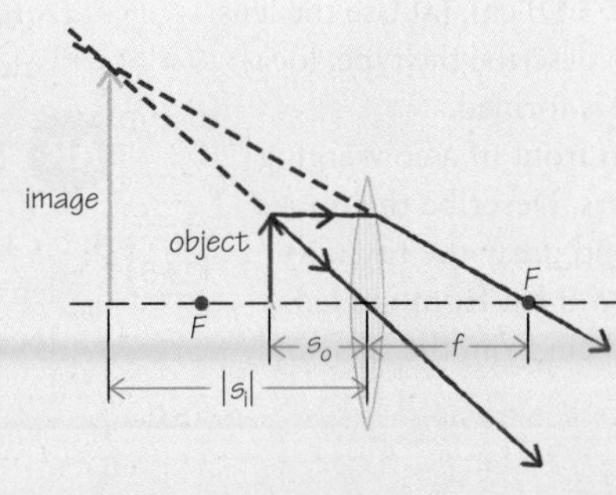

Find the image distance using the lens equation.

We are given $s_o = 9.00$ cm and $f = 15.0$ cm. From Equation 23-11:

$$\frac{1}{s_i} = \frac{1}{f} - \frac{1}{s_o} = \frac{1}{15.0\text{ cm}} - \frac{1}{9.00\text{ cm}}$$
$$= 0.06667\text{ cm}^{-1} - 0.111\text{ cm}^{-1}$$
$$= -0.0444\text{ cm}^{-1}$$
$$s_i = \frac{1}{(0.0444\text{ cm}^{-1})} = -22.5\text{ cm}$$

The negative value of s_i means that the image is on the same side of the lens as the object, just as the ray diagram shows. It must therefore be a virtual image.

Find the image height using the magnification equation.

From Equation 23-14:

$$M = -\frac{s_i}{s_o} = -\frac{(-22.5\text{ cm})}{9.00\text{ cm}} = +2.50$$

The plus sign means the image is upright, in agreement with the ray diagram. The image is 2.50 times as large as the object.

Reflect

Although we didn't try to draw the ray diagram to exact scale, by eye the ratio of the focal length to the object distance looks to be about 2 to 1, which is certainly consistent with the actual values $f = 15.0$ cm and $s_o = 9.00$ cm. We would also expect that the image distance in the diagram is about one and a half times the focal length (22.5 cm compared to 15.0 cm)—and it is. Can you verify that the height of the image in the diagram is about 2.5 times the height of the object?

NOW WORK Problems 1, 2, 4, and 6 from The Takeaway for Section 23-8.

THE TAKEAWAY for Section 23-8

✔ The focal length of a thin lens is determined by the index of refraction of the lens material and the radii of curvature of the front and back surfaces of the lens. These radii can be positive or negative.

✔ The more sharply curved the surfaces of a lens, the smaller the magnitudes of their radii of curvature and the smaller the magnitude of the focal length of the lens.

✔ The same equation that relates object distance, image distance, and focal length for a spherical mirror also applies to thin lenses. The magnification equation is also the same for spherical mirrors and thin lenses.

✔ Image distance s_i is positive when the image is real and negative when the image is virtual. The focal length f is positive for a thin, convex lens and negative for a thin, concave lens.

Prep for the AP® Exam

AP® Building Blocks

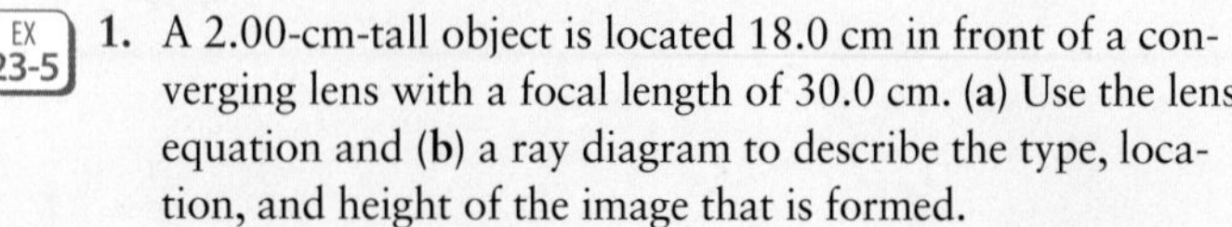

EX 23-5 1. A 2.00-cm-tall object is located 18.0 cm in front of a converging lens with a focal length of 30.0 cm. **(a)** Use the lens equation and **(b)** a ray diagram to describe the type, location, and height of the image that is formed.

EX 23-5 2. A 10.0-cm-tall object is located in front of a converging lens with a power of 5.00 diopters. Describe the image created (type, location, height) and draw the ray diagrams if the object is located **(a)** 5.0 cm from the lens, **(b)** 10.0 cm from the lens, **(c)** 20.0 cm from the lens, and **(d)** 50.0 cm from the lens.

EX 23-3 3. A converging lens is formed from a plastic material that has an index of refraction of 1.55. If the radius of curvature of one surface is 1.25 m and the radius of curvature of the other surface is 1.75 m, use the lensmaker's equation to calculate the focal length and the power of the lens.

AP® Skill Builders

EX 23-5 4. A 10.0-cm-tall object is located in front of a diverging lens with a power of –5.00 diopters. Describe the type, location, and height of the image created, and draw the ray diagrams if the object is located **(a)** 5.0 cm from the lens, (b) 10.0 cm from the lens, (c) 20.0 cm from the lens, and (d) 50.0 cm from the lens.

EX 23-4 5. A glass lens ($n = 1.60$) has a focal length of –31.8 cm and a plano-concave shape. (a) Calculate the radius of curvature of the concave surface. (b) If a lens is constructed from the same glass to form a plano-convex shape with the same radius of curvature, what will the focal length be?

EX 23-5 6. A 2.00-cm-tall object is 30.0 cm in front of a converging lens that has a focal length of 18.0 cm. (a) Use the lens equation and (b) a ray diagram to describe the type, location, and height of the image that is formed.

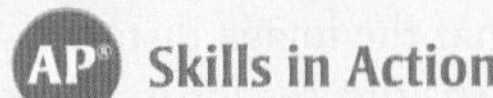

7. You are given a converging lens with focal length f. You place the same object at each of the following object distances. Rank the images that are produced in order of their lateral magnification, from most positive (that is, the largest upright image) to most negative (that is, the largest inverted image). (A) $s_o = 3f$; (B) $s_o = 3f/2$; (C) $s_o = f/2$; (D) $s_o = f/4$.

EX 23-3 8. A thin lens made of glass that has a refractive index equal to 1.60 has surfaces with radii of curvature that have magnitudes equal to 12.0 and 18.0 mm. These surfaces could be either convex or concave. What are the possible values for its focal length? Sketch a cross-sectional view of the lens for each possible combination, making sure to label the radii of curvature of each surface of the lens and the associated focal length of the entire lens.

23-9 A camera and the human eye use different methods to focus on objects at various distances

We can now apply the ideas of the last two sections to two important optical devices: a camera and the human eye. Both of these devices use refraction to form a real, inverted image of an object. In a digital camera (including the camera in a smartphone) the image is formed on a light-sensitive sensor and stored electronically; in the human eye the image is formed on the light-sensitive retina and sent via the optic nerve to the brain. As we will see, however, there are essential differences in how a camera and the human eye form images and how they focus on objects at different distances.

The Camera

Figure 23-36a shows a simplified cross section of a typical camera. Unlike the thin lenses we discussed in Sections 23-7 and 23-8, the lens of a camera can be relatively thick and made up of several individual pieces or *elements*. (Figure 23-36a shows a two-element lens; many smartphone cameras have five or more elements, and lenses used by professional photographers may have more than 15 elements.) The elements typically have different shapes and are made of different kinds of glass. The shapes are chosen to minimize spherical aberration (see Section 23-7) and to provide a sharp image at all points on the light-sensitive sensor.

Unfortunately, glass suffers from dispersion (Section 22-5): The index of refraction is slightly different for light of different wavelengths. The focal length of a lens depends on the index of refraction (Equation 23-24), so if all of the elements of a camera lens were made of the same kind of glass, light of different wavelengths would be brought to a focus at different distances behind the lens. As a result, if the yellow colors of an object formed a sharp image on the sensor,

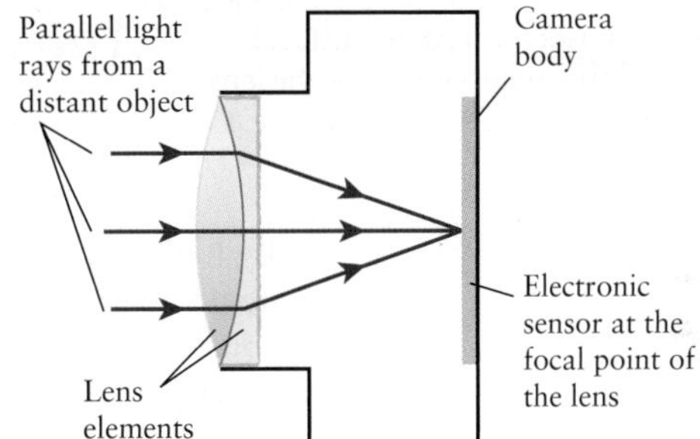

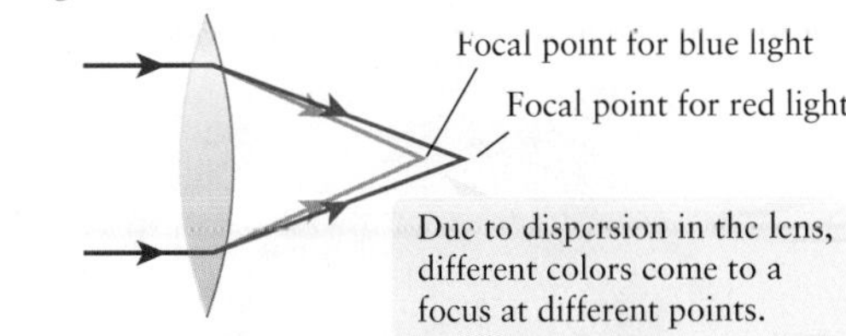

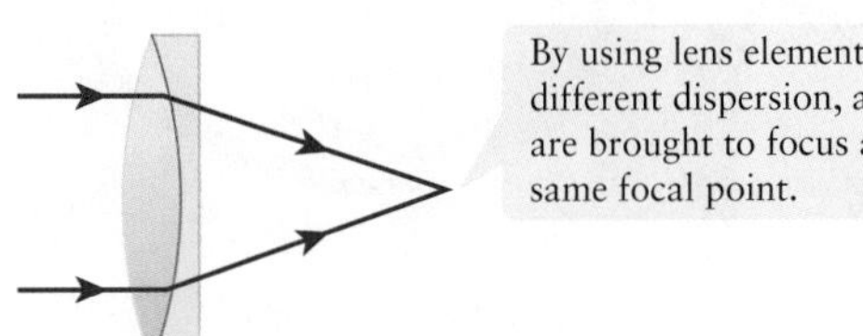

(c) Correcting chromatic aberration.
By using lens elements with different dispersion, all colors are brought to focus at the same focal point.

Figure 23-36 A camera and its lens (a) The lens of a camera is composed of two or more elements. (b) A single-element lens produces chromatic aberration. (c) A lens with more than one element of different indices of refraction can largely eliminate chromatic aberration.

(a) A camera focused on a distant object

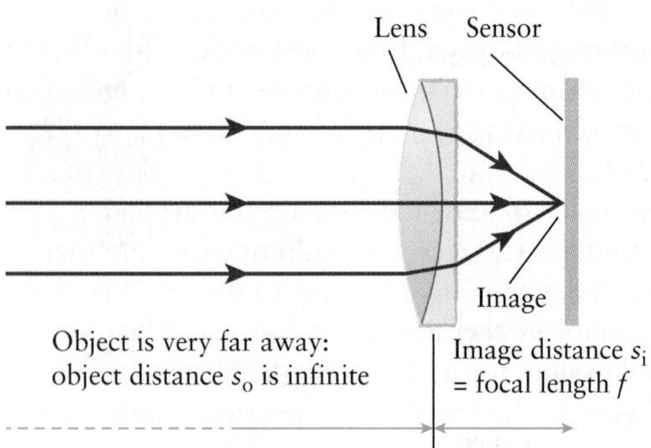

(b) The same camera focused on a nearby object

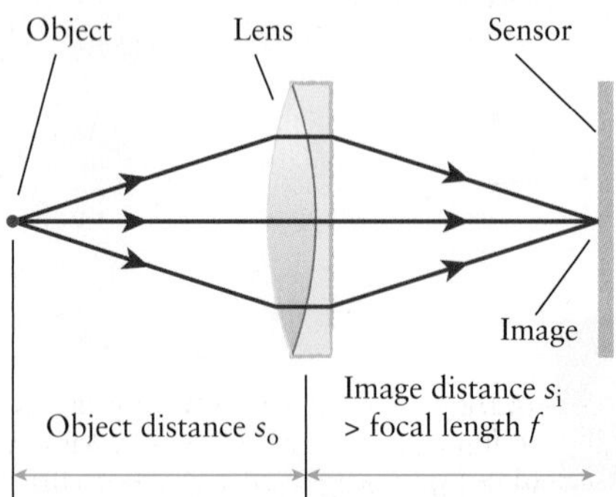

Figure 23-37 Focusing a camera (a) To produce a sharp image of a distant object on the sensor, the distance from lens to sensor must equal the focal length of the lens. (b) If the object is nearby, the lens must be moved away from the sensor to produce a sharp image.

the blue and red colors would be slightly blurred (**Figure 23-36b**). To avoid such *chromatic aberration*, lens designers use different kinds of glass for different lens elements. These kinds of glass are chosen so that the different dispersion of the individual elements largely cancel out. The result is that light of all colors comes to the same focus (**Figure 23-36c**).

The lens equation, Equation 23-11, tells us that for a camera lens of a given focal length f, changing the distance s_o from lens to object will change the distance s_i from lens to image:

$$\frac{1}{s_o} + \frac{1}{s_i} = \frac{1}{f} \tag{23-11}$$

If the object is very distant, s_o is very large and $1/s_o$ is nearly zero. Then Equation 23-11 becomes $1/s_i = 1/f$, which tells us that the image distance s_i equals the focal length f. Hence the camera's light-sensitive sensor is placed a distance f behind the lens so that distant objects will be in sharp focus (**Figure 23-37a**).

As the object approaches the lens, the object distance s_o decreases, so the image distance s_i increases. Hence the sensor must be farther from the lens to record a sharp image. In most professional cameras *focusing* the camera on a nearby object is accomplished by moving the lens farther away from the sensor (**Figure 23-37b**). On some digital cameras you can actually see the lens move outward from the camera body to focus on a nearby object.

In the camera on a smartphone, a different method is used to focus on nearby objects. The front element of the lens remains fixed, but one or more of the rear elements move to decrease the effective focal length of the elements acting in combination.

The Human Eye

Like a camera lens, the human eye is composed of different elements with different indices of refraction (**Figure 23-38**). The *cornea* is the front surface of the eye and has an index of refraction of $n = 1.376$, which is substantially higher than that of the air outside the eye ($n = 1.000$). Behind the cornea is the *aqueous humor*, a fluid that is 98% water and whose index of refraction ($n = 1.336$) is consequently very close to that of pure water ($n = 1.330$). Because the indices of refraction of the cornea and aqueous humor have fairly similar values, most of the refraction caused by the cornea takes place at its front surface rather than its rear surface.

Figure 23-38 The human eye The elements of the human eye work together to bring parallel light rays from a distant object to a focus on the light-sensitive retina.

At the rear of the aqueous humor is the *iris*. This is a diaphragm that opens and closes to regulate the amount of light falling on the sensitive retina. After passing through the aperture of the iris, light rays undergo further refraction when they enter and leave the crystalline *lens*. This is a converging lens about 9 mm in diameter and about 4 mm thick. It is made up of transparent cells called lens fibers arranged in concentric layers rather like the layers of an onion.

Unlike the lenses we studied in Sections 23-7 and 23-8, the index of refraction of the lens is *not* the same throughout its volume: Its value varies from about $n = 1.406$ at the center of the lens to about $n = 1.386$ near its outer rim. This actually enhances the ability of the lens to make light rays converge, and helps to minimize spherical aberration.

The index of refraction of the lens is only slightly greater than that of the aqueous humor in front of the lens ($n = 1.336$) or that of the *vitreous humor* ($n = 1.337$), the transparent gel that lies behind the lens and fills most of the volume of the eye. Hence the lens causes less convergence of light rays than does the cornea. Roughly 70% of the refraction needed to bend parallel light rays coming from a distant object and focus them on the retina comes from the cornea; the remaining 30% comes from the refraction provided by the lens (**Figure 23-39a**).

Unlike in a camera lens, the human eye does not focus on nearby objects by moving its elements forward or back: The distance from the cornea to the center of the lens remains the same, as does the distance from the center of the lens to the retina. Instead, the lens (which is made of flexible biological material) changes shape! To focus on a close object, the *ciliary muscles* that hold the lens in place squeeze the lens around its rim. This causes the lens to deform, as shown in **Figure 23-39b**. The front and back surfaces of the lens become more sharply curved, so their radii of curvature are reduced in magnitude. Hence the focal length f of the lens decreases, the power $P = 1/f$ of the lens increases (see Equation 23-25), and light rays are bent more sharply as they traverse the lens. Hence an image of the nearby object can be brought to a focus on the retina.

(a) An eye focused on a distant object

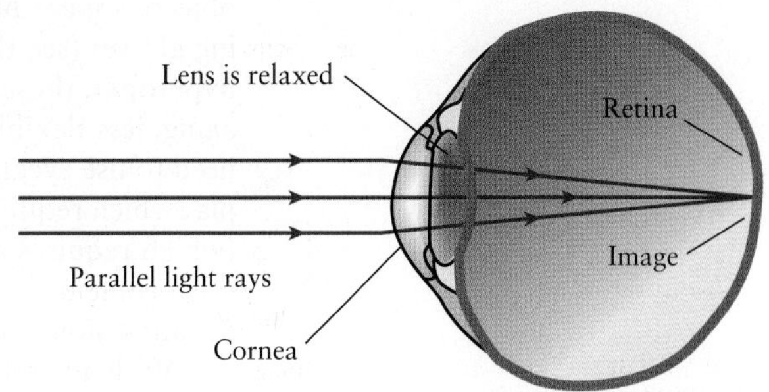

(b) The same eye focused on a nearby object

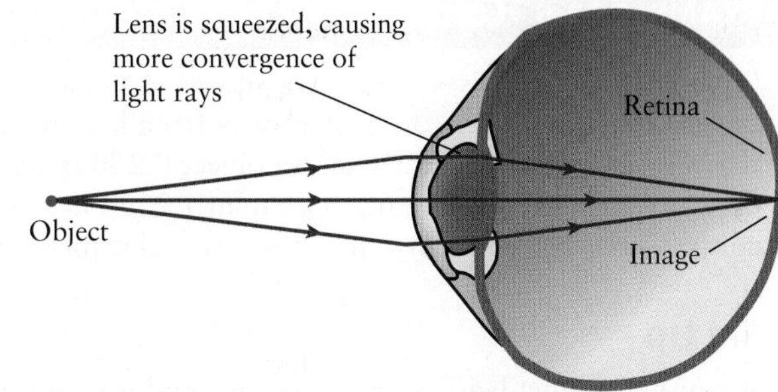

Figure 23-39 Focusing the human eye (a) When the lens is relaxed, parallel light rays from a distant object come to a focus on the light-sensitive retina. (b) To make a sharp image on the retina of a nearby object, the lens must change shape.

In many people the focusing mechanism shown in Figure 23-39 does not work perfectly. One common example is the condition known as *hyperopia* or *farsightedness*, in which the cornea and lens provide too little bending of light rays. As a result, parallel light rays that enter the eye have not yet come to a focus when they reach the retina, so objects appear blurred (**Figure 23-40a**). The lenses of a person with hyperopia must be squeezed, as in Figure 23-39b, to focus both on distant objects and on nearby ones, which causes eye strain and headaches. Even with the lenses squeezed as much as possible, the nearest object that can be seen in sharp focus can still be quite far away. Hyperopia can occur if the eye is too short from front to back, if the cornea has the wrong shape, or if the ciliary muscles are too weak. To compensate for the inadequate convergence of light rays provided by a hyperopic eye, the treatment is to wear converging (convex) eyeglasses or contact lenses.

More common than hyperopia is *myopia* or *nearsightedness*. In this condition the cornea and lens actually do too good of a job of bending light rays and making them converge. As a result, parallel light rays from a distant object come to a focus at a point in front of the retina (**Figure 23-40b**). Nearby objects do appear in focus, however. Myopia can occur if the eye is too long from front to back, or if the cornea or lens has an incorrect shape. The treatment is to wear diverging (concave) eyeglasses or contact lenses. Then parallel light rays coming from a distant object will be diverging when they enter the cornea, and this divergence compensates for the excess convergence caused by the cornea and lens.

Even if you have eyes without hyperopia or myopia, you may need eyeglasses by the time you reach middle age. The reason is that the lenses in our eyes lose some of their flexibility as we age. As a result, the lens is unable to squeeze as in

(a) Hyperopia (farsightedness): Light rays converge too little.

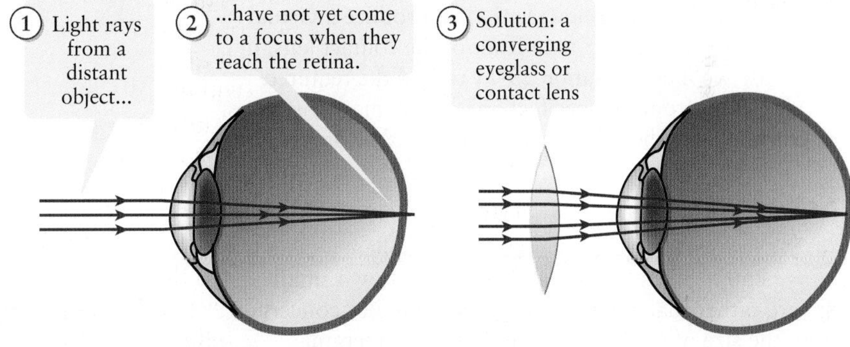

(b) Myopia (nearsightedness): Light rays converge too much.

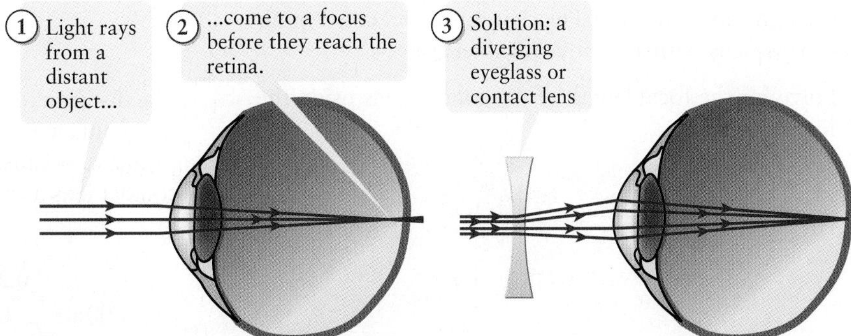

Figure 23-40 Correcting deficiencies of vision (a) Hyperopia is corrected with a converging lens; (b) myopia is corrected with a diverging lens.

Figure 23-39b, so light from nearby objects cannot be focused on the retina and these objects appear blurred. This condition, called *presbyopia*, is treated by wearing reading glasses (see the photograph that opens this chapter) with convex lenses. Just as in hyperopia, these eyeglasses compensate for the inadequate converging power of the aging, less flexible lens. Unlike hyperopia, however, a person with presbyopia may not need to use eyeglasses to see distant objects clearly. Many older people have both myopia (which requires concave, diverging eyeglasses to see distant objects) *and* presbyopia (which requires convex, converging eyeglasses to see nearby objects). The solution to this problem is *bifocal* eyeglasses, which have a concave shape in their upper part for distant vision and a convex shape in their lower part for reading.

With age the lenses of the eyes can also become clouded by cataracts. In this case the lenses are surgically removed and replaced by artificial lenses (see Figure 23-1b).

EXAMPLE 23-6 Correcting for Farsightedness and Nearsightedness

(a) Even with the lenses of her eyes squeezed as much as possible, a certain farsighted person has sharp vision only for objects that are 3.00 m or farther away from her corneas. What should be the focal length and power of a contact lens that will allow her to clearly see an object 0.250 m away from her corneas? (b) A certain nearsighted person can only see objects sharply if they are no more than 1.25 m in front of his corneas. What should be the focal length and power of a contact lens that will allow him to see very distant objects clearly?

Set Up

Because a contact lens sits on the cornea and is very thin, the distance s_o from the contact lens to the object in each case is essentially the same as from the cornea to the object [0.250 m in (a), infinity in (b)]. In part (a) the contact lens must make the eye "think" that an object 0.250 m away is actually 3.00 m away. So this contact lens must produce a virtual image of the object that's 3.00 m in front of the contact lens. Similarly, in part (b) the contact lens must make the eye "think" that a distant object is only 1.25 m away. To do this, the contact lens must produce a virtual image of the object 1.25 m in front of the contact lens. In each case we'll use Equation 23-11 to find the required focal length f from the specified object and image distances. Equation 23-25 then tells us the power of the contact lens.

Lens equation:

$$\frac{1}{s_o} + \frac{1}{s_i} = \frac{1}{f} \quad (23\text{-}11)$$

Lens power:

$$P = \frac{1}{f} \quad (23\text{-}25)$$

Solve

(a) We draw a ray diagram with a small sphere a distance $s_o = 0.250$ m from the contact lens. (We draw a small sphere rather than an arrow because we're not concerned with the size of the image.) The light rays emerging from this contact lens must appear to come from a point 3.00 m in front of the lens, so the contact lens must make the light rays converge. Hence it must be a converging contact lens with a positive focal length.

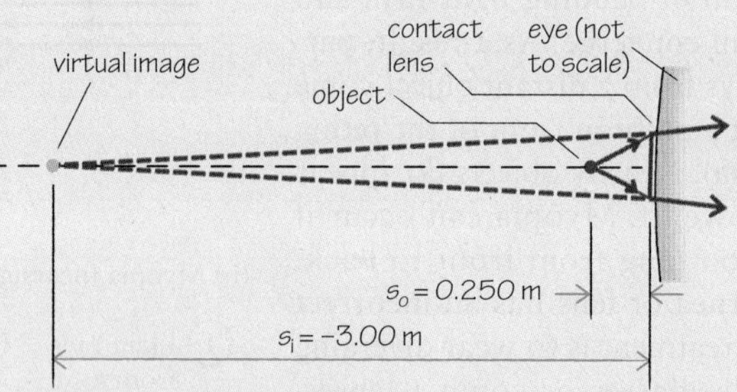

Solve for the focal length of the contact lens using the lens equation.

The distance to the object is $s_o = 0.250$ m, and the distance to the image formed by the contact lens is $s_i = -3.00$ m (negative because the image is on the same side of the contact lens as the object). From Equation 23-11:

$$\frac{1}{f} = \frac{1}{s_o} + \frac{1}{s_i} = \frac{1}{0.250\text{ m}} + \frac{1}{(-3.00\text{ m})}$$

$$= 4.00\text{ m}^{-1} - 0.333\text{ m}^{-1}$$

$$= 3.67\text{ m}^{-1}$$

$$f = \frac{1}{3.67\text{ m}^{-1}} = 0.273\text{ m}$$

The positive focal length means that as predicted, this is a converging contact lens. The power of this contact lens is

$$P = \frac{1}{f} = 3.67\ \text{m}^{-1} = 3.67\ \text{diopters}$$

(b) We again draw a ray diagram, now with parallel light rays from a distant object entering the contact lens. The light rays emerging from the contact lens must appear to come from a point 1.25 m in front of the lens. Because the contact lens makes parallel light rays diverge, it must be a diverging contact lens with a negative focal length.

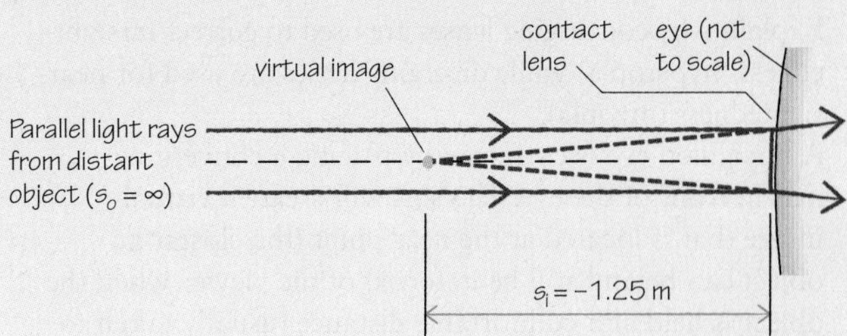

Again, use the lens equation to solve for the focal length of the contact lens.

The distance s_o to the object is infinite (so $1/s_o = 0$), and the image formed by the contact lens is at $s_i = -1.25$ m. (This is negative because the image is on the same side of the contact lens as the object.) From Equation 23-11:

$$\frac{1}{f} = \frac{1}{s_o} + \frac{1}{s_i} = 0 + \frac{1}{(-1.25\ \text{m})}$$

$$\frac{1}{f} = \frac{1}{(-1.25\ \text{m})}$$

Take the reciprocal of both sides:

$$f = -1.25\ \text{m}$$

The negative value means that this is a diverging contact lens, as we expect. The power of this contact lens is

$$P = \frac{1}{f} = \frac{1}{(-1.25\ \text{m})} = -0.800\ \text{m}^{-1} = -0.800\ \text{diopter}$$

Reflect

This example reinforces the idea that you should draw a ray diagram for *any* problem that involves image formation by lenses or mirrors. Note that in part (a) the converging contact lens produces a virtual image. This can happen only if the object is placed inside the focal point of the contact lens, so that the object distance is less than the focal length. This is indeed the case in part (a): The object distance was $s_o = 0.250$ m, and we found that the focal length was $f = 0.273$ m.

NOW WORK Problem 5–7 and 10 from The Takeaway for Section 23-9.

THE TAKEAWAY for Section 23-9

✔ A camera lens refracts parallel light rays from a distant object so that they come to a focus on a light-sensitive sensor. To focus on nearby objects, either the lens is moved away from the sensor, or the focal length of the lens is decreased.

✔ In the human eye, refraction takes place in both the cornea and the lens to focus light rays on the retina. Adjusting the focus from distant to nearby objects is done by reshaping the lens.

✔ Corrective lenses (eyeglasses or contact lenses) are used to either increase or decrease the amount by which light rays are forced to converge within the eye on their way to the retina.

Prep for the AP Exam

AP Building Blocks

1. The image focused on your retina is actually inverted (sketch a simple ray diagram showing this observation). What does this fact say about our definitions of "right side up" and "upside down"?

2. Experimental subjects who wear inverting lenses (glasses that invert all images) for several days adapt to their new perception of the world so well that they can even ride a bicycle. Several days after the glasses are removed, their perceptions return to normal. Discuss this phenomenon and comment.

3. Explain why the lens in a digital camera must move away from the light sensor in order to focus on a nearby object.

AP® Skill Builders

4. Explain why converging lenses are used to correct farsightedness (hyperopia) while diverging lenses are used for nearsightedness (myopia).
5. EX 23-6 A farsighted eye is corrected by placing a converging lens in front of the eye. The lens will create a virtual image that is located at the near point (the closest an object can be and still be in focus) of the viewer when the object is held at a comfortable distance (usually taken to be 25 cm). If a person has a near point of 75 cm, what power reading glasses should be prescribed to treat this hyperopia?
6. EX 23-6 Suppose a given cell phone camera has a single lens and a light sensor that can move to change its distance from the lens. If the camera can focus only on an object 6.5 cm or farther away and the lens has a focal length of 4.3 mm, what are the maximum and minimum distances of the light sensor from the lens?
7. EX 23-6 Andrea, who is nearsighted, wears glasses with lenses that have a power of –1.60 diopters. If the glasses Andrea wears sit 1.40 cm in front of her eyes, what is the farthest distance that objects can be for Andrea to see them clearly without her glasses?

AP® Skills in Action

8. A patient develops cataracts in her myopic eye. As an ophthalmic surgeon, your task is to choose a replacement artificial lens that will also correct for her myopia. To accomplish this, how should the artificial lens compare to the patient's natural lens?
(A) Same shape but a higher index of refraction
(B) Same shape but a lower index of refraction
(C) Same index of refraction but with more sharply curved surfaces
(D) This is not possible.
9. Many older adults can see distant objects clearly but must wear corrective eyeglasses to see nearby objects (for example, to read a book). The lenses of these eyeglasses are
(A) thicker at the middle.
(B) thicker at the edges.
(C) of uniform thickness.
(D) unknown. Not enough information is given to decide.
10. EX 23-6 At a distance of 7.0 cm, a 51 mm by 89 mm business card fills the screen of a cell phone in camera mode. (a) If the focal length of the camera lens is 4.3 mm, what are the length and height of the camera's light sensor? (b) The light sensor is made up of 16×10^6 individual picture elements, or pixels (so this is a 16-megapixel sensor). What is the area of each pixel? If the pixels are square, what is the length of each side of a pixel?

23-10 The concept of angular magnification plays an important role in several optical devices

(a) When an object is far from the eye it subtends a small angle θ.

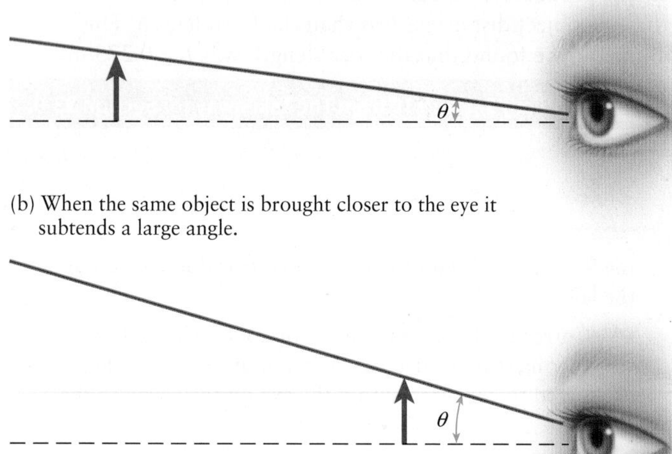

θ
(b) When the same object is brought closer to the eye it subtends a large angle.
θ

Figure 23-41 For a better view, an object should subtend a larger angle Moving an object (a) so that is closer to the eye (b) means that it subtends a larger angle θ.

We've used the idea of *lateral* magnification to compare the *height* of the image made by a mirror or lens to the height of the object. But in many situations the kind of magnification that is most relevant is *angular* magnification. As an example, suppose there's an object you want to inspect that's some distance from your eye (**Figure 23-41a**). To get a better view of this object, you bring it closer to your eye so that it subtends a larger angle θ (**Figure 23-41b**). The greater the value of θ, the more detail you can see. However, some objects (like the wings of a fly or human blood cells) are so small that their details cannot be seen by the naked eye. And some objects (like the Moon and planets) are so distant that it's impractical to bring your eye any closer to them. Let's look at three optical devices that address these limitations: the *magnifying glass*, the *microscope*, and the *telescope*.

The Magnifying Glass

The closest that any object can be brought to a normal, relaxed eye and be in focus is about 25 cm. This is called the *near* point of the eye. (You can see an object at a closer distance, but that requires focusing the eye as in Figure 23-39. This can be tiring.) If the height of the

(a) Without a magnifying lens, the closest an object can be to a relaxed eye and be in focus is 25 cm.

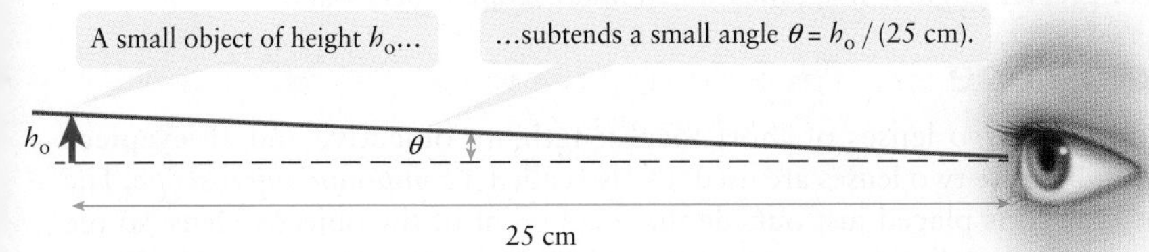
A small object of height h_o...
...subtends a small angle $\theta = h_o/(25 \text{ cm})$.
h_o
θ
25 cm

(b) The same object placed at the focal point F of a magnifying glass (a converging lens of short focal length f)

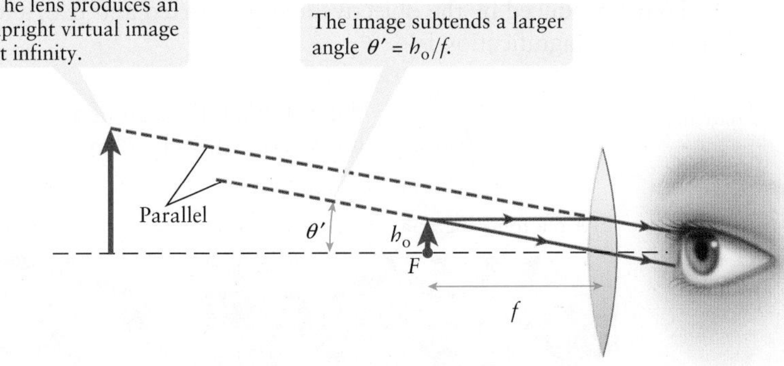
The lens produces an upright virtual image at infinity.
The image subtends a larger angle $\theta' = h_o/f$.
Parallel
θ'
h_o
F
f

(c) The magnified view through a magnifying glass

MRC
M–
M+
ON CE
%
OFF
1
2
3
+
0
=

Paul Michael Hughes/Alamy

Figure 23-42 A magnifying glass (a) An object viewed at the near point of a normal relaxed eye. (b) The same object viewed through a magnifying glass. (c) The image of this object seen through a magnifying glass is upright, is virtual, and subtends a larger angle.

object is h_o, **Figure 23-42a** shows that the angle θ subtended by this object at a 25-cm distance from the eye is given by $\tan\theta = h_o/(25 \text{ cm})$. As we mentioned in Section 23-6, the tangent of a small angle is approximately equal to the angle in radians, so this expression becomes

$$\theta = \frac{h_o}{25 \text{ cm}} \quad \text{(angle subtended by an object at the near point of the eye)} \qquad (23\text{-}27)$$

If the object is small (h_o has a small value), the angle θ is small and it will be difficult to see any detail. The purpose of a **magnifying glass** or *magnifier* is to overcome this limitation. A magnifying glass is simply a converging lens with a focal length f that is shorter than 25 cm. If you place an object at the focal point F on one side of the lens, the light rays emerging from the other side are parallel (**Figure 23-42b**). What you see is an upright virtual image that is infinitely far away (and so appears in focus to the relaxed eye) and that subtends an angle θ', where $\tan\theta' = h_o/f$. Using the same small-angle approximation as in Equation 23-27, we can write the angle subtended by the virtual image as

$$\theta' = \frac{h_o}{f} \quad \text{(angle subtended by the image of an object viewed through a magnifying glass)} \qquad (23\text{-}28)$$

The **angular magnification** M of the magnifying glass is the ratio of the angle θ' (with the magnifying glass, Equation 23-28) to the angle θ (without the magnifying glass, Equation 23-27):

$$M_\theta = \frac{\theta'}{\theta} = \frac{h_o/f}{h_o/(25 \text{ cm})} = \frac{25 \text{ cm}}{f} \quad \text{(angular magnification of a magnifying glass)} \qquad (23\text{-}29)$$

For example, if the magnifying glass has a focal length of 10 cm, the angular magnification is $M_\theta = (25 \text{ cm})/(10 \text{ cm}) = 2.5$. As viewed through the magnifying lens, the object will subtend an angle 2.5 times larger than with the naked eye, and so will appear 2.5 times larger (**Figure 23-42c**).

The smaller the focal length f of the magnifying glass, the greater the angular magnification M_θ given by Equation 23-29. Making f smaller means using more sharply curved lens surfaces, which increases spherical aberration (Section 23-7) and decreases image quality. In practice the largest value of M_θ obtainable with a magnifying glass is about 15. For greater angular magnification, a *microscope* is necessary.

The Microscope

A **microscope** uses *two* lenses of short focal length, an **objective** and an **eyepiece** (**Figure 23-43a**). Because two lenses are used it's also called a *compound microscope*. The object to be viewed is placed just outside the focal point of the objective lens, so the object distance s_o is just slightly larger than the objective focal length $f_{objective}$. The image distance s_i is therefore much larger than either s_o or $f_{objective}$, as **Figure 23-43b** shows. (You can also see this from Equation 23-11, the lens equation: $1/s_o + 1/s_i = 1/f_{objective}$, which we can rewrite as $1/s_i = 1/f_{objective} - 1/s_o$. Because s_o is almost the same as $f_{objective}$, the difference $1/f_{objective} - 1/s_o$ is very small. Hence $1/s_i$ is very small and its reciprocal s_i is large.) The image produced by the objective lens is inverted and real, and from Equation 23-14 the *lateral* magnification is

$$M_{objective} = \frac{h_i}{h_o} = -\frac{s_i}{s_o} \approx -\frac{s_i}{f_{objective}} \quad (23\text{-}30)$$

(lateral magnification of a microscope objective lens)

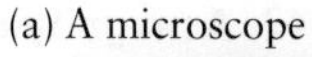
(a) A microscope

(c) A photomicrograph

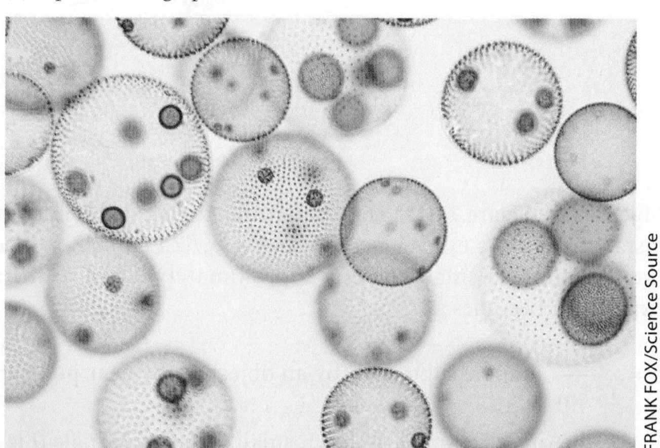

(b) How a microscope works

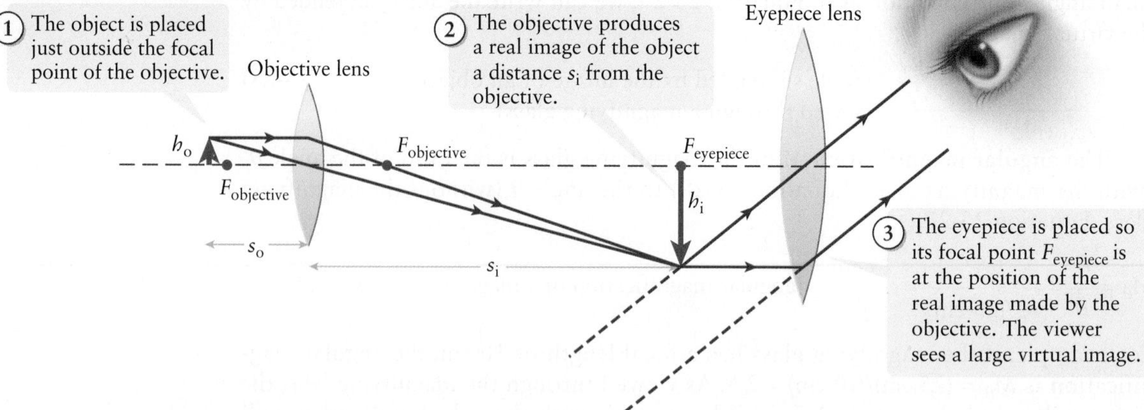

1 The object is placed just outside the focal point of the objective.
Objective lens
2 The objective produces a real image of the object a distance s_i from the objective.
Eyepiece lens
h_o
$F_{objective}$
$F_{objective}$
$F_{eyepiece}$
h_i
s_o
s_i
3 The eyepiece is placed so its focal point $F_{eyepiece}$ is at the position of the real image made by the objective. The viewer sees a large virtual image.

Figure 23-43 A microscope (a) A microscope uses two lenses, an objective and an eyepiece. (b) How a microscope forms an image. (c) To record this microscope image on a light sensor, the eyepiece was not used; instead the sensor was placed at the position of the real image made by the objective shown in (b). The algal colonies shown here (*Volvox*) are about 0.35 mm in diameter and are just visible to the unaided eye; the microscope reveals their detailed structure.

On the far right-hand side of Equation 23-30 we've used the idea that the object distance s_o is almost the same as the objective focal length $f_{objective}$. A typical objective lens for a microscope such as you might use in a biology lab course has a focal length of from 2 to 40 mm.

The second lens, or eyepiece, is just a magnifying glass. The idea is that the real *image* made by the objective lens acts as the *object* for the eyepiece. Just as in Figure 23-42c, we place the eyepiece so that its focal point $F_{eyepiece}$ is at the position of the image formed by the objective. The viewer then sees an enlarged virtual image at infinity. Because the objective produces an inverted image and the eyepiece does not further invert the image, the viewer sees an image that is inverted relative to the object.

The net *angular* magnification of the image delivered to the viewer is the product of the lateral magnification $M_{objective}$ from the objective, Equation 23-30, and the angular magnification $M_{\theta,eyepiece}$ from the eyepiece, Equation 23-29. It's customary to always treat angular magnification as positive, so we use the absolute value of $M_{objective}$:

$$M_{microscope} = |M_{objective}|\,M_{\theta,eyepiece} = \left(\frac{s_i}{f_{objective}}\right)\left(\frac{25\text{ cm}}{f_{eyepiece}}\right) = \frac{s_i(25\text{ cm})}{f_{objective}f_{eyepiece}} \qquad (23\text{-}31)$$

(angular magnification of a microscope)

Equation 23-31 shows that the smaller the focal lengths of the objective and eyepiece lens, the greater the angular magnification. Even relatively inexpensive microscopes can achieve angular magnifications of 2500, called 2500 ×.

Note that if you want to take a photograph using a microscope, the eyepiece is *not* used. Instead a light sensor like that used in a digital camera is placed at the position of the real image made by the objective lens, so that this image is focused on the sensor. **Figure 23-43c** shows an example of a photograph made in this way, called a *photomicrograph*.

The Telescope

Like a microscope, a **telescope** has an objective lens that makes a real image of an object. It also has an eyepiece lens that increases the angular size of that image for the viewer. The difference is that a telescope is used for looking at large, distant objects rather than small objects close at hand.

Figure 23-44 shows a telescope used for astronomy (an *astronomical telescope*). Because astronomical objects are very far away, the angle θ that they subtend is very small. (The full Moon subtends an angle of only about 0.5° as seen from Earth.) The objective lens of focal length $f_{objective}$ makes a real, inverted image of the object. Because the object is essentially infinitely far away, the light rays from the object that enter

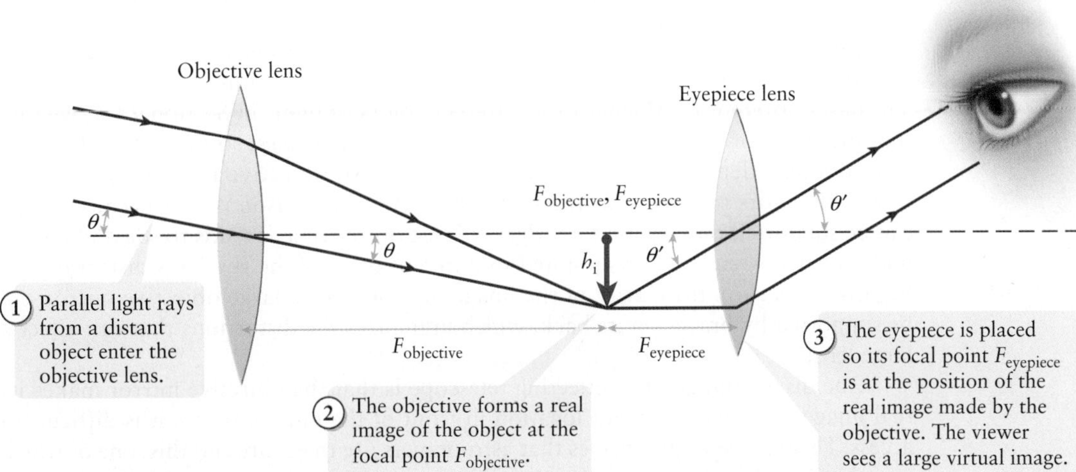
Objective lens
Eyepiece lens
θ
θ
$F_{objective}$, $F_{eyepiece}$
h_i
θ′
θ′
1 Parallel light rays from a distant object enter the objective lens.
$F_{objective}$
$F_{eyepiece}$
2 The objective forms a real image of the object at the focal point $F_{objective}$.
3 The eyepiece is placed so its focal point $F_{eyepiece}$ is at the position of the real image made by the objective. The viewer sees a large virtual image.

Figure 23-44 A refracting telescope In a refracting telescope a converging objective lens makes a real image of a distant object. The eyepiece lens provides a view with an enhanced angular size.

the objective are parallel and the resulting real image is formed at the focal point F, a distance $f_{\text{objective}}$ behind the objective lens. The image has height h_i. As **Figure 23-44a** shows, the angle θ subtended by the distant astronomical object is the same as the angle subtended by the real image as seen from the objective lens. Again using the approximation for small angles that θ in radians is equal to tan θ, we have $\theta = h_i/f_{\text{objective}}$.

To give this image a larger angular size as seen by the viewer, we place the eyepiece lens of focal length f_{eyepiece} so that its focal point is at the same point as the image made by the objective lens. As for the microscope, the image made by the objective lens acts as the object for the eyepiece lens. Figure 23-44 shows that the angle subtended by the final image is $\theta' = h_i/f_{\text{eyepiece}}$. The angular magnification of the telescope as a whole is the ratio of θ' (the angular size of the distant object as seen through the eyepiece) to θ (the angular size of the object as seen by the unaided eye), or

$$M_{\text{telescope}} = \frac{\theta'}{\theta} = \frac{h_i/f_{\text{eyepiece}}}{h_i/f_{\text{objective}}} = \frac{f_{\text{objective}}}{f_{\text{eyepiece}}} \quad \text{(angular magnification of a telescope)} \tag{23-32}$$

Equation 23-32 shows that you can increase the magnification of the telescope by using an objective lens with a *greater* focal length, an eyepiece lens with a *smaller* focal length, or both. A typical telescope used by beginning amateur astronomers can give angular magnifications from about 35 × to about 100 ×. Note that just as for a microscope, if you want to take a photograph using a telescope, you remove the eyepiece and place a light sensor at the point F where the objective makes the real image.

WATCH OUT !

Magnification is not the most important aspect of an astronomical telescope.

It's a common misconception that the primary purpose of a telescope is to magnify images. But in fact magnification is not the most important aspect of a telescope. The reason is that there are limits to how sharp any astronomical image can be. One limit is due to the blurring caused by Earth's atmosphere: As light rays from space travel through our moving atmosphere, they are randomly refracted as they pass through one air mass to the next. This causes stars to twinkle as seen by the naked eye, and it also causes astronomical images to blur and "wobble" as seen through a telescope. A second limit is due to diffraction, which as we saw in Section 23-9 poses fundamental limits on the angular resolution of an image. Magnifying an image that's blurred by the atmosphere, by diffraction, or a combination may make it look bigger, but will not make it any clearer. Thus, beyond a certain point, there's nothing to be gained by further magnification. Astronomers actually place more priority on the *light-gathering power* of a telescope, which is a measure of how much light the telescope can collect from a distant object. This is important because astronomical objects are generally quite dim. For this reason telescopes have large-diameter objectives to gather as much light as possible. This means brighter images, which makes it easier to see faint details.

The telescope shown in Figure 23-44 is called a **refracting telescope** because it uses an objective *lens*. Almost all telescopes used in astronomy research are **reflecting telescopes** that use a converging *mirror* for an objective instead of a converging lens (**Figure 23-45a**). One important reason is chromatic aberration: A simple objective lens has a different focal length for different colors of light. In Section 22-5, we saw that the speed of light was different for different frequencies, so the index of refraction of a lens depends on the color of the light. An objective mirror, by contrast, brings all colors to a focus at the same point. Another reason is that because a lens can be supported only around its edges, a large objective lens tends to sag and distort under its own weight as it is pointed to different parts of the sky. This distortion has negative effects on the clarity of the image. By contrast, a large objective mirror can be supported by braces on its back, which minimizes the distortions that can affect a large lens.

One disadvantage of a reflecting telescope is that the objective mirror makes its real image at an inconvenient location in front of the mirror, where it is difficult to access. There are several schemes that astronomers use to get around this, one of which was developed by Isaac Newton. In a *Newtonian telescope* (**Figure 23-45b**) a small, flat mirror is placed at a 45° angle in front of the focal point. This secondary mirror

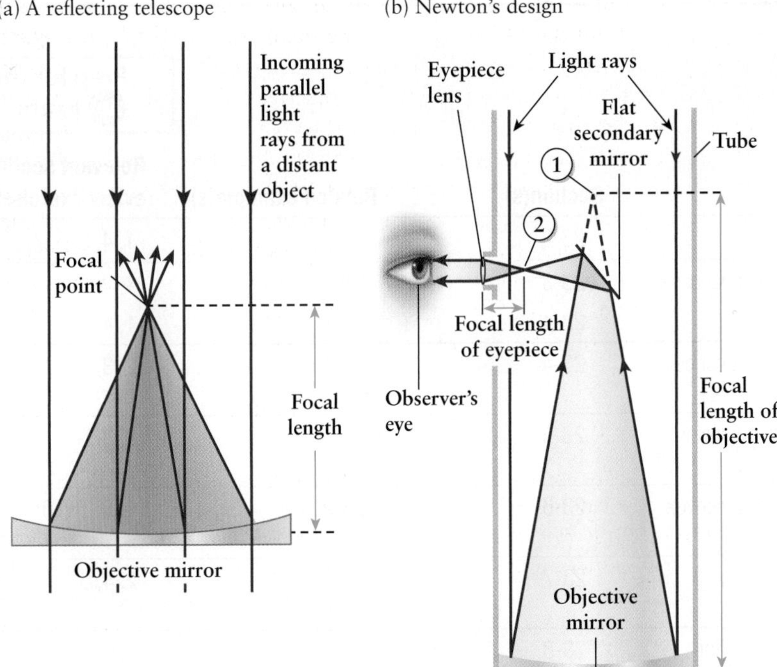

Figure 23-45 A reflecting telescope (a) In a reflecting telescope a concave mirror is used as the objective. (b) Newton's seventeenth-century design (still used in telescopes today) uses a flat 45° mirror to move the image made by the objective mirror from point 1 to point 2, where it can be viewed through the eyepiece.

deflects the light rays and the real image to one side, and the eyepiece is mounted on the side of the telescope. The overall magnification is given by Equation 23-32, the same as for a refracting telescope.

Just as for a microscope, an astronomical telescope produces a final image that is inverted. This is fine for astronomy, but can be distracting if you are using the telescope to look at objects on Earth like a bird in a tree or a distant mountain. A *terrestrial telescope* has an additional lens or prisms between the objective and eyepiece that inverts the image again, giving a final image that is upright. A pair of binoculars is essentially two terrestrial telescopes side by side.

THE TAKEAWAY for Section 23-10

✔ Angular magnification refers to the change in the angle subtended by an object as seen by the viewer caused by an optical system.

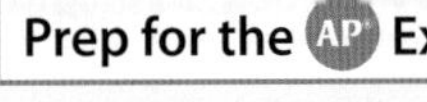

1. You want to use a lens of focal length 9.00 cm that you just happen to have around to examine the hairy details of your favorite pet caterpillar. With the lens close to your eye and the animal at the lens's focal point, what angular magnification do you achieve? Assume that your near point is at 25.0 cm.
2. To view the craters of the Moon, you construct a refracting telescope using a 95.0-cm focal length lens as its objective and a 12.5-cm focal length lens as its eyepiece.
   (a) Determine the angular magnification of your telescope when you look at the Moon.
   (b) Is the image you see upright or inverted with respect to the object?

AP Skill Builders

3. A certain microscope and a certain refracting telescope both advertise an angular magnification of 50 ×. For which of these two devices does the objective lens produce a real image with the greater absolute value of *lateral* magnification?
   (A) The microscope
   (B) The telescope
   (C) The lateral magnification is the same for both devices.
   (D) Not enough information is given to decide.

WHAT DID YOU LEARN?

Prep for the AP® Exam

Chapter learning goals	Section(s)	Related example(s)	Relevant section review exercise(s)
Describe how a plane mirror forms an image.	23-2		1, 4, 5, 6, 7, 9
Use ray diagrams to explain how the image formed by a concave mirror depends on the position of the object.	23-3	23-1	1, 2
	23-4		1, 4, 8
Calculate the position and height of an image made by a concave mirror.	23-4	23-1	2, 3, 5, 9
Explain the differences between the images made by a convex mirror and a concave mirror.	23-5		2, 3, 5
Calculate the position and height of an image made by a convex mirror.	23-6	23-2	1–5, 7–10
Describe how the curved surfaces of a lens make light rays converge or diverge.	23-7		2, 4, 6
Calculate the focal length of a lens based on its composition and shape and the position and magnification of an image made by a lens.	23-8	23-3	3, 8
		23-4	5
		23-5	1, 2, 4, 6
Explain how the eye forms images on the retina, and find the needed focal length for corrective lenses used to compensate for deficiencies in vision.	23-9	23-6	1, 4, 5, 7–9
Calculate the angular magnification of a magnifying glass, a microscope, and a telescope.	23-10		1, 2, 3

Chapter 23 Review

Key Terms

All the Key Terms can be found in the Glossary/Glosario on page G1 in the back of the book.

Chapter Summary

Topic	Equation or Figure	
Plane mirrors: A plane mirror makes an upright, virtual image of any object. The image is the same size as the object, so the lateral magnification is $m = 1$. The image is reversed back to front, not side to side.	The **image distance** is negative. The **object distance** is positive. $$s_i = -\lvert s_i \rvert = -s_o$$ The negative value of s_i indicates that the image is on the opposite side of the mirror from the object.	(23-1)
	Lateral magnification; Image height; Object height $$M = \frac{h_i}{h_o}$$	(23-2)
Spherical mirrors: If we consider only parallel light rays that are close to the principal axis of a spherical concave mirror, the reflected rays all converge at the focal point. The distance from the center of the mirror to the focal point is the focal length. If the mirror is convex rather than concave, parallel light rays diverge rather than converge after reflection. The radius of curvature and the focal length are both negative for a convex mirror.	**Focal length** of a spherical mirror; **Radius of curvature** of the mirror $$f = \frac{r}{2}$$	(23-3)
Image formation by a concave mirror: We can locate the image made by a concave mirror by drawing a ray diagram. If the object is outside the focal point, the image is real and inverted; its size depends on how far the object is from the mirror. If the object is inside the focal point, the image is virtual, upright, and larger than the object.	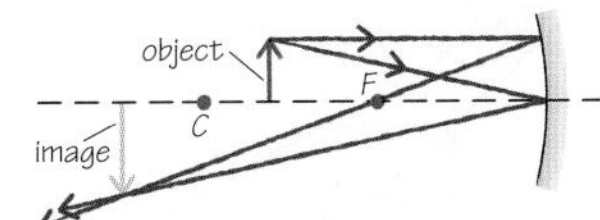	(Example 23-1, figure 1)
	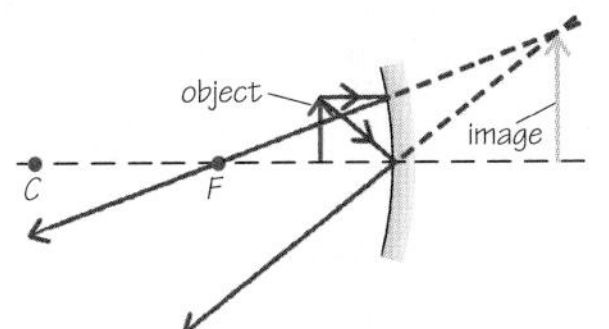	(Example 23-1, figure 2)
Image formation by a convex mirror: A ray diagram also helps us locate the image made by a convex mirror. No matter where the object is placed, the image is virtual, upright, and smaller than the object.	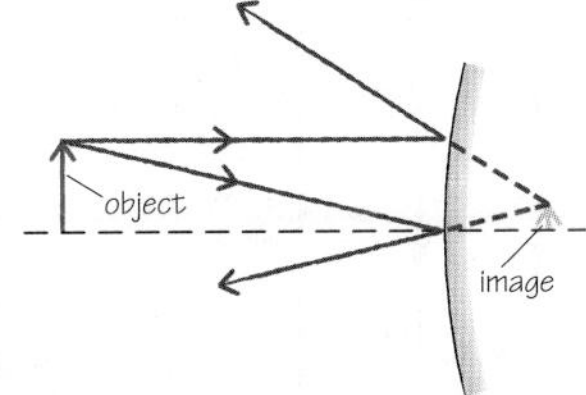	(Example 23-2, figure 1)
The mirror equation and lens equation: The mirror equation and lens equation relate the object distance, image distance, and focal length for either a concave or convex mirror or a converging or diverging lens. The lateral magnification depends on the image, and object distances and can be positive (if the image is upright) or negative (if the image is inverted).	$$\frac{1}{s_o} + \frac{1}{s_i} = \frac{1}{f}$$ Focal length; Object distance; Image distance	(23-11)
	Lateral magnification; Image height; Image distance $$M = \frac{h_i}{h_o} = -\frac{s_i}{s_o}$$ Object height; Object distance	(23-14)

Lenses: Due to refraction, light rays converge as they enter or exit a converging glass lens and diverge as they enter or exit a diverging glass lens. A converging lens behaves similarly to a concave mirror, and a diverging lens behaves similarly to a convex mirror. The focal length of a lens is given by the lensmaker's equation, which involves the index of refraction of the lens material and the radii of curvature of the lens surfaces. This equation is valid if the lens is thin.

These are converging lenses: Parallel light rays that enter either of these lenses will exit the lens converging toward each other.

These are diverging lenses: Parallel light rays that enter either of these lenses will exit the lens diverging away from each other.

Convex Plano-convex Concave Plano-concave

(Figure 23-28)

Focal length of a lens surrounded by air

Index of refraction of the lens material

$$\frac{1}{f} = (n-1)\left(\frac{1}{R_1} - \frac{1}{R_2}\right) \quad (23\text{-}24)$$

Radius of curvature of lens surface 1 (the surface closer to the object)

Radius of curvature of lens surface 2 (the surface farther from the object)

Image formation by lenses: A converging lens produces a real, inverted image if the object is outside the focal point and a virtual, upright, enlarged image if the object is inside the focal point. A diverging lens always produces a virtual, upright, reduced image. The same equation that relates object distance s_o, image distance s_i, and focal length f for a spherical mirror also applies to thin lenses, as does the equation for lateral magnification.

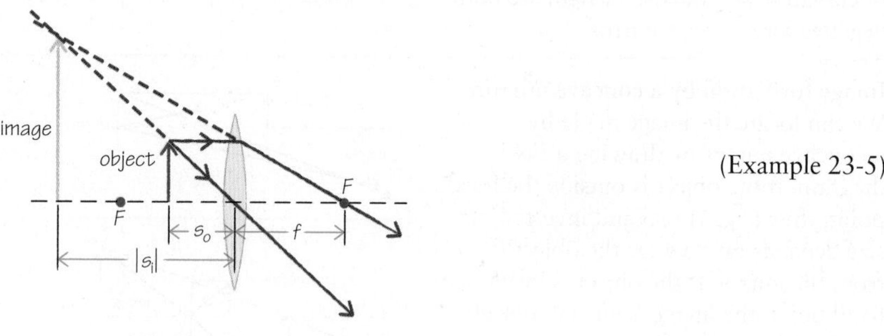

(Example 23-5)

The magnifying glass, microscope, and telescope: A magnifying glass is used for viewing nearby objects. It produces a virtual image that has an angular magnification compared to the object. The objective of a microscope makes a real image of a nearby object, and the objective of a telescope makes a real image of a distant object; in both devices a magnifying lens is used as an eyepiece to provide angular magnification.

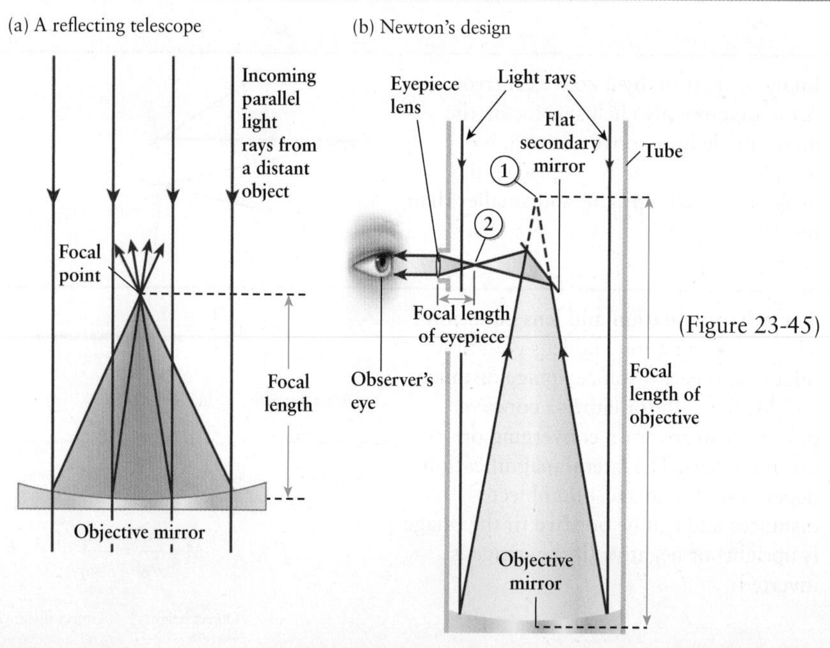

(Figure 23-45)

Chapter 23 Review Problems

Note: In these problems *infinity* means a very large distance compared to the focal length of a lens.

1. A laptop computer is connected to a video projector that projects an image on a screen. If the lens of the projector is half covered, what happens to the image? Explain your answer.
2. Discuss why nearsightedness is not found in all people, but virtually everyone eventually has difficulty focusing on nearby objects as they age.
3. When you view a car's side mirror, you see a smaller image than you would if the mirror were flat. Is the mirror concave or convex? Explain your answer.
4. Explain why looking through a small opening often provides visual acuity even to an extremely nearsighted person.
5. If the angle of incidence on a flat mirror is 0°, what is the angle of reflection?
6. Using a ruler, a protractor, and the law of reflection, show the location of the image of your face when you stand a short distance in front of a plane mirror, as shown in the figure.

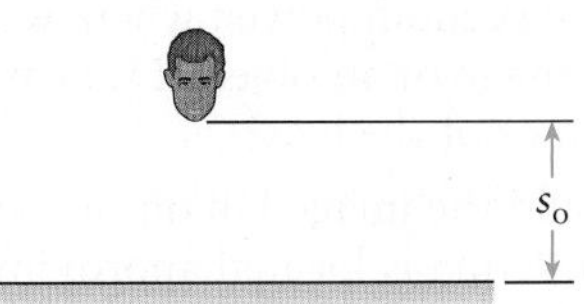

7. At which of the points A through E will the image of the face be visible in the plane mirror of length L as shown in the figure? The distance x equals $L/2$. Points A through E are collinear and separated by a distance of $3L/4$, and point C lies on a line bisecting the mirror. Assume the face lies directly on point C.

A B C D E

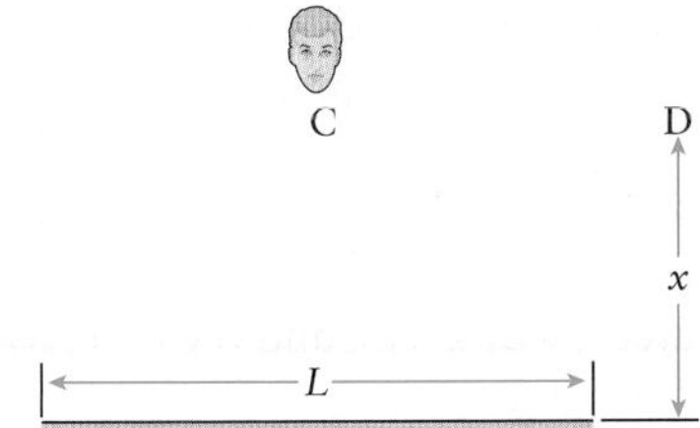

8. One person is looking in a plane mirror at the image of a second person as shown in the figure. Using a ruler, a protractor, and the law of reflection, show the location of the image as seen by the first person.

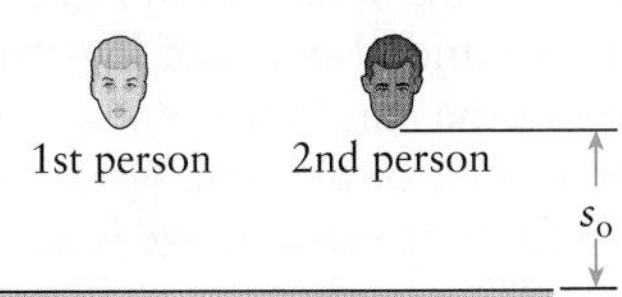

9. You hold an autofocusing camera 2 m in front of a plane mirror and take a picture of the reflection. You discover that the picture of the reflection is out of focus. An autofocus camera determines the distance to the object by measuring the time it takes an infrared pulse to leave the camera, bounce off the subject, and return to the camera. Knowing this, explain what must have happened.
10. Determine the focal length for an unknown lens with the following object and image distances:

Object distance (cm)	Image distance (cm)	Object distance (cm)	Image distance (cm)
30	98	60	37
35	67	65	35
40	53	70	34
45	47	75	33
50	42	80	32
55	38		

11. A square plane mirror of side length s hangs on a wall such that its bottom edge is a height h above the floor. The wall opposite the mirror is a distance d away. Marco, whose eyes are at the exact height of the center of the mirror, stands directly in front of the mirror at the center of its width, as shown in the figure. Derive an expression for the maximum distance x that Marco can stand from the mirror and still see the reflection of the bottom of the wall behind him. Assume that $s < 2h$.

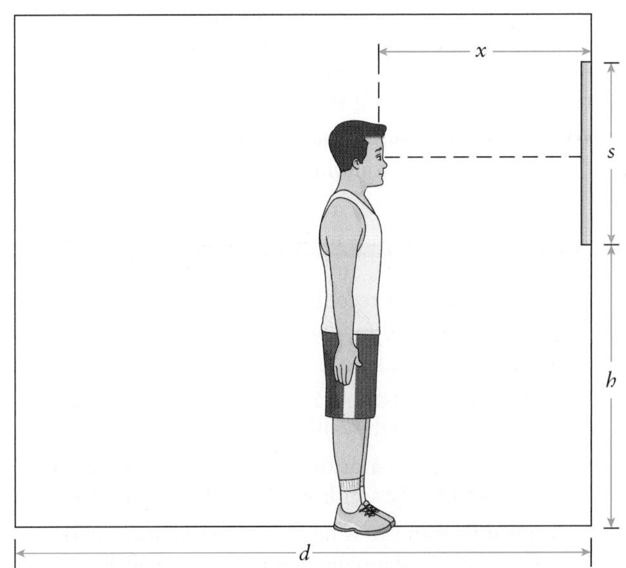

12. You are designing lenses that consist of small double convex pieces of plastic having surfaces with radii

of curvature of magnitudes 3.50 cm on one side and 4.25 cm on the other side. You want the lenses to have a focal length of 1.65 cm in air. What should be the index of refraction of the plastic to achieve the desired focal length?

13. You desire to observe details of the *Statue of Freedom*, the sculpture by Thomas Crawford that is the crowning feature of the dome of the U.S. Capitol in Washington, D.C. For this purpose, you construct a refracting telescope, using as its objective a lens with a focal length of 90.0 cm. To achieve an angular magnification of magnitude 6.0, what focal length should the eyepiece have?

14. A certain lens element inside a camera is a converging lens that has symmetric convex sides ($R_1 = 4.300$ cm and $R_2 = -4.300$ cm) and can be modeled as a thin lens. The lens's index of refraction for blue light is $n_{blue} = 1.588$ and for red light is $n_{red} = 1.582$. Calculate the magnitude of the separation distance of the lens's focal lengths for blue and red light, $|f_{blue} - f_{red}|$.

15. A compound microscope has a barrel length of 160.0 mm and an objective with a 4.500-mm focal length. The total angular magnification of the microscope is –400.0. Find the angular magnification of the eyepiece.

16. A refracting astronomical telescope, or refractor, consists of an eyepiece lens at one end of a cylindrical tube and an objective lens at the other end. The objective lens gathers light from a distant object (such as a planet) and focuses it at the focal point of the eyepiece lens. The eyepiece basically acts as a magnifying lens to create a virtual image of the objective's image. The overall magnification, M, is found to be $M = -f_o/f_e$, where f_o is the focal length of the objective and f_e is the focal length of the eyepiece.
    (a) Calculate the magnification of the 36-in. refractor at the University of California's Lick Observatory on Mount Hamilton near San Jose, California. The focal length of the objective lens is 17.37 m, and the focal length of the eyepiece is 22 mm.
    (b) What is the significance of the negative sign in the magnification equation?

17. The opposite walls of a barber shop are covered by plane mirrors, so that multiple images arise from multiple reflections, and you see many reflected images of yourself receding to infinity. The width of the shop is 6.50 m, and you are standing 2.00 m from the north wall.
    (a) How far apart are the first two images of you behind the north wall?
    (b) What is the separation of the first two images of you behind the south wall? Explain your answer.

18. A typical human eye is nearly spherical and usually about 2.5 cm in diameter. Suppose a person first looks at a coin that is 2.3 cm in diameter, located 30.0 cm from her eye, and then looks up at her friend who is 1.8 m tall and 3.25 m away.
    (a) Find the approximate size of each image (coin and friend) on her retina. (*Hint:* Just consider rays from the top and bottom of the object that pass through the center of the lens.)
    (b) Are the images in part (a) upright or inverted, and are they real or virtual?

19. A geneticist looks through a microscope to determine the phenotype of a fruit fly. The microscope is set to an overall magnification of 400 × with an objective that has a focal length of 0.60 cm. The distance between the eyepiece and objective is 16 cm. Find the focal length of the eyepiece lens assuming a near point of 25 cm (the closest an object can be to the eye and still be seen in focus).

20. A typical human lens has an index of refraction of 1.41. The lens has a double convex shape, but its curvature can be varied by the ciliary muscles acting around its rim. At minimum power, the radius of the front of the lens is 10.0 mm, while that of the back is 6.00 mm. At maximum power the radii are 6.00 and 5.50 mm, respectively. (The numbers can vary somewhat.) The lens is in air.
    (a) Find the ranges of the lens' focal length and its power (in diopters).
    (b) At maximum power, where would the lens form an image of an object 25.0 cm in front of the front surface of the lens?
    (c) Would the image fall on the retina of a human eye? The retina is located approximately 2.5 cm from the lens.

21. A typical person's eye is 2.5 cm in diameter and has a near point (the closest an object can be and still be seen in focus) of 25 cm, and a far point (the farthest an object can be and still be in focus) of infinity.
    (a) What is the range of the effective focal lengths of the focusing mechanism (lens plus cornea) of the typical eye?
    (b) Is the equivalent focusing mechanism of the eye a diverging or a converging lens? Justify your answer without using any mathematics, and then see if your answer is consistent with your result in part (a).

22. The eyepiece and objective in a compound microscope are separated by 20.0 cm, and the objective has a focal length of 8.0 mm.
    (a) If it is to have a magnifying power of 200 ×, what should be the focal length of the eyepiece?
    (b) If the final image is viewed at infinity, how far from the objective should the object be placed?

23. When you place a bright light source 36.0 cm to the left of a lens, you obtain an upright image 14.0 cm from the lens and also a faint inverted image 13.8 cm to the left of the lens that is due to reflection from the front surface of the lens. When the lens is turned around, a faint inverted image is 25.7 cm to the left of the lens. What is the index of refraction of the material?

24. You may have noticed that the eyes of cats appear to glow green in low light. This effect is due to the reflection of light by the *tapetum lucidum*, a highly reflective membrane just behind the retina of the eye (see Figure 22-2a). Light that has passed through the retina without hitting photoreceptors is reflected back to the retina, thus enabling the animal to see much better than humans in low light. The eye of a typical cat is about 1.25 cm in diameter. Assume that the light enters the eye traveling parallel to the principal axis of the lens.
   (a) If some of the light reflected off the *tapetum lucidum* escapes being absorbed by the retina, where will it be focused?
   (b) The refractive index of the liquid in the eye is about 1.4. How does this affect the location of the image in part (a)?

25. A nearsighted eye is corrected by placing a diverging lens in front of the eye. The lens will create a virtual image of a distant object at the far point (the farthest an object can be and still be in focus) of the myopic viewer where it will be clearly seen. In the traditional treatment of myopia, an object at infinity is focused to the far point of the eye. If an individual has a far point of 70 cm, prescribe the correct power of the lens that is needed.

26. Without glasses, a certain person needs to have his eyes 15.0 cm from a book to read comfortably and can focus clearly only on distant objects up to 2.75 m away, but no farther. A typical normal eye should be able to focus on objects that are between 25.0 cm (the near point) and infinity (the far point) from the eye.
   (a) What type of correcting lenses does the person need: single focal length or bifocals? Why?
   (b) What should an optometrist specify as the focal length(s) of the correcting contact lens or lenses?
   (c) What is the power (in diopters) of the correcting lens or lenses?

27. One of the inevitable consequences of aging is a decrease in the flexibility of the lens. This leads to the farsighted condition called *presbyopia* (elder eye). Almost every aging human will experience it to some extent. However, for the myopic person, at some point, it is possible that far vision will be limited by a subpar far point and near vision will be hampered by an expanding near point. One solution is to wear bifocal lenses that are diverging in the upper half to correct the nearsightedness and converging in the lower half to correct the farsightedness.

   Suppose one such individual asks for your help. The patient complains that she can't see far enough to safely drive (her far point is 112 cm) and she can't read the font of her smartphone without holding it beyond arm's length (her near point is 83 cm). Prescribe the bifocals that will correct the visual issues for your patient.

28. A common zoom lens for a digital camera covers a focal length range of 18 to 200 mm. For the purposes of this problem, treat the lens as a thin lens. The lens is zoomed out to 200 mm and is focused on a petroglyph that is 15.0 m away and 38 cm wide.
   (a) How far is the lens from the light sensor of the camera?
   (b) How wide is the image of the petroglyph on the sensors?
   (c) If the closest that the lens can get to the sensor at its 18 mm focal length is 5.2 cm, what is the closest object it can focus on at that focal length?

29. A macro lens is designed to take very close-range photographs of small objects such as insects and flowers. At its closest focusing distance, a certain macro lens has a focal length of 35.0 mm and forms an image on the light sensor of the camera that is 1.09 times the size of the object.
   (a) How close must the object be to the lens to achieve this maximum image size?
   (b) What is the magnification if the object is twice as far from the lens as in part (a)? For this problem, treat the lens as a thin lens.

Prep for the AP Exam

AP® Group Work

Directions: The problem below is designed to be done as group work in class.

Assume you are given two thin lenses of focal lengths f_1 and f_2.

(a) Prove that when the two lenses are pressed next to one another, the effective focal length $f_{combined}$ of the two lenses acting together is given by

$$\frac{1}{f_{combined}} = \frac{1}{f_1} + \frac{1}{f_2}$$

(b) Describe how this relates to a prescription for a contact lens that is placed directly on the eye. (Assume there is no significant separation between the contact lens and the lens of the eye.)

(c) Why would an eyeglass prescription that is identical to a contact lens prescription give a very subtle difference in the image seen by the wearer?

AP® PRACTICE PROBLEMS

Prep for the AP® Exam

Multiple-Choice Questions

Directions: The following questions have a single correct answer.

1. Which is true when an object is moved farther from a plane mirror?
   (A) The height of the image decreases, and the image moves farther from the mirror.
   (B) The height of the image stays the same, and the image moves farther from the mirror.
   (C) The height of the image increases, and the image moves farther from the mirror.
   (D) The height of the image stays the same, and the image moves closer to the mirror.
2. A real image of a real object can form in front of
   (A) no mirror of any type.
   (B) a concave mirror.
   (C) a convex mirror.
   (D) any type of mirror.
3. When an object is placed a little farther from a concave mirror than the focal length, the image is
   (A) magnified and real.
   (B) magnified and virtual.
   (C) smaller and real.
   (D) smaller and virtual.
4. If you want to start a fire using sunlight, which kind of mirror would be most efficient?
   (A) A plane mirror
   (B) A concave mirror
   (C) A convex mirror
   (D) It is not possible to start a fire using sunlight and a mirror; you must use a concave lens.
5. An object is placed at the center of curvature of a concave mirror. The image is
   (A) real and upright.
   (B) real and inverted.
   (C) virtual and upright.
   (D) virtual and inverted.
6. When an object is placed farther from a convex mirror than the absolute value of the focal length, the image is
   (A) larger and real.
   (B) larger and virtual.
   (C) smaller and real.
   (D) smaller and virtual.
7. When a dentist needs a mirror to see an enlarged, upright image of a patient's tooth, what kind of mirror should she use?
   (A) A plane mirror
   (B) A concave mirror
   (C) A convex mirror
   (D) This would require a lens and a mirror.
8. A magnifying lens allows one to look at a very near object by forming an image of it farther away. The object appears larger. To create a magnifying lens, one would use a
   (A) short focal length ($f < 1$ m) converging lens.
   (B) short focal length ($|f| < 1$ m) diverging lens.
   (C) long focal length ($f > 1$ m) converging lens.
   (D) long focal length ($|f| > 1$ m) diverging lens.
9. A compound microscope is a two-lens system used to look at very small objects. Which of the following statements is correct?
   (A) The objective and the eyepiece both have the same focal length, and both serve as magnifying lenses.
   (B) The objective is a short focal length, converging lens and the eyepiece functions as a magnifying lens.
   (C) The objective is a long focal length, converging lens and the eyepiece functions as a magnifying lens.
   (D) The objective is a short focal length, diverging lens and the eyepiece functions as a magnifying lens.
10. An engineer would like to modify a consumer camera to better capture images in the near infrared, just beyond the visible reds. To save money, the decision is made to keep the same lens for producing the image. The index of refraction of the lens is smaller for infrared radiation than for visible light. Which of the following changes could be made to produce a sharp image in the near infrared?
    (A) Relocate the detector closer to the lens
    (B) Relocate the detector farther from the lens
    (C) A sharp image cannot be produced with the same lens system.
    (D) There is no need to change anything.

(Continued)

Free-Response Question

A thin, converging lens having a focal length of magnitude 25.00 cm is placed 1.000 m from a plane mirror that is oriented perpendicular to the principal axis of the lens. A flower, 8.400 cm tall, is 1.450 m from the mirror on the principal axis of the lens.

(a) Draw a ray diagram for the image of the flower formed by just the lens.

(b) Where is the final image of the flower produced by the lens–mirror combination? Is it real or virtual? Upright or inverted? How tall is the image?

(c) If the converging lens is replaced by a diverging lens having a focal length of the same magnitude as the original lens, what will be the answer to part (b)?

Modern Physics

Chapter 24 Quantum Physics and Atomic Structure

Chapter 25 Nuclear Physics

pidjoe/Getty Images

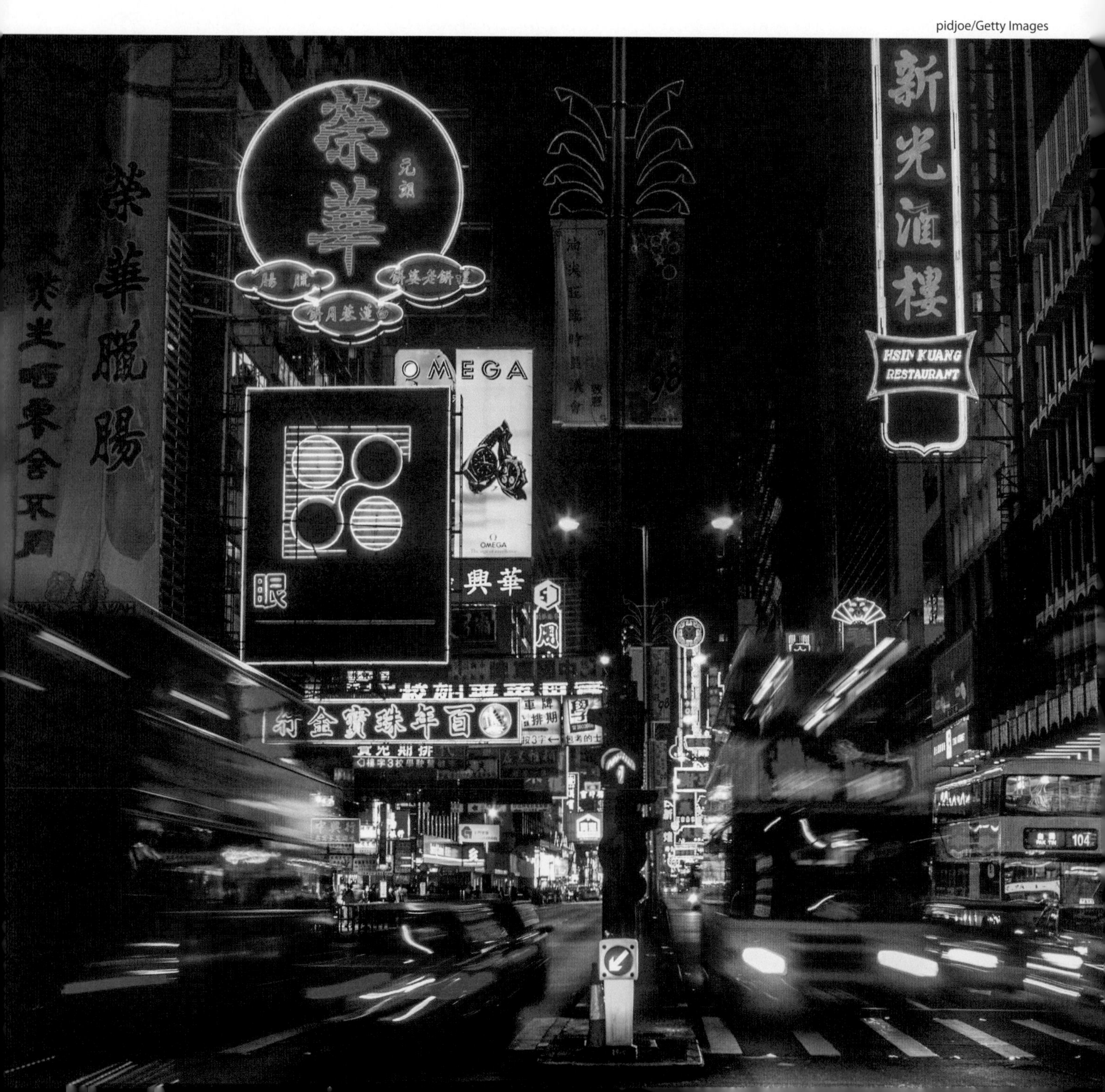

Case Study: What gives a neon sign its reddish glow?

As we saw in Chapter 22, a beam of white light hitting a prism is spread out into the beautiful spectrum of the rainbow due to the phenomenon of dispersion. Light of different wavelengths travels through the glass of the prism at slightly different speeds, resulting in a different angle of refraction for each color of the rainbow. The spectrum we see is continuous; that is, the colors blend together, one into the next, with no gaps.

But what happens when you shine light from a different source on a prism, like the light of a flame, an incandescent gas, or fireworks? Consider, for example, a glowing neon sign. When this light is passed through a prism, the resulting spectrum is not continuous. Rather, we see a series of distinct colored bands, separated by dark gaps. In other words, the atomic spectrum of the glowing neon gas is discrete—but why?

This question puzzled physicists, and even more mysterious was the observation that light sources involving different elements each produce their own unique set of discrete bands. In Chapter 24, we find the resolution of this mystery led to fundamental insights into the nature of atoms and subatomic particles. A key piece of the puzzle was proposed by physicist Niels Bohr, who suggested that the electrons circling the nucleus of an atom might be confined to particular orbits. If this were true, it would mean that electrons can only possess specific amounts of energy, occupying one particular energy state or another, but never falling in between.

Does this explain the mysterious colored bands observed in atomic spectra? Yes! Returning to our neon sign, an electron in a neon atom can gain energy when the gas is heated, temporarily exciting it to a higher energy state. When it falls back down to a lower energy state, that energy is released as a photon. Because the two states always differ by the same discrete amount of energy, such a photon can only carry away that specific energy. As we saw in Chapter 22, the energy of a photon is directly related to its frequency, which determines its color. Therefore, when we look at the distinct colored bands in atomic spectra we see only photons with those particular colors emitted when electrons in excited atoms go down to the allowed levels.

Because those specific energies vary in atoms of different elements, atomic spectra are very useful to scientists, acting as "fingerprints" to help identify the presence of different elements. For example, astronomers study the composition of stars and planets by analyzing their spectra. Spectra are also responsible for the beautiful colors we enjoy in a fireworks show, as different chemical compounds are ignited to produce various hues.

By the end of these chapters, you should be able to describe the absorption or emission of photons by atoms; properties of an atom; properties and behavior of objects exhibiting both particle and wave behavior; electromagnetic radiation emitted by an object due to its temperature; physical properties that constrain the behavior of interacting nuclei, subatomic particles, and nucleons; and interactions between photons and matter using the photoelectric Compton effects.

Quantum Physics and Atomic Structure

NASA, ESA, and T. Brown (STScI)

YOU WILL LEARN TO:

- Recognize the limitations of classical physics for explaining the properties of light and matter.
- Calculate the energy density and intensity of an electromagnetic wave, and the energy of a photon.
- Describe how the photoelectric effect and blackbody radiation provide evidence for the photon picture of light and calculate properties related to these processes.
- Explain why the wavelength of a photon increases if it scatters from an electron, and calculate the change in wavelength for a given scattering angle.
- Calculate the wavelength of a particle such as an electron given its momentum.
- Describe why atoms absorb light at only certain wavelengths, describe why they emit and absorb light at the same wavelengths, and calculate these wavelengths for a hydrogen atom.
- Explain how the Bohr model of the atom explains the spectrum of hydrogen, and calculate predicted allowed energies and angular momenta.
- Describe the Heisenberg uncertainty principle and use it in calculations.

24-1 Experiments that probe the nature of light and matter reveal the limits of classical physics

Up until now, we have avoided using the word *particle* for systems where we could describe their behavior by their center-of-mass motion. Instead, we carefully used the term *object*. In the remainder of this textbook, we will begin to explore the physics of fundamental particles and as we work in this world of truly tiny objects, it is appropriate to use the word *particle*. Although we will move well away from the everyday, classical, physics we have been studying, we won't begin to reach the limits of physics that would require us to consider the internal structure of these particles. We will also see that this amazing world of modern physics builds upon the foundations we established in our study of classical physics, with some fascinating twists.

We first introduce the idea that light and other electromagnetic waves come in particle-like packets of energy called photons. Is there solid evidence for this? Can a photon strike a particle such as an electron and "bounce" or scatter the way a cue ball

Night vision goggles "amplify light" using the photoelectric effect. Even faint light causes a surface to emit electrons, and the current of these electrons can be amplified in a circuit to generate a brighter image that the wearer of the goggles can see.

(a)

Archive Image/Alamy

The characteristic color of neon lights is due to the structure of the neon atom. When excited by an electric current, neon atoms in the light tube jump to a higher energy level. When they jump down to a lower energy level, they emit photons of a specific frequency, wavelength, and color.

(b)

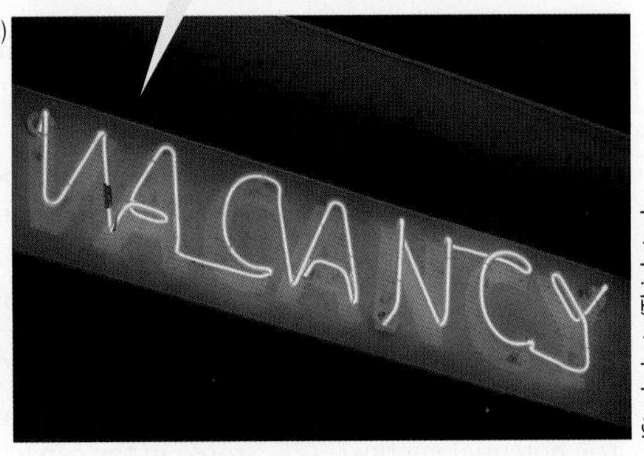

iStockphoto/Thinkstock

Figure 24-1 Photons, electrons, and atoms Two examples of the interaction between light and matter. These can only be understood if we use the ideas that light has particle aspects and matter has wave aspects.

does when it strikes a billiard ball? And if waves have a particle-like nature, do particles ever exhibit properties we associate with waves?

The photoelectric effect (**Figure 24-1a**), in which light striking a surface causes the surface to eject electrons, can be understood only if light comes in the form of photons. The same is true for blackbody radiation, the light emitted by an object as a consequence of its temperature. We'll see direct evidence that photons really do behave like tiny "cue balls" when they scatter from electrons. And we'll learn that particles like electrons also have a wavelike aspect. This wave nature of matter will help us understand the structure of atoms and the manner in which atoms absorb and emit light (**Figure 24-1b**). We'll find that the energies of atoms are *quantized*—that is, they can have only certain very definite values. The lesson of this chapter is that the microscopic world of atoms and light is very different in character from the macroscopic world of ordinary-sized objects that we see around us.

THE TAKEAWAY for Section 24-1

✔ Light waves have some of the characteristics of particles, and particles such as electrons have some of the characteristics of waves.

✔ The energies of atoms are quantized (restricted to certain specific values).

Building Blocks

1. Name, and describe the quantization of a quantity we have already learned about in classical physics.

24-2 Electromagnetic waves carry both electric and magnetic energy, and come in packets called photons

Electromagnetic waves carry energy. You can feel the energy delivered to your skin by sunlight (one kind of electromagnetic wave) on a sunny day. A microwave oven is useful for cooking because the water molecules found in food of all kinds absorb the energy in microwaves, which are at wavelengths between infrared and radio

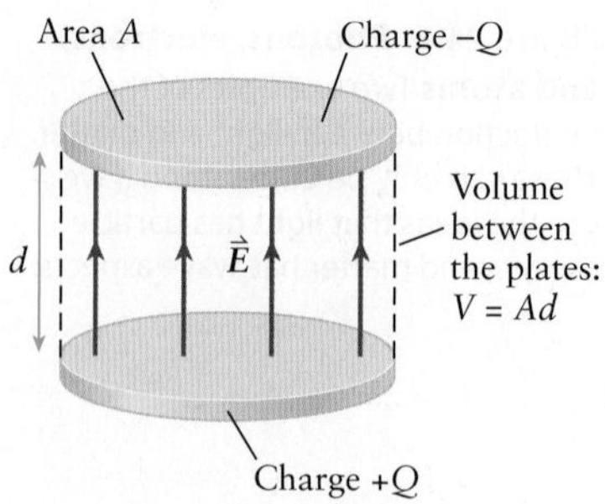

Figure 24-2 Calculating electric energy density The energy stored in this charged capacitor can be thought of as residing in the electric field $\vec{E}$ between the plates.

(Figure 22-3). And the energy in a laser beam is so tightly concentrated that it can be used as a surgical tool (Figure 22-1b).

The energy in an electromagnetic wave is actually contained within the electric and magnetic fields themselves. To see what this means, let's return to the physics of capacitors (Chapter 17) to gain insight into the energy content of electromagnetic waves.

Energy in Electric and Magnetic Fields

Figure 24-2 shows a parallel-plate capacitor with plates of area A separated by a distance d. If the two plates are closely spaced, the electric field between the plates is approximately uniform and fills the volume between the plates (that is, there is very little field outside this volume). With a charge $+Q$ on one plate and a charge $-Q$ on the other plate, the magnitude of this field is

$$E = \frac{Q}{\varepsilon_0 A} \qquad (17\text{-}10)$$

> **AP® Exam Tip**
>
> On the AP® Physics 2 Exam you are not expected to calculate the energy density of electromagnetic waves. However, you should know how to relate the intensity of an electromagnetic wave to the number of photons. This section presents the calculation to help you avoid common misconceptions about this relationship.

It takes work to separate the charges $+q$ and $-q$, and this work goes into the electric potential energy U_{electric} stored in the capacitor. The amount of stored energy is

$$\Delta U_{\text{electric}} = \frac{Q^2}{2C} \qquad (17\text{-}16)$$

where C, the capacitance of the capacitor, is

$$C = \frac{\varepsilon_0 A}{d} \qquad (17\text{-}13)$$

We can think of the electric energy in the capacitor as being stored in the electric field itself. To motivate this idea, let's substitute Equation 17-13 into Equation 17-16 and rearrange:

$$\Delta U_{\text{electric}} = \frac{Q^2}{2}\left(\frac{d}{\varepsilon_0 A}\right) = \frac{1}{2}\varepsilon_0\left(\frac{Q^2}{\varepsilon_0^2 A^2}\right)Ad = \frac{1}{2}\varepsilon_0\left(\frac{Q}{\varepsilon_0 A}\right)^2 Ad \qquad (24\text{-}1)$$

The quantity $Q/(\varepsilon_0 A)$ in parentheses on the far right-hand side of Equation 24-1 is just the electric field magnitude E from Equation 17-10. The quantity Ad is the volume V that the electric field occupies. So we can rewrite Equation 24-11 as

$$\Delta U_{\text{electric}} = \left(\frac{1}{2}\varepsilon_0 E^2\right)V \qquad (24\text{-}2)$$

Equation 24-2 tells us that the energy stored in the capacitor is equal to the volume V occupied by the electric field multiplied by a quantity $(1/2)\varepsilon_0 E^2$ that depends on the electric field magnitude E. This quantity has units of energy per volume (J/m^3) and is called the **electric energy density**:

EQUATION IN WORDS
Electric energy density

$$u_E = \frac{1}{2}\varepsilon_0 E^2 \qquad (24\text{-}3)$$

Electric energy density: u_E; Electric field magnitude: E; Permittivity of free space: ε_0

Equation 24-3 shows that wherever there is an electric field, there is energy. This motivates the idea that the energy stored in a capacitor is stored in the field itself. We derived Equation 24-3 for the special case of a parallel-plate capacitor, but it turns out to be valid in *any* situation where an electric field is present.

We can come to a similar conclusion about the magnetic energy stored in an inductor like the one shown in **Figure 24-3**. Although it is not within the scope of the course, it can be found that the energy stored in an inductor (the magnetic equivalent of a capacitor) of inductance L carrying a current I is $U_{\text{magnetic}} = \frac{1}{2}LI^2$.

For an inductor like that in Figure 24-3, which is a long solenoid (Section 19-7) with N turns of wire, length ℓ, and cross-sectional area A, the inductance is $L = \frac{\mu_0 N^2 A}{\ell}$. Like the electric field inside the parallel-plate capacitor in Figure 24-2, the magnetic field inside the solenoid in Figure 24-3 is nearly uniform and confined to the volume inside the solenoid. The magnitude of this magnetic field is $B = \frac{\mu_0 NI}{\ell}$. Substituting the expression for L into the expression for magnetic energy in a solenoid and rearranging:

$$U_{\text{magnetic}} = \frac{1}{2}\left(\frac{\mu_0 N^2 A}{\ell}\right)I^2 = \frac{1}{2\mu_0}\left(\frac{\mu_0^2 N^2 I^2}{\ell^2}\right)A\ell = \frac{1}{2\mu_0}\left(\frac{\mu_0 NI}{\ell}\right)^2 A\ell \qquad (24\text{-}4)$$

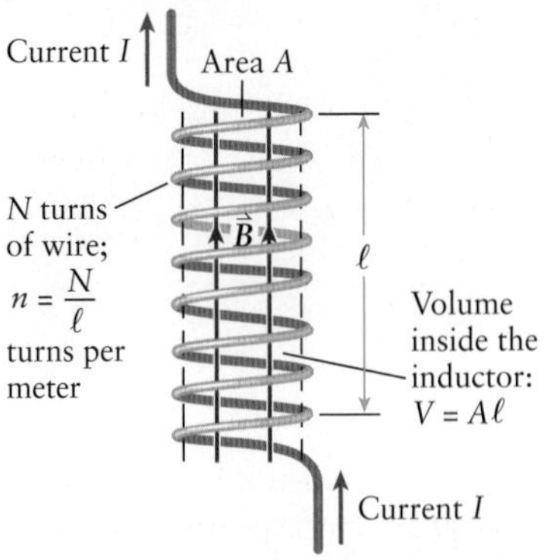

Figure 24-3 Calculating magnetic energy density The energy stored in this current-carrying inductor can be thought of as residing in the magnetic field $\vec{B}$ inside the inductor.

The quantity $\mu_0 NI/\ell$ in parentheses on the far right of Equation 24-4 is the magnitude B of the magnetic field inside the solenoid. The quantity $A\ell$ is the volume V inside the solenoid, which is also the volume that the magnetic field occupies. So Equation 24-4 becomes

$$U_{\text{magnetic}} = \left(\frac{B^2}{2\mu_0}\right)V \qquad (24\text{-}5)$$

Compare this to Equation 24-2 for the electric energy U_{electric}. We see that according to Equation 24-5, the energy stored in the inductor equals the volume V occupied by the magnetic field multiplied by a quantity $B^2/(2\mu_0)$ that depends on the magnetic field magnitude B and has units J/m^3. We call $B^2/(2\mu_0)$ the **magnetic energy density:**

Magnetic energy density | Magnetic field magnitude

$$u_B = \frac{B^2}{2\mu_0} \qquad (24\text{-}6)$$

Permeability of free space

EQUATION IN WORDS
Magnetic energy density

Just as Equation 24-3 tells us that there is electric energy wherever there is an electric field, Equation 24-6 shows that there is magnetic energy wherever there is a magnetic field. This is true for an inductor, so we can think of the energy stored in a current-carrying inductor as being stored in the magnetic field within the inductor. But Equation 24-6 is valid *wherever* there is a magnetic field.

Energy in an Electromagnetic Plane Wave

Let's apply Equations 24-3 and 24-6 for the electric and magnetic energy densities to the sinusoidal electromagnetic plane wave that we introduced in Section 22-2. From Equation 22-3 the electric field of this plane wave has only a y component, and the magnetic field has only a z component:

$$\begin{aligned} E_y(x, t) &= E_0\cos(kx - \omega t + \phi) \\ B_z(x, t) &= B_0\cos(kx - \omega t + \phi) \end{aligned} \qquad (22\text{-}3)$$

Substituting these into Equations 24-3 and 24-6, we find that the electric and magnetic energy densities in the plane wave are

$$\begin{aligned} u_{\text{electric}} &= \frac{1}{2}\varepsilon_0 E^2 = \frac{1}{2}\varepsilon_0 E_0^2\cos^2(kx - \omega t + \phi) \\ u_{\text{magnetic}} &= \frac{B^2}{2\mu_0} = \frac{B_0^2}{2\mu_0}\cos^2(kx - \omega t + \phi) \end{aligned} \qquad (24\text{-}7)$$

Note that u_{electric} and u_{magnetic} both depend on position x and time t in the same way. It may appear from Equation 24-7 that there are different amounts of energy in the electric and magnetic forms, since the coefficients $(1/2)\varepsilon_0 E_0^2$ and $B_0^2/2\mu_0$ are different. But we know from Equations 22-4 and 22-5 that $B_0 = E_0/c$ and $c = 1/\sqrt{\mu_0\varepsilon_0}$, so

$$\frac{B_0^2}{2\mu_0} = \frac{(E_0/c)^2}{2\mu_0} = \frac{\left(E_0\sqrt{\mu_0\varepsilon_0}\right)^2}{2\mu_0} = \frac{1}{2}\left(\frac{\mu_0\varepsilon_0}{\mu_0}\right)E_0^2 = \frac{1}{2}\varepsilon_0 E_0^2 \tag{24-8}$$

In other words, the coefficients of u_{electric} and u_{magnetic} in Equation 24-7 are equal. This means that at any position x and at any time t, an electromagnetic wave in a vacuum has *equal* amounts of electric energy density and magnetic energy density:

$$u_{\text{electric}} = u_{\text{magnetic}} = \frac{1}{2}\varepsilon_0 E_0^2 \cos^2(kx - \omega t + \phi) = \frac{B_0^2}{2\mu_0}\cos^2(kx - \omega t + \phi) \tag{24-9}$$

The *total* energy density u in the wave is the sum of u_{electric} and u_{magnetic}. Equation 24-9 shows that $u_{\text{electric}} = u_{\text{magnetic}}$, so u is equal to $2u_{\text{electric}}$ or $2u_{\text{magnetic}}$:

$$u = u_{\text{electric}} + u_{\text{magnetic}} = \varepsilon_0 E_0^2 \cos^2(kx - \omega t + \phi) = \frac{B_0^2}{\mu_0}\cos^2(kx - \omega t + \phi) \tag{24-10}$$

The value of u at any position varies with time, so it's often more useful to state its *average* value. The average value of the cosine function squared is $\frac{1}{2}$, so

EQUATION IN WORDS
Average energy density in an electromagnetic wave

$$u_{\text{average}} = \frac{1}{2}\varepsilon_0 E_0^2 = \varepsilon_0 E_{\text{rms}}^2 = \frac{B_0^2}{2\mu_0} = \frac{B_{\text{rms}}^2}{\mu_0} \tag{24-11}$$

Average energy density in an electromagnetic wave; Electric field magnitude; Electric field rms value; Magnetic field magnitude; Magnetic field rms value; Permittivity of free space; Permeability of free space

In Equation 24-11 we've rewritten the equation to use the root-mean-square (rms) values of the oscillating electric and magnetic fields, $E_{\text{rms}} = E_0/\sqrt{2}$ and $B_{\text{rms}} = B_0/\sqrt{2}$, replacing the $\frac{1}{2}$ that came from taking the average.

An even more useful way to express the energy carried by an electromagnetic wave is in terms of the wave intensity, or average power per unit area. We introduced this in Chapter 22, as a measure of brightness of light, and discussed that it was proportional to the square of the electric field. We can now see why. **Figure 24-4** shows a portion of a wave that has cross-sectional area A and length ℓ. The energy in this portion of the wave is u_{average} from Equation 24-11 multiplied by the volume $A\ell$ of this portion. This entire portion of the wave moves at speed c through the cross-sectional area A in a time t. The power equals the energy in the wave portion divided by the time $t = \ell/c$ that it takes this portion of the wave to travel at speed c through the cross-sectional area A. The intensity, to which we give the symbol S_{average}, equals the power divided by the area A. So

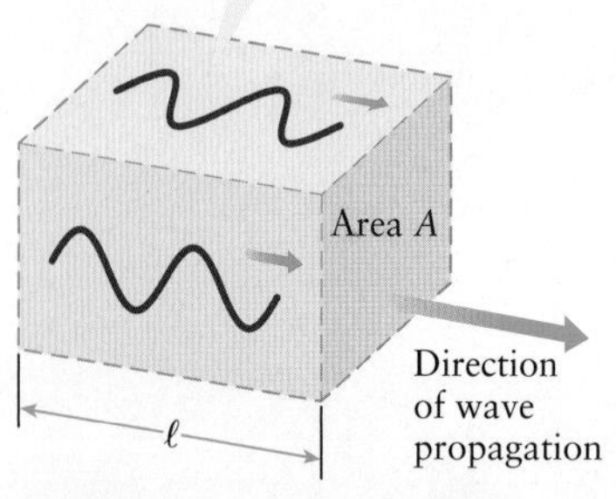
The electromagnetic wave energy in this volume moves with the wave at speed c.
Area A
Direction of wave propagation
ℓ

Figure 24-4 Calculating wave intensity The intensity of an electromagnetic wave equals the amount of wave energy that crosses an area A per unit time, divided by the area A.

$$S_{\text{average}} = \frac{\text{energy}}{\text{time}} \times \frac{1}{\text{area}} = \frac{(u_{\text{average}}A\ell)}{\ell/c} \times \frac{1}{A} = u_{\text{average}}c$$

Using Equation 24-11, we can rewrite this as

EQUATION IN WORDS
Intensity of an electromagnetic wave

$$S_{\text{average}} = c\varepsilon_0 E_{\text{rms}}^2 = \frac{cB_{\text{rms}}^2}{\mu_0} = \frac{E_{\text{rms}}B_{\text{rms}}}{\mu_0} \tag{24-12}$$

Intensity of an electromagnetic wave; Speed of light; Magnetic field rms value; Permittivity of free space; Electric field rms value; Permeability of free space

(To write the last expression in Equation 24-12, we used the result that, because $B_0 = E_0/c$, it follows that $B_{rms} = E_{rms}/c$.)

In Chapter 22 we introduced intensity, which is a measure of power per unit area. Its units are thus watts per square meter, W/m^2. The intensity of sunlight that reaches Earth is about 1.36×10^3 W/m^2. As the following example shows, the intensities of other common electromagnetic waves can be much smaller.

EXAMPLE 24-1 Energy Density and Intensity in a Radio Wave

In Example 22-1 we considered an FM radio wave of frequency 98.7 MHz with an electric field amplitude 1.20×10^{-1} V/m. We found that the magnetic field has amplitude 4.00×10^{-10} T. Calculate the rms values of the electric and magnetic fields, the average energy density in the wave, and the wave intensity.

Set Up

Each rms value is just equal to the amplitude divided by $\sqrt{2}$. Given the rms values, we'll use Equation 24-11 to calculate the average energy density and Equation 24-12 to calculate the intensity.

Root-mean-square values:

$$E_{rms} = \frac{E_0}{\sqrt{2}}$$

$$B_{rms} = \frac{B_0}{\sqrt{2}}$$

Average energy density in an electromagnetic wave:

$$u_{average} = \varepsilon_0 E_{rms}^2 = \frac{B_{rms}^2}{\mu_0} \qquad (24\text{-}11)$$

Intensity of an electromagnetic wave:

$$S_{average} = c\varepsilon_0 E_{rms}^2 = \frac{cB_{rms}^2}{\mu_0} = \frac{E_{rms}B_{rms}}{\mu_0} \qquad (24\text{-}12)$$

Permittivity of free space:

$$\varepsilon_0 = 8.85 \times 10^{-12} \frac{\text{C}}{\text{V} \cdot \text{m}}$$

Permeability of free space:

$$\mu_0 = 4\pi \times 10^{-7}\ \text{T} \cdot \text{m/A}$$

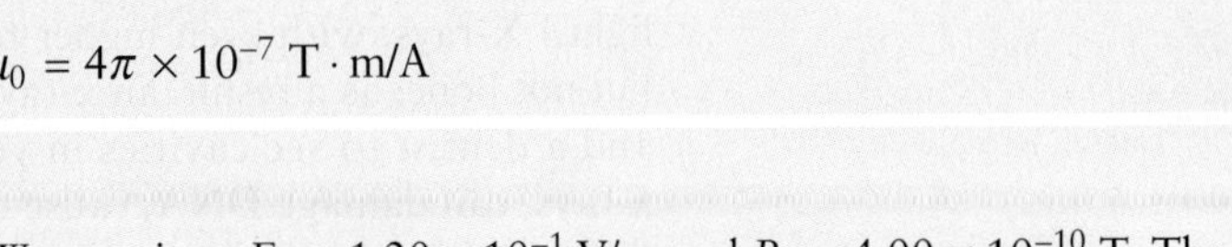

Solve

Calculate the rms values of the electric and magnetic fields.

We are given $E_0 = 1.20 \times 10^{-1}$ V/m and $B_0 = 4.00 \times 10^{-10}$ T. The corresponding rms values are

$$E_{rms} = \frac{E_0}{\sqrt{2}} = \frac{1.20 \times 10^{-1}\ \text{V/m}}{\sqrt{2}} = 8.49 \times 10^{-2}\ \text{V/m}$$

$$B_{rms} = \frac{B_0}{\sqrt{2}} = \frac{4.00 \times 10^{-10}\ \text{T}}{\sqrt{2}} = 2.83 \times 10^{-10}\ \text{T}$$

Use the value of E_{rms} to calculate the average energy density in the wave.

From Equation 24-11:

$$u_{average} = \varepsilon_0 E_{rms}^2 = \left(8.85 \times 10^{-12} \frac{\text{C}}{\text{V} \cdot \text{m}}\right)\left(8.49 \times 10^{-2} \frac{\text{V}}{\text{m}}\right)^2$$

$$= 6.37 \times 10^{-14} \frac{\text{C} \cdot \text{V}}{\text{m}^3}$$

One coulomb times one volt is one joule: $1\ \text{C} \cdot \text{V} = 1$ J. So

$$u_{average} = 6.37 \times 10^{-14}\ \text{J/m}^3$$

Find the intensity of the wave.

Comparing Equations 24-11 and 24-12 shows that the wave intensity is c times the average energy density:

$$S_{\text{average}} = c\varepsilon_0 E_{\text{rms}}^2 = cu_{\text{average}}$$
$$= (3.00 \times 10^8 \text{ m/s})(6.37 \times 10^{-14} \text{ J/m}^3)$$
$$= 1.91 \times 10^{-5} \frac{\text{J}}{\text{m}^2 \cdot \text{s}}$$

One joule per second is one watt: 1 J/s = 1 W. So

$$S_{\text{average}} = 1.91 \times 10^{-5} \text{ W/m}^2$$

Reflect

The average energy density and intensity are both very small quantities. It's a testament to the sensitivity of radio receivers that such a wave is quite easy to detect.

We can check our results by using the alternative expressions for u_{average} and S_{average} given in Equations 24-11 and 24-12. As an example, here's a check on the value of S_{average}.

From Equation 24-12:

$$S_{\text{average}} = \frac{E_{\text{rms}} B_{\text{rms}}}{\mu_0} = \frac{(8.49 \times 10^{-2} \text{ V/m})(2.83 \times 10^{-10} \text{ T})}{4\pi \times 10^{-7} \text{ T} \cdot \text{m/A}}$$
$$= 1.91 \times 10^{-5} \frac{\text{V} \cdot \text{A}}{\text{m}^2}$$

One volt times one ampere is one watt (1 V · A = 1 W), so

$$S_{\text{average}} = 1.91 \times 10^{-5} \text{ W/m}^2$$

This agrees with our calculation above, as it must.

NOW WORK Problems 6–8 from the Takeaway 24-2.

Photons

In Equation 24-12 we see that the intensity of an electromagnetic wave depends on the strength of the electric and magnetic fields that make up the wave but not on the wave frequency. (The frequency f doesn't appear anywhere in this equation.) But everyday experience suggests that wave frequency *does* play a role in the energy carried by an electromagnetic wave. As an example, ultraviolet light can trigger a chemical reaction in the skin that causes a suntan or sunburn, but visible light cannot. (That's why sunscreen contains a substance that allows visible light to pass but blocks ultraviolet light.) X-rays, with even higher frequency than ultraviolet light, can pierce soft tissue but not bone; as a result, an x-ray image allows a physician to diagnose a broken bone and a dentist to see cavities in your teeth. Gamma rays, with higher frequency than x-rays, can damage DNA, cause cancer, and even kill cells. How can we explain these differences?

The explanation is that the energy of an electromagnetic wave propagates as small packets called **photons**. The energy of an individual photon is proportional to the wave frequency, and the proportionality constant h is called **Planck's constant:**

EQUATION IN WORDS
Energy of a photon

Energy of a photon — Wave frequency

$$E = hf \quad (24\text{-}13)$$

Planck's constant = $6.62607015 \times 10^{-34}$ J • s

To three significant digits $h = 6.63 \times 10^{-34}$ J · s.

Equation 24-13 explains why ultraviolet light causes a suntan or sunburn, but light in the visible spectrum does not. For tanning or burning to take place in your skin, individual molecules must absorb a certain minimum amount of energy from light to trigger a chemical change. A given molecule must absorb this energy in the form of a single photon, and a visible-light photon lacks sufficient energy to trigger this chemical change. An ultraviolet photon, by contrast, can trigger the change because it has a shorter wavelength, a higher frequency, and more energy per photon. An x-ray photon is more energetic still, which is why it is able to penetrate soft tissue.

Gamma-ray photons; x-ray photons; and short-wavelength, high-frequency ultraviolet photons have enough energy that they can dislodge an electron from an atom. Such **ionizing radiation** breaks apart molecules by pulling electrons from the chemical bonds that hold atoms together. Although ionizing radiation can directly break DNA molecules in living tissue, it's more likely to disrupt some other, more common molecule, such as water, to create highly reactive free radicals that then damage DNA. Depending on the severity of the damage, the cell may be able to recover. However, if the damage cannot be repaired, or if it is repaired incorrectly, the resulting mutations may be lethal. These effects generally go unnoticed until the next time the cell tries to divide. Because cancerous cells divide more frequently than most other cells in the body, they are more susceptible to radiation damage than are most healthy cells.

Planck's constant is very small, so a single photon carries only a miniscule amount of energy. As an example, a photon of red light with wavelength $\lambda = 750\text{ nm} = 7.50 \times 10^{-7}\text{ m}$ has frequency $f = c/\lambda = (3.00 \times 10^8\text{ m/s})/(7.50 \times 10^{-7}\text{ m}) = 4.00 \times 10^{14}\text{ Hz}$ and energy $E = hf = (6.63 \times 10^{-34}\text{ J} \cdot \text{s})(4.00 \times 10^{14}\text{ Hz}) = 2.65 \times 10^{-19}\text{ J}$. (Recall that $1\text{ Hz} = 1\text{ s}^{-1}$.) That's so small that you don't notice individual photons in the light from a lamp, just as you don't notice individual air molecules in a breeze against your face.

It's common to express photon energies in electron volts (eV). We introduced this unit in Section 17-3: $1\text{ eV} = 1.60 \times 10^{-19}\text{ J}$. For a red photon of wavelength 750 nm, the energy is $E = (2.65 \times 10^{-19}\text{ J})/(1.60 \times 10^{-19}\text{ J/eV}) = 1.66\text{ eV}$. As the following example shows, radio photons have even less energy.

EXAMPLE 24-2 Photons in a Radio Wave

For the radio wave of Examples 22-1 and 24-1, calculate (a) the energy per photon, (b) the number of photons per cubic meter, and (c) the number of photons per second that strike a receiver antenna of area 10.0 cm^2.

Set Up

The radio wave has frequency 98.7 MHz = 98.7×10^6 Hz. We'll use this and Equation 24-13 to determine the energy of a single radio photon. From Example 24-1 we know that the average energy density (energy per unit volume) of the wave is $u_{\text{average}} = 6.37 \times 10^{-14}\text{ J/m}^3$ and the intensity (energy per area per time) is $S_{\text{average}} = 1.91 \times 10^{-5}\text{ W/m}^2$. We'll use these and our calculated value of the photon energy to determine the number of photons per unit volume and the number of photons striking the antenna per time.

Energy of a photon:

$$E = hf \qquad (24\text{-}13)$$

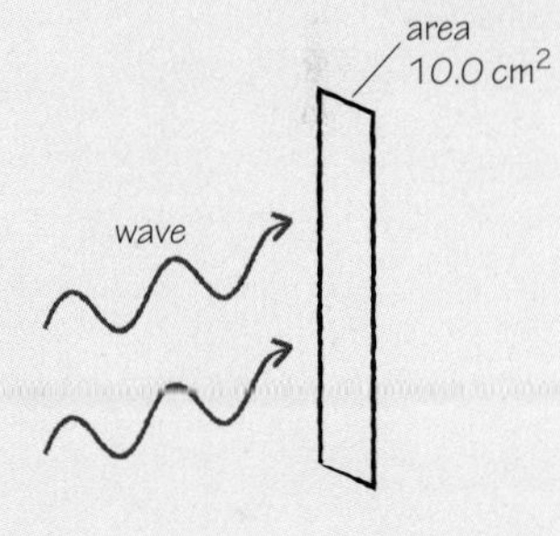

Solve

(a) Calculate the energy of an individual photon of frequency 98.7 MHz. Because the rest of the values we use are 3 significant digits, we will round our constants to 3 significant digits as well.

From Equation 24-13:

$$E = hf = (6.63 \times 10^{-34}\text{ J} \cdot \text{s})(98.7 \times 10^6\text{ Hz}) = 6.54 \times 10^{-26}\text{ J} \cdot \text{s} \cdot \text{Hz}$$

Because $1\text{ Hz} = 1\text{ s}^{-1}$, $1\text{ J} \cdot \text{s} \cdot \text{Hz} = 1\text{ J}$

and so

$$E = 6.54 \times 10^{-26}\text{ J or}$$

$$E = \frac{6.54 \times 10^{-26}\text{ J}}{1.60 \times 10^{-19}\text{ J/eV}} = 4.09 \times 10^{-7}\text{ eV}$$

This is much smaller than the energy of a visible-light photon (2.65×10^{-19} J = 1.66 eV for red light) because the frequency of the radio wave is much less than the frequency of visible light (4.00×10^{14} Hz for red light).

(b) Calculate the photon density (number of photons per unit volume). Because this is a new example, we will use the final numbers we got for our previous example in the calculations here.

The average energy density in the wave is $u_{\text{average}} = 6.37 \times 10^{-14}\ \text{J/m}^3$ and the energy per photon is $E = 6.54 \times 10^{-26}$ J/photon. The number of photons per unit volume is

$$\frac{\text{photons}}{\text{volume}} = \frac{\text{energy}}{\text{volume}} \times \frac{\text{photon}}{\text{energy}} = \frac{\left(\dfrac{\text{energy}}{\text{volume}}\right)}{\left(\dfrac{\text{energy}}{\text{photon}}\right)}$$

$$= \frac{6.37 \times 10^{-14}\ \text{J/m}^3}{6.54 \times 10^{-26}\ \text{J/photon}} = 9.73 \times 10^{11}\ \text{photons/m}^3$$

Each cubic meter of this wave contains 9.73×10^{11} (973 billion) photons.

(c) Calculate the number of photons that strike an area of $10.0\ \text{cm}^2$ in 1.00 s.

The intensity (energy per area per time) of the wave is $S_{\text{average}} = 1.91 \times 10^{-5}\ \text{W/m}^2$, so the rate at which energy arrives at the antenna of area $A = 10.0\ \text{cm}^2$ is

$$\frac{\text{energy}}{\text{area} \cdot \text{time}} \times \text{area} = S_{\text{average}} A$$

$$= (1.91 \times 10^{-5}\ \text{W/m}^2)(10.0\ \text{cm}^2)\left(\frac{1\ \text{m}}{100\ \text{cm}}\right)^2$$

$$= 1.91 \times 10^{-8}\ \text{W} = 1.91 \times 10^{-8}\ \text{J/s}$$

The rate at which photons arrive at the antenna is

$$\frac{\text{photons}}{\text{time}} = \frac{\text{energy}}{\text{time}} \times \frac{\text{photon}}{\text{energy}} = \frac{\left(\dfrac{\text{energy}}{\text{time}}\right)}{\left(\dfrac{\text{energy}}{\text{photon}}\right)}$$

$$= \frac{1.91 \times 10^{-8}\ \text{J/s}}{6.54 \times 10^{-26}\ \text{J/photon}} = 2.92 \times 10^{17}\ \text{photons/s}$$

In one second 2.92×10^{17} (292 quadrillion) photons arrive at the antenna.

Reflect

Even this relatively low-intensity wave contains a tremendous number of photons per cubic meter and delivers an astronomical number of photons per second to a receiver.

NOW WORK Problems 1, 2, 4, and 5 from the Takeaway 24-2.

WATCH OUT !

Photons behave as both particles and waves.

A common *incorrect* way to think about photons is to visualize them as small particles like miniature marbles and to imagine that a large number of photons acting together behave like a wave. The reality is far different! Each individual photon has aspects of *both* wave and particle, and those particles are very different in character from ordinary objects such as marbles. We'll learn more about the curious properties of photons throughout this chapter.

THE TAKEAWAY for Section 24-2

✔ Energy is associated with both electric and magnetic fields. An electromagnetic wave in a vacuum has equal amounts of electric energy and magnetic energy per volume.

✔ The intensity of an electromagnetic wave is the average power per unit area. In a vacuum, the intensity equals the average energy per volume in the wave multiplied by the speed of light *c*.

✔ The energy of an electromagnetic wave comes in packets called photons. The energy of a single photon is proportional to the wave frequency.

Prep for the AP Exam

AP Building Blocks

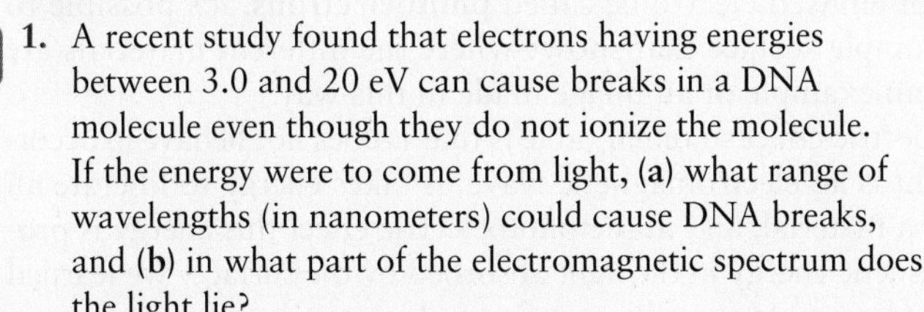

EX 24-2 **1.** A recent study found that electrons having energies between 3.0 and 20 eV can cause breaks in a DNA molecule even though they do not ionize the molecule. If the energy were to come from light, (a) what range of wavelengths (in nanometers) could cause DNA breaks, and (b) in what part of the electromagnetic spectrum does the light lie?

EX 24-2 **2.** Calculate the wavelengths and frequencies of the photons that have the following energies.
(a) $E_{photon} = 3.45 \times 10^{-19}$ J
(b) $E_{photon} = 4.80 \times 10^{-19}$ J
(c) $E_{photon} = 1.28 \times 10^{-18}$ J
(d) $E_{photon} = 4.33 \times 10^{-20}$ J
(e) $E_{photon} = 931$ MeV
(f) $E_{photon} = 2.88$ keV
(g) $E_{photon} = 7.88$ eV
(h) $E_{photon} = 13.6$ eV

3. Describe how the frequency of an electromagnetic wave is related to the energy of the photons of that wave.

EX 24-2 **4.** Calculate the wavelengths and frequencies of photons that have the following energies.
(a) $E_{photon} = 2.33 \times 10^{-19}$ J
(b) $E_{photon} = 4.50 \times 10^{-19}$ J
(c) $E_{photon} = 3.20 \times 10^{-19}$ J
(d) $E_{photon} = 8.55 \times 10^{-19}$ J
(e) $E_{photon} = 63.3$ eV
(f) $E_{photon} = 8.77$ eV
(g) $E_{photon} = 1.98$ eV
(h) $E_{photon} = 4.55$ eV

AP Skill Builders

EX 24-2 **5.** A dental x-ray typically affects 200 g of tissue and delivers about 4.0 μJ of energy using x-rays that have wavelengths of 0.025 nm. What is the energy (in electron volts) of such x-ray photons, and how many photons are absorbed during the dental x-ray? Assume the body absorbs all of the incident x-rays.

EX 24-1 **6.** An electromagnetic plane wave has an intensity $S_{average} = 200$ W/m^2.
(a) Calculate the rms values of the electric and magnetic fields.
(b) Find the amplitudes of the electric and magnetic fields.

EX 24-1 **7.** The amplitude of an electromagnetic wave's electric field is 200 V/m. Calculate (a) the amplitude of the wave's magnetic field and (b) the intensity of the wave.

EX 24-1 **8.** The rms value of an electromagnetic wave's magnetic field is 400 T. Calculate (a) the amplitude of the wave's electric field and (b) the intensity of the wave.

AP Skills in Action

9. Three lasers have equal power output. The first emits a pure violet light, the second emits a pure green light, and the third emits a pure red light. Which laser emits the greater number of photons per second?
(A) The violet laser
(B) The green laser
(C) The red laser
(D) All emit the same number of photons per second.

10. The energy of an ultraviolet light photon is unrelated to the speed of the fundamental electromagnetic waves that make up such radiation. Explain, using physics principles, why this must be true.

24-3 The photoelectric effect and blackbody radiation show that light is absorbed and emitted in the form of photons

By the end of the nineteenth century, it was recognized that light is an electromagnetic wave. Maxwell's equations (Sections 20-5 and 22-2) describe the properties of these waves in great detail, and so physicists were confident that they had a deep understanding of the nature of light. But experimental studies of two very different phenomena showed that the true nature of light is actually more complex. In one of these

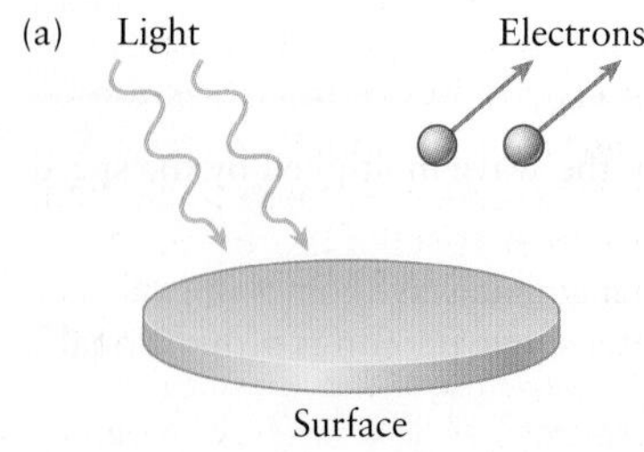

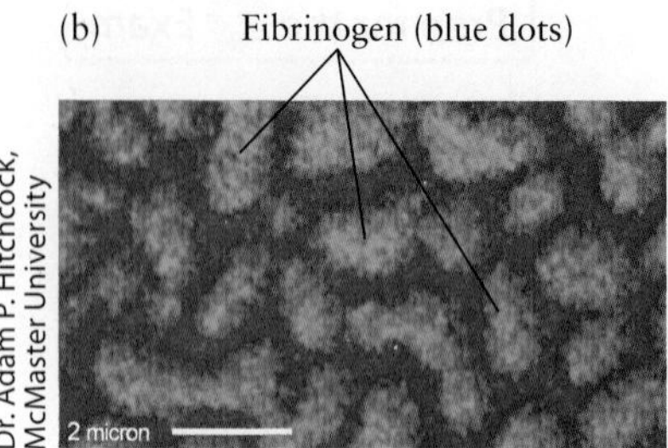

Dr. Adam P. Hitchcock, McMaster University

Figure 24-5 The photoelectric effect (a) In the photoelectric effect electrons escape from a surface when the surface is illuminated with light. (b) This image uses photoelectron electron microscopy (PEEM) to map the spatial distribution of fibrinogen, a blood protein, on a sheet of two different types of polymer. This image could be made because fibrinogen (shown as blue dots) and the two polymers (shown in green and red) emit different numbers of electrons when illuminated with ultraviolet light. Data were measured at the Advanced Light Source, Lawrence Berkeley National Laboratory.

phenomena, the *photoelectric effect*, a material absorbs light and the absorbed energy is used to eject electrons; in the other, called *blackbody radiation*, a material emits electromagnetic radiation when it is heated. The discoveries made by studying these two phenomena radically altered how physicists answered the question "What is light?"

The Photoelectric Effect

When light strikes certain materials, electrons can be ejected from the surface of those materials (**Figure 24-5a**). This **photoelectric effect**, discovered in 1886, plays an important role in biological research through a technique called *photoemission electron microscopy* or PEEM. In this technique a biological sample is illuminated with ultraviolet light or x-rays. Different materials in the surface layer of the sample will emit fewer or greater numbers of electrons in response to this light. By recording the differences in the number of emitted electrons, called **photoelectrons**, it's possible to construct an image of the sample surface that shows where the different materials are located. **Figure 24-5b** shows an example of an image made in this way.

What makes the photoelectric effect so remarkable is that it does not behave in accordance with the idea that light is an electromagnetic wave. It takes energy to liberate an electron from the surface of a material, and in the photoelectric effect this energy is provided by the electric and magnetic energy in the light absorbed by the surface. We learned in the last section that according to Maxwell's equations, the intensity of such a wave depends on the rms values of the electric and magnetic fields in that wave, but not on the frequency of the wave. Equation 22-12 suggests that light of *any* frequency should be able to liberate an electron from the surface of a material, provided the light wave is sufficiently intense. Experiment shows that this is not the case. For example, if the biological sample shown in Figure 24-5b is illuminated with red light, no electrons are ejected no matter how intense the light. But if instead we illuminate the sample with x-rays, which have a higher frequency than red light, electrons *are* ejected. (Figure 24-5b was made by using x-rays.) This is impossible to understand on the basis of Maxwell's equations.

In 1905, the same year that he published his special theory of relativity, Albert Einstein proposed a simple but radical explanation for the strange behavior of the photoelectric effect. Equation 24-12 in the last section is this explanation: Light of frequency f comes in small packets, each with an energy E that is directly proportional to the frequency. (These packets are photons.) We'll see shortly how the value of Planck's constant h is determined from the photoelectric effect. Later in this section we'll see why the constant is named for the German physicist Max Planck.

Let's see how Einstein's idea explains the properties of the photoelectric effect. The minimum amount of energy required to remove a single electron from a material is called the **work function** Φ_0 of the material (Φ is the uppercase Greek letter "phi"). The value of Φ_0 varies from one material to another; it is small if electrons are easy to remove and large if electrons are hard to remove. In a given material, some electrons will be more difficult to remove, but Φ_0 represents the energy required to remove the most easily dislodged electron. Einstein proposed that an electron can absorb only a single photon at a time. So for even the most easily dislodged electron to be ejected from the material, it must absorb a photon with an energy equal to or greater than Φ_0. If the energy of the absorbed photon is greater than Φ_0, the energy that remains after the electron is ejected goes into the kinetic energy of the electron as it flies away from the material. Electrons that require more energy to be ejected will emerge from the material with less kinetic energy, but those with *maximum* kinetic energy will be those that were the easiest to dislodge. So in Einstein's picture the most energetic electrons ejected from the material will emerge with kinetic energy K_{max} given by

EQUATION IN WORDS
Maximum kinetic energy of an electron in the photoelectric effect

Maximum kinetic energy of an electron ejected from a material by the **photoelectric effect**

Work function of the material

$$K_{max} = hf - \Phi_0 \qquad (24\text{-}14)$$

Planck's constant

Frequency of the light used to illuminate the material

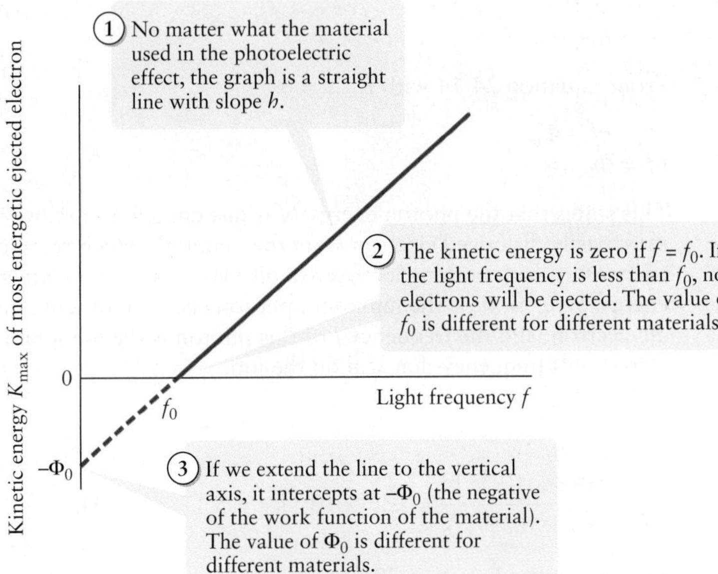
1 No matter what the material used in the photoelectric effect, the graph is a straight line with slope h.
2 The kinetic energy is zero if $f = f_0$. If the light frequency is less than f_0, no electrons will be ejected. The value of f_0 is different for different materials.
3 If we extend the line to the vertical axis, it intercepts at $-\Phi_0$ (the negative of the work function of the material). The value of Φ_0 is different for different materials.
Kinetic energy K_{max} of most energetic ejected electron
0
$-\Phi_0$
f_0
Light frequency f

Figure 24-6 Electron kinetic energy versus frequency in the photoelectric effect The kinetic energy of the most energetic electron ejected in the photoelectric effect depends on the frequency of the light. This cannot be explained using the wave model of light but can be explained by the photon concept.

Equation 24-14 tells us that that a graph of K_{max} as a function of the light frequency f should be a straight line of slope h (see **Figure 24-6**). Because kinetic energy can never be negative, Equation 24-14 also tells us that electrons will be emitted only if $hf - \Phi_0 > 0$, which we can rewrite $f > f_0 = \frac{\Phi_0}{h}$, which shows that electrons will be emitted from the surface only if the light frequency is greater than a threshold frequency f_0 equal to the work function Φ_0 divided by Planck's constant h. This agrees with the observation that no electrons can be ejected from a surface by light of too low a frequency, no matter how intense the light.

The graph shown in Figure 24-6 turns out to be an excellent match to experimental measurements of the maximum kinetic energy of ejected electrons for different light frequencies f. The slope of the graph tells us the value of Planck's constant h. Equation 24-14 also shows that if we extend the graph of K_{max} as a function of f to (unphysical) values of K_{max} less than zero, the graph intercepts the vertical axis at $-\Phi_0$. If we repeat the experimental measurements for a second material with a different work function Φ_0, the straight line intercepts the vertical axis at a different point but has the same slope h as for the first material. This reinforces the idea that Planck's constant is a universal constant; its value does not depend on the properties of the material that absorbs the photons.

The remarkable fit of Einstein's theory to experiment is powerful evidence that light is indeed absorbed in the form of photons with energy $E = hf$ as given by Equation 24-13. Einstein was awarded the Nobel Prize in Physics in 1921 for his explanation of the photoelectric effect.

WATCH OUT !

Some electrons require more energy than Φ_0 to be ejected.

The work function Φ_0 is the smallest amount of energy required to eject an electron from a given material under the most favorable conditions. Other electrons in the material require more energy to eject and will emerge from the material with less kinetic energy than the value K_{max} given by Equation 24-14.

EXAMPLE 24-3 The Photoelectric Effect with Cesium

The work function for a sample of cesium is 3.43×10^{-19} J. (a) What is the minimum frequency of light that will result in electrons being ejected from this sample by the photoelectric effect? (b) What is the maximum wavelength of light that will result in electrons being ejected from this sample by the photoelectric effect?

Set Up

Equation 24-14 tells us that the minimum energy required to eject an electron corresponds to having $K_{max} = 0$, so the electrons just barely make it out of the cesium. We'll use this to find the threshold frequency f_0 that just barely allows an electron to be ejected. We'll find the corresponding wavelength using Equation 22-2.

Maximum kinetic energy of an electron in the photoelectric effect:

$$K_{max} = hf - \Phi_0 \qquad (24\text{-}14)$$

Propagation speed, frequency, and wavelength of an electromagnetic wave:

$$c = f\lambda \qquad (22\text{-}2)$$

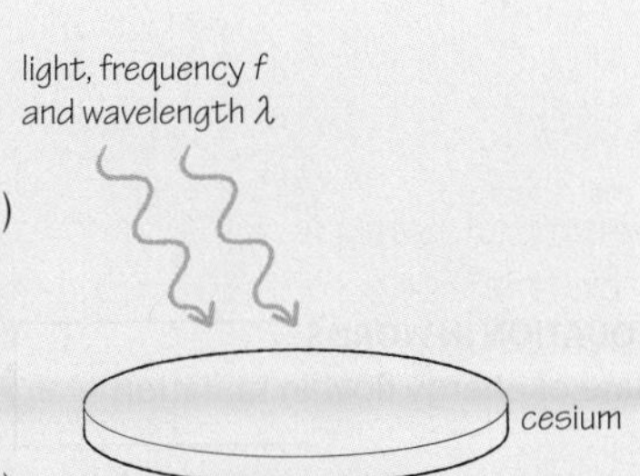
light, frequency f and wavelength λ
cesium

Solve

(a) Use Equation 24-14 to calculate the frequency f that corresponds to $K_{max} = 0$.

From Equation 24-14 with $K_{max} = 0$:

$$0 = hf - \Phi_0$$
$$hf = \Phi_0$$

This states that the photon energy hf is just enough to remove the most easily dislodged electron from the material (which requires energy Φ_0), with nothing left over to give the electron any kinetic energy. So $hf = \Phi_0$ is the minimum photon energy that will eject an electron, and the frequency f of this photon is the minimum (threshold) frequency that will do the job:

$$f_0 = f = \frac{\Phi_0}{h}$$
$$= \frac{3.43 \times 10^{-19}\ \text{J}}{6.63 \times 10^{-34}\ \text{J} \cdot \text{s}} = 5.17 \times 10^{14}\ \text{s}^{-1} = 5.17 \times 10^{14}\ \text{Hz}$$

(b) Find the wavelength that corresponds to the frequency that we calculated in part (a).

We can rewrite Equation 22-2 as

$$\lambda = \frac{c}{f}$$

In words, this says that wavelength is inversely proportional to frequency. So the minimum frequency of light that will eject an electron corresponds to the maximum wavelength that will eject an electron:

$$\lambda_{max} = \frac{c}{f_0} = \frac{3.00 \times 10^8\ \text{m/s}}{5.17 \times 10^{14}\ \text{Hz}}$$
$$= 5.80 \times 10^{-7}\ \text{m} = 580\ \text{nm}$$

(Recall that 1 nm = 10^{-9} m.)

Reflect

Figure 22-3 shows that a wavelength of 580 nm is in the yellow-green part of the visible spectrum. If we illuminate cesium with light of higher frequency and shorter wavelength than this (for example, blue or violet light), the photons will have more energy than the minimum and electrons will be ejected from the cesium. If instead we illuminate cesium with light of lower frequency and longer wavelength (for example, orange or red light), the photons will have less energy than the required minimum and no electrons will be ejected.

NOW WORK Problems 1–3, 8, and 10 from the Takeaway 24-3.

Blackbody Radiation

The photoelectric effect shows that light is *absorbed* in the form of photons. If we are to fully believe the photon concept, however, it must also be true that light is *emitted* in the form of photons. We learned in Section 14-6 that ordinary objects emit electromagnetic radiation as a result of their temperature. If we can find evidence that this emission is in the form of photons, we will have further evidence that the photon description of light is the correct one. Let's take a closer look at radiation of this kind.

Experiment shows that the rate at which an object emits radiation is proportional to its surface area A and to the fourth power of its Kelvin temperature T:

EQUATION IN WORDS
Rate of energy flow in radiation

$$P = e\sigma AT^4 \qquad (14\text{-}20)$$

Rate at which an object emits energy in the form of radiation

Emissivity of the object (a number between 0 and 1)

Temperature of the object on the Kelvin scale

Stefan–Boltzmann constant = $5.6704 \times 10^{-8}\ \text{W} \cdot \text{m}^{-2} \cdot \text{K}^{-4}$

Surface area of the object

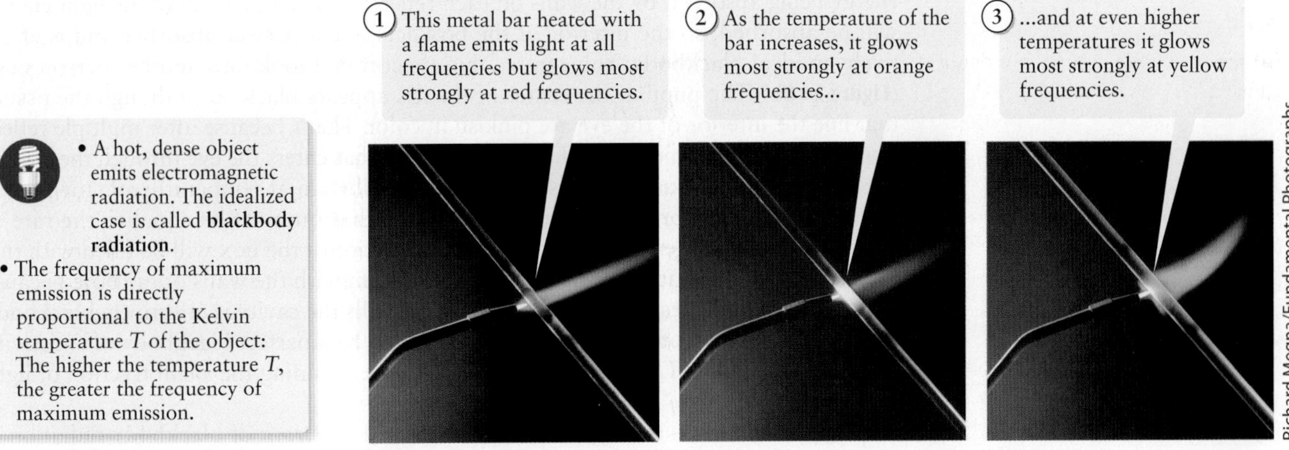

1 This metal bar heated with a flame emits light at all frequencies but glows most strongly at red frequencies.
2 As the temperature of the bar increases, it glows most strongly at orange frequencies...
3 ...and at even higher temperatures it glows most strongly at yellow frequencies.
• A hot, dense object emits electromagnetic radiation. The idealized case is called **blackbody radiation.**
• The frequency of maximum emission is directly proportional to the Kelvin temperature T of the object: The higher the temperature T, the greater the frequency of maximum emission.
Richard Megna/Fundamental Photographs

Figure 24-7 Radiation from heated objects The color of the light from a heated object depends on its temperature.

The higher the temperature of an object of a given size, the greater the radiated power P and so the more brightly it glows.

Experiment also shows that the *color* of the radiation emitted by an object depends on its temperature T (**Figure 24-7**). A heated object emits light at all wavelengths, but emits most strongly at a particular frequency called the *frequency of maximum emission*. As the temperature increases, the frequency of maximum emission increases.

Equation 14-20 shows that the radiated power also depends on a quantity e called the *emissivity*, which depends on the properties of the object's surface. This has its greatest value ($e = 1$) for an idealized type of dense object called a **blackbody**. An ideal blackbody does not reflect any light at all but absorbs all radiation falling on it. If a blackbody is in thermal equilibrium with its surroundings, it must emit energy at the same rate that it absorbs it in order for its temperature T to remain constant. So in addition to being a perfect absorber of energy, an ideal blackbody in thermal equilibrium with its surroundings is also a perfect emitter of energy because it emits as much energy as it absorbs.

Ordinary objects, such as tables, textbooks, and people, are not ideal blackbodies; they reflect light, which is why they are visible. (Even a piece of wood darkened with soot or painted a dull black reflects *some* light.) But it is possible to make a nearly ideal blackbody simply by building a box and drilling a small hole in one side (**Figure 24-8a**). Light that enters the hole will reflect around inside the box, with part of the light

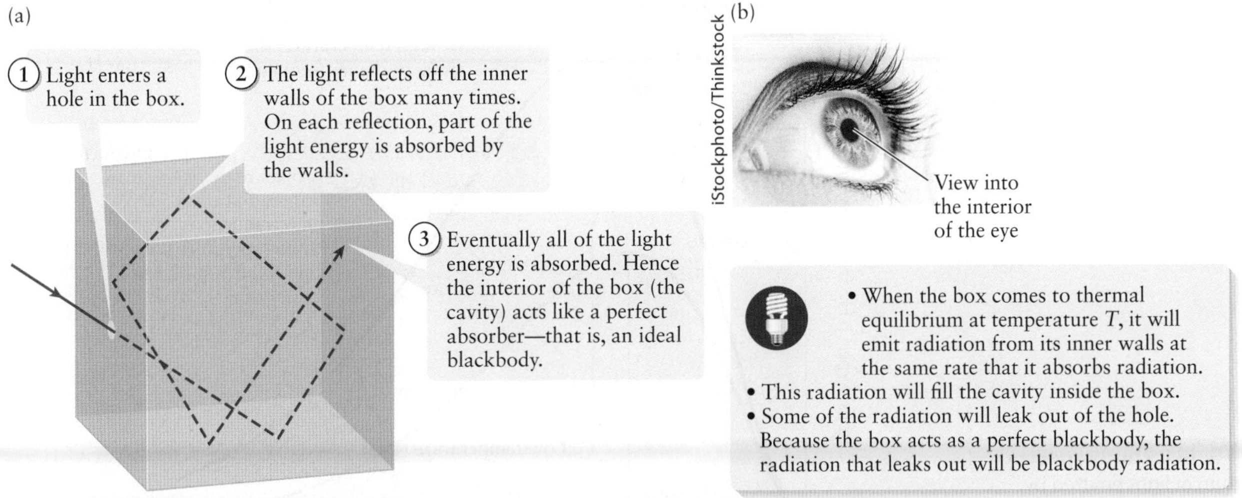

1 Light enters a hole in the box.

Figure 24-8 A blackbody cavity (a) A box with a small hole in one side is a good approximation to an ideal blackbody. (b) The interior of the eye has properties very similar to the box in (a).

energy being absorbed by the walls on each reflection. Eventually all of the light energy will be absorbed, so the interior of the box acts like a perfect absorber and is effectively an ideal blackbody. You can see this effect if you look into another person's eye (**Figure 24-8b**). The pupil at the center of the iris appears black, even though the tissues that line the interior of the eye are pinkish in color. That's because after multiple reflections, those tissues almost completely absorb light that enters the eye through the pupil.

If the box in Figure 24-8a is in thermal equilibrium at temperature T, the rate at which the walls absorb energy in the form of radiation must be equal to the rate at which the walls emit energy. The cavity in the interior of the box will be filled with this radiation, which will itself be in thermal equilibrium with the walls of the box. Because the walls act as an ideal blackbody, the light that fills the cavity is effectively **blackbody radiation**—the kind of light that would be emitted by a perfect blackbody of emissivity $e = 1$ at temperature T. We can study this light by examining the small fraction of light that emerges from the hole in Figure 24-8a.

Figure 24-9 shows the experimentally observed *spectrum* of blackbody radiation—that is, the relative amount of light energy present at different frequencies—for two different temperatures. The high-temperature curve lies above the low-temperature curve, which tells us that the higher the temperature of a blackbody, the greater the amount of radiation at all frequencies. Furthermore, as the blackbody temperature increases, the peak of the curve shifts to a higher frequency. This agrees with our observation about how the frequency of maximum emission varies with an object's temperature (Figure 24-7). Figure 24-9 explains why we can't see the radiation from objects at room temperature, about $T = 300$ K. At this relatively low temperature, the frequency of maximum emission is in the infrared, which our eyes cannot see. There is some emission at visible frequencies (in Figure 24-9, to the right of the peak of the curve), but the amount of emission at $T = 300$ K is so low that our eyes can't detect it.

In the late nineteenth century physicists tried to understand the shape of the blackbody spectrum shown in Figure 24-9 using their knowledge of thermodynamics and electromagnetic waves. Their efforts ended in failure. To understand how they failed, we begin by noting that the electromagnetic waves inside the cavity in Figure 24-8a should be in the form of *standing* waves. That's because the waves will bounce back and forth between the walls of the cavity. We learned in Chapter 21 that when waves bounce back and forth between the ends of a string that's tied down at both ends, steady wave patterns arise for waves of certain wavelengths and frequencies. The same is true for electromagnetic waves in a cavity. The standing wave patterns for electromagnetic waves in a cavity are more complex than for waves on a string: Electromagnetic waves are three-dimensional, not just one-dimensional like those along the length

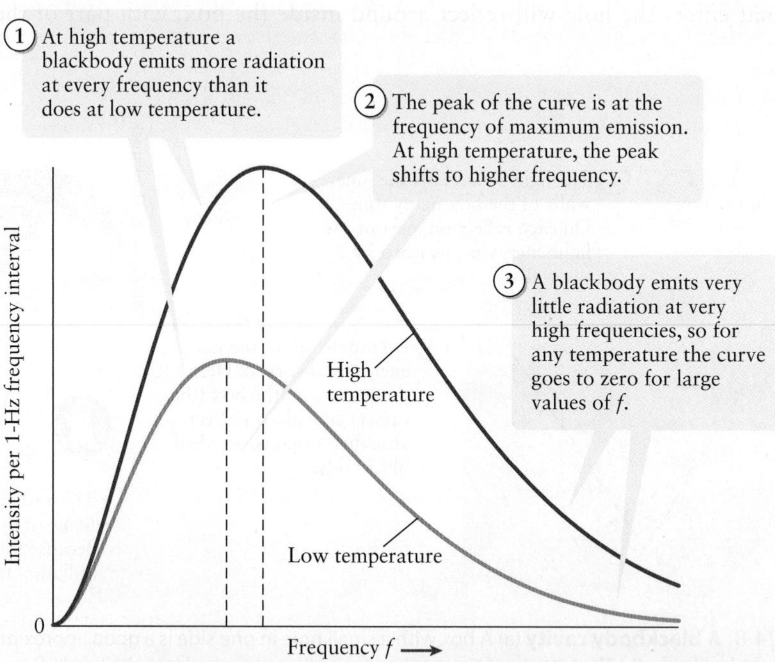

1 At high temperature a blackbody emits more radiation at every frequency than it does at low temperature.
2 The peak of the curve is at the frequency of maximum emission. At high temperature, the peak shifts to higher frequency.
3 A blackbody emits very little radiation at very high frequencies, so for any temperature the curve goes to zero for large values of f.
Intensity per 1-Hz frequency interval
High temperature
Low temperature
0
Frequency f

Figure 24-9 Blackbody spectra The spectrum of light emitted by an ideal blackbody depends on the temperature of the blackbody.

of a string, and involve two varying quantities, the electric field and the magnetic field. For our purposes, however, all we need to know is that these standing waves exist.

The shape of the blackbody spectrum in Figure 24-9 indicates how much energy is present in the standing waves in each frequency range: There is relatively little energy at very low frequencies, more energy at frequencies near the frequency of maximum emission, and again relatively little energy at high frequencies. However, nineteenth-century physics suggested that *every* possible standing wave in the cavity should contain on average the same amount of energy. This conclusion came from the equipartition theorem, which we first encountered in Section 14-3. This theorem states that a molecule in a gas at a Kelvin temperature T has, on average, an amount of energy $(1/2)kT$ for each degree of freedom of the molecule, where $k = 1.381 \times 10^{-23}$ J/K is the Boltzmann constant. Arguments from thermodynamics suggest that for the same reason a standing wave inside a cavity in thermal equilibrium at temperature T should also possess an average amount of energy equal to $(1/2)kT$ per degree of freedom. There are two degrees of freedom per standing wave, one for the electric field and one for the magnetic field, so the total average energy per standing wave should be kT. It turns out that the number of standing waves in a given frequency interval increases with increasing frequency. So according to nineteenth-century physics, the total energy per frequency interval (equal to the energy kT per standing wave multiplied by the number of standing waves per frequency interval) should increase with increasing frequency and *never* decrease (**Figure 24-10**). This is in profound disagreement with the experimentally observed shape of the spectrum.

Theoretical prediction using the equipartition theorem
Experimental data for the same temperature *and* theoretical prediction using the photon model
Intensity per 1-Hz frequency interval
0
Frequency f →

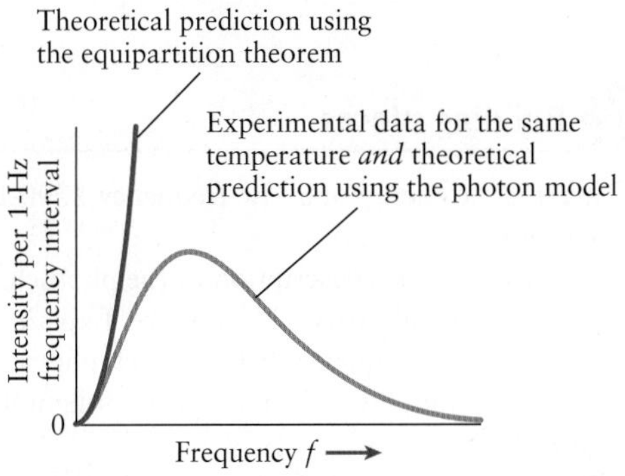

• A photon model accurately describes the spectrum of blackbody radiation.
• The model that does not use the photon concept fails to describe the spectrum.

Figure 24-10 The photon model explains blackbody radiation A model for blackbody radiation that does not use the photon concept predicts (incorrectly) that the intensity should increase without limit as the frequency increases.

If we now introduce the photon concept, however, we *can* match the experimentally observed spectrum. We still use the idea that the average energy available for a standing wave of frequency f is kT, but now this energy goes into photons of that frequency, which each have energy $E = hf$. This energy fundamentally comes from the walls of the box in Figure 24-8a, since these walls emit the photons. At low frequencies the photon energy hf is small compared to the available energy kT, so there will be many photons present for a low-frequency standing wave. But at high frequencies hf is much larger than kT, which means that the energy required to produce a photon is larger than the average energy available to produce one. Hence the average number of photons present for that standing wave will be very small. (It need not be zero, because kT is only the *average* energy available. From time to time the available energy will be greater than kT, and some photons can be produced.) So even though energy is *available* for a high-frequency standing wave, that energy can't be used to create photons, so the amount of energy *present* is quite small. That's just the effect we need to make the theoretical curve in Figure 24-10 decline at large frequencies.

Using a somewhat different version of this photon argument, in 1900 the German physicist Max Planck was able to make a theoretical prediction for the blackbody spectrum that was in excellent agreement with the experimental spectrum shown in Figure 24-10. Planck's theoretical formula was the first to involve the new quantity h that now bears his name, and the value of h given in Equation 24-13 is the one that gives the best match between this formula and the experimental data. Planck was awarded the 1918 Nobel Prize in Physics for his achievement.

The explanation of blackbody spectra in terms of photons is the evidence we were seeking that light is emitted in the form of photons. In the following section we'll see even more compelling evidence for the photon picture of light.

THE TAKEAWAY for Section 24-3

✔ In the photoelectric effect, an electron in a material can absorb light energy that strikes the surface and as a result be ejected from the surface.

✔ For any material, the light must have a certain minimum frequency for electrons to be ejected. This is evidence that light comes in the form of photons, with an energy proportional to their frequency.

✔ A perfect blackbody is an ideal absorber of light and also an ideal emitter of light. The spectrum of light emitted by a blackbody depends on its temperature.

✔ The spectrum of light emitted by a blackbody can be understood only if we use the idea that light is emitted in the form of photons.

Prep for the AP® Exam

AP® Building Blocks

1. What is the energy of a low-frequency 2000-Hz radio photon?

2. Prior to Einstein's description of the photoelectric effect, light was thought to act like a wave. Explain why the existence of a frequency below which photoelectrons are not emitted favors a description of light as a particle instead.

3. What is the minimum frequency of light required to eject electrons from a metal with a work function of 6.53×10^{-19} J?
4. What is the energy of a 0.200-nm x-ray photon?
5. What are the wavelength and frequency of a 3.97×10^{-19}-J photon?

AP® Skill Builders

6. Calculate the range of photon frequencies and energies in the visible spectrum of light (approximately 380–750 nm).
7. Under most conditions, the human eye will respond to a flash of light if 100 photons hit photoreceptors at the back of the eye. Determine the total energy of such a flash if the wavelength is 550 nm (green light).

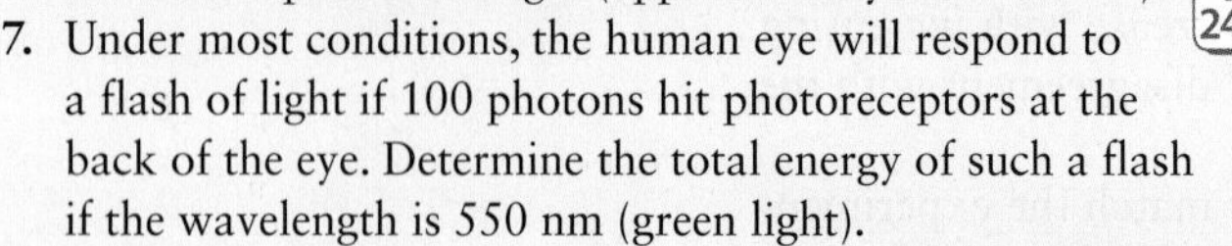

8. The work functions of aluminum, calcium, potassium, and cesium are 6.54×10^{-19} J, 4.65×10^{-19} J, 3.57×10^{-19} J, and 3.36×10^{-19} J, respectively. For which of the metals will photoelectrons be emitted when irradiated with visible light (wavelengths from 380 to 750 nm)?

AP® Skills in Action

9. Two objects of the same size are both perfect blackbodies. One is at a temperature of 3000 K, so its frequency of maximum emission is in the infrared part of the electromagnetic spectrum; the other is at a temperature of 12,000 K, so its frequency of maximum emission is in the ultraviolet part of the spectrum. Compared to the object at 3000 K, the object at 12,000 K
   (A) emits more infrared light.
   (B) emits more visible light.
   (C) emits more ultraviolet light.
   (D) all of the above.

EX 24-3
10. Light that has a 195-nm wavelength strikes a metal surface, and photoelectrons are produced moving as fast as $0.004c$.
   (a) What is the work function of the metal?
   (b) What is the threshold wavelength for the metal above which no photoelectrons will be emitted?

24-4 As a result of its photon character, light changes wavelength when it is scattered

Blackbody radiation and the photoelectric effect suggest that photons can be treated like tiny bundles of energy, and so have a particle-like nature. In the photoelectric effect, a photon strikes an electron and is *absorbed*. But if a photon is like a particle, is it possible that, as in the collisions we studied in Chapter 9, a photon could strike an electron and bounce off? If so, based on our experience with collisions, linear momentum would be conserved in such an interaction and the momentum of the photon should change as a result. This effect, called *Compton scattering*, is further evidence that light does indeed come in the form of photons.

Let's first see how to express the momentum of a photon. Relativity is not part of the curriculum for AP® Physics (but is included in Chapter 26 if you are interested). So, for now, we simply state that for a particle of mass m, we can write the kinetic energy and momentum as

EQUATION IN WORDS
Einsteinian expressions for kinetic energy and momentum

Kinetic energy of a particle of mass m and velocity $\vec{v}$

$$K = (\gamma - 1)mc^2$$
$$\vec{p} = \gamma m\vec{v} \qquad (24\text{-}15)$$

Relativistic gamma for speed v

Momentum of a particle of mass m and velocity $\vec{v}$

At speeds that are a small fraction of the speed of light c, these are approximately equal to the Newtonian expressions:

$$K = \frac{1}{2}mv^2 \text{ and } \vec{p} = m\vec{v}$$

In these expressions the quantity γ (relativistic gamma) is

Relativistic gamma for a particle moving at speed v

$$\gamma = \frac{1}{\sqrt{1 - \frac{v^2}{c^2}}} \qquad (24\text{-}16)$$

Speed of the particle

Speed of light in a vacuum

EQUATION IN WORDS
Relativistic gamma

In Chapter 1, we mentioned that an object of mass m has a rest energy E_0 that is present even when it is not moving (and will dig into this a bit more in Chapter 26):

Rest energy of a particle

Mass of the particle

$$E_0 = mc^2 \qquad (1\text{-}1)$$

Speed of light in a vacuum

EQUATION IN WORDS
Rest energy

If we combine the first of Equations 24-15 with Equations 24-16 and 1-1, we find that the total energy of a particle (kinetic energy plus rest energy) is

$$E = K + E_0 = (\gamma - 1)mc^2 + mc^2 = \gamma mc^2 = \frac{mc^2}{\sqrt{1 - \frac{v^2}{c^2}}} \qquad (24\text{-}17)$$

From the second of Equations 24-15, the magnitude of the momentum of a particle is

$$p = \gamma mv = \frac{mv}{\sqrt{1 - \frac{v^2}{c^2}}} \qquad (24\text{-}18)$$

Comparing Equations 24-17 and 24-18, we see that

$$p = \frac{Ev}{c^2} \text{ (momentum of a particle of total energy } E) \qquad (24\text{-}19)$$

A photon has zero mass, which is why it can travel at the speed of light. (In Chapter 26, we will see that any object with nonzero mass would require an infinite amount of kinetic energy to reach $v = c$.) As such, we can't apply Equation 24-17 or Equation 24-18 directly to photons. But we can use the combination in Equation 24-19, in which mass does not appear explicitly. Setting $v = c$ in Equation 24-19, we get the following relationship for the momentum of a photon:

$$p = \frac{Ec}{c^2} = \frac{E}{c} \text{ (momentum of a photon)} \qquad (24\text{-}20)$$

From Equation 24-13 we can write $E = hf$, and we know that for light waves $c = f\lambda$ (Equation 22-2). If we substitute these into Equation 24-20, we get an alternative expression for the momentum of a photon:

$$p = \frac{hf}{f\lambda}$$

or, simplifying,

Magnitude of the **momentum** of a photon

Energy of the photon

$$p = \frac{E}{c} = \frac{h}{\lambda} \qquad (24\text{-}21)$$

Planck's constant

Speed of light in a vacuum

Wavelength of the photon

EQUATION IN WORDS
Momentum and energy of a photon

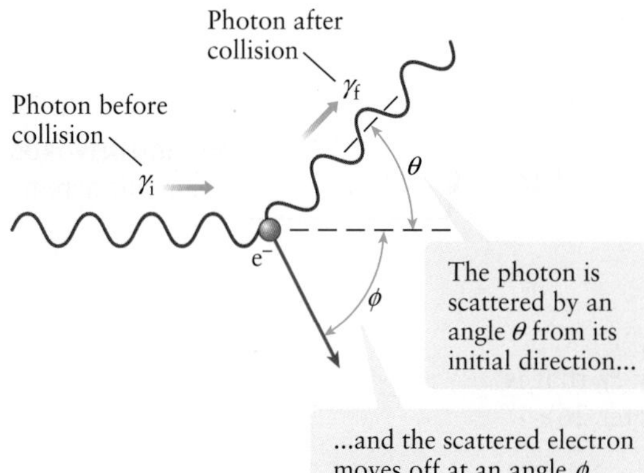

Figure 24-11 Compton scattering A photon undergoes a wavelength shift when it scatters from an electron that is initially at rest.

A photon's momentum is directly proportional to the energy that it carries and inversely proportional to its wavelength. Thus a violet photon of wavelength 400 nm has twice the momentum, as well as twice the energy, of an infrared photon of wavelength 800 nm.

In the early 1920s American physicist Arthur Compton showed conclusively that photons have momentum and that the momentum is inversely proportional to the wavelength, as stated in Equation 24-21. In his experiments, an x-ray photon collided with an electron in a carbon atom, a process now called **Compton scattering.** Compton detected both the electron, which is knocked out of the atom, and the scattered photon. Compton could account for the directions and energies of the electron and the scattered photon by requiring that the total momentum of the electron and the photon be conserved in the collision. In other words, he showed that the photon description of light applies not just to the absorption and emission of light but also to what happens to light when it is scattered. For revealing this fundamental aspect of light, Compton was awarded the Nobel Prize in Physics in 1927.

Figure 24-11 shows the situation that Compton studied, the collision of a photon (symbol γ_i) and a stationary electron (symbol e^-). The electron is scattered at angle ϕ relative to the initial direction of the photon, and the photon scatters at angle θ relative to its initial direction. Although we have labeled the photon γ_f after the collision, the scattered photon is the same photon that collided with the electron. The subscripts "i" for "initial" and "f" for "final" instead imply that the energy, wavelength, and other quantities associated with the photon have changed as a result of the collision. For example, because energy is conserved during the collision and because some of the photon's energy is almost always transferred to the electron, the outgoing photon carries less energy than it had initially. From Equation 24-21 the energy of a photon is

$$E = pc = \frac{hc}{\lambda}$$

Because the photon has less energy after the collision than it had initially even though it still travels at the same speed, E_f is less than E_i, and the final wavelength λ_f is greater than the initial wavelength λ_i. In other words, as the energy of the photon decreases, its wavelength increases. Compton found that the increase in wavelength $\Delta\lambda$ from λ_i to λ_f is a function of the angle θ at which the photon scatters:

EQUATION IN WORDS
Compton scattering equation

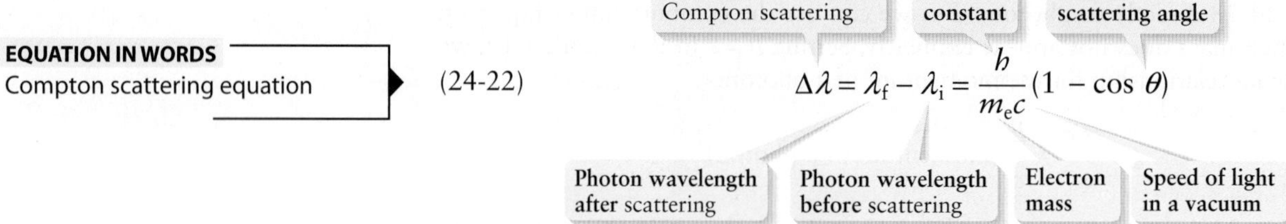

$$\Delta\lambda = \lambda_f - \lambda_i = \frac{h}{m_e c}(1 - \cos\theta) \tag{24-22}$$

The proportionality constant in Equation 24-22 is known as the *Compton wavelength* λ_C:

$$\lambda_C = \frac{h}{m_e c} \tag{24-23}$$

NEED TO REVIEW?
Turn to page M12 in the Math Tutorial in the back of the book for more information on trigonometry.

The scattering angle θ of the photon ranges from 0° (straight forward) to 180° (straight back). When the photon continues in the same direction after the collision, so that θ equals 0° and $\cos\theta$ equals 1, $\Delta\lambda$ equals zero. In other words, if $\theta = 0°$ there is no change in the photon's wavelength and no change in the photon's energy.

The maximum change in a photon's wavelength and energy when it undergoes Compton scattering occurs when it scatters straight back, in the direction opposite to the one in which it approached the electron. In this case $\theta = 180°$, so $\cos\theta = -1$ and the term in parentheses in Equation 24-22 equals 2. The maximum possible change in the photon's wavelength is therefore $2\lambda_C$, or twice the Compton wavelength. Using the known values of h, m_e, and c, we find

$$\lambda_C = \frac{h}{m_e c} = \frac{6.63 \times 10^{-34}\ \mathrm{J \cdot s}}{(9.11 \times 10^{-31}\ \mathrm{kg})(3.00 \times 10^{8}\ \mathrm{m/s})}$$
$$= 2.43 \times 10^{-12}\ \mathrm{m} = 2.43 \times 10^{-3}\ \mathrm{nm} = 0.00243\ \mathrm{nm}$$

(Recall that 1 nm = 10^{-9} m.) By comparison, the wavelength of visible light ranges between about 380 and 750 nm. So if a photon of wavelength 400 nm is scattered by an electron its wavelength changes by a value on the order of 0.00243 nm, a change of less than one part in 10^5. Such a tiny change is very difficult to measure, so we conclude that a visible-light photon undergoes a negligible shift in wavelength due to Compton scattering. But for an x-ray photon with a wavelength on the order of 10^{-11} m = 0.01 nm, a wavelength shift of 0.00243 nm corresponds to a large percentage of the initial wavelength. That's why Compton first noticed this effect in the scattering of x-rays rather than visible light.

If light did not have a photon aspect, we would expect *no* wavelength change on scattering. A light wave of frequency f and wavelength λ encountering an electron would make the electron oscillate at the same frequency f, and the electron would emit radiation with the same frequency f and hence the same wavelength λ as the initial light wave. The change in wavelength that Compton observed is unambiguous evidence that light does indeed have a particle character.

EXAMPLE 24-4 Compton Scattering

A photon carries 2.00×10^{-14} J of energy. It undergoes Compton scattering in a block of carbon. What is the largest fractional change in energy the photon can undergo as a result?

Set Up

Given the initial photon energy $E_i = 2.00 \times 10^{-14}$ J, we can calculate its wavelength λ_i using Equation 24-21. We use Equation 24-22 to calculate the change in wavelength due to scattering; this will be maximum if $\theta = 180°$, so $\cos\theta = -1$. Once we know the final wavelength λ_f, we can use Equation 24-21 again to find the final photon energy. Comparing this to the initial photon energy tells us the fractional change in energy.

Momentum and energy of a photon:

$$p = \frac{E}{c} = \frac{h}{\lambda} \qquad (24\text{-}21)$$

Compton scattering equation:

$$\Delta\lambda = \lambda_f - \lambda_i = \frac{h}{m_e c}(1 - \cos\theta) \qquad (24\text{-}22)$$

Before:
photon
electron at rest
After:
scattered photon
recoiling electron

Solve

First calculate the wavelength of the initial photon using Equation 24-21.

From Equation 24-21 the wavelength of the initial photon is

$$\lambda_i = \frac{hc}{E_i} = \frac{(6.63 \times 10^{-34}\ \mathrm{J \cdot s})(3.00 \times 10^{8}\ \mathrm{m/s})}{2.00 \times 10^{-14}\ \mathrm{J}}$$
$$= (9.95 \times 10^{-12}\ \mathrm{m})\left(\frac{1\ \mathrm{nm}}{10^{-9}\ \mathrm{m}}\right) = 9.95 \times 10^{-3}\ \mathrm{nm}$$

Calculate the wavelength shift using Equation 24-22.

The maximum wavelength shift is with $\theta = 180°$ and $\cos\theta = -1$:

$$\Delta\lambda_{max} = \lambda_f - \lambda_i = \frac{h}{m_e c}(1 - \cos 180°)$$
$$= (2.43 \times 10^{-3}\ \mathrm{nm})[1 - (-1)] = 4.85 \times 10^{-3}\ \mathrm{nm}$$

The wavelength of the final photon equals the wavelength of the initial photon plus the shift $\Delta\lambda$. Use this to find the energy of the final photon.

The final wavelength is

$$\lambda_f = \lambda_i + \Delta\lambda = 9.94 \times 10^{-3} \text{ nm} + 4.85 \times 10^{-3} \text{ nm}$$
$$= 1.480 \times 10^{-2} \text{ nm} = 1.480 \times 10^{-11} \text{ m}$$

The energy of the final photon is

$$E_f = \frac{hc}{\lambda_f} = \frac{(6.63 \times 10^{-34} \text{ J}\cdot\text{s})(3.00 \times 10^8 \text{ m/s})}{1.480 \times 10^{-11} \text{ m}}$$
$$= 1.34 \times 10^{-14} \text{ J}$$

This is less than the energy of the initial photon. The lost energy has gone into the kinetic energy of the scattered electron.

Express the energy change as a fraction of the initial photon energy.

The fractional energy change is the energy change $E_f - E_i$ divided by the initial energy E_i:

$$\text{fractional energy change} = \frac{E_f - E_i}{E_i}$$
$$= \frac{1.34 \times 10^{-14} \text{ J} - 2.00 \times 10^{-14} \text{ J}}{2.00 \times 10^{-14} \text{ J}}$$
$$= -0.328 = -32.8\%$$

Reflect

An initial photon of this high energy and short wavelength can lose as much as 32.8% (nearly one-third) of its initial energy when it undergoes Compton scattering.

NOW WORK Problems 1 and 4–10 from the Takeaway 24-4.

Example 24-4 suggests why x-rays are useful in cancer radiation therapy. If an x-ray photon strikes an electron in a water molecule within a cancerous cell, the photon can scatter and transfer a substantial amount of energy to the electron. This transferred energy is great enough that the electron escapes from the molecule, leaving the water molecule in an ionized state. These ionized water molecules damage the DNA of the cancerous cell and cause cell death.

THE TAKEAWAY for Section 24-4

✔ A photon has zero mass but does have momentum. The magnitude of the momentum is proportional to the photon energy and inversely proportional to the wavelength.

✔ In Compton scattering, a photon scatters from an electron. The photon loses energy and momentum, and these are transferred to the electron. The change in wavelength of the photon depends on the angle through which it is scattered.

Prep for the AP® Exam

1. Calculate the momentum of photons of wavelength (a) 550 nm and (b) 0.0711 nm.
2. The quantity $\lambda_C = h/(mc)$ is called the Compton wavelength. Calculate the numerical value of this quantity for (a) an electron, (b) a proton, and (c) a pi meson (which has a mass of 2.50×10^{-28} kg).
3. The Compton effect is practically unobservable for visible light. Justify this using physics principles.

4. A 0.0750-nm photon Compton scatters off a stationary electron. Determine the maximum speed of the scattered electron.

EX 24-4

5. X-ray photons that have a wavelength of 0.140 nm are scattered off carbon atoms (which possess essentially stationary electrons in their valence shells). What are the wavelengths of the Compton-scattered photons and the

kinetic energies of the scattered electrons if the photons are scattered at angles of (a) 0.00°, (b) 30.0°, (c) 45.0°, (d) 60.0°, (e) 90.0°, and (f) 180°?

EX 24-4 6. Photons that have a wavelength of 0.00225 nm are Compton scattered off stationary electrons at 45.0°. What is the energy of the scattered photons?

EX 24-4 7. If a photon undergoes Compton scattering from a stationary electron and experiences a fractional wavelength change of +7.25%, calculate the angle at which the scattered photons are directed if the original photons have a wavelength of 0.00335 nm.

EX 24-4 8. X-rays that have wavelengths of 0.125 nm are scattered off stationary electrons at an angle of 30.0°.

(a) Calculate the wavelength of the scattered electromagnetic radiation.

(b) Calculate the fractional wavelength change ($\Delta\lambda/\lambda_i = (\lambda_f - \lambda_i)/\lambda_i$) for the scattered x-rays.

Skills in Action

EX 24-4 9. Suppose a photon has a wavelength equal to the Compton wavelength λ_C. If this photon collides with an electron, and the photon is scattered through an angle of 90°, what will be the wavelength of the photon after the collision?

(A) Zero
(B) $\lambda_C/2$
(C) λ_C
(D) $2\lambda_C$

EX 24-4 10. An x-ray source is incident on a collection of stationary electrons. The electrons are scattered with a speed of 4.50×10^5 m/s, and the photon scatters at an angle of 60.0° from the incident direction of the photons. Determine the wavelength of the x-ray source.

24-5 Matter, like light, has aspects of both waves and particles

We have seen that light has a dual nature, with attributes of both waves and particles. Light comes in the form of particles (photons), but these particles have a wave aspect: Associated with them is a frequency f, which determines the photon energy $E = hf$, and a wavelength λ, which determines the photon momentum $p = h/\lambda$. Is it possible that ordinary matter, which we know is made of particles such as electrons, protons, and neutrons, also has a dual nature? Could these particles also have a wave aspect?

In 1924 French graduate student Louis de Broglie (pronounced "de broy") proposed precisely that idea. In particular, he suggested that the relationship $p = h/\lambda$ between momentum p and wavelength λ that applies to photons should also apply to particles such as electrons (**Figure 24-12**). The wavelength of a particle is called its **de Broglie wavelength:**

A particle has a **de Broglie wavelength...** ...equal to **Planck's constant...**

$$\lambda = \frac{h}{p} \tag{24-24}$$

...divided by the **momentum of the particle.** The greater the momentum, the shorter the de Broglie wavelength.

EQUATION IN WORDS
de Broglie wavelength

How large should we expect the wavelength of an electron to be? Let's examine the case of an electron of charge $q = -e$ that gains its momentum by moving through a potential difference of ΔV, so the electron starts at position a and moves to a position b where the potential has a larger positive value (remember from Chapter 17 that a negative charge "falls" toward higher potential). The electron potential energy then changes from $U_a = qV_a$ to $U_b = qV_b = (-e)\Delta V = -e\Delta V$. The change in electric potential energy is

$$\Delta U = U_b - U_a = -e(V_b - V_a) = -e\Delta V < 0$$

The electric potential energy decreases by an amount $e\Delta V$. Mechanical energy is conserved if the only force exerted on the electron is the (conservative) electric force,

Particle description	Wave description
Particle, mass m; Speed v_1; Momentum $p_1 = mv_1$	Wavelength $\lambda_1 = \dfrac{h}{p_1}$
Particle, mass m; Speed $v_2 = 2v_1$; Momentum $p_2 = mv_2 = 2p_1$	Wavelength $\lambda_2 = \dfrac{h}{p_2} = \dfrac{h}{2p_1} = \dfrac{\lambda_1}{2}$

Figure 24-12 Wave-particle duality Matter has both particle aspects (speed and momentum) and wave aspects (wavelength).

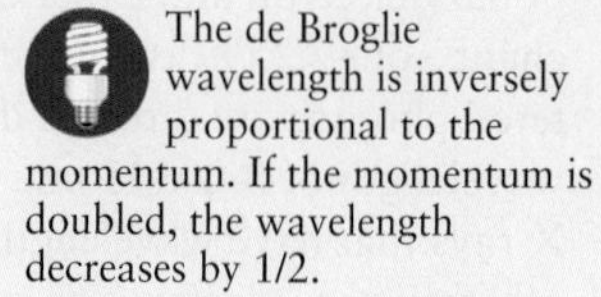

The de Broglie wavelength is inversely proportional to the momentum. If the momentum is doubled, the wavelength decreases by 1/2.

so the decrease in electric potential energy equals the gain in kinetic energy of the electron:

$$\Delta K = -\Delta U = -(-e\Delta V) = +e\Delta V > 0 \qquad (24\text{-}25)$$

If the electron starts at rest, its initial kinetic energy is zero and its final kinetic energy is $K = (1/2)mv^2$, so $\Delta K = (1/2)mv^2 - 0 = (1/2)mv^2$. (We're assuming that the electron is moving at a speed much slower than the speed of light c, so we don't have to use the Einsteinian expression for kinetic energy.) Then, from Equation 24-25, the final kinetic energy of the electron is

$$K = \frac{1}{2}mv^2 = e\Delta V$$

Solve for the final speed of the electron:

$$v^2 = \frac{2K}{m} = \frac{2e\Delta V}{m} \text{ so } v = \sqrt{v^2} = \sqrt{\frac{2e\Delta V}{m}} \qquad (24\text{-}26)$$

The final momentum of the electron is $p = mv$, and its final wavelength is $\lambda = h/p = h/mv$ from Equation 24-24. If we substitute v from Equation 24-26 into the formula for the wavelength of the electron, we get

$$\lambda = \frac{h}{mv} = \frac{h}{m}\sqrt{\frac{m}{2e\Delta V}} = \frac{h}{\sqrt{2me\Delta V}} \qquad (24\text{-}27)$$

Suppose that the electron is accelerated through a potential difference $\Delta V = 50.0$ V. Substituting this into Equation 24-27 along with the values of Planck's constant h, the electron mass m, and the magnitude e of the electron charge, we get

$$\lambda = \frac{6.63 \times 10^{-34}\ \text{J}\cdot\text{s}}{\sqrt{2(9.11 \times 10^{-31}\ \text{kg})(1.60 \times 10^{-19}\ \text{C})(50.0\ \text{V})}}$$
$$= 1.74 \times 10^{-10}\ \text{m} = 0.174\ \text{nm}$$

To see how to measure such a short electron wavelength, note that a photon with this wavelength is in the x-ray region of the electromagnetic spectrum (see Figure 22-3). It was known in the 1920s that x-rays show interference effects when they reflect from adjacent atoms in a crystal: At certain angles waves that reflect from one atom will interfere constructively (so the reflected intensity is high) with those that reflect from neighboring atoms, while at other angles they interfere destructively (so the reflected intensity is near zero). This can happen because the spacing between adjacent atoms in a crystal is around 0.1 nm, comparable to the wavelength of the x-rays. So if electrons have a wave aspect, we expect that a beam of electrons that have been accelerated from rest through 50.0 V should display the same kind of interference effects as a beam of x-rays.

In 1927 the American physicists Clinton Davisson and Lester Germer performed precisely this kind of experiment using a beam of electrons directed at a target of crystalline nickel. They found that the intensity of reflected electrons was greater for certain angles, just as for x-rays. What's more, the angles at which this maximum intensity occurred were precisely those expected if the wavelength of electrons was given by the de Broglie relation, Equation 24-24. This groundbreaking result was quickly confirmed in experiments carried out by the British physicist G. P. Thomson. These results resoundingly confirmed de Broglie's remarkable hypothesis and showed that matter does indeed have a wave aspect. (The 1927 Nobel Prize in Physics went to de Broglie; the 1937 prize was shared by Davisson and Thomson.) The dual character of *both* light and matter, which have both wave and particle characteristics, is called **wave-particle duality.**

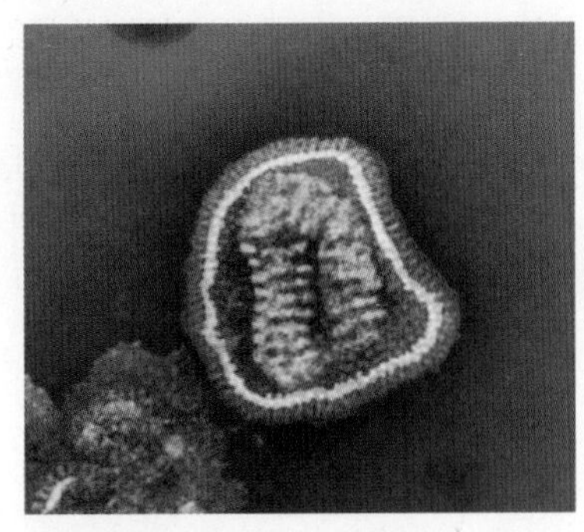

Figure 24-13 An electron micrograph When a beam of low-energy electrons is shot through a thin slice of a specimen in a transmission electron microscope (TEM), the pattern formed by the diffracted electrons forms an image. A TEM captured this (false-color) image of an influenza virus particle, which is only about 100 nm in diameter. Because the wavelengths of low-energy electrons are so much shorter than those of light, a TEM is capable of significantly better resolution than light microscopes (better than 0.005 nm for a TEM, compared to about 0.2 μm with the most powerful optical microscopes).

Wave-particle duality is both surprising and counterintuitive. It is also of tremendous practical use. One important application that has revolutionized biology is the *electron microscope*. A major limitation of ordinary microscopes is that the smallest detail that can be resolved is about the size of a wavelength of visible light (about 380 to 750 nm). This makes microscopes useless for studying the structure of viruses, for example, which range in size from 5 to 300 nm. But as our above example of the 50.0-V electron shows, the wavelength of electron waves can be a fraction of a nanometer. So images made with an *electron microscope* can reveal details that are forever hidden from an ordinary visible-light microscope. **Figure 24-13** is an electron microscope image of an influenza virus, in which details smaller than a nanometer across can be seen.

If particles have wave aspects, why don't we notice these aspects for objects around us? The answer is that any object large enough to be seen by the naked eye has a relatively large momentum, so its de Broglie wavelength (which is inversely proportional to momentum) is infinitesimal. As an example, a dust mote floating in the air (such as you might see when a shaft of sunlight comes through the window) has a mass of about 8×10^{-10} kg. If it drifts at a speed of 1 mm/s = 10^{-3} m/s, its momentum is

$$p = mv = (8 \times 10^{-10} \text{ kg})(10^{-3} \text{ m/s}) = 8 \times 10^{-13} \text{ kg} \cdot \text{m/s}$$

and its de Broglie wavelength is

$$\lambda = \frac{h}{p} = \frac{6.63 \times 10^{-34} \text{ J} \cdot \text{s}}{8 \times 10^{-13} \text{ kg} \cdot \text{m/s}} = 8 \times 10^{-22} \text{ m}$$

That's about 10^{-6} of the diameter of a proton! It's impossible to see wave effects from a wave with such a tiny wavelength. To see diffraction of such a wave, we would have to create a slit whose width is much smaller than the width of a single proton. Wave effects are even smaller for larger objects (greater mass m) moving faster (greater speed v). So the wave aspect of matter is generally noticeable only on the atomic or subatomic scale. Particles such as electrons exhibit noticeable wave properties; objects such as dust motes, baseballs, and humans do not.

EXAMPLE 24-5 Finding the Wavelength of a Room-Temperature Neutron

A nuclear reactor emits *thermal neutrons*. These are neutrons that behave as though they were particles of an ideal gas at Kelvin temperature T. We learned in Section 14-3 that the average kinetic energy of a particle in an ideal gas at temperature T is $(3/2)kT$, where $k = 1.381 \times 10^{-23}$ J/K is the Boltzmann constant. Calculate the de Broglie wavelength of an average neutron at 293 K (room temperature). The mass of a neutron is 1.67×10^{-27} kg.

Set Up

We'll first calculate the kinetic energy of an average neutron using Equation 14-13 (Section 14-3). From this we can find the speed and magnitude of momentum of an average neutron. Equation 24-24 will then allow us to calculate the de Broglie wavelength of such a neutron.

de Broglie wavelength:

$$\lambda = \frac{h}{p} \qquad (24\text{-}24)$$

Temperature and average translational kinetic energy of an ideal gas particle:

$$K_{\text{translational, average}} = \frac{1}{2}m(v^2)_{\text{average}} = \frac{3}{2}kT \qquad (14\text{-}13)$$

Solve

Calculate the translational kinetic energy, speed, and momentum of an average neutron at $T = 293$ K.

From Equation 14-13:

$$K_{\text{translational, average}} = \frac{3}{2}kT = \frac{3}{2}(1.381 \times 10^{-23}\ \text{J/K})(293\ \text{K})$$
$$= 6.07 \times 10^{-21}\ \text{J}$$

Calculate the speed of a neutron with this kinetic energy:

$$K_{\text{translational, average}} = \frac{1}{2}mv^2 \text{ so}$$

$$v = \sqrt{\frac{2K_{\text{translational, average}}}{m}} = \sqrt{\frac{2(6.07 \times 10^{-21}\ \text{J})}{1.67 \times 10^{-27}\ \text{kg}}}$$
$$= 2.70 \times 10^3\ \text{m/s}$$

The magnitude of momentum of a neutron with this speed is

$$p = mv = (1.67 \times 10^{-27}\ \text{kg})(2.70 \times 10^3\ \text{m/s})$$
$$= 4.50 \times 10^{-24}\ \text{kg} \cdot \text{m/s}$$

Calculate the de Broglie wavelength of such a neutron.

From Equation 24-24:

$$\lambda = \frac{h}{p} = \frac{6.63 \times 10^{-34}\ \text{J} \cdot \text{s}}{4.50 \times 10^{-24}\ \text{kg} \cdot \text{m/s}} = 1.47 \times 10^{-10}\ \text{m} = 0.147\ \text{nm}$$

(Recall that 1 J = 1 $\text{kg} \cdot \text{m}^2/\text{s}^2$ and 1 nm = 10^{-9} m.)

Reflect

The de Broglie wavelength of a thermal neutron is about 0.147 nm, a distance that is typical of the size of atoms and of the spacing between atoms within a molecule. For this reason thermal neutrons are useful for studying the structure of complex molecules such as proteins. When the neutrons scatter from a protein molecule, they diffract and produce a diffraction pattern that is characteristic of the particular arrangement of atoms in the molecule. X-rays can have the same wavelength, but they interact with the charges within atoms and so scatter only weakly from relatively small atoms (with a small amount of internal charge) such as hydrogen, carbon, nitrogen, and oxygen found in proteins. Neutrons, by contrast, are electrically neutral and actually scatter more strongly from smaller atoms than from larger ones. This makes neutrons superior to x-rays for studies of protein structure.

NOW WORK Problems 4–10 from the Takeaway 24-5.

THE TAKEAWAY for Section 24-5

✔ Particles can exhibit wave properties such as diffraction.

✔ The wavelength of a particle is inversely proportional to its momentum. Hence wave effects are noticeable only for very small particles such as electrons, for which the momentum is very small.

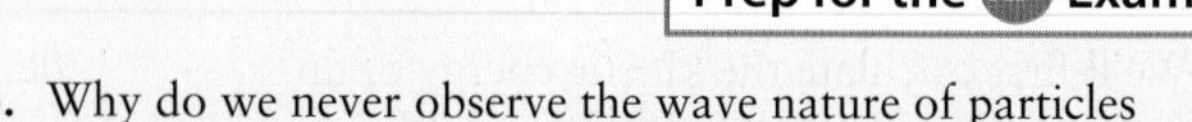

1. An electron and a proton have the same kinetic energy. Which has the longer wavelength?
2. Is the wavelength of an electron the same as the wavelength of a photon if both particles have the same total energy?
3. Why do we never observe the wave nature of particles for everyday objects such as birds or bumblebees, for example?
4. EX 24-5 Calculate the de Broglie wavelength of a 0.150-kg ball moving at 40.0 m/s. Comment on the significance of the result.
5. EX 24-5 Calculate the de Broglie wavelength of an electron that has a speed of 0.00730*c*.

6. Calculate the de Broglie wavelength of a proton ($m = 1.67 \times 10^{-27}$ kg) moving at 4.00×10^5 m/s.

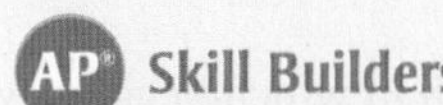
Skill Builders

7. Calculate the de Broglie wavelength for electrons with the following kinetic energies: (a) 1.60×10^{-19} J, (b) 1.60×10^{-18} J, (c) 1.60×10^{-17} J, and (d) 1.60×10^{-16} J.
8. Calculate the de Broglie wavelength of alpha particles ($m_\alpha = 6.64 \times 10^{-27}$ kg) that have a kinetic energy of (a) 1.60×10^{-13} J and (b) 8.00×10^{-13} J.

Skills in Action

9. Rank the following objects in order of their de Broglie wavelength, from longest to shortest.
   (A) A proton moving at 2.00×10^3 m/s
   (B) A proton moving at 4.00×10^3 m/s
   (C) An electron moving at 2.00×10^3 m/s
   (D) An electron moving at 4.00×10^3 m/s

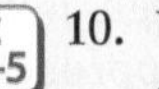
10. Write an equation that relates the Newtonian kinetic energy ($K = (1/2)mv^2$) and mass of a nonrelativistic particle to its de Broglie wavelength. (That is, complete the equation $\lambda(K, m) = ?$)

24-6 The spectra of light emitted and absorbed by atoms show that atomic energies are quantized

We have seen that the late nineteenth and early twentieth centuries were years of tremendous change in the study of physics. Studying the photoelectric effect and blackbody radiation led to the revolutionary concept that light has particle aspects, and de Broglie introduced the no less revolutionary idea that matter has wave aspects. During this same time a key set of experiments radically transformed our understanding of the nature of atoms.

The Nuclear Atom

The early Greeks introduced the idea of the atom, a unit of matter so small that it could not be subdivided. (The word *atom* is derived from the Greek term for indivisible.) In 1897 the British physicist J. J. Thomson discovered that atoms are not in fact indivisible but have an internal structure: All atoms contain negatively charged particles (electrons) that can be removed from the atom. It was known that atoms are electrically neutral, so there must also be positively charged material inside an atom. But what form does this positive charge take?

Thomson proposed that most of the mass of the atom is in the form of electrons and that the positively charged material is a low-density sort of jelly in which the electrons are embedded. This model is sometimes called the plum pudding model, since Thomson envisioned electrons scattered throughout the positive charge much like raisins in the traditional English dessert. (If you're not familiar with plum pudding, think of electrons as pieces of fruit embedded in a gelatin dessert or salad.)

A crucial experiment that tested this model was carried out in 1909 at the University of Manchester in England by the New Zealand–born chemist and physicist Ernest Rutherford with his colleagues Hans Geiger and Ernest Marsden. They fired subatomic particles called *alpha particles* (which were known to be positive helium ions, thousands of times more massive than an electron) at a very thin gold foil. Rutherford expected that if the gold atoms had the structure described in Thomson's model, the alpha particles would be only slightly deflected from their initial direction as a result of passing through the diffuse, positive "pudding" of the atoms. Instead, Rutherford was startled to discover that alpha particles were sometimes scattered at large angles with respect to the initial direction, occasionally leaving the gold foil directly *backward*. In reflecting on this experiment Rutherford later said, "It was quite the most incredible event that ever happened to me in my life. It was as incredible as if you fired a 15-in. shell [a large projectile fired from a military weapon] at a piece of tissue paper and it came back and hit you. On consideration, I realized that this scattering backwards must be the result of a single collision, and when I made calculations I saw that it was impossible to get anything of that order of magnitude unless you took a system in which the mass of the atom was concentrated in a minute nucleus."

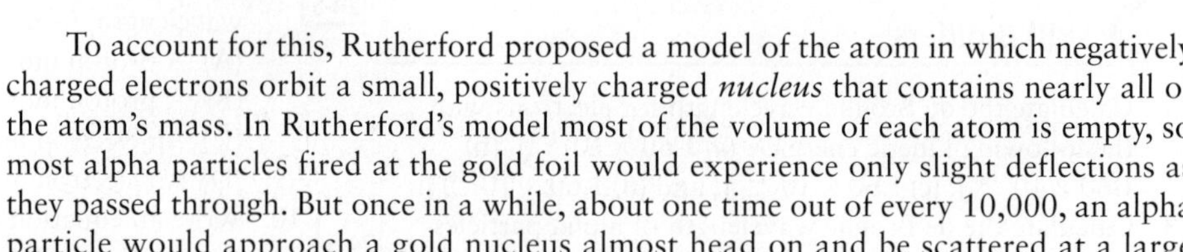

Figure 24-14 The absorption spectrum of the Sun The dark lines in the spectrum of sunlight indicate that certain wavelengths are absorbed when light from the solar interior passes through the Sun's atmosphere.

To account for this, Rutherford proposed a model of the atom in which negatively charged electrons orbit a small, positively charged *nucleus* that contains nearly all of the atom's mass. In Rutherford's model most of the volume of each atom is empty, so most alpha particles fired at the gold foil would experience only slight deflections as they passed through. But once in a while, about one time out of every 10,000, an alpha particle would approach a gold nucleus almost head on and be scattered at a large angle, sometimes directly backward.

In Rutherford's model electrons orbit the nucleus like a satellite orbiting Earth. But while we can place an Earth satellite in an orbit of any size we wish, electrons in an atom behave differently: They can move only in very specific orbits around the nucleus. In this section we'll explore how studies of the spectrum of light emitted and absorbed by atoms led physicists to this discovery. In the following section we'll explore the theories that were developed to explain why only certain electron orbits are allowed.

The Discovery of Atomic Spectra

In the early part of the nineteenth century, the English scientist William Hyde Wollaston and the German physicist Joseph von Fraunhofer independently discovered dark lines in the spectrum of visible light coming from the Sun (**Figure 24-14**). These lines are always in the same locations within the spectrum. Some years later Gustav Kirchhoff, the same physicist we encountered in our study of electric circuits, was able to reproduce these same dark lines in the laboratory. He passed light from a lamp (made to simulate sunlight) through vapors created by heating sodium. The light from the lamp itself had a continuous spectrum like that of a blackbody, but the spectrum of light that had passed through the sodium vapor had two dark lines. Kirchhoff concluded that the dark lines result from certain specific colors of light being absorbed by the sodium vapor. What is more, these lines were at the same position as the closely spaced pair of lines in the yellow-orange region of the Sun's spectrum. (You can easily find these lines in Figure 24-14.) The same mechanism must therefore be happening with sunlight: The light coming from the solar interior has a continuous, blackbody-like spectrum, but certain wavelengths of that light are absorbed by atoms in the Sun's atmosphere. The Sun's atmosphere must contain sodium atoms identical to those in Kirchhoff's laboratory. In light of Kirchhoff's discovery, a spectrum like that shown in Figure 24-14 is called an **absorption spectrum**, and the dark lines are called **absorption lines**.

Scientists soon discovered that each element produces its own characteristic absorption lines when light passes through a vapor containing atoms of that element. Thus an absorption spectrum acts as a "fingerprint" of the chemical composition of the vapor that produced the absorption lines. This is how we know the chemical composition of the Sun's atmosphere. It's also how we determine the chemical composition of the atmospheres of distant stars like those in the photograph that opens this chapter, and how we know that all stars have basically the same chemical makeup (**Figure 24-15**). What was not understood was *why* atoms should selectively absorb only light of certain wavelengths and why the absorbed wavelengths should be different for atoms of different elements.

Science History Images/Alamy

Figure 24-15 Cecilia Payne-Gaposchkin and the composition of stars In her 1925 doctoral thesis, British-born astronomer Cecilia Payne (later Payne-Gaposchkin) was the first to use absorption spectra like that in Figure 24-14 to show that all stars are made primarily of hydrogen and helium. She later became professor of astronomy at Harvard University and the first woman at Harvard to chair a department.

An important clue about the mystery of absorption spectra came from studying the light *emitted* by atoms. Physicists of the nineteenth century discovered that light created by heating a vapor gives rise to an **emission spectrum**, a spectrum that consists only of specific emitted wavelengths. What is more, if the vapor contains atoms of a certain element, the wavelengths in the emission spectrum from those atoms (the **emission lines**) are precisely the same as the wavelengths in the absorption spectrum of that same element (**Figure 24-16**). To explore this, we'll concentrate on the absorption and emission spectra of hydrogen, which has the simplest set of absorption and emission lines of any element. But the underlying physics applies to all elements.

Johann Balmer, a Swiss mathematician, made an analysis of the lines in the absorption and emission spectra of hydrogen. He devised a formula that both reproduced the

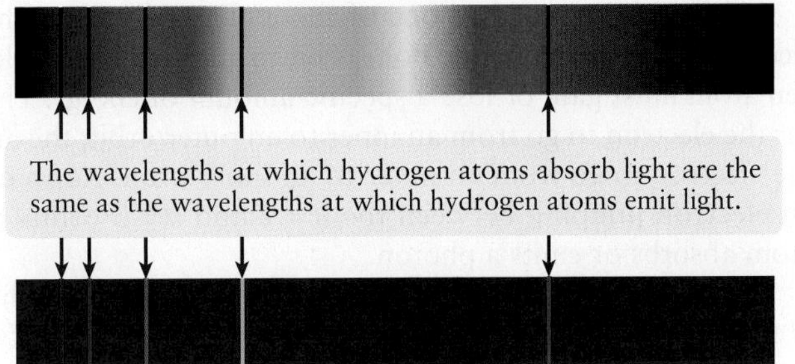

(a) The absorption spectrum produced by passing white light through a gas of hydrogen atoms
The wavelengths at which hydrogen atoms absorb light are the same as the wavelengths at which hydrogen atoms emit light.
(b) The emission spectrum produced by a heated gas of hydrogen atoms

Figure 24-16 Absorption and emission spectra of atomic hydrogen (a) When light passes through a gas, light of specific wavelengths is absorbed, forming dark lines. (b) When a gas is made to glow by passing an electric current through it, it emits only specific wavelengths of light.

- Atoms of each element absorb and emit light at wavelengths that are characteristic of that element.
- The characteristic wavelengths differ from one element to another.
- Hydrogen has the simplest arrangement of characteristic wavelengths of any element.

wavelengths of lines that had been reported and correctly predicted the wavelengths of spectral lines that had not yet been observed. This was later extended by the Swedish physicist Johannes Rydberg. The *Rydberg formula* for the hydrogen spectral lines is

Wavelength of an absorption or emission line in the spectrum of atomic hydrogen

$$\frac{1}{\lambda} = R_H\left(\frac{1}{n^2} - \frac{1}{m^2}\right) \quad (24\text{-}28)$$

Rydberg constant = $1.09737 \times 10^7\ \text{m}^{-1}$

n and m are integers: n can be 1, 2, 3, 4,... and m can be any integer greater than n.

EQUATION IN WORDS
Rydberg formula for the spectral lines of hydrogen

The value of the constant R_H in Equation 24-28, called the *Rydberg constant*, is chosen to match the experimental data. To four significant digits $R_H = 1.097 \times 10^7\ \text{m}^{-1}$. As an example, the hydrogen absorption and emission lines shown in Figure 24-16 all correspond to $n = 2$ in Equation 24-28. The series of wavelengths for which $n = 2$ is called the *Balmer series*. For example, to get the wavelength of the red spectral line in Figure 24-16, set $n = 2$ and $m = 3$ in Equation 24-28 and then take the reciprocal:

$$\frac{1}{\lambda} = R_H\left(\frac{1}{2^2} - \frac{1}{3^2}\right) = (1.097 \times 10^7\ \text{m}^{-1})\left(\frac{1}{4} - \frac{1}{9}\right) = 1.524 \times 10^6\ \text{m}^{-1}$$

$$\lambda = \frac{1}{1.524 \times 10^6\ \text{m}^{-1}} = 6.563 \times 10^{-7}\ \text{m} = 656.3\ \text{nm}$$

The spectral line to the left of this one in Figure 24-16 (in the blue-green part of the spectrum) corresponds to $n = 2$ and $m = 4$; you can show that for this spectral line, $\lambda = 486.2$ nm. The wavelengths with $n = 1$ are all in the ultraviolet and are called the *Lyman series*; the wavelengths with $n = 3$ are all in the infrared and are called the *Paschen series*.

What Balmer and Rydberg did not know was *why* the spectral hydrogen lines were given by this relatively simple formula. It was left to the Danish physicist Niels Bohr to provide the explanation.

Energy Quantization

Bohr realized that to fully understand the structure of the hydrogen atom, he had to be able to derive Balmer's formula using the laws of physics. He first made the rather wild assumption that the electron in a hydrogen atom can orbit the nucleus only in certain specific orbits. (This was a significant break with the ideas of Newton, in whose mechanics any orbit should be possible.) **Figure 24-17** shows the four smallest of these **Bohr orbits**, labeled by the numbers $n = 1$, $n = 2$, $n = 3$, and so on.

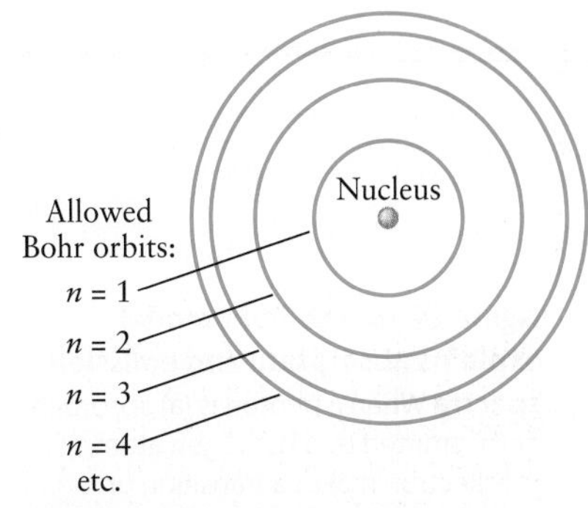

Figure 24-17 Bohr orbits in the hydrogen atom In the model devised by Niels Bohr, electrons in the hydrogen atom are allowed to be in certain orbits only. (The radii of the orbits are not shown to scale.)

Although confined to one of these allowed orbits while circling the nucleus, an electron can jump from one Bohr orbit to another. For an electron to do this, the hydrogen atom must gain or lose a specific amount of energy. The atom must absorb energy for the electron to go from an inner to an outer orbit; the atom must release energy for the electron to go from an outer to an inner orbit. As an example, **Figure 24-18** shows an electron jumping between the $n = 2$ and $n = 3$ orbits of a hydrogen atom as the atom absorbs or emits a photon.

When the electron jumps from one orbit to another, the energy of the photon that is emitted or absorbed equals the difference in energy between these two orbits. This energy difference, and hence the photon energy, is the same whether the jump is from a low orbit to a high orbit (**Figure 24-18a**) or from the high orbit back to the low one (**Figure 24-18b**). According to Einstein, if two photons have the same energy E, the relationship $E = hf$ (Equation 24-13) tells us that they must also have the same frequency f and hence the same wavelength $\lambda = c/f$. It follows that if an atom can emit photons of a given energy and wavelength, it can also absorb photons of precisely the same energy and wavelength. Thus Bohr's picture explains Kirchhoff's observation that atoms emit and absorb the same wavelengths of light.

The Bohr picture also helps us visualize what happens to produce an emission spectrum. When a gas is heated its atoms move around rapidly and can collide with each other. These energetic collisions excite the atoms' electrons into high orbits. The electrons then cascade back down to the innermost possible orbit, emitting photons whose energies are equal to the energy differences between different Bohr orbits. In this fashion, a hot gas produces an emission line spectrum with a variety of different wavelengths.

To produce an absorption spectrum, begin with a relatively cool gas so that the electrons in most of the atoms are in inner, low-energy orbits. If a beam of light with a continuous spectrum is shone through the gas, most wavelengths will pass through undisturbed. Only those photons will be absorbed whose energies are just right to excite an electron to an allowed outer orbit. Hence only certain wavelengths will be absorbed, and dark lines will appear in the spectrum at those wavelengths.

As in Figure 24-18, the energy of the photon that is absorbed or emitted in a jump between orbits must be equal to the *difference* between the energy of the atom with the electron in the larger-radius, higher-energy orbit and the energy of the atom with the electron in the smaller-radius, lower-energy orbit. Bohr concluded that the numbers n and m in the Rydberg formula correspond to the numbers of the orbits between which an electron jumps as it absorbs or emits a photon. The value of n is the number of the lower orbit, and the value of m is the number of the upper orbit. For example, the jump shown in Figure 24-18 corresponds to $n = 2$ and $m = 3$.

We can better understand Bohr's idea by combining the Rydberg formula, Equation 24-26, with the expressions $E = hf$ and $f = c/\lambda$ for a photon. Together these

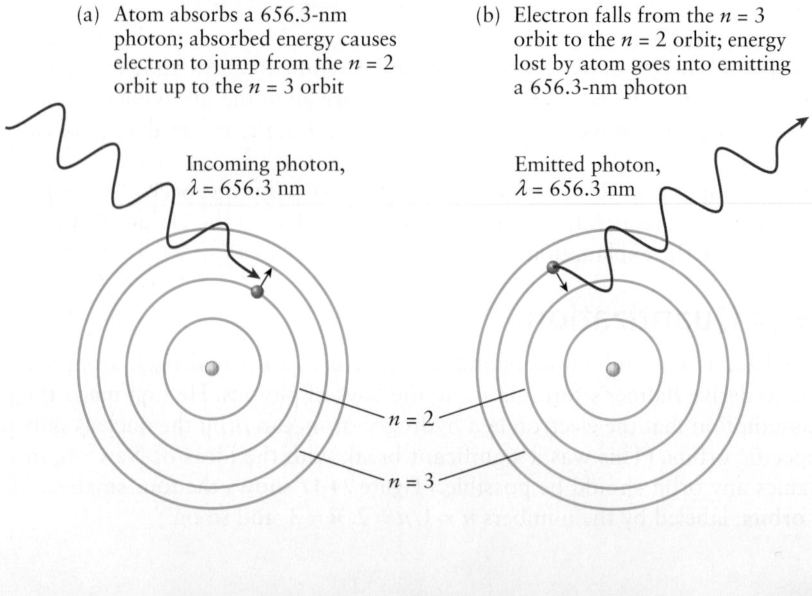

(a) Atom absorbs a 656.3-nm photon; absorbed energy causes electron to jump from the $n = 2$ orbit up to the $n = 3$ orbit

(b) Electron falls from the $n = 3$ orbit to the $n = 2$ orbit; energy lost by atom goes into emitting a 656.3-nm photon

Incoming photon, $\lambda = 656.3$ nm

Emitted photon, $\lambda = 656.3$ nm

$n = 2$

$n = 3$

Figure 24-18 The Bohr model explains absorption and emission spectra When a photon is (a) absorbed or (b) emitted by a hydrogen atom, the electron makes a transition or jump between two allowed orbits. The photon energy equals the difference in energy between the upper and lower electron orbits. (As in Figure 24-17, the radii of the orbits are not shown to scale.)

latter two expressions say that the energy of a photon of wavelength λ is $E = hc/\lambda$. So if we multiply Equation 24-28 by hc, we get an expression for the energy of a photon absorbed or emitted by a hydrogen atom:

$$E_{\text{photon}} = \frac{hc}{\lambda} = hcR_{\text{H}}\left(\frac{1}{n^2} - \frac{1}{m^2}\right) = \frac{hcR_{\text{H}}}{n^2} - \frac{hcR_{\text{H}}}{m^2} = \left(-\frac{hcR_{\text{H}}}{m^2}\right) - \left(-\frac{hcR_{\text{H}}}{n^2}\right) \quad (24\text{-}29)$$

If we say that a hydrogen atom has energy $E_{\text{atom},n} = -hcR_{\text{H}}/n^2$ when the electron is in the nth orbit and has energy $E_{\text{atom},m} = -hcR_{\text{H}}/m^2$ when the electron is in the mth orbit, where m is greater than n, then we can rewrite Equation 24-29 as

$$E_{\text{photon}} = E_{\text{atom},m} - E_{\text{atom},n} \quad (24\text{-}30)$$

Equation 24-30 uses the idea that the energy of a hydrogen atom is **quantized**; that is, the energy can only have certain values. These energies are given by

$$E_{\text{atom},n} = -\frac{hcR_{\text{H}}}{n^2} \quad \text{where } n = 1, 2, 3, 4, \ldots \quad (24\text{-}31)$$

Note that the energy of the atom is negative, and greater values of n correspond to energies that are less negative (that is, closer to zero). This agrees with the idea that the larger the orbit and the greater the value of n for that orbit, the higher the energy.

Each quantized value of the energy is called an **energy level. Figure 24-19** shows several of the energy levels of the hydrogen atom, along with vertical arrows that show the energy of the photon that must be absorbed or emitted in a jump or transition between levels. The transitions that correspond to the Lyman series involve a photon with a very large amount of energy, so these photons have a high frequency and short wavelength: They are all in the ultraviolet part of the spectrum. By contrast, the transitions that correspond to the Paschen series involve a photon with a very small amount of energy, which is why these photons have a low frequency and long wavelength and are in the infrared part of the spectrum. The Balmer series is intermediate between these two; the wavelengths are in either the visible range (380 to 750 nm) or the ultraviolet range.

Elements other than hydrogen also absorb and emit light at specific wavelengths, although those wavelengths do not follow a simple mathematical pattern like the characteristic wavelengths of hydrogen given by Equation 24-28. The conclusion is

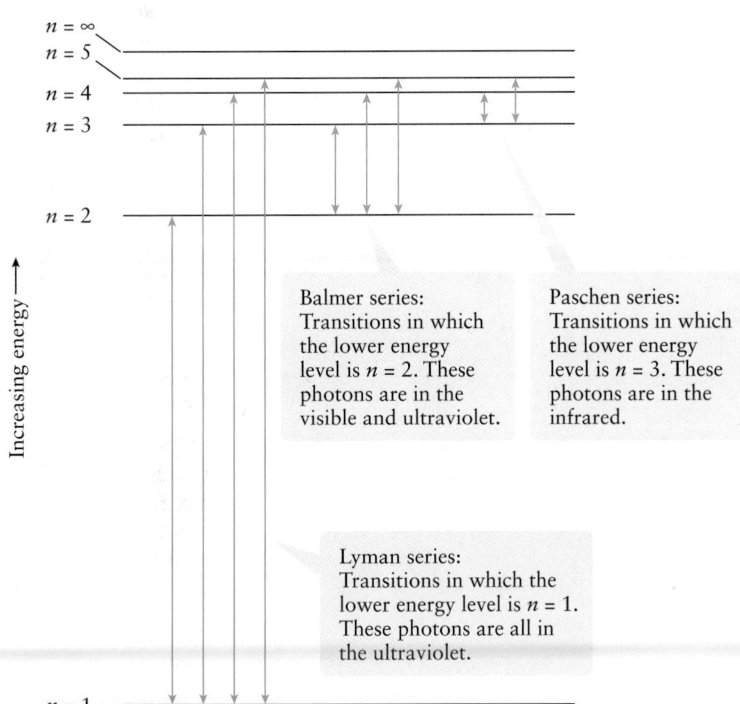

n = ∞
n = 5
n = 4
n = 3
n = 2
n = 1
Increasing energy
Balmer series: Transitions in which the lower energy level is n = 2. These photons are in the visible and ultraviolet.
Paschen series: Transitions in which the lower energy level is n = 3. These photons are in the infrared.
Lyman series: Transitions in which the lower energy level is n = 1. These photons are all in the ultraviolet.

Figure 24-19 Hydrogen energy levels This figure shows some of the lower-lying energy levels of the hydrogen atom and possible transitions between those levels. Note that the energy difference is greatest between the $n = 1$ and $n = 2$ levels, less between the $n = 2$ and $n = 3$ levels, and even less between the $n = 3$ and $n = 4$ levels.

that there are quantized energy levels for atoms of other elements, but the arrangement of energy levels is more complex than for hydrogen. In the following section we'll see how Niels Bohr justified the quantization of energy for the relatively simple case of the hydrogen atom, and how he was able to reproduce Equation 24-31 for the energy levels. We'll then use these ideas to gain insight into the structure of other atoms.

WATCH OUT !

Energy quantization is not just an obscure effect in atomic physics.

Discrete energy levels play an important role in modern technology. One example is a *laser*, a device that emits an intense beam of light of a very specific wavelength. This is possible because the laser contains a material with two distinct energy levels. Light of the laser's characteristic wavelength is emitted when transitions take place from the upper level to the lower one. The laser is unique because these transitions occur coherently rather than at random. Energy is added to the material to pump its molecules into excited states. The excited molecules naturally want to transition to a lower state, but if left on their own, would do so at random times. However, when a photon of energy equal to the difference between two states is sent into the material, it can *stimulate* this transition and cause a second photon of that same energy to be released. This photon can stimulate further emission, so the number of emitted photons increases. The emitted photons are all in phase and all travel in the same general direction. The net result is an intense beam. If materials did not have quantized energy levels, lasers could not exist.

Fluorescent light bulbs also depend on energy quantization. The material inside the bulb emits ultraviolet photons when an electric current passes through it. These photons are absorbed by the white coating on the inner surface of the bulb. Because ultraviolet photons are very energetic, this excites the material of the coating to very high energy levels. The coating then drops down to its initial energy level in a series of small steps. (It's like taking a big leap to the top of a staircase, then coming carefully down the staircase one step at a time.) Each small step between closely spaced energy levels emits a low-energy, visible-light photon. There are so many such energy levels, with a variety of spacing between them, that the net result is that a mixture of almost all visible colors—that is, white light—is emitted from the bulb.

EXAMPLE 24-6 Photon Possibilities

A collection of hydrogen atoms is excited to the $n = 3$ energy level. What are the possible wavelengths that these atoms could emit as they return to the lowest energy ($n = 1$) level?

Set Up

There are two routes that an atom can take from the $n = 3$ level to the $n = 1$ level. One, it could drop down to the $n = 1$ level in a single step by emitting a single photon whose energy is equal to the difference between the energies of the $n = 3$ and $n = 1$ levels. The wavelength of this photon is given by Equation 24-28 with $n = 1$ and $m = 3$. Two, the atom could first drop to the $n = 2$ level by emitting a photon (with a wavelength given by Equation 24-28 with $n = 2$ and $m = 3$), then emit a second photon as it drops from the $n = 2$ level to the $n = 1$ level (with a wavelength given by Equation 24-28 with $n = 1$ and $m = 2$). So three different wavelengths can be emitted by the excited atoms. We'll calculate each of these in turn.

Rydberg formula for the spectral lines of hydrogen:

$$\frac{1}{\lambda} = R_H\left(\frac{1}{n^2} - \frac{1}{m^2}\right) \quad (24\text{-}28)$$

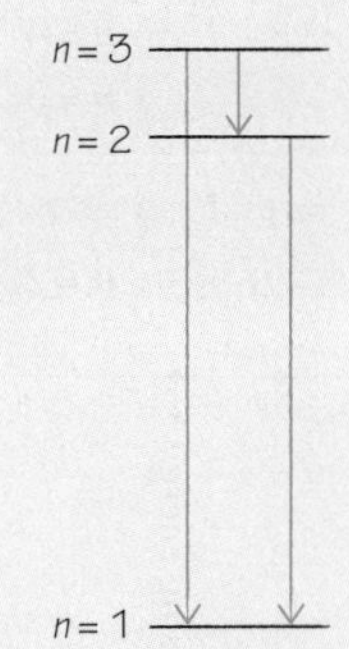

Solve

Use Equation 24-28 to calculate each of the three possible wavelengths.

For the $n = 3$ to $n = 1$ transition:

$$\frac{1}{\lambda} = R_{\text{H}}\left(\frac{1}{1^2} - \frac{1}{3^2}\right) = (1.097 \times 10^7 \text{ m}^{-1})\left(1 - \frac{1}{9}\right)$$

$$= 9.751 \times 10^6 \text{ m}^{-1}$$

$$\lambda = \frac{1}{9.751 \times 10^6 \text{ m}^{-1}} = 1.026 \times 10^{-7} \text{ m} = 102.6 \text{ nm}$$

This is an ultraviolet wavelength.

For the $n = 3$ to $n = 2$ transition:

$$\frac{1}{\lambda} = R_{\text{H}}\left(\frac{1}{2^2} - \frac{1}{3^2}\right) = (1.097 \times 10^7 \text{ m}^{-1})\left(\frac{1}{4} - \frac{1}{9}\right)$$

$$= 1.524 \times 10^6 \text{ m}^{-1}$$

$$\lambda = \frac{1}{1.524 \times 10^6 \text{ m}^{-1}} = 6.563 \times 10^{-7} \text{ m} = 656.3 \text{ nm}$$

This is a visible wavelength (in the red part of the spectrum).

For the $n = 2$ to $n = 1$ transition:

$$\frac{1}{\lambda} = R_{\text{H}}\left(\frac{1}{1^2} - \frac{1}{2^2}\right) = (1.097 \times 10^7 \text{ m}^{-1})\left(1 - \frac{1}{4}\right)$$

$$= 8.228 \times 10^6 \text{ m}^{-1}$$

$$\lambda = \frac{1}{8.228 \times 10^6 \text{ m}^{-1}} = 1.215 \times 10^{-7} \text{ m} = 121.5 \text{ nm}$$

This is another ultraviolet wavelength.

Reflect

Comparing with Figure 24-19 shows that the 656.3-nm wavelength represents an emission line of the Balmer series, while the 102.6- and 121.5-nm wavelengths represent emission lines of the Lyman series.

NOW WORK Problems 1–6 and 10 from the Takeaway 24-6.

THE TAKEAWAY for Section 24-6

✔ Atoms are composed of electrons orbiting a positively charged nucleus that has most of the mass of the atom.

✔ In the Bohr model an electron's orbits around the nucleus can have only certain well-defined energies. Thus the energy of the atom is quantized and can have only certain values. The atom cannot have energies intermediate between those values.

✔ Electrons that make a transition from one allowed orbit to another either radiate or absorb a photon of a well-defined energy. This gives rise to the lines in emission and absorption spectra.

Prep for the AP® Exam

AP® Building Blocks

1. The Balmer formula can be written as follows:

$$\lambda = (364.51 \text{ nm})\left(\frac{m^2}{m^2 - 4}\right)$$

where m is equal to any integer larger than 2. This represents the wavelengths of visible colors that are emitted from the hydrogen atom. Calculate the first four colors (wavelengths) that are observed in the spectrum of hydrogen due to the Balmer series.

2. A hypothetical atom has four unequally spaced energy levels in which a single electron can be found. Suppose a collection of the atoms is excited to the highest of the four levels.
   (a) What is the maximum number of unique spectral lines that could be measured as the atoms relax and return to the lowest, ground state?

(b) Suppose the previous hypothetical atom has 10 energy levels. Now what is the maximum number of unique spectral lines that could be measured in the emission spectrum of the atom?

EX 24-6

3. The Lyman series results from transitions of the electron in hydrogen in which the electron ends at the $n = 1$ energy level. Using the Rydberg formula for the Lyman series, calculate the wavelengths of the photons emitted in the transitions that end in the $n = 1$ level and start in the energy levels that correspond to n equal to 2 through 6, and indicate the initial and final levels of the transition corresponding to each wavelength. State whether each wavelength is visible (380 to 750 nm), ultraviolet (shorter than 380 nm), or infrared (longer than 750 nm).

4. The Balmer series results from transitions of the electron in hydrogen in which the electron ends at the $n = 2$ energy level. Using the Rydberg formula for the Balmer series, calculate the wavelengths of the photons emitted in the transitions that end in the $n = 2$ level and start in the energy levels that correspond to n equal to 3 through 6, and indicate the initial and final levels of the transition corresponding to each wavelength. State whether each wavelength is visible (380 to 750 nm), ultraviolet (shorter than 380 nm), or infrared (longer than 750 nm).

5. The Paschen series results from transitions of the electron in hydrogen in which the electron ends at the $n = 3$ energy level. Using the Rydberg formula for the Paschen series, calculate the wavelengths of the photons emitted in the transitions that end in the $n = 3$ level and start in the energy levels that correspond to n equal to 4 through 6, and indicate the initial and final levels of the transition corresponding to each wavelength. State whether each wavelength is visible (380 to 750 nm), ultraviolet (shorter than 380 nm), or infrared (longer than 750 nm).

AP® Skill Builders

EX 24-6

6. Hydrogen atoms in interstellar gas clouds emit electromagnetic radiation at a wavelength of 21 cm when an electron in the ground state of hydrogen switches spin states. Determine the energy difference between the two spin states in this transition.
7. Express the Balmer formula (see problem 1) in terms of the *frequency* of the photons that are emitted (rather than the wavelength). Extend all numerical values out to five significant digits.
8. Express the Rydberg formula in terms of the *frequency* of the photons that are emitted (rather than the wavelength). Extend all numerical values out to five significant digits.

AP® Skills in Action

9. Rank the following transitions between hydrogen energy levels in terms of the energy of the photon involved, from highest to lowest energy.
(A) An atom drops from the $n = 4$ level to the $n = 2$ level.
(B) An atom rises from the $n = 3$ level to the $n = 5$ level.
(C) An atom drops from the $n = 3$ level to the $n = 1$ level.
(D) An atom drops from the $n = 4$ level to the $n = 3$ level.

EX 24-6

10. Calculate the shortest wavelength (and the highest energy) associated with emitted photons in the following series: (a) Lyman, (b) Balmer, and (c) Paschen (see problems 3, 4, and 5).

24-7 Models by Bohr and Schrödinger give insight into the intriguing structure of the atom

The force that pulls the Moon toward Earth is directed toward Earth's center, yet the Moon does not fall into Earth. In the same way, the negatively charged electron in an atom experiences a Coulomb force that points directly toward the positively charged protons at the center of the atom. Why does the electron orbit rather than fall into the proton? You likely have an intuitive answer that both the Moon and the electron *orbit* rather than fall in. (In a real sense each *is* falling in, but it is always missing the target!) Niels Bohr based his model of the hydrogen atom, for which he received the Nobel Prize in Physics in 1922, on the physics of atomic orbits. The **Bohr model** provides a theoretical foundation for the physics of atomic spectra that we encountered in the previous section.

The Bohr Model

Bohr made two fundamental assumptions in describing the orbit of an electron around a positive atomic nucleus. First he modeled the orbit as uniform circular motion, that is, as an electron moving at constant speed in a circular path. This is not exactly correct but more than satisfactory to provide a broad understanding of the atom. Bohr's second assumption was that only specific values are allowed for the angular momentum of an orbiting electron. We will return to this second assumption shortly.

Bohr considered a single electron orbiting a nucleus of charge $+Ze$, where Z, the *atomic number* of an atom, is the number of protons in the nucleus. The orbiting electron experiences only one force, the Coulomb attraction between it and the protons in the atomic nucleus. The magnitude of the Coulomb force on an electron orbiting in a circle of radius r is, from Equation 16-1,

$$F = \frac{k(Ze)(e)}{r^2} = \frac{kZe^2}{r^2}$$

Here $k = 8.99 \times 10^9 \text{ N} \cdot \text{m}^2/\text{C}^2$ is the Coulomb constant. Newton's second law requires that this force equal the mass of the electron m_e multiplied by the acceleration it experiences. For an object in uniform circular motion at speed v, the magnitude of the acceleration is, from Equation 6-4,

$$a = \frac{v^2}{r}$$

Combining the above two equations into Newton's second law gives

$$\frac{kZe^2}{r^2} = \frac{m_e v^2}{r}$$

If we multiply both sides of this equation by r, we get

$$\frac{kZe^2}{r} = m_e v^2 \qquad (24\text{-}32)$$

Let's set Equation 24-32 aside for a moment and examine Bohr's second assumption—a requirement that only specific values are allowed for the angular momentum associated with the electron. Bohr recognized that the dimensions of Planck's constant h are those of angular momentum, so he constrained the electron's angular momentum to be only multiples of h. Specifically, this requirement is

$$L = n\left(\frac{h}{2\pi}\right) = n\hbar \qquad (24\text{-}33)$$

(orbital angular momentum in the Bohr model)

where L is the electron's orbital angular momentum and n is any integer starting from 1. The constant $\hbar$ (pronounced "h bar") is defined to be h divided by 2π.

To relate angular momentum to Equation 24-32, we express L in terms of the mass of the electron m_e, its speed v, and its distance from the center of the atom r using Equation 8-23:

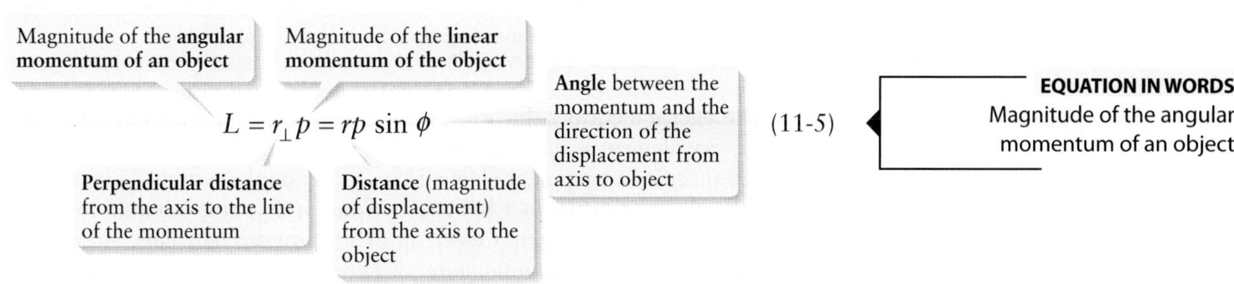

The magnitude of the electron's linear momentum is $p = m_e v$. Because the electron moves in a circle, the vector from the rotation axis to the particle always has the same radius r and is always perpendicular to the momentum vector. So $\phi = 90°$, $\sin\ \phi = 1$, and

$$L = rm_e v \qquad (24\text{-}34)$$

We can now rewrite Equation 24-18 in terms of the angular momentum L by multiplying the right-hand side by 1 in the form of $(r^2/r^2)(m_e/m_e)$:

$$\frac{kZe^2}{r} = \frac{r^2 m_e^2 v^2}{m_e r^2} = \frac{L^2}{m_e r^2}$$

Bohr's requirement that the electron's angular momentum is an integer multiple of $\hbar$ then gives

$$\frac{kZe^2}{r} = \frac{(n\hbar)^2}{m_e r^2}$$

or

$$r_n = \frac{n^2\hbar^2}{m_e kZe^2} \quad \text{where } n = 1, 2, 3, \ldots \tag{24-35}$$

(orbital radii in the Bohr model)

We add the subscript n to the variable r to indicate that the radius can take on only specific values and that the allowed values of radius depend on n. Because the values of r_n are proportional to the square of an integer, the orbital radii of electrons in an atom are quantized.

Notice that both n and Z in Equation 24-35 are dimensionless. Because r_n is a distance, all of the other terms on the right-hand side, taken as they appear in the equation, must have dimensions of distance as well. This distance, usually written as a_0 and called the *Bohr radius*, is

EQUATION IN WORDS
Bohr radius

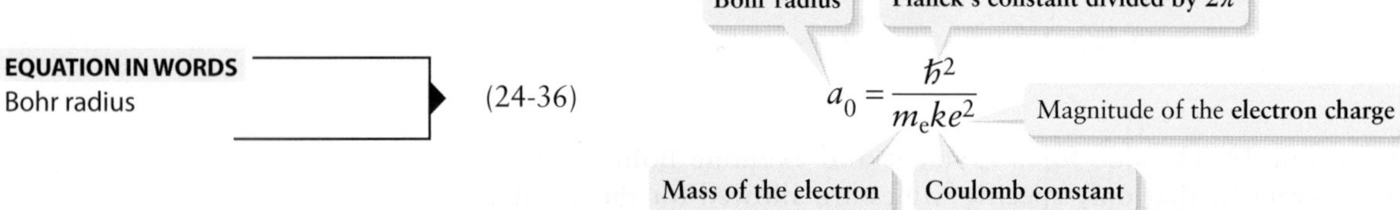

$$a_0 = \frac{\hbar^2}{m_e ke^2} \tag{24-36}$$

Using the values for $\hbar$, m_e, k, and e, we find that the value of the Bohr radius a_0 is approximately

$$a_0 = 0.529 \times 10^{-10} \text{ m} = 0.0529 \text{ nm}$$

In terms of a_0 the quantized radii of the electron orbits (Equation 24-21) are

EQUATION IN WORDS
Orbital radii in the Bohr model

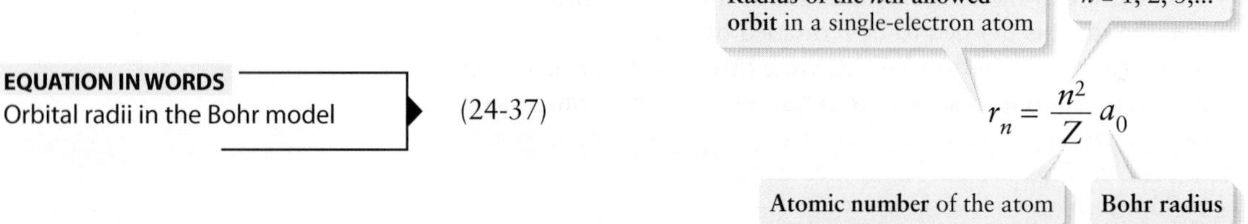

$$r_n = \frac{n^2}{Z} a_0 \tag{24-37}$$

The integer n identifies the orbit, where the $n = 1$ orbit is the closest to the nucleus. Because every element is distinguished by the number of protons it carries, the atomic number Z specifies a particular element. So, for example, setting Z equal to 1 gives the radii of the electron orbits in a hydrogen atom, and setting Z equal to 1 and n equal to 1 gives the radius of the first electron orbit in hydrogen. This is the normal state, usually referred to as the ground state, of hydrogen. Moreover, notice that the radius of the ground state orbit of hydrogen is the Bohr radius. In other words, the radius of a typical hydrogen atom is about 0.05 nm. We can also conclude from Equation 24-37 together with the value of the Bohr radius that, in general, atoms are no more than a few nanometers in radius. The Bohr model sets the scale for atomic sizes.

What is the energy of an electron orbiting an atomic nucleus according to the Bohr model? This is the sum of its kinetic energy and its electric potential energy. The electric potential energy for two point charges q_1 and q_2 is $U_{\text{electric}} = kq_1q_2/r$ (Equation 17-3), so

$$E = \frac{1}{2}m_e v^2 + \frac{k(-e)(Ze)}{r_n} \tag{24-38}$$

We can write the kinetic energy term in terms of angular momentum in a way that is similar to our approach when developing the relationship for r_n. Using Equation 24-34 shows that the kinetic energy term becomes

$$\frac{1}{2}m_e v^2 = \frac{1}{2}\frac{L^2}{m_e r_n^2} = \frac{n^2\hbar^2}{2m_e r_n^2}$$

so the total electron energy from Equation 24-38 is

$$E = \frac{n^2\hbar^2}{2m_e r_n^2} - \frac{kZe^2}{r_n}$$

Substituting Equation 24-37 for r_n and Equation 24-36 for a_0 gives the energy in terms of only the physical constants and the counting integer n:

$$E = \frac{n^2\hbar^2}{2m_e}\left(\frac{m_e kZe^2}{n^2\hbar^2}\right)^2 - kZe^2\frac{m_e kZe^2}{n^2\hbar^2}$$

Simplifying this is not too difficult if you notice that the numerator of both terms is $m_e(kZe^2)^2$ and that both terms have $n^2\hbar^2$ in the denominator:

$$E = \frac{m_e(kZe^2)^2}{2n^2\hbar^2} - \frac{m_e(kZe^2)^2}{n^2\hbar^2}$$

or

$$E_n = -\frac{m_e(kZe^2)^2}{2n^2\hbar^2}, \quad n = 1, 2, 3, \ldots \tag{24-39}$$

(electron energies in the Bohr model)

We add the subscript n to the variable E to indicate that the orbital energy of the electron can take on only specific values that depend on n. Because the values of E_n are proportional to $1/n^2$, the orbital energy of electrons in an atom is quantized. This is in agreement with our conclusion in the previous section that for atomic spectral lines to occur only at specific wavelengths, electrons must orbit the hydrogen nucleus with specific, well-defined energies. Note also that the energy is equal to a negative constant divided by n^2, exactly in accordance with Equation 24-31. (We'll see below that the numerical value of the constant is the same in Equation 24-39 as in Equation 24-31.)

The value of energy of an electron in orbit around an atomic nucleus is negative but closer and closer to zero for increasing values of n. In other words, the lowest energy orbit is the one closest to the nucleus, as we would expect. That the energy is negative emphasizes that the electron is bound to the nucleus, and that energy must be supplied to either move the electron to a higher orbit or to break the electron free from the nucleus altogether.

Because E_n is an energy, and because both n and Z are dimensionless, the combination of other quantities on the right-hand side of Equation 24-39 must have dimensions of energy as well. This energy, usually written as E_0, is called the *Rydberg energy*:

Coulomb constant — Magnitude of the electron charge — Mass of the electron — Rydberg energy — Planck's constant divided by 2π

$$E_0 = \frac{m_e(ke^2)^2}{2\hbar^2} = 2.18 \times 10^{-18}\ \text{J} = 13.6\ \text{eV} \tag{24-40}$$

EQUATION IN WORDS
Rydberg energy

In Equation 24-40 we're given the value of the Rydberg energy in joules and in electron volts (eV). The energy that an electron acquires when it moves through a potential difference of 1 V is 1 eV, or approximately 1.602×10^{-19} J. To give you an idea of the amount of energy 1 eV represents, a photon of visible light carries between about 1.5 and 3 eV.

Using Equation 24-40 we can write Equation 24-39 for the quantized electron orbital energy in terms of the Rydberg energy E_0:

EQUATION IN WORDS
Electron energies in the Bohr model

Electron energy for the *n*th allowed orbit in a single-electron atom

Atomic number of the atom

$$E_n = -\frac{Z^2}{n^2} E_0 \qquad (24\text{-}41)$$

$n = 1, 2, 3,...$

Rydberg energy = 13.6 eV

Again the integer n identifies the orbit, and the atomic number Z specifies a particular element. Setting Z equal to 1 and n equal to 1 therefore tells us that the energy of the ground state of hydrogen is –13.6 eV. We can also conclude from Equation 24-27 that, in general, the energy of electrons in orbit around an atomic nucleus is between about –10 eV and, for the largest elements (for which Z is about 100), -10^5 eV. The Bohr model sets the scale for atomic electron energies.

What does the Bohr model predict for the atomic spectral lines of hydrogen? Every line results from a transition between two electron orbits; that is, the energy of the emitted photon is the energy difference ΔE between two allowed orbits (see Equation 24-30). Since the energy of a photon is hf and its frequency $f = c/\lambda$, the wavelength λ of the photon is then

$$\lambda = \frac{hc}{\Delta E}$$

and the reciprocal of the wavelength (which is what appears in the Rydberg formula) is

$$\frac{1}{\lambda} = \frac{\Delta E}{hc} \qquad (24\text{-}42)$$

Let's consider the transition of an electron from a higher orbit m down to a lower orbit n. We can determine the energy difference between these two orbits by applying Equation 24-41. Atomic number Z equals 1 for hydrogen, so

$$\Delta E = -\frac{1}{m^2} E_0 - \left(-\frac{1}{n^2}\right) E_0$$

or

$$\Delta E = \left(\frac{1}{n^2} - \frac{1}{m^2}\right) E_0$$

Substituting this into Equation 24-42 yields

$$\frac{1}{\lambda} = \frac{E_0}{hc}\left(\frac{1}{n^2} - \frac{1}{m^2}\right)$$

Compare this to the Rydberg formula, Equation 24-28. It has exactly the same form!

How does the value of E_0/hc compare to the value of R_H, which equals approximately $1.10 \times 10^7 \text{ m}^{-1}$? In SI units

$$\frac{E_0}{hc} = \frac{2.18 \times 10^{-18} \text{ J}}{(6.63 \times 10^{-34} \text{ J} \cdot \text{s})(3.00 \times 10^8 \text{ m/s})} = 1.10 \times 10^7 \text{ m}^{-1}$$

The ratio E_0/hc equals R_H, so $E_0 = hcR_H$. That's just what we expect if we compare Equation 24-31 (in which we deduced the energies from the Rydberg equation) and Equation 24-41 (in which we derived the energies from the Bohr assumptions). We conclude that the Bohr model is entirely consistent with the observed spectral lines of hydrogen as described by the Rydberg formula.

EXAMPLE 24-7 Lowest Energy Level of a Lithium Ion

Find the (a) radius, (b) energy, and (c) speed of an electron in the lowest energy level of the doubly ionized lithium ion. An atom of lithium ($Z = 3$) has three electrons, so this ion has just one electron.

Set Up

For this ion, $Z = 3$; for the lowest energy level, $n = 1$. We'll use Equation 24-37 to calculate the radius of this orbit, Equation 24-41 to calculate the energy, and Equation 24-33 for angular momentum to determine the electron speed.

Orbital radii in the Bohr model:

$$r_n = \frac{n^2 a_0}{Z} \quad \text{where } n = 1, 2, 3, \ldots \qquad (24\text{-}37)$$

Electron energies in the Bohr model:

$$E_n = -\frac{Z^2}{n^2} E_0 \quad \text{where } n = 1, 2, 3, \ldots \qquad (24\text{-}41)$$

Orbital angular momentum in the Bohr model:

$$L = n\left(\frac{h}{2\pi}\right) = n\hbar \qquad (24\text{-}33)$$

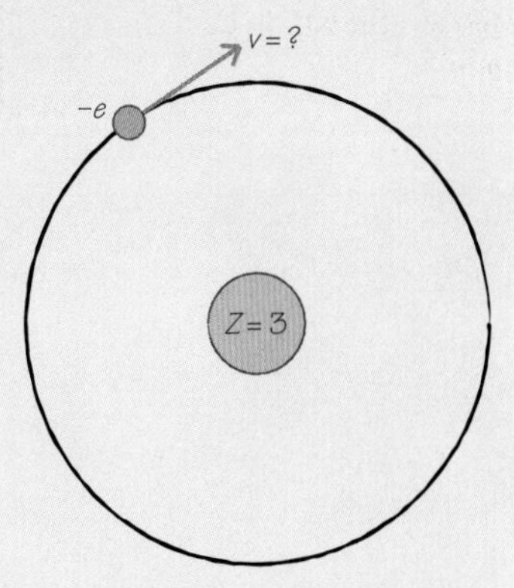

Solve

(a) Calculate the radius of the orbit.

From Equation 24-37 with $Z = 3$ and $n = 1$:

$$r_1 = \frac{1^2 a_0}{3} = \frac{a_0}{3} = \frac{0.0529 \text{ nm}}{3} = 0.0176 \text{ nm}$$

(b) Calculate the energy.

From Equation 24-41 with $Z = 3$ and $n = 1$:

$$E_1 = -\frac{3^2}{1^2} E_0 = -9E_0 = -9(13.6 \text{ eV}) = -122 \text{ eV}$$

(c) To find an expression for the electron speed, combine Equation 24-33 with the equation for the angular momentum of a particle moving in a circular orbit.

Equation 24-34 tells us that the angular momentum of an electron of mass m_e moving in an orbit of radius r at speed v is $L = rm_e v$. Set this equal to the expression for L in Equation 24-33:

$$rm_e v = n\hbar \text{ so}$$

$$v = \frac{n\hbar}{rm_e}$$

We are interested in the case of $n = 1$ and the r_1 we calculated in part (a). Substituting the values for $\hbar = h/(2\pi)$ and the mass m_e of the electron gives

$$v = \frac{1(6.63 \times 10^{-34} \text{ J} \cdot \text{s})}{2\pi\,(1.76 \times 10^{-11} \text{ m})(9.11 \times 10^{-31} \text{ kg})}$$
$$= 6.58 \times 10^6 \text{ m/s}$$

Reflect

The radius of an electron in the lowest level of the doubly ionized lithium ion ($Z = 3$) is one-third the radius of an electron in the ground state of hydrogen. Because there are three protons in the lithium nucleus, compared to one for hydrogen, the Coulomb force on the electron is greater, so it is reasonable that the orbital radius is smaller. Also, notice that the speed of the electron is about 2% of the speed of light in a vacuum.

Why did we insist that the problem be about a doubly ionized lithium ion, rather than a lithium atom? The difference is that when more than one electron is present, each electron is affected not only by the electric force from the nucleus but also by the electric force from the other electrons. We haven't taken the interaction between electrons into account in any of our calculations: The Bohr model is for *single-electron* atoms and ions only.

NOW WORK Problems 1, 4, 5, 8, and 9 from the Takeaway 24-7.

Although the Bohr model successfully predicts atomic spectra and other phenomena associated with hydrogen atoms, it nevertheless leaves us with an outstanding question: Why should the electron orbits be quantized in multiples of $\hbar$? For the answer we turn back to Louis de Broglie. Recall from Section 24-5 that de Broglie postulated that particles, such as electrons, have a wavelike nature. For a nonrelativistic electron of mass m_e moving at speed v, the momentum has magnitude $p = m_e v$. From the de Broglie relation, Equation 24-24, the wavelength of such an electron is

$$\lambda = \frac{h}{m_e v}$$

For an electron orbiting an atomic nucleus we express this in terms of angular momentum by making use of Equation 24-34:

$$\lambda = \frac{hr}{m_e vr} = \frac{hr}{L} = \frac{2\pi\hbar r}{L}$$

(We used $\hbar = h/2\pi$, so $h = 2\pi\hbar$.) Now let L be an integer multiple of $\hbar$, as Bohr required. Then

$$\lambda = \frac{2\pi\hbar r}{n\hbar}$$

or

$$n\lambda = 2\pi r$$

We recognize $2\pi r$ as the circumference of the orbital path of the electron. So Bohr's requirement that L be an integer multiple of $\hbar$ is equivalent to demanding that an integer number of electron wavelengths fit into the circumference of the orbit so that the wave joins smoothly onto itself. In a real sense, it is the wavelike nature of particles that results in the quantization of the energy of atomic electrons.

Confirming Energy Quantization

Bohr's model of the atom works relatively well for hydrogen, for a singly ionized helium ion (an atom with Z equal to 2 but only one electron), and for a doubly ionized lithium ion (an atom with Z equal to 3 but only one electron, as in Example 24-7). For atoms with more than one electron, it isn't possible to make calculations of energy levels using Bohr's physics. However, the general picture of the atom it provides, with electrons in quantized energy states, applies to all atoms. This was confirmed experimentally in 1914 by the German physicists James Franck and Gustav Hertz. (Franck and Hertz were awarded the 1925 Nobel Prize in Physics for their work. Gustav Hertz was a nephew of Heinrich Hertz, whom we encountered earlier.)

Franck and Hertz used a device similar to the one shown schematically in **Figure 24-20** to measure the effect of bombarding atoms in a gas with electrons. The cathode is heated to give off electrons, which are accelerated by a variable potential difference toward the mesh grid. Some electrons pass through the grid and arrive at the anode, where the electron current is detected by the ammeter. Notice that a potential difference is also applied between the grid and the anode, which acts against the electrons; only electrons that carry sufficient energy as they pass the grid will make it to the anode. These electrical components sit inside a tube filled with low-pressure mercury vapor, so collisions between an electron and a mercury (Hg) atom can occur. Franck and Hertz used mercury vapor because the spectral lines of low-pressure mercury gas were well studied; one prominent ultraviolet spectral line has a wavelength $\lambda = 253.7$ nm. The energy of this photon is

$$E = hf = \frac{hc}{\lambda} = \frac{(6.63 \times 10^{-34}\ \text{J}\cdot\text{s})(3.00 \times 10^{8}\ \text{m/s})}{253.7 \times 10^{-9}\ \text{m}}$$

$$= (7.84 \times 10^{-19}\ \text{J})\left(\frac{1\ \text{eV}}{1.60 \times 10^{-19}\ \text{J}}\right) = 4.90\ \text{eV}$$

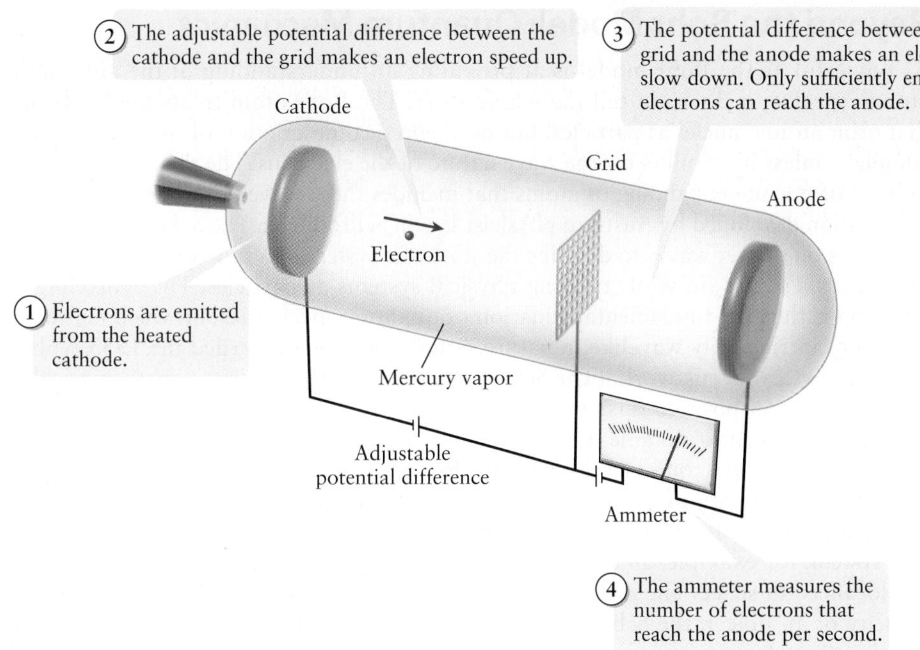

(2) The adjustable potential difference between the cathode and the grid makes an electron speed up.
(3) The potential difference between the grid and the anode makes an electron slow down. Only sufficiently energetic electrons can reach the anode.
Cathode
Grid
Anode
Electron
(1) Electrons are emitted from the heated cathode.
Mercury vapor
Adjustable potential difference
Ammeter
(4) The ammeter measures the number of electrons that reach the anode per second.

Figure 24-20 The Franck–Hertz experiment In this experiment, electrons are accelerated from a cathode toward a mesh grid in a tube filled with mercury vapor. Some electrons pass through the grid, and the current at the anode is measured by the ammeter.

This is the energy emitted when an atomic electron falls from an excited energy level down to a lower energy level. Now consider what happens in the collision of an electron and a mercury atom in the Franck–Hertz tube. In general, the more kinetic energy an electron carries as it approaches the grid, the more likely it will make it through. So the anode current should grow as the potential difference between the cathode and the grid is increased. However, if the energy levels of electrons in Hg (and all) atoms are quantized, when the kinetic energy of the incident electron equals the energy difference between two Hg energy levels, the Hg atom can absorb the electron's energy. When this happens it is less likely that the electron will reach the anode, so the anode current should decrease. **Figure 24-21** shows a typical curve of current versus potential difference from the Franck–Hertz experiment. The general trend, as we would expect, is that the anode current increases as the potential difference is increased. But the current drops dramatically at certain values of accelerating potential difference, in this case, at 4.9 V, and at 9.8 V and 14.7 V, which are multiples of 4.9 V—the energy of the mercury spectral line.

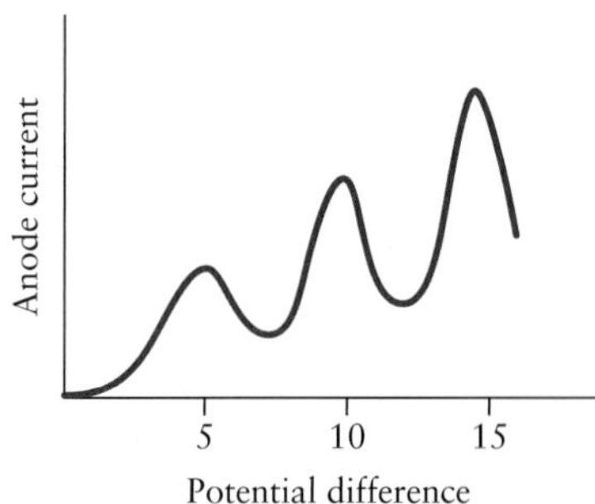

Figure 24-21 Evidence of energy quantization As the potential difference between the cathode and the mesh grid in the Franck–Hertz experiment increases, the anode current increases. When the kinetic energy of the accelerated electrons equals the energy required to excite an electron in a mercury atom to a higher energy level, accelerated electrons lose energy. Not as many electrons reach the anode at the corresponding cathode-grid potential differences, so the anode current decreases.

When the cathode-grid potential difference is set to 4.9 V, the kinetic energy of electrons that leave the cathode reaches 4.9 eV just as they approach the mesh grid. (Recall that the unit of energy eV is the amount of energy gained by an electron when it experiences a potential difference of 1 V.) For this reason electrons that collide with a mercury atom in close proximity to the grid give up their energy in the process of causing an electron in the atom to jump to a higher energy level. Because the collision occurs close to the grid, there is no opportunity for the electron to undergo another acceleration; in other words, there is no opportunity for the electron to acquire enough energy to reach the anode. For this reason when the cathode-grid potential difference is set to 4.9 V, so that the collisions between electrons and Hg atoms occur near the grid with electron kinetic energy equal to 4.9 eV, the number of electrons reaching the anode decreases. In addition, when the potential difference is set to 9.8 V, accelerated electrons that collide with an Hg atom halfway between the cathode and the grid lose their energy and then accelerate up to a kinetic energy of 4.9 eV again by the time they reach the grid. A collision there once again results in the electron transferring its energy to an Hg atom and being unable to reach the anode. A similar phenomenon occurs when the potential difference is set to any integer multiple of 4.9 V. This rise and fall in anode current versus cathode-grid potential difference is shown dramatically in Figure 24-21. The underlying explanation of this phenomenon is the quantization of electron energy levels in the atom, as predicted by Bohr.

Beyond the Bohr Model: Quantum Mechanics

As powerful as the Bohr model is at providing an understanding of the atom and atomic spectra, it does not tell the whole story. The Bohr atom treats the electrons that orbit atomic nuclei as particles. But no theoretical description of the atom can be complete unless it accounts for the wave nature of the electrons. The theoretical underpinning of our understanding of atoms that includes these wave properties is found in an equation developed by Austrian physicist Erwin Schrödinger. The *Schrödinger equation* relies on matter waves to describe the state of a system as a function of time, much like Newton's laws do while treating physical systems as particles. The Schrödinger equation is thus the fundamental equation of **quantum mechanics**, in which matter is treated as intrinsically wavelike in nature. Schrödinger was awarded the 1933 Nobel Prize in Physics for his work. (The Schrödinger equation itself is too mathematically ornate for the purposes of this book.)

Perhaps the most notable difference between Schrödinger's quantum-mechanical description and Newton's classical description of physics is the ability to specify the position and velocity of objects. In Newton's description at any instant of time we can identify a specific position in space and a specific velocity vector for any particle in a system, for example, an electron orbiting an atomic nucleus. A wave, however, is not localized in space. The result is that while Newton's laws predict the position and velocity of an object, the Schrödinger equation predicts the *probability* of finding a certain value of position or velocity. For this reason, electrons in the quantum model of the atom are described not as tiny marbles orbiting the nucleus at fixed radii but rather as a charge distribution. This distribution, sometimes called a *probability cloud*, gives the probability of finding the electron at any given position; the denser the cloud in some region, the more likely it is that the electron will be found there.

Figure 24-22a shows the probability cloud associated with the ground state of hydrogen. The more dense the color in any region in this figure, the more likely it is that the electron will be found in that region. This probability distribution is spherically symmetric; that is, it varies only as a function of radius from the center of the nucleus. For that reason we can also express the same information in a curve of probability versus radius, as in **Figure 24-22b**. Notice that the most probable radius of an electron in the ground state of a hydrogen atom is a_0, the Bohr radius.

The Bohr model employs a single integer, or **quantum number**, to describe electron states. For the Bohr atom this integer is n, which determines the energy level of the electron. In the fully quantum-mechanical view of the atom, *four* quantum numbers are required. These are n, the principal quantum number; ℓ, the angular momentum (or orbital) quantum number; m_ℓ, the magnetic quantum number; and m_s, the electron spin quantum number. The specific values of each of these four quantum numbers completely describe the state of an electron in an atom.

The principal quantum number n plays a role in the quantum atom similar to that which n plays in the classical Bohr atom; in particular, it specifies the energy level or electron shell. The lowest energy state, or ground state, corresponds to n equal to 1.

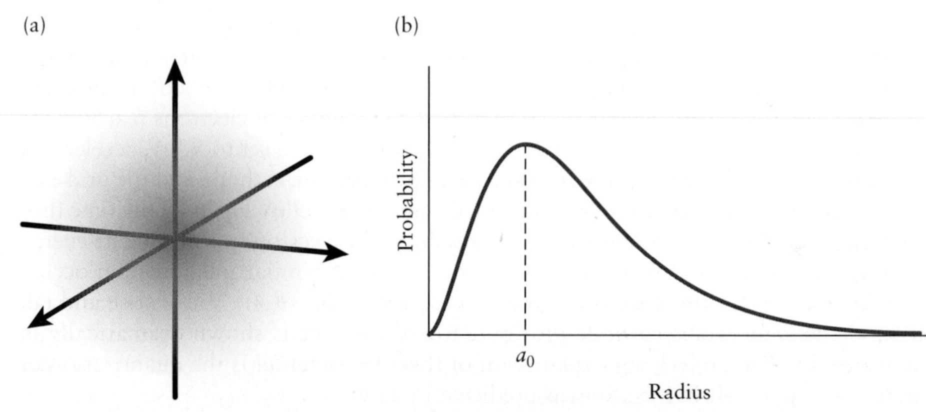

Figure 24-22 The Schrödinger picture of the hydrogen atom (a) In the state of lowest energy, an electron in a hydrogen atom can be found anywhere relative to the nucleus. The darker (denser) the color in a region of the probability cloud, the more likely it is that the electron will be found in that region. (b) The probability cloud in part (a) is spherically symmetric, so we can represent the probability of finding the electron as a function of radius from the center of the nucleus. The probability peaks at a distance from the nucleus equal to a_0, the Bohr radius.

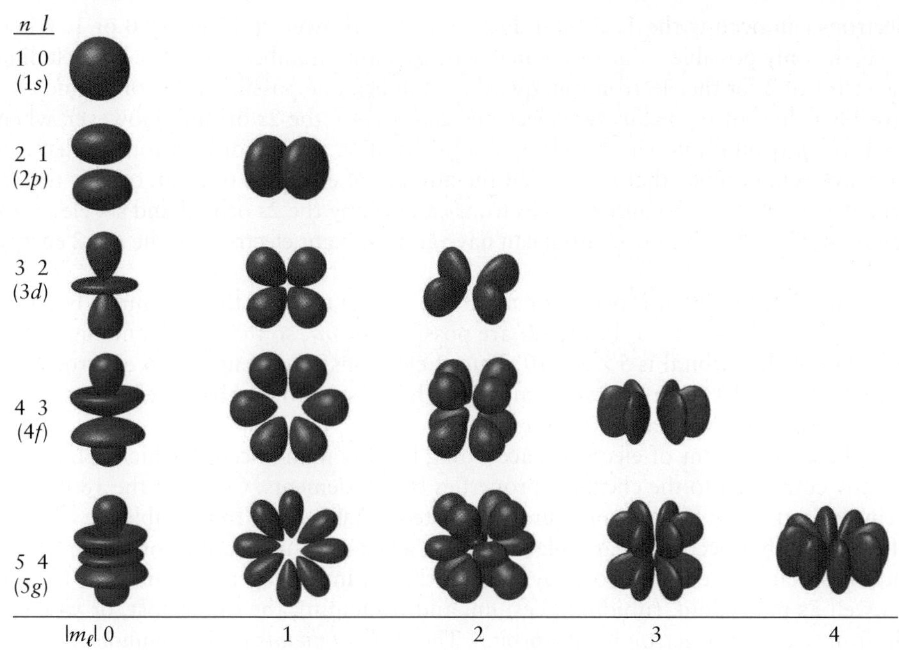

Figure 24-23 Probability clouds for different quantum numbers Different atomic orbitals are specified by specific values of the quantum numbers n, ℓ, and m_ℓ. Each orbital has a unique shape.

The angular momentum quantum number ℓ is a measure of the angular momentum the electron carries. For any value of n, ℓ varies in integer steps from 0 to $n - 1$; each value of ℓ specifies a subshell, or electron orbital, within the energy level specified by n. The shape of each orbital is different. By convention we refer to the orbitals with letters rather than integer numbers; the values of ℓ equal to 0, 1, 2, 3, 4, and 5 correspond to the orbitals s, p, d, f, g, and h, respectively. It is also standard to refer to an electron subshell by giving both n and this orbital letter code together. For example, an electron in the p subshell of the $n = 2$ energy level is said to be in the $2p$ subshell. The electron's energy is slightly dependent on the value of ℓ, an effect called *fine structure* that is not found in the Bohr model.

The magnetic quantum number m_ℓ specifies an orientation of an electron's subshell. It is so called because its value determines the (very small) energy associated with the interaction of a moving charge—the electron—with the magnetic field of the nucleus. The larger the value of ℓ, the more orientations are allowed; m_ℓ can take integer values between $-\ell$ and $+\ell$, including 0. **Figure 24-23** shows the shapes of a number of electron orbitals.

The fourth quantum number m_s involves a new feature of the electron that was not discovered until the 1920s. Electrons have an intrinsic characteristic called **spin**, which is akin to the angular momentum of a rotating sphere. Even electrons that do not orbit an atomic nucleus possess spin, which can take on one of two values, often called spin "up" and spin "down." To fully describe an atomic electron, then, we must also specify its spin state. This is described by the electron spin quantum number m_s, which for an electron can be equal to either $+1/2$ or $-1/2$.

These quantum numbers play an important role in multi-electron atoms. Each electron in a multi-electron atom has a specific value of n, ℓ, m_ℓ, and m_s. (The details of the orbitals are affected by the presence of other electrons, but the same four quantum numbers still apply.) What is more, it turns out that there can be only *one* electron with a specific combination of these four quantum numbers. This fundamental restriction on electrons is called the **Pauli exclusion principle**, after the Swiss physicist Wolfgang Pauli, who deduced this principle in 1925. (Pauli received the 1945 Nobel Prize in Physics for his work.) Let's see what the Pauli exclusion principle tells us about the structure of multi-electron atoms.

For the $n = 1$ shell, only one value of ℓ (equal to 0, which is the s orbital) and therefore only one value of m_ℓ, is allowed. Two values of m_s are always possible, so the maximum number of electrons that can occupy the $n = 1$ shell is $1 \times 2 = 2$, the product of the number of possible m_ℓ values and the number of possible m_s values. That is, two electrons can occupy the 1s orbital. For $n = 2$, ℓ is allowed to be either 0 or 1. When $\ell = 0$, the only possible value of the magnetic quantum number is $m_\ell = 0$. So including the factor of 2 for the electron spin quantum number, one possible value of m_ℓ and two possible values of m_s means two electrons can occupy the 2s orbital. However, when $\ell = 1$ (the p orbital), m_ℓ can be −1, 0, or +1. Including the factor of 2 for the electron spin quantum number, that means that the number of electrons that can occupy the 2p orbital is 3×2, or 6. Because two electrons can occupy the 2s orbital and six electrons can occupy the 2p orbital, an atom can have at most eight electrons in the $n = 2$ energy level.

The pattern above repeats for $n = 3$, up through $\ell = 2$, the d orbital. For this orbital, five values (−2, −1, 0, +1, +2) are possible for m_ℓ, so the maximum number of electrons in this orbital is $5 \times 2 = 10$. Thus 2 electrons can occupy 3s, 6 electrons can occupy 3p, and 10 electrons can occupy 3d. The maximum number of electrons in the $n = 3$ energy shell of an atom is therefore 18.

The arrangement of electrons, according to how many occupy which orbitals, is directly correlated to the chemical properties of the elements. Consider the 18 lightest elements and their electron configurations, listed in **Table 24-1**. (In the table the number of electrons that occupy a particular orbital is given as a superscript, for example, $2p^4$ indicates that four electrons occupy the 2p orbital.) In hydrogen, lithium, and sodium (as well as potassium, rubidium, cesium, and francium), the outermost, or valence, electron is a single electron in an s orbital. These *alkali metals* share common chemical properties and occupy a single column in the periodic table.

Three of the *noble gases* are listed in Table 24-1. These are helium, neon, and argon; in each the outermost subshell is completely full. As a result, all electrons are relatively tightly bound, so these elements do not easily gain, lose, or share electrons. For that reason, the noble gases are relatively inert. Helium, neon, and argon, as well as the other noble gases, occupy a single column in the periodic table.

TABLE 24-1 Electron Configurations of Light Elements

Atomic number	Element	Electron configuration
1	Hydrogen (H)	$1s^1$
2	Helium (He)	$1s^2$
3	Lithium (Li)	$1s^2 2s^1$
4	Beryllium (Be)	$1s^2 2s^2$
5	Boron (B)	$1s^2 2s^2 2p^1$
6	Carbon (C)	$1s^2 2s^2 2p^2$
7	Nitrogen (N)	$1s^2 2s^2 2p^3$
8	Oxygen (O)	$1s^2 2s^2 2p^4$
9	Fluorine (F)	$1s^2 2s^2 2p^5$
10	Neon (Ne)	$1s^2 2s^2 2p^6$
11	Sodium (Na)	$1s^2 2s^2 2p^6 3s^1$
12	Magnesium (Mg)	$1s^2 2s^2 2p^6 3s^2$
13	Aluminum (Al)	$1s^2 2s^2 2p^6 3s^2 3p^1$
14	Silicon (Si)	$1s^2 2s^2 2p^6 3s^2 3p^2$
15	Phosphorus (P)	$1s^2 2s^2 2p^6 3s^2 3p^3$
16	Sulfur (S)	$1s^2 2s^2 2p^6 3s^2 3p^4$
17	Chlorine (Cl)	$1s^2 2s^2 2p^6 3s^2 3p^5$
18	Argon (Ar)	$1s^2 2s^2 2p^6 3s^2 3p^6$

WATCH OUT !

As atomic number increases, electron orbitals do not fill in a continuous fashion.

The configuration of the 18 electrons of argon, the heaviest element listed in Table 24-1, is $1s^2 2s^2 2p^6 3s^2 3p^6$. The next heaviest element is potassium, for which the configuration of the 19 electrons is $1s^2 2s^2 2p^6 3s^2 3p^6 4s^1$. Notice that the additional electron does not occupy the $3d$ orbital, although d follows p in our ordering (s, p, d, f, and so on). Orbitals fill according to the increase in energy required, and for that reason do not fill according to counting up linearly in ℓ (orbital letters) and m_ℓ. Instead, orbitals fill in this order: $1s$, $2s$, $2p$, $3s$, $3p$, $4s$, $3d$, $4p$, $5s$, $4d$, $5p$, $6s$, $4f$, $5d$, $6p$, $7s$, $5f$, $6d$, $7p$. Notice that $4s$, not $3d$, follows the $3p$ orbital in this sequence, which is why the valence electron in potassium is in a $4s$ orbital.

Halogens are elements that are highly reactive; that is, they easily form bonds with certain other elements, especially the alkali metals, to form molecules. Two halogens, fluorine and chlorine, are listed in Table 24-1. The outermost shell in both is one electron short of being full, which means that an atom of one of these elements can readily share an electron with another atom. This is particularly true for an atom of an alkali metal, which has one valence electron that is easily shared. So bring a sodium atom near a chlorine atom, and they will readily bond to form NaCl (table salt).

The ideas of quantum mechanics find applications on scales even smaller than that of the atom. In the final two chapters of this book we will see how quantum-mechanical ideas help us understand the nature of the atomic nucleus and of the fundamental particles that are the essential building blocks of all ordinary matter.

THE TAKEAWAY for Section 24-7

✔ In the Bohr model of the atom, electrons follow Newtonian orbits around the nucleus but with quantized values of orbital angular momentum. As a result, only certain orbital radii and orbital energies are allowed.

✔ The Franck–Hertz experiment confirmed that atomic energies are quantized.

✔ The Schrödinger equation explains the hydrogen atom by describing the electron as a wave. It predicts the probability of finding the electron at a particular location within the atom.

✔ The Pauli exclusion principle allows us to understand the structure of multi-electron atoms.

1. What is the shortest wavelength of electromagnetic radiation that can be emitted by a hydrogen atom?
2. According to classical electromagnetic theory an accelerated charge emits electromagnetic radiation. What would this mean for the electron in the Bohr atom? What would happen to its orbit?
3. Why do you think Bohr's model was originally developed for the element hydrogen?

EX 24-7

4. A hydrogen atom that has an electron in the $n = 2$ state absorbs a photon.
   (a) What wavelength must the photon possess to send the electron to the $n = 4$ state?
   (b) What possible wavelengths would be detected in the spectral lines that result from the de-excitation of the atom as it returns from $n = 4$ to the ground state?

5. Find the amount of energy needed to ionize a hydrogen atom that starts in the Bohr orbit represented by $n = 3$. If an atom is ionized, its outer electron is no longer bound to the atom.

AP® Skill Builders

6. In the Bohr model of hydrogen, there is just one possible state of the electron in the lowest energy level ($n = 1$) and just one possible state of the electron in the next energy level ($n = 2$). In the more accurate Schrödinger picture of the hydrogen atom, in which the electron is described by the four quantum numbers n, ℓ, m_ℓ, and m_s, there are two possible states in the $n = 1$ level and eight possible states in the $n = 2$ level.
   (a) Give the values of n, ℓ, m_ℓ, and m_s for each of the two possible states in the $n = 1$ energy level.
   (b) Give the values of n, ℓ, m_ℓ, and m_s for each of the eight possible states in the $n = 2$ energy level.

7. For carbon (Z = 6) the frequencies of spectral lines resulting from a single electron transition are increased over those for hydrogen by the ratio of the Rydberg constants:

$$\frac{R_C}{R_H} = \frac{1 - m_e/m_C}{1 - m_e/m_H}$$

The mass of a hydrogen atom (m_H) is 1837 times greater than the mass of the electron (m_e), and the mass of the carbon atom (m_C) is 12 times greater than the mass of the hydrogen atom. Find the difference in frequency between the spectral lines for the carbon 272α transition and the hydrogen 272α transition. The 272α transition is from $n_i = 273$ to $n_f = 272$. Express your answer in megahertz (MHz).

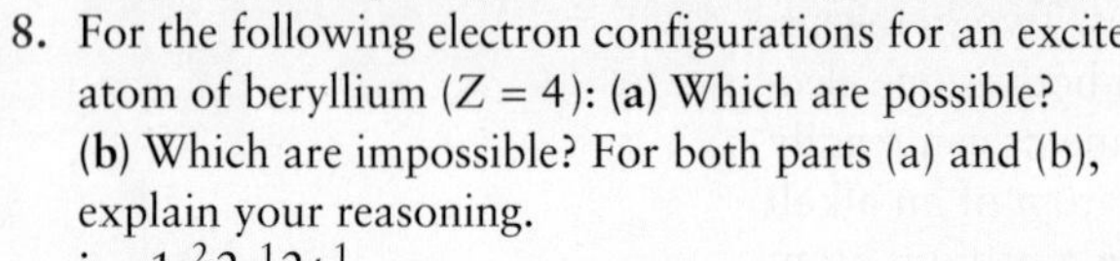

EX 24-7 8. For the following electron configurations for an excited atom of beryllium (Z = 4): **(a)** Which are possible? **(b)** Which are impossible? For both parts (a) and (b), explain your reasoning.

i. $1s^2 2s^1 2p^1$
ii. $1s^1 2s^3$
iii. $1s^1 2p^3$
iv. $1s^3 2s^1$

AP® Skills in Action

EX 24-7 9. The *ionization energy* of an atom is the energy required to remove an electron from the atom. In terms of the Bohr model, it is equal to the energy difference between the $n = 1$ energy level (the level of lowest energy, in which the electron is closest to the nucleus) and the $n = \infty$ energy level (which has an infinite radius, so the electron has moved infinitely far away). Compared to the ionization energy of hydrogen, the energy required to remove the last electron from doubly ionized lithium is

(A) 3 times greater.
(B) 9 times greater.
(C) 27 times greater.
(D) 81 times greater.

10. Derive simple equations that allow you to calculate
(a) the angular momentum and
(b) the speed of an electron in the *n*th Bohr orbit.
(c) Using the equations that you derived, calculate the speed and the angular momentum of an electron in the 10th Bohr orbit of a hydrogen atom.

24-8 In quantum mechanics, it is impossible to know precisely both a particle's position and its momentum

We saw in the previous section that in quantum mechanics, we describe the position of an electron in a hydrogen atom in terms of probabilities rather than certainties. We can calculate the probability that we will find the electron in a certain region around the nucleus, but we cannot say with certainty whether or not we will find it there. This fundamental uncertainty is a direct consequence of the wave nature of the electron. To gain more insight into this, let's investigate a fundamental principle of quantum mechanics called the *Heisenberg uncertainty principle*. We can understand this principle using ideas about sinusoidal waves that we introduced in Chapter 21.

In Section 21-3 we wrote the following wave function to describe a sinusoidal wave with wavelength λ and frequency f:

EQUATION IN WORDS
Wave function for a sinusoidal wave propagating in the $+x$ direction

Wave function for a **sinusoidal wave** propagating in the $+x$ direction
Angular wave number of the wave $= 2\pi/\lambda$

$$y(x,t) = A\cos(kx - \omega t + \phi) \qquad (21\text{-}6)$$

Phase angle
Amplitude of the wave
Angular frequency of the wave $= 2\pi f$

In Equation 21-6 the angular wave number of the wave is $k = 2\pi/\lambda$ and the angular frequency of the wave is $\omega = 2\pi f$. Note that the function in Equation 13-6 is valid for any value of x from $x = -\infty$ to $x = +\infty$. This means that a wave with a single definite wavelength has an infinite extent (**Figure 24-24a**). So the answer to the question "Where is this wave?" is "Everywhere!" A more realistic description of a wave is one that has a finite extent in space; for example, if you make a sound with your voice that lasts 1.00 s, the sound wave that emanates from your mouth at a speed of 343 m/s (the speed of sound in dry air at 20°C) will be (343 m/s)(1.00 s) = 343 m in extent from its leading edge to its trailing edge. Mathematically, a wave of a finite spatial extent Δx can be expressed as a sum of waves of infinite extent like the one shown in Figure 24-24a

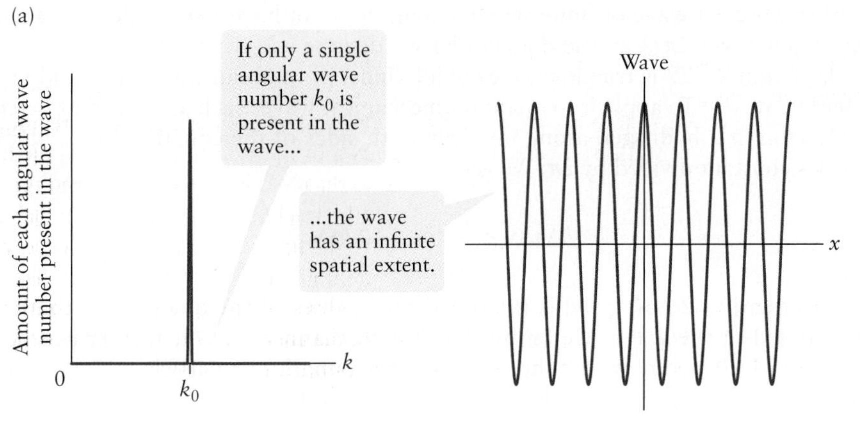
(a)
If only a single angular wave number k_0 is present in the wave...
...the wave has an infinite spatial extent.
Amount of each angular wave number present in the wave
0
k_0
k
Wave
x

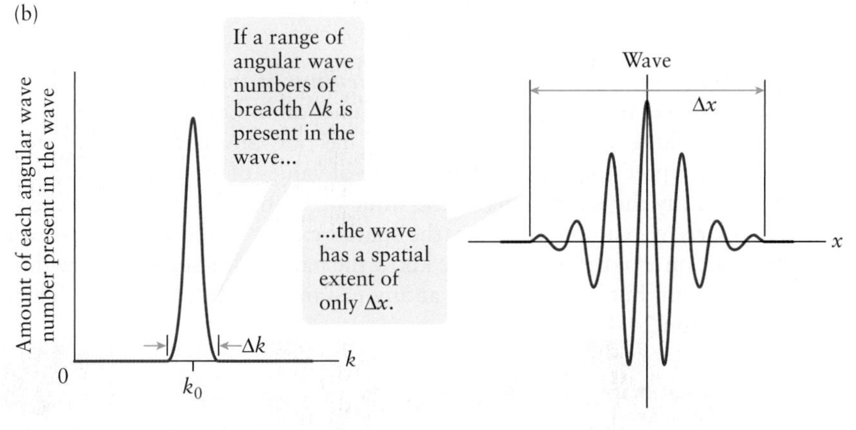
(b)
If a range of angular wave numbers of breadth Δk is present in the wave...
...the wave has a spatial extent of only Δx.
Amount of each angular wave number present in the wave
Δk
0
k_0
k
Wave
Δx
x

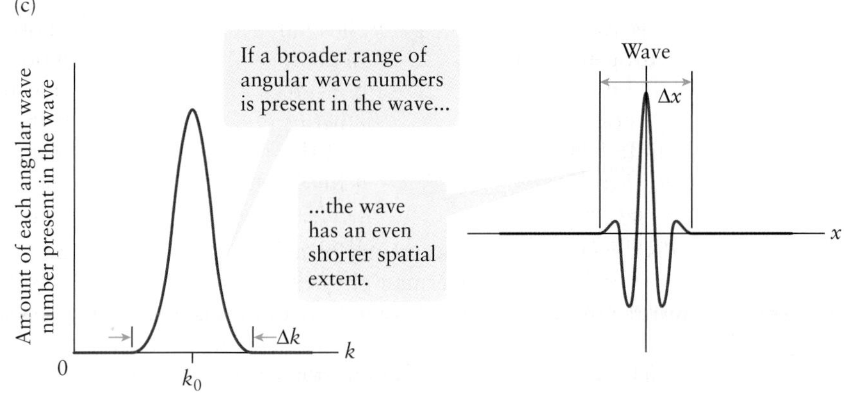
(c)
If a broader range of angular wave numbers is present in the wave...
...the wave has an even shorter spatial extent.
Amount of each angular wave number present in the wave
Δk
0
k_0
k
Wave
Δx
x

Figure 24-24 Angular wave number and the spatial extent of a wave (a) A wave of a single, discrete wavelength λ and hence a single, discrete angular wave number $k = 2\pi/\lambda$ has an infinite spatial extent. (b), (c) Combining waves over a breadth of angular wave number yields a total wave that has a finite spatial extent. The shorter the spatial extent of the wave, the greater the range of angular wave numbers present in the wave.

but with a range of angular wave numbers of breadth Δk (**Figure 24-24b**). To make a wave of shortened spatial extent, we have to add together infinite-extent waves from a broader range of angular wave numbers (**Figure 24-24c**). We can express the relationship between Δx (the spatial extent of a wave) and Δk (the breadth of angular wave numbers that go into that wave) as

$$\Delta k \Delta x \geq \frac{1}{2} \tag{24-43}$$

Equation 24-34 shows that the product of the wave spatial extent Δx and the angular wave number breadth Δk cannot be less than 1/2. (This number arises from the specific way in which Δx and Δk are defined mathematically.) It says that to minimize the spatial extent of a wave necessarily means increasing the range of angular wave numbers that make up the wave. So the shorter the spatial extent of a wave, the less precisely we can answer the question "What is the angular wave number of the wave?"

In other words, a wave of finite spatial extent does not have a single definite angular wave number $k = 2\pi/\lambda$ and so does not have a definite wavelength λ.

Equation 24-29 is true for waves of all kinds, from ocean waves to sound waves to seismic waves. To apply it to quantum-mechanical waves such as those that describe an electron in a hydrogen atom, multiply both sides of the equation by $\hbar$, which is Planck's constant divided by 2π. We get

$$\hbar \Delta k \Delta x \geq \frac{\hbar}{2} \quad \text{or} \quad \Delta(\hbar k)\Delta x \geq \frac{\hbar}{2} \tag{24-44}$$

In Equation 24-30 $\Delta(\hbar k)$ is the breadth of values of the quantity $\hbar k$ that must be included in the wave. We know that $\hbar = h/2\pi$ and $k = 2\pi/\lambda$; furthermore from Equation 24-10 we know that the de Broglie wavelength of a particle of momentum p is $\lambda = h/p$, so $p = h/\lambda$. Putting these together, we find that

$$\hbar k = \left(\frac{h}{2\pi}\right)\left(\frac{2\pi}{\lambda}\right) = \frac{h}{\lambda} = p$$

So the quantity $\hbar k$ is just equal to the *momentum* of the particle that the wave describes. We'll relabel this as p_x since the wave function in Equation 21-6 describes a wave propagating along the x axis, so the particle has only an x component of momentum. So we can interpret $\Delta(\hbar k)$ as Δp_x, the breadth of values of momentum preset in the wave, or equivalently the *uncertainty* in the value of p_x for the particle that the wave describes. Similarly we can interpret Δx, the spatial extent of the wave, as the *uncertainty* in the position x of the particle. (We know the particle lies somewhere within the wave's spatial extent, but we cannot say at any instant exactly where it is.) Then we can rewrite Equation 24-44 as

EQUATION IN WORDS
Heisenberg uncertainty principle for momentum and position

Uncertainty in the x component of momentum of a particle

$\hbar = \dfrac{h}{2\pi}$ = Planck's constant divided by 2π

$$\Delta p_x \Delta x \geq \frac{\hbar}{2} \tag{24-45}$$

Uncertainty in the x coordinate of the particle

Equation 24-45 is the **Heisenberg uncertainty principle for momentum and position**, which was first expressed by the German physicist Werner Heisenberg in 1927. (Heisenberg was awarded the 1932 Nobel Prize in Physics for his work in the development of quantum mechanics.) It states that for any particle, the product of the uncertainty in momentum Δp_x and the uncertainty in position Δx can never be less than $\hbar/2$. This places a fundamental limitation on how well you can simultaneously know the momentum and position of a particle. For example, suppose you know the position of a particle to within an uncertainty Δx. It follows from Equation 24-45 that the the minimum uncertainty in how well you can know the momentum of that particle is $\Delta p_x = \hbar/2\Delta x$. In other words, it's fundamentally impossible to measure the monentum of the particle with an uncertainty less than $\hbar/2\Delta x$. In Newtonian physics it's possible in principle to measure exactly both the momentum and position of a particle; the Heisenberg uncertainty principle says this is impossible, no matter how refined your measuring techniques. Note that the Heisenberg uncertainty principle also applies to the uncertainties in momentum and position for the y and z directions:

$$\Delta p_y \Delta y \geq \frac{\hbar}{2}$$

$$\Delta p_z \Delta z \geq \frac{\hbar}{2}$$

The value of $\hbar/2$ is very small: $\hbar/2 = h/4\pi = 5.27 \times 10^{-35}$ J·s. Hence the Heisenberg uncertainty principle has little consequence for measurements of relatively large objects like baseballs or insects. But on the atomic scale and smaller the uncertainty principle is tremendously important, as the following example shows.

EXAMPLE 24-8 Applying the Heisenberg Uncertainty Principle

(a) If an electron is in the lowest ($n = 1$) energy level of the hydrogen atom, its uncertainty in position is approximately equal to the Bohr radius $a_0 = 0.0529$ nm. Find the minimum uncertainty in its x component of momentum, and find the kinetic energy of an electron with this magnitude of momentum. (b) Repeat these calculations for an electron whose uncertainty in position is one-tenth of a Bohr radius.

Set Up

We use the Heisenberg uncertainty principle for momentum and position, Equation 24-45. For a given position uncertainty Δx, the minimum value for momentum uncertainty Δp_x corresponds to using the equals sign in this equation. If we set Δp_x equal to the momentum p, we can find the speed and hence the kinetic energy of the electron.

Heisenberg uncertainty principle for momentum and position:

$$\Delta p_x \Delta x \geq \frac{\hbar}{2} \tag{24-45}$$

Momentum of a particle of mass m and speed v:

$$p = mv$$

Kinetic energy of a particle of mass m and speed v:

$$K = \frac{1}{2}mv^2$$

Solve

(a) Calculate the minimum momentum uncertainty.

Using the equals sign in Equation 24-45:

$$\Delta p_x \Delta x = \frac{\hbar}{2}$$

$$\Delta p_x = \frac{\hbar}{2\Delta x} = \frac{h}{4\pi\Delta x}$$
$$= \frac{6.63 \times 10^{-34}\ \text{J}\cdot\text{s}}{4\pi\left(0.0529 \times 10^{-9}\ \text{m}\right)}$$
$$= 9.97 \times 10^{-25}\ \frac{\text{J}\cdot\text{s}}{\text{m}} = 9.97 \times 10^{-25}\ \text{kg}\cdot\text{m/s}$$

Find the speed and kinetic energy of an electron with this magnitude of momentum.

If we let $p = \Delta p_x$ and use $p = mv$, the speed of the electron is

$$v = \frac{p}{m} = \frac{9.97 \times 10^{-25}\ \text{kg}\cdot\text{m/s}}{9.11 \times 10^{-31}\ \text{kg}} = 1.09 \times 10^{6}\ \text{m/s}$$

The kinetic energy of an electron moving at this speed is

$$K = \frac{1}{2}mv^2 = \frac{1}{2}(9.11 \times 10^{-31}\ \text{kg})(1.09 \times 10^{6}\ \text{m/s})^2$$
$$= 5.46 \times 10^{-19}\ \text{J}$$
$$= 5.46 \times 10^{-19}\ \text{J}\left(\frac{1\ \text{eV}}{1.602 \times 10^{-19}\ \text{J}}\right) = 3.41\ \text{eV}$$

(b) The Heisenberg uncertainty principle states that the product of the minimum uncertainty in momentum Δp_x and the minimum uncertainty in position Δx is $\hbar/2$. So if we make Δx one-tenth as large, Δp_x must become 10 times larger.

If we reduce the value of Δx by one-tenth from a_0 to $a_0/10$, Δp_x will be 10 times greater:

$$\Delta p_x = 10(9.97 \times 10^{-25}\ \text{kg}\cdot\text{m/s})$$
$$= 9.97 \times 10^{-24}\ \text{kg}\cdot\text{m/s}$$

If this is the magnitude of momentum, then $p = mv$ is 10 times greater than in part (a) and the speed v is also 10 times greater:

$$v = 10(1.09 \times 10^6 \text{ m/s})$$
$$= 1.09 \times 10^7 \text{ m/s}$$

In this case the kinetic energy is

$$K = \frac{1}{2}mv^2 = \frac{1}{2}(9.11 \times 10^{-31} \text{ kg})(1.09 \times 10^7 \text{ m/s})^2$$
$$= 5.46 \times 10^{-17} \text{ J}$$
$$= 5.46 \times 10^{-17} \text{ J}\left(\frac{1 \text{ eV}}{1.602 \times 10^{-19} \text{ J}}\right) = 341 \text{ eV}$$

This is 100 times greater than the kinetic energy we found in part (a).

Reflect

The Heisenberg uncertainty principle for momentum and position tells us why an electron in an atom has kinetic energy and does not simply remain at rest. Because it is confined to a very small volume and so has a very small uncertainty in position, an electron necessarily has a relatively large uncertainty in momentum and hence in velocity. Consequently we can expect it to have on average a nonzero kinetic energy. For such a rough calculation our result of 3.41 eV is not too different from the actual average kinetic energy of an electron in the $n = 1$ energy level, which is 13.6 eV.

NOW WORK Problems 1, 3, 4, and 5 from the Takeaway 24-8.

The Heisenberg uncertainty principle also tells us why a negatively charged electron does not simply fall into the positively charged nucleus. By bringing the electron to one-tenth of its original distance from the nucleus, its kinetic energy increased by a factor of 100. Decreasing the distance r from the electron to the nucleus also decreases its (negative) electron potential energy $U_{\text{electric}} = k(Ze)(-e)/r = -kZe^2/r$, where Z is the number of protons in the nucleus (for hydrogen, $Z = 1$). But because U_{electric} is inversely proportional to r, decreasing r from a_0 to $a_0/10$ will only make the value of U_{electric} more negative by a factor of 10. Because the positive kinetic energy increases by a much larger factor, the result is that it would take a tremendous increase in energy to move the electron closer to the nucleus. Hence the electron cannot fall into the nucleus. So the Heisenberg uncertainty principle prevents atoms from collapsing, and so makes it possible for atoms to exist.

If you continue your study of physics, you will find there is a second form of the Heisenberg uncertainty principle that relates the uncertainty in *energy* of a phenomenon to the *duration* of that phenomenon. The startling result is that it's possible to violate the law of conservation of energy, provided the duration of time during which the law is violated is sufficiently short. This result will helps us understand the properties of the fundamental forces of nature.

THE TAKEAWAY for Section 24-8

✔ A wave has a definite wavelength and angular wave number only if it has infinite spatial extent. The shorter the spatial extent of the wave, the greater the breadth of angular wave numbers there must be in the wave.

✔ The Heisenberg uncertainty principle for momentum and position states that the product of the uncertainty in a particle's momentum and the uncertainty in its position must be greater than a certain minimum value. This is true no matter how these quantities are measured.

Prep for the AP® Exam

AP® Building Blocks

1. Consider the following particles.
   (a) What is the minimum uncertainty in an electron's velocity if the position is known within 1.5 nm?
   (b) What is the minimum uncertainty in a helium atom's velocity if the position is known within 0.10 nm? The mass of a helium atom is 6.646×10^{-27} kg.
2. The Heisenberg uncertainty principle applies not only to very small objects, but also to objects in our everyday world. Why, then, do we not worry about it when simultaneously measuring the position and momentum of a car?

AP® Skill Builders

3. The Heisenberg uncertainty principle for momentum and position plays an important role in the field of microscopy. The simple act of observing a sample involves bombarding the sample with photons. This bombardment results in a change to the physical state of the sample. Consider a 600-nm photon that scatters off an atom located within a crystalline lattice. Using the uncertainty principle, determine the minimum position uncertainty of the atom, assuming all of the momentum of the photon is transferred to the atom (this is an estimate of the upper limit).

4. Consider a 1350-kg automobile clocked by law-enforcement radar at a speed of 38.0 m/s.
   (a) If the position of the car is known to within 1.50 m at the time of the measurement, what is the uncertainty in the velocity of the car?
   (b) If the speed limit is 36.5 m/s, could the driver of the car reasonably evade a speeding ticket by invoking the Heisenberg uncertainty principle?

AP® Skills in Action

EX 24-8

5. A neutron has 1839 times the mass of an electron. If a neutron and an electron both have the same uncertainty in position, the minimum uncertainty in momentum of the neutron will be
   (A) 1839 times the minimum uncertainty in momentum of the electron.
   (B) 1/1839 the minimum uncertainty in momentum of the electron.
   (C) the same as the minimum uncertainty in momentum of the electron.
   (D) impossible to calculate, because it is a neutron.

WHAT DID YOU LEARN?

Chapter learning goals	Section(s)	Related example(s)	Prep for the AP® Exam: Relevant section review exercise(s)
Recognize the limitations of classical physics for explaining the properties of light and matter.	24-1		
	24-3		2
	24-8		2
Calculate the energy density and intensity of an electromagnetic wave, and the energy of a photon.	24-2	24-1	6-8
		24-2	1, 2, 4, 5
Describe how the photoelectric effect and blackbody radiation provide evidence for the photon picture of light and calculate properties related to these processes.	24-3	24-3	1-3, 8, 10
Explain why the wavelength of a photon increases if it scatters from an electron and calculate the change in wavelength for a given scattering angle.	24-4	24-4	1, 4, 10
Calculate the wavelength of a particle such as an electron given its momentum.	24-5	24-5	4-8, 10
Describe why atoms absorb light at only certain wavelengths, describe why they emit and absorb light at the same wavelengths, and calculate these wavelengths for a hydrogen atom.	24-6	24-6	1-6, 10
Explain how the Bohr model of the atom explains the spectrum of hydrogen and calculate predicted allowed energies and angular momenta.	24-7	24-7	1, 5, 8
Describe the Heisenberg uncertainty principle and use it in calculations.	24-8	24-8	1-5

Chapter 24 Review

Key Terms

All the Key Terms can be found in the Glossary/Glosario on page G1 in the back of the book.

absorption line 1214
absorption spectrum 1214
blackbody 1201
blackbody radiation 1202
Bohr model 1220
Bohr orbit 1215
Compton scattering 1206
de Broglie wavelength 1209
electric energy density 1190
emission line 1214
emission spectrum 1214
energy level 1217
Heisenberg uncertainty principle for momentum and position 1234
ionizing radiation 1195
magnetic energy density 1191
Pauli exclusion principle 1229
photoelectric effect 1198
photoelectron 1198
photon 1194
Planck's constant 1194
quantized 1217
quantum mechanics 1228
quantum number 1228
spin 1229
wave-particle duality 1211
work function 1198

Chapter Summary

Topic	Equation or Figure	
Electric and magnetic field energy in electromagnetic waves: The energy density (energy per volume) associated with an electric field is $u_E = (1/2)\varepsilon_0 E^2$, and the energy density associated with a magnetic field is $u_B = B^2/2\mu_0$. In an electromagnetic wave in a vacuum, there are equal amounts of electric energy and magnetic energy. The average energy density in an electromagnetic wave can be expressed in terms of either the field amplitudes or their rms values. The intensity of the wave is the average power per unit area of the wave.	Intensity of an electromagnetic wave; Speed of light; Magnetic field rms value $S_{\text{average}} = c\varepsilon_0 E_{\text{rms}}^2 = \dfrac{cB_{\text{rms}}^2}{\mu_0} = \dfrac{E_{\text{rms}}B_{\text{rms}}}{\mu_0}$ Permittivity of free space; Electric field rms value; Permeability of free space	(24-12)
Photons: The energy of an electromagnetic wave comes in packets called photons that have properties of both wave and particle. The higher the wave frequency, the more energy there is per photon.	$v^2 = \dfrac{2K}{m} = \dfrac{2e\Delta V}{m}$ so $v = \sqrt{v^2} = \sqrt{\dfrac{2e\Delta V}{m}}$	(24-26)
The photoelectric effect: When light shines on a surface, the surface can emit electrons. However, no electrons are emitted if the frequency of the light is below a certain critical value. Einstein showed that this could be explained if light is absorbed in the form of photons whose energy is proportional to their frequency: $E = hf$, where h is Planck's constant.	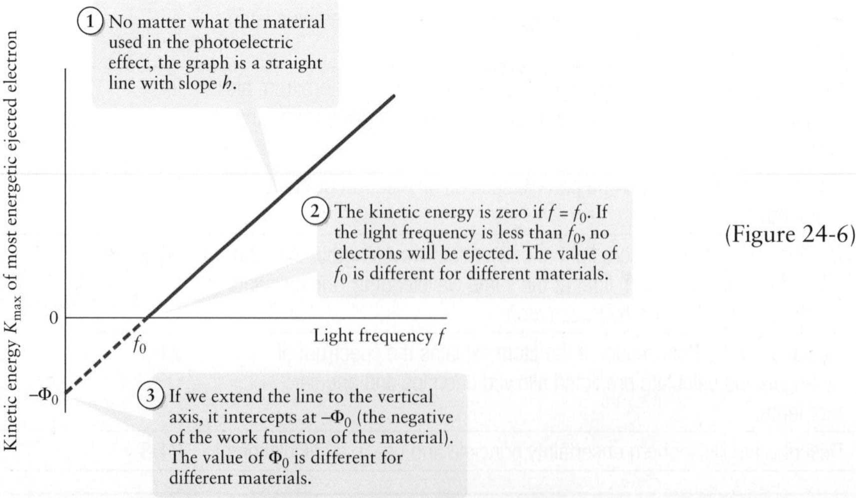 	(Figure 24-6)

Blackbody radiation: Objects emit light due to their temperature. An ideal blackbody (one that does a perfect job of absorbing light) is also a perfect emitter of light. The details of the spectrum of a blackbody can be understood only if a blackbody emits light in the form of photons.

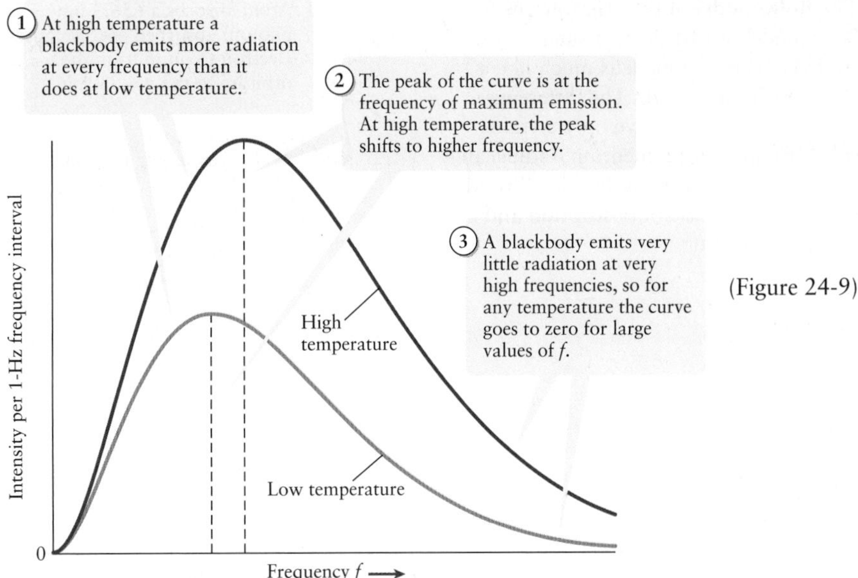

1 At high temperature a blackbody emits more radiation at every frequency than it does at low temperature.
2 The peak of the curve is at the frequency of maximum emission. At high temperature, the peak shifts to higher frequency.
3 A blackbody emits very little radiation at very high frequencies, so for any temperature the curve goes to zero for large values of f.
High temperature
Low temperature
Intensity per 1-Hz frequency interval
0
Frequency f

(Figure 24-9)

Compton scattering: Photons have momentum in inverse proportion to their wavelength. This is demonstrated by Compton scattering, in which a photon undergoes an increase in wavelength when it scatters from an electron.

Magnitude of the **momentum** of a photon | **Energy** of the photon | **Planck's constant** | **Speed of light in a vacuum** | **Wavelength** of the photon

$$p = \frac{E}{c} = \frac{h}{\lambda} \qquad (24\text{-}21)$$

Wavelength change in Compton scattering | **Planck's constant** | Photon **scattering angle** | **Photon wavelength after** scattering | **Photon wavelength before** scattering | **Electron mass** | **Speed of light in a vacuum**

$$\Delta\lambda = \lambda_f - \lambda_i = \frac{h}{m_e c}(1 - \cos\theta) \qquad (24\text{-}22)$$

Wave-particle duality: Just as photons have particle aspects, matter has wave aspects. The de Broglie wavelength of a particle is inversely proportional to the momentum (the same relationship between wavelength and momentum as for a photon).

A particle has a **de Broglie wavelength...** | ...equal to **Planck's constant...**

$$\lambda = \frac{h}{p} \qquad (24\text{-}24)$$

...divided by the **momentum of the particle.** The greater the momentum, the shorter the de Broglie wavelength.

Atomic structure and atomic spectra: Alpha particle scattering shows that the atom is made up of a small positive nucleus surrounded by electrons. The energy of an atom is quantized. It can have only certain definite values. The evidence for this comes from the spectra of atoms, which show that atoms absorb only specific wavelengths of light and that they emit the same wavelengths that they absorb.

(a) The absorption spectrum produced by passing white light through a gas of hydrogen atoms

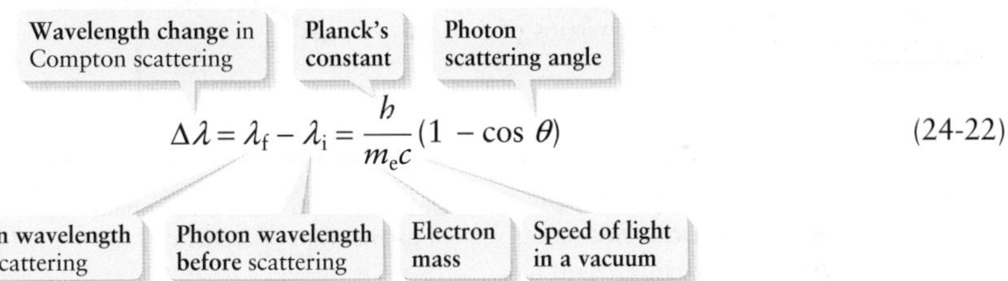

The wavelengths at which hydrogen atoms absorb light are the same as the wavelengths at which hydrogen atoms emit light.

(b) The emission spectrum produced by a heated gas of hydrogen atoms

- Atoms of each element absorb and emit light at wavelengths that are characteristic of that element.
- The characteristic wavelengths differ from one element to another.
- Hydrogen has the simplest arrangement of characteristic wavelengths of any element.

(Figure 24-16)

The Bohr model of the atom: In the Bohr model of hydrogen, a single electron orbits the nucleus much like a satellite orbiting Earth. The difference is that the orbit can have only certain values of angular momentum, radius, and energy. Transitions between the allowed orbits are the cause of absorption and emission spectra.

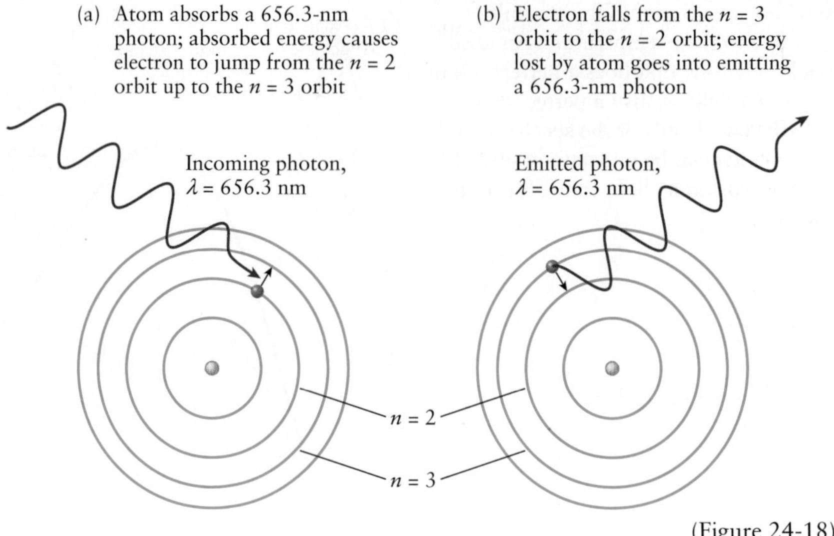

(a) Atom absorbs a 656.3-nm photon; absorbed energy causes electron to jump from the $n = 2$ orbit up to the $n = 3$ orbit
(b) Electron falls from the $n = 3$ orbit to the $n = 2$ orbit; energy lost by atom goes into emitting a 656.3-nm photon
Incoming photon, $\lambda = 656.3$ nm
Emitted photon, $\lambda = 656.3$ nm
$n = 2$
$n = 3$

(Figure 24-18)

The Schrödinger equation and the Pauli exclusion principle: In the more complete Schrödinger description of the atom, the electron is described by a wave. Four quantum numbers describe the state of the electron, and a probability cloud describes the probability of finding an electron at different positions within the atom. The Pauli exclusion principle states that there can be only one electron per state; this helps explain the properties of multi-electron atoms.

(a)

(Figure 24-22a)

The Heisenberg uncertainty principle for momentum and position: Because particles have wave properties, there is a fundamental limit to how well both the momentum and the position of a particle can be known simultaneously. This helps to explain why atoms do not collapse.

Uncertainty in the x component of momentum of a particle

$\hbar = \dfrac{h}{2\pi}$ = Planck's constant divided by 2π

$$\Delta p_x \Delta x \geq \frac{\hbar}{2} \quad (24\text{-}45)$$

Uncertainty in the x coordinate of the particle

Chapter 24 Review Problems

1. Is it possible for photoelectrons to be emitted from a metal plate with relativistic speeds?
2. Describe how the number of photoelectrons emitted from a metal plate in the photoelectric effect would change if
   (a) the intensity of the incident radiation were increased,
   (b) the wavelength of the incident radiation were increased, and
   (c) the work function of the metal were increased.
3. A markedly nonclassical feature of the photoelectric effect is that the energy of the emitted electrons doesn't increase as you increase the intensity of the light striking the metal surface. What change does occur as the intensity is increased?
4. Consider the photoelectric emission of electrons induced by incident light of a single wavelength. The incoming photons all have the same energy, but the emitted electrons have a range of kinetic energies. Why?
5. Does the de Broglie wavelength of a particle increase or decrease as its kinetic energy increases?

6. How does the intensity of light from a blackbody change when its temperature is increased? What changes occur in the body's radiation spectrum?

7. Vitamin D is produced in the skin when 7-dehydrocholesterol reacts with UVB rays (ultraviolet B) having wavelengths between 270 and 300 nm. What is the energy range of the UVB photons?

8. Which of the two Compton scattering experiments more clearly demonstrates the particle nature of electromagnetic radiation: a collision of the photon with an electron or a collision with a proton? Explain your answer.

9. Some of the stars in the photo that opens this chapter are blue, while others are red. Based on the color alone, you can conclude that the blue stars are
   (A) hotter than the red stars.
   (B) cooler than the red stars.
   (C) made of different materials than the red stars.
   (D) at a different temperature and made of different materials than the red stars.

10. Why do you think that Bohr's model of the hydrogen atom is still taught in introductory physics classes?

11. In 2009 astronomers detected gamma-ray photons having energy ranging from 700 GeV to around 5 TeV coming from supernovae (exploding giant stars) in the galaxy M82. (1 GeV = 10^9 eV, 1 TeV = 10^{12} eV.)
   (a) What is the range of wavelengths of the gamma-ray photons detected from M82?
   (b) Calculate the ratio of the energy of a 5-TeV photon to the energy of a visible-light photon having a wavelength of 500 nm.

12. For an electron in the nth state of the hydrogen atom, write expressions for
   (a) the angular momentum of the electron,
   (b) the radius of the electron's orbit,
   (c) the kinetic energy of the electron,
   (d) the total energy of the electron, and
   (e) the speed of the electron.

13. Radio astronomers use radio frequency waves to identify the elements in distant stars. One of the standard lines that is often studied is designated the 272α line. This spectral line refers to the transition in hydrogen from $n_i = 273$ to $n_f = 272$. Calculate the wavelength and frequency of the electromagnetic radiation that is emitted for the 272α transition.

14. Set up a chart for the five quantities listed in problem 12 and calculate the values for $n = 1, 2, 3, 4$, and 5 (four significant digits, SI units). See the following table.

n	L_n	r_n	K_n	E_n	v_n
1					
2					
3					
4					
5					

15. Arthur Compton scattered photons that had a wavelength of 0.0711 nm off a block of carbon during his famous experiment of 1923 at Washington University in St. Louis, Missouri.
   (a) Calculate the frequency and energy of the photons.
   (b) What is the wavelength of the photons that are scattered at 90.0°?
   (c) What is the energy of the photons that are scattered at 90.0°?
   (d) What is the energy of the electrons that recoil from the Compton scattering with $\theta = 90.0°$?

16. The Lyman series ($n_f = 1$), the Balmer series ($n_f = 2$), and the Paschen series ($n_f = 3$) are commonly studied in basic chemistry and physics classes (see problems 3–5 in the Takeaway 24-6). The Brackett series ($n_f = 4$) and the Pfund series ($n_f = 5$) are not so well known.
   (a) Calculate the shortest and longest wavelengths for the spectral lines that are part of the Brackett series. State whether each wavelength is visible (380 to 750 nm), ultraviolet (shorter than 380 nm), or infrared (longer than 750 nm).
   (b) Calculate the shortest and longest wavelengths for the spectral lines that are part of the Pfund series. State whether each wavelength is visible (380 to 750 nm), ultraviolet (shorter than 380 nm), or infrared (longer than 750 nm).

17. Find the energies of the first 10 energy levels (sketch and label an energy-level diagram) for singly ionized helium, He^+.

18. A photon of green light has a wavelength of 525 nm.
   (a) What is the energy of the photon? Give your answer in joules and electron volts.
   (b) What is the wave number of the photon?

19. The threshold wavelength for the photoelectric effect for silver is 262 nm.
   (a) Determine the work function for silver.
   (b) Calculate the maximum kinetic energy of an electron emitted from silver if the incident light has a wavelength of 222 nm.

20. Wien's displacement law states that the wavelength of maximum emission for a blackbody is given by the following formula:

$$\lambda_{max} = \frac{0.290 \text{ K} \cdot \text{cm}}{T}$$

   (a) If the human body acts like a blackbody, use Wien's displacement law to calculate the body's wavelength of maximum emission. Assume an average skin temperature of 34°C.
   (b) In what part of the electromagnetic spectrum is this light?

21. Suppose a blackbody at 400 K radiates just enough heat in 15 min to boil water for a cup of tea. How long will it take to boil the same water if the temperature of the radiator is 500 K?

22. Prove that the Balmer formula is a special case of the Rydberg formula with n set equal to 2.

Rydberg formula: $\frac{1}{\lambda} = R_H\left(\frac{1}{n^2} - \frac{1}{m^2}\right)$

$R_H = 1.09737 \times 10^7\ m^{-1}$

Balmer formula: $\lambda = b\left(\frac{m^2}{m^2 - 4}\right)$ $b = 364.51$ nm

23. A photon Compton scatters off a stationary electron at an angle of 60.0°. The electron moves away with 1.28×10^{-17} J of kinetic energy. Determine the initial wavelength of the photon.

24. A helium-neon laser produces light of wavelength 632.8 nm. The laser beam carries a power of 0.50 mW and strikes a target perpendicular to the beam.
(a) How many photons per second strike the target?
(b) At what rate does the laser beam deliver linear momentum to the target if the photons are all absorbed by the target?

25. The Sun has a wavelength of maximum emission of 475 nm. Using Wien's displacement law (see problem 20), determine the corresponding temperature of the outer layer of the Sun, assuming it is a blackbody.

26. A photon of frequency 4.81×10^{19} Hz scatters off a free stationary electron. Careful measurements reveal that the photon goes off at an angle of 125° with respect to its original direction.
(a) How much energy does the electron gain during the collision?
(b) What percent of its original energy does the photon lose during the collision?

27. Calculate the de Broglie wavelength of a thermal neutron that has a kinetic energy of about 6.41×10^{-21} J.

28. A laboratory oven that contains hydrogen molecules H_2 and oxygen molecules O_2 is maintained at a constant temperature T. Each oxygen molecule is 16 times as massive as a hydrogen molecule. Find the ratio of the de Broglie wavelength of the hydrogen molecule to that of the oxygen molecule, assuming that each molecule has kinetic energy $(5/2)kT$.

29. A hydrogen atom makes a transition from the $n = 5$ state to the ground state and emits a single photon of light in the process. The photon then strikes a piece of silicon, which has a photoelectric work function of 4.8 eV. Is it possible that a photoelectron will be emitted from the silicon? If not, why not? If so, find the maximum possible kinetic energy of the photoelectron.

30. Suppose the electron in the hydrogen atom were bound to the proton by gravitational forces (rather than electrostatic forces). Find
(a) the radius and
(b) the energy of the first orbit.

31. The *E. coli* bacterium is about 2.0 μm long. Suppose you want to study it using photons of that wavelength or electrons having that de Broglie wavelength.
(a) What is the energy of the photon and the energy of the electron?
(b) Which one would be better to use, the photon or the electron? Explain your prediction using physics principles.

32. When x-ray photons are aimed at a carbon target, a photon can undergo Compton scattering from one of the electrons in a carbon atom. But a photon can undergo Compton scattering from any charged particle, and so can be scattered by a carbon nucleus.
(a) Calculate the Compton wavelength $\lambda_C = h/mc$ for a carbon nucleus, which has a mass of 1.99×10^{-26} kg. Express your answer in nanometers (nm).
(b) Find the wavelength shift $\Delta\lambda$ for a photon that undergoes a 180° scattering from a carbon nucleus. Express your answer in nm. Compare your answer to the wavelength shift of 0.00485 nm for a photon that undergoes a 180° scattering from an electron.
(c) The x-ray photons that Arthur Compton used in his 1923 experiment had a wavelength of 0.0711 nm.
 i. Find the fractional changes in wavelength $\frac{\Delta\lambda}{\lambda}$ such a photon undergoes due to Compton scattering from an electron and from a carbon nucleus.
 ii. Explain why Compton was only able to measure the Compton scattering of photons from electrons, not from nuclei.
(d) Suppose you want a photon to undergo a 1.00% increase in wavelength (which is more easily measured) as a result of a 180° Compton scattering from a carbon nucleus. What would be the wavelength and energy of such a photon before scattering? (This is a gamma-ray photon, and such photons were not available to Compton when he did his experiments.)

33. The Franck-Hertz experiment demonstrated energy quantization as electrons with sufficient kinetic energy to excite an atom did not pass through to the anode at the rate expected, apparently having had their kinetic energy absorbed by an atom instead. Figure 24-21 shows that in a Franck–Hertz experiment using mercury vapor, the anode current drops off for cathode-grid potential differences of 4.9 V, 2×4.9 V = 9.8 V, and 3×4.9 V = 14.7 V.
(a) When a helium atom goes from its first excited energy level (electron configuration $1s^1 2p^1$) to its lowest energy level (electron configuration $1s^2$), it emits an ultraviolet photon of wavelength 58.4 nm. What is the energy of such a photon in eV?

(b) If you repeat the Franck–Hertz experiment using helium gas rather than mercury vapor, what are the first three cathode-grid potential differences at which you would expect to see a drop in the anode current?

34. A relativistic electron has a de Broglie wavelength of 346 fm (1 fm = 10^{-15} m). Determine its speed. (You will need first to restate de Broglie's formula for particle waves using the relativistic relationship between speed and momentum shown in Equation 24-15.)

35. In 1913, Niels Bohr proposed a model of the hydrogen atom in which the electron could exist only in specific circular orbits. He did not offer any explanation as to why only those particular orbits would be allowed. In 1924, Louis de Broglie suggested that the electron could only exist in orbits corresponding to certain kinds of wave patterns. The figure here shows an electron orbit that has been "cut and unwrapped" to better show the electron wave. Classify the wave patterns according to whether or not they are compatible with de Broglie's description of the allowed Bohr orbits of the hydrogen atom.

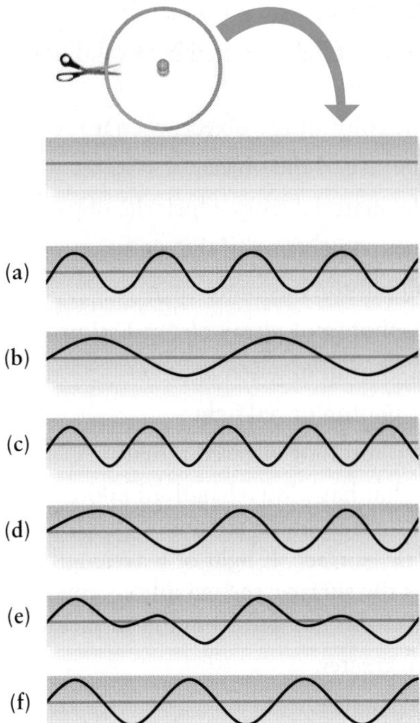

36. Derive an expression for the change in photon *energy* in Compton scattering (that is, an expression for the difference between the final and initial photon energies, E_f and E_i) as a function of the photon scattering angle.

Prep for the AP Exam

AP® Group Work

Directions: The problem below is designed to be done as group work in class.

You want to use a microscope to study the structure of a mitochondrion about 1 μm in size. To be able to observe small details within the mitochondrion, you want to use a wavelength of 0.0500 nm.

(a) If your microscope uses light of this wavelength, what is the momentum p of a photon? What is the energy E of a photon?

(b) If instead, your microscope uses electrons of this de Broglie wavelength, what is the momentum p of an electron? What is the speed v of an electron? (Recall $p = mv$.) What is the kinetic energy $K = (1/2)mv^2$ of an electron?

(c) Comparing your results for the photon momentum and energy in (a) to your results for the electron momentum and kinetic energy in (b), what advantages can you see to using electrons rather than photons?

AP® PRACTICE PROBLEMS

Prep for the AP Exam

Multiple-Choice Questions

Directions: The following questions have a single correct answer.

1. Atoms of an element emit a spectrum that
   (A) is the same as all other elements.
   (B) is evenly spaced.
   (C) is unique to that element.
   (D) is evenly spaced and unique to that element.

2. An ideal blackbody is an object that
   (A) absorbs most of the energy that strikes it and emits a little of the energy it absorbs.
   (B) absorbs all the energy that strikes it and emits all the energy it absorbs.
   (C) absorbs a little of the energy that strikes it and emits most of the energy it absorbs.
   (D) neither absorbs nor emits energy except at ultraviolet ("black light") wavelengths.

(Continued)

3. The color of light emitted by a solid hot object depends on
   (A) the color of the object when it is cold.
   (B) the size of the object.
   (C) the material from which the object is made.
   (D) the temperature of the object.
4. Which photon has more energy?
   (A) A photon of ultraviolet radiation
   (B) A photon of green light
   (C) A photon of red light
   (D) A photon of infrared radiation
5. Light that has a wavelength of 600 nm strikes a metal surface, and a stream of electrons is ejected from the surface. If light of wavelength 500 nm strikes the surface, the maximum kinetic energy of the electrons emitted from the surface will
   (A) be greater.
   (B) be smaller.
   (C) be the same.
   (D) be 5/6 smaller.
6. In the Compton effect experiment, the change in a photon's wavelength depends on
   (A) the scattering angle.
   (B) the initial wavelength of the photon.
   (C) the final wavelength of the photon.
   (D) the density of the scattering material.
7. The maximum change in a photon's energy when it undergoes Compton scattering occurs when its scattering angle is
   (A) 0°. (C) 90°.
   (B) 45°. (D) 180°.
8. As the scattering angle in the Compton effect increases, the energy of the scattered photon
   (A) increases.
   (B) stays the same.
   (C) decreases.
   (D) is proportional to sin θ.
9. The de Broglie wavelength of an object depends only on
   (A) the object's mass.
   (B) the object's speed.
   (C) the object's energy.
   (D) the object's momentum.

Free-Response Question

1. Assume you are to be given an opportunity to explore the photoelectric effect in a lab at a nearby college.
   (a) Describe what quantities you would need to be able to detect and control to experimentally determine the threshold frequency of silver.
   (b) How could you analyze the quantities you described in part (a) to determine the threshold frequency of silver?
   (c) In a different experiment exploring the photoelectric effect with silver, students measure the values in the table below.

Frequency of incident radiation (10^{15} s^{-1})	Maximum kinetic energy (10^{-19} J)
2.00	5.90
2.50	9.21
3.00	12.52
3.50	15.84
4.00	19.15

   i. How could these data be graphed to experimentally determine the value of Planck's constant h and the threshold frequency f_0 for silver?
   ii. Create the graph you described in (i).

   (d) Use your graph to find the value of Planck's constant h and the threshold frequency f_0 for silver.
   (e) In another trial, the graph produced from the data has a positive y intercept. Is this possible? Justify your response using physics principles.

25 Nuclear Physics

imageBROKER/Alamy

YOU WILL LEARN TO:

- Explain how conservation laws play an important role in nuclear physics.
- Describe how we know that the force that holds the nucleus together is both strong and of short range and relate this concept to nuclear density.
- Explain how and why the binding energy per nucleon in a nucleus depends on the size of the nucleus and calculate that energy.
- Describe the properties of nuclear fission and calculate the energy released in nuclear fission.
- Describe the properties of nuclear fusion, compare them to those of nuclear fission, and calculate the energy released in nuclear fission.
- Describe radioactive decay and the processes by which it occurs and calculate the decay constant and half-life for decay processes from given data.

25-1 The quantum concepts that help explain atoms are essential for understanding the nucleus

We learned in Chapter 24 that a handful of radical ideas—the notion that energy is quantized, that electromagnetic waves come in packets called photons, and that the laws of quantum mechanics can specify only the probabilities that particles will behave in certain ways—are essential for understanding the nature and behavior of the atom. These same ideas apply on the even smaller scale of the atomic nucleus.

We'll see that unlike atoms, in which negatively charged electrons are bound to the positively charged nucleus by the attractive electric force, nuclei are bound by a *strong nuclear force*, called the **strong force**, which keeps the protons and neutrons together. This strong force is one of the four fundamental forces. We'll also see that like electrons, nuclei can have an intrinsic angular momentum or *spin*. Magnetic resonance imaging, or MRI, makes use of how nuclear spins respond to an external magnetic field (**Figure 25-1a**). The interplay between the strong force and the electric force (which makes all the protons in a nucleus try to repel each other) means that nuclei of different sizes are more tightly or loosely bound. We'll learn that as a result, the largest nuclei are prone to break into smaller fragments through a process called *fission*. We'll also learn that the smallest nuclei can release energy by *fusion*, in which two nuclei join

Figure 25-1 Applications of nuclear physics The properties of atomic nuclei explain (a) how magnetic resonance imaging (MRI) works, (b) how the Sun provides energy for life on Earth, and (c) how Earth sustains its internal energy.

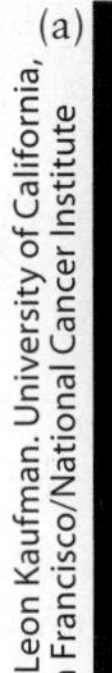

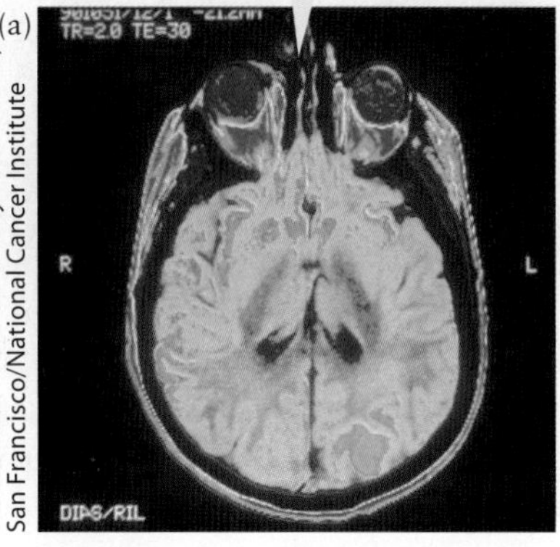

together to form a larger one. Fusion reactions make the Sun shine, and the sunlight that they produce makes life on Earth possible (**Figure 25-1b**).

More than 50% of the energy that powers our planet's geological activity, including volcanic eruptions, comes from the radioactive decay of unstable nuclei in Earth's interior.
(c)
iStockphoto/Thinkstock

Many types of nuclei are **radioactive:** They spontaneously release energy and emit radiation in the form of either particles or photons. Understanding the binding energies of nuclei will help us also understand the three main types of radioactive decay: alpha, beta, and gamma. The energy released by alpha and beta decays in our planet's interior helps power geologic activity such as volcanic eruptions (**Figure 25-1c**). We'll find that the concept of *half-life* will help us understand the nature of radioactive decays of all kinds.

AP® Exam Tip

There are four fundamental forces (interactions), but in AP® Physics we only need to know about three of them, the strong force, gravitational force, and electromagnetic force (which encompasses both the electric and magnetic forces). The interaction that causes radioactive decay is the weak force, the fourth fundamental force. As it is not part of the AP® exam, and we do not need to understand it to learn what we need about radioactive decay, we will not cover the weak force.

THE TAKEAWAY for Section 25-1

✔ Many of the same concepts that help us understand atomic physics are also important in nuclear physics.

✔ The balance between the strong force and the electric force determines the stability of nuclei.

Prep for the AP® Exam

AP® Building Blocks

1. In an atomic nucleus, the strong force binds which of the following particles together?
(A) Neutrons
(B) Protons
(C) Neutrons and protons
(D) Neutrons, protons, and electrons

25-2 The strong force holds nuclei together

As we learned in Chapter 24, an atom has a small, positively charged nucleus at its center. There are two ways you can see that the nucleus must be positively charged. First, because atoms are neutral, positive charge is required to balance the negative charge of the electrons. Second, the Coulomb attraction between the nucleus and the negatively charged electrons provides the force that holds the atom together.

While the nucleus is positively charged, not all of its constituents carry a charge. Nuclei contain both protons, each of which has charge $+e = 1.602 \times 10^{-19}$ C, and neutrons (we mentioned these briefly in Chapter 16) that have zero charge (**Figure 25-2**). Protons and neutrons share many similar properties, including similar masses: The masses of the proton and neutron are 1.6726×10^{-27} kg and 1.6749×10^{-27} kg, respectively. Collectively, we refer to protons and neutrons as **nucleons**.

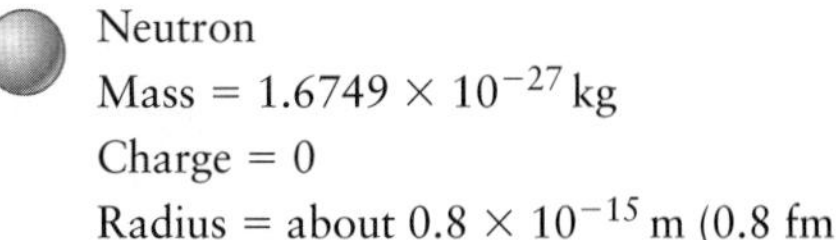

Figure 25-2 Nucleons All atomic nuclei are composed of protons and neutrons, collectively known as nucleons. (The only nucleus that lacks neutrons is the simplest type of hydrogen nucleus, which is made up of a single proton.) The radii of protons and neutrons are very difficult to measure, which is why their values are relatively uncertain.

What holds the protons and neutrons together in the nucleus? It cannot be the electric force, because neutrons do not exert an electric force and protons (which all have the same positive charge $+e = 1.602 \times 10^{-19}$ C) exert *repulsive* electric forces on each other. These forces are very large because protons are close together inside the nucleus. To see just how large these forces are, note that a typical radius of a nucleus is about 5 fm (1 fm = 1 femtometer = 10^{-15} m). The electric force between two protons separated by a distance $r = 5.00$ fm is

$$F = \frac{k(+e)(+e)}{r^2} = \frac{(8.99 \times 10^9 \text{ N} \cdot \text{m}^2/\text{C}^2)(1.602 \times 10^{-19} \text{ C})^2}{(5.00 \times 10^{-15} \text{ m})^2}$$
$$= 9.21 \text{ N}$$

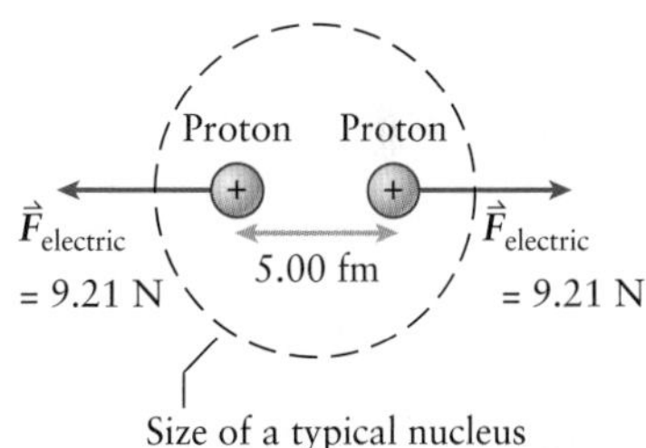

Figure 25-3 Electric repulsion between protons Because protons are so close together inside the nucleus, they exert very substantial electric forces on each other that tend to push them apart. To stabilize the nucleus against these repulsive forces, a much stronger attractive force between nucleons—called the strong force—must be present to hold the nucleus together.

While this force may not seem large, it is exerted on a proton, which has a mass of only 1.6726×10^{-27} kg. If the electric force were the only force exerted on the protons in a nucleus, the resulting acceleration would be enormous, and the nucleus would simply fly apart (**Figure 25-3**).

We conclude that there must be an additional *attractive* force that is exerted on all nucleons (both protons and neutrons) and that binds them together in the nucleus. This attractive force must be stronger than the repulsive electric force between protons, so we call it the strong force. Over short distances the strong force is hundreds of times stronger than the electric force.

WATCH OUT!

The nucleus is not held together by gravitational forces.

It's true that nucleons (protons and neutrons) attract each other through the gravitational force. But the masses of nucleons are so small that the gravitational forces between them are entirely negligible. For the two protons in Figure 25-3, the repulsive electric forces between them are stronger than the attractive gravitational forces by a factor of 10^{46}! We conclude that gravitational forces play no role on the scale of the atomic nucleus.

If the strong force is so strong compared to the electric force, and if protons attract other protons by this force, why are neutrons necessary to help overcome the Coulomb repulsion between protons? The answer lies in the *range* of the strong force, which is the distance beyond which one nucleon no longer experiences a force due to another. Experiments show that the strong force between two nucleons has a range of only about 2.0 fm. The radius of a proton or neutron is about 0.8 fm, so two nucleons must almost be touching to experience the strong force. As a very rough analogy, you can think of protons and neutrons as tiny spheres coated with very strong Velcro, which makes the nucleons stick together if they are brought close enough to each other.

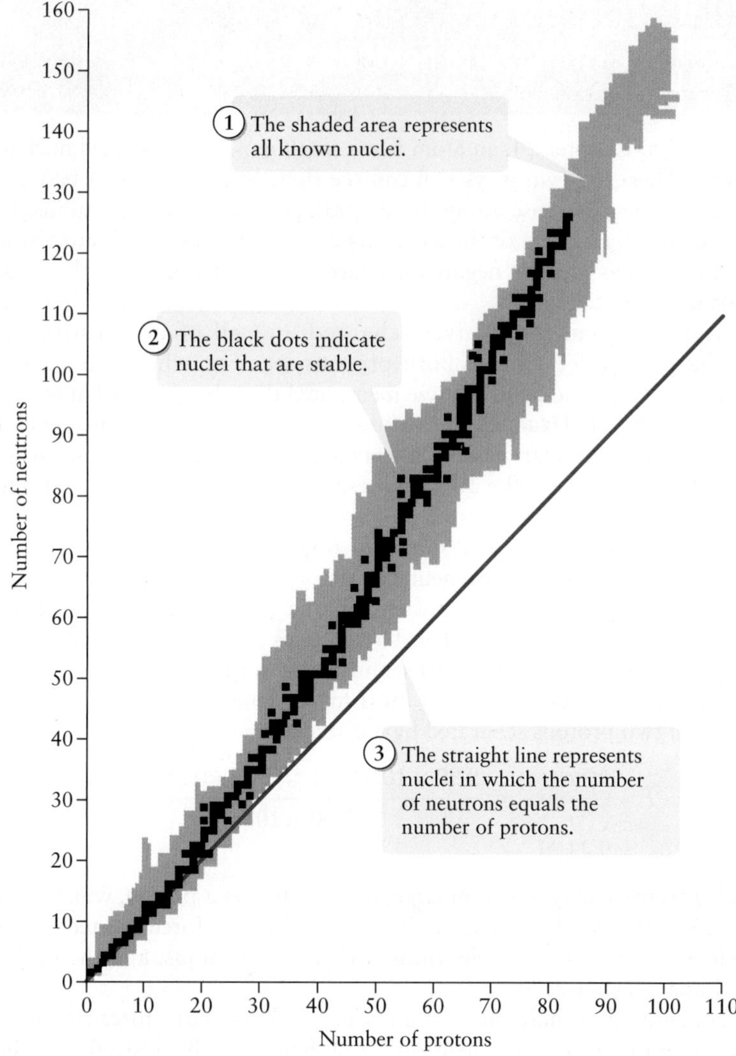
1 The shaded area represents all known nuclei.
2 The black dots indicate nuclei that are stable.
3 The straight line represents nuclei in which the number of neutrons equals the number of protons.
Number of neutrons
Number of protons
0
10
20
30
40
50
60
70
80
90
100
110
120
130
140
150
160

Figure 25-4 Neutrons versus protons in nuclei The interplay between the strong force and the electric force explains the relationship between the numbers of protons and neutrons in nuclei.

For most nuclei, more neutrons than protons must be present to prevent the protons from flying apart due to electrostatic repulsion.

By contrast, the electric repulsion between protons separated by a distance r is proportional to $1/r^2$, so is present even if the distance r is very large. We say that the strong force is a *short-range* force, whereas the electric force is a *long-range* force.

Because the range of the strong force is smaller than the diameter of most nuclei (a few to perhaps 15 fm), each nucleon exerts an attractive force only on its nearest neighbors. Each proton, however, exerts a repulsive force on *every other* proton in the nucleus. As a result, the strong force between neighboring protons cannot overcome the Coulomb repulsion between all the protons. To prevent a nucleus from spontaneously breaking apart (that is, for it to be *stable*), the nucleus must also contain neutrons. You can think of the neutrons as "spacers" that increase the average distance between protons and so decrease their mutual Coulomb repulsion while at the same time contributing to the attractive strong force.

In the smallest stable nuclei, the number of neutrons is about the same as the number of protons; in nuclei with more than about 20 protons, there must be more neutrons than protons for the nucleus to be stable. An unstable nucleus will eventually undergo a spontaneous transformation, termed a *decay*, in which it either splits apart or gives off energy in some other way. **Figure 25-4** shows the number of neutrons versus the number of protons in known atomic nuclei. Stable nuclei are shown in black. Notice that as the number of protons increases, more additional neutrons are required for stability.

Nuclides, Isotopes, and Nuclear Sizes

Atoms of each element have a unique number of protons and the same number of electrons: Hydrogen has one, helium two, lithium three, and so on. Many properties of atoms are related to this number, the **atomic number** Z (in Figure 25-4, Z is indicated as *Number of protons*). Many properties of nuclei arise from *both* Z and the value of the **neutron number** N, which is the number of neutrons in the nucleus (in Figure 25-4, N is indicated as *Number of neutrons*). Nuclei of a given element all have the same number of protons and so have the same value of Z (changing the number of protons changes the element). However, different nuclei of the same element can have different numbers of neutrons and so different values of N. Each combination of Z and N specifies a **nuclide**. For each element, the most common configuration of Z and N corresponds to the most stable nuclide; nuclei of that element with a different number of neutrons are termed **isotopes**. For example, potassium has 19 protons (and 19 electrons). The most stable and most common nuclide of potassium has 20 neutrons, $N = 20$. We denote this nuclide with the symbol ^{39}K, commonly referred to as "potassium-39." The number of protons ($Z = 19$) is understood from the symbol K for potassium, and the total number of protons and neutrons, termed the **mass number** A, is given in the pre-superscript, 39. Note that the mass number of any nucleus equals the atomic number plus the neutron number: $A = Z + N$.

More than 93% of all potassium atoms have a ^{39}K nucleus. However, about 7% of potassium atoms have a ^{41}K nucleus with 19 protons and 22 neutrons, and 0.012% have a ^{40}K nucleus with 19 protons and 21 neutrons. We say that ^{39}K, ^{40}K, and ^{41}K are three isotopes of potassium.

The *size* of a nucleus (its radius and volume) is related to the mass number A. Experiments show that all nuclei are approximately spherical and have radii proportional to the cube root of A, or $A^{1/3}$:

Radius of a nucleus

$$r = r_0 A^{1/3} \quad (25\text{-}1)$$

$r_0 = 1.2 \pm 0.2$ fm

Mass number of the nucleus (total number of protons and neutrons)

EQUATION IN WORDS
Radius of a nucleus

This equation states that if we increase the number of nucleons in a nucleus by a factor of 10, say from $A = 20$ to $A = 200$, the radius of the nucleus will increase by a factor of $10^{1/3} = 2.2$ (**Figure 25-5**).

The volume of a sphere is proportional to r^3. Hence the volume of a nucleus is directly proportional to $(A^{1/3})^3 = A$, which is the mass number of the nucleus and the number of nucleons within that nucleus. This is consistent with our model of nucleons as spheres covered with Velcro. In this model, increasing the number of nucleons by a factor of 10 would result in a ball of nucleons (that is, a nucleus) with 10 times the volume. Because the attractive force is short range, there is no tendency for the nucleons to be compressed together as the size of the nucleus increases.

It's useful to contrast nuclei with planets, which are held together by the *long*-range gravitational force of attraction between all the parts of the planet. As an example, compare the planets Jupiter and Saturn (which have the same chemical composition): Jupiter has 3.3 times more mass than Saturn, but Jupiter has only 1.7 times greater volume because it is more highly compressed and has a greater density. Nuclei of different sizes do *not* behave like planets of different sizes: All

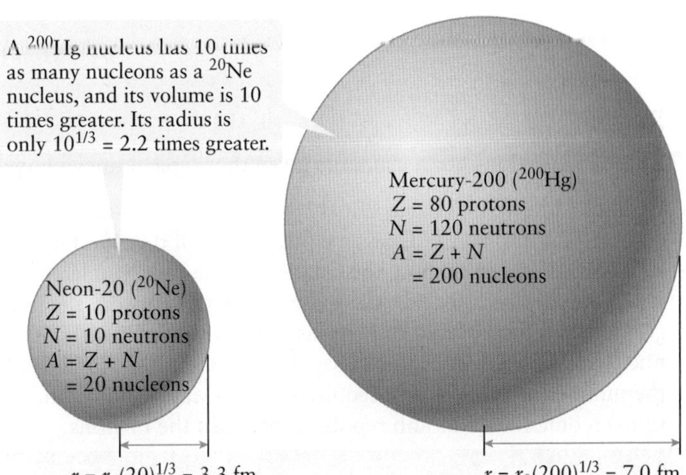

Figure 25-5 Nuclear sizes The volume of an atomic nucleus is proportional to the number of nucleons it contains (the mass number A). The nuclear radius is proportional to the cube root of A.

nuclei have basically the same density (see Example 25-2 below). This is further evidence that the strong force is short-range rather than long-range.

Can we simply add more protons and more neutrons to make larger and larger nuclei? The answer is "no," and for the same reason that neutrons are required for nuclear stability. Figure 25-4 shows that as we work our way up the periodic table to atoms that have more and more protons, more *additional* neutrons are required for stability. Each additional proton exerts a repulsive force on all the others, but neutrons and protons can only attract their nearest neighbors. Those additional neutrons cause the size of the nucleus to grow so that eventually too many neutrons are near the surface of the nucleus and therefore not completely surrounded by neighboring nucleons. At that point, the forces holding the nucleus together are not large enough to overcome the Coulomb repulsion between the protons, and the nucleus cannot be stable. The largest stable nuclide is lead-208 (^{208}Pb), which has 82 protons and 126 neutrons.

EXAMPLE 25-1 Nuclear Radii

Estimate the radius of the nucleus of ^{12}C, a relatively small nucleus; ^{118}Sn, a nucleus of medium size; and ^{236}U, a relatively large nucleus. The nuclides ^{12}C and ^{118}Sn are stable; ^{236}U, like all other isotopes of uranium, is unstable.

Set Up

In each case we use Equation 25-1 to calculate the radius of the nucleus. The value of A for each nucleus is given by the pre-superscript: $A = 12$ for carbon (C), $A = 118$ for tin (Sn), and $A = 236$ for uranium (U).

Radius of a nucleus:

$$r = r_0 A^{1/3} \quad (25\text{-}1)$$

Solve

Apply Equation 25-1 to each nuclide.

We'll use $r_0 = 1.2$ fm in our calculations.

For ^{12}C, which has 6 protons and 6 neutrons,

$$r(^{12}\text{C}) = (1.2 \text{ fm})(12)^{1/3} = 2.7 \text{ fm}$$

For ^{118}Sn, which has 50 protons and 68 neutrons,

$$r(^{118}\text{Sn}) = (1.2 \text{ fm})(118)^{1/3} = 5.9 \text{ fm}$$

For ^{236}U, which has 92 protons and 144 neutrons,

$$r(^{236}\text{U}) = (1.2 \text{ fm})(236)^{1/3} = 7.4 \text{ fm}$$

Reflect

These calculations are reasonable given that typical nuclei have radii of just a few femtometers.

Extend: Although ^{236}U has 236/12 = 19.7 times as many nucleons as ^{12}C, it is only larger in radius by a factor of (7.4 fm)/(2.7 fm) = 2.7. That's a consequence of the $A^{1/3}$ factor in Equation 25-1. Note that $(19.7)^{1/3} = 2.7$.

Notice that in carbon, the smallest of the three nuclei, the number of neutrons equals the number of protons. In tin, which has almost 10 times the mass number of carbon, the number of neutrons required for nuclear stability is 36% greater than the number of protons. And even with nearly 60% more neutrons than protons, the relatively large ^{236}U nucleus is not stable. Thus, we see direct evidence that in larger nuclei more and more additional neutrons, compared to the number of protons, are required for nuclear stability, and that at some size a nucleus is too large for the strong force to overcome the Coulomb repulsion between the protons.

NOW WORK Problem 4 from The Takeaway 25-2.

EXAMPLE 25-2 Nuclear Density

Estimate the density (in kg/m^3) of a nucleus that has mass number A.

Set Up

The density of an object is its mass m divided by its volume V. We'll find the volume of a nucleus from Equation 25-1 and the formula for the volume of a sphere. To estimate the mass of a nucleus, we'll multiply the mass number A (the number of nucleons) by the average mass of a nucleon.

Radius of a nucleus:

$$r = r_0 A^{1/3} \qquad (25\text{-}1)$$

Definition of density:

$$\rho = \frac{m}{V} \qquad (11\text{-}1)$$

Volume of a sphere of radius r:

$$V = \frac{4}{3}\pi r^3$$

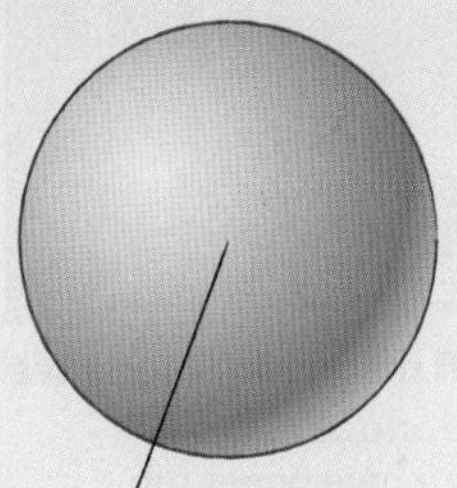

Solve

Find the volume of a nucleus of mass number A.

Substitute Equation 25-1 into the expression for the volume of a sphere of radius r:

$$V = \frac{4}{3}\pi r^3 = \frac{4}{3}\pi(r_0 A^{1/3})^3 = \frac{4}{3}\pi r_0^3 (A^{1/3})^3$$
$$= \frac{4}{3}\pi r_0^3 A$$

Use $r_0 = 1.2\ \text{fm} = 1.2 \times 10^{-15}\ \text{m}$, as in Example 25-1:

$$V = \frac{4}{3}\pi(1.2 \times 10^{-15}\ \text{m})^3 A$$
$$= (7.2 \times 10^{-45}\ \text{m}^3)A$$

Take the average mass of a nucleon to be the average of the proton mass m_p and the neutron mass m_n. Use this to write an expression for the mass of a nucleus of mass number A.

The average mass of a nucleon is

$$m_{avg} = \frac{m_p + m_n}{2} = \frac{(1.6726 \times 10^{-27}\ \text{kg}) + (1.6749 \times 10^{-27}\ \text{kg})}{2}$$
$$= 1.6738 \times 10^{-27}\ \text{kg}$$

A nucleus with A nucleons then has mass

$$m = m_{avg}A = (1.6738 \times 10^{-27}\ \text{kg})A$$

Calculate the density of the nucleus.

The density of the nucleus is

$$\rho = \frac{m}{V} = \frac{(1.6738 \times 10^{-27}\ \text{kg})A}{(7.2 \times 10^{-45}\ \text{m}^3)A} = \frac{1.6738 \times 10^{-27}\ \text{kg}}{7.2 \times 10^{-45}\ \text{m}^3}$$
$$= 2.3 \times 10^{17}\ \text{kg/m}^3$$

Reflect

Our final expression for the density ρ does not depend on A, the mass number of the nucleus. So the density of *all* nuclei is about the same. This agrees with our statements about the short-range character of the strong force. Note also that a block of solid iridium, the densest of all stable elements, has a density of 22,650 kg/m^3 (22.65 times the density of water).

Extend: Our calculation shows that nuclei are 10^{13} times denser than iridium. Nuclei are *extremely* dense! This makes sense: Most of the mass of an atom is concentrated in its nucleus, which has a far smaller volume than the atom as a whole. So nuclear density (mass divided by volume) must be far greater than what we think of as the "ordinary" density of matter such as water or iridium.

NOW WORK Problems 7, 8, and 10 from The Takeaway 25-2.

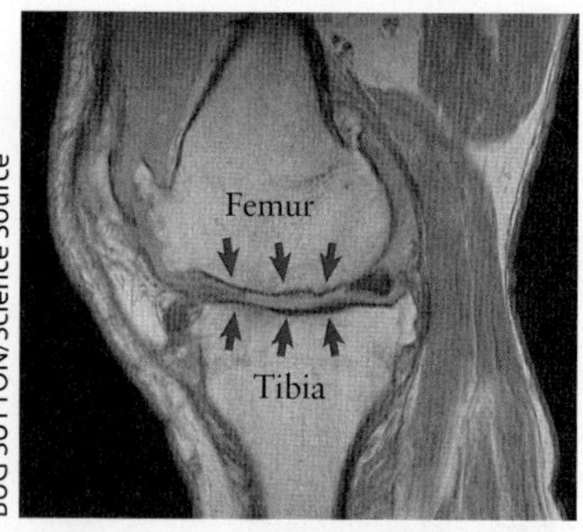

Figure 25-6 A magnetic resonance image The red arrows in this MRI image indicate where bones rub together in the knee of a patient suffering from osteoarthritis.

Nuclear Spin and Magnetic Resonance Imaging

We learned in Section 24-7 that electrons have a type of intrinsic angular momentum called *spin*. (This is something of a misnomer because this angular momentum does not correspond directly to a spinning motion of the electrons.) Protons and neutrons also have spin, and like an electron, the spin of a proton or neutron can take on one of two values, often referred to as "spin up" and "spin down." In nuclei with an even number of nucleons, typically there are as many spin up nucleons as there are spin down, and most such nuclei have zero net spin. (The orbital angular momentum of nucleons moving inside the nucleus can also contribute to the net spin of the nucleus.) But a nucleus with an odd number of nucleons must have a nonzero net spin. In particular, the nucleus of hydrogen—a single proton—has a net spin.

Measurements that make use of the spin of hydrogen nuclei enable us to localize hydrogen in a system, as well as to get information about the material in which those hydrogen atoms are embedded. That's the principle of **magnetic resonance imaging** (MRI). In an MRI scanner, spin information is used to form a three-dimensional map of the density of hydrogen atoms in a body. Living organisms are composed largely of water—our bodies, for example, are 60 to 70% water—and each water molecule contains two hydrogen atoms. As such, MRI is an ideal way to probe the internal structures in the body. The MRI scan in **Figure 25-6** shows the leg bones rubbing against each other in the knee of a person with osteoarthritis. Figure 25-1a shows an MRI scan of a patient's head.

Protons and neutrons also have associated magnetic fields. This field is a dipole field like that of a current-carrying loop of wire (see Section 19-7). The line connecting the north and south poles of a proton or neutron is along the direction of its spin angular momentum vector. As we saw in Section 19-6, a current loop placed in an external magnetic field tends to align with its normal along the direction of the external field. In the same way, when a single proton or neutron is placed in a uniform external magnetic field, the magnetic force tends to align the spin direction either generally parallel to or antiparallel to the external field. The same is true for a nucleus with an odd number A of nucleons. To be precise, the spin direction aligns so that its component along the direction of the external field has a fixed positive or negative value. As such, the spin direction can rotate, or *precess*, around an axis defined by the field direction. **Figure 25-7a** shows this precession for a nucleus with spin aligned with the external field, and **Figure 25-7b** shows this precession for a nucleus with spin anti-aligned with (that is, opposed to) the external field.

(a) A nucleus with its spin aligned with an external magnetic field

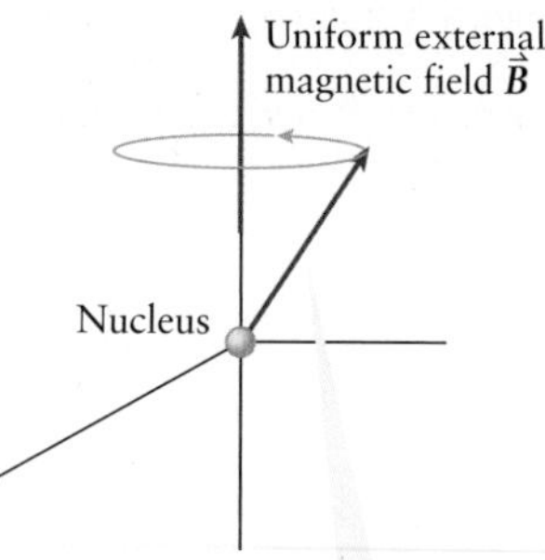

The spin angular momentum vector of the nucleus traces out a cone, but is generally aligned in the same direction as $\vec{B}$.

(b) A nucleus with its spin anti-aligned with an external magnetic field

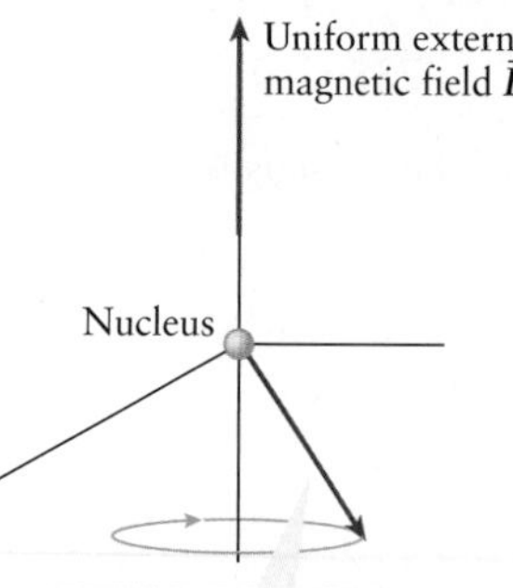

The spin angular momentum vector of the nucleus traces out a cone, but is generally aligned in the direction opposite to $\vec{B}$.

Figure 25-7 Nuclear spin When a nucleus is placed in an external magnetic field, the component of the spin along the direction of the external field is either (a) aligned with the external field or (b) anti-aligned with the external field.

Consider a large number of hydrogen atoms placed in a uniform magnetic field. For each of these atoms the nucleus is a single proton. About half of the protons will end up with spin aligned with the field and half with spin anti-aligned. Now, the energy of the spin-aligned orientation of a nucleus in an external magnetic field (Figure 25-7a) is lower than the energy of the anti-aligned orientation (Figure 25-7b) by an amount ΔE that depends on the magnetic field strength. Suppose we now bathe the atoms in an additional alternating magnetic field with frequency f. The alternating field is made up of photons of frequency f and energy hf, where h is Planck's constant (see Section 24-3). If we choose the frequency f so that hf is equal to ΔE, the photon energy is just equal to the energy difference between the two spin states. As a result, protons that are initially in the lower energy state, with their spin aligned with the field, can absorb a photon of energy $\Delta E = hf$ and flip their spin to be anti-aligned with the external field. What is more, protons that are initially in the higher energy state, with their spin anti-aligned with the field, can be stimulated by the alternating field to emit a photon of energy $\Delta E = hf$ and so flip their spin to the lower energy, aligned state. (A similar sort of *stimulated emission* also occurs in a laser; see Section 24-6.)

The difference in energy between the two spin states of a hydrogen atom depends on the strength of the external field. With the high field strength required to be able to clearly observe the spin-flipping phenomenon, typically around 3 T, the energy difference ΔE is about 5×10^{-7} eV. (As we have learned before, one eV, or one electron volt, is the amount of energy acquired by an electron when it moves through a potential difference of one volt: $1 \text{ eV} = 1.60 \times 10^{-19}$ J. We saw in Section 24-7 that the difference in energy between adjacent energy levels of a hydrogen atom is around 1 to 10 eV. The energy difference ΔE for this spin flip is more than a million times smaller.) From Equation 24-13 the frequency of a photon of this energy is

$$f = \frac{\Delta E}{h} = \frac{5 \times 10^{-7}\ \text{eV}}{4.14 \times 10^{-15}\ \text{eV} \cdot \text{s}} = 1.2 \times 10^{8}\ \text{s}^{-1} = 120\ \text{MHz}$$

(We used the value of h in eV · s.) So, the frequency of the oscillating field used to induce the hydrogen atoms to flip their spins in a strong magnetic field is around 100 MHz, which is in the radio-frequency part of the electromagnetic spectrum. We use the term *resonance* because the spins flip only if the field frequency has just the right value, similar to the behavior of a sound waves in a pipe (Figure 21-18).

Figure 25-8 shows how we "see" this spin flipping and so locate the hydrogen atoms. Initially there will be more atoms in the lower energy state than the higher energy state (lower energy is more likely for a system than higher energy), so when the hydrogen atoms in an external magnetic field are exposed to radio waves with the right frequency f, there is a net absorption of the electromagnetic energy. The radio-frequency signal from the MRI device then stimulates the excited protons to return to their initial (lower energy) state, and in so doing they generate a radio-frequency photon of their own. The MRI device detects that radio-frequency photon and uses it to map the density of hydrogen atoms in the body.

As we have mentioned, the difference in energy ΔE between the two hydrogen spin states depends on the strength of the external magnetic field. In an MRI device the magnetic field is made to vary over a body's volume, so the energy absorbed and then re-emitted in the spin–flip process also varies in different parts of the body. The exact frequencies of the radio energy detected by the MRI device therefore provide the information necessary to create images of high spatial resolution. In addition, the time it takes for the spins of the hydrogen nuclei to return to their equilibrium state depends on the particular molecules in the tissue. Thus, the timing information in an MRI device provides the means to differentiate one type of tissue from another.

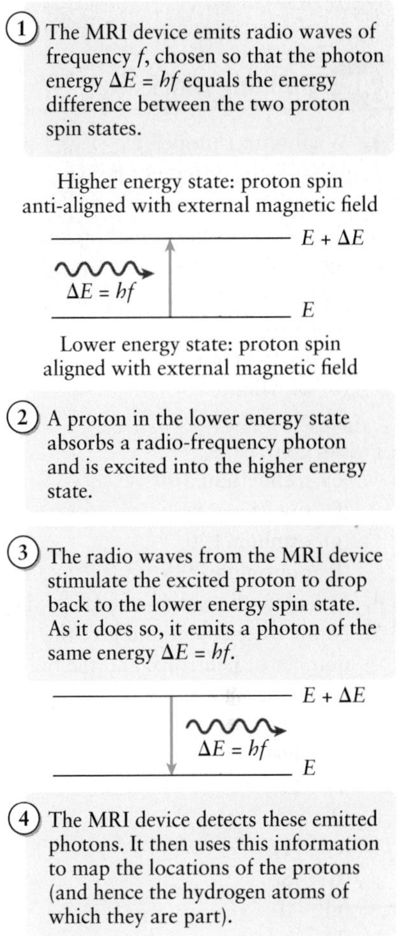

(1) The MRI device emits radio waves of frequency f, chosen so that the photon energy $\Delta E = hf$ equals the energy difference between the two proton spin states.
Higher energy state: proton spin anti-aligned with external magnetic field
$E + \Delta E$
$\Delta E = hf$
E
Lower energy state: proton spin aligned with external magnetic field
(2) A proton in the lower energy state absorbs a radio-frequency photon and is excited into the higher energy state.
(3) The radio waves from the MRI device stimulate the excited proton to drop back to the lower energy spin state. As it does so, it emits a photon of the same energy $\Delta E = hf$.
$E + \Delta E$
$\Delta E = hf$
E
(4) The MRI device detects these emitted photons. It then uses this information to map the locations of the protons (and hence the hydrogen atoms of which they are part).

Figure 25-8 How to make proton spins flip In a magnetic resonance imaging (MRI) scan, the patient is placed in a strong magnetic field. This causes a difference in energy ΔE between the two spin states of the nucleus (a proton) in a hydrogen atom in the patient. Radio waves from the MRI device excite these protons into the higher energy state, flipping their spins; when the proton spin flips back to return to the lower energy state, the proton emits a radio-frequency photon that the MRI device detects.

THE TAKEAWAY for Section 25-2

✔ The strong force binds nucleons (protons and neutrons) together in the nucleus of an atom.

✔ The volume of a nucleus is proportional to its mass number (the total number of nucleons in the nucleus).

✔ To be stable a light nucleus must have about as many neutrons as protons. More massive nuclei with more than about 20 protons require more neutrons than protons to be stable. Very large nuclei with mass number greater than 208 are always unstable.

✔ Protons and neutrons have spin. Magnetic resonance imaging uses the difference in energy between a state in which a nuclear spin is aligned with an external magnetic field and the state in which the spin is anti-aligned with the field.

AP® Building Blocks

1. What is an isotope?
2. What is the difference between atomic number and mass number?
3. Provide the elemental abbreviation (for example, ^{16}O for oxygen-16) and give the number of protons, the number of neutrons, and the mass number for each of the following isotopes.
   (a) hydrogen-3
   (b) beryllium-8
   (c) aluminum-26
   (d) gold-197
   (e) technetium-100
   (f) tungsten-184
   (g) osmium-190
   (h) plutonium-239

EX 25-1 4. Calculate the radius of each of the nuclei in problem 3.

5. Name the element and give the number of protons, the number of neutrons, and the mass number for each of the following nuclei.
   (a) ^{2}H
   (b) ^{4}He
   (c) ^{6}Li
   (d) ^{12}C
   (e) ^{56}Fe
   (f) ^{90}Sr
   (g) ^{131}I
   (h) ^{235}U

AP® Skill Builders

6. Nuclei contain protons and neutrons.
   (a) Describe what is meant by the phrase "larger nuclei are neutron rich."
   (b) Using physics principles, explain why nuclei contain at least as many neutrons as protons.

EX 25-2 7. If our Sun (mass = 1.99×10^{30} kg, radius = 6.96×10^{8} m) were to collapse into a neutron star (an object composed of tightly packed neutrons with roughly the same density as a nucleus), what would the new radius of our "neutron-sun" be?

EX 25-2 8. Given that a nucleus is approximately spherical and has a radius $r = r_0 A^{1/3}$ (where r_0 is about 1.2 fm), determine its approximate mass density. Express your answer in SI units and convert to tons per cubic inch, units that might be used in a news report.

Prep for the AP® Exam

AP® Skills in Action

9. A ^{20}Ne nucleus has 10 protons and 10 neutrons for a total of 20 nucleons, and a ^{160}Dy nucleus has 66 protons and 94 neutrons for a total of 160 nucleons. Compared to a ^{20}Ne nucleus, a ^{160}Dy nucleus has
   (A) double the radius and a greater density.
   (B) 8 times the radius and a greater density.
   (C) double the radius and the same density.
   (D) 8 times the radius and the same density.
   (E) double the radius and a lower density.

EX 25-2 10. Two protons are separated by a distance of 4.0×10^{-15} m inside a nucleus. The proton mass is 1.67×10^{-27} kg.
   (a) Calculate the magnitude of the repulsive electric force between the two protons.
   (b) Calculate the magnitude of the attractive gravitational force between the same two protons.
   (c) Can gravity hold a nucleus together?

25-3 Some nuclei are more tightly bound and more stable than others

Release a ball at the top of a hill, and it rolls down the hill. Pull an object attached to the free end of a spring away from its equilibrium position, and it tends to return to equilibrium. In both cases the systems are finding their way to a more stable configuration. All physical systems do the same; if a more stable configuration exists for a system, it will eventually find itself in that configuration as long as nature provides a mechanism for the transition to take place.

In this context, consider the nucleus of a helium atom, which consists of two protons and two neutrons. This configuration of these four nucleons must be more stable than when they are separate; otherwise, the helium nucleus would end up broken apart. This stability results because the attraction of the strong force between the four nucleons overwhelms the electric repulsive force between the two protons.

Let's be quantitative about *how* stable a given nucleus is. The total mass M_{tot} of the two protons and two neutrons in the nucleus of ^{4}He equals the sum of two proton masses (m_p) and two neutron masses (m_n):

$$\begin{aligned} M_{tot} &= 2m_p + 2m_n \\ &= 2(1.6726 \times 10^{-27}\ \text{kg}) + 2(1.6749 \times 10^{-27}\ \text{kg}) = 6.695 \times 10^{-27}\ \text{kg} \end{aligned}$$

But the actual mass of a helium nucleus is 6.645×10^{-27} kg, which is *less* than the total mass of the two protons and two neutrons (**Figure 25-9**). How is this possible? The answer is the key to nuclear stability; the energy equivalent of the difference in mass is tied up in binding the nucleons together. In Section 24-4 we showed that energy and mass are equivalent, and that an object of mass m has a rest energy $E_0 = mc^2$ (Equation 1-1). Because a ^{4}He nucleus has a smaller mass than its constituent nucleons, it also has a smaller rest energy. The difference between the rest energy of the ^{4}He nucleus and the rest energy of its constituent nucleons is the **binding energy** E_B. You can think of this as the energy that is released when the four nucleons come together to form a ^{4}He nucleus. Alternatively, you can think of the binding energy as the energy that would be required to separate a ^{4}He nucleus into two protons and two neutrons. The greater the binding energy *per nucleon* in a nucleus, the more tightly the nucleus is bound and therefore the more stable it is.

Combined mass of two protons and two neutrons: 6.695×10^{-27} kg

Mass of two protons and two neutrons bound into a ^{4}He nucleus: 6.645×10^{-27} kg

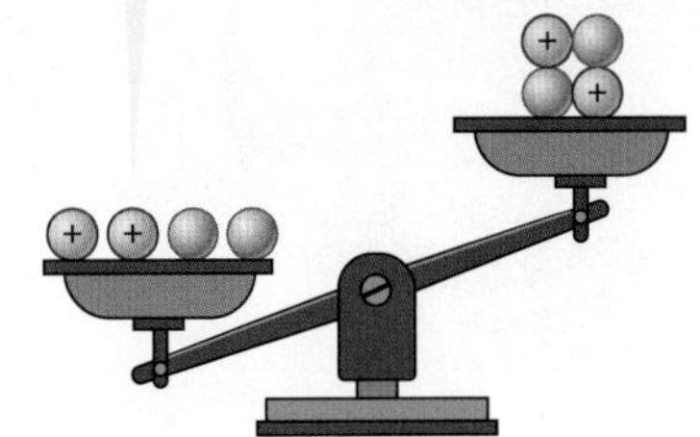

• The mass of a nucleus is less than the sum of the masses of its component nucleons.
• The difference in masses is due to the binding energy of the nucleus.

Figure 25-9 Binding energy Nucleons are bound together in a nucleus by the strong force. Hence the rest energy of any nucleus made up of two or more nucleons is less than the sum of the rest energies of the constituent nucleons, and the same is true of the masses.

Figure 25-10 is a graph of the binding energy per nucleon (E_B/A) for different nuclides as a function of their mass number A. The energy values on the vertical axis are given in MeV, where 1 MeV = 10^6 eV. Figure 25-10 shows that the value of E_B/A for all nuclides is in the range from 1 to 9 MeV, which is why these units are convenient. (By comparison, the binding energy of a hydrogen *atom*—the energy required to separate the single electron in a hydrogen atom from the proton—is 13.6 eV, about 10^{-5} as great as the binding energy of even the most weakly bound nucleus. This indicates how strong the forces are within the nucleus compared to the forces on electrons within atoms.)

Figure 25-10 shows that the nuclide with the smallest binding energy per nucleon is ^{2}H, an isotope of hydrogen with one proton and one neutron. (Just 0.0115% of hydrogen atoms are of this isotope.) As A increases, E_B/A increases rapidly because

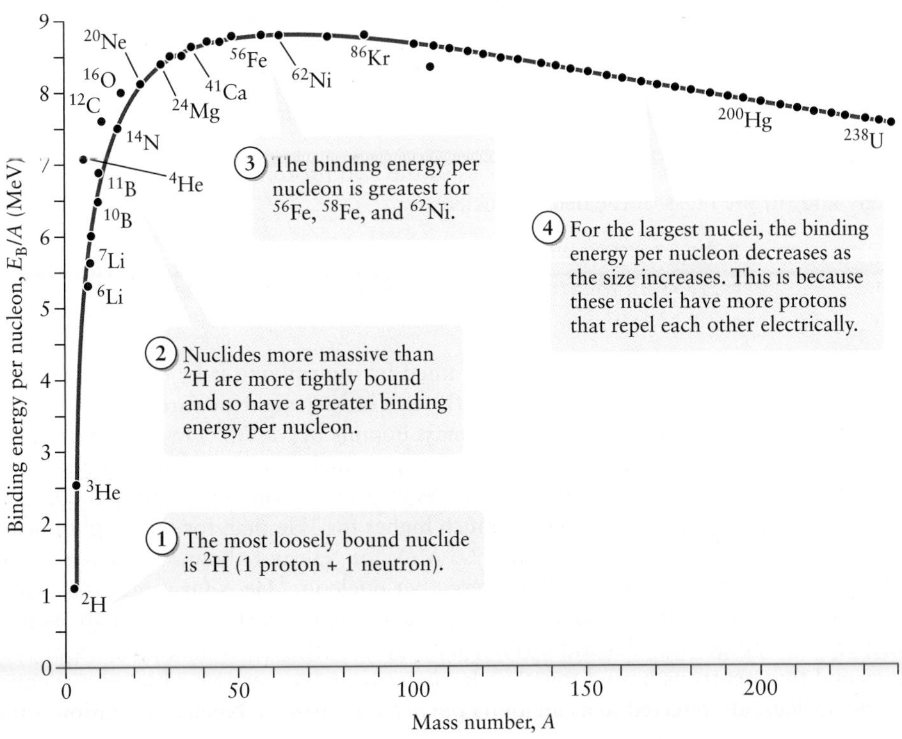
Binding energy per nucleon, E_B/A (MeV)
Mass number, A
^{20}Ne
^{16}O
^{12}C
^{56}Fe
^{41}Ca
^{24}Mg
^{62}Ni
^{86}Kr
^{200}Hg
^{238}U
^{14}N
^{11}B
^{4}He
^{10}B
^{7}Li
^{6}Li
^{3}He
^{2}H
3 The binding energy per nucleon is greatest for ^{56}Fe, ^{58}Fe, and ^{62}Ni.
4 For the largest nuclei, the binding energy per nucleon decreases as the size increases. This is because these nuclei have more protons that repel each other electrically.
2 Nuclides more massive than ^{2}H are more tightly bound and so have a greater binding energy per nucleon.
1 The most loosely bound nuclide is ^{2}H (1 proton + 1 neutron).

Figure 25-10 The curve of binding energy The binding energy per nucleon in a nucleus depends on the mass number A.

Felix Choo/Alamy

Figure 25-11 Five cents' worth of the most stable nuclei A Canadian five-cent coin is made of steel (mostly iron) plated with nickel. Naturally occurring iron is 5.85% ^{54}Fe, 91.75% ^{56}Fe, 2.12% ^{57}Fe, and 0.28% ^{58}Fe; naturally occurring nickel is 68.08% ^{58}Ni, 26.22% ^{60}Ni, 1.14% ^{61}Ni, 3.64% ^{62}Ni, and 0.93% ^{64}Ni. All these isotopes are stable and have very large values of the binding energy per nucleon E_B/A.

more nucleons are surrounded by other nucleons to which they are attracted and so are more tightly bound. The value of E_B/A peaks at about 8.8 MeV for A in the range of 56 to 62 and then decreases slowly for higher and higher values of A. This decrease happens because the number of protons Z also increases as A increases, and the electric repulsion between protons destabilizes the nucleus. The most stable nuclei are the nickel isotope ^{62}Ni and the two iron isotopes ^{56}Fe and ^{58}Fe (**Figure 25-11**). The binding energy per nucleon of these nuclei places them near the peak of the E_B/A curve in Figure 25-10.

Whenever possible nuclei will rearrange themselves to maximize their stability and so maximize their binding energy per nucleon. Figure 25-10 shows that nuclei with relatively *large* values of mass number A (at the far right of the graph) can become more stable by *decreasing* the number of nucleons. One process of this kind is **nuclear fission**, in which nuclei split into smaller pieces. For example, atoms of curium-244 can spontaneously fission into xenon-135 and molybdenum-109. Curium has 96 protons, so ^{244}Cm has 148 neutrons. There are 54 protons in xenon and 42 in molybdenum or 96 total. The total number of nucleons in ^{135}Xe and ^{109}Mo, 244, equals the number of nucleons in ^{244}Cm. In other words, ^{135}Xe and ^{109}Mo contain the same 96 protons and 148 neutrons as in the original ^{244}Cm; the curium atom has split into two fragments. We'll discuss fission in more detail in Section 25-4.

Figure 25-10 also shows that nuclei with relatively *small* values of A (at the far left of the graph) can become more stable by *increasing* the number of nucleons. One way to do this is by **nuclear fusion**, in which two small nuclei join to make a larger one. For example, the 12 protons and 12 neutrons in two carbon-12 nuclei can fuse to form a magnesium-24 nucleus. More than one nucleus can also be formed in a fusion process. For example, when a helium-3 nucleus fuses with a lithium-6 nucleus, the reaction forms two helium-4 nuclei and one hydrogen-1 nucleus. (Count the nucleons: ^{3}He has two protons and one neutron, and ^{6}Li has three protons and three neutrons, for a total of five protons and four neutrons. The two ^{4}He nuclei have two protons and two neutrons each, leaving the one remaining proton as a ^{1}H nucleus.) In Section 25-5 we'll discuss nuclear fusion more carefully.

Let's see how to calculate the binding energy per nucleon of a nucleus such as ^{4}He. We'll do this by finding the total mass of the protons and neutrons that make up the nucleus and comparing it to the mass of the actual nucleus. We then convert that difference to the equivalent energy.

Instead of measuring mass in kilograms, it's most convenient to use units that take advantage of the equivalence between mass and energy. Because rest energy equals mass multiplied by c^2, we'll measure masses in units of MeV/c^2. Note that 1 MeV/c^2 = 1.7827×10^{-30} kg. In these units, the mass of a proton is 938.27 MeV/c^2, the mass of a neutron is 939.57 MeV/c^2, and the mass of a ^{4}He nucleus is 3727.4 MeV/c^2.

The difference Δ between (i) the mass of the two protons and two neutrons separately and (ii) the mass of the helium nucleus is

$$\begin{aligned}\Delta &= 2m_p + 2m_n - m_{He} \\ &= 2(938.27 \text{ MeV}/c^2) + 2(939.57 \text{ MeV}/c^2) - 3727.4 \text{ MeV}/c^2 \\ &= 28.3 \text{ MeV}/c^2\end{aligned}$$

The energy equivalent of any mass is obtained by multiplying it by c^2, so the energy equivalent of this difference is 28.3 MeV. (Notice how straightforward it is to find the energy equivalent of mass when we write mass in units of MeV/c^2.) In other words, the binding energy of ^{4}He is 28.3 MeV. There are four nucleons in the helium nucleus, so E_B/A is about 7.1 MeV. You can verify this result from the curve in Figure 25-10.

The binding energy per nucleon is much higher for ^{4}He than for other light nuclei. (Figure 25-10 shows that E_B/A is about 2.5 MeV for ^{3}He and about 5.3 MeV for ^{6}Li.) Thanks to its relatively high binding energy per nucleon, ^{4}He is far more stable than other light nuclei. For this reason, when large nuclei break apart to transform to a more stable, more energetically favorable state, in many cases they do so by emitting two protons and two neutrons bound together. In such processes these four bound nucleons, a ^{4}He nucleus, are referred to as an **alpha particle** (α particle). **Nuclear radiation** is the emission by a nucleus of either energy (in the form of a photon) or particles, such as an α particle. We'll discuss nuclear radiation in more detail in Section 25-6.

To find the binding energy of ^{4}He, we subtracted the mass of the nucleus from the mass of the two protons and two neutrons separately and then multiplied by c^2 to find the equivalent energy. In general, for a nucleus consisting of N neutrons and Z protons, E_B is

$$E_B = (Nm_n + Zm_p - m_{nucleus})c^2$$

where m_n is the mass of a neutron, m_p is the mass of a proton, and $m_{nucleus}$ is the mass of the nucleus. In practice it's easier to measure the masses of neutral *atoms* (including their electrons) than the masses of isolated atomic nuclei. In terms of these masses, we can write the binding energy of a nucleus as

Binding energy of a nucleus | **Number of neutrons** in the nucleus | **Number of protons** in the nucleus | **Mass of a neutral atom** (including electrons) containing that nucleus

$$E_B = (Nm_n + Zm_{1_H} - m_{atom})c^2 \qquad (25\text{-}2)$$

Mass of a neutron | **Mass of a neutral hydrogen atom** (1 proton + 1 electron) | **Speed of light in a vacuum**

EQUATION IN WORDS
Binding energy of a nucleus

The terms Zm_{1_H} and m_{atom} each include the mass of Z electrons, so the electron masses cancel. The masses of neutral atoms can be found online and are given in the Atomic Numbers and Atomic Weights table in Appendix C. Note that values given in atomic mass units (u or amu) can be expressed as MeV/c^2, where

$$1 \text{ u} = 931.494 \text{ MeV}/c^2$$

EXAMPLE 25-3 The Binding Energy of ^{4}He

Previously, we determined the binding energy per nucleon in the ^{4}He nucleus using the mass of the nucleus. Use the following values to determine the binding energy per nucleon (in MeV/c^2) of the ^{4}He nucleus using the *atomic* mass of ^{4}He: neutron mass = 1.008665 u, atomic mass of ^{1}H = 1.007825 u, and atomic mass of ^{4}He = 4.002602 u.

Set Up

We'll use Equation 25-2 and the values given for the neutron mass, the atomic mass of ^{1}H, and the atomic mass of ^{4}He.

Binding energy of nucleus:

$$E_B = (Nm_n + Zm_{1_H} - m_{atom})c^2 \qquad (25\text{-}2)$$

neutron mass = m_n = 1.008665 u
atomic mass of ^{1}H = m_{1_H} = 1.007825 u
atomic mass of ^{4}He = $m_{4_{He}}$ = 4.002602 u

binding energy of ^{4}He = rest energy of two neutrons + rest energy of two ^{1}H atoms − rest energy of a ^{4}He atom

Solve

Calculate the binding energy of the ^{4}He nucleus, which has two neutrons ($N = 2$) and two protons ($Z = 2$).

Substitute the values of m_n, m_{1_H}, and $m_{4_{He}}$ into Equation 25-2, with $m_{atom} = m_{4_{He}}$:

$$E_B = [2(1.008665 \text{ u}) + 2(1.007825 \text{ u}) - 4.002602 \text{ u}]c^2$$
$$= 0.030378 \text{ u}c^2$$

Substitute $1\ \text{u} = 931.494\ \text{MeV}/c^2$,

$$E_B = 0.030378\ \text{u}c^2\left(\frac{931.494\ \text{MeV}/c^2}{1\ \text{u}}\right)$$
$$= 28.297\ \text{MeV}$$

The binding energy per nucleon equals the binding energy of the nucleus divided by the number of nucleons in the nucleus.

The ^{4}He nucleus has four nucleons (two neutrons and two protons), so the binding energy per nucleon is

$$\frac{E_B}{A} = \frac{28.297\ \text{MeV}}{4} = 7.0742\ \text{MeV}$$

Reflect

Previously, we calculated $E_B/A = 7.1$ MeV to two significant figures using the mass of the ^{4}He nucleus; our new calculation is consistent with this.

Extend: Why would we do this kind of calculation using atomic masses rather than nuclear masses? The reason is that in general the masses of neutral atoms have been well measured, but precise measurements of the masses of atomic nuclei in isolation are difficult to obtain.

NOW WORK Problems 2–8 from The Takeaway for Section 25-3.

THE TAKEAWAY for Section 25-3

✔ The binding energy of a nucleus is the energy that would be required to separate it into its individual nucleons.

✔ The greater the binding energy per nucleon in a nucleus, the more tightly the nucleus is bound and therefore the more stable it is.

✔ The most stable nuclides are ^{56}Fe, ^{58}Fe, and ^{62}Ni. Smaller and larger nuclei have a lower binding energy per nucleon.

Prep for the AP® Exam

AP® Building Blocks

1. A simple idea of nuclear physics can be stated as follows: The whole nucleus weighs less than the sum of its parts. Explain why.

EX 25-3 2. Calculate the atomic mass of each of the isotopes listed below. Give your answer in atomic mass units (u) and in grams (g). The values will include the mass of Z electrons.
(a) ^{1}H
(b) ^{4}He
(c) ^{9}Be
(d) ^{12}C
(e) ^{56}Fe
(f) ^{90}Sr
(g) ^{238}U

EX 25-3 3. What is the binding energy of carbon-12? Give your answer in MeV.

EX 25-3 4. What is the binding energy per nucleon for the following isotopes?
(a) ^{2}H
(b) ^{6}Li
(c) ^{12}C
(d) ^{56}Fe

EX 25-3 5. What minimum energy is needed to remove a neutron from ^{40}Ca, so as to convert it to ^{39}Ca? The atomic masses of the two isotopes are 39.96259098 and 38.97071972 u, respectively.

EX 25-3 6. What is the binding energy of the last neutron of carbon-13? The atomic mass of carbon-13 is 13.003355 u.

AP® Skill Builders

EX 25-3 7. Iodine-131 is a radioactive isotope that is used in the treatment of cancer of the thyroid. The natural tendency of the thyroid to take up iodine creates a pathway for radiation (β^- and γ) that is emitted from this unstable nucleus to be directed into the cancerous tumor with very little collateral damage to surrounding healthy tissue. Another advantage of the isotope is its relatively short half-life (8 days). Calculate the binding energy of iodine-131 and the binding energy per nucleon. The mass of iodine-131 is 130.906124 u.

8. The semi-empirical binding energy formula is given as follows:

$$E_B = (15.8\text{ MeV})A - (17.8\text{ MeV})A^{2/3} - (0.71\text{ MeV})\frac{Z(Z-1)}{A^{1/3}} - (23.7\text{ MeV})\frac{(N-Z)^2}{A}$$

where A is the mass number, N is the number of neutrons, and Z is the number of protons. Using the formula, calculate the binding energy per nucleon for fermium-252 (the mass of fermium-252 is 252.08247 u). Compare your answer with the standard common expression for the binding energy: $E_B = (Nm_n + Zm_p - m_{nucleus})c^2$.

Skills in Action

9. Rank these nuclides in order from most stable to least stable. (A) ^{11}B; (B) ^{20}Ne; (C) ^{86}Kr; (D) ^{200}Hg.
10. Describe two characteristics of the binding energy that are comparable to the work function (from the photoelectric effect) and two characteristics that are dissimilar to the concept of the work function.

25-4 The largest nuclei can release energy by undergoing fission and splitting apart

As we saw in the previous section, nuclei with higher values of binding energy per nucleon (E_B/A) are more stable than those with lower values. Figure 25-10, a plot of the binding energy E_B per nucleon in nuclei versus mass number A, shows that E_B/A decreases as A increases beyond 60 or so. In other words, large nuclei are less stable than smaller ones for A greater than about 60. As a consequence of this instability, these large nuclei can undergo fragmentation or *fission* into smaller nuclei. Fission of a large nucleus can happen spontaneously, or it can be induced by imparting energy to the nucleus through a collision. In either case the smaller fragments have a higher value of E_B/A and are more stable.

Let's look at one of the most important processes of this kind, called **neutron-induced fission**. As an example, the collision of a neutron with a ^{235}U nucleus will cause it to fission. **Figure 25-12** shows one possible result. For a brief time the neutron and ^{235}U nucleus remain stuck together as $^{236}U^*$. (The asterisk indicates that

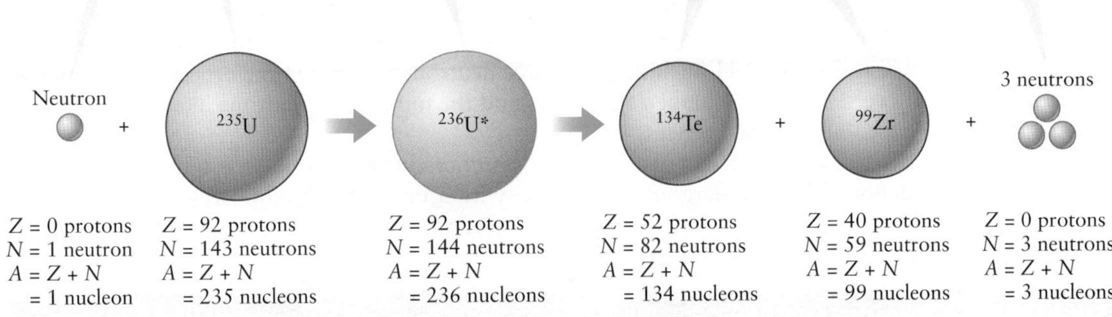

Energy is released in this fission reaction: The total kinetic energy of the fission fragments is much greater than the total kinetic energy of the initial neutron and ^{235}U nucleus.

Figure 25-12 Neutron-induced fission When one of the largest nuclei absorbs a slow-moving neutron, it can fission into smaller, more stable fragments.

this is an excited and short-lived state of ^{236}U.) This excited nucleus quickly fissions into fragments. In this particular reaction the fragments are an isotope of tellurium (^{134}Te), an isotope of zirconium (^{99}Zr), and three neutrons. Because these fragments are all more stable than the original nucleus, energy is released. The process described by Figure 25-12 occurs even when the colliding neutron is moving very slowly and so has essentially zero kinetic energy. The released energy is almost entirely due to the change in binding energy between the initial ^{235}U nucleus and the fission products.

We can estimate the energy released in the process shown in Figure 25-12 by comparing the binding energy of the ^{235}U nucleus to the binding energies of the ^{134}Te and ^{99}Zr fragments. (There is no binding energy associated with the initial or final neutrons; they are not bound to any other particle.) If you make measurements on the graph in Figure 25-10, you'll see that the binding energy per nucleon E_B/A is about 7.6 MeV for $A = 235$, about 8.4 MeV for $A = 134$, and about 8.7 MeV for $A = 99$. The binding energy of each nucleus (E_B) equals the binding energy per nucleon (E_B/A) multiplied by the number of nucleons (A). The energy released in the fission reaction equals the difference between the total binding energy of the fragments and the binding energy of the initial ^{235}U nucleus:

$$\begin{aligned}&(\text{energy released})\\&\quad= (\text{binding energy of } ^{134}\text{Te}) + (\text{binding energy of } ^{99}\text{Zr})\\&\qquad - (\text{binding energy of } ^{235}\text{U})\\&\quad= (134)(8.4\ \text{MeV}) + (99)(8.7\ \text{MeV}) - (235)(7.6\ \text{MeV})\\&\quad= 200\ \text{MeV}\end{aligned}$$

We've given our result to just one significant figure because the values of E_B/A that we measured from Figure 25-10 are just estimates. The actual amount of energy released during this process is about 185 MeV, which is quite close to our estimate. This illustrates the tremendous amount of energy released in fission. By contrast, combustion (a chemical process that involves the electrons in molecules, not the nuclei of atoms) yields only a few eV for every molecule of fuel consumed. The energy release in fission is greater by a factor of several million!

When a heavy nucleus like ^{235}U undergoes fission, a wide variety of fragments can result. Figure 25-12 shows one possible result. Two others are

$$\text{n} + {}^{235}\text{U} \rightarrow {}^{236}\text{U}^* \rightarrow {}^{143}\text{Ba} + {}^{90}\text{Kr} + 3\text{n}$$
$$\text{n} + {}^{235}\text{U} \rightarrow {}^{236}\text{U}^* \rightarrow {}^{140}\text{Xe} + {}^{92}\text{Sr} + 4\text{n}$$

In each case the total number of protons and the total number of neutrons both remain the same. The ^{235}U nucleus has 92 protons and $235 - 92 = 143$ neutrons. In the first of these two processes, ^{143}Ba has 56 protons and 87 neutrons, and ^{90}Kr has 36 protons and 54 neutrons. The total number of protons after the fission has occurred is then $56 + 36 = 92$. The total number of neutrons before the fission is $143 + 1 = 144$ (including the neutron that starts the process). After the fission the number of neutrons is $87 + 54 + 3 = 144$ (including the three neutrons released in the process). You can easily verify that the number of protons and the number of neutrons likewise remain the same in the second process above.

Uranium-235 can also undergo *spontaneous* fission, in which the nucleus fragments without undergoing a collision with a neutron. The following example shows how to calculate the energy released in this process.

EXAMPLE 25-4 Spontaneous Uranium Fission

Determine the energy released when a ^{235}U nucleus spontaneously undergoes fission to ^{140}Xe, ^{92}Sr, and three neutrons. The binding energies per nucleon in the nuclei of ^{235}U, ^{140}Xe, and ^{92}Sr are 7.59 MeV, 8.29 MeV, and 8.65 MeV, respectively.

Set Up

The energy released during the fission process is the difference between the binding energy of the ^{140}Xe and ^{92}Sr nuclei and the binding energy of the ^{235}U nucleus. (There is no binding energy associated with the three unbound neutrons.) The binding energy for each nucleus equals the binding energy per nucleon for that nucleus multiplied by the number of nucleons A.

(energy released)
= (total binding energy of fragments)
− (binding energy of original ^{235}U nucleus)

Solve

Calculate the binding energy for each of the nuclei, and then use these values to calculate the energy released.

The binding energies of the individual nuclei are

For ^{235}U: (235)(7.59 MeV) = 1784 MeV

For ^{140}Xe: (140)(8.29 MeV) = 1161 MeV

For ^{92}Sr: (92)(8.65 MeV) = 796 MeV

The energy released in the spontaneous fission is then

$$\begin{aligned}\text{(energy released)} &= (\text{binding energy of } ^{140}\text{Xe}) + (\text{binding energy of } ^{92}\text{Sr}) \\ &\quad - (\text{binding energy of } ^{235}\text{U}) \\ &= 1161\ \text{MeV} + 796\ \text{MeV} - 1784\ \text{MeV} \\ &= 173\ \text{MeV}\end{aligned}$$

Reflect

This result is consistent with our earlier claim that a typical amount of energy released in the fission of ^{235}U is around 200 MeV.

Extend: Spontaneous fission of ^{235}U is a *very* unlikely process: A given ^{235}U nucleus has a 50% chance of decaying in a period of 7.04×10^8 years, and the probability that it will decay by spontaneous fission is only 7.0×10^{-11}. (Here 7.04×10^8 y is the *half-life* of ^{235}U. We'll discuss this concept more carefully in Section 25-6. In that section we'll see that ^{235}U almost always decays by a different process called *alpha emission*.) By contrast, once a ^{235}U nucleus merges with a neutron to form an excited ^{236}U* nucleus, as in Figure 25-12, it typically undergoes fission within a fraction of a second.

NOW WORK Problems 1–8 in the Takeaway 25-4.

All the examples of fission that we've described result in the release of neutrons. Imagine what can happen if many ^{235}U atoms are close to each other. Should one nucleus be struck by a neutron and fission, as shown in Figure 25-12, there would then be three neutrons moving among the atoms. Should each of these neutrons strike a ^{235}U nucleus and start a fission process, there would be nine neutrons, so possibly nine more fissions. With a sufficient number of ^{235}U atoms present, this *chain reaction* quickly grows, with an accompanying rapid increase in energy released.

Isotopes that are capable of sustaining a fission chain reaction are used as nuclear fuels. Such isotopes are termed *fissile*, and the most common fissile nuclear fuels are ^{233}U, ^{235}U, ^{239}Pu, and ^{241}Pu. In the fission reactions they undergo, typically two or three neutrons are released in addition to larger fragments. These released neutrons don't have to be moving rapidly to trigger additional fission reactions: A chain reaction in a fissile material can be induced by a neutron carrying essentially zero kinetic energy. In some fuels such as ^{235}U, slower, less energetic neutrons are more efficiently absorbed by the fissile nuclei.

Most nuclear reactors use the energy released in a fission chain reaction to heat water and produce steam to drive an electric generator (see Section 20-4). There are several challenges to producing energy in this way. First, a minimum amount, or *critical mass*, of fissile material must be present to sustain a chain reaction. As it happens, only about 0.7% of the naturally occurring uranium in the world is ^{235}U, and the other three commonly used fissile isotopes do not occur naturally. Most naturally occurring uranium is ^{238}U, which is not fissile. To use uranium as a nuclear fuel, then, it is necessary to separate the ^{235}U atoms from the ^{238}U, a costly and difficult process known as *enrichment*. In addition, the ^{238}U atoms that inevitably remain tend to absorb free neutrons and thereby inhibit a chain reaction. Once the critical mass of ^{235}U has been assembled, controlling the chain reaction is another challenge. If too many of the neutrons produced in the fissions result in a second fission, the energy released increases so rapidly that the fuel and whatever vessel is used to contain it can be damaged or even melt. To control a fission chain reaction in a nuclear reactor, control rods made of a substance that is a good absorber of neutrons are inserted between pieces of fuel.

Operating a fission nuclear reactor safely is a significant challenge. First, the fission fragments are radioactive. Many of these fragments, or the fragments produced when they decay, are long lived and tend to produce dangerous radiation for years, centuries, or even longer. In addition, reactors commonly use water as a *moderator*, a material that tends to slow the free neutrons (to make them more easily absorbed by a ^{235}U nucleus). Should the containment vessel rupture, this hot water can be released into the atmosphere in the form of steam carrying radioactive particles. Perhaps the most significant nuclear accident occurred at the Chernobyl nuclear power plant in Ukraine in 1986, in which an uncontrolled chain reaction caused a catastrophic power increase, leading to a series of explosions and the release of large quantities of radioactive steam, fuel, and smoke into the environment. However, with appropriate care, many of the safety concerns around nuclear energy can be addressed. Nuclear energy provides a significant advantage in that it puts no carbon dioxide into the environment. For instance, although nuclear energy is not a large portion of the U.S. energy portfolio, the energy provided by nuclear reactors in 2019 emitted almost 500 million metric tons *less* carbon dioxide into the environment than more traditional production facilities do for the same amount of energy. Active research on the next generation of nuclear power plants is looking to create plants that would operate on used fuel from the current generation of power plants, further reducing safety concerns.

THE TAKEAWAY for Section 25-4

✔ The binding energy per nucleon of nuclides with more than about 60 neutrons and protons decreases with increasing values of *A*. For this reason, the fission process, in which a nucleus breaks up into smaller fragments, leads to more stable configurations of the nucleons in bigger nuclei.

✔ Fission can be triggered by allowing a slow-moving neutron to merge with a fissile nucleus.

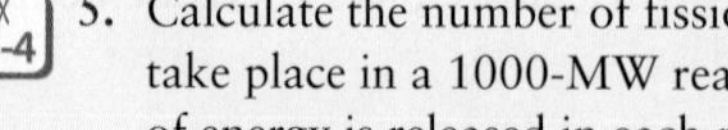

Prep for the AP Exam

AP® Building Blocks

EX 25-4 **1.** Complete the following fission reactions.
(a) $^{242}\text{Am} + \underline{\ ?\ } \rightarrow {}^{90}\text{Sr} + {}^{149}\text{La} + 4\text{n}$
(b) $^{244}\text{Pa} + \text{n} \rightarrow \underline{\ ?\ } + {}^{131}\text{Sb} + 12\text{n}$
(c) $\underline{\ ?\ } + \text{n} \rightarrow {}^{92}\text{Se} + {}^{153}\text{Sm} + 6\text{n}$
(d) $^{262}\text{Fm} + \text{n} \rightarrow {}^{112}\text{Rh} + \underline{\ ?\ } + 9\text{n}$

EX 25-4 **2.** Calculate the energy released in the following nuclear fission reaction.

$$^{239}\text{Pu} + \text{n} \rightarrow {}^{98}\text{Tc} + {}^{138}\text{Sb} + 4\text{n}$$

The atomic masses are $^{239}\text{Pu} = 239.052157$ u, $^{98}\text{Tc} = 97.907215$ u, and $^{138}\text{Sb} = 137.940793$ u.

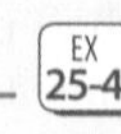

EX 25-4 **3.** Complete the following nuclear fission reaction of thorium-232 and calculate the energy released in the reaction: $^{232}\text{Th} + \text{n} \rightarrow {}^{99}\text{Kr} + {}^{124}\text{Xe} + \underline{\ ?\ }$. The atomic masses are $^{232}\text{Th} = 232.038051$ u, $^{99}\text{Kr} = 98.957606$ u, and $^{124}\text{Xe} = 123.905894$ u.

EX 25-4 **4.** Complete the following fission reactions.
(a) $^{235}\text{U} + \text{n} \rightarrow {}^{128}\text{Sb} + {}^{101}\text{Nb} + \underline{\ ?\ }$
(b) $^{235}\text{U} + \text{n} \rightarrow \underline{\ ?\ } + {}^{116}\text{Pd} + 4\text{n}$
(c) $^{238}\text{U} + \text{n} \rightarrow {}^{99}\text{Kr} + \underline{\ ?\ } + 11\text{n}$
(d) $\underline{\ ?\ } + \text{n} \rightarrow {}^{101}\text{Rb} + {}^{130}\text{Cs} + 8\text{n}$

EX 25-4 **5.** Calculate the number of fission reactions per second that take place in a 1000-MW reactor. Assume that 200 MeV of energy is released in each reaction.

AP® Skill Builders

EX 25-4 **6.** Calculate the energy (in MeV) released in the following nuclear fission reaction.

$$^{242}\text{Am} + \underline{\ ?\ } \rightarrow {}^{90}\text{Sr} + {}^{149}\text{La} + 4\text{n}$$

Start by completing the reaction and use the following nuclear masses: ^{242}Am = 242.059549 u, ^{90}Sr = 89.9077387 u, and ^{149}La = 148.934733 u

EX 25-4 **7.** Assuming that in a fission reactor a neutron loses half its energy in each collision with an atom of the moderator, determine how many collisions are required to slow a 200-MeV neutron to an energy of 0.04 eV.

AP® Skills in Action

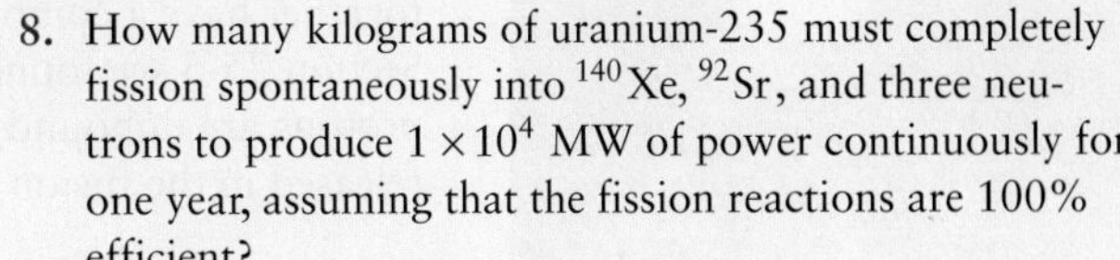

EX 25-4 **8.** How many kilograms of uranium-235 must completely fission spontaneously into ^{140}Xe, ^{92}Sr, and three neutrons to produce 1×10^4 MW of power continuously for one year, assuming that the fission reactions are 100% efficient?

9. Consider the spontaneous fission process $^{20}\text{Ne} \rightarrow {}^{10}\text{B} + {}^{10}\text{B}$. This process does not occur in nature. Why not?
(A) The number of protons does not remain constant.
(B) The number of neutrons does not remain constant.
(C) Both (a) and (b).
(D) This process would absorb energy, not release it.

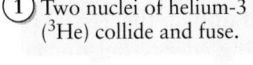

25-5 The smallest nuclei can release energy if they are forced to fuse together

As we saw in the previous section, Figure 25-10 explains why the largest nuclei can undergo fission: The binding energy per nucleon E_B/A is maximum for mass numbers A around 60. By breaking into smaller fragments, nuclei with values of A much greater than 60 can therefore increase their binding energy per nucleon E_B/A and so become more stable. The opposite is true for the lightest nuclei, with A much less than 60. These nuclei can increase the value of E_B/A and become more stable by becoming *larger*. Processes in which two small nuclei combine to form a larger one are called *fusion* processes.

Figure 25-13 shows how two ^{3}He nuclei (each with two protons and one neutron) fuse together to form a ^{4}He nucleus (with two protons and two neutrons). Two protons are left over, and without more neutrons there is no way for the protons to be bound together in a single nucleus. As a result, these protons fly off separately. Energy is released during this process because the final configuration of the protons and neutrons is more stable than the initial configuration. A photon carries away the energy released.

How much energy is released in the process shown in Figure 25-13? As in fission, the energy released in fusion is the difference between the total binding energy of the final

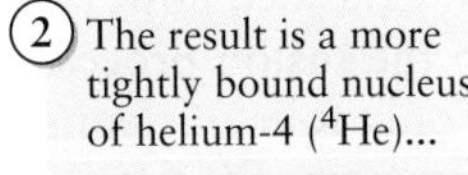
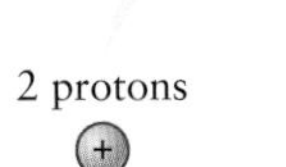
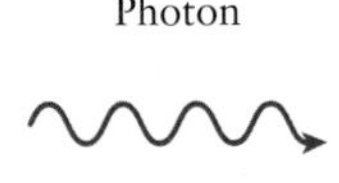

Energy is released in this fusion reaction: The total kinetic energy of the ^{4}He nucleus and protons, plus the energy of the photon, is much greater than the total kinetic energy of the initial ^{3}He nuclei.

Figure 25-13 Nuclear fusion The fusion of two ^{3}He nuclei to make a ^{4}He nucleus is one of the energy-releasing reactions that take place in the core of the Sun.

Norm Lane/Alamy

Figure 25-14 The Sun: A nuclear fusion reactor All of the energy released by the Sun is a result of fusion reactions in its core that convert hydrogen nuclei into helium nuclei.

nuclei and the total binding energy of the original nuclei. From Figure 25-10, the binding energy per nucleon for ^{3}He is approximately 2.5 MeV. Each ^{3}He has three nucleons, so the total binding energy of ^{3}He is 3(2.5 MeV) = 7.5 MeV, and the two ^{3}He nuclei together have a combined binding energy equal to 7.5 MeV + 7.5 MeV = 15.0 MeV. In Section 25-3 we found that a ^{4}He nucleus has a binding energy of 28.3 MeV. The two protons are unbound, so they make no contribution to the binding energy. The energy released in the fusion process is therefore

$$\begin{aligned}\text{(energy released)} &= (\text{binding energy of } ^4\text{He nucleus}) \\ &\quad - (\text{binding energy of two } ^3\text{He nuclei}) \\ &= (28.3 \text{ MeV}) - (15.0 \text{ MeV}) = 13.3 \text{ MeV}\end{aligned}$$

The actual value, found using more accurate values of the binding energies, is closer to 12.86 MeV.

The process shown in Figure 25-13 is the final step in the *proton–proton cycle*, the fusion process that takes place near the center of the Sun and is the source of the Sun's energy (**Figure 25-14**). The cycle begins with the fusing of two protons (the nuclei of ^{1}H) to form ^{2}H. This nucleus fuses with another proton, forming ^{3}He, and finally, two ^{3}He nuclei fuse to form ^{4}He. We can summarize these three steps as

$$\text{Step 1: } ^1\text{H} + {}^1\text{H} \to {}^2\text{H} + \text{e}^+ + \nu_\text{e}$$

$$\text{Step 2: } ^2\text{H} + {}^1\text{H} \to {}^3\text{He} + \gamma$$

$$\text{Step 3: } ^3\text{He} + {}^3\text{He} \to {}^4\text{He} + {}^1\text{H} + {}^1\text{H} + \gamma$$

In step 1 the e^+ particle is a **positron**, a particle with the same mass as an electron but with positive charge $+e$. The particle named ν_e (the Greek letter "nu" with a subscript "e") is a **neutrino**, a nearly massless, neutral particle. Also note that this step involves a proton being converted into a neutron. We'll discuss this conversion, called beta-plus decay, in Section 25-6. Step 1 and the subsequent interaction of the positron with an electron in the Sun (in which the two particles annihilate each other and convert into photons) release 1.44 MeV of energy. The energy released in step 2 is 5.49 MeV. Both these steps must occur twice before step 3 can occur (because step 3 requires two ^{3}He nuclei). This means six protons are used in steps 1 and 2, and in step 3 two of those protons are returned along with a ^{4}He nucleus. The net result is therefore that four ^{1}H nuclei disappear and are replaced by one ^{4}He nucleus. The net energy release is 2 × 1.44 MeV from step 1 happening twice, plus 2 × 5.49 MeV from step 2 happening twice, plus 12.86 MeV from the fusion of two ^{3}He nuclei in step 3. The sum of these is a net energy release of 26.7 MeV as four ^{1}H nuclei are transformed into a ^{4}He nucleus.

Fission processes release more energy than the proton–proton fusion cycle, around 200 MeV compared to 26.7 MeV. Therefore, it might seem that fission is a more effective way to convert fuel to energy. Consider, however, that while 235 nucleons in ^{235}U are spent to release 200 MeV, in the fusion process described above, the 26.7 MeV released come at the expense of only four nucleons. Comparing energy per nucleon (think miles per gallon or kilometers per liter), fission provides less than 1 MeV per nucleon, while proton–proton fusion gives 26.7 MeV divided by 4, or nearly 7 MeV per nucleon. Fusion processes are *much* more efficient at releasing nuclear binding energy.

Hydrogen makes up about 75% of the Sun's mass of approximately 2×10^{30} kg. If the proton–proton cycle leads to more stability, why doesn't all that hydrogen quickly fuse to form ^{4}He, leaving the Sun a gigantic (and cool) ball of helium gas? The answer lies in the same forces at play within nuclei: the Coulomb force that repels protons from each other and the short-ranged strong force that draws them together. For two protons to fuse to form ^{2}H, they must come within a few femtometers of each other, at which point the strong attraction is able to overcome the Coulomb repulsion. This requires the protons to have considerable kinetic energy, which can come from being at high temperature. A temperature of more than 4×10^6 K is required for the proton–proton cycle to start. The temperature at the core of the Sun is around 15×10^6 K, so the proton–proton cycle can and does occur there. Even at that temperature, however, the probability that two nearby protons will fuse is small. This means that only a small fraction—about 4×10^{-19}—of the hydrogen in the Sun is undergoing fusion at any one time. The Sun won't burn out for a long time.

As a star ages, its core temperature increases and additional fusion reactions become possible. For example, three ^{4}He nuclei can fuse to form a ^{12}C nucleus. This requires a temperature of about 10^8 K because the ^{4}He nuclei are more massive than protons and repel each other more strongly due to their greater charge. At even higher temperatures a ^{4}He nucleus can fuse with a ^{12}C nucleus to form a ^{16}O nucleus, a ^{4}He nucleus can fuse with a ^{16}O nucleus to form a ^{20}Ne nucleus, and so on. So as stars age they manufacture heavier and heavier chemical elements. In the most massive stars, which have the highest core temperatures, so much kinetic energy is available at the very end of the star's evolution that fusion processes can produce even the heaviest nuclei up to uranium. Making these massive nuclei by fusion absorbs rather than releases energy, which is why it can happen only in very special circumstances.

These fusion reactions in stars make life on Earth possible. Here's why: When the universe first originated some 13.8 billion years ago, almost all ordinary matter was in the form of hydrogen or helium. All heavier elements had to be manufactured by fusion within stars. After an aging star produces elements heavier than helium and goes through its final stages of evolution, it ejects much of its material into interstellar space (**Figure 25-15**). This material, which is enriched in heavy elements, can then be incorporated into a later generation of stars. Our Sun is such a "second-generation" star, with an elevated abundance of elements heavier than helium. As part of the process by which the Sun formed, some of these elements went into forming Earth and the Sun's other planets. This means that all the nuclei of carbon in the organic compounds that make up your body, the nuclei in the oxygen that you breathe, and the nuclei of iron in the hemoglobin in your blood were manufactured in stars that died billions of years ago. This is one of the great lessons of nuclear physics: You are made of star stuff.

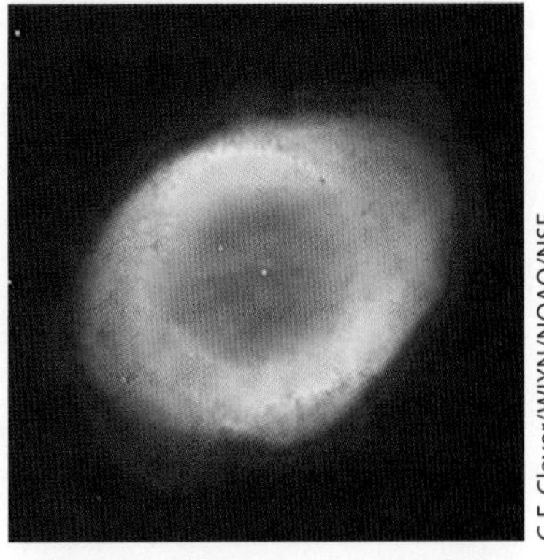
C.F. Claver/WIYN/NOAO/NSF

Figure 25-15 Seeding space with fusion products The Ring Nebula is a cloud of gas emitted by an aging star. The cloud includes nitrogen (outermost edge, shown in red) and oxygen (inner region, shown in green) produced by fusion reactions within the star.

THE TAKEAWAY for Section 25-5

✔ The binding energy per nucleon of nuclides that have fewer than about 60 neutrons and protons is larger for increasing values of mass number A. For this reason, fusion of two small nuclei to form a larger one leads to more stable configuration of the nucleons.

✔ Fusion can take place only if the fusing nuclei come very close to each other to overcome their mutual electric repulsion. This means very high temperatures are required to give nuclei the necessary initial kinetic energies.

AP® Building Blocks

1. Of the elements in the periodic table:
   (a) Which are more likely to undergo nuclear fission?
   (b) Which are more likely to undergo nuclear fusion?
2. What is the difference between fission and fusion?

3. Complete the following fusion reactions.
   (a) $^{2}H + ^{3}H \rightarrow ^{4}He + \underline{?}$
   (b) $^{4}He + ^{4}He \rightarrow ^{7}Be + \underline{?}$
   (c) $^{2}H + ^{2}H \rightarrow ^{3}He + \underline{?}$
   (d) $^{2}H + ^{1}H \rightarrow \gamma + \underline{?}$
   (e) $^{2}H + ^{2}H \rightarrow ^{3}H + \underline{?}$

AP® Skill Builders

4. Calculate the energy released in the fusion reaction in problem 3, part (b). Give your answer in MeV.

5. Each fusion reaction of deuterium (^{2}H) and tritium (^{3}H) releases about 20 MeV. What mass of tritium is needed to create 10^{14} J of energy, the same as that released by exploding 25,000 tons of TNT? Assume that an endless supply of deuterium is available.

AP® Skills in Action

6. How many fusion reactions per second must be sustained to operate a deuterium–tritium fusion power plant that outputs 1000 MW, operating at 33% efficiency?
7. Ordinary hydrogen gas is in the form of diatomic hydrogen (H_2), and more than 99.9% of the atoms in hydrogen gas have a nucleus that is a single proton (^{1}H). Why don't the two hydrogen nuclei in an H_2 molecule spontaneously undergo fusion as in step 1 of the proton–proton cycle: $^{1}H + ^{1}H \rightarrow ^{2}H + e^{+} + \nu_e$?
   (A) The nuclei are too far apart.
   (B) The nuclei are moving too slowly relative to each other.
   (C) The nuclei in an atom are different from those outside an atom.
   (D) Both (A) and (B) are correct.

25-6 Unstable nuclei may emit alpha, beta, or gamma radiation

For many people phrases such as *radioactivity* and *nuclear radiation* are synonymous with danger. Not all nuclear radiation is dangerous, however! Several types of nuclear radiation have important practical applications. In this section we'll explore some aspects of nuclear radiation.

All naturally occurring nuclear processes take place because the final state is more energetically favorable than the initial state. We saw this is true of fission and fusion, and it is true of radiation processes, too. A relatively few nuclides—266 out of more than 3000—are stable. All the rest are radioactive; that is, they decay into another nuclide by radiating away one or more particles. It's also possible for a nucleus in an excited state to radiate energy in the form of a photon as the nucleus transitions to a less excited state.

The three most common kinds of radiation are *alpha*, *beta*, and *gamma* radiation. The terms were coined by Ernest Rutherford, who in his research between 1899 and 1903 classified radiation according to the depth that a radiation particle was able to penetrate other objects. Alpha particles penetrated the least, beta particles more, and gamma particles the most. (Rutherford received the 1908 Nobel Prize in Chemistry for this research, which was the first to show that one element can change into another through radioactive processes.) We now know that alpha particles are ^{4}He nuclei; beta particles are electrons; and gamma particles are photons with energies that can be millions of times greater than the energies of visible-light photons.

Before we consider the properties of these specific types of radiation, let's look at a concept that is common to all of them—*half-life*.

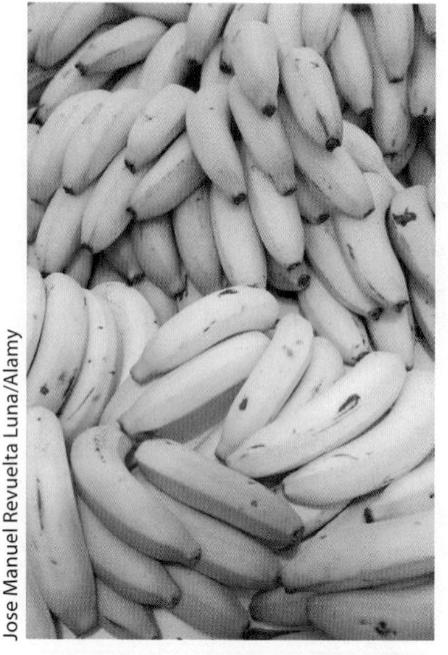

Jose Manuel Revuelta Luna/Alamy

Figure 25-16 Radioactive decays in bananas Bananas are a good source of potassium. About 0.0117% of the potassium atoms in bananas have the radioactive nucleus ^{40}K, for which the decay constant (the probability per second that a nucleus will decay) is very small: $\lambda = 1.76 \times 10^{-17}$ decays/s $= 1.76 \times 10^{-17}$ Bq. The number N of ^{40}K atoms in a typical 0.15-kg banana is about 9×10^{17} (with a combined mass of about 6×10^{-8} kg), so the number of ^{40}K nuclei that decay per second in a banana is $\lambda N = (1.76 \times 10^{-17} \text{ decays/s})(9 \times 10^{17})$, or about 15 decays/s. This is extraordinarily small and poses no health hazard.

Radioactive Decay and Half-Life

Nuclear radiation of all kinds involves physics on the very small scale of the nucleus, so is governed by quantum mechanics. As we learned in the last chapter, quantum mechanics cannot tell us the position or velocity of a particular object at any time. It can, however, tell us the *probability* that an object will be at a particular place or have a particular velocity at any given instant. In the same way, quantum mechanics cannot tell us when a given radioactive nucleus will decay, but it can predict the probability that this nucleus will decay within a given time interval.

As an example, consider the emission of a beta particle in the decay of ^{137}Cs (cesium), in which a neutron is converted into a proton:

$$^{137}\text{Cs} \to {}^{137}\text{Ba} + \text{e}^- + \bar{\nu}_\text{e}$$

The beta particle is the electron (e^-), and $\bar{\nu}_\text{e}$ is an **antineutrino**, related to the light, neutral neutrino particle we discussed in the context of fusion in the previous section. Every ^{137}Cs nucleus can, and eventually will, decay radioactively to a barium nucleus (^{137}Ba). Experiment shows that if a sample of ^{137}Cs contains, say, 10,000 atoms, after 30.17 years only about 5000 will be left. However, if we select any individual ^{137}Cs atom of those 10,000, we have no idea when its nucleus will decay. Perhaps it will decay in the next second or perhaps not for thousands of years.

While we can't make a definitive claim about when any particular atomic nucleus will decay, we can quantify the *rate* at which a group of radioactive atoms decays, that is, the number of decays per second. The probability λ that any one nucleus of a given type will decay in the next second is the same for all such nuclei. The quantity λ is called the **decay constant.** It has units of s^{-1}, because it refers to a probability per second. The value of λ is different for different radioactive nuclides: It is greater for nuclides that decay rapidly and smaller for nuclides that decay slowly.

If we have a sample of N such nuclei, the total number of decays that take place in the next second—that is, the *decay rate*—will be equal to the product of the decay constant λ (the probability that any one nucleus decays in the next second) and N (the number of radioactive nuclei present) (**Figure 25-16**). So the decay rate of a radioactive sample is greater if the sample is larger (the number of nuclei N in the sample is greater) or the nuclei have a higher decay constant λ. The SI unit of decay rate is the **becquerel** (Bq), after the nineteenth-century French physicist Antoine Henri Becquerel, who along with Marie Skłodowska-Curie and Pierre Curie won the 1903 Nobel Prize in Physics for their discovery of radioactivity. The becquerel is equivalent to one radioactive decay per second. Physicists also commonly use units of curies (Ci) for decay rate; 1 Ci = 3.7×10^{10} Bq = 3.7×10^{10} decays/s.

After an elapsed time of Δt seconds, the total number of decays will be equal to $\lambda N \Delta t$. This means that in a time Δt the number of radioactive nuclei decreases by $\lambda N \Delta t$. So the *change* in the number of radioactive nuclei present is

$$\Delta N = -\lambda N \Delta t \qquad (25\text{-}3)$$

The minus sign in Equation 25-3 indicates that the number of nuclei decreases as a result of the decays. If we divide both sides of Equation 25-3 by the elapsed time Δt, we get an expression for the rate of change $\Delta N/\Delta t$ of the number of radioactive nuclei:

In a sample of radioactive material, the **rate of change of the number of radioactive nuclei...**

... is negative because the number of nuclei decreases due to decay...

$$\frac{\Delta N}{\Delta t} = -\lambda N \qquad (25\text{-}4)$$

EQUATION IN WORDS
Radioactive decay equation

...is proportional to the **decay constant** (the probability that a given nucleus decays in a one-second interval)...

...and is proportional to the **number of radioactive nuclei that remain.**

Equation 25-4 tells us that as time goes by the decay rate will decrease because the number of radioactive nuclei will decrease. Using the tools of calculus we can solve Equation 25-4 to find the number of nuclei present as a function of time $N(t)$. The result is

$$N(t) = N_0 e^{-\lambda t} \qquad (25\text{-}5)$$

In Equation 25-5 N_0 is the number of nuclei present at a specific time that we choose to call $t = 0$. This equation tells us that the number of nuclei present decreases exponentially, as **Figure 25-17** shows. The number of decays per second at time t is equal to

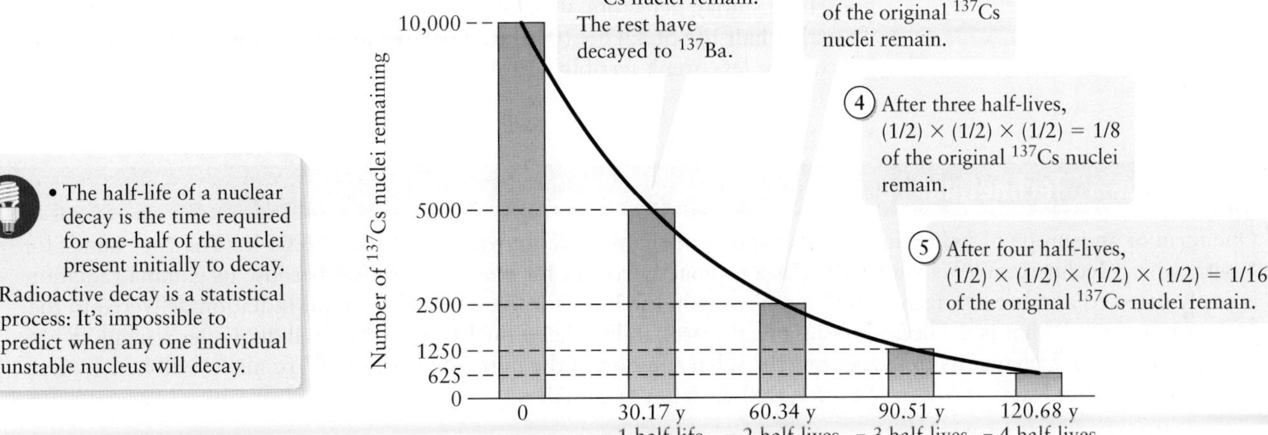

1 Initially we have 10,000 ^{137}Cs nuclei. These nuclei are unstable and decay to ^{137}Ba.
2 After one half-life, 1/2 of the original ^{137}Cs nuclei remain. The rest have decayed to ^{137}Ba.
3 After two half-lives, (1/2) × (1/2) = 1/4 of the original ^{137}Cs nuclei remain.
4 After three half-lives, (1/2) × (1/2) × (1/2) = 1/8 of the original ^{137}Cs nuclei remain.
5 After four half-lives, (1/2) × (1/2) × (1/2) × (1/2) = 1/16 of the original ^{137}Cs nuclei remain.
Number of ^{137}Cs nuclei remaining
10,000
5000
2500
1250
625
0
0
30.17 y = 1 half-life
60.34 y = 2 half-lives
90.51 y = 3 half-lives
120.68 y = 4 half-lives
Time

Figure 25-17 Nuclear decay and half-life This exponential curve shows the evolution of a sample that originally contains 10,000 cesium-137 (^{137}Cs) nuclei, which decay to barium-137 (^{137}Ba).

- The half-life of a nuclear decay is the time required for one-half of the nuclei present initially to decay.
- Radioactive decay is a statistical process: It's impossible to predict when any one individual unstable nucleus will decay.

WATCH OUT !

Be careful not to misuse the concept of half-life.

It's important to note that the graph of nuclear decay in Figure 25-17 is for a *large number* of radioactive nuclei. Statistically, half of the nuclei present at any time will have decayed after one half-life. For a *single* nucleus half-life has a different interpretation: Over one half-life, a given radioactive nucleus has a 50% chance of not decaying (and so a 50% chance of decaying). If it does decay, it can happen at any time during that half-life.

the decay constant λ (the decay probability per second per nucleus), multiplied by $N(t)$, the number of nuclei remaining at time t. We can write this as

$$R(t) = \lambda N(t) = \lambda N_0 e^{-\lambda t}$$

In this equation λN_0 is equal to the decay rate at $t = 0$, which we call R_0. So the decay rate as a function of time is

$$R(t) = R_0 e^{-\lambda t} \tag{25-6}$$

Equation 25-6 tells us that the decay rate, too, will decrease exponentially as time goes by. With the passage of time, a radioactive sample will undergo fewer and fewer decays per second.

Figure 25-17 shows that the number of ^{137}Cs nuclei remaining decreases by one-half every 30.17 years due to beta decay. Because the decay rate is proportional to the number of ^{137}Cs nuclei remaining, the decay rate also decreases by one-half every 30.17 years. This time is called the **half-life** of the radioactive decay, to which we give the symbol $\tau_{1/2}$ (the Greek letter "tau"). If there are N_0 radioactive nuclei present at $t = 0$, there will be $N_0/2$ present at $t = \tau_{1/2}$. If we substitute this into Equation 25-5, we get

$$\frac{N_0}{2} = N_0 e^{-\lambda \tau_{1/2}}$$

Divide both sides of this equation by N_0, then take the natural logarithm of both sides:

$$\ln\left(\frac{1}{2}\right) = \ln(e^{-\lambda \tau_{1/2}}) \tag{25-7}$$

Why do we do this? The reason is that the natural logarithm "undoes" the exponential function: For any x, $\ln(e^x) = x$. Furthermore, $\ln(1/x) = -\ln x$ for any x. If we apply these to Equation 25-7, we get

$$-\ln 2 = -\lambda \tau_{1/2}$$

or

$$\tau_{1/2} = \frac{\ln 2}{\lambda} \tag{25-8}$$

Equation 25-8 says that the half-life is inversely proportional to the decay constant λ, the probability that a given nucleus of a certain type will decay in a 1-s interval. The greater the decay constant, the shorter the half-life and the more rapidly a sample of that nucleus will decay.

The half-life of radioactive sources varies widely, from far less than 1 s to billions of years or more. Common radioactive sources include ^{32}P, a beta emitter used in DNA research, which has a half-life of 14.3 days; ^{241}Am, an alpha emitter often found in household smoke detectors, which has a half-life of 432.2 y; and ^{238}U, an alpha emitter with a half-life of 4.47×10^9 y. Radioactive isotopes with half-lives on the order of a second or less aren't terribly useful because they don't stay around long enough.

EXAMPLE 25-5 Technetium

One form of an isotope of technetium, ^{99m}Tc, undergoes gamma decay with a half-life of 6.01 h. (The "m" stands for "metastable." Technetium-99 is widely used as a radioactive tracer for medical purposes because its gamma radiation is easily detected and because technetium doesn't stay in the body for long. Hence the total radiation delivered to the patient is low.) (a) What is the decay constant λ, the probability that a nucleus of ^{99m}Tc will decay per second, in this sample? Does λ change as time goes on? (b) What fraction of the initial number of ^{99m}Tc nuclei will be left after 1.00 day? (c) What fraction will be left after 4.00 days have elapsed?

Set Up

We'll use Equation 25-8 to relate the decay constant λ to the half-life. Equation 25-5 will tell us the number of ^{99m}Tc nuclei remaining after a time t in terms of the initial number of nuclei N_0.

Half-life of a radioactive substance:

$$\tau_{1/2} = \frac{\ln 2}{\lambda} \qquad (25\text{-}8)$$

Number of radioactive nuclei present at time t:

$$N(t) = N_0 e^{-\lambda t} \qquad (25\text{-}5)$$

Solve

(a) Determine the decay constant for ^{99m}Tc.

We can rewrite Equation 25-8 as

$$\lambda = \frac{\ln 2}{\tau_{1/2}}$$

Substitute $\tau_{1/2} = 6.01$ h:

$$\lambda = \frac{\ln 2}{(6.01\text{ h})}\left(\frac{1\text{ h}}{60\text{ min}}\right)\left(\frac{1\text{ min}}{60\text{ s}}\right) = 3.20 \times 10^{-5}\text{ s}^{-1}$$

The probability that a given ^{99m}Tc nucleus will decay in a 1-s interval is 3.20×10^{-5}, corresponding to odds of 1 in $(1/3.20 \times 10^{-5}) = 31{,}200$. This value depends only on the half-life, so it does not vary with time.

(b) To find the fraction of nuclei remaining after $t = 1.00$ d, substitute this value of t into Equation 25-5.

The fraction of ^{99m}Tc nuclei remaining after a time t equals the number remaining $N(t)$ divided by the number N_0 present initially:

$$\frac{N(t)}{N_0} = \frac{N_0 e^{-\lambda t}}{N_0} = e^{-\lambda t}$$

From part (a) we know the value of λ in s^{-1}, so we need to express t in seconds:

$$t = (1.00\text{ d})\left(\frac{24\text{ h}}{1\text{ d}}\right)\left(\frac{60\text{ min}}{1\text{ h}}\right)\left(\frac{60\text{ s}}{1\text{ min}}\right) = 8.64 \times 10^{4}\text{ s}$$

The fraction of nuclei remaining is then

$$\frac{N(1.00\text{ d})}{N_0} = e^{-(3.20 \times 10^{-5}\text{ s}^{-1})(8.64 \times 10^{4}\text{ s})}$$
$$= e^{-2.77} = 0.0628$$

(c) Repeat part (b) with $t = 4.00$ d.

The elapsed time is now

$$t = (4.00\text{ d})\left(\frac{8.64 \times 10^{4}\text{ s}}{1\text{ d}}\right) = 3.46 \times 10^{5}\text{ s}$$

and the fraction of nuclei remaining is

$$\frac{N(4.00\text{ d})}{N_0} = e^{-(3.20 \times 10^{-5}\text{ s}^{-1})(3.46 \times 10^{5}\text{ s})}$$
$$= e^{-11.1} = 1.55 \times 10^{-5}$$

Reflect

We can check our results by noting that $t = 1.00$ d is almost exactly 4 times the 6.01-h half-life of ^{99m}Tc, and $t = 4.00$ d is almost exactly 16 times the half-life. This check agrees with our calculations, as it should.

After each half-life, the number of ^{99m}Tc nuclei remaining decreases by one-half. So after four half-lives, the fraction of ^{99m}Tc nuclei remaining will be

$$\frac{1}{2} \times \frac{1}{2} \times \frac{1}{2} \times \frac{1}{2} = \frac{1}{2^4} = \frac{1}{16} = 0.0625$$

The fraction actually remaining at $t = 1.00$ d (slightly less than four 6.01-h half-lives) is 0.0628, very close to our estimate.

After 16 half-lives, the fraction of ^{99m}Tc remaining will be

$$\frac{1}{2^{16}} = \frac{1}{65{,}536} = 1.53 \times 10^{-5}$$

The fraction actually remaining at $t = 4.00$ d (slightly less than 16 times the 6.01-h half-life) is 1.55×10^{-5}, which again is very close to our estimate.

NOW WORK Problems 2, 6, and 7 from The Takeaway 25-6.

Alpha Radiation

We saw in Section 25-4 that a large nucleus can increase the binding energy per nucleon E_B/A by breaking into smaller fragments. The most likely decay products are those that are more stable, that is, those with larger binding energy per nucleon. As Figure 25-10 shows, ^{4}He has a greater binding energy per nucleon than any other nucleus with a small value of A. So the nucleus of the ^{4}He atom is far more stable than other small nuclei and therefore a far more probable decay product of large nuclei. For this reason, the radioactive emission of a ^{4}He nucleus is the most likely decay process for a large nucleus. We learned that an alpha particle is a ^{4}He nucleus; the emission of an alpha particle is called **alpha decay**.

The alpha radiation process reduces the number of protons Z of the initial, or parent, nucleus by two and reduces the number of neutrons of the parent by two. The result is a daughter nucleus that has an atomic number $Z - 2$ and a mass number $A - 4$, accompanied by an alpha particle with two protons and two neutrons (**Figure 25-18**).

The daughter nucleus and the alpha particle both carry away the energy released in alpha decay. Because of momentum conservation, the kinetic energy of the alpha particle is far greater than the kinetic energy of the daughter. We can compare these energies by finding the ratio of the kinetic energy K_α of the alpha particle to the kinetic

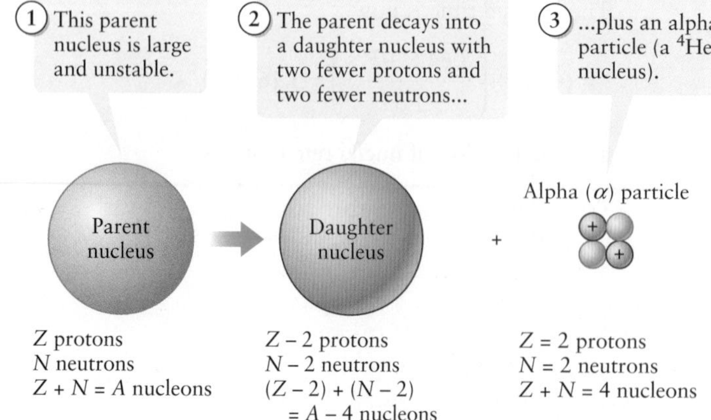

Figure 25-18 Alpha decay Large nuclei can increase their stability by emitting an alpha particle and becoming a smaller, more stable daughter nucleus.

Energy is released in this decay: The daughter nucleus is more tightly bound (has a greater binding energy) than the parent.

energy K_D of the daughter. The alpha particle has mass m_α and is emitted with speed v_α, and the daughter nucleus has mass m_D and is emitted with speed v_D. The kinetic energies of the alpha particle and the daughter are then

$$K_\alpha = \frac{1}{2} m_\alpha v_\alpha^2$$

$$K_D = \frac{1}{2} m_D v_D^2$$

The ratio of these kinetic energies is

$$\frac{K_\alpha}{K_D} = \frac{(1/2) m_\alpha v_\alpha^2}{(1/2) m_D v_D^2} = \frac{m_\alpha}{m_D}\left(\frac{v_\alpha}{v_D}\right)^2 \tag{25-9}$$

We can relate the speeds v_α and v_D as momentum must be constant in the alpha decay because no external forces are exerted on the parent nucleus as it decays. If the parent nucleus is at rest, the total momentum is zero before the decay and so must be zero after the decay. The alpha particle and daughter nucleus must therefore fly off in opposite directions, and each must have the same magnitude of momentum:

$$m_\alpha v_\alpha = m_D v_D \quad \text{so} \quad \frac{v_\alpha}{v_D} = \frac{m_D}{m_\alpha}$$

If we substitute this into Equation 25-9, we find that the ratio of the alpha particle's kinetic energy to that of the daughter nucleus is

$$\frac{K_\alpha}{K_D} = \frac{m_\alpha}{m_D}\left(\frac{v_\alpha}{v_D}\right)^2 = \frac{m_\alpha}{m_D}\left(\frac{m_D}{m_\alpha}\right)^2 = \frac{m_D}{m_\alpha}$$

The mass of the daughter nucleus is much larger than the mass of the alpha particle, so the fraction m_D/m_α is much greater than one and K_α is large compared to K_D.

All nuclei with more than 82 protons are unstable and have some probability of alpha decay. As an example, the element thorium ($Z = 90$ protons) undergoes alpha decay to radium, which contains two fewer protons ($Z = 88$). The α decay of ^{228}Th, for example, is

$$^{228}\text{Th} \rightarrow {}^{224}\text{Ra} + \alpha$$

Just as ^{228}Th decays to ^{224}Ra, ^{224}Ra undergoes alpha decay to ^{220}Rn. This process continues until the daughter nucleus has $Z = 82$ or less. In this case the final alpha decay is to lead, with $Z = 82$:

$$^{228}\text{Th} \rightarrow {}^{224}\text{Ra} + \alpha \quad (\tau_{1/2\ \text{Th}} = 1.91\ \text{y})$$

$$\hookrightarrow {}^{220}\text{Rn} + \alpha \quad (\tau_{1/2\ \text{Ra}} = 3.63\ \text{d})$$

$$\hookrightarrow {}^{216}\text{Po} + \alpha \quad (\tau_{1/2\ \text{Rn}} = 55.6\ \text{s})$$

$$\hookrightarrow {}^{212}\text{Pb} + \alpha \quad (\tau_{1/2\text{Po}} = 0.145\ \text{s})$$

Note that in each alpha decay the daughter nucleus has two fewer protons and two fewer neutrons than its parent.

Another nucleus that decays by alpha radiation is ^{235}U, which has a half-life $\tau_{1/2} = 7.04 \times 10^8$ y. (In Example 25-4 in Section 25-4, we looked at the spontaneous fission of ^{235}U. This is a very rare decay mode; ^{235}U undergoes alpha decay rather than spontaneous fission almost 100% of the time.) Substantial amounts of ^{235}U, ^{238}U ($\tau_{1/2} = 4.47 \times 10^9$ y), and ^{232}Th ($\tau_{1/2} = 1.40 \times 10^{10}$ y) are present in Earth's core, and the energy released by the alpha decay of these isotopes helps to sustain our planet's high internal temperatures. All of Earth's geologic activity, including earthquakes, volcanic eruptions (Figure 25-1c), and the drifting of continents, is powered by the motions of our planet's interior. So alpha decay plays an important role in Earth's dynamic geology.

EXAMPLE 25-6 Alpha Decay of ^{238}U

The uranium isotope ^{238}U undergoes alpha decay to ^{234}Th. The binding energy per nucleon is 7.570 MeV in ^{238}U and 7.597 MeV in ^{234}Th. Find the energy released in the process $^{238}U \rightarrow {}^{234}Th + \alpha$.

Set Up

We'll use the same principle that we used in Example 25-4 (Section 25-4) to find the energy released in fission: The released energy equals the total binding energy of the nuclei present after the decay, minus the binding energy of the original (parent) nucleus. As in that example, we'll find the binding energy for each nucleus by multiplying the binding energy per nucleon times the number of nucleons A.

(energy released)
= (total binding energy of daughter plus alpha particle)
– (binding energy of original ^{238}U nucleus)

Solve

From Example 25-3 (Section 25-3) the binding energy of a ^4He nucleus (an alpha particle) is

$$E_B(^4\text{He}) = 28.297 \text{ MeV}$$

The binding energy of a ^{234}Th nucleus ($A = 234$) is

$$E_B(^{234}\text{Th}) = (234)(7.597 \text{ MeV}) = 1777.7 \text{ MeV}$$

and the binding energy of a ^{238}U nucleus ($A = 238$) is

$$E_B(^{238}\text{U}) = (238)(7.570 \text{ MeV}) = 1801.7 \text{ MeV}$$

The released energy is then

$$\begin{aligned} E_{\text{released}} &= E_B(^{234}\text{Th}) + E_B(^4\text{He}) - E_B(^{238}\text{U}) \\ &= 1777.7 \text{ MeV} + 28.297 \text{ MeV} - 1801.7 \text{ MeV} \\ &= 4.3 \text{ MeV} \end{aligned}$$

Reflect

This energy seems to be of reasonable size, given the binding energy per nucleon. (The alpha particles emitted by radioactive isotopes with long half-lives, such as ^{238}U, tend to have kinetic energies in the 4 to 5 MeV range.)

NOW WORK Problem 3 from The Takeaway 25-6.

Beta Radiation

For most possible mass numbers A, there exist a number of nuclides with that same value of A. For example, molybdenum, technetium, ruthenium, and rhodium each have an isotope with 99 nucleons: ^{99}Mo has 42 protons and 57 neutrons, ^{99}Tc has 43 protons and 56 neutrons, ^{99}Ru has 44 protons and 55 neutrons, and ^{99}Rh has 45 protons and 54 neutrons. (It is also possible to create isotopes of other elements with A equal to 99.) The binding energy per nucleon in each is slightly different, however, so only one of them is the most stable: For $A = 99$, the most stable isotope is ^{99}Ru. If there were a process whereby ^{99}Tc could convert one of its neutrons into a proton, or ^{99}Rh could convert one of its protons into a neutron, either of these nuclei could transform into the more stable ^{99}Ru.

The process that makes this possible is called **beta decay**. There are actually two varieties of beta decay. In **beta-minus decay** a neutron (charge zero) changes into a proton (charge $-e$). The net charge cannot change, so an electron (charge $+e$), also known as a beta-minus (β^-) particle, is also produced and escapes from the nucleus. To account for other conservation requirements a third particle, the neutral and nearly massless antineutrino ($\bar{\nu}_e$), is also created in this process. The full process, then, is

$$n \rightarrow p + e^- + \bar{\nu}_e \quad \text{(beta-minus decay)}$$

In **beta-plus decay** a proton (charge $+e$) changes into a neutron (charge zero). To conserve charge a positively charged electron or *positron*, also called a beta-plus (β^-) particle, is also produced and escapes from the nucleus, along with a neutral and nearly massless neutrino (which, for our purposes, is essentially the same particle as an antineutrino). This process is

$$\mathrm{p} \rightarrow \mathrm{n} + \mathrm{e}^+ + \nu_\mathrm{e} \quad \text{(beta-plus decay)}$$

Figure 25-19a depicts the beta-minus decay of a nucleus with too many neutrons such as $^{99}\mathrm{Tc}$. We can write this process as

$$^{99}\mathrm{Tc} \rightarrow {}^{99}\mathrm{Ru} + \mathrm{e}^- + \bar{\nu}_\mathrm{e}$$

The number of nucleons ($A = 99$) is the same before and after the decay, but the number of protons Z has increased by 1 (from 43 to 44) and the number of neutrons N has decreased by 1 (from 56 to 55). The decay of $^{137}\mathrm{Cs}$ to $^{137}\mathrm{Ba}$, which we discussed at the beginning of this section, is another example of beta-minus decay. In this case Z increases from 55 to 56 and N decreases from 82 to 81.

One particularly important example of beta-plus decay is the decay of potassium-40 ($^{40}\mathrm{K}$) to argon-40 ($^{40}\mathrm{Ar}$) (see Figure 25-16). This process has a very long half-life of 1.25×10^9 y. Because potassium is abundant in Earth's interior, the energy released by this decay makes a substantial contribution to keeping our planet's interior in a fluid state and powering its geologic activity.

Figure 25-19b depicts the beta-plus decay of a nucleus with too many protons such as $^{99}\mathrm{Rh}$. We can write this process as

$$^{99}\mathrm{Rh} \rightarrow {}^{99}\mathrm{Ru} + \mathrm{e}^+ + \nu_\mathrm{e}$$

(a) Nuclei with too many neutrons can undergo beta-minus (β^-) decay.

(1) The daughter nucleus has one more proton and one fewer neutron than the parent.

(2) The nucleus also emits an electron (so electric charge is conserved) and an antineutrino.

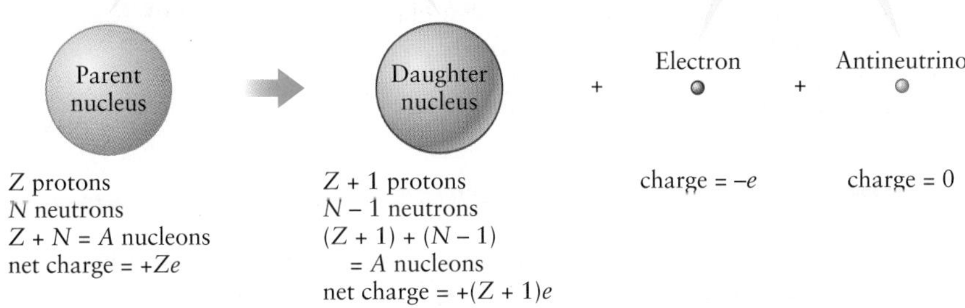

(b) Nuclei with too many protons can undergo beta-plus (β^+) decay.

(1) The daughter nucleus has one fewer proton and one more neutron than the parent.

(2) The nucleus also emits a positron (so electric charge is conserved) and a neutrino.

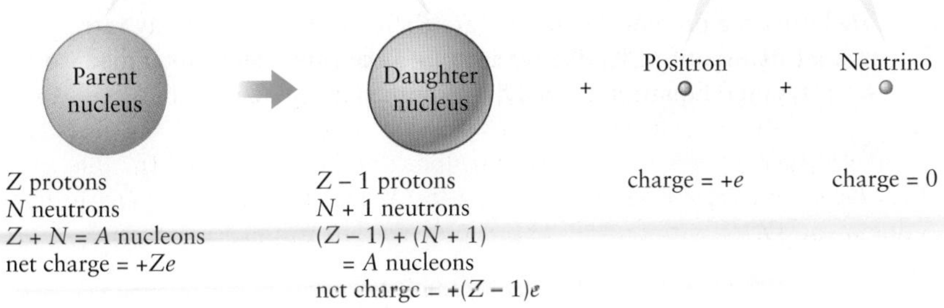

Figure 25-19 Beta decay Nuclei can increase their stability by (a) converting one neutron into a proton or (b) converting one proton into a neutron.

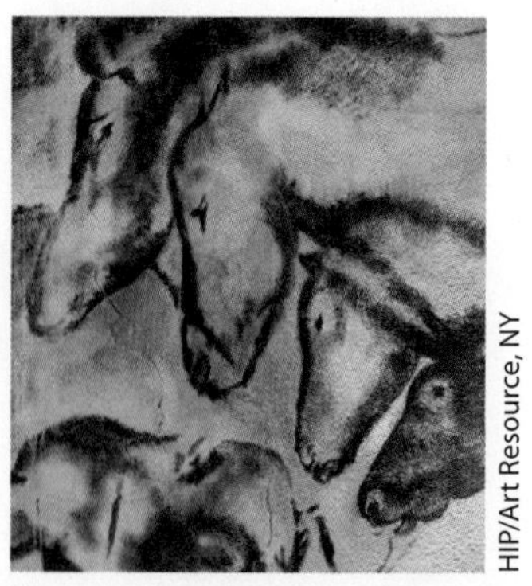

Figure 25-20 Carbon-14 dating and human prehistory These cave paintings in France were created by a prehistoric artist. The amount of ^{14}C present in smoke stains on the walls is less than 3% of what would be expected from smoke produced by burning freshly cut firewood. Detailed analysis shows that between 5.2 and 5.8 half-lives of ^{14}C have elapsed since the smoke was produced, indicating an age of the smoke stains (and the paintings) between 30,000 and 33,000 years.

As for beta-minus decay, the number of nucleons remains the same (in this case $A = 99$). But now the number of protons Z decreases by 1 (from 45 to 44), and the number of neutrons increases by 1 (from 54 to 55).

An important application of beta decay is in carbon-14 dating, a technique used to measure the age of objects that are composed, or partially composed, of organic matter. Almost all carbon atoms in Earth's atmosphere—for example, the carbon in carbon dioxide, CO_2—have a nucleus with a stable isotope of carbon, either ^{12}C (98.9%) or ^{13}C (1.1%). But about 1 in 10^{12} of those carbon atoms has a ^{14}C nucleus. This radioactive isotope of carbon is a β^- emitter with a half-life of 5730 y:

$$^{14}\text{C} \rightarrow {}^{14}\text{N} + \text{e}^- + \bar{\nu}_\text{e}$$

Carbon-14 is constantly produced in the atmosphere by cosmic rays slamming into ^{14}N nuclei. As a result, even though ^{14}C undergoes radioactive decays, the ratio of ^{14}C to ^{12}C in the atmosphere has remained relatively constant for at least tens of thousands of years. The $^{14}C/^{12}C$ ratio is the same in living organisms—for example, plants that breathe in CO_2—as it is in the atmosphere. However, once an organism dies, it no longer replenishes its supply of carbon, so the $^{14}C/^{12}C$ ratio decreases as the ^{14}C decays. As an example, in the Chauvet-Pont-d'Arc Cave in southern France (**Figure 25-20**) there are smoke stains whose $^{14}C/^{12}C$ ratio has been measured. This makes it possible to determine how much time has elapsed since the firewood that created the smoke was part of a living tree. This carbon-14 dating technique tells us that the smoke stains are 30,000 to 33,000 years old—which must be the age of the cave paintings made by the light of that burning firewood.

EXAMPLE 25-7 Ötzi the Iceman

In 1991, two German hikers discovered a human corpse in the Ötztal Alps on the border between Austria and Italy. The remains proved to be a well-preserved natural mummy of a man who lived during the last ice age. The rate of radioactive decay of ^{14}C in the mummy of "Ötzi the Iceman" was measured to be 0.121 Bq per gram (Bq/g). In a living organism, the rate of radioactive decay of ^{14}C is 0.231 Bq/g. How long ago did Ötzi the Iceman live?

Set Up

Once Ötzi died, his body stopped taking in ^{14}C. After that time the number $N(t)$ of ^{14}C nuclei in his body decreased due to beta-minus decay. The ^{14}C decay rate $R(t)$, which is proportional to $N(t)$, decreased in the same manner. We are given $R(t) = 0.121$ Bq/g for the present-day decay rate and $R_0 = 0.231$ Bq/g (the decay rate for a living organism and hence the decay rate at time $t = 0$, the last date on which Ötzi was still alive). We'll solve Equation 25-6 for the present time t (the elapsed time since Ötzi died). We'll also have to use Equation 25-8 to find the decay constant λ from the known half-life $\tau_{1/2} = 5730$ y of ^{14}C.

Decay rate as a function of time:

$$R(t) = R_0 e^{-\lambda t} \qquad (25\text{-}6)$$

Half-life of a radioactive substance:

$$\tau_{1/2} = \frac{\ln 2}{\lambda} \qquad (25\text{-}8)$$

Solve

Rearrange Equations 25-6 and 25-8 to find an expression for the time t since Ötzi died.

We know the present-day decay rate $R(t)$ and the initial decay rate in a living organism R_0. We want to find the time t since Ötzi died, so we rearrange Equation 25-6. Divide both sides by R_0:

$$\frac{R(t)}{R_0} = e^{-\lambda t}$$

Take the natural logarithm of both sides and recall that $\ln e^x = x$:

$$\ln\left(\frac{R(t)}{R_0}\right) = \ln e^{-\lambda t} = -\lambda t$$

Divide both sides by $-\lambda$:

$$t = -\frac{1}{\lambda}\ln\left(\frac{R(t)}{R_0}\right)$$

To get an expression for $1/\lambda$, divide both sides of Equation 25-8 by ln 2:

$$\frac{1}{\lambda} = \frac{\tau_{1/2}}{\ln 2}$$

Putting everything together, the time t since Ötzi died is

$$t = -\frac{\tau_{1/2}}{\ln 2}\ln\left(\frac{R(t)}{R_0}\right)$$

Substitute the given values into the expression for t.

We are given $\tau_{1/2} = 5730$ y for ^{14}C, $R(t) = 0.121$ Bq/g, and $R_0 = 0.231$ Bq/g:

$$t = -\frac{(5730\text{ y})}{\ln 2}\ln\left(\frac{0.121\text{ Bq/g}}{0.231\text{ Bq/g}}\right) = -\left(\frac{5730\text{ y}}{0.693}\right)\ln 0.524$$

$$= -\frac{(5730\text{ y})}{0.693}(-0.647) = 5350\text{ y}$$

Reflect

Ötzi died 5350 y ago, which seems reasonable given his existence in the last ice age. His mummy thus gives us a unique look into life in prehistoric Europe.

Extend: Carbon-14 dating can be used only on objects less than about 50,000 years old, or about 8 to 10 half-lives of ^{14}C. For older objects the decay rate of ^{14}C has decreased to such a small value that it is hard to measure accurately, so any determination of age with this technique becomes difficult. For much older objects such as rocks, a similar approach is used but with isotopes with much longer half-lives. For example, the age of meteorites that fall to Earth is determined by looking at the ratio of uranium to lead (the endpoint of a series of alpha decays that starts with uranium); the oldest meteorites are more than 4.5×10^9 years old.

NOW WORK Problem 8 from The Takeaway 25-6.

Gamma Radiation

A nucleus in an excited state radiates energy in a way analogous to the emission of a photon when an electron in an excited atomic state falls to a state of lower energy (see Section 24-6). Just like atoms, nuclei have excited states of definite energy, and when they transition from an excited state to a less excited one, they will emit a photon of energy equal to the difference in energy between the initial and final states.

The most common way that a nucleus can become excited is following an alpha decay or a beta decay. Although these decay processes result in a more stable configuration of the nucleons, the nucleons that remain in the daughter nucleus may not be, initially, in the most stable arrangement for that particular nuclide. This excited daughter nucleus decays to a more stable configuration, giving off energy in the form of a gamma (γ) ray (**Figure 25-21**). This process is called **gamma decay**.

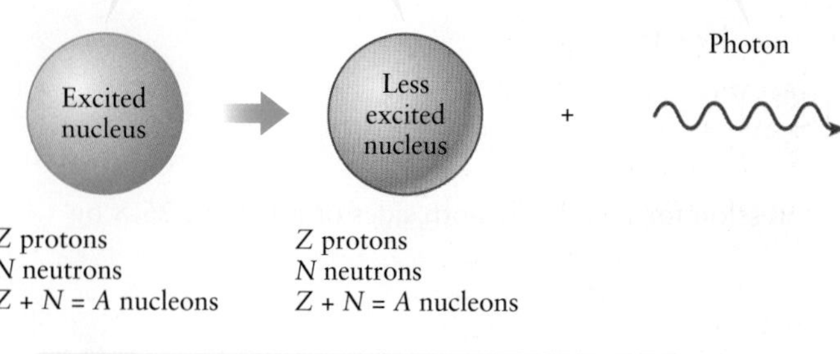
(1) In gamma (γ) decay a nucleus in an excited state drops into a less excited state.
(2) The energy lost by the nucleus goes into a high-energy photon (gamma ray).
Photon
Excited nucleus
Less excited nucleus
+
Z protons
N neutrons
Z + N = A nucleons
Z protons
N neutrons
Z + N = A nucleons

In gamma decay there is no change in the number of protons or the number of neutrons in the nucleus.

Figure 25-21 Gamma decay An excited nucleus can lower its energy by emitting a photon.

The energy carried away when a nucleus in an excited state decays is on the order of 1 MeV. From Equation 22-26 in Section 22-4, $E = hf$, the frequency and wavelength of a photon of energy $E = 1.00$ MeV $= 1.00 \times 10^6$ eV are

$$f = \frac{E}{h} = \frac{1.00 \times 10^6 \text{ eV}}{4.14 \times 10^{-15} \text{ eV} \cdot \text{s}} = 2.42 \times 10^{20} \text{ Hz}$$

$$\lambda = \frac{c}{f} = \frac{3.00 \times 10^8 \text{ m/s}}{2.42 \times 10^{20} \text{ Hz}}$$

$$= 1.24 \times 10^{-12} \text{ m} = 1.24 \times 10^{-3} \text{ nm}$$

Recall that the wavelength of visible light photons is in the range from 380 to 750 nm; a photon emitted by an excited nucleus is in the gamma radiation range, far from the visible part of the spectrum.

Gamma radiation does not change the atomic number Z or the neutron number N of a nucleus; that is, the number of protons and the number of neutrons remain the same after a γ ray is emitted. As an example, earlier we considered the beta-minus decay of ^{137}Cs to ^{137}Ba. In 95% of those decays the ^{137}Ba nucleus is formed in an excited state that we denote as ^{137}Ba*:

$$^{137}\text{Cs} \to {}^{137}\text{Ba}^* + e^- + \bar{\nu}_e$$

The excited barium nucleus then decays to its ground state by emission of a photon of energy 0.662 MeV:

$$^{137}\text{Ba}^* \to {}^{137}\text{Ba} + \gamma$$

The values of $Z = 56$ and $N = 81$ for the barium nucleus do not change in this second step of the radiation process.

Biological Effects of Radiation

We'll finish this section with a discussion of how radiation from nuclei affects living organisms, including humans. As they pass through living tissue, energetic particles cause ionization in molecules such as water, creating free radicals (highly reactive molecules and atoms) that can damage DNA and cause potentially lethal mutations to cells. This is the principle of using radiation to destroy cancerous tumors. The cells in these tumors reproduce more rapidly than healthy cells, and so are more quickly destroyed by the mutations produced by ionizing radiation.

Charged particles—such as the α particles produced in alpha decay and the electrons and positrons produced in beta-minus and beta-plus decays—exert electric forces on molecular electrons to cause ionization. High-energy photons, such as those produced in gamma decay, by contrast cause ionization either by colliding with electrons and knocking them out of molcules (Compton scattering) or by having their energy absorbed by molecular electrons (the photoelectric effect).

In each of these cases the amount of damage depends on the amount of energy absorbed by the tissue from the radiation that passes through it. The energy from radiation absorbed per unit mass of tissue is called the *absorbed dose.* The unit of absorbed dose is the *gray* (Gy), equal to 1 joule per kilogram: 1 Gy = 1 J/kg. However, experiment shows that different kinds of radiation produce different amounts of biological effect for the same absorbed dose. This is described by a factor called the relative biological effectiveness (RBE), which has a different value for different types of radiation. The value of the RBE is 1 for x rays, gamma rays, electrons, and positrons, but about 20 for alpha particles. The amount of biological effect of a given amount of radiation, or *equivalent dose*, is the product of the RBE and the absorbed dose. The unit of absorbed does is the sievert (Sv):

$$\text{equivalent dose (Sv)} = \text{RBE} \times \text{absorbed dose (Gy)}$$

Because alpha particles have such a large RBE, for a given absorbed dose they produce a much greater equivalent dose and have much greater biological effect than does beta or gamma radiation. However, alpha particles are not very penetrating (a piece of paper is enough to stop the alpha particles from most alpha-emitting nuclei), so they do not pose much radiation hazard if they are coming from outside the body.

Because radioactive materials occur naturally in the environment, we are continually exposed to radiation. In one year, a person at sea level has an equivalent dose of about 3 mSv (1 mSv = 10^{-3} Sv), most of which comes from the trace amounts of radon gas found inside homes and other buildings. (Radon, which has no stable isotopes, is a decay product of naturally occurring uranium and thorium. Because it is a gas, radon can seep into buildings, concentrate there, and be inhaled by the occupants. The longest-lived isotope of radon, ^{222}Rn, is an alpha emitter; as alpha particles have a large RBE, even small amounts of inhaled radon have a measurable effect on a person's annual equivalent dose.) At high altitude there is greater exposure to cosmic rays coming from space (which are screened by the atmosphere); the additional equivalent dose from a round-trip airline flight between the west and east coasts of North America is about 0.03 mSv, and is about 1.5 mSv per year for a person living at an elevation of 1500 m (5000 ft). A chest x ray adds an additional 0.1 mSv, a mammogram an additional 0.4 mSv, and a CT scan of the abdomen and pelvis an additional 10 mSv.

To put these numbers in perspective, a whole-body equivalent dose up to about 200 mSv (0.2 Sv) causes no immediately detectable effect. To have fatal consequences a short-term whole-body dose would have to be about 5 Sv, equivalent to having 50,000 simultaneous chest x rays.

THE TAKEAWAY for Section 25-6

✔ Radioactive decay is a statistical process. The number of nuclei and the decay rate both decrease exponentially with time, and both decrease by one-half in a time equal to one half-life.

✔ The three most common modes by which radiation occurs are alpha, beta, and gamma radiation.

✔ Alpha particles are ^{4}He nuclei and are emitted by large nuclei with $Z > 82$. In alpha emission the proton number and neutron number each decrease by 2.

✔ Beta particles are either negatively charged electrons or positively charged positrons. In beta-minus emission a neutron in the nucleus changes into a proton; in beta-plus emission a proton in the nucleus changes into a neutron.

✔ Gamma particles are high-energy photons, typically with energies of about 1 MeV. They are emitted when a nucleus decays from an excited state to a less excited one.

✔ Alpha, beta, and gamma radiation can all damage living tissue by producing ionization.

Prep for the AP® Exam

AP® Building Blocks

1. Complete the following conversions.
   (a) 100 μCi = ____________ Bq
   (b) 1500 decays/min = ____________ Bq
   (c) 16,500 Bq = ____________ Ci
   (d) 7.55×10^{10} Bq = ____________ decays/min

EX 25-5 2. The curie unit is defined as 1 Ci = 3.7×10^{10} Bq, which is about the rate at which radiation is emitted by 1.00 g of radium. From the definition, calculate the half-life of radium.

EX 25-6 3. Complete the following alpha decays.
   (a) ${}^{238}\text{U} \to \alpha + \underline{\;?\;}$
   (b) ${}^{234}\text{Th} \to \underline{\;?\;} + \underline{\;?\;}\,\text{Ra}$
   (c) $\underline{\;?\;} \to \alpha + {}^{236}\text{U}$
   (d) ${}^{214}\text{Bi} \to \alpha + \underline{\;?\;}$

4. Complete the following beta decays.
   (a) ${}^{14}\text{C} \to e^- + \bar{\nu}_e + \underline{\;?\;}$
   (b) ${}^{239}\text{Np} \to e^- + \bar{\nu}_e + \underline{\;?\;}$
   (c) $\underline{\;?\;} \to e + \nu_e + {}^{60}\text{Ni}$
   (d) ${}^{3}\text{H} \to e^- + \bar{\nu}_e + \underline{\;?\;}$
   (e) ${}^{13}\text{N} \to e^+ + \underline{\;?\;} + \underline{\;?\;}$

5. Complete the following gamma decays.
   (a) ${}^{131}\text{I}^* \to \gamma + \underline{\;?\;}$
   (b) $\underline{\;?\;} \to \gamma + {}^{24}\text{Na}$

AP® Skill Builders

EX 25-5 6. Given a radioactive sample:
   (a) Calculate the fraction of the sample that will be left after 6 half-lives.
   (b) Calculate the fraction of the sample that will be left after 7.5 half-lives.

EX 25-5 7. A patient is injected with 7.88 μCi of radioactive iodine-131 that has a half-life of 8.02 days. Assuming that 90% of the iodine ultimately finds its way to the thyroid, what decay rate do you expect to find in the thyroid after 30 days?

AP® Skills in Action

EX 25-7 **8.** The ratio of carbon-14 to carbon-12 in living wood is 1.3×10^{-12}. How many decays per second are there in 550 g of wood?

9. A certain radioactive isotope has a half-life of 5 days. You are given a sample containing a number of nuclei of this isotope. About how long would you have to wait before about 1/1000 of the initial number of nuclei remained?
(A) 20 days
(B) 40 days
(C) 50 days
(D) 1000 days

10. Nickel-64 has an excited state 1.34 MeV above the ground state. The atomic mass of the ground state of this isotope of nickel is 63.927967 u.
(a) What is the mass of the atom when the nucleus is in this excited state?
(b) What is the wavelength of the gamma ray that is emitted when the nucleus decays to the ground state?

WHAT DID YOU LEARN?

Chapter learning goals	Section(s)	Related example(s)	Prep for the AP Exam: Relevant section review exercise(s)
Explain how conservation laws play an important role in nuclear physics.	25-1		4, 5, 9, 10
Describe how we know that the force that holds the nucleus together is both strong and of short range and relate this concept to nuclear density.	25-2	25-1	4, 6
		25-2	7, 8, 10
Explain how and why the binding energy per nucleon in a nucleus depends on the size of the nucleus and calculate that energy.	25-3	25-3	2-8
Describe the properties of nuclear fission, and calculate the energy released in nuclear fission.	25-4	25-4	1-9
Describe the properties of nuclear fusion, compare them to those of nuclear fission, and calculate the energy released in nuclear fission.	25-5	25-4	1-7
Describe radioactive decay and the processes by which it occurs and calculate the decay constant and half-life for decay processes from given data.	25-6	25-5	2, 6, 7
		25-6	3
		25-7	8

Chapter 25 Review

Key Terms

All the Key Terms can be found in the Glossary/Glosario on page G1 in the back of the book.

alpha decay 1270
alpha particle 1256
antineutrino 1266
atomic number 1249
becquerel 1267
beta decay 1272
beta-minus decay 1272
beta-plus decay 1273
binding energy 1255
decay constant 1266
gamma decay 1275
half-life 1268
isotope 1249
magnetic resonance imaging 1252
mass number 1249
neutrino 1264
neutron-induced fission 1259
neutron number 1249
nuclear fission 1256
nuclear fusion 1256
nuclear radiation 1256
nucleon 1247
nuclide 1249
positron 1264
radioactive 1246
strong force 1245

Chapter Summary

Topic	Equation or Figure

The strong force and nuclear sizes: The strong force is a short-range attractive force nucleons (protons or neutrons) exert on each other. Because this force has a short range, the volume of a nucleus is proportional to the mass number (number of nucleons). In larger nuclei, the number of neutrons must exceed the number of protons to counterbalance the electric repulsion between protons.

For most nuclei, more neutrons than protons must be present to prevent the protons from flying apart due to electrostatic repulsion.

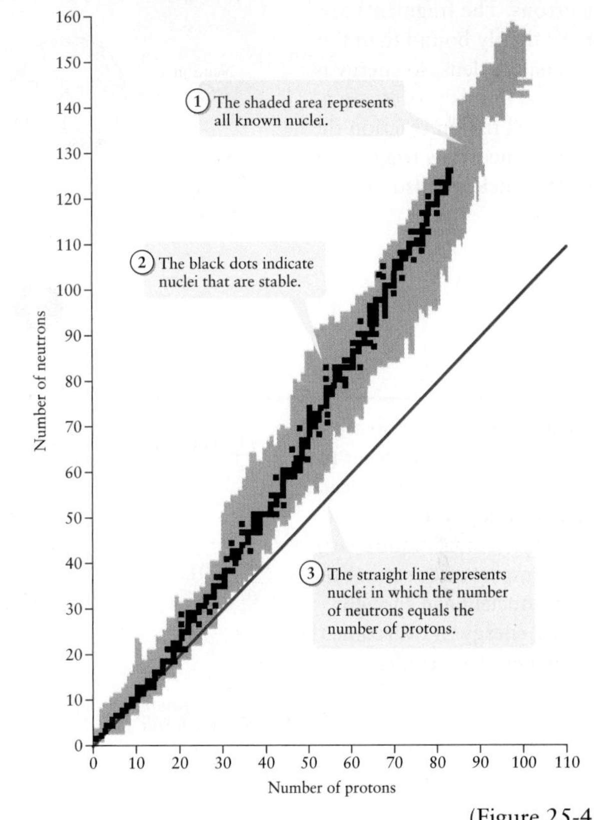

Number of neutrons
Number of protons
1 The shaded area represents all known nuclei.
2 The black dots indicate nuclei that are stable.
3 The straight line represents nuclei in which the number of neutrons equals the number of protons.

(Figure 25-4)

Nuclear binding energy: The binding energy of a nucleus is the energy required to separate it into its constituent nucleons. The binding energy per nucleon is greatest for nuclei with around 60 nucleons; for larger nuclei the electric repulsion between protons makes nuclei less stable. The binding energy of a particular isotope can be calculated from the mass of a neutral atom containing that isotope.

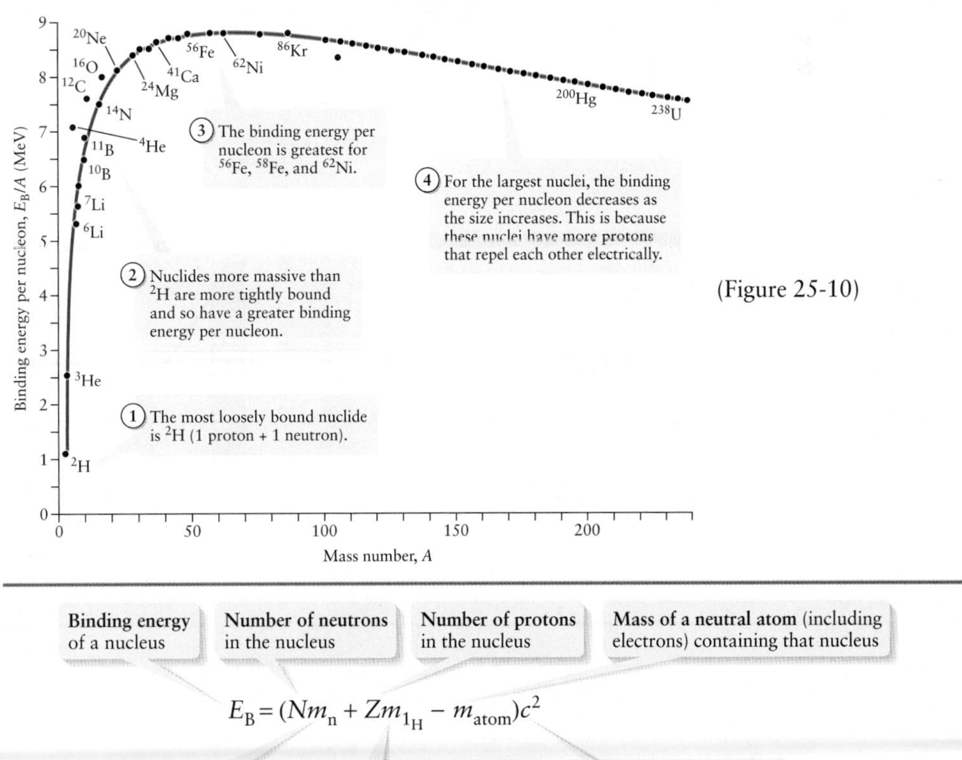

Binding energy per nucleon, E_B/A (MeV)
Mass number, A
^{20}Ne
^{16}O
^{12}C
^{56}Fe
^{41}Ca
^{24}Mg
^{62}Ni
^{86}Kr
^{14}N
^{4}He
^{11}B
^{10}B
^{7}Li
^{6}Li
^{3}He
^{2}H
^{200}Hg
^{238}U
1 The most loosely bound nuclide is ^{2}H (1 proton + 1 neutron).
2 Nuclides more massive than ^{2}H are more tightly bound and so have a greater binding energy per nucleon.
3 The binding energy per nucleon is greatest for ^{56}Fe, ^{58}Fe, and ^{62}Ni.
4 For the largest nuclei, the binding energy per nucleon decreases as the size increases. This is because these nuclei have more protons that repel each other electrically.
Binding energy of a nucleus
Number of neutrons in the nucleus
Number of protons in the nucleus
Mass of a neutral atom (including electrons) containing that nucleus
Mass of a neutron
Mass of a neutral hydrogen atom (1 proton + 1 electron)
Speed of light in a vacuum

(Figure 25-10)

$$E_B = (Nm_n + Zm_{^1H} - m_{atom})c^2 \quad (25\text{-}2)$$

Nuclear fission: When the largest nuclei absorb a neutron, they fragment (fission) into two smaller nuclei plus a few neutrons. The fragments are more tightly bound than the original nucleus, so energy is released in this process. In a sustained fission reaction the released neutrons trigger other, nearby nuclei to also undergo fission.

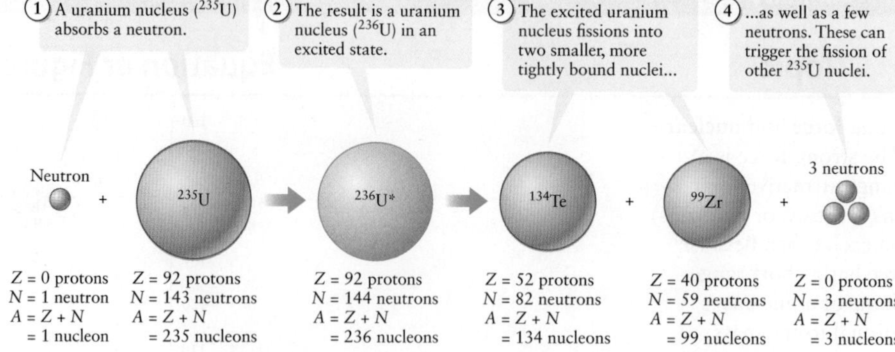

Energy is released in this fission reaction: The total kinetic energy of the fission fragments is much greater than the total kinetic energy of the initial neutron and ^{235}U nucleus.

(Figure 25-12)

Nuclear fusion: The smallest nuclei can merge together to form a larger nucleus, releasing energy in the process. These fusion processes require very high temperatures so that the fusing nuclei have enough kinetic energy to overcome their mutual electric repulsion.

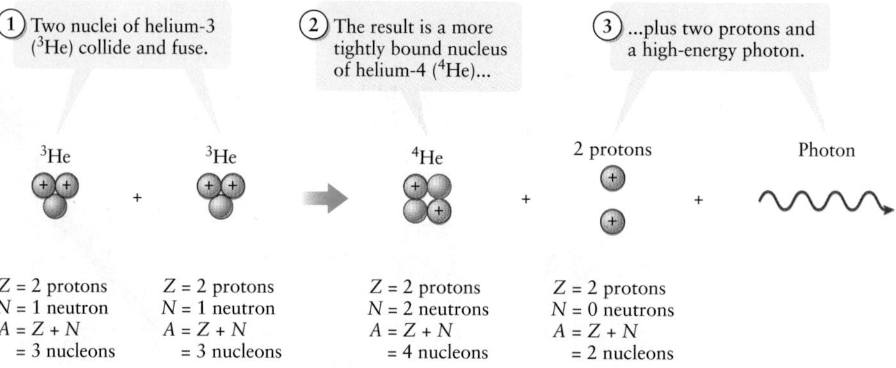

Energy is released in this fusion reaction: The total kinetic energy of the ^{4}He nucleus and protons, plus the energy of the photon, is much greater than the total kinetic energy of the initial ^{3}He nuclei.

(Figure 25-13)

Nuclear decay and half-life: The decay of unstable nuclei is a statistical process. This means that the rate of decay is proportional to the number of unstable nuclei present. As a result the number of nuclei and the decay rate both decline in an exponential manner. The time for the number of nuclei and the decay rate to decrease by one-half is called the half-life.

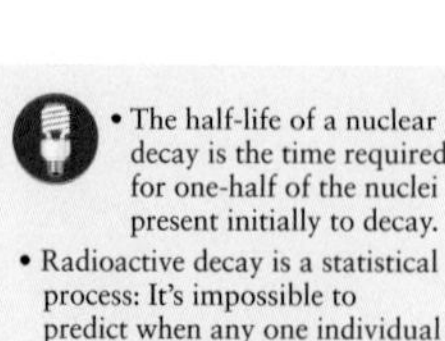

- The half-life of a nuclear decay is the time required for one-half of the nuclei present initially to decay.
- Radioactive decay is a statistical process: It's impossible to predict when any one individual unstable nucleus will decay.

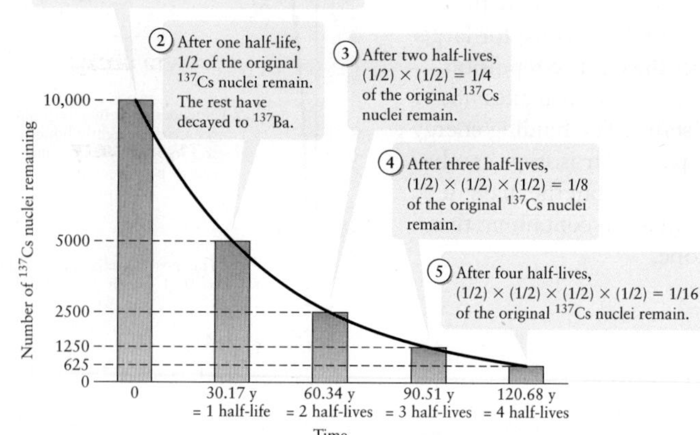

(Figure 25-17)

Alpha, beta, and gamma decays: Large nuclei release energy and become more stable by emitting an alpha particle (a ^{4}He nucleus). Other nuclei with too many neutrons undergo beta-minus decay, in which one neutron changes into a proton; those with too many protons undergo beta-plus decay, in which one proton changes into a neutron. In gamma decay an excited state of a nucleus (indicated by an asterisk) transitions to a less excited state and emits a gamma-ray photon (γ).	An alpha decay: $^{228}\text{Th} \rightarrow {}^{224}\text{Ra} + \alpha$ A beta-minus decay: $^{99}\text{Tc} \rightarrow {}^{99}\text{Ru} + \text{e}^- + \bar{\nu}_\text{e}$ A beta-plus decay: $^{99}\text{Rh} \rightarrow {}^{99}\text{Ru} + \text{e}^+ + \nu_\text{e}$ A gamma decay: $^{137}\text{Ba}^* \rightarrow {}^{137}\text{Ba} + \gamma$

Chapter 25 Review Problems

1. The stenciled outlines of human hands in the chapter opening photo are found in the Cave of the Hands in the Patagonia region of Argentina. They are known to be 9500 to 13,000 years old, an age determined from the radioactive decay of carbon-14. This type of carbon has 6 protons and 8 neutrons in its nucleus and has a half-life of 5730 years. Of the carbon-14 that was present when these outlines were made, the fraction that remains today is closest to

   (A) ½. (C) ⅛.
   (B) ¼. (D) 1⁄16.

2. Describe the basic characteristics of the strong force that exists between nucleons. What other competing force between nucleons is present in the nucleus?
3. If atomic masses are used, explain why the mass of a beta particle is *not* accounted for in the basic beta decay

$$\text{n} \rightarrow \text{p} + \text{e}^- + \bar{\nu}_\text{e}$$

   Assume that the mass of the antineutrino ($\bar{\nu}_\text{e}$) is very small and can be neglected.

4. Our Sun is a nuclear reactor.
   (a) Describe the nuclear reactions that occur in our Sun.
   (b) Discuss how the equilibrium state of the Sun is not permanent and discuss the eventual future of our solar system.
5. An old wooden bowl unearthed in an archeological dig is found to have one-fourth of the amount of carbon-14 present in a similar sample of fresh wood. Determine the age of the bowl.
6. The decay constant of a radioactive nucleus is just that, *constant*. It does not depend on the size of the nuclear sample, the temperature, or any external fields (such as gravity, electricity, or magnetism). Define the decay constant and comment on how nuclear radioactivity would change if the quantity were dependent on temperature.
7. Explain how conservation of energy and momentum would be violated if a neutrino were not emitted in beta decay.
8. Radioactive ^{14}C is used to determine the age of some ancient artifacts.
   (a) Explain how this is done.
   (b) Which types of artifacts can have their age determined in this way and which types cannot?
9. A certain radioactive isotope has a decay constant of 0.00334 s^{-1}. Find the half-life in seconds and days.
10. Describe, in broad terms, the health risks associated with the three major forms of radioactivity: alpha, beta, and gamma. Focus on the dangers due to inherent health risks and the ability of each to penetrate shielding material.
11. A friend says he found a radioactive rock and asks you to carry it to the physics lab in a thin backpack. Is this safe? Explain.
12. Knowing that the binding energy per nucleon for uranium-235 is about 7.6 MeV/nucleon and the binding energy per nucleon for typical fission fragments is about 8.5 MeV/nucleon, find an average energy release per uranium-235 fission reaction in MeV.
13. What fraction of a sample of ^{32}P will be left after 4 months? Its half-life is 14.3 days. Assume 30 days to a month.

14. The masses given below are atomic masses.

(a) Using a spreadsheet or programmable calculator, calculate the binding energy per nucleon for the following isotopes of the five least massive elements.

Hydrogen-1: 1.007825 u	Lithium-8: 8.022486 u	Boron-8: 8.024605 u
Hydrogen-2: 2.014102 u	Lithium-9: 9.026789 u	Boron-10: 10.012936 u
Hydrogen-3: 3.016049 u	Lithium-11: 11.043897 u	Boron-11: 11.009305 u
Helium-3: 3.016029 u	Beryllium-7: 7.016928 u	Boron-12: 12.014352 u
Helium-4: 4.002602 u	Beryllium-9: 9.012174 u	Boron-13: 13.017780 u
Helium-6: 6.018886 u	Beryllium-10: 10.013534	Boron-14: 14.025404 u
Helium-8: 8.033922 u	Beryllium-11: 11.021657 u	Boron-15: 15.031100 u
Lithium-6: 6.015121 u	Beryllium-12: 12.026921 u	
Lithium-7: 7.016003 u	Beryllium-14: 14.024866 u	

(b) Now calculate the binding energy per nucleon for the following isotopes of the five most massive naturally occurring elements.

Radium-221: 221.01391 u	Thorium-228: 228.028716 u	Uranium-232: 232.037131 u
Radium-223: 223.018499 u	Thorium-229: 229.031757 u	Uranium-233: 233.039630 u
Radium-224: 224.020187 u	Thorium-230: 230.033127 u	Uranium-234: 234.040946 u
Radium-226: 226.025402 u	Thorium-231: 231.036299 u	Uranium-235: 235.043924 u
Radium-228: 228.031064 u	Thorium-232: 232.038051 u	Uranium-236: 236.045562 u
Actinium-227: 227.027749 u	Protactinium-231: 231.035880 u	Uranium-238: 238.050784 u
Actinium-228: 228.031015 u	Protactinium-234: 234.043300 u	Uranium-239: 239.054290 u
Thorium-227: 227.027701 u	Uranium-231: 231.036264 u	

(c) Compare your results and comment on any patterns or trends that are obvious.

15. The *fissionability parameter* is defined as the atomic number squared divided by the mass number for any given nucleus (Z^2/A). It can be shown that when this parameter is less than 44, a nucleus will be stable against small deformation; essentially, the nucleus will be stable against spontaneous fission. Calculate the value of this parameter for (a) ^{235}U, (b) ^{238}U, (c) ^{239}Pu, (d) ^{240}Pu, (e) ^{246}Cf, and (f) ^{254}Cf.

16. The stable isotope of sodium is ^{23}Na. What kind of radioactivity would be expected from (a) ^{22}Na and (b) ^{24}Na?

17. Estimate the number of nuclei that are present in a 50-kg human body.

18. At any given instant a sample of radioactive uranium contains many, many different isotopes of atoms that are *not* uranium. Explain why.

19. Iodine-125 is used to treat, among other things, brain tumors and prostate cancer. It decays by gamma decay with a half-life of 59.4 days. Patients who fly soon after receiving ^{125}I implants are given medical statements from the hospital verifying such treatment because their radiation could set off radiation detectors at airports. If the initial decay rate was 525 μCi, (a) what would the rate be at the end of the first year, and (b) how many months after the treatment would the decay rate be reduced by 90%?

20. A friend suggests that the world's energy problems could be solved if only physicists were to pursue the fusion of *heavy* nuclei rather than the fusion of *light* nuclei. To prove his point he suggests that the following fusion reaction should be considered:

$$^{157}\text{Nd} + {}^{80}\text{Ge} \rightarrow {}^{235}\text{U} + 2\text{n}$$

Using the insights that you have acquired in this chapter, show that his argument is flawed. The atomic mass of neodymium-157 is 156.939032 u, and the mass of germanium-80 is 79.925373 u.

21. In an attempt to determine the age of the cave paintings in Chauvet-Pont-d'Arc Cave in France, scientists used carbon-14 dating to measure the age of bones of bears found in the cave. The bears are depicted in the paintings, so presumably the bones are approximately the same age as the paintings. The results showed that the level of ^{14}C was reduced to 2.35% of its present-day level. How old were the bones (and presumably the paintings)?

22. In March 2011, a giant tsunami struck the Fukushima nuclear reactor in Japan, resulting in very large radiation leaks, including cesium-137. The isotope has a 30-y half-life and is a β^- emitter.

(a) What daughter nucleus is left after cesium-137 decays?

(b) How long after the release will it take for the decay rate of the cesium-137 to be reduced by 99%?

23. In one common type of household smoke detector, the radioactive isotope americium-241 decays by alpha emission. The alpha particles produce a small electric current because they are charged. If smoke enters the detector, it blocks the alpha particles, which reduces the current and causes the alarm to go off. The half-life of

^{241}Am is 433 y, and its atomic weight is 241 g/mol. Typical decay rates in smoke detectors are 690 Bq.

(a) Write the alpha decay reaction of ^{241}Am and identify the daughter nucleus.

(b) By how much does the alpha particle current decrease in 1.0 y due to the decay of the americium? How much in 50 y?

(c) How many grams of ^{241}Am are there in a typical smoke detector?

24. In February 2010, it was announced that water containing the carcinogen tritium (^{3}H) was leaking from aging pipes at 27 U.S. nuclear reactors. In one well in Vermont, contaminated water registered 70,500 pCi/L; the federal safety limit was 20,000 pCi/L. Tritium is a β^- emitter with a half-life of 12.3 y.

(a) How many protons and neutrons does the tritium nucleus contain?

(b) Write out the decay equation for tritium and identify the daughter nucleus.

(c) If the leak at the Vermont site is stopped, how long will it take for the water in the contaminated well to reach the federal safety level?

25. You take a course in archaeology that includes field work. An ancient wooden totem pole is excavated from your archaeological dig. The beta decay rate is measured at 150 decays/min. If the totem pole contains 225 g of carbon and the ratio of carbon-14 to carbon-12 in living trees is 1.3×10^{-12}, what is the age of the pole?

26. Determine the decay rate for 500 g of carbon from a tree limb 12 centuries after it is cut off.

27. Consider the ^{238}U nucleus.

(a) Find the approximate radius of the ^{238}U nucleus.

(b) What electric force do two protons on opposite ends of this nucleus exert on each other?

(c) If the electric force in part (b) were the only force exerted on the protons, what would be their acceleration just as they left the nucleus?

(d) Why do the protons in part (b) not accelerate apart?

28. The ages of rocks that contain fossils can be determined using the isotope ^{87}Rb. This isotope of rubidium undergoes beta decay with a half-life of 4.75×10^{10} y. Ancient samples contain a ratio of ^{87}Sr to ^{87}Rb of 0.0225. Given that ^{87}Sr is a stable product of the beta decay of ^{87}Rb, and there was originally no ^{87}Sr present in the rocks, calculate the age of the rock sample. Assume that the decay rate is constant over the relatively short lifetime of the rock compared to the half-life of ^{87}Rb.

29. A chest x ray imparts an equivalent radiation dose of about 1×10^{-3} Sv to a person. Approximately how many radon atoms (which emit alpha particles that have an energy of approximately 5 MeV) do you need to inhale before you expose yourself to a radiation dose equal to that of one chest x ray? Assume the body absorbs x-ray radiation and alpha radiation with the same efficiency, and that all inhaled radon atoms will decay and their emitted radiation will be absorbed by the body.

30. In 2010, physicists first created element number 117 (tennessine, Ts) by colliding ^{48}Ca and ^{249}Bk nuclei. The result was two isotopes of the new element, one of which had a half-life of 14 ms and contained 176 neutrons.

(a) What is the radius of the nucleus of the new element 117?

(b) What percent of the newly created isotope was left 1.0 s after its creation?

31. Three isotopes of aluminum are given in the following table:

Isotope	Atomic mass (u)	E_B/nucleon	Decay process
^{26}Al	25.986892		
^{27}Al	26.981538		Stable
^{28}Al	27.981910		

Calculate the binding energy per nucleon for each isotope and make a prediction of the decay processes for the unstable isotopes aluminum-26 and aluminum-28.

32. Ruthenium-106 is used to treat melanoma in the eye. This isotope decays by β^- emission with a half-life of 373.59 days. One source of the isotope is reprocessed nuclear reactor fuel.

(a) How many protons and neutrons does the ^{106}Ru nucleus contain?

(b) Could we expect to find significant amounts of ^{106}Ru in ore mined from the ground? Why or why not?

(c) Write the decay equation for ^{106}Ru and identify the daughter nucleus.

(d) How many years after ^{106}Ru is implanted in the eye does it take for its decay rate to be reduced by 75%?

33. You are asked to prepare a sample of ruthenium-106 for a radiation treatment. Its half-life is 373.59 days, it is a beta emitter, its atomic weight is 106 g/mol, and its density at room temperature is 12.45 g/cm^3.

(a) How many grams will you need to prepare of a sample having an activity rate of 125 μCi?

(b) If the sample in part (a) is a spherical droplet, what will be its radius?

34. Electron capture by an isolated proton is *not* allowed in nature. Explain why using physics principles. (Specifically, describe why the following nuclear reaction does *not* occur, and for good reason!)

$$e^- + p \nrightarrow n + \nu_e$$

Prep for the AP® Exam

AP® Group Work

Directions: The following problem is designed to be done as group work in class.

Taking into account the recoil (kinetic energy) of the daughter nucleus, calculate the kinetic energy of the alpha particle in the following decay of a ^{235}U nucleus at rest:

$$^{235}\text{U} \rightarrow \alpha + {}^{231}\text{Th}$$

AP® PRACTICE PROBLEMS

Prep for the AP® Exam

Multiple-Choice Questions

Directions: The following questions have a single correct answer.

1. The mass of a nucleus is ____________ the sum of the masses of its nucleons.
   (A) always less than
   (B) sometimes less than
   (C) always more than
   (D) sometimes equal to

2. Assuming that the final particles in each case are more stable than the initial particles, which of the statements regarding fission and fusion reactions is true?
   (A) Fusion absorbs energy and fission releases energy.
   (B) Fusion releases energy and fission absorbs energy.
   (C) Both fusion and fission absorb energy.
   (D) Both fusion and fission release energy.

3. In fission processes, which of the following statements is true?
   (A) Only the total number of protons remains the same.
   (B) Only the total number of neutrons remains the same.
   (C) The total number of protons and the total number of nuclei both remain the same.
   (D) The total number of protons and the total number of neutrons both remain the same.

4. In a spontaneous fission reaction the total mass of the products is
   (A) greater than the mass of the original element.
   (B) less than the mass of the original element.
   (C) the same as the mass of the original element.
   (D) one-half the mass of the original element.

5. What is the source of the Sun's energy?
   (A) Chemical reactions
   (B) Fission reactions
   (C) Fusion reactions
   (D) Gravitational collapse

6. In a fusion reaction the total mass of the products is ____________ the mass of the original elements.
   (A) greater than
   (B) less than
   (C) the same as
   (D) one-half

7. The decay constant λ depends only on
   (A) the number of atoms at the initial time.
   (B) the initial decay rate.
   (C) the half-life.
   (D) whether the decay is alpha, beta, or gamma.

8. The number of polonium-210 atoms in a radioactive sample of ^{210}Po
   (A) decreases linearly with time.
   (B) increases linearly with time.
   (C) decreases exponentially with time.
   (D) remains constant.

9. The decay rate for a sample of any isotope
   (A) decreases linearly with time.
   (B) increases linearly with time.
   (C) decreases exponentially with time.
   (D) increases exponentially with time.

(Continued)

Free-Response Question

1. A radioactive sample is monitored with a radiation detector that registers 5640 counts per minute. Twelve hours later, the detector reads 1410 counts per minute.

(a) Sketch a graph of the number of radioactive particles in the sample as a function of time.

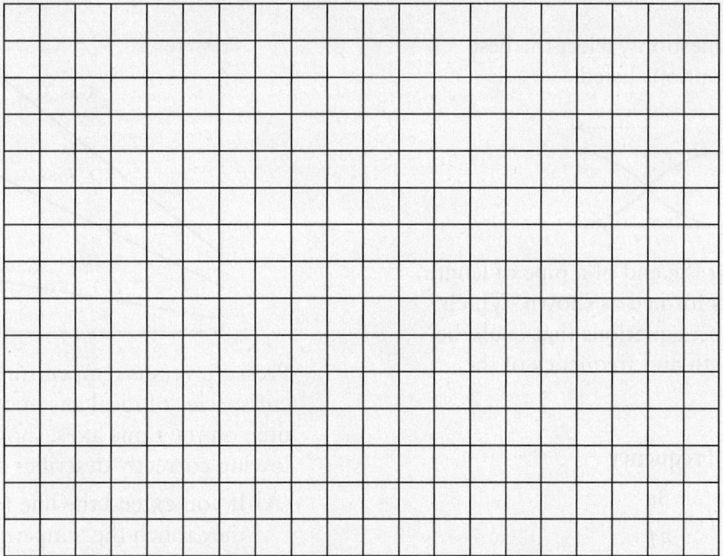

(b) Calculate the decay constant and the half-life of the sample.

(c) Relate the quantities you calculated to the graph you sketched in (a).

AP® Physics 2 Practice Exam

Prep for the AP® Exam

Multiple-Choice Questions

Directions: In the following questions, select the best answer choice from the four options listed.

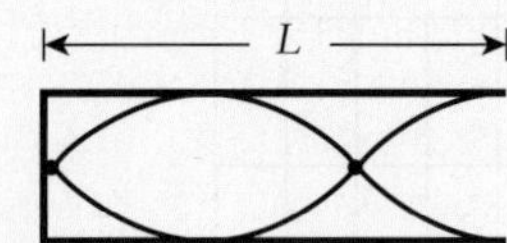

1. A tuning fork is held near the end of a pipe of length, L, and a standing wave is formed as shown. Which of the following are correct equations that could be used to find the wavelength and frequency of the tuning fork?

	Wavelength	Frequency
(A)	$\lambda = \frac{1}{2}L$	$f = \frac{3v}{4L}$
(B)	$\lambda = \frac{3}{2}L$	$f = \frac{3v}{4L}$
(C)	$\lambda = L$	$f = \frac{v}{2L}$
(D)	$\lambda = \frac{4}{3}L$	$f = \frac{3v}{4L}$

2. Eyeglasses for nearsighted people use a diverging lens to form a virtual image of a distant object close to the focal point of the lens. Next, the lens of the eye places an image of the diverging lens image on the back of the retina. What physics principle explains why the image from the diverging lens will be very close to the focal point?
   (A) The images for all diverging lenses are located on the focal point.
   (B) Diverging lenses have negative focal points.
   (C) The object distance is much greater than the focal point, so the fraction of $\frac{1}{s_o}$ approaches 0.
   (D) The object distance is much greater than the focal point, so the fraction of $\frac{1}{s_o}$ approaches infinity.

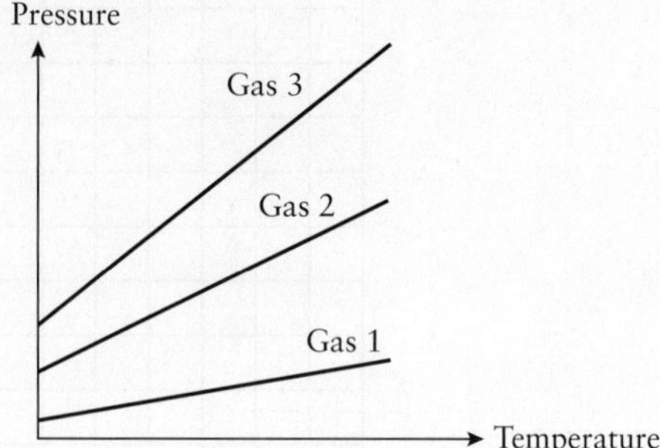

3. Pressure versus temperature is graphed for three different species of ideal gas in containers of constant volume on the same axes, shown above. Which of the following correctly describes a property of these graphs?
   (A) If you extend the line for each species of gas until they touch the temperature axis, they must meet at the same place.
   (B) Because the gases are different species, the points at which their temperatures approach zero are different.
   (C) The specific heat of all three gases depends on their temperature.
   (D) Gas 3 has the highest average kinetic energy at any given temperature.

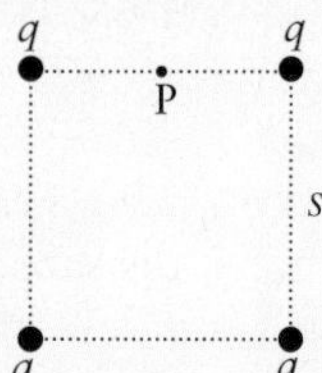

4. Four positive point charges are arranged in a square of side length s as shown above. Which of the following would not change the potential at point P?
   (A) Move one of the point charges adjacent to P to the center of the square
   (B) Move one of the point charges from the bottom corner of the square to the center of the square
   (C) Exchange one of the point charges adjacent to P for a negative point charge
   (D) Remove any of the point charges

5. Two wires are separated by a distance d, and carry currents in the same direction. Wire A has a current I_A and wire B has a current I_B. What is the magnitude of the net magnetic field at distances of $d/4$ from wire A and $3d/4$ from wire B?

(A) $\frac{4\mu_0(3I_A - I_B)}{2\pi d}$

(B) $\frac{2\mu_0(3I_A - I_B)}{3\pi d}$

(C) $\frac{\mu_0(I_A I_B)}{3\pi d}$

(D) $\frac{\mu_0(I_A - I_B)}{2\pi d}$

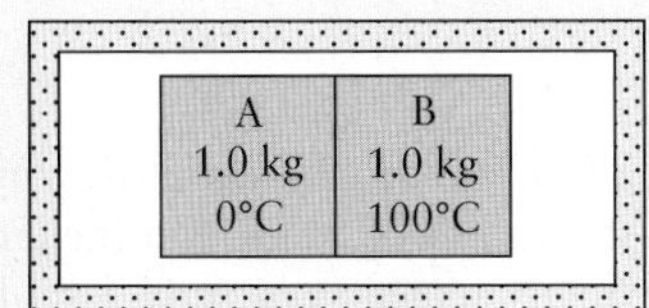

6. Isolated identical solid objects A and B have different initial temperatures as shown. The objects are placed in thermal contact in an insulated chamber and allowed to come to equilibrium. Which statement is correct about the amount of energy that each object transfers, Q_A and Q_B, and the rate that the energy was transferred to or from each object, R_A and R_B?

(A) $Q_A > Q_B$ and $R_A > R_B$, because A is at a higher temperature than B.

(B) $Q_A < Q_B$ and $R_A < R_B$, because B receives more energy.

(C) $Q_A = Q_B$ and $R_A > R_B$, because although the rate of energy transferred is different, each object ends up receiving the same amount of energy.

(D) $Q_A = Q_B$ and $R_A = R_B$, because both objects transfer energy at the same rate and receive the same amount of energy.

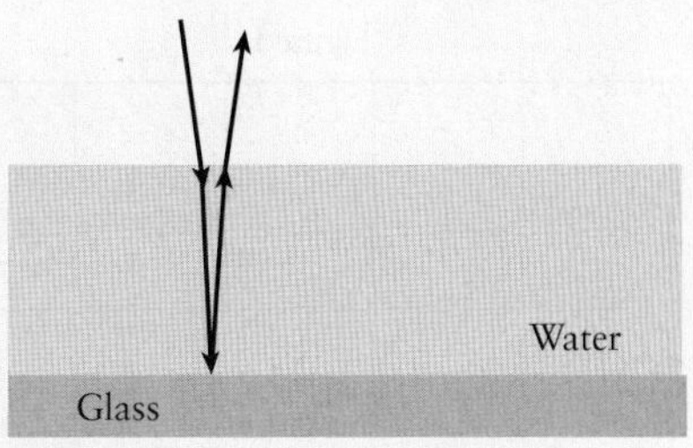

7. Light undergoes thin-film interference due to a layer of water on glass. Light is first in air, then refracts into water ($n = 1.33$), reflects off a piece of glass ($n = 1.62$), and returns to air as shown for a ray of light in the figure above. What is the minimum thickness of the water, in terms of the wavelength of the light in water, that will cause constructive interference?

(A) $\frac{\lambda}{4}$

(B) $\frac{\lambda}{2}$

(C) λ

(D) 4λ

8. Which statement best explains why electron orbitals may be modeled as having an integer number of electron wavelengths in each electron's circular orbit?

(A) Only one electron may occupy each possible electron state.

(B) Incident photons are absorbed to move an electron to a higher energy state.

(C) Electron angular momentum is quantized.

(D) Electron velocities are lower than photon velocities.

9. Rank the categories of electromagnetic waves, according to their wavelengths, from largest to smallest.

G – gamma rays

M – microwaves

R – radio waves

V – visible light

(A) R > M > V > G

(B) V > G > M > R

(C) M > R > G > V

(D) G > V > M > R

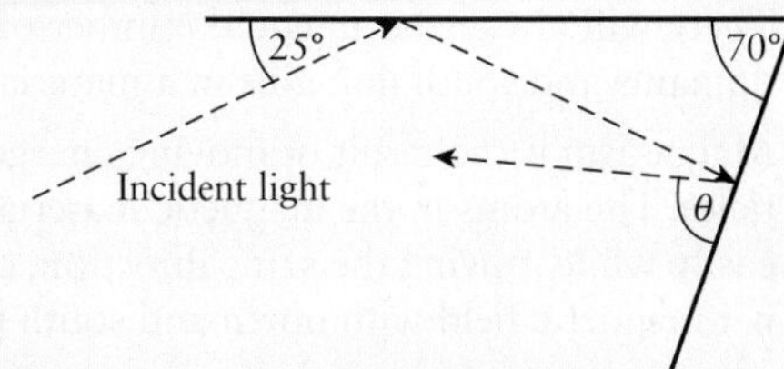

10. Light is reflected from two surfaces as shown. What is the angle θ?

(A) 5°

(B) 25°

(C) 65°

(D) 85°

Figure I

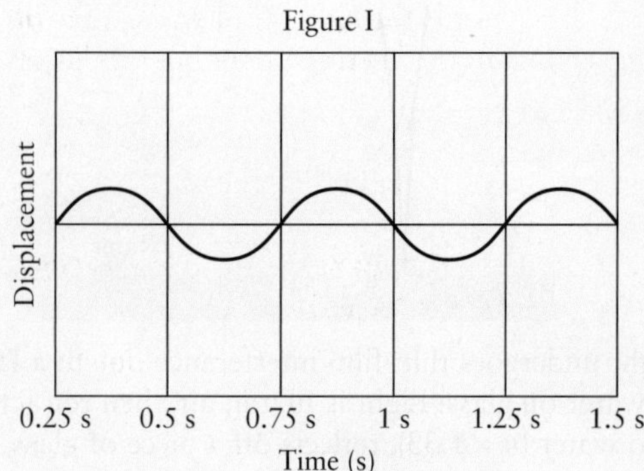

Figure II

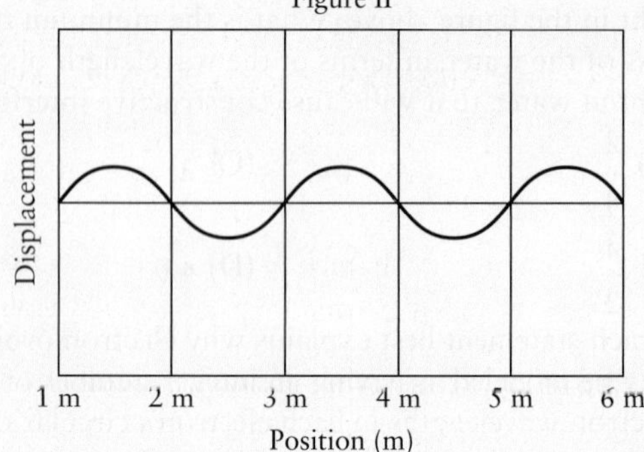

11. Figure I represents the displacement of one point on a wave as a function of time. Figure II represents the displacement of the wave as a function of position at one instant in time. The speed of the wave in the figures shown is

(A) 6.0 m/s.

(B) 4.0 m/s.

(C) 3.0 m/s.

(D) 1.33 m/s.

12. Which is the best explanation for why all magnets have both a north and a south pole?

(A) Magnetism is the result of the alignment of atoms within a material. Because the electrons and protons have opposite polarity, they repel and separate within the material, creating two magnetic poles.

(B) Magnetism is the result of magnetic domains, and each domain has either a north or a south pole. There will always be an equal number of north domains and south domains in a material.

(C) Magnetism is the result of moving charged particles. The atoms in the magnetic material collide easily while moving the same direction, creating a net magnetic field with north and south poles.

(D) Magnetic dipoles result from electric charge moving in a circle. Such motion results in both a north and a south pole, which cannot be separated.

Questions 13–14 refer to the following material.

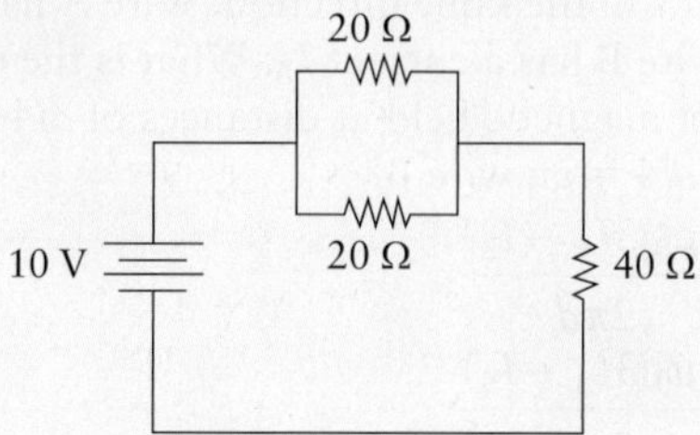

Three resistors are connected to a battery in the circuit shown.

13. The current through each 20-Ω resistor is

(A) 0.05 A.

(B) 0.10 A.

(C) 0.20 A.

(D) 0.50 A.

14. One of the 20-Ω resistors is removed, while the other 20-Ω resistor remains connected in the circuit. Which of the following statements correctly describes the changes to the potential difference across and current through the remaining resistor, if any?

	Potential difference	**Current**
(A)	Increases	Increases
(B)	Remains the same	Increases
(C)	Remains the same	Remains the same
(D)	Decreases	Decreases

15. The half-life of uranium-231 is 4.1 days. A scientist has a sample of mass 3.2×10^{-6} kg. After 20.5 days, approximately what percentage of the sample is still uranium-231?

(A) 0%

(B) 3%

(C) 5%

(D) 20%

16. A converging lens, with a focal length f, is placed at a distance of s_o from an object and a magnified image is observed on the same side of the lens as the object. The lens is replaced with a diverging lens of the same focal length. How will the new values for height of the image and distance from the lens to the image compare?

	h_i	s_i
(A)	Increased	Increased
(B)	Increased	Decreased
(C)	Decreased	Decreased
(D)	Decreased	Increased

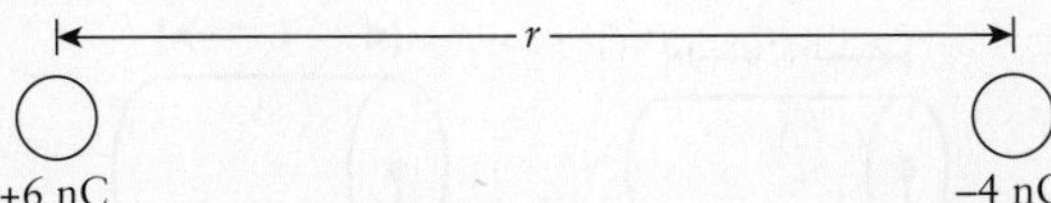

17. Two charged spheres of equal size carrying charges of +6 nC and −4 nC, respectively, are separated by a distance r, as shown. The separation of the spheres is much larger than the size of the spheres. The electric force exerted by the sphere on the left on the sphere on the right has a magnitude F_o. The spheres are brought in contact with one another for a time sufficient to allow them to reach equilibrium. They are then separated to the same distance r. In terms of the magnitude of the original force, F_o, what is the new electric force exerted by the sphere on the left on the sphere on the right?

(A) $\frac{F_o}{24}$

(B) $\frac{F_o}{12}$

(C) $\frac{F_o}{6}$

(D) F_o

Questions 18–20 use the following material.

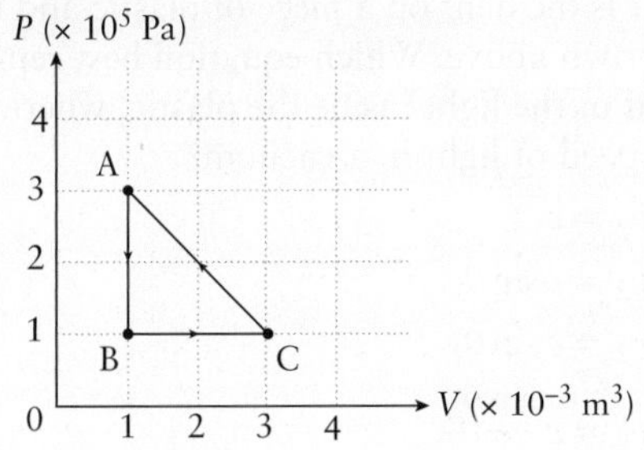

An ideal gas in a cylinder with a movable piston undergoes processes that take the gas from state A to state B to state C as shown above on the graph of pressure P as a function of volume V.

18. In state A the temperature is 300 K. What are the temperatures in states B and C, respectively?

(A) $T_B = 100$ K; $T_C = 300$ K

(B) $T_B = 300$ K; $T_C = 100$ K

(C) $T_B = 300$ K; $T_C = 600$ K

(D) $T_B = 400$ K; $T_C = 600$ K

19. The net work done in the cycle ABCA is

(A) 200 J by the gas.

(B) 200 J on the gas.

(C) 400 J by the gas.

(D) 400 J on the gas.

20. In which process is the amount of work done on or by the gas equal to the heating or cooling of the gas?

(A) AB

(B) BC

(C) CA

(D) The internal energy of the gas remains constant throughout the cycle.

21. The isolines of electric potential due to a negatively charged point charge

(A) are equally spaced circles concentric with the point charge.

(B) are concentric circles, spaced close together near the point charge, with every subsequent isoline farther away.

(C) are equally spaced, but not concentric; nor are they circles.

(D) are noncircular shapes, spaced close together near the point charge, with every subsequent isoline farther away.

Questions 22–24 refer to the following material.

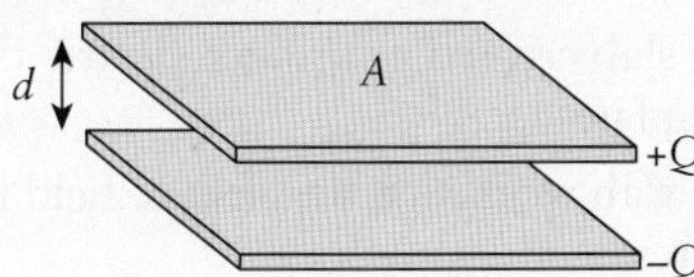

A parallel-plate capacitor consists of two conducting plates of area A separated by a distance d, as shown in the figure.

22. Which of the following statements best describes the electric field E between the plates when a net charge $+Q$ is placed on the top plate and an opposite net charge of equal magnitude $-Q$ is placed on the bottom plate and d is much smaller than the length of the side of one of the plates?

(A) There is no field between the plates because the charges are equal and opposite.

(B) The field points upward and vectors drawn to represent the field would be equally spaced.

(C) The field points downward and vectors drawn to represent the field would be equally spaced.

(D) The field points downward, but because the plates are conducting, the magnitudes and direction of the vectors used to represent the field would vary.

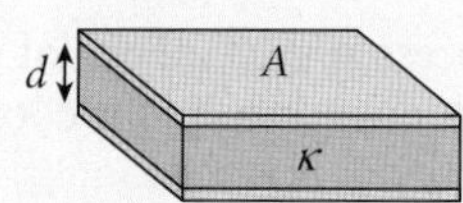

23. A neutral dielectric slab, also of area A and of dielectric constant κ, is inserted between the plates. The net charges on the plates remain the same. Which of the following expressions could represent the magnitude of the net charge Q on the top plate?

(A) $\kappa\varepsilon_0 AE$

(B) $\dfrac{E}{\kappa\varepsilon_0 A}$

(C) $\dfrac{\kappa\varepsilon_0 AE}{d}$

(D) $\dfrac{\kappa\varepsilon_0 d}{AE}$

24. When the dielectric slab was placed between the plates, the capacitance of the device increased. This is because

(A) the slab increased the area of the plates, so that some of the charge Q is transferred to the dielectric.

(B) the slab created a stronger electric field between the plates.

(C) the slab weakened the electric field between the plates.

(D) the potential difference of the slab is larger than that of the plates.

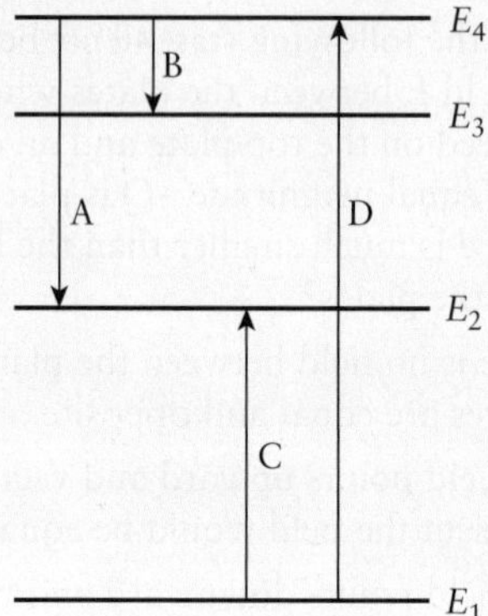

25. The diagram shows energy levels for a hypothetical atom. Four electron transitions are shown on the diagram and include the emission of red light and the emission of blue light. Which transition corresponds to the emission of a photon of blue light?

(A) A

(B) B

(C) C

(D) D

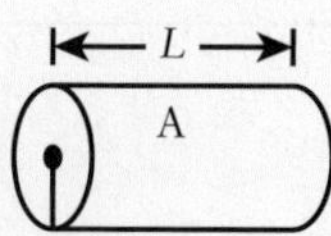

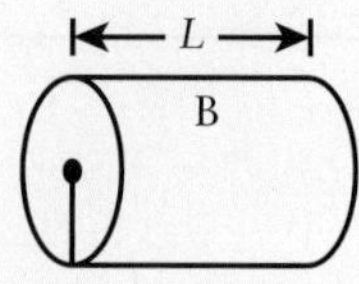

26. Two resistors, A and B, are made from the same material; have equal lengths; and have different radii, where $r_B = 2r_A$ as shown in the figure above. Which of the following equations correctly compares the resistance of B to the resistance of A?

(A) $R_B = 4R_A$

(B) $R_B = 2R_A$

(C) $R_B = \dfrac{1}{2}R_A$

(D) $R_B = \dfrac{1}{4}R_A$

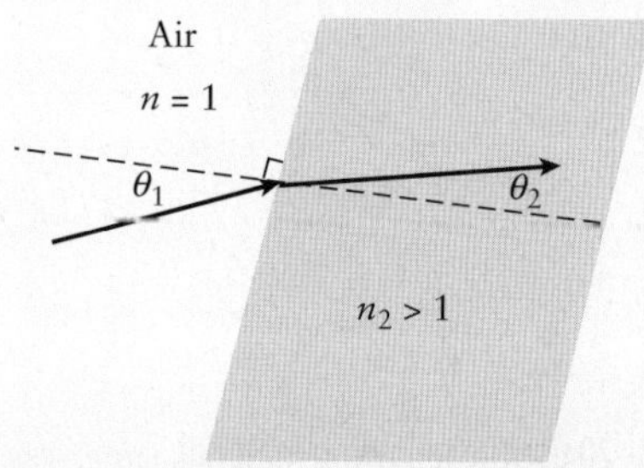

27. Light is incident on a piece of plastic and is refracted, as shown above. Which equation best represents the speed of the light inside the plastic, where c represents the speed of light in a vacuum?

(A) $v_2 = c$

(B) $v_2 = c\sin\theta_1$

(C) $v_2 = c\sin\theta_2$

(D) $v_2 = c\dfrac{\sin\theta_2}{\sin\theta_1}$

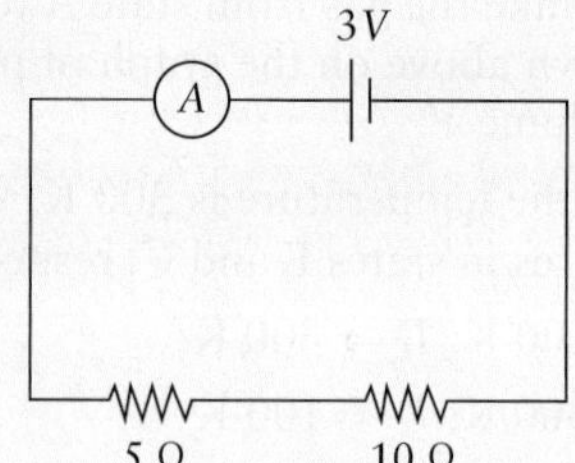

28. An ammeter placed in the circuit shown measures a current of 0.15 A. The resistance of the ammeter, battery, and wires in the circuit is most nearly

(A) zero.

(B) 2 Ω.

(C) 5 Ω.

(D) 15 Ω.

29. The electric field of Earth in a particular region is approximately 120 V/m. A spider of mass 0.1 mg emits electrically charged threads that cause the spider to rise upward. Assume the spider's threads have a total charge of magnitude 100 nC. What is the net work done on the spider after it has ascended 2 m?

(A) 22 μJ

(B) 26 μJ

(C) 2.0 mJ

(D) 22 mJ

r

$+2Q$ $-4Q$

30. Two point charges $+2Q$ and $-4Q$ are placed on a horizontal axis, with $+2Q$ placed at $x = 0$ and $-4Q$ placed at $x = r$, where Q is a positive number. There exists a distance x that a third point charge $+Q$ could be placed on the horizontal axis so that there would be no net force exerted on it by the other point charges. An equation which could be solved for x, where x is the position on the axis, is

(A) $8x^2 = x^2 + r^2$.

(B) $4x^2 = (x + r)^2$.

(C) $\frac{1}{x^2 + r^2} = \frac{2}{x^2}$.

(D) $\frac{2}{(x + r)^2} = \frac{1}{x^2}$.

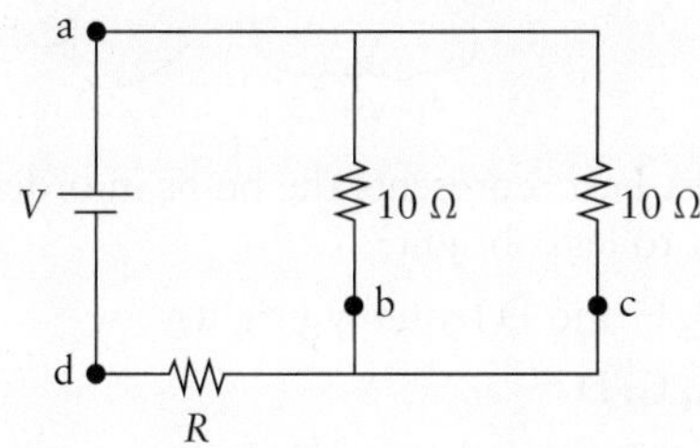

31. The circuit shown has two 10-Ω resistors, an unknown resistor R, and a battery with unknown potential difference V. Which of the following placements of meters will allow a student to make the necessary measurements to determine the value of the unknown resistance R?

(A) Connect a voltmeter between points a and d, and an ammeter at point c

(B) Connect an ammeter between points a and d, and a voltmeter at point b

(C) Connect an ammeter between points a and b, and a voltmeter at point c

(D) Connect a voltmeter between points b and c, and an ammeter at point c

32. A sound wave resonates at the second harmonic with a frequency f in an open pipe of length L.

The pipe is now closed at one end. In terms of the frequency, f, of the sound wave used to produce the resonance pattern in the open pipe, the lowest frequency that would produce resonance in a closed pipe of length L is

(A) $\frac{f}{4}$.

(B) $\frac{f}{2}$.

(C) $2f$.

(D) $4f$.

33. An atom of actinium (Ac-225) undergoes alpha decay and becomes francium (Fr-221) spontaneously through the alpha decay described the by the following equation:

$$^{225}_{89}\text{Ac} \rightarrow {}^{221}_{87}\text{Fr} + {}^{4}_{2}\text{He}$$

The masses of the three isotopes involved in the decay are given below.

$$M_{\text{Ac-225}} = 225.023229 \text{ u}$$

$$M_{\alpha} = 4.001506 \text{ u}$$

$$M_{\text{Fr-221}} = 221.014255 \text{ u}$$

How do the momentum and kinetic energy of the actinium atom compare to the total momentum and kinetic energy of the decay products?

	Actinium momentum	Actinium kinetic energy
(A)	Different	Different
(B)	Different	The same
(C)	The same	Different
(D)	The same	The same

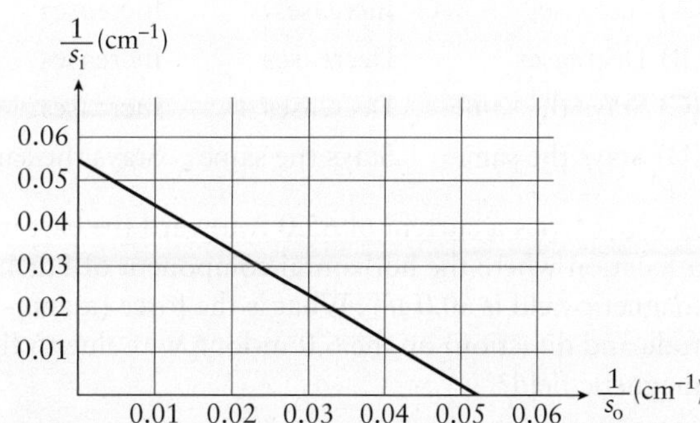

34. The graph above of $\frac{1}{s_i}$ as a function of $\frac{1}{s_o}$ represents a mirror with a focal length of most nearly

(A) 0.5 mm

(B) 50 mm

(C) 100 mm

(D) 200 mm

35. A sample of an ideal monatomic gas is confined in a container of fixed volume. Its temperature increases due to the spontaneous heating of the gas by the surroundings. Which of the following best explains why the pressure the gas exerts on the container increases as the temperature increases?

(A) The molecules are always moving faster when they collide with the container and always exert more force.

(B) The average speed of the molecules increases, causing most collisions to exert more force than before.

(C) The fastest molecules get faster and have more collisions.

(D) The speeds of the molecules are more nearly the same value, even though the average speed may slightly decrease.

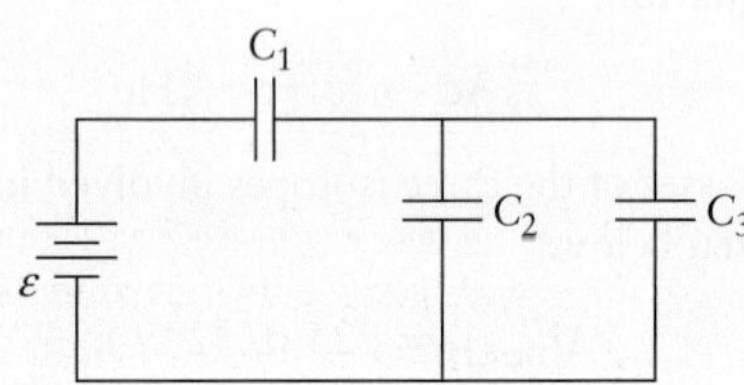

36. A circuit is constructed with three capacitors as shown above. Initially, the capacitances are $C_1 = 2C$, and $C_2 = C_3 = C$. The battery is disconnected and the spacing of the plates of C_1 and C_3 are adjusted. The capacitance of the capacitors is now $C_1 = 1.5C$, $C_2 = C$, and $C_3 = 2C$. The battery is connected and the circuit reaches steady state. How does the charge stored on each capacitor compare to the charge stored on each capacitor in the initial configuration?

	Charge on C_1	Charge on C_2	Charge on C_3
(A)	Increases	Increases	Increases
(B)	Decreases	Decreases	Increases
(C)	Stays the same	Decreases	Increases
(D)	Stays the same	Stays the same	Stays the same

37. A wire carries a current of 65.0 A toward the east in a location where the horizontal component of Earth's magnetic field is 60.0 μT. What is the force (magnitude and direction) on the 6.0-m-long wire due to this magnetic field?

(A) 2.3×10^{-2} N, upward

(B) 2.3×10^{-2} N, downward

(C) 2.2×10^{-6} N, upward

(D) 2.2×10^{-6} N, north

38. Which of the following statements is correct?

(A) Only longitudinal waves can be polarized.

(B) Polarization can occur when longitudinal waves pass through slits.

(C) Polarization blocks specific frequencies of waves.

(D) Polarization can occur when waves are reflected.

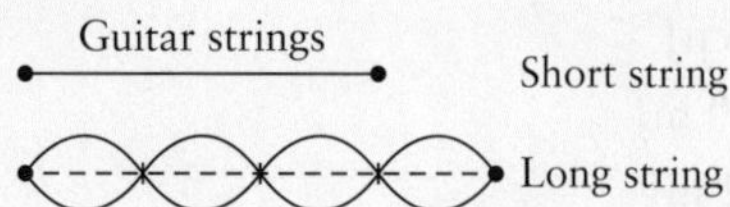

39. A long guitar string and a short guitar string are near one another. The two strings are made of different materials. The long guitar string is plucked, creating a standing wave in the long string with four antinodes. Soon afterward, the short string begins to vibrate with a sympathetic vibration. Which of the following statements must be true?

(A) The frequency of vibration of both strings is the same.

(B) The amplitude of both strings is the same.

(C) The short string must also have four antinodes.

(D) The speed of the waves in both strings is the same.

Questions 40–41 refer to the following material.

Four identical bulbs, A, B, C, and D, are connected to a battery in the circuit shown.

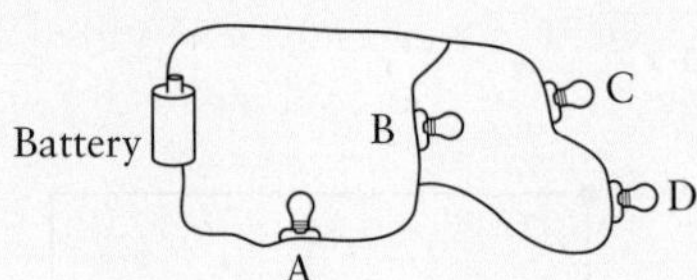

40. Which list best represents the bulbs in order from brightest to least bright?

(A) A, B, C and D (equally bright)

(B) B, A, C, D

(C) C and D (equally bright), B, A

(D) A and B (equally bright), C and D (equally bright)

41. Bulb C burns out, but bulbs A, B, and D remain connected in the circuit, which is otherwise unchanged. How does the brightness of bulb A change after bulb C burns out compared to its brightness before bulb C burns out?

(A) Bulb A is now brighter.

(B) Bulb A is now less bright.

(C) The brightness of bulb A is unchanged.

(D) Bulb A will now have no brightness.

42. A tuning fork of frequency 300 Hz is activated and sends a sound wave toward a classroom wall. The distance between the tuning fork and the wall is equal to one wavelength of the sound wave. The speed of sound at the temperature in the classroom is 343 m/s. The time it takes for one sound wave to return to the tuning fork is most nearly

(A) 0.001 s.
(B) 0.003 s.
(C) 0.007 s.
(D) 0.013 s.

43. A sealed container holds a mixture of two types of ideal gas, Y and Z. The molecules of gas Z have twice the mass of the gas Y molecules. The container and the gas molecules are at thermal equilibrium with the room. Which of the following best describes a relationship between the molecules of the two gases?

(A) There must be twice as many molecules of Y as there are of Z.
(B) On average, molecules of Y are moving twice as fast as molecules of Z.
(C) On average, molecules of Z have twice the kinetic energy as do molecules of Y.
(D) A molecule of Y may have the same kinetic energy as a molecule of Z.

44. Diagrams A–D each show the velocity vector of a charged object in a uniform magnetic field. Which object has the greatest acceleration?

(A) Electron

⊗ B = 4.0 mT into the page
v = 2.0 × 10^3 m/s

(B) Proton

⊗ B = 4.0 mT into the page
v = 3.0 × 10^3 m/s

(C) Electron

⊗ B = 4.0 mT into the page
v = 3.0 × 10^3 m/s

(D) Proton

⊙ B = 4.0 mT into the page
v = 2.0 × 10^3 m/s into the page

Questions 45–46 refer to the following material.

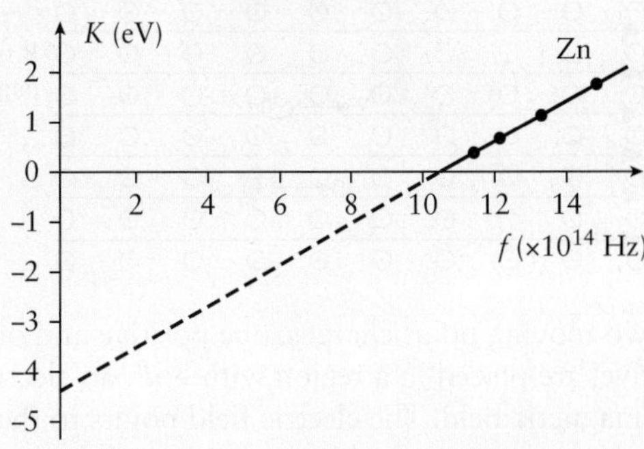

The graph above shows the maximum kinetic energy of electrons ejected from a sample of zinc metal when light of different frequencies interacts with the sample.

45. The work function of zinc is

(A) 0 eV.
(B) 4.3 eV.
(C) any amount of energy between 0 and −4.3 eV.
(D) any amount of energy greater than 4.3 eV.

46. Visible light (of wavelength approximately 400 to 700 nm) interacting with this sample would cause the zinc to

(A) possibly emit electrons, depending on the intensity of the light.
(B) definitely emit electrons, no matter the intensity of the light.
(C) emit electrons if the wavelength approaches the 700-nm wavelength and the intensity is great enough.
(D) never emit electrons, no matter the intensity.

47. A concave mirror produces an image of an object that is real, inverted, and magnified. If the object is moved toward the mirror at half the original distance to the mirror, describe the new image.

(A) Real, inverted, magnified
(B) Real, inverted, reduced
(C) Virtual, upright, magnified
(D) Virtual, upright, reduced

E field

B field out of page

48. Two moving point charges, one positive and one negative, are placed in a region with both an electric and a magnetic field. The electric field points to the left and the magnetic field is coming out of the page, as shown in the figure above. What will be the difference between the change in motion of the point charges, assuming they have the same mass?

(A) The negative point charge will curve upward, and the positive point charge will curve downward.

(B) The positive point charge will accelerate to the left, and the negative point charge will accelerate to the right.

(C) The positive point charge will increase speed and change direction, while the negative point charge will decrease speed and change direction.

(D) The relative magnitude of the fields is necessary to determine the change in motion of the point charges.

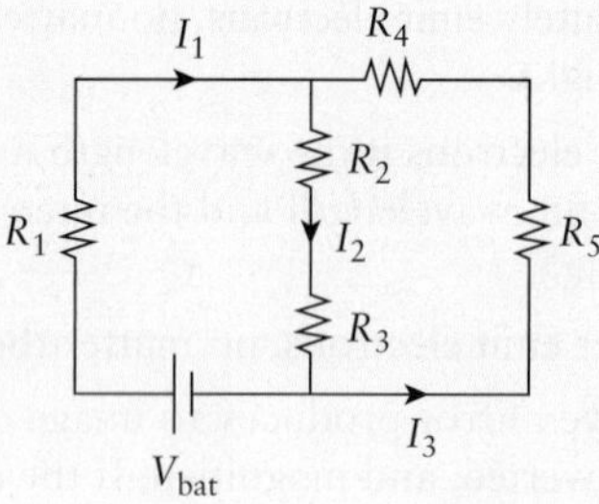

49. A multiloop circuit is shown in the figure. Which of the following equations represent the total electric potential difference for a complete loop?

(A) $V_{bat} + I_1R_1 + I_2(R_2 + R_3) = 0$

(B) $V_{bat} - I_1R_1 - I_3R_4 + I_3R_5 = 0$

(C) $I_3R_4 - I_3R_5 - I_2(R_2 + R_3) = 0$

(D) $V_{bat} - I_1R_1 - I_2(R_2 + R_3) = 0$

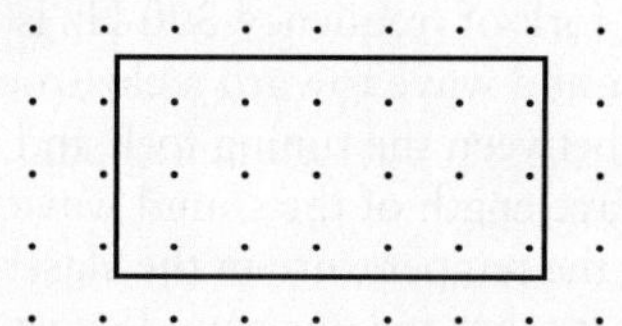

50. A coil is placed inside and perpendicular to a changing magnetic field, as shown. If the magnitude of the magnetic field is decreasing with time, identify the direction of the current in the coil and the reasoning behind your answer.

(A) Clockwise, because the induced magnetic field is into the page

(B) Clockwise, because the induced magnetic field is out of the page

(C) Counterclockwise, because the induced magnetic field is into the page

(D) Counterclockwise, because the induced magnetic field is out of the page

Free-Response Questions

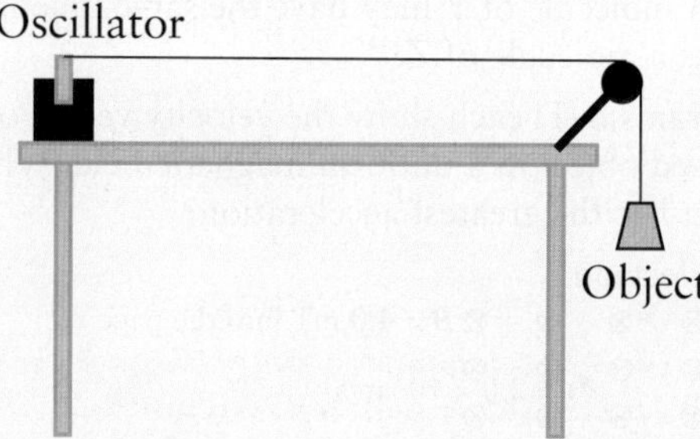

Question 1

Students are given a thin string attached to a single-frequency oscillator on one end. The string passes over a pulley and is attached to a hanging object on the other end, as shown. Students are asked to determine the velocity of the waves on the string.

(a) Design an experimental procedure that the group could use to find the speed of the waves on a string.

i. In the table, list the quantities that would be measured in the group's experiment. Define a symbol to represent each quantity and list the equipment that would be used to measure each quantity. You do not need to fill in every row. If you need additional rows, you may add them.

Quantity to be measured	Symbol for quantity	Equipment for measurement

ii. Describe the overall procedure that could be used to find the speed of the waves on the string. Provide enough detail so that other students could replicate the experiment, including any steps necessary to reduce experimental uncertainty.

(b) Describe how the group can use the quantities measured to determine the speed of the waves on the string.

(c) In another experiment, students determine the length of the string to be 2.4 m and they collect the following data.

Harmonic	Frequency (Hz)		
1st	24.0		
2nd	47.9		
3rd	72.5		
4th	99.6		

i. On the lines below, identify two quantities (from the table above or calculated values) that could be plotted to produce a linear graph that could be used to calculate the speed of the waves on the string. You may use the remaining columns in the table, as needed, to record any quantities (including units) that are not already in the table.

Horizontal axis ____________ Vertical axis ____________

ii. Plot the indicated quantities on the following axes. Clearly scale and label all axes, including units as appropriate.

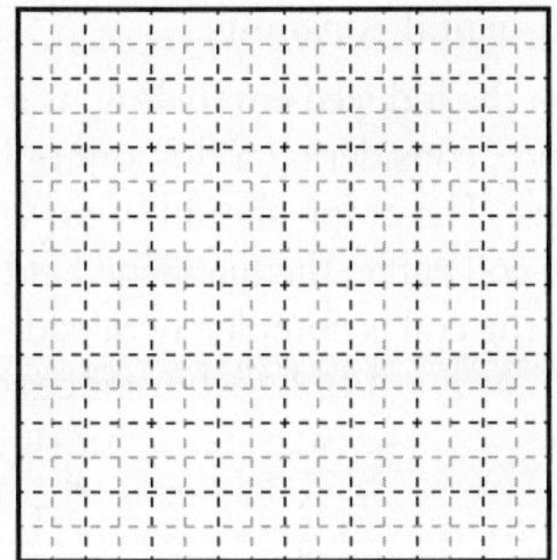

iii. Draw a line of best fit to the data graphed in part (c) ii.

(d) Use the graph in part (c) to calculate the speed of the waves on the string.

(e) Another lab group finds an equation for the speed of waves on a string, $v = \sqrt{\frac{F_T}{m/L}}$, where F_T is the tension in the string and m/L is the linear density of the string. Using this expression, identify two elements of your experimental set up that, if changed, would change the speed of waves on the string. Explain how the change in each element would affect the speed of the waves in the string.

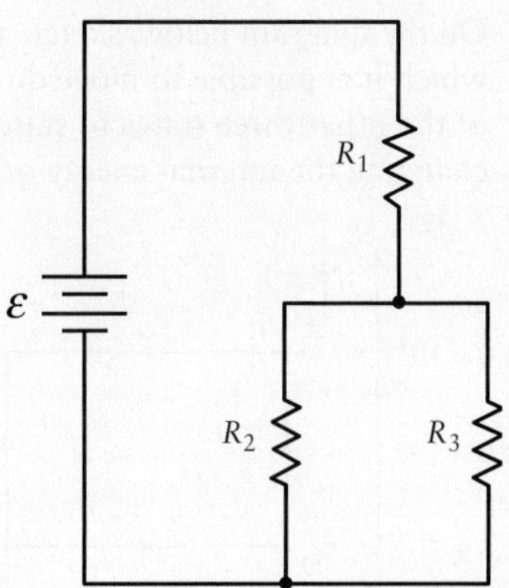

Question 2

Consider the circuit shown above, with three resistors, R_1, R_2, and R_3, connected to a battery of emf $\mathcal{E}$. Let $R_1 = R_3 = R$ and $R_2 = 2R$. Assume the battery is ideal and all three resistors are ohmic. Let the currents in resistors R_1, R_2, and R_3 be I_1, I_2, and I_3, respectively, and the potential differences across these resistors be ΔV_1, ΔV_2, and ΔV_3, respectively.

(a) Rank the currents in the resistors in order from greatest to least. If any of the currents are the same, indicate this explicitly.

Ranking: ________________________

Briefly explain your answer.

(b) Derive an equation for the current through resistor R_3 in terms of $\mathcal{E}$, R_1, R_2, R_3, and fundamental constants.

(c) Which resistor dissipates energy at the greatest rate? Support your answer using physics principles.

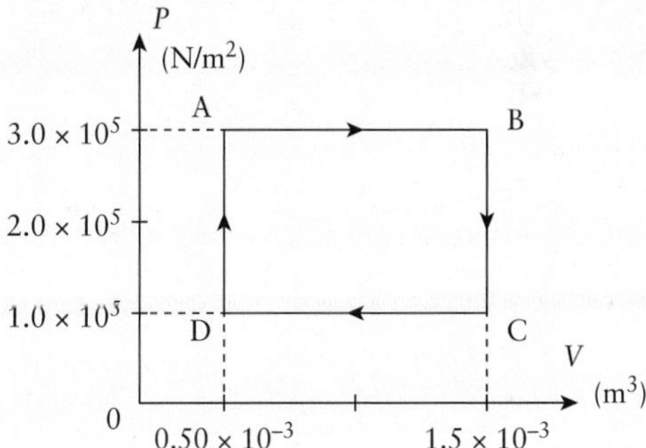

Question 3

The graph above shows the pressure P and volume V relationships for a 0.25-mol sample of an ideal gas confined within a cylinder with a movable piston of negligible friction. The gas is taken through the process A–B–C–D.

(a) i. Calculate the temperature of the gas in state A.

ii. Calculate the heat added or removed in process A–B and explain its sign.

iii. Calculate the net work done in the process A–B–C–D.

(b) i. On the diagram below, sketch a process in which it is possible to move directly from one of the other three states to state A without changing the internal energy of the gas.

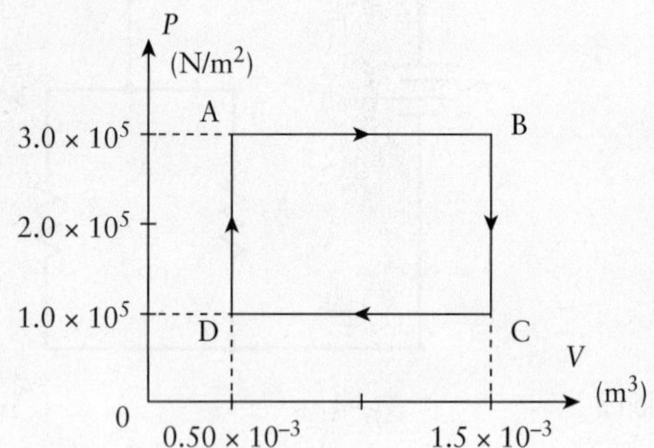

ii. Describe a plausible method for accomplishing this process.

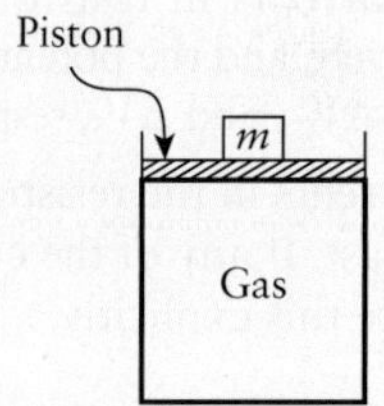

Note: Figure not drawn to scale.

(c) The cylinder is placed upright and a small object of mass m is placed on the piston, causing the piston to move downward. Using a microscopic model, explain why the piston moves down, and why it eventually stops. Does the internal energy of the gas change during this process?

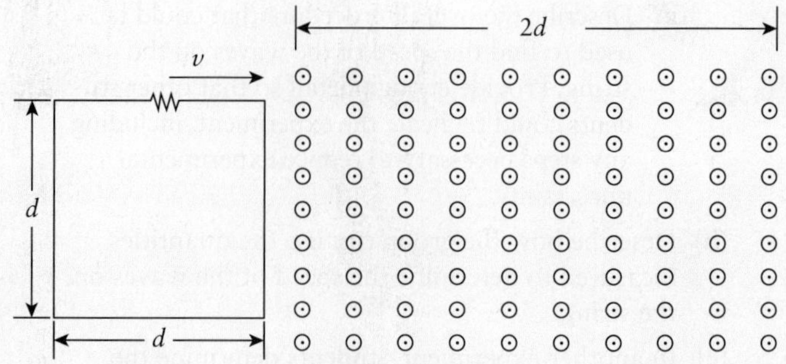

Question 4:

A square coil of wire with a resistance of R is moved at a constant velocity, v, through a region with a constant magnetic field of strength $\mathbf{B}$. As it passes through the field, a current is induced in the coil. The coil of wire has sides of length d.

(a) On the sketches below indicate the direction of the current in the coil when it is in the position shown. If no current is induced, state $I = 0$. Justify your answer for each section.

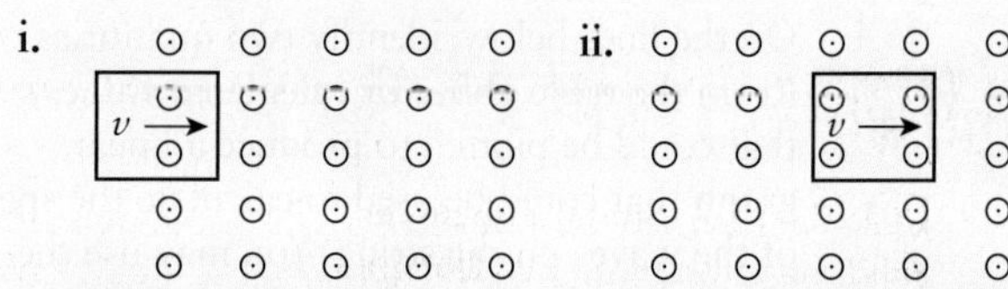

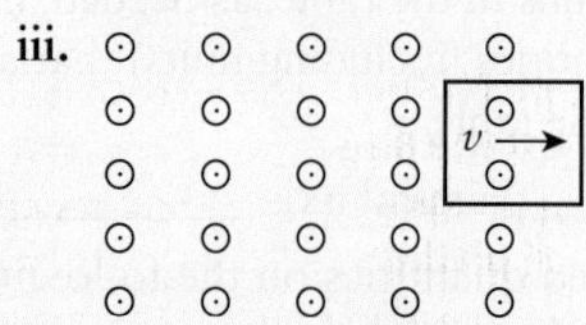

(b) Derive an equation for the magnitude of the maximum current in the loop, using R, B, d, and fundamental constants.

(c) Graph the induced emf below, as a function of the distance, considering a flux out of the page to be positive. The graph should begin when the right edge of the coil enters the magnetic field ($x = 0$) and end when the coil is completely outside the field (its right edge is at $x = 3d$). Label distances on your graph.

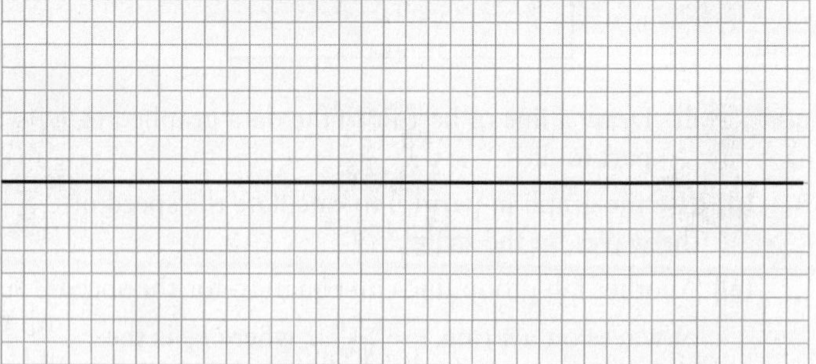

(d) Which features of the equation from part (b) support your answers for parts (a) and (c)? Support your answer using physics principles.

26 Relativity and an Introduction to Particle Physics

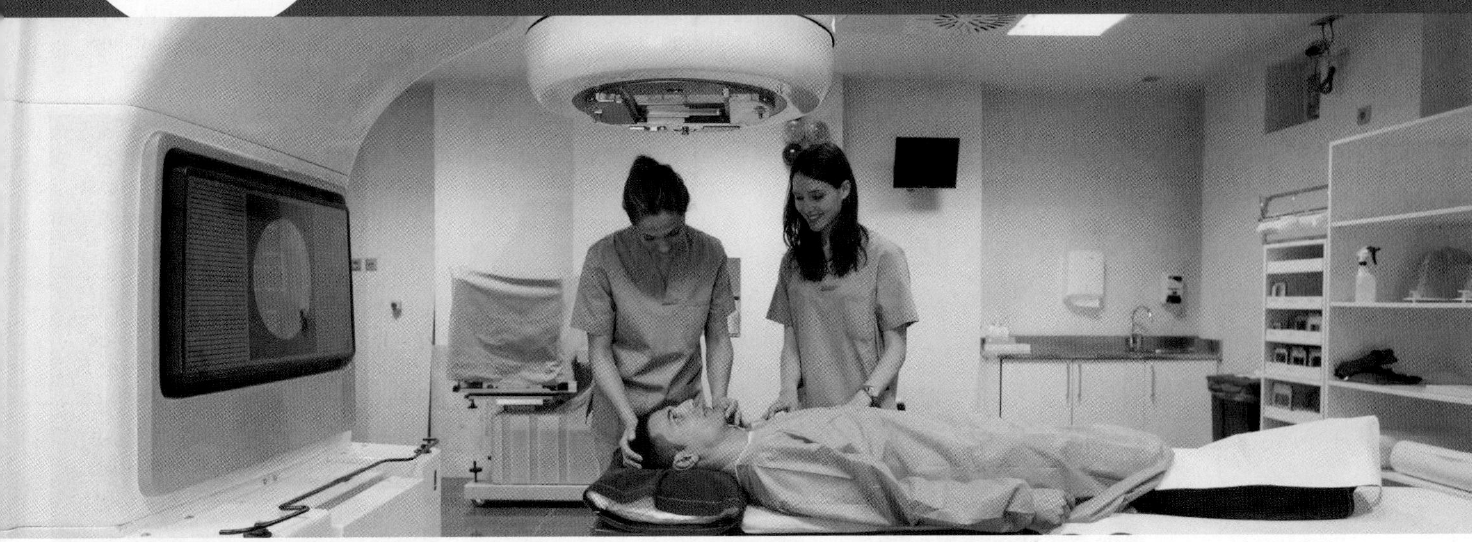

age fotostock/Alamy

YOU WILL LEARN TO:

- Describe how different observers view the same motion in Newtonian physics.
- Explain how the Michelson–Morley experiment helped rule out the ether model of the propagation of light and use the concept in calculations.
- Describe how the time interval between two events can have different values in different reference frames and calculate that the time interval for different reference frames.
- Calculate how the dimensions of an object change when it is in motion.
- Explain why the speed of light in a vacuum is an ultimate speed limit and use this concept to calculate relative velocities.
- Calculate the rest energy of an object with mass, and relativistic momentum and energy.
- Explain what the principle of equivalence tells us about the nature of gravity.
- Describe the difference between hadrons and leptons, and of what hadrons are composed, and use conservation of baryon and lepton numbers to predict allowable reactions.
- Explain how fundamental particles interact with each other, and the differences among the four fundamental forces.

26-1 The concepts of relativity and elementary particles may seem exotic, but they're part of everyday life

Most of our everyday experiences involve objects that move at speeds that are slow compared to the speed of light in a vacuum. In the late nineteenth century, scientists began to realize that the laws of physics developed up to that point (and covered in this book before this chapter) don't properly describe light or objects moving at speeds near the speed of light. Albert Einstein first understood the way to extend physics into this regime of extremely high speed.

In this final chapter we'll first focus on Einstein's *special theory of relativity*. We'll see how discoveries concerning the nature of light helped motivate the central ideas

(a) Relativity tells us that time flows at a different rate aboard an orbiting GPS satellite than on Earth. A GPS receiver must account for this to accurately determine its position.

iStockphoto/Thinkstock

(b) Sunlight is a result of reactions deep inside the Sun that convert mass into energy—a process predicted by the special theory of relativity.

Comstock/Thinkstock

(c) At the Large Hadron Collider near Geneva, Switzerland, protons are accelerated to 99.99999999% of the speed of light. The special theory of relativity is needed to explain how particles behave at such speeds.

James King-Holmes/Alamy

Figure 26-1 Relativity in your world (a) A GPS-equipped mobile phone and (b) the light from the Sun illustrate applications of the theory of relativity. (c) Subatomic particles are regularly accelerated to just below the speed of light. Their behavior cannot be understood without the theory of relativity.

of this theory. We'll also see how a simple postulate—that the speed of light does not depend on the motion of either the emitter or the observer of the light—leads to a radical transformation of our understanding of space and time. We'll discover that the speed of light is an ultimate speed limit, and it's impossible to accelerate an object beyond that speed. We'll also see that mass is simply another form of energy. We'll conclude with a look at Einstein's *general theory of relativity*, which provides new insights into the nature of gravity.

The effects of the special theory of relativity are present in the world around you. Your mobile phone probably has the ability to determine its location using the Global Positioning System, or GPS (**Figure 26-1a**). A GPS receiver detects signals from a collection of satellites that orbit Earth and calculates its position by timing those signals. However, the satellites move at about 28,000 km/h (18,000 mi/h) relative to Earth, and we shall see that special relativity tells us that a moving clock or timekeeper runs at a different rate than a stationary one. Your mobile phone has to be able to correct for this in order to give you accurate positioning information. Ordinary sunlight is also a consequence of relativity: The Sun shines by converting a fraction of its mass into electromagnetic energy, a direct application of the idea that objects have energy as a consequence of having mass (**Figure 26-1b**). And while even the fastest spacecraft travel at only a tiny fraction of the speed of light, subatomic particles can move much faster (**Figure 26-1c**).

"What is the world made of?" "Where did I come from?" You probably asked questions like these when you were a small child. Physicists and astronomers ask these questions throughout their professional lives, in search of ever more sophisticated answers about the nature and origin of the physical universe. We will take a brief look at our present understanding of **particle physics**, the branch of physics that concerns the fundamental constituents of matter and how they interact.

We will conclude with a look at the fundamental particles that make up all the ordinary matter that you see around you, including the matter that makes up your body. These include *leptons*, of which the electron is the best-known example, and *quarks*, which have the curious property that their charges are a fraction of the fundamental charge e. We'll see that these fundamental particles interact with each other by exchanging special fundamental particles, in a process that actually involves violating the law of conservation of energy (but in a way that's nonetheless compatible with the laws of physics). These interactions help explain the strong force that holds the nucleus together, the electric force that keeps electrons in the atom, and the weak force that causes radioactive beta decay. We will find out that such interactions involve antimatter and are mediated by exchange particles (**Figure 26-2**).

Potatoes are a good source of potassium. A small fraction of the potassium is radioactive ^{40}K, which can decay by emitting a positron—a bit of antimatter. This decay also involves a quark, a neutrino, and an exchange particle called a W^+.

Bombaert Patrick/Alamy

Figure 26-2 Particle physics Ordinary objects and common technology connect us to particle physics, the study of fundamental particles and their interactions. The description of the pictured fried potatoes includes many terms we will meet in learning about particle physics in this chapter!

THE TAKEAWAY for Section 26-1

✔ The physics we have learned so far must be modified for objects moving at speeds comparable to the speed of light.

✔ These modifications lead us to new ideas about space, time, and energy.

✔ Ordinary matter such as atoms is composed of a small variety of different fundamental particles.

AP Exam Tip

The topics in Chapter 26 are beyond the scope of the AP® Physics 2 course, although they help support understanding for several of the more recent topics. Thus, there are no AP® Building Blocks, Skill Builders, or Skills in Action, but some problems are provided after sections and at the end of the chapter to help you build and test your understanding of these concepts.

Section 26-1 Takeaway Problems

Note: When working the following problems, ignore relativity. The next section will describe how to include relativity in your calculations.

1. A typical jet airliner has a cruise airspeed—that is, its speed relative to the air through which it is flying—of 900 km/h. If the wind at the airliner's cruise altitude is blowing at 100 km/h from west to east, what is the speed of the airliner relative to the ground if the airplane is flying from west to east? From east to west?
(A) 800 km/h west to east, 1000 km/h east to west
(B) 1000 km/h west to east, 800 km/h east to west
(C) 800 km/h in both directions
(D) 1000 km/h in both directions
2. A bicyclist rides at 8.00 m/s toward the north. A car is moving at 25.0 m/s, also toward the north, and is initially behind the rider. A truck is moving at 15.0 m/s toward the south, approaching the bicycle and the car. **(a)** Make a sketch of the three vehicles and label their respective velocity vectors, relative to the ground. **(b)** Calculate the relative velocities of each vehicle compared to the other two.
3. A reference frame, S, is fixed on the surface of Earth with the x axis pointing toward the east, the y axis pointing toward the north, and the z axis pointing up. A second reference frame, S', is moving at a constant 4.00 m/s toward the east. At time $t = t' = 0$ the origins of both reference frames coincide.
(a) Describe mathematically the relationships between x' and x, y' and y, and z' and z.
(b) A picture is taken in the S reference frame at $t = 4.00$ s at the point (2 m, 1 m, 0 m). Calculate the corresponding values of x', y', and z' for the same event in the S' frame.

Figure 26-3 Relative motion This is the view of one train as seen from the window of another. Which train is moving?

26-2 Newton's mechanics includes some ideas of relativity

In Section 3-4, we introduced the idea of relative **reference frames**. Imagine instead of standing on the platform as in Figure 3-16 that you're on a train, looking out the window at a second train right next to yours. One of the trains is moving (**Figure 26-3**). Is your train moving and the other one stationary, or vice versa? Perhaps both are moving. How can you tell?

To address this question let's return to Newton's first law, which we introduced in Section 4-3:

EQUATION IN WORDS
Newton's first law of motion

If the net external force exerted on an object is zero... ...the object does not accelerate...

$$\text{If } \sum \vec{F}_{\text{ext}} = 0, \text{ then } \vec{a} = 0 \text{ and } \vec{v} = \text{constant} \quad (4\text{-}5)$$

...and the velocity of the object remains constant. If the object is at rest, it remains at rest; if it is in motion, it continues in motion in a straight line at a constant speed.

The first law states that an object on which no net force is exerted could *either* be at rest *or* in motion at a constant velocity. So although we might make a distinction between an object at rest and one in (uniform) motion with respect to us, the laws of physics do not. In the case of the two trains, this tells us something profound: If you have no reference to the ground or the tracks on which the trains move, there is *no* experiment or measurement that would tell you whether the other train is moving at a constant velocity with respect to yours or your train is moving at a constant velocity with respect to the other one.

In Chapter 3, we also introduced the concept of relative velocity. Equation 3-9 showed a relationship that allowed us to convert the velocity as seen from an observer in one reference frame into the one that would be observed in another reference frame. To develop this basic concept enough to serve as a platform for our further study of relativity, we need to introduce more formalism. **Figure 26-4** recasts Figure 3-16. Instead of referring to the two different reference frames as *platform* and *train*, we say that the person standing on the train platform is in frame S and measures the positions of objects using the coordinates x, y, and z. On this person's watch, the time of a certain **event**—that is, something that happens at a particular location at a particular moment of time—is measured as t. The person riding on the train is in frame S' and measures the positions of objects using the coordinates x', y', and z' and the time of an event, t'. If neither reference frame is accelerating, the message of Newton's first law is that *both*

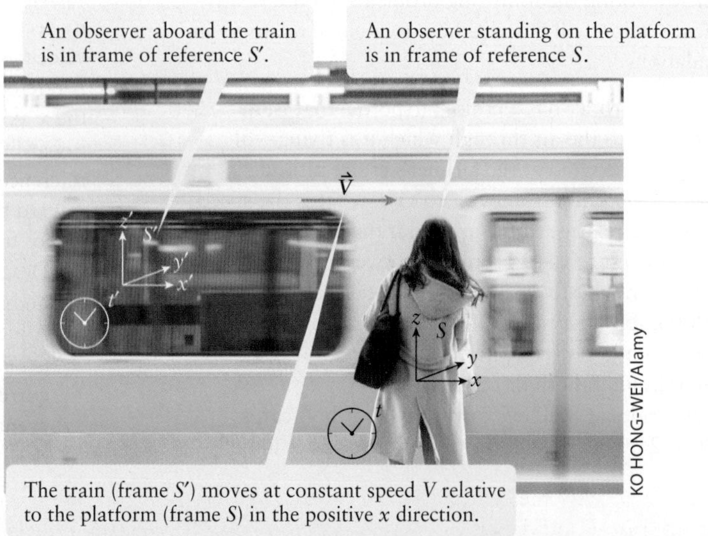

An observer aboard the train is in frame of reference S'.
An observer standing on the platform is in frame of reference S.
$\vec{V}$
z' S' y' x' t'
z S y x t
The train (frame S') moves at constant speed V relative to the platform (frame S) in the positive x direction.
KO HONG-WEI/Alamy

Figure 26-4 Two observers An observer on the platform in a train station uses coordinates x, y, and z and measures time t. An observer aboard the train uses coordinates x', y', and z' and measures time t'.

frames are equally good for making measurements. The velocity of the train's reference frame relative to the platform is now denoted V.

As an example, let's once again consider the child riding on a train (**Figure 26-5**). The train is on a straight track and is moving relative to the platform at a constant speed V. As measured by a person on the platform (reference frame S), the child is moving at a constant velocity with zero acceleration, so the net force on the child must be zero: The upward normal force exerted by the seat exactly balances the downward gravitational force that Earth exerts on the child. As measured by another passenger seated on board the train (reference frame S'), the child is at rest (the child isn't moving relative to that passenger). So as measured in S', the net force on the child must again be zero. The two observers—one in reference frame S and one in reference frame S'—disagree about *how* the child moves. But they agree that the child's motion is in accordance with Newton's first law, Equation 4-5.

It's not just Newton's first law that applies in both reference frames depicted in Figure 26-4. Newton's *second* law also applies in both frames.

As an illustration, suppose the child riding on the train is tossing a ball up and down (**Figure 26-6**). A passenger seated on the train (reference frame S') sees the motion of the ball as purely vertical. By contrast, a person on the platform (reference frame S) sees the ball following a parabolic path: The ball has the same horizontal component of velocity V as the train, and it maintains that horizontal velocity during its flight. As for the child in Figure 26-4, the observers in the two reference frames disagree about how the ball moves. Both observers, however, agree that the ball obeys Newton's second law: When the ball is in flight, only the gravitational force is exerted on it, so the acceleration is downward and has magnitude g. In frame S' the straight up-and-down motion is free fall, as we described in Section 2-6; in frame S the ball is in projectile motion, as we described in Section 3-6. Each description is correct for the reference frame in which the motion is observed.

We refer to a reference frame attached to an object that does not accelerate as an inertial reference frame. An **inertial reference frame** is one in which Newton's first law is valid: If the net force exerted on an object is zero, it either remains at rest or moves with a constant velocity relative to an observer in that reference frame. If one reference frame S is inertial, a second reference frame S' is also inertial if it moves at a constant velocity relative to S. That's the case for the two reference frames depicted in Figure 26-4.

By contrast, a reference frame attached to an accelerated object is a **noninertial frame**. To an observer in a noninertial frame, Newton's first law does not hold true. An example is a reference frame attached to a car that is accelerating forward. A ball sitting on the floor of this car has zero net force on it (the upward normal force exerted by the floor balances the downward gravitational force), yet in this reference frame, the ball accelerates toward the back of the car. A rotating reference frame, such as a carnival merry-go-round, is also noninertial because an object that follows a circular path is accelerating. Just like a ball in a car that accelerates forward, a ball placed on the merry-go-round floor has zero net force exerted on it. Because there is no force pulling it toward the middle of the circular path, as it travels at constant velocity, this ball will roll to the outside of the merry-go-round. In the reference frame of a person riding on the merry-go-round, Newton's first law does not hold true.

Strictly speaking, an observer at rest on Earth's surface, such as the person in frame S standing on the platform in Figure 26-4, is in a noninertial frame. That's because Earth rotates on its axis like a merry-go-round and also moves along a roughly circular orbit around the Sun. However, even though the velocities involved are quite large, the accelerations involved with those motions are so small (each is a small fraction of g) that for many purposes we can ignore them. As a result, we can safely regard the reference frame S in Figure 26-4 as an effectively inertial one, and likewise for the frame S' attached to the train.

We've seen that Newton's first and second laws of motion work equally well in both inertial reference frame S and inertial reference frame S'. The same should be true for any inertial reference frame. This statement is called the **principle of Newtonian relativity**:

The laws of motion are the same in all inertial reference frames.

The word *relativity* means that measurements made relative to one inertial reference frame are just as valid as those made relative to another inertial reference frame.

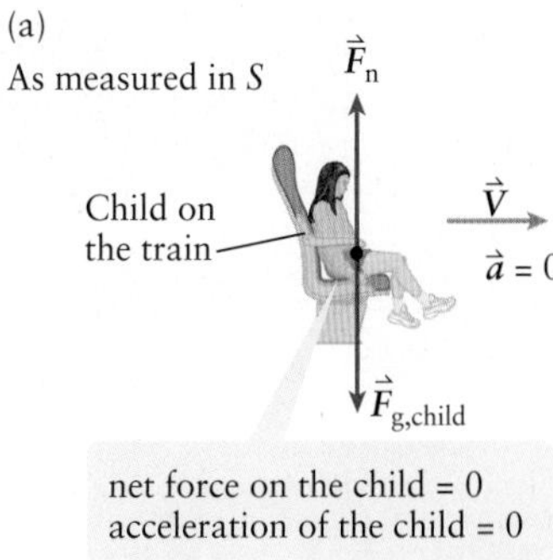

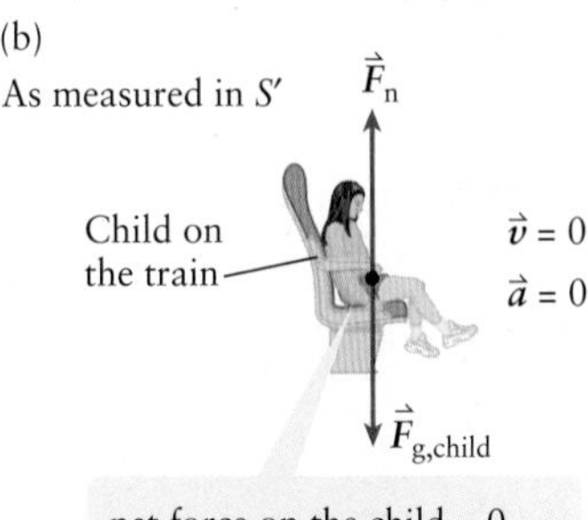

Figure 26-5 Newton's first law in two reference frames The free-body diagram for a child on the train as measured in the two reference frames shown in Figure 26-4.

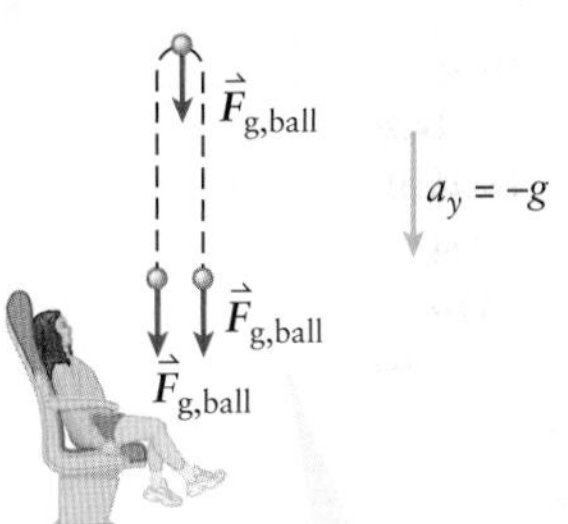

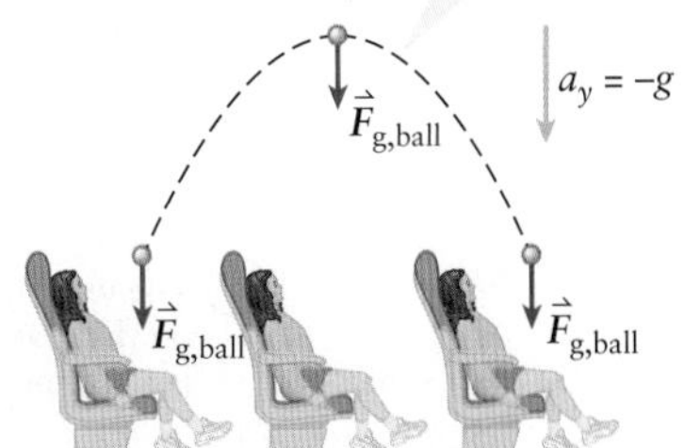

Figure 26-6 A tossed ball in two reference frames The child on the train from Figure 26-5 tosses a ball straight up and down relative to her reference frame S'. (We draw the ball's motion in S' with a slight horizontal displacement to distinguish the up-and-down motion.)

Because the laws of motion don't distinguish between two inertial reference frames S and S', it's meaningless to ask which frame is "really" moving and which frame is "really" at rest. (You may think that frame S is the one that's really at rest because it's stationary with respect to the platform. But remember that the platform is on Earth and that our entire planet is in motion through the solar system.) So another way to express the principle of Newtonian relativity is

> *There is no way to detect absolute motion. Only motion relative to a selected reference frame can be detected.*

Let's apply the principle of Newtonian relativity to the ball shown in Figure 26-6. Suppose you are standing on the platform as the train goes by, so you are in reference frame S. You see the ball following a parabolic path, and you make observations of the x, y, and z coordinates of the ball as functions of time t. The child riding in the train is in reference frame S' and sees the ball moving straight up and down relative to her. As the ball moves she measures the x', y', and z' coordinates of the ball as functions of time t'. The two sets of coordinates have the same orientation, as **Figure 26-7** shows. To calibrate the two clocks—one in frame S, the other in frame S'—you and the child both set your clocks to read zero at the instant that you pass each other, when the origins of your reference frames coincide. So at this instant $t = t' = 0$.

Suppose you and the child both measure the same event—that is, the ball being at a certain point in its motion, such as the high point in its path. In your frame S the coordinates of this event in space and time are x, y, z, and t, and the coordinates of this same event as measured in frame S' are x', y', z', and t'. (Note that we are thinking of time as a fourth coordinate of an event.) As Figure 26-7 shows, a simple set of equations relates the coordinates of the same event in the two frames:

Inertial frame of reference S' moves at speed V in the positive x direction relative to inertial frame of reference S.

EQUATION IN WORDS
Galilean transformation

$$
\begin{aligned}
x' &= x - Vt \\
y' &= y \\
z' &= z \\
t' &= t
\end{aligned}
\tag{26-1}
$$

Coordinates of an event as measured in frame S'

Coordinates of the same event as measured in frame S

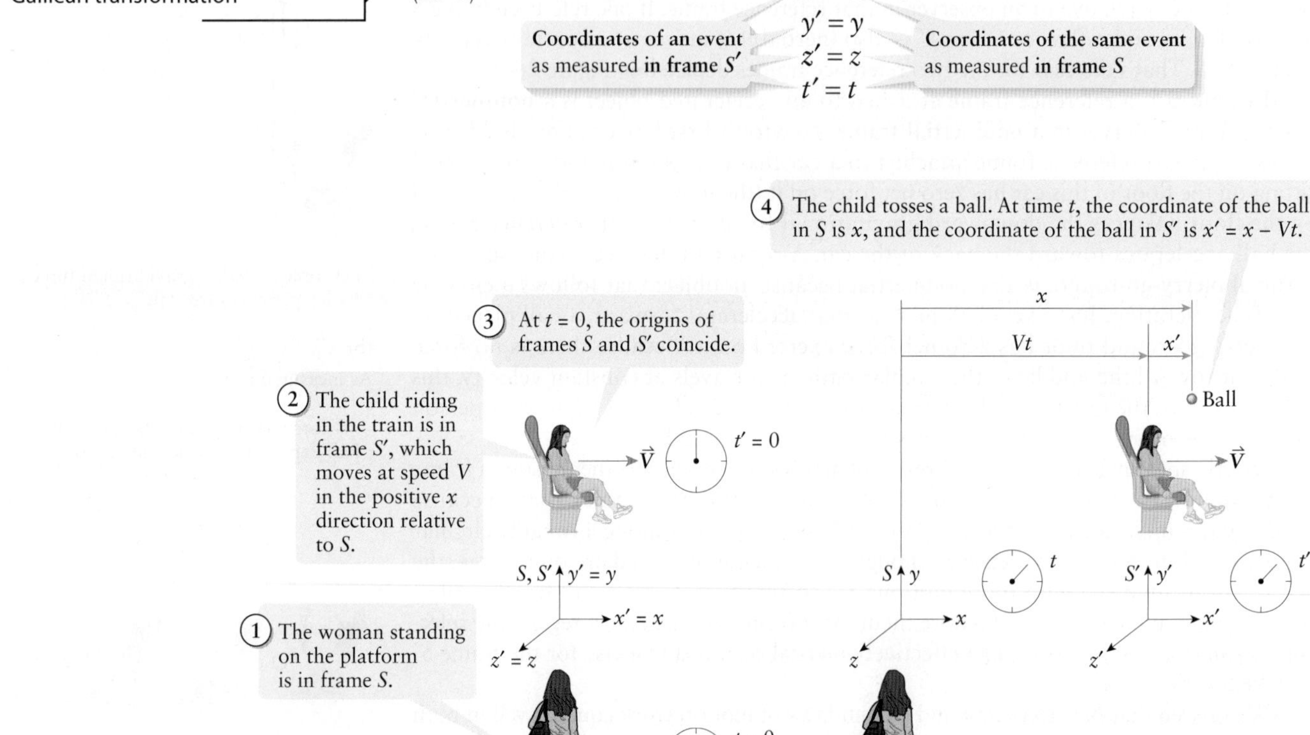

4 The child tosses a ball. At time t, the coordinate of the ball in S is x, and the coordinate of the ball in S′ is x′ = x − Vt.
3 At t = 0, the origins of frames S and S′ coincide.
2 The child riding in the train is in frame S′, which moves at speed V in the positive x direction relative to S.
1 The woman standing on the platform is in frame S.
x
Vt
x′
Ball
V
t′ = 0
t = 0
S, S′ y′ = y
x′ = x
z′ = z
S y
x
z
t
S′ y′
x′
z′
t′

Figure 26-7 Comparing coordinates in two reference frames The coordinates of a ball as measured in the two reference frames shown in Figure 26-4.

This set of equations is known as the **Galilean transformation.** Note that the relative motion of the two reference frames along the positive x axis does not affect the y and z coordinates, which are the same in both reference frames.

We can use the Galilean transformation to compare the velocity of the ball as measured in frame S' to its velocity as measured in frame S. To do this we'll look at two events in the motion of the ball separated by a short time interval from t_1 to t_2 as measured in frame S. During this time interval the coordinates of the ball as measured in frame S change from x_1, y_1, and z_1 to x_2, y_2, and z_2. The components of the ball's velocity $\vec{v}$ in frame S are

$$\begin{aligned} v_x &= \frac{\Delta x}{\Delta t} = \frac{x_2 - x_1}{t_2 - t_1} \\ v_y &= \frac{\Delta y}{\Delta t} = \frac{y_2 - y_1}{t_2 - t_1} \\ v_z &= \frac{\Delta z}{\Delta t} = \frac{z_2 - z_1}{t_2 - t_1} \end{aligned} \tag{26-2}$$

For the same two events as measured in frame S', the time interval is from t_1' to t_2', and the coordinates change from x_1', y_1', and z_1' to x_2', y_2', and z_2'. So in frame S' the components of the object's velocity $\vec{v}$ are

$$\begin{aligned} v_x' &= \frac{\Delta x'}{\Delta t'} = \frac{x_2' - x_1'}{t_2' - t_1'} \\ v_y' &= \frac{\Delta y'}{\Delta t'} = \frac{y_2' - y_1'}{t_2' - t_1'} \\ v_z' &= \frac{\Delta z'}{\Delta t'} = \frac{z_2' - z_1'}{t_2' - t_1'} \end{aligned} \tag{26-3}$$

Now substitute the expressions for x', y', z', and t' from Equations 26-1 into Equations 26-3. We get

$$\begin{aligned} v_x' &= \frac{(x_2 - Vt_2) - (x_1 - Vt_1)}{t_2 - t_1} = \frac{(x_2 - x_1) - V(t_2 - t_1)}{t_2 - t_1} = \frac{(x_2 - x_1)}{t_2 - t_1} - V \\ v_y' &= \frac{y_2 - y_1}{t_2 - t_1} \\ v_z' &= \frac{z_2 - z_1}{t_2 - t_1} \end{aligned} \tag{26-4}$$

If we now compare Equations 26-4 for the velocity components in frame S' to Equations 26-2 for the velocity components in frame S, we see that

Inertial frame of reference S' moves at speed V in the positive x direction relative to inertial frame of reference S.

$$\begin{aligned} v_x' &= v_x - V \\ v_y' &= v_y \\ v_z' &= v_z \end{aligned} \tag{26-5}$$

Velocity components of an object as measured in frame S'

Velocity components of the same object as measured in frame S

EQUATION IN WORDS
Galilean velocity transformation

Equations 26-5, which relate the velocity of an object in frame S' to the velocity of the same object in frame S, are called the Galilean velocity transformation.

As an example, think again of the ball shown in Figure 26-6. If the ball moves straight up and down as measured in frame S', then in that frame the ball is in free fall with only a y component of velocity. The other two components are zero: $v_x' = v_z' = 0$. As measured in frame S, the velocity of the ball has components

$$\begin{aligned} v_x &= v_x' + V = V \\ v_y &= v_y' \\ v_z &= v_z' = 0 \end{aligned}$$

As measured in frame S, the ball moves up and down along the y direction with the same velocity as measured in frame S': $v_y = v'_y$. At the same time, as measured in frame S, the ball maintains a constant velocity V in the x direction. That's just the behavior we expect for a projectile: Its motion is a combination of up-and-down free fall and constant-velocity horizontal motion (see Section 3-6).

WATCH OUT !

There is nothing special about either the *S* frame or the *S′* frame.

In Figures 26-5 and 26-6 we've chosen to think of frame S as at rest and frame S' as moving. This selection is purely our choice because the principle of Newtonian relativity says that *all* inertial reference frames are equivalent. We could just as well say that frame S' is stationary and frame S is moving with speed V in the negative x direction.

You may wonder why we've spent so much time and effort explaining motion as seen from two different inertial reference frames. As we'll discover in the next few sections, the reason is that something remarkable happens when the relative speed V of the two frames is comparable to c, the speed of light in a vacuum. In that case we'll find that the Galilean transformation equations do *not* hold true: As measured in the two different frames, the time interval between events can be different and extended objects can have different dimensions. These remarkable observations will radically transform our notions of the nature of time and space themselves.

EXAMPLE 26-1 Two Cars

You observe two race cars approaching you. A red car is in one lane moving at 24 m/s relative to you. In a second lane a blue car is moving at 36 m/s relative to you. (a) What is the velocity of the red car as measured by the driver of the blue car? (b) What is the velocity of the blue car as measured by the driver of the red car?

Set Up

We'll use Equations 26-5 to transform the velocity of a car as measured in one reference frame to the velocity of the same car as measured in a different reference frame. Note that these equations assume that frame S' is moving relative to frame S at speed V in the positive x direction. We'll use this to decide which reference frame corresponds to S and which to S'.

Galilean velocity transformation:

$$\begin{aligned} v_x &= v'_x - V \\ v_y &= v'_y \\ v_z &= v'_z \end{aligned} \quad (26\text{-}5)$$

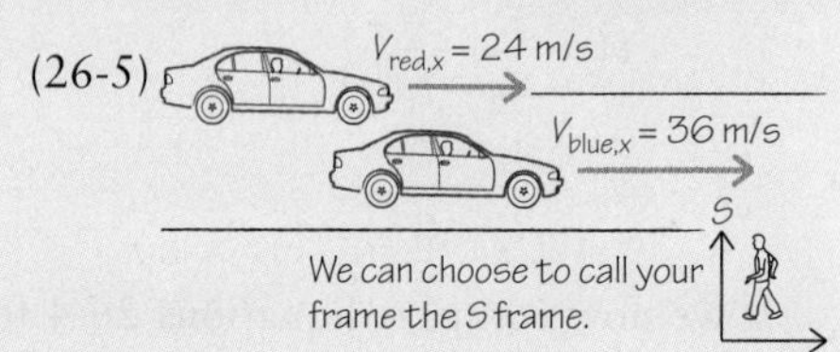
The given speeds of both cars are measured relative to you, that is, in your frame.
$v_{red,x} = 24$ m/s
$v_{blue,x} = 36$ m/s
S
We can choose to call your frame the S frame.

Solve

(a) Let's take the positive x direction to be in the direction both cars are moving relative to you. Then we'll take S to be your reference frame and S' to be the reference frame of the driver of the blue car. The relative speed of these two frames is $V = 36$ m/s (the speed of the blue car relative to you). All the motions in this example are along the x axis, so we don't need the y or z members of Equations 26-5.

Use the x equation from Equations 26-5 to relate the velocity of the red car as measured by you ($v_{red,x} = +24$ m/s) to its velocity as measured by the driver of the blue car ($v'_{red,x}$):

$$\begin{aligned} v'_{red,x} &= v_{red,x} - V \\ &= +24 \text{ m/s} - 36 \text{ m/s} \\ &= -12 \text{ m/s} \end{aligned}$$

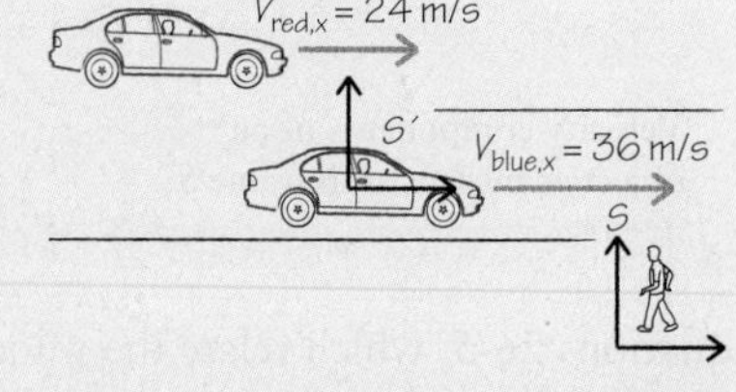
$v_{red,x} = 24$ m/s
S′
$v_{blue,x} = 36$ m/s
S

As measured by the driver of the blue car, the red car is moving in the negative x direction—that is, backward—at 12 m/s.

(b) Again we take S to be your reference frame, but now S' is the reference frame of the driver of the red car. The relative speed of these two frames is $V = 24$ m/s (the speed of the red car relative to you).

Use the x equation from Equations 26-5 to relate the velocity of the blue car as measured by you ($v_{\text{blue},x} = +36$ m/s) to its velocity as measured by the driver of the red car ($v'_{\text{blue},x}$):

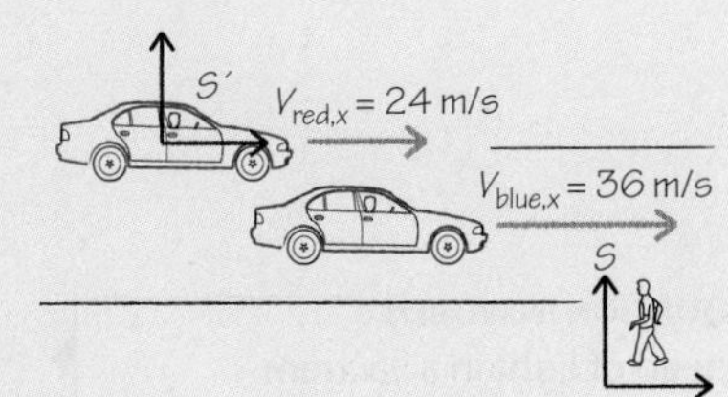

$$\begin{aligned} v'_{\text{blue},x} &= v_{\text{blue},x} - V \\ &= +36 \text{ m/s} - 24 \text{ m/s} \\ &= +12 \text{ m/s} \end{aligned}$$

As measured by the driver of the red car, the blue car is moving in the positive x direction—that is, forward—at 12 m/s.

Reflect

The driver of the blue car sees the red car falling farther and farther behind her at a rate of 12 m/s, while the driver of the red car sees the blue car moving farther and farther in front of him at a rate of 12 m/s. Note that the *magnitude* of the two answers is the same: Both drivers agree that the other driver is moving at a relative speed of 12 m/s.

NOW WORK Problems 1–3 from the Takeaway for Section 26-2.

THE TAKEAWAY for Section 26-2

✔ Newton's laws of motion treat all nonaccelerating objects identically, whether the objects are in motion or at rest.

✔ The laws of motion are the same in all inertial reference frames. There is no way to detect absolute motion; only motion relative to a selected reference frame can be detected.

✔ The Galilean transformations allow you to convert the coordinates and velocity of an object observed in one inertial reference frame to the coordinates and velocity of the same object observed in a different inertial reference frame.

Takeaway Problems for Section 26-2

1. Assume that the origins of S and S' coincide at $t = t' = 0$ s. An observer in inertial reference frame S measures the space and time coordinates of an event to be $x = 750$ m, $y = 250$ m, $z = 250$ m, and $t = 2.0$ μs. What are the space coordinates of the event in inertial reference frame S', which is moving in the $+x$ direction at a speed of $0.01c$ relative to S?

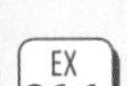

2. At time $t' = 4.00 \times 10^{-3}$ s, as measured in S', a particle is at the point $x' = 10$ m, $y' = 4$ m, and $z' = 6$ m. Compute the corresponding values of x, y, and z, as measured in S, for (a) a relative velocity between S' and S of +500 m/s and (b) a relative velocity between S' and S of −500 m/s. Assume that the origins of S and S' coincide at $t = 0$ and that the motion lies along the x and x' axes.

EX 26-1

3. Suppose that at $t = 6.00 \times 10^{-4}$ s, the space coordinates of a particle are $x = 100$ m, $y = 10$ m, and $z = 30$ m according to coordinate system S. Compute the corresponding values as measured in the frame S' if the relative velocity between S' and S is 150,000 m/s along the x and x' axes. The origins of the reference frames coincide at $t = 0$.

26-3 The Michelson–Morley experiment shows that light does not obey Newtonian relativity

We saw in the preceding section that Newton's laws of motion are the same in all inertial reference frames. Is the same true for the other laws of physics? Physicists asked this very question during the second half of the nineteenth century, specifically about the laws of electromagnetism.

We learned in Section 22-3 that the laws of electromagnetism explain how electromagnetic waves, including visible light, are possible. These laws also predict that the speed of electromagnetic waves in a vacuum is related to the constants ε_0 and μ_0 that appear in the equations of electromagnetism:

EQUATION IN WORDS
Speed of light in a vacuum

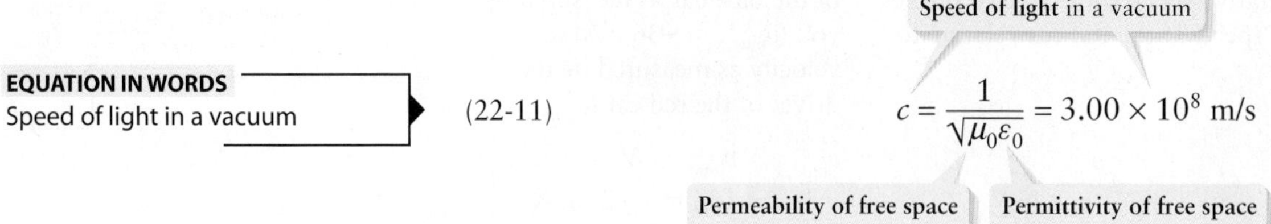

$$c = \frac{1}{\sqrt{\mu_0 \varepsilon_0}} = 3.00 \times 10^8 \text{ m/s} \tag{22-11}$$

Our discussion in Section 26-2 tells us that we can measure relative motion but not absolute motion. So we are forced to ask this question: "Relative to what inertial reference frame is the speed of light in a vacuum equal to c?"

In the nineteenth century the most common answer to this question was to imagine a substance that fills all space. This substance, which was called the luminiferous ether, was thought to be the medium for electromagnetic waves, just as air is the medium for sound waves in our atmosphere. This substance must be of extraordinarily low density so that its presence is almost undetectable. (The word *luminiferous* comes from the Latin for "light-bearing." *Ether* in this phrase has nothing to do with the organic compounds of the same name.) In this model c is the speed of electromagnetic waves relative to the frame in which the luminiferous ether is at rest.

How can we test whether this model is correct? To see the answer, note that the speed of sound waves in dry air is 343 m/s, but you will measure a different speed if a wind is blowing (that is, the air is moving relative to you). If a sound wave is traveling from west to east and a wind is blowing past you at 10 m/s from west to east, the sound will move relative to you at 343 m/s + 10 m/s = 353 m/s. If the sound wave is traveling from west to east and a 10-m/s wind is blowing from east to west, the speed of the sound wave relative to you will be 343 m/s – 10 m/s = 333 m/s. The same should be true for light waves if the luminiferous ether is moving past you so that there is an "ether wind." If a light wave is traveling from west to east and the ether is moving from west to east relative to you at 10 m/s, you would measure the speed of the wave to be $c + 10$ m/s; if the ether is instead moving from east to west relative to you at 10 m/s, you would measure the speed of the wave to be $c - 10$ m/s. If we could detect these small changes in the speed of light, it would be evidence that the luminiferous ether really exists.

Nineteenth-century scientists looked to Earth's motion around the Sun as a source of "ether wind." Our planet moves around its orbit at an average speed of 29.8 km/s = 2.98×10^4 m/s, or about $10^{-4}c$. If the luminiferous ether is at rest relative to the solar system as a whole, we should experience an "ether wind" that blows past our moving planet at $10^{-4}c$. Depending on the direction of that "ether wind" relative to the direction of light propagation, we would expect the speed of light to vary between $(1 + 10^{-4})c$ and $(1 - 10^{-4})c$. The challenge is to design an experiment that can detect such small changes in the speed of light.

In 1887 the American scientists Albert Michelson and Edward Morley carried out the first definitive experiment of this kind. **Figure 26-8** shows a simplified version of the **Michelson–Morley experiment.** Their apparatus split a beam of light into two, sent the two beams along perpendicular paths, and then allowed them to recombine at a viewing screen. What is seen on the viewing screen is an interference pattern between the waves in the two beams. The nature of this pattern depends on the difference in length between the two paths and on whether the speeds at which light travels along each path are the same or different. If there is an "ether wind" that is more nearly aligned with one leg of the interferometer than the other, the speed of light should indeed be different along the two legs. Their apparatus was sensitive enough that Michelson and Morley should have been able to measure the effects of an "ether wind" due to Earth's motion around the Sun.

Michelson and Morley found *no* effect due to an "ether wind." They and other scientists refined and repeated the experiment many times, but the results were always the same: There is no evidence for the existence of the luminiferous ether. The conclusion

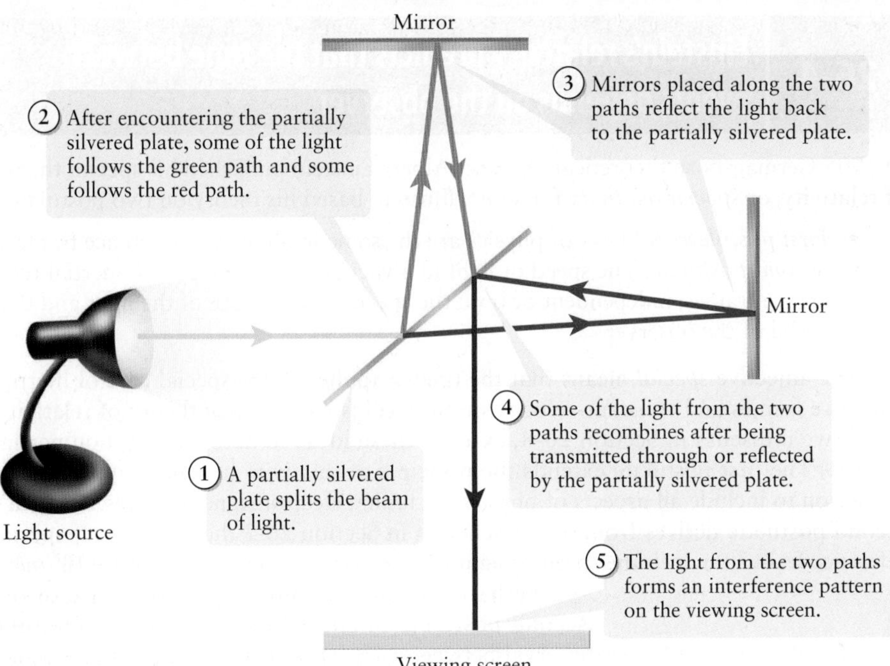

Mirror
(3) Mirrors placed along the two paths reflect the light back to the partially silvered plate.
(2) After encountering the partially silvered plate, some of the light follows the green path and some follows the red path.
Mirror
Light source
(1) A partially silvered plate splits the beam of light.
(4) Some of the light from the two paths recombines after being transmitted through or reflected by the partially silvered plate.
(5) The light from the two paths forms an interference pattern on the viewing screen.
Viewing screen

Figure 26-8 The Michelson–Morley experiment simplified If the luminiferous ether exists and is moving relative to this apparatus, its presence will be apparent in the interference pattern on the viewing screen.

from this and other experiments is that electromagnetic waves do not require the presence of a material medium but can propagate in a complete vacuum.

Under Newtonian relativity, to explain a constant value of the speed of light in a vacuum required that the equations of electromagnetism hold true only in a specific reference frame, one that is at rest relative to the medium for electromagnetic waves. But if there is no luminiferous ether, there is no such medium and hence no such special reference frame. So physicists had no choice but to conclude that Newtonian relativity does not apply to electromagnetic waves. It was left to Albert Einstein to modify the ideas of Newtonian relativity and find a new way to look at the relationships between measurements made in different inertial reference frames. We'll explore Einstein's simple yet radical ideas in the following section.

THE TAKEAWAY for Section 26-3

✔ Experiment shows that electromagnetic waves do not require a material medium; they can easily propagate in a perfect vacuum.

✔ Because electromagnetic waves do not require a material medium, we conclude that electromagnetic waves do not obey Newtonian relativity.

Takeaway Problems for Section 26-3

1. A bottlenose dolphin (*Tursiops truncates*) can swim at speeds up to about 10 m/s. These dolphins produce sounds used for social communication and for echolocation (using sound to detect objects around them while swimming in dark or murky waters). If a swimming dolphin travels at top speed and produces a sound wave that propagates forward, how fast does that wave travel relative to the dolphin? The speed of sound in water is 1500 m/s.
(A) 10 m/s
(B) 1490 m/s
(C) 1500 m/s
(D) 1510 m/s

2. The Michelson–Morley experiment was performed hundreds of times in a futile attempt to find the luminiferous ether. Why was the experiment performed on an enormous slab of marble that was floated in a pool of mercury?

3. If we used radio waves to communicate with an alien space ship approaching Earth at 10% of the speed of light, we would receive their signals at a speed of
(A) $0.10c$.
(B) $0.90c$.
(C) $1.00c$.
(D) $1.10c$.

26-4 Einstein's relativity predicts that the time between events depends on the observer

In 1905 German-born theoretical physicist Albert Einstein published his **special theory of relativity**, or *special relativity* for short. Einstein based his theory on two postulates:

- *First postulate*: All laws of physics are the same in all inertial reference frames.
- *Second postulate*: The speed of light in a vacuum is the same in all inertial reference frames, independent of both the speed of the source of the light and the speed of the observer.

The adjective *special* means that the theory applies to the special case of inertial reference frames and constant-velocity motion. (Einstein's *general* theory of relativity, which we'll discuss in Section 26-8, extends these ideas to accelerating, noninertial frames.) The first postulate extends the principle of relativity beyond Newton's laws of motion to include all aspects of physics, including electromagnetic waves. Einstein's second postulate derives from our conclusion in Section 26-3 that there is no special reference frame in which the speed of light in a vacuum is equal to $c = 3.00 \times 10^8$ m/s.

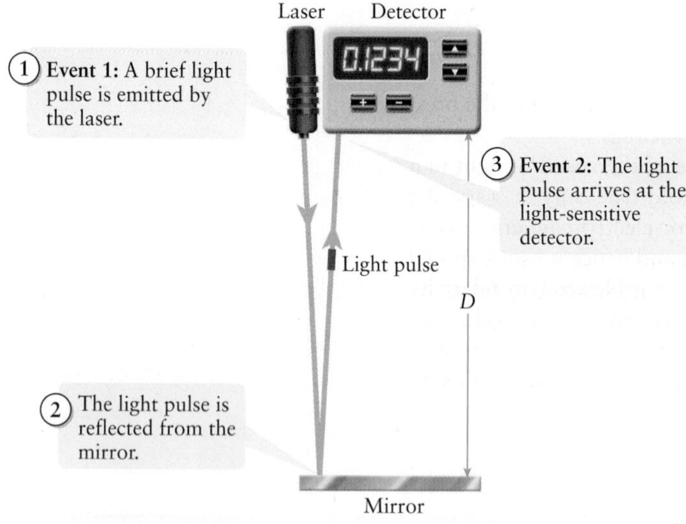

Laser
Detector
0.1234
1 Event 1: A brief light pulse is emitted by the laser.
3 Event 2: The light pulse arrives at the light-sensitive detector.
Light pulse
D
2 The light pulse is reflected from the mirror.
Mirror

Figure 26-9 A light clock The duration of each "tick" of this clock is the time between the start of the pulse and the time when the reflected light arrives at the detector.

Perhaps the most astounding consequence of the second postulate of special relativity is that the time interval between two events is *not* an absolute. Rather, this time interval depends on the motion of the reference frame from which time is measured. To see how this comes about, we'll do a thought experiment.

Imagine a *light clock*, a special clock that uses light to measure intervals of time (**Figure 26-9**). A laser fires an extremely brief burst, or pulse, of light straight downward. This is reflected by a mirror and arrives at a light-sensitive detector next to the laser. For simplicity we can treat the laser and detector as being at the same position. One tick of the clock is the time interval between the pulse leaving the laser (event 1) and the pulse arriving at the detector (event 2). In the reference frame of the clock, these two events occur at the same point in space. We use the term **proper time** for the time interval between two events that occur at the same place, and we denote it by the symbol Δt_{proper}. Another way to think of proper time is that it is the time interval as measured in a reference frame in which the clock is at rest. You can think of the rest frame of the clock as being "attached" to the clock.

WATCH OUT !

Every object—including a moving object—has a rest frame.

An object is always at rest with respect to itself, which means that an object is not moving in a reference frame that is attached to it. Notice, however, that the statement is a relative one—the object is not moving *relative* to its own rest frame. We can still define any number of other frames with respect to which the object *is* in motion. A person standing on the sidewalk is at rest in her own rest frame but is moving as measured from the reference frame of a car driving past.

During one tick of our light clock, a light pulse travels the distance D to the mirror and then the same distance back to the detector, for a total distance of $2D$. We imagine that the clock is placed in a vacuum so that the speed of the light pulse is c. The time interval for the tick is then

$$\Delta t_{\text{proper}} = \frac{2D}{c} \qquad (26\text{-}6)$$

We now let the light clock move at speed V relative to us. In keeping with the way we named frames in discussing Newtonian relativity, we attach a frame S' to the clock and consider ourselves in the S frame. The S' frame therefore moves at speed V relative to the S frame. **Figure 26-10** shows the process of a single tick of the clock as observed from our S frame. The entire clock moves during the time that it takes for the light pulse to move from the laser to the mirror and from the mirror to the detector. The total distance L that the clock moves during this time equals the product of the speed V and the time interval of one clock tick. Be careful, however: We *cannot* assume that the time interval for one tick as measured in S is Δt_{proper} because Δt_{proper} is measured in frame S' at rest with respect to the clock. Instead we use the symbol Δt for the time interval of one tick as observed from the S frame. So

$$L = V\Delta t \tag{26-7}$$

From our vantage point in the S frame, the path followed by the light pulse during time Δt traces out two sides of a triangle (Figure 26-10). The time interval measured in S is therefore equal to the sum of the lengths of these two sides—the distance the light pulse travels—divided by the speed at which the light pulse travels in frame S. The second postulate of special relativity assures us that the speed of the light pulse is c in *all* inertial reference frames, and so is the same in frame S as in frame S'. Thus

$$\Delta t = \frac{2\sqrt{D^2 + \left(\dfrac{L}{2}\right)^2}}{c} \tag{26-8}$$

We now have two expressions for the time interval between the light pulse leaving the laser and the same pulse arriving at the detector. As measured in the clock rest frame S', this time interval is the proper time Δt_{proper} as given by Equation 26-6; as measured in our rest frame S, the time interval is Δt as given by Equation 26-8. To compare these two time intervals, first note that the distance L does not appear in the

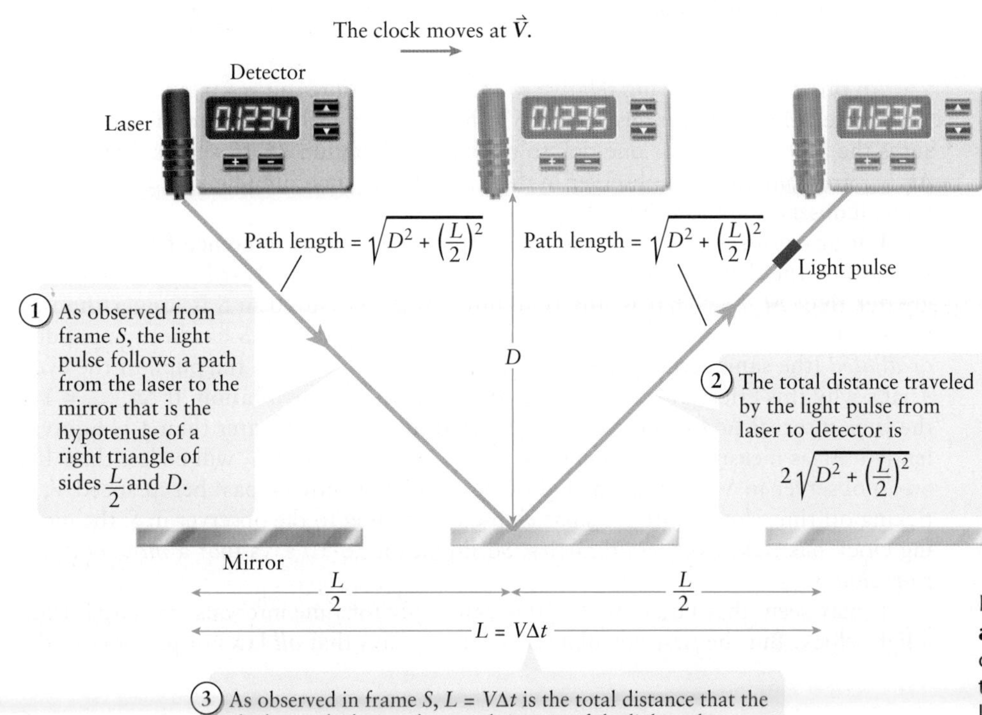

The clock moves at $\vec{V}$.
Detector
Laser
0.1234
0.1235
0.1236
Path length = $\sqrt{D^2 + \left(\frac{L}{2}\right)^2}$
Path length = $\sqrt{D^2 + \left(\frac{L}{2}\right)^2}$
Light pulse
1 As observed from frame S, the light pulse follows a path from the laser to the mirror that is the hypotenuse of a right triangle of sides $\frac{L}{2}$ and D.
D
2 The total distance traveled by the light pulse from laser to detector is $2\sqrt{D^2 + \left(\frac{L}{2}\right)^2}$
Mirror
$\frac{L}{2}$
$\frac{L}{2}$
$L = V\Delta t$
3 As observed in frame S, $L = V\Delta t$ is the total distance that the clock travels during the travel time Δt of the light pulse.

Figure 26-10 A moving light clock and time dilation When the light clock depicted in Figure 26-9 moves, the distance the light pulse travels is longer than twice the distance from the laser to the mirror.

expression for Δt_{proper}. To eliminate L from the expression for Δt, substitute Equation 26-7, $L = V\Delta t$, into Equation 26-8:

$$\Delta t = \frac{2\sqrt{D^2 + \left(\frac{V\Delta t}{2}\right)^2}}{c}$$

Square both sides of this equation and rearrange to bring the two Δt terms together:

$$c^2\Delta t^2 = 4\left(D^2 + \left(\frac{V\Delta t}{2}\right)^2\right) = 4D^2 + V^2\Delta t^2$$

$$c^2\Delta t^2 - V^2\Delta t^2 = 4D^2 \quad \text{or} \quad (c^2 - V^2)\Delta t^2 = 4D^2$$

Rearrange this equation to solve for Δt^2, then take the square root of both sides to solve for Δt:

$$\Delta t^2 = \frac{4D^2}{c^2 - V^2} = \frac{4D^2}{c^2\left(1 - \frac{V^2}{c^2}\right)}$$

(26-9)
$$\Delta t = \frac{2D}{c\sqrt{1 - \frac{V^2}{c^2}}}$$

If we compare Δt (Equation 26-9) to Δt_{proper} (Equation 26-6), we get

EQUATION IN WORDS
Time dilation

(26-10)
$$\Delta t = \frac{\Delta t_{proper}}{\sqrt{1 - \frac{V^2}{c^2}}}$$

- Δt_{proper}: Time interval between two events as measured in frame S', in which those events occur at the same place
- Δt: Time interval between the same two events as measured in frame S
- V: Speed of S' relative to S
- c: Speed of light in a vacuum

An observer in S moving relative to the clock measures a time interval Δt for one tick of the light clock. An observer in S' who is at rest with respect to the clock measures the time interval for one tick to be Δt_{proper}. Equation 26-10 shows that Δt and Δt_{proper} are *not* equal: An observer in the S frame sees time running at a *different* rate than an observer in the S' frame!

For any nonzero value of the speed V of one inertial reference frame relative to another, the denominator in Equation 26-10 is less than 1. It follows that Δt is greater than Δt_{proper}. That is, the time interval as measured in S is longer than as measured in S'. We say that the time interval as measured in S has been expanded or *dilated* (the same term used to refer to an increase in size of the pupil of the eye). That's why this effect of special relativity is known as **time dilation**. If $\Delta t_{proper} = 1$ s, the time interval Δt for one tick as measured in S will be greater than 1 s. Equivalently if Δt as measured in S equals 1 s, Δt_{proper} as measured in S' will be less than 1 s. So an observer in S says that the light clock, which is moving past her at speed V, is ticking off time slowly: After 1 s has elapsed according to the observer in S, the moving clock has ticked off less than 1 s. So Equation 26-10 says that a *moving clock runs slowly*.

It may seem that Equation 26-10 is valid only for time intervals measured using a light clock. But the first postulate of relativity says that *all* laws of physics are the same in every inertial reference frame. This implies that time dilation is also valid for time intervals measured using a mechanical clock, such as a grandfather clock that keeps time with an oscillating pendulum or a wristwatch that uses an oscillating piece of quartz (Figure 12-1b). Imagine a pendulum clock on your desk, with its pendulum swinging back and forth once every second. If the clock is moving, then the swing of the pendulum is slower, so the clock ticks more slowly. As we will see, however, the time dilation effect is very small unless V is very large.

WATCH OUT !

Time dilation occurs whether or not a clock is present.

You don't need to have a clock per se for the effects of time dilation to occur. Imagine, say, that you can place a *Caenorhabditis elegans* worm (**Figure 26-11**) on a spacecraft that will fly past Earth at high speed. *C. elegans* is popular among geneticists because it grows to adulthood in a series of easily identifiable developmental stages that all together take less than 3 days. Were you to observe a *C. elegans* worm as it moved past you at high speed, the growth to adulthood might take weeks or even years, measured on a clock at rest with respect to you.

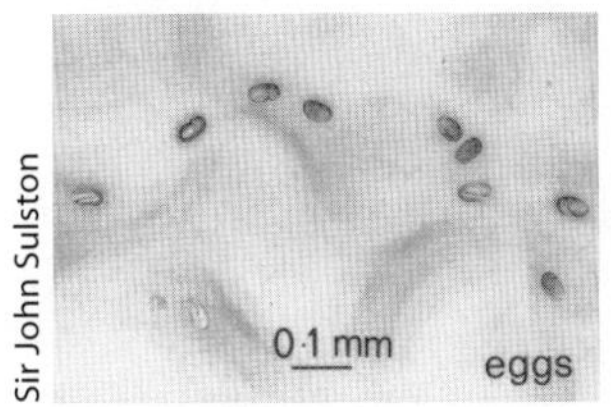

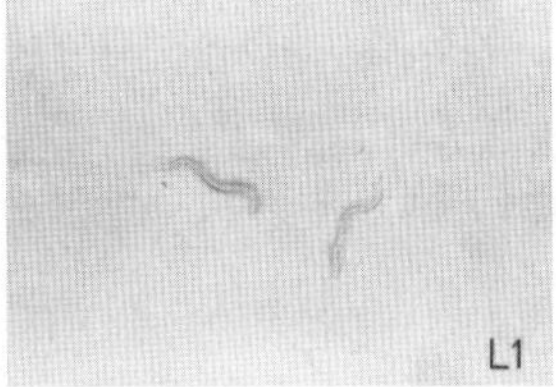

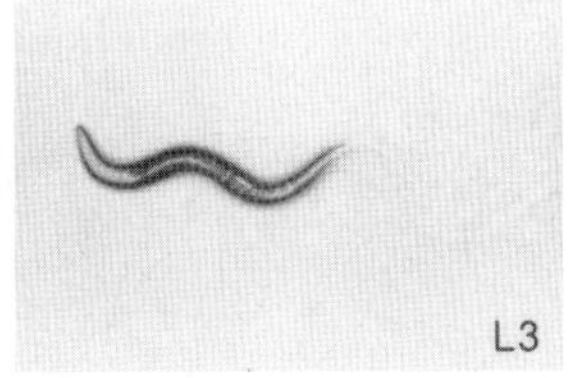

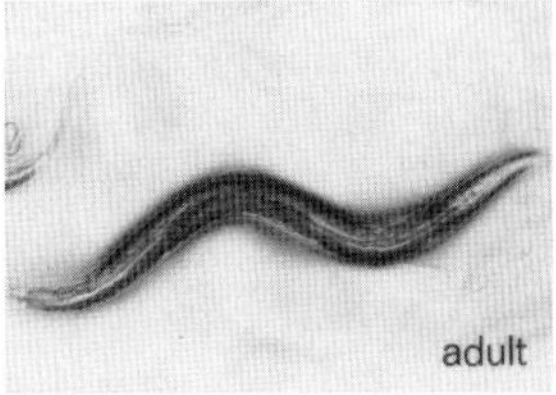

Time ⟶

Figure 26-11 A worm "clock" The worm *Caenorhabditis elegans* grows to adulthood in a series of easily identifiable developmental stages, each of which lasts a well-defined period of time.

WATCH OUT !

The effect of time dilation arises only when events in one frame are viewed from a second frame in motion relative to the first.

We imagined, above, a pendulum clock that ticks once every time the pendulum makes one full oscillation. If the clock sits on your desk, you see it swinging back and forth once every second. If the clock moves relative to you, you see the pendulum swing more slowly. But if you and your clock are moving *together* relative to some other specific object or frame, you will see *no* time dilation effect on your clock, regardless of how fast you and the clock are moving relative to that other object or frame. You will still see the pendulum swinging back and forth once every second. There is no time dilation because you and the clock are not moving relative to each other.

We do not observe time dilation in everyday life because most objects travel very slowly compared to the speed of light. As an example, the greatest launch speed ever given to a spacecraft is 16,260 m/s = 58,536 km/h = 36,373 mi/h relative to Earth. Even this tremendous speed is only 5.42×10^{-5} of the speed of light c, and the factor $1/\sqrt{1 - V^2/c^2}$ in Equation 26-10 is larger than 1 by only 1.47×10^{-9}. Due to time dilation, a clock on board the spacecraft does indeed run slowly as measured from Earth, but by only one second every 21.5 years! As the following two examples show, however, time dilation can be substantial if the speeds involved are **relativistic speeds**—that is, an appreciable fraction of the speed of light.

EXAMPLE 26-2 A Moving Clock

A clock moves past you. What must be the speed of the clock relative to you so that you see it as running at one-half (0.500) the rate of the clock on your cell phone in your hand?

Set Up

We want a moving clock to be observed as running at 0.500 the rate of a clock at rest with respect to you. This means making Δt (the time interval measured by you) equal to twice Δt_{proper} (the time interval measured in the rest frame of the moving clock). That implies the factor $1/\sqrt{1-V^2/c^2}$ in Equation 26-10 must be equal to $1/0.500 = 2.00$. We'll use this to solve for the speed V of the moving clock relative to you.

Time dilation:

$$\Delta t = \frac{\Delta t_{\text{proper}}}{\sqrt{1-\dfrac{V^2}{c^2}}} \qquad (26\text{-}10)$$

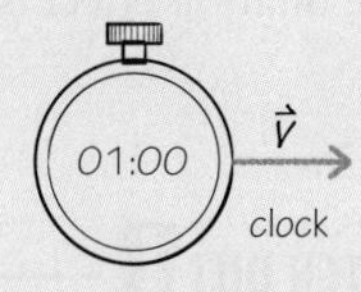

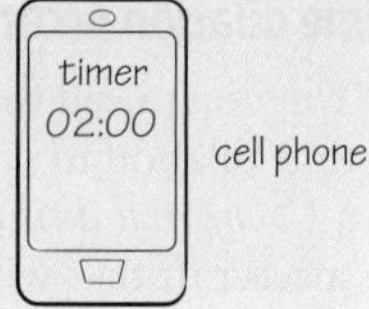

Solve

Determine the value of V that makes $\Delta t = 2.00\Delta t_{\text{proper}}$.

From Equation 26-10 we want

$$\Delta t = \frac{\Delta t_{\text{proper}}}{\sqrt{1-\dfrac{V^2}{c^2}}} = 2.00\Delta t_{\text{proper}} \quad \text{so} \quad \frac{1}{\sqrt{1-\dfrac{V^2}{c^2}}} = 2.00$$

Solve for the value of the speed V:

$$\sqrt{1-\frac{V^2}{c^2}} = \frac{1}{2.00} = 0.500$$

$$1-\frac{V^2}{c^2} = (0.500)^2 = 0.250$$

$$\frac{V^2}{c^2} = 1 - 0.250 = 0.750$$

$$\frac{V}{c} = \sqrt{0.750} = 0.866$$

$$V = 0.866c$$

$$= 0.866(3.00 \times 10^8 \text{ m/s}) = 2.60 \times 10^8 \text{ m/s}$$

Reflect

For you to observe the moving clock running a factor of 2.00 slower than the clock at rest with respect to you, the relative speed between you and the clock would need to be 86.6% of the speed of light in a vacuum. This is more than 10^4 times faster than the top speed of any craft built by humans.

NOW WORK Problem 2 from the Takeaway for Section 26-4.

EXAMPLE 26-3 Muon Decay

Our planet is continually bombarded by fast-moving subatomic particles from space (mostly protons). When these particles collide with the atoms and molecules that make up Earth's upper atmosphere, they can produce a new subatomic particle called the *muon*. These muons travel at high speed, around $0.994c$. Muons naturally decay, however; if you create a number of muons in the lab, the half-life of the muon is 1.56 μs (1 μs = 10^{-6} s). If 1.00×10^6 muons are created at an altitude of 15.0 km, (a) how many would you expect to strike Earth if time-dilation effects were ignored? (b) How many would you expect to strike Earth when time dilation is taken into account?

Set Up

The production and decay of the muon occur at the same place in the rest frame of the muon, so the half-life of 1.56 μs is a proper time interval in the muon frame. We'll call this frame S'. This frame moves at speed $V = 0.994c$ relative to our frame on Earth, so we measure a different half-life Δt as given by Equation 26-10. We'll use the idea that the muon population decreases by 1/2 in each half-life to determine how many reach our planet's surface.

Time dilation:

$$\Delta t = \frac{\Delta t_{\text{proper}}}{\sqrt{1-\frac{V^2}{c^2}}} \qquad (26\text{-}10)$$

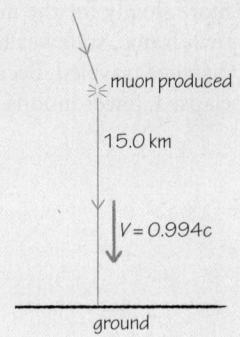

muon produced

15.0 km

$V = 0.994c$

ground

Solve

(a) If there were no time dilation, the half-life of the muon in the Earth frame S would be $\Delta t = \Delta t_{\text{proper}} = 1.56\ \mu\text{s}$, the same as in the muon frame S'. Use this to calculate the number of muons that successfully reach Earth's surface.

Time for a muon to travel a distance $d = 15.0$ km at $V = 0.994c$:

$$T = \frac{d}{V} = \frac{15.0\ \text{km}}{0.994(3.00 \times 10^8\ \text{m/s})}\left(\frac{10^3\ \text{m}}{1\ \text{km}}\right)$$

$$= 5.03 \times 10^{-5}\ \text{s}\left(\frac{1\ \mu\text{s}}{10^{-6}\ \text{s}}\right) = 50.3\ \mu\text{s}$$

Express this as a multiple of the half-life:

$$\frac{T}{\Delta t_{\text{proper}}} = \frac{50.3\ \mu\text{s}}{1.56\ \mu\text{s}} = 32.2\ \text{half-lives}$$

After one half-life, the number of muons has decreased to 1/2 of its initial value; after two-half-lives, to $(1/2) \times (1/2) = 1/2^2$ of its initial value; after three half-lives, to $(1/2) \times (1/2) \times (1/2) = 1/2^3$ of its initial value; and so on. So after 32.2 half-lives, the number of muons remaining would be the original number of 1.00×10^6 multiplied by $1/2^{32.2}$:

$$\text{muons remaining} = (1.00 \times 10^6)\left(\frac{1}{2^{32.2}}\right)$$

$$= (1.00 \times 10^6)(1.97 \times 10^{-10})$$

$$= 1.97 \times 10^{-4}$$

Fewer than 1 muon, on average, survive the trip to Earth's surface.

(b) With time dilation, we first calculate the half-life in the Earth frame using Equation 26-10. We then use the same method as in part (a) to calculate the number of muons that reach Earth's surface.

Accounting for time dilation, the half-life as measured in the Earth frame S is the half-life measured in the muon frame S' divided by $\sqrt{1-V^2/c^2}$:

$$\Delta t = \frac{\Delta t_{\text{proper}}}{\sqrt{1-\frac{V^2}{c^2}}} = \frac{1.56\ \mu\text{s}}{\sqrt{1-\frac{(0.994c)^2}{c^2}}} = \frac{1.56\ \mu\text{s}}{\sqrt{1-(0.994)^2}}$$

$$= \frac{1.56\ \mu\text{s}}{\sqrt{0.0120}} = \frac{1.56\ \mu\text{s}}{0.109}$$

$$= 14.3\ \mu\text{s}$$

Due to time dilation the half-life as measured in the Earth frame is about 9 times longer than the half-life as measured in the muon frame.

The effect of time dilation causes time to run more slowly for the muon than we measure in our own frame, so fewer half-lives elapse for a given distance traveled. Because fewer half-lives have elapsed, fewer muons have decayed.

From part (a), the time for a muon to travel to the surface is 50.3 μs. Expressed as a multiple of the time-dilated half-life, this is

$$\frac{T}{\Delta t} = \frac{50.3\ \mu\text{s}}{14.3\ \mu\text{s}} = 3.53 \text{ half-lives}$$

The number of muons remaining at the surface is the original number of 1.00×10^6 multiplied by $1/2^{3.53}$:

$$\text{muons remaining} = (1.00 \times 10^6)\left(\frac{1}{2^{3.53}}\right)$$
$$= (1.00 \times 10^6)(0.0868)$$
$$= 8.68 \times 10^4$$

About 1 in 11 of the muons produced in the upper atmosphere reaches Earth's surface, far more than would be the case were there no time dilation.

Reflect

One of the earliest confirmations of Einstein's special theory of relativity was a 1941 experiment that compared the number of muons observed in an hour at two elevations in Colorado, 3240 and 1616 m above sea level. Without time dilation, the number of muons observed at the lower elevation would have been about 9% of the number at the higher elevation. The measured number at the lower elevation was close to 80%, in accordance with Einstein's prediction. This experiment provided dramatic evidence that the bizarre phenomenon of time dilation is very real.

NOW WORK Problem 3 from the Takeaway for Section 26-4.

The Twin Paradox

Imagine two twins named Bertha and Eartha. Bertha takes a round trip from Earth to another star and back on a fast space ship that travels at $v = 0.866c$, while her twin Eartha stays at home on Earth. Eartha's clock ticks off 10 years during the time that Bertha is gone. As we saw in Example 26-2, if a clock moves past you at this speed, you will observe it as running at one-half the rate of a clock in your hand. So Eartha expects that when her twin returns, Bertha's clock will have ticked off only 5 years, not 10, and that Bertha will have aged only 5 years. So Eartha predicts that *Bertha* will actually be younger than Eartha after the round trip is done.

But from Bertha's reference frame, Eartha's clock is the one that's moving at $0.866c$, so Bertha observes Eartha's clock to be running slow. Hence Bertha expects that when she returns to Earth, she will find that Eartha's clock has ticked off half as much time as Bertha's clock on the space ship. Therefore, Bertha predicts that Eartha will have aged only half as much as Bertha and that *Eartha* will be the younger twin after the round trip is complete.

It seems like the experiences of the two twins are exactly symmetrical: Each twin sees the other one moving, and hence aging more slowly, and predicts that the other one will have aged less. But clearly both twins can't be correct. So how can we resolve this *twin paradox*?

The explanation is that the two observers, Eartha and Bertha, are *not* perfectly symmetrical. Eartha remained in the same inertial reference frame (Earth) at all times during Bertha's round trip, but Bertha did not: On the outbound leg Bertha was in a frame moving at $0.866c$ in the direction from Earth toward the other star, while on the return leg Bertha was moving at $0.866c$ in the *opposite* direction from the star back toward Earth. It turns out that because Bertha changed her reference frame and her velocity—that is, because she *accelerated*—she is the twin who ages less during the round trip. So Eartha is correct, and after the round trip Eartha will have aged 10 years, while Bertha will have aged only 5 years.

This effect has been experimentally verified many times. If an unstable particle like a muon (Example 26-3) is produced in the laboratory and sent at relativistic speed along a circular round-trip path, its half-life is longer than if the particle is produced at rest; in other words, the particle that takes the round trip ages more slowly. The effect has also been tested using very precise clocks, one that remained in the laboratory and one that was flown on a round trip in a fast-moving airplane. The clock that traveled out and back ticked off a fraction of a second less time than did the clock that stayed at rest, just as predicted by special relativity.

THE TAKEAWAY for Section 26-4

✔ Einstein based his special theory of relativity on two postulates: first that all laws of physics are the same in all inertial reference frames and second that the speed of light in a vacuum is the same in all frames and independent of both the speed of the source of the light and the speed of the observer.

✔ If two events happen at the same place in one frame, the time interval between these events as measured in that frame is called the proper time. As measured from a second frame moving relative to the first one, the time interval between those two events is longer than the proper time. This is called time dilation.

Takeaway Problems for Section 26-4

1. A clock is placed aboard Starship *Alpha*, which Albert flies past Earth at half the speed of light relative to Earth. Barbara flies Starship *Beta* alongside *Alpha* at the same velocity. George pilots Starship *Gamma* past Earth at half the speed of light relative to Earth, but in the opposite direction. Elena observes from Earth. For which of these observers is the time interval between ticks of the clock equal to the proper time?
(A) Albert
(B) Barbara and Albert
(C) Elena
(D) George

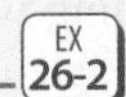

2. An observer in reference frame S observes that a lightning bolt strikes the origin, and 10^{-4} s later a second lightning bolt strikes the same location. What is the time separation between the two lightning bolts determined by a second observer in reference frame S' moving at a speed of $0.8c$ along the collinear x–x' axis?

3. A radioactive particle travels at $0.80c$ relative to the laboratory observers who are performing research. Calculate the half-life of the particle as measured in the laboratory frame compared to the half-life according to the proper frame of the particle.

26-5 Einstein's relativity also predicts that the length of an object depends on the observer

We learned in the preceding section that the time interval between two events is not an absolute but depends on the motion of the observer. As we will see in this section, the *length* of an object is also not an absolute: Different observers will measure the same object as having different dimensions. So the nature of space and time is very different from what had been thought previous to the work of Einstein.

We'll conclude this section by looking at the *Lorentz transformation*, a set of equations that allows us to convert the space and time coordinates of an event in one inertial reference frame to the coordinates in a second inertial reference frame. This transformation is an extension of the Galilean transformation that we explored in Section 26-2. We'll see that the Galilean transformation is actually just a special case of the Lorentz transformation valid for two frames that are moving with respect to each other at a speed far less than c.

Length Contraction

Figure 26-12 shows a thought experiment in which we look at the same motion in two different reference frames, much as we did in Figures 26-9 and 26-10. In **Figure 26-12a** a rod is moving to the right at speed V relative to frame S, which you can think of as our reference frame. The right-hand end of the rod is at $x = 0$ at time $t = 0$, and the left-hand end of the rod is at $x = 0$ at a later time $t = \Delta t_S$. The length L of the rod in our frame is therefore the speed V of the rod multiplied by the time Δt_S needed to travel its own length:

(26-11)
$$L = V\Delta t_S$$
(length of the rod in frame S)

Figure 26-12b shows the same process as observed in the reference frame S' in which the rod is at rest. In this frame the rod has length L_{rest}, and the frame S moves to the left at speed V. The point $x = 0$ on frame S travels from one end of the rod to the other in a time $\Delta t_{S'}$. The length of the rod equals the speed V of frame S multiplied by the time $\Delta t_{S'}$ that frame S needs to travel this length:

(26-12)
$$L_{rest} = V\Delta t_{S'}$$
(length of the rod in its own rest frame S')

To see how the lengths as measured in the two frames compare, note that the length measurements in S and S' both involve the same pair of events: the right-hand end of the rod coinciding with the point $x = 0$ in frame S, and the left-hand end of the rod coinciding with that same point (events 1 and 2 in Figure 26-12). These two events happen at the same location in frame S but at different locations in frame S'. So Δt_S is the proper time interval between these events. The time interval $\Delta t_{S'}$ measured in frame

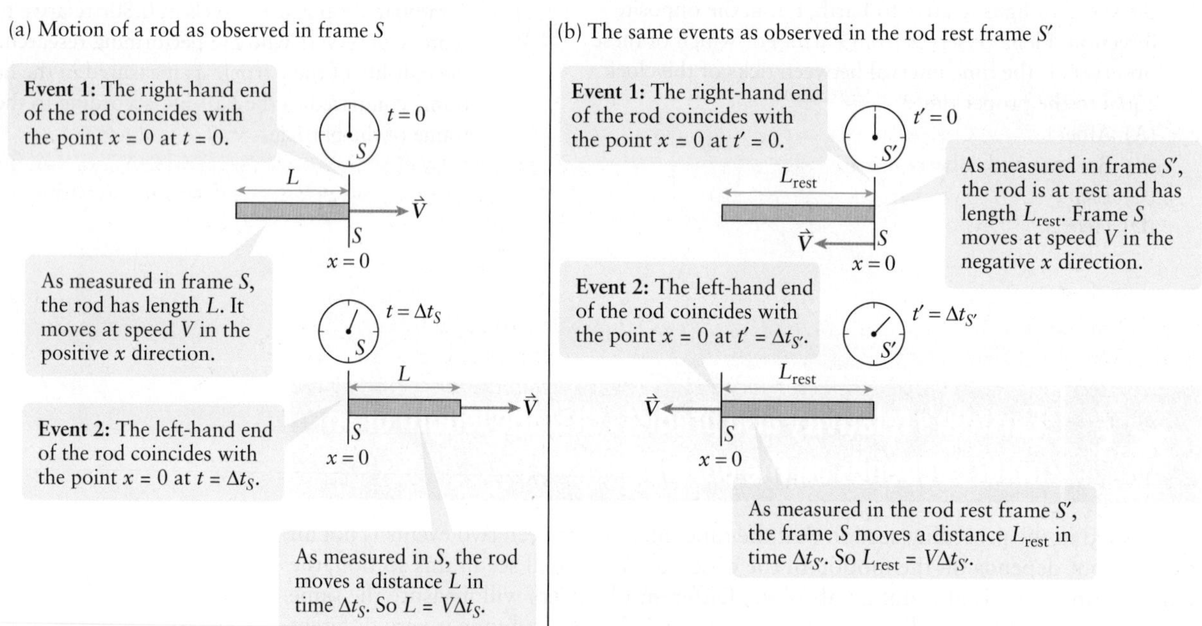

(a) Motion of a rod as observed in frame S
Event 1: The right-hand end of the rod coincides with the point x = 0 at t = 0.
t = 0
S
L
V
x = 0
As measured in frame S, the rod has length L. It moves at speed V in the positive x direction.
t = Δt_S
Event 2: The left-hand end of the rod coincides with the point x = 0 at t = Δt_S.
As measured in S, the rod moves a distance L in time Δt_S. So L = VΔt_S.
(b) The same events as observed in the rod rest frame S′
Event 1: The right-hand end of the rod coincides with the point x = 0 at t′ = 0.
t′ = 0
S′
L_rest
As measured in frame S′, the rod is at rest and has length L_rest. Frame S moves at speed V in the negative x direction.
Event 2: The left-hand end of the rod coincides with the point x = 0 at t′ = Δt_S′.
t′ = Δt_S′
As measured in the rod rest frame S′, the frame S moves a distance L_rest in time Δt_S′. So L_rest = VΔt_S′.

- The two events are (1) the right-hand end of the rod coinciding with $x = 0$ and (2) the left-hand end of the rod coinciding with $x = 0$.
- These two events happen at the same place in frame S, so Δt_S is the proper time interval between these events. The time interval $\Delta t_{S'}$ must therefore be greater than Δt_S (time dilation).
- The distance L_{rest} must therefore be greater than the distance L. So the length L of the moving rod is less than the length L_{rest} of the rod in a frame where it is at rest. This is length contraction.

Figure 26-12 A moving rod and length contraction A rod as observed in (a) a frame in which the rod is moving along its length and (b) a frame in which the rod is at rest.

S' is therefore longer ("dilated") compared to the time interval Δt_S measured in frame S, and the two time intervals are related by Equation 26-10:

$$\Delta t_{S'} = \frac{\Delta t_S}{\sqrt{1 - \frac{V^2}{c^2}}} \tag{26-13}$$

Multiply both sides of Equation 26-13 by V then replace $V\Delta t_S$ by L in accordance with Equation 26-11 and replace $V\Delta t_{S'}$ with L_{rest} according to Equation 26-12:

$$V\Delta t_{S'} = \frac{V\Delta t_S}{\sqrt{1 - \frac{V^2}{c^2}}} \quad \text{or} \quad L_{\text{rest}} = \frac{L}{\sqrt{1 - \frac{V^2}{c^2}}}$$

We can rewrite this as

$$L = L_{\text{rest}}\sqrt{1 - \frac{V^2}{c^2}} \tag{26-14}$$

Length of an object in frame S', in which it is at rest

Length of the same object in frame S, in which the object is moving along its length

Speed of S' relative to S

Speed of light in a vacuum

EQUATION IN WORDS
Length contraction

If the rod is moving, V is greater than zero and the factor $\sqrt{1 - V^2/c^2}$ is less than 1. So the length L of the moving rod is less than the length L_{rest} of the rod at rest. This is called **length contraction:** *A moving object is shortened along the direction in which it is moving.*

WATCH OUT!

Length contraction occurs only along the direction of motion.

There is *no* change in length of a moving object in any direction other than the direction of motion. For example, if the rod in Figure 26-12 is moving in the x direction relative to frame S, an observer in S will measure the rod as having a shorter length in the x direction than will an observer in S' at rest with respect to the rod. But both observers will agree about the height and width of the rod (its dimensions in the y and z directions).

Length contraction gives us an alternative way to understand the results of Example 26-3 in the preceding section, in which we used time dilation to explain how short-lived muons produced at an altitude of 15.0 km are able to survive their trip to Earth's surface. Imagine a vertical rod that extends 15.0 km upward from the ground to where the muons are produced. This rod is stationary relative to Earth, so its length as measured by you on the ground is its rest length: $L_{\text{rest}} = 15.0 \text{ km} = 1.50 \times 10^4 \text{ m}$. But as seen from the frame of a descending muon, this rod is moving upward along its length at $V = 0.994c$. As measured by a muon, the length of this rod is contracted in accordance with Equation 26-14:

$$\begin{aligned} L &= L_{\text{rest}}\sqrt{1 - \frac{V^2}{c^2}} = (15.0 \text{ km})\sqrt{1 - \frac{(0.994c)^2}{c^2}} \\ &= (15.0 \text{ km})\sqrt{1 - (0.994)^2} \\ &= (15.0 \text{ km})\sqrt{0.0120} = (15.0 \text{ km})(0.109) \\ &= 1.64 \text{ km} = 1.64 \times 10^3 \text{ m} \end{aligned}$$

Moving at $V = 0.994c$, this contracted rod travels past the muon in a time

$$\frac{1.64 \times 10^3 \text{ m}}{0.994c} = \frac{1.64 \times 10^3 \text{ m}}{(0.994)(3.00 \times 10^8 \text{ m/s})} = 5.50 \times 10^{-6} \text{ s} = 5.50 \text{ μs}$$

The half-life of the muon in its own rest frame is 1.56 μs, so from the muon's perspective it takes just (5.50 μs)/(1.56 μs) = 3.53 half-lives for the contracted rod to move past it—that is, for the muon to move from where it is produced to Earth's surface. That's exactly the result that we found in Example 26-3 using the ideas of time dilation. The ideas of length contraction and time dilation are mutually consistent!

EXAMPLE 26-4 A Flying Meter Stick I

A meter stick (length 1.00 m) hurtles through space at a speed of $0.800c$ relative to you, with its length aligned with the direction of motion. What do you measure as the length of the meter stick?

Set Up

The meter stick's length along its direction of motion is 1.00 m as measured in its own rest frame S', so $L_{rest} = 1.00$ m. Your frame is frame S, and the two frames are moving relative to each other at $V = 0.800c$. We'll use Equation 26-14 to find the length L as measured in your frame.

Length contraction:

$$L = L_{rest}\sqrt{1-\frac{V^2}{c^2}} \quad (26\text{-}14)$$

Solve

Substitute $L_{rest} = 1.00$ m and $V = 0.800c$ into Equation 26-14 and calculate L.

From Equation 26-14:

$$L = L_{rest}\sqrt{1-\frac{V^2}{c^2}} = (1.00\text{ m})\sqrt{1-\frac{(0.800c)^2}{c^2}}$$

$$= (1.00\text{ m})\sqrt{1-(0.800)^2} = (1.00\text{ m})\sqrt{1-0.640}$$

$$= (1.00\text{ m})\sqrt{0.360} = (1.00\text{ m})(0.600)$$

$$= 0.600\text{ m} = 60.0\text{ cm}$$

Reflect

As measured in your reference frame, the meter stick is only 60.0 cm in length. This is not an optical illusion; the meter stick really is only 60.0% as long in your reference frame as in the rest frame of the meter stick.

NOW WORK Problem 1 from the Takeaway for Section 26-5.

EXAMPLE 26-5 A Flying Meter Stick II

The meter stick from the preceding example again hurtles through space at a speed of $0.800c$ relative to you, but now it is tilted at an angle of 30.0° as measured in its rest frame with respect to the direction of motion. Now what do you measure as the length of the meter stick?

Set Up

As measured in the meter stick's rest frame S', the meter stick is inclined at an angle $\theta' = 30.0°$ to its direction of motion relative to frame S. The x dimension of the stick undergoes length contraction given by Equation 26-14, but the y dimension does not. We'll calculate the x and y dimensions of the stick in your frame S, then use the Pythagorean theorem to find the length of the stick in frame S.

Length contraction:

$$L = L_{rest}\sqrt{1-\frac{V^2}{c^2}} \quad (26\text{-}14)$$

Solve

First calculate the x and y dimensions of the meter stick in its rest frame.

In the rest frame S' of the stick, its dimensions are

$$L_{\text{rest},x} = L_{\text{rest}}\cos\theta' = (1.00\text{ m})\cos 30.0° = 0.866\text{ m}$$
$$L_{\text{rest},y} = L_{\text{rest}}\sin\theta' = (1.00\text{ m})\sin 30.0° = 0.500\text{ m}$$

Calculate the x and y dimensions of the meter stick in your frame, in which the stick is moving at $V = 0.800c$ along the x direction.

In your frame S, the x dimension of the stick is contracted:

$$L_x = L_{\text{rest},x}\sqrt{1-\frac{V^2}{c^2}} = (0.866\text{ m})\sqrt{1-\frac{(0.800c)^2}{c^2}}$$
$$= (0.866\text{ m})\sqrt{1-0.640} = (0.866\text{ m})\sqrt{0.360}$$
$$= 0.520\text{ m}$$

Length contraction occurs only along the direction of motion, so the y dimension of the stick is *not* contracted:

$$L_y = L_{\text{rest},y} = 0.500\text{ m}$$

Use the Pythagorean theorem to calculate the length of the meter stick in your frame.

From the Pythagorean theorem, the length of the meter stick in your frame S is

$$L = \sqrt{L_x^2 + L_y^2}$$
$$= \sqrt{(0.520\text{ m})^2 + (0.500\text{ m})^2}$$
$$= \sqrt{(0.520\text{ m})^2 + (0.500\text{ m})^2}$$
$$= \sqrt{0.520\text{ m}^2} = 0.721\text{ m}$$

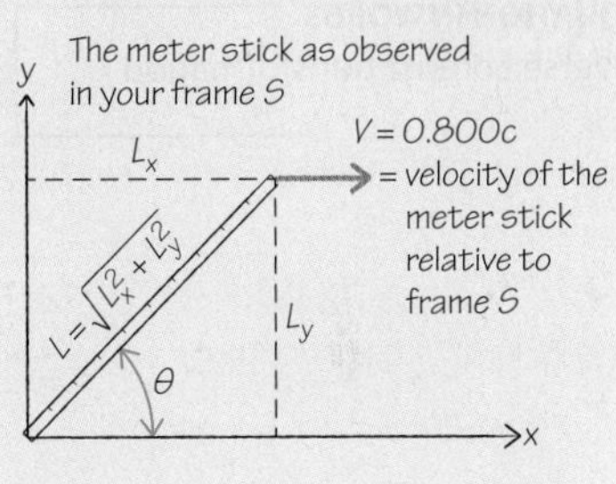

Reflect

Because the meter stick is not aligned with the direction of motion, the amount of contraction is not as great as in Example 26-4, in which the meter stick was completely aligned with the direction of motion. Notice also that the angle θ of the meter stick in your frame is different from the angle $\theta' = 30.0°$ in the stick's rest frame. Can you show that $\theta = 43.9°$?

NOW WORK Problem 2 from the Takeaway for Section 26-5.

The Lorentz Transformation

The Galilean transformation that we presented in Section 26-2 is not consistent with the postulates of special relativity. A set of transformation equations that *is* consistent with relativity is the **Lorentz transformation.** As in Figure 26-4, we take frame S' to be moving at speed V in the positive x direction relative to frame S. The origins of the two frames coincide at $t = 0$ in frame S and $t' = 0$ in frame S'. If an event takes place at coordinates x, y, z, and t in frame S, the coordinates of that same event in frame S' are

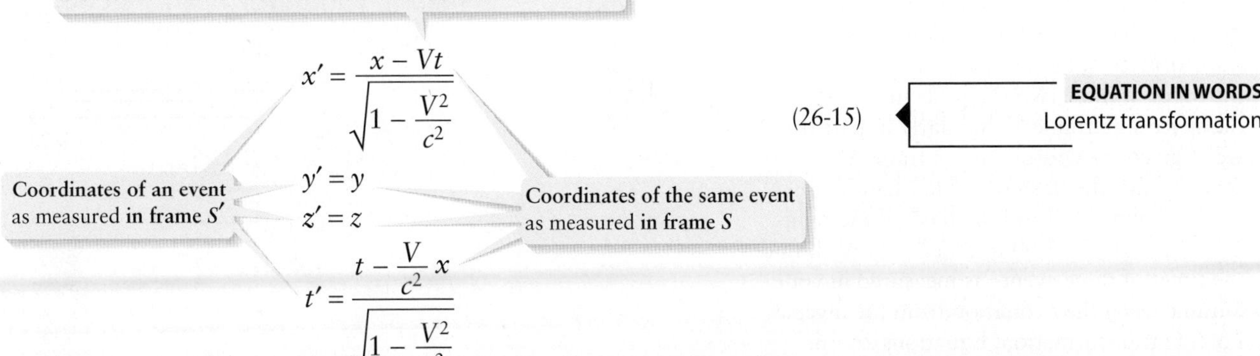

$$x' = \frac{x - Vt}{\sqrt{1-\frac{V^2}{c^2}}}$$
$$y' = y$$
$$z' = z \qquad (26\text{-}15)$$
$$t' = \frac{t - \frac{V}{c^2}x}{\sqrt{1-\frac{V^2}{c^2}}}$$

EQUATION IN WORDS
Lorentz transformation

Note that at speeds that are far less than the speed of light, the ratio V/c is very small and can be treated as essentially zero. Then $\sqrt{1-V^2/c^2}$ is essentially equal to 1, and Equations 26-15 become

$$x' = x - Vt, \quad y' = y, \quad z' = z, \quad t' = t$$

These are just the Galilean transformation equations that we presented in Section 26-2. So at speeds far slower than the speed of light, the equations of Einstein's special theory of relativity reduce to the equations of Newtonian relativity.

Equations 26-15 are useful for finding the coordinates of an event in frame S' if we know the event's coordinates in frame S. If we instead want to determine the coordinates of an event in frame S from the coordinates in frame S', it's most convenient to use the *inverse* Lorentz transformation:

EQUATION IN WORDS
Inverse Lorentz transformation

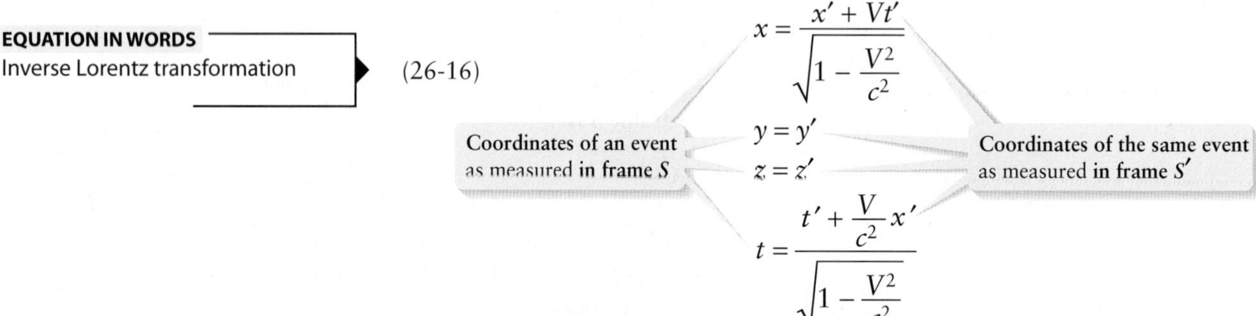
Inertial frame of reference S′ moves at speed V in the positive x direction relative to inertial frame of reference S.
Coordinates of an event as measured in frame S
Coordinates of the same event as measured in frame S′

$$x = \frac{x' + Vt'}{\sqrt{1 - \frac{V^2}{c^2}}} \qquad y = y' \qquad z = z' \qquad t = \frac{t' + \frac{V}{c^2}x'}{\sqrt{1 - \frac{V^2}{c^2}}} \tag{26-16}$$

It's possible to derive the equations for time dilation (Equation 26-10) and length contraction (Equation 26-14) from the Lorentz transformation equations. Instead let's look at another remarkable result of the special theory of relativity that we can deduce from the Lorentz transformation equations.

EXAMPLE 26-6 Simultaneity Is Relative

In your reference frame you have a meter stick that is oriented along the x axis, with one end at $x = 0$ and the other end at $x = 1.00$ m. There is a light bulb at each end of the stick, and you make the two bulbs flash simultaneously (as measured by you) at $t = 0$. A spacecraft flies past you at $V = 0.800c$ in the positive x direction. (a) According to an observer in the spacecraft, are the two light flashes simultaneous? If not, which flash happens first as measured by her? (b) On board the spacecraft is an identical meter stick with a light bulb at each end. The observer on board the spacecraft places the two ends of the stick at $x' = 0$ and $x' = 1.00$ m, and she makes the two bulbs flash simultaneously (as measured by her) at $t' = 0$. According to you, are the two light flashes simultaneous? If not, which flash happens first as measured by you?

Set Up

In part (a) we're given the coordinates in frame S of two events, the flash of the left-hand bulb at $x = 0$ and $t = 0$ and the flash of the right-hand bulb at $x = 1.00$ m and $t = 0$. We'll use the t' equation from the Lorentz transformation, Equations 26-15, to determine the times of these two events as measured in the space ship frame S'. Similarly in part (b) we're given the coordinates in frame S' of two other events, the flash of the left-hand bulb at $x' = 0$ and $t' = 0$ and the flash of the right-hand bulb at $x' = 1.00$ m and $t' = 0$. We'll find the times of these events as measured in your S frame using the t equation from the inverse Lorentz transformation, Equations 26-16.

Time equation from the Lorentz transformation:

$$t' = \frac{t - \frac{V}{c^2}x}{\sqrt{1 - \frac{V^2}{c^2}}} \tag{26-15}$$

(a) Meter stick at rest in frame S:
What does an observer in frame S′ measure?

x = 0
t = 0

x = 1.00 m
t = 0

x

y′
V = 0.800c
S′
x′
z′

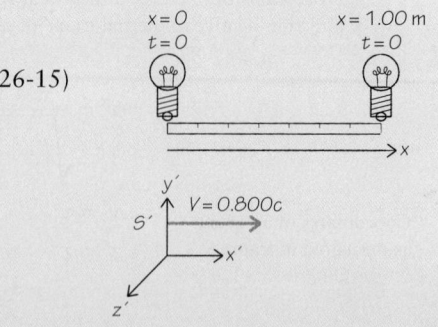

Time equation from the inverse Lorentz transformation:

$$t = \frac{t' + \frac{V}{c^2}x'}{\sqrt{1 - \frac{V^2}{c^2}}} \qquad (26\text{-}16)$$

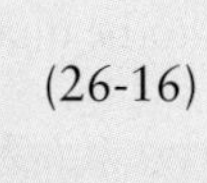

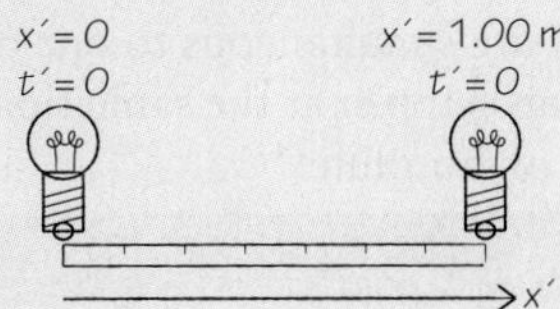

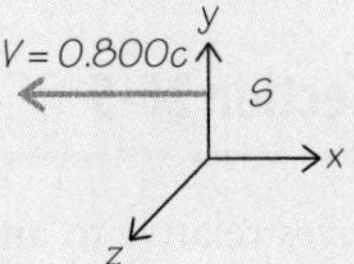

(b) Meter stick at rest in frame S': What does an observer in frame S measure?

Solve

(a) Use the Lorentz transformation to calculate the times of the simultaneous flashes in S as measured in S'.

The left-hand bulb on the meter stick at rest in frame S flashes at $x = 0$, $t = 0$. In frame S' this bulb flashes at

$$t'_{\text{left}} = \frac{(0) - \frac{V}{c^2}(0)}{\sqrt{1 - \frac{V^2}{c^2}}} = 0$$

The right-hand bulb on the meter stick at rest in frame S flashes at $x = 1.00$ m, $t = 0$. In frame S' this bulb flashes at

$$t'_{\text{right}} = \frac{(0) - \frac{V}{c^2}(1.00\text{ m})}{\sqrt{1 - \frac{V^2}{c^2}}} = \frac{-\frac{(0.800c)}{c^2}(1.00\text{ m})}{\sqrt{1 - \frac{(0.800c)^2}{c^2}}}$$

$$= \frac{-0.800\text{ m}}{c\sqrt{1 - (0.800)^2}} = \frac{-0.800\text{ m}}{(3.00 \times 10^8\text{ m/s})(0.600)}$$

$$= -4.44 \times 10^{-9}\text{ s}$$

As observed from the space ship frame S', the two events are *not* simultaneous: The right-hand bulb flashes 4.44×10^{-9} s *before* the left-hand bulb.

(b) Use the inverse Lorentz transformation to calculate the times of the simultaneous flashes in S' as measured in S.

The left-hand bulb on the meter stick at rest in frame S' flashes at $x' = 0$, $t' = 0$. In frame S this bulb flashes at

$$t_{\text{left}} = \frac{(0) + \frac{V}{c^2}(0)}{\sqrt{1 - \frac{V^2}{c^2}}} = 0$$

The right-hand bulb on the meter stick at rest in frame S' flashes at $x' = 1.00$ m, $t' = 0$. In frame S this bulb flashes at

$$t_{\text{right}} = \frac{(0) + \frac{V}{c^2}(1.00\text{ m})}{\sqrt{1 - \frac{V^2}{c^2}}} = \frac{+\frac{(0.800c)}{c^2}(1.00\text{ m})}{\sqrt{1 - \frac{(0.800c)^2}{c^2}}}$$

$$= \frac{+0.800\text{ m}}{c\sqrt{1 - (0.800)^2}} = \frac{+0.800\text{ m}}{(3.00 \times 10^8\text{ m/s})(0.600)}$$

$$= +4.44 \times 10^{-9}\text{ s}$$

As observed from your frame S, the two events are *not* simultaneous: The right-hand bulb flashes 4.44×10^{-9} s *after* the left-hand bulb.

Reflect

This example illustrates yet another counterintuitive consequence of Einstein's special theory of relativity: Two events that are simultaneous to one observer need not be simultaneous to another observer. So even the simple statement "Two things happen at the same time" has to be qualified by stating in which reference frame it holds true. Time in relativity is not an absolute!

NOW WORK Problem 3 from the Takeaway for Section 26-5.

THE TAKEAWAY for Section 26-5

✔ When an object moves relative to an observer, its length in the direction of motion is contracted compared to the length measured in the object's rest frame. This is called length contraction.

✔ The Lorentz transformation allows you to calculate the coordinates of an event measured in one inertial reference frame based on the coordinates of that event measured in another inertial reference frame. Unlike the Galilean transformation, the Lorentz transformation is consistent with the postulates of relativity.

Takeaway Problems for Section 26-5

1. A stick moves past an observer at a speed of $0.44c$. According to the observer the stick is oriented parallel to the direction of motion and is 0.88 m long. Determine the proper length of the stick.

2. A standard tournament domino is 1.5 in. wide and 2.5 in. long. Describe how you might orient a domino so that it will measure 1.5 in. by 1.5 in. as it moves by. What relative speed is required?

3. A spacecraft passes you traveling at $0.600c$. Your alien friend Gaar on the space ship measures the length of the ship as 60 m from front to back. He stands in the middle of the ship and fires one photon backward and one photon forward, and each photon strikes a detector at each end of the ship.
   (a) According to Gaar, what is the travel time of each photon? Verify that he reports the detection events as simultaneous.
   (b) According to you, which photon registers first?

   (c) Use the Lorentz transformation, and the fact that the detection events are simultaneous according to Gaar, to find the time difference according to you.

26-6 The speed of light is the ultimate speed limit

Suppose you are an outfielder running to catch a batted baseball (**Figure 26-13a**). The baseball is traveling at 30.0 m/s relative to the ground, and you are running toward the ball at 10.0 m/s relative to the ground. The Galilean velocity transformation that we learned in Section 26-2 says that, relative to you, the baseball travels at 30.0 m/s plus 10.0 m/s, or 40.0 m/s; the velocities simply add. But suppose instead that you are an astronaut flying in your space ship at 1.00×10^8 m/s, as in **Figure 26-13b**, and you are moving toward a light beam aimed at you by a stationary astronaut. The same idea that we applied to the baseball predicts that relative to you the light beam travels at $c = 3.00 \times 10^8$ m/s (the speed of the light beam relative to the other astronaut) plus 1.00×10^8 m/s (the speed of your space ship relative to the other astronaut), or 4.00×10^8 m/s. But that *cannot* be correct: The second postulate of the special theory of relativity says that the speed of light in a vacuum is the same to all inertial observers. So the light beam must also travel at speed $c = 3.00 \times 10^8$ m/s relative to you. Clearly the Galilean transformation for velocities is inadequate. In this section we'll explore the *Lorentz velocity transformation*, which allows us to combine velocities in a way that is consistent with Einsteinian relativity.

It's possible to derive the transformation of velocities from the Lorentz transformation of coordinates that we introduced in Section 26-5 (see Equations 26-15 and 26-16). We'll skip over the derivation and just present the result for the special case in which all motions are along the same line, as in Figure 26-13. Suppose that inertial reference frame S' is moving at speed V in the positive x direction relative to inertial reference frame S.

(a)

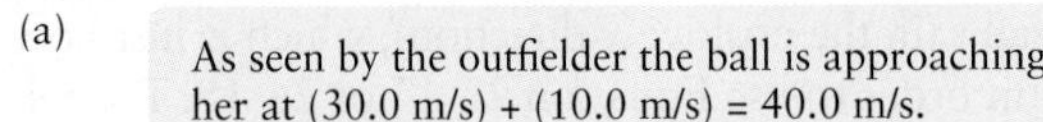

As seen by the outfielder the ball is approaching her at (30.0 m/s) + (10.0 m/s) = 40.0 m/s.

10.0 m/s
30.0 m/s

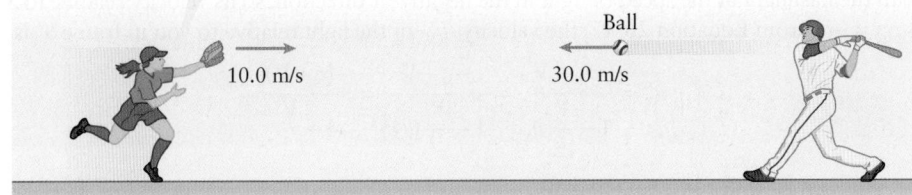

(b)

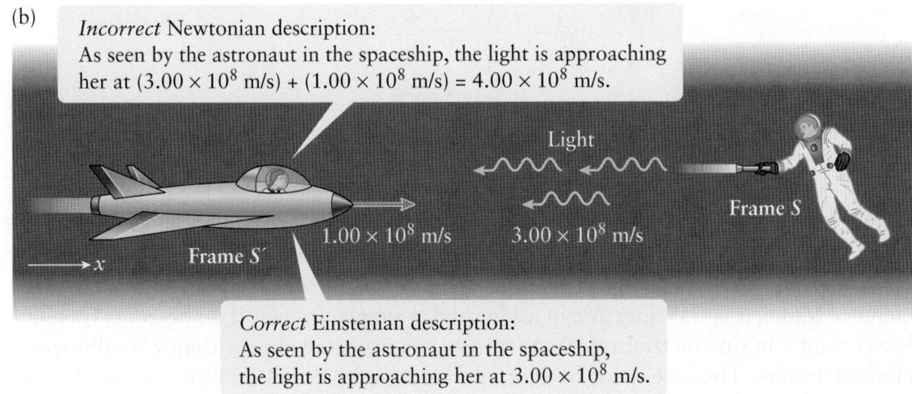

Incorrect Newtonian description:
As seen by the astronaut in the spaceship, the light is approaching her at (3.00 × 10^8 m/s) + (1.00 × 10^8 m/s) = 4.00 × 10^8 m/s.

Correct Einstenian description:
As seen by the astronaut in the spaceship, the light is approaching her at 3.00 × 10^8 m/s.

Figure 26-13 Velocity addition—Newtonian and Einsteinian (a) An outfielder running toward a batted baseball. (b) An astronaut in her space ship flying toward a light beam.

An object is moving relative to frame S along the x direction, with an x component of velocity v_x. In frame S' the same object has an x component of velocity given by

Inertial frame of reference S' moves at speed V in the positive x direction relative to inertial frame of reference S.

$$v_x' = \frac{v_x - V}{1 - \dfrac{V}{c^2}v_x} \qquad (26\text{-}17)$$

x component of velocity of an object moving along the x axis as measured in frame S'

x component of velocity of the same object as measured in frame S

EQUATION IN WORDS
Lorentz velocity transformation

This is called the **Lorentz velocity transformation**. Notice that if the speed V of one frame relative to the other is small compared to c, the ratio V/c is much less than 1 and the denominator is essentially equal to 1. The same thing happens if the velocity v_x of the object relative to frame S is small compared to c. So if either of the speeds involved is a small fraction of the speed of light, Equation 26-17 reduces to $v_x' = v_x - V$, which is just the Galilean velocity transformation (the first of Equations 26-5). This justifies our use of the Galilean velocity transformation for slow-moving objects.

Equation 26-17 allows us to find the object's velocity relative to frame S' if we know its velocity relative to frame S. If instead we know the object's velocity relative to frame S' and want to calculate its velocity relative to frame S, we use the *inverse* Lorentz velocity transformation:

Inertial frame of reference S' moves at speed V in the positive x direction relative to inertial frame of reference S.

$$v_x = \frac{v_x' + V}{1 + \dfrac{V}{c^2}v_x'} \qquad (26\text{-}18)$$

x component of velocity of an object moving along the x axis as measured in frame S

x component of velocity of the same object as measured in frame S'

EQUATION IN WORDS
Inverse Lorentz velocity transformation

1.00 × 10^8 m/s
3.00 × 10^8 m/s

Let's see what the Lorentz velocity transformation tells us about the situation shown in Figure 26-13b. The astronaut with the flashlight is in frame S, and you and your space ship are in frame S'. You are moving to the right (in the positive x direction), which is just how frame S' must move relative to frame S in order to use Equation 26-17 or 26-18. The relative speed of the two frames is V. We know that the light travels relative to the astronaut with the flashlight at the speed of light in the negative x direction, so its velocity relative to S is $v_x = -c$. From Equation 26-17, the velocity v'_x of the light relative to you in frame S' is

$$v'_x = \frac{v_x - V}{1 - \frac{V}{c^2} v_x} = \frac{-c - V}{1 - \frac{V}{c^2}(-c)} = \frac{-(c + V)}{1 + \frac{V}{c}}$$

We can simplify this by factoring c out of the numerator:

$$v'_x = \frac{-c\left(1 + \frac{V}{c}\right)}{1 + \frac{V}{c}} = -c$$

This says that as measured by you in frame S', the light travels at the speed of light c in the negative x direction just as in frame S. Note that while $V = 1.00 \times 10^8$ m/s in Figure 26-13b, our result doesn't depend on the value of V: No matter what the relative speed of the two frames, each observer will see light propagating in a vacuum at the same speed c.

A direct consequence of this calculation is that *no object can move faster than c in any inertial reference frame*. If an object (light) is traveling at speed c in one inertial reference frame, it is traveling at c in all inertial reference frames; if an object is traveling slower than c in one inertial reference frame, it is traveling slower than c in all inertial reference frames. Thus the speed of light in a vacuum represents an ultimate speed limit.

EXAMPLE 26-7 Baseball for Superheroes

Suppose the baseball game in Figure 26-13 is being played by superheroes. The outfielder has super speed and can run at $0.300c$ relative to the ground. The batter has super strength and can bat the ball with such force that the ball ends up traveling horizontally at $0.900c$ relative to the ground. (The bat and ball are made of super materials that can withstand the tremendous forces required.) How fast is the ball moving relative to the outfielder?

Set Up

This is nearly the same situation that we discussed above with the two astronauts and the beam of light, except that the astronauts have been replaced by (super) baseball players and the beam of light has become a baseball. We use Equation 26-17 to calculate the velocity of the ball relative to the outfielder in frame S'.

Lorentz velocity transformation:

$$v'_x = \frac{v_x - V}{1 - \frac{V}{c^2} v_x} \quad (26\text{-}17)$$

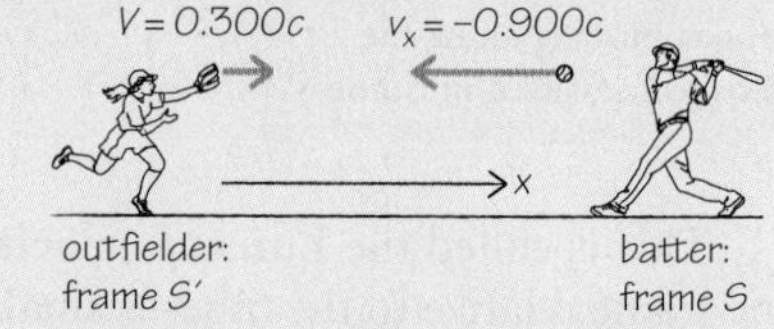

Solve

Use the velocity of the ball relative to the batter (v_x) and the velocity of the outfielder relative to the batter (V) to find the velocity of the ball relative to the outfielder (v'_x).

The outfielder (frame S') is moving in the positive x direction relative to the batter (frame S) at $V = 0.300c$. The velocity of the ball relative to the batter (frame S) is $v_x = -0.900c$ (negative because the ball is moving in the negative x direction). From Equation 26-17 the velocity of the ball relative to the outfielder is

$$v'_x = \frac{v_x - V}{1 - \frac{V}{c^2} v_x} = \frac{-0.900c - 0.300c}{1 - \frac{0.300c}{c^2}(-0.900c)}$$

$$= \frac{-1.200c}{1 + (0.300)(0.900)} = \frac{-1.200c}{1.27}$$

$$= -0.945c$$

Relative to the outfielder the ball is moving at $0.945c$ in the negative x direction (to the left).

Reflect

In the Galilean velocity transformation the velocity of the ball relative to the outfielder would have been $v'_x = v_x - V = -0.900c - 0.300c = -1.200c$, which is faster than c. Thanks to the $1 - (V/c^2)v_x$ term in the denominator of Equation 26-17, the actual speed of the ball relative to the outfielder is slower than c. The batter may be super-powered, but the ball cannot exceed the speed of light in a vacuum as measured by any observer.

NOW WORK Problems 2 and 3 from the Takeaway for Section 26-6.

THE TAKEAWAY for Section 26-6

✔ At speeds that are an appreciable fraction of the speed of light, we must use the Lorentz velocity transformation to calculate relative velocities. This transformation respects the rule that the speed of light in a vacuum is the same in all inertial reference frames.

✔ No object can move faster than c, the speed of light in a vacuum, in any inertial reference frame.

Takeaway Problems for Section 26-6

1. Is it possible to accelerate an object to the speed of light in a real situation? Explain your answer.

EX 26-7 2. Space ship A moves at $0.80c$ toward the right, while space ship B moves in the opposite direction at $0.70c$ (both speeds are measured relative to Earth).
   (a) Calculate the velocity of Earth relative to space ship A.
   (b) Calculate the velocity of Earth relative to space ship B.
   (c) Calculate the velocity of space ship A relative to space ship B.

3. Suppose the super outfielder in Example 26-7 catches the ball and throws it back toward the super batter, who is still standing at home plate. If the outfielder is still running at $0.300c$ and she throws the ball at $0.700c$ relative to her, what is the speed of the ball relative to the batter? (A) $0.331c$; (B) $0.506c$; (C) $0.826c$; (D) $1.00c$; (E) $1.26c$.

26-7 The equations for kinetic energy and momentum must be modified at very high speeds

We have seen that the speed of light in a vacuum c is an ultimate speed limit: No object can travel faster than c in any inertial reference frame. As we'll see in this section, this tells us that the expressions we learned earlier in this book for kinetic energy K (Chapter 7) and momentum $\vec{p}$ (Chapter 9) *cannot* be entirely correct. They fail at speeds that are a reasonable fraction of c. We'll see why this must be so and encounter a new kind of energy called *rest energy* that is intrinsic to any object with mass.

In Chapter 7, we introduced kinetic energy through the work-energy theorem for an object, which allowed us to find the change in kinetic energy as the work done on an object (the force exerted on the object times the displacement of its center of mass in the direction of that force). This says that if an object starts at rest so that its initial kinetic energy K_i is zero, the greater the work done on the object, the greater its final kinetic energy K_f. In principle there is no limit to how much kinetic energy an object can acquire. There is a problem, however: Using Newtonian physics we found that the expression for the kinetic energy K of an object of mass m moving at speed v is $K = \frac{1}{2}mv^2$. Because the speed of light in a vacuum c is the maximum speed an object can acquire, this expression limits the maximum kinetic energy that an object can acquire to $(1/2)mc^2$. This contradicts the idea that there should be no limit on an object's kinetic energy. Clearly, we need an improved equation for kinetic energy.

There is a similar problem with the Newtonian expression for momentum. We saw in Chapter 9 that the change in momentum of an object, the impulse, is determined by the external forces on the object and the duration of the time interval over which the forces are exerted. If a net force is exerted on an object, the longer the time interval Δt

that the force is exerted, the greater the momentum of the object. In principle there is no limit on how long the force can be exerted, so there should be no upper limit on an object's momentum. However, this can't be reconciled with the Newtonian expression for the momentum of an object of mass m with velocity $\vec{v}$:

$$\vec{p} = m\vec{v}$$

According to this expression, the maximum magnitude of momentum that an object of mass m can have is $p = mc$, which would be attained only when an object is moving at the speed of light. This directly contradicts the notion that there should be no upper limit on momentum. Just as for kinetic energy, we need a new expression for momentum that's consistent with the special theory of relativity.

It's possible to derive the correct expressions for K and $\vec{p}$ by analyzing an elastic collision in which both mechanical energy and momentum are constant. The derivation is beyond our scope, so we'll just look at the results. For a particle moving with velocity $\vec{v}$, both the expression for the kinetic energy and the expression for the momentum involve a dimensionless quantity called **relativistic gamma**. We introduced these equations in Chapter 24, in our discussion of conservation of momentum for particle-photon collisions (Compton scattering).

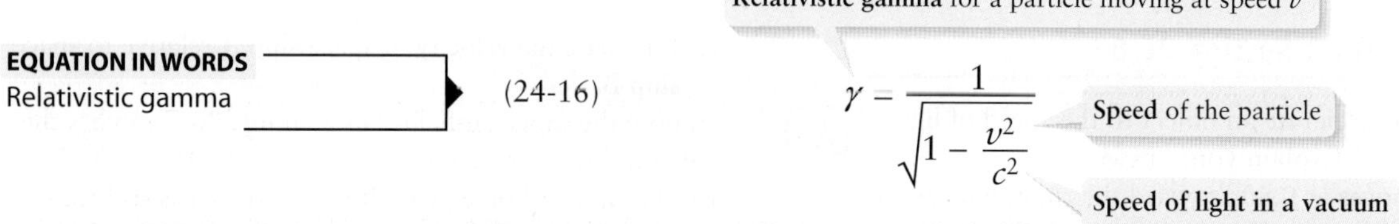

Relativistic gamma is equal to 1 when $v = 0$ and becomes infinitely large as v approaches c (**Figure 26-14**). In terms of relativistic gamma, we can write the correct expressions for kinetic energy and momentum as

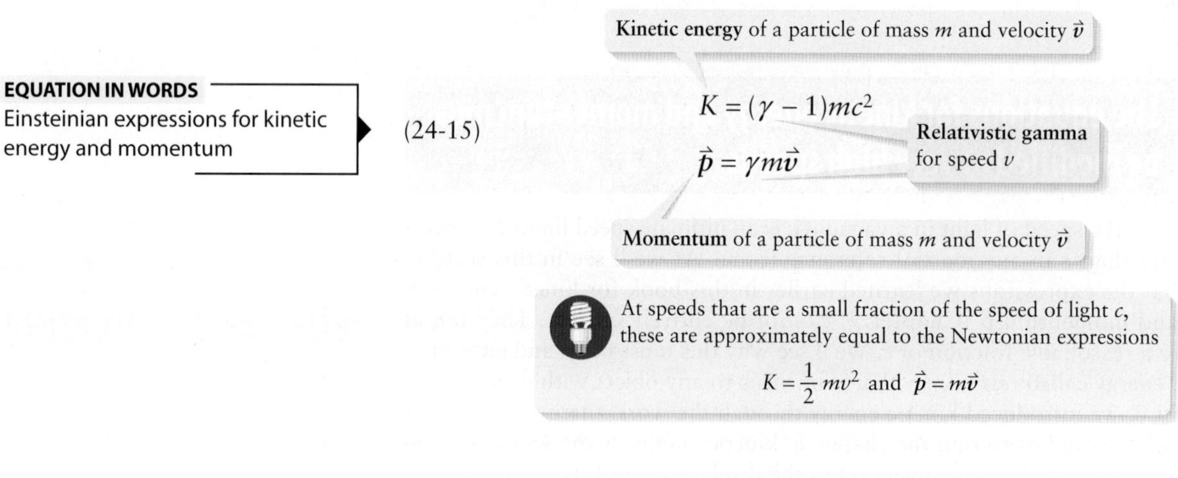

Figure 26-14 Relativistic gamma The quantity $\gamma = 1/\sqrt{1-(v^2/c^2)}$ increases dramatically toward infinity as speed v approaches the speed of light.

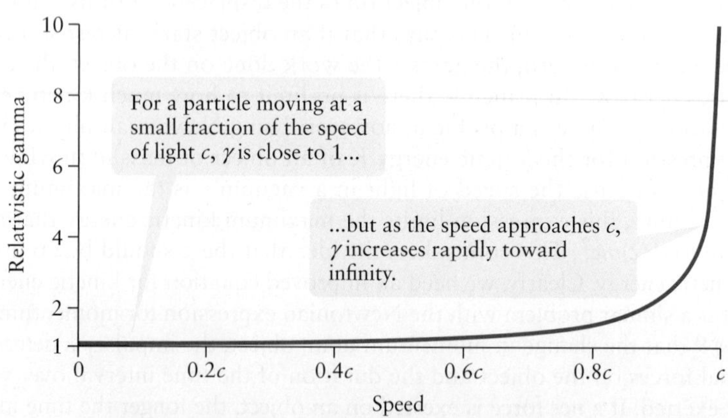

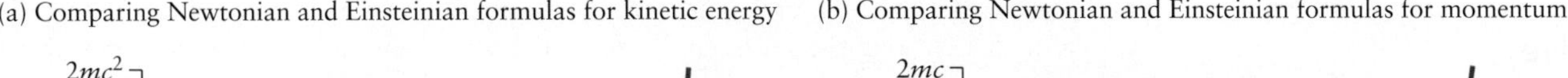

(a) Comparing Newtonian and Einsteinian formulas for kinetic energy
(b) Comparing Newtonian and Einsteinian formulas for momentum

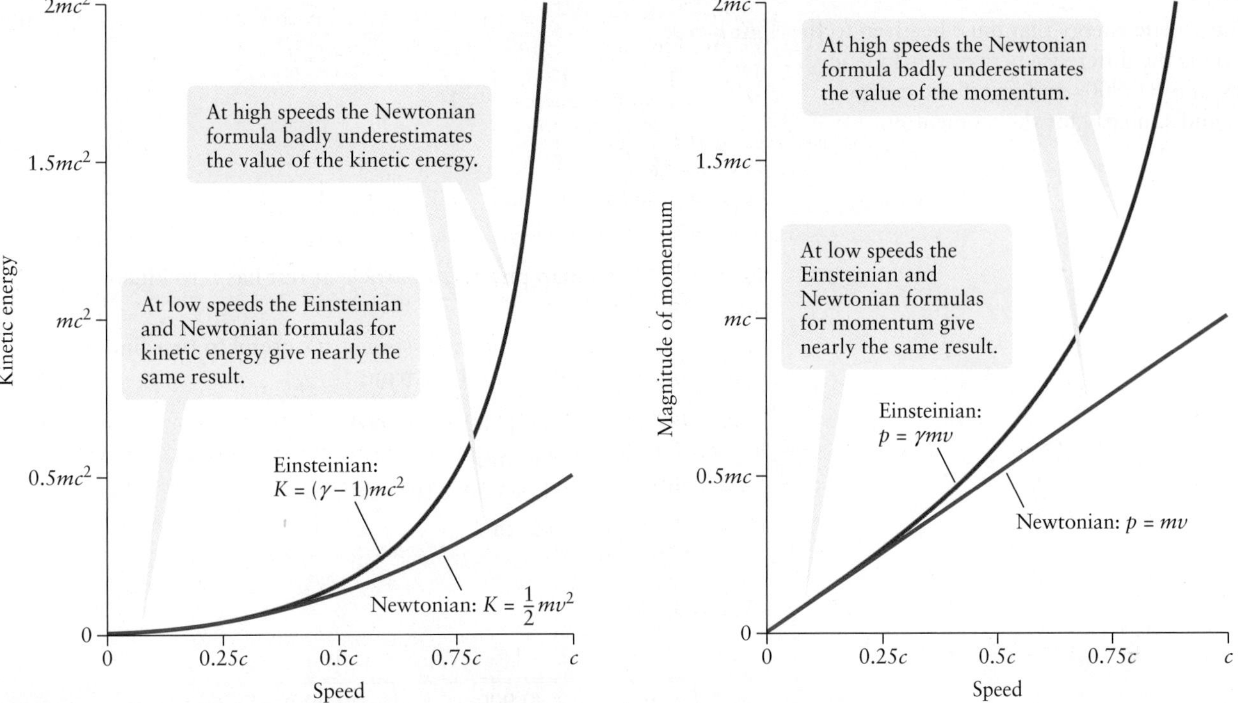

Figure 26-15 Kinetic energy and momentum in relativity These graphs show how the Newtonian and Einsteinian expressions for (a) the kinetic energy of a particle and (b) the momentum of a particle depend on the speed of the particle.

Figure 26-15 compares the kinetic energy K and the magnitude of momentum p from Equations 24-15 to the Newtonian expressions for these quantities. If v is a small fraction of the speed of light, the Einsteinian and Newtonian expressions give essentially identical results. This justifies our use of these expressions in earlier chapters, in which we considered only relatively slow-moving objects. But as the speed v approaches c, the Einsteinian expressions from Equations 24-15 approach infinity. So even though the speed of light is an upper limit to the speed of an object, there is *no* upper limit on the kinetic energy or momentum of an object.

EXAMPLE 26-8 The Energy and Momentum Costs of High Speed

Electrons can be accelerated to speeds very close to the speed of light. The mass of an electron is 9.11×10^{-31} kg. (a) How much kinetic energy must be given to an electron to accelerate it from rest to $0.900c$? How much momentum? (b) How much additional kinetic energy must be given to the electron to accelerate it from $0.900c$ to $0.990c$? How much additional momentum?

Set Up

We use Equations 24-15 to find the kinetic energy K and magnitude of momentum p for each speed. Equation 24-16 gives the value of relativistic gamma for each speed.

Relativistic gamma:

$$\gamma = \frac{1}{\sqrt{1-\dfrac{v^2}{c^2}}} \qquad (24\text{-}16)$$

Einsteinian kinetic energy and momentum:

$$K = (\gamma - 1)mc^2$$
$$\vec{p} = \gamma m\vec{v} \qquad (24\text{-}15)$$

Solve

(a) The kinetic energy that must be given to the electron is the difference between its kinetic energy at $v = 0.900c$ and its kinetic energy at $v = 0$, and similarly for the momentum.

At $v = 0$:

$$\gamma = \frac{1}{\sqrt{1-\frac{v^2}{c^2}}} = \frac{1}{\sqrt{1-0}} = \frac{1}{1} = 1$$

$$K = (\gamma - 1)mc^2 = (1-1)mc^2 = 0$$
$$p = \gamma mv = (1)m(0) = 0$$

Just as in Newtonian physics, a particle at rest has zero kinetic energy and zero momentum.

To calculate K and p at nonzero speeds, it's useful to first find the values of mc and mc^2 for an electron:

$$\begin{aligned} mc &= (9.11 \times 10^{-31}\ \text{kg})(3.00 \times 10^8\ \text{m/s}) \\ &= 2.73 \times 10^{-22}\ \text{kg} \cdot \text{m/s} \\ mc^2 &= (9.11 \times 10^{-31}\ \text{kg})(3.00 \times 10^8\ \text{m/s})^2 \\ &= 8.20 \times 10^{-14}\ \text{kg} \cdot \text{m}^2/\text{s}^2 \\ &= 8.20 \times 10^{-14}\ \text{J} \end{aligned}$$

At $v = 0.900c$:

$$\gamma = \frac{1}{\sqrt{1-\frac{v^2}{c^2}}} = \frac{1}{\sqrt{1-\frac{(0.900c)^2}{c^2}}} = \frac{1}{\sqrt{1-(0.900)^2}}$$
$$= \frac{1}{\sqrt{0.190}} = 2.29$$

$$\begin{aligned} K &= (\gamma - 1)\, mc^2 = (2.29 - 1)(8.20 \times 10^{-14}\ \text{J}) \\ &= 1.06 \times 10^{-13}\ \text{J} \end{aligned}$$

$$\begin{aligned} p &= \gamma mv = (2.29)m(0.900c) = (2.29)(0.900c)mc \\ &= (2.29)(0.900)(2.73 \times 10^{-22}\ \text{kg} \cdot \text{m/s}) \\ &= 5.64 \times 10^{-22}\ \text{kg} \cdot \text{m/s} \end{aligned}$$

The electron begins with $K = 0$ and $p = 0$, so it must be given 1.06×10^{-13} J of kinetic energy and 5.64×10^{-22} kg · m/s of momentum to accelerate it from rest to $0.900c$.

(b) Repeat the calculation in part (a) for the additional kinetic energy and momentum that must be given to the electron to accelerate it from $0.900c$ to $0.990c$.

At $v = 0.990c$:

$$\gamma = \frac{1}{\sqrt{1-\frac{v^2}{c^2}}} = \frac{1}{\sqrt{1-\frac{(0.990c)^2}{c^2}}} = \frac{1}{\sqrt{1-(0.990)^2}}$$
$$= \frac{1}{\sqrt{0.0199}} = 7.09$$

$$\begin{aligned} K &= (\gamma - 1)mc^2 = (7.09 - 1)(8.20 \times 10^{-14}\ \text{J}) \\ &= 4.99 \times 10^{-13}\ \text{J} \end{aligned}$$

$$\begin{aligned} p &= \gamma mv = (7.09)m(0.990c) = (7.09)(0.990)mc \\ &= (7.09)(0.990)(2.73 \times 10^{-22}\ \text{kg} \cdot \text{m/s}) \\ &= 1.92 \times 10^{-21}\ \text{kg} \cdot \text{m/s} \end{aligned}$$

The difference between these values and the values of K and p at $v = 0.900c$ from part (a) tells us the additional kinetic energy and momentum that must be given to the electron:

$$\begin{aligned}\Delta K &= 4.99 \times 10^{-13}\ \text{J} - 1.06 \times 10^{-13}\ \text{J}\\ &= 3.93 \times 10^{-13}\ \text{J}\\ \Delta p &= 1.92 \times 10^{-21}\ \text{kg}\cdot\text{m/s} - 5.64 \times 10^{-22}\ \text{kg}\cdot\text{m/s}\\ &= 1.35 \times 10^{-21}\ \text{kg}\cdot\text{m/s}\end{aligned}$$

Reflect

Compared to accelerating an electron from rest to $0.900c$, accelerating that same electron from $0.900c$ to $0.990c$ requires 3.7 times as much additional kinetic energy and 2.4 times as much additional momentum. As the speed gets closer and closer to c, it requires ever-greater amounts of kinetic energy and momentum to cause an ever-smaller speed increase. You can see that an object can *never* be accelerated from rest to the speed of light. This would require adding infinite amounts of kinetic energy and momentum to the object.

Note that there are objects within your body that are moving at a respectable fraction of the speed of light c, though not as fast as in this example (**Figure 26-16**). Every one of your red blood cells contains molecules of hemoglobin, and every hemoglobin molecule includes four iron atoms. The two innermost (and fastest-moving) electrons in an iron atom have an average speed of about $0.19c$; at this speed γ is about 1.02, and the Einsteinian kinetic energy of the electrons $K = (\gamma - 1)mc^2$ is about 3% greater than predicted by the Newtonian expression $K = \frac{1}{2}mv^2$.

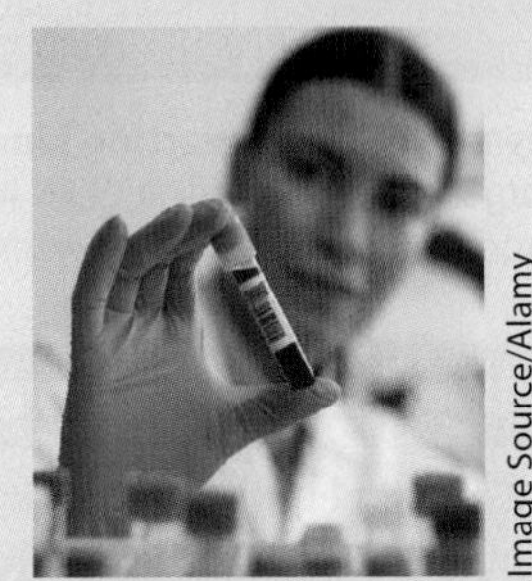

Image Source/Alamy

Figure 26-16 A sample of Einsteinian particles This sample of a patient's blood includes particles moving at about 19% of the speed of light: the innermost electrons in the iron atoms found in hemoglobin.

NOW WORK Problem 2 from the Takeaway for Section 26-7.

We mentioned above that the Einsteinian expressions for kinetic energy and momentum, Equations 24-15, come from an analysis of elastic collisions. In particular we must demand that if energy and momentum are constant in one inertial reference frame, they must be constant in all inertial reference frames. That's required if energy and momentum are to be consistent with the first of the postulates of special relativity. It turns out that for this to be the case, we must also include a term mc^2 in the total energy of a particle of mass m. This quantity is called the **rest energy** of a particle, because it is present even when the particle is not in motion.

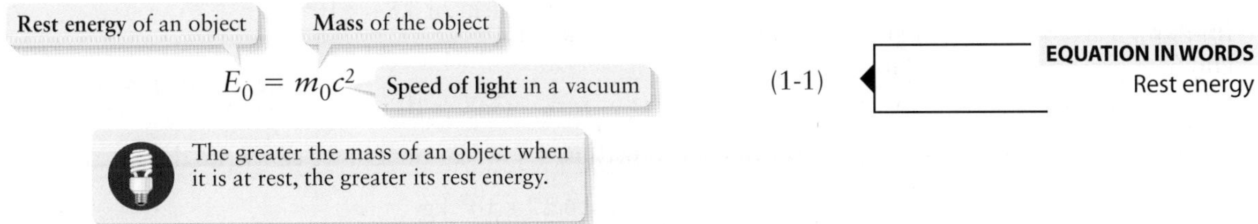

Rest energy of an object
Mass of the object
Speed of light in a vacuum
The greater the mass of an object when it is at rest, the greater its rest energy.

$$E_0 = m_0c^2 \quad (1\text{-}1)$$

EQUATION IN WORDS
Rest energy

Equation 1-1 is one of the most famous in science, and one of the most misunderstood. Rest energy is not potential energy; potential energy is associated with a conservative interaction between objects or systems. (For example, gravitational potential energy is associated with the gravitational interaction between on object and Earth.) It is not kinetic energy; kinetic energy is associated with motion. Rest energy is energy that is intrinsic to an object because of its mass.

Equation 1-1 tells us that mass is simply one possible manifestation of energy. Stated another way, the mass of an object is a measure of its rest energy content. **Figure 26-17** is visual evidence of the equivalence of mass and energy. This image shows the result of a head-on collision between two protons, each of which was moving at just under the speed of light. Dozens of new particles appear after the collision. These

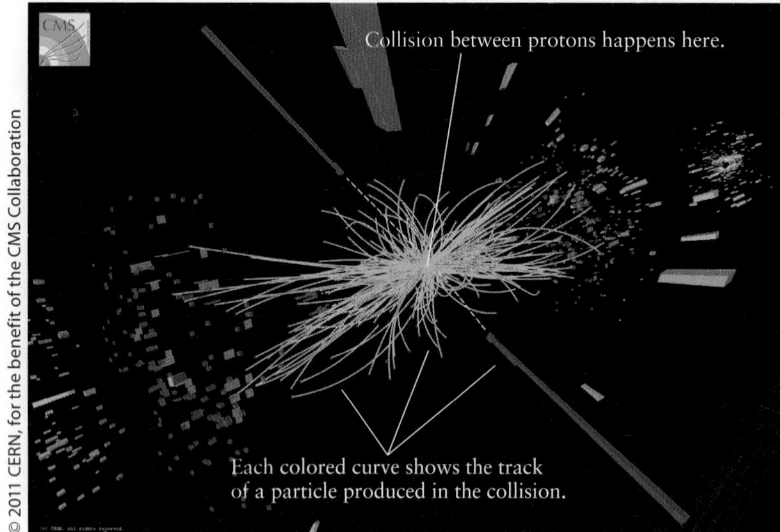
CMS
Collision between protons happens here.
Each colored curve shows the track
of a particle produced in the collision.

Figure 26-17 Converting kinetic energy to new particles This image from the Large Hadron Collider at CERN in Geneva, Switzerland, shows dozens of new particles produced by a collision between energetic protons.

are not fragments of the colliding protons but rather new particles that were created in the collision. This is possible because some of the kinetic energy of the colliding protons was converted into the rest energy of these new particles; another portion of this kinetic energy went into the kinetic energies of the new particles, which fly away from the collision site at high speeds.

You can observe the equivalence of mass and energy in action whenever you go outside on a sunny day or look up at the stars on a clear night. The Sun and stars shine thanks to a process occurring in their interiors in which hydrogen nuclei are fused together to form nuclei of helium. The mass of the products of such a reaction is slightly less than the mass of the hydrogen nuclei present before the reaction. The "lost" mass is converted to energy in the form of electromagnetic radiation, and it is that radiation that we see in the form of sunlight and starlight.

EXAMPLE 26-9 Converting Mass to Energy in the Sun

The Sun emits 3.84×10^{26} J of energy every second. (a) At what rate (in kg/s) is the Sun's mass decreasing? (b) How much mass has the Sun lost since it was formed 4.56×10^9 years ago? Assume that it has emitted energy at the same rate over its entire history. Compare this to the present-day mass of the Sun, 1.99×10^{30} kg.

Set Up

The emitted energy comes from the rest energy of mass that is "lost" in nuclear reactions in the Sun's interior. We use Equation 1-1 to find the amount of mass equivalent to the energy emitted in one second. The total mass lost over the Sun's lifetime equals this amount of mass multiplied by the number of seconds that have elapsed since the Sun formed.

Rest energy:

$$E_0 = m_0 c^2 \qquad (1\text{-}1)$$

Solve

(a) Calculate the amount of mass lost by the Sun per second.

Mass equivalent of 3.84×10^{26} J:

$$m_0 = \frac{E_0}{c^2} = \frac{3.84 \times 10^{26}\ \text{J}}{(3.00 \times 10^8\ \text{m/s})^2} = 4.27 \times 10^9\ \text{J} \cdot \text{s}^2/\text{m}^2$$

Because $1\ \text{J} = 1\ \text{kg} \cdot \text{m}^2/\text{s}^2$, we can write this as

$$m_0 = 4.27 \times 10^9\ \text{kg}$$

The mass of the Sun decreases at a rate of 4.27×10^9 kg/s.

(b) Calculate the total amount of mass lost by the Sun in its history.

The age of the Sun in seconds is

$$(4.56 \times 10^9\ \text{y})\left(\frac{365.25\ \text{d}}{1\ \text{y}}\right)\left(\frac{24\ \text{h}}{1\ \text{d}}\right)\left(\frac{60\ \text{min}}{1\ \text{h}}\right)\left(\frac{60\ \text{s}}{1\ \text{min}}\right) = 1.44 \times 10^{17}\ \text{s}$$

In this number of seconds the total mass lost by the Sun is

$$\left(4.27 \times 10^9\ \frac{\text{kg}}{\text{s}}\right)(1.44 \times 10^{17}\ \text{s}) = 6.14 \times 10^{26}\ \text{kg}$$

As a fraction of the Sun's present-day mass, the amount of mass lost is $(6.14 \times 10^{26}\ \text{kg})/(1.99 \times 10^{30}\ \text{kg}) = 3.09 \times 10^{-4} = 0.0309\%$ of the total mass.

Reflect

In more than 4 billion years of producing energy at a prodigious rate, the Sun has lost only a tiny fraction of its total mass. This is a testament to how much energy can be released by converting even a small amount of mass.

NOW WORK Problem 3 from the Takeaway for Section 26-7.

THE TAKEAWAY for Section 26-7

✔ The mathematical expressions for kinetic energy and momentum have to be modified to be consistent with special relativity. Both the kinetic energy and momentum of an object increase without limit as the object's speed approaches c.

✔ Any object with mass has a kind of energy called rest energy.

Takeaway Problems for Section 26-7

1. The total amount of electric energy produced per year in the United States from all sources is about 1.5×10^{19} J. You propose to use all of this energy to accelerate a spacecraft to $0.990c$ in order to travel to other stars. About how massive could your proposed starship be?
(A) About 3×10^6 kg (the mass of an ocean liner)
(B) About 3×10^5 kg (the mass of a large airliner)
(C) About 3×10^3 kg (the mass of a sport utility vehicle)
(D) About 30 kg (the mass of a kayak or canoe)

2. An electron travels at $0.444c$. Calculate **(a)** its Einsteinian momentum, **(b)** its Einsteinian kinetic energy, **(c)** its rest energy, and **(d)** the total energy of the electron.

EX 26-9

3. Recent home energy bills indicate that a household used 411 kWh of electrical energy and 201 therms for gas heating and cooking in a period of one month (1.0 therm = 29.3 kWh). How many milligrams of mass would need to be converted directly to energy each month to meet the energy needs for the home?

26-8 Einstein's general theory of relativity describes the fundamental nature of gravity

Sitting in a chair in the patent office in Bern, Switzerland, in 1907, a young Albert Einstein had what he would call "the happiest thought of my life." He imagined a man falling freely from the roof of a house and realized that "at least in his immediate surroundings—there exists no gravitational field." If the man released an object, for example, it would accelerate at the same rate as the man accelerated, and because the man would "not feel his own weight," it would appear to him that neither he nor the object was experiencing a gravitational force.

Einstein's thought experiment led him to postulate a new principle called the **principle of equivalence.** This principle states that

A gravitational field is equivalent to an accelerated reference frame in the absence of gravity.

The principle of equivalence dictates that it is not possible to distinguish experimentally between a system in an accelerating frame and a system under the influence of gravity. In other words if you were to drop a ball in a windowless elevator car that makes no noise and doesn't shake, you could not tell from the motion of the ball whether the elevator car were sitting stationary on the surface of a planet and experiencing its gravity or accelerating in empty space, far from sources of gravity.

Let's explore physics in this imaginary elevator car further. In **Figure 26-18a** a ball is thrown horizontally while the elevator car is stationary near Earth's surface. Due to the force of gravity, the ball accelerates downward, following a familiar parabolic arc. The figure shows the position of the ball at five instants, spanning four equal time intervals. What if the stationary elevator car were far from Earth and from any other massive object that could exert a noticeable gravitational force on the ball? In that case the ball would travel along a straight line, as shown in **Figure 26-18b**.

(a)

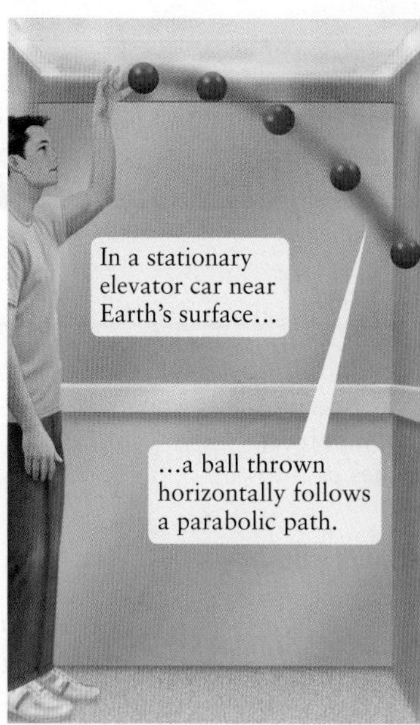
In a stationary elevator car near Earth's surface...
...a ball thrown horizontally follows a parabolic path.

(b)

In a stationary elevator car far from Earth or any other object...
...a ball thrown horizontally travels in a straight line.

Figure 26-18 Elevator cars on Earth and in space A ball is thrown horizontally (a) in an elevator car on Earth's surface and (b) in an elevator car far from any massive object.

Now consider the situation shown in **Figure 26-19**. The elevator car is again far from object that could exert a noticeable gravitational force on the ball, but now the car is accelerating "upward" (toward the top of the page). Because there is no discernible gravitational force, the ball travels in a straight line, as in Figure 26-18b. However, because the elevator car is accelerating, an observer *in the car* sees the ball trace out a parabolic arc with respect to the walls and floor of the car. Remember that there are no windows in the elevator car, and it makes no noise and does not vibrate as it moves. An observer in the car cannot, therefore, detect its motion. The observer observes the effect of the principle of equivalence: It appears that the ball is falling under the influence of gravity.

Einstein's theory of gravitation is therefore a theory of accelerating reference frames. Because this is more general than the case of inertial reference frames, the type that was at the heart of the special theory of relativity, this expanded theory is called the **general theory of relativity**.

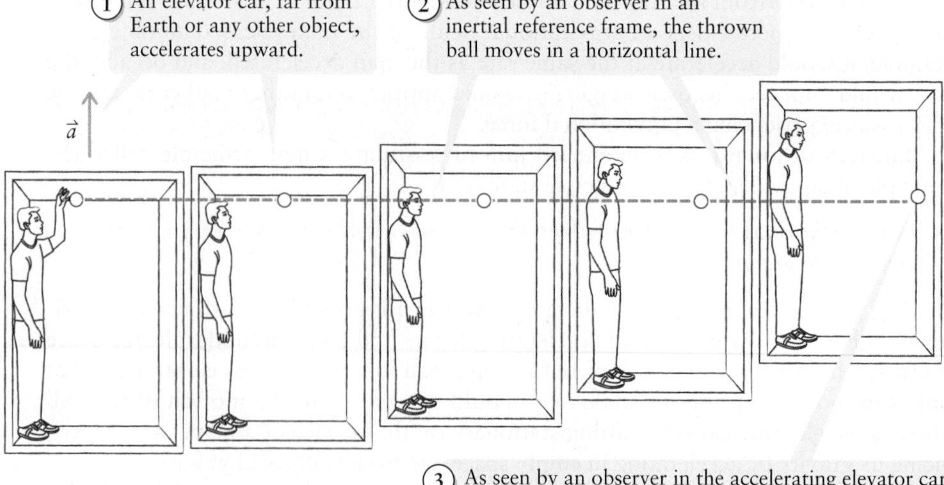
1 An elevator car, far from Earth or any other object, accelerates upward.
2 As seen by an observer in an inertial reference frame, the thrown ball moves in a horizontal line.
$\vec{a}$
3 As seen by an observer in the accelerating elevator car, the ball follows a *curved path*: Where it hits the far wall is lower than where it left the thrower's hand.

Figure 26-19 An accelerating elevator car in space When a ball is thrown horizontally in an elevator car, which is both far from any massive object and also accelerating, it follows a straight path. An observer in the elevator, however, would see the ball follow a parabolic path with respect to the floor and walls of the elevator car.

The trajectory of the ball in the accelerated frame of reference of the elevator car is identical to that of a ball in a uniform gravitational field. This agrees with Einstein's principle of equivalence.

Predictions of General Relativity

Einstein's general theory of relativity does more than give us a way to think about accelerating reference frames. It also makes remarkable predictions about the behavior of light and the nature of time and space, predictions that have been verified by careful experiment and observation. Let's look at a few of these.

Gravity bends light. Imagine that in Figure 26-19 we replace the thrown ball with a beam of light fired horizontally. As seen by an outside observer in an inertial reference frame, the light beam moves in a horizontal straight line toward the far wall. But because the elevator car is accelerating upward, an observer in the car will see the light beam follow a curved path and hit the far wall below the height from which it started. (The curvature of the beam's path will be very much less than that of the trajectory of the ball in Figure 26-19 because the light travels so much faster. Nevertheless, it *will* curve.)

According to the principle of equivalence, the effects of acceleration are indistinguishable from the effects of gravitation. So we conclude that a light beam fired horizontally on Earth will curve downward, just as the path of a ball thrown horizontally curves. This effect is called the **gravitational bending of light**.

Because the speed of light is so great, the gravitational bending of light predicted by Einstein is too small to measure on Earth. For example, a light beam traversing the width of a typical elevator would bend downward by less than 10^{-15} m, about the diameter of a proton. But the bending is measurable if a light beam is acted on by a much stronger gravitational field that acts over a much greater distance. The first measurement of gravitational bending of light was made in 1919 during a total solar eclipse. During totality, when the Moon blocked out the Sun's disk, astronomers photographed the stars around the Sun. These stars were shifted from their usual positions by 4.86×10^{-4} of a degree, consistent with Einstein's predictions.

A more stunning phenomenon associated with gravitational bending of light occurs when multiple images of a distant star form as light is bent by a closer, massive celestial object. **Figure 26-20** depicts such gravitational lensing, in which a massive object is positioned directly between Earth and a distant star. Light from the star cannot reach Earth directly, but the lensing effect results in light that was initially not propagating toward Earth being bent back toward us. In this way light from the star can approach Earth from many directions—for example, the two directions shown in the figure. **Figure 26-21a** presents an example of *gravitational lensing* showing four distinct images. When light is bent around the intervening object and reaches Earth from a full circle around it, the star's light is spread out into a ring around the lensing object, as in **Figure 26-21b**.

Space is curved, and gravitational waves propagate through space. A way to interpret the gravitational bending of light is that light in fact travels in a straight line in empty space but that *space itself is curved* by the presence of a massive object. Indeed, Einstein envisioned gravity as being caused by a curvature of space. In this picture, a massive object like Earth curves the space around it, and a light beam bends because it follows that curvature. Furthermore, an object like a ball falls toward Earth because it responds to the curvature of space that Earth produces. (A 0.1-kg ball and a 10-kg ball sense the same curvature of space due to Earth, which explains why gravity produces the same acceleration on objects of different mass.) Einstein's full mathematical formulation of the general theory of relativity is a set of equations that describe how the curvature of space and time, collectively referred to as spacetime, are affected by the presence of mass and energy.

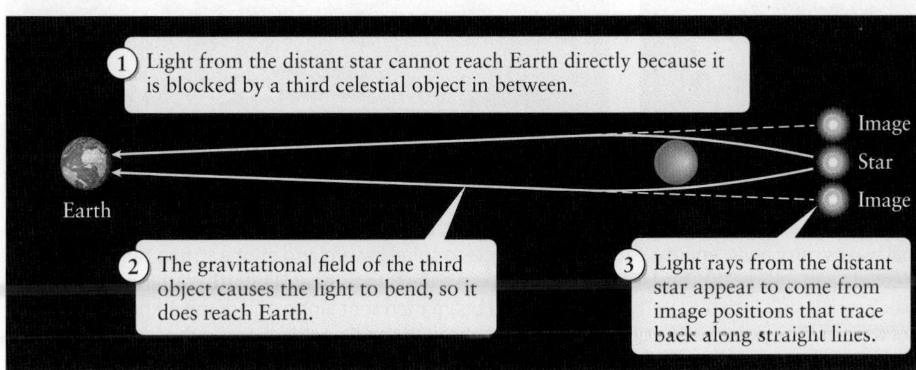

1 Light from the distant star cannot reach Earth directly because it is blocked by a third celestial object in between.
Image
Star
Image
Earth
2 The gravitational field of the third object causes the light to bend, so it does reach Earth.
3 Light rays from the distant star appear to come from image positions that trace back along straight lines.

Figure 26-20 Gravity deflects light A massive object positioned between Earth and a distant star bends light from the star so that it can be seen on Earth. This is called gravitational lensing.

(a)
This galaxy acts as a gravitational lens...
Image 1
Image 2
Image 3
Image 4
...and makes four images of an even more distant supernova.

NASA, ESA, and S. Rodney (JHU) and the FrontierSN team; T. Treu (UCLA), P. Kelly (UC Berkeley), and the GLASS team; J. Lotz (STScI) and the Frontier Fields team; M. Postman (STScI) and the CLASH team; and Z. Levay (STScI)

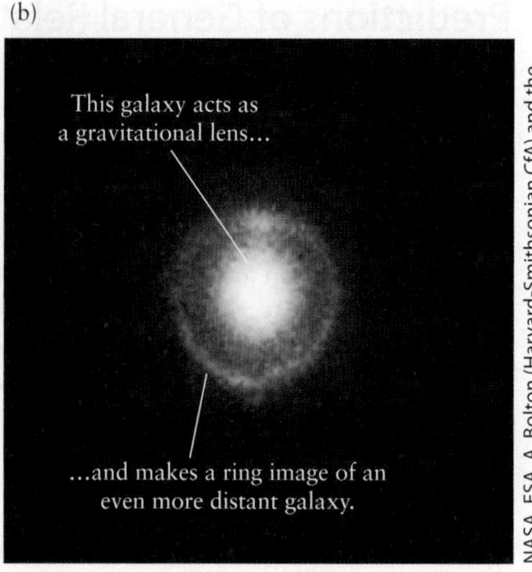
(b)
This galaxy acts as a gravitational lens...
...and makes a ring image of an even more distant galaxy.

NASA, ESA, A. Bolton (Harvard-Smithsonian CfA) and the SLACS Team

Figure 26-21 Gravitational lenses Two examples of a massive, distant galaxy acting as a gravitational lens. (a) This gravitational lens makes four images of an even more distant supernova (an exploding star). (b) A more distant galaxy is located directly along our line of sight to the gravitational lens. The resulting image of the more distant galaxy is a nearly perfect circular ring.

How can we test this idea? The general theory of relativity predicts not only that spacetime curves in response to the presence of massive objects, but also that ripples in spacetime called **gravitational waves**—that is, small variations in the curvature of spacetime—should spread away from massive objects that are oscillating in a certain way (**Figure 26-22a**). (Newton's theory of gravitation makes no such prediction.) Such gravitational waves were detected for the first time in 2015 by the Laser Interferometer Gravitational-Wave Observatory (LIGO), a set of two detectors located 3000 km apart in the U.S. states of Washington and Louisiana (**Figure 26-22b**).

The operating principle of each LIGO detector is the same as that of the Michelson–Morley apparatus shown schematically in Figure 26-8. A light beam is split in two, sent along two perpendicular arms of the detector, reflected from mirrors at the end of each arm (in the case of the LIGO detectors, 4 km away), and then recombined. If a gravitational wave from space passes through a LIGO detector, the change in the curvature of spacetime will slightly change the length of each arm of the detector and hence the distance that each light beam travels, and this changes the interference pattern formed when the beams recombine. Local disturbances such as an earthquake could also trigger a change in these distances, which would produce a false signal. That's why LIGO has two detectors located 3000 km apart: Only a disturbance coming from space will produce the same signal in both detectors.

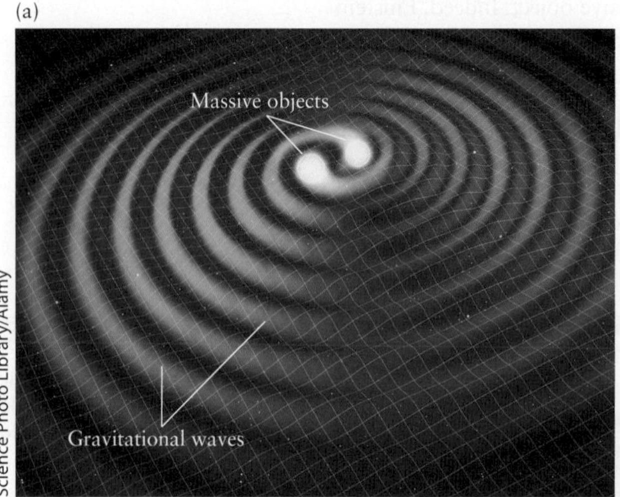
(a)
Massive objects
Gravitational waves

Science Photo Library/Alamy

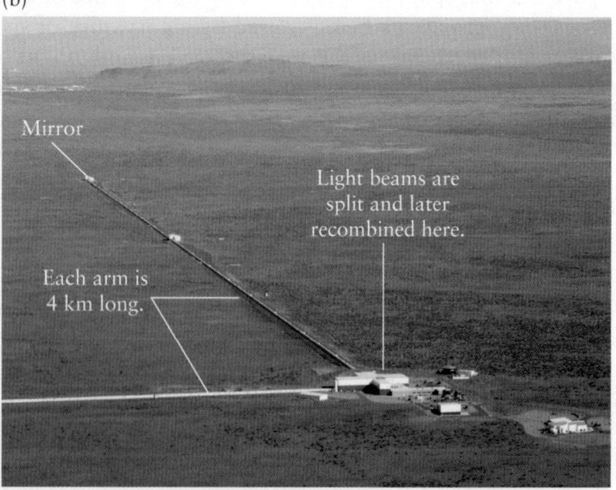
(b)
Mirror
Light beams are split and later recombined here.
Each arm is 4 km long.

Xinhua/Alamy

Figure 26-22 Gravitational waves (a) As two massive celestial objects orbit each other, they produce ripples in spacetime called gravitational waves. These gravitational waves carry away energy, causing the objects to spiral inward and emit a strong burst of gravitational waves when they finally collide and merge. (b) This LIGO gravitational wave detector is near Hanford, Washington. (The other is in Livingston, Louisiana.) A laser beam is split into two, with each individual beam then sent along one of the 4-km arms and reflected back by a mirror at the end of the arm. The passage of a gravitational wave through the detector changes the lengths of each of the arms, which causes a change in the interference pattern formed when the two light beams are recombined.

As of this writing, LIGO has made 90 confirmed detections of gravitational wave events, most of them caused by a pair of objects many times more massive than the Sun colliding with each other. (These objects are thought to be black holes. In a black hole the material has been so compressed that the gravitational field near the black hole is extremely strong, so much so that nothing, not even light, can escape. *Black holes* are another prediction of general relativity.) Although such a cataclysm emits an immense amount of gravitational wave energy (equivalent to converting several times the mass of the Sun completely into energy), the signal detected by LIGO is miniscule: In each detection each 4-km arm of the detector changed in length by only about 10^{-18} m, about 10^{-3} as large as the radius of a proton. The 2017 Nobel Prize in Physics was awarded to Rainer Weiss, Barry Barish, and Kip Thorne for their work on LIGO.

Gravity affects time. We saw in Sections 26-4 and 26-5 that in the special theory of relativity, both time intervals and distances are affected by motion. Similarly, in the general theory of relativity, a massive object such as Earth affects time and curves space. Einstein predicted that clocks on the ground floor of a building should tick slightly more slowly than clocks on the top floor, which are farther from Earth. This **gravitational slowing of time** has been measured even for differences in height as small as 33 cm (1 ft): An experiment in 2010 using extremely precise clocks showed that the lower clock would fall behind by about 9×10^{-8} s over a 79-year human lifetime, in agreement with Einstein's prediction.

An important application of the gravitational slowing of time is the Global Positioning System, or GPS (Figure 26-1a). A GPS receiver uses signals from orbiting satellites, each of which carries extremely accurate clocks, to triangulate its position on Earth. The effects of gravity slow down time on Earth's surface compared to the satellites' clocks, so the general theory of relativity *must* be taken into account for an accurate GPS result. If the gravitational slowing of time were not taken into account, a GPS receiver would accumulate errors of more than 10 km per day!

The general theory of relativity has never made an incorrect prediction. It now stands as our most accurate and complete description of gravity.

THE TAKEAWAY for Section 26-8

✔ The principle of equivalence, a central postulate of Einstein's general theory of relativity, states that a gravitational field is equivalent to an accelerated reference frame in the absence of gravity.

✔ It is not possible to distinguish between a system in an accelerating frame and a system under the influence of gravity.

✔ General relativity predicts the gravitational bending of light, the existence of gravitational waves, and the gravitational slowing of time. All of these have been observed.

Takeaway Problems for Section 26-8

1. What is the fundamental postulate of the general theory of relativity?
2. Consider two atomic clocks, one at the GPS ground control station near Colorado Springs (elevation 1830 m) and the other one in orbit in a GPS satellite (altitude 20,200 km). According to the general theory of relativity, which atomic clock runs slow?
   (A) The clock in Colorado runs slow.
   (B) The clock in orbit runs slow.
   (C) The clocks keep identical time.
   (D) The orbiting clock is 95% slower than the clock in Colorado.
3. What would an observer inside an elevator measure for the free-fall acceleration near the surface of Earth if the elevator accelerates upward at 18.0 m/s^2?

26-9 Most forms of matter can be explained by just a handful of fundamental particles

By the late nineteenth century scientists had concluded that all matter was composed of atoms and that atoms could not be subdivided into more fundamental particles. It was thought that a hydrogen atom was intrinsically different from a carbon atom, which in turn was intrinsically different from an iron atom, and so on.

Electron
Mass = 9.1094×10^{-27} kg
Charge = $-e = -1.60 \times 10^{-19}$ C

Proton
Mass = 1.6726×10^{-27} kg
Charge = $+e = 1.60 \times 10^{-19}$ C

Neutron
Mass = 1.6749×10^{-27} kg
Charge = 0

Figure 26-23 The first three subatomic particles to be discovered All atoms are combinations of electrons (discovered 1897), protons (discovered 1917), and neutrons (discovered 1932). While the electron is a truly fundamental particle, the proton and neutron are not: They are composed of other, more fundamental particles called quarks and gluons.

As we learned in Section 26-5, this conclusion was quite incorrect. In 1897 J. J. Thomson discovered the electron, a particle of charge $-e = -1.602 \times 10^{-19}$ C, which turned out to be a constituent of the atoms of every element. In 1909 Ernest Rutherford discovered the atomic nucleus, and in 1917 he found evidence that all nuclei contain a positively charged particle of charge $+e$ that is identical to a hydrogen nucleus—that is, what we now call a proton. In 1932 the English physicist James Chadwick discovered the neutron, which has zero charge. As we learned in Chapter 25, all nuclei are composed of protons and neutrons (referred to collectively as nucleons), and all atoms are made of nuclei plus electrons. So by 1932 it seemed that these three subatomic particles—electron, proton, and neutron (**Figure 26-23**)—were the truly fundamental building blocks of matter.

That conclusion, too, turned out to be wildly incorrect. Since 1932 literally hundreds of other subatomic particles have been discovered. Almost all of these are unstable and decay to other particles with a radioactive half-life of a fraction of a second. (In this aspect they resemble the neutron: A free neutron that is not incorporated into a nucleus undergoes beta decay with a half-life of about 15 minutes.) But none of these additional particles can be regarded as simple combinations of protons, neutrons, and electrons. They include the neutrino, which has no charge, interacts hardly at all with other particles, and has a mass so close to zero that it has yet to be accurately measured; the muon, which resembles an electron in almost every way except that it is 207 times more massive; the pion, which like the proton and neutron interacts through the strong force but has only about one-seventh the mass of a proton; and the delta, which resembles a proton or neutron but is about 30% more massive and comes in four varieties, with charges $+2e$, $+e$, 0, and $-e$. The discovery of these additional particles forced physicists to once again ask the question, "What *are* the fundamental building blocks of matter?"

Hadrons and Quarks

To answer this question, it's useful to distinguish between particles that interact through the strong force, including the proton and neutron, and those that do not, such as the electron. Because protons and neutrons are relatively heavy and electrons are relatively light, we use the term **hadrons** (from the Greek word for stout or thick) for particles that interact through the strong force and the term **leptons** (from the Greek word for small or delicate) for those that do not. (The photon is considered to be in a special category of its own, to which we will return later.) The vast majority of new particles discovered since 1932 are hadrons, so we'll look at these first.

In 1964 the American physicists Murray Gell-Mann and George Zweig independently proposed that all hadrons are made of more fundamental entities that Gell-Mann whimsically named **quarks**. The first evidence that quarks really exist came from experiments carried out at the Stanford Linear Accelerator Center, or SLAC, in 1967 (**Figure 26-24a**). These experiments were the same in principle as Rutherford's

(a)

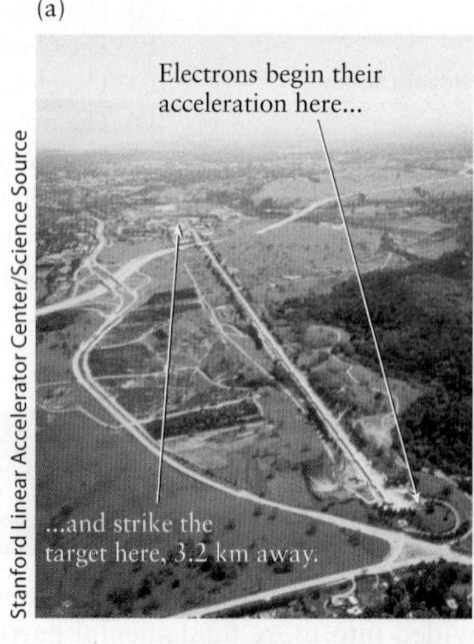
Electrons begin their acceleration here...
...and strike the target here, 3.2 km away.
Stanford Linear Accelerator Center/Science Source

(b)

When high-energy electrons scatter from protons, a substantial number scatter backward.

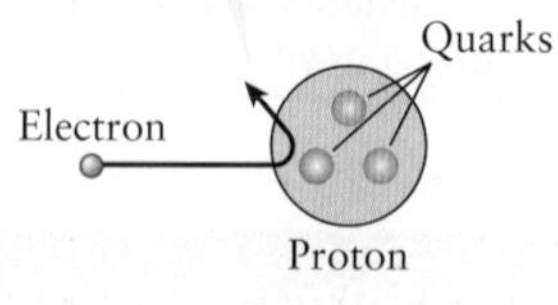
Quarks
Electron
Proton

This is evidence that there are small charged objects (quarks) inside the proton.

Figure 26-24 Discovering quarks (a) In a seminal series of experiments at the Stanford Linear Accelerator Center, electrons were accelerated to a kinetic energy of 20 GeV = 2×10^4 MeV and fired into a target containing protons. (b) Measuring how the electrons scattered from the protons provided evidence of the existence of quarks.

1909 experiment that led to the discovery of the atomic nucleus. As we saw in Section 24-6, Rutherford aimed a beam of alpha particles at a target of gold. Had the charge inside the atom been distributed more or less uniformly, the alpha particles would have undergone only gentle deflections as they passed through the gold atoms. Instead, Rutherford found that some alpha particles were scattered by very large angles. This was evidence that charge was highly concentrated into a very small object within the atom—namely, the nucleus. The experiment at SLAC used electrons instead of alpha particles and aimed these electrons at a target of protons (the nuclei of hydrogen atoms inside a tank of liquid hydrogen). The 3.2-km-long accelerator gave each electron a tremendous kinetic energy—some 20,000 MeV, compared to the 7 MeV of Rutherford's alpha particles—and a correspondingly large momentum. This was done so that the electron would have a de Broglie wavelength much smaller than the size of the proton and so would be sensitive to fine details of the proton's internal structure as it passed through the proton.

Much as in the Rutherford experiment six decades before, many physicists expected that the electrons would undergo only small-angle deflections because they thought the charge inside the proton was distributed uniformly over its volume. And just as in the Rutherford experiment, what they found was that a substantial number of electrons were scattered by very large angles (**Figure 26-24b**). The conclusion was that the proton's charge is carried by smaller entities inside the proton, which are the quarks.

These experiments and a host of others confirm that Gell-Mann and Zweig were correct and that quarks are the fundamental building blocks of all hadrons. To explain all of the hundreds of hadron varieties currently known, we need six varieties or *flavors* of quarks. These are known as the up (u), down (d), charm (c), strange (s), top (t), and bottom (b) quarks. **Table 26-1** lists the six quarks, along with their masses and charges. The quarks are divided into three groups, or *generations*, of two quarks: u and d in the first generation, c and s in the second generation, and t and b in the third generation. As Table 26-1 shows, the quarks in each generation are more massive than those in the previous generation. Physicists usually write the quark generations as

$$\begin{pmatrix} \mathrm{u} \\ \mathrm{d} \end{pmatrix}\begin{pmatrix} \mathrm{c} \\ \mathrm{s} \end{pmatrix}\begin{pmatrix} \mathrm{t} \\ \mathrm{b} \end{pmatrix}$$

The quarks along the top row (u, c, and t) all have the same charge, as do the quarks along the bottom row (d, s, and b). However, all six quarks differ not only in mass but also in other subtle properties. (These properties are beyond our scope in this brief introduction.) All attempts to find evidence of a fourth-generation quark have failed; as best we know, there are only three generations of quarks.

Note that each flavor of quark has a charge that is a *fraction* of e, either $+2e/3$ or $-e/3$. The explanation is that the proton, neutron, and other related particles are actually combinations of three quarks. **Figure 26-25** shows four examples of such combinations. Any hadron that can be made up of three quarks is called a **baryon**. Baryons always have a net charge that is an integer multiple of e, between $-2e$ and $2e$. Two examples are the least massive baryons, the proton (**Figure 26-25a**) of net charge $+e$ and the neutron (**Figure 26-25b**) with net charge zero. This picture explains why the neutron, which has zero net charge, nonetheless produces a magnetic field of its own: The u quark and two d quarks within the neutron are each charged, and the motions of these charged particles within the neutron generate a magnetic field.

Note that in Table 26-1 we list only an *approximate* mass for each quark. We know quite precise values of the masses of other particles; for example, the mass of the proton is known to seven significant figures. By contrast, we know only rough values for the masses of the quarks. That's because, unlike protons, neutrons, or electrons, quarks

TABLE 26-1 The Six Quarks

Quark	Symbol	Charge	Approximate mass
Up	u	$+\frac{2}{3}e$	2.16 MeV/c^2
Down	d	$-\frac{1}{3}e$	4.67 MeV/c^2
Charm	c	$+\frac{2}{3}e$	1.27 GeV/c^2
Strange	s	$-\frac{1}{3}e$	93 MeV/c^2
Top	t	$+\frac{2}{3}e$	173 GeV/c^2
Bottom	b	$-\frac{1}{3}e$	4.18 GeV/c^2

Note: 1 GeV/c^2 = 10^3 MeV/c^2 = 10^9 eV/c^2.

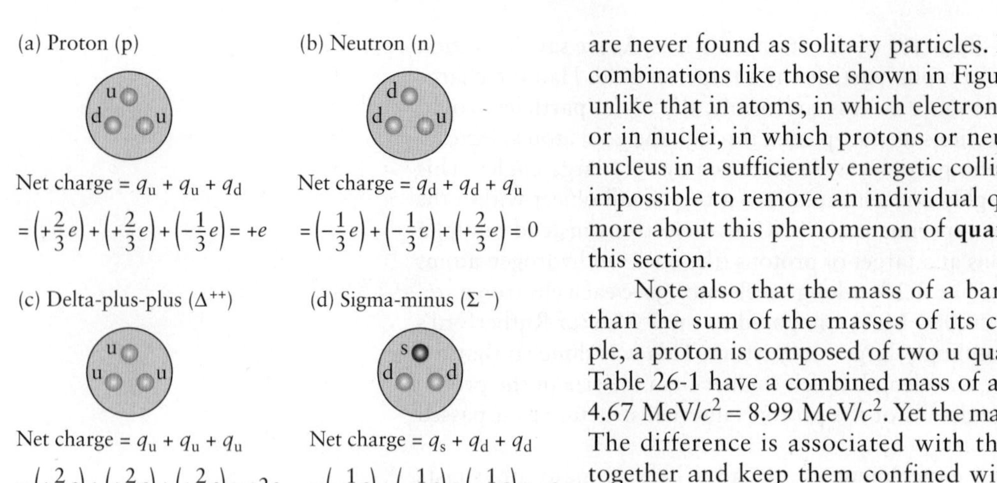

(a) Proton (p)
u d u
Net charge = $q_u + q_u + q_d$
$= \left(+\frac{2}{3}e\right) + \left(+\frac{2}{3}e\right) + \left(-\frac{1}{3}e\right) = +e$
(b) Neutron (n)
d d u
Net charge = $q_d + q_d + q_u$
$= \left(-\frac{1}{3}e\right) + \left(-\frac{1}{3}e\right) + \left(+\frac{2}{3}e\right) = 0$
(c) Delta-plus-plus (Δ^{++})
u u u
Net charge = $q_u + q_u + q_u$
$= \left(+\frac{2}{3}e\right) + \left(+\frac{2}{3}e\right) + \left(+\frac{2}{3}e\right) = +2e$
(d) Sigma-minus (Σ^-)
s d d
Net charge = $q_s + q_d + q_d$
$= \left(-\frac{1}{3}e\right) + \left(-\frac{1}{3}e\right) + \left(-\frac{1}{3}e\right) = -e$

Figure 26-25 Baryons All baryons, including (a) the proton and (b) the neutron, are made of combinations of three quarks. The proton is the only stable baryon; the neutron and all others, like (c) and (d), decay into other particles.

are never found as solitary particles. Instead, they are found only in combinations like those shown in Figure 26-25. This situation is quite unlike that in atoms, in which electrons can be removed by ionization, or in nuclei, in which protons or neutrons can be removed from a nucleus in a sufficiently energetic collision. By contrast, it seems to be impossible to remove an individual quark from a baryon. We'll say more about this phenomenon of **quark confinement** a little later in this section.

Note also that the mass of a baryon is generally much greater than the sum of the masses of its constituent quarks. For example, a proton is composed of two u quarks and a d quark, which from Table 26-1 have a combined mass of approximately $2(2.16 \text{ MeV}/c^2) + 4.67 \text{ MeV}/c^2 = 8.99 \text{ MeV}/c^2$. Yet the mass of the proton is $938.3 \text{ MeV}/c^2$. The difference is associated with the forces that bind the quarks together and keep them confined within the proton. We'll explore these forces in the next section.

Other hadrons are made up of a quark and an **antiquark**, the **antimatter** version of a quark. For every type of matter particle there is an antimatter partner that is identical to it in every way except that it is oppositely charged. We've already encountered one example of antimatter, the positron (e^+) produced in β^+ decay (Section 26-5); the positron and electron have the same mass and are identical except for the sign of their charges ($-e$ for the electron, $+e$ for the positron). Each of the six quarks has an antiquark associated with it. Antiquarks are signified by adding "bar" to the name or placing a bar over the quark's symbol. For example, the antiquark associated with the up quark (u) is the up-bar or $\bar{u}$. The $\bar{u}$ antiquark has charge $-2e/3$, the opposite of the $+2e/3$ charge of the u quark. In the same way, the $\bar{d}$ antiquark carries charge $+e/3$, the opposite of the $-e/3$ charge of the d quark.

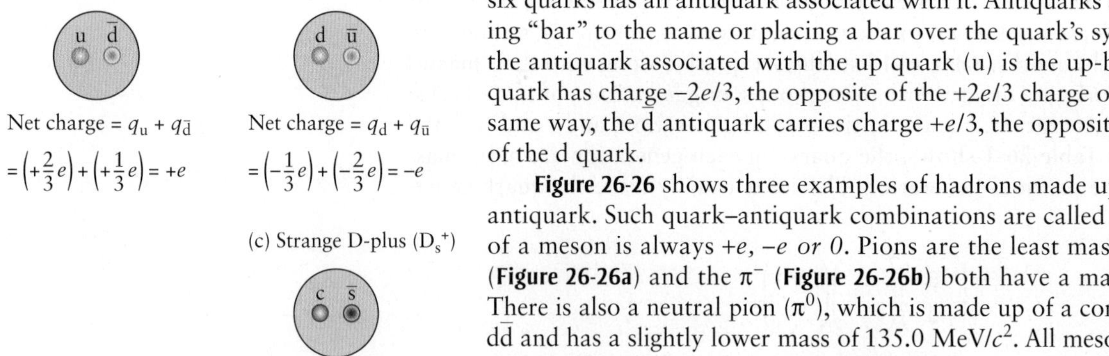

Figure 26-26 Mesons All mesons are made of a quark and an antiquark.

Figure 26-26 shows three examples of hadrons made up of a quark and an antiquark. Such quark–antiquark combinations are called **mesons**. The charge of a meson is always $+e$, $-e$ *or 0*. Pions are the least massive mesons; the π^+ (**Figure 26-26a**) and the π^- (**Figure 26-26b**) both have a mass of $139.6 \text{ MeV}/c^2$. There is also a neutral pion (π^0), which is made up of a combination of $u\bar{u}$ and $d\bar{d}$ and has a slightly lower mass of $135.0 \text{ MeV}/c^2$. All mesons, including pions, are unstable; they decay with a half-life that is a small fraction of a second.

Just as three quarks make up a baryon, three antiquarks make up an *antibaryon*. For each variety of baryon, there is a corresponding variety of antibaryon. For example, the proton has quark content uud and charge $(+2e/3) + (+2e/3) + (-e/3) = +e$; the corresponding antibaryon is the antiproton, which has quark content $\bar{u}\bar{u}\bar{d}$ and charge $(-2e/3) + (-2e/3) + (+e/3) = -e$.

When a particle and its antimatter partner collide, their total mass can be converted to its equivalent energy. For example, the collision of an electron and a positron results in two or more photons that carry the total energy of the two particles, a process known as *annihilation*. A quark and an antiquark can also annihilate each other; we'll see examples of these processes in the next section. In the reverse process, energy can be converted into a particle and its corresponding antiparticle. This helps us understand what happens when we try to remove a single quark from a hadron such as a proton (**Figure 26-27**). Experiment shows that unlike the electric force holding an electron inside an atom, the force that holds a quark inside a hadron does *not* decrease with increasing distance but instead increases as the quarks are separated. So it would take an infinite amount of work (force times distance) to remove a quark from a hadron, which is why we saw that quarks are *confined* inside hadrons. If we do enough work on a quark while trying to remove it, the energy added by this work can produce a new quark–antiquark pair. These combine with the original quarks so that we end up with two hadrons, as Figure 26-7 shows.

The force that binds quarks together is the strong force. When three quarks are bound together in a proton or a neutron, the strong force outside the nucleon is very small and behaves somewhat differently because of how it binds the quarks together.

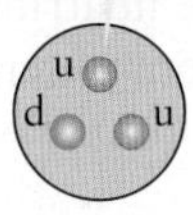
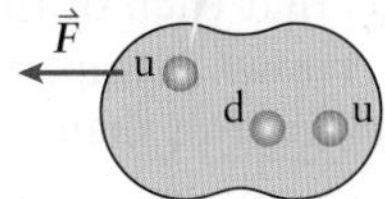
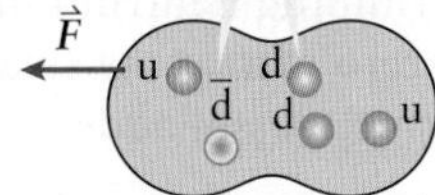
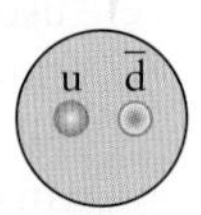
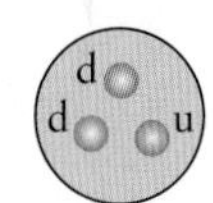

1 What happens if we attempt to remove a quark (in this case, a u quark) from a hadron (in this case, a proton)?
2 In our attempt we exert a force over a distance, which means we do work on the u quark and so add energy to it.
3 The force holding the u quark inside the proton does not decrease with distance, so we cannot remove this quark. Instead, the energy we add goes into creating a new quark–antiquark pair (in this case, a d quark and a $\bar{d}$ antiquark).
4 Rather than removing a quark from the original hadron, the net result is that we end up with *two* hadrons (in this case, a positive pion and a neutron).
Proton (p)
Positive pion (π^+)
Neutron (n)

Figure 26-27 Quark confinement Attempting to remove a quark from a hadron just results in producing additional hadrons.

However, this tiny residual interaction from the force does is what holds protons and neutrons together despite the electric repulsion between like charged protons. Because neutrons interact through this same force, which is always attractive, neutrons help hold a nucleus against repulsion. We will learn why the strong force between nucleons behaves differently in Section 26-10.

In practice the way that physicists experiment with adding energy to quarks is by colliding hadrons with each other; the kinetic energy of the collision goes into producing new hadrons as in Figure 26-27. Recall Figure 26-17, which shows the result of a collision between two protons at the Large Hadron Collider at CERN in Switzerland. The kinetic energy in this collision was so great that multiple baryons, antibaryons, and mesons were produced; the paths of these after the collision are shown by the many yellow tracks in Figure 26-17.

Leptons

While all hadrons are composed of quarks or antiquarks, there are other particles that are not composed of quarks at all. The *leptons* are another category of particles that are as fundamental as the quarks; to the best of our knowledge, leptons are not made up of smaller constituents. The electron is a lepton, as is the muon that we encountered in Section 26-4 (see Example 26-3). The electron neutrino that is created in hydrogen fusion (Section 25-5) and in beta decay (Section 25-6) is also a lepton. There are six leptons, listed in **Table 26-2**, each of which has an antimatter partner.

Notice that in Table 26-2 we have listed the leptons in three groups of two. As is the case for the quarks, leptons form three generations, which we usually write as

$$\begin{pmatrix} e^- \\ \nu_e \end{pmatrix} \begin{pmatrix} \mu^- \\ \nu_\mu \end{pmatrix} \begin{pmatrix} \tau^- \\ \nu_\tau \end{pmatrix}$$

As for the quarks listed in Table 26-1, the leptons in each successive generation are more massive than their counterparts in the preceding generation. In many ways, the muon and tau are more massive versions of the electron, so they share many properties and interact with other particles in similar ways. One way in which electrons, muons, and tau particles *are* significantly different (in addition to the differences in mass) is that while the electron is stable and does not decay, the other two are unstable: The muon has a half-life of 1.56 μs (see Example 26-3), and the tau has a half-life of about 2.0×10^{-13} s.

Each lepton also has an antimatter particle associated with it. We have already encountered the electron and positron. In the same way, the antimuon (μ^+) and antitau (τ^+) have the same masses as the muon and tau, respectively, but the μ^+ and τ^+

TABLE 26-2 The Six Leptons

Lepton	Symbol	Charge	Mass
Electron	e^-	$-e$	0.5110 MeV/c^2
Electron neutrino	ν_e	0	<1.1 eV/c^2
Muon	μ^-	$-e$	105.7 MeV/c^2
Muon neutrino	ν_μ	0	<0.19 MeV/c^2
Tau	τ^-	$-e$	1.777 GeV/c^2
Tau neutrino	τ_ν	0	<18.2 MeV/c^2

have positive charge $+e$ rather than negative charge $-e$. Like the neutrinos ν_e, ν_μ, and ν_τ, the antineutrinos $\bar{\nu}_e$, $\bar{\nu}_\mu$, and $\bar{\nu}_\tau$ have zero charge; they differ from the neutrinos in another way that we'll describe below.

Notice that in Table 26-2 we've given only upper limits for the masses of the three types of neutrino. That's because neutrinos have zero charge, so the methods physicists normally employ to determine the masses of particles (which involve measuring how a particle responds to electric and magnetic forces) can't be used. The indirect methods used for determining neutrino masses tell us that each of the three neutrinos has a small but nonzero mass; how small is not yet known, except we are pretty certain that all three neutrino masses added together would be less than a *millionth* of the mass of an electron!

Unlike quarks, which appear in groups of three to form baryons or in quark–antiquark combinations to form mesons, leptons do not group together to form other particles. In the following section we'll see the reason for this. We'll also discover an essential third class of particles in addition to hadrons and leptons, the *exchange particles*.

Conservation Laws for Hadrons and Leptons

In earlier chapters we encountered a number of important conservation laws, including the conservation of energy and the conservation of momentum. There are several additional conservation laws that govern the behavior of hadrons and leptons.

- *Baryon number is conserved.* Every baryon (composed of three quarks) is assigned *baryon number* B equal to +1, and every antibaryon is assigned B equal to −1. Mesons are assigned $B = 0$, as are leptons. Experiments show that in every process that involves baryons, the sum of the values of B for all particles present before the process equals the sum of the values of B for all particles present after the process. Consider, for example, the β decay of a neutron:

(26-19)
$$\begin{array}{rcccc} & \mathrm{n} \rightarrow & \mathrm{p} + & \mathrm{e}^- + & \bar{\nu}_\mathrm{e} \\ B = & 1 & 1 & 0 & 0 \end{array}$$

The neutron and proton are both baryons, so each has $B = +1$. The electron and the antineutrino are leptons, each with $B = 0$. The total baryon number equals 1 before the decay and equals $1 + 0 + 0 = 1$ after the decay, so baryon number is conserved. Quarks have a fractional baryon number: Every quark has $B = +1/3$, and every antiquark has $B = -1/3$. That's consistent with a baryon with three quarks having $B = 3(+1/3) = +1$ and a meson with a quark and an antiquark having $B = (+1/3) + (-1/3) = 0$.

- *Lepton number is conserved.* Every electron (e^-) and electron neutrino (ν) is assigned *electron-lepton number* $L_e = +1$, and every positron (e^+) and electron antineutrino ($\bar{\nu}_e$) is assigned Le equal to −1. Muons, muon neutrinos, tau particles, and tau neutrinos each have a similarly defined *muon–lepton number* L_μ or a *tau-lepton number* L_τ. A particle that is not a lepton is assigned L_e, L_μ, and L_τ equal to zero. Experiment shows that each of these lepton numbers is separately conserved.

As an example, again consider the β^- decay of a neutron from Equation 26-19:

(26-20)
$$\begin{array}{rcccc} & \mathrm{n} \rightarrow & \mathrm{p} + & \mathrm{e}^- + & \bar{\nu}_\mathrm{e} \\ L_\mathrm{e} = & 0 & 0 & 1 & -1 \end{array}$$

Both the neutron and the proton are baryons, so each has $L_e = 0$; the electron has $L_e = +1$, and the electron antineutrino has $L_e = -1$. The total value of L_e before the decay is zero, and afterward it is $0 + 1 + (-1) = 0$. The electron–lepton number is conserved in β decay. A second example is the decay of a tau into an electron, an electron antineutrino, and a tau neutrino:

(26-21)
$$\begin{array}{rcccc} & \tau^- \rightarrow & \mathrm{e}^- + & \bar{\nu}_\mathrm{e} + & \nu_\tau \\ L_e = & 0 & 1 & -1 & 0 \\ L_\tau = & 1 & 0 & 0 & 1 \end{array}$$

The tau (τ) and tau neutrino (ν_τ) each have $L_e = 0$ and $L_\tau = 1$, the electron (e^-) has $L_e = 1$ and $L_\tau = 0$, and the electron antineutrino ($\bar{\nu}_e$) has $L_e = -1$ and $L_\tau = 0$. You can see that the total electron-lepton number L_e equals 0 before and after the decay, and the total tau-lepton number L_τ equals 1 before and after the decay. Both of these lepton numbers are separately conserved in the decay of the tau.

Physicists have searched for processes in which either baryon number or one of the lepton numbers is not conserved. No such process has ever been observed. It appears that these four laws—conservation of baryon number, electron-lepton number, muon-lepton number, and tau-lepton number—are as universal and fundamental as the law of conservation of electric charge (which says that the net electric charge has the same value before and after any process).

THE TAKEAWAY for Section 26-9

✔ Quarks are the constituents of hadrons. Quarks interact through the strong force and the hadrons interact through the residual strong force. There are six varieties of quarks, each of which has a charge that is a fraction of *e*.

✔ Baryons such as the proton are made up of three quarks. Mesons are made up of a quark and an antiquark. Quarks are never found in isolation, but are always confined inside hadrons.

✔ Leptons, which include electrons and neutrinos, have no constituent particles. They do not interact through the strong force.

✔ In the interactions of subatomic particles, baryon number and the three lepton numbers are each individually conserved.

Takeaway Problems for Section 26-9

1. A neutral η (eta) meson at rest decays into two photons:

$$\eta \rightarrow \gamma + \gamma$$

Calculate the energy, momentum, and wavelength for each of the identical photons. The mass of the eta particle is 547 MeV/c^2.

2. What is the charge of the particle that is composed of (a) the quark combination uds? (b) the quark combination uss? (c) Which particle is likely to be more massive? Explain all your answers.

3. Two protons with the same speed collide head-on in a particle accelerator, causing the reaction

$$\mathrm{p} + \mathrm{p} \rightarrow \mathrm{p} + \mathrm{p} + \pi^0$$

Calculate the minimum kinetic energy of each of the incident protons.

26-10 Four fundamental forces describe all interactions between material objects

Since early in our study of physics we have seen the importance of *forces*, the interactions that can be described as pushes and pulls that one object or system exerts on another. We've encountered three fundamental kinds of forces: the gravitational force, the electromagnetic force, and the strong force. *Gravity* is an attractive force that draws objects closer together. The gravitational force attracts you to Earth and keeps Earth in orbit around the Sun. Electric and magnetic forces, two manifestations of the *electromagnetic force*, cause charged objects to accelerate. As a result, electrons can be bound to atomic nuclei, and atoms can bond together. Note that all contact forces, including the normal force between two objects that touch or the force of friction, are electromagnetic in nature; they arise from the electromagnetic interactions between the atoms of the two surfaces in contact. As we explored in Chapter 25, the *strong force* binds protons and neutrons together to form atomic nuclei. To this list we will add a force that we have not yet discussed but the effects of which we encountered in Section 25-6; this **weak force** is at the heart of the interaction that governs beta decay.

Now that we are exploring the fundamental constituents of matter, we are in a position to ask a central question about force: How do objects exert forces on each other? And what is fundamentally different about the four different kinds of forces: gravitational, electromagnetic, strong, and weak? If we can understand how fundamental particles such as quarks and leptons exert forces of different kinds on each other, we will be closer to answering these questions.

As we will see, the manner in which fundamental particles exert forces on each other—that is, how they *interact*—comes from a remarkable aspect of nature: It is possible to *violate* the law of conservation of energy, provided we do it for a sufficiently short time. We can see this by returning to the *Heisenberg uncertainty principle*, which we introduced in Chapter 24.

The Heisenberg Uncertainty Principle Revisited

In Section 24-8 we learned that due to the wave nature of matter, there is a fundamental limitation on how well you can simultaneously know the momentum and position of a particle. This is the *Heisenberg uncertainty principle for momentum and position*:

EQUATION IN WORDS
Heisenberg uncertainty principle for momentum and position

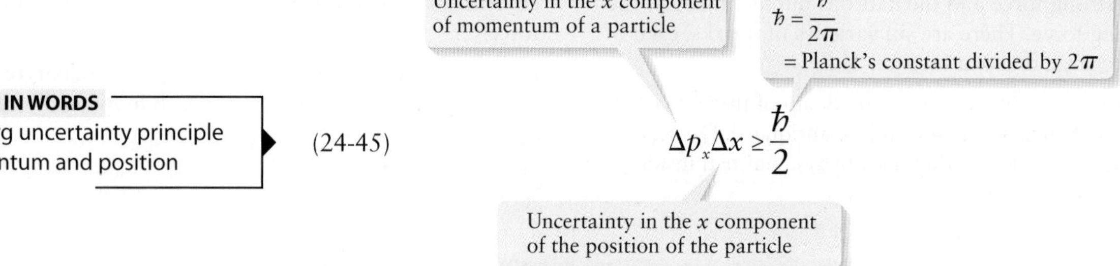

(24-45) $$\Delta p_x \Delta x \geq \frac{\hbar}{2}$$

We deduced this from the properties of waves. To make a wave that propagates in the x direction and that has a small spatial extent—equivalent to a small uncertainty Δx in the x component of the position of the wave—we have to combine sinusoidal waves with a range of wavelengths λ and hence a range of angular wave numbers $k = 2\pi/\lambda$ (see Figure 24-24). In quantum mechanics the x component of momentum of a particle is given by $p_x = h/\lambda = \hbar k$. So a particle—represented by a wave with a small spatial extent—will necessarily include a range of values of momentum p_x. Because its momentum does not have a single unique value, the particle has a momentum uncertainty Δp_x. Equation 24-45 indicates that the smaller the spatial extent given by Δx, the greater the range of angular wave numbers and hence the greater the momentum uncertainty Δp_x.

There is a similar relationship between the *duration in time* of a wave and the range of *frequencies* that the wave includes. To see this, recall from Section 21-3 the expression for a sinusoidal wave propagating in the $+x$ direction:

EQUATION IN WORDS
Wave function for a sinusoidal wave propagating in the $+x$ direction

(21-6) $$y(x,t) = A\cos(kx - \omega t + \phi)$$

Wave function for a **sinusoidal wave** propagating in the $+x$ direction
Angular wave number of the wave $= 2\pi/\lambda$
Phase angle
Amplitude of the wave
Angular frequency of the wave $= 2\pi f$

In Equation 21-6 the angular wave number of the wave is $k = 2\pi/\lambda$ and the angular frequency of the wave is $\omega = 2\pi f$, where f is the wave frequency. Note that the function in Equation 21-6 is valid for any value of x from $-\infty$ to $+\infty$ *and* for any time t from $-\infty$ to $+\infty$. So a wave with a single definite wavelength and a single frequency has an infinite spatial extent *and* an infinite duration. This means that the wave has always been present and will always be present. A more realistic description of a wave is one that has a finite duration; for example, the wave produced when you turn a source of waves (like a laser pointer) on and then off again.

Note that for waves of all kinds, including the quantum-mechanical waves that describe particles, changing the wavelength also changes the frequency. Note also in

Equation 21-6 that k and x appear together in the same way as do ω and t. So if we combine sinusoidal waves of infinite extent and infinite durattion with a range of values of k to create a wave with a small *spatial extent* (that is, one that it is present over only a limited region of space), that wave will also have a range of values of ω and will have a short *duration* (that is, one that is present for only a limited length of time). Hence a wave of short duration necessarily includes a range of frequencies—that is, its frequency does not have a single definite value, but necessarily has some uncertainty.

To see how the frequency uncertainty of a wave is related to its duration in time, recall from Section 24-8 this relationship between the angular wave number uncertainty Δk and the spatial extent Δx (that is, the uncertainty in position):

$$\Delta k \Delta x \geq \frac{1}{2} \qquad (24\text{-}43)$$

Because in Equation 21-6 k and x appear in the same combination as angular frequency ω and time t, it follows that the relationship between the angular frequency uncertainty $\Delta\omega$ and the time duration Δt has the same form as Equation 24-42:

$$\Delta\omega \Delta t \geq \frac{1}{2} \qquad (26\text{-}22)$$

Equation 26-22 says that the product of $\Delta\omega$, the range of angular frequencies present in the wave, and the duration Δt of the wave cannot be less than ½. (This number arises from the specific way in which $\Delta\omega$ and Δt are defined mathematically.) It says that to minimize the duration of a wave necessarily means increasing the range of frequencies that make up the wave. So the shorter the duration of a wave, the less precisely we can answer the question, "What is the frequency of the wave?"

How does Equation 26-22 make it possible to violate energy conservation? To see the answer, let's consider the photon, the particle associated with electromagnetic waves. We learned in Section 24-2 that the energy of a photon is proportional to its frequency:

Energy of a photon — Wave **frequency**

$$E = hf \qquad (24\text{-}11)$$

Planck's constant = $6.62607015 \times 10^{-34}$ J • s

EQUATION IN WORDS
Energy of a photon

We can rewrite this in terms of the angular frequency ω of the photon. Because $\omega = 2\pi f$, it follows that $f = 2\pi/\omega$ and the energy of a photon is

$$E = h\left(\frac{\omega}{2\pi}\right) = \left(\frac{h}{2\pi}\right)\omega = \hbar\omega \qquad (26\text{-}23)$$

(Recall that $\hbar$ is Planck's constant divided by $2\pi : \hbar = h/2\pi$.) Let's now take Equation 26-22 and multiply both sides of the equation by $\hbar$. We get

$$\hbar\Delta\omega\Delta t \geq \frac{\hbar}{2} \quad \text{or} \quad \Delta(\hbar\omega)\Delta t \geq \frac{\hbar}{2} \qquad (26\text{-}24)$$

Here $\Delta(\hbar\omega)$ is the breadth of values of the quantity $\hbar\omega$ that must be included in the wave—in other words, it is the uncertainty in the value of $\hbar\omega$. But Equation 26-23 tells us that $E = \hbar\omega$ is the energy of a photon associated with the wave. So $\Delta(\hbar\omega)$ is the uncertainty in photon energy ΔE, and Equation 26-24 becomes

$$\Delta E \Delta t \geq \frac{\hbar}{2} \text{ for a photon} \qquad (26\text{-}25)$$

This means that just as a wave of finite duration does not have a single definite frequency, a photon of finite duration does not have a single definite energy: It includes energy values that extend over a range of breadth ΔE. We can think of ΔE as the *uncertainty* in the energy of the photon. Equation 26-25 says that if a photon has a duration Δt, the product of Δt and the uncertainty in energy ΔE of the photon cannot be less than $\hbar/2$ and so ΔE cannot be less than $\hbar/(2\Delta t)$. This energy uncertainty is *not* a result of the limitations of an experimental apparatus that we might use to measure the energy of a photon. Rather, it is intrinsic to photons because of their wave nature.

We have used the equations $E = hf$ and $E = \hbar\omega$ to apply to photons only. But this relationship applies to particles of *all* kinds: We can think of anything with an energy E as having an associated angular frequency ω given by $E = \hbar\omega$. Then Equation 26-25 applies to phenomena in general, not just to photons. This implies that any physical phenomenon that has a finite duration Δt will necessarily have an uncertainty ΔE in energy given by Equation 26-25. The minimum value of this uncertainty is found by replacing the $\geq$ sign (greater than or equal to) in Equation 26-25 with an equals sign. The shorter the duration Δt of the phenomenon, the greater the minimum value of the energy uncertainty ΔE and the more uncertain the energy of the phenomenon. This is known as the **Heisenberg uncertainty principle for energy and time:**

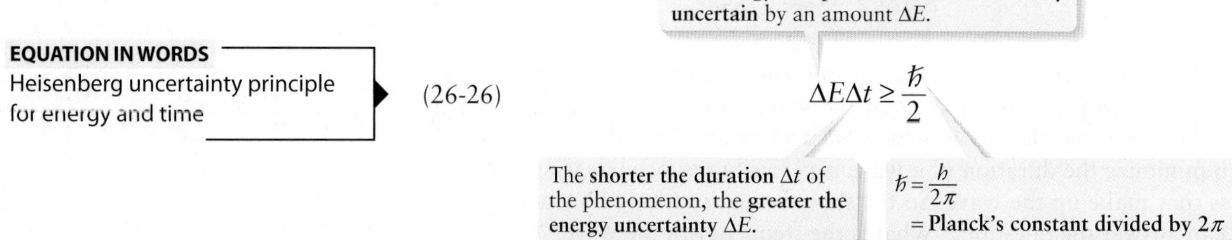
EQUATION IN WORDS
Heisenberg uncertainty principle for energy and time
(26-26)
$\Delta E \Delta t \geq \frac{\hbar}{2}$
The energy of a phenomenon is necessarily uncertain by an amount ΔE.
The shorter the duration Δt of the phenomenon, the greater the energy uncertainty ΔE.
$\hbar = \frac{h}{2\pi}$
= Planck's constant divided by 2π

Another way to interpret Equation 26-26 is to say that it places limits on how precisely the law of conservation of energy must be obeyed. Suppose a system undergoes some kind of process that lasts for a time Δt. Equation 26-6 shows that the energy of the system is necessarily uncertain during that process, and the minimum energy uncertainty is given by $\Delta E \Delta t = \hbar/2$ or $\Delta E = \hbar/(2\Delta t)$. So it's fundamentally impossible to measure the energy of the system with an uncertainty less than $\hbar/(2\Delta t)$. This means that during the process the energy E of the system could actually vary by as much as $\hbar/(2\Delta t)$, and there would be no way that we could tell that the energy had changed value—that is, that the energy was not conserved. The shorter the duration Δt of the process, the greater the amount $\hbar/(2\Delta t)$ by which energy conservation can be (temporarily) violated during that duration of time. Stated another way, it's acceptable to violate the law of conservation of energy by an amount ΔE, provided the duration of time during which the law is violated is no more than $\Delta t = \hbar/(2\Delta E)$.

Exchange Particles: The Electromagnetic Force

The Heisenberg uncertainty principle for energy and time helps us understand the following bold statement: *All forces result from the exchange of particles.* To see what we mean by this statement, first consider the electromagnetic force. Up to this point we've used the idea that charged particles exert electromagnetic forces on each other even at a distance, with no physical contact required. But a more sophisticated way to look at this force is to envision that when two charged particles exert an electric or magnetic force on each other, they do so by exchanging a photon: One of the charged particles emits the photon, and the other absorbs it. The exchanged photon has energy, and it violates the law of conservation of energy for this photon to spontaneously appear and be emitted by one of the charged particles. But as we have seen, it's perfectly acceptable to violate the law of conservation of energy by an amount ΔE, provided we do so for a time no longer than $\Delta t = \hbar/(2\Delta E)$. So the uncertainty principle proves that what we've described can take place provided the exchanged photon is absorbed by the second particle (and so disappears) within a time $\hbar/(2\Delta E)$ after the photon of energy ΔE was emitted by the first particle.

In this picture we say that the electromagnetic force is *mediated* by the exchange of a photon and that the exchanged photon is the *mediator* of the force. The particles that mediate forces are called **exchange particles**. You should try to envision a continuous stream of photons going back and forth between charged particles so that the two particles continuously exert an electromagnetic force on each other. Any charged particle can and does emit and absorb photons exchanged in this way. These photons are not the same as the photons emitted by a light bulb or a laser, however; they can exist for only a finite time before they must disappear. We call them **virtual particles.**

We can represent the exchange process by a diagram such as the one shown in **Figure 26-28**. In this slightly simplified version of a *Feynman diagram*, invented by American physicist Richard Feynman to visualize and analyze processes that involve fundamental particles, time runs from left to right. The lines associated with each particle do not represent actual paths that the particles take through space but only indicate which particles interact with which other particles. In Figure 26-28 two electrons exchange a photon and in so doing each exerts an electromagnetic force on the other. A particle such as an electron is drawn as a solid, straight line in Feynman diagrams. An exchange particle is drawn as either a wavy line (as for the photon) or a spiral line.

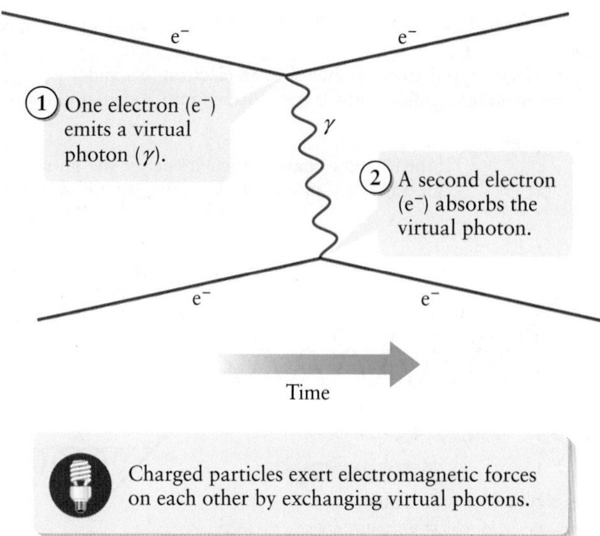
e^-
e^-
1 One electron (e^-) emits a virtual photon (γ).
γ
2 A second electron (e^-) absorbs the virtual photon.
e^-
e^-
Time
Charged particles exert electromagnetic forces on each other by exchanging virtual photons.

Figure 26-28 Photon exchange and the electromagnetic interaction All electric and magnetic interactions between charged particles involve the exchange of virtual photons.

This picture helps us understand Coulomb's law, which tells us that the electric force between two charged particles separated by a distance r decreases in proportion to $1/r^2$. In other words, the electric force goes to zero only when the charges are infinitely far apart. This is possible because the photon has zero mass, so its minimum energy $E = hf$ is zero. (If the photon did have a mass m, its minimum energy would be its rest energy mc^2.) If one charged particle violates conservation of energy by creating and emitting a photon of energy ΔE, the uncertainty principle says that this photon can exist no longer than $\Delta t = \hbar/(2\Delta E)$. Even traveling at the speed of light c, this photon can travel no farther than $c\Delta t = c\hbar/(2\Delta E)$, so that distance is the maximum separation at which two charged particles can interact by exchanging photons of energy ΔE. But because the lower limit on the energy of a photon is zero, the amount ΔE by which the particle violates conservation of energy can be as small as we like. So the distance $r = c\Delta t = c\hbar/(2\Delta E)$ can be arbitrarily large, and two charged particles can interact via the electromagnetic force at any distance out to infinity. But because only low-energy photons (with small ΔE) can be exchanged between charged particles separated by great distances r, the force is quite weak if r is large—just as in Coulomb's law.

Exchange Particles: The Strong Force

We use a similar picture to explain the strong interaction between quarks. The Feynman diagram in **Figure 26-29** shows two quarks that exert forces on each other by exchanging a particle called the **gluon**. This rather whimsical name expresses the idea that the exchange of gluons provides the force that confines, or glues, quarks inside baryons. Gluon exchange is also how the quark and antiquark inside a meson interact with each other and how the antiquarks inside an antibaryon interact.

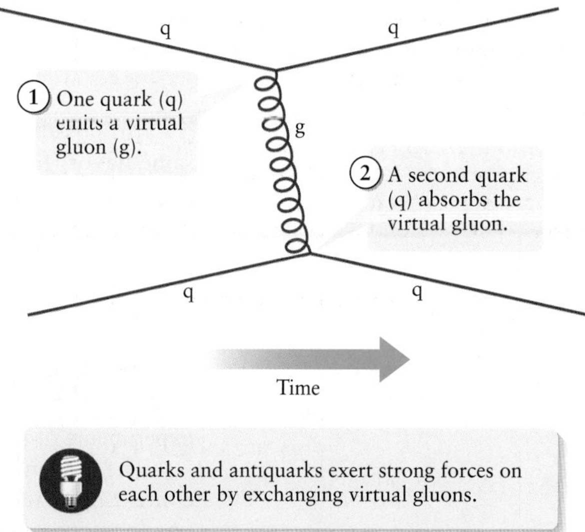
q
q
1 One quark (q) emits a virtual gluon (g).
g
2 A second quark (q) absorbs the virtual gluon.
q
q
Time
Quarks and antiquarks exert strong forces on each other by exchanging virtual gluons.

Figure 26-29 Gluon exchange and the strong interaction between quarks The forces that bind quarks together inside baryons and mesons are the result of the exchange of virtual gluons.

Like the photon the gluon has zero mass, so the lower limit on the energy of a gluon is zero. Using the same argument we made above for photons, it follows that quarks or antiquarks separated by any distance, no matter how large, can interact by exchanging gluons. But unlike photons, the kinds of particles that emit and absorb gluons are not simply those with electric

(a)

① In electron–positron annihilation, an electron (e⁻) and a positron (e⁺) collide and are transformed into a virtual photon.

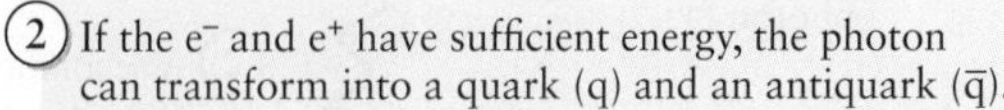
② If the e⁻ and e⁺ have sufficient energy, the photon can transform into a quark (q) and an antiquark (q̄).

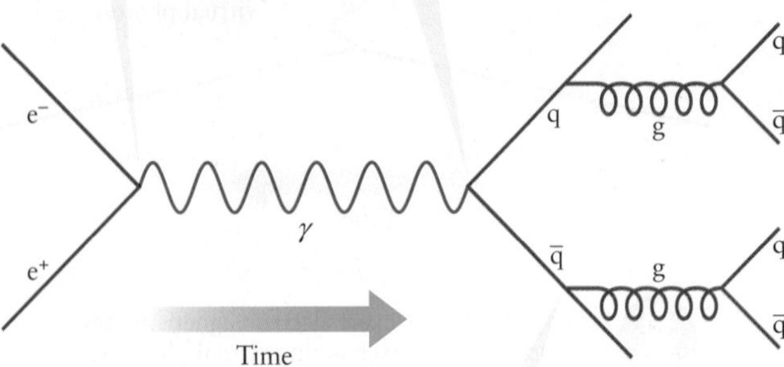
e⁻
e⁺
γ
Time
q
q̄
g
g
q
q̄
q
q̄

③ The quark and antiquark can emit gluons, which transform into quark–antiquark pairs. The shower of quarks and antiquarks coalesces into hadrons, including baryons (three quarks), antibaryons (three antiquarks), and/or mesons (quark–antiquark pairs).

(b)

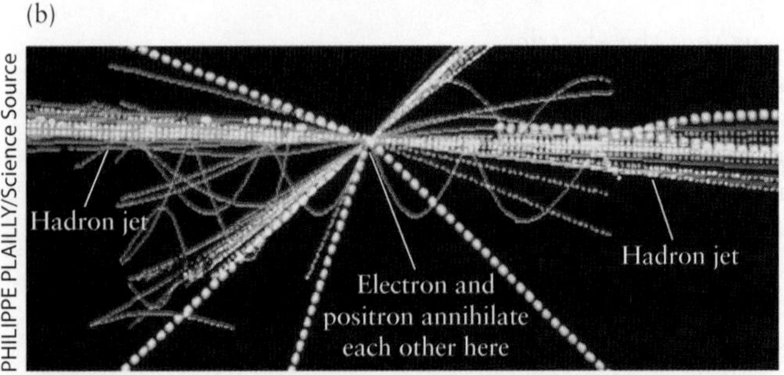
Hadron jet
Hadron jet
Electron and
positron annihilate
each other here

PHILIPPE PLAILLY/Science Source

Figure 26-30 Electron–positron annihilation and hadron production (a) When an electron and a positron collide, the result can be a shower of hadrons produced via virtual photons and gluons. (b) This image shows the result of a high-energy electron–positron collision at CERN in Geneva, Switzerland. Two jets of hadrons are produced along with other particles.

charge: They are particles that have a different attribute called *color*. (This is yet another of the light-hearted names associated with quark physics. It has nothing to do with the wavelength of light or the perception of color by the human eye.) Quarks and antiquarks carry color, as do gluons themselves. So unlike photons, gluons can interact with each other by exchanging other gluons. (By contrast, photons have no electric charge and cannot interact directly with other photons.) As a consequence of this curious aspect of gluons, the force that gluons mediate between quarks does *not* get weaker as the quarks move farther apart. This helps to explain why this force makes it impossible to remove an isolated quark from a hadron.

As we mentioned in Section 26-9, the mass of a nucleon (a proton or neutron) is much greater than the sum of the masses of the three quarks that constitute the nucleon. The gluon picture helps us understand why this is: Part of the additional mass is associated (through the relationship $E = mc^2$) with the energy of the virtual gluons that are continuously exchanged between the constituent quarks. The remaining mass is associated with the kinetic energies of the constituent quarks, which are in high-speed motion inside the nucleon.

Because quarks are charged, they also interact electromagnetically by exchanging photons. But these interactions have a relatively small effect compared to the dominant strong interaction mediated by gluons.

As we mentioned in the previous section, when an electron and a positron (matter and antimatter) collide, they annihilate each other. The Feynman diagram in **Figure 26-30a** shows one possible outcome of such an annihilation. The electron of charge $-e$ and the positron of charge $+e$ disappear and are replaced by a single photon. This *virtual* photon does not have the proper relationship between energy and momentum that a real photon must have, so it violates the conservation laws that govern the electron–positron collision. Like an exchange photon, this virtual photon can exist for only a finite time before it must disappear. In some cases the virtual photon will transform back into an electron–positron pair. But as Figure 26-30a shows, it's also possible for the virtual photon to become a quark and an antiquark of the same flavor (for instance, a u quark and a ū antiquark, or an s quark and an s̄ antiquark). The quark and antiquark can themselves emit virtual gluons, which can in turn transform into quark–antiquark pairs. Depending on how much energy is available in the original collision between electron and positron, many such quark–antiquark pairs can be produced. These will sort themselves into combinations of quarks (baryons), combinations of antiquarks (antibaryons), and quark–antiquark pairs (mesons). The result will be a shower of hadrons emanating from the site of the electron–positron collision. **Figure 26-30b** shows "jets" of hadrons emerging from the site of just such a collision. Many varieties of hadrons were first identified in electron–positron collision experiments of this sort.

The notion of gluon exchange also helps us understand the strong force in the context in which we first encountered it in Section 25-2: a force that attracts nucleons (protons or neutrons) to each other. **Figure 26-31** shows how this force between nucleons arises. The meson that is produced in this way acts as an exchange particle between the nucleons, and this exchange gives rise to the attractive force between nucleons. The meson exchange particle has a substantial mass m_{exchange}, so the minimum energy that an exchanged meson can have is its rest energy $m_{\text{exchange}}c^2$.

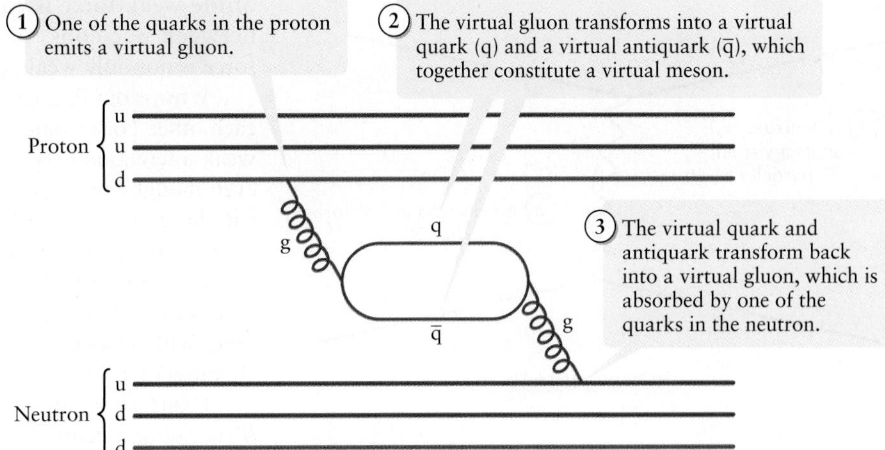

1 One of the quarks in the proton emits a virtual gluon.
2 The virtual gluon transforms into a virtual quark (q) and a virtual antiquark ($\bar{q}$), which together constitute a virtual meson.
3 The virtual quark and antiquark transform back into a virtual gluon, which is absorbed by one of the quarks in the neutron.
Proton
u
u
d
g
q
$\bar{q}$
g
Neutron
u
d
d

Figure 26-31 The strong force between nucleons The force that binds protons and neutrons together has its origin in the interaction between quarks and gluons.

- Nucleons (protons and neutrons) exert the strong force on each other by exchanging virtual mesons.
- This exchange is actually due to interactions between quarks and gluons.

Producing such a meson means violating the conservation of energy by an amount of at least $\Delta E = m_{\text{exchange}}c^2$, which means that the meson can exist for a time no longer than $\Delta t = \hbar/(2\Delta E) = \hbar/2m_{\text{exchange}}c^2$. Even moving at the speed of light, the maximum distance that this exchange meson can travel during a time $\Delta t = \hbar/2m_{\text{exchange}}c^2$ is

$$c\Delta t = \frac{c\hbar}{2m_{\text{exchange}}c^2} = \frac{\hbar}{2m_{\text{exchange}}c} \tag{26-27}$$

(range of a force mediated by an exchange particle of mass m_{exchange})

Equation 26-27 tells us the *range* of the force mediated by the exchange meson. There are many types of mesons, but the type whose exchange will have the longest range is the one with the smallest mass. (That's because the range in Equation 26-27 is inversely proportional to the mass of the exchange particle.) The mass of the lightest meson, the neutral pion (π^0), is 135.0 MeV/c^2 = 135.0×10^6 eV/c^2, so the maximum range of the strong force between nucleons should be

$$\frac{\hbar}{2m_{\text{neutral pion}}c} = \frac{\hbar c}{2m_{\text{neutral pion}}c^2} = \frac{hc}{4\pi m_{\text{neutral pion}}c^2}$$

$$= \frac{(4.136 \times 10^{-15}\ \text{eV} \cdot \text{s})(3.00 \times 10^8\ \text{m/s})}{4\pi(135.0 \times 10^6\ \text{eV}/c^2)c^2}$$

$$= 7.31 \times 10^{-16}\ \text{m} = 0.731\ \text{fm}$$

Because this range is so small, the strong force between nucleons is a short-range force, just as we discussed in Section 25-2. Indeed the rough value of 0.731 fm that we calculated here is of the same order of magnitude as the value we gave in Section 25-2 for the range of the strong force between nucleons.

Exchange Particles: The Weak Force and the Gravitational Force

The weak force is mediated by *three* different exchange particles. These are the neutral Z_0 (charge zero) and the positively and negatively charged W particles, W^+ (charge $+e$) and W^- (charge $-e$). These particles are not massless: The Z^0 has mass 91.2 GeV/c^2, about 100 times the mass of a proton, and the W^+ and W^- both have mass 80.4 GeV/c^2. The calculation above shows that the range of the force mediated by an exchange particle of mass m_{exchange} is inversely proportional to the mass. So because the Z^0, W^+, and W^- have roughly 600 times the mass of the neutral pion (135.0 MeV/c^2), the range

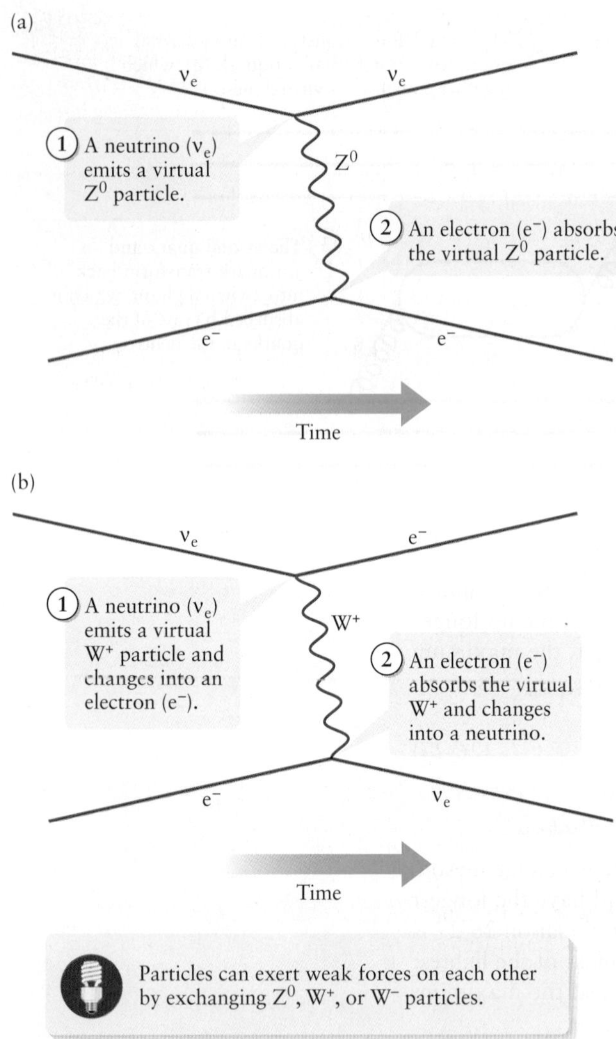

Figure 26-32 The weak interaction The weak force between particles involves the exchange of massive virtual particles. These come in both (a) neutral and (b) charged varieties.

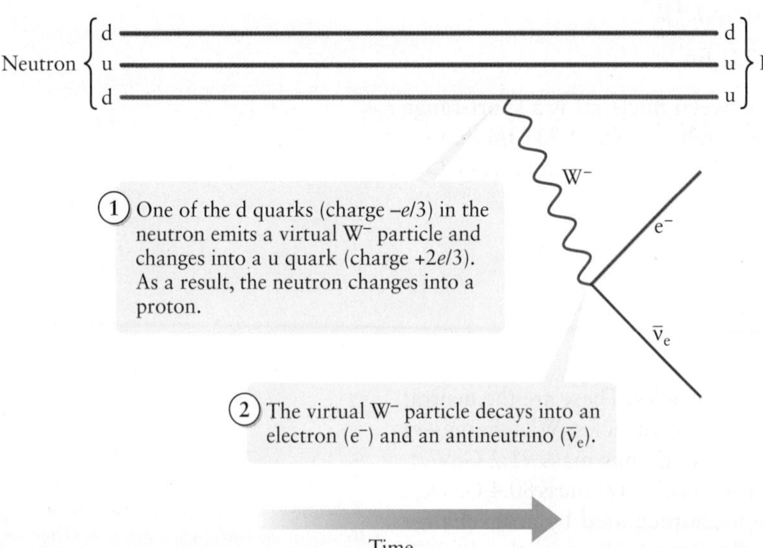

Neutron
d
u
d
d
u
u
Proton
W^-
e^-
$\bar{\nu}_e$
1 One of the d quarks (charge $-e/3$) in the neutron emits a virtual W^- particle and changes into a u quark (charge $+2e/3$). As a result, the neutron changes into a proton.
2 The virtual W^- particle decays into an electron (e^-) and an antineutrino ($\bar{\nu}_e$).
Time

Figure 26-33 Beta decay of a neutron The weak interaction describes how a free neutron decays into a proton, an electron, and an antineutrino.

of the weak force is roughly 1/600 that of the strong force between nucleons, or about 10^{-18} m = 10^{-3} fm. The weak force is not only weak, it is of *extremely* short range!

It turns out that any two particles can exert a weak force on each other. For instance, **Figure 26-32** shows two examples of the weak interaction between a neutrino and an electron. Note that even though the weak interaction does not directly involve electric charge, charge is still conserved. In **Figure 26-32a** the neutrino remains neutral, and the electron retains its charge of $-e$. In **Figure 26-32b** the neutrino emits a W^+ of charge $+e$ and becomes an electron of charge $-e$, so charge is conserved. The W^+ combines with an electron of charge $-e$ and becomes a neutrino of charge zero, and again charge is conserved.

Figure 26-33 shows how the weak force gives rise to the β^- decay of a neutron (quark content udd) to a proton (quark content uud):

$$n \rightarrow p + e^- + \bar{\nu}_e$$

Compared to a u quark, a d quark is substantially more massive (see Table 26-1) and so has a substantially greater rest energy. So the d quark can lower its energy by becoming a u quark. It does this by emitting a W^- particle, as shown. This is a very short-lived virtual particle: The rest energy of the W^- (80.4 GeV/c^2) is far greater than the rest energy of the original d quark (about 4.67 MeV/c^2), so emitting a W^- means violating energy conservation by quite a bit. The W^- of charge $-e$ quickly decays into an electron of charge $-e$ and a neutral antineutrino $\bar{\nu}_e$. (It has to be an *anti*neutrino to conserve electron–lepton number L_e: The W^- is not a lepton and therefore has $L_e = 0$, so the W^- must decay into particles with a net value of L_e equal to zero. The electron, with $L_e = +1$, must therefore be accompanied by a neutral particle with $L_e = -1$, which means an antineutrino rather than a neutrino.)

As we discussed in Section 25-6, in some nuclei it's energetically favorable for β^+ decay to take place, in which a proton changes to a neutron. This is similar to the process shown in Figure 26-33, except that one of the u quarks in a proton changes to a d quark by emitting a W^+ rather than a W^-. The W^+ then decays into a positron (e^+) and a neutrino (ν_e). One nucleus in which this can happen is the potassium isotope ^{40}K, which is found in foods such as potatoes. The antimatter produced in this way—the positron—quickly encounters an atomic electron, and the positron and electron annihilate each other. Happily, this happens very infrequently in your food (and releases very little energy when it does happen). So you needn't worry about biting into a bit of antimatter when you eat a potato!

A particle named the graviton is thought to mediate the gravitational force. So far no experimental evidence has confirmed the existence of this particle, although no experimental evidence has excluded it either. Because the gravitational force between two masses separated by a distance r is proportional to $1/r^2$, just like the electric force between two charged particles, the graviton is thought to have the same mass as the photon: zero. The graviton also carries no charge.

Table 26-3 summarizes the four fundamental forces and the exchange particles that mediate these forces.

TABLE 26-3 The Four Fundamental Forces

Force	Range	Mediator(s)	Strength relative to the strong force
Gravity	Infinite	Graviton	10^{-40}
Weak	~10^{-3} fm	Z^0, W^+, W^-	10^{-6}
Electromagnetic	Infinite	Photon (γ)	10^{-2}
Strong*	~1 fm	Gluon (g)	1

*Gluons mediate the strong force between quarks, and this force has infinite range. As described in the text, the strong force between *nucleons* (protons and neutrons) has a range of only about 1 fm because it involves not just gluons but also virtual meson exchange.

The Standard Model and the Higgs Particle

The picture that Table 26-3 summarizes—in which two classes of fundamental particles, the six quarks and the six leptons, interact by means of six exchange particles, the graviton, the Z^0, the W^+, the W^-, the photon, and the gluon—is called the **Standard Model**. What this very simple table does not reflect is the decades of experimental and theoretical effort (and several Nobel Prizes in Physics) that have gone into constructing this model. One long-standing question about the Standard Model that has recently been answered is this: Of the exchange particles listed in Table 26-3, why do the Z^0, W^+, and W^- all have substantial masses (91.2, 80.4, and 80.4 GeV/c^2, respectively), while the graviton, photon, and gluon each have zero mass? The answer proposed in the 1960s was that there is a field that fills all of space, now called the **Higgs field**, and the interactions of the Z^0, W^+, and W^- particles with this field give these particles their masses. Unlike electric and magnetic fields, which are vectors with a magnitude and direction, the Higgs field is a scalar that has no direction. Because the Higgs field has the same nonzero value everywhere in space, the masses of the Z^0, W^+, and W^- have the same values no matter where in space these particles are found.

This proposal may remind you of a debunked idea that we encountered in Section 26-3. In the nineteenth century, it was thought that there was a substance called the luminiferous ether that filled all space and that acted as a medium for the propagation of light waves. The Michelson–Morley experiment demonstrated that no such luminiferous ether exists. To determine whether the Higgs field exists, twenty-first-century physicists used the idea that for each kind of field there is a corresponding particle. For example, the particle that corresponds to the electric and magnetic fields is the photon. The electric and magnetic forces are mediated by virtual photon exchange, and an electromagnetic wave is made up of photons. To explain the observed masses of the Z^0, W^+, and W^-, the **Higgs particle** that corresponds to the Higgs field would have to have zero charge, have a mass about 50% greater than that of the W^+ or W^-, and be unstable. If such a particle could be produced in a high-energy collision of other particles, it would suggest that the Higgs field does indeed exist.

In 2012, after decades of experimental effort, the Higgs particle was discovered in collisions between high-energy protons at the Large Hadron Collider in Switzerland (**Figure 26-34**). The observed Higgs particle mass of 125 GeV/c^2 is within the expected range of values, and as predicted the Higgs particle has zero charge and a half-life of about 10–22 s. This discovery provided strong experimental support that unlike the luminiferous ether, the Higgs field actually does exist throughout space. This field is now regarded as an essential part of the Standard Model.

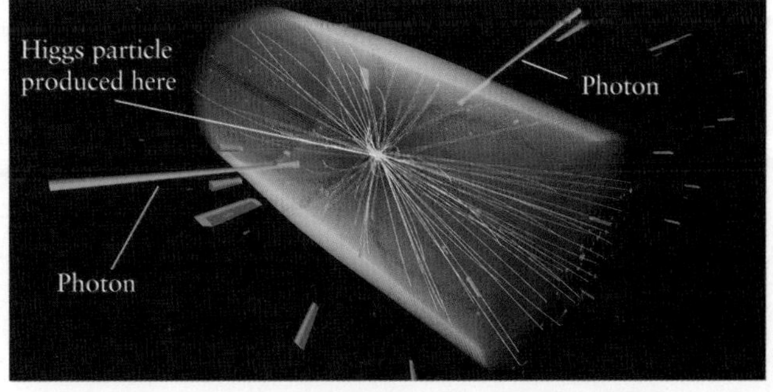

Figure 26-34 Detecting the Higgs particle A Higgs particle was produced in the collision shown here between high-energy protons at the Large Hadron Collider at CERN. The Higgs particle then decayed into two photons. By measuring the energy of these photons, physicists can determine the mass of the Higgs particle.

The Higgs field is thought to explain not just the masses of the Z^0, W^+, and W^-, but also the masses of the leptons and the quarks. In this picture, the reason the muon has a greater mass than the electron is that muons interact more strongly with the Higgs field than electrons do. In future experiments, physicists will investigate how particles of various kinds interact with the Higgs *field* by measuring how these particles interact with Higgs *particles*. These and other aspects of the Standard Model are the subject of active research by physicists around the globe.

THE TAKEAWAY for Section 26-10

✔ Particles exert forces on each other through the exchange of virtual particles that are emitted by one particle and absorbed by another.

✔ The photon is the exchange particle that mediates the electromagnetic force; the gluon mediates the strong force; and the Z^0, W^+, and W^- particles mediate the weak force. The graviton, which has not yet been detected, is thought to mediate the gravitational force.

✔ The Standard Model is our picture of six types of quarks and six types of leptons that interact by means of these various exchange particles. A key part of the Standard Model is the Higgs field, a field that exists throughout all space and explains the large masses of the Z^0, W^+, and W^-.

Takeaway Problems for Section 26-10

1. A c quark can decay into an s quark via the weak force. In this process, what does the c quark emit?
   (A) A W^+, which decays into a positron and a neutrino
   (B) A W^+, which decays into a positron and an antineutrino
   (C) A W^-, which decays into an electron and a neutrino
   (D) A W^-, which decays into an electron and an antineutrino.
2. Discuss the similarities and differences between a photon and a neutrino.
3. Two protons with enough energy to possibly produce the Higgs particle are smashed together in a series of experiments. Each of these experiments produces a unique collection of signals corresponding to the presence of particles produced directly or indirectly as a result of the proton–proton collision. One experiment produces signals indicating the presence of an electron, two muons, two high-energy photons, and a Z^0 boson. Could this collection of particles have been produced by the decay of a Higgs particle? Explain your answer.

WHAT DID YOU LEARN?

Chapter learning goals	Section(s)	Related example(s)	Relevant section review exercise(s)
Describe how different observers view the same motion in Newtonian physics.	26-2	26-1	1-3
Explain how the Michelson–Morley experiment helped rule out the ether model of the propagation of light and use the concept in calculations.	26-3		1-3
Describe how the time interval between two events can have different values in different reference frames and calculate that time interval for different reference frames.	26-4	26-2 26-3	1-3
Calculate how the dimensions of an object change when it is in motion.	26-5	26-4 26-5 26-6	1-3
Explain why the speed of light in a vacuum is an ultimate speed limit and use this concept to calculate relative velocities.	26-6	26-7	1-3
Calculate the rest energy of an object with mass, and relativistic momentum and energy.	26-7	26-8 26-9	1-3
Explain what the principle of equivalence tells us about the nature of gravity.	26-8		1-3
Describe the difference between hadrons and leptons, and of what hadrons are composed, and use conservation of baryon and lepton numbers to predict allowable reactions.	26-9		1-3
Explain how fundamental particles interact with each other, and the differences among the four fundamental forces.	26-10		1-3

Chapter 26 Review

Key Terms

All the Key Terms can be found in the Glossary/Glosario on page G1 in the back of the book.

Chapter Summary

Topic	Equation or Figure
Newtonian relativity: The principle of Newtonian relativity states that the laws of motion are the same in all inertial reference frames. The Galilean transformation relates the coordinates of an event in one frame to the coordinates of the same event in another frame, and the Galilean velocity transformation relates the velocity of an object relative to one frame to its velocity relative to another frame. These transformations are valid only for speeds that are small compared to the speed of light c.	Inertial frame of reference S' moves at speed V in the positive x direction relative to inertial frame of reference S. $x' = x - Vt$, $y' = y$, $z' = z$, $t' = t$ (26-1) Coordinates of an event as measured in frame S' / Coordinates of the same event as measured in frame S Inertial frame of reference S' moves at speed V in the positive x direction relative to inertial frame of reference S. $v'_x = v_x - V$, $v'_y = v_y$, $v'_z = v_z$ (26-5) Velocity components of an object as measured in frame S' / Velocity components of the same object as measured in frame S

The Michelson–Morley experiment: In the nineteenth century it was hypothesized that space was filled with a material medium, called the luminiferous ether, that was required for light to propagate through space. The Michelson–Morley experiment provided evidence that the ether does not exist and motivated Einstein's special theory of relativity.

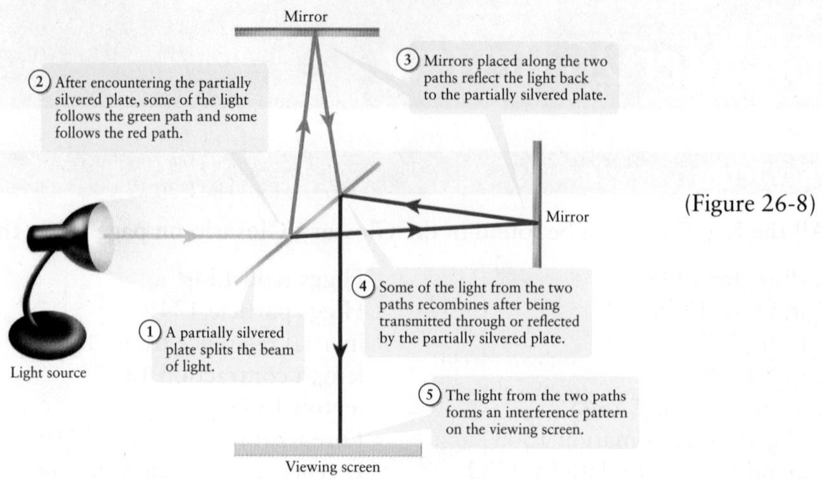

Mirror
② After encountering the partially silvered plate, some of the light follows the green path and some follows the red path.
③ Mirrors placed along the two paths reflect the light back to the partially silvered plate.
Mirror
Light source
① A partially silvered plate splits the beam of light.
④ Some of the light from the two paths recombines after being transmitted through or reflected by the partially silvered plate.
⑤ The light from the two paths forms an interference pattern on the viewing screen.
Viewing screen

(Figure 26-8)

The special theory of relativity: Einstein's special theory of relativity is based on the postulates that all laws of physics are the same in all inertial reference frames, and the speed of light in a vacuum is the same in all inertial reference frames. Two consequences are that moving clocks run slow, and moving extended objects are shortened along their direction of motion.

Time interval between two events as measured in frame S', in which those events occur at the same place

$$\Delta t = \frac{\Delta t_{\text{proper}}}{\sqrt{1 - \dfrac{V^2}{c^2}}} \tag{26-10}$$

Time interval between the same two events as measured in frame S

Speed of S' relative to S

Speed of light in a vacuum

Length of an object in frame S', in which it is at rest

$$L = L_{\text{rest}}\sqrt{1 - \frac{V^2}{c^2}} \tag{26-14}$$

Length of the same object in frame S, in which the object is moving along its length

Speed of S' relative to S

Speed of light in a vacuum

The Lorentz transformation: The Lorentz transformation is a generalization of the Galilean transformation that is valid for all speeds. The Lorentz velocity transformation is a generalization of the Galilean velocity transformation: It shows that the speed of light c is an ultimate speed limit and that an object at rest can never be accelerated to c.

Inertial frame of reference S' moves at speed V in the positive x direction relative to inertial frame of reference S.

$$x' = \frac{x - Vt}{\sqrt{1 - \dfrac{V^2}{c^2}}} \tag{26-15}$$

$$y' = y$$

$$z' = z$$

$$t' = \frac{t - \dfrac{V}{c^2}x}{\sqrt{1 - \dfrac{V^2}{c^2}}}$$

Coordinates of an event as measured in frame S'

Coordinates of the same event as measured in frame S

Inertial frame of reference S' moves at speed V in the positive x direction relative to inertial frame of reference S.

x component of velocity of an object moving along the x axis as measured in frame S'

$$v_x' = \frac{v_x - V}{1 - \frac{V}{c^2} v_x} \quad (26\text{-}17)$$

x component of velocity of the same object as measured in frame S

Kinetic energy, momentum, and rest energy: The formulas for kinetic energy and momentum must be modified to account for c being the ultimate speed limit. The relativity postulates also show that an object with mass m has a rest energy $E_0 = mc^2$ even when it is not in motion.

Kinetic energy of a particle of mass m and velocity $\vec{v}$

$$K = (\gamma - 1)mc^2$$

$$\vec{p} = \gamma m\vec{v} \quad (24\text{-}15)$$

Relativistic gamma for speed v

Momentum of a particle of mass m and velocity $\vec{v}$

At speeds that are a small fraction of the speed of light c, these are approximately equal to the Newtonian expressions

$$K = \frac{1}{2}mv^2 \text{ and } \vec{p} = m\vec{v}$$

The general theory of relativity: The principle of equivalence states that a gravitational field is equivalent to an accelerated reference frame in the absence of gravity. This principle shows that light is affected by gravity just as objects with mass are.

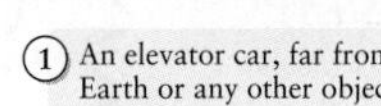

1 An elevator car, far from Earth or any other object, accelerates upward.
2 As seen by an observer in an inertial reference frame, the thrown ball moves in a horizontal line.

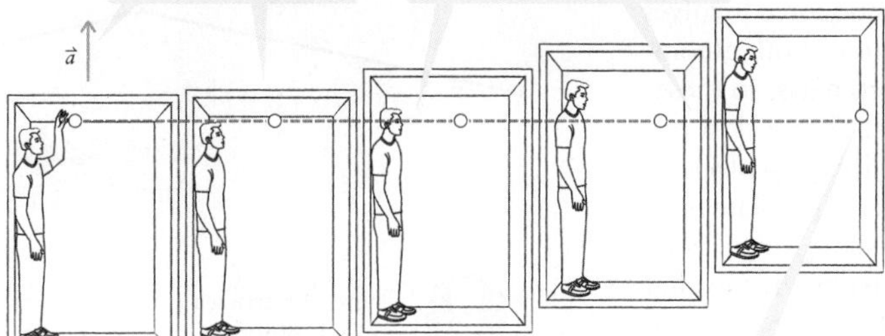

$\vec{a}$
3 As seen by an observer in the accelerating elevator car, the ball follows a *curved path*: Where it hits the far wall is lower than where it left the thrower's hand.

The trajectory of the ball in the accelerated frame of reference of the elevator car is identical to that of a ball in a uniform gravitational field. This agrees with Einstein's principle of equivalence.

(Figure 26-19)

Hadrons and leptons: Hadrons, including the proton and neutron, are particles that interact through the strong force; leptons, including the electron and neutrino, are particles that do not. Hadrons are composed of more fundamental particles called quarks, which are confined to the interior of hadrons and cannot exist in isolation. Hadrons include baryons, which are made of three quarks, and mesons, which are made of a quark and an antiquark. Leptons are themselves fundamental particles; they are not believed to be made of anything simpler.

(a) Proton (p)

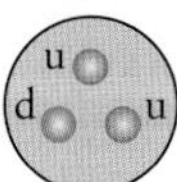

Net charge = $q_u + q_u + q_d$

$= \left(+\frac{2}{3}e\right) + \left(+\frac{2}{3}e\right) + \left(-\frac{1}{3}e\right) = +e$

(b) Neutron (n)

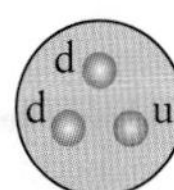

Net charge = $q_d + q_d + q_u$

$= \left(-\frac{1}{3}e\right) + \left(-\frac{1}{3}e\right) + \left(+\frac{2}{3}e\right) = 0$

(c) Delta-plus-plus (Δ^{++})

Net charge = $q_u + q_u + q_u$

$= \left(+\frac{2}{3}e\right) + \left(+\frac{2}{3}e\right) + \left(+\frac{2}{3}e\right) = +2e$

(d) Sigma-minus (Σ^-)

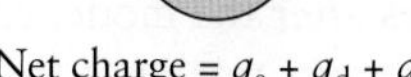

Net charge = $q_s + q_d + q_d$

$= \left(-\frac{1}{3}e\right) + \left(-\frac{1}{3}e\right) + \left(-\frac{1}{3}e\right) = -e$

(Figure 26-25)

(a) Positive pion (π^+)

u d̄

Net charge = $q_u + q_{\bar{d}}$

$= \left(+\frac{2}{3}e\right) + \left(+\frac{1}{3}e\right) = +e$

(b) Negative pion (π^-)

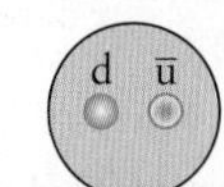
d ū

Net charge = $q_d + q_{\bar{u}}$

$= \left(-\frac{1}{3}e\right) + \left(-\frac{2}{3}e\right) = -e$

(c) Strange D-plus (D_s^+)

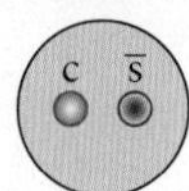
c s̄

Net charge = $q_c + q_{\bar{s}}$

$= \left(+\frac{2}{3}e\right) + \left(+\frac{1}{3}e\right) = +e$

(Figure 26-26)

The four forces and exchange particles: All interactions between particles can be understood in terms of four basic forces. Each of these forces involves the exchange of virtual particles, which is permitted by the Heisenberg uncertainty principle. The strong force between quarks involves the exchange of gluons, and the electromagnetic force involves the exchange of photons; both gluons and photons have zero mass, so these are long-range forces. The weak force responsible for beta decay involves the exchange of massive particles, so this force has a very short range. The gravitational force involves the exchange of gravitons, which have not yet been detected experimentally. Because the gravitational force is long range, gravitons are expected to have zero mass just like photons.

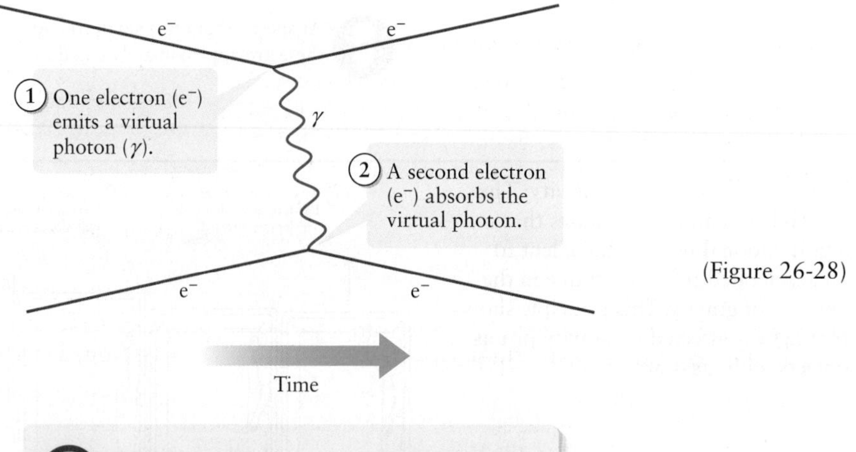
e⁻
e⁻
1 One electron (e⁻) emits a virtual photon (γ).
γ
2 A second electron (e⁻) absorbs the virtual photon.
e⁻
e⁻
Time

Charged particles exert electromagnetic forces on each other by exchanging virtual photons.

(Figure 26-28)

The energy of a phenomenon is necessarily uncertain by an amount ΔE.

$$\Delta E \Delta t \geq \frac{\hbar}{2} \qquad (26\text{-}26)$$

The shorter the duration Δt of the phenomenon, the greater the energy uncertainty ΔE.

$\hbar = \frac{h}{2\pi}$ = Planck's constant divided by 2π

Chapter 26 Review Problems

1. A radio-controlled model car travels at 15.0 m/s, to the right, relative to the parking lot that it is driving on. A girl on her scooter chases after the model car at 4.00 m/s. The parking lot represents reference frame S, the model car is in frame S', and the girl is described by frame S''. Assume that the car and the girl are both at the origin of the parking lot at $t = 0$ s. Write the Galilean transformation equations between (a) S and S', (b) S and S'', and (c) S' and S''.

2. A floatplane lands on a river. The velocity of the plane relative to the air is 30 m/s, due east; the velocity of the wind is 20 m/s, due north; and the current in the river is 5 m/s, due south. Calculate the velocity of the plane relative to the water just before it lands.

3. Time dilation means that
   (A) the slowing of time in a moving reference frame is only an illusion resulting from motion.
   (B) time really does pass more slowly in a reference frame moving relative to a reference frame at relative rest.
   (C) time really does pass more slowly in a reference frame at rest relative to a reference frame that is moving.
   (D) time is unchanging regardless of the reference frame.

4. Which of the following statements is/are true?
   (A) The laws of motion are the same in all inertial reference frames.
   (B) The laws of motion are the same in all reference frames.
   (C) There is no way to detect absolute motion.
   (D) Two of the above statements are true.

5. A meter stick hurtles through space at a speed of $0.95c$ with its length perpendicular to the direction of motion. You measure its length to be equal to
   (A) 0.05 m.
   (B) 0.95 m.
   (C) 1.00 m.
   (D) 1.05 m.

6. Suppose that the distance in the Michelson–Morley experiment from the partially silvered plate to either mirror is 20,000 m.
   (a) How much time would light, moving at speed c, need to travel from the partially silvered plate to one mirror and back again?

   How much *additional* time would be required *if the ether existed*, if the Galilean velocity transformation were valid, and if Earth moved relative to the ether along the direction from the partially silvered plate to this mirror at the following speeds: **(b)** at $0.01c$, **(c)** at $0.1c$, **(d)** at $0.5c$, and **(e)** at $0.9c$?

7. How fast must a positive pion (an unstable particle) be moving to travel 100 m (according to the laboratory frame) before it has a 50% chance of decaying? The half-life, at rest, of a positive pion is 1.80×10^{-8} s. Give your answer in units of meters per second (m/s) and as a fraction of the speed of light (in other words, find both v and v/c) to four decimal places.

8. A space ship flies by Earth at $0.92c$. It fires a rocket at $0.75c$ in the forward direction, relative to the space ship. What is the velocity of the rocket relative to Earth?

9. Suppose the space ship in the previous problem continues to fly by Earth at $0.920c$. This time, however, it fires a rocket at $0.750c$ in the backward direction relative to the space ship. What is the velocity of the rocket relative to Earth?

10. A clock is placed aboard Starship *Alpha*, which Albert flies past Earth at half the speed of light relative to Earth. Elena observes from Earth, where she has an identical clock. Which pair of words correctly fills in the blanks in this statement: "Elena measures Albert's clock as running ______, and Albert measures Elena's clock as running ______."
    (A) slow; fast
    (B) slow; slow
    (C) fast; fast
    (D) fast; slow

11. The United States used approximately 4.12×10^{12} kWh of electrical energy during 2019.
    (a) If we could generate all the energy using a matter–antimatter reactor, how many kilograms of fuel would the reactor need, assuming 100% efficiency in the annihilation?
    (b) Suppose the fuel consisted of iron and anti-iron, each of density 7800 kg/m^3. If the iron and anti-iron were each stored as a cube, what would be the dimensions of each cube?

12. A car is driving along the freeway. Estimate the ratio of its kinetic energy to its rest energy.

13. A beam of positive pions (unstable particles) has a speed of $0.88c$. Their half-life, as measured in the reference frame of the laboratory, is 1.80×10^{-8} s. What is the distance traveled by the laboratory, as measured by the pion, during one half-life?

14. Suppose a jet plane flies at 300 m/s relative to an observer on the ground. Using only special relativity, determine how far the plane must travel, as measured by an observer on the ground, before the clocks aboard the plane are 10 s behind clocks on the ground. Assume the two clocks were originally synchronized to start.

15. A proton ($m = 1.673 \times 10^{-27}$ kg) is traveling at $0.50c$. Calculate its **(a)** Einsteinian momentum and its **(b)** Einsteinian kinetic energy.

16. We know 1 kg of trinitrotoluene (TNT) yields an energy of 4.2 MJ. The energy released comes from the chemical bonds in the material. How much mass would be required to create an explosion equivalent to 1.8×10^9 kg TNT? Assume all the energy released comes simply from the rest energy of the material.

17. At the end of the linear accelerator at the SLAC National Accelerator laboratory, electrons have a speed of $0.99999999995c$.
    (a) Calculate the value of γ for an electron with this speed.
    (b) What time interval would an observer at rest relative to the accelerator measure for a time interval of 1.66 μs measured from the electron's perspective?

18. The half-life of a certain subatomic particle at rest is about 1×10^{-8} s. How fast is a beam of these particles moving if one-half of them decay in 6×10^{-8} s as measured in the laboratory?

19. Based on experiments in your lab, you know that a certain radioactive isotope has a half-life of 2.25 μs. As a high-speed space ship passes your lab, you measure that the same isotope at rest inside the space ship takes 3.15 μs for one-half of it to decay.
    (a) What is the half-life of the isotope as measured by an astronaut working inside the space ship?
    (b) How fast is the space ship traveling relative to Earth?

20. High-speed cosmic rays strike atoms in Earth's upper atmosphere and create secondary showers of particles. Suppose a particle in one of the showers is created 25.0 km above the surface and travels downward at 90.0% the speed of light. Consider the following two events: "A particle is created in the upper atmosphere" and "a particle strikes the ground." We can view the events from two reference frames, one fixed on Earth and one traveling with the created particle.
    (a) In which of the reference frames are the proper time and the rest length between the two events measured? Explain your reasoning.
    (b) In the particle's reference frame how long after its creation does it take it to reach the ground?
    (c) In Earth's reference frame how long after creation does it take the particle to reach the ground?
    (d) Show that the times in parts (b) and (c) are consistent with time dilation.

21. Twin astronauts Anselma and Baraka have identical pulse rates of 70 beats/min on Earth. Anselma remains on Earth, but Baraka is assigned to a space voyage during which he travels at $0.75c$ relative to Earth. What will be Baraka's pulse rate as measured by (a) Anselma on Earth and (b) the physician in Baraka's spacecraft?

22. A rocket is traveling at speed v relative to Earth. Inside a lab on the rocket, two laser beams are turned on, one pointing in the forward direction and the other pointing in the backward direction relative to the rocket's velocity.
    (a) What is the speed of each laser beam relative to the laboratory in the rocket?
    (b) Use the Lorentz velocity transformation to find the velocity of each laser beam as measured by an observer at rest on Earth.
    (c) As observed from Earth, how fast are the two laser beams separating from *each other*?

23. A 1.0×10^3-kg rocket is flying at $0.90c$ relative to your lab. Calculate the kinetic energy of the rocket using the Einsteinian formula and the ordinary Newtonian formula. What is the percent error if we use the Newtonian formula? Does the Newtonian formula overestimate or underestimate the kinetic energy?

24. In cancer radiotherapy, such as that in the cover photograph for this chapter, a beam of electrons is accelerated to nearly the speed of light. The kinetic energy of the electrons is used to create intense x-ray beams that can be accurately targeted on the location of the cancerous tissue. Which requires more energy: (A) accelerating an electron from rest to 90% of the speed of light, or (B) further accelerating that electron from 90 to 99% of the speed of light?

25. The kinetic energy of a neutral pion (π^0) is 860 MeV. This pion decays to two photons, one of which has energy 640 MeV.
    (a) Calculate the energy of the other photon.
    (b) Draw the Feynman diagram for the process.

26. Which of the following reactions are possible? If a reaction is not possible, tell which conservation law(s) is/are violated.
    (A) $n \rightarrow p + e^- + \bar{\nu}_e$
    (B) $\mu^- \rightarrow e^- + \bar{\nu}_e + \nu_\mu$
    (C) $\pi^- \rightarrow \mu^- + \bar{\nu}_\mu$

Group Work

Directions: The problem below is designed to be done as group work in class.

Muons have a proper half-life of 1.56×10^{-6} s. Suppose a muon is formed at an altitude of 3000 m and travels at a speed of $0.950c$ straight toward Earth.

(a) Does the muon reach Earth's surface within one half-life? Complete the problem from both perspectives: the muon's point of view and Earth's reference frame.

(b) Calculate the minimum speed of the muon so that it *just barely* reaches Earth's surface after one half-life.

Math Tutorial

In this tutorial, we review some of the basic results of algebra, geometry, trigonometry, and calculus. In many cases, we merely state results without proof. **Table M-1** lists some mathematical symbols.

M-1 Significant digits

Many numbers we work with in science are the result of measurement and are therefore known only within a degree of uncertainty. You will notice this if you check the values for constants from different sources! This uncertainty should be reflected in the number of digits used. For example, if you have a 1-meter-long rule with scale spacing of 1 cm, you know that you can measure the height of a box to within a fifth of a centimeter or so. Using this rule, you might find that the box height is 27.0 cm. If there is a scale with a spacing of 1 mm on your rule, you might perhaps measure the box height to be 27.03 cm. However, if there is a scale with a spacing of 1 mm on your rule, you might not be able to measure the height more accurately than 27.03 cm because the height might vary by 0.01 cm or so, depending on where you measure the height of the box. When you write down that the height of the box is 27.03 cm, you are stating that your best estimate of the height is 27.03 cm, but you are not claiming that it is exactly 27.030000...cm high. The four digits in 27.03 cm are called **significant digits**. Your measured length, 27.03 cm, has four significant digits. Significant digits are also called significant figures.

The number of significant digits in an answer to a calculation will depend on the number of significant digits in the given data. When you work with numbers that have uncertainties, you should be careful not to include more digits than the certainty of measurement warrants. *Approximate* calculations (order-of-magnitude estimates) always result in answers that have only one significant digit or none (a 0 with no decimals is considered no significant figures). When you multiply, divide, add, or subtract numbers, you must consider the accuracy of the results. Listed below are some rules that will help you determine the number of significant digits of your results.

(1) When multiplying or dividing quantities, the number of significant digits in the final answer is no greater than that in the quantity with the fewest significant digits.

(2) When adding or subtracting quantities, the number of decimal places in the answer should match that of the term with the smallest number of decimal places.

(3) Exact values have an unlimited number of significant digits. For example, a value determined by counting, such as 2 tables, has no uncertainty and is an exact value. In addition, the conversion factor 0.0254000...m/in. is an exact value because 1.000...inches is exactly equal to 0.0254000...meters. (The yard is,

TABLE M-1 Mathematical Symbols

Symbol	Meaning
$=$	is equal to
$\neq$	is not equal to
$\approx$	is approximately equal to
$\sim$	is of the order of
$\propto$	is proportional to
$>$	is greater than
$\geq$	is greater than or equal to
$>>$	is much greater than
$<$	is less than
$\leq$	is less than or equal to
$<<$	is much less than
Δx	change in x
$\lvert x \rvert$	absolute value of x
$n!$	$n(n-1)(n-2)\ldots1$
Σ	sum

by definition, equal to exactly 0.9144 m, and 0.9144 divided by 36 is exactly equal to 0.0254.)

(4) Sometimes zeros are significant and sometimes they are not. If a zero is before a leading nonzero digit, then the zero is not significant. For example, the number 0.00890 has three significant digits. The first three zeroes are not significant digits but are merely markers to locate the decimal point. Note that the zero after the nine is significant.

(5) Zeros that are between nonzero digits are significant. For example, 5603 has four significant digits.

(6) The number of significant digits in numbers with trailing zeros and no decimal point is ambiguous. For example, 31,000 could have as many as five significant digits or as few as two significant digits. To prevent ambiguity, you should report numbers by using scientific notation or by using a decimal point.

EXAMPLE M-1 Finding the Average of Three Numbers

Find the average of 19.90, –7.524, and –11.8179.

Set Up

You will be adding 3 numbers and then dividing the result by 3. The first number has four significant digits, the second number has four, and the third number has six.

Solve

Sum the three numbers.

$$19.90 + (-7.524) + (-11.8179) = 0.55\textit{81}$$

If the problem only asked for the sum of the three numbers, we would round the answer to the least number of decimal places among all the numbers being added—the answer would be 0.56 (0.5581 rounds up to 0.56 to two significant digits). However, we must divide this intermediate result by 3, so we use the intermediate answer with the two extra digits (italicized and red).

$$\frac{0.55\textit{81}}{3} = 0.18\textit{60333}\ldots$$

Only two of the digits in the intermediate answer, 0.55*81*..., are significant digits, so we must round the final number to get our final answer. The number 3 in the denominator is a whole number and has an unlimited number of significant digits. Thus, the final answer has the same number of significant digits as the numerator, which is 2.

The final answer is 0.19.

Reflect

The sum in step 1 has two significant digits following the decimal point, the same as the number being summed with the least number of significant digits after the decimal point.

M-2 Equations

An equation is a statement written using numbers and symbols to indicate that two quantities, written on either side of an equal sign (=), are equal. The quantity on either side of the equal sign may consist of a single term, or of a sum, difference, product, or quotient of two or more terms. For example, the equation $x = 1 - (ay + b)/(cx - d)$ contains three terms, x, 1, and $(ay + b)/(cx - d)$.

You can perform the following operations on equations:

(1) The same quantity can be added to or subtracted from each side of an equation.

(2) Each side of an equation can be multiplied or divided by the same quantity.

(3) Each side of an equation can be raised to the same power.

These operations are meant to be applied to each *side* of the equation rather than each term in the equation. (Because multiplication is distributive over addition, operation 2—and only operation 2—of the preceding operations also applies term by term.)

Caution: *Division by zero is forbidden at any stage in solving an equation; results (if any) would be invalid. This includes dividing by something that could be zero for some value of what you are solving for, like* $\sin\theta$*, when you are solving for* θ*.*

Adding or Subtracting Equal Amounts

To find x when $x-3=7$, add 3 to both sides of the equation: $(x-3)+3=7+3$; thus, $x=10$.

Multiplying or Dividing by Equal Amounts

If $3x=17$, solve for x by dividing both sides of the equation by 3; thus, $x=\frac{17}{3}$, or 5.7.

EXAMPLE M-2 Simplifying Reciprocals in an Equation

Solve the following equation for x:

$$\frac{1}{x}+\frac{1}{4}=\frac{1}{3}$$

Equations containing reciprocals of unknowns occur in many circumstances in physics. Two instances of this are geometric optics and electric circuit analysis.

Set Up

In this equation, the term containing x is on the same side of the equation as a term not containing x. Furthermore, x is found in the denominator of a fraction. We'll start by isolating the $1/x$ term, find common denominators, and then multiply both sides of the equation by appropriate quantities.

Solve

Subtract $\frac{1}{4}$ from each side.

$$\frac{1}{x}=\frac{1}{3}-\frac{1}{4}$$

Simplify the right side of the equation by using the lowest common denominator.

Begin by multiplying both terms on the right-hand side by appropriate forms of 1.

$$\frac{1}{x}=\frac{1}{3}\frac{4}{4}-\frac{1}{4}\frac{3}{3}=\frac{4}{12}-\frac{3}{12}$$

$$=\frac{4-3}{12}=\frac{1}{12} \quad \text{so} \quad \frac{1}{x}=\frac{1}{12}$$

Multiply both sides of the equation by $12x$ to determine the value of x.

$$12\cancel{x}\frac{1}{\cancel{x}}=\cancel{12}x\frac{1}{\cancel{12}}$$

$$12=x$$

Reflect

To check our answer, substitute 12 for x in the left side of original equation.

$$\frac{1}{x}+\frac{1}{4}=\frac{1}{12}+\frac{3}{12}=\frac{4}{12}=\frac{1}{3}$$

M-3 Direct and inverse proportions

When we say variable quantities x and y are directly proportional, we mean that as x and y change, the ratio x/y is constant. To say that two quantities are proportional is to say that they are directly proportional. When we say variable quantities x and y are inversely proportional, we mean that as x and y change, the ratio xy is constant.

Relationships of direct and inverse proportion are common in physics. Objects moving at the same velocity have momenta directly proportional to their masses.

The ideal gas law ($PV = nRT$) states that pressure P is directly proportional to (absolute) temperature T, when volume V remains constant, and is inversely proportional to volume, when temperature remains constant. Ohm's law ($\Delta V = IR$) states that the potential difference, ΔV, across a resistor is directly proportional to the electric current in the resistor when the resistance remains constant.

Constant of Proportionality

When two quantities are directly proportional, the two quantities are related by a *constant of proportionality*. If you are paid for working at a regular rate R in dollars per day, for example, the money m you earn is directly proportional to the time t you work; the rate R is the constant of proportionality that relates the money earned in dollars to the time worked t in days:

$$\frac{m}{t} = R \quad \text{or} \quad m = Rt$$

If you earn \$400 in 5 days, the value of R is \$400/(5 days) = \$80/day. To find the amount you earn in 8 days, you could perform the calculation

$$m = (\$80/\text{day})(8 \text{ days}) = \$640$$

Sometimes the constant of proportionality can be ignored in proportion problems. Because the amount you earn in 8 days is $\frac{8}{5}$ times what you earn in 5 days, this amount is

$$m_{8 \text{ days}} = 8 \text{ days} \frac{\$400}{5 \text{ days}} = \$640$$

EXAMPLE M-3 Painting Cubes

You need 15.4 mL of paint to cover one side of a cube. The area of one side of the cube is 426 cm^2. What is the relation between the volume of paint needed and the area to be covered? How much paint do you need to paint one side of a cube on which one side has an area of 503 cm^2?

Set Up

To determine the amount of paint for the side whose area is 503 cm^2 we will set up a proportion.

Solve

The volume V of paint needed increases in proportion to the area A to be covered.

V and A are directly proportional.
That is, $\frac{V}{A} = k$ or $V = kA$
where k is the proportionality constant

Determine the value of the proportionality constant using the given values $V_1 = 15.4$ mL and $A_1 = 426$ cm^2.

$$k = \frac{V_1}{A_1} = \frac{15.4 \text{ mL}}{426 \text{ cm}^2} = 0.0362 \text{ mL/cm}^2$$

Determine the volume of paint needed to paint a side of a cube whose area is 503 cm^2 using the proportionality constant in step 1.

$$V_2 = kA_2 = (0.0362 \text{ mL/cm}^2)(503 \text{ cm}^2) = 18.2 \text{ mL}$$

Reflect

Our value for V_2 is greater than the value for V_1, as expected. The amount of paint needed to cover an area equal to 503 cm^2 should be greater than the amount of paint needed to cover an area of 426 cm^2 because 503 cm^2 is larger than 426 cm^2.

M-4 Linear equations

A linear equation is an equation of the form $x + 2y - 4z = 3$. That is, an equation is linear if each term either is constant or is the product of a constant and a variable raised to the first power. Such equations are said to be linear because the plots of these equations form lines or planes. The equations of direct proportion between two variables are linear equations.

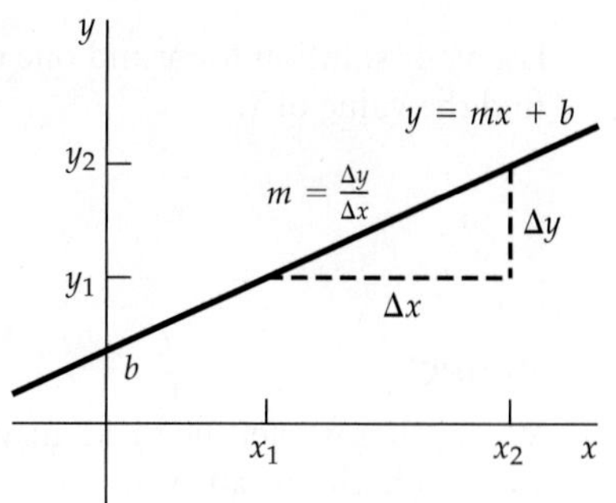

Figure M-1 Graph of the linear equation $y = mx + b$, where b is the y intercept and $m = \Delta y/\Delta x$ is the slope.

Graph of a Straight Line

A linear equation relating y and x can always be put into the standard form

$$y = mx + b \qquad \text{(M-1)}$$

where m and b are constants that may be either positive or negative. **Figure M-1** shows a graph of the values of x and y that satisfy Equation M-1. The constant b, called the y intercept, is the value of y at $x = 0$. The constant m is the slope of the line, which equals the ratio of the change in y to the corresponding change in x. In the figure, we have indicated two points on the line, (x_1, y_1) and (x_2, y_2), and the changes $\Delta x = x_2 - x_1$ and $\Delta y = y_2 - y_1$. The slope m is then

$$m = \frac{y_2 - y_1}{x_2 - x_1} = \frac{\Delta y}{\Delta x}$$

If x and y are both unknown in the equation $y = mx + b$, there are no unique values of x and y that are solutions to the equation. Any pair of values (x_1, y_1) on the line in Figure M-1 will satisfy the equation. If we have two equations, each with the same two unknowns x and y, the equations can be solved simultaneously for the unknowns. Example M-4 shows two methods for simultaneously solving two linear equations.

EXAMPLE M-4 Using Two Equations to Solve for Two Unknowns

Find any and all values of x and y that simultaneously satisfy

$$3x - 2y = 8 \qquad \text{(M-2)}$$

and

$$y - x = 2 \qquad \text{(M-3)}$$

Set Up

Graph the two equations (**Figure M-2**). At the point where the lines intersect, the values of x and y satisfy both equations.

While the graph is one complete solution, we can also solve two simultaneous equations by first solving either equation for one variable in terms of the other variable and then substituting the result into the second equation.

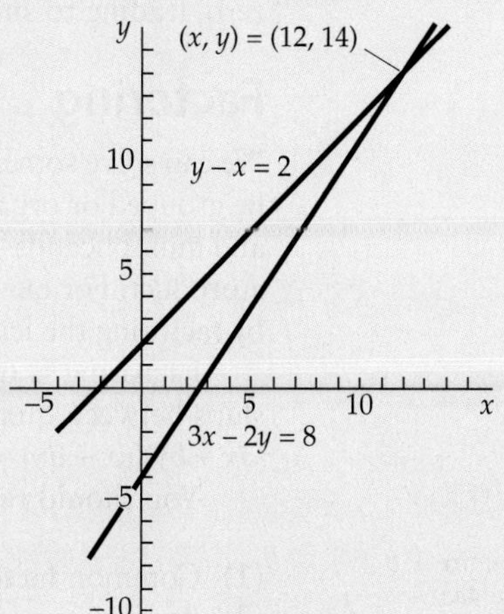

Figure M-2 Graph of Equations M-2 and M-3. At the point where the lines intersect, the values of x and y satisfy both equations.

Solve

Solve Equation M-3 for y. $\qquad y = x + 2$

Substitute this value for y into Equation M-2. $\qquad 3x - 2(x + 2) = 8$

Simplify the equation and solve for x. $\qquad 3x - 2x - 4 = 8$

$$x - 4 = 8$$

$$x = 12$$

Use your solution for x and one of the given equations to find the value of y.

Return to Equation M-3 and substitute $x = 12$.

$$y - x = 2, \text{ where } x = 12$$
$$y - 12 = 2$$
$$y = 2 + 12 = 14$$

Reflect

We see that we get the same answer using the graphical solution or the algebraic one as we must. This gives us a good way to check our answer.

Extend: An alternative method is to multiply one equation by a constant such that one of the unknown terms is eliminated when the equations are added or subtracted.

We can multiply through Equation M-3 by 2.

$$2(y - x) = 2(2)$$
$$2y - 2x = 4$$

Add the result to Equation M-2 and solve for x:

$$\cancel{2y} - 2x = 4$$
$$3x - \cancel{2y} = 8$$
$$3x - 2x = 12 \Rightarrow x = 12$$

Substitute into Equation M-3 and solve for y:

$$y - 12 = 2 \Rightarrow y = 14$$

M-5 Quadratic equations and factoring

A common quadratic equation is of the form

$$ax^2 + bx + c = 0 \qquad \text{(M-4)}$$

where a, b, and c are constants. The quadratic equation has two solutions or roots—values of x for which the equation is true. In each term of the equation the powers of the variables are integers that sum to 2, 1, or 0.

A full quadratic equation is of the form $ax^2 + bxy + cy^2 + ex + fy + g = 0$, where x and y are variables and a, b, c, e, f, and g are constants. Any of these constants can be zero, leading to simpler quadratic equation forms.

Factoring

We can solve some quadratic equations by factoring. Very often terms of an equation can be grouped or organized into other terms. When we factor terms, we look for multipliers and multiplicands—which we now call factors—that will yield two or more new terms as a product. For example, we can find the roots of the quadratic equation $x^2 - 3x + 2 = 0$ by factoring the left side to get $(x - 2)(x - 1) = 0$. The roots are $x = 2$ and $x = 1$.

Factoring is useful for simplifying equations and for understanding the relationships between quantities. You should be familiar with the multiplication of the factors $(ax + by)(cx + dy) = acx^2 + (ad + bc)xy + bdy^2$.

You should readily recognize some typical factorable combinations:

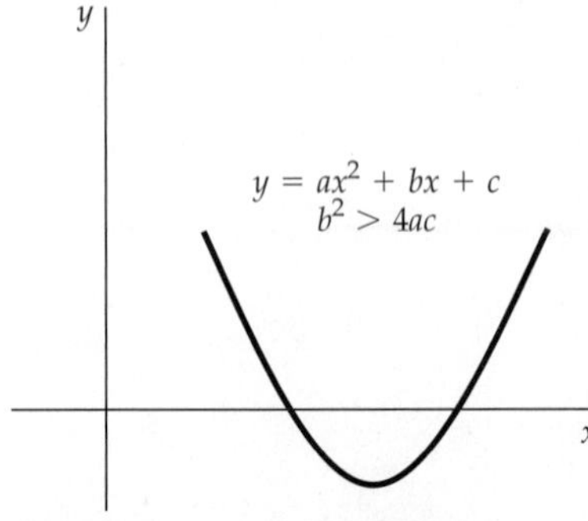

Figure M-3 Graph of y versus x when $y = ax^2 + bx + c$ for the case $b^2 > 4ac$. The two values of x for which $y = 0$ satisfy the quadratic equation (Equation M-4).

(1) Common factor: $2ax + 3ay = a(2x + 3y)$
(2) Perfect square: $x^2 - 2xy + y^2 = (x - y)^2$ (If the expression on the left side of a quadratic equation in standard form is a perfect square, the two roots will be equal.)
(3) Difference of squares: $x^2 - y^2 = (x + y)(x - y)$

Also, look for factors that are prime numbers (2, 5, 7, etc.) because these factors can help you simplify terms quickly. For example, the equation $98x^2 - 140 = 0$ can be simplified because 98 and 140 share the common factor 2. That is, $98x^2 - 140 = 0$ becomes $2(49x^2 - 70) = 0$, so we have $49x^2 - 70 = 0$.

This result can be further simplified because 49 and 70 share the common factor 7. Thus, $49x^2 - 70 = 0$ becomes $7(7x^2 - 10) = 0$, so we have $7x^2 - 10 = 0$.

The Quadratic Formula

Not all quadratic equations can be solved by factoring. However, *any* quadratic equation in the standard form $ax^2 + bx + c = 0$ can be solved by the quadratic formula,

$$x = \frac{-b \pm \sqrt{b^2 - 4ac}}{2a} = -\frac{b}{2a} \pm \frac{1}{2a}\sqrt{b^2 - 4ac} \qquad \text{(M-5)}$$

When b^2 is greater than $4ac$, there are two solutions corresponding to the + and − signs, respectively. **Figure M-3** shows a graph of y versus x where $y = ax^2 + bx + c$. The curve, a parabola, crosses the x axis twice. [The simplest representation of a parabola in (x, y) coordinates is an equation of the form $y = ax^2 + bx + c$.] The two roots of this equation are the values for which $y = 0$; that is, they are the x *intercepts*.

When b^2 is less than $4ac$, the graph of y versus x does not intersect the x axis, as is shown in **Figure M-4**; there are still two roots, but they are not real numbers. When $b^2 = 4ac$, the graph of y versus x is tangent to the x axis at the point $x = -b/2a$; the two roots are each equal to $-b/2a$.

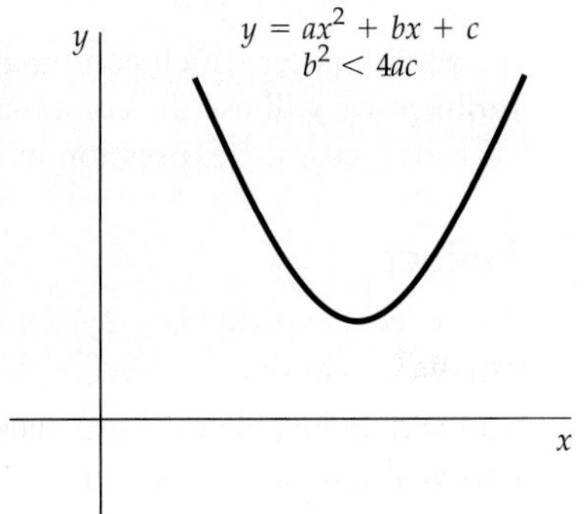

Figure M-4 Graph of y versus x when $y = ax^2 + bx + c$ for the case $b^2 < 4ac$. In this case, there are no real values of x for which $y = 0$.

EXAMPLE M-5 Factoring a Second-Degree Polynomial

Factor the expression $6x^2 + 19xy + 10y^2$.

Set Up

We examine the coefficients of the terms to see whether the expression can be factored without resorting to more advanced methods. Remember that the multiplication $(ax + by)(cx + dy) = acx^2 + (ad + bc)xy + bdy^2$.

Solve

The coefficient of x^2 is 6, which can be factored two ways.

$ac = 6$

$3 \cdot 2 = 6$ or $6 \cdot 1 = 6$

The coefficient of y^2 is 10, which can also be factored two ways.

$bd = 10$

$5 \cdot 2 = 10$ or $10 \cdot 1 = 10$

List the possibilities for a, b, c, and d in a table. Include a column for $ad + bc$.

If $a = 3$, then $c = 2$, and vice versa. In addition, if $a = 6$, then $c = 1$, and vice versa. For each value of a there are four values for b.

a	b	c	d	$ad + bc$
3	5	2	2	16
3	2	2	5	19
3	10	2	1	23
3	1	2	10	32
2	5	3	2	19
2	2	3	5	16
2	10	3	1	32
2	1	3	10	23
6	5	1	2	17
6	2	1	5	32
6	10	1	1	16
6	1	1	10	61
1	5	6	2	32
1	2	6	5	17
1	10	6	1	61
1	1	6	10	16

Find a combination such that $ad + bc = 19$. As you can see from the table there are two such combinations.

$ad + bc = 19$

$3 \cdot 5 + 2 \cdot 2 = 19$ and

$2 \cdot 2 + 5 \cdot 3 = 19$

It doesn't matter which combination we choose. To finish this problem we will use the combination in the second row of the table to factor the expression in question.

$$6x^2 + 19xy + 10y^2 = (3x + 2y)(2x + 5y)$$

Reflect

As a check, expand $(3x + 2y)(2x + 5y)$ to see if we return to the original equation.

$$(3x + 2y)(2x + 5y) = 6x^2 + 15xy + 4xy + 10y^2$$
$$= 6x^2 + 19xy + 10y^2$$

You should be able to show that the combination in the fifth row is also an acceptable factoring.

M-6 Exponents and logarithms

Exponents

The notation x^n stands for the quantity obtained by multiplying x by itself n times. For example, $x^2 = x \cdot x$ and $x^3 = x \cdot x \cdot x$. The quantity n is called the **power**, or the exponent, of x (the base). Listed below are some rules that will help you simplify terms that have exponents.

(1) When two powers of x are multiplied, the exponents are added:

(M-6) $$(x^m)(x^n) = x^{m+n}$$

Example: $x^2 x^3 = x^{2+3} = (x \cdot x)(x \cdot x \cdot x) = x^5$.

(2) Any number (except 0) raised to the 0 power is defined to be 1:

(M-7) $$x^0 = 1$$

(3) Based on rule 2,

$$x^n x^{-n} = x^0 = 1$$

(M-8) $$x^{-n} = \frac{1}{x^n}$$

(4) When two powers are divided, the exponents are subtracted:

(M-9) $$\frac{x^n}{x^m} = x^n x^{-m} = x^{n-m}$$

(5) When a power is raised to another power, the exponents are multiplied:

(M-10) $$(x^n)^m = x^{nm}$$

(6) When exponents are written as fractions, they represent the roots of the base. For example,

$$x^{1/2} \cdot x^{1/2} = x$$

so

$$x^{1/2} = \sqrt{x} \quad (x > 0)$$

EXAMPLE M-6 Simplifying a Quantity That Has Exponents

Simplify $\dfrac{x^4 x^7}{x^8}$.

Set Up

According to rule 1, when two powers of x are multiplied, the exponents are added.

$$(x^m)(x^n) = x^{m+n} \qquad \text{(M-6)}$$

Rule 4 states that when two powers are divided, the exponents are subtracted.

$$\frac{x^n}{x^m} = x^n x^{-m} = x^{n-m} \tag{M-9}$$

Solve

Simplify the numerator x^4x^7 using rule 1.

$$x^4x^7 = x^{4+7} = x^{11}$$

Simplify $\dfrac{x^{11}}{x^8}$ using rule 4.

$$\frac{x^{11}}{x^8} = x^{11}x^{-8} = x^{11-8} = x^3$$

Reflect

Use the value $x = 2$ to test our answer.

$$\frac{2^4 2^7}{2^8} = 2^3 = 8$$

$$\frac{2^4 2^7}{2^8} = \frac{(16)(128)}{256} = \frac{2048}{256} = 8$$

Logarithms

Any positive number can be expressed as some power of any other positive number except one. If y is related to x by $y = a^x$, then the number x is said to be the logarithm of y to the base a, and the relation is written

$$x = \log_a y$$

Thus, logarithms are *exponents*, and the rules for working with logarithms correspond to similar laws for exponents. Listed below are some rules that will help you simplify terms that have logarithms.

(1) If $y_1 = a^n$ and $y_2 = a^m$, then

$$y_1y_2 = a^n a^m = a^{n+m}$$

Correspondingly,

$$\log_a y_1y_2 = \log_a a^{n+m} = n + m = \log_a a^n + \log_a a^m = \log_a y_1 + \log_a y_2 \tag{M-11}$$

It then follows that

$$\log_a y^n = n \log_a y \tag{M-12}$$

(2) Because $a^1 = a$ and $a^0 = 1$,

$$\log_a a = 1 \tag{M-13}$$

and

$$\log_a 1 = 0 \tag{M-14}$$

There are two bases in common use: Logarithms to base 10 are called common logarithms, and logarithms to base e (where $e = 2.718\ldots$) are called natural logarithms.

In this text, the symbol ln is used for natural logarithms and the symbol log, without a subscript, is used for common logarithms. Thus,

$$\log_e x = \ln x \quad \text{and} \quad \log_{10} x = \log x \tag{M-15}$$

and $y = \ln x$ implies

$$x = e^y \tag{M-16}$$

Logarithms can be changed from one base to another. Suppose that

$$z = \log x \tag{M-17}$$

Then

$$10^z = 10^{\log x} = x \tag{M-18}$$

Taking the natural logarithm of both sides of Equation M-18, we obtain

$$z \ln 10 = \ln x$$

Substituting $\log x$ for z (see Equation M-17) gives

$$\ln x = (\ln 10)\log x \qquad \text{(M-19)}$$

EXAMPLE M-7 Converting Between Common Logarithms and Natural Logarithms

The steps leading to Equation M-19 show that, in general, $\log_b x = (\log_b a)\log_a x$, and thus that conversion of logarithms from one base to another requires only multiplication by a constant. Describe the mathematical relation between the constant for converting common logarithms to natural logarithms and the constant for converting natural logarithms to common logarithms.

Set Up

We have a general mathematical formula for converting logarithms from one base to another. We look for the mathematical relation by exchanging a for b and vice versa in the formula.

Solve

We have a formula for converting logarithms from base a to base b.

$$\log_b x = (\log_b a)\log_a x$$

To convert from base b to base a, exchange all a for b and vice versa.

$$\log_a x = (\log_a b)\log_b x$$

Divide both sides of the first equation by $\log_a x$.

$$\frac{\log_b x}{\log_a x} = \log_b a$$

Divide both sides of the second equation by $(\log_a b)\log_a x$.

$$\frac{1}{\log_a b} = \frac{\log_b x}{\log_a x}$$

The results show that the conversion factors $\log_b a$ and $\log_a b$ are reciprocals of one another.

$$\frac{1}{\log_a b} = \log_b a$$

Reflect

We saw that these values must be reciprocals. For the value of $\log_{10} e$, your calculator will give 0.43429. For ln 10, your calculator will give 2.3026. Multiply 0.43429 by 2.3026 and you will get 1.0000. So our answer makes sense.

M-7 Geometry

The properties of the most common geometric figures—bounded shapes in two or three dimensions whose lengths, areas, or volumes are governed by specific ratios—are a basic analytical tool in physics. For example, the characteristic ratios within triangles give us the laws of *trigonometry* (see Section M-8), which in turn give us the theory of vectors, essential in analyzing motion in two or three dimensions. Circles and spheres are essential for understanding, among other concepts, angular momentum and the probability densities of quantum mechanics. Geometry is also helpful in analyzing graphs to find the "area under the curve."

Basic Formulas in Geometry

Circle The ratio of the circumference of a circle to its diameter is a number π, which has the approximate value

$$\pi = 3.141\,592$$

The circumference C of a circle is thus related to its diameter d and its radius r by

$$C = \pi d = 2\pi r \quad \text{circumference of circle} \tag{M-20}$$

The area of a circle is (**Figure M-5**)

$$A = \pi r^2 \quad \text{area of circle} \tag{M-21}$$

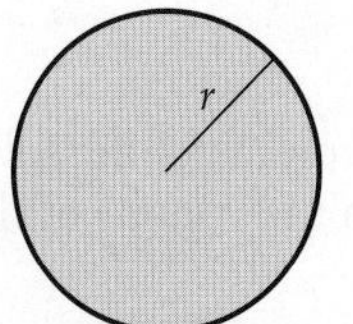

Area of a circle $A = \pi r^2$

Figure M-5 Area of a circle.

Parallelogram The area of a parallelogram is the base b multiplied by the height h (**Figure M-6**):

$$A = bh$$

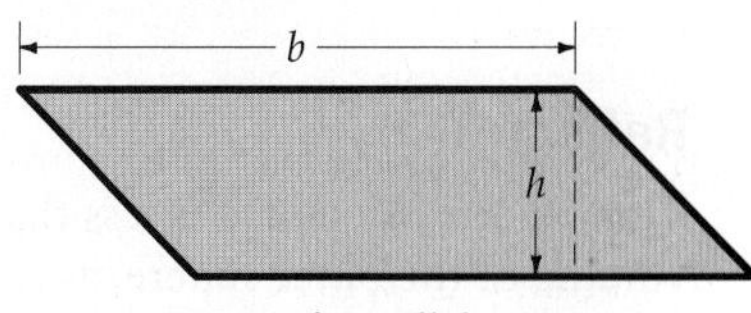

Area of parallelogram
$A = bh$

Figure M-6 Area of a parallelogram.

Triangle The area of a triangle is one-half the base multiplied by the height (**Figure M-7**):

$$A = \frac{1}{2}bh$$

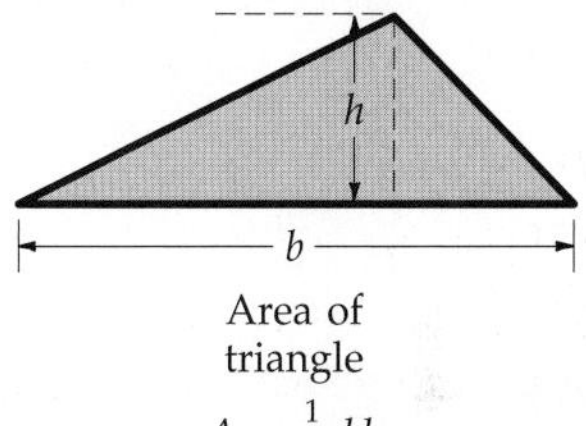

Area of triangle
$A = \frac{1}{2}bh$

Figure M-7 Area of a triangle.

Sphere A sphere of radius r (**Figure M-8**) has a surface area given by

$$A = 4\pi r^2 \quad \text{surface area of sphere} \tag{M-22}$$

and a volume given by

$$V = \frac{4}{3}\pi r^3 \quad \text{volume of sphere} \tag{M-23}$$

Cylinder A cylinder of radius r and length L (**Figure M-9**) has a surface area (not including the end faces) of

$$A = 2\pi rL \quad \text{surface of cylinder} \tag{M-24}$$

and volume of

$$V = \pi r^2 L \quad \text{volume of cylinder} \tag{M-24}$$

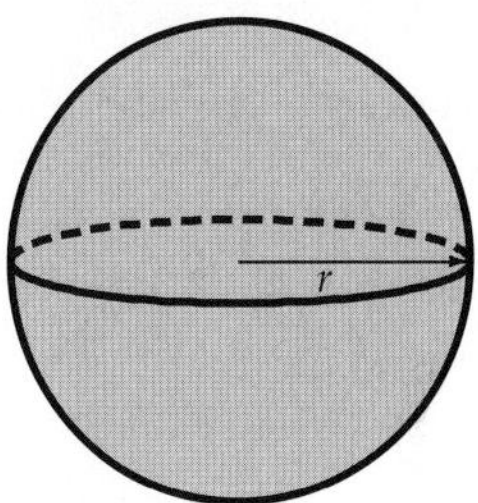

Spherical surface area
$A = 4\pi r^2$
Spherical volume
$V = \frac{4}{3}\pi r^3$

Figure M-8 Surface area and volume of a sphere.

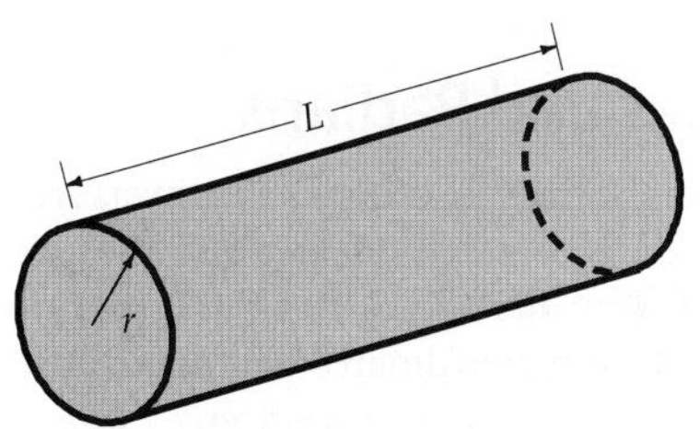

Cylindrical surface area
$A = 2\pi rL$
Cylindrical volume
$V = \pi r^2 L$

Figure M-9 Surface area (not including the end faces) and the volume of a cylinder.

EXAMPLE M-8 Calculating the Volume of a Spherical Shell

An aluminum spherical shell has an outer diameter of 40.0 cm and an inner diameter of 38.0 cm. What is the volume of the aluminum in this shell?

Set Up

The volume of the aluminum in the spherical shell is the volume that remains when we subtract the volume of the inner sphere having $d_i = 2r_i = 38.0$ cm from the volume of the outer sphere having $d_o = 2r_o = 40.0$ cm.

Spherical volume:

$$V = \frac{4}{3}\pi r^3 \tag{M-23}$$

Solve

Subtract the volume of the sphere of radius r_i from the volume of the sphere of radius r_o.

$$V = V_o - V_i = \frac{4}{3}\pi r_o^3 - \frac{4}{3}\pi r_i^3 = \frac{4}{3}\pi(r_o^3 - r_i^3)$$

Substitute 20.0 cm for r_o and 19.0 cm for r_i.

$$V = \frac{4}{3}\pi\left[(20.0\text{ cm})^3 - (19.0\text{ cm})^3\right]$$
$$= 4.78 \times 10^3\text{ cm}^3$$

Reflect

The volume calculated is less than the volume of the outer sphere, as it must be.

$$V_o = \frac{4}{3}\pi r_o^3 = \frac{4}{3}\pi(20.0\text{ cm})^3$$
$$= 3.35 \times 10^4\text{ cm}^3$$

M-8 Trigonometry

Trigonometry, which gets its name from Greek roots meaning "triangle" and "measure," is the study of some important mathematical functions, called trigonometric functions. These functions are most simply defined as ratios of the sides of right triangles. However, these right-triangle definitions are of limited use because they are valid only for angles between zero and 90°. However, the validity of the right-triangle definitions can be extended by defining the trigonometric functions in terms of the ratio of the coordinates of points on a circle of unit radius drawn centered at the origin of the x–y plane.

In physics, we first encounter trigonometric functions when we use vectors to analyze motion in two dimensions and they are also useful in analyzing components on free-body diagrams. Trigonometric functions are also essential in the analysis of any kind of periodic behavior, such as circular motion, oscillatory motion, and wave mechanics.

Angles and Their Measure: Degrees and Radians

The size of an angle formed by two intersecting straight lines is known as its measure. The standard way of finding the measure of an angle is to place the angle so that its vertex, or point of intersection of the two lines that form the angle, is at the center of a circle located at the origin of a graph that has Cartesian coordinates and one of the lines extends rightward on the positive x axis. The distance traveled *counterclockwise* on the circumference from the positive x axis to reach the intersection of the circumference with the other line defines the measure of the angle. (Traveling clockwise to the second line would simply give us a negative measure; to illustrate basic concepts, we position the angle so that the smaller rotation will be in the counterclockwise direction.) The measure of the angle in **Figure M-10** is θ, illustrated in radians.

One of the most familiar units for expressing the measure of an angle is the degree, which equals 1/360 of the full distance around the circumference of the circle. For greater precision, or for smaller angles, we either show degrees plus minutes (′) and seconds (″), with 1′ = 1°/60 and 1″ = 1′/60 = 1°/3600; or show degrees as an ordinary decimal number.

For scientific work, a more useful measure of an angle is the radian (rad). Again, place the angle with its vertex at the center of a circle and measure counterclockwise rotation around the circumference. The measure of the angle in radians is then defined as the length of the circular arc from one line to the other divided by the radius of the circle (Figure M-10). If s is the arc length and r is the radius of the circle, the angle θ measured in radians is

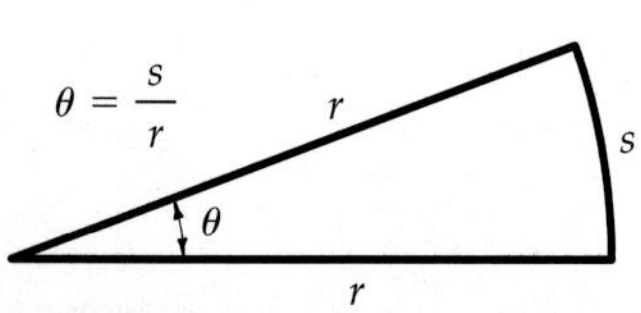

$$\theta = \frac{s}{r} \tag{M-26}$$

$\theta = \frac{s}{r}$ r s θ r

Figure M-10 The angle θ in radians is defined to be the ratio s/r, where s is the arc length intercepted on a circle of radius r.

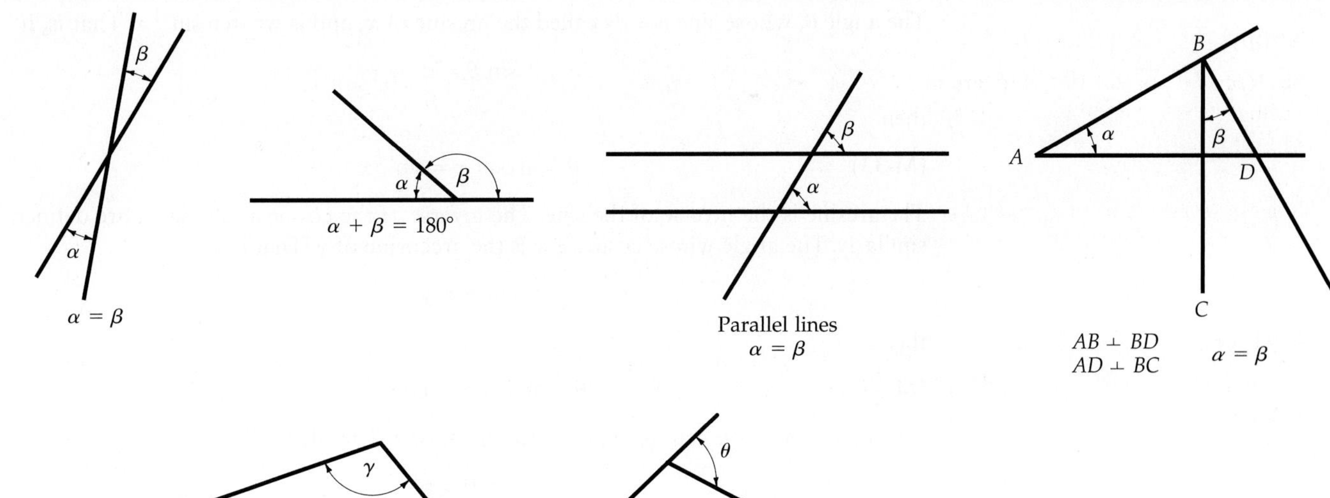

Figure M-11 Some useful relations for angles.

Because the angle measured in radians is the ratio of two lengths, it is dimensionless. The relation between radians and degrees is

$$360° = 2\pi \text{ rad}$$

or

$$1 \text{ rad} = \frac{360°}{2\pi} = 57.3°$$

Figure M-11 shows some useful relations for angles.

The Trigonometric Functions

Figure M-12 shows a right triangle formed by drawing the line segment BC perpendicular to AC. The lengths of the sides are labeled a, b, and c. The right-triangle definitions of the trigonometric functions $\sin\theta$ (the sine), $\cos\theta$ (the cosine), and $\tan\theta$ (the tangent) for an acute angle θ are

$$\sin\theta = \frac{a}{c} = \frac{\text{opposite side}}{\text{hypotensue}} \tag{M-27}$$

$$\cos\theta = \frac{b}{c} = \frac{\text{adjacent side}}{\text{hypotensue}} \tag{M-28}$$

$$\tan\theta = \frac{a}{b} = \frac{\text{opposite side}}{\text{adjacent side}} = \frac{\sin\theta}{\cos\theta} \tag{M-29}$$

(Acute angles are angles whose positive rotation around the circumference of a circle measures less than 90° or $\pi/2$.) Three other trigonometric functions—the secant (sec), the cosecant (csc), and the cotangent (cot), defined as the reciprocals of these functions—are

$$\csc\theta = \frac{c}{a} = \frac{1}{\sin\theta} \tag{M-30}$$

$$\sec\theta = \frac{c}{b} = \frac{1}{\cos\theta} \tag{M-31}$$

$$\cot\theta = \frac{b}{a} = \frac{1}{\tan\theta} = \frac{\cos\theta}{\sin\theta} \tag{M-32}$$

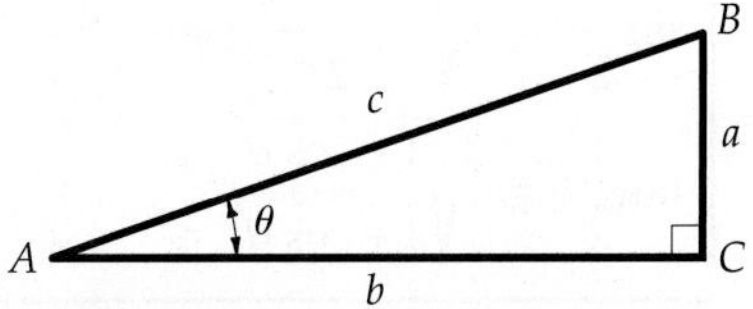

Figure M-12 A right triangle with sides of length *a* and *b* and a hypotenuse of length *c*.

The angle θ, whose sine is x, is called the arcsine of x, and is written $\sin^{-1} x$. That is, if

$$\sin \theta = x$$

then

(M-33) $$\theta = \arcsin x = \sin^{-1} x$$

The arcsine is the inverse of the sine. The inverse of the cosine and tangent are defined similarly. The angle whose cosine is y is the arccosine of y. That is, if

$$\cos \theta = y$$

then

(M-34) $$\theta = \arccos y = \cos^{-1} y$$

The angle whose tangent is z is the arctangent of z. That is, if

$$\tan \theta = z$$

then

(M-35) $$\theta = \arctan z = \tan^{-1} z$$

Trigonometric Identities

We can derive several useful formulas, called trigonometric identities, by examining relationships between the trigonometric functions. Equations M-30 through M-32 list three of the most obvious identities, formulas expressing some trigonometric functions as reciprocals of others. Almost as easy to discern are identities derived from the Pythagorean theorem,

(M-36) $$a^2 + b^2 = c^2$$

Simple algebraic manipulation of Equation M-36 gives us three more identities. First, if we divide each term in Equation M-36 by c^2, we obtain

$$\frac{a^2}{c^2} + \frac{b^2}{c^2} = 1$$

or, from the definitions of $\sin \theta$ (which is a/c) and $\cos \theta$ (which is b/c),

(M-37) $$\sin^2 \theta + \cos^2 \theta = 1$$

Similarly, we can divide each term in Equation M-36 by a^2 or b^2 and obtain

(M-38) $$1 + \cot^2 \theta = \csc^2 \theta$$

and

(M-39) $$1 + \tan^2 \theta = \sec^2 \theta$$

Table M-2 lists these last three and many more trigonometric identities. Notice that they fall into four categories: functions of sums or differences of angles, sums or differences of squared functions, functions of double angles (2θ), and functions of half angles $\left(\frac{1}{2}\theta\right)$. Notice that some of the formulas contain paired alternatives, expressed with the signs $\pm$ and $\mp$; in such formulas, remember to always apply the formula with either all the upper or all the lower alternatives.

TABLE M-2 Trigonometric Identities

$$\sin(A \pm B) = \sin A \cos B \pm \cos A \sin B$$

$$\cos(A \pm B) = \cos A \cos B \mp \sin A \sin B$$

$$\tan(A \pm B) = \frac{\tan A \pm \tan B}{1 \mp \tan A \tan B}$$

$$\sin A \pm \sin B = 2 \sin\left[\frac{1}{2}(A \pm B)\right]\cos\left[\frac{1}{2}(A \mp B)\right]$$

$$\cos A + \cos B = 2 \cos\left[\frac{1}{2}(A + B)\right]\cos\left[\frac{1}{2}(A - B)\right]$$

$$\cos A - \cos B = 2 \sin\left[\frac{1}{2}(A + B)\right]\sin\left[\frac{1}{2}(B - A)\right]$$

$$\tan A \pm \tan B = \frac{\sin(A \pm B)}{\cos A \cos B}$$

$$\sin^2 \theta + \cos^2 \theta = 1; \sec^2 \theta - \tan^2 \theta = 1;$$

$$\csc^2 \theta - \cot^2 \theta = 1$$

$$\sin 2\theta = 2 \sin \theta \cos \theta$$

$$\cos 2\theta = \cos^2 \theta - \sin^2 \theta = 2 \cos^2 \theta - 1 = 1 - 2 \sin^2 \theta$$

$$\tan 2\theta = \frac{2 \tan \theta}{1 - \tan^2 \theta}$$

$$\sin \frac{1}{2}\theta = \pm\sqrt{\frac{1 - \cos \theta}{2}}; \cos \frac{1}{2}\theta = \pm\sqrt{\frac{1 + \cos \theta}{2}};$$

$$\tan \frac{1}{2}\theta = \pm\sqrt{\frac{1 - \cos \theta}{1 + \cos \theta}}$$

Some Important Values of the Functions

Figure M-13 is a diagram of an *isosceles* right triangle (an isosceles triangle is a triangle with two equal sides), from which we can find the sine, cosine, and tangent of 45°. The two acute angles of this triangle are equal. Because the sum of the three angles in a triangle must equal 180° and the right angle is 90°, each acute angle must be 45°. For convenience, let us assume that the equal sides each have a length of 1 unit. The Pythagorean theorem gives us a value for the hypotenuse of

$$c = \sqrt{a^2 + b^2} = \sqrt{1^2 + 1^2} = \sqrt{2} \text{ units}$$

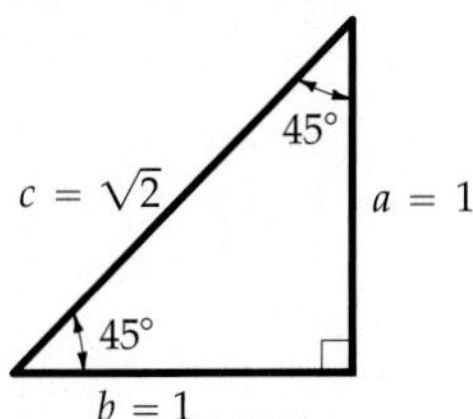

Figure M-13 An isosceles right triangle.

We calculate the values of the functions as follows:

$$\sin 45° = \frac{a}{c} = \frac{1}{\sqrt{2}} = 0.707 \qquad \cos 45° = \frac{b}{c} = \frac{1}{\sqrt{2}} = 0.707 \qquad \tan 45° = \frac{a}{b} = \frac{1}{1} = 1$$

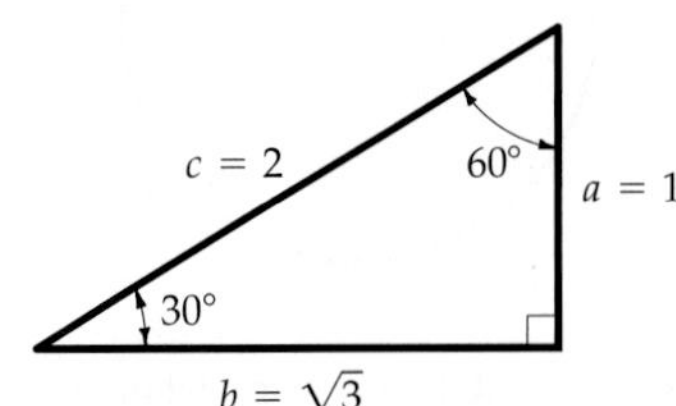

Figure M-14 A 30°–60°–90° right triangle.

Another common triangle, a 30°–60°–90° right triangle, is shown in **Figure M-14**. Because this particular right triangle is in effect half of an *equilateral triangle* (a 60°–60°–60° triangle or a triangle having three equal sides and three equal angles), we can see that the sine of 30° must be exactly 0.5 (**Figure M-15**). The equilateral triangle must have all sides equal to c, the hypotenuse of the 30°–60°–90° right triangle. Thus, side a is one-half the length of the hypotenuse, and so

$$\sin 30° = \frac{1}{2}$$

To find the other ratios within the 30°–60°–90° right triangle, let us assign a value of 1 to the side opposite the 30° angle. Then

$$c = \frac{1}{0.5} = 2 \qquad b = \sqrt{c^2 - a^2} = \sqrt{2^2 - 1^2} = \sqrt{3}$$

$$\cos 30° = \frac{b}{c} = \frac{\sqrt{3}}{2} = 0.866 \qquad \tan 30° = \frac{a}{b} = \frac{1}{\sqrt{3}} = 0.577$$

$$\sin 60° = \frac{b}{c} = \cos 30° = 0.866 \qquad \cos 60° = \frac{a}{c} = \sin 30° = \frac{1}{2}$$

$$\tan 60° = \frac{b}{a} = \frac{\sqrt{3}}{1} = 1.732$$

(a)

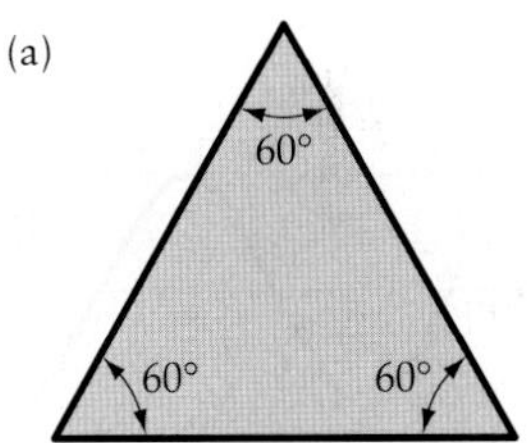

(b)

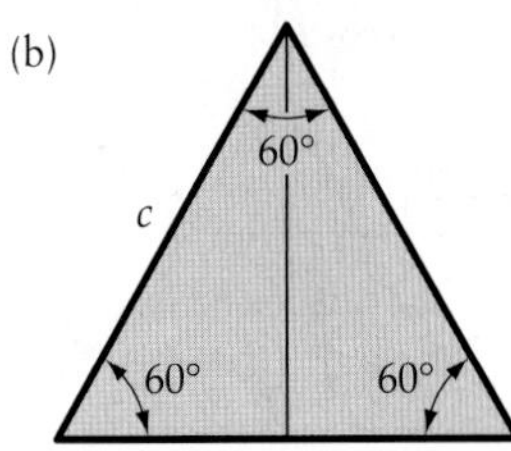

Figure M-15 (a) An equilateral triangle. (b) An equilateral triangle that has been bisected to form two 30°–60°–90° right triangles.

Small-Angle Approximation

For small angles, the length a is nearly equal to the arc length s, as can be seen in **Figure M-16**. The angle $\theta = s/c$ is therefore nearly equal to $\sin\theta = a/c$:

$$\sin\theta \approx \theta \quad \text{for small values of } \theta \tag{M-40}$$

Similarly, the lengths c and b are nearly equal, so $\tan\theta = a/b$ is nearly equal to both θ and $\sin\theta$ for small values of θ:

$$\tan\theta \approx \sin\theta \approx \theta \quad \text{for small values of } \theta \tag{M-41}$$

Equations M-40 and M-41 hold only if θ is measured in radians. Because $\cos\theta = b/c$, and because these lengths are nearly equal for small values of θ, we have

$$\cos\theta \approx 1 \quad \text{for small values of } \theta \tag{M-42}$$

Figure M-17 shows graphs of θ, $\sin\theta$, and $\tan\theta$ versus θ for small values of θ. If accuracy of a few percent is needed, small-angle approximations can be used only for angles of about a quarter of a radian (or about 15°) or less. Below this value, as the angle becomes smaller, the approximation $\theta \approx \sin\theta \approx \tan\theta$ is even more accurate.

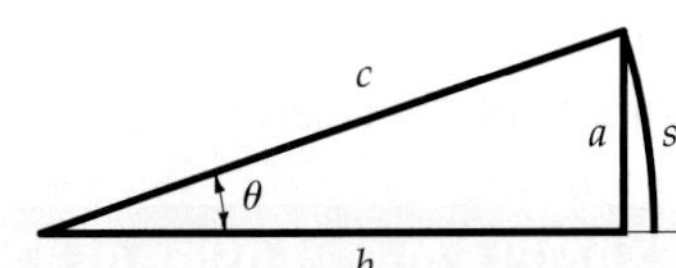

Figure M-16 For small angles, $\sin\theta = a/c$, $\tan\theta = a/b$, and the angle $\theta = s/c$ are all approximately equal.

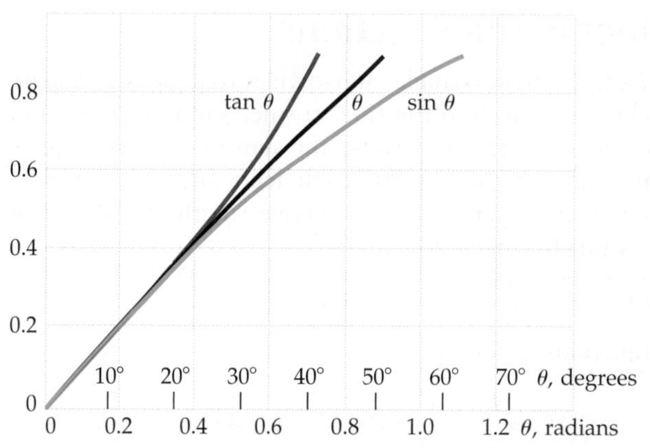

Figure M-17 Graphs of *tan θ*, *θ*, and *sin θ* versus *θ* for small values of *θ*.

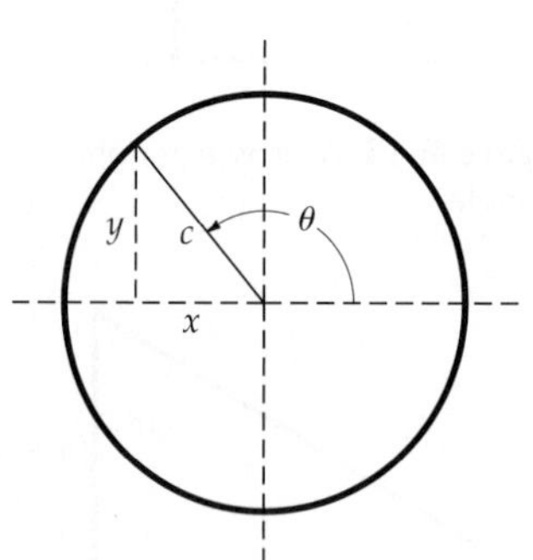

Figure M-18 Diagram for defining the trigonometric functions for an obtuse angle.

Trigonometric Functions as Functions of Real Numbers

So far we have illustrated the trigonometric functions as properties of angles. **Figure M-18** shows an *obtuse* angle with its vertex at the origin and one side along the x axis. The trigonometric functions for a "general" angle such as this are defined by

$$\sin\theta = \frac{y}{c} \tag{M-43}$$

$$\cos\theta = \frac{x}{c} \tag{M-44}$$

$$\tan\theta = \frac{y}{x} \tag{M-45}$$

It is important to remember that values of x to the left of the vertical axis and values of y below the horizontal axis are negative; c in the figure is always regarded as positive since it is the square root of squared numbers. **Figure M-19** shows plots of the general sine, cosine, and tangent functions versus θ. The sine and cosine functions have a period of 2π rad. Thus, for any value of θ, $\sin(\theta + 2\pi) = \sin\theta$, and so forth. That is, when an angle changes by 2π rad, the function returns to its original value. The tangent function has a period of π rad. Thus, $\tan(\theta + \pi) = \tan\theta$, and so forth. Some other useful relations are

$$\sin(\pi - \theta) = \sin\theta \tag{M-46}$$

$$\cos(\pi - \theta) = -\cos\theta \tag{M-47}$$

$$\sin\left(\frac{1}{2}\pi - \theta\right) = \cos\theta \tag{M-48}$$

$$\cos\left(\frac{1}{2}\pi - \theta\right) = \sin\theta \tag{M-49}$$

Because the radian is dimensionless, it is not hard to see from the plots in Figure M-19 that the trigonometric functions are functions of all real numbers.

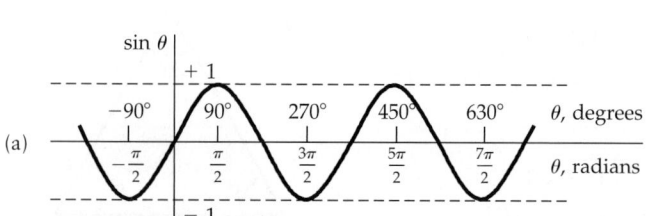

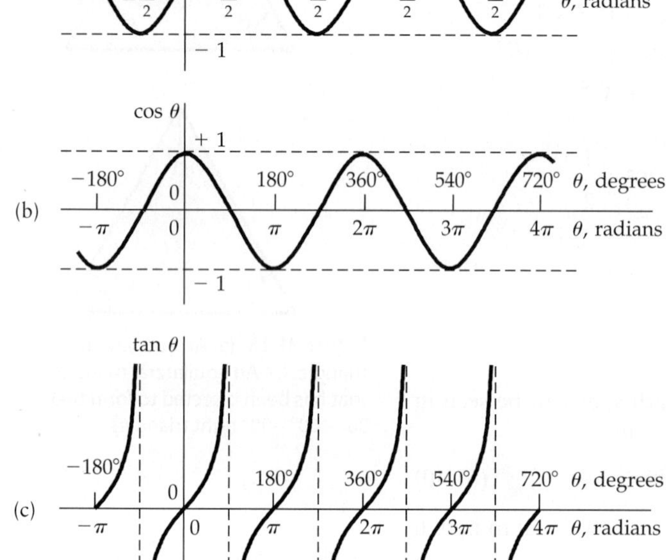

Figure M-19 The trigonometric functions *sin θ*, *cos θ*, and *tan θ* versus *θ*.

EXAMPLE M-9 Cosine of a Sum

Using the suitable trigonometric identity from Table M-2, find $\cos(135° + 22°)$. Give your answer with four significant figures.

Set Up

As long as all angles are given in degrees, there is no need to convert to radians, because all operations are numerical values of the functions. Be sure, however, that your calculator is in degree mode. The suitable identity is $\cos(A \pm B) = \cos A \cos B \mp \sin A \sin B$, where the upper signs are appropriate.

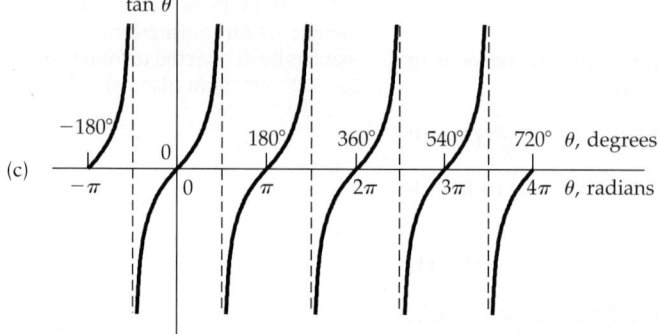

Solve

Write the trigonometric identity for the cosine of a sum, with $A = 135°$ and $B = 22°$.	$\cos(135° + 22°) = (\cos 135°)(\cos 22°) - (\sin 135°)(\sin 22°)$
Using a calculator, find cos 135°, sin 135°, cos 22°, and sin 22°.	$\cos 135° = -0.7071$ $\cos 22° = 0.9272$ $\sin 135° = 0.7071$ $\sin 22° = 0.3746$
Enter the values in the formula and calculate the answer.	$\cos(135° + 22°) = (-0.7071)(0.9272) - (0.7071)(0.3746) = -0.9205$

Reflect

We can see in Figure M-18 that, since each quarter of a circle is 90°, 157° must have a negative value for a cosine, one larger in magnitude than 0.5, so this answer is plausible. We can also check with a calculator that $\cos(135° + 22°) = \cos(157°) = -0.9205$, and it does.

M-9 The dot product

For two vectors $\vec{A}$ and $\vec{B}$ separated by angle θ, as shown in **Figure M-20**, their dot product C is defined as

$$C = \vec{A} \cdot \vec{B} = AB \cos\theta \tag{M-50}$$

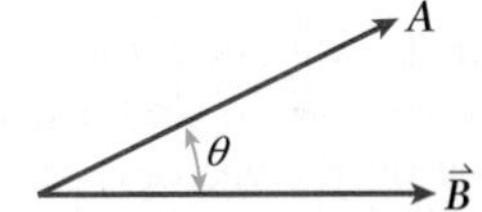

Figure M-20 Two vectors separated by an angle θ.

which you can read as, "C equals A dot B." In Equation M-50, A and B are the magnitudes of vectors $\vec{A}$ and $\vec{B}$, respectively. As a result, the dot product of two vectors is a scalar quantity. This is why $\vec{A} \cdot \vec{B}$ is also called the scalar product of $\vec{A}$ and $\vec{B}$.

Physically, the dot product $\vec{A} \cdot \vec{B}$ is a measure of how parallel the two vectors are. We can think of it as the magnitude of vector $\vec{A}$ multiplied by the component of vector $\vec{B}$ that is parallel to $\vec{A}$. Referring to **Figure M-21a**, we see that $B \cos\theta$ is the component of $\vec{B}$ that is parallel to $\vec{A}$. That is, $B \cos\theta$ tells us how much of $\vec{B}$ points in the direction of $\vec{A}$. Alternatively, the dot product $\vec{A} \cdot \vec{B}$ can be thought of as the magnitude of vector $\vec{B}$ multiplied by the component of vector $\vec{A}$ parallel to $\vec{B}$ (**Figure M-21b**).

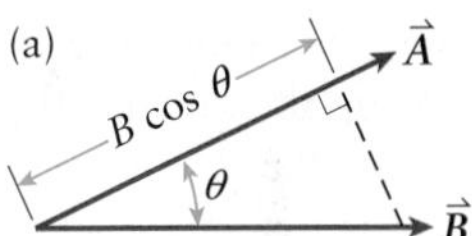

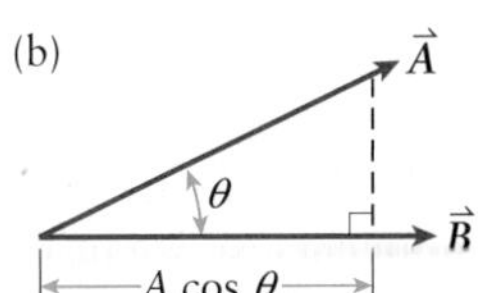

Figure M-21 The dot product is a measure of how parallel two vectors are. (a) $B \cos\theta$ is the component of $\vec{B}$ that is parallel to $\vec{A}$. (b) $A \cos\theta$ is the component of $\vec{A}$ that is parallel to $\vec{B}$.

The dot product is commutative; the order of the vectors in a dot product does not affect the result:

$$\vec{A} \cdot \vec{B} = \vec{B} \cdot \vec{A}$$

The dot product is also distributive, which means

$$\vec{A} \cdot (\vec{B} + \vec{C}) = \vec{A} \cdot \vec{B} + \vec{A} \cdot \vec{C}$$

Three special cases of the dot product are particularly important in physics. First, the dot product of two vectors $\vec{A}$ and $\vec{B}$ that point in the same direction (so $\theta = 0$ and $\cos\theta = \cos 0 = 1$) equals the product of their magnitudes:

$$\vec{A} \cdot \vec{B} = AB \cos 0 = AB \tag{M-51}$$

(if $\vec{A}$ and $\vec{B}$ point in the same direction)

(As an example, the dot product of a vector $\vec{A}$ with itself is equal to the square of its magnitude: $\vec{A} \cdot \vec{A} = AA \cos 0 = A^2$.)

Second, the dot product of two perpendicular vectors $\vec{A}$ and $\vec{B}$ (so $\theta = 90°$ and $\cos\theta = \cos 90° = 0$ is zero:

$$\vec{A} \cdot \vec{B} = AB \cos 90° = 0 \tag{M-52}$$

(if $\vec{A}$ and $\vec{B}$ are perpendicular)

Third, if two vectors $\vec{A}$ and $\vec{B}$ point in opposite directions (so $\theta = 180°$ and $\cos\theta = \cos 180° = -1$), their dot product equals the *negative* of the product of their magnitudes:

$$\vec{A} \cdot \vec{B} \cos 180° = -AB \tag{M-53}$$

(if $\vec{A}$ and $\vec{B}$ point in opposite directions)

Finally, it's useful to know how to calculate the dot product of two vectors $\vec{A}$ and $\vec{B}$ that are expressed in terms of their components A_x, A_y, A_z, and B_x, B_y, and B_z:

$$\vec{A}\cdot\vec{B} = A_xB_x + A_yB_y + A_zB_z \tag{M-54}$$

You can verify that Equation M-54 is correct by thinking of $\vec{A}$ as the sum of three vectors: $\vec{A}_1$, which has only an x-component A_x; $\vec{A}_2$, which has only a y-component A_y; and $\vec{A}_3$, which has only a z-component A_z. From the definition of the dot product, $\vec{A}_1\cdot\vec{B}$ is equal to A_x multiplied by the component of $\vec{B}$ in the direction of $\vec{A}_1$, or $\vec{A}_1\cdot\vec{B} = A_xB_x$. Similarly, $\vec{A}_2\cdot\vec{B} = A_yB_y$ and $\vec{A}_3\cdot\vec{B} = A_zB_z$. Since $\vec{A} = \vec{A}_1 + \vec{A}_2 + \vec{A}_3$ and the dot product is distributive, it follows that

$$\vec{A}\cdot\vec{B} = (\vec{A}_1 + \vec{A}_2 + \vec{A}_3)\cdot\vec{B} = \vec{A}_1\cdot\vec{B} + \vec{A}_2\cdot\vec{B} + \vec{A}_3\cdot\vec{B} = A_xB_x + A_yB_y + A_zB_z$$

That's the same as Equation M-54. If the vectors have only x and y components, Equation M-54 simplifies to $\vec{A}\cdot\vec{B} = A_xB_x + A_yB_y$.

EXAMPLE M-10 The Dot Product

(a) Calculate the dot product of vector $\vec{A}$ with magnitude 5.00 pointed in a horizontal direction 36.9° north of east and vector $\vec{B}$ of magnitude 1.50 pointed in a horizontal direction 53.1° south of west. (b) What is the dot product of vector $\vec{C}$ with components $C_x = 4.00$, $C_y = 3.00$ and vector $\vec{D}$ with components $D_x = -0.900$, $D_y = -1.20$?

Set Up

In part (a) we know the magnitude and direction of the vectors, so we'll use Equation M-50. In part (b) the vectors are given in terms of components, so we'll evaluate the dot product using Equation M-54.

$$\vec{A}\cdot\vec{B} = AB\cos\theta \tag{M-50}$$

Dot product of two vectors in terms of components:

$$\vec{A}\cdot\vec{B} = A_xB_x + A_yB_y + A_zB_z \tag{M-54}$$

Solve

(a) The drawing shows that the angle between $\vec{A}$ and $\vec{B}$ is $\theta = 163.8°$. We use this in Equation M-50 to evaluate the dot product.

$$\begin{aligned}\vec{A}\cdot\vec{B} &= AB\cos\theta\\ &= (5.00)(1.50)\cos 163.8°\\ &= (5.00)(1.50)(-0.960)\\ &= -7.20\end{aligned}$$

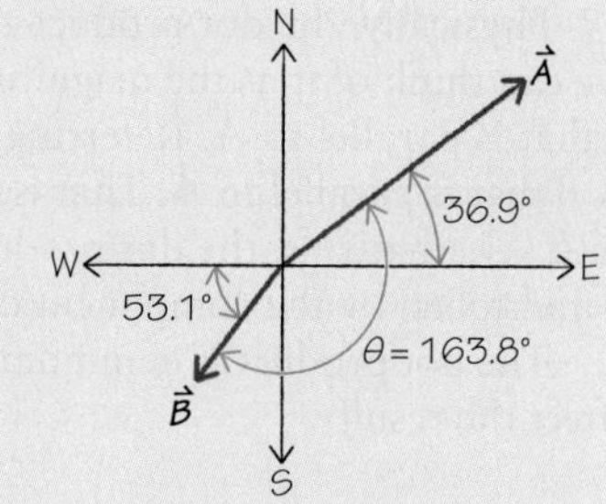

(b) Both $\vec{C}$ and $\vec{D}$ are in the x–y plane and have no z components, so we just need the first two terms in Equation M-54 to calculate their dot product.

$$\begin{aligned}\vec{C}\cdot\vec{D} &= C_xD_x + C_yD_y\\ &= (4.00)(-0.900) + (3.00)(-1.20)\\ &= -7.20\end{aligned}$$

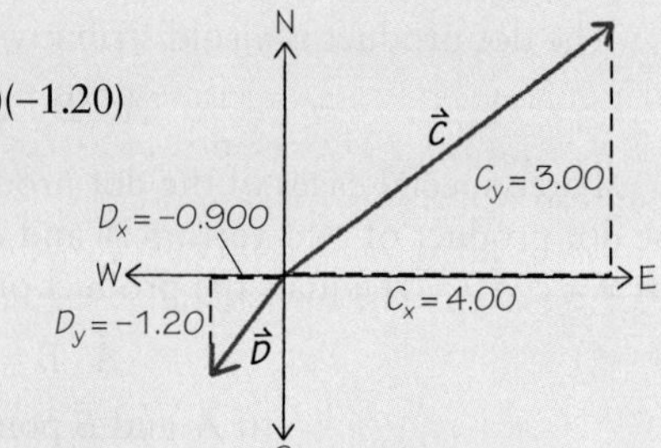

Reflect

It's not a coincidence that we got the same result in part (b) as in part (a): Vectors $\vec{A}$ and $\vec{C}$ are the same, as are vectors $\vec{B}$ and $\vec{D}$. [You can verify this by using the techniques from Chapter 3 to calculate the components of the vectors $\vec{A}$ and $\vec{B}$ in part (a). You'll find that the components are the same as those of $\vec{C}$ and $\vec{D}$ in part (b).] This should give you confidence that the method of calculating the dot product using components gives you the same result as the method that involves the magnitudes and directions of the vectors.

Notice that the angle between vectors $\vec{A}$ and $\vec{B}$ is between 90° and 180°, and the dot product is negative.

M-10 The cross product

The dot product, described in section M-9, is only one way to multiply two vectors. We can also multiply two vectors $\vec{A}$ and $\vec{B}$ using the cross product

$$\vec{C} = \vec{A} \times \vec{B} \tag{M-55}$$

The symbol "×" represents the mathematical operation known as the cross product. As you can see from Equation M-55, the result of taking the cross product of two vectors is also a vector, which is why it is also called a vector product. The magnitude of the resulting vector is the product of the magnitudes of the two vectors and the sine of the angle between them. That is, the magnitude of the cross product of $\vec{A}$ and $\vec{B}$ is

$$\vec{C} = \left|\vec{A} \times \vec{B}\right| = AB \sin\phi \tag{M-56}$$

where according to convention ϕ is defined as the angle that goes from $\vec{A}$ to $\vec{B}$. $\vec{C}$ points in the direction perpendicular to both $\vec{A}$ and $\vec{B}$ as shown in **Figure M-22.**

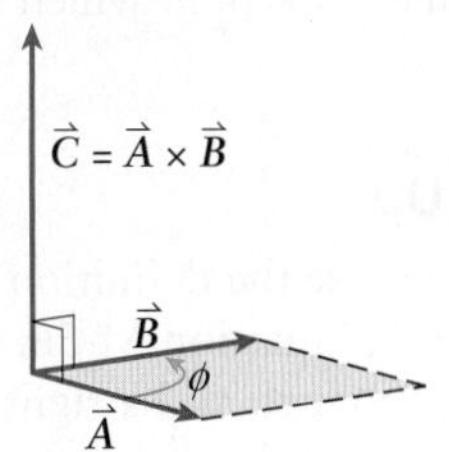

Figure M-22 The cross product is a vector $\vec{C}$ that is perpendicular to both $\vec{A}$ and $\vec{B}$, and has a magnitude $AB \sin\phi$, which equals the area of the parallelogram shown.

The magnitude of the cross product $\vec{A} \times \vec{B}$ can be interpreted as the magnitude of vector $\vec{A}$ multiplied by the component of vector $\vec{B}$ perpendicular to $\vec{A}$, or the magnitude of vector $\vec{B}$ multiplied by the component of vector $\vec{A}$ perpendicular to $\vec{B}$.

Note that the order of the two vectors in a cross product makes a difference. The cross product of $\vec{B}$ and $\vec{A}$ is the negative of the cross product of $\vec{A}$ and $\vec{B}$ or

$$\vec{A} \times \vec{B} = -\vec{B} \times \vec{A} \tag{M-57}$$

This results from the definition of the angle ϕ in Equation M-56. Since ϕ is directed from the first vector to the second vector, if you travel the angle from the second vector to the first—in reverse direction—ϕ becomes negative. And the sine of a negative angle is also negative.

In addition, the cross product obeys the distributive law under addition:

$$\vec{A} \times (\vec{B} + \vec{C}) = \vec{A} \times \vec{B} + \vec{A} \times \vec{C} \tag{M-58}$$

To determine the direction of the cross product $\vec{C} = \vec{A} \times \vec{B}$, you can use a right-hand rule. To apply this rule, point the fingers of your right hand in the direction of the first vector of the cross product (in this case $\vec{A}$). Then curl your fingers toward the second vector, $\vec{B}$. If you stick your thumb straight out, it points in the direction of the cross product, vector $\vec{C}$ (**Figure M-23a**). If you instead want to find the direction of the cross product $\vec{B} \times \vec{A}$, begin by pointing the fingers of your right hand in the direction of vector $\vec{B}$. Then curl them toward vector $\vec{A}$. Your thumb again points in the direction of the cross product (**Figure M-23b**). Note that because you must curl your fingers in the opposite direction as for $\vec{C} = \vec{A} \times \vec{B}$, the cross product of $\vec{B} \times \vec{A}$ points in the opposite direction of $\vec{A} \times \vec{B}$, which is just what we stated in Equation M-58.

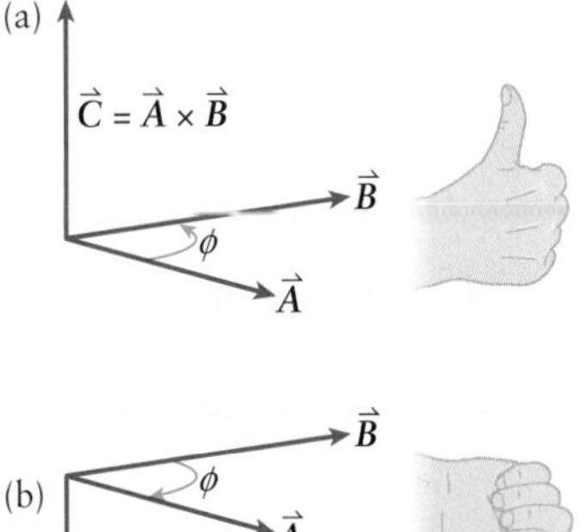

Figure M-23 (a) To find the direction of the cross product, point the fingers of your right hand in the direction of vector $\vec{A}$, then curl them toward vector $\vec{B}$. Your thumb points in the direction of the cross product. (b) The direction of $\vec{B} \times \vec{A}$ points in the opposite direction of.

There are two special cases of the cross product that are worth pointing out. The first is the cross product for two perpendicular vectors, for which $\phi = 90°$, so $\sin\phi = 1$.

$$\left|\vec{A} \times \vec{B}\right| = AB \sin 90° = AB(1) = AB$$

(magnitude of the cross product of two perpendicular vectors)

The second special case is the cross product of two parallel vectors, for which $\phi = 0$, so $\sin\phi = 0$.

$$\left|\vec{A} \times \vec{B}\right| = AB \sin 0 = AB(0) = 0$$

(magnitude of the cross product for two parallel vectors)

One example of a cross product of two parallel vectors is the cross product of a vector with itself: $\vec{A} \times \vec{A} = 0$.

EXAMPLE M-11 The Cross Product

Evaluate $\vec{A} \times \vec{B}$, in which the components of vector $\vec{A}$ are $A_x = 5$, $A_y = 0$, and the components of vector $\vec{B}$ are $B_x = 9$, $B_y = 7$.

Set Up

We will use the definition of the magnitude of the cross product, Equation M-56, to find the magnitude of the cross product, and a right-hand rule to determine the direction of the cross product.

We will have to use the components of vector $\vec{B}$ to determine its magnitude and the angle it makes with the x axis and vector $\vec{A}$.

$$C = \left|\vec{A} \times \vec{B}\right| = AB \sin \phi \quad \text{(M-56)}$$

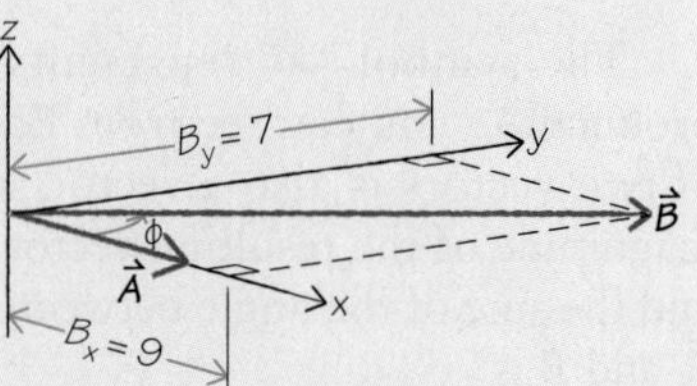

Finding vector magnitude and direction from vector components:

$$A = \sqrt{A_x^2 + A_y^2}$$

$$\tan \theta = \frac{A_y}{A_x} \quad \text{(3-2)}$$

Solve

Begin by determining the magnitude and direction of vector $\vec{B}$ using its components and Equations 3-2.

Determine the magnitude of vector $\vec{B}$ from its components:

$$B = \sqrt{B_x^2 + B_y^2} = \sqrt{9^2 + 7^2} = 11.4$$

Determine the angle $\vec{B}$ makes with the x axis (and vector $\vec{A}$) from its components:

$$\tan \phi = \frac{B_y}{B_x} = \frac{7}{9} = 0.778, \text{ so}$$

$$\phi = \arctan 0.778 = 37.9°$$

Because vector $\vec{A}$ has only an x component, its magnitude is equal to its x component, and the angle it makes with the x axis is 0.

Determine the magnitude of vector $\vec{A}$ from its components:

$$A = \sqrt{A_x^2 + A_y^2} = \sqrt{5^2 + 0^2} = 5$$

Because $\vec{A}$ has only an x component, it makes an angle of zero degrees with the x axis.

Now that we know both the magnitude and direction of the vectors, we can use Equation M-56 to determine the magnitude of the cross product.

Apply Equation M-56 to the two vectors:

$$\begin{aligned}\left|\vec{A} \times \vec{B}\right| &= AB \sin \phi \\ &= (5.00)(11.4) \sin 37.9° \\ &= (5.00)(11.4)(0.614) \\ &= 35.0\end{aligned}$$

Use a right-hand rule to determine the direction of the cross product.

From the figure, if we first point the fingers of our right hand in the direction of $\vec{A}$ (along the x axis), and then curl them toward $\vec{B}$, we see that the thumb points in the positive z direction. So the cross product $\vec{A} \times \vec{B}$ has a magnitude of 35 in the $+z$ direction.

Reflect

The vectors $\vec{A}$ and $\vec{B}$ lie in the x–y plane, so the cross product, which must be perpendicular to both vectors, should point along the z axis, which is just what we found.

Appendix A SI Units and Conversion Factors

Base Units*	
Length	The *meter* (m) is the distance traveled by light in a vacuum in 1/299,792,458 s.
Time	The *second* (s) is the duration of 9,192,631,770 periods of the unperturbed ground-state hyperfine transition frequency of the ^{133}Cs atom.
Mass	The *kilogram* (kg) is defined by taking the fixed numerical value of the Planck constant h to be $6.626\ 070\ 15 \times 10^{-34}$ when expressed in the units J s, which is equal to kg m^2 s^{-1}, where the meter and the second are defined as described above.
Mole	The *mole* (mol) is the amount of substance. One mole contains exactly $6.022\ 140\ 76 \times 10^{23}$ elementary entities. This number is the fixed numerical value of the Avogadro constant, N_A, when expressed in the unit mol^{-1} and is called the Avogadro number. The amount of substance, symbol n, of a system is a measure of the number of specified elementary entities. An elementary entity may be an atom, a molecule, an ion, an electron, or any other particle or specified group of particles.
Current	The *ampere* (A) is the SI unit of electric current, defined by taking the fixed numerical value of the elementary charge e to be $1.602\ 176\ 634 \times 10^{-19}$ when expressed in the unit C, which is equal to A s, where the second is defined as described above.
Temperature	The *kelvin* (K) is the SI unit of thermodynamic temperature. It is defined by taking the fixed numerical value of the Boltzmann constant k to be $1.380\ 649 \times 10^{-23}$ when expressed in the unit J K^{-1}, which is equal to kg m^2 s^{-2} K^{-1}, where the kilogram, meter, and second are the SI units of thermodynamic temperature.
Luminous intensity	The *candela* (cd) is the luminous intensity in a given direction, it is defined in terms of a source that emits monochromatic radiation of frequency 540×10^{12} Hz and that has a radiant intensity, in that direction of a fixed numerical value of 1/683 W/steradian.

* These definitions are found on the Internet at http://physics.nist.gov/cuu/Units/current.html.

Derived Units		
Force	newton (N)	$1\ N = 1\ kg \cdot m/s^2$
Work, energy	joule (J)	$1\ J = 1\ Nm$
Power	watt (W)	$1\ W = 1\ J/s$
Frequency	hertz (Hz)	$1\ Hz = cy/s$
Electric charge	coulomb (C)	$1\ C = 1\ A \cdot s$
Electric potential difference	volt (V)	$1\ V = 1\ J/C$
Resistance	ohm (Ω)	$1\ \Omega = 1\ V/A$
Capacitance	farad (F)	$1\ F = 1\ C/V$
Magnetic field	tesla (T)	$1\ T = 1\ N/(A \cdot m)$
Magnetic flux	weber (Wb)	$1\ Wb = 1\ T \cdot m^2$
Inductance	henry (H)	$1\ H = 1\ J/A^2$

Conversion Factors

Conversion factors are written as equations for simplicity; relations marked with an asterisk are exact.

Length

1 km = 0.6214 mi

1 mi = 1.609 km

1 m = 1.0936 yard = 3.281 ft = 39.37 in.

*1 in. = 2.54 cm

*1 ft = 12 in. = 30.48 cm

*1 yard = 3 ft = 91.44 cm

1 light-year = 1 $c \cdot$ y = 9.461×10^{15} m

*1Å = 0.1 nm

Area

*1 $m^2 = 10^4$ cm^2

1 km^2 = 0.3861 mi^2 = 247.1 acres

*1 $in.^2$ = 6.4516 cm^2

1 $ft^2 = 9.29 \times 10^{-2}$ m^2

1 m^2 = 10.76 ft^2

*1 acre = 43560 ft^2

1 mi^2 = 640 acres = 2.590 km^2

Volume

*1 $m^3 = 10^6$ cm^3

*1 L = 1000 $cm^3 = 10^{-3}$ m^3

1 gal = 3.785 L

1 gal = 4 qt = 8 pt = 128 oz = 231 $in.^3$

1 $in.^3$ = 16.39 cm^3

1 ft^3 = 1728 $in.^3$ = 28.32 L

= 2.832×10^4 cm^3

Time

*1 h = 60 min = 3.6 ks

*1 d = 24 h = 1440 min = 86.4 ks

1 y = 365.25 day = 3.156×10^7 s

Speed

*1 m/s = 3.6 km/h

1 km/h = 0.2778 m/s = 0.6214 mi/h

1 mi/h = 0.4470 m/s = 1.609 km/h

1 mi/h = 1.467 ft/s

Angle and Angular Speed

*π rad = 180°

1 rad = 57.30°

1° = 1.745×10^{-2} rad

1 rev/min = 0.1047 rad/s

1 rad/s = 9.549 rev/min

Mass

*1 kg = 1000 g

*1 tonne = 1000 kg = 1 Mg

1 u = 1.6605×10^{-27} kg
931.49 MeV/c^2

1 kg = 6.022×10^{26} u

1 slug = 14.59 kg

1 kg = 6.852×10^{-2} slug

Density

*1 g/cm^3 = 1000 kg/m^3 = 1 kg/L

(1 g/cm^3)g = 62.4 lb/ft^3

Force

1 N = 0.2248 lbf = 10^5 dyn

*1 lbf = 4.448222 N

(1 kg)g = 2.2046 lbf

Pressure

*1 Pa = 1 N/m^2

*1 atm = 101.325 kPa = 1.01325 bar

1 atm = 14.7 lbf/$in.^2$ = 760 mmHg

= 29.9 in.Hg = 33.9 ftH_2O

1 lbf/$in.^2$ = 6.895 kPa

1 torr = 1 mmHg = 133.32 Pa

1 bar = 100 kPa

Energy

*1 kW · h = 3.6 MJ

*1 cal = 4.186 J

1 ft · lbf = 1.356 J = 1.286×10^{-3} BTU

*1 L · atm = 101.325 J

1 L · atm = 24.217 cal

1 BTU = 778 ft · lbf = 252 cal = 1054.35 J

1 eV = 1.602×10^{-19} J

1 u · c^2 = 931.49 MeV

*1 erg = 10^{-7} J

Power

1 horsepower = 550 ft · lbf/s = 745.7 W

1 BTU/h = 2.931×10^{-4} kW

1 W = 1.341×10^{-3} horsepower

= 0.7376 ft · lbf/s

Magnetic Field

*1 T = 10^4 G

Thermal Conductivity

1 W/(m · K) = 6.938 BTU · in./(h · ft^2 · °F)

1 BTU · in./(h · ft^2 · °F) = 0.1441 W/(m · K)

Appendix B Numerical Data

Terrestrial Data	
Free-fall acceleration g	
Standard value (at sea level at 45° latitude)*	9.806 65 m/s^2; 32.1740 ft/s^2
At equator*	9.7804 m/s^2
At poles*	9.8322 m/s^2
Mass of Earth M_E	5.97×10^{24} kg
Radius of Earth R_E, mean equatorial value	6.38×10^6 m; 3960 mi
Escape speed	1.12×10^4 m/s; 6.96 mi/s
Solar constant†	1.36 kW/m^2
Standard temperature and pressure (STP):	
Temperature	273.15 K
Pressure	101.3 kPa (1.00 atm)
Molar mass of air	28.97 g/mol
Density of air (273.15 K, 101.3 kPa), ρ_{air}	1.29 kg/m^3
Speed of sound (273.15 K, 101.3 kPa)	331 m/s
Latent heat of fusion of H_2O (0°C, 1 atm)	334 kJ/kg
Latent heat of vaporization of H_2O (100°C, 1 atm)	2.26 MJ/kg

* Measured relative to Earth's surface.

† Average power incident normally on 1 m^2 outside Earth's atmosphere at the mean distance from Earth to the Sun.

Astronomical Data*	
Earth	
Distance to the Moon, mean†	3.844×10^8 m; 2.389×10^5 mi
Distance to the Sun, mean†	1.496×10^{11} m; 9.29×10^7 mi; 1.00 AU
Orbital speed, mean	2.98×10^4 m/s
Moon	
Mass	7.35×10^{22} kg
Radius (mean)	1.737×10^6 m
Period	27.32 days
Acceleration of gravity at surface	1.62 m/s^2
Sun	
Mass	1.99×10^{30} kg
Radius	6.96×10^8 m

* Additional solar system data are available from NASA at http://nssdc.gsfc.nasa.gov/planetary/planetfact.html.

† Center to center.

Physical Constants*

Newtonian constant of gravitation	G	$6.674\,30(15) \times 10^{-11}$ Nm^2/kg^2
Speed of light in vacuum (exact)	c	$2.997\,924\,58 \times 10^8$ m/s
Elementary charge (exact)	e	$1.602\,176\,634 \times 10^{-19}$ C
Avogadro's constant (exact)	N_A	$6.022\,140\,76 \times 10^{23}$ particles/mol
Molar gas constant	R	$8.314\,462\,618\ldots$ J/(mol · K)
Boltzmann constant (exact)	$k = R/N_A$	$1.380\,649 \times 10^{-23}$ J/K
Stefan-Boltzmann constant	$\sigma = (\pi^2/60)k^4/(\hbar^3 c^2)$	$5.670\,374\,419\ldots \times 10^{-8}$ W/(m^2 · K^4)
Atomic mass constant	$m_u = (1/12)m(^{12}\text{C})$	$1.660\,539\,066\,60(50) \times 10^{-27}$ kg = 1 u
Permeability of free space	μ_0	$4\pi \times 10^{-7}$ N/A^2
		$1.256\,637\,062\,12(19) \times 10^{-6}$ N/A^2
Permittivity of free space	$\varepsilon_0 = 1/(\mu_0 c^2)$	$8.854\,187\,8128(13)\ldots \times 10^{-12}$ C/(Nm^2)
Coulomb constant†	$k = 1/(4\pi\varepsilon_0)$	$8.987\,551\,7923(14) \times 10^9$ N · m^2/C^2
Planck's constant (exact)	h	$6.626\,070\,15 \times 10^{-34}$ J · s
		$4.135\,667\,696\ldots \times 10^{-15}$ eV · s
	$\hbar = h/(2\pi)$	$1.054\,571\,817\ldots \times 10^{-34}$ J · s
		$6.582\,119\,569\ldots \times 10^{-16}$ eV · s
Mass of electron	m_e	$9.109\,383\,7015(28) \times 10^{-31}$ kg
		$0.510\,998\,950\,00(15)$ MeV/c^2
Mass of proton	m_p	$1.672\,621\,923\,69(51) \times 10^{-27}$ kg
		$938.272\,088\,16(29)$ MeV/c^2
Mass of neutron	m_n	$1.674\,927\,498\,04(95) \times 10^{-27}$ kg
		$939.565\,420\,52(54)$ MeV/c^2
Bohr magneton	$m_B = e\hbar/(2m_e)$	$9.274\,010\,0783(28) \times 10^{-24}$ J/T
		$5.788\,381\,8060(17) \times 10^{-5}$ eV/T
Nuclear magneton	$m_n = e\hbar/(2m_p)$	$5.050\,783\,7461(15) \times 10^{-27}$ J/T
		$3.152\,451\,258\,44(96) \times 10^{-8}$ eV/T
Magnetic flux quantum (exact)	$\phi_0 = h/(2e)$	$2.067\,833\,848\ldots \times 10^{-15}$ T · m^2
Rydberg constant	R_∞	$1.097\,373\,156\,8160(21) \times 10^7$ m^{-1}
Josephson constant (exact)	$K_J = 2e/h$	$4.835\,978\,484\ldots \times 10^{14}$ Hz/V
Compton wavelength	$\lambda_C = h/(m_e c)$	$2.426\,310\,238\,67(73) \times 10^{-12}$ m

* Updated values for these and other constants may be found on the Internet at https://www.nist.gov/pml/fundamental-physical-constants. The numbers in parentheses represent the uncertainties in the last two digits. (For example, 2.044 43(13) stands for 2.044 43 ± 0.000 13.) Values without uncertainties are exact, including those values with ellipses (. . .).

† The Coulomb constant is no longer exactly defined since the redefinition of SI base units and is subject to the measurement error in the fine structure constant.

Appendix C Periodic Table of Elements*

1	2	3	4	5	6	7	8	9	10	11	12	13	14	15	16	17	18
1 H																	2 He
3 Li	4 Be											5 B	6 C	7 N	8 O	9 F	10 Ne
11 Na	12 Mg											13 Al	14 Si	15 P	16 S	17 Cl	18 Ar
19 K	20 Ca	21 Sc	22 Ti	23 V	24 Cr	25 Mn	26 Fe	27 Co	28 Ni	29 Cu	30 Zn	31 Ga	32 Ge	33 As	34 Se	35 Br	36 Kr
37 Rb	38 Sr	39 Y	40 Zr	41 Nb	42 Mo	43 Tc	44 Ru	45 Rh	46 Pd	47 Ag	48 Cd	49 In	50 Sn	51 Sb	52 Te	53 I	54 Xe
55 Cs	56 Ba	57–71 Lanthanoids	72 Hf	73 Ta	74 W	75 Re	76 Os	77 Ir	78 Pt	79 Au	80 Hg	81 Tl	82 Pb	83 Bi	84 Po	85 At	86 Rn
87 Fr	88 Ra	89–103 Actinoids	104 Rf	105 Db	106 Sg	107 Bh	108 Hs	109 Mt	110 Ds	111 Rg	112 Cn	113 Nh	114 Fl	115 Mc	116 Lv	117 Ts	118 Og

Lanthanoids	57 La	58 Ce	59 Pr	60 Nd	61 Pm	62 Sm	63 Eu	64 Gd	65 Tb	66 Dy	67 Ho	68 Er	69 Tm	70 Yb	71 Lu
Actinoids	89 Ac	90 Th	91 Pa	92 U	93 Np	94 Pu	95 Am	96 Cm	97 Bk	98 Cf	99 Es	100 Fm	101 Md	102 No	103 Lr

* From https://iupac.org/what-we-do/periodic-table-of-elements/.

Atomic Numbers and Atomic Weights*

Atomic Number	Name	Symbol	Weight	Atomic Number	Name	Symbol	Weight
1	Hydrogen	H	[1.007 84; 1.008 11]	60	Neodymium	Nd	144.242(3)
2	Helium	He	4.002602(2)	61	Promethium	Pm	
3	Lithium	Li	[6.938; 6.997]	62	Samarium	Sm	150.36(2)
4	Beryllium	Be	9.0121831(5)	63	Europium	Eu	151.964(1)
5	Boron	B	[10.806; 10.821]	64	Gadolinium	Gd	157.25(3)
6	Carbon	C	[12.009 6; 12.011 6]	65	Terbium	Tb	158.925354(8)
7	Nitrogen	N	[14.006 43; 14.007 28]	66	Dysprosium	Dy	162.500(1)
8	Oxygen	O	[15.999 03; 15.999 77]	67	Holmium	Ho	164.930328(7)
9	Fluorine	F	18.998403163(6)	68	Erbium	Er	167.259(3)
10	Neon	Ne	20.1797(6)	69	Thulium	Tm	168.934218(6)
11	Sodium	Na	22.98976928(2)	70	Ytterbium	Yb	173.045(10)
12	Magnesium	Mg	[24.304, 24.307]	71	Lutetium	Lu	174.9668(1)
13	Aluminum	Al	26.9815384(3)	72	Hafnium	Hf	178.486(6)
14	Silicon	Si	[28.084; 28.086]	73	Tantalum	Ta	180.94788(2)
15	Phosphorus	P	30.973761998(5)	74	Tungsten	W	183.84(1)
16	Sulfur	S	[32.059; 32.076]	75	Rhenium	Re	186.207(1)
17	Chlorine	Cl	[35.446; 35.457]	76	Osmium	Os	190.23(3)
18	Argon	Ar	[39.792, 39.963]	77	Iridium	Ir	192.217(2)
19	Potassium	K	39.0983(1)	78	Platinum	Pt	195.084(9)
20	Calcium	Ca	40.078(4)	79	Gold	Au	196.966570(4)
21	Scandium	Sc	44.955908(5)	80	Mercury	Hg	200.592(3)
22	Titanium	Ti	47.867(1)	81	Thallium	Tl	[204.382; 204.385]
23	Vanadium	V	50.9415(1)	82	Lead	Pb	[206.14, 207.94]
24	Chromium	Cr	51.9961(6)	83	Bismuth	Bi	208.98040(1)
25	Manganese	Mn	54.938043(2)	84	Polonium	Po	
26	Iron	Fe	55.845(2)	85	Astatine	At	
27	Cobalt	Co	58.933194(3)	86	Radon	Rn	
28	Nickel	Ni	58.6934(4)	87	Francium	Fr	
29	Copper	Cu	63.546(3)	88	Radium	Ra	
30	Zinc	Zn	65.38(2)	89	Actinium	Ac	
31	Gallium	Ga	69.723(1)	90	Thorium	Th	232.0377(4)
32	Germanium	Ge	72.630(8)	91	Protactinium	Pa	231.03588(1)
33	Arsenic	As	74.921595(6)	92	Uranium	U	238.02891(3)
34	Selenium	Se	78.971(8)	93	Neptunium	Np	
35	Bromine	Br	[79.901, 79.907]	94	Plutonium	Pu	
36	Krypton	Kr	83.798(2)	95	Americium	Am	
37	Rubidium	Rb	85.4678(3)	96	Curium	Cm	
38	Strontium	Sr	87.62(1)	97	Berkelium	Bk	
39	Yttrium	Y	88.90584(1)	98	Californium	Cf	
40	Zirconium	Zr	91.224(2)	99	Einsteinium	Es	
41	Niobium	Nb	92.90637(1)	100	Fermiun	Fm	
42	Molybdenum	Mo	95.95(1)	101	Mendelevium	Md	
43	Technetium	Tc		102	Nobelium	No	
44	Ruthenium	Ru	101.07(2)	103	Lawrencium	Lr	
45	Rhodium	Rh	102.90549(2)	104	Rutherfordium	Rf	
46	Palladium	Pd	106.42(1)	105	Dubnium	Db	
47	Silver	Ag	107.8682(2)	106	Seaborgium	Sg	
48	Cadmium	Cd	112.414(4)	107	Bohrium	Bh	
49	Indium	In	114.818(1)	108	Hassium	Hs	
50	Tin	Sn	118.710(7)	109	Meitnerium	Mt	
51	Antimony	Sb	121.760(1)	110	Darmstadtium	Ds	
52	Tellurium	Te	127.60(3)	111	Roentgenium	Rg	
53	Iodine	I	126.90447(3)	112	Copernicium	Cn	
54	Xenon	Xe	131.293(6)	113	Nihonium	Nh	
55	Cesium	Cs	132.90545196(6)	114	Flerovium	Fl	
56	Barium	Ba	137.327(7)	115	Moscovium	Mc	
57	Lanthanum	La	138.90547(7)	116	Livermorium	Lv	
58	Cerium	Ce	140.116(1)	117	Tennessine	Ts	
59	Praseodymium	Pr	140.90766(1)	118	Oganesson	Og	

* Some weights are listed as intervals ([*a*; *b*]; $a \le$ atomic weight $\le b$) because these weights are not constant but depend on the physical, chemical, and nuclear histories of the samples used. Atomic weights are not listed for some elements because these elements do not have stable isotopes. Exceptions are thorium, protactinium, and uranium. From http://www.ciaaw.org/atomic-weights.htm.

Glossary/Glosario

absolute pressure: Total pressure at a point in a fluid, equal to the sum of the gauge and the atmospheric pressures. (p. 621)

presión absoluta: La presión total en un punto de un fluido, que es igual a la suma de las presiones manométricas y atmosféricas. (pág. 621)

absolute zero: The lowest temperature that is theoretically possible, at which the motion of particles is at a minimum; 0 on the Kelvin scale; –273.15°C or –459.67°F. (p. 687)

cero absoluto: Temperatura más baja teóricamente posible, donde el movimiento de partículas es mínimo; 0 en la escala Kelvin; –273.15°C o –459.67°F. (pág. 687)

absorption line: Dark lines in an otherwise continuous spectrum. These indicate certain wavelengths of light that are absorbed by the atoms of the intervening medium. (p. 1214)

líneas de absorción: Líneas oscuras en un espectro continuo. Indican ciertas longitudes de onda de luz que absorben los átomos del medio interviniente. (pág. 1214)

absorption spectrum: A continuous spectrum, broken by a specific pattern of dark lines or bands, observed when light traverses a particular absorbing medium. (p. 1214)

espectro de absorción: Espectro continuo, interrumpido por un patrón específico de líneas o bandas oscuras, que se observa cuando la luz atraviesa un medio absorbente particular. (pág. 1214)

AC (alternating current): An electric current that reverses its direction many times a second at regular intervals. (p. 882)

AC (corriente alterna): Corriente eléctrica que invierte su dirección muchas veces por segundo a intervalos regulares. (pág. 882)

ac generator: Alternating-current generator; as its coil rotates in a magnetic field, an oscillating emf is generated in the turns of wire that make up the coil. (p. 992)

generador ac: Generador de corriente alterna; a medida que gira su bobina en un campo magnético, se genera un campo electromagnético oscilante en el alambre que forma la bobina. (pág. 992)

acceleration: Acceleration is the time rate of change of the velocity of an object or center of mass of a system. In one dimension, although only one component is needed to describe acceleration, it is still a vector. (p. 45)

aceleración: La aceleración es la tasa de cambio de velocidad de un objeto o centro de masa de un sistema. En una dimensión, si bien solo se necesita un componente para describir la aceleración, sigue siendo un vector. (pág. 45)

acceleration due to gravity: The magnitude of the average acceleration due to gravity near Earth's surface is typically denoted by the symbol *g*. While its value varies slightly with location on Earth's surface, it is approximately 9.8 m/s^2. This is a magnitude, so is always positive. The sign depends on the coordinate system chosen. (p. 71)

aceleración por gravedad: La magnitud de la aceleración promedio por gravedad cerca de la superficie de la Tierra. Se denota con el símbolo *g*. Si bien su valor varía ligeramente según su posición en la superficie de la Tierra, es aproximadamente 9.8 m/s^2. Es una magnitud y por ello siempre es positiva. Su signo depende del sistema de coordenadas elegido. (pág. 71)

adiabatic process: In thermodynamics, a process that occurs without transfer of energy by thermal processes or matter into or out of a system. (p. 734)

proceso adiabático: En termodinámica, proceso que ocurre sin transferencia de energía por procesos térmicos o sin entrada o salida de materia de un sistema. (pág. 734)

alpha decay: Type of radioactive decay in which an atomic nucleus emits an alpha particle (helium nucleus) and thereby transforms into a nucleus with a mass number that is reduced by four and an atomic number that is reduced by two. (p. 1270)

desintegración alfa: Tipo de desintegración radiactiva en la que un núcleo atómico emite una partícula alfa (núcleo de helio) y, por lo tanto, se transforma en un núcleo cuyo número de masa se reduce en cuatro unidades y cuyo número atómico se reduce en dos unidades. (pág. 1270)

alpha particle: A positively charged particle, indistinguishable from a helium nucleus and consisting of two protons and two neutrons. (p. 1270)

partícula alfa: Partícula con carga positiva que no es distinguible de un núcleo de helio y que consta de dos protones y dos neutrones. (pág. 1270)

alternating current (AC): *See* AC (alternating current). (p. 882)

corriente alterna (AC): *Ver* AC (corriente alterna). (pág. 882)

Ampere: The SI unit of electric current, equal to a flow of one coulomb per second. (p. 877)

amperio: La unidad de corriente eléctrica del SI, que equivale a un flujo de un culombio por segundo. (pág. 877)

Ampère's law: For any closed loop path, the circulation of a magnetic field created by an electric current is equal to the size of that electric current times the permeability of free space; discovered by French physicist André-Marie Ampère. (p. 957)

ley de Ampère: En cualquier trayectoria de contorno cerrado, la circulación de un campo magnético creado por una corriente eléctrica es igual al tamaño de esa corriente eléctrica por la permeabilidad del espacio libre; descubierta por el físico francés André-Marie Ampère. (pág. 957)

Amperian loop: An imaginary closed path in space around a current-carrying conductor. (p. 957)

bucle amperiano: Trayectoria cerrada imaginaria en el espacio que rodea a un conductor de corriente. (pág. 957)

amplitude: The maximum displacement of an oscillation, measured from equilibrium. (p. 557)

angular acceleration: The rate at which the angular velocity is changing with respect to time. (p. 470)

angular displacement: The angle through which an object or a point on a rotating extended object moves on a circular path in a given time interval. (p. 451)

angular frequency: A scalar measure of rotation rate; the angular displacement per unit time describing a periodic process (such as an oscillation) expressed as radians per second. (p. 566)

angular magnification: The ratio of the angular size of an object viewed through an optical device (such as a magnifying glass, a microscope, or a telescope) to its angular size viewed with the unaided eye. (p. 1173)

angular momentum: The product of an object's or system's rotational inertia and its angular velocity. A measure of an object's or system's ability to set another object or system into rotational motion. A vector quantity, its direction is determined by a right-hand rule. (p. 506)

angular position: The angle of a line to the point of interest on an extended object, measured from a reference axis. (p. 471)

angular resolution: The angle that separates two objects that are just barely resolved through a circular aperture. (p. 1117)

angular speed: How fast the angular position is changing with respect to time. As long as a system does not change its direction of rotation, angular speed is the magnitude of the angular velocity. (p. 452)

angular velocity: The angular displacement of a point on a rotating extended object divided by the time interval for the point to undergo that displacement. A vector quantity, its direction is determined by a right-hand rule. (p. 452)

angular wave number: The reciprocal of wavelength multiplied by 2π. (p. 1025)

antimatter: Particles with the same properties as ordinary particles, but with the opposite electric charge. (p. 1338)

antineutrino: The antiparticle of a neutrino. (p. 1266)

antinode: Positions along a standing wave at which the oscillation is maximal. (p. 1038)

antiquark: The antiparticle of a quark. (p. 1338)

apparent weight (effective weight): The normal force exerted on an object. It is what a scale placed under the object would measure. (p. 175)

apparent weightlessness (effective weightlessness): The appearance of floating of an object or system when no force of support is needed because the object or system and its surroundings are all in free fall (accelerating at the same rate). (p. 272)

Archimedes' principle: The buoyant force exerted on an object immersed in a fluid by the fluid is equal to the weight of the fluid that the object displaces. (p. 630)

amplitud: Desplazamiento máximo de una oscilación. Se mide desde el punto de equilibrio. (pág. 557)

aceleración angular: Tasa de cambio de velocidad angular con respecto al tiempo. (pág. 470)

desplazamiento angular: Ángulo sobre el cual un objeto o un punto sobre un objeto extendido en rotación se mueve sobre un trayecto circular en un intervalo de tiempo determinado. (pág. 451)

frecuencia angular: Medida escalar de la tasa de rotación; el desplazamiento angular por unidad de tiempo que describe un proceso periódico (como una oscilación) se expresa en radianes por segundo. (pág. 566)

aumento angular: Relación entre el tamaño angular de un objeto al verlo a través de un dispositivo óptico (una lupa, un microscopio o un telescopio) y su tamaño angular a simple vista. (pág. 1173)

momento angular: El producto de la inercia rotacional de un objeto o sistema y su velocidad angular. Medida de la capacidad de un objeto o sistema para generar movimiento rotacional en otro objeto o sistema. Un vector de cantidad cuya dirección está determinada por una regla de la mano derecha. (pág. 506)

posición angular: Ángulo de una línea hasta el punto de interés sobre un objeto extendido, que se mide desde un eje de referencia. (pág. 471)

resolución angular: Ángulo que separa dos objetos que apenas se logran distinguir a través de una apertura circular. (pág. 1117)

velocidad angular: Tasa de cambio de posición angular con respecto al tiempo. Siempre y cuando el sistema no cambie la dirección de su rotación, la velocidad angular es la magnitud de la velocidad angular. (pág. 452)

velocidad angular: Desplazamiento angular de un punto sobre un objeto extendido en rotación dividido por el intervalo de tiempo requerido para que el punto complete ese desplazamiento. Un vector de cantidad cuya dirección está determinada por una regla de la mano derecha. (pág. 452)

número de onda angular: Longitud de onda recíproca multiplicada por 2π. (pág. 1025)

antimateria: Partículas con las mismas propiedades que las partículas ordinarias, pero con carga eléctrica opuesta. (pág. 1338)

antineutrino: Antipartícula de un neutrino. (pág. 1266)

antinodo: Posiciones a lo largo de una onda estacionaria donde la oscilación es máxima. (pág. 1038)

anticuark: Antipartícula de un cuark. (pág. 1338)

peso aparente (peso efectivo): La fuerza normal que se ejerce sobre un objeto. Es lo que se mediría al poner una balanza bajo el objeto. (pág. 175)

ingravidez aparente (ingravidez efectiva): Un objeto o sistema que parece flotar al no requerir una fuerza de apoyo porque el objeto o el sistema y sus alrededores están en caída libre (con la misma tasa de aceleración). (pág. 272)

principio de Arquímedes: La fuerza de flotación ejercida sobre un objeto sumergido en un fluido por éste es igual al peso del fluido que el objeto desplaza. (pág. 630)

atmosphere: A unit of pressure. The average value of atmospheric pressure at sea level; equal to 1.01325×10^5 Pa or about 14.7 pounds per square inch. (p. 613)

atomic number: The number of protons in an atomic nucleus; determines the chemical properties of an element and its place in the periodic table. (p. 1249)

average velocity: Average velocity is a vector quantity, the displacement of an object divided by the time interval for the object to undergo that displacement. (p. 36)

baryon: Any hadron that can be made up of three quarks. (p. 1337)

battery: An electrochemical cell which is designed to use a chemical reaction to produce an electric potential difference. (p. 876)

beat frequency: Rate of the periodic variations of amplitude when two waves of different frequencies interfere. (p. 1051)

beats: Periodic variations in amplitude when two waves of different frequency interfere. (p. 1050)

becquerel: The SI unit of radioactive decay rate, equal to one disintegration per second. (p. 1267)

Bernoulli's equation: An expression of the work-energy theorem, which provides the relationship among pressure, speed, and height in an ideal fluid in motion. (p. 649)

Bernoulli's principle: In a moving fluid, the pressure is lower where the fluid is moving rapidly. (p. 647)

beta decay: A type of radioactive decay in which a beta particle (an electron or a positron) is emitted from an atomic nucleus. (p. 1272)

beta-minus decay: When a neutron decays into a proton, an electron, and an electron antineutrino. (p. 1272)

beta-plus decay: When a proton decays into a neutron, a positron, and an electron neutrino. (p. 1273)

binding energy: The energy required to disassemble an atomic nucleus into its component protons and neutrons. (p. 1255)

blackbody: A system that does not reflect any light at all but absorbs all radiation falling on it. (p. 1201)

blackbody radiation: Light emitted by a perfect blackbody of emissivity = 1. (p. 1202)

Bohr model: Theory of atomic structure in which a small, positively charged nucleus is surrounded by electrons that travel in circular orbits around that nucleus. (p. 1220)

Bohr orbit: In the Bohr model, one of the orbits in which electrons in an atom travel around the nucleus. (p. 1215)

Boltzmann constant: Physical constant in the ideal gas law which has the same value for all gases and relates the average kinetic energy of the particles in a gas to the temperature of the gas; equal to 1.38065×10^{-23} J/K. (p. 690)

atmósfera: Unidad de presión. El valor promedio de la presión atmosférica a nivel del mar; equivale a 1.01325×10^5 Pa o a aproximadamente 14.7 libras por pulgada cuadrada. (pág. 613)

número atómico: Número de protones en un núcleo atómico; determina las propiedades químicas de un elemento y su lugar en la tabla periódica. (pág. 1249)

velocidad promedio: La velocidad promedio es una cantidad vectorial, es el desplazamiento de un objeto dividido por el intervalo de tiempo necesario para que el objeto complete dicho desplazamiento. (pág. 36)

barión: Cualquier hadrón compuesto de tres cuarks. (pág. 1337)

batería: Celda electroquímica diseñada para albergar una reacción que produce una diferencia de potencial eléctrico. (pág. 876)

frecuencia de batimiento: Tasa de variaciones periódicas de amplitud cuando interfieren dos ondas de frecuencia diferente. (pág. 1051)

batimientos: Variaciones periódicas de amplitud cuando interfieren dos ondas de frecuencia diferente. (pág. 1050)

becquerel: Unidad del SI para la tasa de desintegración radiactiva, equivale a una desintegración por segundo. (pág. 1267)

ecuación de Bernoulli: Una expresión del teorema del trabajo y la energía, que muestra la relación entre presión, velocidad y altura de un fluido ideal en movimiento. (pág. 649)

principio de Bernoulli: En un fluido en movimiento, la presión es mas baja donde el fluido se mueve rápidamente. (pág. 647)

desintegración beta: Tipo de desintegración radiactiva en la que una partícula beta (un electrón o positrón) se emite desde un núcleo atómico. (pág. 1272)

desintegración beta menos: Cuando un neutrón se desintegra en un protón, un electrón y un antineutrino electrónico. (pág. 1272)

desintegración beta más: Cuando un neutrón se desintegra en un neutrón, un positrón y un neutrino electrónico. (pág. 1273)

energía de enlace: Energía requerida para descomponer un núcleo atómico en los protones y neutrones que lo componen. (pág. 1255)

cuerpo negro: Sistema que no refleja luz pero absorbe toda la radiación que cae sobre él. (pág. 1201)

radiación de cuerpo negro: Luz emitida por un cuerpo negro perfecto de emisividad = 1. (pág. 1202)

modelo de Bohr: Teoría de la estructura atómica en la que un pequeño núcleo que tiene carga positiva está rodeado de electrones que viajan en órbitas circulares alrededor de él. (pág. 1220)

órbita de Bohr: En el modelo de Bohr, una de las órbitas sobre las cuales los electrones de un átomo viajan alrededor del núcleo. (pág. 1215)

constante de Boltzmann: Constante física en la ley de los gases ideales que tiene el mismo valor para todos los gases y relaciona la energía cinética promedio de las partículas en un gas con la temperatura del gas; es igual a 1.38065×10^{-23} J/K. (pág. 690)

boundary layer: In a fluid, the layer next to a solid surface within which the fluid speed increases from zero at the surface to full speed at the edge of the layer. (p. 641)

Brewster's angle: An angle of incidence at which light with a particular polarization is perfectly transmitted through a transparent dielectric surface, with no reflection. (p. 1094)

British thermal unit: The quantity of energy required to increase the temperature of one pound (1 lb) of pure water from 63°F to 64°F. (p. 704)

buoyant force: The upward force exerted by a fluid on an object placed in it. (p. 630)

calorie: The quantity of energy required to increase the temperature of one gram (1 g) of pure water from 14.5°C to 15.5°C. (p. 704)

capacitance: The ability of a system to store electric charge. (p. 844)

capacitor: A system or device that can store positive and negative charge, consisting of one or more pairs of conductors that may be separated by an insulator or a single isolated conductor separated from ground. (p. 841)

Carnot cycle: An ideal, reversible thermodynamic cycle consisting of two isothermal processes and two adiabatic processes; the most efficient cycle in a heat engine; first proposed in the early 1800s by Sadi Carnot. (p. 756)

Cavendish experiment: Experiment used to determine the value of the gravitational constant. (p. 262)

Celsius scale: The most common temperature scale; based on the work of eighteenth-century Swedish astronomer Anders Celsius. In this scale the freezing point of water is approximately 0°C, and the boiling point is approximately 100°C. (p. 686)

center of curvature: (of a mirror) The center of the sphere defined by the surface of a concave or convex mirror. (p. 1136)

center of mass: A point representing the average position of the mass in a system; the system moves as though all of the mass were concentrated at that point and all external forces exerted on it. (p. 162)

centripetal acceleration: An acceleration that points toward the inside of an object's curving trajectory perpendicular to its instantaneous motion; changes the object's direction, but not its speed. (p. 239)

centripetal force: The force that points toward the inside of an object's curving trajectory and produces a centripetal acceleration. (p. 246)

charged: Possessing a net electric charge. (p. 782)

circuit: The complete path of an electric current, including sources of emf, resistors, or other circuit elements. (p. 877)

circuit branch: A single pathway through a circuit between two junctions. (p. 894)

circuit element: Any component of a circuit, such as a battery, resistor, inductor, or capacitor. (p. 890)

capa límite o fronteriza: En un fluido, la capa junto a una superficie sólida en la cual la velocidad del fluido aumenta desde cero en la superficie hasta la velocidad máxima en el borde de la capa. (pág. 641)

ángulo de Brewster: Ángulo de incidencia en el que la luz con una polarización particular se transmite a través de una superficie dieléctrica, sin reflexión. (pág. 1094)

unidad térmica británica: Cantidad de energía requerida para aumentar la temperatura de una libra (1 lb) de agua purificada de 63°F a 64°F. (pág. 704)

fuerza de flotación: La fuerza ascendente que ejerce un fluido sobre un objeto colocado en él. (pág. 630)

caloría: Cantidad de energía requerida para aumentar la temperatura de un gramo (1 g) de agua purificada de 14.5°C a 15.5°C. (pág. 704)

capacidad eléctrica: Capacidad de un sistema para almacenar una carga eléctrica. (pág. 844)

condensador eléctrico: Sistema o dispositivo que puede almacenar cargas positivas o negativas, y consiste en uno o más pares de conductores que pueden estar separados por un aislador o por un conductor aislado separado de la tierra. (pág. 841)

ciclo de Carnot: Ciclo termodinámico reversible ideal que consta de dos procesos isotérmicos y dos procesos adiabáticos; el ciclo más eficiente en un motor térmico; propuesto por primera vez a principios del siglo XIX por Sadi Carnot. (pág. 756)

experimento de Cavendish: Experimento que se usa para determinar el valor de la constante gravitacional. (pág. 262)

escala Celsius: Escala de temperatura más común; basada en el trabajo del astrónomo sueco del siglo XVIII Anders Celsius. En esta escala, el punto de congelación del agua es de aproximadamente 0°C y el punto de ebullición es de aproximadamente 100°C. (pág. 686)

centro de curvatura: (de un espejo) El centro de la esfera definido por la superficie de un espejo cóncavo o convexo. (pág. 1136)

centro de masa: Punto que representa la posición promedio de la masa en un sistema; el sistema se mueve como si toda la masa estuviera concentrada en ese punto y todas las fuerzas externas se ejercieran sobre él. (pág. 162)

aceleración centrípeta: Aceleración dirigida hacia el interior de la trayectoria curvilínea de un objeto perpendicular a su movimiento instantáneo; cambia la dirección del objeto pero no su velocidad. (pág. 239)

fuerza centrípeta: Fuerza dirigida hacia el interior de la trayectoria curvilínea de un objeto que produce aceleración centrípeta. (pág. 246)

cargado: Que posee una carga eléctrica neta. (pág. 782)

circuito: Trayectoria completa de una corriente eléctrica, incluyendo fuentes de fem, resistencias u otros elementos del circuito. (pág. 877)

rama de un circuito: Trayectoria única de un circuito entre dos empalmes. (pág. 894)

elemento de un circuito: Cualquier componente de un circuito, como una batería, una resistencia, un inductor o un condensador. (pág. 890)

circulation (of a magnetic field): For an Amperian loop, the sum of products of the component of magnetic field parallel to each loop segment multiplied by the segment length along the loop. (p. 956)

circulación (de un campo magnético): En un bucle amperiano, la suma de los productos del componente del campo magnético paralelo a cada segmento del bucle multiplicado por la longitud del segmento a lo largo del bucle. (pág. 956)

closed pipe: A pipe which is open at one end and blocked at the other. (p. 1044)

tubo cerrado: Tubo que está abierto en un extremo y bloqueado en el otro. (pág. 1044)

closed, isolated system: System where no energy is transferred to or from the system and there are no interactions between objects in the system and objects outside of the system. (p. 287)

sistema cerrado, aislado: Aquellos sistemas donde no hay transferencia energética, es decir, que no entra ni sale energía y no hay interacciones entre el sistema y los objetos que están fuera del sistema. (pág. 287)

coefficient of kinetic friction (μ_k): The ratio of the force of friction exerted on an object sliding on a surface to the normal force exerted by the surface on the object; depends on the properties of the surfaces in contact. (p. 204)

coeficiente de fricción cinética (μ_k): La relación entre la fuerza de fricción ejercida sobre un objeto que se desliza sobre una superficie y la fuerza normal ejercida por la superficie sobre el objeto; depende de las propiedades de las superficies en contacto. (pág. 204)

coefficient of performance: (of a heat pump or refrigerator) The ratio of useful heating or cooling provided to work required. (p. 761)

coeficiente de rendimiento: (de una bomba térmica o refrigerador) Relación entre la calefacción o refrigeración útil proporcionada y el trabajo que requiere. (pág. 761)

coefficient of rolling friction (μ_r): The ratio of the force of friction to the normal force exerted by a surface on an object. The force of friction resists the motion of an extended object and is exerted at the point of contact on the extended object rolling on a surface. μ_r depends on properties of the surfaces in contact and also on deformation of the surfaces, so is always neglected in pure rolling motion. (p. 208)

coeficiente de rodadura (μ_r): La razón de la fuerza de fricción a la fuerza normal ejercida por una superficie sobre un objeto. La fuerza de fricción se resiste al movimiento de un cuerpo extenso y se ejerce en el punto de contacto con el cuerpo extenso que rueda por la superficie. μ_r depende de las propiedades de las superficies en contacto, así como de la deformación de las superficies, de modo que se puede despreciar en un movimiento de rodadura. (pág. 208)

coefficient of static friction (μ_s): The ratio of the force of friction exerted on an object that is stationary on a surface to the normal force exerted by the surface on the object; depends on the properties of the surfaces in contact. (p. 197)

coeficiente de fricción estática (μ_s): La relación entre la fuerza de fricción ejercida sobre un objeto estacionario sobre una superficie y la fuerza normal que ejerce la superficie sobre el objeto; depende de las propiedades de las superficies en contacto. (pág. 197)

collision: An interaction between objects or systems that occurs over a short period of time and can be described by internal forces that are much larger than the net external force exerted on the objects or systems so that the net external force can be neglected during the interaction. In a collision, momentum is constant. (p. 386)

choque: La interacción entre dos objetos y sistemas que sucede en un período de tiempo corto y que se describe por fuerzas internas que son mucho más fuertes que la fuerza externa neta ejercida sobre los objetos o sistemas, de modo que la fuerza externa neta es despreciable durante la interacción. En las colisiones, el momento se mantiene constante. (pág. 386)

completely inelastic collision: A collision in which the interacting objects or systems stick together after they collide. Momentum is constant, but any kinetic energy over that required to conserve momentum is converted to internal energy. (p. 401)

choque perfectamente inelástico: Choque en el cual los objetos o sistemas que interactúan se pegan después de chocar. El momento es constante, pero toda la energía cinética que no es necesaria para conservar el momento se convierte en energía interna. (pág. 401)

component: The part of a vector along a single coordinate axis. Components define a rectangle (or rectangular prism in three dimensions) and are a common way to specify a vector. The tail and tip of the vector lie at opposite ends of the rectangle or prism and the components of the vector are the lengths and directions of each side of the rectangle or prism along an axis in the coordinate system. (p. 95)

componente: La parte de un vector a lo largo de un único eje de coordenadas. Las componentes definen un rectángulo (o un prisma rectangular en tres dimensiones) y son una manera común de especificar un vector. La cola y la punta del vector se encuentran en extremos opuestos del rectángulo o prisma y las componentes del vector son las longitudes y las direcciones de cada lado del rectángulo o prisma sobre un eje en el sistema de coordenadas. (pág. 95)

component method: The concept that you can resolve vectors into two or three independent (perpendicular) vectors (for example, in the x and y directions). (p. 95)

método de las componentes: El concepto según el cual es posible resolver vectores en dos o tres vectores (por ejemplo, en las direcciones x e y) independientes (perpendiculares). (pág. 95)

compressible fluid: A fluid that can be easily compacted by squeezing. (p. 606)

fluido compresible: Un fluido que se puede compactar fácilmente al comprimirlo. (pág. 606)

Compton scattering: Elastic scattering of a photon by a free charged particle, usually an electron. (p. 1206)

dispersión de Compton: Dispersión elástica de un fotón por una partícula de carga libre, generalmente un electrón. (pág. 1206)

concave: Curved inward, such as for a mirror or lens surface. (p. 1135)

condensation: The phase change from gas to liquid. (p. 706)

conduction: Energy transfer by the collision of molecules in one system with the molecules in another. (p. 712)

conductor: A substance in which charge can move freely. (p. 787)

conservation: A quantity that is conserved can be transferred between objects or systems, or converted from one form to another, but is neither created or destroyed. (p. 286)

conservation law: A statement that a measurable physical quantity of a system does not change, except through transfers of the quantity, as the system evolves over time. (p. 287)

conservation of angular momentum: If there is no net external torque on a system, the angular momentum of the system is constant. (p. 520)

conservative force: A model for an interaction associated only with the configuration of objects or a system. Examples include the gravitational force or the force exerted by an ideal spring. If the interaction is completely within the system, the interaction can instead be modeled by a potential energy. (p. 316)

constant velocity: Movement at a steady speed in the same direction. (p. 33)

constructive interference: The mutual reinforcement of waves of the same frequency and in phase such that the amplitude of the total wave is the sum of the amplitudes of the individual waves. (p. 1030)

contact forces: Forces describing the interaction of one object or system touching (in contact with) another that are the result of a large number of electric forces between the atoms and molecules in the objects; friction can also depend on surface features. (p. 147)

contact time: The amount of time that colliding objects or systems are in contact and hence the amount of time that they exert forces on each other. (p. 415)

continuity equation: An expression of conservation of mass in fluid dynamics. In a steady flow, the mass of fluid that flows into and out of a region must be the same, thus $A_1v_1 = A_2v_2$ (the product of a pipe's cross-sectional area A and the flow speed has the same value at point 1 as at point 2). (p. 642)

convection: Energy transfer by the motion of a liquid or gas (such as air) caused by the tendency of hotter, less dense material to rise and colder, denser material to sink under the influence of gravity. (p. 712)

converging lens: Lens that brings incoming parallel light rays to a focus at the focal point on the transmitted side of the lens. (p. 1155)

convex: Curved outward, such as for a mirror or lens surface. (p. 1135)

cóncavo: Con curva hacia el interior, por ejemplo en la superficie de un espejo o lente. (pág. 1135)

condensación: Cambio de estado de gas a líquido. (pág. 706)

conducción: Transferencia de energía por la colisión de moléculas de un sistema con las moléculas de otro. (pág. 712)

conductor: Sustancia en la cual las cargas se mueven libremente. (pág. 787)

conservación: Una cantidad que se conserva puede ser transferida entre objetos o sistemas, o se puede convertir de una forma a otra, pero nunca se crea ni se destruye. (pág. 286)

ley de la conservación: Ley según la cual una cantidad física medible de un sistema no cambia, salvo que sea a través de transferencias de la cantidad conforme el sistema evoluciona con el tiempo. (pág. 287)

conservación del momento angular: Si no hay fuerza neta externa sobre un sistema, el movimiento angular del sistema es constante. (pág. 520)

fuerza conservativa: Modelo de una interacción asociada sólo con la configuración de objetos o con un sistema. Algunos ejemplos son la fuerza gravitacional o la fuerza ejercida por un resorte ideal. Si la interacción ocurre completamente dentro del sistema, la interacción puede estar modelada en cambio por una energía potencial. (pág. 316)

velocidad constante: Movimiento con rapidez fija en una misma dirección. (pág. 33)

interferencia constructiva: Refuerzo mutuo de ondas de la misma frecuencia y en fase tal que la amplitud total de la onda es la suma de las amplitudes de las ondas individuales. (pág. 1030)

fuerzas de contacto: Fuerzas que describen la interacción de un objeto o sistema, al tocar (entrar en contacto con) otro objeto como resultado de un gran número de fuerzas eléctricas entre los átomos y moléculas de los objetos; la fricción también varía según las características de la superficie. (pág. 147)

tiempo de contacto: La cantidad de tiempo que los objetos o sistemas en colisión se mantienen en contacto y la cantidad de tiempo durante el cual ejercen fuerzas entre sí. (pág. 415)

ecuación de continuidad: Una expresión de la conservación de la masa en la dinámica de los fluidos. En un flujo constante, la masa de fluido que entra y sale de una región debe ser la misma, asi que $A_1v_1 = A_2v_2$ (el producto del área de la sección transversal A de una tubería y la velocidad del flujo tiene el mismo valor en el punto 1 que en el punto 2). (pág. 642)

convección: Transferencia de energía como consecuencia del movimiento de un líquido o gas (por ejemplo, el aire) causada por la tendencia del material más cálido y menos denso a elevarse mientras que el material más denso y frío se hunde ante la influencia de la gravedad. (pág. 712)

lente convergente: Lente que enfoca los rayos de luz paralelos entrantes en el punto focal en el lado transmitido de la lente. (pág. 1155)

convexo: Con curva hacia el exterior, por ejemplo en la superficie de un espejo o lente. (pág. 1135)

cooling: Energy is transferred out of a system by a thermal process. (p. 700)

coordinate system: A three-dimensional reference system, for example, three perpendicular number lines (axes). For physical quantities (such as position), these axes will have units of measurement as well (such as meters). (p. 34)

coulomb: The unit of electric charge, equal to the amount of charge conveyed in one second by a current of one ampere; named after eighteenth-century French physicist Charles-Augustin de Coulomb, who uncovered the fundamental law that governs the interaction of charged objects. (p. 784)

Coulomb's constant: A proportionality constant used to determine the magnitude of the electric forces between two point charges; equal to $8.99 \times 10^9\ \text{N} \cdot \text{m}^2/\text{C}^2$. (p. 790)

Coulomb's law: The magnitude of the force of electrostatic attraction or repulsion exerted by one electrically charged object on another is directly proportional to the product of their charges and inversely proportional to the square of the distance between them. (p. 790)

critical angle: The angle of incidence beyond which total internal reflection occurs. (p. 1084)

critical point: A point on a phase diagram at which both the liquid and gas phases of a substance have the same density and are therefore indistinguishable. (p. 710)

cross product (vector product): The product of two vectors in three dimensions that is itself a vector at right angles to both the original vectors, with a direction given by the right-hand rule and a magnitude equal to the product of the magnitudes of the original vectors and the sine of the angle between their directions. (p. 481)

current: The rate at which charge moves past any point in a circuit. (p. 877)

current loop: A single loop that carries a current provided by a source of emf. (p. 950)

cycle: A complete series of occurrences that repeats. (p. 553)

DC (direct current): An electric current that does not change direction in a circuit. (p. 882)

de Broglie wavelength: The wavelength of an object, given by Planck's constant divided by the momentum of the object. (p. 1209)

decay constant: Proportionality between the size of a population of radioactive nuclei and the rate at which the population decreases because of radioactive decay. (p. 1266)

degree of freedom: In thermodynamics, a possible form of motion of an object. (p. 697)

density: The mass of a substance divided by the volume that it occupies. (p. 606)

deposition: The phase change from gas to solid. (p. 706)

destructive interference: The partial or complete cancelation of two waves of the same frequency that are 180° out of phase such that the amplitude of the total wave is the difference of the amplitudes of the individual waves. (p. 1031)

enfriamiento: Energía que se transmite hacia el exterior de un sistema por un proceso termal. (pág. 700)

sistema de coordenadas: Sistema de referencia tridimensional, por ejemplo, tres líneas numéricas perpendiculares (ejes). Para cantidades físicas (por ejemplo, posición), estos ejes tendrán también unidades de medida (por ejemplo, metros). (pág. 34)

culombio: Unidad de carga eléctrica, equivalente a la cantidad de carga que transmite una corriente de un amperio en un segundo; su nombre viene del físico francés Charles-Augustin de Coulomb, que descubrió la ley fundamental que gobierna la interacción de objetos con carga. (pág. 784)

constante de Coulomb: Constante de proporcionalidad que determina la magnitud de fuerzas eléctricas entre dos cargas puntuales; es igual a $8.99 \times 10^9\ \text{N} \cdot \text{m}^2/\text{C}^2$. (pág. 790)

ley de Coulomb: La magnitud de la fuerza de la atracción electrostática o repulsión ejercida por un objeto con carga eléctrica sobre otro es directamente proporcional al producto de sus cargas e inversamente proporcional al cuadrado de la distancia entre ellos. (pág. 790)

ángulo crítico: Ángulo de incidencia más allá del cual ocurre la reflexión interna total. (pág. 1084)

punto crítico: Punto en un diagrama de fase en el que la fase líquida y gaseosa de una sustancia tienen la misma densidad y, por lo tanto, son indistinguibles. (pág. 710)

producto cruz (producto vectorial): El producto de dos vectores en tres dimensiones, que es en sí un vector en ángulo recto con los dos vectores originales, cuya dirección es dada por la regla de la mano derecha y cuya magnitud es igual al producto de las magnitudes de los vectores originales y el seno del ángulo entre sus direcciones. (pág. 481)

corriente: La velocidad a la que una carga pasa por cualquier punto de un circuito. (pág. 877)

bucle de corriente Bucle sencillo que carga una corriente proporcionada por una fuente de fem. (pág. 950)

ciclo: Serie completa de sucesos que se repite. (pág. 553)

CC (corriente continua): Corriente eléctrica que no cambia de dirección en un circuito. (pág. 882)

longitud de onda de Broglie: Longitud de onda de un objeto, proporcionada por la constante de Planck dividida por el momento del objeto. (pág. 1209)

constante de desintegración: Proporción entre el tamaño de una población de núcleos radiactivos y la tasa a la que la población disminuye debido a la desintegración radiactiva. (pág. 1266)

grado de libertad: En la termodinámica, una forma posible de movimiento de un objeto. (pág. 697)

densidad: La masa de una sustancia dividida por el volumen que ocupa. (pág. 606)

deposición: Cambio de fase de gas a sólido. (pág. 706)

interferencia destructiva: Cancelación parcial o completa de dos ondas de la misma frecuencia que están desfasadas 180°, de tal manera que la amplitud total de la onda es la diferencia de las amplitudes de las ondas individuales. (pág. 1031)

diamagnetic: A material that tends to become magnetized in a direction opposite to that of the applied magnetic field. (p. 962)

diamagnetismo: Material con tendencia a ser magnetizado en dirección opuesta a la dirección del campo magnético aplicado. (pág. 962)

dielectric: A material that is both an insulator and polarizable. (p. 857)

dieléctrico: Material que es tanto aislante como polarizable. (pág. 857)

dielectric constant: A measure of polarizability of a material. The greater the dielectric constant of a material, the more the material is polarized when it is placed in an electric field. (p. 859)

constante dieléctrica: Medida de la polarizabilidad de un material. Entre mayor sea la constante dieléctrica de un material, más se polariza el material cuando es colocado en un campo eléctrico. (pág. 859)

diffraction: The bending of light around an obstacle or aperture. (p. 1109)

difracción: Curvatura de la luz alrededor de un obstáculo o apertura. (pág. 1109)

diffraction maxima: Locations in a diffraction pattern where waves interfere constructively, producing a bright fringe. (p. 1111)

difracción máxima: Ubicaciones en un patrón de difracción donde las ondas interfieren de manera constructiva y producen una franja brillante. (pág. 1111)

diffraction minima: Locations in a diffraction pattern where waves interfere destructively, producing a dark fringe. (p. 1111)

difracción mínima: Ubicaciones en un patrón de difracción donde las ondas interfieren destructivamente y producen una franja oscura. (pág. 1111)

diffraction pattern: The distinctive pattern of bright and dark fringes caused when light is diffracted through a slit or aperture. (p. 1111)

patrón de difracción: Patrón distintivo de franjas oscuras y brillantes que se produce cuando se difracta la luz por una ranura o apertura. (pág. 1111)

diffuse reflection: Reflection of light from a surface that is not perfectly smooth, resulting in incident light bending in many random directions. (p. 1131)

reflexión difusa: Reflexión de luz de una superficie que no es perfectamente lisa, lo cual resulta en que la luz incidente se desvíe en muchas direcciones aleatorias. (pág. 1131)

dimension: A measure of a physical quantity. (p. 22)

dimensión: Una medida de una cantidad física. (pág. 22)

dimensional analysis: The process of ensuring that in an equation the dimensions are the same on both sides. (p. 22)

análisis dimensional: El proceso de asegurarse de que, en una ecuación, las dimensiones sean iguales de ambos lados. (pág. 22)

diopter: A unit of optical power for a lens or mirror that is equal to the reciprocal of the focal length (in meters). (p. 1161)

dioptría: Unidad de potencia óptica de un lente o espejo que es igual al recíproco de la distancia focal (en metros). (pág. 1161)

direct current (DC): *See* DC. (p. 882)

corriente continua (CC): *Ver* CC. (pág. 882)

direction: Information needed to define a vector. The direction of the vector is from its tail to its tip, expressed in terms of the coordinate system chosen. For one-dimensional motion, direction is completely specified by + or –. (p. 34)

dirección: Información necesaria para definir un vector. La dirección del vector se mide de punta a punta y se expresa en términos del sistema de coordenadas elegido. Para el movimiento unidimensional, la dirección se especifica con + o –. (pág. 34)

dispersion: The separation of light according to wavelength due to differing propagation speeds of different wavelengths of light; a prism separates white light into its component colors because light of each color travels at a different speed through the material. (p. 1088)

dispersión: Separación de la luz por longitud de onda debida a las diferentes tasas de propagación de las diferentes longitudes de onda de la luz; un prisma separa la luz blanca en los colores que la componen porque la luz de cada color atraviesa el material a una velocidad diferente. (pág. 1088)

displacement: The change in the position of an object, independent of the path taken between where it starts and ends. In working in one dimension, we use just a single component, but displacement is always a vector. (p. 34)

desplazamiento: El cambio de posición de un objeto, independientemente del trayecto recorrido entre el punto de inicio y el final. Al trabajar en una dimensión, solo usamos una componente, pero el desplazamiento es siempre un vector. (pág. 34)

displacement current: A quantity appearing in Maxwell's equations that is defined in terms of the rate of change of the electric flux and creates a magnetic field in the same way as a conventional current. (p. 1001)

corriente de desplazamiento: Cantidad que aparece en las ecuaciones de Maxwell que se define en términos de la tasa de cambio del flujo eléctrico y crea un campo magnético de la misma manera que una corriente convencional. (pág. 1001)

distance: The length of a specific path between two points in space. (p. 35)

distancia: La longitud de un trayecto específico entre dos puntos en el espacio. (pág. 35)

diverging lens: Lens that causes incoming parallel light rays to spread away from the principal axis on the transmitted side of the lens as if they had come from the focal point on the incident side of the lens. (p. 1155)

lente divergente: Lente a través de la cual los rayos de luz paralelos entrantes se separan del eje principal del lado transmitido de la lente como si se originaran en el punto focal del lado incidente de la lente. (pág. 1155)

Doppler effect: The change in frequency of a wave caused by an observer moving relative to its source or its source moving relative to the observer. (p. 1053)

efecto Doppler: Cambio de frecuencia de una onda causado por un observador que se mueve en relación con la fuente de la onda, o cuando una fuente se mueve en relación al observador. (pág. 1053)

double-slit interference: The pattern of bright and dark fringes resulting when monochromatic light illuminates a pair of closely spaced narrow slits. (p. 1105)

interferencia de doble rendija: Patrón de franjas brillantes y oscuras que resultan cuando la luz monocromática ilumina un par de ranuras estrechas cercanas. (pág. 1105)

drag force: Force that resists the motion of an object through a liquid such as water or a gas such as air (p. 218)

fuerza de arrastre: Fuerza que resiste el movimiento de un objeto a través de un líquido, por ejemplo, el agua, o un gas, como el aire. (pág. 218)

drift speed: The average speed at which charge carriers move through a conductor. (p. 879)

velocidad de deriva: Velocidad promedio a la que los portadores de carga se mueven por un conductor. (pág. 879)

eccentricity: Parameter determining the circularity of an ellipse; a perfect circle has zero eccentricity. (p. 532)

excentricidad: Parámetro que determina cuánto se desvía una elipse de una circunferencia; cuando un círculo es perfecto, su excentricidad es cero. (pág. 532)

effective weight (apparent weight): *See* apparent weight. (p. 175)

peso efectivo (peso aparente): *Ver* peso aparente. (pág. 175)

effective weightlessness (apparent weightlessness): *See* apparent weightlessness. (p. 272)

ingravidez efectiva (ingravidez aparente): *Ver* ingravidez aparente. (pág. 272)

efficiency: The useful work done by an engine in one cycle divided by the amount of energy transferred to the engine during the cycle to allow it to do that amount of work. (p. 754)

eficiencia: El trabajo útil llevado a cabo por un motor en un ciclo que se divide por la cantidad de energía transmitida al motor durante el ciclo para permitir que realice esa cantidad de trabajo. (pág. 754)

elastic collision: A collision in which the interacting objects or systems undergo a conservative interaction. Momentum and mechanical energy are constant. (p. 398)

choque elástico: Choque en el cual los objetos o sistemas interactúan de manera conservativa. El momento y la energía mecánica se mantienen constantes. (pág. 398)

electric charge: The fundamental property of an object or system that determines how it interacts with other objects or systems containing charge. Electric charge is conserved. (p. 782)

carga eléctrica: Propiedad fundamental de un objeto o sistema que determina cómo interactúa con otros objetos o sistemas que contienen cargas. La carga eléctrica se conserva. (pág. 782)

electric dipole: A combination of two point charges of the same magnitude but opposite signs separated by a small distance. (p. 802)

dipolo eléctrico: Combinación de dos o más cargas de la misma magnitud, pero de signos opuestos separadas por una distancia corta. (pág. 802)

electric field lines: Lines indicating the direction and magnitude of an electric field. (p. 802)

líneas de campo eléctrico: Líneas que indican la dirección y magnitud de un campo eléctrico. (pág. 802)

electric force: The force between electrically charged objects. (p. 783)

fuerza eléctrica: Fuerza entre objetos con carga eléctrica. (pág. 783)

electric potential: The electric potential energy when a charged object is at a given position in a system of charged objects or an electric field divided by the value of the charge at that position. (p. 829)

potencial eléctrico: Energía eléctrica potencial de un objeto cuando está en una posición dada dentro de un sistema de objetos cargados o un campo eléctrico dividido por el valor de la carga en esa posición. (pág. 829)

electric potential difference: The difference in electric potential between two locations. (p. 830)

diferencia de potencial eléctrico: Diferencia de potencial eléctrico entre dos ubicaciones. (pág. 830)

electric potential energy: Potential energy due to the relative positions of a configuration of charged objects; the equivalent of the work required to assemble the charged objects in that system into this configuration from initial positions at infinity. (p. 820)

energía potencial eléctrica: Energía potencial debida a las posiciónes relativa de una configuración de objetos con carga; el equivalente al trabajo requerido para ensamblar los objetos cargados de ese sistema en esta configuración desde posiciones iniciales en el infinito. (pág. 820)

electromagnetic induction: The process whereby a changing magnetic field induces an electric field. (p. 979)

inducción electromagnética: Proceso a través del cual un campo magnético cambiante induce un campo eléctrico. (pág. 979)

electromagnetic spectrum: The range of electromagnetic waves according to wavelength. (p. 1071)

espectro electromagnético: Rango de ondas electromagnéticas de acuerdo a su longitud de onda. (pág. 1071)

electromagnetic wave: Wave that propagates by simultaneous periodic variations of electric and magnetic fields. These include radio waves, infrared, visible light, ultraviolet, x-rays, and gamma rays. (p. 1068)

onda electromagnética: Onda que se propaga a través de variaciones periódicas simultáneas de campos eléctricos y magnéticos. Estas incluyen ondas de radio, infrarrojas, de luz visible, ultravioleta, rayos x y rayos gama. (pág. 1068)

electromagnetism: An umbrella term that covers both electricity and magnetism because both involve interactions between charged objects. A fundamental force; when relativistic effects are considered, the electric and magnetic forces are shown to both be manifestations of the electromagnetic force. (p. 936)

electromagnetismo: Término general que cubre tanto electricidad como magnetismo porque ambos involucran interacciones entre objetos con carga. Una fuerza fundamental; cuando se consideran los efectos relativistas, se demuestra que tanto las fuerzas eléctricas como las magnéticas son manifestaciones de la fuerza electromagnética. (pág. 936)

electron volt: A unit of energy equal to the work done on an electron in accelerating it through a potential difference of one volt. (p. 834)

electronvoltio: Unidad de energía que equivale al trabajo llevado a cabo sobre un electrón al acelerarlo a través de una diferencia potencial de un voltio. (pág. 834)

emission lines: Bright lines in the emission spectrum of a gas. (p. 1214)

líneas de emisión: Líneas brillantes en el espectro de emisión de un gas. (pág. 1214)

emission spectrum: A spectrum that consists only of specific emitted wavelengths. (p. 1214)

espectro de emisión: Espectro que consiste únicamente de una longitud de onda emitida específica. (pág. 1214)

emissivity: How well or how poorly a surface radiates. (p. 713)

emisividad: Qué tan bien o mal emite radiaciones una superficie. (pág. 713)

energy: A scalar quantity used to measure the state (that is, speed, temperature, configuration) of an object or system or, equivalently, its capacity to cause motion. (p. 287)

energía: La cantidad escalar que se usa para medir el estado (es decir, velocidad, temperatura, configuración) de un objeto o sistema, o, de igual manera, su capacidad para causar movimiento. (pág. 287)

energy level: Quantized value of energy, for example, of electrons in an atom. (p. 1217)

nivel de energía: Valor cuantificado de energía, por ejemplo, de electrones en un átomo. (pág. 1217)

entropy: A measure of the amount of disorder of a system. (p. 764)

entropía: Medida de la cantidad de desorden en un sistema. (pág. 764)

equation of hydrostatic equilibrium: The equation $p = p_0 + \rho g d$, which must be satisfied for a fluid to remain at rest. (p. 618)

ecuación de equilibrio hidrostático: La ecuación $p = p_0 + \rho g d$, que debe cumplirse para que un fluido permanezca en reposo. (pág. 618)

equation of state: A relationship among the quantities of pressure, volume, and temperature of a system. (p. 713)

ecuación de estado: Relación entre las cantidades de presión, volumen y temperatura en un sistema. (pág. 713)

equilibrium: State in which the net external force on an object is zero. (p. 158)

equilibrio: Estado en el cual la fuerza externa neta sobre un objeto es cero. (pág. 158)

equipartition theorem: Principle stating that the energy of a molecule is shared equally among each degree of freedom. (p. 697)

teorema de equipartición: Principio que establece que la energía de una molécula se reparte en partes iguales entre cada grado de libertad. (pág. 697)

equipotential: A curve or line along which the electric potential has the same value at all points. (p. 837)

equipotencial: Curva o linea a lo largo de la cual el potencial eléctrico tiene el mismo valor en todos los puntos. (pág. 837)

equipotential surface: A surface on which the electric potential has the same value at all points. (p. 837)

superficie equipotencial: Superficie en la cual el potencial eléctrico tiene el mismo valor en todos los puntos. (pág. 837)

equipotential volume: Space in which the electric potential has the same value everywhere. The inside of a conductor in equilibrium is always an equipotential volume. (p. 838)

volumen equipotencial: Espacio en el cual el potencial eléctrico tiene el mismo valor en todas partes. El interior de un conductor en equilibrio siempre es un volumen equipotencial. (pág. 838)

equivalent capacitance: Effective capacitance of an arrangement of two or more connected capacitors. (p. 851)

capacidad equivalente: Capacidad efectiva de dos o más condensadores conectados. (pág. 851)

equivalent resistance: Effective resistance of an arrangement of two or more connected resistors. (p. 893)

resistencia equivalente: Capacidad efectiva de dos o más resistencias conectadas. (pág. 893)

escape speed: The speed at which a projectile must be launched from a planet, moon, or other body in space so that it never falls back to that body. (p. 365)

velocidad de escape: La velocidad que necesita un proyectil para ser lanzado de un planeta, luna u otro cuerpo espacial y nunca volver a caer sobre ese cuerpo. (pág. 365)

event: Something that happens at a certain point in time. (p. 1300)

evento: Algo que sucede en cierto punto del tiempo. (pág. 1300)

exchange particle: A virtual particle that interacts with ordinary particles to mediate forces, producing the effects of attraction and repulsion. (p. 1345)

partícula de intercambio: Partícula virtual que interactúa con partículas ordinarias para mediar fuerzas, de tal manera que produce los efectos de atracción y repulsión. (pág. 1345)

explosive collision: A collision in which internal energy is converted to mechanical energy during the interaction; momentum is constant. (p. 399)

choque explosivo: Un choque donde la energía interna se convierte en energía mecánica durante la interacción; el momento es constante. (pág. 399)

exponent: A quantity representing the power to which a given number or expression is to be raised, usually expressed as a raised symbol beside the number or expression. (p. 13)

exponente: Cantidad que representa el poder al cual un número o expresión debe ser elevado. Normalmente se expresa como un símbolo elevado al lado del número o expresión. (pág. 13)

exponential function: The irrational number *e* raised to a power. (p. 914)

función exponencial: Número irracional *e* elevado a una potencia. (pág. 914)

extended free-body diagram (force diagram): A free-body diagram modified to show where on an extended object—with respect to the rotational axis—the forces are exerted. This type of diagram aids in the calculation of torque in cases where there is rotational motion. (p. 487)

diagrama de cuerpo libre extenso (diagrama de fuerzas): Un diagrama de cuerpo libre modificado para mostrar el punto en un cuerpo extenso—con respecto al eje de rotación—donde se ejercen las fuerzas, para ayudar al cálculo de torque en aquellos casos donde hay movimiento rotacional. (pág. 487)

extended object: A system made up of connected objects that may change position with respect to one another when the system moves. (p. 449)

objeto extendido: Sistema de objetos conectados que pueden cambiar de posición entre sí cuando se mueva el sistema. (pág. 449)

external forces: Forces exerted on an object in a system by objects outside of the system. For an object not in a system, all other objects or systems are external and so all forces on it are external. (p. 148)

fuerzas externas: Fuerzas ejercidas sobre un objeto en un sistema por objetos externos al sistema. Para objetos que no están en un sistema, todos los demás objetos o sistemas son externos y, por tanto, todas las fuerzas en los mismos son externas. (pág. 148)

eyepiece: A magnifying glass used to give an enlarged view of the image produced by the objective lens of a microscope or telescope. (p. 1174)

ocular: Lupa que da una perspectiva amplificada de la imagen producida por la lente objetivo de un microscopio o telescopio. (pág. 1174)

Fahrenheit scale: The official temperature scale used in the United States and its territories, the Bahamas, Palau, Belize, the Cayman Islands, the Federated States of Micronesia, and the Marshall Islands; originated by the Dutch-German-Polish physicist Daniel Gabriel Fahrenheit in 1724. On this scale water freezes at 32°F and boils at 212°F. (p. 686)

escala Fahrenheit: Escala oficial de temperatura que se usa en los Estados Unidos y sus territorios, las Bahamas, Palau, Belice, las Islas Caimán, la Federación de Estados de Micronesia y las Islas Marshall; fue originada por el físico holandés-alemán-polaco Daniel Gabriel Fahrenheit en 1724. En esta escala el agua se congela a 32°F y hierve a 212°F. (pág. 686)

farad: The SI unit of electrical capacitance, equal to the capacitance of a capacitor in which one coulomb of charge causes a potential difference of one volt; named after English physicist and chemist Michael Faraday. (p. 844)

faradio: Unidad de capacidad eléctrica del sistema internacional, que equivale a la capacidad de un condensador en el cual un culombio de carga genera una diferencia potencial de un voltio; recibe su nombre por el físico y químico inglés Michael Faraday. (pág. 844)

Faraday's law (of induction): An emf is induced in a loop if the magnetic flux through that loop changes. (p. 983)

ley de Faraday (de inducción): Se induce una fem en un bucle cuando cambia el flujo magnético del bucle. (pág. 983)

ferromagnetic: A ferromagnetic material has a high susceptibility to magnetization, the strength of which depends on that of the applied magnetizing field and that may persist after removal of the applied field. (p. 962)

ferromagnético: Un material ferromagnético tiene una alta susceptibilidad a la magnetización, cuya fuerza depende de la fuerza del campo magnetizante aplicado, y que puede permanecer después de retirar el campo aplicado. (pág. 962)

field: Associates some quantity with every point in space. When modeling forces with fields, these fields must be vectors, because forces are vectors. Fields can also be used to model scalar quantities such as temperature that vary in space. (p. 264)

campo: Asocia alguna cantidad con cada punto en el espacio. Al modelar fuerzas con campos, estos campos deben ser vectores, porque las fuerzas son vectores. Los campos también sirven para modelar cantidades escalares que varían en el espacio, como la temperatura. (pág. 264)

first law of thermodynamics: The change in the internal energy of a system equals the sum of the energy transferred into the system by thermal processes and the work done on the system during the processes. This is the work-energy theorem for a system, expanded to include heating and cooling. (p. 733)

primera ley de la termodinámica: El cambio en la energía interna de un sistema es igual a la suma de energía transferida al interior del sistema mediante procesos térmicos y el trabajo realizado en el sistema durante el proceso. Este es el teorema de trabajo-energía de un sistema, ampliado para incluir calefacción y enfriamiento. (pág. 733)

fluid: A substance (a gas or a liquid) that can flow because its molecules can move freely with respect to each other. (p. 605)

fluido: Una sustancia (un gas o un líquido) que puede fluir porque sus moléculas se mueven libremente unas respecto a otras. (pág. 605)

fluid resistance: The resistance to motion experienced by an object as it moves through a fluid (*see* drag force). (p. 218)

resistencia fluida: La resistencia al movimiento que experimenta un objeto al atravesar un fluido (*ver* fuerza de arrastre). (pág. 218)

focal length: The distance from the focal point to the center of a mirror or lens. (p. 1136)

focal point: The point along the principal axis of a mirror or lens at which incident rays parallel to the principal axis converge and come to a common focus or from which they appear to diverge for a diverging lens or mirror. (p. 1136)

force: A way to describe interactions of an object or system with other objects and systems. (p. 144)

force diagram (extended free-body diagram): *See* extended free-body diagram. (p. 487)

force pair: In an interaction between objects A and B, the forces exerted by A on B and by B on A are always the same magnitude and in opposite directions. (p. 166)

frame of reference: *See* reference frame. (p. 103)

free fall: Any motion of an object or system where gravity is the only force exerted on it. (p. 69)

free-body diagram: A graphical representation of all of the external forces exerted on an object or the center of mass of a system. (p. 161)

freezing: The phase change from liquid to solid. (p. 706)

frequency (*f*): The number of cycles of an oscillation per unit of time. (p. 556)

fringe: The bright or dark regions in an interference or diffraction pattern. (p. 1104)

front: (of a wave) Wave crest. (p. 1077)

fundamental frequency: The lowest natural frequency of an oscillating system. (p. 1039)

fundamental mode: For an oscillating system, the standing wave mode corresponding to the fundamental frequency. (p. 1039)

fundamental particles: Particles such as electrons and quarks that we believe have no internal structure. (p. 8)

fundamental units: One of a set of unrelated units of measurement, adopted for measurement of a chosen subset of physical quantities. All other units of measurement can be expressed in terms of combinations of fundamental units. (p. 11)

fusion: The change from solid to liquid; melting. (p. 706)

Galilean transformation: Set of equations used to translate between the coordinates of two reference frames which differ only by constant relative motion. (p. 1303)

gamma decay: A radioactive process in which an atomic nucleus loses energy by emitting a gamma ray (photon) without a change in its atomic number or mass number. (p. 1275)

gas: A fluid that expands to fill whatever volume is available to it. (p. 605)

gauge pressure: The amount by which pressure exceeds atmospheric pressure. (p. 621)

Gauss's law for the electric field: The net electric flux through a closed surface (called a Gaussian surface) equals the net charge enclosed by that surface divided by the permittivity. Charges outside the surface have no effect on the net electric flux through the surface; named after nineteenth-century German mathematician and physicist Carl Friedrich Gauss. (p. 995)

distancia focal: Distancia desde el punto focal hasta el centro de un espejo o lente. (pág. 1136)

punto focal: Punto sobre el eje principal de un espejo o lente en el que los rayos incidentes paralelos al eje principal convergen y llegan a un foco común o del cual parecen divergir en una lente o espejo divergente. (pág. 1136)

fuerza: Una manera de describir las interacciones de un objeto o sistema con otros objetos o sistemas. (pág. 144)

diagrama de fuerzas (diagrama de cuerpo libre extenso): *Ver* diagrama de cuerpo libre extenso. (pág. 487)

par de fuerzas: En una interacción entre objetos A y B, las fuerzas ejercidas por A sobre B y por B sobre A siempre tienen la misma magnitud y direcciones opuestas. (pág. 166)

sistema de referencia: *Ver* marco de referencia. (pág. 103)

caída libre: Cualquier movimiento de un objeto o sistema donde la gravedad es la única fuerza ejercida sobre ellos. (pág. 69)

diagrama de cuerpo libre: Representación gráfica de todas las fuerzas externas ejercidas sobre un objeto o en el centro de masa de un sistema. (pág. 161)

congelamiento: Cambio de fase de líquido a sólido. (pág. 706)

frecuencia (*f*): El número de ciclos de una oscilación por unidad de tiempo. (pág. 556)

franja: Regiones oscuras o brillantes en una interferencia o patrón de difracción. (pág. 1104)

frente: (de una onda) Cresta de una onda. (pág. 1077)

frecuencia fundamental: Frecuencia natural más baja de un sistema oscilante. (pág. 1039)

modo fundamental: Para un sistema oscilante, el modo de una onda estacionaria que corresponde a la frecuencia fundamental. (pág. 1039)

partículas fundamentales: Partículas como electrones y quarks que creemos que no tienen estructura interna. (pág. 8)

unidades fundamentales: Una unidad de medida de un conjunto no relacionado, adoptada para medir un subconjunto de cantidades físicas. Todas las demás unidades de medida se expresan en términos de combinaciones de unidades fundamentales. (pág. 11)

fusión: Cambio de sólido a líquido; cuando algo se derrite. (pág. 706)

transformación de Galileo: Conjunto de ecuaciones que se usa para traducir entre las coordenadas de dos puntos de referencias que difieren solo por un movimiento constante relativo. (pág. 1303)

desintegración gamma: Proceso radiactivo donde un núcleo atómico pierde energía al emitir un rayo gamma (fotón) sin un cambio en su número atómico o número de masa. (pág. 1275)

gas: Un fluido que se expande hasta llenar cualquier volumen que tenga disponible. (pág. 605)

presión manométrica: La cantidad en la que la presión supera a la presión atmosférica. (pág. 621)

ley de Gauss del campo eléctrico: El flujo eléctrico neto a través de una superficie (llamada superficie gaussiana) es igual a la carga neta que encierra esa superficie dividida por la permitividad. Los cambios al exterior de la superficie no afectan el flujo eléctrico neto de la superficie; recibe su nombre del matemático y físico alemán del siglo XIX, Carl Friederich Gauss. (pág. 995)

general theory of relativity: Albert Einstein's theory which provides a unified description of gravity as a geometric property of space and time, or spacetime. (p. 1332)

geometric optics: The science of mirrors and lenses. (p. 1131)

global warming: An increase in the global average surface temperature caused by the greenhouse effect. (p. 714)

gluon: A subatomic particle of a class that is thought to bind quarks together. (p. 1345)

gravitational bending of light: Effect in which light passing near a massive object will curve under the influence of the gravitational force exerted by the object. (p. 1333)

gravitational constant: The constant involved in the calculation of gravitational force between two objects; equal to $6.67 \times 10^{-11}\ \text{Nm}^2/\text{kg}^2$, and commonly referred to as the Newtonian constant of gravitation or the universal gravitational constant. (p. 259)

gravitational force: The force of attraction between all objects or systems with mass in the universe; most commonly used in this class for the attraction of Earth's mass for objects near its surface. The gravitational force is one of the four fundamental forces. (p. 146)

gravitational mass: Property of an object that determines the amount of force exerted on the object through a gravitational interaction. (p. 149)

gravitational potential energy: The potential energy stored in the configuration of a system because of a gravitational interaction (*mgh* near the surface of Earth). (p. 318)

gravitational slowing of time: Effect in which gravity influences the rate of a ticking clock; clocks on the ground floor of a building tick more slowly than clocks on the top floor, which are farther from Earth's center. (p. 1335)

gravitational waves: Small variations in the curvature of spacetime that spread away from moving massive objects. (p. 1334)

greenhouse effect: Warming effect in which the atmosphere prevents some of the radiation emitted by Earth's surface from escaping into space. (p. 713)

greenhouse gas: One of several gases in the atmosphere that are transparent to visible light but not to infrared radiation. (p. 713)

hadron: A particle made up of quarks. (p. 1336)

half-life: The time taken for the rate of radioactive decays of a specified isotope to fall to half its original value. (p. 1268)

Hall effect: The creation of a potential difference across a conductor perpendicular to the direction of a current in the conductor due to the force exerted on the charge carriers by an external magnetic field that has a component perpendicular to the direction of motion of the charge carriers. (938)

teoría general de la relatividad: Teoría de Albert Einstein que ofrece una descripción unificada de la gravedad como una propiedad geométrica del espacio y tiempo, o espacio-tiempo. (pág. 1332)

óptica geométrica: Ciencia de espejos y lentes. (pág. 1131)

calentamiento global: Aumento de la temperatura superficial global promedio causado por el efecto invernadero. (pág. 714)

gluon: Partícula subatómica de una clase que se cree que une a los cuarks. (pág. 1345)

curvatura gravitacional de la luz: Efecto en el que un rayo luz que pasa junto a un objeto enorme se curva bajo la influencia de la fuerza gravitacional que ejerce el objeto. (pág. 1333)

constante gravitacional: La constante para el cálculo de la fuerza gravitacional entre dos objetos; es igual a $6.67 \times 10^{-11}\ \text{Nm}^2/\text{kg}^2$, y comúnmente conocida como la constante gravitacional newtoniana o constante de gravitación universal. (pág. 259)

fuerza gravitacional: La fuerza de atracción entre todos los objetos o sistemas en el universo que tienen masa; su uso más común en esta clase es como referencia para la atracción de la masa de la Tierra con los objetos cercanos a su superficie. La fuerza gravitacional es una de las cuatro fuerzas fundamentales. (pág. 146)

masa gravitacional: Propiedad de un objeto que determina la cantidad de fuerza ejercida sobre ese objeto a través de una interacción gravitacional. (pág. 149)

energía potencial gravitacional: La energía potencial almacenada en la configuración de un sistema por causa de una interacción gravitacional (*mgh* cerca de la superficie de la Tierra). (pág. 318)

dilatación gravitacional del tiempo: Efecto en el que la gravedad influye en la velocidad de las manecillas del reloj; las manecillas de los relojes que se encuentran en la planta baja de un edificio giran más lento que los relojes de la planta más alta, que se encuentran más lejos del centro de la Tierra. (pág. 1335)

ondas gravitacionales: Pequeñas variaciones en la curvatura del espacio-tiempo que se dispersan en dirección opuesta de objetos enormes. (pág. 1334)

efecto invernadero: Efecto de calentamiento en el que la atmósfera impide que parte de la radiación de la superficie de la Tierra escape al espacio. (pág. 713)

gas invernadero: Uno de varios gases en la atmósfera que son transparentes con la luz visible pero no con la radiación infrarroja. (pág. 713)

hadrón: Partícula compuesta de cuarks. (pág. 1336)

semivida: Tiempo que toma la tasa de desintegración radiactiva de un isótopo específico en caer a la mitad de su valor original. (pág. 1268)

efecto Hall: Creación de una diferencia de potencial a lo largo de un conductor perpendicular a la dirección de la corriente en un conductor como resultado de la fuerza ejercida sobre los portadores de carga por un campo magnético externo que tiene un componente perpendicular a la dirección del movimiento de los portadores de carga. (pág. 938)

harmonic property: Characteristic of simple harmonic motion in which the angular frequency, period, and frequency of an oscillation are independent of the amplitude; result of the restoring force obeying Hooke's law. (p. 567)

propiedad armónica: Las características de un movimiento armónico simple en el cual la frecuencia angular, período y frecuencia de una oscilación son independientes de la amplitud; es el resultado de la fuerza restauradora que establece la ley de Hooke. (pág. 567)

heat: The amount of energy that is transferred by a thermal process. (p. 700)

calor: Cantidad de energía transferida por un proceso térmico. (pág. 700)

heat engine: A device that converts the energy transfer due to a temperature difference between systems into work. (p. 753)

motor térmico: Dispositivo que convierte la transferencia de energía en trabajo gracias a una diferencia de temperaturas entre dos sistemas. (pág. 753)

heating: Any thermal process in which energy is transferred into a system. (p. 700)

calefacción: Cualquier proceso térmico en el que se transfiere energía a un sistema. (pág. 700)

Heisenberg uncertainty principle for energy and time: The shorter the duration of a phenomenon, the greater the uncertainty in the energy of that phenomenon. This principle is only noticeable in quantum mechanical systems. (p. 1344)

principio de incertidumbre de la energía y el tiempo de Heisenberg: Entre menos dure un fenómeno, mayor será la incertidumbre de la energía en ese fenómeno. Este principio solo se puede observar en sistemas mecánicos cuánticos. (pág. 1344)

hertz (Hz): The SI unit of frequency, equal to one cycle per second. (p. 556)

hercio (Hz): La unidad de frecuencia del Sistema Internacional. Es igual a un ciclo por segundo. (pág. 556)

Higgs field: The theoretical field that gives fundamental particles their mass. (p. 1349)

campo de Higgs: El campo teórico del cual las partículas fundamentales reciben su masa. (pág. 1349)

Higgs particle: Subatomic particle associated with the Higgs field. (p. 1349)

partícula de Higgs: Partícula subatómica asociada con el campo de Higgs. (pág. 1349)

Hooke's law: To stretch or compress an ideal spring (or a real spring by a relatively small amount) the force that must be exerted is directly proportional to the amount of the change in the length of the spring from its relaxed equilibrium length, and in the opposite direction. (p. 224)

ley de Hooke: Para estirar o comprimir ligeramente un resorte ideal (o un resorte real) la fuerza que debe ejercerse es directamente proporcional a la cantidad de cambio de longitud del resorte cuando está en reposo o en equilibrio, y debe ser ejercida en la dirección opuesta. (pág. 224)

Huygens' principle: The wave front at a later time is the superposition of all of the wavelets emitted at the starting time and is tangent to the leading edges of the wavelets. (p. 1077)

principio de Huygens: El frente de onda en un momento posterior es la superposición de todas las ondas pequeñas emitidas en el momento inicial y es tangente a los bordes de las ondas pequeñas que van al frente. (pág. 1077)

hydrostatic equilibrium: State in which a fluid is at rest. (p. 616)

equilibrio hidrostático: El estado en el cual un fluido está en reposo. (pág. 616)

ideal battery: A battery for which internal resistance may be neglected; acts only as a perfect source of emf. (p. 893)

batería ideal: Batería en la cual la resistencia interna es insignificante; actúa únicamente como fuente perfecta de fem. (pág. 893)

ideal gas: A theoretical gas composed of a set of randomly moving, noninteracting molecules. (p. 690)

gas ideal: Gas teórico compuesto de un conjunto de moléculas que se mueven al azar y no interactúan entre sí. (pág. 690)

ideal gas constant: A constant in the ideal gas law, expressed in units of energy per temperature increment per mole; equal to 8.314 J/(mol • K). (p. 690)

constante de los gases ideales: Constante de la ley de gases ideales, expresada en unidades de energía por aumento de temperatura por mol; igual a 8.314 J/(mol • K). (pág. 690)

ideal gas law: Equation of the state of a hypothetical ideal gas. (p. 690)

ley de los gases ideales: Ecuación del estado de un gas ideal hipotético. (pág. 690)

ideal spring: A system that has negligible mass and always exerts a force proportional to the distance that it is stretched or compressed. While few systems are ideal springs overall, many systems behave as ideal springs over small enough extensions or compressions. (p. 224)

resorte ideal: Un sistema que tiene una masa despreciable y siempre ejerce una fuerza proporcional a la distancia que se estira o se comprime. Si bien pocos sistemas son resortes ideales en general, muchos sistemas se comportan como resortes ideales si las extensiones o compresiones son lo suficientemente pequeñas. (pág. 224)

image: An appearance of an object formed by light rays reflected by a mirror or focused by a lens. (p. 1133)

imagen: Aspecto de un objeto formado por rayos de luz reflejados por un espejo o enfocados por una lente. (pág. 1133)

image distance: The distance from a mirror to an object's reflected image, or the distance from a lens to the image. (p. 1133)

distancia de la imagen: Distancia de un espejo a la imagen reflejada del objeto, o la distancia de una lente a la imagen. (pág. 1133)

image height: The height of an object's image reflected in a mirror or focused by a lens. (p. 1133)

altura de la imagen: Altura de la imagen de un objeto reflejado en un espejo o enfocado por una lente. (pág. 1133)

impulse: The transfer of momentum through an external force exerted on a system for a time interval Δt. (p. 415)

impulse-momentum theorem: Momentum is always conserved. Any change in momentum of a system is equal to the transfer of momentum into or out of that system, called impulse. (p. 415)

incident light: Incoming light that strikes a surface. (p. 1076)

incompressible fluid: A fluid for which volume and density change very little when squeezed. (p. 606)

index of refraction: A measure of the speed of light traveling in a medium; equal to the speed of light in a vacuum divided by the speed of light in the medium. (p. 1079)

induced emf: An emf induced around a loop by a changing magnetic flux through the loop. (p. 981)

induced magnetic field: A magnetic field produced by the current in a loop that is caused by an induced emf in that loop. (p. 987)

inelastic collision: A collision in which the interacting objects or systems undergo a nonconservative interaction. Momentum is constant, but some mechanical energy is converted to internal energy. (p. 399)

inertia: Resistance to change in motion. (p. 149)

inertial frame of reference: A frame of reference that does not accelerate. In an inertial reference frame Newton's laws are observed to hold true; in a noninertial frame of reference the frame itself is accelerating, creating the appearance of fictitious forces. (p. 173)

inertial mass: The property of an object or system that determines how its motion changes when it interacts with other objects and systems. This property is sometimes referred to as inertia. (p. 149)

inertial reference frame: A frame of reference attached to an object that does not accelerate. (p. 1301)

instantaneous: When used to describe a quantity, means the value of that quantity at a particular instant in time, where an instant represents a very short time interval, approaching $\Delta t = 0$. (p. 47)

insulator: Substance in which charge carriers are not able to move freely. (p. 787)

intensity: Average wave power per unit area. (p. 1084)

interference: The combination of two or more waves to form a resultant wave in which the displacement is either reinforced or canceled. (p. 1031)

interference maxima: Locations in an interference pattern where waves interfere constructively, producing a bright fringe. (p. 1105)

interference minima: Locations in an interference pattern where waves interfere destructively, producing a dark fringe. (p. 1105)

interference pattern: The pattern that results when two or more waves interfere with each other, showing distinct regions of constructive and destructive interference. (p. 1032)

impulso: La transferencia de impulso a través de una fuerza externa ejercida en un sistema durante un intervalo de tiempo Δt. (pág. 415)

teorema del impulso y el momento: El momento siempre se conserva. Cualquier cambio en el momento de un sistema es equivalente a la transferencia de momento hacia dentro o fuera de ese sistema, lo cual se conoce como impulso. (pág. 415)

luz incidente: Luz entrante que impacta una superficie. (pág. 1076)

fluido incompresible: Un fluido cuyo volumen y densidad cambian muy poco cuando se lo comprime. (pág. 606)

índice de refracción: Medida de la velocidad de la luz que viaja a través de un medio; equivale a la velocidad de la luz en un vacío dividida por la velocidad de la luz en el medio. (pág. 1079)

fem inducida: fem inducida alrededor de un bucle al cambiar el flujo magnético en el bucle. (pág. 981)

campo magnético inducido: Campo magnético producido por la corriente en un bucle que es causada por una fem inducida en ese bucle. (pág. 987)

choque inelástico: Choque en el cual los objetos o sistemas interactúan de manera no conservativa. El momento es constante, pero cierta energía mecánica es convertida en energía interna. (pág. 399)

inercia: Resistencia al cambio en movimiento. (pág. 149)

sistema de referencia inercial: Un sistema de referencia que no acelera. En un sistema de referencia inercial, las leyes de Newton son verdaderas; en un sistema de referencia no inercial, el sistema en sí está en aceleración, lo cual crea la apariencia de fuerzas ficticias. (pág. 173)

masa inercial: La propiedad de un objeto o sistema que determina la manera en que cambia su movimiento cuando interacciona con otros objetos y sistemas. Esta propiedad también es conocida como inercia. (pág. 149)

sistema de referencia inercial: Sistema de referencia unido a un objeto que no acelera. (pág. 1301)

instantáneo: Cuando se usa para describir una cantidad, significa el valor de una cantidad en un instante particular en el tiempo, donde un instante representa un intervalo de tiempo muy corto en el que $\Delta t = 0$. (pág. 47)

aislante: Sustancia en la cual los portadores de carga no se mueven libremente. (pág. 787)

intensidad: Potencia promedio de una onda por unidad de área. (pág. 1084)

interferencia: Combinación de dos o más ondas para formar una onda resultante en la cual el desplazamiento se refuerza o cancela. (pág. 1031)

interferencia máxima: Puntos en un patrón de interferencia donde las ondas interfieren de manera constructiva, de tal manera que se produce una franja brillante. (pág. 1105)

interferencia mínima: Puntos en un patrón de interferencia donde las ondas interfieren de manera destructiva, de tal manera que se produce una franja oscura. (pág. 1105)

patrón de interferencia: Patrón que resulta cuando dos o más ondas interfieren entre sí, de tal manera que muestran distintas regiones de interferencia constructiva y destructiva. (pág. 1032)

internal energy: In mechanics, internal energy is sometimes used to represent all the types of energy internal to a system. In thermodynamics, internal energy specifically refers to the kinetic and potential energies of the individual molecules making up the system. This definition depends on temperature and phase. (p. 730)

internal forces: Interactions within a system. These are negligible for any system that can be modeled as an object, and cannot change the motion of the center of mass of a system. (p. 148)

internal resistance: The resistance to current passing through a real battery. (p. 891)

inverted image: Image (produced by a mirror or lens) that is flipped upside down relative to the corresponding object. (p. 1137)

inviscid flow: Flow of a fluid that is assumed to have no viscosity. (p. 641)

ionizing radiation: Photons with enough energy to dislodge an electron from an atom. (p. 1195)

irreversible process: A process that cannot return both the system and the surroundings to their original conditions. (p. 756)

irrotational flow: Flow in which the speed varies gradually from one part of the fluid to another, with no abrupt changes. (p. 641)

isobaric process: Process in which the pressure of the system remains constant. (p. 734)

isotherm: Contours of constant temperature on a *PV* diagram. (p. 737)

isothermal process: Process in which the temperature of the system remains constant. (p. 734)

isotope: Variants of a particular chemical element that share the same number of protons in the nucleus of each atom but differ in neutron numbers. (p. 1249)

isovolumetric process: Process in which the volume of the system remains constant. (p. 734)

joule: The SI unit of energy; 1 J = (1 N)(1 m). (p. 290)

junction: Points in a circuit where the current splits to follow multiple pathways or currents in multiple pathways come together into one; sometimes referred to as nodes. (p. 894)

kelvin: The SI unit of thermodynamic temperature. A degree kelvin is equal in magnitude to a degree Celsius. (p. 687)

Kelvin scale: Temperature scale based on the relationship between pressure and temperature of low-density gases; first proposed by nineteenth-century Scottish physicist and engineer William Thomson (1st Baron Kelvin). (p. 686)

kilogram (kg): SI unit of mass. (p. 149)

kinematics: The branch of physics concerned with the motion of objects or systems without reference to the causes of that motion. (p. 32)

energía interna: En la mecánica, la energía interna sirve a veces para representar todos los tipos de energía que son internas en un sistema. En termodinámica, la energía interna se refiere específicamente a la energía potencial y cinética de las moléculas individuales que componen el sistema. Esta definición depende de la temperatura y la fase. (pág. 730)

fuerzas internas: Interacciones dentro de un sistema. Son despreciables para cualquier sistema que se modela como objeto y no pueden cambiar el movimiento en el centro de masa de un sistema. (pág. 148)

resistencia interna: Resistencia a la corriente que atraviesa una batería real. (pág. 891)

imagen invertida: Imagen (producida por un espejo o lente) que se voltea boca abajo en relación al objeto correspondiente. (pág. 1137)

flujo no viscoso: El flujo de un fluido que se considera que no tiene viscosidad. (pág. 641)

radiación ionizante: Fotones con suficiente energía para desalojar un electrón de un átomo. (pág. 1195)

proceso irreversible: Proceso que no puede reproducir las condiciones originales de un sistema o sus alrededores. (pág. 756)

flujo irrotacional: El flujo en el que la velocidad varía gradualmente de una parte a otra del fluido, sin cambios bruscos. (pág. 641)

proceso isobárico: Proceso en el que la presión del sistema permanece constante. (pág. 734)

isoterma: Contornos de una temperatura constante en un diagrama *PV*. (pág. 737)

proceso isotérmico: Proceso en el que la temperatura de un sistema permanece constante. (pág. 734)

isótopo: Variantes de un elemento químico particular que comparten el mismo número de protones en el núcleo de cada átomo, pero difieren en número de neutrones. (pág. 1249)

proceso isovolumétrico: Proceso en el cual el volumen del sistema permanece constante. (pág. 734)

joule: La unidad de energía del Sistema Internacional; 1 J = (1 N)(1 m). (pág. 290)

empalme: Puntos en un circuito en el que la corriente se divide para seguir varias trayectorias o donde corrientes en distintas trayectorias se unen; a veces se conocen como nodos. (pág. 894)

kelvin: La unidad del SI para temperatura termodinámica. Un grado kelvin es igual en magnitud a un grado Celsius. (pág. 687)

escala Kelvin: Escala de temperatura basada en la relación entre la presión y la temperatura de gases de baja densidad; originalmente postulada por el físico e ingeniero escocés del siglo IXI, William Thomson (primer barón Kelvin). (pág. 686)

kilogramo (kg): Unidad de masa del Sistema Internacional. (pág. 149)

cinemática: La rama de la física que estudia el movimiento de objetos o sistemas sin hacer referencia a las causas de ese movimiento. (pág. 32)

kinetic energy: The energy that an object has due to its motion. (p. 288)

kinetic friction: An interaction between two surfaces that are moving relative to each other that resists that relative motion; the result of intermolecular forces, and surface roughness and defects, it is always parallel to the surface exerting it at the point of contact. (p. 145)

kinetic friction force: Force used to model the effect of the kinetic friction interaction on an object or system. (p. 147)

Kirchhoff's junction rule: The sum of the currents into a junction equals the sum of the currents out of it; an application of the law of conservation of charge. (p.894)

Kirchhoff's loop rule: The sum of the changes in electric potential around a closed loop in a circuit must equal zero; an application of the law of conservation of energy. (p. 892)

laminar flow: Smooth fluid flow in which each parcel of fluid follows the same trajectory as the one directly in front of it. (p. 639)

latent heat (of fusion or vaporization): The amount of energy per unit mass transferred out of or into a substance to cause the phase of the substance to change. (p. 706)

lateral magnification: The ratio of the height of an image to the height of the corresponding object. (p. 1133)

law: A summary of the relationships between variables for some aspect of the natural world or universe that has been repeatedly tested and verified in accordance with the scientific method, using accepted protocols of observation, measurement, and evaluation of results. (p. 4)

law of areas: A line joining the Sun and a planet sweeps out equal areas in equal intervals of time, regardless of the position of the planet in the orbit (Kepler's second law). (p. 533)

law of conservation of energy: Energy can be converted from one type into another or transferred between objects or systems, but never created or destroyed. (p. 287)

law of conservation of momentum: If the net external force on a system of objects is zero, then the total momentum of the system does not change. Momentum can be transferred between systems, but is never created or destroyed. (p. 380)

law of orbits: The orbit of each planet is an ellipse with the Sun located at one focus of the ellipse (Kepler's first law). (p. 532)

law of periods: The square of the period of a planet's orbit is proportional to the cube of the semimajor axis of the orbit (Kepler's third law). (p. 535)

law of reflection: The angle of the reflected light from a normal to the surface is the same magnitude as the angle of the incident light from the normal. (p. 1076)

law of universal gravitation: Any two objects exert a gravitational force of attraction on each other in a direction along the line joining the objects, with a magnitude proportional to the product of the masses of the objects and inversely proportional to the square of the distance between them. (p. 236)

energía cinética: La energía de un objeto debida a su movimiento. (pág. 288)

fricción cinética: Una interacción entre dos superficies que se mueven una con relación a otra que resiste ese movimiento relativo; es el resultado de fuerzas intermoleculares y de la rugosidad y los defectos de la superficie, y siempre es paralela a la superficie que la ejerce en el punto de contacto. (pág. 145)

fuerza de fricción cinética: Fuerza que modela el efecto de la interacción de la fricción cinética sobre un objeto o sistema. (pág. 147)

regla de nodos de Kirchhoff: La suma de las corrientes que entran en un nodo es igual a la suma de las corrientes que salen de él; una aplicación de la ley de conservación de la carga. (pág.894)

regla de bucles de Kirchhoff: La suma de los cambios en el potencial eléctrico alrededor de un bucle cerrado en un circuito debe ser igual a cero; aplicación de la ley de conservación de la energía. (pág. 892)

flujo laminar: El flujo suave de un fluido en el que cada lámina de fluido sigue la misma trayectoria que la lámina que está directamente enfrente. (pág. 639)

calor latente (de fusión o vaporización): Cantidad de energía por unidad de masa transferida hacia dentro o fuera de una sustancia para que la fase de la sustancia cambie. (pág. 706)

aumento lateral: Relación entre la altura de una imagen y la altura del objeto correspondiente. (pág. 1133)

ley: Resumen de las relaciones entre variables de ciertos aspectos del mundo natural o del universo, que se ha examinado y verificado repetidamente según el método científico, siguiendo protocolos establecidos para la observación, medida y evaluación de los resultados. (pág. 4)

ley de áreas: Una línea entre el Sol y un planeta recorre áreas iguales en intervalos de tiempo iguales sin importar la posición del planeta en la órbita (segunda ley de Kepler). (pág. 533)

ley de conservación de la energía: La energía se convierte de un tipo de energía en otro o se transfiere entre objetos o sistemas, pero nunca se crea ni se destruye. (pág. 287)

ley de conservación del momento: Si la fuerza externa ejercida sobre un sistema de objetos es cero, el momento del sistema no cambia. El momento se puede transferir entre sistemas, pero nunca se crea ni destruye. (pág. 380)

ley de órbitas: La órbita de cada planeta es una elipse en la cual el Sol está ubicado en uno de los focos de la elipse (primera ley de Kepler). (pág. 532)

ley de los períodos: El cuadrado del período de la órbita de un planeta es proporcional al cubo del eje semimayor de la órbita (tercera ley de Kepler). (pág. 535)

ley de la reflexión: El ángulo de la luz reflejada desde una normal a la superficie tiene la misma magnitud que el ángulo de la luz incidente desde la normal. (pág. 1076)

ley de gravitación universal: Dos objetos cualesquiera ejercen una fuerza gravitacional de atracción entre sí en dirección de una línea que une los objetos, con magnitud proporcional al producto de la masa de los objetos e inversamente proporcional al cuadrado de la distancia entre ellos. (pág. 236)

length contraction: The shortening of a system or extended object along the direction of motion; important at speeds close to that of light, predicted by the theory of special relativity. (p. 1317)

lens: A piece of glass or other transparent material with a curved surface for concentrating or dispersing light rays. (p. 1154)

lens equation: Expression relating object distance and image distance to focal length. (p. 1142)

lensmaker's equation: Expression relating the focal length of a lens in air to its index of refraction and the curvature of its surfaces. (p. 1160)

Lenz's law: The direction of the magnetic field induced within a conducting loop opposes the change in magnetic flux that created it; named after Russian physicist Heinrich Lenz. (p. 988)

lepton: A particle not made of quarks and so which does not interact through the strong force. (p. 1336)

lever arm: The perpendicular distance from the rotation axis to the line of action of a force exerted on an extended object. (p. 478)

line of action: An extension of the force vector $\vec{F}$ through the point where the force is exerted on an extended object. (p. 478)

linear mass density: Mass per unit length. (p. 1017)

linear momentum (momentum): The product of an object's mass and its velocity. A measure of an object's ability to set another object in motion. (p. 383)

linear motion: Motion in a line. (p. 32)

linearly polarized (wave): A transverse wave for which the amplitude of the disturbance is along a single direction. An example of a linearly polarized mechanical wave is a wave on a string for which the disturbance is set up by always moving the end up and down, producing a vertically linearly polarized wave. For electromagnetic waves, the direction of the electric field determines the direction of the polarization. (p. 1092)

liquid: A fluid that maintains the same volume regardless of the shape and size of its container. (p. 605)

longitudinal wave: A traveling disturbance or vibration in which the individual parts of the wave medium move in the direction parallel to the direction of wave propagation. (p. 1016)

Lorentz transformation: Set of equations used to translate between the coordinates of two reference frames moving relative to one another, consistent with special relativity. (p. 1319)

Lorentz velocity transformation: Set of equations relating the velocity of an object in two reference frames moving relative to one another, consistent with special relativity. (p. 1323)

Mach cone: The conical pressure wave front produced by an object moving at a speed greater than that of sound. (p. 1056)

contracción de longitud: Reducción de un sistema u objeto extendido sobre la dirección del movimiento; es importante a velocidades cercanas a la velocidad de la luz, como lo predice la teoría de la relatividad especial. (pág. 1317)

lente: Pieza hecha de vidrio u otro material transparente con una superficie curva para concentrar o dispersar rayos de luz. (pág. 1154)

ecuación de la lente: Expresión que relaciona la distancia del objeto y la distancia de la imagen con la distancia focal. (pág. 1142)

ecuación del fabricante de lentes: Expresión que relaciona la distancia focal de una lente en el aire con su índice de refracción y la curvatura de sus superficies. (pág. 1160)

ley de Lenz: La dirección del campo magnético inducido dentro de un bucle conductor es opuesta al cambio en el flujo magnético que lo creó; lleva el nombre del físico ruso Heinrich Lenz. (pág. 988)

leptón: Partícula que no está hecha de cuarks y por lo tanto no experimenta interacción fuerte. (pág. 1336)

brazo de palanca: La distancia perpendicular del eje de rotación a la línea de acción de una fuerza ejercida sobre un objeto extendido. (pág. 478)

línea de acción: Extensión del vector de fuerza $\vec{F}$ a través del punto donde se ejerce la fuerza sobre un objeto extendido. (pág. 478)

densidad lineal de masa: Masa por unidad de longitud. (pág. 1017)

momento lineal (momento): El producto de la masa de un objeto y su velocidad. Es una medida de la capacidad de un objeto para iniciar el movimiento de otro objeto. (pág. 383)

movimiento lineal: Movimiento en línea. (pág. 32)

(onda) polarizada linealmente: Onda transversal para la cual la amplitud de la perturbación es a lo largo de una sola dirección. Un ejemplo de una onda mecánica polarizada linealmente es una onda en una cuerda en la cual se mueve un extremo hacia arriba y abajo para iniciar la perturbación, lo cual produce una onda polarizada linealmente en sentido vertical. En las ondas electromagnéticas, la dirección del campo eléctrico determina la dirección de la polarización. (pág. 1092)

líquido: Un fluido que mantiene el mismo volumen independientemente de la forma y el tamaño del recipiente que lo contiene. (pág. 605)

onda longitudinal: Perturbación o vibración en movimiento en la que las partes individuales del medio ondulatorio se mueven en dirección paralela a la dirección de propagación de la onda. (pág. 1016)

transformación de Lorentz: Conjunto de ecuaciones que se usan para traducir entre las coordenadas de dos marcos de referencia que se mueven entre sí, de manera consistente con la relatividad especial. (pág. 1319)

transformación de la velocidad de Lorentz: Conjunto de ecuaciones que relacionan la velocidad de un objeto en dos marcos de referencia que se mueven entre sí, de manera consistente con la relatividad especial. (pág. 1323)

cono de Mach: Frente de onda de presión cónica producido por un objeto que viaja a una velocidad mayor que la del sonido. (pág. 1056)

Mach number: The ratio of an object's speed to the speed of sound. (p. 1056)

macroscopic: Physical objects that are measurable and can be seen by the naked eye. (p. 196)

magnet: A material or an object that produces a magnetic field. (p. 932)

magnetic dipole: A magnet that produces a magnetic field consistent with equal and opposite magnetic poles (that may be located at the same point in space); the most fundamental is a current loop. The magnetic field caused by a magnetic dipole points away from the magnet's north pole and toward the magnet's south pole. (p. 935)

magnetic energy density: The energy per unit volume stored in a magnetic field, such as in an inductor. (p. 1191)

magnetic flux: The area of a surface multiplied by the component of the magnetic field that is perpendicular to that surface. (p. 982)

magnetic force: Force of attraction or repulsion that arises between magnets, or between electrically charged particles because of their motion. (p. 933)

magnetic poles: Regions near a magnet that have the strongest external magnetic field, which points away from or into the magnet at these points. A dipole magnet has two poles. (p. 934)

magnetic resonance imaging (MRI): A form of medical imaging that measures the response of atomic nuclei in body tissues to radio waves when placed in a strong magnetic field, producing detailed images of internal organs. (p. 1152)

magnetism: The interaction between magnets or between electrically charged particles due to their motion. (p. 933)

magnifying glass: A converging lens used to view a nearby object and produce a virtual image with large angular magnification. (p. 1173)

magnitude: Size. The magnitude of a vector is the "length" of a vector from its tail to its tip, in whatever units are appropriate for the quantity the vector represents (for instance, m for displacement, or m/s for velocity). (p. 34)

mass: A measure of the amount of material in an object or system. (p. 10) A measure of how much matter an object has; SI unit is the kilogram. (p. 149)

mass number: The total number of protons and neutrons in an atomic nucleus. (p. 1249)

mass spectrometer: A device used to determine the masses of individual atoms and molecules. (p. 941)

Maxwell's equations: Four basic equations that describe all electromagnetic phenomena. (p. 995)

Maxwell-Ampère law: Magnetic fields can be generated in two ways: by electrical current and by changing electric fields. (p. 1001)

mean free path: The average distance that a molecule travels from the time at which it collides with one molecule to when it collides with another molecule. (p. 697)

número de Mach: Relación entre la velocidad de un objeto y la velocidad del sonido. (pág. 1056)

macroscópico: Objetos físicos que se pueden medir y que son perceptibles a simple vista. (pág. 196)

imán: Material u objeto que produce un campo magnético. (pág. 932)

dipolo magnético: Imán que produce un campo magnético consistente con polos magnéticos iguales y opuestos (pueden estar en el mismo punto en el espacio); el más fundamental es un bucle de corriente. El campo magnético causado por un dipolo magnético apunta en dirección opuesta del polo norte del imán y hacia el polo sur del imán. (pág. 935)

densidad de energía magnética: Energía por unidad de volumen almacenada en un campo magnético, como en un inductor. (pág. 1191)

flujo magnético: Área de una superficie multiplicada por el componente del campo magnético que es perpendicular a esa superficie. (pág. 982)

fuerza magnética: Fuerza de atracción o repulsión que surge entre imanes o entre partículas con carga eléctrica debido a su movimiento. (pág. 933)

polos magnéticos: Regiones cercanas a un imán que tienen el campo magnético externo más fuerte, el cual apunta en dirección opuesta del imán o hacia el imán en estos puntos. Un imán dipolar tiene dos polos. (pág. 934)

imagen por resonancia magnética (IRM): Forma de tomar imágenes médicas que mide la respuesta de los núcleos atómicos en los tejidos del cuerpo ante las ondas de radio al colocarlas en un fuerte campo magnético, lo cual produce imágenes detalladas de los órganos internos. (pág. 1152)

magnetismo: Interacción entre imanes o partículas con carga eléctrica causada por su movimiento. (pág. 933)

lupa: Lente convergente que se usa para ver un objeto cercano y producir una imagen virtual con gran aumento angular. (pág. 1173)

magnitud: Tamaño. La magnitud de un vector es la "longitud" de un vector de punta a punta, en la unidad que sea apropiada para la cantidad que representa el vector (por ejemplo, m para desplazamiento o m/s para velocidad). (pág. 34)

masa: Medida de la cantidad de material en un objeto o sistema. (pág. 10) Medida de la cantidad de materia en un objeto; la unidad de masa del Sistema Internacional es el kilogramo. (pág. 149)

número de masa: Número total de protones y neutrones en un núcleo atómico. (pág. 1249)

espectrómetro de masas: Dispositivo que determina la masa de átomos y moléculas individuales. (pág. 941)

ecuaciones de Maxwell: Cuatro ecuaciones básicas que describen todos los fenómenos electromagnéticos. (pág. 995)

ley de Maxwell-Ampère: Los campos magnéticos se pueden generar de dos formas: por corriente eléctrica y por cambios en los campos eléctricos. (pág. 1001)

camino libre medio: La distancia promedio que recorre una molécula desde el momento en que choca con una molécula hasta que choca con otra molécula. (pág. 697)

mechanical energy: Energy related to motion, including reversible changes in configuration. (p. 334)

mechanical energy transfer (work): Transfer of energy to or from an object or system through a mechanical process. (p. 335)

mechanical wave: A disturbance that propagates through a material that we call the medium for the wave. The medium can be a solid, liquid, or gas. (p. 1014)

medium: The substance through which a mechanical wave propagates. (p. 1014)

meson: A hadron composed of two quarks. (p. 1338)

Michelson–Morley experiment: An experiment performed in 1887 attempting to detect the velocity of Earth with respect to the hypothetical luminiferous ether; found no evidence for the ether's existence. (p. 1306)

microscope: An optical device used to make enlarged images of small, nearby objects. (p. 1174)

millimeters of mercury: A unit of pressure. The height of a column of mercury in a barometer used to measure pressure; 760 mmHg equals 1 atm. (p. 621)

mirror equation: Expression relating object distance and image distance to the focal length of a mirror. (p. 1142)

model: Conceptual or mathematical models identify the most important characteristics of a phenomenon or system to simplify analysis. (p. 4)

molar specific heat: The energy required to raise the temperature of one mole of substance by one kelvin. (p. 745)

molar specific heat at constant volume: The energy required to raise the temperature of one mole of substance by one kelvin if the volume is held constant. (p. 745)

mole: Unit measuring the quantity of a substance; one mole equals the number of atoms in exactly 12 grams of carbon-12, given by Avogadro's number: 6.022×10^{22}. (p. 690)

momentum (linear momentum): *See* linear momentum. (p. 380)

monochromatic light: Light that has a single definite wavelength. (p. 1097)

motion in a plane: Another name for two-dimensional motion, also sometimes referred to as planar motion. (p. 103)

motional emf: emf generated in a conductor due to its motion relative to a magnetic field. (p. 980)

multiloop circuit: A circuit with more than one complete closed pathway that current can take to return to a point in the circuit. (p. 894)

net external force (net force): The vector sum of all external forces exerted on the object or system of interest. (p. 145)

net force (net external force): *See* net external force. (p. 145)

neutral: Having a net charge of zero; could contain equal amounts of positive and negative charge. (p. 782)

neutrino: A nearly massless, neutral particle. (p. 1264)

neutron number: The number of neutrons in the nucleus of an atom. (p. 1249)

energía mecánica: Energía relacionada con el movimiento, incluyendo cambios de configuración reversibles. (pág. 334)

transferencia de energía mecánica (trabajo): Transferencia de energía hacia o desde un objeto o sistema a través de un proceso mecánico. (pág. 335)

onda mecánica: Perturbación que se propaga a través de un material, que se conoce como el medio de la onda. El medio puede ser un sólido, un líquido o un gas. (pág. 1014)

medio: Sustancia a través de la cual se propaga una onda mecánica. (pág. 1014)

mesón: Hadrón compuesto de dos cuarks. (pág. 1338)

Experimento de Michelson-Morley: Experimento llevado a cabo en 1887 con el fin de detectar la velocidad de la Tierra con respecto al hipotético éter luminífero; no se encontró evidencia de la existencia del éter. (pág. 1306)

microscopio: Dispositivo óptico que amplifica la imagen de objetos pequeños y cercanos. (pág. 1174)

milímetros de mercurio: Unidad de presión. La altura de una columna de mercurio de un barómetro que mide la presión; 760 mmHg equivalen a 1 atm. (pág. 621)

ecuación del espejo: Expresión que relaciona la distancia del objeto y la distancia de la imagen con la distancia focal de un espejo. (pág. 1142)

modelo: Los modelos conceptuales o matemáticos identifican las características más importantes de un fenómeno o sistema para simplificar los análisis. (pág. 4)

calor específico molar: Energía requerida para elevar en un kelvin la temperatura de un mol de sustancia. (pág. 745)

calor específico molar a volumen constante: Energía requerida para elevar la temperatura de un mol de sustancia en un kelvin si el volumen se mantiene constante. (pág. 745)

mol: Unidad de medida de la cantidad de una sustancia; un mol es igual al número de átomos en exactamente 12 gramos de carbono-12, de acuerdo con el número de Avogadro: 6.022×10^{22}. (pág. 690)

momento (momento lineal): *Ver* momento linear. (pág. 380)

luz monocromática: Luz que tiene una longitud de onda única. (pág. 1097)

movimiento en un plano: Nombre alterno para el movimiento en dos dimensiones, a veces conocido como movimiento plano. (pág. 103)

fem de movimiento: fem generada en un conductor debido a su movimiento en relación a un campo magnético. (pág. 980)

circuito de bucles múltiples: Circuito que tiene más de un camino cerrado completo por el que la corriente puede regresar a un punto del circuito. (pág. 894)

fuerza externa neta (fuerza neta): La suma vectorial de todas las fuerzas externas ejercidas sobre el objeto o sistema de interés. (pág. 145)

fuerza neta (fuerza externa neta): *Ver* fuerza externa neta. (pág. 145)

neutral: Tener una carga neta de cero; puede contener cantidades iguales de carga positiva y negativa. (pág. 782)

neutrino: Partícula neutra que casi no tiene masa. (pág. 1264)

número de neutrones: Número de neutrones en el núcleo de un átomo. (pág. 1249)

neutron-induced fission: The radioactive decay of an atomic nucleus initiated by the collision of a neutron. (p. 1259)

newton (N): The SI unit of force; equivalent to the force that would give a one-kilogram object an acceleration of one meter per second per second. (p. 150)

Newton's first law: The law of inertia: An object at rest stays at rest, and an object in motion stays in motion with the same speed and in the same direction, unless a net force is exerted on it. (p. 158)

Newton's laws of motion: Isaac Newton's three fundamental relationships between force and motion. (p. 148)

Newton's second law: If a net external force is exerted on an object, the object accelerates. The net external force is equal to the product of the object's mass and the object's acceleration. This is later refined to be the net external force is equal to the object's change in momentum. (p. 148)

Newton's third law: If object A exerts a force on object B, object B exerts a force on object A that has the same magnitude but is in the opposite direction. (p. 166)

node: (of a wave) Any point where the displacement of a wave is always zero. (p. 1037)

nonconservative force: A force describing a dissipative interaction for which work done between two points depends on the path taken between those points, such as friction. (p. 316)

noninertial frame: A frame of reference moving with an accelerated object. (p. 1301)

normal: A line that is perpendicular to a surface. (p. 951)

normal force: Force that is exerted on any object placed in contact with another solid object or surface; the result of interatomic forces, it is always perpendicular to the surface exerting it at the point of contact. (p. 147)

no-slip condition: Requirement that the velocity of a fluid be zero next to a solid surface. (p. 640)

nuclear fission: A nuclear reaction in which a heavy nucleus splits spontaneously or on impact with another particle, releasing energy. (p. 1256)

nuclear fusion: A nuclear reaction in which atomic nuclei of low atomic number fuse to form a heavier nucleus, releasing energy. (p. 1256)

nuclear radiation: The emission by a nucleus of either energy (in the form of a photon) or particles (such as an alpha particle). (p. 1256)

nucleon: A proton or neutron. (p. 1247)

nuclide: An atomic species characterized by the specific constitution of its nucleus, that is, by its number of protons and its number of neutrons. (p. 1249)

object: Something for which you can use the object model. (p. 8)

fisión inducida por neutrones: Desintegración radiactiva de un núcleo atómico que inicia por la colisión de un neutrón. (pág. 1259)

newton (N): La unidad de fuerza del Sistema Internacional; equivale a la fuerza que causaría una aceleración de un metro por segundo por segundo a un objeto de un kilogramo. (pág. 150)

primera ley de Newton: Ley de la inercia: Un objeto en reposo permanecerá en reposo, y un objeto en movimiento se mantendrá en movimiento con la misma velocidad y en la misma dirección, salvo que se ejerza una fuerza neta sobre él. (pág. 158)

leyes de Newton sobre el movimiento: Las tres relaciones fundamentales de Isaac Newton entre fuerza y movimiento. (pág. 148)

segunda ley de Newton: Si se ejerce una fuerza externa neta sobre un objeto, el objeto acelera. La fuerza externa neta es igual al producto de la masa del objeto y la aceleración del objeto. Posteriormente fue refinada para decir que la fuerza externa neta es igual al cambio de momento del objeto. (pág. 148)

tercera ley de Newton: Si un objeto A ejerce una fuerza sobre un objeto B, el objeto B ejerce una fuerza sobre el objeto A con la misma magnitud pero en dirección opuesta. (pág. 166)

nodo: (de una onda) Cualquier punto donde el desplazamiento de una onda siempre es cero. (pág. 1037)

fuerza no conservativa: Una fuerza que describe una interacción disipativa para la cual el trabajo realizado entre dos puntos depende de la trayectoria tomada entre aquellos puntos, por ejemplo, la fricción. (pág. 316)

marco no inercial: Marco de referencia que se mueve con un objeto acelerado. (pág. 1301)

normal: Línea perpendicular a una superficie. (pág. 951)

fuerza normal: Fuerza que se ejerce sobre cualquier objeto que está en contacto con otro objeto sólido o con una superficie; como resultado de fuerzas interatómicas, siempre es perpendicular a la superficie que la ejerce en el punto de contacto. (pág. 147)

condición antideslizante: El requisito de que la velocidad de un fluido que está junto a una superficie sólida sea cero. (pág. 640)

fisión nuclear: Reacción nuclear en la que un núcleo pesado se divide espontáneamente o al impactar con otra partícula, de tal manera que libera energía. (pág. 1256)

fusión nuclear: Reacción nuclear en la que núcleos atómicos de bajo número atómico se fusionan para formar un núcleo más pesado, de tal manera que se libera energía. (pág. 1256)

radiación nuclear: Emisión por un núcleo de energía (en forma de fotón) o por partículas (como una partícula alfa). (pág. 1256)

nucleón: Un protón o neutrón. (pág. 1247)

nucleido: Especie atómica caracterizada por la constitución específica de su núcleo, es decir, por su número de protones y su número de neutrones. (pág. 1249)

objeto: Algo para lo cual se usa el modelo de objeto. (pág. 8)

object distance: An object's distance from a mirror or lens. (p. 1133)

object height: The vertical extent of an object that is reflected in a mirror or refracted by a lens. (p. 1133)

object model: Using the object model for a system means that, for the purposes of the analysis, you can ignore any internal structure, and that every point in the system moves at the same velocity, so that the motion of the system may be described by the motion of its center of mass. (p. 8)

objective: In a microscope, the converging lens used to produce a magnified real image of a small object. In a telescope, the converging lens or mirror used to produce a small real image of a distant object. (p. 1174)

ohm: The SI unit of electrical resistance; the resistance in a circuit element that carries a current of one ampere when a potential difference of one volt is placed across it. (p. 884)

one-dimensional wave: A wave that propagates along a single dimension of space. (p. 1019)

open pipe: A pipe that is unblocked on both ends. (p. 1047)

optical device: An instrument that changes the direction of light rays in a regular way. (p. 1130)

orbital period: The time required for one object to complete an orbit about another object. The square of the orbital period is proportional to the cubed value of the distance (the length of the semimajor axis of the orbit) between them and inversely proportional to the product of the masses of the orbited and orbiting object. (p. 268)

origin: (also reference position) The center of the coordinate system (where the axes intersect), defined to be the 0 position along each axis. (p. 34)

oscillation: Periodic motion of an object, or disturbance of a medium, around an equilibrium. (p. 552)

parabola: A symmetrical open plane curve formed by the intersection of a cone with a plane parallel to its side. The path of a projectile under the influence of gravity ideally follows a curve of this shape. (p. 117)

parallel capacitors: An arrangement of capacitors connected along multiple paths (not in series), resulting in a multiloop circuit; the total capacitance is equal to the sum of all the individual capacitances. (p. 853)

parallel resistors: An arrangement of resistors connected along multiple paths (not in series), resulting in a multiloop circuit; the reciprocal of the total resistance is equal to the sum of the reciprocals of all the individual resistances. (p. 894)

parallel-axis theorem: Expression determining the rotational inertia of a rigid body about a given axis in terms of the rotational inertia about a parallel axis running through its center of mass and the distance between the two axes. (p. 463)

parallel-plate capacitor: A capacitor made of two parallel conducting plates separated from each other. (p. 841)

distancia del objeto: Distancia de un objeto a un espejo o lente. (pág. 1133)

altura del objeto: Extensión vertical de un objeto reflejado en un espejo o refractado por una lente. (pág. 1133)

modelo de objeto: Usar el modelo de objeto para un sistema significa que, para propósitos del análisis, se puede ignorar cualquier estructura interna y que cada punto del sistema se mueve a la misma velocidad, de modo que el movimiento del sistema se puede describir por el movimiento de su centro de masa. (pág. 8)

objetivo: En un microscopio, la lente convergente que produce una imagen real ampliada de un objeto pequeño. En un telescopio, la lente o espejo convergente que produce una pequeña imagen real de un objeto distante. (pág. 1174)

ohm: Unidad del SI de resistencia eléctrica; la resistencia en un elemento de un circuito que transporta una corriente de un amperio con una diferencia de potencial de un voltio. (pág. 884)

onda unidimensional: Onda que se propaga en una sola dimensión del espacio. (pág. 1019)

tubo abierto: Tubo desbloqueado en ambos extremos. (pág. 1047)

dispositivo óptico: Instrumento que cambia la dirección de los rayos de luz de manera regular. (pág. 1130)

período orbital: El tiempo necesario para que un objeto complete una órbita alrededor de otro objeto. El cuadrado del período orbital es proporcional al cubo del valor de la distancia (la distancia del eje semimayor de la órbita) entre ellos y es inversamente proporcional al producto de las masas del objeto que orbita y del objeto orbitado. (pág. 268)

origen: (también conocido como posición de referencia) Es el centro del sistema de coordenadas (donde los ejes se intersectan), definido como la posición 0 a lo largo de cada eje. (pág. 34)

oscilación: Movimiento periódico de un objeto, o perturbación de un medio, alrededor de un equilibrio. (pág. 552)

parábola: Una curva simétrica de plano abierto formada por la intersección de un cono con un plano paralelo a su lado. La trayectoria de un proyectil bajo la influencia de la gravedad sigue idealmente una curva con esta forma. (pág. 117)

condensadores en paralelo: Disposición de condensadores conectados a lo largo de varios caminos (no en serie), lo que da como resultado un circuito de bucles múltiples; la capacidad total es igual a la suma de todas las capacidades individuales (pág. 853)

resistencias en paralelo: Disposición de resistencias conectadas a lo largo de varios caminos (no en serie), lo que da como resultado un circuito de bucles múltiples; el recíproco de la resistencia total es igual a la suma de los recíprocos de todas las resistencias individuales. (pág. 894)

teorema de los ejes paralelos: Expresión que determina la inercia rotacional de un cuerpo rígido con respecto a un eje determinado en términos de la inercia rotacional sobre un eje paralelo que atraviesa su centro de masa y la distancia entre los dos ejes. (pág. 463)

condensador de placas paralelas: Condensador formado por dos placas conductoras paralelas separadas entre sí. (pág. 841)

paramagnetic: A substance that is weakly attracted by the poles of a magnet but which does not retain any permanent magnetization; if the external magnetic field is removed, random thermal motion will cause the atomic current loops in the substance to return to their original, nonaligned orientations. (p. 962)

paramagnético: Sustancia débilmente atraída por los polos de un imán pero que no retiene una magnetización permanente; si se elimina el campo magnético externo, el movimiento térmico aleatorio hará que los bucles de corriente atómica en la sustancia vuelvan a sus orientaciones originales no alineadas. (pág. 962)

partially polarized light: Light in which the orientation of the electric field changes randomly but is more likely to be in one orientation than in other orientations. (p. 1092)

luz parcialmente polarizada: Luz en la cual la orientación del campo eléctrico cambia de manera aleatoria, aunque es más probable que esté en una orientación que en otras orientaciones. (pág. 1092)

particle physics: The branch of physics concerned with the fundamental constituents of matter and how they interact. (p. 1298)

física de partículas: Rama de la física que se ocupa de los constituyentes fundamentales de la materia y la manera en que interactúan. (pág. 1298)

pascal: The SI unit for pressure, equal to one newton per square meter. (p. 613)

pascal: La unidad del Sistema Internacional para la presión, que es igual a un newton por metro cuadrado. (pág. 613)

Pascal's principle: Pressure applied to a confined, static fluid is transmitted undiminished to every part of the fluid as well as to the walls of the container. (p. 628)

principio de Pascal: La presión aplicada a un fluido estático confinado se transmite equitativamente a cada parte del fluido, así como a las paredes del recipiente. (pág. 628)

path length difference: The difference in the distance traversed by two waves traveling to the same point from different locations. (p. 1032)

diferencia de longitud de camino: Diferencia en la distancia que recorren dos ondas que viajan al mismo punto desde diferentes lugares. (pág. 1032)

Pauli exclusion principle: The quantum mechanical principle that no two electrons may occupy the same quantum state simultaneously. (p. 1229)

principio de exclusión de Pauli: Principio de la mecánica cuántica según el cual dos electrones no pueden ocupar el mismo estado cuántico de manera simultánea. (pág. 1229)

pendula: Systems that oscillate back and forth due to the restoring force of gravity. (p. 582)

péndulos: Sistemas que oscilan hacia adelante y hacia atrás gracias a la fuerza restauradora de la gravedad. (pág. 582)

period (T): The time for one complete cycle of an object in circular motion or for a system to complete one full cycle of an oscillation. (p. 553)

periodo (T): Tiempo para un ciclo completo de un objeto en movimiento circular o para que un sistema complete un ciclo de una oscilación. (pág. 553)

permeability of free space: A constant involved in the relationship between an electric current and the magnetic field that it produces in a vacuum. (p. 956)

permeabilidad del espacio libre: Constante involucrada en la relación entre una corriente eléctrica y el campo magnético que produce en el vacío. (pág. 956)

permittivity of free space: A constant involved in the relationship between an electric charge and the electric field that it produces in a vacuum. (p. 806)

permitividad del espacio libre: Constante involucrada en la relación entre una carga eléctrica y el campo eléctrico que produce en el vacío. (pág. 806)

phase: A physically distinct state of matter: solid, liquid, gas, or plasma. (p. 706)

fase: Estado físicamente distintivo de la materia: sólido, líquido, gas o plasma. (pág. 706)

phase angle: The angular position indicating where in the oscillation cycle a point of interest is at for $t = 0$. (p. 569)

ángulo de fase: La posición angular que indica en qué parte de un ciclo de oscilación un punto de interés es $t = 0$. (pág. 569)

phase change: The transformation from one state of matter to another. (p. 706)

cambio de fase: Transformación de un estado de la materia a otro. (pág. 706)

phase diagram: A graph of pressure P versus temperature T for a substance, showing the values of P and T for each phase of the substance. (p. 710)

diagrama de fase: Gráfico de presión P contra temperatura T de una sustancia, que muestra los valores de P y T en cada fase de la sustancia. (pág. 710)

photoelectric effect: The emission, or ejection, of electrons from the surface of a material in response to incident light. (p. 1198)

efecto fotoeléctrico: Emisión, o expulsión, de electrones de la superficie de un material en respuesta a una luz incidente. (pág. 1198)

photoelectrons: Electrons emitted through the photoelectric effect. (p. 1198)

fotoelectrones: Electrones emitidos a través del efecto fotoeléctrico. (pág. 1198)

photon: The quantum of electromagnetic energy, regarded as a discrete particle having zero mass, no electric charge, and an indefinitely long lifetime. In some models, the photon is the exchange particle of the electromagnetic force. (p. 1194)

fotón: Cuanto de la energía electromagnética, considerado una partícula discreta que tiene masa cero, sin carga eléctrica y una vida útil indefinidamente larga. En algunos modelos, el fotón es la partícula de intercambio de la fuerza electromagnética. (pág. 1194)

physical pendulum: A pendulum whose mass is distributed throughout its volume. (p. 588)

péndulo físico: Un péndulo cuya masa está distribuida a lo largo de su volumen. (pág. 588)

Planck's constant: A physical constant relating the ratio of the energy of a photon to its frequency; equal to 6.626×10^{-34} J · s. (p. 1194)

constante de Planck: Constante física que relaciona la correspondencia entre la energía de un fotón y su frecuencia; es igual a 6.626×10^{-34} J · s. (pág. 1194)

plane mirror: A flat, reflecting surface. (p. 1132)

espejo plano: Superficie plana y reflectante. (pág. 1132)

plate: Piece of metal used in a parallel-plate capacitor to store charge. (p. 841)

placa: Pieza de metal que se usa en condensadores de placas paralelas para almacenar carga. (pág. 841)

point charge: A model that can be used when a charged object is much smaller than the distance between it and other charged objects, no matter how large or small the object. (p. 790)

carga puntual: Modelo que se usa cuando un objeto cargado es mucho más pequeño que la distancia entre él y otros objetos cargados, sin importar el tamaño del objeto. (pág. 790)

polarization: The orientation of the electric field in a light wave. (p. 1092)

polarización: Orientación del campo eléctrico en una onda de luz. (pág. 1092)

polarizing filter: A transparent sheet which contains long-chain molecules that are all oriented in the same direction, used to polarize the light passing through it by absorbing energy from the electric field along the direction of the molecule. (p. 1092)

filtro polarizador: Hoja transparente que contiene moléculas de cadena larga orientadas en la misma dirección, que se utiliza para polarizar la luz que la atraviesa al absorber energía del campo eléctrico en la dirección de la molécula. (pág. 1092)

position: The vector that represents the location of the object or center of mass of a system. In one dimension, although only one component is needed to describe position, it is still a vector. (p. 34)

posición: El vector que representa la ubicación del objeto o centro de masa de un sistema. En una dimensión, si bien solo se necesita una componente para describir la ubicación, sigue siendo un vector. (pág. 34)

positron: A subatomic particle with the same mass as an electron and a numerically equal but positive (opposite) charge. (p. 1264)

positrón: Partícula subatómica con la misma masa que un electrón y cuya carga es numéricamente igual pero positiva (opuesta). (pág. 1264)

potential energy: The energy stored in the configuration of a system because of a conservative interaction. (p. 288)

energía potencial: La energía almacenada en la configuración de un sistema a causa de una interacción conservativa. (pág. 288)

pound-force (lbf): The English unit of force. (p. 150)

libra-fuerza (lbf): Unidad inglesa de fuerza. (pág. 150)

power: The rate at which energy is transferred into or out of a system, or converted from one type of energy to another. The SI unit of power is the joule per second, or watt (abbreviated W): 1 W = 1 J/s. (p. 354)

potencia: La velocidad a la que la energía se transfiere desde o hacia un sistema, o se convierte de un tipo de energía en otro. La unidad de potencia en el Sistema Internacional es el joule por segundo, o vatio (abreviado con una W): 1 W = 1 J/s. (pág. 354)

power (of a lens or mirror): The reciprocal of the focal length (in meters). (p. 905)

potencia (de una lente o espejo): Recíproco de la distancia focal (en metros). (pág. 905)

pressure: The magnitude of the force per unit area exerted perpendicular to the surface of an object or system. (p. 612)

presión: La magnitud de la fuerza por unidad de área que se ejerce en dirección perpendicular a la superficie de un objeto o sistema. (pág. 612)

pressure wave: A mechanical wave consisting of periodic variations in pressure in the medium carrying it. A pressure wave is always longitudinal. The most common pressure wave is sound. (p. 1016)

onda de presión: Onda mecánica que consiste en variaciones periódicas de presión en el medio que la transporta. Las ondas de presión siempre son longitudinales. El sonido es la onda de presión más común. (pág. 1016)

principal axis: A line passing through the geometric center of the surface of a lens or spherical mirror and through the centers of curvature of both surfaces of the lens. (p. 1136)

eje principal: Línea que atraviesa el centro geométrico de la superficie de una lente o espejo esférico y que pasa por los centros de curvatura de ambas superficies de la lente. (pág. 1136)

principle of equivalence: The effect of a gravitational field on an object or system in a stationary reference frame is equivalent to that of an accelerated reference frame in the absence of gravity. (p. 1331)

principio de equivalencia: El efecto de un campo gravitacional sobre un objeto o sistema en un sistema de referencia estacionario es equivalente al de un sistema de referencia en aceleración en ausencia de gravedad. (pág. 1331)

principle of Newtonian relativity: The laws of motion are the same in all inertial reference frames. (p. 1301)

principio de relatividad de Newton: Las leyes del movimiento son las mismas en todos los sistemas de referencia inercial. (pág. 1301)

projectile: An object undergoing projectile motion. (p. 114)

proyectil: Un objeto en movimiento parabólico. (pág. 114)

projectile motion: Free-fall motion for an object that was launched with an initial velocity so that its motion has both vertical and horizontal components; for an object not launched this is simply free fall. (p. 114)

movimiento parabólico: Movimiento en caída libre de un objeto lanzado con una velocidad inicial tal su movimiento incluye componentes verticales y horizontales; para un objeto que no fue lanzado, es simplemente una caída libre. (pág. 114)

propagation speed: The rate at which a wave travels in a medium. (p. 1022)

velocidad de propagación: Tasa a la que viaja una onda por un medio. (pág. 1022)

proper time: The time interval between two events in a reference frame in which the events occur at the same place. (p. 1308)

tiempo propio: Intervalo de tiempo entre dos eventos en un sistema de referencia en el que los eventos ocurren en el mismo lugar. (pág. 1308)

***PV* diagram:** A graph that plots the pressure of a system on the vertical axis versus the volume of the system on the horizontal axis. (p. 735)

diagrama *PV*: Gráfica que traza la presión de un sistema en el eje vertical contra el volumen del sistema en el eje horizontal. (pág. 735)

quantized: Restricted to only certain values, such as charge. (p. 1217)

cuantificar: Restringido a ciertos valores, como la carga. (pág. 1217)

quantum mechanics: Branch of physics that deals with the motions and interactions of atoms and subatomic particles, incorporating the concepts of quantization of energy, wave-particle duality, and the uncertainty principle. (p. 1228)

mecánica cuántica: Rama de la física que se ocupa del movimiento e interacción de los átomos y las partículas subatómicas, e incorpora los conceptos de cuantización de la energía, dualidad onda-partícula y el principio de incertidumbre. (pág. 1228)

quantum number: Number describing the value of a physical quantity in a quantum mechanical system such as an atom. (p. 1228)

número cuántico: Número que describe el valor de una cantidad física en un sistema mecánico cuántico, por ejemplo, un átomo. (pág. 1228)

quark: Any of a number (currently only six are known to exist) of fundamental subatomic particles which have a fractional electric charge, and are postulated as the building blocks of hadrons. (p. 1336)

cuark: Cualquier partícula entre un número de partículas subatómicas fundamentales (actualmente solo se conoce la existencia de seis) que tienen una carga eléctrica fraccionaria y se postulan como los componentes básicos de los hadrones. (pág. 1336)

quark confinement: The phenomenon wherein quarks can never be isolated. (p. 1338)

confinamiento de cuarks: Fenómeno en el que los cuarks nunca pueden estar aislados. (pág. 1338)

radiate: A method of energy transfer through a thermal process, where the energy either warms or cools the surroundings or generates light. (p. 906)

irradiar: Método de transferencia de energía por medio de un proceso térmico en el cual la energía calienta o enfría el entorno o genera luz. (pág. 906)

radiation: Energy transfer by the emission (or absorption) of electromagnetic waves. (p. 712)

radiación: Transferencia de energía por la emisión (o absorción) de ondas electromagnéticas. (pág. 712)

radioactive: A radioactive nuclide is one that decays into another nuclide by emitting ionizing radiation or particles. (p. 1246)

radiactivo: Un nucleido radiactivo es aquel que se desintegra en otro nucleido al emitir radiación o partículas ionizantes. (pág. 1246)

radius of curvature: The distance from a curved surface (such as a mirror) to its center of curvature along the optical axis. (p. 1136)

radio de curvatura: La distancia desde una superficie curva (como un espejo) hasta el centro de su curvatura a lo largo del eje óptico. (pág. 1136)

ratio of specific heats: For a given substance, the ratio of molar specific heat at constant pressure to the molar specific heat at constant volume. (p. 748)

relación de calores específicos: En una sustancia, la relación entre el calor específico molar a presión constante y el calor específico molar a volumen constante. (pág. 748)

ray: An arrow that points in the direction of light propagation. (p. 1077)

rayo: Flecha que apunta en la dirección de propagación de la luz. (pág. 1077)

ray diagram: Drawing used to determine the position of an image made by a mirror or lens. (p. 1133)

diagrama de rayos: Dibujo que sirve para determinar la posición de una imagen producida por un espejo o lente. (pág. 1133)

reaction force: The force exerted by object A on object B in reaction to a force exerted by object B on object A (*see* force pair). (p. 166)

fuerza de reacción: La fuerza ejercida por el objeto A sobre el objeto B en reacción a una fuerza ejercida por el objeto B sobre el objeto A (*ver* par de fuerzas) (pág. 166)

real image: An image formed by light rays coming together. (p. 1133)

imagen real: Imagen formada por rayos de luz que se unen. (pág. 1133)

reference frame: A coordinate system with respect to which we can make observations or measurements. (p. 103)

marco de referencia: Un sistema de coordenadas con respecto al cual podemos hacer observaciones o mediciones. (pág. 103)

reflecting telescope: A telescope in which the objective is a converging mirror. (p. 1176)

telescopio reflector: Telescopio en el cual el objetivo es un espejo convergente. (pág. 1176)

refracting telescope: A telescope in which the objective is a converging lens. (p. 1176)

telescopio refractante: Telescopio en el cual el objetivo es una lente convergente. (pág. 1176)

refraction: The change in direction of a ray of light that travels from one medium into another; the angle that the refracted light makes to the normal is not equal to the angle of the incident light to the normal. (p. 1076)

refrigerator: A device that takes in energy and uses it to transfer energy through a thermal process from a system at low temperature to a system at high temperature. (p. 760)

relative velocity: The velocity of one object relative to—and as measured by—another object or an observer. (p. 105)

relativistic gamma: A dimensionless quantity that is equal to 1 when an object is at rest and becomes infinitely large as the speed of the object approaches the speed of light. (p. 1326)

relativistic speed: A speed that is a significant proportion of the speed of light. (p. 1311)

reservoir: A part of a system large enough to provide either heating or cooling without a change in its temperature. (p. 753)

resistance: The resistance of a resistor is proportional to its length and inversely proportional to its cross-sectional area. The constant of proportionality is the resistivity of the material. (p. 884)

resistivity: A fundamental property of a material that quantifies how strongly it opposes electric current; depends on its molecular and atomic structure, and the temperature of the material. (p. 884)

resistor: A circuit component intended to add resistance to current in the circuit. (p. 885)

resolve: To optically distinguish; as in telling two closely spaced objects apart. (p. 1117)

rest energy: Energy of a particle that is not in motion, due to its mass. (p. 1329)

restoring force: A force that tends to bring an object back toward equilibrium. (p. 555)

reversible process: An ideal process that can return both the system and the surroundings to their original conditions without increasing entropy; throughout an entire reversible process the system is in thermodynamic equilibrium with its surroundings. (p. 756)

right-hand rule: A rule that uses the right hand to determine the orientation of vector quantities normal to a plane; for example, used to find the direction of the angular momentum vector around an axis of rotation. (p. 481)

rigid body: An extended object whose shape doesn't change as it rotates. (p. 449)

rolling friction: Force exerted on a rolling extended object by the surface upon which it rolls (due to small deformations of the surface and of the extended object), directed opposite to the direction of motion of the center of mass of the rolling extended object. (p. 208)

rolling without slipping: The state in which a rigid body rolls uniformly across a surface without sliding. (p. 511)

root-mean-square speed (rms speed): (of molecules) A measure of how fast gas molecules move; the square root of the average value of the square of individual speeds. (p. 694)

refracción: Cambio de dirección de un rayo de luz que viaja de un medio a otro; el ángulo que forma la luz refractada con la normal no es igual al ángulo de la luz incidente con la normal. (pág. 1076)

refrigerador: Dispositivo que consume energía y la usa para transferirla a través de un proceso térmico de un sistema a baja temperatura a un sistema a alta temperatura. (pág. 760)

velocidad relativa: La velocidad de un objeto en relación con—y medida a partir de—otro objeto u observador. (pág. 105)

gamma relativista: Cantidad sin dimensiones que es igual a 1 cuando un objeto está en reposo y se vuelve infinitamente grande cuando la velocidad del objeto se acerca a la velocidad de la luz. (pág. 1326)

velocidad relativista: Velocidad que es proporcional a la velocidad de la luz. (pág. 1311)

reservorio: Parte de un sistema suficientemente grande para proporcionar calefacción o enfriamiento sin cambiar su temperatura. (pág. 753)

resistencia: La resistencia de un resistor es proporcional a su longitud e inversamente proporcional a su área transversal. La constante de proporcionalidad es la resistividad del material. (pág. 884)

resistividad: Propiedad fundamental de un material que cuantifica su oposición de corriente eléctrica; depende de su estructura molecular y atómica, y de la temperatura del material. (pág. 884)

resistor: Componente de un circuito que aumenta la resistencia a la corriente del circuito. (pág. 885)

resolución: Distinguir ópticamente; poder discernir entre dos objetos cercanos. (pág. 1117)

energía en reposo: Energía de una partícula que no está en movimiento, debido a su masa. (pág. 1329)

fuerza restauradora: Una fuerza que devuelve a un objeto hacia el equilibrio. (pág. 555)

proceso reversible: Proceso ideal que devuelve tanto al sistema como al entorno a sus condiciones originales sin aumentar la entropía; en un proceso reversible, el sistema está en equilibrio termodinámico con su entorno. (pág. 756)

regla de la mano derecha: Regla que usa la mano derecha para determinar la orientación de las cantidades vectoriales normales en un plano; por ejemplo, se utiliza para encontrar la dirección del vector de momento angular alrededor de un eje de rotación. (pág. 481)

cuerpo rígido: Un objeto extendido cuya forma no cambia al rotar. (pág. 449)

fricción de rodadura: Fuerza que ejerce una superficie sobre un cuerpo extenso que rueda sobre ella debido a pequeñas deformaciones tanto en el cuerpo extenso como en la superficie, y que se dirige en sentido opuesto a la dirección del movimiento del centro de masa del cuerpo extenso rodante. (pág. 208)

rodar sin deslizarse: Estado en el cual un cuerpo rígido rueda uniformemente por una superficie sin deslizarse. (pág. 511)

velocidad media cuadrática (velocidad rms): (de moléculas) Medida de qué tan rápido se mueven las moléculas de gas; la raíz cuadrada del valor medio del cuadrado de las velocidades individuales. (pág. 694)

rotation: Motion in which an extended object rotates about an axis. (p. 449)

rotational equilibrium: A state of zero angular acceleration, resulting from a rigid body on which no net torques are exerted. (p. 480)

rotational inertia: (also moment of inertia) A property of a body that defines its resistance to a change in angular velocity about an axis of rotation. (p. 455)

rotational kinematics: The study of rotational motion, including angular velocities and angular acceleration, without reference to the causes of that motion. (p. 470)

rotational kinetic energy: Kinetic energy associated with the rotation of a system around an axis. (p. 455)

scalar: Quantities that can be described simply by stating a number and a unit (that is, a quantity with no direction). (p. 35)

scientific notation: A way of precisely and compactly expressing a number so that its significant digits are clear. Numbers (the first factor) are usually written as a number between 1 and 10 multiplied by the appropriate power of ten (the second factor). (p. 13)

scientific practice: The practice used by scientists to carry out their work, which is an iterative process of observation, analysis, and logical reasoning that links evidence and claims to explain phenomena and to make predictions about what will happen in the context of observations. (p. 4)

scientific questioning: A scientific question has a definite answer that is testable (you can design an experiment to test the answer) and the hypothesis upon which it is based is falsifiable (you could test to see if the hypothesis can be ruled out). (p. 4)

second law of thermodynamics: The amount of disorder in an isolated system either always increases or, if the system is in equilibrium, stays the same. (p. 752)

semiconductors: Substances with electrical properties that are intermediate between those of insulators and conductors. (p. 788)

semimajor axis: Half of the distance of the longest diameter of an ellipse (the major axis; a line segment that runs through the center and both foci, with ends at the widest points of the perimeter). (p. 532)

series (capacitors): An arrangement of capacitors connected along a single path (not in parallel); the reciprocal of the total capacitance is equal to the sum of the reciprocals of the individual capacitances. (p. 852)

shock wave: A sudden pressure increase in a narrow region of a medium (for example, air), such as that caused by an object moving faster than the speed of sound in that medium. (p. 1056)

significant digits (also significant figures): When a number is expressed in scientific notation, the number of significant digits (or significant figures) is the number of single digits (0 through 9 inclusive) needed to express the number to within the uncertainty of calculation or measurement. (p. 16)

rotación: Movimiento en el cual un objeto extendido rota alrededor de un eje. (pág. 449)

equilibrio rotacional: Estado de aceleración angular nula, que resulta de un cuerpo rígido sobre el cual no se ejerce ningún par neto. (pág. 480)

inercia rotacional: (también conocida como momento de inercia) Propiedad de un cuerpo que define su resistencia a un cambio de velocidad angular alrededor de un eje de rotación. (pág. 455)

cinemática rotacional: El estudio del movimiento rotacional, que incluye velocidades angulares y aceleración angular, sin referirse a las causas de dicho movimiento. (pág. 470)

energía cinética rotacional: La energía cinética asociada con la rotación de un sistema alrededor de un eje. (pág. 455)

escalar: Cantidades que se describen con un número y una unidad (es decir, una cantidad sin dirección). (pág. 35)

notación científica: Una forma de expresar números de forma precisa y compacta para que sus cifras significativas sean claras. Los números (el primer factor) generalmente se escriben como un número entre 1 y 10 multiplicado por la potencia correspondiente de diez (el segundo factor). (pág. 13)

práctica científica: La práctica que hacen los científicos para llevar a cabo su trabajo. Es un proceso iterativo de observación, análisis y razonamiento lógico que vincula la evidencia con afirmaciones para explicar fenómenos y hacer predicciones sobre lo que sucederá en el contexto de las observaciones. (pág. 4)

cuestionamiento científico: Las preguntas científicas tienen respuestas definitivas que se pueden probar (se puede diseñar un experimento para probar la respuesta) y la hipótesis en la que se basa es falsable (se puede realizar una prueba para ver si se puede descartar la hipótesis). (pág. 4)

segunda ley de la termodinámica: La cantidad de desorden en un sistema aislado siempre aumenta o, si el sistema está en equilibrio, permanece igual. (pág. 752)

semiconductores: Sustancias con propiedades eléctricas que están entre los aislantes y los conductores. (pág. 788)

eje semimayor: La mitad de la distancia del diámetro más largo de una elipse (el eje mayor; un segmento que atraviesa el centro y ambos focos, con extremos en los puntos más anchos del perímetro). (pág. 532)

series (condensadores): Disposición de condensadores conectados a lo largo de un solo camino (no en paralelo); el recíproco de la capacitancia total es igual a la suma de los recíprocos de las capacitancias individuales. (pág. 852)

onda de choque: Aumento repentino de la presión en una región estrecha de un medio (por ejemplo, el aire), como aquel que causa un objeto al moverse a una velocidad mayor que la del sonido en ese medio. (pág. 1056)

dígitos significativos: (también conocidos como cifras significativas) Cuando se expresa un número con notación científica, el número de dígitos significativos (o cifras significativas) es el número de dígitos únicos (de 0 a 9 inclusive) requeridos para expresar el número dentro de la incertidumbre de cálculo o medición. (pág. 16)

simple harmonic motion (SHM): Oscillatory motion under a restoring force that is proportional to the displacement from the equilibrium position (a linear restoring force, such as Hooke's law). (p. 567)

movimiento armónico simple (MAS): Un movimiento oscilatorio bajo una fuerza restauradora que es proporcional al desplazamiento desde el punto de equilibrio (una fuerza restauradora lineal, como la ley de Hooke). (pág. 567)

simple pendulum: A model of a pendulum in which all of the mass is concentrated at a single point. (p. 582)

péndulo simple: Modelo de un péndulo en el cual toda la masa se concentra en un punto único. (pág. 582)

single-loop circuit: A circuit with only one complete closed pathway that current can take around the circuit. (p. 891)

circuito de bucle único: Circuito con un solo camino cerrado que la corriente puede tomar alrededor del circuito. (pág. 891)

single-slit diffraction: Experiment in which a wave passes through a single narrow opening, producing a pattern of bright and dark fringes. (p. 1111)

difracción de una sola rendija: Experimento en el que una onda cruza una sola abertura estrecha, de tal manera que produce un patrón de franjas brillantes y oscuras. (pág. 1111)

sinusoidal function: A mathematical curve that describes a smooth repetitive oscillation, such as the sine and cosine functions. (p. 569)

función sinusoidal: Una curva matemática que describe una oscilación suave repetitiva, como las funciones de seno y coseno. (pág. 569)

sinusoidal plane wave: An idealized kind of wave that has all the key properties of a more general electromagnetic wave where the disturbance of the wave oscillates in a sinusoidal fashion. A plane wave describes a physical quantity whose value, at any moment, is constant through any plane that is perpendicular to a fixed direction in space. (p. 1072)

onda plana sinusoidal: Tipo idealizado de onda que tiene todas las propiedades clave de una onda electromagnética general y en el cual la perturbación de la onda oscila de forma sinusoidal. Una onda plana describe una cantidad física cuyo valor, en cualquier momento, es constante a lo largo de cualquier plano que sea perpendicular a una dirección fija en el espacio. (pág. 1072)

sinusoidal wave: A wave in which the wave pattern at any instant is a sinusoidal function. (p. 1019)

onda sinusoidal: Onda en la que el patrón de onda en cualquier instante es una función sinusoidal. (pág. 1019)

Snell's law of refraction: Expression describing the relationship between the angles of incidence and refraction when referring to light or other waves passing through a boundary between two different media, such as water, glass, or air. (p. 1080)

Ley de refracción de Snell: Expresión que describe la relación entre los ángulos de incidencia y refracción de la luz u otras ondas que cruzan el límite entre dos medios diferentes, como el agua, el vidrio o el aire. (pág. 1080)

solenoid: A straight helical coil of wire. (p. 962)

solenoide: Bobina helicoidal recta de alambre. (pág. 962)

solid: A substance whose individual molecules cannot move freely and thus remain in essentially fixed positions relative to one another. (p. 604)

sólido: Una sustancia cuyas moléculas individuales no se mueven libremente y, por lo tanto, permanecen en una posición esencialmente fija en relación con las demás moléculas. (pág. 604)

sonic boom: A loud, explosive noise caused by the shock wave from an object traveling faster than the speed of sound in a medium. (p. 1056)

boom sónico: Ruido fuerte y explosivo producido por la onda de choque de un objeto que viaja a una velocidad mayor que la velocidad del sonido en un medio. (pág. 1056)

source of emf: A source of electric energy which can be used to provide a potential difference, such as a battery. (p. 875)

fuente de fem: Fuente de energía eléctrica que proporciona una diferencia de potencial, como una batería. (pág. 875)

special theory of relativity: Theory developed by Albert Einstein that states: (1) All laws of physics are the same in all inertial frames, and (2) the speed of light in a vacuum is the same in all inertial frames, independent of both the speed of the source of the light and the speed of the observer. (p. 1308)

teoría especial de la relatividad: Teoría desarrollada por Albert Einstein que establece: (1) todas las leyes de la física son las mismas en todos los marcos inerciales, y (2) la velocidad de la luz en el vacío es la misma en todos los marcos inerciales, independientemente de la velocidad de la fuente de luz y de la velocidad del observador. (pág. 1308)

specific gravity: The density of a substance divided by the density of 4°C liquid water. (p. 607)

gravedad específica: La densidad de una sustancia dividida por la densidad del agua líquida a 4°C. (pág. 607)

specific heat: The amount of energy per unit mass required to raise the temperature of a substance by one kelvin. (p. 701)

calor específico: Cantidad de energía por unidad de masa requerida para aumentar la temperatura de una sustancia en un kelvin. (pág. 701)

specular reflection: Type of surface reflection in which light rays striking a smooth surface in a single direction reflect from the surface in a single outgoing direction. (p. 1132)

reflexión especular: Tipo de reflexión superficial en la cual los rayos de luz que impactan una superficie lisa en una sola dirección se reflejan desde la superficie en una sola dirección saliente. (pág. 1132)

speed: The distance traveled divided by the time it took to move that distance. (p. 34)

rapidez: La distancia recorrida por el tiempo que tomó recorrer esa distancia. (pág. 34)

speed of light: The speed at which all electromagnetic waves—including radio waves, x-rays, and others—travel in a vacuum. (p. 1070)

velocidad de la luz: Velocidad a la que todas las ondas electromagnéticas, incluyendo las ondas de radio, los rayos X y otras, viajan en el vacío. (pág. 1070)

spin: An intrinsic characteristic of electrons, protons, and neutrons akin to the angular momentum of a rotating sphere. (p. 1229)

espín: Característica intrínseca de los electrones, protones y neutrones similar al momento angular de una esfera giratoria. (pág. 1229)

spring constant: The ratio of the force exerted on a spring to the change in the length of the spring caused by it; determined by the material of which the spring is made, and the shape of the spring. (p. 224)

constante elástica: La relación entre la fuerza ejercida sobre un resorte y el cambio en la longitud del resorte causado por dicha fuerza; depende del material del resorte y su forma. (pág. 224)

spring potential energy: Potential energy due to the configuration of a spring (how far it is compressed or extended from its relaxed equilibrium length). (p. 322)

energía potencial elástica: Energía potencial debido a la configuración de un resorte (cuánto se comprime o se extiende de su longitud de equilibrio). (pág. 322)

Standard Model: A mathematical description of elementary particles and the electromagnetic, weak, and strong forces by which they interact. (p. 1349)

Modelo estándar: Descripción matemática de las partículas elementales y las fuerzas electromagnéticas, débiles y fuertes mediante las cuales interactúan. (pág. 1349)

standing wave: A wave in which each point in the medium has a constant amplitude, giving it the appearance of being stationary. (p. 1037)

onda estacionaria: Onda en la que cada punto del medio tiene una amplitud constante, lo cual le da la apariencia de ser estacionaria. (pág. 1037)

standing wave mode: The conditions under which a standing wave is possible; for example, a whole number of half-wavelengths must fit onto a string fixed at both ends. (p. 1037)

modo de onda estacionaria: Condiciones bajo las cuales es posible una onda estacionaria; por ejemplo, en una cuerda fija en ambos extremos debe caber un número entero de medias longitudes de onda. (pág. 1037)

state variables: Quantities such as volume, temperature, pressure, and internal energy that depend on the state or condition of a system. (p. 730)

variables de estado: Cantidades como volumen, temperatura, presión y energía interna que dependen del estado o condición de un sistema. (pág. 730)

static friction force: Force exerted on a stationary object by the surface on which it is placed that opposes any relative motion of the object and surface; varies between zero and a maximum value that depends on the properties of the object and surface. (p. 147)

fuerza de fricción estática: La fuerza que ejerce una superficie en un objeto estacionario colocado sobre esa superficie y que se opone a cualquier movimiento relativo del objeto y la superficie; varía entre cero y un valor máximo que depende de las propiedades del objeto y la superficie. (pág. 147)

steady flow: Type of fluid flow in which the flow pattern does not change with time. (p. 639)

flujo constante: El tipo de flujo de un fluido en el que el patrón de flujo no cambia con el tiempo. (pág. 639)

Stefan–Boltzmann constant: The constant of proportionality in the Stefan–Boltzmann law which states that the total energy radiated per unit surface area of a blackbody per unit time is proportional to the fourth power of the thermodynamic temperature; equal to 5.670×10^{-8} W/(m$^2 \cdot$ K^4). (p. 713)

constante de Stefan-Boltzmann: Constante de proporcionalidad en la ley de Stefan-Boltzmann que establece que la energía total radiada por unidad de superficie de un cuerpo negro por unidad de tiempo es proporcional a la cuarta potencia de la temperatura termodinámica; es igual a 5.670×10^{-8} W/(m$^2 \cdot$ K^4). (pág. 713)

streamline: The path followed by parcels of fluid in laminar flow. (p. 639)

línea de corriente: La trayectoria de las láminas de fluido durante el flujo laminar. (pág. 639)

strong force: An attractive force between protons and neutrons that is stronger than the repulsive electric force between protons; one of the four fundamental forces. (p. 1245)

fuerza potente: Fuerza de atracción entre protones y neutrones que es más fuerte que la fuerza eléctrica de repulsión entre protones; una de las cuatro fuerzas fundamentales. (pág. 1245)

sublimation: The phase change from solid directly to gas. (p. 706)

sublimación: Cambio de fase de sólido directamente a gas. (pág. 706)

surface tension: The behavior of a fluid, as though its surface were a membrane being stretched. To minimize potential energy due to the attractive forces between molecules, the liquid assumes a shape that minimizes its surface area. (p. 656)

tensión superficial: El comportamiento de un fluido, como si su superficie fuera una membrana estirada. Para minimizar la energía potencial que resulta de las fuerzas de atracción entre las moléculas, el líquido toma una forma que minimiza su superficie. (pág. 656)

surface wave: A wave that propagates along the interface between two media (for example, a seismic wave that travels along the surface of the Earth or a wave on the surface of water). (p. 1016)

onda superficial: Onda que se propaga sobre una interfaz entre dos medios (por ejemplo, una onda sísmica que viaja a lo largo de la superficie de la Tierra o una onda en la superficie del agua). (pág. 1016)

system: Something for which you must use the system model; a system for which internal structure or interactions cannot be ignored. (p. 8)

sistema: Algo que requiere el uso del modelo de sistema; un sistema en el cual no se puede ignorar la estructura interna ni las interacciones. (pág. 8)

system model: We use the system model when we cannot ignore internal structure. (p. 8)

modelo de sistema: Usamos el modelo de sistema cuando no podemos ignorar la estructura interna. (pág. 8)

Système International (SI): The international standard decimal system of weights and measures derived from and extending the metric system of units. (p. 10)

telescope: An optical device used to make images of distant objects. (p. 1175)

temperature: A measure of the kinetic energy associated with molecular motion. (p. 684)

tesla: The SI unit of magnetic field strength. (p. 937)

test charge: A point charge with such a small magnitude of charge that its effect on the electric field in which it is placed is negligible. (p. 829)

theory: An explanation of an aspect of the natural world or universe that has been repeatedly tested and verified in accordance with the scientific method, using accepted protocols of observation, measurement, and evaluation of results. (p. 4)

thermal conductivity: A measure of how easily energy passes through a specified material by conduction. (p. 717)

thermal contact: A state in which two or more systems can exchange energy through thermal processes. (p. 684)

thermal energy: Energy associated with the random motion of atoms and molecules in a system. (p. 342)

thermal equilibrium: The condition in which two systems in thermal contact exchange no energy through thermal processes; in thermal equilibrium the objects are said to be at the same temperature. (p. 684)

thermal expansion: The tendency of matter to increase in length, area, or volume in response to an increase in temperature. (p. 688)

thermal processes: Processes related to or involving temperature. (p. 4)

thermodynamic process: Any process that changes the state of a system. (p. 730)

thermodynamics: The branch of physics that describes the behavior of measurable macroscopic physical quantities of a system (such as temperature, pressure, and volume) which depend on the microscopic objects making up the system. (p. 682)

thermometer: Instrument used to measure temperature. (p. 684)

thin film: A very fine layer of a substance on a supporting material; for example, the thin lining behind the retina of the eyes of some animals or a thin layer of oil on water. (p. 1097)

thin lens: A lens with a thickness that is negligible compared to the radii of curvature of the lens surfaces. (p. 1156)

third law of thermodynamics: It is possible for the temperature of a system to be arbitrarily close to absolute zero, but it can never reach absolute zero. (p. 759)

time constant: (for an RC circuit) The product of resistance multiplied by capacitance. (p. 915)

time dilation: A difference of elapsed time between two events as measured by observers either moving relative to each other or located at different distances from large gravitational masses. (p. 1310)

Système International (SI): El sistema decimal estándar internacional de pesos y medidas derivado del sistema métrico de unidades. (pág. 10)

telescopio: Dispositivo óptico que se usa para crear imágenes de objetos distantes. (pág. 1175)

temperatura: Medida de la energía cinética asociada con el movimiento molecular. (pág. 684)

tesla: Unidad del SI de fuerza de campo magnético. (pág. 937)

carga de prueba: Carga puntual con una magnitud de carga tan pequeña que su efecto sobre el campo eléctrico en el que se encuentra es insignificante. (pág. 829)

teoría: Explicación de un aspecto del mundo natural o del universo, que se ha examinado y verificado repetidamente según el método científico, siguiendo protocolos establecidos para la observación, medida y evaluación de los resultados. (pág. 4)

conductividad térmica: Medida de qué tan fácilmente pasa la energía a través de un material específico por conducción. (pág. 717)

contacto térmico: Estado en el que dos sistemas o más intercambian energía por medio de procesos térmicos. (pág. 684)

energía térmica: Energía asociada con el movimiento aleatorio de átomos y moléculas en un sistema. (pág. 342)

equilibrio térmico: Condición en la que dos sistemas en contacto térmico no intercambian energía a través de procesos térmicos; cuando dos objetos están en equilibrio térmico se dice que están a la misma temperatura. (pág. 684)

expansión térmica: Tendencia de la materia a aumentar en longitud, área o volumen en respuesta a un aumento de la temperatura. (pág. 688)

procesos térmicos: Procesos que involucran o están relacionados con la temperatura. (pág. 4)

proceso termodinámico: Cualquier proceso que cambie el estado de un sistema: (pág. 730)

termodinámica: Rama de la física que describe el comportamiento de las cantidades físicas macroscópicas que se pueden medir en un sistema (como la temperatura, la presión y el volumen) que dependen de los objetos microscópicos que componen el sistema. (pág. 682)

termómetro: Instrumento que mide la temperatura. (pág. 684)

lámina delgada: Capa muy fina de una sustancia sobre un material de soporte; por ejemplo, la capa delgada detrás de la retina de los ojos de algunos animales o una capa fina de aceite sobre el agua. (pág. 1097)

lente delgada: Lente con un grosor insignificante en comparación con los radios de curvatura de las superficies de la lente. (pág. 1156)

tercera ley de la termodinámica: La temperatura de un sistema puede estar arbitrariamente cerca del cero absoluto, pero nunca puede llegar al cero absoluto. (pág. 759)

constante de tiempo: (para un circuito RC) El producto de la resistencia multiplicado por la capacitancia. (pág. 915)

dilatación del tiempo: Diferencia del tiempo transcurrido entre dos eventos, medida por observadores que se mueven entre sí o que se encuentran a diferentes distancias de grandes masas gravitacionales. (pág. 1310)

torque: The tendency of a force to rotate an extended object. (p. 477)

torr: A unit of pressure. Named after the inventor of the barometer, equivalent to a millimeter of mercury; equal to 1/760 atm or 133.32 pascals. (p. 621)

total internal reflection: The phenomenon occurring when a wave strikes a boundary from a medium of higher index of refraction into a medium of lower index of refraction at an angle larger than the critical angle with respect to the normal; 100% of the incident light is reflected back into the first medium. (p. 1084)

total mechanical energy: The sum of kinetic energy and potential energy; constant in closed, isolated systems made up of objects experiencing only conservative interactions. (p. 335)

total momentum: The vector sum of the momenta of a system of objects. (p. 387)

trajectory: The path that a moving object follows through space as a function of time. (p. 103)

translation: Motion in which the center of mass of a system moves from one point in space to another. (p. 448)

translational kinetic energy: The kinetic energy of an object or system associated with translational motion. (p. 300)

transverse wave: A traveling disturbance or vibration in which the individual parts of the wave medium move in a direction perpendicular to the direction of wave propagation. (p. 1015)

traveling wave: A wave of the form $y(x, t) = A \cos (kx - \omega t)$ in which a disturbance propagates from one location to another. (p. 1037)

triple point: The particular combination of pressure and temperature of a material at which the solid, liquid, and vapor phases all coexist. (p. 687)

turbulent flow: Type of fluid motion in which the velocity of the flow at any point is continuously undergoing changes in both magnitude and direction. (p. 640)

uniform circular motion: The motion of an object going around a circular path at a constant speed. (p. 238)

uniform density: Constant mass density throughout a volume. (p. 617)

unit: A magnitude of a physical quantity, defined and adopted by convention or by law, that is used as a standard for measurement of the same kind of quantity. (p. 10)

unpolarized light: Light resulting from waves in which the electric fields are randomly oriented. Common light sources such as the Sun or fluorescent lights produce unpolarized light. (p. 1092)

unsteady flow: Fluid motion in which the velocity changes with time. (p. 639)

upright image: Image (produced by a lens or mirror) that has the same orientation as the corresponding object. (p. 1139)

vacuum: A space containing no matter, including air. (p. 156)

momento de fuerza: La tendencia de una fuerza a rotar un objeto extendido. (pág. 477)

torr: Unidad de presión. La unidad, que lleva el nombre del inventor del barómetro, equivale a un milímetro de mercurio; es igual a 1/760 atm o 133.32 pascales. (pág. 621)

reflexión interna total: Fenómeno que ocurre cuando una onda impacta un límite desde un medio de índice de refracción más alto hacia un medio de índice de refracción más bajo en un ángulo mayor que el ángulo crítico respecto a la normal; 100% de la luz incidente se refleja de nuevo en el primer medio. (pág. 1084)

energía mecánica total: La suma de la energía cinética y la energía potencial; es constante en sistemas cerrados y aislados compuestos de objetos que únicamente experimentan interacciones conservativas. (pág. 335)

momento total: La suma vectorial de los momentos de un sistema de objetos. (pág. 387)

trayectoria: El camino que sigue un objeto en movimiento a través del espacio como función del tiempo. (pág. 103)

traslación: Movimiento en el cual el centro de masa de un sistema se mueve de un punto en el espacio a otro. (pág. 448)

energía cinética de traslación: La energía cinética de un objeto o sistema asociada con el movimiento traslacional. (pág. 300)

onda transversal: Perturbación o vibración en movimiento en la cual las partes individuales del medio ondulatorio se mueven en una dirección perpendicular a la dirección de propagación de la onda. (pág. 1015)

onda viajera: Onda en la forma $y(x, t) = A \cos (kx - \omega t)$, donde una perturbación se propaga de un lugar a otro. (pág. 1037)

punto triple: Combinación particular de presión y temperatura de un material en la cual coexisten las fases sólida, líquida y de vapor. (pág. 687)

flujo turbulento: El tipo de movimiento de fluidos en el que, en cualquier punto, la velocidad del flujo experimenta cambios continuos tanto de magnitud como de dirección. (pág. 640)

movimiento circular uniforme: El movimiento de un objeto sobre una trayectoria circular a una velocidad constante. (pág. 238)

densidad uniforme: La densidad de masa constante en todo un volumen. (pág. 617)

unidad: Magnitud física de cantidad, definida y adoptada por convención o por ley, que se usa como estándar para medir un tipo de cantidad. (pág. 10)

luz no polarizada: Luz resultante de ondas en las que la orientación de los campos magnéticos es aleatoria. Las fuentes de luz comunes, como el sol o las luces fluorescentes, producen luz no polarizada. (pág. 1092)

flujo inestable: El movimiento del fluido en el que la velocidad cambia con el tiempo. (pág. 639)

imagen derecha: Imagen (producida por una lente o espejo) que tiene la misma orientación que el objeto correspondiente. (pág. 1139)

vacío: Un espacio que no contiene materia, incluyendo el aire. (pág. 156)

vapor: A substance in the gas phase at a temperature lower than its critical temperature. (p. 687)

vaporization: The phase change from solid to liquid. (p. 706)

vector: A quantity having direction and magnitude. A vector represents the coordinates of a point in a coordinate system, or the separation between two points in a coordinate system, and so vectors have both magnitude and direction. (p. 34)

vector addition: An operation in which two or more vectors are added to find the vector sum. (p. 91)

vector difference: An operation in which one vector is subtracted from another, or the negative of the vector is added in finding a vector sum. (p. 92)

vector product (cross product): *See* cross product. (p. 481)

velocity: The change in position (including direction) of an object or the center of mass of a system divided by the time it took to make that change in position. In one dimension, although only one component is needed to describe velocity, it is still a vector. (p. 34)

virtual image: An image at a location from which rays of reflected or refracted light only appear to diverge; for example, the image seen in a plane mirror seems to be behind the mirror as the light diverges from the surface as if it came from the point behind the mirror. (p. 1133)

virtual particle: A particle whose existence is allowed by the uncertainty principle and that exhibits many of the characteristics of an ordinary particle, but that exists only for a limited time. (p. 1345)

viscosity: A measure of the resistance to flow of a fluid. (p. 640)

visible light: The range of wavelengths visible to the human eye. (p. 1071)

volt: The SI unit of electromotive force or electric potential; the emf required to drive one ampere of current through one ohm of resistance. (p. 830)

volume flow rate: Volume of fluid per unit time passing a given point. (p. 643)

watt: The SI unit of power; equal to one joule per second. (p. 354)

wave: A disturbance that transports energy and momentum through space without transporting matter. (p. 1012)

wave function: A mathematical description of the properties of a wave, expressing the displacement of the wave medium at every position and at every time. (p. 1024)

wave pulse: A single sudden change in the position of a medium that propagates through the medium as a wave but does not repeat. (p. 1012)

wavelength: The distance between successive crests of a wave. (p. 1020)

wavelet: A tiny segment of a larger wave. (p. 1077)

wave-particle duality: The dual character of both light and matter, which have both wave and particle characteristics. (p. 1211)

vapor: Sustancia en fase gaseosa que está a una temperatura inferior a su temperatura crítica. (pág. 687)

vaporización: Cambio de fase de sólido a líquido. (pág. 706)

vector: Una cantidad que tiene dirección y magnitud. Los vectores representan las coordenadas de un punto en un sistema de coordenadas, o la separación de dos puntos en un sistema de coordenadas. Por tanto, los vectores tienen magnitud y dirección. (pág. 34)

suma de vectores: Una operación en la cual se suman dos o más vectores para encontrar el total de esos vectores. (pág. 91)

diferencia de vectores: Una operación en la cual se resta un vector de otro, o el vector negativo se suma al vector para encontrar el total. (pág. 92)

producto vectorial (producto cruz): *Ver* producto cruz. (pág. 481)

velocidad: El cambio de posición (y de dirección) de un objeto o del centro de masa de un sistema dividido por el tiempo que tomó ese cambio de posición. En una dimensión, si bien solo se necesita una componente para describir la velocidad, sigue siendo un vector. (pág. 34)

imagen virtual: Imagen en un lugar desde el cual los rayos de luz reflejada o refractada solo parecen divergir; por ejemplo, la imagen que se ve en un espejo plano parece estar detrás del espejo ya que la luz diverge de la superficie como si viniera del punto detrás del espejo. (pág. 1133)

partícula virtual: Partícula cuya existencia entera está permitida por el principio de incertidumbre y que exhibe muchas de las características de una partícula ordinaria, pero que existe solo por un tiempo limitado. (pág. 1345)

viscosidad: Una medida de la resistencia de un fluido a fluir. (pág. 640)

luz visible: Rango de longitudes de onda visibles para el ojo humano. (pág. 1071)

voltio: Unidad del SI de fuerza electromotriz o potencial eléctrico; la fem que se requiere para impulsar un amperio de corriente a través de un ohm de resistencia. (pág. 830)

caudal volumétrico: El volumen de un fluido por unidad de tiempo al pasar por un punto determinado. (pág. 643)

vatio: La unidad de potencia del Sistema Internacional, igual a un joule por segundo. (pág. 354)

onda: Perturbación que transporta energía y momento por el espacio sin transportar materia. (pág. 1012)

función de onda Descripción matemática de las propiedades de una onda, que expresa el desplazamiento del medio de la onda en cada posición y en cada momento. (pág. 1024)

pulso de onda: Cambio repentino de la posición de un medio, que se propaga por el medio a manera de onda, pero no se repite. (pág. 1012)

longitud de onda: Distancia entre las crestas sucesivas de una onda. (pág. 1020)

ondícula: Segmento diminuto de una onda más grande. (pág. 1077)

dualidad onda-partícula: Carácter dual tanto de la luz como de la materia, que tienen características de ondas y partículas. (pág. 1211)

weak force: An interaction between elementary particles, often involving neutrinos or antineutrinos, that is responsible for certain kinds of radioactive decay; one of the four fundamental forces. (p. 1341)

weight: The magnitude of the gravitational force exerted on an object. (p. 156)

weighted average: A mean calculated by giving values in a data set more influence according to some attribute of the data, such as how often a given value appears in the data set. (p. 420)

work (mechanical energy transfer): *See* mechanical energy transfer. (p. 287)

work function: The minimum amount of energy required to remove a single electron from a material. (p. 1198)

work-energy theorem: The sum of the work done on a system by external forces exerted along the motion of their points of contact is equal to the change in the energy of the system. For an object, the motion of the point of contact and the center of mass are equivalent so all the energy change is kinetic. (p. 298)

zeroth law of thermodynamics: If two systems are each in thermal equilibrium with a third system, they are also in thermal equilibrium with each other. (p. 685)

fuerza débil: Interacción entre partículas elementales, a veces con neutrinos o antineutrinos, que genera ciertos tipos de desintegración radiactiva; una de las cuatro fuerzas fundamentales. (pág. 1341)

peso: La magnitud de la fuerza gravitacional ejercida sobre un objeto. (pág. 156)

media ponderada: Una media que se calcula otorgando mayor influencia a los valores en un conjunto de datos de acuerdo a algún atributo, por ejemplo, con qué frecuencia aparece cierto valor dentro del conjunto. (pág. 420)

trabajo (transferencia de energía mecánica): *Ver* transferencia de energía mecánica. (pág. 287)

función de trabajo: Cantidad mínima de energía requerida para retirar un solo electrón de un material. (pág. 1198)

teorema del trabajo y la energía: La suma del trabajo llevado a cabo en un sistema por fuerzas externas ejercidas a lo largo del movimiento de sus puntos de contacto es igual al cambio en la energía del sistema. Para un objeto, el movimiento en el punto de contacto y el centro de masa son equivalentes y, por tanto, todo el cambio de energía es cinético. (pág. 298)

ley cero de la termodinámica: Si dos sistemas están en equilibrio con un sistema, también están en equilibrio térmico entre sí. (pág. 685)

ANSWERS TO ODD PROBLEMS

Chapter 1

1-1 Scientists use special practices to understand and describe the natural world

1. (a) Choosing the right tool is as important as being able to apply the tool. A screw can be hammered into the board, ruining the head of the screw and making it impossible to remove.
   (b) Using a formula without understanding its limitations is like asking a question without knowing that the context matters or using a memorized phrase to answer the question you think you are being asked when you are being asked something quite different. For example, asking for directions to *the* hotel in Paris can mean that you want to be directed to the building where the mayor of Paris works, City Hall, Hotel de Ville.

1-2 Success in physics requires well-developed problem solving using mathematical, graphical, and reasoning skills

1. Without a documented design the architect of my new home has no way to completely communicate the design to me, without which I cannot be sure that the outcome will conform to my "mental model." Without an opportunity to visualize the new home I may not actually know what it is that I want as an outcome. Similarly, in solving a physics problem the drawing both communicates my strategy and helps me to clarify that strategy and its outcome.

1-3 Scientists use simplifying models to make it possible to solve problems; an "object" will be an important model in your studies

1. (a) The motion can be described without reference to the Sun except as an origin in a coordinate frame. The overall orbital motion in a year is described simply by the path of a single point. Hence the object model would be sufficient. If we want to understand the "day" we have to worry about the rotation of Earth about its own axis, so not all parts of Earth are moving in the same direction at the same speed, so an object model is no longer sufficient.
   (b) For an explanation of this motion we need to consider the interactions between two objects—Earth and the Sun. A system is needed because both Earth and the Sun were highlighted and we are being asked to describe this in terms of the interaction between them.
   (c) Often when a person is inserted into a physics problem there is a tendency to look for intention—the girl interacts with the bicycle intentionally—and then assume there is an interaction between the two objects. But the girl could be inanimate and the girl–bike object would still move down the hill, so the interaction of the girl and the bike is not important to the problem and we can treat the girl–bike as an object.
   (d) The property of mass requires no structure to understand, so the object model can be used.
   (e) The explanation for the shape of the water requires the description of the interaction between the water and the bowl. All of the particles in the water do not move in exactly the same way, so an object model cannot be used for the water.
   (f) The ice skater is moving in a straight line, so his or her motion can be described by a single position; therefore an object model is sufficient.
   (g) The ice skater is spinning: All points on the ice skater are not moving in the same direction at the same speed during this motion, so he or she must be described as a system.

   If we must include interactions between objects, which affect the behavior of the system, or if the internal structure of the system is demonstrated because not all points move in exactly the same way, we must use a system model. When the behavior we are interested in can be sufficiently described by the motion and position of a single point in space, the object model is appropriate.
3. The object model simplifies and so makes it possible to solve problems. The actual plane that you board was probably selected by the airline only a few days before your departure. The selection would have been made based on many interactions within the system of planes owned by the carrier. You are not aware of events near the time of your departure that may have slightly altered the flight time. Delays on the part of any of the ground crew or boarding and luggage processes could delay your flight; to account for these you would need to consider the system. The abstract "plane" that will fly you to your destination has eliminated all of these details and made it possible for you to solve the problem.

   Strong winds as you are en route could slow or speed up your plane, or weather could cause your plane to take a different route. Such issues affect the plane as a whole, so the plane can still be modeled as an object if their effects can be described externally, like a change in an object's direction or speed.

1-4 Measurements in physics are based on standard units of time, length, and mass

1. time: SI unit second; length: SI unit meter; mass: SI unit kilogram
3. The distance in meters can be calculated:

$$(5420\ \cancel{\text{km}})\left(\frac{1000\text{ m}}{1\ \cancel{\text{km}}}\right) = 5.42 \times 10^6\text{ m}$$

The time can be converted to seconds:

$$(6\ \cancel{h})\left(\frac{3600\ \text{s}}{\cancel{h}}\right)+(40\ \cancel{\text{min}})\left(\frac{60\ \text{s}}{\cancel{\text{min}}}\right)=2.40\times10^{4}\ \text{s}$$

$$\text{speed} = (5.42\times10^{6}\ \text{m})/(2.40\times10^{4}\ \text{s}) = 2.26\times10^{2}\ \text{m/s}$$

5. From smallest to largest, the order is (e), (b), (a), (d), (c). Some examples of objects of about the size of the five distances include 200 nanometers (e) is the diameter of a typical virus; 7 μm (b) is the diameter of a human red blood cell; 0.1 mm (a) is the diameter of a typical human hair; 165 cm (d) is the average height of a woman in the United States; and 6380 km (c) is the equatorial radius of Earth.

7. **Set Up**

What is the problem asking for? I need to figure out, from the numbers given, how many neurons are on the surface of the cerebral cortex.

Known quantities are the surface area of the cerebral cortex and the number of neurons on a cm^2 of it. The only unknown is the answer.

This is really an estimation problem using unit conversions.

A picture that captures the geometry is needed to check that my calculation shows the total area and the relative size of a single cm^2. I choose A to represent the area of the cortex, and a to be the area of a single cm^2 sample.

This square represents the area of the cerebral cortex if it were smooth, $A = 1\ \text{m}^2$.

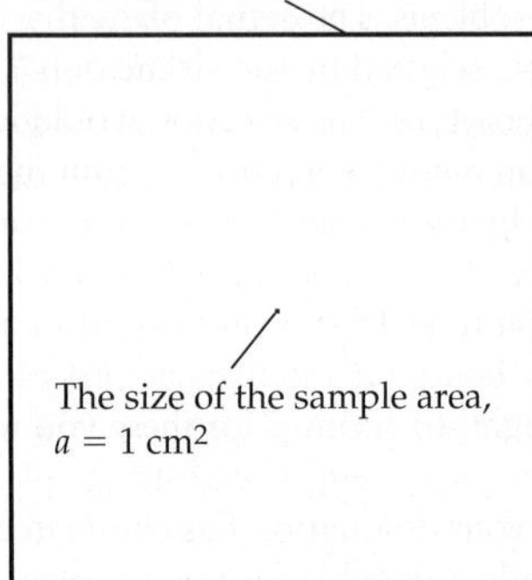

Solve

$$N \geq (1\ \text{m}^2)\times(100\ \text{cm/m})^2\times(1.2\times10^{5}\ \text{neurons/cm}^2)$$
$$= 1.2\times10^{9}\ \text{neurons, and}$$
$$N \leq (1\ \text{m}^2)\times(100\ \text{cm/m})^2\times(3.2\times10^{6}\ \text{neurons/cm}^2)$$
$$= 3.2\times10^{10}\ \text{neurons}$$

Reflect

Does it make sense? There are A/a little squares (10^4 small squares) and each has between 1.2×10^5 and 3.2×10^6 neurons, so 10^{10} is the right size. The units in the conversion factors cancel, so our answer is reasonable and passes the first check that we did the math right.

1-5 Correct use of significant digits helps keep track of uncertainties in numerical values

1. (a) $m_A = (1.395)(1.989\times10^{30}\ \text{kg}) = 2.775\times10^{30}\ \text{kg}$
(b) $m_B = (0.3180)(1.989\times10^{30}\ \text{kg}) = 6.325\times10^{29}\ \text{kg}$
(c) $m_A + m_B = (1.395+0.3180)(1.989\times10^{30}\ \text{kg}) = 3.407\times10^{30}\ \text{kg}$
[*Note:* If the answers for parts (a) and (b) are simply added, the result 3.408×10^{30} kg is slightly different because those answers were individually rounded.]

3. The number we started with here, 10, had two significant digits, so at best we would expect two in our answer, 22,000. The upper and lower errors are not symmetric in such a system, with 9000 being adequate on the lower end, and over 14,000 on the high end! Obviously, there is no simple rule. You will find the same sort of variability in other sorts of functions. You will not be expected to determine uncertainty in such calculations in this textbook, but you should realize that making certain types of calculations even from relatively certain numbers can cause large uncertainties.

5. B is correct. There are two significant digits in 0.028.

7. Converting 128.01 g to kg:

$$(128.01\ \text{g})\left(\frac{1\ \text{kg}}{1000\ \text{g}}\right) = 0.12801\ \text{kg}$$

A measurement of mass with a balance whose precision is 0.005 kg means that we know only that the value of the mass of the object lies in the interval $0.125\ \text{kg} \leq m_2 < 0.135\ \text{kg}$. The measurement of the mass using the higher precision balance lies within this interval, so we cannot claim that the two masses are different.

9. You may have predicted 30 inches, give or take 1 inch; that is, 30 ± 1 inches. Converting that statement to centimeters gives 30 inches $\left(\frac{2.54\ \text{cm}}{1\ \text{inch}}\right) = 76$ cm, and one inch is about 3 cm. We then have 76 ± 3 cm. If when you measured the box with a meter stick (a more precise instrument, so we will consider this our theoretical or expected value) you got 71 cm, then your percent error would be found by applying the given formula with 76 cm as your measured value and 71 cm as your expected value.

$$\text{percent error: }\left(\frac{76\ \text{cm} - 71\ \text{cm}}{71\ \text{cm}}\right)\times100$$
$$= \frac{5\ \text{cm}}{71\ \text{cm}}\times100 \approx 7\%$$

11. Count the number of fists needed to cover an arc of 90°. Could you have been off by a whole fist? Yes, probably. Assuming you counted between 11 and 12 fists in the 90° arc, then you could report an interval

$$\frac{90^\circ}{12\ \text{fists}} < \frac{\theta}{\text{fist}} < \frac{90^\circ}{11\ \text{fists}} \Rightarrow 7^\circ < \frac{\theta}{\text{fist}} < 8^\circ.$$

A simpler way to report this measurement is

$$\frac{\theta}{\text{fist}} = 7.5^\circ \pm 0.5^\circ.$$

13. This convention is based on the reasonable assumption that every observer can visually divide the smallest increment into two halves and assign a value as less than, greater than, or equal to the midpoint of the interval. Someone using these measurements then does not need to make assumptions about the visual acuity of the observer.

1-6 Dimensional analysis is a powerful way to check the results of a physics calculation

1. The ratio 10^2 cm/1 m is equal to 1. Multiplying by 1 does not change a value.

$$1\ \text{m}^3\left(\frac{10^2\ \text{cm}}{1\ \text{m}}\right)^3 = 10^6\ \text{cm}^3$$

3. A convenient notation that is often used is to specify the dimension of a quantity x is $[x]$. Then,

$$[a] = \frac{[\Delta v]}{[\Delta t]} = \frac{1}{\text{T}}\frac{\text{L}}{\text{T}} = \frac{\text{L}}{\text{T}^2}.$$

So the SI units of acceleration are m/s^2.

5. (a) $[c^2] = \text{L}^2/\text{T}^2$. So $[E] = \text{M L}^2/\text{T}^2$.
(b) The product mc^2 has units of joules. Evaluating the units of the right-hand side of the equation, the SI units of the joule are $\text{kg}\cdot\text{m}^2/\text{s}^2$.
7. 2π is dimensionless, so dimensions of the right-hand side of the equation $T = 2\pi\sqrt{m/k}$ are

$$\sqrt{\frac{\text{mass}}{\text{mass}/(\text{time})^2}} = \sqrt{\text{mass} \times \frac{(\text{time})^2}{\text{mass}}}$$
$$= \sqrt{(\text{time})^2} = \text{time}$$

Both sides have dimensions of time, so this relationship could be correct. Because the constants in the equation are not checked, we cannot say the equation is definitely correct. If the dimensional analysis did not work, we could say the equation is definitely not correct.

Chapter 1 Review Problems

1. (E) > (B) = (C) > (D) > (A). It is easiest to answer this question by first converting all the choices into meters.
3. (D) (2.5). When dividing quantities, the quantity with the fewest significant digits, which is 0.28 in this case, dictates the number of significant digits in the final answer.
5. (C) (1810). Both 25.8 and 70.0 have three significant digits. When multiplying quantities the quantity with the fewest significant digits dictates the number of significant digits in the final answer.
7. (a) 3×10^5 m
(b) 3.37×10^{-5} m
(c) 7.75×10^{10} W
9. For numbers smaller than 1 it is customary to include the 0 before the decimal place.
(a) 0.00442
(b) 0.00000709
(c) 828
(d) 6,020,000
(e) 456,000
(f) 0.0224
(g) 0.0000375
(h) 0.000138
11.

$$\left(7.9\frac{\text{km}}{\text{kg}}\right)\left(\frac{1\ \text{mile}}{1.609\ \text{km}}\right)\left(\frac{0.729\ \text{kg}}{\text{L}}\right)\left(\frac{3.785\ \text{L}}{\text{gal}}\right) = 14\ \text{miles/gal}$$

In the conversion of km to miles we take the precision to be fixed by the denominator so that 1 is exact. Then there are just two significant digits in the answer fixed by the precision of the metric efficiency.
13. $[x] = \text{L}$, $[a] = \text{L}/\text{T}^2$, $[t^2] = \text{T}^2$; $(\text{L}/\text{T}^2)(\text{T}^2) = \text{L}$
15. To see if an equation is dimensionally consistent we need to make sure the dimensions on the left side equal the dimensions on the right side. Position has the dimension of length; speed has dimensions of length per time; and time has the dimension of time.

$$x = vt + x_0$$
$$\text{L} \stackrel{?}{=} \frac{\text{L}}{\text{T}}\text{T} + \text{L}$$
$$\text{L} = \text{L} + \text{L}$$

17. To check that the equation is dimensionally correct, we substitute the dimensions of period (time), length (length), and acceleration due to gravity (length per time squared) into the equation. The 2π is dimensionless. The results are T = T, so this is dimensionally correct.
19. $$\frac{30.0\ \text{miles}}{1\ \text{hour}} \times \frac{1\ \text{hour}}{3600\ \text{seconds}} \times \frac{1609\ \text{meters}}{1\ \text{mile}} = 13.4\ \text{m/s}$$
21. First convert the given quantities into the basic SI units. Report the correct number of significant digits. When dividing quantities, the number with the fewest significant digits dictates the number of significant digits in the answer; 37.1 has three significant digits.

$$v = \frac{\Delta x}{\Delta t} = \frac{11.12\ \text{km}}{37.1\ \mu\text{s}} = \frac{1.112 \times 10^4\ \text{m}}{3.71 \times 10^{-5}\ \text{s}} = 3.00 \times 10^8\ \frac{\text{m}}{\text{s}}$$

23. $$\frac{1.33\ \text{euros}}{\text{liter}} \times \frac{3.79\ \text{liters}}{\text{gallon}} \times \frac{1.00\ \text{dollar}}{0.84\ \text{euro}} - \$6.00\ \text{per gallon}$$

Chapter 1 AP® Practice Problems

1. (A) $3\text{t} = 2\text{c} \Rightarrow 1 = \frac{2\text{c}}{3\text{t}}$ $\quad V = \ell wh$
3. (D) x is in meters so all of the terms on the right must also come out in meters:

$$\text{m} = \text{p}_{\text{units}}\ \text{s} \Rightarrow \text{p}_{\text{units}} = \frac{\text{m}}{\text{s}}; \quad \text{m} = \text{q}_{\text{units}}\ \text{s}^{-1}$$
$$\Rightarrow \text{q}_{\text{units}} = \text{m}\cdot\text{s}; \quad \text{m} = \text{r}_{\text{units}}\ \text{s}^2 \Rightarrow \text{r}_{\text{units}} = \frac{\text{m}}{\text{s}^2}$$

5. (D) The slope of a line is the numerical value of its steepness, the rise over the run, or the rate of change of the vertical value with respect to the horizontal value. In this case, the ranking of the lines based on slope is $B > (A = D) > C$.

Chapter 2

2-1 Studying linear motion is the first step in understanding physics

1. The aluminum bar attachment allows the connection of the bicycle and wagon to remain rigid so they can be represented as a single object in (a) or (b). Only when the rope is taut does the bicycle connected to the wagon by a rope remain rigid. When the rider of the bicycle applies the brakes, the distance between the bicycle and the wagon changes when connected by the rope, and so they do not travel at exactly the same speed; therefore, the object model would only be valid for (a).

2-2 Constant velocity means moving at a constant speed without changing direction

1. Displacement is the difference between the positions of an object at two different times. If the coordinate origin lies 3 m to my left and to the right is considered positive, my position is +3 m. If I later move 1 m further to the right from the coordinate origin, my position is +4 m and my displacement between the two times is 1m (4 m – 3 m). If instead I considered my initial position the origin, my final position would be +1 m, but my displacement would still be 1m (1 m – 0 m), independent of the origin.

3. Sentence(s) should be complete; for example, **(a)** A round trip has a zero displacement, **(b)** Displacement will be the same as the distance traveled if direction does not change, and **(c)** The displacement can never be greater than distance, because displacement is a measure of how much your position changed: a single line, the shortest distance between two points.

5. $$\left(1\frac{\text{km}}{\text{h}}\right)\left(\frac{1000\text{ m}}{\text{km}}\right)\left(\frac{1\text{ h}}{3600\text{ s}}\right) = 0.278\text{ m/s}$$

$$\left(1\frac{\text{mi}}{\text{h}}\right)\left(\frac{1\text{ h}}{3600\text{ s}}\right)\left(\frac{1610\text{ m}}{\text{mi}}\right) = 0.447\text{ m/s}$$

7. Because we are told the tracks are parallel, we may assume they are straight. For straight tracks, we can calculate the displacements from the given information.

When the trains are passing, the magnitude of the sum of the magnitudes of the displacements (the distances) of the train from Boston, d_A and the the train headed to Boston, d_B is the total distance from Boston to New York City, D. So we can find one of the distances in terms of the other: $D = d_A + d_B$.

Because the time interval, Δt_{meet}, is the same for both for the time from when they leave until when they meet, the distances traveled, found from $v_{average}\Delta t_{meet}$ have the same ratio as their average speeds. The average speeds can be found from the total distance, D and the time the train takes for each route, t_A = time from Boston to NYC = 220 minutes and t_B = time from NYC to Boston = 245 minutes:

$$d_A/d_B = (v_{average,A}\Delta t_{meet})/(v_{average,B}\Delta t_{meet}) = (D/t_A)/(D/t_B) = t_B/t_A = 245\text{ min}/220\text{ min} = 1.11364\text{, so } d_A = 1.11364 d_B$$

Eliminating d_A:

$$D = 221\text{ mi} = d_B + 1.11364 d_B$$

Solving, $d_B = 105$ mi and $d_A = 116$ mi. To get displacement we must include direction, $\Delta x_A = 116$ miles from Boston toward NYC, and $\Delta x_B = 105$ miles from NYC toward Boston.

9. **(a)** We are not told the path is straight, but we are asked about distance and speed so direction does not matter and we can represent it as a straight path.

Hiker 1:

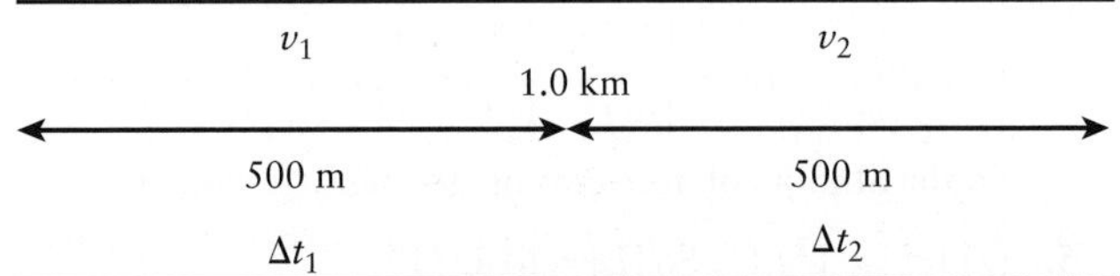

Hiker 2:

(b) Hiker 1 goes 500 m at a speed of v_1. So, the time for the first segment is 500 m/v_1. Likewise, the time that it takes to go the 500 m of the second segment is 500 m/v_2. The total time is the sum

$$\text{hiker}_1\text{ time} = \frac{500\text{ m}}{v_1} + \frac{500\text{ m}}{v_2}$$

and

$$\text{hiker}_2\text{ time} = \Delta t_1 + \Delta t_2$$

(c) The average speed is the distance divided by the elapsed time. So,

$$\text{hiker}_1\text{ average speed} = \frac{1000\text{ m}}{\dfrac{500\text{ m}}{v_1} + \dfrac{500\text{ m}}{v_2}}$$

$$\text{hiker}_2\text{ average speed} = \frac{1000\text{ m}}{\Delta t_1 + \Delta t_2}$$

11. **(a)**

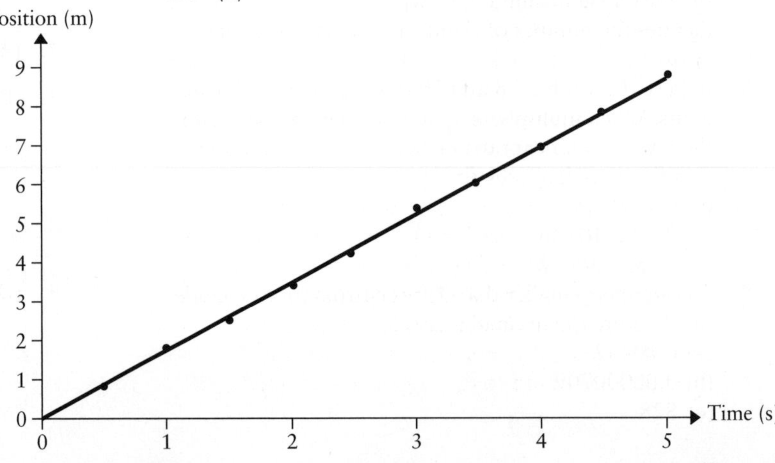

(b) The data are consistent with a constant velocity in that the deviations of the positions from the constant slope lie both above and below the line.
(c) Average velocity is calculated from taking two points on the line of best fit. The line passes through the origin, and we can read 4.8 m at 2.8 s pretty clearly on the graph so $\Delta x/\Delta t = (4.8 \text{ m} - 0)/(2.8 \text{ s} - 0) = 1.7$ m/s. Because the graph is consistent with a constant velocity motion, the value of the constant velocity is the average velocity. The limitation of the significant digits gives an estimate of uncertainty. The best answer should be expressed as 1.7 ± 0.1 m/s.

13. Possible questions include:
Was the average speed of this trip nonzero, and if so was the average velocity of this trip nonzero?
Was the average velocity equal to zero?

15. The displacement can be as small as 3408 m and as large as 3410 m. The time can be as short as 68.5 s and as long as 69.5 s. The smallest ratio and largest ratio of these variables is therefore 3408 m/69.5 s and 3410 m/68.5 s, respectively. So, the answer is the interval [49.0 m/s, 49.8 m/s]. This can be reported as 49.4 m/s ± 0.4 m/s.

17. Average speed for each interval: **(a)** 18 m/s, **(b)** 16 m/s, and **(c)** 10 m/s. The average speed is greater at earlier times—nearly 20 m/s at the start of the race and down to 10 m/s at the end. Training might guide the horse to maintain a steadier pace. For example, with an average of 15 m/s the horse would have gone 750 m in the same total time.

2-3 Velocity is the rate of change of position, and acceleration is the rate of change of velocity

1. The slopes of these graphs have units of m/s and m/s^2.
3. Because the graph is a line, we can choose any two points from which to calculate the slope.

$$a_{\text{average}} = \frac{8 \text{ m/s} - (-4 \text{ m/s})}{9 \text{ s}} = \frac{12}{9} \text{ m/s}^2 = \frac{4}{3} \text{ m/s}^2$$

5. (a) In situations labeled (i) and (iv) the signs of the velocity and acceleration are the same, so the speed of the object will increase.
(b) In situations labeled (ii) and (iii) the signs of the velocity and acceleration are opposite, so the speed of the object will decrease—at least initially. If the motion continues long enough, the object will come to a stop and then begin to speed up in the direction of the acceleration.
(c) When velocity and acceleration have opposite signs, (ii) and (iii), the speed of the object decreases, so the object eventually comes to a stop. When both properties have the same sign the speed increases, and the object does not come to a stop.
(d) The object begins at −20 m, so when the velocity and acceleration are both positive (i) the object moves toward the origin, and must eventually pass through the origin. When the object is moving away from the origin and the acceleration is positive (iii), it will eventually come to a stop and must eventually return to the origin, given enough time.
(e) When the object is moving away from the origin and the acceleration is negative (iv), the object will not return to the origin. The case in which the velocity is positive and the acceleration is negative (ii) is uncertain, and depends on the relative magnitudes of these properties to determine if the object reaches the origin before it changes direction.

7. The simplest test is "Were all the values of velocity measured the same to within the uncertainty in the value of the velocity?" To test the claim that the motion is constant-velocity motion using acceleration, the average acceleration can be calculated between the first and second times and between the second and third times. If the calculated accelerations are different from zero by an amount less than the uncertainty in the calculated values at each time, then the possibility that the motion is at constant velocity cannot be ruled out.

9. (a) First, a calculation of the average speed of the Hyperloop One system is required. We want it to be in units of miles per hour to make the comparison.

$$\text{speed} = \frac{\Delta x}{\Delta t} = \left(\frac{139 \text{ km}}{13 \text{ min}}\right)\left(\frac{60 \text{ min}}{1 \text{ h}}\right)\left(\frac{1 \text{ mi}}{1.609 \text{ km}}\right) = 4.0 \times 10^2 \text{ mi/h}$$

To reach a nearby city at a distance of 45 mi would require 1 h by car at 45 mi/h. Using the Hyperloop with this targeted speed, just 6.8 minutes would be needed.
(b) Because the Hyperloop has to speed up from rest and come to rest at the destination, some time will be spent traveling at less than the average speed. In order to compensate, the maximum speed must be greater than the targeted speed. The average speed would stay the same in both situations.

11. (a) The object slows at a constant acceleration, momentarily stops, and then speeds up at a constant acceleration. The direction of motion of the object as it slows down is opposite to the direction of the motion as it speeds up. This might be a lab cart rolling up and then back down an inclined plane or a ball thrown up into the air that then comes back down.
(b) The acceleration is the change in velocity divided by the change in time. Because the acceleration is constant, any pair of points on the graph can be used to calculate the acceleration. For example,

$$a = \frac{-4 \text{ m/s} - 4 \text{ m/s}}{8 \text{ s}} = -1 \text{ m/s}^2$$

(c) $$v_{\text{average}} = \frac{4 \text{ m/s} + 0 \text{ m/s}}{2} = 2 \text{ m/s}$$

(d) The displacement is the product of average velocity and elapsed time, Equation 2-1. The object will be displaced in the positive direction by 8 m.

2-4 Tools for describing constant acceleration motion

1. (a) (D)
(b) (B)
3. "Slowing down" describes a change in speed, whereas "going slow" describes speed. The ambiguity of common language occurs in statements such as "go

slower," which could mean "go slower than you are now" (slow down), or it could mean "when you drive you should go slower than some given speed" (go slow).

5. Leon's difficulty is in differentiating between speed and acceleration, revealed by asking questions such as, "How long does it take an Eclipse to reach its maximum speed?"
7. (a) Same sign: The magnitude of the velocity of the object will continue to increase in the initial direction.
   (b) Opposite signs: The velocity will change sign, so the magnitude of the velocity initially decreases and then increases in the direction opposite the initial motion.
9. (D) The velocity will decrease in magnitude until it becomes momentarily zero and then increase in magnitude in the opposite direction.

2-5 Solving linear motion problems: constant acceleration

1. (C) The rest have something wrong with the mathematical routine.
3. (a) $\Delta x = v_0 t + \frac{1}{2}at^2 = -10 \text{ m/s } t + 5t^2$
   (b) $5t^2 - 10t - 15 = 0 \Rightarrow t^2 - 2t - 3 = 0$
      i. $t = \dfrac{-b \pm \sqrt{b^2 - 4ac}}{2a} = \dfrac{2 \pm \sqrt{4+12}}{2} = 1 \text{ s} \pm 2 \text{ s}$
      ii. $(t-3)(t+1) = 0 \Rightarrow t = 3 \text{ s} \quad t = -1 \text{ s}$. The negative time is meaningless in the context of the problem; something happened at $t = 0$, so 3 s is the only correct answer.
5.
$$a < 0.72 \text{ m/s}^2 \Rightarrow \Delta t > \frac{\Delta v}{a} = \left(\frac{270 \text{ km}}{\text{hr}}\right)\left(\frac{1 \text{ hr}}{3600 \text{ s}}\right)\left(\frac{1000 \text{ m}}{\text{km}}\right)$$
$$\times \frac{1}{0.72 \text{ m/s}^2} = 104 \text{ s} = 1.74 \text{ min}$$
   To clearly express the significant digits of the answer we could write 1.0×10^2 s or 1.7 min.
7. (a) This is the time at which the velocity goes to zero from its initial value at $t = 0, v_0$. Taking the initial direction of motion to be positive, the acceleration will be negative.
$$v = v_0 + a\Delta t_{stop} = 0 \Rightarrow \Delta t_{stop} = -\frac{\Delta v_0}{a} = \frac{20 \text{ m/s}}{2 \text{ m/s}^2} = 10 \text{ s}$$
   (b) Use the relationship between velocity and time at the two other times:
$$v = v_0 + at = 20 \text{ m/s} - 2 \text{ m/s}^2(5 \text{ s}) = 10 \text{ m/s}$$
$$v = v_0 + at = 20 \text{ m/s} - 2 \text{ m/s}^2(20 \text{ s}) = -20 \text{ m/s}$$
   10 m/s west and later 20 m/s east makes sense: Initial velocity is west so the velocity before reaching zero will be west, and the final will be in the direction of the acceleration, east.
9. (a)
   $a = \Delta v/\Delta t = (135 \text{ mm/s} - 75 \text{ mm/s})(2.4 \text{ s} - 1.2 \text{ s})$
   $= 5.0 \times 10^{-2}$ m/s
   (b)
$$5.0 \times 10^{-2} \text{ m/s}^2 \left(\frac{1 \text{ body length}}{0.3 \text{ mm}}\right)\left(\frac{1000 \text{ mm}}{\text{m}}\right)$$
   $= 170$ body lengths/s^2
   The average speed of *Daphnia* from the graph is 100 mm/s, ~300 body lengths/second to one significant digit.
   (c)
      i. $v_{average} = \dfrac{100 \text{ m}}{9.6 \text{ s}} = 10 \text{ m/s}$
      ii. $\dfrac{(11.4 \text{ m/s})^2 - (5.6 \text{ m/s})^2}{2(20 \text{ m})} = 2.5 \text{ m/s}^2$
      iii. In units of a 2.0-m body length these are 5.2 body lengths/s and 1.2 body lengths/s^2.
   (d) The graph shows that the average velocity during a stroke is roughly 100 mm/s. Humans that compete as Olympic sprinters have maximum velocities that are around 10 m/s. In units of body lengths per second the water flea has a speed that is roughly 66 times greater than the human. The water flea has an acceleration that is about 160 times greater than the human when expressed in body lengths/s^2.

   Questions might be:
   "How are the joints in the limbs of water fleas different from the joints in humans?"
   "How are the muscles in the two species different?"
   "Does the movement of water around the water flea propel rather than resist motion?"
11.
$$a = \frac{\Delta v}{\Delta t} \Rightarrow \Delta t = \frac{\Delta v}{a} = \frac{0 - 4.0 \text{ m/s}}{-3.0 \text{ m/s}^2} = 1.3 \text{ s}$$
$$\Delta x = v_0\Delta t + \frac{a}{2}\Delta t^2 = 4.0 \text{ m/s } (1.3 \text{ s}) - \frac{3}{2} \text{ m/s}^2 (1.3 \text{ s})^2 = 2.7 \text{ m}$$

2-6 Objects falling freely near Earth's surface have constant acceleration

1. As the ball rises the speed decreases from the initial value, reaching zero at the highest point of its rise. As it falls from the highest point the speed increases. Taking up to be the positive direction, the velocity during the rise is positive and the velocity during the descent is negative. The acceleration during the entire flight has a constant magnitude, g. With the upward direction as positive the downward acceleration vector is negative.
3. (B) The ball comes instantaneously to rest at the top of its flight, while the acceleration is causing its direction to change.
5. (a) Taking downward to be positive y:
$$v = v_0 + gt \Rightarrow t = \frac{v - v_0}{g} = \frac{0 - (-18.0 \text{ m/s})}{9.80 \text{ m/s}^2} = 1.84 \text{ s}$$
   The path is symmetric, with the ball taking as long to fall as it did to rise. (From the highest point where the velocity is 0 m/s the ball falls until it has a speed of 18.0 m/s.)

(b) $y = y_0 + v_{0y}t + \frac{1}{2}gt^2 = 0 - 18.0 \text{ m/s}(1.84 \text{ s})$
$+ \frac{1}{2}(9.80 \text{ m/s}^2)(1.84 \text{ s})^2 = -16.5 \text{ m}$

(c) $v = v_0 - gt \Rightarrow t = \frac{v - v_0}{-g} = \frac{0 - 18.0 \text{ m/s}}{-9.80 \text{ m/s}^2} = 1.84 \text{ s}$

$y = y_0 + v_{0y}t - \frac{1}{2}gt^2 = 0 + 18.0 \text{ m/s } (1.84 \text{ s})$
$- \frac{1}{2}(9.80 \text{ m/s}^2)(1.84 \text{ s})^2 = 16.5 \text{ m}$

(d)

$$\Delta v = g\Delta t \Rightarrow \Delta t = \frac{0 - (-18.0 \text{ m/s})}{9.80 \text{ m/s}^2} = 1.84 \text{ s}$$

$$v^2 = v_0^2 + 2g\Delta y \Rightarrow \Delta y = \frac{0 - (-18.0 \text{ m/s})^2}{2\left(9.80 \text{ m/s}^2\right)} = -16.5 \text{ m}$$

7. If the ball is not going too fast, then there is little resistance to its motion from air, and only the constant downward acceleration due to gravity is influencing the motion. To argue this mathematically, consider the top of the motion to be $t = 0$. Then the equation of motion is completely symmetric about this time and so you cannot tell which way it is going in time because $(-t)^2 = t^2$. You could determine the direction of time from the motion of a ball only with cues from the environment. For example, if the ball were thrown into the air but fell into a pile of sand the behavior of the sand would indicate the direction of time.
9. The speed of the ball goes to zero at the maximum height reached, in a time $\Delta v/a$. The same time interval elapses as the ball returns to the height of the cliff. The speed of the ball as it falls from rest increases as $v = at = av_0/a = v_0$. Because the speed of the ball at the instant that it returns to the height of the cliff is identical to the launch speed, v_0, the final velocities in both cases are the same.

Chapter 2 Review Problems

1. (B) The slope of an $x - t$ plot gives information regarding the speed of the object. At point B the slope is 0, which means the object is momentarily stationary.
3. (C) Both balls are undergoing free fall, which means they accelerate at the acceleration due to gravity. The mass of the object does not factor into this calculation.
5. (a) Displacement = 0 so average velocity = 0.
   (b) Average speed takes the total distance covered into account.

$$v_{\text{average},x} = \frac{\Delta x}{\Delta t} = \left(\frac{4000 \text{ m}}{1.00 \text{ h}}\right)\left(\frac{1 \text{ km}}{1000 \text{ m}}\right) = 4.00 \frac{\text{km}}{\text{h}}$$

   (c) $v_{\text{average},x} = \frac{\Delta x}{\Delta t} = \left(\frac{25.0 \text{ m}}{9.27 \text{ s}}\right)\left(\frac{1 \text{ km}}{1000 \text{ m}}\right)\left(\frac{3600 \text{ s}}{1 \text{ h}}\right)$
   $= 9.71 \frac{\text{km}}{\text{h}}$
7. The total distance is twice the number of laps multiplied by the length of the pool. The displacement for each lap is zero. For the pool described, the total distance is 1 km assuming that the swimmer moves in a straight line.
9. In both cases the car is initially traveling at 90.0 km/h. Assuming it takes the same distance to come to a stop once the brakes are applied, the drunk driver travels for an extra 0.680 s at 90.0 km/h before hitting the brakes. Use the definition of average speed to calculate the extra distance the drunk driver travels.
   $v_{\text{relative},x} = \frac{\Delta x}{\Delta t} = 17.0$ m, which could put the car in the middle of an intersection!
11. In one-dimensional motion, the object's direction changes whenever the sign of its velocity changes. The slope of a line tangent to the curve of a v–t graph is equal to the object's acceleration. Average acceleration is the time rate of change of velocity.
   (a) Maximum velocity is the maximum value of the curve on the v–t graph, 2 m/s, which occurs between 0 and 2 s.
   (b) Velocity is positive between 0 and approximately 2.5 s, negative between approximately 2.5 and 4.75 s, and positive again between approximately 4.75 and 6 s. The object changes direction twice, at approximately 2.5 and 4.75 s.
   (c) The curve is a straight line between 2 and 3 s, which indicates that acceleration is constant during this time interval. The constant acceleration is thus equal to average acceleration here:

$$a_{\text{average}} = \frac{\Delta v}{\Delta t} = \frac{-2\frac{\text{m}}{\text{s}} - 2\frac{\text{m}}{\text{s}}}{3 \text{ s} - 2 \text{ s}} = -4\frac{\text{m}}{\text{s}^2}$$

   (d) $a_{\text{average}} = \frac{\Delta v}{\Delta t} = \frac{1\frac{\text{m}}{\text{s}} - 2\frac{\text{m}}{\text{s}}}{6 \text{ s} - 0 \text{ s}} = -0.167\frac{\text{m}}{\text{s}^2} = -0.2\frac{\text{m}}{\text{s}^2}$
13. Use the constant acceleration equations of motion to determine position and velocity as a function of time. The car's initial position at time $t = 0$ is $x = 0$. Its initial velocity is also 0 at this time, and its acceleration is 3 m/s^2:

$$x = \frac{1}{2}\left(3\frac{\text{m}}{\text{s}^2}\right)t^2 = \left(1.5\frac{\text{m}}{\text{s}^2}\right)t^2$$

   Between 5 and 6 s, the car's acceleration is 0, and its speed is constant at 15 m/s. Its displacement between 5 and 6 s is:

$$x = \left(15\frac{\text{m}}{\text{s}}\right)(1 \text{ s}) = 15 \text{ m}$$

The car's position at $t = 6$ s is then $37.5 + 15 = 52.5$ m. The truck is at $x = 0$ at $t = 2$ s. Its velocity is constant at 20 m/s:

Time (s)	Car position (m)
0	0
1	1.5
2	6
3	13.5
4	24
5	37.5
6	52.5

Time (s)	Truck position (m)
2	0
3	20
4	40
5	60
6	80

15. During the first interval (interval A), the car starts at 35.0 km/h and accelerates up to 45.0 km/h. In the second interval (interval B), the car starts at 65.0 km/h and accelerates up to 75.0 km/h. We are told the accelerations are constant, which means the average acceleration is equal to the instantaneous acceleration.

(a)

$$a_{Ax} = \frac{\Delta v_{Ax}}{\Delta t} = \frac{\left(45.0\frac{\text{km}}{\text{h}} - 35.0\frac{\text{km}}{\text{h}}\right)\left(\frac{1\text{ h}}{3600\text{ s}}\right)\left(\frac{1000\text{ m}}{1\text{ km}}\right)}{5.00\text{ s}}$$

$$= \frac{\left(2.78\frac{\text{m}}{\text{s}}\right)}{5.00\text{ s}} = 0.556\frac{\text{m}}{\text{s}^2}$$

$$a_{Bx} = \frac{\Delta v_{Bx}}{\Delta t} = \frac{\left(75.0\frac{\text{km}}{\text{h}} - 65.0\frac{\text{km}}{\text{h}}\right)\left(\frac{1\text{ h}}{3600\text{ s}}\right)\left(\frac{1000\text{ m}}{1\text{ km}}\right)}{5.00\text{ s}}$$

$$= \frac{\left(2.78\frac{\text{m}}{\text{s}}\right)}{5.00\text{ s}} = 0.556\frac{\text{m}}{\text{s}^2}$$

(b)

$$x_A = x_{0A} + v_{0A}t + \frac{1}{2}a_{Ax}t^2 = \left(35.0\frac{\text{km}}{\text{h}} \times \frac{1\text{ h}}{3600\text{ s}} \times \frac{1000\text{ m}}{1\text{ km}}\right)(5.00\text{ s})$$

$$+\frac{1}{2}\left(0.556\frac{\text{m}}{\text{s}^2}\right)(5.00\text{ s})^2 = (48.6\text{ m}) + (6.96\text{ m}) = 55.6\text{ m}$$

$$x_B = x_{0B} + v_{0B}t + \frac{1}{2}a_{Bx}t^2 = \left(65.0\frac{\text{km}}{\text{h}} \times \frac{1\text{ h}}{3600\text{ s}} \times \frac{1000\text{ m}}{1\text{ km}}\right)(5.00\text{ s})$$

$$+\frac{1}{2}\left(0.556\frac{\text{m}}{\text{s}^2}\right)(5.00\text{ s})^2 = (90.3\text{ m}) + (6.95\text{ m}) = 97.2\text{ m}$$

17. A graph of this motion could be broken up into three areas, for the time periods $t_1, 4t_1$, and t_1. In the first and last time periods, the areas between the graph and the time axis, A_1 and A_3, are those of triangles as the elevator accelerates. A_2 is a rectangle as the velocity is constant. The total distance of 400 meters is the sum of the areas A_1, A_2, and A_3. The areas of the triangles are $A_1 = A_3 = \frac{1}{2}t_1\left(8.00\frac{\text{m}}{\text{s}}\right)$. The area of the rectangle is $A_2 = 4t_1\left(8.00\frac{\text{m}}{\text{s}}\right)$. Adding the areas, we get $\left(40.0\frac{\text{m}}{\text{s}}\right)t_1 = 400$ m, so $t_1 = 10$ s and the total time is 60 s.

19. If we know the vertical velocity of the jump, the time for the ascent can be calculated:

$$\Delta v = 0 - v_0 = -g\Delta t$$

and the time in the air is $2\Delta t$.

21. (a) When speeding up, the magnitude of the velocity is increasing in the direction of motion.

Because acceleration is defined as the rate of change of velocity it must be in the same direction as the velocity or it would decrease the magnitude of the velocity. Therefore, when speeding up velocity and acceleration are parallel.

(b) Velocity is a vector in the direction of motion. When slowing down, the magnitude of the velocity in the direction of motion is decreasing, and because acceleration is defined as the rate of change of velocity, it is in the opposite direction of the motion to cause the magnitude of the velocity to decrease. Therefore, when slowing down velocity and acceleration are antiparallel.

(c) Displacement is a vector in the direction of motion. Acceleration and displacement are antiparallel for one-dimensional motion that is slowing down because the magnitude of the velocity is decreasing in the direction of motion, and because acceleration is defined as the rate of change of velocity, it opposes the direction of motion.

(d) Displacement is a vector in the direction of motion. Acceleration and displacement are parallel for one-dimensional motion that is speeding up because the magnitude of the velocity is increasing in the direction of motion, and because acceleration is defined as the rate of change of velocity, it is in the direction of motion.

23. The design should explicitly identify the dependent (acceleration) and independent (propeller turns) variables. The way that the acceleration is obtained from measurements of speed should be described. This description should use as much of the data as possible rather than just a pair of points. For example, a graph of the velocity as a function of time for a particular number of turns of the rubber band would show a segment (which may be the entire track length or only a part of the track) where the acceleration is approximately constant and the velocity is increasing linearly. Best-fit lines through these data would make use of a larger sample of the motion than just the last value of the speed along this line and the slope of this line is an estimate of the average acceleration. Then a graph of the average acceleration versus the number of turns of the rubber band, n, should be constructed. If the graph is simple, then a best-fit line should be constructed and the equation of this line should be reported. An uncertainty estimate should be made for n, and then this uncertainty should be used to describe the interval of the measurement.

25. (**a**) The initial speed of the ball is 0 ($v_{0y} = 0$), $y_0 = 25.0$ m, $y = 10.0$ m, and $a_y = -g$.

$v_y^2 = v_{0y}^2 + 2a_y(y - y_0)$

$$v_y = \sqrt{v_{0y}^2 + 2a_y(\Delta y)} = \sqrt{0^2 + 2\left(-9.80\frac{\text{m}}{\text{s}^2}\right)((10.0\text{ m}) - (25.0\text{ m}))}$$

$$= 17.1\frac{\text{m}}{\text{s}}$$

(**b**) $y = y_0 + v_{0y}t + \frac{1}{2}a_y t^2$

$$t = \sqrt{\frac{2\Delta y}{a_y}} = \sqrt{\frac{2(-25.0\text{ m})}{\left(-9.80\frac{\text{m}}{\text{s}^2}\right)}} = 2.26\text{ s}$$

27. The fox's speed is 0 at its maximum height. The total time the fox is in the air is twice the time it takes for the fox to jump from the ground to $y = 0.850$ m, because it takes the same amount of time to come back down to the ground.

(**a**) $v_y^2 = v_{0y}^2 + 2a_y(y - y_0)$

$$v_{0y} = \sqrt{v_y^2 + 2g(\Delta y)} = \sqrt{0 + 2\left(9.80\frac{\text{m}}{\text{s}^2}\right)(0.850\text{ m})} = 4.08\frac{\text{m}}{\text{s}}$$

(**b**) $v_y = v_{0y} + a_y t = v_{0y} + (-g)t$

$$t = \frac{v_y - v_{0y}}{-g} = \frac{0 - \left(4.08\frac{\text{m}}{\text{s}}\right)}{-9.80\frac{\text{m}}{\text{s}^2}} = 0.416\text{ s}$$

This is the time it takes for the fox to jump 0.850 m in the air from the ground. The total amount of time the fox is in the air is $2t = 0.833$ s.

29. We can calculate the velocity of the ball using $v_y(t) = v_{0y} + a_y t = v_{0y} - gt$. "How fast" is magnitude.

$$|v_y(t = 1\text{ s})| = \left|\left(18.0\frac{\text{m}}{\text{s}}\right) - \left(9.80\frac{\text{m}}{\text{s}^2}\right)(1.00\text{ s})\right| = 8.2\frac{\text{m}}{\text{s}}$$

$$|v_y(t = 2\text{ s})| = \left|\left(18.0\frac{\text{m}}{\text{s}}\right) - \left(9.80\frac{\text{m}}{\text{s}^2}\right)(2.00\text{ s})\right| = 1.6\frac{\text{m}}{\text{s}}$$

$$v_y(t = 5\text{ s})| = \left|\left(18.0\frac{\text{m}}{\text{s}}\right) - \left(9.80\frac{\text{m}}{\text{s}^2}\right)(5.00\text{ s})\right| = 31.0\frac{\text{m}}{\text{s}}$$

Using the same equation, we can find the time for the maximum height; that is, the time when $v = 0$.

$0 = v_{0y} - gt \Rightarrow 0 = 18.0\text{ m/s} - (9.80\text{ m/s}^2)t \Rightarrow$

$$t = \frac{18.0\text{ m/s}}{9.80\text{ m/s}^2} = 1.84\text{ s}$$

31. Use kinematics to determine the falling time of the first stone. Choose downward as the positive x direction.

$$\Delta x = v_{ox}t + \frac{1}{2}a_x t^2$$

$$50.0\text{ m} = \frac{1}{2}\left(9.80\frac{\text{m}}{\text{s}^2}\right)t^2;\ t = 3.19\text{ s}$$

Thus, the second stone is only falling for 1.19 seconds. Using the same kinematics equation,

$$50.0\text{ m} = v_{ox}(1.19\text{ s}) + \frac{1}{2}\left(9.80\frac{\text{m}}{\text{s}^2}\right)(1.19\text{ s})^2$$

Thus, $v_0 = |v_{0x}| = 36.0\frac{\text{m}}{\text{s}}$.

33. We can calculate the velocity of the ball when it is at the top of the window from the length of the window, the acceleration due to gravity, and the time it takes to pass by the window. Once we have the velocity at that point, we can determine the height the ball needed to fall to achieve that speed, assuming that its initial velocity was 0. Throughout the problem, we will define *down* to be negative.

Speed of ball at top of window:

$$y = y_0 + v_{0y}t + \frac{1}{2}a_y t^2$$

$$\Delta y = v_{0y}t + \frac{1}{2}(-g)t^2$$

$$v_{0y} = \frac{\Delta y + \frac{1}{2}gt^2}{t} = \frac{(-1.50\text{ m}) + \frac{1}{2}\left(9.80\frac{\text{m}}{\text{s}^2}\right)(0.180\text{ s})^2}{(0.180\text{ s})}$$

$$= -7.45\frac{\text{m}}{\text{s}}$$

Distance the ball drops to achieve a speed of 7.45 m/s:

$v_y^2 = v_{0y}^2 + 2a_y(y - y_0)$

$v_y^2 = v_{0y}^2 + 2(-g)(\Delta y)$

$$\Delta y = \frac{v_y^2 - v_{0y}^2}{2(-g)} = \frac{\left(-7.45\frac{\text{m}}{\text{s}}\right)^2 - 0}{2\left(-9.80\frac{\text{m}}{\text{s}^2}\right)} = -2.83\text{ m}$$

The ball started at a distance of 2.83 m above your window.

35. We know $v_{\text{average},x} = \frac{\Delta x}{\Delta t}$, and we know the distance $\Delta x = 1000\text{ m} - 750\text{ m} = 250$ m, so all we need to know is time for the last portion.

For the entire race

$$v_{\text{average},x} = \frac{\Delta x}{\Delta t} \Rightarrow \Delta t = \frac{\Delta x}{v_{\text{averge},x}} = \frac{1000\text{ m}}{8.00\text{ m/s}} = 125\text{ s}.$$

For the completed portion of the race

$$v_{\text{average},x} = \frac{\Delta x}{\Delta t} \Rightarrow \Delta t = \frac{\Delta x}{v_{\text{averge},x}} = \frac{750\text{ m}}{7.20\text{ m/s}} = 104\text{ s}.$$

The remaining time is then

$$\Delta t = 125\text{ s} - 104\text{ s} = 21\text{ s} \Rightarrow v_{\text{average},x} = \frac{\Delta x}{\Delta t} = \frac{250\text{ m}}{21\text{ s}}$$

$$= 12.0\text{ m/s}$$

37. We will take the upward direction as positive. We can divide the motion into three parts, each with a different acceleration. Then the positive displacement for each part is determined. The positive displacements are added together to obtain the maximum altitude.

(a) Stage 1:

$$\Delta y_1 = v_0 t + \frac{1}{2}a_1 t^2 = \frac{1}{2}(2.00\text{ m/s}^2)(15.0\text{ s})^2 = 225\text{ m}$$

Stage 2: We still need the initial velocity for this part. At the end of stage 1 the rocket will have a velocity of

$$v = v_0 + a_3 t = 0\text{ m/s} + (2.00\text{ m/s}^2)(15.0\text{ s}) = 30.0\text{ m/s}.$$

$$\Delta y_2 = v_0 t + \frac{1}{2}a_2 t^2 = (30.0\text{ m/s})(12.0\text{ s}) + \frac{1}{2}(3.00\text{ m/s}^2)(12.0\text{ s})^2 = 576\text{ m}$$

Stage 3: At the end of stage 2 the rocket will have achieved a velocity of

$$v = v_0 + a_2 t = 30.0\text{ m/s} + (3.00\text{ m/s}^2)(12.0\text{ s}) = 66.0\text{ m/s}$$

At its highest point the velocity will be zero. The acceleration due to gravity then gives us the remaining time to altitude:

$$0 = v_0 + a_3 t \Rightarrow 66.0\text{ m/s} + (-9.80\text{ m/s}^2)t \Rightarrow t = 6.73\text{ s}$$

And the distance traveled in that time is

$$\Delta y_3 = v_{ave}\,\Delta t = \frac{v_1 + v_2}{2}\Delta t = \frac{66.0\text{ m/s} + 0\text{ m/s}}{2}(6.73\text{ s}) = 222\text{ m}$$

giving us a maximum altitude for the sum of the three stages of 225 m + 576 m + 222 m = 1023 m = 1.02×10^3 m.

(b) i. To get the average speed we need to know the distance traveled and the time required. The distance round trip is simply 2(1023 m) = 2046 m. We can find the total time to reach the maximum altitude from the given values and calculations for the third stage in part (a).

$t_{total} = 15.0\text{ s} + 12.0\text{ s} + 6.7\text{ s} = 33.7\text{ s}$, so we just need to calculate the time it takes the rocket to fall back down to Earth.

$$\Delta y = v_{0y} t + \frac{1}{2}a_y t^2 \Rightarrow -1023\text{ m} = 0\text{ m/s }(t) + \frac{1}{2}(-9.80\text{ m/s}^2)t^2 \Rightarrow t = 14.4\text{ s}$$

The total time for the flight is then 33.7 s + 14.4 s = 48.1 s.

The average speed is then equal to the distance divided by the time.

$$\frac{2046\text{ m}}{48.1\text{ s}} = 42.5\text{ m/s}$$

ii. The average velocity is zero because the displacement is zero.

Chapter 2 AP® Practice Problems

1. (A) $v_{average} = \frac{\Delta x}{\Delta t}$

For each of these segments, change in position/change in time is the same, 1 m/s. This is found by looking at the chords drawn between the ends of the curve for each segment and observing that they are parallel to the straight line for the other object.

3. (D) $\Delta y = y_0 + v_{y0}t + \frac{1}{2}a_y t^2$

Because we do not know the initial v_y, we start at the top of the motion where $v_y = 0$. From the graph given, the object falls 18 m in 1.5 s.

$$\Delta y = \frac{1}{2}a_y t^2 \Rightarrow -18\text{ m} = \frac{1}{2}a_y(1.5\text{ s})^2 \Rightarrow |a_y| = 16\text{ m/s}^2$$

5. (D) On a $v - t$ graph the displacement is represented by the area bounded by the graph.

7. (C) The magnitude of the acceleration is the absolute value of the slope of the $v - t$ graph. A and B are both on the steepest section of the graph, thus the acceleration at those two points has the greatest magnitude. The slope at point D is negative, and has a magnitude less than the magnitude at A and B. The slope at point C is zero, so the acceleration at point C is zero.

Chapter 3

3-1 The ideas of linear motion help us understand motion in two or three dimensions

1. (a) Let the time that Angelo's opponent takes be T_0. Angelo has a shorter distance to run because he will run along the diagonal that is a distance of $\sqrt{2}L$. The difference between the times that it will take the two runners to reach the far corner is Δt, and that is how long Angelo can delay his start.

$$T_0 = \frac{2L}{v} \quad T_A = \frac{\sqrt{2}L}{v} \quad \Delta t = T_0 - T_A = \frac{L}{v}\left(2 - \sqrt{2}\right) = 24\text{ s}\left(2 - \sqrt{2}\right) = 14\text{ s}$$

(b) The two-dimensional motion of the opponent in the plane of the block can be thought of as two one-dimensional motions to find total distance traveled. Then, because distance and speed are scalars, no vector math is needed.

3-2 A vector quantity has both a magnitude and a direction

1. A scalar, such as distance or speed, has only a magnitude. A vector, such as displacement or velocity, has both magnitude and direction.

3. (a)–(e) These will vary based on choice of scale. For (a), the eastward vector $\vec{B}$ should be 1.5 times the length of the northward vector $\vec{A}$. You can choose to represent either east or north as up, but the two directions must be perpendicular and to this scale or the representation is not accurate and the rest of the problem will not work. The magnitude of the resultant vector should be given as 18 km, and the scale found by dividing 18 km by the length in cm measured for the resultant vector, giving a scale in km/cm.

(f) $\vec{A} = \vec{C} - \vec{B}$. So, the canteen is at Dora's current location plus 15 km west. Or, "The canteen is at my current location minus 15 km east."

(g) The scale of the map is a scalar. It tells you by what factor to multiply the vectors.

5. The location of each part in the machine can be described by one or more vectors. Multiplying each of the vectors that describe the machine by the same constant used to "scale up" the design does not change the shape of any part, and so there is no change in the shape of the entire machine or the interactions among the parts of the machine. For example, multiplying each side of a triangle by the same constant produces a triangle that is similar; the shape is not changed. The real-world concern that would arise is that the larger parts are heavier and may be more susceptible to bending, so there are still things the engineer needs to take into consideration beyond just factors of scale.

3-3 Vectors can be described in terms of components

1. $r = \sqrt{r_x^2 + r_y^2} = \sqrt{25 + 6.25} = 5.6$

3. (b), (d), (e)

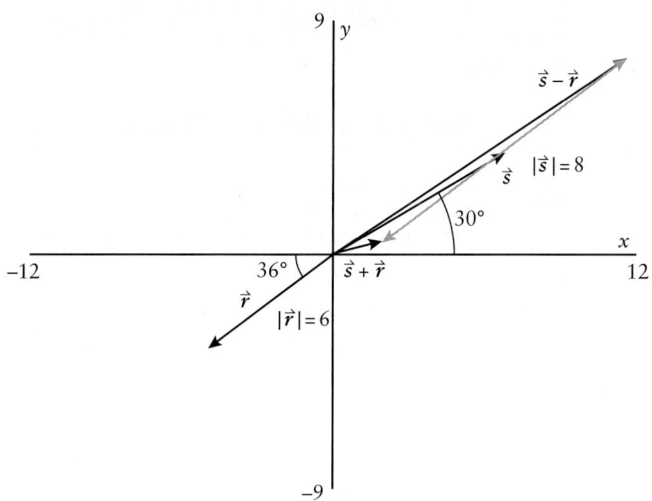

(a) $r_x = 6.0 \cos(216°) = -4.9 \quad r_y = 6.0 \sin(216°) = -3.5$

(c) $s_x = 8.0 \cos(30°) = 6.9 \quad s_y = 8.0 \sin(30°) = 4.0$

(f) $s_x + r_x = 2.0$, $s_y + r_y = 0.5$, $s_x - r_x = 11.8$, and $s_y - r_y = 7.5$. These values should be easily confirmed within the precision of the graph.

5. The x component of the sum of vectors is the sum of the x components of those vectors. Similar for the y component. Do this for each part. Denoting the vector sum as $\vec{R}$:

(a) For $\vec{A} + \vec{B} = \vec{R}$, $R_x = A_x + B_x = 6 + 7 = 13$;

$R_y = A_y + B_y = 9 + 23 = 32$

(b) For $\vec{A} - 2\vec{C} = \vec{R}$, $R_x = A_x - 2C_x = 6 - 0 = 6$;

$R_y = A_y - 2C_y = 9 - (2 \times 26) = -43$

(c) For $\vec{A} + \vec{B} - \vec{C} = \vec{R}$, $R_x = A_x + B_x - C_x = 6 + 7 - 0 = 13$; $R_y = A_y + B_y - C_y = 9 + 23 - 26 = 6$

(d) For $\vec{A} + \frac{1}{2}\vec{B} - 3\vec{C} = \vec{R}$, $R_x = A_x + \frac{1}{2}B_x - 3C_x = 6 + 3.5 - 0 = 9.5$; $R_y = A_y + \frac{1}{2}B_y - 3C_y = 9 + 11.5 - 78 = -57.5$

7. (a) and (b) A unit vector has a magnitude of 1. The solution should include a vector of length 1 for both 30° and 60° angles. The x component of the unit vector is the length of the vector times the cosine of the angle between the vector and the x axis. The y component is found using the sine of the angle. Whereas for this problem you should read them from your graph, we can calculate:

For 60 degrees: x is $1 \cos 60 = \frac{1}{2}$ and y is $1 \sin 60 = (3)^{\frac{1}{2}}/2$.

For 30 degrees: x is $1 \cos 30 = (3)^{\frac{1}{2}}/2$ and y is $1 \sin 30 = \frac{1}{2}$.

(c) For each of these we can calculate that $x^2 + y^2 = 1^2 = 1$. Reading it off the graph you may get slightly different answers, but it should still give you close to 1.

(d) The ratio of the x component and the hypotenuse is the cosine of the angle. The ratio of the y component and the hypotenuse is the sine of the angle. The sum of squares is consistent with the identity $\cos^2(\theta) + \sin^2(\theta) = 1$.

9. (a) and (b) The magnitude of the total displacement is found by summing the squared components given, because we are told that one is directed across the river and one is along the river (we will assume the river is straight over these distances). $\theta = \tan^{-1}\dfrac{300 \text{ m}}{200 \text{ m}} = 56°$

$$d = \sqrt{d_x^2 + d_y^2} = \sqrt{(200 \text{ m})^2 + (300 \text{ m})^2} = 360 \text{ m}$$

(c) To get the speed of the river we need to use the given displacement across the river (200 m, (0.20 km) and the time for the trip. The motor supplied a velocity component of 2.0 km/h across the river. The boat traveled 0.30 km down river. For the displacement across the river, calculate time:

$$v_y = \frac{\Delta y}{\Delta t} \Rightarrow \Delta t = \frac{\Delta y}{v_y} = \frac{0.20 \text{ km}}{2.0 \text{ km/h}} = 0.10 \text{ h}$$

Then the speed of the water current,

$$v_x = \frac{\Delta x}{\Delta t} = \frac{0.30 \text{ km}}{0.10 \text{ h}} = 3.0 \text{ km/h}$$

(d) We need the time to go twice as far, and are asked for T in minutes, so we must convert from hours to minutes for that answer. For help with conversions see Chapter 1.

$$T = \frac{400 \text{ m}}{v_{\text{boat}}} = \frac{400 \text{ m}}{2.0 \text{ km/h}} = 12 \text{ min}$$

3-4 Motion in a plane: reference frames, velocity, and relative motion

1. (a) $\vec{v}_{bw} + \vec{v}_{ws} = \vec{v}_{bs}$; b = boat, w = water, s = shore

$$\vec{v}_{bs} = 8.00\frac{m}{s} + \left(-3.00\frac{m}{s}\right) = 5.00\frac{m}{s}; \Delta x_{bs} = v_{bs}t$$

$$1.00 \times 10^3 \text{ m} = \left(5.00\frac{m}{s}\right)t \text{ and } t = 200 \text{ s}$$

(b) $\vec{v}_{bw} + \vec{v}_{ws} = \vec{v}_{bs}$

$$\vec{v}_{bs} = 8.00\frac{m}{s} + \left(3.00\frac{m}{s}\right) = 11.00\frac{m}{s}; \Delta x_{bs} = v_{bs}t$$

$$1.00 \times 10^3 \text{ m} = \left(11.00\frac{m}{s}\right)t \text{ and } t = 90.9 \text{ s}$$

3. (a) We are given the components of the initial and final position vectors, from which we can find the displacement. Divide that displacement by the time taken to get velocity. Defining west to be negative x and south to be negative y:

$$x_i = 6.0 \text{ km} \quad y_i = -4.0 \text{ km} \quad x_f = -10.0 \text{ km} \quad y_f = 6.0 \text{ km}$$

$$v_x = \frac{-10.0 \text{ km} - 6.0 \text{ km}}{4.0 \text{ h}} = -4.0 \text{ km/h}$$

$$v_y = \frac{6.0 \text{ km} - (-4.0 \text{ km})}{4.0 \text{ h}} = 2.5 \text{ km/h}$$

$$v = \sqrt{\frac{(-4.0)^2 \text{ km}^2}{\text{h}^2} + \frac{(2.5)^2 \text{ km}^2}{\text{h}^2}} = 4.7 \text{ km/h}$$

$$\theta = \tan^{-1}\left(\frac{2.5 \text{ km/h}}{-4.0 \text{ km/h}}\right) = 180° + (-32°) = 148° \text{ north of east}$$

(b) If he traveled in a straight line, we could find the speed as the magnitude of the velocity vector, 4.7 km/h. However, it is likely that his average speed is *greater than* the magnitude of his average velocity, as he would probably have deviated from a straight-line path, and so traveled a greater distance!

(c) The average velocity depends only on the displacement and time, and is independent of the other details of the trip, *so they are the same.*

5. $\vec{v}_{\text{spider relative to river bank}} = \vec{v}_{sb} = \vec{v}_{sl} + \vec{v}_{lb}$

s = spider, l = leaf, b = riverbank

$\vec{v}_{sl} = 0.10$ m/s toward the river bank

$\vec{v}_{lb} = 0.50$ m/s parallel to the riverbank

$$|\vec{v}_{sb}| = \sqrt{0.50^2 \text{ m}^2/\text{s}^2 + 0.10^2 \text{ m}^2/\text{s}^2} = 0.51 \text{ m/s}$$

at $\theta = \tan^{-1}\dfrac{0.50 \text{ m/s}}{0.10 \text{ m/s}} = 79°$, or 11° toward the riverbank from the current

7. $v = \sqrt{(70 \text{ m/s})^2 + (-18 \text{ m/s})^2} = 72 \text{ m/s}$

$$\theta = \tan^{-1}\frac{70 \text{ m/s}}{-18 \text{ m/s}} = -76° + 180° = 104° \text{ north of east,}$$

or with respect to the plane, 14° west of north

9. The velocity of the tomato in the frame of the victim (g = ground) is the velocity of the tomato in the frame of the biker plus the velocity of the biker in the frame of the victim.

(a) $v_{tg,ix} = (4.0 \text{ m/s}) \cos 45° = 2.83 \text{ m/s} \sim 2.8 \text{ m/s}$

$v_{tg,iy} = (4.0 \text{ m/s}) \sin 45° + v_{bg,y} = 8.83 \text{ m/s} \sim 8.8 \text{ m/s}$

$|\vec{v}_{tg,i}| = \sqrt{(2.83 \text{ m/s})^2 + (8.83 \text{ m/s})^2} = 9.3 \text{ m/s}$

The direction of the velocity is

$$\theta = \tan^{-1}\left(\frac{2.83 \text{ m/s}}{8.83 \text{ m/s}}\right) = 18° \text{ with respect to the forward}$$

motion of the bike.

(b) $t_{flight} = \dfrac{3.0 \text{ m}}{v_{tg,ix}} = \dfrac{3.0 \text{ m}}{2.83 \text{ m/s}} = 1.1 \text{ s}$

(c) 18° with respect to the edge of the sidewalk (that is, the forward motion of the bike)

(d) For example, how far it falls due to gravity, or how far it travels in the bike's direction

3-5 Motion in a plane: acceleration and projectile motion

1. The downward motion of the thrown ball is the motion of an object in free fall with no initial downward velocity. This is the same as for the dropped ball.

3. The x component of velocity and the (downward) acceleration will be the same at all points. The vertical component of the velocity will be upward to start, decrease to zero at point 3, and then increase downward (with the same magnitude at times 1 and 5, and 2 and 4).

5.

$$a_x = \frac{v_{xf} - v_{xi}}{t_f - t_i} = \frac{(15.0 \text{ m/s}) \cos(30°) - (30.0 \text{ m/s}) \cos(180° + 45°)}{5.0 \text{ s}} = 6.8 \text{ m/s}^2$$

$$a_y = \frac{v_{yf} - v_{yi}}{t_f - t_i} = \frac{(15.0 \text{ m/s}) \sin(30°) - (30.0 \text{ m/s}) \sin(180° + 45°)}{5.0 \text{ s}} = 5.7 \text{ m/s}^2$$

$|\vec{a}| = \sqrt{a_x^2 + a_y^2} = \sqrt{(6.8 \text{ m/s}^2)^2 + (5.7 \text{ m/s}^2)^2} = 8.9 \text{ m/s}^2$ at an angle with the x axis of

$$\theta = \tan^{-1}\frac{a_y}{a_x} = \tan^{-1}\frac{5.7 \text{ m/s}^2}{6.8 \text{ m/s}^2} = 40°$$

Because we do not know exactly what happened between the initial and final states, we cannot determine if the acceleration was constant.

7. (a) The acceleration is only in the $-x$ direction, so the y component of the velocity does not change and the x component of the velocity will become increasingly negative.

$$v_{f,x} = v_{i,x} + a_x\Delta t \quad v_{f,y} = v_{i,y}$$

(b) $v_{f,x} = v_{f,y} = 16.0 \text{ m/s} \cdot \cos(45.0°) = 11.3 \text{ m/s}$

$a_x\Delta t = -2.00 \text{ m/s}^2 \cdot 2.70 \text{ s} = -5.40 \text{ m/s} \quad a_y\Delta t = 0$

$v_{i,x} = 11.3 \text{ m/s} - (-5.40 \text{ m/s}) = 16.7 \text{ m/s} \quad v_{i,y} = 11.3 \text{ m/s}$

(c) $v_i = \sqrt{11.3^2 \text{ m}^2/\text{s}^2 + 16.7^2 \text{ m}^2/\text{s}^2} = 20.2 \text{ m/s}$

3-6 You can solve projectile motion problems using techniques learned for straight-line motion

1. (a) $v_{0x} = 14.2 \text{ m/s} \cos(60.0°) = 7.10 \text{ m/s}$

$v_{0y} = 14.2 \text{ m/s} \sin(60.0°) = 12.3 \text{ m/s}$

(b) $x = v_{0x}t + \frac{1}{2}a_x t^2 = 7.10 \text{ m/s } t$

$y = y_0 + v_{0y}t + \frac{1}{2}a_y t^2 = 30.0 \text{ m} + 12.3 \text{ m/s} \times t - 4.90 \text{ m/s}^2 t^2$

(c) In the second equation in (b), the unit m cancels from each term. Multiplying through by the unit s^2 and rearranging, we get: $4.90\, t_{\text{flight}}^2 - 12.3 \text{ s } t_{\text{flight}} - 30.0 \text{ s}^2 = 0 \Rightarrow$

$$t_{\text{flight}} = \frac{12.3 \text{ s} \pm \sqrt{(-12.3 \text{ s})^2 - 4(4.90)(-30.0 \text{ s}^2)}}{9.80} = 4.03 \text{ s}$$

as the positive root for time

(d) and (i) $x = (7.10 \text{ m/s})\, t_{\text{flight}} = 29$ m (28.6 m if you don't prematurely round)

(e) $v = v_{0y} - gt \Rightarrow t_{\text{max height}} = \frac{v_{0y}}{g} = \frac{12.3 \text{ m/s}}{9.80 \text{ m/s}^2} = 1.26 \text{ s}$

(f) $y_{\text{max}} = 30.0 \text{ m} + 12.3 \text{ m/s } t_{\text{max height}} - 4.90 \text{ m/s}^2 \times t_{\text{max height}}^2 = 30.0 \text{ m} + 12.3 \text{ m/s} \times (1.26 \text{ s}) - 4.90 \text{ m/s}^2 (1.26 \text{ s})^2 = 37.7 \text{ m}$

(g) $y = 0 = y_0 + v_{0y}t + \frac{1}{2}a_y t^2 = 37.7 \text{ m} - 4.90 \text{ m/s}^2 \times t^2 \Rightarrow t_f = \sqrt{\frac{37.7 \text{ m}}{4.90 \text{ m/s}^2}} = 2.77 \text{ s}$

(h) $t_{\text{flight}} = 1.26 \text{ s} + 2.77 \text{ s} = 4.03 \text{ s}$

3. The prey has a path between zags with an elapsed time greater than the lower limit and less than the upper limit, so a throw at the beginning of the path that anticipates the prey's position along the same vector can hit the target. This means that the time of flight must lie in this interval and we can use these limits to find the limits on the interval of the velocity. From the velocities and the times we can find the distances.

(a) and (b)

$$1 \text{ s} \le t_{\text{flight}} \le 2.5 \text{ s} \Rightarrow 1 \text{ s} \le \frac{2v \sin\theta}{g} \le 2.5 \text{ s} \Rightarrow$$

$$\frac{g}{\sqrt{2}} \text{s} \le v \le \frac{5\,g}{2\sqrt{2}} \text{s} \Rightarrow 7 \text{ m/s} \le v \le 17 \text{ m/s} \Rightarrow$$

$$5 \text{ m} \le x \le 30 \text{ m}$$

(c) The predicted speed is smaller than that of an Olympic thrower (as expected), given that athletes use several seconds running up to the release point. The 7.8-m range falls within our calculated range, so this is also reasonable.

5. (B) In (C), the acceleration and the initial y component of the velocity have the same sign, indicating a is positive. In (A), it is assumed that the initial and final heights are the same. In (D) the vertical height is added to the horizontal position.

7. (a) (D) The remainder of the problem develops evidence.

(b) A dart with a very high speed might be expected to hit the monkey close to h. The acceleration due to gravity on another planet could be different. If the acceleration due to gravity is large, then the velocity of the dart would need to increase to hit the monkey near h. A potential question is: Does the ratio of the acceleration due to gravity and the initial velocity determine where the collision occurs? A dart that is shot straight up will have a long time of flight and cannot hit the target. Another potential question then is: Is there a connection between the time of flight for a successful shot and the time that it takes the monkey to hit the ground that depends on the angle?

(c) d and v_{0x} should be in the same direction, and h and v_{0y} should be in the same direction, perpendicular to d.

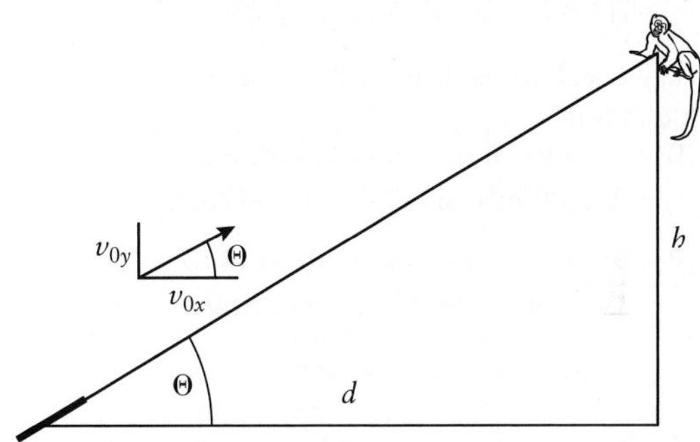

(d) $T = \frac{d}{v_{0x}}$

(e) $y_p(T) = v_{0y}T - \frac{1}{2}gT^2 = \frac{v_{0y}d}{v_{0x}} - \frac{1}{2}gT^2 = \frac{v_{0y}d}{v_{0x}} - \frac{g}{2}\left(\frac{d}{v_{0x}}\right)^2$

(f) $y_m(T) = h - \frac{1}{2}gT^2 = d\tan\theta - \frac{1}{2}gT^2 = \frac{v_{0y}d}{v_{0x}} - \frac{1}{2}g\left(\frac{d}{v_{0x}}\right)^2$

(g) and (h) At the collision time T, $y_p(T) = y_m(T)$. Because the only difference is the first term, the ratio of the components of the initial velocity must equal the ratio of h/d. So, the zookeeper should aim at the monkey.

(i) i. When v_{0x} is very large the projectile will travel d in a very short time, and it and the monkey will have fallen a very small distance, so the two will collide near h.

ii. To obtain the minimum v_{0x} for the projectile to reach the distance d when the monkey is still in the air, we can set $y_m = 0$:

$$\frac{v_{0y}d}{v_{0x}} = h = \frac{1}{2}gT^2 = \frac{g}{2}\left(\frac{d}{v_{0x}}\right)^2 \Rightarrow v_{0x} = \sqrt{\frac{gd^2}{2h}}$$

9. (B) The magnitude of the velocity is the square root of the sum of squares of its components. The x component never changes and the y component becomes zero at the highest point.

Chapter 3 Review Problems

1. During the motion of a projectile, only v_x, $a_x = 0$, and $a_y = -g$ (if up is positsive) are constant. The position (x, y) of the projectile is constantly changing while it is moving. Acceleration in the y direction is nonzero, which means v_y also changes.

3. (a) $v_y = (34.00\text{ m/s})\sin 210.0° = -17.00\text{ m/s}$
$v_x = (34.00\text{ m/s})\cos 210.0° = -29.44\text{ m/s}$
(b) $v_x = (1200.0\text{ m/s})\cos(43.0°) = 878\text{ m/s}$
$v_y = (1200.0\text{ m/s})\sin(43.0°) = 818\text{ m/s}$
(c) $R = \sqrt{R_x^2 + R_y^2} = \sqrt{120^2\text{ m}^2 + 345^2\text{ m}^2} = 365\text{ m}$
(d) $\alpha = \tan^{-1}\left(\dfrac{120\text{ m}}{345\text{ m}}\right) = 19.2°$
(e) $v_x = (15.0\text{ m/s})\cos(12.0°) = 14.7\text{ m/s}$
$v_y = (15.0\text{ m/s})\sin(12.0°) = 3.12\text{ m/s}$

5. You can check to see which of the vectors are in the right quadrant.
(a) (B) $v_x = 240.00\text{ m/s} \cdot \cos(330.0°) = 207.85\text{ m/s}$
$v_y = 240.00\text{ m/s} \cdot \sin(330.0°) = -120.00\text{ m/s}$
(b) (B) $v_x = 15.00\text{ m/s} \cdot \cos(12.0°) = 14.67\text{ m/s}$
$v_y = 15.00\text{ m/s} \cdot \sin(12.0°) = 3.12\text{ m/s}$
(c) (B) $v = \sqrt{v_x^2 + v_y^2} = 365.27\text{ m/s}$
(d) (A) $\tan(\alpha) = \dfrac{31.0\text{ m/s}}{8.0\text{ m/s}} \Rightarrow \alpha = 76°$
(e) (C) $\tan(\alpha) = \dfrac{12.0\text{ m/s}^2}{-15.0\text{ m/s}^2} \Rightarrow \alpha = -38.7°$ or
$\alpha = 180° - 38.7° = 141°$
The ratio 12/(−15) is the same as −12/15, so taking the inverse tangent on your calculator gives an angle in the fourth quadrant. You must rotate by 180° to get the correct answer.
(f) (B) $v = \sqrt{v_x^2 + v_y^2} = \sqrt{(31.0\text{ m/s})^2 + (8.0\text{ m/s})^2}$
$= 32.0\text{ m/s}$

7. (a) $\sqrt{13} = 3.6$ at 326° (b) $\sqrt{8} = 2.8$ at 135°
(c) $\sqrt{4} = 2.0$ at 270°

9. Rather than calculate, reason. The vector $\vec{D}$ has no x component so it lies on the negative y axis, 90° below the x axis, and has a length of 16.00.

11. The rock's speed will be greater than the speed with which it was thrown. Gravity is accelerating the rock in the y direction, so the magnitude of the y component of its velocity will increase.

13. (a)

−x $v_f = -40$ m/s

$v_i = -30$ m/s

at

−y

(b) $at = \sqrt{(-30\text{ m/s})^2 + (-40\text{ m/s})^2} = 50\text{ m/s}$
$|\vec{a}| = \dfrac{\Delta v}{\Delta t} = \dfrac{50\text{ m/s}}{10\text{ s}} = 5\text{ m/s}^2$
(c) $\theta = \tan^{-1}\dfrac{-30\text{ m/s}}{-40\text{ m/s}} = 37°$ north of west, or
$180° - 37° = 143°$ north of east
$\Rightarrow a_x = 5\text{ m/s}^2(\cos 143°) = -4\text{ m/s}^2,$
$a_y = 5\text{ m/s}^2(\sin 143°) = 3\text{ m/s}^2$

15. $v = \sqrt{v_x^2 + v_y^2} = \sqrt{11.3^2 + 3.5^2}\text{ m/s} = 11.8\text{ m/s}$
$\theta = \tan^{-1}\dfrac{3.5\text{ m/s}}{11.3\text{ m/s}} = 17°$

17. $L_x = 450\text{ m} \cdot \cos(20°) + 250\text{ m} \cdot \cos(270°) + 630\text{ m} \cdot \cos(70°) = 423\text{ m} + 0\text{ m} + 215\text{ m} = 638\text{ m} = 6.4 \times 10^2\text{ m}$
$L_y = 450\text{ m} \cdot \sin(20°) + 250\text{ m} \cdot \sin(270°) + 630\text{ m} \cdot \sin(70°) = 154\text{ m} - 250\text{ m} + 592\text{ m} = 496\text{ m} = 5.0 \times 10^2\text{ m}$
$L = \sqrt{L_x^2 + L_y^2} = \sqrt{(638\text{ m})^2 + (496\text{ m})^2} = 810\text{ m} = 8.1 \times 10^2\text{ m}$
$\tan(\alpha) = \dfrac{496\text{ m}}{638\text{ m}} \Rightarrow \alpha = 37.8° = 38°$

19. (a)

t (s)	v_x (m/s)	v_y (m/s)	x (m)	y (m)
0.0	2	0	0	0
0.2	2	−0.8	0.4	−0.1
0.4	2	−1.6	0.8	−0.3
0.6	2	−2.4	1.2	−0.7
0.8	2	−3.2	1.6	−1.3
1.0	2	−4.0	2.0	−2.0

(b)

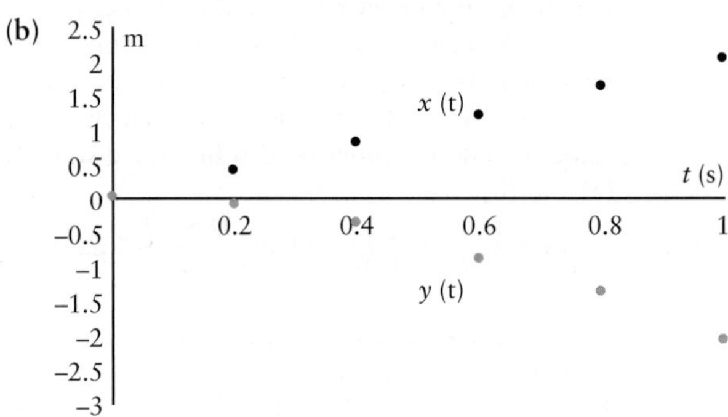

(c) The y–t and y–x graphs will have the same shape because x–t is a linear relationship.

21. We will assume the football is traveling in the positive x direction and take up to be the positive y direction. The x component of velocity is constant because $a_x = 0$, so
$v_x = v_{0x} = v_0\cos(30.0°) = \left(25.0\dfrac{\text{m}}{\text{s}}\right)\cos(30.0°) = 21.7\dfrac{\text{m}}{\text{s}}$
The y component is
$v_y^2 = v_{0y}^2 + 2a_y(y - y_0)$

$$v_y = \pm\sqrt{v_{0y}^2 + 2a_y(\Delta y)} = \pm\sqrt{(25.0 \text{ m/s})^2 \sin^2(30.0°) + 2(-9.80 \text{ m/s}^2)(4.00 \text{ m})}$$
$$= \pm 8.82 \frac{\text{m}}{\text{s}}$$

The plus or minus depends on whether the ball is before or after the highest point in its trajectory.

23. First, find Δt, from $x = v_x \Delta t$. Substitute this value into the equation for $\Delta y = 1/2\ a_y\ \Delta t$
$= 0.5 \times 9.80 \text{ m/s}^2 \times (1.0 \text{ m}/3.0 \times 10^6 \text{ m/s})^2$
$= 5.4 \times 10^{-13}$ m

25. (a) First, consider the situation on the left. You are given the magnitude and direction of the initial velocity.

$v_{i,y} = 420$ m/s sin (60.0°) = 364 m/s and

$v_{i,x} = 420$ m/s cos (60.0°) = 210 m/s

Using the y component, the time to the highest point of the trajectory, where the vertical speed is momentarily zero, can be calculated:

$$t_{y,\text{max}} = \frac{v_{i,y}}{g} = \frac{364 \text{ m/s}}{9.80 \text{ m/s}^2} = 37.1 \text{ s}$$

$$y_{\text{max}} = 0 + v_{i,y}t_{y,\text{max}} - \frac{1}{2}gt_{y,\text{max}}^2 = 6.76 \text{ km}$$

Then from this maximum height the projectile falls to h in a time t_{fall}:

$$6760 \text{ m} - \frac{g}{2}t_{\text{fall}}^2 = 400 \text{ m} \Rightarrow t_{\text{fall}} = 36.0 \text{ s}$$

The time of flight of the projectile is the sum of the rising and falling times, 73.1 s.
In a time of 73.1 s, the projectile travels a horizontal distance:

$$x = v_{i,x}t_{\text{flight}} = 73.1 \text{ s } (210 \text{ m/s}) = 1.54 \times 10^4 \text{ m}$$

In the situation on the right, the initial velocity is the same. The time that it takes the missile to rise to its maximum height and vertical distance traveled in that time are the same. But the missile now ends at $-h$ from the height at which it was launched, so it falls 7160 m.

$$-7.16 \times 10^3 \text{ m} = -\frac{1}{2}gt_{\text{fall}}^2 \Rightarrow t_{\text{fall}} = 38.2 \text{ s}$$

The time of flight now is 75.3 s and the horizontal distance that the missile travels is 1.58×10^4 m. Firing from the higher elevation adds another 460 m to the range.

(b) The difference in the time of flight is equal to the difference in the time for the missile to fall from its maximum height. We found the times from $y_{\text{final}} = y_{\text{max}} - 1/2gt^2{}_{\text{fall}}$.
Solving for each time and then taking the difference:

$$t_{\text{fall,high}} - t_{\text{fall,low}} = \Delta t = \sqrt{\frac{2(y_{\text{max}} + h)}{g}} - \sqrt{\frac{2(y_{\text{max}} - h)}{g}}$$

Multiply Δt by the x component of the velocity to find the added range, Δx:

$$\Delta x = v_{i,x}\left[\sqrt{\frac{2(h + y_{\text{max}})}{g}} - \sqrt{\frac{2(y_{\text{max}} - h)}{g}}\right]$$

27. Find the time to the top of the flight (zero y velocity after leaving the water) and multiply by 2 to get the total time of flight (0.826 s). Time of flight × the x velocity component gives $x = 3.99$ m.

29. Take the hikers' starting point as the origin, east to point toward positive x and north to point toward positive y. Measured from the positive x axis, Cassie's direction is 180° – 45.0° = 135°. Represent the first leg of each hike as a vector, then calculate the vector representing the second leg because the two legs of each journey need to add to 12.0 km north and 0 km east. The distance that each needs to walk is the magnitude of the second-leg vector.

$x_{\text{M1}} = (10.0 \text{ km}) \cos 30.0° = 8.66$ km

$y_{\text{M1}} = (10.0 \text{ km}) \sin 30.0° = 5.00$ km

Marcus's remaining walk must then have components of –8.66 km in the x direction and 12.00 km – 5.00 km = 7.00 km in the y direction.

$$r_{\text{M2}} = \sqrt{x_{\text{M2}}^2 + y_{\text{M2}}^2} = \sqrt{(-8.66 \text{ km})^2 + (7.00 \text{ km})^2} = 11.1 \text{ km}$$

$x_{\text{CI}} = (15.0 \text{ km}) \cos 135° = -10.6$ km

$y_{\text{CI}} = (15.0 \text{ km}) \sin 135° = 10.6$ km

Cassie's remaining walk must then have components of +10.6 km in the x direction and 12.00 km – 10.6 km = 1.39 km in the y direction.

$$r_{\text{C2}} = \sqrt{x_{\text{C2}}^2 + y_{\text{C2}}^2} = \sqrt{(10.6 \text{ km})^2 + (1.39 \text{ km})^2} = 10.7 \text{ km}$$

31. To determine which balloon hits the ground first, we need to calculate the time it takes each balloon to hit the ground; whichever balloon has the shorter time will hit the ground first. The difference is the initial velocity, as both are only accelerated by gravity. Velocity in the horizontal (x) direction does not affect time of fall.
Take y positive as up.
Time for balloon 1 to hit the ground:

$$y = y_0 + v_{0y}t + \frac{1}{2}a_yt^2 = y_0 + 0 + \frac{1}{2}a_yt^2$$

$$t = \sqrt{\frac{2(\Delta y)}{a_y}} = \sqrt{\frac{2(-6.00 \text{ m})}{\left(-9.80 \frac{\text{m}}{\text{s}^2}\right)}} = 1.11 \text{ s}$$

Time for balloon 2 to hit the ground:

$$\frac{1}{2}a_yt^2 + v_{0y}t - \Delta y = 0$$

$$t = \frac{-v_{0y} \pm \sqrt{v_{0y}^2 - 4\left(\frac{1}{2}a_y\right)(-\Delta y)}}{2\left(\frac{1}{2}a_y\right)}$$

$$= \frac{-\left(-2.00 \frac{\text{m}}{\text{s}}\right) \pm \sqrt{\left(-2.00 \frac{\text{m}}{\text{s}}\right)^2 + 2\left(-9.80 \frac{\text{m}}{\text{s}^2}\right)(-6.00 \text{ m})}}{\left(-9.80 \frac{\text{m}}{\text{s}^2}\right)}$$

$$= \frac{2.00 \pm 11.03}{-9.80} \text{ s}$$

Taking the positive solution: $t = 0.921$ s
Balloon 2 lands 0.19 s before balloon 1.

33. We will set the origin to be at the surface of the water directly beneath his starting position. The positive x direction is the same direction as his initial velocity and up will be the positive y direction. Once the boy leaves the platform, he will undergo projectile motion. His initial velocity is solely in the x direction.

(a) $v_y^2 = v_{0y}^2 + 2a_y(y - y_0) = 0 + 2a_y(\Delta y)$

$$v_y = \sqrt{2a_y(\Delta y)} = \sqrt{2\left(-9.80\frac{\text{m}}{\text{s}}\right)(-10.0\text{ m})} = 14.0\frac{\text{m}}{\text{s}}$$

$$v = \sqrt{v_x^2 + v_y^2} = \sqrt{\left(5.00\frac{\text{m}}{\text{s}}\right)^2 + \left(14.0\frac{\text{m}}{\text{s}}\right)^2} = 14.9\frac{\text{m}}{\text{s}}$$

(b) $y = y_0 + v_{0y}t + \frac{1}{2}a_yt^2 = y_0 + 0 + \frac{1}{2}a_yt^2$

$$= \sqrt{\frac{2(\Delta y)}{a_y}} = \sqrt{\frac{2(-10.0\text{ m})}{\left(-9.80\frac{\text{m}}{\text{s}^2}\right)}} = 1.43\text{ s}$$

(c) $x = x_0 + v_{0x}t + \frac{1}{2}a_xt^2 = 0 + \left(5.00\frac{\text{m}}{\text{s}}\right)(1.43\text{ s}) + 0$

$= 7.15\text{ m}$

35. (a) The ball is launched horizontally, so the initial vertical velocity is zero. Applying $\Delta x = v_0t + \frac{1}{2}at^2$ to the vertical motion:

$-1.5\text{ m} = \frac{1}{2}\left(-9.8\frac{\text{m}}{\text{s}^2}\right)t^2 \Rightarrow t = 0.55$ seconds. The horizontal speed $v_x = \frac{15\text{ m}}{0.55\text{ s}} = 27\text{ m/s}$.

(b) Because the ball is caught at the same height as thrown, the maximum range is obtained when the ball is thrown at a 45° angle above the horizontal. The time in the air can be calculated using the vertical motion.

$$\Delta y = v_{0y}t + \frac{1}{2}a_yt^2 \Rightarrow 0 = \left(27\frac{\text{m}}{\text{s}}\right)\sin(45°)t + \frac{1}{2}\left(-9.8\frac{\text{m}}{\text{s}^2}\right)t^2$$

$t = 3.9\text{ s}$

The horizontal range can then be found.

$$\Delta x = v_{0x}t = \left(27\frac{\text{m}}{\text{s}}\right)\cos(45°)\ (3.9\text{ s}) = 75\text{ m}$$

37. (a) i. Time upstream: $t_u = \frac{D/2}{v_{BW} - v_{WS}}$.

Time downstream: $t_d = \frac{D/2}{v_{BW} + v_{WS}}$.

Thus the total time $= t_1 = t_u + t_d = \frac{D/2}{v_{BW} - v_{WS}} + \frac{D/2}{v_{BW} + v_{WS}} = \frac{D}{2}\left(\frac{2v_{BW}}{v_{BW}^2 - v_{WS}^2}\right) = \left(\frac{Dv_{BW}}{v_{BW}^2 - v_{WS}^2}\right)$

ii. $v = \sqrt{v_{BW}^2 - v_{WS}^2}$ is the same speed for both directions, across and back, and thus for the total distance, D. So the time is $t_2 = \frac{D}{\sqrt{v_{BW}^2 - v_{WS}^2}}$.

(b) To compare the two values, we put them over the same common denominator.

So $t_1 = D\frac{v_{BW}}{v_{BW}^2 - v_{WS}^2}$ and $t_2 = D\frac{1}{\sqrt{v_{BW}^2 - v_{WS}^2}}\left(\frac{\sqrt{v_{BW}^2 - v_{WS}^2}}{\sqrt{v_{BW}^2 - v_{WS}^2}}\right) = D\frac{\sqrt{v_{BW}^2 - v_{WS}^2}}{v_{BW}^2 - v_{WS}^2}$. Because the numerator is greater for t_1 than it is for $t_2\left(v_{BW} > \sqrt{v_{BW}^2 - v_{WS}^2}\right)$, the time to go upstream and back is greater than the time to go across and back.

39. The horizontal range can be found by $x = v_{0x}t$, recognizing the time to the maximum range is twice the time to the maximum height, assuming it lands at the same height as it was thrown.
To solve for the time to the maximum height, we recognize that $v_y = 0$ and $v_{0y} = v_0\sin\theta$ so

$$y = v_{\text{average}}t = \frac{1}{2}(v_{0y} + v_f)t$$

$$h = \frac{1}{2}(v_0\sin\theta + 0)t$$

$$t = \frac{2h}{v_0\sin\theta}$$

And substituting back into the equation in the horizontal direction, we find

$$5h = (v_0\cos\theta)2\left(\frac{2h}{v_0\sin\theta}\right)$$

$$\theta = \tan^{-1}\left(\frac{4}{5}\right) = 39°$$

Chapter 3 AP® Practice Problems

1. (B) vector addition

$\vec{v}_{\text{obj,obs}} = \vec{v}_{\text{obj,belt}} + \vec{v}_{\text{belt,obs}} \Rightarrow \vec{v}_{\text{obj,belt}} = \vec{v}_{\text{obj,obs}} - \vec{v}_{\text{belt,obs}}$

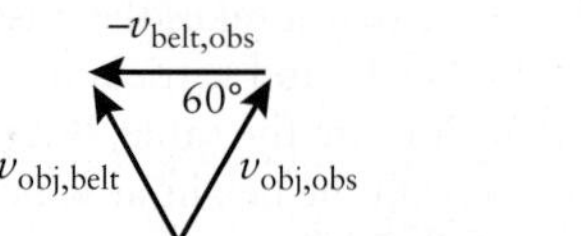

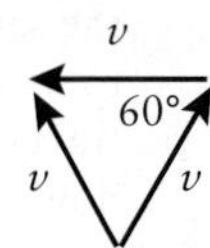

An equilateral triangle; therefore the answer is v.

3. (B) *A quick solution* comes from the fact that (C) and (D) cannot be correct from dimensional analysis, and the acceleration in the y direction is greater than in the x, so the magnitude of y is greater than the magnitude of x. *The mathematical solution is:*

$x = x_0 + v_{0x}t + \frac{1}{2}a_xt^2 \Rightarrow x = \frac{1}{2}\left(\frac{g}{2}\right)t^2$ and

$y = y_0 + v_{0y}t + \frac{1}{2}a_yt^2 \Rightarrow y = \frac{1}{2}(-g)t^2 = -\frac{1}{2}gt^2$

From the x equation:

$\frac{1}{2}gt^2 = 2x$ and substituting in the y equation $y = -2x$

5. (D) $x = x_0 + v_{0x}t + \frac{1}{2}a_x t^2$
There is no acceleration in the x direction, so the object would undergo equal changes in the x position in equal times. During the interval 7.5 m – 10.0 m the change in x clearly differs from the other intervals by more than the experimental error indicated by the circle size, so even though the x and y positions appear to be on a correct trajectory, the recording of the time may have been in error.
7. (A) The vertical component determines the time the object is in flight. The more positive the vertical component, the longer the time the object is in the air. The vertical component of velocity is positive for object A, zero for object B, and negative for object C.

Chapter 4

4-1 How objects move is determined by their interactions with other objects, which can be described by forces

1. The more mass the ball has, the more difficult it is to change its direction or speed. A greater force must be exerted on a ball as the mass of the ball increases to accelerate it.

4-2 If a net external force is exerted on an object, the object accelerates

1. The units of acceleration are m/s^2. The units of the right-hand side of this equation are kg · m/s^2. So the SI units of the newton (N) are kg · m/s^2.
3. (D) A net force of zero results in an acceleration of zero, so the object's velocity stops changing.
5. (A) For a given force, acceleration decreases as mass increases: $a_x = \sum F_{\text{ext},x}/m$.
7. $\sum F_{\text{ext},x} = ma_x = 8.00 \times 10^3$ N
9. The horizontal forces are F_2 and the x component of F_1. If $F_{1x} - F_2 = F_1 \cos\theta - F_2 > 0$, there is a horizontal acceleration in the $+x$ direction. If $F_{1x} - F_2 = F_1 \cos\theta - F_2 < 0$, a is in the $-x$ direction.
11. The slope of the velocity-versus-time graph is the acceleration, which is proportional to the net force exerted on the box. If $a_x = 0$, $F_{\text{ext},x} = 0$, so there is a force of 60 N in the opposite direction and the velocity–time graph is a horizontal line. Doubling the magnitude of the applied force to $2F$ will cause the net force to equal F and the slope of the velocity–time graph will be F/m.
13. The forward displacement of the body relative to the head shows that the relative velocity of the body was greater in the direction the rider was facing. So, the car was struck from behind, which caused a forward acceleration. Because the body is more connected to the car, the body achieved the same acceleration as the car, but the head, pulled forward only by the neck, did not initially accelerate as quickly.
15. (a) To hold the boat in a fixed position, $\sum \vec{F}_{\text{ext}} = 0$, so vector addition is done to find the third force.

(b)

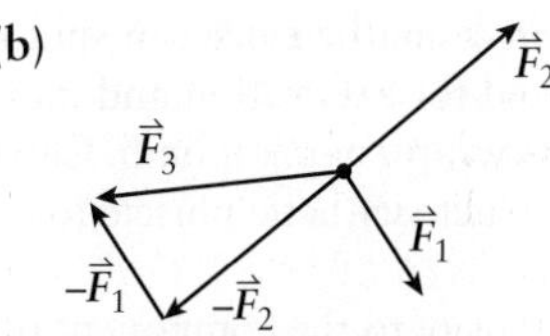

(c) x: $F_1\cos\theta_1 + F_2\cos\theta_2 + F_3\cos\theta_3 = 0$
y: $-F_1\sin\theta_1 + F_2\sin\theta_2 + F_3\sin\theta_3 = 0$

(d) $500\text{ N} + 1597\text{ N} + F_3\cos\theta_3 = 0$
$-866\text{ N} + 1204\text{ N} + F_3\sin\theta_3 = 0$

(e) $F_3\cos\theta_3 = -2097\text{ N} \quad F_3\sin\theta_3 = -338\text{ N} \Rightarrow \theta_3$
$= \tan^{-1}\frac{338}{2097} = 9.16° + 180° = 189° \quad F_3 = 2120\text{ N}$

(f) Given that the net force on the boat is zero when the engines are on, the accelerations due to the sum of the other forces should both be positive, because the components of F_3 are both negative. The angle is given relative to the positive x axis.

$$a_x = \frac{1000\text{ N}\cos(60.0°) + 2000\text{ N}\cos(37.0°)}{m}$$
$$= 0.932\text{ m/s}^2 \Rightarrow \Delta x = \frac{0.932}{2}\Delta t^2 = 46.6\text{ m}$$
$$a_y = \frac{-1000\text{ N}\sin(60.0°) + 2000\text{ N}\sin(37.0°)}{m}$$
$$= 0.150\text{ m/s}^2 \Rightarrow \Delta y = \frac{0.150}{2}\Delta t^2 = 7.50\text{ m}$$

and the displacement magnitude and direction are

$$\Delta d = \sqrt{\Delta x^2 + \Delta y^2} = 47.2\text{ m} \quad \theta = \tan^{-1}\frac{7.50\text{ m}}{46.6\text{ m}} = 9.14°$$

17. The maximum force that accelerates the 1300-kg car is exerted on the total mass of the system of two vehicles:
$$a = \frac{F}{m} = \frac{1300\text{ kg }(5.0\text{ N/kg})}{1300\text{ kg} + 1700\text{ kg}} = 2.2\text{ N/kg}$$

4-3 Mass and weight are distinct but related concepts

1. $\sum \vec{F}_{\text{ext}} = 0$. The scale initially had zero velocity, and continues to have zero velocity, and the floor supports the additional force the person exerts on the scale as well as the weight of the scale. If the weight of the person is 75 kg · g downward, then the normal force of the scale on the person, which is what the scale displays, is 75 kg · g upward.
3. $F_g = mg = 120\text{ kg}\frac{9.8\text{ N}}{\text{kg}} = 1180\text{ N} = 1200\text{ N}$
5. $F_g = mg_{\text{Mars}} = m(0.380)g_{\text{Earth}} = 80.0\text{ kg}(0.380)\frac{9.80\text{ N}}{\text{kg}}$
$= 298\text{ N}$
7. That the object doesn't accelerate downward under the influence of gravity indicates there must be an upward force from the surface.

9. A single sheet placed in a small crevice can split rock. Can these sheets be used for excavation and mining? They can be shaped to wrap a human limb. Can these sheets be used to construct synthetic muscle for prosthetic limbs?
11. (C) F_n is equal and opposite to the component of F_g perpendicular to the ramp's surface.

4-4 A free-body diagram is essential in solving any problem involving forces

1. One thing is touching the crate, so there is one contact force exerted on the crate. We will assume this is on Earth, so there is also a downward force of gravity. Draw these two vectors on a simple dot to represent the crate. Because the velocity is constant the net force on the crate is zero, so the lengths of the vectors must be equal and in opposite directions.
3. F_g is exerted on the center of mass of the system. For the seesaw to be horizontal the center of mass needs to be above the pivot so the larger child should sit closer to the pivot. The gravitational force on the board is exerted at the center of the board, so the gravitational force on the board need not be considered if the board is centered at the pivot.
5. (a) In A, there is no component of the motion along y, so the acceleration, and thus the sum of the forces in that direction will be zero.
   (b) For the block, there are three forces, the two strings that exert contact forces on the block and F_g. In C, the coordinate system is chosen so that at least one unknown force lies along an axis, making the problem easier to analyze because it reduces the number of unknowns to only the one component for that vector.
7. (a) For example: With the pencil in a vertical position move the horizontal shape around until the shape is balanced on the sharp point. That will be the location of the center of mass.
   (b) The procedure presented does not work because there is no cardboard that can make contact with the pencil when the center of mass of the object is located in a "hole."
   (c) This time place the pencil in a horizontal position and balance the shape on it. Mark the position of the length of the pencil across the shape. Rotate the shape and again balance it on the pencil and mark the new line showing the direction of the pencil. These two drawn lines, if projected across the shape, will cross at the center of mass even if it is in the hole.
9. (a)

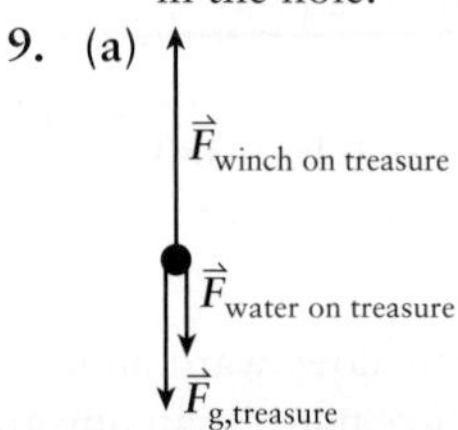

   (b) The net force is 540 N upward. So, a is 2.7 m/s^2 upward. $\Delta y = 1.35\ \text{m/s}^2\ t^2$.

4-5 Newton's third law relates the forces that two objects exert on each other

1. We see the tide rise in response to the force of gravity exerted by the Moon on the sea.
3. If the boxer and his opponent were two objects, then the boxer is correct because the forces are always equal in magnitude. But, the boxer is incorrect because they do have internal structures. A gloved fist and an unprotected jaw are two components of this internal structure and they differ in their response to external forces.
5. (a) The forces on the object are the two tensions, directed along the ropes, and the force of gravity, downward.
   (b) $F_{T1,y} = F_{g,\text{object}}$; $F_{T1} \sin (60°) = (5.0\ \text{kg})\left(9.8 \dfrac{\text{N}}{\text{kg}}\right)$;

   $F_{T1} = 57$ N

   (c) $F_{T1,x} = F_{T2}$; $F_{T1} \cos (60°) = F_{T2}$; (57 N) $\cos (60°) = F_{T2}$; $F_{T2} = 28$ N
   (d) The vertical component of the tension in the diagonal rope must always equal the force of gravity, so the magnitude of the tension must increase for the vertical component to remain the same. Therefore, the horizontal component of that force must also increase. And thus, tension in the horizontal rope must also increase.
7. (a) Amy's claim is that by Newton's third law, the force of the fan on the sail is equal and opposite to the force of the sail on the fan. Because they are both attached to the cart, the cart experiences a zero net force.
   (b) The cart does not move.
   (c) The sheet of paper is blown out horizontally or falls off of the system and the cart moves. This is a test of Amy's claim because she predicted the outcome if the force of the sail on the cart were removed from the system, so that the fan is exerting a force on the air, which is not attached to the cart, and so the air is exerting an oppositely directed force of equal magnitude on the fan, and thus the cart.

4-6 All problems involving forces can be solved using the same series of steps

1. (a)

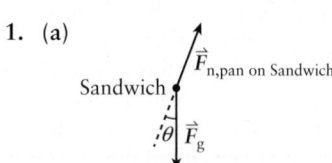

Sandwich $\vec{F}_{\text{n,pan on Sandwich}}$ θ $\vec{F}_g$

   (b) $F_n = mg \cos \theta = 0.200\ \text{kg}\ (9.8\ \text{N/kg}) \cos (20°) = 1.8\ \text{N}$

   The normal force exerted by the sandwich on the pan is equal in magnitude to the force exerted by the pan on the sandwich and in the opposite direction.

   (c) $a_x = F_g \sin \theta / m = g \sin \theta = 3.4\ \text{m/s}^2$ in the $+x$ direction

3. (a) The bar exerts a force on the building and the building exerts an equal and opposite force on the bar. If the balcony is not to collapse, there must be an upward component of the tension force on the bar to balance its weight.
(b) $mg - F_T \sin\theta = 0 \quad F_{\text{building on bar}} - F_T \cos\theta = 0$

$$F_T = \frac{mg}{\sin\theta} = \frac{F_{\text{building on bar}}}{\cos\theta} \Rightarrow F_{\text{building on bar}} = \frac{mg}{\tan\theta}$$

(c) $F_{\text{building on bar}} = \dfrac{mg}{\tan\theta} = \dfrac{3\text{ m}^3(2500\text{ kg/m}^3)(9.8\text{ N/kg})}{\tan(3°)}$

$$= 1.4 \times 10^6\text{ kg}\cdot\text{m/s}^2 \approx 1\text{ MN}$$

Although 1.4 is less than 1.5, and so is rounded down, the physics would suggest that the force on the bar would be rounded up for safety reasons.

5. (b) (also describes (a)) The problem is in one dimension, so the magnitudes of the forces are the magnitudes for this single component. The cable exerts a force upward and Earth is pulling downward on the combined mass of the elevator and zombies. As long as the elevator accelerates only downward, the zombies are secure by Newton's second law:

$(m_{\text{zombies}} + m_{\text{elevator}})a_{\text{elevator+zombies}} = F_{\text{cable on elevator}} - (m_{\text{zombies}} + m_{\text{elevator}})g$

$F_{\text{cable on elevator}} = (m_{\text{zombies}} + m_{\text{elevator}})a_{\text{elevator+zombies}} + (m_{\text{zombies}} + m_{\text{elevator}})g$

$F_{\text{cable on elevator}} = (m_{\text{zombies}} + m_{\text{elevator}})(-g/4) + (m_{\text{zombies}} + m_{\text{elevator}})g$

$= 3/4(m_{\text{zombies}} + m_{\text{elevator}})g = 8.0 \times 10^3\text{ N}$

(c) Coming to a stop at the 5th floor, the elevator accelerates upward. The cable snaps, because the force the cable exerts supporting both the zombies and elevator causes this acceleration. The "fatal" acceleration can be determined.

$F_{\text{cable on elevator}} - (m_{\text{zombies}} + m_{\text{elevator}})g = (m_{\text{zombies}} + m_{\text{elevator}})a$ and solve

$$1 + \frac{a}{g} = \frac{12{,}000\text{ N}}{10{,}600\text{ N}} \Rightarrow a = 1.3\text{ m/s}^2$$

(d) After the 5th floor we have a new episode: "Weightless Dead in Free Fall." $F_n = 0$.

7. (a) Because the novice and the abbot are connected by the rope, one will move up when the other moves down, so if we take the upward direction as positive y for the novice and negative y for the abbot, we can assign them the same $\boldsymbol{a_y}$. The tension forces are the same size. The force of gravity on the abbot is larger.
(b) and (c)

$$F_{g,\text{abbot}} - F_T = \frac{F_{g,\text{abbot}}}{g}a_y \Rightarrow F_T = F_{g,\text{abbot}}(1 - a_y/g)$$

$$F_T - F_{g,\text{novice}} = \frac{F_{g,\text{novice}}}{g}a_y \Rightarrow F_T = F_{g,\text{novice}}(1 + a_y/g)$$

$$a_y = \frac{F_{g,\text{abbot}} - F_{g,\text{novice}}}{F_{g,\text{abbot}} + F_{g,\text{novice}}}g = \frac{F_{g,\text{abbot}}(1 - 1/2)}{F_{g,\text{abbot}}(1 + 1/2)}g = \frac{g}{3}$$

(d) $\Delta y = \dfrac{1}{2}a_y\Delta t^2 \Rightarrow \Delta t_{\text{abbot}} = \sqrt{\dfrac{2\Delta y}{a_y}} = \sqrt{\dfrac{15}{g}} = 1.2\text{ s}$

(e) If the abbot lets go of the rope the instant he touches the ground, the novice will still have an upward velocity.

$v_y = v_{0y} + a_y t = 0 + 3.3\text{ m/s}^2(1.2\text{ s}) = 4.0\text{ m/s}$

After the abbot reaches the ground, the novice, now at 2.5 m, would continue upward an additional distance given by

$v_f^2 = v_i^2 + 2(-g)\Delta y = (4.0\text{ m/s})^2 + 2(-9.80\text{ m/s}^2)\Delta y \Rightarrow$
$\Delta y = 0.81\text{ m}$

So the novice's time to free fall the total distance to the ground is

$$\Delta y_{\text{total}} = \frac{1}{2}a_y t^2 \Rightarrow t = \sqrt{\frac{2\Delta y_{\text{total}}}{a_y}} = \sqrt{\frac{6.6\text{ m}}{9.8\text{ m/s}^2}} = 0.82\text{ s}$$

9. (a) Take upward to be positive; barbell and scale both touch the lifter. Gravity is exerted on both. Lifter exerts force on barbell.
(b) $F_{n,\text{scale on lifter}} - m_{\text{lifter}}g - F_{\text{barbell on lifter}} = 0$

$F_{\text{lifter on barbell}} - m_{\text{barbell}}g = m_{\text{barbell}}a$

Adding these together and applying Newton's third law to the interaction between the lifter and the barbell:

$F_{n,\text{scale on lifter}} = m_{\text{barbell}}g + m_{\text{lifter}}g + m_{\text{barbell}}a$

(c) When the barbell accelerates upward (after a), the normal force is greater than the sum of the weights. Between b and c, the acceleration is changing direction from upward to downward. When the barbell accelerates downward (toward c), the normal force is less than the sum of the weights.
(d) When the acceleration is zero, the normal force is equal to the sum of the weights. If the barbell has a weight of $9.8\text{ N/kg}\cdot 25\text{ kg} = 240\text{ N}$ and the total normal force is approximately equal to 840 N when $a = 0$, then the weight of the lifter is approximately 600 N, which equals a mass of approximately 61 kg.

11. (a) Taking downward as positive: $ma = mg - F_T \Rightarrow$ $F_T = mg - ma = 850\text{ N} - ma \le 800\text{ N}$ so a must be greater than (downward)

$$a \ge \frac{F_{\text{net}}}{m} = \frac{F_g - F_{T,\text{rope}}}{\left(\dfrac{F_g}{g}\right)} = \frac{850\text{ N} - 800\text{ N}}{\left(\dfrac{850\text{ N}}{9.8\text{ m/s}^2}\right)} = 0.58\text{ m/s}^2$$

(b) The start of the descent is the most hazardous. You must start over the edge with a sufficient downward acceleration. For example, the person holding the rope should walk forward as the descent begins, accelerating to not let the rope become taut.

13. (a) Write the equations for each diagram, choosing the direction of acceleration as positive.
$M_1g - F_T = M_1a \quad F_T - M_2g\sin\theta = M_2a$
$M_2g\cos\theta - F_{n,2} = 0$

(b) Solve each for F_T and set equal. Then set $a = 0$.

$M_1 g - M_1 a = M_2 a + M_2 g \sin\theta \Rightarrow M_1 g$

$= M_2 g \sin(30.0°) \Rightarrow M_1 = \dfrac{M_2}{2}$

(c) $M_1 g - M_1 a = M_2 a + M_2 g \sin\theta \Rightarrow a = 0.11\ g$

Block 1 falls and block 2 rises with a displacement that depends on time as

$$\Delta x_1 = \frac{1}{2}(0.11\ \text{g})\Delta t^2$$

15. (A) Violations of the third law have never been observed.

Chapter 4 Review Problems

1. (a) To reduce the net force on the line, the line should be reeled in when the fish is accelerating toward the boat, and reeled out when the fish accelerates away from the boat.

   (b) When the fish is lifted out of the water the tension in the line is equal to the weight of the fish plus the product of the acceleration and mass of the fish. The net provides a force opposite gravity, keeping the line from snapping.

3. For example: Muscles, guitar strings, and backpack straps are under tension. You increase the tension in shoestrings when you lace them tighter. The threads in your clothes are under tension near your shoulders, supporting the weight of cloth hanging down.

5. Claim: The larger and smaller blocks, neglecting air resistance, remain at the same separation. Evidence: The equation that describes the fall of each block is $\Delta x = gt^2/2$ (assuming they start from rest when the rope breaks). This is the same for both blocks. So, the distance fallen, Δx, is the same for both blocks in the same time interval, and the distance between them never changes. If air resistance is considered, there may be a different answer, depending on the distance of the fall, the cross-sectional area of the blocks, and the masses of the blocks.

7. The student is neglecting the fact that the action and reaction forces in Newton's third law are for two objects exerting forces of equal magnitude on each other in opposite directions. Net force, on the other hand, is the vector sum of forces exerted on a single object, and is described by Newton's second law of motion.

9. A child is pulling a 4.0-kg sled up a snowy hill that is inclined at 20° with the horizontal. The pulling force is exerted parallel to the hill and has a magnitude of 40 N. Friction can be considered negligible. The equation describes the magnitude of the sled's acceleration.

11. Because the driver is wearing her seat belt, we can model her motion as the motion of the car to which she is attached. Assuming the acceleration of the driver is constant, we can calculate this acceleration's magnitude and direction from the constant acceleration equations. (Take the initial direction of motion to be positive x.)

Assume the only force on the driver in the x direction is the contact force exerted by the seat belt.

Acceleration of the driver:

$$v_x^2 = v_{0x}^2 + 2a_x(x - x_0)$$

$$a_x = \frac{v_x^2 - v_{0x}^2}{2(\Delta x)} = \frac{0 - \left(14.0\ \frac{\text{m}}{\text{s}}\right)^2}{2(3.00\ \text{m})} = -32.7\ \frac{\text{m}}{\text{s}^2}$$

Then, Newton's second law for the driver:

$$\sum F_{\text{ext},x} = F_{\text{belt on driver}} = m_{\text{driver}} a_x$$

$$F_{\text{belt on driver}} = m_{\text{driver}} a_x = (52.0\ \text{kg})\left(-32.7\ \frac{\text{m}}{\text{s}^2}\right) = -1.70 \times 10^3\ \text{N}$$

13. (a) The bird is at rest on the wire, so the net force on the bird must be zero. Because the bird lands exactly in the wire's center, the magnitude of the tension forces on the left and right are the same because the horizontal components cancel. The wire is touching the bird, pulling it upward, and gravity is pulling down on the bird. A tension force is always along the string, wire, or rope causing the tension force.

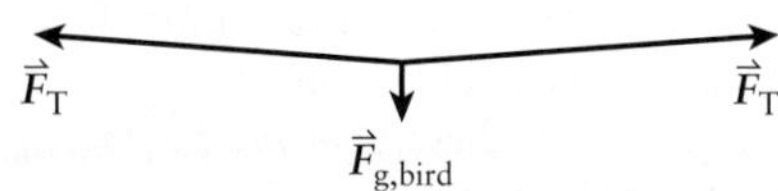

(b) and (c) We know the x and y components of the net force are both zero. We can actually solve this with just the y component and using the approximation that $\sin\theta \approx \tan\theta$.

$2F_{Ty} - m_{\text{bird}} g = 0 \quad F_{Ty} = F_T \sin\theta \approx F_T \Delta y/(L/2)$

where θ is the angle that the wire makes with the horizontal at the point where the bird sits. If we can neglect the mass of the wire, the tension is the same throughout the wire, so solving for

$$F_T\text{: } F_T = \frac{m_{\text{bird}} g\ (L/2)}{2\Delta y} = \frac{0.500\ \text{kg}\ (9.8\ \text{N/kg})25.0\ \text{m}}{2(0.15\ \text{m})} = 410\ \text{N}$$

15. (a) $F_1 \sin\theta_1 - F_2 \sin\theta_2 = Ma_y$

$F_1 \cos\theta_1 + F_2 \cos\theta_2 = Ma_x$

(b) Solving the x equation: $F_2 \cos\theta_2 = Ma_x - F_1 \cos\theta_1$

$F_2 \cos\theta_2 = \left(2.4 \times 10^{-8}\ \text{kg}\right)\left(1.7 \times 10^3\ \text{N/kg}\right)$

$-4.8 \times 10^{-5}\ \text{N} \cos(17°) = -5.1 \times 10^{-6}\ \text{N}$

$Ma_y = 0$, so $F_1 \sin\theta_1 - F_2 \sin\theta_2 = 0 \Rightarrow F_2 \sin\theta_2$

$= 1.4 \times 10^{-5}\ \text{N}$

$$\frac{F_2 \sin\theta_2}{F_2 \cos\theta_2} = \tan\theta_2 = \frac{1.4 \times 10^{-5}\ \text{N}}{-5.1 \times 10^{-6}\ \text{N}} \Rightarrow$$

$\theta = 70°$ below the $+x$ axis, so

$$F_2 = \frac{4.8 \times 10^{-5}\ \text{N} \sin(17°)}{\sin(70°)} = 1.5 \times 10^{-5}\ \text{N}$$

17. Assume initial direction of motion is positive x, so the passenger's displacement will be positive. The only force exerted on the passenger in the x direction is the contact force exerted by the air bag. Calculate the acceleration: $v_f^2 = v_i^2 + 2a_x\Delta x$. Use the acceleration and Newton's second law to solve for the force the air bag exerts on the passenger.

$$a_x = \frac{v_f^2 - v_i^2}{2(\Delta x)} = \frac{0 - \left(28.0\frac{\text{m}}{\text{s}}\right)^2}{2(1.5 \text{ m})} = -261\frac{\text{m}}{\text{s}^2}$$

$$\sum F_{\text{ext},x} = F_{\text{airbag on passenger}} = m_{\text{passenger}}a_x$$

$$F_{\text{airbag on passenger}} = m_{\text{passenger}}a_x = -1.2 \times 10^4 \text{ N}$$

We chose the initial direction of motion to be positive, so the negative tells us this force is opposite the direction of motion, which makes sense because it slows the passenger down. This is a catastrophic acceleration. Crumple zones and airbags increase stopping distance, which decreases acceleration and so decreases the force.

19. (a) The gravitational force exerted on the fish is exerted at the center of mass of the fish. This point must, therefore, align along the string in whatever orientation is chosen for the fish. The center of mass of the fish in the x–y plane must be at the intersection of these lines because the center of mass must lie somewhere along both of them.
(b) To find the actual center of mass of the fish, which is not two-dimensional, the fish should be hung so that a third line along the ventral (looking down) view of the fish can be placed.

21. We can use the constant acceleration kinematics equation $v_x^2 = v_{0x}^2 + 2a_x\Delta x$ to determine the required initial velocity for the given final velocity and displacement. We'll find the acceleration along the ramp by constructing a free-body diagram of the object and applying Newton's second law. We won't need to analyze forces perpendicular to the ramp, because no motion takes place in that direction. Choose up the incline to be the positive x direction. Our chosen equation will not tell us the direction of the velocity, but we know it must be positive x, given our choice to move up the incline.

$$\sum F_{\text{ext},x} = ma_x \quad \text{gives} \quad -mg\sin\theta = ma_x$$

$$a_x = -g\sin\theta = -\left(9.80\frac{\text{m}}{\text{s}^2}\right)\sin(9.00°) = -1.533\frac{\text{m}}{\text{s}^2}$$

$$v_{0x} = \pm\sqrt{v_x^2 - 2a_x\Delta x}$$

$$v_{0x} = \pm\sqrt{\left(0\frac{\text{m}}{\text{s}}\right)^2 - 2\left(-1.533\frac{\text{m}}{\text{s}^2}\right)(4.00 \text{ m})} = \pm 3.50\frac{\text{m}}{\text{s}}$$

$$v_{0x} = 3.50\frac{\text{m}}{\text{s}}$$

23. There is no acceleration perpendicular to the slope, so the net force in that direction is zero and we can use one-dimensional kinematics parallel to the slope. To determine the skier's final speed, we can calculate her final velocity with the kinematic equation $v_x = v_{0x} + a_x t$. Find her acceleration along the ski slope by drawing a free-body diagram of the skier (normal force perpendicular to the slope, gravity downward) and applying Newton's second law:

$$a_x = g\sin\theta = \left(9.80\frac{\text{m}}{\text{s}^2}\right)\sin(18.0°) = 3.03\frac{\text{m}}{\text{s}^2}$$

$$v_x = \left(0\frac{\text{m}}{\text{s}}\right) + \left(3.03\frac{\text{m}}{\text{s}^2}\right)(25.0 \text{ s}) = 75.7\frac{\text{m}}{\text{s}}$$

This incredible speed is almost twice the usual speed for a downhill racer. Neglecting friction and air drag is not truly viable in this case.

25. (a) The pulley is not accelerating, so for the forces on it to be balanced, there must be an upward net force on box B. Because the pulley and rope are assumed to be of negligible mass, the tension must be the same size throughout the rope. Choose the direction of acceleration as positive for each block, but to find the sum of the tensions on the pulley choose down to be positive (direction assignment must be consistent within any given calculation).

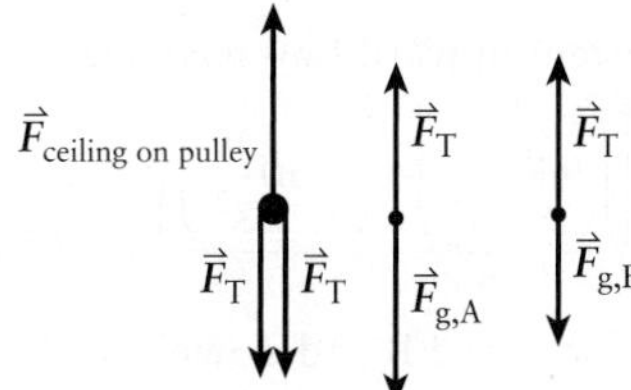

(b) $F_T - m_B g = m_B a \Rightarrow F_T =$

$2.00 \text{ kg } (9.80 \text{ N/kg} + 3.00 \text{ N/kg}) = 25.6 \text{ N}$

(c) $m_A g - F_T = m_A a \Rightarrow m_A = \dfrac{F_T}{g - a} = \dfrac{25.6 \text{ N}}{6.80 \text{ N/kg}}$
$= 3.76 \text{ kg}$

(d) The sum of the two tensions downward on the pulley is 51.2 N. The pulley is not accelerating so the ceiling must pull upward on the pulley at 51.2 N. By Newton's third law, the pulley must pull downward on the ceiling by this same amount.

27. We will use a coordinate system where the y axis is perpendicular to the ground and up is considered positive. The effective weight of the person is equal in magnitude to the normal force of the scale on the person. Use the y component of Newton's second law to calculate this magnitude because the net force on the person in the y direction must give the y component of the acceleration (the x component is balanced by a force of friction or the person would slide sideways). The other force with a vertical component is gravity.

$$\sum F_{\text{ext},y} = F_n - F_{\text{g,rider}} = F_n - m_{\text{rider}}g = m_{\text{rider}}a_y = m_{\text{rider}}\, a\sin(39.0°)$$

$$F_n = m_{\text{rider}}g + m_{\text{rider}}a\sin(39.0°) = m_{\text{rider}}g\left(1 + \left(\frac{a}{g}\right)\sin(39.0°)\right)$$

$$= (588 \text{ N})\left(1 + \left(\frac{1.25\frac{\text{m}}{\text{s}^2}}{9.80\frac{\text{m}}{\text{s}^2}}\right)\sin(39.0°)\right) = 635 \text{ N}$$

29. The man, chair, and bucket move together, so we can treat them as one object. The force that the man exerts on the rope is equal and opposite to the force the rope exerts upward on the man. The rope passes over the fixed pulley and the other end is attached to the bosun's chair, exerting the same amount of upward force. In a free-body diagram, the forces on the object are gravity downward and the two tension forces upward. We'll use Newton's second law for the y components for the combined object to relate the forces to acceleration. Let upward be the positive y direction:

$$\sum F_{\text{ext},y} = 2F_T - F_g = 2F_T - m_{\text{obj}}g = m_{\text{obj}}a_y$$

$$F_T = \frac{m_{\text{obj}}(g + a_y)}{2}$$

Part (a), at constant upward speed:

$$F_T = \frac{(95.0\text{ kg})\left[\left(9.80\frac{\text{m}}{\text{s}^2}\right)+0\right]}{2} = 466\text{ N}$$

Part (b), at constant upward 1.50 m/s^2 acceleration:

$$F_T = \frac{(95.0\text{ kg})\left[\left(9.80\frac{\text{m}}{\text{s}^2}\right)+\left(1.50\frac{\text{m}}{\text{s}^2}\right)\right]}{2} = 537\text{ N}$$

31. (a) Take right for A and B and down for C as positive.

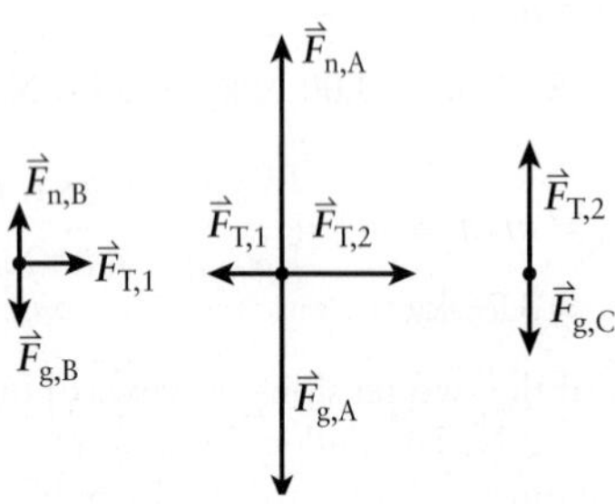

(b) $m_C g - F_{T,2} = m_C a \quad F_{T,2} - F_{T,1} = m_A a \quad F_{T,1} = m_B a$ and $F_{n,B} = F_{g,B} \quad F_{n,A} = F_{g,A}$

(c) $F_{T,2}$ must be greater than $F_{T,1}$ because it must accelerate both blocks A and B if $a \neq 0$ and m_C, m_B and $m_A \neq 0$.

(d) $m_C g = (m_C + m_B + m_A)a \Rightarrow a = \dfrac{m_C g}{(m_C + m_B + m_A)}$, so

$$F_{T,1} = m_B \frac{m_C g}{(m_A + m_B + m_C)}$$

$$F_{T,2} = (m_B + m_A)\frac{m_C g}{(m_A + m_B + m_C)}$$

33. Although the sizes are different, we can reuse the free-body diagrams and initial work from problem 31.

(a) $a = \dfrac{m_C g}{(m_C + m_B + m_A)} = \dfrac{4}{7}g$

(b) $F_{T,2} = (m_A + m_B)a = (1.00\text{ kg} + 2.00\text{ kg})\dfrac{4}{7}g = \left(\dfrac{12}{7}\text{ kg}\right)g$

(c) $F_{T,1} = m_A a = (1.00\text{ kg})(4/7)g = (4/7\text{ kg})g$

35. First, calculate the speed with which the diver enters the water using the constant acceleration equations. The acceleration necessary for her to come to rest in 5.00 m will be the minimum average acceleration needed to stop within this distance. Before the diver enters the water, a free-body diagram only includes the downward force of gravity. In the water, that force is still present, but there is also an upward force due to the water. Because she must slow down, the upward force must be greater. We're asked for the average force of the water on the diver, so we can treat it as a constant acceleration and use Newton's second law.

Speed entering the water, taking upward as positive:

$$v_y^2 = v_{0y}^2 + 2a_y(y - y_0)$$

$$v_y = \sqrt{v_{0y}^2 + 2a_y(\Delta y)} = \sqrt{0 + 2\left(-9.80\frac{\text{m}}{\text{s}^2}\right)(-10.0\text{ m})} = 14.0\frac{\text{m}}{\text{s}}$$

Acceleration necessary to stop the diver in the water:

$$a_y = \frac{v_y^2 - v_{0y}^2}{2(\Delta y)} = \frac{0 - \left(14.0\frac{\text{m}}{\text{s}}\right)^2}{2(-5.00\text{ m})} = 19.6\frac{\text{m}}{\text{s}^2}$$

Average force of the water on the diver:

$$\sum F_{\text{ext},y} = F_{\text{water on diver}} - F_{\text{g,diver}} = F_{\text{water on diver}} - m_{\text{diver}}g = m_{\text{diver}}a_y$$

$$F_{\text{water on diver}} = m_{\text{diver}}(g + a_y) = (62.0\text{ kg})\left(9.80\frac{\text{m}}{\text{s}^2} + 19.6\frac{\text{m}}{\text{s}^2}\right)$$

$$= 1.82 \times 10^3\text{ N}$$

Chapter 4 AP® Practice Problems

1. (B) $\vec{F}_{\text{A on B}} = -\vec{F}_{\text{B on A}}$
The forces must represent the interaction between a pair of objects. (A) and (C) both have the flaw that the forces are exerted on the same object.

3. (C) $\vec{a} = \dfrac{\sum \vec{F}_{\text{ext}}}{m}$

$F = M(2.0\text{ m/s}^2) = (M + 1.0\text{ kg})(0.50\text{ m/s}^2)$

5. (A) $\vec{a} = \dfrac{\sum \vec{F}_{\text{ext}}}{m}$, and for a_{cm}, we can treat it as a single $5M$ object.

7. (A) Box 2 is at rest. There are only two forces exerted on it, the force of gravity, $F_{g,2}$, and the upward normal force exerted by box 1, $F_{n,2}$. So those two forces must be equal in magnitude and opposite in direction. The force of gravity, $F_{g,1}$, on box 1 is less than the force of gravity on box 2, because box 1 has less mass than box 2. There are three forces on box 1: the force of gravity exerted by box 1 (downward), the normal force exerted by the floor (upward), and the contact force exerted by box 2 (downward). Because box 1 is at rest, the net force is zero, so the force exerted by the floor must have the same magnitude as the sum of the other two forces. Thus $F_{n,1}$ must be bigger than $F_{n,2}$.

Chapter 5

5-1 We can use Newton's laws in situations beyond those we have already studied

1. An object on an inclined plane, being pulled by a cord at a slightly upward angle on a horizontal floor, on a surface at the top of a vertical circle (like on a roller coaster, being at the top of a loop and upside down) or in an elevator that is accelerating either up or down, are examples of the normal force not being equal to the object's weight.

5-2 The static friction force changes magnitude to offset other forces being exerted on a system

1. On a horizontal surface when no other forces are exerted on the box, the normal force will be the same magnitude as the weight of the box, so to just overcome friction

$$F_{ext} = \mu_S mg = 0.67(5.00 \text{ kg})(9.8 \text{ N/kg}) = 33 \text{ N}$$

3. (D) The direction of the friction force on an object is always that which would oppose the relative motion between two surfaces that would occur if there were no friction. The static friction force is only directly proportional to the normal force when it is maximum.
5. (a) At $\theta = 0°$, the force exerted by the person on the table does not contribute to the normal force exerted on the table, the horizontal force on the table is greatest, and the normal forces exerted by the floor on both the person and table are the same because their masses are equal. Because the coefficient of friction between the person and the floor is smaller, and the contact force exerted by the person on the table is equal in magnitude and opposite in direction to the contact force that the table exerts on the person, the acceleration of the person is always greater. Therefore, there is no angle at which the person can exert a force for which the table will slip first.
   (b) At $\theta = 0°$, the normal forces are equal to the weights, so the friction forces exerted on the person and the table are equal in magnitude when the product of friction coefficient and weight are equal. This occurs when the mass of the person is five times greater than the mass of the table.
   (c) The vertical forces must cancel, because the table is not accelerating in that direction. Given (except for $\theta = 0°$) that there is a *downward* component to the force of the person on the table, the normal force of the floor on the table will be *increased* by the amount of the vertical component of the force of the person on the table, which also increases the magnitude of the maximum static friction force.

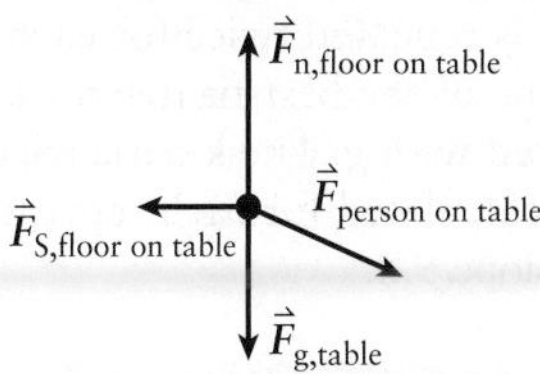

   (d) The vertical forces must cancel, because the person is not accelerating in the vertical direction. Given (except for $\theta = 0°$) that there is an *upward* component to the force of the table on the person, the normal force of the floor on the person will be decreased by the amount of the vertical component of the force of the table on the person.

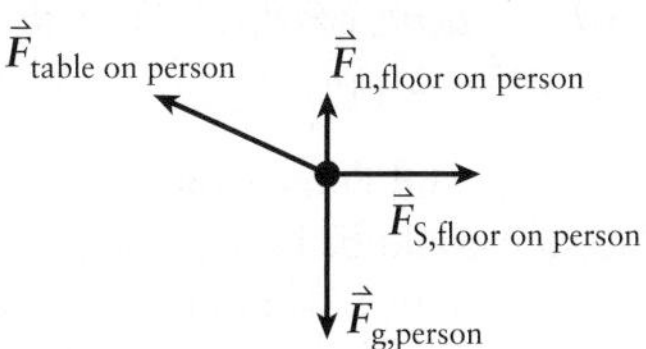

   (e) Solve for the case of zero force (no sliding) and then the exerted contact force must be less than or equal to that force or the table will slide. In both diagrams take up (y) and to the right (x) to be positive. Write equations for x and y components separately:

$$F_{\text{person on table},x} - F_{S,\text{floor on table}} = 0$$
$$F_{g,\text{table}} + F_{\text{person on table},y} - F_{n,\text{floor on table}} = 0$$
$$F_{\text{table on person},x} - F_{S,\text{floor on person}} = 0$$
$$F_{g,\text{person}} - F_{\text{table on person},y} - F_{n,\text{floor on person}} = 0$$

So,

$$F_{\text{person on table}} \le \frac{\mu_{S,\text{table}} F_{g,\text{table}}}{\cos\theta - \mu_{S,\text{table}} \sin\theta}, \text{ or table slides}$$

$$F_{\text{table on person}} \le \frac{\mu_{S,\text{person}} F_{g,\text{person}}}{\cos\theta + \mu_{S,\text{person}} \sin\theta}, \text{ or person slides}$$

   (f) As the angle decreases, more of the force exerted by the person lies in the x direction and less of the force pushes down on the table. Therefore, as the angle increases, the normal force on the table increases and the friction force that resists the motion of the table increases. For some ratio of weights of the person and the table both the person and table slide simultaneously, and the values of this ratio are graphed as the curve. For any angle, if the ratio of weights lies above the point on this curve the table will slide first and for ratios below this value the person will slide first.
7. (a) The block on the plane would be experiencing a force of gravity exerted by Earth, a normal force exerted by the plane, and a force of friction opposing motion along the plane. The static friction force is zero when the plane is horizontal, as there is nothing causing the block to try to move on the plane. The greater the angle, the larger the component of the gravitational force down the plane, so the bigger the friction force up the plane. As the component of the gravitational force down the plane increases, the component of the gravitational force perpendicular to the plane decreases, so the normal force is decreasing. The block will not move until the force of gravity

exceeds the limit of the friction force, $\mu_S F_n$. At the angle just before it begins to slip, you could set the component of the force due to gravity down the incline equal to the friction force and solve for the coefficient of static friction.

(b) Apply the second law at the angle identified in part (a):

$$mg \sin\theta_{slip} - \mu_S mg \cos\theta_{slip} = 0 \Rightarrow \mu_S = \tan\theta_{slip}$$

$$\theta_{slip} = \tan^{-1}\mu_S = 11°$$

(c) Lubricants A and D have measured intervals that overlap so it could be impossible to tell them apart. Lubricants B and E, similarly, have measured intervals that overlap. Lubricant C (grease G with molybdenum disulfide added) has the lowest static friction coefficient, and so would be the best lubricant.

5-3 The kinetic friction force on a sliding object has a constant magnitude

1. The friction force between an object and a surface is $\mu_k F_n$. Because the surface is horizontal, normal and gravitational forces balance, so the friction force is equal to $\mu_k mg$, $m = 12.7\ \text{N}/(0.37 \cdot 9.8\ \text{N/kg}) = 3.5\ \text{kg}$.
3. First, draw a free-body diagram showing the forces exerted on the mop head. Use the free-body diagram to write equations using Newton's second law. From the sum of forces in the vertical direction (0), find the normal force on the mop head, and then use that to find the friction force exerted by the floor on the mop head. The floor exerts two forces on the mop head: F_n upward and F_k opposing the direction of motion. The other two forces on the mop head are mg and the force due to the handle, F. Taking downward to be positive:

$$F_{net,y} = F\sin\theta + mg - F_n = 0 \Rightarrow F_n = F\sin\theta + mg$$
$$= 50.0\ \text{N}\sin(50.0°) + 3.75\ \text{kg}(9.80\ \text{N/kg}) = 75.1\ \text{N}$$

and

$$F_{net,x} = F\cos\theta - F_k = F\cos\theta - \mu_k F_n$$

$$a_x = \frac{F_{n,x}}{m} = \frac{2.1\ \text{N}}{3.75\ \text{kg}} = 0.56\ \text{N/kg} = 0.56\ \text{m/s}^2$$

5. $v_{fx}^2 = v_{ix}^2 + 2a_x\Delta x = 0 \quad a_x = -g\sin\theta - \mu_k g\cos\theta$

$$\mu_k = \frac{-\left(-\dfrac{v_{ix}^2}{2\Delta x}\right)}{g\cos\theta} - \tan\theta \Rightarrow \mu_k = \frac{v_{ix}^2}{2g\Delta x\cos\theta} - \tan\theta$$

$$\mu_k = -\tan\theta + \frac{(4.5\ \text{m/s})^2}{2g\cos\theta\ \Delta x} = -\tan(10°)$$
$$+ \frac{(4.5\ \text{m/s})^2}{2(9.8\ \text{m/s}^2)\cos(10°)(2.1\ \text{m})} = 0.32$$

While in the range reported, many additional trials are needed to consider the range verified.

7. (a) Three forces: a normal force and force of friction exerted by the ramp and force of gravity exerted by Earth. The normal force and the perpendicular-to-the-ramp component of the gravitational force cancel, and the component of the gravitational force along the plane is bigger than the friction force.

(b) The gravitational force along the incline is increasing, too. So, just the increase in the acceleration is not sufficient to justify the claim.

(c) Assuming that the friction force is proportional to the normal force, apply Newton's second law perpendicular to the ramp to get the normal force, and then along the ramp (taking the direction of motion to be positive):

$$mg\sin\theta - \mu_k mg\cos\theta = ma$$

(d) To allow a graph with a line of best fit, which we can use to find the variable of interest, define variables: $x = 1/\tan\theta$ and $y = a/\sin\theta$:

$$g\left(1 - \mu_k\frac{1}{\tan\theta}\right) = \frac{a}{\sin\theta} \Rightarrow y = g(1 - \mu_k x)$$

(e)

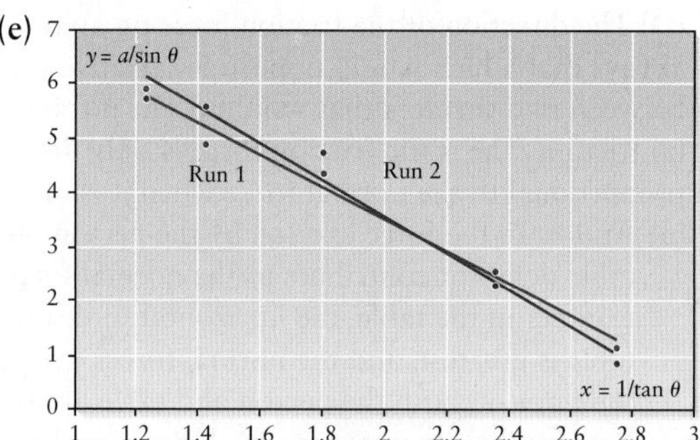

Best fit lines from run 1 and run 2 give intercepts of 9.4 and 10.3 m/s^2, respectively, and 9.8 m/s^2 is in this interval, so the data do not disprove the claim.

(f) To test the claim, we must change the surface area while maintaining the mass, which can be achieved by laying the two blocks on the board in two ways: in contact side by side or stacked. They remain in contact as they descend, so the side-by-side blocks form a single object with the same mass and twice the contacting surface area as the stacked blocks. The motion can then be analyzed for differences.

Independent and dependent variables: surface area and acceleration.

First a pair of stacked blocks is considered. For each of five inclinations, runs are made while velocity data are collected. Best fit lines through the velocities for each run are used to determine acceleration. The graph of [all accelerations]/ sin θ versus 1/tan θ is made from these data. The slope and intercept are determined. This process is repeated twice for each inclination. The slope of the best-fit line for each graph can be compared with g. The kinetic friction coefficient can be calculated for each repetition as $\mu_k = -\text{slope}/9.8$.

The same data collection process described is carried out for the side-by-side blocks.

9. (A) Each term is proportional to mass.

$$mg\sin\theta - \mu_k mg\cos\theta = ma$$

5-4 Problems involving friction are solved like any other force problems

1. (a) (A) The horizontal force is less than the minimum of 14.7 N required to overcome the static friction force.
   (b) (D) Static friction force is overcome, and the constant kinetic friction force opposes the motion.

$$a = \frac{F - F_k}{m} = \frac{20.0\text{ N} - (0.45)(2.0\text{ kg})(9.8\text{ N/kg})}{2.0\text{ kg}}$$
$$= 5.6\text{ m/s}^2$$

3. (a)

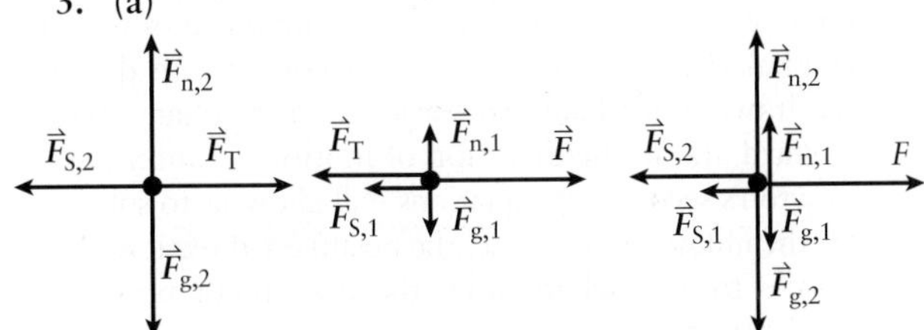

   (b) $F_T - F_{S,2} = M_2 a \quad F - F_T - F_{S,1} = M_1 a$
   $F - F_{S,2} - F_{S,1} = (M_1 + M_2)a$
   (c) At the minimum $F_T = \mu_{S,2} M_2 g = 0.43(12.0\text{ kg})(9.8\text{ N/kg}) = 51\text{ N}$ But this requires that the force F consistent with this tension is large enough to cause the first block to slide.
   $F \geq F_{S,2} + F_{S,1} = \mu_{S,2} M_2 g + \mu_{S,1} M_1 g = 78\text{ N}$

5. (a) The normal force of the large block on the smaller block is the net force on the small block in the horizontal direction: $F = F_n = ma$. The upward friction force on the small block is therefore $\mu_S F_n = \mu_S ma$. For the small block not to slide, the upward friction force must balance the downward force of gravity.
   (b) Setting the magnitudes equal: $F_S = F_g$,

$$F_n = ma \rightarrow \mu_S ma = mg \rightarrow \mu_s = \frac{g}{a} = \frac{9.8\text{ m/s}^2}{15\text{ m/s}^2} = 0.65$$

7. (a) For the system to be treated as a single object, the accelerations of the blocks must be equal.

$$F - F_{S,1\text{ on }2} = M_2 a_2 \quad F_{S,2\text{ on }1} = M_1 a_1 \quad a_1 = a_2 \Rightarrow$$
$$F = (M_1 + M_2)a$$

So

$$a = \frac{F_{S,2\text{ on }1}}{M_1} \Rightarrow F = \frac{(M_1 + M_2)}{M_1} F_{S,2\text{ on }1}$$
$$\leq \frac{(M_1 + M_2)}{M_1}\mu_S g M_1 = \mu_S (M_1 + M_2) g$$

   (b) Friction forces are internal, so

$$a = \frac{F}{M_1 + M_2} \leq \mu_S g$$

   (c) The friction between the blocks is now kinetic.

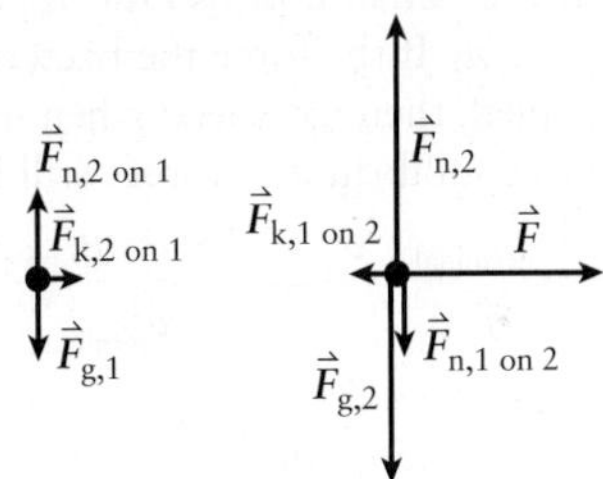

   (d) $F - F_{k,1\text{ on }2} = M_2 a_2 \quad F_{k,2\text{ on }1} = F_{k,1\text{ on }2} = M_1 a_1$
   $= \mu_k M_1 g \quad a_1 = \mu_k g$

$$a_2 = \frac{F - F_{k,1\text{ on }2}}{M_2} = \frac{F - M_1\mu_k g}{M_2}$$

   (e) a_2 is the ratio of the difference between the applied force and kinetic friction force and the larger mass. For a large force F, the friction contribution will probably be negligible, and can be ignored. If this is the case, the smaller block, which accelerated only due to kinetic friction, will not accelerate at all, but will be "left behind" as the larger block slides beneath it at approximately F/M_2.

9. Examples: Does liquid water form a film that lubricates the boundary between skate blades and ice in the same way that it does between the tires and the road? Is the water filling pits in the ice surface as it does pits in the road surface, smoothing the surface as in the model? Do copper skates have more pitting than mild steel or plastic, or less pitting? When skates slide over ice, does this melting occur less with copper than with mild steel or plastic?

11. Lower block: Forces are kinetic friction force due to table to the left, kinetic friction force due to upper block to the left, and tension to the right. Upper block: same tension, to the right, kinetic friction force due to lower block to the right, and the force pulling to the left.

    Vertical second law equations are used to find the normal forces needed for the friction forces in the horizontal equations. The acceleration is zero because the blocks are moving at constant velocity.

$$F - M\mu_k g - F_T = 0$$
$$F_T - (M + M_1)\mu_k g - M\mu_k g = 0$$

The objects cannot be treated as a single object given that not each point on each block is going through the same displacement. Set these two equations equal (because they both equal zero), cancel the tension force, and solve for M.

$$F - (3M + M_1)\mu_k g = 0 \Rightarrow M = \frac{F - M_1\mu_k g}{3\mu_k g}$$

5-5 An object moving through air or water experiences a drag force

1. The net force is zero for terminal speed ($a = 0$).

$$F_{net} = F_{propulsion} - F_{drag} = 0 \Rightarrow F_{propulsion} = F_{drag}$$
$$= cv^2 = (0.31\text{ Ns}^2/\text{m}^2)(1.5 \times 10^{-4}\text{ m/s})^2 = 7.0 \times 10^{-9}\text{ N}$$

3. (B) is correct while the speed is increasing.
5. (C) This is a very small body, so the appropriate drag force is $F_{drag} = bv$. If the force the bacteria exerts on the water is doubled, then the speed when the bacterium again reaches equilibrium of forces will be $2v_i$.
7. (a) $a = g - \frac{cv^2_{terminal}}{m} = 0 \Rightarrow c = \frac{mg}{v^2_{terminal}}$

$$= \frac{4\pi r^3 \rho}{3} \frac{g}{v^2_{terminal}} = 33.5\ \text{mm}^3 \cdot \left(\frac{1\ \text{m}}{1.00 \times 10^3\ \text{mm}}\right)^3$$

$$\cdot \frac{1.00 \times 10^3\ \text{kg}}{\text{m}^3} \cdot \frac{9.80\ \text{m/s}^2}{72.3\ \text{m}^2/\text{s}^2} = 4.55 \cdot 10^{-6}\ \text{kg/m}$$

(b) As the cross-sectional area increases, the drag force increases. When the radius of the drop is doubled the cross-sectional area of the drop (and thus c) increases by a factor of 4. When the radius of the drop is doubled the volume of the drop increases by a factor of 8, and so does the mass. Given that

$$c = \frac{mg}{v^2_{terminal}} \Rightarrow v_{terminal} = \sqrt{\frac{mg}{c}}$$

So, $c_{8\,mm} = 4c_{4\,mm} \quad m_{8\,mm} = 8m_{4\,mm} \quad v^2_{terminal} = \frac{mg}{c}$

$$\Rightarrow v_{terminal,8\,mm} = \sqrt{2}v_{terminal,4\,mm} = 12\ \text{m/s}$$

9. No violation. The law relates acceleration to net force.
11. The terminal speed of the rescuer can be larger than that of the rescued if the rescuer reduces her drag coefficient by reducing her cross-sectional area.

5-6 An ideal spring force can be used to model many interactions

1. In equilibrium, the forces on the object will sum to zero. For positive y upward, $F_s - mg = 0$ so $\Delta y = mg/k = 0.0588\ \text{m} = 5.88\ \text{cm}$.
3. (a) Plot the data given with units.
(b) The net force on the object in each case is zero so $F_{sx} - mg = 0 \Rightarrow k\Delta x = mg$. Comparing to the equation for a straight line, the theoretical y intercept should be 0, and the slope k. From the best fit line, $k = 0.96$ N/m.

Chapter 5 Review Problems

1. (A) The static friction coefficient is the dominant one in pure rolling motion. Because at 45° sine and cosine are equal, and the cosine value decreases and sine value increases for angles ≥ 45°, the static friction coefficient must be ≥1 for the wheel to accelerate up the incline.

$$\sum F_{ext,x} = mg\mu_S \cos\theta - mg \sin\theta \geq 0 \quad \mu_S \geq \sin 45°/\cos 45° = 1$$

3. Because the book is not changing speed, $F_{net} = 0$ so

$$\mu_k F_n = \mu_k mg = F_{applied} = mg/2 \Rightarrow \mu_k = 0.5$$

5. (a) Does not change. It is an intrinsic property of the book–Earth system.
(b) Decreases. The normal force that the wall exerts on the book is equal and opposite to the force exerted by the tip of your finger.
(c) Does not change. The friction force exerted by the wall on the book is equal and opposite to the gravitational force exerted on the book given that the book remains at rest. Because the weight of the book does not change, neither does the friction force that opposes the gravitational force.
(d) Does not change. The maximum static friction force is a constant limiting value. If the weight of the book exceeds this maximum force, the book will slide.
7. First construct a free-body diagram for the dog. The forces exerted on the dog are the dog's weight, the normal force, kinetic friction exerted by the floor, and the tension in the rope toy. The normal force and the upward component of tension will balance gravity. Because velocity is constant, the horizontal component of tension will balance the friction force. The components do not get drawn individually, so tension is the resultant vector upward and in the direction of motion. Combining Newton's second law equations will allow us to solve for the unknown force. Let the positive x direction be parallel to the level ground in the direction of motion, and let the positive y direction be upward.

Newton's second law y components:

$$\sum F_{ext,y} = F_n + F_{T,y} - F_{g,dog} = F_n + F_T \sin\theta - m_{dog}g = m_{dog}a_y = 0$$
$$F_n = m_{dog}g - F_T \sin\theta$$

Newton's second law x components:

$$\sum F_{ext,x} = F_{T,x} - F_k = F_T \cos\theta - \mu_k F_n = m_{dog}a_x = 0$$

Combining equations and solving for tension:

$$F_T \cos\theta - \mu_k(m_{dog}g - F_T \sin\theta) = 0$$

$$F_T = \frac{\mu_k m_{dog} g}{\cos\theta + \mu_k \sin\theta} = \frac{(0.367)(18.5\ \text{kg})\left(9.80\ \frac{\text{m}}{\text{s}^2}\right)}{\cos(23.0°) + (0.367)\sin(23.0°)}$$

$$F_T = 62.5\ \text{N}$$

9. Only (D) is supported by the evidence. As velocity increases, the drag coefficient increases for these measurements near the melting point of ice. A linear increase in the friction force as velocity increases is a dependence that is consistent with a drag force at low velocities through a highly resistant medium.
11. To remove the book, the librarian must overcome the static friction forces that the other books in the stack exert on the third book. The four books above the book in question can be treated as a single object with mass $4M$, because we are not interested in the forces among them, only in the force they exert on the third book. The same is true for the two books below. Construct a free-body diagram of the book in question and the collective object above and below the third book. Applying Newton's second law to these objects will allow us to calculate the minimum necessary yanking force, which will have to be just bigger than one that gives a net horizontal force of

zero when the static friction force on the third book is maximum. The kinetic friction force will be less, but the book won't start moving unless we overcome static friction. Let the direction of the yanking force be the positive x direction, and let upward be the positive y direction.

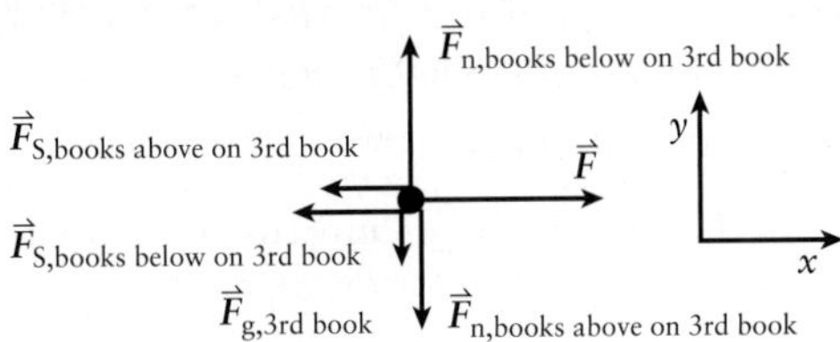

First find the normal forces.

$$\sum F_{\text{ext},y} = F_{\text{n,3rd book on books above}} - F_{\text{g,books above}} = m_{\text{books above}} a_y = 0$$

$$F_{\text{n,3rd book on books above}} = m_{\text{books above}} g = 4\ Mg$$

$$\sum F_{\text{ext},y} = F_{\text{n,books below on 3rd book}} - F_{\text{n,books above on 3rd book}} - F_{\text{g,3rd book}} = m_{\text{3rd book}} a_y$$

$$F_{\text{n,books below on 3rd book}} - 4\ Mg - Mg = m_{\text{3rd book}} a_y = 0$$

$$F_{\text{n,books below on 3rd book}} = 5\ Mg$$

Newton's second law x components for third book, substituting previous results:

$$0 = \sum F_{\text{ext},x} = F - F_{\text{S,books above on 3rd book}} - F_{\text{S,books below on 3rd book}}$$

$$0 = F - \mu_S F_{\text{n,books above on 3rd book}} - \mu_S F_{\text{n,books below on 3rd book}}$$

$$0 = F - 9\mu_S Mg$$

This was the minimum condition, so $F > 9\mu_S Mg$

13. The furniture is moving at constant velocity so the external force you are exerting on the object must be equal and opposite the friction force that the floor exerts on the object. The magnitude of this force is the weight of the object multiplied by the coefficient of kinetic friction. When the weight is suddenly increased the friction force is increased, producing a net force directed opposite to its motion, so it begins to slow down. When the speed reaches zero, the object is at rest and the magnitude of the static friction force will be equal in magnitude to the force you are exerting, so $F_{\text{net}} = 0$.

15. Because her speed is constant, the magnitude of the drag force is equal to the magnitude of the component of her weight parallel to the incline. This allows us to solve for the drag coefficient c, which depends on only her shape and the air, not the speed. In order to travel down the hill at a faster speed, an external force in the same direction as the parallel component of her weight must be exerted on her to accelerate her to a new terminal speed, in which case the net force will again be zero, with a larger drag force. Define the positive x axis down the incline.

$$F_{\text{applied}} = m_{\text{girl+scooter}} g \sin(10.0°) - cv_{1,x}^2 = m_{\text{girl+scooter}} a_x = 0$$

$$c = \frac{m_{\text{girl+scooter}} g \sin(10.0°)}{v_{1,x}^2}$$

This coefficient remains the same for the higher speed, when an external force must be exerted on the girl and scooter in the direction of motion (while the other forces in the free-body diagram remain the same):

$$\sum F_{\text{ext},x} = 0 = F_{\text{applied}} + F_{\text{g,girl+scooter},x} - F_{\text{drag}} = F_{\text{applied}} + m_{\text{girl+scooter}}\, g \sin(10.0°) - cv_{2,x}^2$$

$$F_{\text{applied}} = cv_{2,x}^2 - m_{\text{girl+scooter}}\, g\sin(10.0°) = cv_{2,x}^2 - cv_{1,x}^2 = cv_{1,x}^2\left(\frac{v_{2,x}^2}{v_{1,x}^2} - 1\right)$$

$$= m_{\text{girl+scooter}} g \sin(10.0°) \times \left(\frac{v_{2,x}^2}{v_{1,x}^2} - 1\right) = 151\ \text{N}$$

17. Relate the acceleration to the net force on the package and then calculate the coefficient of kinetic friction between the package and the ramp. The free-body diagram will be the same as the one in Figure 5-11b. Define positive x as down the incline and the y axis perpendicular to and positive above the incline.

Finding the acceleration:

$$v_x^2 = v_{0x}^2 + 2a_x(x - x_0)$$

$$a_x = \frac{v_x^2 - v_{0x}^2}{2(\Delta x)} = \frac{0 - \left(2.00\ \frac{\text{m}}{\text{s}}\right)^2}{2(12.0\ \text{m})} = -0.17\ \frac{\text{m}}{\text{s}^2}$$

Newton's second law:

$$\sum F_{\text{ext},y} = F_n - F_{g,y} = F_n - mg\cos(20.0°) = ma_y = 0$$

$$F_n = mg\cos(20.0°)$$

$$\sum F_{\text{ext},x} = F_{g,x} - F_k = mg\sin(20.0°) - \mu_k F_n = mg\sin(20.0°) - \mu_k mg\cos(20.0°) = ma_x$$

$$\mu_k = \frac{g\sin(20.0°) - a_x}{g\cos(20.0°)} = \frac{\left(9.80\ \frac{\text{m}}{\text{s}^2}\right)\sin(20.0°) - \left(-0.167\ \frac{\text{m}}{\text{s}^2}\right)}{\left(9.80\ \frac{\text{m}}{\text{s}^2}\right)\cos(20.0°)}$$

$$= 0.38$$

19. In the case of static friction, both blocks are stationary, and the tension is equal to the weight of block 2. We'll assume that block 1 would move down the ramp in the case of kinetic friction. Static friction needs to counteract the difference between the parallel component of gravity and the tension. If static friction is large enough to hold the blocks in place, then we've already calculated the tension. If the static friction is smaller, then the blocks will accelerate and kinetic friction will take over. Then we need to calculate the acceleration of the two blocks in order to determine the tension. For block 1 we'll define positive x down the ramp and positive y as extending out from the ramp perpendicularly. For block 2 we'll define the x axis in the vertical direction, where upward is positive.

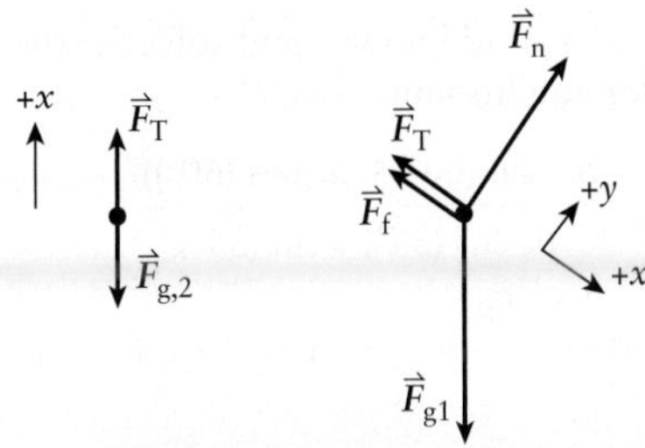

Newton's second law:

$$\sum F_{\text{ext on }2,x} = F_T - F_{g,2} = F_T - m_2 g = m_2 a_x$$

$$\sum F_{\text{ext on }1,y} = F_n - F_{g,1y} = F_n - m_1 g\cos(30.0°) = m_1 a_y = 0$$

$$F_n = m_1 g\cos(30.0°)$$

$$\sum F_{\text{ext on }1,x} = F_{g,1x} - F_T - F_f = m_1 a_x$$

Testing static friction:

$$F_{S,max} = \mu_S F_n = \mu_S m_1 g\cos(30.0°)$$

$$= (0.500)(1.00\text{ kg})\left(9.80\frac{\text{m}}{\text{s}^2}\right)\cos(30.0°) = 4.24\text{ N}$$

$$F_{g,1x} - F_T = m_1 g\sin(30.0°) - m_2 g = g(m_1\sin(30.0°) - m_2)$$

$$= \left(9.80\frac{\text{m}}{\text{s}^2}\right)((1.00\text{ kg})\sin(30°) - (0.400\text{ kg})) = 0.980\text{ N}$$

Because the maximum magnitude of static friction is larger than the forces opposing it, the blocks will remain at rest and the tension in the string is equal to the weight of m_2: $F_T = m_2 g = 3.92$ N

21. To determine the range of values m_2 can take such that the system remains at rest, use Newton's second law to calculate the value of mass that gives the maximum magnitude of static friction in the system. Consider two cases: (a) static friction prevents block 2 from sliding up the ramp and (b) static friction prevents block 2 from sliding down the ramp. Static friction will be exerted in different directions on each block in each case, so we need to redraw the free-body diagrams for each case. Coordinate systems for each block are shown.

Assuming block 2 would slide up the ramp:

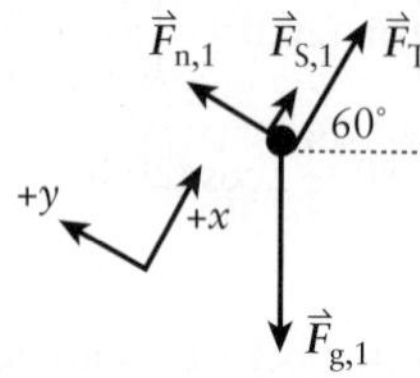

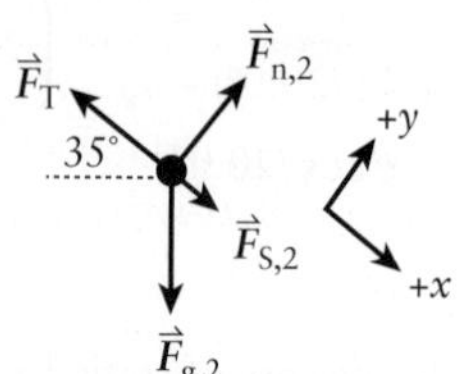

$$\sum F_{\text{ext on }1,y} = F_{n,1} - F_{g,1y} = F_{n,1} - m_1 g\cos(60°) = m_1 a_y = 0$$

$$F_{n,1} = m_1 g\cos(60°)$$

$$\sum F_{\text{ext on }1,x} = F_T + F_{S,1} - F_{g,1x} = F_T + \mu_S F_{n,1} - m_1 g\sin(60°)$$

$$= F_T + \mu_S m_1 g\cos(60°) - m_1 g\sin(60°) = m_1 a_x = 0$$

$$F_T = m_1 g\sin(60°) - \mu_S m_1 g\cos(60°)$$

$$= m_1 g(\sin(60°) - \mu_S\cos(60°))$$

$$\sum F_{\text{ext on }2,y} = F_{n,2} - F_{g,2y} = F_{n,2} - m_2 g\cos(35°) = m_2 a_y = 0$$

$$F_{n,2} = m_2 g\cos(35°)$$

$$\sum F_{\text{ext on }2,x} = F_{S,2} + F_{g,2x} - F_T = \mu_S F_{n,2} + m_2 g\sin(35°) - F_T$$

$$0 = \mu_S m_2 g\cos(35°) + m_2 g\sin(35°) - F_T$$

Substitute F_T and knowns and solve for the minimum value for equilibrium:

$$m_{2,min} = \frac{m_1(\sin(60°) - \mu_S\cos(60°))}{\mu_S\cos(35°) + \sin(35°)}$$

$$= 3.51\text{ kg}$$

Assuming block 2 would slide down the ramp, the friction force would reverse and the free-body diagrams would have these relationships between the forces:

$$\sum F_{\text{ext on }1,y} = F_{n,1} - F_{g,1y} = F_{n,1} - m_1 g\cos(60°) = m_1 a_y = 0$$

$$F_{n,1} = m_1 g\cos(60°)$$

$$\sum F_{\text{ext on }1,x} = F_T - F_{S,1} - F_{g,1x} = F_T - \mu_S F_{n,1} - m_1 g\sin(60°)$$

$$= F_T - \mu_S m_1 g\cos(60°) - m_1 g\sin(60°) = m_1 a_x = 0$$

$$F_T = m_1 g\sin(60°) + \mu_S m_1 g\cos(60°) = m_1 g(\sin(60°) + \mu_S\cos(60°))$$

$$\sum F_{\text{ext on }2,y} = F_{n,2} - F_{g,2y} = F_{n,2} - m_2 g\cos(35°) = m_2 a_y = 0$$

$$F_{n,2} = m_2 g\cos(35°)$$

$$\sum F_{\text{ext on }2,x} = -F_{S,2} + F_{g,2x} - F_T = -\mu_S F_{n,2} + m_2 g\sin(35°) - F_T$$

$$0 = -\mu_S m_2 g\cos(35°) + m_2 g\sin(35°) - F_T$$

Substitute F_T and knowns and solve for the maximum value for equilibrium:

$$m_{2,max} = \frac{m_1(\sin(60°) + \mu_S\cos(60°))}{-\mu_S\cos(35°) + \sin(35°)}$$

$$= 52.6\text{ kg}$$

23. Example application: While moving into her new home, a student is pushing or pulling a 2.0-kg box up an incline of 30° with a force of 20 N. The component of gravity pulls the box down the incline, and the friction force resists the student's pushing the box up the incline. The coefficient of kinetic friction between the block and the incline is 0.40.

25. (a) $F_{S,max} = F_T$; $F_T = \mu_S F_n$; $F_n = mg$; so $F_T = \mu_S mg$. M is not moving and thus not accelerating so, $F_n = Mg$. Setting the two expressions for tension equal, $Mg = \mu_S mg$, and solving $M = \mu_S m$. $M = 0.52(2.5\text{ kg}) = 1.3$ kg.

(b) For the block on the table: $a = \dfrac{F_T - F_k}{m}$, where $F_k = \mu_k mg$.

For the hanging block: $a = \dfrac{Mg - F_T}{M}$.

Solving each expression for tension: $F_T = ma + \mu_k mg$ and $F_T = Mg - Ma$. Combining and solving for acceleration:

$$a = \frac{Mg - \mu_k mg}{m + M} = \frac{(1.5\text{ kg})\left(9.8\frac{\text{N}}{\text{kg}}\right) - (0.32)(2.5\text{ kg})\left(9.8\frac{\text{N}}{\text{kg}}\right)}{(1.5\text{ kg} + 2.5\text{ kg})}$$

$$= 1.7\frac{\text{m}}{\text{s}^2}.$$

Plugging back in: $F_T = (1.5\text{ kg})\left(9.8\frac{\text{N}}{\text{kg}} - 1.7\frac{\text{m}}{\text{s}^2}\right) = 12\text{ N}.$

Chapter 5 AP® Practice Problems

1. (D) Draw a free-body diagram (FBD) and apply Newton's second law:

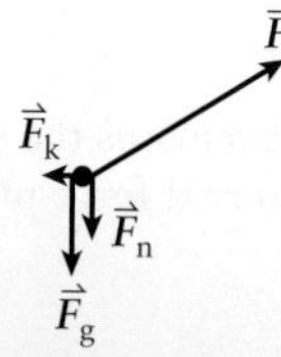

$$\sum F_y = ma_y = 0 \quad \sum F_x = ma_x = F\cos\theta - F_k = ma$$
$$F_k = \mu_k F_n = \mu_k(F\sin\theta - mg)$$

3. (D) $F_k = \mu_k F_n$ and $a_x = \dfrac{\sum F_{ext,x}}{m}$

$$F_k = \mu_k F_n = \mu_k mg = ma_x \Rightarrow \mu_k = \frac{a_x}{g} = \frac{2\ \text{m/s}^2}{10\ \text{m/s}^2} = 0.2$$

5. (A) $F_k = \mu_k F_n$ and $F_S \le \mu_S F_n$
The block slides up the incline with constant kinetic friction directed down the incline, until the applied force up the incline is sufficiently reduced so that the block stops. Then static friction takes over and decreases as the applied force is further reduced. As the applied force and the component of the gravitational force on the block along the incline cancel, the static friction force goes instantaneously to zero, and then becomes positive as it is resisting the net force down the incline once the component of the gravitational force is larger than the magnitude of the applied force. When the net force down the incline due to the gravitational force and the applied force exceeds the static friction force, the box then slides down the incline with a constant kinetic friction force, which has the same magnitude as it originally had, but is oppositely directed.

7. (B) The weight of the rope on the incline is $\dfrac{x}{L}mg$.
Acceleration (and net force) is along the incline. The normal force is the magnitude of the component of the weight perpendicular to the incline.

$$F_{\text{down indine}} = \frac{x}{L}mg\sin\theta = \frac{x}{L}mg\frac{0.75\ \text{m}}{1.25\ \text{m}} = \frac{x}{L}mg\frac{3}{5} = \frac{3x}{5L}mg;$$

$$F_k = \mu_k F_n = \mu_k\frac{x}{L}mg\cos\theta = \mu_k\frac{x}{L}mg\frac{1.00\ \text{m}}{1.25\ \text{m}} = \mu_k\frac{x}{L}mg\frac{4}{5} = 0.25\frac{x}{L}mg\frac{4}{5} = \frac{x}{5L}mg$$

$$F_{net} = \frac{3x}{5L}mg - \frac{x}{5L}mg = ma \Rightarrow a = \frac{2x}{5L}g$$

9. (D) Acceleration is constant, so average speed can be calculated from initial and final speeds, and $v_f = 0$.

11. (C) The maximum static friction force possible is $F_{S,max} = \mu_S F_n = \mu_S mg = 0.5(200\ \text{N}) = 100\ \text{N}$. The kinetic friction force is $F_k = \mu_k F_n = \mu_k mg = (0.3)(200\ \text{N}) = 60\ \text{N}$. In case *A*, the force of 75 N is not enough to get the block moving, so the friction force is equal to the pulling force, and thus is 75 N. This is also true in case *C*, so the friction force is 50 N. In cases *B* and *D*, the friction force is greater than the maximum static friction force, so the block moves and thus is subject to a constant kinetic friction force of 60 N. Thus, $A > (B = D) > C$.

Chapter 6

6-1 Gravitation is a force of universal importance; add circular motion and you start explaining the motion of the planets

1. The loudness of a sound and the apparent brightness of a light become weaker as the distance between the ear or eye and the source increases. The light or sound spread out from the source. One can visualize this as a sphere of increasing radius with the source of the signal at the center of the sphere and the signal spreading out over a surface that increases with the square of the radius. The gravitational force also becomes weaker as distance from the source increases proportional to the surface area of a sphere at that radius.

6-2 An object moving in a circle is accelerating even if its speed is constant

1. When stopping and starting the velocity is decreasing (acceleration in opposite direction as the instantaneous direction of the velocity) and then increasing (acceleration in same direction as the instantaneous direction of the velocity). When riding in a circle, the velocity is constantly changing direction. The acceleration associated with this change in direction is always directed toward the center of the circular path. When the speed along the circular path is constant, then there is no acceleration in the instantaneous direction of the velocity, which is always tangent to the circular path.
3. The acceleration is perpendicular to the velocity and points in toward the center of the circular motion. In the situation described, the acceleration will initially be pointed east. Its direction will rotate clockwise from initially pointing to the east to pointing toward the south to zero when the velocity is straight again. See Figure 6-4.
5. As it travels once around the circle, the distance traveled is equal to the circumference. Time can be found by dividing the total distance traveled by the speed with which the object covered that distance.

$$2\pi r = 2\pi(1.25\ \text{m}) = 7.854\ \text{m}$$

$$\text{time} = \frac{\text{distance}}{\text{speed}} = \frac{7.854\ \text{m}}{2.25\ \text{m/s}} = 3.49\ \text{s}$$

7. (a) In terms of the components of acceleration, velocity, and displacement perpendicular and parallel to Earth's surface:
   - In the model describing projectile motion there is no horizontal component of the acceleration, so the horizontal component of the velocity is constant. There is a vertical component of the acceleration, which is a constant in both magnitude and direction, downward, so that the perpendicular component of the velocity varies with time.
   - In the model describing uniform circular motion there is also no parallel component of the accelera-tion and there is a perpendicular component with a constant magnitude, but it has a changing

direction so that it always points downward toward the surface. Like projectile motion, the parallel component of the velocity remains constant but in this case the perpendicular component of the velocity is zero because the perpendicular component of the acceleration is just large enough to cause the change in velocity direction needed for uniform circular motion.

- The displacement in projectile motion changes in both the vertical and horizontal directions, and the horizontal component is a linear function of time, whereas the vertical component is a quadratic function of time. The vertical component changes direction at the highest point during the projectile motion. In uniform circular motion, there is also a linear time dependence in the parallel displacement but there is no perpendicular displacement; the height remains constant.

(b)

Vector components	Projectile motion	Uniform circular motion
Displacement perpendicular to the surface	Changes with time as t^2	Zero
Displacement parallel to the surface	Changes with time as t	Changes with time as t
Velocity perpendicular to the surface	Changes with time as t	Zero
Velocity parallel to the surface	Constant	Constant
Acceleration perpendicular to the surface	Constant downward	Constant downward
Acceleration parallel to the surface	Zero	Zero

This table suggests a connection between projectile motion and uniform circular motion.

9. (a) As the two runners are rounding the semicircular segment their velocities are changing direction and so they are accelerating. The runner in the outside lane has a larger radius; $r_{\text{Kelly}} = r_{\text{Mary}} + \Delta r$. Because they enter and exit the turn in the same time, the periods are equal. Because the radius of Kelly's motion is larger and the period is the same, Kelly must have a larger acceleration.

$$a_{\text{cent}} = v^2/r = (2\pi r/T)^2/r = \frac{4\pi^2 r}{T^2} \Rightarrow a_{\text{Kelly,cent}}$$

$$= \frac{r_{\text{Kelly}}}{r_{\text{Mary}}} a_{\text{Mary,cent}} = \frac{r_{\text{Mary}} + \Delta r}{r_{\text{Mary}}} a_{\text{Mary,cent}}$$

$$a_{\text{Kelly,cent}} = (1 + \Delta r/r_{\text{Mary}}) a_{\text{Mary,cent}}$$

(b) If the width of a lane is taken to be 1.50 m and Mary is running in the middle of her lane, then $r_{\text{Mary}} = 20.0\text{ m} - 0.75\text{ m}$ and

$$\frac{a_{\text{Kelly,cent}}}{a_{\text{Mary,cent}}} = (1 + 1.50\text{ m}/(20.0\text{ m} - 0.75\text{ m})) \approx 1.08$$

(c) We can find Mary's acceleration using the relationship between centripetal acceleration, velocity, and radius for uniform circular motion. The ratio of their accelerations was found in part (b), so we can use Mary's acceleration to find Kelly's speed.

$$a_{\text{Mary,cent}} = \frac{v_{\text{Mary}}^2}{r_{\text{Mary}}} = \frac{8.33^2\text{ m}^2/\text{s}^2}{19.25\text{ m}} = 3.60\text{ m/s}^2 \Rightarrow$$

$$a_{\text{Kelly,cent}} = (1.078)3.60\text{ m/s}^2 = 3.89\text{ m/s}^2$$

$$\frac{v_{\text{Kelly}}^2}{r_{\text{Kelly}}} = a_{\text{Kelly,cent}} \Rightarrow \frac{v_{\text{Kelly}}^2}{(20.0\text{ m} + 0.75)\text{ m}} = 3.89\text{ m/s}^2 \Rightarrow$$

$$v_{\text{Kelly}} = \sqrt{(20.75\text{ m})(3.89\text{ m/s}^2)} = 8.98\text{ m/s}$$

6-3 For an object in uniform circular motion, the net force exerted on the object points toward the center of the circle

1. (a) When an object moves with constant speed in a circle, the sum of all the forces exerted on it will point toward the center of the circle. The normal force exerted by the surface will balance the downward force of gravity, leaving just the force of tension from the string as the net force exerted on the washer. Thus, for the given speed, washer mass, and string length, the tension that keeps the washer in a horizontal circle is

$$F_{\text{T}} = m\frac{v^2}{r} = 0.0250\text{ kg}\left(\frac{36\text{ m}^2/\text{s}^2}{0.600\text{ m}}\right) = 1.5\text{ N}$$

(b) At the top of the circle, there are two forces exerted on the washer: the tension of the string and the weight of the washer. Weight is always downward and tension is always directed along the string causing the force, so both forces point toward the center of the circle.

$$F_{\text{T}} + mg = m\frac{v^2}{r} \Rightarrow F_{\text{T}} = m\frac{v^2}{r} - mg$$

$$= 0.0250\text{ kg}\left(\frac{36\text{ m}^2/\text{s}^2}{0.600\text{ m}}\right) - 0.0250\text{ kg}(9.8\text{ N/kg}) = 1.3\text{ N}$$

(c) When the washer is at the bottom of the circle, the tension due to the string is now upward, in the direction needed to cause the centripetal acceleration (and the weight of the washer is still downward). Taking up to be positive:

$$F_{\text{T}} - mg = m\frac{v^2}{r} \Rightarrow F_{\text{T}} = m\frac{v^2}{r} + mg$$

$$= 0.0250\text{ kg}\left(\frac{36\text{ m}^2/\text{s}^2}{0.600\text{ m}}\right) + 0.0250\text{ kg}(9.8\text{ N/kg}) = 1.7\text{ N}$$

3. (a) If the child lets go of the railing, the only force holding the standing child in the circular path is the friction force.

$$\frac{mv^2}{r} = F_{\text{S}} \le \mu_{\text{S}} gm \Rightarrow v^2/r = \mu_{\text{S}} g$$

If the period is 1.0 s, then the speed on the edge of the merry-go-round is

$$v = \frac{2\pi r}{T} = 4\pi \text{ m/s}$$

Standing closer to the middle, the speed will be reduced by 4, because the distance traveled is reduced by 4.

Yet even for the child wearing sneakers at both distances,

$$r = 2.0 \text{ m}: \frac{v^2}{r} = 8\pi^2 \text{ N/kg} > \mu_S g = 0.8 \text{ (9.8 N/kg)} = 7.8 \text{ N/kg}$$

$$r = 0.5 \text{ m}: \frac{v^2}{r} = 2\pi^2 \text{ N/kg} > 7.8 \text{ N/kg}$$

(b) For her charge wearing socks,

$$\mu_S g \geq \frac{1}{r}\left(\frac{2\pi r}{T}\right)^2 = \frac{4\pi^2 r}{T^2} \Rightarrow T = \pi\sqrt{\frac{4r}{\mu_S g}}$$

$$= \pi\sqrt{\frac{8.0 \text{ m}}{0.1(9.8 \text{ N/kg})}} = 9 \text{ s}$$

If they wear sneakers,

$$\mu_S g \geq \frac{1}{r}\left(\frac{2\pi r}{T}\right)^2 = \frac{4\pi^2 r}{T^2} \Rightarrow T = \pi\sqrt{\frac{4r}{\mu_S g}}$$

$$= \pi\sqrt{\frac{8.0 \text{ m}}{0.8(9.8 \text{ N/kg})}} = 3.2 \text{ s}$$

5. Representing the time to complete one revolution as T:

$$T = \frac{1 \text{ min}}{1.20 \times 10^3 \text{ rev}} = \frac{60 \text{ s}}{1.20 \times 10^3 \text{ rev}} = 5.00 \times 10^{-2} \text{ s}$$

The speed is the displacement, $2\pi r$, divided by the period. Substituting this relationship into the equation for centripetal acceleration:

$$a_{\text{cent}} = \frac{v^2}{r} = \frac{(2\pi r)^2}{T^2 r} = \frac{4\pi^2 r}{T^2} = \frac{4\pi^2(0.25 \text{ m})}{(5.00 \times 10^{-2})^2} = 3900 \text{ N/kg}$$

7. (a) (D) Evaluate the inverse tangent of the ratio on the right-hand side of the expression to calculate the bank angle. The elevation of the outer edge divided by the width of the road is equal to the sine of that angle. To evaluate the right-hand side of the expression the radius of curvature, speed limit, and probable static friction coefficient for slippery conditions with tires that are well worn would need to be known.

(b) (B) or (A) The angle of the banked turn and radius of curvature are already fixed by the existing construction. Given a lower estimate of the coefficient of static friction for slippery conditions, the speed limit can be calculated.

(c) The radius of curvature can be evaluated from expression C if the expected speed limit, bank angle, and a lower estimate of the static friction coefficient for slippery conditions are known. The area over which the new construction must be completed is determined by the radius of curvature.

9. (a) The tension force in the string supplies the centripetal force on block 1 (on the table), and must balance the force of gravity on the hanging block 2, given that we are told it remains stationary. The table will exert a normal force on block 1 that balances the force of gravity on it since it is not accelerating vertically, and the tension provides the centripetal force.

(b) Apply Newton's second law to the vertical forces on the hanging object, where $F_{\text{net}} = 0$: $F_T = m_2 g = 0.225 \text{ kg}(9.8 \text{ N/kg}) = 2.2 \text{ N}$

(c) Apply Newton's second law to the centripetal motion of object 1:

$$F_T = m_1\frac{v^2}{r} \Rightarrow v = \sqrt{\frac{rF_T}{m_1}} = \sqrt{\frac{1.0 \text{ m}(2.21 \text{ N})}{0.125 \text{ kg}}} = 4.2 \text{ m/s}$$

6-4 Newton's law of universal gravitation explains the orbit of the Moon, and introduces us to the concept of field

1. Acceleration due to gravity would increase more if the radius were cut in half and the mass kept the same, because

$$g = \frac{F_g}{m} = \frac{GM_{\text{Earth}}}{r_{\text{Earth}}^2}$$

$$g \to 2g \quad \text{if} \quad M_{\text{Earth}} \to 2M_{\text{Earth}}$$

and

$$g \to 4g \quad \text{if} \quad r_{\text{Earth}} \to r_{\text{Earth}}/2$$

The forces the two objects exert on each other have the same magnitude. Because acceleration is F/m, the ratio of the accelerations is the inverse of the ratio of the masses. Because the mass of a person is so much less than the mass of Earth, the acceleration of Earth due to a person would be negligible.

3. To get the distance from the center of the Moon to the apple, subtract the radius of Earth from the Earth–Moon distance. Because we know things fall downward and the force of the Moon's gravity would be upward, the ratio had better be pretty small!

$$\frac{F_{\text{Moon on apple}}}{F_{\text{Earth on apple}}} = \frac{GM_{\text{Moon}}m_{\text{apple}}}{r_{\text{Moon–apple}}^2}\frac{r_{\text{Earth–apple}}^2}{Gm_{\text{apple}}M_{\text{Earth}}}$$

$$= \frac{M_{\text{Moon}}}{M_{\text{Earth}}}\frac{r_{\text{Earth–apple}}^2}{r_{\text{Moon–apple}}^2} = \frac{7.348 \times 10^{22} \text{ kg}}{5.972 \times 10^{24} \text{ kg}}\left[\frac{6.371 \times 10^6 \text{ m}}{3.780 \times 10^8 \text{ m}}\right]^2$$

$$= 3.495 \times 10^{-6}$$

5. (a) $$g_{452b} = \frac{GM_{452b}}{r_{452b}^2} = G\frac{5M_{\text{Earth}}}{(1.6)^2 R_{\text{Earth}}^2} = \frac{5}{(1.6)^2}g_{\text{Earth}}$$

$$= 19 \text{ m/s}^2 \approx 2g_{\text{Earth}}$$

(b) An object with a weight of 10 N on Earth has a weight of 20 N.

7. Use the equation for uniform circular motion, using the distance traveled in one orbit divided by the period of the orbit for the velocity:

$$a_{\text{cent}} = \left(\frac{2\pi r}{T}\right)^2\frac{1}{r} = \frac{4\pi^2 r}{T^2} = 4\pi^2(3.84 \times 10^8 \text{ m})$$

$$\times\left[\frac{1}{27.3 \text{ days}}\left(\frac{1 \text{ day}}{24 \text{ h}}\right)\left(\frac{1 \text{ h}}{3600 \text{ s}}\right)\right]^2 = 2.72 \times 10^{-3} \text{ m/s}^2$$

9. (a)

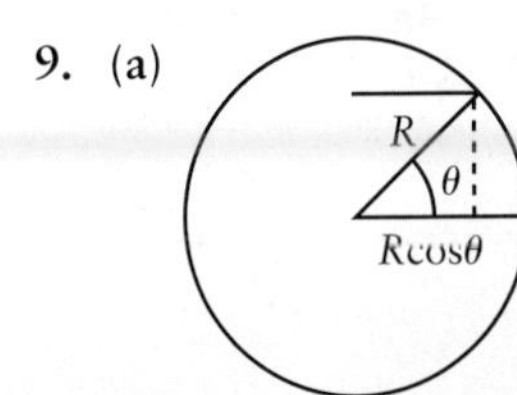

(b) i. At latitude 0° N

$$\frac{v^2}{R} = \frac{4\pi^2 R\cos\theta}{T^2} = \frac{4\pi^2(6.371\times10^6\text{ m})\cos(0°)}{(86400\text{ s})^2} = 0.0337\text{ m/s}^2$$

ii. At latitude 40.0° N

$$\frac{v^2}{R} = \frac{4\pi^2 R\cos\theta}{T^2} = \frac{4\pi^2(6.371\times10^6\text{ m})\cos 40.0°}{(86400\text{ s})^2} = 0.0258\text{ m/s}^2$$

6-5 Newton's law of universal gravitation begins to explain the orbits of planets and satellites

1. (a) Average speed is distance divided by time, and period is the time to complete the circular path:

$$v = \frac{2\pi r}{T} = \frac{2\pi(13\text{ m})}{26\text{ s}} = 3.1\text{ m/s}$$

(b) $v = \frac{2\pi r}{T} = \frac{2\pi(79.25\text{ m})}{26\text{ s}} = 19\text{ m/s}$

(c) The period is the time it takes to travel through one complete circle, and so is the circumference of the circle divided by the speed:

$$v = \frac{48\text{ km}}{\text{h}}\frac{1\text{ h}}{3600\text{ s}}\frac{1000\text{ m}}{\text{km}} = 13.3\text{ m/s} = \frac{2\pi r}{T} \Rightarrow$$

$$T = \frac{2\pi(79.25\text{ m})}{13.3\text{ m/s}} = 37\text{ s}$$

3. Express the second law in terms of the period. Note all vectors are centripetal.

$$F_{\text{net}} = \frac{GMm}{r^2} = ma = \frac{4\pi^2 rm}{T^2} \Rightarrow T^2 = \frac{4\pi^2 r^3}{GM}$$

The period increases as the radius increases.

5. (A) The force is proportional to the product of masses. Both masses are doubled.
7. (D) The period squared is proportional to the orbital radius cubed. A's orbital radius is 4 times larger than that of planet B, and $4^3 = 64$, so the period of planet A's orbit is 8 times that of planet B.
9. (a) With a coordinate system in which the direction toward the Sun is negative:

$$\frac{F_{\text{net L1}}}{m} = -\frac{GM_{\text{Sun}}}{r^2_{\text{Sun–Ll}}} + \frac{GM_{\text{Earth}}}{r^2_{\text{Earth–L1}}} = a_{\text{L1,cent}}$$

$$\frac{F_{\text{net L2}}}{m} = -\frac{GM_{\text{Sun}}}{r^2_{\text{Sun–L2}}} - \frac{GM_{\text{Earth}}}{r^2_{\text{Earth–L2}}} = a_{\text{L2,cent}}$$

(b) Expressing the centripetal acceleration in terms of the period of L_1:

$$\frac{F_{\text{net L1}}}{m} = -\frac{GM_{\text{Sun}}}{r^2_{\text{Sun–L1}}} + \frac{GM_{\text{Earth}}}{r^2_{\text{Earth–L1}}} = a_{\text{L1,cent}} = -\frac{v^2}{r_{\text{Sun–L1}}} = -\frac{4\pi^2 r_{\text{Sun–L1}}}{T^2_{\text{L1}}}$$

So,

$$-\frac{GM_{\text{Sun}}}{r^3_{\text{Sun–L1}}} + \frac{GM_{\text{Earth}}}{r^2_{\text{Earth–L1}} r_{\text{Sun–L1}}} = -\frac{4\pi^2}{T^2_{\text{L1}}}$$

And

$$-\frac{4\pi^2 r_{\text{Earth–Sun}}}{T^2_{\text{Earth}}} = -\frac{GM_{\text{Sun}}}{r^2_{\text{Earth–Sun}}} \Rightarrow -\frac{4\pi^2}{T^2_{\text{Earth}}} = -\frac{GM_{\text{Sun}}}{r^3_{\text{Earth–Sun}}}$$

If the periods of the orbits of L_1 and Earth are equal:

$$-\frac{4\pi^2}{T^2_{\text{L1}}} = -\frac{4\pi^2}{T^2_{\text{Earth}}} \Rightarrow -\frac{GM_{\text{Sun}}}{r^3_{\text{Sun–L1}}} + \frac{GM_{\text{Earth}}}{r^2_{\text{Earth–L1}} r_{\text{Sun–L1}}} = -\frac{GM_{\text{Sun}}}{r^3_{\text{Earth–Sun}}}$$

(c) An object at L_3 has a gravitational force due to the Sun with the same magnitude exerted on it as an object at the position of Earth. The additional force due to Earth would be in the same direction as the Sun's, but much smaller because of the much larger radius at L_3 so reasonable to neglect. They then must have the same centripetal acceleration, and so the same period.

6-6 Apparent weight and what it means to be "weightless"

1. The net force exerted on you in an elevator is the sum of the downward gravitational force exerted on you by Earth and the upward contact force exerted on you by the floor of the elevator. If the elevator is not in free fall, then because your acceleration is less than g, the elevator must exert an upward contact force on you. If, for example, the cable supporting the elevator breaks, then both you and the elevator car are in free fall and the floor of the car no longer exerts an upward normal force on you and you would feel weightless.
3. (C) The gravitational field at a distance r from the center of Earth is $g = \frac{GM_{\text{Earth}}}{r^2}$. Although for small changes in elevation the effect is not noticeable, as the climber ascends the mountain the value of r increases and the gravitational field strength decreases.

Chapter 6 Review Problems

1. $v = \frac{2\pi r}{T} = \frac{2\pi(38\text{ m})}{20\text{ min}}\frac{1\text{ min}}{60\text{ s}} = 0.2\text{ m/s}$

$$a_{\text{cent}} = \frac{v^2}{r} = \frac{(0.20\text{ m/s})^2}{38\text{ m}} = 0.001\text{ m/s}^2$$

3. The speed of a point on the spinner's edge is the circumference divided by the period. The period is the inverse of the frequency, and we can find the frequency because we know it rotates 4000 times every minute:

$$v = \frac{\pi d}{T} = \pi(0.060\text{ m})\frac{4000\text{ rev}}{1\text{ min}}\frac{1\text{ min}}{60\text{ s}} = 12.57\text{ m/s} = 13\text{ m/s}$$

$$a_{\text{cent}} = \frac{v^2}{r} = \frac{158.0\text{ m}^2/\text{s}^2}{0.030\text{ m}} = 5300\text{ m/s}^2$$

5. The pitcher rotates her arm about her shoulder in a circle of radius r = the length of her arm. Calculate the length of her arm directly from the definition:

$$a_{\text{cent}} = \frac{v^2}{r}$$

$$r = \frac{v^2}{a_{\text{cent}}} = \frac{\left(34.3\frac{\text{m}}{\text{s}}\right)^2}{1960\frac{\text{m}}{\text{s}^2}} = 0.600\text{ m}$$

7. Assuming that the protons are moving at c,

$$F_{\text{cent}} = m\frac{v^2}{r} = 1.673 \times 10^{-27}\ \text{kg}\frac{9.0 \times 10^{16}\ \text{m}^2/\text{s}^2}{(27{,}000\ \text{m}/2\pi)} = 3.5 \times 10^{-14}\ \text{N}$$

9. The friction force must provide the needed centripetal acceleration if the coin does not slip. We can find the speed from the radius and the period, and the period from revolutions per minute.

$$v = \frac{2\pi(0.120\ \text{m})}{\left(\frac{1\ \text{min}}{78.0\ \text{rev}}\right)\left(\frac{60\ \text{s}}{1\ \text{min}}\right)} = 0.980\ \text{m/s}$$

So

$$m\frac{v^2}{r} = \mu_{\text{S}} mg \Rightarrow \mu_{\text{S}} = \frac{v^2}{gr} = \frac{(0.980\ \text{m/s})^2}{(9.80\ \text{m/s}^2)(0.120\ \text{m})} = 0.817$$

11. **(a)** The release is like cutting a string, the ball continues in its direction of motion tangential to the circular path when the centripetal force is removed so at the bottom of the arc it is moving forward.

(b) A free-body diagram for the release point shows the force of her hand on the ball pointing straight upward, opposing the smaller force of gravity. Using Newton's second law and the definition of centripetal acceleration:

$$\sum F_{\text{ext},y} = F_{\text{hand on ball}} - F_{\text{g,ball}} = F_{\text{hand on ball}} - m_{\text{ball}}g =$$

$$m_{\text{ball}}\,a_y = m_{\text{ball}}\left(\frac{v^2}{r}\right)$$

$$F_{\text{hand on ball}} = m_{\text{ball}}\left(g + \frac{v^2}{r}\right) =$$

$$(0.19\ \text{kg})\left(\left(9.8\frac{\text{m}}{\text{s}^2}\right) + \frac{\left(33.0\frac{\text{m}}{\text{s}}\right)^2}{0.600\ \text{m}}\right) = 350\ \text{N}$$

13. **(a)**

0.2 m

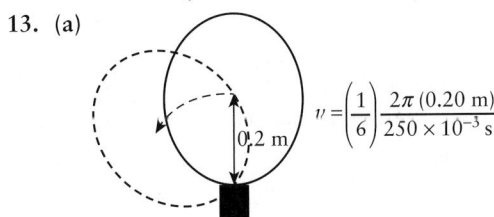

$$v = \left(\frac{1}{6}\right)\frac{2\pi\,(0.20\ \text{m})}{250 \times 10^{-3}\ \text{s}}$$

(b) 60° = 1/6 of a complete circle. Find velocity from 1/6 of a circle in the given time and centripetal acceleration is

$$v^2/r = ((2\pi r/6)/T)^2/r = 4\pi^2 r/36T^2$$

$$= 4\pi^2 r/36T^2 = 4\pi^2(0.20\ \text{m})/36(250 \times 10^{-3}\ \text{s})^2 = 3.5\ \text{m/s}^2$$

(c) $F = 0.06(75\ \text{kg})(3.5\ \text{N/kg}) = 16\ \text{N}$

15.

$$F_{\text{Sun on Earth}} = \frac{Gm_{\text{Earth}}m_{\text{Sun}}}{r^2_{\text{Sun to Earth}}}$$

$$= \frac{\left(6.67 \times 10^{-11}\frac{\text{Nm}}{\text{kg}^2}\right)(5.97 \times 10^{24}\ \text{kg})(1.99 \times 10^{30}\ \text{kg})}{(1.50 \times 10^{11}\ \text{m})^2}$$

$$= 3.52 \times 10^{22}\ \text{N}$$

17. **(a)**

$$v = \frac{2\pi r}{T} = \frac{2\pi(1.5 \times 10^{11}\ \text{m})}{365.25\ \text{day}}\frac{1\ \text{day}}{86{,}400\ \text{s}}$$

$$= 29{,}900\ \text{m/s} = 3.0 \times 10^4\ \text{m/s}$$

(b)

$$a = \frac{F}{m} = \frac{GM_{\text{Sun}}}{r^2_{\text{Earth–Sun}}} = \frac{v^2}{r} = \frac{(29{,}900\ \text{m/s})^2}{r_{\text{Earth–Sun}}} \Rightarrow$$

$$M_{\text{Sun}} = \frac{(29{,}900\ \text{m/s})^2}{6.67 \times 10^{-11}\ \text{Nm}^2/\text{kg}^2}(1.5 \times 10^{11}\ \text{m})$$

$$= 2.0 \times 10^{30}\ \text{kg}$$

19. **(a)**

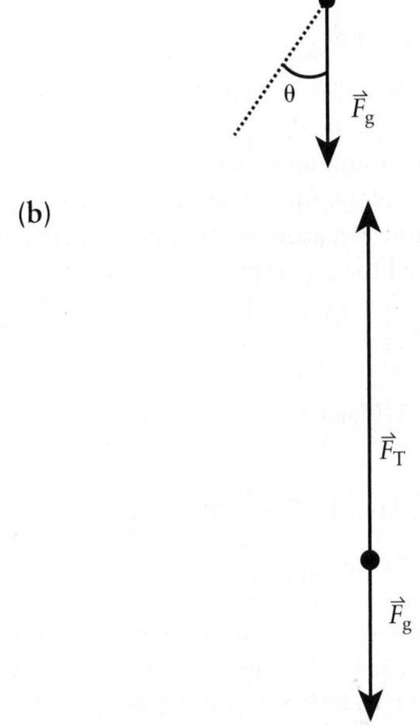

(b)

$\vec{F}_{\text{T}}$

$\vec{F}_{\text{g}}$

(c) Draw the coordinate system at each instant with one axis along the string, we'll call it y, and the x axis in the direction of motion. Define θ as the angle between the y axis aligned with $\vec{F}_{\text{T}}$ and $\vec{F}_{\text{g}}$. So the x component of gravity is $F_{\text{g}} \sin\theta$ and the y component of gravity is $-F_{\text{g}} \cos\theta$.

For part (a): $F_{\text{T}} - F_{\text{g}}\cos\theta = ma_y = 0$; $F_{\text{g}}\sin\theta = ma_x$

For part (b): $F_{\text{T}} - F_{\text{g}} = ma_y$; $ma_x = 0$

(d) Because the magnitude of the velocity is approximately constant as it moves through this point the path of the ball is well approximated as uniform circular motion at that instant. This means the acceleration and the velocity are perpendicular so only the vertical component of the acceleration is nonzero.

21. **(a)** Assuming the speed is constant, we can calculate the centripetal acceleration directly from the linear speed (circumference/period) and the radius of the device.

$$v = 30.6\frac{\text{rev}}{\text{min}} \times \frac{1\ \text{min}}{60\ \text{s}} \times \frac{2\pi\left(\frac{13.4\ \text{m}}{2}\right)}{1\ \text{rev}} = 21.47\frac{\text{m}}{\text{s}}$$

$$a_{\text{cent}} = \frac{v^2}{r} = \frac{\left(21.47\frac{\text{m}}{\text{s}}\right)^2}{\left(\frac{13.4\ \text{m}}{2}\right)} = 68.8\frac{\text{m}}{\text{s}^2} = 7\ g$$

(b) To decrease the acceleration, we need to increase the period, T. The new speed is the circumference of the circle divided by T_{new}. We can calculate the ratio to T_{old} for the original speed. Rewrite centripetal acceleration in terms of period, then solve that for the period:

$$a_{cent} = \frac{4\pi^2 r}{T^2} \qquad T = \sqrt{\frac{4\pi^2 r}{a_{cent}}}$$

$$\frac{T_{new}}{T_{old}} = \frac{\sqrt{\dfrac{4\pi^2 r}{a_{cent,new}}}}{\sqrt{\dfrac{4\pi^2 r}{a_{cent,old}}}} = \sqrt{\frac{a_{cent,old}}{a_{cent,new}}} = \sqrt{\frac{a_{cent,old}}{0.750 a_{cent,old}}} =$$

$1.15 \Rightarrow T_{new} = 1.15 T_{old}$

The time for the pilot to make one spin must be increased by 15%.

23. The truck is undergoing uniform circular motion and that static friction is the force responsible for the truck's acceleration. Because we are given the friction force, we only need the x components (defined as toward the center of the circle) of the forces.
Newton's second law:

$$\sum F_{ext,x} = F_{S,max} = m_{truck} a_{cent} = m_{truck}\left(\frac{v^2}{r}\right)$$

$$r = \frac{m_{truck} v^2}{F_{S,max}} = \frac{(1.50 \times 10^3 \text{ kg})\left(20.0 \dfrac{\text{m}}{\text{s}}\right)^2}{8.00 \times 10^3 \text{ N}} = 75.0 \text{ m}$$

25. The normal force required to hold the person in the circular path must result in a great enough upward static friction force to balance gravity. We can use Newton's second law and the centripetal acceleration to calculate the minimum coefficient of static friction between the wall and the passenger.

Free-body diagram of the passenger:

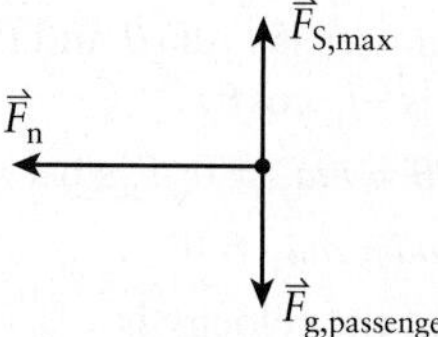

Newton's second law:

$$\sum F_{ext,x} = F_n = m_{passenger} a_{cent} = m_{passenger}\left(\frac{v^2}{r}\right)$$

$$\sum F_{ext,y} = F_{S,max} - F_{g,passenger} = \mu_S F_n - m_{passenger} g =$$

$$\mu_S\left(\frac{m_{passenger} v^2}{r}\right) - m_{passenger} g = m_{passenger} a_y = 0$$

$$\mu_S = \frac{gr}{v^2} = \frac{\left(9.80 \dfrac{\text{m}}{\text{s}^2}\right)(3.50 \text{ m})}{\left(25.0 \dfrac{\text{rev}}{\text{min}} \times \dfrac{1 \text{ min}}{60 \text{ s}} \times \dfrac{2\pi(3.50 \text{ m})}{1 \text{ rev}}\right)^2} = 0.409$$

27. The radius of the orbit is the radius of Earth plus the altitude h. *Sputnik*'s weight at any point is equal to its mass multiplied by the acceleration due to gravity at that point.

(a) Solve Equation 6-10 for the radius:

$$T^2 = \frac{4\pi^2 r^3}{G m_{Earth}} = \frac{4\pi^2 (R_{Earth} + h)^3}{G m_{Earth}}$$

$$h = \sqrt[3]{\frac{G m_{Earth} T^2}{4\pi^2}} - R_{Earth}$$

$$= \sqrt[3]{\frac{\left(6.67 \times 10^{-11} \dfrac{\text{Nm}}{\text{kg}^2}\right)(5.97 \times 10^{24} \text{ kg})\left(96.2 \text{ min} \times \dfrac{60 \text{ s}}{1 \text{ min}}\right)^2}{4\pi^2}}$$

$$-(6.37 \times 10^6 \text{ m}) = 5.8 \times 10^5 \text{ m} = 5.8 \times 10^2 \text{ km}$$

(b) Weight in orbit:

$$F_{\text{Earth on Sputnik}} = m_{Sputnik}\frac{G m_{Earth}}{R^2}$$

$$= (83.6 \text{ kg})\frac{\left(6.67 \times 10^{-11} \dfrac{\text{Nm}}{\text{kg}^2}\right)(5.97 \times 10^{24} \text{ kg})}{((6.37 \times 10^3 \text{ km}) + (5.8 \times 10^5 \text{ m}))^2} = 6.89 \times 10^2 \text{ N}$$

Weight on Earth's surface:

$$F_{g,Sputnik} = m_{Sputnik} g = (83.6 \text{ kg})\left(9.80 \frac{\text{m}}{\text{s}^2}\right) = 8.19 \times 10^2 \text{ N}$$

29. (a) $a = \dfrac{F_{net}}{m}$; $a_{cent} = \dfrac{v^2}{R}$. Therefore, $\dfrac{v^2}{R} = \dfrac{mg - F_n}{m}$.

Because $F_n = mg - m\dfrac{v^2}{R}$ and the speed of the ride is

$$v = \frac{2\pi R}{T} = \frac{2\pi(13.0 \text{ m})}{26.0 \text{ s}} = 3.14 \frac{\text{m}}{\text{s}}.$$

Thus,

$$F_n = (71.0 \text{ kg})\left(9.80 \frac{\text{m}}{\text{s}^2}\right) - (71.0 \text{ kg})\frac{\left(3.14 \dfrac{\text{m}}{\text{s}}\right)^2}{13.0 \text{ m}} = 642 \text{ N}.$$

(b) $a = \dfrac{F_{net}}{m}$; $a_{cent} = \dfrac{v^2}{R}$. Therefore, $\dfrac{v^2}{R} = \dfrac{F_n - mg}{m}$. Because

$$F_n = mg + m\frac{v^2}{R},$$

$$F_n = (71.0 \text{ kg})\left(9.80 \frac{\text{m}}{\text{s}^2}\right) + (71.0 \text{ kg})\frac{\left(3.14 \dfrac{\text{m}}{\text{s}}\right)}{13.0 \text{ m}} = 750 \text{ N}.$$

31. The satellite is in circular motion, and the net centripetal force is supplied by Earth's gravity.

$a = \dfrac{F_{net}}{m}$; $a_{cent} = \dfrac{v^2}{R}$. Therefore, $\dfrac{v^2}{R} = \dfrac{F_G}{m}$;

$F_G = \dfrac{G M_{Earth} m}{R^2}$. Thus, $v^2 = \dfrac{G M_{Earth}}{R}$. Because $v = \dfrac{2\pi R}{T}$,

$\left(\dfrac{2\pi R}{T}\right)^2 = \dfrac{G M_{Earth}}{R}$; $4\pi^2 R^3 = G M_{Earth} T^2$.

$T = 12 \text{ h} = 43{,}200 \text{ s}$

$$4\pi^2 R^3 = \left(6.67 \times 10^{-11} \frac{\text{Nm}^2}{\text{kg}^2}\right)(5.97 \times 10^{24} \text{ kg})(43{,}200 \text{ s})^2$$

$R = 2.66 \times 10^7 \text{ m}$

$h = R - R_{\text{Earth}} = 2.66 \times 10^7 \text{ m} - 6.37 \times 10^6 \text{ m}$

$h = 2.02 \times 10^7 \text{ m}$

33. **(a)** The free-body diagrams are the same shape for both, with the lift force on the plane and the normal force on the person, and look like the diagram in Example 6-4.

(b) $F_{Lx} = F_L \sin\theta = m_{\text{plane}}\dfrac{v^2}{r}$ $\quad F_{L,y} - F_{g,\text{plane}} =$

$F_L \cos\theta - F_{g,\text{plane}} = 0$

$$F_L = \frac{m_{\text{plane}}g}{\cos\theta} \Rightarrow F_L \sin\theta = m_{\text{plane}}g\tan\theta = m_{\text{plane}}\frac{v^2}{r}$$

$a_{\text{cent}} = g\tan\theta$

(c) $F_n \cos\theta = F_{g,\text{person}}$

(d) The apparent weight of the person is the normal force exerted on the person, and so is limited to 1.2 times the person's weight.

$$F_n = \frac{F_{g,\text{person}}}{\cos\theta} \le 1.2F_{g,\text{person}} \Rightarrow \theta \le \cos^{-1}\left(\frac{1}{1.2}\right) = 33.56° \approx 34°$$

(e) Using our expression for centripetal acceleration from part (b):

$$g\tan\theta = \frac{v^2}{r} \Rightarrow r = \frac{v^2}{g\tan\theta} = \frac{(75 \text{ m/s})^2}{(9.8 \text{ m/s}^2)\tan(33.56°)}$$

$$= 865.2 \text{ m} = 870 \text{ m}$$

The turn is made through 90° or $\pi/2$ radians.

$$v = \frac{(\pi/2)r}{T} \Rightarrow T = \frac{\pi r/2}{v} = \frac{\pi(865 \text{ m})/2}{75 \text{ m/s}} = 18 \text{ s}$$

(f) To make the turn in less time at the same speed, the bank angle must be increased. First the radius of the turn can be calculated.

$$T = \frac{\pi r/2}{v} = 15 \text{ s} \Rightarrow r = \frac{15 \text{ s} \times 2}{\pi}(75 \text{ m/s}) = 716 \text{ m} \approx 720 \text{ m}$$

And from that the bank angle can be calculated.

$$r = \frac{v^2}{g\tan\theta} \Rightarrow \theta = \tan^{-1}\frac{v^2}{gr} = \tan^{-1}\frac{(75 \text{ m/s})^2}{(9.8 \text{ m/s}^2)716 \text{ m}} = 38.7° \approx 39°$$

And from that the ratio of apparent and actual weights can be obtained.

$$\frac{F_n}{F_{g,\text{person}}} = \frac{1}{\cos\theta} = \frac{1}{\cos(38.7°)} = 1.3$$

(g) The only variables that can be altered to meet the reduced time limit are the plane's speed or the radius of the turn, both of which will increase the acceleration required to maintain circular motion. Centripetal acceleration depends on the gravitational constant and the tangent of the bank angle θ, so any change in centripetal acceleration will require a change in bank angle.

Chapter 6 AP® Practice Problems

1. (A) Two forces: (1) the downward one caused by gravity and (2) the normal force exerted on the object by the track.

At the top and bottom of the track both gravity and the normal force by the track are along a vertical line. At the top, both forces are downward, toward the center of the circle—the acceleration is centripetal and the object's speed is not changing at this point, just its direction. At the bottom, the normal force is upward and the gravitational force is downward; the acceleration is centripetal so the net force and acceleration are upward and speed is not changing at this point.

On the sides of the track the normal forces are directed toward the center. On the right side we then have two components of the acceleration: that normal to the track and that of gravity downward. They add to a net acceleration downward to the left. On the left side we have the same two components but the normal force is to the right so the net acceleration is downward toward the right.

3. (A) Because there is no atmosphere to keep you moving with the surface of the planet, there must be a net force on you directed toward the center of the planet to cause the centripetal acceleration. The gravitational force exerted on you does not change, so the normal force (apparent weight) will be less, such that the net force will be toward the center.

$F_{\text{net}} = mg_p - F_n = m\dfrac{v^2}{R}$. For this problem the apparent weight is $F_n = \dfrac{3}{4}mg_p$.

Substituting that gives us $mg_p - \dfrac{3}{4}mg_p = \dfrac{1}{4}mg_p = m\dfrac{v^2}{R}$.

Solving for v gives $v = \dfrac{\sqrt{g_pR}}{2}$.

5. (A) $v = \dfrac{2\pi r}{T}$; $a = \dfrac{v^2}{r}$ and $\left|\vec{F}_g\right| = G\dfrac{m_1m_2}{r^2}$

The basic relation for circular orbits comes from combining the law of gravitation with the relation for centripetal force: $m\dfrac{v^2}{r} = G\dfrac{Mm}{r^2} \Rightarrow v^2 = \dfrac{GM}{r}$. To find the period as a function of velocity, use $v = \dfrac{2\pi r}{T}$ to eliminate r from our equation, which gives $v^2 = \dfrac{2\pi GM}{vT}$ or $v^3 = \dfrac{2\pi GM}{T}$. Solve for period, $T = \dfrac{2\pi GM}{v^3}$. From this it can be seen that doubling the speed will reduce the period by a factor of 8, or to one-eighth of its original value.

7. (D) Only when the traveler is located on the line connecting the centers, and the magnitudes of the forces are the same, will the net force be zero. Along a line perpendicular to the line that connects the centers of the two stars, the component of the force is nonzero.

9. (C) The mass increases proportionally with density. For F_2 the spheres are most dense and for F_3 the least dense.

Chapter 7

7-1 The ideas of work and energy are intimately related, and this relationship is based on a conservation principle

1. When wind blows across a field of tall grass, the grass bends in the wind. This bending is a displacement in the direction the wind is pushing the grass, a signature of work. The wind must have energy to be able to do work.
3. The exchange of matter between the surroundings and the system always involves exchange of energy because it involves the exchange of atoms or molecules that are always in motion, and so have kinetic energy.

7-2 The work done by a constant force exerted on a moving object depends on the magnitude of the force and the distance the object moves in the direction of the force

1. (a) The net force is zero because the apple moves at constant speed (acceleration is zero). The upward force you exert must therefore be equal in magnitude to the downward gravitational force, *mg*.
   (b) The work done by you is equal to the force that you exert on the apple multiplied by the distance that the apple is displaced in the same direction as this force.
   (c) The sign is positive, even though the convention that we use for the positive direction is arbitrary, as the force and the displacement are in the same direction.
   (d) We know that work is done because we observe an acceleration as the apple moves downward, from which we infer a force. We know that the work done is the product of the force exerted on the object and displacement of the object in the direction of the force.
   (e) The force of gravity by Earth on the apple does work on the apple as it falls. This force can be expressed as *mg* downward and if the displacement is a distance *d* downward, then the work done is *mgd*.
   (f) As you lifted the apple, the force you exerted was always in equilibrium with the gravitational force exerted on the apple by Earth and in the direction of the apple's motion. This means the work done by you on the apple as it is raised is identical to the work done by Earth on the apple as it falls and the work done by Earth as the apple was raised was the negative of this value. This assumes no other forces, such as air resistance.
   (g) In part (b), we considered only the work you did on the apple. The net work would be zero because the apple (an object) did not accelerate. In part (d) only one force was exerted on the apple as it accelerated, so the net work (that is, the energy transferred to it) is *mgd*.
   (h) The work done on the apple as it falls is *mgd*. When it hits the ground, the ground pushes up on the apple, doing enough negative work on it to bring it to a stop. The apple does positive work on the ground, probably resulting in a dent in the dirt.

   The forces involved (equal magnitude and opposite direction) have to be much bigger than *mg* because the change in the apple's energy ($-mg \times 1$ m) occurs over a much smaller distance.
3. The work done by the force *F* decreases as the angle increases: the vector component of the force that is exerted in the same direction as the displacement *d* gets smaller as the angle increases.
5. $W = Fd = mgd = (446\text{ kg})(9.8\text{ N/m})(2.0\text{ m}) = 8700\text{ J}$. We are forced to use the precision of measurement implied by the height.
7. (a) and (b)

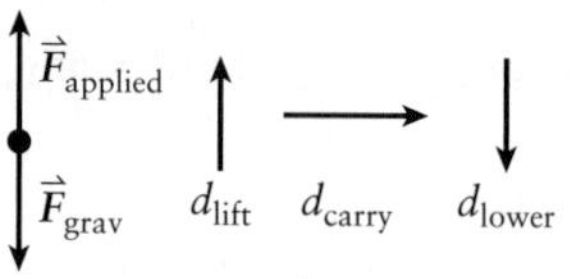

   $\vec{F}_{\text{applied}}$ $\vec{F}_{\text{grav}}$ d_{lift} d_{carry} d_{lower}

   (c) i. $W_{\text{lift}} = F_{\text{applied}} d_{\text{lift}} \cos(0°) = F_{\text{applied}} d_{\text{lift}}$
   ii. $W_{\text{carry}} = F_{\text{applied}} d_{\text{carry}} \cos(90°) = 0$
   iii. $W_{\text{lower}} = F_{\text{applied}} d_{\text{lower}} \cos(180°) = -F_{\text{applied}} d_{\text{lower}}$
   Because $d_{\text{lift}} = d_{\text{lower}}$, $W_{\text{lower}} = -F_{\text{applied}} d_{\text{lift}}$.
   $W_{\text{total}} = W_{\text{lift}} + W_{\text{carry}} + W_{\text{lower}} = W_{\text{lift}} - W_{\text{lift}} = 0$
   (d) To hold the box at a constant height, muscles convert chemical energy even though gravity is not doing work on the box. You feel tired as a result of this energy consumption. The fact that there is a change in your internal energy without you doing work on the box creates a conceptual dissonance. Your energy was not transferred to the box, however, because its speed did not change!
   (e) Your answers for the net work done on each segment and overall do not change. When you lift the box, you do positive work to get it moving in the direction you are lifting it. However, when you stop it going upward, you do an equal amount of negative work, because you remove the same amount of kinetic energy you gave it. You do positive work on the box to get it going in the horizontal direction and an equal amount of negative work to stop its motion in the horizontal direction. You do negative work on the box as you begin to set it down (you are still lifting upward if it isn't going to hit the floor too hard) but Earth is doing positive work that is larger in magnitude than this negative work, so the box begins to accelerate downward. Then either you or the ground must do negative work to bring the box to a stop. The total work done on the box as it moved downward is still zero, if it starts and stops at rest and is well-modeled as an object. If you don't slow the box and it falls hard enough to break, then the object model no longer fits, and some energy is transferred internally to the box to bend or break it.

7-3 Newton's second law applied to an object allows us to determine a formula for kinetic energy and state the work-energy theorem for an object

1. $K = \dfrac{mv^2}{2} = 0.250\text{ g}\dfrac{1\text{ kg}}{1000\text{ g}}\dfrac{100\text{ m}^2/\text{s}^2}{2} = 12.5\text{ mJ}$

3. $W = \Delta K = \dfrac{mv_f^2}{2} - \dfrac{mv_i^2}{2}$

$= 0.5 \times 2100 \text{ kg} \times [12.0^2 \text{ m}^2/\text{s}^2 - 22.0^2 \text{ m}^2/\text{s}^2] = -357 \text{ kJ}$

$= -360 \text{ kJ}$

5. Substituting $a_x = \Sigma F_{\text{ext},x}/m$ into the equation, multiplying both sides by m, and dividing by 2 leads to the relationship between work and change in kinetic energy for an object. The equality only holds if the object model holds, because all points in the system must travel the same distance as the center of mass.

$$v_{fx}^2 = v_{ix}^2 + 2\frac{F_{\text{ext},x}}{m}(x_f - x_i) \Rightarrow \frac{mv_{fx}^2}{2} - \frac{mv_{ix}^2}{2} = F_{\text{ext},x}(x_f - x_i)$$

7. (a) As the coefficient of kinetic friction decreases, the velocity of the sled as it descends the slope increases unless the rider exerts other forces on the surface. As the coefficient of kinetic friction decreases, the distance required for the sled to stop increases, making the possibility of the sled and rider going over the cliff much greater.

(b) $K_f = 0 \quad K_f - K_i = -F_k\Delta x$

$= -\mu_k mg\Delta x \Rightarrow \mu_k \geq \dfrac{v_{ix}^2}{2g\Delta x}$

7-4 The work-energy theorem can simplify many physics problems

1. The same work is done by the launchers, so the kinetic energies of the cars will be the same. Kinetic energy is proportional to mass multiplied by velocity squared. So the speed of the car with more mass is decreased in proportion to the inverse of the square root of the ratio of the masses.

3. (a)

$\vec{F}_{\text{wind on sail}}$, $\vec{F}_{\text{drag on hull}}$, $\vec{F}_{\text{water on keel}}$

(b) The northern component of the force exerted on the sail by the wind, $F_{\text{wind on sail}} \sin(25.0°) = 0.782$ N, is canceled by the force exerted by the water on the keel. The net force to the east is $F_{\text{wind on sail}} \cos(25.0°) - F_{\text{drag on hull}} = 0.927$ N. So the work done by the net force in 3.55 m is 3.29 J. With a mass of 0.325 kg, the speed of the boat at 3.55 m is calculated to be 4.50 m/s.

(c) $v_{fx} = v_{ix} + (\Sigma F_{\text{ext},x}/m)\,\Delta t \Rightarrow v_x(t) = v_{0x}$

$+ 2.85 \text{ m/s}^2 t = 2.85 \text{ m/s}^2 t$

(d) Newton's second law is equivalent to the work-energy theorem for an object when the time-independent form of the kinematic equations is used. If time is a variable of interest, you should use kinematics.

5. (a) The gravitational force does positive work on the woman as she falls 10 m through the air and 4 m through the water because the force of gravity and the displacement are in the same direction.

$$W = Fd = mgd = 65.0 \text{ kg } (9.8 \text{ N/kg})(14.0 \text{ m}) = 8920 \text{ J}$$

The net work done on the woman over the 14 meters is zero as her net change in kinetic energy is zero, so the work done to the woman by the water is −8920 J.

(b) The kinetic energy of the woman when she enters the water is equal to the work done on her by the gravitational force as she fell 10 m through the air.

$$K = Fd = mgd = 65.0 \text{ kg } (9.80 \text{ N/kg})(10.0 \text{ m}) = 6370 \text{ J}$$

Because she must lose this much kinetic energy in the water, the angle between the displacement and average net force (magnitude F_{average}) must be 180°. In this case, $d = 4$ m.

$W = \Delta K = K_{\text{final}} - K_{\text{initial}} = 0 - 6370 \text{ J} = -6370 \text{ J}$

$W = F_{\text{average}} d \cos 180° = -6370 \text{ J} = F_{\text{average}}(4 \text{ m})(-1)$

$F_{\text{average}} = 1590 \text{ N}$

(c) The force of gravity exerted on the woman is still downward, so the force exerted solely by the water is larger in magnitude than the average net force on the woman and in the opposite direction.

7-5 The work-energy theorem is also valid for curved paths and varying forces, and, with a little more information, systems as well as objects

1. (a)

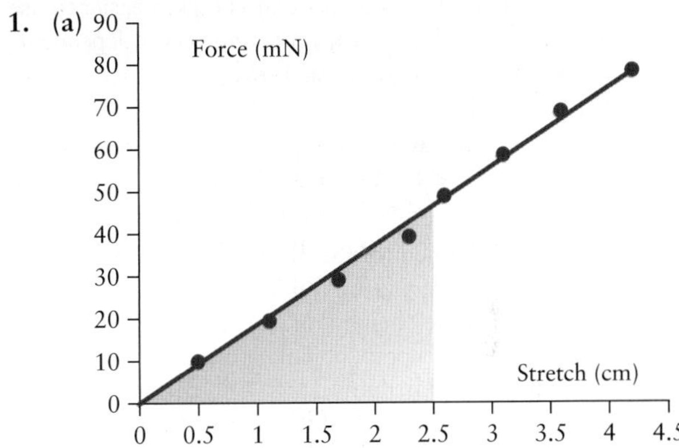

(b) Linear regression (line of best fit) of the data gives a slope $= k = 1.86$ N/m $\cong 1.9$ N/m to correct significant digits.

(c) $W = \text{area} = F\Delta x/2 = \dfrac{k\Delta x^2}{2}$

$= \dfrac{1}{2}(1.9 \text{ N/m})(2.5 \times 10^{-2} \text{ m})^2 = 0.59 \text{ mJ}$

3. (D) The tension force is always perpendicular to the displacement.

5. Algebraically, the work done by the pulling force to stretch the spring and move the object is:

$W = F\Delta x/2 = \dfrac{k}{2}(\Delta x_f^2 - \Delta x_i^2)$

$= \dfrac{450 \text{ N/m}}{2}(0.18^2 \text{ m}^2 - 0.12^2 \text{ m}^2) = 4.1 \text{ J}$

So, the work done on the object by the spring is −4.1 J, given that the kinetic energy of the object does not change.

Graphically:

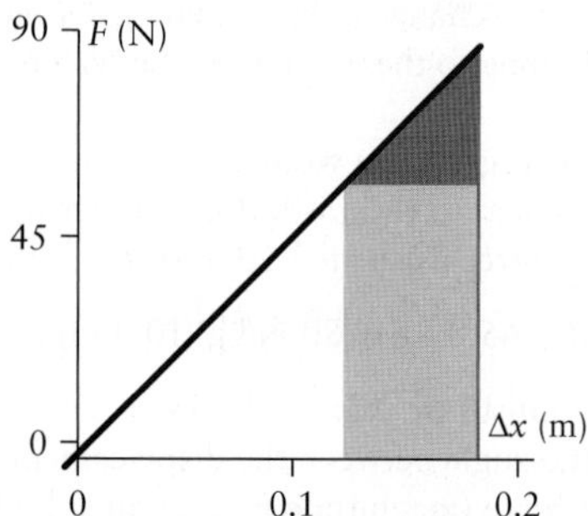

Calculate the area under the force-displacement graph to find the work done by the pulling force, which is the opposite of the work done by the spring.

$$W = 0.12\text{ m }(450\text{ N/m}) \times (0.18\text{ m} - 0.12\text{ m}) + (0.18\text{ m }(450\text{ N/m}) - 0.12(450\text{ N/m}))/2 \times (0.18\text{ m} - 0.12\text{ m}) = 4.1\text{ J}$$

7. (a) The forces, such as the force exerted by the flick of the finger; the component of the gravitational force that depends on the local shape of the bowl; and the component of the normal force are not easily modeled. As the cube slides up the bowl, its direction of motion is always changing. So, an initial and final state analysis based on scalar energy is easier to use than an analysis of the time- and vector-dependent motion up the side of the bowl.

(b) and (d)

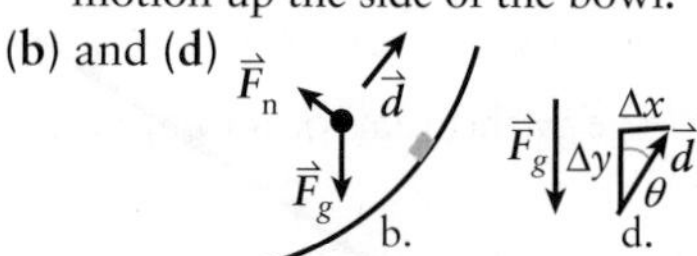

The component of the displacement vector that is along the gravitational field lies along a line that measures the "height" of the bowl. The other component is perpendicular to that line and so makes no contribution.

(c) The normal force is always perpendicular to the displacement of the ice cube and so does not do work on the ice cube.

(e) By adding all of these vertical components, we find the final position reached that is important is the vertical distance traveled by the ice cube as it slides up the side of the bowl. If we apply the work-energy theorem for an object to the ice cube, then the sum of the work done on the ice cube by the gravitational force over each of these increments is the negative of the initial kinetic energy of the ice cube (the energy of the ice cube is zero at its maximum height). If we are considering the Earth–cube system, then the sum of these incremental displacements multiplied by mg is the change in potential energy, which is equal to the initial kinetic energy of the ice cube.

(f) Using either the object model or the cube-Earth system, the sum of terms involving the displacement along the field is proportional to the mass of the ice cube. The initial kinetic energy is also proportional to the mass of the ice cube, and so the mass of the ice cube cancels out and makes no contribution to the calculated height.

The predicted height is very sensitive to the estimate of the initial velocity because height is proportional to the square of the velocity.

9.

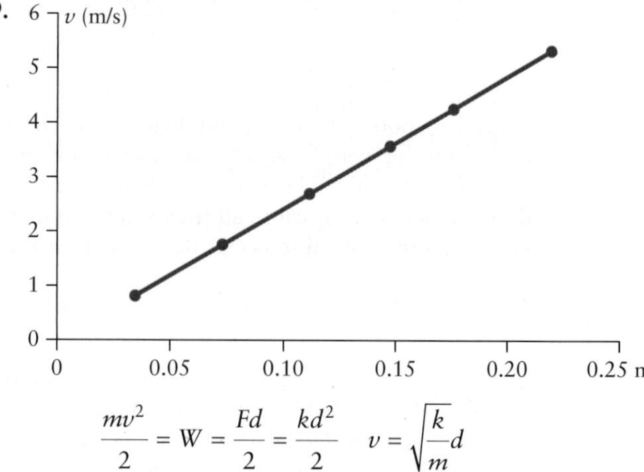

$$\frac{mv^2}{2} = W = \frac{Fd}{2} = \frac{kd^2}{2} \qquad v = \sqrt{\frac{k}{m}}d$$

The best fit line is $v = 24\text{ s}^{-1}d$. So $k = 120$ N/m.

7-6 Potential energy is energy related to reversible changes in a system's configuration

1. A force is conservative if the work done by the force on an object or a system does not depend on the path of the displacement of the point at which that force is exerted. For a conservative force, work done over a closed path is zero. Conservative forces can always be expressed in terms of a configuration; for example, the spring force is expressed in terms of how far the end of the spring is extended from its equilibrium position and gravity depends on the distance from the center of Earth to the point at which it is exerted.

3. $$U_{\text{spring}} = \frac{kx^2}{2} = \frac{k(0.125\text{ m})^2}{2} = 3.33\text{ J}$$

$$\Rightarrow k = \frac{6.66\text{ J}}{0.0156\text{ m}^2} = 426\text{ kg/s}^2 = 426\text{ N/m}$$

5. (a)

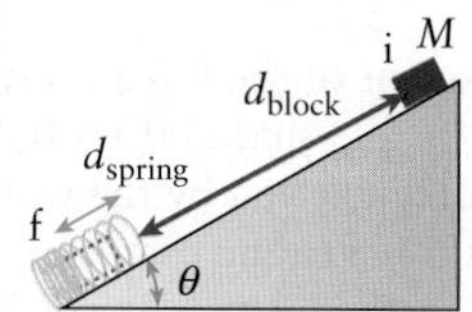

where d_{block} is the displacement of the block prior to collision with the spring and d_{spring} is the maximum compression of the spring after collision. So, the total displacement of the block during the motion is $d_{\text{block}} + d_{\text{spring}}$.

(b) Here are three alternatives for the selection of the system:

i. The block, M, could be treated using an object model with external forces exerted by Earth and the spring (not recommended because it requires using the average force = $F/2$ trick).

ii. The block and the spring could be treated as a system and work would be done by the external gravitational force.

iii. Block–spring–Earth could be treated as a system and no work would be done.

(c) Looking at system choices ii and iii:

ii. $K_f = K_i = 0 \Rightarrow \Delta K = 0$

$U_i = 0 \quad U_f = kd^2_{spring}/2 \quad \Delta U = kd^2_{spring}/2$

$W = Mg(d_{block} + d_{spring})\cos(\pi/2 - \theta)$

$= Mg\sin(\theta)(d_{block} + d_{spring}) = \Delta U = kd^2_{spring}/2$

iii. $K_f = K_i = 0 \Rightarrow \Delta K = 0$

Taking the origin of the coordinate along the ramp to be at the top of the ramp (and the spring is relaxed):

$$U_i = 0 \quad \Delta U = kd^2_{spring}/2 - Mg\sin(\theta)(d_{block} + d_{spring}) = 0$$
$$W = 0$$

(d) Each route leads to the same result.

$$Mg\sin(\theta)(d_{block} + d_{spring}) = kd^2_{spring}/2$$

from which we obtain

$$d_{block} = \frac{kd^2_{spring}/2}{Mg\sin(\theta)} - d_{spring}$$

$$= \frac{(1.35\times10^4\ \text{N/m})(5.50\times10^{-2}\ \text{m})^2}{2(12.0\ \text{kg})(9.80\ \text{N/kg})\sin 28.0°} - 5.50\times10^{-2}\ \text{m} = 0.315\ \text{m}$$

(e) See part (c).

7. (a) Negative changes in a quantity can easily be confused with values of the quantity. It is relative, rather than absolute, values of quantities that are of interest in applications of energy conservation. Andrea could point out that kinetic energy is proportional to the product of mass and velocity squared, neither of which can be negative. She could recognize that Carrie is confusing absolute and relative quantities and agree that the change in kinetic energy can be negative. Carrie might also be neglecting the idea that both force and displacement are vectors that have signs.

(b) Carrie is applying the work-energy theorem for an object to a system with components that can change relative positions within the system—the cart is nearer to or farther from the center of Earth—so Carrie is neglecting the potential energy change of the cart–Earth system. By neglecting this term, Carrie arrives at $\Delta K = 0$ and then to a negative kinetic energy, which is not possible.

Chapter 7 Review Problems

1. The gravitational force exerted by Earth on any object is directed from the object toward the center of Earth. In its circular orbit, the displacement of the satellite is always perpendicular to the direction of the force of gravity, so there can be no work done by this force.

3. The net work required to increase the car's speed is equal to the change in the car's kinetic energy.

$$W_{net} = \Delta K = K_f - K_i = \frac{1}{2}mv_f^2 - \frac{1}{2}mv_i^2 = \frac{1}{2}m(v_f^2 - v_i^2)$$
$$= \frac{1}{2}(1250\ \text{kg})\left(\left(30.0\frac{\text{m}}{\text{s}}\right)^2 - \left(20.0\frac{\text{m}}{\text{s}}\right)^2\right)$$
$$= 3.13\times10^5\ \text{J}$$

5. We can apply the work-energy theorem for an object because we are not changing the shape of the train. Because the train is on a level track, gravity does no work.

Train mass in kg:

$$m_{train} = (8500\ \text{metric tons})\left(\frac{10^3\ \text{kg}}{1\ \text{metric ton}}\right) = 8.5\times10^6\ \text{kg}$$

Train's initial speed in m/s:

$$v_i = \left(90\frac{\text{km}}{\text{h}}\right)\left(\frac{10^3\ \text{m}}{1\ \text{km}}\right)\left(\frac{1\ \text{h}}{3600\ \text{s}}\right) = 25\frac{\text{m}}{\text{s}}$$

Initial kinetic energy of train (final is zero):

$$K_i = \frac{1}{2}m_{train}v_i^2 = \frac{1}{2}(8.5\times10^6\ \text{kg})\left(25\frac{\text{m}}{\text{s}}\right)^2 = 2.66\times10^9\ \text{J}$$

$$W_{net} = \Delta K = K_f - K_i = 0\ \text{J} - 2.66\times10^9\ \text{J} = -2.7\times10^9\ \text{J}$$

7. To decrease the ball's kinetic energy the force of the glove on the ball points in the opposite direction to the ball's displacement. We can use the work-energy theorem for an object to calculate the magnitude of the average force the glove exerts on the ball during the catch.

$$\Delta K = \frac{1}{2}mv_f^2 - \frac{1}{2}mv_i^2 = 0 - \frac{1}{2}mv_i^2$$
$$W_{net} = W_{\text{glove on ball}} = F_{\text{glove on ball}}d\cos(180°) = -F_{\text{glove on ball}}d$$

Using the work-energy theorem for an object:

$$-\frac{1}{2}mv_i^2 = -F_{\text{glove on ball}}d$$

$$F_{\text{glove on ball}} = \frac{mv_i^2}{2d} = \frac{(0.145\ \text{kg})\left(44.0\frac{\text{m}}{\text{s}}\right)^2}{2(0.125\ \text{m})} = 1.12\times10^3\ \text{N}$$

The force of the glove on the hand will be slightly less than the force of the glove on the ball because the glove will compress somewhat and its internal energy will increase.

9. (a) $W_{you} = F_{applied}d\cos(\theta) = (120\ \text{N})(20.0\ \text{m})\cos(20°) = 2255\ \text{J} = 2300\ \text{J}$

(b) $W_{grav} = F_g d\cos(\theta) = mgd\cos(\theta) = (18.0\ \text{kg})(9.8\ \text{N/kg})(20.0\ \text{m})\cos(100°) = -613\ \text{J} = -610\ \text{J}$

(c) $W_n = F_n d\cos(\theta) = F_n d\cos(90°) = 0\ \text{J}$

(d) $W_{total} = W_{you} + W_{grav} + W_n = 2255\ \text{J} + (-613\ \text{J}) + 0 = 1642\ \text{J} = 1600\ \text{J}$

11. Even though the path is irregular, the force is directed parallel to each incremental element of displacement, so the entire 910 m is along the direction of the force.

$$W_{drag} = F_{drag}d\cos\theta = F_{drag}d\cos(0°)$$
$$W_{drag} = F_{drag}d\cos\theta = (625\ \text{N})(910\ \text{m}) = 568{,}750\ \text{J}$$
$$= 5.69\times10^5\ \text{J}$$

13. (a)

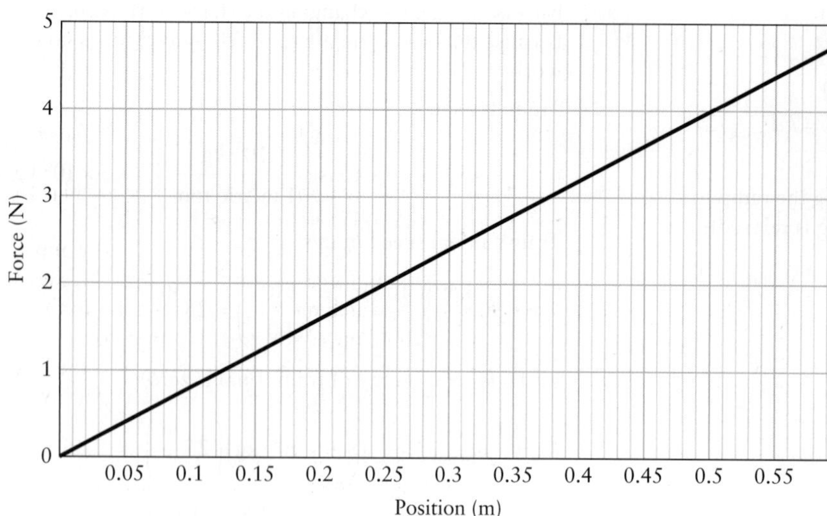

(b) $W = \text{area} = \frac{1}{2}bh = \frac{1}{2}(0.5\text{ m})(4.0\text{ N}) = 1.0\text{ J}$

(c) $U_{s,i} = K_f; 1.0\text{ J} = \frac{1}{2}(0.750\text{ kg})v_f^2;\ v_f = 1.6\frac{\text{m}}{\text{s}}$

15. $$\Delta U_{grav} = mg\Delta y = (40.0\text{ kg})\left(9.80\frac{\text{m}}{\text{s}^2}\right)(-4.35\text{ m})$$
$$= -1.71\times10^3\text{ J} = -1.71\text{ kJ}$$

17. In equilibrium, the net force on the object is zero. This means the upward force exerted by the spring on the object is equal in magnitude to the downward gravitational force on it. We can use this to find the amount by which the spring compresses, which we can then use to calculate the potential energy stored in the spring.

(a) $\sum F_{ext,y} = F_s - F_g = k\Delta y - mg = ma_y = 0$

$$k\Delta y - mg = 0$$

$$\Delta y = \frac{mg}{k} = \frac{(2.00\text{ kg})\left(9.80\frac{\text{m}}{\text{s}^2}\right)}{\left(2.00\times10^2\frac{\text{N}}{\text{m}}\right)} = 9.80\times10^{-2}\text{ m}$$

(b) $\Delta U_s = \frac{1}{2}ky_f^2 - \frac{1}{2}ky_i^2 = \frac{1}{2}ky_f^2 - 0$

$$= \frac{1}{2}\left(2.00\times10^2\frac{\text{N}}{\text{m}}\right)(9.80\times10^{-2}\text{ m})^2 = 0.960\text{ J}$$

19. The only nonzero work is the work due to kinetic friction, which is equal to the change in the truck's kinetic energy. We can use the work-energy theorem for an object to calculate the coefficient of kinetic friction between the gravel in the truck lane and the truck's tires.

$$\Delta K = W_{net} = F_k d\cos(180°) = -\mu_k F_n d$$
$$= -\mu_k(mg)d = -\frac{1}{2}mv_i^2$$

$$\mu_k = \frac{v_i^2}{2gd} = \frac{\left(24.6\frac{\text{m}}{\text{s}}\right)^2}{2\left(9.80\frac{\text{m}}{\text{s}^2}\right)(35.0\text{ m})} = 0.882$$

21. If the wheels were free to roll, the person and wheelchair would accelerate down the ramp. Because we do not yet know how to work with rolling motion, we will instead look at the changes in kinetic energy of the center of mass, given that we do know how to do that, and it does not require understanding of rolling motion. We cannot just use work, because the point of application of the force due to the ramp is on the edge of a wheel, and so the chair will not move the same way as the center of mass. Identifying the forces: The net force on the wheelchair is the normal force exerted by the ramp on the wheelchair perpendicular to the surface, the downward gravitational force exerted by Earth on the wheelchair, and a friction force exerted by the ramp on the wheels. The downward gravitational force has a component perpendicular to the surface, which is balanced by the normal force, and a component down the ramp. To go up or down the ramp at constant speed, the friction force on the wheels must be directed up the ramp to balance the force of gravity down the ramp. The engineer would need data on the ratio of the range of forces per unit mass that a typical wheelchair occupant could exert, and the range of coefficients of friction. The selection of 12:1 indicates that, because $\theta = \tan^{-1}(1/12) \sim 5°$, a lower limit of that force per unit mass is $g\sin(5°) = 0.8$ N/kg .

23. (a) (D) Because the initial and final heights are the same, no work is done by the gravitational force and the tension force is always perpendicular to the displacement.
(b) (C) The ball has moved in the same direction as the force of gravity exerted on it between its initial and final heights so $W = mgd = 6$ J.
(c) (D) The tension force is always perpendicular to the displacement.

25. (a) The fact that the gravitational force is directed downward is not a convention. But the signs of displacement uphill and the sign of the force are both conventions that must be chosen consistently. The displacement and force for this situation are always oppositely directed, so the work is always negative.
(b) The work done by the foot on the pedal is positive because force and the displacement are in the same direction.
(c) The force of the pedal on the foot is equal and opposite to the force of the foot on the pedal. The work is negative because the force and displacement will be in opposite directions.

Chapter 7 AP® Practice Problems

1. (C) $W = Fd\cos\theta$
At C the force of gravity is perpendicular to the displacement of the ball as it moves through this point;

therefore, the work done on the ball as it moves through this point is zero.

3. (C) $W_{net} = \Delta K, W = Fd\cos\theta$
$W_{net} = W_{grav} + W_{string} = \Delta K$; therefore
$W_{string} = K_f - K_i - mgh \Rightarrow W_{string} = \frac{1}{2}mv_f^2 - mgs$

5. (A) $W_{net} = \Delta K$
Because the block was initially at rest, the net work is the final kinetic energy.

7. (D) The potential energy stored in the spring is proportional to $1/2k\Delta x^2$. We are told the compression doesn't change, so the spring energy increases by a factor that is the same as the increase in the spring constant, which is 4. The gravitational potential energy is proportional to H, so the height reached must also increase by a factor of 4.

9. (D)

$$\Delta K = W = \Sigma F_{ext,x}D = (F_{app} - \mu m_{block}g)D$$
$$W_Q = (F - \mu mg)D$$
$$W_R = (2F - \mu mg)D$$
$$W_S = (F - \mu(2m)g)D$$
$$W_T = (2F - \mu(2m)g)D$$

Chapter 8

8-1 Total energy is always conserved, but it is constant only for a closed, isolated system

1. Enrique is confusing total energy with mechanical energy. There is a friction force, which is nonconservative, exerted on the block, causing the block to come to a rest. If we treat friction as an external force, he is neglecting the transfer of energy out of the system, causing the table and block to become warm. If we think of the friction force as internal, then energy is converted from kinetic to internal. Total energy includes all these quantities and is conserved.

3. (C) When an external force is exerted on a system, mechanical energy is not constant. Neglecting air resistance, there are no nonconservative forces within the system until the rock strikes the water. As the rock falls through the water, mechanical energy is not constant because the drag force exerted by the water on the rock does work on the system. Total energy is conserved; the mechanical energy gets dissipated in warming the water and rock.

8-2 Choosing systems and considering multiple interactions, including nonconservative ones

1. (A) is not correct. Friction and air resistance can be neglected. Because there are no nonconservative forces exerted on either boy, the total mechanical energy of each boy is constant and independent of the path taken.
(B) is not correct. The work done in both cases is equal to the product of the gravitational force and the displacement in the direction of the force. The displacement in the direction of the force (straight downward) is the same for both boys.
(C) is correct. There are no nonconservative forces exerted on either boy so the total mechanical energy of each boy–Earth system is constant and independent of the path taken. The change in the gravitational potential energy for each boy–Earth system is the same, so the change in the kinetic energy of both boys is the same.
(D) is not correct. The normal force is perpendicular to the displacement and therefore does no work.

3. (a) (B) The net force accelerates the box up the ramp (also the friction force cannot be greater than the normal force, which must be perpendicular to the surface), and F_g must be directly downward.
(b) In each case, there are three interactions with the box:
   1. The second box is pulling on it connected by the string.
   2. Earth is interacting with it through gravity.
   3. The surface is interacting with it through friction.

If your system is the box m_2, you would need to use a force description for the interactions with Earth and the other box.

If your system is the two boxes and Earth, you would describe the system by the changes in gravitational potential energy of the two boxes with Earth.

If your system is the box m_2 and Earth, the other box, m_1, is external to the system. Then, you would talk about gravitational potential energy to describe the interaction of the box (m_2) with Earth, as well as using this box–Earth system model to describe changes in kinetic energy, but you would use a force description to describe the interaction with the second box.

In each of these descriptions, the surface was external to the system, so you would use a force description to describe the energy dissipated by the friction interaction with the surface. You could put this interaction inside the system in any of the three situations listed, in which case you would describe the friction interaction by the change in internal energy of the system that the interaction would cause.

(c) The friction force exerted on the box by the ramp is nonconservative and mechanical energy will be dissipated by the interaction. Because this is friction, the transfer is not cleanly defined: Some of the mechanical energy dissipated will end up heating or damaging the box and some the ramp. The energy transfer is negative (mechanical energy decreases).

5. (a) At all stages, there is a gravitational interaction between Earth and the ball, but given that Earth is in the system, the interaction is internal and we only consider the interaction as a source of potential energy. In stage 1, an external force is exerted by the student that compresses a spring. Because the spring is held in place by the cocking mechanism, it exerts only a normal force on the plastic ball. The normal force exerted by the spring and the barrel of the launcher together must balance the weight of the ball, but individually these forces vary, depending on the orientation of the launcher. Stage 2: Once the

student pulls the trigger, a spring force is exerted on the ball, moving it in the direction of the spring extension, along the barrel of the launcher. The barrel may exert a friction force on the ball. Stage 3: The only external force on the ball is air resistance.

(b) Stage 1: Potential energy stored in the spring increases as positive work is done on the spring by the student cocking the launcher. Stage 2: Assuming that the mass of the spring can be neglected, the spring will do an amount of work on the ball–Earth system equal to the potential energy of the compressed spring at the end of stage 1 as the spring relaxes. This work equals the increase in the ball's kinetic energy plus the change in the gravitational potential energy of the ball–Earth system, minus any energy lost to friction within the barrel. At the beginning of this step, the velocity of the ball was zero, so the change in kinetic energy at any time is the ball's kinetic energy at that time. During stage 3, the ball–Earth system has the total energy it had when it left the barrel, and the ball moves upward until its kinetic energy (except for that energy given up to friction with the air) is all converted into additional gravitational potential energy of the ball–Earth system. When the energy is all potential, the ball will come to a stop, and then begin falling toward Earth, converting the potential energy of the ball–Earth system into kinetic energy and any energy lost to air resistance, until it hits the ground and either comes to a stop or bounces.

(c) The work done by the student is only as the launcher is cocked and the trigger is pulled. Pulling the trigger does very little work; it just releases the spring, starting stage 2. Cocking the launcher does positive work on the spring in stage 1. If friction with the barrel and air could be neglected, the spring would transfer this energy and this would be the total mechanical energy of the ball–Earth system from launch until the ball returned to the same height from which it was launched (a level we will define as 0 of gravitational potential energy for the ball–Earth system). If we cannot neglect the friction forces as the ball moves through stages 2 and 3, the total mechanical energy will be reduced by these friction interactions, reducing the height to which the ball flies and its speed upon returning to Earth.

8-3 Energy conservation is an important tool for solving a wide variety of problems

1. Energy is conserved, so $\Delta K + \Delta U = W$. Because we know that $K_f = 0$, $K_i = \dfrac{mv_i^2}{2}$

   and $\Delta U = 0$, so $W = -\mu_k mgd$

   $$-\frac{mv_i^2}{2} = -\mu_k mgd$$

   $$v_i = \sqrt{2\mu_k gd} = \sqrt{2(0.480)(9.80 \text{ N/kg})(88.0 \text{ m})} = 28.8 \text{ m/s}$$

3. The block starts and stops at rest, so initial and final K are zero. When the block is at the top of the ramp the spring is not compressed, so there is only gravitational potential energy. At the bottom of the ramp, there is both gravitational potential energy and energy stored in the compressed spring.

   $$K_i = 0 \quad K_f = 0 \quad U_i = mgL\sin\theta$$

   $$U_f = mg\sin\theta(d - \Delta d) + \frac{k}{2}\Delta d^2$$

5. (a) Taking $U_i = 0$, $\Delta U = U_f$; Because speed does not change, $K_i = K_f$, so $\Delta K = 0$.

   This means the change in the energy, which is the energy dissipated to the environment, is

   $$W = \Delta E = \Delta U = U_f$$
   $$= m_2 g\Delta y \sin\theta - m_1 g\Delta y$$

   (b) This energy is dissipated through friction, $W = -F_k\Delta y$, where

   $$F_k = \mu_k F_n = \mu_k m_2 g \cos\theta$$

   Setting the two expressions for work equal and canceling Δy gives

   $$m_2 g \sin\theta - m_1 g = -F_k = -\mu_k m_2 g \cos\theta$$

   So, $m_2 \sin\theta + \mu_k m_2 \cos\theta = m_1$.

   (c) For static friction, you would have the same equilibrium condition, but it would be a static equilibrium: nothing moving, instead of dynamic equilibrium, where velocity is constant. So, the same expression as we found in part (b) will answer this question but with the coefficient of static friction, which involves an inequality. The expression obtained in part (b) shows that only for a single value of the inclination can the system be in equilibrium with the maximum coefficient of static friction. For an angle smaller than this, the system will still be in equilibrium, but the static friction needed will be less. So, the experimental design involves slowly adjusting the angle of inclination until the system is stationary. The design should include statements making it clear that the masses are already known and include multiple trials. Finally, an expression that can be used to evaluate the friction coefficient should be given.

   $$\mu_{S,max} = \frac{m_1 - m_2\sin\theta}{m_2\cos\theta}$$

7. We are told that the system is closed and isolated and are given a potential energy function.

   (a) (D) $\Delta K + \Delta U = 0 \Rightarrow K(X) = U_a - U(X)$. So when $U(X) = U_a$ the kinetic energy is 0. Potential energy will then be converted to kinetic energy and motion within the system will reverse when kinetic energy again reaches zero. These turning points occur at $X = 0$ and $X = X_4$ and repeat continuously. As the potential energy of the system changes in between these points, the object will slow down or speed up, but it does not come to a stop, and so does not change direction of motion between these points.

   (b) (C) It will lose an amount of mechanical energy U_b; so it will not be able to make it out of the minimum at X_3 but would still have some energy at that point, and thus would be in motion if it had made it into this well before the work was completed.

(c) (C) $\Delta K + \Delta U = 0 \Rightarrow K(X) = U_c - U(X)$, so when $X = X_3$ the kinetic energy will be $U_c - U_d$.

9. (a) Convenient locations for initial and final states are where the ball is released and where the maximum velocity of the ball is predicted, when the string is vertical (minimum potential energy).

(b) We are told that the string is very light, so we neglect the mass of the string. We set the gravitational potential energy of the object–Earth system at zero for the lowest point to make the math easier.

$$U_i = mg(L - L\cos\theta) = mgL(1 - \cos\theta),\ K_i = 0,\ U_f = 0,\ K_f = \frac{1}{2}mv_f^2$$

(c) We must assume that there are no nonconservative forces, such as interaction with the support or friction or drag that would remove mechanical energy from the system.

(d) $K_f - K_i + U_f - U_i = 0 \Rightarrow \dfrac{mv_f^2}{2} - 0 + 0 - U_i$

$$\frac{1}{2}mv_f^2 - mgL(1 - \cos\theta) = 0$$

(e) $v_f = \sqrt{2gL(1 - \cos\theta)} = 1.81\ \text{m/s}$

8-4 Power is the rate at which energy is transferred into or out of a system or converted within a system

1. Let the system be the cyclist and the Earth. The power is a transfer of energy out of the system (negative). Assume the average velocity is a constant velocity.

(a) Because the kinetic and the potential energies of the system do not change, the power is equal to the change in internal energy of the cyclist.

$$\Delta K + \Delta U + \Delta E_{\text{internal}} = -P\Delta t \qquad \Delta K = \Delta U = 0$$

$$\Delta E_{\text{internal}} = -P\Delta t = -350\ \text{W} \times \frac{12{,}300\ \text{m}}{22{,}500\ \text{m/h}}\ \frac{3600\ \text{s}}{\text{h}} = -690\ \text{kJ}$$

(b) Independent of the speed of the cyclist there is a change in the gravitational potential energy of the cyclist–Earth system. So now the internal energy of the cyclist must provide both the energy expended on the flat ride and the change in the gravitational potential energy of the system:

$$\Delta U_{\text{grav}} = mg\Delta h = 70\ \text{kg}(9.8\ \text{N/kg})(12300\ \text{m})\sin(3.9^\circ) = 0.574\ \text{MJ}$$

So

$$\Delta K + \Delta U + \Delta E_{\text{internal}} = -P\Delta t \qquad \Delta K = 0$$

$$\Delta E_{\text{internal}} = -P\Delta t - \Delta U_{\text{grav}} = -0.689\ \text{MJ} - 0.574\ \text{MJ} = -1.263\ \text{MJ} \approx -1.3\ \text{MJ}$$

3. Solar energy received can increase the total mechanical energy of the Earth system, increase the internal energy of the system, or be transferred out of the system. The Earth system would be warming excessively if this much energy were accumulating each year during the last 3.8 billion years. Most of this energy must eventually be transferred out of the system. Some of this energy allows life to develop and become more organized within the system. Some of this energy gets converted within the system to do things like help plants (and therefore animals) grow (chemical) and move clouds around (kinetic), and some of the resulting chemical energy can be stored for a long time (such as in oil and coal stored in the ground), but ultimately all these types of energy end up as random motion of molecules that warm up the planet, unless the energy is transferred out of the Earth system.

5. (a) $P = Fv = 120\ \text{N}(10\ \text{m/s}) = 1.2 \times 10^3\ \dfrac{\text{Nm}}{\text{s}} \approx 1\ \text{kW}$

(b) $P = c\sin\theta\cos\theta\, v = c\sin(25^\circ)\cos(25^\circ)v$

$$= 120\ \text{kg ms}^{-2}v$$

$$c = \frac{120\ \text{kg}\cdot\text{m/s}^2}{\sin 25^\circ \cos 25^\circ} = 310\ \text{kg}\cdot\text{m/s}^2$$

(c) $P = 310\ \text{kg ms}^{-2}\sin\theta_{\text{match}}\cos\theta_{\text{match}}v$

$P = 0.373\ \text{kW}$

$0.373\ \text{kW} = 310\ \text{kg m/s}^2\sin\theta_{\text{match}}\cos\theta_{\text{match}}v$

(d) It is important not to confuse a description with an explanation. An explanation seeks the reason. Wind speed increases with height. So, increasing the height of the modern turbine increases the power output. Blades on modern turbines are massive and as mass increases, stability decreases. A smaller number of blades has less mass than a large number. As rotational speed increases, stability decreases. So, the design of modern turbines involves greater height for increased power output, and fewer blades and a lower rotation speed to improve stability.

The water pump was built originally with a humbler technology—ropes and ladders—and a larger number of lighter blades has a low burden. The pump runs at a higher speed, and no gears are needed to increase the speed if the blades are turning rapidly.

7. Assume that the elevator both reaches the average speed and stops instantaneously. This is a good approximation since any perception of the acceleration, even in a very fast elevator, is very brief. Accelerating the elevator does require energy, which is neglected to simplify the problem. We find the magnitude of the power:

$$\Delta K + \Delta U = W \qquad \Delta K = 0 \qquad W = mg\Delta h$$
$$= 1600\ \text{kg}(9.8\ \text{N/kg})(382.2\ \text{m}) = 6.0\ \text{MJ}$$

$$P = \frac{W}{\Delta t} = \frac{W}{(\Delta h/v)} = \frac{6.0\ \text{MJ}}{(382.2\ \text{m}/16.83\ \text{m/s})} = 0.26\ \text{MW}$$

Alternatively we could make the assumption that the force exerted on the elevator by the cables that raise and lower it is constant and equal to mg—assuming constant velocity motion. Then we can express the magnitude of the power in terms of the constant force and speed.

$$P = Fv = 1600\ \text{kg}(9.8\ \text{N/kg})(16.83\ \text{m/s}) = 0.26\ \text{MW}$$

If we had considered the accelerations, we would find a greater power output needed during the initial upward acceleration, to increase the kinetic energy, and then less power output when the elevator comes to a halt as the kinetic energy decreases.

8-5 Gravitational potential energy is much more general, and profound, than our near-Earth approximation

1. $\Delta U_{grav} = U_{grav,f} - U_{grav,i} = -\frac{GM_{Earth}m}{r_{Earth}+h} - \left[-\frac{GM_{Eath}m}{r_{Earth}}\right]$

   Because the leading term has a larger denominator the magnitude is smaller than the second term. Therefore, the difference is positive.

3. $\Delta E = 0 = K_f - K_i + U_f - U_i = K_f - \frac{m}{2}\left[\frac{2\pi r_i}{T}\right]^2$

   $$+\left[-\frac{GM_{Earth}\,m}{r_f} + \frac{GM_{Earth}\,m}{r_i}\right]$$

   $$K_f = \frac{1000\text{ kg}}{2}\left[\frac{2\pi\left(5.0\times10^5\text{ m}+6.37\times10^6\text{ m}\right)}{1.5(3600\text{ s})}\right]^2$$

   $$-(6.67\times10^{-11}\text{ m}^3\text{ kg}^{-1}\text{ s}^{-2})(5.97\times10^{24}\text{ kg})(1000\text{ kg})$$

   $$\left[-\frac{1}{5.4\times10^5\text{ m}+6.37\times10^6\text{ m}} + \frac{1}{5.0\times10^5\text{ m}+6.37\times10^6\text{ m}}\right]$$

   The numerical value of the final kinetic energy is 3.16×10^{10} J.

5. (a)

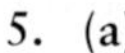

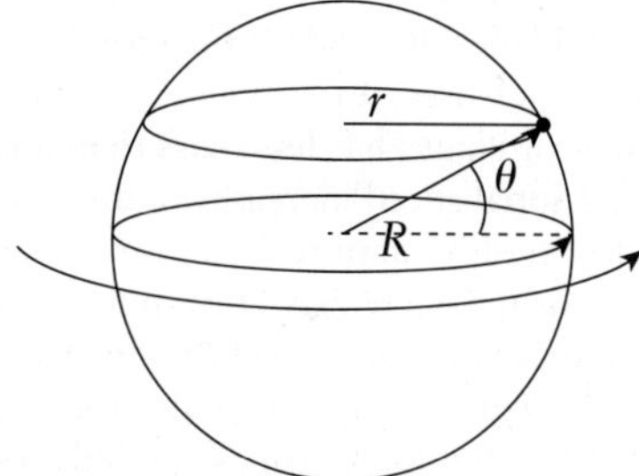

   $$v = \frac{2\pi r}{24\text{ h}} = \frac{2\pi R\cos(\theta)}{24\text{ h}} =$$

   $$\frac{2\pi\cdot6.37\cdot10^6\text{ m}\cos(28.6°)}{24\text{ h}}\frac{1\text{ h}}{3600\text{ s}} = 407\text{ m/s}$$

   (b) The final state is defined at a distance from Earth where the potential energy is negligible and where the rocket is at rest.

   $$K_f = 0 \quad U_f = 0 \Rightarrow -\frac{mv_{escape}^2}{2} + \frac{Gm_{Earth}m}{r_{Earth}} = 0$$

   (c) $$v_{escape} = \sqrt{\frac{2GM_{Earth}}{r_{Earth}}}$$

   $$= \sqrt{\frac{2\left(6.67\times10^{-11}\text{ Nm}^2/\text{kg}^2\right)\left(5.97\times10^{24}\text{ kg}\right)}{6.37\times10^6\text{ m}}}$$

   $$= 11{,}200\text{ m/s}$$

   So, the kinetic energy of the motion of the launch site is only a small fraction of the kinetic energy of the escape velocity.

   $$\frac{v_{surface}^2}{v_{escape}^2} < 0.1\%$$

   (d) Rockets are not designed for escape but for orbital distances. For the approximate parameters of the ISS orbit provided, the final kinetic energy per unit mass is

   $$\frac{K_i}{m} = \frac{1}{2}\left[\frac{2\pi\left(6.77\times10^6\text{ m}\right)}{90\text{ min (60 s/min)}}\right]^2$$

   $$-6.67\times10^{-11}\text{ Nm}^2/\text{kg}^2\left(5.97\times10^{24}\text{ kg}\right)\times$$

   $$\left[\frac{1}{6.77\times10^6\text{ m}} - \frac{1}{6.37\times10^6\text{ m}}\right] = 3.5\times10^7\text{ J}$$

   The ratio of kinetic energies is still very small, ~0.24%.

7. (a) If we want to save fuel, we would want to go by a path where conservation of mechanical energy would be as close to keeping us on our desired path as possible. If we fly through the point of maximum potential energy, we continue in a straight line. If we fly a little off the maximum toward either star, the potential energy is decreasing toward that star, so our ship would start to fall toward it and we would have to fire our engines to get back on course.

   (b) As the mass of one of two interacting objects increases, the gravitational potential energy for the same separation of those two objects is increasingly negative. So if the ship passes closer to the more massive star, the potential energy of the system decreases more than if it were nearer to the less massive star. The total gravitational potential energy is the sum of contributions caused by the interactions with each star. Because the gravitational potential energy function falls off as $1/r$, the closer star has a bigger effect.

   (c) The potential energy is at a maximum at a point where the net force is zero. The gravitational force exerted by each star on the rocket ship pulls the ship toward that star, so the forces have opposite directions between the two stars, and so the net force is smaller than either individual force exerted on the rocket ship. That is where the maximum potential energy would be. If the stars have the same mass, the maximum is located at the midpoint between the two stars.

   $$F_{net} = Gm\left[\frac{M_1}{(d/2)^2} - \frac{M_2}{(d/2)^2}\right] = 0 \text{ if } M_1 = M_2$$

Chapter 8 Review Problems

1. (a) The instant before it strikes the ground

   (b) Probably just before it strikes the ground, although it may have been close to that speed for a while

3. Because the book is not accelerating in the vertical direction, the magnitude of the normal force is equal to the weight of the book. These forces are perpendicular to the displacement, so do no work. The only work is done by kinetic friction and so is equal to the change in the book's kinetic energy.

   $$W_{net} = W_{nonconservative} = F_k d\cos(180°) = -(\mu_k F_n)d = -\mu_k mgd$$

   $$\Delta K = \frac{1}{2}mv_f^2 - \frac{1}{2}mv_i^2 = 0 - \frac{1}{2}mv_i^2$$

Using the work-energy theorem for an object:

$$-\mu_k mgd = -\frac{1}{2}mv_i^2$$

$$\mu_k = \frac{v_i^2}{2gd} = \frac{\left(4.00\frac{m}{s}\right)^2}{2\left(9.80\frac{m}{s^2}\right)(3.25\ m)} = 0.251$$

5. Use conservation of energy to calculate the speed of the balloon right before it hits the ground. Defining the ground to be the zero of potential energy for the Earth–balloon system,

$$mgy_i + \frac{1}{2}mv_i^2 = 0 + \frac{1}{2}mv_f^2$$

$$v_f = \sqrt{2gy_i + v_i^2} = \sqrt{2\left(9.80\frac{m}{s^2}\right)(5.00\ m) + \left(12.0\frac{m}{s}\right)^2} = 15.6\frac{m}{s}$$

7. Use conservation of mechanical energy to first determine the skier's final speed in the absence of air resistance and use this to calculate the skier's final speed taking air resistance into account.

$$\frac{1}{2}mv_i^2 + mgy_i = \frac{1}{2}mv_{f,ideal}^2 + mgy_f$$

$$v_{f,ideal}^2 = v_i^2 + 2g(y_i - y_f)$$

With air resistance: $K_{f,actual} = \frac{1}{2}K_{f,ideal}$

$$\frac{1}{2}mv_{f,actual}^2 = \frac{1}{2}\left(\frac{1}{2}mv_{f,ideal}^2\right)$$

$$v_{f,actual} = \sqrt{\frac{v_{f,ideal}^2}{2}} = \sqrt{\frac{v_i^2 + 2g(y_i - y_f)}{2}}$$

$$= \sqrt{\frac{\left(4.00\frac{m}{s}\right)^2 + 2\left(9.80\frac{m}{s^2}\right)((4212\ m) - (4039\ m))}{2}}$$

$$= 41.3\frac{m}{s}$$

9. Set the initial height equal to zero for both. Ignoring any air resistance, total energy is constant in each case. Energy for each includes kinetic energy, spring potential energy, and gravitational potential energy of the Earth–marble system. Set initial and final total energies equal in each case, and solve for the maximum height for each when the final speed equals zero (that is, at the highest point in the motion).

Conservation of energy for Neil:

$$E_{N,f} = E_{N,i}$$

$$\frac{1}{2}mv_{N,f}^2 + \frac{1}{2}k_N y_{N,f}^2 + mgh_{N,f} = \frac{1}{2}mv_{N,i}^2 + \frac{1}{2}k_N y_{N,i}^2 + mgh_{N,i}$$

$$mgh_{N,f} = \frac{1}{2}k_N y_{N,i}^2$$

$$h_{N,f} = \frac{k_N y_{N,i}^2}{2\ mg} = \frac{\left(50.8\frac{N}{m}\right)(0.140\ m)^2}{2(0.0045\ kg)\left(9.80\frac{m}{s^2}\right)} = 11.3\ m$$

Conservation of energy for Gus:

$$mgh_{G,f} = \frac{1}{2}k_G y_{G,i}^2$$

$$h_{G,f} = \frac{k_G y_{G,i}^2}{2\ mg} = \frac{\left(12.7\frac{N}{m}\right)(0.270\ m)^2}{2(0.0045\ kg)\left(9.80\frac{m}{s^2}\right)} = 10.5\ m$$

Neil wins, with a height of 11.3 m above the starting point.

11. Use conservation of mechanical energy to calculate the dolphin's initial speed. Set the zero of gravitational potential energy for the dolphin–Earth system at the water's surface.

$$0 + \frac{1}{2}mv_i^2 = mgy_f + 0$$

$$v_i = \sqrt{2gy_f} = \sqrt{2\left(9.80\frac{m}{s^2}\right)(2.50\ m)} = 7.00\frac{m}{s}$$

13. Use conservation of mechanical energy to calculate the initial speed v_i of the froghopper. From this initial speed, we can calculate the initial kinetic energy of the insect, which is equal to the energy stored in its legs that is used for the jump. At the top of its jump it will still have its initial horizontal velocity, so it will have kinetic energy at the top.

$$0 + \frac{1}{2}mv_i^2 = mgh_f + \frac{1}{2}mv_f^2$$

$$\frac{1}{2}v_i^2 = gh_f + \frac{1}{2}(v_i\cos(58.0°))^2$$

$$v_i = \sqrt{\frac{2gh_f}{1-\cos^2(58.0°)}} = \sqrt{\frac{2\left(9.80\frac{m}{s^2}\right)(0.290\ m)}{1-\cos^2(58.0°)}} = 2.81\frac{m}{s}$$

15. The work done by kinetic friction is equal to the change in the mechanical energy of the system; from this we can calculate the distance up the ramp, d. The force of kinetic friction, which is exerted opposite to the displacement of the block, has a magnitude of $F_k = \mu_k F_n = \mu_k mg\cos(37.0°)$. Define the bottom of the ramp to have a gravitational potential energy of zero.

$$W_{nonconservative} = \Delta K + \Delta U = \frac{1}{2}m(v_f^2 - v_i^2) + mg(\Delta y)$$

$$-(\mu_k mg\cos(37°))d = \frac{1}{2}m(0 - v_i^2) + mg(d\sin(37°) - 0)$$

$$\mu_k gd\cos(37°) + gd\sin(37°) = \frac{1}{2}v_i^2$$

$$d = \frac{v_i^2}{2g(\sin(37°) + \mu_k\cos(37°))} =$$

$$\frac{\left(20.0\frac{m}{s}\right)^2}{2\left(9.8\frac{m}{s^2}\right)(\sin(37°) + (0.50)\cos(37°))} = 20\ m$$

17. (a) $K_i = 0 \quad K_f = \dfrac{mv_f^2}{2} \quad U_f = U_i = 0$

$$W = K_f - K_i = \frac{mv_f^2}{2}$$

$$W = \frac{0.145\text{ kg}}{2}\left(1600\text{ m}^2/\text{s}^2\right) = 116\text{ J} \approx 120\text{ J}$$

(b) Power is the work done divided by the interval of time during which the work is done.

$$P_{av} = \frac{W}{\Delta t} = \frac{116\text{ J}}{50 \times 10^{-3}\text{ s}} = 2.3\text{ kW}$$

19. Amit's model is $F = \dfrac{k}{x}$. Kamala could clarify by asking if Amit means the length of the spring or the change in length of the spring or where the origin is for the coordinate system in his model.

21. (a) Use the Earth–spring–block system so that spring and gravitational potential energies may be used.

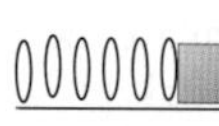

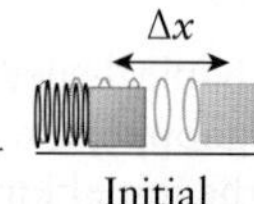

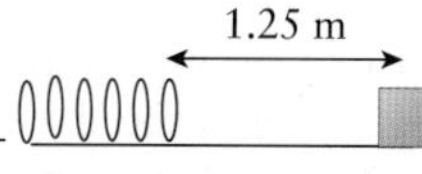

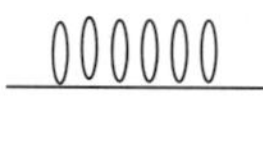

(b) For the part of the motion on the tabletop, setting $U_{grav} = 0$ at that height,

$$K_i = 0,\ U_{s,i} = k\Delta x^2/2,\ U_{s,f} = 0,\ K_f = 1/2mv_x^2$$

where v_x is the velocity as the block leaves the tabletop.
$K_f + U_{s,f} = K_i + U_{s,i}$ so, substituting these values:

$$\frac{mv_x^2}{2} = \frac{k\Delta x^2}{2} \Longrightarrow v_x = \Delta x\sqrt{\frac{k}{m}}$$

(c) Choosing the block as the system, when it leaves the table it is in free fall, so we can calculate the time it takes to hit the floor, t, from the height and g. Then calculate the speed in the horizontal direction, a constant, by dividing the horizontal displacement by t:

$$d = v_x t = v_x\sqrt{\frac{2h}{g}} \Rightarrow v_x = d\sqrt{\frac{g}{2h}}$$

(d) From parts (b) and (c), if there is no friction force on the block as it slides across the table then the two predictions for the velocity should agree. If there is a friction force exerted on the block by the table then the predicted velocity as the block leaves the table would be larger than observed, and the distance, d, would be smaller.

$$v_x = \Delta x\sqrt{\frac{k}{m}} = 0.15\text{ m}\sqrt{\frac{75\text{ N/m}}{0.10\text{ kg}}} = 4.11\text{ m/s} \approx 4.1\text{ m/s}$$

and

$$v_x = d\sqrt{\frac{g}{2h}} = 1.60\text{ m}\sqrt{\frac{9.8\text{ m/s}^2}{2(1.0\text{ m})}} = 3.54\text{ m/s} \approx 3.5\text{ m/s}$$

The fact that the two velocities are different and that the velocity predicted from the conversion of the spring potential energy is greater is evidence of a friction force exerted on the block. (This is assuming your experimental error is small.)

(e) The difference between the kinetic energies of these two velocities is the work done on the block by the table:

$$W = \frac{mv_{x,\text{predicted}}^2}{2} - \frac{mv_{x,\text{observed}}^2}{2}$$

$$\frac{1}{2}mv_{x,\text{predicted}}^2 - \mu_k mgL = \frac{1}{2}mv_{x,\text{observed}}^2 \Rightarrow$$

$$\mu_k = \frac{1}{2}\frac{v_{x,\text{predicted}}^2 - v_{x,\text{observed}}^2}{gL} = \frac{1}{2}\frac{(4.11\text{ m/s})^2 - (3.54\text{ m/s})^2}{\left(9.8\text{ m/s}^2\right)(1.25\text{ m})} = 0.18$$

23. (a) Use energy conservation, choosing the block–spring system:

$$\Delta K + \Delta U = 0 \quad K_f = 0 \quad U_i = 0 \quad K_i = \frac{mv_i^2}{2}$$

$$U_f = \frac{kx_f^2}{2} \quad mv_i^2 = kx_f^2 \Rightarrow x_f = \sqrt{\frac{m}{k}}v_i$$

(b) If the data are graphed, the line of best fit is $x = 0.0088v + 0.003$, with x in units of m. The constant can be calculated from the slope:

$$0.0088\text{ s} = \sqrt{\frac{m}{k}} \Rightarrow k = \frac{m}{7.74 \times 10^{-5}\text{ s}^2} = \frac{0.100\text{ kg}}{7.74 \times 10^{-5}\text{ s}^2}$$
$$= 1290\text{ N/m} \approx 1000\text{ N/m}$$

25. $K_i + U_{grav,i} = K_f + U_{grav,f} = 0 \quad \dfrac{1}{2}mv_E{}^2 - \dfrac{GM_E m}{R_E} = 0$

$$v_E = \sqrt{\frac{2GM_E}{R_E}}$$

$$v_P = \sqrt{\frac{2G(8M_E)}{(2R_E)}} = 2\sqrt{\frac{2GM_E}{R_E}} = 2v_E$$

27. $U_{grav,i} = U_{s,f};\quad mgx_{max} = \dfrac{1}{2}k(x_{max})^2$. Solving this equation for k, we get $k = \dfrac{2mg}{x_{max}}$.

Chapter 8 AP® Practice Problems

1. (D) $W = Fd\cos\theta$

$$W_{ABC} = W_{AB} + W_{BC} = F_y\Delta y + F_x\Delta x$$
$$= \left[\left(2\frac{\text{N}}{\text{m}^2}\right)(0\text{ m})^2\right](2\text{ m}) + \left[\left(2\frac{\text{N}}{\text{m}}\right)(2\text{ m})\right](3\text{ m})$$
$$= 12\text{ Nm} = 12\text{ J} \approx 10\text{ J}$$

$$W_{ADC} = W_{AD} + W_{DC} = F_x \Delta x + F_y \Delta y$$
$$= \left[\left(2\frac{\text{N}}{\text{m}}\right)(0\text{ m})\right](3\text{ m}) + \left[\left(2\frac{\text{N}}{\text{m}^2}\right)(3\text{ m})^2\right](2\text{ m})$$
$$= 36\text{ Nm} = 36\text{ J} \approx 40\text{ J}$$

The force is nonconservative because the work is path dependent.

3. (B) Energy of the system is conserved.

$$\Delta U + \Delta K + \Delta E_{\text{internal}} = 0 \Rightarrow 0 - mgh + K_f - 0 - W_{\text{nonconservative}} = 0$$
$$-mg\frac{L}{2} + \frac{1}{2}mv_f^2 + F_k L = 0;\ \text{solve for } v_f$$
$$= \sqrt{\frac{mgL - 2F_k L}{m}} = \sqrt{\frac{2.0\text{ kg}(9.8\text{ m/s}^2)(4.0\text{ m}) - 2(1.0\text{ N})4.0\text{ m}}{2.0\text{ kg}}}$$
$$= 5.9\text{ m/s}$$

5. (A) Conservation of energy

$$\frac{1}{2}mv^2 - \frac{GM_E m}{R_E} = 0 - \frac{GM_E m}{R}$$
$$\frac{1}{2}\left(\frac{1}{2}\sqrt{\frac{2GM_E}{R_E}}\right)^2 - \frac{GM_E}{R_E} = -\frac{GM_E}{R}$$
$$\frac{1}{4}\left(\frac{1}{R_E}\right) - \frac{1}{R_E} = -\frac{1}{R} \Rightarrow -\frac{3}{4R_E} = -\frac{1}{R} \Rightarrow R = \frac{4}{3}R_E$$

7. (B) $\frac{1}{2}mv^2 = \frac{1}{2}kx^2$, thus, $x = \sqrt{\frac{mv^2}{k}}$. Thus, the speed has a greater effect on the compression than the mass, and compression is greatest in case C and least in case A.

Chapter 9

9-1 Newton's third law will help lead us to the idea of momentum

1. (a) When the line connecting the center of mass and the point where the back legs contact the floor is parallel to the gravitational force, the system is stable. If your center of mass (near your navel) lies to either side of the point of contact between the back legs of the chair and the floor you will tip in that direction. If you tip backward, you tip over, so that is the unstable direction.

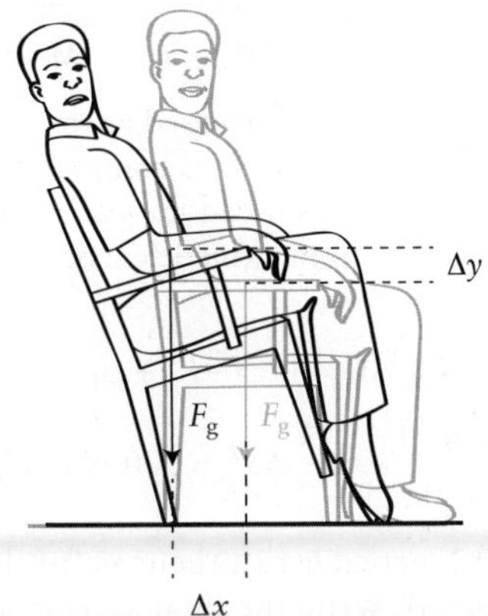

So, the limit of stability is when the gravitational force connects your navel to the point of contact. To represent the angle at this stability limit, imagine sitting on the chair and observing the displacement of your navel as you tilt backward. As you lean back your navel rises by a distance Δy and moves backward by a distance Δx. The tangent of the angle at the limit of stability is the ratio of displacements, $\Delta y/\Delta x$.

(b) To measure the angle, you can have one friend measure the displacements and another catch you when you fall. The front edge of the seat will be displaced in the same way that your navel is. A string taped to the edge of the chair seat can be used to measure the Δx displacement on a ruler resting on the floor.

9-2 Momentum is a vector that depends on an object's mass, speed, and direction of motion

1. $p = mv = (1.00 \times 10^4\text{ kg})(15\text{ m/s}) = 1.5 \times 10^5\text{ kg}\cdot\text{m/s}$

$$K = \frac{1}{2}(1.00 \times 10^4\text{ kg})\left(15\frac{\text{m}}{\text{s}}\right)^2 = 1.1 \times 10^6\text{ J}$$

3. Because we are told to take $+x$ as the forward direction, a negative value is in the backward direction.
   (a) $p_x = 1250\text{ kg}(-5.00\text{ m/s}) = -6250\text{ kg}\cdot\text{m/s}$
   (b) $p_x = 1250\text{ kg}(14.0\text{ m/s}) = 17{,}500\text{ kg}\cdot\text{m/s}$
   (c) $\Delta p_x = p_{fx} - p_{ix} = 17500\text{ kg}\cdot\text{m/s} - (-6250\text{ kg}\cdot\text{m/s}) = 23{,}800\text{ kg}\cdot\text{m/s}$

5. Yes. The momentum is the same when the speed of the tennis ball is 18 times greater than the speed of the basketball and the velocities of both objects are in the same direction.

7. (a) i. The momentum is at an angle of 37° above the horizontal, with a magnitude of $p = 0.145\text{ kg}(34.9\text{ m/s}) \approx 5.1\text{ kg}\cdot\text{m/s}$
   ii. $p_x = 5.06\text{ kg}\cdot\text{m/s}\cos(37°) = 4.04\text{ kg}\cdot\text{m/s} \approx 4.0\text{ kg}\cdot\text{m/s}$
   iii. $p_y = 5.06\text{ kg}\cdot\text{m/s}\sin(37°) = 3.05\text{ kg}\cdot\text{m/s} \approx 3.0\text{ kg}\cdot\text{m/s}$

   (b) $t_{\text{flight}} = 2\frac{v\sin\theta}{g}$ $\quad \Delta x = v_x t_{\text{flight}} = \frac{2v^2\cos\theta\sin\theta}{g}$

$$= \frac{2(34.9\text{ m/s})^2\cos(37°)\sin(37°)}{9.8\text{ m/s}^2} = 1.2 \times 10^2\text{ m}$$

   It does not reach the wall.

9-3 The total momentum of a system is always conserved; it is constant for systems that are closed and isolated

1. To solve this problem, we must assume it is a collision. That is, any external forces are much smaller than the internal forces exerted on the interacting objects. This lets us use constant momentum (assign east as positive x):

$$p_{ix} = 2.00\text{ kg}\ (4.00\text{ m/s})$$
$$p_{fx} = 6.00\text{ kg}\ (2.00\text{ m/s}) + 2.00\text{ kg}(v_x)$$
$$p_{fx} - p_{ix} = 0 \Rightarrow v_x = \frac{8.00\text{ kg}\cdot\text{m/s} - 12.00\text{ kg}\cdot\text{m/s}}{2.00\text{ kg}}$$
$$= -2.00\text{ m/s east, which is the same as } +2.00\text{ m/s west}$$

3. If we assume there are no external forces exerted on the system, then we can use constant momentum to solve the problem. Initial and final momenta of the system are expressed using $+x$ to indicate east and $+y$ to indicate north.

$$p_{\text{total,ix}} = 0.170 \text{ kg } (2.0 \text{ m/s}) \cos(30°)$$
$$p_{\text{total,iy}} = 0.170 \text{ kg } (2.0 \text{ m/s}) \sin(30°)$$
$$p_{\text{total,fx}} = 0.170 \text{ kg } v_{0,\text{f}} \cos(10°)$$
$$p_{\text{total,fy}} = 0.170 \text{ kg } v_{0,\text{f}} \sin(10°) + 0.156 \text{ kg } v_{8\text{f}}$$

The x components can be used to determine the final speed of the cue ball.

$$p_{\text{total,ix}} = 0.170 \text{ kg}(2.0 \text{ m/s})\cos(30°) = p_{\text{total,fx}}$$
$$= 0.170 \text{ kg } v_{0,\text{f}} \cos(10°) \Rightarrow v_{0,\text{f}} = \frac{0.170 \text{ kg } (2.0 \text{ m/s}) \cos(30°)}{0.170 \text{ kg } \cos(10°)}$$
$$= 1.76 \text{ m/s} \approx 1.8 \text{ m/s}$$

The y component can be used to determine the final speed of the 8-ball.

$$p_{\text{total,iy}} = 0.170 \text{ kg}(2.0 \text{ m/s}) \sin(30°) = p_{\text{total,fy}}$$
$$= 0.170 \text{ kg}(1.76 \text{ m/s}) \sin(10°) + 0.156 \text{ kg } v_{8\text{f}}$$
$$v_{8\text{f}} = \frac{0.170 \text{ kg } (2.0 \text{ m/s}) \sin(30°) - 0.170 \text{ kg } (1.76 \text{ m/s}) \sin(10°)}{0.156 \text{ kg}}$$
$$= 0.76 \text{ m/s}$$

5. During the brief time interval, Δt, only two objects in the system collide. On each object, there are external and internal forces and the sum of these is used in Newton's second law to write

$$\vec{F}_{\text{net,A}} = \vec{F}_{\text{A,external}} + \vec{F}_{\text{B on A}} = m_{\text{A}}\vec{a}_{\text{A}}$$
$$\vec{F}_{\text{net,B}} = \vec{F}_{\text{B,external}} + \vec{F}_{\text{A on B}} = m_{\text{B}}\vec{a}_{\text{B}}$$

Adding these equations, and using the fact that the internal forces are equal and opposite, the net force is the sum of external forces:

$$\vec{F}_{\text{A,external}} + \vec{F}_{\text{B,external}} + \vec{F}_{\text{B on A}} + \vec{F}_{\text{A on B}} = \vec{F}_{\text{A,external}} + \vec{F}_{\text{B,external}}$$
$$= m_{\text{A}}\vec{a}_{\text{A}} + m_{\text{B}}\vec{a}_{\text{B}}$$

If there are no external forces exerted on the two objects during the collision, because $\Delta\vec{v}/\Delta t = \vec{a}$, multiplying by Δt results in

$$(m_{\text{A}}\vec{a}_{\text{A}} + m_{\text{B}}\vec{a}_{\text{B}}) \times \Delta t = 0 \times \Delta t \Rightarrow m_{\text{A}}\vec{v}_{\text{A}} + m_{\text{B}}\vec{v}_{\text{B}} = 0$$

The change in momentum of the pair of objects is equal to zero. Therefore, the momentum does not change and is constant.

7. (a) $\vec{F}_{\text{internal}} = (\vec{F}_{1 \text{ on } 2} + \vec{F}_{2 \text{ on } 1}) + (\vec{F}_{1 \text{ on } 3} + \vec{F}_{3 \text{ on } 1}) + (\vec{F}_{2 \text{ on } 3} + \vec{F}_{3 \text{ on } 2}) = 0$
Each pair of forces in parentheses sums to zero because of Newton's third law:

$$\vec{F}_{\text{i on j}} = -\vec{F}_{\text{j on i}}$$

The sum of internal forces exerted between objects in a system is equal to zero.
(b) There are no terms $F_{\text{j on j}}$ because an object does not exert a force on itself.

9. (a) The force exerted on the receiver by the linebacker is equal and opposite to the force exerted by the receiver on the linebacker.
(b) The changes in momentum for each player are equal in magnitude and opposite in direction.
(c) The time interval is the same for both players. Because the receiver has a smaller mass, the change in velocity of the receiver must be larger for the change in momentum to have the same magnitude for both players. Then the change in velocity per unit time, the acceleration, is larger for the receiver. The difference can be big.

9-4 In an inelastic collision some of the mechanical energy is dissipated

1. Because we neglect all other interactions, momentum is constant.

$$3Mv_0 - M2v_0 = -3M\left(\frac{v_0}{5}\right) + Mv \Rightarrow v = \frac{8Mv_0}{5M}$$
$$= 1.6v_0, \text{ in the } +x \text{ direction}$$

To see if the collision is elastic, we check to see if the initial and final kinetic energies are equal.

$$K_{\text{before}} = \frac{3M}{2}v_0^2 + \frac{M}{2}4v_0^2 = \frac{7M}{2}v_0^2$$
$$K_{\text{after}} = \frac{3M}{2}\frac{v_0^2}{25} + \frac{M}{2}\frac{64v_0^2}{25} = \frac{67M}{50}v_0^2$$

The change in kinetic energy is negative, so this is an inelastic collision:

$$\Delta K = \frac{67M}{50}v_0^2 - \frac{175M}{50}v_0^2 = -\frac{108M}{50}v_0^2 = -2.2\ Mv_0^2$$

Roughly 60% of the kinetic energy the system had prior to collision is dissipated into other types of energy, such as warmth, sound, and damage to the colliding objects.

3. This is a totally inelastic collision because the objects stick together. First, the initial momentum is calculated. Then, from that, using conservation of momentum, an expression for the final momentum is calculated. We are working with one vector component and define to the right as positive.

$$p_{\text{before}} = mv_{\text{before}} = (0.0120 \text{ kg})(200 \text{ m/s}) = 2.40 \text{ kg} \cdot \text{m/s}$$
$$\Rightarrow v_{\text{after}} = \frac{p_{\text{before}}}{M + 0.0120 \text{ kg}} = \frac{2.40 \text{ kg} \cdot \text{m/s}}{(M + 0.0120 \text{ kg})}$$

Mechanical energy is dissipated as the bullet lodges in the block, but as the spring is compressed mechanical energy is conserved as no additional energy is dissipated. So expressing the initial kinetic energy in terms of the momentum:

$$\Delta K + \Delta U = 0 \Rightarrow -K_{\text{i}} + U_{\text{f}} = 0$$
$$\frac{(2.40 \text{ kg} \cdot \text{m/s})^2}{2(M + 0.0120 \text{ kg})} = \frac{k}{2}\Delta x^2 = 100 \text{ N/m } (0.300 \text{ m})^2 = 9.00 \text{ J}$$

From that kinetic energy, the mass of the block can be determined. Write the final kinetic energy of the block–bullet system in terms of the mass expression found from momentum and set it equal to 9.00 J.

$$M + 0.0120\text{ kg} = \frac{(2.40\text{ kg}\cdot\text{m/s})^2}{2(9.00\text{ J})} = 0.320\text{ kg} \Rightarrow M = 0.308\text{ kg}$$

5. The total momentum of the system is constant during the interaction, so:

$$\vec{p}_{\text{total,before}} = m_1\vec{v}_1 + m_2\vec{v}_2$$

must be equal to

$$\vec{p}_{\text{total,after}} = (m_1 + m_2)\vec{v}_{\text{after}}$$

The components of the momentum before the collision are calculated:

$$p_{\text{total,before},x} = 105\text{ kg }(6.3\text{ m/s}) + 92\text{ kg }(5.6\text{ m/s})\cos(72°)$$

$$p_{\text{total,before,y}} = 92\text{ kg }(5.6\text{ m/s})\sin(72°)$$

Because the momentum is the same both before and after the collision, we can use these components to find the final velocity components, remembering to use the combined mass:

$$v_{\text{after},x} = \frac{1}{197\text{ kg}}\left[105\text{ kg }(6.3\text{ m/s}) + 92\text{ kg }(5.6\text{ m/s})\cos(72°)\right]$$

$$= 4.2\text{ m/s}$$

$$v_{\text{after,y}} = \frac{1}{197\text{ kg}}\left[92\text{ kg }(5.6\text{ m/s})\sin(72°)\right] = 2.5\text{ m/s}$$

The magnitude of the final speed is the square root of the sum of the squares:

$$v_{\text{after}} = \sqrt{(4.17\text{ m/s})^2 + (2.49\text{ m/s})^2} = 4.9\text{ m/s}$$

The components above completely specify the vector, but if you want to write it as a magnitude and direction, the direction of the final velocity is the inverse tangent of the ratio of the y to the x component.

$$\tan\theta = \frac{v_{\text{after,y}}}{v_{\text{after},x}} = \frac{2.49\text{ m/s}}{4.17\text{ m/s}} = 0.60 \Rightarrow \theta = \tan^{-1}(0.60) = 31°$$

above the $+x$ axis

7. (a) (E) The golf ball. There is nonzero momentum and kinetic energy after the club strikes the ball. Before the interaction, both were zero.
   (b) (A) The two pucks. The friction force exerted by the ice surface is negligible, so that the total momentum does not change. The hockey pucks are rigid, so internal energy changes will be close to zero. If the collision of the hockey pucks is inelastic, then there would be some dissipation of kinetic energy into internal energy, and (B) would be a correct response.
   (c) (B) Bullet–block. The final momentum of the combined block and bullet is equal to the initial momentum of the bullet, and kinetic energy is dissipated into internal energy, given that the bullet sticks in the block.
   (d) (F) Two cars. During the collision, the momentum is constant and not zero. After the collision the magnitude of the momentum of the system is reduced until, when the cars come to a stop, the momentum and kinetic energy of each car is zero.
   (e) (C) The balloon (and the air in it). Internal energy before the balloon pops is converted to positive kinetic energy immediately after the balloon pops. Immediately after the balloon pops, the momentum of the system remains constant (the pieces of the balloon fly off in every direction).
   (f) (D) Cart. The friction force exerted on the gliding cart and the energy dissipated by the rubber band are negligible. The direction of the momentum is reversed (so momentum changes) by the collision, but the change in kinetic energy is negligible.
   (g) (F) Cart. There must be a dissipative interaction. The final magnitude of the momentum and kinetic energy of the cart are both zero, so both momentum and energy decrease.
   (h) (A) Cart–Earth. There is no change in mechanical energy as the cart rises or comes back down. The momentum of the cart changes direction, but the momentum of the cart–Earth system remains constant.

9. (D) The initial momenta of the ball and clay lump are both mv. The final momentum of the ball is $-mv$ so its change in momentum is $-2mv$. The final momentum of the clay lump is 0, so its change in momentum is $-mv$.

11. (D) In a completely inelastic collision the objects fuse into a single object. The initial momentum of the system was not equal to zero, so the final momentum of the system must be nonzero. Therefore, not all of the mechanical energy of the system can be dissipated.

9-5 In an elastic collision both momentum and mechanical energy are constant

1. (a) $v_{2fx} = \dfrac{m_2 - m_1}{m_1 + m_2}v_{2ix} + \dfrac{2m_1}{m_1 + m_2}v_{1ix}$
   (b) The first expression is the one worked out in Example 9-8, with 1 as A and 2 as B. If the masses are the same, $m_1 - m_2 = 0$, so $v_{1fx} = 0$ as we predicted, and we get the velocity of the second from $v_{\text{B}fx} = (m_\text{A}/m_\text{B})(v_{\text{A}ix} - v_{\text{A}fx})$ the same way we did before, $v_{2fx} = (m_1/m_2)(v_{1ix}) = v_{1ix}$.
   (c) $v_{2ix} = 0$ $m_2 = 4.00$ kg $v_{1ix} = 3.00$ m/s
   $m_1 = 2.00$ kg

$$v_{1fx} = \frac{-2.00\text{ kg}}{6.00\text{ kg}}(3.00\text{ m/s}) = -1.00\text{ m/s}$$

$$v_{2fx} = \frac{4.00\text{ kg}}{6.00\text{ kg}}(3.00\text{ m/s}) = 2.00\text{ m/s}$$

   (d) $v_{2ix} = 4.00$ m/s $m_2 = 6.00$ kg $v_{1ix} = 8.00$ m/s
   $m_1 = 10.00$ kg

$$v_{1fx} = \frac{m_1 - m_2}{m_1 + m_2}v_{1ix} + \frac{2m_2}{m_1 + m_2}v_{2ix} = \frac{10.00\text{ kg} - 6.00\text{ kg}}{16.00\text{ kg}}8.00\text{ m/s}$$

$$+\frac{12.00\text{ kg}}{16.00\text{ kg}}4.00\text{ m/s} = 5.00\text{ m/s, east}$$

$$v_{2fx} = \frac{m_2 - m_1}{m_1 + m_2}v_{2ix} + \frac{2m_1}{m_1 + m_2}v_{1ix} = \frac{6.00\text{ kg} - 10.00\text{ kg}}{16.00\text{ kg}}4.00\text{ m/s}$$

$$+\frac{20.00\text{ kg}}{16.00\text{ kg}}8.00\text{ m/s} = 9.00\text{ m/s, east}$$

(e) $v_{2ix} = -2.00$ m/s $\quad m_2 = 0.155$ kg $\quad v_{1ix} = 4.00$ m/s

$m_1 = 0.170$ kg

$$v_{1fx} = \frac{m_1 - m_2}{m_1 + m_2}v_{1ix} + \frac{2m_2}{m_1 + m_2}v_{2ix} = \frac{0.170\text{ kg} - 0.155\text{ kg}}{0.325\text{ kg}}4.00\text{ m/s} + \frac{0.310\text{ kg}}{0.325\text{ kg}}(-2.00\text{ m/s}) = -1.72\text{ m/s}$$

$$v_{2fx} = \frac{m_2 - m_1}{m_1 + m_2}v_{2ix} + \frac{2m_1}{m_1 + m_2}v_{1ix} = \frac{0.155\text{ kg} - 0.170\text{ kg}}{0.325\text{ kg}}(-2.00\text{ m/s}) + \frac{0.340\text{ kg}}{0.325\text{ kg}}4.00\text{ m/s} = 4.28\text{ m/s}$$

3. (a) For each collision momentum is conserved and mechanical energy is constant. No energy is transferred out of the system of balls and Earth and the rigid balls do not acquire internal energy.
   (b) No energy is being transferred to the system, so the total mechanical energy cannot increase. At the instant that the second ball stops, all of the mechanical energy has become gravitational potential, which depends on height. Therefore, because the balls are identical, and the one initially moving is now at the height of the one initially at rest, the height of the ball initially at rest can be no greater than the height of the ball that was raised.
   (c) Momentum is conserved. If the momentum after collision of the raised ball were negative, then the momentum of the ball initially at rest would have to increase by this additional amount. Because of the conclusion reached in part (b) this is not possible, as it would violate the conservation of mechanical energy. The ball initially at rest would have a speed after the collision that would result in a height greater than the height from which the ball initially in motion before the collision fell.
   (d) Because energy is conserved, and the ball that is initially at rest has the full kinetic energy after the collision, the ball that is initially raised cannot have a velocity after collision.
   (e) In the Newton's cradle with at least three balls, you can raise and release two balls simultaneously, and both the momentum and mechanical energy of the system are constant. When two balls are raised in the initial state, two balls must be elevated in the final state. Otherwise, mechanical energy cannot be conserved simultaneously with momentum.
5. Because the internal forces exerted on each sphere in the system have the same magnitude and opposite direction during the same time period, the spheres' changes in momentum are equal in magnitude and opposite in direction. The momentum of the system is constant. When a single sphere is the system, then an external force is exerted on the system by the other sphere and the change in momentum of the system is due to the force exerted by the other sphere. In the first case of two spheres in the system, momentum is conserved and constant. In the second case momentum is conserved but the momentum of the single-object system changes as momentum is transferred to or from the other sphere.

9-6 What happens in a collision is related to the time the colliding objects are in contact

1. Taking the positive direction as the direction of the initial path of the hawk:

$$\vec{F}_{\text{wind on hawk}}\Delta t = \Delta\vec{p} = \vec{p}_f - \vec{p}_i = m(-7.00\text{ m/s}) - m(5.00\text{ m/s}) = -10.0\text{ N}(1.20\text{ s})$$

$$m = \frac{10.0\text{ N}(1.20\text{ s})}{12.00\text{ m/s}} = 1.00\text{ kg}$$

3. $F_{\text{bat on ball}}\Delta t = 2.5 \times 10^4\text{ N }(0.500\text{ ms}) = 1.3 \times 10^4\text{ N}\cdot\text{ms} = 13\text{ Ns}$

   Taking the positive x direction to be the initial direction of the ball and the impulse in the negative x direction:

$$F_{\text{bat on ball},x}\Delta t = \Delta p_x = p_{fx} - p_{ix} = -12.5\text{ Ns} = p_{fx} - (0.145\text{ kg})(40.0\text{ m/s}) \Rightarrow p_{fx} = -12.5\text{ Ns} + 5.80\text{ Ns}$$

$$v_{\text{ball,fx}} = \frac{p_{fx}}{m} = \frac{-6.70\text{ Ns}}{0.145\text{ kg}} = -46.2\text{ m/s},\ v = 46.2\text{ m/s}$$

5. If the arrow is identical but has twice the momentum, then the speed is also doubled. Once we know this, because this problem is one dimensional, we can solve it using Equation 2-11, and the fact that $a_x = F_x/m$, assuming the target exerts the same force on each arrow $v_x^2 = v_{0x}^2 + 2a_x(x - x_0)$ solving for $v_x = 0$ after a displacement $x - x_0$. If v_{0x} is doubled, then the stopping distance increases by a factor of 4.
7. (a) The product $-\vec{F}\Delta t$ is equal to the change in momentum during impact. With a case composed of a deformable material the collision time is extended. As the duration of the collision increases the magnitude of the force decreases for a given momentum change.
   (b) We can use conservation of mechanical energy of the phone–Earth system to find the initial momentum of the phone as it strikes the floor. We are told to assume the final momentum of the phone is zero. We can find the force from the impulse the floor must exert on the phone to bring it to rest and the time interval given.

$$\vec{F}\Delta t = -\vec{p}_i = -m\sqrt{2gh} \qquad \Delta t < 180\text{ ms}$$

$$|F| = \frac{m\sqrt{2gh}}{\Delta t} > \frac{0.174\text{ kg}\sqrt{2(9.80\text{ N/kg})(1.22\text{ m})}}{0.180\text{ s}} = 4.7\text{ N}$$

   So forces greater than 4.7 N must be survivable to pass the test. This is actually a lower limit, because the floor would exert a greater force on the phone if the phone bounced upward after the collision (the phone's change in momentum would be greater). Significant digits may imply more precision than is accurate; there are significant changes in force as time changes slightly!
   (c) We have already found the velocity of a phone dropped this distance. We can find the time needed from the impulse relationship as well, with the force

we are given. Because this problem is one dimensional, we can solve it using Equation 2-11 and the fact that $a_x = F_x/m$. Taking the direction of the falling phone to be positive, the upward force exerted on the phone results in a negative acceleration over the distance of the thickness of the case:

$$v_x^2 = v_{0x}^2 + 2a_x(x - x_0)$$

The falling object converts gravitational potential energy into kinetic energy, so that the initial speed of the object can be expressed:

$$v_{ix}^2 = 2gh$$

The final speed of the phone is zero. So

$$-2gh = 2\frac{F_x}{m}(x - x_0)$$

Then

$$x - x_0 = -\frac{mgh}{F_x} = -\frac{0.174 \text{ kg } (9.8 \text{ N/kg})(1.22 \text{ m})}{-4 \text{ N}} = 0.52 \text{ m} \approx 0.5 \text{ m}$$

The time of impact can be determined from Equation 9-22,

$$F_x\Delta t = mv_{f,x} - mv_{i,x} = 0 - mv_{i,x} = -m\sqrt{2gh}$$

$$\Delta t = \frac{-m\sqrt{2gh}}{F_x} = \frac{-0.174 \text{ kg}\sqrt{2\left(9.80\dfrac{\text{m}}{\text{s}}\right)(1.22 \text{ m})}}{-4 \text{ N}} = 0.21 \text{ s} \approx 0.2 \text{ s}$$

(d) No commercial cell phone case would be expected to have a width of 1 m. Yet 75% of cell phone users have cases and 50% of users have never cracked a screen. So the model given is not appropriate, given that it is not a good predictor of performance. Case materials also have internal structure. The case does not simply function by increasing the distance over which the acceleration occurs, but by absorbing and redistributing some of the energy. Additionally, cracks in the screen are a result of internal rearrangement of the screen material. Internal structure necessitates the use of a system model. In fact, design practices involve a trial and error sequence with extensive testing. The distance in part (b), at a slightly larger force, is not much smaller. We just didn't notice the issue because we weren't considering distance.

9. (a) $\vec{F}_{\text{cart on probe}}\Delta t = -\vec{F}_{\text{probe on cat}}\Delta t = \Delta\vec{p} = \vec{p}_f - \vec{p}_i$
The area can be calculated to find the impulse. There is uncertainty in the measurement of the force, so the best-fit lines give the best representation of the data in the intervals between 1 and 6 ms and 12 and 18 ms. Calculating the heights of these right triangles using the equations of the best-fit line increases the quality of the measurement. So, for the first triangle height = slope × 6 ms, width = 6 ms, and area = 1/2 width times height. For the second triangle the height and area are found in the same way as for the first. For the rectangle, the area is the base times the height. The initial direction of motion is positive.

$$\text{area} = [6 \text{ ms} \times (4.36 \text{ N/ms})(6 \text{ ms})/2] + [6 \text{ ms} \times (76.9 \text{ N} - (4.21 \text{ N/ms})(12 \text{ ms}))/2] + 6 \text{ ms} \times 25 \text{ N} = 308 \text{ N}\cdot\text{ms}$$

$$\vec{p}_i - \vec{p}_f = \vec{F}_{\text{probe on cart}}\Delta t = 0.308 \text{ Ns} \Rightarrow \vec{p}_f = 0.159 \text{ Ns} - 0.308 \text{ Ns}$$

$$= -0.149 \text{ Ns} \Rightarrow v_{\text{cart}} = \frac{p}{m} = -0.79 \text{ m/s}$$

(b) $$\Delta K = \frac{p_f^2}{2 \text{ m}} - \frac{p_i^2}{2 \text{ m}} = \frac{(-0.1487 \text{ Ns})^2}{2(0.1874 \text{ kg})} - \frac{(0.1592 \text{ Ns})^2}{2(0.1874 \text{ kg})}$$

$$= 0.0590 \text{ J} - 0.0676 \text{ J} = -0.0086 \text{ J}$$

9-7 The center of mass of a system moves as though all the system's mass were concentrated there

1. Using Equation 9-25:

$$x_{CM} = \frac{1}{M_{\text{tot}}}\sum_i m_i x_i = \frac{1}{m_{\text{tot}}}[4.00 \text{ kg } (-6.0 \text{ m}) + 2.00 \text{ kg } (2.0 \text{ m}) + 3.00 \text{ kg } (-4.0 \text{ m})] = \frac{-32 \text{ kg}\cdot\text{m}}{9.00 \text{ kg}} = -3.6 \text{ m}$$

and

$$y_{CM} = \frac{1}{M_{\text{tot}}}\sum_i m_i y_i = \frac{1}{m_{\text{tot}}}[4.00 \text{ kg } (2.0 \text{ m}) + 2.00 \text{ kg } (2.0 \text{ m}) + 3.00 \text{ kg } (-4.0 \text{ m})] = 0$$

3. (a) In the coordinate system with its origin, $x = 0$, at the center of mass of the boat, the dog is 1.5 m from the center with a mass of 20 kg and the man is 1.5 m from the center of the boat in the opposite direction with a mass of 65 kg. So the center of mass of the system is closer to the man than it is to the dog. The center of mass, quantitatively, is

$$x_{CM} = \frac{1}{(135 \text{ kg} + 65 \text{ kg} + 20 \text{ kg})}[65 \text{ kg } (-1.5 \text{ m}) + 20 \text{ kg } (1.5 \text{ m})] = -0.31 \text{ m}$$

(b) In the reference frame fixed on the boat, the boat does not move and the man, who remains seated, does not move. Because the dog is moving in the reference frame of the boat the momentum is the product of the velocity and mass of the dog.

(c) Assuming that there is no force in the plane of the surface of the water exerted on the boat by the water, the momentum of the boat–man–dog system must be constant because the momentum of the system is constant in the absence of an external force. From that reference frame, as the dog moves backward, the man and boat move forward and the center of mass does not move.

(d) In the coordinate frame fixed on the shore the dog moves in a direction away from the shore. The center of mass of the system must remain constant in the absence of an external force, so the boat must move toward the shore.

(e)

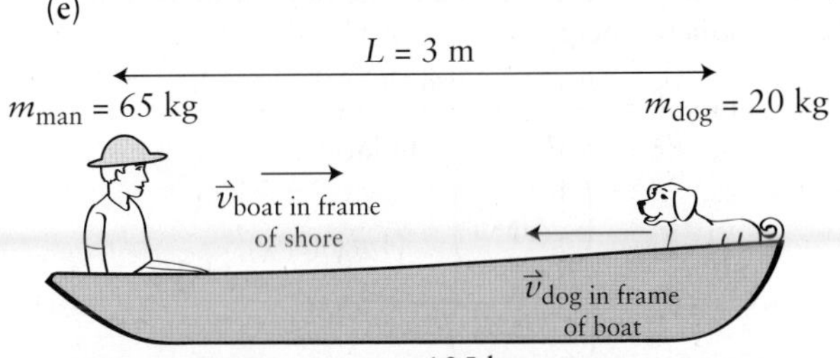

5. (a) $x_{CM} = \dfrac{1}{m_A + m_B}(m_A x_A + m_B x_B) \Rightarrow$

$x_{CM}(m_A + m_B) - (m_A x_A + m_B x_B) = 0$

This can be rearranged:

$m_A(x_{CM} - x_A) - m_B(x_B - x_{CM}) = 0$

This is equivalent to the claim.

(b) $x_{CM} = \dfrac{1}{m_A + m_B}(m_A x_A + m_B x_B) \Rightarrow x_{CM} = \dfrac{m_A x_A}{m_A + m_B}$

7. (a) Yes. No external force is exerted on the system.
(b) Yes. The potential energy stored in the spring is converted to the kinetic energy of the carts.
(c) Because the masses of both carts and their contents are equal, the center of mass of the system will be at the origin if the positions of the carts are symmetric with respect to the origin, as they are.
(d) When $n = 10$ and $m = 0$, the mass of cart 1 is twice the mass of cart 2. The distance from the origin to the location of cart 2 is therefore twice the distance from the origin to cart 1. And because the total separation between the carts is $L + d$:

$x_{1,0} = -\dfrac{L+d}{3} \quad x_{2,0} = x_{1,0} + L + d = \dfrac{2(L+d)}{3}$

(e) Initial momentum is zero, so conservation of momentum means the individual momenta must be oppositely directed, with cart 1 moving off to the left:

$$\left(M + n\left(\frac{M}{10}\right)\bar{v}_1\right) + \left(M + m\left(\frac{M}{10}\right)\bar{v}_2\right) = 0$$

$$\left(M\left(1 + \left(\frac{n}{10}\right)\bar{v}_1\right)\right) + \left(M\left(1 + \left(\frac{m}{10}\right)\bar{v}_2\right)\right) = 0$$

$$(1 + n/10)\bar{v}_1 = -(1 + m/10)\bar{v}_2$$

(f) Conservation of mechanical energy gives

$$(M + nM/10)v_1^2 + (M + mM/10)v_2^2 = k(d_0 - d)^2$$

$$\Rightarrow (1 + n/10)v_1^2 + (1 + m/10)v_2^2 = \frac{k(d_0 - d)^2}{M}$$

(g) The speeds of the carts are measured and the momentum of each cart calculated. If the sum of the momenta remains 0 within the uncertainty of the measurements, then the model is not rejected.
(h) The speeds of the carts are measured and the kinetic energy of each cart calculated. If the sum of the kinetic energies is constant and equal to $k(d_0 - d)^2/2$ within the uncertainty of the measurements, then the model is not rejected.

Chapter 9 Review Problems

1. Label the two balls (below we use A for the lighter ball) and calculate the magnitude of their momentum and kinetic energy.

(a) $\dfrac{p_B}{p_A} = \dfrac{m_B v_B}{m_A v_A} = \dfrac{(4m_A)(2v_A)}{m_A v_A} = 8$

(b) $\dfrac{K_B}{K_A} = \dfrac{\left(\frac{1}{2}m_B v_B^2\right)}{\left(\frac{1}{2}m_A v_A^2\right)} = \dfrac{(4m_A)(2v_A)^2}{m_A v_A^2} = 16$

3. $p = mv \Rightarrow v = \dfrac{p}{m} = \dfrac{550\ \text{kg}\cdot\text{m/s}}{75\ \text{kg}} = 7.3\ \text{m/s}$

5. We can use the impulse equation to determine the final speed of the opponent's head, assuming it is a free object and begins at rest. Assume the direction of the fist is the $+x$ direction. $\Delta p_x = F_x \Delta t$. Because $v_i = 0$, $p_i = 0$, and $\Delta p_x = p_f = mv_{fx}$. This means $v = |F_x \Delta t/m| = 56$ m/s. Although the neck keeps the head from flying off, this huge acceleration can lead to brain damage as the human head is not well modeled as an object; the brain moves inside it.

7. $K_{after} = 9\dfrac{p_{before}^2}{2m} = \dfrac{p_{after}^2}{2m} \Rightarrow \dfrac{p_{after}^2}{p_{before}^2} = 9 \quad \dfrac{p_{after}}{p_{before}} = 3$

9. $K_1 = \dfrac{p_1^2}{2m_1} = K_2 = \dfrac{p_2^2}{2m_2} \Rightarrow |p_2| = \left|p_1\sqrt{\dfrac{m_2}{m_1}}\right|$

So when the masses are not equal, objects with the same kinetic energy will have different momenta and if the masses are equal, then objects with the same kinetic energy must have the same magnitude of momentum.

11. Take right to be positive x.

$$\Delta p_{total} = \Delta p_{1x} + \Delta p_{2x} = p_{2,fx} - p_{2,ix} + p_{1,fx} - p_{1,ix} = 0$$

$$p_{2,fx} = p_{2,ix} - p_{1,fx} + p_{1,ix}$$

$$p_{1,ix} = 3.00\ \text{kg}\ (6.00\ \text{m/s}) \quad p_{2,ix} = -5.00\ \text{kg}\ (4.00\ \text{m/s})$$

$$p_{1,fx} = -3.00\ \text{kg}\ (2.00\ \text{m/s})$$

$$p_{2,fx} = -20.0\ \text{kg}\cdot\text{m/s} + 6.00\ \text{kg}\cdot\text{m/s} + 18.0\ \text{kg}\cdot\text{m/s}$$

$$= 4.0\ \text{kg}\cdot\text{m/s} \Rightarrow v_{2,fx} = \frac{4.0\ \text{kg}\cdot\text{m/s}}{5.00\ \text{kg}} = 0.80\ \text{m/s}$$

13. (B) Motion in the x direction, so ± denotes vector direction

$$\Delta p_{total} = \Delta p_{1x} + \Delta p_{2x} = p_{2,fx} - p_{2,ix} + p_{1,fx} - p_{1,ix} = 0$$

$$p_{1,fx} + p_{2,fx} = p_{2,ix} + p_{1,ix}$$

$$p_{2,ix} = -\frac{m}{2}v \quad p_{1,ix} = mv \quad p_{1,fx} + p_{2,fx} = \frac{3m}{2}v_{fx} = mv - \frac{m}{2}v$$

$$= \frac{m}{2}v \Rightarrow v_{fx} = \frac{v}{3}$$

15. Immediately after the collision the two players move together with the same final velocity. We can use conservation of momentum to calculate their final velocity. The final velocity is (v_{fx}, v_{fy}).
x component:

$$m_A v_{Aix} + m_B v_{Bix} = (m_A + m_B)v_{fx}$$

$$v_{fx} = \frac{m_A v_{Aix} + m_B v_{Bix}}{m_A + m_B} = \frac{0 + (75.0\ \text{kg})\left(9.00\ \frac{\text{m}}{\text{s}}\right)}{(85.0\ \text{kg}) + (75.0\ \text{kg})} = 4.22\ \frac{\text{m}}{\text{s}}$$

y component:

$$m_A v_{Aiy} + m_B v_{Biy} = (m_A + m_B)v_{fy}$$

$$v_{fy} = \frac{m_A v_{Aiy} + m_B v_{Biy}}{m_A + m_B} = \frac{(85.0\ \text{kg})\left(8.00\ \frac{\text{m}}{\text{s}}\right) + 0}{(85.0\ \text{kg}) + (75.0\ \text{kg})} = 4.25\ \frac{\text{m}}{\text{s}}$$

17. The collision can be treated as one dimensional. Define the direction the bat is moving to be the

positive x direction. We can use the definition of impulse to calculate the impulse given to the ball by the bat. Apply conservation of momentum to find the bat's final velocity. The average force the bat exerts on the ball is equal to the ball's change in momentum divided by the contact time.

(a) $m_{\text{ball}}\Delta v_{\text{ball},x} = m_{\text{ball}}(v_{\text{ball,f}x} - v_{\text{ball,i}x})$

$$= (0.145\text{ kg})\left(\left(42.5\frac{\text{m}}{\text{s}}\right)-\left(-31.3\frac{\text{m}}{\text{s}}\right)\right) = 10.7\frac{\text{kg}\cdot\text{m}}{\text{s}}$$

(b) $m_{\text{bat}}v_{\text{bat,i}x} + m_{\text{ball}}v_{\text{ball,i}x} = m_{\text{bat}}v_{\text{bat,f}x} + m_{\text{ball}}v_{\text{ball,f}x}$

$$v_{\text{bat,f}x} = \frac{m_{\text{bat}}v_{\text{bat,i}x} + m_{\text{ball}}v_{\text{ball,i}x} - m_{\text{ball}}v_{\text{ball,f}x}}{m_{\text{bat}}}$$

$$= \frac{(0.850\text{ kg})\left(31.3\frac{\text{m}}{\text{s}}\right) + (0.145\text{ kg})\left(-31.3\frac{\text{m}}{\text{s}}\right) - (0.145\text{ kg})\left(42.5\frac{\text{m}}{\text{s}}\right)}{0.850\text{ kg}}$$

$$= 18.7\frac{\text{m}}{\text{s}}$$

Impulse:

$$m_{\text{bat}}\Delta v_{\text{bat},x} = m_{\text{bat}}(v_{\text{bat,f}x} - v_{\text{bat,i}x})$$

$$= (0.850\text{ kg})\left(\left(18.7\frac{\text{m}}{\text{s}}\right)-\left(31.3\frac{\text{m}}{\text{s}}\right)\right) = -10.7\frac{\text{kg}\cdot\text{m}}{\text{s}}$$

(c) $F_{\text{average},x} = \dfrac{\Delta p_x}{\Delta t} = \dfrac{\left(10.7\dfrac{\text{kg}\cdot\text{m}}{\text{s}}\right)}{1.20\times10^{-3}\text{ s}} = 8.92\times10^3\text{ N}$

(d) Although the force is large, the ball and bat are in contact for only 1.20 ms, so the bat doesn't shatter.

19. Take east to be positive x. Use the work–kinetic energy theorem to calculate the velocity $v_{\text{w,i}x}$ of the wreckage just after the cars collide (this initial velocity for the slide is the final velocity from the collision $v_{\text{f}x}$); the only force doing work on the cars is kinetic friction. The final kinetic energy of the wreck is zero, when they come to rest. Once we have the velocity of the wreckage at the beginning of the slide, which is the same as immediately after the collision, we can use conservation of momentum to calculate the initial velocity of car A (its velocity before the collision). Because momentum only uses mass and velocity, if we keep the units consistent, we can use mph for velocity. We know the velocity following the collision is positive, so we can just find the magnitude of $v_{\text{f}x}$.

Speed following collision:

$$\Delta K = W_{\text{net}} = W_{\text{fric}}$$

$$\frac{1}{2}(m_A + m_B)(0 - v_{\text{f}x}^2) = -\mu_k(m_A + m_B)gd$$

$$0 - v_{\text{f}x}^2 = -2\mu_k gd$$

$$v_{\text{f}x} = \sqrt{2\mu_k gd} = \sqrt{2(0.750)\left(9.80\frac{\text{m}}{\text{s}^2}\right)(6.00\text{ m})} = 9.39\frac{\text{m}}{\text{s}}$$

$$= \left(9.39\frac{\text{m}}{\text{s}}\right)\left(\frac{1\text{ km}}{1000\text{ m}}\right)\left(\frac{0.6215\text{ mi}}{1\text{ km}}\right)\left(\frac{3600\text{ s}}{1\text{ h}}\right) = 21.0\text{ mph}$$

$$v_{\text{Bi}x} = \left(-72.0\frac{\text{km}}{\text{h}}\right)\left(\frac{0.6215\text{ mi}}{1\text{ km}}\right) = -44.7\text{ mph}$$

$p_f = p_i$, so $(m_A + m_B)v_{\text{f}x} = m_A v_{\text{Ai}x} + m_B v_{\text{Bi}x}$. Now solve for the initial velocity of car A:

$$v_{\text{Ai}x} = \frac{(m_A + m_B)v_{\text{f}x} - m_B v_{\text{Bi}x}}{m_A}$$

$$= \frac{(680\text{ kg} + 500\text{ kg})(21.0\text{ mph}) - (500\text{ kg})(-44.7\text{ mph})}{680\text{ kg}}$$

$$= 69.3\text{ mph}$$

21. Use conservation of momentum in order to calculate the velocity of the pendulum + marble after the collision. Use conservation of mechanical energy to calculate the final height y_f that the pendulum + marble reaches. Assume the initial height of the pendulum + marble is zero for the gravitational potential energy of the Earth–pendulum–marble system. Initially the motion of the pendulum (which is always tangent to the string) is purely horizontal, so momentum is in one dimension for the collision. Take the initial direction of motion of the marble to be $+x$.

Conservation of momentum:

$$m_M v_{M,\text{i}x} + m_p v_{P,\text{i}x} = (m_M + m_p)v_{\text{f}x}$$

$$v_{\text{f}x} = \frac{m_M v_{M,\text{i}x} + m_p v_{P,\text{i}x}}{m_M + m_P} = \frac{m_M v_{M,\text{i}x} + 0}{m_M + m_P}$$

$$= \frac{(0.00750\text{ kg})\left(6.00\frac{\text{m}}{\text{s}}\right)}{(0.00750\text{ kg}) + (0.250\text{ kg})} = 0.175\frac{\text{m}}{\text{s}}$$

Conservation of mechanical energy:

$$U_i + K_i = U_f + K_f$$

$$0 + \frac{1}{2}(m_M + m_P)v_{\text{f}x}^2 = (m_M + m_P)gy_f + 0$$

$$y_f = \frac{v_{\text{f}x}^2}{2g} = \frac{\left(0.175\frac{\text{m}}{\text{s}}\right)^2}{2\left(9.80\frac{\text{m}}{\text{s}^2}\right)} = 0.00156\text{ m} = 1.56\text{ mm}$$

23. $p_i = p_f$; $0 = m_1v_{1f} + m_2v_{2f}$; $m_2 = 2m_1$

$$K = \frac{1}{2}m_1v_{1f}^2 + \frac{1}{2}m_2v_{2f}^2;\text{ thus, }K = \frac{1}{2}m_1v_{1f}^2 + \frac{1}{2}(2m_1)\left(\frac{m_1v_{1f}}{2m_1}\right)^2$$

$$K = \frac{1}{2}m_1v_{1f}^2\left(1+\frac{2}{2^2}\right) = \frac{1}{2}m_1v_{1f}^2(1.5) = 1.5\text{ K}_{1f};$$

$6000\text{ J} = 1.5K_{1f}$; $K_{1f} = 4000\text{ J}$ and $K_{2f} = 2000\text{ J}$

25. $\frac{1}{2}(m_{\text{bullet}} + M_{\text{block}})v^2 = (m_{\text{bullet}} + M_{\text{block}})gh$;

$$v = \sqrt{2gh} = \sqrt{2\left(9.80\frac{\text{N}}{\text{kg}}\right)1.30\text{ m}} = 5.05\frac{\text{m}}{\text{s}}.$$

$m_{\text{bullet}}v_i = (m_{\text{bullet}} + M_{\text{block}})v_f$;

$$(0.025\text{ kg})\left(250\frac{\text{m}}{\text{s}}\right) = (0.025\text{ kg} + M_{\text{block}})\left(5.05\frac{\text{m}}{\text{s}}\right);$$

$M_{\text{block}} = 1.21\text{ kg} \approx 1.2\text{ kg}$

Chapter 9 AP® Practice Problems

1. (D) $\vec{p} = m\vec{v}$ and $\Delta\vec{p} = \vec{p}_2 - \vec{p}_1$

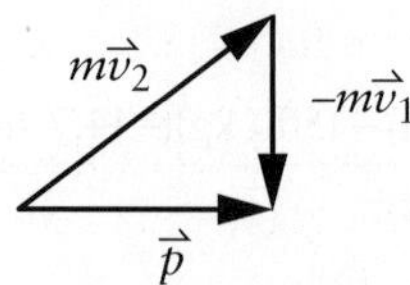

$(mv_2)^2 - (mv_1)^2 = (\Delta p)^2 \Rightarrow$

$\Delta p = \sqrt{(20\ \text{kg}\cdot\text{m/s})^2 - (12\ \text{kg}\cdot\text{m/s})^2} = \sqrt{256\ (\text{kg}\cdot\text{m/s})^2}$
$= 16\ \text{kg}\cdot\text{m/s}$

3. (A) For motion in one dimension: $\Delta p_x = F_x\Delta t$ and $\Delta p_x = mv_{2x} - mv_{1x}$

$\Delta p_x = m\Delta v_x \Rightarrow \Delta v_x = \dfrac{\Delta p_x}{m} = \dfrac{F_x\Delta t}{m}$

Consider the three time intervals 0–1.0 s, 1 s–2.5 s, and 2.5 s–4.0 s. Because they are relatively smooth you can either find the area under the graph or use the average force for each interval multiplied by the interval to estimate the impulse:

$F_x\Delta t = 2\ \text{N}(1\ \text{s}) + 3.5\ \text{N}(1.5\ \text{s}) + 3.5\ \text{N}(1.5\ \text{s}) = 12.5\ \text{Ns}.$

Dividing by the mass of 0.5 kg we get ~25 m/s for the change in velocity.

5. (A) Newton's third law and $\Delta\vec{p} = \vec{F}\Delta t$
The forces that the two objects exert on each other must be the same magnitude and opposite in direction, and the time of the interaction is the same, so their changes in momentum must be equal and opposite. Even though the forces they exerted on each other are the same and for the same time, in general, the two objects could be moving through different displacements during the interaction if the interaction is a long-distance force (such as that using magnetic bumpers). Remember from Chapter 7 that only for contact forces is the work equal in magnitude. Therefore, even though the time of contact is the same, the average displacements of each object during the collision is not necessarily the same. In general, the work done on each by the other will not be the same.

7. (D) $\Delta p = F\Delta t$

$d = \dfrac{1}{2}a(\Delta t)^2$, thus $\Delta t = \sqrt{\dfrac{2d}{a}}$ and $a = \dfrac{F}{m}$ and so

$\Delta p = F\sqrt{\dfrac{2dm}{F}} = \sqrt{2dFm}$

Chapter 10

10-1 Rotation is an important and ubiquitous kind of motion

1. (a) Numbering the images as a time sequence from 1 (leftmost) to 5 (rightmost), the wheel is spinning clockwise. The hub of a wheel spinning clockwise would translate to the right. Numbering the images as a time sequence from 1 (rightmost) to 5 (leftmost), the wheel is spinning counterclockwise. The hub of the wheel would translate to the left.
(b) The circumference of the wheel is $2\pi r = 2\pi$m. The rock moves a distance 2πm around the wheel between the first and last views. If the views are ordered in time from left to right, the sign of the angular displacement is negative. If the views are ordered in time from right to left, the angular displacement is positive.
(c) There are four time intervals between the five subsequent images in the diagram, which shows one complete rotation of the wheel. $4 \times 0.125\ \text{s} = 0.500\ \text{s}$. Thus, the time for one complete revolution is ½ second.
(d) If the wheel does not slip over the surface, then the distance that a point on the rim travels is the same as the distance that the hub travels. If the rim were a ribbon that unrolled as the wheel turns the length of one rotation (2π meters), then the ribbon would now be stretched between the first and last views shown. So the translational speed of the wheel rim is $2\pi\text{m}/1/2\ \text{s} = 4\pi$ m/s. The translational speed of the hub is therefore 4π m/s.
(e) As the wheel turns, the direction the rock is moving relative to the hub when at the top of the wheel is the same direction that the hub is moving. But the direction the rock is moving relative to the hub when at the bottom of the wheel is opposite to the direction that the hub is moving. The wheel pivots at the bottom, so the motion there is nearly zero.
(f) For the spoke that joins the topmost point of the wheel to the hub, the end of the spoke at the hub moves at the speed of the hub. But the end of the spoke at the top of the wheel is moving more quickly, at the speed of the rim, and in the same direction as the hub. Using this information, the translational velocity of the rock at the top of the wheel is greater and in the same direction as the translational velocity of the hub.

10-2 An extended object's rotational kinetic energy is related to its angular velocity and how its mass is distributed

1. (a) $45.0\dfrac{\text{rev}}{\text{min}}\dfrac{2\pi\ \text{rad}}{1\ \text{rev}}\dfrac{1\ \text{min}}{60\ \text{s}} = 4.71\ \text{rad/s}$

(b) $33\dfrac{1}{3}\ \text{rpm}\dfrac{2\pi\ \text{rad}}{1\ \text{rev}}\dfrac{1\ \text{min}}{60\ \text{s}} = 3.49\ \text{rad/s}$

(c) $2\pi\ \text{rev/s}\dfrac{2\pi\ \text{rad}}{1\ \text{rev}} = (2\pi)^2\ \text{rad/s}$

3. $v = r\omega \Rightarrow \omega = \dfrac{v}{r} = \dfrac{380\ \text{m/s}}{7600\ \text{m}} = \dfrac{1}{20}\ \text{rad/s}$

5. $K_{\text{rotational}} = \dfrac{I\omega^2}{2} = \dfrac{2.00\ \text{kg}\cdot\text{m}^2(3.00\ \text{rad/s})^2}{2}$
$= 9.00\ \text{kg}\cdot\text{m}^2/\text{s}^2 = 9.00\ \text{J}$

7. (a) $I = mr^2 = 1.00 \times 10^{-3}\ \text{kg}\ (0.120\ \text{m})^2$
$= 1.44 \times 10^{-5}\ \text{kg}\cdot\text{m}^2$

(b) $K_{\text{rotational}} = \dfrac{I\omega^2}{2} = \dfrac{1.44 \times 10^{-5}\ \text{kg} \cdot \text{m}^2 (300\ \text{rev/s})^2}{2}$

$\times \left(\dfrac{2\pi\ \text{rad}}{\text{rev}}\right)^2 = 25.6\ \text{J}$

9. (a) $a_{\text{cent}} = \dfrac{v^2}{r} = \dfrac{(r\omega)^2}{r} = r\omega^2$

(b) All points in a rotating rigid body have the same angular velocity; ω does not depend on r. Therefore, the result in part (a) shows that the centripetal acceleration is a linearly increasing function of the distance from the center of the disk.

11. $K_{\text{rotational}} = \dfrac{I}{2}\omega^2 \Rightarrow \omega = \sqrt{\dfrac{2K_{\text{rotational}}}{I}} = \sqrt{\dfrac{2(2.75\ \text{J})}{0.330\ \text{kg} \cdot \text{m}^2}}$

$= 4.08\ \text{rad/s}\left(\dfrac{1\ \text{rev}}{2\pi\ \text{rad}}\right)\left(\dfrac{60\ \text{s}}{\text{min}}\right) = 39.0\ \text{rpm}$

13. $\theta = \dfrac{2\pi\ \text{rad}}{\text{year}} \dfrac{1\ \text{year}}{365.25\ \text{days}} = 1.72 \times 10^{-2}\ \text{rad/day}$

$\dfrac{1.72 \times 10^{-2}\ \text{rad}}{\text{day}} \dfrac{360°}{2\pi\ \text{rad}} = 0.986°/\text{day}$

10-3 An extended object's rotational inertia depends on its mass distribution and the choice of rotation axis

1. The rotational inertia increases as the distribution of mass is shifted outward from the axis of rotation. So the hoop has the largest rotational inertia. Next in order of decreasing rotational inertia are the spherical shell, then the solid cylinder and finally the solid sphere. There is more mass closer to the axis of rotation for the solid sphere than for the cylinder.

3. $I = \sum_i m_i x_i^2 = 0.010\ \text{kg}(1\ \text{m})^2 + 0.015\ \text{kg}(3\ \text{m})^2$
$+ 0.020\ \text{kg}(6\ \text{m})^2 = 0.865\ \text{kg} \cdot \text{m}^2 \approx 0.9\ \text{kg} \cdot \text{m}^2$

5. $I = I_{\text{CM}} + MR^2 = \dfrac{MR^2}{2} + MR^2 = \dfrac{3MR^2}{2}$

7. (a) A line is drawn (as shown in the figure) in the direction of the gravitational force through point A. The string is then attached through hole B and hung on the wall. A second line is drawn containing the point B in the direction of the gravitational force. The center of mass of the fish must lie somewhere along both lines, because no matter what point you suspend the fish from, the force of gravity will point straight downward through the center of mass, given that the center of mass is the point at which the force of gravity is always exerted. So where the two lines intersect is the center of mass.

(b) An outline of the fish is drawn on the paper. One of the vertical lines is aligned with the axis of rotation and the line is marked. The distance of the midpoint in each of the vertical strips from the axis of rotation is measured and recorded. The fish is cut out of the sheet of paper, and the mass of the paper fish is determined. The strips are cut from the paper copy of the fish and the mass of each strip is determined. The square of the distance from the axis of rotation to the midpoint of the strip multiplied by the mass of the strip is recorded. The sum of these terms is the rotational inertia of the fish about this axis of rotation, once the mass of the paper is replaced with the mass of the cardboard (once you add them all up, only the total mass matters).

9. To have the smallest rotational inertia, the bead with the largest mass should be closest to the axis of rotation.

$I = I_{\text{rod}} + m_3 L_A^2 + m_2 L_B^2 + m_1 L_C^2$

10-4 The equations for rotational kinematics are almost identical to those for linear motion

1. (a) $\omega_i = 1.0 \dfrac{\text{rev}}{\text{s}} \dfrac{2\pi\ \text{rad}}{\text{rev}} = 2\pi\ \text{rad/s};\ \omega_f = 0$

$\alpha = -0.020\ \text{rev/s}^2 \left(\dfrac{2\pi\ \text{rad}}{\text{rev}}\right) = -0.040\pi\ \text{rad/s}^2$

$\omega_f = \omega_i + \alpha\Delta t \Rightarrow \Delta t = \dfrac{-\omega_i}{\alpha} = \dfrac{-2\pi\ \text{rad/s}}{-0.040\pi\ \text{rad/s}^2} = 5.0 \times 10^1\ \text{s}$

(b) $\omega_i = 1.0\ \text{rev/s}\left(\dfrac{2\pi\ \text{rad}}{\text{rev}}\right) = 2\pi\ \text{rad/s};\ \omega_f = 0$

$\omega_f = \omega_i + \alpha\Delta t \Rightarrow \Delta t = \dfrac{-2\pi\ \text{rad/s}}{-0.020\ \text{rad/s}^2} = 3.1 \times 10^2\ \text{s}$

(c) $\Delta\theta = \omega_i\Delta t + \dfrac{\alpha}{2}\Delta t^2$

$\Delta\theta = 2\pi\ \text{rad/s}(3.1 \times 10^2\ \text{s}) - \dfrac{1}{2}0.020\ \text{rad/s}^2(3.1 \times 10^2\ \text{s})^2$

$= 9.9 \times 10^2\ \text{rad}\left(\dfrac{1\ \text{rev}}{2\pi\ \text{rad}}\right) = 1.6 \times 10^2\ \text{rev}$

The angular displacement of the wheel is 9.9×10^2 rad, which is 1.6×10^2 rev.

3. (a) $\Delta\theta = 1\ \text{h}\dfrac{2\pi\ \text{rad}}{24\ \text{h}} = \dfrac{\pi\ \text{rad}}{12}$

(b) $\omega = \dfrac{2\pi\ \text{rad}}{24\ \text{h}} \dfrac{1\ \text{h}}{3600\ \text{s}} = 7.3 \times 10^{-5}\ \text{rad/s}\left(\dfrac{1\ \text{rev}}{2\pi\ \text{rad}}\right)$

$\times \dfrac{60\ \text{s}}{1\ \text{min}} = 6.9 \times 10^{-4}\ \text{rpm}$

5. (a)

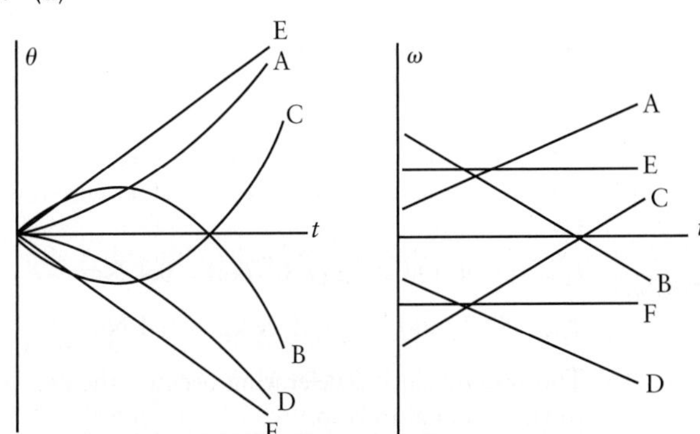

(b) i. C and B have angular velocities that change sign, and at these points the disk momentarily stops.
ii. C and B can return the disk to $\theta = 0$, given enough time.
iii. A, E, D, and F will not return the disk to $\theta = 0$.
iv. B, D, and F result in displacements that are multiples of -2π, given enough time.
v. A, C, and E result in displacements that are multiples of $+2\pi$, given enough time.

7. At any point on the disk, the linear velocity v will equal the radius r at that point multiplied by the angular velocity ω. As the laser scans the disk from close to the center toward the outer edge, r increases, and therefore the speed at which the depressions would be scanned would also increase if the angular velocity were constant. For the sampling rate to remain constant, v must be kept constant; therefore, ω must compensate for the laser's position over the disk. As r increases, ω must decrease to keep v, and thus the sampling rate, constant.

9. (a) $\omega_f = 18.0 \text{ rpm}\left(\frac{2\pi \text{ rad}}{\text{rev}}\right)\left(\frac{1 \text{ min}}{60 \text{ s}}\right) = 1.88 \text{ rad/s}$

$$\alpha_{\text{average}} = \frac{1.88 \text{ rad/s}}{43.0 \text{ s}} = 4.38 \times 10^{-2} \text{ rad/s}^2$$

(b) $\theta = 0 \text{ rad/s } \Delta t + \frac{\alpha}{2}\Delta t^2 = \frac{1}{2}4.38 \times 10^{-2} \text{ rad/s}^2 (43.0 \text{ s})^2$

$= 40.5 \text{ rad}$

(c) $v_{\max} = r\omega_f = 2.00 \text{ m}(1.88 \text{ rad/s}) = 3.76 \text{ m/s}$

11. (a) Plot the data given
(b) A best-fit line gives an angular acceleration of -1.32 rad/s^2.

10-5 Torque is to rotation as force is to translation

1. The force exerted on the barbell by gravity is directly downward, so $\phi = \pi/2$ at a distance of the length of your arm from your shoulder.

$$\tau = Fr \sin\phi = 10.0 \text{ kg}(9.8 \text{ N/kg})(0.75 \text{ m})\sin(\pi/2)$$
$$= 74 \text{ Nm}$$

3. $\tau = rF\sin\phi$, where ϕ is the angle between the lever arm and the force, $\vec{F}$. Here, ϕ is 90° in case (a) and 65°, (90° – 25°), in case (b).

$$\tau_{90} = Fr \sin\phi = 75.0 \text{ N}(0.850 \text{ m}) \sin(90°) = 63.8 \text{ Nm} \approx 64 \text{ Nm}$$

$$\tau_{25} = Fr \sin\phi = 75.0 \text{ N}(0.850 \text{ m}) \sin(65°) = 57.8 \text{ Nm} \approx 58 \text{ Nm}$$

5. (a) The 0.1-kg object exerts a counterclockwise (ccw) torque. The 0.2-kg object exerts a clockwise (cw) torque.

$$\tau_{\text{ccw}} = (0.4 \text{ m})(0.1 \text{ kg})(9.8 \text{ N/kg}) = 0.4 \text{ Nm}$$
$$\tau_{\text{cw}} = (0.2 \text{ m})(0.2 \text{ kg})(9.8 \text{ N/kg}) = 0.4 \text{ Nm}$$

(b) The system is not accelerating because the net torque is equal to zero.
(c) The 0.1-kg object exerts a counterclockwise (ccw) torque. The 0.2-kg object exerts a cw torque. The center of mass of the meter stick exerts a cw torque.

$$\tau_{\text{ccw}} = (0.3 \text{ m})(0.1 \text{ kg})(9.8 \text{ N/kg}) = 0.3 \text{ Nm}$$
$$\tau_{\text{cw}} = (0.3 \text{ m})(0.2 \text{ kg})(9.8 \text{ N/kg}) + (0.1 \text{ m})(0.05 \text{ kg})(9.8 \text{ N/kg}) = 0.6 \text{ Nm}$$

(d) $I = m_L(x_L - x_{\text{fulcrum}})^2 + m_R(x_R - x_{\text{fulcrum}})^2 + m_{\text{meter stick}}(L^2/12 + (x_{\text{fulcrum}} - L/2)^2)$

$$= (0.1 \text{ kg})(0.3 \text{ m})^2 + (0.2 \text{ kg})(0.3 \text{ m})^2 + (0.05 \text{ kg})((1.0 \text{ m})^2/12 + (0.1 \text{ m})^2) = 0.0317 \text{ kg}\cdot\text{m}^2$$

$$\alpha = \frac{\tau_{\text{net}}}{I} = \frac{-0.6 \text{ Nm} + 0.3 \text{ Nm}}{0.0317 \text{ kg}\cdot\text{m}^2} = -9.5 \text{ rad/s}^2$$

(e) The angular acceleration points into the page because the result is negative.

7. $\tau = Fr \le 10 \text{ Nm} \Rightarrow F \le \frac{10 \text{ Nm}}{0.02 \text{ m}} = 500 \text{ N}$ (We took the axis as one edge of the cap.)

9. (a) The total rotational inertia is the sum of the rotational inertias for the rod and the sphere. The center of the sphere is a distance $L + R$ from the axis of rotation. So the parallel-axis theorem can be used to calculate the contribution of the sphere, and this added to that of the rod is $I = M(L+R)^2 + 2MR^2/5 + mL^2/3$.
(b) The forces are exerted at the center of mass of the rod (mg) and of the sphere (Mg). The magnitude of the position $\vec{r}$ from the rotational axis to the point at which the force is exerted for the rod is $L/2$ and for the ball is $L + R$. So

$$\tau_{\text{rod}}(t=0) = -mg(L/2)\sin\theta \quad \tau_{\text{sphere}}(t=0) = -Mg(L+R)\sin\theta$$
$$\sum\tau(t=0) = -[mL/2 + M(L+R)]g \sin\theta$$

(c) The ball–rod system will increase its angular speed until $\theta = 0$, assuming that mechanical energy is conserved. The angular speed will then begin to decrease and finally the sphere will come to a momentary stop at $-\theta$. So the angular speed is a maximum at $\theta = 0$, requiring that the angular acceleration be zero at that point. Also, because the position from the rotational axis to the point at which the force is exerted, $\vec{r}$, and the net force are parallel when $\theta = 0$, the torque is zero. Given $\alpha(\theta = 0) = \tau(\theta = 0)/I, \alpha(\theta = 0)$ is zero.
(d) In terms of the parameters in this system the initial angular acceleration is τ/I:

$$\alpha(t=0) = \frac{\sum\tau(t=0)}{I} = -\frac{[mL/2 + M(L+R)]g \sin\theta}{M(L+R)^2 + 2MR^2/5 + mL^2/3}$$

11. (a) The cable exerts a force along its length so up and to the right from the left end of the beam. The force of gravity is exerted downward on the beam at its center of mass, so 200 N downward at the pivot. Gravity also pulls downward on the person at 600 N. Because the person is not accelerating, there must be an upward normal force exerted on the person by the beam to balance gravity. By Newton's third law, the person exerts a 600-N force downward on the beam, at the point 2.0 m from the left end of the beam. The force of gravity on the beam exerts no torque about the pivot. The perpendicular component of the tension exerts a

clockwise torque on the beam, and the force the person exerts on the beam exerts a counterclockwise torque on the beam. Because the tension in the cable also exerts a force along the beam to the right, the pivot must exert a force to the left as well as upward on the beam.

(b) The torques must sum to zero to maintain rotational equilibrium. The farther the person walks toward the left end, the more torque is exerted in the counterclockwise direction. We are told the tension force can increase to balance this and keep the beam in rotational equilibrium. As the person walks toward the right, the torque caused by the person decreases until it reaches zero when he reaches the center of the beam. At that point, the tension in the cable would also be zero, and the beam would be in rotational equilibrium. If the person walks farther to the right, past the pivot, the torque the person exerts on the beam would be in the clockwise direction and could not be balanced by a torque from the cable, so the person can walk 2 m in either direction from their initial position.

10-6 The techniques used for solving problems with Newton's second law also apply to rotation problems

1. $\sum F_{\text{block}}: ma = mg - F_T; \quad \sum \tau_{\text{pulley}}: F_T r = I\alpha = Ia/r$

$\Rightarrow ma + Ia/r^2 = mg$

$$a = \frac{mg}{m + I/r^2} = \frac{0.250 \text{ kg}(9.80 \text{ N/kg})}{0.250 \text{ kg} + 1.00 \text{ kg}/2} = 3.27 \text{ N/kg}$$

$$\Rightarrow \alpha = \frac{a}{r} = \frac{3.27 \text{ N/kg}}{0.25 \text{ m}} = 13 \text{ rad s}^{-2}$$

$$F_T = \frac{Ia}{r^2} = \frac{Ma}{2} = \frac{(1.00 \text{ kg})(3.267 \text{ N/kg})}{2} = 1.63 \text{ N}$$

3. (a) The average angular acceleration is

$$\alpha = \frac{\Delta\omega}{\Delta t} = \frac{7200 \text{ rpm}}{2.50 \text{ s}}\frac{2\pi \text{ rad}}{\text{rev}}\frac{1 \text{ min}}{60 \text{ s}} = 3.0 \times 10^2 \text{ rad/s}^2$$

The torque on the disk is

$$\tau = I\alpha = \frac{mr^2}{2}\alpha = \frac{(7.50 \times 10^{-3} \text{ kg})(0.03175 \text{ m})^2}{2} 3.0 \times 10^2 \text{ rad/s}^2$$

$= 1.1 \times 10^{-3}$ Nm

(b) Not the final answer, so leave more digits so we get a more accurate power calculation.

$$\Delta K_{\text{rotational}} = \frac{I\omega^2}{2} = \frac{(7.50 \times 10^{-3} \text{ kg})(0.03175 \text{ m})^2}{4}\left(7200 \text{ rpm}\frac{2\pi \text{ rad}}{\text{rev}}\frac{1 \text{ min}}{60 \text{ s}}\right)^2$$

$= 1.07$ J ($= 1.1$ J)

The power is

$$P = \frac{\Delta K_{\text{rotational}}}{\Delta t} = \frac{1.07 \text{ J}}{2.50 \text{ s}} = 0.43 \text{ W}$$

5. (a) Let F_{T1} be the tension in the string attached to the block on the table. Let F_{T2} be the tension in the string attached to the hanging block. The linear acceleration of a point on the rim of the pulley must be the same as the accelerations of the blocks as long as the string does not stretch or slip on the pulley. (The accelerations of the blocks are the same as long as the string does not stretch.) Using Newton's second law, the linear acceleration can be expressed in terms of the tensions. The linear acceleration of a point on the rim of the pulley is related to the angular acceleration by $a = \alpha r$. This will give us three equations, allowing us to solve for our three unknowns.

$$F_{T1} = m_1 a; \quad m_2 g - F_{T2} = m_2 a; \; (F_{T2} - F_{T1})r = \frac{m_{\text{pulley}} r^2}{2}\frac{a}{r}$$

Then

$$F_{T1} = m_1 a; \quad F_{T2} = m_2 g - m_2 a; \; (F_{T2} - F_{T1})r = (m_2 g - m_2 a - m_1 a)r = \frac{m_{\text{pulley}} r^2}{2}\frac{a}{r}$$

which can be solved for a

$$\left(m_2 + m_1 + \frac{m_{\text{pulley}}}{2}\right)a = m_2 g \Rightarrow a = \frac{m_2 g}{\left(m_2 + m_1 + \frac{m_{\text{pulley}}}{2}\right)}$$

$$= \frac{4.00 \text{ kg}(9.80 \text{ N/kg})}{2.00 \text{ kg} + 4.00 \text{ kg} + 0.250 \text{ kg}} = 6.27 \text{ N/kg}$$

From the acceleration, the tensions can be calculated.

$$F_{T1} = (2.00 \text{ kg})(6.27 \text{ N/kg}) = 12.5 \text{ N}$$

$$F_{T2} = 4.00 \text{ kg}(9.80 \text{ N/kg} - 6.27 \text{ N/kg}) = 14.1 \text{ N}$$

(b) $x = \frac{1}{2}at^2 \Rightarrow t = \sqrt{\frac{2x}{a}} = \sqrt{\frac{2(2.25 \text{ m})}{6.27 \text{ m/s}^2}} = 0.847 \text{ s}$

(c) $\omega = \alpha t = \frac{a}{r}t = \frac{6.27 \text{ m/s}^2}{0.0400 \text{ m}}(0.847 \text{ s}) = 133 \text{ rad/s}$

7. The gravitational force is exerted on the center of mass, which is on the axis of rotation. So the lever arm is zero and no torque is exerted. The cylinder will therefore slide down the incline. The cylinder will only roll if there is a friction force exerted on its surface, resulting in a torque.

9. (a) $Mg - F_{TL} = Ma \quad F_{TR} = ma \quad (F_{TL} - F_{TR})R = I\alpha$

$$a = \frac{M}{M + m + I/R^2}g = xg$$

(b)

m (kg)	a (m/s^2)	x
0.35	3.38	0.345
0.37	3.31	0.333
0.39	3.19	0.323
0.41	3.06	0.313
0.43	2.98	0.303
0.45	2.92	0.294
0.47	2.80	0.286
0.49	2.76	0.278

(c) $a = \dfrac{M}{M + m + I/R^2} g$

The slope is 9.7 m/s^2.

(d) By making the term in the denominator smaller by introducing $0 < c < 1$, the ratio (value of x) will increase and the slope will decrease.

(e) She should consider at least these possible sources of error:

- Is the track horizontal? She can place the glider on the track at rest and look for acceleration.
- Does the pulley exert a friction force on the string? She can listen for noise as the pulley turns. She can look to see if the post or screw on which the pulley spins is bent or too tight.
- Is the track straight? She can lay a meter stick on the track and look for gaps between the track and the stick.
- Are the surfaces of the glider and air track smooth? She can run her hand over both surfaces looking for spurs or abrasions.
- Have the measurements of mass been made accurately? She can repeat them.
- Is the measurement of acceleration being made accurately? She can use a stopwatch to measure the time it takes the hanging object to descend a given distance and use these data to confirm the value of the acceleration.

Chapter 10 Review Problems

1. (D) Kinetic energy is proportional to the product of the rotational inertia and the angular speed squared. Multiplying I by 1/5 decreases $K_{\text{rotational}}$ by 1/5, but multiplying ω by 5 increases $K_{\text{rotational}}$ by 25.

3. (A) The rotational inertia is proportional to the product of the mass and the radius squared. The mass (in terms of the density and volume) is proportional to the radius cubed, so the mass term will increase by a factor of 8, and the radius term will increase by a factor of 4, increasing I by a factor of 32.

5. (a) Use the average angular speed to calculate the total number of rotations the turntable undergoes when starting or slowing down. In between, the turntable spins at a constant angular speed for 2827 s (47 min, 7 s); we can determine the number of rotations from the definition of angular speed. The total number of rotations is the sum of these three values.

Starting up:

$$\Delta\theta = \omega_{\text{average},z} t = \left(\frac{\omega_{2z} + \omega_{1z}}{2}\right) t = \left(\frac{\left(\dfrac{100 \text{ rot}}{3 \text{ min}} \times \dfrac{1 \text{ min}}{60 \text{ s}}\right) + 0}{2}\right)$$

$$\times (4.5 \text{ s}) = 1.2 \text{ rot}$$

(b) Spinning at constant speed:

$$\Delta\theta = \omega_{\text{average},z} t = \left(\frac{100 \text{ rot}}{3 \text{ min}} \times \frac{1 \text{ min}}{60 \text{ s}}\right)(2827 \text{ s}) = 1571 \text{ rot}$$

Slowing down:

$$\Delta\theta = \omega_{\text{average},z} t = \left(\frac{\omega_{2z} + \omega_{1z}}{2}\right) t = \left(\frac{0 + \left(\dfrac{100 \text{ rev}}{3 \text{ min}} \times \dfrac{1 \text{ min}}{60 \text{ s}}\right)}{2}\right)$$

$$\times (8 \text{ s}) = 2 \text{ rev}$$

Total number of rotations:

$\Delta\theta_{\text{total}} = (1.2 \text{ rot}) + (1571 \text{ rot}) + (2.2 \text{ rot})$
$= 1574 \text{ rot} \approx 1.6 \times 10^3 \text{ rot}$

7. (a) $I_{\text{CM}} = \dfrac{1}{12} ML^2$. The angular speed should be converted from rev/s to rad/s; thus

$$\omega = 3.00 \frac{\text{rev}}{\text{s}} \times 2\pi \frac{\text{rad}}{\text{rev}} = 18.8 \frac{\text{rad}}{\text{s}}.$$

$$K_{\text{rotationalCM}} = \frac{1}{2} I\omega^2 = \frac{1}{2}\left(\frac{1}{12}(0.103 \text{ kg})(1.00 \text{ m})^2\right)$$

$$\left(\frac{18.8 \text{ rad}}{\text{s}}\right)^2 = 1.52 \text{ J}$$

(b)

$$I_{\text{end}} = \frac{1}{3} ML^2 = \frac{1}{3}(0.103 \text{ kg})(1.00 \text{ m})^2 = 0.0343 \text{ kg} \cdot \text{m}^2$$

$$K_{\text{rotational,end}} = \frac{1}{2} I\omega^2 = \frac{1}{2}(0.0343 \text{ kg} \cdot \text{m}^2)\left(18.8 \frac{\text{rad}}{\text{s}}\right)^2 = 6.09 \text{ J}$$

9. The rotational inertia for a sphere in rotation about an axis that is tangent to its surface can be found from the parallel-axis theorem to be $I_{\text{tangential axis}} = 7MR^2/5$.

$$K_{\text{central axis}} = \frac{2MR^2}{5}\frac{\omega^2}{2} \qquad K_{\text{tangential axis}} = \frac{7MR^2}{5}\frac{\omega^2}{2}$$

$$\frac{K_{\text{central axis}}}{K_{\text{tangential axis}}} = \frac{2}{7}$$

11. (a) $\omega_f = 45.0 \text{ rpm}\left(2\pi \dfrac{\text{rad}}{\text{rev}}\right)\left(\dfrac{1 \text{ min}}{60 \text{ s}}\right) = 4.71 \dfrac{\text{rad}}{\text{s}}$

$\omega_f = \omega_0 + \alpha t$; $4.71 \dfrac{\text{rad}}{\text{s}} = 0 + \alpha(2.50 \text{ s})$; $\alpha = 1.88 \text{ rad/s}^2$

(b) The total angular displacement is the displacement during the acceleration, during the constant angular speed, and during the deceleration.

Speeding up:

$$\Delta\theta_1 = \frac{1}{2}\alpha t^2 = \frac{1}{2}\left(1.88 \frac{\text{rad}}{\text{s}^2}\right)(2.50 \text{ s})^2 = 5.88 \text{ rad}$$

Constant speed:

$$\Delta\theta_2 = \omega t = \left(4.71 \frac{\text{rad}}{\text{s}}\right)(220 \text{ s}) = 1037 \text{ rad}$$

Slowing down:

$$\Delta\theta_3 = \frac{\omega + \omega_0}{2} t = \frac{0 + 4.71 \dfrac{\text{rad}}{\text{s}}}{2}(4.80 \text{ s}) = 11.3 \text{ rad}$$

Total $= 5.88 \text{ rad} + 1037 \text{ rad} + 11.3 \text{ rad} = 1054 \text{ rad}$
$= 1.05 \times 10^3 \text{ rad} = 1054 \text{ rad } (1 \text{ rev}/2\pi \text{ rad})$
$= 168$ rotations

13. (a)

$$\tau_{net} = I\alpha = 0 = F_{ext}r - 4F_k R \Rightarrow F_{ext} = \frac{4F_k R}{r}$$

$$= \frac{4(45.0\ \text{N})(1.00\ \text{m})}{0.60\ \text{m}} = 3.0 \times 10^2\ \text{N}$$

(b)

$$\alpha = \frac{\tau_{net}}{I} = 1\ \text{rad/s}^2 \Rightarrow F_{ext} = \frac{1\ \text{rad/s}^2(119\ \text{kg}\cdot\text{m}^2) + 4F_k R}{r}$$

$$= \frac{1\ \text{rad/s}^2(119\ \text{kg}\cdot\text{m}^2) + 4(45.0\ \text{N})(1.00\ \text{m})}{0.60\ \text{m}} = 5.0 \times 10^2\ \text{N}$$

15. (a) $\frac{v}{r} = \frac{V}{R_{lever}} \Rightarrow V = \frac{R_{lever}}{r}v$

(b) Using conservation of energy, with Earth in the system, the maximum height is when the kinetic energy is zero:

$$\frac{m_{puck}V^2}{2} = m_{puck}gh \Rightarrow V > \sqrt{2gh} = 10.8\ \text{m/s} \approx 11\ \text{m/s}$$

(c) We can find the height we need using the same approach as in part (b). First, we use the fact that the sledge and the puck have the same angular velocity to find the required velocity for the sledge.

$$v = \frac{r}{R_{lever}}V = \frac{0.50\ \text{m}}{1.0\ \text{m}}10.8\ \text{m/s} = 5.4\ \text{m/s} = \sqrt{2gh_{drop}} \Rightarrow h_{drop} = 1.5\ \text{m}$$

The bell could be rung by dropping the head of the sledge from roughly twice the length of the handle.

(d) The sledgehammer is allowed to fall from position 1 until it is vertically downward, as the gravitational force exerted on the center of mass of the sledge is negative (actually in the head, which is much heavier than the handle), rotating the handle in the clockwise direction. Between the positions numbered as 2–5, one hand slides from a point near the head to a point nearer the axis of rotation. The contact force of the hand nearer the head is exerted on a lever arm that is approximately the distance between the hands. This torque is negative, rotating the handle in the clockwise direction. The gravitational force exerted on the center of mass of the sledge is now positive, rotating the handle in the counterclockwise direction. The torque resulting from the contact force must be greater in magnitude than the torque exerted by the gravitational force. Otherwise, the handle will not rotate as shown. As the lever arm during the first half of the motion decreases, so does the torque produced by the force exerted at this lever arm. But the torque produced by the gravitational force also decreases as the handle rotates, because the lever arm for this force is the horizontal component of the distance from the center of mass of the sledgehammer where the vertical gravitational force is exerted to the rotational axis. After the top of the arc, which is about halfway between positions 5 and 6, gravity is again rotating the head in the direction desired, so he can pull his hand back and all of the torque is exerted by the gravitational force, which is greatest just after position 7.

17. $\frac{1}{2}gt_{flight}^2 = 5.0\ \text{m} \Rightarrow t_{flight} = \sqrt{\frac{2h}{g}} = \sqrt{\frac{2(5.0\ \text{m})}{9.80\ \text{m/s}^2}} = 1.01\ \text{s}$

$$\Rightarrow \omega = \frac{2.5(2\pi\ \text{rad})}{1.01\ \text{s}} = 16\ \text{rad/s}$$

19. (a) The acceleration is in the counterclockwise direction.

$$\alpha = \frac{\Delta\omega}{\Delta t} = 5.00 \times 10^2\ \frac{\text{rev}}{\text{min}}\ \frac{2\pi\ \text{rad}}{\text{rev}}\ \frac{1\ \text{min}}{60\ \text{s}}\ \frac{1}{15.0\ \text{s}} = 3.49\ \text{rad/s}^2$$

(b) $\tau = I\alpha \Rightarrow I = \frac{\tau}{\alpha}$

(c) $I = \frac{75.0\ \text{Nm}}{3.49\ \text{rad/s}^2} = 21.5\ \text{kg}\cdot\text{m}^2$

(d) $\alpha = \frac{\Delta\omega}{\Delta t} = \frac{-5.00 \times 10^2\ \text{rev/min}}{2.00 \times 10^2\ \text{s}}\ \frac{2\pi\ \text{rad}}{\text{rev}}\ \frac{1\ \text{min}}{60\ \text{s}}$

$= -0.262\ \text{rad/s}^2$

So

$$\tau = I\alpha = 21.5\ \text{kg}\cdot\text{m}^2(-0.262\ \text{rad/s}^2) = -5.63\ \text{Nm}$$

The direction of the torque is clockwise.

21. (a)

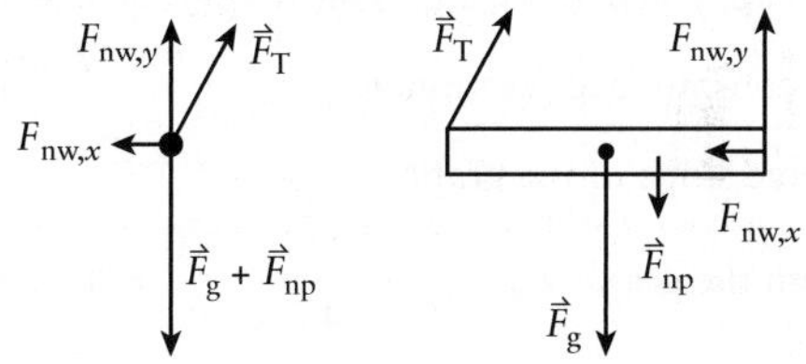

$F_{nw,y}$ $\vec{F}_T$ $F_{nw,x}$ $\vec{F}_g + \vec{F}_{np}$

$\vec{F}_T$ $F_{nw,y}$ $F_{nw,x}$ $\vec{F}_{np}$ $\vec{F}_g$

(b) Because the beam is not accelerating, the forces balance, and so do the torques. We can calculate the torque about any point on the beam, so we are free to choose which point we want to use. We label the normal force the person exerts on the beam (in response to the beam pushing upward on the person) as $\vec{F}_{np}$. We use the subscript nw for the normal force exerted by the bracket on the wall. Because it is attached, it can have both upward and outward components. The sum of the forces in the horizontal direction is (where θ is the angle between the beam and the tension force):

$F_{T,x} - F_{nw,x} = 0$

so $F_{nw,x} = F_{T,x} = F_T \cos\theta$

The sum of the forces in the vertical direction is

$F_{nw,y} + F_{T,y} - F_{np} - F_g = 0$

so $F_{T,y} = F_{np} + F_g - F_{nw,y}$

and as $F_{T,y} = F_T \sin\theta$,

$F_T = (\csc\theta)(F_{np} + F_g - F_{nw,y})$

Choose a point about which to balance torques: $\tau_{net} = \tau_p + \tau_g + \tau_w + \tau_T = 0$. If we calculate torque about the left end of the beam, the torque due to the tension in the cable is zero, because the lever arm is zero. The other forces are all perpendicular or parallel to the beam, so $|\tau| = Fr$ or 0 for each. Choosing counterclockwise positive, the torque due to the person is $\tau_p = -(F_{np})(r_p)$ and if the length of the beam is ℓ, the torque due to gravity on the beam is $\tau_g = -F_g r_g = -F_g(\ell/2)$. The torque due to the wall on the beam is $\tau_w = +(F_{nw,y})(\ell)$. So, $0 = -(F_{np})(r_p) - F_g(\ell/2) + (F_{nw,y})(\ell)$. This simplifies to $F_{nw,y} = (F_{np})(r_p/\ell) + F_g/2$.

We have enough information to solve this:

$$F_{nw,y} = \frac{1}{3.0 \text{ m}}(75.0 \text{ kg})(g)(2.0 \text{ m}) + \frac{1}{2}(50.0 \text{ kg})(g) = 735 \text{ N}$$

and so

$$F_T = (\csc 60°)(m_p g + m_b g - F_{nw,y})$$

$$= (\csc 60°)((75.0 \text{ kg})(g) + (50.0 \text{ kg})(g) - 735 \text{ N}) = 566 \text{ N}$$

and

$$F_{nw,x} = F_T \cos\theta = (566 \text{ N})\cos 60° = 283 \text{ N}$$

Chapter 10 AP® Practice Questions

1. (D) This can be done by using the relationship for constant acceleration rotational motion, where the initial angular speed and initial angular position are both zero, so

   $\theta = 1/2\alpha t^2$

   $2 \text{ rad} = 1/2\alpha(4\text{s})^2$

   $\alpha = 0.25 \text{ rad/s}^2$. Then, $\omega_4 = 0.25 \text{ rad/s}^2 \times 4\text{s} = 1 \text{ rad/s}$

   Instead, you can use a graphical approach, given that for constant acceleration: $\omega_{average} = \dfrac{\omega_2 + \omega_1}{2}$ and the averge slope of the graph.

   From the graph, $\omega_{average} = \dfrac{2.0 \text{ rad} - 0}{4.0 \text{ s} - 0} = 0.50 \text{ rad/s} \Rightarrow$

   $0.50 \text{ rad/s} = \dfrac{\omega_2 - \omega_1}{2} = \dfrac{\omega_2}{2} \Rightarrow \omega_2 = 1.0 \text{ rad/s}$

3. (A) $I = \sum_{i=1}^{N} m_i r_i^2$

   c > a because due to its shape and orientation, more of its mass is beyond the $2r$ radius than is within
   a > d because a large portion of the triangle's mass is closer than the $2r$ radius
   d > b because, even though b has twice the mass, the distance to the center of mass of d is large enough that the distance squared is larger by at least a factor of 2

5. (D) The net external torque on the system (and acceleration) is caused by the force of gravity on the hanging block. Systems A and C have the same mass, and system B has a smaller mass. Systems A and B have the same net accelerating force and C has a greater net force causing the acceleration. The effect of the rotational inertia of the pulley is the same for all three systems, so that does not affect the ranking. Thus, the block has the greatest acceleration in C and has the least acceleration in A.

7. (A) $K_{rotational} = \dfrac{1}{2}\omega^2 = \dfrac{1}{2}\left(\dfrac{1}{12}ML^2\right)\omega^2$

   Simply investigate the ML^2 factor. Halving the original length and doubling the cross-sectional area keeps the mass the same. Then one-half the length squared gives the factor of one-quarter that was needed.

Chapter 11

11-1 Angular momentum and the next conservation law: conservation of angular momentum

1. Momentum depends on an object's mass and velocity. Angular quantities depend on location of mass in reference to a rotation axis, so there must be some dependence on a rotation axis and its perpendicular distance to the direction of the momentum as well.
3. When you pull your arms inward you are bringing mass closer to the rotation axis, or reducing your rotational inertia.

11-2 Conservation of mechanical energy also applies to rotating extended objects

1. Denoting the radius of Earth as R_E and the radius of the orbit of Earth as R_{E-S}:

$$K_{rotational} = \frac{I}{2}\omega^2 = \frac{MR_E^2}{5}\left(\frac{2\pi \text{ rad}}{\text{day}}\right)^2\left(\frac{1 \text{ day}}{86{,}400 \text{ s}}\right)^2$$

$$K_{translational} = \frac{M}{2}v^2 = \frac{MR_{E-S}^2}{2}\omega^2 = \frac{MR_{E-S}^2}{2}\left(\frac{2\pi \text{ rad}}{365.25 \text{ day}}\right)^2\left(\frac{1 \text{ day}}{86{,}400 \text{ s}}\right)^2$$

$$\frac{K_{rotational}}{K_{translational}} = \frac{2}{5}\frac{R_E^2}{R_{E-S}^2}\left(\frac{365.25\text{d}}{1\text{d}}\right)^2 = \frac{2}{5}\left(\frac{6.38 \times 10^6 \text{ m}}{1.5 \times 10^{11} \text{ m}}\right)^2(365.25)^2$$

$$= 9.7 \times 10^{-5}$$

3.

$$K_{rotational} = \frac{I}{2}\omega^2 = \frac{MR^2}{4}\omega^2 = 15.0 \text{ J} \Rightarrow \omega = \sqrt{\frac{60.0 \text{ J}}{60.0 \text{ kg}(0.175 \text{ m})^2}}$$

$$= 5.71 \text{ rad/s}\left(\frac{1 \text{ rev}}{2\pi \text{ rad}}\right)\left(\frac{60 \text{ s}}{\text{min}}\right) = 54.6 \text{ rpm}$$

5. $\Delta U_{grav} = -\Delta K = -K_f + 0 = 0 - mgh$

   $h = L \sin\theta = 2.00 \text{ m} \sin(25.0°) = 0.845 \text{ m}$

$$K_f = \frac{I}{2}\omega_f^2 + \frac{m}{2}v_f^2 = \frac{m}{4}(r\omega_f)^2 + \frac{m}{2}v_f^2 = \left(\frac{1}{4} + \frac{1}{2}\right)mv_f^2 = \frac{3mv_f^2}{4}$$

$$K_f = mgh \Rightarrow v_f^2 = \frac{4gh}{3} = \frac{4(9.80 \text{ m/s}^2)(0.845 \text{ m})}{3}$$

$$= 11.0 \text{ m}^2/\text{s}^2$$

   So $v_f = 3.32$ m/s.

7. (B) There are no forces or torques exerted on the system. So the angular acceleration of the system is zero. The angular speed of the system is constant and mechanical energy is conserved.
9. The radius of the loop is taken as R and the radius of the marble as r. h is taken as the initial height of the center of mass of the marble as measured from the bottom of the loop.

   Taking $U_{grav} = 0$ when the marble is at the bottom of the loop, the conservation of mechanical energy gives us

   $mgh = mg(2R - r) + \dfrac{1}{2}mv_f^2 + \dfrac{1}{2}I\omega_f^2$ where $I = \dfrac{2}{5}mr^2$

   and $r\omega_f = v_f$

   $gh = g(2R - r) + \dfrac{1}{2}v_f^2 + \dfrac{1}{2}\left(\dfrac{2}{5}r^2\right)\omega_f^2 = g(2R - r) + \dfrac{7}{10}v_f^2$

In order to barely make it around the loop

$$a_{\text{cent}} = \frac{v_f^2}{r_{\text{cirde}}} = g = \frac{v_f^2}{R-r}$$

$$gh = g(2R-r) + \frac{7}{10}g(R-r). \text{ Rearranging:}$$

$$h = 2.7R - 1.7r = 2.7(0.200 \text{ m}) - 1.7(0.005 \text{ m})$$

$$= 0.540 \text{ m} - 0.0085 \text{ m} = 0.532 \text{ m}$$

11. $\Delta U = -\Delta K = 0 + K_i = mg\Delta h$

$$mg\Delta h = \frac{m}{2}v_{\text{cm,i}}^2 + \frac{I}{2}\omega_i^2 = \frac{m}{2}v_{\text{cm,i}}^2\left(1 + \frac{I}{mr^2}\right) \Rightarrow \Delta h = \frac{v_{\text{cm,i}}^2}{2g}\left(1 + \frac{I}{mr^2}\right)$$

$$\Delta h = \frac{v_{\text{cm,i}}^2}{2g}(1+c) \Rightarrow \Delta h_{\text{hoop}} = \frac{v_{\text{cm,i}}^2}{g};\ \Delta h_{\text{disk}} = \frac{3v_{\text{cm,i}}^2}{4g};\ \Delta h_{\text{sphere}} = \frac{7v_{\text{cm,i}}^2}{10g}$$

so $\Delta h_{\text{hoop}} > \Delta h_{\text{disk}} > \Delta h_{\text{sphere}}$. This makes sense. Because the hoop has the largest rotational inertia, it has the largest initial kinetic energy, so makes it the farthest up the incline.

11-3 Angular momentum is always conserved; it is constant when there is zero net torque (or angular impulse) exerted on a system

1. The angular momentum's direction would be found by wrapping the fingers of your right hand in the direction of rotation of the tire: So your extended thumb points to the left from your location. As long as you say perpendicular to the tire at its axis of rotation, that is good enough.

3. (a)

$$L = \frac{2mR^2}{5}\omega = \frac{2(5.98\times10^{24}\text{ kg})(6.38\times10^6\text{ m})^2}{5}\frac{2\pi\text{ rad}}{24\text{ h}}$$

$$\times\frac{1\text{ h}}{3600\text{ s}} = 7.08\times10^{33}\text{ kg}\cdot\text{m}^2\text{ s}^{-1}$$

(b)

$$L = mr^2\omega = (5.98\times10^{24}\text{ kg})(1.5\times10^{11}\text{ m})^2\frac{2\pi\text{rad}}{365.3\text{ d}}\frac{1\text{ d}}{24\text{ h}}$$

$$\times\frac{1\text{ h}}{3600\text{ s}} = 2.7\times10^{40}\text{ kg}\cdot\text{m}^2\text{ s}^{-1}$$

5. (a) The direction of the velocity of the wad of gum is perpendicular to the plane of rotation of the turntable so the gum has no angular momentum about the axis of rotation of the turntable. If we define our system as gum-turntable, there is no external torque on the system about the axis of rotation of the turntable, so the angular momentum of the system about that axis does not change. The rotational inertia of the system does increase when the gum sticks to the turntable, however. Because angular momentum is constant and the rotational inertia changes, the angular speed must change.

(b) $$L_i = \frac{mr^2}{2}\omega_i = \frac{(2.150\text{ kg})(0.1600\text{ m})^2}{2}$$

$$\times\ 33.33\frac{\text{rev}}{\text{min}}\frac{2\pi\text{ rad}}{\text{rev}}\frac{1\text{ min}}{60\text{ s}} = 0.09606\text{ kg}\cdot\text{m}^2/\text{s}$$

$$I_f = I_i + m_{\text{gum}}r_{\text{gum}}^2 = \frac{(2.150\text{ kg})(0.1600\text{ m})^2}{2} +$$

$$0.01100\text{ kg }(0.1000\text{ m})^2 = 0.02763\text{ kg}\cdot\text{m}^2$$

$$L_f = I_f\omega_f = I_i\omega_i \Rightarrow \omega_f = \frac{L_i}{I_f} = \frac{0.09606\text{ kg}\cdot\text{m}^2/\text{s}}{0.02763\text{ kg}\cdot\text{m}^2}$$

$$= 3.477\frac{\text{rad}}{\text{s}}\frac{60\text{ s}}{\text{min}}\frac{1\text{ rev}}{2\pi\text{ rad}} = 33.20\text{ rpm}$$

7. Angular momentum is the ability of an object to set another object in rotational motion. If an object moving in a line strikes something constrained to a rotational axis, like an object on the end of a string, where the other end of the string is tied down somewhere, it has the ability to set that second object into rotational motion. Even though the first object is not attached to the center of rotation (the point to which the other end of the string is attached), the object that it strikes cannot just move off in linear motion. The motion of the second object is determined by a combination of the linear momentum transferred to it by the first object and the distance of the object from its rotation axis, so it ends up in rotational motion. So an object traveling in a line can still set another object in rotational motion, and thus it has angular momentum!

9. (a) $$I = 2M\left(\frac{\ell}{4}\right)^2 + \frac{M}{8}\frac{(2\ell)^2}{12} = \frac{M\ell^2}{8} + \frac{M\ell^2}{24} = \frac{M\ell^2}{6}$$

(b) $$K_{\text{rotational,i}} = \frac{M\ell^2\omega_i^2}{12}$$

(c) $$I_f = 2M\ell^2 + \frac{M}{8}\frac{(2\ell)^2}{12} = 2M\ell^2 + \frac{M\ell^2}{24} = \frac{49M\ell^2}{24}$$

$$L_f = L_i \Rightarrow \omega_f = \frac{I_i\omega_i}{I_f} = \frac{M\ell^2}{6}\frac{24}{49\,M\ell^2}\omega_1 = \frac{4}{49}\omega_i$$

(d) $$K_{\text{rotational,f}} = \frac{I_f\omega_f^2}{2} = \frac{49\,M\ell^2}{(2)(24)}\left(\frac{4}{49}\omega_i\right)^2 = \frac{M\ell^2}{147}\omega_i^2$$

$$W = K_{\text{rotational,f}} - K_{\text{rotational,i}} = \frac{M\ell^2}{147}\omega_i^2 - \frac{M\ell^2}{12}\omega_i^2 = -\frac{135\,M\ell^2}{1764}\omega_i^2$$

$$= -\frac{15M\ell^2}{196}\omega_i^2$$

11. The second sentence is corrected by replacing the term "rotational kinetic energy" with "angular momentum," and the conservation principle is the conservation of angular momentum. Because the force is toward the axis of rotation it cannot change the angular momentum, but as it is in the direction of motion of the ball it can do work. Therefore, the phrase "angular momentum will not remain constant" must be changed to "rotational kinetic energy will not remain constant" because the force pulling the ball inward along the radial component of the ball's motion is doing positive work on the ball, increasing its kinetic energy.

11-4 Newton's law of universal gravitation along with gravitational potential energy and angular momentum explains Kepler's laws for the orbits of planets and satellites

1. A is the kinetic energy because the kinetic energy is always positive. B is the total energy because it is equal in magnitude and opposite in sign to the kinetic energy, leaving graph C as the potential energy. The derivation of Equation 11-7 shows that the potential energy is twice as negative as the total energy.

3. (a)

$$T = \sqrt{\frac{4\pi^2 r^3}{GM}} = \sqrt{\frac{4\pi^2(3.00 \times 10^5\ \text{m} + 6.371 \times 10^6\ \text{m})^3}{(6.67 \times 10^{-11}\ \text{m}^3/\text{kg} \cdot \text{s}^2)(5.97 \times 10^{24}\ \text{kg})}}$$
$$= 5.43 \times 10^3\ \text{s} = 1.51\ \text{h}$$

(b) The Sun "rises" every time the shuttle emerges from Earth's shadow; once per orbit.

$$\frac{1\ \text{orbit}}{1.51\ \text{h}}\frac{24\ \text{h}}{\text{day}} = 15.9\ \text{sunrises} \approx 16\ \text{sunrises}$$

(c) $v = \frac{2\pi r}{T} = \frac{2\pi(6.671 \times 10^6\ \text{m})}{5.425 \times 10^3\ \text{s}} = 7.73\ \text{km/s}$

5. (a)

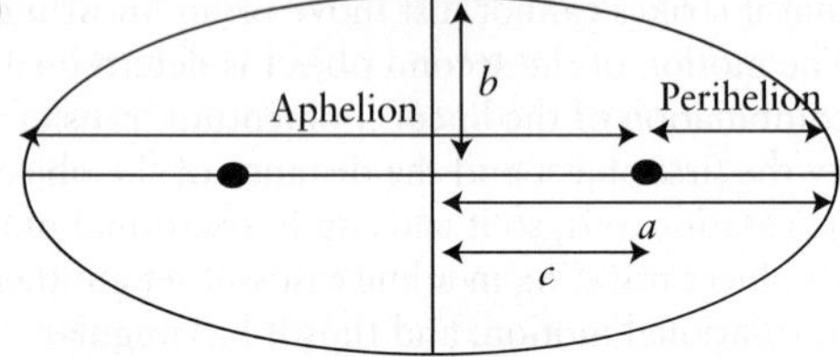

(b) In this figure a is the semimajor axis. The foci are equidistant from the intersection of the minor and major axes. So the distance from the focus on the right to the aphelion is $a + c$ and the distance to the perihelion is $a - c$. So the sum of these distances is $2a$.

(c) Let the aphelion and perihelion distances be r_a and r_p.

$$2a = r_a + r_p \Rightarrow a = \frac{r_a + r_p}{2} = \frac{1.099\ \text{au} + 0.746\ \text{au}}{2}$$
$$= 0.9225\ \text{au}\frac{1.496 \times 10^{11}\ \text{m}}{\text{au}} = 1.380 \times 10^{11}\ \text{m}$$

$$T^2 = \frac{4\pi^2}{GM}a^3 = \frac{4\pi^2(1.380 \times 10^{11}\ \text{m})^3}{(6.674 \times 10^{-11}\ \text{kg}^{-1}\ \text{m}^3\text{s}^{-2})(1.989 \times 10^{30}\ \text{kg})}$$
$$T = 2.796 \times 10^7\ \text{s} = 324\ \text{days}$$

(d) $v_p r_p = v_a r_a \Rightarrow \frac{v_p}{v_a} = \frac{r_a}{r_p}$

The angular momentum of the orbiting object is constant and equal to mrv. As r becomes smaller, v must become larger.

7. (a) A force exerted on an object in the direction of its motion over a displacement changes its kinetic energy. If the force is in the opposite direction to the motion, the change in kinetic energy is negative and the angular momentum at point P is decreased (the force is along the line of motion and so can change the angular momentum). If the spacecraft's angular momentum is decreased at a particular radius, its speed must decrease, and it is no longer going fast enough to maintain a circular orbit at this radius, so it falls slightly toward the planet as it orbits, making point P now the aphelion for the orbit. (Decrease the energy too much and the spacecraft crashes into the planet.) If instead kinetic energy increases, the spacecraft now has more kinetic energy than it can have for a circular orbit at this radius and it travels away from the planet, making point P the perihelion. (Increase the kinetic energy until total energy is no longer negative and the spacecraft escapes orbit.)

(b) The new orbits are ellipses. The semimajor axis in the backward force case is smaller, as shown in the diagram below for orbit 1. The semimajor axis in the forward force case is larger, as shown in the diagram for orbit 2.

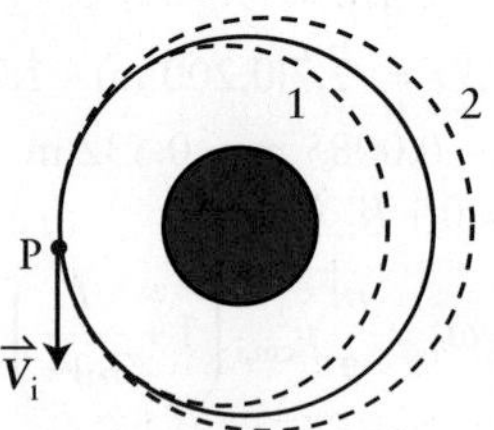

(c) The astronaut could use the thruster to exert the same force for the same time (the same impulse) on the exact opposite side of the orbit, bringing the kinetic energy to the amount needed to establish a circular orbit at the new radius achieved at that point.

9. (a) The magnitude of the gravitational force of Earth on the Moon can be used with the orbital period and radius of the Moon to calculate the mass of Earth.

$$\frac{GM_{\text{Earth}}m_{\text{Moon}}}{r^2_{\text{Earth–Moon}}} = m_{\text{Moon}}\frac{v^2_{\text{Moon}}}{r_{\text{Earth–Moon}}} = m_{\text{Moon}}\frac{4\pi^2 r_{\text{Earth–Moon}}}{T^2_{\text{Earth–Moon}}}$$

Solving:

$$M_{\text{Earth}} = \frac{4\pi^2 r^3_{\text{Earth–Moon}}}{GT^2_{\text{Earth–Moon}}}$$

But the mass of the Moon cancels, so this approach provides no method to calculate the mass of the Moon.

(b) The lunar orbiter does provide the necessary data:

$$M_{\text{Moon}} = \frac{4\pi^2 a^3_{\text{orbiter–Moon}}}{GT^2_{\text{orbiter–Moon}}} = \frac{4\pi^2(1.848 \times 10^6\ \text{m})^3}{G(7130\ \text{s})^2}$$
$$= 7.35 \times 10^{22}\ \text{kg}$$

The percent relative error in Newton's measurement using tides was about 50%.

$$\%\text{rel error} = 100\frac{|81.2 - 39.8|}{81.2} = 51.0\%$$

11. The orbital speed of the satellite is now greater than the rotational speed of Earth. The period has decreased. The evidence of this is that the velocity is inversely proportional to period. Therefore, the orbital radius has decreased. The evidence of this claim is provided by Kepler's third law. The satellite is at a lower Earth orbit.

Chapter 11 Review Problems

1. (a) The radius of the circular motion around the tetherball pole is $R \sin \phi$, where ϕ is the angle between the rope and the pole. So the angular momentum is

$$L = mvr = mvR \sin \phi$$

(b) The angular velocity is the tangential velocity divided by the radius of the circular motion in the plane of the motion:

$$L = mvR \sin\phi = m\omega R^2 \sin^2 \phi$$

(c) $L = 0.300 \text{ kg } (1.25 \text{ m})^2 (\sin 28.0°)^2 \; 41.5 \frac{\text{rev}}{\text{min}} \frac{2\pi\text{rad}}{\text{rev}}$

$$\times \frac{1 \text{ min}}{60 \text{ s}} = 0.449 \text{ kg} \cdot \text{m}^2 \text{ s}^{-1}$$

3. Because the angular momentum L of this electron and the radius of the orbital are given, we can calculate the speed of the electron using $L = m_e v_\perp r$.

$$v_\perp = \frac{L}{m_e r} = \frac{1.055 \times 10^{-34} \text{ J} \cdot \text{s}}{(9.11 \times 10^{-31} \text{ kg})(5.29 \times 10^{-11} \text{ m})} = 2.19 \times 10^6 \frac{\text{m}}{\text{s}}$$

5. The glob falls straight down so no torque is exerted by the collision of the glob and the turntable about the rotational axis. Because there is no external torque, angular momentum of the glob–turntable system is constant. The mass distribution of the rotating system changes and the rotational inertia increases. The angular speed decreases by the ratio of the initial and final rotational inertias because angular momentum is constant. While the angular speed of the disk decreases, the magnitude of its angular acceleration increases because it had no acceleration before. Because the kinetic energy is proportional to the square of the angular momentum divided by the rotational inertia, the numerator in the expression for the rotational kinetic energy remains constant but the denominator increases. Therefore, the kinetic energy of the system is decreased. The internal energy increases by the same amount that the kinetic energy decreases. This addition to the internal energy involves the cohesion of the glob to the surface of the disk.

7. (a) You can make inferences about the relative masses of the two people only if the seesaw has zero angular acceleration, in which case there is no net torque. If they are equidistant from the fulcrum and there is no net torque, then they must each have the same mass.
(b) The effect is a shift in the centers of mass of the riders, so that the rotational inertia is greater on the side where a rider leans back (and reduced by a rider who leans forward). The external torque due to the force of gravity on the two people is then greater on the side where a rider leans back (and reduced on the side where a rider leans forward). The rotation changes to favor the side with the greater torque. By alternating this procedure, the riders can cause an increase in the speed of the rotation.

9. (a) The direction of the angular momentum of a spinning wheel is perpendicular to the plane of the wheel. Looking toward the wheel, the direction of the angular momentum is out of the plane of the wheel (toward you) when the rotation is counterclockwise and into the plane of the wheel (away from you) when the rotation is clockwise.
(b) $\alpha = \dfrac{\Delta\omega}{\Delta t}$
(c) If the wheel is spinning in a counterclockwise direction the angular velocity is out of the plane, and if it is spinning in the clockwise direction the angular velocity points into the plane. If the magnitude of the angular velocity is increasing for clockwise rotations or if the angular velocity is decreasing for counterclockwise rotations, the angular acceleration points into the plane of the wheel—it is negative. If the magnitude of the angular velocity is increasing for counterclockwise rotations or if the angular velocity is decreasing for clockwise rotations, the angular acceleration points out of the plane of the wheel—it is positive. If there is an angular acceleration, then a torque must be exerted. The torque points in the same direction as the angular acceleration. The torque and the angular acceleration are proportional, and the constant of proportionality is the rotational inertia.
(d) When a bike topples over, a torque due to the gravitational force is exerted at the bike's center of mass, with the lever arm being the horizontal displacement of the bike's center of mass from its balanced position of directly over the point where the tires contact the road, so any amount of displacement in either direction will cause the bike to begin to tip.

11. (a) For a star, the only interactions it has are gravitational interactions. Gravitational forces are always exerted at a system's center of mass. Because the star is not held in place by any constraint, its rotational axis must also pass through its center of mass, so there are no torques exerted on the star, and its angular momentum is constant. The rotational inertia is reduced as the star collapses, so its angular speed must increase.
(b) Rotational inertia depends on r^2, so if r decreases by 10, I decreases by 10^2, and angular velocity increases by 100.

13. (a) If the astronaut holds the spinner straight over his head so that its rotational axis is along his long axis, then when he flips it over, his angular momentum will have to change so that the angular momentum of the system remains constant.
(b) When he flips it, its angular momentum becomes the negative of its initial value, so his angular momentum will have to be twice its initial value and the two angular momenta sum to the initial value.

$$I_{\text{fidget}}\omega_{\text{fidget}} = -I_{\text{fidget}}\omega_{\text{fidget}} + I_{\text{astro}}\omega_{\text{astro}}$$

$$2I_{\text{fidget}}\omega_{\text{fidget}} = I_{\text{astro}}\omega_{\text{astro}}$$

$$\omega_{\text{astro}} = 2(I_{\text{fidget}}\omega_{\text{fidget}})/I_{\text{astro}}; I_{\text{astro}} = MR^2/2$$

$$\omega_{\text{astro}} = 2 \times 10^{-3} \text{ kg} \cdot \text{m}^2 \times 3000 \text{ rpm}/7.5 \text{ kg} \cdot \text{m}^2 = 0.8 \text{ rpm}$$

(c) Now, $I_{\text{astro}} = MR^2/4 + M\ell^2/12 = 19.95 \text{ kg} \cdot \text{m}^2$, so $\omega_{\text{astro}} = 0.3$ rpm.

15. Because 1.00 au is also the orbital radius of Earth, the period of the other planet is just scaled by the square root of the ratio of the mass of the star to the mass of the Sun.

$$T = 365.25 \text{ d}\sqrt{\frac{M_{\text{Sun}}}{1.75M_{\text{Sun}}}} = 276 \text{ d}$$

17. (a) For a sphere of mass M and radius R and a ring with the same mass M spinning at the same angular speed ω to have the same rotational kinetic energy, we know they must have the same rotational inertia

because $K_{\text{rotational}} = 1/2I\omega^2$. So they will have different radii. Setting the rotational inertias equal, we can solve for R_{ring} in terms of R.

$$MR_{\text{ring}}^2 = \frac{2}{5}MR^2$$

$$R_{\text{ring}} = R\sqrt{\frac{2}{5}}$$

(b) The magnitude of the angular momentum of an object is equal to the product of the rotational inertia and the angular speed. Because the rotational inertia and the angular speed are equal for the two objects, their angular momenta will also be equal.

19. The Sun is an isolated object, so its angular momentum will remain the same.

(a) $L_{\text{iz}} = L_{\text{fz}}$

$$I_{\text{i}}\omega_{\text{iz}} = I_{\text{f}}\omega_{\text{fz}}$$

$$\left(\frac{2}{5}m_{\text{Si}}R_{\text{Si}}^2\right)\omega_{\text{iz}} = \left(\frac{2}{5}m_{\text{Sf}}R_{\text{Sf}}^2\right)\omega_{\text{fz}}$$

$$m_{\text{Si}}R_{\text{Si}}^2\omega_{\text{iz}} = m_{\text{Sf}}R_{\text{Sf}}^2\omega_{\text{fz}}$$

$$\omega_{\text{fz}} = \frac{R_{\text{Si}}^2}{R_{\text{Sf}}^2}\omega_{\text{iz}} = \frac{\left(6.96\times10^8\ \text{m}\right)^2}{\left(8.00\times10^6\ \text{m}\right)^2}\omega_{\text{iz}} = 7570\ \omega_{\text{iz}}$$

(b) $$\frac{K_{\text{f}}}{K_{\text{i}}} = \frac{\left(\frac{1}{2}I_{\text{f}}\omega_{\text{f}}^2\right)}{\left(\frac{1}{2}I_{\text{i}}\omega_{\text{i}}^2\right)} = \frac{(I_{\text{f}}\omega_{\text{f}})\,\omega_{\text{f}}}{(I_{\text{i}}\omega_{\text{i}})\,\omega_{\text{i}}} = \frac{\omega_{\text{f}}}{\omega_{\text{i}}} = \frac{7570\omega_{\text{i}}}{\omega_{\text{i}}} = 7570$$

The final kinetic energy is roughly 7570 times larger than the initial kinetic energy.

21. Use conservation of energy to find the speed of the ball at point B. Define potential energy of the Earth–ball system at point B as zero. The sum of the force of gravity and the normal force must provide the centripetal acceleration to keep the ball traveling in a circle of radius 60 cm over the hill.

Conservation of energy to find speed of ball:

$$U_{\text{grav,i}} + K_{\text{translational,i}} + K_{\text{rotational,i}} = U_{\text{grav,f}} + K_{\text{translational,f}} + K_{\text{rotational,f}}$$

$$Mgh_1 + \frac{1}{2}Mv_{\text{CM,i}}^2 + \frac{1}{2}I_{\text{CM}}\omega_{\text{i}}^2 = Mgh_{\text{f}} + \frac{1}{2}Mv_{\text{CM,f}}^2 + \frac{1}{2}I_{\text{CM}}\omega_{\text{f}}^2$$

$$Mgh_{\text{i}} + \frac{1}{2}Mv_{\text{CM,i}}^2 + \frac{1}{2}\left(\frac{2}{5}MR_{\text{ball}}^2\right)\left(\frac{v_{\text{CM,i}}^2}{R_{\text{ball}}^2}\right) = 0 + \frac{1}{2}Mv_{\text{CM,f}}^2 + \frac{1}{2}\left(\frac{2}{5}MR_{\text{ball}}^2\right)\left(\frac{v_{\text{CM,f}}^2}{R_{\text{ball}}^2}\right)$$

$$gh_{\text{i}} + \frac{1}{2}v_{\text{CM,i}}^2 + \frac{1}{5}v_{\text{CM,i}}^2 = \frac{1}{2}v_{\text{CM,f}}^2 + \frac{1}{5}v_{\text{CM,f}}^2$$

$$v_{\text{CM,f}}^2 = \frac{10}{7}gh_{\text{i}} + v_{\text{CM,i}}^2 = \frac{10}{7}\left(9.80\frac{\text{m}}{\text{s}^2}\right)(0.10\ \text{m}) + \left(2.0\frac{\text{m}}{\text{s}}\right)^2 = 5.4\frac{\text{m}^2}{\text{s}^2}$$

A free-body diagram of the ball and Newton's second law (where up is $+y$) to determine the magnitude of the normal force on the ball at point B reveals:

$$\sum F_{\text{ext},y} = F_{\text{n}} - F_{\text{g}} = M\left(-\frac{v_{\text{CM,f}}^2}{R_{\text{ball}}}\right)$$

$$F_{\text{n}} = F_{\text{g}} - M\left(\frac{v_{\text{CM,f}}^2}{R_{\text{ball}}}\right) = Mg - M\left(\frac{v_{\text{CM,f}}^2}{R_{\text{ball}}}\right) = M\left(g - \frac{v_{\text{CM,f}}^2}{R_{\text{ball}}}\right)$$

$$= (0.160\ \text{kg})\left(\left(9.8\frac{\text{m}}{\text{s}^2}\right) - \frac{\left(5.4\frac{\text{m}^2}{\text{s}^2}\right)}{(0.60\ \text{m})}\right) = 1.3\times10^{-1}\ \text{N}$$

23. (a) The friction force on the cylinder keeps it from slipping, and the torque from the applied force makes it rotate. The friction force opposes the rotation of the cylinder against the surface.

Apply Newton's second law of motion to translation and rotation. Let $+x$ be to the right, and counterclockwise be positive for rotation.

$$F_{\text{T}} + F_{\text{S}} = ma_{\text{CM},x} = ma_{\text{CM}}$$

$$F_{\text{S}}R - F_{\text{T}}R = I_{\text{CM}}\alpha_{\text{CM}}$$

$$\alpha_{\text{CM}} = -\frac{a_{\text{CM}}}{R}$$

$$F_{\text{S}}R - F_{\text{T}}R = -(mR^2/2)a_{\text{CM}}/R$$

$$-F_{\text{S}} + F_{\text{T}} = ma_{\text{CM}}/2$$

Add this to the equation for forces to eliminate the unknown F_{S}:

$$2F_{\text{T}} = ma_{\text{CM}} + ma_{\text{CM}}/2 = 3ma_{\text{CM}}/2$$

$$a_{\text{CM}} = 4F_{\text{T}}/3m = (4/3)(3.00\ \text{N}/7.00\ \text{kg}) = 0.571\ \text{N/kg}$$

(b) The direction of the angular motion of the cylinder is clockwise. The magnitude of the angular acceleration is $\alpha_{\text{CM}} = a_{\text{CM}}/R = (0.571\ \text{N/kg})/(0.450\ \text{m}) = 1.27\ \text{rad/s}^2$.

(c) $F_{\text{S}} = ma_{\text{CM}} - F_{\text{T}} = (7.00\ \text{kg})(0.571\ \text{N/kg}) - 3.00\ \text{N} = 1.00\ \text{N}$ in the positive x direction.

25. The initial rotational energy of the disk is $\frac{1}{2}I\omega_{\text{lower}}^2 = \frac{1}{2}\left(\frac{1}{2}(2M)(2R)^2\right)\omega_{\text{lower}}^2 = 2MR^2\omega_{\text{lower}}^2$, where ω_{lower} is the initial angular speed of the lower disk. To find the final kinetic energy of the system of two disks, we need to find the common final angular speed using conservation of angular momentum.

$$I_{\text{lower}}\omega_{\text{lower}} = I_{\text{both}}\omega_{\text{both}}$$

$$\frac{1}{2}(2M)(2R)^2\omega_{\text{lower}} = \left[\frac{1}{2}(2M)(2R)^2 + \frac{1}{2}MR^2\right]\omega_{\text{both}}$$

$$4MR^2\omega_{\text{lower}} = \frac{9}{2}MR^2\omega_{\text{both}}$$

$$\omega_{\text{both}} = \frac{8}{9}\omega_{\text{lower}}$$

Substituting, the final kinetic energy is

$\frac{1}{2}I_{\text{both}}\omega_{\text{both}}^2 = \frac{1}{2}\left(\frac{9}{2}MR^2\right)\left(\frac{8}{9}\omega_{\text{lower}}\right)^2 = \frac{64}{36}MR^2\omega_{\text{lower}}^2 = \frac{16}{9}MR^2\omega_{\text{lower}}^2$. Thus, the fraction lost is

$$1 - \frac{K_{\text{f}}}{K_{\text{i}}} = 1 - \frac{\frac{16}{9}MR^2\omega_{\text{lower}}^2}{2MR^2\omega_{\text{lower}}^2} = 1 - \frac{8}{9} = \frac{1}{9}.$$

27. Assume that the acrobats are objects and that each one is the same distance R_{acrobat} from the rotation axis. The merry-go-round and acrobats make up an isolated system, which means angular momentum is constant. Although we are not given the exact radius of the merry-go-round, we can find R_{acrobat} in terms of R_{MGR} to determine how the acrobats should redistribute themselves such that the final angular velocity is $\omega_{iz}/2$.

$$L_{\text{MGR,i}} + L_{\text{acrobat,i}} = L_{\text{MGR,f}} + L_{\text{acrobat,f}}$$
$$I_{\text{MGR}}\omega_{iz} = (I_{\text{MGR}} + I_{\text{acrobat,f}})\omega_{fz}$$
$$I_{\text{MGR}}\omega_{iz} = \left(I_{\text{MGR}} + I_{\text{acrobat,f}}\right)\frac{\omega_{iz}}{2}$$
$$I_{\text{MGR}} = I_{\text{acrobat,f}}$$
$$\frac{1}{2}M_{\text{MGR}}R_{\text{MGR}}^2 = 10M_{\text{acrobat}}R_{\text{acrobat}}^2$$
$$R_{\text{acrobat}} = \sqrt{\frac{M_{\text{MGR}}R_{\text{MGR}}^2}{20M_{\text{acobat}}}} = R_{\text{MGR}}\sqrt{\frac{(1.00 \times 10^3 \text{ kg})}{20(50.0 \text{ kg})}} = R_{\text{MGR}}$$

All 10 acrobats should move to the edge of the merry-go-round.

29. (a) $W = F_{\text{applied}}R_{\text{axle}}\theta_{\text{axle}} = \tau_{\text{applied}}\theta_{\text{axle}}$. Assuming the torque is constant, just the angle is changing, and the change in angle is angular velocity, so

$$P_{\text{input}} = W/\Delta t = \tau_{\text{applied}}\omega_{\text{axle}}$$

(b) If the wheel and axle rotate together, then the angular rotation of the wheel and the axle must be the same. If the angular speed is constant, the total torque on the wheel must be zero. This means $\tau_{\text{load}} = -\tau_{\text{applied}}$ so $P_{\text{ouput}} = \tau_{\text{load}}\omega_{\text{wheel}} = -\tau_{\text{applied}}\omega_{\text{wheel}}(\omega_{\text{axle}}/\omega_{\text{axle}}) = -P_{\text{input}}(\omega_{\text{wheel}}/\omega_{\text{axle}})$. So when the angular speed of the wheel is equal to the angular speed of the axle, as it must be for a rigid body, $P_{\text{input}} = -P_{\text{output}}$ and $\varepsilon = 1$.

Chapter 11 AP® Practice Problems

1. (B) $L = pr_\perp = mvr_\perp$; because the mass, the speed, and the perpendicular distance to the origin are constant, the angular momentum is constant.

3. (B) Because the ring is not pivoted, it will rotate about its center of mass if there is a torque:

$$\Delta L = \tau \Delta t$$
$$\Delta L = L - 0 = 0.40 \text{ N}(0.050 \text{ m})(2.0 \text{ s})$$
$$= 0.040 \text{ kg} \cdot \text{m}^2/\text{s}$$

5. (C) There are no external torques, so $L_f = L_i$. $L = I\omega$, and $K_{\text{rotational}} = \dfrac{1}{2}I\omega^2$:

$$L_f = L_i \Rightarrow 2I\omega = I\omega_0 \Rightarrow \omega = \frac{\omega_0}{2} \Rightarrow K_{\text{rotational}} = \frac{1}{2}2I\left(\frac{\omega_0}{2}\right)^2 = \frac{I\omega_0^2}{4}$$

7. (A) Use conservation of energy to figure out the height to which each disk will rise, including Earth in the system:

$$K_{\text{translational}} + K_{\text{rotational}} = U_{\text{grav}}$$
$$\frac{1}{2}mv_0^2 + \frac{1}{2}I\omega^2 = mgh$$
$$v_0 = \omega r; \frac{1}{2}mv_0^2 + \frac{1}{2}\left(\frac{1}{2}mr^2\right)\left(\frac{v_0}{r}\right)^2 = mgh$$

$\dfrac{3}{4}mv_0^2 = mgh$; $h = \dfrac{3v_0^2}{4g}$. Thus, the height to which each disk rises is independent of its mass and radius, and all three disks will rise to the same maximum height.

Chapter 12

12-1 We live in a world of oscillations

1. You can control your breathing rate by voluntary muscular contractions. Contractions of the heart are thought to be controlled by the involuntary (autonomic) nervous system. You may find that the period of the heartbeat becomes a multiple of the period of the breathing rate. A factor of 5 is commonly observed. You may find that your heart rate slows while you pay attention to your breathing rate. Possible questions are
   - Does the awareness of your heartbeat control your breath rate and if so is there a voluntary control of your heartbeat?
   - Does the awareness of your breath rate control your heart rate and if so is there a path of communication between the breath rate and the heartbeat?
   - If heart rate is controlled only by the autonomic nervous system, is there signaling between thought and the autonomic system, such as the release of signaling molecules?

3. (a) The equilibrium is the "zero" point of an oscillation. In this case it would be the average ocean temperature as measured over a period of years.
   (b) The mechanism that causes a periodic event, such as day following night, can sometimes be found by detecting other processes with the same period. In this case, Earth rotates once on its axis each day, turning toward or away from the Sun. It is more difficult to find the causes of events that are seemingly random because they may involve several causes, and although each mechanism may be periodic, when they are combined they appear random.
   (c) What are the possible cyclic processes that could cause earthquakes and how could the periods of these cycles interact to produce the time period observed in our data?

12-2 Oscillations are caused by the interplay between a restoring force and inertia

1. When the inertia of the system causes it to pass through the equilibrium point, the direction of the restoring force reverses, causing a sequence of oscillations. In an ideal case, a slowly moving system would just mean a very slow or very small oscillation. If the motion is slow enough, in most real-world systems there is some resistance to motion and the inertia of the system will not carry it through the equilibrium point. For any system to oscillate, the object must have enough kinetic energy when it passes through the equilibrium position to overshoot equilibrium and thus oscillate.

3. In both sequences, the restoring force is the gravitational force. Although in both sequences the restoring force is the same, oscillations of the bead occur only in the lower sequence. In the upper sequence the tube is rotated very slowly. In the upper sequence the bead gradually moves in the tube while remaining at approximately the same height above the table and returns to rest at the equilibrium position. The bead does not overshoot the equilibrium position at the bottom of the U-tube and there are no oscillations of the bead. In the lower sequence the tube is rotated quickly and the bead is quickly elevated to a height within an arm of the tube. When it moves back down the tube, it overshoots the equilibrium position, causing an oscillation around the equilibrium position, returning repeatedly to nearly the initial height. Alternatively, we will see we could describe this by energy conservation. In the first case, where the tube is rotated slowly, the bead never changes its separation with Earth significantly over time, and so does not have enough kinetic energy to overshoot equilibrium, whereas in the second case it does.
5. (a) $f = \dfrac{1250}{T} = \dfrac{1250}{20.0\text{ min}}\dfrac{1\text{ min}}{60\text{ s}} = 1.04\text{ Hz}$

   (b) $f = \dfrac{1}{T} \Rightarrow T = \dfrac{1}{f} = \dfrac{1}{1.04\text{ s}^{-1}} = 0.960\text{ s}$

12-3 An object changes length when under tensile or compressive stress; Hooke's law is a special case

1. (C) < (A) = (D) < (B) The easier a rod is to stretch, the smaller its spring constant k. We know that stiffness depends directly on the area and inversely on the length for a given material stiffness. All four rods are made of the same material. So the rod with the smallest value of k is the one with the smallest ratio A/L_0—that is, the one with the smallest cross-sectional area and the longest length. The values of this ratio for each of the four rods are
   (A) $(2.00\text{ mm}^2)/(5.00\text{ cm}) = 0.400\text{ mm}^2/\text{cm}$
   (B) $(2.00\text{ mm}^2)/(2.50\text{ cm}) = 0.800\text{ mm}^2/\text{cm}$
   (C) $(1.00\text{ mm}^2)/(5.00\text{ cm}) = 0.200\text{ mm}^2/\text{cm}$
   (D) $(1.00\text{ mm}^2)/(2.50\text{ cm}) = 0.400\text{ mm}^2/\text{cm}$.
3. Among questions that might be fruitful:
   - What are the mechanisms by which skin cells exert a restoring force?
   - Do the contents of skin cells, such as the fraction of solids or dissolved ions, change over time?
   - Do mechanisms, such as cell surface transport systems that move water across the cell boundary, become less effective with time?
   - Do skin cells reproduce less frequently with increasing age?
5. These data show that as the angle of the force exerted on the bone changes from the transverse orientation to the longitudinal direction the stiffness generally decreases. When a force with a magnitude F is exerted on a bone, a larger stiffness results in a smaller displacement. A break occurs when there is a displacement that is larger than the material can respond to elastically. So because the bone is stiffer in the direction perpendicular to the long axis of the bone than parallel to the long axis (so it will displace less for the same force in the perpendicular direction), a larger force is needed to break a bone than is needed to split a bone.

12-4 The simplest form of oscillation occurs when the restoring force obeys Hooke's law

1. (a) The units of frequency f are cycles per second or simply per second, s^{-1}, because it can be assumed that a cycle is being described. The hertz (Hz) is equivalent to s^{-1}.
   (b) The angular frequency is 2π times the frequency, which can be written as $\omega = 2\pi f$.
   (c) The period is the time elapsed between periodically recurring events or the time of a cycle. So because frequency is the number of cycles per second, the period is the reciprocal of the frequency, which can be written as $T = 1/f$.
   (d) The angular frequency is 2π multiplied by the frequency, and the frequency is the reciprocal of the period. So the period is 2π divided by the angular frequency, which can be written as $T = 2\pi/\omega$.
3. (a) $F_{s,x} = -kx$; $F_x = ma_x$ Using Equations 12-14 and 12-18 to find another relationship for a,

   $$a_x = -\omega^2 A\cos(\omega t) = -\omega^2 x \Rightarrow \omega^2 = \frac{k}{m}$$

   (b) The period is used to calculate the angular frequency:

   $$T = \frac{2\pi}{\omega} = \frac{16\text{ s}}{14\text{ cycles}} \Rightarrow \omega = \frac{28\pi}{16\text{ s}} = 5.5\text{ s}^{-1}$$

   The angular frequency is used to calculate the mass:

   $$m = \frac{k}{\omega^2} = \frac{200\text{ N/m}}{5.5^2\text{ s}^{-2}} = 6.6\text{ Ns}^2/\text{m} = 6.6\text{ kg}$$

5. (a) The object is initially at $x = 0$ and then its position increases. So a sine function without a phase angle can be used to express its position. The amplitude of the motion is 4 cm and the period is 0.4 s. So by comparison with Equation 12-14, where $\omega = 2\pi/T$

   $$x = (4\text{ cm})\sin\left(\frac{2\pi}{0.4\text{ s}}t\right)$$

   (b) $T = 0.4\text{ s}\quad \omega = \dfrac{2\pi}{T} = \dfrac{2\pi}{0.4\text{ s}} = 15.7\text{ rad/s} = 2\times10^1\text{ rad/s}$

   $$f = \frac{1}{T} = \frac{1}{0.4\text{ s}} = 3\text{ s}^{-1}$$

   (c) Because the mass and period are known and the period can be expressed in terms of m and k,

   $$T = 2\pi\sqrt{\frac{m}{k}} \Rightarrow k = \frac{4\pi^2 m}{T^2} = \frac{4\pi^2(1.2\text{ kg})}{0.4^2\text{ s}^2} = 3\times10^2\text{ N/m}$$

   (d) From our comparison to circular motion, we know the maximum speed is $A\omega$. From Equation 12-16,

   $$v_{\max} = A\omega = 4\text{ cm}\frac{15.7\text{ rad}}{\text{s}} = 63\text{ cm/s} \approx 60\text{ cm/s} = 0.6\text{ m/s}$$

Alternatively, the speed can be estimated from the slope of the graph at times when the position is zero. Looking at a change of 0.02 m, the time to make that change is somewhere between 0.03 and 0.04 second (a little smaller than the smallest division but more than half), so the speed is somewhere between 0.5 m/s and 0.7 m/s.

(e) From our comparison to circular motion, we know the maximum magnitude of the acceleration is $A\omega^2$. From Equation 12-17,

$$a_{\max} = A\omega^2 = 4\text{ cm}\frac{15.7^2\text{ rad}^2}{\text{s}^2} = 986\text{ cm/s}^2$$

$$= 1\times10^3\text{ cm/s}^2 = 1\times10^1\text{ m/s}^2$$

7. (a) Using

$$\omega = 2\pi/T$$

Mass (g)	Period (ms)	Angular frequency (s^{-1})
10.0	1.5	4200
30.0	2.6	2400
60.0	3.7	1700
90.0	4.5	1400

(b) Plot the square of the angular frequency versus $1/m$. The graph should be linear with a slope of k.

Mass (kg)	Angular frequency squared (s^{-2}) × 10^6	1/mass (kg^{-1})
0.010	17.5	100
0.030	5.84	33.3
0.060	2.88	16.7
0.090	1.95	11.1

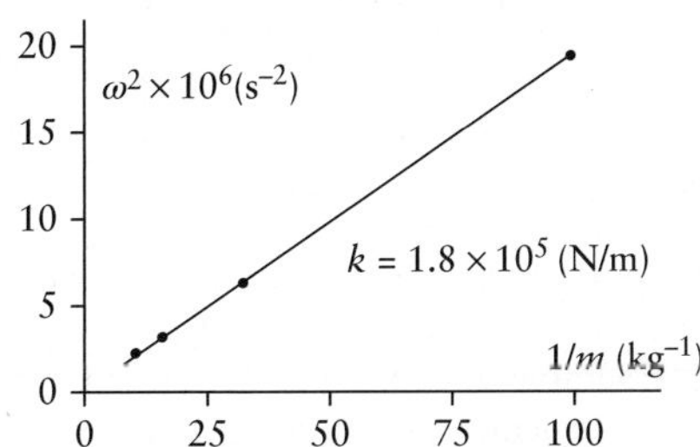

9. The angular frequency of the oscillation depends on the ratio of the spring constant and the mass and is inversely proportional to the period.

$$\omega^2 = \frac{k}{m} = \left(\frac{2\pi}{T}\right)^2 \Rightarrow T = 2\pi\sqrt{\frac{m}{k}} = 2\pi\sqrt{\frac{0.200\text{ kg}}{55.0\text{ N/m}}} = 0.379\text{ s}$$

12-5 Mechanical energy is constant in simple harmonic motion

1. (A) The kinetic energy is proportional to the product of mass and velocity squared. The velocity is proportional to $A\omega$, where A is the amplitude of the oscillation and ω is the angular frequency. If the mass and spring constant do not change, the angular frequency does not change. So increasing the amplitude increases the energy of the system. Therefore, (A) is correct while (B) and (C) are incorrect.

In a simple harmonic motion the mechanical energy is conserved. The potential energy and the velocity vary with time but the sum of the kinetic and potential energies is constant. (D) is incorrect.

3. (a) At the instant described the total mechanical energy is 4.0 J.

$$E = U(t) + K(t) = 2.4\text{ J} + 1.6\text{ J} = 4.0\text{ J}$$

At maximum amplitude, the velocity is zero and the potential energy is 4.0 J. So

$$E = U_{\max} = 4.0\text{ J} = \frac{A^2k}{2} = 0.200^2\text{ m}^2\frac{k}{2}$$

$$k = 2.0\times10^2\text{ J/m}^2 = 2.0\times10^2\text{ N/m}$$

(b) The maximum displacement is the amplitude, and the force exerted on the object at that instant is in the negative direction and its magnitude is the product of the spring constant and the displacement: 2.0×10^2 N/m (0.200 m) = 40 N.

5. (a) The sum of the forces at the new equilibrium must be zero (spring force up, gravity down), so $k\Delta x_{eq} = mg$, and then solving, $mg/k = \Delta x_{eq}$.

(b) Defining the reference for potential energy as $U_{total}(y=0) = 0$ means that the gravitational potential energy must be the negative of the spring potential energy at this point:

$$U_{total} = 0 = U_{grav} + 1/2k\Delta x_{eq}{}^2 \Rightarrow U_{grav} = -1/2k\Delta x_{eq}{}^2$$
$$= -1/2k(mg/k)^2$$

(c) For the highest point of the oscillation, $y = A$ and the extension of the spring from equilibrium is

$A - \dfrac{mg}{k}$, so

$U_s(y=A) = 1/2k(A-(mg/k))^2 = 1/2kA^2 - mgA + 1/2k(mg/k)^2$, and

$U_{grav}(y=A) = U_{grav}(y=0) + mgA = -1/2k(mg/k)^2 + mgA$

Adding both types of potential energy we get:
$U_{total}(y=A) = 1/2kA^2$
For the lowest point of the oscillation, $y = -A$ and the extension of the spring from equilibrium is

$-A - \dfrac{mg}{k}$

Following the same steps as for the highest point,
$U_s(y=-A) = 1/2k(A+(mg/k))^2$ and
$U_{grav}(y=-A) = -1/2k(mg/k)^2 - mgA$, so
$U_{total}(y=-A) = 1/2kA^2$

7. When $U = K$, $K = U = 1/2\ E_{total} = 1/2(kA^2/2)$.
Potential energy depends on x^2, and x depends on $A\cos(2\pi ft)$ (Equation 12-8). Kinetic energy depends on velocity squared, which has exactly the same form, except it is a sine if displacement is a cosine, as we can see from Equation 12-9. So we need the $\sin^2$ or $\cos^2$ term to equal ½. This happens at $\pi/4$, $3\pi/4$, and so on (any odd number multiple of $\pi/4$).
Because $f = 1/T$, and the motion goes through 2π in each full period, $2\pi f$ = odd factors of $\pi/4$ happens at odd factors of 1/8 T so $U = K$ when $t = T/8$, $3T/8$, $5T/8$, and $7T/8$.

12-6 The motion of a pendulum is approximately simple harmonic

1. (a) The period increases when the length of the string increases, in proportion to the square root of the string length.

$$T = 2\pi\sqrt{\frac{L}{g}}$$

(b) It decreases. The angular frequency is proportional to the reciprocal of the period. It decreases in proportion to the reciprocal of the square root of the string length:

$$\omega = \frac{2\pi}{T} = \sqrt{\frac{g}{L}}$$

(c) The linear speed increases. The linear speed is a maximum when the object is at the lowest point in its arc. As the pendulum bob falls from the highest point at either end of its arc, potential energy is converted to kinetic energy. If the length of the string is increased, the distance that the pendulum bob falls increases and the linear speed increases.

(d) The linear speed v of the pendulum can be expressed using the law of energy conservation (and noticing that U_{grav} is defined as 0 at the top of the string):

$$U_f = -mgL \quad U_i = -mgL\cos\theta_0$$
$$\Delta U = U_f - U_i = -mgL(1-\cos\theta_0)$$

and

$$K_f = -\Delta U = mgL(1-\cos\theta_0) = \frac{mv_f^2}{2} = \frac{mL^2\omega_f^2}{2}$$

The replacement of v with $L\omega$ uses the relationship between the linear and angular speeds defined in Chapter 10, Equation 10-1. So

$$\omega_f = \sqrt{\frac{2g(1-\cos\theta_0)}{L}} \quad \omega_f' = \sqrt{\frac{2g(1-\cos\theta_0)}{L'}}$$

(e) The definition of angular frequency in part (b) and the expression for the angular speed in part (d) share the same symbol. However, the angular frequency is a constant of the motion; it is the number of cycles completed each second. The angular speed of a pendulum changes continuously. In uniform circular motion angular speed does not change, and so is essentially identical to angular frequency.

3. $T = 2\pi\sqrt{\frac{L}{g}} = 2\pi\sqrt{\frac{1.24 \text{ m}}{9.80 \text{ m/s}^2}} = 2.24 \text{ s} \quad \omega = \frac{2\pi}{T} = \sqrt{\frac{g}{L}}$
$= 2.81$ rad/s

5. (a) The origin for the y axis might be taken at the point of attachment of the string to the support (shown in the following figure, on the left) or at the location of the pendulum bob at the lowest point in its arc (shown on the right). In either case the change in potential energy is the same.

(b) $\Delta U_{conservation} + \Delta K_{measured} = 0 \Rightarrow \Delta U_{conservation} = -\frac{mv_f^2}{2}$

(c) By comparing the difference between the final and initial potential energies calculated from the change in heights and the difference calculated from measurements of the final speed the principle of energy conservation is tested.

(d)

Approximate angle from vertical	Period (s)	Elapsed time at photogate (s)	Speed of object at $\theta = 0°$ (m/s)	Kinetic energy of object at $\theta = 0°$ (J)
5°	2.694	0.0273	0.366	0.003
10°	2.702	0.0136	0.735	0.011
15°	2.715	0.0090	1.111	0.025
20°	2.733	0.0069	1.449	0.042
25°	2.757	0.0055	1.818	0.066
30°	2.787	0.0046	2.174	0.095

(e)

Approximate angle from vertical	ΔU(J) from part (a)	$\Delta U_{conservation}$ (J) from part (b)
5°	−0.0027	−0.0027
10°	−0.011	−0.011
15°	−0.024	−0.025
20°	−0.043	−0.042
25°	−0.066	−0.066
30°	−0.095	−0.095

(f) There is no difference between the two methods of calculating the potential energy of the system attributable to consistent changes in initial displacement. Slight differences are from rounding or measurement errors. The total energy is constant, independent of the initial displacement, and there is no violation of the energy conservation.

(g) The graph (shown on next page) of the percent relative error of the function of initial angular displacement shows that the period does depend on the initial angular displacement. Because the majority of these angles were not small enough for the motion to be modeled as simple harmonic motion, we cannot use this result to draw any conclusions about whether the configuration of the pendulum was truly simple. The percent relative error was small at 5°, which is a safe angle for the small angle approximation.

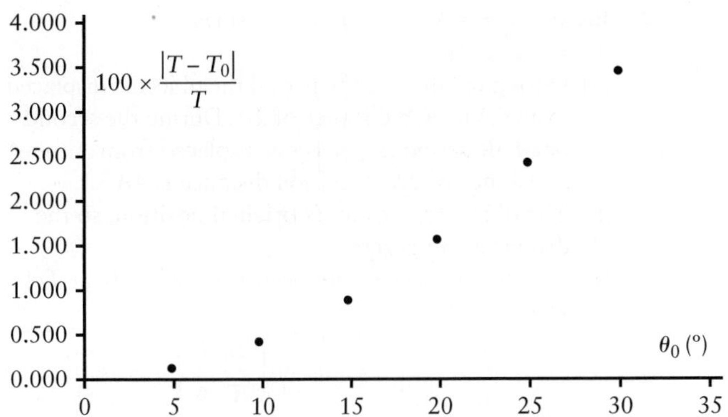

7. For parts (a) and (c) the initial gravitational potential energy is the same, U_i. The total energy of the oscillator is the initial potential energy, since it is released from rest. Let $c = x/L$

$$U_i = mg(L - L\cos\theta_0) \quad U_f = mg(L - x)(1 - \cos\theta).$$

So equating the initial and final potential energies by energy conservation, substituting $c = x/L$, and dividing out the mg,

$mg(L - x)(1 - \cos\theta) = mgL(1 - \cos\theta_0) \Rightarrow (1 - c)\cos\theta$ $= \cos\theta_0 - c$; dividing both sides by $(1 - c)$ gives the answer.

(b) Energy conservation requires that, when dropped from a greater initial displacement, the bob will have a greater final displacement.

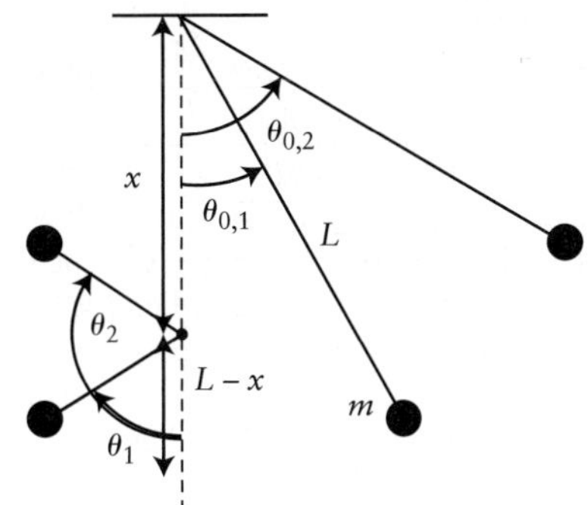

(d) $\cos 180° = -1 = \frac{\cos\theta_c - c}{(1 - c)} \Rightarrow \cos\theta_c - c = c - 1 \Rightarrow \theta_c$ $= \cos^{-1}(2c - 1)$

For initial angular displacements greater than θ_c, the bob will rotate around the peg while the string is wrapping around the peg. Right at the critical angle, the string will go slack as the bob approaches vertical, because the pendulum will not be swinging fast enough for circular motion.

(e) The pendulum swings through one-half of a period when it has length L and then swings through one-half of a different period when it has length $L - x$.

$$T = T_{\text{before peg}} + T_{\text{after peg}} = 2\pi\frac{1}{2}\sqrt{\frac{L}{g}} + 2\pi\frac{1}{2}\sqrt{\frac{L - x}{g}}$$

(f) The independent variable is x and the dependent variable is the total period. The variable x is measured with a meter stick and T is measured with a stopwatch. Measurement of the elapsed time over multiple cycles will improve precision. It is important the peg be located along the dashed line. A measurement of uncertainty could be made with multiple trials at the same value of x. A discussion of the possible dependence of the period on the initial angular displacement should be included. Although the initial displacement should not affect the energy, it will affect the period.

9. On the Moon, the restoring force on the pendulum would be smaller, because the acceleration due to gravity is smaller. Both the pendulum on the Moon and the pendulum on Earth are on the same length string, so they will have the same amplitude. The pendulum on Earth has a greater restoring force, and so will have a greater acceleration returning it back to equilibrium; it therefore takes less time to return to equilibrium and so will take less time to complete a cycle of motion. Because a period is a measure of how long a cycle of motion takes, this means the pendulum on Earth has a smaller period than the pendulum on the Moon.

12-7 A physical pendulum has its mass distributed over its volume

1. (B)

$$T = 2\pi\sqrt{\frac{I}{mgh}} = 2\pi\sqrt{\frac{mL^2}{3}\frac{1}{mg(L/2)}} = 2\pi\sqrt{\frac{2L}{3g}}$$

3. The center of mass of the system is located at the center of the sphere, so the distance $d = L + R = 1.5\text{ m} + 0.5\text{ m} = 2.0\text{ m}$.

$$I = \frac{2mR^2}{5} + md^2 = \frac{2(1.0\text{ kg})(0.50\text{ m})^2}{5} + 1.0\text{ kg}(2.0\text{ m})^2$$

$$= 4.1\text{ kg}\cdot\text{m}^2$$

$$T = 2\pi\sqrt{\frac{I}{mgh}} = 2\pi\sqrt{\frac{4.1\text{ kg}\cdot\text{m}^2}{1.0\text{ kg }(9.80\text{ m/s}^2)(2.0\text{ m})}} = 2.9\text{ s}$$

5. (a) The center of mass lies somewhere on the x axis. A second hole is punched at a point off the x axis and the cardboard is hung at this new point. The center of mass will also lie on a vertical line that contains this point. So the center of mass lies at the intersection of the two lines.

(b) $T_n = 2\pi\sqrt{\frac{I_n}{mgx_n}}$

Squaring both sides of this equation and solving the result for the rotational inertia,

$$I_n = \frac{mgx_nT_n^2}{4\pi^2}$$

(c) Using the parallel-axis theorem,

$$\frac{mgx_nT_n^2}{4\pi^2} = I_n = mx_n^2 + I_{CM} \Rightarrow \frac{x_ngT_n^2}{4\pi^2} = x_n^2 + \frac{I_{CM}}{m}$$

Define a new variable y:

$$y = \frac{x_ngT_n^2}{4\pi^2} = x_n^2 + \frac{I_{CM}}{m}$$

So a plot of y versus x^2 should have a slope of 1 and an intercept equal to I_{CM}/m. When the mass of the piece of cardboard is known using a pan balance, the rotational inertia about the center of mass can be determined.

(d) The best fit to these data is $y = 1.06x^2 + 0.0048\ \text{m}^2$. So $I_{CM}/m = 0.0048\ \text{m}^2$ and $I_{CM} = 1.2 \times 10^{-4}\ \text{kg} \cdot \text{m}^2$.

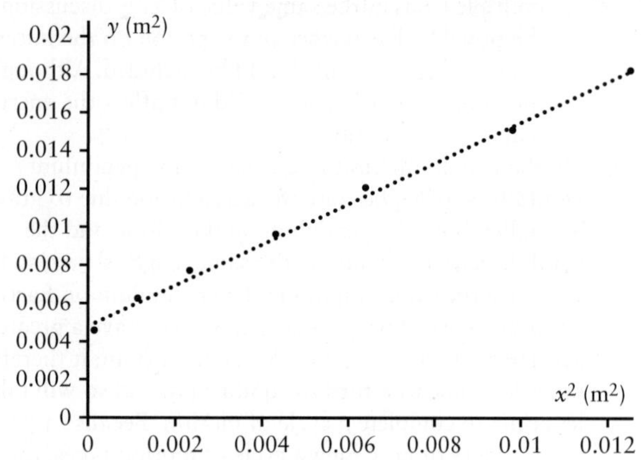

Chapter 12 Review Problems

1. (a) $T = \frac{1}{f} = \frac{1}{15\ \text{s}^{-1}} = 0.067\ \text{s}$

(b) Let n be the number of oscillations:

$$n = 120\ \text{s} \frac{1}{0.067\ \text{s}} = 1800$$

3. (a) Substituting the given values into Equation 12-16:

$$v_x = -0.15\ \text{m}\pi \sin(\pi t + \pi/3)$$
$$v_x(1.0\ \text{s}) = -0.15\ \text{m}\pi\text{s}^{-1} \sin(4\pi/3) = 0.41\ \text{m/s}$$

(b) Substituting the given values into Equation 12-17:

$$a_x = -0.15\ \text{m}\pi^2 \cos(\pi t + \pi/3)$$
$$a_x(2.0\ \text{s}) = -0.15\ \text{m}\pi^2 \cos(2\pi + \pi/3) = -0.74\ \text{m/s}^2$$

5. The choice of cosine or sine is based on the initial conditions of the motion. Because cos (0) = 1 and the position is equal to A when $t = 0$, the cosine is used:

$$x(t) = A\cos(2\pi t/T)$$

Because sin (0) = 0 and the position is equal to 0 when $t = 0$, the sine is used. The oscillation has amplitude A so

$$x(t) = \pm A \sin(2\pi t/T)$$

This is incomplete. When the motion starts at A, then displacement has to be decreasing—there is only one way the object can be going. When the motion starts at $x = 0$, the object could be moving in either direction, so we don't have enough information. We need a plus or minus sign in front of the A to describe the direction.

7. (C) $T = 2\pi\sqrt{\frac{m}{k}} \Rightarrow f = \frac{1}{T} = \frac{1}{2\pi}\sqrt{\frac{k}{m}}$

When $m' = m/4$

$$f' = \frac{1}{2\pi}\sqrt{\frac{k}{m'}} = \frac{1}{2\pi}\sqrt{\frac{4k}{M}} = 2f$$

9. Because $x = -A\cos\omega t$, $v_x = A\omega \sin\omega t$ and $a_x = A\omega^2 \cos\omega t$.

(a) During one-half of the period the object is displaced from $-A$ to A, a distance of $2A$. During the second one-half period the object is displaced from A to $-A$, a distance of $2A$. The total distance is $4A$.

(b) The object returns to its original position, so the displacement is zero.

(c) The maximum velocity occurs when $x = 0$, at $T/4$, where

$$v_x(t = T/4) = A\omega \sin\left(\frac{2\pi}{T}\frac{T}{4}\right) = A\sqrt{k/m}$$

Using energy would find the speed, but would not tell you the direction.

The maximum magnitude of the velocity (maximum speed) occurs when $x = 0$ at $T/4$ and $3T/4$, because at $3T/4$,

$$v_x(t = 3T/4) = A\omega \sin\left(\frac{2\pi}{T}\frac{3T}{4}\right) = -A\omega$$

(d) The average velocity is the displacement divided by the period and is equal to zero.

(e) The average speed is the distance traveled divided by the period, which is $4A/T$.

(f) The maximum magnitude of the acceleration occurs when $t = 0$ and $t = T/2$. The maximum in the vector acceleration occurs when $t = 0$, where $a_x = +A\omega^2$. When $t = T/2$, $a_x = -A\omega^2$. The maximum force is when the acceleration is maximum, and is equal to $ma_x = Ak$.

11. $T = 2\pi\sqrt{\frac{1.00\ \text{kg}}{16.0\ \text{N/m}}} = \frac{\pi}{2}\ \text{s}$; $\omega = \frac{2\pi}{T} = 4.00\ \text{s}^{-1}$

$x = 0.200\ \text{m} \cos(4.00\ \text{s}^{-1}t)$; $v_x = -0.800\ \text{m/s} \sin(4.00\ \text{s}^{-1}t)$

(a) x is at the equilibrium position $x = 0$ when $\cos(4.00\ \text{s}^{-1}t) = 0$. This happens at the time when $4t = \pi/2, 3\pi/2, 5\pi/2 \ldots$ or when $t = (2n+1)\pi/8$ s. In terms of the period these are $t = T/4, 3T/4, 5T/4, 7T/4, \ldots (2n+1)T/4$.

(b) x is first equal to -0.100 m when

$$-0.100\ \text{m} = 0.200\ \text{m} \cos(4.00\ \text{s}^{-1}t) \Rightarrow \cos(4.00\ \text{s}^{-1}t) = -0.500 \Rightarrow$$
$$t = \frac{1}{4.00}\cos^{-1}(-0.500)\ \text{s} = 0.5236\ \text{s}$$

The oscillator passes $x = -0.100$ m for the first time when $t = 0.5236\ \text{s} = \pi/6$ s, where $4t = 120°$ or $2\pi/3$ rad. Then the oscillator again passes $x = -0.100$ m when $4t = 240°$ or $4\pi/3$ rad and $t = \pi/3$ s. In terms of the period, $T = \pi/2$ s, these times are $t = T/3$ and $2T/3$.

(c) The oscillator passes $x = 0.100$ m for the first time $\cos(4t) = 0.5$, so $4t = 60° = \pi/3$, and $t = \pi/12$ s. The $\cos(4t) = 0.5$ again at $4t = 300°$, and then any time you add $n \times 360°$ to either of these angles. But at half of these occurrences the oscillator is moving in the positive direction, toward $x = 0.200$ m. Those times when the oscillator is at $x = 0.100$ m and moving to the left occur where $\sin(4t) > 0$, in the first quadrant. These occur when $4t = 60°$ and

420°, and in general at times when $4t = 60°$ plus a multiple of 360°. In radians these are where $4t = \pi/3 + 2\pi n$, where n is a positive integer 1, 2, 3... The times at which these occur are $t = \pi/12 + n\pi/2$ s.
In terms of the period, $T = \pi/2$, the first two times are $t = T/6$ and $7T/6$.

13. Use the definition of the period of a simple pendulum $T = 2\pi\sqrt{\frac{L}{g}}$ and solve for its length.

$$L = g\left(\frac{T}{2\pi}\right)^2 = \left(9.80\frac{\text{m}}{\text{s}^2}\right)\left(\frac{\left(\frac{25.0 \text{ s}}{14 \text{ cycles}}\right)}{2\pi}\right)^2 = 0.792 \text{ m}$$

15. (a) $T = 2\pi\sqrt{\frac{L}{g}} = 2\pi\sqrt{\frac{0.50 \text{ m}}{9.80 \text{ m/s}^2}} = 1.4 \text{ s}$

(b) (i) We know from part (a) the period for a full swing of the pendulum is 1.4 s, so for this pendulum to swing from –8° to +8° is about 0.7 s or one-half of the period.
(ii) We are asked to compare the result for (i) with the time interval to move from –4° to +4°, or for positions at one-half the amplitude to either side of equilibrium. That gives us

$$\theta = \theta_{\text{max}} \cos\left(\sqrt{\frac{g}{L}}t\right) \Rightarrow \pm 0.50 = \cos\left(\sqrt{\frac{g}{L}}t\right) \Rightarrow \cos^{-1}(\pm 0.50)$$

$$= \sqrt{\frac{g}{L}}t \Rightarrow t = \sqrt{\frac{L}{g}}\cos^{-1}(\pm 0.50)$$

The first time this occurs is the result of the + value and occurs at 0.237 s, and the next time is for the negative value, and is at 0.473 s. The time interval is then 0.237 s = 0.24 s. This is 0.24 s/0.7 s = 0.3, or about one-third of the time taken by the motion from –8° to +8°.

17. $T = 2\pi\sqrt{\frac{L}{g}}$ $\quad g = 9.8 \text{ m/s}^2 \quad 2.0 \text{ s} = 2\pi\sqrt{\frac{L}{9.8\frac{\text{m}}{\text{s}^2}}}$, so

$L = 0.99$ m

$$g_{\text{Mars}} = \frac{GM_{\text{Mars}}}{R^2_{\text{Mars}}}$$

$$g_{\text{Mars}} = \frac{\left(6.67 \times 10^{-11}\frac{\text{Nm}^2}{\text{kg}^2}\right)(6.4 \times 10^{23} \text{ kg})}{(3.4 \times 10^6 \text{ m})^2} = 3.7\frac{\text{m}}{\text{s}^2}$$

$$T_{\text{Mars}} = 2\pi\sqrt{\frac{L}{g_{\text{Mars}}}} = 2\pi\sqrt{\frac{0.99 \text{ m}}{3.7\frac{\text{m}}{\text{s}^2}}} = 3.3 \text{ s}$$

19. (a) The total energy of an oscillating spring–object system is equal to $\frac{1}{2}kA^2$. It is also equal to the elastic potential energy stored in the spring plus the kinetic energy of the moving object. Setting these equal, we can solve for an algebraic expression for the speed v as a function of x.

$$\frac{1}{2}kx^2 + \frac{1}{2}mv^2 = \frac{1}{2}kA^2$$

$$v = \sqrt{\frac{k}{m}\left(A^2 - x^2\right)}$$

(b)

$$v(0 \text{ m}) = \sqrt{\frac{\left(85\frac{\text{N}}{\text{m}}\right)}{(0.250 \text{ kg})}\left((0.100 \text{ m})^2 - (0 \text{ m})^2\right)} = 1.84\frac{\text{m}}{\text{s}} = 1.8\frac{\text{m}}{\text{s}}$$

$$v(0.020 \text{ m}) = \sqrt{\frac{\left(85\frac{\text{N}}{\text{m}}\right)}{(0.250 \text{ kg})}\left((0.100 \text{ m})^2 - (0.020 \text{ m})^2\right)}$$
$$= 1.81\frac{\text{m}}{\text{s}} = 1.8\frac{\text{m}}{\text{s}}$$

$$v(0.050 \text{ m}) = \sqrt{\frac{\left(85\frac{\text{N}}{\text{m}}\right)}{(0.250 \text{ kg})}\left((0.100 \text{ m})^2 - (0.050 \text{ m})^2\right)}$$
$$= 1.60\frac{\text{m}}{\text{s}} = 1.6\frac{\text{m}}{\text{s}}$$

$$v(0.080 \text{ m}) = \sqrt{\frac{\left(85\frac{\text{N}}{\text{m}}\right)}{(0.250 \text{ kg})}\left((0.100 \text{ m})^2 - (0.080 \text{ m})^2\right)}$$
$$= 1.11\frac{\text{m}}{\text{s}} = 1.1\frac{\text{m}}{\text{s}}$$

$$v(0.100 \text{ m}) = \sqrt{\frac{\left(85\frac{\text{N}}{\text{m}}\right)}{(0.250 \text{ kg})}\left((0.100 \text{ m})^2 - (0.100 \text{ m})^2\right)} = 0\frac{\text{m}}{\text{s}}$$

21. (a) $f_1 = \frac{1}{2\pi}\sqrt{\frac{k}{M}}$ and $f_2 = \frac{1}{2\pi}\sqrt{\frac{k}{2M}}$, thus $f_2 = \frac{f_1}{\sqrt{2}}$.

(b) Because the two systems have the same total energy and spring constant, they must have the same amplitude of motion. $v_{\text{max}} = \omega A$ and $\omega = \sqrt{\frac{k}{m}}$.
Thus, the maximum speed depends inversely on the square root of the mass of the system, because A and k are the same.

So $v_{\text{max},2} = \frac{v_{\text{max},1}}{\sqrt{2}}$.

(c) According to Equation 12-10, the magnitude of the maximum acceleration is $\omega^2 A$, as the cosine function varies from –1 to 1. And $\omega^2 = \frac{k}{m}$.

$a_{\text{max},1} = \frac{k}{M}A$ and $a_{\text{max},2} = \frac{k}{2M}A$. Thus,

$a_{\text{max},2} = \frac{a_{\text{max},1}}{2}$.

23. (a) $\dfrac{f_{Earth}}{f_{moon}} = \dfrac{3.50\ s^{-1}}{1.82\ s^{-1}} = \sqrt{\dfrac{g_{Earth}}{g_{moon}}} \Rightarrow g_{moon} =$

$$g_{Earth}\left(\frac{1.82}{3.50}\right)^2 = 2.65\ m/s^2$$

$$g_{moon} = \frac{GM_{moon}}{R_{moon}^2} \Rightarrow$$

$$M_{moon} = \frac{2.65\ m/s^2 (2.740\times10^6\ m)^2}{6.67\times10^{-11}\ m^3\ s^{-2}\ kg^{-1}} = 2.98\times10^{23}\ kg$$

(b) The object–spring system's frequency does not depend on the gravitational acceleration of the moon, and so does not provide the benefit of a direct relationship between the measurement of the pendulum's frequency and the computation of the moon's gravitational acceleration. Therefore, it would not serve as a means for determining the moon's mass.

25. Use conservation of linear momentum to determine the speed of the bullet–block system immediately after the completely inelastic collision of the bullet and the block. This speed at the equilibrium is the maximum system speed, which will allow us to determine the amplitude of the simple harmonic motion, and subsequently to find the position function for the postcollision bullet–block system.

Conservation of system linear momentum (constant):

$$m_{bullet}v_{bullet,ix} + m_{block}v_{block,ix} = m_{bullet+block}v_{bullet+block,fx}$$

$$v_{bullet+block,fx} = \frac{m_{bullet}v_{bullet,ix} + m_{block}\ v_{block,ix}}{m_{bullet+block}}$$

$$v_{bullet+block,fx} = \frac{(7.50\times10^{-3}\ kg)\left(3.55\times10^2\ \frac{m}{s}\right) + (0.750\ kg)(0)}{\left(7.50\times10^{-3}\ kg + 0.750\ kg\right)}$$

$$= 3.5149\ \frac{m}{s} \approx 3.51\ \frac{m}{s}$$

Simple harmonic motion angular frequency:

$$\omega = \sqrt{\frac{k}{m_{bullet+block}}} = \sqrt{\frac{9.80\times10^3\ \frac{N}{m}}{0.7575\ kg}} = 113.74\ \frac{rad}{s} \approx 114\ \frac{rad}{s}$$

Simple harmonic motion amplitude:

$$\frac{1}{2}kA^2 = \frac{1}{2}m_{bullet+block}v_{max}^2$$

$$A = v_{max}\sqrt{\frac{m_{bullet+block}}{k}} = \frac{v_{max}}{\omega} = \frac{\left(3.5149\ \frac{m}{s}\right)}{\left(113.74\ \frac{rad}{s}\right)} = 0.0309\ m$$

$$= 3.09\ cm$$

Initial conditions of the simple harmonic motion indicate the appropriate function and phase angle. At time $t = 0$, the position of the system is $x = 0$ and the system is moving in the positive direction. This suggests that a simple sine function with zero phase angle best describes the motion:

$$x(t) = A\sin(\omega t) = (3.09\ cm)\sin\left(\left(114\ \frac{rad}{s}\right)t\right)$$

Alternatively, the cosine function could be used to model the motion in the form of Equation 12-12, using a phase angle of $-\pi/2$ radians, by recognizing that $\sin(\omega t) = \cos\left(\omega t - \left(\frac{\pi}{2}\ rad\right)\right)$.

27. Use a phase angle of zero because the object starts at its maximum amplitude at $t = 0$. Its velocity will be zero to start and acceleration (maximum magnitude to get it going) will be negative, given that its position has to start decreasing as soon as it starts moving. The angular frequency of the motion is $\omega = \sqrt{\frac{k}{m}}$. Write $x(t)$ in terms of givens and compare it with Equations 12-14 and 12-15 to determine the functional forms for the velocity and acceleration of the object with respect to time. The maximum values of the speed and acceleration are equal to the absolute values of the amplitudes of these functions. To find the velocity at a given time, plug that time directly into the equation for v_x.

General form of the velocity as a function of time for simple harmonic motion:

$$v_x = -\omega A\sin(\omega t + \phi)$$

General form of the acceleration as a function of time for simple harmonic motion:

$$a_x = -\omega^2 A\cos(\omega t + \phi)$$

By comparison, the amplitude of the velocity expression is equal to the amplitude of the position, A, multiplied by $-\omega$; the amplitude of the acceleration expression is equal to the amplitude of the position multiplied by $-\omega^2$; and the phase angle is equal to 0 in all cases. Therefore,

$$v_x = -A\omega\sin(\omega t)$$
$$a_x = -\omega^2 A\cos(\omega t)$$

(a) $v_{max} = |-A\omega| = A\sqrt{\dfrac{k}{m}} = (0.100\ m)\sqrt{\dfrac{\left(75\ \frac{N}{m}\right)}{2.0\ kg}} = 0.61\ \dfrac{m}{s}$

(b) $a_{max} = |-A\omega^2| = A\left(\sqrt{\dfrac{k}{m}}\right)^2 = (0.100\ m)\dfrac{\left(75\ \frac{N}{m}\right)}{2.0\ kg} = 3.8\ \dfrac{m}{s^2}$

(c) $v_x = -A\omega\sin(\omega t)$

$$v_x(t = 5.0\ s) = -(0.100\ m)\left(\sqrt{\frac{\left(75\ \frac{N}{m}\right)}{2.0\ kg}}\right)\sin\left[\left(\left(\sqrt{\frac{\left(75\ \frac{N}{m}\right)}{2.0\ kg}}\right)(5.0\ s)\right)\right]$$

$$= 0.44\ \frac{m}{s}$$

Chapter 12 AP® Practice Problems

1. (D) $U = 1/2kx^2$, and $K = E - U$.
3. (C) One-half of a full cycle of an oscillation is 1 s, so the period of the pendulum is 2(1 s) = 2 s, regardless of amplitude, for small angles. We are told the period of the circular motion is approximately the same period, so the speed of the pendulum is $v = \frac{2\pi r}{t} = \frac{2\pi(0.60 \text{ m})}{2.0 \text{ s}} = 1.9$ m/s.
5. (A) The point furthest from equilibrium because for SHM the restoring force, and therefore the acceleration, are directly proportional to the displacement from equilibrium.
7. (A) Amplitude is measured from the equilibrium position.
   From the graph, the equilibrium position is at 6 cm and the maximum displacement is at 11 cm, so the amplitude is 11 cm – 6 cm = 5 cm.

Chapter 13

13-1 Liquids and gases are both examples of fluids

1. (a) For a thrown ball, if there were no air resistance, the ball would not slow down in the horizontal direction as it flew, and would gain speed more quickly in the downward direction, because there would be no acceleration in the horizontal direction and no force opposing the downward acceleration due to gravity.
   (b) The ball would slow down more quickly so would not travel as high. It would be going slower when it came back to the initial height. If fast enough, the ball could reach terminal velocity coming down, with the fluid resistance balancing the force of gravity on the ball.
   (c) The car has no net force exerted on it to travel at constant velocity, but the engine must be pushing the tires onto the road harder than if there were just rotational friction, so the road is pushing the car harder to overcome the force opposite the car's motion, which is caused by the air resistance.
3. The solid rigid body maintains its shape (is incompressible) and does not absorb any of the liquid.

13-2 Density measures the amount of mass per unit volume

1. $\rho = \frac{m}{V}$ We are given the 5.98×10^{24} kg. The volume is $\frac{4}{3}\pi R^3$ and R is given as 6.38×10^6 m so $\rho = 5.50 \times 10^3$ kg/m^3.
3. We find density of aluminum in Table 13.1. Then we know $\rho = \frac{m}{V}$ and the volume is $\frac{4}{3}\pi R^3$. Solving for R,
   $$\frac{4}{3}\pi R^3 = \frac{m}{\rho}, \text{ so } R = \left(\left(\frac{3}{4\pi}\right)\times\left(\frac{24.8 \text{ kg}}{2.70\times10^3 \text{ kg/m}^3}\right)\right)^{1/3}$$
   $R = 0.130$ m
5. We find density of ice in Table 13.1. $\rho = \frac{m}{V}$. For a cube: $V = L \times L \times L = L^3$. Substitute into density and solve for L. $V = L^3 = m/\rho$, so $L = \sqrt[3]{m/\rho} = 0.0725 \text{ m} \approx 7.3$ cm.
7. From the appendix, the mass of the Sun is 1.99×10^{30} kg. We are given mass and radius, and volume is $\frac{4}{3}\pi R^3$. We can use the density equation to solve for the volume of the Sun given the mass, setting it equal to the density of a neutron.
   $$\rho_{\text{neutron}} = \frac{m_{\text{neutron}}}{V_{\text{neutron}}} = \frac{m_{\text{star}}}{V_{\text{star}}} \Rightarrow \frac{m_{\text{neutron}}}{\frac{4}{3}\pi R^3_{\text{neutron}}} = \frac{m_{\text{star}}}{\frac{4}{3}\pi R^3_{\text{star}}}$$
   $$R_{\text{star}} = \left(\frac{m_{\text{star}}}{m_{\text{neutron}}}\right)^{1/3} R_{\text{neutron}} = 12.6 \text{ km} \approx 13 \text{ km}$$
9. (D) Iron has a much greater density than does ice, but the mass m and so weight of each block depends on its volume $V(m = \rho V)$. Because we do not know the volume, we cannot tell.

13-3 Pressure in a fluid is caused by the impact of molecules

1. Pressure is the perpendicular component of force per unit area.
   (a) The force in this case is her weight, distributed over the area of all four feet:
   $$P_{1\text{ foot}} = \frac{F_\perp}{A} = \left(\frac{3000 \text{ kg} \times g}{4 \times \pi(0.25 \text{ m})^2}\right) = 3.7 \times 10^4 \text{ Pa}$$
   (b) The force in this case is your weight, distributed over the area of two much smaller feet (mine are about 0.1 m wide and 0.28 m long):
   $$P_{1\text{ foot}} = \frac{F_\perp}{A} = \left(\frac{75 \text{ kg} \times g}{2 \times (0.1 \text{ m})(0.28 \text{ m})}\right) = 1.3 \times 10^4 \text{ Pa}$$
3. In a submarine, the pressure inside is approximately 1 atm to make the people comfortable, which is less than the pressure outside, since pressure increases with depth. Because the pressure inward on the door would be greater than the pressure outward, the net force on the door would be inward, and the door would have to be reinforced against caving in.
5. (D) Unlike force, pressure is *not* a vector and so has no direction, so any statement that refers to the direction of pressure is incorrect.

13-4 In a fluid at rest pressure increases with increasing depth

1. Because pressure increases with depth, raising your hand will decrease the blood pressure and, thus, slow the bleeding.
3. We are given a 25-cm-tall tube with half the height mercury and half water so the depth of each is 12.5 cm. Since we are asked for absolute pressure we must remember to include atmospheric pressure: $P_0 = 1 \text{ atm} = 1.01 \times 10^5 \text{ Pa} = 1.01 \times 10^5 \text{ N/m}^2$.
   $$\rho_{\text{Hg}} = 1.3595 \times 10^4 \text{ kg/m}^3$$
   $$\rho_{\text{water}} = 1.000 \times 10^3 \text{ kg/m}^3$$

$P = P_0 + \rho_{Hg} g d_{Hg} + \rho_{water} g d_{water}$
$P = 1.19 \times 10^5$ Pa

5. (D) The gauge pressure doubles but the atmospheric pressure will not double, so the overall pressure will be the doubled gauge pressure plus atmospheric pressure; so it increases, but does not double.
7. Pressure increases linearly with water depth; take an average pressure at mid-depth of the water. Roughly approximate the dam as a flat surface with dimensions 221 m by 379 m, but since 50 m is above water, 221 m – 50 m = 171 m is the height of our rectangle, and area is approximately 6.5×10^4 m^2. This also means the depth of water halfway down the height of the dam will be about 85 m, so we will use this depth to find our "average" pressure.

$$P = (1.0100 \times 10^5 \text{ Pa}) + (998 \text{ kg/m}^3)(9.80 \text{ m/s}^2)(85 \text{ m})$$
$$= 1.0100 \times 10^5 \text{ Pa} + 8.3 \times 10^5 \text{ kg/(m}\cdot\text{s}^2) \approx 9 \times 10^5 \text{ Pa}$$

Take our average value and multiply by the area we found. The total hydrostatic force (that is, total force on the water side of the dam) is about 6×10^{10} N.
9. An atmosphere is about 760 mmHg so add the atmosphere to the gauge pressure: 760 mmHg + 120 mmHg = 880 mmHg.
11. The difference in pressure is the difference caused by a depth of 1.75 m, converted to mmHg.

$$\Delta P_{blood} = (1.06 \times 10^3 \text{ kg/m}^3)(9.80 \text{ m/s}^2)(1.75 \text{ m}) = 1.82 \times 10^4 \text{ Pa}$$
$$\Delta P_{blood} = \left(\frac{760 \text{ mmHg}}{1.0100 \times 10^5 \text{ Pa}}\right) 1.82 \times 10^4 \text{ Pa} = 137 \text{ mmHg}$$

13-5 A difference in pressure on opposite sides of an object produces a net force on the object

1. The net force perpendicular to a surface is found from the difference in pressure between its two sides.

$$F_{net} = (P_{in} - P_{out})A = (0.95 \text{ atm} - 0.85 \text{ atm})A$$
$$F_{net} = (0.10 \text{ atm})\left(\frac{1.01 \times 10^5 \text{ N/m}^2}{1 \text{ atm}}\right)(1000 \text{ cm}^2)\left(\frac{1 \text{ m}}{100 \text{ cm}}\right)^2$$
$$F_{net} = 1.0 \text{ kN, directed outward}$$

3. Because pressure increases with increasing depth in a fluid, including the atmosphere, the air pressure on the bottom of the paper is slightly greater than the pressure on the top, and the net force due to these pressures is upward (although very small). The sum of *all* forces exerted on the paper is zero, because the paper is at rest, but that's not the question.
5. The force of the water will be larger than the atmospheric pressure inside the drum so there will be an inward force. The depth is 250 m and the area of the sides of a cylinder are found from circumference times height. The absolute pressure the water exerts on the drum at this depth is $P_{water} = P_0 + \rho g d$ but P_{air} inside the drum that we need to subtract to find the net force is P_0, so $F_{net} = (P_{water} - P_{air})A = (\rho g d)A = [(1.025 \times 10^3 \text{ kg/m}^3)(9.80 \text{ m/s}^2)(250 \text{ m})] \times [\pi(0.549 \text{ m})(0.876 \text{ m})] = 3.79$ MN toward the inside of the drum

13-6 A pressure increase at one point in a fluid causes a pressure increase throughout the fluid

1. Because we are going to take the ratio of the areas, we don't need to convert units:

$$F_2 = F_1\left(\frac{A_2}{A_1}\right) = 150 \text{ N}\left(\frac{750 \text{ cm}^2}{8.0 \text{ cm}^2}\right) = 1.4 \times 10^4 \text{ N}$$

3. (C) Because the cap on side 2 has 5 times the radius and so $A_2/A_1 = 25$, from Equation 13-11 the force on side 2 (F_2) is 25 times greater than the force on side 1 (F_1). From Equation 13-12, the displacement of the cap on side 2 (d_2) is only 1/25 as much as the displacement of the cap on side 1 (d_1). Work is the product of force and displacement in the direction of that force, so the work done on side 2 is the *same* as the work done on side 1, because 25 × (1/25) = 1. The hydraulic jack doesn't increase the amount of work that you can do but does make it possible for you to do that work by applying a smaller force.
5. (a) $F_2 = F_1\left(\frac{A_2}{A_1}\right) = 16 \text{ N}\left(\frac{4.0 \text{ m}^2}{0.033 \text{ m}^2}\right) = 1.9 \times 10^3 \text{ N}$
   (b) The same volume of liquid must go up as goes down, so the changes in height will be inversely proportional to the areas: $\Delta y_2 = \left(\frac{0.033 \text{ m}^2}{4.0 \text{ m}^2}\right)\Delta y_1 = 8.25 \times 10^{-3} \Delta y_1$.
   (c) The work must be equal. In each case it is the force exerted by the displacement over which that force is exerted (the Δy that is on the same side as the force). The results of (b) compared to our calculation in (a) show the ratio of the displacements is the inverse of the ratio of the forces, so when we calculate work as force times displacement for each piston $W_1 = 3.2 \text{ J} = W_2 = 3.2$ J.

13-7 Archimedes' principle helps us understand buoyancy

1. It will sit higher. Regardless of whether the coin is on the top or bottom of the cube, since the cube–coin combination is floating, the buoyant force must equal the weight of the cube plus the weight of the coin for the cube and coin not to sink. So the amount of water displaced is the same in both cases. When the coin is on the top, only the cube is displacing water, but when the coin is on the bottom, both the cube and the coin displace water. When the volume of the coin is contributing to the buoyant force, the cube itself won't displace as much water as when the buoyant force was due totally to the amount of the cube that was submerged. For that reason, more of the cube will be above the surface of the water when the coin is taped to the bottom.
3. When the bottle was floating it displaced a volume of water equal to its weight, because to float, the buoyant force must equal the bottle's weight, and according to Archimedes' principle, the buoyant force is equal to the weight of the displaced water. However, when the bottle has sunk below the surface of the water and filled with water, the volume of water it displaces is

equal only to the volume of the glass that forms the bottle. Because the bottle sank, we know that the glass must be denser than water, which means the glass itself takes up less volume than the equivalent weight of water (the acceleration is downward, so the force of gravity is greater than the buoyant force now, so the buoyant force is less than it was when the bottle floated). The net result is that when the bottle sinks, the water level goes down.

5. **(a)** The free-body diagram will have two forces, the downward force of gravity exerted on the block by Earth and the upward buoyant force. The two forces will be the same size.
   (b) Find mass from its volume and density, which is 0.6 that of water:

$$F_b = mg = (0.00600\ \text{m}^3)(0.600)(1000\ \text{kg/m}^3)(9.80\ \text{m/s}^2) = 35.3\ \text{N}$$

   (c) The fraction is the ratio of the density of the object to water, which is what the specific gravity is, so $V_{\text{submerged}}/V_{\text{block}} = 0.600$.
   (d) The weight of the water must be exactly the same as the buoyant force. $F_{g,\text{water}} = 35.3$ N.

7. We use V for the volume of the person. Other volumes will be denoted by subscripts.
   (a) Use $F_{g,\text{apparent}} = F_g - \rho_{\text{fluid}}Vg$ to find his volume. With his weight and volume find his average density, ρ: $F_g = \rho V g$.

$$19.7\ \text{N} = 833\ \text{N} - 1.00\times10^3\ \text{kg/m}^3 V(9.8\ \text{m/s}^2)$$
$$V = 8.299\times10^{-2}\ \text{m}^3$$

Then:

$$833\ \text{N} = \rho(8.299\times10^{-2}\ \text{m}^3)(9.8\ \text{m/s}^2)$$
$$\rho = 1.02\times10^3\ \text{kg/m}^3$$

   (b) Use the average density, and those of fat and muscle, to find the percentages.
   The total volume needs to add up as well so we can find the volume of muscle in terms of the volume we want to find, the fat, and $\frac{V_{\text{fat}}}{V}$ is the % we want.

total weight = weight fat + weight muscle

$$\rho_{\text{average}}V = \rho_{\text{fat}}V_{\text{fat}} + \rho_{\text{muscle}}V_{\text{muscle}}$$
$$V_{\text{muscle}} = V - V_{\text{fat}}$$
$$\rho_{\text{average}} = \rho_{\text{fat}}\frac{V_{\text{fat}}}{V} + \rho_{\text{muscle}}\frac{V - V_{\text{fat}}}{V}$$
$$1.024\times10^3\ \text{kg/m}^3 = 9.30\times10^2\ \text{kg/m}^3\frac{V_{\text{fat}}}{V} + 1.06\times10^3\ \text{kg/m}^3\left(1 - \frac{V_{\text{fat}}}{V}\right)$$
$$\frac{V_{\text{fat}}}{V} = 0.277 \text{ or } 28\%$$

9. **(a)** Apparent weight will decrease by an amount proportional to its density. We know its specific gravity, and that it is completely submerged:

$$F_{g,\text{apparent}} = \rho_{\text{object}}V_{\text{object}}g - \rho_{\text{fluid}}V_{\text{object}}g$$
$$F_{g,\text{apparent}} = 19.3\rho_{\text{fluid}}V_{\text{object}}g - \rho_{\text{fluid}}V_{\text{object}}g = 18.3\rho_{\text{fluid}}V_{\text{object}}g$$

To find the volume, we can use the measured weights:

$$4.88\ \text{N} = 5.15\ \text{N} - \rho_{\text{fluid}}V_{\text{object}}g$$
$$V_{\text{object}} = 2.755\times10^{-5}\ \text{m}^3$$
$$F_{g,\text{apparent}} = 18.3\rho_{\text{fluid}}V_{\text{object}}g = 18.3\times(1000\ \text{kg/m}^3)\times2.755\times10^{-5}\ \text{m}^3(9.8\ \text{m/s}^2) = 4.94\ \text{N}$$

   (b) The crown is most likely made of gold as the apparent weight is within 1.3% of that of gold, and this is an experiment.

11. **(a)** At the top of the cube $P_{\text{water}} = P_0 + \rho g d$, where d is the distance below the surface of the water where the top of the cube is located. At the bottom of the cube $P_{\text{water}} = P_0 + \rho g(d+s)$. Taking the difference, $\Delta P = \rho_{\text{water}}gs$.
   (b) The only forces exerted on the cube are the gravitational force downward and the buoyant force upward. The cube is completely submerged, so if we take upward to be positive, the net force on the cube is: $\rho_{\text{water}}gs^3 - m_{\text{cube}}g$.
   (c) We find the weight from the density of the displaced water, the volume, and g: $\rho_{\text{water}}s^3g$.

13-8 Fluids in motion: a more robust definition of an ideal fluid, and application of conservation of mass

1. If we assume no water is lost, the continuity equation applies because mass is conserved. The speed will increase when the cross-sectional area decreases: The speed of the water in the valley will be slower than that in the narrow channel.

3. The speed of the water times the cross-sectional area of the flow is equal to the volume flow rate, so $4.5\times10^{-1}\ \text{m}^3/\text{s} = v[\pi(0.0075\ \text{m}/2)^2] \rightarrow v = 1.0\times10^4$ m/s.

5. It will decrease. According to the continuity equation, because mass is conserved, the product of the cross-sectional area of a flow of fluid and the speed of flow is constant. For that reason, as the speed of the flow increases, the cross-sectional area must decrease. The water speeds up as it falls due to the acceleration due to gravity.

7. The diameter is decreased by 60%, so $r_2 = 0.4r_1$ and we are given $v_1 = v_0$.

$$v_2 = \frac{A_1}{A_2}v_1$$
$$v_2 = \frac{(\pi r_1^2)}{[\pi(0.4r_1)^2]}(v_0) = 6.25v_0$$

9. Flow rate multiplied by time is volume. 1.35 billion gallons is the volume, and looking it up on the Internet, 1 cubic meter is 264.17 gallons. From the endpaper, 3.281 ft = 1 m.

$$1.35\times10^9\ \text{gallons}\times\left(\frac{1\ \text{m}^3}{264.17\ \text{gallons}}\right) = 40{,}000\ \text{ft}^3/\text{s} \times\left(\frac{1\ \text{m}}{3.281\ \text{ft}}\right)^3\times\Delta t$$

Solving, $\Delta t = 4512\ \text{s} = 1.25$ h.

13-9 Bernoulli's equation, an expression of the work-energy theorem, helps us relate pressure and speed in fluid motion

1. (A) By Bernoulli's principle, when two regions in a flowing fluid have different flow speed, the pressure must be lower in the region in which the flow is faster. When you blow between two papers the air flows faster

between the papers than in the region surrounding the papers, so the pressure between them is lower than in the surrounding region, causing the papers to be drawn together.

3. The pressure difference between the water in the container and the outside air will cause the water to shoot out in the horizontal direction. It will then start to fall due to gravity.
5. Start with conservation of energy: Bernoulli's equation. With the atmospheric pressure lowering, we can guess that the water will go further up, because the pressure giving the water the initial energy is the same. Bernoulli's equation must hold for all locations, so the total for the lower atmospheric pressure condition for the water's maximum height will be equal to the total when the pressure is normal at water's maximum height, given that this must equal the initial condition when the water is ejected from the pipe. The speed of the water at the top of its flight will be zero, so kinetic energy = 0 at the top.

$$P_1 + \frac{1}{2}\rho v_1^2 + \rho g y_1 = P_2 + \frac{1}{2}\rho v_2^2 + \rho g y_2$$

$$101.3 \times 10^3 \text{ Pa} + (998 \text{ kg/m}^3)(9.8 \text{ m/s}^2)(5.00 \text{ m}) = (0.877)(101.3 \times 10^3 \text{ Pa}) + (998 \text{ kg/m}^3)(9.8 \text{ m/s}^2)y_2$$

$$y_2 = 6.27 \text{ m}$$

7. Bernoulli's equation can be used to find the pressure required to give the water the necessary energy, after we use the continuity equation to get the speed at which the water leaves the truck's tank. That will be the speed the water has when it gets to the nozzle. The water is at rest in the tank, and we will assume that the tank is the full 5.00 m below the level of the nozzle on the end of the hose and atmospheric pressure is the same at both ends of the hose. So we will just include the initial pressure due to the truck, as the atmospheric pressure would cancel.
First:

$$v_{nozzle} = \frac{A_{hose}}{A_{nozzle}} v_{hose}$$

Substituting values:

$$20 \text{ m/s} = \frac{\pi\left(\frac{0.117 \text{ m}}{2}\right)^2}{\pi\left(\frac{0.020 \text{ m}}{2}\right)^2}(v_{hose})$$

$$v_{hose} = 0.584 \text{ m/s}$$

1 = the tank and 2 = the end of the hose right before the nozzle. Apply conservation of energy to find P_1:

$$P_1 + \frac{1}{2}\rho v_1^2 + \rho g y_1 = P_2 + \frac{1}{2}\rho v_2^2 + \rho g y_2$$

$$P_1 + 0 + 0 = 0 + \frac{1}{2}(998 \text{ kg/m}^3)(0.584 \text{ m/s})^2 + (998 \text{ kg/m}^3)(9.80 \text{ m/s}^2)(5.00 \text{ m})$$

$$P_1 = 4.91 \times 10^4 \text{ Pa}$$

13-10 Surface tension explains the shape of raindrops and how respiration is possible

1. The raindrop must be considered a system. The internal interactions within the raindrop are the ones that cause the potential energy that is minimized by changing the shape of the raindrop. When considering the change in speed of the raindrop, we treat it as an object, and the interaction that is the source of the potential energy is in the Earth–raindrop system.

Chapter 13 Review Problems

1. No. The pieces could be of different sizes.
3. We know that the pressure increases as you go deeper in a fluid, to support the weight of the fluid above, if it is in hydrostatic equilibrium. Air is light, so we don't expect a big change.

$$P_0 = 1 \text{ atm} = 1.01 \times 10^5 \text{ Pa} = 1.01 \times 10^5 \text{ N/m}^2$$

$$\rho_{air} = 1.23 \text{ kg/m}^3$$

$$P = P_0 + \rho_{air} g d_{air} = 1.02 \times 10^5 \text{ Pa}$$

5. To find absolute pressure, we must add atmospheric pressure. Convert psi to pascals. The table of conversion factors in the back of the book gives 101.3 kPa (1 atmosphere of pressure) = 14.70 lbf/in.^{2}. So, the tire is about 4 atm, so we should get something close to 5 atm, and we do: 65 lbf/in.^{2} × (101.3 kPa/14.70 lbf/in.^{2}) + 101.3 kPa = 550 kPa = 5.5×10^5 Pa
7. This is a unit conversion problem:

$$\left(\frac{50.0 \text{ mi}}{\text{gallon}}\right) \times \left(\frac{1 \text{ gallon}}{3.788 \text{ L}}\right) \times \left(\frac{1 \text{ L}}{10^{-3} \text{ m}^3}\right) \times \left(\frac{1 \text{ m}^3}{737 \text{ kg}}\right) = 17.9 \text{ mi/kg}$$

9. The volume of the ball multiplied by the density of the compressed air = mass of the compressed air. The mass of the air + the shell divided by the volume will give us the average density:

$$V = (4\pi/3)R^3 = \frac{4\pi}{3}\left(\frac{0.75 \text{ m}}{2\pi}\right)^3 = 7.124 \times 10^{-3} \text{ m}^3$$

$$M_{air} = 7.124 \times 10^{-3} \text{ m}^3 \times 1.89 \text{ kg/m}^3$$

So,

$$M_{total} = 0.62369 \text{ kg} + M_{air}$$

$$\rho_{average} = \frac{M_{total}}{V} = 8.9 \times 10^1 \text{ kg/m}^3$$

11. The water level will fall because the boat displaces a volume of water equal to the weight of the boat. If the boat is removed from the water, the water will no longer be displaced.
13. Landing airplanes should approach from the east (into the wind). When flying into the wind, the speed of the airplane relative to the ground is reduced and the airplane can stop in a shorter distance.
15. While jumping up, the person will have less blood pumped to the brain, as more blood will be in the lower body due to gravity. This may result in a fainting spell. Once the person is horizontal,

the blood will evenly distribute, and consciousness will be regained.

17. (A) The pipe has ½ the diameter, so the area of the pipe will be ¼ the size of the larger pipe. The volume flow rate must be the same because mass is conserved, so the speed in the larger pipe will be ¼ that of the speed in the smaller pipe.

19. $P_{absolute} = P_0 + \rho gd$ and $P_{gauge} = \rho gd$. We are told to assume the same atmospheric pressure as Earth. We also assume the same density for sea water:
(a) $P_0 = 101.3 \times 10^3$ Pa, $\rho = 1.025 \times 10^3$ kg/m^3, $g = 0.379(9.80$ m/s$^2)$, $d = 0.50 \times 10^3$ m:

Gauge: 1.9×10^6 Pa Absolute: 2.0×10^6 Pa

(b) Calculate the pressure, or realize that the product of the acceleration due to gravity and the depth needs to be the same: $d_{Earth} = d_{Mars}\dfrac{g_{Mars}}{g_{Earth}} = 0.379 d_{Mars} = 190$ m

21. (a) The force the nurse exerts on the plunger will cause a pressure in the fluid. To get the fluid to go into the patient from the needle, it must be at a greater gauge pressure than the blood, so we will use 140 mmHg as our value for the pressure of the liquid in the needle. We are going to neglect the fact that the fluid should change in pressure as it goes from the broader syringe into the narrower needle as we will assume that the speed of the fluid is very, very slow.

$$A_{syringe} = \pi(0.30 \times 10^{-3} \text{ m})^2 = 2.827 \times 10^{-7} \text{ m}^2$$

$$A_{needle} = \pi(0.125 \times 10^{-3} \text{ m})^2 = 4.909 \times 10^{-8} \text{ m}^2$$

$$\Delta P_{fluid} = \left(\frac{1.0100 \times 10^5 \text{ Pa}}{760 \text{ mmHg}}\right) 140 \text{ mmHg} = 1.8605 \times 10^4 \text{ Pa}$$

$$F_{net} = (\Delta P_{fluid})A = (1.8605 \times 10^4 \text{ Pa})(2.827 \times 10^{-7} \text{m}^2) = 5.3 \times 10^{-3} \text{ N}$$

(b) Use Bernoulli's equation, neglecting any change in height.

$$P_1 + \frac{1}{2}\rho v_1^2 + \rho g y_1 = P_2 + \frac{1}{2}\rho v_2^2 + \rho g y_2$$

$$P_1 + \frac{1}{2}\rho v_1^2 = P_2 + \frac{1}{2}\rho v_2^2$$

From the continuity equation:

$v_2 = \dfrac{A_1}{A_2} v_1$ so taking 1 to be the needle and 2 the syringe:

$$P_{needle} - P_{syringe} = \frac{1}{2}\rho_{liquid}(v_{syringe}^2 - v_{needle}^2) = \frac{1}{2}\rho_{liquid}\left[\left(\frac{A_{needle}}{A_{syringe}}\right)^2 v_{needle}^2 - v_{needle}^2\right]$$

$$= \frac{1}{2}\rho_{liquid} v_{needle}^2 \left[\left(\frac{A_{needle}}{A_{syringe}}\right)^2 - 1\right]$$

$$= \frac{1}{2}(998 \text{ kg/m}^3)(5 \times 10^{-3} \text{ m/s})^2 \left[\left(\frac{4.909 \times 10^{-8} \text{ m}^2}{2.827 \times 10^{-7} \text{ m}^2}\right)^2 - 1\right]$$

$$= -1.2 \times 10^{-2} \text{ kg/(m} \cdot \text{s}^2) = -1.2 \times 10^{-2} \text{ N/m}^2$$

(c) Yes, at these slow speeds (which seem reasonable given how fast the nurse pushes on the plunger of the needle), the pressure difference due to the change in speed is very tiny.

23. This is a Torricelli variation of Bernoulli's equation, so assume atmospheric pressure both at the top of the tank and at the opening of the tube, and the tank is big enough that we can neglect any change in the height of the top of the water, or the speed of the water at the top. The only thing we need the diameter of the pipe for is to tell that this is a small opening, but not so small we need to worry about nonideal flow.

$$\frac{1}{2}\rho v_{opening}^2 + \rho g y_{opening} = \frac{1}{2}\rho v_{surface}^2 + \rho g y_{surface}$$

$$\frac{1}{2}v_{opening}^2 + g(0 \text{ m}) = 0 + g(18.0 \text{ m})$$

$$v_{opening} = 18.8 \text{ m/s}$$

25. (A) If the boxes have the same mass, the one with the larger density will be smaller and will have a smaller buoyant force exerted on it, if they are both held submerged. So when they are released there is a larger upward force on the less dense box so it will have a greater acceleration because the force exerted on it is greater and the masses are the same.

Chapter 13 AP® Practice Problems

1. (B) The buoyant force is the mass of the fluid that the submerged part of the object has displaced, so it depends on the volume of the object that is under the water.

3. (D) Volume flow rate is the volume of water per unit time. To fill a $V = 0.1$ m $\times$ 0.1 m $\times$ 0.1 m container in 90 seconds, divide the volume of the container (the volume of liquid that will be required to fill it) in m^3 by time in seconds. Rate = $(10.0 \times 10^{-2}$ m$)^3/90$ s $= 1.11 \times 10^{-5}$ m^3/s.

5. (A) Because mass is conserved, the speed × cross-sectional area must be constant. Because the cross-sectional area increases by a factor of 100, the speed must decrease by a factor of 100.

Chapter 14

14-1 A knowledge of thermodynamics is essential for understanding almost everything around you—including your own body

1. When we studied pressure in fluids, we described the force on an object placed in the fluid as being due to the sum of many tiny forces exerted by the molecules in the fluid colliding with the object. So, pressure is the macroscopic measurable quantity associated with the effect of the motion of the microscopic objects making up the fluid.

14-2 Temperature is a measure of the energy within a system

1. The temperature at which the pressure of an ideal gas would become zero is –273.15°C. This is the zero point of the Kelvin temperature scale.
3. (a) 82
   (b) 14
   (c) –229
   (d) 299
   (e) 99
5. Thermometer A (200–270 K) can be used as a freezer gauge. Thermometer B (230–270 K) can be used as a meteorological gauge (for outdoor temperatures). Thermometer C (300–550 K) can be used as an oven gauge. Thermometer D (300–315 K) can be used as a medical thermometer for humans.
7. (c) < (b) < (a)
9. Place the thermometer in an ice bath and let it come to equilibrium. The level of the mercury corresponds to 0°C. Then boil some water, place the thermometer in it, and let it reach equilibrium. The level of the mercury in this case corresponds to 100°C. Then we can divide the distance between the 0 mark and the 100 mark by 10 to get the marks for 10°C, 20°C, and so on. Dividing these regions into 10 again will give the individual degree markings.

14-3 In a gas, temperature and molecular kinetic energy are directly related

1. 17.0 L
3. 872 m/s
5. 6×10^3 J
7. (B)
9. (a) The formula is derived using the sum of the square of the speed in each dimension. In a one-dimensional system, the 3 should be replaced with 1: $K_{\text{translational,average}} = (1/2)kT$.
   (b) $K_{\text{translational,average}} = (10/2)kT = 5kT$

14-4 Heat is the amount of energy that is transferred in a thermal process

1. 78.5 kJ
3. (D)
5. (a) The final temperature is closer to the initial temperature of A. (b) For the two systems, $Q_A = m_A c_A \Delta T_A$ and $Q_B = m_B c_B \Delta T_B$. If the two systems exchange energy with each other only, $Q_A = -Q_B$ (see Example 14-5), so $m_A c_A \Delta T_A = -m_B c_B \Delta T_B$. Two systems are made of the same substance, and so have the same specific heat ($c_A = c_B$); therefore, our equation simplifies to $m_A \Delta T_A = -m_B \Delta T_B$.
7. 35.0°C
9. The calculated number of strokes to produce the amount of energy is greater than 18, even if only by a small amount, so 19.

14-5 Heating and cooling do not always result in a temperature change

1. First, the temperature of the ice rises to 0°C. Second, all the ice melts. Third, the water warms up to 100°C. Fourth, all the water boils. Finally, the temperature of the steam rises.
3. 79 kJ
5. (a) They are the same temperature.
   (b) The steam can cause a more severe burn because of its latent heat of vaporization.
7. (d) > (b) > (c) > (a) > (e)
9. (a) 35.7 g
   (b) two cubes

14-6 Thermal processes of energy transfer are radiation, convection, and conduction

1. (B)
3. a factor of 18.32 times
5. 6.12 kW
7. (D)
9. 2.9×10^6 J

Chapter 14 Review Problems

1. (a) Fahrenheit: $T_{F,\text{low}} = 98.1°\text{F}$, $T_{F,\text{high}} = 98.6°\text{F}$
   Kelvin: $T_{K,\text{low}} = 309.8$ K, $T_{K,\text{high}} = 310.2$ K
   (b) the Kelvin temperature scale
3. (a) $T_B < T_A$ (b) $T_B > T_A$
5. The length of a thermometer is based on the maximum and minimum values it is designed to read, not the temperature scale it employs.
7. If A is in thermal equilibrium with B, and B is in thermal equilibrium with C, then A is in thermal equilibrium with C. This is used to establish that temperature is actually a property of systems.
9. 1.3×10^9 m
11. 8.20×10^{-21} J
13. Heated air expands and becomes more buoyant. Cooler, denser air descends, displacing the warmer air. Thus, the warm air rises, and the cooler air sinks.
15. 3.85×10^5 J
17. 8.12×10^9 kg
19. 1110 K
21. 248 kJ [164.7 kJ (to raise the temperature) + 83.6 kJ (to melt the sample)]
23. The two gases are in thermal contact for a long time, so they reach thermal equilibrium: $T_A = T_B$. $PV = NkT$, so $P_A V_A = P_B V_B$. We cannot say anything about the individual values of P and V, only their product.
25. (a) $PV = NkT$ is most appropriate if counting molecules; $PV = nRT$ is most appropriate if counting moles.
    (b) P is the pressure of the gas, V is its volume, N is the number of molecules, n is the number of moles of

molecules, k is the Boltzmann constant, R is the universal gas constant, and T is the temperature of the gas.

27. The iron melts 7.87 g of ice, the copper melts 10.1 g of ice, and the water melts 14.6 g of ice.
29. $c = 0.384$ J/g · K ≈ 400 J/kg · K. Therefore, it is most likely copper or iron/steel.
31. (a) 26.7°C
    (b) yes
    (c) Her body prevents this increase by radiating energy and evaporation of sweat.
33. $P = 3.9 \times 10^{26}$ W
35. $H = 3.0 \times 10^{2}$ W
37. (a) $E_{\text{treadmill}}/E_{\text{person}} = 35.9\%$, remainder of energy goes into heating the body: 64.1%
    (b) $m = 0.230$ kg
39. (a) $m_{\text{water}} = 2.4 \times 10^{16}$ kg
    (b) $L = 2.9 \times 10^{4}$ m or 18 mi
41. 0.235 kg

Chapter 14 AP® Practice Problems

1. (C)
3. (C)

Chapter 15

15-1 The laws of thermodynamics involve energy and entropy

1. Yes, for example, when the heating of the system is equal to the work done by the system.

15-2 The first law of thermodynamics applies conservation of energy to thermal processes

1. +300 J
3. Yes, there is no change in internal energy for an isothermal process, so the work done by the gas must equal the heating of the system.
5. The temperature will remain constant when the work done by (on) the system is equal to the heating (cooling) of the system.

15-3 A graph of pressure versus volume helps to describe what occurs in a thermodynamic process

1. +8.80 kJ
3. −800 kJ
5. −9.22 kJ
7. 0 J
9. (a) (i) > (ii) > (iii) = (iv)
   (b) (iv) > (iii) > (ii) > (i)

15-4 The concept of molar specific heat helps us understand isobaric, isovolumetric, and adiabatic processes for ideal gases

1. 2.9 kJ
3. 27.5
5. −321 K
7. 19.0 g
9. 1.4
11. 1.25

15-5 The second law of thermodynamics describes why some processes are impossible

1. 40.0%
3. 14.8%
5. 3.41
7. 13.7 kJ
9. (a) 5.43
   (b) 225 J
   (c) −24°C

15-6 The entropy of a system is a measure of its disorder

1. −16.9 J/K
3. 245 J/K
5. Energy is transferred from the water to its environment, decreasing its temperature and so its entropy.
7. (B)
9. 1.48 kJ/K
11. 25.9 J/K

Chapter 15 Review Problems

1. (B)
3. (C)
5. There is just one configuration of the fragments that will reassemble the cup, compared to countless configurations that will not.
7. When quickly compressed there is insufficient time for the system to transfer an amount of energy equal to the work done on the system out of the system by thermal processes, increasing internal energy of the system.
9. High efficiency is a primary objective for a steam-electric generating plant. The temperature of the low-temperature reservoir is usually fixed by the environment, so increasing the feed-steam temperature is the only way to increase the Carnot efficiency limit.
11. No. Parts of the system may still be able to exchange energy with each other, which could increase entropy within the system.
13. Energy is transferred to the refrigerator's working substance both through heating from the things inside the box and as work done on it by the compressor. All this energy then heats the room air. With the door open, the compressor works harder. This extra energy also heats the air. Because no heat engine is 100% efficient, this extra heating is greater than the cooling due to the open refrigerator.
15. longer, shorter, longer
17. Entropy is a state function that does not depend on the history of changes to the system. So we can calculate

the change in entropy by assuming that the final state was achieved by isothermal expansion in which energy must have been transferred into the system to cause the increase in the system's volume; so the entropy of the final state will be greater than the entropy of the initial state.

19. –7500 J
21. 300 J
23. The water's entropy increases by 24.7 J/K.
25. –602 J/K
27. 2.88 J/K
29. (a) 38.0%
    (b) 0.288 J/K
    (c) 78.6 J
31. The efficiency drops from 10.5% to 8.9%.
33. (a) 47°C
    (b) 180 J
35. (a) 10.7%
    (b) 2.02 TJ
37. 25.5 kJ/K; increases

Chapter 15 AP® Practice Problems

1. (A)
3. (C)

Chapter 16

16-1 Electric forces and electric charge are all around you—and within you

1. A normal force opposes the attractive gravitational force exerted on you by a surface; therefore, charged objects must exert repulsive forces on other charged objects. The force exerted on a stretched string is opposed by attractive forces between charged objects within the string. When you slide an object over a surface, a force between charged objects is exerted in a direction that opposes your push.

16-2 Matter contains objects with positive and negative electric charge

1. 6×10^{12}.
3. (a) 3×10^{13}
   (b) 3×10^{13}
5. $+2e$
7. (a) The hanging rod will be rotated clockwise because of the greater repulsion caused by proximity of the like charge to the right of its lower end.
   (b) The hanging rod will be rotated counterclockwise because of the greater attraction caused by the proximity of the opposite charge to the right of its lower end.
9. If a charged insulating object repels *both* charged glass (positive) and charged rubber (negative), then you may have discovered a new kind of charge.

16-3 Charge moves freely in a conductor but not in an insulator

1. (a) 0.223 C/m^3
   (b) The charge is uniformly distributed within the sphere, rather than concentrated only on the outer surface.
3. (a) -1.56×10^{-3} C/m^2
   (b) Independent variables: amount of charge, Q, radius, R, of a metal sphere. Dependent variable: measured surface charge, σ. A graph of σ versus $x = 1/R^2$ should be linear, with a slope that is proportional to Q.
5. (B)
7. (a) This evidence does not support the claim because there are no significant differences reported here.
   (b) The increase in the threshold charge as area increases is not significant, so the claim is not well supported by the data.
9. The plastic comb gains a net negative charge from your hair, repelling the free electrons in the aluminum. Some of the can's free electrons move away from the comb, leaving a positive charge near the comb and creating a net attractive force on the can. When the comb touches the can, it transfers some electrons. Even though the can now has a net negative charge, the force is again attractive because the attractive force on the nearby positive charge exceeds the repulsive force on the larger, but considerably more distant, negative charge.

16-4 Coulomb's law describes the force between charged objects

1. (a) $r > 2000\ R$
   (b) 0.2%
   (c) If two spherical charged objects are far enough away, they behave as if they are points with no volume. As the distance separating them decreases, the accuracy of this representation decreases.
3. (a) 5.08 m
   (b) No. Equating the gravitational and Coulomb forces exerted between the charges results in the false equation

$$\frac{G}{k} = \frac{q_e^2}{m_e^2}$$

5. (a) 1.1×10^{-8} N
   (b) 40°

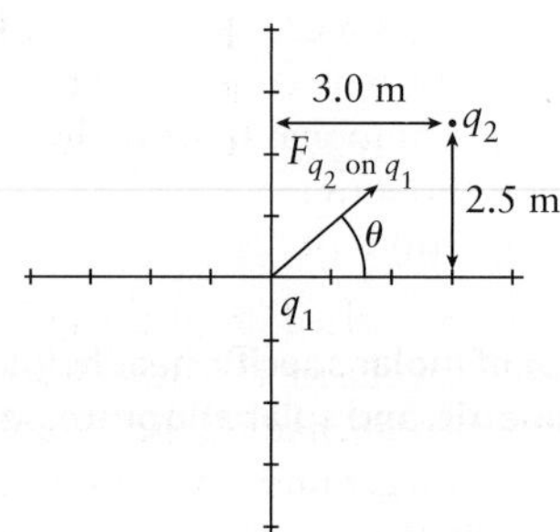

7. (a) The protons and electrons exert attractive and repulsive forces of equal magnitude on the point charge, so the net force is 0.

(b) When the atom is ionized, the net force exerted on the point charge is the sum of two forces with the same magnitude and direction, so it is nonzero.

9. (a) There are two equations (electric force relationship, ratio of the two charges) and two unknowns (separation distance, charge magnitudes), so a unique solution exists.
   (b) To find the solution, let $q_A = 2q_B$ and substitute this into the electric force law. Solve for q_B, and then determine q_A.
   (c) $q_B = \pm 6.0$ nC, $q_A = \pm 12.0$ nC
   (d) The force is repulsive, so the charges must have the same sign.

16-5 Electric forces are the true cause of many other forces you experience, and electric fields can help model electric forces

1. (a) Because the sign of the point charge q is positive, the force is attractive and the force vector points toward Q.
   (b) Arrows represent the direction of the force. The circle represents all points the same distance from $-Q$, represented by the dot at the center of the circle.

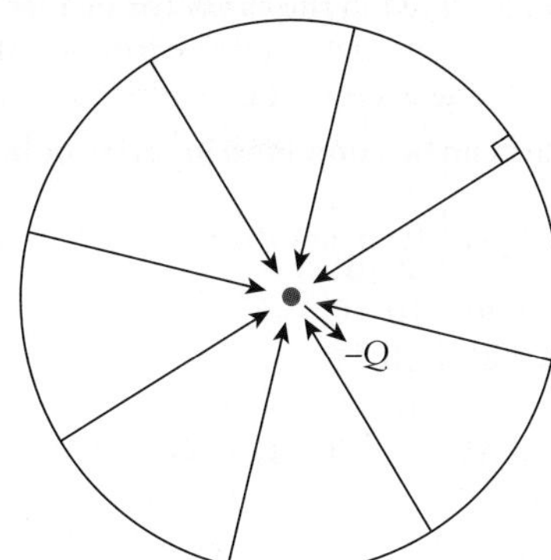

   (c) the same direction
   (d) Each force depends on the product of the two interacting quantities (charge or mass), and the inverse of the square of the distance separating their centers. The gravitational force is always attractive, whereas the electric force can be repulsive or attractive.
   (e) The electric forces near the large, planar surface will appear to be parallel.
3. (B)
5. The electric force is more short ranged because charges of each sign over some region tend to cancel each other, whereas the gravitational force is always attractive.
7. (a) $q = \dfrac{4\pi\rho g}{3E} r^3$
   (b) Many different-sized droplets can be suspended, each having a charge that is an integer multiple of the fundamental charge e. In a graph of the charge versus the integer number, the slope is the charge of the electron e.
9. (a) -4.3×10^4 N/C
   (b) -4.3×10^{-3} N
   (c) 260 nC

Chapter 16 Review Problems

1. 4.646×10^{-18} C
3. Electrons are transferred to the comb from your hair, polarizing the paper, which is then attracted to the comb. When they touch, some charge is transferred to the paper. The similarly charged paper and comb repel each other.
5. (a) invalid
   (b) invalid
   (c) invalid
   (d) incomplete
   (e) invalid
7. (C)
9. The electric forces are equal in magnitude but opposite in direction, downward for the proton and upward for the electron.
11. (B)
13. (a) 1.6×10^7 electrons
    (b) no
15. (a) 1.73×10^{-6} N
    The magnitudes of the forces are equal, but directed away from each other: The force on q_1 is in the $-x$ direction, and the force on q_2 is in the $+x$ direction.
    (b) (i) The direction of the force exerted on each is reversed.
    (ii) The direction of each force is reversed and the magnitude is increased by a factor of 4 to 6.90×10^{-6} N.
17. (a)

q_1 3.00 nC; $E = 0$; $q_2 > 0$
$x_1 = 0$; x_q; x_2

    (b) The electric force F_q exerted on a positive charge q at x_q must be zero. Therefore, to cancel the repulsive force exerted by charge q_1, $x_2 > x_q$.
    (c) $x_2 = 3x_q$
19. (a) 5.333×10^3 N/C in the direction of the force
    (b) -1.173×10^{-4} N directed opposite to the electric field
21. (a) $F_{\text{net on }3,x} = 0.00321$ N in the $+x$ direction, $F_{\text{net on }3,y} = -0.00797$ N in the $-y$ direction
    (b) $E_x = 2.14 \times 10^4$ N/C, $E_y = -5.31 \times 10^4$ N/C
23. (a) -1.71×10^3 N/C in the $-x$ direction
    (b) $t = 3.73 \times 10^{-11}$ s
25. 14.2 N/C in the direction of travel of the electron
27. (a) Down. The electric force opposing gravity requires a downward electric field because the drop is negatively charged.
    (b) 9060 N/C
    (c) seven electrons

Chapter 16 AP® Practice Problems

Multiple Choice

1. (B)
3. (B)
5. (C)

Chapter 17

17-1 Electric energy is important in nature and for technology; electric and gravitational potential energy have similar forms

1. (a) It decreases.
   (b) negative
   (c) The potential energy must decrease, because this is the same as an object falling toward Earth.

17-2 Electric potential energy of a system changes when a charged object moves in an electric field

1. (a) 8.99×10^{-9} J
   (b) 4.10×10^{-4} m/s
3. -3.33×10^{-7} J
5. $2kQ^2\left(-\frac{1}{x}-\frac{1}{y}+\frac{1}{\sqrt{x^2+y^2}}\right)$
7. (a) $v = 3.2 \times 10^2$ m/s
   (b) $a = 1.4 \times 10^{10}$ m/s^2
9. (a) 4.5×10^{-12} J
   (b) 3.7×10^7 m/s
   (c) Momentum, 0.017. The error is small.

17-3 Electric potential difference relates to the change in electric potential energy

1. (a) $\Delta V_q = \frac{4kq}{r} = 4kq \text{ m}^{-1}$
   (b) Electric potential, a scalar property, is unchanged, whereas electric field, a vector property, changes in direction.
3. 8.8 μJ
5. -2.85×10^3 V
7. (a) 6.35×10^2 J/C
   (b) 73 J/C; the amount of electric potential energy it would take to move a unit of positive charge from the initial to final positions
   (c) -1.2×10^{-17} J
   (d) The change in electric potential would be zero, so the net work done is also zero.
9. (C)

17-4 The electric potential has the same value everywhere on an equipotential surface

1.

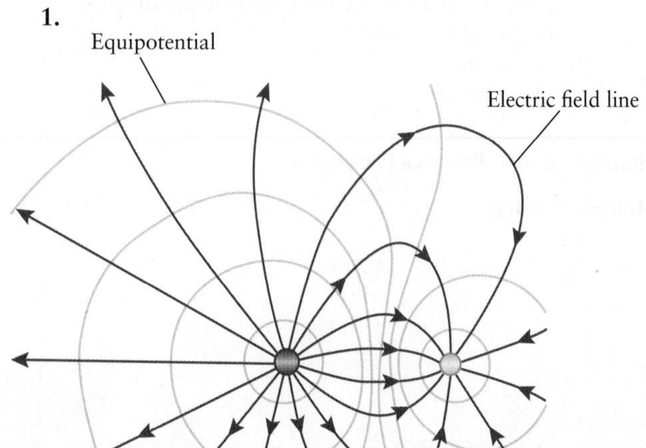

3. (a) 0.5 V/m pointing to the right
   (b) 0.86 V/m ≈ 0.9 V/m pointing to the right
5. Zero. The electric field is perpendicular to an equipotential.
7. very far from the two point charges and anywhere on the line perpendicular to the line separating them equidistant from both charges
9. (B)

17-5 A capacitor stores equal amounts of positive and negative charge

1. $C = 1.00 \times 10^{-6}$ F
3. $A = 2.26 \times 10^{-4}$ m^2
5. $d_2 = 0.250$ mm
7. increase plate area, decrease separation, increase the dielectric constant
9. (D)

17-6 A capacitor is a storehouse of electric potential energy

1. $C = 2.00 \times 10^{-6}$ F
3. $\Delta U_{electric} = 8 \times 10^{-3}$ J
5. (a) $C = 10^{-7}$ F
   (b) $C = 2 \times 10^{-7}$ F
7. The energy stored in the capacitor increases.
9. (C)

17-7 Capacitors can be combined in series or in parallel

1. $C_{equiv} = 0.2$ μF
3. (a) $C_{equiv} = 150.0$ μF
   (b) $q_1 = 1.20 \times 10^{-4}$ C
   $q_2 = 4.80 \times 10^{-4}$ C
   $q_3 = 1.20 \times 10^{-3}$ C
   (c) $\Delta V_1 = \Delta V_2 = \Delta V_3 = 12.0$ V
5. (A)
7. (a) 55.0 μF
   (b) 5.00 μF
9. (B)

17-8 Placing a dielectric between the plates of a capacitor increases the capacitance

1. $C = 4.8 \times 10^{-12}$ F
3. 802
5. $U_{electric} = 3.7 \times 10^{-6}$ J
7. $C_{equiv} = 4.6 \times 10^{-13}$ F
9. (D)

Chapter 17 Review Problems

1. Electric potential is the electric potential energy per unit charge caused by a configuration of charged objects. Electric field is the electric force per unit charge at a point in space due to a configuration of charged objects.
3. If only one charged object exists, then there is no electric force exerted on it. If there is no force, then that force is not available to do any work, so there is no potential energy.

5. This statement makes sense only if the zero point of the electric potential has been defined previously.
7. (a) Yes, a region of constant potential must have zero electric field because potential decreases in the direction of a field, so for it not to change there must not be a field.
   (b) No, if the electric field is zero, the potential need only be constant.
9. For a given potential difference, capacitors in parallel store more charge than any one of the individual capacitors.
11. For a given potential difference across them, the capacitors in parallel have a greater total area and so can store a greater amount of charge, thus having a larger capacitance.
13. $\Delta V = 0.094$ V
15. (a) $V = 3.60 \times 10^3$ V
    (b) $V = -3.60 \times 10^3$ V
17. 7.14 cm and 16.7 cm
19. (a) $V_A = -2.73 \times 10^3$ V
    (b) $V_B - V_A = -2.67 \times 10^4$ V
    (c) $W_{required} = 4.28 \times 10^{-15}$ J
21. (a) 1.12×10^{-20} J; (b) increases
23. (a) $V = 27.2$ V
    (b) $E_{ionization} = 2.18 \times 10^{-18}$ J
25. $Q_1 = Q_2 = 1.8 \times 10^{-5}$ C
27. $Q_2 = 8.00 \times 10^{-6}$ C; $Q_1 = 4.00 \times 10^{-6}$ C
29. $q_{equiv} = 551 \times 10^{-6}$ C
    $U_{electric} = 2.07 \times 10^{-2}$ J
31. (a) 6×10^8 J
    (b) 2.65×10^2 kg
33. 14 μF
35. $C = 1 \times 10^{-6}$ F
37. (a) 1/2
    (b) energy doubles, charge remains constant
39. (a) 6.3×10^{-10} J
    (b) 1.4×10^{-7}
41. 1×10^5 m/s

Chapter 17 AP® Practice Problems

1. (C)
3. (B)
5. (C)

Chapter 18

18-1 Life on Earth and our technological society are possible only because of charge in motion

1. You would be using your internal energy to return the box–Earth system to a state of higher gravitational potential energy. Conversion of your internal energy would result in hot, tired muscles, and eventually you would slow down.

18-2 Electric current equals the rate at which charge moves

1. (a) 2.2×10^{17} electrons/s
   (b) opposite
3. (a) 1.1×10^{20} electrons
   (b) 3.683×10^{-6} s
5. There is no contradiction.
7. (a) 4.82×10^3 A
   (b) same direction
9. (A)

18-3 The resistance to current through an object depends on the object's resistivity and dimensions

1. $\dfrac{d_1}{d_2} \approx 10$
3. 0.50 A
5. 1.4 Ω·m
7. (a) $\Delta V_{ab} = IR = I\dfrac{\rho L}{A}$
   L is the length of the separation between points a and b, over which the electric potential difference is measured. A is constant in these measurements. A graph of I versus ΔV_{ab} should be linear with a slope $A/\rho L$.
   The goal is to determine the resistivity ρ as a function of distance d relative to the soil–air surface. If the depth d is the midpoint between points a and b, the locations can be adjusted, always keeping point d at the midpoint, to obtain an estimate of the resistivity at the midpoint.
   (b) Charge is conserved, so the current into the core is equal to the current out of the core.
9. (D)

18-4 Conservation of energy and conservation of charge make it possible to analyze electric circuits

1. (a) 2100 V
   (b) –2100 V
   (c) 30 V
   (d) –30 V
3. 4.50 Ω
5. (a)

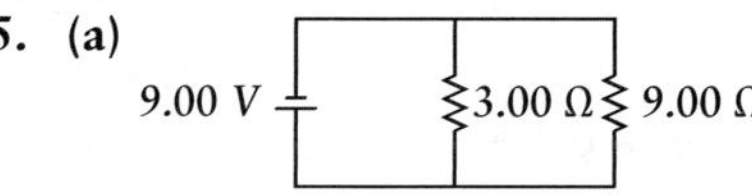

   (b) 2.25 Ω
   (c) 4.00 A
   (d) 9.00 V
   (e) $I_1 = 1.00$ A, $I_2 = 3.00$ A
   (f) 1.00 A. The potential difference across the resistor would be the same. The current through the resistor is the same.
7. (a) Taking the current in 1 to be into the junction (to the right):
   (i) $-I_1 = I_2 + I_3$ (ii) $I_1 = I_2 - I_3$
   Taking the current in 1 to be out of the junction (to the left):
   (i) $I_1 = I_2 + I_3$ (ii) $I_1 = I_3 - I_2$
   (b) Current (0.1A) is leaving the junction in branch 1.
9. $I_1 = 63.75$ mA, $I_2 = 5.625$ mA, $I_3 = 58.13$ mA

18-5 The rate at which energy is transferred by a circuit element depends on the current through the element and the electric potential difference across it

1. $[I][\Delta V] = \frac{C}{s}\frac{J}{C} = \frac{J}{s}$; $[I^2][R] = \frac{C^2}{s^2}\frac{J \cdot s}{C^2} = \frac{J}{s}$;

   $\frac{[\Delta V]^2}{[R]} = \frac{J^2}{C^2}\left(\frac{J \cdot s}{C^2}\right)^{-1} = \frac{J}{s}$
3. 58 MW
5. (a) Applying the loop rule to the right-hand portion of the circuit: $-\Delta V_{R2} + \Delta V_{R3} = 0 \Rightarrow \Delta V_{R3} = \Delta V_{R2}$. Given that the power for both resistors is the same, and the potential difference across both resistors is the same:

   $$\frac{\Delta V^2}{P} = R_2 = R_3$$

   (b) 0.500 A
   (c) $I_1 = I_{bat} = 0.500\text{ A}, I_2 = I_3 = \frac{I_{bat}}{2} = 0.250\text{ A}$
   (d) $R_1 = 6.00\ \Omega, R_2 = R_3 = 24.0\ \Omega$
   (e) $R_{equiv} = 18.0\ \Omega$
7. (a) 50.0 Ω
   (b) 0.200 A
   (c) $\Delta V_1 = 2.00\text{ V}, \Delta V_2 = 8.00\text{ V}$
   (d) $P_1 = 0.400\text{ W}, P_2 = 1.60\text{ W}$
   (e) $P = I\Delta V = 2.00\text{ W}$
9. 60-W bulb: 27 W, 120-W bulb: 13 W

18-6 A circuit containing a resistor and a capacitor has a current that varies with time

1. 12.0 s
3. The resistor in the circuit limits the current. Because the current is finite, not infinite, it requires time for the charge to move onto and off of the capacitor plates.
5. 38.8 kΩ
7. $C \approx 33\ \mu F$
9. (C)

Chapter 18 Review Problems

1. This energy is converted to warmth in the wire.
3. 1.1×10^{22}
5. 18J
7. (a) 0.48 nA
   (b) In the direction of the potassium ion motion
   (c) 0.96 nA
9. 1/3
11. $R_{circuit} = 16\ \Omega, I_{total} = 0.32\text{ A}$
    $I_{8\Omega} = 0.32\text{ A}, I_{12\Omega} = 0.20\text{ A}, I_{20\Omega} = 0.12\text{ A}$
    $P_{8\Omega} = 0.83\text{ W} \approx 0.8\text{ W}, P_{12\Omega} = 0.49\text{ W} \approx 0.5\text{ W},$
    $P_{20\Omega} = 0.29\text{ W} \approx 0.3\text{ W}$
13. $R_A = 6.0\ \Omega, R_B = 3.0\ \Omega$
15. $I_{6\Omega} = 0.60\text{ A}, I_{12\Omega} = 0.30\text{ A}, I_{total} = 1.5\text{ A}, V = 3.6\text{ V}$ for each resistor
17. The resistance changes.
19. 0.943 m
21. ≈\$1.30
23. 0.0282 Ω
25. 12 Ω
27. 4.0 V
29. $R_x = 3\ \Omega$
31. $270\ \Omega \cdot m$
33. $P_{5\Omega} = 45\text{ W}, P_{8\Omega} = 28\text{ W}, P_{10\Omega} = 90\text{ W}, P_{16\Omega} = 56\text{ W}$
35. (a) 4.50×10^3 C; total charge that the battery can deliver as current
    (b) 6.8×10^3 J
    (c) 3 h
    (d) 1×10^{-5} J; the battery stores 600 million times as much energy as the typical capacitor.

Chapter 18 AP® Practice Problems

Multiple Choice

1. (C)
3. (B)
5. (D)
7. (D)
9. (D)

Chapter 19

19-1 The magnetic force, like the electric force, is a long-range force

1. There must be moving charge carriers inside the material from which the refrigerator is made.

19-2 Magnetism is an interaction between moving charged objects

1. (B)
3. geographic east

19-3 A moving point charge can experience a magnetic force

1. 5×10^{-10} N
3. 0
5. west
7. 5.8×10^{-18} N to the left
9. (A)

19-4 A mass spectrometer uses magnetic forces to differentiate atoms of different masses

1. 5×10^{-3} T in the $+x$ direction
3. (a) 4.26×10^{-5} m
   (b) 5.60×10^5 Hz

19-5 Magnetic fields exert forces on current-carrying wires

1. 2.5 T
3. 0.74 A
5. 1.5 N
7. 16 N to the left
9. (D)

19-6 A magnetic field can exert a torque on a current loop

1. 0.0160 T
3. $0\ N \cdot m$
5. (A)

7. (A)
9. 1.44×10^{-5} N · m

19-7 Current-carrying wires create magnetic fields

1. (C)
3. B_O and B_P point in the $+z$ direction (out of the page). B_Q and B_R point in the $-z$ direction (into the page).
5. 1×10^{-8} T = 10 nT
7. $B = 4 \times 10^{-6}$ T = 0.08 B_{Earth}
9. (a) The clockwise current results in a magnetic dipole pointing into the page.
   (b) The dipole will align with the external field.
   (c) Some of the dipoles within the material will align with the external field.
   (d) More of the dipoles within the materials will align with the external field.
   (e) In the paramagnetic material, the dipoles will become random again. Some dipoles will stay in the orientation of the external field in the ferromagnetic material.

19-8 Two current-carrying wires exert magnetic forces on each other

1. 1.03×10^{-8} N away from wire 2
3. 0.313 N
5. (B)
7. 2.40×10^{-4} N to the left
9. (D)

Chapter 19 Review Problems

1. 0.3 fN (3×10^{-16} N) vertically downward
3. (a) 5.00×10^4 G
   (b) 2.5 T
   (c) 0.743 μT
   (d) 18.8 G
5. 0.0109 T
7. 2.00×10^{-5} T and points into the page
9. 2.1×10^8 m/s^2; highly dangerous
11. 42.8
13. 0.40 N
15. (a) 298 A ≈ 3.0×10^2 A
    (b) 0.62 T
    (c) increase the number of windings, N
17. $B = \dfrac{\mu_0 I}{2\sqrt{2}\pi d}$ and makes an angle of 315° with the $+x$ axis
19. 9.12×10^{-31} kg
21. (a) $B = 0.7155\dfrac{\mu_0 NI}{R}$ in the $+x$ direction
    (b) $B = 0.221\dfrac{\mu_0 NI}{R}$ in the $+x$ direction
23. 0.386
25. (a) Current is out of page in the wire, into page in the outer shell. There is a counterclockwise magnetic field within the wire and in the gap between the wire and the shell. There is no magnetic field outside the shell.
    (b) $r < R_i$: $B = \dfrac{\mu_0 I r}{2\pi R_i^2}$; $R_i \le r < R_o$: $B = \dfrac{\mu_0 I}{2\pi r}$; $r \ge R_o$: $B = 0$
    (c) (i) $B_1 = 1.3 \times 10^{-4}$ T (ii) $B_2 = 1.0 \times 10^{-4}$ T

Chapter 19 AP® Practice Problems

1. (D)
3. (D)
5. (B)

Chapter 20

20-1 The world runs on electromagnetic induction

1. The changing magnetic flux through the electronic circuitry will induce an emf in the device.

20-2 A changing magnetic flux creates an electric field

1. 1.3×10^{-2} V
3. (a) 0.204 Wb
   (b) 0.220 V
5. 2.66×10^{-5} V

20-3 Lenz's law describes the direction of the induced emf

1. (a) clockwise
   (b) counterclockwise
   (c) counterclockwise
   (d) clockwise
3. 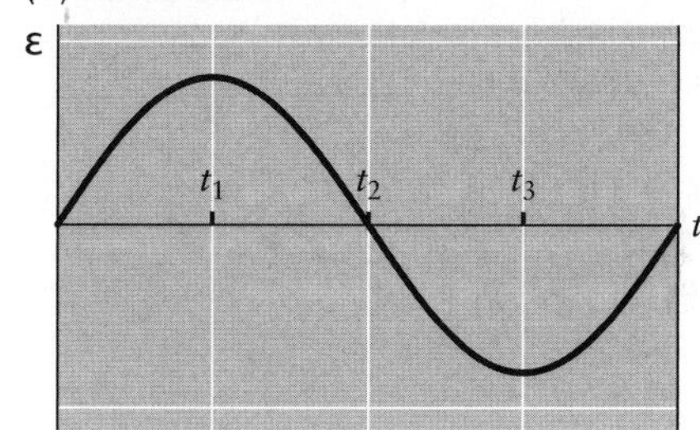

5. (a) (i) $I = 0$
   (ii) 4.39×10^{-2} A
   (iii) $I = 0$
   (b) 0.220 C
   (c) I is zero except from 0 to 10 cm, where it is 4.39×10^{-2} A.
7. (C)

20-4 Faraday's law explains how alternating currents are generated

1. ω = 15.2 rad/s ≈ 15 rad/s
3. $P_{average}$ = 9.3 mW
5. (C)

20-5 Maxwell's equations tie electricity and magnetism together

1. (C)
3. (D)
5. 1.0×10^{-5} T

Chapter 20 Review Problems

1. (B)
3. The induced emf will create an induced current in the clockwise direction, which—interacting with the magnetic field—will exert a force to the left on the sliding rod. This will gradually slow the rod until it comes to a stop.
5. 1.2×10^{-3} T/s
7. 2.1 V
9. (a) The current will pass up through the ammeter.
   (b) The current would pass down through the ammeter.
11. 7.3 mW
13. 9.28 s

Chapter 20 AP® Practice Problems

1. (C)
3. (A)
5. (B)

Chapter 21

21-1 Waves transport ene rgy and momentum from place to place without transporting matter

1. No. The molecules of air that leave your mouth collide with others, transferring their momentum, and causing a disturbance in the medium. This compression travels through the air, and transfers momentum to your friend's eardrum.

21-2 Mechanical waves can be transverse, longitudinal, or a combination of these; their speed depends on the properties of the medium

1. (a) vertical, gravitational
   (b) radial
   (c) (a) transverse, (b) longitudinal
3. Stiffness is a measure of the restoring force. Density is a measure of the inertial resistance to deformation. So in two materials with the same stiffness, the one with smaller density has larger wave speed.
5. (D)

21-3 Sinusoidal waves are related to simple harmonic motion

1. 1200 m
3. $\omega = 2\pi f = \dfrac{2\pi}{T} = 2.00\pi \times 10^2\ \text{s}^{-1} = 628$ rad/s

   $f = \dfrac{1}{T} = 1.00 \times 10^2\ \text{s}^{-1}$
5. $\omega = 1.5 \times 10^2$ rad/s, $\lambda = 0.92$ m
7. (a) $y(x,t) = (0.30\ \text{m}) \sin\left(\dfrac{2\pi x}{0.70\ \text{m}} - 4.0\ \text{s}^{-1}\pi t\right)$

   (b) $y(x,t) = (0.30\ \text{m}) \cos\left(\dfrac{2\pi x}{0.70\ \text{m}} - 4.0\ \text{s}^{-1}\pi t\right)$

   (c) $y(x,t) = (0.30\ \text{m}) \sin\left(\dfrac{2\pi x}{0.70\ \text{m}} + 4.0\ \text{s}^{-1}\pi t\right)$

   (d) $y(x,t) = (0.30\ \text{m}) \cos\left(\dfrac{2\pi x}{0.70\ \text{m}} + 4.0\ \text{s}^{-1}\pi t\right)$
9. (a) If $x > x_0$ and $x < x_0 + \lambda$ then the value of y from the table of values in the interval from 0 to λ can be obtained by calculating $x - \lambda$. If the value of $x \pm \lambda$ is not in the interval, then the wavelength is subtracted (or added) again until a value in the interval $[0, \lambda]$ is obtained.
   (b) The algorithm showed that as the wave propagates in the positive direction, the cosine (which is really a shorthand expression for a table lookup) is obtaining values from behind (– or to the left of) the current position. If the propagation is in the negative direction then the table lookup is made from points that are in front of (+ or to the right of) the current position.
11. (a) $T = 0.40\pi$ s = 1.3 s
    (b)

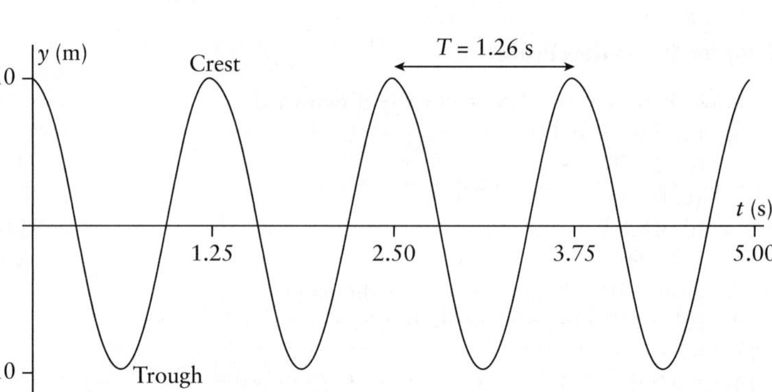

13. (a) $0.38\ \text{s} < \Delta t < 0.45\ \text{s}$
    (b) shortened
    (c) $3.6\ \text{cm} < L < 5.0\ \text{cm}$
    (d) The time interval will tend to be increased by the response time, so the speed will tend to be decreased.

21-4 Waves pass through each other without changing shape; while they overlap, the net displacement is just the sum of the displacements of the individual waves

1.

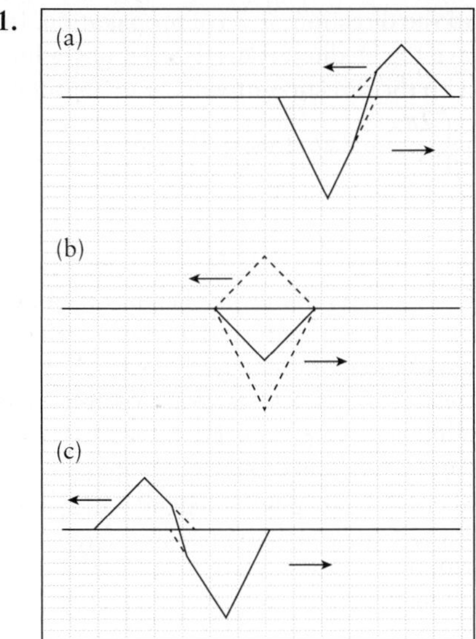

3.

B. $\pi/3$ rad $= 60°$ $f(x) = \cos\left(\frac{2\pi x}{L} + \frac{\pi}{3}\right)$

C. $\pi/2$ rad $= 90°$ $f(x) = \cos\left(\frac{2\pi x}{L} - \frac{\pi}{2}\right)$

D. $3\pi/5$ rad $= 108°$ $f(x) = \cos\left(\frac{2\pi x}{L} + \frac{3\pi}{5}\right)$

E. $5\pi/6$ rad $= 300°$ $f(x) = \cos\left(\frac{2\pi x}{L} + \frac{5\pi}{6}\right)$

F. π rad $= 180°$ $f(x) = \cos\left(\frac{2\pi x}{L} - \pi\right)$ **or**

$f(x) = \cos\left(\frac{2\pi x}{L} + \pi\right)$

5. **(a)** is incorrect. If the difference between the distances from the point to each speaker is equal to a whole number of wavelengths plus one-half of a wavelength, then the interference is destructive. This is not always the case. For example, at the midpoint the distances are equal, so that the difference is 0 and the interference is constructive.
   (b) is incorrect. If the difference between the distances from the point to each speaker is equal to a whole number of wavelengths, then the interference is constructive. At other points, it is not.
   (c) is correct. If the difference between the distances from the point to each speaker is equal to neither a whole number of wavelengths nor a whole number of wavelengths plus one-half of a wavelength, then the interference is neither constructive nor destructive. But because of the fact that between two whole numbers of wavelengths there is a distance that is one half- wavelength larger than the smaller of the two, constructive and destructive interferences alternate.

   Having demonstrated that C is correct, (d) is incorrect.

7. **(a)** $2v_p$
   (b) $t = t_0 + w/4v_P$
   - The leading edges of the pulses first meet:
     A, B, C: no overlap between pulses, width of combined pulse is $2w$, no change in amplitude
   - One-half of each pulse is yet to overlap:
     A: pulses overlap on each side by $w/2$, width of combined pulse is $3w/2$, center of combined pulse has amplitude of 2
     B: pulses overlap on each side by $w/2$, width of the combined pulse is $3w/2$, center of combined pulse has amplitude of 0
     C: pulses overlap on each side by $w/2$, width of combined pulse is $3w/2$, center of combined pulse has amplitude of 3
   - The pulses are fully aligned.
     A: pulses completely overlap, width of combined pulse is w, amplitude of combined pulse is 2
     B: pulses completely overlap, width of combined pulse is w, amplitude of combined pulse is 0
     C: pulses completely overlap, width of combined pulse is w, amplitude of combined pulse is 3
   - One-half of each pulse still overlaps.
     A: pulses overlap on each side by $w/2$, width of combined pulse is $3w/2$, center of combined pulse has amplitude of 2
     B: pulses overlap on each side by $w/2$, width of combined pulse is $3w/2$, center of combined pulse has amplitude of 0
     C: pulses overlap on each side by $w/2$, width of combined pulse is, $3w/2$, center of combined pulse has amplitude of 3
   - The trailing edges of the pulses are aligned.
     A, B, C: no overlap between the pulses, width of the combined pulse is $2w$, no change in amplitude

21-5 A standing wave is caused by interference between waves traveling in opposite directions

1. **(a)** 4.00 m
   (b)

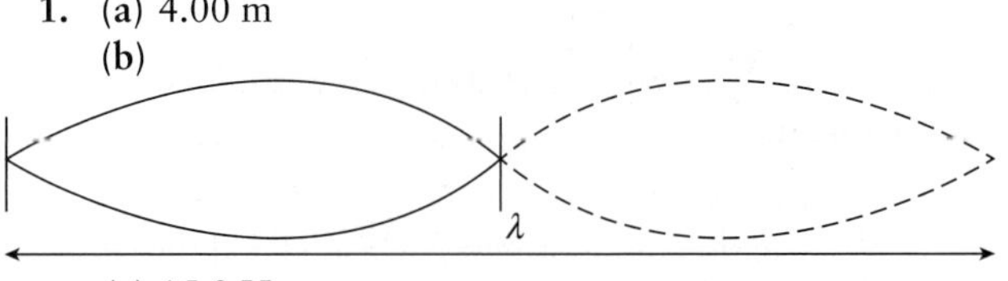

   (c) 15.0 Hz
3. Permitted wavelengths for this string are those that have zero wave heights (nodes) at the fixed points at 0 and L, in addition to any other nodes between 0 and L. The longest wavelength that satisfies this condition is $2L$.
5. **(a)** The algorithm adds a phase of π to the waveforms of the environmental acoustic data and then adds the result to the signal sent to the speakers.
   (b) Microphones could be placed on the phone case to gather acoustic data from all directions except front-facing to the phone. The environmental wave data collected could be phase-shifted and added to the waveform containing the voice and the environmental noise to cancel the noise. If the voice waveform was phase shifted and added to the

original voice waveform, and then broadcast into the environment, the voice waveform could be masked by destructive interference.

7. (C)

21-6 Wind instruments, the human voice, and the human ear use standing sound waves

1. The fundamental frequency is that of the longest wavelength with a displacement antinode at each open end. That is a wavelength of $2L$. As shown below, the fundamental and second harmonics have displacement antinodes at both ends. The first (fundamental) harmonic, with a wavelength of $2L$, can be constructed with one node at the midpoint of the tube, and the second harmonic, with a wavelength of L, can be constructed with two nodes within the tube. So, standing waves with wavelengths $2L/n$ for all integer values n greater than zero can be supported by a tube open at both ends.

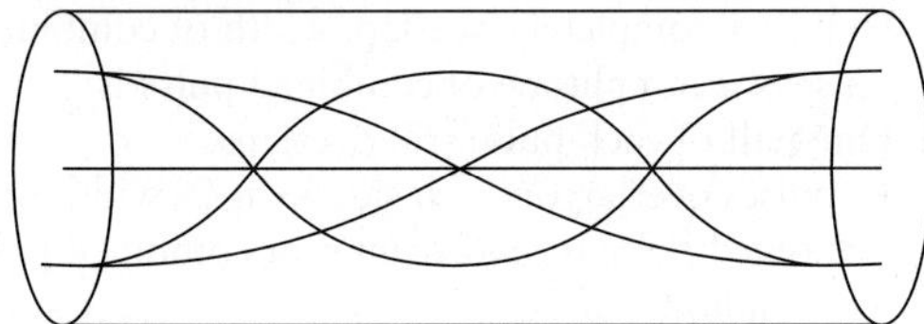

3. (D)
5. 350 m/s
7. The air at the open end of the tube is restricted to atmospheric pressure. When an increased pressure wave arrives at the open end of the tube, the air at the end of the tube responds with a lowered density so as to maintain the pressure at the air pressure of the room. When a decreased pressure wave arrives at the open end, the density of the air at the open end must increase. The density variation is large because the pressure variation is restricted, producing a displacement antinode.
9. The tube must be open.

21-7 Two sound waves of slightly different frequencies produce beats

1. either 436 or 444 Hz
3. If the tension of the string is increased and the beat frequency decreases, then continue increasing the tension until the beat frequency is zero. If the beat frequency increases with increased tension, then decrease the tension until the beat frequency is zero.

21-8 The frequency of a sound depends on the motion of the source and the listener

1. (C)
3. (a) 637 Hz
   (b) 567 Hz
5. 2.1 m/s

Chapter 21 Review Problems

1. $[v_p] = \text{m/s}; [\lambda][f] = \text{ms}^{-1}$
3. In a longitudinal wave the medium is displaced along a line parallel to the direction of wave propagation. In a transverse wave the medium is displaced along a line perpendicular to the direction of wave propagation.
5. $y(x,t) = A\cos\left(\frac{2\pi x}{\lambda} - 2\pi ft + \phi\right) = (0.10\text{ m})$
   $\cos(10\pi x - \pi t - \pi/2)$
   or $y(x,t) = (0.10\text{ m})\sin(10\pi x - \pi t)$
7. (a) The pulse has a maximum wave height of $2A$ and the same width because of constructive interference.
   (b) The pulse has a maximum wave height of $1.5A$ and the same width because of constructive interference.
   (c) The pulse has zero amplitude because of destructive interference.
   (d) The pulse has a negative wave height, $-A/2$, and the same width because of destructive interference.
9. (a) 8.85 Hz
   (b) 2.06 cm
11. (C)
13. 2.2 kg/m
15. 1.04×10^2 N
17. 6.55×10^2 N
19. (a) 30 Hz
    (b) 90 Hz
21. 3.00 m, open at both ends

Chapter 21 AP® Practice Problems

1. (B)
3. (B)
5. (A)

Chapter 22

22-1 Light is one example of an electromagnetic wave, and its wave nature explains much about how light behaves

1. (B)

22-2 In an electromagnetic plane wave, electric and magnetic fields both oscillate

1. (a) 7.25×10^{-8} m; ultraviolet light
   (b) 4.29×10^{-7} m; violet visible light
   (c) 3.75×10^{-9} m; x ray
   (d) 1.00×10^{-5} m; infrared light
   (e) 3.33×10^{-5} m; infrared light
   (f) 8.72×10^{-10} m; x ray
   (g) 3.65×10^{-8} m; ultraviolet light
   (h) 5.00×10^{-8} m; ultraviolet light
3. (a) 4.29×10^{14} Hz
   (b) 5.00×10^{14} Hz
   (c) 6.00×10^{14} Hz
   (d) 7.50×10^{14} Hz
   (e) 3.00×10^{15} Hz
   (f) 9.01×10^{18} Hz
   (g) 6.00×10^{11} Hz
   (h) 4.74×10^{18} Hz

5. between 2.8 and 3.4 m
7. 1.0×10^6 Hz = 1.0 MHz
9. (D)

22-3 Huygens' principle explains the reflection and refraction of light

1. 1.52
3. 1.00029
5. (a) 1.41
   (b) 1.46
   (c) 1.17
   (d) 1.00
7. 763 km
9. (C)

22-4 In some cases light undergoes total internal reflection at the boundary between media

1. example: fiber optics used for communication and laparoscopic surgery
3. (a) 41.8°
   (b) 48.8°
   (c) 58.5°
   (d) no critical angle
5. 34.4°
7. 59.1°
9. (A)

22-5 The dispersion of light explains the colors from a prism or a rainbow

1. 1.50 or 1.47
3. 0.1136 cm ≈ 0.11 cm
5. (a) 1.58 (rounded to three significant digits), 1.50
   (b) 18.5°, 19.5°
   (c) 33.3°, 35.3°
7. (a) 59.4°
   (b) 7.95×10^{-3} m

22-6 In a polarized light wave the electric field vector points in a specific direction

1. vertically polarized and has an intensity of 39 W/m^2
3. 30°, 150°
5. (a) $\theta_B = 36.9°$
   (b) $\theta_c = 48.8°$
7. polarized 75.0° from the vertical and has an intensity of 1.5×10^2 W/m^2
9. (D)

22-7 Light waves reflected from the surfaces of a thin film can interfere with each other, producing dazzling effects

1. soap bubbles reflecting different colors, nonreflective coatings (for example, thin film coatings on photographic lenses), colors on oil floating on a puddle of water
3. (a) shifted by half a wavelength
   (b) shifted by half a wavelength
5. 648 nm
7. $t = 93.8$ nm; $t = 282$ nm; $t = 469$ nm; $t = 657$ nm
9. $\lambda = 653$ nm and $\lambda = 435$ nm are not seen in the film.

22-8 Interference can occur when light passes through two narrow, parallel slits

1. (a) 3.90 mm
   (b) 5.85 mm
3. 5.6 mm
5. 540 nm
7. (A)

22-9 Diffraction is the spreading of light when it passes through a narrow opening

1. $m = 1$: 3.15°; $m = 2$: 6.32°; $m = 3$: 9.50°
3. $m = 2$
5. 4.06×10^{-5} m ≈ 41 μm
7. 8.25×10^{-6} m ≈ 8.25 μm
9. (A)

22-10 The diffraction of light through a circular aperture is important in optics

1. 1.4×10^{-4} rad
3. 2.7×10^{-7} rad
5. 6.9×10^{-5} m
7. (B), (A), (C)
9. (a) Bright light: 3.4×10^{-4} rad, dim light: 8.4×10^{-5} rad. The angular resolution is better in dim light than in bright light because the angular resolution is smaller for dim light.
   (b) By squinting, we partially close our eye, thereby limiting the light entering it. This causes the pupil to enlarge, which improves our angular resolution.

Chapter 22 Review Problems

1. $\Delta x = 3$ m
3. $\Delta t = 1.28$ s
5. $c = f\lambda = \left(\frac{\omega}{2\pi}\right)\left(\frac{2\pi}{k}\right) = \frac{\omega}{k}$
7. Yes, diffraction occurs for all wavelike phenomena. Sound waves are diffracted through an open window and water waves are diffracted around an obstacle in the water, for example.
9. No, the critical angle depends only on the indices of refraction of the water and the air.
11. Most glare is created by light that is reflected from horizontal surfaces, so it is advantageous to block out the light that is horizontally polarized.
13. 4 m
15. (a) $\theta_2 = 25.4°$
    (b) $\theta_3 = 28.0°$
17. 36.9°
19. $n_2/n_1 = 1.54$
21. (a) $n_2 = 2.0$
    (b) $BC = 1.0$ cm

23. $n_1 \sin(\theta_1) = n_n \sin(\theta_n)$ is true, so if the last medium is the same as the first, the angle of refraction is identical to the angle of incidence.
25. (a) $\theta = 49.8°$
    (b) $\theta = 30.9°$
27. $\theta_C = 24.2°$
29. $n = 1.60$
31. 29° or 61°
33. 188 μm
35. Small-angle approximation: 5.81×10^{-3} m ≈ 6 mm (rounded to one significant digit)
37. (a) $L = 8.8 \times 10^{15}$ m = 0.93 ly
    (b) $D = 9.1 \times 10^2$ m = 9×10^2 m (assuming precision of only one significant digit)
39. (a) maximum angular resolution (at shortest wavelength, or 55 μm):

$$\theta_R = 1.9 \times 10^{-5} \text{ rad} = 4.0''$$

    (b) $D = 0.035$ m
    (c) $x = 2.7 \times 10^{13}$ m
41. (a) 25
    (b) 25

Chapter 22 AP® Practice Problems

1. (B)
3. (D)
5. (D)
7. (B)
9. (A)

Chapter 23

23-1 Mirrors or lenses can be used to form images

1. (A)

23-2 A plane mirror produces an image that is reversed back to front

1. Your image in a plane mirror is reversed back to front.
3. A real image forms where light rays come together. No light rays actually meet where a virtual image forms.
5. 1.8 m
7. (A)
9. (D)

23-3 A concave mirror can produce an image of a different size than the object

1. upside-down and the same size
3. Yes. $s_o > f$
5. (B)

23-4 Simple equations give the position and magnification of the image made by a concave mirror

1. (a)

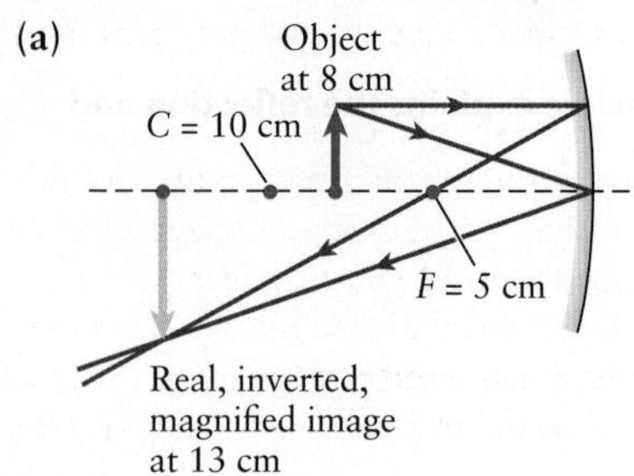

   (b) $s_i = 13.3$ cm, $M = -1.67$. The image is real and inverted.
3. (a)

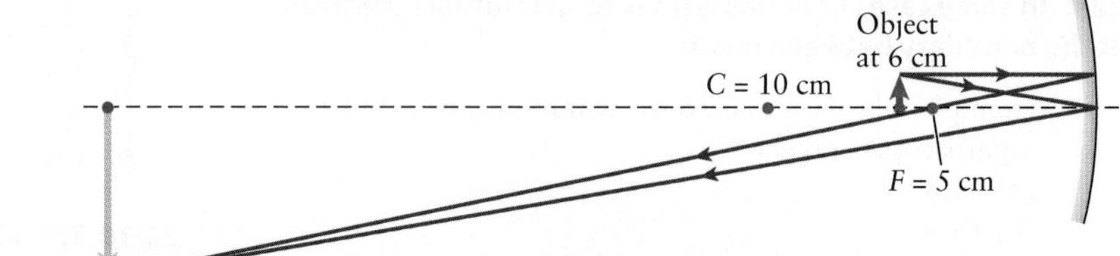

   (b) $s_i = 30$ cm, $h_i = -5.0$ cm
5. (a) $s_i = -10$ cm, $h_i = 20$ cm. The image is virtual and upright.
   (b) $s_i = 20$ cm, $h_i = -10$ cm. The image is real and inverted.
   (c) $s_i = 12.5$ cm ≈ 13 cm, $h_i = -2.5$ cm. The image is real and inverted.
   (d) $s_i = 11.1$ cm ≈ 11 cm, $h_i = -1.1$ cm. The image is real and inverted.
7. (a) $f = 13.3$ cm
   (b) The image is real.
   (c) $h_i = -12.5$ cm
9. (B)

23-5 A convex mirror always produces an image that is smaller than the object

1. In both cases, the images are virtual. The image of an "up close" object is larger than the image of the same object "far away." In both cases the image is smaller than the object.
3. A convex mirror is the opposite of a concave mirror. Use a phrase such as "The cave goes in" to remember the shape of a concave mirror.
5. (B)

23-6 The same equations used for concave mirrors also work for convex mirrors

1. 0.862 m behind the mirror
3. $s_i = -4.29$ m, $h_i = 0.429$ m, virtual, upright.
5. $s_i = -0.93$ cm, length = 0.93 cm, virtual, upright.
7. $s_i = -6.32$ cm, $h_i = 0.947$ cm, virtual, upright.
9. (C)

23-7 Convex lenses form images like concave mirrors and vice versa

1. No, a real image in a converging lens occurs on the opposite side of the lens from the object.
3. Only converging lenses can create real images. Diverging lenses produce only virtual images. A converging lens will produce a real image when the object is outside the focal point (that is, $s_o > f$).
5. Two. Images form where all the refracted rays (or their backward extensions) intersect. Therefore, wherever two rays intersect, the rest of the rays also intersect.
7. (C)

23-8 The focal length of a lens is determined by its index of refraction and the curvature of its surfaces

1. (a) $s_i = -45.0$ cm, $h_i = 5.00$ cm, virtual, upright.
   (b)

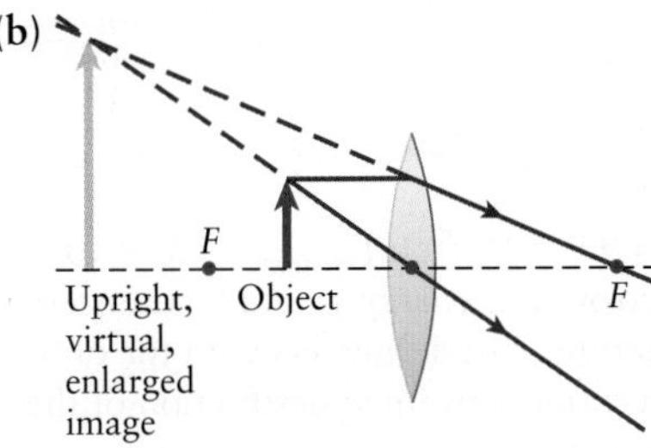

3. $P = 0.7543\ \text{m}^{-1}$; $f = 1.33$ m
5. (a) 19.1 cm = 0.191 m
   (b) +31.8 cm
7. (C) > (D) > (A) > (B)

23-9 A camera and the human eye use different methods to focus on objects at various distances

1. Our brain interprets an inverted image on the retina as a physically upright object.
3. f is fixed so s_i must increase as s_o decreases.
5. $P = 2.7$ diopters
7. 63.9 cm
9. (A)

23-10 The concept of angular magnification plays an important role in several optical devices

1. 2.78
3. (A)

Chapter 23 Review Problems

1. The full image is formed, but it is dimmer than before because it is formed with half as much light.
3. Only convex mirrors produce images that are upright and smaller than the object.
5. $\theta_{\text{reflected}} = \theta_{\text{incident}} = 0°$
7. B, C, and D
9. The camera will focus the lens on the surface of the mirror
11. $x = sd/2h$
13. 15 cm
15. −11.25
17. (a) 9.00 m
    (b) 4.00 m
19. $f_e = -2$ cm (rounded to one significant figure)
21. (a) 2.3 cm $\leq f \leq$ 2.5 cm
    (b) converging lens
23. 1.52
25. −1.4 diopters
27. −0.893 diopters; 2.80 diopters
29. (a) 67.1 mm
    (b) −0.353

Chapter 23 AP® Practice Problems

Multiple-Choice

1. (B)
3. (A)
5. (B)
7. (B)
9. (B)

Chapter 24

24-1 Experiments that probe the nature of light and matter reveal the limits of classical physics

1. In classical (pre-Bohr) physics, the only quantity known to be quantized without respect to its system was electric charge, which was found only in multiples of e.

24-2 Electromagnetic waves carry both electric and magnetic energy, and come in packets called photons

1. (a) 62 to 4.1×10^2 nm
   (b) It extends from violet visible light into the UV.
3. The energy of a photon is proportional to the wave frequency.
5. 5.0×10^4 eV; 5.0×10^8 photons
7. (a) $6.67 \times 10^{-7}\ \text{T} \approx 7 \times 10^{-7}\ \text{T}$
   (b) $53.1\ \text{W/m}^2 \approx 5 \times 10^1\ \text{W/m}^2$
9. (C)

24-3 The photoelectric effect and blackbody radiation show that light is absorbed and emitted in the form of photons

1. $1.33 \times 10^{-30}\ \text{J} \approx 1 \times 10^{-30}\ \text{J}$
3. 9.85×10^{14} Hz
5. 5.01×10^{-7} m = 501 nm; 5.99×10^{14} Hz
7. 3.62×10^{-17} J
9. (D)

24-4 As a result of its photon character, light changes wavelength when it is scattered

1. (a) $1.21 \times 10^{-27}\ \text{kg} \cdot \text{m/s} = 2.26\ \text{eV/c}$
   (b) $9.32 \times 10^{-24}\ \text{kg} \cdot \text{m/s} = 17{,}500\ \text{eV/c}$
3. Because the Compton wavelength (0.0024 nm) is very small compared to the wavelength of visible light, the change in the wavelength of a visible photon would be negligibly small.

5. (a) $\lambda = 0.140$ nm; $K = 0$ eV
   (b) $\lambda = 0.14033$ nm; $K = 20.6$ eV
   (c) $\lambda = 0.14071$ nm; $K = 44.8$ eV
   (d) $\lambda = 0.1412$ nm; $K = 76.2$ eV
   (e) $\lambda = 0.14243$ nm; $K = 151$ eV
   (f) $\lambda = 0.14485$ nm; $K = 297$ eV
7. 25.8°
9. (D)

24-5 Matter, like light, has aspects of both waves and particles

1. electron
3. Because the mass of a macroscopic object is so large, the de Broglie wavelength is too small to observe.
5. 3.32×10^{-10} m
7. (a) 1.23×10^{-9} m
   (b) 3.88×10^{-10} m
   (c) 1.23×10^{-10} m
   (d) 3.88×10^{-11} m
9. (C), (D), (A), (B)

24-6 The spectra of light emitted and absorbed by atoms show that atomic energies are quantized

1. 656.11 nm; 486.01 nm; 433.94 nm; 410.07 nm
3. From $m = 2$ to $n = 1$: 1.215×10^{-7} m = 121.5 nm
   From $m = 3$ to $n = 1$: 1.026×10^{-7} m = 102.6 nm
   From $m = 4$ to $n = 1$: 9.723×10^{-8} m = 97.23 nm
   From $m = 5$ to $n = 1$: 9.496×10^{-8} m = 94.96 nm
   From $m = 6$ to $n = 1$: 9.376×10^{-8} m = 93.76 nm
   These are all ultraviolet wavelengths.
5. From $m = 4$ to $n = 3$: 1.875×10^{-6} m = 1875 nm
   From $m = 5$ to $n = 3$: 1.282×10^{-6} m = 1282 nm
   From $m = 6$ to $n = 3$: 1.094×10^{-6} m = 1094 nm
   These are all infrared wavelengths.
7. $(8.2303 \times 10^{14}\text{ Hz})\left(\dfrac{m^2 - 4}{m^2}\right)$
9. (C) > (A) > (B) > (D)

24-7 Models by Bohr and Schrödinger give insight into the intriguing structure of the atom

1. 91.16 nm
3. Hydrogen has only one electron.
5. 1.51 eV
7. $\Delta f = 0.162$ MHz
9. (B)

24-8 In quantum mechanics, it is impossible to know precisely both a particle's position and its momentum

1. (a) 3.86×10^4 m/s
   (b) 79.4 m/s
3. 47.7 nm
5. (C)

Chapter 24 Review Problems

1. yes
3. The rate at which the electrons are emitted increases.
5. Momentum increases, so the de Broglie wavelength decreases.
7. 7.37×10^{-19} J; 6.63×10^{-19} J
9. (A)
11. (a) 1.78×10^{-18} m – 2.49×10^{-19} m
    (b) 2×10^{12}
13. $\lambda = 92.2$ cm; $f = 325$ MHz
15. (a) $f = 4.22 \times 10^{18}$ Hz; $E = 1.75 \times 10^4$ eV
    (b) 7.35×10^{-11} m = 0.0735 nm
    (c) 1.69×10^4 eV
    (d) ~580 eV
17. $E_{10} = -0.544$ eV; $E_9 = -0.672$ eV; $E_8 = -0.850$ eV; $E_7 = -1.11$ eV; $E_6 = -1.51$ eV; $E_5 = -2.18$ eV; $E_4 = -3.40$ eV; $E_3 = -6.04$ eV; $E_2 = -13.6$ eV; $E_1 = -54.4$ eV
19. (a) 7.59×10^{-19} J
    (b) 1.37×10^{-19} J
21. 6.1 min
23. 0.137 nm
25. 6110 K
27. 1.43×10^{-10} m
29. 8.3 eV
31. (a) $E_{photon} = 9.9 \times 10^{-20}$ J; $E_{electron} = 6.0 \times 10^{-26}$ J
    (b) The electron; less energy would be transferred to the bacterium, interfering less with the observations and preventing unwanted destruction of the bacterium features.
33. (a) $E = 21.3$ eV
    (b) The first three cathode-grid potential differences at which one expects to see a drop in anode current are the first three whole-number multiples of the potential difference that would result in the energy gain in (a); thus 21.3, 42.6, and 63.9 V.
35. Only choice (b) is compatible.

Chapter 24 AP® Practice Problems

1. (C)
3. (D)
5. (A)
7. (D)
9. (D)

Chapter 25

25-1 The quantum concepts that help explain atoms are essential for understanding the nucleus

1. (C)

25-2 The strong force holds nuclei together

1. Isotopes are nuclides with the same atomic number Z but different numbers of neutrons N. They have different mass numbers $A = Z + N$.
3. (a) ^{3}H: $Z = 1$, $N = 2$, $A = 3$
   (b) ^{8}Be: $Z = 4$, $N = 4$, $A = 8$
   (c) ^{26}Al: $Z = 13$, $N = 13$, $A = 26$
   (d) ^{197}Au: $Z = 79$, $N = 118$, $A = 197$
   (e) ^{100}Tc: $Z = 43$, $N = 57$, $A = 100$
   (f) ^{184}W: $Z = 74$, $N = 110$, $A = 184$

(g) ^{190}Os: $Z = 76$, $N = 114$, $A = 190$
(h) ^{239}Pu: $Z = 94$, $N = 145$, $A = 239$

5. (a) hydrogen: $Z = 1$, $N = 1$, $A = 2$
   (b) helium: $Z = 2$, $N = 2$, $A = 4$
   (c) lithium: $Z = 3$, $N = 3$, $A = 6$
   (d) carbon: $Z = 6$, $N = 6$, $A = 12$
   (e) iron: $Z = 26$, $N = 30$, $A = 56$
   (f) strontium: $Z = 38$, $N = 52$, $A = 90$
   (g) iodine: $Z = 53$, $N = 78$, $A = 131$
   (h) uranium: $Z = 92$, $N = 143$, $A = 235$
7. 1.3×10^4 m
9. (C)

25-3 Some nuclei are more tightly bound and more stable than others

1. Some of the mass is converted to energy when the parts are combined. The converted energy is the binding energy of the nucleus.
3. $E_B = 92.162$ MeV
5. $\Delta E = 15.643$ MeV
7. binding energy: 1103.33 MeV; binding energy per nucleon: 8.42235 MeV
9. (C), (B), (D), (A)

25-4 The largest nuclei can release energy by undergoing fission and splitting apart

1. (a) n
   (b) ^{102}Zr (zirconium-102)
   (c) ^{250}Cm (curium-250)
   (d) ^{142}Cs (cesium-142)
3. The missing product is 10n; $E_{released} = 89.951$ MeV.
5. 3×10^{19} reactions/s
7. 33 collisions
9. (D)

25-5 The smallest nuclei can release energy if they are forced to fuse together

1. (a) Elements with a higher atomic number than iron are more likely to undergo fission.
   (b) Elements with a lower atomic number than iron are more likely to undergo fusion.
3. (a) n
   (b) n
   (c) n
   (d) ^{3}He
   (e) ^{1}H
5. 156 g ≈ 200 g (rounded to one significant digit)
7. (D)

25-6 Unstable nuclei may emit alpha, beta, or gamma radiation

1. (a) 4×10^6 Bq
   (b) 25 Bq
   (c) 4.46×10^{-7} Ci
   (d) 4.53×10^{12} decays/min
3. (a) ^{234}Th (thorium-234)
   (b) α, ^{230}Ra (radium-230)
   (c) ^{240}Pu (plutonium-240)
   (d) ^{210}Tl (thallium-210)
5. (a) ^{131}I (iodine-131)
   (b) ^{24}Na* (sodium-24)
7. $R = 0.5$ μCi
9. (C)

Chapter 25 Review Problems

1. (B)
3. There is one electron mass included in the atomic mass of hydrogen.
5. 1.15×10^4 y
7. Application of conservation of energy and momentum to the two-body decay would require that the β particle be ejected with a single unique energy for any specific recoil momentum of the proton. Instead we observe experimentally that β particles are produced with a range of energies from zero to a maximum value.
9. $\tau_{1/2} = 208$ s $= 0.00240$ d
11. No. All forms of radiation can penetrate materials like fabric.
13. 0.00298 ≈ 0.30% (rounded to two significant digits)
15. (a) 36
    (b) 36
    (c) 37
    (d) 37
    (e) 39
    (f) 38
17. Assuming the elemental composition of the body is 63% H, 26% O, 10% C, and 1% N, there would be about 2×10^{28} nuclei in a 50-kg body.
19. (a) 7.40 μCi
    (b) 197 d × (1 mo)/(30 d) ≈ 6.6 mo
21. 3.10×10^4 y
23. (a) $^{241}\text{Am} \rightarrow {}^{237}\text{Np} + \alpha$
    (b) 0.16%, 8%
    (c) 5.4×10^{-9} g
25. age of the pole: 2.5×10^4 y
27. (a) $r = 7.44$ fm ≈ 7.4 fm
    (b) $F_{electric} = 1.04$ N ≈ 1.0 N
    (c) $a = 6.2 \times 10^{26}$ m/s^2
    (d) The strong nuclear force holds them together.
29. 6.2×10^7 atoms
31. 8.14978 MeV/nucleon; $^{26}\text{Al} \rightarrow e^+ + \nu_e + {}^{26}\text{Mg}$; 8.33159 MeV/nucleon; 8.30992 MeV/nucleon; $^{28}\text{Al} \rightarrow e^- + \bar{\nu}_e + {}^{28}\text{Si}$
33. (a) 3.79×10^{-8} g
    (b) 8.99×10^{-4} cm $= 8.99$ μm

Chapter 25 AP® Practice Problems

Multiple-Choice

1. (A)
3. (D)
5. (C)
7. (C)
9. (C)

Chapter 26

26-1 The concepts of relativity and elementary particles may seem exotic, but they're part of everyday life

1. (B)
3. (a) $x' = x - (4.00\text{ m/s})t$; $y' = y$; $z' = z$
   (b) $x' = -14$ m; $y' = 1$ m; $z' = 0$ m

26-2 Newton's mechanics includes some ideas of relativity

1. $x' = 7.4 \times 10^2$ m; $y' = y = 250$ m; $z' = z = 250$ m; $t' = t = 2.0 \times 10^{-6}$ s
3. $x' = 10$ m; $y' = y = 10$ m; $z' = z = 30$ m; $t' = t = 6.00 \times 10^{-4}$ s

26-3 The Michelson–Morley experiment shows that light does not obey Newtonian relativity

1. (B)
3. (C)

26-4 Einstein's relativity predicts that the time between events depends on the observer

1. (B)
3. 1.67

26-5 Einstein's relativity also predicts that the length of an object depends on the observer

1. 0.98 m
3. (a) 1×10^{-7} s, for both
   (b) The backward photon registers first.
   (c) 1.5×10^{-7} s

26-6 The speed of light is the ultimate speed limit

1. No. Equation $E = mc^2$ shows that as the speed of an object approaches the speed of light, the amount of energy required to further increase the speed approaches infinity.
3. (C)

26-7 The equations for kinetic energy and momentum must be modified at very high speeds

1. (D)
3. $m = 0.252$ mg

26-8 Einstein's general theory of relativity describes the fundamental nature of gravity

1. The equivalence of gravity and acceleration is fundamental to the theory of general relativity.
3. 25.65 8.20 m/s^2 in the upward direction

26-9 Most forms of matter can be explained by just a handful of fundamental particles

1. $E_\gamma = 273.5$ MeV, $\lambda = 4.54 \times 10^{-6}$ nm, and $p = 1.46 \times 10^{-19}$ kg · m/s for each photon
3. $K_p = 67.50$ MeV

26-10 Four fundamental forces describe all interactions between material objects

1. (A)
3. No. The Higgs particle must have zero charge. This collection of particles will have a net negative charge.

Chapter 26 Review Problems

1. (a) $x' = x - (15.0\text{ m/s})t$; $y' = y$; $z' = z$
   (b) $x'' = x - (4.00\text{ m/s})t$; $y'' = y$; $z'' = z$
   (c) $x' = x'' - (11.0\text{ m/s})t$; $y' = y''$; $z' = z''$
3. (B)
5. (C)
7. $V = 2.9956 \times 10^8$ m/s; $V/c = 0.9985$
9. $0.548c$
11. (a) mass necessary to produce this much energy: $m = 165$ kg
    (b) $L = 0.219$ m $= 21.9$ cm
13. $\Delta x = 2.3$ m
15. $p = 2.9 \times 10^{-19}$ kg · m/s
    (a) $K = 2.3 \times 10^{-11}$ J
17. (a) $\gamma = 100{,}000$
    (b) $\Delta t = 0.166$ s
19. (a) $\Delta t_{\text{proper}} \approx 2.25 \times 10^{-6}$ s
    (b) $V = 0.700c$
21. (a) $f = 50$ beats/min
    (b) 70 beats/min
23. $K_{\text{Einsteinian}} = 1.2 \times 10^{20}$ J, $K_{\text{Newtonian}} = 3.6 \times 10^{19}$ J; % error = 70%; The Newtonian formula underestimates the kinetic energy.
25. (a) $K_{\gamma 2} = 3.6 \times 10^2$ MeV
    (b)

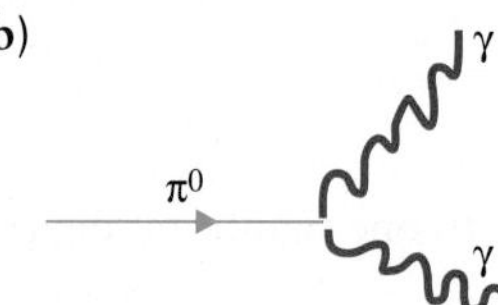

Index

Note: Page numbers preceded by A indicate Appendices; those preceded by M indicate Math Tutorial.

A

B

C

D

E

F

G

H

I

J

K

N

O

P

Q

R

S

T

U

V

W

X

Y

Z

Prefixes for Powers of 10*

Multiple	Prefix	Abbreviation
10^{24}	yotta	Y
10^{21}	zetta	Z
10^{18}	exa	E
10^{15}	peta	P
10^{12}	tera	T
10^{9}	**giga**	G
10^{6}	**mega**	M
10^{3}	**kilo**	k
10^{2}	hecto	h
10^{1}	deka	da
10^{-1}	deci	d
10^{-2}	**centi**	c
10^{-3}	**milli**	m
10^{-6}	**micro**	μ
10^{-9}	**nano**	n
10^{-12}	**pico**	p
10^{-15}	femto	f
10^{-18}	atto	a
10^{-21}	zepto	z
10^{-24}	yocto	y

* Commonly used prefixes are bolded. All prefixes are pronounced with the accent on the first syllable.

The Greek Alphabet

English pronunciation	Uppercase	Lowercase	English pronunciation	Uppercase	Lowercase
alpha	A	α	nu	N	ν
beta	B	β	xi	Ξ	ξ
gamma	G	γ	omicron	O	o
delta	Δ	δ	pi	Π	π
epsilon	E	ϵ, ε	rho	P	ρ
zeta	Z	ζ	sigma	Σ	σ
eta	H	η	tau	T	τ
theta	Θ	θ	upsilon	Υ	υ
iota	I	ι	phi	Φ	ϕ
kappa	K	κ	chi	X	χ
lambda	Λ	λ	psi	Ψ	ψ
mu	M	μ	omega	Ω	ω

Vector Products

$\vec{A} \cdot \vec{B} = AB \cos \theta$ $\quad |\vec{A} \times \vec{B}| = AB \sin \theta$

Terrestrial and Astronomical Data*

acceleration due to gravity at Earth's surface	$g^{\dagger}$	9.798 m/s^2
radius of Earth (volumetric mean)	R_E	6.371×10^6 m
mass of Earth	M_E	5.9724×10^{24} kg
mass of the Sun		1.9985×10^{30} kg
mass of the Moon		7.346×10^{22} kg
escape speed at Earth's surface		11.2 km/s = 6.96 mi/s
standard temperature and pressure (STP)		0°C = 273.15 K 1 atm = 101.3 kPa
Earth–Moon distance (mean)‡		3.84×10^8 m
Earth–Sun distance (mean)‡		1.50×10^{11} m
speed of sound in dry air (20°C, 1 atm)		343 m/s
density of dry air (STP)		1.29 kg/m^3
density of dry air (20°C, 1 atm)		1.20 kg/m^3
density of water (4°C, 1 atm)		1000 kg/m^3

* Additional data on the solar system can be found in Appendix B and at http://nssdc.gsfc.nasa.gov/planetary/planetfact.html.

† NASA cites a single average value for the planet, but this value changes slightly with location so see Appendix B for more detail.

‡ Center to center.

Mathematical Symbols

$=$	is equal to
$\equiv$	is identical to (equivalent)
$\neq$	is not equal to
$\approx$	is approximately equal to
$\sim$	is of the order of
$\propto$	is proportional to
$>$	is greater than
$\geq$	is greater than or equal to
$>>$	is much greater than
$<$	is less than
$\leq$	is less than or equal to
$<<$	is much less than
Δx	change in x
$\lvert x \rvert$	absolute value of x
$n!$	$n(n-1)(n-2)\ldots1$
Σ	sum
$\Delta t \to 0$	Δt approaches zero